The Inhibitor Index

A Desk Reference on Enzyme Inhibitors, Receptor Antagonists, Drugs, Toxins, Poisons & Therapeutic Leads

Daniel L. Purich

Department of Biochemistry & Molecular Biology
University of Florida College of Medicine
Gainesville, Florida U.S.A.

CRC Press
Taylor & Francis Group
Boca Raton London New York

CRC Press is an imprint of the
Taylor & Francis Group, an **informa** business

Cover Image: Crystal structure of the ABL kinase (semi-transparent silvery gray surface), with imatinib (Gleevec™) bound to an inactive conformation, such that the activation-loop blocks the substrate binding site, holding the enzyme's DFG motif in its out-conformation, exposing a critical hydrophobic pocket, and inducing downward folding of the Gly-rich loop. Imatinib's success as a first-line treatment for Philadelphia chromosome-positive Chronic Myelogenous Leukemia spurred an industry-wide search for other oncokinase inhibitors. **Lower Right:** Chemical structure of imatinib, permitting the reader to comprehend precisely how the inhibitor is lodged within its high-affinity site. Many thanks the author's colleague Professor Linda Bloom for her rendering of imatinib's binding pose, based on published atomic coordinates (PDB ID: 2hyy) from Cowan-Jacob, Fendrich, Floersheimer, *et al.* (2007) *Acta Crystallog.* (*D Biol. Crystallogr.*) **63**, 80-93.

CRC Press
Taylor & Francis Group
6000 Broken Sound Parkway NW, Suite 300
Boca Raton, FL 33487-2742

© 2017 by Taylor & Francis Group, LLC
CRC Press is an imprint of Taylor & Francis Group, an Informa business

No claim to original U.S. Government works

Printed in Canada on acid-free paper

International Standard Book Number-13: 978-1-138-73921-5 (Hardback)

Visit the Taylor & Francis Web site at
http://www.taylorandfrancis.com

and the CRC Press Web site at
http://www.crcpress.com

PREFACE

The roots of this reference book first took hold in 1967, when the purchase of a small notebook permitted me – then a first-year graduate student at Iowa State University – to record the names and properties of every enzyme and metabolic inhibitor then in use in the Department of Biochemistry & Biophysics. That practice was sparked by my graduate work in the laboratory of Professor Herbert J. Fromm, the enzyme kineticist who pioneered the use of dead-end competitive inhibitors, alternative substrate inhibitors, and multisubstrate geometrical analogues to define the steady-state behavior of multisubstrate enzyme-catalyzed reactions. Although a famously demanding taskmaster, Herb never discouraged his students from visiting the Physical Sciences Reading Room, and I spent many an hour reading research journals in quiet comfort, learning more and more about the primary modes of action of countless inhibitors. During a visit to Ames several years later, the celebrated enzymologist Ephraim Racker presented a seminar, in which he restated his oft-quoted and playful adage: "An inhibitor is most specific on the day of its discovery." Little did Racker know how he had inspired me to pay more attention to the off-target and toxic effects of inhibitors.

This systematic effort continued, when, as a NIH Staff Fellow, I joined Earl Stadtman in the NIH Laboratory of Biochemistry to study the adenylylation/deadenylylation cascade that controls the *Escherichia coli* glutamine synthetase, an enzyme that is unrivaled in the complexity of its feedback inhibition, posttranslational modification, and transcriptional regulation. As a mecca for biomedical research, Earl and Terry Stadtman's lab attracted many famous researchers, who were always eager to share their latest findings. The Library in nearby Building 10 was also an ideal setting for me to add new compounds and to gather details about the actions of inhibitors acting on all manner of biochemical and physiological processes.

This practice continued with heightened interest, when I moved to the Department of Chemistry at the University of California Santa Barbara, where I filled the position and occupied the office previously occupied by the late Professor B. R. Baker, a pioneer in the synthesis of puromycin and anti-folates. Baker was the author of *The Design of Active Site-Designed Irreversible Enzyme Inhibitors*. That seminal monograph described the logic of his determined efforts to synthesize and screen many hundreds of potential anticancer agents.

In the early 1970s, the field of enzyme chemistry was also abuzz with new ideas about the nature of enzyme intermediates, the action of mechanism-based inhibitors, the theory and design of first-generation transition-state inhibitors, as well as first recognition of slow, tight-binding inhibitors. The nature of allosteric feedback inhibition was yet another focus of enzyme science in that era.

Upon becoming the Chair of Biochemistry & Molecular Biology at the University of Florida College of Medicine in 1984, I had the opportunity to attend numerous pharmacology, physiology, and neuroscience seminars, a practice that fostered greater awareness of the clinical significance of receptor antagonists. A few years later, the department was fortunate to hire Dr. R. Donald Allison as an instructor. Previously my doctoral student at UC Santa Barbara, Don was keenly aware of the "Inhibitor Project", and, in 2004, offered to consolidate my inhibitor database and to collect additional information. Tragically, that effort was cut short in 2010, when Don perished from injuries from an auto accident. I dedicate this monograph to Don Allison in honor of our 37-year friendship. A better friend no one can have!

Now, fifty years after this project began, the first edition of *The Inhibitor Index* will provide readers with a single-source reference citing over 50,000 original research publications on the actions of nearly 8000 enzyme inhibitors, receptor antagonists, drugs, toxins, poisons, and/or therapeutic leads. Many are widely used in research and medicine; others should stimulate novel approaches and further research; still others are offered to acknowledge their historical significance. Future editions will include additional entries from an already compiled master source list of over 12,000 compounds. Readers are advised that suggestions for new entries and corrections will always be welcomed.

Lastly, I thank Fiona Macdonald at CRC Press for making the publication of this book so pleasant, even as she deftly applied her considerable chemical prowess to detect and correct my errors.

Dan Purich
Gainesville, Florida
March, 2017

INTRODUCTION

In the molecular life sciences, metabolic inhibitors and receptor antagonists are indispensable tools of the trade. Because enzymes carry out their tasks with fidelity, catalysis underlies most cellular processes, from facilitating central pathway metabolism to driving signal transduction (*e.g.,* 3',5'-cyclic-AMP, 3',5'-cyclic- GMP, and nitric oxide stimulation of protein kinases). Receptors are no less remarkable, and such considerations explain why cellular processes are exquisitely sensitive to inhibitors and antagonists. By blocking specific enzymes and/or receptor-mediated signaling, inhibitors and antagonists simplify the analysis of complex cellular processes. They are also incredibly useful in demonstrating that a process is functionally linked to an inhibitor-sensitive enzyme or an antagonist-sensitive receptor. Indeed, such target- specific agents provide a straightforward way to minimize the influence of competing pathways that might otherwise obscure mechanistic features of a reaction or process of experimental interest.

Inhibitors are incredibly versatile in the analysis of metabolite and signal transduction pathways. Although mutations can be used to block specific enzymes and receptors, an inhibitor's potency (a) can be conveniently ramped from low to high, based on the chosen value of $[I]/K_i$, where $[I]$ is the inhibitor concentration and K_i is the inhibition constant; (b) can be modulated by changes in substrate concentration, which can often displace or compete with the inhibitor; and (b) can be attenuated or completely reversed by dilution or dialysis. Enzyme inhibitors and receptor antagonists also allow one to retard processes that might otherwise occur on a timescale too rapid for convenient observation.

Through the use of microbe-killing antibiotics, cholesterol-reducing statins, mood-controlling agents, etc., modern medicine has benefitted from enzyme inhibitors and receptor antagonists. Chief limitations on their use concern bioavailability, potency, specificity, and above all – safety. Although improved mainly by trial and error, bioavailability is often predicted by *Lipinski's Rule of Fives*, which suggests that druglike properties of small molecules can achieved if the molecular mass is less than *five* hundred daltons, if the $\log_{10}$ value of the *n*-octanol/H_2O partition coefficient P exceeds *five*, if there are fewer than *five* hydrogen-bond donors, and if there are fewer than twice *five* (*i.e.,* 10) hydrogen-bond acceptor groups. That said, more and more drugs are administered intravenously or orally as membrane-penetrating prodrugs. Improvements in potency are driven by greater awareness of the atomic-level structure of enzymes and receptors and by the successful characterization of the underlying mechanisms of catalysis and signal transduction. Specificity is always a challenge, because off-target effects are not always evident, especially when agents are used persistently to treat chronic ailments. And, consistent with *Rackers's Rule*, asserting that an inhibitor is most specific on the day of its discovery, one must await further investigation to uncover off-target effects of inhibitors and antagonists. The advent of metabolomics, pharmacogenomics, and toxicogenomics offer hope that off-target effects can be detected or predicted, thus providing a wider margin of safety.

Beyond human medicine, enzyme inhibitors and receptor antagonists improve the lives of pets and show animals as well as the health and productivity of livestock. And no one can dispute the impact, albeit controversial, of glyphosate, the broad-spectrum systemic herbicide that provides farmers a way to control weeds, and, when coupled with GM glyphosate-resistant seeds, a way to improve per-acre crop yields.

How Compounds Were Selected for The Inhibitor Index – This reference describes the inhibitory action of nearly 8000 biologically significant agents, ranging from natural products to those now synthesized on the basis of structural biology and fragment-based approaches. Included are natural antibiotics, ultrahigh-affinity transition-state analogues, toxins, and receptor-neutralizing antibodies, ranging from molecules consisting of only a few atoms to those with 30,000 atoms. The goal of *The Inhibitor Index* is to describe both their primary mode of action as well as their off-target effects, simply because lack of specificity most often compromises the unambiguous use of inhibitors and antagonists.

The challenge of selecting compounds is illustrated with inhibitors of angiotensinogen-converting enzyme (ACE). Well over one-million candidate compounds have been tested for their antihypertensive effects, making it abundantly clear that no book or database can include all of the known ACE inhibitors. The same can be said for the myriad of receptor ligands that have been tested for action as agonists, partial agonists, inverse agonists, and antagonists or as direct and indirect inhibitors of receptor-mediated autophosphorylation and/or G-protein-coupled processes. Such considerations drive the logic for focusing mainly on: (a) enzyme inhibitors that molecular life scientists reliably use to impede/block cellular processes (*e.g.,* transporters, ion channels, metabolic pathways, chromosome remodeling, replication, transcription, translation, protein folding, cytoskeletal processes, or distinct organelle functions, etc.); (b) those receptor antagonists that are well-characterized in terms of their ability to reduce signal transduction processes vitalto cell and/or organ function; and (c) inhibitors that have already become drugs or are valuable as "lead molecules" for the development of novel drug-based therapies.

Of the 12,000 molecules currently in *The Inhibitor Index* master file, it seemed prudent to begin with approximately 7800 molecules, a number large enough to represent the inhibitors frequently used by practicing molecular life scientists, while including other drugs, antibiotics, and poisons used less often. The goal was to provide a one-volume compendium with adequate documentation of each entry's class, structure, formula weight, solubility, mode of action, off-target effects, and pertinent literature citations. Target selectivity and frequency of citation have and will always commend a molecule for inclusion in the first and subsequent editions. Even so, other notable compounds, having been historically significant by serving as prototypes for developing later-generation agents. Their mechanisms of action are milestones in the molecular life sciences. Other less frequently use toxins and poisons are also included, as are many drug leads, to foster broader awareness and encourage future development.

How Inhibitor Index Entries Are Curated – Every effort has been made to confirm the accuracy of the structures, stereochemistry, physical and chemical properties as well as other details for listed compounds. In most cases, the original research paper was examined, and, when conflicts arose, they were often resolved by contacting by one or more authors or by consulting other experts. Given the ongoing need to rename certain enzymes, alternate names were also used to search for inhibitor information. It has proved infeasible, however, to specify the organism or tissue of origin for every enzyme/receptor target; however, the literature citations should permit interested readers with ready access to these and other details. Because this book is intended for practicing molecular life scientists, no information concerning chemical syntheses and/or patents is provided. And, given the rapid pace of biomedical research, the latest findings on currently listed compounds will be incorporated into future editions, and every effort will be made to add promising inhibitors and antagonists in a timely manner.

Making Best Use of The Inhibitor Index – Experiments with enzyme inhibitors and receptor antagonists are least ambiguous when researchers maximize target specificity/selectivity. This can be best achieved by employing the electronic version to identify all agents having a particular enzyme or receptor as the primary target. When affinity data are available, one should favor agents having the lowest K_i or IC_{50} values, thereby minimizing the occurrence of spurious off-target effects, as the concentration of inhibitor or antagonist reaches the oversaturation range (*e.g.,* when [Inhibitor] or [Antagonist] $\gg K_i$ or IC_{50}). When multiple agents are available, experiments should be conducted in parallel using each inhibitor. Greatest confidence arises when the target enzyme or receptor is inhibited by two or more structurally unrelated, target-specific inhibitors/antagonists, each exhibiting a unique set of off-target effects.

Disclaimers – The information in this book is solely intended for laboratory use and *NOT* for clinical purposes. Physicians or veterinarians should instead consult *The Physicians Desk Reference* or *The Veterinarians Desk Reference*, as well as the manufacturer's FDA-approved package insert describing a drug's pharmacologic action, approved indications and dosages. The author also makes no direct, or implied, warranty as to the safety of any listed practices; nor should the reader infer that information provided here is completely current. Because *ALL* listed substances inhibit one or more vitally important enzymes or receptors, they should be treated as dangerous or potentially so. Nor should the absence a specific cautionary be taken to imply that a listed substance is harmless. Recommendations provided here are based on information contained in source publications; however, best-practice standards evolve, when previously unknown hazards are discovered. Readers seeking specific advice concerning the safe handling of agents listed in *The Inhibitor Index* should consult additional sources, including *Chem Sources* or specific-use data sheets supplied by manufacturers and distributors. In the absence of such information, it would be wise to treat all listed agents as potentially hazardous. Good laboratory practice demands that experimenters avoid contacting, swallowing, or inhaling chemical fluids, fumes and particulates. One must likewise minimize exposure to skin, eyes, and nasal passages by appropriate use of gloves, glasses, masks, laboratory garb, closed-toe shoes, as well as a chemical fume hood. Note also that static electricity frequently disperses finely divided powders during weighing, with consequential risk of inhalation. Protein and peptide inhibitors may also be allergenic. Above all, one must proceed with an abundance of caution. The reader accepts full responsibility for the safe use of agents mentioned in this book. The author and publisher bear no responsibility for any loss or damage, arising from use of information provided in *The Inhibitor Index*, whether such loss is direct, indirect, special, or incidental.

ABOUT THE AUTHOR

Dan Purich earned his Ph.D. for ground-breaking work on brain hexokinase, adenine nucleotide control, and multisubstrate inhibitors under Herbert Fromm at Iowa State University. After investigating the glutamine synthetase adenylylation cascade as an NIH Staff Fellow under Earl Stadtman, he joined the Chemistry Department at UC Santa Barbara. Awarded a Sloan Fellowship in Chemistry and NIH Research Career Development Award, Purich rose to full professor before moving to the University of Florida College of Medicine to become Chair of Biochemistry & Molecular Biology. He served on the *Journal of Biological Chemistry, Archives of Biochemistry & Biophysics,* and *Biophysical Chemistry* editorial boards. He also edited the six-volume *Enzyme Kinetics & Mechanism* sub-series of Methods in Enzymology and wrote The Handbook of Biochemical Kinetics (2000), The Enzyme Reference (2002), and Enzyme Kinetics: Catalysis & Control (2010). Author of over 170 peer-reviewed papers, chapters and reviews on enzyme action and cytoskeleton self-assembly, Purich discovered the actoclampins, the (+)-end-tracking motors that power actin-based motility by harnessing the Gibbs energy of on-filament ATP hydrolysis, while catalyzing monomer insertion. For fun, he travels and writes mystery novels.

– A –

A, *See Adenine; Alanine*

A3, *See N-(2-Aminoethyl)-5-chloronaphthalene-1-Sulfonamide*

A20 Protein

This novel zinc finger protein (MW = 89614 g; Isoelectric Point = 8.62) is a key player in the negative feedback regulation of the NF-κB pathway in response to multiple stimuli. Tumor necrosis factor alpha (TNFα) dramatically increases A20 expression in all tissues. A20 drives the adipogenic capacity of human mesenchymal stromal cells (MSCs), even in the presence of TNFα. **1**. Dorronsoro, Lang, Jakobsson, *et al.* (2013) *Cell Death Dis.* **4**, e972.

A46 Protein

This receptor signal-interference protein (MW = 27622 g; GenBank: CAM58344.1) targets the Toll-like receptor/ interleukin-1R (TIR) domain of Toll-like receptors (TLRs) and intracellular adaptor proteins, such as MAL, TRAM, TRIF and MyD88, promoting vaccinia virus replication in host cells. Full-length A46 behaves as a tetramer, whereas variants lacking the N-terminal 80 residues are dimeric. Both forms bind to the TLR domains of MAL and MyD88 with K_d values in the low μM range. A46 also shows a Bcl-2 like fold, but differences from that of A52, another Vaccinia signaling interference protein. A46 employs helices α-4 and α-6 to dimerize, whereas to the α1-α6 face is used by A52 and other Bcl-2 like Vaccinia proteins. Comparisons of A46 and A52 exemplify how subtle changes in viral proteins with the same fold can lead to crucial differences in biological activity. **1**. Fedosyuk, Grishkovskaya, De Almeida Ribeiro & Skern (2013) *J. Biol. Chem.* **289**, 3749.

A66

This potent and highly selective p110α inhibitor (FW = 393.53; CAS 1166227-08-2; Solubility: 79 mg/mL DMSO, 1 mg/mL H_2O), also known as (2*S*)-N^1-(5-(2-*tert*-butylthiazol-4-yl)-4-methylthiazol-2-yl)pyrrolidine-1, 2-dicarboxamide, has a IC_{50} value of 32 nM for the wild-type phosphoinositide-3 kinase p110 α as well as oncogenic forms of p110α such as p110α E545K (IC_{50} = 43 nM) and p110α H1047R (IC_{50} = 43 nM) (1). A66 inhibition also demonstrated that cancer-specific mutations in the iSH_2 (inter-SH_2) and nSH_2 (N-terminal SH_2) domains of p85α, the regulatory subunit of phosphatidylinositide 3-kinase (PI3K), show gain of function, inducing oncogenic cellular transformation, stimulating cellular proliferation, and enhancing PI3K signaling (2). **1**. Jamieson, *et al.* (2011) *Biochem. J.* **438**, 53. **2**. Sun, Hillmann, Hofmann, Hart & Vogt (2010) *Proc. Natl. Acad. Sci. USA* 107, 15547.

A 83-01

This TGFβ inhibitor (FW = 421.52 g/mol; CAS 909910-43-6; Solubility: 50 mM in DMSO), named 3-(6-methyl-2-pyridinyl)-*N*-phenyl-4-(4-quinolinyl)-1*H*-pyrazole-1-carbothioamide, selectively targets TGF-β type I receptor ALK5 kinase (IC_{50} = 12 nM), Type I Activin/Nodal receptor ALK4 (IC_{50} = 45 nM), and type I nodal receptor ALK7 (IC_{50} = 7.5 nM), blocking the phosphorylation of Smad2 and inhibiting TGF-β-induced epithelial-to-mesenchymal transitions. TGF-β signaling inhibitors represent a useful strategy for treating patients with tumor growth and metastasis in advanced cancer. A-83-01 only weakly inhibits ALK-1, ALK-2, ALK-3, ALK-6 and MAPK. It also has little or no effect on bone morphogenetic protein type I receptors, p38 mitogen-activated protein kinase, or extracellular regulated kinase (1). A-83-01 inhibits Smad signaling and epithelial-to-mesenchymal transition by transforming growth factor-β (1). Its use also instrumental in demonstrating that attachment to Laminin-111 facilitates TGF-β-induced expression of matrix metalloproteinase-2 in synovial fibroblasts (2). **1**. Tojo, *et al.* (2005) *Cancer. Sci.* **96**, 791. **2**. Hoberg, *et al.* (2008) *Ann. Rheum. Dis.* **66**, 446.

A-100

This novel exosite-directed inhibitor (*Sequence*: EEWEVLCWTWET CERGEG-NH₂; MW = 2242.42 g/mol), first identified and affinity-matured using naive and partially randomized peptide phage libraries against the immobilized tissue factor x Factor VIIa (TF x FVIIa) complex, is an exosite inhibitor of human coagulation factor VIIa (IC_{50} = 1.5 nM). This so-called A-series peptide represents a new class of peptide exosite inhibitors that are capable of attenuating, rather than completely inhibiting, the activity of TF x FVIIa, commending the development anticoagulants with an increased therapeutic window. **1**. Dennis, Roberge, Quan & Lazarus (2001) *Biochemistry* **40**, 9513.

A-132

This antibiotic (FW = 728.71 g/mol) is a chartreusin derivative that has been used in the treatment of breast cancer (1). A pharmacokinetic and pharmacodynamic (PK/PD) study of 6-*O*-(3-ethoxypropionyl)-3',4'-*O*-exo-benzylidene-chartreusin (IST-622) demonstrated that the latter is converted to A-132 and 3''-demethyl-3',4'-*O*-exo-benzylidene-chartreusin (A-132M) (2), **Target(s):** DNA topoisomerase II (1). **1**. Christmann-Franck, Bertrand, Goupil-Lamy, *et al.* (2004) *J. Med. Chem.* **47**, 6840. **2**. Asai, Yamamoto, Toi, *et al.* (2002) *Cancer Chemother. Pharmacol.* 49, 468.

A-183

This pentadecapeptide (MW = 1999.2 g/mol; *Sequence*: EEWEVLCWT WETCER, with disulfide linkage between Cys-7 and Cys-13) is an exosite inhibitor of human coagulation factor VIIa, IC_{50} = 1.6 nM (1-4). A-183 was a partial (hyperbolic) mixed-type inhibitor of FX activation having a K_i of 200 pM as well as acting as a partial competitive inhibitor of amidolytic activity. The location of the exosite was refined by a series of FVIIa alanine mutants, which showed that proximal residues Trp-61 and Leu-251 were critical for binding. **1**. Eigenbrot, Kirchhofer, *et al.* (2001) *Structure* **9**, 627. **2**. Dennis, Roberge, Quan & Lazarus (2001) *Biochemistry* **40**, 9513. **3**. Roberge, Santell, Dennis, *et al.* (2001) *Biochemistry* **40**, 9522. **4**. Maun, Eigenbrot & Lazarus (2003) *J. Biol. Chem.* **278**, 21823.

A238L Protein

This 226-residue immunomodulatory protein (MW = 26205) from African Swine Fever Virus inhibits calcineurin phosphatase activity and down-regulates the activation of the NF-κB and NF-AT transcription factors. A238L competitively inhibits calcineurin by occupying a substrate recognition site, while at the same time leaving the catalytic center fully accessible (1). The 1.7 Å structure of the A238L-calcineurin complex reveals how calcineurin recognizes residues in A238L that are analogous to the substrate motif, "LxVP." The high-resolution structure enabled modeling of a peptide substrate bound to calcineurin, even predicting substrate interactions beyond the catalytic center. Importantly, "LxVP" sequences in immunosuppressants (like Cyclosporin A, FK506, and A238L) all bind to the same site on calcineurin, allowing the latter to prevent "LxVP"-mediated substrate recognition and underscoring the critical role of this interaction for substrate dephosphorylation. (*See Cyclosporin A; FK506*) **1**. Grigoriu, Bond, Cossio, *et al.* (2013) *PLoS Biol.* **11**, e1001492.

A 779

This peptidomimetic Mas antagonist (FW = 872.98 g/mol; CAS 159432-28-7; Sequence: Asp-Arg-Val-Tyr-Ile-His-D-Ala (*or* DRVYIH-D-A); Solubility: 1 mg/mL H_2O) targets the Mas receptor, *or* Ang-(1-7) Receptor, the Class I seven-transmembrane G-protein-coupled angiotensin-(1-7) receptor that preferentially couples to the G_q protein, thereby activating the phospholipase C signaling pathway, with roles in hypotension, smooth muscle relaxation and cardio-protection by mediating the effects of angiotensin-(1-7). A-779 exhibits no significant affinity for AT_1 or AT_2 receptors at a concentration of 1 μM. It also inhibits antidiuretic effect of Ang-(1-7) in water-loaded rats. A779 attenuates monocrotaline-induced pulmonary fibrosis in rats. **1**. Bruce, *et al* (2015) *Brit. J. Pharmacol.* **172**, 2219. **2**. Becker, *et al* (2005) *Brain Res.* **1040**, 121. **3**. Santos, *et al* (1994) *Brain Res. Bull.* **35**, 293.

A-3253, *See* 4-Aminocatechol

A9415C. *See* Dehydrosinefungin

A21978C

This set of branched peptidolactone antibiotics (CAS 74754-47-5), formed nonribosomally by *Streptomyces roseosporus*, are antibacterial 13-residue peptides containing a medium-chain-acylated amino terminus and a ten-residue lactone ring (1-4). These antibiotics are closely related to daptomycin and A54145. **1**. Debono, Barnhart, Carrell, *et al.* (1987) *J. Antibiot.* **40**, 761. **2**. Eliopoulos, Thauvin, Gerson & Moellering, Jr. (1985) *Antimicrob. Agents Chemother.* **27**, 357. **3**. Wessels, von Dohren & Kleinkauf (1996) *Eur. J. Biochem.* **242**, 665. **4**. Lakey, Maget-Dana & Ptak (1989) *Biochim. Biophys. Acta* **985**, 60.

A23187

This Ca^{2+}/Mg^{2+} ionophore and antibiotic (FW = 523.62 g/mol; CAS 52665-69-7), also known as calcimycin and calcium ionophore III, from *Streptomyces chartreusensis*, forms high-affinity, membrane-permeant one-to-one complexes with calcium or magnesium ions at pH 7.4 (1-3). A23187 also stimulates nitric oxide production in cell cultures via the calmodulin-dependent nitric oxide synthase. **Metal Ion Selectivity:** This ionophore is highly selective for divalent cations over monovalent cations. Selectivity for Ca^{2+} is higher than for Mg^{2+}, as reflected by a 2.6-fold difference in binding constants. A23187 stock solutions (10 mM) can be prepared and stored almost indefinitely in dimethyl sulfoxide at –20°C. Other divalent cations that bind include Mn^{2+}, Sr^{2+}, and Ba^{2+} (3). A23187 also transports certain tervalent cations: efficiency for $Ca^{2+} > Nd^{3+} > La^{3+} > Eu^{3+} > Gd^{3+} > Er^{3+} > Yb^{3+} > Lu^{3+}$ (2,3). Over the 10–250 mM concentration range, La^{3+} is transported by an electroneutral mechanism, presumably as a $(A23187)_2$:La(III)-OH ternary complex. The equilibrium constants for protonation of A23187 and formation of complexes with monovalent and divalent cations have been determined in methanol-water mixtures and in suspensions of phospholipid vesicles (4,5). When compared as a function of ionic radius and cation charge, the binding constant data and associated free energies of complex formation indicate that lack of conformational mobility in A23187 is responsible for its high cation size selectivity. The relative complex stabilities, as determined by organic phase-transfer methods, are Mn^{2+} (210) »» Ca^{2+} (2.6) ≈ Mg^{2+} (1.0) »» Sr^{2+} › Ba^{2+} (3). **Spectrophotometry:** In ethanol, the free acid displays absorption peaks at 378, 278, 225, and 204 nm, with respective molar extinction values of 8200, 18200, 26200, and 28,200 $M^{-1}cm^{-1}$ (3). Upon combination with a divalent metal ion, the UV/Visible and fluorescence spectra are altered dramatically. Moreover, difference spectral titrations demonstrate that A23187 forms a 1:2 metal-to-ligand complex with Mn^{2+} and other alkaline earth divalent cations, as confirmed by X-ray crystallography (6). **Primary Mode of Action:** A23187 catalytically shuttles metal ions forth and back across the membrane bilayer, and the effective concentration range for its ionophoric action is 5–15 μM. To explain A23187's transport selectivity for divalent cations, Pfeiffer & Lardy (2) proposed that this ionophore forms a family of five complexes, of which only charge-neutral species are membrane-permeant. Although calcium ionophores are assumed to directly facilitate the transport of Ca^{2+} across the plasma membrane, A23187-mediated increases in cytosolic Ca^{2+} concentration in different Fura-2-loaded cells gave evidence of three ion-transfer components: (a) activation of Ca^{2+} influx through native Ca^{2+} channels, which was sensitive to drugs which inhibited the receptor-operated Ca^{2+} influx; (b) phospholipase C-dependent mobilization of Ca^{2+} from intracellular stores; and (c) a small ionophoric contribution at low A23187 concentrations (1). **Depletion of Cell Calcium Ion:** To deplete cells of calcium ions, suspend cells in a Ca^{2+}-free medium, supplemented with about 0.1–0.2 mM EGTA (pH 7.6–8.0). Because the extracellular Ca^{2+} concentration is approximately 5 mM in human fluids, addition of 10 μM A23187 to intact cells will greatly increase the intracellular calcium ion concentration, and the reader should consult the entry on calcium ions to identify likely inhibitory targets. **Depletion of Mitochondrial Magnesium Ion:** When employed to modify the mitochondrial free Mg^{2+} concentration, one typically uses 0.8 nmol A23187 per mg mitochondrial protein, along with sufficient EDTA to chelate all the Mg^{2+} present in the incubation medium. *Note*: A23187 uncouples oxidative phosphorylation and will inhibit ATPase activity. (*See also Bromo-A23187; Iophorous Antibiotics*) **Target(s):** mitochondrial ATPase (4,5); oxidative phosphorylation (5); Ca^{2+}-transporting ATPase (7); [pyruvate dehydrogenase (acetyl-transferring)]-phosphatase (8). **1**. Dedkova, Sigova & Zinchenko (2000) *Membr. Cell Biol.* **13**, 357. **2**. Pfeiffer & Lardy (1976), *Biochemistry* **15**, 935. **3**. Pfeiffer, Reed & Lardy (1974) *Biochemistry* **19**, 4007. **4**. Wang, Taylor & Pfeiffer (1998) *Biophys. J.* **75**, 1244. **5**. Taylor, Pfeiffer, Chapman, Craig & Thomas (1993) *Pure & Applied Chem.* **65**, 579. **6**. Smith & Duax (1976) *J. Am. Chem. Soc.* **98**, 6. **7**. Almeida, Benchimol, De Souza & Okorokov (2003) *Biochim. Biophys. Acta* **1615**, 60. **8**. Severson, Denton, Pask & Randle (1974) *Biochem. J.* **140**, 225.

A23871, *See* Efrapeptin

A-53612

This phenidone homologue (FW = 176.22 g/mol), also known as 1-phenyl[2H]tetrahydropyradazin-3-one, inhibits erythrocyte 5-lipoxygenase, or arachidonate 5-lipoxygenase (human), IC_{50} = 0.7 μM (1,2). **1**. Brooks, Albert, Dyer, et al. (1992) Bioorg. Med. Chem. Lett. **2**, 1353. **2**. Young (1999) Eur. J. Med. Chem. **34**, 671.

A54145

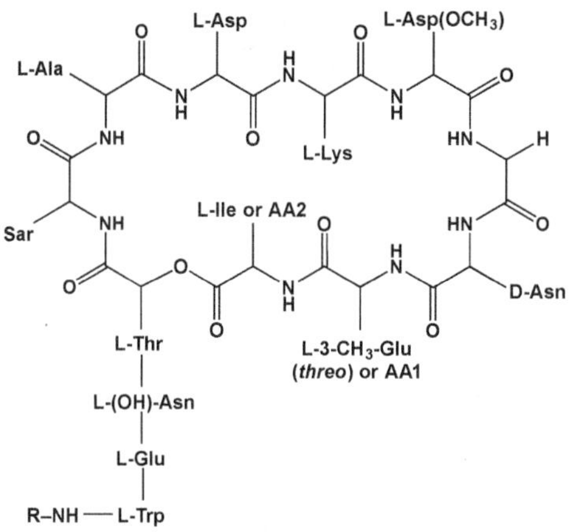

These 13-residue branched peptidolactone antibiotics from Streptomyces fradiae, possessing a medium-chain-acylated amino terminus and a 10 amino-acid residue lactone ring, are closely related to daptomycin and A21978C. A54145 forms (substituents) are Form A (R = 8-methylnonanoyl-; AA_1 = L-Glu); Form B (R = decanolyl-); Form B_1 (R = 8-methylnonanoyl-); Form C (R = 8-methyldecanoyl-; AA_2 = L-Val); Form D (R = 8-methyldecanoyl-; AA_1 = L-Glu); Form E (R = 8-methyldecanoyl-); Form F (R = 8-methyldecanoyl-; AA_1 = L-Glu; AA1 = L-Glu; AA_2 = L-Val). $MIC_{0.25}$ is approximately 32 μg/mL against Gram-positive aerobic organisms. **Target(s):** peptidoglycan synthesis (1-6). **1**. Wessels, von Dohren & Kleinkauf (1996) Eur. J. Biochem. **242**, 665. **2**. Counter, Allen, Fukuda, Hobbs, Ott, Ensminger, Mynderse, Preston & Wu (1990) J. Antibiot. **43**, 616. **3**. Boeck & Wetzel (1990) J. Antibiot. **43**, 607. **4**. Fukuda, Debono, Molloy & Mynderse (1990) J. Antibiot. **43**, 601. **5**. Fukuda, Du Bus, Baker, Berry & Mynderse (1990) J. Antibiot. **43**, 594. **6**. Boeck, Papiska, Wetzel, Mynderse, Fukuda, Mertz & Berry (1990) J. Antibiot. **43**, 58.

A-63124

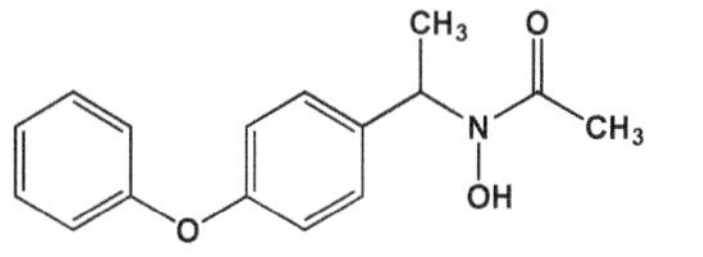

This hydroxyamide (FW = 271.32 g/mol) inhibits 5-lipoxygenase, or arachidonate 5-lipoxygenase, the enzyme that catalyzes two steps in biosynthesis of leukotrienes (LTs). Because leukotrienes are lipid mediators of inflammation, LT antagonism is used in treatment of asthma and atherosclerosis. **See** Zileuton **1**. Young (1999) Eur. J. Med. Chem. **34**, 671.

A-64077, See Zileuton

A-70450

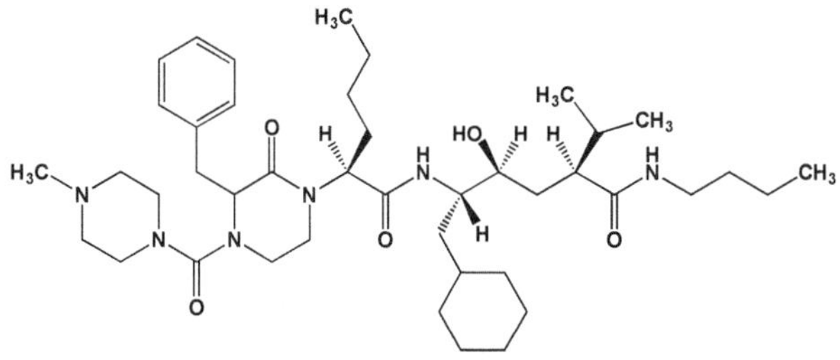

This peptidomimetic (FW = 739.06 g/mol; CAS 142928-23-2) is a strong inhibitor of the secreted aspartic proteinase of Candida albicans: IC_{50} = 1.4 nM as well as renin, IC_{50} = 7.1 nM, and cathepsin D, IC_{50} = 770 nM. **Target(s):** candidapepsin, or secreted aspartic proteinase of Candida albicans (1-4,6); renin (4,6); cathepsin D (4,6); candidapepsin, or secreted aspartic proteinase 2X of Candida tropicalis (5). **1**. Cutfield, Dodson, Anderson, et al. (1995) Structure **3**, 1261. **2**. Lerner & Goldman (1993) J. Gen. Microbiol. **139**, 1643. **3**. Abad-Zapatero, Goldman, et al. (1998) Adv. Exp. Med. Biol. **436**, 297. **4**. Leung, Abbenante & Fairlie (2000) J. Med. Chem. **43**, 305. **5**. Symersky, Monod & Foundling (1997) Biochemistry **36**, 12700. **6**. Pranav-Kumar & Kulkarni (2002) Bioorg. Med. Chem. **10**, 1153.

A-71561, See 1-(3-Phenylpropionyl)piperidine-3-(R,S)-carboxylic Acid [4-Amino-1(S)-(benzothiazole-2-carbonyl)butyl] Amide

A-72517, See Zankiren

A-76745, See Fenleuton

A-77003

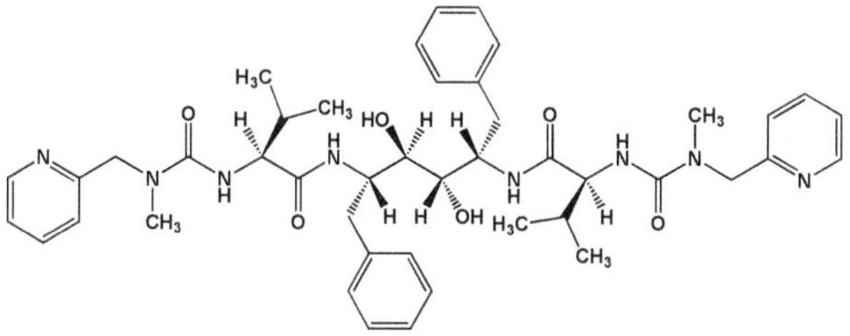

This rotationally symmetric peptidomimetic (FW = 794.99 g/mol) inhibits HIV-1 protease, K_i = 1.4 nM (1-3). When tested in asymptomatic HIV patients, plasma A77003 concentrations reached the in vitro IC_{50} (0.16 μg/mL) in the 0.28-mg/kg/h dosage group, but never attained the IC_{90} (0.48 μg/mL). No statistically significant effect on CD4 cell numbers occurred in any of the groups, and there was no evidence of antiviral activity, as determined by HIV-1 p24 antigen level, quantitative plasma and cell culture, and quantitation of viral RNA in plasma (2). The lack of antiviral activity observed in the study may be a consequence of the low concentrations in plasma in all groups. **1**. Wilson, Phylip, Mills, et al. (1997) Biochim. Biophys. Acta **1339**, 113. **2**. Reedijk, Boucher, van Bommel, et al. (1995) Antimicrob. Agents Chemother. **39**, 1559. **3**. Kageyama, Weinstein, Shirasaka, et al. (1992) Antimicrob. Agents Chemother. **36**, 926

A-79175, See A-78733

A-79285

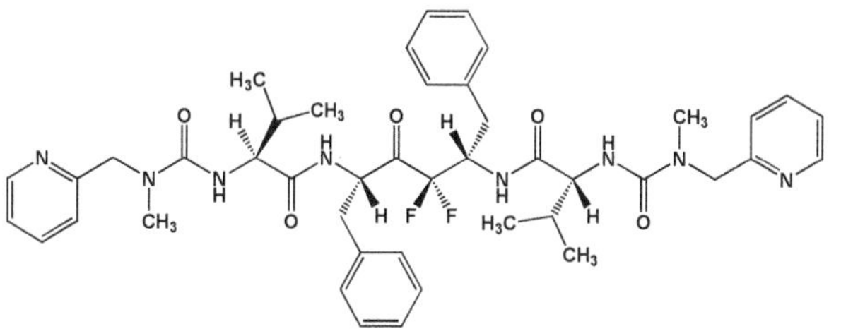

This pseudo-C_2 symmetric difluoroketone (FW = 812.96 g/mol) inhibits HIV-1 protease, or HIV-1 retropepsin (1-2). In the water, the ketone quickly converted to a gem-diol. Refinement of the 1.7-Å structure of the EI complex revealed a unique, and almost symmetric, set of interactions between the geminal hydroxyl groups, the geminal fluorine atoms, and the active-site aspartate residues, with only one active-site aspartate protonated. **1**. Silva, Cachau, Sham & Erickson (1996) J. Mol. Biol. **255**, 321. **2**. Silva, Cachau, Baldwin, Gulnik, Sham & Erickson (1995) Adv. Exp. Med. Biol. **362**, 451.

A-85380, See 3-(2(S)-Azetidinylmethoxy)pyridine

A-85783

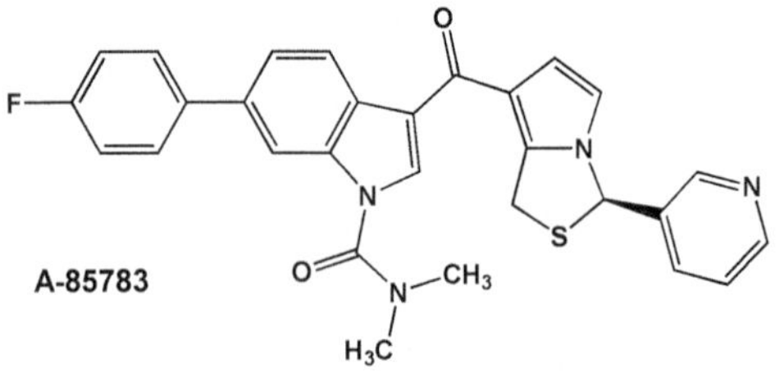

A-85783

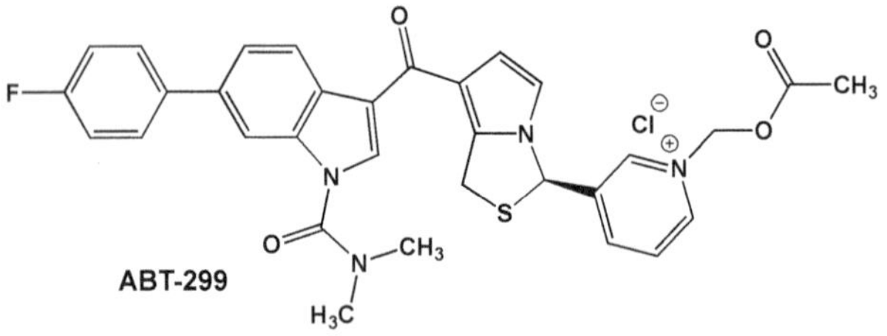

ABT-299

This substituted indole (FW$_{\text{free-base}}$ = 510.59 g/mol), also known as (R)-6-(4-fluorophenyl)-N,N-dimethyl-3-[3-(3-pyridinyl)-1H,3H-pyrrolo[1,2-c]thia-zol-7-yl]carbonyl-1H-indole-1-carboxamide, is a potent platelet-activating factor receptor antagonist (1,2). Given that the lipid mediator platelet-activating factor (PAF) plays an important role in cutaneous inflammation, pretreatment of tissue with A-85783 results in an inhibition of PAF-induced inflammation (1). ABT-299 is the water-soluble prodrug of A-85783 (**See also ABT-299**). **1**. Travers, Pei, Morin & Hood (1998) *Arch. Dermatol. Res.* **290**, 569. **2**. Albert, Conway, Magoc, *et al.* (1996) *J. Pharmacol. Exp. Ther.* **277**, 1595.

A-86719.1, See ABT-719

A-134974

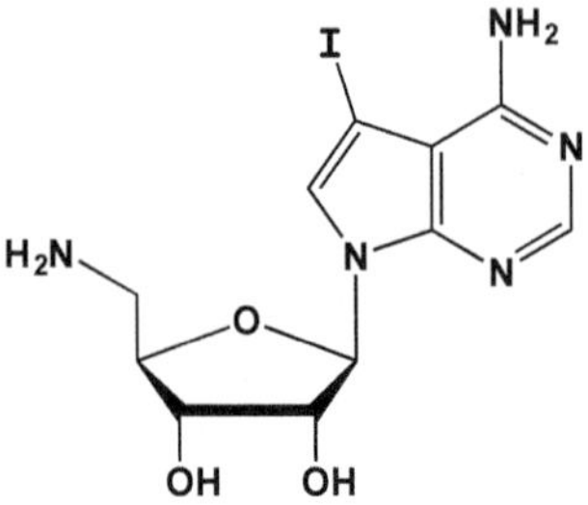

This adenosine and 5-iodotubercidin analogue (FW$_{\text{free-base}}$ = 375.17 g/mol), also known as [(1'R,2'S,3'R,4'S)-2',3' dihydroxy-4'-aminocyclopentyl]-4-amino-5-iodopyrrolopyrimidine, is a potent adenosine kinase inhibitor, IC$_{50}$ = 60 pM (1-3). Because the use of adenosine receptor agonists is plagued by dose-limiting cardiovascular side effects, there is great interest in developing adenosine kinase inhibitors (AKIs) as a means for raising steady-state adenosine concentrations. A-134974 reduces thermal hyperalgesia and depresses locomotor activity. **1**. McGaraughty, Chu, Wismer, *et al.* (2001) *J. Pharmacol. Exp. Ther.* **296**, 501. **2**. Zhu, Mikusa, Chu, *et al.* (2001) *Brain Res.* **905**, 104. **3**. Iqbal, Burbiel & Müller (2006) *Electrophoresis* **27**, 2505.

A-170634

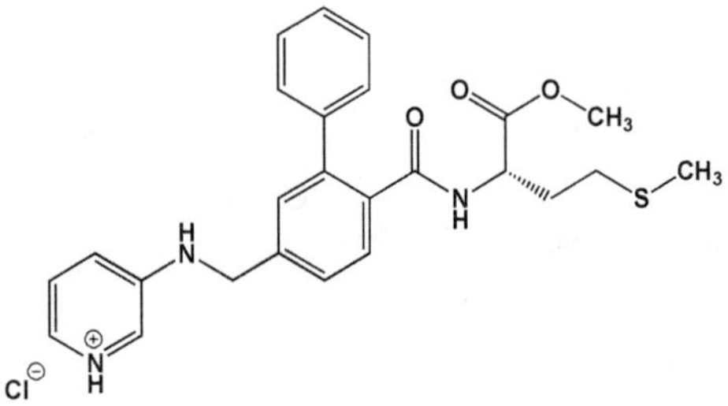

This CAAX-directed prenylation inhibitor (FW$_{\text{free-base}}$ = 449.57 g/mol) targets protein farnesyl-transferase: IC$_{50}$ = 120 nM (1), inhibiting Ras processing, blocking anchorage-dependent and anchorage-independent growth of HCT116 K-ras-mutated cells, and decreasing human umbilical vein endothelial cell capillary structure formation as well as vascular endothelial growth factor secretion from tumor cells (2). As a potent and selective peptidomimetic CAAX FTI with anti-angiogenic properties, A-170634 may affect tumor growth in vivo by one or more antitumor pathways. **1**. Gibbs (2001) *The Enzymes*, 3rd ed., **21**, 81. **2**. Gu, Tahir, Wang, *et al.* (1999) *Eur. J. Cancer* **35**, 1394.

A-315456

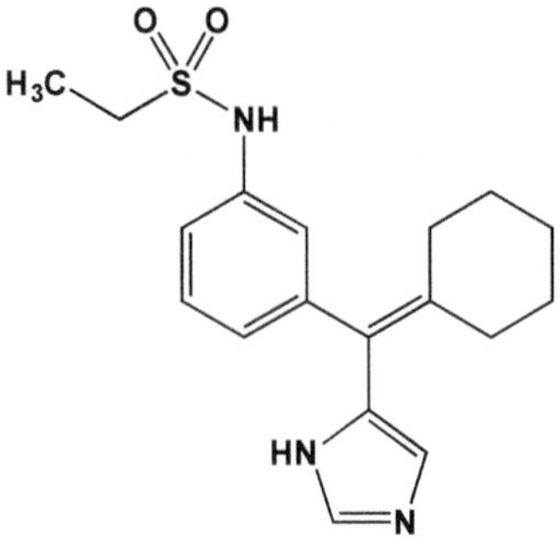

This selective α$_{1D}$-adrenergic receptor antagonist (FW$_{\text{free-base}}$ = 345.47 g/mol; CAS 258527-07-0), also known systematically as N-[3-(cyclo-hexylidene-(1H)-imidazol-4-ylmethyl)-phenyl]ethanesulfonamide, exhibits minimal dopamine D$_2$ and 5-HT$_{1A}$ receptor affinity. **1**. Buckner, Milicic, Daza, *et al.* (2001) *Eur. J. Pharmacol.* **433**, 123.

A-317491

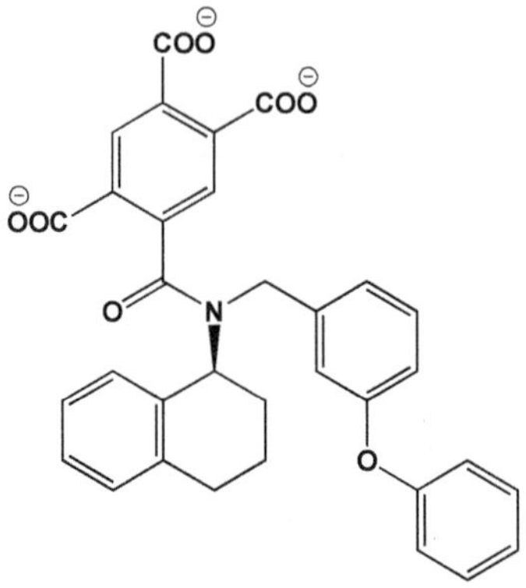

This tricarboxylic acid (FW$_{\text{free-acid}}$ = 565.58 g/mol) is a P2X$_3$-selective and P2X$_2$/P2X$_3$ receptor antagonist that reduces chronic inflammatory and neuropathic pain. A-317491 inhibits nNOS activation, a process requiring translocation of nNOS from the cytosol to the membrane after protein kinase C-mediated activation of purinergic P2X and P2Y receptors (2). A-317491 blocks both hyperalgesia and allodynia in different animal models of pathological pain. A-317344 is the R-enantiomer (1). **1**. Jarvis, Burgard, McGaraughty, *et al.* (2002) *Proc. Natl. Acad. Sci. U.S.A.* **99**, 17179. **2**. Ohnishi, Matsumura & Ito (2009) *Mol. Pain* **5**, 40.

A-331440

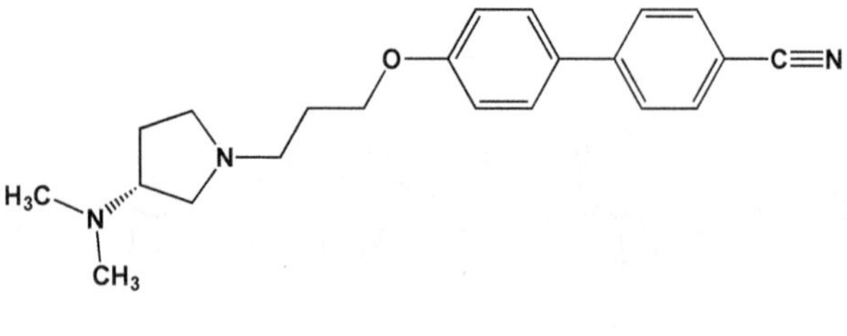

This pyrrolidine (FW$_{\text{free-base}}$ = 349.48 g/mol) is a strong, non-imidazole H$_3$ histamine receptor antagonist, K$_i$ = 25 nM. At higher doses, A-331440

reduced body fat and normalized an insulin tolerance test, showing that this histamine receptor antagonist has potential as an antiobesity agent. **1**. Faghih, Esbenshade, Krueger, *et al.* (2004) *Inflamm. Res.* **53** Suppl. 1, S79. **2**. Hancock, Bennani, Bush, *et al.* (2004) *Eur. J. Pharmacol.* **487**, 183.

A-425619

This TRPV1 antagonist (FW = 345.32 g/mol; CAS 581809-67-8; Soluble to 100 mM in DMSO), also named *N*-5-isoquinolinyl-*N*′-[[(4-(trifluoromethyl)phenyl]methyl]urea, targets (IC_{50} = 3-9 nM) Transient Receptor Potential cation channel subfamily V member 1 (also known as the capsaicin receptor and the vanilloid receptor 1), a nonselective cation channel that integrates responses to multiple stimuli, such as capsaicin, acid, heat, endovanilloids,and allyl isothiocyanate (the pungent compound present in mustard and wasabi). TRPV1 plays a key role in the transmission of inflammatory pain. A-425619 potently blocks capsaicin-evoked increases in intracellular calcium concentrations in HEK293 cells expressing recombinant human TRPV1 receptors, IC_{50} = 5 nM (1). It shows similar potency (IC_{50} = 3-4 nM) in blocking TRPV1 receptor activation by anandamide and *N*-arachidonoyl-dopamine (1). Electrophysiological experiments show that A-425619 also potently blocks the activation of native TRPV1 channels in rat dorsal root ganglion neurons (IC_{50} = 9 nM). Compared with other known TRPV1 antagonists, A425619 exhibits superior potency in blocking both naive and phorbol ester-sensitized TRPV1 receptors. Like capsazepine, A-425619 demonstrates competitive antagonism (pA_2 = 2.5 nM) of capsaicin-evoked calcium flux (1). Systemic injection of A-425619 (3-30 micromol/kg, iv) reduces neuronal wide dynamic range (*or* WDR) responses to thermal stimulation in both complete Freund's adjuvant- (*or* CFA-) inflamed (47 °C) and uninjured (52 degrees C) rats. However, the efficacy of A-425619 to attenuate thermally-evoked WDR activity is significantly greater (P < 0.01) in CFA-treated rats. Both intradorsal root ganglion (DRG; L5; 20 nmol) and intraplantar (30-300 nmol) injection of A-425619 reduces WDR responses to thermal stimulation. While the effectiveness of A-425619 is similar between CFA-inflamed and uninjured rats after intraplantar injection, the effects of A-425619 after intra-DRG injection are enhanced in the inflamed rats (compared with the uninjured rats). Spontaneous WDR discharges are unaltered by systemic or site-specific injections of A-425619. **1**. El Kouhen, Surow, Bianchi *et al.* (2005) *J. Pharmacol. Exp. Ther.* **314**, 400. **2**. McGaraughty, Chu, Faltynek & Jarvis (2006) *J. Neurophysiol.* 95, 18.

A-674563

This ATP-competitive protein kinase inhibitor (FW = 358.44; CAS 552325-73-2; Solubility (25°C): 70 mg/mL DMSO; 70 mg/mL H_2O), systematically named (2*S*)-1-(5-(3-metsignalhyl-1*H*-indazol-5-yl)pyridin-3-yloxy)-3-phenylpropan-2-amine, shows slightly greater inhibition of Akt1, as compared to protein kinase A (PKA) and cyclin-dependent kinase-2 (CDK2), as indicated by K_i values of 11 nM, 16 nM, and 46 nM. A-674563 does not inhibit Akt phosphorylation itself, but blocks the phosphorylation of Akt's downstream targets in a dose-dependent manner. **Cyclin Target Selectivity:** Cdk1 (weak, if any), Cdk2 (++), Cdk3 (weak, if any), Cdk4 (weak, if any), Cdk5 (weak, if any), Cdk6 (weak, if any), Cdk7 (weak, if any), Cdk8 (weak, if any), Cdk9 (weak, if any), Cdk10 (weak, if any), Akt1 (++), PKA (++).**1**. Luo, *et al.* (2005) *Mol. Cancer Ther.* **4**, 977.

2. Zhu, *et al.* (2008) *Cancer Res.* **68**, 2895. **3**. Okuzumi, *et al.* (2010) *Mol. Biosyst.* **6**, 1389.

A-769662

This potent, reversible AMPK activator (FW = 360.39 g/mol; CAS 844499-71-4) targets AMP-stimulated protein kinase (EC_{50} = 0.8 μM), a key sensor and regulator of intracellular and whole-body energy metabolism. (***See also*** *Acadesine*) AMPK activation has been shown to alleviate symptoms of type 2 diabetes, making it a druggable target. Short-term treatment of normal Sprague Dawley rats with A-769662 decreases liver malonyl CoA levels and the respiratory exchange ratio, V_{CO2}/V_{O2}, indicating increased whole-body fatty acid oxidation (1). Treatment (30 mg/kg b.i.d.) of ob/ob mice decreases hepatic expression of PEPCK, G6Pase, and FAS, lowered plasma glucose (by 40%), reduces body weight gain, and significantly decreases both plasma and liver triglycerides (1). A-769662 activates AMPK, both allosterically and by inhibiting AMPK dephosphorylation (at Thr-172), similar to the effects of the natural effector, 5'-AMP (2). It also activates AMPK harboring a γ-subunit mutation that abolishes AMP activation. An AMPK complex that lacks the glycogen binding domain of the β-subunit abolishes the allosteric effect of A-769662, but not allosteric activation by AMP (2). AMPK activation by A-769662 results in the inhibition of endothelial cell proliferation, by elevating p21 and p27 expression, show that AMPK regulates endothelial cell migration and differentiation (3). **Off-Target Effect(s):** Contrary to the mechanism of action of most proteasome inhibitors, A769662 inhibits the 26S proteasome by an AMPK-independent mechanism that does not affect the proteolytic activities of the 20S core subunit (4). This novel mechanism of inhibition of 26S proteasome activity is reversible and results in cell-cycle arrest. Such side effects of this new activator of AMPK should be considered when this compound is used as an alternative activator of the kinase (4). Other AMPK activators also induce nucleoli reorganization, with attendant changes in cell proliferation. Among the compounds tested, phenformin and resveratrol had the most pronounced impact on nucleolar organization (5). **1**. Cool, et al, (2006) *Cell Metab.* **3**, 403. **2**. Sanders (2007) *J. Biol. Chem.* **282**, 32539. **3**. Peyton, Liu, Yu, Yates & Durante (2012) *J. Pharmacol. Exp. Ther.* **342**, 827. **4**. Moreno, et al, (2008) *FEBS Lett.* **583**, 2650. **5**. Kodiha, Salimi, Wang & Stochaj (2014) *PLoS One.* **9**, e88087.

A771726, *See Leflunomide*

A-803467

This potent sodium channel blocker (FW = 357.79 g/mol; CAS 944261-79-4; Solubility: 72 mg/mL DMSO; <1 mg/mL H_2O), also named 5-(4-chlorophenyl)-*N*-(3,5-dimethoxyphenyl)furan-2-carboxamide, selectively targets $Na_V1.8$ channels (IC_{50} = 8 nM), blocking tetrodotoxin-resistant currents and exhibiting >100x selectivity against human $Na_V1.2$, $Na_V1.3$, $Na_V1.5$, and $Na_V1.7$. A-803467 potently blocks tetrodotoxin-resistant currents (IC_{50} = 140 nM), generating spontaneous and electrically evoked action potentials *in vitro* in rat dorsal root ganglion neurons (1). Later studies reported the additive antinociceptive effects of the selective $Na_V1.8$ blocker A-803467 and selective TRPV1 antagonists in rat inflammatory and neuropathic pain models (2). **Inhibition of Drug Efflux:** ATP-binding cassette (ABC) multidrug transporters (such as ABCB1, *or* MDR1/P-glycoprotein); ABCC1 (MRP1) and ABCG2 (BCRP/MXR)) mediate drug efflux in human cancers (3). ABCG2 is a 72-kDa half transporter that is specifically localized at the apical surface of enterocytes, the luminal surface of liver canaliculi, the luminal surface of the proximal convoluted tubule of the kidneys, the blood–brain barrier (BBB), blood–testis barrier (BTB), blood–placental and blood–retinal barriers. Because of its localization on the secretory surface of the major organs involved in drug transport, ABCG2 alters absorption, distribution, metabolism and elimination of its substrate drugs. A-803467 significantly increases the cellular sensitivity to ABCG2 substrates in drug-resistant cells overexpressing either wild-type or mutant ABCG2. At 7.5 μM, A-803467 significantly increases the intracellular accumulation of [^{3}H]-mitoxantrone by inhibiting the transport activity of ABCG2, without altering its expression levels. In addition, A-803467 stimulates the ATPase activity in membranes overexpressed with ABCG2. In a murine model system, combination treatment of A-803467 (35 mg/kg) and topotecan (3 mg/kg) significantly inhibits the tumor growth in mice xenografted with ABCG2-overexpressing cancer cells. Such findings indicate that a combination of A-803467 and ABCG2 substrates may potentially be a novel therapeutic treatment in ABCG2-positive drug resistant cancers. **1**. Jarvis, Honore, Shieh, *et al.* (2007) *Proc. Natl. Acad. Sci. U.S.A.* **104**, 8520. **2**. Joshi, Honore, Hernandez, *et al.* (2009) *J. Pain* **10**, 306. **3**. Anreddy, Patel, Zhang, *et al.* (2015) *Oncotarget* **6**, 39276.

A 939572

This potent oleate synthesis inhibitor (FW = 387.86 g/mol; CAS 1032229-3-6; Soluble to 100 mM in DMSO), also named 4-(2-chlorophenoxy)-*N*-[3-[(methylamino)carbonyl]phenyl]-1-piperidine carboxamide, selectively targets stearoyl-CoA desaturase 1 (SCD1) inhibitor (IC$_{50}$ = 0.4 nM), showing selectivity over a range of kinases and hERG channels. The utility of human pluripotent stem cells (hPSCs) in cell therapy is hindered by the tumorigenic risk from residual undifferentiated cells. A-939572 was identified among 15 pluripotent cell-specific inhibitors (PluriSINs), of which nine share a common structural moiety. It selectively eliminates hPSCs, while sparing a large array of progenitor and differentiated cells, doing so by inducing ER stress, protein synthesis attenuation, and apoptosis (1). Further examination identified A-939572 as an inhibitor of stearoyl-CoA desaturase (SCD1), revealing a unique role for lipid metabolism in hPSCs. A-939572 is also cytotoxic to mouse blastocysts, indicating that the dependence on oleate is inherent to the pluripotent state. Application of A-939572 also prevents teratoma formation from tumorigenic undifferentiated cells. **1**. Xin, Zhao, Serby, *et al.* (2008) *Bioorg. Med. Chem. Lett.* **18**, 4298. **2**. Ben-David, Gan, Golan-Lev, *et al.* (2013) *Cell Stem Cell.* **12**, 167.

A- 971432

This orally active potent and selective S1P$_5$ agonist (FW = 366.24 g/mol; CAS 1240308-45-5; Soluble to 100 mM in 1 eq. NaOH and to 100 mM in 1eq. HCl), also named 1-[[4-[(3,4-dichlorophenyl)-methoxy]phenyl] methyl]-3-azetidinecarboxylic acid, selectively targets Sphingosine-1-phosphate receptor-5, a receptor that is highly expressed on endothelial cells within the blood-brain barrier, where it helps to maintain barrier integrity in *in vitro* models. S1P acts as an extracellular ligand for a family of G-protein coupled receptors that are crucial in cell migration. S1P5 is exclusively expressed in oligodendrocytes and oligodendrocyte precursor cells (OPCs), which migrate considerable distances during brain development. [For reference, S1P binding to the S1P5 receptor blocks OPC migration (IC$_{50}$ = 30 nM).] EC$_{50}$ values for A-971432 activation of S1P5 receptor are 4.1 and 5.7 nM in cAMP and GTPγS assays, respectively. A-971432 is highly efficacious in reversing lipid accumulation and age-related cognitive decline in rats. It also improves blood-brain barrier integrity *in vitro* and reverses age-related cognitive decline in mice. Such

results suggest S1P5 agonism may prove beneficial in treating neurodegenerative disorders involving lipid imbalance and/or compromised blood-brain barrier. **1**. Hobson, Harris, van der Kam, *et al.* (2015) *J. Med. Chem.* **58**, 9154.

A-1070722

This potent, brain-penetrating GSK inhibitor (FW = 362.31 g/mol; CAS 1384424-80-9), also named 1-(7-methoxyquinolin-4-yl)-3-[6-(trifluoro methyl)pyridin-2-yl]urea, targets glycogen synthase kinase-3, with K_i = 0.6 nM for both GSK-3α and GSK-3β, with >50x weaker action on other kinases tested, including CDK family members. Its high degree of selectivity was confirmed using a biochemical affinity approach employing ATP-binding in rat brain lysates. In cells, A-1070722 decreases phosphorylation of the microtubule-associated protein Tau, as well as accumulation of cellular β-catenin. In addition, A-1070722 protected rat primary cortical neurons against a β-amyloid or glutamate challenge. **1**. Bakker, van Gaalen, Bitner, *et al.* (2011) Society for Neuroscience National Meeting, Washington, DC, November 12-16, *Abstract* 568.04/Z7.

AA-861, See *2-(12-Hydroxydodeca-5,10-diynyl)-3,5,6-trimethyl-p-benzoquinone*

AAL Toxins

These sphingoid-like alkaloids (AAL Toxin TA$_1$; FW = 521.64 g/mL; CAS 79367-52-5; AAL Toxin TA$_2$; FW = 521.65 g/mol; CAS 79367-51-4) from the ascomycete fungus *Alternaria alternata* inhibit ceramide synthase, an early biochemical event in the animal diseases associated with consumption of *Fusarium moniliforme*-contaminated corn. Because free sphingoid bases are precursors to plant "ceramides," their accumulation suggests that the primary biochemical lesion is inhibition of *de novo* ceramide synthesis and re-acylation of free sphingoid bases. **Target(s):** ceramide synthase, *or* sphingosine *N*-acyltransferase (2,3). **1**. Caldas, Jones, Ward, Winter & Gilchrist (1994) *J. Agri. Food Chem.* **42**, 327. **2**. Wang & Merrill, Jr. (2000) *Meth. Enzymol.* **311**, 15. **3**. Merrill, Jr., Wang, Gilchrist & Riley (1993) *Adv. Lipid Res.* **26**, 215.

AB1010, See *Masitinib*

ABA, See *Abscisic Acid*

***p*ABA, See** *4-Aminobenzoic Acid*

Abacavir

This carbocyclic diaminopurine nucleoside analogue and antiviral prodrug(.FW = 286.34 g/mol; CAS 136470-78-5), named systematically as (1*S*,4*R*)-4-(2-amino-6-(cyclopropylamino)-9*H*-purin-9-yl)-2-cyclopentene-

1-methanol, is metabolized to carbovir triphosphate (*See Carbovir*), latter inhibiting viral reverse transcriptase (1,2). When incorporated into the growing viral DNA strand, abacavir blocks strand elongation, leading to a premature halt in viral DNA synthesis and chain termination. It is typically used in combination with other antiretroviral agents in the treatment of HIV infection. **Systemic Abacavir Hypersensitivity:** Abacavir induces powerful immunolosensitivity, which is driven by activation of HLA-B*5701 in $CD8^+$ T cells, resulting in secretion of inflammatory mediators (*e.g.*, TNF-α and IFN-γ). Delayed hypersensitivity reactions, including systemic syndromes and organ-specific toxicities, can be triggered by abacavir (3,4), and there is a strong genetic association with the human leukocyte antigen (HLA)-B*57:01 allele. Abacavir is a prodrug, and one its metabolites binds to an antigen-presenting cleft unique to HLA-B*5701, explaining why the drug *per se* does not cause a hypersensitivity syndrome in carriers of other HLA-B alleles and why compounds similar to abacavir fail to react with HLA-B*5701. Patients should be tested for hypersensitivity prior to using abacavir. **Target(s):** HIV-1 reverse transcriptase (1,2); RNA-directed DNA polymerase (1,2). **1.** Tisdale, Alnadaf & Cousens (1997) *Antimicrob. Agents Chemother.* **41**, 1094. **2.** El Safadi, Vivet-Boudou & Marquet (2007) *Appl. Microbiol. Biotechnol.* **75**, 723. **3.** Mallal, Nolan, Witt, *et al.* (2002) *Lancet* **359**. 727. **4.** Chessman, Kostenko, Lethborg, Purcell, *et al.* (2008) *Immunity* **28**, 822.

Abamectin

These macrocyclic metabolites (FW$_{Abermectin-B1a}$ = 873.09 g/mol, R = –CH_2CH_3; FW$_{Abermectin-B1b}$ = 859.06, R = –CH_3; CAS 71751-41-2; Trade Names: Agri-Mek®, Avid®, Avert®), obtained from *Streptomyces avermitilis*, are natural insecticides, acaricides, and antihelmitics that stimulate γ-aminobutyrate release. Abamectin is more effective when consumed orally, and irreversible paralysis is the main effect. By the 1985 convention, commercial sources contain 80% B$_{1a}$ and 20 % B$_{1b}$. *Note:* Abamectin is toxic (in rats, LD$_{50}$(oral) = 10 mg/kg; LD$_{50}$(dermal) = >20000 mg/kg). Abamectin is safest when applied to mature plants, because harder leaves are more resistant to abamectin penetration. **1.** Wislocki, *et al.* "Environmental Aspects of Abamectin Use in Crop Protection" *in* Campbell (ed.). 1989. *Ivermectin and Abamectin.* Springer-Verlag, NY.

Abamine

This tertiary amine (FW = 373.42 g/mol) inhibits 9-*cis*-epoxycarotenoid dioxygenase, with respective K_i values of 132 μM and 39 μM for the *Arabdiposis thaliana* and spinach enzymes (1-3). Abscisic acid (ABA) is a major regulator in the adaptation of plants to environmental stresses, plant growth, and development. In higher plants, the ABA biosynthesis pathway involves the oxidative cleavage of 9-*cis*-epoxycarotenoids, the key regulatory step in the pathway catalyzed by 9-cis-epoxycarotenoid dioxygenase. **1.** Boyd, Gai, Nelson, *et al.* (2009) *Bioorg. Med. Chem.* **17**, 2902. **2.** Han, Kitahata, Saito, *et al.* (2004) *Bioorg. Med. Chem. Lett.* **14**, 3033. **3.** Han, Kitahata, Sekimata, *et al.* (2004) *Plant Physiol.* **135**, 1574.

Abatacept

This recombinant anti-inflammatory protein (MW = 92 kDa; CAS 213252-14-3), marketed under the trade name Orencia®, inhibits T-cell activation in rheumatoid arthritis and reduces the B-cell production of rheumatoid factor. **Primary Mode of Inhibitory Action:** Abatacept is a CTLA4-Ig fusion protein, consisting of the extracellular domain of the cytotoxic T lymphocyte-associated antigen 4-immunoglobulin CTLA-4, along with the hinge, CH2, and CH3 domains of IgG1. It selectively modulates co-stimulation by interrupting the CD28:CD80/86 pathway, resulting in downregulation of T-cell activation (1-3). Presentation of antigen to the T-cell receptor without co-stimulation results in antigen-specific unresponsiveness upon re-challenge, a phenomenon known as anergy *in vitro* and tolerance *in vivo*. **Use in Therapy:** Abatacept is used to treat arthritic inflammation (*e.g.*, pain, swelling and prolonged joint stiffness). Abatacept is a second-line drug that is mainly used, when other anti-rheumatic drugs, such as methotrexate, fail to control inflammation. When used in combination with methotrexate, abatacept prevents the progression of joint damage and improves physical function in rheumatoid arthritis patients. **1.** Allison (2005) *Issues Emerg. Health Technol.* (73), 1. **2.** Kremer (2005) *J. Clin. Rheumatol.* **11** (3 Suppl.), S55. **3.** Teng, Turkiewicz & Moreland (2005) *Expert Opin. Biol. Ther.* **5**, 1245.

Abciximab

This antithrombotic (MW = 47455 g/mol; CAS 143653-53-6; also known as Reopro®) consists of an immunoglobulin G$_1$ Fab fragment from a chimeric human-murine monoclonal antibody directed against the human platelet glycoprotein IIb/IIIa receptor (1,2). Abciximab also facilitates the dispersal of newly formed platelet aggregates *in vitro*, by partially displacing fibrinogen from activated GPIIb/IIIareceptors. *In vivo*, abciximab may destabilize coronary thrombi by preventing aggregate formation and dispersing mural thrombi (3). **Target(s):** coagulation factor Xa, in presence of heparin (1); glycoprotein IIb/IIIa receptor (1,2). **1.** Li, Spencer, Ball & Becker (2000) *J. Thromb. Thrombolysis* **10**, 69. **2.** Knight, Wagner, Jordan, *et al.* (1995) *Mol. Immunol.* **32**, 1271. **3.** Marciniak, Mascelli, Furman, *et al.* (2002) *Thrombosis & Haemostasis* **87**, 928.

Abemaciclib

This oral cell cycle inhibitor (FW$_{free-base}$ = 506.61 g/mol; FW$_{mesylate-salt}$ = 602.70 g/mol; CAS 1231930-82-7 (mesylate salt)), also known as LY2835219 and *N*-[5-[(4-ethyl-1-piperazinyl)methyl]-2-pyridinyl]-5-fluoro-4-[4-fluoro-2-methyl-1-(1-methylethyl)-1*H*-benzimidazol-6-yl]-2-pyrimidinamine, targets the cyclin-dependent kinase CDK4, *or* cyclin D$_1$ (IC$_{50}$ = 2 nM) and CDK6, *or* cyclin D$_3$ (IC$_{50}$ = 6 nM), inhibiting retinoblastoma (Rb) protein phosphorylation in early G$_1$, thereby arresting the cell cycle in the G$_1$, suppressing DNA synthesis, and inhibiting cancer cell growth (1). LY2835219 inhibits activation of AKT and ERK, but not mTOR (2). **1.** Sanchez-Martinez, Gelbert & Shannon (2011) *Molec. Cancer Therapeut.* **10**, (*Supplement* 1) Abstract Number B234. **2.** Ku, Yi, Koh, *et al.* (2016) *Oncotarget* **12**, 14803.

Abietamide

This diterpenoid amide (FW = 301.47 g/mol; CAS 40713-30-2) inhibits [pyruvate dehydrogenase (acetyl-transferring)] kinase, IC$_{50}$ = 6.7 μM. Abietamide, its oxime, and dehydroabietyl amine-2 exhibit a blood glucose lowering effect in the diabetic *ob/ob* mouse after a single oral dose of 100 micromol/kg, but the mechanism of this effect is unlikely to be related to PDK inhibition. **1**. Aicher, Damon, Koletar, *et al.* (1999) *Bioorgan. Med. Chem. Lett.* **9**, 2223.

Abietic Acid

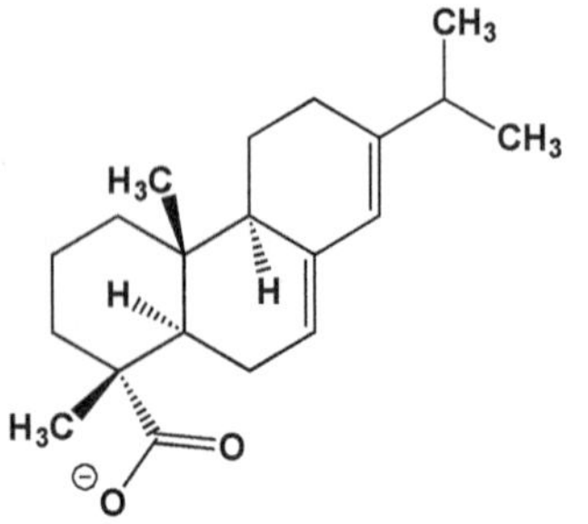

This diterpenoid resin acid (FW$_{free-acid}$ = 302.46 g/mol; CAS 514-10-3; M.P. = 172-175°C), a major component of rosin, is also present in a many plants and amber. Abietic acid has a low solubility in water and is soluble in organic solvents. Note: Abietic acid readily isomerizes to neoabietic acid. **Target(s):** 5-lipoxygenase (1); myeloperoxidase (2); [pyruvate dehydrogenase (acetyl transferring)] kinase, IC$_{50}$ = 20 μM (3). **1**. Ulusu, Ercil, Sakar & Tezcan (2002) *Phytother. Res.* **16**, 88. **2**. Fernandez, Tornos, Garcia, *et al.* (2001) *J. Pharm. Pharmacol.* **53**, 867. **3**. Aicher, Damon, Koletar, *et al.* (1999) *Bioorgan. Med. Chem. Lett.* **9**, 2223.

Abiraterone Acetate

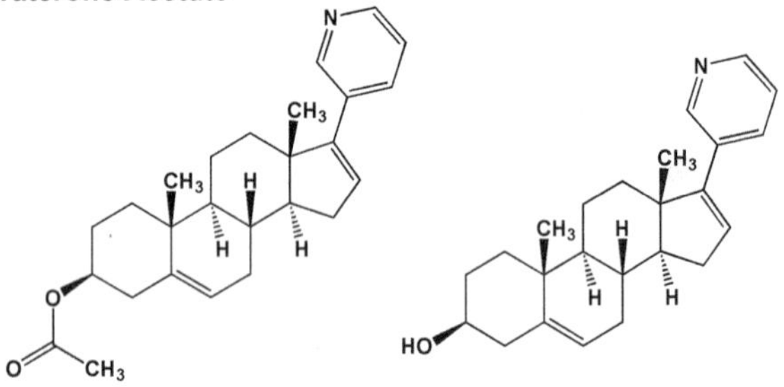

This steroid-based antineoplastic pro-drug (FW = 391.55 g/mol; CAS 154229-18-2), also known as Zytiga™, CB-7630, and systematically as (3β)-17-(pyridin-3-yl)androsta-5,16-dien-3-ol, is converted *in vivo* to abiraterone (FW = 349.51 g/mol; CAS 154229-19-3), an androgen biosynthesis inhibitor that targets 17-α-hydroxylase/C17,20-lyase, *or* CYP17 (K_i = 2 nM), an enzyme expressed in testicular, adrenal, and prostatic tumor tissues and required for androgen biosynthesis (1-3). CYP17A1 catalyzes the conversion of pregnenolone (and progesterone) to 17-α-hydroxy derivatives *via* its 17 α-hydroxylase activity, as well as the subsequent formation of dehydroepiandrosterone, or DHEA, (and androstenedione) *via* its C17,20 lyase activity. Zytiga is effective for the treatment of castration-resistant prostate cancer (previously known as hormone-resistant prostate cancer). A sensitive and selective LC-MS/MS method is available for the quantification of abiraterone acetate and abiraterone (4). The major circulating metabolites of abiraterone in humans are abiraterone sulfate and the *N*-oxide of abiraterone sulfate. Each of these metabolites occur at >100 times the respective concentration of abiraterone. The sulfate metabolite is approximately 7x more potent than abiraterone against rat testicular microsomal CYP17 (5). **1**. Barrie, Potter, Goddard, *et al.* (1994) *J. Steroid Biochem. Mol. Biol.* **50**, 267. **2**. Haidar, Ehmer, Barassin, Batzl-Hartmann, and Hartmann (2003) *J. Steroid Biochem. Mol. Biol.* **84**, 555. **3**. O'Donnell, Judson, Dowsett, *et al.* (2004) *Br. J. Cancer* **90**, 2317. **4**. Martins, Asad, Wilsher & Raynaud (2006) *J. Chromatogr. B Analyt. Technol. Biomed. Life Sci.* **843**, 262. **5**. *Australian Public Assessment Report for Abiraterone acetate (Zytiga)* (2012) Sponsor: Janssen-Cilag, Ltd.

ABM-1/2 Inhibitors of Actin-Based Motors

ABM-1 Inhibitor: FEFPPPPTDEC

ABM-2 Inhibitor: GPPPPPGPPPPPGPPPPP

These actin-based motility inhibitors (***See also*** *ActA FEFPPPPTDE Fragment; VASP GPPPPP Trimer*) are based respectively on (a) the consensus oligoproline docking sites within *Listeria monocytogenes* ActA surface protein, which binds to human platelet vasodilator-stimulated phospho-protein (VASP) and related proteins containing an EVH1 pleckstrin homology-like domain (1,2) and (b) profilin-binding domains in Ena/VASP proteins. Analysis of known actin regulatory proteins led to the identification of distinct Actin-Based Motility homology sequences:

Consensus ABM-1 Sequence: (D/E)FPPPPX(D/E) [where X = P or T]

Consensus ABM-2 Sequence: XPPPPP [where X = G, A, L, P, or S]

Actin-based motility involves a cascade of binding interactions designed to assemble actin regulatory proteins into functional locomotory units. *Listeria* ActA surface protein contains a series of nearly identical EFPPPPTDE-type oligoproline sequences for binding vasodilator-stimulated phosphoprotein (VASP). The latter, a tetrameric protein with 20-24 GPPPPP docking sites, binds numerous molecules of profilin, a 15-kDa regulatory protein known to promote actin filament assembly. Laine *et al.* (4) demonstrated that proteolysis of the focal contact component vinculin unmasks an ActA homologue for actin-based *Shigella* motility. The ABM-1 sequence (PDFPPPPPDL) is located at or near the C-terminus of the p90 proteolytic fragment of vinculin. Unmasking of this site serves as a molecular switch that initiates assembly of an actin based motility complex containing VASP and profilin. Another focal adhesion protein zyxin (4) contains several ABM-1 homology sequences that are also functionally active in reorganizing the actin cytoskeletal network (1-6). **FPPPP-mito**: Actin-based motility of lammellipodia and filopodia may be suppressed by transfecting target cells with a vector containing a cDNA coding for a chimeric protein possessing a EVH1 ligand as well as a mitochondria-targeting sequence (7). In this case Ena/VASP proteins are misdirected from their sites at the leading edge to motilityidling sites on the outer surface of mitochondria. **FPPPP-CAAX**: Actin-based motility of lammellipodia and filopodia may also be suppressed by transfecting target cells with a vector containing a cDNA coding for a chimeric protein possessing an EVH1 ligand as well as the CAAX peripheral membrane-targeting sequence (7). The latter is modified by farnesyltransferase (*Reaction*: Farnesyl-PP + Protein-CAAX ⇌ Protein-C(*S*-Farnesyl)AAX + PP$_i$). In this way, Ena/VASP proteins are misdirected from their sites at the leading edge to motility-idling sites on the inner surface of the peripheral membrane. **1**. Purich (2016) *Biophys. Chem.* **209**, 41. **2**. Purich & Southwick (1997) *Biochem. Biophys. Res. Comm.* **231**, 686. **3**. Southwick & Purich (1996) *New Engl. J. Med.* **334**, 770. **4**. Laine, Zeile, Kang, Purich & Southwick (1997) *J. Cell Biol.* **138**, 1255. **5**. Golsteyn, Beckerle, Koay & Friedrich (1997) *J. Cell Sci.* **110**, 1893. **6**. Purich & Southwick (1999) *Mol. Cell Biol. Res. Commun.* **1**, 176. 7. Bear, Loureiro, Libova, *et al.* (2000) Cell **102**, 717.

abn-cbd

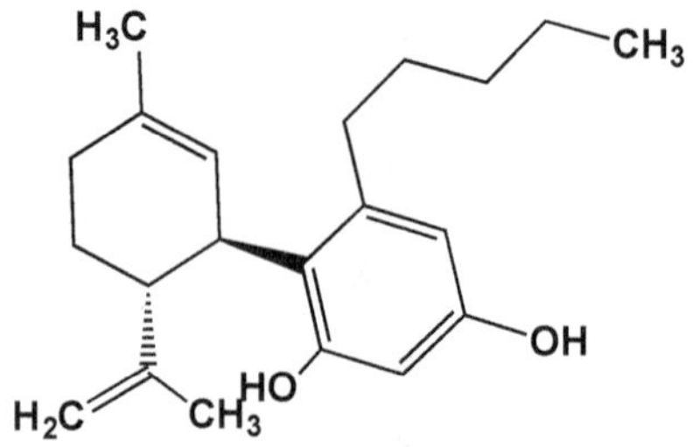

This synthetic cannabidiol regioisomer (FW = 314.47 g/mol; CAS 22972-55-0), also called Abnormal Cannabidiol and 4-[(1*R*,6*R*)-3-methyl-6-(1-methylethenyl)-2-cyclohexen-1-yl]-5-pentyl-1,3-benzenediol, produces vasodilator effects, lowering blood pressure and inducing cell migration and proliferation, as well as mitogen-activated protein kinase activation in microglia, without psychoactive effects (1-4). Abnormal Cannabidiol binds to a site (possibly the "orphan" receptor GPR18) that is distinct from the CB$_1$ and CB$_2$ receptors. Abnormal Cannabidiol relaxes human pulmonary arteries in an endothelium-independent and endothelium-dependent manner. The latter component is probably mediated via the putative endothelial cannabinoid receptor, activation of which may release endothelium-derived hyperpolarizing factor, which in turn acts via calcium-activated potassium channels. Abnormal cannabidiol is behaviorally inactive; it may have a therapeutic implication in vascular diseases, especially in the treatment of pulmonary hypertension (5). 1. Jarai, *et al.*

(1999) *Proc. Natl.Acad. Sci. U.S.A.* **96**, 14136. **2**. Ho & Hiley (2003) *Brit. J. Pharmacol.* **138**, 1320. **3**. Mo, *et al.* (2004) *Eur. J. Pharmacol.* **489**, 21. **4**. Ryberg, *et al.* (2007) *Brit. J. Pharmacol.* **152**, 1092. **4**. Kozłowska, Baranowska, Schlicker, *et al.* (2007) *J. Hypertens.* **25**, 2240.

ABP 688

This high-affinity mGluR antagonist (FW = 240.31 g/mol; CAS 924298-51-1; Soluble to 100 mM in DMSO; 50 mM in 1 eq. HCl), also named 3-[2-(6-methyl-2-pyridinyl)ethynyl]-2-cyclohexene-1-one *O*-methyloxime, targets the human metabolotrophic glutamate receptor-5 (mGlu$_5$) (K_i = 1.7 nM) and inhibits glutamate-induced calcium release from Ltk-cells expressing human mGlu$_5$ receptors (IC$_{50}$ = 2.3 nM) (1). Indeed, [^{11}C]-ABP688 is a useful PET tracer for *in vivo* imaging of the mGluR5 in rodents (2). **1**. Hintermann, Vranesic, Allgeier, *et al.* (2007) *Bioorg. Med. Chem.* **15**, 903. **2**. Ametamey, Kessler, Honer, *et al.* (2006) *J. Nucl. Med.* **47**, 698.

Abrin

This type-II ribosome-inactivating glycoprotein toxin (MW = 65 kDa; CAS 1393-62-0) from the seeds of *Abrus precatorius* (common names: rosary pea and jequirity) is composed of an acidic A chain and a neutral B chain joined by disulfide bonds. The A chain exhibits the toxic rRNA *N*-glycosidase activity, catalyzing the endohydrolysis of a specific *N*-glycosidic bond at an adenosine on rRNA, thereby irreversibly inactivating protein synthesis (**See** *Ricin*). The B chain has lectin-like properties that assist A chain entry into cells. Chewing jequirity seeds has proved fatal, because each molecule of abrin can inactivate as many as 1,500 ribosomes per second (~75x more active than ricin). *Note*: Abrin should not be confused with abrine, or N^α-methyl-L-tryptophan. **Mode of Apoptotic Action:** The abrin A chain also promotes apoptosis, which is preceded by mitochondrial cytochrome *c* release, followed by caspase-3 and -9 activation (4). The broad-spectrum caspase inhibitor, benzyloxycarbonyl-Val-Ala-Asp-fluoromethyl ketone (zVADfmk), prevents abrin-triggered caspase activation and reduces apoptotic cell death, but is without effect on cytochrome *c* release. Such results suggest that the sequential caspase activation is crucial for abrin-induced apoptosis (4). Abrin also inhibits the mitochondrial antioxidant protein AOP-1 (5). Recent studies suggest that abrin-induced apoptosis depends on p38 MAPK (but not JNK), attended by activation of caspase-2 and caspase-8, triggering Bid cleavage and and to subsequent mitochondrial membrane potential loss (6). Such a pathway connects the signaling events from ER stress to mitochondrial death machinery. **Target(s):** protein biosynthesis (1-3); ribosome (1-3); antioxidant protein AOP-1 (4). **1**. Jiménez (1976) *Trends Biochem. Sci.* **1**, 28. **2**. Benson, Olnes, Phil, Skorve & Abraham (1975) *Eur. J. Biochem.* **59**, 573. **3**. Olsnes, Fernandez-Puentes, Carrasco & Vazquez (1975) *Eur. J. Biochem.* **60**, 281. **4**. Qu and Qing (2004) *J. Biochem. and Molec. Biol.* **37**, 445. **5**. Shih, Wu, Hung, Yang & Lin (2001) *J. Biol. Chem.* **276**, 21870. **6**. Mishra & Karande (2014) **9**, e92586.

cis,trans-Abscisate (Abscisic Acid)

This plant metabolite (FW$_{\text{free-acid}}$ = 264.32 g/mol; CAS 14375-45-2; M.P. 183-185 °C; Soluble in DMSO; *Abbreviation*: ABA), known also as (+/-)-abscisic acid and named systematically as (2-*cis*,4-*trans*)-5-(1-hydroxy-2,6,6-trimethyl-4-oxo-2-cyclohexen-1-yl)-3-methyl-2,4-pentadienoic acid, is a component of a abscisoyl-β-D-glucopyranoside, which plays a profound role in plant growth and development, promotes abscission, and regulates seed maturation, germination, and adaptation to environmental stresses (*e.g.*, dehydration, high salt concentrations, and low temperature) (1-5). ABA is found in most vascular plants, green algae, certain fungi, and mosses. Note that *trans,trans* abscisic acid, which is also commercially available, has been reported in low concentrations in a few organisms. ABA also inhibits DNA replication in the rice shoot apical meristem and

inhibits shoot growth in rice by inducing expression of OsKRP4, OsKRP5, and OsKRP6, which collectively inhibit the G$_1$/S transition (6). Salicylate blocks these effects, overcoming ABA-induced cell-cycle arrest by blocking expression of OsKRP genes. **Target(s):** coleoptile growth (1); seed germination (1,4,5); inward K$^+$ currents (2); indole-3 acetate β-glucosyltransferase, weakly inhibited (3). **1**. Grill & Himmelbach (1998) *Curr. Opin. Plant Biol.* **1**, 412. **2**. Schwartz, Wu, Tucker & Assmann (1994) *Proc. Natl. Acad. Sci. U.S.A.* **91**, 4019. **3**. Leznicki & R. S. Bandurski (1988) *Plant Physiol.* **88**, 1481 **4**. Cutler, Rose, Squires, *et al.* (2000) *Biochemistry* **39**, 13614. **5**. Ueno, Araki, Hirai, *et al.* (2005) *Bioorg. Med. Chem.* **13**, 3359. **6**. Meguro & Sato (2014) *Sci. Rep.* **4**, 4555.

Abscisic Aldehyde

This aldehyde (FW = 248.32 g/mol; CAS 41944-86-9) is the immediate biosynthetic precursor of abscisic acid. The short-chain alcohol dehydrogenase ABA2 catalyzes the conversion of xanthoxin to abscisic aldehyde, with the latter undergoing selective oxidation (*Abscisic Aldehyde Oxygenase Reaction*: Abscisic aldehyde + O$_2$ + H$_2$O $\rightleftharpoons$ Abscissic Acid + H$_2$O$_2$). **Target(s):** (+)-abscisate 8'-hydroxylase, mildly inhibitory. **1**. Ueno, Araki, Hirai, *et al.* (2005) *Bioorg. Med. Chem.* **13**, 3359.

ABT

This cytochrome P450 suicide inhibitor (FW = 134.14 g/mol; CAS 1614-12-6; Soluble to 100 mM in water and to 100 mM in DMSO), also named 1*H*-benzotriazol-1-amine, inhibits the synthesis of eicosanoid 20-hydroxyeicosa-tetraenoate (*or* 20-HETE). ABT reduces intimal hyperplasia and vascular remodeling following endothelial injury in rat carotid arteries (1,2). Microsomal benzphetamine *N*-demethylase (BND) and 7-ethoxyresorufin O-deethylation (ERF) activities (catalyzed by P-450 isozymes 2 and 6, respectively) and specific P-450 content were determined after incubation with ABT (1). In vitro destruction of P-450 was dependent on ABT concentration and required NADPH. Significant losses in both activities are observed only at ABT concentrations >10 μM (1). **1**. Mathews, Dostal & Bend (1985). *J. Pharmacol. Exp. Ther.* **235**, 186. **2**. Orozco, Liu, Perkins, *et al.* (2013) *J Pharmacol. Exp. Ther.* **346**, 67.

ABT-199, See *Venetoclax*

ABT-263, See *Navitoclax*

ABT-267, See *Ombitasvir*

ABT-299, See *A-85783*

ABT-378, See *Lopinavir*

ABT-450, See *Paritaprevir*

ABT-491

This substituted indole (FW$_{\text{free-base}}$ = 479.52 g/mol; IUPAC: 4-ethynyl-*N,N*-dimethyl-3-[3-fluoro-4-[(2-methyl-1*H*-imidazo[4,5-*c*]pyridin-1-yl)methyl].benzoyl]-1*H*-indole-1-carboxamide, is a strong human platelet-activating factor receptor antagonist, K_i = 0.6 nM (1-3). Its binding kinetics to the

PAF receptor is consistent with a slower off-rate for the antagonist compared to PAF. Inhibition of PAF binding is selective and correlates with functional antagonism of PAF-mediated cellular responses (*e.g.*, Ca^{2+} mobilization, priming, and degranulation). Administration of ABT-491 *in vivo* results in potent inhibition of PAF-induced inflammatory responses (*e.g.*, increased vascular permeability, hypotension, and edema) and PAF-induced lethality. Oral potency (ED_{50}) was between 0.03 and 0.4 mg/kg in rat, mouse, and guinea pig. When administered intravenously in these species, ABT-491 exhibited ED_{50} values between 0.005 and 0.016 mg/kg. In the rat, an oral dose of 0.5 mg/kg provides > 50% protection for 8 hours against a cutaneous PAF challenge. When administered orally, ABT-491 is also effective in inhibiting lipopolysaccharide-induced hypotension ($ED_{50} =$ 0.04 mg/kg), gastrointestinal damage (0.05 mg/kg, 79% inhibition), and lethality (1 mg/kg, 85% vs. 57% survival). **1**. Albert, Magoc, Tapang, *et al.* (1997) *Eur. J. Pharmacol.* **325**, 69. **2**. Albert, Malo, Tapang, *et al.* (1998) *J. Pharmacol. Exp. Ther.* **284**, 83. **3**. Curtin, Davidsen, Heyman, *et al.* (1998) *J. Med. Chem.* **41**, 74.

ABT-538, *See* Ritonavir

ABT-578, *See* Zotarolimus

ABT-702

This potent and orally effective adenine-like non-nucleoside inhibitor ($FW_{free-base} =$ 463.34 g/mol; CAS 214697-26-4), also known as 4-amino-5-(3-bromophenyl)-7-(6-morpholinopyridin 3-yl)pyrido[2,3-*d*]pyrimidine), targets cytosolic adenosine kinase ($IC_{50} =$ 1.7 nM), thereby increasing adenosine and providing anti-seizure activity akin to that of adenosine receptor agonists, but with fewer dose-limiting cardiac side effects (1-5). ABT-702 shows analgesic effect in animal models of pain via nociceptive and motor reflex pathways in the isolated spinal cord of neonatal rats (6). It inhibits slow Ventral Root Potentials (sVRPs) in the nociceptive pathway more potently than monosynaptic reflex potentials (MSRs) in the motor reflex pathway. ABT-702's inhibitory effects are mimicked by exogenously applied adenosine, blocked by the adenosine A_1 receptor antagonist 8-cyclopentyl-1,3-dipropylxanthine, and augmented by the adenosine deaminase (ADA) inhibitor EHNA (*erythro*-9-(2-hydroxy-3-nonyl) adenine). Equilibrative nucleoside transporter (ENT) inhibitors reverse the effects of ABT-702, but not those of adenosine. ABT-702 stimulates adenosine release from the spinal cord, an effect that is also reversed by ENT inhibitors. ABT-702-facilitates release of adenosine by way of ENTs inhibits nociceptive pathways more potently than motor reflex pathways in the spinal cord via activation of A_1 receptors (6). **1**. Jarvis, Yu, Kohlhaas, *et al.* (2000) *J. Pharmacol. Exp. Ther.* **295**, 1156. **2**. Lee, Jiang, Cowart, *et al.* (2001) *J. Med. Chem.* **44**, 2133. **3**. Matulenko, Paight, Frey, *et al.* (2007) *Bioorg. Med. Chem.* **15**, 1586. **4**. Kowaluk, Mikusa, Wismer, *et al.* (2000) *J. Pharmacol. Exp. Ther.* **295**, 1165. **5**. Iqbal, Burbiel & Müller (2006) *Electrophoresis* **27**, 2505. **6**. Otsuguro, Tomonari, Otsuka, *et al.* (2015) *Neuropharmacol.* **97**, 160.

ABT-719

This first-in-class 2-pyridone-containing bacterial antibiotic (FW = 345.37 g/mol), also named 8-(3-(*S*)-aminopyrrolidin-1-yl)-1-cyclopropyl-7-fluoro-

9-methyl-4*H*-4-oxoquinolizine-3-carboxylic acid, inhibits DNA topoisomerase II. ABT-719 was more effective than ciprofloxacin, sparfloxacin, and clinafloxacin against Gram-positive bacteria. It was particularly active against *Staphylococcus aureus* (MIC_{90} = 0.015 μg/mL) and *Streptococcus pneumoniae* (MIC_{90} = 0.03 μg/mL). ABT-719 was also highly active of the compounds tested against ciprofloxacin-resistant *S. aureus* isolates (MIC_{90} = 0.25 μg/mL), compared with 64 μg/mL for ciprofloxacin. Against Gram-negative organisms, ABT-719 was as active as, or slightly more active, than ciprofloxacin and was the most active compound against ciprofloxacin-resistant *Pseudomonas aeruginosa* (MIC_{90} = 2 μg/mL). ABT-719 was also the most active compound against both Gram-positive and Gram-negative anaerobes, with MIC_{90} values ranging from 0.12 to 0.25 μg/mL. **1**. Flamm, Vojtko, Chu, *et al.* (1995) *Antimicrob. Agents Chemother.* **39**, 964.

ABT-737

This small-molecule inhibitor (F.W. = 813.43; CAS 852808-04-9; Solubility: 160 mg/mL DMSO, <1 mg/mL H_2O), also known as 4-[4-[(4'-chloro[1,1'-biphenyl]-2-yl)methyl]-1-piperazinyl]-*N*-[[4-[[(1*R*)-3-(dimethyl amino)-1-[(phenylthio)methyl]propyl]amino]-3-nitrophenyl]sulfonyl] benzamide, is a BH3 (Bcl-2 homology domain) mimetic that targets the apoptosis regulatory protein Bcl-xL (B-cell lymphoma-extra large), Bcl-2 (B-cell lymphoma 2) and pro-survival protein Bcl-w with EC_{50} values of 79 nM, 30 nM and 198 nM, respectively (1-7). Significantly, ABT-737 induces autophagy through a BAX- and BAK1-independent mechanism that is likely disrupt BECN1 binding to antiapoptotic BCL2 family members (8). When administered along with Sorafenib (an inhibitor of VEGFR and PDGFR tyrosine protein kinases), ABT-737 induces apoptosis of targeted tumor cells. ABT-737 sensitizes androgen-dependent castration-resistant prostate cancer PC3 cells to ENMD-1198-mediated, caspase-dependent apoptosis (9). While showing demonstrated efficacy in several forms of hematological and non-hematological cancers, ABT-737's efficacy is diminished in cancer cells producing high levels of Mcl-1, another anti-apoptotic member of the Bcl-2 family (10). When used in combination with the Mcl-1-selective inhibitor maritoclax (***See*** *Maritoclax*), however, ABT-737 greatly decreases the survival of human melanoma cells (10). **1**. Konopleva, *et al.* (2006) *Cancer Cell* **10**, 375. **2**. van Delft, *et al.* (2006) *Cancer Cell* **10**, 389. **3**. Del Gaizo Moore, *et al.* (2007) *J. Clin. Invest.* **117**, 112. **4**. Tahir, *et al.* (2007) *Cancer Res.* **67**, 1176. **5**. Ishitsuka, *et al.* (2012) *Cancer Lett.* **317**, 218. **6**. Hikita, *et al.* (2010) *Hepatology* **52**, 1310-21. **7**. Whitecross, *et al.* (2009) *Blood* **113**, 1982. **8**. Pedro, Wei, Sica, *et al.* (2015) *Autophagy* **11**, 452. **9**. Parrondo, de Las Pozas, Reiner & Perez-Stable (2013) *Peer J.* **1**, e144. **10**. Pandey, Gowda, Doi, *et al.* (2013) *PLoS One* **8**, e78570.

ABT-761, *See* Atreleuton

ABT 888, *See* Veliparib

ABT-980, *See* Fiduxosin

Abu, *See* 2-Aminobutyric Acid

A2bu, *See* L-2,4-Diaminobutyric Acid

Ac2-12

This annexin/lipocortin-1 peptidomimetic (FW = 1351.58 g/mol; Sequence: Acetyl-AMVSEFLKQAW; CAS 256447-08-2; Soluble to 1 mg/mL in 20% (vol/vol) acetonitrile-H_2O) markedly reduces the degree of neutrophil adhesion and emigration across mouse mesenteric postcapillary venules, when administered at 13 mg/kg (s.c.), but does not modify cell rolling, when assessed directly by intravital microscopy (1). Ac2-12 also promotes neutrophil detachment from mesenteric endothelium after activation by intraperitoneal injection of 1 mg zymosan in mice (*i.e.*, zymosan peritonitis). Ac2-12 phenocopies the effects observed in Formylated

Peptide Receptor (FPR) knock-out mice (2). **1**. Lim, Solito, Russo-Marie, Flower & Perretti (1998) *Proc. Natl. Acad. Sci. U.S.A.* **95**, 14535. **2**. Perretti, Getting, Solito, Murphy & Gao (2001) *Am. J. Pathol.* **158**, 1969.

AC220, *See Quizartinib*

Ac2-26

This anti-inflammatory agent (FW = 3089.46 g/mol; Acetyl-AMVSEF LKQAWFIENEEQEYVQTVK), corresponding to N-terminal residues 2-26 of steroid-inducible lipocortin-1; CAS 151988-33-9; Soluble to 1 mg/mL in PBS Buffer) inhibits leukocyte extravasation, reducing neutrophil adhesion and emigration and promoting neutrophil detachment from activated mesenteric endothelium in mice *in vivo* (1). Such observations show that peptides derived from LC1 mimic the activity of the full-length molecule, pointing the way for development of new anti-inflammatory inhibitors of leukocyte trafficking. **1**. Perretti, Ahluwalia, Harris, Goulding & Flower (1993) *J. Immunol.* **151**, 4306. **2**. Lim, Solito, Russo-Marie, Flower & Perretti (1998) *Proc. Natl. Acad. Sci. U.S.A.* **95**, 14535.

AC-4

This photoswitchable TRPV1 channel blocker (FW = 514.56 g/mol; CAS 1459809-09-6; Solubility: 100 mM in DMSO; 20 mM in H_2O, with gentle warming), also known as 1,3,4,5-tetrahydro-7,8-dihydro-*N*-[2-[4-[2-[4-(trifluoromethyl)phenyl]diazenyl]phenyl]ethyl]-2*H*-2-benzazepine-2-carbothioamide, acts an antagonist of <u>T</u>ransient <u>R</u>eceptor <u>P</u>otential cation channel, subfamily <u>V</u> member <u>1</u>, when in its *trans*-conformation during voltage-gated activation, but as an antagonist when in the *cis*-formation on capsaicin-induced TRPV1 currents. (AC4 switches conformation from *cis* to *trans* at 440 nm and *trans* to *cis* at 360 nm.) TRPV1 (also known as the Vanilloid Receptor 1) is a nonselective cation channel that is permeable to Ca^{2+} ions, but shows little discrimination between mono- and divalent cations. It is abundantly expressed in nociceptive neurons (*e.g.*, dorsal root ganglion (DRG), trigeminal ganglion (TG), as well as spinal and peripheral nerve terminals and cornea. This channel is located in the plasma membrane but occasionally is found on sarco-/endo-plasmic reticulum membranes, where it may function in intracellular Ca^{2+} release. TRPV1 activation gives rise to burning and pain sensations, making it a target for developing analgesics. TRPV1 is activated by several chemical and physical stimuli (*e.g.*, voltage, heat, capsaicin, spider toxins, low pH, and the endocannabinoid anandamide). AC4 acts as a cis-antagonist of CAP-induced TRPV1 currents but becomes a trans-antagonist upon voltage activation, in good agreement with the observation that agonists interact with TRPV1 channels in a state-dependent fashion and that specific antagonists differ in their ability to block distinct agonists. **1**. Stein, Breit, Fehrentz, Gudermann & Trauner (2013) *Angew. Chem. Int. Ed. Engl.* **52**, 9845.

AC 187

This orally active receptor antagonist (FW = 2890.25 g/mol; Sequence: Acetyl-VLGKLSQELHKLQTYPRTNTGSNTY-NH$_2$; CAS 151804-77-2) targets (IC$_{50}$ = 0.5 nM) receptors for amylin, the 37-residue hormone (Sequence: KCNTATCATQRLANFLVHSSNNFGAILSSTNVGSNTY, with a single disulfide bridge) co-secreted with the pancreatic β-cells with insulin (1-3). Amylin slows gastric emptying, thereby promoting satiety and preventing post-prandial spikes in blood glucose levels. Amylin is postulated to act as a hormonal signal from pancreas to brain, inhibiting food intake and regulating energy reserves by reducing food intake, body weight, and adiposity, when administered systemically or into the brain (2). AC-187 displays 38x and 400x selectivity over calcitonin and CGRP receptors, respectively. By attenuating the activation of initiator and effector caspases, AC-187 also blocks amyloid β-induced neurotoxicity *in vitro*. Other effects include increasing glucagon secretion, accelerating gastric emptying, altering plasma glucose levels, and increasing food intake *in vivo*. **1**. Jhamandas & MacTavish (2004) *J. Neurosci.* **24**, 5579. **2**. Reidelberger, Haver, Arnelo, *et al.* (2004) *Am. J. Physiol. Inter. Comp. Physiol.* **287**, R568. **3**. Gedulin, Jodka, Herrmann & Young. (2006) *Regul. Pept.* **137**, 121.

AC 222164

This herbicide (FW = 260.29 g/mol) uncompetitively inhibits acetolactate synthase, K_i = 1.7 μM (1) as well as branched-chain amino acids (2). Often used as an herbicide in the cultivation of soybeans, Scepter® has been known to inhibit the growth of corn, and this herbicide is surprisingly long-lived in event of drought. **1**. Shaner, Anderson & Stidham (1984) *Plant Physiol.* **76**, 545. **2**. Hawkes, Howard & Pontin (1990) in *Herbicides and Plant Metabolism* (Dodge, ed.), p. 313, Soc. Exp. Biol. Sem. Ser., Cambridge University Press, UK.

ACA, *See* 2-(p-Amylcinnamoyl)-amino-p-chlorobenzoate

AC 243997, *See* Imazapyr

AC 252214, *See* Imazaquin

Acacetin

This naturally occurring flavone (FW = 286.27 g/mol; CAS 480-44-4), also known as 5,7-dihydroxy-4'-methoxyflavone and 7-*O*-methylapigenin, and systematically named as 5,7-dihydroxy-2-(4-methoxyphenyl)-chromen-4-one, from the black locust *Robinia pseudoacacia* is the aglycon of linarin and acaciin and is biosynthesized by apigenin 4'-O-methyltransferase from *S*-adenosyl-methionine and 5,7,4'-trihydroxyflavone (apigenin), yielding *S*-adenosylhomocysteine and acacetin. The chemical synthesis of acacetin was accomplished by Robert Robinson, who was awarded the 1947 Nobel Prize in Chemistry for his work on alkaloids and organic synthesis. **Target(s):** glutathione *S*-transferase (1); xanthine oxidase, K_i = 0.11 μM (2); CYP1A (3); CYP1B1 (3); glutathione-disulfide reductase (4); DNA topoisomerase I (5); [myosin light-chain] kinase (6); and protein-tyrosine kinase, *or* non-specific protein-tyrosine kinase (7). **1**. Zhang & Das (1994) *Biochem. Pharmacol.* **47**, 2063. **2**. Nguyen, Awale, Tezuka, Ueda, Tran & Kadota (2006) *Planta Med.* **72**, 46. **3**. Doostdar, Burke & Mayer (2000) *Toxicology* **144**, 31. **4**. Zhang, Yang, Tang, Wong & Mack (1997) *Biochem. Pharmacol.* **54**, 1047. **5**. Boege, Straub, Kehr, *et al.* (1996) *J. Biol. Chem.* **271**, 2262. **6**. Jinsart, Ternai & Polya (1991) *Biol. Chem. Hoppe-Seyler* **372**, 819. **7**. Hollósy & Kéri (2004) *Curr. Med. Chem. Anticancer Agents* **4**, 173.

Acadesine

This dephosphorylated form of a key purine nucleotide biosynthetic intermediate (FW = 258.23 g/mol; CAS 2627-69-2), also known as 5-aminoimidazole-4-carboxamide 1-β-D-ribofuranoside and 5-amino-1-((2R,3R,4S,5R)-3,4-dihydroxy-5-(hydroxymethyl)tetrahydro-furan-2-yl)-1H-imidazole-4-carboxamide, is phosphorylated to aminoimidazole carboxamide ribonucleotide (or AICAR, or ZMP), which mimics AMP's activating effect on AMP-dependent protein kinase, or AMPK (*See also A-769662*). **Tool for Exploring AMPK Signal Transduction Pathways:** The latter is believed to protect cells against environmental stress (*e.g.* heat shock) by switching off biosynthetic pathways, the key signal being elevation of AMP. Identification of novel targets for the kinase cascade are thus facilitated by development of acadesine as a specific agent for activating the kinase in intact cells (1). ZMP mimics both activating effects of AMP on AMPK (*i.e.*, by direct allosteric activation and by promotion of phosphorylation by AMPK kinase). Unlike existing methods for activating AMPK in intact cells (*e.g.* fructose, heat shock), AICAR does not perturb the cellular contents of ATP, ADP or AMP. Acadesine (500 μM) causes a transient 12x activation of AMPK within 15 minutes in rat hepatocytes and 2-3x activation of AMPK in adipocytes, again without affecting levels of ATP, ADP or AMP. Acadesine activates AMPK and induces apoptosis in B-cell chronic lymphocytic leukemia cells but not in T lymphocytes (2). **Target(s):** At 500 μM, acadesine dramatically inhibits both fatty acid and sterol synthesis. It likewise phosphorylates and inactivates HMG-CoA reductase (1). Although acadesine kills chronic myelogenous leukemia (CML) cells, AMPK knock-down by Sh-RNA fail to prevent the effect of acadesine, indicating an AMPK-independent mechanism (3). Surprisingly, acadesine triggers relocation and activation of several PKC isoforms (3). Acadesine inhibits neutrophil CD11b up-regulation *in vitro* and during *in vivo* cardiopulmonary bypass (4). Other AMPK activators also induce nucleoli re-organization, with attendant changes in cell proliferation. Among the compounds tested, phenformin and resveratrol had the most pronounced impact on nucleolar organization (5). Notably, AICA riboside uptake into cells, as mediated by the adenosine transport system, is blocked by a number of protein kinase inhibitors, such that ZMP does not accumulate to sufficient levels to stimulate AMPK (6). Such findings demonstrate that careful interpretation is required when using AICA riboside in conjunction with protein kinase inhibitors to investigate the physiological role of AMPK. 1. Corton, *et al.* (1995) *Eur. J. Biochem.* **229**, 558. 2. Campàs, *et al.* (2003) *Blood* **101**, 3674. 3. Robert, *et al.* (2009) *PLoS One* **4**, e7889. 4. Mathew, *et al.* (1995) *J. Thorac. Cardiovasc. Surg.* **109**, 448. 5. Kodiha, Salimi, Wang & Stochaj (2014) *PLoS One.* **9**, e88087. 6. Fryer, Parbu-Patel & Carling (2002) *FEBS Lett.* **531**, 189.

Acamprosate

This antimetabolite (FW = 181.21 g/mol; CAS 77337-76-9; Symbol = AC), also known as *N*-acetylhomotaurate, decreases voluntary ethanol drinking in laboratory animals and relapse behavior in human alcoholics. Although glutamatergic mechanisms have been implicated in several recent studies on the action of acamprosate, its underlying mechanism of action remains unknown. While AC appears to bind to *N*-methyl-D-aspartic acid (NMDA) targets (1), acamprosate strongly inhibits the binding of taurine to taurine receptors. It has little effect on the binding of glutamate to glutamate receptors or muscimol to GABA_A receptors (2). AC is neurotoxic, at least in culture, triggering neuronal damage at 1 mM. The underlying mechanism of AC-induced neuronal injury appears to be due to its action in increasing intracellular calcium ion levels (2). 1. Tomek, Lacrosse, Nemirovsky & Olive (2013) *Pharmaceuticals* (Basel) **6**, 251. 2. Wu, Jin, Schloss, *et al.* (2001) *J. Biomed. Sci.* **8**, 96.

Acanthoate (Acanthoic Acid)

This diterpenoid (FW = 302.46 g/mol), also named 1R,4aR,7S, 8aS,10aS)-1,4a,7-trimethyl-7-vinyl-1,2,3,4,4a,6,7,8,8a,9,10,10a-dodecahydrophenanthrene-1-carboxylic acid, from the roots of *Acanthopanax koreanum*, inhibits human protein tyrosine-phosphatase 1B (IC_50 = 24 μM) and cyclooxygenases-1 and -2, with IC_50 values of 116 and 790 μM, respectively. **Target(s):** cyclooxygenases-1 and -2 (1,2); protein-tyrosine-phosphatase (3). 1. Jung, Shim, Suh, *et al.* (2007) *Cancer Sci.* **98**, 1943. 2. Suh, Kim, Park, *et al.* (2001) *Bioorg. Med. Chem. Lett.* **11**, 559. 3. Na, Oh, Kim, *et al.* (2006) *Bioorg. Med. Chem. Lett.* **16**, 3061.

Acarbose

This pseudotetrasaccharide (FW = 645.61 g/mol; CAS 56180-94-0), also named *O*-[4,6-dideoxy-4-[1S-(1α,4α,5β,6α)-4,5,6-trihydroxy-3-hydroxy methyl-2-cyclohexen-1-yl]amino-α-D-glucopyranosyl]-(1→4)-*O*-α-D-gluco pyranosyl-(1→4)-D-glucopyranose, inhibits many α-glucosidases, including pig intestinal sucrase (IC_50 = 0.5 μM). Acarbose also reduces sugar absorption in the gastrointestinal tract as well as the level of circulating glycosylated hemoglobin (*i.e.*, HbA_1c). First marketed in Germany in 1990 as Glucobay™ and in the United States in 1996 as Precose™ for the treatment of Type 2 diabetes, acarbose enables patients to achieve better control of blood sugar levels, while on starch-containing diets. **Target(s):** cyclomaltodextrin glucano-transferase (1,15-18,45-47); *O*-glycosyl glycosidase (2); α-glucosidase, *or* maltase (3,4,8,9,18,29-31,48); α-amylase (3,6,7,10,11, 18,20,36-40); sucrose α-glucosidase, *or* sucrase (2,3,12, 28); glycogen phosphorylase *a* (3,14); glucan 1,4-α-glucosidase, *or* glucoamylase (3,19,20,32-35); 4-α-glucanotransferase (5,42-44); maltodextrin phosphorylase (13,20); exo(1→4)-α-D-glucan lyase, K_i = 20 nM (21-23); maltose-transporting ATPase (24); glucan 1,4-α maltohydrolase (25); cyclomaltodextrinase (26,27); oligosaccharide 4-α-glucosyltransferase (41); amylomaltase (42,44); and maltodextrin phosphorylase, poorly inhibited (49). 1. Nakamura, Haga & Yamane (1993) *Biochemistry* **32**, 6624. 2. Schmidt, Frommer, Junge, *et al.* (1977) *Naturwissenschaften* **64**, 535. 3. Mooser (1992) *The Enzymes*, 3rd ed., **20**, 187. 4. Matsui, Kobayashi, Hayashida & Matsumoto (2002) *Biosci. Biotechnol. Biochem.* **66**, 689. 5. Roujeinikova, Raasch, Sedelnikova, Liebl & Rice (2002) *J. Mol. Biol.* **321**, 149. 6. Talamond, Desseaux, Moreau, Santimone & Marchis-Mouren (2002) *Comp. Biochem. Physiol. B Biochem. Mol. Biol.* **133**, 351. 7. Yoon & Robyt (2003) *Carbohydr. Res.* **338**, 1969. 8. Martin & Montgomery (1996) *Amer. J. Health Syst. Pharm.* **53**, 2277. 9. Walton, Sherif, Noy & Alberti (1979) *Brit. Med. J.* **1**, 220. 10. Mukherjee, Chatterjee & Mukherjee (1982) *Indian J. Biochem. Biophys.* **19**, 288. 11. Burrill, Brannon & Kretchmer (1981) *Anal. Biochem.* **117**, 402. 12. Hanozet, Pircher, Vanni, Oesch & Semenza (1981) *J. Biol. Chem.* **256**, 3703. 13. O'Reilly, Watson & Johnson (1999) *Biochemistry* **38**, 5337. 14. Goldsmith, Fletterick & Withers (1987) *J. Biol. Chem.* **262**, 1449. 15. Strokopytov, Penninga, Rozeboom, *et al.* (1995) *Biochemistry* **34**, 2234. 16. Villette, Helbecque, Albani, Sicard & Bouquelet (1993) *Biotechnol. Appl. Biochem.* **17**, 205. 17. Mosi, Sham, Uitdehaag, Ruiterkamp, Dijkstra & Withers (1998) *Biochemistry* **37**, 17192. 18. Kim, Lee, Lee, *et al.* (1999) *Arch. Biochem. Biophys.* **371**, 277. 19. De Mot, Van Oudendijck & Verachtert (1985) *Antonie Van Leeuwenhoek* **51**, 275. 20. De Mot & Verachtert (1987) *Eur. J. Biochem.* **164**, 643. 21. Lee, Yu & Withers (2002) *J. Amer. Chem. Soc.* **124**, 4948. 22. Yu, Christensen, Kragh, Bojsen & Marcussen (1997) *Biochim. Biophys. Acta* **1339**, 311. 23. Lee, Yu & Withers (2003) *Biochemistry* **42**, 13081. 24. Hulsmann, Lurz, Scheffel & Schneider (2000) *J. Bacteriol.* **182**, 6292. 25. Oh, Kim, Kim, *et al.* (2005) *FEMS Microbiol. Lett.* **252**, 175. 26. Kaulpiboon & Pongsawasdi (2005) *Enzyme Microb. Technol.* **36**, 168. 27. Galvin, Kelly & Fogarty (1994) *Appl. Microbiol. Biotechnol.* **42**, 46. 28. Karley, Ashford, Minto, Pritchard & Douglas (2005) *J. Insect Physiol.* **51**, 1313. 29. Rowe & Margaritis (2004) *Biochem. Eng. J.* **17**, 121. 30. Saha & Zeikus (1991) *Appl. Microbiol. Biotechnol.* **35**, 568. 31. Frandsen,

Lok, Mirgorodskaya, Roepstorff & Svensson (2000) *Plant Physiol.* **123**, 275. **32.** Fierobe, Clarke, Tull & Svensson (1998) *Biochemistry* **37**, 3753. **33.** Sauer, Sigurskjold, Christensen, *et al.* (2000) *Biochim. Biophys. Acta* **1543**, 275. **34.** Ono, Shigeta & Oka (1988) *Agric. Biol. Chem.* **52**, 1707. **35.** Aleshin, Feng, Honzatko & Reilly (2003) *J. Mol. Biol.* **327**, 61. **36.** Nahoum, Roux, Anton, *et al.* (2000) *Biochem. J.* **346**, 201. **37.** Li, Begum, Numao, *et al.* (2005) *Biochemistry* **44**, 3347. **38.** Kang, Lee, Kim, *et al.* (2004) *FEMS Microbiol. Lett.* **233**, 53. **39.** Funke & Melzig (2005) *Pharmazie* **60**, 796. **40.** Oudjeriouat, Moreau, Santimone, *et al.* (2003) *Eur. J. Biochem.* **270**, 3871. **41.** Nebinger (1986) *Biol. Chem. Hoppe-Seyler* **367**, 169. **42.** Kaper, Talik, Ettema, Bos, van der Maarel & Dijkhuizen (2005) *Appl. Environ. Microbiol.* **71**, 5098. **43.** Imamura, Fushinobu, Yamamoto, *et al.* (2003) *J. Biol. Chem.* **278**, 19378. **44.** Kaper, Leemhuis, Uitdehaag, *et al.* (2007) *Biochemistry* **46**, 5261. **45.** Bovetto, JVillette, Fontaine, Sicard & Bouquelet (1992) *Biotechnol. Appl. Biochem.* **15**, 59. **46.** Nakamura, Haga & Yamane (1994) *Biochemistry* **33**, 9929. **47.** Kelly, Leemhuis, Gätjen & Dijkhuizen (2008) *J. Biol. Chem.* **283**, 10727. **48.** Li, Lu, Su, *et al.* (2008) *Planta Med.* **74**, 287. **49.** Watson, McCleverty, Geremia, *et al.* (1999) *EMBO J.* **18**, 4619.

Acarviosine-glucose

This pseudo-trisaccharide (FW = 475.49 g/mol) is a truncated acarbose analogue that inhibits α-amylase (1,2) and cyclomaltodextrin glucanosyltransferase (1). IsoAca and AcvGlc are competitive inhibitors that allow one to classify alpha-glucosidases into two groups (3). Enzymes of the first group were strongly inhibited by AcvGlc and weakly by IsoAca, in which the K_i values of AcvGlc (0.35-3.0 μM) were 21- to 440-fold smaller than those of IsoAca. However, the second group of enzymes showed similar K_i values, ranging from 1.6 to 8.0 μM for both compounds. This classification for α-glucosidases is in total agreement with that based on the similarity of their amino acid sequences (family I and family II). This indicated that the α-glucosidase families I and II could be clearly distinguished based on their inhibition kinetic data for IsoAca and AcvGlc. **1.** Kim, Lee, Lee, *et al.* (1999) *Arch. Biochem. Biophys.* **371**, 277. **2.** Li, Begum, Numao, *et al.* (2005) *Biochemistry* **44**, 3347. **3.** Kimura, Lee, Lee, *et al.* (2004) Carbohydr Res. **339**, 1035.

ACEA

This synthetic CB₁ receptor agonist (FW = 365.99 g/mol; CAS 220556-69-4), also known as arachidonyl-2'-chloroethylamide and *N*-(2-chloroethyl)-5Z,8Z,11Z,14Z-eicosatetraenamide, selectively targets the CB₁ cannabinoid receptor (K_i = 1.4 nM) with >1400-fold selectivity over CB₂ receptors. Two subtypes of the cannabinoid receptor (CB₁ and CB₂) are expressed in mammalian tissues. Although selective antagonists are available for each of the subtypes, most of the available cannabinoid agonists bind to both CB₁ and CB₂ with similar affinities (1). ACEA possesses the characteristics of CB₁ receptor agonists, inhibiting forskolin-induced cAMP accumulation in Chinese hamster ovary cells expressing the human CB₁ receptor. It also increases the binding of [³⁵S]GTPγS to cerebellar membranes and inhibits electrically evoked contractions of the mouse *vas deferens* (1). ACEA produces hypothermia in mice, an effect that is inhibited by co-administration of the CB₁ receptor antagonist SR141716A (1). Like other CB₁ receptor agonists (*e.g.*, CP55940, Win 55212-2, Δ⁹-tetrahydrocannabinol, arachidonylethanolamide (AEA)), arachidonyl-2-chloroethylamide produces increases in gait width, a measure of truncal ataxia (2). All of these CB₁ agonists significantly increased the number of slips on the bar cross test, which is consistent with loss of motor coordination. Pretreatment with the CB₁ receptor antagonist SR141716 attenuated both the change in gait width and number of slips induced by

CP55940 and AEA (2). *See also* **ACPA** **1.** Hillard, Manna, Greenberg, *et al.* (1999) *J. Pharmacol. Exp. Ther.* **289**, 1427. **2.** Patel & Hillard (2001) *J. Pharmacol. Exp. Ther.* **297**, 629.

ACEA-1011

This quinoxaline-2,3-dione (FW = 264.59 g/mol), also known as 5-chloro-7-trifluoromethyl-2,3-quinoxalinedione, is an AMPA (*i.e.*, α-amino-3-hydroxy-5-methylisoxazole-4-propionic acid) receptor antagonist. ACEA-1011 also blocks the currents in DRG cells with a K_b of approximately 1 μM, but in cortical neurons the K_b for this drug was 10-12 μM. **1.** Wilding & Huettner (1996) *Mol. Pharmacol.* **49**, 540.

Acebutolol

This lipophilic, cardioselective β-adrenergic blocker (FW$_{free-base}$ = 336.43 g/mol; CAS Reg. Nos 37517-30-9 and 34381-68-5 (HCl), also known as Sectral™ and named systematically as (*RS*)-*N*-{3-acetyl-4-[2-hydroxy-3-(propan-2-ylamino)-propoxy]phenyl}butanamide, displays intrinsic sympathomimetic activity (ISA) and is used to treat hypertension and angina. Acebutolol is metabolized to diacetolol, which is as active as acebutolol and shows a same pharmacologic profile. **Target(s):** β-adrenergic receptor (1,2). **1.** Cuthbert & Owusu-Ankomah (1971) *Brit. J. Pharmacol.* **43**, 639. **2.** Basil, Jordan, Loveless & Maxwell (1973) *Brit. J. Pharmacol.* **48**, 198.

Acein-1, *See* L-Tyrosyl-L-leucyl-L-tyrosyl-L-glutamyl-L-isoleucyl-L-alanyl-L-arginine

Acein-2, *See* L-Leucyl-L-isoleucyl-L-tyrosine

Acenocoumarol

This warfarin analogue (FW = 353.33 g/mol; CAS 152-72-7), also known as acenocoumarin and (*RS*)-4-hydroxy-3-[1-(4-nitrophenyl)-3-oxobutyl]-2*H*-chromen-2-one, is a short-lived oral anti-coagulant that functions by inhibiting vitamin K epoxide reductase, with higher intrinsic anticoagulant potency *in vitro* than warfarin. Acenocoumarol also inhibits rat liver NAD(P)H dehydrogenase competitively with respect to NADH, K_i = 1.7 μM (2) and galactose transport (3). **1.** Ufer (2005) *Clin. Pharmacokinet.* **44**, 1227. **2.** Almeda, Bing, Laura & Friedman (1981) *Biochemistry* **20**, 3731. **3.** Larralde, Ruano, Bolufer & Jordana (1975) *Arch. Int. Physiol. Biochim.* **83**, 271.

ACES

This alkyl-sulfonic acid (FW = 182.20 g/mol; CAS 7365-82-4; pK_2 = 6.88 at 20°C, with dpK_2/dT = –0.020 pH unit/°C), named systematically as *N*-(2-acetamido)-2-aminoethanesulfonic acid, is frequently used as a neutral pH buffer. *Note*: Chloroethane sulfonate is an essential ingedient in the chemical synthesis of ethane sulfonic acid-containing Good's Buffers, giving rise to variable contaminating levels of oligovinylsulfonate (OVS) that inhibit ribonuclease (1) and are likely to act similarly with other enzymes that bind polyanions. **Target(s):** GABA (γ-aminobutyrate) receptor (2). 1. Smith, Soellner & Raines (203) *J. Biol. Chem.* **278**, 20934. **2.** Tunnicliff & Smith (1981) *J. Neurochem.* **36**, 1122.

Acetaldehyde

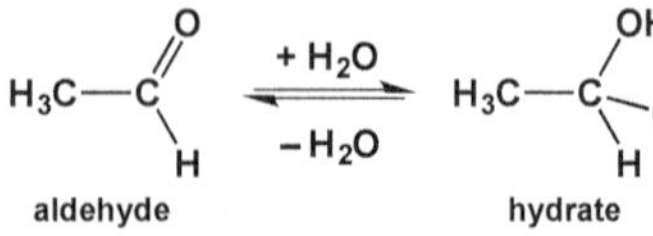

aldehyde ⇌ hydrate

This simple aldehyde (FW = 44.053 g/mol; CAS 75-07-0; Miscible with water and ethanol), is a metabolite produced during carbohydrate degradation. The pure aldehyde is a flammable (M.P. = –123.5°C; B.P. = 21°C) and is best stored in a sealed glass ampoule which should be cooled on ice prior to opening. Storage in a glass ampoule also suppresses spontaneous polymerization at room temperature. As illustrated above, the carbonyl of acetaldehyde readily undergoes hydration to form $CH_3CH(OH)_2$ in aqueous solution, especially in the presence of imidazole or phosphate. Indeed, reversible hydration of acetaldehyde is a general acid- and general base-catalyzed reaction, wherein acidic and basic components in the reaction medium independently contribute to the overall reaction rate. Acetaldehyde reacts with protein amino and sulfhydryl groups to form a variety of adducts, of which are most are Schiff bases and hemiacetals that are readily reversible. The ε-amino groups of certain lysine residues and free N-terminal residues bind acetaldehyde and can be reduced by sodium cyanoborohydride to produce ethylated amino groups formed. **Target(s):** trimethylamine dehydrogenase (1); pyruvate carboxylase (2-5); pyruvate decarboxylase (6,7,13,14,42,43,47); sarcosine oxidase (8); 2-aminoadipate-6-semialdehyde dehydrogenase (9); lipase (10); xanthine oxidase (11,21,52); pyrophosphatase, *or* inorganic diphosphatase (12); monoamine oxidase (15); proteasome (16); pyruvate dehydrogenase (17,37,41); chymotrypsin (18,27); Ca^{2+} channels, voltage-dependent (19); glutamyl aminopeptidase, *or* aminopeptidase A (20); membrane alanyl aminopeptidase, *or* aminopeptidase M (20); lysozyme (22); peptidyl-dipeptidase A, *or* angiotensin-converting enzyme (23); tubulin polymerization, *or* microtubule self-assembly (24,29,30); porphobilinogen synthase, *or* δ-aminolevulinate dehydratase (25); carboxypeptidase C, *or* cathepsin A (26); trypsin (27); methylated-DNA[protein]-cysteine *S* methyltransferase, *or* O^6-methylguanine transferase (28); dihydroxyacetone-phosphate acyltransferase (31); ribonuclease A (32,33); glucose-6-phosphate dehydrogenase (33); nitrogenase (34); hydrogenase (34); Na^+/K^+-exchanging ATPase (35,36,39,40); Mg^{2+}-ATPase (35,40); 5'-nucleotidase (35); succinate dehydrogenase (36); cytochrome *c* oxidase (36); phosphoenolpyruvate carboxykinase (38); 4 hydroxy-2-oxoglutarate aldolase (44); hydroxynitrilase, *or* hydroxynitrile lyase (45); acetoacetate decarboxylase (46); amidase (48,49); malate synthase, weakly inhibited (50); glycerone-phosphate *O*-acyltransferase, *or* dihydroxyacetone-phosphate *O*-acyltransferase (51); and xanthine dehydrogenase (53). **1.** McIntire (1990) *Meth. Enzymol.* **188**, 250. **2.** Karrer & Visconti (1947) *Helv. Chim. Acta* **30**, 268. **3.** Massart (1950) *The Enzymes*, 1st ed., **1** (part 1), 307. **4.** Vennesland (1951) *The Enzymes*, 1st ed., **2** (part 1), 183. **5.** Utter (1961) *The Enzymes*, 2nd ed., **5**, 319. **6.** Bauchop & Dawes (1959) *Biochim. Biophys. Acta* **36**, 294. **7.** Webb (1966) *Enzyme and Metabolic Inhibitors*, vol. 2, p. 432, Academic Press, New York. **8.** Suzuki (1981) *J. Biochem.* **89**, 599. **9.** Calvert & Rodwell (1966) *J. Biol. Chem.* **241**, 409. **10.** Weinstein & Wynne (1936) *J. Biol. Chem.* **112**, 649. **11.** Ball (1939) *J. Biol. Chem.* **128**, 51. **12.** Naganna (1950) *J. Biol. Chem.* **183**, 693. **13.** Gruber & Wesselius (1962) *Biochim. Biophys. Acta* **57**, 171. **14.** Gruber & Wassenaar (1960) *Biochim. Biophys. Acta* **38**, 355. **15.** Vitek, Rysanek & Svehla (1968) *Act. Nerv. Super. (Praha)* **10**, 328. **16.** Rouach, Andraud, Aufrere & Beauge (2005) *Alcohol. Alcohol* **40**, 359. **17.** Hard, Raha, Spino, Robinson & Koren (2001) *Alcohol* **25**, 1. **18.** Brecher & Yang (1998) *J. Investig. Med.* **46**, 146. **19.** Morales, Ram, Song & Brown (1997) *Toxicol. Appl. Pharmacol.* **143**, 70. **20.** Brecher, Stauffer & Knight (1996) *Alcohol* **13**, 125. **21.** Aversa & Meany (1996) *Physiol. Chem. Phys. Med NMR* **28**, 153. **22.** Brecher, Riley & Basista (1995) *Alcohol* **12**, 169. **23.** Thevananther & Brecher (1993) *Alcohol* **10**, 545. **24.** Smith, Jennett, Sorrell & Tuma (1992) *Biochem. Pharmacol.* **44**, 65. **25.** Solomon & Crouch (1990) *J. Lab. Clin. Med.* **116**, 228. **26.** Ostrowska & Kaminski (1990) *Acta Biochim. Pol.* **37**, 141. **27.** Skrzydlewska, Worowski, Zakrzewska, *et al.* (1989) *Mater. Med. Pol.* **21**, 225. **28.** Espina, Lima,

Lieber & Garro (1988) *Carcinogenesis* **9**, 761. **29.** McKinnon, de Jersey, Shanley & Ward (1987) *Neurosci. Lett.* **79**, 163. **30.** McKinnon, Davidson, De Jersey, Shanley & Ward (1987) *Brain Res.* **416**, 90. **31.** Dobrowsky & Ballas (1987) *J. Biol. Chem.* **262**, 3136. **32.** Mauch, Tuma & Sorrell (1987) *Alcohol* **22**, 103. **33.** Mauch, Donohue, Jr., Zetterman, Sorrell & Tuma (1986) *Hepatology* **6**, 263. **34.** Slatyer, Daday & Smith (1983) *Biochem. J.* **212**, 755. **35.** Gonzalez-Calvin, Saunders & Williams (1983) *Biochem. Pharmacol.* **32**, 1723. **36.** Lieberthal, Oldfield & Shanley (1980) *Adv. Exp. Med. Biol.* **132**, 797. **37.** Alkonyi, Bolygo, Gyocsi & Sumegi (1978) *Acta Biochim. Biophys. Acad. Sci. Hung.* **13**, 143. **38.** Baxter (1976) *Biochem. Biophys. Res. Commun.* **70**, 965. **39.** Williams, M. Tada, Katz & Rubin (1975) *Biochem. Pharmacol.* **24**, 27. **40.** Tabakoff (1974) *Res. Commun. Chem. Pathol. Pharmacol.* **7**, 621. **41.** Blass & Lewis (1973) *Biochem. J.* **131**, 415. **42.** Singer & J. Pinsky (1952) *J. Biol. Chem.* **196**, 375. **43.** King & Cheldelin (1954) *J. Biol. Chem.* **208**, 821. **44.** Grady, Wang & Dekker (1981) *Biochemistry* **20**, 2497. **45.** Chueskul & Chulavatnatol (1996) *Arch. Biochem. Biophys.* **334**, 401. **46.** Autor & Fridovich (1970) *J. Biol. Chem.* **245**, 5214. **47.** Goetz, Iwan, Hauer, Breuer & Pohl (2001) *Biotechnol. Bioeng.* **74**, 317. **48.** Woods, Findlater & Orsi (1979) *Biochim. Biophys. Acta* **567**, 225. **49.** Gregoriou & Brown (1979) *Eur. J. Biochem.* **96**, 101. **50.** Miernyk & Trelease (1981) *Phytochemistry* **20**, 2657. **51.** Dobrowsky & Ballas (1987) *J. Biol. Chem.* **262**, 3136. **52.** Morpeth & Bray (1984) *Biochemistry* **23**, 1332. **53.** Pérez-Vicente, Alamillo, Cárdenas & Pineda (1992) *Biochim. Biophys. Acta* **1117**, 159.

Acetaldol, *See Aldol*

Acetal Phosphatide, *See specific plasmalogen*

Acetamide

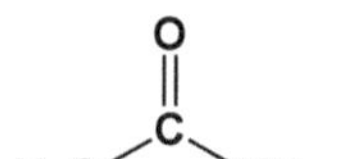

This simple amide (FW = 59.07 g/mol; CAS 142-26-7; M.P. = 81°C; B.P. = 222° C), Deliquescent solid; Solubility: 97 g/100 mL H_2O). **Target(s):** thermopsin (1,2); β-*N*-acetylglucosaminidase (3); urease, weakly inhibited (4); alcohol dehydrogenase, weakly inhibited (5); carbonic anhydrase (6,10); acetylcholinesterase, weakly inhibited (7,8); Na^+/K^+-exchanging ATPase, weakly inhibited (9); urea carboxylase (11); amidase, also alternative substrate (12); pyroglutamyl-peptidase I (13); soluble epoxide hydrolase, K_i = 37 mM (14); glucan 1,3-β-glucosidase (15); and β-*N*-acetylhexosaminidase (16,17). **1.** Tang & Lin (1998) in *Handbook of Proteolytic Enzymes* (Barrett, Rawlings & Woessner, eds), p. 980, Academic Press, San Diego. **2.** Fusek, Lin & Tang (1990) *J. Biol. Chem.* **265**, 1496. **3.** Pugh, Leaback & Walker (1957) *Biochem. J.* **65**, 464. **4.** Kistiakowsky & Shaw (1953) *J. Amer. Chem. Soc.* **75**, 866. **5.** Winer & Theorell (1960) *Acta Chem. Scand.* **14**, 1729. **6.** Whitney, Nyman & Malmström (1967) *J. Biol. Chem.* **242**, 4212. **7.** Butler, Jackson & Polya (1962) *Enzymologia* **24**, 52. **8.** Bergmann, Wilson & Nachmansohn (1950) *J. Biol. Chem.* **186**, 693. **9.** Pisareva (1976) *Tsitologiia* **18**, 1403. **10.** Liang & Lipscomb (1989) *Biochemistry* **28**, 9724. **11.** Castric & Levenberg (1976) *Biochim. Biophys. Acta* **438**, 574. **12.** Woods, Findlater & Orsi (1979) *Biochim. Biophys. Acta* **567**, 225. **13.** Bharadwaj, Sinha Roy, Saha & Hati (1992) *Indian J. Biochem. Biophys.* **29**, 442. **14.** Rink, Kingma, Lutje Spelberg & Janssen (2000) *Biochemistry* **39**, 5600. **15.** Pitson, Seviour, McDougall, Woodward & Stone (1995) *Biochem. J.* **308**, 733. **16.** Barber & Ride (1989) *Plant Sci.* **60**, 163. **17.** Li & Li (1970) *J. Biol. Chem.* **245**, 5153.

N-(2-Acetamido)-2-aminoethanesulfonate, *See ACES*

3-[(6-Acetamido-2-amino-3-pyridinyl)methyl-5-(2-hydroxyethyl)-4-methylthiazolium Pyrophosphate

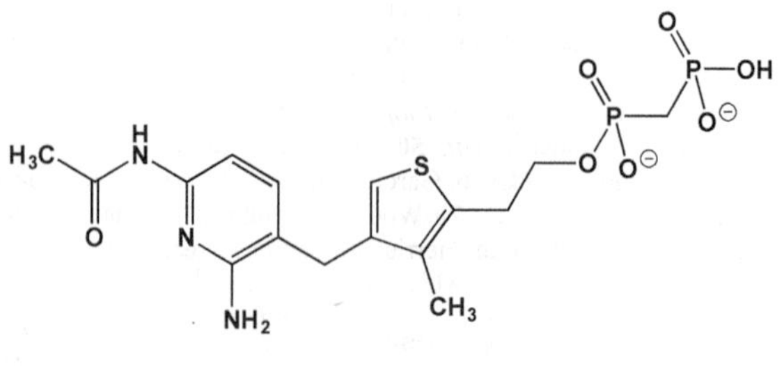

This pyridine analogue ($FW_{hydrochloride}$ = 502.81 g/mol) of thiamin pyrophosphate inhibits transketolase, K_i = 23 nM. A key enzyme in the synthesis of ribose, and hence nucleotide biosynthesis, transketolase is a druggable target for treatment of cancer. In nude mice with xenografted HCT-116 tumors, this inhibitor almost completely suppressed transketolase activity in blood, spleen, and tumor cells, but there was little effect on other thiamine-utilizing enzymes α-ketoglutarate dehydrogenase or glucose-6-phosphate dehydrogenase. **1.** Thomas, Le Huerou, De Meese, *et al.* (2008) *Bioorg. Med. Chem. Lett.* **18**, 2206.

3-[(2-Acetamido-4-amino-5-pyrimidinyl)methyl]-5-(2-hydroxyethyl)-4-methylthiophene Pyrophosphate

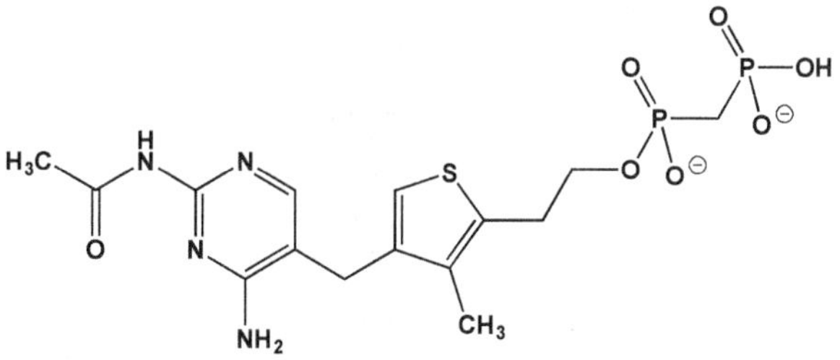

This desamino thiamin pyrophosphate analogue (FW = 462.05 g/mol) inhibits transketolase (TK), K_i = 3.0 nM (1). The depyrophosphorylated derivative is a weaker inhibitor (K_i = 1.1 μM). Most thiamine antagonists have a permanent +1 charge on the B-ring, but the affinity of this inhibitor shows that this charge is not required for binding. **1.** Thomas, De Meese, Le Huerou, *et al.* (2008) *Bioorg. Med. Chem. Lett.* **18**, 509.

rel-(3*aS*,6*S*,6*aR*)-4-(4-Acetamidobenzenesulfonyl)-6-methyl-5-oxohexahydro-pyrrolo[3,2-*b*]pyrrole-1-carboxylic Acid Benzyl Ester

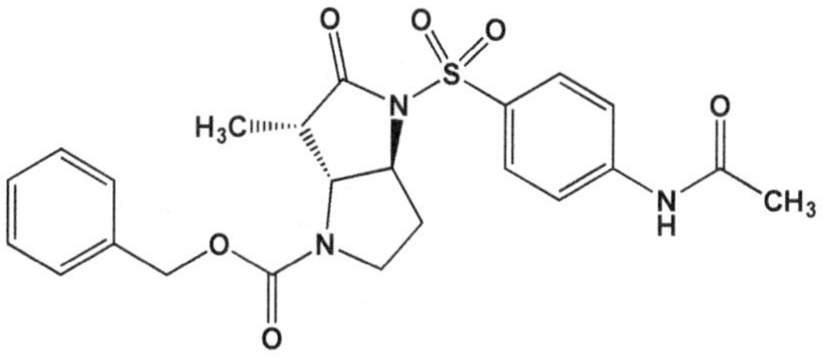

This mechanism-based HCMV protease inhibitor (FW = 459.52 g/mol) targets human cytomegalovirus assemblin. ESI/MS studies have shown that this and related inhibitors can bind covalently and reversibly to the viral enzyme in a time-dependent manner by a mechanism which is consistent with acylation of HCMV δAla protease at the active site nucleophile Ser-132. **1.** Borthwick, Angier, Crame, *et al.* (2000) *J. Med. Chem.* **43**, 4452.

2-Acetamido-2-deoxy-D-galactonolactone

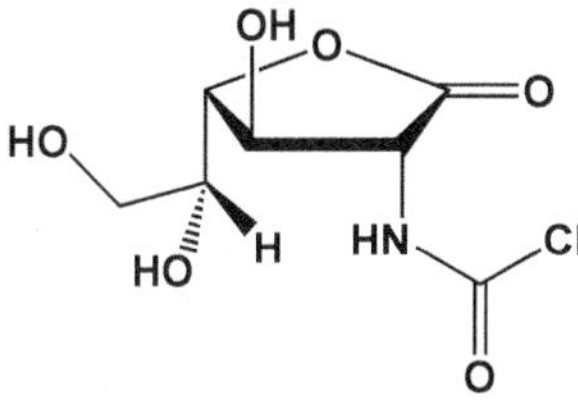

This lactone (FW = 219.19 g/mol) inhibits β-*N*-acetylgalactosaminidase (1,2), β-*N*-acetylglucosaminidase (1,2), and β-*N* acetylhexosaminidase (3-8). Note that the 1,4-lactone is shown above; however, in some cases, the 1,5-lactone may be the true inhibitor. **1.** Walker & Nicholas (1961) *Biochem. J.* **78**, 10p. **2.** Webb (1966) *Enzyme and Metabolic Inhibitors*, vol. **2**, p. 419-420, Academic Press, New York. **3.** Pócsi, Kiss, Zsoldos-Mádi & Pintér (1990) *Biochim. Biophys. Acta* **1039**, 119. **4.** Pokorny & Glaudemans (1975) *FEBS Lett.* **50**, 66. **5.** Calvo, Reglero & Cabezas (1978) *Biochem. J.* **175**, 743. **6.** Garcia-Alonso, Reglero & Cabezas (1990) *Int. J. Biochem.* **22**, 645. **7.** Woollen, Heymorth & Walker (1961) *Biochem. J.* **78**, 111. **8.** Mian, Herries, Cowen & Batte (1979) *Biochem. J.* **177**, 319.

2-Acetamido-2-deoxy-D-glucose, *See* N-Acetyl-D-glucosamine

2-Acetamido-2-deoxy-D-mannopyranose, *See* N-Acetyl-D-mannosamine

2-Acetamido-2,3-dideoxy-3-*C*-(2',6'-anhydro-7'-deoxy-D-*glycero*-D-*manno*-heptitol-7'-*C*-yl)-D-galactose, *See* α-*C*(1→3)-D-Mannopyranoside of N-Acetylgalactosamine

6-Acetamido-1,6-dideoxy-D-mannojirimycin

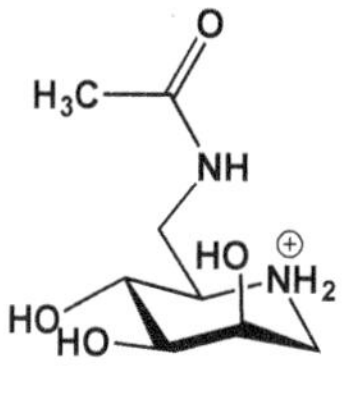

This mannosamine analogue (FW = 204.23 g/mol) potently inhibits α-L-fucosidase (1). When the ring nitrogen is protonated, this analogue mimics the charge of oxa-carbenium ion intermediate thought to form transiently in many glycosidase reactions. **1.** Berteau, McCort, Goasdoué, Tissot & Daniel (2002) *Glycobiology* **12**, 273.

N-(2-Acetamido)iminodiacetic Acid, *See* ADA

2-Acetamido-1,5-imino-1,2,5-trideoxy-D-glucitol

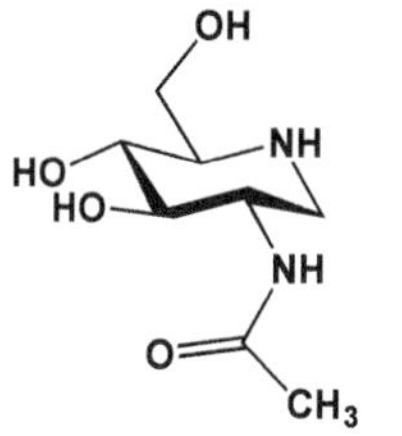

This *N*-acetylglucosamine analogue (FW = 204.23 g/mol), also known as 2-acetamido-1,2-dideoxynojirimycin, inhibits β-*N*-acetylhexosamindase (1,2) and β-*N*-acetylglucosaminidase (2). When the ring nitrogen is protonated, this analogue mimics the charge of oxa-carbenium ion intermediate that forms transiently in many glycosidase reactions. This property results in tight binding. **1.** Nagahashi, Tu, Fleet & Namgoong (1990) *Plant Physiol.* **92**, 413. **2.** Woynarowska, Wikiel, Sharma, *et al.* (1992) *Anticancer Res.* **12**, 161.

4-Acetamido-4'-isothiocyanatostilbene-2,2'-disulfonate

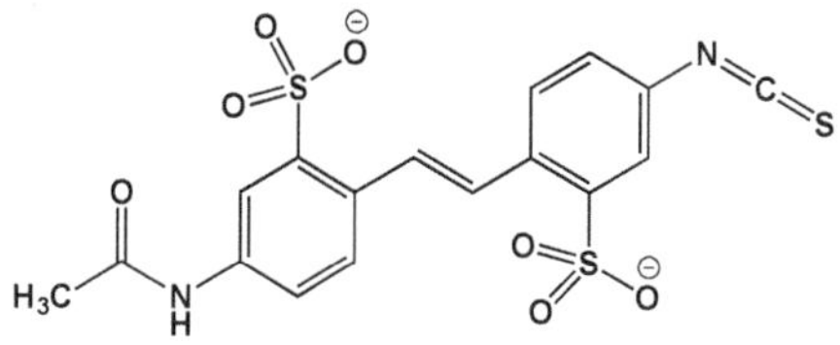

This reagent ($FW_{disodium salt}$ = 498.47 g/mol), also known bythe abbreviation SITS, inhibits anion permeability of membranes (1). It exists primarily in the *trans* geometric isomer. SITS also can function as a fluorescent marker for membranes. **Target(s):** anion transport (1); Cl⁻ channel (2); Na⁺/K⁺-exchanging ATPase (3); F₁ ATPase (4); and thiamin-triphosphatase (5). **1.** Zoccoli & Karnovsky (1980) *J. Biol. Chem.* **255**, 1113. **2.** Tauskela, Mealing, Comas, *et al.* (2003) *Eur. J. Pharmacol.* **464**, 17. **3.** Pedemonte, Kirley, Treuheit & Kaplan (1992) *FEBS Lett.* **314**, 97. **4.** Sakai, Kanazawa, Tsuda & Tsuchiya (1990) *Biochim. Biophys. Acta* **1018**, 18. **5.** Matsuda, Tonomura, Baba & Iwata (1991) *Int. J. Biochem.* **23**, 1111.

N-(4-Acetamidophenylsulfonyl)-*N*-(4-nitrobenzyl)-β-alanine Hydroxamate

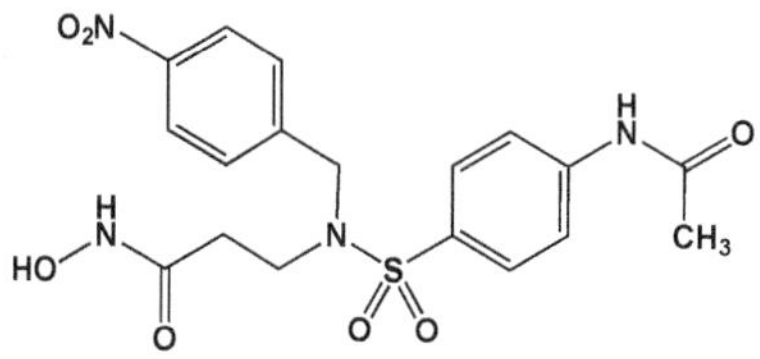

This hydroxamate (FW = 435.44 g/mol) inhibits interstitial collagenase (K_i = 13 nM), gelatinase A (K_i = 9 nM), neutrophil collagenase (K_i = 8 nM), gelatinase B (K_i = 11 nM), and microbial collagenase (K_i = 13 nM). **Target(s):** gelatinase A (matrix metalloproteinase-2); gelatinase B (matrix metalloproteinase-9); interstitial collagenase (matrix metalloproteinase-1); microbial collagenase; neutrophil collagenase (matrix metalloproteinase-8). **1.** Scozzafava, Ilies, Manole & Supuran (2000) *Eur. J. Pharm. Sci.* **11**, 69.

5-AcetamidosulfonyAcid1-(3,4-dichlorobenzyl)-isatin 3-(*O*-a cetyl)oxime

This acylated isatin oxime (FW = 484.32 g/mol) inhibits ubiquitin thiolesterase, IC_{50} = 15 μM. Neuronal ubiquitin C-terminal hydrolase (UCH-L1) has been linked to Parkinson's disease (PD) as well as the progression of certain nonneuronal tumors, and neuropathic pain. **1.** Liu, Lashuel, Choi, *et al.* (2003) *Chem. Biol.* **10**, 837.

2-Acetamido-1,2,5-trideoxy-1,5-imino-D-glucitol, *See* 2-Acetamido-1,5-imino-1,2,5-trideoxy-D-glucitol

5-Acetamidouracil

This xanthine/urate analogue (FW = 169.14 g/mol) inhibits *Streptococcus faecalis* xanthine phosphoribosyltransferase. Inhibition studies with 42 purine and purine analogues indicated that oxo groups at positions-2 and-6 of the purine ring were required for optimal binding. The substitution of thio for oxo reduced binding to the enzyme ca. 20-fold. In contrast to its rigid specificity with respect to the 2,6-dioxo substituents, the enzyme bound a variety of 4,5-condensed pyrimidine systems containing a nitrogen at the position corresponding to the *N*-7 of xanthine. At concentrations of 1 mM, hypoxanthine, adenine, and 4,6-dihydroxypyrazolo[3,4-*d*]pyrimidine were converted to their corresponding ribonucleotides at rates approximately 0.1% of the rate for xanthine. **1.** Miller, Adamczyk, Fyfe & Elion (1974) *Arch. Biochem. Biophys.* **165**, 349.

Acetaminophen

This centrally acting analgesic and antipyretic agent (FW = 151.17 g/mol; CAS 103-90-2; Solubility: 100 mM in DMSO), also known as *N*-acetyl-*p*-aminophenol, paracetamol, Tylenol, and Panadol, inhibits prostaglandin biosynthesis through its action against prostaglandin-endoperoxide synthase (EC 1.14.99.1). (**See also** *Phenacetin*) **Cyclooxygenase Inhibition:** Acetaminophen is somewhat more active against COX-3 (IC_{50} = 460 μM) *versus* COX-1 (IC_{50} > 1000 μM) and COX-2 (IC_{50} > 1000 μM), but its true selectivity and mechanism of therapeutic action remains to be fully determined. Under most circumstances, acetaminophen is a relatively weak cyclooxygenase inhibitor; however, its potency against both purified ovine cyclooxygenase-1 (oCOX-1) and human cyclooxygenase-2 (hCOX-2) increases approximately 30x in the presence of glutathione peroxidase and glutathione, yielding IC_{50} values of 33 μM and 980 μM, respectively (1). Acetaminophen is a good reducing agent of both oCOX-1 and hCOX-2, suggesting an inhibitory mechanism in which it acts to reduce the active oxidized form of COX to the resting form (1). This results in inhibition of phenoxyl radical formation from a critical tyrosine residue essential for the cyclooxygenase activity of COX-1 and COX-2 and prostaglandin (PG) synthesis (2). Acetaminophen shows selectivity for inhibition of the synthesis of PGs and related factors, when low levels of arachidonic acid and peroxides are available (2). Conversely, it has little activity at substantial levels of arachidonic acid and peroxides. Furthermore, inhibition would therefore be expected to be more effective under conditions of low peroxide concentration, a prediction that is consistent with the known tissue selectivity of acetaminophen (2). Although the exact site and mechanism of acetaminophen analgesia remains to be established, it appears to elevate the pain threshold by inhibiting the nitric oxide pathway, as mediated by a variety of neurotransmitter receptors, including *N*-methyl-D-aspartate and substance P. **Antipyretic Effects:** Acetaminophen is thought to inhibit the action of endogenous pyrogens on the heat-regulating centers in the brain by blocking the formation and release of prostaglandins in the central nervous system. **Pharmacokinetics:** Oral acetaminophen is rapidly and almost completely absorbed from the GI tract, primarily in the small intestine by passive transport. Relative bioavailability is 85-98%. Acetaminophen appears to be widely distributed throughout most body fluids, except fat. The apparent volume of distribution of acetaminophen is 0.95 L/kg. A small fraction (10-25%) of acetaminophen is bound to plasma proteins. The biologic half-life of acetaminophen in normal adults is approximately 2-3 hours in the usual dosage range, and the half-life for CSF clearance is 3 hours. **Key Pharmacokinetic Parameters:** *See* Appendix II in Goodman & Gilman's *THE PHARMACOLOGICAL BASIS OF THERAPEUTICS*, 12th Edition (Brunton, Chabner & Knollmann, eds.) McGraw-Hill Medical, New York (2011). **Toxicity:** After FDA approval in 1961, acetaminophen quickly became one of the most widely used analgesics. Often taken in excess by those ignoring clearly stated warnings, acute overdoses can result in potentially fatal liver damage. Harmful effects of acetaminophen have been attributed to the production of its toxic metabolite, *N*-acetyl-*p*-benzoquinone imine (*or* NAPQI), a topoisomerase II poison. (**See** *N-Acetyl-p-benzoquinone imine*) Acetylcysteine is often effective in the treatment of acetaminophen overdose, acting, in part, by inhibiting cytochrome P-448. **Target(s):** cyclooxygenase, *or* prostaglandin synthase (1-3,6); peroxynitrite-mediated chemiluminescence (4); glucosamine synthase (5); phenylpyruvate tautomerase (7); adenosine deaminase, K_i = 0.13 mM (8); β-glucosidase (9); arylamine *N*-acetyltransferase (10); phosphoribosylamino-imidazole carboxamide formyltransferase (11); thiopurine *S* methyltransferase (12); linoleate diol synthase (13); tryptophan 2,3-dioxygenase (14); 3-hydroxyanthranilate 3,4-dioxygenase (15); myeloperoxidase [where inhibition involves acetaminophen oxidation and concomitant decreased formation of halogenating oxidants (e.g. hypochlorous acid, hypobromous acid) that may be associated with multiple inflammatory pathologies including atherosclerosis and rheumatic diseases (2).] **1.** Ouellet & Percival (2001) *Arch. Biochem. Biophys.* **387**, 273. **2.** Graham, Davies, Day, Mohamudally, Scott (2013) *Inflammopharmacol.* **21**, 201. **3.** Flower & Vane (1972) *Nature* **240**, 410. **4.** Rork, Van Dyke, Spiler & Merrill (2004) *Exp. Biol. Med. (Maywood)* **229**, 1154. **5.** Goodman, Kent & Truelove (1977) *Arch. Int. Pharmacodyn. Ther.* **226**, 4. **6.** Humes, Winter, Sadowski & Kuehl (1981) *Proc. Natl. Acad. Sci. U.S.A.* **78**, 2053. **7.** Molnar & Garai (2005) *Int. Immunopharmacol.* **5**, 849. **8.** Ataie, Safarian, Divsalar, *et al.* (2004) *J. Enzyme Inhib. Med. Chem.* **19**, 71. **9.** Dale, Ensley, Kern, Sastry & Byers (1985) *Biochemistry* **24**, 3530. **10.** Makarova (2008) *Curr. Drug Metab.* **9**, 538. **11.** Ha, Morgan, Vaughn, Eto & Baggott (1990) *Biochem. J.* **272**, 339. **12.** Oselin & Anier (2007) *Drug Metab. Dispos.* **35**, 1452. **13.** Brodowsky, Hamberg & Oliw (1994) *Eur. J. Pharmacol.* **254**, 43. **14.** Daya & Anoopkumar-Dukie (2000) *Life Sci.* **67**, 235. **15.** Maharaj, Maharaj & Daya (2006) *Metab. Brain Dis.* **21**, 189.

Acetanilide

This acetylated aniline (FW = 135.17 g/mol; CAS 103-84-4 M.P. = 113-115°C) is a component in the synthesis of numerous pharmaceuticals, inhibitors, and dyes. Acetanilide is toxic *in vivo*, owing in part to its hydrolysis to aniline, with consequential methemoglobinemia. **Target(s):** chymotrypsin (1,2); D-3-hydroxybutyrate dehydrogenase (3); and amidase, also as an alternative substrate (4). **1.** Baker (1967) *Design of Active-Site-Directed Irreversible Enzyme Inhibitors*, Wiley, New York. **2.** Wallace,

Kurtz & Niemann (1963) *Biochemistry* **2**, 824. **3**. Gotterer (1969) *Biochemistry* **8**, 641. **4**. Alt, Heymann & Krisch (1975) *Eur. J. Biochem.* **53**, 357.

Acetarsone

This organoarsenical (FW$_{free-acid}$ = 275.09 g/mol; CAS 97-44-4), also known as Stovarsol, Spirozid, Ehrlich 594, and [3-(acetylamino)-4-hydroxyphenyl]arsonate, is an antiprotozoal agent that was first prepared by Paul Ehrlich in 1909 during his classic campaign to identify a "silver bullet" for treating syphilis (1). **Target(s):** γ-butyrobetaine dioxygenase, mildly inhibited (2). **1**. Mitchell (1935) *Can. Med. Assoc. J.* **33**, 377. **2**. Lindstedt & Lindstedt (1970) *J. Biol. Chem.* **245**, 4178.

Acetate (Acetic Acid)

This two-carbon alkyl acid (FW = 60.05 g/mol; CAS 64-19-7) is the most common carboxylic acid in biological systems. Acetate is the anion form or conjugate base of acetic acid, which has a pK_a value of 4.76 at 25°C (the dpK_a/dT value is +0.0002). The acid is a colorless liquid (M.P. = 16.7°C; B.P. = 118°C) with a distinctively strong and pungent odor. Acetic acid is miscible with water, ethanol, diethyl ether, glycerol, and carbon tetrachloride. The pure acid is often referred to as glacial acetic acid because, when poured on a surface, it resembles ice in appearance. Note that, just as water has an autoprotolysis constant of 10^{-14} M^2 at 24°C, the autoprotolysis constant for glacial acetic acid is 10^{-13} M^2 at 25°C. Salts of acetic acid (*e.g.*, sodium acetate, ammonium acetate, etc.) are very hygroscopic and are very soluble in water; for example, 119 g of sodium acetate dissolve in 100 mL water at 0°C and 253 g potassium acetate dissolve in 100 mL water at 20°C. **Target(s):** sarcosine dehydrogenase (1); β-*N*-acetylhexosaminidase (2,7,25,26,43,89-101); membrane dipeptidase (3); (*S*)-2-hydroxy-acid oxidase (4); 4-aminobutyrate aminotransferase (5,46,120); penicillo-carboxypeptidase (6); hexose oxidase (8); diacetyl reductase, *or* acetoin dehydrogenase (9); 3-dehydroquinate dehydratase (10,62); methylamine oxidase (11); glutamyl endopeptidase (12); endopeptidase Clp (13); glutathione *S*-transferase (14); catalase (15); alanine aminotransferase (16); L-lactate dehydrogenase (17,23); carbonic anhydrase (18,28,50,63-66); triose-phosphate isomerase (19); phosphoglucomutase (20); aspartate aminotransferase (21); rat liver mitochondrial ATPase (22); galactose oxidase (24); glutamate decarboxylase (25,33); β-*N*-acetyl-glucosaminidase (25,26,43,94); pantothenate synthetase (27); aspartate 4-decarboxylase (29,30); myosin ATPase (31); estradiol-17β dehydrogenase (31); β-galactosidase (31); angiotensin-converting enzyme, *or* peptidyl-dipeptidase A (34); aldose reductase, *or* aldehyde reductase (35); alcohol dehydrogenase, *or* weakly inhibited (36); *N*-methylglutamate dehydrogenase (37); creatine kinase (38); sarcosine oxidase (39); sepiapterin reductase (40); rhodanese, *or* thiosulfate sulfurtransferase (41); hyaluronoglucosaminidase, *or* hyaluronidase (42); cystathionine γ-lyase, *or* cysteine desulfhydrase (44); pyrophosphatase, *or* inorganic diphosphatase (45); cutinase (47); laccase (48); urease (49); pyridoxamine:pyruvate aminotransferase (51); pepsin (52); xanthine oxidase (53); UDP-*N*-acetylmuramoyl-tripeptide:D-alanyl-D-alanine ligase (54); asparaginyl-tRNA synthetase (55); aspartyl-tRNA synthetase (55); bisphosphoglycerate mutase (56); isopentenyl-diphosphate Δ-isomerase (57); alanine racemase (58); pectin lyase (59); cyanate hydratase, *or* cyanase (60); nitrile hydratase (61); mandelonitrile lyase (67); glutamate decarboxylase (68); oxaloacetate decarboxylase (69,70); haloacetate dehalogenase (71); 1,5-anhydro-D-fructose reductase, *or* 1,5-anhydro-D-mannitol forming (72); succinate:3-hydroxy-3-methylglutarate CoA-transferase (73); theanine hydrolase (74); N^4-(β-*N*-acetyl-glucosaminyl)-L-asparaginase (75); pantothenase (76); β-ureidopropionase (77); amidase (78-80); carboxypeptidase A, inhibited by concentrated

acetic acid (81); carboxypeptidase B, inhibited by concentrated acetic acid (81); carboxypeptidase A (82,83); peptidyl-dipeptidase A, *or* angiotensin I converting enzyme (84); aminopeptidase B (85); chitosanase (86,87); mannosyl-glycoprotein endo-β-*N* acetylglucosaminidase (88); α-L-rhamnosidase, weakly inhibited (102); α,α-trehalase (103,104); poly(A) specific ribonuclease (105); 2',3'-cyclic-nucleotide 3'-phosphodiesterase (106,107); cephalosporin-C deacetylase (108-111); lipase, *or* triacylglycerol lipase (112); [β-adrenergic-receptor] kinase (113; [branched-chain α-keto-acid dehydrogenase] kinase, *or* [3-methyl-2-oxobutanoate dehydrogenase (acetyl-transferring)] kinase (114); thymidine kinase (115); glucokinase (116); hexokinase IV (116); serine:glyoxylate aminotransferase (117); 2-aminoethylphosphonate:pyruvate aminotransferase (118); L-lysine 6-aminotransferase (119); protein geranylgeranyltransferase type II (121); methionine *S*-adenosyltransferase, weakly inhibited (122); nicotinate-nucleotide diphosphorylase (carboxylating) (123); starch phosphorylase, weakly inhibited (124); peptide α-*N* acetyltransferase (125); amino-acid *N*-acetyltransferase (126); acetolactate synthase (127); aspartate carbamoyl-transferase (128); glutamate formimidoyltransferase (129,130); glycine/sarcosine *N*-methyltransferase (131); tRNA (cytosine-5-)-methyltransferase (132); glycine *N*-methyltransferase (133); superoxide reductase (134); phenol 2-monooxygenase, weakly inhibited (135); 4-hydroxybenzoate 3-monooxygenase, weakly inhibited (136); lactate 2-monooxygenase, product inhibition (137,138); lysine 2-monooxygenase (139); protocatechuate 3,4-dioxygenase (140); manganese peroxidase (141,142); and laccase (143). **1**. Mackenzie & Hoskins (1962) *Meth. Enzymol.* **5**, 738. **2**. Frohwein & Gatt (1969) *Meth. Enzymol.* **14**, 161. **3**. Campbell (1970) *Meth. Enzymol.* **19**, 722. **4**. Jorns (1975) *Meth. Enzymol.* **41**, 337. **5**. Kraus & Noack (1996) *J. Enzyme Inhib.* **10**, 125. **6**. Hofmann (1976) *Meth. Enzymol.* **45**, 587. **7**. Sarber, Distler & Jourdian (1978) *Meth. Enzymol.* **50**, 520. **8**. Ikawa (1982) *Meth. Enzymol.* **89**, 145. **9**. Martín Sarmiento & Burgos (1982) *Meth. Enzymol.* **89**, 516. **10**. Chaudhuri, Duncan & Coggins (1987) *Meth. Enzymol.* **142**, 320. **11**. McIntire (1990) *Meth. Enzymol.* **188**, 227. **12**. Birktoft & Breddam (1994) *Meth. Enzymol.* **244**, 114. **13**. Maurizi, Thompson, Singh & Kim (1994) *Meth. Enzymol.* **244**, 314. **14**. J. Dierickx (1984) *Res. Commun. Chem. Pathol. Pharmacol.* **44**, 327. **15**. Nicholls & Schonbaum (1963) *The Enzymes*, 2nd. ed., **8**, 147. **16**. Saier & Jenkins (1967) *J. Biol. Chem.* **242**, 101. **17**. Schwert & Winer (1963) *The Enzymes*, 2nd ed., **7**, 142. **18**. Lindskog, Henderson, Kannan, *et al.* (1971) *The Enzymes*, 3rd ed., **5**, 587. **19**. Noltmann (1972) *The Enzymes*, 3rd ed., **6**, 271. **20**. Ray & Peck (1972) *The Enzymes*, 3rd ed., **6**, 407. **21**. Braunstein (1973) *The Enzymes*, 3rd ed., **9**, 379. **22**. Penefsky (1974) *The Enzymes*, 3rd ed., **10**, 375. **23**. Holbrook, Liljas, Steindel & Rossmann (1975) *The Enzymes*, 3rd ed., **11**, 191. **24**. Malmström, Andréasson & Reinhammar (1975) *The Enzymes*, 3rd ed., **12**, 507. **25**. Webb (1966) *Enzyme and Metabolic Inhibitors*, vol. **2**, Academic Press, New York. **26**. Pugh, Leaback & Walker (1957) *Biochem. J.* **65**, 464. **27**. Van Oorschot & Hilton (1963) *Arch. Biochem. Biophys.* **100**, 289. **28**. Verpoorte, Mehta & Edsall (1967) *J. Biol. Chem.* **242**, 4221. **29**. Novogrodsky & Meister (1964) *J. Biol. Chem.* **239**, 879. **30**. Soda, Novogrodsky & Meister (1964) *Biochemistry* **3**, 1450. **31**. Warren & Cheatum (1966) *Biochemistry* **5**, 1702. **32**. Kolbe (1845) *Annalen* **54**, 145. **33**. Susz, Haber & Roberts (1966) *Biochemistry* **5**, 2870. **34**. Oshima & Nagasawa (1977) *J. Biochem.* **81**, 57. **35**. Hayman & Kinoshita (1965) *J. Biol. Chem.* **240**, 877. **36**. Winer & Theorell (1960) *Acta Chem. Scand.* **14**, 1729. **37**. Hersh, Stark, Worthen & Fiero (1972) *Arch. Biochem. Biophys.* **150**, 219. **38**. Noda, Nihei & Morales (1960) *J. Biol. Chem.* **235**, 2830. **39**. Frisell & Mackenzie (1955) *J. Biol. Chem.* **217**, 275. **40**. Katoh (1971) *Arch. Biochem. Biophys.* **146**, 202. **41**. Alexander & Volini (1987) *J. Biol. Chem.* **262**, 6595. **42**. Linker (1984) in *Methods of Enzymatic Analysis* (Bergmeyer, ed.), vol. **4**, 256. **43**. Frohwein & Gatt (1967) *Biochemistry* **6**, 2775. **44**. Fromageot & Grand (1942) *Enzymologia* **11**, 81. **45**. Naganna (1950) *J. Biol. Chem.* **183**, 693. **46**. Sytinsky & Vasilijev (1970) *Enzymologia* **39**, 1. **47**. Maeda, Yamagata, Abe, *et al.* (2005) *Appl. Microbiol. Biotechnol.* **67**, 778. **48**. Bertrand (1907) *Ann. Institut Pasteur* **21**, 673. **49**. Onodera (1915) *Biochem. J.* **9**, 544. **50**. Pocker & Stone (1968) *Biochemistry* **7**, 2936. **51**. Kolb, Cole & Snell (1968) *Biochemistry* **7**, 2946. **52**. Hollands, Voynick & Fruton (1969) *Biochemistry* **8**, 575. **53**. Booth (1935) *Biochem. J.* **29**, 1732. **54**. Dementin, Bouhss, Auger, *et al.* (2001) *Eur. J. Biochem.* **268**, 5800. **55**. Lea & Fowden (1973) *Phytochemistry* **12**, 1903. **56**. Rose (1973) *Arch. Biochem. Biophys.* **158**, 903. **57**. Banthorpe, Doonan & Gutowski (1977) *Arch. Biochem. Biophys.* **184**, 381. **58**. Morollo, Petsko & Ringe (1999) *Biochemistry* **38**, 3293. **59**. Van Houdenhoven (1975) *Mededelingen Landbouwhogeschool, Communications Agricultural*

University **75-13**, 1. **60**. Anderson & Little (1986) *Biochemistry* **25**, 1621. **61**. Amarant, Vered & Bohak (1989) *Biotechnol. Appl. Biochem.* **11**, 49. **62**. Chaudhuri, Lambert, McColl & Coggins (1986) *Biochem. J.* **239**, 699. **63**. Yachandra, Powers & Spiro (1983) *J. Amer. Chem. Soc.* **105**, 6596. **64**. Mananes, Daleo & Vega (1993) *Comp. Biochem. Physiol. B* **105**, 175. **65**. Kimber & Pai (2000) *EMBO J.* **19**, 1407. **66**. Baird, Waheed, Okuyama, Sly & Fierke (1997) *Biochemistry* **36**, 2669. **67**. Jorns (1980) *Biochim. Biophys. Acta* **613**, 203. **68**. Spink, Porter, Wu & Martin (1985) *Biochem. J.* **231**, 695. **69**. Benziman, Russo, Hochmann & Weinhouse (1978) *J. Bacteriol.* **134**, 1. **70**. Labrou & Clonis (1999) *Arch. Biochem. Biophys.* **365**, 17. **71**. Goldman (1965) *J. Biol. Chem.* **240**, 3434. **72**. Dambe, Kühn, Brossette, Giffhorn & Scheidig (2006) *Biochemistry* **45**, 10030. **73**. Francesconi, Donella-Deana, Furlanetto, Cavallini, Palatini & Deana (1989) *Biochim. Biophys. Acta* **999**, 163. **74**. Tsushida & Takeo (1985) *Agric. Biol. Chem.* **49**, 2913. **75**. Mahadevan & Tappel (1967) *J. Biol. Chem.* **242**, 4568. **76**. Airas (1976) *Biochim. Biophys. Acta* **452**, 201. **77**. Wasternack, Lippmann & Reinbotte (1979) *Biochim. Biophys. Acta* **570**, 341. **78**. Woods, Findlater & Orsi (1979) *Biochim. Biophys. Acta* **567**, 225. **79**. Maestracci, Bui, Thiery, Arnaud & Galzy (1984) *Biotechnol. Lett.* **6**, 149. **80**. Maestracci, Thiery, Bui, Arnaud & Galzy (1984) *Arch. Microbiol.* **138**, 315. **81**. Gates & Travis (1973) *Biochemistry* **12**, 1867. **82**. Banci, Schröder & Kollman (1992) *Proteins* **13**, 288. **83**. Auld, Bertini, Donaire, Messori & Moratal (1992) *Biochemistry* **31**, 3840. **84**. Oshima & Nagasawa (1977) *J. Biochem.* **81**, 57. **85**. Hopsu, Mäkinen & Glenner (1966) *Arch. Biochem. Biophys.* **114**, 567. **86**. Alfonso, Martinez & Reyes (1992) *FEMS Microbiol. Lett.* **95**, 187. **87**. Somashekar & Joseph (1996) *Biores. Technol.* **55**, 35. **88**. DeGasperi, Li & Li (1989) *J. Biol. Chem.* **264**, 9329. **89**. Calvo, Reglero & Cabezas (1978) *Biochem. J.* **175**, 743. **90**. Barber & Ride (1989) *Plant Sci.* **60**, 163. **91**. Wheeler, Bharadwaj & Gregory (1982) *J. Gen. Microbiol.* **128**, 1063. **92**. Li & Li (1970) *J. Biol. Chem.* **245**, 5153. **93**. Geimba, Riffel & Brandelli (1998) *J. Appl. Microbiol.* **85**, 708. **94**. Khar & Anand (1977) *Biochim. Biophys. Acta* **483**, 141. **95**. Stirling (1972) *Biochim. Biophys. Acta* **271**, 154. **96**. Verpoorte (1974) *Biochemistry* **13**, 793. **97**. Potier, Teitelbaum, Melancon & Dallaire (1979) *Biochim. Biophys. Acta* **566**, 80. **98**. Jin, Jo, Kim, *et al.* (2002) *J. Biochem. Mol. Biol.* **35**, 313. **99**. Bedi, Shah & Bahl (1984) *Arch. Biochem. Biophys.* **233**, 237. **100**. Sakai, Nakanishi & Kato (1993) *Biosci. Biotechnol. Biochem.* **57**, 965. **101**. Ohtakara, Yoshida, Murakami & Izumi (1981) *Agric. Biol. Chem.* **45**, 239. **102**. Mutter, Beldman, Schols & Voragen (1994) *Plant Physiol.* **106**, 241. **103**. Biswas & Ghosh (1996) *Biochim. Biophys. Acta* **1290**, 95. **104**. Londesborough & Varimo (1984) *Biochem. J.* **219**, 511. **105**. Korner, Wormington, Muckenthaler, *et al.* (1998) *EMBO J.* **17**, 5427. **106**. Sprinkle (1989) *CRC Crit. Rev. Clin. Neurobiol.* **4**, 235. **107**. Sims & Carnegie (1978) *Adv. Neurochem.* **3**, 1. **108**. Abbott & Fukuda (1975) *Appl. Microbiol.* **30**, 413. **109**. Abbott, Cerimele & Fukuda (1976) *Biotechnol. Bioeng.* **18**, 1033. **110**. Abbott & Fukuda (1978) *Meth. Enzymol.* **43**, 731. **111**. Takimoto, Mitsushima, Yagi & Sonoyama (1994) *J. Ferment. Bioeng.* **77**, 17. **112**. Muderhwa, Ratomahenina, Pina, Graille & Galzy (1986) *Appl. Microbiol. Biotechnol.* **23**, 348. **113**. Benovic, Mayor, Staniczewski, Lefkowitz & Caron (1987) *J. Biol. Chem.* **262**, 9026. **114**. Paxton & Harris (1984) *Arch. Biochem. Biophys.* **231**, 58. **115**. Rohde & Lezius (1971) *Hoppe-Seyler's Z. Physiol. Chem.* **352**, 1507. **116**. Vandercammen & Van Schaftingen (1991) *Eur. J. Biochem.* **200**, 545. **117**. Smith (1973) *Biochim. Biophys. Acta* **321**, 156. **118**. Dumora, Lacoste & Cassaigne (1983) *Eur. J. Biochem.* **133**, 119. **119**. Yoshimura, Tanizawa, Tanaka & Soda (1984) *J. Biochem.* **95**, 559. **120**. Schousboe, Wu & Roberts (1974) *J. Neurochem.* **23**, 1189. **121**. Seabra, Goldstein, Sudhof & Brown (1992) *J. Biol. Chem.* **267**, 14497. **122**. Luo, Yuan, Luo & Zhao (2008) *Biotechnol. Prog.* **24**, 214. **123**. Shibata & Iwai (1980) *Biochim. Biophys. Acta* **611**, 280. **124**. Schwarz, Brecker & Nidetzky (2007) *FEBS J.* **274**, 5105. **125**. Yamada & Bradshaw (1991) *Biochemistry* **30**, 1010. **126**. Qu, Morizono, Shi, Tuchman & Caldovic (2007) *BMC Biochem.* **8**, 4. **127**. Kaushal, Pabbi & Sharma (2003) *World J. Microbiol. Biotechnol.* **19**, 487. **128**. Jacobson & Stark (1975) *J. Biol. Chem.* **250**, 6852. **129**. Miller & Waelsch (1957) *J. Biol. Chem.* **228**, 397. **130**. Tabor & Wyngarden (1959) *J. Biol. Chem.* **234**, 1830. **131**. Waditee, Tanaka, Aoki, *et al.* (2003) *J. Biol. Chem.* **278**, 4932. **132**. Keith, Winters & Moss (1980) *J. Biol. Chem.* **255**, 4636. **133**. Heady & Kerr (1973) *J. Biol. Chem.* **248**, 69. **134**. Brines & Kovacs (2007) *Eur. J. Inorg. Chem.* **2007**, 29. **135**. Neujahr (1983) *Biochemistry* **22**, 580. **136**. Shoun, Arima & Beppu (1983) *J. Biochem.* **93**, 169. **137**. Durfor & Cromartie (1981) *Arch. Biochem. Biophys.* **210**, 710. **138**. Ghisla & Massey (1975) *J. Biol. Chem.* **250**, 577. **139**. Vandecasteele & Hermann (1972) *Eur. J. Biochem.* **31**, 80. **140**. Bull & Ballou (1981) *J. Biol. Chem.* **256**, 12673. **141**. Urek & Pazarlioglu (2004) *Proc. Biochem.* **39**, 2061. **142**. Airken & Irvine (1990)

Arch. Biochem. Biophys. **276**, 405. **143**. Junghanns, Pecyna, Böhm, *et al.* (2009) *Appl. Microbiol. Biotechnol.* **84**, 1095.

Acetazolamide

This universally used diuretic (FW = 222.25 g/mol; CAS 59-66-5), also known as Diamox™, *N*-(5-sulfamoyl-1,3,4-thiadiazol-2-yl)acetamide, and 2-acetamido-1,3,4-thiadiazole-5-sulfonamide, is a potent and reversible noncompetitive carbonic anhydrase inhibitor. Although aqueous solutions are stable for up to week, the recommended practice is to freshly prepare acetazolamide solutions and use them within 24 hours. **Primary Mode of Action:** Carbonic anhydrase catalysis may be depicted as the following multi-step reaction sequence: *Step*-1, $E–Zn^{2+}(H_2O) \rightleftharpoons E–Zn^{2+}(^-OH) + H^+$; *Step*-2, $E–Zn^{2+}(^-OH) + O=C=O \rightleftharpoons E–Zn^{2+}(^-OC(=O)OH)$; *Step*-3, $E–Zn^{2+}(^-OC(=O)OH) + H_2O \rightleftharpoons HCO_3^- + E–Zn^{2+}(H_2O)$, where the latter is the transient E·H⁺ species. Acetazolamide binds to E·H⁺, and, because E·H⁺ must be deprotonated to regenerate the active catalyst, acetazolamide effectively blocks catalysis. Note that the sulfonamide moiety is believed to resemble enzyme-bound carbon dioxide, thus accounting for its high affinity ($K_i \approx 10^{-8}$ M). **Medical Uses:** Because hemoglobin's O_2 affinity is reduced at lower blood pH and because carbonic anhydrase inhibition tends to lower blood pH, acetazolamide is frequently used to reduce the severity and duration of symptoms of "mountain sickness" (*e.g.*, upset stomach, headache (and, in severe cases, cerebral edema), shortness of breath (often attending pulmonary edema), dizziness, drowsiness, and fatigue). Acetazolamide also reduces hydrogen ion secretion in the renal tubule, increasing renal excretion of sodium, potassium, bicarbonate, and resulting in net water loss. Acetazolamide is also used to treat glaucoma (*See also Dorzolamide*), to reduce edema, and to control seizures in certain types of epilepsy. Acetazolamide also decreases renal ammonia production. Because there are multiple isozyme forms of carbonic anhydrase, including some ectoplasmic isozymes, acetazolamide is likely to have metabolic consequences beyond its effects on diuresis. **Target(s):** carbonic anhydrase (1-36); chloride-transporting ATPase (37); 3,5'-cyclic-nucleotide phosphodiesterase (38); dihydroorotase (K_i = 4.2 mM (39); glutaminase (40); γ-glutamyl transpeptidase, *or* γ-glutamyltransferase (41-44).

1. Davis (1961) *The Enzymes*, 2nd ed., **5**, 545. **2**. Wells, Kandel, Kandel & Gornall (1975) *J. Biol. Chem.* **250**, 3522. **3**. Whitney, Folsch, Nyman & Malmstrom (1967) *J. Biol. Chem.* **242**, 4206. **4**. Lindskog, Henderson, Kannan, *et al.* (1971) *The Enzymes*, 3rd. ed., **5**, 587. **5**. Pocker & Stone (1967) *Biochemistry* **6**, 668. **6**. Pocker & Meany (1965) *Biochemistry* **4**, 2535. **7**. Eriksson, Kylsten, Jones & Liljas (1988) *Proteins* **4**, 283. **8**. Bülbül, Hisar, Beydemir, Çiftçi & Küfrevioglu (2003) *J. Enzyme Inhib. Med. Chem.* **18**, 371. **9**. Olander & Kaiser (1970) *J. Amer. Chem. Soc.* **92**, 5758. **10**. Maren, Parcell & Malik (1960) *J. Pharmacol. Exp. Ther.* **130**, 389. **11**. Maren (1963) *J. Pharmacol. Exp. Ther.* **139**, 129 and 140. **12**. Whitney & Briggle (1982) *J. Biol. Chem.* **257**, 12056. **13**. Maren, Mayer & Wadsworth (1954) *Bull. Johns Hopkins Hosp.* **95**, 199. **14**. Davis (1959) *J. Amer. Chem. Soc.* **81**, 5674. **15**. Waygood (1955) *Meth. Enzymol.* **2**, 836. **16**. Pocker & Meany (1965) *Biochemistry* **4**, 2535. **17**. Pocker & Storm (1968) *Biochemistry* **7**, 1202. **18**. Pocker & Dickerson (1968) *Biochemistry* **7**, 1995. **19**. Pocker & Stone (1968) *Biochemistry* **7**, 2936. **20**. Taylor, King & Burgen (1970) *Biochemistry* **9**, 2638. **21**. Winum, Vullo, Casini, *et al.* (2003) *J. Med. Chem.* **46**, 2197. **22**. Chirica, Elleby & Lindskog (2001) *Biochim. Biophys. Acta* **1544**, 55. **23**. Rumeau, Cuine, Fina, Gault, Nicole & Peltier (1996) *Planta* **199**, 79. **24**. Brundell, Falkbring & Nyman (1972) *Biochim. Biophys. Acta* **284**, 311. **25**. Pocker & Ng (1973) *Biochemistry* **12**, 5127. **26**. Yachandra, Powers & Spiro (1983) *J. Amer. Chem. Soc.* **105**, 6596. **27**. Mananes, Daleo & Vega (1993) *Comp. Biochem. Physiol. B* **105**, 175. **28**. Engberg, Millqvist, Pohl & Lindskog (1985) *Arch. Biochem. Biophys.* **241**, 628. **29**. Sanyal, Pessah & Maren (1981) *Biochim. Biophys. Acta* **657**, 128. **30**. Bernstein & Schraer (1972) *J. Biol. Chem.* **247**, 1306. **31**. Wingo, Tu, Laipis & Silverman (2001) *Biochem. Biophys. Res. Commun.* **288**, 666. **32**. Walk & Metzner (1975) *Hoppe-Seyler's Z. Physiol. Chem.* **356**, 1733. **33**. Heck, Tanhauser, Manda, *et al.* (1994) *J. Biol. Chem.* **269**, 24742. **34**. Earnhardt, Qian, Tu, *et al.* (1998) *Biochemistry* **37**, 10837. **35**. Kisiel & Graf (1972) *Phytochemistry* **11**, 113.

36. Baird, Waheed, Okuyama, Sly & Fierke (1997) *Biochemistry* **36**, 2669. **37.** Gerencser (1996) *CRC Crit. Rev. Biochem.* **31**, 303. **38.** Drummond & Yamamoto (1971) *The Enzymes*, 3rd, **4**, 355. **39.** Pradhan & Sander (1973) *Life Sci.* **13**, 1747. **40.** Beaton (1961) *Can. J. Med. Sci.* **39**, 663. **41.** Thornley-Brown, Bass & Welbourne (1982) *Biochem. Pharmacol.* **31**, 3347. **42.** Allison (1985) *Meth. Enzymol.* **113**, 419. **43.** Yasumoto, Iwami, Fushiki & Mitsuda (1978) *J. Biochem.* **84**, 1227. **44.** Rambabu & Pattabiraman (1982) *J. Biosci.* **4**, 287.

Acetic Anhydride

This carboxylic anhydride (FW = 102.09 g/mol; CAS 108-24-7; M.P. = –73°C; B.P. = 139°C) reacts with amino groups in peptides/proteins under slightly alkaline conditions. If the amino group is essential for biological or catalytic activity, its acetylation will often inhibit the modified peptide/protein. Tyrosyl residues are also acetylated; however, under alkaline condition, *O*-acetyltyrosyl residues hydrolyze readily. The latter can occasionally be trapped as the *N*-acetohydroxamate after reaction with neutral hydroxylamine. Acetic anhydride hydrolyzes slowly ($t_{1/2}$ at 0°C = 50 min; $t_{1/2}$ at 25°C = 4.5 min) to yield two molecules of acetic acid, the presence of which imparts a distinct odor. ***CAUTION***: This liquid is a skin and eye irritant. Avoid breathing acetic anhydride vapor. In view of its relatively low flash point of 54°C, acetic anhydride is a combustible liquid requiring appropriate caution. **Target(s):** glucose dehydrogenase (1); citrate (*si*)-synthase (2); thrombin (3,5); carboxypeptidase A (4,6,22); chymotrypsin (7); acetoacetate decarboxylase (8); trypsin (9); crotoxin B (10); dendrotoxin I (11); cAMP-dependent protein kinase (12); Na$^+$/K$^+$-exchanging ATPase (13,14); collagenase (15); alkaline phosphatase (16,24); fructose-1,6-bisphosphatase (17); arylsulfatase (18); lysozyme (19); glutamine synthetase (20,25); glutamate dehydrogenase (21); hexokinase (23); phenylalanine ammonia-lyase (26,27); 3-oxoacid CoA-transferase, *or* 3-ketoacid CoA-transferase (28); aryl-acylamidase, but stimulated at low concentrations (29,30); butyrylcholinesterase, but stimulated at low concentrations (29); acetylcholinesterase (30); leukotriene-A4 hydrolase (31); endo-1,4-β-xylanase (32); ribonuclease T$_1$ (33); poly(3-hydroxyoctanoate) depolymerase (34,35); phospholipase A$_2$ (36); sterol *O*-acyltransferase, *or* cholesterol *O*-acyltransferase, *or* ACAT (37,38); and glycerol-3-phosphate *O*-acyltransferase (39). **1.** Sadoff (1966) *Meth. Enzymol.* **9**, 103. **2.** Srere (1969) *Meth. Enzymol.* **13**, 3. **3.** Magnusson (1970) *Meth. Enzymol.* **19**, 157. **4.** Pétra (1970) *Meth. Enzymol.* **19**, 460. **5.** Lundblad, Kingdon & Mann (1976) *Meth. Enzymol.* **45**, 156. **6.** Sigman & Mooser (1975) *Ann. Rev. Biochem.* **44**, 889. **7.** Oppenheimer, Labouesse & Hess (1966) *J. Biol. Chem.* **241**, 2720. **8.** O'Leary & Westheimer (1968) *Biochemistry* **7**, 913. **9.** Chevallier, Yon & Labouesse (1969) *Biochim. Biophys. Acta* **181**, 73. **10.** Soares, Mancin, Cecchini, *et al.* (2001) *Int. J. Biochem. Cell Biol.* **33**, 877. **11.** Harvey, Rowan, Vatanpour, *et al.* (1997) *Toxicon* **35**, 1263. **12.** Buechler, Vedvick & Taylor (1989) *Biochemistry* **28**, 3018. **13.** Xu & Kyte (1989) *Biochemistry* **28**, 3009. **14.** De Pont, Van Emst-De Vries & Bonting (1984) *J. Bioenerg. Biomembr.* **16**, 263. **15.** Mookhtiar, Wang & Van Wart (1986) *Arch. Biochem. Biophys.* **246**, 645. **16.** Reid & Wilson (1971) *The Enzymes*, 3rd ed., **4**, 373. **17.** Pontremoli & Horecker (1971) *The Enzymes*, 3rd ed., **4**, 611. **18.** Nicholls & Roy (1971) *The Enzymes*, 3rd ed., **5**, 21. **19.** Imoto, Johnson, North, Phillips & Rupley (1972) *The Enzymes*, 3rd ed., **7**, 665. **20.** Meister (1974) *The Enzymes*, 3rd ed., **10**, 699. **21.** Smith, Austen, Blumenthal & Nyc (1975) *The Enzymes*, 3rd ed., **11**, 293. **22.** Riordan, Vallee & Saunders (1963) *Biochemistry* **2**, 1460. **23.** Grillo (1968) *Enzymologia* **34**, 7. **24.** Thomas & Moss (1972) *Enzymologia* **42**, 65. **25.** Wilk, Meister & Haschemeyer (1969) *Biochemistry* **8**, 3168. **26.** Hodgins (1972) *Arch. Biochem. Biophys.* **149**, 91. **27.** Jorrin, Lopez-Valbuena & Tena (1988) *Biochim. Biophys. Acta* **964**, 73. **28.** Hersh & Jencks (1967) *J. Biol. Chem.* **242**, 3468. **29.** Boopathy & Balasubramanian (1985) *Eur. J. Biochem.* **151**, 351. **30.** Majumdar & Balasubramanian (1984) *Biochemistry* **23**, 4088. **31.** Mueller, Samuelsson & Haeggström (1995) *Biochemistry* **34**, 3536. **32.** Gupta, Bhushnan & Hoondal (2000) *J. Appl. Microbiol.* **88**, 325. **33.** Takahashi (1966) *J. Biochem.* **60**, 239. **34.** Kim, Kim, Kim & Rhee (2005) *J. Microbiol.* **43**, 285. **35.** Kim, Kim, Nam, Bae & Rhee (2003) *Antonie Leeuwenhoek* **83**, 183. **36.** Shashidharamurthy & Kemparaju (2006) *Toxicon* **47**, 727. **37.** Kinnunen, DeMichele & Lange (1988) *Biochemistry* **27**, 7344. **38.** Chang, Chang & Cheng (1997) *Ann. Rev. Biochem.* **66**, 613. **39.** Negrel, Ailhaud & Mutaftschiev (1973) *Biochim. Biophys. Acta* **291**, 635.

Acetoacetate (Acetoacetic Acid)

This β-keto acid (FW$_{free-acid}$ = 102.09 g/mol; CAS 541-50-4; M.P. = 36-37°C; pK_a = 3.58 at 18°C), also known as 3-oxobutanoic acid, is one of the three ketone bodies (*i.e.*, fatty acid-derived metabolites) formed during ketogenesis (***See also*** *Acetone and D-3-Hydroxybutyrate*). **Formation:** When carbohydrates are scarce, metabolic energy is obtained from fatty acid degradation, and ketone bodies are produced in hepatocytes from acetyl-CoA within the mitochondrial matrix. High acetyl CoA concentrations (from β-oxidation of fatty acids) inhibit pyruvate dehydrogenase and also activate pyruvate carboxylase allosterically. High ATP and NADH inhibit isocitrate dehydrogenase, decreasing TCA cycle flux, resulting in an increase malate concentration (derived from oxaloacetate). Malate leaves the mitochondrion to undergo gluconeogenesis. Those on low-carbohydrate diets also develop ketosis (called nutritional ketosis), where ketone body concentrations typically range between 0.5 to 5 mM. Ketone bodies are taken up by various cells, whereupon they reconverted to acetyl-CoA, which enters the tricarboxylic acid cycle and oxidized in the mitochondria for energy. By providing an alternative substrate to fuel oxidative phosphorylation, ketone bodies support mammalian cell survival during periods of energy insufficiency. The free acid is unstable to heat, decomposing to acetone and CO$_2$, and is miscible with water and ethanol. Neutral solutions are stable for two-to-three weeks at –15°C. In the pathological ketoacidosis, however, they can rise to 15-25 mM, lowering blood pH considerably. Ketoacidosis is a complication of untreated Type I diabetes and is occasionally in end-stage Type II diabetes (known as diabetic ketoacidosis). **Acetoacetate-Requiring & -Utizing Enzymes:** Acetoacetate is the direct product of reactions catalyzed by 3-hydroxybutyrate dehydrogenase, hydroxyacid:oxoacid transhydrogenase, 2-oxopropyl-CoM reductase, butyrate:acetoacetate CoA-transferase, acetoacetyl-CoA hydrolase, fumarylacetoacetase, hydroxymethylglutaryl-CoA lyase, acetylene-carboxylate hydratase, and acetone carboxylase. It is a substrate for acetoacetate decarboxylase, 3-oxoacid CoA-transferase, and acetoacetate:CoA ligase. **Target(s):** succinate dehydrogenase (1-3); aldehyde reductase, *or* aldose reductase (4); succinate:3-hydroxy-3-methylglutarate CoA-transferase (5,11); papain (6); monoamine oxidase (6); urease (6); F$_1$ ATPase (7); 3-methylcrotonyl-CoA carboxylase (8); malate dehydrogenase (9); oxaloacetate tautomerase (10); allophanate hydrolase (12); prostaglandin-endoperoxide synthase, *or* cyclooxygenase, weakly inhibited (13). **1.** Veeger, DerVartanian & Zeylemaker (1969) *Meth. Enzymol.* **13**, 81. **2.** Hatefi (1978) *Meth. Enzymol.* **53**, 27. **3.** Hatefi & Stiggall (1976) *The Enzymes*, 3rd ed., **13**, 175. **4.** Hayman & Kinoshita (1965) *J. Biol. Chem.* **240**, 877. **5.** Deana, Rigoni, Deana & Galzigna (1981) *Biochim. Biophys. Acta* **662**, 119. **6.** Nath & Sivakumar (1969) *Enzymologia* **36**, 86. **7.** Reddi, Bhai, Sivakumar, Nath & M. Nath (1969) *Enzymologia* **37**, 227. **8.** Landaas (1977) *Scand. J. Clin. Lab. Invest.* **37**, 411. **9.** Green (1936) *Biochem. J.* **30**, 2095. **10.** Annett & Kosicki (1969) *J. Biol. Chem.* **244**, 2059. **11.** Francesconi, Donella-Deana, Furlanetto, *et al.* (1989) *Biochim. Biophys. Acta* **999**, 163. **12.** Maitz, Haas & Castric (1982) *Biochim. Biophys. Acta* **714**, 486. **13.** Ryan & Davis (1988) *Biochem. Soc. Trans.* **16**, 398.

Acetoacetate Ethyl Ester, *See Ethyl Acetoacetate*

Acetoacetyl-CoA

This β-ketoacyl-CoA (FW = 851.61 g/mol; CAS 1420-36-6), also known as acetoacetyl-*S*-CoA and 3-oxobutyroyl-CoA, is an intermediate in fatty acid β-oxidation and ketogenesis. Note that acetoacetyl-CoA is rapidly hydrolyzed in alkaline solutions. Acetoacetyl-CoA is a product of the reactions catalyzed by 3-hydroxyacyl-CoA dehydrogenase, 3-hydroxybutyryl-CoA dihydrogenase, acetyl-CoA *C*-acetyltransferase, butyrate: acetoacetate CoA-transferase, and acetoacetate:CoA ligase. It is a substrate for β-ketoacyl-CoA thiolase, hydroxy-methylglutaryl-CoA synthase, and acetoacetyl-CoA hydrolase. Note that it is also a strong substrate inhibitor of 3-hydroxy-3-methylglutaryl-CoA synthase. **Target(s):**

pyruvate carboxylase (1,9), 3-methylcrotonoyl-CoA carboxylase (2,7,15,18,27); enoyl-CoA hydratase, *or* crotonase, the enol tautomer is the inhibitory species (3-5,8,19,20,26,29); [branched-chain α-keto-acid dehydrogenase] kinase, *or* 3-methyl-2-oxobutanoate dehydrogenase (acetyl transferring)] kinase (6,11,22,32); 3-ketoacid CoA-transferase (10); 3-methylcrotonoyl-CoA carboxylase (12); 3-hydroxy-3-methylglutaryl-CoA reductase (13,28); citrate synthase (14); short-chain acyl-CoA dehydrogenase (16); L-3-hydroxyacyl-CoA dehydrogenase (17); acyl-CoA dehydrogenase (21); 2-methyl branched-chain acyl-CoA dehydrogenase (23); succinate:3-hydroxy-3 methylglutarate CoA-transferase (24); glutaryl-CoA dehydrogenase (25); 3-hydroxyisobutyryl-CoA hydrolase (30,31); pantothenate kinase (33); [acyl-carrier-protein] *S*-malonyltransferase (34); acetyl-CoA *C*-acyltransferase (35,36); and arylamine *N*-acetyltransferase (37). **1.** Scrutton & Fatebene (1976) *FEBS Lett.* **62**, 220. **2.** Diez, Wurtele & Nikolau (1994) *Arch. Biochem. Biophys.* **310**, 64. **3.** Waterson & Hill (1972) *J. Biol. Chem.* **247**, 5258. **4.** Steinman & Hill (1975) *Meth. Enzymol.* **35**, 136. **5.** Fong & Schulz (1981) *Meth. Enzymol.* **71**, 390. **6.** Paxton (1988) *Meth. Enzymol.* **166**, 313. **7.** Wurtele & Nikolau (2000) *Meth. Enzymol.* **324**, 280. **8.** Hill & Teipel (1971) *The Enzymes*, 3rd. ed., **5**, 539. **9.** Scrutton & Young (1972) *The Enzymes*, 3rd ed., **6**, 1. **10.** Jencks (1973) *The Enzymes*, 3rd ed., **9**, 483. **11.** Randle, Patston & Espinal (1987) *The Enzymes*, 3rd ed., **18**, 97. **12.** Chen, Wurtele, Wang & Nikolau (1993) *Arch. Biochem. Biophys.* **305**, 103. **13.** Kirtley & Rudney (1967) *Biochemistry* **6**, 230. **14.** Srere & Matsuoka (1972) *Biochem. Med.* **6**, 262. **15.** Chen, Wurtele, Wang & Nikolau (1993) *Arch. Biochem. Biophys.* **305**, 103. **16.** Battaile, Molin-Case, Paschke, *et al.* (2002) *J. Biol. Chem.* **277**, 12200. **17.** Schifferdecker & Schulz (1974) *Life Sci.* **14**, 1487. **18.** Landaas (1977) *Scand. J. Clin. Lab. Invest.* **37**, 411. **19.** Furuta, Miyazawa, Osumi, Hashimoto & Ui (1980) *J. Biochem.* **88**, 1059. **20.** Fujita, Shimakata & Kusaka (1980) *J. Biochem.* **88**, 1045. **21.** Mizzer & Thorpe (1981) *Biochemistry* **20**, 4965. **22.** Paxton & Harris (1984) *Arch. Biochem. Biophys.* **231**, 48. **23.** Komuniecki, Fekete & Thissen-Parra (1985) *J. Biol. Chem.* **260**, 4770. **24.** Francešconi, Donella-Deana, Furlanetto, *et al.* (1989) *Biochim. Biophys. Acta* **999**, 163. **25.** Byron, Stankovich & Husain (1990) *Biochemistry* **29**, 3691. **26.** Engel, Mathieu, Zeelen, Hiltunen & Wierenga (1996) *EMBO J.* **15**, 5135. **27.** Maier & Lichtenthaler (1998) *J. Plant Physiol.* **152**, 213. **28.** Kawachi & Rudney (1970) *Biochemistry* **9**, 1700. **29.** Fong & Schulz (1977) *J. Biol. Chem.* **252**, 542. **30.** Shimomura, Murakami, Nakai, *et al.* (2000) *Meth. Enzymol.* **324**, 229. **31.** Shimomura, Murakami, Fujitsuka, *et al.* (1994) *J. Biol. Chem.* **269**, 14248. **32.** Espinal, Beggs & Randle (1988) *Meth. Enzymol.* **166**, 166. **33.** Vallari, Jackowski & Rock (1987) *J. Biol. Chem.* **262**, 2468. **34.** Guerra & Ohlrogge (1986) *Arch. Biochem. Biophys.* **246**, 274. **35.** Yamashita, Itsuki, Kimoto, Hiemori & Tsuji (2006) *Biochim. Biophys. Acta* **1761**, 17. **36.** Pantazaki, Ioannou & Kyriakidis (2005) *Mol. Cell. Biochem.* **269**, 27. **37.** Kawamura, Graham, Mushtaq, *et al.* (2005) *Biochem. Pharmacol.* **69**, 347.

Acetobixin

This cellulose biosynthesis inhibitor, *or* CBI (FW = 287.38 g/mol) reduces incorporation [^{14}C]-labeled glucose into crystalline cellulose. Cellulose is biosynthesized at the plasma membrane by the multi-enzyme/multi-component cellulose synthase A (CesA) complex. In living *Arabidopsis* cells, acetobixan treatment caused CesA particles localized at the plasma membrane (PM) to rapidly re-localize to cytoplasmic vesicles. Moreover, cortical microtubule dynamics were not disrupted by acetobixan, suggesting specific activity towards cellulose synthesis. **1.** Xia, Lei, Brabham, *et al.* (2014) *PLoS One* **9**, e95245.

Acetochlor

This toluidine derivative (FW = 269.77 g/mol; CAS 34256-82-1; low water solubility (200 ppm)), also known as MON-097, Harness™, Surpass™, and Trophy™, and named systematically as 2-chloro-*N*-(ethoxymethyl)-*N*-(2-ethyl-6-methylphenyl)acetamide, is a pre-emergent herbicide that acts on most annual grasses and certain broadleaf weeds. Acetochlor inhibits human multidrug resistance transporters MDR1, as measured by ATPase activity, inhibition of fluorescent dye efflux and vesicular transport. This herbicide also inhibits synthesis of very-long-chain fatty acids in plants. *Caution:* Acetoclor is classified as a probable human carcinogen.

Acetone

This common organic solvent (FW = 58.08 g/mol; CAS 67-64-1; M.P. = – 94.7 °C; B.P. = 56.3 °C; flammable liquid; freely soluble in water, ethanol, chloroform, and other polar organic solvents; dipole moment of 2.69 D at 20°C; dielectric constant of 20.7 at 25°C), inhibits many enzymes by forming reversible adducts with active-site amino and thiol groups. Acetone is also a nonspecific protein denaturant at high concentrations. A highly effective lipid-extracting solvent, ice-cold acetone is frequently used to prepare tissue powders that are both long-lived and well suited for the extraction and purification of both water-soluble and membrane-associated proteins (1). **Target(s):** catechol 2,3-dioxygenase (2,17,71-73); subtilisin (3); lipase, *or* triacylglycerol lipase (4,28,54-56); acetyl-CoA carboxylase (5); L-glycol dehydrogenase (6,21); sterol *O* acyltransferase, *or* cholesterol *O*-acyltransferase, *or* ACAT (7,63,64); microsomal glycerol-3-phosphate *O*-acyltransferase (8,15); β-fructofuranosidase, *or* invertase (9); α-glucosidase, *or* maltase (10); β-lactamase (11); carbonic anhydrase, weakly inhibited (12,25,34); DNA polymerase (13); camphor 5-monooxygenase, *or* CYP101 (14); chymotrypsin (16); Ca^{2+}-dependent ATPase (18); epoxide hydrolase (19); acetoacetate decarboxylase (20); enolase, in presence of ammonium sulfate and EDTA (22); diacetyl reductase, *or* α-diketone reductase, NAD$^+$-dependent (23); adenylate cyclase, *or* adenyl cyclase (24); esterase *or* lipase (26); asparaginase (27); cystathionine γ-lyase, *or* cysteine desulfurase (29); dihydropyrimidinase, *or* hydantoinase (30); pyrophosphatase, *or* inorganic diphosphatase (31); aminoacylase (32); ribonuclease A, in the presence of sodium borohydride (33); 3-hydroxybutyrate dehydrogenase (35); hydroxynitrilase, *or* hydroxynitrile lyase (36,37); uroporphyrinogen decarboxylase (38); myosin ATPase, strongly inhibited under alkaline conditions (39); thiamin-triphosphatase (40); cytidine deaminase (41); pantothenase (42); arylformamidase (43); coagulation factor IXa (44); pyroglutamyl-peptidase I (45); cholesterol-5,6-oxide hydrolase (46); soluble epoxide hydrolase (47-49); poly(ADP-ribose) glycohydrolase (50); β-D-fucosidase (51); chitinase (52); retinyl-palmitate esterase (53); carboxylesterase, weakly inhibited (57,58); diacylglycerol cholinephosphotransferase (59); ethanolaminephospho-transferase (59); *N* acetyllactosamine synthase (60); lactose synthase (60); glucuronosyltransferase (61); 2,3 diaminopropionate *N*-oxalyltransferase (62); 2-acylglycerol *O*-acyltransferase, *or* monoacylglycerol *O*-acyltransferase, weakly inhibited (65); steroid 11β-monooxygenase (66); bacterial luciferase, *or* alkanal monooxygenase (FMN-linked), weakly inhibited (67); phenylacetone monooxygenase (68); nitronate monooxygenase (69); biphenyl-2,3-diol 1,2-dioxygenase (70); peroxidase (74); and laccase (75). **1.** Dixon & Webb (1960) Enzymes, Academic Press. **2.** Nozaki (1970) *Meth. Enzymol.* **17A**, 522. **3.** Ottesen & Svendsen (1970) *Meth. Enzymol.* **19**, 199. **4.** Khoo & Steinberg (1975) *Meth. Enzymol.* **35**, 181. **5.** Mishina, Kamiryo & Numa (1981) *Meth. Enzymol.* **71**, 37. **6.** Burgos & Martín-Sarmiento (1982) *Meth. Enzymol.* **89**, 523. **7.** Billheimer (1985) *Meth. Enzymol.* **111**, 286. **8.** Haldar & Vancura (1992) *Meth. Enzymol.* **209**, 64. **9.** Neuberg & Mandl (1950) *The Enzymes*, 1st ed., **1** (part 1), 527. **10.** Gottschalk (1950) *The Enzymes*, 1st ed., **1** (part 1), 551. **11.** Abraham (1951) *The Enzymes*, 1st ed., **1** (part 2), 1170. **12.** Lindskog, Henderson, Kannan, *et al.* (1971) *The Enzymes*, 3rd. ed., **5**, 587. **13.** Loeb (1974) *The Enzymes*, 3rd ed., **10**, 173. **14.** Ullrich & Duppel (1975) *The Enzymes*, 3rd ed., **12**, 253. **15.** Bell & Coleman (1983) *The Enzymes*, 3rd ed., **16**, 87. **16.** Clement & Bender (1963) *Biochemistry* **2**, 836. **17.** Zhang, Yin, Zheng, *et al.* (1998) *Sheng Wu Hua Xue Yu Sheng Wu Wu Li Xue Bao (Shanghai)* **30**, 579. **18.** Kosterin, Bratkova, Slinchenko & Zimina (1998) *Biofizika* **43**, 1037. **19.** Schladt, Hartmann, Worner, Thomas & Oesch (1988) *Eur. J. Biochem.* **176**, 31. **20.** Lopez-Soriano, Alemany & Argiles

(1985) *Int. J. Biochem.* **17**, 1271. **21.** Gonzalez Prieto, Martin-Sarmiento & Burgos (1983) *Arch. Biochem. Biophys.* **224**, 372. **22.** Winstead & Wold (1965) *Biochemistry* **4**, 2145. **23.** Gonzalez, Vidal, Bernardo & Martin (1988) *Biochimie* **70**, 1791. **24.** Huang, Smith & Zahler (1982) *J. Cyclic Nucleotide Res.* **8**, 385. **25.** Whitney, Nyman & Malmström (1967) *J. Biol. Chem.* **242**, 4212. **26.** Falk (1913) *J. Amer. Chem. Soc.* **35**, 616. **27.** Geddes & Hunter (1928) *J. Biol. Chem.* **77**, 197. **28.** Weinstein & Wynne (1936) *J. Biol. Chem.* **112**, 649. **29.** Laskowski & Fromageot (1941) *J. Biol. Chem.* **140**, 663. **30.** Eadie, Bernheim & Bernheim (1949) *J. Biol. Chem.* **181**, 449. **31.** Balocco (1962) *Enzymologia* **24**, 275. **32.** Tosa, Mori & Chibata (1971) *Enzymologia* **40**, 49. **33.** Means & Feeney (1968) *Biochemistry* **7**, 2192. **34.** Pocker & Stone (1968) *Biochemistry* **7**, 2936. **35.** Gotterer (1969) *Biochemistry* **8**, 641. **36.** Chueskul & Chulavatnatol (1996) *Arch. Biochem. Biophys.* **334**, 401. **37.** Zuegg, Gruber, Gugganig, Wagner & Kratky (1999) *Protein Sci.* **8**, 1990. **38.** Kawanishi, Seki & Sano (1983) *J. Biol. Chem.* **258**, 4285. **39.** Kameyama, Ichikawa, Sunaga, *et al.* (1985) *J. Biochem.* **97**, 625. **40.** Iwata, Baba, Matsuda & Terashita (1975) *J. Neurochem.* **24**, 1209. **41.** Wang, Sable & Lampen (1950) *J. Biol. Chem.* **184**, 17. **42.** Airas (1983) *Anal. Biochem.* **134**, 122. **43.** Santti (1969) *Hoppe-Seyler's Z. Physiol. Chem.* **350**, 1279. **44.** Stürzebecher, Kopetzki, Bode & Hopfner (1997) *FEBS Lett.* **412**, 295. **45.** Exterkate (1979) *FEMS Microbiol. Lett.* **5**, 111. **46.** Levin, Michaud, Thomas & Jerina (1983) *Arch. Biochem. Biophys.* **220**, 485. **47.** Morisseau, Archelas, Guitton, *et al.* (1999) *Eur. J. Biochem.* **263**, 386. **48.** Meijer & Depierre (1985) *Eur. J. Biochem.* **150**, 7. **49.** Schladt, Hartman, Wörner, Thomas & Oesch (1988) *Eur. J. Biochem.* **176**, 31. **50.** Hatakeyama, Nemoto, Ueda & Hayaishi (1986) *J. Biol. Chem.* **261**, 14902. **51.** Nunoura, Ohdan, Yano, Yamamoto & Kumagai (1996) *Biosci. Biotechnol. Biochem.* **60**, 188. **52.** Wang, Hsiao & Chang (2002) *J. Agric. Food Chem.* **50**, 2249. **53.** Mahadevan, Ayyoub & Roels (1966) *J. Biol. Chem.* **241**, 57. **54.** Quyen, Le, Nguyen, Oh & Lee (2005) *Protein Expr. Purif.* **39**, 97. **55.** Quyen, Schmidt-Dannert & Schmid (2003) *Protein Expr. Purif.* **28**, 102. **56.** Sharma, Chisti & Banerjee (2001) *Biotechnol. Adv.* **19**, 627. **57.** Choi, Kim, Kim, Ryu & Kim (2003) *Protein Expr. Purif.* **29**, 85. **58.** Junge & Heymann (1979) *Eur. J. Biochem.* **95**, 519. **59.** Coleman & Bell (1977) *J. Biol. Chem.* **252**, 3050. **60.** Pisvejcova, Rossi, Husakova, *et al.* (2006) *J. Mol. Catal. B* **39**, 98. **61.** Uchaipichat, Mackenzie, Guo, *et al.* (2004) *Drug Metab. Dispos.* **32**, 413. **62.** Malathi, Padmanaban & Sarma (1970) *Phytochemistry* **9**, 1603. **63.** Erickson, Shrewsbury, Brooks & Meyer (1980) *J. Lipid Res.* **21**, 930. **64.** Taketani, Nishino & Katsuki (1979) *Biochim. Biophys. Acta* **575**, 148. **65.** Coleman & Haynes (1986) *J. Biol. Chem.* **261**, 224. **66.** Watanucki, Tilley & Hall (1978) *Biochemistry* **17**, 127. **67.** Curry, Lieb & Franks (1990) *Biochemistry* **29**, 4641. **68.** De Gonzalo, Ottolina, Zambianchi, Fraaije & Carrea (2006) *J. Mol. Catal. B* **39**, 91. **69.** Kido, Soda & Asada (1978) *J. Biol. Chem.* **253**, 226. **70.** Vaillancourt, Han, Fortin, Bolin & Eltis (1998) *J. Biol. Chem.* **273**, 34887. **71.** Kobayashi, Ishida, Horiike, *et al.* (1995) *J. Biochem.* **117**, 614. **72.** Bertini, Briganti & Scozzafava (1994) *FEBS Lett.* **343**, 56. **73.** Kachhy & Modi (1976) *Indian J. Biochem. Biophys.* **13**, 234. **74.** Marzouki, Limam, Smaali, Ulber & Marzouki (2005) *Appl. Biochem. Biotechnol.* **127**, 201. **75.** Junghanns, Pecyna, Böhm, *et al.* (2009) *Appl. Microbiol. Biotechnol.* **84**, 1095.

S-Acetonyl-CoA

This coenzyme A derivative (FW = 821.59 g/mol), also known as 1-[(S-(coenzyme A)]propan-2-one, is a nonreactive acetyl-CoA analogue and is a potent competitive inhibitor of citrate synthase, phosphotransacetylase, and carnitine acetyltransferase. **Target(s):** citrate (*si*)-synthase (1,3,5); protein acetyltransferase (2,4); phosphate acetyltransferase, *or* phosphotransacetylase (1); carnitine acetyltransferase (1); malate synthase (6); and histone acetyltransferase (7). **1.** Rubenstein & Dryer (1980) *J. Biol. Chem.* **255**, 7858. **2.** Redman & Rubenstein (1984) *Meth. Enzymol.* **106**, 179. **3.** Wiegand, Remington, Deisenhofer & Huber (1984) *J. Mol. Biol.* **174**, 205. **4.** Rubenstein, Smith, Deuchler & Redman (1981) *J. Biol. Chem.* **256**, 8149. **5.** Kurz, Shah, Crane, *et al.* (1992) *Biochemistry* **31**, 7899. **6.** Miernyk & Trelease (1981) *Phytochemistry* **20**, 2657. **7.** Cullis, Wolfenden, Cousens & Alberts (1982) *J. Biol. Chem.* **257**, 12165.

Acetophenone

This aromatic ketone (FW = 120.15 g/mol; CAS 98-86-2; M.P. = 20.5°C; B.P. = 202°C; slightly soluble in water) inhibits carboxylase (1,2), aldehyde dehydrogenase (NAD$^+$), K_i = 48 μM (3,4), chymotrypsin (5), esterase (6), lipase (7), NAD$^+$ ADP-ribosyltransferase, poly(ADP-ribose) polymerase, IC$_{50}$ = 2.3 mM (8), tyrosinase, *or* monophenol monooxygenase, IC$_{50}$ = 0.85 mM (9), catechol 2,3-dioxygenase (10). **1.** Karrer & Visconti (1947) *Helv. Chim. Acta* **30**, 268. **2.** Massart (1950) *The Enzymes*, 1st ed., **1** (part 1), 307. **3.** Weiner, Freytag, Fox & Hu (1982) *Prog. Clin. Biol. Res.* **114**, 91. **4.** Deitrich & Hellerman (1963) *J. Biol. Chem.* **238**, 1683. **5.** Wildnauer & Canady (1966) *Biochemistry* **5**, 2885. **6.** Murray (1929) *Biochem. J.* **23**, 292. **7.** Weinstein & Wynne (1936) *J. Biol. Chem.* **112**, 649. **8.** Banasik, Komura, Shimoyama & Ueda (1992) *J. Biol. Chem.* **267**, 1569. **9.** Liu, Yi, Wan, Ma & Song (2008) *Bioorg. Med. Chem.* **16**, 1096. **10.** Bertini, Briganti & Scozzafava (1994) *FEBS Lett.* **343**, 56.

Acetosyringone

This acetophenone-based, plant-pathogen recognition substance (FW = 196.20 g/mol; CAS 2478-38-8) is a volatile antioxidant produced within the wounds of certin plants. Acetosyringone triggers the *VirA* gene in *Agrobacterium tumefaciens* to promote more efficient transformation. Acetosyringone is also produced by the male leaffooted bug *Leptoglossus phyllopus* for use as a pheromone. **Target(s):** α-glucosidase (1). **1.** Berthelot & Delmotte (1999) *Appl. Environ. Microbiol.* **65**, 2907.

(3*aS*,6*S*,6*aR*)-4-Acetoxyacetyl-6-methyl-5-oxohexahydro-pyrrolo[3,2-*b*]pyrrole-1-carboxylic Acid Benzyl Ester

This pyrrolidine-5,5-*trans*-lactam (FW = 374.39 g/mol) inhibits human cytomegalovirus assemblin, IC$_{50}$ = 6 μM, and the *rel*-(3*aS*,6*R*,6*aR*)-stereoisomer is a weaker inhibitor (IC$_{50}$ = 805 μM). **Target(s):** assemblin (1,2); elastase (2); and thrombin (2). **1.** Borthwick, Crame, Ertl, *et al.* (2002) *J. Med. Chem.* **45**, 1. **2.** Borthwick, Angier, Crame, *et al.* (2000) *J. Med. Chem.* **43**, 4452.

3β-Acetoxy-17-(4'-imidazolyl)androsta-5,16-diene

This synthetic steroid (FW = 380.25 g/mol) inhibits human steroid 17α-monooxygenase, *or* CYP17, (IC$_{50}$ = 75 nM), a key enzyme in androgen biosynthesis. The partially reduced 3β-acetoxy-17β-(4'-imidazolyl)androst-

5-ene is a slightly weaker inhibitor (human IC_{50} = 200 nM). **1**. Ling, Li, Liu, *et al.* (1997) *J. Med. Chem.* **40**, 3297.

o-(Acetoxyphenyl)hept-2-ynyl Sulfide

This aspirin analogue (FW = 262.37 g/mol; Abbreviated APHS) preferentially acetylates and irreversibly inactivates cyclooxygenase. Relative to aspirin, APHS was 60 times as reactive against COX-2 and 100 times as selective for its inhibition; it also inhibited COX-2 in cultured macrophages and colon cancer cells and in the rat air pouch *in vivo*. **Target(s):** cyclooxygenase-1, IC_{50} = 10 μM (1); and cyclooxygenase-2, IC_{50} = 0.8 μM (2). **1**. Hochgesang, Nemeth-Cawley, Rowlinson, Caprioli & Marnett (2003) *Arch. Biochem. Biophys.* **409**, 127. **2**. Kalgutkar, Crews, Rowlinson, *et al.* (1998) *Science* **280**, 1268.

Acetylacetone

This β-diketone (FW = 100.12 g/mol; flammable liquid; M.P. = –23°C; B.P. = 140.5°C; moderately soluble in water, 0.1 g/mL), also known as 2,4-pentanedione and acetoacetone, reacts with amino groups and also chelates metal ions. **Target(s):** acetoacetate decarboxylase, slow-binding inhibition (1-3,8,14,15); NAD^+ glycohydrolase, *or* NADase (4,13); aspartate aminotransferase (5); peroxidase (6); lipase, *or* triacylglycerol lipase (7,17); actin interaction with tropomyosin (9); malic enzyme, *or* malate dehydrogenase (10); pyridoxamine-5'-phosphate oxidase (11); estradiol 17β-dehydrogenase (12); acid phosphatase (16); aminolevulinate aminotransferase (18); tryptophan dimethylallyltransferase (19); and tyrosinase, *or* monophenol monooxygenase (20). **1**. Neece & Fridovich (1967) *J. Biol. Chem.* **242**, 2939. **2**. Morrison (1982) *Trends Biochem. Sci.* **7**, 102. **3**. Fridovich (1972) *The Enzymes*, 3rd ed., **6**, 255. **4**. Yost & Anderson (1982) *J. Biol. Chem.* **257**, 767. **5**. Gilbert & O'Leary (1977) *Biochim. Biophys. Acta* **483**, 79. **6**. Gemant (1977) *Mol. Biol. Rep.* **3**, 283. **7**. Weinstein & Wynne (1936) *J. Biol. Chem.* **112**, 649. **8**. Fridovich (1968) *J. Biol. Chem.* **243**, 1043. **9**. El-Saleh, Thieret, Johnson & Potter (1984) *J. Biol. Chem.* **259**, 11014. **10**. Chang & Huang (1981) *Biochim. Biophys. Acta* **660**, 341. **11**. Choi & McCormick (1981) *Biochemistry* **20**, 5722. **12**. Inano, Ohba & Tamaoki (1983) *J. Steroid Biochem.* **19**, 1617. **13**. Verhaeghe, De Wolf, Lagrou, *et al.* (1990) *Int. J. Biochem.* **22**, 197. **14**. Kimura, Yasuda, Tanigaki-Nagae, *et al.* (1986) *Agric. Biol. Chem.* **50**, 2509. **15**. Autor & Fridovich (1970) *J. Biol. Chem.* **245**, 5214. **16**. Sugiura, Kawabe, Tanaka, Fujimoto & Ohara (1981) *J. Biol. Chem.* **256**, 10664. **17**. Muderhwa, Ratomahenina, Pina, Graille & Galzy (1986) *Appl. Microbiol. Biotechnol.* **23**, 348. **18**. Neuberger & Turner (1963) *Biochim. Biophys. Acta* **67**, 342. **19**. Cress, Chayet & Rilling (1981) *J. Biol. Chem.* **256**, 10917. **20**. Oda, Kato, Isoda, *et al.* (1989) *Agric. Biol. Chem.* **53**, 2053.

N-Acetyl-Adhesin Amide Fragment 1025-1044

This acetylated eicosapeptide amide (MW = 2161.44 g/mol; *Sequence*: N-Acetyl-QLKTADLPAGRDETTSFVLV-NH$_2$) inhibits *Streptococcus mutans* adhesion to salivary receptors. *S. mutans* is the main etiological agent of dental caries. The earliest step in microbial infection is adherence by specific microbial adhesins to the mucosa of the oro-intestinal, nasorespiratory, or genito-urinary tract. The binding of a cell surface adhesin of *Streptococcus mutans* to salivary receptors can be inhibited *in vitro*, as measured by surface plasmon resonance, using a synthetic peptide (p1025) corresponding to residues 1025-1044 of the adhesin. **1**. Kelly, Younson, Hikmat, *et al.* (1999) *Nat. Biotechnol.* **17**, 42.

N-Acetyl-L-alanine

This acetylated amino acid ($FW_{free-acid}$ = 131.13 g/mol) is frequently used in peptide synthesis. N-Acetyl-L-alanine, together with N-acetyl-L-leucine, was the first amino acid mixture to be separated with partition chromatography. **Target(s):** acylaminoacyl-peptidase (1-3). **1**. Kobayashi & Smith (1987) *J. Biol. Chem.* **262**, 11435. **2**. Sharma & Orthwerth (1993) *Eur. J. Biochem.* **216**, 631. **3**. Raphel, Giardina, Guevel, *et al.* (1999) *Biochim. Biophys. Acta* **1432**, 371.

N-Acetyl-L-alanine Chloromethyl Ketone

This halomethyl ketone (FW = 163.60 g/mol), also known as N-acetyl-3-amino-1-chloro-2-butanone and 3 acetamido-1-chloro-2-butanone, irreversibly inactivates elastase (1) and acylaminoacyl-peptidase (2). **1**. Powers (1977) *Meth. Enzymol.* **46**, 197. **2**. Kobayashi & Smith (1987) *J. Biol. Chem.* **262**, 11435.

N-Acetyl-L-alanine 2-(N-(p-nitrophenoxy)carbonylsarcosyl)-2-methylhydrazide

This azapeptide (FW = 395.37 g/mol) inhibits human leukocyte elastase, which, in addition to catalyzing elastin proteolysis, also disrupts tight junctions, damages certain tissues, breaks down cytokines and alpha proteinase inhibitor, cleaves IgA and IgG, cleave complement component C3bi, a component and the CR1, receptor that triggers phagocytosis in neutrophils. **1**. Powers, Boone, Carroll, *et al.* (1984) *J. Biol. Chem.* **259**, 4288.

N-Acetyl-L-alanyl-L-alanyl-L-alanine Chloromethyl Ketone

This halomethyl ketone (FW = 305.76 g/mol) irreversibly inactivates elastase, which, in addition to catalyzing elastin proteolysis, also disrupts tight junctions, damages certain tissues, breaks down cytokines and alpha proteinase inhibitor, cleaves IgA and IgG, cleave complement component C3bi, a component and the CR1, receptor that triggers phagocytosis in neutrophils. **Target(s):** elastase (1,2); leukocyte elastase (3); myeloblastin, weakly inhibited (3); on pancreatic endopeptidase E (4). **1**. Powers (1977) *Meth. Enzymol.* **46**, 197. **2**. Galdston & Carter (1975) *Amer. Rev. Respir. Dis.* **111**, 873. **3**. Kam, Kerrigan, Dolman, *et al.* (1992) *FEBS Lett.* **297**, 119. **4**. Mallory & Travis (1975) *Biochemistry* **14**, 722.

N-Acetyl-L-alanyl-L-alanyl-azanorleucine 2,2,2-Trifluoroethyl Ester

This azapeptide ester (FW = 398.38 g/mol) inhibits a number of serine proteinases, including cathepsin G (1) and human leukocyte elastase (2). **1**. Gupton, Carroll, Tuhy, Kam & Powers (1984) *J. Biol. Chem.* **259**, 4279. **2**. Powers, Boone, Carroll, *et al.* (1984) *J. Biol. Chem.* **259**, 4288.

N-Acetyl-L-alanyl-L-alanyl-L-prolyl-L-alanine Chloromethyl Ketone

This halomethyl ketone (FW = 402.88 g/mol) irreversibly inactivates several serine proteinases, including cathepsin B (1), cathepsin H (2), pancreatic elastase (3,6,7), leukocyte elastase (4,6,7), myeloblastin, weakly inhibited (4), pancreatic endopeptidase E (5), cathepsin G (7), and chymotrypsin (7). **1**. Olstein & Liener (1983) *J. Biol. Chem.* **258**, 11049. **2**. Raghav, Kamboj, Parnami & Singh (1995) *Indian J. Biochem. Biophys.* **32**, 279. **3**. Powers (1977) *Meth. Enzymol.* **46**, 197. **4**. Kam, Kerrigan, Dolman, *et al.* (1992) *FEBS Lett.* **297**, 119. **5**. Mallory & Travis (1975) *Biochemistry* **14**, 722. **6**. Starkey & Barrett (1976) *Biochem. J.* **155**, 265. **7**. Powers, Gupton, Harley, Nishino & Whitley (1977) *Biochim. Biophys. Acta* **485**, 156.

N^α-Acetyl-L-alanyl-L-cysteinyl-L-arginyl-L-alanyl-L-threonyl-L-lysyl-L-methionyl-L-leucinamide

This acetylated octapeptide amide (FW = 936.18 g.mol; *Sequence*: Ac-ACRATKML-NH₂) inhibits *Clostridium botulinum* bontoxilysin, K_i = 0.15 nM (1). The D-cysteinyl analogue is slightly stronger (K_i = 0.026 nM). **1**. Schmidt, Stafford & Bostian (1998) *FEBS Lett.* **435**, 61.

N^α-Acetyl-D-alanyl-D-glutamic Acid

This acetylated dipeptide (FW = 261.26 g/mol; Sequence: Acetyl-AE) is a competitive inhibitor of muramoylpentapeptide carboxypeptidase (1,2) and zinc D-Ala-D-Ala carboxypeptidases (3). **1**. Nieto, Perkins, Leyh-Bouille, Frère & Ghuysen (1973) *Biochem. J.* **131**, 163. **2**. DeCoen, Lamotte-Brasseur, Ghuysen, Frère & Perkins (1981) *Eur. J. Biochem.* **121**, 221. **3**. Joris, Van Beeumen, Casagrande, *et al.* (1983) *Eur. J. Biochem.* **130**, 53.

N^α-Acetyl-Acetyl-L-alanyl-L-glutamyl-L-valyl-L-aspartal, *See N^α-(t-Butoxycarbonyl)-L-alanyl-L-glutamyl-L-valyl-L-aspartate α-Semialdehyde*

N-Acetyl-L-alanylglycyl-L-phenylalanine Chloromethyl Ketone

This halomethyl ketone (FW = 367.83 g/mol) irreversibly inactivates chymotrypsin (1), chymase (2) tryptase (2,3), and cathepsin G (4). **1**. Segal, Powers, Cohen, Davies & Wilcox (1971) *Biochemistry* **10**, 3728. **2**. Powers, Tanaka, Harper, *et al.* (1985) *Biochemistry* **24**, 2048. **3**. Yoshida, Everitt, Neurath, Woodbury & Powers (1980) *Biochemistry* **19**, 5799. **4**. Powers, Gupton, Harley, Nishino & Whitley (1977) *Biochim. Biophys. Acta* **485**, 156.

N-Acetyl-L-alanyl-L-leucyl-L-cysteinyl-L-aspartyl-L-aspartyl-L-prolyl-L-arginyl-L-valyl-L-aspartyl-L-arginyl-L-tryptophanyl-L-tyrosyl-L-cysteinyl-L-glutaminyl-L-phenylalanyl-L-valyl-L-glutamylglycinamide (3-13 Disulfide Bridge)

This octadecapeptide amide (*Sequence*: Acetyl-ALCDDPRDRWYCQF VEG-NH₂; MW = 2215.47 g/mol), with an intrapeptide disulfide bond between Cys3 and Cys13), inhibits tissue-factor-dependent clotting by blocking the activation of coagulation factors VIIIa and X, the latter with an IC₅₀ of 1 nM). **1**. Dennis, Eigenbrot, Skelton, *et al.* (2000) *Nature* **404**, 465.

N-Acetyl-L-alanyl-L-phenylalanine Chloromethyl Ketone

This halomethyl ketone (FW = 310.78 g/mol) irreversibly inactivates protease IV (1), chymotrypsin (2,3), tryptase (4), and cathepsin G (5). **1**. Miller (1998) in *Handbook of Proteolytic Enzymes* (Barrett, Rawlings & Woessner, eds), p. 1589, Academic Press, San Diego. **2**. Shaw (1972) *Meth. Enzymol.* **25**, 655. **3**. Segal, Powers, Cohen, Davies & Wilcox (1971) *Biochemistry* **10**, 3728. **4**. Yoshida, Everitt, Neurath, Woodbury & Powers (1980) *Biochemistry* **19**, 5799. **5**. Powers, Gupton, Harley, Nishino & Whitley (1977) *Biochim. Biophys. Acta* **485**, 156.

N-Acetyl-β-aletheine

This pantothenate analogue (FW = 190.27 g/mol) inhibits 3-oxoacid CoA-transferase, *or* 3-ketoacid CoA-transferase, which catalyzes the reversible conversion of succinyl-CoA and acetoacetate into acetoacetyl-CoA and succinate through a covalent enzyme thiol ester intermediate, E-CoA. Substrate analogue binding studies with *N*-Acetyl-β-aletheine helped to clarify how this enzyme exploits binding energy from noncovalent interactions with CoA to bring about an increase in k_{cat}/K_m of ~10^{10} $M^{-1}s^{-1}$. **1**. Whitty, Fierke & Jencks (1995) *Biochemistry* **34**, 11678.

2-(Acetylamino)-N-[4-[(aminoiminomethyl)amino]-1-(2-benzothiazolyl-carbonyl)butyl]acetamido

This benzothiazole derivative of glycylarginine (FW = 391.47 g/mol) is a transition-state analogue that potently and reversibly inhibits trypsin (K_i = 260 nM) and tryptase (K_i = 470 nM), the latter a human mast cell enzyme (EC 3.4.21.59) with therapeutic potential for the treatment pf allergic or inflammatory disorders. The 1.9-Å resolution X-ray structure of the enzyme-inhibitor complex depicts a hemiketal adduct involving Ser-189, and hydrogen bonds with His-57 and Gln-192. **1**. Costanzo, Yabut, Almond, *et al.* (2003) *J. Med. Chem.* **46**, 3865.

8-(N-Acetyl-4-aminobenzyl)-1-benzyl-3-butylxanthine

This 3-alkyl-1,8-dibenzylxanthine (FW = 445.52 g/mol) inhibits human GTP-dependent phosphoenol-pyruvate carboxykinase, *or* PEPCK (IC₅₀ = 19 μM), a druggable target for improving glucose homeostasis in Type 2 diabetes. This agent is the first nonsubstrate-like inhibitor of human cytosolic PEPCK that is competitive with GTP. **1**. Foley, Wang, Dunten, *et al.* (2003) *Bioorg. Med. Chem. Lett.* **13**, 3607.

8-(N-Acetyl-4-aminobenzyl)-3-butyl-1-(2-fluorobenzyl)xanthine

This GTP-competitive synthetic purine (FW = 463.51 g/mol) inhibits human phosphoenolpyruvate carboxykinase (GTP-requiring), IC₅₀ = 2.1 μM. Increased hepatic gluconeogenesis is thought to lead to fasting hyperglycemia in patients with Type 2 diabetes. Because cytosolic phosphoenolyruvate carboxyinase (PEPCK) catalyzes the rate-limiting step in gluconeogenesis, it is deemed a druggable target for controlling Type 2 diabetes. **1**. Foley, Wang, Dunten, *et al.* (2003) *Bioorg. Med. Chem. Lett.* **13**, 3607.

S-Acetyl-2(S)-amino-3-[4-(benzyloxy)phenyl]-1-propanethiol

This thioamine (FW_{free-base} = 315.43 g/mol), *S*-acetyl-3-[4-(benzyloxy) phenyl-2-amino-1-propanethiol, inhibits leukotriene A₄ hydrolase, IC₅₀ = 1 μM. The *R*-enantiomer is a weaker inhibitor (IC₅₀ = 8 μM). A free thiol is needed for sub-μM binding, with the enzyme preferring the *R*-enantiomer over the *S*-enantiomer, in contrast to the stereoselectivity displayed towards bestatin, an inhibitor of somewhat similar structure. Substitution of acid moieties around the periphery of the benzyloxyphenyl portion results in substantially weaker binding, suggesting this group resides within a large hydrophobic pocket when bound to the enzyme. **1**. Ollmann, Hogg, Muñoz, *et al.* (1995) *Bioorg. Med. Chem.* **3**, 969.

[3-(*N*-Acetylamino)-3-(carboxymethyl)propanoyl]-7-glutamate

This acetylated dipeptide (FW = 306.27 g/mol) inhibits glutamate carboxypeptidase II, *or N*-acetylated α-linked acidic dipeptidase, K_i = 3.2 μM. These results suggest that the relative spacing between the side chain carboxyl and the alpha-carboxyl group of the C-terminal residue may be important for binding to the active site of the enzyme. They also indicate that the χ_1 torsional angle for the aspartyl residue is ~0°. **1**. Subasinghe, Schulte, Chan, *et al.* (1990) *J. Med. Chem.* **33**, 2734.

N-Acetyl-DL-aminocarnitine

This carnitine analogue (FW$_{inner-salt}$ = 202.25 g/mol), also known as 3-acetamino-4-(trimethylammonio)butyric acid, is a strong inhibitor of carnitine *O*-palmitoyltransferase (IC$_{50}$ for carnitine *O*-palmitoyltransferase-2 is 0.8 μM) and carnitine *O*-acetyltransferase (K_i = 24 μM) (1,2). Moreover, this compound readily undergoes metabolic hydrolysis to form aminocarnitine, a good inhibitor of fatty acid oxidation. Aminocarnitine also acts as a hypoglycemic and antiketogenic compound and alters lipidic metabolism by inhibiting carnitine acetyltransferase (CAT) and carnitine palmitoyltransferase (CPT). **1**. Jenkins & Griffith (1985) *J. Biol. Chem.* **260**, 14748. **2**. Brass, Gandour & Griffith (1991) *Biochim. Biophys. Acta* **1095**, 17.

2-Acetylaminoethyltrimethylammonium Iodide

This nonhydrolyzable acetylcholine analogue (FW = 272.13 g/mol), also known as acetylaminocholine, inhibits choline *O*-acetyltransferase competitively with respect to choline and noncompetitive with respect to acetyl-CoA, while product inhibition by choline is competitive with respect to acetylcholine and noncompetitive with respect to coenzyme A. These and related kinetic data are not consistent with the previously proposed ordered Theorell-Chance reaction mechanism, and have been interpreted in terms of a random binding mechanism. **1**. Hersh & Peet (1977) *J. Biol. Chem.* **252**, 4796.

4-(Acetylamino)-3-guanidinobenzoic Acid

This benzoic acid derivative (FW = 236.23 g/mol), also known as BCX 140, inhibits influenza virus neuraminidase, *or* sialidase, with an IC$_{50}$ value of 2.5-10 μM, but did not exhibit *in vivo* activity in a mouse model of influenza (1,2). **1**. Singh, Jedrzejas, Air, *et al.* (1995) *J. Med. Chem.* **38**, 3217. **2**. Chand, Babu, Bantia, *et al.* (1997) *J. Med. Chem.* **40**, 4030.

(2*S*,4*R*)-1-Acetyl-*N*-[(1*S*)-4-[(aminoiminomethyl)-amino]-1-(2-benzothiazolylcarbonyl)-butyl]-4-hydroxy-2-pyrrolidine carboxamide

This arginine derivative (FW = 430.53 g/mol), also known as RWJ-56423, inhibits several serine proteinases (1), including trypsin and tryptase, for which K_i = 8.1 and 10 nM, respectively). The (2*R*)-stereoisomer is a weaker inhibitor. **Target(s):** coagulation factor Xa, weakly inhibited; kallikrein; plasmin; thrombin (weakly inhibited); trypsin; tryptase; and u-plasminogen activator, *or* urokinase. **1**. Costanzo, Yabut, Almond, *et al.* (2003) *J. Med. Chem.* **46**, 3865.

5-(Acetylamino)-4-oxo-6-phenyl-2-hexenoate and Methyl Ester

This acid (FW = 261.28 g/mol; CAS 477769-02-1; Abbreviation: AOPHA) and its methyl ester (FW = 275.30 g/mol; Abbreviation: AOPHA-Me) are both mechanism-based inactivators of peptidylglycine monooxygenase (PAM). Significantly, AOPHA-Me also inhibits carrageenan-induced edema in rats in a dose-dependent manner. Neither 4-phenyl-3-butenoic acid nor AOPHA-Me exhibits significant cyclooxygenase (COX) inhibition *in vitro*, indicating that the anti-inflammatory activities of PBA and AOPHA-Me are unlikely to be the consequence of COX inhibition. **1**. Bauer, Sunman, Foster, *et al.* (2007) *J. Pharmacol. Exp. Ther.* **320**, 1171.

N$^\alpha$-Acetyl-L-arginine

This acetylated amino acid (FW$_{inner-salt}$ = 216.24 g/mol; CAS 155-84-0; IUPAC Name: (2*S*)-2-acetamido-5-(diaminomethylideneamino)pentanoate) inhibits carboxypeptidase B (1). **Target(s):** carboxypeptidase B (1,2); actinidain (3); lysine carboxypeptidase, *or* lysine(arginine) carboxypeptidase; carboxypeptidase N (4). **1**. Folk (1971) *The Enzymes*, 3rd ed., **3**, 57. **2**. Zisapel, Navon & Sokolovsky (1975) *Eur. J. Biochem.* **52**, 487. **3**. J. Boland & Hardman (1972) *FEBS Lett.* **27**, 282. **4**. Juillerat-Jeanneret, Roth & Bargetzi (1982) *Hoppe-Seyler's Z. Physiol. Chem.* **363**, 51.

N-Acetyl-L-arginyl-L-arginyl-L-lysyl-L-arginyl-L-arginyl-L-arginamide

This acetylated hexapeptide amide (MW = 970.18 g/mol; *Sequence:* Ac-RRKRRR-NH$_2$; Isoelectric Point = 12.6) inhibits furin and proprotein convertase-2 (1). Significantly, proteolytic activation of the spike protein at a novel RRRR/S motif is implicated in furin-dependent entry, syncytium formation, and infectivity of coronavirus infectious bronchitis virus in cultured cells (2). **1**. Cameron, Appel, Houghten & Lindberg (2000) *J. Biol. Chem.* **275**, 36741. **2**. Yamada & Liu (2009) *J. Virol.* **83**, 8744.

N-Acetyl-L-arginyl-L-arginyl-L-prolyl-L-tyrosyl-L-isoleucyl-L-leucine and N-Acetyl-L-arginyl-L-arginyl-L-prolyl-L-tyrosyl-L-isoleucyl-L-leucineamide

This acetylated hexapeptide (FW = 860.05 g/mol; *Sequence*: Ac-RRPYIL) and its corresponding amide (FW = 859.97 g/mol; *Sequence*: Ac-RRPYIL-NH$_2$), where the sequence corresponds to the C-terminus of neurotensin, inhibit neurolysin, *or* endopeptidase-24.16 (1,2). Both are poor alternative substrates, suggesting that slow catalysis account for their inhibitory effects on natural neurolysin substrates (1,2). **1**. Barelli, Vincent & Checler (1988) *Eur. J. Biochem.* **175**, 481. **2**. Vincent, Dauch, Vincent & Checler (1997) *J. Neurochem.* **68**, 837.

N$^\alpha$-Acetyl-L-arginyl-L-cysteinylglycyl-L-valyl-L-prolyl-L-aspartate α-Amide

This pentapeptide amide (FW$_{free-base}$ = 686.79 g/mol; *Sequence*: N^α-Acetyl-RCGVPD-NH$_2$) inhibits matrix metalloproteinase-3, *or* stromelysin I, IC$_{50}$ = 5 μM (1,2). **1**. Hanglow, Lugo, Walsky, *et al.* (1993) *Agents Actions* **39**, C148. **2**. Fotouhi, Lugo, Visnick, *et al.* (1994) *J. Biol. Chem.* **269**, 30227.

{N-[N-Acetyl-L-asparaginyl]-(5S-amino-4S-hydroxy-2-isopropyl-7-methyloctanoyl)}-L-isoleucinamide

This pseudopeptide amide (FW = 613.78 g/mol; *Sequence*: Acetyl-Asn-Leuψ[CH(OH)CH$_2$]Val-Ile-NH$_2$), also known as U92516E, inhibits rhizopuspepsin, K_i = 510 nM (1) and HIV-1 protease, K_i = 90 nM (2). **1**. Lowther, Sawyer, Staples, *et al.* (1992) in *Peptides: Chemistry and Biology Proc. of the 12th Amer. Peptide Symp.* (June 16-21, 1991) (Smith & Rivier, eds), pp. 413-414. **2**. Sawyer, Staples, Liu, *et al.* (1992) *Int. J. Pept. Protein Res.* **40**, 274.

N$^\alpha$-Acetyl-L-asparaginyl-L-cysteinyl-L-arginyl-L-alanyl-L-threonyl-L-lysyl-L-methionyl-L-leucinamide

This acetylated octapeptide (FW = 993.23 g/mol; *Sequence*: Acetyl-QCRATKML-NH$_2$) amide inhibits *Clostridium botulinum* bontoxilysin, K_i = 0.50 nM (1). The D-cysteinyl analogue is a slightly better inhibitor (K_i = 0.15 nM). **1**. Schmidt, Stafford & Bostian (1998) *FEBS Lett.* **435**, 61.

N-Acetyl-L-aspartic Acid

This acetylated amino acid (FW$_{free-acid}$ = 175.14 g/mol; M.P. = 137-140°C), which reportedly makes up slighly less than 1% of the dry weight of human brain (1), plays roles in myelin biosynthesis and osmotic regulation. It is also an alternative activator of carbamoyl-phosphate synthetase I. **Target(s):** *Escherichia coli* aspartate ammonia-lyase, K_i = 6.7 mM (2); 2-(acetamidomethylene)succinate hydrolase, inhibited by *N*-acetyl-DL-aspartate (3); glutamate carboxypeptidase II, *or* N-acetylated-γ-linked acidic dipeptidase (4); amino-acid *N*-acetyltransferase (5). **1**. Birken & Oldendorf (1989) *Neurosci. Biobehav. Rev.* **13**, 23. **2**. Falzone, Karsten, Conley & Viola (1988) *Biochemistry* **27**, 9089. **3**. Huynh & Snell (1985) *J. Biol. Chem.* **260**, 2379. **4**. Robinson, Blakely, Couto & Coyle (1987) *J. Biol. Chem.* **262**, 14498. **5**. Shigesada & Tatibana (1978) *Eur. J. Biochem.* **84**, 285.

N-Acetyl-L-aspartyl-L-aspartyl-L-aspartyl-L-tryptophanyl-L-aspartyl-L-phenylalanine

This acetylated hexapeptide (*Sequence*: Ac–DDDWDF; FW = 854.22 g/mol), corresponding the C-terminus of the small subunit of *Mycobacterium tuberculosis* ribonucleoside diphosphate reductase, inhibits

ribonucleoside-diphosphate reductase, IC$_{50}$ = 100 μM (1,2). **1**. Nurbo, Roos, Muthas, *et al.* (2007) *J. Pept. Sci.* **13**, 822. **2**. Yang, Curran, Li, *et al.* (1997) *J. Bacteriol.* **179**, 6408.

N-Acetyl-L-aspartyl-D-aspartyl-L-isoleucyl-L-valyl-L-prolyl-L-cysteine, See N-Acetyl-L-aspartyl-L-aspartyl-L-isoleucyl-L-valyl-L-prolyl-L-cysteine

N-Acetyl-L-aspartyl-L-aspartyl-L-isoleucyl-L-valyl-L-prolyl-L-cysteine

This acetylated hexapeptide (*Sequence*: Ac-DDIVPC; FW$_{free-acid}$ = 686.78 g/mol), whose sequence corresponds to the N-terminal cleavage product of an HCV dodecapeptide substrate derived from the NS5A/5B cleavage site, inhibits hepacivirin, *or* hepacivirin (hepatitis C virus NS3 serine proteinase, IC$_{50}$ = 28 μM (1). **1**. Llinàs-Brunet, Bailey, Fazal, *et al.* (1998) *Bioorg. Med. Chem. Lett.* **8**, 1713.

N-Acetyl-L-aspartyl-D-aspartyl-L-isoleucyl-L-valyl-L-prolyl-DL-norvaline Benzylamide

This acetylated hexapeptide (FW$_{free-acid}$ = 844.96 g/mol; *Sequence*: Ac-DDIVP-norV-benzylamide), whose sequence corresponds to the N-terminal cleavage product of an HCV dodecapeptide substrate derived from the NS5A/5B cleavage site, inhibits hepacivirin, *or* hepacivirin (hepatitis C virus NS3 serine proteinase, IC$_{50}$ = 0.7 μM (1). **1**. Llinàs-Brunet, Bailey, Fazal, *et al.* (1998) *Bioorg. Med. Chem. Lett.* **8**, 1713.

N-Acetyl-L-aspartyl-D-(γ-carboxy)glutamyl-L-leucyl-L-isoleucyl-3-cyclohexyl-L-alanyl-L-cysteine

This acetylated hexapeptide (FW = 788.96 g/mol; *Sequence*: Ac-Asp-(γ)D-Glu-Leu-Ile-cyclohex-Ala-Cys) inhibits hepacivirin, *or* hepatitis C virus NS3 serine proteinase, IC$_{50}$ = 1.5 nM (1,2). The strategy began by optimizing S'-binding in the context of noncleavable decapeptides spanning the P$_6$-P$_4'$ region. Binding was sequentially increased by starting the previously optimized P-region (1), by changing the P$_4'$ residue, and by combinatorially optimizing positions P$_2'$-P$_3'$ (2) That effort led to an increase in binding of more than three orders of magnitude, with the best decapeptides showing IC$_{50}$ values below 200 pM. The binding mode of the decapeptide inhibitors shares features with the binding mode of the natural substrates, but incorporates novel interactions in the S' subsite (2). Similar inhibitor potency was observed with full-length Hepatitis C NS3 protein, expressed by viral mRNA extracted from the serum of a patient infected with HCV Strain-1a. Sequencing of that protease gene identified the virus to be a new variant closely related to Strain H77. The NS3 protein from Strain-1a differs in 15 out of 631 amino acid residues, of which none are predicted to be directly involved in catalysis or binding of substrate or cofactor (3). **1**. Ingallinella, *et al.* (1998) Biochemistry **37**, 8906. **2**. Ingallinella, Bianchi, Ingenito, *et al.* (2000) *Biochemistry* **39**, 12898. **3**. Poliakov, Hubatsch, Shuman, Stenberg & Danielson (2002) *Protein Expr. Purif.* **25**, 363.

N-Acetyl-p-benzoquinone imine

This toxic acetaminophen metabolite (FW = 149.15g/mol; CAS 50700-49-7; Symbol: NAPQI), also systematically named *N*-(4-oxo-1-cyclohexa-2,5-dienylidene)acetamide, causes severe liver damage, usually within 3–4 days after acetaminophen (paracetamol) overdose (typically 4-6 g) (**See** *Acetaminophen*). Severe NAPQI toxicity is known to lead to death from fulminant liver failure. NAPQI induces the depletion of reduced glutathione (GSH), the peroxidation of lipids, as well as cytotoxicity in hepatocytes to the same extent as excess acetaminophen (1). The preferred antidote for NAPQI poisoning is acetylcysteine, which can be given in large amounts without side effects (**See** *Acetylcysteine*). Notably, NAPQI is a strong topoisomerase II poison, increasing levels of enzyme-mediated DNA cleavage >5-fold at 100 μM (2). It induces scission at a number of DNA sites similar to those observed in the presence of etoposide. NAPQI strongly impairs the ability of topoisomerase IIα to reseal cleaved DNA molecules, suggesting inhibition of DNA religation is the primary mechanism for enhancing cleavage. NAPQI also increases levels of DNA scission mediated by human topoisomerase IIα in CEM leukemia cells (2). In mice, knockout of Cu,Zn-SOD superoxide dismutase (SOD1) alone, or in combination with selenium-dependent glutathione peroxidase-1 (GPX1) blocks acetaminophen-induced liver toxicity (3). **1.** van de Straat, de Vries, Debets & Vermeulen (1987) *Biochem. Pharmacol.* **36**, 2065. **2.** Bender, Lindsey, Burden & Osheroff (2004) *Biochemistry* **43**, 3731. **3.** Lei, Zhu, McClung, Aregullin & Roneker (2006) *Biochem J.* **399**, 455.

Acetyl Ceramide, See *D-erythro-N-Acetylsphingosine*

10-Acetyl-2-chloroacetylphenothiazine

This phenothiazine (FW = 317.80 g/mol) inhibits trypanothione-disulfide reductase (IC$_{50}$ = 440 μM) and glutathione-disulfide reductase (1,2). The corresponding *O*-benzyloxime derivative also inhibits. **1.** Iribarne, Paulino, Aguilera & Tapia (2009) *J. Mol. Graph. Model.* **28**, 371. **2.** Chan, Yin, Garforth, *et al.* (1998) *J. Med. Chem.* **41**, 148.

Acetylcholine

This prototypical neurotransmitter (FW$_{chloride}$ = 181.66 g/mol; CAS 51-84-3; aqueous solutions, relatively stable at pH 4, but decompose under alkaline conditions; IUPAC: 2-Acetoxy-*N,N,N*-trimethylethanaminium), acts (a) *peripherally* as a neurotransmitter at the neuromuscular junction (*i.e.*, its release by motor neurons activates muscle contraction; (b) *autonomicallly* (*i.e.*, functionally independent; not under voluntary control), both as an internal transmitter for the sympathetic nervous system and as the final product released by the parasympathetic nervous system; and (c) *centrally*, as a neuromodulator that allows the brain to process information rather than to transmit information from point to point, also affecting arousal, attention, and motivation. Acetylcholine-containing synaptic vesicles (30-nm diameter; containing ~5000 acetylcholine molecules, the net positive charge of which is neutralized by the high molar content of GTP) are formed presynaptically. Upon fusion with the axon terminus, each vesicle spills its contents, causing a miniature end-plate potential (*or* MEPPs) that is <1 mV in amplitude and insufficiernt to reach threshold. However, once an action potential is strong enough to bring about the release of many acetylcholine-containing vesicles, acetylcholine diffuses across the neuromuscular junction, binds to ligand-gated nicotinic receptors on the muscle fiber, increasing sodium and potassium conductance and depolarizing the muscle cell membrane (sarcolemma).

Nicotinic Receptors	N1 *or* Nm (neuromuscular junction)
	N2 *or* Nn (autonomic ganglia, CNS)

Muscarinic Receptors	M1 (striatum, cortex, hippocampus)
	M2 (forebrain, thalamus, heart, pupil, spinal cord, exocrine)
	M3 (brain, hypothalamus, exocrine, pupil, peripheral arteries)
	M4 (striatum, cortex, hippocampus, spinal cord)
	M5 (dopaminergic neurons, basal ganglia, brain vasculature)

Acetylcholine binds at (a) *Nicotinic Acetylcholine Receptors*, which muscle-type (located on muscle end-plates and blocked by curare) and neuronal-type (located in autonomic ganglia (both sympathetic and parasympathetic) and in central nervous system and blocked by hexamethonium) ligand-gated ion channels that, when activated, become permeable to sodium, potassium, and calcium ions; (b) *Muscarinic Acetylcholine Receptors*, which have a more complex mechanism that affects target cells over a longer timescale. In mammals, there are five muscarinic receptor subtypes (*e.g.*, M$_1$, M$_2$, M$_3$, M$_4$ and M$_5$) that function as G-protein-coupled receptors (GPCRs), exerting their effects via a second messenger system. M$_1$, M$_3$, and M$_5$ are Gq-coupled subtypes that increase intracellular levels of Inositol trisphosphate, *or* InsP$_3$ (*or* Inositol 1,4,5-trisphosphate (Ins3P) and triphosphoinositol (IP$_3$)) and Ca^{2+} through its activation of phospholipase C. Their effect on target cells is usually excitatory. M$_2$ and M$_4$ subtypes are G$_i$/G$_o$-coupled and decrease intracellular levels of cAMP by inhibiting adenylate cyclase. Muscarinic acetylcholine receptors are found in both CNS and peripheral neurons in the heart, lungs, upper gastrointestinal tract, and sweat glands. **Target(s):** Acetylcholine inhibits several enzymes *in vitro*, but its rapid hydrolysis *in vivo* probably limits its effectiveness as an inhibitor. Aqueous solutions are relatively stable at pH 4, but acetylcholine decomposes in alkaline pH conditions. Its targets include: aryl-acylamidase (1,2); choline kinase (3,11); glutamate dehydrogenase (4); Na$^+$/K$^+$-exchanging ATPase (5); diisopropylfluorophosphatase (6); glycerophosphocholine choline phosphodiesterase (7); choline sulfotransferase (8); diacylglycerol choline-phosphotransferase (9); ethanolamine phosphotransferase (9); ethanolamine kinase (10); and carnitine *O*-acetyltransferase (12). **1.** Oommen & Balasubramanian (1979) *Eur. J. Biochem.* **94**, 135. **2.** George & Balasubramanian (1980) *Eur. J. Biochem.* **111**, 511. **3.** Ishidate & Nakazawa (1992) *Meth. Enzymol.* **209**, 121. **4.** Smith, Austen, Blumenthal & Nyc (1975) *The Enzymes*, 3rd ed., **11**, 293. **5.** Takachuk, Lopina & Boldyrev (1975) *Biokhimiia* **40**, 1032. **6.** Cohen & Warringa (1957) *Biochim. Biophys. Acta* **26**, 29. **7.** Sok (1998) *Neurochem. Res.* **23**, 1061. **8.** Orsi & Spencer (1964) *J. Biochem.* **56**, 81. **9.** Strosznajder, Radominska-Pyrek & Horrocks (1979) *Biochim. Biophys. Acta* **574**, 48. **10.** Sung & Johnstone (1967) *Biochem. J.* **105**, 497. **11.** Brostrom & Browning (1973) *J. Biol. Chem.* **248**, 2364. **12.** White & Wu (1973) *Biochemistry* **12**, 841. **Target(s):** aryl-acylamidase (1,2); choline kinase (3,11); glutamate dehydrogenase (4); Na$^+$/K$^+$-exchanging ATPase (5); diisopropylfluorophosphatase (6); glycerophosphocholine cholinephosphodiesterase (7); choline sulfotransferase (8); diacylglycerol choline-phosphotransferase (9); ethanolamine phosphotransferase (9); ethanolamine kinase (10); and carnitine *O*-acetyltransferase (12). **1.** Oommen & Balasubramanian (1979) *Eur. J. Biochem.* **94**, 135. **2.** George & Balasubramanian (1980) *Eur. J. Biochem.* **111**, 511. **3.** Ishidate & Nakazawa (1992) *Meth. Enzymol.* **209**, 121. **4.** Smith, Austen, Blumenthal & Nyc (1975) *The Enzymes*, 3rd ed., **11**, 293. **5.** Takachuk, Lopina & Boldyrev (1975) *Biokhimiia* **40**, 1032. **6.** Cohen & Warringa (1957) *Biochim. Biophys. Acta* **26**, 29. **7.** Sok (1998) *Neurochem. Res.* **23**, 1061. **8.** Orsi & Spencer (1964) *J. Biochem.* **56**, 81. **9.** Strosznajder, Radominska-Pyrek & Horrocks (1979) *Biochim. Biophys. Acta* **574**, 48. **10.** Sung & Johnstone (1967) *Biochem. J.* **105**, 497. **11.** Brostrom & Browning (1973) *J. Biol. Chem.* **248**, 2364. **12.** White & Wu (1973) *Biochemistry* **12**, 841.

Acetyl-Coenzyme A (*or* Acetyl-CoA)

This key thiolester (FW = 809.57 g/mol; CAS 72-89-9) is the main C$_2$ acyl donor in biosynthetic reactions in all living organisms. Acetyl-CoA is also an immediate product of fatty acid β-oxidation. The standard Gibbs free energy of hydrolysis (ΔG°') of its thioester linkage is about −32.2 kJ/mol. Acetyl-CoA is water-soluble and stable in acid conditions (*e.g.*, resisting hydrolysis for 15 minutes at pH 3.5-5 at 100°C), but readily hydrolyzed in alkaline solutions. Acetyl-S-CoA should be stored in a desiccator at +4°C. Its λ$_{max}$ at 260 nm (ε = 16,400 M^{-1}cm^{-1}) at pH 7 is chiefly attributable to its adenine moiety. As is true for other thiolesters, the α-carbon of the acetyl moiety is a weak carbon acid, forming reactive carbanion species $^-$CH$_2$C(=O)-S-CoA. **Target(s):** glucose-6-P dehydrogenase (1,2); pyruvate kinase (3); NAD(P)$^+$ transhydrogenase, AB-specific (4); enoyl-[acyl-carrier protein] reductase (5); pantothenate kinase (6,31-40); 3-hydroxy-3-methylglutaryl-CoA reductase (7,13); malonyl-CoA decarboxylase (8);

oxaloacetate decarboxylase (9); methylmalonyl-CoA epimerase (11); pyruvate deydrogenase complex (12); succinyl-CoA synthetase (14); malyl-CoA lyase (15); phosphoketolase (16); aspartate 4 decarboxylase (17); succinate:3-hydroxy-3-methylglutarate CoA-transferase (18); [3-methyl-2 oxobutanoate dehydrogenase (2-methylpropanoyl-transferring)] phosphatase, or [branched-chain α-keto-acid dehydrogenase] phosphatase (19-21); phosphoprotein phosphatase (19-22); palmitoyl CoA hydrolase, or acyl-CoA hydrolase (23,24); (S)-methylmalonyl-CoA hydrolase (25); 3 hydroxyisobutyryl-CoA hydrolase (26,27); [protein-PII] uridylyltransferase (28); adenylate kinase (29); glucose-1,6-bisphosphate synthase (30); cysteine synthase (41); 4-hydroxybenzoate nonaprenyltransferase, weakly inhibited (42); 2-methylcitrate synthase (43); decylcitrate synthase (44); anthocyanin 5-O-glucoside 6'''-O-malonyltransferase (45); anthocyanin 6''-O-malonyl-transferase (46); propionyl-CoA C_2-trimethyltridecanoyl-transferase (47); β-ketoacyl-[acyl-carrier-protein] synthase I (48); [acyl-carrier-protein] S-malonyltransferase (49); and carnitine O-palmitoyltransferase (50-52). **1**. Olive & Levy (1975) *Meth. Enzymol.* **41**, 196. **2**. Miethe & Babel (1990) *Meth. Enzymol.* **188**, 346. **3**. Kayne (1973) *The Enzymes*, 3rd ed., **8**, 353. **4**. Rydström, Hoek & Ernster (1976) *The Enzymes*, 3rd ed., **13**, 51. **5**. Wakil & Stoops (1983) *The Enzymes*, 3rd ed., **16**, 3. **6**. Halvorsen & Skrede (1982) *Eur. J. Biochem.* **124**, 211. **7**. Kirtley & Rudney (1967) *Biochemistry* **6**, 230. **8**. Buckner, Kolattukudy & Poulose (1976) *Arch. Biochem. Biophys.* **177**, 539. **9**. Horton & Kornberg (1964) *Biochim. Biophys. Acta* **89**, 381. **10**. Lipmann, Kaplan, Novelli, Tuttle & Guirard (1947) *J. Biol. Chem.* **167**, 869. **11**. Stabler, Marcell & Allen (1985) *Arch. Biochem. Biophys.* **241**, 252. **12**. Schwartz & Reed (1970) *Biochemistry* **9**, 1434. **13**. Kawachi & Rudney (1970) *Biochemistry* **9**, 1700. **14**. Leitzmann, Wu & Boyer (1970) *Biochemistry* **9**, 2338. **15**. Hersh (1974) *J. Biol. Chem.* **249**, 5208. **16**. Whitworth & Ratledge (1977) *J. Gen. Microbiol.* **102**, 397. **17**. Rathod & Fellman (1985) *Arch. Biochem. Biophys.* **238**, 447. **18**. Francesconi, Donella Deana, Furlanetto, *et al.* (1989) *Biochim. Biophys. Acta* **999**, 163. **19**. Damuni & Reed (1987) *J. Biol. Chem.* **262**, 5129. **20**. Damuni, Merryfield, Humphreys & Reed (1984) *Proc. Natl. Acad. Sci. U.S.A.* **81**, 4335. **21**. Reed & Damuni (1987) *Adv. Protein Phosphatases* **4**, 59. **22**. Damuni & Reed (1987) *J. Biol. Chem.* **262**, 5133. **23**. Sanjanwala, Sun & MacQuarrie (1987) *Arch. Biochem. Biophys.* **258**, 299. **24**. Berge (1979) *Biochim. Biophys. Acta* **574**, 321. **25**. Kovachy, Copley & Allen (1983) *J. Biol. Chem.* **258**, 11415. **26**. Shimomura, Murakami, Nakai, *et al.* (2000) *Meth. Enzymol.* **324**, 229. **27**. Shimomura, Murakami, Fujitsuka, *et al.* (1994) *J. Biol. Chem.* **269**, 14248. **28**. Engleman & Francis (1978) *Arch. Biochem. Biophys.* **191**, 602. **29**. McKellar, Charles & Butler (1980) *Arch. Microbiol.* **124**, 275. **30**. Rose, Warms & Kaklij (1975) *J. Biol. Chem.* **250**, 3466. **31**. Vallari, Jackowski & Rock (1987) *J. Biol. Chem.* **262**, 2468. **32**. Calder, Williams, Ramaswamy, *et al.* (1999) *J. Biol. Chem.* **274**, 2014. **33**. Leonardi, Rock, Jackowski & Zhang (2007) *Proc. Natl. Acad. Sci. U.S.A.* **104**, 1494. **34**. Zhang, Rock & Jackowski (2005) *J. Biol. Chem.* **280**, 32594. **35**. Rock, Calder, Karim & Jackowski (2000) *J. Biol. Chem.* **275**, 1377. **36**. Lehane, Marchetti, Spry, *et al.* (2007) *J. Biol. Chem.* **282**, 25395. **37**. Fisher, Robishaw & Neely (1985) *J. Biol. Chem.* **260**, 15745. **38**. Halvorsen & Tverdal (1986) *Scand. J. Clin. Lab. Invest.* **46**, 67. **39**. Zhang, Rock & Jackowski (2006) *J. Biol. Chem.* **281**, 107. **40**. Kotzbauer, Truax, Trojanowski & Lee (2005) *J. Neurosci.* **25**, 689. **41**. Cook & Wedding (1978) *J. Biol. Chem.* **253**, 7874. **42**. Kawahara, Koizumi, Kawaji, *et al.* (1991) *Agric. Biol. Chem.* **55**, 2307. **43**. Maerker, Rohde, Brakhage & Brock (2005) *FEBS J.* **272**, 3615. **44**. Måhlén (1971) *Eur. J. Biochem.* **22**, 104. **45**. Suzuki, Nakayama, Yonekura-Sakakibara, *et al.* (2001) *J. Biol. Chem.* **276**, 49013. **46**. Suzuki, Sawada, Yonekura-Sakakibara, *et al.* (2003) *Plant Biotechnol.* **20**, 229. **47**. Bun-Ya, Maebuchi, Hashimoto, Yokota & Kamiryo (1997) *Eur. J. Biochem.* **245**, 252. **48**. Toomey & Wakil (1966) *J. Biol. Chem.* **241**, 1159. **49**. Joshi (1972) *Biochem. J.* **128**, 43P. **50**. Kashfi, Mynatt & Cook (1994) *Biochim. Biophys. Acta* **1212**, 245. **51**. Zierz & Engel (1987) *Biochem. J.* **245**, 205. **52**. McCormick, Notar-Francesco & Sriwatanakul (1983) *Biochem. J.* **216**, 499.

1-Acetyl-7-cyano-2,3-diphenylindolizine

This cyano-indolizine (FW = 336.13 g/mol) inhibits (IC_{50} = 23 μM) arachidonate 15-lipoxygenase, or 15-lipoxygenase, an enzyme that has been implicated in oxidation of low-density lipoproteins (LDL) and may be involved in the development of atherosclerosis. **1**. Gundersen, Malterud, Negussie, *et al.* (2003) *Bioorg. Med. Chem.* **11**, 5409.

cyclic-[N-Acetyl-Cys³,Nle⁴,Arg⁵,³-(2'-Naphthyl)-D-Ala⁷,Cys¹¹]-α-Melanocyte-Stimulating Hormone Fragment 3-11 Amide

This cyclic nonapeptide amide (FW = 1266.50 g/mol; *Sequence*: CH_3CO-Cys-Nle-Arg-His-D-2-Nal-Arg-Trp-Gly-Cys-NH_2, where Nle refers to L-norleucine, and D-2-Nal refers to 3-(2 naphthyl)-D-alanine), is a modified fragment of α-melanocyte-stimulating hormone (α-MSH) that acts as a melanocortin-4 receptor antagonist, K_d = 0.29 nM. This agent antagonizes an aMSH-induced cAMP response in cells expressing the human MC1, MC3, MC4, or MC5 receptor and also causes a dose-dependent increase in food intake, with a maximum response when a 1-nmol dose injected intracerebroventricularly in freely feeding rats. It does not show evidence of affecting emotionality or locomotor activity, suggesting that the MC4 receptor does not mediate anxiogenic-like and locomotor effects related to the melanocortic peptides. **1**. Kask, Mutulis, Muceniece, *et al.* (1998) *Endocrinology* **139**, 5006.

N-Acetyl-L-cysteine & N-Acetyl-L-cysteine amide

These neuroprotective antioxidants and copper chelators (*N*-Acetyl-L-cycteine: $FW_{free-acid}$ = 163.20 g/mol; CAS 616-91-1; Abbreviation: NAC; (*N*-Acetyl-L-cycteine amide: $FW_{free-acid}$ = 162.21 g/mol; CAS 38520-57-9; Abbreviation: NACA) are metabolizable cysteine derivatives that are frequently used as an antidote in acetaminophen (Tylenol) and carbon monoxide (CO) poisoning. After oral administration of a 200–400 mg dose, NAC enters circulation, where it is rapidly oxidized to its homo-disulfide, reaching peak plasma concentrations of 0.35–40 mg/L in 1–2 hours. Its volume of distribution is 0.33–0.47 L/kg, with significant protein binding of 50%. Upon transporter-independent cellular uptake, *N*-acetyl-L-cysteine is deacetylated, giving rise to elevated intracellular cysteine concentrations and thereby enhancing glutathione production. While NAC inhibits neuronal cell apoptosis, it induces apoptosis in smooth muscle cells. NAC inhibits homocysteine-enhanced expression of an ox-LDL receptor. *N*-acetylcysteine amide crosses the blood-brain barrier, chelating Cu^{2+} (which is known to catalyze free radical formation) and preventing ROS-induced activation of JNK, p38 and MMP-9. *Note*: Mercapturic acids are *S*-substituted derivatives of NAC (**1**). **Target(s):** N^4-(β-*N*-acetylglucosaminyl)-L-asparaginase, K_i = 3.2 mM (2,3); cysteine dioxygenase (4,5); alkaline phosphatase (6); HIV replication (7);

prolyl/cysteinyl-tRNA synthetase, *Methanocaldococcus jannaschii* (8); alliin lyase (9); and 5'-nucleotidase (10); serine *O*-acetyltransferase (11,12). **1**. Offen, Gilgun-Sherki, Barhum, *et al.* (2004) *J. Neurochem.* **89**, 1241. **2**. Dugal & Stromme (1977) *Biochem. J.* **165**, 497. **3**. Dugal (1978) *Biochem. J.* **171**, 799. **4**. Yamaguchi & Hosokawa (1987) *Meth. Enzymol.* **143**, 395. **5**. Yamaguchi, Hosokawa, Kohashi, *et al.* (1978) *J. Biochem.* **83**, 479. **6**. Agus, Cox & Griffin (1966) *Biochim. Biophys. Acta* **118**, 363. **7**. Roederer, Staal, Raju, *et al.* (1990) *Proc. Natl. Acad. Sci. U.S.A.* **87**, 4884. **8**. Ambrogelly, Ahel, Polycarpo, *et al.* (2002) *J. Biol. Chem.* **277**, 34749. **9**. Jansen, Muller & Knobloch (1989) *Planta Med.* **55**, 440. **10**. Liu & Sok (2000) *Neurochem. Res.* **25**, 1475. **11**. Nozaki, Shigeta, Saito-Nakano, Imada & Kruger (2001) *J. Biol. Chem.* **276**, 6516. **12**. Kredich & Tomkins (1966) *J. Biol. Chem.* **241**, 4955.

N^{α}-Acetyl-L-cysteinyl-L-arginyl-L-alanyl-L-threonyl-L-lysyl-L-methionyl-L-leucinamide

This acetylated heptapeptide amide (*Sequence*: Ac-CRATKML-NH$_2$; FW = 865.10 g/mol) inhibits *Clostridium botulinum* bontoxilysin: K_i = 1.9 pM. The D-cysteinyl analogue has a K_i value of 1.8 pM (1,2). The D-arginine-containing analogue is considerably weaker (K_i = 0.41 nM). **1**. Schmidt, Stafford & Bostian (1998) *FEBS Lett.* **435**, 61. **2**. Ahmed, Byrne, Jensen, *et al.* (2001) *J. Protein Chem.* **20**, 221.

N^{α}-Acetyl-L-cysteinyl-D-cysteinyl-L-arginyl-L-alanyl-L-threonyl-L-lysyl-L-methionyl-L-leucinamide

This acetylated octapeptide amide (*Sequence*: Ac-C-DC-RATKML-NH$_2$; FW = 968.24 g/mol) inhibits *Clostridium botulinum* botoxilysin (K_i = 0.11 nM), taking advantage of the fact that Botox A's requirement for arginine as the P$_1$' inhibitor residue is dissimilar to other zinc metalloproteases. Substrate analog peptides with P$_4$, P$_3$, P$_2$' or P$_3$' cysteine are readily hydrolyzed by the toxin, but those with P$_1$ or P$_2$ cysteine were not cleaved and were inhibitors. Peptides with either D- or L-cysteine as the N-terminus, followed by the last six residues of the substrate, are the most effective inhibitors, each with a Ki value of 2 microM. Elimination of the cysteine sulfhydryl group results in much less effective inhibitors, suggesting that inhibition is due primarily to binding of the active-site zinc by the sulfhydryl group. **1**. Schmidt, Stafford & Bostian (1998) *FEBS Lett.* **435**, 61.

N-Acetyl-L-cysteinyl-L-glutaminyl-L-arginyl-L-alanyl-L-threonyl-L-lysyl-L-methionyl-L-leucinamide

This acetylated octapeptide amide (FW = 994.22 g/mol; *sequence*; Ac-CERATKML-NH$_2$) inhibits (K_i = 0.19 nM) *Clostridium botulinum* bontoxilysin, the zinc metalloprotease/neurotoxin (*or* botox A) that cleaves only one peptide bond in the synaptosomal protein, SNAP-25. The D-cysteinyl analogue is a slightly better inhibitor, K_i = 0.14 nM. Botox A displays an unusual requirement for arginine as the P$_1$' inhibitor residue, demonstrating that botox A differs from most other zinc metalloproteases and offering a strategy for achieving inhibitor selectivity. **1**. Schmidt, Stafford & Bostian (1998) *FEBS Lett.* **435**, 61.

N^2-Acetyl-2'-deoxyguanosine 5'-Monophosphate

This acetylated dGMP (FW$_{free-acid}$ = 389.26 g/mol) competitively inhibits a bifunctional enzyme exhibiting IMP cyclohydrolase (and 5-aminoimidazole-4-carboxamide ribotide (AICAR) transformylase) activity in human CCRF-CEM leukemia cells. Other competitive inhibitors of IMP cyclohydrolase include: 2-mercaptoinosine 5'-monophosphate (K_i = 0.09 μM), xanthosine 5'-monophosphate (K_i = 0.12 μM), 2-fluoroadenine arabinoside 5'-monophosphate (K_i = 0.16 μM), 6-mercaptopurine riboside 5'-monophosphate (K_i = 0.20 μM), adenosine N^1-oxide 5'-monophosphate (K_i = 0.28 μM), and N^6-(carboxymethyl)adenosine 5'-monophosphate (K_i = 1.7 μM). **1**. Szabados, Hindmarsh, Phillips, Duggleby & Christopherson (1994) *Biochemistry* **33**, 14237.

6-*S*-Acetyldihydrolipoamide

This acetylated dihydrolipoamide (FW = 381.02 g/mol) is thought to mimic an intermediate in the reaction catalyzed by the dihydrolipoyllysine-residue acetyltransferase component within the pyruvate dehydrogenase complex. **1**. Butterworth, Tsai, Eley, Roche & Reed (1975) *J. Biol. Chem.* **250**, 1921.

DL-*erythro-N*-Acetyldihydrosphingosine

D-isomer

This ceramide analogue (FW = 343.55 g/mol), also known as DL-*erythro*-C^2-dihydroceramide and DL-*erythro-N*-acetylsphinganine, is an inactive form of C^2 ceramide that inhibits acyl-CoA:cholesterol acyltransferase, *or* ACAT, as does DL-*erythro-N*-acetylsphingosine. **1**. Ridgway (1995) *Biochim. Biophys. Acta* **1256**, 39.

7-Acetyl-2,3-diphenyl-1-(*p*-tosyloxy)indolizine

This indolizine (FW = 481.57 g/mol) inhibits 15-lipoxygenase, *or* arachidonate 15-lipoxygenase, IC$_{50}$ = 23 μM. The enzyme inhibition was not affected by preincubation or the presence of a detergent and no significant particle formation was observed. Hence, inhibition from aggregates of indolizines, promiscuous inhibition, is highly unlikely. **1**. Teklu, Gundersen, Larsen, Malterud & Rise (2005) *Bioorg. Med. Chem.* **13**, 3127.

Acetyldithio-CoA

This readily enolizable acetyl-CoA analogue (FW = 814.64 g/mol), a product analogue of the reaction catalyzed by 3-hydroxy-3-methylglutaryl-CoA, supports enzyme-catalyzed exchange of the methyl protons of the acetyl group with the solvent. Citrate synthase catalyzes the slow condensation of acetyldithio-CoA with oxaloacetate to form thiocitrate. **Target(s)**: 3-hydroxy-3-methylglutaryl-CoA lyase (1); 3-hydroxy-3-methylglutaryl-CoA synthase (2). **1**. Tuinstra, Wang, Mitchell & Miziorko (2004) *Biochemistry* **43**, 5287. **2**. Wang, Misra & Miziorko (2004) *J. Biol. Chem.* **279**, 40283.

N-Acetyldopamine

This metabolite (FW = 195.22 g/mol) is a relatively nontoxic analogue and precursor of dopamine that has shown significant antitumor activity. **Target(s):** aromatic L-amino-acid decarboxylase, *or* L-dopa decarboxylase (1,2) and sepiapterin reductase (3,4). **1.** Fragoulis & Sekeris (1975) *Arch. Biochem. Biophys.* **168**, 15. **2.** Clark, Pass, Venkataraman & Hodgetts (1978) *Mol. Gen. Genet.* **162**, 287. **3.** Ruiz-Vazquez, Silva & Ferre (1996) *Comp. Biochem. Physiol. B Biochem. Mol. Biol.* **113**, 131. **4.** Smith, Duch, Edelstein & Bigham (1992) *J. Biol. Chem.* **267**, 5599.

N-Acetyl-[D-Trp16]-Endothelin 1 Fragment 16-21, See *N-Acetyl-D-tryptophanyl-L-leucyl-L-aspartyl-L-isoleucyl-L-isoleucyl-L-tryptophan*

N-Acetyl-Eglin c

This acetylated polypeptide contains 70-residue (MW = 8.1 kDa; *Sequence*: Ac-TEFGSELKSFPEVVGKTVDCAREYFTLHYPQYDVYFLPEGSPVT LDLRYNRVRVFYNPGTNVVNHVPHVG) from the medicinal leech (*Hirudo medicinalis*) inhibits cathepsin G and elastase (1). *See also Eglin* **1.** Seemuller, Eulitz, Fritz & Strobl (1980) *Hoppe-Seyler's Z. Physiol. Chem.* **361**, 1841.

Acetylene

This industrial gas H–C≡C–H (FW = 26.038 g/mol; CAS 74-86-2; IUPAC Name: Ethyne) sublimes at –31°C) serves as an alternative substrate for nitrogenase. It also alkylates the prosthetic heme group of many cytochrome P450 enzymes, especially those acting on olefinic π-bond systems to form epoxides. Acetylene is also formed from 1,1,1-trichloroethane (TCE), a widely used industrial solvent that exhibits low acute toxicity. **Target(s):** methane monooxygenase (1,2,18-22); cytochrome P450 (3,6); hydrogenase (4,7,11); nitrous oxide reductase (5,9); ammonia monooxygenase (8); alkene monooxygenase (10,17); nitrogenase, as alternative substrate (12-16); and L-ascorbate peroxidase (23,24). **1.** Pilkington & Dalton (1990) *Meth. Enzymol.* **188**, 181. **2.** Colby, Dalton & Whittenbury (1975) *Biochem. J.* **151**, 459. **3.** Ator & Ortiz de Montellano (1990) *The Enzymes*, 3rd ed., **19**, 213. **4.** Smith, Hill & Yates (1976) *Nature* **262**, 209. **5.** Kristjansson & Hollocher (1980) *J. Biol. Chem.* **255**, 704. **6.** Ortiz de Montellano, Kunze, Beilan & Wheeler (1982) *Biochemistry* **21**, 1331. **7.** Tibelius & Knowles (1984) *J. Bacteriol.* **160**, 103. **8.** Hyman & Wood (1985) *Biochem. J.* **227**, 719. **9.** Teraguchi & Hollocher (1989) *J. Biol. Chem.* **264**, 1972. **10.** Hartmans, Weber, Somhorst & de Bont (1991) *J. Gen. Microbiol.* **137**, 2555. **11.** Sun, Hyman & Arp (1992) *Biochemistry* **31**, 3158. **12.** Seefeldt, Dance & Dean (2004) *Biochemistry* **43**, 1401. **13.** Igarashi & Seefeldt (2003) *Crit. Rev. Biochem. Mol. Biol.* **38**, 351. **14.** Mortenson & Thorneley (1979) *Ann. Rev. Biochem.* **48**, 387. **15.** Christiansen, Cash, Seefeldt & Dean (2000) *J. Biol. Chem.* **275**, 11459. **16.** Erickson, Nyborg, Johnson, *et al.* (1999) *Biochemistry* **38**, 14279. **17.** Fosdike, Smith & Dalton (2005) *FEBS J.* **272**, 2661. **18.** Stirling & Dalton (1979) *Eur. J. Biochem.* **96**, 205. **19.** Zahn & DiSpirito (1996) *J. Bacteriol.* **178**, 1018. **20.** Shiemke, Cook, Miley & Singleton (1995) *Arch. Biochem. Biophys.* **321**, 421. **21.** Murrell, Gilbert & McDonald (2000) *Arch. Microbiol.* **173**, 325. **22.** Lontoh, DiSpirito & Semrau (1999) *Arch. Microbiol.* **171**, 301. **23.** Dalton, Hanus, Russell & Evans (1987) *Plant Physiol.* **83**, 789. **24.** Caldwell, Turano & McMahon (1998) *Planta* **204**, 120.

(+)-8'-Acetyleneabscisic Acid

This seed germination retardant (FW = 276.33 g/mol; CAS 191610-47-6; IUPAC Name: (2Z,4E)-*rel*-5-[(1R,6R)-6-ethynyl-1-hydroxy-2,6-dimethyl-4-oxo-2-cyclohexen-1-yl]-3-methyl-2,4-pentadienoic Acid Methyl Ester, is a mechanism-based inhibitor of (+)-abscisate 8'-hydroxylase (K_i = 19 µM; k_{inact} = 0.053 min^{-1} (1,2). **1.** Cutler, Rose, Squires, *et al.* (2000) *Biochemistry* **39**, 13614. **2.** Rose, Cutler, Irvine, *et al.* (1997) *Bioorg. Med. Chem. Lett.* **7**, 2543.

Acetylene Dicarboxylic Acid

This toxic dicarboxylic acid (FW$_{free-acid}$ = 114.06 g/mol), also known as butynedioic acid, inhibits fumarase (1,2), succinate dehydrogenase (3,4,7), glutamate decarboxylase (5), succinate oxidase (6), argininosuccinate lyase (8), maleate hydratase (9), and 3-oxoacid CoA-transferase, the latter weakly (10). **1.** Hill & Bradshaw (1969) *Meth. Enzymol.* **13**, 91. **2.** Massey (1953) *Biochem. J.* **55**, 172. **3.** Webb (1966) *Enzyme and Metabolic Inhibitors*, vol. 2, pp. 36 & 239, Academic Press, New York. **4.** Dietrich, Monson, Williams & Elvehjem (1952) *J. Biol. Chem.* **197**, 37. **5.** Fonda (1972) *Biochemistry* **11**, 1304. **6.** Thomson (1959) *Proc. Soc. Exp. Biol. Med.* **101**, 589. **7.** Hellerman, Reiss, Parmar, Wein & Lasser (1960) *J. Biol. Chem.* **235**, 2468. **8.** Garred, Mathis & Raushel (1983) *Biochemistry* **22**, 3729. **9.** Ueda, Yamada & Asano (1994) *Appl. Microbiol. Biotechnol.* **41**, 215. **10.** Fenselau & Wallis (1974) *Biochemistry* **13**, 3884.

8-Acetylenyl-6-indolin-N-yl-9-(β-D-ribofuranosyl)-purine

This modified nucleoside (FW = 381.39 g/mol) inhibits human adenosine kinase, IC$_{50}$ = 1 µM (1). Because the use of adenosine receptor agonists is plagued by dose-limiting cardiovascular side effects, there is great interest in developing adenosine kinase inhibitors (AKIs) as a means for raising steady-state adenosine concentrations. **1.** Bookser, Matelich, Ollis & Ugarkar (2005) *J. Med. Chem.* **48**, 3389.

Acetyl-1,N⁶-EthenoCoA, See *1,N⁶-Etheno-acetyl-CoA*

N-Acetyl-S-trans,trans-farnesyl-L-cysteine

This N- and S-acylated L-cysteine (FW$_{free-acid}$ = 367.55 g/mol) is an alternative substrate (K_m = 20 µM) of S-farnesylcysteine methyltransferase (*i.e.*, protein-S-isoprenylcysteine O-methyltransferase) and thus inhibits posttranslational modification of proteins (1-5) as well as chemotaxis (3). **1.** Volker & Stock (1995) *Meth. Enzymol.* **255**, 65. **2.** Choi, Niedbala, Lynch, *et al.* (2001) *Meth. Enzymol.* **332**, 103. **3.** Young, Ambroziak, Kim & Clarke (2001) *The Enzymes*, 3rd. ed., **21**, 155. **4.** Volker, Miller, McCleary, *et al.* (1991) *J. Biol. Chem.* **266**, 21515. **5.** Papaharalambus, Sajjad, Syed, *et al.* (2005) *J. Biol. Chem.* **280**, 18790.

N-[N-Acetyl-L-(N-indole-formyl)tryptophanyl-L-prolyl-L-phenylalanyl-L-histidyl-(2S-amino-3-phenylpropyl)]-L-phenylalaninamide

This pseudopeptide amide (FW$_{free-base}$ = 935.10 g/mol; *Sequence*: Ac-Ftr-Pro-Phe-His-Pheψ[CH$_2$-NH]Phe-NH$_2$, where Ftr refers to an N-indole-formyltryptophanyl residue and a reduced peptide bond between two phenylalanyl residues), also known as U-71908E, inhibits renin, IC$_{50}$ = 0.56

nM. **1**. Sawyer, Maggiora, Liu, *et al.* (1990) in *Peptides: Chemistry, Structure, and Biology* (Rivier & Marshall, eds), p. 46, Escom Sci. Publ.

N-[*N*-Acetyl-L-(*N*-indole-formyl)tryptophanyl-L-prolyl-L-phenylalanyl-L-histidyl-(2S-amino-3-phenylpropyl)]-L-phenylalanyl-L-valyl-L-tyrosinamide

This renin-inhibiting pseudopeptide amide (FW$_{\text{free-base}}$ = 1197.40 g/mol; *Sequence*: Ac-Ftr-Pro-Phe-His-Pheψ[CH$_2$-NH]Phe-Val-Tyr-NH$_2$, where Ftr refers to an *N*-indole-formyltryptophanyl residue and a reduced peptide bond between two phenylalanyl residues) inhibits renin, IC$_{50}$ = 3.0 nM (1). **1**. Sawyer, Maggiora, Liu, *et al.* (1990) in *Peptides: Chemistry, Structure, and Biology* (Rivier & Marshall, eds), p. 46, Escom Sci. Publ.

N-Acetyl-D-galactosamine

This acetylated aminosugar (FW = 221.21 g/mol), also known as 2-acetamido-2-deoxy-D-galactopyranose and abbreviated D-Gal*p*NAc and NAGA, is a component of many glycoproteins, chondroitin, and blood-group substances. *N*-Acetyl-D-galactosamine will also inhibit the binding of ligands to many lectins. **Target(s)**: β-*N*-acetylhexosaminidase (1,2,13-15,17,26-43); α-*N*-acetylgalactosaminidase (3,12,19,44-46); discoidins I and II, *or Dictyostelium discoideum* agglutinins (4); *Bauhinia purpurea* agglutinin (5); lima bean (*Phaseolus lunatus*) lectin (6); *Wistaria floribunda* phytomitogen (7); horse gram (*Dolichos biflorus*) lectin (8); *Sophora japonica* hemagglutinin (9); soybean (*Glycine max*) agglutinin (10,16); snail (*Helix pomatia*) hemagglutinin (11); α-galactosidase B (12); β-*N*-acetylglucosaminidase (13-15,35); β-*N*-acetylgalactosaminidase (13-15,17,24,25); galactose oxidase, also alternative substrate (18); keratan sulfotransferase, weakly inhibited (20); aralin (21); rRNA *N*-glycosylase (21); peptidoglycan β-*N*-acetylmuramidase (22,23); polypeptide *N* acetylgalactosaminyltransferase (47,48). **1**. Frohwein & Gatt (1969) *Meth. Enzymol.* **14**, 161. **2**. Johnson, Mook & Brady (1972) *Meth. Enzymol.* **28**, 857. **3**. Weissmann (1972) *Meth. Enzymol.* **28**, 801. **4**. Barondes, Rosen, Frazier, Simpson & Haywood (1978) *Meth. Enzymol.* **50**, 306. **5**. Osawa, Irimura & Kawaguchi (1978) *Meth. Enzymol.* **50**, 367. **6**. Galbraith & Goldstein (1972) *Meth. Enzymol.* **28**, 318. **7**. Osawa & Toyoshima (1972) *Meth. Enzymol.* **28**, 328. **8**. Etzler (1972) *Meth. Enzymol.* **28**, 340. **9**. Poretz (1972) *Meth. Enzymol.* **28**, 349. **10**. Lis & Sharon (1972) *Meth. Enzymol.* **28**, 360. **11**. Hammerström (1972) *Meth. Enzymol.* **28**, 368. **12**. Brady (1983) *The Enzymes*, 3rd ed., **16**, 409. **13**. Webb (1966) *Enzyme and Metabolic Inhibitors*, vol. 2, p. 419, Academic Press, New York. **14**. Watkins (1959) *Biochem. J.* **71**, 261. **15**. Walker & Nicholas (1961) *Biochem. J.* **78**, 10p. **16**. Lis, Sela, Sachs & Sharon (1970) *Biochim. Biophys. Acta* **211**, 582. **17**. Frohwein & Gatt (1967) *Biochemistry* **6**, 2775. **18**. Yip & Dain (1968) *Enzymologia* **35**, 368. **19**. Weissmann & Hinrichsen (1969) *Biochemistry* **8**, 2034. **20**. Yamamoto, Takahashi, Ogata & Nakazawa (2001) *Arch. Biochem. Biophys.* **392**, 87. **21**. Tomatsu, Kondo, Yoshikawa, *et al.* (2004) *Biol. Chem.* **385**, 819. **22**. del Rio & Berkeley (1976) *Eur. J. Biochem.* **65**, 3. **23**. del Rio, Berkeley, Brewer & Roberts (1973) *FEBS Lett.* **37**, 7. **24**. Tanaka & Ozaki (1997) *J. Biochem.* **122**, 330. **25**. Izumi & Suzuki (1983) *J. Biol. Chem.* **258**, 6991. **26**. Calvo, Reglero & Cabezas (1978) *Biochem. J.* **175**, 743. **27**. Garcia-Alonso, Reglero & Cabezas (1990) *Int. J. Biochem.* **22**, 645. **28**. Woollen, Heymorth & Walker (1961) *Biochem. J.* **78**, 111. **29**. Edwards, Thomas & Westwood (1975) *Biochem. J.* **151**, 145. **30**. Barber & Ride (1989) *Plant Sci.* **60**, 163. **31**. Berkeley, Brewer, Ortiz & Gillespie (1973) *Biochim. Biophys. Acta* **309**, 157. **32**. Mommsen (1980) *Biochim. Biophys. Acta*

612, 361. **33**.-Li & Li (1970) *J. Biol. Chem.* **245**, 5153. **34**. Geimba, Riffel & Brandelli (1998) *J. Appl. Microbiol.* **85**, 708. **35**. Khar & Anand (1977) *Biochim. Biophys. Acta* **483**, 141. **36**. Stirling (1972) *Biochim. Biophys. Acta* **271**, 154. **37**. Potier, Teitelbaum, Melancon & Dallaire (1979) *Biochim. Biophys. Acta* **566**, 80. **38**. Gers-Barlag, Bartz & Ruediger (1988) *Phytochemistry* **27**, 3739. **39**. Sakai, Nakanishi & Kato (1993) *Biosci. Biotechnol. Biochem.* **57**, 965. **40**. Mian, Herries, Cowen & Batte (1979) *Biochem. J.* **177**, 319. **41**. Eriquez & Pisano (1979) *J. Bacteriol.* **137**, 620. **42**. Sakai, Narihara, Kasama, Wakayama & Moriguchi (1994) *Appl. Environ. Microbiol.* **60**, 2911. **43**. Poulton, Thomas, Ottwell & McCormick (1985) *Plant Sci.* **42**, 107. **44**. Uda, Li & Li (1977) *J. Biol. Chem.* **252**, 5194. **45**. Chien (1986) *J. Chin. Biochem. Soc.* **15**, 86. **46**. Itoh & Uda (1984) *J. Biochem.* **95**, 959. **47**. Tenno, Saeki, Kézdy, Elhammer & Kurosaka (2002) *J. Biol. Chem.* **277**, 47088. **48**. Wandall, Irazoqui, Tarp, *et al.* (2007) *Glycobiology* **17**, 374.

N-Acetyl-D-glucosamine

This acetylated aminosugar (FW = 221.21 g/mol; CAS 7512-17-6), also known as β-acetamido-2-deoxy-D-glucopyranose and abbreviated D-Glc*p*NAc and NAG, is a component of many glycoproteins. NAG also inhibits the binding of ligands to many lectins. **Target(s)**: glucokinase (1,5,12,19,66-69,71); β-*N*-acetylhexosaminidase (2,3,8,12,14-16,21,27-48); α-*N*-acetylglucosaminidase (4,18); hexokinase (6,19,20,69-72); acetyl-CoA:α-glucosaminide *N*-acetyltransferase, *or* heparan-α-glucosaminide *N*-acetyltransferase (7); chitinase (9,58-60); concanavalin A (10); lysozyme, both anomers (11,22,50-57); fructokinase (12,13); α-galactosidase (12,14); β-*N*-acetylgalactosaminidase (12,14,16,21); β-*N*-acetylglucosaminidase (12,14-16,34,37,48); glucosamine kinase (17); keratan sulfotransferase (23); *N*-acetylmuramoyl-L-alanine amidase (24); peptidoglycan β-*N*-acetylmuramidase (25,26); α-*N*-acetylglucosaminidase (49); *N*-acetylglucosamine-1-phosphodiester α-*N*-acetylglucosaminidase (61-64); acyl-phosphate:hexose phosphotransferase, also weak alternative substrate (65); α-1,6-mannosyl-glycoprotein 4-β-*N*-acetylglucosaminyl transferase, weakly inhibited (73); β-1,4-mannosyl-glycoprotein 4-β-*N*-acetylglucosaminyltransferase (74); and β-*N*-acetylglucosaminyl-glycopeptide β-1,4-galactosyl-transferase (75,76). **1**. Walker & Parry (1966) *Meth. Enzymol.* **9**, 381. **2**. Frohwein & Gatt (1969) *Meth. Enzymol.* **14**, 161. **3**. Johnson, Mook & Brady (1972) *Meth. Enzymol.* **28**, 857. **4**. Weissmann (1972) *Meth. Enzymol.* **28**, 796. **5**. Maitra (1975) *Meth. Enzymol.* **42**, 25. **6**. Fornaini, Dachà, Magnani & Stocchi (1982) *Meth. Enzymol.* **90**, 3. **7**. Bame & Rome (1987) *Meth. Enzymol.* **138**, 607. **8**. Ohtakara (1988) *Meth. Enzymol.* **161**, 462. **9**. Boller, Gehri, Mauch & Vögeli (1988) *Meth. Enzymol.* **161**, 479. **10**. Goldstein, Hollerman & Smith (1965) *Biochemistry* **4**, 876. **11**. Imoto, Johnson, North, Phillips & Rupley (1972) *The Enzymes*, 3rd ed., **7**, 665. **12**. Webb (1966) *Enzyme and Metabolic Inhibitors*, vol. 2, Academic Press, New York. **13**. Maley & Lardy (1955) *J. Biol. Chem.* **214**, 765. **14**. Watkins (1959) *Biochem. J.* **71**, 261. **15**. Pugh, Leaback & Walker (1957) *Biochem. J.* **65**, 464. **16**. Walker &. Nicholas (1961) *Biochem. J.* **78**, 10p. **17**. Harpur & Quastel (1949) *Nature* **164**, 693. **18**. Weissmann, Rowan, Marshall & Friederici (1967) *Biochemistry* **6**, 207. **19**. González, Ureta, Babul, Rabajille & Niemeyer (1967) *Biochemistry* **6**, 460. **20**. Ning, Purich & Fromm (1969) *J. Biol. Chem.* **244**, 3840. **21**. Frohwein & Gatt (1967) *Biochemistry* **6**, 2775. **22**. Dahlquist & Raftery (1968) *Biochemistry* **7**, 3269. **23**. Yamamoto, Takahashi, Ogata & Nakazawa (2001) *Arch. Biochem. Biophys.* **392**, 87. **24**. Van Heijenoort, Parquet, Flouret & Van Heijenoort (1975) *Eur. J. Biochem.* **58**, 611. **25**. del Rio & Berkeley (1976) *Eur. J. Biochem.* **65**, 3. **26**. del Rio, Berkeley, Brewer & Roberts (1973) *FEBS Lett.* **37**, 7. **27**. Tomiya, Narang, Park, *et al.* (2006) *J. Biol. Chem.* **281**, 19545. **28**. Calvo, Reglero & Cabezas (1978) *Biochem. J.* **175**, 743. **29**. Garcia-Alonso, Reglero & Cabezas (1990) *Int. J. Biochem.* **22**, 645. **30**. Woollen, Heymorth & Walker (1961) *Biochem. J.* **78**, 111. **31**. Edwards, Thomas & Westwood (1975) *Biochem. J.* **151**, 145. **32**. Barber & Ride (1989) *Plant Sci.* **60**, 163. **33**. Berkeley, Brewer, Ortiz & Gillespie (1973) *Biochim. Biophys. Acta* **309**, 157. **34**. Mommsen (1980) *Biochim. Biophys. Acta* **612**, 361. **35**. Li & Li (1970) *J. Biol. Chem.* **245**, 5153. **36**. Geimba, Riffel

& Brandelli (1998) *J. Appl. Microbiol.* **85**, 708. **37**. Khar & Anand (1977) *Biochim. Biophys. Acta* **483**, 141. **38**. Potier, Teitelbaum, Melancon & Dallaire (1979) *Biochim. Biophys. Acta* **566**, 80. **39**. Jones & Kosman (1980) *J. Biol. Chem.* **255**, 11861. **40**. Gers-Barlag, Bartz & Ruediger (1988) *Phytochemistry* **27**, 3739. **41**. Sakai, Nakanishi & Kato (1993) *Biosci. Biotechnol. Biochem.* **57**, 965. **42**. Mian, Herries, Cowen & Batte (1979) *Biochem. J.* **177**, 319. **43**. Koga, Iwamoto, Sakamoto, *et al.* (1991) *Agric. Biol. Chem.* **55**, 2817. **44**. Eriquez & Pisano (1979) *J. Bacteriol.* **137**, 620. **45**. Sakai, Narihara, Kasama, Wakayama & Moriguchi (1994) *Appl. Environ. Microbiol.* **60**, 2911. **46**. Sone & Misaki (1978) *J. Biochem.* **83**, 1135. **47**. Ohtakara, Yoshida, Murakami & Izumi (1981) *Agric. Biol. Chem.* **45**, 239. **48**. Cohen (1986) *Plant Sci.* **43**, 93. **49**. Nok, Shuaibu, Choudhry, *et al.* (2001) *J. Biochem. Mol. Toxicol.* **15**, 221. **50**. Bernard, Canioni, Cozzone, Berthou & Jolles (1990) *Int. J. Protein Res.* **36**, 46. **51**. Croux, Canard, Goma & Sucaille (1992) *Appl. Environ. Microbiol.* **58**, 1075. **52**. Glazer, Barel, Howard & Brown (1969) *J. Biol. Chem.* **244**, 3583. **53**. Perin & Jolles (1973) *Mol. Cell. Biochem.* **2**, 189. **54**. Howard & Glazer (1967) *J. Biol. Chem.* **242**, 5715. **55**. Pahud & Widmer (1982) *Biochem. J.* **201**, 661. **56**. Bernier, van Leemputten, Horisberger, Bush & Jolles (1971) *FEBS Lett.* **14**, 100. **57**. Gayen, Som, Sinha & Sen (1977) *Arch. Biochem. Biophys.* **183**, 432. **58**. Bhushan (2000) *J. Appl. Microbiol.* **88**, 800. **59**. Molano, Polacheck, Duran & Cabib (1979) *J. Biol. Chem.* **254**, 4901. **60**. Boller, Gehri, Mauch & Vögeli (1983) *Planta* **157**, 22. **61**. Page, Zhao, Tao & Miller (1996) *Glycobiology* **6**, 619. **62**. Mullis, Huynh & Kornfeld (1994) *J. Biol. Chem.* **269**, 1718. **63**. Varki & Kornfeld (1981) *J. Biol. Chem.* **256**, 9937. **64**. Lee & Pierce (1995) *Arch. Biochem. Biophys.* **319**, 413. **65**. Casazza & Fromm (1977) *Biochemistry* **16**, 3091. **66**. Goward, Hartwell, Atkinson & Scawen (1986) *Biochem. J.* **237**, 415. **67**. Xu, Harrison, Weber & Pilkis (1995) *J. Biol. Chem.* **270**, 9939. **68**. Klein & Charles (1986) *Can. J. Microbiol.* **32**, 937. **69**. Monasterio & Cardenas (2003) *Biochem. J.* **371**, 29. **70**. Racagni & Machado de Domenech (1983) *Mol. Biochem. Parasitol.* **9**, 181. **71**. Vandercammen & Van Schaftingen (1991) *Eur. J. Biochem.* **200**, 545. **72**. Aleshin, Zeng, Bourenkov, *et al.* (1998) *Structure* **6**, 39. **73**. Brockhausen, Hull, Hindsgaul, *et al.* (1989) *J. Biol. Chem.* **264**, 11211. **74**. Ikeda, Koyota, Ihara, *et al.* (2000) *J. Biol. Chem.* **128**, 609. **75**. Bella, Whitehead & Kim (1977) *Biochem. J.* **167**, 621. **76**. Rao, Garver & Mendicino (1976) *Biochemistry* **15**, 5001.

N-Acetyl-D-glucosamine-1,5-lactone

This *N*-acetylglucosamine derivative (FW = 219.19 g/mol; deliquescent), inhibits α-*N*-acetylglucosaminidase (1,5), mannosyl-glycoprotein endo-β-*N* acetylglucosaminidase (2,4,6,7,9-11), lysozyme (3), β-*N*-acetylgalactos-aminidase (4,7), chitinase (8), and β-*N* acetylhexosaminidase (12-22). *Note:* Several literature citations do not distinguish between the 1,5- or the 1,4-lactone. **1**. Weissmann (1972) *Meth. Enzymol.* **28**, 796. **2**. Horsch, Hoesch, Fleet & Rast (1993) *J. Enzyme Inhib.* **7**, 47. **3**. Imoto, Johnson, North, Phillips & Rupley (1972) *The Enzymes*, 3rd ed., **7**, 665. **4**. Webb (1966) *Enzyme and Metabolic Inhibitors*, vol. **2**, Academic Press, New York. **5**. Weissmann, Rowan, Marshall & Friederici (1967) *Biochemistry* **6**, 207. **6**. Levvy & Conchie (1966) *Meth. Enzymol.* **8**, 571. **7**. Walker & Nicholas (1961) *Biochem. J.* **78**, 10p. **8**. Pegg (1988) *Meth. Enzymol.* **161**, 474. **9**. Couling & Goodey (1970) *Biochem. J.* **119**, 303. **10**. Horsch, Hoesch, Vasella & Rast (1991) *Eur. J. Biochem.* **197**, 815. **11**. Koide & Muramatsu (1974) *J. Biol. Chem.* **249**, 4897. **12**. Maier, Strater, Schuette, *et al.* (2003) *J. Mol. Biol.* **328**, 669. **13**. Calvo, Reglero & Cabezas (1978) *Biochem. J.* **175**, 743. **14**. Garcia-Alonso, Reglero & Cabezas (1990) *Int. J. Biochem.* **22**, 645. **15**. Woollen, Heymorth & Walker (1961) *Biochem. J.* **78**, 111. **16**. Berkeley, Brewer, Ortiz & Gillespie (1973) *Biochim. Biophys. Acta* **309**, 157. **17**. Mommsen (1980) *Biochim. Biophys. Acta* **612**, 361. **18**. Wheeler, Bharadwaj & Gregory (1982) *J. Gen. Microbiol.* **128**, 1063. **19**. Li & Li (1970) *J. Biol. Chem.* **245**, 5153. **20**. Jones & Kosman (1980) *J. Biol. Chem.* **255**, 11861. **21**. Mian, Herries, Cowen & Batte (1979) *Biochem. J.* **177**, 319. **22**. Reyes & Byrde (1973) *Biochem. J.* **131**, 381.

N-Acetyl-D-glucosamine 1-Phosphate

This phosphorylated metabolite (FW$_{free-acid}$ = 301.19 g/mol; hygroscopic; pK_a values of < 1.4 and 6.0), als named 2-acetamido-2-deoxy-D-glucose 1-phosphate, is a product of the reaction catalyzed by glucosamine-1 phosphate *N*-acetyltransferase and is a substrate for *N*-acetylglucosamine-1-phosphate uridyltransferase (*or* UDP-*N*-acetylglucosamine diphos-phorylase) and phosphoacetylglucosamine mutase. The α-anomer is more stable in acids than the β-epimer; however, both are completely hydrolyzed in ten minutes in 0.1 M HCl at 100°C. The β-anomer is completely hydrolyzed in 0.67 M H$_2$SO$_4$ and 26°C in 45 min, whereas only 5-6% of the α-anomer has undergone hydrolysis. **Target(s):** *N*-acetylglucosamine-1-phosphodiester α-*N*-acetylglucosaminidase, inhibited by the α-anomer (1-4); β-1,4-mannosyl-glycoprotein 4-β-*N*-acetylglucosaminyl-transferase, inhibited by the α-anomer (5). **1**. Page, Zhao, Tao & Miller (1996) *Glycobiology* **6**, 619. **2**. Mullis, Huynh & Kornfeld (1994) *J. Biol. Chem.* **269**, 1718. **3**. Varki & Kornfeld (1981) *J. Biol. Chem.* **256**, 9937. **4**. Lee & Pierce (1995) *Arch. Biochem. Biophys.* **319**, 413. **5**. Ikeda, Koyota, Ihara, *et al.* (2000) *J. Biochem.* **128**, 609.

N-Acetyl-D-glucosamine 6-Phosphate

This phosphorylated metabolite (FW$_{free-acid}$ = 301.19 g/mol), also known as 2-acetamido-2-deoxy-D-glucose 6-phosphate, is hygroscopic. About 60% will be hydrolyzed in three minutes in 0.2 M alkali at 100°C. A product of the reaction catalyzed by glucosamine 6-phosphate *N*-acetyltransferase and *N*-acetylglucosamine kinase, *N*-acetylglucosamine 6-P inhibits glucosamine-6-phosphate deaminase. **1**. Weidanz, Campbell, DeLucas, *et al.* (1995) *Brit. J. Hematol.* **91**, 72.

O-[*N*-Acetyl-D-glucosaminyl-1,2-(α-6-deoxy-D-mannosyl)-1,6-(β-D-glucosyl)]-8-octanol

This substrate analogue (FW = 624.70 g/mol) inhibits α-1,6-mannosyl-glycoprotein 6-β-*N*-acetylglucosaminyl-transferase, K_i = 30 μM (1-3). **1**. Khan, Crawley, Kanie & Hindsgaul (1993) *J. Biol. Chem.* **268**, 2468. **2**. Palcic, Ripka, Kaur, *et al.* (1990) *J. Biol. Chem.* **265**, 6759. **3**. Hindsgaul, Kaur, Srivastava, *et al.* (1991) *J. Biol. Chem.* **266**, 17858.

O-[*N*-Acetyl-D-glucosaminyl-1,2-(α-4-*O*-methyl-D-mannosyl)-1,6-(β-D-glucosyl)]-8-octanol

This substrate analogue (FW = 641.58 g/mol) inhibits α-1,6-mannosyl-glycoprotein 6-β-N-acetylglucosaminyl-transferase, K_i = 14 μM. NMR data show no evidence of important conformational differences between the trisaccharide analogs and the tetrasaccharides, and kinetic experiments detected no differences for the binding of UDP-GlcNAc in their presence. These findings suggest that the large methyl group introduced on O-4' sterically prevented the formation of product even though both potential substrates were bound by the enzyme. This "steric exclusion" strategy offers potential f designing inhibitors for that class of glycosyltransferases in which the reactive hydroxyl group is also an essential recognition element. **1.** Khan, Crawley, Kanie & Hindsgaul (1993) *J. Biol. Chem.* **268**, 2468.

N-Acetyl-L-glutamic Acid

This acetylated amino acid (FW$_{free-acid}$ = 189.17 g/mol), itself an essential activator for the mammalian urea cycle enzyme carbamoyl-phosphate synthetase I, *or* carbamoyl-phosphate synthetase (ammonia), is an intermediate in the biosynthesis of arginine in microorganisms, including *Escherichia coli*, *Neurospora crassa*, *Pisum sativum*, *Mycobacterium smegmatis*, *Pseudomonas aeruginosa*, and *P. putida*). N-Acetyl-L-glutamic acid is the product of the reactions catalyzed by amino-acid N-acetyltransferase and glutamate N-acetyltransferase (*Reaction:* L-Glutamate + Acetyl-CoA ⇌ N-Acetyl-L-glutamate + CoA–SH), the latter activated by arginine. **Target(s):** N-formylglutamate deformylase (1); 4-methyleneglutaminase (2); glutamate carboxypeptidase II (3); glutamate carboxypeptidase (4); and glutamate formimidoyltransferase (5). **1.** Hu, Mulfinger & Phillips (1987) *J. Bacteriol.* **169**, 4696. **2.** Ibrahim, Lea & Fowden (1984) *Phytochemistry* **23**, 1545. **3.** Robinson, Blakely, Couto & Coyle (1987) *J. Biol. Chem.* **262**, 14498. **4.** Goldman & Levy (1967) *Proc. Natl. Acad. Sci. U.S.A.* **58**, 1299. **5.** Miller & Waelsch (1957) *J. Biol. Chem.* **228**, 397.

N$^\alpha$-Acetyl-L-glutamine

This acetylated amino acid (FW$_{free-acid}$ = 188.18 g/mol) inhibits asparagine synthetase (1), N-formylglutamate deformylase, weakly (2), 4-methyleneglutaminase (3), and amino-acid N-acetyltransferase (4). **1.** Lea & Fowden (1975) *Proc. R. Soc. Lond. B Biol. Sci.* **192**, 13. **2.** Hu, Mulfinger & Phillips (1987) *J. Bacteriol.* **169**, 4696. **3.** Ibrahim, Lea & Fowden (1984) *Phytochemistry* **23**, 1545. **4.** Shigesada & Tatibana (1978) *Eur. J. Biochem.* **84**, 285.

{N-[N-Acetyl-L-glutaminyl-L-asparaginyl]-(5S-amino-4S-hydroxy-2-isopropyl-7-methyloctanoyl)}-L-isoleucinamide

This pseudopeptide amide (*Sequence:* Ac-Gln-Asn-Leuψ[CH(OH)CH$_2$]Val-Ile-NH$_2$; FW = 627.77 g/mol), also known as U92517E, inhibits rhizopuspepsin, K_i = 1.36 nM (1) and HIV-1 protease, K_i = 28 nM (2). **1.** Lowther, Sawyer, Staples, *et al.* (1992) in *Peptides: Chemistry and Biology Proc. of the 12th Amer. Peptide Symp.* (June 16-21, 1991) (Smith & Rivier, eds), p. 413. **2.** Sawyer, Staples, Liu, *et al.* (1992) *Int. J. Pept. Protein Res.* **40**, 274.

N-Acetyl-L-glutamyl-L-aspartyl-L-aspartyl-L-aspartyl-L-tryptophanyl-L-aspartyl-L-phenylalanine

This acetylated heptapeptide (*Sequence:* Ac-QDDDWDF; FW = 982.94 g/mol) inhibits ribonucleoside-diphosphate reductase. The sequence is derived from the C-terminus of the small subunit of *Mycobacterium tuberculosis* ribonucleoside diphosphate reductase, IC$_{50}$ = 20 or 135 μM (1,2). Substitution of an alanyl residue at each position weakens the inhibitory effect. **1.** Nurbo, Roos, Muthas, *et al.* (2007) *J. Pept. Sci.* **13**, 822. **2.** Yang, Curran, Li, *et al.* (1997) *J. Bacteriol.* **179**, 6408.

N$^\omega$-Acetylhistamine

This acetylated histamine (FW$_{free-base}$ = 153.18 g/mol) inhibits β-glucosidase (1), cytochrome P450 (2), and glutaminyl-peptide cyclotransferase (3-6). **1.** Field, Haines, Chrystal & Luszniak (1991) *Biochem. J.* **274**, 885. **2.** Morris, Tucker, Crewe, *et al.* (1989) *Biochem. Pharmacol.* **38**, 2639. **3.** Schilling, Niestroj, Rahfeld, *et al.* (2003) *J. Biol. Chem.* **278**, 49773. **4.** Schilling, Lindner, Koch, *et al.* (2007) *Biochemistry* **46**, 10921. **5.** Huang, Liu, Cheng, Ko & Wang (2005) *Proc. Natl. Acad. Sci. U.S.A.* **102**, 13117. **6.** Huang, Liu & Wang (2005) *Protein Expr. Purif.* **43**, 65.

N-Acetylhomocysteine Thiolactone

This reagent (FW = 159.21 g/mol), also called citiolone, has been used to covalently modify proteins, thereby introducing a sulfhydryl group into the protein. It has also been used to treat a number of liver disorders. **Target(s):** ribonuclease (1,2); phenylalanine monooxygenase (3); and insulin (4). **1.** Shall & Barnard (1969) *J. Mol. Biol.* **41**, 237. **2.** Richards & Wyckoff (1971) *The Enzymes*, 3rd ed., **4**, 647. **3.** Weiss, Hui & Lajtha (1979) *Res. Commun. Chem. Pathol. Pharmacol.* **25**, 153. **4.** Virupaksha & Tarver (1964) *Biochemistry* **3**, 1507.

N-Acetylhydantocidin 5'-Monophosphate

This nucleoside derivative (FW$_{free-acid}$ = 340.18 g/mol), also known as N-acetyl-5'-phosphohydantocidin, is the metabolically phosphorylated form of N-acetyl-5'-phosphohydantocidin, and, when tested in *Arabidopsis thaliana*, potently inhibits adenylosuccinate synthetase, but not adenylosuccinate lyase. Such findings explain the observation that its herbicidal effects of N-acetyl-5'-phosphohydantocidin are reversed, when the agar growth medium is supplemented with AMP, but not IMP or GMP, suggesting that hydantocidin blocked the two-step conversion of IMP to AMP in *de novo* purine biosynthesis pathway. **1.** Siehl, Subramanian, Walters, *et al.* (1996) *Plant Physiol.* **110**, 753.

3-(N-Acetyl-N-hydroxyamino)-2-benzylpropanoic Acid

This hydroxamate (FW = 189.26 g/mol) inhibits bovine carboxypeptidase A, K_i = 8.06 μM. Its hydroxamate anion displaces a relatively acidic $_{H2O}$ ligand (pK_a = 6) from the active-site zinc ion of carboxypeptidase A. In solution, metal ion-coordinating, N-substituted hydroxamic acid groups often exist as a mixture of *syn* and *anti* rotamers, with relative abundances depending on their pK_a. Pyrrolidinone analogues that are conformationally in the *syn* form are not especially potent inhibitors. Structure-activity relationships thus define criteria for designing hydroxamic acid inhibitors of metalloenzymes. **1**. Mock & Cheng (2000) *Biochemistry* **39**, 13945.

N-Acetyl-5-hydroxytryptamine, *See* *N-Acetylserotonin*

N-Acetylimidazole

This relatively indiscriminant acetylating reagent (FW = 110.12 g/mol; Most stable at pH 7.5) modifies both amino groups and tyrosyl hydroxyl groups within proteins, but the latter are preferred under normal reaction conditions. When utilizing N-acetylimidazole, fresh solutions must always be used, because the reagent is relatively unstable in water. Above pH 8.0, the O-acetyltyrosyl products are readily deacetylated. When residues essential for catalysis are modified, treatment with N-acetylimidazole can lead to loss of enzyme activity. **Target(s):** mouse submandibular renin (1); aspergillopepsin I (2); carboxypeptidase A, peptidase activity is inhibited, while esterase activity is enhanced (3-5,37-39,58); thrombin (6); phosphofructokinase (7); fructose-1,6-bisphosphate aldolase (8,32,41); nitric-oxide synthase (9); prostaglandin endoperoxide synthase (10); H^+-importing ATPase (11); Na^+/K^+-exchanging ATPase (12,18); transketolase (13); ribulose-1,5-bisphosphate carboxylase (14); neutrophil collagenase (15); succinate dehydrogenase (16); cytochrome P450LM2 (17); diamine oxidase (19); angiotensin I-converting enzyme, *or* peptidyl-dipeptidase A (20); trypsin (21); 3-hydroxyisobutyrate dehydrogenase (22); D-β hydroxybutyrate dehydrogenase (23); α-amylase (24,25); microbial collagenase (26); fructose-1,6 bisphosphatase (27,29); lysozyme (28,33); arylsulfatase (30); D-lysine 5,6-aminomutase (31,45); arginine kinase (34); glutamine synthetase (35,51); filamentous phage gene-5 DNA binding protein (36); ribonuclease A (40); protein-L-isoaspartate(D-aspartate) O-methyltransferase (42); galactolipase (43); ornithine carbamoyltransferase (44); aldose 1-epimerase, *or* mutarotase (46); lactoylglutathione lyase, *or* glyoxalase I (47); alliin lyase, slightly inhibited (48); phenylalanine ammonia-lyase (49); heparin lyase (50); aminocarboxymuconate-semialdehyde decarboxylase (52); thiamin-triphosphatase (53); 3-oxoacid CoA-transferase, *or* 3-ketoacid CoA-transferase (54); allantoicase (55); N-acyl-D-amino-acid deacylase (56); aryl-acylamidase, but stimulated at low concentrations (57); butyrylcholinesterase, but stimulated at low concentrations (57); membrane dipeptidase, *or* renal dipeptidase, *or* dehydropeptidase I (59); soluble epoxide hydrolase, weakly inhibited (60); leukotriene-A4 hydrolase (61); glucan endo-1,6-β-glucosidase (62); β-N-acetylhexosaminidase (63); α-N acetylgalactosaminidase (64); glucan endo-1,3-β-D-glucosidase (65); β-D-fucosidase66; β-glucosidase (67); chitinase (68); endo-1,4-β-xylanase (69); α-amylase (70); fructose-1,6-bisphosphatase (71); protein glutamate methylesterase (72); galactolipase (73); gluconolactonase (74); serine-phosphoethanolamine synthase (75); sulfate adenylyltransferase (76); aspartate kinase (77); N-acetyllactosamine synthase (78); β-N-acetylglucosaminylglycopeptide β-1,4-galactosyltransferase (78); lactose synthase (79); starch phosphorylase (80); glutaminyl-peptide cyclotransferase (81); N-acetylneuraminate 7-O- (or 9-O-) acetyltransferase (82); dihydrolipoyllysine-residue acetyltransferase (83); ornithine carbamoyltransferase (84) **1**. Suzuki, Murakami, Nakamura & Inagami (1998) in *Handbook of Proteolytic Enzymes* (Barrett, Rawlings & Woessner, eds), p. 856, Academic Press, San Diego. **2**. Ichishima (1970) *Meth. Enzymol.* **19**, 397. **3**. Pétra (1970) *Meth. Enzymol.* **19**, 460. **4**. Riordan & Vallee (1972) *Meth. Enzymol.* **25**, 500. **5**. Sigman & Mooser (1975) *Ann. Rev. Biochem.* **44**, 889. **6**. Lundblad, Kingdon & Mann (1976) *Meth. Enzymol.* **45**, 156. **7**. Chapman, Sanner & Pihl (1969) *Eur. J. Biochem.* **7**, 588. **8**. Pugh & Horecker (1967) *Arch. Biochem. Biophys.* **122**, 196. **9**. Amin, Vyas, Attur, *et al.* (1995) *Proc. Natl. Acad. Sci. U.S.A.* **92**, 7926. **10**. Scherer, Karthein, Strieder & Ruf (1992) *Eur. J. Biochem.* **205**, 751. **11**. Kulanthaivel, Simon, Burckhardt, *et al.* (1990) *Biochemistry*

29, 10807. **12**. Arguello & Kaplan (1990) *Biochemistry* **29**, 5775. **13**. Kuimov, Kovina & Kochetov (1988) *Biochem. Int.* **17**, 517. **14**. Purohit & Bhagwat (1988) *Indian J. Biochem. Biophys.* **25**, 313. **15**. Mookhtiar, Wang & Van Wart (1986) *Arch. Biochem. Biophys.* **246**, 645. **16**. Nakae & Shono (1986) *Histochem. J.* **18**, 169. **17**. Janig, Friedrich, Smettan, *et al.* (1985) *Biomed. Biochim. Acta* **44**, 1071. **18**. Masiak & D'angelo (1975) *Biochim. Biophys. Acta* **382**, 83. **19**. Bieganski, Osinska & Maslinski (1982) *Agents Actions* **12**, 41. **20**. Weare, Stewart, Gafford & Erdos (1981) *Hypertension* **3**, 150. **21**. Houston & Walsh (1970) *Biochemistry* **9**, 156. **22**. Hawes, Crabb, Chan, *et al.* (2000) *Meth. Enzymol.* **324**, 218. **23**. el Kebbaj, Gaudemer & Latruffe (1986) *Arch. Biochem. Biophys.* **244**, 671. **24**. Kochhar & Dua (1985) *Arch. Biochem. Biophys.* **240**, 757. **25**. Connellan & Shaw (1970) *J. Biol. Chem.* **245**, 2845. **26**. Nordwig (1971) *Adv. Enzymol.* **34**, 155. **27**. Pontremoli, Grazi & Accorsi (1966) *Biochemistry* **5**, 3568. **28**. Cohen (1970) *The Enzymes*, 3rd ed., **1**, 147. **29**. Pontremoli & Horecker (1971) *The Enzymes*, 3rd ed., **4**, 611. **30**. Nicholls & Roy (1971) *The Enzymes*, 3rd ed., **5**, 21. **31**. Stadtman (1972) *The Enzymes*, 3rd ed., **6**, 539. **32**. Horecker, Tsolas & Lai (1972) *The Enzymes*, 3rd ed., **7**, 213. **33**. Imoto, Johnson, North, Phillips & Rupley (1972) *The Enzymes*, 3rd ed., **7**, 665. **34**. Morrison (1973) *The Enzymes*, 3rd ed., **8**, 457. **35**. Meister (1974) *The Enzymes*, 3rd ed., **10**, 699. **36**. Kowalczykowski, Bear & von Hippel (1981) *The Enzymes*, 3rd ed., **14**, 373. **37**. Simpson, Riordan & Vallee (1963) *Biochemistry* **2**, 616. **38**. Whitaker, Menger & Bender (1966) *Biochemistry* **5**, 386. **39**. Coleman, Pulido & Vallee (1966) *Biochemistry* **5**, 2019. **40**. Simpson & Vallee (1966) *Biochemistry* **5**, 2531. **41**. Pontremoli, Grazi & Accorsi (1966) *Biochemistry* **5**, 3072. **42**. Polastro, Deconinck, Devogel, *et al.* (1978) *Biochem. Biophys. Res. Commun.* **81**, 920. **43**. Hirayama, Matsuda, Takeda, Maenaka & Takatsuka (1975) *Biochim. Biophys. Acta* **384**, 127. **44**. Grillo & Coghe (1969) *Ital. J. Biochem.* **18**, 133. **45**. Morley & Stadtman (1972) *Biochemistry* **11**, 600. **46**. Fishman, Pentchev & Bailey (1973) *Biochemistry* **12**, 2490. **47**. Baskaran & Balasubramanian (1987) *Biochim. Biophys. Acta* **913**, 377. **48**. Jin, Choi & Yang (2001) *Bull. Korean Chem. Soc.* **22**, 68. **49**. El-Shora (2002) *Plant Sci.* **162**, 1. **50**. Yapeng, Ningguo, Xiulan, Jing, Shijun & Shuzheng (2003) *J. Biochem.* **134**, 365. **51**. Wilk, Meister & Haschemeyer (1970) *Biochemistry* **9**, 2039. **52**. Egashira, Kouhashi, Ohta & Sanada (1996) *J. Nutr. Sci. Vitaminol.* **42**, 173. **53**. Lakaye, Makarchikov, Wins, *et al.* (2004) *Int. J. Biochem. Cell Biol.* **36**, 1348. **54**. Hersh & W. Jencks (1967) *J. Biol. Chem.* **242**, 3468. **55**. Piedras, Munoz, Aguilar & Pineda (2000) *Arch. Biochem. Biophys.* **378**, 340. **56**. Wakayama, Yada, Kanda, *et al.* (2000) *Biosci. Biotechnol. Biochem.* **64**, 1. **57**. Boopathy & Balasubramanian (1985) *Eur. J. Biochem.* **151**, 351. **58**. Peterson, Sokolovsky & Vallee (1976) *Biochemistry* **15**, 2501. **59**. Adachi, Katayama, Nakazato & Tsujimoto (1993) *Biochim. Biophys. Acta* **1163**, 42. **60**. Dietze, Stephens, Magdalou, *et al.* (1993) *Comp. Biochem. Physiol. B* **104**, 299. **61**. Mueller, Samuelsson & Haeggström (1995) *Biochemistry* **34**, 3536. **62**. Pitson, Seviour, McDougall, Stone & Sadek (1996) *Biochem. J.* **316**, 841. **63**. Jin, Jo, Kim, *et al.* (2002) *J. Biochem. Mol. Biol.* **35**, 313. **64**. Bakunina, Kuhlmann, Likhosherstov, *et al.* (2002) *Biochemistry (Moscow)* **67**, 689. **65**. Moore & Stone (1972) *Biochim. Biophys. Acta* **258**, 248. **66**. Colas (1981) *Biochim. Biophys. Acta* **657**, 535. **67**. Pitson, Seviour & McDougall (1997) *Enzyme Microb. Technol.* **21**, 182. **68**. Chang, Hsueh & Sung (1996) *Biochem. Mol. Biol. Int.* **40**, 417. **69**. Kubackova, Karacsonyi, Bilisics & Toman (1978) *Folia Microbiol.* **23**, 202. **70**. Chang, Tang & Lin (1995) *Biochem. Mol. Biol. Int.* **36**, 185. **71**. Zimmermann, Kelly & Latzko (1978) *J. Biol. Chem.* **253**, 5952. **72**. Snyder, Stock & Koshland (1984) *Meth. Enzymol.* **106**, 321. **73**. Hirayama, Matsuda, Takeda, Maenaka & Takatsuka (1975) *Biochim. Biophys. Acta* **384**, 127. **74**. Bailey, Roberts, Buess & Carper (1979) *Arch. Biochem. Biophys.* **192**, 482. **75**. Allen & Rosenberg (1968) *Biochim. Biophys. Acta* **151**, 504. **76**. Farley, Christie, Seubert & Segel (1979) *J. Biol. Chem.* **254**, 3537. **77**. Keng & Viola (1996) *Arch. Biochem. Biophys.* **335**, 73. **78**. Chandler, J. C. Silvia & K. E. Ebner (1980) *Biochim. Biophys. Acta* **616**, 179. **79**. Hill & Brew (1975) *Adv. Enzymol. Relat. Areas Mol. Biol.* **43**, 411. **80**. Hamdan & Diopoh (1991) *Plant Sci.* **76**, 1. **81**. Schilling, Niestroj, Rahfeld, *et al.* (2003) *J. Biol. Chem.* **278**, 49773. **82**. Vandamme-Feldhaus & Schauer (1998) *J. Biochem.* **124**, 111. **83**. Schwartz & Reed (1969) *J. Biol. Chem.* **244**, 6074. **84**. Marshall & Cohen (1972) *J. Biol. Chem.* **247**, 1669.

N-Acetyl-L-isoleucyl-L-alanyl-L-phenylalaninal

This peptide aldehyde (*Sequence*: Ac-IAF-CHO; FW = 375.47 g/mol), which as a hydrate adopts a structure resembling the tetrahedral intermediate-like oxyanion in protease catalysis, inhibits the pepstatin-insensitive carboxyl protease from *Pseudomonas* sp., K_i = 0.6 μM, and kumamolysin, K_i = 0.9 μM at 22.4°C. **Target(s):** kumamolysin (1,2); pseudomonalisin, *or* pseudomonapepsin (3), and tripeptidyl peptidase I (4). **1**. Oyama, Hamada, Ogasawara, *et al.* (2002) *J. Biochem.* **131**, 757. **2**. Comellas-Bigler, Fuentes-Prior, Maskos, *et al.* (2002) *Structure* **10**, 865. **3**. Wlodawer, Li, Gustchina, *et al.* (2001) *Biochemistry* **40**, 15602. **4**. Oyama, Fujisawa, Suzuki, *et al.* (2005) *J. Biochem.* **138**, 127.

N^α-Acetyl-L-isoleucyl-L-glutamyl-L-threonyl-L-aspartate Semialdehyde

This tetrapeptide aldehyde (*Sequence*: Ac-IETD-CHO; $FW_{free-acid}$ = 502.52 g/mol), is a strong inhibitor of caspase-8 and granzyme B. Ac-IETD-CHO also blocks the cleavage of the pro-form of caspase 3 and prevents the formation of active caspase-3. *Note*: A cell-permeable derivative (*Sequence*: Ac-AAVALLPAVLLALLAPIETD-CHO; $FW_{free-acid}$ = 2000.45 g/mol) is also commercially available. This eicosapeptide consists of the N-terminal sequence (residues 1-16) of the signal peptide of Kaposi fibroblast growth factor linked to IETD-CHO. **Target(s):** caspase-8 (1-4); caspase-6 (1); caspase-9 (1); caspase-10 (1); granzyme B (1); caspase-3 activation (2). **1**. Roy & Nicholson (2000) *Meth. Enzymol.* **322**, 110. **2**. Han, Hendrickson, Bremner & Wyche (1997) *J. Biol. Chem.* **272**, 13432. **3**. Gao, Ren, Zhang, *et al.* (2001) *Exp. Cell. Res.* **265**, 145. **4**. Watt, Koeplinger, Mildner, *et al.* (1999) *Structure Fold. Des.* **7**, 1135.

3-*O*-Acetyl-11-keto-β-boswellic Acid

This naturally occurring pentacyclic triterpene ($FW_{free\ acid}$ = 512.73 g/mol) from the gum resin exudate of stems of the tree *Boswellia serrata* (the source of frankincense) is an orally active, non-redox and non-competitive inhibitor of 5-lipoxygenase and leukotriene biosynthesis. **Target(s):** DNA topoisomerase I (1,5); DNA topoisomerase II (1); elastase (2); 5-lipoxygenase, *or* arachidonate 5-lipoxygenase (2,4,6-10); 12-lipoxygenase, *or* arachidonate 12-lipoxygenase (11); and leukotriene biosynthesis (3). **1**. Syrovets, Buchele, Gedig, Slupsky & Simmet (2000) *Mol. Pharmacol.* **58**, 71. **2**. Safayhi, Rall, Sailer & Ammon (1997) *J. Pharmacol. Exp. Ther.* **281**, 460. **3**. Wildfeuer, Neu, Safayhi, *et al.* (1998) *Arzneimittelforschung* **48**, 668. **4**. Bishnoi, Patil, Kumar & Kulkarni (2006) *Indian J. Exp. Biol.* **44**, 128. **5**. Hoernlein, Orlikowsky, Zehrer, *et al.* (1999) *J. Pharmacol. Exp. Ther.* **288**, 613. **6**. Sailer, Schweizer, Boden, Ammon & Safayhi (1998) *Eur. J. Biochem.* **256**, 364. **7**. Sailer, Subramanian, Rall, *et al.* (1996) *Brit. J. Pharmacol.* **117**, 615. **8**. Safayhi, Sailer & Ammon (1995) *Mol. Pharmacol.* **47**, 1212. **9**. Safayhi, Mack, Sabieraj, *et al.* (1992) *J. Pharmacol. Exp. Ther.* **261**, 1143. **10**. Schneider & Bucar (2005) *Phytother. Res.* **19**, 81. **11**. Poeckel, Tausch, Kather, Jauch & Werz (2006) *Mol. Pharmacol.* **70**, 1071.

6-Acetylkynurenate, *See 6-Acetyl-4-oxo-4H-quinoline-2-carboxylate*

7-Acetylkynurenate *See 7-Acetyl-4-oxo-4H-quinoline-2-carboxylate*

8-Acetylkynurenate, *See 8-Acetyl-4-oxo-4H-quinoline-2-carboxylate*

N-Acetyllactosamine

This modified disaccharide (FW = 383.35 g/mol), often abbreviated Gal-(β1,4)-GlcNAc, is a common component of glycoproteins and an alternative acceptor substrate for many sialyltransferases. *N*-Acetyllactosamine is also a substrate for β-galactosidases. **Target(s):** electrolectin (1); *Bauhinia purpurea* agglutinin (2); calf heart lectin (3); *Rana catesbiana* lectin (4); *Datura stramonium* lectin (5); *Achatina fulica* agglutinin (6); galaptin, human spleen (7); *Arum maculatum* lectin (8); *Phalera flavescens* agglutinin (9); rabbit serum lectin (10); *Octopus vulgaris* lectin (11); β-galactoside α-2,6-sialyltransferase, also an alternative substrate (12); and glucuronosyltransferase (13). **1**. Teichberg (1978) *Meth. Enzymol.* **50**, 291. **2**. Osawa, Irimura & Kawaguchi (1978) *Meth. Enzymol.* **50**, 367. **3**. Childs & Feizi (1979) *FEBS Lett.* **99**, 175. **4**. Sakakibara, Takayanagi, Kawauchi, Watanabe & Hakomori (1976) *Biochim. Biophys. Acta* **444**, 386. **5**. Crowley, Goldstein, Arnarp & Lonngren (1984) *Arch. Biochem. Biophys.* **231**, 524. **6**. Sarkar, Bachhawat & Mandal (1984) *Arch. Biochem. Biophys.* **233**, 286. **7**. Lee, Ichikawa, Allen & Lee (1990) *J. Biol. Chem.* **265**, 7864. **8**. Allen (1995) *Biochim. Biophys. Acta* **1244**, 129. **9**. Umetsu, Yamashita, Suzuki, Yamashita & Suzuki (1993) *Arch. Biochem. Biophys.* **301**, 200. **10**. Hosomi, Takeya & Yazawa (1993) *Biochim. Biophys. Acta* **1157**, 45. **11**. Rogener, Renwrantz & Uhlenbruck (1985) *Dev. Comp. Immunol.* **9**, 605. **12**. Paulson, Rearick & Hill (1977) *J. Biol. Chem.* **252**, 2363. **13**. Kakuda, Oka & Kawasaki (2004) *Protein Expr. Purif.* **35**, 111.

N-Acetyl-L-leucine Chloromethyl Ketone

This halomethyl ketone (FW = 205.68 g/mol) inhibits acylaminoacyl-peptidase, *or* acylpeptide hydrolase, and induces apoptosis of U937 cells (1,2). **1**. Yamaguchi, Kambayashi, Toda, *et al.* (1999) *Biochem. Biophys. Res. Commun.* **263**, 139. **2**. Scaloni, Jones, Barra, *et al.* (1992) *J. Biol. Chem.* **267**, 3811.

N-Acetyl-L-leucyl-L-alanyl-L-alanyl-L-(*N,N*-dimethylglutaminal)

This peptide aldehyde (FW = 200.24 g/mol) irreversibly inhibits picornain 3C proteases (1), enzymes that adopt a chymotrypsin-like fold and display an active-site configuration like those of the serine proteinases (2,3). Peptide-aldehydes based on the preferred peptide substrates for hepatitis A virus (HAV) 3C proteinase were synthesized by reduction of a thioester precursor. Acetyl-Leu-Ala-Ala-(*N,N*-dimethylglutaminal) is a reversible, slow-binding inhibitor for HAV 3C with a K_i^* of (4.2 x 10^{-8}) M. This inhibitor shows 50x less activity against the highly homologous human Rhinovirus (Strain 14) 3C proteinase, whose peptide substrate specificity is slightly different, suggesting a high degree of selectivity. NMR spectrometry of the adduct of the ^{13}C-labeled inhibitor with the HAV-3C proteinase indicate that a thiohemiacetal is formed between the enzyme and the aldehyde carbon, as previously noted for peptide-aldehyde inhibitors of papain (4,5). **1**. Malcolm, Lowe, Shechosky, *et al.* (1995) *Biochemistry* **34**, 8172. **2**. Allaire, *et al.* (1994) *Nature* **369**, 72. **3**. Matthews, *et al.* (1994) *Cell* **77**, 761. **4**. Lewis & Wolfenden (1977) *Biochemistry* **16**, 4890. **5**. Gamcsik, *et al.* (1983) *J. Am. Chem. Soc.* **105**, 6324.

3-[N^1-(*N*-Acetyl-L-leucyl-L-alanyl-L-alanyl)-N^2-(*o*-nitrophenyl-sulfenyl)hydrazino]-*N,N*-(dimethyl)propanamide

This azaglutamine derivative (FW = 568.68 g/mol) inactivates picornain 3C, a cysteine proteinase essential for cleavage of the initially synthesized viral polyprotein precursor into mature fragments. Because picornain 3C is essential for viral replication *in vivo*, it is an obvious druggable target for developing antivirals. **1.** Huang, Malcolm & Vederas (1999) *Bioorg. Med. Chem.* **7**, 607.

N-Acetyl-L-leucyl-L-arginal

This dipeptide aldehyde (FW$_{free-base}$ = 313.40 g/mol) inhibits dipeptidyl-peptidase III, most likely by forming a tetrahedral transition-state mimic. (1,2). Note that this is a shorter analogue of leupeptin, a broad-specificity serine/cysteine protease inhibitor. **See also** *Leupeptin* **1.** McDonald (1998) in *Handbook of Proteolytic Enzymes* (Barrett, Rawlings & Woessner, eds), p. 536, Academic Press, San Diego. **2.** Nishikiori, Kawahara, Naganawa, *et al.* (1984) *J. Antibiot.* **37**, 680.

N-{*N*-[*N*-Acetyl-L-leucyl-L-asparaginyl]-(3*S*-amino-2*S*-hydroxy-4-phenylbutyl)}-L-prolyl-L-isoleucyl-L-valine Methyl Ester and Amide

This peptide analogue (*Sequence*: Ac-Leu-Asn-[Phe-HEA-Pro]-Ile-Val-Ome; FW = 773.97 g/mol), inhibits HIV-1 protease, *or* HIV-1 retropepsin, K_i = 21 nM (1-3). The corresponding amide has a K_i value of 3 nM. **1.** Rich, Green, Toth, *et al.* (1990) *J. Med. Chem.* **33**, 1285. **2.** March, Abbenante, Bergman, *et al.* (1996) *J. Amer. Chem. Soc.* **118**, 3375. **3.** Abbenante, March, Bergman, *et al.* (1995) *J. Amer. Chem. Soc.* **117**, 10220.

N^α-Acetyl-L-leucyl-L-glutamyl-L-histidyl-L-aspartate-α-semialdehyde

This tetrapeptide aldehyde (FW = 636.57 g/mol; Sequence: Ac-LEHD-CHO) inhibits caspase-9 (1,2). Acetyl-AAVALLPAVLLALLAPLEHD-CHO is the commercially available cell-permeable derivative of theis tetrapeptide, and this eicosapeptide aldehyde consists of the N-terminal sequence (residues 1-16) of the cell-penetrating signal peptide of Kaposi fibroblast growth factor covalently linked to LEHD-CHO. **1.** Thornberry & Lazebnik (1998) *Science* **281**, 1312. **2.** Thornberry, Rano, Peterson, *et al.* (1997) *J. Biol. Chem.* **272**, 17907.

N^α-Acetyl-L-leucyl-L-glutamyl-L-valyl-L-aspartate-α-semialdehyde

This tetrapeptide aldehyde (Sequence: Ac-LEVD-CHO; FW$_{free-acid}$ = 500.55 g/mol), inhibits caspase-4 (1). The commercially available cell-penetrating derivative (*N*^{α}-Acetyl-AAVALLPAVLLALLAPLEVD-CHO) is an eicosapeptide, consisting of the N-terminal sequence (residues 1-16) of the signal peptide of Kaposi fibroblast growth factor covalently linked to LEVD-CHO. **1.** Talanian, Quinlan, Trautz, *et al.* (1997) *J. Biol. Chem.* **272**, 9677.

N-Acetyl-L-leucyl-L-leucyl-L-arginal, See *Leupeptin*

N-Acetyl-L-leucyl-L-leucyl-L-arginyl-L-valyl-L-lysyl-L-arginine

This acetylated hexapeptide (FW = 827.07 g/mol; *Sequence*: Ac-LLRVKR) and its amide Ac-LLRVKR-NH$_2$ (FW = 826.06 g/mol) inhibit furin and proprotein convertase 2, respectively (1,2). **1.** Apletalina, Muller & Lindberg (2000) *J. Biol. Chem.* **275**, 14667. **2.** Cameron, Appel, Houghten & Lindberg (2000) *J. Biol. Chem.* **275**, 36741.

N-Acetyl-L-leucyl-L-leucyl-L-methioninal

This membrane-permeable, transition state-mimicking endopeptidase inhibitor (FW = 401.57 g/mol; Sequence: Ac-LLM-CHO) targets calpain I, K_i = 120 nM, calpain II, K_i = 230 nM, cathepsin B, K_i = 100 nM, and cathepsin L, K_i = 600 pM. **Target(s):** calpain-1 (1,3); calpain-2, *or* μ-calpain (1,3,6,7); cathepsin L (1); cathepsin B (1); proteasome (chymotrypsin and trypsin-like activity, K_i = 33 (2); HIV retropepsin, *or* HIV protease, IC$_{50}$ = 5 μM (4); peptide deformylase (5); and cathepsin K (8). **1.** Sasaki, Kishi, Saito, *et al.* (1990) *J. Enzyme Inhib.* **3**, 195. **2.** Vinitsky, Michaud, Powers & Orlowski (1992) *Biochemistry* **31**, 9421. **3.** Saito & Nixon (1993) *Neurochem. Res.* **18**, 231. **4.** Sarubbi, Seneci, Angelastro, *et al.* (1993) *FEBS Lett.* **319**, 253. **5.** Durand, Green, O'Connell & Grant (1999) *Arch. Biochem. Biophys.* **367**, 297. **6.** Ladrat, Verrez-Bagnis, Noel & Fleurence (2002) *Mar. Biotechnol.* **4**, 51. **7.** Wang, Ma, Su, Chen & Jiang (1993) *J. Agric. Food Chem.* **41**, 1379. **8.** Aibe, Yazawa, Abe, *et al.* (1996) *Biol. Pharm. Bull.* **19**, 1026.

N-Acetyl-L-leucyl-L-leucyl-L-norleucinal

This acetylated tripeptide aldehyde (FW = 383.53 g/mol), also known as Calpain Inhibitor I and ALLN, inhibits many proteinases by forming a tightly bound, reversible tetrahedral hemiacetal adduct that resembles a key intermediate (shown in the inset above) in the catalytic reaction cycles of the above enzymes. ALLN also inhibits cyclin B degradation and alters the rate of processing and secretion of the β-amyloid precursor protein. ALLN also reportedly inhibits nitric oxide production by interfering with transcription of the inducible nitric oxide synthase gene. **Target(s):** calpain-1, K_i = 0.19 μM; (1,6,7,10); *Thermoplasma* proteasome (Archaean proteasome) (2); HslVU protease (3); proteasome endopeptidase complex, K_i = 5.7 μM (4,5,13,14); calpain-2, *or* μ-calpain, K_i = 0.22 μM (1,10,15,16); cathepsin L, K_i = 0.5 nM (1); cathepsin B, K_i = 0.15 μM (1); γ-secretase (8,9); HIV retropepsin, IC$_{50}$ = 5.2 μM (11); pepsin, IC$_{50}$ = 24 μM (11); cathepsin D, IC$_{50}$ = 35 μM (11); and peptide deformylase (12). Like its substrate counterparts, this inhibitor enters through a narrow side entrance to the active site of the PRE2 subunit of the 20S proteasome from *Saccharomyces cerevisiae*, whereupon it binds and forms the hemiacetal (17). **1.** Sasaki, Kishi, Saito, *et al.* (1990) *J. Enzyme Inhib.* **3**, 195. **2.** Lupas, Kania & Baumeister (1998) in *Handbook of Proteolytic Enzymes* (Barrett, Rawlings & Woessner, eds), p. 486, Academic Press, San Diego. **3.** Rohrwild, Huang, Goldberg, Yoo & Chung (1998) in *Handbook of Proteolytic Enzymes* (Barrett, Rawlings & Woessner, eds), p. 492, Academic Press, San Diego. **4.** Seemüller, Dolenc & Lupas (1998) in *Handbook of Proteolytic Enzymes* (Barrett, Rawlings & Woessner, eds), p. 495, Academic Press, San Diego. **5.** Vinitsky, Michaud, Powers & Orlowski (1992) *Biochemistry* **31**, 9421. **6.** Rami & Krieglstein (1993) *Brain Res.* **609**, 67. **7.** Squier, Miller, Malkinson & Cohen (1994) *J. Cell Physiol.* **159**, 229. **8.** Zhang, Song & Parker (1999) *J. Biol. Chem.* **274**, 8966. **9.** Klafki, Abramowski, Swoboda, Paganetti & Staufenbiel (1996) *J. Biol. Chem.* **271**, 28655. **10.** Saito & Nixon (1993) *Neurochem. Res.* **18**, 231. **11.** Sarubbi, Seneci, Angelastro, *et al.* (1993) *FEBS Lett.* **319**, 253. **12.** Durand, Green, O'Connell & Grant (1999) *Arch. Biochem. Biophys.* **367**, 297. **13.** Drexler (1997) *Proc. Natl. Acad. Sci. U.S.A.* **94**, 855. **14.** Wojcikiewicz, Xu, Webster, Alzayady & Gao (2003) *J. Biol. Chem.* **278**, 940. **15.** Ladrat, Verrez-Bagnis, Noel & Fleurence (2002) *Mar. Biotechnol.* **4**, 51. **16.** Wang, Ma, Su, Chen & Jiang (1993) *J. Agric. Food Chem.* **41**, 1379. **17.** Groll, Ditzel, Lowe, *et al.* (1997) *Nature*, **386**, 463.

N^ε-Acetyl-L-lysine

This posttranslationally derived amino acid (FW = 188.23 g/mol), first isolated from calf thymus histone IV in the late 1960s (1-3) and formed by histone acetyltransferase or synthetically by acetylation of the copper complex of L-lysine (4), inhibits D-lysine 5,6-aminomutase, by the racemic mixture (5,6), histone acetyltransferase (7), and lysyl-tRNA synthetase (8). **1.** J. DeLange, Smith, Fambrough & Bonner (1968) *Proc. Natl. Acad. Sci. U.S.A.* **61**, 1145. **2.** DeLange, Fambrough, Smith & Bonner (1969) *J. Biol. Chem.* **244**, 319. **3.** Gershey, Vidali & Allfrey (1968) *J. Biol. Chem.* **243**, 5018. **4.** Benoiton (1963) *Can. J. Chem.* **41**, 1718. **5.** Morley & Stadtman (1970) *Biochemistry* **9**, 4890. **6.** Stadtman (1972) *The Enzymes*, 3rd ed., **6**, 539. **7.** Wiktorowicz, Campos & Bonner (1981) *Biochemistry* **20**, 1464. **8.** Takita, Ohkubo, Shima, *et al.* (1996) *J. Biochem.* **119**, 680.

N^α-Acetyllysyltyrosylcysteine Amide

This novel inhibitor (FW = 453.55 g/mol; Sequence: Ac-KYC-NH$_2$) targets myeloperoxidase, a heme peroxidase that is formed in and released by activated neutrophils, macrophages and monocytes, playing important roles in host defense. Ferric MPO reacts with hydrogen peroxide to form compound I, an oxy-ferryl-cation (P'Fe^{4+}=O) radical intermediate that oxidizes a wide variety of substrates to form an equally wide variety of toxic oxidants and free radicals that kill invading bacteria. Ac-KYC-NH$_2$ reduces production of hypochlorous acid, lowers the peroxidation of LDL lipids, suppresses protein nitration, and inhibits tryptophan oxidation. Ac-KYC-NH$_2$ specifically inhibits the myeloperoxidase activity released from PMA-activated HL-60 cells, but not the NOX activity essential for HL-60 cells to generate O$_2$$^{\cdot-}$ that can dismutate to hydrogen peroxide to activate MPO. While Ac-KYC-NH$_2$ is cytotoxic to bovine aortic endothelial cell cultures, even when incubated at 4 mM, it protects these cells from MPO-induced injury and death at 100x lower concentrations. Support for intramolecular electron transfer playing a role in the mechanisms by which Ac-KYC-NH$_2$ detoxifies MPO activity comes from data showing that MPO-dependent Ac-KYC-NH$_2$ oxidation forms exclusively a disulfide, not dityrosine, since the latter is formed only when one Tyr$^{\cdot}$ free radical condenses with another Tyr$^{\cdot}$ free radical. Notably, Ac-KYC-NH$_2$ enjoys great advantage over azides, hydrazides and hydroxamic acids, which are MPO suicide inhibitors that irreversibly modify the iron heme site of MPO, but lack specificity and are inherently toxic. **1.** Zhang, Jing, Shi, *et al.* (2013) *J. Lipid Res.* **54**, 3016.

N-Acetyl-L-lysyl-(*N*-methyl)-L-leucyl-L-valyl-(*N*-methyl)-L-phenylalanyl-L-phenylalaninamide

This acetylated peptapentide amide (FW$_{free-base}$ = 721.93 g/mol; water-soluble), itself containing *N*-methylated amino acids is homologous to the central core domain of the Alzheimer's β-amyloid peptide and adopts the structure of an extended β-strand. **Target(s):** amyloid β–40 fibrillogenesis. **1.** Gordon, Tappe & Meredith (2002) *J. Pept. Res.* **60**, 37.

N^1-[*N*-Acetyl-D-lysyl-L-phenylalanyl-L-phenylalanyl-L-prolyl-L-leucyl-L-glutamic Acid α-Amide

This substituted hexapeptide amide (FW = 822.98 g/mol; *Sequence:* Ac-KFFPLE-NH$_2$) inhibits human tissue kallikrein, thrombin, coagulation factor Xa, plasmin, cathepsin G, and plasma kallikrein, with K_i values of 29, 700, 130, 170, 170, and 8 nM, respectively. If inhibition of kinin release is the main goal, acetyl-Lys-Phe-Phe-Pro-Leu-Glu-NH$_2$ may be a useful lead molecule for designing specific inhibitors for human tissue kallikrein (hK1), plasma kallikrein, or for both at same time. **1.** Pimenta, S. Fogaça, Melo, Juliano & Juliano (2003) *Biochem. J.* **371**, 1021.

N-Acetyl-D-mannosamine

This acylamino sugar (FW = 221.21 g/mol), also known as 2-acetamido-2-deoxy-D-mannose, inhibits α-*N*-acetylglucosaminidase (1,2), β-*N*-acetylhexosaminidase (3,4), peptidoglycan β-*N*-acetylmuramidase (5,6), *N*-acetylglucosamine kinase (7), mannokinase (8), and hexokinase (9). **1.** Weissmann (1972) *Meth. Enzymol.* **28**, 796. **2.** Weissmann, Rowan, Marshall & Friederici (1967) *Biochemistry* **6**, 207. **3.** Mommsen (1980) *Biochim. Biophys. Acta* **612**, 361. **4.** Geimba, Riffel & Brandelli (1998) *J. Appl. Microbiol.* **85**, 708. **5.** del Rio & Berkeley (1976) *Eur. J. Biochem.* **65**, 3. **6.** del Rio, Berkeley, Brewer & Roberts (1973) *FEBS Lett.* **37**, 7. **7.** Allen & Walker (1980) *Biochem. J.* **185**, 577. **8.** Sabater, Sebastián & Asensio (1972) *Biochim. Biophys. Acta* **284**, 406. **9.** E. Machado de Domenech & Sols (1980) *FEBS Lett.* **119**, 174.

S-Acetylmercaptosuccinic Anhydride

This reagent (FW = 174.18 g/mol), also known as 2-(acetylthio)succinic anhydride, has been used to acylate proteins and introduce sulfhydryl groups (1). *Note:* Care must be exercised when working with this moisture-sensitive agent, which is also skin irritant. **Target(s):** phenylalanine monooxygenase (2). **1.** Klotz & Heiney (1962) *Arch. Biochem. Biophys.* **96**, 605. **2.** Weiss, Hui & Lajtha (1979) *Res. Commun. Chem. Pathol. Pharmacol.* **25**, 153.

3-(*N*-Acetyl-L-methionyl)pyruvic Acid

This methionine derivative (FW$_{free-acid}$ = 261.30 g/mol) inhibits peptidylamidoglycolate lyase, a bifunctional enzyme catalyzing C-terminal amidation, as seen in the essential post-translational modification of many neuropeptides. C-terminal amidation entails sequential action by peptidylglycine mono-oxygenase (EC 1.14.17.3) and peptidylamidoglycolate lyase (EC 4.3.2.5), with the mono-oxygenase catalysing conversion of a glycine-extended pro-peptide into the corresponding α-hydroxyglycine derivative, which is then converted by the lyase into amidated peptide plus glyoxylate. Since the mono-oxygenase and lyase reactions exhibit tandem reaction stereospecificities, the possibility of channelling of the α-hydroxy intermediate was investigated. Selective inhibition of the mono-oxygenase by competitive ester inhibitors, as well as mechanism-based mono-oxygenase inactivation by the novel olefinic inhibitor 5-acetamido-4-oxo-6-phenylhex-2-enoate (*N*-acetylphenylalanyl acrylate), was found to exert little to no effect on the kinetic parameters of the lyase domain of the AE from *Xenopus laevis*. Similarly, inhibition of the lyase domain by the potent dioxo inhibitor 2,4-dioxo-5-acetamido-6-phenylhexanoate had little effect on the activity of the monooxygenase domain in the bifunctional enzyme. A assessment of intermediate accumulation and conversion, bolstered by kinetic investigations of the reactivities of the monofunctional and bifunctional enzyme forms towards substrates and inhibitors, provide strong evidence for the kinetic independence of the mono-oxygenase and lyase, absent any evidence for substrate channelling. **1.** Moore & May (1999) *Biochem. J.* **341**, 33.

N-Acetyl-5-methoxytryptamine, *See Melatonin*

Acetyl-β-methylcholine Chloride

This acetylcholine analogue (FW = 195.69 g/mol), also known as methacholine chloride, is photosensitive, very deliquescent, and hygroscopic (the bromide and iodide salts are less hygroscopic). Acetyl-β-methylcholine chloride is a muscarinic receptor agonist and parasympathomimetic bronchoconstrictor (the (+)-enantiomer is the more active stereoisomer). Note that it is a weaker acetylcholinesterase substrate than acetylcholine; hence, its physiological effects are longer lasting. **Target(s):** aryl-acylamidase (1); carnitine acetyltransferase (2); ethanolamine kinase (3). **1.** Oommen & Balasubramanian (1979) *Eur. J. Biochem.* **94**, 135. **2.** Fritz & Schultz (1965) *J. Biol. Chem.* **240**, 2188. **3.** Sung & Johnstone (1967) *Biochem. J.* **105**, 497.

N-Acetyl-α-methyl-L-phenylalanine Methyl Ester

While this ester (FW = 235.28 g/mol) is a weak alternative substrate for chymotrypsin, both stereoisomers are effective competitive inhibitors when chymotrypsin operates on better substrates (1-3). **1.** Almond, Manning & Niemann (1962) *Biochemistry* **1**, 243. **2.** Baker (1967) *Design of Active-Site-Directed Irreversible Enzyme Inhibitors*, Wiley, New York. **3.** Hein & Niemann (1962) *J. Amer. Chem. Soc.* **84**, 4495.

N-Acetylmuramic Acid

This modified *N*-acetyl-D-glucosamine (FW$_{\text{free-acid}}$ = 293.27 g/mol; CAS 10597-89-4), also known as 2-acetamido-3-*O*-(D-1-carboxyethyl)-2 deoxy-D-glucose, is a component of many peptidoglycans and inhibits lysozyme and peptidoglycan β-*N*-acetylmuramidase. In bacterial cell wall peptidoglycan, alternating units of *N*-acetylglucosamine (GlcNAc) and *N*-acetylmuramic acid (MurNAc) form a polymer that is cross-linked with oligopeptides at the lactic acid residue of MurNAc. **Target(s):** lysozyme (1,6); peptidoglycan β-*N*-acetylmuramidase (2,5); alkaline phosphatase (3); *N*-acetylmuramoyl-L-alanine amidase (4). **1.** Imoto, Johnson, North, Phillips & Rupley (1972) *The Enzymes*, 3rd ed., **7**, 665. **2.** del Rio & Berkeley (1976) *Eur. J. Biochem.* **65**, 3. **3.** Komoda & Sakagishi (1977) *Biochim. Biophys. Acta* **482**, 79. **4.** Van Heijenoort, Parquet, Flouret & Van Heijenoort (1975) *Eur. J. Biochem.* **58**, 611. **5.** del Rio, Berkeley, Brewer & Roberts (1973) *FEBS Lett.* **37**, 7. **6.** Pahud & Widmer (1982) *Biochem. J.* **201**, 661.

N-Acetylmuramoyl-L-alanyl-D-isoglutamine

This peptidoglycan derivative (FW$_{\text{free-acid}}$ = 492.48 g/mol), also known as muramyl-dipeptide and MDP, is a water-soluble, synthetic immunoadjuvant (1) that enhances the antigenicity of weak antigens. **Target(s):** HIV replication (2). **1.** Lefrancier, Choay, Derrien & Lederman (1977) *Int. J. Pept. Protein Res.* **9**, 249. **2.** Masihi, Lange, Rohde-Schulz & Chedid (1990) *AIDS Res. Hum. Retroviruses* **6**, 393.

N¹-Acetyl-N⁴-(Nᵃ-2-naphthalenesulfonyl-L-3-amidinophenylalanyl)piperazine

This substituted phenylalanine (FW = 506.61 g/mol) inhibits thrombin and trypsin, the latter with a K_i value of 44 nM. Isothermal calorimetry and protein crystallographic analysis demonstrated changes of protonation states attend inhibitor binding, resulting from induced pK_a shifts that depend on interactions with protein functional groups. A strong negative heat capacity change ΔC_p is detected suggests binding becomes more exothermic and entropically less favorable with increasing temperature. Due to a mutual compensation, the Gibbs free energy of inhibitor binding remains virtually unchanged. The strongly negative ΔC_p value cannot solely be explained on the basis of the removal of hydrophobic surface portions of the protein or ligand from water exposure, suggesting additional contributions from modulations of the local water structure, changes in vibrational modes, or other ordering parameters. **1.** Dullweber, Stubbs, Musil, Stürzebecher & Klebe (2001) *J. Mol. Biol.* **313**, 593.

N-Acetyl-β-(2-naphthyl)-D-alanyl-4-chloro-D-phenylalanyl-β-(3-pyridyl)-D-alanylglycyl-L-arginyl-L-prolyl-D-alaninamide
This modified heptapeptide amide is a luteinizing hormone-releasing agent (*i.e.*, a gonadotropin releasing hormone) and has a high receptor binding affinity (IC$_{50}$ = 7 nM), as compared to that of gonadotropin itself (IC$_{50}$ = 2 nM). **1.** Yahalom, Rahimipour, Koch, Ben-Aroya & Fridkin (2000) *J. Med. Chem.* **43**, 2831.

N-Acetyl-β-(1-naphthyl)-L-alanyl-L-valylstatinyl-L-glutamyl-β-(1-naphthyl)-L-alaninamide

This peptidomimetic (FW$_{\text{free-acid}}$ = 839.00 g/mol), also known as LP-149, is a strong inhibitor of feline immunodeficiency virus retropepsin and HIV retropepsin (K_i = 260 and 1.7 nM, respectively). **Target(s):** feline immunodeficiency virus retropepsin (1,2); HIV retropepsin, *or* HIV protease (1-3). **1.** Dunn (1998) in *Handbook of Proteolytic Enzymes*, p.

949, Academic Press, San Diego. **2**. Wlodawer, Gustchina, Reshetnikova, *et al* (1995) *Nature Struct. Biol.* **2**, 480. **3**. Dunn, Pennington, Frase & Nash (1999) *Biopolymers* **51**, 69.

N-Acetylneuraminic Acid

This acetylated sugar acid (FW$_{free\text{-}acid}$ = 309.27 g/mol; frequently abbreviated NANA), also known as 5-(acetylamino)-3,5-dideoxy-D-*glycero*-D-*galacto*-2-nonulosonic acid, is a building block for many complex carbohydrates, glycoproteins, and proteoglycans. Although often regarded to be synonymous with sialate, this term actually refers to a class of acylated neuraminates and esters. Other sialates include *N*-glycolylneuraminate and *N,O*^{9}-diacetylneuraminate. Soluble in water and methanol, NANA is more stable than neuraminate, its deacetylated analogue. Greatest stability is observed at pH 4.5, but *N*-acetylneuraminic acid is moderately stable in alkaline solutions. NANA decomposes at 185-187°C and has a pK_a value of 2.60 at 25°C. **Target(s):** alkaline phosphatase (1); exo-α-sialidase, *or* neuraminidase; sialidase (2-7); limulin, *or Limulus polyphemus* agglutinin (8); phospholipase A$_2$ (9); neolactotetrasylceramide α-2,3 sialyltransferase (10); and ganglioside galactosyltransferase (11). **1**. Komoda & Sakagishi (1977) *Biochim. Biophys. Acta* **482**, 79. **2**. Gottschalk & Bhargava (1971) *The Enzymes*, 3rd ed., **5**, 321. **3**. Sastre, Cobaleda, Cabezas & Villar (1991) *Biol. Chem. Hoppe-Seyler* **372**, 923. **4**. Miyagi, Hata, Hasegawa & Aoyagi (1993) *Glycoconjugate J.* **10**, 45. **5**. Wang, Tanenbaum & Flashner (1978) *Biochim. Biophys. Acta* **523**, 170. **6**. Baricos, Cortez-Schwartz & Shah (1986) *Biochem. J.* **239**, 705. **7**. Burg & Muthing (2001) *Carbohydr. Res.* **330**, 335. **8**. Barondes & Nowak (1978) *Meth. Enzymol.* **50**, 302. **9**. Yang, Farooqui & Horrocks (1996) *Adv. Exp. Med. Biol.* **416**, 309. **10**. Dasgupta, Chien & Hogan (1986) *Biochim. Biophys. Acta* **876**, 363. **11**. Yip & Dain (1970) *Biochem. J.* **118**, 247.

N-Acetyl-[Nle^4,Asp^5,D-2-Nal^7,Lys^{10}]-Cyclo-α-Melanocyte Stimulating Hormone Amide Fragment 4-10

This peptide lactam (*Sequence*: CH$_3$CO-Nle-*cyclo*(β-Asp-His-D-2-Nal-Arg-Trp-ε-Lys-NH$_2$, where Nle refers to L-norleucine and D-2-Nal refers to D-2'-naphthylalanine) is an analogue of α-melanocyte stimulating hormone (α-MSH) and a potent antagonist of the melanocortin-4 receptor. It is also a weaker antagonist at the melanocortin-3 receptor. **1**. Hruby, Lu, Sharma, *et al.* (1995) *J. Med. Chem.* **38**, 3454.

O-Acetyloctahydropiericidin B$_1$

This reduced antibiotic (FW = 479.70 g/mol) and its deacetylated derivative are structural analogues of ubiquinone and moderately inhibit NADH dehydrogenase and electron transport. **1**. Jeng, Hall, Crane, *et al.* (1968) *Biochemistry* **7**, 1311.

3-(Acetyl)oxanilic Acid

This arylated oxamic acid (FW$_{free\ acid}$ = 207.19 g/mol), also called *N*-(3-acetylphenyl)oxamic acid, inhibits lactate and glutamate dehydrogenase

(1,2). **1**. Baker (1967) *Design of Active-Site-Directed Irreversible Enzyme Inhibitors*, Wiley, New York. **2**. Baker, Lee, Tong, Ross & Martinez (1962) *J. Theoret. Biol.* **3**, 446.

3-Acetyl-4-oxo-1,7-heptanedioic Acid

This porphobilinogen analogue (FW = 216.19 g/mol) inhibits porphobilinogen synthase, *or* δ-aminolevulinate dehydratase, with the formation of a Schiff base complex between the inhibitor and an active-site lysine. Comparative studies of pea ALAD with *E. coli* ALAD using the inhibitors 3-acetyl-4-oxoheptane-1,7-dioic acid (AOHD) and succinylacetone (SA) indicated similar modes of inhibition, Studies with the ALA homologue, 4-amino-3-oxobutanoic acid (AOB), revealed that it is specific for the A-site of both the pea and *E. coli* ALADs. An interesting difference exists between the enzymes, however, with pea ALAD far more susceptible to inhibition with AOB than the bacterial enzyme. **1**. Cheung, Spencer, Timko & Shoolingin-Jordan (1997) *Biochemistry* **36**, 1148.

N-Acetyl-4-oxo-D-neuraminic Acid

This neuraminic acid derivative (FW$_{free\ acid}$ = 307.26 g/mol) inhibits *N*-acetylneuraminate lyase (K_i = 25 μM and 150 μM for the *Clostridum perfringens* and *Escherichia coli* enzymes, respectively) (1,2). **1**. Groß & Brossmer (1988) *FEBS Lett.* **232**, 145. **2**. Schauer, Wember, Wirtz-Peitz & Ferreira do Amaral (1971) *Hoppe-Seyler's Z. Physiol. Chem.* **352**, 1073.

[(*S*)-1-((3*R,S*)-1-Acetyl-4-oxopiperidin-3-ylcarbamoyl)-3-methylbutyl]carbamic Acid Benzyl Ester

This leucine derivative (FW = 415.49 g/mol) inhibits cathepsin K, ($K_{i,app}$ = 230 nM), a cysteine protease in the papain superfamily that is predominantly expressed in osteoclasts and a drug tareget in the treatment of osteoporosis. **1**. Marquis, Ru, Zeng, *et al.* (2001) *J. Med. Chem.* **44**, 725.

2-Acetyloxybenzoic Acid, *See Acetylsalicylic Acid*

4-Acetyloxybenzoyl Chloride Phenylhydrazone

This hydrazone (FW = 288.73 g/mol) inhibits soybean lipoxygenase and human platelet 12-lipoxygenase (K_i = 11 and 29 nM, respectively, using measurements of oxygen consumption. **1**. Wallach & Brown (1981) *Biochim. Biophys. Acta* **663**, 361.

Acetyl Pepstatin

This peptide (FW$_{free\ acid}$ = 643.82 g/mol; CAS 11076-29-2; Soluble to 2 mg/mL in phosphate-buffered saline), also known as *N*-acetyl-L-valyl-L-valyl-Sta-L-alanyl-Sta, where Sta refers to (3*S*,4*S*)-4-amino-3-hydroxy-6-methylheptanoic acid, inhibits a number of proteases. **Target(s):** cathepsin E (1); HIV-1 retropepsin, *or* human immunodeficiency virus 1 protease, K_i = 20 nM (2,3,5,6,10,12); HIV-2 retropepsin, *or* human immunodeficiency virus 2 protease, K_i = 5 nM (3,9,14); pepsin (4,7,8); gastricsin (4,8); cathepsin D (4,8); chymosin (4,7,8); penicillopepsin (4); mucorpepsin, *or Mucor pusillus* pepsin (4); endothiapepsin, *or Endothia parasitica* proteinase (4); saccharopepsin, *or* yeast proteinase A (4); rhizopuspepsin, K_i = 10.3 nM (11,15); and plasmepsin II (13). **1**. Kay & Tatnell (1998) in *Handbook of Proteolytic Enzymes*, p. 819, Academic Press, San Diego. **2**. Richards, Roberts, Dunn, Graves & Kay (1989) *FEBS Lett.* **247**, 113. **3**. Dunn (1998) in *Handbook of Proteolytic Enzymes*, p. 929, Academic Press, San Diego. **4**. Valler, Kay, Aoyagi & Dunn (1985) *J. Enzyme Inhib.* **1**, 77. **5**. Ringe (1994) *Meth. Enzymol.* **241**, 157. **6**. Vacca (1994) *Meth. Enzymol.* **241**, 311. **7**. Nakatani, Tsuchiya, Morita, *et al.* (1980) *Biochim. Biophys. Acta* **614**, 144. **8**. Kay, Valler & Dunn (1983) in *Proteinase Inhibitors: Medical and Biological Aspects*, p. 201, Springer-Verlag, Berlin. **9**. Richards, Broadhurst, Ritchie, *et al.* (1989) *FEBS Lett.* **253**, 214. **10**. Pennington, Festin, Maccecchini, *et al.* (1991) *Peptides 1990*, ESCOM Publ., Proc. 21st Eur. Peptide Sympos., pp. 787. **11**. Lowther, Sawyer, Staples, *et al.* (1992) in *Peptides: Chemistry and Biology, Proc. of the 12th Amer. Peptide Symp.* (June 16-21, 1991), p. 413. **12**. Prashar & Hosur (2004) *Biochem. Biophys. Res. Commun.* **323**, 1229. **13**. Hill, Tyas, Phylip, *et al.* (1994) *FEBS Lett.* **352**, 155. **14**. Gustchina, Weber & Wlodawer (1991) *Adv. Exp. Med. Biol.* **306**, 549. **15**. Nakatani, Hiromi & Kitagishi (1988) *Arch. Biochem. Biophys.* **263**, 311.

N-Acetylphenethylamine, *See* N-β-Phenethylacetamide

N-Acetylphenothiazine

This acylated phenothiazine (FW = 241.31 g/mol), also known as 10-acetyl-10*H*-phenothiazine, inhibits acetylcholinesterase and butyrylcholinesterase, with K_i values of 39 and 36 μM, respectively, for the human enzymes. **1**. Darvesh, McDonald, Penwell, *et al.* (2005) *Bioorg. Med. Chem.* **13**, 211.

*N*¹-Acetyl-*N*⁴-[*N*-(phenoxycarbonyl)-L-valyl-L-prolyl]-1,4(*S*)-diamino-2,2-difluoro-5-methyl-3-oxohexane

This peptide analogue (FW = 526.58 g/mol) inhibits leukocyte elastase (K_i = 2.3 nM), which, in addition to catalyzing elastin proteolysis, also disrupts tight junctions, damages certain tissues, breaks down cytokines and alpha proteinase inhibitor, cleaves IgA and IgG, cleave complement component C3bi, a component and the CR1 receptor that triggers phagocytosis in neutrophils. **1**. Veale, Bernstein, Bohnert, *et al.* (1997) *J. Med. Chem.* **40**, 3173.

N-Acetyl-DL-phenylalanine Trifluoromethyl Ketone

This halomethyl ketone (FW = 259.23 g/mol), also known as *N*-(1-benzyl-2-oxo-3,3,3-trifluoropropyl)acetamide, inhibits chymase and chymotrypsin (K_i = 22 and 24 μM, respectively). **Target(s):** chymase (1); chymotrypsin (1,2). **1**. Akahoshi, Ashimori, Yoshimura, *et al.* (2001) *Bioorg. Med. Chem.* **9**, 301. **2**. Lin, Cassidy & Frey (1998) *Biochemistry* **37**, 11940.

N-(*N*-Acetyl-L-phenylalanyl)aminoacetonitrile

This peptidonitrile (FW = 245.28 g/mol) is a reversible inhibitor of papain. ¹³C NMR spectroscopy demonstrated covalent adduct formation, the reversibility of which was shown by displacement with an aldehyde inhibitor, regenerating the nitrile. No hydrolysis of the nitrile was observed. The covalent adduct is most likely a thioimidate formed between the catalytically essential thiol and nitrile. Based on kinetic studies with several *p*-nitroanilide substrates and their corresponding nitrile inhibitors, correlation was found between K_i and k_{cat}/K_m was observed. The finding that the pH-dependence of K_i parallels that of k_{cat}/K_m suggests that the interaction of nitriles and papain has considerable transition-state character. **1**. Liang & Abeles (1987) *Arch. Biochem. Biophys.* **252**, 626.

(2*S*,6*RS*)-2-(*N*-(*N*-Acetyl-L-phenylalanyl)amino)-6-hydrazinoheptane-1,7-dioate

This slow, tight-binding inhibitor (FW = 392.42 g/mol) targets *Escherichia coli* succinyldiaminopimelate aminotransferase (K_i = 0.9 μM). These dipeptide substrates were converted to hydrazines by treatment with hydrazine, followed by reduction with sodium cyanoborohydride. **1**. Cox, Schouten, Stentiford & Wareing (1998) *Bioorg. Med. Chem. Lett.* **8**, 945.

N-Acetyl-L-phenylalanyl-azaalanine Phenyl Ester

This azapeptide ester (FW = 355.39 g/mol) inactivates papain and cathepsin B. The ester derivatives inactivated papain and cathepsin B at rates which increased dramatically with leaving group hydrophobicity and electronegativity. The authors stress that the ease of synthesis coupled with good solution stability suggests that azapeptide esters may be useful as active site titrants of cysteine proteinases and probes of their biological function *in vivo*. **1**. Xing & Hanzlik (1998) *J. Med. Chem.* **41**, 1344.

N-Acetyl-L-phenylalanylglycyl-L-alanyl-L-leucine Chloromethyl Ketone

This halomethyl ketone (FW = 480.99 g/mol) inhibits subtilisin, protein-tyrosine kinase, cathepsin G4, and rat mast cell proteinase II. **1**. Powers (1977) *Meth. Enzymol.* **46**, 197. **2**. Navarro, Abdel-Ghany & Racker

(1982) *Biochemistry* **21**, 6138. **3.** Yoshida, Everitt, Neurath, Woodbury & Powers (1980) *Biochemistry* **19**, 5799. **4.** Powers, Gupton, Harley, Nishino & Whitley (1977) *Biochim. Biophys. Acta* **485**, 156.

N-Acetyl-L-phenylalanyl-L-histidyl-(4-amino-5-cyclohexyl-3-hydroxypentanoic acid) *N*-(2(*S*)-Methylbutyl)amide

This peptide amide analogue (FWfree base = 610.80 g/mol) inhibits a number of aspartic proteinases. The IC_{50} value for renin was 42 nM whereas gastricsin and pepsin had weak affinities (IC_{50} = 15 and 16 μM, respectively). **Target(s):** cathepsin D; cathepsin E; endothiapepsin; gastricsin; pepsin; renin. **1.** Jupp, Dunn, Jacobs, *et al.* (1990) *Biochem J.* **265**, 871.

N-Acetyl-L-phenylalanyl-L-histidyl-(4-amino-5-cyclohexyl-3-hydroxypentanoyl)-L-leucyl-L-phenylalaninamide

This peptide amide analogue (FW$_{free-base}$ = 801.00 g/mol) inhibits a number of aspartic proteinases, including cathepsin D, cathepsin E, endothiapepsin, gastricsin, pepsin, and renin. The IC_{50} value for renin was 3 nM, whereas endothiapepsin and pepsin had weak affinities (IC_{50} = 2.7 and 1.5 μM, respectively). **1.** Jupp, Dunn, Jacobs, *et al.* (1990) *Biochem J.* **265**, 871.

N-[*N*-Acetyl-L-phenylalanyl-L-histidyl]-*N*-(2*S*-amino-3-phenylpropyl)-L-phenylalaninamide

This pseudopeptide amide (FW$_{free-base}$ = 623.76 g/mol; Ac-Phe-His-Pheψ[CH₂-NH]Phe-NH₂, where ψ indicates a pseudopeptide linkage, with a reduced peptide bond) inhibits renin with IC_{50} = 450 nM. **1.** Sawyer, Maggiora, *et al.* (1990) in *Peptides: Chemistry, Structure, and Biology* (Rivier & Marshall, eds), pp. 46-48, Escom Sci. Publ.

N-Acetyl-D-phenylalanyl-L-prolyl-L-boroarginine

This boroarginine-containing tripeptide (FW$_{HCl}$ = 496.80 g/mol), also known as DuP714, is a potent inhibitor of thrombin (K_i = 41 pM). It also inhibits, albeit with a weaker affinity, other proteinases, including plasma kallikrein (K_i = 1.9 nM) and coagulation factor Xa. **1.** Weber, Lee, Lewandowski, *et al.* (1995) *Biochemistry* **34**, 3750.

N-Acetyl-D-phenylalanyl-L-prolyl-L-boroarginine-pinanediol

This peptide analogue (FW$_{free-base}$ = 594.56 g/mol) inhibits thrombin (K_i = 0.07 nM). It is a slightly weaker inhibitor of trypsin and plasmin (K_i = 0.14 and 9.9 nM, respectively). **1.** Tapparelli, Metternich, Ehrhardt, *et al.* (1993) *J. Biol. Chem.* **268**, 4734. **2.** Kettner & Shenvi (1988) European Patent Application 0293-881 (1/6/1988).

N-Acetyl-L-phenylalanyl-L-threonyl-L-leucyl-L-aspartyl-L-alanyl-L-aspartyl-L-phenylalanine

This acetylated heptapeptide (FW$_{free-acid}$ = 869.93 g/mol) inhibits mammalian ribonucleotide reductase. The sequence Ac-FTLDADF corresponds to the C-terminus of mR₂ subunit of the reductase and disrupts the enzyme's quaternary structure. **1.** Tan, Gao, Kaur & Cooperman (2004) *Bioorg. Med. Chem. Lett.* **14**, 5301. **2.** Cooperman, Gao, Tan, Kashlan & Kaur (2005) *Adv. Enzym. Regul.* **45**, 112. **3.** Yang, Spanevello, Celiker, Hirschmann, Rubin & Cooperman (1990) *FEBS Lett.* **272**, 61. **4.** Xu, Fairman, Wijerathna, *et al.* (2008) *J. Med. Chem.* **51**, 4653.

Acetyl Phosphate

This acyl phosphate (FW$_{free-acid}$ = 140.03 g/mol; CAS 590-54-5; pK_a values: 1.2 and 4.8; Most stable as dilithium salt), first identified by Nobelist Fritz Lipmann, led to the development of his concept of group transfer potential. Depending upon the nature of the enzyme-catalyzed reaction, acetyl-P can serve as a substrate for either acetyl or phosphoryl transfer. Acetyl phosphate is a labile metabolite that is rapidly hydrolyzed in acidic or alkaline conditions: approximately one-fifth is hydrolyzed after thirty minutes at pH 7 and room temperature (45% is hydrolyzed in ten minutes in 0.5 M HCl). Solutions should always be freshly prepared (aqueous neutral solutions are relatively stable when stored at −30°C). Acetyl phosphate is a product of the reactions catalyzed by acetate kinase, acetate kinase (diphosphate), pyruvate oxidase, glycine reductase, sarcosine reductase, betaine reductase, phosphate acetyltransferase, phosphoketolase, and fructose-6-phosphate phosphoketolase. It is an alternative product for propionate kinase. It is a substrate for sulfoacetaldehyde acetyltransferase and is an alternative substrate for several phosphotransferases and ATPases (*e.g.*, a substrate for the Na⁺/K⁺-exchanging ATPase). **Target(s):** triose-phosphate isomerase (1,2); thiamin-phosphate diphosphorylase (3,4,12); glyceraldehyde-3-phosphate dehydrogenase (5); pyruvate dehydrogenase (8); formyltetrahydrofolate synthetase (9); acetyl-CoA synthetase, *or* acetate:CoA ligase (10); acetoacetate decarboxylase, inhibited by the monoanion (11); aspartate carbamoyltransferase (13,14). **1.** Krietsch (1975) *Meth. Enzymol.* **41**, 434. **2.** Krietsch (1975) *Meth. Enzymol.* **41**, 438. **3.** Penttinen (1979) *Meth. Enzymol.* **62**, 68. **4.** Kawasaki & Esaki (1970) *Biochem. Biophys. Res. Commun.* **40**, 1468. **5.** Harris & Waters (1976) *The Enzymes*, 3rd ed., **13**, 1. **6.** Lipmann (1940) *J. Biol. Chem.* **134**, 463. **7.** Lipmann (1944) *J. Biol. Chem.* **155**, 55. **8.** Carlsson, Kujala & Edlund (1985) *Infect. Immun.* **49**, 674. **9.** Buttlaire, Balfe, Wendland & Himes (1979) *Biochim. Biophys. Acta* **567**, 453. **10.** Preston, Wall & Emerich (1990) *Biochem. J.* **267**, 179. **11.** Kluger & Nakaoka (1974)

Biochemistry **13**, 910. **12**. Kayama & Kawasaki (1973) *Arch. Biochem. Biophys.* **158**, 242. **13**. Burns, Mendz & Hazell (1997) *Arch. Biochem. Biophys.* **347**, 119. **14**. Savithri, Vaidyanathan & Rao (1978) *Proc. Indian Acad. Sci. Sect. B* **87B**, 81.

Acetylphosphinate (Acetylphosphinic Acid)

This phosphinic acid (FW$_{\text{free-acid}}$ = 108.03 g/mol; CAS 50654-76-7) is a strong inhibitor of the pyruvate dehydrogenase complex (requiring the presence of thiamin pyrophosphate) and inactivates formate *C*-acetyltransferase. Acetylphosphinate is the transamination product of 1-aminoethylphosphinate. **Target(s):** formate *C*-acetyltransferase, *or* pyruvate formate-lyase (1-3); pyruvate dehydrogenase (4,5). **1**. Brush & Kozarich (1992) *The Enzymes*, 3rd ed., **20**, 317. **2**. Parast, Wong, Lewisch, Kozarich, Peisach & Magliozzo (1995) *Biochemistry* **34**, 2393. **3**. Ulissi-DeMario, Brush & Kozarich (1991) *J. Amer. Chem. Soc.* **113**, 4341. **4**. Laber & Amrhein (1987) *Biochem. J.* **248**, 351. **5**. Schonbrunn-Hanebeck, Laber & Amrhein (1990) *Biochemistry* **29**, 4880.

Acetylphosphonate (Acetylphosphonic Acid)

This pyruvate analogue (FW$_{\text{free-acid}}$ = 124.03 g/mol; CAS 6881-54-5) inhibits pyruvate (1), pyruvate dehydrogenase (2,3), pyruvate oxidase (3), and phosphoenolpyruvate:protein phosphotransferase (4). **1**. Kluger (1977) *J. Amer. Chem. Soc.* **99**, 4504. **2**. Dixon, Giddens, Harrison, *et al.* (1991) *J. Enzyme Inhib.* **5**, 111. **3**. O'Brien, Kluger, Pike & Gennis (1980) *Biochim. Biophys. Acta* **613**, 10. **4**. Saier, Schmidt & Lin (1980) *J. Biol. Chem.* **255**, 8579.

N-[*N*-Acetyl-L-prolyl-L-phenylalanyl-L-histidyl]-*N*-(2*S*-amino-3-cyclohexylpropyl)-L-phenylalaninamide

This pseudopeptide amide (FW = 726.92 g/mol), also known as U-79465E and Ac-Pro-Phe-His-Chaψ[CH$_2$ NH]Phe-NH$_2$ (where ψ indicates a reduced peptide bond between a cyclohexylalanyl and a phenylalanyl residue), potently inhibits renin, IC$_{50}$ = 1.7 nM, and weakly inhibits rhizopuspepsin, K_i = 218 μM (2). **1**. Sawyer, Maggiora, Liu, *et al.* (1990) in *Peptides: Chemistry, Structure, and Biology* (Rivier & Marshall, eds), pp. 46-48, Escom Sci. Publ. **2**. Lowther, Sawyer, Staples, *et al.* (1992) in *Peptides: Chemistry and Biology Proc. of the 12th Amer. Peptide Symp.* (June 16-21, 1991) (Smith & Rivier, eds), pp. 413.

{*N*-[*N*-Acetyl-L-prolyl-L-phenylalanyl-L-histidyl]-(5*S*-amino-4*S*-hydroxy-2-isopropyl-7-methyloctanoyl)}-L-isoleucinamide

This pseudopeptide amide (FW$_{\text{free-base}}$ = 674.93 g/mol), also known as U77647E and Ac-Pro-Phe-His Leuψ[CH(OH)CH$_2$]Val-Ile-NH$_2$ (where ψ indicates a reduced peptide bond, inhibits rhizopuspepsin: K_i = 54 nM. **1**. Lowther, Sawyer, Staples, *et al.* (1992) in *Peptides: Chemistry and Biology Proc. of the 12th Amer. Peptide Symp.* (June 16-21, 1991) (Smith & Rivier, eds), pp. 413.

N-[*N*-Acetyl-L-prolyl-L-phenylalanyl-L-histidyl]-*N*-(2*S*-amino-3-phenylpropyl)-L-(4-chlorophenyl)alaninamide

This pseudopeptide amide (FW$_{\text{free-base}}$ = 755.32 g/mol), also known as U-80011E and Ac-Pro-Phe-His-Pheψ[CH$_2$NH]Phe(*p*-Cl)-NH$_2$ (where ψ indicates a reduced peptide bond between a phenylalanyl and a *p*-chlorophenylalanyl residue), inhibits renin, IC$_{50}$ = 38 nM (1) and rhizopuspepsin, K_i = 13.3 μM (2). **1**. Sawyer, Maggiora, Liu, *et al.* (1990) in *Peptides: Chemistry, Structure, and Biology* (Rivier & Marshall, eds), pp. 46-48, Escom Sci. Publ. **2**. Lowther, Sawyer, Staples, *et al.* (1992) in *Peptides: Chemistry and Biology Proc. of the 12th Amer. Peptide Symp.* (June 16-21, 1991) (Smith & Rivier, eds), pp. 413.

N-[*N*-Acetyl-L-prolyl-L-phenylalanyl-L-histidyl]-*N*-(2*S*-amino-3-phenylpropyl)-L-tyrosinamide

This pseudopeptide amide (FW$_{\text{free-base}}$ = 736.87 g/mol), also known as Ac-Pro-Phe-His-Pheψ[CH$_2$-NH]Tyr-NH$_2$, containing a reduced pseudopeptide bond between a phenylalanyl and a tyrosyl residue, inhibits renin with IC$_{50}$ = 25 nM. **1**. Sawyer, Maggiora, Liu, *et al.* (1990) in *Peptides: Chemistry, Structure, and Biology* (Rivier & Marshall, eds), pp. 46-48, Escom Sci. Publ.

N-Acetyl-purinomycin

This puromycin analogue (FW = 513.55 g/mol; CAS 22852-13-7; Solubility: 100 mM in DMSO; 100 mM HCl salt in H$_2$O) downregulates SnoN and Ski protein expression and promotes TGF-β signaling, the latter independent of MAPK activation. Unlike puromycin, *N*-acetyl-purinomycin does not bind to ribosomes or block protein synthesis (1). (***See** Puromycin*) SnoN and Ski proteins function as Smad transcriptional co-repressors and are implicated in the regulation of diverse cellular processes such as proliferation, differentiation and transformation (1). Transforming growth factor-β (TGF-β) signaling causes SnoN and Ski protein degradation via proteasome with the participation of phosphorylated R-Smad proteins. Intriguingly, the antibiotics anisomycin (ANS) and puromycin (PURO) are also able to downregulate Ski and SnoN proteins via proteasome. By controlling Ski and SnoN protein levels, antibiotic analogues that lack ribotoxic effects are promising tools for studying TGF-β signaling. **1.** Hernández-Damián, Tecalco-Cruz, Ríos-López, *et al.* (2013) *Biochim. Biophys. Acta* **1830**, 5049.

3-Acetylpyridine

This substituted nicotinamide analogue (FW = 121.14 g/mol; Boiling Point = 220°C), also known as methyl 3-pyridyl ketone, inhibits many NAD^+-dependentreactions. 3-Acetylpyridine is often an alternative substrate for nicotinamide-utilizing enzymes. **Target(s):** nicotinamidase (1,8,10,11); NAD^+ ADP-ribosyltransferase (2); NAD^+ nucleosidase, *or* NADase (3-5,13); cAMP phosphodiesterase (6); acetylcholinesterase (7); NAD^+ diphosphatase (9); nicotinamide phosphoribosyltransferase, weakly inhibited (13); nicotinate phosphoribosyltransferase (14). **1.** Joshi, Calbreath & Handler (1971) *Meth. Enzymol.* **18B**, 180. **2.** Nishizuka, Ueda & Hayaishi (1971) *Meth. Enzymol.* **18B**, 230. **3.** Anderson & Yuan (1980) *Meth. Enzymol.* **66**, 144. **4.** McIlwain (1950) *Biochem. J.* **46**, 612. **5.** Webb (1966) *Enzyme and Metabolic Inhibitors*, vol. **2**, p. 491, Academic Press, New York. **6.** Shimoyama, Kawai, Nasu, Shioji & Hoshi (1975) *Physiol. Chem. Phys.* **7**, 125. **7.** Bergmann, Wilson & Nachmansohn (1950) *J. Biol. Chem.* **186**, 693. **8.** Grossowicz & Halpern (1956) *Biochim. Biophys. Acta* **20**, 576. **9.** Anderson & Lang (1966) *Biochem. J.* **101**, 392. **10.** Johnson & Gadd (1974) *Int. J. Biochem.* **5**, 633. **11.** Tanigawa, Shimiyama, Dohi & Ueda (1972) *J. Biol. Chem.* **247**, 8036. **12.** Nakazawa, Ueda, Honjo, *et al.* (1968) *Biochem. Biophys. Res. Commun.* **32**, 143. **13.** Dietrich & O. Muniz (1972) *Biochemistry* **11**, 1691. **14.** Imsande (1964) *Biochim. Biophys. Acta* **85**, 255.

3-Acetylpyridine Adenine Dinucleotide

This NAD^+ analogue (FW = 662.44 g/mol; CAS 86-08-8), also known as 3-acetyl-NAD^+, is an alternative coenzyme substrate for many oxidoreductases. **Target(s):** UDP-glucuronate decarboxylase (1); nicotinamide phosphoribosyltransferase (2,12); NADase (3,4); 1-pyrroline-5-carboxylate reductase (5); lactate dehydrogenase (6,7); ADP-ribose diphosphatase (8); NADH kinase (9); NAD^+ kinase (10,11). **1.** Ankel & Feingold (1966) *Meth. Enzymol.* **8**, 287. **2.** Dietrich (1971) *Meth. Enzymol.* **18B**, 144. **3.** Everse, JEverse & Simeral (1980) *Meth. Enzymol.* **66**, 137. **4.** Anderson & Yuan (1980) *Meth. Enzymol.* **66**, 144. **5.** Krishna, Beilstein & Leisinger (1979) *Biochem. J.* **181**, 223. **6.** Okabe, Hayakawa, Hamada & Koike (1968) *Biochemistry* **7**, 79. **7.** Ciaccio (1966) *J. Biol. Chem.* **241**, 1581. **8.** Wu, Lennon & Suhadolnik (1978) *Biochim. Biophys. Acta* **520**, 588. **9.** Iwahashi, Hitoshio, Tajima & Nakamura (1989) *J. Biochem.* **105**, 588. **10.** Chung (1967) *J. Biol. Chem.* **242**, 1182. **11.** Lerner, Niere, Ludwig & Ziegler (2001) *Biochem. Biophys. Res. Commun.* **288**, 69. **12.** Dietrich & Muniz (1972) *Biochemistry* **11**, 1691.

3-Acetylpyridine Adenine Dinucleotide Phosphate

This $NADP^+$ analogue (FW$_{free-acid}$ = 742.42; CAS 150729-98-9 (free acid)), also known as 3-acetyl-$NADP^+$, is a poor substrate for many $NADP^+$-dependent enzymes, including glucose-6-phosphate dehydrogenase (1); 4-hydroxyacetophenone monooxygenase (2); phenylacetone monooxygenase (3). **1.** Noltmann & Kuby (1963) *The Enzymes*, 2nd ed., **7**, 223. **2.** Kamerbeek, Fraaije & Janssen (2004) *Eur. J. Biochem.* **271**, 2107. **3.** Fraaije, Kamerbeek, Heidekamp, Fortin & Janssen (2004) *J. Biol. Chem.* **279**, 3354.

Acetylsalicylate (Acetylsalicylic Acid & Aspirin)

This first-in-class pain-relieving and fever-lowering anti-inflammatory agent (FW$_{free-acid}$ = 180.16 g/mol; CAS 50-78-2; pK_a = 3.49 at 25°C; Solubility = 3.3 mg/mL at 25°C; Symbol: ASA) is the original synthetic nonsteroidal anti-inflammatory drug, *or* NSAID. Aspirin exerts its effects by inactivating cyclooxygenase (*or* prostaglandin G/H synthase, *or* PGHS). It shows somewhat greater selectivity toward COX-1, the type-1 isozyme, with attendent inhibition of prostaglandin formation. Acetylsalicylic acid also inhibits platelet thromboxane synthesis as well as the ADP- and collagen-induced platelet release reaction. Low-dose aspirin is effective in preventing ~20% of atherothrombotic complications (including non-fatal myocardial infarction, non-fatal stroke, or vascular death) in patients with prior evidence of myocardial infarction, stroke, or transient cerebral ischemia. **Mechanism of Action:** That aspirin targets prostaglandin formation was first demonstrated by Piper & Vane (1), a finding that earned Vane the Nobel Prize in Physiology and Medicine in 1982. Aspirin selectively acetylates the hydroxyl group of Ser-530 in COX-1, irreversibly inhibiting the enzyme (2). This outcome requires *de novo* enzyme synthesis before prostanoid synthesis can resume. When purified PGHS is acetylated, its COX activity is inhibited, but *not* its hydroperoxidase activity. One acetyl group is incorporated per monomer of the dimeric enzyme. At low concentrations, aspirin acetylates PGHS rapidly (usually within minutes) in a highly selective reaction that prevents arachidonate binding. At higher concentrations, aspirin non-specifically acetylates a variety of proteins and even nucleic acids. Upon modification of COX-2, the latter produces lipoxins, of which most are anti-inflammatory. **Pharmacokinetics & Metabolism:** Aspirin is both a drug and prodrug, depending on its target. It is rapidly absorbed from the stomach and intestine by passive diffusion. Pertinent PK parameters for acetylsalicylate are: Oral Bioavailability = 68%; Clearance = 39 L/hour; Volume of Distribution = 10.5 L; and $t_{1/2}$ = 0.25 hours. Acetylsalicylate is transformed into salicylate in stomach, intestinal mucosa, blood, and liver, the latter being the main site. Salicylate distributes rapidly into the body fluid compartments, binding to albumin within plasma. With increasing total plasma salicylate concentrations, the unbound fraction increases. Pertinent PK parameters for salicylate are: Clearance = 0.6-3.6 L/hour, depending on dose; Volume of Distribution = 11.5 L; and $t_{1/2}$ = 2-30 hours, depending on aspirin dose. Salicylate also activates AMP-stimulated protein kinase (AMPK), a property likely to account for some of aspirin's anti-cancer and anti-inflammatory effects (3). ASA can also be used during a heart attack to reduce the risk of dying from the heart attack. **Chemical Synthesis:** Acetylsalicylate was first synthesized from salicylate and acetyl chloride by Felix Hofmann at Friedrich Bayer & Co. in 1897, and its name universally known name, "aspirin", was coined by then supervising pharmacologist, Heinrich Dreser, to avoid confusion with salicylic acid. The famously high purity and stability of the Bayer product is evident by the absence of a vinegary odor, long after the packaged product is first opened. Prolonged exposure to moisture, however, generates both acetate and salicylate, with the latter inhibiting some 20-30 enzymes (*See Salicylate*). **Nonenzymatic Aspirin Hydrolysis:** Hydrolysis of acetylsalicylic acid presents the interesting case where the reaction rate is pH-independent over range from 4 to 8. As shown nearly a half century ago (4), aspirin undergoes general base catalysis. Later work suggests that aspirin anion exists only fleetingly, rearranging by acetyl transfer to the *ortho*-carboxylate group, as indicated by IR, UV and NMR (5). The mixed anhydride then cyclizes to the more stable bicyclic orthoacetate isomer, a process facilitated by time and increasing pH. **Target(s):** prostaglandin-endoperoxide synthase, *or* cyclooxygenase (6,7,10,11,16,28-31); ζ-crystallin (8); creatine kinase (9), 15-

hydroxyprostaglandin dehydrogenase (10); cyclooxygenase-16 (16); cyclooxygenase-26 (16); hemoglobin S polymerization (12), however, *See* Reference 13; glutamine:D-fructose-6-phosphate aminotransferase (14); ATPase, mitochondrial (15); aryl-acylamidase (17); palmitoyl-CoA hydrolase, *or* acyl-CoA hydrolase (18); phospholipase A$_2$ (19); estrone sulfotransferase (20); phenol sulfotransferase, *or* aryl sulfotransferase (20); IκB kinase (21,22); procollagen glucosyltransferase (23); *N*-hydroxyarylamine *O*-acetyltransferase (24); 1-alkylglycerophosphocholine *O*-acetyltransferase (25,26); thiopurine *S*-methyltransferase (32); 3-hydroxyanthranilate 3,4 dioxygenase (32). **1**. Piper & Vane (1969) *Nature* **223**, 29. **2**. Roth & Majerus (1975) *J. Clin. Invest.* **56**, 624. **3**. Hardie (2014) *J. Internal Med.* **276**, 543. **4**. Fersht & Kirby (*J. Am. Chem. Soc.* (1967) **89**, 4857. **5**. Chandrasekhar & Kumar (2011) *Tet. Lett.* **52**, 3561. **6**. van der Ouderaa & Buytenhek (1982) *Meth. Enzymol.* **86**, 60. **7**. Roth (1982) *Meth. Enzymol.* **86**, 392. **8**. Bazzi, Rabbani & Duhaiman (2002) *Int. J. Biochem. Cell Biol.* **34**, 70. **9**. Watts (1973) *The Enzymes*, 3rd ed., **8**, 383. **10**. Pace-Asciak & Smith (1983) *The Enzymes*, 3rd ed., **16**, 543. **11**. Kalgutkar, Crews, Rowlinson, Garner, Seibert & Marnett (1998) *Science* **280**, 1268. **12**. Klotz & Tam (1973) *Proc. Natl. Acad. Sci. U.S.A.* **70**, 1313. **13**. de Furia, Cerami, Bunn, Lu & Peterson (1973) *Proc. Natl. Acad. Sci. U.S.A.* **70**, 3707. **14**. Chatterjee & Stefanovich (1976) *Arzneimittel-forschung.* **26**, 502. **15**. Chatterjee & Stefanovich (1976) *Arzneimittelforschung* **26**, 499. **16**. Lecomte, Laneuville, Ji, DeWitt & Smith (1994) *J. Biol. Chem.* **269**, 13207. **17**. Khanna, Kaur, Sanwal & Ali (1992) *J. Pharmacol. Exp. Ther.* **262**, 1225. **18**. Dixon, Osterloh & Becker (1990) *J. Pharm. Sci.* **79**, 103. **19**. Hendrickson, Trygstad, Loftness & Sailer (1981) *Arch. Biochem. Biophys.* **212**, 508. **20**. King, Ghosh & Wu (2006) *Curr. Drug Metab.* **7**, 745. **21**. Burke (2003) *Curr. Opin. Drug Discov. Devel.* **6**, 720. **22**. Luo, Kamata & Karin (2005) *J. Clin. Invest.* **115**, 2625. **23**. Barber & Jamieson (1971) *Biochim. Biophys. Acta* **252**, 533. **24**. Yamamura, Sayama, Kakikawa, Mori, Taketo & Kodaira (2000) *Biochim. Biophys. Acta* **1475**, 10. **25**. Hurst & Bazan (1997) *J. Ocul. Pharmacol. Ther.* **13**, 415. **26**. Yamazaki, Sugatani, Fujii, *et al.* (1994) *Biochem. Pharmacol.* **47**, 995. **27**. Zhou & Chowbay (2007) *Curr. Pharmacogenomics* **5**, 103. **28**. Meade, Smith & DeWitt (1993) *J. Biol. Chem.* **268**, 6610. **29**. Miyamoto, Ogino, Yamamoto & Hayaishi (1976) *J. Biol. Chem.* **251**, 2629. **30**. Roth, Ciok & Ozols (1980) *J. Biol. Chem.* **255**, 1301. **31**. Rogge, Ho, Liu, Kulmacz & Tsai (2006) *Biochemistry* **45**, 523. **32**. Maharaj, Maharaj & Daya (2006) *Metab. Brain Dis.* **21**, 189.

N-Acetyl-seryl-aspartyl-lysyl-proline

This human plasma hemoregulatory peptide (FW = 487.51 g/mol; *Sequence*: AcSDKP), also known by the trade name Seraspenide®, is a bone marrow-derived negative regulator that inhibits pluripotent hematopoietic stem cell entry into the S-phase of the cell cycle (1-4). It increases the survival of mice treated with chemotherapeutic agents, such that the agents are selectively cytotoxic to cancer, but not untransformed, cells. Ac-SDKP inhibits endothelin-1-induced activation of p44/42 mitogen-activated protein kinase and collagen production in cultured rat cardiac fibroblasts by preserving SHP-2 activity and thereby preventing p44/42 MAPK activation (5). Incubation of normal human mononuclear cells with AcSDKP (10^{-10} to 10^{-9} M) significantly inhibits GMCF- and erythroid burst-forming growth, but no effect is seen at higher AcSDKP concentrations (6). **Metabolism:** AcSDKP is most likely derived from thymosin-β4 (7), the human actin monomer-sequestering protein (*Sequence*: MSDKPDMAEIEKFDKSKLKKTETQEKNPLPSKETIEQEK QAGES), suggesting the likely involvement of proteases and an acetylase. AcSDKP is completely degraded ($t_{1/2}$ = 80 min) in human plasma (8). After 128 nmol/kg was given as a 12-hr intravenous infusion, NAc-SDKP was quickly eliminated, with a mean $t_{1/2}$ of 4.5 min (9). While insensitive to leupeptin. AcSDKP cleavage is sensitive to metalloprotease inhibitors (8). The finding that cleavage is blocked by specific angiotensin I-converting enzyme (*or* ACE) inhibitors, demonstrated that the first step is catalyzed by this hydrolase. Hydrolysis of AcSDKP by commercial rabbit lung ACE generated the C-terminal dipeptide Lys-Pro. Depending on the kinetics, an alternative substrate like AcSDKP may well be an endogenous ACE modulator (8). **1**. Frindel & Guigon (1977) *Exp. Hematol.* 5, 74. **2**. Lord, Mori, Wright & Lajtha (1976) *Brit. J. H/ematol.* 34, 441. **3**. Lenfant, Wdzieczak-Bakala, Guittet, *et al.* (1989) *Proc. Natl. Acad. Sci. U.S.A.* **86**, 779. **4**. Wdzieczak-Bakala, Fache, Lenfant, *et al.* (1990) *Leukemia* 4, 235. **5**. Peng, Carretero, Peterson, *et al.* (2012) *Pflugers Arch.* **464**, 415. **6**. Guigon, Bonnet, Lemoine, *et al.* (1990) *Exp. Hematol.* **18**, 1112. **7**. Grillon, Rieger, Bakala, *et al.* (1990) *FEBS Lett.* **274**, 30. **8**. Rieger, Saez-Servent, Papet, *et al.* (1993) *Biochem. J.* **296**, 373. **9**. Ezan, Carde, Le Kerneau, *et al.* (1994) *Drug Metab. Dispos.* **22**, 843.

D-*erythro*-*N*-Acetylsphingosine

This short-chain ceramide (FW = 341.53 g/mol), also known as D-*erythro*-C2-ceramide, is a cell-permeable, biologically active ceramide that induces differentiation and apoptosis. It also stimulates certain protein phosphatases as well as phosphatidylglycerolphosphate synthase. **Target(s):** acyl-CoA:cholesterol acyltransferase, or ACAT (1); Ca^{2+} channels, L-type (2,3); choline-phosphate cytidylyltransferase, *or* CTP:phosphocholine cytidylyltransferase (4,9,10); Na$^+$/Ca^{2+} exchange (5); phospholipase D (6,7); ubiquinol:cytochrome *c* reductase, complex III (8); ceramide kinase, as weak alternative substrate (11,12). **1**. Ridgway (1995) *Biochim. Biophys. Acta* **1256**, 39. **2**. Chik, Li, Karpinski & Ho (2004) *Mol. Cell. Endocrinol.* **218**, 175. **3**. Chik, Li, Negishi, Karpinski & Ho (1999) *Endocrinology* **140**, 5682. **4**. Ramos, El Mouedden, Claro & Jackowski (2002) *Mol. Pharmacol.* **62**, 1068. **5**. Condrescu & Reeves (2001) *J. Biol. Chem.* **276**, 4046. **6**. Singh, Stromberg, Bourgoin, *et al.* (2001) *Biochemistry* **40**, 11227. **7**. Albert, Piper & Large (2005) *J. Physiol.* **566**, 769. **8**. Gudz, Tserng & Hoppel (1997) *J. Biol. Chem.* **272**, 24154. **9**. Awasthi, Vivekananda, Awasthi, Smith & King (2001) *Amer. J. Physiol.* **281**, L108. **10**. Jackowski & Fagone (2005) *J. Biol. Chem.* **280**, 853. **11**. Baumruker, Bornancin & Billich (2005) *Immunol. Lett.* **96**, 175. **12**. Mitsutake, Kim, Inagaki, *et al.* (2004) *J. Biol. Chem.* **279**, 17570.

N-Acetyltryptamine

This partial receptor agonist (FW = 202.26 g/mol; CAS 1016-47-3; Soluble to 100 mM in DMSO) is a melatonin (*or* 5-methoxy-*N*-acetyltryptamine) mimic and is more potent than melatonin in inhibiting dopamine release fom presynaptic melatonin receptor site of rabbit retina labeled *in vitro* with [^{3}H]dopamine. **1**. Dubocovich (1988) *J. Pharmacol. Exp. Ther.* **246**, 902.

N-Acetyl-L-tryptophan 3,5-bis(trifluoromethyl)-benzyl Ester

This ester (FW = 472.39 g/mol) is a potent substance P receptor antagonist of the human substance P receptor, *or* neurokinin-1 (NK-1) receptor, with 230- and 10-fold reduced affinity for mutant NK-1 receptors in which histidine 265 or histidine 197, respectively, are replaced by alanine. These residues play key roles in the binding of quinuclidine antagonists to the NK-1 receptor. **1**. MacLeod, Merchant, Cascieri, *et al.* (1993) *J. Med. Chem.* **36**, 2044. **2**. Cascieri, Macleod, Underwood, *et al.* (1994) *J. Biol. Chem.* **269**, 6587.

N^{α}-Acetyl-L-tryptophanyl-L-glutamyl-L-histidyl-L-aspartate α-Semialdehyde

This tetrapeptide aldehyde (MW = 585.57), abbreviated Ac-WEHD-CHO, strongly inhibits caspase-1, K_i = 56 pM (1-4), caspase-3 (4); caspase-4 (1,4); caspase-5 (1,4); caspase-6 (4); caspase-8, K_i = 21.1 nM (4); caspase-9 (4); caspase-10 (4). **1**. Roy & Nicholson (2000) *Meth. Enzymol.* **322**, 110. **2**. Rano, Timkey, Peterson, *et al.* (1997) *Chem. Biol.* **4**, 149. **3**. Hayashi, Jikihara, Yagi, *et al.* (2001) *Brain Res.* **893**, 113. **4**. Garcia-Calvo, Peterson, Leiting, *et al.* (1998) *J. Biol. Chem.* **273**, 32608.

N-Acetyl-D-tryptophanyl-L-leucyl-L-aspartyl-L-isoleucyl-L-isoleucyl-L-tryptophan

This acetylated hexapeptide (FW$_{free-acid}$ = 887.05 g/mol), corresponding to residues 16-21 of human, pig, and dog endothelin-1 (except that D-Trp16 is in place of L-His16), inhibits endothelin ETA receptor. Endothelins are 21-amino acid vasoconstricting peptides mainly produced in the endothelium and exerting a key role in vascular homeostasis. **1**. Cody, Doherty, He, *et al.* (1992) *J. Med. Chem.* **35**, 3301.

N-Acetyl-L-tyrosine Ethyl Ester

This ester (FW = 251.28 g/mol; Abbreviation ATEE) is a substrate for chymotrypsin that is frequently used in assays of that enzyme. The reaction can be followed by observing the production of free carboxyl groups or by the decrease in absorbance at 237 nm and pH 7. Supersaturated stock solutions (*i.e.*, 10-20 mM) can be prepared by dissolving ATEE in water at 85°C, followed by rapidly cooling to room temperature. These solutions are typically stable for six to twelve hours. *N*-Acetyl-L-tyrosine ethyl ester inhibits protein-disulfide reductase, glutathione-requiring (1); thiolsubtilisin (2); complement component C' (3). **1**. Tomizawa (1971)

Meth. Enzymol. **17B**, 515. **2**. Philipp, Polgar & Bender (1970) *Meth. Enzymol.* **19**, 215. **3**. Shin & Mayer (1968) *Biochemistry* **7**, 3003.

N^{α}-Acetyl-L-tyrosyl-L-valyl-L-alanyl-L-aspartate Chloromethyl Ketone

This halomethyl ketone (FW$_{free-acid}$ = 541.00 g/mol), often abbreviated Ac-YVAD-*CMK* and referred to as interleukin-1β converting enzyme inhibitor II, inhibits caspase-1 (K_i = 760 pM) and caspase-4. It also protects against neurodegeneration by inhibition of cathepsin B. **Target(s):** caspase-1 (1-4); caspase-4 (3); cathepsin B (4). **See also** *L-Tyrosyl-L-valyl-L-alanyl-L-aspartic Chloromethyl Ketone* **1**. Walker, Talanian, Brady, *et al.* (1994) *Cell* **78**, 343. **2**. Enari, Hug & Nagata (1995) *Nature* **375**, 78. **3**. Garcia-Calvo, Peterson, Leiting, *et al.* (1998) *J. Biol. Chem.* **273**, 32608. **4**. Gray, Haran, Schneider, *et al.* (2001) *J. Biol. Chem.* **276**, 32750.

N^{α}-Acetyl-L-tyrosyl-L-valyl-L-alanyl-L-aspartate α-Semialdehyde

This tetrapeptide aldehyde (FW$_{free-acid}$ = 492.53 g/mol), often abbreviated Ac-YVAD-CHO, inhibits caspase-1 (K_i = 0.2 nM) and caspase-4. A cell-permeable derivative of this inhibitor is Ac-AAVALLPAVLLALLA PYVAD-CHO, in which the N-terminus of YVAD-CHO has been linked to the N-terminal sequence (residues 1-16) of the membrane-penetrating signal peptide of the Kaposi fibroblast growth factor. **Target(s):** caspase-1 (1-7,10); caspase-4 (3,6,9,10); caspase-3 (5); caspase-5 (3,6,9); caspase-8 (3); caspase-9 (3); caspase-10 (3,8). **1**. Thornberry, Bull, Calaycay, *et al.* (1992) *Nature* **356**, 768. **2**. Wilson, Black, Thomson, *et al.* (1994) *Nature* **370**, 270. **3**. Garcia-Calvo, Peterson, Leiting, Ruel, Nicholson & Thornberry (1998) *J. Biol. Chem.* **273**, 32608. **4**. Chapman (1992) *Bioorgan. Med. Chem. Chem., Lett.* **2**, 613. **5**. Thornberry (1998) in *Handbook of Proteolytic Enzymes* (Barrett, Rawlings & Woessner, eds), p. 732, Academic Press, San Diego. **6**. Roy & Nicholson (2000) *Meth. Enzymol.* **322**, 110. **7**. Thornberry (1994) *Meth. Enzymol.* **244**, 615. **8**. Garcia-Calvo, Peterson, Leiting, *et al.* (1998) *J. Biol. Chem.* **273**, 32608. **9**. Fassy, Krebs, Rey, *et al.* (1998) *Eur. J. Biochem.* **253**, 76. **10**. Margolin, Raybuck, Wilson, *et al.* (1997) *J. Biol. Chem.* **272**, 7223.

N^{α}-Acetyl-L-valyl-L-aspartyl-L-valyl-L-alanyl-L-aspartate α-Semialdehyde

This pentapeptide aldehyde (FW$_{free-acid}$ = 543.57 g/mol), often abbreviated Ac-VDVAD-CHO, inhibits caspase-2, K_i = 3.5 nM (1), caspase-3, K_i = 1.0

nM (1), caspase-7, K_i = 7.5 nM (1), and an ecdysone-inducible *Drosophila* caspase, DRONC (2). **1.** Talanian, Quinlan, Trautz, *et al.* (1997) *J. Biol. Chem.* **272**, 9677. **2.** Dorstyn, Colussi, Quinn, Richardson & Kumar (1999) *Proc. Natl. Acad. Sci. U.S.A.* **96**, 4307.

{*N*-[*N*-Acetyl-L-valyl-L-valyl]-(5*S*-amino-4*S*-hydroxy-2-isopropyl-7-methyloctanoyl)}-L-isoleucine 2-Pyridylmethylamide

This pseudopeptide amide (FW = 674.93 g/mol), also named U85964E and Ac-Val-Val-Leuψ[CH(OH)CH₂]Val-Ile-Amp (where ψ indicates a reduced peptide linkage), inhibits rhizopuspepsin, K_i = 0.12 nM (1,2), HIV-1 protease, K_i = 19 nM (1), HIV-2 protease, K_i = 12 nM), and human pepsin, K_i = 0.1 nM (1). **1.** Lowther, Sawyer, Staples, *et al.* (1992) in *Peptides: Chemistry and BiologyL Proc. of the 12th Amer. Peptide Symp.* (June 16-21, 1991) (Smith & Rivier, eds), p. 413. **2.** Sawyer, Staples, Liu, *et al.* (1992) *Int. J. Pept. Protein Res.* **40**, 274.

ACHP

This novel IKK inhibitor (FW = 364.44 g/mol; CAS 406208-42-2; Soluble to 20 mM in DMSO), also named 2-amino-6-[2-(cyclopropylmethoxy)-6-hydroxyphenyl]-4-(4-piperidinyl)-3-pyridinecarbonitrile, targets IκB kinases, with respective IC₅₀ values of 8.5 and 250 nM for IKKβ and IKKα. ACHP's selectivity is indicated by its > 20-μM IC₅₀ values toward IKK3, Syk, and MAPKKK4. Inhibition of the TNFα-mediated gene expression could occur at low ACHP concentration (<1 μmol/L), with higher concentrations (>10 μmol/L) required to inhibit the constitutive phosphorylation of p65, expression of NF-κB-mediated genes (*e.g.*, *CYCLIN D₁*, *BCL-xₗ*, *XIAP*, *c-IAP₁*, and *IL-6*), ultimately achieving myeloma cytostasis. Such findings indicate that ACHP's growth inhibitory effects may be mediated through inhibition of both IKKα and IKKβ. By inhibiting NF-κB's DNA binding activity, ACHP blocks the NF-κB pathway in multiple myeloma cell lines, inducing growth arrest and apoptosis. **1.** Sanda, Iida, Ogura, *et al.* (2005) *Clin. Cancer Res.* **11**, 1974.

Acitretin

This orally active, synthetic retinoid (FW = 326.43 g/mol; CAS 55079-83-9), systematically named as (2*E*,4*E*,6*E*,8*E*)-9-(4-methoxy-2,3,6-trimethyl-phenyl)-3,7-dimethylnona-2,4,6,8-tetraenoic acid, inhibits calcineurin, weakly (1), retinol dehydrogenase (2), retinyl-palmitate esterase (3), and ribonuclease P (4,5). Acitretin is used to treat psoriasis by binding to nuclear receptors regulating gene transcription and inducing keratinocyte

differentiation, thereby reducing epidermal hyperplasia, 1. Spannaus-Martin & Martin (2000) *Biochem. Pharmacol.* **60**, 803. **2.** Karlsson, Vahlquist, Kedishvili & Törmä (2003) *Biochem. Biophys. Res. Commun.* **303**, 273. **3.** Ritter & Smith (1996) *Biochim. Biophys. Acta* **1291**, 228. **4.** Papadimou, Georgiou, Tsambaos & Drainas (1998) *J. Biol. Chem.* **273**, 24375. **5.** Papadimou, Pavlidou, Séraphin, Tsambaos & Drainas (2003) *Biol. Chem.* **384**, 457.

Acivicin

This cytotoxic isoxazole and copper chelator (FW = 178.57 g/mol; CAS 42228-92-2; Source: *Streptomyces sviceus*), also known as L-(α*S*,5*S*)-α-amino-3-chloro-4,5-dihydro-5-isoxazoleacetic acid, AT-125, and NSC-165301, strongly inhibits γ-glutamyl transpeptidase (1-5,9,10,18,19,26-30). Its clinical application in cancer treatment failed as a consequence of unacceptable toxicity, and the cause(s) of the desired and undesired biological effects have never been elucidated and only limited information about acivicin-specific targets is available (32). Target deconvolution by quantitative mass spectrometry (MS) has now revealed acivicin's preference for the specific aldehyde dehydrogenase known as ALDH4A1 by binding to the catalytic site (32). Moreover, siRNA-mediated downregulation of ALDH4A1 results in a severe inhibition of cell growth, a finding that may explain acivicin's cytotoxicity (32). **Targets:** Acivicin is thought to be a glutamine analogue, an assumption that is amply supported by its ability to inhibit the following enzymes that possess essential amidohydrolase activities that generate nascent ammonia: asparagine synthetase (6,24); carbamoyl-phosphate synthetase (7,8,12,13,23); anthranilate synthase (11); glutamate synthase (11); CTP symthetase (13,15,25); amidophospho-ribosyltransferase (13); glutamin-(asparagin)ase, *or* glutaminase-asparaginase (14); GMP synthase, glutamine-dependent (15,20,22); phosphoribosyl-formylglycinamidine synthetase (formylglycinamidine ribonucleotide synthetase (16); thiol oxidase (17,31); γ-glutamyl hydrolase (21); imidazole-glycerol phosphate synthetase (22). **1.** Buchanan (1982) *Meth. Enzymol.* **87**, 76. **2.** Griffith & Meister (1980) *Proc. Natl. Acad. Sci. U.S.A.* **77**, 3384. **3.** Allison (1985) *Meth. Enzymol.* **113**, 419. **4.** Meister, Tate & (1981) *Meth. Enzymol.* **77**, 237. **5.** Schasteen, Curthoys & Reed (1983) *Biochem. Biophys. Res. Commun.* **112**, 564. **6.** Pinkus (1977) *Meth. Enzymol.* **46**, 414. **7.** Casey & Anderson (1983) *J. Biol. Chem.* **258**, 8723. **8.** Miles, Thoden, Holden & Raushel (2002) *J. Biol. Chem.* **277**, 4368. **9.** Allen, Meck & Yunis (1980) *Res. Commun. Chem. Pathol. Pharmacol.* **27**, 175. **10.** Reed, Ellis & Meck (1980) *Biochem. Biophys. Res. Commun.* **94**, 1273. **11.** Tso, Bower & Zalkin (1980) *J. Biol. Chem.* **255**, 6734. **12.** Aoki, Sebolt & Weber (1982) *Biochem. Pharmacol.* **31**, 927. **13.** Denton, Lui, Aoki, Sebolt & Weber (1982) *Life Sci.* **30**, 1073. **14.** Steckel, Roberts, Philips & Chou (1983) *Biochem. Pharmacol.* **32**, 971. **15.** Achleitner, Lui & Weber (1985) *Adv. Enzyme Regul.* **24**, 225. **16.** Elliott & Weber (1985) *Biochem. Pharmacol.* **34**, 243. **17.** Lash & Jones (1986) *Arch. Biochem. Biophys.* **247**, 120. **18.** Stole, Smith, Manning & Meister (1994) *J. Biol. Chem.* **269**, 21435. **19.** Smith, Ikeda, Fujii, Taniguchi & Meister (1995) *Proc. Natl. Acad. Sci. U.S.A.* **92**, 2360. **20.** Nakamura, Straub, Wu & Lou (1995) *J. Biol. Chem.* **270**, 23450. **21.** Waltham, Li, Gritsman, Tong & Bertino (1997) *Mol. Pharmacol.* **51**, 825. **22.** Chittur, Klem, Shafer & Davisson (2001) *Biochemistry* **40**, 876. **23.** Aoki & Oya (1987) *Mol. Biochem. Parasitol.* **23**, 173. **24.** Milman & Cooney (1979) *Biochem. J.* **181**, 51. **25.** Hofer, Steverding, Chabes, Brun & Thelander (2001) *Proc. Natl. Acad. Sci. U.S.A.* **98**, 6412. **26.** Kwiecien, Rokita, Lorenc-Koci, Sokolowska & Wlodek (2007) *Fundam. Clin. Pharmacol.* **21**, 95. **27.** Suzuki, Kumagai & Tochikura (1986) *J. Bacteriol.* **168**, 1325. **28.** Minami, Suzuki & Kumagai (2003) *Enzyme Microb. Technol.* **32**, 431. **29.** Nakano, Okawa, Yamauchi, Koizumi & Sekiya (2006) *Biosci. Biotechnol. Biochem.* **70**, 369. **30.** Hussein & Walter (1996) *Mol. Biochem. Parasitol.* **77**, 41. **31.** Lash, Dones & Orrenius (1984) *Biochim. Biophys. Acta* **779**, 191. **32.** Kreuzer, Bach, Forler & Sieber (2014) *Chem. Sci.* **6**, 237.

Aclacinomycin A

This non-peptidic aclacinomycin antibiotic (FW = 811.88 g/mol; CAS CAS 57576-44-0; *Source*: strain of *Streptomyces galilaeus*), also known as aclarubicin, induces DNA strand scission. **Target(s):** nitric oxide synthase (1); RNA biosynthesis (2); DNA polymerase I (3,5); RNA polymerase, *Escherichia coli* (3); reverse transcriptase, avian myeloblastosis virus (3); Na$^+$/K$^+$-exchanging ATPase (4); Ca^{2+}-transporting ATPase (4); cyclic-nucleotide phosphodiesterase (6); electron transport and oxidative phosphorylation, mitochondrial (7); DNA helicase (8); DNA topoisomerase II (9,13-15); 20S proteasome, chymotrypsin-like activity (10); DNA topoisomerase I (11); 3'-5' DNA helicase, *Plasmodium falciparum* (12). **1.** Luo & Vincent (1994) *Biochem. Pharmacol.* **47**, 2111. **2.** DuVernay & Crooke (1980) *J. Antibiot.* **33**, 1048. **3.** Komiyama, Oki & Inui (1983) *Biochim. Biophys. Acta* **740**, 80. **4.** Kitao & Hattori (1983) *Experientia* **39**, 1362. **5.** Fritzsche, Wahnert, Chaires, Dattagupta, Schlessinger & Crothers (1987) *Biochemistry* **26**, 1996. **6.** Nwankwoala & West (1988) *Clin. Exp. Pharmacol. Physiol.* **15**, 805. **7.** Shinozawa, Gomita & Araki (1991) *Physiol. Chem. Phys. Med. NMR* **23**, 101. **8.** Bachur, Yu, Johnson, Hickey, Wu & Malkas (1992) *Mol. Pharmacol.* **41**, 993. **9.** Holm, Jensen, Sehested & Hansen (1994) *Cancer Chemother. Pharmacol.* **34**, 503. **10.** Figueiredo-Pereira, Chen, Li & Johdo (1996) *J. Biol. Chem.* **271**, 16455. **11.** Nitiss, Pourquier & Pommier (1997) *Cancer Res.* **57**, 4564. **12.** Suntornthiticharoen, Petmitr & Chavalitshewinkoon-Petmitr (2006) *Parasitology*, **133**, 389. **13.** Andersen, Bendixen & Westergaard (1996) *DNA Replication in Eucaryotic Cells, Cold Spring Harbor Laboratory Press*, p. 587. **14.** Insaf, Danks & Witiak (1996) *Curr. Med. Chem.* **3**, 437. **15.** Christmann-Franck, Bertrand, Goupil-Lamy, *et al.* (2004) *J. Med. Chem.* **47**, 6840.

cis-Aconitate (*cis*-Aconitic Acid)

This tricarboxylic acid (FW$_{free-acid}$ = 174.11 g/mol; CAS 585-84-2), systematically referred to as (Z)-prop-1-ene-1,2,3-tricarboxylic acid, is an intermediate in the reversible conversion of citrate to isocitrate: the enzyme aconitate hydratase catalyzes the interconversion of citrate and *cis*-aconitate as well as *cis*-aconitate with isocitrate. *cis*-Aconitate is also a product of the reactions catalyzed by citrate dehydratase and aconitate Δ-isomerase. It is a substrate for aconitate decarboxylase and an alternative substrate for *trans*-aconitate 2-methyltransferase and *trans*-aconitate 3-methyltransferase. **Target(s):** phosphofructokinase (1,2,4); glutamate decarboxylase (2,5,7); isocitrate dehydrogenase, NADP$^+$-dependent (3,11); 3-carboxy-*cis,cis*-muconate cyclase (6); chorismate mutase/prephenate dehydratase (8,10); α-ketoglutarate dehydrogenase (9); isocitrate lyase (12); phosphoglucomutase, weakly inhibited (13); prephenate dehydratase (8); fumarase, *or* fumarate hydratase (15); glucose-1,6-bisphosphate synthase (16). **Note:** Aqueous solutions are relatively stable between pH 7 and 10 at temperatures up to 90°C; however, isomerization to *trans*-aconitate occurs at pH values outside of that range (69% *cis* aconitate is recovered after 96 hours at pH 3 and 25°C [43.7% after 72 hours at pH 3 and 65°C] and 73.6 % is recovered after 96 hours at pH 13 and 25°C [48.5% after 72 hours at pH 13 and 65°C] (14). **1.** Bloxham & Lardy (1973) *The Enzymes*, 3rd ed., **8**, 239. **2.** Webb (1966) *Enzyme and Metabolic Inhibitors*, vol. **2**, Academic Press, New York. **3.** Ochoa & Weisz-Tabori (1948) *J. Biol. Chem.* **174**, 123. **4.** Passonneau & Lowry (1963) *Biochem. Biophys. Res. Commun.* **13**, 372. **5.** Gerig & Kwock (1979) *FEBS Lett.* **105**, 155. **6.**

Thatcher & Cain (1975) *Eur. J. Biochem.* **56**, 193. **7.** Endo & Kitahara (1977) *Hoppe-Seyler's Z. Physiol. Chem.* **358**, 1365. **8.** Baldwin & Davidson (1983) *Biochim. Biophys. Acta* **742**, 374. **9.** Meixner-Monori, Kubicek, Habison, Kubicek-Pranz & Rohr (1985) *J. Bacteriol.* **161**, 265. **10.** Ma & Davidson (1985) *Biochim. Biophys. Acta* **827**, 1. **11.** Shikata, Ozaki, Kawai, Ito & Okamoto (1988) *Biochim. Biophys. Acta* **952**, 282. **12.** Ko, Vanni & McFadden (1989) *Arch. Biochem. Biophys.* **274**, 155. **13.** Popova, Matasova & Lapot'ko (1998) *Biochemistry (Moscow)* **63**, 697. **14.** Ambler & Roberts (1948) *J. Org. Chem.* **13**, 399. **15.** Flint (1994) *Arch. Biochem. Biophys.* **311**, 509. **16.** Rose, Warms & Wong (1977) *J. Biol. Chem.* **252**, 4262.

trans-Aconitate (*trans*-Aconitic Acid)

This tricarboxylic acid (FW$_{free-acid}$ = 174.11 g/mol; CAS 585-84-2), named systematically as (E)-prop-1-ene-1,2,3-tricarboxylic acid, is the geometric isomer of the intermediate in the reversible conversion of citrate to isocitrate. *trans*-Aconitate occurs naturally and is a substrate for aconitate Δ-isomerase, *trans*-aconitate 2-methyltransferase, and *trans*-aconitate 3-methyltransferase. **Target(s):** aconitase, *or cis*-aconitate hydratase (1,4-8,10,13,14,18,19); fumarase, *or* fumarate hydratase (2,3,9,10,15,20); glutamate decarboxylase (10,11); chorismate mutase/prephenate dehydratase (12); carboxy-*cis,cis*-muconate cyclase (16); phosphoenolpyruvate mutase (17); prephenate dehydratase (12); aconitate decarboxylase (21); β-fructofuranosidase, *or* invertase (22); glucose-1,6 bisphosphate synthase (23). **1.** Anfinsen (1955) *Meth. Enzymol.* **1**, 695. **2.** Hill & Bradshaw (1969) *Meth. Enzymol.* **13**, 91. **3.** Massey (1953) *Biochem. J.* **55**, 172. **4.** Ochoa (1951) *The Enzymes*, 1st ed., **1** (part 2), 1217. **5.** Saffran & Prado (1949) *J. Biol. Chem.* **180**, 1301. **6.** Dickman (1961) *The Enzymes*, 2nd ed., **5**, 495. **7.** Morrison & Still (1947) *Aust. J. Sci.* **9**, 150. **8.** Glusker (1971) *The Enzymes*, 3rd ed., **5**, 413. **9.** Hill & Teipel (1971) *The Enzymes*, 3rd. ed., **5**, 539. **10.** Webb (1966) *Enzyme and Metabolic Inhibitors*, vol. **2**, Academic Press, New York. **11.** Endo & Kitahara (1977) *Hoppe-Seyler's Z. Physiol. Chem.* **358**, 1365. **12.** Baldwin & Davidson (1983) *Biochim. Biophys. Acta* **742**, 374. **13.** Lauble, Kennedy, Beinert & Stout (1992) *Biochemistry* **31**, 2735. **14.** Lauble, Kennedy, Beinert & Stout (1994) *J. Mol. Biol.* **237**, 437. **15.** Rebholz & Northrop (1994) *Arch. Biochem. Biophys.* **312**, 227. **16.** Thatcher & Cain (1975) *Eur. J. Biochem.* **56**, 193. **17.** Seidel & Knowles (1994) *Biochemistry* **33**, 5641. **18.** Eprintsev, Semenova & Popov (2002) *Biochemistry (Moscow)* **67**, 795. **19.** Uhrigshardt, Walden, John & Anemuller (2001) *Eur. J. Biochem.* **268**, 1760. **20.** Flint (1994) *Arch. Biochem. Biophys.* **311**, 509. **21.** Bentley & Thiessen (1957) *J. Biol. Chem.* **226**, 703. **22.** Prado, Vattuone, Fleischmacher & Sampietro (1985) *J. Biol. Chem.* **260**, 4952. **23.** Rose, Warms & Wong (1977) *J. Biol. Chem.* **252**, 4262.

Aconitine

This violently poisonous alkaloid (FW$_{free-base}$ = 645.75 g/mol; CAS 302-27-2), obtained from *Aconitum napellus* (wolf's bane or monk's hood) is a neurotoxin that activates tetrodotoxin-sensitive sodium channels (1), inducing presynaptic depolarization and blocking nerve action potential. In addition, aconitine blocks norepinephrine reuptake (K_i = 230 nM) and inhibits pig heart aconitase (K_i = 0.11 mM). In humans, the lethal dose can be <2 mg. In ancient times, aconitine preparations were used as arrow poisons. **Target(s):** aconitase (2); neurotransmitter release (3); norepinephrine reuptake (4). **1.** Conn (1983) *Meth. Enzymol.* **103**, 401. **2.**

Hernanz & Silio (1983) *Comp. Biochem. Physiol. C* **76**, 335. **3**. Onur, Bozdagi & Ayata (1995) *Neuropharmacology* **34**, 1139. **4**. Seitz & Ameri (1998) *Biochem. Pharmacol.* **55**, 883.

ACPA

This highly selective, synthetic CB_1 receptor agonist (FW = 343.55 g/mol; CAS 229021-64-1), also known as arachidonylcyclopropylamide and *N*-(cyclopropyl)-5*Z*,8*Z*,11*Z*,14*Z*-eicosatetraenamide, targets CB_1 cannabinoid receptor agonist (K_i = 2.2 nM) with >325-fold selectivity over CB_2 receptors. Two subtypes of the cannabinoid receptor (CB_1 and CB_2) are expressed in mammalian tissues. Although selective antagonists are available for each of the subtypes, most of the available cannabinoid agonists bind to both CB_1 and CB_2 with similar affinities (1). ACPA possesses the characteristics of CB_1 receptor agonists, inhibiting forskolin-induced cAMP accumulation in Chinese hamster ovary cells expressing the human CB_1 receptor. It also increases the binding of [^{35}S]GTPγS to cerebellar membranes and inhibits electrically evoked contractions of the mouse *vas deferens* (1). **See also** *ACEA* 1. Hillard, Manna, Greenberg, *et al.* (1999) *J. Pharmacol. Exp. Ther.* **289**, 1427. **2**. Patel & Hillard (2001) *J. Pharmacol. Exp. Ther.* **297**, 629.

Acridine Orange

This cell-permeable dye (FW$_{free-base}$ = 265.36 g/mol; CAS 65-61-2), also known as *N,N,N',N'*-tetramethylacridine-3,6-diamine, is an intercalator that binds tightly to nucleic acids. Binding to DNA is characterized by excitation maximum at 502 nm (cyan) and an emission maximum at 525 nm, producing a green fluorescence. (A 120 µM solution of acridine orange will detect 25-50 ng DNA per band in an agarose or polyacrylamide gel.) Binding to RNA is characterized by an excitation maximum at 460 nm and the emission maximum is located at 650 nm, yielding a reddish-orange fluorescence. Acridine orange also penetrates acidic compartments (*e.g.,* lysosomes and phagolysosomes); upon protonation, the dye emits orange light when excited by blue light. **1**. Ranadive & Korgaonkar (1960) *Biochim. Biophys. Acta* **39**, 547. **2**. Boyle, Nelson, Dollish & Olsen (1962) *Arch. Biochem. Biophys.* **96**, 47.

Acrivastine

This oral, nonsedating antihistamine (FW$_{free-acid}$ = 348.44 g/mol; CAS 87848-99-5), also named Semprex and (*E*)-3-{6-[(*E*)-1-(4-methylphenyl)-3-pyrrolidin-1-yl-prop-1-enyl]-pyridin-2-yl}prop-2-enoic acid, is a histamine H_1 receptor antagonist, K_i = 0.63 nM. **1**. Hill, Ganellin, Timmerman, *et al.* (1997) *Pharmacol. Rev.* **49**, 253.

Acrolein

This α,β-unsaturated aldehyde and thiol-reactive reagent (FW = 56.06 g/mol; CAS 107-02-8; M.P. = –88°C; B.P. = 52.5°C), also known as propenal, is an unstable liquid that polymerizes readily in the presence of light, alkali, or strong acid. Acrolein is soluble in water and ethanol. It is highly toxic, reactive, and irritating aldehyde that is often a product of organic pyrolysis (it has been identified in automobile exhaust, cigarette smoke, and burnt eggs). Acrolein has also been used as a herbicide. In addition, acrolein concentrations are reportedly elevated in the brains of individuals with Alzheimer's disease. Note also that acrolein is a reactive metabolite of both cyclophosphamide and allylamine. Acrolein is an irreversible active-site inhibitor of deoxyribose 5-phosphate aldolase. Treatment of carbonic anhydrase with acrolein results predominately in the alkylation of a single histidyl residue, which is implicated in the transfer of a proton from zinc-bound water to buffer in solution. **Target(s):** 21-hydroxysteroid dehydrogenase, NAD$^+$-dependent (1,9); carboxylase (2,3); aspartate aminotransferase (4); lecithin:cholesterol acyltransferase, *or* LCAT (5,25); succinate-semialdehyde dehydrogenase (6); glutathione *S*-transferase (7,18,21,27); deoxyribose-5-phosphate aldolase (8); carbonic anhydrase, *or* carbonate dehydratase (10,11,28); asparaginase12; NADPH:cytochrome *c* reductase (13,17); aldehyde dehydrogenase (14,19); *O*6-methylguanine-DNA methyl-transferase (15); 15 hydroxyprostaglandin dehydrogenase, NAD$^+$-dependent (16); DNA methyltransferase (20); electron transport (22); glucose-6-phosphate dehydrogenase (23); DNA polymerase α (24); DNA-directed DNA polymerase (24); glutathione-disulfide reductase (26); alcohol dehydrogenase (29); protein phosphatase type 1 (30); pyruvate dehydrogenase complex (31); α-ketoglutarate dehydrogenase complex (31); protein-disulfide isomerase (32); phospholipid-translocating ATPase, *or* flippase (33); glycerol-2 phosphatase (34). **1**. Monder & Furfine (1969) *Meth. Enzymol.* **15**, 667. **2**. Karrer & Visconti (1947) *Helv. Chim. Acta* **30**, 268. **3**. Massart (1950) *The Enzymes*, 1st ed., **1** (part 1), 307. **4**. Southwell, Yeargans, Kowalewski & Seidler (2002) *J. Enzyme Inhib. Med. Chem.* **17**, 19. **5**. Chen & Loo (1995) *J. Biochem. Toxicol.* **10**, 121. **6**. Nguyen & Picklo, Sr. (2003) *Biochim. Biophys. Acta* **1637**, 107. **7**. van Iersel, Ploemen, Lo Bello, Federici & van Bladeren (1997) *Chem. Biol. Interact.* **108**, 67. **8**. Wilton (1976) *Biochem. J.* **153**, 495. **9**. Monder & White (1962) *Biochem. Biophys. Res. Commun.* **8**, 383. **10**. Tu, Wynns & Silverman (1989) *J. Biol. Chem.* **264**, 12389. **11**. Khalifah & Silverman (1991) in *The Carbonic Anhydrases* (Dodgson, Tashian, Gros & Carter, eds), pp. 49-70, Plenum Publ., New York. **12**. Bilimoria & Nisbet (1971) *Proc. Soc. Exp. Biol. Med.* **136**, 698. **13**. Patel, Ortiz, Kolmstetter & Leibman (1984) *Drug Metab. Dispos.* **12**, 460. **14**. Nagasawa, Elberling & DeMaster (1984) *J. Med. Chem.* **27**, 1335. **15**. Krokan, Grafstrom, Sundqvist, Esterbauer & Harris (1985) *Carcinogenesis* **6**, 1755. **16**. Liu & Tai (1985) *Biochem. Pharmacol.* **34**, 4275. **17**. Cooper, Witmer & Witz (1987) *Biochem. Pharmacol.* **36**, 627. **18**. Ansari, Singh, Gan & Awasthi (1987) *Toxicol. Lett.* **37**, 57. **19**. Mitchell & Petersen (1988) *Drug Metab. Dispos.* **16**, 37. **20**. Cox, Goorha & Irving (1988) *Carcinogenesis* **9**, 463. **21**. Scott & Kirsch (1988) *Biochem. Int.* **16**, 439. **22**. Biagini, Toraason, Lynch & Winston (1990) *Toxicology* **62**, 95. **23**. Trieff, Ficklen & Gan (1993) *Toxicol. Lett.* **69**, 121. **24**. Catalano & Kuchta (1995) *Biochem. Biophys. Res. Commun.* **214**, 971. **25**. McCall, Tang, Bielicki & Forte (1995) *Arterioscler. Thromb. Vasc. Biol.* **15**, 1599. **26**. Vander Jagt, Hunsaker, Vander Jagt, *et al.* (1997) *Biochem. Pharmacol.* **53**, 1133. **27**. van Iersel, Ploemen, Lo Bello, Federici & van Bladeren (1997) *Chem. Biol. Interact.* **108**, 67. **28**. Earnhardt, Qian, Tu, *et al.* (1998) *Biochemistry* **37**, 1083. **29**. Trivic & Leskovac (1999) *Biochem. Mol. Biol. Int.* **47**, 1. **30**. Witt, Schultz, Dolker & Eger (2000) *Bioorg. Med. Chem.* **8**, 807. **31**. Pocernich & Butterfield (2003) *Neurotox. Res.* **5**, 515. **32**. Liu & Sok (2004) *Biol. Chem.* **385**, 633. **33**. Castegna, Lauderback, Mohmmad-Abdul & Butterfield (2004) *Brain Res.* **1004**, 193. **34**. Sexton, Cronshaw & Hall (1971) *Protoplasma* **73**, 417.

Acrylamide

This common laboratory reagent and neurotoxin (FW = 71.079 g/mol; CAS 79-06-1) reacts with sulfhydryl and amino groups. Acrylamide readily polymerizes in the presence of UV light or free radicals (such as those generated by ammonium persulfate) or when heated at the melting point (84.5°C). When copolymerized with a bifunctional cross-linker, acrylamide polymerization forms a well-hydrated gel suitable for many forms of protein and nucleic acid electrophoresis. It is conveniently recrystallized by dissolving in hot chloroform and cooling ovenight. **Caution:** Acrylamide is absorbed through unbroken skin, and the powdery solid is easily inhaled. Extreme care must be exercised when working with this reagent. **Target(s):** creatine kinase (1,2,12); cytochrome P450 C-2 (13); mitochondrial respiration (4); protein-glutamine γ-glutamyltransferase, *or* transglutaminase (5); enolase (6,9); glyceraldehyde-3 phosphate dehydrogenase (6-8); alcohol dehydrogenase (10); cytochrome b_2 (11). **1.** Meng, Zhou & Zhou (2001) *Int. J. Biochem. Cell Biol.* **33**, 1064. **2.** Matsuoka, Igisu, Lin & Inoue (1990) *Brain Res.* **507**, 351. **3.** Narasimhulu (1991) *Biochemistry* **30**, 9319. **4.** Medrano & LoPachin (1989) *Neurotoxicology* **10**, 249. **5.** Signorini, Dallocchio & Bergamini (1988) *Biochim. Biophys. Acta* **957**, 168. **6.** Sakamoto & Hashimoto (1985) *Arch. Toxicol.* **57**, 276. **7.** Ross, Sabri & Spencer (1985) *Brain Res.* **340**, 189. **8.** Vyas, Lowndes & Howland (1985) *Neurotoxicology* **6**, 123. **9.** Tanii & Hashimoto (1984) *Experientia* **40**, 971. **10.** Dixit, Mukhtar & Seth (1981) *Toxicol. Lett.* **7**, 487. **11.** Kulis, Shvirmitskas, Antanavichius & Vaitkiavichius (1982) *Biokhimiia* **47**, 582. **12.** Kohriyama, Matsuoka & Igisu (1994) *Arch. Toxicol.* **68**, 67.

ACT 335827

This brain-penetrant and orally available orexin receptor antagonist (FW = 518.64 g/mol; CAS 1354039-86-3; Soluble to 100 mM in DMSO, also known as (a*R*,1*S*)-1-[(3,4-Dimethoxyphenyl)methyl]-3,4-dihydro-6,7-dimethoxy-*N*-(1-methylethyl)-a-phenyl-2(1*H*)-isoquinoline acetamide, elicits its anxiolytic effects *in vivo* by targeting OX_1 (K_i = 41 nM) and OX_2 (K_i = 560 nM) receptors. ACT-335827 decreases fear, compulsive behaviors, and autonomic stress reactions in rats. **1.** Steiner, Gatfield, Brisbare-Roch, *et al.* (2013) *Chem. Med. Chem.* **8**, 898.

ACT-064992, *See Macitentan*

ACT-129968, *See Setipiprant*

ActA Fragment 269-278 or 304-313

This decapeptide FEFPPPPTDE (MW = 1175.26 g/mol), corresponding the second oligoproline repeat in *Listeria monocytogenese* ActA surface protein, blocks actin-based locomotion *in vitro* or when micro-injected into *Listeria*-infected or *Shigella*-infected host cells, typically at submicromolar concentrations. FEFPPPPTDE inhibits actin-based motility by disrupting the binding of vasodilator stimulated phosphoprotein (VASP) and other members of the Ena/VASP superfamily from binding to ActA on the surface of motile *Listeria* or proteolytically cleaved vinculin in the case of *Shigella flexnerii*. **See also ABM-1 & ABM-2 Sequences in Actin-Based Motors Target(s):** *Listeria* locomotion (1); vasodilator-stimulated phosphoprotein (VASP) binding to *Listeria* ActA, vinculin p90, and zyxin (1,2); *Shigella* locomotion (1). **1.** Zeile, Purich & Southwick (1996) *J. Cell Biol.* **133**, 49. **2.** Purich & F. Southwick (1997) *Biochem. Biophys. Res. Commun.* **231**, 686.

Acteoside

This plant glycoside and antioxidant (FW = 624.60 g/mol; CAS 61276-17-3), also known as verbascoside, inhibits TNFα-mediated apoptosis in LPS-stimulated macrophages (1) as well as 1-methyl-4-phenylpyridinium-induced apoptosis in cerebellar granule neurons (2). It also inhibits proliferation of human promyelocytic HL-60 leukemia cells by inducing cell cycle arrest at the G_0/G_1 phase, followed by differentiation into

monocytes (3). Acteoside inhibits PMA-induced invasion and migration of human fibrosarcoma cells by Ca^{2+}-dependent CaMK/ERK and JNK/NF-κB-signaling pathways. Acteoside thus shows the potential as antimetastatic agent (4). Another intriguing finding is that acteoside inhibits amyloid-β aggregation (5). **Target(s):** aldose reductase (6); protein kinase C (7); integrase, HIV-1 (8); peptidyl-dipeptidase A, *or* angiotensin-converting enzyme (9,10); tyrosinase, *or* monophenol monooxygenase (11); 5-lipoxygenase, *or* arachidonate 5-lipoxygenase (12,13). **1.** Xiong, Hase, Tezuka, Namba & Kadota (1999) *Life Sci.* **65**, 421. **2.** Pu, Song, Li, Tu, & Li (2003) *Planta Med.* **69**, 65. **3.** Lee, Kim, Lee, *et al.* (2007) *Carcinogenesis* **28**, 1928. **4.** Hwang, Kim, Choi, *et al.* (2011) *Mol. Nutr. Food Res.* **55**, Supplement-1, S103. **5.** Kurisu M, Miyamae Y, Murakami, *et al.* (2013) *Biosci. Biotechnol. Biochem.* **77**, 1329. **6.** Kohda, Tanaka, Yamaoka, *et al.* (1989) *Chem. Pharm. Bull. (Tokyo)* **37**, 3153. **7.** Herbert, Maffrand, Taoubi, *et al.* (1991) *J. Nat. Prod.* **54**, 1595. **8.** Kong, Wolfender, Cheng, Hostettmann & Tan (1999) *Planta Med.* **65**, 744. **9.** Oh, Kang, Kwon, *et al.* (2003) *Phytother. Res.* **17**, 811. **10.** Kang, Lee, Kim, Lee & Lee (2003) *J. Ethnopharmacol.* **89**, 151. **11.** Karioti, Protopappa, Magoulas & Skaltsa (2007) *Bioorg. Med. Chem.* **15**, 2708. **12.** Kimura, Okuda, Nishibe & Arichi (1987) *Planta Med.* **53**, 148. **13.** Schneider & Bucar (2005) *Phytother. Res.* **19**, 81.

Actinomycin D

This widely used member (FW = 1255.44 g/mol; CAS 50-76-0; where Sar = sarcosine; mV = *N*-methyl-valine) of the actinomycin family of structurally related antibiotics was isolated from *Streptomyces* strains by Selman Waksman, who was awarded the 1952 Nobel Prize in Physiology or Medicine for his codiscovery of streptomycin. All actinomycins share a substituted phenoxazine ring and contain two cyclic heterodetic peptides differing only with respect to side-chain groups. Also known as actinomycin C1, cosmegen, and dactinomycin, actinomycin D is a potent transcription inhibitor in eukaryotes, prokaryotes, and chloroplasts, acting by intercalating into duplex DNA (reportedly at GpC sequences), blocking the latter's ability to serve as a template for DNA-directed RNA polymerases. Use of actinomycin D actually assisted in the discovery of messenger RNA. Actinomycin D does not greatly alter DNA replication and does not bind to RNA or single-stranded DNA and will inhibit the replication of DNA viruses. DNA polymerase can be inhibited at elevated concentrations. Despite its low therapeutic index, actinomycin D is used as an antineoplastic agent (under the generic name Dactomycin) to treat various sarcomas, carcinomas, and adenocarcinomas. In 1943, it became the first antibiotic discovered to have antitumor activity. **Properties:** Unstable in dilute base or bright light, actinomycins form bright red-colored crystals that are highly soluble in benzene and chloroform but only slightly soluble in water and methanol. The UV/visible spectrum of actinomycin D exhibits λmax values (in methanol) at 244 and 441 nm. **Target(s):** DNA-directed RNA polymerase (1-4,12,15,16,27,28); viral RNA polymerase (5,14); deoxyribonuclease I (6); DNA-directed DNA polymerase (7,18); RNA-directed DNA polymerase (8); deoxyribonuclease II (spleen acid deoxyribonuclease)10; DNA methyltransferases (11); *Eco*RI (13); *Eco*RII restriction endonuclease (13); *Hin*DII restriction endonuclease (13); *Hin*DIII restriction endonuclease (13); *Hpa*I restriction endonuclease)13; *Hpa*II restriction endonuclease (13); exodeoxyribonuclease I (17); DNA ligase (ATP), Vaccinia virus (19); DNA topoisomerase I (20-22); DNA topoisomerase II (23,24); P-glycoprotein (25); xenobiotic-transporting ATPase, *or* multidrug-resistance protein (25); thermitase, inhibited by actinomycins D and SG3 (26); subtilisin (26); peptidase K, *or* proteinase K (26); tryptophan 2,3-dioxygenase, inhibited by actinomycin D (29). **1.** Hurwitz (1963) *Meth. Enzymol.* **6**, 23. **2.** Weiss (1968) *Meth. Enzymol.* **12B**, 555. **3.** Krakow & Horsley (1968) *Meth. Enzymol.* **12B**, 566. **4.** Scott & Tomkins (1975) *Meth. Enzymol.* **40**, 273. **5.** Gershon & Moss (1996) *Meth. Enzymol.* **275**, 208. **6.** Eron & McAuslan (1966) *Biochim. Biophys. Acta* **114**, 633. **7.** Honikel, Sippel & Hartmann (1968) *Z. physiol. Chem.* **349**, 957. **8.** Gurgo,

Ray, Thiry & Green (1971) *Nature New Biol.* **229**, 111. **9**. Baltimore (1970) *Nature* **226**, 1209. **10**. Bernardi (1971) *The Enzymes*, 3rd ed., **4**, 271. **11**. Kerr & Borek (1973) *The Enzymes*, 3rd ed., **9**, 167. **12**. Chamberlin (1974) *The Enzymes*, 3rd ed., **10**, 333. **13**. Wells, Klein & Singleton (1981) *The Enzymes*, 3rd ed., **14**, 157. **14**. Ho & Walters (1966) *Biochemistry* **5**, 231. **15**. Kersten, Kersten, Szybalski & Fiandt (1966) *Biochemistry* **5**, 236. **16**. Harbers & Müller (1962) *Biochem. Biophys. Res. Commun.* **7**, 107. **17**. Blakesley, Dodgson, Nes & Wells (1977) *J. Biol. Chem.* **252**, 7300. **18**. Müller, Yamazaki & Zahn (1972) *Enzymologia* **43**, 1. **19**. Shuman (1995) *Biochemistry* **34**, 16138. **20**. Alkorta, Park, Kong, *et al.* (1999) *Arch. Biochem. Biophys.* **362**, 123. **21**. Kohsaka (1989) *Agric. Biol. Chem.* **53**, 3357. **22**. Vosberg (1985) *Curr. Top. Microbiol. Immunol.* **114**, 19. **23**. Insaf, Danks & Witiak (1996) *Curr. Med. Chem.* **3**, 437. **24**. Saijo, Enomoto, Hanaoka & Ui (1990) *Biochemistry* **29**, 583. **25**. Shapiro & Ling (1997) *Eur. J. Biochem.* **250**, 130. **26**. Betzel, Rachev, Dolashka & Genov (1993) *Biochim. Biophys. Acta* **1161**, 47. **27**. Sethi (1971) *Prog. Biophys. Mol. Biol.* **23**, 67. **28**. Allan & Kropinski (1987) *Biochem. Cell Biol.* **65**, 776. **29**. Hitchcock & Katz (1988) *Arch. Biochem. Biophys.* **261**, 148.

Actinonin

This antibiotic (FW = 385.50 g/mol; CAS 13434-13-4), also named (2*R*)-*N*⁴-hydroxy-*N*¹-{(2*S*)-1-[(2*S*)-2-(hydroxymethyl)pyrrolidin-1-yl]-3-methyl-1-oxobutan-2-yl}-2-pentylbutanediamide, is an actinomycete metabolite that potently inhibits peptide deformylase as well as other peptidases. **Target(s):** leucyl aminopeptidase, *Leishmania* sp. (8); membrane alanyl aminopeptidase, *or* aminopeptidase N (1,2,27-32); meprin A, K_i = 0.135 μM (3,4,11); astacin, weakly inhibited (4,11,24); acyl amidase-like leucyl aminopeptidase (5); neprilysin (6); neutrophil leucyl aminopeptidase, weakly inhibited (7); collagenase (9,10); interstitial collagenase, *or* fibroblast collagenase (12); stromelysin 1 (12); peptide deformylase (13,15-22); gelatinase A (14); ADAMTS-4 endopeptidase, *or* aggrecanase (23); meprin B (24); meprin A (11,24); cytosol alanyl aminopeptidase (25); aminopeptidase B (26). **1**. Umezawa, Aoyagi, Tanaka, *et al.* (1985) *J. Antibiot.* **38**, 1629. **2**. Turner (1998) in *Handbook of Proteolytic Enzymes* (Barrett, Rawlings & Woessner, eds), p. 996, Academic Press, San Diego. **3**. Johnson & Bond (1998) in *Handbook of Proteolytic Enzymes* (Barrett, Rawlings & Woessner, eds), p. 1222, Academic Press, San Diego. **4**. Wolz & Bond (1995) *Meth. Enzymol.* **248**, 325. **5**. Matsuda, Katayama, Hara, *et al.* (2000) *Arch. Androl.* **44**, 1. **6**. Hachisu, Hiranuma, Shibazaki, *et al.* (1987) *Eur. J. Pharmacol.* **137**, 59. **7**. Matsuda, Katsuragi, Saiga, Tanaka & Nakamura (1988) *Biochem. Int.* **16**, 383. **8**. Morty & Morehead (2002) *J. Biol. Chem.* **277**, 26057. **9**. Faucher, Lelievre & Cartwright (1987) *J. Antibiot.* **40**, 1757. **10**. Lelievre, Bouboutou, Boiziau & Cartwright (1989) *Pathol. Biol. (Paris)* **37**, 43. **11**. Wolz (1994) *Arch. Biochem. Biophys.* **310**, 144. **12**. Ho, Qoronfleh, Wahl, *et al.* (1994) *Gene* **146**, 297. **13**. Chen, Patel, Hackbarth, *et al.* (2000) *Biochemistry* **39**, 1256. **14**. Lee, Jang, Kim, *et al.* (2002) *FEBS Lett.* **519**, 147. **15**. Lee, Antczak, Li, *et al.* (2003) *Biochem. Biophys. Res. Commun.* **312**, 309. **16**. Van Aller, Nandigama, Petit, *et al.* (2005) *Biochemistry* **44**, 253. **17**. Apfel, Banner, Bur, *et al.* (2000) *J. Med. Chem.* **43**, 2324. **18**. Dirk, Williams & Houtz (2001) *Plant Physiol.* **127**, 97. **19**. Giglione, Pierre & Meinnel (2000) *Mol. Microbiol.* **36**, 1197. **20**. Serero, Giglione & Meinnel (2001) *J. Mol. Biol.* **314**, 695. **21**. Guilloteau, Mathieu, Giglione, *et al.* (2002) *J. Mol. Biol.* **320**, 951. **22**. Lee, Antczak, Li, *et al.* (2003) *Biochem. Biophys. Res. Commun.* **312**, 309. **23**. Sugimoto, Takahashi, Yamamoto, Shimada & Tanzawa (1999) *J. Biochem.* **126**, 449. **24**. Bertenshaw, Turk, Hubbard, *et al.* (2001) *J. Biol. Chem.* **276**, 13248. **25**. Yamamoto, Li, Huang, Ohkubo & Nishi (1998) *Biol. Chem.* **379**, 711. **26**. Petrov, Fagard & Lijnen (2004) *Cardiovasc. Res.* **61**, 724. **27**. Xu & Li (2005) *Curr. Med. Chem. Anticancer Agents* **5**, 281. **28**. Bauvois & Dauzonne (2006) *Med. Res. Rev.* **26**, 88. **29**. Hua, Tsukamoto, Taguchi, *et al.* (1998) *Biochim. Biophys. Acta* **1383**, 301. **30**. Huang, Takahara, Kinouchi, *et al.* (1997) *J. Biochem.* **122**, 779. **31**. Terashima & Bunnett (1995) *J. Gastroenterol.* **30**, 696. **32**.

Thielitz, Bukowska, Wolke, *et al.* (2004) *Biochem. Biophys. Res. Commun.* **321**, 795.

Aculeacin A

This antifungal antibiotic (FW = 1036.23 g/mol; CAS 58814-86-1) is a cyclopeptide-containing long-chain fatty acid that inhibits cell wall biosynthesis by targetting 1,3-β-D-glucan synthase. **1**. Taft, Zugel & Selitrennikoff (1991) *J. Enzyme Inhib.* **5**, 41. **2**. Beaulieu, Tang, Zeckner & Parr, Jr. (1993) *FEMS Microbiol. Lett.* **108**, 133.

ACV 1

GCCSDPRCNYDHPEIC-NH₂

This neuronal nicotinic acetylcholine receptor antagonist (FW = 1806.98 g/mol; CAS 740980-24-9), also known as Conotoxin Vc1.1, is produced from mRNA in venom duct of *Conus victoriae* (1). ACV 1 selectively targets the $\alpha_9\alpha_{10}$ nAChR subtype (IC₅₀ = 19 nM), with weaker action against $\alpha_6/\alpha_3\beta_2\beta_3$ (IC₅₀ = 140 nM), $\alpha_6/\alpha_3\beta_4$ (IC₅₀ = 980 nM), $\alpha_3\beta_4$ (IC₅₀ = 4200 nM), and $\alpha_3\beta_2$ (IC₅₀ = 7300 nM) nAChR subtypes (2). ACV-1 suppresses vascular responses to unmyelinated sensory nerve C-fiber activation in rats, as related to pain transmission (1). ACV-1 and its posttranslationally modified analogues vc1a, [P6O]Vc1.1, and [E14γ]Vc1.1 are equally potent at inhibiting ACh-evoked currents mediated by $\alpha_9\alpha_{10}$ nAChRs (3). **1**. Sandall, Satkunanathan, Keays, *et al.* (2003) *Biochemistry* **42**, 6904. **2**. Vincler, Wittenauer, Parker, *et al.* (2006) *Proc. Natl. Acad. Sci. USA.* **103**, 17880. **3**. Nevin, Clark, Klimis, *et al.* (2007) *Mol. Pharmacol.* **72**, 1406.

Acyclovir

This guanosine analogue (FW = 225.21 g/mol; CAS 59277-89-3), also known as acycloguanosine and 9-(2 hydroxyethoxymethyl)guanine, is a widely used and highly effective pro-drug for the treatment of herpes infections (1-5). Once transported in cells, where it is metabolically phosphorylated by viral thymidine kinases to the monophosphate and then to to the di- and triphosphate by host cell guanylate kinase and nucleoside 5'-diphosphokinase, acyclovir becomes a potent inhibitor of viral DNA polymerases, with minimal host toxicity. The triphosphate halts viral replication via inhibition of viral DNA polymerases, attended by incorporation and termination of growing viral DNA chains. The selectivity of acyclovir action was investigated by Gertrude Elion and coworkers (Elion shared the 1988 Nobel Prize in Physiology or Medicine). **Target(s):** viral DNA polymerases, via the 5'-triphosphate (1-4); HSV-1 replication (2); varicella zoster virus replication (2); murine cytomegalovirus DNA polymerase (5); purine-nucleoside phosphorylase (6,8-11); thymidine kinase (7); glycine *N*-methyltransferase, weakly inhibited (12); tryptophan 2,3-dioxygenase (13). **Key Pharmacokinetic Parameters:** After intravenous dosing of patients with normal renal function, 8–14% of the dose is recovered in the urine as the metabolite 9-carboxymethoxymethylguanine (14). Adequate distribution of acyclovir occurs in the cerebrospinal fluid, vesicular fluid, vaginal secretions and tissues. The low plasma protein binding of acyclovir precludes drug interactions involving binding displacement. When IV doses in the range of 2.5–15 mg/kg are given every 8 h to adult patients, dose-independent kinetics iss observed. Continuous infusions of acyclovir over an equivalent daily dose range have achieved predictable plasma levels. Acyclovir half-life and total body clearance are influenced significantly by renal function, and dosage adjustments should be made for patients with impaired renal function (14). ***See also*** Appendix II in Goodman & Gilman's THE PHARMACOLOGICAL BASIS OF THERAPEUTICS, 12th Edition (Brunton, Chabner & Knollmann, eds.) McGraw-Hill Medical, New York (2011). For addition information on acyclovir's PK properties. **1**. Wilson, Porter & Reardon (1996) *Meth. Enzymol.* **275**, 398. **2**. Prisbe & Chen (1996) *Meth. Enzymol.* **275**, 425. **3**. Ator & Ortiz de Montellano (1990) *The Enzymes*, 3rd ed., **19**, 213. **4**. Elion, Furman, Fyfe, *et al.* (1977) *Proc. Natl. Acad. Sci. U.S.A.* **74**, 5716. **5**. Ochiai, Kumura & YMinamishima (1992) *Antiviral Res.* **17**, 1. **6**. dos Santos, Canduri, Pereira, *et al.* (2003) *Biochem. Biophys. Res. Commun.* **308**, 553. **7**. Rechtin, Black, Mao, Lewis & Drake (1995) *J. Biol. Chem.* **270**, 7055. **8**. Kulikowska, Bzowska, Wierzchowski & Shugar (1986) *Biochim. Biophys. Acta* **874**, 355. **9**. Bzowska, Kulikowska & Shugar (2000) *Pharmacol. Ther.* **88**, 349. **10**. Wielgus-Kutrowska, Tebbe, Schroder, *et al.* (1998) *Adv. Exp. Med. Biol.* **431**, 259. **11**. Canduri, Fadel, Basso, *et al.* (2005) *Biochem. Biophys. Res. Commun.* **327**, 646. **12**. Kloor, Karnahl & Kömpf (2004) *Biochem. Cell Biol.* **82**, 369. **13**. Müller & Daya (2008) *Metab. Brain Dis.* **23**, 351. **14**. de Miranda & Blum (1983) *J. Antimicrob. Chemother.* **12** (*Suppl* B) 29.

Acyclovir Triphosphate

This nucleotide analogue, also known as acycloguanosine triphosphate and 9-([2'-hydroxyethoxy]methyl)guanine 2'-triphosphate (FWfree acid = 465.15 g/mol), halts viral replication via inhibition of viral DNA polymerases and incorporation and termination of growing viral DNA chains. The selectivity of its action was investigated by Nobelist Gertrude Elion and coworkers. **Target(s):** DNA polymerases, viral (1-4); mRNA (nucleoside-2'-*O*-)-methyltransferase (5). **1**. Ator & Ortiz de Montellano (1990) *The Enzymes*, 3rd ed., **19**, 213. **2**. Elion, Furman, Fyfe, de Miranda, Beauchamp & Schaeffer (1977) *Proc. Natl. Acad. Sci. U.S.A.* **74**, 5716. **3**. Brown & Wright (1995) *Meth. Enzymol.* **262**, 202. **4**. Wilson, Porter & Reardon (1996) *Meth. Enzymol.* **275**, 398. **5**. Benarroch, Egloff, Mulard, Guerreiro, Romette & Canard (2004) *J. Biol. Chem.* **279**, 35638.

ACZ885, *See Canakinumab*

Adalimumab

This widely used human monoclonal antibody (Mol. Wt. = 144,190.3 kDa; CAS 331731-18-1; bioavailability = 64% Subcutaneous; Half-life = 10-20 days), known commercially as Humira, is used in the treatment of autoimmune diseases by blocking the action of the soluble pro-inflammatory cytokine, tumor necrosis factor, or TNF (1). Adalimumab and four other anti-TNF agents (infliximab, etanercept, golimumab and certolizumab pegol) are approved worldwide for the treatment of rheumatoid arthritis. These anti-TNF agents bind to and neutralize soluble TNF-alpha, exerting different effects on transmembrane TNF-alpha-expressing cells. Differences in affinity and avidity for soluble and transmembrane TNF-alpha have been demonstrated (2). **1**. Kempeni (1999) *Ann. Rheum. Dis.* **58**, 170. **2**. Benucci, Saviola, Manfredi, Sarzi-Puttini & Atzeni (2012) *Acta Biomed.* **83**, 72.

2-((3R,5R,7R)-Adamantan-1-yl)-N-(2-(4-benzylpiperazin-1-yl)-2-oxoethyl)acetamide

This novel Ebola virus entry inhibitor (FW = 409.57 g/mol) targets Niemann-Pick C₁ (NPC₁), an endosomal membrane protein that plays an essential role in Ebola infection by binding to the virus glycoprotein (GP). When combined with earlier findings, an emerging model of EboV infection involves cleavage of the GP₁ subunit by endosomal cathepsin

proteases, removing heavily glycosylated domains and exposing an N-terminal ligand for binding to NPC₁. **1**. Côté, Misasi, Ren, *et al.* (2011) *Nature* **477**, 344.

N-[*N*-(3-Adamant-1-yl)carbamoyl)]-3-amidino-D-phenylalanine *N*-Phenethylamide

This phenylalanine derivative (FW = 457.58 g/mol), also known as (*R*)-2-(3-adamant-1-ylureido)-3-(3-carbimidoylphenyl)-*N*-benzylpropionamide, inhibits serine proteases with moderately high affinity. Other amides also inhibit (*e.g.*, *N*-(4-aminophenyl)ethylamide, *N*-(2-chlorophenyl)ethylamide, *N*-(3-chlorophenyl)ethylamide, and *N*-(4-*tert*-butylphenyl)ethylamide). **Target(s):** coagulation factor Xa (K_i = 0.025 μM); thrombin (K_i = 0.9 μM); trypsin (K_i = 7.0 μM). **1**. Mueller, Sperl, Stürzebecher, Bode & Moroder (2002) *Biol. Chem.* **383**, 1185.

Adapalene

This synthetic retinoid (FW = 412.52 g/mol; CAS 106685-40-9), also known by the trade name Differin® Gel – 0.1% and systematically as 6-[3-(1-adamantyl)-4-methoxyphenyl] naphthalene-2-carboxylic acid, received FDA approval in 1996 for the treatment of acne. **Primary Mode of Action:** Adapalene does not bind to cytosolic retinoic acid binding proteins, binding instead to nuclear retinoic acid receptor (RAR) subtypes RARβ and RARγ, thereby inhibiting keratinocyte differentiation. This inhibitory action on keratinocyte differentiation/ proliferation underlies adapalene's ability to resolve blemishes that form when oil becomes trapped in pore of the dermis. **Target(s):** triacylglycerol biosynthesis (1); inflammation (2); cell proliferation (3); induction of apoptosis in colorectal cancer cells *in vitro* (3). **1**. Sato, Akimoto, Kitamura, *et al.* (2013) *J. Dermatol. Sci.* **70**, 204. **2**. Akdeniz, Calka, Ozbek & Metin (2005) *Clin. Exp. Dermatol.* **30**, 570. **3**. Ocker, Herold, Ganslmayer, Hahn & Schuppan (2003) *Int. J. Cancer* **107**, 453.

Adaptaquin

This brain-penetrant PHD2 inhibitor and antioxidant (FW = 377.82 g/mol; CAS 385786-48-1; Soluble to 100 mM in DMSO), also named 7-[(4-chlorophenyl)[(3-hydroxy-2-pyridinyl)amino]methyl]-8-quinolinol, targets Hypoxia-Inducible Factor (*or* HIF) prolyl hydroxylase-2 (1), a critical regulator of hematopoietic stem cell (HSC) maintenance during steady-state and stress (2). Indeed, this branched oxyquinoline molecule was named adaptaquin (AQ) to signify its ability to inhibit the HIF-PHDs and activate *adaptive* responses to hypoxia (1). Hypoxia is a prominent feature in the maintenance of HSC quiescence and multipotency. Adaptaquin also exhibits neuroprotective effects, enhancing functional recovery in rodent intracerebral hemorrhage models, via inhibition of ATF4 dependent genes (3). **1**. Neitemeier, *et al.* (2016) *Cell Death Dis.* **7**, e2214. **2**. Singh, Franke, Kalucka, *et al.* (2013) *Blood* **121**, 5158. **3**. Karuppagounder, *et al* (2016) *Sci.Transl.Med.* **8**, 328.

Adderall

This proprietary mixture of amphetamine and dextroamphetamine salts (CAS Reg. Nos = 300-62-9 and 51-64-9; NLM Unique ID = C090411) is a CNS stimulant that is widely used to treat Attention Deficit Hyperactivity Disorder (ADHD) and narcolepsy. These agents act by blocking the reuptake of dopamine and norepinephrine into presynaptic neurons and by stimulating the release of these monoamines into the synaptic cleft (*See Amphetamine*). Adderall is also used as a performance and cognitive enhancer, and recreationally as an aphrodisiac and euphoriant.

Adecypenol

This compound (FW = 280.28 g/mol; CAS Number = 104493-13-2) from *Streptomyces* sp. OM-3223 potently inhibits adenosine deaminase. The K_i value for the calf intestinal enzyme is 4.7 nM, showing slow, tight-binding characteristics.   **1.** Omura, Ishikawa, Kuga, *et al.* (1986) *J. Antibiot. (Tokyo)* **39**, 1219. **2.** Omura, Tanaka, Kuga & Imamura (1986) *J. Antibiot. (Tokyo)* **39**, 309.

Adefovir & Adefovir Dipivoxil

Adefovir

Adefovir dipivoxil

This nucleotide analogue (FW = 273.19 g/mol; CAS 106941-25-7; IUPAC: {[2-(6-amino-9*H*-purin-9-yl)ethoxy]methyl}phosphonic acid) and its membrane-permeant prodrug (FW = 501.48 g/mol; CAS 142340-99-6; IUPAC: [2-(6-aminopurin-9-yl)ethoxymethyl-(2,2-dimethylpropanoyloxy methoxy)phosphoryl]oxymethyl 2,2-dimethylpropanoate; Trade Names: Hepsera® and Preveon®) are technically both prodrugs, inasmuch as they are metabolically transformed to adefovir triphosphate, which is the active reverse transcriptase inhibitor. It is approved for treating chronic hepatitis B in adults with evidently active viral replication as well as either evidence of persistent elevations in serum aminotransferases (primarily ALT) or histologically active disease. Adefovir also offers the advantage that resistance develops more slowly than Lamivudine. (*See also Entecavir; Lamivudine; Telbivudine; Tenofovir; Emtricitabine*) **1.** De Clercq, Holý & Rosenberg (1989) *Antimicrob. Agents Chemother.* **33**, 185. **2.** Barditch-Crovo, Toole, Hendrix *et al.* (1997) *J. Infect. Dis.* **176**, 406.

Adenanthin

This diterpenoid (FW = 484.63 g/mol; CAS 111917-59-0) from the leaves of *Rabdosia adenantha*, induces differentiation of acute promyelocytic leukemia (APL) cells. Adenanthin targets the conserved cysteine thiols of peroxiredoxins I and II (*or* Prx I and Prx II), inhibiting their peroxidase activities and consequently increasing cellular hydrogen peroxide. The latter leads to the activation of Extracellular signal-Regulated Kinases (ERKs) and increased transcription of CCAAT/enhancer-binding protein β (*or* C/EBPs), further contributing to differentiation. Adenanthin induces APL-like cell differentiation, represses tumor growth in vivo and prolongs the survival of mouse APL models that are sensitive and resistant to retinoic acid. **1.** Liu, Yin, Zhou, *et al.* (2012) *Nature Chem. Biol.* **8**, 486.

Adenine 9-β-D-Arabinofuranoside

This ribo- and 2'-deoxyriboadenosine analogue and anti-viral (FW = 267.24 g/mol; CAS 58-61-7; λ_{max} = 259 nm; ε_M = 15400 M^{-1}cm^{-1}; Source: *Streptomyces antibioticus* strain), also known as Ara-A, arabinofuranosyladenine, vidarabine, spongoadenosine and (2*R*,3*S*,4*S*,5*R*)-2-(6-amino-9*H*-purin-9-yl)-5-(hydroxymethyl)oxolane -3,4-diol, undergoes metabolic phosphorylation to form Ara-ATP. The latter inhibits viral DNA replication by sequence-specific inhibition of strand elongation, attended by incorporation of this purine nucleoside (2). Vidarabine, the very first synthetic purine nucleoside anti-viral, was synthesized by B. R. Baker and his associates in 1960 (1). **Target(s):** adenosylhomo-cysteinase, *or S*-adenosylhomocysteine hydrolase (3-10,16,19-22); adenylate cyclase (11,18); DNA polymerase, via the 5'-triphosphate (12-14); double-strand DNA breaks (15); herpes virus replication (14); GMP synthetase (17); [3-hydroxy-3-methylglutaryl-CoA reductase (NADPH)] kinase (23); deoxyguanosine kinase (24); deoxyadenosine kinase (25); deoxycytidine kinase (26); adenosine kinase (27); *S*-methyl-5'-deoxyadenosine phosphorylase, weakly (28). **1.** Lee, Benitez, Goodman & Baker (1960) *J. Amer. Chem. Soc.* **82**, 2648. **2.** Ohno, Spriggs, Matsukage, Ohno & Kufe (1989) *Cancer Res.* **49**, 2077. **3.** Thong, Coombs & Sanderson (1985) *Mol. Biochem. Parasitol.* **17**, 35. **4.** Magnuson, Perryman, Decker & Magnuson (1984) *Int. J. Biochem.* **16**, 1163. **5.** Ueland & Helland (1983) *J. Biol. Chem.* **258**, 747. **6.** Chiang, Guranowski & Segall (1981) *Arch. Biochem. Biophys.* **207**, 175. **7.** Helland & Ueland (1981) *Cancer Res.* **41**, 673. **8.** Chiang (1987) *Meth. Enzymol.* **143**, 377. **9.** Ator & Ortiz de Montellano (1990) *The Enzymes*, 3rd ed., **19**, 213. **10.** Fabianowska-Majewska, Duley & Simmonds (1994) *Biochem. Pharmacol.* **48**, 897. **11.** Onoda, Braun & Wrenn (1987) *Biochem. Pharmacol.* **36**, 1907. **12.** Chadwick, Bassendine, Crawford, Thomas & Sherlock (1978) *Brit. Med. J.* **2**, 531. **13.** Mustafi, Heaton, Brinkman & Schwartz (1994) *Int. J. Radiat. Biol.* **65**, 675. **14.** Prisbe & Chen (1996) *Meth. Enzymol.* **275**, 425. **15.** Bryant & Blocher (1982) *Int. J. Radiat. Biol. Relat. Stud. Phys. Chem. Med.* **42**, 385. **16.** Schanche, Schanche, Ueland & Montgomery (1984) *Cancer Res.* **44**, 4297. **17.** Spector & Beecham (1975) *J. Biol. Chem.* **250**, 3101. **18.** Jayaswal, Bressan & Handa (1985) *FEMS Microbiol. Lett.* **27**, 313. **19.** Jarvi, McCarthy, Mehdi, *et al.* (1991) *J. Med. Chem.* **34**, 647. **20.** Shimizu, Shiozaki, Ohshiro & Yamada (1984) *Eur. J. Biochem.* **141**, 385. **21.** Hershfield, Aiyar, Premakumar & Small (1985) *Biochem. J.* **230**, 43. **22.** Minotto, Ko, Edwards & Bagnara (1998) *Exp. Parasitol.* **90**, 175. **23.** Musi, Hayashi, Fujii, *et al.* (2001) *Amer. J. Physiol.* **280**, E677. **24.** Yamada, Goto & Ogasawara (1982) *Biochim. Biophys. Acta* **709**, 265. **25.** Deibler, Reznik & Ives (1977) *J. Biol. Chem.* **252**, 8240. **26.** Yamada, Goto & Ogasawara (1983) *Biochim. Biophys. Acta* **761**, 34. **27.** Miller, Adamczyk, Miller, *et al.* (1979) *J. Biol. Chem.* **254**, 2346. **28.** Koszalka & Krenitsky (1986) *Adv. Exp. Med. Biol.* **195B**, 559.

Adenine 9-β-D-Arabinofuranoside 5'-Monophosphate

This AMP and dAMP analogue (FWfree acid = 347.22 g/mol), also known as Ara-AMP, is metabolically transformed into Ara-ATP, which can be incorporated into viral DNA. Ara-AMP is also an alternative substrate of adenylate kinase. **Target(s):** adenine phosphoribosyl-transferase (1); fructose-1,6-bisphosphatase (2); 5' phosphoribosyl-1-pyrophosphate synthetase (1); replication, viral (3); sphingomyelin phosphodiesterase, *or* sphingomyelinase (4,5,9); adenylosuccinate lyase (7,8); ADP:thymidine kinase (6); AMP:thymidine kinase (6). **1.** Becher & Schollmeyer (1983) *Klin. Wochenschr.* **61**, 751. **2.** Marcus (1976) *Cancer Res.* **36**, 1847. **3.** Weller, Bassendine, Craxi, *et al.* (1982) *Gut* **23**, 717. **4.** Quintern & Sandhoff (1991) *Meth. Enzymol.* **197**, 536. **5.** Quintern, Weitz, Nehrkorn, *et al.* (1987) *Biochim. Biophys. Acta* **922**, 323. **6.** Labenz, Müller & Falke (1984) *Arch. Virol.* **81**, 205. **7.** Spector, Jones & Elion (1979) *J. Biol. Chem.* **254**, 8422. **8.** Spector (1977) *Biochim. Biophys. Acta* **481**, 741. **9.** Quintern, Weitz, Nehrkorn, *et al.* (1987) *Biochim. Biophys. Acta* **922**, 323.

Adenine 9-β-D-Arabinofuranoside 5'-Triphosphate

This ATP and dATP analogue (FWfree-acid = 507.18 g/mol; CAS 3714-60-1), also known as 9-(5-O-(hydroxy(hydroxy(phosphonooxy) phosphinyl)-phosphinyl)-β-D-arabinofuranosyl)-9H-purin-6-amine,Ara-ATP, adenine arabinoside 5'-triphosphate, and vidarabine 5'-triphosphate, is incorporated into viral DNA, thus inhibiting viral polymerase-mediated DNA synthesis via chain termination. *Note:* Some DNA polymerases are known to extend chains ending in the ara-nucleotide. Ara-ATP is available commercially for *in vitro* experimentation. **Target(s):** DNA primase (1,2,28); mRNA translocation (3); DNA-directed DNA polymerase (4,5,7,10-12,14-16,18-20,28); DNA polymerase α (4,5,11,12,14,28); DNA polymerase β (4,5,11,12,14); DNA polymerase γ (11,14); ribonucleotide reductase (4,9); S-adenosylhomocysteine hydrolase (4,6); RNA polymerase (8); polynucleotide adenylyltransferase (13,17); GMP synthetase (21); lysyl-tRNA synthetase (22); leucyl-tRNA synthetase (23); adenylate cyclase (24); [hydroxymethylglutaryl-CoA reductase (NADPH)] kinase, possible alternative substrate (25); duck hepatitis B virus reverse transcriptase (26); RNA-directed DNA polymerase (26); DNA nucleotidyl exotransferase, *or* terminal deoxyribonucleotidyl transferase (27). **1.** Kuchta & Willhelm (1991) *Biochemistry* **30**, 797. **2.** Suzuki, Masaki, Koiwai & Yoshida (1986) *J. Biochem.* **99**, 1673. **3.** Schroder, Nitzgen, Bernd, *et al.* (1984) *Cancer Res.* **44**, 3812. **4.** White, Shaddix, Brockman & Bennett (1982) *Cancer Res.* **42**, 2260. **5.** Allaudeen, Kozarich & Sartorelli (1982) *Nucl. Acids Res.* **10**, 1379. **6.** Chiang, Guranowski & Segall (1981) *Arch. Biochem. Biophys.* **207**, 175. **7.** Hess, Arnold & Meyer zum Buschenfelde (1981) *Antimicrob. Agents Chemother.* **19**, 44. **8.** Chanda & Banerjee (1980) *Virology* **107**, 562. **9.** Chang & Cheng (1980) *Cancer Res.* **40**, 3555. **10.** Lee, Byrnes, Downey & So (1980) *Biochemistry* **19**, 215. **11.** Ono, Ohashi, Yamamoto, *et al.* (1979) *Cancer Res.* **39**, 4673. **12.** Okura & Yoshida (1978) *J. Biochem.* **84**, 727. **13.** Rose & Jacob (1978) *Biochem. Biophys. Res. Commun.* **81**, 1418. **14.** Reinke, Drach, Shipman & Weissbach (1978) *IARC Sci. Publ.* **24**, Pt. 2, 999. **15.** Dicioccio & Srivastava (1977) *Eur. J. Biochem.* **79**, 411. **16.** Rashbaum & Cozzarelli (1976) *Nature* **264**, 679. **17.** Edmonds (1982) *The Enzymes*, 3rd ed., **15**, 217. **18.** Brown & Wright (1995) *Meth. Enzymol.* **262**, 202. **19.** Prisbe & Chen (1996) *Meth. Enzymol.* **275**, 425. **20.** Weissbach (1981) *The Enzymes*, 3rd ed., **14**, 67. **21.** Spector (1978) *Meth. Enzymol.* **51**, 219. **22.** Freist, Sternbach, von der Haar & Cramer (1978) *Eur. J. Biochem.* **84**, 499. **23.** Marutzky, Flossdorf & Kula (1976) *Nucl. Acids Res.* **3**, 2067. **24.** Yang & Epstein (1983) *J. Biol. Chem.* **258**, 3750. **25.** Henin, Vincent & Van den Berghe (1996) *Biochim. Biophys. Acta* **1290**, 197. **26.** Offensperger, Walter, Offensperger, *et al.* (1988) *Virology* **164**, 48. **27.** Müller, Zahn &

Arendes (1978) *FEBS Lett.* **94**, 47. **28.** Kuchta, Ilsley, Kravig, Schubert & Harris (1992) *Biochemistry* **31**, 4720.

Adeninylpentylcobalamin

This vitamin B₁₂ analogue (FW = 1534.69 g/mol) inhibits Class-II ribonucleoside-triphosphate reductase (K_i = 1.3 μM), thereby generating a radical intermediate by mechanisms involving 5'-deoxyadenosy-cobalamin. In 5'-deoxyadenosylcobalamin-dependent eliminases, the γ-band of bound cob(II)alamin is shifted from the 473 nm for free cob(II)alamin to longer wavelengths, 475–480 nm. However, in mutases, the γ-band of bound cob(II)alamin is shifted to shorter wavelengths, 465–470 nm. Such results provide strong evidence that two subclasses of AdoCbl-dependent enzymes exist, and they give insight into the probable post-homolysis state in RTPR and other eliminases. They likewise identify the C₅-analog as the tightest-binding analogue for crystallization and other biophysical studies. 1. Suto, Poppe, Rétey & Finke (1999) *Bioorg. Chem.* **27**, 451.

Adenosine 2',5'-Bisphosphate

This rare ribonucleotide (FWfree-acid = 427.20 g/mol; λ_{max} = 259 nm; ε_M = 15400 M⁻¹cm⁻¹), a structural element of NADP⁺ and NADPH, inhibits erythrulose reductase (1); isocitrate dehydrogenase, NADP⁺-requiring (2); RNA ligase (ATP) (3); NAD⁺ diphosphatase (4); ferredoxin:NADP⁺ reductase (5,6,11); methenyl-tetrahydrofolate cyclohydrolase (7); barley nuclease (8); thiol sulfotransferase (9); estrone sulfotransferase (10); and [methionine synthase] reductase (12). **1.** Uehara & Hosomi (1982) *Meth. Enzymol.* **89**, 232. **2.** Seelig & Colman (1978) *Arch. Biochem. Biophys.* **188**, 394. **3.** Sugiura, Suzuki, Ohtsuka, *et al.* (1979) *FEBS Lett.* **97**, 73. **4.** Nakajima, Fukunaga, Sasaki & Usami (1973) *Biochim. Biophys. Acta* **293**, 242. **5.** Serrano & Rivas (1982) *Anal. Biochem.* **126**, 109. **6.** Batie & Kamin (1986) *J. Biol. Chem.* **261**, 11214. **7.** Pelletier & MacKenzie (1994) *Biochemistry* **33**, 1900. **8.** Sasakuma & Oleson (1979) *Phytochemistry* **18**, 1873. **9.** Schmidt & Christen (1978) *Planta* **140**, 239. **10.** Horwitz, Misra, Rozhin, *et al.* (1980) *Biochim. Biophys. Acta* **613**, 85. **11.** Maeda, Lee, Ikegami, *et al.* (2005) *Biochemistry* **44**, 10644. **12.** Wolthers, Lou, Toogood, Leys & Scrutton (2007) *Biochemistry* **46**, 11833.

Adenosine 3',5'-Bisphosphate

This ribonucleotide (FWfree-acid = 427.20 g/mol; *Symbol*: pAp), also known as 3'-phosphoadenosine 5'-phosphate and adenosine 3' phosphate 5'-phosphate, is produced in mammalian cells, primarily by the transfer of the sulfate group of PAPS (phosphoadenosine 5′ phosphosulfate) to various acceptor molecules. pAp is recycled into AMP and P_i by pApphosphatases, hydrolases that are competitively inhibited by sub-mM concentrations of lithium (1,2). **Target(s):** bile-salt sulfotransferase (3,23-26); ribonuclease A, *or* pancreatic ribonuclease (4); Rat1p nuclease (5); sulfate adenylyltransferase, *or* ATP sulfurylase (6,41-43); phenol sulfotransferase, *or* aryl sulfotransferase; as product inhibitor (7,29,36,37); adenylosuccinate synthetase (6); 4-chlorobenzoyl-CoA dehalogenase, weakly inhibited (9); nucleotide diphosphatase (10,11); sphingomyelin phosphodiesterase (12); glycochenodeoxycholate sulfotransferase (13); scymnol sulfotransferase (14,15); flavonol 3-sulfotransferase (16); deoxysulfoglucosinolate sulfotransferase (17-19); cortisol sulfotransferase (20); thiol sulfotransferase (21); steroid sulfotransferase (22); galactosylceramide sulfotransferase (27); tyrosine-ester sulfotransferase, *or* aryl sulfotransferase IV (28,29); [heparan sulfate]-glucosamine *N*-sulfotransferase (30); choline sulfotransferase (31); estrone sulfotransferase (32,33); alcohol sulfotransferase (34,35); holo-[acyl-carrier-protein] synthase, as product inhibitor (36-40); *N*-acetylneuraminate 4-*O*-acetyltransferase (44); arylamine *N*-acetyltransferase (45); and adenylyl-sulfate reductase (46). **1.** York, Ponder & Majerus (1995) *Proc. Natl. Acad. Sci. U.S.A.* **92**, 5149. **2.** Yenush., Bellés, López-Coronado, et al. (2000) *FEBS Lett.* **467**, 321. **3.** Chen (1981) *Meth. Enzymol.* **77**, 213. **4.** Russo, Acharya & Shapiro (2001) *Meth. Enzymol.* **341**, 629. **5.** Johnson (2001) *Meth. Enzymol.* **342**, 260. **6.** Peck (1974) *The Enzymes*, 3rd ed., **10**, 651. **7.** Sekura & Jakoby (1979) *J. Biol. Chem.* **254**, 5658. **8.** Van der Weyden & Kelly (1974) *J. Biol. Chem.* **249**, 7282. **9.** Luo, Taylor, Xiang, *et al.* (2001) *Biochemistry* **40**, 15684. **10.** Grobben, Claes, Roymans, *et al.* (2000) *Brit. J. Pharmacol.* **130**, 139. **11.** Grobben. Anciaux, Roymans, *et al.* (1999) *J. Neurochem.* **72**, 826. **12.** Quintern, Weitz, Nehrkorn, *et al.* (1987) *Biochim. Biophys. Acta* **922**, 323. **13.** Falany, Wheeler, Coward, Keehan, Falany & Barnes (1992) *J. Biochem. Toxicol.* **7**, 241. **14.** Pettigrew, Wright & Macrides (1998) *Comp. Biochem. Physiol. B* **121B**, 243, 299, and 341. **15.** Macrides, Faktor, Kalafatis & Amiet (1994) *Comp. Biochem. Physiol. B* **107B**, 461. **16.** Varin & Ibrahim (1992) *J. Biol. Chem.* **267**, 1858. **17.** Glendening & Poulton (1990) *Plant Physiol.* **94**, 811. **18.** Jain, Wassink, Kolenovsky & Underhill (1990) *Phytochemistry* **29**, 1425. **19.** Klein, Reichelt, Gershenzon & Papenbrock (2006) *FEBS J.* **273**, 122. **18.** Singer & Bruns (1980) *Can. J. Biochem.* **58**, 660. **20.** Schmidt & Christen (1978) *Planta* **140**, 239. **21.** Falany, Vasquez & Kalb (1989) *Biochem. J.* **260**, 641. **22.** Chen & Segel (1985) *Arch. Biochem. Biophys.* **241**, 371. **23.** Chen (1982) *Biochim. Biophys. Acta* **717**, 316. **24.** Chen, Bolt & Admirand (1977) *Biochim. Biophys. Acta* **480**, 219. **25.** Chen, Imperato & Bolt (1978) *Biochim. Biophys. Acta* **522**, 443. **26.** Tennekoon, Aitchison & M. Zabura (1985) *Arch. Biochem. Biophys.* **240**, 932. **27.** Duffel & Jakoby (1981) *J. Biol. Chem.* **256**, 11123. **28.** Saidha & Schiff (1994) *Biochem. J.* **298**, 45. **29.** Wei, Swiedler, Ishihara, Orellana & Hirschberg (1993) *Proc. Natl. Acad. Sci. U.S.A.* **90**, 3885. **30.** Renosto & Segel (1977) *Arch. Biochem. Biophys.* **180**, 416. **31.** Horwitz, Misra, Rozhin, *et al.* (1978) *Biochim. Biophys. Acta* **525**, 364. **32.** Horwitz, Misra, Rozhin, *et al.* (1980) *Biochim. Biophys. Acta* **613**, 85. **33.** Rearick & Calhoun (2001) *Biochim. Cell Biol.* **79**, 499. **34.** Ryan & Carroll (1976) *Biochim. Biophys. Acta* **429**, 391. **35.** Novakova, Van Dyck, Glatz, Van Schepdael & Hoogmartens (2004) *J. Chromatogr. A* **1032**, 319. **36.** Whittemore, Pearce & Roth (1985) *Biochemistry* **24**, 2477. **37.** Yang, Fernandez & Lamppa (1994) *Eur. J. Biochem.* **224**, 743. **38.** Elhussein, Miernyk & Ohlrogge (1988) *Biochem. J.* **252**, 39. **39.** McAllister, Peery, Meier, Fischl & Zhao (2000) *J. Biol. Chem.* **275**, 30864. **40.** Renosto, Patel, Martin, *et al.* (1993) *Arch. Biochem. Biophys.* **307**, 272. **41.** Renosto, Martin, Wailes, Daley & Segel (1990) *J. Biol. Chem.* **265**, 10300. **42.** MacRae, Segel & Fisher (2002) *Nat. Struct. Biol.* **9**, 945. **43.** Iwersen, Dora, Kohla, Gasa & Schauer (2003) *Biol. Chem.* **384**, 1035. **44.** Andres, Kolb, Schreiber & Weiss (1983) *Biochim. Biophys. Acta* **746**, 193. **45.** Setya, Murillo & Leustek (1996) *Proc. Natl. Acad. Sci. U.S.A.* **93**, 13383.

Adenosine-5′-(β-bromoethyl Phosphonate)

This nucleotide analogue (FW$_{free-acid}$ = 438.17 g/mol; λ_{max} = 259 nm; ε_M = 15400 $M^{-1}cm^{-1}$), also known as AMP-CH_2-CH_2-Br and adenosine 5′-(2

bromoethyl)phosphate, inhibits tryptophanyl-tRNA synthetase (1,2), isocitrate dehydrogenase (3), and phosphorylase kinase (4). **1.** Kisselev, Favorova & Kolaleva (1979) *Meth. Enzymol.* **59**, 174. **2.** Kovaleva, Ivanov, Madoian, *et al.* (1978) *Biokhimiia* **43**, 525. **3.** Colman (1990) *The Enzymes*, 3rd ed., **19**, 283. **4.** Guliaeva, Vul'fson & Severin (1978) *Biokhimiia* **43**, 373.

Adenosine 5′-Carboxamide

This nucleoside analogue and coronary dilator (FW = 280.24 g/mol; λ_{max} = 259 nm; ε_M = 15400 $M^{-1}cm^{-1}$) inhibits adenosylhomocysteinase (1) and 5′-methylthioadenosine nucleosidase (2-4). **1.** Wnuk, Liu, Yuan, Borchardt & Robins (1996) *J. Med. Chem.* **39**, 4162. **2.** Guranowski, Chiang & Cantoni (1983) *Meth. Enzymol.* **94**, 365. **3.** Guranowski, Chiang & Cantoni (1981) *Eur. J. Biochem.* **114**, 293. **4.** Yamakawa & Schweiger (1979) *Dev. Cell Biol.* **3**, 241.

Adenosine-5′-(β-chloroethyl Phosphate)

This nucleotide analogue (FW$_{free-acid}$ = 409.72 g/mol), often abbreviated as AMP-O-CH_2-CH_2-Cl inhibits tryptophanyl-tRNA synthetase (1,2), phosphorylase kinase (3), and leucyl-tRNA synthetase (4). **1.** Kisselev, Favorova & Kolaleva (1979) *Meth. Enzymol.* **59**, 174. **2.** Kovaleva, Ivanov, Madoian, *et al.* (1978) *Biokhimiia* **43**, 525. **3.** Guliaeva, Vul'fson & Severin (1978) *Biokhimiia* **43**, 373. **4.** Krauspe, Kovaleva, Guliaev, Baranova & Agalarova (1978) *Biokhimiia* **43**, 656.

Adenosine 2′,3′-Cyclic Monophosphate

This adenosine phosphoric diester (FW$_{free-acid}$ = 329.21 g/mol; λ_{max} = 259 nm; ε_M = 15400 $M^{-1}cm^{-1}$), also known as 2′,3′-cyclic AMP and 2′,3′-cAMP, is produced nonenzymatically as an intermediate in the alkaline hydrolysis of RNA as well as an enzymatically as an intermediate by several ribonucleases (*e.g.*, pancreatic ribonuclease, ribonuclease T₁, ribonuclease T₂, ribonuclease U₂, *Bacillus subtilis* ribonuclease, *etc.*). 2′,3′-cyclic AMP is also a substrate for 2′,3′-cyclic-nucleotide 2′-phosphodiesterase (EC 3.1.4.16) and 2′,3′-cyclic nucleotide 3′-phosphodiesterase (EC 3.1.4.37). A 2′,3′-cyclic phosphate termini is produced in the endonucleolytic cleavage of pre-tRNA by tRNA-intron endonuclease (EC 3.1.27.9). This cyclic nucleotide can also be used in the synthesis of oligonucleotides of the form ApX using the reversible reaction catalyzed by *Ustilago sphaerogena* ribonuclease U2: 2′,3′-cAMP is reacted with an acceptor substrate containing a free 5′-hydroxyl group (this reactant can be a single nucleotide or nucleoside or an oligonucleotide). *Note:* Hydrolysis by dilute acid or base produces a mixture of 2′-AMP and 3′-AMP (often referred to as 2′(3′)-AMP). Do not confuse this mixture with 2′,3′-cAMP; both are commercially available. **Target(s):** choline kinase (1); 3′-exonuclease, salmon testes (2); nucleotide diphosphatase, weakly inhibited (3); and poly(ADP-ribose) glycohydrolase (4,5). **1.** Ishidate & Nakazawa (1992) *Meth. Enzymol.* **209**, 121. **2.** Menon & Smith (1970) *Biochemistry* **9**, 1584. **3.** Bartkiewicz, Sierakowska & Shugar (1984) *Eur. J. Biochem.* **143**, 419. **4.** Sugimura, Yamada, Miwa, *et al.*

(1973) *Biochem. Soc. Trans.* **1**, 642. **5**. Miwa, Tanaka, Matsushima & Sugimura (1974) *J. Biol. Chem.* **249**, 3475.

Adenosine 3',5'-Cyclic Monophosphate

This prototypical second messenger (FW$_{free\text{-}acid}$ = 329.21 g/mol; λ_{max} = 259 nm; ε_M = 15400 M^{-1}cm^{-1}; Abbreviation: 3',5'-cyclic AMP, 3',5'-cAMP, opr cAMP), the isolation of which in the late 1950s earned Earl Sutherland his share of the 1971 Nobel Prize in Medicine & Physiology, activates susceptible protein kinases in signal transduction cascades. Elevated cAMP concentrations also activate cyclicAMP-gated ion channels; Exchange Proteins Activated by cAMP (or EPAC), such as RAPGEF3; and popeye domain-containing proteins (or Popdc). While many enzymes are weakly inhibited by cAMP (with inhibition constants typically in the 0.1 mM range or above), the peak physiologic concentration of cAMP rarely exceeds 10-20 μM. **Target(s):** phosphate-activated glutaminase (1); glyceraldehyde-3-phosphate dehydrogenase, K_i = 0.11 mM for the yeast enzyme (2,16,17); choline kinase (3); *Bacillus subtilis* ribonuclease (7); NAD(P)$^+$ transhydrogenase, AB-specific (8); phosphofructokinase, weakly inhibited (9,10); hexokinase (12); adenosine deaminase (13); NAD$^+$ nucleosidase, *or* NADase (14,29,30); poly(ADP-ribose) glycohydrolase (15,31-37); GMP synthetase (18); 5-phosphoribosylamine synthetase, *or* ribose-5-phosphate:ammonia ligase (19); tyrosyl-tRNA synthetase, weakly inhibited (20); adenylylsulfatase (21); CDP-diacylglycerol diphosphatase (22); nucleotide diphosphatase, weakly inhibited (23); ATP diphosphatase, slight inhibition (24); keratan sulfotransferase, weakly inhibited (25); adenosylhomocysteinase (26-28); 6-phospho-β-glucosidase (38); prenyl-diphosphatase (39); CMP-*N*-acylneuraminate phosphodiesterase (40,41); 2',3'-cyclic-nucleotide 3'-phosphodiesterase, *or* cCMP phosphodiesterase (42,43); 3',5'-cyclic-GMP phospho-diesterase, *or* cGMP phosphodiesterase (44-47); sphingomyelin phosphodiesterase (48); phosphoprotein phosphatase (49); 5'-nucleotidase (50); acid phosphatase (51); diacylglycerol cholinephosphotransferase (52); ethanolamine-phosphotransferase (52); glucose-1-phosphate adenylyltransferase (53); nucleoside diphosphate kinase (54,55); AMP:thymidine kinase (56); diacylglycerol kinase (57); nucleoside phosphotransferase (58); 1-phosphatidylinositol 4-kinase (59-62); NAD$^+$ kinase (63; procollagen glucosyltransferase64; chitin synthase, weakly inhibited (65); glycogen phosphorylase, weakly inhibited (66); luciferase, firefly (*Photinus*-luciferin 4-monooxygenase, ATP-hydrolyzing (67,68).

1. Kvamme, Torgner & Svenneby (1985) *Meth. Enzymol.* **113**, 241. **2**. Brosemer (1975) *Meth. Enzymol.* **41**, 273. **3**. Ishidate & Y. Nakazawa (1992) *Meth. Enzymol.* **209**, 121. **4**. Rall, Sutherland & Berthet (1957) *J. Biol. Chem.* **224**, 463. **5**. Rall & Sutherland (1958) *J. Biol. Chem.* **232**, 1065. **6**. Sutherland & Rall (1958) *J. Biol. Chem.* **232**, 1077. **7**. Kerjan & Szulmajster (1976) *Biochimie* **58**, 533. **8**. Rydström, Hoek & Ernster (1976) *The Enzymes*, 3rd ed., **13**, 51. **9**. Mansour (1963) *J. Biol. Chem.* **238**, 2285. **10**. Webb (1966) *Enzyme and Metabolic Inhibitors*, vol. 2, p. 474, Academic Press, New York. **11**. Sutherland & Rall (1957) *J. Amer. Chem. Soc.* **79**, 3608. **12**. Ning, Purich & Fromm (1969) *J. Biol. Chem.* **244**, 3840. **13**. Van Heukelom, Boom, Bartstra & Staal (1976) *Clin. Chim. Acta* **72**, 109. **14**. Ueda, Fukushima, Okayama & Hayaishi (1975) *J. Biol. Chem.* **250**, 7541. **15**. Tanuma, Kawashima & Endo (1986) *J. Biol. Chem.* **261**, 965. **16**. Yang & Deal (1969) *Biochemistry* **8**, 2806. **17**. DeMark, Benjamin & Fife (1983) *Arch. Biochem. Biophys.* **223**, 360. **18**. Hirai, Matsuda & Nakagawa (1987) *J. Biochem.* **102**, 893. **19**. Westby & Tsai (1974) *J. Bacteriol.* **117**, 1099. **20**. Santi & Peña (1973) *J. Med. Chem.* **16**, 273. **21**. Stokes, Denner & Dodgson (1973) *Biochim. Biophys. Acta* **315**, 402. **22**. Raetz, Hirschberg, Dowhan, Wickner & Kennedy (1972) *J. Biol. Chem.* **247**, 2245. **23**. Bartkiewicz, Sierakowska & Shugar (1984) *Eur. J. Biochem.* **143**, 419. **24**. Kawamura, Tonotsuka & Nagano (1976) *Biochim. Biophys. Acta* **421**, 195. **25**. Keller, Driesch, Stein, *et al.* (1983) *Hoppe-Seyler's Z. Physiol. Chem.* **364**, 239. **26**. Kloor, Kurz, Fuchs, Faust & Osswald (1996) *Kidney Blood Press. Res.* **19**, 100. **27**. Hohman, Guitton & Veron (1984) *Arch. Biochem. Biophys.* **233**, 785. **28**. Knudsen & Yall (1972) *J. Bacteriol.* **112**, 569. **29**. Schuber, Pascal & Travo (1978) *Eur. J. Biochem.* **83**, 205. **30**. Travo, Muller & Schuber (1979) *Eur. J. Biochem.* **96**, 141.

31. Tavassoli, Tavassoli & Shall (1983) *Eur. J. Biochem.* **135**, 449. **32**. Hatakeyama, Nemoto, Ueda & Hayaishi (1986) *J. Biol. Chem.* **261**, 14902. **33**. Tanuma, Sakagami & Endo (1989) *Biochem. Int.* **18**, 701. **34**. Sugimura, Yamada, Miwa, *et al.* (1973) *Biochem. Soc. Trans.* **1**, 642. **35**. Miwa, Tanaka, Matsushima & Sugimura (1974) *J. Biol. Chem.* **249**, 3475. **36**. Maruta, Inageda, Aoki, Nishina & Tanuma (1991) *Biochemistry* **30**, 5907. **37**.Tanuma & Endo (1990) *Eur. J. Biochem.* **191**, 57. **38**. Wilson & Fox (1974) *J. Biol. Chem.* **249**, 5586. **39**. Tsai & Gaylor (1966) *J. Biol. Chem.* **241**, 4043. **40**. Kean & Bighouse (1974) *J. Biol. Chem.* **249**, 7813. **41**. van Dijk, Maier & van den Eijnden (1976) *Biochim. Biophys. Acta* **444**, 816. **42**. Sprinkle (1989) *CRC Crit. Rev. Clin. Neurobiol.* **4**, 235. **43**. Sims & Carnegie (1978) *Adv. Neurochem.* **3**, 1. **44**. Morishima (1975) *Biochim. Biophys. Acta* **410**, 310. **45**. Hwang, Clark & Bernlohr (1974) *Biochem. Biophys. Res. Commun.* **58**, 707. **46**. Methven, Lemon & Bhoola (1980) *Biochem. J.* **186**, 491. **47**. Bosgraaf, Russcher, Snippe, *et al.* (2002) *Mol. Biol. Cell* **13**, 3878. **48**. Testai, Landek, Goswami, Ahmed & Dawson (2004) *J. Neurochem.* **89**, 636. **49**. Nakai & Thomas (1974) *J. Biol. Chem.* **249**, 6459. **50**. Carter & Tipton (1986) *Phytochemistry* **25**, 33. **51**. Leelapon, Sarath & Staswick (2004) *Planta* **219**, 1071. **52**. Strosznajder, Radominska-Pyrek & Horrocks (1979) *Biochim. Biophys. Acta* **574**, 48. **53**. Sowokinos & Preiss (1982) *Plant Physiol.* **69**, 1459. **54**. Tiwari, Kishan, Chakrabarti & Chakraborti (2004) *J. Biol. Chem.* **279**, 43595. **55**. Brodbeck, Rohling, Wohlleben, Thompson & Süsstrunk (1996) *Eur. J. Biochem.* **239**, 208. **56**. Grivell & Jackson (1976) *Biochem. J.* **155**, 571. **57**. Kato & Takenawa (1990) *J. Biol. Chem.* **265**, 794. **58**. Billich & Witzel (1986) *Biol. Chem. Hoppe-Seyler* **367**, 291. **59**. Vogel & Hoppe (1986) *Eur. J. Biochem.* **154**, 253. **60**. Buckley (1977) *Biochim. Biophys. Acta* **498**, 1. **61**. Steinert, Wissing & Wagner (1994) *Plant Sci.* **101**, 105. **62**. Yamakawa & Takenawa (1988) *J. Biol. Chem.* **263**, 17555. **63**. Blomquist (1973) *J. Biol. Chem.* **248**, 7044. **64**. Barber & Jamieson (1971) *Biochim. Biophys. Acta* **252**, 533. **65**. Jan (1974) *J. Biol. Chem.* **249**, 1973. **66**. Spearman, Khandelwal & Hamilton (1973) *Arch. Biochem. Biophys.* **154**, 306. **67**. Lee, Denburg & McElroy (1970) *Arch. Biochem. Biophys.* **141**, 38. **68**. Leach (1981) *J. Appl. Biochem.* **3**, 473.

Adenosine 2',3'-Dialdehyde

This nucleoside derivative (FW = 265.23 g/mol) alkylates susceptible enzymes, most often by combining with an ε-amino group of an active-site lysine residue. Because the dialdehyde is of limited stability, it is usually prepared by treating adenosine with a slight molar excess of sodium periodate, followed by a larger molar excess of ethylene glycol to reduce ant unreacted periodate. **Target(s):** adenosine kinase (1); histamine *N*-methyltransferase (2); *S*-adenosylhomocysteinase, *or* *S*-AdoHcy hydrolase, IC$_{50}$ = 2-3 μM (3,4); ribonucleotide reductase (5); histone-arginine *N*-methyltransferase (6).

1. Pak, Haas, Decking & Schrader (1994) *Neuropharmacology* **33**, 1049. **2**. Borchardt, Wu & Wu (1978) *Biochemistry* **17**, 4145. **3**. Bartel & Borchardt (1984) *Mol. Pharmacol.* **25**, 418. **4**. Patel-Thombre & Borchardt (1985) *Biochemistry* **24**, 1130. **5**. Cory & Mansell (1975) *Cancer Res.* **35**, 390. **6**. Herrmann, Lee, Bedford & Fackelmayer (2005) *J. Biol. Chem.* **280**, 38005.

Adenosine 5'-Diphosphate

Adenosine 5'-Diphosphate Dianion

⬡ = H$_2$O

Me^{2+} Complex with Adenosine 5'-Diphosphate

This adenine nucleotide (FW$_{free-acid}$ = 504.16 g/mol; CAS 58-64-0; Molar Absorptivity = 15,400 M^{-1}cm^{-1}, λ = 259 nm) is a product in ATP-dependent transphosphorylases, phosphohydrolases, and molecular motors; as such, ADP often inhibits these enzymes. **Enzymatic Phosphorylation:** ADP is a substrate for adenylate kinase (*Reaction*: ADP^{2-} + MeADP $\rightleftharpoons$ MeATP^{2-} + AMP) and other enzymes that stabilize ATP concentrations in prokaryotes [*e.g.*, acetate kinase (*Reaction*: MgADP + Acetyl-phosphate $\rightleftharpoons$ MeATP^{2-} + Acetate)] and eukaryotes [*e.g.*, pyruvate kinase (*Reaction*: MgADP + Phosphoenolpyruvate $\rightleftharpoons$ MgATP^{2-} + Pyruvate), creatine kinase (*Reaction*: MgADP + Creatine-phosphate $\rightleftharpoons$ MgATP^{2-} + Creatine), arginine kinase (*Reaction*: MgADP + Arginine-phosphate $\rightleftharpoons$ MgATP^{2-} + Arginine), and nuclecleotide diphosphate kinase (*Reaction*: ADP^{2-} + MgGTP^{2-} $\rightleftharpoons$ MgATP^{2-} + GDP^{2-})]. **ATP Synthase:** ADP is a primary substrate for the F$_O$F$_1$ ATP synthase (*Reaction*: MgADP + P$_i$ + High Chemiosmotic Gradient Energization State $\rightleftharpoons$ MgATP^{2-} + Low Chemiosmotic Gradient Energization State). ADP can also become entrapped within a catalytic site of the rotary motor, when proton motive is low, absent, or uncoupled, and its inhibitory action under such conditions is believed to prevent wasteful hydrolysis of ATP (*Reaction*: MgATP^{2-} + H$_2$O $\rightarrow$ MgADP + P$_i$). **Metal Ion Binding Properties:** As a polyanion, ADP not only binds physiologic divalent cations Mg^{2+} and Ca^{2+}, but also forms reversible complexes with Mn^{2+} and Co^{2+}. For reversible complexation of ADP^{2-} with a metal ion Me^{2+}, (*Reaction*: ADP^{2-} + Me^{2+} $\rightleftharpoons$ MeADP), $K_{formation}$ = [MeADP]/[ADP^{2-}]$_{free}$[Me^{2+}]$_{free}$, indicating that [MeADP]/[ADP^{2-}]$_{free}$ = $K_{formation}$ × [Me^{2+}]$_{free}$. In many cases, metal-free ADP is not a substrate and instead acts as a revesible inhibitor. Good experimental design therefore demands rigorous control of free metal ion concentration to control the ratio of metal-bound and metal-free forms (1,2). When exposed to Cr(III) at elevated temperature, ADP also forms ligand exchange-inert complexes with Cr^{3+} (2). **Platelet Aggregation:** ADP is also a well-known activator of platelet aggregation, as mediated by the ADP receptors P2Y1, P2Y12 and P2X1. Upon conversion to adenosine by ecto-ADPases, platelet activation is inhibited by means of adenosine receptors. **Target(s):** *Hydrogenomonas facilis* ribulosediphosphate (RuDP) carboxylase and NADH-, ATP-dependent CO$_2$ fixation (3); platelet (Na$^+$/K$^+$)-ATPase (4); hydrogen-ion transport in chloroplasts (5); pyruvate dehydrogenase kinase (6); 5-oxo-L-prolinase, *or* L-pyroglutamate hydrolase (7); α-NADH-dependent reductase, rat liver microsomes (8); nitrogenase (9); *Trypanosoma cruzi* hexokinase (10); maize leaf acetyl-coenzyme A carboxylase (11); rat brain mitochondrial calcium-efflux (12); sarcoplasmic reticulum Ca^{2+} ATPase (13); Na$^+$-Na$^+$ exchange mediated by (Na$^+$/K$^+$)-ATPase reconstituted into liposomes (14); nitrate and nitrite assimilation in *Zea mays* under dark conditions (15); PGE$_1$-activated platelet adenylate cyclase in rats and rabbits (16); mitochondrial F$_1$-ATPase, inactive complex formed upon binding ADP at a catalytic site (1,17,18); ATP-sensitive K$^+$ channels, frog skeletal muscle (19); human 5-phosphoribosyl-1-pyrophosphate synthetase (20); *Crithidia fasciculata* glutathionyl-spermidine synthetase (21); myosin V ATPase (22); cystic fibrosis transmembrane conductance regulator (ABC transporter) via its adenylate kinase activity (23); V type ATPase/synthase (24). **1.** Fromm (1975) *Initial Rate Enzyme Kinetics*, Springer. **2.** Purich (2010) *Enzyme Kinetics: Catalysis & Control*, Academic Press. **3.** McFadden & Tu (1967) *J. Bacteriol.* **93**, 886. **4.** Moake, Ahmed & Bachur (1970) Biochim. Biophys. Acta. **219**, 484. **5.** McCarty, Fuhrman & Tsuchiya (1971) *Proc. Natl. Acad. Sci. U.S.A.* **68**, 2522. **6.** Roche & Reed (1974) *Biochem. Biophys. Res. Commun.* **59**, 1341. **7.** Van Der Werf, Griffith & Meister (1975) *J. Biol. Chem.* **250**, 6686. **8.** Miyake, Nakamura, Takayama & Horiike (1975) *J. Biochem.* **78**, 773. **9.** Weston, Kotake & Davis (1983) *Arch. Biochem. Biophys.* **225**, 809. **10.** Racagni & Machado de Domenech (1983) *Mol. Biochem. Parasitol.* **9**, 181. **11.** Nikolau & Hawke (1984) *Arch. Biochem. Biophys.* **228**, 86. **12.** Vitorica & Satrústegui (1985) *Biochem. J.* **225**, 41. **13.** Coll & Murphy (1985) *FEBS Lett.* **187**, 131. **14.** Cornelius & Skou (1985) *Biochim. Biophys. Acta.* **818**, 211. **15.** Watt, Gray & Cresswell (1987) *Planta* **172**, 548. **16.** Defreyn, Gachet, Savi *et al.* (1991) *Thromb. Haemost.* **65**, 186. **17.** Chernyak & Cross (1992) *Arch. Biochem. Biophys.* **295**, 247. **18.** Walker (1994) *Curr. Opin. Struct. Biol.* **4**, 912. **19.** Forestier & Vivaudou (1993) *J. Membr. Biol.* **132**, 87. **20.** Fry, Becker & Switzer (1995) *Mol. Pharmacol.* 47, 810. **21.** Koenig, Menge, Kiess, Wray & Flohé (1997) *J. Biol. Chem.* **272**, 11908 (Erratum in: *J. Biol. Chem.* **280**, 7407). **22.** De La Cruz, Sweeney & Ostap (2000) *Biophys J.* 79, 1524. **23.** Randak & Welsh (2005) *Proc. Natl. Acad. Sci. U.S.A.* **102**, 2216. **24.** Kishikawa, Nakanishi, Furuike, Tamakoshi & Yokoyama (2013) *J. Biol. Chem.* **289**, 403.

Adenosine 5'-(β,γ-Imido)triphosphate

This ATP analogue (FW$_{free-acid}$ = 506.20 g/mol), also called AMP-PNP and β,γ-imidoadenosine 5'-triphosphate, is an inhibitor of many ATP-dependent enzymes. It can also serve as an alternative substrate for enzymes that hydrolyze ATP between the α and β phosphorus atoms. Snake venom phosphodiesterase hydrolyzes AMP-PNP to produce AMP and imidodiphosphate (PNP). *Escherichia coli* alkaline phosphatase will hydrolyze the nucleotide to orthophosphate and the corresponding nucleoside diphosphate derivative. AMP-PNP is very soluble in water and tends to bind metal ions with a greater affinity than ATP. It is unstable in acidic solutions, producing the corresponding phosphoramidate and orthophosphate. **Target(s):** choline kinase (1); formyltetrahydrofolate synthetase (2); methionine *S* adenosyltransferase, also as alternative substrate (3,91); chaperonin ATPase, *or* chaperonin 60 (4,56); 5'→3' exoribonuclease (5); mitogen-activated protein kinase activated protein kinase (26); pantothenate kinase (7); adenosine-5'-phosphosulfate kinase (8); ecto-ATPase (9); ATP synthase, *or* F$_1$ ATPase, *or* H$^+$-transporting two-sector ATPase (10,17,19,20,24,25,63); DNA helicase (11,29,34,64); DNA gyrase (12,33); protein kinase C (13); kinesin (14,57,59); glycerol kinase (15); myosin ATPase (16,30); cAMP dependent protein kinase (18); K$^+$/H$^+$-ATPase (21); Na$^+$/K$^+$-exchanging ATPase (22); T$_4$ polynucleotide kinase (23); glutamine synthetase (26); [myosin light-chain] kinase (27); DNA topoisomerase II (28,36,51,52); type I site-specific deoxyribonucleases (32); phosphoprotein phosphatase types 1 and 2A (35); dynein ATPase (37,38); GTP diphosphokinase, *or* GTP pyrophosphokinase (39); cobaltochelatase (40); RNA ligase (ATP) (41); DNA ligase (ATP), bacteriophage T$_7$ (42); asparagine synthetase (43); CTP synthetase (44,45); dethiobiotin synthetase (46); tetrahydrofolate synthetase (47); phosphoribosylamino-imidazole-succinocarboxamide synthetase (48); glycyl-tRNA synthetase (49); leucyl-tRNA synthetase (50); guanylate cyclase (53); adenylate cyclase (54); protein-synthesizing GTPase, elongation factor (55); minus-end-directed kinesin ATPase (57,58); centromere binding protein E58; plus-end-directed kinesin ATPase (14,59); channel-conductance-controlling ATPase, cystic-fibrosis membrane-conductance-regulating protein (60); polar-amino-acid-transporting ATPase, histidine permease (61,62); nucleoside-triphosphatase (64); nucleotide diphosphatase (65); apyrase (66); endopeptidase La (67); nucleotidases (68); 5'-nucleotidase (69); estrone sulfotransferase (70); alcohol sulfotransferase (70); triphosphate:protein phosphotransferase (71); histidine kinase (72); [tau protein] kinase, *or* glycogen synthase kinase 3β (73); phosphorylase kinase (74); cAMP-dependent protein kinase, *or* protein kinase A (75); [pyruvate dehydrogenase (acetyl-transferring)] kinase (76,77); non-specific serine/threonine protein kinase (78); PknB serine/threonine protein kinase (78); pyruvate,orthophosphate dikinase (79); GTP diphosphokinase (80); ribose-phosphate diphosphokinase, *or* phosphoribosyl-pyrophosphate synthetase (81); phosphomevalonate kinase (82); 3-phosphoglycerate kinase (83); polo kinase (84); polynucleotide 5'-hydroxyl-kinase (85,86); homoserine kinase (87); xylulokinase (88); ribulokinase (89); glucokinase (90); hexokinase IV (90). **1.** Ishidate & Nakazawa (1992) *Meth. Enzymol.* **209**, 121. **2.** Buttlaire (1980) *Meth. Enzymol.* **66**, 585. **3.** Markham, Hafner, Tabor & Tabor (1983) *Meth. Enzymol.* **94**, 219. **4.** Torres-Ruiz & McFadden (1998) *Meth. Enzymol.* **290**, 147. **5.** Slobin (2001) *Meth. Enzymol.* **342**, 282. **6.** Schindler, Godbey, Hood, *et al.* (2002) *Biochim. Biophys. Acta* **1598**, 88. **7.** Yun, Park, Kim, *et al.* (2000) *J. Biol. Chem.* **275**, 28093. **8.** Satishchandran & Markham (2000) *Arch. Biochem. Biophys.* **378**, 210. **9.** Chen & Lin (1997) *Biochem. Biophys. Res. Commun.* **233**, 442. **10.** Pedersen, Hullihen, Bianchet, Amzel & Lebowitz (1995) *J. Biol. Chem.* **270**, 1775. **11.** Poll, Harrison, Umthun, Dobbs & Benbow (1994) *Biochemistry* **33**, 3841. **12.** Tamura, Bates & Gellert (1992) *J. Biol. Chem.* **267**, 9214. **13.** Leventhal & Bertics (1991) *Biochemistry* **30**, 1385. **14.** Schnapp, Crise, Sheetz, Reese & Khan (1990) *Proc. Natl. Acad. Sci. U.S.A.* **87**, 10053. **15.** Pettigrew, Yu & Liu (1990) *Biochemistry* **29**, 8620. **16.** Sleep & Glyn (1986) *Biochemistry* **25**, 1149. **17.** Gresser, Beharry & Moennich (1984) *Curr.*

Top. Cell. Regul. **24**, 365. **18.** Whitehouse, Feramisco, Casnellie, Krebs & Walsh (1983) *J. Biol. Chem.* **258**, 3693. **19.** Belda, Carmona, Canovas, Gomez-Fernandez & Lozano (1983) *Biochem J.* **210**, 727. **20.** Baubichon, Godinot, Di Pietro & Gautheron (1981) *Biochem. Biophys. Res. Commun.* **100**, 1032. **21.** van de Ven, Schrijen, de Pont & Bonting (1981) *Biochim. Biophys. Acta* **640**, 487. **22.** Robinson (1980) *J. Bioenerg. Biomembr.* **12**, 165. **23.** Lillehaug (1978) *Biochim. Biophys. Acta* **525**, 357. **24.** Schuster, Ebel & Lardy (1975) *J. Biol. Chem.* **250**, 7848. **25.** Lardy, Schuster & Ebel (1975) *J. Supramol. Struct.* **3**, 214. **26.** Liaw, Jun & Eisenberg (1994) *Biochemistry* **33**, 11184. **27.** Tokimasa (1995) *Neurosci. Lett.* **197**, 75. **28.** Bojanowski, Maniotis, Plisov, Larsen & Ingber (1998) *J. Cell Biochem.* **69**, 127. **29.** Kopel, Pozner, Baran & Manor (1996) *Nucl. Acids Res.* **24**, 330. **30.** Yount, Ojala & Babcock (1971) *Biochemistry* **10**, 2490. **31.** Yount, Babcock, Ballantyne & Ojala (1971) *Biochemistry* **10**, 2484. **32.** Endlich & Linn (1981) *The Enzymes*, 3rd ed., **14**, 137. **33.** Gellert (1981) *The Enzymes*, 3rd ed., **14**, 345. **34.** Gefter (1981) *The Enzymes*, 3rd ed., **14**, 367. **35.** Ballou & Fischer (1986) *The Enzymes*, 3rd ed., **17**, 311. **36.** Goto, Laipis & Wang (1984) *J. Biol. Chem.* **259**, 10422. **37.** Terry & Purich (1982) *Adv. Enzymol. Relat. Areas Mol. Biol.* **53**, 113. **38.** Penningroth & Wittman (1978) *J. Cell Biol.* **79**, 827. **39.** Sy & Akers (1976) *Biochemistry* **15**, 4399. **40.** Debussche, Couder, Thibaut, *et al.* (1992) *J. Bacteriol.* **174**, 7445. **41.** Juodka & Labeikyte (1991) *Nucleosides Nucleotides* **10**, 367. **42.** Doherty, Ashford, Subramanya & Wigley (1996) *J. Biol. Chem.* **271**, 11083. **43.** Boehlein, Stewart, Walworth, *et al.* (1998) *Biochemistry* **37**, 13230. **44.** Willemoës & Sigurskjold (2002) *Eur. J. Biochem.* **269**, 4772. **45.** Kizaki, Ohsaka & Sakurada (1982) *Biochem. Biophys. Res. Commun.* **108**, 286. **46.** Alexeev, Baxter, Campopiano, *et al.* (1998) *Tetrahedron* **54**, 15891. **47.** Cichowicz & Shane (1987) *Biochemistry* **26**, 513. **48.** Nelson, Binkowski, Honzatko & Fromm (2005) *Biochemistry* **44**, 766. **49.** Dignam, Nada & Chaires (2003) *Biochemistry* **42**, 5333. **50.** Marutzky, Flossdorf & Kula (1976) *Nucl. Acids Res.* **3**, 2067. **51.** Gellert (1981) *Ann. Rev. Biochem.* **50**, 879. **52.** Melendy & Ray (1989) *J. Biol. Chem.* **264**, 1870. **53.** Gorczyca, Van Hooser & Palczewski (1994) *Biochemistry* **33**, 3217. **54.** Yang & Epstein (1983) *J. Biol. Chem.* **258**, 3750. **55.** Uritani & Miyazaki (1988) *J. Biochem.* **103**, 522. **56.** Torres-Ruiz & McFadden (1992) *Arch. Biochem. Biophys.* **295**, 172. **57.** Walker, Salmon & Endow (1990) *Nature* **347**, 780. **58.** Thrower, Jordan, Schaar, Yen & Wilson (1995) *EMBO J.* **14**, 918. **59.** Vale, Reese & Sheetz (1985) *Cell* **42**, 39. **60.** Weinreich, Riordan & Nagel (1999) *J. Gen. Physiol.* **114**, 55. **61.** Nikaido, Liu & Ames (1997) *J. Biol. Chem.* **272**, 27745. **62.** Liu, Liu & Ames (1997) *J. Biol. Chem.* **272**, 21883. **63.** Lo Piero & Petrone (1992) *Comp. Biochem. Physiol. B* **103**, 235. **64.** Borowski, Niebuhr, Schmitz, *et al.* (2002) *Acta Biochim. Pol.* **49**, 597. **65.** Rossomando & Jahngen (1983) *J. Biol. Chem.* **258**, 7653. **66.** Knowles, Isler & Reece (1983) *Biochim. Biophys. Acta* **731**, 88. **67.** Waxman & Goldberg (1982) *Proc. Natl. Acad. Sci. U.S.A.* **79**, 4883. **68.** Nikbakht & Stone (2000) *Brain Res.* **860**, 161. **69.** Newby, Luzio & Hales (1975) *Biochem. J.* **146**, 625. **70.** Allali-Hassani, Pan, Dombrovski, *et al.* (2007) *PLoS Biol.* **5**, e97. **71.** Tsutsui (1986) *J. Biol. Chem.* **261**, 2645. **72.** Gilmour, Foster, Sheng, *et al.* (2005) *J. Bacteriol.* **187**, 8196. **73.** Aoki, Yokota, Sugiura, *et al.* (2004) *Acta Crystallogr. Sect. D* **60**, 439. **74.** Farrar & Carlson (1991) *Biochemistry* **30**, 10274. **75.** Johnson, Noble & Owen (1996) *Cell* **85**, 149. **76.** Mann, Dragland, Vinluan, *et al.* (2000) *Biochim. Biophys. Acta* **1480**, 283. **77.** Bao, Kasten, Yan & Roche (2004) *Biochemistry* **43**, 13432. **78.** Ortiz-Lombardia, Pompeo, Boitel & Alzari (2003) *J. Biol. Chem.* **278**, 13094. **79.** Ye, Wei, McGuire, *et al.* (2001) *J. Biol. Chem.* **276**, 37630. **80.** Sy & Akers (1976) *Biochemistry* **15**, 4399. **81.** Sonoda, Kita, Ishijima, *et al.* (1997) *J. Biochem.* **122**, 635. **82.** Pilloff, Dabovic, Romanowski, *et al.* (2003) *J. Biol. Chem.* **278**, 4510. **83.** Kóvári, Flachner, Náray-Szabó & Vas (2002) *Biochemistry* **41**, 8796. **84.** Kothe, Kohls, Low, *et al.* (2007) *Biochemistry* **46**, 5960. **85.** Mani, Karimi-Busheri, Fanta, Cass & Weinfeld (2003) *Biochemistry* **42**, 12077. **86.** Lillehaug (1978) *Biochim. Biophys. Acta* **525**, 357. **87.** Shames & Wedler (1984) *Arch. Biochem. Biophys.* **235**, 359. **88.** Di Luccio, Petschacher, Voegtli, *et al.* (2007) *J. Mol. Biol.* **365**, 783. **89.** Lee, Gerratana & Cleland (2001) *Arch. Biochem. Biophys.* **396**, 219. **90.** Kim, Kalinowski & Marcinkeviciene (2007) *Biochemistry* **46**, 1423. **91.** Markham, Hafner, Tabor & Tabor (1980) *J. Biol. Chem.* **255**, 9082.

Adenosine 5'-(α,β-Methylene)diphosphate, *See* α,β-Methyleneadenosine 5'-Diphosphate

Adenosine 5'-(α,β-Methylene)-γ-thio-triphosphate, *See* α,β-Methyleneadenosine 5'-Triphosphate

Adenosine 5'-(α,β-Methylene)triphosphate, *See* α,β-Methyleneadenosine 5'-Triphosphate

Adenosine 5'-Monophosphate

This adenine nucleotide (FW$_{free-acid}$ = 346.06 g/mol; CAS 61-19-8; Molar Absorptivity = 15,400 M^{-1}cm^{-1}, λ = 259 nm) is formed by the action of adenylosuccinate lyase in the *de novo* purine nucleotide biosynthetic pathway. AMP is a feedback inhibitor of adenylosuccinate synthase. AMP is also a substrate for adenylate kinase (*Reaction*: MeATP^{2-} + AMP ⇌ ADP^{2-} + MeADP). **Target(s):** skeletal muscle fructose 1,6-bisphosphatase by adenosine monophosphate (1-4); pea seed phosphofructokinase by phosphoenolpyruvate (5); *Escherichia coli* ATP phosphoribosyltransferase (6); glutamine 5-phosphoribosyl-1-pyrophosphate amidotransferase (7); ADP-glucose synthetase (8); brown adipocyte nonselective cation channel (9). **1.** Opie & Newsholme (1967) *Biochem. J.* **104**, 353. **2.** Stone & Fromm (1980) *Biochemistry* **19**, 620. **3.** Liang, Zhang, Huang, Lipscomb (1993) *Proc. Natl. Acad. Sci. U.S.A.* **90**, 2132. **4.** Kurbanov, Choe, Honzatko & Fromm (1998) *J. Biol. Chem.* **273**, 17511. **5.** Kelly & Turner (1969) *Biochem. J.* **115**, 481. **6.** Tebar, Leyva, Laynez & Ballesteros (1978) *Rev Esp Fisiol.* **34**, 159. **7.** Tsuda, Katunuma & Weber (1979) *J. Biochem.* **85**, 1347. **8.** Leckie, Porter, Roth, Tieber & Dietzler (1984) *Arch. Biochem. Biophys.* **235**, 493. **9.** Halonen & Nedergaard (2002) *J. Membr. Biol.* **188**, 183. **10.**

Adenosine 5'-Monophosphate 2',3'-Dialdehyde

This nucleotide analogue (FW$_{free-acid}$ = 345.21 g/mol), also known as dial AMP and oAMP, alkylates susceptible enzymes, most often combining with an ε-amino group of an active-site lysine residue. Because the dialdehyde is of limited stability, it is usually prepared by treating adenosine with a slight molar excess of sodium periodate, followed by a larger molar excess of ethylene glycol to reduce any unreacted periodate. **Target(s):** F$_1$ ATPase (1,2); pyruvate,orthophosphate dikinase (3-5); ribonucleotide reductase (6); fructose-1,6-bisphosphatase (7); sphingomyelin phosphodiesterase (8). **1.** Satre, Lunardi, Dianoux, *et al.* (1986) *Meth. Enzymol.* **126**, 712. **2.** de Melo, Satre & Vignais (1984) *FEBS Lett.* **169**, 101. **3.** Colman (1990) *The Enzymes*, 3rd ed., **19**, 283. **4.** Phillips (1988) *Biochemistry* **27**, 3314. **5.** Evans, Goss & Wood (1980) *Biochemistry* **19**, 5809. **6.** Cory & Mansell (1975) *Cancer Res.* **35**, 390. **7.** Sakai, Kawashima, Suzuki & Imahori (1987) *J. Biochem.* **102**, 377. **8.** Quintern, Weitz, Nehrkorn, *et al.* (1987) *Biochim. Biophys. Acta* **922**, 323.

Adenosine 5'-Monosulfate

This AMP analogue (FW$_{free-acid}$ = 347.31 g/mol; λ$_{max}$ = 259 nm; ε$_M$ = 15400 M^{-1}cm^{-1}) competes with many adenine-containing nucleotides and coenzymes, including AMP, ADP, ATP, NAD(H), NADP(H), FAD(H), *etc.* **Target(s):** D-amino-acid oxidase (1-3; adenylate kinase (4); rhodopsin kinase (5); adenosine 5' phosphosulfate kinase, moderately inhibited (6); adenosine deaminase (7); sulfate adenylyltransferase (8); 5'-nucleotidase (9); nucleotide diphosphatase (10,11); estrone sulfotransferase (12). **1.** Yagi (1971) *Meth. Enzymol.* **17B**, 608. **2.** Meister & Wellner (1963) *The Enzymes*, 2nd ed., 7, 609. **3.** Yagi & Ozawa (1959) *Nature* **184** (Suppl. 16), 1227. **4.** Callaghan & Weber (1959) *Biochem. J.* **73**, 473. **5.**

Palczewski, McDowell & Hargrave (1988) *J. Biol. Chem.* **263**, 14067. **6.** Renosto, Martin & Segel (1991) *Arch. Biochem. Biophys.* **284**, 30. **7.** Challa, Johnson, Robertson & Gunasekaran (1999) *J. Basic Microbiol.* **39**, 97. **8.** Renosto, Patel, Martin, *et al.* (1993) *Arch. Biochem. Biophys.* **307**, 272. **9.** Carter & Tipton (1986) *Phytochemistry* **25**, 33. **10.** Kahn & Anderson (1996) *J. Biol. Chem.* **261**, 6016. **11.** Jacobson & Kaplan (1957) *J. Biol. Chem.* **226**, 427. **12.** Horwitz, Misra, Rozhin, *et al.* (1978) *Biochim. Biophys. Acta* **525**, 364.

P^1-(Adenosine-5')-P^3-(nicotinamide riboside-5')triphosphate, *Similar in action to P^1-(Adenosine-5')-P^4-(nicotinamide riboside-5')tetraphosphate*

P^1-(Adenosine-5')-P^4-(nicotinamide riboside-5')tetraphosphate

This multisubstrate geometric analogue (FW$_{\text{free-acid}}$ = 816.4 g/mol) targets nicotinamide-nucleotide adenylyltransferase, most likely occupying both the NMN and ATP substrate subsites (1,2). (***See** Adenylate Kinase Inhibitors for discussion of the related agents, P^1,P^4-di-(adenosine-5')tetraphosphate; P^1,P^5-di-(adenosine-5')pentaphosphate*) **1.** Franchetti, Cappellacci, Pasqualini, *et al.* (2003) *Nucleosides Nucleotides Nucleic Acids* **22**, 865. **2.** Sorci, Cimadamore, Scotti, *et al.* (2007) *Biochemistry* **46**, 4912.

Adenosine N^1-Oxide

This oxidized purine-containing nucleoside (FW = 283.24 g/mol), also called adenosine 1-*N*-oxide, inhibits adenosylhomocysteinase (1,2), dihydroorotase (3), GMP synthetase (4), and adenosine kinase (also alternative substrate) (5). **1.** Guranowski, Montgomery, Cantoni & P. K. Chiang (1981) *Biochemistry* **20**, 110. **2.** Shimizu, Shiozaki, Ohshiro & Yamada (1984) *Eur. J. Biochem.* **141**, 385. **3.** Bresnick & Blatchford (1964) *Biochim. Biophys. Acta* **81**, 150. **4.** Spector & Beecham III (1975) *J. Biol. Chem.* **250**, 3101. **5.** Long & Parker (2006) *Biochem. Pharmacol.* **71**, 1671.

Adenosine 5'-(β,γ-Peroxytriphosphate)

This ATP analogue (FW = 523.18 g/mol), abbreviated AMP-PO$_2$P, is a competitive inhibitor of a number of kinases. It is also an alternative substrate for NAD pyrophosphorylase. **Target(s):** glycerokinase (K_i = 140 µM)1; hexokinase (K_i = 90 µM); nucleoside diphosphate kinase (K_i = 3.1 mM); phosphofructokinase (K_i = 200 µM); phosphoglycerate kinase (K_i = 400 µM); phosphoribosyl-pyrophosphate synthetase (K_i = 33 µM). NAD$^+$ pyrophosphorylase is the only enzyme, among 13 tested, that uses adenosine 5'-(β,γ-peroxytriphosphate) as a substrate. The peroxy compounds tested inactive with *Escherichia coli* RNA polymerase and DNA polymerase I, as well as with wheat germ RNA polymerase II. **1.** Gibson & Leonard (1984) *Biochemistry* **23**, 78.

Adenosine 3'-Phosphate 5'-Phosphosulfate

This nucleotidyl-sulfate (FW$_{\text{free-acid}}$ = 507.27 g/mol; *Symbol*: PAPS), also known as 3'-phosphoadenosine-5'-phosphosulfate and 3' phospho-adenylylsulfate, is an important metabolite in sulfate reduction and in the formation of sulfate esters. Note that PAPS is an unstable derivative; the solid decomposes at 15-20% per day at 37°C. Commercial preparations can often contain substantial amounts of adenosine 3',5'-bisphosphate and sulfate as impurities. AMPS is even more unstable under acidic conditions; the half-life at 37°C in 0.1 M HCl is 6 minutes. Aqueous solutions are more stable in alkaline conditions. Solutions requiring storage should be kept frozen at pH 8.0. In addition, solutions should be kept free of ammonia-free, because NH$_3$ readily reacts with PAPS to produce the 5' phosphoramidate. PAPS is readily prepared in the laboratory via the action of ribonuclease T$_2$ on adenosine 2',3'-cyclic phosphate 5'-phosphosulfate (1). **Target(s):** sulfate adenylyltransferase, *or* ATP Sulfurylase (2,3); threonine synthase, weakly inhibited (4). **1.** Horwitz, Neenan, Misra, *et al.* (1977) *Biochim. Biophys. Acta* **480**, 376. **2.** Peck (1974) *The Enzymes*, 3rd ed., **10**, 651. **3.** Hanna, Ng, MacRae, *et al.* (2004) *J. Biol. Chem.* **279**, 4415. **4.** Giovanelli, Mudd, Datko & Thompson (1986) *Plant Physiol.* **81**, 577.

Adenosine 5'-*O*-(γ-Thio)triphosphate

This ATP analogue (FW$_{\text{free-acid}}$ = 523.25 g/mol), also known as adenosine 5'-(3-thiotriphosphate) and ATPγ-S, is an inhibitor and/or alternative substrate of a number of ATP-dependent systems. ATPγ-S is a P2 purinergic agonist and will increase the activity of Ca^{2+}-activated K$^+$ channels. (P1 and P2 purinergic receptor refers to classes of membrane receptors that mediate relaxation of gut smooth muscle as a response to the release of adenosine and ATP, respectively. **Target(s):** diphosphomevalonate decarboxylase (1,12); folylpolyglutamate synthetase (2,7); protein-tyrosine phosphatase (3); recA ATPase (4); pyruvate dehydrogenase complex (5); succinyl CoA synthetase (6); RNA ligase, ATP-dependent (8); glutaminyl-tRNA synthetase, glutamine-hydrolyzing (9); CTP synthetase (10); tetrahydrofolate synthetase (11); chaperonin ATPase (13); microtubule-severing ATPase, *or* katanin (14,15); polar-amino-acid-transporting ATPase, *or* histidine permease (16,17); Na$^+$-transporting two-sector ATPase (18); Na$^+$-exporting ATPase (18); DNA helicase (19); nucleoside-truphosphatase (19); insulysin (20); endopeptidase La (21); protein-tyrosine kinase (22); Src protein tyrosine kinase (22); selenide,water dikinase (23); acetate kinase (24); diacylglycerol kinase (25,26); 1-phosphatidylinositol 4-kinase (27,28); galactokinase (29). **1.** Cardemil & Jabalquinto (1985) *Meth. Enzymol.* **110**, 86. **2.** Bognar & Shane (1986) *Meth. Enzymol.* **122**, 349. **3.** Zhao (1996) *Biochem. Biophys. Res. Commun.* **218**, 480. **4.** Weinstock, McEntee & Lehman (1981) *J. Biol. Chem.* **256**, 8850. **5.** Radcliffe, Kerbey & Randle (1980) *FEBS Lett.* **111**, 47. **6.** Nishimura & Mitchell (1984) *J. Biol. Chem.* **259**, 9642. **7.** Cichowicz & Shane (1987) *Biochemistry* **26**, 513. **8.** Juodka & Labeikytė (1991) *Nucleosides Nucleotides* **10**, 367. **9.** Horiuchi, Harpel, Shen, *et al.* (2001) *Biochemistry* **40**, 6450. **10.** Willemoës & Sigurskjold (2002) *Eur. J. Biochem.* **269**, 4772. **11.** Cichowicz & Shane (1987) *Biochemistry* **26**, 513. **12.** Jabalquinto & Cardemil (1989) *Biochim. Biophys. Acta* **996**, 257. **13.** Meyer, Gillespie, Walther, *et al.* (2003) *Cell* **113**, 369. **14.** McNally & Vale (1993) *Cell* **75**, 419. **15.** Lohret, McNally & Quarmby (1998) *Mol. Biol.* **9**, 1195. **16.** Nikaido, Liu & Ames (1997) *J. Biol. Chem.* **272**, 27745. **17.** Liu, Liu & Ames (1997) *J. Biol. Chem.* **272**, 21883. **18.** Murata, Igarashi, Kakinuma & Yamato (2000) *J. Biol. Chem.* **275**, 13415. **19.** Borowski, Niebuhr, Schmitz, Hosmane, Bretner, Siwecka & Kulikowski (2002) *Acta Biochim. Pol.* **49**, 597. **20.** Camberos, Pérez,

Udrisar, Wanderley & Cresto (2001) *Exp. Biol. Med. (Maywood)* **226**, 334. **21**. Larimore, Waxman & Goldberg (1982) *J. Biol. Chem.* **257**, 4187. **22**. Nam, Lee, Ye, Sun & Parang (2004) *Bioorg. Med. Chem.* **12**, 5753. **23**. Veres, Kim, Scholz & Stadtman (1994) *J. Biol. Chem.* **269**, 10597. **24**. Gorrell, Lawrence & Ferry (2005) *J. Biol. Chem.* **280**, 10731. **25**. Kato & Takenawa (1990) *J. Biol. Chem.* **265**, 794. **26**. Wissing & Wagner (1992) *Plant Physiol.* **98**, 1148. **27**. Scholz, Barritt & Kwok (1991) *Eur. J. Biochem.* **201**, 249. **28**. Flanagan & Thorner (1992) *J. Biol. Chem.* **267**, 24117. **29**. Lavine, Cantlay, Roberts & Morse (1982) *Biochim. Biophys. Acta* **717**, 76.

Adenosine 5'-Triphosphate

Adenosine 5'-Triphosphate Tri-anion

⬤ = H_2O

Me^{2+} Complex with Adenosine 5'-Triphosphate

This adenine nucleotide ($FW_{\text{free-acid}}$ = 507.18 g/mol; CAS 56-65-5; Molar Absorptivity = 15,400 $M^{-1}cm^{-1}$, λ = 259 nm) is the phosphoryl-donor substrate for numerous transphosphorylases, phosphohydrolases, and molecular motors; as such, ADP often inhibits these enzymes. **Enzymatic Phosphorylation:** ATP is a substrate for adenylate kinase reaction: $MeATP^{2-} + AMP \rightleftharpoons ADP^{2-} + MeADP$).It is also the product of other reactions that stabilize ATP concentrations in prokaryotes [*e.g.*, acetate kinase (*Reaction*: $MgADP$ + Acetyl-phosphate $\rightleftharpoons MeATP^{2-}$ + Acetate)] and eukaryotes [*e.g.*, pyruvate kinase (*Reaction*: $MgADP$ + Phosphoenolpyruvate $\rightleftharpoons MgATP^{2-}$ + Pyruvate), creatine kinase (*Reaction*: $MgADP$ + Creatine-phosphate $\rightleftharpoons MgATP^{2-}$ + Creatine), arginine kinase (*Reaction*: $MgADP$ + Arginine-phosphate $\rightleftharpoons MgATP^{2-}$ + Arginine), and nuclecleotide diphosphate kinase (*Reaction*: $ADP^{2-} + MgGTP^{2-} \rightleftharpoons MgATP^{2-} + GDP^{2-}$)]. **ATP Synthase:** ADP is a primary product of the F_OF_1 and CF_OCF_1 ATP synthases (*Reaction*: $MgADP + P_i$ + High Chemiosmotic Gradient Energization State $\rightleftharpoons MgATP^{2-}$ + Low Chemiosmotic Gradient Energization State). When the proton motive is low, absent, or uncoupled, the synthase catalyzes hydrolysis of ATP (*Reaction*: $MgATP^{2-} + H_2O \rightarrow MgADP + P_i$). **Metal Ion Binding Properties:** As a polyanion, ATP not only binds physiologic divalent cations Mg^{2+} and Ca^{2+}, but also forms reversible complexes with Mn^{2+} and Co^{2+}. For reversible complexation of ATP^{4-} with a metal ion Me^{2+}, (*Reaction*: $ATP^{4-} + Me^{2+} \rightleftharpoons MeATP^{2-}$), $K_{\text{formation}}$ = $[MeATP^{2-}]/[ATP^{4-}]_{\text{free}}[Me^{2+}]_{\text{free}}$, indicating that $[MeATP^{2-}]/[ATP^{4-}]_{\text{free}}$ = $K_{\text{formation}} \times [Me^{2+}]_{\text{free}}$. Formation constants typically range from 10^4–10^5 M^{-1}. In many cases, metal-free ATP is not a substrate and instead acts as a reversible inhibitor. In the case of bovine brain hexokinase, metal-free ATP^{4-} is a linear competitive inhibitor relative to $MgATP^{2-}$ complex (1). $HATP^{3-}$ also inhibits this enzyme. Good experimental design therefore requires rigorous control of free metal ion concentration to control the ratio of metal-bound and metal-free forms (2,3). When exposed to Cr(III) at elevated temperature, ATP also forms ligand exchange-inert complexes with Cr^{3+} (3). **False Inhibition by Metal Ion Chelation:** Many enzymes require divalent or trivalent metal ions for activity, and the addition of metal ion-free ATP inhibits such enzymes by sequestering metal ions. When ATP is found to inhibit an enzyme of interest, it is advisable to determine whether the observed effect remains after ATP is augmented by sufficient Mg^{2+} to form $MgATP^{2-}$, thereby minimizing nonphysiologic effects of ATP^{4-}. The simplest way to test for this possibility is to add equimolar amounts of ATP and $MgCl_2$ to a buffer already supplemented with 1-3 mM additional

$MgCl_2$. Under such conditions, >95% of the ATP will be $MgATP^{2-}$ complex (1-3). **Target(s):** Phosphofructokinase, liver – low ATP concentrations inhibit PFK by decreasing its affinity for fructose 6-phosphate (citrate and other tricarboxylic acid cycle intermediates also inhibit, and this inhibition was relieved by either AMP or fructose 1,6-diphosphate; however, higher concentrations of ATP decrease and finally remove the effect of these activators (4,5); brain hexokinase (by metal-free ATP^{4-}) – the apparently sigmoidal saturation curves for $MgATP^{2-}$ observed by Bachelard (6) studies can be corrected to hyperbolic curves by use of a stability constant for $MgATP^{2-}$ complex formation, especially when corrected for the presence of the inhibitory free uncomplexed ATP^{4-} concentration (1), which presents a simple model demonstrating how enzymes obeying Michaelis-Menten kinetics can falsely demonstrate sigmoidal velocity responses, if the true substrate is the metal-substrate complex (1); phosphorylase *b* (7); locust pyruvate kinase (8); skeletal muscle pyruvate kinase reaction, dependence on the total magnesium ion concentration (9); glucose-6-phosphate dehydrogenase, variability in ATP inhibition (10); *Rhodopseudomonas spheroides* aminolevulinate (ALA) synthetase activity (11); phosphorylase *b*, effect on AMP activation (12); leaf adenosine diphosphate sulphurylase (13); *Escherichia coli* phosphofructokinase, ATP-sensitive and ATP-insensitive forms (14); cardiac mitochondrial pyruvate dehydrogenase (15); phosphofructokinase, skeletal muscle – citrate, 3-P-glycerate and P-enolpyruvate each act synergistically in the presence of ATP to inhibit the PFK reaction, and the sensitivity to these inhibitors is more pronounced at elevated ATP concentrations; moreover, these metabolites do not inhibit, when ITP is used as the phosphoryl donor (16); *Escherichia coli* glutaminase, metal ion effects on (17); terminal deoxynucleotidyltransferase, mechanism of ATP (18); *Kluyveromyces lactis* phosphofructokinase, ATP inhibition and Pasteur Effect (19); human phosphofructokinase isozymes, ATP inhibition (20); NAD-linked glutamate dehydrogenase (21); liver phosphofructokinase, lessened ATP inhibition after hormonally stimulated phosphorylation (22); muscle AMP deaminase, lowered ATP inhibition at higher (physiologic) ATP concentration (23); mammalian brain Na^+/K^+-ATPase (24); ATP synthase, catalytic and regulatory features of ATP inhibition (25); maize leaf acetyl-coenzyme A carboxylase,inhibition by ATP^{4-} but not $MgATP^{2-}$ (26); bovine heart NAD-specific isocitrate dehydrogenase, activation and inhibition by ATP (27); ATP-inhibited K+ channels, paradoxical role of ATP in maintain its channel-open state (28); human red cell sugar transporter (29); adult and fetal myocardial phosphofructokinase, relief of cooperativity and competition between fructose 2,6-bisphosphate, ATP, and citrate (30); muscle phosphofructokinase, desensitization to ATP inhibition by removal of a carboxyl-terminal heptadecapeptide (31); soluble 5'-nucleotidase of rat kidney (32); DNA topoisomerase, type I (33); modulation of nucleotide-sensitive K^+ channel gating in insulin-secreting cells by changes in the ratio ATP^{4-}/ADP^{3-} and by nonhydrolyzable derivatives of both ATP and ADP (34); mitochondrial NAD(P)-malic enzyme, from herring skeletal muscle (35); 6-Phosphofructo-2-kinase and fructose-2,6-bisphosphatase, *Saccharomyces cerevisiae* (36); 5-nucleotidase, rat heart (37); CTP:cholinephosphate cytidylyltransferase, castor bean endosperm (38); *Phycomyces* pyruvate kinase (39); NAD(P)$^+$-malic enzyme, ATP inhibition competes with activating cations in modulating the activity in the mitochondrial matrix of *Xenopus laevis* oocytes (40); neutrophil migration (41); *Escherichia coli* phosphofructo-1-kinase (42); ventricular myocyte calcium current, neuromodulation by extracellular ATP (43); Ca^{2+}-dependent K^+ channels, in soma membrane of cultured leech *Retzius* neurons (44); ADP-ribosylation of membrane-bound actin and actin-binding to membranes (45); agonist-induced mobilization of internal calcium in human platelets by extracellular ATP (46); *Saccharomyces cerevisiae* glycine amide ribonucleotide synthetase (47); *Saccharomyces cerevisiae* aminoimidazole ribonucleotide synthetase (47); adrenergic agonist-induced hypertrophy of neonatal cardiac myocytes, inhibition by extracellular ATP (48); hepatic fatty acid metabolism, inhibition by extracellular ATP (49); Cytochrome *c* oxidase, allosteric ATP inhibition in eukaryotes but not prokaryotes (50); ATP-inhibitable K^+ channels in insulin secreting cells (51); ATP inhibition of K_{ATP} channels, PIP$_2$ and PIP as determinants of (52); kidney phosphofructokinase, relief from ATP inhibition by ribose 1,5-bisphosphate (53,54); glycolytic flux in *Saccharomyces cerevisiae*, targeting PFK and PK (55); inositol 1,4,5-trisphosphate receptor type-1 and type-3, differential modulation by ATP (56); 2-oxoglutarate dehydrogenase complex (57); Mg^{2+} uptake in MDCT cells via P2X purinoceptors (58); NMDA receptors in cultured hippocampal neurons (59); endothelin-1 production by rat inner medullary collecting duct cells (60); UDP-glucuronosyltransferase, *or* UGT (61);

Ins(1,4,5)P₃-evoked Ca²⁺ release in smooth muscle via P2Y1 receptors (62). **1**. Purich & Fromm (1972) *Biochem. J.* **130**, 63. **2**. Fromm (1975) *Initial Rate Enzyme Kinetics*, Springer. **3**. Purich (2010) *Enzyme Kinetics: Catalysis & Control*, Acaademic Press. **4**. Underwood & Newsholme (1965) *Biochem. J.* **95**, 868. **5**. Underwood & Newsholme (1967) *Biochem. J.* **104**, 296. **6**. Newsholme, Rolleston & Taylor (1968) *Biochem. J.* **106**, 193. **7**. Damjanovich, Sümegi, Tóth (1968) *Experientia* **24**, 351. **8**. Bailey & Walker (1969) *Biochem. J.* **111**, 359. **9**. Holmsen & Storm (1969) *Biochem. J.* **112**, 303. **10**. Smith & Anwer (1971) *Experientia* **27**, 835. **11**. Fanica-Gaignier & Clement-Metral (1971) *Biochem. Biophys. Res. Commun.* **44**, 192. **12**. Scopes (1973) *Biochem. J.* **134**, 197. **13**. Burnell & Anderson (1973) *Biochem. J.* **133**, 417. **14**. Doelle (1975) *Eur. J. Biochem.* **50**, 335. **15**. Chiang & Sacktor (1975) *J. Biol. Chem.* **250**, 3399. **16**. Colombo, Tate, Girotti & Kemp (1975) *J. Biol. Chem.* **250**, 9404. **17**. Prusiner & Stadtman (1976) *J. Biol. Chem.* **251**, 3463. **18**. Modak (1978) *Biochemistry* **17**, 3116. **19**. Royt & MacQuillan (1979) Antonie Van Leeuwenhoek. **45**, 241. **20**. Kahn, Meienhofer, Cottreau, Lagrange & Dreyfus (1979) *Human Genet.* **48**, 93. **21**. Tokushige, Miyamoto & Katsuki (1979) *J. Biochem.* **85**, 1415. **22**. Kagimoto & Uyeda (1979) *J. Biol. Chem.* **254**, 5584. **23**. Wheeler & Lowenstein (1979) *J. Biol. Chem.* **254**, 8994. **24**. Swann (1983) *Arch. Biochem. Biophys.* **221**, 148. **25**. Boyer (2000) *Biochim. Biophys. Acta* **1458**, 252. **26**. Nikolau & Hawke (1984) *Arch. Biochem. Biophys.* **228**, 86. **27**. Gabriel, Milner & Plaut (1985) *Arch. Biochem. Biophys.* **240**, 128. **28**. Findlay & Dunne (1986) *Pflugers Arch.* **407**, 238. **29**. Carruthers (1986) *J. Biol. Chem.* **261**, 11028. **30**. Bristow, Bier, Lange (1987) *J. Biol. Chem.* **262**, 2171. **31**. Valaitis, Foe & Kemp (1987) *J. Biol. Chem.* **262**, 5044. **32**. Le Hir & Dubach (1988) *Am. J. Physiol.* **254** (Pt 2), F191. **33**. Chen & Castora (1988) *Biochemistry* **27**, 4386. **34**. Dunne, West-Jordan, Abraham, Edwards & Petersen (1988) *J. Membr. Biol.* **104**, 165. **35**. Skorkowski & Storey (1988) *Fish Physiol. Biochem.* **5**, 241. **36**. Hofmann, Bedri, Kessler, Kretschmer & Schellenberger (1989) *Adv. Enzyme Regul.* **28**, 283. **37**. Headrick & Willis (1989) *Biochem. J.* **261**, 541. **38**. Wang & Moore (1989) *Arch. Biochem. Biophys.* **274**, 338. **39**. Del Valle, Busto, De Arriaga & Soler (1990) *J. Enzyme Inhib.* **3**, 219. **40**. Petrucci & Cesare (1990) *Int. J. Biochem.* **22**, 137. **41**. Boonen, van St.eveninck, de Koster & Elferink (1991) *Agents Actions* **32**, 100. **42**. Zheng & Kemp (1992) *J. Biol. Chem.* **267**, 23640. **43**. Qu, Campbell, Himmel & Strauss (1993) *Adv. Exp. Med. Biol.* **346**, 11. **44**. Frey, Hanke & Schlue (1993) *J. Membr. Biol.* **134**, 131. **45**. Schroeder, Just & Aktories (1994) *Eur. J. Cell Biol.* **63**, 3. **46**. Soslau, McKenzie, Brodsky & Devlin (1995) *Biochim. Biophys. Acta* **1268**, 73. **47**. Tret'iakov, Ryzhova, Velichutina, et al. (1995) *Biokhimiia* **60**, 2011. **48**. Zheng, Boluyt, Long, et al. (1996) *Circ. Res.* **78**, 525. **49**. Guzmán, Velasco & Castro (1996) *Am. J. Physiol.* **270**, G701. **50**. Follmann, Arnold, Ferguson-Miller & Kadenbach (1998) *Biochem. Mol. Biol. Int.* **45**, 1047. **51**. Dzeja, Zeleznikar & Goldberg (1998) *Mol. Cell Biochem.* **184**, 169. **52**. Baukrowitz, Schulte, Oliver, et al. (1998) *Science* **282**, 1141. **53**. Ozeki, Mitsui, Sugiya & Furuyama (1999) *Comp. Biochem. Physiol. B, Biochem. Mol. Biol.* **124**, 327. **54**. Sawada, Mitsui, Sugiya & Furuyama (2000) *Int. J. Biochem. Cell Biol.* **32**, 447. **55**. Larsson, Påhlman & Gustafsson (2000) *Yeast* **16**, 797. **56**. Maes, Missiaen, De Smet, et al. (2000) *Cell Calcium* **27**, 257. **57**. Rodríguez-Zavala, Pardo & Moreno-Sánchez (2000) *Arch. Biochem. Biophys.* **379**, 78. **58**. Dai, Kang, Kerstan, Ritchie & Quamme (2001) *Am. J. Physiol. Renal Physiol.* **281**, F833. **59**. Ortinau, Laube & Zimmermann (2003) *J. Neurosci.* **23**, 4996. **60**. Hughes, Stricklett, Kishore & Kohan (2006) *Exp. Biol. Med.* **231**, 1006. **61**. Ishii, An, Nishimura & Yamada (2012) *Drug Metab. Dispos.* **40**, 2081. **62**. MacMillan, Kennedy & McCarron (2012) *J. Cell Sci.* **125**, 5151.

[5-(Adenosin-5'-O-yl)pentyl]cobalamin

This vitamin B₁₂ analogue (FW = 1665.74 g/mol) inhibits ribonucleoside-triphosphate reductase (K_i = 7.7 µM). Note that the propyl, butyl, hexyl, and heptyl homologues also inhibit (K_i = 55.8, 18.9, 24.6, and 12.8 µM, respectively). In addition, all five (adenosin-5'-O-ylalkyl)cobalamins also inhibit glycerol dehydratase (K_i = 10.5, 9.7, 5.9, 15.1, and 11.7 nM for the C3, C4, C5, C6, and C7 analogues, respectively), diol dehydratase (K_i = 0.77, 0.86, 0.50, 0.83, and 0.63 µM, respectively), and methylmalonyl-CoA mutase (K_i = 2.48, 1.45, 1.13, 0.77, and 2.10 µM, respectively). **Target(s):** diol dehydratase (1,2); glycerol dehydratase (1,2); methylmalonyl-CoA mutase (1,3); ribonucleoside-triphosphate reductase (1). **1**. Suto, Poppe, Rétey & Finke (1999) *Bioorg. Chem.* **27**, 451. **2**. Poppe & Rétey (1997) *Eur. J. Biochem.* **245**, 398. **3**. Poppe & Rétey (1995) *Arch. Biochem. Biophys.* **316**, 541.

S-(5'-Adenosyl)-L-cysteine

This substituted cysteine (FW = 370.39 g/mol; CAS 35899-53-7) is an alternative substrate for S-adenosylhomocysteinases, often acting as a potent competitive inhibitor that protracts each catalytic round. **Target(s):** acyl-homoserine-lactone synthase (1); spermidine synthase (2); mRNA (guanine-N⁷-)-methyltransferase (3); mRNA (nucleoside-2'-O-)-methyltransferase (3); histone-lysine N methyltransferase (4); tRNA (adenine-N¹-)-methyl-transferase (5); tRNA (guanine-N⁷-)methyltransferase (6). **1**. Parsek, Val, Hanzelka, Cronan & Greenberg (1999) *Proc. Natl. Acad. Sci. U.S.A.* **96**, 4360. **2**. Hibasami, Borchardt, Chen, Coward & Pegg (1980) *Biochem. J.* **187**, 419. **3**. Pugh & Borchardt (1982) *Biochemistry* **21**, 1535. **4**. Tuck, Farooqui & Paik (1985) *J. Biol. Chem.* **260**, 7114. **5**. Salas & Sellinger (1978) *J. Neurochem.* **31**, 85. **6**. Paolella, Ciliberto, Traboni, Cimino & Salvatore (1982) *Arch. Biochem. Biophys.* **219**, 149.

S-Adenosyl-D-homocysteine

This epimer (FW = 384.41 g/mol; CAS 979-92-0) of S-adenosyl-L-homocysteine inhibits acyl-homoserine-lactone synthase. **Target(s):** acyl-homoserine-lactone synthase (1,2); [cytochrome c]-arginine N-methyltransferase (3); [cytochrome c]-methionine S-methyltransferase (3); histone-arginine N-methyltransferase (4); [cytochrome c]-lysine N-methyltransferase (5); mRNA (guanine-N⁷-)-methyltransferase (6,7); mRNA (nucleoside-2'-O-)-methyltransferase (6); tRNA (adenine-N¹-)-methyltransferase (8); tRNA (uracil-5-)-methyltransferase (9); tRNA (guanine-N⁷-)-methyltransferase (10); tRNA (guanine-N²-)-methyltrans-ferase (11). **1**. Parsek, Val, Hanzelka, Cronan, & Greenberg (1999) *Proc. Natl. Acad. Sci. U.S.A.* **96**, 4360. **2**. Hanzelka, Parsek, Val, et al. (1999) *J. Bacteriol.* **181**, 5766. **3**. Farooqui, Tuck & Paik (1985) *J. Biol. Chem.* **260**, 537. **4**. Gupta, Jensen, Kim & Paik (1982) *J. Biol. Chem.* **257**, 9677. **5**. DiMaria, S. Kim & Paik (1982) *Biochemistry* **21**, 1036. **6**. Pugh & Borchardt (1982) *Biochemistry* **21**, 1535. **7**. Pugh, Borchardt & Stone (1977) *Biochemistry* **16**, 3928. **8**. Salas & Sellinger (1978) *J. Neurochem.* **31**, 85. **9**. Shugart & Chastain (1979) *Enzyme* **24**, 353. **10**. Paolella, Ciliberto, Traboni, Cimino & Salvatore (1982) *Arch. Biochem. Biophys.* **219**, 149. **11**. Hildesheim, Hildesheim, Blanchard, Farrugia & Michelot (1973) *Biochimie* **55**, 541.

S-Adenosyl-L-homocysteine

This S-substituted homocysteine (FW = 384.42 g/mol; CAS 979-92-0; Abbreviation: SAH) is an L-cysteine synthesis intermediate and potent product inhibitor of numerous SAM-dependent methyltransferases. SAH is also a substrate for adenosylhomocysteine nucleosidase, adenosylhomocysteinase, and S-adenosylhomocysteine deaminase. S-adenosyl-L-homocysteine is relatively stable in neutral or alkaline solutions; however, oxidation is more rapid in alkaline conditions. Because solid SAH slowly oxidizes to its sulfoxide, it should be stored under nitrogen and at a low temperature. 2,2'-Thioethanol (thiodiglycol) is often added to protect against oxidation. Treatment of SAH in 0.1 M HCl at 100°C results in the formation of S-ribosyl-L-homocysteine in ninety minutes. **Target(s):** L-lysine 2,3-aminomutase (1,15); S-adenosylmethionine hydrolase (2,18); formate C-acetyltransferase, or pyruvate formate lyase (3,37); spermidine synthase, or putrescine aminopropyltransferase (4); 5'-methylthioadenosine nucleosidase (5,19); 1-aminocyclopropane-1 carboxylate synthase (6,16,17); O-acetylhomoserine aminocarboxy-propyltransferase, or O-acetyl-homoserine sulfhydrylase (7,26,27); type I site-specific Deoxyribonuclease (8,21); valyl-tRNA synthetase (9); cyclopropane-fatty-acyl-phospholipid synthase (10,93,94); homoserine O-acetyltransferase (11); guanidinoacetate methyltransferase (12); leucine transport (13); S-adenosylmethionine cyclotransferase (14); ribonuclease H (20); S-methyl-5-thioribose kinase (22); adenosine kinase (23); adenosyl-methionine:8-amino-7-oxononanoate transaminase or diaminopelargonate aminotransferase (24); adenosyl-fluoride synthase (25); cystathionine γ synthase (28); spermine synthase, weakly inhibited (29); methionine S-adenosyltransferase (30,31); S-adenosylmethionine cyclotransferase (32,33); S-methyl-5'-thioadenosine phosphorylase, weakly inhibited (34); acyl-homoserine-lactone synthase (35,36); homoserine O-acetyltransferase, (38); glycine/sarcosine/ dimethylglycine N-methyl-transferase (39); dimethylglycine N-methyl-transferase (40); sarcosine/ dimethylglycine N-methyltransferase (40,41); glycine/sarcosine N-methyltransferase (40,41); kaempferol 4'-O-methyltransferase (42); isoliquiritigenin 2'-O methyltransferase (43); vitexin 2''-O-rhamnoside 7-O-methyltransferase (44); corydaline synthase (45); 3'-demethylstaurosporine O-methyltransferase (47); chlorophenol O-methyltransferase (48); inositol 4-methyltransferase (49); (RS)-norcoclaurine 6-O-methyltransferase (50,51); [myelin basic protein] arginine N-methyltransferase (52,53); histone-arginine N-methyltransferase (54-56); [cytochrome c] arginine N-methyltransferase (57); [cytochrome c]-methionine S-methyltransferase (57); (S)-tetrahydroprotoberberine N-methyltransferase (58,59); 6-O-methylnorlaudanosoline 5'-O methyltransferase (60); coilumbamine O-methyltransferase (61); site-specific DNA-methyltransferase, cytosine-N4-specific, M.BamHI (62); site-specific DNA-methyltransferase adenine-specific (62,99-103); sterigmatocystin 8-O-methyltransferase (63); 6-hydroxymellein O-methyltransferase (64,65); uroporphyrinogen-III C-methyltransferase (66); tryptophan 2-C methyltransferase, weakly inhibited (67); caffeoyl-CoA O-methyltransferase (68); phosphoethanolamine N-methyltransferase (69-71); demethylmacrocin O-methyltransferase (72); macrocin O-methyltransferase (73); protein-S-isoprenylcysteine O-methyltransferase (74-82); thioether S-methyltransferase (83); tocopherol methyltransferase (84); tabersonine 16-O-methyltransferase (85); isobutyraldoxime O-methyltransferase (86); 8-hydroxyquercetin 8-O methyltransferase (87); pyridine N-methyltransferase (88); methylquercetagetin 6-O-methyltransferase (89); 3,7-dimethyl-quercetin 4'-O-methyltransferase (89); protein-glutamate O-methyltransferase (90-92); protein-L-isoaspartate (D-aspartate) O-methyltransferase (95-97); quercetin 3-O-methyltransferase (89); apigenin 4'-O-methyltransferase (98); phosphatidyl-N methylethanolamine N-methyltransferase (104); caffeate O-methyltransferase (105-108); thiopurine S methyltransferase (109,110); rRNA (adenosine-2'-O-)-methyltransferase (111); calmodulin-lysine N-methyltransferase (112,113); [cytochrome c]-lysine N-methyltransferase (114-118); mRNA (guanine N7-)-methyltransferase (119,123-126); mRNA (nucleoside-2'-O-)-methyltransferase (119-122); putrescine N-methyltransferase (127,128); rRNA (adenine-N6-)-methyltransferase (129); isoflavone 4' O-methyltransferase (130);

dimethylhistidine N-methyltransferase (131); histone-lysine N methyltransferase (132-134); luteolin O-methyltransferase (135); sterol 24-C-methyltransferase, or cycloartenol 24-C-methyltransferase (46,136); DNA (cytosine-5-)-methyltransferase (137-140); HaeIII DNA methyltransferase (137); HhaI DNA methyltransferase (137); tRNA (adenine-N1-) methyltransferase (141-146); tRNA (uracil-5-)-methyltransferase (147-152); tRNA guanosine-2'-O methyltransferase (153); tRNA (guanine-N7-)-methyltransferase (150,154); tRNA (guanine-N2-) methyltransferase (144,155-158); tRNA (guanine-N1-)-methyltransferase (150,158-161); tRNA (cytosine 5-)-methyltransferase (150,162,163); phenylethanolamine N-methyltransferase (164-166); glycine N methyltransferase (167-169); phosphatidylethanolamine N-methyltransferase (170-173); fatty-acid O-methyltransferase (174,175); magnesium protoporphyrin IX methyltransferase (176-181); homocysteine S-methyltransferase (182); thiol S-methyltransferase (183-185); histamine N-methyltransferase (186-190); nicotinate N-methyltransferase (191); catechol O-methyltransferase (192-196); betaine:homocysteine S-methyltransferase (197); acetylserotonin O-methyltransferase, or hydroxyindole O-methyltransferase (198-200); guanidinoacetate N-methyltransferase (201-203); nicotinamide N methyltransferase (204). **1.** Chirpich & Barker (1971) Meth. Enzymol. **17B**, 215. **2.** Gefter (1971) Meth. Enzymol. **17B**, 406. **3.** Knappe & Blaschkowski (1975) Meth. Enzymol. **41**, 508. **4.** Pegg (1983) Meth. Enzymol. **94**, 294. **5.** Guranowski, Chiang & Cantoni (1983) Meth. Enzymol. **94**, 365. **6.** Adams & Yang (1987) Meth. Enzymol. **143**, 426. **7.** Yamagata (1987) Meth. Enzymol. **143**, 465. **8.** Endlich & Linn (1981) The Enzymes, 3rd ed., **14**, 137. **9.** Jakubowski (1982) Biochim. Biophys. Acta **709**, 325. **10.** Chung & Law (1964) Biochemistry **3**, 1989. **11.** Shiio & Ozaki (1981) J. Biochem. **89**, 1493. **12.** Im, Chiang & Cantoni (1979) J. Biol. Chem. **254**, 11047. **13.** Law & Ferro (1980) J. Bacteriol. **143**, 427. **14.** Mudd (1959) J. Biol. Chem. **234**, 87. **15.** Chirpich, Zappia, Costilow & Barker (1970) J. Biol. Chem. **245**, 1778. **16.** Jakubowicz (2002) Acta Biochim. Pol. **49**, 757. **17.** Nakajima & Imaseki (1986) Plant Cell Physiol. **27**, 969. **18.** Spoerel & Herrlich (1979) Eur. J. Biochem. **95**, 227. **19.** Yamakawa & Schweiger (1979) Dev. Cell Biol. **3**, 241. **20.** Stavrianopoulos, Gambino-Giuffrida & Chargraff (1976) Proc. Natl. Acad. Sci. U.S.A. **73**, 1087. **21.** Janscak, Abadjieva & Firman (1996) J. Mol. Biol. **257**, 977. **22.** Guranowski (1983) Plant Physiol. **71**, 932. **23.** Palella, Andres & Fox (1980) J. Biol. Chem. **255**, 5264. **24.** Stoner & Eisenberg (1975) J. Biol. Chem. **250**, 4037. **25.** Schaffrath, Deng & O'Hagan (2003) FEBS Lett. **547**, 111. **26.** Yamagata (1984) J. Biochem. **96**, 1511. **27.** Kerr (1971) J. Biol. Chem. **246**, 95. **28.** Kerr & Flavin (1969) Biochim. Biophys. Acta **177**, 177. **29.** Hibasami, Borchardt, Chen, Coward & Pegg (1980) Biochem. J. **187**, 419. **30.** Reguera, Balaña-Fouce, et al. (2002) J. Biol. Chem. **277**, 3158. **31.** Yarlett, Garofalo, Goldberg, et al. (1993) Biochim. Biophys. Acta **1181**, 68. **32.** Mudd (1959) J. Biol. Chem. **234**, 87. **33.** Swiatek, Simon & Chao (1973) Biochemistry **12**, 4670. **34.** Ferro, Wrobel & Nicolette (1979) Biochim. Biophys. Acta **570**, 65. **35.** Parsek, Val, Hanzelka, Cronan & Greenberg (1999) Proc. Natl. Acad. Sci. U.S.A. **96**, 4360. **36.** Hanzelka, Parsek, Val, et al. (1999) J. Bacteriol. **181**, 5766. **37.** Wood & Jungermann (1972) FEBS Lett. **27**, 49. **38.** Yamagata (1987) J. Bacteriol. **169**, 3458. **39.** Lai, Wang, Chuang, Wu & Lee (2006) Res. Microbiol. **157**, 948. **40.** Waditee, Tanaka, Aoki, et al. (2003) J. Biol. Chem. **278**, 4932. **41.** Nyyssola, Reinikainen & Leisola (2001) Appl. Environ. Microbiol. **67**, 2044. **42.** Curir, Lanzotti, Dolci, et al. (2003) Eur. J. Biochem. **270**, 3422. **43.** Maxwell, Edwards & Dixon (1992) Arch. Biochem. Biophys. **293**, 158. **44.** Knogge & Weissenböck (1984) Eur. J. Biochem. **140**, 113. **45.** Rueffer, Bauer & Zenk (1994) Can. J. Chem. **72**, 170. **46.** Nes, Song, Dennis, et al. (2003) J. Biol. Chem. **278**, 34505. **47.** Weidner, Kittelmann, Goeke, Ghisalba & Zähner (1998) J. Antibiot. **51**, 697. **48.** Coque, Alvarez-Rodríguez & Larriba (2003) Appl. Environ. Microbiol. **69**, 5089. **49.** Wanek & Richter (1995) Planta **197**, 427. **50.** Rueffer, Nagakura & Zenk (1983) J. Med. Plant Res. **49**, 131. **51.** Sato, Tsujita, Katagiri, Yoshida & Yamada (1994) Eur. J. Biochem. **225**, 125. **52.** Ghosh, Paik & Kim (1988) J. Biol. Chem. **263**, 19024. **53.** Park, Greenstein, Paik & Kim (1989) J. Mol. Neurosci. **1**, 151. **54.** Gupta, Jensen, Kim & Paik (1982) J. Biol. Chem. **257**, 9677. **55.** Rajpurohit, Lee, Park, Paik & Kim (1994) J. Biol. Chem. **269**, 1075. **56.** Lee, Kim & Paik (1977) Biochemistry **16**, 78. **57.** Farooqui, Tuck & Paik (1985) J. Biol. Chem. **260**, 537. **58.** O'Keefe & Beecher (1994) Plant Physiol. **105**, 395. **59.** Rueffer, Zumstein & Zenk (1990) Phytochemistry **29**, 3727. **60.** Rueffer, Nagakura & Zenk (1983) Planta Med. **49**, 196. **61.** Rueffer, Amann & Zenk (1986) Plant Cell Rep. **3**, 182. **62.** Malygin, Zinoviev, Evdokimov, et al. (2003) J. Biol. Chem. **278**, 15713. **63.** Liu, Bhatnagar & Chu (1999) Nat. Toxins **7**, 63. **64.** Kurosaki, Kizawa & Nishi (1989) Phytochemistry **28**, 1843. **65.** Kurosaki

(1996) *Phytochemistry* **41**, 1023. **66**. Blanche, Debussche, Thibaut, Crouzet & Cameron (1989) *J. Bacteriol.* **171**, 4222. **67**. Frenzel, Zhou & Floss (1990) *Arch. Biochem. Biophys.* **278**, 35. **68**. Pakusch & Matern (1991) *Plant Physiol.* **96**, 327. **69**. Smith, Summers & Weretilnyk (2000) *Physiol. Plant.* **108**, 286. **70**. Brendza, Haakenson, Cahoon, *et al.* (2007) *Biochem. J.* **404**, 439. **71**. Pessi, Kociubinski & Mamoun (2004) *Proc. Natl. Acad. Sci. U.S.A.* **101**, 6206. **72**. Kreuzman, Turner & W. Yeh (1988) *J. Biol. Chem.* **263**, 15626. **73**. Bauer, Kreuzman, Dotzlaf & Yeh (1988) *J. Biol. Chem.* **263**, 15619. **74**. Pillinger, Volker, Stock, Weissmann & Philips (1994) *J. Biol. Chem.* **269**, 1486. **75**. De Busser, Van Dessel & Lagrou (2000) *Int. J. Biochem. Cell Biol.* **32**, 1007. **76**. Shi & Rando (1992) *J. Biol. Chem.* **267**, 9547. **77**. Hasne & Lawrence (1999) *Biochem. J.* **342**, 513. **78**. Baron & Casey (2004) *BMC Biochem.* **5**, 19. **79**. Giner & Rando (1994) *Biochemistry* **33**, 15116. **80**. Klein, Ben-Baruch, Marciano, *et al.* (1994) *Biochim. Biophys. Acta* **1226**, 330. **81**. Stephenson & Clarke (1990) *J. Biol. Chem.* **265**, 16348. **82**. Li, Kowluru & Metz (1996) *Biochem. J.* **316**, 345. **83**. Mozier, McConnell & Hoffman (1988) *J. Biol. Chem.* **263**, 4527. **84**. Koch, Lemke, Heise & Mock (2003) *Eur. J. Biochem.* **270**, 84. **85**. Levac, Murata, Kim & De Luca (2007) *Plant J.* **53**, 225. **86**. Harper & Kennedy (1985) *Biochem. J.* **226**, 147. **87**. Jay, De Luca & Ibrahim (1985) *Eur. J. Biochem.* **153**, 321. **88**. Damani, Shaker, Crooks, Godin & Nwosu (1986) *Xenobiotica* **16**, 645. **89**. De Luca & Ibrahim (1985) *Arch. Biochem. Biophys.* **238**, 606. **90**. Kim (1984) *Meth. Enzymol.* **106**, 295. **91**. Rollins & Dahlquist (1980) *Biochemistry* **19**, 4627. **92**. Burgess-Cassler, Ullah & Ordal (1982) *J. Biol. Chem.* **257**, 8412. **93**. Taylor & Cronan (1979) *Biochemistry* **18**, 3292. **94**. Wang, Grogan & Cronan (1992) *Biochemistry* **31**, 11020. **95**. Thapar, Griffith, Yeates & Clarke (2002) *J. Biol. Chem.* **277**, 1058. **96**. Ota, Gilbert & Clarke (1988) *Biochem. Biophys. Res. Commun.* **151**, 1136. **97**. Gingras, Ménard & Béliveau (1991) *Biochim. Biophys. Acta* **1066**, 261. **98**. Kuroki & Poulton (1981) *Z. Naturforsch. C* **36c**, 916. **99**. Kossykh, Schlagman & Hattman (1995) *J. Biol. Chem.* **270**, 14389. **100**. Mashhoon, Carroll, Pruss, *et al.* (2004) *J. Biol. Chem.* **279**, 52075. **101**. Bheemanaik, Chandrashekaran, Nagaraja & Rao (2003) *J. Biol. Chem.* **278**, 7863. **102**. Kossykh, Schlagman & Hattman (1997) *J. Bacteriol.* **179**, 3239. **103**. Evdokimov, Zinoviev, Malygin, Schlagman & Hattman (2002) *J. Biol. Chem.* **277**, 279. **104**. Schneider & Vance (1979) *J. Biol. Chem.* **254**, 3886. **105**. Poulton & V. S. Butt (1975) *Biochim. Biophys. Acta* **403**, 301. **106**. Poulton, K. Hahlbrock & H. Grisebach (1976) *Arch. Biochem. Biophys.* **176**, 449. **107**. Edwards & Dixon (1991) *Arch. Biochem. Biophys.* **287**, 372. **108**. Vance & Bryan (1981) *Phytochemistry* **20**, 41. **109**. Woodson & Weinshilboum (1983) *Biochem. Pharmacol.* **32**, 819. **110**. Szumlanski, Honchel, Scott & Weinshilboum (1992) *Pharmacogenetics* **2**, 148. **111**. Thompson & Cundliffe (1981) *J. Gen. Microbiol.* **124**, 291. **112**. Pech & Nelson (1994) *Biochim. Biophys. Acta* **1199**, 183. **113**. Wright, Bertics & Siegel (1996) *J. Biol. Chem.* **271**, 12737. **114**. Park, Frost, Tuck, *et al.* (1987) *J. Biol. Chem.* **262**, 14702. **115**. DiMaria, Kim & Paik (1982) *Biochemistry* **21**, 1036. **116**. Durbin, Nochumson, Kim, Paik & Chan (1978) *J. Biol. Chem.* **253**, 1427. **117**. Durbin, Kim, Jun & Paik (1983) *Korean J. Biochem.* **15**, 19. **118**. DiMaria, Polastro, DeLange, Kim & Paik (1979) *J. Biol. Chem.* **254**, 4645. **119**. Pugh & Borchardt (1982) *Biochemistry* **21**, 1535. **120**. Barbosa & Moss (1978) *J. Biol. Chem.* **253**, 7698. **121**. Pugh, Borchardt & Stone (1978) *J. Biol. Chem.* **253**, 4075. **122**. Langberg & Moss (1981) *J. Biol. Chem.* **256**, 10054. **123**. Locht, Beaudart & Delcour (1983) *Eur. J. Biochem.* **134**, 117. **124**. Martin & Moss (1975) *J. Biol. Chem.* **250**, 9330. **125**. Pugh, Borchardt & Stone (1977) *Biochemistry* **16**, 3928. **126**. Zheng, Hausmann, Liu, *et al.* (2006) *J. Biol. Chem.* **281**, 35904. **127**. Walton, Peerless, Robins, *et al.* (1994) *Planta* **193**, 9. **128**. Biastoff, Teuber, Zhou & Dräger (2006) *Planta Med.* **72**, 1136. **129**. Denoya & Dubnau (1989) *J. Biol. Chem.* **264**, 2615. **130**. Wengenmayer, Ebel & Grisebach (1974) *Eur. J. Biochem.* **50**, 135. **131**. Ishikawa & Melville (1970) *J. Biol. Chem.* **245**, 5967. **132**. Venkatesan & McManus (1979) *Biochemistry* **18**, 5365. **133**. Tuck, Farooqui & Paik (1985) *J. Biol. Chem.* **260**, 7114. **134**. Lobet, Lhoest & Colson (1989) *Biochim. Biophys. Acta* **997**, 224. **135**. Poulton, Hahlbrock & Grisebach (1977) *Arch. Biochem. Biophys.* **180**, 543. **136**. Ator, Schmidt, Adams & Dolle (1989) *Biochemistry* **28**, 9633. **137**. Cohen, Griffiths, Tawfik & Loakes (2005) *Org. Biomol. Chem.* **3**, 152. **138**. Theiss, Schleicher, Schimpff-Weiland & Follmann (1987) *Eur. J. Biochem.* **167**, 89. **139**. Gold & Hurwitz (1964) *J. Biol. Chem.* **239**, 3858. **140**. Simon, Grunert, von Acken, Döring & Kröger (1978) *Nucl. Acids Res.* **5**, 2153. **141**. Brahmachari & Ramakrisnan (1984) *Arch. Microbiol.* **140**, 91. **142**. Mutzel, Malchow, Meyer & Kersten (1986) *Eur. J. Biochem.* **160**, 101. **143**. Glick & Leboy (1977) *J. Biol. Chem.* **252**, 4790. **144**. Glick, Ross & Leboy (1975) *Nucl. Acids Res.* **2**, 1639. **145**. Salas & Sellinger (1978) *J. Neurochem.* **31**, 85. **146**. Hurwitz, Gold & Anders (1964) *J. Biol. Chem.* **239**, 3474. **147**. Santi & Hardy (1987) *Biochemistry* **26**, 8599. **148**. Ny, Lindström, Hagervall & Björk (1988) *Eur. J. Biochem.* **177**, 467. **149**. Shugart (1978) *Biochemistry* **17**, 1068. **150**. Hurwitz, Gold & Anders (1964) *J. Biol. Chem.* **239**, 3474. **151**. Greenberg & Dudock (1980) *J. Biol. Chem.* **255**, 8296. **152**. Shugart & Chastain (1979) *Enzyme* **24**, 353. **153**. Kumagai, Watanabe, Oshima (1982) *J. Biol. Chem.* **257**, 7388. **154**. Paolella, Ciliberto, Traboni, Cimino & Salvatore (1982) *Arch. Biochem. Biophys.* **219**, 149. **155**. Pierré, Berneman, Vedel, Robert-Géro & Vigier (1978) *Biochem. Biophys. Res. Commun.* **81**, 315. **156**. Hildesheim, Hildesheim, Blanchard, Farrugia & Michelot (1973) *Biochimie* **55**, 541. **157**. Taylor & Gantt (1979) *Biochemistry* **18**, 5253. **158**. Glick, Averyhart & Leboy (1978) *Biochim. Biophys. Acta* **518**, 158. **159**. Redlak, Andreos-Selim. Giece, Florentz & Holmes (1997) *Biochemistry* **36**, 8699. **160**. Gabryszuk & Holmes (1997) *RNA* **3**, 1327. **161**. Hjalmarsson, Byström & Björk (1983) *J. Biol. Chem.* **258**, 1343. **162**. Kahle & Kröger (1975) *Biochem. Soc. Trans.* **3**, 908. **163**. Keith, Winters & Moss (1980) *J. Biol. Chem.* **255**, 4636. **164**. Wu, Gee, Lin, *et al.* (2005) *J. Med. Chem.* **48**, 7243. **165**. Lee, Schulz & Fuller (1978) *Arch. Biochem. Biophys.* **185**, 239. **166**. Wong, Yamasaki & Ciaranello (1987) *Brain Res.* **410**, 32. **167**. Kloor, Karnahl & Kömpf (2004) *Biochem. Cell Biol.* **82**, 369. **168**. Heady & Kerr (1973) *J. Biol. Chem.* **248**, 69. **169**. Takata, Huang, Komoto, *et al.* (2003) *Biochemistry* **42**, 8394. **170**. Gaynor & Carman (1990) *Biochim. Biophys. Acta* **1045**, 156. **171**. Makishima, Toyoshima & Osawa (1985) *Arch. Biochem. Biophys.* **238**, 315. **172**. Tahara, Ogawa, Sakakibara & Yamada (1987) *Agric. Biol. Chem.* **51**, 1425. **173**. Tahara, Ogawa, Sakakibara & Yamada (1986) *Agric. Biol. Chem.* **50**, 257. **174**. Akamatsu & Law (1970) *J. Biol. Chem.* **245**, 709. **175**. Safayhi, Anazodo & Ammon (1991) *Int. J. Biochem.* **23**, 769. **176**. Gibson, Neuberger & Tait (1963) *Biochem. J.* **88**, 325. **177**. Sawicki & Willows (2007) *Biochem. J.* **406**, 469. **178**. Henchigeri, Chan & Richards (1981) *Photosynthetica* **15**, 351. **179**. Henchigeri & Richards (1982) *Photosynthetica* **16**, 554. **180**. Henchigeri, Nelson & Richards (1984) *Photosynthetica* **18**, 168. **181**. Shepherd, Reid & Hunter (2003) *Biochem. J.* **371**, 351. **182**. Balish & Shapiro (1967) *Arch. Biochem. Biophys.* **119**, 62. **183**. Borchardt & Cheng (1978) *Biochim. Biophys. Acta* **522**, 340. **184**. Weisiger & Jacoby (1979) *Arch. Biochem. Biophys.* **196**, 631. **185**. Weinshilboum, Sladek & Klumpp (1979) *Clin. Chim. Acta* **97**, 59. **186**. Francis, Thompson & Greaves (1980) *Biochem. J.* **187**, 819. **187**. Harvima, Kajander, Harvima & Fraki (1985) *Biochim. Biophys. Acta* **841**, 42. **188**. Matuszewska & Borchardt (1983) *J. Neurochem.* **41**, 113. **189**. Borchardt & Matuszewska (1985) *Adv. Biosci.* **51**, 163. **190**. Matuszewska & Borchardt (1985) *Prep. Biochem.* **15**, 145. **191**. Upmeier, Gross, Köster & Barz (1988) *Arch. Biochem. Biophys.* **262**, 445. **192**. Ball, Knuppen, Haupt & Breuer (1972) *Eur. J. Biochem.* **26**, 560. **193**. Veser (1987) *J. Bacteriol.* **169**, 3696. **194**. Borchardt & Cheng (1978) *Biochim. Biophys. Acta* **522**, 49. **195**. Dhar & Rosazza (2000) *Appl. Environ. Microbiol.* **66**, 4877. **196**. Coward, Slisz & Wu (1973) *Biochemistry* **12**, 2291. **197**. Finkelstein, Kyle & Harris (1974) *Arch. Biochem. Biophys.* **165**, 774. **198**. Karahasanoglu & Ozand (1972) *J. Neurochem.* **19**, 411. **199**. Tedesco, Morton & Reiter (1994) *J. Pineal Res.* **16**, 121. **200**. Kuwano & Takahashi (1984) *Biochim. Biophys. Acta* **787**, 1. **201**. Im, Chiang & Cantoni (1979) *J. Biol. Chem.* **254**, 11047. **202**. Komoto, Yamada, Takata, *et al.* (2004) *Biochemistry* **43**, 14385. **203**. Takata & Fujioka (1992) *Biochemistry* **31**, 4369. **204**. Rini, Szumlanski, Guerciolini & Weinshilboum (1990) *Clin. Chim. Acta* **186**, 359.

Adenylate Kinase Inhibitors

Most cells contain adenylate kinase (*Reaction:* ADP + MgADP $\rightleftharpoons$ MgATP + AMP) in abundance, maintaining adenine nucleotide concentrations at, or very near, equilibrium. Even after extensive purification, many proteins and enzymes are contaminated by trace amounts of adenylate kinase, often frustrating experiments requiring the presence of one or more adenine nucleotide. The presence of this kinase can confound efforts to quantify processes requiring ADP or any process requiring the presence of both ATP and AMP. To minimize the effects by contaminating levels of adenylate kinase, one can utilize the bisubstrate geometrical analogues Ap$_5$A and Ap$_4$A that simultaneously occupy both nucleotide-binding pockets within the active site of this kinase (1,2). Effective inhibition by Ap$_5$A and Ap$_4$A occurs at 0.3 and 1 mM, respectively. Powers *et al.* (3) also demonstrated the inhibitory effects of a series of di(adenosine-5')-polyphosphates (*e.g.*, Ap$_n$A, where n = 2, 3, 4, 5 and 6) on carbamyl phosphate synthetase. Only Ap$_5$A was found to be an effective inhibitor of the overall reaction. *See* P^1, P^4-Di(adenosine 5') tetraphosphate; P^1, P^5-Di(adenosine 5') pentaphosphate **1**. Purich & Fromm (1972) *Biochim. Biophys. Acta* **276**, 563. **2**. Leinhard & Secemski (1973) *J. Biol. Chem.* **248**, 1121. **3**. Powers, Griffith & Meister (1977) *J. Biol. Chem.* **252**, 3558.

Adenylomalic Acid

This adenylosuccinate analogue (FW$_{\text{free-acid}}$ = 479.30 g/mol), also known as N^6-(DL-1,2-dicarboxy-*threo*-2-hydroxyethyl)adenosine 5'-monophosphate, inhibits (K_i = 1.2 μM) chicken liver adenylosuccinate lyase (*Reaction*: Adenyosuccinate $\rightleftharpoons$ AMP + Fumarate). Adenylomalixcacid therefore represents a lead molecule for the development of novel inhibitors of AMP synthesis. 1. Brand & Lowenstein (1978) *Biochemistry* **17**, 1365.

Adenylosuccinate

This purine nucleotide biosynthesis intermediate (FW$_{\text{free-acid}}$ = 463.30 g/mol; CAS 19240-42-7; Symbols: AMP-Succ, Succ-AMP or S-AMP), also named succinyl-AMP and N^6-(1,2-dicarboxyethyl)AMP, is the immediate product of the adenylosuccinate synthetase (*Raction*: IMP + Aspartate + GTP $\rightleftharpoons$ Adenylosuccuccinate + GDP + P$_i$) and the substrate for adenylosuccinate lyase, *or* ADSL (*Reaction*: Adenylosuccinate $\rightleftharpoons$ AMP + Fumarate). In ADSL deficiency, biologic fluids accumulate succinyladenosine (S-Ado), which is formed by enzymatic hydrolysis of AMP-Succ (1). Adenylosuccinate also appears to be an insulin secretagogue that is formed during glucose-induced purine metabolism and may play a role in loss of glucose-stimulated insulin secretion (GSIS) from islet β-cells, a transition that heralds the onset of type 2 diabetes, *or* T2D (2). Indeed, exposure of the robustly glucose responsive INS-1-derived insulinoma cell line 832/13 to stimulatory glucose concentrations results in a striking increase in adenylosuccinate levels, and molecular or pharmacological manipulation of S-AMP levels results in alterations in insulin secretion that are consistent with a regulatory role for this metabolite (2). Addition of S-AMP to the interior of patch-clamped human β-cells amplifies exocytosis, an effect that depends upon expression of Sentrin/SUMO-specific protease-1 (SENP1). S-AMP also overcomes a defect in glucose-induced exocytosis in β-cells from a human donor with T2D. S-AMP is therefore an insulin secretagogue capable of reversing β cell dysfunction in T2D (2). The N^6-(D-1,2-dicarboxyethyl)adenosine 5'-monophosphate epimer inhibits adenylosuccinate lyase (K_i = 100 μM). Adenylosuccinate also inhibits IMP cyclohydrolase, which in humans is actually a bifunctional enzyme consisting of 5-aminoimidazole-4-carboxamide ribotide (AICAR) transformylase and IMP cyclohydrolase activities (3). **See also** *Hydantocidin* 1. Brand & Lowenstein (1978) *Biochemistry* **17**, 1365. 2. Gooding, Jensen, Dai, *et al.* (2015) *Cell Rep.* **13**, 157. 3. Szabados, Hindmarsh, Phillips, Duggleby & Christopherson (1994) *Biochemistry* **33**, 14237.

9-*O*-(Adenylyl)spectinomycin

This alternative product (FW$_{\text{free-acid}}$ = 642.54 g/mol), which is formed by the *Escherichia coli* streptomycin/spectinomycin resistance factor, inhibits streptomycin 3''-adenylyltransferase, the newly discovered bifunctional enzyme and aminoglycoside antibiotic resistance factor in *Serratia marcescens*. 1. Kim, Hesek, Zajicek, Vakulenko & Mobashery (2006) *Biochemistry* **45**, 8368.

Adipic Acid

This six-carbon dicarboxylic acid (FW$_{\text{free-acid}}$ = 146.14 g/mol; CAS 124-04-9; M.P. = 152°C; pK_1 = 4.43 and pK_2 = 5.41) derives its name from *adipis* (Greek for "fat"), because it was first prepared by oxidation of various lipids. It is often employed in controlled-release formulations that facilitate pH-independent release for weakly basic and weakly acidic drugs. **Target(s):** glutamate decarboxylase (1,10); kynurenine:oxoglutarate aminotransferase (2,5,11,18,27,30,31); carbamoyl-phosphate synthetase (3); fumarase (4,6); aspartate aminotransferase (7-10,17,33); succinate dehydrogenase (10,12); γ-butyrobetaine dioxygenase, *or* γ-butyrobetaine hydroxylase, K_i = 0.15 mM (13); D-aspartate oxidase, weakly inhibited (14); pyrophosphatase, *or* inorganic diphosphatase (15); 4-aminobutyrate aminotransferase (16,29); glutamate dehydrogenase (19); carboxy-*cis,cis*-muconate cyclase (20); nitrile hydratase (21); aconitase, *or* aconitate hydratase (22); diaminopimelate decarboxylase (23); cytosol alanyl aminopeptidase (24); 3-oxoadipate CoA-transferase, activated at 2 mM, but inhibited at higher concentrations (25); 2-aminoadipate aminotransferase (26,27); pyridoxamine:oxaloacetate aminotransferase, weakly inhibited at pH 8 (28); leucine aminotransferase (32); procollagen-lysine 5-dioxygenase (34); procollagen-proline 3 dioxygenase (34). **1.** Fonda (1972) *Biochemistry* **11**, 1304. **2.** Tanizawa, Asada & Soda (1985) *Meth. Enzymol.* **113**, 90. **3.** Jones (1962) *Meth. Enzymol.* **5**, 903. **4.** Hill & Bradshaw (1969) *Meth. Enzymol.* **13**, 91. **5.** Tobes (1987) *Meth. Enzymol.* **142**, 217. **6.** Massey (1953) *Biochem. J.* **55**, 172. **7.** Jenkins, Yphantis & Sizer (1959) *J. Biol. Chem.* **234**, 51. **8.** Velick & Vavra (1962) *The Enzymes*, 2nd ed., **6**, 219. **9.** Braunstein (1973) *The Enzymes*, 3rd ed., **9**, 379. **10.** Webb (1966) *Enzyme and Metabolic Inhibitors*, vol. **2**, Academic Press, New York. **11.** Mason (1959) *J. Biol. Chem.* **234**, 2770. **12.** Potter & Elvehjem (1937) *J. Biol. Chem.* **117**, 341. **13.** Ng, Hanauske-Abel & England (1991) *J. Biol. Chem.* **266**, 1526. **14.** Dixon & Kenworthy (1967) *Biochim. Biophys. Acta* **146**, 54. **15.** Naganna (1950) *J. Biol. Chem.* **183**, 693. **16.** Sytinsky & Vasilijev (1970) *Enzymologia* **39**, 1. **17.** Martinez-Carrion, Barber & Pazoles (1977) *Biochim. Biophys. Acta* **482**, 323. **18.** Asada, Sawa, Tanizawa & Soda (1986) *J. Biochem.* **99**, 1101. **19.** Bonete, Perez-Pomares, Ferrer & Camacho (1996) *Biochim. Biophys. Acta* **1289**, 14. **20.** Thatcher & Cain (1975) *Eur. J. Biochem.* **56**, 193. **21.** Moreau, Azza, Arnaud & Galzy (1993) *J. Basic Microbiol.* **33**, 323. **22.** Treton & Heslot (1978) *Agric. Biol. Chem.* **42**, 1201. **23.** Ray, Bonanno, Rajashankar, *et al.* (2002) *Structure* **10**, 1499. **24.** Garner & Behal (1977) *Arch. Biochem. Biophys.* **182**, 667. **25.** MacLean, MacPherson, Aneja & Finan (2006) *Appl. Environ. Microbiol.* **72**, 5403. **26.** Deshmukh &

Mungre (1989) *Biochem. J.* **261**, 761. **27**. Tobes & Mason (1975) *Biochem. Biophys. Res. Commun.* **62**, 390. **28**. Wu & Mason (1964) *J. Biol. Chem.* **239**, 1492. **29**. Schousboe, Wu & Roberts (1974) *J. Neurochem.* **23**, 1189. **30**. Takeuchi, Otsuka & Shibata (1983) *Biochim. Biophys. Acta* **743**, 323. **31**. Mawal, Mukhopadhyay & Deshmukh (1991) *Biochem. J.* **279**, 595. **32**. Pathre, Singh, Viswanathan & Sane (1987) *Phytochemistry* **26**, 2913. **33**. Tanaka, Tokuda, Tachibana, Taniguchi & Oi (1990) *Agric. Biol. Chem.* **54**, 625. **34**. Majamaa, Turpeenniemi-Hujanen, Latipää, *et al.* (1985) *Biochem. J.* **229**, 127.

Adiphenine

This anticholinergic agent (FW$_{free-base}$ = 311.42 g/mol; CAS 50-42-0), also named Diphacil and 2-(diethylamino)ethyl 2,2-diphenylacetate, targets both muscarinic (1,2) and nicotinic (3) acetylcholine receptors (3). Adiphenine is a fat-soluble compound that freely crosses the blood brain barrier. In rats, adiphenine was extensively metabolized by hydrolysis of the ester bond into diethylaminoethanol, diphenylacetic acid, diphenylacetic acid glucuronide and, in small quantities, the corresponding glycine and glutamine conjugates (4). **1**. Eglen & Whiting (1987) *Brit. J. Pharmacol.* **90**, 701. **2**. Witkin, Gordon & Chiang (1987) *J. Pharmacol. Exp. Ther.* **242**, 796. **3**. Gentry & Lukas (2001) *J. Pharmacol. Exp. Ther.* **299**, 1038. **4**. Michelot, Madelmont, Jordan, *et al.* (1981) *Xenobiotica* **11**, 123. **5**. Michelot, Madelmont, Rousset, *et al.* (1982) *Xenobiotica* **12**, 457.

Adiponectin

This 244-residue anti-diabetic hormone (MW$_{monomer}$ = 30 kDa; Symbol: Ad) is a adipocyte-derived polypeptide that regulates energy homeostasis and glucose and lipid metabolism, mainly by increasing the insulin responsiveness of susceptible cells and by stimulating the phosphorylation/activation of the 5'-AMP-activated protein kinase, *or* AMPK (1-3). Adiponectin circulates in human plasma mainly as a 180-kDa middle molecular weight (MMW) hexamer and a high molecular weight (HMW) multimer of approximately 360 kDa. It has four topologically distinct regions: a short signal sequence targeting the hormone for secretion; a short species-variant region; a 65-residue region resembling collagens; and a globular domain. This adipokine, which bears similarities in domain structure to the complement protein C1q, is processed into at least three homomeric complexes, trimer, hexamer, and a high-molecular-weight (HMW) octadecamer. In skeletal muscle. AMPK is stimulated with globular and full-length Ad, whereas AMPK is only stimulated by full-length Ad in liver. Ad stimulates phosphorylation of acetyl coenzyme A carboxylase (ACC), fatty-acid oxidation, glucose uptake and lactate production in myocytes, phosphorylation of ACC and reduction of molecules involved in gluconeogenesis in the liver, and reduction of glucose levels *in vivo*. Inhibition of AMPK activation (**See Dominant Negative Mutants**) blocks each of these effects, indicating that stimulation of glucose utilization and fatty-acid oxidation by Ad occurs through activation of AMPK (4). **1**. Nawrocki, Rajala, Tomas, *et al.* (2006) *J. Biol. Chem.* **281**, 2654. **2**. Fruebis, Tsao, Javorschi, *et al.* (2001) *Proc. Natl. Acad. Sci. USA* **98**:2005. **3**. Yamauchi, Kamon, Minokoshi, et al. (2002) *Nature Med.* **8**, 1288. **4**. Woods, Azzout-Marniche, Foretz, *et al.* (2000) *Mol. Cell Biol.* **20**, 6704.

AdipoRon

This orally bioavailable adiponectin receptor agonist (FW = 428.52 g/mol; CAS 924416-43-3; Solubility: 100 mg/mL DMSO; <1 mg/mL H$_2$O), also known as 2-(4-benzoylphenoxy)-*N*-(1-benzylpiperidin-4-yl)acetamide,

stimulates AMPK-mediated phosphorylation and activation of AdipoR1 (K_d = 1.8 µM) and AdipoR1 (K_d = 3.1 µM), which are orientated oppositely to most GPCRs within the membrane (*i.e.*, cytoplasmic N-terminus, extracellular C-terminus), such that they cannot associate with G-proteins. In wild type mice, AdipoRon (orally administered at 50 mg/kg) improves insulin resistance, glucose intolerance, and dyslipidemia by activating AdipoR1–AMPK–PGC-1α pathways in skeletal muscle, while by activating AdipoR2–PPAR-α (Peroxisome Proliferator-Activated Receptor) pathways in the liver. **1**. Okada-Iwabu, Yamauchi, Iwabu, *et al.* (2013) *Nature* **503**, 493.

Adociasulfate-2

This marine natural product (FW$_{disodium-salt}$ = 738.92 g/mol; CAS 208181-29-7; Abbreviated: AS-2) from the sponge *Haliclona* sp. (also known as *Adocia* sp.) specifically inhibits kinesin activity by interfering with microtubule binding. This mechanism is unlike that of any known kinesin (or other motor) inhibitor, and AS-2 is thought to mimic tubulin binding to a region within kinesin's microtubule-binding site.1 This potent inhibitor (IC$_{50}$ = 2.7 µM) may generally arrest transport by other members of the kinesin superfamily. **1**. Sakowicz, Berdelis, Ray, *et al.* (1998) *Science* **280**, 292.

Ado-Trastuzumab Emtansine

This antibody-drug conjugate *or* ADC (FW = 146.3 kDa; CAS 400010-39-1; *Abbreviation*: T-DM1, marketed under the name Kadcyla™) is a novel cancer therapeutic that attaches to surface determinants on targeted cancer cells and, upon internalization, releases the powerful anti-microtubule drug mertansine, thereby killing them. T-DM1 is a humanized epidermal growth factor receptor (HER2)-targeted antibody-drug conjugate consisting of trastuzumab (MW = 145,500; CAS 180288-69-1; also known as Herclon™ and Herceptin™) joined by means of a stable, 4-mercaptovalerate thioether linker to the potent cytotoxic agent DM1 *or* Derivative of Maytansine-1 (*or* Mertansine, FW = 739.29 g/mol). Kadcyla is the first FDA-approved ADC for treating HER2-positive mBC, an aggressive form of the disease. To generate thioTMAb-mpeo-DM1, mertansine was conjugated to an engineered cysteine residue at Ala114 (Kabat numbering) on each trastuzumab-heavy chain, resulting in two DM1 molecules per antibody (1). ThioTMAb-mpeo-DM1 retained similar in vitro anti-cell proliferation activity and human epidermal growth factor receptor 2 (HER2) binding properties to that of the conventional ADC. Furthermore, it showed improved efficacy over the conventional ADC at DM1-equivalent doses (µg/m^2) and retained efficacy at equivalent antibody doses (mg/kg). Moreover, its pharmacokinetic profile is unaffected by circulating levels of HER2 extracellular domain or residual trastuzumab (2). **1**. Junutula, Flagella, Graham, *et al.* (2010) *Clin. Cancer Res.* **16**, 4769. **2**. Girish, Gupta, Wang, *et al.* (2012) *Cancer Chemother Pharmacol.* **69**, 1229.

ADP-D-glucose

This starch biosynthesis intermediate (FW$_{\text{free-acid}}$ = 589.35 g/mol; CAS 2140-58-1), also known as adenosine (5')diphospho(1)-α-D-glucose, has the following inhibitory targets: UDP-glucuronate 4-epimerase (1,7); starch phosphorylase (2,15,16,19-26,28,29,32); UDP-glucose pyrophosphorylase, or glucose-1-phosphate uridylyltransferase (3); alcohol dehydrogenase, K_i = 0.48 mM (4); ADP-glyceromanno-heptose 6-epimerase (5,6); UDP-sugar diphosphatase (8); pyruvate,water dikinase, weakly inhibited (9); glucose-1-phosphate cytidylyltransferase, weakly inhibited (10); β-1,4-mannosyl-glycoprotein 4-β-N acetylglucosaminyltransferase (11); maltose synthase (12); α,α-trehalose-phosphate synthase, or UDP forming (13); glycogen synthase, alternative substrate (14); glycogen phosphorylase (17,18,27,30,31,33). **1.** Gaunt, Ankel & Schutzbach (1972) *Meth. Enzymol.* **28**, 426. **2.** Mu, Yu, Wasserman & Carman (2001) *Arch. Biochem. Biophys.* **388**, 155. **3.** Kornfeld (1965) *Fed. Proc.* **24**, 536. **4.** Weiner (1969) *Biochemistry* **8**, 526. **5.** Ding, Seto, Ahmed & Coleman (1994) *J. Biol. Chem.* **269**, 24384. **6.** Deacon, Ni, W. Coleman & Ealick (2000) *Structure Fold Des.* **8**, 453. **7.** Gaunt, Maitra & Ankel (1974) *J. Biol. Chem.* **249**, 2366. **8.** Glaser, Melo & Paul (1967) *J. Biol. Chem.* **242**, 1944. **9.** Chulavatnatol & Atkinson (1973) *J. Biol. Chem.* **248**, 2712. **10.** Kimata & Suzuki (1966) *J. Biol. Chem.* **241**, 1099. **11.** Ikeda, Koyota, Ihara, *et al.* (2000) *J. Biochem.* **128**, 609. **12.** Schilling (1982) *Planta* **154**, 87. **13.** Londesborough & Vuorio (1991) *J. Gen. Microbiol.* **137**, 323. **14.** Zea, MacDonell & Pohl (2003) *J. Amer. Chem. Soc.* **125**, 13666. **15.** Lee & Braun (1973) *Arch. Biochem. Biophys.* **156**, 276. **16.** Nakamura & Imamura (1983) *Phytochemistry* **22**, 835. **17.** Schultz & Ankel (1970) *Biochim. Biophys. Acta* **215**, 39. **18.** Chen & Segel (1968) *Arch. Biochem. Biophys.* **127**, 175. **19.** Kokesh, Stephenson & Kakuda (1977) *Biochim. Biophys. Acta* **483**, 258. **20.** Oluoha & Ugochukwu (1991) *Biol. Plant.* **33**, 249. **21.** Matheson & Richardson (1973) *Phytochemistry* **17**, 195. **22.** Oluoha (1990) *Biol. Plant.* **32**, 64. **23.** Oluoha & Ndukwu (1999) *Biokemistry* **9**, 37. **24.** Dauvillée, Chochois, Steup, *et al.* (2006) *Plant J.* **48**, 274. **25.** Hsu, Yang, Su & Lee (2004) *Bot. Bull. Acad. Sin.* **45**, 187. **26.** Nakamura & Imamura (1983) *Phytochemistry* **22**, 2395. **27.** Robson & Morris (1974) *Biochem. J.* **144**, 513. **28.** Chang & Su (1986) *Plant Physiol.* **80**, 534. **29.** Yu & Pedersen (1991) *Physiol. Plant.* **81**, 149. **30.** Takata, Takaha, Okada, Takagi & Imanaka (1998) *J. Ferment. Bioeng.* **85**, 156. **31.** Thomas & Wright (1976) *J. Biol. Chem.* **251**, 1253. **32.** Kruger & Rees (1983) *Phytochemistry* **22**, 1891. **33.** Spearman, Khandelwal & Hamilton (1973) *Arch. Biochem. Biophys.* **154**, 306.

ADP-D-ribose

This sugar nucleotide (FW$_{\text{free-acid}}$ = 559.32 g/mol; CAS 68414-18-6), often referred to as ADPR and adenosine 5'-diphosphoribose, is a primary 5'-phosphoribosylation substrate in both purine and pyrimidine nucleotide biosynthesis. It is also component of NAD$^+$ and inhibits a number of NAD$^+$-dependent enzymes. ADPR binds to and activates the Transient Receptor Potential cation, subfamily M, member 2 (or TRPM2) channel, a non-selective calcium-permeable cation channel. **Target(s):** formaldehyde dehydrogenase (glutathione) (1,18); S-(hydroxymethyl)-glutathione dehydrogenase (1,18); nicotinamide-nucleotide adenylyltransferase (2,11,33,34); alcohol dehydrogenase, K_d = 9 μM at pH 6 (3,4,12-14,19); NAD$^+$ kinase (5,9,35); isocitrate dehydrogenase, NAD$^+$ (6,8); malate dehydrogenase (7); poly(ADP-ribose) glycohydrolase (10); DNA-(apurinic or apyrimidinic site) lyase (15); ADP-ribosyl-[dinitrogen reductase]

hydrolase (16); nucleotide diphosphatase (17); [protein ADP-ribosylarginine] hydrolase (20-24); NMN nucleosidase (25); NAD(P)$^+$ nucleosidase, product inhibition (26); NAD$^+$ nucleosidase, NADase, product inhibition (27-30); phosphodiesterase I, or 5' exonuclease (31); ribose-5-phosphate adenylyltransferase (32); NAD$^+$:protein-arginine ADP ribosyltransferase (36); glycogen phosphorylase (37); procollagen-proline 4-dioxygenase, weakly inhibited (38); lactate dehydrogenase (39). **1.** Uotila & Koivusalo (1981) *Meth. Enzymol.* **77**, 314. **2.** Magni, Emanuelli, Amici, Raffaelli & Ruggieri (1997) *Meth. Enzymol.* **280**, 241. **3.** Li & Vallee (1964) *J. Biol. Chem.* **239**, 792. **4.** Brändén, Jörnvall, Eklund & Furugren (1975) *The Enzymes*, 3rd ed., **11**, 103. **5.** Wang (1955) *Meth. Enzymol.* **2**, 652. **6.** Plaut (1969) *Meth. Enzymol.* **13**, 34. **7.** Banaszak & Bradshaw (1975) *The Enzymes*, 3rd ed., **11**, 369. **8.** Chen & Plaut (1963) *Biochemistry* **2**, 1023. **9.** Wang & Kaplan (1954) *J. Biol. Chem.* **206**, 311. **10.** Tanuma, Kawashima & Endo (1986) *J. Biol. Chem.* **261**, 965. **11.** Emanuelli, Natalini, Raffaelli, Ruggieri, Vita & Magni (1992) *Arch. Biochem. Biophys.* **298**, 29. **12.** Weiner (1969) *Biochemistry* **8**, 526. **13.** Dalziel (1963) *J. Biol. Chem.* **238**, 1538. **14.** Yonetani (1963) *Acta Chem. Scand.* **17**, S96. **15.** Kane & Linn (1981) *J. Biol. Chem.* **256**, 3405. **16.** Nielsen, Bao, Roberts & Ludden (1994) *Biochem. J.* **302**, 801. **17.** Jacobson & Kaplan (1957) *J. Biol. Chem.* **226**, 427. **18.** Uotila & Mannervik (1979) *Biochem. J.* **177**, 869. **19.** Asante-Appiah & Chan (1996) *Biochem. J.* **320**, 17. **20.** Moss, Oppenheimer, West & Stanley (1986) *Biochemistry* **25**, 5408. **21.** Takada, Okazaki & Moss (1994) *Mol. Cell. Biochem.* **138**, 119. **22.** Moss, Zolkiewska & Okazaki (1997) *Adv. Exp. Med. Biol.* **419**, 25. **23.** Maehama, Nishina & Katada (1994) *J. Biochem.* **116**, 1134. **24.** Kim & Graves (1990) *Anal. Biochem.* **187**, 251. **25.** Foster (1981) *J. Bacteriol.* **145**, 1002. **26.** Berthelier, Tixier, Muller-Steffner, Schuber & Deterre (1998) *Biochem. J.* **330**, 1383. **27.** Travo, Muller & Schuber (1979) *Eur. J. Biochem.* **96**, 141. **28.** Yuan & Anderson (1972) *J. Biol. Chem.* **247**, 515. **29.** De Wolf, van Dessel, Lagrou, Hilderson & Dierick (1985) *Biochem. J.* **226**, 415. **30.** Kim, Kim, Kim, *et al.* (1993) *Arch. Biochem. Biophys.* **305**, 147. **31.** Picher & Boucher (2000) *Amer. J. Respir. Cell Mol. Biol.* **23**, 255. **32.** Evans & Pietro (1966) *Arch. Biochem. Biophys.* **113**, 236. **33.** Magni, Amici, Emanuelli, *et al.* (2004) *Curr. Med. Chem.* **11**, 873. **34.** Barile, Passarella, Danese & Quagliariello (1996) *Biochem. Mol. Biol. Int.* **38**, 297. **35.** Blomquist (1973) *J. Biol. Chem.* **248**, 7044. **36.** Zheng, Morrison, Chung, Moss & Bortell (2006) *J. Cell. Biochem.* **98**, 851. **37.** Robson & Morris (1974) *Biochem. J.* **144**, 513. **38.** Hussain, Ghani & Hunt (1989) *J. Biol. Chem.* **264**, 7850. **39.** Subramanian & Ross (1978) *Biochemistry* **17**, 2193.

ADP[S] or ADPβS, *See Adenosine 5'-(2-Thiodiphosphate)*

Adrafinil

This hydroxamate-containing pro-drug (FW = 289.35 g/mol; CAS 63547-13-7), also known by its code name CRL-40028 and trade name Olmifon®, is a mild CNS stimulant used to relieve excessive sleepiness and inattention in elderly patients. Adrafinil is mainly metabolized to modafinil (*See Modafinil*), with which it shares many pharmacologic properties. Given its abuse by athletes in training, the World Anti-Doping Agency (WADA) banned adrafinil and modafinil in sports since 2004.

Adrenate

This long-chain, unsaturated fatty acid (FW$_{\text{free-acid}}$ = 332.53 g/mol; CAS 28874-58-0), also known as adrenic acid and (7Z,10Z,13Z,16Z)-docosatetraenoic acid, inhibits arachidonyl-CoA synthetase, or arachidonate:CoA ligase (1) and prostaglandin biosynthesis (2). **1.** Taylor, Sprecher & Russel (1985) *Biochim. Biophys. Acta* **833**, 229. **2.** Cagen & Baer (1980) *Life Sci.* 26, 765.

Aducanumab

This high-affinity, fully human IgG1 monoclonal antibody (MW = 149.5 kDa; CAS 1384260-65-4), also known by the developmental code name BIIB037, selectively targets aggregated forms of β-amyloid protein (EC$_{50}$ = 0.1 nM), while showing weak binding to Aβ monomer. In the brains of transgenic mice, aducanumab preferentially binds to parenchymal Aβ over vascular Aβ deposits, consistent with the lack of effect on vascular Aβ following chronic dosing. Aducanumab dose-dependently reduces amyloid deposition in six cortical regions of the brain. [ch]Aducanumab, a murine IgG2a/κ chimeric analogue, dose-dependently reduces Aβ measured in brain homogenates by up to 50% relative to the vehicle control in the diethylamine fraction that extracted soluble monomeric and oligomeric forms of Aβ$_{40}$ and Aβ$_{42}$, and in the guanidine hydrochloride fraction that extracted insoluble Aβ fibrils. The clearance of Aβ deposits was accompanied by enhanced recruitment of microglia. Together with the reduced potency of the aglycosylated form of [ch]aducanumab and the *ex vivo* phagocytosis data, such findings suggest that FcγR-mediated microglial recruitment and phagocytosis played an important role in Aβ clearance in these models. Activated microglia appeared to encapsulate the remaining central dense core of plaques in treated animals, possibly isolating them from the surrounding neurophil. **1.** Sevigny, Chiao, Bussière, *et al.* (2016) *Nature* **537**, 50.

ADX 71743

This brain-penetrant mGlu$_7$ allosteric negative modulator (FW = 269.34 g/mol; CAS 1431641-29-0; Solubility: 100 mM in DMSO), also named 6-(2,4-dimethylphenyl)-2-ethyl-6,7-dihydro-4(5H)-benzoxazolone, potently and selectively targets the metabotropic glutamate receptor-7, itself a promising novel target for treatment of anxiety, post-traumatic stress disorder, depression, drug abuse, and schizophrenia. When tested *in vitro*, Schild plot analysis and reversibility tests at the target confirmed the NAM properties of the compound and attenuation of L-(+)-2-amino-4-phosphonobutyric acid (*or* L-AP4) -induced synaptic depression confirmed activity at the native receptor. When tested *in vitro*, ADX-71743 gives a IC$_{50}$ value *versus* EC$_{80}$ of glutamate = 22 nM at human mGlu$_7$ in a 3',5'-cyclic-AMP assay, with IC$_{50}$ values versus EC$_{80}$ of L-AP4 are 63 and 125 nM at human mGlu$_7$ in Ca^{2+} and cAMP assays respectively; IC$_{50}$ value versus EC$_{80}$ of L-AP4 = 88 nM at rmGlu$_7$ in a Ca^{2+}assay). Selective for mGlu$_7$ over all other mGlu receptor subtypes and a panel of relevant GPCRs. Exhibits an anxiolytic-like profile *in vivo*. The Schild plot experiments clearly indicate the noncompetitive nature of ADX71743 inhibition, whereas simple wash experiments seem to indicate that its binding to the receptor is readily reversible. **Other Targets:** When tested for its action at other mGlu-expressing cells in series of FLIPR experiments, ADX71743 showed no detectable activity (either agonist or allosteric effects) in cell lines expressing hmGlu$_3$, hmGlu$_4$, rmGlu$_5$, hmGlu$_6$, and hmGlu$_8$. A negligible inhibition of rmGlu$_1$ (32% at 30 μM) and a weak positive allosteric modulator effect on hmGlu$_2$ (EC$_{50}$ = 11 μM) was measured. When further tested in a functional GPCR screen against 27 targets in agonist and antagonist modes, ADX71743 exhibited no stimulation or inhibition above 27% was observed. **1.** Kalinichev, Rouillier, Girard, *et al.* (2013) *J. Pharmacol. Exp. Ther.* **344**, 624.

AEBSF, *See 4-(2-Aminoethyl)benzenesulfonyl Fluoride*

Aedesin

This cecropin-like antimicrobial (FW = 3676.54 g/mol; Sequence: GGLKKLGKKLEGAGKRVFKASEKALPVVVGIKAIGK, matching residues 26-61 in the longer polypeptide AAEL000598, *or* Q17NR1), is a pore-forming, helix-bend-helix peptide that kill drug-resistant Gram-negative bacterial strains displaying resistance to carbapenems, aminoglycosides, cephalosporins, antifolates, fourth generation fluoroquinolones, and monobactams (**See also** *Cecropins*). Aedesin is isolated from infected salivary glands from the mosquito *Aedes aegypti*, the well-known dengue virus (*or* DV) vector. Its anti-bacterial activity is salt-

resistant, indicating it remains active in physiological fluids. Aedesin has two amphipatic α-helices within the Lys^{30}-Lys^{48} and Val^{52}-Ile^{59} regions at the N- and C-terminal part of the peptide. While the N-terminal region contains a large stretch of positively charged residues including six Lys residues, the C-terminal helix is clearly hydrophobic, with two Ile and three Val residues, separated by a Lys residue. Gram-positive strains are insensitive to Aedesin's lytic effects. Significantly, Aedesin plays a role in protecting mosquitoes from infection by dengue virus, which is not simply passively transported to humans by its vector *A. egypti*. Viral replication in the mosquito salivary gland is essential for the later injection of infectious saliva into the human host and continuation of the dengue virus transmission cycle. Aedesin may offer a measure of protection against overgrowth of virus in the salivary gland. In this respect, the mosquito vector is likely to have various ways to assure its own survival while harboring a variety of pathogens. **1.** Godreuil, Leban, Padilla, *et al.* (2014) *PLoS One.* **9**, e105441.

Aerothrinicin 3

This cyclic lipopeptide lactone (FW = 1474.72 g/mol), produced by *Deuteromycotina* spp. NR7379, inhibits *Saccharomyces cerevisiae* 1,3-β-glucan synthase, which contains two catalytic subunits (Fks1p and Fks2p) and makes a main component of the cell wall. Although highly homologous (88.1% identity), Fks2p is more sensitive than Fks1p to inhibition by aerothricin-3. By constructing a series of chimeric FKS genes and examining their sensitivity to aerothricin-3, it was shown that a region around the fourth extracellular domain of Fks2p, containing 10 different amino acid residues from those of Fks1p, conferred Fks1p aerothricin-3 sensitivity, when the region was replaced with a corresponding region of Fks1p. To identify essential amino acid residue(s) responsible for the sensitivity, each of the 10 non-conserved amino acids of Fks1p was substituted into the corresponding amino acid of Fks2p by site-directed mutagenesis. Only one amino acid substitution of Fks1p (Lys-1336-Ile) was needed to confer Fks1p hypersensitivity to aerothricin-3. On the other hand, reverse substitution of the corresponding amino acid of Fks2p (Ile-1355-Lys) resulted in loss of hypersensitivity to aerothricin-3. Such findings results suggest this isoleucine of Fks2p plays a key role in aerothricin-3 sensitivity. **1.** Kondoh, Takasuka, Arisawa, Aoki & Watanabe (2002) *J. Biol. Chem.* **277**, 41744.

AF 802, *See Alectinib*

Afatinib

This oral quinazoline derivative and EGFR/HER2-directed protein kinase inhibitor (FW = 485.94; CASs = 439081-18-2 (free base), 936631-70-8 (maleic acid salt), 1254955-21-9 (HCl salt); Solubility (at 25°C): 197 mg/mL DMSO, 1 mg/mL Water), also known by its code name BIBW2992, its trade names Gilotrif® Tomtovok®, Tovok®, and its systematic name (S,E)-N-(4-(3-chloro-4-fluorophenyl-amino)-7-(tetra-hydrofuran-3-yloxy)quinazolin-6-yl)-4-(dimethylamino)but-2-enamide, irreversibly inactivates EGFR^{wt}, EGFR^{L858R}, EGFR$^{L858R/T790M}$ and HER2 with IC_{50} values of 0.5 nM, 0.4 nM, 10 nM and 14 nM, respectively (1). **Mode of Action:** The irreversible binding of afatinib to HER2 inactivates its interactions with a preferred partner of EGFR, and blocking the HER2-EGFR heterodimerization reduces their intrinsic tyrosine kinase activities. Irreversible inhibitors (such as afatinib and dacomitinib) that target all ErbB family receptor tyrosine kinases are intended to confer sustained disease control in ErbB-dependent cancers. Because nearly all EGFR-mutated patients eventually develop resistance to reversible EGFR-TKIs after a median of 14 months, tafatinib's irreversible action is thought to be a promising feature of its mode of action. **Pharmacokinetics:** Afatinib's PK profile is best described by a two-compartment disposition model, with first-order absorption and linear elimination. There was a slightly more than proportional increase in exposure with increasing dose, most likely due to dose-dependent relative bioavailability. For the therapeutic dose of 40 mg, the estimated apparent total clearance rate at steady state was 734 mL/min (2). **1.** Li, Ambrogio, Shimamura *et al.* (2008) *Oncogene* **27**, 4702. **2.** Freiwald, Schmid, Fleury, *et al.* (2014) *Cancer Chemother. Pharmacol.* **73**, 759.

Aflatoxins

Aflatoxin B₁   Aflatoxin B₂
Aflatoxin G₁   Aflatoxin G₂
Aflatoxin M₁   Aflatoxin M₂

These mycotoxins (FW_{B1} = 312.30 g/mol; CAS 1162-65-8; FW_{B2} = 314.29 g/mol; CAS 7220-81-7; FW_{G1} = 328.27 g/mol; CAS 1165-39-5; FW_{G2} = 330.29 g/mol; CAS 7241-98-77; FW_{M1} = 328.27 g/mol; CAS 6795-23-9; FW_{M2} = 330.29 g/mol; CAS 6885-57-0), produced by the fungus *Aspergillus flavus*, *A. parasiticus*, *A. oryzae*, and some *Penicillium* strains, are a set of closely related metabolites that are known carcinogens. The B and G designations in the name refer to the blue or green fluorescence, respectively, that are exhibited by the mycotoxins (afloxins B₂ and G₂ are dihydro derivatives of B₁ and G₁ that are less toxic). The M series of aflatoxins are hydroxylated derivatives of the B series toxins isolated from the milk of cows that ingested aflatoxin-contaminated feed. Initially innocuous forms are bioactivated by members of the cytochrome p450 family into mutagenic and carcinogenic intermediates. Aflatoxin-B₁, for example, is converted into aflatoxin B₁-8,9 *exo*-epoxide, which is in turn converted into the 8,9-dihydroxy-8-N^7-guanyl-9-hydroxy aflatoxin B₁ adduct. This adduct is metabolized into aflatoxin B₁ formaminopyrimidine adduct. It is these adducts that prove to be strongly mutagenic and often carcinogenic. Chronic hepatitis B virus (HBV) infection and persistent AFB_1 exposure are major risk factors for developing hepatocellular carcinogenesis, suggesting synergy in the action of these agents. **Nucleic Acid Interactions:** Aflatoxins have dimensions and so-called bay region aromatic characteristics that allow them to intercalate between base-pairs in DNA and double-stranded segments of RNA. Major groove recognition of DNA by proteins also relies the variation in hydrogen-bond donor/acceptor content to makes DNA base-pairs distinguishable from one another, and aflatoxins may mimic major groove-binding proteins. Aflatoxins also inhibit RNA biosynthesis, transcription, and protein biosynthesis. **CAUTION**: Aflatoxin B₁ is a highly potent mutagen and carcinogen. Avoid direct exposure to the skin or eyes as well as inhalation. **Target(s):** RNA polymerase (1-4,6,8,9); DNase-I (deoxyribonuclease I), by Aflatoxin B₂ₐ > Aflatoxin G₂ₐ; (5); electron transport (7,11); nucleoside transport (10); oxidative phosphorylation, uncoupled (11); β-galactoside α-2,6-sialyltransferase, inhibited by Aflatoxins B₁, B₂ₐ, and G₁ (12); glycoprotein sialyltransferase (12); acetylcholinesterase, inhibited by Aflatoxin B₁ (13); aminopeptidase B (14). **1.** Gelboin, Wortham, Wilson, Friedman & Wogan (1966) *Science* **154**, 1205. **2.** Sporn, Dingman, Phelps & Wogan (1966) *Science* **151**, 1539. **3.** Yu (1977) *J. Biol. Chem.* **252**, 3245. **4.** Yu (1981) *J. Biol. Chem.* **256**, 3292. **5.** Schabort & Pitout (1971) *Enzymologia* **41**, 201. **6.** King & Nicholson (1969) *Biochem. J.* **114**, 679. **7.** Doherty & Campbell (1972) *Res. Commun. Chem. Pathol. Pharmacol.* **3**, 601. **8.** Saunders, Barker & Smuckler (1972) *Cancer Res.* **32**, 2487. **9.** Akinrimisi, Benecke & Seifart (1974) *Eur. J. Biochem.* **42**, 333. **10.** Kunimoto, Kurimoto, Aibara & Miyaki (1974) *Cancer Res.* **34**, 968. **11.** Ramachandra Pai, Jayanthi Bai & Venkitasubramanian (1975) *Chem. Biol. Interact.* **10**, 123. **12.** Bernacki & Gurtoo (1975) *Res. Commun. Chem. Pathol. Pharmacol.* **10**, 681. **13.** Cometa, Lorenzini, Fortuna, Volpe, Meneguz & Palmery (2005) *Toxicology* **206**, 125. **14.** Sharma, Padwal-Desai & Ninjoor (1989) *Biochem. Biophys. Res. Commun.* **159**, 464.

AFN-1252

This selective, orally available, first-in-class antibiotic (FW = 375.43 g/mol; CAS 620175-39-5; Formulated in 1% Poloxamer 407) is a potent inhibitor of *Staphylococcus aureus* enoyl-acyl carrier protein reductase (FabI), thereby preventing fatty acyl chain elongation and disrupting biosynthesis of both saturated and unsaturated fatty acids needed for bacterial growth. It demonstrates exceptional potency and specificity against staphylococcal isolates, with typical *S. aureus* MIC ranges of 0.002–0.12 µg/mL and MIC_{90}s of ≤ 0.015 µg/mL (1). **Mode of Action:** FabI is the only enoyl-ACP reductase in *S. aureus*, *S. epidermidis*, and related staphylococci, with no alternative enzyme or rescue pathway, including processing of exogenously supplied fatty acids (1,2). Such metabolic features suggest FabI may be essential to cell viability in *Staphylococcus* spp. The inhibitor's target was convincingly confirmed (a) by using direct enzyme assays (IC_{50} = 14 nM, with no inhibition of the human enzyme, even at 67 µM); (b) by employing radiolabeled acetate to demonstrate that AFN-1252 greatly reduced incorporation into lipids; (c) by co-crystallizing AFN-1252 and FabI in the presence of the unnatural substrate 3′-NADPH, and (d) by genetically determining AFN-1252's spontaneous resistance frequency and characterizing AFN-1252-resistant FabI mutants (2). In so-called time-kill assays, AFN-1252 caused a time-dependent reduction in the viability of *S. aureus* ATCC 29213 (MSSA) and *S. aureus* ATCC 43300 (MRSA), with similar magnitudes and rates of killing were obtained at 4 and 128 times the MIC (*i.e.*, 2.0- and 2.9-$\log_{10}$ reductions) (2). **1.** Karlowsky, Kaplan, Hafkin, *et al.* (2009) *Antimicrob. Agents Chemother.* **53**, 3544. **2.** Parsons, Kukula, Jackson, *et al.* (2013) *Antimicrob. Agents Chemother.* **57**, 2182.

Afoxolaner

This orally active isoxazoline-based pesticide (FW = 901.18 g/mol; CAS 1093861-60-9; IUPAC Name: 4-[5-[3-chloro-5-(trifluoromethyl)phenyl]-4,5-dihydro-5-(trifluoromethyl)-3-isoxazolyl]-*N*-[2-oxo-2-[(2,2,2-trifluoro-ethyl)amino]ethyl]-1-naphthalenecarboxamide, is highly active against fleas (*Ctenocephalides felis*) and ticks (*Dermacentor variabilis*) in dogs. Afoxolaner is more potent, when tested *in vitro*, than any other compound ever tested in the membrane feeding system, including the avermectins. At an oral dose of 2.5 mg/kg, afoxolaner reaches pharmacologically effective plasma concentrations (0.1–0.2 µg/mL), blocking native and expressed insect GABA-gated chloride channels with nanomolar potency. Afoxolaner has comparable potency between wild-type channels as well as channels possessing the dieldrin resistance-conferring Alanine-302-Serine mutation. Lack of cyclodiene cross-resistance for afoxolaner suggests that afoxolaner blocks GABA-gated chloride channels by binding at a site topologically distinct from the cyclodienes. **1**. Shoop, Hartline, Gould, *et al.* (2014) *Vetern. Parisitol.* **201**, 179.

Afuresertib

This potent and orally bioavailable pan-AKT inhibitor (FW$_{\text{free-base}}$ = 427.32 g/mol; CAS 1047644-62-1; Soluble in DMSO) is ATP-competitive and targets AKT1 (K_i = 0.08 nM), AKT1$^{\text{E17K}}$ (K_i = 0.2 nM), AKT2 (K_i = 2 nM), and AKT3 (K_i = 2.6 nM), a family of serine-threonine kinase in the PI3K-AKT pathway, which is often disregulated in many human malignancies, leading to increased cell survival, growth and proliferation (1). PIP3 lipids tether these kinases to the membrane via their plextrin homology domain, enabling activation by Thr308 phosphorylation by PDK1 and Ser473 phosphorylation by the mTORC2 complex. Activated AKT in turn phosphorylates FOXO, TSC1/2, PRAS40, and GSK3β. Combination of Lapatinib with Afuresertib (or Uprosertib) is synergistic in HER2$^+$/PIK3CA^{mut} cell lines, but not in HER2$^+$/PIK3CA$^{\text{wild-type}}$ cell lines (2). Measured changes in phosphoprotein levels in 15 cell revealed that p-S6RP levels were less well attenuated by lapatinib in HER2$^+$/PIK3CA^{mut} cells compared to HER2$^+$/PIK3CA$^{\text{wild-type}}$ cells and that Lapatinib plus Afuresertib (or Uprosertib) reduced p-S6RP levels to those achieved in HER2$^+$/PIK3CA$^{\text{wild-type}}$ cells with lapatinib alone. There is also compensatory up-regulation of p-HER3, and p-HER2 is blunted in PIK3CA(mut) cells, following Lapatinib plus Afuresertib (or Uprosertib) treatment (2). ***See also Uprosertib* Other Targets:** P70SK6 (IC$_{50}$ = 251 nM), PKA (IC$_{50}$ = 1.3 nM), PKCα (IC$_{50}$ = >1000 nM), PKCβ1 (IC$_{50}$ = 430 nM), PKCβ2 (IC$_{50}$ = >1000 nM), PKCδ (IC$_{50}$ = 1000 nM), PKCγ (IC$_{50}$ = >1000 nM), PKCε (IC$_{50}$ = >1000 nM), PKCθ (IC$_{50}$ = 510 nM), PKCη (IC$_{50}$ = 210 nM), PKCξ (IC$_{50}$ = >1000 nM), PKG1α (IC$_{50}$ = 0.9 nM), PKG1β (IC$_{50}$ = 4 nM), ROCK (IC$_{50}$ = 100 nM), RSK1 (IC$_{50}$ = 320 nM). **1**. Dumble, Crouthamel, Zhang, *et al.* (2014) *PLoS One* **9**, e100880. **2**. Korkola, Collisson, Heiser, *et al.* (2015) *PLoS One* **10**, e0133219.

AG013736, *See* Axitinib
AG 9, *See* Tyrphostin 1
AG 10, *See* Tyrphostin 10
AG 17, *See* Tyrphostin A9
AG 18, *See* Tyrphostin 23
AG 30, *See* Tyrphostin AG30
AG 43, *See* Tyrphostin 63
AG 82, *See* Tyrphostin 25
AG 99, *See* Tyrphostin 46
AG 112, *See* Tyrphostin 48
AG-120, *See* Ivosidenib
AG 126, *See* Tyrphostin AG126
AG 183,*See* Tyrphostin 51
AG 213, *See* Tyrphostin 47
AG 370, *See* Tyrphostin AG370
AG 490, *See* Tyrphostin AG490

AG957

This tyrosine kinase inhibitor (FW = 273.28 g/mol; CAS 140674-76-6; Soluble in 5 mg/mL DMSO) targets BCR/ABL tyrosine kinase activity, restores β1 integrin-mediated cell adhesion, and alters inhibitory signaling in chronic myelogenous leukemia hematopoietic progenitors (1-3). AG957 selectively blocks the tyrosine kinase activity of human p210$_{\text{bcr-abl}}$ (K_i = 750 nM) versus p140$_{\text{c-abl}}$ (K_i = 10 μM). It also decreases p210$_{\text{bcr-abl}}$ phosphorylation in viable K562 cells at concentrations and durations of exposure that likewise inhibit cell proliferation. AG957 altesr the physical state of p210$_{\text{bcr-abl}}$ and the association of p210$_{\text{bcr-abl}}$ with Shc and Grb2 in K562 cells. **1**. Bhatia, Munthe & Verfaillie (1998) *Leukemia* **12**, 1708. **2**. Kaur & Sausville (1996) *Anticancer Drugs* **7**, 815. **3**. Anafi., *et al.* (1992) *J. Biol. Chem.* **267**, 4518.

AG 1343, *See* Nelfinavir
AG 1346, *See* Nelfinavir
AG 1433, *See* Tyrphostin AG1433
AG-1749, *See* Lansoprazole

AG 1776

This peptidomimetic inhibitor (FW = 574.72 g/mol), also known as KNI-764 and JE-2147, targets HIV-1 protease and plasmepsin IV, the latter with a K_i of 110 nM (3-5). **1**. Clemente, Moose, Hemrajani, *et al.* (2004) *Biochemistry* **43**, 12141. **2**. Clemente, Hemrajani, Blum, Goodenow & Dunn (2003) *Biochemistry* **42**, 15029. **3**. Clemente, Govindasamy, Madabushi, *et al.* (2006) *Acta Crystallogr. D Biol. Crystallogr.* **62**, 246. **4**. Madabushi, Chakraborty, Fisher, *et al.* (2005) *Acta Crystallogr. Sect. F Struct. Biol. Cryst. Commun.* **61**, 228. **5**. Gutiérrez-de-Terán, Nervall, Dunn, Clemente & Åqvist (2006) *FEBS Lett.* **580**, 5910.

AG 7088, *See* Rupintrivir

AG013958

This novel and potent protein kinase inhibitor (FW = 305.28 g/mol; CAS 900515-16-4; Solubility: 60 mg/mL DMSO), also known as (*Z*)-5-((5-(4-fluoro-2-hydroxyphenyl)furan-2-yl)methylene)thiazolidine-2,4-dione, targets PI3Kγ (IC$_{50}$ = 30 nM) in a cell-free assay, showing 30x greater selectivity toward PI3Kγ versus PI3Kα and much lower inhibitory activity toward PI3Kβ and PI3Kδ. **1**. Pomel, Klicic, Covini, *et al.* (2006) *J. Med. Chem.* **49**, 3857. **2**. Edling, Selvaggi, Buus, *et al.* (2010) *Clin. Cancer Res.* **16**, 4928.

AG-014699, *See* Rucaparib
AGA, *See* N-Acetyl-L-glutamate and N-Acetyl-L-glutamic Acid
Agenerase, *See* Amprenavir
Agent Orange, *See* (2,4,5-Trichlorophenoxy)acetate and (2,4,5-Trichlorophenoxy)acetic Acid; TCDD

AGI-5198

This isocitrate dehydrogenase-1 inhibitor (FW = 462.56 g/mol; CAS 1355326-35-0; Solubility: 24 mg/mL DMSO; <1 mg/mL H$_2$O; Prepared as 10 mM stock in DMSO and then dilute 50x to final concentration), also named *N*-[2-(cyclohexylamino)-1-(2-methylphenyl)-2-oxoethyl]-*N*-(3-fluorophenyl)-2-methyl-1*H*-imidazole-1-acetamide, potently and selectively targets IDH1^{R132H} (IC$_{50}$ = 0.07 μM) and IDH1^{R132C} (IC$_{50}$ = 0.16 μM) mutants, blocking production of the epigenetically active oncometabolite *R*-2-hydroxyglutarate. Near-complete inhibition of *R*-2-hydroxyglutarate formation induces histone H3K9me3 demethylation and expression of genes associated with gliogenic differentiation, thereby impairing growth of mutant-containing cells without appreciable changes in genome-wide DNA methylation. **1.** Rohle, Popovici-Muller, Palaskas, *et al.* (2013) *Science* **340**, 626.

AGM-1470, *See TNP-470*

Agmatine

This arginine metabolite (FW$_{free-base}$ = 130.19 g/mol), also known as 4-(aminobutyl)guanidine, is a putative endogenous agonist at imidazoline receptors and a ligand for α2-adrenergic receptors. Agmatine is also a precursor in the biosynthesis of a phosphagen (*i.e.*, phosphoagmatine) in certain invertebrates. In addition, agmatine blocks NMDA-activated channels in hippocampal neurons. Agmatine is a product of the reactions catalyzed by arginine decarboxylase, diguanidinobutanase, and acylagmatine amidase. **Target(s):** GfMEP thermostable lysine-specific metalloendopeptidase (1); polyamine oxidase (2,18); trypsin (2); nitric-oxide synthase (2,13,14,34); ancrod (3); diamine oxidase, weakly inhibited (4); Arginase (5,16,25); arginine 2-monooxygenase (7); guanidinobutyrase (8,22); *N*-methyl-D-aspartate (NMDA) receptors (10,19); *S*-adenosylmethionine decarboxylase (11,12); cyanophycin synthetase (15); voltage-gated Ca^{2+} channels (10,17,19); arginine decarboxylase (11); ornithine decarboxylase (20,21); arginine deaminase (23,24); peptidyl-Lys metalloendopeptidase (26); venombin A$_3$; arginine kinase (27,28); guanidinoacetate kinase (29); deoxyhypusine synthase (30); spermidine synthase (31); arginine *N*-succinyltransferase (32); ornithine carbamoyltransferase (33). **1.** Takio (1998) in *Handb. Proteolytic Enzymes* (Barrett, Rawlings & Woessner, eds), p. 1538, Academic Press, San Diego. **2.** Federico, Leone, Botta, *et al.* (2001) *J. Enzyme Inhib.* **16**, 147. **3.** Nolan, Hall & Barlow (1976) *Meth. Enzymol.* **45**, 205. **4.** Webb (1966) *Enzyme and Metabolic Inhibitors*, vol. 2, p. 363, Academic Press, New York. **5.** Colleluori & Ash (2001) *Biochemistry* **40**, 9356. **7.** Olomucki, Dangba & Nguyen Van Thoai (1964) *Biochim. Biophys. Acta* **85**, 480. **8.** Chou & Rodwell (1972) *J. Biol. Chem.* **247**, 4486. **9.** Li, Regunathan, Barrow, Eshraghi, Cooper & Reis (1994) *Science* **263**, 966. **10.** Yang & Reis (1999) *J. Pharmacol. Exp. Ther.* **288**, 544. **11.** Yang & Cho (1991) *Biochem. Biophys. Res. Commun.* **181**, 1181. **12.** Choi & Cho (1994) *Biochim. Biophys. Acta* **1201**, 466. **13.** Auguet, Viossat, Marin & Chabrier (1995) *Jpn. J. Pharmacol.* **69**, 285. **14.** Galea, Regunathan, Eliopoulos, Feinstein & Reis (1996) *Biochem. J.* **316**, 247. **15.** Aboulmagd, Oppermann-Sanio & Steinbuchel (2001) *Appl. Environ. Microbiol.* **67**, 2176. **16.** Hwang, Kim & Cho (2001) *Phytochemistry* **58**, 1015. **17.** Zheng, Weng, Gai, Li & Xiao (2004) *Acta Pharmacol. Sin.* **25**, 281. **18.** Cona, Manetti, Leone, *et al.* (2004) *Biochemistry* **43**, 3426. **19.** Askalany, Yamakura, Petrenko, *et al.* (2005) *Neurosci. Res.* **52**, 387. **20.** Lee & Cho (2001) *J. Biochem. Mol. Biol.* **34**, 408. **21.** Schaeffer & Donatelli (1990) *Biochem. J.* **270**, ¡599. **22.** Yorifuji, Shimizu, Hirata, *et al.* (1992) *Biosci. Biotechnol. Biochem.* **56**, 773. **23.** Knodler, Sekyere, Stewart, Schofield & Edwards (1998) *J. Biol. Chem.* **273**, 4470. **24.** Smith, Ganaway & Fahrney (1978) *J. Biol. Chem.* **253**, 6016. **25.** Patchett, Daniel & Morgan (1991) *Biochim. Biophys. Acta* **1077**, 291. **26.** Nonaka, Ishikawa, Tsumuraya, Hashimoto, Dohmae & Takio (1995) *J. Biochem.* **118**, 1014. **27.** Pereira, Alonso, Ivaldi, *et al.* (2003) *J. Eukaryot. Microbiol.* **50**, 132. **28.** Pereira, Alonso, Paveto, *et al.* (2000) *J. Biol. Chem.* **275**, 1495. **29.** Shirokane, Nakajima & Mizusawa (1991) *Agric. Biol. Chem.* **55**, 2235. **30.** Jakus, Wolff, Park & Folk (1993) *J. Biol. Chem.* **268**, 13151. **31.** Graser & Hartmann (2000) *Planta* **211**, 239. **32.** Tricot, Vander Wauven, Wattiez, Falmagne & Stalon (1994) *Eur. J. Biochem.* **224**, 853. **33.** Baker & Yon (1983) *Phytochemistry* **22**, 2171. **34.** Demady, Jianmongkol, Vuletich, Bender & Osawa (2001) *Mol. Pharmacol.* **59**, 24.

AGN 1135, *See Rasagiline*

AH6809

This substituted xanthene (FW$_{free-acid}$ = 298.30 g/mol; CAS 33458-93-4), named systematically as 6-isopropoxy-9-oxoxanthene-2-carboxylic acid, is an antagonist that binds to EP$_1$ and EP$_2$ receptors (1,3) and DP receptors (2-4). AH6809 inhibits prostaglandin E$_2$ and D$_2$-stimulated increases in cyclic AMP. **1.** Woodward, Pepperl, Burkey & Regan (1995) *Biochem. Pharmacol.* **50**, 1731. **2.** Crider, Griffin & Sharif (1999) *Brit. J. Pharmacol.* **127**, 204. **3.** Chan, Jones & Lau (2000) *Brit. J. Pharmacol.* **129**, 589. **4.** Keery & Lumley (1988) *Brit. J. Pharmacol.* **94**, 745.

AH 23848

This prostaglandin analogue (FW$_{free-acid}$ = 477.60 g/mol; CAS 81496-19-7), also named (4*Z*)-*rel*-7-[(1*R*,2*R*,5*S*)-5-([1,1'-biphenyl]-4-ylmethoxy)-2-(4-morpholinyl)-3-oxocyclopentyl]-4-heptenoic acid, is a potent, specific thromboxane receptor-blocking drug that is orally active and has a long duration of action (1-4). It also blocks the action of Prostaglandin E$_2$ (PGE$_2$), which activates four E-prostanoid receptors, EP$_1$, EP$_2$, EP$_3$, and EP$_4$. AH-23848 is a dual antagonist of TP^1 and EP$_4$ receptors (5). It inhibits TXA$_2$-induced platelet aggregation, IC$_{50}$ = 0.26 μM (1) and the contraction of human bronchial smooth muscle induced by the TP agonist U-46619 (6). AH-23848 also impairs PGE$_2$-mediated relaxation of piglet saphenous vein by antagonizing the PGE$_2$ receptor EP$_4$ (5) By inhibiting EP$_4$, it likewise suppresses serum-induced cAMP generation, cyclin A synthesis, and the poliferation of fibroblasts (7). Accumulation at G$_0$/G$_1$ after AH-6809 treatment appears to depend on intracellular Ca^{2+} concentration, because a 6-h treatment with thapsigargin (1 μM) allowed G$_0$/G$_1$-arrested cells to enter S phase. Similarly, treatment with 20 μM forskolin for 6 h allowed S-phase and G$_2$/M progression of AH-23848B-treated cells. Antagonism of the EP4 receptor with either AH23848 or ONO-AE3-208 reduced breast cancer metastasis as compared with vehicle-treated controls (8). The therapeutic effect was comparable to that observed with the dual COX-1/COX-2 inhibitor indomethacin. EP3 antagonism had no effect on tumor metastasis. Mammary tumor cells migrated in vitro in response to PGE and this chemotactic response was blocked by EP receptor antagonists (8). **1.** Brittain, Boutal, Carter, *et al.* (1985) *Circulation* **72**, 1208. **2.** Reeves & Stables (1987) *Prostaglandins* **34**, 829. **3.** Davis & Sharif (2000) *Brit. J. Pharmacol.* **130**, 1919. **4.** Crider, Griffin & Sharif (2000) *Prostaglandins Leukot. Essent. Fatty Acids* **62**, 21. **5.** Coleman, Grix, Head, *et al.* (1994) *Prostaglandins* **47**, 151. **6.** Coleman & Sheldrick (1989) *Brit. J. Pharmacol.* **96**, 688. **7.** Sanchez & Moreno (2002) *Am. J. Physiol. Cell Physiol.* **282**, C280. **8.** Ma, Kundu, Rifat, Walser & Fulton (2006) *Cancer Res.* **66**, 2923.

AICA, *See 5-Aminoimidazole-4-carboxamide*

AICAR, *See 5-Aminoimidazole-4-carboxamide-1-β-D-ribofuranosyl 5'-Monophosphate*

AIP Peptide, *See Autocamtide-2 Related Inhibitory Peptide*

AIR, *See 5-Amino-1-(5-phospho-D-ribosyl)imidazole*

Ajmalicine, *See Raubasine*

Ajoene

This asymmetric disulfide (FW = 234.41 g/mol; CAS 92285-01-3), the major sulfur-containing compound purified from garlic, exhibits antiplatelet activity and has a possible role in the prevention and treatment of cancer. It has also been reported to inhibit prenylation of proteins. **Target(s):** lipases (1); trypsin-like activity of the 20S proteasome (2); glutathione-disulfide reductase (3); trypanothione reductase (3); protein-tyrosine phosphatase (4); soybean lipoxygenase (5). **1**. Ransac, Gargouri, Marguet, *et al.* (1997) *Meth. Enzymol.* **286**, 190. **2**. Xu, Monsarrat, Gairin & Girbal-Neuhauser (2004) *Fundam. Clin. Pharmacol.* **18**, 171. **3**. Gallwitz, Bonse, Martinez-Cruz, *et al.* (1999) *J. Med. Chem.* **42**, 364. **4**. Villar, Alvarino & Flores (1997) *Biochim. Biophys. Acta* **1337**, 233. **5**. Belman, Solomon, Segal, Block & Barany (1989) *Biochem. Toxicol.* **4**, 151.

AK 295

This unusual α-ketoamide (FW$_{\text{free-base}}$ = 504.63 g/mol) inhibits calpain (K_i = 41 μM), providing a significant dose-dependent neuroprotective effect after focal brain ischemia in rats. **Target(s):** calpain-1 (1,2); calpain-2 (2); cathepsin B, weakly inhibited (2). **1**. Bartus, Hayward, Elliott, *et al.* (1994) *Stroke* **25**, 2265. **2**. Li, Ortega-Vilain, Patil, *et al.* (1996) *J. Med. Chem.* **39**, 4089.

Akt Inhibitor, *See specific inhibitor; e.g., 1-L-6-Hydroxymethyl-chiro-inositol 2-(R)-2-O methyl-3-O-octadecylcarbonate*

Akti-1/2

This dual Akt1/Akt2 inhibitor (FW = 551.64 g/mol; CAS 612847-09-3; Soluble to 20 mM in DMSO with gentle warming), also named 1,3-dihydro-1-[1-[[4-(6-phenyl-1*H*-imidazo[4,5-*g*]quinoxalin-7-yl)phenyl]-methyl]-4-piperidinyl]-2*H*-benzimidazol-2-one, targets Protein kinase B (PKB) isoforms Akt1 (IC$_{50}$ = 50 nM) and Akt2 (IC$_{50}$ = 210 nM), serine/threonine-specific protein kinases that play a key role in multiple cellular processes such as glucose metabolism, apoptosis, cell proliferation, transcription and cell migration. Akti-1/2 is selective for Akt1 and 2 over a panel of other tyrosine and serine/threonine kinases (1-3). It also sensitizes LnCaP cells to TRAIL (TNF-related apoptosis-inducing ligand) induced apoptosis. Complete inhibition is achieved in liver cells treated with 1-10 μM (4). Akti-1/2 also blocks insulin regulation of PEP carboxykinase and glucose 6-phosphatase expression (4). **1**. Barnett, *et al.* (2005) *Biochem. J.* **385**, 399. **2**. DeFeo-Jones, *et al.* (2005) *Mol. Cancer Ther.* **4**, 271. **3**. Lindsley, Zhao, Leister, *et al.* (2005) *Bioorg. Med. Chem. Lett.* **15**, 761. **4**. Logie, Ruiz-Alcaraz, Keane, *et al.*(2007) *Diabetes* **56**, 2218.

Akt Inhibitor II

This phosphatidylinositol analogue (FW$_{\text{free-acid}}$ = 598.76 g/mol), also known as SH-5, inhibits the proto-oncogenic serine/threonine kinase Akt (protein kinase B), inducing apoptosis. Specific inhibition of Akt by these compounds validates ligand design targeted to the PH domains of crucial signaling proteins, thus providing a unique class of possible cancer therapeutics. **1**. Kozikowski, Sun, Brognard & Dennis (2003) *J. Amer. Chem. Soc.* **125**, 1144.

Akt Inhibitor III, *Similar in action to Akt Inhibitor II*

Akt-I-1

This reversible, cell-active protein kinase inhibitor (FW = 378.51 g/mol; CAS 473382-39-7) targets Akt1, IC$_{50}$ = 4.6 μM, showing linear mixed-type inhibition against ATP and peptide substrate. Akt/PKB (protein kinase B) is a serine/threonine kinase that plays a key role in regulating cell survival and proliferation. Akt1, Akt2 and Akt3 possess an N-terminal PH (pleckstrin homology) domain as well as a kinase domain, separated by a 39-amino-acid hinge region. Akt-I-1 is virtually inactive against Akt2 (IC$_{50}$ > 250 μM) or Akt3 (IC$_{50}$ > 250 μM). It also blocks the phosphorylation and activation of Akt1 by PDK1 (Phosphoinositide-Dependent Kinase-1). Akt-I-1 blocks Akt phosphorylation at Thr-308 and Ser-473 and reduces the levels of active Akt within cells, as well as blocking phosphorylation of known Akt substrates and promoting TRAIL- (or Tumour-necrosis-factor-Related Apoptosis-Inducing Ligand-) induced apoptosis in LNCap prostate cancer cells. This inhibitor is weakly active (IC$_{50}$ > 250 μM) against ΔPH-Akt1, a form lacking its pleckstrin homology domain. In fact, antibodies to the Akt PH domain or hinge region block Akt-I-1 inhibition of Akt. **1**. Barnett, Defeo-Jones, Fu, *et al.* (2005) *Biochem. J.* **385** (*Part 2*), 399.

Al, *See Aluminum and Aluminum Ions*

AL-8810

This prostaglandin PGF2α analogue (FW$_{\text{free-acid}}$ = 402.51 g/mol; CAS 246246-19-5), also named (5*Z*,13*E*)-(9*S*,11*S*,15*R*)-9,15-dihydroxy-11-fluoro-15-(2-indanyl)-16,17,18,19,20-pentanor-5,13-prostadienoate, is a strong and selective FP prostanoid receptor antagonist: K_i = 400-500 nM. AL 8810 is an 11β-fluoro analogue of PGF$_{2\alpha}$ which acts as a potent and selective antagonist at the FP receptor (1). AL 8810 exhibits weak intrinsic agonist activity on FP receptor preparations in the 200-300 nM range, but fully antagonizes the activity of the potent FP receptor agonist fluprostenol (EC$_{50}$ = 430 nM). AL-8810 fully antagonizes the bimatoprost-induced calcium mobilization in Swiss 3T3 fibroblasts at 100 μM, indicating that bimatoprost acts as an FP agonist in this preparation (2). Its K_i for inhibiting several potent agonists at the cloned human ciliary body FP receptor is in the range of 1-2 μM (3). **1**. Griffin, Klimko, Crider & Sharif (1999) *J. Pharmacol. Exp. Ther.* **290**, 1278. **2**. Sharif, Williams & Kelly (2001) *Eur. J. Pharmacol.* **432**, 211. **3**. Sharif, Kelly & Crider (2002) *J. Ocul. Pharmacol* **18**, 313.

[Ala286]-Calmodulin-Dependent Protein Kinase II Fragment 281-301 Amide, *See Calmodulin-Dependent Protein Kinase II Fragment 281-301*

Alamethicin

These voltage-dependent, channel-forming, peptaibol antibiotics (MW$_{\text{Alamethicin F-30}}$ = 1964.40; CAS 27061-78-5) from the soil fungus *Trichoderma viride* NRRL 3199 are membrane-active oligopeptides isolated from that exhibit anti-bacterial and anti-fungal properties (1,2). Peptaibols are amphipathic, usually highly helical in structure, and typically form voltage dependent ion channels that uncouple oxidative phosphorylation (3,4), often attended by in bacterial and fungal cell death. Alamethicin F-30 has the sequence Ac-Aib-Pro-Aib-Ala-Aib-Ala-Gln-Aib-Val-Aib-Gly-Leu-Aib-Pro-Val-Aib-Aib-Glu-Gln-Phe-OH; Alamethicin F-50 has the sequence Ac-Aib-Pro-Aib-Ala-Aib-Ala-Gln-Aib-Val-Aib-Gly-Leu-Aib-Pro-Val-Aib Aib-Gln-Gln-Phe-OH; and Alamethicin II has the sequence Ac-Aib-Pro-Aib-Ala-Aib-Aib-Gln-Aib-Val-Aib-Gly-Leu-Aib-Pro-Val-Aib-Aib-Glu-Gln-Phe-OH. *Note*: The name "peptaibol" is derived from the prefix "*pep-*", the central syllable "*aib*", and the suffix "*-ol*" to designate its <u>pep</u>tide structure, the presence of an α-<u>a</u>mino<u>i</u>so<u>b</u>utyryl unit, and the C-terminal alco<u>hol</u>. **1.** Rinehart, Gaudioso, Moore, *et al.* (1981) *J. Amer. Chem. Soc.* **103**, 6517. **2.** Brueckner & Przybylski (1984) *J. Chrom.* **296**, 263. **3.** Reed (1979) *Meth. Enzymol.* **55**, 435. **4.** Das, Basu & Balaram (1985) *Biochem. Int.* **11**, 357.

L-Alanosine

This antibiotic and aspartate/asparagine analogue (FW = 149.11 g/mol; CAS 5854-93-3), also named L-2-amino-3-(*N*-hydroxy-*N*-nitrosoamino) propionic acid, isolated from fermentation broths of *Streptomyces alanosinicus* (1,2), inhibits adenylosuccinate synthetase, disrupting *de novo* synthesis of AMP in both malignant and normal cells. That alanosine inhibits the adenylosuccinate synthetase reaction, the first step in the conversion of IMP to AMP, is evidenced by the accumulation of IMP, but not adenylosuccinate, in treated cells (3). L-Alanosine's action is potentiated in methylthioadenosine phosphorylase (MTAP) deficiency. Clinical use is limited by its toxicity. At a concentration as low as 2.7 μM, L-alanosine completely inhibits the incorporation of hypoxanthine into adenosine triphosphate by cultured Novikoff rat hepatoma cells. A related agent, L-alanosyl-5-aminoimidazole-4-carboxylic acid ribonucleotide (*or* alanosyl-AICOR) has been synthesized enzymatically using 4-(*N*-succino)-5-aminoimidazole-4-carboxamide ribonucleotide (*or* SAICAR) synthetase in conjunction with 5-aminoimidazole-4-carboxylic acid ribonucleotide and alanosine (4). Alanosyl-AICOR is not a substrate of adenylosuccinate lyase from rat skeletal muscle, but it was an apparent competitive inhibitor in both reactions catalyzed by the enzyme (4). **1.** Murthy, Thiemann, Coronelli, Sensi (1966) *Nature* **211**, 1198. **2.** Thiemann & Beretta (1966) *J. Antibiot.* (Tokyo) **19**, 155. **3.** Graff & Plagemann (1976) *Cancer Res.* **36**, 1428. **4.** Casey & Lowenstein (1987) *Biochem. Pharmacol.* **36**, 705.

1-Alaninechlamydocin

This β-amino acid-containing cyclic tetrapeptide (FW = 427.62 g/mol; CAS 141446-96-0), from a Great Lakes-derived *Tolypocladium* sp. fungal isolate, showed potent antiproliferative/cytotoxic activities in a human MIA PaCa-2 pancreatic cancer cell line (GI$_{50}$ = 5.3 nM, GI$_{100}$ = 8.8 nM, LC$_{50}$ 22 nM). L-Alanine-chlamydocin induced G$_2$/M cell-cycle arrest and apoptosis. Its inhibitory effects are believed to result from inhibition of histone deacetylase (HDAC). **1.** Du, Risinger, King, Powell & Cichewicz (2014) *J. Nat. Prod.* **77**, 1753.

Alaproclate

This ester (FW$_{free-base}$ = 255.74 g/mol; CASs = 60719-82-6 (free base) and 60719-83-7 (hydrochloride)), also known as GAE-654 and DL-alanine 2-(4-chlorophenyl 0-1,1-dimethylethyl ester, is a serotonin-uptake inhibitor (1), an NMDA (*N*-methyl-D-aspartate) receptor antagonist (2,4), and a K$^+$ channel blocker (3,4). **1.** Lindberg, Thorberg, Bengtsson, *et al.* (1978) *J. Med. Chem.* **21**, 448. **2.** Wilkinson, Courtney, Westlind-Danielsson, Hallnemo & Akerman (1994) *J. Pharmacol. Exp. Ther.* **271**, 1314. **3.**

Hedlund (1987) *Neuropharmacology* **26**, 1535. **4.** Svensson, Werkman & Rogawski (1994) *Neuro-pharmacology* **33**, 795.

Alatrofloxacin

This fourth-generation fluoroquinolone-class antibiotic (FW = 558.51 g/mol; CAS 146961-76-4; quickly hydrolyzes to form trovafloxacin), also named 7-[(1*R*,5*S*)-6-{[(2*S*)-1-{[(2*S*)-2-aminopropanoyl]amino}-1-oxopro-pan-2-yl]amino}-3-azabicyclo[3.1.0]hexan-3-yl]-1-(2,4-difluorophenyl)-6-fluoro-4-oxo-1,8-naphthyridine-3-carboxylic acid, is a systemic anti-bacterial that targets Type II DNA topoisomerases (gyrases) required for bacterial replication and transcription. This fluoroquinolone shows an extensive *in vitro* antibiotic spectrum, with high activity against Gram-positive coccus, anaerobic and atypical pneumonia-producing bacteria (1). (For the prototypical member of this antibiotic class, **See** *Ciprofloxacin*) Intravenous alatrofloxacin, followed by oral trovafloxacin (TVX), is safe and well tolerated. Alatrofloxacin also lacks the phototoxicity, cardiovascular toxicity, and hemolytic anemia associated with earlier fluoroquinolones. Given its metabolism to trovafloxacin, it should be noted that the latter significantly increases the formation of nitrotyrosine in mice that are heterozygous for mitochondrial superoxide dismutase-2 (2). Using the NO-selective probe DAF-2, TVX also increases the production of mitochondrial NO in immortalized human hepatocytes. Similarly, mitochondrial Ca^{2+} is increased by TVX, suggesting calcium ion-dependent activation of mitochondrial NOS activity. The relationship between these observations and trovafloxacin-induced leukopenia remains to be explored. **1.** Hsiao, Younis & Boelsterli (2010) *Chem. Biol. Interact.* **188**, 204.

Albiglutide

This GLP-1 agonist and anti-diabetic drug (MW = 72.9 kDa; CAS 782500-75-8), marketed as Eperzan® and Tanzeum®, by GlaxoSmithKline (GSK) is a dipeptidyl peptidase-4-resistant glucagon-like peptide-1 dimer fused to human albumin. Albiglutide has a half-life of 4-7 days, commending it for biweekly or weekly administration in the treatment of Type 2 diabetes. The half-life is considerably longer than exenatide (Byetta®) and liraglutide (Victoza®).

L-Albizziin

This L-glutamine analogue (FW = 147.13 g/mol; CAS 1483-07-4; IUPAC: L-(−)-2-amino-3-ureidopropionic acid; CAS: 1483-07-4), also known as L-2-amino-3-ureidopropionic acid, from *Albizzia julibrissin* seeds inhibits many glutamine-dependent enzymes. **Target(s):** asparaginase (1); glutaminase (1); asparagine synthetase (2,13,14); glutamate synthase (3,4); glutaminase (5,6); glutamine: fructose-6-phosphate aminotransferase (isomerizing), *or* glucosamine-6 phosphate synthase (7); glutaminyl-tRNA synthetase (8); phosphoribosyl-formylglycinamidine synthetase, *or* formylglycinamide ribotide amidotransferase (9-12,15); carbamoyl-serine ammonia lyase, weakly inhibited (16); 4-methyleneglutaminase (17,18). **1.** Kovalenko, Tsvetkova & Nikolaev (1977) *Vopr. Med. Khim.* **23**, 618. **2.** Horowitz & Meister (1972) *J. Biol. Chem.* **247**, 6708. **3.** Meister (1985) *Meth. Enzymol.* **113**, 327. **4.** Masters & Meister (1982) *J. Biol. Chem.* **257**, 8711. **5.** Gmelin, Strauss & Hasenmaier (1958) *Z. Naturforsch.* **13B**, 252. **6.** Dura, Flores & Toldra (2002) *Int. J. Food Microbiol.* **76**, 117. **7.** Chmara, Andruszkiewicz & Borowski (1985) *Biochim. Biophys. Acta* **870**, 357. **8.** Lea & Fowden (1973) *Phytochemistry* **12**, 1903. **9.** Buchanan, Ohnoki & Hong (1978) *Meth. Enzymol.* **51**, 193. **10.** Buchanan (1982) *Meth. Enzymol.* **87**, 76. **11.** Pinkus (1977) *Meth. Enzymol.* **46**, 414. **12.** Schroeder, Allison & Buchanan (1969) *J. Biol. Chem.* **244**, 5856. **13.** Lea

& Fowden (1975) *Proc. R. Soc. Lond. B Biol. Sci.* **192**, 13. **14.** Pike & Beevers (1982) *Biochim. Biophys. Acta* **708**, 203. **15.** Schendel, Mueller, Stubbe, Shiau & Smith (1989) *Biochemistry* **28**, 2459. **16.** Cooper & Meister (1973) *Biochem. Biophys. Res. Commun.* **55**, 780. **17.** Ibrahim, Lea & Fowden (1984) *Phytochemistry* **23**, 1545. **18.** Winter & Dekker (1991) *Plant Physiol.* **95**, 206.

Albumin, Bovine Serum

This abundant blood protein (MW = 66,463 Da; CAS 9048-46-8) regulates osmotic pressure and serves as a binding protein for fatty acid hormone, Ca^{2+}, K^+, Na^+, bilirubin, and many hydrophobic and amphipathic pharmaceuticals. The ease of purification made led to its use as an protein standard in biochemistry. BSA is frequently used (at 1–2 mg/mL) to stabilize proteins and enzymes by exerting what is called the protective colloid effect. BSA often protects against dilution-induced dissociation/inactivation of susceptible. BSA also reduces enzyme adsorption to the surfaces of plastic labware. Because BSA binds a wide variety of substances (including substrates, cofactors, or other components of the reaction mixture), it may inhibit enzyme activity, and one should run control experiments before using BSA indiscriminately. **1.** Purich (2010) *Enzyme Kinetics: Catalysis & Control*, Academic Press (Elsevier), New York.

Albuterol, *See Salbutamol*

Alcapton, *See Homogentisate and Homogenisic Acid*

Aldehyde-Containing Protease Inhibitors

The catalytic mechanisms of serine and cysteine proteases are multi-step catalytic process, of which a key step is the formation of oxygen- or thiol-ester intermediate. Nested around that intermediate is a manifold of transitory species, including any tetrahedral adducts leading to and from that covalent acyl-enzyme. Papain, for example, is a thiol proteinase that is strongly inhibited by aldehydes resembling carboxylic acids that are released upon enzyme-catalyzed hydrolysis of specific substrates. Westerik & Wolfenden (1) found that at pH 5.5, approximately one mol acetyl-L-phenylalanylamino-acetaldehyde is bound per mol papain (K_i = 4.6 x 10^{-8} M). These aldehydes protect the Enzyme-SH against inactivation by the thiol-modifying reagent *N*-ethylmaleimide, presumably as a consequence of the enzyme forming an unusually stable thiohemiacetal with stereoelectronic features resembling tetrahedral intermediates in enzyme-catalyzed substrate hydrolysis. Such ideas have guided the design of countless peptide aldehydes that have proved to be especially effective inhibitors. The objective is to achieve specificity by capitalizing on knowledge of a target enzyme's preference for side-chains in the peptide substrate, while relying on reversible formation of hemiacetal or hemithioacetal adducts to gain potency. **1.** Westerik &Wolfenden (1972) *J. Biol. Chem.* **247**, 8195.

Aldicarb

This insecticide (FW = 190.26 g/mol; CAS 116-06-3), also known systematically as 2-methyl-2-(methylthio)-propanal *O*-(*N*-methyl-carbamoyl)oxime, inhibits acetyl-cholinesterase (1-6), aryl acylamidase (7), and carboxylesterase (8). It also binds to the nicotinic acetylcholine receptor (1). **CAUTION:** Exposure to high amounts of aldicarb can cause muscle weakness, blurred vision, headache, nausea, tearing, sweating, and tremors. High doses can be fatal to humans as a consequence of its ability to paralyze the respiratory system. Application of aldicarb should be undertaken only by qualified operators, heeding all guidelines and safety devices. **1.** Smulders, Bueters, Van Kleef & Vijverberg (2003) *Toxicol. Appl. Pharmacol.* **193**, 139. **2.** Payne, Stansbury & Weiden (1966) *J. Agric. Food Chem.* **14**, 356. **3.** Bakry, Sherby, Eldefrawi & Eldefrawi (1986) *Neurotoxicology* **7**, 1. **4.** Ahmed (1991) *J. Egypt. Soc. Parasitol.* **21**, 283. **5.** Singh & Spassova (1998) *Comp. Biochem. Physiol. C*

Pharmacol. Toxicol. Endocrinol. **119**, 97. **6.** Perkins & Schlenk (2000) *Toxicol. Sci.* **53**, 308. **7.** Engelhardt & Wallnöfer (1975) *Appl. Microbiol.* **29**, 717. **8.** McGhee (1987) *Biochemistry* **26**, 4101.

Alectinib

This potent, selective, and orally available ALK inhibitor (FW = 482.62 g/mol; CAS 1256580-46-7; Soluble in DMSO, not in H_2O), also known as AF 802, CH5424802, RO5424802, and 9-ethyl-6,6-dimethyl-8-(4-morpholinopiperidin-1-yl)-11-oxo-6,11-dihydro-5*H*-benzo[*b*]carbazole-3-carbonitrile, exhibits antitumor activity against cancers with ALK (IC_{50} = 1.9 nM) and mutantforms, as seen in nonsmall cell lung cancer (NSCLC) cells expressing EML4-ALK fusion, and anaplastic large-cell lymphoma (ALCL) cells expressing NPM-ALK fusion *in vitro* and *in vivo* (1,2). Alectinib inhibits ALK^{L1196M} (IC_{50} = 1.56 nM), the common resistance-conferring gatekeeper mutation that defeats the action of many kinase inhibitors, and blocks EML4-ALK^{L1196M}-driven cell growth. **1.** Sakamoto, Tsukaguchi, Hiroshima, *et al.* (2011) *Cancer Cell* **19**, 679. **2.** Kinoshita, Asoh, Furuichi, *et al.* (2012) *Bioorg. Med. Chem.* **20**, 1271. **3.** Yang (2013) *Lancet Oncol.* **14**, 564. **4.** Seto, Kiura, Nishio, *et al.* (2013) *Lancet Oncol.* **14**, 590. **5:** Latif, Saeed & Kim (2013) *Arch Pharm Res.* **36**, 1051. **6.** Solomon, Wilner & Shaw (2014) *Clin. Pharmacol. Ther.* **95**, 15.

Alefacept

This recombinant immunosuppressive protein (FW = 51800 g/mol; CAS 222535-22-0), marketed under the trade name Amevive®, reduces the number of psoriasis-associated T lymphocytes, thereby alleviating an underlying cause of the disorder. Alefacept is indicated in patients with moderate to severe chronic plaque psoriasis. The dimeric fusion protein consists of the extracellular CD2-binding portion of the human Leukocyte Function Antigen-3 (*or* LFA-3) linked to the Fc (hinge, C_H2 and C_H3 domains) portion of human IgG_1 (1). Amevive is a fully humanized toxin that binds to CD2 and blocks costimulatory signaling, selectively inducing apoptosis of activated memory T cells involved in the pathogenesis of psoriasis. Alefacept has a slow onset of action, peaking approximately 18 weeks after the first injection of a 12-week course (2). Because psoriasis is a chronic disorder, patients require lifelong access to injectable biologic agents targeting specific immune mediators, raising concerns about their longterm safety. Alefacept is only effective in a small proportion of patients, and its maximal efficacy is not achieved until 4-6 weeks after a treatment course. **1.** Cooper, Morgan & Harding (2003) *Eur. J. Immunol.* **33**, 666. **2.** Hodak & David (2004) *Dermatol. Ther.* **17**, 383.

Alendronate

This widely used methylenebisphosphonate bone resorption inhibitor ($FW_{free-acid}$ = 249.10 g/mol; CAS 66376-36-1 (disodium salt)), also known by the trade name Fosamax® and systematically named 4-amino-1-hydroxybutylidene-1,1-bisphosphonic acid, remains the drug of choice for treating osteoporosis. Alendronate shows preferential uptake at sites of bone resorption, where it blocks osteoclastic activity by inhibiting ruffled border formation (1). Alendronate is also the treatment of choice for hypercalcemia of malignancy, where a single infusion will normalize serum calcium in 80% of the patients. Paget's disease also shows an excellent long-term response to alendronate. **Inhibition of Bone Resorption:** Alendronate is a polyanion that binds firmly to positively charged calcium ions at active resorption sites of bone surfaces. Unlike pyrophosphate, however, bisphosphonates cannot be hydrolyzed by inorganic pyrophosphatase to orthophosphate. Instead, alendronate ion forms a multidendrate Ca^{2+} complex that binds tenaciously at physiologic pH. Under therapy, normal bone mineralization is unimpeded, with alendronate deposited onto and incorporated within the bone-matrix in a pharmacologically latent form. During osteoclast-mediated bone resorption, actin filament-rich podosomes allow osteoclasts to contact the bone surface, allowing the local hydrogen ion concentration to increase, lowering the pH to ~5.5. This process facilitates local hydroxyapatite hydrolysis, with release of Ca^{2+}, phosphate, and alendronate. Allendronate uptake by osteoclasts requires fluid-phase endocytosis and is enhanced by Ca^{2+} ions. Subsequent transfer of alendronate from endocytic vesicles to the cytosol requires endosomal acidification. Liberated in this manner, the drug retards further resorption by inhibiting farnesyl-diphosphate synthase as well as prenylation of Rac and Rho. Posttranslational modification of these small GTPases is required for tether these regulators to the peripheral membrane, where they activate actin polymerization and promote podosome formation. By inhibiting podosome assembly, alendronate suppresses the activation of vacuolar ATPase (or H^+-transporting proton pump), thus preventing proton export to the bone surface and further hydrolytic erosion of hydroxyapatite (1,2). **Pharmacokinetics:** After oral administration, systemic alendronate bioavailability is low, amounting to ~0.6% under fasting conditions. Once absorbed, about half binds to exposed bone surfaces, with the remainder excreted unchanged within 4-6 hours by the kidney. When taken with meals and beverages (other than water), alendronate's bioavailability is greatly reduced. When administered at the directed amount, it inhibits bone resorption without affecting bone mineralization. Alendronate is not cytotoxic for osteoclasts. Alendronate also appears to be without effect on the dynamics of vascular calcification. **Key Pharmacokinetic Parameters:** *See* Appendix II in Goodman & Gilman's *THE PHARMACOLOGICAL BASIS OF THERAPEUTICS*, 12th Edition (Brunton, Chabner & Knollmann, eds.) McGraw-Hill Medical, New York (2011). **Amelioration of Elastase-induced Emphysema:** In mice, delivery of the alendronate via aerosol inhalation (not orally administered) relieves elastase-induced emphysema, inhibiting airspace enlargement after elastase instillation (3). Aerosol inhalation induces macrophage apoptosis in bronchoalveolar fluid via caspase-3- and mevalonate-dependent pathways. Cytometry indicates that the $F4/80^+CD11b^{high}CD11c^{mild}$ population characterizing inflammatory macrophages, and the $F4/80^+CD11b^{mild}CD11c^{high}$ population defining resident alveolar macrophages take up substantial amounts of the bisphosphonate imaging agent OsteoSense680 after aerosol inhalation (3). Alendronate also inhibits macrophage migratory and phagocytotic activities and blunts the inflammatory response of alveolar macrophages by inhibiting nuclear factor-κB (NFκB) signalling. **Acid/Base Properties:** $pK_1 = 2.6$ (at 298 K) and 2.76 (at 310 K), $pK_2 = 6.7$ (at 298 K) and 6.8 (at 310 K), $pK_3 = 11.5$ (at 298 K) and 11.3 (at 310 K), and $pK_4 = 12.4$ (at 298 K) and 11.8 (at 298 K). **Target(s):** protein-tyrosine phosphatase (4,6-8); geranyl*trans*-transferase, *or* farnesyl diphosphate synthase (9-13,17-19); alkaline phosphatase (14); interstitial collagenase, *or* MMP-1 (15); gelatinase A, *or* MMP-2 (15); stromelysin-1, *or* MMP-3 (15); neutrophil collagenase, *or* MMP-8 (15); gelatinase B, *or* MMP-9 (15); macrophage elastase, *or* MMP-12 (15); collagenase-3, *or* MMP-13 (15); enamelysin, *or* MMP-20 (15); pyrophosphate-dependent phosphofructo-kinase, *or* diphosphate: fructose-6-phosphate 1-phosphotransferase) (16); geranyl-diphosphate synthase (20); protein-tyrosine phosphatases (PTPs), most likely as an alendronate-metal ion complex, inasmuch as alendronate inhibition is suppressed by EDTA (21). **1.** Rodan & Balena (1993) *Ann. Med.* **25**, 373. **2.** Russell, Watts, Ebetino & Rogers (2008) *Osteoporosis Internat.* **19**, 733. **3.** Ueno, Maeno, Nishimura, *et al.* (2015) *Nature Commun.* **6**, 6332. **4.** Skorey, Ly, Kelly, *et al.* (1997) *J. Biol. Chem.* **272**, 22472. **5.** Reszka & Rodan (2004) *Mini Rev. Med. Chem.* **4**, 711. **6.** Endo, Rutledge, Opas, *et al.* (1996) *J. Bone Miner. Res.* **11**, 535. **7.** Schmidt, Rutledge, Endo, *et al.* (1996) *Proc. Natl. Acad. Sci. U.S.A.* **93**, 3068. **8.** Opas, Rutledge, Golub, *et al.* (1997) *Biochem. Pharmacol.* **54**, 721. **9.** van Beek, Pieterman, Cohen, Lowik & Papapoulos (1999) *Biochem. Biophys. Res. Commun.* **264**, 108. **10.** Keller & Fliesler (1999) *Biochem. Biophys. Res. Commun.* **266**, 560. **11.** Bergstrom, Bostedor, Masarachia, Reszka & Rodan (2000) *Arch. Biochem. Biophys.* **373**, 231. **12.** Dunford, Thompson, Coxon, *et al.* (2001) *J. Pharmacol. Exp. Ther.* **296**, 235. **13.** Montalvetti, Bailey, Martin, *et al.* (2001) *J. Biol. Chem.* **276**, 33930. **14.** Vaisman, McCarthy & Cortizo (2005) *Biol. Trace Elem. Res.* **104**, 131. **15.** Heikkila, Teronen, Moilanen, *et al.* (2002) *Anticancer Drugs* **13**, 245. **16.** Bruchhaus, Jacobs, Denart & Tannich (1996) *Biochem. J.* **316**, 57. **17.** Sanders, Song, Chan, *et al.* (2005) *J. Med. Chem.* **48**, 2957. **18.** Glickman & Schmid (2007) *Assay Drug Dev.*

Technol. **5**, 205. **19.** Dunford, Kwaasi, Rogers, *et al.* (2008) *J. Med. Chem.* **51**, 2187. **20.** Burke, Klettke & Croteau (2004) *Arch. Biochem. Biophys.* **422**, 52. **21.** Skorey, Li, Kelly, *et al.* (1997) *J. Biol. Chem.* **272**, 22472.

Alfentanil

This potent analgesic (FW = 416.52 g/mol; CAS 71195-58-9), also known by its code name R-39209 and trade names Alfenta® and Rapifen®, is a synthetic, short-acting μ-opioid agonist (1). Rat liver alfentanil oxidation was increased threefold by treatment with pregnenolone 1-α-carbonitrile, a CYP3A inducer, but not by treatment with phenobarbital, β-naphthoflavone, or pyrazole, which are CYP2B, CYP1A, and CYP2E1 inducers, respectively (2). **Key Pharmacokinetic Parameters:** *See* Appendix II in Goodman & Gilman's *THE PHARMACOLOGICAL BASIS OF THERAPEUTICS*, 12th Edition (Brunton, Chabner & Knollmann, eds.) McGraw-Hill Medical, New York (2011). **1.** Komatsu, Turan & Orhan-Sungur (2007) *Anaesthesia* **62**, 1266. **2.** Kharasch & Thummel (1993) *Anesth. Analg.* **76**, 1033.

Alirocumab

This humanized anti-PCSK9 monoclonal antibody (MW = 146 kDa; CAS 1245916-14-6), developed as REGN727 (Regeneron Pharmaceuticals) and SAR236553 and Praluent® in cooperation with Sanofi, inhibits atherosclerosis, improves plaque morphology, and enhances statin effects (1-4). Its target, Pro-protein Convertase Subtilisin/Kexin type 9 (or PCSK9), is a secreted pro-protein convertase. Inhibition of PCSK9 increases hepatic Low-Density Lipoprotein Receptors (LDLRs), thereby enhancing hepatic LDL clearance. PCSK9 undergoes autocatalytic processing of its pro-domain within the endoplasmic reticulum, and its inhibitory pro-peptide segment remains associated with it following subsequent secretion. PCSK9 phosphorylation at Ser-47 in its pro-peptide and Ser-688 in its C-terminal domain by a Golgi casein kinase-like protein kinase appears to stabilize secreted PCSK9. The level of circulating PCSK9 appears to be inversely related to blood HDL levels, and it is understandable that cholesterol-lowering statins have the effect of increasing PCSK9 gene expression as well as the circulating level of PCSK9. Alirocumab binds to circulating PCSK9, blocking the latter's action on surface LDLR. Alirocumab dose-dependently decreases plasma lipids and atherosclerosis progression, and it enhances the beneficial effects of atorvastatin in ApoE3-Leiden transgenic mice. (*See also Evolocumab*) **1.** McKenney, Koren, Kereiakes, *et al.* (2012) *J. Am. Coll. Cardiol.* **59**, 2344. **2.** Kastelein, Robinson, Farnier, *et al.* (2014) *Cardiovasc. Drugs Ther.* **28**, 281. **3.** Kastelein, Robinson, Farnier, *et al.* (2014) *Cardiovasc Drugs Ther.* **28**, 281. **4.** Kühnast, van der Hoorn, Pieterman, *et al.* (2014) *J. Lipid Res.* **55**, 2103.

Alisertib

This orally available protein kinase inhibitor (FW = 518.92 g/mol; CAS 1028486-01-2; Solubility: 1 mg/mL DMSO, <1mg/mL H_2O), also known by its code name MLN8237 and its systematic name 4-[[9-chloro-7-(2-fluoro-6-methoxyphenyl)-5*H*-pyrimido[5,4-*d*][2]benzazepin-2-yl]amino]-2-methoxybenzoic acid, targets Aurora A, *or* AAK (IC$_{50}$ = 1.2 nM), showing >200x selectivity over Aurora B, *or* ABK (1). **Mode of Inhibitory Action:** Alisertib produces a dose-dependent decrease in bipolar and aligned chromosomes in the HCT-116 xenograft model, a drug-induced phenotype consistent with AAK inhibition (1). Targeting Aurora A kinase activity with the investigational agent alisertib likewise increases the efficacy of cytarabine through a FOXO-dependent mechanism (2). MLN8237 likewise induces polyploidization and expression of mature megakaryocyte markers in acute megakaryocytic leukemia (AMKL) blasts and displayed potent anti-AMKL activity *in vivo* (3). **Drug Formulation:** Typically prepared in an inert solubilizing carrier (2-hydroxypropyl-β-cyclodextrin, 10% weight/volume) in sodium bicarbonate, 1% weight/volume). **1.** Manfredi, Ecsedy, Chakravarty, *et al.* (2011) *Clin. Cancer Res.* **17**, 7614. **2.** Kelly, Nawrocki, Espitia, *et al.* (2012) *Int. J. Cancer* **131**, 2693. **3.** Wen, Goldenson, Silver, *et al.* (2012) *Cell* **150**, 575.

Aliskiren

This first-in-class oral renin inhibitor (FW = 551.76 g/mol; CAS 173334-57-1) also known by its alternate code name SPP100, its trade names: Tekturna™ and Rasilez™; and its IUPAC name, (2*S*,4*S*,5*S*,7*S*)-5-amino-*N*-(2-carbamoyl-2,2-dimethylethyl)-4-hydroxy-7-{[4-methoxy-3-(3-methoxy propoxy)phenyl]methyl}-8-methyl-2-(propan-2-yl)nonanamide, targets renin (IC$_{50}$ = 0.6 nM; *in vivo* $t_{1/2}$ = 40 hours; bioavailability = 2.6%) and is effective in treating hypertension. Aliskiren not only blocks renin but also binds to prorenin, inducing a conformational change. In this respect, aliskiren can be employed to titrate prorenin concentrations. Among renin/angiotensinogen blocking agents, direct renin inhibitors show excellent efficacy in hypertension control and also exhibit pharmacologic tolerance comparable with other renin-angiotensin suppressors. Aliskiren is effective in controlling blood pressure as monotherapy or in combination with other antihypertensive drugs, irrespective of patient's age, ethnicity or sex (2). (***See*** *VTP-27999*) **Key Pharmacokinetic Parameters:** *See* Appendix II in Goodman & Gilman's THE PHARMACOLOGICAL BASIS OF THERAPEUTICS, 12th Edition (Brunton, Chabner & Knollmann, eds.) McGraw-Hill Medical, New York (2011). **1.** Nussberger, Wuerzner, Jensen, *et al.* (2003) *Biochem. Biophys. Res. Commun.* **308**, 698. **2.** Juncos (2013) *Ther. Adv. Cardiovasc. Dis.* **7**, 153.

Alizarin

This anthraquinone dye (FW = 240.22 g/mol; CAS 72-48-0), also called Mordant Red 11, is a natural product from the madder plant (*Rubia tinctorium*), where it is glycosidically linked to primeverose. **Target(s):** steroid 5α-reductase (1); CYP1A1 (2); CYP1A2 (2); CYP1B1 (2); HIV-1 proteinase (3); glutathione *S*-transferase (4,5,8); cystathionine β-synthase (6); [myosin light-chain] kinase (7); procollagen-proline 4-dioxygenase, IC$_{50}$ ≈ 79 μM (9). **1.** Hiipakka, Zhang, Dai, Dai & Liao (2002) *Biochem. Pharmacol.* **63**, 1165. **2.** Takahashi, Fujita, Kamataki, Arimoto-Kobayashi, Okamoto & Negishi (2002) *Mutat. Res.* **508**, 147. **3.** Brinkworth & Fairlie (1995) *Biochim. Biophys. Acta* **1253**, 5. **4.** Das, Singh, Mukhtar & Awasthi (1986) *Biochem. Biophys. Res. Commun.* **141**, 1170. **5.** Das, Bickers & Mukhtar (1984) *Biochem. Biophys. Res. Commun.* **120**, 427. **6.** Walker & Barrett (1992) *Exp. Parasitol.* **74**, 205. **7.** Jinsart, Ternai & Polya (1992) *Biol. Chem. Hoppe-Seyler* **373**, 903. **8.** Liebau, Eckelt, Wildenburg, *et al.* (1997) *Biochem. J.* **324**, 659. **9.** Cunliffe & Franklin (1986) *Biochem. J.* **239**, 311.

Allicin

This naturally occurring sulfoxide and antibacterial and antifungal agent (FW = 162.28 g/mol; CAS 539-86-6), also known as diallyl thiosulfinate, and 2-propene-1-sulfinothioic acid *S*-2-propenyl ester, has the fragrance of freshly crushed garlic (*Allium sativum*). **Target(s):** urease (1-3); papain (2,4,10); amylase (2); succinate dehydrogenase (4); xanthine oxidase (4); glyceraldehyde-3-phosphate dehydrogenase (4); choline oxidase (4); hexokinase (4); cholinesterase (4,13); glyoxylase (4); alcohol dehydrogenase (4,9,10); lactate dehydrogenase (4); tyrosinase (4); alkaline phosphatase (2); RNA biosynthesis (5); prostaglandin-E synthase, *or* prostaglandin-H$_2$ isomerase (6); acetyl-CoA synthetase, *or* acetate:CoA ligase (7); lanosterol 14α-demethylase (8); histolysain and/or amoebapain (9); alcohol dehydrogenase, NADP$^+$-dependent (10); protein farnesyltransferase (11); protein geranylgeranyltransferase (11); CYP1A2 (12); CYP2C9 (12); CYP2C19 (12); CYP2D6 (12); CY3A4 (12); acetylcholinesterase (13); 5-lipoxygenase (14); arachidonate 5-lipoxygenase (14). **1.** Juszkiewicz, Zaborska, Sepiol, Gora & Zaborska (2003) *J. Enzyme Inhib. Med. Chem.* **18**, 419. **2.** Rao, Rao & Venkataraman (1946) *J. Sci. Industr. Res. B* **5**, 31. **3.** Agarwala, Murti & Shrivastava (1952) *J. Sci. Industr. Res. B* **11**, 165. **4.** Wills (1956) *Biochem. J.* **63**, 514. **5.** Feldberg, Chang, Kotik, *et al.* (1988) *Antimicrob. Agents Chemother.* **32**, 1763. **6.** Shalinsky, McNamara & Agrawal (1989) *Prostaglandins* **37**, 135. **7.** Focke, Feld & Lichtenthaler (1990) *FEBS Lett.* **261**, 106. **8.** Gebhardt, Beck & Wagner (1994) *Biochim. Biophys. Acta* **1213**, 57. **9.** Ankri, Miron, Rabinkov, Wilchek & Mirelman (1997) *Antimicrob. Agents Chemother.* **41**, 2286. **10.** Rabinkov, Miron, Konstantinovski, *et al.* (1998) *Biochim. Biophys. Acta* **1379**, 233. **11.** Lee, Park, Oh & Yang (1998) *Planta Med.* **64**, 303. **12.** Zou, Harkey & Henderson (2002) *Life Sci.* **71**, 1579. **13.** Millard, Shnyrov, Newstead, *et al.* (2003) *Protein Sci.* **12**, 2337. **14.** Schneider & Bucar (2005) *Phytother. Res.* **19**, 81.

L-Alloisoleucine

This isoleucine isomer (FW = 131.17 g/mol; CAS 1509-34-8; Symbol, L-alle), named systematically as (2*S*,3*R*)-2-amino-3-methylpentanoic acid, is a diastereoisomer of the proteogenic amino acid, L-isoleucine. Alloisoleucine is observed in the blood plasma and urine of patients with maple syrup urine disease and other disorders associated with metabolic pathways of the branched chain amino acids (1,2). Alloisoleucine is also a component of the antibiotics malformin B$_1$ and globomycin. **Target(s):** threonine deaminase (3). **1.** Schadewaldt, Bodner-Leidecker, Hammen & Wendel (1999) *Clin. Chem.* **45**, 1734. **2.** Matthews, Ben-Galim, Haymond & Bier (1980) *Pediatr. Res.* **14**, 854. **3.** Ahmed, Bollon, Rogers & Magee (1976) *Biochimie* **58**, 225.

Allopurinol

major resonance species

This purine analogue (FW = 136.11 g/mol; CAS 315-30-0), also known by the trade names Zyloprim™ and Lopurin™ and by its systematic name 4-hydroxypyrazolo(3,4-*d*)pyrimidine, is used mainly to treat gout by inhibiting xanthine oxidase, thereby limiting the build-up of uric acid.

Allopurinol and oxipurinol are metabolically converted by purine nucleoside phosphorylase to their corresponding L-ribosyl derivatives that are then phosphorylated. **History:** Seeking to retard the breakdown of the anticancer drug mercaptopurine, which is used to treat acute lymphoblastic leukemia, Nobelist Gertrude Elion and coworkers synthesized allopurinol to enhance mercaptopurine's action. Allopurinol is a slowly catalyzed alternative substrate for xanthine oxidase, which produces oxipurinol (**See** *Oxipurinol*). Because oxipurinol dissociates extremely slow ($t_{1/2}$ = 300 min) from the enzyme, oxipurinol is a potent xanthine oxidase product inhibitor. **Therapeutic Use:** Allopurinol is dispensed as 30-mg and 100-mg tablets. Its long-term use is generally unremarkable, but both allopurinol and oxipurinol cause orotic aciduria (from < 2 mg/day to 6-30 mg/day) and orotidinuria (from < 3 mg/day to 24-55 mg/day), while also increasing orotidine:PRPP transferase and orotidylate decarboxylase activity, for which they are inhibitors. **Allopurinol Hypersensitivity:** Allopurinol-derived oxypurinol is a common cause of severe cutaneous adverse drug reactions (SCAR), including Stevens-Johnson Syndrome (SJS) and Toxic Epidermal Necrolysis (TEN). A strong association between "allopurinol"-induced SCAR and the Human Leukocyte Antigen HLA-B*5801 was observed in a Han Chinese population, with high frequency of this allele, whereas there is only a moderate association was observed in European and Japanese populations, where this allele has a low frequency. Although genotyping is now required to identify individuals susceptible to SJS prior to starting abacavir and carbamazepine therapy (**See** *Abacavir and Carbamazepine*), no such screening is advocated for allopurinol, in part due to the lack of rapid/inexpensive screening for the HLA-B*58:01 allele. **Pharmacokinetics:** After oral dosage, the PK parameters of allopurinol include an oral bioavailability of 79%, an elimination $t_{1/2}$ of 1.2 hours, an apparent oral clearance of 15.8 mL/min/kg, and an apparent volume of distribution of 1.31 L/kg (1). Assuming that 90 mg of oxipurinol is formed from every 100mg of allopurinol, the pharmacokinetic parameters of oxipurinol in subjects with normal renal function are a $t_{1/2}$ of 23 hours, CL/F of 0.31 mL/min/kg, V_d/F of 0.59 L/kg, and renal clearance (relative to creatinine) of 0.19. Oxipurinol is cleared almost entirely by urinary excretion (1). **Key Pharmacokinetic Parameters:** *See* Appendix II in Goodman & Gilman's *THE PHARMACOLOGICAL BASIS OF THERAPEUTICS*, 12th Edition (Brunton, Chabner & Knollmann, eds.) McGraw-Hill Medical, New York (2011). **Target(s):** xanthine oxidase, slow, tight-binding inhibitor (1-5,8,9,14,19-23); orotate phosphoribosyl-transferase (6); purine nucleoside phosphorylase (7,15); tryptophan pyrrolase, *or* tryptophan 2,3-dioxygenase (10,16); salicylhydroxamic acid reductase (11); guanine deaminase (12); glutamine:D-fructose-6-phosphate aminotransferase (13); xanthine phosphoribosyl-transferase (17,18); hypoxanthine(guanine) phosphoribosyl-transferase (18); xanthine dehydrogenase (24-29); orotidine:PRPP transferase and orotidylate decarboxylase (30). **1.** Day, Graham, Hicks, *et al.* (2007) *Clin. Pharmacokinet.* **46**, 623. **2.** Kelley & Beardmore (1970) *Science* **169**, 388. **3.** Terada, Leff & Repine (1990) *Meth. Enzymol.* **186**, 651. **4.** Cha, Agarwal & Parks (1975) *Biochem. Pharmacol.* **24**, 2187. **5.** Bray (1975) *The Enzymes*, 3rd ed., **12**, 299. **6.** Jones, Kavipurapu & Traut (1978) *Meth. Enzymol.* **51**, 155. **7.** Parks & Agarwal (1972) *The Enzymes*, 3rd ed., **7**, 483. **8.** Webb (1966) *Enzyme and Metabolic Inhibitors*, vol. 2, p. 283, Academic Press, New York. **9.** Elion, Callahan, Nathan, *et al.* (1963) *Biochem. Pharmacol.* **12**, 85. **10.** Becking & Johnson (1967) *Can. J. Biochem.* **45**, 1667. **11.** Katsura, Kitamura & Tatsumi (1993) *Arch. Biochem. Biophys.* **302**, 356. **12.** Martinez-Farnos, Gubert & Bozal (1978) *Rev. Esp. Fisiol.* **34**, 295. **13.** Chatterjee & Stefanovich (1976) *Arzneimittelforschung* **26**, 502. **14.** Watts, Watts & Seegmiller (1965) *J. Lab. Clin. Med.* **66**, 688. **15.** Ray, Olson & Fridland (2004) *Antimicrob. Agents Chemother.* **48**, 1089. **16.** Badawy & Evans (1973) *Biochem. J.* **133**, 585. **17.** Miller, Adamczyk, Fyfe & Elion (1974) *Arch. Biochem. Biophys.* **165**, 349. **18.** Naguib, Iltzsch, el Kouni, Panzica & el Kouni (1995) *Biochem. Pharmacol.* **50**, 1685. **19.** Gupta, Rodrigues, Esteves, *et al.* (2008) *Eur. J. Med. Chem.* **43**, 771. **20.** Hsieh, Wu, Yang, Choong & Chen (2007) *Bioorg. Med. Chem.* **15**, 3450. **21.** Chang, Lee, Chen, *et al.* (2007) *Free Rad. Biol. Med.* **43**, 1541. **22.** Lin, Tsai, Chen, *et al.* (2008) *Biochem. Pharmacol.* **75**, 1416. **23.** Dew, Day & Morgan (2005) *J. Agric. Food Chem.* **53**, 6510. **24.** Lyon & Garrett (1978) *J. Biol. Chem.* **253**, 2604. **25.** Ziang & Edmondson (1996) *Biochemistry* **35**, 5441. **26.** Schräder, Rienhöfer & Andreesen (1999) *Eur. J. Biochem.* **264**, 862. **27.** Nguyen & Feierbend (1978) *Plant Sci. Lett.* **13**, 125. **28.** Sauer, Frebortova, Sebela, *et al.* (2002) *Plant Physiol. Biochem.* **40**, 393. **29.** Atmani, Baghiani, Harrison & Benboubetra (2005) *Int. Dairy J.* **15**, 1113. **30.** Suttle, Becroft & Webster (1989) in *The Metabolic Basis of Inherited Disease*, 6th ed., p. 1103.

Allopurinol Ribonucleoside 5'-Monophosphate

This unnatural nucleotide ($FW_{free-acid}$ = 348.21 g/mol) is an AMP and IMP analogue that inhibits orotidylate decarboxylase, *or* orotidine-5'-phosphate decarboxylase (1-5); GMP reductase (6), adenylosuccinate synthetase (7), and amidophosphoribosyl-transferase (8). **1.** Silva & Hatfield (1978) *Meth. Enzymol.* **51**, 143. **2.** Jones, Kavipurapu & Traut (1978) *Meth. Enzymol.* **51**, 155. **3.** Porter & Short (2000) *Biochemistry* **39**, 11788. **4.** Fyfe, Miller & Krenitsky (1973) *J. Biol. Chem.* **248**, 3801. **5.** Miller & Wolfenden (2002) *Ann. Rev. Biochem.* **71**, 847. **6.** Spector, Jones, LaFon, *et al.* (1984) *Biochem. Pharmacol.* **33**, 1611. **7.** Spector & R. L. Miller (1976) *Biochim. Biophys. Acta* **445**, 509. **8.** Holmes, McDonald, McCord, Wyngaarden & Kelley (1973) *J. Biol. Chem.* **248**, 144.

Allosamidin

This glycoside (FW = 622.63 g/mol; CAS 103782-08-7), first isolated from a species of *Streptomyces* in 1987, is the first naturally occurring chitinase inhibitor identified. It inhibits insect class-18 chitinases as well as chitinase CiX1 of *Coccidioides immitis*, K_i = 60 nM (1-7). **Target(s):** chitinase (1-7); chitotriosidase (8); hevamine, a chitinase from the rubber tree *Hevea brasiliensis* (5). **1.** Milewski, O'Donnell & Gooday (1992) *J. Gen. Microbiol.* **138**, 2545. **2.** Sakuda, Isogai, Matsumoto & Suzuki (1987) *J. Antibiot.* **40**, 296. **3.** Bortone, Monzingo, Ernst & Robertus (2002) *J. Mol. Biol.* **320**, 293. **4.** Dickinson, Keer, Hitchcock & Adams (1989) *J. Gen. Microbiol.* **135**, 1417. **5.** Bokma, Barends, van Scheltingab, Dijkstra & Beintema (2000) *FEBS Lett.* **478**, 119. **6.** Sakurada, Morgavi, Komatani, Tomita & Onodera (1997) *Curr. Microbiol.* **35**, 48. **7.** Sakurada, Morgavi, Ushirone, Komatani, Tomita & R. Onodera (1998) *Curr. Microbiol.* **37**, 60. **8.** Fusetti, von Moeller, Houston, *et al.* (2002) *J. Biol. Chem.* **277**, 25537.

L-Allothreonine

This nonproteogenic diastereoisomer of L-threonine (FW = 119.12 g/mol; CAS 28954-12-3) is found in the antibiotics telomycin, syringostatin, globomycin, enduracidins A and B, and lysobactin. **Target(s):** serine hydroxymethyl-transferase, *or* glycine Hydroxymethyltransferase (1,6); L-threonine 3-dehydrogenase, also an alternative substrate (2); threonine dehydratase, inhibited by racemic mixture (3); L-serine dehydratase, inhibited by racemic mixture (4); and kynureninase, weakly inhibited by racemic mixture (5); 3-phosphoglycerate dehydrogenase (6). **1.** Liu, Reig, Nasrallah & Stover (2000) *Biochemistry* **39**, 11523. **2.** Green & Elliott (1964) *Biochem. J.* **92**, 537. **3.** Maeba & Sanwal (1966) *Biochemistry* **5**, 525. **4.** Ramos & Wiame (1982) *Eur. J. Biochem.* **123**, 571. **5.** Jakoby & Bonner (1953) *J. Biol. Chem.* **205**, 709. **6.** Schirch & Gross (1968) *J. Biol. Chem.* **243**, 5651.

Alloxan

This sulfhydryl reagent (FW = 142.07 g/mol; CASs = 50-71-5 (free base) and 2244-11-3 (monohydrate)), also named as 2,4,5,6-tetraoxopyrimidine, is an unstable tetraoxopyrimidine that induces diabetes by selectively destroying pancreatic β-cells in laboratory animals upon selective uptake via the GLUT2 transporter. **Protocol for Using Alloxan to Induce Diabetes:** *Step*-1: Weigh and measure baseline blood glucose level of mice (Age: 6 weeks or older), adhering to the IACUC Guidelines for Glucose Monitoring of Blood. *Step*-2: Inject i.v. or i.p. with alloxan (70-90 mg/kg). *Step*-3: Beginning the day after the alloxan injection, blood glucose levels of each mouse should be checked daily or every other day for 5-7 days. Within 2-3 days after onset of diabetes (as indicated by blood glucose levels), mice should be implanted with an insulin-secreting pellet, when long-term study is necessary. Pellets typically release 0.1 U each 24 hr for >30 days. Mice weighing less than 25 grams require one insulin pellet; larger mice typically receive two insulin pellets to restore normoglycemia. **Target(s):** hexokinase (1,5,24-28,59); glucokinase (2,60,63,64, 66); succinate dehydrogenase (3,4,41,56); phosphofructokinase (6,9,26); alcohol dehydrogenase, yeast (7); glucose-6 phosphatase (8,23); aconitase (10,57,65); ATPase (11,12); arylamine acetyltransferase (13,14); carbonic anhydrase, weakly inhibited (15); choline acetyltransferase (16); choline dehydrogenase (17); choline oxidase (18); cytochrome oxidase (19,20); fructose-1,6-bisphosphatase (21); galactonolactone dehydrogenase (22); alkaline phosphatase (29,30,50); phosphoglucomutase (23,31); glyceraldehyde-3-phosphate dehydrogenase (32); phosphoprotein phosphatase (33,52); pyrophosphatase (11,12,34,49,51); pyruvate decarboxylase (36-38); rhodanese, *or* thiosulfate sulfurtransferase (39); sedoheptulokinase (40); succinate oxidase (42,43); urease (44); xanthine oxidase (45,73); adenylate cyclase, *Phycomyces blakesleeanus* (47); stearate dehydrogenase (48); acid phosphatase (50,72); allantoinase, weakly inhibited (53); oxidative phosphorylation, moderately inhibited (54); protein *O*-linked *N*-acetylglucosaminyl-transferase (55); Na$^+$/K$^+$-exchanging ATPase (56); calmodulin-dependent protein kinase (58,67); Ca^{2+}-dependent ATPase (61,62); isocitrate dehydrogenase, NAD (65); glutamate dehydrogenase (65); α-ketoglutarate dehydrogenase (65); succinyl-CoA synthetase (65); fumarase (65); citrate synthase, moderately inhibited (65); isocitrate dehydrogenase, NADP (65); phosphate transport (68); xanthine dehydrogenase (69); papain (70); cathepsin (70); barbiturase (71). **1.** Crane & Sols (1955) *Meth. Enzymol.* **1**, 277. **2.** Lenzen, Brand & Panten (1988) *Brit. J. Pharmacol.* **95**, 851. **3.** Hopkins, Morgan & Lutwak-Mann (1938) *Biochem. J.* **32**, 1829. **4.** Hopkins & Morgan (1938) *Biochem. J.* **32**, 611. **5.** Colowick (1951) *The Enzymes*, 1st ed., **2** (part 1), 114. **6.** Lardy (1962) *The Enzymes*, 2nd ed., **6**, 67. **7.** Sund & Theorell (1963) *The Enzymes*, 2nd ed., **7**, 25. **8.** Nordlie (1971) *The Enzymes*, 3rd ed., **4**, 543. **9.** Bloxham & Lardy (1973) *The Enzymes*, 3rd ed., **8**, 239. **10.** Krebs & Eggleston (1944) *Biochem. J.* **38**, 426. **11.** Gore (1951) *Biochem. J.* **50**, 18. **12.** Gordon (1953) *Biochem. J.* **55**, 812. **13.** Wrenshall (1957) *Ann. N. Y. Acad. Sci.* **71**, 164. **14.** Cooperstein & Lazarow (1958) *J. Biol. Chem.* **232**, 695. **15.** Chiba, Kawai & Kondo (1954) *Bull. Res. Inst. Food Sci., Kyoto Univ.* **13**, 53. **16.** Torda & Wolff (1946) *Amer. J. Physiol.* **147**, 384. **17.** Gordon & Quastel (1948) *Biochem. J.* **42**, 337. **18.** Rothschild, Cori & Barron (1954) *J. Biol. Chem.* **208**, 41. **19.** Becker & Rauschke (1951) *Z. Ges. Exptl. Med.* **117**, 374. **20.** Robuschi (1952) *Boll. Soc. Ital. Biol. Sper.* **28**, 624. **21.** Walsh & Walsh (1948) *Nature* **161**, 976. **22.** Mapson, Isherwood & Chen (1954) *Biochem. J.* **56**, 21. **23.** Broh-Kahn, Mirsky, Perisutti & Brand (1948) *Arch. Biochem.* **16**, 87. **24.** Sato, Takemori & Ebata (1956) *J. Biochem.* **43**, 623. **25.** Saltman (1953) *J. Biol. Chem.* **200**, 145. **26.** Griffiths (1949) *Arch. Biochem.* **20**, 451. **27.** Bhattacharya (1959) *Nature* **184**, 1638. **28.** Villar-Palasi, Carballido, Sols & Arteta (1957) *Nature* **180**, 387. **29.** Burgen & Lorch (1947) *Biochem. J.* **41**, 223. **30.** Larralde & Pons (1949) *Rev. Espan. Fisiol.* **5**, 37. **31.** Lehmann (1939) *Biochem. J.* **33**, 1241. **32.** Rapkine, Rapkine & Trpinac (1939) *Compt. Rend.* **209**, 253. **33.** Sundararajan & Sarma (1954) *Biochem. J.* **56**, 125. **34.** Gordon (1950) *Biochem. J.* **46**, 96. **35.** Naganna & Menon (1948) *J. Biol. Chem.* **174**, 501. **36.** Kensler, Young & Rhoads (1942) *J. Biol. Chem.* **143**, 465. **37.** Kuhn & Beinert (1947) *Ber.* **80**, 101. **38.** Stoppani, Actis, Deferrari & Gonzalez (1952) *Nature* **170**, 842. **39.** Sörbo (1951) *Acta Chem. Scand.* **5**, 724. **40.** Ebata, Sato & Bak (1955) *J. Biochem.* **42**, 715. **41.** Klebanoff (1955) *Can. J. Biochem. Physiol.* **33**, 780. **42.** Bhattacharya (1954) *Science* **120**, 841. **43.** Hirade (1952) *J. Biochem.* **39**, 165. **44.** Gray, Brooke & Gerhart (1959) *Nature* **184**, 1936. **45.** Bruns (1954) *Naturwissenschaften* **41**, 360. **46.** Schönberg & R. Moubacher (1952) *Chem. Rev.* **50**, 261. **47.** Cohen, Ness & Whiddon (1980) *Phytochemistry* **19**, 1913. **48.** Singer & Barron (1945) *J. Biol. Chem.* **157**, 241. **49.** Naganna & Menon (1948) *J. Biol. Chem.* **174**, 501. **50.** Ponz &

Larralde (1951) *Enzymologia* **14**, 325. **51.** Naganna (1950) *J. Biol. Chem.* **183**, 693. **52.** Bargoni, Fossa & Sisini (1963) *Enzymologia* **26**, 65. **53.** Franke, Thiemann, Remily, Möchel & Heye (1965) *Enzymologia* **29**, 251. **54.** Younathan (1962) *J. Biol. Chem.* **237**, 608. **55.** Konrad, Zhang, Hale, *et al.* (2002) *Biochem. Biophys. Res. Commun.* **293**, 207. **56.** Mishra, Das, Routray & Behera (1993) *Indian J. Physiol. Pharmacol.* **37**, 151. **5S.** Lenzen & Mirzaie-Petri (1992) *Naunyn Schmiedebergs Arch. Pharmacol.* **346**, 532. **58.** Kloepper, Norling, McDaniel & Landt (1991) *Cell Calcium* **12**, 351. **59.** Lenzen, Freytag, Panten, Flatt & Bailey (1990) *Pharmacol. Toxicol.* **66**, 157. **60.** Lenzen, Freytag & Panten (1988) *Mol. Pharmacol.* **34**, 395. **61.** Kwan (1988) *Biochem. J.* **254**, 293. **62.** Kwan & Beazley (1988) *J. Bioenerg. Biomembr.* **20**, 517. **63.** Lenzen, Tiedge & Panten (1987) *Acta Endocrinol. (Copenhagen)* **115**, 21. **64.** Hara, Miwa & Okuda (1986) *Chem. Pharm. Bull. (Tokyo)* **34**, 4731. **65.** Boquist & Ericsson (1984) *FEBS Lett.* **178**, 245. **66.** Miwa, Hara, Matsunaga & Okuda (1984) *Biochem. Int.* **9**, 595. **67.** Colca, Kotagal, Brooks, *et al.* (1983) *J. Biol. Chem.* **258**, 7260. **68.** Nelson & Boquist (1982) *Acta Diabetol. Lat.* **19**, 319. **69.** Mitidieri & Affonso (1979) *An. Acad. Bras. Cienc.* **51**, 753. **70.** Purr (1935) *Biochem. J.* **29**, 5. **71.** Soong, Ogawa, Sakuradani & Shimizu (2002) *J. Biol. Chem.* **277**, 7051. **72.** Chen & Chen (1988) *Arch. Biochem. Biophys.* **262**, 427. **73.** De Renzo (1956) *Adv. Enzymol. Relat. Subj. Biochem.* **17**, 293.

Alogliptin

This orally available DPP-IV inhibitor (FW = 339.39 g/mol; CAS 850649-62-6), also known as 2-({6-[(3*R*)-3-aminopiperidin-1-yl]-3-methyl-2,4-dioxo-3,4-dihydropyrimidin-1(2*H*)-yl}methyl)benzonitrile and by its trade name Nesina® targets dipeptidyl peptidase IV, *or* DPP-4, thereby retarding the inactivation of incretin hormones GLP-1 (glucagon-like peptide-1) and GIP (glucose-dependent insulinotropic peptide), both of which play a role in regulating blood glucose levels (1). Alogliptin inhibition is competitive, K_i = 24 nM (2). In 2013, alogliptin received FDA approval for the treatment of type II diabetes mellitus. DPP-4 inhibitors have become widely accepted in clinical practice because of their low risk of hypoglycemia, favorable adverse-effect profile, and once-daily dosing. The only reported side-effect is mild hypoglycemia, suggesting that some individuals may benefit from a slightly lower dose. Other DPP-4 inhibitors include sitagliptin phosphate (Januvia®) and saxagliptin (Onglyza®). DPP-4 inhibition has been associated with enhanced β-cell survival and neogenesis in streptozotocin-treated diabetic rats (*See also Streptozotocin*) (3). Significantly, the addition of alogliptin resulted in clinically significant reductions in HbA$_{1c}$ (typically from 1.4 to 1%), without increased incidence of hypoglycemia, in type 2 diabetes patients who are inadequately controlled by glyburide monotherapy (4). **1.** Feng, Zhang, Wallace, *et al.* (2007) *J. Med. Chem.* **50**, 2297. **2.** Thomas, Eckhardt, Langkopf, *et al.* (2008) *J. Pharmacol. Exp. Ther.* **325**, 175. **3.** Pospisilik, Stafford, Demuth, McIntosh & Pederson (2002) *Diabetes* **51**, 2677. **4.** Marino & Cole (2014) *J. Pharm. Pract.* **28**, 99.

Allyl (5*S*,6*S*)-6-[(*R*)-Acetoxyethyl]penem-3-carboxylate

This penicillin-like ester (FW = 297.33 g/mol), containing an 2,3-unsaturated thiazolidine ring, is a potent inhibitor of *Escherichia coli* leader peptidase I (1-3). Signal peptidase (SPase) I is responsible for the cleavage of signal peptides of many secreted proteins in bacteria. Because of its unique physiological and biochemical properties, it serves as a potential target for development of novel antibacterial agents. SPase I belongs to a novel class of serine protease that utilize a serine and a lysine to form a unique catalytic dyad for peptide hydrolysis (2). Because of this unique catalytic mechanism, they are not sensitive to the classic protease inhibitors. (5S,6S)-penem is covalently bound as an acyl-enzyme intermediate to the γ-oxygen of a serine residue at position-90, demonstrating that this residue acts as the nucleophile in the hydrolytic mechanism of signal-peptide cleavage (2). **1**. Carlos, Paetzel, Klenotic, Strynadka & Dalbey (2002) *The Enzymes*, 3rd ed., **22**, 27. **2**. Paetzel, Dalbey & Strynadka (1998) *Nature* **396**, 186. **3**. Allsop, Brooks, Edwards, Kaura & Southgate (1996) *J. Antibiot.* **49**, 921.

Allyl Alcohol

This extremely hazardous hydroxyalkene (FW = 58.08 g/mol; CAS 107-18-6; M.P. = –50°C; B.P. = 96-97°C; Solubility: Miscible with water and ethanol; Lachrymator), also known as 2-propen-1-ol, is a mechanism-based irreversible inhibitor of yeast alcohol dehydrogenase. In mammals, allyl alcohol requires activation by alcohol dehydrogenases (ADH) to form the highly reactive and toxic metabolite acrolein, which shows similar toxicity in zebrafish embryos and adults. *Note:* Allyl alcohol is more toxic than related alcohols, and its threshold limit value (TLV) is 2 ppm. **1**. Rando (1977) *Meth. Enzymol.* **46**, 28. **2**. Rando (1974) *Biochem. Pharmacol.* **23**, 2328. **3**. Trivic & Leskovac (1999) *Biochem. Mol. Biol. Int.* **47**, 1.

Allylamine

This toxic aminoalkene (FW$_{free-base}$ = 57.095 g/mol; B.P. = 55-58°C; Miscible with Water and Ethanol), also known as 2-propen-1-amine targets monoamine oxidase (1-4,8,9), methylamine dehydrogenase (5,11), diamine oxidase, *or* amine oxidase (copper-containing) (6), alanine aminotransferase (7), aspartate aminotransferase (7), mitochondrial electron transport, weakly inhibited (10). **1**. Rando (1977) *Meth. Enzymol.* **46**, 28. **2**. Maycock (1980) *Meth. Enzymol.* **66**, 294. **3**. Ator & Ortiz de Montellano (1990) *The Enzymes*, 3rd ed., **19**, 213. **4**. Brush & Kozarich (1992) *The Enzymes*, 3rd ed., **20**, 317. **5**. Kiriukhin, Chistoserdov & Tsygankov (1990) *Meth. Enzymol.* **188**, 247. **6**. Jeon & Sayre (2003) *Biochem. Biophys. Res. Commun.* **304**, 788. **7**. Kuzuya, Kitagawa & Yamada (1967) *J. Nutr.* **93**, 280. **8**. Rando & Eigner (1977) *Mol. Pharmacol.* **13**, 1005. **9**. Silverman, Hiebert & Vazquez (1985) *J. Biol. Chem.* **260**, 14648. **10**. Biagini, Toraason, Lynch & Winston (1990) *Toxicology* **62**, 95. **11**. Davidson, Graichen & Jones (1995) *Biochem. J.* **308**, 487.

2-(3-Allylamino-5-chloro-2-oxo-6-phenyl-2H-pyrazin-1-yl)-N-(4-[amino(imino)-methyl]benzyl)acetamido

This substituted pyrazinone (FW = 451.93 g/mol) inhibits coagulation factor VIIa (IC$_{50}$ = 0.7 μM) and thrombin (IC$_{50}$ = 8.7 μM). The pyrazinone core orients the substituents in the correct spatial arrangement to probe the S$_1$, S$_2$, and S$_3$ pockets within the active site of the enzyme. **1**. Parlow, Case, Dice, *et al.* (2003) *J. Med. Chem.* **46**, 4050.

4-(Allylamino)-3-hydroxy-2,2-dimethyl-3,4-dihydro-2H-benzo[g]chromene-5,10-dione

This quinone (FW = 313.35 g/mol) inhibits indoleamine 2,3-dioxygenase, with an IC$_{50}$ value of 0.186 μM for the *cis* isomer and 0.183 μM for the *trans*. Indoleamine 2,3-dioxygenase (IDO) is emerging as an important new therapeutic target for the treatment of cancer, chronic viral infections, and other diseases characterized by pathological immune suppression. **1**. Kumar, Malachowski, DuHadaway, *et al.* (2008) *J. Med. Chem.* **51**, 1706.

1-Allyl-2-[3-(benzylamino)propoxy]-9H-carbazole

This derivative (FW = 382.51 g/mol), also known as YM 75440, inhibits squalene synthase (IC$_{50}$ = 150 and 63 nM for the rat and human enzymes, respectively). It significantly reduced both plasma total cholesterol and plasma triglyceride levels following oral dosing to rats with a reduced tendency to elevate plasma transaminase levels. **1**. Ishihara, Kakuta, Moritani, Ugawa & Yanagisawa (2004) *Bioorg. Med. Chem.* **12**, 5899.

3(S)-Allyl-4(S)-benzyl-2-oxoazetidine-1-carboxylic Acid (4-Pyridinylmethyl)amide

This β-lactam (FW = 335.42 g/mol) inhibits human pancreatic elastase, human leukocyte elastase, and chymotrypsin, with IC$_{50}$ values of 4.7, 0.3, and 6 μM, respectively. It also targets human cytomegalovirus protease. Modest antiviral activity was found in a plaque reduction assay. **1**. Yoakim, Ogilvie, Cameron, *et al.* (1998) *J. Med. Chem.* **41**, 2882.

L-C-Allylglycine

This amino acid (FW = 115.13 g/mol), also known as L-2-amino-4-pentenoic acid, is a convulsant and inhibits glutamate decarboxylase. **Target(s):** cystathionine γ-lyase (1); cysteine desulfurase (2); cysteine sulfinate decarboxylase, *or* sulfinoalanine decarboxylase, weakly inhibited (3); glutamate decarboxylase (4-7); protein biosynthesis (8); UDP-N-acetylmuramate:L-alanine ligase, weakly inhibited (9). **1**. Beeler & Churchich (1978) *Biochim. Biophys. Acta* **522**, 251. **2**. Zheng, White, Cash & Dean (1994) *Biochemistry* **33**, 4714. **3**. Heinamaki, Peramaa & Piha (1982) *Acta Chem. Scand. B* **36**, 287. **4**. Alberici, de Lores Arna & De Robertis (1969) *Biochem. Pharmacol.* **18**, 137. **5**. Fisher & Davies (1974) *J. Neurochem.* **23**, 427. **6**. Orlowski, Reingold & Stanley (1977) *J. Neurochem.* **28**, 349. **7**. Blindermann, Maitre, Ossola & Mandel (1978) *Eur. J. Biochem.* **86**, 143. **8**. De Canal & De Lores Arnaiz (1972) *Biochem. Pharmacol.* **21**, 133. **9**. Liger, Masson, Blanot, van Heijenoort & Parquet (1995) *Eur. J. Biochem.* **230**, 80.

3-Allyl-4-[3-(isopropylamino)propoxy]biphenyl

This biphenyl derivative (FW = 308.45 g/mol) inhibits squalene synthase (IC$_{50}$ = 93 nM for the rat enzyme), significantly reducing both plasma total cholesterol and plasma triglyceride levels following oral dosing to rats with a reduced tendency to elevate plasma transaminase levels. **1**. Ishihara, Kakuta, Moritani, Ugawa & Yanagisawa (2004) *Bioorg. Med. Chem.* **12**, 5899.

1-Allyl-2-[3-(isopropylamino)propoxy]-9H-carbazole

This carbazole derivative (FW = 308.45 g/mol) inhibits squalene synthase (IC$_{50}$ = 66 nM for the rat enzyme), significantly reducing both plasma total cholesterol and plasma triglyceride levels following oral dosing to rats with a reduced tendency to elevate plasma transaminase levels. **1**. Ishihara, Kakuta, Moritani, Ugawa & Yanagisawa (2004) *Bioorg. Med. Chem.* **12**, 5899.

4-Allyl-2-methoxyphenyl Benzoate

This eugenol ester (FW = 268.31 g/mol), also known as eugenyl benzoate, inhibits 15-lipoxygenase (IC$_{50}$ = 4.4 μM). Other esters also inhibit (*e.g.*, 2-pyridinecarboxylate (IC$_{50}$ = 33.3 μM), nicotinate (IC$_{50}$ = 1.7 μM), isonicotinate (IC$_{50}$ = 2.3 μM), 2- fluorobenzoate (IC$_{50}$ = 6.7 μM), 3-fluorobenzoate (IC$_{50}$ = 2.1 μM), and 4-fluorobenzoate (IC$_{50}$ = 7.2 μM), 2-chlorobenzoate (IC$_{50}$ = 168 μM), 3-chlorobenzoate (IC$_{50}$ = 5.2 μM), and 4-chlorobenzoate (IC$_{50}$ = 134.3 μM), 2-methylbenzoate (IC$_{50}$ = 77 μM), 3-methylbenzoate (IC$_{50}$ = 11 μM), and 4-methylbenzoate (IC$_{50}$ = 15 μM), 3- (IC$_{50}$ = 23 μM), 4-methoxybenzoate (IC$_{50}$ = 19 μM), cyclohexanecarboxylate (IC$_{50}$ = 2.2 μM), and adamantanecarboxylate (IC$_{50}$ = 0.017 μM). **1**. Sadeghian, Seyedi, Saberi, Arghiani & Riazi (2008) *Bioorg. Med. Chem.* **16**, 890.

[4-[6-(Allylmethylamino)hexyloxy]phenyl](4-nitrophenyl) methanone

This orally active non-terpenoic inhibitor (FW$_{free-base}$ = 396.49 g/mol) targets human lanosterol synthase (IC$_{50}$ = 1.9 nM), with demonstrated efficacy in lowering plasma total cholesterol (TC) and plasma low-density lipoprotein cholesterol (LDL-C) in hyperlipidemic hamsters. **1**. Dehmlow, Aebi, Jolidon, *et al.* (2003) *J. Med. Chem.* **46**, 3354.

N-(α(R)-Allyl-4-methylbenzylaminocarbonyl) 4(S)-(4-Carboxymethylphenoxy)-3,3-diethylazetidin-2-one

This alkylazetidinone (FW$_{free-acid}$ = 464.56 g/mol), also known as L-680755, inactivates human, green monkey, dog, and rat leukocyte elastase (1). Its specificity toward human pancreatic elastase versus porcine pancreatic elastase is consistent with the differences in substrate specificity reported for these enzymes (2). **1**. Knight, Green, Chabin, *et al.* (1992) *Biochemistry* **31**, 8160. **2**. Zimmerman & Ashe (1977) *Biochim. Biophys. Acta* **480**, 241

1-Allyl-2-thiourea

This substituted thiourea (FW = 116.19 g/mol), also known as thiosinamine, is a noncompetitive inhibitor of ammonia monooxygenase. **Target(s):** methane monooxygenase (1); photosynthesis (2); tyrosinase (3); cytochrome *c* oxidase (3); succinate dehydrogenase (3); ammonia monooxygenase (4,5); alkane monooxygenase (6). **1**. Tonge, Harrison & Higgins (1977) *Biochem. J.* **161**, 333. **2**. Green, McCarthy & King (1939) *J. Biol. Chem.* **128**, 447. **3**. DuBois & Erway (1946) *J. Biol. Chem.* **165**, 711. **4**. Rasche, Hyman & Arp (1991) *Appl. Environ. Microbiol.* **57**, 2986. **5**. Juliette, Hyman & Arp (1993) *Appl. Environ. Microbiol.* **59**, 3728. **6**. Hamamura, Yeager & Arp (2001) *Appl. Environ. Microbiol.* **67**, 4992.

Almorexant

This orally available, dual orexin (*or* hypocretin) antagonist (FW = 549.02 g/mol; CAS 913358-93-7; Solubility: 79 mg/mL DMSO; <1 mg/mL H$_2$O; Dose: ~300 mg/kg body weight, administered in polyethylene glycol 400 or 0.25% methylcellulose in water), also named (*R*)-2-((*S*)-1-(4-(trifluoromethyl)phenethyl)-6,7-dimethoxy-3,4-dihydroisoquinolin-2(1*H*)-yl)-*N*-methyl-2-phenylacetamide·HCl, targets for OX$_1$ (IC$_{50}$ = 6.6 nM) and OX$_2$ (IC$_{50}$ = 3.4 nM) receptors, decreased cell proliferation and neurogenesis specifically in the ventral hippocampus (1,2). Mounting evidence suggests that an increase of orexin receptor signaling is likely to underlie the pathophysiology of major depression. These receptors (**See also** *Orexins A & B; Fluoxetine; IPSU*) **1**. Brisbare-Roch, Dingemanse, Koberstein, *et al.* (2007) *Nature Med.* **13**, 150. **2**. Nollet, Gaillard, Tanti, *et al.* (2012) *Neuropsycho-pharmacol.* **37**, 2210.

Alpelisib

This orally bioavailable phosphatidylinositol 3-kinase (PI3K) inhibitor (FW = 441.47 g/mol; CAS 1217486-61-7; Soluble in DMSO), also named BYL719 and (S)-N^1-(4-methyl-5-(2-(1,1,1-trifluoro-2-methylpropan-2-yl)pyridin-4-yl)thiazol-2-yl)pyrrolidine-1,2-dicarboxamide, targets PIK3 in the PI3K/AKT kinase (or Protein Kinase B) signaling pathway, thereby blocking the activation of the PI3K signaling pathway and impeding tumor cell growth/survival in susceptible tumor cells, including PIK3CA-mutant breast cancer. Dysregulated PI3K signaling is believed to be a contributing factor in tumor resistance to antineoplastic agents. **1**. Azab, Vali, Abraham, et al. (2014) Brit. J. Haematol. **165**, 89. **2**. Brady, Zhang, Seok, Wang & Yu (2013) Mol Cancer Ther. **13**, 60. **3**. Garrett, Sutton, Kurupi, et al. (2013) Cancer Res. **73**, 6013. **4**. Elkabets, Vora, Juric, et al. (2013) Sci. Transl. Med. 5, 196. **5**. Furet, Guagnano, Fairhurst, et al. (2013) Bioorg. Med. Chem. Lett. 23, 3741. **6**. Young, Pfefferle, Owens, et al. (2013) Cancer Res. **73**, 4075.

Alprazolam

This potent benzodiazepine-class depressant and anxiolytic drug (FW = 308.77 g/mol; CAS 28981-97-7), known by the trade name Xanax® and by its IUPAC name, 8-chloro-1-methyl-6-phenyl-4H-[1,2,4]triazolo[4,3-a][1,4]benzodiazepine, exhibit wide-ranging properties that commend its use as a highly effective sedative, hypnotic, skeletal muscle relaxant, anticonvulsant, and amnestic properties. When given daily at a dosage of 0.5 to 4.0 mg, it is as effective an anxiolytic agent as diazepam and chlordiazepoxide. Protracted use, however, results in tolerance, requiring higher doses to maintain efficacy. Among the most frequently abused CNS depressants, alprazolam is classified as a controlled substance (Schedule IV) by the U.S. Drug Enforcement Agency. **Primary Mode of Action:** Alprazolam targets the GABA_A receptor, of which there are numerous subtypes, each showing varying interactions with alprazolam. Located different regions of the brain, diazepam-type GABA_A receptor agonists cause hyperpolarization and inhibition of neuronal excitability as a consequence of chloride ion influx. **Pharmacokinetics:** When given orally or intravenously, only 1% of alprazolam reaches the brain, as detected with [^{11}C]-alprazolam and positron emission tomography (2). The mean kinetic parameters (1) for intravenous and oral alprazolam administration, respectively, were: volume of of distribution of 0.72 and 0.84 L/kg; elimination $t_{1/2}$ of 11.7 and 11.8 hour; clearance rate of 0.74 and 0.89 mL/min/kg (2). There were no significant differences between intravenous and oral alprazolam in terms of V_d, $t_{1/2}$, or AUC. The mean fraction absorbed after oral administration was 0.92. With the exception of rapidity of onset, the pharmacodynamic profiles of intravenous and oral alprazolam are similar for a 1-mg dose (2). In humans, alprazolam is metabolized by CYP3A4 to 4-hydroxyalprazolam and α-hydroxyalprazolam. **Key Pharmacokinetic Parameters:** See Appendix II in Goodman & Gilman's THE PHARMACOLOGICAL BASIS OF THERAPEUTICS, 12th Edition (Brunton, Chabner & Knollmann, eds.) McGraw-Hill Medical, New York (2011). **1**. Dobbs, Banks, Fleishaker, et al. (1995) Nucl. Med. Biol. **22**, 459. **2**. Smith, Kroboth, Vanderlugt, Phillips & Juhl (1984) Psychopharmacol. (Berlin) **84**, 452.

Alrestatin

This aldose reductase inhibitor and drug (FW_free-acid = 255.23 g/mol; CAS 51411-04-2), also known as 1,3-dioxo-1H-benz[de]isoquinoline-2(3H)-acetic acid) suppresses diabetes-associated, osmotic cell and tissue damage by inhibiting aldose reduction and thereby reducing the accumulation intracellular sorbitol. **Primary Mode of Action of Aldose Reductase Inhibitors:** A major cause of diabetic neuropathy is the intraneural osmotic pressure that builds up as a consequence of the over-accumulation of sorbitol, a polyol formed by aldose reductase (Reaction: Glucose + NADPH $\rightleftharpoons$ Sorbitol + NADP$^+$ + H$^+$). Similar considerations apply to cataract formation in the lens, another tissue rich in aldose reductase. In diabetes, aldose reductase activity increases as the concentration of glucose rises in the lens, peripheral nerves and glomerulus (tissues that are insulin-insensitive); because sorbitol lacks a membrane carrier, its contributes to intracellular osmotic pressure, disrupting cell-cell interactions (especially synapses), eventually leading to retinopathy and neuropathy. The additive effects of aldose reductase (AR) and polyol dehydrogenase in producing sorbitol from glucose and fructose, acting in combination with age-dependent decreased hexokinase is believed to account for diabetic cataract formation in human lenses under high glucose stress. AR's K_m for glucose of AR is roughly 200 mM, whereas its K_m for NADPH is 0.06 mM. NADP inhibits human lens AR noncompetitively and has a K_i that is roughly equal to the K_m for NADPH. Notably, the K_m for fructose is 40 mM and that for NADH is 0.02 mM in the polyol dehydrogenase (PD) reaction. Therefore, although sorbitol formation is modest during normoglycemia, such is not the case for diabetic hyperglycemia. Moreover, the recent increased reliance on high-fructose corn syrup as a sweetener is problematic, in that glucose-sensing mechanisms in humans are largely unresponsive to fructose. Because sorbitol is not transported out of the lens, any increase in intracellular sorbitol must be compensated osmotically by the considerable uptake of water, a well-characterized cataractogenic event. By inhibiting sorbitol dehydrogenase, alrestatin lowers tha undesirable net accumulatrion of sorbitol during hyperglycemic episodes. At high enough concentrations, alrestatin also inhibits PD. **Target(s):** aldose reductase, or aldehyde reductase (1-3,6-10); 4-aminobutyrate aminotransferase (4); carbonyl reductase (5); succinate-semialdehyde dehydrogenase (4); polyol dehydrogenase, weakly inhibited, except at elevated concentrations (6); hexonate dehydrogenase, or glucuronate reductase (8). **1**. Wermuth & von Wartburg (1982) Meth. Enzymol. **89**, 181. **2**. Whittle & Turner (1981) Biochem. Pharmacol. **30**, 1191. **3**. Davidson & Murphy (1985) Prog. Clin. Biol. Res. **174**, 251. **4**. Whittle & Turner (1978) J. Neurochem. **31**, 145. **5**. Wermuth (1981) J. Biol. Chem. **256**, 1206. **6**. Jedziniak, Chylack, Cheng, et al. (1981) Invest. Ophthalmol. Vis. Sci. **20**, 314. **7**. Srivastava, Petrash, Sadana, et al. (1982) J. Neurochem. **39**, 810. **9**. Cromlish & Flynn (1983) J. Biol. Chem. **258**, 3416. **10**. Petrash & Srivastava (1982) Biochim. Biophys. Acta **707**, 105.

Alsterpaullone

This cell-permeable GSK3β (FW = 293.28 g/mol; CAS 237430-03-4), also named NSC 705701 and 9-nitro-7,12-dihydroindolo[3,2-d][1]benzazepin-6(5H)-one, targets cyclin-dependent kinases (CDKs) and glycogen synthase kinase 3β potently, reversibly, and ATP competitively. Displays remarkable antitumor activity in vitro. Alsterpaullone inhibits growth of the colon cancer cell line HCT-116, with an IC_{50} in the nanomolar range. **Target(s):** cyclin-dependent kinase, or cyclin-dependent protein kinase 2/cyclin A, IC_{50} = 0.08 μM (1,4); glycogen synthase kinase 3β, or [tau protein] kinase; IC_{50} = 0.11 μM (1,3,5); lymphocyte kinase, IC_{50} = 0.47 μM (1); cyclin-dependent protein kinase 1/cyclin B, IC_{50} = 0.035 μM (2,5); cyclin-dependent protein kinase 5/p25 (3,5); mitogen-activated protein kinase (1); protein kinase C, weakly inhibited (1); cAMP-dependent protein kinase, or protein kinase A; weakly inhibited (1); non-specific serine/threonine protein kinase (1); protein-tyrosine kinase, or non-specific protein-tyrosine kinase (1), Lck protein-tyrosine kinase (1). **1**. Bain, McLauchlan, Elliott & Cohen (2003) Biochem. J. **371**, 199. **2**. Schultz, Link, Leost, et al. (1999) J. Med. Chem. **42**, 2909. **3**. Leost, Schultz, Link, et al. (2000) Eur. J. Biochem. **267**, 5983. **4**. Woodard, Li, Kathcart, et al.

(2003) *J. Med. Chem.* **46**, 3877. **5**. Gompel, Soulié, Ceballos-Picot & Meijer (2004) *Neurosignals* **13**, 134.

(−)-Altenuene

This mycotoxin (FW = 292.28 g/mol; CAS 889101-41-1) from *Alternaria tenuis* (an endophytic fungus associated with *Vinca rosea*, or *Catharanthus roseus*) partially inhibits acetylcholinesterase (maximal inhibition = 78%) and butyrylcholinesterase (maximal inhibition = 73%). Other *Alternaria* mycotoxins include tenuazonic acid, alternariol monomethyl ether, alternariol, and altertoxin I. **See** *Tenuazonic Acid* **1**. Bhagat, Kaur, Kaur, *et al.* (2016) *J. Appl. Microbiol.* **121**, 1015.

Altropane

This phenyltropane derivative (FW = 429.27 g/mol; CAS 180468-34-2), also known as O-587, IACFT, and 2β-carbomethoxy-3β-(4-fluorophenyl)-*N*-((*E*)-3-iodoprop-2-enyl)tropane, is a dopamine reuptake inhibitor that displays high affinity and specificity both *in vitro* and *in vivo* in laboratory animals (1). The favorable binding properties of altropane, together with its rapid entry into primate brain and highly localized distribution in dopamine-rich brain regions, suggest it is a suitable iodinated probe for monitoring the dopamine transporter *in vitro* and *in vivo* by SPECT or PET imaging (2). Altropane also shows promise for treating attention deficit hyperactivity disorder (ADHD). **1**. Elmaleh, Fischman, Shoup, *et al.* (1996) *J. Nucl. Med.* **37**, 1197. **2**. Madras, Meltzer, Liang, *et al.* (1998) *Synapse* **29**, 93.

Aluminum Chloride, *See Aluminum and Aluminum Ions*

Aluminum(III)-ATP Complex

This one-to-one, exchange-stable complex, forms as a consequence of the much tighter binding of tervalent Al^{3+} to nucleoside triphosphates than with divalent magnesium or calcium ions. Al(III)ATP inhibits a number of ATP-dependent enzymes (1-6) and is likely to act in a similar manner with many more. Solheim & Fromm (7) describe a simple method for removing aluminum from ATP. **Target(s):** hexokinase, slow-binding inhibition (1-6); glycerokinase (2,4). **1**. Morrison (1982) *Trends Biochem. Sci.* **7**, 102. **2**. Viola, Morrison & Cleland (1980) *Biochemistry* **19**, 3131. **3**. Macdonald & Martin (1988) *Trends Biochem. Sci.* **13**, 15. **4**. Furumo & Viola (1989) *Inorg. Chem.* **28**, 820. **5**. Womack & Colowick (1979) *Proc. Natl. Acad. Sci. U.S.A.* **76**, 5080. **6**. Solheim & Fromm (1980) *Biochemistry* **19**, 6074. **6**. Solheim & Fromm (1980) *Analyt. Biochem.* **109**, 266.

Aluminum Ion (Al³⁺)

This tervalent inorganic ion, derived from its parent Group IIIB metal, readily forms the amphoteric species aluminum hydroxide [Al(OH)₃] that hydrates to the insoluble hexacoordinate species [Al(OH)₃(H₂O)₃]. Insoluble aluminum hydroxide can be formed by the addition of hydroxide ion, OH⁻ to form a soluble salt of Al³⁺ (*Reaction*: [Al(H₂O)₆]³⁺(aq) + 3 OH⁻(aq) → Al(H₂O)₃(OH)₃(s) + 3 H₂O(l)). The insoluble metal hydroxide can act as a base, since it can be redissolved upon reaction with an acid (*Reaction*: Al(OH)₃(H₂O)₃(s) + 3 H₃O⁺(aq) → [Al(H₂O)₆]³⁺(aq) + 3 H₂O(l)). Alternatively, the metal hydroxide can act as an acid, since it can react with a base (*Reaction*: Al(OH)₃(H₂O)₃(s) + OH⁻(aq) → [Al(OH)₄(H₂O)₂](aq) + H₂O(l)). pH therefore strongly influences the aqueous species that form, as does temperature and ionic strength. **Nature of Enzyme-Bound Aluminum Ions:** The amphoteric nature of aluminum ions is strongly influenced by the presence of other ligands, including amines, carboxylic acids, and phosphoric acid species. For this reason, the best way

to determine the enzyme-bound species is through the use of X-ray crystallography or NMR structures of the enzyme-aluminum species in a buffer that approximates the solution properties used in assays of enzymatic activity. Because the enzyme is likely to provide one or more ligands, the precise solution species that first combines with the enzyme is necessarily conjectural. **Aluminum Exposure & Neurodegenerative Disease:** Various lines of evidence, mostly conjectural, have associated aluminum exposure to such disorders as dialysis encephalopathy, amyotrophic lateral sclerosis, Parkinson Disease in the Kii Peninsula and Guam, and Alzheimer's disease (AD). The complex characteristics of aluminum bioavailability make it difficult to evaluate its toxicity, and a direct causal relationship remains to be established. Given that protein misfolding and associated oligomerization and/or polymerization appears to be an emerging theme in protein folding disorders, one cannot discount a role of aluminum ions in promoting such protein structural rearrangements. **Sources:** Anhydrous AlCl₃ (Sigma-Aldrich 563919 and Fisher-Acros AC36481-0100 (both are 99.999% trace metals basis); CAS 7446-70-0; MW(anhydrous) = 133.34 g/mol); AlCl₃ Hydrate (Sigma-Aldrich 229393 (99.999% on trace metals basis); CAS 10124-27-3). **Caution:** Anhydrous AlCl₃ *reacts violently* with water, generating considerable heat. The recommended practice for hydrolysis is to allow anhydrous AlCl₃ to hydrate slowly (over a few days) in a moist atmosphere. This can be accomplished by placing two beakers, one containing the anhydrous powder and the other a few mL of water in an otherwise empty desiccator. (The anhydrous powder essentially serves as the desiccant.) Another simple method is to expose a weighed sample of anhydrous powder to the ambient atmosphere for a few days, allowing it to absorb ambient humidity. Unless highly experienced, it is advisable to prepare only a small quantity (0.2-0.5 g) at a time. **Target(s):** phospholipase C (10,43,121,122); phosphorylase phosphatase (2); Ferroxidase (3,42,152); acetylcholinesterase (14,42,52); apyrase, starting from aluminum chloride (5); H⁺ ATPase (6); K⁺ channels (7); porphobilinogen synthase, *or* δ-aminolevulinate dehydratase; starting with AlCl₃ (8,14,18); hexokinase (9,16,34,40,41,42); phospholipase D (10,119,120); butyrylcholinesterase (11,42); ras p21 (12); catalase (13); nitrogenase, starting with aluminum fluoride (15); trypsin (17,53); chymotrypsin (17); isocitrate dehydrogenase, NAD (19); isocitrate dehydrogenase, NADP (19,20); acid phosphatase (21,22,126-128); alkaline phosphatase, starting from aluminum chloride (22,35,42); protein kinase C (23); calpain (24); H⁺-translocating ATPase (25); superoxide dismutase (26); glucose-6-phosphate dehydrogenase (27,28); transducin (29,30); GTPase (29,30); adenylate cyclase, activated and inhibited under different conditions (31,32); porphobilinogen deaminase (33); mannan endo-1,4-β mannosidase (34); aquacobalamin reductase (38); aquacobalamin reductase, NADPH (37,38,153); secretory tomato ribonucleases (39); glycerokinase (41); 3′,5′-cyclic-nucleotide phosphodiesterase (42); catechol *O*-methyltransferase (42); amidase (44); phospholipase A₂ (45); ascorbate peroxidase (46); tyrosinase (47); TF₁ ATPase (48,49); nucleotide diphosphatase, nucleotide pyrophosphatase (50); aminoacylase (51); pyruvate decarboxylase (52); 6-carboxyhexanoyl-CoA synthetase, *or* 6-carboxyhexanoate: CoA ligase (54); mannose isomerase (55); indoleacet-aldoxime dehydratase (56); isocitrate lyase (57); acetylenedicarboxylate decarboxylase, starting from Al(NO₃)₃ (58); uroporphyrinogen decarboxylase (59); aromatic-L-amino-acid decarboxylase, *or* dopa decarboxylase (60,61); phloretin hydrolase (72); Mg²⁺-importing ATPase (63); NAD diphosphatase (64); exopolyphosphatase (65); GTP cyclohydrolase I (66); guanosine deaminase (67); D-arginase (68); arginine deaminase (69); arginase II (70); calpain-2, *or* m-calpain (71); calpain-1, μ-calpain (71); IgA-specific serine endopeptidase (72); prolyl oligopeptidase (73); cystinyl aminopeptidase, oxytocinase (74); leucyl aminopeptidase (75); microsomal epoxide hydrolase (weakly inhibited (76); soluble epoxide hydrolase (76); α-glucuronidase (77); glucan 1,4-α-maltohexaosidase, weakly inhibited (78); fructan β-fructosidase (79); inulinase (79); galacturan 1,4-α-galacturonidase (80); levanase (81); glucan 1,4-α-maltotetraohydrolase (82); β-*N*-acetylhexosaminidase (83,84); pullulanase (85); xylan 1,4-β-xylosidase (86); hyaluronoglucosaminidase, *or* hyaluronidase (87); xylan endo-1,3-β-xylosidase (88,89); α,α-trehalase (90,91); β-galactosidase (92); β glucosidase (93); α-glucosidase (94,95); chitinase, starting from by AlCl₃ (96); endo-1,4-β-xylanase (97,98); glucan 1,4-α-glucosidase, glucoamylase (99,100); β-amylase (101,102); α-amylase (103-108); *Basidobolus haptosporus* nuclease (nuclease Bh1) (109,111); ribonuclease IX (110); Arylsulfatase, starting from by AlCl₃ (112,113); 3′,5′-cyclic-nucleotide phosphodiesterase (114-117); sphingomyelin phospho-diesterase, *or* neutral sphingomyelinase (118); 4-phytase (123,125); 3- or 4-phytase (124); phospholipase A₁ (129);

Lysophospholipase (129,132,133); 1,4-lactonase (130); pectinesterase (131); lipase, or triacylglycerol lipase (134); acetate kinase (135); pantothenate kinase, starting from by AlCl₃ (136); dihydroxyphenylalanine aminotransferase (137); alanine aminotransferase (138); hydroxymethylbilane synthase (139); 3-phosphoshikimate 1-carboxyvinyl-transferase, or 3-enolpyruvoylshikimate-5 phosphate synthase (140); galactose-6-sulfurylase (141); 1,4-β-D-xylan synthase (142); nicotinate nucleotide diphosphorylase, carboxylating (143,144); anthocyanin 3'-O-β-glucosyl-transferase (145); xyloglucan:xyloglucosyl transferase (146); sucrose:sucrose fructosyltransferase (147); galactinol: sucrose galactosyltransferase, or raffinose synthase (148); cyclomaltodextrin glucanotransferase (149,150); dextransucrase (151); polyphenol oxidase (154); tyrosinase, monophenol monooxygenase (154); stizolobate synthase (155); stizolobinate synthase (155); protocatechuate 3,4 dioxygenase (156); catechol 1,2-dioxygenase (157); L-ascorbate peroxidase (158,159); peroxidase (160,161); laccase (162,163); catechol oxidase (164); polyphenol oxidase (164). **1.** Hayaishi (1955) *Meth. Enzymol.* **1**, 660. **2.** Rall & Sutherland (1962) *Meth. Enzymol.* **5**, 377. **3.** Huber & Frieden (1970) *J. Biol. Chem.* **245**, 3979. **4.** Marquis & Lerrick (1982) *Biochem. Pharmacol.* **31**, 1437. **5.** Schetinger, Wyse, Da Silva, *et al.* (1995) *Biol. Trace Elem. Res.* **50**, 209. **6.** Ahn, Sivaguru, Osawa, Chung & Matsumoto (2001) *Plant Physiol.* **126**, 1381. **7.** Liu & Luan (2001) *Plant Cell* **13**, 1453. **8.** Pimentel Vieira, Rocha, *et al.* (2000) *Toxicol. Lett.* **117**, 45. **9.** Socorro, Olmo, Teijon, Blanco & Teijon (2000) *J. Protein Chem.* **19**, 199. **10.** Li & Fleming (1999) *FEBS Lett.* **461**, 1. **11.** Sarkarati, Cokugras & Tezcan (1999) *Comp. Biochem. Physiol. C Pharmacol. Toxicol. Endocrinol.* **122**, 181. **12.** Landino & Macdonald (1997) *J. Inorg. Biochem.* **66**, 99. **13.** Chainy, Samanta & Rout (1996) *Res. Commun. Mol. Pathol. Pharmacol.* **94**, 217. **14.** Schroeder & Caspers (1996) *Biochem. Pharmacol.* **52**, 927. **15.** Renner & Howard (1996) *Biochemistry* **35**, 5353. **16.** Exley, Price & Birchall (1994) *J. Inorg. Biochem.* **54**, 297. **17.** Zatta, Bordin & Favarato (1993) *Arch. Biochem. Biophys.* **303**, 407. **18.** Zaman, Zaman, Dabrowski & Miszta (1993) *Comp. Biochem. Physiol. C* **104**, 269. **19.** Yoshino, Yamada & Murakami (1992) *Int. J. Biochem.* **24**, 1615. **20.** Yoshino & Murakami (1992) *Biometals* **5**, 217. **21.** Domenech, Lisa, Salvano & Garrido (1992) *FEBS Lett.* **299**, 96. **22.** Levy, Schoen, Flowers & Staelin (1991) *J. Biomed. Mater. Res.* **25**, 905. **23.** Katsuyama, Saijoh, Inoue & Sumino (1987) *Arch. Toxicol.* **63**, 474. **24.** Nixon, JClarke, Logvinenko, Tan, Hoult & Grynspan (1990) *J. Neurochem.* **55**, 1950. **25.** Sturr & Marquis (1990) *Arch. Microbiol.* **155**, 22. **26.** Shainkin-Kestenbaum, Adler, Berlyne & Caruso (1989) *Clin. Sci. (Lond.)* **77**, 463. **27.** Cho & Joshi (1989) *J. Neurochem.* **53**, 616. **28.** Cho & Joshi (1989) *Toxicol. Lett.* **47**, 21. **29.** Miller, Hubbard, Litman & Macdonald (1989) *J. Biol. Chem.* **264**, 243. **30.** Kanaho, Moss & Vaughan (1985) *J. Biol. Chem.* **260**, 11493. **31.** Bellorin-Font, Weaver, Stokes, McConkey, Slatopolsky & Martin (1985) *Endocrinology* **117**, 1456. **32.** Sharp & Rosenberry (1985) *Biophys. Chem.* **21**, 261. **33.** Farmer & Hollebone (1984) *Can. J. Biochem. Cell Biol.* **62**, 49. **34.** Solheim & Fromm (1980) *Biochemistry* **19**, 6074. **35.** Levy, Schoen, Flowers & Staelin (1991) *J. Biomed. Mater. Res.* **25**, 905. **36.** Kusakabe & Takahashi (1988) *Meth. Enzymol.* **160**, 611. **37.** Watanabe & Nakano (1997) *Meth. Enzymol.* **281**, 289. **38.** Watanabe & Nakano (1997) *Meth. Enzymol.* **281**, 295. **39.** Abel & Köck (2001) *Meth. Enzymol.* **341**, 351. **40.** Morrison (1982) *Trends Biochem. Sci.* **7**, 102. **41.** Viola, Morrison & Cleland (1980) *Biochemistry* **19**, 3131. **42.** Macdonald & Martin (1988) *Trends Biochem. Sci.* **13**, 15. **43.** Zeller (1951) *The Enzymes*, 1st ed., **1** (part 2), 986. **44.** Varner (1960) *The Enzymes*, 2nd ed., **4**, 243. **45.** Hanahan (1971) *The Enzymes*, 3rd ed, **5**, 71. **46.** Shigeoka, Nakano & Kitaoka (1980) *Arch. Biochem. Biophys.* **201**, 121. **47.** Bodine & Tahmisian (1943) *Arch. Biochem.* **2**, 403. **48.** Dou, Grodsky, Matsui, Yoshida & Allison (1997) *Biochemistry* **36**, 371. **49.** Grodsky, Dou & Allison (1998) *Biochemistry* **37**, 1007. **50.** Kornberg & Pricer (1950) *J. Biol. Chem.* **182**, 763. **51.** Tosa, Mori, Fuse & Chibata (1967) *Enzymologia* **32**, 153. **52.** Singer & Pinsky (1952) *J. Biol. Chem.* **196**, 375. **53.** Clifford (1933) *Biochem. J.* **27**, 326. **54.** Izumi, Morita, Tani & Ogata (1974) *Agric. Biol. Chem.* **38**, 2257. **55.** Hirose, Maeda, Yokoi & Takasaki (2001) *Biosci. Biotechnol. Biochem.* **65**, 658. **56.** Shulka & Mahadevan (1968) *Arch. Biochem. Biophys.* **125**, 873. **57.** Tanaka, Yoshida, Watanabe, Izumi & Mitsunaga (1997) *Eur. J. Biochem.* **249**, 820. **58.** Yamada & Jakoby (1958) *J. Biol. Chem.* **233**, 706. **59.** Straka & Kushner (1983) *Biochemistry* **22**, 4664. **60.** Fragoulis & Sekeris (1975) *Arch. Biochem. Biophys.* **168**, 15. **61.** Bossinakou & Fragoulis (1996) *Comp. Biochem. Physiol. B* **113**, 213. **62.** Minamikawa, Jayasankar, Bohm, Taylor & Towers (1970) *Biochem. J.* **116**, 889. **63.** Li, Tutone, Drummond, Gardner & Luan (2001) *Plant Cell* **13**, 2761. **64.** Davies & King (1978) *Biochem. J.* **175**, 669. **65.** Afansieva & Kulaev (1973) *Biochim. Biophys. Acta* **321**, 336. **66.** Kobashi, Hariu &

Iwai (1976) *Agric. Biol. Chem.* **40**, 1597. **67.** Ishida, Shirafuji, Kida & Yoneda (1969) *Agric. Biol. Chem.* **33**, 384. **68.** Nadai (1958) *J. Biochem.* **45**, 1011. **69.** Shibatani, Kakimoto & Chibata (1975) *J. Biol. Chem.* **250**, 4580. **70.** Tormanen (2001) *J. Enzyme Inhib.* **16**, 443. **71.** Sazontova, Matskevich & Arkhipenko (1999) *Pathophysiology* **6**, 91. **72.** Bleeg, Reinholdt & Kilian (1985) *FEBS Lett.* **188**, 357. **73.** Mizutani, Sumi, Suzuki, Narita & Tomoda (1984) *Biochim. Biophys. Acta* **786**, 113. **74.** Hayashi & Oshima (1976) *J. Biochem.* **80**, 389. **75.** Nampoothiri, Nagy, Kovacs, Szakacs & Pandey (2005) *Lett. Appl. Microbiol.* **41**, 498. **76.** Draper & Hammock (1999) *Toxicol. Sci.* **52**, 26. **77.** Uchida, Nanri, Kawabata, Kusakabe & Murakami (1992) *Biosci. Biotechnol. Biochem.* **56**, 1608. **78.** Nakakuki, Hayashi, Monma, Kawashima & Kainuma (1983) *Biotechnol. Bioeng.* **25**, 1095. **79.** Zhang, Zhao, Zhu, Ohta & Wang (2004) *Protein Expr. Purif.* **35**, 272. **80.** Celestino, de Freitas, Javier *et al.* (2006) *J. Biotechnol.* **123**, 33. **81.** Kang, Lee, Lee & Lee (1999) *Biotechnol. Appl. Biochem.* **29**, 263. **82.** Sakano, Kashiyama & Kobayashi (1983) *Agric. Biol. Chem.* **47**, 1761. **83.** Jin, Jo, Kim, *et al.* (2002) *J. Biochem. Mol. Biol.* **35**, 313. **84.** Lisboa De Marco, Valadares-Inglis & Felix (2004) *Appl. Microbiol. Biotechnol.* **64**, 70. **85.** Ohba & Ueda (1975) *Agric. Biol. Chem.* **39**, 967. **86.** Andrade, Polizeli, Terenzi & Jorge (2004) *Process Biochem.* **39**, 1931. **87.** Ozegowski, Gunther & Reichardt (1994) *Zentralbl. Bakteriol.* **280**, 497. **88.** Araki, Inoue & Morishita (1998) *J. Gen. Appl. Microbiol.* **44**, 269. **89.** Araki, Tani, Maeda, *et al.* (1999) *Biosci. Biotechnol. Biochem.* **63**, 2017. **90.** De Almeida, Polizeli, Terenzi & Jorge (1999) *FEMS Microbiol. Lett.* **171**, 11. **91.** Kadowaki, de L. Polizeli, Terenzi & Jorge (1996) *Biochim. Biophys. Acta* **1291**, 199. **92.** Choi, Kim, Lee & Lee (1995) *Biotechnol. Appl. Biochem.* **22**, 191. **93.** Patchett, Daniel & Morgan (1987) *Biochem. J.* **243**, 779. **94.** Yamasaki & Suzuki (1978) *Agric. Biol. Chem.* **42**, 971. **95.** Yamasaki & Suzuki (1980) *Agric. Biol. Chem.* **44**, 707. **96.** Skujins, Pukite & McLaren (1970) *Enzymologia* **39**, 353. **97.** Khasin, Alchanati & Shoham (1993) *Appl. Environ. Microbiol.* **59**, 1725. **98.** Frederick, Kiang, Frederick & Reilly (1985) *Biotechnol. Bioeng.* **27**, 525. **99.** Yamasaki, Suzuki & Ozawa (1977) *Agric. Biol. Chem.* **41**, 2149. **100.** Tsao, Hsu, Chao & Jiang (2004) *Fish. Sci.* **70**, 174. **101.** Srivastava (1987) *Enzyme Microb. Technol.* **9**, 749. **102.** Ray (2000) *Acta Microbiol. Immunol. Hung.* **47**, 29. **103.** Krishnan & Chandra (1983) *Appl. Environ. Microbiol.* **46**, 430. **104.** Hayashida, Teramoto & Inoue (1988) *Appl. Environ. Microbiol.* **54**, 1516. **105.** Aguilar, Morlon-Guyot, Trejo-Aguilar & Guyot (2000) *Enzyme Microb. Technol.* **27**, 406. **106.** Tsao, Hsu, Chao & Jiang (2004) *Fish. Sci.* **70**, 174. **107.** Sakano, Hiraiwa, Fukushima & T. Kobayashi (1982) *Agric. Biol. Chem.* **46**, 1121. **108.** Nirmala & Muralikrishna (2003) *Phytochemistry* **62**, 21. **109.** Desai & Shankar (2001) *J. Biochem. Mol. Biol. Biophys.* **5**, 267. **110.** Laliloti, Sideris & Fragoulis (1992) *Insect Biochem. Mol. Biol.* **22**, 125. **111.** Desai & Shankar (2000) *Eur. J. Biochem.* **267**, 5123. **112.** Ueki, Sawada, Fukagawa & Oki (1995) *Biosci. Biotechnol. Biochem.* **59**, 1062 and 1069. **113.** Sakurai, Isobe & Shiota (1980) *Agric. Biol. Chem.* **44**, 1. **114.** Lim, Thannimali & Ong (2000) *World J. Microbiol. Biotechnol.* **16**, 19. **115.** Lim, Palanisamy & Ong (1986) *Arch. Microbiol.* **146**, 142. **116.** Lim, Woon, Tan & Ong (1989) *Int. J. Biochem.* **21**, 909. **117.** Lim, Krishnan & Ong (1989) *Biochim. Biophys. Acta* **991**, 353. **118.** Lister, Crawford-Redick & Loomis (1993) *Biochim. Biophys. Acta* **1165**, 314. **119.** Kokusho, Kato, Machida & Iwasaki (1987) *Agric. Biol. Chem.* **51**, 2515. **120.** Okawa & Yamaguchi (1974) *J. Biochem.* **78**, 363. **121.** Okawa & Yamaguchi (1975) *J. Biochem.* **78**, 537. **122.** Piña-Chable & Hernández-Sotomayor (2001) *Prostaglandins* **65**, 45. **123.** Sutardi & Buckle (1986) *J. Food Biochem.* **10**, 197. **124.** Vats & Banerjee (2005) *J. Ind. Microbiol. Biotechnol.* **32**, 141. **125.** Hubel & Beck (1996) *Plant Physiol.* **112**, 1429. **126.** Andrews & Pallavicini (1973) *Biochim. Biophys. Acta* **321**, 197. **127.** Tejera Garcia, Olivera, Iribarne & Lluch (2004) *Plant Physiol. Biochem.* **42**, 585. **128.** Shirai, Takenouchi, Yamashita & Wakabayashi (1970) *Enzymologia* **39**, 125. **129.** Uehara, Hasegawa & Iwai (1979) *Agric. Biol. Chem.* **43**, 517. **130.** Kataoka, Nomura, Shinohara, *et al.* (2000) *Biosci. Biotechnol. Biochem.* **64**, 1255. **131.** Versteeg (1979) *Versl. Landbouwkd. Onderz. (Agric. Res. Rep.)* **892**, 1. **132.** Doi & Nojima (1975) *J. Biol. Chem.* **250**, 5208. **133.** Ichimasa, Morooka & Niimura (1984) *J. Biochem.* **95**, 137. **134.** Kumar, Kikon, Upadhyay, Kanwar & Gupta (2005) *Protein Expr. Purif.* **41**, 38. **145.** Gorrell, Lawrence & Ferry (2005) *J. Biol. Chem.* **280**, 10731. **136.** Shimizu, Kubo, Tani & Ogata (1973) *Agric. Biol. Chem.* **37**, 2863. **137.** Nagasaki, Sugita & Fukawa (1973) *Agric. Biol. Chem.* **37**, 1701. **138.** Agarwal (1985) *Indian J. Biochem. Biophys.* **22**, 102. **139.** Farmer & Hollebone (1984) *Can. J. Biochem. Cell Biol.* **62**, 49. **140.** Steinrücken & Amrhein (1984) *Eur. J. Biochem.* **143**, 341. **141.** Rees (1961) *Biochem. J.* **80**, 449. **142.** Urahara, Tsuchiya, Kotake, *et al.* (2004) *Physiol. Plant.* **122**, 169. **143.** Taguchi & Iwai (1975) *Agric. Biol. Chem.*

39, 1599. **144**. Taguchi & Iwai (1974) *J. Nutr. Sci. Vitaminol.* **20**, 283. **145**. Fukuchi-Mizutani, Okuhara, Fukui, *et al.* (2003) *Plant Physiol.* **132**, 1652. **146**. Takeda & Fry (2004) *Planta* **219**, 722. **147**. Yun (1996) *Enzyme Microb. Technol.* **19**, 107. **148**. Lehle & Tanner (1973) *Eur. J. Biochem.* **38**, 103. **149**. Akimaru, Yagi & Yamamoto (1991) *J. Ferment. Bioeng.* **71**, 322. **150**. Cao, Jin, Wang & Chen (2005) *Food Res. Int.* **38**, 309. **151**. Yalin, Jin, Jianhua, Da & Zigang (2008) *J. Biotechnol.* **133**, 505. **152**. Frieden & Hsieh (1976) *Adv. Enzymol. Rel. Areas Mol. Biol.* **44**, 187. **153**. Watanabe, Oki, Nakano & Kitaoka (1987) *J. Biol. Chem.* **262**, 11514. **154**. Ayaz, Demir, Torun, Kolcuoglu & Colak (2007) *Food Chem.* **106**, 291. **155**. Saito & Komamine (1978) *Eur. J. Biochem.* **82**, 385. **156**. Chen, Dilworth & Glenn (1984) *Arch. Microbiol.* **138**, 187. **157**. Chen, Glenn & Dilworth (1985) *Arch. Microbiol.* **141**, 225. **158**. Sharma & Dubey (2004) *Plant Sci.* **167**, 541. **159**. Shigeoka, Nakano & Kitaoka (1980) *Arch. Biochem. Biophys.* **201**, 121. **160**. Bhatti, Najma, Asgher, Hanif & Zia (2006) *Protein Pept. Lett.* **13**, 799. **161**. Koga, Ogawa, Choi & Shimizu (1999) *Biochim. Biophys. Acta* **1435**, 117. **162**. Park & Park (2008) *J. Microbiol. Biotechnol.* **18**, 670. **163**. Ben Younes, Mechichi & Sayadi (2007) *J. Appl. Microbiol.* **102**, 1033. **164**. Motoda (1979) *J. Ferment. Technol.* **57**, 79.

Aluminum Tetrafluoride

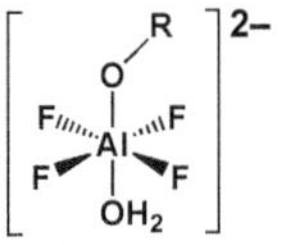

This aqueous aluminum fluoride anion is a presumptive geometric analogue of an associated intermediate formed in S_N2-type nucleophilic mechanisms, as observed with many phosphoryl transferases and phosphate ester hydrolases. Tightly bound fluoride substituents exhibit exceptional hydrogen-bonding capacity, a property that is likely to explain why AlF_4^- species interact as strongly as they do with enzyme-bound water molecules and other hydrogen-donating groups within the active sites of enzymes binding nucleoside 5'-triphosphates. This fluoroaluminate complex is readily formed by mixing aqueous solutions of sodium fluoride and aluminum chloride (*see below*). The dihydrate $[AlF_4(H_2O)_2]^-$ is the most likely species in water. Either of the two water molecules may be displaced by a nucleophile, R–O$^-$ in the structure shown above, thereby yielding a dianion. Analogous tightly bound inorganic fluorides include beryllium fluoride (BeF_n), scandium fluoride (ScF_n), and gallium fluoride (GaF_n). AlF_4^- activates members of the heterotrimeric G-protein ($G\alpha\beta\gamma$) family by binding to inactive $G\alpha$·GDP complex near the site occupied by the γ-phosphate in $G\alpha$·GTP complex (1). Recent work (2), however, illustrates the need for caution in interpreting the mode of AlF_4^- action. These investigators found that p190 RhoGAP forms a high-affinity complex with Rho GTPases in the presence of fluoride ions *without* any requirement for aluminum ions or even guanine nucleotide. Thus, fluoride may play a role other than as a γ-phosphate mimic. This finding supports the results of an earlier study by Murphy & Coll (3), who found that fluoride is a slow, tight-binding inhibitor of the calcium ATPase of sarcoplasmic reticulum. Any role for aluminum was ruled out by the observation that the addition of EGTA to 10 mM or aluminum sulfate to 0.2 mM or of deferoxamine to 0.5 mM produced no significant change in k_{obs} for fluoride inactivation of the pump ATPase. Investigators should carefully consider alternative modes of action when a requirement for aluminum or guanine nucleotide has not yet been demonstrated. One should also question any assumption that enzymes passively bind AlF_4^- as a tightly bound competitive inhibitor. The phosphoryl oxygens of active-site nucleotides may combine with AlF_4^- to form species resembling the structure presented above. Other active-site changes can occur. Yousafzai & Eady (4), for example, found that incubation of MoFe (Kp1) and Fe (Kp2) component proteins of *Klebsiella pneumoniae* nitrogenase are incubated with MgADP and AlF_4^- in the presence of the reducing agent dithionite produces a stable transition-state complex. The EPR signal associated with reduced Kp2 was not detectable; yet, Kp1 retained the S = 3/2 EPR signal arising from the dithionite reduced state of the protein's MoFe cofactor center. This observation suggested that the [Fe$_4$S$_4$] center was somehow oxidized. No satisfactory explanation for the fate of the electrons lost by Kp2 has been forthcoming. In a later report, Yousafzai and Eady (5) demonstrated that H$_2$ was evolved; likewise, they observed the formation of H$_2$ by nitrogenase in the absence of AlF_4^- in reaction mixtures containing MgADP, but not MgATP. **Preparation:** AlF_4^- is generated by combining aqueous Al2(SO$_4$)$_3$ and NaF solutions at a 4:1

fluorine:aluminum ratio. If a pH buffer is required, sulfonic acid buffers, such as HEPES or MES, should be used. Avoid Tris buffer, because primary amines may interact with aluminum fluoride. **Target(s):** F$_1$ ATPase, via MgADP-fluoroaluminate complex (6-10); RecA ATPase (11); Ca^{2+}-transporting ATPase (12); myosin ATPase (13,18); nitrogenase (14,16); CF$_1$ ATPase (15); dynamin GTPase (17); H$^+$-exporting ATPase (19); exopolyphosphatase (20); acetate kinase (21). **1**. Sondek, Lambright, Noel, Hamm, & Sigler (1994) *Nature* **372**, 276. **2**. Vincent, Brouns, Hart, Settleman (1998) *Proc. Natl. Acad. Sci. U.S.A.* **95**, 2210. **3**. Murphy & Coll (1992) *J. Biol. Chem.* **267**, 5229. **4**. Yousafzai & Eady (1997) *Biochem. J.* **326**, 637. **5**. Yousafzai & Eady (1999) *Biochem. J.* **339**, 511. **6**. Ren, Dou, Stelzer & Allison (1999) *J. Biol. Chem.* **274**, 31366. **7**. Bandyopadhyay, Muneyuki & Allison (2005) *Biochemistry* **44**, 2441. **8**. Allison, Ren & Dou (2000) *J. Bioenerg. Biomembr.* **32**, 531. **9**. Lunardi, Dupuis, Garin, *et al.* (1988) *Proc. Natl. Acad. Sci. U.S.A.* **85**, 8958. **10**. Issartel, Dupuis, Lunardi & P. V. Vignais (1991) *Biochemistry* **30**, 4726. **11**. Moreau & Carlier (1989) *J. Biol. Chem.* **264**, 2302. **12**. Troullier, Girardet & Dupont (1992) *J. Biol. Chem.* **267**, 22821. **13**. Maruta, Henry, Sykes & Ikebe (1993) *J. Biol. Chem.* **268**, 7093. **14**. Renner & Howard (1996) *Biochemistry* **35**, 5353. **15**. Gunther & Huchzermeyer (1998) *Eur. J. Biochem.* **258**, 710. **16**. Igarashi & Seefeldt (2003) *Crit. Rev. Biochem. Mol. Biol.* **38**, 351. **17**. Shpetner & Vallee (1992) *Nature* **355**, 733. **18**. Park, Ajtai & Burghardt (1999) *Biochim. Biophys. Acta* **1430**, 127. **19**. Rapin-Legroux, Troullier, Dufour & Dupont (1994) *Biochim. Biophys. Acta* **1184**, 127. **20**. Bolesch & Keasling (2000) *Biochem. Biophys. Res. Commun.* **274**, 236. **21**. Gorrell, Lawrence & Ferry (2005) *J. Biol. Chem.* **280**, 10731.

Alvespimycin

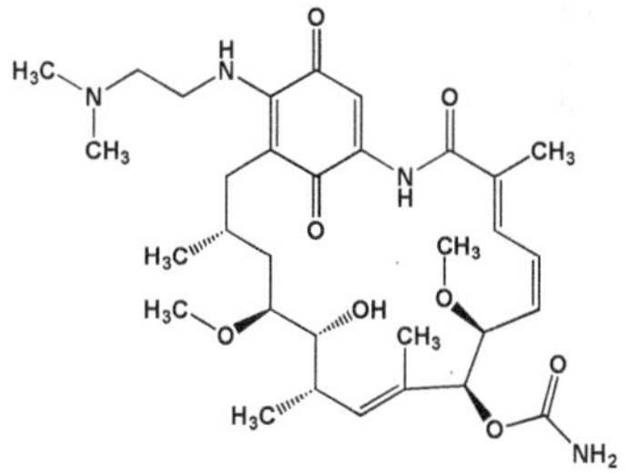

This potent Hsp90 inhibitor (FW$_{HCl-Salt}$ = 653.21 g/mol; CAS 467214-21-7), also named 17-DMAG and 17-demethoxy-17-[[2-(dimethylamino) ethyl]amino]geldanamycin, targets Heat Shock Protein-90 (IC$_{50}$ = 62 nM in a cell-free assay). (*See also* Ansamycin; Deguelin; Derrubone; Ganetespib; Geldanamycin; Herbimycin; Macbecin; Radicicol; Tanespimycin; and the nonansamycin Hsp90 inhibitor KW-2478) **1**. Ge, Normant, Porter, *et al.* (2006) *J. Med. Chem.* **49**, 4606. **2**. Lang, Klein, Moser, *et al.* (2007) *Mol. Cancer Ther.* **6**, 1123.

Alvocidib, *See Flavopiridol*

AM251

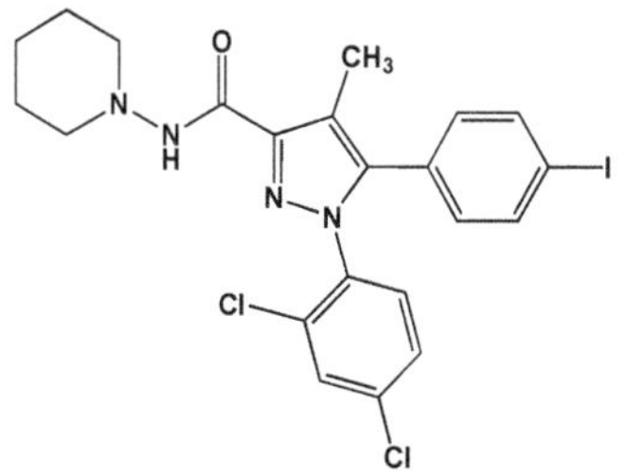

This CB$_1$ inverse agonist (FW = 555.24 g/mol; CAS 183232-66-8; Solubility: 100 mM in DMSO; Fluorescent Derivative: Tocrifluor T1117), also named 1-(2,4-dichlorophenyl)-5-(4-iodophenyl)-4-methyl-*N*-(1-piperidyl)pyrazole-3-carboxamide, binds to the cannabinoid-1 receptor, potently blocking (K_i = 7.5 nM; IC$_{50}$ = 8 nM) the action of endocannabinoids (*See Anandamide*). AM251 potentiates the forskolin-stimulated accumulation of cAMP, and it reduces the basal levels of inositol phosphate production in cells expressing the CB$_2$ receptors. Such findings demonstrate that AM251 acts as inverse agonists at the human CB$_2$ receptor acting via pathways coupled to members of the G$_\alpha$ family of GPCRs. **1**. Lan, Liu, Fan, *et al.* (1999) *J. Med. Chem.* **42**, 769.

AM803, *See GSK2190915*

AM-1155, *See* Gatifloxacin

α-Amanitin

This bicyclic octapeptide toxin (FW = 918.98 g/mol; CAS 23109-05-9; LD_{50} = 0.1 mg/kg *i.p* in albino mice) from the the death-cap fungus *Amanita phalloides* binds very tightly to eukaryotic RNA polymerase II (K_i ≈ 10 nM) and less tightly to eukaryotic RNA polymerase III (K_i ≈ 1 μM), blocking mRNA synthesis and inducing cytolysis of hepatocytes and kidney cells. α-Amanitin istable at room temperature, soluble in water and ethanol, and melts with decomposition at 254-255°C. **Caution:** Gloves should be worn, and measures should be taken to avoid inhalation. Symptoms of a toxic dose may not appear for 6-15 hours after exposure. More than 90% of deaths caused by mushroom poisonings are due to *A. phalloides* and related species. **Target(s):** RNA polymerase II (1-4,8); RNA polymerase III (1-3,7); yeast RNA polymerase B (2); RNA polymerase I (7); certain plant RNA polymerases (*e.g.*, plant RNA polymerase II ()2); DNA-directed DNA polymerase (5); DNA-directed RNA polymerase (1-4,6-8). **1**. Scott & Tomkins (1975) *Meth. Enzymol.* **40**, 273. **2**. Chambon (1974) *The Enzymes*, 3rd ed., **10**, 261. **3**. Lewis & Burgess (1982) *The Enzymes*, 3rd ed., **15**, 109. **4**. Preston, Hencin & Gabbay (1981) *Arch. Biochem. Biophys.* **209**, 63. **5**. Spampinato, Pairoba, Colombo, Benediktsson & Andreo (1994) *Biosci. Biotechnol. Biochem.* **58**, 822. **6**. Sethi (1971) *Prog. Biophys. Mol. Biol.* **23**, 67. **7**. Köck & Cornelissen (1991) *Mol. Microbiol.* **5**, 835. **8**. Sadhukhan, Chakraborty, Dasgupta & Majumder (1997) *Mol. Cell. Biochem.* **171**, 105.

Amastatin

This peptide analogue (FW = 474.56 g/mol; CAS 100938-10-1), also known as (2S,3R)-3-amino-2-hydroxy-5-methylhexanoyl-L-valyl-L-valyl-L-aspartate, is a high-affinity, reversible competitive inhibitor of human serum and porcine kidney aminopeptidase (K_i ≈ 1–2 μM). Prepare as a 1 mM stock solution in methanol or ethanol, store at –20 ° C for one month, and use at final concentrations of 2–20 μM. At room temperature, aqueous solutions are stable for one day. **Target(s):** leucyl aminopeptidase (1,9,17,22,27,55,66-74); membrane alanyl aminopeptidase, *or* aminopeptidase M, *or* aminopeptidase N (1,2,19, 22,28,48,60-65); glutamyl aminopeptidase, *or* aminopeptidase A (3,18,42-48); cystinyl aminopeptidase, *or* oxytocinase (4,54-59); insulin-regulated membrane aminopeptidase (5); aminopeptidase PS (6); aminopeptidase B (7,49-52); nardilysin (8,15,24,25); aminopeptidase Y, *or* aminopeptidase Co (10,11,37); *Vibrio* aminopeptidase (12); aminopeptidase T (13); Xaa-Trp aminopeptidase, *or* aminopeptidase W (14,34-36); acyl amidase-like leucyl aminopeptidase (16); L-to-D-amino-acid-residue isomerase (20,21); bacterial leucyl aminopeptidase (12,22); biotinidase (23); mitochondrial intermediate peptidase (26); pyroglutamyl peptidase I, weakly inhibited (27); dipeptidyl-peptidase III, moderately inhibited (29); cytosol nonspecific dipeptidase (30); aminopeptidase I (31-33); cytosol alanyl aminopeptidase (38); Xaa-Pro aminopeptidase, *or* aminopeptidase P (39-41); prolyl aminopeptidase, *Aneurinibacillus* sp. (53); leucyl aminopeptidase/cystinyl aminopeptidase (55). **1**. Rich, Moon & Harbeson (1984) *J. Med. Chem.* **27**, 417. **2**. Turner (1998) in *Handb. Proteolytic Enzymes* (Barrett, Rawlings & Woessner, eds), p. 996, Academic Press,

San Diego. **3**. Wang & Cooper (1998) in *Handb. Proteolytic Enzymes* (Barrett, Rawlings & Woessner, eds), p. 1002, Academic Press, San Diego. **4**. Mizutani, Tsujimoto & Nakazato (1998) in *Handb. Proteolytic Enzymes* (Barrett, Rawlings & Woessner, eds), p. 1008, Academic Press, San Diego. **5**. Keller (1998) in *Handb. Proteolytic Enzymes* (Barrett, Rawlings & Woessner, eds), p. 1011, Academic Press, San Diego. **6**. Dando & Barrett (1998) in *Handb. Proteolytic Enzymes* (Barrett, Rawlings & Woessner, eds), p. 1013, Academic Press, San Diego. **7**. Foulon, Cadel & Cohen (1998) in *Handb. Proteolytic Enzymes* (Barrett, Rawlings & Woessner, eds), p. 1026, Academic Press, San Diego. **8**. Prat, Chesneau, Foulon & Cohen (1998) in *Handb. Proteolytic Enzymes* (Barrett, Rawlings & Woessner, eds), p. 1370, Academic Press, San Diego. **9**. Sträter & Lipscomb (1998) in *Handb. Proteolytic Enzymes* (Barrett, Rawlings & Woessner, eds), p. 1384, Academic Press, San Diego. **10**. Wolf (1998) in *Handb. Proteolytic Enzymes* (Barrett, Rawlings & Woessner, eds), p. 1427, Academic Press, San Diego. **11**. Yasuhara (1998) in *Handb. Proteolytic Enzymes* (Barrett, Rawlings & Woessner, eds), p. 1429, Academic Press, San Diego. **12**. Chevier & D'Orchymont (1998) in *Handb. Proteolytic Enzymes* (Barrett, Rawlings & Woessner, eds), p. 1433, Academic Press, San Diego. **13**. Motoshima & Kaminogawa (1998) in *Handb. Proteolytic Enzymes* (Barrett, Rawlings & Woessner, eds), p. 1452, Academic Press, San Diego. **14**. Hooper (1998) in *Handb. Proteolytic Enzymes* (Barrett, Rawlings & Woessner, eds), p. 1513, Academic Press, San Diego. **15**. Cohen, Pierotti, Chesneau, Foulon & Prat (1995) *Meth. Enzymol.* **248**, 703. **16**. Matsuda, Katayama, Hara, *et al.* (2000) *Arch. Androl.* **44**, 1. **17**. Kim & Lipscomb (1993) *Biochemistry* **32**, 8465. **18**. Aoyagi, Tobe, Kojima, *et al.* (1978) *J. Antibiot.* **31**, 636. **19**. Tokioka-Terao, Hiwada & Kokubu (1984) *Enzyme* **32**, 65. **20**. Torres, Tsampazi, Kennett, *et al.* (2007) *Amino Acids* **32**, 63. **21**. Torres, Tsampazi, Tsampazi, *et al.* (2006) *FEBS Lett.* **580**, 1587. **22**. Wilkes & Prescott (1985) *J. Biol. Chem.* **260**, 13154. **23**. Oizumi & Hayakawa (1991) *Biochim. Biophys. Acta* **1074**, 433. **24**. Chesneau, Pierotti, Barre, *et al.* (1994) *J. Biol. Chem.* **269**, 2056. **25**. Foulon, Cadel, Prat, *et al.* (1997) *Ann. Endocrinol.* **58**, 357. **26**. Kalousek, Isaya & Rosenberg (1992) *EMBO J.* **11**, 2803. **27**. Mantle, Lauffart & Gibson (1991) *Clin. Chim. Acta* **197**, 35. **28**. Vanha-Perttula (1988) *Clin. Chim. Acta* **177**, 179. **29**. Abramic, Schleuder, Dolovcak, *et al.* (2000) *Biol. Chem.* **381**, 1233. **30**. Lenney (1990) *Biol. Chem. Hoppe-Seyler* **371**, 433. **31**. Tisljar & Wolf (1993) *FEBS Lett.* **322**, 191. **32**. Desimone, Kruger, Wessel, *et al.* (2000) *J. Chromatogr. B* **737**, 285. **33**. Noguchi, Nagata, Koganei, *et al.* (2002) *J. Agric. Food Chem.* **50**, 3540. **34**. Gee & Kenny (1987) *Biochem. J.* **246**, 97. **35**. Tieku & Hooper (1992) *Biochem. Pharmacol.* **44**, 1725. **36**. Tieku & Hooper (1993) *Biochem. Soc. Trans.* **21**, 250S. **37**. Yasuhara, Nakai & Ohashi (1994) *J. Biol. Chem.* **269**, 13644. **38**. Yamamoto, Li, Huang, Ohkubo & Nishi (1998) *Biol. Chem.* **379**, 711. **39**. Orawski, Susz & Simmons (1987) *Mol. Cell. Biochem.* **75**, 123. **40**. McDonnell, Fitzgerald, Fhaoláin, Jennings & O'Cuinn (1997) *J. Dairy Res.* **64**, 399. **41**. Harbeck & Mentlein (1991) *Eur. J. Biochem.* **198**, 451. **42**. Goto, Hattori, Ishii, Mizutani & Tsujimoto (2006) *J. Biol. Chem.* **281**, 23503. **43**. Petrovic & Vitale (1990) *Comp. Biochem. Physiol. B* **95**, 589. **44**. Iturrioz, Reaux-Le Goazigo & Llorens-Cortes (2004) in *Aminopeptidases in Biology and Disease* (Hooper & Lendaechel, eds), vol. **2**, 229, Springer. **45**. Tobe, Morishima, Aoyagi, *et al.* (1982) *Agric. Biol. Chem.* **46**, 1865. **46**. Nagatsu, Nagatsu, Yamamoto, Glenner & Mehl (1970) *Biochim. Biophys. Acta* **198**, 255. **47**. Tobe, Kojima, Aoyagi & Umezawa (1980) *Biochim. Biophys. Acta* **613**, 459. **48**. Lalu, Lampelo & Vanha-Perttula (1986) *Biochim. Biophys. Acta* **873**, 190. **49**. Tanioka, Hattori, Masuda, *et al.* (2003) *J. Biol. Chem.* **278**, 32275. **50**. Cadel, Pierotti, Foulon, *et al.* (1995) *Mol. Cell. Endocrinol.* **110**, 149. **51**. Goldstein, Nelson, Kordula, Mayo & Travis (2002) *Infect. Immun.* **70**, 836. **52**. Petrov, Fagard & Lijnen (2004) *Cardiovasc. Res.* **61**, 724. **53**. Murai, Tsujimoto, Matsui & Watanabe (2004) *J. Appl. Microbiol.* **96**, 810. **54**. Itoh, Watanabe, Nagamatsu, *et al.* (1997) *Biol. Pharm. Bull.* **20**, 20. **55**. Matsumoto, Rogi, Yamashiro, *et al.* (2000) *Eur. J. Biochem.* **267**, 46. **56**. Itoh & Nagamatsu (1995) *Biochim. Biophys. Acta* **1243**, 203. **57**. Krishna & Kanagasabapathy (1989) *J. Endocrinol.* **121**, 537. **58**. Burbach, De Bree, Terwel, *et al.* (1993) *Peptides* **14**, 807. **59**. Tisljar & Wolf (1993) *FEBS Lett.* **322**, 191. **60**. Xu & Li (2005) *Curr. Med. Chem. Anticancer Agents* **5**, 281. **61**. Bauvois & Dauzonne (2006) *Med. Res. Rev.* **26**, 88. **62**. Hua, Tsukamoto, Taguchi, *et al.* (1998) *Biochim. Biophys. Acta* **1383**, 301. **63**. Yoshimoto, Tamesa, Gushi, Murayama & Tsuru (1988) *Agric. Biol. Chem.* **52**, 217. **64**. Huang, Takahara, Kinouchi, *et al.* (1997) *J. Biochem.* **122**, 779. **65**. Terashima & N. W. Bunnett (1995) *J. Gastroenterol.* **30**, 696. **66**. Lauffart, McDermott, Gibson & Mantle (1988) *Biochem. Soc. Trans.* **16**, 850. **67**. Kohno, Kanda & Kanno (1986) *J. Biol. Chem.* **261**, 10744. **68**. Gu, Holzer & Walling (1999) *Eur. J. Biochem.* **263**, 726. **69**.

Xu, Shawar & Dresden (1990) *Exp. Parasitol.* **70**, 124.   **70**. Karadzic, Izrael, Gojgic-Cvijovic & Vujcic (2002) *J. Biosci. Bioeng.* **94**, 309.   **71**. Kumagai, Watanabe & Fujimoto (1991) *Biochem. Med. Metab. Biol.* **46**, 110.   **72**. Gibson, Biggins, Lauffart, Mantle & McDermott (1991) *Neuropeptides* **19**, 163.   **73**. Hattori, Kitatani, Matsumoto, *et al.* (2000) *J. Biochem.* **128**, 755.   **74**. Sträter & Lipscomb (2004) in *Handbook of Metalloproteins* (Messerschmidt, Huber, Wieghardt & Poulos, eds) **3**, 199, Wiley, New York.

Ambrisentan

This orally active endothelin receptor antagonist (FW = 378.42 g/mol; CAS 177036-94-1), also known by its code names LU-208075, BSF-208075, its trade names Letairis™ and Volibris™, and by its IUPAC name (2*S*)-2-[(4,6-dimethylpyrimidin-2-yl)oxy]-3-methoxy-3,3-diphenylpropanoic acid, is a FDA-approved, once-daily drug for treating pulmonary hypertension. Because endothelin constricts blood vessels and elevates blood pressure, endothelin receptor antagonists (ETRAs) prevent constriction/narrowing of blood vessels, thereby enhancing blood flow. There are no known interactions between ambrisentan and cytochrome P450 isoenzymes (metabolism, induction or inhibition) that might alter the activity of P450-metabolized drugs. **Key Pharmacokinetic Parameters:** *See* Appendix II in Goodman & Gilman's THE PHARMACOLOGICAL BASIS OF THERAPEUTICS, 12th Edition (Brunton, Chabner & Knollmann, eds.) McGraw-Hill Medical, New York (2011).   **1**. Vatter, Zimmermann, Jung, *et al.* (2002) *Clin Sci (Lond)* **103**, 408S.   **2**. Witzigmann, Ludwig, Escher, *et al.* (2002) *Transplant Proc.* **34**, 2387.   **3**. Galié, Badesch, Oudiz, *et al.* (2005) *J. Amer. Coll. Cardiol.* **46**, 529.

AMD3100, *See* Plerixafor

Amdinocillin

This amidino-penicillin (FW = 325.43 g/mol), also known as mecillinam, is a semi-synthetic antibacterial that inhibits certain penicillin-binding proteins, indirectly blocking cell division. **Target(s):** penicillin-binding protein 5, a serine-type D-Ala-D-Ala carboxypeptidase (1); penicillin-binding protein 4 (2); penicillin-binding protein 1 (3); penicillin-binding protein 2 (4-7,9); *Actinomadura* R39 D-alanyl-D-alanine-cleaving serine peptidase (8). Under laboratory selection conditions, frequencies of mutation to amdinocillin resistance in *Escherichia coli* varied from 8×10^{-8} to 2×10^{-5} per cell, depending on the amdinocillin concentration (9). Of the genes demonstrated to confer amdinocillin resistance, eight novel genes in amdinocillin resistance encode functions involved in the respiratory chain, the ribosome, cysteine biosynthesis, tRNA synthesis, and pyrophosphate metabolism.   **1**. Wilkin (1998) in *Handb. Proteolytic Enzymes*, p. 418, Academic Press, San Diego.   **2**. Wilkin (1998) in *Handb. Proteolytic Enzymes*, p. 435, Academic Press, San Diego.   **3**. Storey & Chopra (2001) *Antimicrob. Agents Chemother.* **45**, 303.   **4**. Vinella & D'Ari (1994) *J. Bacteriol.* **176**, 966.   **5**. Vinella, Joseleau-Petit, Thevenet, Bouloc & R. D'Ari (1993) *J. Bacteriol.* **175**, 6704.   **6**. Licht, Gally, Henderson, Young & Cooper (1993) *Res. Microbiol.* **144**, 423.   **7**. Gutmann, Vincent, Billot-Klein, *et al.* (1986) *Antimicrob. Agents Chemother.* **30**, 906.   **8**. Kelly, Frere, Klein & Ghuysen (1981) *Biochem. J.* **199**, 129.   **9**. Wientjes & Nanninga (1991) *Res. Microbiol.* **142**, 333.   **9**. Thulin, Sundqvist & Andersson (2015) *Antimicrob. Agents Chemother.* **59**, 1718.

Amentoflavone

This biflavonoid (FW = 538.47 g/mol; CAS 1617-53-4), also known as 3',8"-biapegenin, from *Ginkgo biloba*, *Taxodium mucronatum*, and St. John's wort (*Hypericum perforatum*), scavenges superoxide radicals, inhibits lipid peroxidation, and has anti-inflammatory properties. **Target(s):** CYP1A1 (1); CYP1B1 (1); 3',5'-cyclic-nucleotide phosphodiesterase, *or* cAMP phosphodiesterase; IC_{50} = 22 μM (2,6); RNA-directed DNA polymerase (3); avian myeloblastosis reverse transcriptase (3); Rous-associated virus-2 reverse transcriptase (3); Maloney murine leukemia virus reverse transcriptase (3); phospholipase Cγ1 (4); cyclooxygenase (however, see ref. 11) (5,11); α-glucosidase (7); α-amylase (7); phospholipase A₂ (8); DNA topoisomerase I (9); CYP2C9 (10); cathepsin B (12); 3',5'-cyclic-GMP phosphodiesterase (13); phosphodiesterase 5 (13); inositol-trisphosphate 3 kinase (14).   **1**. Chaudhary & Willett (2006) *Toxicology* **217**, 194.   **2**. Beretz, Briancon-Scheid, Stierle, *et al.* (1986) *Biochem. Pharmacol.* **35**, 257.   **3**. Spedding, Ratty & Middleton (1989) *Antiviral Res.* **12**, 99.   **4**. Lee, Oh, Kim, *et al.* (1996) *Planta Med.* **62**, 93.   **5**. Kim, Mani, Iversen & Ziboh (1998) *Prostaglandins Leukot. Essent. Fatty Acids* **58**, 17.   **6**. Saponara & Bosisio (1998) *J. Nat. Prod.* **61**, 1386.   **7**. Kim, Kwon & Son (2000) *Biosci. Biotechnol. Biochem.* **64**, 2458.   **8**. Kim, Pham & Ziboh (2001) *Prostaglandins Leukot. Essent. Fatty Acids* **65**, 281.   **9**. Grynberg, Carvalho, Velandia, *et al.* (2002) *Braz. J. Med. Biol. Res.* **35**, 819.   **10**. von Moltke, Weemhoff, Bedir, *et al.* (2004) *J. Pharm. Pharmacol.* **56**, 1039.   **11**. Seaver & Smith (2004) *J. Herb. Pharmacother.* **4**, 11.   **12**. Pan, Tan, Zeng, Zhang & Jia (2005) *Bioorg. Med. Chem.* **13**, 5819.   **13**. Dell'Agli, Galli & Bosisio (2006) *Planta Med.* **72**, 468.   **14**. Mayr, Windhorst & Hillemeier (2005) *J. Biol. Chem.* **280**, 13229.

Amezinium

This sympathomimetic and antihypotensive drug (FW = 313.32 g/mol; CAS 30578-37-1), also named 4-amino-6-methoxy-1-phenyl-pyridazinium methyl sulfate, amezinium-metilsulfate, LU 1631, and Regulton®, increases the arterial blood pressure and heart rate of anaesthetized animals and of pithed rats by stimulating vascular α-adrenoceptors and cardiac β₁-adrenoceptors (1). This action is unaffected by ganglionic blockade with hexamethonium. The α-adrenergic blocking drug phentolamine antagonizes the blood pressure increasing effect of amezinium, and the β-adrenergic blocking drug propranolol antagonizes the heart rate increasing effect. Noradrenaline depletion by pretreatment with reserpine reduces the pressor effect of amezinium to approximately the same extent as it reduces the effect of tyramine. It completely abolished the heart rate increasing effect. Under such conditions, high doses of amezinium reduced the heart rate (1). Amezinium is taken up by adrenergic neurons. Amezinium also inhibits intraneuronal monoamine oxidase, with IC_{50} values of 5 μM and >1mM for MAO-A and MAO-B, respectively (1,2).   **1**. Araújo, Caramona & Osswald (1983) *Eur. J. Pharmacol.* **90**, 203.   **2**. Lenke, Gries & Kretzschmar (1981) *Arzneimittel-Forschung* **31**, 1558.

AMF-26

This novel anticancer agent (FW = 403.55 g/mol), also known as (2E,4E)-5-((1S,2S,4aR,6R,7S,8S,8aS)-7-hydroxy-2,6,8-trimethyl-1,2,4a,5,6,7,8,8a-octahydronaphthalen-1-yl)-2-methyl-N-(pyridin-3-ylmethyl)penta-2,4-dienamide), is a powerful cell growth inhibitor (GI$_{50}$ = 12 nM) and equally potent Golgi disruptor (EC$_{50}$ = 27 nM), inhibiting activation of ADP-ribosylation factors (or Arfs) by its guanine nucleotide exchange factors (GEFs). Arfs are Ras superfamily small GTPases that play a major role in mediating vesicular transport in the secretory and endocytic pathways. Molecular dynamics (MD) simulations suggest that AMF-26 binds to the contact surface of the Arf1-Sec7 domain near the site for brefeldin A. AMF-26 strongly affected membrane traffic, including the cis-Golgi and trans-Golgi networks, and the endosomal systems. When tested with a panel of 39 cancer cell lines, the GI$_{50}$ was consistently below 40 nM. The total synthesis of AMF-26 has also been described (2). **1**. Ohashi, Iijima, Yamaotsu, et al. (2012) J. Biol. Chem. **287**, 3885. **2**. Shiina, Umezaki, Ohashi, et al. (2013) J. Med. Chem. **56**, 150.

AMG-1

This protein kinase inhibitor (FW = 556.58 g/mol; CAS 913376-84-8), targets human c-Met (or hepatocyte growth factor receptor (HGFR)) and RON (or Recepteur d'Origine Nantais) kinase, with IC$_{50}$ values of 4 nM and 9 nM respectively. **1**. Araki, Karasawa, Kawashima, et al. (1989) Meth. Find. Exp. Clin. Pharmacol. **11**, 731.

AMG-47a

This selective protein kinase inhibitor (FW = 535.60 g/mol; CAS 882663-88-9; Soluble in DMSO), also named 3-(2-(2-morpholinoethylamino) quinazolin-6-yl)-N-(3-(trifluoromethyl)phenyl)-4-methylbenzamide, targets Lck (or Lymphocyte-specific kinase), showing a sub-nM IC$_{50}$, and low (<10 nM) inhibition of so-called hard-to-inhibit kinases (e.g., KDR, Src, and MAPKα (or p38α). AMG-47 is effective against the Lck mediated anti-inflammatory activity (ED$_{50}$ = 11 mg/kg, or 630nM) in vivo. **1**. Carver, Dexheimer, Hsu, et al. (2014) PLoS One **9**, e103836.

AMG-51

This protein kinase inhibitor (FW = 629.65 g/mol; CAS 890019-63-3; Solubility: 15 mM in DMSO), also known as 5-(3-fluoro-4-(6-methoxy-7-(3-morpholinopropoxy)quinolin-4-yloxy)phenyl)-2-(4-fluorophenylamino)-

3-methylpyrimidin-4(3H)-one, targets c-Met, or hepatocyte growth factor receptor (HGFR), a protein tyrosine kinase that plays a key role in several cellular processes and is overexpressed or mutated in different human cancers. AMG-51 shows the enzyme selectivity of c-Met with a (K_i = 4.9 nM), with off-targets, such as IGFR (K_i = 22nM), Ron or Recepteur d'Origine Nantais kinase (K_i = 28nM), and KDR (K_i = 139 nM). **1**. D'Angelo, Bellon, Booker, et al. (2008) J. Med. Chem. **51**, 5766.

AMG-145, *See Evolocumab*

AMG 162, *See Denosumab*

AMG-319

This PI3Kδ inhibitor (FW = 385.40 g/mol; CAS 1608125-21-8; Solubility: 77 mg/mL DMSO; <1 mg/mL H$_2$O), also named N-((S)-1-(7-fluoro-2-(pyridin-2-yl)quinolin-3-yl)ethyl)-9H-purin-6-amine, potently and selectively targets phosphatidylinositol-4,5-bisphosphate 3-kinase (IC$_{50}$ = 16 nM in whole blood), an intracellular signal-transducing enzyme that phosphorylates the position-3 hydroxyl in phosphatidylinositol (PtdIns). **Other Target(s):** PI3Kγ (IC$_{50}$ = 850 nM); PI3Kβ (IC$_{50}$ = 2.7μM); and PI3Kα (IC$_{50}$ = 33 μM). **1**. Cushing, Hao, Shin, et al. (2015) J. Med. Chem. **58**, 480.

AMG-458

This potent c-Met protein kinase inhibitor (FW = 539.58 g/mol; CAS 913376-83-7; Solubility: 21 mg/mL DMSO; ~1 mg/mL H$_2$O), also named 1-(2-hydroxy-2-methylpropyl)-N-(5-(7-methoxyquinolin-4-yloxy)pyridin-2-yl)-5-methyl-3-oxo-2-phenyl-2,3-dihydro-1H-pyrazole-4-carboxamide, targets the *MET* proto-oncogene product c-Met, which is the hepatocyte growth factor receptor possessing tyrosine-kinase activity. The primary single-chain precursor protein is posttranslationally cleaved to produce the α and β subunits that are disulfide linked to form the mature receptor. Various mutations in the MET gene are associated with papillary renal carcinoma. AMG-458 preferentially inhibits c-Met (K_i = 1.2 nM), showing ~350-fold greater potency than VEGFR2 in cells (1). AMG-458 significantly inhibited tumor growth in the NIH3T3/TPR-Met and U-87 MG xenograft models with no adverse effect on body weight (1). AMG 458 binds covalently to liver microsomal proteins from rats and humans, even in the absence of NADPH. When [^{14}C]AMG-458 is incubated with liver microsomes in the presence of glutathione or N-acetyl cysteine, quinolone-type thioether adducts can be detected by radiochromatography and LC/MS/MS analysis (2). AMG-458 was more effective in cells that expressed higher levels of c-Met/p-Met, suggesting that higher levels of c-Met and p-Met in non-small cell lung cancer (NSCLC) tissue may classify a subset of tumors that are more sensitive to molecular therapies against this receptor (3). **1**. Liu, Siegmund, Xi *et al.* (2008) *J. Med. Chem.* **51**, 3688. **2**. Teffera, Colletti, Harmange, *et al.* (2008) *Chem. Res. Toxicol.* **21**, 2216. **3**. Li, Torossian, Sun, *et al.* (2012) *Int. J. Radiat. Oncol. Biol. Phys.* **84**, e525.

AMG 706, *See Motesanib*

AMG-900

This potent and highly selective mitotic protein kinase inhibitor (FW = 503.58 g/mol; CAS 945595-80-2; Solubility: 100 mg/mL DMSO), also named *N*-(4-(3-(2-aminopyrimidin-4-yl)pyridin-2-yloxy)phenyl)-4-(4-methylthiophen-2-yl)phthalazin-1-amine, targets Aurora A (IC$_{50}$ = 5 nM), Aurora B (IC$_{50}$ = 4 nM), and Aurora C (IC$_{50}$ = 1 nM) protein kinases, with >10-fold selectivity versus p38α, Tyk2, JNK2, Met and Tie2 (1). The modal tumor cell response to AMG 900 treatment is aborted cell division without prolonged mitotic arrest, ultimately resulting in cell death (1). AMG 900 exhibits acceptable PK properties in preclinical species and is predicted to have low clearance in humans (2). Male rats metabolize AMG 900 primarily through hydroxylation with subsequent sulfate conjugation on the pyrimidinyl-pyridine side-chain, whereas female rats favor oxidation on the thiophene ring's methyl group, which is then metabolized to a carboxylic acid, attended by conjugation to an acyl glucuronide (3). At low-nM concentrations, AMG 900, whether administered alone or in combination with microtubule-targeting drugs (paclitaxel or ixabepilone), may be an effective intervention strategy for the treatment of metastatic breast cancer and provide potential therapeutic options for patients with multidrug-resistant tumors (4). It is, in fact, also active against taxane-resistant tumor cell lines (5). **1**. Payton, Bush, Chung, *et al.* (2010) *Cancer Res.* **70**, 9846. **2**. Huang, Be, Berry, *et al.* (2011) *Xenobiotica* **41**, 400. **3**. Waldon, Berry, Lin, *et al.* (2011) *Drug Metab. Lett.* **5**, 290. **4**. Bush, Payton, Heller, *et al.* (2013) *Mol Cancer Ther.* **12**, 2356. **5**. Payton, Bush, Chung, *et al.* (2010) *Cancer Res.* **70**, 9846.

AMG-Tie2-1

This highly potent signal transduction inhibitor (FW = 479.75 g/mol; CAS 870223-96-4; Solubility: 15 mM in DMSO), also named 3-(3-(2-(methylamino)pyrimidin-4-yl)pyridin-2-yloxy)-*N*-(3-(trifluoromethyl)-phenyl)-4-methylbenzamide, targets the intrinsic protein tyrosine kinase of Tie2, a receptor that binds angiopoietin-1 and mediates signaling pathway(s) in embryonic vascular development. Tie-2 possesses a unique extracellular region that contains two immunoglobulin-like domains, three epidermal growth factor (EGF)-like domains and three fibronectin type III repeats. AMG-Tie2-1 exhibits >30-fold selectivity over a panel of kinases, good oral exposure, and *in vivo* inhibition of Tie-2 autophosphorylation. **1**. Hodous, Geuns-Meyer, Hughes, *et al.* (2007) *J. Med. Chem.* **50**, 611.

Amicetins

Amicetin A

Amicetin B

These pyrimidine-containing glycoside antibiotic (FW$_{Amicetin-A}$ = 617.69 g/mol (free base); CAS 17650-86-1; FW$_{Amicetin-B}$ = 532.59 g/mol (free base)) from *Streptomyces vinaceus-drappus* and *S. fasciculatus* inhibits protein biosynthesis, exerting its bacteriostatic most potently on Gram-positive bacteria. Amicetin B, also called plicacetin, was isolated from *Streptomyces plicatus* and has the identical structure minus the α-methylseryl residue. It is weaker in action compared to amicetin A. **Target(s):** peptidyltransferase (1-5,9-11,14-17); peptidyl-tRNA hydrolase, *or* aminoacyl-tRNA hydrolase (6); ribonuclease P (7,8); protein biosynthesis (1-5,9-14).**1**. Fernandez-Muñoz, Monro & Vazquez (1971) *Meth. Enzymol.* **20**, 481. **2**. Gottesman (1971) *Meth. Enzymol.* **20**, 490. **3**. Pestka (1974) *Meth. Enzymol.* **30**, 261. **4**. Carrasco, Battaner & Vazquez (1974) *Meth. Enzymol.* **30**, 282. **5**. Jiménez (1976) *Trends Biochem. Sci.* **1**, 28. **6**. Tate & Caskey (1974) *The Enzymes*, 3rd ed., **10**, 87. **7**. Kalavrizioti, Vourekas, Tekos, *et al.* (2003) *Mol. Biol. Rep.* **30**, 9. **8**. Stathopoulos, Tsagla, Tekos & Drainas (2000) *Mol. Biol. Rep.* **27**, 107. **9**. Gu & Lovett (1995) *J. Bacteriol.* **177**, 3616. **10**. Leviev, Rodriguez-Fonseca, Phan, *et al.* (1994) *EMBO J.* **13**, 1682. **11**. Theocharis, Synetos, Kalpaxis, Drainas & Coutsogeorgopoulos (1992) *Arch. Biochem. Biophys.* **292**, 266. **12**. Pestka (1974) *Meth. Enzymol.* **30**, 261. **13**. Bloch & Coutsogeorgopoulos (1966) *Biochemistry* **5**, 3345. **14**. Coutsogeorgopoulos (1967) *Biochemistry* **6**, 1704. **15**. Spirin & Asatryan (1976) *FEBS Lett.* **70**, 101. **16**. Cerná, Rychlík & Lichtenthaler (1973) *FEBS Lett.* **30**, 147. **17**. Lichtenthaler, Cerná & Rychlík (1975) *FEBS Lett.* **53**, 184.

Amifloxacin

This fluoroquinolone-class antibiotic (FW = 334.31 g/mol; CAS 86393-37-5) is a broad-spectrum systemic agent that targets Type II DNA topoisomerases (gyrases) required for bacterial replication and transcription (1). Ninety percent of *Escherichia coli*, *Klebsiella* species, *Aeromonas*, *Salmonella*, *Shigella*, *Citrobacter*, *Enterobacter* species, *Proteus mirabilis*, *Serratia marcescens*, and *Morganella morganii* were inhibited by ≤ 0.5 µg/mL (2). Amifloxacin inhibited *Branhamella*, *Haemophilus*, and *Neisseria* at ≤ 0.25 µg/mL, and 90% of *Pseudomonas aeruginosa*, including gentamicin and carbenicillin-resistant isolates, at 4 ≤g/mL (2). It also inhibited *Staphylococci*, including methicillin-resistant isolates, but was less active against *Streptococci* and *Bacteroides* species. For mechanism of action, see Ciprofloxacin, which is the prototypical member of this antibiotic class. **1**. Johnson & Benzinger (1985) *Antimicrob. Agents Chemother.* **27**, 774. **2**. Neu & Labthavikul (1985) *Diagnost. Microbiol. Infect. Dis.* **3**, 469.

Amikacin

This aminoglycoside-class antibiotic (FW = 585.60 g/mol; CAS 37517-28-5) is a broad-spectrum semisynthetic derivative of kanamycin A that has become a go-to drug for treating hospital-acquired infections by multidrug-resistant, Gram-negative bacteria (*e.g.*, *Pseudomonas aeruginosa*, *Acinetobacter*, *Enterobacter*, as well as *Serratia marcescens* and *Providencia stuartii*). (**For additional mechanistic details, See** *Neomycins*) Because amikacin is not absorbed orally, it must be delivered intravenously or by an intranasal using liposomes as carriers. Test results revealed that 60.5% of the isolates were resistant to tobramycin (at 8 µg/mL), 67.1% to gentamicin (at 8 µg/mL), 86.2% to kanamycin A (at 20 µg/mL), but only 8.6% to amikacin (at 20 µg/mL). Amikacin's broad spectrum comports with the fact that it is a poor substrate for most enzymes that inactivate other aminoglycosides through *O*-phosphorylation, *O*-adenylylation, or *N*-acetylation (1). **Pharmacokinetics:** A dose of 300 mg/m^2 intramuscularly produced a highest mean serum concentration of 25.4 µg/mL with a mean serum concentration of 3.1 µg/mL at 8 hours (2). The same dose intravenously produced a mean serum concentration of 52 µg/mL with a mean serum concentration of 2 µg/mL at 8 hours. The mean urinary excretion during the first 6 hours was 75 and 66%, respectively. When administered at 150 mg/m^2 every 6 hours, there was evidence of some drug accumulation (2). A loading dose of 150 mg/m^2 administered IV over 30 min, followed by a continuous infusion at 200 mg/m^2 every 6 hours to maintain serum concentrations of 8 µg/mL. No major toxicity was observed with any of these drug regimens (2). **Key Pharmacokinetic Parameters:** *See* Appendix II in Goodman & Gilman's THE PHARMACOLOGICAL BASIS OF THERAPEUTICS, 12th Edition (Brunton, Chabner & Knollmann, eds.) McGraw-Hill Medical, New York (2011). **1.** Price, Pursiano & DeFuria (1974) *Antimicrob. Agents Chemother.* **5**, 143. **2.** Bodey, Valdivieso, Feld & Rodriguez (1974) *Antimicrob. Agents Chemother.* **5**, 508.

Amiloride

This potassium-sparing diuretic (FW = 229.63 g/mol; CAS 2016-88-8; Absorbance at 10 mg/mL (water) = 642 at 212 nm, 555 at 285 nm, and 617 at 362 nm; pK_a = 8.7; Soluble in hot water (50 mg/mL), yielding a clear, yellow-green solution), also known as MK 870, Midamor®, and 3,5-diamino-6-chloro-*N*-(diaminomethylene)pyrazine-2-carboxamide, is an epithelial sodium channel (*or* ENaC) blocker used to manage hypertension and congestive heart failure (1-4). Two classes of Na$^+$ transporters are sensitive to this drug: (a) the conductive Na$^+$ entry pathway found in electrically high resistance epithelia and (b) a Na$^+$-H$^+$ electroneutral exchange system found in certain leaky epithelia, such as the renal proximal tubule. Amiloride exhibits a K_i value of <1 µM for the kidney transporter and approximately 1 mM in the colon. Midamor is a potassium-conserving (antikaliuretic) drug that possesses weak (compared with thiazide diuretics) natriuretic, diuretic, and antihypertensive activity. Like other potassium-conserving agents, amiloride may cause hyperkalemia (*i.e.*, serum K$^+$ levels >5.5 mEq/L), which, if uncorrected, can be fatal. Amiloride-induced hyperkalemia occurs in ~10% of patients, when used without a kaliuretic diuretic. This complication is more frequent in patients showing evidence of renal impairment and diabetes mellitus. When used along with a thiazide diuretic in patients without these complications, the risk of Amiloride-induced hyperkalemia drops to 1-2%. Amiloride is not metabolized by the liver and is instead excreted unchanged by the kidney. Amiloride displaces both adenosine A$_1$ receptor agonist and antagonist binding with a K_i value in the low micromolar range, when assayed in calf brain membranes (5). Inhibition is counteracted by NaCl and protons. Amiloride (IC$_{50}$ = 0.25 mM) and 5-(*N*-ethyl-*N*-isopropyl)amiloride (IC$_{50}$ = 0.11 mM) also inhibit coxsackievirus B$_3$ RNA polymerase in a single-nucleotide incorporation assay (6,7), although the argument that amiloride

competes with incoming nucleoside triphosphates is unconvincing. A high-fidelity RNA dependent CVB$_3$ RNA polymerase (RdRp) variant is amiloride-resistant (7). **1.** Baer, Jones, Spitzer & Russo (1967) *J. Pharm. Exper. Ther.* **157**, 435. **2.** Baba, Lant, Smith, Townsend & Wilson (1968) *Clin. Pharmacol. Therapeut.* 9, 318. **3.** Smith & Smith (1973) *Brit. J. Pharmacol.* **48**, 646. **4.** Garritsen, Ijzerman, Beukers, Cragoe & Soudijn (1990) *Biochem. Pharmacol.* **40**, 827. **5.** Benos (1982) *Am. J. Physiol.* 242, C131. **6.** Harrison, Gazina, Purcell, Anderson & Petrou (2008) *J. Virol.* 82, 1465. **6.** Gazina, Smidansky, Holien, *et al.* (2011) *J. Virol.* **85**, 10364. **7.** Levi, Gnädig, Beaucourt, *et al.* (2010) *PLoS Pathog.* **6**, e1001163.

Amineptine

This synthetic dopamine reuptake inhibitor (FW = 338.47 g/mol; CAS 57574-09-1), also known by the trade names Survector®, Maneon®, Directim®, Neolior®, Provector®, and Viaspera® as well as its IUPAC name, 7-[(10,11-dihydro-5*H*-dibenzo[a,d]-cyclohepten-5-yl)amino]heptanoic acid, significantly reduces the effect of 6-hydroxy-dopamine on brain dopamine, increases striatal homovanillic acid concentrations, and shows cross-tolerance (1). Despite these amphetamine-like properties, amineptine does not alter brain noradrenaline or acetylcholine concentrations, suggesting amineptine is a new type of antidepressant with a brain biochemical profile differing from that of other drugs used in depressive disorders. Amineptine is also a norepinephrine reuptake inhibitor (2). This agent is also ineffective in reducing amphetamine withdrawal symptoms or craving (3). Amineptine failed to secure FDA approval and is not a duly authorized drug in the U.S. **1.** Samanin, Jori, Bernasconi, Morpugo & Garattini (1977) *J. Pharm. Pharmacol.* 29, 555. **2.** Invernizzi & Garattini (2004) *Prog. Neuropsychopharmacol. Biol. Psychiatry* **28**, 819. **3.** Shoptaw, Kao, Heinzerling & Ling (2009) *Cochrane Database Syst. Rev.* CD003021.

4-Aminobutyraldehyde, *See* *4-Aminobutanal*

2-Aminobutyrate (2-Aminobutyric Acid)

This water-soluble amino acid (FW = 103.12 g/mol), also called α-aminobutyric acid and 2-aminobutanoic acid, is an alanine homologue that has often been used to probe substrate specificity. **Target(s):** tryptophanase, by the L-enantiomer (1); 4-aminobutyrate aminotransferase (2); arginase, weakly inhibited (3); L-alanine aminotransferase, by the L-enantiomer (4,10); homoserine kinase (5,9,23,24); D-alanyl-poly(phosphoribitol) synthetase, *or* D-alanine:poly(phosphoribitol) ligase, inhibited by the D-enantiomer (6); saccharopine dehydrogenase (7); methionyl-tRNA synthetase (8); kynureninase, by DL-2-aminobutyrate (11); UDP-*N*-acetylmuramate:L-alanine ligase (inhibited by the L-enantiomer (12,13); cysteinyl-tRNA synthetase, also weak alternative substrate (14); D-lysine 5,6-aminomutase (15); 2-aminohexano-6-lactam racemase, inhibited by both D- and L-isomers (16); methionine γ-lyase (17); 1-aminocyclopropane-1-carboxylate deaminase, inhibited by the L-enantiomer (18,19); arginine deaminase, inhibited by the L-enantiomer (20); [glutamine-synthetase] adenylyltransferase (21); aspartate kinase, inhibited by the L-enantiomer (22); (*R*)-3-amino-2 methylpropionate:pyruvate aminotransferase, by L-2-aminobutyrate (25); asparagine:oxo-acid aminotransferase, weakly (26); acetolactate synthase , inhibited by the L-enantiomer (27); ornithine carbamoyltransferase, inhibited by the L-enantiomer (28-31); aminocyclopropane-carboxylate oxidase, inhibited by D-isomer (32). **1.** Morino & Snell (1970) *Meth. Enzymol.* **17A**, 439. **2.** Der Garabedian, Lotti & Vermeersch (1986) *Eur. J. Biochem.* **156**, 589. **3.** Hunter & Downs (1945) *J. Biol. Chem.* **157**, 427. **4.** Bulos & Handler (1965) *J. Biol. Chem.* **240**, 3283. **5.** Webb (1966) *Enzyme and Metabolic Inhibitors*, vol. **2**, p. 357, Academic Press, New York. **6.** Reusch & Neuhaus (1971) *J. Biol. Chem.* **246**, 6136. **7.** Fujioka & Nakatani (1972) *Eur. J. Biochem.* **25**, 301. **8.** Hahn & Brown (1967) *Biochim. Biophys.*

Acta **146**, 264. **9**. Wormser & Pardee (1958) *Arch. Biochem. Biophys.* **78**, 416. **10**. Jakoby & Bonner (1953) *J. Biol. Chem.* **205**, 709. **11**. Saier & Jenkins (1967) *J. Biol. Chem.* **242**, 101. **12**. Liger, Blanot & van Heijenoort (1991) *FEMS Microbiol. Lett.* **80**, 111. **13**. Liger, Masson, Blanot, van Heijenoort & Parquet (1995) *Eur. J. Biochem.* **230**, 80. **14**. Burnell & Whatley (1977) *Biochim. Biophys. Acta* **481**, 266. **15**. Morley & Stadtman (1970) *Biochemistry* **9**, 4890. **16**. Ahmed, Esaki, Tanaka & Soda (1985) *Agric. Biol. Chem.* **49**, 2991. **17**. Lockwood & Coombs (1991) *Biochem. J.* **279**, 675. **18**. Minami, Uchiyama, Murakami, *et al.* (1998) *J. Biochem.* **123**, 1112. **19**. Honma, Shimomura, Shiraishi, Ichihara & Sakamura (1979) *Agric. Biol. Chem.* **43**, 1677. **20**. Smith, Ganaway & Fahrney (1978) *J. Biol. Chem.* **253**, 6016. **21**. Ebner, Wolf, Gancedo, Elsässer & Holzer (1970) *Eur. J. Biochem.* **14**, 535. **22**. Keng & Viola (1996) *Arch. Biochem. Biophys.* **335**, 73. **23**. Burr, Walker, Truffa-Bachi & Cohen (1976) *Eur. J. Biochem.* **62**, 519. **24**. Thèze, Kleidman & Saint Girons (1974) *J. Bacteriol.* **118**, 577. **25**. Kontani, Kaneko, Kikugawa, Fujimoto & Tamaki (1993) *Biochim. Biophys. Acta* **1156**, 161. **26**. Cooper (1977) *J. Biol. Chem.* **252**, 2032. **27**. Oda, Nakano & Kitaoka (1982) *J. Gen. Microbiol.* **128**, 1211. **28**. Marshall & Cohen (1972) *J. Biol. Chem.* **247**, 1654. **29**. Lusty, Jilka & Nietsch (1979) *J. Biol. Chem.* **254**, 10030. **30**. Pierson, Cox & Gilbert (1977) *J. Biol. Chem.* **252**, 6464. **31**. Szilagyi, Vargha & Szabo (1987) *FEMS Microbiol. Lett.* **48**, 115. **32**. Charng, Chou, Jiaang, Chen & Yang (2001) *Arch. Biochem. Biophys.* **385**, 179.

4-Aminobutyrate (4-Aminobutyric Acid)

This neurotransmitter (FW = 103.12 g/mol; CAS 56-12-2; pK_1 = 4.03 and pK_2 =10.56; Very Soluble in H_2O), commonly referred to as γ-aminobutyric acid (GABA), is synthesized by pyridoxal 5'-phosphate-dependent decarboxylation of L-glutamate, stored in vesicles within a neuron's synaptic terminal, and, upon release, is an agonist for the $GABA_A$ and $GABA_B$ receptors. $GABA_A$ receptors are ligand-activated chloride channels: when activated by GABA, they allow the flow of chloride ions across the membrane of the cell. GABA is s the chief inhibitory neurotransmitter in the mammalian Central Nervous System, where its principal role is to reduce neuronal excitability throughout the nervous system. In humans, GABA is also directly responsible for the regulation of muscle tone. **Target(s):** ornithine carbamoyltransferase (1); pyrroline-5-carboxylate dehydrogenase (2,6); pantothenate synthetase, *or* pantoate: β-alanine ligase (3,11); putrescine carbamoyltransferase (4); carboxypeptidase B (5); L-carnitine uptake (7); D-alanyl-D-alanine synthetase (D-alanine:D-alanine ligase (8); methionyl-tRNA synthetase (9); pyridoxal kinase (10); lysyl-tRNA synthetase (12); D-lysine 5,6-aminomutase (13); ornithine decarboxylase (14); guanidinopropionase (15); guanidinobutyrase (15); acetylspermidine deacetylase, weakly inhibited (16); β-ureidopropionase (17); 2 aminoethylphosphonate:pyruvate aminotransferase, weakly inhibited (18); ornithine δ aminotransferase (19,20); carnitine acetyltransferase (21). **1**. Nakamura & Jones (1970) *Meth. Enzymol.* **17A**, 286. **2**. Strecker (1971) *Meth. Enzymol.* **17B**, 262. **3**. Miyatake, Nakano & Kitaoka (1979) *Meth. Enzymol.* **62**, 215. **4**. Stalon (1983) *Meth. Enzymol.* **94**, 339. **5**. Folk & Gladner (1958) *J. Biol. Chem.* **231**, 379. **6**. Webb (1966) *Enzyme and Metabolic Inhibitors*, vol. 2, p. 336, Academic Press, New York. **7**. Hannuniemi & Kontro (1988) *Neurochem. Res.* **13**, 317. **8**. Lacoste, Poulsen, Cassaigne & Neuzil (1979) *Curr. Microbiol.* **2**, 113. **9**. Hahn & Brown (1967) *Biochim. Biophys. Acta* **146**, 264. **10**. Kerry & Kwok (1986) *Prep. Biochem.* **16**, 199. **11**. Miyatake, Nakano & Kitaoka (1978) *J. Nutr. Sci. Vitaminol.* **24**, 243. **12**. Levengood, Ataide, Roy & Ibba (2004) *J. Biol. Chem.* **279**, 17707. **13**. Morley & Stadtman (1970) *Biochemistry* **9**, 4890. **14**. Guirard & Snell (1980) *J. Biol. Chem.* **255**, 5960. **15**. Yorifuji, Sugai, Matsumoto & Tabuchi (1982) *Agric. Biol. Chem.* **46**, 1361. **16**. Huang, Dredar, Manneh, Blankenship & Fries (1992) *J. Med. Chem.* **35**, 2414. **17**. Matthews & Traut (1987) *J. Biol. Chem.* **262**, 7232. **18**. Dumora, Lacoste & Cassaigne (1983) *Eur. J. Biochem.* **133**, 119. **19**. Strecker (1965) *J. Biol. Chem.* **240**, 1225. **20**. Kalita, Kerman & Strecker (1976) *Biochim. Biophys. Acta* **429**, 780. **21**. Fritz & Schultz (1965) *J. Biol. Chem.* **240**, 2188.

γ-Aminobutyrine, *See 2,4-Diaminobutyrate and 2,4-Diaminobutyric Acid*

γ-Aminobutyryl-L-alanylglycine

This glutathione analogue (FW = 231.25 g/mol) weakly inhibits glutathionylspermidine synthetase from *Crithidia fasciculata*, which, like other trypanosome, synthesizes trypanothione enzymatically synthesized from spermidine and glutathione through the consecutive action of the ATP-dependent carbon-nitrogen ligases, glutathionylspermidine synthetase and trypanothione synthetase. **1**. De Craecker, Verbruggen, Rajan, *et al.* (1997) *Mol. Biochem. Parasitol.* **84**, 25.

α-Aminocaproate (α-Aminocaproic Acid), *See Norleucine*

ε-Amino-*n*-caproate (ε-Amino-*n*-caproic Acid), *See 6-Aminohexanoate (6-Aminohexanoic Acid)*

ε-Aminocaproic Acid Hydrochloride *p*-Carbethoxy-phenyl Ester, *See p-Carbethoxyphenyl ε-Amino-caproate*

***N*-(ε-Aminocaproyl)-α-galactopyranosylamine**, *See N-(6-Aminohexanoyl)-α-galactopyranosylamine*

DL-α-Aminocaprylate (DL-α-Aminocaprylic Acid), *See DL-2-Aminooctanoate (DL-2-Aminooctanoic Acid)*

***N*-(Aminocarbonyl)-L-phenylalanine**, *See N-Carbamoyl-L-phenylalanine*

N-(3-Aminocarbonylphenyl)-4-(2-chlorophenoxy)piperidine-1-carboxamide

This orally bioavailable piperidine derivative (FW = 372.83 g/mol) inhibits stearoyl-CoA 9-desaturase (mouse, IC_{50} = 10 nM), a well-known druggable target in the management of fat metabolism and weight-gain. **1**. Xin, Zhao, Serby, *et al.* (2008) *Bioorg. Med. Chem. Lett.* **18**, 4298.

2(*S*)-Amino-3-[4-(4-carboxybenzyl)oxyphenyl]-1-propanethiol

This amino thiol (FW = 317.41 g/mol) inhibits leukotriene-A_4 hydrolase (IC_{50} = 12 μM). The *m*-carboxy analogue has an IC_{50} value of 6 μM. The free thiol group is necessary for sub-micromolar binding and the enzyme prefers the *R*-enantiomer over the *S*-enantiomer. **1**. Ollmann, Hogg, Muñoz, *et al.* (1995) *Bioorg. Med. Chem.* **3**, 969.

2-(4-Amino-4-carboxybutyl)aziridine-2-carboxylate

This metabolically activated species (FW = 202.21 g/mol), formed by spontaneous hydrolysis of α-(halomethyl)diaminopimelic acids, is a potent irreversible inhibitor of *Escherichia coli* diaminopimelate epimerase (EC 5.1.1.7). Under physiological conditions, 2-(4-amino-4-carboxybutyl) aziridine-2-carboxylate (aziridino-DAP) reacts quickly with and inactivates this non-pyridoxal enzyme. **1**. Gerhart, Higgins, Tardif & Ducep (1990) *J. Med. Chem.* **33**, 2157. **2**. Scapin & Blanchard (1998) *Adv. Enzymol. Relat. Areas Mol. Biol.* **72**, 279.

(4R*,5S*,6S*)-4-Amino-5-[(1-carboxyethenyl)oxy]-6-hydroxycyclohex-1-enecarboxylate

This substrate analogue (FW = 243.22 g/mol) inhibits *Escherichia coli* *p*-aminobenzoate synthase, anthranilate synthase, and isochorismate synthase (K_i = 2.3, 0.38, and 0.027 µM, respectively, when normalized for axial and equatorial conformations). **Target(s):** *p*-aminobenzoate synthase (1); anthranilate synthase (1); isochorismate synthase1; 4-amino-4-deoxychorismate synthase, *K*i = 5.6 µM (2). **1.** Kozlowski, Tom, Seto, Sefler & Bartlett (1995) *J. Amer. Chem. Soc.* **117**, 2128. **2.** He, Stigers Lavoie, Bartlett & Toney (2004) *J. Amer. Chem. Soc.* **126**, 2378.

(R)-S-(2-Amino-2-carboxyethyl)-L-homocysteine, *See* L-Cystathionine

Nᵉ-(DL-2-Amino-2-carboxyethyl)-L-lysine

This nephrotoxic amino acid analogue (FW = 233.27 g/mol), also known as lysinoalanine, is also a metal ion chelator. This property demands the experimenter to carry out measurements at a constant free metal ion concentration to differentiate enzyme inhibition effects from metal ion chelation effects. **Target(s):** lysyl-tRNA synthetase (1); carboxypeptidase A (2,3); carboxypeptidase B (2); alcohol dehydrogenase, yeast (2); cytochrome *c* oxidase (3). **1.** Lifsey, Farkas & Reyniers (1988) *Chem. Biol. Interact.* **68**, 241. **2.** Hayashi (1982) *J. Biol. Chem.* **257**, 13896. **3.** Friedman, Grosjean & Zahnley (1986) *Adv. Exp. Med. Biol.* **199**, 531.

2-Amino-4-{O-[4-(carboxymethyl)phenyl]-O-methylphosphono}butanoic Acid

This γ-phosphono diester analogue (FW = 329.29 g/mol) of glutamate inactivates *Escherichia coli* and human γ-glutamyl transpeptidase, *or* γ-glutamyltransferase, *or* GGT (EC 2.3.2.2). By detoxifying xenobiotics and reactive oxygen species, GGT plays a role in cancer cell drug resistance and metastasis. It is also implicated in Parkinson's Disease, neurodegenerative disease, diabetes, and cardiovascular diseases. **1.** Han, Hiratake, Kamiyama & Sakata (2007) *Biochemistry* **46**, 1432.

(R,S)-2-Amino-4-[2-(carboxymethylsulfanyl)ethylsulfanyl] butyrate

This homocysteine derivative (FW = 221.35 g/mol) inhibits betaine: homocysteine *S*-methyltransferase, *or* BHMT (IC₅₀ = 96 nM). The 2-(carboxymethylsulfoxide)ethyl analogue is a weaker inhibitor. BHMT is very sensitive to the structure of substituents on the sulfur atom of homocysteine. The *S*-carboxybutyl and *S*-carboxypentyl derivatives are the most potent inhibitors, and an additional sulfur atom in the alkyl chain is well tolerated. The respective (*R,S*)-5-(3-amino-3-carboxy-propylsulfanyl)-pentanoic, (*R,S*)-6-(3-amino-3-carboxy-propylsulfanyl)-hexanoic, and (*R,S*)-2-amino-4-(2-carboxymethylsulfanylethylsulfanyl)-butyric acids are also very potent inhibitors and among the strongest ever reported. **1.** Jiracek, Collinsova, Rosenberg, *et al.* (2006) *J. Med. Chem.* **49**, 3982.

3(S)-[[(1S)-5-Amino-1-carboxypentyl]amino]-2,3,4,5-tetrahydro-2-oxo-1H-1-benzazepine-1-acetic Acid

This benzazepineone derivative (FW = 363.41 g/mol) is a potent inhibitor of peptidyl-dipeptidase A, *or* human angiotensin I-converting enzyme (IC₅₀ = 7 nM) and is especially potent, when tested in dogs for inhibition of angiotensin I pressor response, having an ID₅₀ of 0.07 mg/kg orally. **1.** Stanton, Watthey, Desai, *et al.* (1985) *J. Med. Chem.* **28**, 1603.

(3R,4S)-4-(3'(S)-Amino-3'-carboxypropyl-thiomethyl)-1-[(9-deazaadenin-9-yl)-methyl]-3-hydroxypyrrolidine

This immucillin-based *S*-adenosylhomocysteine analogue (FW = 369.47 g/mol) inhibits (K_i' = 6 pM) 5'-methylthioadenosine/*S*-adenosyl-homocysteine nucleosidase, which recycles 5'-methylthio-adenosine from the polyamine pathway via adenine phosphoribosyl-transferase as well as recycling adenine and 5-methylthioriboseto methionine. **1.** Singh, Evans, Lenz, *et al.* (2005) *J. Biol. Chem.* **280**, 18265.

N-{[(2S,3R)-3-Amino-5-carboxyl-2-sulfhydryl]-pentanoyl}-L-isoleucyl-L-aspartate

This peptide analogue (FW = 421.47 g/mol), also known as *N*-{[(2*S*,3*R*)-3-amino-5-carboxy-2-mercapto]pentanoyl}-L-isoleucyl-L-aspartate, inhibits glutamyl aminopeptidase (K_i = 3.56 nM), neprilysin (K_i = 4 nM), and peptidyl-dipeptidase A (K_i = 1800 nM). The (2*R*,3*S*)-, (2*S*,3*S*)-, and (2*R*,3*R*)-diastereoisomers also inhibit glutamyl aminopeptidase (K_i = 16, 13 , 267 nM, respectively). The (2*R*,3*S*)- and (2*S*,3*S*)-isomers also inhibit membrane alanyl aminopeptidase (K_i = 3900 and 120 nM, respectively). **Target(s):** glutamyl aminopeptidase, *or* aminopeptidase A; membrane alanyl aminopeptidase; neprilysin; peptidyl-dipeptidase A (angiotensin I-converting enzyme). **1.** David, Bischoff, Meudal, *et al.* (1999) *J. Med. Chem.* **42**, 5197.

Aminocarnitine

This carnitine isostere and fatty acid oxidation inhibitor (FW = 161.23 g/mol; CAS 98063-21-9), also known as 3-amino-4-(trimethyl-ammonio)-butyric acid, is a strong inhibitor of carnitine *O*-palmitoyltransferase. Aminocarnitine is a hypoglycemic and antiketogenic agent that alters lipid metabolism by inhibiting both carnitine acetyltransferase (CAT) and carnitine palmitoyltransferase (CPT). The (*R*)-stereoisomer, also known as emeriamine, has been isolated from a fungus. **Target(s):** carnitine *O*-

palmitoyltransferase (1,2,4-7); carnitine *O*-octanoyltransferase, moderately inhibited (3). **1**. Jenkins & Griffith (1985) *J. Biol. Chem.* **260**, 14748. **2**. Kanamaru, Shinagawa, Asai, *et al.* (1985) *Life Sci.* **37**, 217. **3**. Chung & Bieber (1993) *J. Biol. Chem.* **268**, 4519. **4**. Hertel, Gellerich, Hein & Zierz (1999) *Adv. Exp. Med. Biol.* **466**, 87. **5**. Traufeller, Gellerich & Zierz (2004) *Biochim. Biophys. Acta* **1608**, 149. **6**. Murthy, Ramsay & Pande (1990) *Biochem. Soc. Trans.* **18**, 604. **7**. Murthy, Ramsay & Pande (1990) *Biochem. J.* **267**, 293.

7-Aminocephalosporanic Acid

This β-lactam (FW = 272.28 g/mol; CAS 957-68-6), obtained by mild hydrolysis of cephalosporin C, is a precursor in the preparation of synthetic cephalosporins. **Target(s):** D-stereospecific aminopeptidase (1,8); γ-aminobutyrate aminotransferase (2); serine-type D-Ala-D-Ala carboxypeptidase (3); zinc D-Ala-D-Ala carboxypeptidase (3); glutathione S-transferase π (4); β-lactamase, also weak alternative substrate (5); glutaryl-7-aminocephalosporanic acid acylase, product inhibition (6,7). **1**. Asano (1998) in *Handb. Proteolytic Enzymes* (Barrett, Rawlings & Woessner, eds.), p. 430, Academic Press, San Diego. **2**. Hopkins & Silverman (1992) *J. Enzyme Inhib.* **6**, 125. **3**. Kelly, Frère, Klein & Ghuysen (1981) *Biochem. J.* **199**, 129. **4**. Adams & Sikakana (1992) *Biochem. Pharmacol.* **43**, 1757. **5**. Dryjanski & Pratt (1995) *Biochemistry* **34**, 3561. **6**. Aramori, Fukagawa, Tsumura, *et al.* (1992) *J. Ferment. Bioeng.* **73**, 185. **7**. Lee, Chang, Liu & Chu (1998) *Biotechnol. Appl. Biochem.* **28** (Pt. 2), 113. **8**. Asano, Kato, Yamada & Kondo (1992) *Biochemistry* **31**, 2316.

(2*R*,4*R*)-2-Amino-4-[(3-chloro)-2-benzo[*b*]thienyl]methyl-pentanedioate

This D-glutamate derivative (FW = 292.34 g/mol), also known as 4*R*-[(3-chloro)-2-benzo[*b*]thienyl]methyl-D-glutamic acid, potently inhibits glutamate racemase (K_i = 16 nM). It shows potent whole cell antibacterial activity against *S. pneumoniae* PN-R6, with good correlation between minimal inhibitory concentration and inhibition of enzyme assay. **1**. de Dios, Prieto, Martín, *et al.* (2002) *J. Med. Chem.* **45**, 4559.

(*R*)-7-Amino-2-(4-(4-chlorobiphenyl)ylsulfonyl)-1,2,3,4 tetrahydroisoquinolin-3-carboxylic Acid

This isoquinoline derivative (FW = 442.92 g/mol) inhibits human neutrophil collagenase, *or* matrix metalloproteinase 8, IC$_{50}$ = 7 nM. Docking studies of a reference compound are based on crystal structures of MMP-8 complexed with peptidic inhibitors to propose a model of its bioactive conformation, and this model was validated by a 1. 7-Å crystal structure of MMP-8's catalytic domain. The weaker zinc ion-binding properties of carboxylates was compensated by introducing an optimally fitting P1' residue. **1**. Matter, Schwab, Barbier, *et al.* (1999) *J. Med. Chem.* **42**, 1908.

2-Amino-3-chlorobutyrate

This valine analogue (FW = 137.57 g/mol; CAS 14561-56-9), also known as *threo*-α-amino-β-chlorobutyric acid and 2-amino-3-chlorobutanoic acid, is an alternative substrate and competitive inhibitor of valyl-tRNA synthetase (1). β-Chlorovaline is also incorporated into the hemoglobin α- and β-chains in rabbit reticulocytes, reducing the protein's half-life to ten minutes, instead of its normal value of 120 days. **Target(s):** valyl-tRNA synthetase, inhibited by DL- and D-*allo*-isomers (1); aspartate 4-decarboxylase, inhibited by the L-*threo*-isomer (3). **1**. Bergmann, Berg & Dieckmann (1961) *J. Biol. Chem.* **236**, 1735. **2**. Rabinovitz & Fisher (1964) *Biochim. Biophys. Acta* **91**, 313. **3**. Tate, Relyea & Meister (1969) *Biochemistry* **8**, 5016.

7-Amino-4-chloro-3-ethoxyisocoumarin

This substitiuted isocoumarin (FW = 239.66 g/mol) is a strong inactivator of human leukocyte elastase. The half-life of this isocoumarin is about 216 minutes at pH 7.5 and 25°C. **Target(s):** leukocyte elastase (1-3); pancreatic elastase (1); cathepsin G (1); chymotrypsin (1); chymase (1); tryptase (1); streptogrisin A (1); thrombin (1); coagulation factor XIIa (1); plasma kallikrein (1); trypsin (1); myeloblastin (3). **1**. Harper & Powers (1985) *Biochemistry* **24**, 7200. **2**. Kerrigan, Oleksyszyn, Kam, Selzler & Powers (1995) *J. Med. Chem.* **38**, 544. **3**. Kam, Kerrigan, Dolman, *et al.* (1992) *FEBS Lett.* **297**, 119.

4-Amino-*N*-(3-chloro-4-fluorophenyl)-*N*'-hydroxy-1,2,5-oxadiazole-3-carboximidamide

This carboximidamide (FW = 271.64 g/mol) is a competitive inhibitor of human indoleamine 2,3-dioxygenase (IC$_{50}$ = 67 nM). In mice, this agent demonstrated pharmacodynamic inhibition of the dioxygenase, as measured by decreased plasma kynurenine levels and dose-dependent efficacy against GM-CSF-secreting B16 melanoma tumors. **1**. Yue, Douty, Wayland, *et al.* (2009) *J. Med. Chem.* **52**, 7364.

7-Amino-4-chloro-3-methoxyisocoumarin

This yellow isocoumarin (FW = 225.63 g/mol; Melting Point = 163-165°C), also known as γ secretase inhibitor XI and JLKG, inhibits a number of proteases as well as γ-secretase and blocks the production of both amyloid-β40 and -β42 (IC$_{50}$ < 100 μM). The half-life for this isocoumarin is about 200 minutes at 25°C in 100 μM HEPES, 500 μM NaCl, 10% dimethyl sulfoxide, and pH 7.5. **Target(s):** leukocyte elastase (1-3); pancreatic elastase (2); cathepsin G (1,2); chymotrypsin (1,2); chymase (2); tryptase (2); streptogrisin A (2); γ-secretase (4); granzyme B (5); complement factor D (6). **1**. Powers & Kam (1994) *Meth. Enzymol.* **244**, 442. **2**. Harper & Powers (1985) *Biochemistry* **24**, 7200. **3**. Kerrigan, Oleksyszyn, Kam, Selzler & Powers (1995) *J. Med. Chem.* **38**, 544. **4**. Petit, Bihel, Alves da Costa, *et al.* (2001) *Nat. Cell Biol.* **3**, 507. **5**. Odake, Kam, Narasimhan, *et al.* (1991) *Biochemistry* **30**, 2217. **6**. Kam, Oglesby, Pangburn, Volanakis & Powers (1992) *J. Immunol.* **149**, 163.

1-(4-{5-[6-Amino-5-(4-chlorophenyl)pyrimidin-4-ylethynyl]pyridin-2-yl}piperazin-1-yl)ethanone

This pyrimidine (FW = 432.92 g/mol) inhibits (IC_{50} = 1 nM) cytosolic adenosine kinase, a key intracellular enzyme regulating intracellular and extracellular concentrations of adenosine (ADO), itself a powerful endogenous modulator of intercellular signalling that reduces cell excitability during tissue stress and trauma. Adenosine levels increase at seizure foci as part of a negative feedback mechanism that controls seizure activity through adenosine receptor signaling. Agents that inhibit adenosine kinases increase/sustain site-/event-specific adenosine surges, thereby providing anti-seizure activity akin to that of adenosine receptor agonists, but with fewer dose-limiting cardiac side effects. **1**. Matulenko, Paight, Frey, *et al.* (2007) *Bioorg. Med. Chem.* **15**, 1586.

(R)-α-Amino-5-chloro-1-(phosphonomethyl)-1H-benzimidazole-2-propanoic Acid

This structurally complex amino acid (FW = 333.67 g/mol), also known as (R)-α-amino-5-chloro-1-(phosphonomethyl)-1H-benzimidazole-2-propanoic acid hydrochloride and EAB-318, is an antagonist for NMDA (N-methyl-D-aspartate) and AMPA (α-amino-3-hydroxy-5-methyl-4-isoxazolepropionic acid) receptors. EAB-318 also protects neurons against ischemia induced cell death. **1**. Sun, Chiu, Kowal, *et al.* (2004) *J. Pharmacol. Exp. Ther.* **310**, 563.

2-Amino-5-chlorotetralin

This conformationally rigid phenylethylamine substrate analogue (FW = 181.67 g/mol) competitively inhibits (K_i = 0.2 μM) phenylethanolamine N-methyltransferase, *or* norepinephrine N-methyltransferase (IC_{50} = 5.2 μM). (Its 6-, 7-, and 8-chlorinated analogues are slightly weaker inhibitors.) 2-Amino-5-chlorotetralin inhibits the reuptake of serotonin and norepinephrine, and likely induces their release. **1**. Fuller & Molloy (1977) *Biochem. Pharmacol.* **26**, 446.

22-Aminocholesterol

This cholesterol derivative ($FW_{free-base}$ = 401.67 g/mol; CAS 50921-65-8) inhibits cholesterol monooxygenase (side-chain-cleaving), *or* CYP11A1,

where the (22R)-stereoisomer (IC_{50} = 25 nM) is the stronger inhibitor than the (22S)-isomer (IC_{50} = 13 μM). **1**. Vickery (1991) *Meth. Enzymol.* **206**, 548. **2**. Nagahisa, Foo, Gut & Orme-Johnson (1985) *J. Biol. Chem.* **260**, 846. **3**. Jacobs, Singh & Vickery (1987) *Biochemistry* **26**, 4541.

1-Aminocyclobutane-1-carboxylate

This NMDA receptor partial agonist (FW = 115.13 g/mol; CAS 22264-50-2; Symbol: ACBC; Soluble to 100 mM in water) is a specific antagonist of the N-methyl-D-aspartate receptor-coupled glycine receptor (1,2). Unfortunately, ACBC is rapidly eliminated ($t_{1/2}$ = 5 min in mouse brain; $t_{1/2}$ = 5 4 min in rat cerebrospinal fluid), rendering it of questionable utility *in vivo*. **1**. Hood, Sun, Compton & Monahan (1989) *Eur. J. Pharmacol.* **161**, 281. **2**. Rao, Cler, Compton, *et al.* (1990) *Neuropharmacol.* **29**, 305.

cis-1-Aminocyclobutane-1,3-dicarboxylate

This potent, competitive and selective transport inhibitor (FW = 159.14 g/mol; CAS 73550-55-7; Symbol: cis-ACBD; Soluble to 100 mM with addition of 1 equivalent of NaOH) targets synaptosomal glutamate transporters (K_i = 8 μM), reducing uptake of glutamic acid. **1**. Allan, *et al.* (1990) *J. Med. Chem.* **33**, 2905. **2**. Fletcher, *et al.* (1991) *Neurosci. Lett.* **121**, 133. **3**. Koch, *et al.* (1999) *Mol. Pharmacol.* **56**, 1095.

cis-3-Aminocyclohex-4-ene-1-carboxylate

This conformationally constrained amino-acid (FW = 140.16 g/mol) is a mechanism-based enzyme inhibitor that is an analogue of the antiepilepsy drug vigabatrin, inactivating porcine 4-aminobutyrate aminotransferase, a key enzyme in γ-aminobutyrate metabolism. An energy-minimized computer model of vigabatrin bound to PLP indicated that the major Michael addition pathway could only occur if the vinyl group could rotate by 180°. Further study revealed that the agent inactivates by an enamine pathway. (*See* Vigabatrin) **1**. Choi, Storici, Schirmer & Silverman (2002) *J. Amer. Chem. Soc.* **124**, 1620.

(2S)-1-[(2S)-2-Amino-2-cyclopentylacetyl]-2-cyanopyrrolidine

This pyrrolidone (FW = 220.30 g/mol), also known as (2S)-1-[(αS)-α-cyclopentylglycyl]-2-cyanopyrrolidine, inhibits human dipeptidyl-peptidase IV (K_i = 1 nM), a target for diabetes therapy due to its indirect regulatory role in modulating plasma glucose concentration. Dipeptidyl peptidase IV (DPP-IV) belongs to a family of serine peptidases. Its indirect regulatory role in plasma glucose modulation makes it an attractive druggable target for diabetes therapy. DPP-IV inactivates the glucagon-like peptide (GLP-1) and several other naturally produced bioactive peptides that contain preferentially a proline or alanine residue in the second amino acid sequence position by cleaving the N-terminal dipeptide. **1**. Longenecker, Stewart, Madar, *et al.* (2006) *Biochemistry* **45**, 7474.

1-Amino-1-cyclopropanecarboxylate

This unusual conformationally confined amino acid (FW = 101.11 g/mol; CAS 68781-13-5) is an intermediate in the production of ethene from L-methionine.It is a product of the reaction catalyzed by 1-aminocyclopropane-1-carboxylate synthase and is a substrate for aminocyclopropanecarboxylate oxidase and 1 aminocyclopropane-1-carboxylate deaminase. 1-Amino-1-cyclopropanecarboxylate is also an NMDA (N-methyl-D-aspartate) receptor agonist, binding at the glycine co-agonist site. **Target(s):** 2,2-dialkylglycine decarboxylase (pyruvate) (1-3). 1. Malashkevich, Strop, Keller, Jansonius & Toney (1999) *J. Mol. Biol.* **294**, 193. 2. Zhou, Kay & Toney (1998) *Biochemistry* **37**, 5761. 3. Sun, Bagdassarian & Toney (1998) *Biochemistry* **37**, 3876.

1-Aminocyclopropanol

This cycloalkyl-hemiaminal ($FW_{free-base}$ = 73.09 g/mol) inhibits aldehyde dehydrogenases. The active agent is cyclopropanone hydrate, formed by hydrolysis of 1-aminocyclopropanol ($t_{1/2}$ = 30 min at 27°C and pH 7.4). 1. Tottmar & Lindberg (1977) *Acta Pharmacol. Toxicol. (Copenhagen)* **40**, 476. 2. Marchner & Tottmar (1978) *Acta Pharmacol. Toxicol. (Copenhagen)* **43**, 219. 3. Wiseman & Abeles (1979) *Biochemistry* **18**, 427.

1-Amino-2-cyclopropene-1-carboxylate

This substrate analogue (FW = 84.10 g/mol) inactivates 1-aminocyclopriopanecarboxylate (ACC) oxidase (*Reaction*: ACC + Ascorbate + O_2 ⇌ Ethylene + Cyanide + Dehydroascorbate + CO_2 + 2 H_2O). Ring-opening of enzyme-bound ACC is believed to result in the elimination of ethylene together with an unstable intermediate, cyanoformate ion, which then decomposes to cyanide ion and carbon dioxide. 1-Amino-2-cyclopropene-1-carboxylate most likely intercepts an active-site nucleophile. 1. Pirrung (1999) *Acc. Chem. Res.* **32**, 711.

3-Amino-1-(cyclopropylamino)heptan-2-one

This transition-state analogue ($FW_{free-base}$ = 184.28 g/mol) inhibits *Staphylococcus aureus* methionyl aminopeptidase (IC_{50} = 7 µM). by mimicking the tetrahedral intermediate for amide bond hydrolysis. 1. Douangamath, Dale, D'Arcy, *et al.* (2004) *J. Med. Chem.* **47**, 1325.

1-Amino-2-cyclopropylcyclopropane-1-carboxylate

This substrate analogue (FW = 141.17 g/mol) is a mechanism-based inactivator of aminocyclopropanecarboxylate oxidase (Reaction: 1-Aminocyclopropane-1-carboxylate + Ascorbate + O_2 ⇌ Ethylene + Cyanide + Dehydroascorbate + CO_2 + 2 H_2O). 1. Pirrung (1999) *Acc. Chem. Res.* **32**, 711.

2-Amino-2-deoxy-D-glucitol 6-Phosphate

$$CH_2OH$$
$$HC-NH_2$$
$$HO-CH$$
$$HC-OH$$
$$HC-OH$$
$$CH_2OPO_3H^-$$

This reduced aminosugar-phosphate (FW = 261.17 g/mol) inhibits glutamine:fructose-6-phosphate aminotransferase (isomerizing), with a K_i value of 0.2 µM for the *Escherichia coli* enzyme. **Target(s):** glutamine:fructose-6-phosphate aminotransferase (isomerizing), *or* glucosamine 5-phosphate deaminase, *or* glucosamine-6-phosphate isomerase (1-11); and glucose-6-phosphate isomerase (7). 1. Montero-Morán, Lara-González, Alvarez-Añorve, Plumbridge & Calcagno (2001) *Biochemistry* **40**, 10187. 2. Lara-Lemus & Calcagno (1998) *Biochim. Biophys. Acta* **1388**, 1. 3. Midelfort & Rose (1977) *Biochemistry* **16**, 1590. 4. Bustos-Jaimes, Sosa-Peinado, Rudiño-Piñera, Horjales & Calcagno (2002) *J. Mol. Biol.* **319**, 183. 5. Steimle, Lindmark & Jarroll (1997) *Mol. Biochem. Parasitol.* **84**, 149. 6. Olchowy, Jedrzejczak, Milewski & Rypniewski (2005) *Acta Crystallogr. Sect. F* **61**, 994. 7. Milewski, Janiak & Wojciechowski (2006) *Arch. Biochem. Biophys.* **450**, 39. 8. Floquet, Richez, Durand, *et al.* (2007) *Bioorg. Med. Chem. Lett.* **17**, 1966. 9. Wojciechowski, Milewski, Mazerski & Borowski (2005) *Acta Biochim. Pol.* **52**, 647. 10. Badet-Denisot, Leriche, Massiere & Badet (1995) *Bioorg. Med. Chem. Lett.* **5**, 815. 11. Bearne (1996) *J. Biol. Chem.* **271**, 3052.

(3S)-3-Amino-1-diaminophosphinyl-2-pyrrolidinone

This sulphostin analogue ($FW_{free-base}$ = 166.14 g/mol) inhibits dipeptidyl-peptidase IV (IC_{50} = 11 nM). All the functional groups on the piperidine ring appear to be crucial for DPP-IV inhibition. 1. Abe, Akiyama, Umezawa, *et al.* (2005) *Bioorg. Med. Chem.* **13**, 785.

4-Amino-4,5-dideoxy-L-arabinose Imine

This L-rhamnose analogue (FW = 115.13 g/mol) noncompetitively inhibits α-L-rhamnosidase (K_{is} = 0.14 µM, K_{ii} = 1.2 µM). Substitution on the nitrogen atom shifted the inhibition mechanism from mixed to competitive. 1. Provencher, Steensma & Wong (1994) *Bioorg. Med. Chem.* **2**, 1179.

2-Amino-4,5-dihydro-8-phenylimidazole[2,1-b]thiazolo[5,4-g]benzothiazole

This benzothiazole ($FW_{free-base}$ = 324.43 g/mol), also known as YJA20379-1, inhibits gastric H^+/K^+ ATPase, with IC_{50} values of 21 and 24 µM at pH 6.4 and 7.4, respectively. Oral administration of YJA20379-1 also prevented the formation of ethanol-, indomethacin-, and water immersion stress-induced gastric lesions as well as mepirizole-induced duodenal ulcers in rats. YJA20379-1 accelerated the healing of acetic acid-induced chronic gastric ulcers in rats. Such results suggest YJA20379-1 is a potent inhibitor of gastric H⁺-K⁺ ATPase, with much shorter duration of antisecretory action than omeprazole. **1**. Sohn, Chang, Choi, *et al.* (1999) *Can. J. Physiol. Pharmacol.* **77**, 330.

(2S,3R,12RS)-2-Amino-3,12-dihydroxy-2-hydroxymethylstearate

This sphingosine analogue (FW = 365.51 g/mol) inhibits serine *C*-palmitoyltransferase reaction (IC$_{50}$ = 3.2 nM), the first step of the biosynthesis of sphingolipids that modulate various cellular proliferation, differentiation, and apoptosis. **1**. Yamaji-Hasegawa, Takahashi, Tetsuka, Senoh & Kobayashi (2005) *Biochemistry* **44**, 268.

1-[4-Amino-2,6-di(isopropyl)phenyl]-3-[1-butyl-4-(3-methoxyphenyl)-2(1H)-oxo-1,8-naphthyridin-3-yl]urea

This orally available naphthyridinylurea (FW = 541.70 g/mol), also known as SMP-797, inhibits rat cholesterol *O*-acyltransferase, *or* ACAT (IC$_{50}$ = 8.3 nM), with inducible effect on the expression of low-density lipoprotein receptor. SMP-797 also increased low-density lipoprotein receptor expression in HepG2 cells like atorvastatin, an HMG-CoA reductase inhibitor, although other acyl-coenzyme A: cholesterol acyltransferase inhibitor had no effect, suggesting that the increase of low-density lipoprotein receptor expression by SMP-797 was independent of its acyl-coenzyme A: cholesterol acyltransferase inhibitory action and did not result from the inhibition of hepatic cholesterol synthesis. The 3-amino analogue also inhibits (IC$_{50}$ = 14 nM). **1**. Ban, Muraoka, Ioriya & Ohashi (2006) *Bioorg. Med. Chem. Lett.* **16**, 44.

(S)-2-Amino-3-(4-{5-[3,4-dimethoxybenzylamino]pyrazin-2-yl}phenyl)propanoic Acid

This amino acid (FW = 406.44 g/mol) inhibits (IC$_{50}$ = 69 nM) tryptophan 5-monooxygenase, thereby providing the promise of novel treatments for gastrointestinal disorders associated with dysregulation of the serotonergic system, such as chemotherapy-induced emesis and irritable bowel syndrome. **1**. Shi, Devasagagayaraj, Gu, *et al.* (2008) *J. Med. Chem.* **51**, 3684.

2-Amino-5-[(2,5-dimethoxyphenyl)sulfanyl]-6-methylthieno[2,3-d]pyrimidin-4(3H)-one

This thienopyrimidine (FW = 337.42 g/mol) is a dual inhibitor of dihydrofolate reductase (IC$_{50}$ = 22 µM) and human thymidylate synthase (IC$_{50}$ = 4.6 µM). The IC$_{50}$ value for *Toxoplasma gondii* dihydrofolate reductase is 56 nM. **1**. Gangjee, Qiu, Li & Kisliuk (2008) *J. Med. Chem.* **51**, 5789.

4-(2-Aminoethyl)benzenesulfonyl Fluoride

This yellow sulfonyl fluoride (FW$_{hydrochloride}$ = 239.70 g/mol; CAS 30827-99-7; Solubility: 100 mM in H$_2$O; 100 mM in DMSO), also named 4-(2-aminoethyl)benzenesulfonyl fluoride hydrochloride, often abbreviated AEBSF and Pefabloc SC hydrochloride, is a potent and irreversible inhibitor of many serine proteases and esterases. It is more soluble in water than phenylmethanesulfonyl fluoride (PMSF) and less toxic (aqueous solutions are stable for one-to-two months at 4°C). AEBSF slowly hydrolyzes above pH 8. **Target(s):** α-lytic protease (1); kedarcidin chromoprotein (2); chymotrypsin (17); kallikrein (15,25); plasmin (16); trypsin (3,20); platelet-activating-factor:sphingosine acetyltransferase (4); proteasome (5); thrombin (6,7); chymase (8); t-plasminogen activator (9,38); *Pseudoalteromonas* protease CP1 (10); *Sarcocystis neurona* merozoite serine protease (11); platelet activating factor-degrading acetylhydrolase (12); phospholipase A$_2$ (12,18); cinnamoyl esterase (13); NADPH oxidase (14); Z-Pro prolinal-insensitive Z-Gly-Pro-, *or* 7-amino-4-methylcoumarin-hydrolyzing peptidase (19); cinnamoyl ester hydrolase (21); dipeptidyl-peptidase II (22,47,48); dipeptidyl-peptidase IV (23,46); complement component C1a (24); aromatic-L-amino-acid decarboxylase, filarial nematode (*Dirofilaria immitis*) (26); *N*-acyl-D-amino-acid deacylase (27); amidase (28); envelysin (29); site-1 protease (30); mannan-binding lectin-associated serine protease-2 (31); hepacivirin, *or* hepatitis C virus NS3 serine proteinase (32); oligopeptidase B (33,34); myeloblastin (35-37); complement factor I (39); cucumisin (40); pyroglutamyl-peptidase II (41); cyanophycinase (42); Xaa-Xaa-Pro tripeptidyl peptidase, *or* prolyltripeptidyl aminopeptidase (43-45); aminopeptidase B (49,50); membrane alanyl aminopeptidase, *or* aminopeptidase N (51); calciciviron (52); feruloyl esterase (53); 1-alkyl-2 acetylglycerophosphocholine esterase, *or* platelet-activating-factor acetylhydrolase (54,55); vinorine synthase (56); platelet-activating factor acetyltransferase (57); phosphatidylcholine:sterol *O*-acyltransferase, *or* LCAT (58). **1**. Rader & Agard (1998) in *Handb. Proteolytic Enzymes* (Barrett, Rawlings & Woessner, eds.), p. 251, Academic Press, San Diego. **2**. Zein (1998) in *Handb. Proteolytic Enzymes* (Barrett, Rawlings & Woessner, eds.), p. 1604, Academic Press, San Diego. **3**. Basir, van der Burg, Scheringa, Tons & Bouwman (1997) *Transplant Proc.* **29**, 1939. **4**. Lee (2000) *Meth. Enzymol.* **311**, 117. **5**. Rivett, Savory & Djaballah (1994) *Meth. Enzymol.* **244**, 331. **6**. Markwardt, Hoffmann & Körbs (1973) *Thromb. Res.* **2**, 343. **7**. Lawson, Valenty, Wos & Lobo (1982) *Folia Haematol. Int. Mag. Klin. Morphol. Blutforsch.* **109**, 52. **8**. Su, Bochan, Hanna, Froelich & Brahmi (1994) *Eur. J. Immunol.* **24**, 2073. **9**. Herbert, Lamarche, Prabonnaud, Dol & Gauthier (1994) *J. Biol. Chem.* **269**, 3076. **10**. Sanchez-Porro, Mellado, Bertoldo, Antranikian & Ventosa (2003) *Extremophiles* **7**, 221. **11**. Barr & Warner (2003) *J. Parasitol.* **89**, 385. **12**. Dentan, Tselepis, M. J. Chapman & E. Ninio (1996) *Biochim. Biophys. Acta* **1299**, 353. **13**. Kroon, Faulds & Williamson (1996) *Biotechnol. Appl. Biochem.* **23**, 255. **14**. Diatchuk, Lotan, Koshkin, Wikstroem & Pick (1997) *J. Biol. Chem.* **272**, 13292. **15**. Wirth, Fink, Rudolphi, Heitsch, Deutschlander & Wiemer (1999) *Eur. J. Pharmacol.* **382**, 27. **16**. Chu & Kawinski (1998) *Biochem. Biophys. Res. Commun.* **253**, 128. **17**. Rose, Palcic, Helms & Lakey (2003) *Transplantation* **75**, 462. **18**. Carpenter, Dennis, Challis, *et al.* (2001) *FEBS Lett.* **505**, 357. **19**. Birney & O'Connor (2001) *Protein Expr. Purif.* **22**, 286. **20**. Lam, Coast & Rayne (2000) *Insect Biochem. Mol. Biol.* **30**, 85. **21**. Fillingham, Kroon, Williamson, Gilbert & Hazlewood (1999) *Biochem. J.* **343**, 215. **22**. Huang, Takagaki, Kani & Ohkubo (1996) *Biochim. Biophys. Acta* **1290**, 149. **23**. Ohkubo, Huang, Ochiai, Takagaki & Kani (1994) *J. Biochem.* **116**, 1182. **24**. Baker & M. Cory (1971) *J. Med. Chem.* **14**, 119. **25**. Markwardt, Drawert & Walsmann (1974) *Biochem. Pharmacol.* **23**, 2247. **26**. Tang & Frank (2001) *Biol. Chem.* **382**, 115. **27**. Wakayama, Yada, Kanda, *et al.* (2000) *Biosci. Biotechnol. Biochem.* **64**, 1. **28**. Hirrlinger, Stolz & Knackmuss (1996) *J. Bacteriol.* **178**, 3501. **29**. D'Aniello, Denuce, de Vincentiis, di Fiore & Scippa (1997) *Biochim. Biophys. Acta* **1339**, 101. **30**. Cheng, Espenshade, Slaughter, *et al.* (1999) *J. Biol. Chem.* **274**, 22805. **31**. Wong, Kojima, Dobo, Ambrus & Sim (1999) *Mol. Immunol.* **36** 853. **32**. Mori, Yamada, Kimura, *et al.* (1996) *FEBS Lett.* **378**, 37. **33**. Morty, Fülöp & Andrews (2002) *J. Bacteriol.* **184**, 3329. **34**. Caldas, Cherqui, Pereira & Simoes (2002) *Appl. Environ. Microbiol.* **68**, 1297. **35**. Hirche, Crouch, Espinola, *et al.* (2004) *J. Biol. Chem.* **279**, 27688. **36**. Witko-Sarsat, Canteloup, Durant, *et al.* (2002) *J. Biol. Chem.* **277**, 47338. **37**. Dublet, Ruello, Pederzoli, *et al.*

(2005) *J. Biol. Chem.* **280**, 30242. **38**. Fredriksson, Li, Fieber, Li & Eriksson (2004) *EMBO J.* **23**, 3793. **39**. Tsiftsoglou & Sim (2004) *J. Immunol.* **173**, 367. **40**. Kaneda, Yonezawa & Uchikoba (1997) *Biosci. Biotechnol. Biochem.* **61**, 2100. **41**. Gallagher & O'Connor (1998) *Int. J. Biochem. Cell Biol.* **30**, 115. **42**. Richter, Hejazi, Kraft, Ziegler & Lockau (1999) *Eur. J. Biochem.* **263**, 163. **43**. Umezawa, Yokoyama, Kikuchi, *et al.* (2004) *J. Biochem.* **136**, 293. **44**. Banbula, Mak, Bugno, *et al.* (1999) *J. Biol. Chem.* **274**, 9246. **45**. Fujimura, Ueda, Shibata & Hirai (2003) *FEMS Microbiol. Lett.* **219**, 305. **46**. Davy, Thomsen, Juliano, *et al.* (2000) *Plant Physiol.* **122**, 425. **47**. Araki, Li, Yamamoto, *et al.* (2001) *J. Biochem.* **129**, 279. **48**. Huang, Takagaki, Kani & Ohkubo (1996) *Biochim. Biophys. Acta* **1290**, 149. **49**. Mercado-Flores, Noriega-Reyes, Ramirez-Zavala, Hernandez-Rodriguez & Villa-Tanaca (2004) *FEMS Microbiol. Lett.* **234**, 247. **50**. Goldstein, Nelson, Kordula, Mayo & Travis (2002) *Infect. Immun.* **70**, 836. **51**. Contreras-Rodriguez, Ramirez-Zavala, Contreras, Schurig, Sriranganathan & Lopez-Merino (2003) *Infect. Immun.* **71**, 5238. **52**. Zeitler, Estes & Prasad (2006) *J. Virol.* **80**, 5050. **53**. Kroon, Faulds & Williamson (1996) *Biotechnol. Appl. Biochem.* **23**, 255. **54**. Tsoukatos, Liapikos, Tselepis, Chapman & Ninio (2001) *Biochem. J.* **357**, 457. **55**. Tselepis, Dentan, Karabina, Chapman & Ninio (1995) *Arterioscler. Thromb. Vasc. Biol.* **15**, 1764. **56**. Bayer, Ma & Stöckigt (2004) *Bioorg. Med. Chem.* **12**, 2787. **57**. Tsoukatos, Liapikos, Tselepis, Chapman & Ninio (2001) *Biochem. J.* **357**, 457. **58**. Nakamura, Kotite, Gan, *et al.* (2004) *Biochemistry* **43**, 14811.

[1(*RS*)-Aminoethyl][2(*RS*)-benzyl-2-carboxy-1-ethyl]phosphinate

This pentavalent phosphinate-containing transition-state analogue ($FW_{hydrochloride}$ = 307.71 g/mol) inhibits D-alanyl-D-alanine synthetase, *or* D-alanine:D-alanine ligase. A mechanism for the inhibition of D-Ala-D-Ala ligase by these compounds is proposed to involve an ATP-dependent formation of phosphorylated inhibitor within the enzyme's active site. The antibacterial activities of these compounds are modest although their spectra include both Gram-positive and Gram-negative susceptible organisms. **1**. Parsons, Patchett, Bull, *et al.* (1988) *J. Med. Chem.* **31**, 1772.

(*S*)-2-Amino-3-(4-{5-[(9-ethyl-9*H*-carbazol-3-yl)methylamino]pyrazin-2-yl}phenyl)propanoic Acid

This amino acid (FW = 465.56 g/mol) inhibits tryptophan hydroxylase (TPH), *or* tryptophan 5-monooxygenase (IC_{50} = 31 nM). This class of TPH inhibitors exhibits excellent potency in *in vitro* biochemical and cell-based assays, and it selectively reduces serotonin levels in the murine intestine after oral administration without affecting levels in the brain. Such inhibitors may provide novel treatments for gastrointestinal disorders associated with dysregulation of the serotonergic system, such as chemotherapy-induced emesis and irritable bowel syndrome. **1**. Shi, Devasagayaraj, Gu, *et al.* (2008) *J. Med. Chem.* **51**, 3684.

[1(*S*)-Aminoethyl][2-carboxy-2(*RS*)-heptyl-1-ethyl]phosphinate

This pentavalent phosphinate-containing transition-state analogue ($FW_{hydrochloride}$ = 315.76 g/mol) is a strong inhibitor of D-alanyl-D-alanine synthetase, *or* D-alanine:D-alanine ligase. **1**. McDermott, Creuzet, Griffin, *et al.* (1990) *Biochemistry* **29**, 5767. **2**. Duncan & Walsh (1988) *Biochemistry* **27**, 3709. **3**. Parsons, Patchett, Bull, *et al.* (1988) *J. Med. Chem.* **31**, 1772.

[1(*RS*)-Aminoethyl][2(*RS*)-carboxy-1-hexyl]phosphinate

This pentavalent phosphinate-containing transition-state analogue ($FW_{hydrochloride}$ = 273.70 g/mol) inhibits D-alanyl-D-alanine synthetase, *or* D-alanine:D-alanine ligase. A mechanism for the inhibition of D-Ala-D-Ala ligase by these compounds is proposed to involve an ATP-dependent formation of phosphorylated inhibitor within the enzyme's active site. The antibacterial activities of these compounds are modest although their spectra include both Gram-positive and Gram-negative susceptible organisms. **1**. Parsons, Patchett, Bull, *et al.* (1988) *J. Med. Chem.* **31**, 1772.

[1(*S*)-Aminoethyl][2-carboxy-2(*R*)-methyl-1-ethyl]phosphinate

This aminoalkyl dipeptide analogue and pentavalent phosphinate-containing transition-state analogue ($FW_{hydrochloride}$ = 231.62 g/mol) is an ATP-dependent, slow-binding inhibitor of the D-Ala:D-Ala ligase (*i.e.*, D-alanyl-D-alanine synthetase) from *Salmonella typhimurium*. After ATP-dependent phosphorylation, dissociation of the enzyme-bound inhibitor has a half-life of 17 days at 37°C. The mechanism of inhibition is analogous to that of glutamine synthetase by methionine sulfoximine and phosphinothricin. McDermott *et al.* (1) used rotational resonance, then a newly developed solid-state NMR method, for structural studies of an inhibited complex formed by reaction of the enzyme with ATP and the aminoalkyl dipeptide analogue. The measured NMR coupling properties indicate that the two species are bridged in a P–O–P linkage, with a P–P through-space distance of 2.7 ± 0.2 Å, unambiguously demonstrating that the inactivation mechanism involves phosphorylation of enzyme-bound inhibitor by ATP to form a phosphoryl-phosphinate adduct. **1**. McDermott, Creuzet, Griffin, *et al.* (1990) *Biochemistry* **29**, 5767. **2**. Parsons, Patchett, Bull, *et al.* (1988) *J. Med. Chem.* **31**, 1772. **3**. Fan, Park, Walsh & Knox (1997) *Biochemistry* **36**, 2531. **4**. Shi & Walsh (1995) *Biochemistry* **34**, 2768.

N-(2-Aminoethyl)-5-chloroisoquinoline-8-sulfonamide

This arylsulfonamide protein kinase inhibitor ($FW_{free-base}$ = 285.75 g/mol), also known as CKI-7, targets casein kinase 1 (K_i = 8.5 μM) and casein kinase 2 (K_i = 70 μM). **Target(s):** casein kinase 1 (1-9); casein kinase 2 (1-3); non-specific serine/threonine protein kinase (1-9). **1**. Hidaka, Watanabe & Kobayashi (1991) *Meth. Enzymol.* **201**, 328. **2**. Eriksson, Toivola, Sahlgren, Mikhailov & Härmälä-Braskén (1998) *Meth. Enzymol.* **298**, 542. **3**. Chijiwa, Hagiwara & Hidaka (1989) *J. Biol. Chem.* **264**, 4924.

4. Xu, Carmel, Kuret & Cheng (1996) *Proc. Natl. Acad. Sci. U.S.A.* **93**, 6308. **5**. Fish, Cegielska, Getman, Landes & Virshup (1995) *J. Biol. Chem.* **270**, 14875. **6**. Barik, Taylor & Chakrabarti (1997) *J. Biol. Chem.* **272**, 26132. **7**. Graves, Haas, Hagedorn, DePaoli-Roach & Roach (1993) *J. Biol. Chem.* **268**, 6394. **8**. Price (2006) *Genes Dev.* **20**, 399. **9**. Zhai, Graves, Robinson, *et al.* (1995) *J. Biol. Chem.* **270**, 12717.

S-(β-Aminoethyl)-L-cysteine

This L-lysine analogue (FW$_{dihydrochloride}$ = 237.14 g/mol), also known as L-4-thialysine, inhibits a number of lysine-dependent enzymes. This thioether is also formed upon acid hydrolysis of thiol-containing proteins previously treated with ethyleneimine to introduce additional trypsin-cleavage sites. **See** *Ethyleneimine* **Target(s):** L-lysine 2,3-aminomutase (1,7,11); peptidyl-Lys metalloendopeptidase, *or Armillaria mellea* protease (2,18); α-aminoadipate-semialdehyde dehydrogenase (3); D-lysine 5,6 aminomutase (4,10); lysyl-tRNA synthetase (5,8,9); aspartate kinase (5,6,19,20); dihydrodipicolinate synthase (12-15); diaminopimelate decarboxylase (16); kynureninase (17); hydroxylysine kinase (21); homocitrate synthase (22-25). **1**. Chirpich & Barker (1971) *Meth. Enzymol.* **17B**, 215. **2**. Takio (1998) in *Handb. Proteolytic Enzymes* (Barrett, Rawlings & Woessner, eds.), p. 1538, Academic Press, San Diego. **3**. Schmidt, Bode & Birnbaum (1990) *FEMS Microbiol. Lett.* **58**, 41. **4**. Stadtman (1972) *The Enzymes*, 3rd ed., **6**, 539. **5**. Coles & Brenchley (1976) *Biochim. Biophys. Acta* **428**, 647. **6**. Di Girolamo, De Marco, Busiello & Cini (1982) *Mol. Cell Biochem.* **49**, 43. **7**. Miller, Bandarian, Reed & Frey (2001) *Arch. Biochem. Biophys.* **387**, 281. **8**. Levengood, Ataide, Roy & Ibba (2004) *J. Biol. Chem.* **279**, 17707. **9**. Ataide & Ibba (2004) *Biochemistry* **43**, 11836. **10**. Morley & Stadtman (1970) *Biochemistry* **9**, 4890. **11**. Chirpich, Zappia, Costilow & Barker (1970) *J. Biol. Chem.* **245**, 1778. **12**. Laber, Gomis-Rueth, Romao & R. Huber (1992) *Biochem. J.* **288**, 691. **13**. Kumpaisal, Hashimoto & Yamada (1987) *Plant Physiol.* **85**, 145. **14**. Wallsgrove & Mazelis (1981) *Phytochemistry* **20**, 2651. **15**. Frisch, Gengenbach, Tommy, *et al.* (1991) *Plant Physiol.* **96**, 444. **16**. Rosner (1975) *J. Bacteriol.* **121**, 20. **17**. Tanizawa & Soda (1979) *J. Biochem.* **86**, 499. **18**. Lewis, Basford & Walton (1978) *Biochim. Biophys. Acta* **522**, 551. **19**. Hamano, Nicchu, Shimizu, *et al.* (2007) *Appl. Microbiol. Biotechnol.* **76**, 873. **20**. Ferreira, Meinhardt & Azevedo (2006) *Ann. Appl. Biol.* **149**, 77. **21**. Hiles & Henderson (1972) *J. Biol. Chem.* **247**, 646. **22**. Schmidt, Bode, Lindner & Birnbaum (1985) *J. Basic Microbiol.* **25**, 675. **23**. Takenouchi, Tanaka & Soda (1981) *J. Ferment. Technol.* **59**, 429. **24**. Gray & Bhattacharjee (1976) *J. Gen. Microbiol.* **97**, 117. **25**. Shimizu, Yamana, Tanaka & Soda (1984) *Agric. Biol. Chem.* **48**, 2871.

N-(2-Aminoethyl)-5-isoquinolinesulfonamide

This substituted isoquinoline (FW$_{free-base}$ = 251.31 g/mol), commonly referred to as H-9, inhibits a number of protein kinases: *e.g.*, protein kinase G (K_i = 0.87 μM), protein kinase A (Ki = 1.9 μM), protein kinase C (K_i = 18 μM), and myosin light-chain kinase (IC$_{50}$ = 70 μM). **1**. Inagaki, Watanabe & Hidaka (1985) *J. Biol. Chem.* **260**, 2922.

S-(2-Aminoethyl)isothiourea & S-(2-Aminoethyl)isothiouronium Cation

This substituted isothiourea (FW$_{free-base}$ = 119.19 g/mol; *Abbreviation*: AET), inhibits nitric-oxide synthase, with respective K_d values of 12 μM and 3 μM for the endothelial and inducible enzymes. **Target(s):** nitric oxide synthase (1,2,4); arginase (3); glutamate decarboxylase (5);

creatinase (6).　　**1**. Southan, Szabo & Thiemermann (1995) *Brit. J. Pharmacol.* **114**, 510. **2**. Southan, Zingarelli, O'Connor, Salzman & Szabo (1996) *Brit. J. Pharmacol.* **117**, 619. **3**. Colleluori & Ash (2001) *Biochemistry* **40**, 9356. **4**. Handy, Wallace & Moore (1996) *Pharmacol. Biochem. Behav.* **55**, 179. **5**. Prabhakaran, Harris & Kirchheimer (1983) *Arch. Microbiol.* **134**, 320. **6**. Coll, Knof, Ohga, *et al.* (1990) *J. Mol. Biol.* **214**, 597.

N,N-(2-Aminoethyl)-D-mannoamidine

This half-chair mannosidase transition-state mimic (FW = 219.24 g/mol) inhibits several glycosidases: α-mannosidase (K_i = 6 nM), β-mannosidase (K_i = 9 nM), α glucosidase (K_i = 81 μM), β-glucosidase (K_i = 6600 μM), α-galactosidase (K_i = 218 μM), β-galactosidase (K_i = 82 μM), α-fucosidase (K_i = 6 μM), and β-N acetylhexosaminidase (K_i = 99 μM). **1**. Heck, Vincent, Murray, *et al.* (2004) *J. Amer. Chem. Soc.* **126**, 1971.

4-(5-(3-Amino-3-ethylpentyl)thiophen-2-yl)-N-[4-([4-pyrrolidin-1-yl]piperidin-1-yl]carbonyl)phenyl]pyrimidin-2-amine

This substituted pyrimidine (FW = 546.78 g/mol) inhibits human IκB kinase 1 and 2, with IC$_{50}$ values of 0.03 and 0.06 μM, respectively, demonstrating significant *in vivo* activity in an acute model of cytokine release. **1**. Waelchli, Bollbuck, Bruns, *et al.* (2006) *Bioorg. Med. Chem. Lett.* **16**, 108.

D-3-[(1-Aminoethyl)phosphinyl]-D-2-ethylpropionic Acid

This phosphinate-containing dipeptide analogue (FW = 209.18 g/mol) is a slow-binding inhibitor of *Enterococcus faecium* D-Ala-D-Ala dipeptidase (K_i^* = 90 nM) a D-,D-dipeptidase required for vancomycin resistance in *Enterococcus faecium*. This inhibitor showed a time-dependent onset of inhibition of VanX and a time-dependent return to uninhibited steady-state rates upon dilution of the enzyme/inhibitor mixture. **1**. Wu & Walsh (1995) *Proc. Natl. Acad. Sci. U.S.A.* **92**, 11603.

D-3-[(1-Aminoethyl)phosphinyl]-2-heptylpropionic Acid

This phosphinate-containing dipeptide analogue (FW$_{hydrochloride}$ = 315.78 g/mol) is converted by enzymatic phosphorylation from a low-affinity inhibitor form to an extremely tightly bound analogue of a D-alanyl-D-alanine synthetase reaction intermediate. This compound is a potent active site-directed inhibitor and is competitive with D-alanine (K_i = 1.2 μM), exhibiting time-dependent inhibition in the presence of ATP. Kinetic analysis revealed a rapid onset of steady-state inhibition (k_{on} = 1.35 x 10^4 M^{-1} s^{-1}), followed by very slow dissociation of inhibitory complex(es) with a half-life of 8.2 h. **1**. Duncan & Walsh (1988) *Biochemistry* **27**, 3709.

(R)-(+)-trans-4-(1-Aminoethyl)-N-4-pyridyl)cyclohexane carboxamide

This pyridine derivative (FW$_{free-base}$ = 247.34 g/mol; CAS 146986-50-7; Water-soluble; λ_{max} = 269.6 nm), also known as Y-27632, binds specifically to the ROCK family of protein kinases and inhibits their activity (1-5). Y-27632 is also a smooth muscle relaxant, inhibits stress fiber formation, and inhibits lysophosphatidate-induced neurite retraction (2). **Target(s):** Rho-dependent protein kinase (ROCK-I and ROCK-II) (1-5); mitogen- and stress- activated protein kinase 1, mildly inhibited (3); [mitogen-activated protein kinase]-activated protein kinase-1, weakly inhibited (3); phosphorylase kinase, weakly inhibited (3). **1.** Amano, Fukata, Shimokawa & Kaibuchi (2000) *Meth. Enzymol.* **325**, 149. **2.** Narumiya, Ishizaki & Uehata (2000) *Meth. Enzymol.* **325**, 273. **3.** Davies, Reddy, Caivano & Cohen (2000) *Biochem. J.* **351**, 95. **4.** Niggli, Schmid & Nievergelt (2006) *Biochem. Biophys. Res. Commun.* **343**, 602. **5.** Grimm, Haas, Willipinski-Stapelfeldt, *et al.* (2005) *Cardiovasc. Res.* **65**, 211.

4-Amino-5-fluoropentanoate

This γ-amino acid (FW = 135.14 g/mol) is a mechanism-based inhibitor that inactivates γ-aminobutyrate aminotransferase (1-8) by an enamine mechanism first suggested by Metzler and co-workers (9). **Target(s):** γ-aminobutyrate aminotransferase (1-9,12); hypotaurine aminotransferase (2); glutamate 1-semialdehyde 2,1-aminomutase, *or* glutamate-1-semialdehyde aminotransferase (10,11). **1.** Silverman (1995) *Meth. Enzymol.* **249**, 240. **2.** Fellman (1987) *Meth. Enzymol.* **143**, 183. **3.** Silverman & Invergo (1986) *Biochemistry* **25**, 6817. **4.** Silverman & Levy (1981) *Biochemistry* **20**, 1197. **5.** Silverman & Levy (1980) *Biochem. Biophys. Res. Commun.* **95**, 250. **6.** Silverman & Levy (1980) *J. Org. Chem.* **45**, 815. **7.** Silverman, Levy, Muztar & Hirsch (1981) *Biochem. Biophys. Res. Commun.* **102**, 520. **8.** Silverman, Muztar, Levy & Hirsch (1983) *Life Sci.* **32**, 2717. **9.** Likos, Ueno, Feldhaus & Metzler (1982) *Biochemistry* **21**, 4377. **10.** Bishop, Gough, Mahoney, Smith & Rogers (1999) *FEBS Lett.* **450**, 57. **11.** Bull, Breu, Kannangara, Rogers & Smith (1990) *Arch. Microbiol.* **154**, 56. **12.** Wang & Silverman (2006) *Bioorg. Med. Chem.* **14**, 2242.

2-[5-Amino-2-(4-fluorophenyl)-6-oxo-1,6-dihydro-1-pyrimidinyl]-N-[2-methyl-1-(4,4,5,5-tetramethyl-1,3,2-dioxaborolan-2-yl)propyl]acetamide

This substituted 5-aminopyridin-6-one (FW = 444.31 g/mol) inhibits human leukocyte elastase, K_i = 6.2 nM. This compound contains a boronic acid moiety that binds covalently to the Ser-195 hydroxyl of the enzyme. **1.** Veale, Bernstein, Bryant, *et al.* (1995) *J. Med. Chem.* **38**, 98

8-Amino-9-(2-furylmethyl)guanine

This synthetic purine nucleoside (FW = 246.23 g/mol) inhibits calf spleen purine nucleoside phosphorylase (IC$_{50}$ = 92 nM). **See also Immucillins 1.** Castilho, Postigo, de Paula, *et al.* (2006) *Bioorg. Med. Chem.* **14**, 516. **2.** Bzowska, Kulikowska & Shugar (2000) *Pharmacol. Ther.* **88**, 349.

3-Aminoglutarate, See β-Glutamate

DL-Aminoglutethimide

This substituted glutarimide (FW = 232.28 g/mol), also known as 2-(4-aminophenyl)-2-ethylglutarimide, inhibits many steroid monooxygenases. Aminoglutethimide blocks adrenal steroid biosynthesis and the production of estrogens in extraglandular tissues. **Target(s):** aromatase, *or* CYP19, IC$_{50}$ values for the (+)-R- and (−)-S-stereoisomers are 0.86 and 23.15 μM, respectively (1-3,7,10,16,19); steroid 11β-monooxygenase (4,9); calcidiol 1-monooxygenase, *or* 25-hydroxycholecalciferol 1α-monooxygenase (5); cholesterol monooxygenase, side-chain-cleaving (6,9-11,17,18); ecdysone 20-monooxygenase (8); steroid 7α-monooxygenase, *or* cholesterol 7α-monooxygenase (12,13); estradiol 2-monooxygenase (14,16); estradiol 16α-monooxygenase (14); cholestanetriol 26-monooxygenase, *or* 5β-cholestane-3α,7α,12α-triol 27-hydroxylase, weakly inhibited (15). **1.** Rowlands, Davies, Shearer & Dowsett (1991) *J. Enzyme Inhib.* **4**, 307. **2.** Coulson, King & Wiseman (1984) *Trends Biochem. Sci.* **9**, 446. **3.** Ogbunude & Aboul-Enein (1994) *Chirality* **6**, 623. **4.** Faglia, Gattinoni, Travaglini, *et al.* (1971) *Metabolism* **20**, 266. **5.** Ghazarian, Jefcoate, Knutson, Orme-Johnson & DeLuca (1974) *J. Biol. Chem.* **249**, 3026. **6.** Uzgiris, Whipple & Salhanick (1977) *Endocrinology* **101**, 89. **7.** Graves & Salhanick (1979) *Endocrinology* **105**, 52. **8.** Smith, Bollenbacher, Cooper, *et al.* (1979) *Mol. Cell Endocrinol.* **15**, 111. **9.** Whipple, Grodzicki, Hourihan & Salhanick (1981) *Steroids* **37**, 673. **10.** Salhanick (1982) *Cancer Res.* **42** (8 Suppl.), 3315s. **11.** Sheets & Vickery (1982) *Naunyn Schmiedeberg's Arch. Pharmacol.* **321**, 70. **12.** Schwartz & Margolis (1983) *J. Lipid Res.* **24**, 28. **13.** Ogishima, Deguchi & Okuda (1987) *J. Biol. Chem.* **262**, 7646. **14.** Numazawa & Satoh (1988) *J. Steroid Biochem.* **29**, 221. **15.** Okuda, Masumoto & Ohyama (1988) *J. Biol. Chem.* **263**, 18138. **16.** Purba, King, Richert & Bhatnagar (1994) *J. Steroid Biochem. Mol. Biol.* **48**, 215. **17.** Patte-Mensah, Kappes, Freund-Mercier, Tsutsui & Mensah-Nyagan (2003) *J. Neurochem.* **86**, 1233. **18.** Ahmed (2000) *Biochem. Biophys. Res. Commun.* **274**, 821. **19.** Maiti, Reddy, Sturdy, *et al.* (2009) *J. Med. Chem.* **52**, 1873.

Aminoguanidine

This substituted hydrazine (FW$_{hydrochloride}$ = 110.55 g/mol; CAS 79-17-4), also named pimagedine, is a strong inhibitor of diamine oxidases as well as nitric oxide synthase. Note that aminoguanidine also inhibits the formation of advanced glycosylation end-products, reportedly by reacting with and trapping Amadori rearrangement-derived fragmentation products in solution. In view of this property, aminoguanidine has often been used to treat complications associated with chronic diabetes. **Target(s):** amine oxidase (copper-containing), *or* diamine oxidase (1,2,6,8,9,13,18,19,25); methylamine dehydrogenase (3); tryptophan tryptophylquinone enzymes (3); nitric-oxide synthase (4,10,14-16,20,26,27); β-fructofuranosidase, *or* invertase (5); monoamine oxidase, *or* amine oxidase (7,10); histidine decarboxylase (9,12,25); aldose reductase (10); catalase (11); formation of advanced glycosylation end products (17); adenosylmethionine decarboxylase (21); arginine deaminase (22); arginine kinase (23); aspartate aminotransferase (24); histamine *N*-methyltransferase (25). **1.** Luhova, Slavik, Frebort, *et al.* (1996) *J. Enzyme Inhib.* **10**, 251. **2.** Padiglia, Medda, Lorrai, *et al.* (2000) *J. Enzyme Inhib.* **15**, 91. **3.** Davidson, Brooks, Graichen, Jones & Hyun (1995) *Meth. Enzymol.* **258**, 176. **4.** Corbett & McDaniel (1996) *Meth. Enzymol.* **268**, 398. **5.** Neuberg & Mandl (1950) *The Enzymes*, 1st ed., **1** (part 1), 527. **6.** Zeller (1963) *The Enzymes*, 2nd ed., **8**, 313. **7.** Blaschko (1963) *The Enzymes*, 2nd ed., **8**, 337. **8.** Malmström, Andréasson & Reinhammar (1975) *The Enzymes*, 3rd ed., **12**, 507. **9.** Webb (1966) *Enzyme and Metabolic Inhibitors*, vol. **2**, Academic Press, New York. **10.** Griffith & Stuehr (1995) *Ann. Rev. Physiol.* **57**, 707. **11.** Feinstein, Seaholm & Ballonoff (1964) *Enzymologia* **27**, 30. **12.** Kobayashi & Maudsley (1971) *Brit. J. Pharmacol.* **43**, 426P. **13.** Crabbe, Childs & Bardsley (1975) *Eur. J. Biochem.* **60**, 325. **14.** Corbett, Tilton, Chang, *et al.* (1992) *Diabetes* **41**, 552. **15.** Misko, Moore, Kasten, *et al.* (1993) *Eur. J. Pharmacol.* **233**, 119. **16.** Griffiths, Messent, MacAllister & Evans (1993) *Brit. J. Pharmacol.* **110**, 963. **17.** Edelstein & Brownlee (1992) *Diabetes* **41**, 26. **18.** Schayer, Kennedy & Smiley (1953) *J. Biol. Chem.* **205**, 739. **19.** Schuler (1952) *Experientia* **8**, 230. **20.** Hasan, Heesen, Corbett, *et al.* (1993) *Eur. J. Pharmacol.* **249**, 101. **21.** Svensson, Mett & Persson (1997) *Biochem. J.* **322**, 297. **22.** Smith, Ganaway & Fahrney (1978) *J. Biol. Chem.* **253**, 6016. **23.** Pereira, Alonso, Ivaldi, *et al.* (2003) *J. Eukaryot. Microbiol.* **50**, 132. **24.** Okada & Ayabe (1995) *J. Nutr. Sci. Vitaminol.* **41**, 43. **25.** Beaven & Roderick (1980) *Biochem. Pharmacol.* **29**, 2897. **26.** Chandok, Ytterberg, van Wijk & Klessig (2003) *Cell* **113**, 469. **27.** Hong, Kim, Choi, *et al.* (2003) *FEMS Microbiol. Lett.* **222**, 177.

6-Aminohexanoate

This ε-amino acid (FW = 131.17 g/mol; CAS 60-32-2; *Abbreviation*: εAhx), also known as ε-amino-*n*-caproic acid, is a lysine derivative/analogue that inhibits many peptidases and likewise retards fibrinolysis. 6-Aminohexanoate is soluble in water and has pK_a values of 4.43 and 10.75. Note that 6-aminohexanoic acid is the subunit of the polymer nylon 6. **Target(s):** plasmin (1,2,10,19,25,40); pyrroline-5-carboxylate dehydrogenase (3); u-plasminogen activator, *or* urokinase (4,5,10,14,19,23,24); lysine carboxypeptidase, *or* lysine(arginine) carboxypeptidase, *or* carboxypeptidase N (6,7,42,44-47); carboxypeptidase B (8,9,11-13,42,48-50); coagulation factor VIIa (15); leukocyte phagocytosis and clumping (16,17); penicillopepsin (18); crayfish carboxypeptidase (20,21); lysyl endopeptidase (22); trypsin (25); complement component C_1 (26); pantothenate synthetase, *or* pantoate:β-alanine ligase, slightly inhibited (27); lysyl-tRNA synthetase (28); D-lysine 5,6-aminomutase (29); lysine decarboxylase (30); ornithine decarboxylase (31); guanidinobutyrase (32); acetylspermidine deacetylase (33); peptidyl-Lys metalloendopeptidase (34); procollagen C endopeptidase (35); cathepsin D (36); t-plasminogen activator (37); complement factor I, weakly inhibited (38); acrosin (39); carboxypeptidase U (41-43); carboxypeptidase M (42); carboxypeptidase E (42); cytosol alanyl aminopeptidase (51); membrane alanyl aminopeptidase (52); 2,3,4,5 tetrahydropyridine-2,6-dicarboxylate *N*-succinyltransferase (53); ceruloplasmin (54); ferroxidase (54); lysine 2-monooxygenase (55). **1.** Steffen & Steffen (1976) *Clin. Chem.* **22**, 381. **2.** Robbins & Summaria (1970) *Meth. Enzymol.* **19**, 184. **3.** Strecker (1971) *Meth. Enzymol.* **17B**, 262. **4.** White & Barlow (1970) *Meth. Enzymol.* **19**, 665. **5.** Barlow (1976) *Meth. Enzymol.* **45**, 239. **6.** Plummer & Erdös (1981) *Meth. Enzymol.* **80**, 442. **7.** Maderazo, Woronick & Ward (1988) *Meth. Enzymol.* **162**, 223. **8.** Folk & Gladner (1958) *J. Biol. Chem.* **231**, 379. **9.** Neurath (1960) *The Enzymes*, 2nd ed., **4**, 11. **10.** Ablondi & Hagan (1960) *The Enzymes*, 2nd ed., **4**, 175. **11.** Folk (1971) *The Enzymes*, 3rd ed., **3**, 57. **12.** Wintersberger, Cox & Neurath (1962) *Biochemistry* **1**, 1069. **13.** Webb (1966) *Enzyme and Metabolic Inhibitors*, vol. **2**, p. 367, Academic Press, New York. **14.** Lorand & Condit (1965) *Biochemistry* **4**, 265. **15.** Radcliffe & Heinze (1978) *Arch. Biochem. Biophys.* **189**, 185. **16.** Allison, Lancaster & Crosthwaite (1963) *Amer. J. Path.* **43**, 775. **17.** Allison &

Lancaster (1964) *Ann. N. Y. Acad. Sci.* **116**, 936. **18.** Hofmann & Shaw (1964) *Biochim. Biophys. Acta* **92**, 543. **19.** Johnson, Skoza & Tse (1969) *Thromb. Diath. Haemorrh. Suppl.* **32**, 105. **20.** Zwilling & Neurath (1981) *Meth. Enzymol.* **80**, 633. **21.** Zwilling, Jakob, Bauer, Neurath & Enfield (1979) *Eur. J. Biochem.* **94**, 223. **22.** Sakiyama & Masaki (1994) *Meth. Enzymol.* **244**, 126. **23.** Propping, Zaneveld, Tauber & Schumacher (1978) *Biochem. J.* **171**, 435. **24.** Alkjaersig, Fletcher & Sherry (1959) *J. Biol. Chem.* **234**, 832. **25.** Ablondi, Hagan, Philips & De Renzo (1959) *Arch. Biochem. Biophys.* **82**, 153. **26.** Soter, Austen & Gigli (1975) *J. Immunol.* **114**, 928. **27.** Miyatake, Nakano & Kitaoka (1978) *J. Nutr. Sci. Vitaminol.* **24**, 243. **28.** Takita, Ohkubo, Shima, *et al.* (1996) *J. Biochem.* **119**, 680. **29.** Morley & Stadtman (1970) *Biochemistry* **9**, 4890. **30.** Yamamoto, Imamura, Kusaba & Shinoda (1991) *Chem. Pharm. Bull.* **39**, 3067. **31.** Guirard & Snell (1980) *J. Biol. Chem.* **255**, 5960. **32.** Yorifuji, Shimizu, Hirata, *et al.* (1992) *Biosci. Biotechnol. Biochem.* **56**, 773. **33.** Huang, Dredar, Manneh, Blankenship & Fries (1992) *J. Med. Chem.* **35**, 2414. **34.** Nonaka, Ishikawa, Tsumuraya, *et al.* (1995) *J. Biochem.* **118**, 1014. **35.** Kessler, Takahara, Biniaminov, Brusel & Greenspan (1971) *Science* **271**, 360. **36.** Woessner (1973) *J. Biol. Chem.* **248**, 1634. **37.** Johnsen, Ravn, Berglund, *et al.* (1998) *Biochemistry* **37**, 12631. **38.** Tsiftsoglou & Sim (2004) *J. Immunol.* **173**, 367. **39.** Polakoski & McRorie (1973) *J. Biol. Chem.* **248**, 8183. **40.** Rickli & Otavsky (1975) *Eur. J. Biochem.* **59**, 441. **41.** Bouma, Marx, Mosnier & Meijers (2001) *Thromb. Res.* **101**, 329. **42.** Mao, Colussi, Bailey, *et al.* (2003) *Anal. Biochem.* **319**, 159. **43.** Boffa, Wang, Bajzar & Nesheim (1998) *J. Biol. Chem.* **273**, 2127. **44.** Erdös (1979) *Fed. Proc.* **38**, 2774. **45.** Jeanneret, Roth & Bargetzi (1976) *Hoppe-Seyler's Z. Physiol. Chem.* **357**, 867. **46.** Schweisfurth (1984) *Dtsch. Med. Wochenschr.* **109**, 1254. **47.** Juillerat-Jeanneret, Roth & Bargetzi (1982) *Hoppe-Seyler's Z. Physiol. Chem.* **363**, 51. **48.** Gates & Travis (1973) *Biochemistry* **12**, 1867. **49.** Bradley, Naudé, Muramoto, Yamauchi & Oelofsen (1996) *Int. J. Biochem. Cell Biol.* **28**, 521. **50.** Marinkovic, Marinkovic, Erdös & Robinson (1977) *Biochem. J.* **163**, 253. **51.** Garner & Behal (1977) *Arch. Biochem. Biophys.* **182**, 667. **52.** Lalu, Lampelo & Vanha-Perttula (1986) *Biochim. Biophys. Acta* **873**, 190. **53.** Berges, DeWolf, Dunn, *et al.* (1986) *J. Biol. Chem.* **261**, 6160. **54.** Frieden & Hsieh (1976) *Adv. Enzymol. Rel. Areas Mol. Biol.* **44**, 187. **55.** Flashner & Massey (1974) *J. Biol. Chem.* **249**, 2587.

N-(6-Aminohexyl)-5-chloro-1-naphthalenesulfonamide

This naphthylsulfonamide (FW$_{hydrochloride}$ = 377.33 g/mol), often referred to as W-7, is a calmodulin antagonist that inhibits a number of calmodulin-dependent reactions. **Target(s):** 3',5'-cyclic-nucleotide phosphodiesterase (1-4,9,11), [myosin light-chain] kinase (1-4,6,13,14); calmodulin (1-3); protein kinase C (3,5,8,10); Ca^{2+}-exporting ATPase (3); cAMP-dependent protein kinase (3,4); cGMP-dependent protein kinase (3,4); casein kinase I (3); casein kinase II (3); pyruvate dehydrogenase, pea (7); [3-hydroxy-3-methylglutaryl-CoA reductase (NADPH)] kinase (12); Ca^{2+}/calmodulin-dependent protein kinase (15-19); diphosphoinositol-pentakisphosphate kinase (20); NAD$^+$ kinase (21); thymidine kinase (22); fatty-acid *O*-methyltransferase (23); nitric-oxide synthase (24,25). **1.** Hidaka, Sasaki, Tanaka, *et al.* (1981) *Proc. Natl. Acad. Sci. U.S.A.* **78**, 4354. **2.** Hidaka & Tanaka (1983) *Meth. Enzymol.* **102**, 185. **3.** Hidaka, Inagaki, Nishikawa & Tanaka (1988) *Meth. Enzymol.* **159**, 652. **4.** Hidaka, Watanabe & Kobayashi (1991) *Meth. Enzymol.* **201**, 328. **5.** Kikkawa & Nishizuka (1986) *The Enzymes*, 3rd ed., **17**, 167. **6.** Asano (1989) *J. Pharmacol. Exp. Ther.* **251**, 764. **7.** Miernyk, Fang & Randall (1987) *J. Biol. Chem.* **262**, 15338. **8.** Schatzman, Raynor & Kuo (1983) *Biochim. Biophys. Acta* **755**, 144. **9.** Itoh & Hidaka (1984) *J. Biochem.* **96**, 1721. **10.** O'Brian & Ward (1989) *Biochem. Pharmacol.* **38**, 1737. **11.** Ryan & Toscano (1985) *Arch. Biochem. Biophys.* **241**, 403. **12.** Beg, Stonik & Brewer (1987) *J. Biol. Chem.* **262**, 13228. **13.** Niggli, Schmid & Nievergelt (2006) *Biochem. Biophys. Res. Commun.* **343**, 602. **14.** Yanase, Ikeda, Ogata, *et al.* (2003) *Biochem. Biophys. Res. Commun.* **305**, 223. **15.** Rodriguez-Mora, LaHair, Howe, McCubrey & Franklin (2005) *Exp. Opin. Ther. Targets* **9**, 791. **16.** Davletova, Mészáros, Miskolczi, *et al.* (2001) *J. Exp. Bot.* **52**, 215. **17.** Dhillon, Sharma & Khuller (2003) *Mol. Cell. Biochem.* **252**, 183. **18.** Pan,

Means & Liu (2005) *EMBO J.* **24**, 2104. **19**. Calderilla-Barbosa, Ortega & Cisneros (2006) *J. Neurochem.* **98**, 713. **20**. Safrany (2004) *Mol. Pharmacol.* **66**, 1585. **21**. Muto (1983) *Z. Pflanzenphysiol.* **109**, 385. **22**. O'Day, Chatterjee-Chakraborty, Wagler & Myre (2005) *Biochem. Biophys. Res. Commun.* **331**, 1494. **23**. Safayhi, Anazodo & Ammon (1991) *Int. J. Biochem.* **23**, 769. **24**. Bodnárová, Martásek & Moroz (2005) *J. Inorg. Biochem.* **99**, 922. **25**. Côté & Roberge (1996) *Free Radic. Biol. Med.* **21**, 109.

(2S,3R)-2-Amino-3-hydroxy-2-hydroxymethyl-12-oxostearate

This sphingosine analogue (FW = 347.50 g/mol) inhibits serine *C*-palmitoyltransferase (IC_{50} = 3.5 nM), the enzyme that catalyzes the first step of the biosynthesis of all sphingolipids. The methyl ester is a slightly weaker inhibitor (IC_{50} = 17 nM). **1**. Yamaji-Hasegawa, Takahashi, Tetsuka, Senoh & Kobayashi (2005) *Biochemistry* **44**, 268.

2-Amino-4-hydroxypteridine-6-aldehyde

This yellow pteridine (FW = 191.15 g/mol), also known as 6-pteridylaldehyde, pterin-6-aldehyde, and 2-amino-4-hydroxy-6-pteridinecarboxaldehyde, is a photoproduct that inhibits xanthine oxidase, for which it is a slow substrate, producing the 6-carboxylate. It also exhibits a superoxide anion radical scavenging activity (7). **Note:** This derivative is a frequent contaminant in folic acid solutions, as a consequence of unrecognized photodegradation. **Target(s):** xanthine oxidase, *or* pterin oxidase (1-12,15); guanine deaminase, *or* guanase (4,14); quinine oxidase (6,13); xanthine dehydrogenase (16). **1**. Theorell (1951) *The Enzymes*, 1st ed., **2** (part 1), 335. **2**. Bray (1963) *The Enzymes*, 2nd ed., **7**, 533. **3**. Bray (1975) *The Enzymes*, 3rd ed., **12**, 299. **4**. Webb (1966) *Enzyme and Metabolic Inhibitors*, vol. **2**, Academic Press, New York. **5**. Terada, Leff & Repine (1990) *Meth. Enzymol.* **186**, 651. **6**. Kalckar, Kjeldgaard & Klenow (1948) *J. Biol. Chem.* **174**, 771. **7**. Watanabe, Arai, Mori, *et al.* (1997) *Biochem. Biophys. Res. Commun.* **233**, 447. **8**. Hille, George, Eidsness & Cramer (1989) *Inorg. Chem.* **28**, 4018. **9**. Hofstee (1949) *J. Biol. Chem.* **179**, 633. **10**. H. Lowry, Bessey & Crawford (1949) *J. Biol. Chem.* **180**, 399. **11**. Dietrich, Monson, Williams & Elvehjem (1952) *J. Biol. Chem.* **197**, 37. **12**. Litwack, Bothwell, Williams & Elvehjem (1953) *J. Biol. Chem.* **200**, 303. **13**. Villela (1963) *Enzymologia* **25**, 261. **14**. Dietrich & Shapiro (1953) *J. Biol. Chem.* **203**, 89. **15**. De Renzo (1956) *Adv. Enzymol. Relat. Subj. Biochem.* **17**, 293. **16**. Yen & Glassman (1967) *Biochim. Biophys. Acta* **146**, 35.

5-Amino-1H-imidazole-4-carboxamide

This free-base of a key purine nucleotide biosynthesis intermediate (FW$_{free-base}$ = 126.12 g/mol; λ_{max} = of 266 nm at pH 7, with ε = 12700 $M^{-1}cm^{-1}$; λ_{max} = 277 nm at pH 13, with ε = 12500 $M^{-1}cm^{-1}$), also known as 4-amino-5-imidazolecarboxamide, enzymatically undergoes PRPP-dependent conversion into AICAR, the actual purine nucleotide biosynthetic pathway intermediate. (**See next entry**). **Target(s):** guanine deaminase (1-4,7,8,10-15); adenosine deaminase (5); adenine deaminase (6); NAD(P)H dehydrogenase (quinone), *or* DT diaphorase (9); xanthine phosphoribosyltransferase (16,17); hypoxanthine (guanine) phosphoribosyltransferase (17); urate-ribonucleotide phosphorylase, weakly inhibited (18); glutaminyl-peptide cyclotransferase, weakly inhibited (1).9 **1**. Glantz &

Lewis (1978) *Meth. Enzymol.* **51**, 512. **2**. Baker (1967) *Design of Active-Site-Directed Irreversible Enzyme Inhibitors*, Wiley, New York. **3**. Mandel (1957) *Biochim. Biophys. Acta* **25**, 402. **4**. Martinez-Farnos, Gubert & Bozal (1978) *Rev. Esp. Fisiol.* **34**, 295. **5**. Agarwal & Parks (1978) *Meth. Enzymol.* **51**, 502. **6**. Hartenstein & Fridovich (1967) *J. Biol. Chem.* **242**, 740. **7**. Miyamoto, Ogawa, Shiraki & Nakagawa (1982) *J. Biochem.* **91**, 167. **8**. Gupta & Glantz (1985) *Arch. Biochem. Biophys.* **236**, 266. **9**. Roberts, Marchbank, Kotsaki-Kovatsi, *et al.* (1989) *Biochem. Pharmacol.* **38**, 4137. **10**. Pugh & Bieber (1984) *Comp. Biochem. Physiol. B* **77**, 619. **11**. Kim & Kimm (1982) *Korean J. Biochem.* **14**, 77. **12**. Fogle & Bieber (1975) *Prep. Biochem.* **5**, 59. **13**. Kimm, Park & Lee (1985) *Korean J. Biochem.* **17**, 139. **14**. Kimm, Park & Kim (1987) *Korean J. Biochem.* **19**, 39. **15**. Lewis & Glantz (1974) *J. Biol. Chem.* **249**, 3862. **16**. Miller, Adamczyk, Fyfe & Elion (1974) *Arch. Biochem. Biophys.* **165**, 349. **17**. Naguib, Iltzsch, el Kouni, Panzica & el Kouni (1995) *Biochem. Pharmacol.* **50**, 1685. **18**. Laster & Blair (1963) *J. Biol. Chem.* **238**, 3348. **19**. Schilling, Niestroj, Rahfeld, *et al.*h (2003) *J. Biol. Chem.* **278**, 49773.

5-Amino-1H-imidazole-4-carboxamide riboside

This key purine nucleotide precursor and Type-2 diabetes prodrug (FW$_{free-base}$ = 126.12 g/mol; λ_{max} = of 266 nm at pH 7, with ε = 12700 $M^{-1}cm^{-1}$; λ_{max} = 277 nm at pH 13, with ε = 12500 $M^{-1}cm^{-1}$; Abreviation: AICAriboside), also known as Acadesine® and 4-amino-5-imidazolecarboxamide ribonucleoside, is phosphorylated to form 5-Amino-1*H*-imidazole-4-carboxamide ribonucleoside 5'-phosphate (AICAR), an allosteric activator of AMP-stimulated protein kinase (AMPK). (Other reports attribute AMPK activation by AICAR to the latter's conversion to XMP.) Addition of 50-500 µM acadesine to suspensions of isolated rat hepatocytes results in the accumulation of millimolar concentrations of the AICAR (1). AMP-activated protein kinase is a multisubstrate protein kinase that, in liver, inactivates both acetyl-CoA carboxylase, the rate-limiting enzyme of fatty acid synthesis, and 3-hydroxy-3-methyl-glutaryl-CoA reductase, the rate-limiting enzyme of cholesterol synthesis. AICAR stimulates rat liver AMP-activated protein kinase up to 10-fold, with a half-maximal effect at approximately 5 mM (1). This was accompanied by a dose-dependent inactivation of both acetyl-CoA carboxylase and 3-hydroxy-3-methylglutaryl-CoA reductase. Addition of 50-500 µM acadesine to hepatocyte suspensions incubated in the presence of various substrates, including glucose and lactate/pyruvate, caused a parallel inhibition of both fatty acid and cholesterol synthesis (1). Acadesine also stimulates glucose uptake and increases the activity of p38 mitogen-activated protein kinases α and β in skeletal muscle tissue. 5-Aminoimidazole-4-carboxamide riboside is also a competitive inhibitor of adenosine deaminase, K_i = 0.36 mM (2). AICAR activation of the p53 pathway is attenuated by rapamycin, an immunosuppressant that inhibits mTOR kinase. **1**. Henin, Vincent, Gruber & Van den Berghe (1985) *FASEB J.* **9**, 541. **2**. Baggott, Vaughn & Hudson (1986) *Biochem. J.* **236**, 193.

6-[(2-Aminoimidazol-1-yl)methyl]-5-bromouracil

This substituted uracil (FW = 287.10 g/mol) is a potent inhibitor of thymidine phosphorylase (IC50 = 20 nM). The 4-amino analogue is a weaker inhibitor. **1**. Reigan, Gbaj, Stratford, Bryce & Freeman (2008) *Eur. J. Med. Chem.* **43**, 1248. **2**. Reigan, Edwards, Gbaj, *et al.* (2005) *J. Med. Chem.* **48**, 392. **3**. Gbaj, Edwards, Reigan, *et al.* (2006) *J. Enzyme Inhib. Med. Chem.* **21**, 69.

N-(2-((4-((Amino(imino)methyl)amino)-1(S)-carboxybutyl)amino)propyl)-L-histidine

This histargin homologue and ACE1 inhibitor (FW = 358.42 g/mol), consisting of L-histidine and L-arginine linked by a propylene bridge to their α-amino groups, targets carboxypeptidase B, lysine carboxypeptidase (*or* lysine(arginine) carboxypeptidase, carboxypeptidase N, and peptidyl-dipeptidase A (*or* angiotensin I-converting enzyme). **1**. Moriguchi, Umeda, Miyazaki, *et al.* (1988) *J. Antibiot.* **41**, 1823.

2-Aminoindan-2-phosphonate

This conformationally constrained phenylalanine analogue (FW$_{hydrochloride}$ = 249.63 g/mol) is a slow-binding, inhibitor of phenylalanine ammonia-lyase, with a K_i of 7 nM for the *Petroselinum crispum* enzyme). The corresponding carboxylic acid is a weak alternative substrate and competitive inhibitor. **1**. Appert, Zon & Amrhein (2003) *Phytochemistry* **62**, 415. **2**. Dubery & Smit (1994) *Biochim. Biophys. Acta* **1207**, 24. **3**. Campbell & Ellis (1992) *Plant Physiol.* **98**, 62. **4**. Kim, Kronstad & Ellis (1996) *Phytochemistry* **43**, 351. **5**. Hisaminato, Murata & Homma (2001) *Biosci. Biotechnol. Biochem.* **65**, 1016.

4-Amino-5-iodo-7-(β-D-*erythro*-furanosyl)pyrrolo[2,3-*d*]pyrimidine

This tubercidin analogue (FW = 362.12 g/mol) inhibits human adenosine kinase (IC$_{50}$ = 6 nM), a key intracellular enzyme regulating intracellular and extracellular concentrations of adenosine (ADO), itself an powerful endogenous modulator of intercellular signalling that reduces cell excitability during tissue stress and trauma. **1**. Boyer, Ugarkar, Solbach, *et al.* (2005) *J. Med. Chem.* **48**, 6430.

5-Aminolevulinate (5-Aminolevulinic Acid)

This key metabolic precursor (FW = 131.13 g/mol; CAS 106-60-5; pK_a values = 4.05 and 8.90 at 25°C; Symbol: ALA), also known as δ-aminolevulinic acid, is essential for the biosynthesis of metal ion-binding tetrapyrrole ring systems (porphyrins, chlorophylls, and cobalamins). In non-photosynthetic eukaryotes (animals, insects, fungi, protozoa, and alphaproteobacteria), δ-aminolevulinic acid is produced by the enzyme ALA synthase, using glycine and succinyl CoA as substrates. In plants, algae, bacteria, and archaea, it is produced from glutamyl-tRNA and glutamate-1-semialdehyde. 5-Aminolevulinic acid inhibits (*R*)-3-amino-2-methylpropionate:pyruvate aminotransferase. **ALA Phototherapy:** Protoporphyrin IX, the immediate heme precursor is a highly effective tissue photosensitizer that is synthesized in four steps from 5-aminolevulinic acid. ALA synthesis is regulated via a feedback inhibition and gene repression mechanism linked to the concentration of free heme. In certain cell and tissue types, addition of exogenous ALA bypasses these regulation mechanisms, inducing uptake and synthesis of photosensitizing concentrations of Protoporphyrin IX, *or* PpIX. Topical application of ALA to certain malignant and non-malignant skin lesions, for example, can induce a clinically useful degree of lesion-specific photosensitization (*e.g.*, superficial basal cell carcinomas show high response rate (~79%) after a single phototherapy treatment). ALA also induces localized tissue-specific photosensitization, when injected intradermally. In this sense, ALA and its methyl ester (methyl aminolevulinate, *or* MAL; trade name: Metvix®) are prodrugs that increase the amounts of the active drug (PpIX). **1**. Kontani, Kaneko, Kikugawa, Fujimoto & Tamaki (1993) *Biochim. Biophys. Acta* **1156**, 161.

Aminomalonate (Aminomalonic Acid)

This amino dicarboxylic acid (FW$_{hydrochloride}$ = 155.54 g/mol), which readily undergoes decarboxylation in aqueous solutions, inhibits δ-aminolevulinate synthase (1-3,14,15), aspartate 4-decarboxylase, as an alternative substrate (4,8); asparagine synthetase (5,6,9,10), serine hydroxymethyl-transferase (*or* glycine hydroxymethyltransferase) (1,7); glycine acyltransferase, weakly inhibited (1), phosphoribosylglycinamide synthetase (phosphoribosylamine:glycine ligase, weakly inhibited (1), aspartyl-tRNA synthetase (11), serine-sulfate ammonia-lyase (13), and glycine *C*-acetyltransferase (16). **1**. Matthew & Neuberger (1963) *Biochem. J.* **87**, 601. **2**. Jordan & Laghai-Newton (1986) *Meth. Enzymol.* **123**, 435. **3**. Jordan & Shemin (1972) *The Enzymes*, 3rd ed., **7**, 339. **4**. Tate, Novogrodsky, Soda, Miles & Meister (1970) *Meth. Enzymol.* **17A**, 681. **5**. Meister (1974) *The Enzymes*, 3rd ed., **10**, 561. **6**. Horowitz & Meister (1972) *J. Biol. Chem.* **247**, 6708. **7**. Webb (1966) *Enzyme and Metabolic Inhibitors*, vol. 2, p. 239, Academic Press, New York. **8**. Palekar, Tate & Meister (1970) *Biochemistry* **9**, 2310. **9**. Milman, Muth & Cooney (1979) *Enzyme* **24**, 36. **10**. Milman & Cooney (1979) *Biochem. J.* **181**, 51. **11**. Lea & Fowden (1973) *Phytochemistry* **12**, 1903. **12**. Thanassi (1970) *Biochemistry* **9**, 525. **13**. Tudball & Thomas (1973) *Eur. J. Biochem.* **40**, 25. **14**. Varadharajan, Dhanasekaran, Bonday, Rangarajan & Padmanaban (2002) *Biochem. J.* **367**, 321. **15**. Nandi (1978) *J. Biol. Chem.* **253**, 8872. **16**. Mukherjee & Dekker (1987) *J. Biol. Chem.* **262**, 14441.

N-[2(S)-[2(R)-Amino-3-mercaptopropylamino]-3-methylbutyl]-L-phenylalanyl-L-methionine

This cell-permeable Cys-Val-Phe-Met peptidomimetic (FW = 470.70 g/mol), also known as FTase inhibitor I, is a potent inhibitor of protein farnesyltransferase (IC$_{50}$ = 21 nM). It will also inhibit protein geranylgeranyltransferase at higher concentrations (IC$_{50}$ = 0.79 μM). **1**. Garcia, Rowell, Ackermann, Kowalczyk & Lewis (1993) *J. Biol. Chem.* **268**, 18415. **2**. Cox, Garcia, Westwick, *et al.* (1994) *J. Biol. Chem.* **269**, 19203.

N-4-[2(R)-Amino-3-mercaptopropyl]amino-2-naphthylbenzoyl-L-leucine

This peptidomimetic (FW = 465.62 g/mol), often called GGTI-297, inhibits protein geranylgeranyltransferase type I, K_i = 56 nM (1-5) and protein farnesyltransferase, K_i = 200 nM (6). **1.** Sebti & Hamilton (2000) *Meth. Enzymol.* **325**, 381. **2.** Gibbs (2001) *The Enzymes*, 3rd ed., **21**, 81. **3.** Yokoyama & Gelb (2001) *The Enzymes*, 3rd ed., **21**, 105. **4.** Sun, Qian, Hamilton & Sebti (1998) *Oncogene* **16**, 1467. **5.** Wilson, Erdman, Castellano & Maltese (1998) *Biochem. J.* **333**, 497. **6.** Yokoyama, Trobridge, Buckner, *et al.* (1998) *J. Biol. Chem.* **273**, 26497.

N-4-[2(R)-Amino-3-mercaptopropyl]amino-2-phenylbenzoyl-L-leucine

This protein farnesyltransferase inhibitor (FW = 415.56 g/mol), often referred to as GGTI-287, is a leucine-containing L-cysteinyl-L-valyl-L-leucyl-L-leucine peptidomimetic, in which the two central aminoacyl residues are replaced with 2-phenyl-4-aminobenzoic acid. It inhibits protein geranylgeranyltransferase I (IC$_{50}$ = 5 nM) and protein farnesyltransferase (IC$_{50}$ = 25 nM). **1.** Lerner, Qian, Hamilton & Sebti (1995) *J. Biol. Chem.* **270**, 26770. **2.** Qian, Vogt, Vasudevan, Sebti & Hamilton (1998) *Bioorg. Med. Chem.* **6**, 293.

N-[3-(Aminomethyl)benzyl]acetamidine

This slow, tight-binding inhibitor (FW$_{dihydrochloride}$ = 250.17 g/mol), targets the inducible nitric-oxide synthase, *or* iNOS (K_d = 7 nM). The K_d values for the human brain and endothelial enzymes are 2 and 50 μM, respectively. **1.** Kankuri, Vaali, Knowles, *et al.* (2001) *J. Pharmacol. Exp. Ther.* **298**, 1128. **2.** Thomsen, Scott, Topley, *et al.* (1997) *Cancer Res.* **57**, 3300. **3.** Laszlo & Whittle (1997) *Eur. J. Pharmacol.* **334**, 99. **4.** Garvey, Oplinger, Furfine, *et al.* (1997) *J. Biol. Chem.* **272**, 4959.

2-[[(1-Amino-3-methylbutyl)hydroxyphosphinyl]methyl]-4-methylpentanoate

This peptide analogue (FW = 279.32 g/mol), also known as LeuP[CH$_2$]Leu, inhibits leucyl aminopeptidase (K_i = 65 nM) and membrane alanyl aminopeptidase, *or* aminopeptidase N (K_i = 1.02 μM). **1.** Grembecka, Mucha, Cierpicki & Kafarski (2003) *J. Med. Chem.* **46**, 2641.

trans-1-Amino-4-methylcyclohexane

This sterically constrained cycloalkylamine (FW = 113.20 g/mol; CAS 2523-55-9), also known as *trans*-4-methylcyclohexylamine, inhibits spermidine synthase (*Reaction*: *S*-Adenosyl 3-(methylthio)propylamine + Putrescine $\rightleftharpoons$ 5'-*S*-Methyl-5'-thioadenosine + Spermidine), with the *cis* isomer a weaker inhibitor as is 1-amino-3-methylcyclohexane. **1.** Ikeguchi, Bewley & Pegg (2006) *J. Biochem.* **139**, 1. **2.** Shirahata, Takahashi, Beppu, Hosoda & Samejima (1993) *Biochem. Pharmacol.* **45**, 1897. **3.** Goda, Watanabe, Takeda, *et al.* (2004) *Biol. Pharm. Bull.* **27**, 1327. **4.** Haider, Eschbach, de Soudas Dias, *et al.* (2005) *Mol. Biochem. Parasitol.* **142**, 224. **5.** Enomoto, Nagasaki, Yamauchi, *et al.* (2006) *Anal. Biochem.* **351**, 229. **6.** Dufe, Lüersen, Eschbach, *et al.* (2005) *FEBS Lett.* **579**, 6037. **7.** Shirahata, Morohoshi & Samejima (1988) *Chem. Pharm. Bull. (Tokyo)* **36**, 3220. **8.** Shirahata, Morohohi, Fukai, Akatsu & Samejima (1991) *Biochem. Pharmacol.* **41**, 205. **9.** Burger, Birkholtz, Joubert, *et al.* (2007) *Bioorg. Med. Chem.* **15**, 1628.

1-Amino-3-methylcyclopentane

A sterically constrained cycloalkylamine (FW = 99.18 g/mol), consisting of a racemic mixture of *cis* and *trans* isomers, inhibits spermidine synthase (IC$_{50}$ = 15 μM). The active site has a relatively large hydrophobic cavity adjacent to a negatively charged binding site, with which a protonated amino group of putrescine interacts, while putrescine's other amino group is situated in the hydrophobic cavity as a free form to be aminopropylated by decarboxylated *S*-adenosylmethionine. **1.** Shirahata, Morohohi, Fukai, Akatsu & Samejima (1991) *Biochem. Pharmacol.* **41**, 205.

Aminomethylenebisphosphonate

This pyrophosphate analogue (FW = 190.01 g/mol) inhibits vacuolar H$^+$-translocating inorganic pyrophosphatase. Proton-translocating pyrophosphatases of from bacteria and plants are competitively inhibited by aminomethylene-diphosphonate and have inhibition constants below 2 μM, whereas liver mitochondrial pyrophosphatase is two orders of magnitude less sensitive to this compound, but extremely sensitive to imidodiphosphate. Aminomethylene-bisphosphonic acid derivatives also have herbicidal properties. **1.** Zhen, Baykov, Bakuleva & Rea (1994) *Plant Physiol.* **104**, 153. **2.** Gordon-Weeks, Parmar, Davies & Leigh (1999) *Biochem. J.* **337**, 373. **3.** Drozdowicz, Shaw, Nishi, *et al.* (2003) *J. Biol. Chem.* **278**, 1075. **4.** McIntosh & Vaidya (2002) *Int. J. Parasitol.* **32**, 1.

1-Amino-S-methylisothiourea

This thiourea derivative, (FW$_{free-base}$ = 105.16 g/mol; frequently supplied as the *p*-toluenesulfonate salt, with FW = 277.36 g/mol; *Abbreviation*: AMITU), also known as *S*-methylisothiosemicarbazide, is a strong, irreversible inhibitor of neuronal nitric-oxide synthase (IC$_{50}$ = 3 μM). It is a somewhat weaker inhibitor of the inducible (IC$_{50}$ = 24 μM) and endothelial (IC$_{50}$ = 103 μM) isozymes. **Target(s):** catalase (1); nitric-oxide synthase (2). **1.** Feinstein, Seaholm & Ballonoff (1964) *Enzymologia* **27**, 30. **2.** Wolff, Gauld, Neulander & Southan (1997) *J. Pharmacol. Exp. Ther.* **283**, 265.

2-Amino-6-methyl-5-(2-naphthylsulfanyl)thieno[2,3-*d*]pyrimidin-4(3*H*)-one

This dual TS-DHFR inhibitor (FW = 339.44 g/mol) inhibits human thymidylate synthase (IC$_{50}$ = 0.12 µM) and dihydrofolate reductase (IC$_{50}$ = 2.9 µM). The IC$_{50}$ value for *Toxoplasma gondii* dihydrofolate reductase is 44 nM. 1. Gangjee, Qiu, Li & Kisliuk (2008) *J. Med. Chem.* **51**, 5789.

N-{4-[(2-Amino-6-methyl-4-oxo-3,4-dihydro-thieno[2,3-*d*]pyrimidin-5-yl)sulfanyl]benzoyl}-L-glutamate

This dual TS-DHFR inhibitor (FW = 462.51 g/mol) inhibits dihydrofolate reductase (human, IC$_{50}$ = 20 nM) and thymidylate synthase (human, IC$_{50}$ = 40 nM). The IC$_{50}$ value for *Toxoplasma gondii* dihydrofolate reductase is 8 nM. Such dual inhibitors of the one-carbon folate cycle promise to exert enhanced pharmacologic potency as antineoplastic agents. 1. Gangjee, Qiu, Li & Kisliuk (2008) *J. Med. Chem.* **51**, 5789.

(*S*)-2-Amino-2-methyl-4-phosphonobutyrate

This glutamate isostere and presynaptic depressant (FW = 197.13 g/mol; FW$_{hydrochloride}$ = 232. g/mol; CAS 157381-42-5; Solubility: 100 mM in H$_2$O), also known as MAP4 and L-2-amino-4-phosphonobutyrate (L-AP4), is a selective Group III metabotropic glutamate receptor antagonist in some electrophysiological systems, but also acts as Group II/Group III agonist in certain neurochemical systems. MAP4 selectively and competitively antagonizes the depression of monosynaptic excitation produced by L-AP4, K_d = 22 µM (1). At 10x higher concentrations, MAP4 also antagonizes synaptic depression produced by L-CCG-I, but in an apparently non-competitive manner. MAP4 was virtually without effect on depression produced by (1*S*,3*R*)- or (1*S*,3*S*)-1-aminocyclopentane-1,3-dicarboxylate (4). Given the fact that mammalian glutamine synthetases bind both D- and L-glutamate as well as accommodate bulky side-chains on the α-carbon (5), one must consider the likelihood that (S)-2-amino-2-methyl-4-phosphonobutyrate will inhibit glutamine synthetase. Similar considerations may be at play for enzymes catalyzing the oligo-glutamylation of folates. 1. Knöpfel, Lukic, Leonard, *et al.* (1995) *Neuropharmacology* **34**, 1099. 2. Salt & Eaton (1995) *Neuroscience* **65**, 5. 3. Jane, Jones, Pook, Tse & Watkins (1994) *Brit. J. Pharmacol.* **112**, 809. 4. 1. Jane, Jones, Pook, Tse & Watkins (1994) *Br. J. Pharmacol.* **112**, 809. 5. Purich (1998) *Adv. Enzymol.* **72**, 9.

2-Amino-2-methyl-1,3-propanediol

This pH-buffer compound (FW$_{free-base}$ = 105.14 g/mol; Solubility: 250 g per 100 mL at 20°C; Abbreviation: AMPD) is a structural analogue of the common laboratory buffer known as Tris and is itself an effective buffer (pK_a = 8.79 at 25°C; dpK_a/dT = –0.029). **Target(s):** D-arabinose isomerase (1,2); L-fucose isomerase (1,2); sucrose α-glucosidase, *or* sucrase-isomaltase (3,7); oligo-1,6-glucosidase, *or* isomaltase (7); choline oxidase (4,5); D-serine ammonia lyase (6); amylo-α-1,6-glucosidase/4-α-glucanotransferase, *or* glycogen debranching enzyme (8); acid phosphatase (9); lysine 2-monooxygenase (10). 1. Yamanaka & Izumori (1975) *Meth. Enzymol.* **41**, 462. 2. Yamanaka & Izumori (1976) *Agric. Biol. Chem.* **40**, 439. 3. Vasseur, Frangne, Cauzac, Mahmood & Alvarado (1990) *J. Enzyme Inhib.* **4**, 15. 4. Webb (1966) *Enzyme and Metabolic Inhibitors*, vol. 2, p. 289, Academic Press, New York. 5. Wells (1954) *J. Biol. Chem.* **207**, 575. 6. Federiuk & Shafer (1981) *J. Biol. Chem.* **256**, 7416. 7. Kano, Usami, Adachi, Tatematsu & Hirano (1996) *Biol. Pharm. Bull.* **19**, 341. 8. Gillard & Nelson (1977) *Biochemistry* **16**, 3978. 9. Garrido, Lisa & Domenech (1988) *Mol. Cell. Biochem.* **84**, 41. 10. Flashner & Massey (1974) *J. Biol. Chem.* **249**, 2579.

2-Amino-2-methyl-1-propanol

This α-amino alcohol (FW$_{free-base}$ = 89.137 g/mol; CAS 124-68-5; M.P.= 30-31°C; Buffer Properties: pK_a = 9.69 at 25°C; dpK_a/dT = –0.032) is an analogue of Tris, choline, and ethanolamine. **Target(s):** sucrose α-glucosidase, *or* sucrase-isomaltase (1,7); ethanolamine ammonia-lyase (2,6); choline oxidase (3,4); ethanolamine kinase (5); acid phosphatase (8). 1. Vasseur, Frangne, Cauzac, Mahmood & Alvarado (1990) *J. Enzyme Inhib.* **4**, 15. 2. Stadtman (1972) *The Enzymes*, 3rd ed., **6**, 539. 3. Webb (1966) *Enzyme and Metabolic Inhibitors*, vol. 2, p. 289, Academic Press, New York. 4. Wells (1954) *J. Biol. Chem.* **207**, 575. 5. Weinhold & Rethy (1972) *Biochim. Biophys. Acta* **276**, 143. 6. Kaplan & Stadtman (1968) *J. Biol. Chem.* **243**, 1787. 7. Kano, Usami, Adachi, Tatematsu & Hirano (1996) *Biol. Pharm. Bull.* **19**, 341. 8. Garrido, Lisa & Domenech (1988) *Mol. Cell. Biochem.* **84**, 41.

2-Amino-4-methyl-6-propylpyridine

This substituted pyridine (FW = 150.23 g/mol) is a non-selective inhibitor of nitric-oxide synthase (human inducible, endothelial, and neuronal IC$_{50}$ = 0.01, 0.05, and 0.01 µM, respectively), demonstrating that flipping of the pyridine ring in these new inhibitors allows the piperidine to interact with different residues and confer excellent selectivity. 1. Connolly, Aberg, Arvai, *et al.* (2004) *J. Med. Chem.* **47**, 3320.

4-Amino-*N*^{10}-methylpteroyl-L-aspartate

This folic acid analogue (FW$_{free-acid}$ = 440.42 g/mol), which is identical to methotrexate, albeit with an aspartate residue in place of a glutamate substituent, is a weak inhibitor of many folate-dependent enzymes. **See also** Methotrexate 1. Baker (1967) *Design of Active-Site-Directed Irreversible Enzyme Inhibitors*, Wiley, New York. 2. Mead, Greenberg, Schrecker, Seeger & Tomcufcik (1965) *Biochem. Pharmacol.* **14**, 105.

4-Amino-N^{10}-methylpteroyl-L-glutamate, See *Methotrexate*

2-Amino-4-methylpyridine

This substituted pyridine, (FW = 108.14 g/mol), also known as 2-amino-4-picoline, is a strong inhibitor of inducible (K_d = 59 nM), neuronal (K_d = 111 nM), and endothelial (K_d = 136 nM) nitric oxide synthase. The inducible isozyme within RAW 264.7 cells has an IC_{50} value of 6 nM. **1**. Faraci, Nagel, Verdries, *et al.* (1996) *Brit. J. Pharmacol.* **119**, 1101. **2**. Boer, Ulrich, Klein, *et al.* (2000) *Mol. Pharmacol.* **58**, 1026. **3**. Connolly, Aberg, Arvai, *et al.* (2004) *J. Med. Chem.* **47**, 3320.

3-[(2-Amino-6-methyl-3-pyridinyl)methyl-5-(2-hydroxyethyl)-4-methylthiazolium Pyrophosphate

This thiamin pyrophosphate analogue ($FW_{hydrochloride}$ = 459.78 g/mol) selectively inhibits transketolase, K_i = 33 nM (1). When tested in nude mice xenografted with HCT-116 tumor cells and dosed with this inhibitor, transketolase activity is almost totally suppressed in blood, spleen, and tumor cells, with little effect on α-ketoglutarate dehydrogenase or glucose-6-phosphate dehydrogenase, two other thiamine-utilizing enzymes (2). 6-Trifluoromethyl, 6-cyano, 6-ethyl, 6-chloro, and 6-demethyl analogues also inhibit, with K_i values of 104, 57, 10, 46, and 36 nM, respectively (1). 4-Ethylthiazolium, 4-(hydroxymethyl)-thiazolium, 2-methylthiazolium, 2-ethyl-4-methyl-thiazolium, and 2,4-dimethylthiazolium derivatives show K_i values of 40, 25, 36, 75, and 25 nM, respectively (1). The 1-(R) stereoisomer is only weakly inhibitory. **1**. Esakova, Meshalkina, Golbik, Hübner & Kochetov (2004) *Eur. J. Biochem.* **271**, 4189. **2**. Thomas, Le Huerou, De Meese, *et al.* (2008) *Bioorg. Med. Chem. Lett.* **18**, 2206.

4-Amino-1,8-naphthalimide

This PARP inhibitor (FW = 212.21 g/mol) is a strong inhibitor of NAD:ADP-ribosyltransferase *or* poly(ADP-ribose) polymerase (IC_{50} = 0.18 μM). In addition, 4-Amino-1,8-naphthalimide is a radiation sensitizer at non-toxic and low concentrations. **1**. Thiemermann, Bowes, Myint & Vane (1997) *Proc. Natl. Acad. Sci. U.S.A.* **94**, 679. **2**. Banasik, Komura, Shimoyama & Ueda (1992) *J. Biol. Chem.* **267**, 1569. **3**. Schlicker, Peschke, Burkle, Hahn & Kim (1999) *Int. J. Radiat. Biol.* **75**, 91.

4-Amino-N-[4-(3-nitrophenyl)thiazol-2-yl]benzenesulfonamide

This sulfonamide (FW = 348.40 g/mol) inhibits kynurenine 3-monooxygenase (IC_{50} = 40 nM), blocking rat and gerbil kynurenine 3-hydroxylase after oral administration, with an ED_{50} value of 3-5 μmol/kg in gerbil brain. **1**. Röver, Cesura, Huguenin, Kettler & Szente (1997) *J. Med. Chem.* **40**, 4378.

8-Amino-2'-nordeoxyguanosine

This nucleoside analogue (FW = 270.25 g/mol), also known as 9-(1,3-dihydroxy-2-propoxymethyl)-8-methylguanine, 2,8-dimethyl-9-(1,3-dihydroxy-2-propoxymethyl)-6-hydroxypurine and 8-aminoganciclovir, is a moderately strong inhibitor of human purine-nucleoside phosphorylase: K_i = 0.26 μM. **1**. Stein, Stoeckler, Li, *et al.* (1987) *Biochem. Pharmacol.* **36**, 1237.

L-2-Amino-4-oxo-5-chloropentanoate

This chloromethyl ketone (FW = 165.58 g/mol), also known as 5-chloro-4-oxo-L-norvaline (CONV), is a glutamine and asparagine analogue that inactivates many glutamine-dependent biosynthetic enzymes by alkylating an active-site cysteinyl residue in the glutamine-hydrolyzing subunit. **Target(s):** carbamoyl-phosphate synthetase II, glutamine-hydrolyzing (1,5,8,10,15,19); glutamate synthase (2,5); γ-glutamylcysteine synthetase (3,5,6,12,14,26); asparagine synthetase (4 7,16,17,20-23); anthranilate synthetase (5,18,27); glutaminase (5,11); phosphoribosylformyl-glycinamidine synthetase (5); CTP synthetase (5); glutamine transport (9); aspartate-β semialdehyde dehydrogenase (13); GMP synthetase (24); carbamoyl-phosphate synthetase I (25); asparaginase (28); amidophospho-ribosyl-transferase (29). **1**. Kaseman & Meister (1985) *Meth. Enzymol.* **113**, 305. **2**. Meister (1985) *Meth. Enzymol.* **113**, 327. **3**. Seelig & Meister (1985) *Meth. Enzymol.* **113**, 379. **4**. Horowitz, Nelson & Meister (1970) *Meth. Enzymol.* **17A**, 726. **5**. Pinkus (1977) *Meth. Enzymol.* **46**, 414. **6**. Meister (1995) *Meth. Enzymol.* **252**, 26. **7**. Meister (1974) *The Enzymes*, 3rd ed., **10**, 561. **8**. Khedouri, Anderson & Meister (1966) *Biochemistry* **5**, 3552. **9**. Novogrodsky, Nehring, Jr., & A. Meister (1979) *Proc. Natl. Acad. Sci. U.S.A.* **76**, 4932. **10**. Lusty & Liao (1993) *Biochemistry* **32**, 1278. **11**. Shapiro, Morehouse & Curthoys (1982) *Biochem. J.* **207**, 561. **12**. Sekura & Meister (1977) *J. Biol. Chem.* **252**, 2606. **13**. Biellmann, Eid, Hirth & Jornvall (1980) *Eur. J. Biochem.* **104**, 59. **14**. Beamer, Griffith, Gass, Anderson & Meister (1980) *J. Biol. Chem.* **55**, 11732. **15**. Casey & Anderson (1983) *J. Biol. Chem.* **258**, 8723. **16**. Horowitz & Meister (1972) *J. Biol. Chem.* **247**, 6708. **17**. Jayaram, Cooney, Milman, Homan & Rosenbluth (1976) *Biochem. Pharmacol.* **25**, 1571. **18**. Tamir & Srinivasan (1970) *Meth. Enzymol.* **17A**, 401. **19**. Shaw (1970) *The Enzymes*, 3rd ed., **1**, 91. **20**. Lea & Fowden (1975) *Proc. R. Soc. Lond. B Biol. Sci.* **192**, 13. **21**. Mehlhaff & Schuster (1991) *Arch. Biochem. Biophys.* **284**, 143. **22**. Milman & Cooney (1979) *Biochem. J.* **181**, 51. **23**. Reitzer & Magasanik (1982) *J. Bacteriol.* **151**, 1299. **24**. Zalkin & Truitt (1977) *J. Biol. Chem.* **252**, 5431. **25**. Lusty (1978) *Eur. J. Biochem.* **85**, 373. **26**. Griffith & Mulcahy (1999) *Adv. Enzymol. Relat. Areas Mol. Biol.* **73**, 209. **27**. Goto, Zalkin, Keim & Heinrikson (1976) *J. Biol. Chem.* **251**, 941. **28**. Lea, Fowden & Miflin (1978) *Phytochemistry* **17**, 217. **29**. Messenger & Zalkin (1979) *J. Biol. Chem.* **254**, 3382.

N-{4-[4-(2-Amino-4-oxo-4,7-dihydro-3H-pyrrolo[2,3-d]pyrimidin-6-yl)butyl]benzoyl}-L-glutamate

This glutamate-containing folate/pemetrexed analogue (FW = 453.46 g/mol) inhibits the *de novo* purine biosynthesis by targeting phosphoribosyl glycinamide .formyltransferase (IC_{50} = 6.8 nM). Its combined properties (*e.g.*, selective FR targeting, lack of transport by the reduced folate carrier, and GARFTase inhibition) making it a promising antitumor agent. **1**. Deng, Wang, Cherian, *et al.* (2008) *J. Med. Chem.* **51**, 5052.

N-[3-[4-Amino-2-oxo-1-(1*H*-pyrazol-5-ylmethyl)-1,2-dihydropyrimidin-5-yl]prop-2-yn-1-yl]ethanesulfonamide

This substituted pyrimidine (FW = 337.38 g/mol) inhibits 4-(cytidine-5'-diphospho)-2-*C*-methyl-D-erythritol kinase, an enzyme in the non-mevalonate pathway for isoprenoid biosynthesis (K_i = 1.6 µM for the *Mycobacterium tuberculosis* enzyme). **1**. Hirsch, Lauw, Gersbach, *et al.* (2007) *ChemMedChem* **2**, 806.

Aminooxyacetate

This amino acid analogue (FW$_{free-acid}$ = 91.07 g/mol; CAS 645-88-5), also known as aminooxyacetic acid *O*-(carboxymethyl)hydroxylamine, is a good nucleophile that readily reacts with the aldehyde functional group in the vitamin B$_6$ cofactor pyridoxal 5-phosphate (PLP), thereby inhibiting numerous PLP-dependent systems, especially transaminases (aminotransferases). Like methoxyhydroxylamine, this hydroxylamine analogue also reacts rapidly (<10 seconds) at pH 5 to form complexes having absorption maxima in the 370-390 nm range. **1**. Roberts (1952) *J. Biol. Chem.* **198**, 95. **2**. Roberts & Simonsen (1963) *Biochem. Pharmacol.* **12**, 113.

6-Aminopenicillanate

This β-lactam (FW = 216.26 g/mol; CAS 551-16-6; Abbreviation: 6-APS), also known as 6-aminopenicillanic acid and (2*S*,5*R*,6*R*)-6-amino-3,3-dimethyl-7-oxo-4-thia-1-azabicyclo[3.2.0]heptane-2-carboxylic acid, is a common starting reagent in the preparation of semi-synthetic penicillins and as an intermediate in the biosynthesis of naturally occurring penicillins. Note that 6-APS possesses no antibiotic activity on its own. 6-Aminopenicillanate is a product of the reaction catalyzed by penicillin amidase. The structure and chemistry of this metabolite was first determined by Ernst Chain, who shared the Nobel Prize in Physiology or Medicine in 1945 for his work on penicillins. **Target(s):** serine-type D-Ala-D-Ala carboxypeptidase, *or* penicillin-binding protein-5 (1,12); D-stereospecific aminopeptidase (2,13); penicillin-binding protein-4 (3); γ-aminobutyrate aminotransferase (4); cytosol alanyl aminopeptidase (5); D-alanine carboxypeptidase (6,8); penicillin amidase (7,9-11). **1**. Wilkin (1998) in *Handb. Proteolytic Enzymes* (Barrett, Rawlings & Woessner, eds.), p. 418, Academic Press, San Diego. **2**. Asano (1998) in *Handb. Proteolytic Enzymes* (Barrett, Rawlings & Woessner, eds.), p. 430, Academic Press, San Diego. **3**. Wilkin (1998) in *Handb. Proteolytic Enzymes* (Barrett, Rawlings & Woessner, eds.), p. 435, Academic Press, San Diego. **4**. Hopkins & Silverman (1992) *J. Enzyme Inhib.* **6**, 125. **5**. Starnes, Szechinski & Behal (1982) *Eur. J. Biochem.* **124**, 363. **6**. Storm, Blumberg & Strominger (1974) *J. Bacteriol.* **117**, 783. **7**. Chiang & Bennett (1967) *J. Bacteriol.* **93**, 302. **8**. Blumberg & Strominger (1971) *Proc. Natl. Acad. Sci. U.S.A.* **68**, 2814. **9**. Savidge & Cole (1975) *Meth. Enzymol.* **43**, 705. **10**. Warburton, Balasingham, Dunnil & Lilly (1972) *Biochim. Biophys. Acta* **284**, 278. **11**. Balasingham, Warburton, Dunnil & Lilly (1972) *Biochim. Biophys. Acta* **276**, 250. **12**. Frère, Kelly, Klein &

Ghuysen (1982) *Biochem. J.* **203**, 223. **13**. Asano, Kato, Yamada & Kondo (1992) *Biochemistry* **31**, 2316.

(*S*)-2-Aminopentane-1,5-dithiol

This dithiol (FW = 151.30 g/mol), also known as EC27, inhibits membrane alanyl aminopeptidase (K_i = 32 nM). **1**. Zini, Fournie-Zaluski, Chauvel, *et al.* (1996) *Proc. Natl. Acad. Sci. U.S.A.* **93**, 11968. **2**. Llorens-Cortes (1998) *C. R. Seances Soc. Biol. Fil.* **192**, 607. **3**. Bauvois & Dauzonne (2006) *Med. Res. Rev.* **26**, 88.

p-Aminophenylarsine Oxide

This toxic thiol-reactive arsenical, (FW = 183.04 g/mol; CAS 73791-39-6 (for the dehydrate)), also known as reduced Atoxyl, inhibits many enzymes. In 1905, Thomas and Breinl reported the successful use of Atoxyl in the treatment of trypanosomiasis, albeit at high doses. **CAUTION:** Avoid inhalation of dry powder. **Target(s):** D-amino-acid oxidase (1-3); β-amylase (4); choline oxidase (5); cytochrome *c* oxidase (6); lipase (2,3); monoamine oxidase (3); nicotinamide methyltransferase (7); papain (8); pyruvate decarboxylase (3); succinate oxidase (3); urease (9); protein-disulfide isomerase (10); pantetheine hydrolase (11); amylo-α-1,6-glucosidase/4-α-glucano-transferase, *or* glycogen debranching enzyme (12); arginyltransferase (13). **1**. Krebs (1951) *The Enzymes*, 1st ed., **2** (part 1), 499. **2**. Singer (1948) *J. Biol. Chem.* **174**, 11. **3**. Singer & Barron (1945) *J. Biol. Chem.* **157**, 241. **4**. Ghosh (1958) *Proc. Intern. Symp. Enzyme Chem., Tokyo Kyoto 1957*, p. 532, Maruzen, Tokyo. **5**. Rothschild, Cori & Barron (1954) *J. Biol. Chem.* **208**, 41. **6**. Kreke, Schaefer, Seibert & Cook (1950) *J. Biol. Chem.* **185**, 469. **7**. Cantoni (1951) *J. Biol. Chem.* **189**, 745. **8**. Bersin (1934) *Z. Physiol. Chem.* **222**, 177. **9**. Yall & Green (1952) *Proc. Soc. Exptl. Biol. Med.* **79**, 306. **10**. Gallina, Hanley, Mandel, *et al.* (2002) *J. Biol. Chem.* **277**, 50579. **11**. Ricci, Nardini, Chiaraluce, Dupre & Cavallini (1986) *Biochim. Biophys. Acta* **870**, 82. **12**. Liu, de Castro, Takrama, Bilous, Vinayagamoorthy, Madsen & Bleackley (1993) *Arch. Biochem. Biophys.* **306**, 232. **13**. Li & Pickart (1995) *Biochemistry* **34**, 139.

m-Aminophenylboronic Acid

This arylboronic acid (FW = 136.95 g/mol; CAS 280563-63-5), also known as *m*-aminobenzeneboronic acid and *m*-aminophenylboron dihydroxide, inhibits many enzymes, including microbial β-lactamases. Affinity chromatography on *m*-aminophenylboronic acid agarose columns also affords a simple way to measure glycosylated albumin. **Target(s):** lecithin:cholesterol acyltransferase (1); lipase, *or* triacylglycerol lipase (2,3); β-lactamase (4,7,9-11); lipoprotein lipase (5); pantothenase (6,12); ADP-ribosyltransferase (8); pancreatic elastase (13); NAD nucleosidase, *or* NADase (8); peptidyltransferase (14); glycerol-3-phosphate *O*-acyltransferase (15). **1**. Jauhiainen, Ridgway & Dolphin (1987) *Biochim. Biophys. Acta* **918**, 175. **2**. Uusi-Oukari, Ehnholm & Jauhiainen (1996) *J. Chromatogr. B Biomed. Appl.* **682**, 233. **3**. Garner (1980) *J. Biol. Chem.* **255**, 5064. **4**. Usher, Blaszczak, Weston, Shoichet & Remington (1998) *Biochemistry* **37**, 16082. **5**. Jackson (1983) *The Enzymes*, 3rd ed., **16**, 141. **6**. Airas (1988) *Biochem. J.* **250**, 447. **7**. Amicosante, Felici, Segatore, *et al.* (1989) *J. Chemother.* **1**, 394. **8**. Penyige, Deak, Kalmanczhelyi & Barabas (1996) *Microbiology* **142**, 1937. **9**. Voladri, Lakey, Hennigan, *et al.* (1998) *Antimicrob. Agents Chemother.* **42**, 1375. **10**. Tondi, Powers,

Caselli, *et al.* (2001) *Chem. Biol.* **8**, 593. **11**. Bauvois, Ibuka, Celso, *et al.* (2005) *Antimicrob. Agents Chemother.* **49**, 4240. **12**. Airas (1978) *Biochemistry* **17**, 4932. **13**. Smoum, Rubinstein & Srebnik (2003) *Bioorg. Chem.* **31**, 464. **14**. Cerná & Rychlík (1980) *FEBS Lett.* **119**, 342. **15**. Ganesh Bhat, Wang, *et al.* (1999) *Biochim. Biophys. Acta* **1439**, 415.

2-[[(1-Amino-3-phenylpropyl)hydroxyphosphinyl]methyl]-3-phenylpropionate

This new class of highly potent transition-state analogue (FW = 361.38 g/mol), also known as hPheP[CH₂]Phe and 2-[[(1-amino-3-phenylpropyl)hydroxyphosphinyl]methyl]-3-phenylpropionic acid, most likely mimics a catalytically required tetrahedral adduct, thereby inhibiting leucyl aminopeptidase (K_i = 66 nM) and aminopeptidase N (K_i = 276 nM). The X-ray structure of bovine lens leucine aminopeptidase complexed with the phosphonic acid analogue of leucine (LeuP) guided the structure-based design of novel inhibitors and for the analysis of their interactions with the substrate binding site. These inhibitors were designed by modification of phosphonic group in the LeuP structure to account for the docking of substituents located at the S' side of the enzyme. **1**. Grembecka, Mucha, Cierpicki & Kafarski (2003) *J. Med. Chem.* **46**, 2641.

4-(2-Aminophenylthio)butylphosphonate

This phosphonate (FW$_{free-acid}$ = 261.28 g/mol), also known as 4-(2-aminophenylthio)butylphosphonic acid, is a transition-state analogue that inhibits *Salmonella typhimurium* tryptophan synthase (IC$_{50}$ = 178 nM). This and related inhibitors were designed to mimic the transition state formed during the α-reaction of the enzyme and, as expected, have affinities much greater than that of the natural substrate indole-3-glycerol phosphate or its nonhydrolyzable analogue indole propanol phosphate (IPP). **1**. Sachpatzidis, Dealwis, Lubetsky, *et al.* (1999) *Biochemistry* **38**, 12665.

DL-*E*-2-Amino-5-phosphono-3-pentenoate

This substituted α-amino acid (FW$_{hydrochloride}$ = 231.57 g/mol) is an irreversible inhibitor of threonine synthase, K_i = 0.4 mM, k_{inact} = 0.25 min^{-1} (1) and a strong competitive inhibitor of cystathionine γ-synthase, K_i = 1.1 µM (2-4). The *Z*-isomer is a stronger inhibitor of *Escherichia coli* threonine synthase. **1**. Laber, Gerbling, Harde, *et al.* (1994) *Biochemistry* **33**, 3413. **2**. Kreft, Townsend, Pohlenz & Laber (1994) *Plant Physiol.* **104**, 1215. **3**. Clausen, Wahl, Messerschmidt, *et al.* (1999) *Biol. Chem.* **380**, 1237. **4**. Steegborn, Laber, Messerschmidt, Huber & Clausen (2001) *J. Mol. Biol.* **311**, 789.

N-{[(2S,3R)-3-Amino-5-phosphono-2-sulfhydryl]pentanoyl}-L-isoleucyl-L-aspartate

This phosphonopeptide (FW = 457.44 g/mol), also known as *N*-{[(2S,3R)-3-amino-2-mercapto-5-phosphono]pentanoyl}-L-isoleucyl-L-aspartate, inhibits glutamyl aminopeptidase (K_i = 12 nM), neprilysin (K_i = 5.4 nM), and peptidyl-dipeptidase A (K_i = 2200 nM). The (2R,3R) diastereoisomer also inhibits glutamyl aminopeptidase (K_i = 910 nM). **1**. David, Bischoff, Meudal, *et al.* (1999) *J. Med. Chem.* **42**, 5197.

1-Aminoproline

D-enantiomer   **L-enantiomer**

These *N*-substituted amino acids (FW = 130.15 g/mol), a component of the naturally occurring vitamin B₆ antagonist linatine (*N*-γ-L-glutamyl)amino-D-proline; CAS 10139-06-7), inactivates D-lysine 5,6-aminomutase, a pyridoxal 5-P/adenosylcobalamin-derpendent enzyme that catalyzes the reversible and nearly isoenergetic transformation of D-lysine into 2,5-diaminohexanoate (2,5-DAH) and of l-β-lysine into 3,5-diaminohexanoate (3,5-DAH). **1**. Morley & Stadtman (1972) *Biochemistry* **11**, 600. **2**. Stadtman (1972) *The Enzymes*, 3rd ed., **6**, 539.

β-Aminopropionitrile

This veterinary antirheumatic drug (FW = 70.09 g/mol; CAS 151-18-8; Symbol: BAPN), also known as 2-cyanoethylamine, targets lysyl oxidase (LOX; IC$_{50}$ = 10 µM), a Cu(II)-containing amine oxidase that catalyzes the oxidation of the ε-amino group of peptidyl lysine to peptidyl aldehyde, consuming O₂ and H₂O, and producing ammonia and H₂O₂ (1). β-Aminopropionitrile also decreases tissue colonization by MDA-MB-231-Luc2 cells following their intracardiac administration (2). BAPN pretreatment significantly reduces the number of metastases that develop as well as the whole animal tumor burden. The former result suggests that the numbers of cells 'seeding' a given tissue site is reduced by BAPN pre-treatment (2). When BAPN initiated 3 hours after tumor cell injection, there is less of an inhibition on metastasis development, suggesting that this agent is most effective when it acts on tumor cells and/or host tissues during the initial phase of the extravasation process (2). β-aminopropionitrile also attenuates body weight increase and fat mass increase otherwise observed in obese animals, shifting adipocyte size toward smaller adipocytes (3). Reference (4) describes prodrugs that can release BAPN selectively under hypoxic conditions. *Note*: Because BAPN metabolism yields cyanide, the specificity of this agent may be compromised (***See** Cyanide*). **1**. Narayanan, Siegel & Martin (1972) *Biochem. Biophys. Res. Commun.* **46**, 745. **2**. Bondareva, Bondy, Ayres, *et al.* (2009) *PLoS One* **4**, e5620. **3**. Miana, Galán, Martínez-Martínez, *et al.* (2015) *Dis. Model Mech.* **8**, 543. **4**. Granchi, Funaioli, Erler, (2009) *ChemMedChem.* **4**, 1590.

N-(3-Aminopropyl)-*exo*-2-aminonorbornane

This substrate analogue (FW = 169.29 g/mol) inhibits spermine synthase (IC$_{50}$ = 0.27 µM). The *endo* analogue is a much weaker inhibitor (IC$_{50}$ = 10 µM). The active site possesses a relatively large hydrophobic cavity lying adjacent to a negatively charged site, to which a protonated amino group of putrescine binds, with another amino group of putrescine being situated in the hydrophobic cavity as a free form to be aminopropylated by decarboxylated S-adenosylmethionine. **1**. Shirahata, Morohohi, Fukai, Akatsu & Samejima (1991) *Biochem. Pharmacol.* **41**, 205. **2**. Shirahata, Takahashi, Beppu, Hosoda & Samejima (1993) *Bochem. Pharmacol.* **45**, 1897.

2-[({2-[(3-Aminopropyl)amino]phenyl}sulfonyl)amino]-5,6,7,8-tetrahydro-1-naphthalenecarboxylic Acid

This orally active anthranilic acid sulfonamide (FW = 403.50 g/mol) reversibly inhibits methionyl aminopeptidase (IC$_{50}$ = 13 nM), showing good cellular activity in both proliferation and methionine processing assays. **1**. Sheppard, Wang, Kawai, *et al.* (2006) *J. Med. Chem.* **49**, 3832.

N-(3-Aminopropyl)-1,4-diaminobut-2-ene

trans-isomer

cis-isomer

This spermidine analogue (FW$_{free-base}$ = 144.24 g/mol) inhibits human deoxyhypusine synthase (EC 2.5.1.46), the enzyme that catalyzes the posttranslational modification of in Eukaryotic Translation Initiation Factor 5A-1, converting a lysyl residue into deoxyhypusine [*or* N^ε-(4-amino-2-hydroxybutyl)-lysine]. The synthase transfers the 4-aminobutyl moiety from spermidine to a specific lysine residue in the eIF5A precur   sor protein, forming the deoxyhypusine-containing eIF5A intermediate. Both the *cis* and *trans* isomers of *N*-(3-aminopropyl)-1,4-diaminobut-2-ene are inhibitory. **1**. Park, Wolff, Folk & Park (2003) *J. Biol. Chem.* **278**, 32683.

N,N-(3-Aminopropyl)-D-glucoamidine

This glucose analogue (FW = 232.26 g/mol) inhibits α-mannosidase (K_i = 90 nM), β-mannosidase (K_i = 0.41 μM), α-glucosidase (K_i = 14 μM), β-glucosidase (K_i = 71 μM), α-galactosidase (K_i = 44 μM), β-galactosidase (K_i = 44 μM), α-fucosidase (K_i = 91 μM), and β-*N* acetylhexosaminidase (K_i = 252 μM). Its likely mode of action is that protonation of the ring nitrogen forms a positively charged mimic of oxacarbenium ion intermediates formed in enzyme-catalyzed hydrolysis of glycosides. **1**. Heck, Vincent, Murray, *et al.* (2004) *J. Amer. Chem. Soc.* **126**, 1971.

5-(2-Aminopropyl)indole

This psychoactive stimulant (FW = 174.25 g/mol; CAS 3784-30-3; *Symbol*: 5-IT) is a recreationally abused substance associated with fatal and non-fatal intoxications, showing acute symptoms of monoaminergic (*e.g.* serotonin) toxicity, resulting from increased circulating serotonin levels (1,2). 5-IT targets the type A isozyme of human monoamine oxidase, *or* MAO-A (IC$_{50}$ = 1.6 μM and K_i = 0.25 μM), whereas MAO-B inhibition is not observed, even at 500 μM (2). For reference, clorgyline, harmaline, toloxatone, and moclobemide have IC$_{50}$ values of 16 nM, 20 nM, 6.7 μM and >500 μM. In Sweden, 5-IT is scheduled as a hazardous substance (September, 2012), with prohibitions on possession and sale. Its isomer, 3-(2-aminopropyl)indole (*or* α-methyltryptamine, AMT) may represent yet another factor in 5-IT toxicity. **1**. Coppola & Mondola (2013) *Am. J. Psychiatry* **170**, 226. **2**. Herraiz & Brandt (2013) *Drug Test Anal.* **6**, 607.

6-Amino-9-D-psicofuranosylpurine, *See* Psicofuranine

Aminopterin

This folate antagonist (FW$_{free-acid}$ = 440.42 g/mol; CAS 54-62-6; Solubility: 50 mg/mL in 2 M NaOH (store for 1-2 days at 4°C)), also known as 4-amino-4-deoxyfolate and 4-aminofolate, was the first antifolate drug used clinically in the treatment of acute lymphoblastic leukemia, beginning in the late 1940s. (*See also Methotrexate*) Upon transport into cells via the folate transporter, aminopterin is polyglutamylated, which in turn, inhibits dihydrofolate reductase. Although considerably more potent than methotrexate, aminopterin is also far more toxic. Indeed, aminopterin has been used as a rodenticide. Aminopterin-polyglutamate is degraded intracellularly by γ-glutamyl hydrolase. **Target(s):** dihydrofolate reductase (1-6,23,32); dihydropterine reductase (7,9); dihydropterine reductase, NADPH-dependent (8); GTP cyclohydrolase I, weakly inhibited (10); NAD(P)H dehydrogenase, quinone-dependent (11); NADPH dehydrogenase, quinone-dependent (11); glutamate dehydrogenase (12); lactate dehydrogenase (12); malate dehydrogenase (12); serine hydroxymethyltransferase, *or* glycine hydroxymethyl-transferase (13,31); mandelate 4-monooxygenase (14); phenylalanine 4-monooxygenase (15); xanthine oxidase (16,17,21); choline oxidase (19); purine-nucleoside phosphorylase (22); formimidoyltetrahydrofolate cyclodeaminase (24); sepiapterin deaminase (25,26); methenyltetrahydrofolate cyclohydrolase, weakly inhibited (27); methionyl-tRNA formyltransferase (28); deoxycytidylate 5-hydroxymethyltransferase (29); glutamate formimidoyltransferase (30); thymidylate synthase (32-36); anthranilate 3-monooxygenase (37); benzoate 4-monooxygenase, weakly inhibited (38). **1**. Baker (1959) *Cancer Chemother. Rep.* **4**, 1. **2**. Mathews, Scrimgeour & Huennekens (1963) *Meth. Enzymol.* **6**, 364. **3**. Baker (1967) *Design of Active-Site-Directed Irreversible Enzyme Inhibitors*, Wiley, New York. **4**. Futterman (1957) *J. Biol. Chem.* **228**, 1031. **5**. Bertino, Gabrio & Huennekens (1960) *Biochem. Biophys. Res. Commun.* **3**, 461. **6**. Osborn, Freeman & Huennekens (1958) *Proc. Soc. Exptl. Biol. Med.* **97**, 429. **7**. Hasegawa & Nakanishi (1987) *Meth. Enzymol.* **142**, 103. **8**. Hasegawa & Nakanishi (1987) *Meth. Enzymol.* **142**, 111. **9**. Shen, Smith, Davis, Brubaker & Abell (1982) *J. Biol. Chem.* **257**, 7294. **10**. Shen, Alam & Zhang (1988) *Biochim. Biophys. Acta* **965**, 9. **11**. Koli, Yearby, Scott & Donaldson (1969) *J. Biol. Chem.* **244**, 621. **12**. Vogel, Snyder & Schulman (1963) *Biochem. Biophys. Res. Commun.* **10**, 97. **13**. Ramesh & Appaji Rao (1980) *Biochem. J.* **187**, 623. **14**. Bhat & Vaidyanathan (1976) *Arch. Biochem. Biophys.* **176**, 314. **15**. Kaufman & Levenberg (1959) *J. Biol. Chem.* **234**, 2683. **16**. Hofstee (1949) *J. Biol. Chem.* **179**, 633. **17**. Williams (1950) *J. Biol. Chem.* **187**, 47. **18**. Farber, Diamond, Mercer, Sylvester & Wolff (1948) *N. Engl. J. Med.* **238**, 787. **19**. Dinning, Keith, Davis & Day (1950) *Arch. Biochem.* **27**, 89. **20**. Osborn, Freeman & Huennekens (1958) *Proc. Soc. Exp. Biol. Med.* **97**, 429. **21**. Robinson, Pilot & Meany (1990) *Physiol. Chem. Phys. Med. NMR* **22**, 95. **22**. Lewis (1978) *Arch. Biochem. Biophys.* **190**, 662. **23**. Erickson & Mathews (1972) *J. Biol. Chem.* **247**, 5661. **24**. Uyeda & Rabinowitz (1967) *J. Biol. Chem.* **242**, 24. **25**. Tsusue (1971) *J. Biochem.* **69**, 781. **26**. Tsusue & Mazda (1977) *Experientia* **33**, 854. **27**. Suzuki & Iwai (1973) *Plant Cell Physiol.* **14**, 319. **28**. Gambin, Crosti & Bianchetti (1980) *Biochim. Biophys. Acta* **613**, 73. **29**. Lee, Gautam-Basak, Wooley & Sander (1988) *Biochemistry* **27**, 1367. **30**. Tabor & Wyngarden (1959) *J. Biol. Chem.* **234**, 1830. **31**. Rao & Rao (1982) *Plant Physiol.* **69**, 11. **32**. McCuen & Sirotnak (1975) *Biochim. Biophys. Acta* **384**, 369. **33**. Chalabi & Gutteridge (1977) *Biochim. Biophys. Acta* **481**, 71. **34**. Bisson & Thorner (1981) *J. Biol. Chem.* **256**, 12456. **35**. So, Wong & Ko (1994) *Exp. Parasitol.* **79**, 526. **36**. Dolnick & Cheng (1978) *J. Biol. Chem.* **253**, 3563. **37**. Nair & Vaidyanathan (1965) *Biochim. Biophys. Acta* **110**, 521. **38**. McNamee & Durham (1985) *Biochem. Biophys. Res. Commun.* **129**, 485.

2-Aminopurine

This substituted purine (FW = 135.13 g/mol) is a mutagen for many organisms. It is metabolized to 2-aminopurine-2'-deoxyriboside 5'-triphosphate and incorporated into DNA. Aminopurine is also a competitive inhibitor with respect to ATP of double-stranded RNA-dependent protein kinase. **Target(s):** adenosine deaminase (1); double-stranded RNA-dependent protein kinase (2-5); Ca^{2+}/calmodulin-dependent protein kinase (6); casein kinase (7); nonspecific serine/threonine protein kinase (7); methionine S-adenosyltransferase, weakly inhibited (8); hypoxanthine(guanine) phosphoribosyltransferase (9); xanthine phosphoribosyltransferase (9). **1**. Akedo, Nishihara, Shinkai, Komatsu & Ishikawa (1972) *Biochim. Biophys. Acta* **276**, 257. **2**. De Benedetti, Williams & Baglioni (1985) *J. Virol.* **54**, 408. **3**. De Benedetti & Baglioni (1983) *J. Biol. Chem.* **258**, 14556. **4**. Farrell, Balkow, Hunt, Jackson & Trachsel (1977) *Cell* **11**, 187. **5**. Hu & Conway (1993) *J. Interferon Res.* **13**, 323. **6**. Boulton, Gregory & Cobb (1991) *Biochemistry* **30**, 278. **7**. Mottet, Ruys, Demazy, Raes & Michiels (2005) *Int. J. Cancer* **117**, 764. **8**. Berger & Knodel (2003) *BMC Microbiol.* **3**, 12. **9**. Naguib, Iltzsch, el Kouni, Panzica & el Kouni (1995) *Biochem. Pharmacol.* **50**, 1685.

2-Aminopurine 9-β-D-Ribofuranoside

This adenosine analogue (FW = 267.24 g/mol), also known as 2-amino-9-β-D-ribofuranosylpurine, inhibits adenosine deaminase, K_i = 4 μM (1) and adenosine kinase (2). The latter is a key intracellular enzyme regulating intracellular and extracellular concentrations of adenosine (ADO), itself an powerful endogenous modulator of intercellular signalling that reduces cell excitability during tissue stress and trauma. **1**. Simon, Bauer, Tolman & Robins (1970) *Biochemistry* **9**, 573. **2**. Miller, Adamczyk, Miller, *et al.* (1979) *J. Biol. Chem.* **254**, 2346.

8-Aminopurine 9-β-D-Ribofuranoside

This nucleoside (FW = 267.24 g/mol) inhibits adenosine kinase, a purine salvage enzyme that phosphorylates adenosine to AMP. In *Mycobacterium tuberculosis*, adenosine kinase also catalyzes an essential step in the conversion of 2-methyl-Ado to a compound with selective antimycobacterial activity. This study revealed the presence of a hydrophobic pocket near the N^6- and N^1-positions that can accommodate substitutions at least as large as a benzyl group. **1**. Long & Parker (2006) *Biochem. Pharmacol.* **71**, 1671.

N-{(E)-3-[(2R,3S,4R,5R)-5-(6-Amino-6H-purin-9-yl)-3,4-dihydroxytetrahydrofuran-2-yl]prop-2-enyl}-2,3-dihydroxy-5-nitrobenzene-1-carboxamide

This bisubstrate analogue (FW = 473.40 g/mol) inhibits catechol O-methyltransferase (IC_{50} = 9 nM), a druggable target for the treatment of Parkinson's Disease. There is a large dependence of binding affinity on inhibitor preorganisation and the length of the linker between nucleoside and catechol moieties. This bisubstrate analogue exhibits competitive kinetics for the SAM and mixed inhibition kinetics for the catechol binding site. Its bisubstrate binding mode was confirmed by X-ray structure analysis of the ternary complex formed by the inhibitor, COMT and a Mg^{2+} ion. **1**. Lerner, Masjost, Ruf, *et al.* (2003) *Org. Biomol. Chem.* **1**, 42.

Aminopurvalanol A

This cell-permeable purvalanol derivative (FW = 403.92 g/mol), also named (2R)-2-[[6-[(3-Amino-5-chlorophenyl)amino]-9-(1-methylethyl)-9H-purin-2-yl]amino]-3-methyl-1-butanol, targets cdk1/cyclin B (IC_{50} = 0.033 μM), cdk2/cyclin A (IC_{50} = 0.033 μM), cdk2/cyclin E (IC_{50} = 0.028 μM), and cdk5/p35 (IC_{50} = 0.020 μM). It also inhibits ERK1 (IC_{50} = 12 μM) and ERK2 (IC_{50} = 3.1 μM), with 3000-times greater selectivity versus a range of other protein kinases (IC_{50} > 100 μM). Aminopurvalanol A arrests cell cycle at G_2/M boundary (IC_{50} = 1.25 μM), and induces apoptosis at concentrations > 10 μM (1). **1**. Gompel, Soulié, Ceballos-Picot & Meijer (2004) *Neurosignals* **13**, 134. **2**. Rosania, Merlie, Gray, *et al.* (1999) *Proc. Natl. Acad. Sci. U.S.A.* **96**, 4797. **3**. Lu & Schulze-Gahmen (2006) *J. Med. Chem.* **49**, 3826.

4-Aminopyrazolo[3,4-d]pyrimidine

This adenine/allopurinol analogue (FW = 135.18 g/mol; CAS 2380-63-4) is taken up by many cells, where it is transformed into its corresponding nucleoside 5'-mono-, di, and tri-phosphates. **Target(s):** adenine phosphoribosyltransferase, alternative substrate (1); xanthine oxidase (2,5); ribosomal-inactivating proteins RNA-N-glycosidase activity (*e.g.*, Shiga toxin I, ricin, gelonin, *etc.* (3); adenine deaminase (4). **1**. Dewey & Kidder (1977) *Can. J. Biochem.* **55**, 110. **2**. Sheu, Lin & Chiang (1997) *Anticancer Res.* **17**, 1043. **3**. Brigotti, Rizzi, Carnicelli, L. Montanaro & S. Sperti (2000) *Life Sci.* **68**, 331. **4**. Jun & Sakai (1979) *J. Ferment. Technol.* **57**, 294. **5**. Hsieh, Wu, Yang, Choong & Chen (2007) *Bioorg. Med. Chem.* **15**, 3450.

4-Aminopyridine

This substituted pyridine (FW = 94.116 g/mol; CAS Registry Nuymber = 504-24-5; M.P. = 158-159°C), also known as fampridine, inhibits chymotrypsin and is also a K^+ channel blocker. 4-Aminopyridine is used in the treatment of multiple sclerosis. **Target(s):** chymotrypsin, K_i = 2.9 mM (1); K^+ channels, *or* Kv channels (2,3,5-11); Ca^{2+} transporting ATPase (4); steroid 11β-monooxygenase (12); Na^+ channels (13). **1**. Wallace, Kurtz & Niemann (1963) *Biochemistry* **2**, 824. **2**. Hu, Liu, Zeng, *et al.* (2006) *Neuropharmacology* 51, 737. **3**. Davies, Pettit, Agarwal & Standen (1991)

Pflugers Arch. **419**, 25. **4**. Ishida & Honda (1993) *J. Biol. Chem.* **268**, 4021. **5**. Castle & Slawsky (1993) *J. Pharmacol. Exp. Ther.* **265**, 1450. **6**. Castle, Fadous, Logothetis & Wang (1994) *Mol. Pharmacol.* **45**, 1242. **7**. Robertson & Nelson (1994) *Amer. J. Physiol.* **267**, C1589. **8**. Castle, Fadous, Logothetis & Wang (1994) *Mol. Pharmacol.* **46**, 1175. **9**. Russell, Publicover, Hart, *et al.* (1994) *J. Physiol.* **481**, 571. **10**. Lymangrover & Martin (1981) *Mol. Cell. Endocrinol.* **21**, 199. **11**. Hermann & Gorman (1981) *J. Gen. Physiol.* **78**, 63. **12**. Fraser, Holloway & Kenyon (1986) *J. Steroid Biochem.* **24**, 777. **13**. Lu, Liu, Liao, *et al.* (2005) *Toxicol. Appl. Pharmacol.* **207**, 275.

3-Aminopyridine Adenine Dinucleotide

This NAD$^+$ analogue (FW = 636.43 g/mol) inhibits numerousf dehydrogenases and NAD$^+$-dependent enzymes, including NAD$^+$ nucleosidase, *or* NADase (1,3,4), alcohol dehydrogenase (2,3), lactate dehydrogenase (3), glucose-6-phosphate dehydrogenase, weakly inhibited (3), and malate dehydrogenase (3). **1**. Anderson & Yuan (1980) *Meth. Enzymol.* **66**, 144. **2**. Bränden, Jörnvall, Eklund & Furugren (1975) *The Enzymes*, 3rd ed., **11**, 103. **3**. Fisher, Vercellotti & Anderson (1973) *J. Biol. Chem.* **248**, 4293. **4**. Muller-Steffner, Slama & Schuber (1996) *Biochem. Biophys. Res. Commun.* **228**, 128.

3-Aminopyridine Adenine Dinucleotide Phosphate

This NADP$^+$ analogue (FW = 716.41 g/mol) inhibits a number of NADP$^+$-dependent dehydrogenases, including glucose-6-phosphate dehydrogenase (1), methenyl-tetrahydrofolate cyclohydrolase (2), phenylacetone monooxygenase (3; 4), and hydroxyacetophenone monooxygenase (4,5). **1**. Anderson, Yuan & Vercellotti (1975) *Mol. Cell. Biochem.* **8**, 89. **2**. Pelletier & MacKenzie (1994) *Biochemistry* **33**, 1900. **3**. Fraaije, Kamerbeek, Heidekamp, Fortin & Janssen (2004) *J. Biol. Chem.* **279**, 3354. **4**. van den Heuvel, Tahallah, Kamerbeek, *et al.* (2005) *J. Biol. Chem.* **280**, 32115. **5**. Kamerbeek, Fraaije & Janssen (2004) *Eur. J. Biochem.* **271**, 2107.

6-(6-Aminopyridin-3-ylethynyl)-5-(4-chlorophenyl)-pyrimidin-4-ylamine

This pyrimidine (FW = 321.77 g/mol) inhibits cytosolic adenosine kinase, IC$_{50}$ = 14 nM. Adenosine levels increase at seizure foci as part of a negative feedback mechanism that controls seizure activity through adenosine receptor signaling. Agents that inhibit adenosine kinases increase/sustain site-/event-specific adenosine surges, thereby providing anti-seizure activity akin to that of adenosine receptor agonists, but with fewer dose-

limiting cardiac side effects. **1**. Matulenko, Paight, Frey, *et al.* (2007) *Bioorg. Med. Chem.* **15**, 1586.

3-[(6-Amino-3-pyridinyl)methyl-5-(2-hydroxyethyl)-4-methylthiazolium Chloride Pyrophosphate

This pyridine analogue of thiamin pyrophosphate (FW$_{chloride}$ = 433.75 g/mol) inhibits transketolase (K_i = 7 nM). 2- Amino-3-pyridinyl and 4-amino-3-pyridinyl analogues are slightly weaker inhibitors (K_i = 36 and 64 nM, respectively). Transketolase activity was almost completely suppressed in blood, spleen, and tumor cells, but there was little effect on the activity of the other thiamine-utilizing enzymes α-ketoglutarate dehydrogenase or glucose-6-phosphate dehydrogenase. **1**. Thomas, Le Huerou, De Meese, *et al.* (2008) *Bioorg. Med. Chem. Lett.* **18**, 2206.

6-Amino-2-[(quinolin-4-ylmethyl)amino]-8H-imidazo[4,5-g]quinazolin-8-one

This substrate analogue (FW = 357.38 g/mol) inhibits tRNA-guanine transglycosylase (K_i = 27 nM) a newly recognized target for disease-causing *Shigella* bacteria. Crystallographic and pK_a data suggest that the aminoimidazole moiety of the central *lin*-benzoguanine scaffold is protonated and binding is stabilized by charge-assisted hydrogen bonding. **1**. Hörtner, Ritschel, Stengl, *et al.* (2007) *Angew. Chem.* **46**, 8266.

2-Amino-5-(β-D-ribofuranosyl)-2,5-dihydro[1,5,2] diazaphosphorin-6(1H)-one 2-oxide

This phosphopyrimidine nucleoside (FW = 279.19 g/mol) is a potent, slow-binding inhibitor of *Escherichia coli* cytidine deaminase, most likely by mimicking a pentavalent associative transition-state species bearing both the in-coming nucleophile and its nascent exiphile. **1**. Ashley & Bartlett (1984) *J. Biol. Chem.* **259**, 13621. **2**. Ashley & Bartlett (1982) *Biochem. Biophys. Res. Commun.* **108**, 1467.

4-Aminosalicylate

This aromatic acid (FW$_{free-acid}$ = 153.14 g/mol: CAS 2066-89-9; *Abbreviation*: PAS), also known by the trade name Paser® as well as 4-aminosalicylic acid or *p*-aminosalicylic acid, is an analogue of *p*-aminobenzoate, an essential intermediate in the bacterial folate synthesis. PAS potently inhibits the growth of *Mycobacterium bovi*, *M. smegmatis*, and *M. tuberculosis*. It is also used for the treatment of such inflammatory bowel diseases as ulcerative colitis and Crohn's disease, most likely by inhibiting NF-κB. **Primary Mode of Antimicrobial Action:** Although in use clinically for over 60 years, *p*-aminosalicylate's inhibitory mechanism had proved elusive. Recent studies, however, demonstrated that PAS is prodrug, one that targets dihydrofolate reductase, *or* DHFR (1). PAS is first converted by dihydropteroate synthase (DHPS) a nd dihydrofolate synthase (DHFS) to a hydroxyl-dihydrofolate antimetabolite, which in turn inhibits DHFR. PAS is about as efficient a DHPS substrate as its natural substrate pABA, and inhibition of DHPS or mutation in DHFS prevents the formation of the antimetabolite and confers resistance to PAS.

N-{[(2S,3R)-3-Amino-2-sulfhydryl-5-sulfonate]pentanoyl}-L-isoleucyl-L-aspartate

This peptide analogue (FW = 441.53 g/mol; *N*-{[(2*S*,3*R*)-3-amino-2-mercapto-5-sulfonate]pentanoyl}-L-isoleucyl-L-aspartate) inhibits glutamyl aminopeptidase (K_i = 3.2 nM), neprilysin (K_i = 22 nM), and peptidyl-dipeptidase A (K_i = 1200 nM). The (2*R*,3*S*)-, (2*S*,3*S*)-, and (2*R*,3*R*)-diastereoisomers also inhibit glutamyl aminopeptidase (K_i = 50, 20, and 53 nM, respectively). The (2*S*,3*S*)-isomer also weakly inhibits membrane alanyl aminopeptidase (K_i = 23 μM). Replacing the L-isoleucyl residue with its D-enantiomer weakens the inhibition (K_i = 14.7 nM for the (2*S*,3*R*)-isomer and glutamyl aminopeptidase). The analogue containing D-aspartate is a strong inhibitor of glutamyl aminopeptidase (K_i = 5.36 nM). α-L-Aspartamide analogues are weaker inhibitors (K_i = 15 nM for the (2*S*,3*R*)-isomer and glutamyl aminopeptidase). **1.** David, Bischoff, Meudal, *et al.* (1999) *J. Med. Chem.* **42**, 5197.

5-Aminosulfonyl-1-(3,4-dichlorobenzyl)isatin 3-(O-Acetyl)oxime

This acylated isatin oxime (FW = 442.28 g/mol) inhibits ubiquitin thiolesterase (IC$_{50}$ = 12 μM), an enzyme linked to Parkinson Disease, progression of certain nonneuronal tumors, as well as neuropathic pain. Certain lung-derived tumor cell lines express the enzyme, which it is not expressed in normal lung tissue. **1.** Liu, Lashuel, Choi, *et al.* (2003) *Chem. Biol.* **10**, 837.

9-Amino-1,2,3,4-tetrahydroacridine

This acetylcholinesterase inhibitor (FW$_{free-base}$ = 198.27 g/mol; CAS 321-64-2 and 1684-40-8 (HCl)), also known as Tacrine and 1,2,3,4-tetrahydro-5-aminoacridine, was the first orally available, centrally-acting cholinesterase inhibitor approved for the treatment of Alzheimer's disease. Its use in the U.S. was discontinued out of concerns about its side effects, mainly hepatoxicity and gastrointestinal discomfort. This agent is also a likely DNA intercalator. **Target(s):** acetylcholinesterase, K_i = 25 nM (1-3,6,8,12-17); cholinesterase, *or* butyrylcholinesterase (2,3,5,6,13,15,16); aldehyde oxidase (4); ion channels (7); K$^+$ current (8,9); histamine *N*-methyltransferase (10,18-20); apyrase, *or* ATP diphosphohydrolase (11); aryl-acylamidase (13). **1.** Harel, Schalk, Ehret-Sabatier, *et al.* (1993) *Proc.*

Natl. Acad. Sci. U.S.A. **90**, 9031. **2.** Debord, N'Diaye, Bollinger, *et al.* (1997) *J. Enzyme Inhib.* **12**, 13. **3.** Darvesh, Walsh, Kumar, *et al.* (2003) *Alzheimer Dis. Assoc. Disord.* **17**, 117. **4.** Obach, Huynh, Allen & Beedham (2004) *J. Clin. Pharmacol.* **44**, 7. **5.** Benveniste, Hemmingsen & Juul (1967) *Acta Anaesthesiol. Scand.* **11**, 297. **6.** Bajgar, Fusek, Patocka & Hrdina (1979) *Physiol. Bohemoslov.* **28**, 31. **7.** Rogawski (1987) *Eur. J. Pharmacol.* **142**, 169. **8.** Drukarch, Kits, Van der Meer, Lodder & Stoof (1987) *Eur. J. Pharmacol.* **141**, 153. **9.** Kotake, Hisatome, Matsuoka, *et al.* (1990) *Cardiovasc. Res.* **24**, 42. **10.** Cumming, Reiner & Vincent (1990) *Biochem. Pharmacol.* **40**, 1345. **11.** Bonan, Battastini, Schetinger, *et al.* (1997) *Gen. Pharmacol.* **28**, 761. **12.** Kaul (1962) *J. Pharm. Pharmacol.* **14**, 243. **13.** Costagli & Galli (1998) *Biochem. Pharmacol.* **55**, 1733. **14.** Pietsch & Gütschow (2005) *J. Med. Chem.* **48**, 8270. **15.** Luo, Yu, Zhan, *et al.* (2005) *J. Med. Chem.* **48**, 986. **16.** Giacobini (2003) *Neurochem. Res.* **28**, 515. **17.** Haviv, Wong, Greenblatt, *et al.* (2005) *J. Amer. Chem. Soc.* **127**, 11029. **18.** Graßmann, Apelt, Sippl, *et al.* (2003) *Bioorg. Med. Chem.* **11**, 2163. **19.** Horton, Sawada, Nishibori & Cheng (2005) *J. Mol. Biol.* **353**, 334. **20.** Taraschenko, Barnes, Herrick-Davis, *et al.* (2005) *Methods Find. Exp. Clin. Pharmacol.* **27**, 161.

2-Aminotetralin

This conformationally constrained cyclohexylamine (FW = 147.22 g/mol; CAS 2954-50-9, 1743-01-7 (HCl), 21880-87-5 (*S*-isomer), 21966-60-9 (*R*-isomer)), also called β-tetrahydronaphthylamine, and its many derivatives are inhibitors of serotonin and norepinephrine transport and likely induces their release as well. **Target(s):** monamine oxidase, *or* amine oxidase (1); 5-hydroxytryptamine transport (2); norepinephrine transport (2,3); phenylethanolamine *N*-methyltransferase (4). **1.** Blaschko (1963) *The Enzymes*, 2nd ed., **8**, 337. **2.** Bruinvels (1971) *Brit. J. Pharmacol.* **42**, 281. **3.** Knepper, Grunewald & Rutledge (1988) *J. Pharmacol. Exp. Ther.* **247**, 487. **4.** Fuller & Molloy (1977) *Biochem. Pharmacol.* **26**, 446.

17-(2-Amino-4-thiazolyl)androsta-5,16-dien-3β-ol

This synthetic steroid (FW$_{free-base}$ = 370.56 g/mol) inhibits 17α-hydroxyprogesterone aldolase, *or* steroid C17(20) lyase, or Cytochrome P450 17A1 (IC$_{50}$ = 63 nM (1), key enzyme in the steroidogenic pathway that produces progestins, mineralocorticoids, glucocorticoids, androgens, and estrogens, as well as steroid 17α-monooxygenase (2). These studies confirmed that steroids bearing a heteroaromatic substituent at C-17 were designed as inhibitors of C17(20) lyase. . **1.** Burkhart, Gates, Laughlin, Resvick & Peet (1996) *Bioorg. Med. Chem.* **4**, 1411. **2.** Njar & Brodie (1999) *Curr. Pharm. Des.* **5**, 163.

3-Amino-1-(3-trifluoromethyl)phenyl-2-pyrazoline

This substituted pyrazoline (FW = 229.20 g/mol), also known as 4,5-dihydro-1-[3-(trifluoromethyl)phenyl]-1*H* pyrazol-3-amine and BW755C, inhibits both the cyclooxygenase and lipoxygenase pathways. **Target(s):** 12-lipoxygenase, *or* arachidonate 12-lipoxygenase; (1 3,13,14; however, see 8); cyclooxygenase, *or* prostaglandin-endoperoxide synthase (2,6,8);

lipoxygenase (4,5,15); 5-lipoxygenase, *or* arachidonate 5-lipoxygenase (2,7-12); 15-lipoxygenase, *or* arachidonate 15-lipoxygenase (8). **1.** Koshishara, Murota, Petasis & Nicolaou (1982) *FEBS Lett.* **143**, 13. **2.** Coutts, Khandwala & Vaninwegen (1985) in *Prostaglandins, Leukotrienes, and Lipoxins* (Bailey, ed.), pp. 627-637, Plenum Press, New York. **3.** Higgs, Flower & Vane (1979) *Biochem. Pharmacol.* **28**, 1959. **4.** Salari, Braquet & Borgeat (1984) *Prostaglandins Leukot. Med.* **13**, 53. **5.** Sircar, Schwender & Johnson (1983) *Prostaglandins* **25**, 393. **6.** Miller, Munster, Wasvary, *et al.* (1994) *Biochem. Biophys. Res. Commun.* **201**, 356. **7.** Furakawa, Yoshimoto, Ochi & Yamamoto (1984) *Biochim. Biophys. Acta* **795**, 458. **8.** Brooks, Albert, Dyer, *et al.*r (1992) *Bioorg. Med. Chem. Lett.* **2**, 1353. **9.** Riendeau, Falgueyret, Nathaniel, *et al.* (1989) *Biochem. Pharmacol.* **38**, 2313. **10.** Mulliez, Leblanc, Girard, Rigaud & Chottard (1987) *Biochim. Biophys. Acta* **916**, 13. **11.** Denis, Falgueyret, Riendeau & Abramovitz (1991) *J. Biol. Chem.* **266**, 5072. **12.** Young (1999) *Eur. J. Med. Chem.* **34**, 671. **13.** Yokoyama, Mizuno, Mitachi, *et al.* (1983) *Biochim. Biophys. Acta* **750**, 237. **14.** Yoshimoto, Miyamoto, Ochi & Yamamoto (1982) *Biochim. Biophys. Acta* **713**, 638. **15.** Beneytout, Andrianarison, Rakotoarisoa & Tixier (1989) *Plant Physiol.* **91**, 367.

Amiodarone

This photosensitive benzofuran (FW$_{free-base}$ = 645.32 g/mol; CAS 19774-82-4 (HCl)), also known as Nexterone and systematically as (2-{4-[(2-butyl-1-benzofuran-3-yl)carbonyl]-2,6-diiodophenoxy}ethyl)diethylamine, is a non-selective ion channel blocker and an α- and β-adrenergic receptor agonist and antiarrhythmic agent used for various types of cardiac dysrhythmias, both ventricular and atrial. **Key Pharmacokinetic Parameters:** *See* Appendix II in Goodman & Gilman's *THE PHARMACOLOGICAL BASIS OF THERAPEUTICS*, 12th Edition (Brunton, Chabner & Knollmann, eds.) McGraw-Hill Medical, New York (2011). **Target(s):** Na$^+$ channels (1); phospholipase A$_1$ (2); calmodulin-dependent action (3-5); protein kinase C (4); Ca^{2+} channels (6,10,16); carnitine palmitoyltransferase (17; K+ channels (8,9,15); Na$^+$/K$^+$-exchanging ATPase (11,12); lysosomal phospholipases (13); phospholipase A$_2$ (14,18); phospholipase C (14); cholestenol Δ-isomerase, *or* sterol Δ^{7,8}-isomerase (17); phosphatidylglycerol-selective phospholipase A$_2$ (18); retinyl ester hydrolase (19). **1.** Revenko, Khodorov & Avrutskii (1980) *Biull. Eksp. Biol. Med.* **89**, 702. **2.** Hostetler, Giordano & Jellison (1988) *Biochim. Biophys. Acta* **959**, 316. **3.** Nokin, Blondiaux, Schaeffer, Jungbluth & Lugnier (1989) *Naunyn Schmiedebergs Arch. Pharmacol.* **339**, 367. **4.** Silver, Connell, Dillon, *et al.* (1989) *Cardiovasc. Drugs Ther.* **3**, 675. **5.** Deziel, Davis, Davis, *et al.* (1989) *Arch. Biochem. Biophys.* **274**, 463. **6.** Lubic, Nguyen, Dave & Giacomini (1994) *J. Cardiovasc. Pharmacol.* **24**, 707. **7.** Kennedy, Unger & Horowitz (1996) *Biochem. Pharmacol.* **52**, 273. **8.** Watanabe, Hara, Tamagawa & Nakaya (1996) *J. Pharmacol. Exp. Ther.* **279**, 617. **9.** Holmes, Sun, Porter, *et al.* (2000) *J. Cardiovasc. Electrophysiol.* **11**, 1152. **10.** Ding, Chen, Klitzner & Wetzel (2001) *J. Investig. Med.* **49**, 346. **11.** Broekhuysen, Clinet & Delisee (1972) *Biochem. Pharmacol.* **21**, 2951. **12.** Chatelain, Laruel & Gillard (1985) *Biochem. Biophys. Res. Commun.* **129**, 148. **13.** Heath, Costa-Jussa, Jacobs & Jacobson (1985) *Brit. J. Exp. Pathol.* **66**, 391. **14.** Shaikh, Downar & Butany (1987) *Mol. Cell. Biochem.* **76**, 163. **15.** Haworth, Goknur & Berkoff (1989) *Circ. Res.* **65**, 1157. **16.** Valenzuela & Bennett (1991) *J. Cardiovasc. Pharmacol.* **17**, 894. **17.** Moebius, Reiter, Bermoser, *et al.* (1998) *Mol. Pharmacol.* **54**, 591. **18.** Shinozaki & Waite (1999) *Biochemistry* **38**, 1669. **19.** Schindler (2001) *Lipids* **36**, 543.

Amiton

This toxic organothiophosphate (FW$_{free-base}$ = 269.35 g/mol; CAS 78-53-5), also known as Tetram and *O,O*-diethyl-*S*-(2 diethylaminoethyl) phosphorothiolate, inhibits acetylcholinesterase and has been used as an insecticide. Amiton has also been classified as a chemical warfare agent and has been given the designation VG. **CAUTION:** Extreme care must be exercised in handling this agent. A single drop on the skin is often lethal. There are no confirmed reports of amiton having been used in combat. **Target(s):** acetylcholinesterase (1-6); butyrylcholinesterase (6,7). **1.** Froede & Wilson (1971) *The Enzymes*, 3rd ed., **5**, 87. **2.** Bracha & O'Brien (1968) *Biochemistry* **7**, 1545. **3.** Aharoni & O'Brien (1968) *Biochemistry* **7**, 1538. **4.** Ellison (2000) *Handbook of Chemical and Biological Warfare Agents*, CRC Press, Boca Raton. **5.** Iverson & Main (1969) *Biochemistry* **8**, 1889. **6.** Main (1969) *J. Biol. Chem.* **244**, 829. **7.** Brown, Kalow, Pilz, Whittaker & Woronick (1981) *Adv. Clin. Chem.* **22**, 1.

Amitriptyline

This tricyclic antidepressant (FW$_{free-base}$ = 277.41 g/mol; CAS 50-48-6 (free base) and 549-18-8 (hydrochloride)), marketed under Tryptomer, Elavil, Tryptizol, Laroxyl, Saroten, Sarotex, Lentizol, Endep, and systematically named 3-(10,11-dihydro-5*H*-dibenzo[*a,d*]cycloheptene-5-ylidene)-*N,N*-dimethylpropan-1-amine, is widely used to treat depressive disorders, anxiety disorders, attention deficit hyperactivity disorder, migraine prophylaxis, eating disorders, bipolar disorder, post-herpetic neuralgia, and insomnia. The pK_a = 9.4, and its hydrochloride salt is freely soluble in water. (*See Mianserin*) **Target(s):** Ca^{2+}-dependent cyclic nucleotide phosphodiesterase (1); monoamine oxidase (2,8); norepinephrine transporter (3,4); serotonin reuptake (4-6); K$^+$ channels (7,10,11); Na$^+$ channels, voltage dependent (9,13,14); butyrylcholinesterase, *or* cholinesterase (12); glucuronosyltransferase (15). **1.** Hidaka, Inagaki, Nishikawa & Tanaka (1988) *Meth. Enzymol.* **159**, 652. **2.** Von Voigtlander & E. G. Losey (1976) *Biochem. Pharmacol.* **25**, 217. **3.** Glowinski & Axelrod (1964) *Nature* **204**, 1318. **4.** Sanchez & Hyttel (1999) *Cell Mol. Neurobiol.* **19**, 467. **5.** Todrick & Tait (1969) *J. Pharm. Pharmacol.* **21**, 751. **6.** Ross & Renyi (1969) *Eur. J. Pharmacol.* **7**, 270. **7.** Teschemacher, Seward, Hancox & Witchel (1999) *Brit. J. Pharmacol.* **128**, 479. **8.** Schraven & Reibert (1984) *Arzneimittelforschung* **34**, 1258. **9.** Ishii & Sumi (1992) *Eur. J. Pharmacol.* **221**, 377. **10.** Wooltorton & Mathie (1995) *Brit. J. Pharmacol.* **116**, 2191. **11.** Yoshida, Hisatome, Nawada, *et al.* (1996) *Eur. J. Pharmacol.* **312**, 115. **12.** Cokugras & Tezcan (1997) *Gen. Pharmacol.* **29**, 835. **13.** Bielefeldt, Ozaki, Whiteis & Gebhart (2002) *Dig. Dis. Sci.* **47**, 959. **14.** Wang, Russell & Wang (2004) *Pain* **110**, 166. **15.** Hara, Nakajima, Miyamoto & Yokoi (2007) *Drug Metab. Pharmacokinet.* **22**, 103.

Amlodipine

This long-acting, dihydropyridine-class calcium channel blocker (FW = 408.88 g/mol; CAS 88150-42-9), also known by the trade name Norvasc™ is an antihypertensive agent indicated for the treatment of angina pectoris. **Primary Mode of Action:** Amlodipine acts by relaxing arterial wall smooth muscle, thereby decreasing total resistance in peripheral circulation and lowering blood pressure. Amlopidine toxicity is the leading cause of drug overdose, often leading to profound hypotension and shock. **Pharmacokinetics:** Intravenous amlopidine administration (single dose = 10 mg) results in a mean plasma $t_{1/2}$ of 34 hours, with a mean clearance of 7 mL min^{-1}kg^{-1} and a mean apparent volume of distribution of 21 L/kg (1). Oral administration of single 10-mg dose gives a mean systemic availability of 64% and a mean plasma half-life of 36 hours. **Key Pharmacokinetic Parameters:** *See* Appendix II in Goodman & Gilman's *THE PHARMACOLOGICAL BASIS OF THERAPEUTICS*, 12th Edition (Brunton, Chabner & Knollmann, eds.) McGraw-Hill Medical, New York (2011). **1.** Faulkner, McGibney, Chasseaud, Perry & Taylor (1986) *Brit. J. Clin. Pharmacol.* **22**, 21.

Ammonia (Ammonium Ion)

This important nitrogenous metabolite (FW = 17.031 g/mol) is extremely soluble in water, with which it readily forms hydrogen bonds, often abstracting a proton to become ammonium ion (Ionic Radius = 1.43 Å). Although aqueous ammonia solutions are commonly called ammonium hydroxide solutions, NH$_4$OH is unlikely to exist to any significant degree. In fact, a 1 M ammonia solution only contains 0.0042 M ammonium and hydroxide ions. Because $pK_a = 14 - pK_b$, the pK_a for the reaction of ammonium ion with water to yield ammonia and water is 9.25 ($K_a = 5.62$ x 10^{-10} M). Ammonia protonation and ammonium ion deprotonation occur on the sub-nanosecond timescale, as first demonstrated in the chemical relaxation experiments by Nobelist Manfred Eigen. Ammonia is also an excellent nucleophile, and the rapidity of forming and breaking hydrogen bonds in aqueous solution permits ammonia to react well with electrophilic centers in nucleophilic displacement reactions. Ammonium ion itself is inert in nucleophilic reactions. At normal blood pH ($\approx$ 7.4), nearly 99% of all circulating ammonia is in the ammonium ion form. Ammonium ion is converted to glutamine in the glutamine synthetase reaction, which serves to deprotonate this metabolite at neutral pH. Enzymes requiring reactive ammonia then hydrolyze the γ-amide of glutamine to generate what is called "nascent ammonia". Such enzymes possess an endogenous glutaminase activity that is usually tightly coupled by means of an ammonia transfer tunnel to ammonia consumption in subsequent biosynthetic reaction steps, thereby avoiding futile hydrolysis of this amid nitrogen donor. Protected transfer of nascent ammonia is accomplished by a tunnel that excludes water and proton donors. For humans, ammonia toxicity refers to the effects on the brain that arise due to depletion of α-keto-glutarate (by means of the glutamate dehydrogenase (NADPH-dependent) and glutamine synthetase reactions) and the collapse of certain transmembrane proton gradients when circulating ammonia rises above 0.5 mM. Although ammonia and ammonium ions inhibit over two hundred enzyme-catalyzed reactions, none are listed here. Such effects are generally nonphysiologic, characterized by millimolar inhibition constants.

AMN107, *See* Nilotinib

Amodiaquine

This antimalarial and anti-inflammatory agent (FW = 355.87 g/mol; CAS 86-42-0), also known by the trade names Camoquin® and Flavoquine® as well as its systematic name 4-[(7-chloroquinolin-4-yl)amino]-2-[(diethylamino)methyl]phenol, is more effective than in inhibiting chloroquine-resistant *Plasmodium falciparum* infections. Amodiaquine binds to the parasite DNA, preventing replication, transcription, and translation. It is active against the asexual erythrocytic forms of *Plasmodium* species. While not marketed in the United States, amodiaquine is widely available in Africa. ***See*** *Artemisinin; Chloroquine*

Amoxicillin

This semi-synthetic, moderate-spectrum β-lactam antibiotic (FW = 365.40 g/mol; CAS 26787-78-0; Abbreviation: Amox) also known as 2S,5R,6R)-6-{[(2R)-2-amino-2-(4-hydroxyphenyl)-acetyl]amino}-3,3-dimethyl-7-oxo-4-thia-1-azabicyclo[3.2.0]heptane-24-carboxylate,) is bacteriolytic at 10 μg/mL, or less, for many bacterial pathogens, including *Streptococcus, Bacillus subtilis, Enterococcus, Haemophilus, Helicobacter,* and *Moraxella*. Clinically, amoxicillin is effective in eradicating bacteriuria due to susceptible organisms and is well tolerated (1). Its stability in gastric acid and rapid uptake after oral administration commend its use for treating *Helicobacterium pylori*, the major causative agent in the occurrence of peptic ulcers and gastric cancer in humans. Treatment typically involves a two- or three-antibiotic cocktail that almost always includes amoxicillin. **Mode of Inhibitory Action:** Amoxicillin binds to <u>P</u>enicillin-<u>B</u>inding Protein <u>1A</u>, *or* PBP-1A, which is situated inside the bacterial cell well. Like many other β-lactam antibiotics, amoxicillin acylates the penicillin-sensitive transpeptidase C-terminal domain, a process attended by β-lactam ring opening. Enzyme nactivation blocks cross-link formation between nearby linear peptidoglycan strands, thereby inhibitin bacterial cell wall synthesis, with subsequent lysis mediated by autolysins. **Amoxicillin Resistance:** Certain bacteria (*e.g., Citrobacter, Klebsiella* and *Pseudomonas aeruginosa*) are highly resistant to amoxicillin, and *Escherichia coli* and most clinical isolates of *Staphylococcus aureus* have already developed resistance to amoxicillin to varying degrees. Amox resistance is typically multifactorial, as illustrated for the case of *H. pylori* (2). Progression of spontaneous genetic mutations contributing to *H. pylori* resistance, when exposed to increasing amoxicillin concentrations *in vitro*, revealed that mutations in a number of genes (notably pbp1, pbp2, hefC, hopC, and hofH). Five isolates, each expressing multiple mutated genes, and four transformed strains expressing individually mutated pbp1, hefC, or hofH, were characterized using minimum inhibitory concentrations, amoxicillin uptake, and efflux studies (2). Mutations in pbp1, hefC, hopC, hofH, and possibly pbp2 contribute to high-level amoxicillin resistance in *H. pylori*. Such findings indicate the complex evolution of amoxicillin resistance, with certain families of genes appearing to be more susceptible to resistance mutations (2). **Key Pharmacokinetic Parameters:** *See* Appendix II in Goodman & Gilman's *THE PHARMACOLOGICAL BASIS OF THERAPEUTICS*, 12th Edition (Brunton, Chabner & Knollmann, eds.) McGraw-Hill Medical, New York (2011). **1.** Handsfield, Clark, Wallace, Holmes & Turck (1973) *Antimicrob. Agents Chemother.* **3**, 262. **2.** Qureshi, Gallaher & Schiller (2014) *Microb. Drug Resist.* **20**, 509.

AMPCP, *See* α,β-Methyleneadenosine 5'-Diphosphate

AMPCPP, *See* α,β-Methyleneadenosine 5'-Triphosphate

AMPD, *See* 2-Amino-2-methyl-1,3-propanediol
Amphetamine

Dextrooamphetamine
(2S)-Phenylpropane-2-amine

Levoamphetamine
(2R)-Phenylpropane-2-amine

This psychostimulant and drug (FW$_{free-base}$ = 135.21 g/mol; CAS 300-62-9 (free base) and 60-13-9 (sulfate (2:1)), also known as 1-phenyl-2-aminopropane and by its IUPAC name (2S)-1-phenylpropan-2-amine, increases one's sense of wakefulness and improves focus, with attendant decreases in fatigue and appetite. Amphetamine inhibits catecholamine release, inhibits monoamine oxidase, and acts as an adrenergic receptor agonist. The salt (*e.g.,* the hydrochloride) is very soluble in water. (*Note:* The terms dextroamphetamine and dexamphetamine refer to the (+)-(*S*)-enantiomer, the more physiologically active form of the two stereoisomers.) **Mechanism of Amphetamine Action:** To regulate the extracellular dopamine, the presynaptic dopamine transporter (DAT) utilizes the chemiosmotic energy of the Na^+/Cl^- gradient to drive the reuptake of released dopamine, thereby conserving dopamine and decreasing the need for *de novo* synthesis of dopamine to replenish vesicular dopamine stores. In the Facilitated Exchange Diffusion Model, dopamine efflux results from the translocation of amphetamine into the cell, which is attended by outward movement of dopamine into the extracellular space. In the Vesicle Depletion Model, amphetamine induces dopamine reverse transport by elevating the cytoplasmic its concentration, thus altering the transmembrane dopamine gradient. Amphetamine and similarly acting drugs compete with dopamine for uptake and also induce DAT-mediated dopamine efflux. **Caution:** In view of its so-called performance-enhancing effects, amphetamine has become a recreational drug, the abuse of which has become a matter of great medical and societal concern. Drug tolerance raises the risk for using ever-higher doses, thus increasing the likelihood of amphetamine addiction/dependence. **Key Pharmacokinetic Parameters:** *See* Appendix II in Goodman & Gilman's THE PHARMACOLOGICAL BASIS OF THERAPEUTICS, 12th Edition (Brunton, Chabner & Knollmann, eds.) McGraw-Hill Medical, New York (2011). **Target(s):** alcohol dehydrogenase (1); dopamine β-monooxygenase (2); CYP2D6 (3); hemoglobin S polymerization (4); monoamine oxidase (5,6,8); aldehyde dehydrogenase (7); norepinephrine uptake (9); phenylalanyl-tRNA synthetase, K_i = 12 μM (10); lysine 2-monooxygenase (11). **1.** Brändén, Jörnvall, Eklund & Furugren (1975) *The Enzymes*, 3rd ed., **11**, 103. **2.** Goldstein & Contrera (1962) *J. Biol. Chem.* **327**, 1898. **3.** Wu, Otton, Inaba, Kalow & Sellers (1997) *Biochem. Pharmacol.* **53**, 1605. **4.** Noguchi & Schechter (1978) *Biochemistry* **17**, 5455. **5.** Mantle, Tipton & Garrett (1976) *Biochem. Pharmacol.* **25**, 2073. **6.** Blaschko (1963) *The Enzymes*, 2nd ed., **8**, 337. **7.** Messiha (1981) *Neurotoxicology* **2**, 703. **8.** Blaschko & Stroemblad (1960) *Arzneimittelforschung* **10**, 327. **9.** Haggendal & Hamberger (1967) *Acta Physiol. Scand.* **70**, 277. **10.** Santi & Danenberg (1971) *Biochemistry* **10**, 4813. **11.** Nakazawa, Hori & Hayaishi (1972) *J. Biol. Chem.* **247**, 3439.

Amphomycin

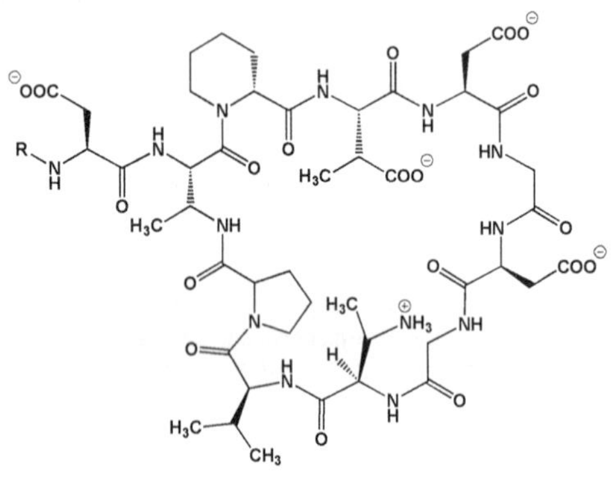

This lipopeptide antibiotic (FW = 1290.43 g/mol; CAS 1402-82-0; Source: *Streptomyces canus*), where the R group in the above structure is (+)-3-anteisotridecenoic acid or (+)-isododecenoic acid, inhibits protein glycosylation. **Target(s):** protein glycosylation (1-5,8-10); dolichyl-phosphate α-*N*-acetylglucosaminyl-transferase (1,2,5,6,10); dolichyl-phosphate β-D-mannosyltransferase (1,2,4-6,10-13); glycosyl-phosphatidylinositol anchor biosynthesis (4,12); dolichyl-phosphate β-D-glucosyltransferase (6,18); dolichyl-phosphate D-xylosyltransferase (6); phospho-*N*-acetylmuramoyl pentapeptide-transferase (7); α-mannosyl transferases (14,15); α1,6-mannosyltransferase (15); UDP-*N*-acetylglucosamine:dolichyl-phosphate *N*-acetylglucos-aminephospho-transferase (16); *N*-acetylglucos-aminyldiphosphodolichol *N*-acetylglucosaminyl-transferase (17); dolichyl-phosphate β-D-mannosyltransferase (19,20); cellulose synthase, UDP-forming (21). **1.** Elbein (1983) *Meth. Enzymol.* **98**, 135. **2.** Elbein (1987) *Meth. Enzymol.* **138**, 661. **3.** Schwarz & Datema (1982) *Meth. Enzymol.* **83**, 432. **4.** Vidugiriene & Menon (1995) *Meth. Enzymol.* **250**, 513. **5.** Schwarz & Datema (1980) *Trends Biochem. Sci.* **5**, 65. **6.** Presper & Heath (1983) *The Enzymes*, 3rd ed., **16**, 449. **7.** Tanaka, Oiwa, Matsukura, Inokoshi & Omura (1982) *J. Antibiot.* **35**, 1216. **8.** Tanaka, Iwai, Oiwa, *et al.* (1977) *Biochim. Biophys. Acta* **497**, 633. **9.** Ericson, Gafford & Elbein (1979) *Arch. Biochem. Biophys.* **191**, 698. **10.** Kang, Spencer & Elbein (1978) *J. Biol. Chem.* **253**, 8860. **11.** Spencer & Elbein (1980) *Proc. Natl. Acad. Sci. U.S.A.* **77**, 252. **12.** Tomavo, Dubremetz & Schwarz (1992) *J. Biol. Chem.* **267**, 21446. **13.** Zhu & Laine (1996) *Glycobiology* **6**, 811. **14.** Brown, Guther, Field & Ferguson (1997)

Glycobiology **7**, 549. **15.** Brown, Field, Barker, *et al.* (2001) *Bioorg. Med. Chem.* **9**, 815. **16.** Kaushal & Elbein (1986) *Plant Physiol.* **82**, 748. **17.** Kean & Niu (1998) *Glycoconjugate J.* **15**, 11. **18.** Arroyo-Flores, Rodríguez-Bonilla, Villagómez-Castro, *et al.* (2000) *Fungal Genet. Biol.* **30**, 127. **19.** Jensen & Schutzbach (1986) *Carbohydr. Res.* **149**, 199. **20.** Zhu & Laine (1996) *Glycobiology* **6**, 811. **21.** Haass, Hackspacher & Franz (1985) *Plant Sci.* **41**, 1.

Amphotericin B

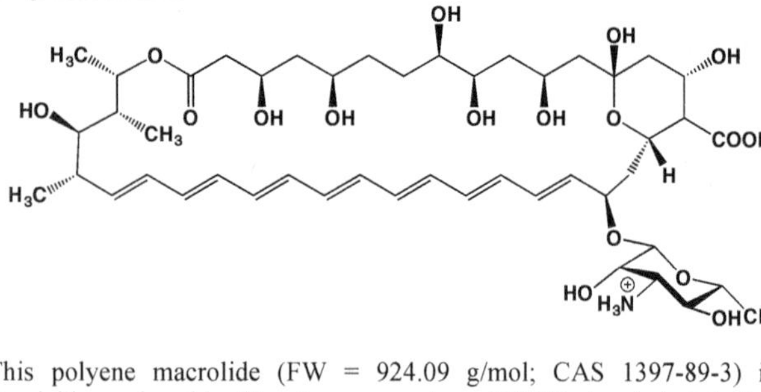

This polyene macrolide (FW = 924.09 g/mol; CAS 1397-89-3) is a fungicide produced by *Streptomyces nodosus* M4575. Amphotericin B acts by permeabilizing sterol-containing membranes; hence, it is not active against mitochondrial membranes. It binds selectively to ergosterol in the cell membrane of susceptible fungi, resulting in cell death. Amphotericin B is used to treat serious systemic fungal infections. Ambisome for Injection is a sterile, non-pyrogenic lyophilized form of amphotericin B suitable for intravenous infusion. **Key Pharmacokinetic Parameters:** *See* Appendix II in Goodman & Gilman's THE PHARMACOLOGICAL BASIS OF THERAPEUTICS, 12th Edition (Brunton, Chabner & Knollmann, eds.) McGraw-Hill Medical, New York. **Target(s):** photosystem 1 (1); photosynthetic electron transport (2,3); chitin synthase, the methyl ester also inhibits (4,7); DNA biosynthesis (5); Na^+/K^+-exchanging ATPase (6); sterol 24-*C* methyltransferase, inhibits at concentrations above 60 nM (8); sterol 14-demethylase, *or* CYP51 (9). **1.** Trebst (1980) *Meth. Enzymol.* **69**, 675. **2.** Bishop (1973) *Biochem. Biophys. Res. Commun.* **54**, 816. **3.** Nolan & Bishop (1975) *Arch. Biochem. Biophys.* **166**, 323. **4.** Rast & Bartnicki-Garcia (1981) *Proc. Natl. Acad. Sci. U.S.A.* **78**, 1233. **5.** Kuwano, Akiyama, Takaki, Okano & Nishimoto (1981) *Biochim. Biophys. Acta* **652**, 266. **6.** Vertut-Doi, Hannaert & Bolard (1988) *Biochem. Biophys. Res. Commun.* **157**, 692. **7.** Haenseler, Nyhlen & Rast (1983) *Exp. Mycol.* **7**, 17. **8.** Mukhtar, Hakkou & Bonaly (1994) *Mycopathologia* **126**, 75. **9.** Mellado, Garcia-Effron, Buitrago, *et al.* (2005) *Antimicrob. Agents Chemother.* **49**, 2536.

Ampicillin

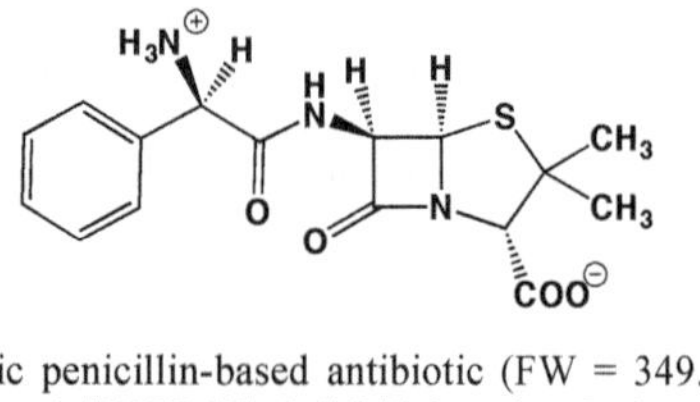

This semisynthetic penicillin-based antibiotic (FW = 349.41 g/mol; CAS 69-53-4), also named (2*S*,5*R*,6*R*)-6-([(2*R*)-2-amino-2-phenylacetyl]amino)-3,3-dimethyl-7-oxo-4-thia-1-azabicyclo[3.2.0]heptane-2-carboxylate, inhibits peptidoglycan synthesis needed for bacterial proliferation. It has an increased resistance to acidic pH and a wider antibacterial spectrum than most penicillins. It is also effective against many Gram-negative species. Ampicillin was the first semisynthetic penicillin analogue to be administered orally in clinical practice. An ampicillin resistance gene (abbreviated *ampR*) is often employed as a selectable marker by molecular biologists. **Target(s):** muramoylpentapeptide carboxypeptidase (1); glycopeptide transpeptidase (1); serine-type D-Ala-D-Ala carboxypeptidase, *or* penicillin-binding protein-5 (2,3,10,11); D-stereospecific aminopeptidase (4,12); penicillin-binding protein-4 (5); neprilysin, *or* enkephalinase (6); β-lactamase, *or* cephalosporinase (7); lysostaphin, *or* glycylglycine endopeptidase (8); acylaminoacyl-peptidase (9); cytosol alanyl aminopeptidase (13); α-amino-acid esterase, weak alternative substrate (14); deacetoxycepholosporin-C synthase (15). **1.** Izaki, Matsuhashi & Strominger (1966) *Meth. Enzymol.* **8**, 487. **2.** Izaki & Strominger (1970) *Meth. Enzymol.* **17A**, 182. **3.** Wilkin (1998) in *Handb. Proteolytic Enzymes* (Barrett, Rawlings & Woessner, eds.), p. 418, Academic Press, San Diego. **4.** Asano (1998) in *Handb. Proteolytic*

Enzymes (Barrett, Rawlings & Woessner, eds.), p. 430, Academic Press, San Diego. **5.** Wilkin (1998) in *Handb. Proteolytic Enzymes* (Barrett, Rawlings & Woessner, eds.), p. 435, Academic Press, San Diego. **6.** Livingston, Smith, Sewell & Ahmed (1992) *J. Enzyme Inhib.* **6**, 165. **7.** Toda, Inoue & Mitsuhashi (1981) *J. Antibiot.* **34**, 1469. **8.** Ramadurai, Lockwood, Nadakavukaren & Jayaswal (1999) *Microbiology* **145**, 801. **9.** Sharma & Orthwerth (1993) *Eur. J. Biochem.* **216**, 631. **10.** Coyette, Ghuysen & Fontana (1978) *Eur. J. Biochem.* **88**, 297. **11.** Marquet, Nieto & Diaz-Mauriño (1976) *Eur. J. Biochem.* **68**, 581. **12.** Asano, Nakazawa, Kato & Kondo (1989) *J. Biol. Chem.* **264**, 14233. **13.** Starnes, Szechinski & Behal (1982) *Eur. J. Biochem.* **124**, 363. **14.** Takahashi, Yamazaki & Kato (1974) *Biochem. J.* **137**, 497. **15.** Dotzlaf & Yeh (1989) *J. Biol. Chem.* **264**, 10219.

AMPPCP, See *β,γ-Methyleneadenosine 5′-Triphosphate*

AMPPNP, See *Adenosine 5′-(β,γ-Imido)triphosphate*

Amprenavir

This peptide analogue (FW = 505.64 g/mol; CAS 161814-49-9), also known as KVX-478, VX-478, and by its trade name as Agenerase™ is an antiviral agent that inhibits human immunodeficiency virus-1 protease (K_i = 0.6 nM). With its relatively long *in vivo* half-life of 7-9 hours, amprenavir requires less frequent administration. **1.** Kim, Baker, Dwyer, *et al.* (1995) *J. Amer. Chem. Soc.* **117**, 1181. **2.** Clemente, Coman, Thiaville, *et al.* (2006) *Biochemistry* **45**, 5468. **3.** Specker, Böttcher, Brass, *et al.* (2006) *ChemMedChem* **1**, 106. **4.** Surleraux, de Kock, Verschueren, *et al.* (2005) *J. Med. Chem.* **48**, 1965. **5.** Shuman, Haemaelaeinen & Danielson (2004) *J. Mol. Recognit.* **17**, 106.

AMP-S, See *Adenosine 5′-Thiomonophosphate*

AMPSF, See *4-Amidinophenylmethanesulfonyl Fluoride*

Ampullosporins

This set of at eight or more peptaibol antibiotics are membrane-active oligopeptides isolated from soil fungi and exhibit anti-bacterial and anti-fungal properties. They contain α-aminoisobutyryl residues; hence, the "aib" in the class name) and usually end in a C-terminal alcohol (thus, the "ol" in the class name). The peptaibols are amphipathic, usually highly helical in structure, and typically form voltage-dependent ion channels. This will often lead to cell death. Ampullosporin has the sequence Ac-Trp-Ala-Aib-Aib-Leu-Aib-Gln-Aib-Aib-Aib-Gln-Leu-Aib-Gln-Leu-OH, ampullosporin B is Ac-Trp-Ala-Aib-Aib-Leu-Aib-Gln-Ala-Aib-Aib-Gln-Leu-Gln-Leu-OH, ampullosporin C is Ac-Trp-Ala-Aib-Aib-Leu-Aib-Gln-Aib-Ala-Aib-Gln-Leu-Aib-Gln-Leu-OH, ampullosporin D is Ac-Trp-Ala-Aib-Aib-Leu-Aib-Gln-Aib-Aib-Ala-Gln-Leu-Aib-Gln-Leu-OH, and ampullosporin E1 has the sequence Ac-Trp-Ala-Aib-Aib-Leu-Aib-Gln-Ala-Aib-Aib-Gln Leu-Ala-Gln-Leu-OH where Ac refers to the *N*-acetyl group, Aib refers to an α-aminoisobutyryl residue, and Leu-OH refers to L-leucinol. The ampullosporins are isolated from *Sepedonium ampullosporum*. **1.** Ritzau, Heinze, Dornberger, *et al.* (1997) *J. Antibiot.* **50**, 722. **2.** Kronen, Kleinwachter, Schlegel, Hartl & Grafe (2001) *J. Antibiot.* **54**, 175.

Amrinone

This 2-pyridinone (FW = 187.20 g/mol; CAS 60719-84-8), also known by the trade name Inocor® and its systematic name, 5-amino-3,4′-bipyridin-6(1*H*)-one, is a noncatechol, nonglycoside phosphodiesterase-3 inhibitor that retards the breakdown of both cAMP and cGMP. The vasodilator mechanisms of amrinone do not involve the opening of either ATP-sensitive K$^+$ (K$_{ATP}$) and large-conductance Ca^{2+}-activated K$^+$ (*or* KCa) channels. **1.** Orito, Satoh & Taira (1994) *Tohoku J. Exp. Med.* **172**, 163.

***m*-AMSA, See** *Amsacrine*

Amsacrine

This 9-anilinoacridine, (FW$_{free-base}$ = 393.47 g/mol; CAS 51264-14-3), also known as *m*-AMSA, is a cytostatic intercalating agent that stimulates and stabilizes the complex formed between DNA and DNA topoisomerase II. The *o*-analogue is a slightly weaker inhibitor of the *Sulfolobus shibatae* enzyme. Amsacrine is used to treat acute myelogenous leukemia. **Target(s):** DNA topoisomerase II (1,2,5-14); aldehyde oxidase (3); HERG (human ether-a-go go-related gene) K+ currents (4). **1.** McDonald, Eldredge, Barrows & Ireland (1994) *J. Med. Chem.* **37**, 3819. **2.** Nelson, Tewey & Liu (1984) *Proc. Natl. Acad. Sci. U.S.A.* **81**, 1361. **3.** Gormley, Rossitch, D'Anna & Cysyk (1983) *Biochem. Biophys. Res. Commun.* **116**, 759. **4.** Thomas, Hammerling, Wu, *et al.* (2004) *Brit. J. Pharmacol.* **142**, 485. **5.** Low, Orton & Friedman (2003) *Eur. J. Biochem.* **270**, 4173. **6.** Sullivan, Latham, Rowe & Ross (1989) *Biochemistry* **28**, 5680. **7.** Melendy & Ray (1989) *J. Biol. Chem.* **264**, 1870. **8.** Vosberg (1985) *Curr. Top. Microbiol. Immunol.* **114**, 19. **9.** Hammonds, Maxwell & Jenkins (1998) *Antimicrob. Agents Chemother.* **42**, 889. **10.** Insaf, Danks & Witiak (1996) *Curr. Med. Chem.* **3**, 437. **11.** Saijo, Enomoto, Hanaoka & M. Ui (1990) *Biochemistry* **29**, 583. **12.** Bergerat, Gadelle & Forterre (1994) *J. Biol. Chem.* **269**, 27663. **13.** Dickey, Choi, Van Etten & Osheroff (2005) *Biochemistry* **44**, 3899. **14.** Dickey & Osheroff (2005) *Biochemistry* **44**, 11546.

Amuvatinib

This potent, multi-target signal transduction inhibitor (FW = 447.51 g/mol; CAS 850879-09-3; Solubility: 32 mg/mL DMSO; <1 mg/mL H$_2$O), also known as MP-470, is highly effective against c-Kit, IC$_{50}$ = 10 nM; PDGFRα IC$_{50}$ = 40 nM; and Flt3, IC$_{50}$ = 81 nM (1-4). Whether alone or in combination with erlotinib, amuvatinib inhibits prostate cancer *in vitro* and *in vivo*, with an IC$_{50}$ in the low-μM range. **1.** Mahadevan, Cooke, Riley, *et al.* (2007) *Oncogene* **26**, 3909. **2.** Qi, Cooke, Stejskal, *et al.* (2009) *BMC Cancer* **9**, 142. **3.** Welsh, Mahadevan, Ellsworth, *et al.* (2009) *Radiat. Oncol.* **4**, 69. **4.** Zhao, Luoto, Meng, Bristow, *et al.* (2011) *Radiother. Oncol.* **101**, 59-65.

Amygdalin

This cyanogenic glycoside (FW = 457.43 g/mol; CAS 29883-15-6) a constituent of bitter almonds as well as the seeds of peaches and apricots, inhibits trehalase. Often called laetrile by the nonscientific community, this substance was the focus of bogus claims in the 1970s and 1980s that it is effective in curing cancer. Numerous clinical investigations discount any idea that amygdalin is effective as a cancer treatment. Hydrolysis of amygdalin is attended by the release of cyanide, a property accounting for its toxicity and mild cytostatic properties. The hydrolysis of amygdalin by emulsin in 1830 was one of the first chemical studies on enzyme catalysis. **Target(s):** α,α-trehalase (1,2); β-galactosidase, *Tenebrio molitor* (3); thioglucosidase, *or* myrosinase, *or* sinigrinase (4-6); cyclomaltodextrin glucanotransferase (7). **1**. Silva, Terra & Ferreira (2006) *Comp. Biochem. Physiol. B Biochem. Mol. Biol.* **143**, 367. **2**. Silva, Terra & Ferreira (2004) *Insect Biochem. Mol. Biol.* **34**, 1089. **3**. Ferreira, Terra & Ferreira (2003) *Insect Biochem. Mol. Biol.* **33**, 253. **4**. Ohtsuru & Hata (1973) *Agric. Biol. Chem.* **37**, 2543. **5**. Ohtsuru, Tsuruo & Hata (1969) *Agric. Biol. Chem.* **33**, 1315. **6**. Tani, Ohtsuru & Hata (1974) *Agric. Biol. Chem.* **38**, 1623. **7**. Bovetto, Villette, Fontaine, Sicard & Bouquelet (1992) *Biotechnol. Appl. Biochem.* **15**, 59.

Amylin

This 37-residue pancreatic β-cell hormone (MW = 3.9 kDa; Sequence: KCNTATCATQRLANFLVHSSNNFGAILSSTNVGSNTY), also called Islet Amyloid Polypeptide (IAPP), is co-secreted with insulin and plays a role in glycemic regulation by retarding gastric emptying and promoting satiety. Amylin helps to prevent post-prandial spikes in blood glucose. Amylin potently inhibits (EC_{50} = 18 pM) amino acid-stimulated glucagon secretion. (*Note*: Human amylin is amyloidogenic, whereas the synthetic hormone known as Pramlintide (MW = 3951.41g; CAS 151126-32-8; Sequence: KCNTATCATQRLANFLVHSSNNFGPILPPTNVGSNTY·NH₂; Tradename: Symlin®) contains three prolyl residues that are found in Rat Amylin (Sequence: KCNTATCATQRLANFLVRSSNNLGPVLPPTNG SNTY) and is nonamyloidogenic.) **1**. Young (2005) *Adv. Pharmacol.* **52**, 151.

Anabasine & Anabaseine

Anabasine   Anabaseine

Anabasine (FW = 162.23 g/mol; CAS 40774-73-0) is a piperidine-containing alkaloid and natural insecticide found in the Tree Tobacco *Nicotiana glauca* (1,2). As a potent partial agonist of nicotinic acetylcholine receptors, it produces a depolarizing block at autonomic ganglionic and neuromuscular synapses, resulting in nicotine-like symptoms. Anabaseine (FW = 160.22 g/mol; CAS 3471-05-4), isolated from the marine worm *Paranemertes peregrina Coe* (3) and *Aphaenogaster* ants (4), interacts preferentially with α_7 nicotinic receptors, inducing half-maximal probability of channel opening at 10 μM (5). Anabaseine undergoes pH-dependent ring opening, and, because only 29% is cyclic at pH 7.4 (6), its closed ring form is likely to be a more potent partial agonist than anabasine. Receptor interactions of Anabasine and Anabaseine inform the design of receptor-selective nicotinic agonists (*See* GTS-21). **1**. Pyriki & Oehler (1954) *Pharmazie* **9**, 685. **2**. Feinstein & McCabe (1950) *Science* **112**, 534. **3**. Kem, Abbott & Coates (1971) *Toxicon* **9**, 15. **4**. Wheeler, Olubajo, Storm & Duffield (1981) *Science* **211**, 1051. **5**. Kem, Mahnir, Papke & Lingle (1997) *J. Pharmacol. Exp. Ther.* **283**, 979. **6**. Zoltewicz, Bloom, Kem (1989) *J. Org. Chem.* **54**, 4462.

Anacardic Acids

Anacardic Acid (2-hydroxy-6-pentadecyl-benzoic acid)

Anacardic Acid (2-hydroxy-6-[(8Z,11Z)-pentadeca-8,11,14-trienyl]benzoic acid)

These urushiol-related phenolic lipids (IUPAC Name: 2-hydroxy-6-pentadecylbenzoic acid; FW = 348.52 g/mol; CAS 16611-84-0; IUPAC Name: 2-hydroxy-6-[(8Z,11Z)-pentadeca-8,11,14-trienyl]benzoic acid; FW = 342.47 g/mol; CAS 11034-77-8), first isolated from the shell of the Indian cashew nut *Anacardium occidentale*, consists of salicylic acid with an 15 or 17 carbon atom alkyl chain. The first is a potent inhibitor of p300 and p300/CBP-associated factor histone acetyltranferases, also showing antimicrobial activity as well as inhibition of prostaglandin synthase, tyrosinase and lipoxygenase. The second is active against *Streptococcus mutans* and methicillin-resistant *Staphylococcus aureus*. **1**. Izzo & Dawson (1949) *J. Org. Chem.* **14**, 1039. **2**. Symes & Dawson (1953) *J. Amer. Chem. Soc.* **75**, 4952. **3**. Paul & Yeddanapalli (1954) *Nature* **174**, 604.

Anacetrapib

This cholesteryl ester transfer protein (*or* CETP) inhibitor (FW = 637.51 g/mol; CAS 875446-37-0), also known by the code name MK-0859, reduces plasma cholesterol concentrations, thereby increasing serum high-density lipoprotein levels and preventing cardiovascular disease associated with hypercholesteremia (1,2). X-ray crystal studies suggest that CETP is a crescent-shaped protein with a 6-nm hydrophobic tunnel traversing its core (3) Both of the tunnel's two entrances face the protein's concave surface, which most likely is the interaction platform for binding HDL particles and engaging its lipoprotein constituents. CETP mediates the reciprocal transfer of neutral lipids (*e.g.*, cholesteryl esters and triglycerides) and phospholipids between different lipoprotein fractions in blood plasma. Anacetrapib is unlike many CETP inhibitors, which increase systolic blood pressure. Anacetrapib also exhibits a low-to-moderate degree of absorption after oral dosing and majority of the absorbed dose is eliminated via oxidation to a series of hydroxylated metabolites that undergo conjugation with glucuronate before excretion into bile (2). Unlike torcetrapib, another CETP inhibitor, anacetrapib does not elevate blood pressure, alter electrolytes, or cause any significant side effects. **1**. Krishna, Anderson, Bergman, *et al.* (2007) *Lancet* **370**, 1907. **2**. Tan, Hartmann, Chen, *et al.* (2010) *Drug Metab. Dispos.* **38**, 459. **3**. Qiu, Mistry, Ammirati, *et al.* (2007) *Nature Struct. Mol. Biol.* **14**, 106.

Anagrelide

This potent platelet aggregation inhibitor (FW = 256.09 g/mol; CAS 68475-42-3), also known as 6,7-dichloro-1,5-dihydroimidazo[2,1-b]quinazolin-2(3H)-one, BL-4162A, and Agrylin®, blocks the action of a variety of aggregating agents added platelet rich plasma, EC_{50} < 1 μg/mL, *or* 4 nM (1). **Primary Mode of Action:** Although the exact mechanism of its selective inhibition of megakaryocyte (MK) production of platelets remains uncertain, anagrelide is known to be a potent inhibitor of phosphodiesterase-II (IC_{50} = 36 nM) and lipoprotein-associated phospholipase A₂ (*or* Lp-PLA₂), the latter also known as platelet-activating factor acetylhydrolase (*or* PAF-AH). PDE-II hydrolyzes both cGMP and cAMP. Binding of cGMP to its regulatory GAF-B domain favors cAMP hydrolysis to 5'-AMP, thereby reducing cGMP hydrolysis to 5'-GMP. This property, which facilitates cross-regulation of the cAMP and cGMP pathways, suggests that a potent PDE-II inhibitor should potentiate the effects of cAMP and/or cGMP, the concentrations of which should increase in anagrelide-sensitive cells. Lp-PLA₂ plays pivotal role in platelet maturation by specifically hydrolyzing Platelet-Activating Factor (PAF = acetyl-glyceryl-ether-phosphorylcholine) as well as other glycerophospholipids containing short, truncated, and/or oxidized fatty acyl groups at the *sn*-2 position of the glycerol backbone. At a final concentration of 100 ng/mL, anagrelide selectively blocks *in vitro* MK maturation, resulting in a 50% decrease in the total number of CD41a$^+$ MKs (2). In humans, anagrelide has the intriguing ability to promote as a species-specific platelet-lowering activity at dose levels lower than those required to inhibit platelet aggregation (3). **Target(s):** collagen- and immune complex-induced platelet aggregation and release (4); suppresses megakaryocytopoiesis by reducing the expression levels of the transcription GATA-1 and FOG-1 via a PDEIII-independent mechanism that is differentiation context-specific but does not involve inhibition of MPL-mediated early signal transduction events (5). **1**. Fleming & Buyniski (1979) *Thromb. Res.* **15**, 373. **2**. Lane, Hattori, Dias, *et al.* (2001) *Exp. Hematol.* **29**, 1417. **3**. Tefferi, Silverstein, Petitt, Mesa & Solberg (1997) *Semin. Thromb. Hemost.* **23**, 379. **4**. Clark, Reid, Tevaarwerk (1981) *Thromb. Res.* **21**, 215. **5**. Ahluwalia, Donovan, Singh, Butcher & Erusalimsky (2010) *J. Thromb. Haemost.* **8**, 2252.

Anandamide

This lipid neurotransmitter and endocannabinoid receptor ligand (FW = 347.54 g/mol; CAS 94421-68-8), also known as arachidonylethanolamide, inhibits calcium ion currents and activates the MAP kinase signaling pathway. Anandamide is a naturally occurring effector of central and peripheral nervous systems, as mediated by CB$_1$ and CB$_2$ cannabinoid receptors, respectively. (These receptors were originally characterized as binding Δ^9-tetrahydrocannabinol, *or* THC, marijuana's major psychoactive ingredient.) Anandamide, 2-arachidonoylglycerol, and congeners (*N*-oleoylethanolamide and *N*-palmitoylethanolamide) are lipophilic signalling molecules with multiple physiological roles in learning and memory, neuroinflammation, oxidative stress, neuroprotection and neurogenesis. Anandamide likewise plays a role in regulating uterine implantation of the early-stage embryo (blastocyst). Anandamide synthesis occurs from *N*-acetylphosphatidylethanolamine by multiple pathways that include phospholipase A$_2$, phospholipase C, and NAPE-hydrolyzing phospholipase D (*or* NAPE-PLD). **Target(s):** adenylate cyclase (1,2); Na$^+$ channels (3,4); α_7-nicotinic acetylcholine receptor (5); 5 HT$_3$ receptors (6); Ca^{2+} channels (7,8,10,11); K$^+$ channels, voltage-gated (9); acylglycerol lipase, *or* monoacylglycerol lipase (12). **1**. Vogel, Barg, Levy, *et al.* (1993) *J. Neurochem.* **61**, 352. **2**. Van der Kloot (1994) *Brain Res.* **649**, 181. **3**. Kim, Kim, Shin, *et al.* (2005) *Brain Res.* **1062**, 39. **4**. Nicholson, Liao, Zheng, *et al.* (2003) *Brain Res.* **978**, 194. **5**. Oz, Ravindran, Diaz-Ruiz, Zhang & Morales (2003) *J. Pharmacol. Exp. Ther.* **306**, 1003. **6**. Oz, Zhang & Morales (2002) *Synapse* **46**, 150. **7**. Chemin, Monteil, Perez-Reyes, J. Nargeot & P. Lory (2001) *EMBO J.* **20**, 7033. **8**. Oz, Tchugunova & Dunn (2000) *Eur. J. Pharmacol.* **404**, 13. **9**. Poling, Rogawski, Salem, & Vicini (1996) *Neuropharmacology* **35**, 983. **10**. Mackie, Devane & Hille (1993) *Mol. Pharmacol.* **44**, 498. **11**. Twitchell, Brown & Mackie (1997) *J. Neurophysiol.* **78**, 43. **12**. Ghafouri, Tiger, *et al.* (2004) *Brit. J. Pharmacol.* **143**, 774.

Anastrozole

This non-steroidal aromatase inhibitor and breast cancer adjuvant (FW = 293.37 g/mol; CAS 120511-73-1), also known by its code name ZD1033, its proprietary name Arimidex™ and its systematic name 2,2'-[5-(1*H*-1,2,4-triazol-1-ylmethyl)-1,3-phenylene]bis(2-methylpropanenitrile), is a FDA-approved drug for treating breast cancer after surgery, with or without radiotherapy. **Mechanism of Inhibitory Action:** Arimidex is an orally available drug for the treatment of postmenopausal women with hormone receptor-positive or hormone receptor-unknown locally advanced or metastatic breast cancer and for the treatment of postmenopausal women with advanced breast cancer that has progressed following treatment with tamoxifen. It is a reversible competitive aromatase inhibitor (IC$_{50}$ of 15 nM) that limits the conversion of androgens to estrogens in extragonadal tissues (1). At 0.5 μM, anastrozole does not elevate plasma 11-deoxycorticosterone in monkeys, and at 1.5 μM, does not affect plasma aldosterone levels or Na$^+$/K$^+$ excretion in rats, plasma K$^+$ concentrations in dogs, or cause adrenal hypertrophy in rats or dogs. Anastrozole is without discernible effect on adrenocorticoid hormone synthesis *in vivo*, even at high multiples of its maximally effective aromatase-inhibiting dose. **Pharmacokinetics:** Anastrozole is rapidly absorbed, with maximum plasma concentrations occurring within 2 hour after oral administration (2). Plasma concentrations of anastrozole rose with increasing doses of the drug. The elimination $t_{1/2}$ of Anastrozole in humans ranged from 30-60 hours. Consistent with anastrozole's long plasma $t_{1/2}$, steady-state plasma concentrations were 3-4x higher than plasma concentrations observed after single administration of 1, 3, 5, or 10 mg (2). **Key Pharmacokinetic Parameters:** *See* Appendix II in Goodman & Gilman's *THE PHARMACOLOGICAL BASIS OF THERAPEUTICS*, 12th Edition (Brunton, Chabner & Knollmann, eds.) McGraw-Hill Medical, New York. **1**. Dukes, Edwards, Large, Smith & Boyle (1996) *J. Steroid Biochem. Mol. Biol.* **58**, 439. **2**. Plourde, Dyroff, Dowsett, *et al.* (1995) *J. Steroid Biochem. Mol. Biol.* **53**, 175.

5α-Androstan-17-one

This steroid (FW = 274.45 g/mol), which, except for the absence of a hydroxyl group, is otherwise identical to 3α-hydroxy-5α-androstan-17-one (*or* androsterone), inhibits rat mammary glucose-6-phosphate dehydrogenase (K_i = 0.3 μM). **1**. Raineri & Levy (1970) *Biochemistry* **9**, 2233.

5-Androstene-3β,17β-diol

This sterol (FW = 290.45 g/mol), also referred to simply as androstenediol, is a metabolite of dehydroepiandrosterone (prasterone) isolated independently in the laboratories of Butenandt and Ruzicka, who were awarded the Nobel Prize in Chemistry in 1939. **Target(s):** glucose-6-phosphate dehydrogenase (1); steroid 16α-monooxygenase (2); 17β hydroxysteroid dehydrogenase, also alternative substrate (3); 17-ketosteroid reductase (4); estrone sulfotransferase (5). **1**. Criss & McKerns (1969) *Biochim. Biophys. Acta* **184**, 486. **2**. Sano, Shibusawa, Yoshida, *et al.* (1980) *Acta Obstet. Gynecol. Scand.* **59**, 245. **3**. Bonney, Reed & James

(1983) *J. Steroid Biochem.* **18**, 59. **4**. MacIndoe, Hinkhouse & Woods (1990) *Breast Cancer Res. Treat.* **16**, 261. **5**. Rozhin, Huo, Zemlicka & Brooks (1977) *J. Biol. Chem.* **252**, 7214.

4-Androstene-3,17-dione

This androgen (FW = 286.41 g/mol), also referred to simply as androstenedione, is synthesized in the adrenal gland, testis, and ovary as a testosterone and estrone precursor. Androstenedione exhibits approximately one-fifth the androgenic activity of testosterone. **Target(s):** steroid 21-monooxygenase (1,14); sterol demethylase (2); glucose-6-phosphate dehydrogenase, K_i = 60.7 μM (3,4,8,9); 3-keto-5β-steroid 4-dehydrogenase (5); glutamate dehydrogenase (6); 3β-hydroxysteroid dehydrogenase (7); 6-phosphogluconate dehydrogenase (8); isocitrate dehydrogenase (8); steroid sulfatase (10); steroid Δ-isomerase (11); sterol *O*-acyltransferase, *or* cholesterol *O*-acyltransferase, *or* ACAT (12); arylamine *N*-acetyltransferase (13). **1**. Sharma (1964) *Biochemistry* **3**, 1093. **2**. Gaylor, Chang, Nightingale, Recio & Ying (1965) *Biochemistry* **4**, 1144. **3**. McKerns & Kaleita (1960) *Biochem. Biophys. Res. Commun.* **2**, 344. **4**. Criss & McKerns (1969) *Biochim. Biophys. Acta* **184**, 486. **5**. Davidson (1969) *Meth. Enzymol.* **15**, 656. **6**. Frieden (1963) *The Enzymes*, 2nd ed., **7**, 3. **7**. Townsley (1975) *Acta Endocrinol. (Copenh.)* **79**, 740. **8**. McKerns (1963) *Biochim. Biophys. Acta* **73**, 507. **9**. Douville & Warren (1968) *Biochemistry* **7**, 4052. **10**. Notation & Ungar (1969) *Biochemistry* **8**, 501. **11**. Weintraub, Vincent, Baulieu & Alfsen (1977) *Biochemistry* **16**, 5045. **12**. Simpson & Burkhart (1980) *Arch. Biochem. Biophys.* **200**, 79. **13**. Kawamura, Westwood, Wakefield, *et al.* (2008) *Biochem. Pharmacol.* **75**, 1550. **14**. Bélanger, Tremblay, Vallée, Provencher & Perron (1995) *Endocr. Res.* **21**, 329.

Androsterone

This steroid (FW = 290.45 g/mol), also known as 5α-androstan-3α-ol-17-one and 3α-hydroxy-5α-androstan-17-one, is a weak androgen (approximately one-tenth that of testosterone). Androsterone was first isolated by Adolph Butenandt and coworkers in 1929-1931. Leopold Ruzicka showed that cholesterol could be converted to androsterone, and the two shared the 1939 Nobel Prize in Chemistry. **Target(s):** 3-keto-5β-steroid 4-dehydrogenase (1); 21-hydroxysteroid dehydrogenase (2,4,7,8); sterol acyltransferase, *or* cholesterol acyltransferase (3); glucose-6-phosphate dehydrogenase (5,6); steroid sulfatase (9); steroid sulfotransferase, also alternative substrate showing substrate inhibition (10); glucuronosyltransferase (11). **1**. Davidson (1969) *Meth. Enzymol.* **15**, 656. **2**. Monder & Furfine (1969) *Meth. Enzymol.* **15**, 667. **3**. Billheimer (1985) *Meth. Enzymol.* **111**, 286. **4**. Talalay (1963) *The Enzymes*, 2nd ed., **7**, 177. **5**. Noltmann & Kuby (1963) *The Enzymes*, 2nd ed., **7**, 223. **6**. McKerns & Kaleita (1960) *Biochem. Biophys. Res. Commun.* **2**, 344. **7**. Monder & White (1962) *Biochem. Biophys. Res. Commun.* **8**, 383. **8**. Monder & White (1965) *J. Biol. Chem.* **240**, 71. **9**. Notation & Ungar (1969) *Biochemistry* **8**, 501. **10**. Chang, Shi, Rehse & Lin (2004) *J. Biol.*

Chem. **279**, 2689. **11**. Yoshigae, Konno, Takasaki & Ikeda (2000) *J. Toxicol. Sci.* **25**, 433.

Anecortave Acetate

This novel angiogenesis pro-drug (FW = 386.48 g/mol; CAS 7753-60-8) is taken up and metabolically de-esterified to form the aqnecortave, a synthetic steroid that suppresses the development and elongation of neovascular sprouts and tip cell motility in a VEGF-treated retinal explant model in a dose-dependent manner (1). Whether used as monotherapy or in combination with sub-therapeutic doses of cisplatin, anecortave acetate significantly controlled tumor burden in a murine retinoblastoma model (2). **1**. Unoki, Murakami, Ogino, Nukada & Yoshimura (2010) *Invest. Ophthalmol. Vis. Sci.* **51**, 2347. **2**. Jockovich, Murray, Escalona-Benz, Hernandez & Feuer (2006) *Invest. Ophthalmol. Vis. Sci.* **47**, 1264.

2,2'-Anhydro-1-(arabinofuranosyl)-5-ethyluracil

2,2'-anhydro-
5-ethyluridine

2,2'-Anhydro-1-(arabinofuranosyl)-
5-ethyluracil

This nucleoside analogue (FW = 253.23 g/mol), misnamed 2,2'-anhydro-5-ethyluridine, inhibits rat uridine phosphorylase (K_i = 25 nM). This anhydropyrimidine nucleoside has little or no flexibility about the *N*-glycosidic bond, as confirmed by NMR spectroscopy. **1**. Veres, Neszmélyi, Szabolcs & Dénes (1988) *Eur. J. Biochem.* **178**, 173. **2**. Veres, Szabolcs, Szinai, *et al.* (1985) *Biochem. Pharmacol.* **34**, 1737. **3**. Drabikowska (1996) *Acta Biochim. Pol.* **43**, 733. **4**. Watanabe, Hino, Wada, Eliason & Uchida (1995) *J. Biol. Chem.* **270**, 12191. **5**. el Kouni, Fardos, Naguib, *et al.* (1988) *J. Biol. Chem.* **263**, 6081.

2,6-Anhydro-3-deoxy-2β-phosphonylmethyl-8-phosphate-D-*glycero*-D-*talo*-octonate

This isosteric phosphonate analogue (FW = 410.20 g/mol) inhibits 3-deoxy-8-phosphooctulonate synthase (K_i = 5 μM). **1**. Baasov, Sheffer-Dee-Noor, Kohen, Jakob & Belakhov (1993) *Eur. J. Biochem.* **217**, 991.

1,5-Anhydro-D-glucitol

This naturally occurring sugar ether (FW = 164.15 g/mol; CAS 154-58-5; Symbol: 1,5-AG), also known as 1-deoxy-D-glucose and GlycoMark®, is a metabolically inert polyol that competes with glucose for reabsorption in the kidneys. 1,5-AG is rapidly depleted as blood glucose levels exceed the renal threshold for glucosuria. 1,5-AG is prepared by LiAlH₄ reduction of 1-bromo-α-D-glucopyranose 2,3,4,6-tetraacetate. **Biomarker for Glycemic Control:** Although still regarded as the gold standard in diabetes management, glycated hemoglobin (*or* HbA1c) is not a direct index of recent glycemic control (*i.e.,* excursions from normoglycemia over immediately preceeding days or weeks). The plasma level of 1,5-AG reflects episodes of acute hyperglycemia more sensitively than HbA1c does and is correlated with fasting plasma glucose (FPG) and postprandial hyperglycemic peaks. In this regard, 1,5-AG levels appear to be a useful index of short-term glucose status, postprandial hyperglycemia, and glycemic variability (1-3). 1,5-AG predicts rapid changes in glycemia more accurately than HbA1c or fructosamine. It is also more tightly associated with glucose fluctuations and postprandial glucose. 1,5-AG provides information that is complementary information to A1C. Guidelines in Diabetes Assessment: >10 µg 1,5-AG per mL (*or* 60 µM) = Well Controlled; 5-10 µg 1,5-AG per mL (*or* 30-60 µM) = Moderately Controlled; 2-5 µg 1,5-AG per mL (*or* 12-30 µM) = Poorly Controlled; <2 µg 1,5-AG per mL (*or* <12 µM) = Very Poorly Controlled. *Note*: Because 1,5-anhydroglucitol is transported by sodium–glucose cotransporter, 1,5-AG assays may provide inaccurate assessment of glycemic control in T2DM patients treated with an (SGLT2) inhibitor, such as canagliflozin (4). **Target(s):** concanavalin A (5); hexokinase, as a weakly inhibiting alternative substrate (6,7); aldose 1-epimerase, *or* mutarotase (8-11); glucose 1-oxidase (11); trehalose (12); glucokinase (11); β-glucosidase (K_i = 60 mM for the sweet almond enzyme (12); α,α-trehalose phosphorylase, configuration-retaining (13); cellobiose phosphorylase (14). **1.** Dungan, Buse, Largay, *et al.* (2006) *Diabetes Care.* **29**, 1214. **2.** McGill, Cole, Nowatzke, *et al.* (2004) *Diabetes Care.* **27**, 1859. **3.** Nguyen, Rodriguez, Mason, *et al.* (2007) *Pediatr Diabetes.* **8**, 214. **4.** Balis, Tong & Meininger (2014) *J. Diabetes.* **6**, 378. **5.** Goldstein, Hollerman & Smith (1965) *Biochemistry* **4**, 876. **6.** Webb (1966) *Enzyme and Metabolic Inhibitors,* vol. **2**, p. 379, Academic Press, New York. **7.** Sols (1956) *Biochim. Biophys. Acta* **19**, 144. **8.** Keston (1964) *J. Biol. Chem.* **239**, 3241. **9.** Keston (1964) *Science* **143**, 698. **10.** Bailey & Pentchev (1964) *Biochem. Biophys. Res. Commun.* **14**, 161. **11.** Taguchi, Haruna & Okuda (1993) *Biotechnol. Appl. Biochem.* **18**, 275. **12.** Dale, Ensley, Kern, Sastry & Byers (1985) *Biochemistry* **24**, 3530. **13.** Eis, Watkins, Prohaska & Nidetzky (2001) *Biochem. J.* **356**, 757. **14.** Nidetzky, Eis & Albert (2000) *Biochem. J.* **351**, 649.

2,5-Anhydro-D-glucitol 1,6-Bisphosphate

This glucitol analogue (FW_free-acid = 324.12 g/mol) of the α-anomer of D-fructose 1,6-bisphosphate and inhibits fructose-1,6-bisphosphate aldolase and fructose-1,6 bisphosphatase, with respective K_i values of 13 and 0.55 µM. **Target(s):** fructose-1,6-bisphosphate aldolase (1); fructose-1,6-bisphosphatase (2,3); pyrophosphate-dependent phospho-fructokinase, *or* diphosphate: fructose-6-phosphate 1-phosphotransferase (4). **1.** Hartman & Barker (1965) *Biochemistry* **4**, 1068. **2.** Marcus (1976) *J. Biol. Chem.* **251**, 2963. **3.** Villeret, Huang, Zhang & Lipscomb (1995) *Biochemistry* **34**, 4307. **4.** Bertagnolli, Younathan, Voll, Pittman & Cook (1986) *Biochemistry* **25**, 4674.

1,5-Anhydro-D-glucitol 6-Phosphate

This glucitol derivative (FW_free-acid = 244.14 g/mol) is an analogue of D-glucose 6-phosphate. **Target(s):** glucose-6-phosphate isomerase (1,2); hexokinase (2,3,4,6-13); phosphoglucomutase (2); glucokinase (4); inositol-3-phosphate synthase, *or* inositol-1-phosphate synthase (5). **1.** Noltmann (1972) *The Enzymes*, 3ʳᵈ., **6**, 271. **2.** Webb (1966) *Enzyme and Metabolic Inhibitors*, vol. **2**, Academic Press, New York. **3.** Kosow & Rose (1968) *J. Biol. Chem.* **243**, 3623. **4.** Taguchi, Haruna & Okuda (1993) *Biotechnol. Appl. Biochem.* **18**, 275. **5.** Barnett, Rasheed & Corina (1973) *Biochem. Soc. Trans.* **1**, 1267. **6.** de Cerqueira Cesar & Wilson (2002) *Arch. Biochem. Biophys.* **397**, 106. **7.** Tsai (2007) *J. Biomed. Sci.* **14**, 195. **8.** Aleshin, Malfois, Liu, *et al.* (1999) *Biochemistry* **38**, 8359. **9.** Sebastian, Wilson, Mulichak & Garavito (1999) *Arch. Biochem. Biophys.* **362**, 203. **10.** Hashimoto & Wilson (2002) *Arch. Biochem. Biophys.* **399**, 109. **11.** White & Wilson (1989) *Arch. Biochem. Biophys.* **274**, 375. **12.** Armstrong, Wilson & Shoemaker (1996) *Protein Expr. Purif.* **8**, 374. **13.** Wilson (2003) *J. Exp. Biol.* **206**, 2049.

2,5-Anhydro-D-mannitol 1,6-Bisphosphate

This mannitol derivative (FW_free-acid = 324.12 g/mol) is an analogue of the β-anomer of D-fructose 1,6-bisphosphate and inhibits fructose-1,6-bisphosphate aldolase and fructose-1,6 bisphosphatase (Ki = 30 µM and 33 nM, respectively). **Target(s):** fructose-1,6-bisphosphatase (1-5); fructose-1,6-bisphosphate aldolase (6); pyrophosphate-dependent phospho-fructokinase, *or* diphosphate: fructose-6-phosphate 1-phosphotransferase (7). **1.** Pilkis, Claus, Kountz & El-Maghrabi (1987) *The Enzymes*, 3rd ed., **18**, 3. **2.** Marcus (1976) *J. Biol. Chem.* **251**, 2963. **3.** Zhang, Liang, Huang, Ke & Lipscomb (1993) *Biochemistry* **32**, 1844. **4.** Hanson, Ho, Wiseberg, *et al.* (1984) *J. Biol. Chem.* **259**, 218. **5.** Ganson & Fromm (1984) *Curr. Top. Cell. Regul.* **24**, 197. **6.** Hartman & Barker (1965) *Biochemistry* **4**, 1068. **7.** Bertagnolli & Cook (1984) *Biochemistry* **23**, 4101.

N-(4-(Anilinocarbonyl)phenylsulfonyl)-2-(benzofuran-2-yl)tryptamine

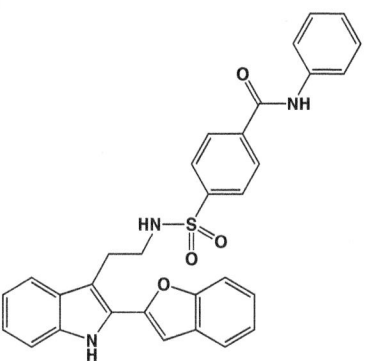

This modified tryptamine (FW = 519.63 g/mol) inhibits 15-lipoxygenase, *or* arachidonate 15-lipoxygenase, IC₅₀ = 15 nM. Compounds with aryl substituents at C-2 of the indole core of tryptamine and homotryptamine sulfonamides proved to be potent inhibitors of the isolated enzyme. **1.** Weinstein, Liu, Gu, *et al.* (2005) *Bioorg. Med. Chem. Lett.* **15**, 1435.

Anipamil

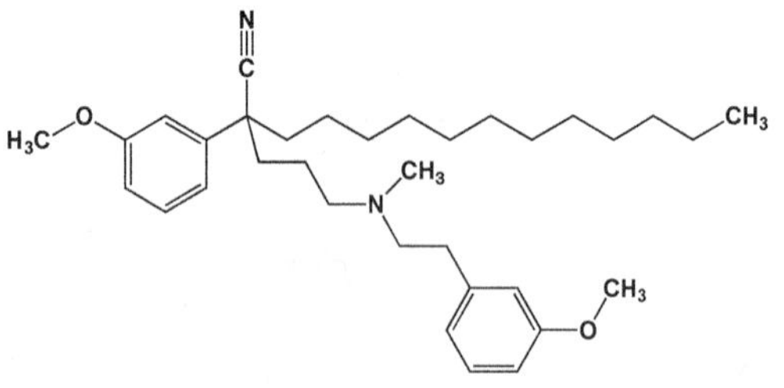

This novel, orally active phenylalkylamine derivative (FW = 520.79 g/mol; CAS 83200-10-6; IUPAC: 2-(3-methoxyphenyl)-2-[3-[2-(3-methoxy-phenyl)ethylmethylamino]-propyl]tetradecanenitrile, exhibits calcium channel-blocking activity that is confined peculiarly to myocardial muscle. 1. Raddino, Poli, Pasini & Ferrari (1992) *Naunyn. Schmiedebergs Arch. Pharmacol.* **346**, 339.

Anisomycin

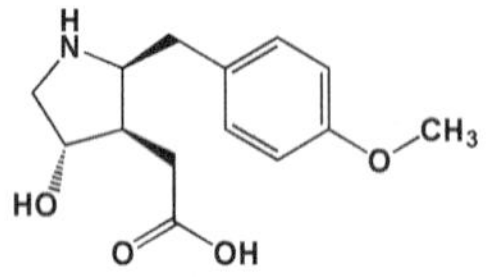

Anisomycin 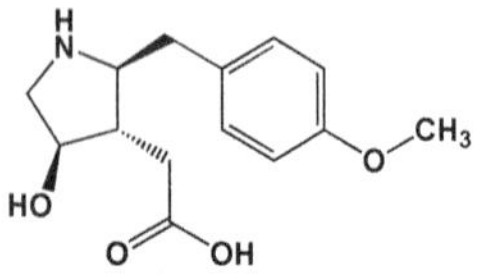   **2S,3S,4S-isomer**

This ribotoxin, antibiotic and famous amnestic agent (FW$_{free-base}$ = 265.31 g/mol; CAS 22862-76-6; alkali-labile; also slowly decomposes in acid), systematically named (2*R*,3*S*,4*S*)-4-hydroxy-2-(4-methoxybenzyl)-pyrrolidin-3-yl acetate, from *Streptomyces griseolus* and *S. roseochromogenes*, inhibits protein biosynthesis by binding to the 3'-end of the large (23S/28S) ribosomal RNA, a region directly involved in the three sequential steps of translational elongation, namely aminoacyl-tRNA binding, peptidyl transfer, and ribosomal translocation (1-8). **Structural Features:** The large ribosomal subunit of *Haloarcula marismortui* has provided 3-Å resolution structural data on antibiotic binding site(s) and the poses taken by anisomycin, blasticidin S, chloramphenicol, sparsomycin, and virginiamycin when they are bound to ribosomes (8). Two hydrophobic crevices, one located at the peptidyl transferase center and the other situated at the entrance to the peptide exit tunnel play roles in antibiotic binding. The aromatic ring of anisomycin binds to the active-site hydrophobic crevice, as does the aromatic ring of puromycin, while the aromatic ring of chloramphenicol binds to the exit tunnel hydrophobic crevice (8). These antibiotics bind at sites that overlap with either peptidyl-tRNA or aminoacyl-tRNA sites, consistent with their functioning as competitive inhibitors of peptide bond formation. *In vitro* **Mechanism of Action:** Anisomycin's inhibitory mechanism has been investigated using an *in vitro* rabbit reticulocyte system, using preformed Complex C (consisting of rabbit reticulocyte ribosomes, *E. coli* AcPhe-tRNA, and polyU) that reacts with excess puromycin and catalyzes peptide bond formation between AcPhe and puromycin in a pseudo-first-order, non-elongating reaction. A detailed kinetic investigation (9) showed that AcPhe-puromycin is produced in a pseudo-first-order reaction from the preformed AcPhe-tRNA/poly(U)/80S ribosome complex (designated Complex C) and excess of puromycin serving as substrate. Anisomycin acts as a mixed noncompetitive inhibitor (K_i = 6 x 10^{-7} M; K_{ii} = 12 x 10^{-7} M), with the product, AcPhe-puromycin, is derived only from Complex C, according to the puromycin reaction. **Amnestic Properties:** One well known characteristic of long-term memory is its apparent sensitivity to anisomycin, when the latter is administered directly to the brain at or around the time of training (10). About an hour or more after ending a training, memory enters a long-term, anisomycin-insensitrive phase. Even beyond this period, however, there appear to be time windows during which later expression of memory is impaired by injection of this inhibitor (11). Anisomycin is also known to disrupt traumatic memory consolidation, attenuating posttraumatic stress response in experimental animals (12). Because anisomycin was first shown to be a ribosome inhibitor that blocks translation, these observations led to the now long-held conclusion that the consolidation of memory somehow requires protein synthesis. More recently, however, anisomycin was also shown to activate certain stress-activated protein kinases (SAPKs), including c-Jun amino-terminal kinases (JNKs), thereby altering the efficiency of their associated signal transduction pathways (12,13). Other aminohexose pyrimidine nucleoside antibiotics that bind to the anisomycin-binding region of ribosomes likewise activate SAPK/JNK1. Consistent with this finding is the separate observation that ribosomes that are active at the time of exposure to Ricin A chain and α-Sarcin can initiate signal transduction from the damaged 28S rRNA to SAPK/JNK1, whereas inactive ribosomes cannot (12,13). Recent studies have shown that anisomycin also significantly inhibits mammalian cell proliferation by activating the P53/P21/P27 signaling to decrease the expression of ICBP90, thereby inhibiting expression of P-CDK2 and blocking cells in S and G$_2$/M phases (14). **Target(s):** ribosomal peptidyltransferase (1-8); tyrosine hydroxylase (15); acetylcholine esterase (16,17); α-L-rhamnosidase, K_i = 52 μM for the 2*S*,3*S*,4*S*-isomer of anisomycin [structure shown above] (18). 1. Fernandez-Muñoz, Monro & Vazquez (1971) *Meth. Enzymol.* **20**, 481. 2. Pestka (1974) *Meth. Enzymol.* **30**, 261. 3. Carrasco, Battaner & Vazquez (1974) *Meth. Enzymol.* **30**, 282. 4. Barbacid & Vazquez (1974) *Meth. Enzymol.* **30**, 426. 5. Scott & Tomkins (1975) *Meth. Enzymol.* **40**, 273. 6. Jiménez (1976) *Trends Biochem. Sci.* **1**, 28. 7. Grollman (1967) *J. Biol. Chem.* **242**, 3226. 8. *Pharmacol.* **53**, 1089. 8. Hansen, Moore & Steitz (2003) *J. Mol. Biol.* **330**, 1061. 9. Ioannou, Coutsogeorgopoulos & Synetos (1998) *Mol. Pharmacol.* **53**, 1089. 10. Rudy, Biedenkapp, Moineau & Bolding (2006) *Learn. Mem.* **13**, 1. 11. Reis, Jarome & Helmstetter (2013) *Front. Behav. Neurosci.* **7**, 150. 12. Cohen, Kaplan, Matar, *et al.* (2006) *Biol. Psychiatry* **60**, 767. 12. Iordanov, Pribnow, Magun, *et al.* (1997) *Mol. Cell Biol.* **17**, 3373. 13. Iordanov, Pribnow, Magun, *et al.* (1998) *J. Biol. Chem.* **273**, 15794. 14. Yu, Xing, Tang, *et al.* (2013) *Pharmacol. Rep.* **65**, 435. 15. Squire, Kuczenski & Barondes (1974) *Brain Res.* **82**, 241. 16. Moss & Fahrney (1976) *J. Neurochem.* **26**, 1155. 17. Zech & Domagk (1976) *Life Sci.* **19**, 157. 18. Kim, Curtis-Long, Seo, *et al.* (2005) *Bioorg. Med. Chem. Lett.* **15**, 4282.

Anthracimycin

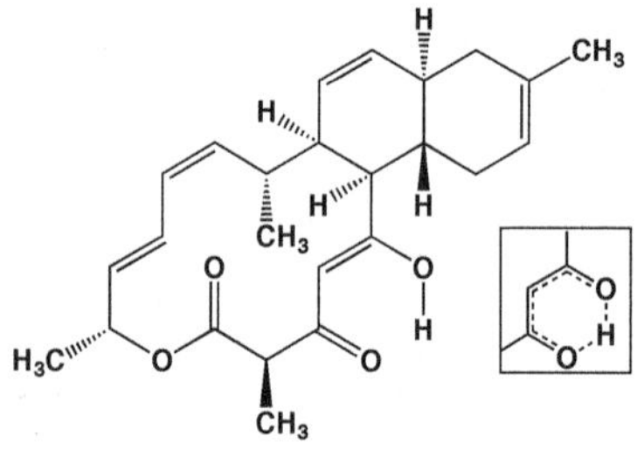

This first-in-class polyketide antibiotic (FW = 396.28 g/mol), isolated from a marine-derived *Streptomyces* species (Strain CNH365) from near-shore sediments off the California coast and cultured in saline, shows significant activity against *Bacillus anthracis* (MIC = 31 ng/mL *in vitro*). Anthracimycin's antibiotic spectrum includes Gram-positive staphylococci, enterococci, and streptococci. Its X-ray structure showed the enol proton to be virtually equidistant between each oxygen atom, suggesting a rapid keto–enol tautomerization, with pseudo-ring formation (See inset in structure above). 1. Jang, Nam, Locke, *et al.* (2013) *Angew. Chem. (Internat. Ed.)* **52**, 7822.

Anthranilate (Anthranilic Acid)

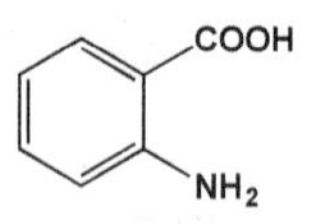

This tryptophan biosynthetic intermediate (FW$_{free-acid}$ = 137.14 g/mol; CAS 118-92-3; pK_1 = 2.05, pK_2 = 4.95), also known as 2-aminobenzoic acid, is a product of the anthanilate synthase reaction (Chorismate + L-glutamine $\rightleftharpoons$ anthranilate + pyruvate + L-glutamate). It is also an alternative substrate for 4-aminobenzoate 1-monooxygenase. **Target(s):** arylformamidase (1,24-26); indole-3-glycerol-phosphate synthase (2); 3-hydroxyanthranilate 3,4-dioxygenase (3); D-amino-acid oxidase (4,10-12,16,18,19); 3-hydroxy-anthranilate oxidase (5); tryptophanase, *or* tryptophan indole-lyase (6,14,20); acetoin dehydrogenase (7); alcohol dehydrogenase, NADP-dependent (7); carbonyl reductase (7); glutamine synthetase, *Neurospora crassa* (8); phenoxazinone synthase, *or* o-aminophenol oxidase (9); esterase (13); alanine aminotransferase (15); L-amino-acid oxidase (17); 4,5-dihydroxyphthalate decarboxylase (21); o-pyrocatechuate decarbox-ylase (22); kynureninase, as product inhibitor (23); kynurenine:

oxoglutarate amino-transferase, weakly inhibited (27); orotate phosphoribosyl-transferase, weakly inhibited (28); 3' dimethyl-staurosporine *O*-methyltransferase (29); kynurenine 3-monooxygenase, weakly inhibited (30); phenylalanine 2-monooxygenase (31); 3-hydroxyanthranilate 3,4-dioxygenase (32); *o*-aminophenol oxidase (33). **1**. Knox (1955) *Meth. Enzymol.* **2**, 242. **2**. Smith & Yanofsky (1962) *Meth. Enzymol.* **5**, 794. **3**. Nishizuka, Ichiyama & Hayaishi (1970) *Meth. Enzymol.* **17A**, 463. **4**. Bright & Porter (1975) *The Enzymes*, 3rd ed., **12**, 421. **5**. Vescia & di Prisco (1962) *J. Biol. Chem.* **237**, 2381. **6**. Hoch, Simpson & DeMoss (1966) *Biochemistry* **5**, 2229. **7**. Hara, Seiriki, Nakayama & Sawada (1985) *Prog. Clin. Biol. Res.* **174**, 291. **8**. Kapoor & Bray (1968) *Biochemistry* **7**, 3583. **9**. Katz & Weissbach (1970) *Meth. Enzymol.* **17A**, 549. **10**. Webb (1966) *Enzyme and Metabolic Inhibitors*, vol. 2, p. 341, Academic Press, New York. **11**. Bartlett (1948) *J. Amer. Chem. Soc.* **70**, 1010. **12**. Nishikimi, Osamura & Yagi (1971) *J. Biochem.* **70**, 457. **13**. Weber & King (1935) *J. Biol. Chem.* **108**, 131. **14**. Fenske & DeMoss (1975) *J. Biol. Chem.* **250**, 7554. **15**. Katsos, Philippidis & Palaiologos (1981) *Horm. Metab. Res.* **13**, 85. **16**. Fitzpatrick & Massey (1982) *J. Biol. Chem.* **257**, 9958. **17**. Soltysik, Byron, Einarsdottir & Stankovich (1987) *Biochim. Biophys. Acta* **911**, 201. **18**. Pollegioni, Buto, Tischer, Ghisla & Pilone (1993) *Biochem. Mol. Biol. Int.* **31**, 709. **19**. Pollegioni, Diederichs, Molla, S. *et al.* (2002) *J. Mol. Biol.* **324**, 535. **20**. Cowell, Moser & DeMoss (1973) *Biochim. Biophys. Acta* **315**, 449. **21**. Nakazawa & Hayashi (1978) *Appl. Environ. Microbiol.* **36**, 264. **22**. Kamath, Dasgupta & Vaidyanathan (1987) *Biochem. Biophys. Res. Commun.* **145**, 586. **23**. Koushik, Moore, Sundararaju & Phillips (1998) *Biochemistry* **37**, 1376. **24**. Serrano & Nagayama (1991) *Comp. Biochem. Physiol. B Comp. Biochem.* **99**, 281. **25**. Shinohara & Ishiguro (1970) *Biochim. Biophys. Acta* **198**, 324. **26**. Menge (1979) *Hoppe-Seyler's Z. Physiol. Chem.* **360**, 185. **27**. Takeuchi, Otsuka & Shibata (1983) *Biochim. Biophys. Acta* **743**, 323. **28**. Victor, Greenberg & Sloan (1979) *J. Biol. Chem.* **254**, 2647. **29**. Weidner, Kittelmann, Goeke, Ghisalba & Zähner (1998) *J. Antibiot.* **51**, 697. **30**. Saito, Quearry, Saito, Nowak, Markey & Heyes (1993) *Arch. Biochem. Biophys.* **307**, 104. **31**. Ida, Kurabayashi, Suguro, *et al.* (2008) *J. Biol. Chem.* **283**, 16584. **32**. Nandi, Lightcap, Koo, *et al.* (2003) *Int. J. Biochem. Cell Biol.* **35**, 1085. **33**. Subba Rao & Vaidyanathan (1967) *Arch. Biochem. Biophys.* **118**, 388.

Anthrax Lethal Toxin

This potentially death-causing, zinc metalloproteinase, (*Abbreviation*: LeTx), produced by *Bacillus anthracis,* is a holocomplex consisting of (a) Anthrax Lethal Factor (*or* ATLF), a zinc-metalloproteinase; (b) Protective Antigen (*or* PA), an 83-kDa bacterial component; and (c) Anthrax Edema Factor, an adenylate cyclase that also associates with PA to form Anthrax Edema Toxin. Upon binding to a cell surface receptor, PA^{83} undergoes proteolysis to PA^{63}, the latter a 63-kDa fragment that then self-assembles into a heptameric membrane channel that mediates the entry of Lethal Factor and/or Edema Factor into the host cell. Once within the cytosol of target cells, ATLF cleaves mitogen-activated protein kinase kinase 1 (MEK1) and other targets. Proteasome inhibitors (*e.g.*, *N*-acetyl-Leu-Leu-Norleucinal, MG132, and Lactacystin) efficiently block ATLF cytotoxicity, whereas other protease inhibitors ineffective (1). Inhibitor concentrations needed to block ATLF cytotoxicity are similar to those required to inhibit the proteasome-dependent IκB-α degradation, as induced by lipopolysaccharide. Proteosome inhibitors do not interfere with the subsequent proteolytic cleavage of MEK1 within LeTx-treated cells, indicating that they do not directly block the protease activity of ATLF. (That said, the proteasome inhibitors also prevent ATP depletion, a well known early effect of LeTx.) LeTx inhibition of actin assembly is mediated by blockade of Hsp27 phosphorylation, and can be reproduced by treating cells with the p38 mitogen-activated protein (MAP) kinase inhibitor SB203580 (2). Nonphosphorylated Hsp27 inhibits actin-based *Listeria* motility in cell extracts, most likely by binding to and sequestering actin monomers. **Structural Features:** Peptide substrates and small-molecule inhibitors bind to ATLF in the space between Domains-3 and -4. Domain-3 is attached on a hinge to Domain-2 and can move through an angular arc of >35° in response to ligand binding (3). Structures of ATLF·Inhibitor complexes reveal three frequently populated ATLF conformational states termed `bioactive', `open' and `tight' (1). The bioactive position is observed with large substrate peptides, with all peptide-recognition subsites open and accessible. The tight state, which occurs in unliganded and small-molecule complex structures, has Domain-3 serving as a clamp, situated over certain substrate subsites and blocking access. The open position appears to be an intermediate state between these extremes and is observed owing to steric constraints imposed by specific bound ligands (1). **1**. Tang & Leppla

(1999) *Infect. Immun.* **67**, 3055. **2**. During, Gibson, Li, *et al.* (2007) *EMBO J.* **26**, 2240. **3**. Maize, Kurbanov, De La Mora-Rey, *et al.* (2014) *Acta Crystallogr. D Biol. Crystallogr.* **70**, 2813. **4**. Sun & Jacquez (2016) *Toxins* (Basel). **8**, E34.

Antiamoebins

This set of at least sixteen peptide antibiotics are peptaibols, membrane-active oligopeptides isolated from soil fungi that exhibit anti-bacterial and anti-fungal properties. They contain α-aminoisobutyryl residues (thus, the "aib" in the class name) and usually end in a C-terminal alcohol (thus, the "ol" in the class name). The peptaibols are amphipathic, usually highly helical in structure, and typically form voltage-dependent ion channels. This will often lead to cell death. Antiamoebin-I sequence: Ac-Phe-Aib-Aib-Aib-Iva-Gly-Leu-Aib-Aib-Hyp-Gln-Iva-Hyp-Aib-Pro-Phe-OH, Antiamoebin-II: Ac-Phe-Aib-Aib-Aib-Iva-Gly-Leu-Aib-Aib-Hyp-Gln-Iva-Pro-Aib-Pro-Phe-OH, Antiamoebin-III: Ac-Phe-Aib-Aib-Aib-Aib-Gly-Leu-Aib-Aib-Hyp-Gln-Iva-Hyp-Aib-Pro-Phe-OH, and Antiamoebin-IV has the sequence Ac-Phe-Aib-Aib-Aib-Iva-Gly-Leu-Aib-Aib-Hyp-Gln-Iva-Hyp-Aib-Pro-Phe-OH, where Ac indicates an *N*-acetyl group, Aib refers to an α-aminoisobutyryl residue, Hyp represents a hydroxyprolyl residue, Iva refers to an isovaleryl residue, and Phe-OH refers to L-phenylalaninol. Antiamoebins-I to-V are isolated from *Emericellopsis poonensis*, *Emericellopsis synnematicola*, and *Cephalosporium pimprina* (1,2). Antiamoebins VI to XVI are isolated from *Stilbella erythrocephal, Stilbella fimetaria*, and *Gliocladium catenulatum* (3). **Likely Structure of the Active Ionophore:** Molecular dynamic simulations were carried out to obtain a structural model consistent with measured conductances (4). By counting ions crossing the channel and by solving the Nernst-Planck equation, the motion of ions inside the channel appears to be diffusive. The calculated conductance of octameric channels is markedly higher than observed in single channel recordings, and the tetramer appears to be nonconducting. The calculated conductance for a hexameric pore was 74 ± 20 pS at 75 mV, agreeing well with the 90-pS value measured at 75 mV. These findings suggest a six-monomer channel can accommodate K$^+$ and Cl$^-$, while keeping their inner shells intact. The free energy barrier for K$^+$ is only 2.2 kcal/mol, and that for Cl$^-$ is nearly 5 kcal/mol, making the channel selective for cations. **1**. Rinehart, Gaudioso, Moore, *et al.* (1981) *J. Amer. Chem. Soc.* **103**, 6517. **2**. Brueckner & Przybylski (1984) *J. Chrom.* **296**, 263. **3**. Jaworski & Brueckner (2000) *J. Pept. Sci.* **6**, 149. **4**. Wilson, Wei, Bjelkmar, Wallace & Pohorille (2010) *Biophys J.* **100**, 2394.

Antibiotic 1037, *See Toyocamycin*

Antibiotic AKD-1C, *See Dinactin*

Antibiotic S-3466A, *See Dinactin*

α₁-Antichymotrypsin

This glycoprotein (MW = 68000 Da) is a serpin that forms very long-lived, enzymatically inactive 1:1 complex (denoted E*I*) with chymotrypsin and other chymotrypsin-like proteases, but not trypsin or leukocyte elastase. **Mechanism of Action:** Serpins share a conserved tertiary structure, in which an exposed region of amino acid residues (called the Reactive Center Loop, or RCL) serves as "bait" for its target proteinase. Within E*I*, the serpin and protease are covalently linked (E*–I*) after nucleophilic attack by Ser-195 on the P₁ residue in the serpin RCL. While formally similar to acyl-enzyme in serine proteinase catalysis, subsequent hydrolysis of E*–I* is extremely slow as a result of structural changes within the enzyme leading to distortion of the active site. There is presently an ongoing debate about whether the enzyme, bound to P₁, maintains its original position at the top of the serpin molecule (1) or translocates across the entire length of the serpin, attended by insertion of RCL residues P₁-P₁₄ within β-sheet A as well as a large separation of the enzyme and RCL residue P₁' (2). **Target(s):** chymotrypsin (3,9,12,14,16,25,30,44); pancreatic elastase I (14,17,19); cathepsin G (5,9,10,12-15,28,29,40-43); chymase (6,9,9, 12,26,36-39); seaprose (7,27); prohormone thiol protease (8,24); DNA polymerase α (18); DNA primase (20,21); clipsin (22); NADPH oxidase-enzyme complex (23); pappalysin-1 (31); kallikrein 8, *or* neuropsin, weakly inhibited (32); stratum corneum chymotryptic enzyme, *or* kallikrein-7 (33); semenogelase (34,35). **1**. O'Malley & Cooperman (2001) *J. Biol. Chem.* **276**, 6631. **2**. Wright & Scarsdale (1995) *Proteins* **22**, 210. **3**. Travis & Salvesen (1983) *Ann. Rev. Biochem.* **52**, 655. **4**. Bieth (1998) in *Handb. Proteolytic Enzymes* (Barrett, Rawlings & Woessner, eds.), p. 46, Academic Press, San Diego. **5**. Salvesen (1998) in *Handb. Proteolytic Enzymes* (Barrett, Rawlings & Woessner, eds.), p. 60, Academic Press, San

Diego. **6.** Caughey (1998) in *Handb. Proteolytic Enzymes* (Barrett, Rawlings & Woessner, eds.), p. 66, Academic Press, San Diego. **7.** Kolattukudy & Sirakova (1998) in *Handb. Proteolytic Enzymes* (Barrett, Rawlings & Woessner, eds.), p. 320, Academic Press, San Diego. **8.** Hook (1998) in *Handb. Proteolytic Enzymes* (Barrett, Rawlings & Woessner, eds.), p. 779, Academic Press, San Diego. **9.** Travis & Morii (1981) *Meth. Enzymol.* **80**, 765. **10.** Rest (1988) *Meth. Enzymol.* **163**, 309. **11.** Carrell & Travis (1985) *Trends Biochem. Sci.* **10**, 20. **12.** Travis, Bowen & Baugh (1978) *Biochemistry* **17**, 5651. **13.** Barrett (1981) *Meth. Enzymol.* **80**, 561. **14.** Travis, Garner & Bowen (1978) *Biochemistry* **17**, 5647. **15.** Laine, Davril & Hayem (1982) *Biochem. Biophys. Res. Commun.* **105**, 186. **16.** Laine, Davril & Hayem (1984) *Eur. J. Biochem.* **140**, 105. **17.** Laine, Davril, Rabaud, Vercaigne-Marko & Hayem (1985) *Eur. J. Biochem.* **151**, 327. **18.** Tsuda, Masuyama & Katsunuma (1986) *Cancer Res.* **46**, 6139. **19.** Davril. Laine & Hayem (1987) *Biochem. J.* **245**, 699. **20.** Takada, Tsuda, Yamamura & Katsunuma (1988) *Biochem. Int.* **16**, 949. **21.** Takada, Tsuda, Matsumoto, *et al.* (1988) *Tokai J. Exp. Clin. Med.* **13**, 321. **22.** Nelson & Siman (1990) *J. Biol. Chem.* **265**, 3836. **23.** Kilpatrick, McCawley, Nachiappan, *et al.* (1992) *J. Immunol.* **149**, 3059. **24.** Hook, Purviance, Azaryan, Hubbard & Krieger (1993) *J. Biol. Chem.* **268**, 20570. **25.** Cooperman, Stavridi, Nickberg, *et al.* (1993) *J. Biol. Chem.* **268**, 23616. **26.** Schechter, Jordan, James, *et al.* (1993) *J. Biol. Chem.* **268**, 23626. **27.** Korzus, Luisetti & Travis (1994) *Biol. Chem. Hoppe-Seyler* **375**, 335. **28.** Patston (1995) *Inflammation* **19**, 75. **29.** Duranton, Adam & Bieth (1998) *Biochemistry* **37**, 11239. **30.** Luo, Zhou & Cooperman (1999) *J. Biol. Chem.* **274**, 17733. **31.** Parker, Gockerman, Busby & Clemmons (1995) *Endocrinology* **136**, 2470. **32.** Shimizu, Yoshida, Shibata, *et al.* (1998) *J. Biol. Chem.* **273**, 11189. **33.** Franzke, Baici, Bartels, Christophers & Wiedow (1996) *J. Biol. Chem.* **271**, 21886. **34.** Christensson, Laurell & Lilja (1990) *Eur. J. Biochem.* **194**, 755. **35.** Malm, Hellman, Hogg & Lilja (2000) *Prostate* **45**, 132. **36.** Schechter, Sprows, Schoenberger, *et al.* (1989) *J. Biol. Chem.* **264**, 21308. **37.** Kido, Fukusen & Katunuma (1985) *Arch. Biochem. Biophys.* **239**, 436. **38.** Katunuma, Fukusen & Kido (1986) *Adv. Enzyme Regul.* **25**, 241. **39.** Pejler & Berg (1995) *Eur. J. Biochem.* **233**, 192. **40.** Stefansson, Yepes, Gorlatova, *et al.* (2004) *J. Biol. Chem.* **279**, 29981. **41.** Ermolieff, Boudier, Laine, Meyer & Bieth (1994) *J. Biol. Chem.* **269**, 29502. **42.** Maison, Villiers & Colomb (1991) *J. Immunol.* **147**, 921. **43.** Mistry, Snashall, Totty, Guz & Tetley (1999) *Arch. Biochem. Biophys.* **368**, 7. **44.** O'Malley & Cooperman (2001) *J. Biol. Chem.* **276**, 6631.

Antihypertensive Agents

These blood pressure-lowering substances are used to prevent the complications of high blood pressure, mainly stroke and myocardial infarction. **DIURETICS**, which facilitate excretion of excess salt and water from body fluids, include: **Loop Diuretics**, such as Bumetanide, Ethacrynic Acid, Furosemide, and Torsemide; **Thiazide Diuretics**, such as Epitizide, Hydrochlorothiazide, Chlorothiazide, and Bendroflumethiazide; **Thiazide-like Diuretics**, including Indapamide, Chlorthalidon, and Metolazone; as well as **Potassium-sparing Diuretics**, such as Amiloride, Spironolactone, and Triamterene. **CALCIUM CHANNEL BLOCKERS**, which block the entry of calcium into muscle cells in artery walls, include: **Dihydropyridines**, such as Amlodipine, Cilnidipine, Felodipine, Isradipine, Lercanidipine, Levamlodipine, Nicardipine, Nifedipine, Nimodipine, Nitrendipine, and **Non-dihydropyridines**, such as Diltiazem and Verapamil. **ACE INHIBITORS**, which target angiotensin-converting enzyme-mediated conversion of angiotensin I into angiotensin II, include Benazepril, Captopril, Enalapril, Fosinopril, Lisinopril, Perindopril, Quinapril, Ramipril, and Trandolapril. **ANGIOTENSIN II RECEPTOR ANTAGONISTS**, which work by inhibiting the activation of angiotensin receptors, include Candesartan, Eprosartan, Irbesartan, Losartan, Olmesartan, Telmisartan, and Valsartan. Adrenergic Receptor Antagonists, of which there are several subtypes: β-**BLOCKERS**, such as Atenolol, Metoprolol, Nadolol, Nebivolol, Oxprenolol, Pindolol, Propranolol, and Timolol; α-**BLOCKERS**, such as Doxazosin, Phentolamine, Indoramin, Phenoxybenzamine, Prazosin, Terazosin, and Tolazoline; **MIXED** α/β-**BLOCKERS**, which Bucindolol, Carvedilol, and Labetalol; **VASODILATORS**, which act directly on the smooth muscle of arteries to relax, include Sodium Nitroprusside and Hydralazine; **BENZODIAZEPINES**, which, as GABA$_A$ receptor agonists in brain, slow down neurotransmission and dilate blood vessels; **RENIN INHIBITORS**, Renin Inhibitors, which include Aliskiren; **ALDOSTERONE RECEPTOR ANTAGONISTS**, which include Eplerenone and Spironolactone; and α$_2$-**ADRENERGIC RECEPTOR AGONISTS**, such as Clonidine, Guanabenz, Guanfacine, Methyl-DOPA, and Moxonidine. *See specific agent for details on mode of action.*

Antimony Sodium Gluconate

This pentavalent antimony complex (FW = 925.90 g/mol; CAS 16037-91-5) has remained a first-line drug for treating leishmaniasis, a disease caused by a protozoan parasite of the genus *Leishmania*, transmitted to humans through the bite of sandflies that previously fed on an infected vertebrate reservoir (1). Entering by means of the phosphate transporter, Sb(V) accumulation is higher in axenic amastigotes than in promastigotes for a number of species. Gluconate competitively inhibits Sb(V) uptake in axenic amastigotes, and Sb(V) is believed to enter into the parasites by means of a protein recognizing a sugar moiety resembling gluconate (2). It is likely that pentavalent antimony is a pro-drug, undergoing metabolic reduction by a parasite-specific enzyme, thiol-dependent reductase to Sb(III), the highly toxic anti-leishmanial agent. The most likely reductant is trypanothione, main intracellular thiol in *Leishmania* parasites, containing two glutathione units joined by a spermidine (3). Potential molecular targets of Sb(III) are trypanothione reductase as well as zinc-finger proteins, the latter involved in DNA replication, structure and repair. Sodium stibogluconate, but not Sb(III), specifically inhibits type I DNA topoisomerase from *Leishmania donovani* (4). Treatment with sodium antimony gluconate at the therapeutic dose may result in minor side effects such as arthralgia, myalgia, transient elevation of hepatocellular enzyme levels, and minor changes in ECG (5). **1.** Haldar, Sen & Roy (2011) *Molec. Biol. Internat.* **2011**, Article ID 571242. **2.** Brochu, Wang, Roy, *et al.* (2003) *Antimicrob. Agents Chemother.* **47**, 3073. **3.** Ferreira Cdos, Martins, Demicheli, *et al.* (2003) *Biometals* **16**, 441. **4.** Chakraborty & Majumder (1988) *Biochem. Biophys. Res. Communs.* **152**, 605. **5.** Sundar & Chakravarty (2010) *Int. J. Environ. Res. Public Health* **7**, 4267.

Antimycin A

This mixture of four antibiotic components from *Streptomyces*, that inhibit electron transport (at complex III), mitochondrial oxidative phosphorylation (1-4,10,12-14,17) as well as photo-phosphorylation (5) Antimycin A$_1$ inhibits electron transfer in complex III by blocking the transfer of electrons between Cyt b_H and Coenzyme Q bound at the Q$_N$ site. The antimycins also induce apoptosis. The two main components of this mixture are antimycin A$_1$ and A$_3$. All four exhibit the same activity with respect to electron transport. The four antimycins differ only in the alkyl chain attached at the 8 position of the nine-membered ring: Antimycin A$_1$ (FW = 548.63 g/mol; CAS 642-15-9; M.P. = 149-150°C; *R* = *n*-hexyl); Antimycin A$_2$ (FW = 534.6 g/mol; CAS 27220-57-1) *R* = *n*-pentyl; Antimycin A$_3$, (FW = 520.58 g/mol; M.P. = 170.5-171.5°C; *R* = *n*-butyl); and Antimycin A$_4$, *R* = *n*-propyl. Antimycin A$_1$ is practically insoluble in water but is soluble in diethyl ether, ethanol, acetone, and chloroform. This is also true for antimycin A$_3$. **Target(s):** ubiquinol:cytochrome-*c* reductase (complex III) (1,2,8-10,12-16,19,29-48); oxidative phosphorylation (1-4,8-10,12-17); aldehyde oxidase (3); sarcosine oxidase (4); mitochondrial electron transport (1,8-10,15,19); chloroplast electron transport (5); photophosphorylation (5); succinate oxidase (6,19,20); NADH oxidase6; very-long-chain fatty acid α-hydroxylase (7); calcidiol 1 monooxygenase, *or* 25-hydroxyvitamin D 1α-hydroxylase (11,23-25); ATP $\rightleftarrows$ P$_i$ exchange activity in *Rhodospirillum rubrum* (18); steroid 21-monooxygenase, *or* CYP21A1, weakly inhibited (21); squalene monooxygenase, *or* squalene epoxidase (22); *trans*-cinnamate 4-monooxygenase (26); procollagen proline 4-dioxygenase, weakly inhibited (27); hydrogen:quinone oxidoreductase (28). *Note*: The antimycin A-like molecules AAL_1 and AAL_2 inhibit cyclic electron transport around Photosystem I in ruptured chloroplasts (49). **1**. Slater (1967) *Meth. Enzymol.* **10**, 48. **2**. Rieske (1967) *Meth. Enzymol.* **10**, 239. **3**. Rajagopalan & Handler (1966) *Meth. Enzymol.* **9**, 364. **4**. Cronin, Frisell & Mackenzie (1967) *Meth. Enzymol.* **10**, 302. **5**. Izawa & Good (1972) *Meth. Enzymol.* **24**, 355. **6**. Kurup, Vaidyanathan & Ramasarma (1966) *Arch. Biochem. Biophys.* **113**, 548. **7**. Kishimoto (1978) *Meth. Enzymol.* **52**, 310. **8**. Hatefi & Stiggall (1978) *Meth. Enzymol.* **53**, 5. **9**. Hatefi (1978) *Meth. Enzymol.* **53**, 35. **10**. Singer (1979) *Meth. Enzymol.* **55**, 454. **11**. Lobaugh, Almond & Drezner (1986) *Meth. Enzymol.* **123**, 159. **12**. Linke & Weiss (1986) *Meth. Enzymol.* **126**, 201. **13**. von Jagow & Link (1986) *Meth. Enzymol.* **126**, 253. **14**. Schägger, Brandt, Gencic & von Jagow (1995) *Meth. Enzymol.* **260**, 82. **15**. Rapoport & Schewe (1977) *Trends Biochem. Sci.* **2**, 186. **16**. Goldberger & Green (1963) *The Enzymes*, 2nd ed., **8**, 81. **17**. Walter & Lardy (1964) *Biochemistry* **3**, 812. **18**. Zaugg & Vernon (1966) *Biochemistry* **5**, 34. **19**. Potter & Reif (1952) *J. Biol. Chem.* **194**, 287. **20**. Reif & Potter (1953) *J. Biol. Chem.* **205**, 279. **21**. Ryan & Engel (1957) *J. Biol. Chem.* **225**, 103. **22**. Ryder & Dupont (1984) *Biochim. Biophys. Acta* **794**, 466. **23**. Gray, Omdahl, Ghazarian & DeLuca (1972) *J. Biol. Chem.* **247**, 7528. **24**. Henry & Norman (1974) *J. Biol. Chem.* **249**, 7529. **25**. Paulson & DeLuca (1985) *J. Biol. Chem.* **260**, 11488. **26**. Billett & Smith (1978) *Phytochemistry* **17**, 1511. **27**. Hutton, Tappel & Udenfriend (1967) *Arch. Biochem. Biophys.* **118**, 231. **28**. Ferber & Maier (1993) *FEMS Microbiol. Lett.* **110**, 257. **29**. Trumpower (1990) *Microbiol. Rev.* **54**, 101. **30**. Hauska, Hurt, Gabellini & Lockau (1983) *Biochim. Biophys. Acta* **726**, 97. **31**. Yang & Trumpower (1986) *J. Biol. Chem.* **261**, 12282. **32**. Gabellini, Bowyer, Hurt, Melandri & Hauska (1982) *Eur. J. Biochem.* **126**, 105. **33**. Gao, Wen, Esser, *et al.* (2003) *Biochemistry* **42**, 9067. **34**. Kriauciunas, Yu, Yu, Wynn & Knaff (1989) *Biochim. Biophys. Acta* **976**, 70. **35**. Yu, Mei & Yu (1984) *J. Biol. Chem.* **259**, 5752. **36**. Wikström, Krab & Saraste (1981) *Ann. Rev. Biochem.* **50**, 623. **37**. Rieske (1976) *Biochim. Biophys. Acta* **456**, 195. **38**. Siedow, Power, de la Rosa & Palmer (1978) *J. Biol. Chem.* **253**, 2392. **39**. Diggens & Ragan (1982) *Biochem. J.* **202**, 527. **40**. Matsuno-Yagi & Hatefi (1997) *J. Biol. Chem.* **272**, 16928. **41**. Matsuno-Yagi & Hatefi (2001) *J. Biol. Chem.* **276**, 19006. **42**. Brasseur (1988) *J. Biol. Chem.* **263**, 12571. **43**. di Rago & Colson (1988) *J. Biol. Chem.* **263**, 12564. **44**. Pember, Fleck, Moberg & Walker (2005) *Arch. Biochem. Biophys.* **435**, 280. **45**. Rotsaert, Ding & Trumpower (2008) *Biochim. Biophys. Acta* **1777**, 211. **46**. Ding, di Rago & Trumpower (2006) *J. Biol. Chem.* **281**, 36036. **47**. Gutierrez-Cirlos, Merbitz-Zahradnik & Trumpower (2004) *J. Biol. Chem.* **279**, 8708. **48**. Covian & Trumpower (2006) *J. Biol. Chem.* **281**, 30925. **49**. Taira, Okegawa, Sugimoto, *et al.* (2013) *FEBS Open Bio.* **3**, 406.

Antipain

This peptide analogue and irreversible serine and cysteine protease inhibitor (FW = 604.70 g/mol; CAS 37691-11-5), also known as *N*-(N^α-carbonyl-L-arginyl-L-valyl-L-arginal)-L-phenylalanine and [(*S*)-1-carboxy-2-phenylethyl]carbamoyl-L-arginyl-L-valyl-L-arginal, is produced by certain *Streptomyces* species (1,2). When bound to wheat carboxypeptidase (3) and *Leishmania* major oligopeptidase B (4), the backbone carbonyl of antipain's terminal arginine forms a covalent bond to the active site serine, giving detailed information on the enzyme active site and extended substrate binding pockets. A 10-mM stock solution is stable for one week at 4°C. **Target(s)**: Over 200 proteinases are inhibited by antipain. **1**. Suda, Aoyagi, Hamada, Takeuchi & Umezawa (1972). *J. Antibiot.* **25**, 263. **2**. Umezawa, Tatsuta, Fujimoto, Tsuchiya & Umezawa (1972). *J. Antibiot.*

25, 267. **3**. Bullock, Breddam & Remington (1996) *J. Mol. Biol.* **255**, 714. **4**. McLuskey, Paterson, Bland, Isaacs & Mottram (2010) *J. Biol. Chem.* **285**, 39249.

α_2-Antiplasmin

This fast-acting lycoprotein serpin (MW = 58 kDa (452 aminoacyl residues); however, on SDS-gel, M_r is 67000) single-chain rapidly binds to plasmin ($k_{on} = 2 \times 10^7$ $M^{-1}s^{-1}$) to form a high-affinity 1:1 complex. Efficient binding requires the presence of a free active site and free lysine-binding site(s) on plasmin (or plasminogen). (*For a discussion of the likely mechanism of inhibitory action, See Serpins; also α_1-Antichymotrypsin*) **Target(s)**: plasmin (1,2,8-10,13-17,19,20,24,25); coagulation factor XIIa (3); coagulation factor Xia, slow interaction (4,17); limulus clotting factor B (5,11,22); limulus clotting enzyme (6,12); limulus coagulation factor G (7); trypsin (8,16,18); chymotrypsin, slow interaction (8,16,18); t-plasminogen activator, slow interaction (8,23); thrombin, slow interaction (16,17); plasma kallikrein, slow interaction (17); matriptase (21). **1**. Wiman (1980) *Biochem. J.* **191**, 229. **2**. Castellino (1998) in *Handb. Proteolytic Enzymes*, p. 190, Academic Press, San Diego. **3**. Ratnoff (1998) in *Handb. Proteolytic Enzymes*, p. 144, Academic Press, San Diego. **4**. Walsh (1998) in *Handb. Proteolytic Enzymes*, p. 153, Academic Press, San Diego. **5**. Kawabata, Muta & Iwanaga (1998) in *Handb. Proteolytic Enzymes*, p. 212, Academic Press, San Diego. **6**. Kawabata, Muta & Iwanaga (1998) in *Handb. Proteolytic Enzymes*, p. 213, Academic Press, San Diego. **7**. Kawabata, Muta & Iwanaga (1998) in *Handb. Proteolytic Enzymes*, p. 214, Academic Press, San Diego. **8**. Wiman (1981) *Meth. Enzymol.* **80**, 395. **9**. Plow, Miles & Collen (1989) *Meth. Enzymol.* **169**, 296. **10**. Aoki, Sumi, Miura & Hirosawa (1993) *Meth. Enzymol.* **223**, 185. **11**. Nakamura, Muta, Oda, Morita & Iwanaga (1993) *Meth. Enzymol.* **223**, 346. **12**. Muta, Nakamura, Hashimoto, Morita & Iwanaga (1993) *Meth. Enzymol.* **223**, 352. **13**. Seegers (1951) *The Enzymes*, 1st ed., **1** (part 2), 1106. **14**. Ablondi & Hagan (1960) *The Enzymes*, 2nd ed., **4**, 175. **15**. Mullertz (1974) *Biochem. J.* **174**, 273. **16**. Edy & Collen (1977) *Biochim. Biophys. Acta* **484**, 423. **17**. Saito, Goldsmith, Moroi & Aoki (1979) *Proc. Natl. Acad. Sci. U.S.A.* **76**, 2013. **18**. Enghild, Valnickova, Thogersen, Pizzo & Salvesen (1993) *Biochem. J.* **291**, 933. **19**. Hall, Humphries & Gonias (1991) *J. Biol. Chem.* **266**, 12329. **20**. Nilsson & Wiman (1982) *FEBS Lett.* **142**, 111. **21**. Szabo, Netzel-Arnett, Hobson, Antalis & Bugge (2005) *Biochem. J.* **390**, 231. **22**. Nakamura, Horiuchi, Morita & Iwanaga (1986) *J. Biochem.* **99**, 847. **23**. Saksela (1985) *Biochim. Biophys. Acta* **823**, 35. **24**. Kolev, Lerant, Tenekejiev & Machovich (1994) *J. Biol. Chem.* **269**, 17030. **25**. Ries & Zenker (2003) *Blood Coagul. Fibrinolysis* **14**, 203.

Antipyretic Agents

These substances have the effect of directly or indirectly overriding prostaglandin-induced increase in temperature. Commonly used antipyretics include: Acetylsalicylic acid (Aspirin) and Salicylate Salts (*e.g.*, Choline Salicylate, Magnesium Salicylate, and Sodium Salicylate); Phenazone; Non-Steroidal Anti-Inflammatory Drugs, *or* NSAIDs (*e.g.*, Ibuprofen, Naproxen, Ketoprofen, Nimesulide); Paracetamol (also known as Acetaminophen); Metamizole (now banned for causing agranulocytosis) and Nabumetone.

Antisauvagine-30

This highly selective CRF_2 antagonist (MW = 3650.29 g/mol; CAS 220673-95-0; Sequence: D-Phe-His-Leu-Leu-Arg-Lys-Met-Ile-Glu-Ile-Glu-Lys-Gln-Glu-Ala-Ala-Asn-Asn-Arg-Leu-Leu-Leu-Asp-Thr-Ile-NH_2; Solubility: 5 mg/mL H_2O; Alternate Name: [D-Phe[11],His[12]]-Sauvagine (11-40)) targets human Corticotropin-Releasing hormone (CRH) Receptor-2 subtypes $CRF_{2\beta}$ (K_i = 1.4 nM), with a much higher value of 154 nM for CRF_1 (1-3). (*See also Astressin; K41448*) Type-2 GPCRs reside in plasma membranes of hormone-sensitive cells and bind CRH, a peptide hormone and neurotransmitter involved in stress responses. Antisauvagine-30 blocks (pA_2 = 8.49) sauvagine-stimulated accumulation of cAMP in HEK-m$CRF_{2\beta}$ cells. It also prevents stress-enhanced fear conditioning and MEK1/2-dependent activation of ERK1/2 in mice and suggesting a role for protein kinase signaling in the hippocampus and its modulation by CRF_2 as a possible link between stress and fear memory. **1**. Ruhmann, *et al.* (1998) Proc. Natl. Acad. Sci. USA **95**, 15264. **2**. Brauns, *et al.* (2001) *Neuropharmacol.* **41**, 507. 3. Sananbenesi, *et al.* (2003) *J.Neurosci.* **23**, 11436.

Antisense Drugs/Effectors

These short DNA- or RNA-like analogues (consisting of 12-21 bases, with molecular weight values that are typically below 2,000) hybridize with specific mRNA molecules, thereby preventing synthesis of a protein (wild type or mutant), whose presence results in or promotes a metabolic disorder, infection, or degenerative disease. Some antisense drugs direct their inhibitory effects on the expression of a particular mRNA splice-variant. Still others combine the extreme sequence-specificity of antisense oligonucleotides with the mRNA-cleaving power of catalytic RNA (ribozymes) in search-and-destroy strategies aimed at degrading target mRNA molecules. To minimize hydrolysis and inactivation, antisense oligonucleotides are synthesized with nuclease-resistant phosphorothioate linkages. Use of suitably derivatized purine and pyrimidine bases can also improve pharmacokinetics. (*See Mipomersen*) **1**. Phillips, ed. (1999) "Antisense Technology, Part A: General Methods, Methods of Delivery, and RNA Studies", *Methods in Enzymology*, vol. 313. **2**. Phillips, ed. (1999) "Antisense Technology, Part B: Applications", *Methods in Enzymology*, vol. 314. **3**. Kurreck, *ed.* (2008) *Therapeutic Oligonucleotides*, RSC Biomolecular Sciences, Oxford, UK.

Anti-Siglec-15 Antibodies

These monoclonal antibodies target Siglec-15, a functionally important, but little-studied, sialic acid-binding receptor located on the outer surface of bone-resorbing osteoclasts. At the molecular level, Siglec-15 interacts with the adapter protein DAP12 (*or* DNAX Activation Protein of 12 kDa) to activate Akt (also known as "protein kinase B", *or* PKB), when clustered on the cell surface of osteoclasts. This key regulatory interaction controls osteoclast differentiation and downstream activity of RANKL (*or* Receptor Activator of Nuclear factor Kappa-B Ligand). Antibody-mediated receptor dimerization is followed by rapid internalization, lysosomal targeting, and degradation, resulting in inhibition of functional osteoclast differentiation *in vitro*. Treatment of mice with these antibodies markedly increases bone mineral density, a finding that is likewise consistent with inhibition of osteoclast activity. Significantly, osteoblast numbers were maintained despite the anti-resorptive activity of these antibodies, supporting the further development of Siglec-15 antibodies as a novel class of bone loss therapeutics. **1**. Stuible, Moraitis, Fortin, *et al.* (2014) *J. Biol. Chem.* **289**, 6498.

Antithrombin

This human plasma serpin (MW = 50.2 kDa; CAS 9000-94-6; *Abbreviation*: AT), also known as human plasma-derived antithrombin, antithrombin III, ATIII, and by its trade name ATryn®, plays a central role in hemostasis, serving as the principal inhibitor of thrombin (*Reaction*: Fibrinogen + H_2O ⇌ Fibrin + Fibrinopeptides *a* and *b*) and Factor Xa (*Reaction*: Prothrombin + H_2O ⇌ Thrombin). Given that AT is one of the major naturally occurring inhibitors of coagulation, acquired or hereditary deficiencies of this protein result in excessive thrombin generation. Use of ATryn is indicated for prophylactic treatment of patients with congenital antithrombin deficiency, a disorder that is manifested by dysregulated coagulation, abnormal blood clotting, and deep vein thrombosis. (*For a discussion of the likely mechanism of inhibitory action, See Serpins; also α_1-Antichymotrypsin*) **Other Target(s):** AT also inhibits Factors IXa, XIa, and XIIab, in addition to kallikrein and plasmin. **Primary Mode of Inhibitory Action:** Antithrombin is a liver-derived glycoprotein that neutralizes the catalytic activities of thrombin and Factor Xa by forming a complex that is rapidly removed from the circulation. It is essentially unreactive with Factor IXa in the absence of heparin (*i.e.*, $k_{on} \approx 10$ $M^{-1}s^{-1}$) but undergoes a remarkable approximately one million-fold enhancement in reactivity with this proteinase to the physiologically relevant range (*i.e.*, $k_{on} \approx 10^7$ $M^{-1}s^{-1}$), when activated by heparin in the presence of physiologic levels of calcium. This enhancement results from: (a) allosteric activation of antithrombin by a sequence-specific heparin pentasaccharide (300-500x enhancement), (b) allosteric activation of factor IXa by calcium ions (4-8x enhancement), and (c) heparin bridging of antithrombin and factor IXa, as augmented by calcium ions (130-1000x, depending on heparin chain length) (1). **Structural Features:** Antithrombin is a 432-residue protein that contains three disulfide bonds and four canonical glycosylation sites. Exosite interactions are responsible for the unusual specificity of antithrombin and heparin for thrombin, Factor IXa, and Factor Xa, each with quite distinct substrate specificities. Antithrombin circulates in blood in two major isoforms, α and β differing in their glycosylation, a property that is linked to differences in their affinities for heparin. Kinetic studies suggest that similar induced-fit mechanisms determine pentasaccharide binding to native and latent antithrombins, and kinetic simulations support a three-step mechanism of allosteric activation of native antithrombin involving successive conformational changes. Equilibrium binding studies of pentasaccharide interactions with native and latent antithrombins and the salt dependence of these interactions suggest that each conformational change is associated

with distinct spectroscopic changes and is driven by a progressively better fit of the pentasaccharide in the binding site (3). **Use of Transgenics in Antithrombin Production:** Given its complex structure, antithrombin III resisted expression by standard bacterial or yeast protocols. ATryn is thus the first-ever transgenically produced therapeutic protein (4) and the first transgenically produced recombinant antithrombin to be approved in the U.S. ATryn is purified from the milk of a genetically modified goat, after introduction of the antithrombin III gene into the nucleus of its fertilized egg. The glycosylation profile differs from plasma-derived antithrombin, resulting in increased heparin affinity and a higher clearance rate. **1**. Bedsted, Swanson, Chuang, *et al.* (2003) *Biochemistry* **42**, 8143. **2**. Fyfe & Tait (2009) *Expert Rev. Hematol.* **2**, 499. **3**. Schedin-Weiss, Richard & Olson (2010) *Arch. Biochem. Biophys.* **504**, 169. **4**. United States Patent 5843705: Transgenically Produced Antithrombin III, teaching methods of transgenic production.

α_1-Antitrypsin

This plasma serpin (MW = 51 kDa; Abbreviation: A1AT; Accession Number: NP 001121173; Gene: *SERPINA1*; Chromosome Location: 14q32.1) chiefly inhibits neutrophil elastase, but also shows broader action (hence its alternative name: α_1-proteinase inhibitor, *or* API). Although elastase is its physiologic target, A1AT was designated as antitrypsin in recognition of its ability to irreversibly inactivate trypsin, the prototypical serine protease. (*For a discussion of the most likely mechanism of action, See Serpins; α_1-Antichymotrypsin*) α_1-Antitrypsin protects the lungs from neutrophil elastase, an enzyme that disrupts connective tissue. α_1-Antitrypsin deficiency, an autosomal codominant hereditary disorder, is characterized by the accumulation of the inactive (misfolded/aggregated) A1AT polypeptide within the liver, leading to chronic uninhibited tissue breakdown and proceeding to cirrhosis of the liver (and, in smokers, pulmonary emphysema). Normal α_1-antitrypsin blood levels range from 1.0 to 2.7 g/L, but, in individuals with PiSS, PiMZ and PiSZ genotypes, circulating A1AT levels can fall by 40–60%. While such lower levels are sufficient to protect the lungs of nonsmokers from the effects of elastase, those who smoke suffer the consequential oxidation of Methionine-382, with loss of serpin activity and progression to emphysema. A naturally occurring Methionine-358-Arginine (M358R) mutant has reduced inhibition of neutrophil elastase, while significantly enhancing inhibition of thrombin, factor XIa, and kallikrein. **Target(s):** trypsin (1,15,18,19,24,31); pancreatic elastase (2,18,23,24,61; pancreatic elastase II (3,49,61); pancreatic endopeptidase E (4,50); leukocyte elastase (5,17,18,26,31, 42,58-60,66,69,73); cathepsin G (6,16,18,26,67-71); myeloblastin (7,28,39-44); sheep mast cell proteinase (18); tissue kallikrein (9,10,18,21,62; however, inhibition is slow, 21); semenogelase (11); coagulation factor Xia (12,20,25); protein C (activated) (13,51-53,63); glutamyl endopeptidase II (14); chymotrypsin (18,26); thrombin (18,32); plasmin (18,73); acrosin (18,72); collagenase (18,23); tryptase (22); *Coccidioses* endopeptidase (27); granzyme A (29); granzyme B (30,38); kexin (by α_1-antitrypsin Pittsburgh) (33); coagulation factor Xa (34); t plasminogen activator (48); u-plasminogen activator (urokinase) (35,47,48); kallikrein 8 (neuropsin), weakly inhibited (36); matriptase (37); furin, inhibited by α1-antitrypsin Pittsburgh, *i.e.*, Met35Arg (45,46); chymase (54-57); plasma kallikrein, inhibited by altered α_1-antitrypsin (63); complement subcomponent C1, inhibited by altered α_1-antitrypsin (63); coagulation factor XIIa, inhibited by altered α_1-antitrypsin (63); brachyuran (64); prolyl oligopeptidase, partially inhibited (65). **1**. Halfon & Craik (1998) in *Handb. Proteolytic Enzymes*, p. 12, Academic Press, San Diego. **2**. Bieth (1998) *ibid*, p. 42. **3**. Bieth (1998) *ibid*, p. 46. **4**. Bieth (1998) *ibid*, p. 48. **5**. Bieth (1998) *ibid*, p. 54. **6**. Salvesen (1998) in *Handb. Proteolytic Enzymes*, p. 60, Academic Press, San Diego. **7**. Hoidal (1998) in *Handb. Proteolytic Enzymes*, p. 62, Academic Press, San Diego. **8**. Pemberton (1998) in *Handb. Proteolytic Enzymes*, p. 76, Academic Press, San Diego. **9**. Chao (1998) in *Handb. Proteolytic Enzymes*, p. 97, Academic Press, San Diego. **10**. Chao (1998) in *Handb. Proteolytic Enzymes*, p. 111, Academic Press, San Diego. **11**. Chao (1998) in *Handb. Proteolytic Enzymes*, p. 102, Acadelic Press, San Diego. **12**. Walsh (1998) in *Handb. Proteolytic Enzymes*, p. 153, Academic Press, San Diego. **13**. Shen & Dahlbäck (1998) in *Handb. Proteolytic Enzymes*, p. 174, Academic Press, San Diego. **14**. Stennicke & Breddam (1998) in *Handb. Proteolytic Enzymes*, p. 246, Academic Press, San Diego. **15**. Dejgaard, Ortapamuk & Ozer (1999) *J. Enzyme Inhib.* **14**, 391. **16**. Barrett (1981) *Meth. Enzymol.* **80**, 561. **17**. Barrett (1981) *Meth. Enzymol.* **80**, 581. **18**. Travis & Johnson (1981) *Meth. Enzymol.* **80**, 754. **19**. Pannell, Johnson & Travis (1974) *Biochemistry* **13**, 5439. **20**. Kurachi & Davie (1981) *Meth. Enzymol.* **80**,

211. **21**. Geiger & Fritz (1981) *Meth. Enzymol.* **80**, 466. **22**. Woodbury, Everitt & Neurath (1981) *Meth. Enzymol.* **80**, 588. **23**. Rest (1988) *Meth. Enzymol.* **163**, 309. **24**. Arnaud & Chapuis-Cellier (1988) *Meth. Enzymol.* **163**, 400. **25**. Walsh, Baglia & Jameson (1993) *Meth. Enzymol.* **222**, 65. **26**. Knight (1995) *Meth. Enzymol.* **248**, 85. **27**. Rawlings & Barrett (1994) *Meth. Enzymol.* **244**, 19. **28**. Hoidel, Rao & Gray (1994) *Meth. Enzymol.* **244**, 61. **29**. Simon & Kramer (1994) *Meth. Enzymol.* **244**, 68. **30**. Peitsch & Tschopp (1994) *Meth. Enzymol.* **244**, 80. **31**. Carrell & Travis (1985) *Trends Biochem. Sci.* **10**, 20. **32**. Magnusson (1971) *The Enzymes*, 3rd ed., **3**, 277. **33**. Rockwell & Fuller (2002) *The Enzymes*, 3rd ed., **22**, 259. **34**. Ellis, Scully, MacGregor & Kakkar (1982) *Biochim. Biophys. Acta* **701**, 24. **35**. Clemmensen & Christensen (1976) *Biochim. Biophys. Acta* **429**, 591. **36**. Shimizu, Yoshida, Shibata, *et al.* (1998) *J. Biol. Chem.* **273**, 11189. **37**. Szabo, Netzel-Arnett, Hobson, Antalis & Bugge (2005) *Biochem. J.* **390**, 231. **38**. Poe, Blake, Boulton, *et al.* (1991) *J. Biol. Chem.* **266**, 98. **39**. Goldmann, Niles & Arnaout (1999) *Eur. J. Biochem.* **261**, 155. **40**. Dolman, van de Wiel, Kam, *et al.* (1992) *FEBS Lett.* **314**, 117. **41**. Dolman, van de Wiel, Kam, *et al.* (1993) *Adv. Exp. Med. Biol.* **336**, 55. **42**. Wiesner, Litwiller, Hummel, *et al.* (2005) *FEBS Lett.* **579**, 5305. **43**. Korkmaz, Attucci, Moreau, *et al.* (2004) *Amer. J. Respir. Cell Mol. Biol.* **30**, 801. **44**. Rao, Wehner, Marshall, *et al.* (1991) *J. Biol. Chem.* **266**, 9540. **45**. Brennan & Nakayama (1994) *FEBS Lett.* **338**, 147. **46**. Rufaut, Brennan, Hakes, Dixon & Birch (1993) *J. Biol. Chem.* **268**, 20291. **47**. Chao (1983) *J. Biol. Chem.* **258**, 4434. **48**. Saksela (1985) *Biochim. Biophys. Acta* **823**, 35. **49**. Davril, Laine & Hayem (1987) *Biochem. J.* **245**, 699. **50**. Mallory & Travis (1975) *Biochemistry* **14**, 722. **51**. Rezaie & Esmon (1993) *J. Biol. Chem.* **268**, 19943. **52**. Heeb & Griffin (1988) *J. Biol. Chem.* **263**, 11613. **53**. Hermans & Stone (1993) *Biochem. J.* **295**, 239. **54**. Newlands, Knox, Pirie-Sheperd & Miller (1993) *Biochem. J.* **294**, 127. **55**. Schechter, Sprows, Schoenberger, *et al.* (1989) *J. Biol. Chem.* **264**, 21308. **56**. Kido, Fukusen & Katunuma (1985) *Arch. Biochem. Biophys.* **239**, 436. **57**. Katunuma, Fukusen & Kido (1986) *Adv. Enzyme Regul.* **25**, 241. **58**. Pacholok, Davies, Dorn, *et al.* (1995) *Biochem. Pharmacol.* **49**, 1513. **59**. Ohlsson & Olsson (1974) *Eur. J. Biochem.* **42**, 519. **60**. Jones, Elphick, Pettitt, Everard & Evans (2002) *Eur. Respir. J.* **19**, 1136. **61**. Largman, Brodrick & Geokas (1976) *Biochemistry* **15**, 2491. **62**. Chan, Springman & Clark (1998) *Protein Expr. Purif.* **12**, 361. **63**. Sulikowski, Bauer & Patston (2002) *Protein Sci.* **11**, 2230. **64**. Roy, Colas & Durand (1996) *Comp. Biochem. Physiol. B* **115**, 87. **65**. Goossens, De Meester, Vanhoof, *et al.* (1995) *Eur. J. Biochem.* **233**, 432. **66**. Stefansson, Yepes, Gorlatova, *et al.* (2004) *J. Biol. Chem.* **279**, 29981. **67**. Duranton, Adam & Bieth (1998) *Biochemistry* **37**, 11239. **68**. Ermolieff, Boudier, Laine, Meyer & Bieth (1994) *J. Biol. Chem.* **269**, 29502. **69**. Mistry, Snashall, Totty, Guz & Tetley (1999) *Arch. Biochem. Biophys.* **368**, 7. **70**. Bjoerk & Ohlsson (1991) *Biol. Chem. Hoppe-Seyler* **372**, 419. **71**. Cavarra, Santucci & Lungarella (1995) *Biol. Chem. Hoppe-Seyler* **376**, 371. **72**. Hermans, Monard, Jones & Stone (1995) *Biochemistry* **34**, 3678. **73**. Kolev, Lerant, Tenekejiev & Machovich (1994) *J. Biol. Chem.* **269**, 17030.

Antofloxacin

This novel 8-amino derivative of levofloxacin (FW = 376.49 g/mol; Symbol: ATFX and AX) is a long-lived (and hence dosed daily) fluoroquinolone antibiotic displaying broad-spectrum *in vitro* activity, with high effectiveness against *Staphylococcus aureus* (1). The fact that cyclosprin A, probenecid, erythromycin and cimetidine all inhibit the biliary excretion of ATFX glucuronide suggests that multiple transporters are likely to mediate the biliary excretion of ATFX (2). Using phenacetin as a microsomal CYP1A2 substrate, antofloxacin was found to require NADPH for greatest inhibition (3). When studied by whole-cell patch-clamp technique in transiently transfected HEK293 cells, the administration of AX caused voltage- and time-dependent inhibition of HERG K$^+$ current (I(HERG/MiRP1)) in a concentration-dependent manner, but does not

markedly modify the properties of channel kinetics, including activation, inactivation, deactivation and recovery from inactivation as well (4). **1**. Xiao & Xiao (2008) *Acta Pharmacol. Sin.* **29**, 1253. **2**. Hu, Liu, Xie, Wang & Liu (2007) *Xenobiotica* **37**, 579. **3**. Zhu, Liao, Xie, Wang & Liu (2009) *Xenobiotica* **39**, 293. **4**. Guo, Han, Liu, *et al.* (2010) *Basic Clin. Pharmacol. Toxicol.* **107**, 643.

AP5

This NMDA antagonist (FW = 197.13 g/mol; CAS 76326-31-3; Solubility: 10 mM in H$_2$O; 100 mM, when combined with 1 equivalent NaOH), also known as APV and (2*R*)-amino-5-phosphonovaleric acid, targets the *N*-methyl-D-aspartate receptor (K_d = 1.4 μM), binding competitively at the ligand (glutamate) binding site. Although AP5 is isosteric with γ-glutamyl-P, an intermediate in the glutamine synthetase catalysis (3), there have been no reports on how AP5 affects glutamine synthetases. **1**. Davies & Watkins (1982) *Brain Res.* **235**, 378. **2**. Evans, Francis, Jones, Smith & Watkins (1982) *Brit. J. Pharmacol.* **75**, 65.

AP22161

This Src SH2 binding disruptor (FW = 577.68 g/mol; CAS 268741-42-0) targets the cysteine residue in the phosphotyrosine-binding pocket of the Src SH2 domain in Src family tyrosine kinases (*e.g.*, Src, Fyn, Yes, Yrk, Lyn, Hck, Fgr, Blk, Lck, Frk/Rak and Iyk/Bsk). A fluorescence-polarization-based competitive binding assay was utilized to determine the IC$_{50}$ of AP-22161 binding to the Src SH2 domain, the Yes SH2 domain and the tandem ZAP SH2 domains. Src has been implicated in the regulation of osteoclast functional activity, and AP 22161 inhibits resorption of dentine. AP22161 demonstrates no sign of toxicity at tested concentrations, as monitored by the presence of tartrate-resistant acid phosphatase (TRAP)-positive cells and surrounding fibroblasts. *See also UCS15A; pYEEI; and ζ-ITAM peptide.* **Target(s):** Src SH2 (IC$_{50}$ = 0.24 μM); YES SH2 (IC$_{50}$ = 29.4 μM); ZAP SH2 (IC$_{50}$ = 422 μM); dentine resorption by rabbit osteoclasts (IC$_{50}$ = 43 μM). **1**. Violette, Shakespeare, Bartlett, *et al.* (2000) **7**, 225.

AP23573, *See Ridaforolimus*

AP26113

This potent ALK inhibitor (FW = 529.02 g/mol; CAS 1197958-12-5; Solubility: 45 mg/mL DMSO; <1 mg/mL H$_2$O), also named 5-chloro-*N*2-

[4-[4-(dimethylamino)-1-piperidinyl]-2-methoxyphenyl]-*N*4-[2-(dimethyl-phosphinyl)phenyl]-2,4-pyrimidinediamine, inhibits wild type Lymphoma Kinase, *or* ALK (K_i = 0.62 nM) and overcomes crizotinib resistance in non-small cell lung cancers harboring the fusion oncogene EML4-ALK. Cells often develop the L1196M gatekeeper mutation within the kinase domain, a changes that renders them resistant to crizotinib, even at higher drug doses (1 µM). AP26113 is highly active against these resistant cancer cells, both *in vitro* and *in vivo*. **1**. Katayama, Khan, Benes, *et al.* (2011) *Proc. Natl. Acad. Sci. U.S.A.* **108**, 7535.

Ap₄A, *See* P^1,P^4-Di(adenosine-5') Tetraphosphate

Ap₅A, *See* P^1,P^5-Di(adenosine-5') Pentaphosphate

Apamin

This bee venom octadecapeptide toxin (FW = 2027.34 g/mol; CAS 24345-16-2; NCBI Reference Sequence: NP_001011612.1) potently blocks the small-conductance Ca^{2+}-activated potassium ion (*or* SK) channels hSK1 as well as rSK2, with IC_{50} values of 3.3 nM and 83 pM. It shows greater effectiveness at the SK2 channel (IC_{50} = 0.06-0.4 nM) than SK1 (IC_{50} = 1-12 nM). SK3 (IC_{50} = 1-13 nM), and SK4 (IC_{50} = 1 µM) channels (1-3), and is active against channels within neurons, vascular endothelium, bladder smooth muscle, and certain cancers (4,5). Apamin does not inhibit human cardiac Na^+ current, L-type Ca^{2+} current or other major K^+ currents (6). Structurally, apamin forms a stable structure, consisting of a C-terminal α-helix and two reverse turns, that is stabilized by two disulfide bonds connecting Cys-1 to Cys-11 and Cys-3 to Cys-15. A minor constituent of venom of the bee (*Apis mellifera*), apamin amounts to only 2-3% (w/w) of its dry venom. The smallest known neurotoxic polypeptide, apamin is derived by proteolytic processing of the 48-residue pre-pro-protein (7-9). The precursors of the bee venom constituents apamin and MCD peptide are encoded by two genes in tandem which share the same 3'-exon (10). **1**. Maylie, Bond, Herson, *et al.* (2004) *J. Physiol.* **554**, 255. **2**. Wulff & Köhler (2013) *J. Cardiovasc. Pharmacol.* **61**, 102. **3**. Triggle, Hollenberg, Anderson, *et al.* (2003) *J. Smooth Muscle Res.* **39**, 249. **4**. Pedarzani & Stocker (2008) *Cellul. Molec. Life Sci.* **65**, 3196 (2008). 5. Alvarez-Fischer, Noelker, Vulinovic, *et al.* (2013) *PLoS One* **8**, 1. **6**. Yu, Ai, Weiss, *et al.* *PLoS One* **9**, 1. **7**. Habermann (1972) *Science* **117**, 314. **8**. Shipolini, *et al.* (1967) *J. Chem. Soc. D. Chem. Commun.* **1967**, 679. **9**. Mourre, *et al.* (1997) *Brain Res.* **778**, 405. **10**. Gmachl & Kreil (1995) *J. Biol. Chem.* **270**, 12704.

APCIN

This Cdc20 inhibitor (FW = 438.65 g/mol; CAS 300815-04-7; Solubility: 100 mM in DMSO), also named 3-(2-methyl-5-nitroimidazol-1-yl)-*N*-(2,2,2-trichloro-1-phenylaminoethyl)propionamide, targets substrate interaction of <u>C</u>ell division <u>c</u>ontrol protein 20, blocking substrate-mediated Cdc20 loading onto the <u>A</u>naphase-<u>P</u>romoting <u>C</u>omplex (1,2). APC is a 13-subunit ubiquitin ligase that initiates the metaphase-anaphase transition and mitotic exit by targeting proteins such as securin and cyclin B_1 for ubiquitin-dependent destruction by the proteasome. APC regulates cell cycle progression by forming APC·Cdc20 and APC·Cdh1, two closely related, but functionally distinct, E_3 ubiquitin ligase sub-complexes. Cdh1 and Cdc20 have opposing functions in tumorigenesis, with the former largely functioning as a tumor suppressor, and the latter exhibiting an oncogenic function. Because blocking mitotic exit is an effective approach for inducing tumour cell death, the APC/C represents a potential novel target for cancer therapy. APC/C activation during mitosis requires Cdc20binding, forming a co-receptor with the APC/C to recognize substrates containing a destruction box (D-box). APC/C-dependent

proteolysis and mitotic exit can be synergistically inhibited by simultaneously disrupting two protein-protein interactions within the APC/C-Cdc20-substrate ternary complex. <u>A</u>PC inhibitor, *or* Apcin, binds to Cdc20, competitively inhibiting ubiquitylation of D-box-containing substrates. Analysis of the crystal structure of the apcin-Cdc20 complex suggests that apcin occupies the D-box-binding pocket on the side face of the WD40-domain. The ability of Apcin to block mitotic exit is enhanced by co-addition of tosyl-l-arginine methyl ester (**See** TAME), a small molecule that blocks the APC/C-Cdc20 interaction (3). Apcin also inhibits APC-dependent ubiquitylation and prolongs mitotic duration in RPE1 cells, when administered in combination with proTAME, a prodrug of TAME *in vitro* (**See** pro-TAME). **1**. Wang, *et al.* (2015) *Pharmacol. Ther.* **151**, 141. **2**. Sackton et al (2014) *Nature* **514**, 646. **3**. Zeng, Sigoillot, Gaur, *et al.* (2010) *Cancer Cell* **18**, 382.

APEX57219

This HIV-1 RT inhibitor (FW = 424.36 g/mol), first identified using a FRET-based high-throughput screening for the AZT-MP excision repair activity, has K_i = 140 nM for wild-type reverse transcriptase, 220 nM for [M41L/L210W/T215Y]-RT, 207 nM for [D67N/K70R]-RT, and 242 nM for [D67N/K70R/T215F/K219Q]-RT. In the RNA-dependent DNA polymerase assay, APEX57219 gave a K_i value of 330 nM for wild-type RT. APEX57219 is a competitive inhibitor with respect to template/primer binding, and a non-competitive inhibitor with respect to dNTP binding. That APEX57219 abrogates the interactions between enzyme and its nucleic acid substrate is indicated by its ability to inhibit all of HIV-1 RT activities (*i.e.*, DNA polymerase, RNase H and NRTI-MP excision), while retaining activity against common NNRTI resistance mutations. APEX57219 also inhibits HIV replication and the level of reverse transcription. **1**. Bago, Malik, Munson, *et al.* (2014) *Biochem. J.* **462**, 425.

Aphidicolin

This photosensitive diterpenoid (FW = 338.49 g/mol; CAS 38966-21-1; M.P. = 227–232°C; IUPAC: (3*R*,4*R*,4*aR*,6*aS*,8*R*,9*R*,11*aS*,11*bS*)tetradeca hydro-3,9-dihydroxy-4,11*b*-dimethyl-8,11*a*-methano-11*aH*-cyclohepta[*a*] naphthalene-4,9-dimethanol) from *Cephalosporium aphidicola* and *Nigrospora oryzae* inhibits many nuclear replicative DNA polymerases, thereby arresting the cell cycle in the S phase. Inhibition of DNA polymerase α is noncompetitive with respect to dATP, dGTP, dTTP as well as DNA, but is competitive relative to dCTP. Certain cells become aphidicolin-resistant by any one of the following adaptations: (a) a mutation in DNA polymerase α that greatly decreases affinity for the drug; (b) overproduction of DNA polymerase α; or (c) increased synthesis of dCTP. Replication of mitochondrial and chloroplast DNA is unaffected by aphidicolin. **Target(s):** DNA polymerase α (1-3,5,7,10,12,17,21-23,26); φ29 DNA polymerase (4,7); DNA polymerase δ (3,6,7,8,21-23,26,27,29,32); DNA polymerase ζ (28); bacteriophage T4 DNA polymerase (7,16); DNA polymerase ε (7,11,12,20-23); *Pyrococcus furiosus* DNA polymerase BI (9); Herpes simplex-induced DNA polymerase (10,12); *Vaccinia*-induced DNA polymerase (10,12); plant cell α-like DNA polymerase (10); yeast DNA polymerase I (10); yeast DNA polymerase II (10). **1**. Hubermann (1981) *Cell* **23**, 647. **2**. Ward & Weissbach (1986) *Meth. Enzymol.* **118**, 97. **3**. Mitsis, Chiang & Lehman (1995) *Meth. Enzymol.* **262**, 62. **4**. Lázaro, Blanco & Salas (1995) *Meth. Enzymol.* **262**, 42. **5**. Wang, Copeland, Rogge & Dong (1995) *Meth. Enzymol.* **262**, 77. **6**. Downey & So (1995) *Meth. Enzymol.* **262**, 84. **7**. Brown & Wright (1995) *Meth. Enzymol.* **262**, 202. **8**. Malkas & Hickey (1996) *Meth. Enzymol.* **275**, 133. **9**. Ishino & Ishino (2001) *Meth. Enzymol.* **334**, 249. **10**. Spadari, Sala & Pedrali-Noy (1982) *Trends Biochem. Sci.* **7**, 29. **11**. Cheng & Kuchta (1993) *Biochemistry* **32**, 8568. **12**. Weissbach (1981) *The Enzymes*, 3rd ed., **14**, 67. **13**. Trost, Nishimura, Yamamoto & McElvaine (1979) *J. Amer. Chem. Soc.* **101**, 1328. **14**. McMurry, Andrus, Ksander, Musser & Johnson (1979) *J. Amer. Chem. Soc.* **101**, 1330. **15**. Corey, Tius & Das (1980) *J. Amer. Chem. Soc.* **102**, 1742. **16**. Khan, Reha-Krantz & Wright (1994) *Nucl. Acids Res.* **22**, 232. **17**. Arabshahi, Brown, Khan & Wright (1988) *Nucl. Acids Res.* **16**, 5107. **18**. Spampinato, Pairoba, Colombo, Benediktsson & Andreo (1994) *Biosci. Biotechnol. Biochem.* **58**, 822. **19**. Oliveros, Yanez, Salas, *et al.* (1997) *J. Biol. Chem.* **272**, 30899. **20**. Niranjanakumari & Gopinathan (1993) *J. Biol. Chem.* **268**, 15557. **21**. Wang (1991) *Ann. Rev. Biochem.* **60**, 513. **22**. Bambara & Jessee (1991) *Biochim. Biophys. Acta* **1088**, 11. **23**. Wright, Hübscher, Khan, Focher & Verri (1994) *FEBS Lett.* **341**, 128. **24**. Burrows & Goward (1992) *Biochem. J.* **287**, 971. **25**. Zhu & Ito (1994) *Biochim. Biophys. Acta* **1219**, 267. **26**. Lehman & Karguni (1989) *J. Biol. Chem.* **264**, 4265. **25**. Aoyagi, Matsuoka, Furunobu, Matsukage & Sakaguchi (1994) *J. Biol. Chem.* **169**, 6045. **28**. Takeuchi, Oshige, Uchida, *et al.* (2004) *Biochem. J.* **382**, 535. **29**. Peck, Germer & Cress (1992) *Nucl. Acids Res.* **20**, 5779. **30**. Weissbach (1979) *Arch. Biochem. Biophys.* **198**, 386. **31**. Chavalitshewinkoon, De Vries, Stam, *et al.* (1993) *Mol. Biochem. Parasitol.* **61**, 243. **32**. Brown, Duncan & Campbell (1993) *J. Biol. Chem.* **268**, 982. **33**. Pisani, De Felice, Manco & Rossi (1998) *Extremophiles* **2**, 171.

Apicidin

This fungal metabolite and potential oral chemotherapeutic agent (FW = 623.79 g/mol; CAS 183506-66-3), also known as [cyclo(*N*-*O*-methyl-L-tryptophanyl-L-isoleucinyl-D-pipecolinyl-L-2-amino-8-oxodecanoyl)], is an antiprotozoal agent. It is cell permeable and is a strong inhibitor of histone deacetylase (IC$_{50}$ = 0.7 nM). Apicidin also inhibits proliferation of tumor cells *via* induction of p21WAF1/Cip1 and gelsolin. Apicidin's low bioavailability of apicidin is mainly due to the P-gycoprotein-mediated efflux (3). **1**. Han, Ahn, Park, *et al.* (2000) *Cancer Res.* **60**, 6068. **2**. Darkin-Rattray, Gurnett, Myers, *et al.* (1996) *Proc. Natl. Acad. Sci. U.S.A.* **93**, 13143. **3**. Shin, Yoo, Kim, *et al.* (2014) *Drug Metab. Dispos.* **42**, 974.

Apixaban

This oral factor Xa inhibitor (FW = 459.51 g/mol; CAS 503612-47-3), also named Eliquis® and 1-(4-methoxyphenyl)-7-oxo-6-[4-(2-oxopiperidin-1-yl)phenyl]-4,5-dihydropyrazolo[5,4-*c*]pyridine-3-carboxamide, directly and selectively targets coagulation Factor Xa. With its extremely low IC$_{50}$ of 80 pM for human FXa, apixaban has >30,000-fold selectivity for FXa over other human coagulation proteases (1). (**See also** *Heparin, Clopidogrel Dabigatran, Rivaroxaban, and Warfarin*) **Mode of Inhibitor Action:** Direct FXa inhibitors constitute a class of anticoagulant drugs acting directly, without requiring antithrombin as a mediator.) With association rate constant of 20 μM^{-1}s^{-1}, apixaban produces rapid-onset FXa inhibition, inhibiting both free as well as prothrombinase- and clot-bound FXa activity *in vitro* (2). When human tissue factor (TF) plus 7.5 mM CaCl$_2$ was added to platelet-rich plasma, aggregometry confirmed that TF-induced platelet aggregation was inhibited by apixaban, IC$_{50}$ = 4 nM (2). **Pharmacodynamics:** Apixaban is FDA-approved for the prevention of venous thromboembolism after elective hip or knee replacement. Apixaban has good bioavailability, low clearance, and a small volume of distribution (2). Although warfarin's anticoagulative action can be reversed (**See** *Warfarin*), there is no known substance that will rapidly reverse apixaban's inhibitory action. Apixaban shows only a low potential for drug-drug interactions, but it is a substrate for both CYP3A4 and P-gp. As such, inhibitors of these enzymes can be expected to alter apixaban exposure. *See other direct Factor Xa inhibitors: Rivaroxaban, or Xarelto®); Betrixaban; Darexaban (discontinued); Edoxaban, or Lixiana®; and Otamixaban injectable (discontinued).* **1**. Pinto, Orwat, Koch, *et al.* (2007) *J. Med. Chem.* **50**, 5339. **2**. Wong, Pinto, Zhang (2011) *J. Thromb. Thrombolysis* **31**, 478. **3**. Wong & Jiang (2010) *Thromb. Haemost.* **104**, 302.

Apocynin

This neutrophil oxidative burst antagonist and potent anti-inflammatory agent (FW = 166.17 g/mol; CAS 4098.02-2; slight vanilla fragrance), also named 1-(4-hydroxy-3-methoxyphenyl)ethanone, from the root of Canadian hemp *Apocynum cannabinum* and the perennial Nepalese herb *Picrorhiza kurroa*, targets NADPH oxidase (IC$_{50}$ = 10-15 μM), which in activated polymorphonuclear (PMN) leukocytes catalyzes NADPH-dependent O$_2$ reduction to superoxide, the latter employed by immune cells to kill bacteria and fungi (1). Apocynin effectively prevents superoxide production in white blood cells or neutrophilic granulocytes, but does not affect phagocytic or other defense roles of granulocytes. Oxygen uptake by neutrophils incubated with 300 μM apocynin was completely inhibited at 7 min upon addition of serum-treated zymosan (STZ), with a lagtime of inhibition of 2-3 min (1). However, the capacity of neutrophils for intracellular killing of *Staphylococcus aureus* was not impaired by apocynin (1). Such selective inhibition makes apocynin a tool for modulating NADPH oxidase without interfering with other immune system actions. Apocynin is rapidly absorbed after oral administration at 50 mg/kg in rats, reaching peak plasma level within 5 min (2). Its presence in plasma is detectable for up to 48 hours. Apocynin bioavailability is 8.3%. When measured *in vitro*, plasma protein binding was found to be 83-86% and 71-73% in rat and human plasma, respectively. Apocynin is stable in gastric (pH 1.2), intestinal (pH 6.8) and physiological (pH 7.4) fluids, including microsomal (rat and human) stability studies (2). IIIHyQ, a related trimer hydroxylated quinone, inhibits endothelial NADPH oxidase, IC$_{50}$ = ~30 nM. *In vitro*, IIIHyQ disrupts the interaction between p47$_{phox}$ and p22$_{phox}$, thereby blocking the activation of the Nox2 isoform. A key cysteine residues in p47$_{phox}$ is the IIIHyQ target (3). **Other Target(s):** Arachidonic acid-induced aggregation of bovine platelets, possibly by inhibiting thromboxane formation (4); angiotensin II-induced endothelin-1 expression (5); ischemia-reperfusion lung injury, but the mechanism of protection remains unclear (6). **1**. Stolk, Hiltermann, Dijkman & Verhoeven (1994) *Am. J. Respir. Cell Mol. Biol.* **11**, 95. **2**. Chandasana, Chhonker, Bala, *et al.* (2015) *J. Chromatogr. B Analyt. Technol. Biomed. Life Sci.* **985C**, 180. **3**. Mora-Pale, Kwon, Linhardt & Dordick (2012) *Free Radic. Biol. Med.* **52**, 962. **4**. Engels, Renirie, Hart, Labadie & Nijkamp (1992) *FEBS Lett.* **305**, 25. **5**. An, Boyd, Zhu, *et al.* (2007) *Cardiovasc. Res.* **75**, 702. **6**. Dodd-O & Pearse (2000) *Am. J. Physiol. Heart Circ. Physiol.* **279**, H303.

Apolipoprotein C-III

This 79-residue lipid-binding protein (MW = 8,750 Da), also known as Apo-CIII and ApoC3, is a major component of HDL, triglyceride-rich lipoproteins (TRL) (*i.e.*, VLDL and chylomicrons), and, to a lesser extent,

LDL. Apo-CIII inhibits lipoprotein lipase and hepatic lipase, thereby delaying the uptake and metabolism of triglyceride-rich particles in liver. While Apo-CIII promotes assembly and secretion of triglyceride-rich VLDL particles from hepatic cells under lipid-rich conditions, its over-expression of human apoC-III in the plasma of transgenic mice results in hypertriglyceridemia. Indeed, high apoC-III levels are linked to hypertriglyceridemia, inflammation, atherosclerosis, and metabolic syndrome. Rare mutations that disrupt *APOC3* function are associated with lower levels of plasma triglycerides and ApoC3 (1,2), and carriers of these mutations show a reduced risk of coronary heart disease. Changes in apoC-III levels are directly associated with changes in cardiovascular risk and the atherogenicity of the lipoproteins on which apoC-III resides (3). Emerging roles of apoC-III include influencing the atherogenicity of high-density lipoprotein, trafficking of intestinal dietary triglycerides, and modulating pancreatic β-cell survival (3). **1.** Do, Willer, Schmidt, *et al.* (2013) *Nature Genetics* **45**, 1345. **2.** TG & HDL Working Group, NHLBI Exome Sequencing Project (2014) *New. Engl. J. Med.* **371**, 22.. **3.** Kohan (2015) *Curr. Opin. Endocrinol. Diabetes Obes.* **22**, 119.

Apoptin

This 121-residue nucleocytoplasmic shuttling protein (MW = 13.6 kDa; GenBank: AAO45416.1), from Chicken Anemia Virus (CAV) and Human Gyrovirus Type 1, associates with APC (anaphase-promoting complex), inducing G_2/M cell-cycle arrest and apoptosis in the absence of p53. When grown *in vitro*, CAV only replicates in transformed chicken cell lines, suggesting that at least a part of the CAV life-cycle requires transformation-like cellular events (1). Apoptin triggers programmed cell death by activating the mitochondrial pathway (*or intrinsic apoptosis*), acting independently of the death receptor pathway (*or extrinsic apoptosis*). Synthesis of apoptin alone induces apoptosis in various human transformed and/or tumorigenic cell lines, but not in normal human diploid cells (1). Apoptin also interacts with the SH3 domain of the BCR-ABL1 fusion protein, inhibiting BCR-ABL1 kinase and its downstream targets (2). A chimeric protein (Dose: 1 µM), consisting of apoptin fsed the Tat cell-penetrating sequence, as well as the apoptin-derived decapeptide (*Sequence*: PKPPSKKRSC; Dose: 1-2 µM), but not the sequence-scrambled peptide PRPPSRSPKC, is effective in treating chronic myelogenous leukemia (CML) by inhibiting the BCR-ABL1 down-stream target, c-Myc, with an efficacy comparable to full-length apoptin and imatinib (**See** *Imatinib*). Synthetic apoptin inhibits cell proliferation in murine 32Dp210 cells and human K562 cells, and is also effective *ex vivo* in both imatinib-resistant and imatinib-sensitive CML patient samples (3). **1.** Noteborn (2004) *Vet. Microbiol.* **98**, 89. **2.** Panigrahi, Stetefeld, Jangamreddy *et al.* (2012) *PLoS One* **7**, e28395. **3.** Jangamreddy, Panigrahi, Lotfi, *et al.* (2014) *Oncotarget* **5**, 7198.

Apoptosis Inhibitor of Macrophages

This novel murine macrophage-derived soluble factor (MW ≈ 40 kDa; *Symbol*: AIM), also known as Spα and Api6, belongs to the scavenger receptor cysteine-rich domain superfamily of glycoproteins secreted by tissue macrophages in the spleen, lymph node, thymus, bone marrow, liver and fetal liver. AIM confers apoptosis resistance to macrophages exposed to various pro-apoptotic signals, including cytotoxic oxidized-low density lipoprotein (oxLDL), dexamethasone, irradiation, Fas/CD95, and fulminant hepatitis (1,2). **1.** Miyazaki, Hirokami, Matsuhashi, Takatsuka & Naito (1999) *J. Exp. Med.* **189**, 413. **2.** Arai, Shelton, Chen, *et al.* (2005) *Cell Metab.* **1**, 201.

Apoptosis Repressor with CARD, See ARC

Apremilast

This thalidomide-like psoriasis drug (FW = 460.50 g/mol; CAS 608141-41-9; Solubility: 90 mg/mL DMSO; <1 mg/mL H₂O), also known as CC-

10004, Otezla® and *N*-[2-[(1*S*)-1-(3-ethoxy-4-methoxyphenyl)-2-(methyl-sulfonyl)ethyl]-2,3-dihydro-1,3-dioxo-1*H*-isoindol-4-yl]acetamide, is a potent, orally active phosphodiesterase inhibitor that targets phosphodiesterase PDE4 (IC_{50} = 74 nM), itself a proinflammatory mediator, and TNF-α, IC_{50} = 77 nM (1-3). Apremilast inhibits PBMC production of the chemokines CXCL9 and CXCL10, cytokines interferon-γ and tumor necrosis factor-α (TNF-α), and interleukins IL-2, IL-12 and IL-23 from human rheumatoid synovial membrane cultures (3). Apremilast significantly reduces clinical scores in both murine models of arthritis over a ten-day treatment period and maintains healthy joint architecture in a dose-dependent manner. Unlike rolipram, however, apremilast demonstrated no adverse behavioral effects in naïve mice (4). Otezla is specifically indicated by the Food & Drug Administration for the treatment of patients with moderate to severe plaque psoriasis who are typically candidates for phototherapy or systemic therapy. That said, the specific mechanism(s) for therapeutic action in psoriatic arthritis patients and psoriasis patients is unclear. **1.** Man, Schafer, Wong, *et al.* (2009) *J. Med. Chem.* **52**, 1522. **2.** Muller, Shire, Wong, *et al.* (1998) *Bioorg. Med. Chem. Lett.* **8**, 2669. **3.** Schafer, Parton, Gandhi, *et al.* (2010) *Brit. J. Pharmacol.* **159**, 842. **4.** McCann, *et al.* (2010) *Arthritis Res. Ther.* **12**, R107.

Aprepitant

This anti-emetic (FW = 534.43 g/mol; CAS 170729-80-3), also known by its code names MK 869 (*or* MK 0869) and L 754030, its trade name Emend® and its systematic name 5-([[(2*R*,3*S*)-2-((*R*)-1-[3,5-bis(trifluoromethyl) phenyl]ethoxy]-3-(4-fluorophenyl)morpholino]methyl) -1*H*-1,2,4-triazol-3(2*H*)-one, is a selective high-affinity Substance P antagonist (*or* SPA) that acts by blocking Substance P binding to the Neurokinin 1 (NK₁) receptor (IC_{50} = 0.12 nM). NK₁ is a G-protein-coupled receptor in the central and peripheral nervous systems, and is especially abundant in the vomiting center. (**See** *prodrug Fosaprepitant*) In preclinical studies, aprepitant did not interact with monoamine systems in the manner seen with established antidepressant drugs (1). Once daily oral administration of MK-869 was effective in reducing delayed emesis and nausea after high-dose Cisplatin (2). At 5-70 µM, aprepitant also elicits growth cell inhibition in a concentration-dependent manner in all tumor cell lines studied (3). Aprepitant moderately inhibits CYP3A4 in humans, as measured with midazolam as a probe drug (4). **See** *Orvepitant; Rolapitant* **Key Pharmacokinetic Parameters:** *See* Appendix II in Goodman & Gilman's *THE PHARMACOLOGICAL BASIS OF THERAPEUTICS*, 12ᵗʰ Edition (Brunton, Chabner & Knollmann, eds.) McGraw-Hill Medical, New York. **1.** Kramer, Cutler, Feighner, *et al.* (1998) *Science* **281**, 1640. **2.** Campos, Pereira, Reinhardt, *et al.* (2001) *J. Clin. Oncol.* **19**, 1759. **3.** Muñoz & Rosso (2010) *Invest. New Drugs* **28**, 187. **4.** Majumdar, McCrea, Panebianco, *et al.* (2003) Clin. Pharmacol. Ther. **74**, 150.

Apricoxib

This oral COX-2 inhibitor (FW = 356.44 g/mol; CAS 162011-90-7), also named CS-706 and 4-[2-(4-ethoxyphenyl)-4-methyl-1*H*-pyrrol-1-yl]-benzene sulfonamide, targets Cyclooxygenase-2 (1), which is overexpressed in advanced stage non-small-cell lung cancer (NSCLC), possibly as a result of elevated levels of COX-2-dependent prostaglandin E_2 (PGE_2). That said, a randomized, double-blind, placebo-controlled, multicenter phase II study (2) found little evidence of efficacy of apricoxib in combination with either docetaxel or pemetrexed in patients with biomarker-selected NSCLC. **See also** Rofecoxib 1. Kirane (2012) *Clin. Cancer Res.* **18**, 5031. **2**. Edelman, Tan, Fidler, *et al.* (2015) J. Clin. Oncol. **33**, 189.

Aprotinin

RPDFCLEPPYTGPCKARIIRYFYNAKAGLCQTFVYGGCRAKRNNFKSAEDCMRTCGGA

This serine protease inhibitor (MW ≈ 6.5 kDa; CASs = 9087-70-1 and 9004-04-0; Solubility: 10 mg/mL or 1.5 mM), also called bovine pancreatic trypsin inhibitor or Trasylol™, is a naturally occurring 58-residue oligopeptide, with disulfides joining Cys-5 to Cys-55, Cys-14 to Cys-38, and Cys-30 to Cys-51) that potently inhibits trypsin, (K_i = 60 fM), chymotrypsin (K_i = 10 nM), kallikrein (K_i = 1-2 nM), plasmin (K_i = 5 nM), elastase (K_i = 4-5 μM), and urokinase (K_i = 8-10 μM). Aprotinin forms extremely stable complexes at neutral pH, but binding is reversible, with most aprotinin-enzyme complexes dissociating slowly at pH >10 or <3. Given the great diversity of proteases and their substrates, aprotinin can be an invaluable tool in unraveling the action of serine proteases in the posttranslational processing of peptide/protein ligands in signal transduction pathways.

AR-A014418

This ATP-competitive protein kinase inhibitor (FW = 308.31 g/mol; CAS 487021-52-3; Solubility: 62 mg/mL DMSO; <1 mg/mL H_2O) selectively targets glycogen synthase kinase 3β (GSK3β),with IC_{50} = 104 nM; K_i = 38 nM. Attesting to its high specificity for GSK3, AR-A014418 does not significantly inhibit Cdk2 or Cdk5 (IC_{50} > 100 μM) or 26 other kinases (1). AR-A014418 inhibits phosphorylation of the microtubule-associated protein Tau at Ser-396 (*i.e.*, the GSK3-specific site) in cells stably expressing human four-repeat Tau (1). AR-A014418 protects N2A neuroblastoma cells against cell death mediated by inhibition of the phosphatidylinositol 3-kinase/protein kinase B survival pathway. AR-A014418 inhibits neurodegeneration mediated by β-amyloid peptide in hippocampal slices (1). Treatment of neuroblastoma cell lines with AR-A014418 reduced the level of GSK-3α phosphorylation at Tyr-279 compared to GSK-3β phosphorylation at Tyr-216, and attenuated growth via the maintenance of apoptosis. (2). AR-A014418-dependent antinociceptive effects are induced by modulation of the glutamatergic system through metabotropic and ionotropic (NMDA) receptors and the inhibition of TNF-α and IL-1β cytokine signaling (3). **1**. Bhat, Xue, Berg, *et al.* (2003) *J. Biol. Chem.* **278**, 45937. **2**. Carter, Kunnimalaiyaan, Chen, Gamblin & Kunnimalaiyaan (2014) *Cancer Biol. Ther.* **15**, 510. **3**. Martins, *et al.* (2011) *J. Pain.* **12**, 315.

Arachidonyl-2'-chloroethylamide

This synthetic CB_1 receptor agonist and anandamide (*N*-arachidonylethanolamide) analogue (FW = 365.99 g/mol; CAS 220556-69-

4; Symbol: ACEA) is a selective agonist of cannabinoid receptor-1 (K_i = 1.4 nM) but has low affinity to the cannabinoid receptor 2 (K_i = 3.1 μM). ACEA inhibits forskolin-induced cAMP accumulation in Chinese hamster ovary cells expressing the human CB_1 receptor. It also increases the binding of [^{35}S]GTPγS to cerebellar membranes and inhibits electrically evoked contractions of the mouse vas deferens. ACEA produces hypothermia in mice, an effect that is inhibited by co-administration of the CB_1 receptor antagonist SR141716A. **1**. Hillard, Manna, Greenberg, *et al.* (1999) *J. Pharmacol. Exp. Ther.* **289**, 1427.

2-Arachidonoylglycerol

This potent endocannabinoid (FW = 378.30 g/mol; CAS 53847-30-6; *Symbol*: 2-AG), also known as 1,3-dihydroxy-2-propanyl (5Z,8Z,11Z,14Z)-5,8,11,14-eicosatetraenoate, is an endogenous agonist of the CB_1, the G-protein-coupled Cannabinoid receptor type-1 found primarily in the central and peripheral nervous system. 2-AG is found at highest concentrations in the CNS, where it exerts its cannabinoid-like neuromodulatory effects (1,2). Found in milk, 2-AG plays a role in sustaining infant suckling, and the selective CB1 receptor antagonist SR141617A (*N*-(piperidin-1-yl)-5-(4-chlorophenyl)-1-(2,4-dichlorophenyl)-4-methyl-1*H*-pyrazole-3-carboxamide) permanently prevents milk ingestion in a dose-dependent manner, when administered to mouse pups, within 1 day of birth (3). 2-AG is formed by phospholipase C (PLC) and diacylglycerol lipase (DAGL) from arachidonic acid-containing diacylglycerol (DAG). In the CNS, three serine-hydrolases, monoacylglycerol lipase (MAGL), α,β-hydrolase-domain-6 (ABHD6) and α,β-hydrolase-domain 12 (ABHD12) are responsible for inactivation of the primary 2-arachidonoylglycerol. Irreversible ABHD6 inhibitors show exceptional potency and selectivity in cells (<5 nM) and, at equivalent doses in mice (1 mg/kg), acting as systemic and peripherally restricted inhibitors, respectively (1). Indeed, selective knockdown of ABHD6 in metabolic tissues protects mice from high-fat-diet-induced obesity, hepatic steatosis, and systemic insulin resistance (2). **1**. Hsu, Tsuboi, Chang, *et al.* (2013) *J Med Chem.* **56**, 8270. **2**. Thomas, Betters, Lord, *et al.* (2013) *Cell Rep.* **5**, 508. **3**. Fride, Foox, Rosenberg, *et al.* (2003) *Eur. J. Pharmacol.* **461**, 27.

Aranidipine

This calcium channel blocker (FW = 388.37 g/mol; CAS 86780-90-7), also named MPC-1303, Sapresta® and the methyl-2-oxopropyl 1,4-dihydro-2,6-dimethyl-4-(2-nitrophenyl)-3,5-pyridine-dicarboxylate, exhibits antihypertensive effects that were more potent than other dihydropyridines, and its pharmacologically active metabolites displayed antihypertensive effects comparable to other dihydropyridines. **1**. Kanda, Haruno, Miyake, & Nagasaka (1992) J. Cardiovasc. Pharmacol. 20, 723.

ARC

This 208-residue skeletal and cardiac muscle protein (MW = 22629.46 g; Protein Accession Number AAC34993; Isoelectric Point = 4.11), known fully as Apoptosis Repressor with CARD (or Caspase recruitment domain), is an endogenous inhibitor of apoptosis that is induced by caspase-8 and CED-3, but not caspase-9. Significantly, the enzymatic activity of caspase-8 is inhibited by ARC in 293T cells (1). Consistent with the inhibition of caspase-8, ARC attenuates apoptosis induced by FADD and TRADD as well as that triggered by stimulation of caspase-8 death receptors (*e.g.*, CD95/Fas, tumor necrosis factor-R1, and TRAMP/DR3) (1). ARC also antagonizes both the extrinsic (death receptor) and intrinsic (mitochondrial/ER) pathways. ARC also inhibits TNFα-induced necrosis when overexpressed in its wild-type form, but not by a CARD mutant defective for inhibiting apoptosis. Conversely, knockdown of ARC exacerbated TNFα-induced necrosis, an effect that was rescued by reconstitution with wild type ARC, but not with CARD-defective ARC. A likely mechanism for these effects involves the interaction of ARC with TNF Receptor-1, interfering with recruitment of RIP1, a critical mediator of TNFα-induced regulated necrosis. Induction of hypoxia-inducible factors (HIFs) is needed for tumor cell to adapt to a low-oxygen environment, and ARC expression is itself induced in a HIF1-dependent manner (3). **1.** Koseki, Inohara, Chen & Núñez (1998) *Proc. Natl. Acad. Sci. U.S.A.* **95**, 5156. **2.** Kung, Dai, Deng & Kitsis (2014) *Cell Death Differ.* **21**, 634. **3.** Razorenova, Castellini, Colavitti, *et al.* (2013) *Mol. Cell Biol.* **34**, 739.

Arctigenin

This plant lignan (FW = 372.41 g/mol; CAS 7770-78-7; IUPAC: (3*R*,4*R*)-4-[(3,4-dimethoxy-phenyl)methyl]-3-[(4-hydroxy-3-methoxyphenyl) methyl]-2-tetrahydrofuranone) is a functional ingredient of several traditional Chinese herbs and dose-dependently reduces proliferation of human dermal microvascular endothelial cells without altering their migratory and tube-forming capacity. Arctigenin treatment also resulted in a decreased cellular expression of phosphorylated serine/threonine protein kinase AKT, vascular EGFR2, and proliferating cell nuclear antigen and inhibited vascular sprouting from aortic rings. **1.** Gu, Scheuer, Feng, Menger & Laschke (2013) *Anticancer Drugs* **24**, 781.

Arecoline

This CNS-active alkaloid (FW = 155.19 g/mol; CAS 63-75-2; $pK_a \sim 6.8$), also named methyl 1-methyl-1,2,5,6-tetrahydropyridine-3-carboxylate, from the areca nut (*Areca catechu*), is a *partial* agonist of muscarinic M_1, M_2, M_3, and M_4 acetylcholine receptors, giving rise to its well-known parasympathetic effects, including pupillary and bronchial constriction (1-3). (*Note*: Some muscarinic agonists also cause cause significant receptor internalization and down-regulation; others are allosteric ligands. In this respect, the action of arecoline may not be that of a classical partial agonist acting at a receptor's primary binding site.) Arecoline induces dose-dependent antinociception (0.3-1 mg/kg, *i.p.*) that can be prevented by muscarinic antagonists: pirenzepine (0.1 µg/mouse, *i.c.v.*) or *S*-(–)-ET-126 (0.01 µg/mouse, *i.c.v.*) (1). Arecoline excites rat *locus coeruleus* neurons by activating the M_2-muscarinic receptor (2). It also excites the colonic smooth muscle motility via M_3 receptors in rabbits (3). In some Asian countries, the areca nut is chewed along with betel leaf for its stimulating effects. This practice can have devastating long-term consequences, including aggressive oral cancer. Oral submucous fibrosis (OSF), a chronic inflammatory disease characterized by the accumulation of excess collagen, is a condition occurring often in individuals who frequently chew areca nuts (4). Notably, arecoline markedly induces morphologic change in HaCaT cells (transformed human epithelial cells derived from adult skin), but has no obvious effect on Hel cells (human embryo fibroblasts) (4). Arecoline also attenuates the 2,3,7,8-tetrachlorodibenzo-*p*-dioxin (TCDD) - induced CYP1A1 activation, mainly via down-regulation of AhR expression in human hepatoma cells, suggesting the possible involvement of arecoline in the AhR-mediated metabolism of environmental toxicants in liver (5). **1.** Ghelardini, Galeotti, Lelli & Bartolini (2001) *Farmaco* **56**, 383. **2.** Yang YR, Chang, Chen & Chiu (2000) *Chin. J. Physiol.* **43**, 23. **3.** Xie, Chen, Liu, *et al.* (2004) *Chin. J. Physiol.* **47**, 89. **4.** Li, Gao, Zhou, *et al.* (2014) *Oncol. Rep.* 31, 2422. **5.** Chang, Miao, Lee, *et al.* (2007) *J. Hazard. Mater.* **146**, 356.

Argadin

This cyclopentapeptide (FW = 674.70 g/mol; CAS 289665-92-5), found in fungal cultures of a species of *Clonostachys*, inhibits insect chitinases (*e.g.*, blowfly [*Lucilia cuprina*] chitinase: IC_{50} = 0.15 µM at 37°C and 3.4 nM at 20°C). The K_i value for chitinase B of *Serratia marcescens* is 20 nM. **1.** Arai, Shiomi, Yamaguchi, *et al.* (2000) *Chem. Pharm. Bull.* **48**, 1442. **2.** Houston, Shiomi, Arai, *et al.* (2002) *Proc. Natl. Acad. Sci. U.S.A.* **99**, 9127.

Argatroban

This anticoagulant (FW = 508.64 g/mol; CAS 74863-84-6), also named (2*R*,4*R*)-4-methyl-1-*N*2-[(3-methyl-1,2,3,4-tetrahydro-8-quinolilyl)sulfon-yl]-L-arginyl-2-piperidine carboxylic acid, Novastan®, and MD-805, is a direct thrombin inhibitor (1-11) that is an effective treatment for heparin-induced thrombocytopenia (HIT), an immunoglobulin-mediated adverse drug reaction characterized by platelet activation, thrombocytopenia, and a high risk of thrombotic complications among patients who are receiving or have recently received heparin. HIT is an autoimmune-like disorder, with the target antigen a multimolecular complex of the "self" protein, Platelet Factor-4, and heparin. Continued use of heparin is ill-advised and dangerous. Argatroban is also indicated for the treatment of heparin-induced thrombocytopenia and thrombosis syndrome (HITTS), an immune-mediated response to the administration of heparin that results in life-threatening thrombosis. Argatroban is one of four FDA-approved parenteral direct thrombin inhibitors, the others being lepirudin, desirudin, and bivalirudin. Argatroban is metabolized in the liver ($t_{1/2}$ = ~50 min). Other DTI's include recombinant hirudins, bivalirudin, and ximelagatran, either alone or in combination with melagatran (2). Argatroban is a significantly weaker inhibitor of trypsin, K_i = 4.3 µM (6,11), plasmin (K_i = 800 µM), coagulation factor Xa, and plasma kallikrein (K_i = 2 mM). Commercial sources often supply a mixture of 21(*R*) and 21(*S*) Diastereoisomers. The latter is more potent. **1.** Koster, Fischer, Harder & Mertzlufft (2007) *Biologics* **1**, 105. **2.** Di Nisio, Middeldorp & Büller (2005) *New Eng. J. Med.* **353**, 10. **3.** Nilsson, Sjoling-Ericksson & Deinum (1998) *J. Enzyme Inhib.* **13**, 11. **4.** Okamoto & Hijikata-Okunomiya (1993) *Meth. Enzymol.* **222**, 328. **5.** Matsuo, Koide & Kario (1997) *Semin. Thromb. Hemost.* **23**, 51. **6.** Tapparelli, Matternich, Ehrardt, *et al.* (1993) *J. Biol. Chem.* **268**, 4734. **7.** Kikumoto, Tamao, Tezuka, *et al.* (1984) *Biochemistry* **23**, 85. **8.** Rawson, VanGorp, Yang & Kogan (1993) *J. Pharm. Sci.* **82**, 672. **9.** Okamoto, Hijikata, Kikumoto, *et al.* (1981) *Biochem. Biophys. Res. Commun.* **101**, 440. **10.** Weitz (2003) *Thromb. Res.* **109** Suppl. 1, S17. **11.** Hijikata-Okunomiya, Tamao, Kikumoto & Okamoto (2000) *J. Biol. Chem.* **275**, 18995.

D-Arginine

This amino acid ($FW_{free-base}$ = 174.20 g/mol), named systematically as (R)-2-amino-5-guanidinopentanoic acid, is the enantiomer of the more abundant and proteogenic amino acid. **Target(s):** arginine N-succinyltransferase (1); ribosomal peptidyltransferase (2); arginine decarboxylase (3,4,10-12); carboxypeptidase B (5,16); clostripain (6); arginine 2-monooxygenase, weakly inhibited (7); acetylcholinesterase, weakly inhibited (8); tyrosine:arginine ligase, or kyotorphin synthetase (9); arginine deaminase (13); Arginase (14); lysine carboxypeptidase, or lysine(arginine) carboxypeptidase, or carboxypeptidase N (15); aminopeptidase B (17); arginine kinase (18-20). **1.** Tricot, Vander Wauven, Wattiez, Falmagne & Stalon (1994) *Eur. J. Biochem.* **224**, 853. **2.** Palacian & Vazquez (1979) *Eur. J. Biochem.* **101**, 469. **3.** Rosenfeld & Roberts (1976) *J. Bacteriol.* **125**, 601. **4.** Smith (1983) *Meth. Enzymol.* **94**, 176. **5.** Folk (1971) *The Enzymes*, 3rd ed., **3**, 57. **6.** Mitchell & Harrington (1971) *The Enzymes*, 3rd. ed., **3**, 699. **7.** Olomucki, Dangba, Nguyen Van Thoai (1964) *Biochim. Biophys. Acta* **85**, 480. **8.** Bergmann, Wilson & Nachmansohn (1950) *J. Biol. Chem.* **186**, 693. **9.** Kawabata, Muguruma, Tanaka & Takagi (1996) *Peptides* **17**, 407. **10.** Smith (1979) *Phytochemistry* **18**, 1447. **11.** Ramakrishna & Adiga (1975) *Eur. J. Biochem.* **59**, 377. **12.** Balbo, Patel, Sell, *et al.* (2003) *Biochemistry* **42**, 15189. **13.** Park, Hirotani, Nakano & Kitaoka (1984) *Agric. Biol. Chem.* **48**, 483. **14.** Patchett, Daniel & Morgan (1991) *Biochim. Biophys. Acta* **1077**, 291. **15.** Juillerat-Jeanneret, Roth & Bargetzi (1982) *Hoppe-Seyler's Z. Physiol. Chem.* **363**, 51. **16.** Bradley, Naudé, Muramoto, Yamauchi & Oelofsen (1996) *Int. J. Biochem. Cell Biol.* **28**, 521. **17.** Kawata, Takayama, Ninomiya & Makisumi (1980) *J. Biochem.* **88**, 1601. **18.** Cheung (1973) *Arch. Biochem. Biophys.* **154**, 28. **19.** Brown & Grossman (2004) *Arch. Insect Biochem. Physiol.* **57**, 166. **20.** Baker (1976) *Insect Biochem.* **6**, 449.

L-Arginine

This basic proteogenic amino acid and nitric oxide precursor ($FW_{monohydrochloride}$ = 210.66 g/mol; $FW_{free-base}$ = 174.20 g/mol; pKa values of 1.82, 8.99, and 12.48 (guanido) at 25°C), also called (S)-2-amino-5-guanidinopentanoate, is symbolized by Arg or R. There are six codons for arginine: AGA, AGG, CGU, CGC, CGA, and CGG. (The first two are not in the mitochondria of mammals, and AGA is also not in the mitochondria of *Drosophila melanogaster*). The guanidino group reacts with a number of reagents (*e.g.*, phenylglyoxal and 1,2-cyclohexanedione), often resulting in loss of catalytic activity for enzymes with active-site arginine residues.. Arginine is also converted into creatine and creatinine. **Target(s):** amino-acid N-acetyltransferase (N-acetylglutamate synthase; arginine is an activator of the mammalian and frog enzymes) (1,79,85-92); N-acetylglutamate kinase (2,7,15,17,28,73,78); omptin (3,11); procollagen C-endopeptidase (4,56,57); peptidyl-Lys metalloendopeptidase, or *Armillaria mellea* protease, weakly inhibited (5,54,55); GfMEP, thermostable lysine-specific metalloendopeptidase (5,55); ancrod (6); ribosomal peptidyltransferase (8); putrescine carbamoyltransferase (9); lysine carboxypeptidase, or lysine(arginine) carboxypeptidase, or carboxypeptidase N (10,64); ornithine carbamoyltransferase (12,93-95); carboxypeptidase B (13); clostripain (14); carbamate kinase (16); homoserine kinase, weakly inhibited (18,22); saccharopine dehydrogenase (19); argininosuccinate synthetase (20,35); Xaa-Arg Dipeptidase, or aminoacyl-lysine Dipeptidase (27); aminopeptidase B (23,66-68); γ-glutamyl transpeptidase, weakly inhibited (24,25); group I self-splicing intron (26); kynureninase (27); asparagine synthetase, weakly inhibited (29); alkaline phosphatase (30); acetylcholinesterase (31); glutamate decarboxylase, weakly inhibited (32); urate oxidase, weakly inhibited (33); carbamoyl-phosphate synthetase (34); glutamine synthetase (36,37); lysyl-tRNA synthetase (38); 2-aminohexano-6-lactam racemase (39); threonine ammonia-lyase, or threonine dehydratase (40); hyaluronate lyase (41); dihydrodipicolinate synthase (42-45); lysine transport (46); 5-guanidino-2-oxopentanoate decarboxylase, or α-ketoarginine decarboxylase; weakly inhibited (47); ornithine decarboxylase (48-50); kynureninase (51); agmatine deaminase (52); agmatinase (53); oligopeptidase B (58); venombin A (60; acrosin (59-62); carboxypeptidase E (63); membrane alanyl aminopeptidase, or aminopeptidase N (65,69,70); [citrate lyase] deacetylase (71); guanidinoacetate kinase, weakly inhibited (78);

homoserine kinase (86); ornithine aminotransferase (87); spermidine synthase (88); homocitrate synthase (89); leucyltransferase. **1.** Powers-Lee (1985) *Meth. Enzymol.* **113**, 27. **2.** Dénes (1970) *Meth. Enzymol.* **17A**, 269. **3.** Toledo & Mangel (1998) in *Handb. Proteolytic Enzymes*, p. 518, Academic Press, San Diego. **4.** Kessler (1998) in *Handb. Proteolytic Enzymes*, p. 1236, Academic Press, San Diego. **5.** Takio (1998) in *Handb. Proteolytic Enzymes*, p. 1538, Academic Press, San Diego. **6.** Nolan, Hall & Barlow (1976) *Meth. Enzymol.* **45**, 205. **7.** Wolf & Weiss (1980) *J. Biol. Chem.* **255**, 9189. **8.** Palacian & Vazquez (1979) *Eur. J. Biochem.* **101**, 469. **9.** Stalon (1983) *Meth. Enzymol.* **94**, 339. /**10.** Ryan (1988) *Meth. Enzymol.* **163**, 186. **11.** Mangel, Toledo, Brown, Worzalla, Lee & Dunn (1994) *Meth. Enzymol.* **244**, 384. **12.** Ahmad, Bhatnagar & Venkitasubramanian (1986) *Biochem. Cell Biol.* **64**, 1349. **13.** Folk (1971) *The Enzymes*, 3rd ed., **3**, 57. **14.** Mitchell & Harrington (1971) *The Enzymes*, 3rd. ed., **3**, 699. **15.** Fernández-Murga, Ramón-Maiques, Gil-Ortiz, Fita & Rubio (2002) *Acta Crystallogr. D Biol. Crystallogr.* **58**, 1045. **16.** Raijman & Jones (1973) *The Enzymes*, 3rd ed., **9**, 97. **17.** Dénes (1973) *The Enzymes*, 3rd ed., **9**, 511. **18.** Webb (1966) *Enzyme and Metabolic Inhibitors*, vol. 2, p. 357, Academic Press, New York. **19.** Fujioka & Nakatani (1972) *Eur. J. Biochem.* **25**, 301. **20.** Takada, Saheki, Igarashi & Katsunuma (1979) *J. Biochem.* **85**, 1309. **21.** Kumon, Matsuoka, Kakimoto, Nakajima & Sano (1970) *Biochim. Biophys. Acta* **200**, 466. **22.** Wormser & Pardee (1958) *Arch. Biochem. Biophys.* **78**, 416. **23.** Kawata, Takayama, Ninomiya & Makisumi (1980) *J. Biochem.* **88**, 1601. **24.** Allison (1985) *Meth. Enzymol.* **113**, 419. **25.** Thompson & Meister (1977) *J. Biol. Chem.* **252**, 6792. **26.** Liu & Leibowitz (1995) *Nucl. Acids Res.* **23**, 1284. **27.** Jakoby & Bonner (1953) *J. Biol. Chem.* **205**, 709. **28.** Haas & Leisinger (1975) *Eur. J. Biochem.* **52**, 377. **29.** Horowitz & Meister (1972) *J. Biol. Chem.* **247**, 6708. **30.** Bodansky & Strachman (1949) *J. Biol. Chem.* **179**, 81. **31.** Bergmann, Wilson & Nachmansohn (1950) *J. Biol. Chem.* **186**, 693. **32.** Roberts & Frankel (1951) *J. Biol. Chem.* **190**, 505. **33.** Bentley & Truscoe (1969) *Enzymologia* **37**, 285. **34.** Llamas, Suarez, Quesada, Bejar & del Moral (2003) *Extremephiles* **7**, 205. **35.** Hilger, Simon & Stalon (1979) *Eur. J. Biochem.* **94**, 153. **36.** Southern, Parker & Woods (1987) *J. Gen. Microbiol.* **133**, 2437. **37.** Blanco, Alana, Llama & Serra (1989) *J. Bacteriol.* **171**, 1158. **38.** Levengood, Ataide, Roy & Ibba (2004) *J. Biol. Chem.* **279**, 17707. **39.** Ahmed, Esaki, Tanaka & Soda (1985) *Agric. Biol. Chem.* **49**, 2991. **40.** Choi & Kim (1995) *J. Biochem. Mol. Biol.* **28**, 118. **41.** Akhtar & Bhakuni (2003) *J. Biol. Chem.* **278**, 25509. **42.** Dereppe, Bold, Ghisalba, Ebert & Schaer (1991) *Plant Physiol.* **98**, 813. **43.** Kumpaisal, Hashimoto & Yamada (1987) *Plant Physiol.* **85**, 145. **44.** Wallsgrove & Mazelis (1981) *Phytochemistry* **20**, 2651. **45.** Mazelis, Watley & Whatley (1977) *FEBS Lett.* **84**, 236. **46.** Maretzki & Thom (1970) *Biochemistry* **9**, 2731. **47.** Vanderbilt, Gaby, Rodwell & Bahler (1975) *J. Biol. Chem.* **250**, 5322. **48.** Guirard & Snell (1980) *J. Biol. Chem.* **255**, 5960. **49.** Schaeffer & Donatelli (1990) *Biochem. J.* **270**, 599. **50.** Pantazaki, Anagnostopoulos, Lioliou & Kyriakidis (1999) *Mol. Cell. Biochem.* **195**, 55. **51.** Jakoby & Bonner (1953) *J. Biol. Chem.* **205**, 709. **52.** Legaz, Iglesias & Vicente (1983) *Z. Pflanzenphysiol.* **110**, 53. **53.** Satishchandran & Boyle (1986) *J. Bacteriol.* **165**, 843. **54.** Lewis, Basford & Walton (1978) *Biochim. Biophys. Acta* **522**, 551. **55.** Nonaka, Ishikawa, Tsumuraya, *et al.* (1995) *J. Biochem.* **118**, 1014. **56.** Hojima, van der Rest & Prockop (1985) *J. Biol. Chem.* **260**, 15996. **57.** Kessler, Takahara, Biniaminov, Brusel & Greenspan (1971) *Science* **271**, 360. **58.** Pacaud & Richaud (1975) *J. Biol. Chem.* **250**, 7771. **59.** Polakoski & McRorie (1973) *J. Biol. Chem.* **248**, 8183. **60.** Brown, Andani & Hartree (1975) *Biochem. J.* **149**, 133. **61.** Polakoski, McRorie & Williams (1973) *J. Biol. Chem.* **248**, 8178. **62.** Schiessler, Fritz, Arnold, Fink & Tschesche (1972) *Hoppe-Seyler's Z. Physiol. Chem.* **353**, 1638. **63.** Hook (1990) *Life Sci.* **47**, 1135. **64.** Juillerat-Jeanneret, Roth & Bargetzi (1982) *Hoppe-Seyler's Z. Physiol. Chem.* **363**, 51. **65.** Lalu, Lampelo & Vanha-Perttula (1986) *Biochim. Biophys. Acta* **873**, 190. **66.** Goldstein, Nelson, Kordula, Mayo & Travis (2002) *Infect. Immun.* **70**, 836. **67.** Lauffart, McDermott, Jones & Mantle (1988) *Biochem. Soc. Trans.* **16**, 849. **68.** Kawata, Takayama, Ninomiya & Makisumi (1980) *J. Biochem.* **88**, 1601. **69.** Bauvois & Dauzonne (2006) *Med. Res. Rev.* **26**, 88. **70.** McCaman & Villarejo (1982) *Arch. Biochem. Biophys.* **213**, 384. **71.** Giffhorn & Gottschalk (1975) *J. Bacteriol.* **124**, 1052. **72.** Shirokane, Nakajima & Mizusawa (1991) *Agric. Biol. Chem.* **55**, 2235. **73.** Maheswaran, Urbanke & Forchhammer (2004) *J. Biol. Chem.* **279**, 55202. **74.** Faragó & Dénes (1967) *Biochim. Biophys. Acta* **136**, 6. **75.** Pauwels, Abadjieva, Hilven, Stankiewicz & Crabeel (2003) *Eur. J. Biochem.* **270**, 1014. **76.** Ramón-Maiques, Fernández-Murga, Gil-Ortiz, *et al.* (2006) *J. Mol. Biol.* **356**, 695. **77.** Chen, Ferrar, Lohmeir-Vogel, *et al.* (2006) *J. Biol. Chem.* **281**, 5726. **78.** Fernández-Murga, Gil-Ortiz, Llácer &/ Rubio (2004) *J. Bacteriol.* **186**, 6142. **79.** Qu,

Morizono, Shi, Tuchman & Caldovic (2007) *BMC Biochem.* **8**, 4.   **80.** Thoen, Rognes & Aarnes (1978) *Plant Sci. Lett.* **13**, 103.   **81.** Yasuda, Tanizawa, Misono, Toyama & Soda (1981) *J. Bacteriol.* **148**, 43.   **82.** Graser & Hartmann (2000) *Planta* **211**, 239.   **83.** Gray & Bhattacharjee (1976) *J. Gen. Microbiol.* **97**, 117.   **84.** Soffer (1973) *J. Biol. Chem.* **248**, 8424.   **85.** Haas & Leisinger (1974) *Biochem. Biophys. Res. Commun.* **60**, 42.   **86.** Errey & Blanchard (2005) *J. Bacteriol.* **187**, 3039.   **87.** Haskins, Panglao, Qu, *et al.* (2008) *BMC Biochem.* **9**, 24.   **88.** Morris & Thompson (1975) *Plant Physiol.* **55**, 960.   **89.** Marvil & Leisinger (1977) *J. Biol. Chem.* **252**, 3295.   **90.** Leisinger & Haas (1975) *J. Biol. Chem.* **250**, 1690.   **91.** Hinde, Jacobson, Weiss & Davis (1986) *J. Biol. Chem.* **261**, 5848.   **92.** Haas, Kurer & Leisinger (1972) *Eur. J. Biochem.* **31**, 290.   **93.** Baker & Yon (1983) *Phytochemistry* **22**, 2171.   **94.** Timm, Van Rompaey, Tricot, *e al.* (1992) *Mol. Gen. Genet.* **234**, 475.   **95.** Abdelal, Kennedy & Nainan (1977) *J. Bacteriol.* **129**, 1387.

L-Arginine Methyl Ester

This amino acid ester ($FW_{dihydrochloride}$ = 261.15 g/mol) has the following inhibitory targets: alkaline phosphatase (1); cyanophycin synthetase (2); hyaluronate lyase (3); arginine decarboxylase (4,5); protein-arginine deiminase (reference probably meant N^{ω}-nitro-L-arginine methyl ester) (6); aminopeptidase B, weakly inhibited (7); [protein ADP-ribosylarginine] hydrolase, weakly inhibited (8); arginine kinase (9); *N*-acetylglutamate kinase (10); leucyltransferase (11).   **1.** Fishman & Sie (1971) *Enzymologia* **41**, 141.   **2.** Aboulmagd, Oppermann-Sanio & Steinbuchel (2001) *Appl. Environ. Microbiol.* **67**, 2176.   **3.** Akhtar & Bhakuni (2003) *J. Biol. Chem.* **278**, 25509.   **4.** Graham, Xu & White (2002) *J. Biol. Chem.* **277**, 23500.   **5.** Balbo, Patel, Sell, *et al.* (2003) *Biochemistry* **42**, 15189.   **6.** McGraw, Potempa, Farley & Travis (1999) *Infect. Immun.* **67**, 3248.   **7.** Kawata, Takayama, Ninomiya & Makisumi (1980) *J. Biochem.* **88**, 1601.   **8.** Kim & Graves (1990) *Anal. Biochem.* **187**, 251.   **9.** Baker (1976) *Insect Biochem.* **6**, 449.   **10.** Fernández-Murga, Gil-Ortiz, Llácer & Rubio (2004) *J. Bacteriol.* **186**, 6142.   **11.** Soffer (1973) *J. Biol. Chem.* **248**, 8424.

Arginine Phosphate, See N^{ω}-Phospho-L-arginine

[8-Arginine]vasopressin, See *Vasopressin*

[8-Arginine]vasotocin, See *Vasotocin*

Argininic Acid

This α-hydroxy arginine analogue (FW = 175.19 g/mol; CAS 157-07-3), also known as α-hydroxy-δ-guanidinovaleric acid, has the following inhibitor targets: carboxypeptidase B (1-4); arginase (5,13); lysine carboxypeptidase, *or* lysine(arginine) carboxypeptidase (6); Na^+/K^+-exchanging ATPase (7,8); arginine decarboxylase, inhibited by the L-enantiomer (9,10,12); 5-guanidino-2-oxopentanoate decarboxylase, *or* α ketoarginine decarboxylase; weakly inhibited, K_i = 6.9 mM (11).   **1.** Folk & Gladner (1958) *J. Biol. Chem.* **231**, 379.   **2.** Neurath (1960) *The Enzymes*, 2nd ed., **4**, 11.   **3.** Folk (1971) *The Enzymes*, 3rd ed., **3**, 57.   **4.** Zisapel, Navon & Sokolovsky (1975) *Eur. J. Biochem.* **52**, 487.   **5.** Colleluori & Ash (2001) *Biochemistry* **40**, 9356.   **6.** Juillerat-Jeanneret, Roth & Bargetzi (1982) *Hoppe-Seyler's Z. Physiol. Chem.* **363**, 51.   **7.** da Silva, Parolo, Streck, Wajner, Wannmacher & Wyse (1999) *Brain Res.* **838**, 78.   **8.** Zugno, Stefanello, Streck, *et al.* (2003) *Int. J. Dev. Neurosci.* **21**, 183.   **9.** Blethen, Boeker & Snell (1968) *J. Biol. Chem.* **243**, 1671.   **10.** Boeker & Snell (1971) *Meth. Enzymol.* **17B**, 657.   **11.** Vanderbilt, Gaby, Rodwell & Bahler (1975) *J. Biol. Chem.* **250**, 5322.   **12.** Balbo, Patel, Sell, *et al.* (2003) *Biochemistry* **42**, 15189.   **13.** Patchett, Daniel & Morgan (1991) *Biochim. Biophys. Acta* **1077**, 291.

L-Argininosuccinate

This Urea Cycle intermediate (FW = 290.27 g/mol; CAS 2387-71-5; pK_a values: 1.62, 2.70, 4.26, 9.58, and >12), also known as (N^{ω}-L-arginino)-succinic acid and N^{ω}-(1,2-dicarboxyethyl)arginine, is a nonproteinogenic L-amino acid formed by argininosuccinate synthetase (*Reaction*: Citrulline + $MgATP^{2-}$ + L-Aspartate $\rightleftharpoons$ L-Argininosuccinate + AMP + PP_i) and is subsequently converted to arginine by argininosuccinate lyase. **Target(s):** arginyl-tRNA synthetase (1,2,9; but see ref. 2); diacylglycerol cholinephosphotransferase (3); nitric oxide synthase (4,5); adenylosuccinate synthetase (6,7); histidyl tRNA synthetase, weakly inhibited (8).   **1.** Nazario (1967) *Biochim. Biophys. Acta* **145**, 146.   **2.** Charlier & Gerlo (1976) *Eur. J. Biochem.* **70**, 137.   **3.** Choy (1993) *Biochem. J.* **289**, 727.   **4.** Hussain (1998) *Comp. Biochem. Physiol. A Mol. Integr. Physiol.* **119**, 191.   **5.** Gold, Wood, Buga, Byrns & Ignarro (1989) *Biochem. Biophys. Res. Commun.* **161**, 536.   **6.** Muirhead & Bishop (1974) *J. Biol. Chem.* **249**, 459.   **7.** Ogawa, Shiraki, Matsuda, Kakiuchi & Nakagawa (1977) *J. Biochem.* **81**, 859.   **8.** Chen & Somberg (1980) *Biochim. Biophys. Acta* **613**, 514.   **9.** Williams, Yem, McGinnis & Williams (1973) *J. Bacteriol.* **115**, 228.

L-Arginyl-L-alanylglycyl-L-phenylalanyl-L-alanyl-L-prolyl-L-phenylalanyl-L-arginine

This octapeptide (FW = 835.96 g/mol; Sequence: RAGFAPFA), also known as [des-Pro3,Ala2,6]-bradykinin, inhibits peptidyl-dipeptidase A, *or* angiotensin I-converting enzyme: (K_i = 30 nM), giving better inhibition than captopril.   **1.** Chaturvedi, Huelar, Gunthorpe, *et al.* (1993) *Pept. Res.* **6**, 308.

D-Arginyl-L-homophenylalanyl-L-arginine 2-(3,3,3-Trifluoro)ethoxymethyl Ketone

This tripeptide analogue (FW = 588.66 g/mol) inhibits plasma kallikrein, *or* PK (K_i = 54 nM, showing good selectivity for PK versus tissue kallikrein, thrombin and plasmin.   **1.** Evans, Jones, Pitt, *et al.* (1996) *Immunopharmacology* **32**, 115.

L-Arginyl-L-lysyl-L-lysyl-L-tyrosyl-L-lysyl-L-tyrosyl-L-arginyl-L-arginyl- L-lysinamide

This nonapeptide amide, also known as Peptide-18 (FW = 1324.62 g/mol; *Sequence*: RKKYKYRRK-NH$_2$), inhibits [myosin light-chain] kinase (K_i = 52 nM).   **1.** Lukas, Mirzoeva, Slomczynska & Watterson (1999) *J. Med. Chem.* **42**, 910.   **2.** Deng, Williams & Schultz (2005) *Dev. Biol.* **278**, 358.

L-Arginyl-D-tryptophanyl-*N*-methyl-L-phenylalanyl-D-tryptophanyl-L-leucyl-L-methioninamide

This hexapeptide amide (FW = 951.21 g/mol), also known as Antagonist G and [Arg6,D-Trp7,9,*N*-Me-Phe8]-Substance P fragment 6-11, is a broad spectrum neuropeptide growth factor antagonist used in the treatment of small-cell lung cancer.   **1.** Jones, Cummings, Langdon, MacLellan & Smyth (1995) *Biochem. Pharmacol.* **50**, 585.   **2.** Laufer, Wormser, Friedman, *et al.* (1985) *Proc. Natl. Acad. Sci. U.S.A.* **82**, 7444.

Argyrol

This silver-gelatin colloid (MW = undefined), prepared by a proprietary process, in which silver nitrate, sodium hydroxide, and gelatin are combined, inhibits the erosive growth of *Neisseria gonorrhoeae* introduced into a neonate's eyes during passage through an infected birth canal. The preparation was perfected and commercialized by Philadelphia physician, Albert Coombs Barnes, who used the proceeds to amass one of the finest private collections of Impressionist paintings, now comprising the world-famous Barnes Collection. **Note:** Despite claims to the contrary, some micoorganisms manage to resist silver toxicity by expressing ATP-dependent cation pumps that export Ag^+.   **1.** Bruns (1906) *Trans. Am. Ophthalmol. Soc.* **11**, 28.   **2.** Lancaster (1920) *Trans. Am. Ophthalmol. Soc.* **18**, 151.

Arietin

This cysteine-rich disintegrin (M_r = 8500) from the snake *Bitis arietans* contains an RGD loop that facilitates tight binding to integrins in a manner that inhibits platelet aggregation stimulated by ADP, thrombin, collagen, and U46619, with IC_{50} = 130-270 nM. At 65 nM, arietin blocks fibrinogen-induced aggregation of elastase-treated platelets. **1**. Huang, Wang, Teng & Ouyang (1991) *Biochim. Biophys. Acta* **1074**, 144. **2**. Huang, Wang, Teng, Liu & Ouyang (1991) *Biochim. Biophys. Acta* **1074**, 136.

Aripiprazole

This antipsychotic drug (FW = 448.39 g/mol; CAS 129722-12-9; Solubility = 90 mg/mL DMSO, <1 mg/mL H_2O), known also by its code name OPC-145977, its trade name Ambilify®, and its systematic name (4-[4-(2,3-dichlorophenyl)-1-piperazinyl]butyloxy)-3,4-dihydro-2(1*H*)-quinolinone, is a dual dopamine autoreceptor agonist and postsynaptic D_2 receptor antagonist (1). (**See Brexpiprazole**) Ambilify is indicated for the treatment of schizophrenia in individuals age 13 years and older and for bipolar disorder in individuals age 10 years and older. It is considered to be safe, effective, and well tolerated for treating both positive and negative symptoms in schizophrenia and schizoaffective disorder. Aripiprazole is also the first non-D_2 receptor antagonist with evident antipsychotic effects, representing a novel treatment for such disorders. Aripiprazole is a partial agonist of human $5\text{-}HT_{1A}$ receptor, K_i = 4.2 nM (2). Coadministration with medications that inhibit (*e.g.*, paroxetine and fluoxetine) or induce (*e.g.*, carbamazepine) CYP2D6 and CYP3A4 has the effect of increasing and decreasing plasma aripiprazole levels, respectively. (**See also** *Amisulpride; Aripiprazole; Olanzapine; Quetiapine; Remoxipride; Risperidone; Sertindole; Ziprasidone; Zotepine*) **Key Pharmacokinetic Parameters:** *See* Appendix II in Goodman & Gilman's THE PHARMACOLOGICAL BASIS OF THERAPEUTICS, 12th Edition (Brunton, Chabner & Knollmann, eds.) McGraw-Hill Medical, New York. **Other Target(s):** $5\text{-}HT_{1B}$ serotonin receptor, K_i = 830 nM; $5\text{-}HT_{1D}$ receptor, K_i = 66 nM; $5\text{-}HT_{2A}$ receptor, K_i = 8.7 nM; $5\text{-}HT_{2B}$ receptor, K_i = 0.36 nM; $5\text{-}HT_{2C}$ receptor, K_i = 22 nM, partial agonist; $5\text{-}HT_3$ receptor, K_i = 630 nM; $5\text{-}HT_{5A}$ receptor, K_i = 1200 nM; $5\text{-}HT_6$ receptor, K_i = 640 nM; $5\text{-}HT_7$ receptor, K_i = 10 nM, weak partial agonist; D_1 receptor, K_i = 1200 nM; D_2 receptor, K_i = 1.6 nM, partial agonist; D_3 receptor, K_i = 5.4 nM, partial agonist; D_4 receptor, K_i = 510 nM, partial agonist; D_5 receptor, K_i = 2100 nM; α_{1A}-adrenergic receptor, K_i = 26 nM; α_{1B}-adrenergic receptor, K_i = 34 nM; α_{2A}-adrenergic receptor, K_i = 74 nM; α_{2B}-adrenergic receptor, K_i = 100 nM; α_{2C}-adrenergic receptor, K_i = 38 nM; β_1-adrenergic receptor, K_i = 140 nM; β_2-adrenergic receptor, K_i = 160 nM; H_1 receptor, K_i = 28 nM; M_1 receptor, K_i = 6800 nM; M_2 receptor, K_i = 3500 nM; M_3 receptor, K_i = 4700 nM; M_4 receptor, K_i = 1500 nM; M_5 receptor, K_i = 2300 nM; SERT, K_i = 1100 nM; NET, K_i = 2100 nM; and DAT, K_i = 320 nM. **1**. Kikuchi, Tottori, Uwahodo, *et al.* (1995) *J. Pharmacol. Exp. Ther.* **274**, 329. **2**. Jordan, Koprivica, Chen, *et al.* (2002) *Eur. J. Pharmacol.* **441**, 137.

Aristeromycin

This adenosine analogue (FW = 265.27 g/mol CAS 19186-33-5), also known as carbocyclic adenosine and 9-[2,3-dihydroxy-4-(hydroxy methyl)cyclopentyl]adenine, inhibits purine biosynthesis as well as *S*-adenosylhomocysteine hydrolase, *or* adenosylhomocysteinase (1-5). **1**. Mizutani, Masuoka, Imoto, Kawada & Umezawa (1995) *Biochem. Biophys. Res. Commun.* **207**, 69. **2**. Chiang (1987) *Meth. Enzymol.* **143**, 377. **3**. Guranowski & Jakubowski (1987) *Meth. Enzymol.* **143**, 430. **4**. Guranowski, Montgomery, Cantoni & Chiang (1981) *Biochemistry* **20**, 110. **5**. Ishikura, Nakamura, Sugawara, *et al.* (1983) *Nucl. Acids Symp. Ser.* (12), 119.

Arjunolate (Arjunolic Acid)

This pentacyclic triterpenoid (FW = 488.70; CAS 465-00-9), also known as 2α,3β,23-trihydroxyolean-12-en-28-oic acid, from several plants, including flowers of *Campsis grandiflora*, inhibits cholesterol *O*-acyltransferase (1) and tyrosinase, the latter with an IC_{50} value of 1 μM (2). The methyl ester also inhibits tyrosinase (IC50 = 2.3 μM). **1**. Kim, Han, Chung, *et al.* (2005) *Arch. Pharm. Res.* **28**, 550. **2**. Ullah, Hussain, Hussain, *et al.* (2007) *Phytother. Res.* **21**, 1976.

Armodafinil

This vigilance-promoting, *or* eugeroic, drug (FW = 287.38 g/mol; CAS 112111-43-0), also known by its trade name Nuvigil® and its systematic name (–)-2-[(*R*)-(diphenylmethyl)sulfinyl]acetamide, is the (*R*)-enantiomer of modafinil and has FDA approval for the treatment of excessive sleepiness caused either by narcolepsy or shift-worker sleep disorder. Armodafinil is a unique dopamine uptake inhibitor that binds to the dopamine transporter (DAT), inhibiting DA uptake less potently than cocaine. *R*-modafinil has ~3x higher affinity for DAT than its enantiomer, a property thought, on the basis molecular docking studies, to arise from differences in binding modes (1). *R*-modafinil was significantly less potent in the DAT-Y156F mutant compared with wild-type DAT, whereas *S*-modafinil was less affected (1). Studies on Y335A DAT also showed that the R- and S-enantiomers tolerated the inwardly facing conformation better than cocaine, a finding that was further supported by the reactivity of [2-(trimethylammonium)ethyl]-methanethio-sulfonate reactivity with the DAT-E2C/I159C double-mutant. Both *R*- and *S*-enantiomers increase extracellular DA concentrations in the nucleus accumbens shell less efficaciously than cocaine, but with a longer duration of action (1). Loland, Mereu, Okunola, *et al.* (2012) *Biol. Psychiatry* **72**, 405.

ARN 272

This FLAT inhibitor (FW = 432.47 g/mol; CAS 488793-85-7), also named 4-[[4-(4-hydroxy-phenyl)-1-phthalazinyl]amino]-*N*-phenylbenzamide, targets the FAAH-Like Anandamide Transporter (IC_{50} = 1.8 μM), a catalytically silent variant of fatty acid amide hydrolase-1 (FAAH-1) that drives anandamide transport. **Mode of Action:** The intensity and duration of anandamide signaling appear to be controlled by a two-step elimination process, in which the anandamide or an anandamide-like substance is first internalized by neurons and astrocytes through the action of FLAT and then undergoes hydrolysis through the action of the intracellular membrane-bound amidases, FAAH-1 and FAAH-2. ARN-272 attenuates anandamide internalization, thereby reducing its deactivation. ARN272 competitively antagonizes [^{3}H]-anandamide binding to FLAT (IC$_{50}$ = 1.8 µM) and inhibits [^{3}H]-anandamide accumulation in both FLAT-expressing Hek293 cells (IC$_{50}$ ≈ 3 µM) and primary cultures of cortical neurons prepared from rats or wild-type mice. By contrast, ARN272 is without significant effect on the residual [^{3}H]-anandamide accumulation observed in cortical transport system may also facilitate the release of this lipid mediator from cells It also exhibits analgesic effects in rodent models of CB$_1$ endocannabinoid receptor-mediated nociceptive and inflammatory pain. **1.** Fu, Bottegoni, Sasso, *et al.* (2011) *Nature Neurosci.* **15**, 64.

ARN398 & ARN14988

ARN398

ARN14988

These sphingolipid-signaling inhibitors (FW$_{ARN398}$ = 307.27 g/mol; FW$_{ARN14988}$ = 373.83 g/mol) target human acid ceramidase, with IC$_{50}$ values of 7.7 and 12.8 nM, respectively, thereby increasing cellular ceramide levels and decreasing sphingosine 1-phosphate, and synergizing with several antitumor agents. Compared to normal melanocytes, ceramide production is suppressed in melanoma cells, and elevation of ceramide levels enhances the cytotoxic effects of chemotherapeutic drugs on proliferative melanoma cells in culture. ARN398 is unstable, showing a plasma half-life of 1 min and an *in vitro* half-life of 120 min at pH 4.5. ARN398 is more stable, showing a plasma half-life of 300 min and an *in vitro* half-life of 140 min at pH 4.5. **1.** Realini, Palese, Pizzirani, *et al.* (2015) *J. Biol. Chem.* **291**, 2422.

Arphamenine A

This dipeptide analogue (FW$_{di-Arphenenine-Sulfate}$ = 738.85 g/mol; CAS 144110-37-2), isolated from *Chromobacterium violaceum*, strongly inhibits aminopeptidase B. Stock solutions are stable for about one month at –20°C. **Target(s):** aminopeptidase B (1,2,9-13); leukotriene A$_4$ hydrolase aminopeptidase activity (3); dipeptide transport (4); leucyl aminopeptidase, *Leishmania* sp. and *Psoroptes* cuniculi (5,6); Xaa-Trp aminopeptidase, *or* aminopeptidase W (7); cytosol alanyl aminopeptidase (8); prolyl aminopeptidase, from *Grifola frondosa* (14); membrane alanyl aminopeptidase, aminopeptidase N (15). **1.** Foulon, Cadel & Cohen (1998) in *Handb. Proteolytic Enzymes* (Barrett, Rawlings & Woessner, eds.), p.

1026, Academic Press, San Diego. **2.** Umezawa, Aoyagi, Ohuchi, *et al.* (1983) *J. Antibiot.* **36**, 1572. **3.** Orning, Gierse & Fitzpatrick (1994) *J. Biol. Chem.* **269**, 11269. **4.** Enjoh, Hashimoto, Arai & Shimizu (1996) *Biosci. Biotechnol. Biochem.* **60**, 1893. **5.** Morty & Morehead (2002) *J. Biol. Chem.* **277**, 26057. **6.** Nisbet & Billingsley (2002) *Insect Biochem. Mol. Biol.* **32**, 1123. **7.** Tieku & Hooper (1992) *Biochem. Pharmacol.* **44**, 1725. **8.** Yamamoto, Li, Huang, Ohkubo & Nishi (1998) *Biol. Chem.* **379**, 711. **9.** Ishiura, Yamamoto, Yamamoto, ei al. (1987) *J. Biochem.* **102**, 1023. **10.** Cadel, Piesse, Gouzy-Darmon, Cohen & Foulon (2004) in *Proteases in Biology and Disease* (Hooper & Lendeckel, eds.) vol. **2**, 113, Springer. **11.** Cadel, Pierotti, Foulon, *et al.* (1995) *Mol. Cell. Endocrinol.* **110**, 149. **12.** Petrov, Fagard & Lijnen (2004) *Cardiovasc. Res.* **61**, 724. **13.** Nagata, Mizutani, Nomura, *et al.* (1991) *Enzyme* **45**, 165. **14.** Hiwatashi, Hori, Takahashi, *et al.* (2004) *Biosci. Biotechnol. Biochem.* **68**, 1395. **15.** Huang, Takahara, Kinouchi, *et al.* (1997) *J. Biochem.* **122**, 779.

ARQ 197, *See Tivantinib*

AR-R17477

This neuronal NOS-specific inhibitor (FW = 369.91 g/mol), also known as *N*-(4-(2-((3-chlorophenylmethyl)amino)ethyl)phenyl)-2-thiophenecarboxamide, inhibits neuronal nitric-oxide synthase (rat IC$_{50}$ = 35 nM). It is a weaker inhibitor of the inducible and endothelial synthases. Crystal structures of iNOS$_{oxy}$ and nNOS$_{oxy}$ complexed with AR-R17447 suggest that specificity is provided by the interaction of the chlorophenyl group with an isoform-unique substrate access channel residue (L337 in rat neuronal NOS, N115 in mouse inducible NOS). This is confirmed by biochemical analysis of site-directed mutants. AR-R 17477 also blocks some effects of phencyclidine, while having no observable behavioural effects when given alone (2). **1.** Federov, Vasan, Ghosh & Schlichting (2004) *Proc. Natl. Acad. Sci. U.S.A.* **101**, 5892. 2. Johansson, Deveney, Reif & Jackson (1999) Pharmacology & Toxicology **84**, 226.

Arrestin

This rod outer-segment protein (MW = 48-kDa) inhibits the dephosphorylation of photolyzed rhodopsin by binding preferentially to the phosphorylated protein and blocking dephosphorylation. Arrestin mediates the quenching of phototransduction. **Target(s):** G protein-coupled receptor/G protein coupling (1); protein phosphatase 2A, *or* opsin phosphatase (2); rhodopsin kinase (3). **1.** Mundell, Orsini & Benovic (2002) *Meth. Enzymol.* **343**, 600. **2.** Palczewski, McDowell, Jakes, Ingebritsen & Hargraves (1989) *J. Biol. Chem.* **264**, 15770. **3.** Doza, Minke, Chorev & Selinger (1992) *Eur. J. Biochem.* **209**, 1035.

ARRY-162, *See Binimetinib*

ARRY-142886, *See Selumetinib*

ARRY 334543

This ErbB1/2 inhibitor (FW = 466.94 g/mol; CAS 845272-21-1), also named (*R*)-*N*^{4}-(3-chloro-4-(thiazol-2-ylmethoxy)phenyl)-*N*^{6}-(4-methyl-4,5-dihydrooxazol-2-yl)quinazoline-4,6-diamine, selectively targets the intrinsic protein kinase activities of ErbB1, *or* Epidermal Growth Factor Receptor (EGFR), IC$_{50}$ = 7 nM and ErbB2, *or* Human Epidermal growth factor Receptor (HER2), IC$_{50}$ = 2 nM (1). ARRY-334543 also reverses multidrug resistance by antagonizing the activity of ATP-binding cassette subfamily G member 2 (2). **1.** Rabindran, Discafani, Rosfjord, *et al.* (2004) *Cancer Res.* **64**, 3958. **2.** Wang, Patel, Sim, *et al.* (2014) *J. Cell Biochem.* **115**, 1381.

ARRY-438162, *See* Binimetinib

p-Arsanilate (*p*-Arsanilic Acid)

This toxic arsenical (FW$_{free-acid}$ = 217.06 g/mol; CAS 202-674-3), also known as *p*-arsanilic acid, (4-aminophenyl)arsonic acid, atoxylic acid, and atoxyl, is a precursor in the chemical synthesis of many pharmaceuticals. The sodium salt is used as an anthelminitic in farm animals. The free acid is slightly soluble in cold water while the sodium salt is soluble in about six parts water. Nobelist Robert Koch reported in 1906 on the effectiveness of atoxyl on human trypanosomiasis, and Nobelist Paul Ehrlich determined the structure of atoxyl. **Target(s):** subtilisin (1,3); esterase (2,4,5); chymotrypsin (3); trypsin (3); triacylglycerol lipase (6,10,12,14-16); cholesterol esterase (7); lipoprotein lipase (8); carboxylesterase (2,4,5,9); amidase (11); aryl esterase (12); cholinesterase (13); thioesterase (19). **1.** Ottesen & Svendsen (1970) *Meth. Enzymol.* **19**, 199. **2.** Juhl & Vilmann (1990) *Scand. J. Dent. Res.* **98**, 286. **3.** Glazer (1968) *J. Biol. Chem.* **243**, 3693. **4.** Nachlas & Seligman (1949) *J. Biol. Chem.* **181**, 343. **5.** Rona & Pavlovic (1922) *Biochem. Z.* **130**, 225. **6.** Bier (1955) *Meth. Enzymol.* **1**, 627. **7.** Vahouny & Treadwell (1969) *Meth. Enzymol.* **15**, 537. **8.** Hofstee (1960) *The Enzymes*, 2nd ed., **4**, 485. **9.** Krisch (1971) *The Enzymes*, 3rd ed., **5**, 43. **10.** Itoh, Kayashima & Fujimi (1939) *J. Biochem.* **30**, 283. **11.** Krisch (1963) *Biochem. Z.* **337**, 546. **12.** Myers, Schotte, Boer & Borsje-Bakker (1955) *Biochem. J.* **61**, 521. **13.** Frommel, Herschberg & Piquet (1944) *Helv. Physiol. Pharmacol. Acta* **2**, 169. **14.** Gordon & Quastel (1948) *Biochem. J.* **42**, 337. **15.** Rona & Bach (1920) *Biochem. Z.* **111**, 166. **16.** Rona & Pavlovic (1922) *Biochem. Z.* **130**, 225. **17.** Rona & Haas (1923) *Biochem. Z.* **141**, 222. **18.** Gyotoku (1930) *Biochem. Z.* **217**, 279. **19.** Suzuoki & Suzuoki (1954) *Nature* **173**, 83.

Arsenate

This inorganic dianion HAsO$_4^{2-}$ (FW$_{dianion}$ = 139.93 g/mol) and conjugate base of arsenic acid, H$_3$AsO$_4$ (FW$_{free-acid}$ = 141.94 g/mol; CAS 7778-39-4), is an environmental hazzard that acts as a phosphate analogue in many enzyme-catalyzed reactions. The similarity of arsenate and phosphate is evident in their pK_a values: 2.25, 6.77, and 11.60 (at 18°C) for the former, and 2.15, 7.20, and 12.33 (at 25°C) for the latter. The As–O bond length and bond angle are 1.6-2.0 Å and 100° (compare to 1.5 Å and 117° for P$_i$). As such, arsenate enters most cells and compartments via phosphate transporters, albeit at somewhat lower rates. Once within, arsenate is incorporated into many metabolites, forming glucose 1-arsenate by the action of glycogen phosphorylase, glyceraldehyde 3-arsenate by glyceraldehyde 3-P dehydrogenase, and ADP-arsenate by ATP synthase. Similarly, reactions like polynucleotide phosphorylase produce arsenate esters. A key property of arsenate esters (*i.e.*, R–O–AsO$_3^{2-}$), however, is their kinetic and thermodynamic instability, readily undergoing hydrolysis in aqueous solutions. As such, arsenate forms product-like compounds that persist within an enzyme's active sites, but, once released from the enzyme's surface, rapidly hydrolyze to ROH and HAsO$_4^{2-}$. One also cannot discount the likelihood thar arsenate is an effective competitive inhibitor for many enzymes, and, given the well-known phosphate attenuation of glucose-6-P inhibition of brain hexokinase, one may also anticipate higher rates of glycolysis in the presence of arsenate. Arsenate contamination in drinking water poses serious worldwide health threats, producing multiple adverse effects in humans. Indeed, certain geographical regions have intolerably high levels of arsenate pollution. (*Note*: Arsenate is also redox-active, with an oxidation-reduction potential of +139 mV (compare to P$_i$ value of –690 mV), a property exploited by some microorganisms as a source of metabolic energy.) **Target(s):** deoxyribonuclease (11,25); glycerol-2-phosphatase (2); exopolyphosphatase (3); ribulose-bisphosphate

carboxylase (4); fatty-acid synthase (5); steroid 11β-monooxygenase (6); aspartate carbamoyltransferase (7); AMP nucleosidase (8,67); nucleoside-diphosphate kinase (9,27); trehalose-phosphatase (10,81); cholesterol esterase (11); photo-phosphorylation uncoupler, in the presence of ADP (12); triose-phosphate isomerase (13,55-57); alkaline phosphatase (14,20,22,24,28,29,32,47,95-98); luciferin, firefly (*Photinus*-luciferin 4-monooxygenase (15,120-123); putrescine carbamoyltransferase (16); flagellar motor (17); arylformamidase (18); Ca^{2+}-transporting ATPase (19); 3-phosphoglycerate kinase (21); acid phosphatase (22,23,30,33,35, 36,38,39,84-94); polynucleotide phosphorylase (26); glucose-6 phosphatase (31,45); carbonic anhydrase, *or* carbonate dehydratase (34); phosphoprotein phosphatase (37); glycerol-1-phosphatase (40); methylglyoxal synthase (41); pyrophosphatase, *or* inorganic diphosphatase (42); phosphoglucomutase (43); xylose oxidase (44); malic enzyme, NAD$^+$-dependent (decarboxylating) (46); oxidative phosphorylation (48); adenylosuccinate synthetase (49); asparagine synthetase (ADP-forming) (50); carboxy-*cis,cis*-muconate cyclase (51); phosphopentomutase (52); protein-disulfide isomerase (53); xylose isomerase (54); pectin lyase (58); xenobiotic-transporting ATPase, multidrug-resistance protein (59,60); trimetaphosphatase, slightly inhibited (61); sorbose reductase (62); cyanoalanine nitrilase (63); Arginase (64); 4-methyleneglutaminase, weakly inhibited (65); aryl-acylamidase (66); glucan 1,3-β-glucosidase (68); β-D-fucosidase (69); inulinase (70); 2.,3'-cyclic-nucleotide 3'-phosphodiesterase (71); phosphodiesterase I, *or* 5'-exonuclease, weakly inhibited (72); protein-tyrosine-phosphatase (73); 3-phosphoglycerate phosphatase (74); 4 phytase (75); 3-phytase (76,77); phosphoprotein phosphatase (78); histidinol-phosphatase (79); methylphosphothioglycerate phosphatase (80); glucose-1-phosphatase (82); 5'-nucleotidase (83); cephalosporin-C deacetylase (99); ribose-5-phosphate adenylyltransferase (100); sulfate adenylyltransferase (ADP) (101,102); ammonia kinase (103); nucleoside phosphotransferase (104); riboflavin phosphotransferase (105); pantothenate kinase (106); hexokinase (107); arylamine glucosyltransferase, weakly inhibited (108); glycogen phosphorylase (109); homocitrate synthase (110); phosphate acety-ltransferase (111-114); transketolase (115); ornithine carbamoyltransferase (116); aspartate carbamoyltransferase (117); glutamate formimidoyl-transferase, weakly inhibited (118); cyclohexanone monooxygenase (119); gentisate 1,2-dioxygenase (124); protocatechuate 3,4 dioxygenase (125); L-ascorbate peroxidase (126); *o*-aminophenol oxidase (127); glutathione:cystine transhydrogenase (128); sulfite oxidase (129,130). **1.** McDonald (1955) *Meth. Enzymol.* **2**, 437. **2.** Morton (1955) *Meth. Enzymol.* **2**, 533. **3.** Mandl & Neuberg (1955) *Meth. Enzymol.* **2**, 577. **4.** Racker (1962) *Meth. Enzymol.* **5**, 266. **5.** Lynan (1962) *Meth. Enzymol.* **5**, 443. **6.** Hayano & Dorfman (1962) *Meth. Enzymol.* **5**, 503. **7.** Jones (1962) *Meth. Enzymol.* **5**, 903. **8.** Heppel (1963) *Meth. Enzymol.* **6**, 117. **9.** Bessman (1963) *Meth. Enzymol.* **6**, 163. **10.** Friedman (1966) *Meth. Enzymol.* **8**, 372. **11.** Vahouny & Treadwell (1969) *Meth. Enzymol.* **15**, 537. **12.** Izawa & Good (1972) *Meth. Enzymol.* **24**, 355. **13.** Gracy (1975) *Meth. Enzymol.* **41**, 442. **14.** Chen & H. M. Zhou (1999) *J. Enzyme Inhib.* **14**, 251. **15.** DeLuca & McElroy (1978) *Meth. Enzymol.* **57**, 3. **16.** Stalon (1983) *Meth. Enzymol.* **94**, 339. **17.** Macnab (1986) *Meth. Enzymol.* **125**, 563. **18.** Katz, Brown & Hitchcock (1987) *Meth. Enzymol.* **142**, 225. **19.** Fassold & Hasselbach (1988) *Meth. Enzymol.* **157**, 220. **20.** Garen & Levinthal (1960) *Biochim. Biophys. Acta* **38**, 470. **21.** Axelrod & Bandurski (1953) *J. Biol. Chem.* **204**, 939. **22.** Roche (1950) *The Enzymes*, 1st ed., **1** (part 1), 473. **23.** Schmidt (1961) *The Enzymes*, 2nd ed., **5**, 37. **24.** Stadtman (1961) *The Enzymes*, 2nd ed., **5**, 55. **25.** Laskowski (1961) *The Enzymes*, 2nd ed., **5**, 123. **26.** Grunberg-Manago (1961) *The Enzymes*, 2nd ed., **5**, 257. **27.** Weaver (1962) *The Enzymes*, 2nd ed., **6**, 151. **28.** Reid & Wilson (1971) *The Enzymes*, 3rd ed., **4**, 373. **29.** Fernley (1971) *The Enzymes*, 3rd ed., **4**, 417. **30.** Hollander (1971) *The Enzymes*, 3rd ed., **4**, 449. **31.** Nordlie (1971) *The Enzymes*, 3rd ed., **4**, 543. **32.** Gerlt (1992) *The Enzymes*, 3rd ed., **20**, 95. **33.** London, McHugh & Hudson (1958) *Arch. Biochem. Biophys.* **73**, 72. **34.** Vullo, Franchi, Gallori, *et al.* (2003) *J. Enzyme Inhib. Med. Chem.* **18**, 403. **35.** Schlosnagle, Bazer, Tsibris & Roberts (1974) *J. Biol. Chem.* **249**, 7574. **36.** Bazer, Chen, Knight, *et al.* (1975) *J. Anim. Sci.* **41**, 1112. **37.** Roberts & Bazer (1976) *Biochem. Biophys. Res. Commun.* **68**, 450. **38.** Ketcham, Baumbach, Bazer & Roberts (1985) *J. Biol. Chem.* **260**, 5768. **39.** Allen, Nuttleman, Ketcham & Roberts (1989) *J. Bone Miner. Res.* **4**, 47. **40.** Sussman & Avron (1981) *Biochim. Biophys. Acta* **661**, 199. **41.** Hopper & Cooper (1972) *Biochem. J.* **128**, 321. **42.** Naganna & Menon (1948) *J. Biol. Chem.* **174**, 501. **43.** Jagannathan & Luck (1949) *J. Biol. Chem.* **179**, 561. **44.** Lyr (1962) *Enzymologia* **24**, 69. **45.** Swanson (1950) *J. Biol. Chem.* **184**, 647. **46.** Korkes, del Campillo &

Ochoa (1950) *J. Biol. Chem.* **187**, 891. **47.** Scutt & Moss (1968) *Enzymologia* **35**, 157. **48.** Crane & Lipmann (1953) *J. Biol. Chem.* **201**, 235. **49.** Muirhead & Bishop (1974) *J. Biol. Chem.* **249**, 459. **50.** Nair (1969) *Arch. Biochem. Biophys.* **133**, 208. **51.** Thatcher & Cain (1975) *Eur. J. Biochem.* **56**, 193. **52.** Barsky & Hoffee (1983) *Biochim. Biophys. Acta* **743**, 162. **53.** De Azevedo, Brockway, Freedman, Greenwell & Roden (1984) *Biochem. Soc. Trans.* **12**, 1043. **54.** Vartak, Srinivasan, Powar, Rele & Khire (1984) *Biotechnol. Lett.* **6**, 493. **55.** Xiang, Jung & Sampson (2004) *Biochemistry* **43**, 11436. **56.** Kurzok & Feierabend (1984) *Biochim. Biophys. Acta* **788**, 214. **57.** Lambeir, Opperdoes & Wierenga (1987) *Eur. J. Biochem.* **168**, 69. **58.** Afifi, Fawzi & Foaad (2002) *Ann. Microbiol.* **52**, 287. **59.** Paul, Breuninger & Kruh (1996) *Biochemistry* **35**, 14003. **60.** Paul, Breuninger, Tew, Shen & Kruh (1996) *Proc. Natl. Acad. Sci. U.S.A.* **93**, 6929. **61.** Meyerhof, Shatas & Kaplan (1953) *Biochim. Biophys. Acta* **12**, 121. **62.** Sugisawa, Hoshino & Fujiwara (1991) *Agric. Biol. Chem.* **55**, 2043. **63.** Yanase, Sakai & Tonomura (1983) *Agric. Biol. Chem.* **47**, 473. **64.** Nakamura, Fujita & Kimura (1973) *Agric. Biol. Chem.* **37**, 2827. **65.** Ibrahim, Lea & Fowden (1984) *Phytochemistry* **23**, 1545. **66.** Engelhardt, Wallnöfer & Plapp (1973) *Appl. Microbiol.* **26**, 709. **67.** Hurwitz, Heppel & Horecker (1957) *J. Biol. Chem.* **226**, 525. **68.** Nagasaki, Saito & Yamamoto (1977) *Agric. Biol. Chem.* **41**, 493. **69.** Giordani & Noat (1988) *Eur. J. Biochem.* **175**, 619. **70.** Hamdy (2002) *Indian J. Exp. Biol.* **40**, 1393. **71.** Sims & Carnegie (1978) *Adv. Neurochem.* **3**, 1. **72.** Udvardy, Marre & Farkas (1970) *Biochim. Biophys. Acta* **206**, 392. **73.** Zhou, Bhattacharjee & Mukhopadhyay (2006) *Mol. Biochem. Parasitol.* **148**, 161. **74.** Randall & Tolbert (1971) *J. Biol. Chem.* **246**, 5510. **75.** Mahajan & Dua (1997) *J. Agric. Food Chem.* **45**, 2504. **76.** Youssef, Ghareib & Nour el Dein (1987) *Zentralbl. Mikrobiol.* **142**, 397. **77.** Ghareib, Youssef & Nour el Dein (1988) *Zentralbl. Mikrobiol.* **143**, 397. **78.** Zhou, Clemens, Hakes, Barford & Dixon (1993) *J. Biol. Chem.* **268**, 17754. **79.** Brady & Houston (1972) *Anal. Chem.* **48**, 480. **80.** Black & Wright (1956) *J. Biol. Chem.* **221**, 171. **81.** Friedman (1960) *Arch. Biochem. Biophys.* **88**, 339. **82.** Faulkner (1955) *Biochem. J.* **60**, 590. **83.** Ikura & Horikoshi (1989) *Agric. Biol. Chem.* **53**, 645. **84.** Shimada, Shinmyo & Enatsu (1977) *Biochim. Biophys. Acta* **480**, 417. **85.** Sugiura, Kawabe, Tanaka, Fujimoto & Ohara (1981) *J. Biol. Chem.* **256**, 10664. **86.** Lawrence & van Etten (1981) *Arch. Biochem. Biophys.* **206**, 122. **87.** Andrews & Pallavicini (1973) *Biochim. Biophys. Acta* **321**, 197. **88.** Reilly, Baron, Nano & Kuhlenschmidt (1996) *J. Biol. Chem.* **271**, 10973. **89.** Giordani, Nari, Noat & Sauve (1986) *Plant Sci.* **43**, 207. **90.** Odds & Hierholzer (1973) *J. Bacteriol.* **114**, 257. **91.** Turner & Plaxton (2001) *Planta* **214**, 243. **92.** Dibenedetto (1972) *Biochim. Biophys. Acta* **286**, 363. **93.** Tominaga & Mori (1974) *J. Biochem.* **76**, 397. **94.** Kawabe, Sugiura, Terauchi & Tanaka (1984) *Biochim. Biophys. Acta* **784**, 81. **95.** Sakurai, Toda & Shiota (1981) *Agric. Biol. Chem.* **45**, 1959. **96.** Fernandez, Gascon & Schwenke (1981) *Curr. Microbiol.* **6**, 121. **97.** Adler (1978) *Biochim. Biophys. Acta* **522**, 113. **98.** Harkness (1968) *Arch. Biochem. Biophys.* **126**, 513. **99.** Cardoza, Velasco, Martin & Liras (2000) *Appl. Microbiol. Biotechnol.* **54**, 406. **100.** Stern & Avron (1966) *Biochim. Biophys. Acta* **118**, 577. **101.** Adams & Nicholas (1972) *Biochem. J.* **128**, 647. **102.** Grunberg-Manago, Del Campillo-Campbell, Dondon & Michelson (1966) *Biochim. Biophys. Acta* **123**, 1. **103.** Dowler & Nakada (1968) *J. Biol. Chem.* **243**, 1434. **104.** Billich & Witzel (1986) *Biol. Chem. Hoppe-Seyler* **367**, 291. **105.** Katagiri, Yamada & Imai (1959) *J. Biochem.* **46**, 1119. **106.** Shimizu, Kubo, Tani & Ogata (1973) *Agric. Biol. Chem.* **37**, 2863. **107.** White & J. E. Wilson (1989) *Arch. Biochem. Biophys.* **274**, 375. **108.** Frear (1968) *Phytochemistry* **7**, 381. **109.** Boeck & Schinzel (1996) *Eur. J. Biochem.* **239**, 150. **110.** Gray & Bhattacharjee (1976) *Can. J. Microbiol.* **22**, 1664. **111.** Lundie & Ferry (1989) *J. Biol. Chem.* **264**, 18392. **112.** Henkin & Abeles (1976) *Biochemistry* **15**, 3472. **113.** Preston, Zeiher, Wall & Emerich (1989) *Appl. Environ. Microbiol.* **55**, 165. **114.** Robinson & Sagers (1972) *J. Bacteriol.* **112**, 465. **115.** Klein & Brand (1977) *Hoppe-Seyler's Z. Physiol. Chem.* **358**, 1325. **116.** Legrain & Stalon (1976) *Eur. J. Biochem.* **63**, 289. **117.** Lowenstein & Cohen (1956) *J. Biol. Chem.* **220**, 57. **118.** Miller & Waelsch (1957) *J. Biol. Chem.* **228**, 397. **119.** Trower, Buckland & Griffin (1989) *Eur. J. Biochem.* **181**, 199. **120.** Leach (1981) *J. Appl. Biochem.* **3**, 473. **121.** Webster, Chang, Manley, Spivey & Leach (1980) *Anal. Biochem.* **106**, 7. **122.** DeLuca, Wannlund & McElroy (1979) *Anal. Biochem.* **95**, 194. **123.** Leach & Webster (1986) *Meth. Enzymol.* **133**, 51. **124.** Suemori, Kurane & Tomizuka (1993) *Biosci. Biotechnol. Biochem.* **57**, 1781. **125.** Kurane, Ara, Nakamura, Suzuki & Fukuoka (1984) *Agric. Biol. Chem.* **48**, 2105. **126.** Sharma & Dubey (2004) *Plant Sci.* **167**, 541. **127.** Nair & Vining (1965) *Biochim. Biophys. Acta* **96**, 318. **128.** Minoda, Kurane & Yamada (1973) *Agric. Biol. Chem.* **37**, 2511. **129.** George, Garrett, Graf, Prince & Rajagopalan (1998) *J.*

Amer. Chem. Soc. **120**, 4522. **130.** D'Errico, Di Salle, La Cara, Rossi & Cannio (2006) *J. Bacteriol.* **188**, 694.

Arsenic Trioxide

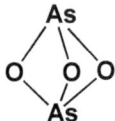

This toxic inorganic oxide As_2O_3 (FW = 197.84 g/mol; CAS 1327-53-3; Solubility = 37 g/L H_2O at 20ºC and 115 g/L H_2O at 100ºC), a commercially useful precursor employed in the synthesis of other arsenic compounds, including organoarsenic compounds, has enjoyed great interest in view of its ability to induce clinical remission in patients with acute promyelocytic leukemia (APL), particularly those who have relapsed after retinoic acid treatment. Although no longer produced in the U.S., arsenic trioxide has been employed as a herbicide, pesticide, and rodenticide. **Mechanism of As_2O_3 Action:** APL is characterized by a specific reciprocal chromosome translocation t(15;17) that generates the *PML-RARα* fusion gene and expression of its oncoprotein PML-RARα. Upon dimerization, (PML-RARα)₂ binds the histone-deacetylase recruiting co-repressor complex with higher affinity than the wild-type RARα. Fortunately, that co-repressor complex can be dissociated by pharmacological doses of all-*trans* retinoic acid (ATRA), providing complete remission for roughly 90% of APL patients, especially when combined with chemotherapy. For the remaining 10%, who are refractory to ATRA treatment, relief has been achieved through arsenic trioxide-induced proteolysis of PML-RARα (1,2). For such purposes, arsenic trioxide is marketed under the trade name Trisenox®. **As_2O_3-Mediated Reactivation of Methylation-Silenced Genes in Human Cancer:** DNA (cytosine-5)-methyltransferases help establish and regulate tissue-specific DNA methylation patterns, thereby maintaining tissue integrity and preventing both tumorigenesis and developmental abnormalities. In cancer, tumor suppressor genes are silenced by oncomethylation, and reactivation of them occurs when 5-azacytidine and 5-aza-2'-deoxycytidine selectively and rapidly induce proteasome-mediated degradation of maintenance DNA methyltransferase-1. Arsenic trioxide inhibits DNA methyltransferase and restores methylation-silenced genes in human liver cancer cells. Upon treatment of HepG2 and Huh-7 cells with 2–10 µM As_2O_3 for 24, 48, and 72 hours, arsenic trioxide decreased DNMT1 mRNA expression of DNMT1 and dose-dependently inhibited catalytic activity of DNMT1 (3). **As_2O_3 Toxicity:** The considerable cytotoxicity of arsenic trioxide appears to be linked to its ability to induce oxidative stress, reducing glutathione stores and generating reactive oxygen species, the latter evidenced by increased malondialdehyde, a marker of lipid peroxidation. As_2O_3 also significantly increases the levels of antioxidant enzymes, among them superoxide dismutase and catalase. Arsenic trioxide-exposed cells show increased levels of reactive oxygen species, lipid peroxidation, suggesting that oxidative stress might be the primary mechanism for arsenic trioxide toxicity in humans (4). **Chemical Properties:** Arsenic trioxide undergoes slow hydrolysis to form As(OH)₃, a pyramidal molecule consisting of three hydroxyl groups bonded sp^2 to arsenic. In humans, As_2O_3 is metabolized to arsenate, mono-methylarsonate and di-methylarsinate, the latter two formed by the action of liver methyltransferases. **Caution:** Care should be exercised in handling all arsenicals, including use of face mask to avoid airborne particulates and chemically resistant rubber gloves with liquids. (In 1858, an estimated 200 people in Bradford, England were poisoned, with twenty succumbing, when arsenic trioxide was accidentally added to some candies. That incident led to the British Pharmacy Act of 1868, recognizing the druggist as the legal custodian of all listed poisons.) **Target(s):** Acetyl-CoA acetyltransferase (5); acetyl-CoA acyltransferase (6); pyruvate dehydrogenase (7); cystathionine γ-lyase, *or* cysteine desulfhydrase (8); catalase (9,10); thiaminase (11); IκB kinase (12); formaldehyde transketolase (13); 2-hydroxypyridine 5-monooxygenase (14); iodide peroxidase, *or* thyroid peroxidase (15). **1.** Gambacorti-Passerini, Grignani, *et al.* (1993) *Blood* **81**, 1369. **2.** Jing (2004) *Leuk. Lymphoma* **45**, 639. **3.** Cui, Wakai, Shirai, *et al.* (2006) **37**, 298. **4.** Alarifi, Ali, Alkahtani, Siddqui & Ali (2013) *Onco Targets Ther.* **6**, 75. **5.** Hartmann & Lynen (1961) *The Enzymes*, 2nd ed., **5**, 381. **6.** Gehring & Lynen (1972) *The Enzymes*, 3rd ed., **7**, 391. **7.** Samikkannu, Chen, Yih, *et al.* (2003) *Chem. Res. Toxicol.* **16**, 409. **8.** Lawrence & Smythe (1943) *Arch. Biochem.* **2**. 225. **9.** Harris & Creighton (1915) *J. Biol. Chem.* **22**, 535. **10.** Senter (1903) *Zeit. für physikol. Chem.* **44**, 257. **11.** Ikehata (1960) *J. Gen. Appl. Microbiol.* **6**, 30.

12. Luo, Kamata & Karin (2005) *J. Clin. Invest.* **115**, 2625. **13**. Kato, Higuchi, Sakazawa, *et al.* (1982) *Biochim. Biophys. Acta* **715**, 143. **14**. Sharma, Kaul & Shukla (1984) *Biol. Mem.* **9**, 43. **15**. Palazzolo & Jansen (2008) *Biol. Trace Elem. Res.* **126**, 49.

Arsphenamine

This organoarsenical (FW$_{disodium-salt}$ = 458.04 g/mol; CAS 1936-28-3; IUPAC Name: 4,4'-(1,2-diarsenediyl)bis[2-aminophenol]), more famously as "Ehrlich's 606" or Salvarsan, was found to be highly effective against syphilis, culminating a determined search by Nobelist Paul Ehrlich for a "magic bullet" against the spirochete recently discovered by Schaudinn and Hoffmann. Ehrlich had his associate, Sahachiro Hata, noted the effectiveness of what was the 606th drug to be tested on syphilis-infected rabbits. Mass spectral analysis (1) later demonstrated that the actual structure is almost certainly a mixture of the cyclic trimer and pentamer, with no evidence of an As=As double bond. Salvarsan is likely to be a slow-release formulation, giving rise to the bioactive species R–As(OH)$_2$ (1). While other agents (notably, penicillin) have long-since supplanted Salvarsan in the treatment of syphilis, its discovery remains a landmark in the history of chemotherapy, one that led to the development of many other antiprotozoal and antibacterial agents (2). **1**. Lloyd, Morgan, Nicholson & Ronimus (2005) *Angew. Chem. Int. Ed. Engl.* **44**, 941. **2**. Purich (2010) *Enzyme Kinetics: Catalysis & Control*, pp. 486-487, Academic Press-Elsevier, New York.

Artemisinin

This first-line antimalarial (FW = 282.34 g/mol; CAS 63968-64-9), also known as quinhaosu, is a endoperoxy sesquiterpene lactone that is active at nanomolar concentrations *in vitro* against both chloroquine-sensitive and resistant strains of *Plasmodium falciparum*. It was isolated from the ancient medicinal herb *Artemisia annua* (or *Qinghao*) by Nobelist Youyou Tu (屠呦呦). The principal advantage of artemisinin over antimalarial drugs is that they accelerate parasite clearance by clearing young, circulating, ring-stage parasites, thereby preventing further maturation and parasite sequestration (1,2). Various mechanisms for its toxic effects on falciparum have been proposed, including heme alkylation and/or inactivation (3). In areas of resistance to mefloquine (the second-line treatment for chloroquine-sensitive or resistant malaria), combination therapy with artemisinin or its derivatives appears to improve sustained parasite clearance compared with either drug alone (4). **Primary Mode of Action:** Artemisinin brings a rapid and dramatic increase reactive oxygen species (ROS) in yeast and malarial mitochondria, but not in mammalian mitochondria (5). The presence of ROS scavengers ameliorates artemisinin's effects. Lacking a crucial endoperoxide bridge, deoxy-artemisinin is without effect on malarial mitochondrial membrane potential of ROS production. A distantly related antimalarial endoperoxide, OZ209, also causes ROS production and depolarizes isolated malarial mitochondria. Interference of the mitochondrial electron transport system (ETS) by desferrioxamine also alters the artemisinin sensitivity of *P. falciparum* and reduces the extent of artemisinin-induced ROS production (5). Such results strongly implicate the parasite mitochondrion is the major, if not the sole, target for artemisinin's therapeutic action on *P. falciparum*. **Artemisinin Resistance:** Various models have been advanced to explain how malarial organisms gain resistance to antimalarial drugs. One idea is that one or more of these drugs is inactivated by the cytochrome P450 (or CYP) system. What seems more likely is that ATP-binding cassette (ABC)-type transporters confer multi-drug resistance (MDR) through their ability to

actively export such antimalarials as chloroquine, mefloquine, and artemisinin. Resistance is, in fact, marked by increased copy numbers of the *P. falciparum* (Pf) multidrug-resistance gene *PfMDR1*. Resistance is manifested as slow *in vivo* clearance of the parasite without evidence of reduction in conventional *in vitro* susceptibility tests (6). **Global Plan for Containing Artemensin Resistance:** Artemisinin-based combination therapies (ACTs) represent the most potent weapon in treating malaria arising from *P. falciparum*. Given that pockets of resistance to artemisinin have emerged along the Cambodia-Thailand border, the World Health Organization (WHO) and Roll Back Malaria (RBM) have advanced a five-step action plan: (*a*) stop the spread of resistant parasites; (*b*) increase monitoring and surveillance for artemisinin resistance; (*c*) improve access to malaria diagnostic testing and rational treatment with ACTs; (*d*) invest in artemisinin resistance-related research; and (*e*) motivate action and mobilize resources. The success of this global plan will depend on a well-coordinated and adequately funded response from many stakeholders acting globally, regionally, nationally, and locally. **Artemisinin Derivatives:** Unlike most other anti-malarials, this conformationally constrained and highly oxygenated sesquiterpene lactone peroxide lacks nitrogen-containing heterocyclic rings. Even so, artemisinin is only sparingly soluble in water or oil as well as poor GI absorption – features emphasizing the opportunity for improvement. Artemisinin has proved to be a remarkably versatile scaffold for the semi-synthesis of many pharmacologically active derivatives. Chaturvedi, *et al.* (7) present a thorough review of the synthesis, chemical and physical properties, and structure–activity relationships of artemisinin derivatives for use as anti-malarial, immunosuppressant, and anti-cancer agents. **Antineoplastic Properties:** Artemisinin has selective anticancer activities in various types of cancers, both *in vitro* and *in vivo*. Artemisinin derivatives induce iron-dependent cell death (ferroptosis) in tumor cells (12). In human cervical cancer cells, artemisinin induces apoptosis and represses cell proliferation by inhibiting the telomerase subunit ERα and downstream components like VEGF (13). **Target(s):** dihydroorotate dehydrogenase (8); glutathione *S*-transferase (9,10); K$^+$ current (11). **1**. Kuile, White, Holloway, Pasvol & Krishna (1993) *Exp. Parasitol.* **76**, 85. **2**. Udomsangpetch, Pipitaporn, Krishna, *et al.* (1996) *J. Infect. Dis.* **173**, 691. **3**. Robert, Dechy-Cabaret, Cazelles & Meunier (2002) *Acc. Chem. Res.* **35**, 167. **4**. McIntosh & Olliaro (2000) *Cochrane Database Syst Rev.* CD000256. **5**. Wang, Huang, Li, *et al.* (2010) *PLoS One* **5**, e9582. **6**. Dondorp, Nosten, Yi, *et al.* (2009) *New Engl. J. Med.* **361**, 4557. **7**. Chaturvedi, Goswami, Pratim Saikia, Barua & Rao (2010) *Chem. Soc. Rev.* (2010) **39**, 435. **8**. Ittarat, Asawamahasakda & Meshnick (1994) *Exp. Parasitol.* **79**, 507. **9**. Mukanganyama, Widersten, Naik, Mannervik & Hasler (2002) *Int. J. Cancer* **97**, 700. **10**. Srivastava, Puri, Kamboj & Pandey (1999) *Trop. Med. Int. Health* **4**, 251. **11**. Yang, Luo, Bao, Zhang & Wang (1998) *Zhongguo Yao Li Xue Bao* **19**, 269. **12**. Ooko, Saeed, Kadioglu, *et al.* (2015) *Phytomedicine* **22**, 1045. **13**. Mondal & Chatterji (2015) *J. Cell Biochem.* **116**, 1968.

Arzanol

This phloroglucinol α-pyrone (FW = 402.44 g/mol; CAS 32274-52-5), also known as 3-[[3-acetyl-2,4,6-trihydroxy-5-(3-methyl-2-buten-1-yl)phenyl] methyl]-6-ethyl-4-hydroxy-5-methyl- 2*H*-pyran-2-one from a Mediterranean plant *Helichrysum italicum* (ssp. *Microphyllum*; family Asteraceae) exhibits anti-inflammatory, anti-HIV, and antioxidant activities (1). Arzanol also protects linoleic acid against free-radical attack in assays of autoxidation and EDTA-mediated oxidation (1). At noncytotoxic concentrations, arzanol also showed a strong inhibition of TBH-induced oxidative stress in VERO cells. It also protects low density lipoprotein (LDL) against the oxidative modification induced by Cu^{2+}, inhibiting the increase of oxidative products, such as conjugated dienes, fatty acid hydroperoxides, 7β-hydroxycholesterol, and 7-ketocholesterol (2). Moreover, its cell-penetrating properties commend its use as a lipid-protecting antioxidant (2). Arzanol inhibited 5-lipoxygenase and related leukotriene formation in neutrophils, as well as the activity of cyclooxygenase-1 (COX-1) and formation of COX-2-derived prostaglandin

PG-E$_2$ *in vitro*, IC$_{50}$ = 2-9µM (3). **1.** Rosa, Deiana, Atzeri, *et al.* (2007) *Chem. Biol. Interact.* **165**, 117. **2.** Rosa, Pollastro, Atzeri, *et al.* (2011) *Chem. Phys. Lipids* **164**, 24. **3.** Bauer, Koeberle, Dehm, *et al.* (2011) *Biochem. Pharmacol.* **81**, 259.

AS252424

This potent and selective PI3K inhibitor (FW = 305.28 g/mol; CAS 900515-16-4), also known as (5*Z*)-5-[[5-(4-fluoro-2-hydroxyphenyl)-2-furanyl]methylene]-2,4-thiazolidinedione, targets human recombinant PI3Kα (IC$_{50}$ = 30 nM), with far weaker inhibition of PI3Kβ (IC$_{50}$ = 940 nM), PI3Kγ (IC$_{50}$ = 20,000 nM), and PI3Kδ (IC$_{50}$ = 20,000 nM). These intracellular signal-transducing enzymes phosphorylate the position-3 hydroxyl in phosphatidylinositol (PtdIns). Oral administration of AS252424 in a mouse model for acute peritonitis led to a significant reduction of leukocyte recruitment. **1.** Pomel, Klicic, Covini, *et al.* (2006) *J. Med. Chem.* **49**, 3857.

AS703569, See *MSC1992371A1*

AS703026

This novel and orally bioavailable MEK1/2 inhibitor (FW = 431.20 g/mol; CAS 1236699-92-5), also known as (S)-*N*-(2,3-dihydroxypropyl)-3-((2-fluoro-4-iodophenyl)amino)-isonicotinamide, selectively targets MEK1/2 (noncompetitive, with IC$_{50}$ values of 5-11 nM). AS703026 reduces the growth and survival of multiple myeloma (MM) cells. It also inhibits cytokine-induced osteoclast differentiation about 10x more potently than AZD6244. Inhibition of proliferation by AS703026 leads to G$_0$-G$_1$ cell cycle arrest, attended by reduction in MAF oncogene expression. AS703026 also induces apoptosis (via Caspase-3) as well as poly(ADP-ribose) polymerase (PARP) cleavage in MM cells, whether in the absence or presence of bone marrow stromal cells (BMSCs). Importantly, AS703026 also sensitizes MM cells to other anti-MM therapies employing dexamethasone, melphalan, lenalidomide, perifosine, bortezomib, or rapamycin. Growth reduction in mice bearing H929 MM xenograft tumors correlates with down-regulated pERK1/2, induced PARP cleavage, and decreased microvasculature. BMSC-induced viability of MM patient cells was similarly blocked within the same dose range. **1.** Kim, Kong, Fulciniti, *et al.* (2010) *Brit. J. Hematol.* **149**, 537.

Ascaris Aspartic Proteinase Inhibitor Protein
This inhibitory protein (MW ~ 17000) from a pig intestinal parasitic worm is highly specific (K$_i$ of 0.5 nM) for pig pepsin, and reacts more weakly with other proteases. **Target(s):** pepsin A, human (1,8); gastricsin, human (1); pepsin, pig (1,3-10); gastricsin, pig (1,6); cathepsin E (2-4,7,11,12); pepsin A, chicken (6). **1.** Valler, Kay, Aoyagi & Dunn (1985) *J. Enzyme Inhib.* **1**, 77. **2.** Kageyama (1995) *Meth. Enzymol.* **248**, 120. **3.** Keilova & Tomasek (1977) *Acta Biol. Med. Ger.* **36**, 1873. **4.** Keilova & Tomasek (1972) *Biochim. Biophys. Acta* **284**, 461. **5.** Peanasky & Abu-Erreish (1971) in *Proc. Int. Res. Conf. Proteinase Inhib., Munich 1970* (Fritz & Tchesche, eds.), p. 281, Walter de Gruyter, Berlin. **6.** Kay, Valler & Dunn (1983) in *Proteinase Inhibitors: Medical and Biological Aspects* (Katunema, Holzer & Umezawa, eds.), pp. 201-210, Springer-Verlag, Berlin. **7.** Jupp, Richards, Kay, *et al.* (1988) *Biochem. J.* **254**, 895. **8.** Petersen, Chernaia, Rao-Naik, *et al.* (1998) *Adv. Exp. Med. Biol.* **436**, 391. **9.** Ng, Petersen, Cherney, *et al.* (2000) *Nat. Struct. Biol.* **7**, 653. **10.** Bhatt & Dunn (2000) *Bioorg. Chem.* **28**, 374. **11.** Hill, Montgomery & Kay (1993) *FEBS Lett.* **326**, 101. **12.** Liu, Tsukuba, Okamoto, Ohishi & Yamamoto (2002) *J. Biochem.* **132**, 493.

Ascomycin

This antifungal antibiotic (FW = 792.02 g/mol; CAS 104987-12-4; Solubility: 50 mM in DMSO), also known as FK520, isolated from *Streptomyces hygroscopicus*, is also a powerful immunosuppressant. Note that ascomycin is a structural analogue of tacrolimus (*or* FK506) and, as such, operates by a similar mechanism of action (*See Tacrolimus*). Ascomycin inhibits allogenic T-lymphocyte proliferation and calcineurin phosphatase activity. **1.** Petros, Kawai, Luly & Fesik (1992) *FEBS Lett.* **308**, 309. **2.** Ivery & Weiler (1997) *Bioorg. Med. Chem.* **5**, 217. **3.** Salowe & Hermes (1998) *Arch. Biochem. Biophys.* **355**, 165.

Ascorbate (Ascorbic Acid)

ascorbic acid   ascorbate

This water-soluble vitamin and anti-Scurvy factor (FW$_{free-acid}$ = 176.13 g/mol; CAS 50-81-7), also known as vitamin C, is structurally the γ-lactone of 2-oxo-L-gulonic acid. It was discovered as a major constituent of Hungarian paprika, affording Nobel Laureate Albert Szent-Györgyi the means for its isolation (1) and subsequent chemical and physiologic analysis. One hydroxyl group of the ene-diol is more acidic than acetic acid (*i.e.,* pK$_a$ of 4.17 for ascorbic acid versus pK$_a$ of 4.76 for acetic acid). Its second pK$_a$ is 11.6, with a redox potential ($E^{\circ\prime}$) of 0.058 V at pH 7. With its reducing and chelating properties, ascorbic acid is also an efficient enhancer of non-heme iron absorption. It protects unsaturated fatty acids (and their esters) from peroxidation and converts lipid peroxyl radicals and alkoxyl radical into lipid hydroperoxides and lipid hydroxides, respectively. **Roles as an Enzyme Cofactor:** Humans cannot synthesize ascorbate, because we lack the enzyme catalyzing the last step of its biosynthesis (*i.e.,* oxidation of l-gulonolactone to L-ascorbic acid by l-gulonolactone oxidase, an endoplasmic reticulum enzyme). Ascorbate is also an electron donor for numerous enzymes and thereby plays an important role in a wide variety of biochemical processes, including collagen formation, norepinephrine biosynthesis, and mitochondrial fatty acid transport. As an essential cofactor in reactions catalyzed by Cu$^+$-dependent monooxygenases and Fe^{2+}-dependent dioxygenases, vitamin C is required by γ-butyrobetaine dioxygenase, procollagen-proline dioxygenase, pyrimidine-deoxynucleoside 2'-dioxygenase, procollagen-lysine 5-dioxygenase, thymine dioxygenase, procollagen-proline 3-dioxygenase, trimethyllysine dioxygenase, flavanone 3-dioxygenase, pyrimidine-deoxynucleoside 1'-dioxygenase, hyoscyamine (6S)-dioxygenase, 6β-hydroxyhyoscyamine epoxidase, gibberellin 3β-dioxygenase, taurine dioxygenase, phytanoyl CoA dioxygenase, leucocyanidin oxygenase, desacetoxyvindoline 4-hydroxylase, flavone synthase, and 2-acetolactate mutase. Ascorbic acid is a required Fe(III)-reducing cofactor for prolyl hydroxylase and lysyl hydroxylase, the enzymes responsible for the hydroxylation of proline and lysine residues in tropocollagen. **Effects on the Efficacy of Anticancer Agents:** Pharmacologic concentrations of intracellular vitamin C antagonize the therapeutic cytotoxic effects of antineoplastic chemotherapeutic agents (2). Given the current widespread use of vitamin C as a nutritional supplement, one may anticpate adverse consequences in patients who are receiving cancer chemotherapy. Vitamin C enters the mitochondria where it may play a role in quenching local reactive oxygen species (ROS). Alternatively, because some tissues cannot absorb ascorbate, but do concentrate dehydroascorbate (DHA), the latter may act as an oxidant, paradoxically increasing the level of ROS in those cells that take up DHA. As the fully oxidized form of vitamin C, dehydroascorbate is reduced spontaneously by glutathione as well as enzymatically in reactions requiring glutathione or NADPH. (*See also Dehydroascorbate for additional details*) **Target(s):** steroid 11β-monooxygenase (3,117); phosphodiesterase I (4); steroid 21 monooxygenase, *or* CYP21A1 (5,44,99,100); *o*-aminophenol oxidase (6); ecarin (7); catalase (8,16,37,38,58); 6 phosphofructokinase (9); lactate dehydrogenase (9); adenylate kinase (9,30,86); β-amylase (10,36,47); papain (11,42); carboxylesterase (12); lipoxygenase (13); urate oxidase, *or* uricase (14); neuraminidase, *or* sialidase (15); GTP cyclohydrolase I (20); hyaluronoglucosaminidase, *or* hyaluronidase (22,34); hyaluronate lyase (22,49); tyrosine 3-monooxygenase (23,25); thyroxine deiodinase, *or* iodothyronine deiodinase (24); dihydrodiol dehydrogenase (2,24); alkaline phosphatase (26); L-galactose dehydrogenase (27); 3-hydroxy-3-methylglutaryl-CoA reductase (29); β-glucuronidase (30); lactoylglutathione lyase (31); NADH oxidase (33,39); pyrophosphatase, *or* inorganic diphosphatase (39); catechol oxidase (40); dipeptidyl-peptidase I, *or* cathepsin C (45); xanthine oxidase (46); GDP-mannose 3,5-epimerase, mildly inhibited (48); isocitrate lyase, in the presence of Fe^{2+} (50); trimethylamine oxide aldolase (51); fructose-1,6-bisphosphate aldolase (52); nucleotide diphosphatase (53); arylacetonitrilase (54); 3-β-hydroxy-5α-steroid dehydrogenase (55); acid phosphatase (56,81); alkaline phosphatase, in the presence of copper (56); 1,5-anhydro-D-fructose reductase (57); arylformamidase (59); cathepsin D (60); glutamate carboxypeptidase (61); thioglucosidase, *or* myrosinase, an activator for several thioglucosidases (62-64); levanase (65,66); β-*N*-acetylhexosaminidase, in the presence of Cu^{2+} (67); β-glucuronidase (68,69); β-fructofuranosidase, *or* invertase (70); α-galactosidase (71); β-glucosidase (72); dextranase (73); cerebroside-sulfatase, *or* arylsulfatase A (73,75); glycerophosphocholine cholinephosphodiesterase (76); phosphodiesterase I, *or* 5'-exonuclease; weakly inhibited (77); 4-phytase (78); 5'-nucleotidase (79); acid phosphatase (80-82); *S*-formylglutathione hydrolase (83); acetyl-CoA hydrolase (84); lipase, *or* triacylglycerol lipase (85); alanine aminotransferase (87); aspartate aminotransferase (88); glutathione *S*-transferase (89); glutaminyl-peptide cyclotransferase (90); naringenin-chalcone synthase (91); 1-alkylglycerophosphocholine *O* acetyltransferase (92); arylamine *N*-acetyltransferase (93); catechol *O*-methyltransferase (94,95); xanthine oxidase (96,97); deoxyhypusine monooxygenase (98); catechol oxidase (102,103,105,106,139-144,146-150); polyphenol oxidase (105-112,115,139-144,146-150); tyrosinase, *or* monophenol monooxygenase (100-105,107-115,145); phenylalanine 4-monooxygenase (116); *trans*-cinnamate 4-monooxygenase (118); phenol 2-monooxygenase (119); salicylate 1-monooxygenase, weakly inhibited (120); clavaminate synthase (121); proline 4-dioxygenase (122); 3-hydroxy-2-methylquinolin-4-one 2,4-dioxygenase, weakly inhibited (123); 3-hydroxy-4-oxoquinoline 2,4-dioxygenase (123); 2,3-dihydroxyindole 2,3 dioxygenase (124); lipoxygenase (125); homogentisate 1,2-dioxygenase (126,127); catechol 1,2 dioxygenase (128); manganese peroxidase (129,130); peroxidase (131-134); peroxidase, horseradish (132); NADPH peroxidase (135); *o*-aminophenol oxidase (136,137); laccase (138); ubiquinol:cytochrome-*c* reductase, complex III (151,152).

1. Szent-Györgyi (1928) *Biochem. J.* **22**, 1387. **2.** Heaney, Gardner, Karasavvas, *et al.* (2008) *Cancer Res.* **68**, 8031. **3.** Hayano & Dorfman (1962) *Meth. Enzymol.* **5**, 503. **4.** Razzell (1963) *Meth. Enzymol.* **6**, 236. **5.** Rosenthal & Narasimhulu (1969) *Meth. Enzymol.* **15**, 596. **6.** Subba Rao & Vaidyanathan (1970) *Meth. Enzymol.* **17A**, 554. **7.** Paine (1998) in *Handb. Proteolytic Enzymes* (Barrett, Rawlings & Woessner, eds.), p. 1283, Academic Press, San Diego. **8.** Orr (1970) *Meth. Enzymol.* **18A**, 59. **9.** Russell, Williams & Austin (2000) *J. Enzyme Inhib.* **15**, 283. **10.** Rowe & Weill (1962) *Biochim. Biophys. Acta* **65**, 245. **11.** Ozawa, Ohnishi & Tanaka (1962) *J. Biochem.* **51**, 372. **12.** Krisch (1963) *Biochem. Z.* **337**, 546. **13.** Holman & Bergström (1951) *The Enzymes*, 1st ed., **2** (part 1), 559. **14.** Mahler (1963) *The Enzymes*, 2nd ed., **8**, 285. **15.** Gottschalk & Bhargava (1971) *The Enzymes*, 3rd ed., **5**, 321. **16.** Davison, Kettle & Fatur (1986) *J. Biol. Chem.* **261**, 1193. **17.** Szent-Györgyi (1928) *Biochem. J.* **22**, 1387. **18.** Haworth & Szent-Györgyi (1933) *Nature* **131**, 24. **19.** Ault, Baird, Carrington, *et al.* (1933) *J. Chem. Soc.*, 1419. **20.** Reichstein, Grüssner & Oppenauer (1933) *Helv. Chim. Acta* **16**, 561 and 1019. **21.** Burg & Brown (1968) *J. Biol. Chem.* **243**, 2349. **22.** Okorukwu & Vercruysse (2003) *J. Enzyme Inhib. Med. Chem.* **18**, 377. **23.** Roskoski, Gahn & Roskoski (1993) *Eur. J. Biochem.* **218**, 363. **24.** Nakagawa & Ruegamer (1967) *Biochemistry* **6**, 1249. **25.** Shinoda, Hara, Nakayama, Deyashiki & Yamaguchi (1992) *J. Biochem.* **112**, 840. **26.** Hara, Shinoda, Kanazu, *et al.* (1991) *Biochem. J.* **275**, 121. **27.** Wilgus & Roskoski (1988) *J. Neurochem.* **51**, 1232. **28.** Miggiano, Mordente, Martorana, Meucci &

Castelli (1984) *Biochim. Biophys. Acta* **789**, 343. **29.** Mieda, Yabuta, Rapolu, *et al.* (2004) *Plant Cell Physiol.* **45**, 1271. **30.** Russell, Williams & Gapuz (1997) *Biochem. Biophys. Res. Commun.* **233**, 386. **31.** Harwood, Greene & Stacpoole (1986) *J. Biol. Chem.* **261**, 7127. **32.** Diez & Cabezas (1979) *Eur. J. Biochem.* **93**, 301. **33.** Iio, Okabe & Omura (1976) *J. Nutr. Sci. Vitaminol. (Tokyo)* **22**, 53. **34.** Beiler & Martin (1947) *J. Biol. Chem.* **171**, 507. **35.** Mayer (1959) *Enzymologia* **20**, 313. **36.** Nadkarni & Sohonie (1963) *Enzymologia* **25**, 337. **37.** Orr (1967) *Biochemistry* **6**, 2995 and 3000. **38.** Foulkes & Lemberg (1948) *Australian J. Exptl. Biol. Med. Sci.* **26**, 307. **39.** Naganna (1950) *J. Biol. Chem.* **183**, 693. **40.** Aerts & Vercauteren (1964) *Enzymologia* **28**, 1. **41.** Suzuki (1966) *Enzymologia* **30**, 215. **42.** Skelton (1968) *Enzymologia* **35**, 275. **43.** Russell, Williams, Abbott, DeRosales & Vargas (2006) *J. Enzyme Inhib. Med. Chem.* **21**, 61. **44.** Kitabaci (1967) *Steroids* **10**, 567. **45.** Tallan, Jones & Fruton (1952) *J. Biol. Chem.* **194**, 793. **46.** Feigelson (1952) *J. Biol. Chem.* **197**, 843. **47.** Hanes (1935) *Biochem. J.* **29**, 2588. **48.** Wolucka & Van Montagu (2003) *J. Biol. Chem.* **278**, 47483. **49.** Li, Taylor, Kelly & Jedrzejas (2001) *J. Biol. Chem.* **276**, 15125. **50.** Rua, Soler, Busto & de Arriaga (2002) *Fungal Genet. Biol.* **35**, 223. **51.** Parkin & Hultin (1986) *J. Biochem.* **100**, 77. **52.** Bais, James, Rofe & Conyers (1985) *Biochem. J.* **230**, 53. **53.** Twu, Haroz & Bretthauer (1977) *Arch. Biochem. Biophys.* **184**, 249. **54.** Nagasawa, Mauger & Yamada (1990) *Eur. J. Biochem.* **194**, 765. **55.** Warneck & Seitz (1990) *Z. Naturforsch. C* **45**, 963. **56.** Giri (1939) *Biochem. J.* **33**, 309. **57.** Sakuma, Kametani & Akanuma (1998) *J. Biochem.* **123**, 189. **58.** Lemberg & Legge (1943) *Biochem. J.* **37**, 117. **59.** Serrano & Nagayama (1991) *Comp. Biochem. Physiol. B Comp. Biochem.* **99**, 281. **60.** von Clausbruch & Tschesche (1988) *Biol. Chem. Hoppe-Seyler* **369**, 683. **61.** Albrecht, Boldizsar & Hutchinson (1978) *J. Bacteriol.* **134**, 506. **62.** Jwanny, El-Sayed, Rashad, Mahmoud & Abdallah (1995) *Phytochemistry* **39**, 1301. **63.** Tani, Ohtsuru T. Hata (1974) *Agric. Biol. Chem.* **38**, 1623. **64.** Pontoppidan, Ekbom, Eriksson & Meijer (2001) *Eur. J. Biochem.* **268**, 1041. **65.** Kang, Lee, Lee & Lee (1999) *Biotechnol. Appl. Biochem.* **29**, 263. **66.** Chaudhary, Gupta, Gupta & Banerjee (1996) *J. Biotechnol.* **46**, 55. **67.** Verpoorte (1974) *Biochemistry* **13**, 793. **68.** Gupta & Singh (1983) *Biochim. Biophys. Acta* **748**, 398. **69.** Diez & Cabezas (1979) *Eur. J. Biochem.* **93**, 301. **70.** Prado, Vattuone, Fleischmacher & Sampietro (1985) *J. Biol. Chem.* **260**, 4952. **71.** Chinen, Nakamura & Fukuda (1981) *J. Biochem.* **90**, 1453. **72.** Dale, Ensley, Kern, Sastry & Byers (1985) *Biochemistry* **24**, 3530. **73.** Madhu & Prabhu (1985) *Enzyme Microb. Technol.* **7**, 279. **74.** Farooqui & Srivastava (1979) *Biochem. J.* **181**, 331. **75.** Farooqui & Bachhawat (1972) *Biochem. J.* **126**, 1025. **76.** Sok (1998) *J. Neurochem.* **70**, 1167. **77.** Lerch & Wolf (1972) *Biochim. Biophys. Acta* **258**, 206. **78.** Sutardi & Buckle (1986) *J. Food Biochem.* **10**, 197. **79.** Marseno, Hori & Miyazawa (1993) *J. Agric. Food Chem.* **41**, 1208. **80.** Hayman, Warburton, Pringle, Coles & Chambers (1989) *Biochem. J.* **261**, 601. **81.** Tominaga & Mori (1974) *J. Biochem.* **76**, 397. **82.** Ching, Lin & Metzger (1987) *Plant Physiol.* **84**, 789. **83.** Uotila & Koivusalo (1974) *J. Biol. Chem.* **249**, 7664. **84.** Nakanishi, Isohashi, Matsunaga & Sakamoto (1985) *Eur. J. Biochem.* **152**, 337. **85.** Sharma, Soni, Vohra, Gupta & Gupta (2002) *Process Biochem.* **37**, 1075. **86.** Russell, Williams, Abbott, DeRosales & Vargas (2006) *J. Enzyme Inhib. Med. Chem.* **21**, 61. **87.** Lain-Guelbenzu, Cárdenas & Muñoz-Blanco (1991) *Eur. J. Biochem.* **202**, 881. **88.** Lain-Guelbenzu, Muñoz-Blanco & Cárdenas (1990) *Eur. J. Biochem.* **188**, 529. **89.** Letelier, Martinez, Gonzalez-Lira, Faundez & Aracena-Parks (2006) *Chem. Biol. Interact.* **164**, 39. **90.** Busby, Quackenbush, Humm, Youngblood & Kizer (1987) *J. Biol. Chem.* **262**, 8532. **91.** Saleh, Fritsch, Kreuzaler & Grisebach (1978) *Phytochemistry* **17**, 183. **92.** Hurst & Bazan (1997) *J. Ocul. Pharmacol. Ther.* **13**, 415. **93.** Makarova (2008) *Curr. Drug Metab.* **9**, 538. **94.** Bonifácio, Palma, Almeida & Soares-da-Silva (2007) *CNS Drug Reviews* **13**, 352. **95.** Veser (1987) *J. Bacteriol.* **169**, 3696. **96.** De Renzo (1956) *Adv. Enzymol. Relat. Subj. Biochem.* **17**, 293. **97.** Machida & Nakanishi (1981) *Agric. Biol. Chem.* **45**, 425. **98.** Abbruzzese, Park & Folk (1986) *J. Biol. Chem.* **261**, 3085. **99.** Ryan & Engel (1957) *J. Biol. Chem.* **225**, 103. **100.** Greenfield, Ponticorvo, Chasalow & Lieberman (1980) *Arch. Biochem. Biophys.* **200**, 232. **101.** Parvez, Kang, Chung & Bae (2007) *Phytother. Res.* **21**, 805. **102.** Kim & Uyama (2005) *Cell. Mol. Life Sci.* **62**, 1707. **103.** Sharma & Ali (1980) *Phytochemistry* **19**, 1597. **104.** Ikediobi & Obasuyi (1982) *Phytochemistry* **21**, 2815. **105.** Kumar, Mohan & Murugan (2008) *Food Chem.* **110**, 328. **106.** Güllçin, Küfrevioglu & Octay (2005) *J. Enzyme Inhib. Med. Chem.* **20**, 297. **107.** Dogan & Salman (2007) *Eur. Food Res. Technol.* **226**, 93. **108.** Ayaz, Demir, Torun, Kolcuoglu & Colak (2007) *Food Chem.* **106**, 291. **109.** Sakiroglu, Oztürk, Pepe & Erat (2008) *J. Enzyme Inhib. Med. Chem.* **23**, 380. **110.** Gawlik-Dziki, Szymanowska & Baraniak (2007) *Food Chem.* **105**, 1047. **111.**

Gawlik-Dziki, Zlotek & Swieca (2007) *Food Chem.* **107**, 129. **112**. Liu, Cao, Xie, *et al.* (2007) *J. Agric. Food Chem.* **55**, 7140. **113**. Orenes-Pinero, Garcia-Carmona & Sanchez-Ferrer (2007) *J. Mol. Catal. B* **47**, 143. **114**. Wichers, Peetsma, Malingre & Hulzing (1984) *Planta* **162**, 334. **115**. Hsu, Kalan & Bills (1984) *Plant Sci. Lett.* **34**, 315. **116**. Martínez, Olafsdottir, Haavik &Flatmark (1992) *Biochem. Biophys. Res. Commun.* **182**, 92. **117**. Yanagibashi, Kobayashi & Hall (1990) *Biochem. Biophys. Res. Commun.* **170**, 1256. **118**. Billett & Smith (1978) *Phytochemistry* **17**, 1511. **119**. Neujahr & Gaal (1973) *Eur. J. Biochem.* **35**, 386. **120**. Yamamoto, Katagiri, Maeno & Hayaishi (1965) *J. Biol. Chem.* **240**, 3408. **121**. Salowe, Marsh & Townsend (1990) *Biochemistry* **29**, 6499. **122**. Lawrence, Sobey, Field, Baldwin & Schofield (1996) *Biochem. J.* **313**, 185. **123**. Bauer, Max, Fetzner & Lingens (1996) *Eur. J. Biochem.* **240**, 576. **124**. Fujioka & Wada (1968) *Biochim. Biophys. Acta* **158**, 70. **125**. Barone, Briante, D'Auria, *et al.* (1999) *J. Agric. Food Chem.* **47**, 1924. **126**. Veldhuizen, Vaillancourt, Whiting, *et al.* (2005) *Biochem. J.* **386**, 305. **127**. Schmidt, Müller & Kress (1995) *Eur. J. Biochem.* **228**, 425. **128**. Itoh (1981) *Agric. Biol. Chem.* **45**, 2787. **129**. Urek & Pazarlioglu (2004) *Proc. Biochem.* **39**, 2061. **130**. Paszczynski, Huynh & Crawford (1986) *Arch. Biochem. Biophys.* **244**, 750. **131**. Martinez, Civello, Chaves & Anon (2001) *Phytochemistry* **58**, 379. **132**. Galzigna, Rizzoli, Schiappelli, *et al.* (1996) *Free Radic. Biol. Med.* **20**, 807. **133**. Serrano-Martinez, Fortea, del Amor & Nunez-Delicado (2007) *Food Chem.* **107**, 193. **134**. Kouakou, Dué, Kouadio, *et al.* (2009) *Appl. Biochem. Biotechnol.* **157**, 575. **135**. Conn, Kraemer, Liu & Vennesland (1952) *J. Biol. Chem.* **194**, 143. **136**. Subba Rao & Vaidyanathan (1967) *Arch. Biochem. Biophys.* **118**, 388. **137**. Nair & Vining (1965) *Biochim. Biophys. Acta* **96**, 318. **138**. Ishigami, Hirose & Yamada (1988) *J. Gen. Appl. Microbiol.* **34**, 401. **139**. Motoda (1979) *J. Ferment. Technol.* **57**, 79. **140**. Paul & Gowda (2000) *J. Agric. Food Chem.* **48**, 3839. **141**. Yamamoto, Yoshitama & Teramoto (2002) *Plant Biotechnol.* **19**, 95. **142**. Sellés-Marchart, Casado-Vela & Bru-Martínez (2006) *Arch. Biochem. Biophys.* **446**, 175. **143**. Erat, Sakiroglu & Küfrevioglu (2006) *Food Chem.* **95**, 503. **144**. Wuyts, De Waele & Sweenen (2006) *Plant Physiol. Biochem.* **44**, 308. **145**. Gandía-Herrero, Escribano & García-Carmona (2007) *J. Agric. Food Chem.* **55**, 1546. **146**. Dogan, Turan, Ertürk & Arslan (2005) *J. Agric. Food Chem.* **53**, 776. **147**. Yang, Fujita, Kohno, *et al.* (2001) *J. Agric. Food. Chem.* **49**, 1446. **148**. Yang, Fujita, Ashrafuzzaman, Nakamura & Hayashi (2000) *J. Agric. Food. Chem.* **48**, 2732. **149**. Wong, Luh & Whitaker (1971) *Plant Physiol.* **48**, 19. **150**. Jang, Shin & Song (2005) *Food Sci. Biotechnol.* **14**, 117. **151**. Matsuno-Yagi & Hatefi (1997) *J. Biol. Chem.* **272**, 16928. **152**. Matsuno-Yagi & Hatefi (2001) *J. Biol. Chem.* **276**, 19006.

Asenapine

This atypical antipsychotic agent (FW = 285.77 g/mol; CAS 65576-45-6), marketed under the trade names Saphris®, and also known as Org 5222 and (3a*RS*,12b*RS*)-*rel*-5-chloro-2,3,3a,12b-tetrahydro-2-methyl-1*H*-dibenz[2,3:6,7]-oxepino-[4,5-*c*]pyrrole, is multi-receptor antagonist with the following spectrum of binding interactions: serotonin 5-HT$_{1A}$ receptor, K_i = 2.5 nM; serotonin 5-HT$_{1B}$ receptor, K_i = 4.0 nM; serotonin 5-HT$_{2A}$ receptor, K_i = 0.06 nM; serotonin 5-HT$_{2B}$ receptor, K_i = 0.16 nM; serotonin 5-HT$_{2C}$ receptor, K_i = 0.03 nM; serotonin 5-HT$_{5A}$ receptor, K_i = 1.6 nM; serotonin 5-HT$_6$ receptor, K_i = 1.5 nM; serotonin 5-HT$_7$ receptor, K_i = 0.13 nM; α_1-Adrenergic receptor, K_i = 1.2 nM; α_{2A}-Adrenergic receptor, K_i = 1.2 nM; α_{2B}-Adrenergic receptor, K_i = 0.25 nM; α_{2C}-Adrenergic receptor, K_i = 1.2 nM; dopamine D$_1$-receptor, K_i = 1.4 nM; dopamine D$_2$-receptor, K_i = 1.3 nM; dopamine D$_3$-receptor, K_i = 0.4 nM; dopamine D$_4$-receptor, K_i = 1.1 nM; histamine H$_1$-receptor, K_i = 1.0 nM; and histamine H$_2$-receptor, K_i = 6 nM (1,2). Like other atypical antipsychotic drugs, asenapine preferentially enhances dopamine and acetylcholine efflux in the rat medial prefrontal cortex and hippocampus. See Reference-x for asenapine's UV, IR, NMR, and mass spectra as well as X-ray analysis, thermal properties, solubilities and partition coefficient (3). **1**. Shahid, Walker, Zorn & Wong (2009) *J.*

Psychopharmacol. **23**, 65. **2**. De Boer T, Tonnaer JA, De Vos CJ, Van Delft AM. (1990) *Arzneimittelforschung.* **40**, 550. **3**. Funke CW, Hindriks H, Sam AP (1990) *Arzneimittelforschung* **40**, 536.

ASP015K, *See Peficitinib*

ASP1941, *See Ipragliflozin*

ASP2151

This structurally novel oxadiazolephenyl-class helicase/ primase inhibitor (FW = 482.16 g/mol; CAS 841301-32-4), also known as amenamevir, is not a nucleoside and, as such, bypasses drug-resistance mechanisms associated with nucleoside transport and phosphorylation as well as the broad spectrum of inhibition that limit the efficacy of nucleoside 5'-di- and tri-phosphate purine and pyrimidine analogues as anticancer agents. ASP2151 possesses potent antiviral activity not only against HSV-1 and HSV-2 but also against VZV. ASP2151 is selective with low cytotoxicity *in vitro*. It is also orally available and well tolerated. ASP2151 inhibited the single-stranded DNA-dependent ATPase, helicase and primase activities associated with the HSV-1 helicase − primase complex. Antiviral assays revealed that ASP2151, unlike other known HSV helicase/primase inhibitors (HPI's), exerts equipotent activity against VZV, HSV-1 and HSV-2 through prevention of viral DNA replication. Further, the anti-VZV activity of ASP2151 (EC$_{50}$ = 0.038 – 0.10 μM) was more potent against all strains tested than that of aciclovir (EC$_{50}$ = 1.3 – 27 μM). ASP2151 was also active against aciclovir- resistant VZV. Amino acid substitutions were found in helicase and primase subunits of ASP2151-resistant VZV. In a mouse zosteriform-spread model, ASP2151 was orally active and inhibited disease progression more potently than valaciclovir. **1**. Hono, Katsumata, Kontani, *et al.* (2010) *J. Antimicrob. Chemother.* **65**, 1733.

ASP3026

This potent signal transduction inhibitor (FW = 580.74 g/mol; CAS 1097917-15-1; Solubility: 14 mg/mL DMSO; <1 mg/mL H$_2$O), also known as N^2-[2-methoxy-4-[4-(4-methyl-1-piperazinyl)-1-piperidinyl]phenyl]-N^4-[2-[(1-methylethyl)sulfonyl]phenyl]-1,3,5-triazine-2,4-diamine, targets ALK, the Activin receptor-Like Kinase activity of Transforming Growth Factor-β (TGF-β) receptors, with IC$_{50}$ of 3.5 nM. ASP3026 inhibits ALK in an ATP-competitive manner with an inhibitory spectrum that differs from crizotinib, a dual ALK/MET inhibitor. In patients with non-small cell lung cancer, a therapeutic target is the EML4-ALK fusion protein, which is formed between echinoderm microtubule-associated protein like 4 and ALK as a consequence of the rearrangement of ALK and EML4 genes. In mice xenografted with EML4-ALK-expressing NCI-H2228 cells, orally administered ASP3026 is well absorbed in tumor tissues, reaching concentrations >10-fold higher than those in plasma, and induces tumor regression with a wide therapeutic margin between efficacious and toxic doses. In the same mouse model, ASP3026 enhances the antitumor activities of paclitaxel and pemetrexed without affecting body weight. ASP3026 also showed potent antitumor activities, including tumor shrinkage to a undetectable level, in hEML4-ALK transgenic mice and prolonged survival in mice with intrapleural NCI-H2228 xenografts. Finally, ASP3026 exhibited potent antitumor activity against cells expressing EML4-ALK^{L1196M}, a mutation conferring resistance to crizotinib. **1**. Mori, Ueno, Konagai, *et al.* (2014) *Mol. Cancer Ther.* **13**, 329.

D-Asparagine

This amino acid (FW = 132.12 g/mol; FW$_{monohydrate}$ = 150.13 g/mol), also called (*R*)-2-amino-3-carbamoylpropanoate, is the enantiomer of the nonproteogenic amino acid L-asparagine. D-Asparagine found in certain bacterial cell wall peptidoglycans as well as in a form of bacitracin A. **Target(s):** L-asparaginase (1-4,8-10); allantoicase (5); D-alanine aminotransferase (6); glutamin (asparagin-)ase (7). **1**. Sobis & Mikucki (1991) *Acta Microbiol. Pol.* **40**, 143. **2**. Varner (1960) *The Enzymes*, 2nd ed., **4**, 243. **3**. Grossowicz & Halpern (1956) *Nature* **177**, 623. **4**. Webb (1966) *Enzyme and Metabolic Inhibitors*, vol. 2, p. 269, Academic Press, New York. **5**. van der Drift & Vogels (1970) *Biochim. Biophys. Acta* **198**, 339. **6**. Martinez-Carrion & Jenkins (1965) *J. Biol. Chem.* **240**, 3547. **7**. Imada & Igarski (1973) *J. Takeda Res. Lab.* **32**, 140. **8**. Chang & Farnden (1981) *Arch. Biochem. Biophys.* **208**, 49. **9**. Raha, Roy, Dey & Chakrabarty (1990) *Biochem. Int.* **21**, 987. **10**. Pritsa & Kyriakidis (2001) *Mol. Cell. Biochem.* **216**, 93.

L-Asparagine

This proteogenic amino acid (FW = 132.12 g/mol; FW$_{monohydrate}$ = 150.13 g/mol) has pK_a values of 2.14 and 8.72 at 25°C), also called (*S*)-2-amino-3-carbamoylpropanoate, has a polar side chain that, when incorporated in a protein, is most often found on the surface. L-asparaginyl residues in proteins can undergo racemization as well as deamidation to produce a mixture of D- and L-aspartyl residues as well as isoaspartyl residues. Asparaginyl residues are also the sites for *N*-glycosylation in the formation of glycoproteins. The asparaginyl residue modified is usually in one of the following sequences: –Asn–Xaa–Cys–, –Asn–Xaa–Ser–, or –Asn–Xaa–Thr–, where Xaa is any residue other than proline. **Target(s):** hemoglobin S polymerization (1); kynureninase, weakly inhibited (2); glutamate synthase (3); acetylcholinesterase (4); glutamine synthetase (5); allantoicase (6); theanine hydrolase (7); N^4-(β-*N*-acetylglucosaminyl)-L-asparaginase, also alternative substrate (8); leucyl aminopeptidase (9); [citrate lyase] deacetylase, weakly inhibited (10); serine:glyoxylate aminotransferase, also alternative substrate (11); ornithine carbamoyltransferase, weakly inhibited (12). **1**. Rumen (1975) *Blood* **45**, 45. **2**. Jakoby & Bonner (1953) *J. Biol. Chem.* **205**, 709. **3**. Boland & Benny (1977) *Eur. J. Biochem.* **79**, 355. **4**. Bergmann, Wilson & Nachmansohn (1950) *J. Biol. Chem.* **186**, 693. **5**. Singh & Singh (1990) *Arch. Int. Physiol. Biochim.* **98**, 95. **6**. Elzainy & El-Awamry (1980) *Egypt. J. Bot.* **23**, 137. **7**. Tsushida & Takeo (1985) *Agric. Biol. Chem.* **49**, 2913. **8**. Noronkoski, Stoineva, Petkov & Mononen (1997) *FEBS Lett.* **412**, 149. **9**. Machuga & Ives (1984) *Biochim. Biophys. Acta* **789**, 26. **10**. Giffhorn & Gottschalk (1975) *J. Bacteriol.* **124**, 1052. **11**. Ireland & Joy (1983) *Arch. Biochem. Biophys.* **223**, 291. **12**. Legrain & Stalon (1976) *Eur. J. Biochem.* **63**, 289.

L-Asparaginyl-L-isoleucyl-L-valyl-L-asparaginyl-L-valyl-L-serinyl-L-leucyl-L-valyl-L-lysine

This non-toxic, nine-residue peptide (FW = 985.19 g/mol; *Sequence*: NIVNVSLVK; Isoelectric Point = 8.75) inhibits insulin fibrillation, a phenomenon limiting long-term insulin storage and reducing insulin efficacy in treating Type II diabetes. Insulin oligomer dissociation into monomers is the key step for onset of fibrillation, and time-course of insulin fibrillation at 62°C (based on the fluorescence-generating binding of the amyloid cross-β-sensitive dye Thioflavin-T) shows an increase in the lag-time from 120 min in the absence peptide, as compared to 236 min in its presence. Transmission electron micrographs show branched insulin fibrils in its absence and less inter-fibril association in its presence. Both size-exclusion chromatography and dynamic light scattering show that insulin primarily exists as trimer, the conversion of which into a monomer is resisted by the peptide. Saturation transfer difference nuclear magnetic resonance confirms that the hydrophobic residues in the peptide are in close contact with an insulin hydrophobic groove. Significantly, insulin aggregation appears to be related to the interplay of B-chain Phe-24 and B-chain Tyr-26. **1**. Banerjee, Kar, Datta, *et al.* (2013) *PLoS One* **8**, e72318.

L-Asparaginyl-L-seryl-L-phenylalanyl-L-threonyl-L-leucyl-L-aspartyl-L-alanyl-L-aspartyl-L-phenylalanine

This acetylated nonapeptide (*Sequence*: Ac-NSFTLDADF; FW = 1072.12 g/mol) inhibits mammalian ribonucleotide reductase. The sequence corresponds to the C-terminus of the R2 subunit of the reductase. **1**. Cory, Cory & Downes (1995) in *Purine and Pyrimidine Metabolism in Man*, 8th ed. (Sohota & Taylor, eds.), p. 631.

Asparagusate (Asparagusic Acid)

This naturally occurring dithiolane (FW = 150.21 g/mol; CAS 2224-02-4), systematically named 1,2-dithiolane-4-carboxylic acid, is present in asparagus and may well be the metabolic precursor to characteristically odorous thiols associated with the consumption of this vegetable. The reduced form is likely to have the same protein disulfide reducing properties as dithiothreitol (DTT) and dithioerythritol (DTE). It also structurally resembles oxidized lipoic acid. Asparagusate's biological function remains uncertain, but it may prevent fungal growth and repel insect attack (1). It also possesses a considerable nematocidal activity (2,3) In plants, asparagusic acid is most abundant in apical regions of the emerging shoots and other vulnerable areas, decreasing markedly in woody parts of plants (4). Other 1,2-dithiolanes, especially those isolated from mangroves (4-hydroxy-1,2-dithiolane, its sulphoxide and sulfone), exhibit potent antibacterial and insecticidal activities. Asparagusic acid is also a growth inhibitor in other higher plants (similar in potency to abscisic acid), perhaps preventing overcrowding and/or competition for scarce nutriments by adjacent plants. **1**. Iriuchijima (1977) *Yuki Gosei Kagaku Kyokaishi* **35**, 394. **2**. Birch, Robertson & Fellows (1993) *Pestic. Sci.* **39**, 141. **3**. Takasugi, Yachida, Anetai, Masamune & Kegawawa (1975). *Chem. Lett.* **4**, 43. **4**. Yanagawa (1979) *Meth. Enzymol.* **62**, 181.

Aspartame

This popular artificial sweetener (FW = 294.31 g/mol), also known as *N*-(L-α-aspartyl)-L-phenylalanine methyl ester and EQUAL®, is a dipeptide ester that is ~160x sweeter than in aqueous sucrose. Aspartame exhibits dose-dependent inhibition of L-glutamate binding to the *N*-methyl-D-aspartame (NMDA) receptor in rat brain synaptosomes (1). **Target(s):** peptidyl-dipeptidase A, *or* angiotensin-converting enzyme; weakly inhibited (2); thrombin, weakly inhibited (3); acetylcholinesterase, weakly inhibited (4). **Note**: L-Aspartyl-L-phenylalanine, formed by the action of esterases, inhibits Angiotensin Converting Enzyme from rabbit lung (K_i = 11 μM, comparable to the IC$_{50}$ of 12 μM for 2-D-methyl-succinyl-L-proline, an orally active antihypertensive agent in rats (5). **1**. Pan-Hou, Suda, Ohe, Sumi & Yoshioka (1990) *Brain Res.* **520**, 351. **2**. Grobelny & Galardy (1985) *Biochem. Biophys. Res. Commun.* **128**, 960. **3**. Scheffler & Berliner (2004) *Biophys. Chem.* **112**, 285. **4**. Tsakiris, Giannoulia-Karantana, Simintzi & Schulpis (2006) *Pharmacol. Res.* **53**, 1. **5**. Grobelny & Galardy (1985) *Biochem. Biophys. Res. Commun.* **128**, 960.

D-Aspartate

This conjugate base (FW$_{\text{free-acid}}$ = 133.10 g/mol; FW$_{\text{monosodium salt}}$ = 155.09 g/mol; pK_a values of 1.99, 3.90 (β-carboxyl), and 9.90 at 25°C; dpK_{a3}/dT = –0.022), also known as (R)-2-aminobutanedioic acid, is the less common enantiomer of the proteogenic amino acid L-aspartate. D-Aspartyl residues can be found in bacitracin A and in bacterial cell walls. **Target(s):** argininosuccinate synthetase (1,5); adenylosuccinate synthetase (2); D-alanine aminotransferase (3); glutamine synthetase, weakly inhibited (4); UDP-N-acetylmuramoyl-L-alanine:D-glutamate ligase, weakly inhibited (6); asparagine synthetase (7); aspartate ammonia lyase (8,9); glutamate decarboxylase, weakly inhibited (10); L-asparaginase (11); γ-D-glutamyl-*meso* diaminopimilate, weakly inhibited (12); tyrosine aminotransferase, weakly inhibited (13). **1.** Ratner (1970) *Meth. Enzymol.* **17A**, 298. **2.** Ratner (1962) *The Enzymes*, 2nd ed., **6**, 495. **3.** Martinez-Carrion & Jenkins (1965) *J. Biol. Chem.* **240**, 3547. **4.** Schou, Grossowicz & Waelsch (1951) *J. Biol. Chem.* **192**, 187. **5** Ratner (1973) *Adv. Enzymol. Relat. Areas Mol. Biol.* **39**, 1. **6.** Michaud, Blanot, Flouret & van Heijenoort (1987) *Eur. J. Biochem.* **166**, 631. **7.** Al Dawody & Varner (1961) *Fed. Proc.* **20**, 10C. **8.** Mizuta & Tokushige (1975) *Biochim. Biophys. Acta* **403**, 221. **9.** Suzuki, Yamaguchi & Tokushige (1973) *Biochim. Biophys. Acta* **321**, 369. **10.** Tunnicliff (1990) *Int. J. Biochem.* **27**, 1235. **11.** Chang & Farnden (1981) *Arch. Biochem. Biophys.* **208**, 49. **12.** Arminjon, Guinand, Vacheron & Michel (1977) *Eur. J. Biochem.* **73**, 557. **13.** Miller & Litwack (1971) *J. Biol. Chem.* **246**, 3234.

L-Aspartate

This conjugate base (FW$_{\text{free-acid}}$ = 133.10 g/mol; FW$_{\text{monosodium salt}}$ = 155.09 g/mol; pKa values of 1.99, 3.90 (β-carboxyl), and 9.90 at 25°C; dpK_{a3}/dT = –0.022). Aspartic acid is soluble in water (0.5 g/100 mL at 25°C and 2.87 g/100 mL at 75°C; the solubility increases with pH) and easily forms supersaturated solutions. It is more soluble in salt solutions. It is slightly soluble in ethanol (0.00016 g/100 mL at 25°C) and insoluble in diethyl ether. It is stable in hot acid solutions. **Target(s):** pyruvate carboxylase (1-3,14,30-34); glutamate dehydrogenase (4); carbamoyl phosphate synthetase(5); phosphoenolpyruvate carboxylase (6,7,9,15,40-59) glycosylasparaginase (8); β-glucuronidase (10); pantothenate synthetase, *or* pantoate:β-alanine ligase 9-11); arginase, weakly inhibited (12); malic enzyme, *or* malate dehydrogenase (decarboxylating) (13); L-lactate dehydrogenase, particularly LDH-X (16); glutamate decarboxylase, weakly inhibited (17,18,19,60); succinate dehydrogenase (19); glutamine synthetase (20,29,36-38); γ-glutamyl transpeptidase, weakly inhibited (21,22); N-methylglutamate dehydrogenase (23); kynureninase, weakly inhibited (34); glutamate synthase (25); alkaline phosphatase (26); pyrophosphatase, *or* inorganic diphosphatase (27); acetylcholinesterase (28); 4-methyleneglutamine synthetase, *or* 4-methyleneglutamate:ammonia ligase (35); serine-sulfate ammonia-lyase (weakly inhibited)48; theanine hydrolase (61); N^4-(β-N-acetylglucosaminyl)-L-asparaginase (62,63); gametolysin (64); glutamate carboxypeptidase II (65); glutamate carboxypeptidase, weakly inhibited (66); dipeptidase E (67); glutamyl aminopeptidase, *or* aminopeptidase A (68); phosphoenolpyruvate phosphatase (69); [citrate lyase] deacetylase (70,71); acetylcholinesterase, weakly inhibited (72); arginine kinase (73,74); pyruvate kinase (75); homoserine kinase, weakly inhibited (76); phosphoribulokinase (77); asparagine:oxo-acid aminotransferase, weakly inhibited (78,79); kynurenine:oxoglutarate aminotransferase (80); tyrosine aminotransferase (81); nicotinate-nucleotide diphosphorylase, carboxylating (82,83); arginyltransferase, weakly inhibited (84); aspartyltransferase (85); glutamate formimidoyltransferase (86); cysteine dioxygenase (87); catalase (88,89). **1.** Osmani & Scrutton (1985) *Eur. J. Biochem.* **147**, 119. **2.** Palacian, de Torrontegui & Losada (1966) *Biochem. Biophys. Res. Commun.* **24**, 644. **3.** Young, Tolbert & Utter (1969) *Meth. Enzymol.* **13**, 250. **4.** Strecker (1955) *Meth. Enzymol.* **2**, 220. **5.** Jones (1962) *Meth.*

Enzymol. **5**, 903. **6.** Maeba & Sanwal (1969) *Meth. Enzymol.* **13**, 283. **7.** Cánovas & Kornberg (1969) *Meth. Enzymol.* **13**, 288. **8.** Risley, Huang, Kaylor, Malik & Xia (2001) *J. Enzyme Inhib.* **16**, 269. **9.** Mori & Shiio (1985) *J. Biochem.* **98**, 1621. **10.** Kreamer, Siegel & Gourley (2001) *Pediatr. Res.* **50**, 460. **11.** Miyatake, Nakano & Kitaoka (1979) *Meth. Enzymol.* **62**, 215. **12.** Hunter & Downs (1945) *J. Biol. Chem.* **157**, 427. **13.** Kun (1963) *The Enzymes*, 2nd ed., **7**, 149. **14.** Scrutton & Young (1972) *The Enzymes*, 3rd ed., **6**, 1. **15.** Utter & Kolenbrander (1972) *The Enzymes*, 3rd ed., **6**, 117. **16.** Holbrook, Liljas, Steindel & Rossmann (1975) *The Enzymes*, 3rd ed., **11**, 191. **17.** Webb (1966) *Enzyme and Metabolic Inhibitors*, vol. 2, p. 327, Academic Press, New York. **18.** Wingo & Awapara (1950) *J. Biol. Chem.* **187**, 267. **19.** Potter & Elvehjem (1937) *J. Biol. Chem.* **117**, 341. **20.** Orr & Haselkorn (1981) *J. Biol. Chem.* **256**, 13099. **21.** Allison (1985) *Meth. Enzymol.* **113**, 419. **22.** Thompson & Meister (1977) *J. Biol. Chem.* **252**, 6792. **23.** Hersh, Stark, Worthen & Fiero (1972) *Arch. Biochem. Biophys.* **150**, 219. **24.** Jakoby & Bonner (1953) *J. Biol. Chem.* **205**, 709. **25.** Boland & Benny (1977) *Eur. J. Biochem.* **79**, 35. **26.** Bodansky & Strachman (1949) *J. Biol. Chem.* **179**, 81. **27.** Naganna (1950) *J. Biol. Chem.* **183**, 693. **28.** Bergmann, Wilson & Nachmansohn (1950) *J. Biol. Chem.* **186**, 693. **29.** Schou, Grossowicz & Waelsch (1951) *J. Biol. Chem.* **192**, 187. **30.** Osman, Marston, Selmes, Chapman & Scrutton (1981) *Eur. J. Biochem.* **118**, 271. **31.** Osmani, Mayer, Marston, Selmes & Scrutton (1984) *Eur. J. Biochem.* **139**, 509. **32.** Modak & Kelly (1995) *Microbiology* **141**, 2619. **33.** Gurr & Jones (1977) *Arch. Biochem. Biophys.* **179**, 444. **34.** Wood, Sundaram, Blackburn & Marsh (1983) *Biochem. Soc. Trans.* **11**, 741. **35.** Winter, Su & Dekker (1983) *Biochem. Biophys. Res. Commun.* **111**, 484. **36.** Blanco, Alana, Llama & Serra (1989) *J. Bacteriol.* **171**, 1158. **37.** Krishnan, Singhal & Dua (1986) *Biochemistry* **25**, 1589. **38.** Singh & Singh (1990) *Arch. Int. Physiol. Biochim.* **98**, 95. **39.** Tudball & Thomas (1973) *Eur. J. Biochem.* **40**, 25. **40.** Schwitzguebel & Ettlinger (1979) *Arch. Microbiol.* **122**, 109. **41.** Iglesias, Gonzalez & Andreo (1986) *Planta* **168**, 239. **42.** Rivoal, Plaxton & Turpin (1998) *Biochem. J.* **331**, 201. **43.** Schuller, Turpin & Plaxton (1990) *Plant Physiol.* **94**, 1429. **44.** Marczewski (1989) *Physiol. Plant* **76**, 539. **45.** Schuller, Plaxton & Turpin (1990) *Plant Physiol.* **93**, 1303. **46.** Gold & Smith (1974) *Arch. Biochem. Biophys.* **164**, 447. **47.** Yoshida, Tanaka, Mitsunaga & Izumi (1995) *Biosci. Biotechnol. Biochem.* **59**, 140. **48.** McDaniel & Siu (1972) *J. Bacteriol.* **109**, 385. **49.** Nakamura, Minoguchi & Izui (1996) *J. Biochem.* **120**, 518. **50.** Chen, Omiya, Hata & Izui (2002) *Plant Cell Physiol.* **43**, 159. **51.** Patel, Kraszewski & Mukhopadhyay (2004) *J. Bacteriol.* **186**, 5129. **52.** Sadaie, Nagano, Suzuki, Shinoyama & Fujii (1997) *Biosci. Biotechnol. Biochem.* **61**, 625. **53.** Daniel, Bryant & Woodward (1984) *Biochem. J.* **218**, 387. **54.** Mares, Barthova & Leblova (1979) *Collect. Czech. Chem. Commun.* **44**, 1835. **55.** Moraes & Plaxton (2000) *Eur. J. Biochem.* **267**, 4465. **56.** Hoban & Lyric (1975) *Can. J. Biochem.* **53**, 875. **57.** Andreo, Gonzalez & Iglesias (1987) *FEBS Lett.* **213**, 1. **58.** Blonde & Plaxton (2003) *J. Biol. Chem.* **278**, 11867. **59.** Tripodi, Turner, Gennidakis & Plaxton (2005) *Plant Physiol.* **139**, 969. **60.** Blindermann, Maitre, Ossola & Mandel (1978) *Eur. J. Biochem.* **86**, 143. **61.** Tsushida & Takeo (1985) *Agric. Biol. Chem.* **49**, 2913. **62.** Risley, Huang, Kaylor, Malik & Xia (2001) *J. Enzyme Inhib.* **16**, 269. **63.** Kohno & Yamashina (1972) *Biochim. Biophys. Acta* **258**, 600. **64.** Matsuda, Saito, Yamaguchi & Kawase (1985) *J. Biol. Chem.* **260**, 6373. **65.** Robinson, Blakely, Couto & Coyle (1987) *J. Biol. Chem.* **262**, 14498. **66.** McCullough, Chabner & Bertino (1971) *J. Biol. Chem.* **246**, 7207. **67.** Conlin, Hakensson, Liljas & Millèr (1994) *J. Bacteriol.* **176**, 166. **68.** Lalu, Lampelo & Vanha-Perttula (1986) *Biochim. Biophys. Acta* **873**, 190. **69.** Duff, Lefebvre & Plaxton (1989) *Plant Physiol.* **90**, 734. **70.** Giffhorn, Rode, Kuhn & Gottschalk (1980) *Eur. J. Biochem.* **111**, 461. **71.** Giffhorn & Gottschalk (1975) *J. Bacteriol.* **124**, 1052. **72.** Tsakiris, Giannoulia-Karantana, Simintzi & Schulpis (2006) *Pharmacol. Res.* **53**, 1. **73.** Pereira, Alonso, Paveto, *et al.* (2000) *J. Biol. Chem.* **275**, 1495. **74.** Tang, Yang & Zhang (2006) *Int. J. Biol. Macromol.* **40**, 15. **75.** Lin, Turpin & Plaxton (1989) *Arch. Biochem. Biophys.* **269**, 228. **76.** Finkelnburg & Klemme (1987) *FEMS Microbiol. Lett.* **48**, 93. **77.** Marsden & Codd (1984) *J. Gen. Microbiol.* **130**, 999. **78.** Maul & Schuster (1986) *Arch. Biochem. Biophys.* **251**, 585. **79.** Cooper (1977) *J. Biol. Chem.* **252**, 2032. **80.** Kocki, Luchowski, Luchowska, *et al.* (2003) *Neurosci. Lett.* **346**, 97. **81.** Miller & Litwack (1971) *J. Biol. Chem.* **246**, 3234. **82.** Shibata & Iwai (1980) *Biochim. Biophys. Acta* **611**, 280. **83.** Iwai, Shibata & Taguchi (1979) *Agric. Biol. Chem.* **43**, 351. **84.** Soffer (1973) *J. Biol. Chem.* **248**, 2918. **85.** Jayaram, Ramakrishnan & Vaidyanathan (1969) *Indian J. Biochem.* **6**, 106. **86.** Miller & Waelsch (1957) *J. Biol. Chem.* **228**, 397. **87.** Chai, Bruyere & Maroney (2006) *J.*

Biol. Chem. **281**, 15774. **88**. Dincer & Aydemir (2001) *J. Enzyme Inhib.* **16**, 165. **89**. Chatterjee & Sanwal (1993) *Mol. Cell. Biochem.* **126**, 125.

L-Aspartate β-Semialdehyde

This semialdehyde (FW = 117.10 g/mol), also known as L-aspartic acid β-semialdehyde, is an intermediate in the biosynthesis of L-lysine, L-threonine, and L-methionine. It is a product of the reaction catalyzed by β-aspartate semialdehyde dehydrogenase and a substrate for dihydropicolinate synthase and homoserine dehydrogenase. *Escherichia coli* asparate ammonia-lyase catalyzes the deamination of the semialdehyde to produce fumarate semialdehyde and ammonium ion; the semialdehyde product partitions between subsequent release or irreversible enzyme inactivation. This semialdehyde is unstable as a solid or in neutral solutions; hence, soutions should be freshly prepared (aqueous solutions are stable for several days at 4°C in 0.25-4 M HCl, and especially in an inert atmosphere). **Target(s):** asparaginase (1); aspartate ammonia-lyase (2-5); aspartate kinase (6); homoserine kinase (7). **1**. Westerik & Wolfenden (1974) *J. Biol. Chem.* **249**, 6351. **2**. Tokushige (1985) *Meth. Enzymol.* **113**, 618. **3**. Giorgianni, Beranova, Wesdemiotis & Viola (1997) *Arch. Biochem. Biophys.* **341**, 329. **4**. Schindler & Viola (1994) *Biochemistry* **33**, 9365. **5**. Yumoto, Okada & Tokushige (1982) *Biochem. Biophys. Res. Commun.* **104**, 859. **6**. Truffa-Bachi (1973) *The Enzymes*, 3rd ed., **8**, 509. **7**. Shames & Wedler (1984) *Arch. Biochem. Biophys.* **235**, 359.

ASS234

The multi-target enzyme inhibitor (FW = g/mol), also known as *N*-((5-(3-(1-benzylpiperidin-4-yl)propoxy)-1-methyl-1*H*-indol-2-yl)methyl)-*N*-methylprop-2-yn-1-amine, bears the MAO-inhibiting propargyl group attached to a cholinesterase-inhibiting donepezil moiety retains the ability to inhibit human acetyl- and butyryl-cholinesterases as well as monamine oxidases. With a k_{inact}/K_I value of 3×10^6 min^{-1} M^{-1}, ASS234 inhibition of MAO A and MAO B by ASS234 is almost as effective as clorgyline. It also forms the same N^5-adduct with the isoalloxazine ring of the FAD cofactor. The kinetic studies demonstrate that ASS234 is not only a reversible inhibitor of both acetyl and butyryl-cholinesterases with μM affinity, but is also a highly potent irreversible inhibitor of MAO A similarly to clorgyline. **1**. Estebana, Allanb, Samadi, *et al.* (2014) *Biochim. Biophys. Acta* **1844**, 1104.

Astemizole

This once-daily, orally bioavailable, second-generation antihistamine (FW = 458.57 g/mol; CAS 68844-77-9), also known as R43512, Hismanal® and 1-[(4-fluorophenyl)methyl]-*N*-[1-[2-(4-methoxyphenyl)ethyl]-4-piperidyl] benzoimidazol-2-amine, is selective H$_1$ receptor inverse agonist, with rapid onset of action. With the withdrawal of both astemizole and terfenadine (which produce unexpected and significant prolongation of the cardiac QT interval, with other more serious cardiac effects) from the U.S. and European markets, current widely available second-generation antihistamines include, loratadine, cetirizine, and fexofenadine (1). Because histamine favors the proliferation of normal and malignant cells, several anti-histamine drugs, including astemizole, inhibit tumor cell proliferation (2). Astemizole has also drawn interest by binding to ether à-go-go 1 (Eag1) and Eag-related gene (Erg) potassium channels, two important protein targets in cancer progression (2). **1**. Van Wauwe, Awouters, Neimegeers, *et al.* (1981) *Arch. Int. Pharmacodyn. Ther.* **251**, 39. **2**. García-Quiroz & Camacho (2011) *Anticancer Agents Med. Chem.* **11**, 307.

Astressin

This novel CRF antagonist (MW = 3663.20 g/mol; CAS 170809-51-5; Sequence: D-Phe-His-Leu-Leu-Arg-Glu-Val-Leu-Glu-Nle-Ala-Arg-Ala-Gln-Gln-Ala-Gln-*cyclo*-(γ-Glu-Ala-His-ε-Lys)Asn-Arg-Leu-Leu-Mle-Glu-Ile-Ile-NH$_2$; Solubility: 1 mg/mL in Acetic acid/water (10% vol/vol); Alternate Name: [D-Phe^{12}, Nle^{21,38}, Glu^{30}, Lys^{33}]-CRF (12-41)) targets human Corticotropin-Releasing hormone (CRH) Receptor subtypes CRF$_{2\alpha}$ (K_i = 1.5 nM) and CRF$_{2\beta}$ (K_i = 1 nM), with a comparable value (2 nM) for CRF$_1$. (**See also** *Antisauvagine-30; K41448*) These type-2 GPCRs reside in plasma membranes of hormone-sensitive cells and bind CRH, a peptide hormone and neurotransmitter involved in stress responses. Astressin is a structurally constrained and competitive antagonists of corticotropin-releasing factor (1). Astressin is neuroprotective in the hippocampus when administered after a seizure (2). Peripheral injection of astressin blocks peripheral CRF- and abdominal surgery-induced delayed gastric emptying in rats (3). A related antagonist is Astressin 2B (FW = 4041.69 g/mol; CAS 681260-70-8; Sequence: Ac-Asp-Leu-Ser-D-Phe-His-α-methyl-Leu-Leu-Arg-Lys-Nle-Ile-Glu-Ile-Glu-Lys-Gln-Glu-Lys-Glu-Lys-Gln-Gln-Ala-*cyclo*-(γ-Glu-Ala-Asn-ε-Lys)Leu-Leu-Leu-Asp-α-methyl-Leu-Ile-NH$_2$). Astressin 2B is a potent and selective antagonist of CRF$_2$ (IC$_{50}$ = 1.3 nM) and CRF$_1$ (IC$_{50}$ > 500 nM) (4-6). **1**. Gulyas, *et al.* (1995) *Proc. Natl. Acad. Sci. U.S.A.* **92**, 10575. **2**. Maecker, *et al.* (1997) *Brain Res.* **744**, 166. **3**. Martinez, *et al.* (1999) *J. Pharmacol. Exp. Ther.* **290**, 629. **4**. Rivier, *et al.* (2002). *J. Med. Chem.* **45**, 4737. **5**. Hoare, *et al.* (2005) *Peptides* **26**, 457. **6**. Henry, *et al.* (2006) *J. Neurosci.* **26**, 9142.

Astressin 2B

This Astressin analogue (FW = 4041.69 g/mol; CAS 681260-70-8; Sequence: Ac-Asp-Leu-Ser-D-Phe-His-α-methyl-Leu-Leu-Arg-Lys-Nle-Ile-Glu-Ile-Glu-Lys-Gln-Glu-Lys-Glu-Lys-Gln-Gln-Ala-*cyclo*-(γ-Glu-Ala-Asn-ε-Lys)Leu-Leu-Leu-Asp-α-methyl-Leu-Ile-NH$_2$) is a potent and selective antagonist of CRF$_2$ (IC$_{50}$ = 1.3 nM) and CRF$_1$ (IC$_{50}$ > 500 nM) (1-3). (**See also** *Antisauvagine-30; Astressin; K41448*) **1**. Rivier, *et al.* (2002). *J. Med. Chem.* **45**, 4737. **2**. Hoare, *et al.* (2005) *Peptides* **26**, 457. **3**. Henry, *et al.* (2006) *J. Neurosci.* **26**, 9142.

Asunaprevir

This highly selective HCV antiviral (FW = 748.29 g/mol; CAS 630420-16-5; Symbol: ASV), also named BMS-650032, targets NS3 protease, a serine proteinase required for Hepatitis C Virus (HCV) polyprotein processing, showing good antiviral activity against replicons based on HCV Genotype 1a (EC$_{50}$ = 4 nM), Genotype 1b (EC$_{50}$ = 1.2 nM), Genotype 4 (EC$_{50}$ = 1.8 nM), Genotype 5 (EC$_{50}$ = 1.7 nM), and Genotype 6 (EC$_{50}$ = 0.9 nM) (1). It is far less effective against Genotype 2 (EC$_{50}$ = 67 nM) and Genotype 3 (EC$_{50}$ = >1100 nM). Asunaprevir is a peptidomimetic that occupies the active site, inhibiting the HCV NS3/4A serine protease, with little or no action against related viruses (1). The average K_i for ASV is approximately 0.4 nM and 0.2 nM for Genotype 1a and Genotype 1b, respectively. Its acylsulfonamide moiety interacts noncovalently with the NS3 protease – unlike telaprevir's α-ketoamide moiety, which is linked covalently to NS3/4A's catalytic serine and reverses slowly over time. ASV exhibits an excellent selectivity index (>40,000x) against all serine/cysteine proteases evaluated, including human leukocyte elastase, porcine pancreatic elastase, and three members of the chymotrypsin family (1). Although ASV shows a low barrier to resistance, it can be combined with daclatasvir to achieve a very high rate of viral eradication in both naïve and treatment-experienced patients, showing a sustained virological response rate of 80%-90% (2,3). **1**. McPhee, Sheaffer, Friborg, *et al.* (2012) *Antimicrob. Agents Chemother.* **56**, 5387. **2**. Lok, Gardiner, Lawitz, *et al.* (2012) *New Engl. J. Med.* **366**, 216. **3**. Gentile, Buonomo, Zappulo, *et al.* (2014) *Ther. Clin. Risk Manag.* **10**, 493.

Asymmetric Dimethylarginine

This endogenous dimethylated amino acid (FW = 202.26 g/mol; CAS 30315-93-6; *Symbol*: ADMA), also named $N^г,N^г$-dimethylarginine and 2-amino-5-[(amino-N,N-dimethylaminomethylene)amino]pentanoic acid, is formed by the proteolysis of posttranslationally $N^г,N^г$-dimethylated arginine-containing proteins, with up to 10 mg excreted daily in the urine. ADMA targets cellular L-arginine uptake and nitric oxide synthase, thereby reducing the rate of NO formation (1). Elevated plasma ADMA levels are detected in patients or experimental animals with hypercholesterolemia, renal failure, atherosclerosis, hypertension, thrombotic microangiopathy, peripheral arterial occlusive disease and in the regenerated endothelial cells after angioplasty (3). Within endothelial cells, NO induces DDAH2 expression *via* a cGMP-mediated process, thereby reducing both ADMA and $N^г$-monomethylarginine (*or* MMA). When acting in concert, these enzymes constitute a positive feedback loop that maintains NO levels in endothelial cells (4). A high circulating ADMA level has also emerged as an independent risk marker predicting future CVD (cardiovascular disease) events (5). ADMA and its metabolizing enzymes also show significant temporal changes after Traumatic Brain Injury (TBI) and may be new targets in TBI treatment (6). **Pharmacologic Manipulation of ADMA Concentration:** PD-404182 potently inhibits dimethylarginine dimethylaminohydrolase isoform-1 (DDAH1), the human enzyme that degrades ADMA (*See PD-404182*) (7). On the other hand, chloroketone-containing peptide C21 irreversibly inhibits protein arginine methyltransferase 1 (PRMT1), the human enzyme that catalyzes the *S*-adenosyl-methionine-dependent post-translational methylation of arginine residues (*See C21*). PRMT1 generates the majority of protein-bound ADMA that represents the ultimate source of ADMA. **1**. Vallance, Leone, Calver, Collier & Moncada (1992). *J. Cardiovasc. Pharmacol* **20**, S60. **2**. Leiper & Vallance (1999) *Cardiovasc. Res.* **43**, 542. **3**. Masuda & Azuma (2002) *Nihon Yakurigaku Zasshi.* **119**, 29. **4**. Sakurada, Shichiri, Imamura, Azuma & Hirata (2008) *Hypertension* **52**, 903. **5**. Jawalekar, Karnik & Bhutey (2013) *Biochem. Res. Int.* **2013**, 189430. **6**. Jung, Wispel, Zweckberger, *et al.* (2014) *Int. J. Mol. Sci.* **15**, 4088. **7**. Ghebremariam, Erlanson & Cooke (2013) *J. Pharmacol. Exp. Ther.* **348**, 69.

AT-125, *See Acivicin*

AT-877, *See Fasudil*

AT-4140, *See Sparfloxacin*

AT7519

This novel small-molecule multi-CDK inhibitor (FW = 382.24 g/mol; CAS 844442-38-2, 902135-91-5 (HCl), 902135-89-1 (methanesulfonate); Solubility: 10 mg/mL DMSO; <1 mg/mL Water; Formulation: Dissolved in 0.9% saline), named systematically as 4-(2,6-dichlorobenzamido)-*N*-(piperidin-4-yl)-1*H*-pyrazole-3-carboxamide, targets cyclin dependent kinases CDK1/cyclin B, CDK2/Cyclin A, CDK3/Cyclin E, CDK4/Cyclin D1, CDK5/p35 and CDK6/Cyclin D3 with IC_{50} values of 210 nM, 47 nM, 360 nM, 100 nM, 13 nM and 170 nM, respectively (1). AT7519 is inactive against all non-CDK kinases with the exception of GSK3β (IC_{50} = 89 nM) (2). AT7519 displays potent cytotoxicity and apoptosis in multiple myeloma (MM) models, resulting *in vivo* tumor growth inhibition and prolonged survival (3). AT7519 inhibits CDK-mediated phosphorylation of RNA polymerase II (RNA pol II), attended by reduced RNA synthesis. AT7519 inhibits glycogen synthase kinase-3β) phosphorylation, and pretreatment with a selective GSK-3 inhibitor as well as shRNA GSK-3β knockdown restores MM survival, suggesting the involvement of GSK-3β in AT7519-induced apoptosis. GSK-3β activation was independent of RNA pol II dephosphorylation, as confirmed by α-amanitin, a specific RNA pol II inhibitor. **1**. Squires, *et al.* (2009) *Mol. Cancer Ther.* **8**, 324. **2**. Santo, *et al.* (2010) *Oncogene* **29**, 2325. **3**. Santo, Vallet, Hideshima, *et al.* (2010) *Oncogene* **29**, 2325.

AT7867

This potent ATP-competitive inhibitor (FW = 337.85; CAS 857531-00-1; Solubility: 68 mg/mL DMSO, <1 mg/mL H_2O), known systematically as 4-(4-(1*H*-pyrazol-4-yl)phenyl)-4-(4-chlorophenyl)piperidine, targets Akt1, Akt2 and Akt3 Protein Kinases B (*or* PKB), with IC_{50} values of 32 nM, 17 nM and 47 nM, respectively. AT7867 also inhibits the structurally related AGC kinases, p70S6K and PKA, with IC_{50} values of 20 nM and 85 nM, respectively. **1**. Grimshaw (2010) *Mol. Cancer Ther.* **9**, 1100.

AT9283

This potent JAK protein kinase inhibitor (FW = 381.43 g/mol; CAS 896466-04-9; Solubility = 76 mg/mL DMSO), also named 1-cyclopropyl-3-(3-(5-(morpholinomethyl)-1*H*-benzo[*d*]imidazol-2-yl)-1*H*-pyrazol-4-yl)urea, targets JAK3 (IC_{50} = 1.1 nM), JAK2 (IC_{50} = 1.2 nM), Aurora A (IC_{50} = 3 nM), Aurora B (IC_{50} = 3 nM), and and Abl (T315I). **1**. Howard, Berdini, Boulstridge, *et al.* (2009) *J. Med. Chem.* **52**, 379. **2**. Qi, Liu, Cooke, *et al.* (2012) *Int. J. Cancer* **130**, 2997.

AT-13387

This selective and potent chaperone inhibitor (FW = 409.53 g/mol; CAS 912999-49-6; Solubility: 25 mg/mL DMSO, <1 mg/mL H_2O), also named 2,4-dihydroxy-5-isopropylphenyl)-[5-(4-methylpiperazin-1-ylmethyl)-1,3-dihydroisoindol-2-yl]methanone, targets Hsp90α, K_i = 18 nM. Like other Hsp90 inhibitors, AT-13387 interacts with the N-terminal ATP-binding site to prevent ATP binding and stop the chaperone cycle, leading to degradation of multiple client proteins involved in tumor progression. AT13387 displays a long duration of action *in vitro* and *in vivo* in non-small cell lung cancer (2). It is also is effective against imatinib-sensitive and -resistant gastrointestinal stromal tumor models (3). The total synthesis of AT13387 has been reported (4). **1**. Woodhead, Angove, Carr, *et al.*

(2010) *J. Med. Chem.* **53**, 5956. **2.** Graham, Curry, Smyth, *et al.* (2012) *Cancer Sci.* **103**, 522. **3.** Smyth, Van Looy, Curry, *et al.* (2012) *Mol. Cancer Ther.* **11**, 1799. **4.** Patel & Barrett (2012) *J. Org. Chem.* **77**, 11296.

Ataluren

This reported translation modulator (FW = 284.24 g/mol; CAS 775304-57-9), also known as PTC124 and 3-[5-(2-fluorophenyl)-[1,2,4]oxadiazol-3-yl]-benzoate, reportedly induces dose- and time-dependent ribosomal readthrough of premature stop codons with greater potency than gentamicin (1,2). There is nonetheless some question as to whether the *Photinus pyralis* firefly luciferase (FLuc) reporter assay is an artifact. Importantly, FLuc inhibition and stabilization results from formation of an inhibitory product of the FLuc-catalyzed reaction between ATP and ataluren (3). A 2.0-Å cocrystal structure reveals the inhibitor to be the acyl-AMP mixed-anhydride adduct ataluren-AMP, which is a high-affinity (K_d = 120 pM) multisubstrate geometrical inhibitor. Biochemical assays, liquid chromatography/mass-spectrometry, and near-attack conformer modeling demonstrate that formation of this inhibitor depends upon precise positioning and reactivity of a key meta-carboxylate of ataluren within the FLuc active site. The inhibitory activity of ataluren-AMP is relieved by free coenzyme A, a component present at high concentrations in luciferase detection reagents used for cell-based assays. This finding explains why ataluren can appear to increase, instead of inhibit, FLuc activity in cell-based reporter gene assays. To test the ability of PTC124 to read through nonsense mutations, McElroy *et al.* (4) comparing ataluren's efficacy with that of the classical aminoglycoside antibiotic read-through agent, geneticin (G418) across a diverse range of *in vitro* reporter assays. That study confirmed the off-target FLuc activity of PTC124, but found that, while G418 exhibits varying activity in every read-through assay, there is no evidence of activity for PTC124. **1.** Welch, Barton, Zhuo, *et al.* (2007) *Nature* **447**, 87. **2.** Hirawat, Welch, Elfring, *et al.* (2007) *J. Clin. Pharmacol.* **47**, 430. **3.** Auld, Lovell, Thorne, *et al.* (2010) *Proc. Natl. Acad. Sci. U.S.A.* **107**, 4878. **4.** McElroy, Nomura, Torrie, *et al.* (2013) *PLoS Biol.* **11**, e1001593.

Atazanavir

This antiviral peptide drug (FW = 704.88 g/mol; CAS 198904-31-3), also known as BMS-232632, Rayataz, and systematically as methyl *N*-[(1*S*)-1-{[(2*S*,3*S*)-3-hydroxy-4-[(2*S*)-2-[(methoxycarbonyl)amino]-3,3-dimethyl-*N'*-{[4-(pyridin-2-yl)phenyl]methyl}butanehydrazido]-1-phenylbutan-2-yl]-carbamoyl}-2,2-dimethyl-propyl]carbamate, is a potent inhibitor of HIV-1 protease, particularly of indinavir- and saquinavir-resistant strains. Proof-of-principle studies indicate BMS-232632 blocks cleavage of viral precursor proteins in HIV-infected cells, demonstrating its function as an HIV Prt inhibitor. BMS-232632 is generally more potent than the five currently approved HIV-1 Prt inhibitors. BMS-232632 is highly selective for HIV-1 Prt and exhibits cytotoxicity only at concentrations 6,500- to 23,000-fold higher than required for anti-HIV activity. One mechanism that contribute to reduced atazanavir concentrations is the expression of ATP-binding cassette drug efflux transporters, especially P-glycoprotein multi-drug resistance factor. Because atazanavir is a substrate of the cytochrome P450 3A4 isoenzyme, it may alter serum concentrations of other drugs metabolized by the CYP3A4 pathway, such as rifabutin, clarithromycin, and lipid-lowering agents. **Key Pharmacokinetic Parameters:** *See* Appendix

II in Goodman & Gilman's THE PHARMACOLOGICAL BASIS OF THERAPEUTICS, 12th Edition (Brunton, Chabner & Knollmann, eds.) McGraw-Hill Medical, New York. **Target(s):** HIV-1 retropepsin (1,4,5); glucuronosyl-transferase, *or* UDPglucuronosyltransferases: UGT1A1, UGT1A3, and UGT1A4 (2); CYP3A (3); P-glycoprotein (3). **1.** Robinson, Riccardi, Gong, *et al.* (2000) *Antimicrob. Agents Chemother.* **44**, 2093. **2.** Zhang, Chando, Everett, *et al.* (2005) *Drug Metab. Dispos.* **33**, 1729. **3.** Perloff, Duan, Skolnik, Greenblatt & von Moltke (2005) *Drug Metab. Dispos.* **33**, 764. **4.** Clemente, Coman, Thiaville, *et al.* (2006) *Biochemistry* **45**, 5468. **5.** Surleraux, de Kock, Verschueren, *et al.* (2005) *J. Med. Chem.* **48**, 1965.

ATB-346

This H$_2$S-liberating naproxen analogue and nonsteroidal anti-inflammatory drug, *or* NSAID (FW = 365.44 g/mol; CAS 1226895-20-0; Soluble in DMSO, not in water), also named 4-carbamothioylphenyl 2-(6-methoxynaphthalen-2-yl)propanoate, is an anti-inflammatory agent with properties similar to naproxen, suppressing gastric prostaglandin E$_2$ synthesis as effectively, but with substantially reduced gastrointestinal toxicity (1,2). NSAID-induced gastroenteropathy is a significant limitation to the use of this class of drugs, and hydrogen sulfide is known to be an important mediator of gastric mucosal defense. ATB-346 inhibits cyclooxygenase-2 (COX2), suppressing leukocyte infiltration more effectively than naproxen. ATB-346 is as effective as naproxen in adjuvant-induced arthritis in rats, with a more rapid onset of activity. Unlike naproxen, ATB-346 does not elevate blood pressure in hypertensive rats. (**See also** Naproxen; NCX 429) **1.** Wallace, Caliendo, Santagada & Cirino (2010) *Brit. J. Pharmacol.* **159**, 1236. **2.** Blackler, Syer, Bolla, Ongini & Wallace (2012) *PLoS One* **7**, e35196.

Atenolol

This selective β$_1$ receptor antagonist, *or* β-blocker (FW = 266.34 g/mol; CAS 29122-68-7), also know as Tenormin® and (*RS*)-2-{4-[2-hydroxy-3-(propan-2-ylamino)propoxy]phenyl}-acetamide, is a widely used drug that slows the heart and reduces its workload. Unlike propranolol, atenolol does not penetrate the brain barrier and thus avoids the latter's well-known CNS side-effects. β-Blockers are no longer the recommended first-line therapy for primary hypertension, in view of data indicating that β-blockers are inferior to other antihypertensives and no better than placebo, in spite of its blood pressure-reducing action (1). Atenolol is useful in the prevention of angina pectoris, particularly in patients in whom there is associated arterial hypertension (2). **Key Pharmacokinetic Parameters:** *See* Appendix II in Goodman & Gilman's THE PHARMACOLOGICAL BASIS OF THERAPEUTICS, 12th Edition (Brunton, Chabner & Knollmann, eds.) McGraw-Hill Medical, New York (2011). **1.** Ripley & Saseen (2014) *Ann. Pharmacother.* **48**, 723. **2.** Hernández-Cañero, González, Cardonne, Pérez-Medina & Garcia-Barreto (1972) *Cor. Vasa.* **20**, 99.

Atomoxetine

This ADHD drug (FW$_{\text{free-base}}$ = 255.36 g/mol; FW$_{\text{hydrochloride}}$ = 291.81 g/mol; CAS 83015-26-3), also known as LY139603, Strattera®, and systematically as (3*R*)-*N*-methyl-3-(2-methylphenoxy)-3-phenylpropan-1-amine; (*R*)-*N*-

methyl-3-phenyl-3-(*o*-tolyloxy)-propan-1-amine, potently inhibits the presynaptic norepinephrine transporter, K_i = 4.5 nM (1-3) and is used to treat depression and attention-deficit/hyperactivity disorder (4). In 2002, Strattera became the first FDA-approved nonstimulant for the treatment of ADHD. The mechanism of action is related to its selective inhibition of presynaptic norepinephrine reuptake in the prefrontal cortex (5). Atomoxetine demonstrates high affinity and selectivity for norepinephrine transporters, but little or no affinity for neurotransmitter receptors. Atomoxetine demonstrates preferential binding to areas of known high distribution of noradrenergic neurons, such as the fronto-cortical subsystem. Atomoxetine undergoes extensive biotransformation, which is affected by poor metabolism by cytochrome P450 (CYP2D6) in a small percentage of the population (5). (Note: Atomoxetine was originally named tomoxetine, but was changed to avoid any potential confusion with tamoxifen.) **Key Pharmacokinetic Parameters:** *See* Appendix II in Goodman & Gilman's THE PHARMACOLOGICAL BASIS OF THERAPEUTICS, 12th Edition (Brunton, Chabner & Knollmann, eds.) McGraw-Hill Medical, New York. **1.** Farid, Bergstrom, Ziege, Parli & L. Lemberger (1985) *J. Clin. Pharmacol.* **25**, 296. **2.** Zerbe, Rowe, Enas, *et al.* (1985) *J. Pharmacol. Exp. Ther.* **232**, 139. **3.** Gehlert, Gackenheimer & Robertson (1993) *Neurosci. Lett.* **157**, 203. **4.** Kratochvil, Vaughan, Harrington & Burke (2003) *Expert Opin. Pharmacother.* **4**, 1165. **5.** Garnock-Jones & Keating (2009) *Pediatr. Drugs* **11**, 203.

Atopaxar

This novel orally active thrombin receptor antagonist (FW = 528.62 g/mol; CAS 751475-53-3), also known as E-5555, reversibly inhibits protease-activated receptor-1, *or* PAR-1 (IC$_{50}$ = 19 nM). The inhibition of thrombin-mediated platelet activation by means of protease-activated receptor-1 inhibitors represents an attractive therapeutic option for patients with atherothrombotic disease processes. (**See** *Vorapaxar*) **1.** Kogushi, *et al.* (2011) *Eur. J. Pharmacol.* **657**, 131. **2.** Marion, et al. (2006) *J. Clin. Pharmacol.* **46**, Abstract 102. **3.** Takeuchi, *et al.* (2007) *Eur. Heart J.* **28** (*Suppl.* 1), Abstract 264. **4.** Goto, *et al.* (2010) *Eur. Heart J.* **31**, 2601.

Atorvastatin

This widely used statin (FW$_{free-acid}$ = 558.65 g/mol; CAS 134523-00-5), also known as Lipitor™ and systematically named (3*R*,5*R*)-7-[2-(4-fluorophenyl)-3-phenyl-4-(phenylcarbamoyl)-5-propan-2-ylpyrrol-1-yl]-3,5-dihydroxy-heptanoic acid, is a potent inhibitor of 3- hydroxy-3-

methylglutaryl-CoA reductase, IC$_{50}$ = 8 nM (1-4). When taken with a low-fat diet, atorvastatin reduces the risk of heart attack, stroke, certain kinds of heart surgeries, and chest pain in patients with heart disease or several common heart disease risk factors (*i.e.*, family history of heart disease, high blood pressure, age, low HDL cholesterol, and smoking). *Note*: Because atorvastatin is a CYP3A4-metabolized statin, co-administration of a CYP3A4 inhibitors can result in adverse drug interactions. (**See** *Statin Side Effects*) **Pharmacokinetics & Metabolism:** Atorvastatin PK and metabolism are well characterized in humans. *Absorption*: Atorvastatin is rapidly absorbed after oral administration, reaching maximum plasma concentrations within 1-2 hours. Its extent of absorption increases in proportion to atorvastatin dose, with an absolute bioavailability of ~14% and a systemic availability of HMG-CoA reductase inhibitory activity at ~30%. The low systemic availability of atorvastatin is attributed to pre-systemic clearance in gastrointestinal mucosa and/or hepatic first-pass metabolism. Although food decreases the rate and extent of drug absorption by approximately 25% and 9%, respectively, as assessed by C$_{max}$ and AUC parameters, LDL-C reduction is similar whether atorvastatin is given with or without food. Plasma atorvastatin concentrations are lower (~30% for C$_{max}$ and AUC) following evening drug administration compared with morning. *Distribution*: The mean volume of distribution of atorvastatin is approximately 381 L. More than 98% of atorvastatin binds to plasma proteins. A blood/plasma ratio of ~0.25 indicates poor drug penetration into red blood cells. *Metabolism*: Atorvastatin is extensively metabolized to ortho- and para-hydroxylated derivatives and various β-oxidation products, but the *in vitro* inhibition of HMG-CoA reductase by these metabolites is equivalent to that of atorvastatin, and ~70% of circulating inhibitory activity for HMG-CoA reductase is attributed to active metabolites. *In vitro* studies suggest the importance of atorvastatin metabolism by cytochrome P450 3A4, consistent with increased plasma concentrations of atorvastatin in humans following coadministration with erythromycin, a known inhibitor of this cytochrome isozyme. *Excretion*: Atorvastatin and its metabolites are eliminated primarily in bile following hepatic and/or extra-hepatic metabolism; however, the drug does not appear to undergo enterohepatic recirculation. The mean plasma elimination half-life of atorvastatin in humans is ~14 hours, but the half-life of inhibitory activity for HMG-CoA reductase is 20-30 hours, mainly the result of the contribution of active metabolites. **Key Pharmacokinetic Parameters:** *See* Appendix II in Goodman & Gilman's THE PHARMACOLOGICAL BASIS OF THERAPEUTICS, 12th Edition (Brunton, Chabner & Knollmann, eds.) McGraw-Hill Medical, New York. **1.** Istvan & Deisenhofer (2001) *Science* **292**, 1160. **2.** Baumann, Butler, Deering, *et al.* (1992) *Tetrahedron Lett.* **33**, 2283. **3.** Kearney, Crawford, Mehta & Radebaugh (1993) *Pharm. Res.* **10**, 1461. **4.** Roth (2002) *Prog. Med. Chem.* **40**, 1.

Atractyloside

This toxic mitochondrial ATP-ADP translocation inhibitor and pro-apoptotic agent (FW = 802.98 g/mol for the dipotassium salt; CAS 102130-43-8), first isolated from the Mediterranean thistle *Atractylis gummifera*, blocks nucleotide binding to the carrier, thereby, inhibiting mitochondrial import of ADP and reducing oxidative phosphorylation. The amount of atractyloside required for achieving half-maximal inhibition is approximately one nmol atractyloside per mg mitochondrial protein. The dissociation constant for atractylate for the ATP-ADP translocase is about 5 x 10^{-8} M. Rat heart studies indicate that Atractyloside inhibits chloride channels within the mitochondrial membrane. (**See** *Bongkrekic Acid; Carboxyatractyloside*) **CAUTION:** Atractyloside poisoning is an infrequent but often fatal form of herbal poisoning, which occurs worldwide but especially in Africa and the Mediterranean regions. **Target(s):** ATP/ADP antiport or translocator, *or* adenine nucleotide translocator (1-8); 2,3' cyclic-nucleotide 3'-phosphodiesterase (9); and oxidative phosphorylation (10). **1.** Slater (1967) *Meth. Enzymol.* **10**, 48. **2.** Vignais, Brandolin, Lauquin & Chabert (1979) *Meth. Enzymol.* **55**, 518. **3.** Klingenberg, Grebe & Scherer (1975) *Eur. J. Biochem.* **52**, 351. **4.** Santi & Luciani, eds. (1978) *Atractyloside: Chemistry, Biochemistry, and Toxicology*, Piccin Medical Books, Padova, Italy. **5.** Kemp & Slater (1964) *Biochim. Biophys. Acta* **92**, 178. **6.** Pfaff, Heldt & Klingenberg (1969) *Eur. J. Biochem.* **10**, 484. **7.** Silva Lima, Denslow & Fernandes de Melo (1977) *Physiol. Plant* **41**, 193. **8.** Lima & Denslow (1979) *Arch. Biochem. Biophys.* **193**, 368. **9.** Dreiling, Schilling & Reitz (1981) *Biochim. Biophys. Acta* **640**, 114. **10.** Vignais & Vignais (1964) *Biochem. Biophys. Res. Commun.* **14**, 559.

Atrazine

This herbicide (FW = 215.68 g/mol; CAS 1912-24-9), also known as 2-chloro-4-(2-propylamino)-6-ethylamino-*s*-triazine, is a chloroplast electron transport inhibitor and popular selective herbicide. Atrazine is used to eliminate pre- and post-emergence broadleaf and grassy weeds from corn fields. **Target(s):** photosynthetic electron transport (1-3); photosystem II (4); *s*-triazine hydrolase (5); cyclic-nucleotide phosphodiesterase (6,7); unspecific monooxygenase (8). **1.** Izawa & Good (1972) *Meth. Enzymol.* **24**, 355. **2.** Dodge (1977) *Spec. Publ., Chem. Soc.* **29**, 7. **3.** Tischer & Strotmann (1977) *Biochim. Biophys. Acta* **460**, 113. **4.** Stemler & Murphy (1985) *Plant Physiol.* **77**, 179. **5.** Mulbry (1994) *Appl. Environ. Microbiol.* **60**, 613. **6.** Roberge, Hakk & Larsen (2004) *Toxicol. Lett.* **154**, 61. **7.** Roberge, Hakk & Larsen (2006) *Food Chem. Toxicol.* **44**, 885. **8.** Gorinova, Nedkovska & Atanassov (2005) *Biotechnol. Biotechnol. Equip.* **19**, 105.

Atropine

This parasympatholytic alkaloid (FW = 289.37 g/mol; CAS 51-55-8; White solid; M.P. = 114-116°C; Solubility: 2.2 mg/mL H_2O; Atropine HCl is highly soluble in water), also called (±)-hyoscamine, can be isolated from the deadly nightshade (*Atropa belladonna*) and related organisms. Atropine is a racemic mixture, with hyoscamine as the 3*S*-isomer. (*See also Hyoscyamine*) When delivered promptly, atropine can be an effective antidote for nerve agent poisoning and other cholinesterase inhibitors (*e.g.*, organophosphorus insecticides). Atropine is a competitive nonselective antagonist at central and peripheral muscarinic acetylcholine receptors. Its ability to block the effects of vagal stimulation was noted in 1867. **Target(s):** acid phosphatase (1); [β-adrenergic-receptor] kinase (2); cholinesterase, *or* butyrylcholinesterase (3); β-fructofuranosidase, *or* invertase (4); muscarinic acetylcholine receptors (5,6); phospholipase A_2 (7); tropinesterase (8); urease, reportedly inhibited by the free base (9). **1.** Lisa, Garrido & Domenech (1984) *Mol. Cell. Biochem.* **63**, 113. **2.** Kwatra, Benovic, Caron, Lefkowitz & Hosey (1989) *Biochemistry* **28**, 4543. **3.** Nagasawa, Sugisaki, Tani & Ogata (1976) *Biochim. Biophys. Acta* **429**, 817. **4.** Neuberg & Mandl (1950) *The Enzymes*, 1st ed., **1** (part 1), 527. **5.** Trovero, Brochet, Breton, *et al.* (1998) *Toxicol. Appl. Pharmacol.* **150**, 321. **6.** Al-Badr & Muhtadi (1985) *Anal. Profiles Drug Subs.* **14**, 325. **7.** Singh, Jabeen, Pal, *et al.* (2006) *Proteins* **64**, 89. **8.** Moog & Krisch (1974) *Hoppe-Seyler's Z. Physiol. Chem.* **355**, 529. **9.** Onodera (1915) *Biochem. J.* **9**, 544.

Auranofin

This gold-containing anti-rheumatic (FW = 679.49 g/mol; CAS 34031-32-8; photosensitive and thermolabile), also known as SK&F D-39162 and 1-thio-β-D-glucopyranosatotriethylphosphine gold-2,3,4,6-tetraacetate, is gold compound intended to improve on gold thioglucose and gold thiomalate as a treatment for rheumatoid arthritis as well as other inflammatory and proliferative diseases. **Mechanism(s) of Action:** Auranofin potently inhibits the release of lysosomal enzymes from leukocytes *in vitro* (1). At 1-10 µM, auranofin produces a dose-dependent reduction in extracellular levels of the lysosomal marker enzymes, β-glucuronidase and lysozyme, which are selectively released from rat leukocytes during phagocytosis of zymosan particles. The reduction in extracellular levels of lysosomal enzymes appears to be caused by inhibition of their selective cellular release, since effective concentrations of auranofin does not produce leukocyte cytotoxicity or inhibition of cell-free lysosomal enzyme activity (1). Morphologic and biochemical evidence indicated that auranofin also interferes with phagocytosis of zymosan particles. Another auranofin target is Vascular Endothelial Growth Factor Receptor-3, an endothelial cell (EC) surface receptor that is essential for angiogiogenesis and lymphangiogenesis (2). In both primary EC as well as EC cell lines, auranofin down-regulates VEGFR3 in a dose-dependent manner. At doses ≥ 1 µM, auranofin decreases the cellular survival protein thioredoxin reductase (TrxR2), TrxR2-dependent Trx2, and transcription factor NF-κB. At low doses (≤ 0.5 µM), however, auranofin specifically down-regulates VEGFR3 and VEGFR3-mediated EC proliferation and migration, two steps that are required for *in vivo* lymphangiogenesis. Auranofin-induced VEGFR3 down-regulation is blocked by antioxidant *N*-acetyl-L-cysteine (NAC) and the lysosome inhibitor chloroquine, but is promoted by proteasomal inhibitor MG132. Such results suggest auranofin induces VEGFR3 degradation through a lysosome-dependent pathway (2). **Effect on Diphtheria Toxin Processing:** Another auranofin target is diphtheria toxin, which delivers its catalytic domain DTA from acidified endosomes into the cytosol in a process that requires reduction of the disulfide bond linking DTA to the transport domain (3). Thioredoxin reduces this disulfide, and thioredoxin reductase (TrxR) is part of a cytosolic complex facilitating DTA-translocation. As a TrxR-selective inhibitor, auranofin prevents DTA delivery into the cytosol and intoxication of HeLa cells by diphtheria toxin (3). **Target(s):** thioredoxin reductase (2-4,9,28); glutathione peroxidase (5,24); glutathione-disulfide reductase (5); nitric-oxide synthase (6); phosphofructo-kinase (7); 5-lipoxygenase (8,21); trypsin-like neutral protease (10,11); cathepsins (11,26,27); histamine release (12,20); Na^+/K^+-exchanging ATPses (13); neutrophil collagenase (14,15); bile acid uptake (16); protein kinase C (17-19); phospholipase C, moderately inhibited (22); adenylate cyclase (23); IκB kinase (25). **1.** Dimartino & Walz (1977) *Inflammation* **2**, 131. **2.** Chen, Zhou, Huang, Lu & Min (2014) *Anticancer Agents Med. Chem.* **14**, 946. **3.** Schnell, Dmochewitz-Kück, Feigl, Montecucco & Barth (2015) *Toxicon.* **16**, 123. **4.** Kim, Kang, Lee, Choe & Hwang (2001) *Brit. J. Pharmacol.* **134**, 571. **5.** Gromer, Arscott, Williams, Schirmer & Becker (1998) *J. Biol. Chem.* **273**, 20096. **6.** Yamashita, Niki, Yamada, Mue & Ohuchi (1997) *Eur. J. Pharmacol.* **338**, 151. **7.** Anderson, Van Rensburg, Joone & Lessing (1991) *Mol. Pharmacol.* **40**, 427. **8.** Peters-Golden & Shelly (1989) *Biochem. Pharmacol.* **38**, 1589. **9.** Gromer, Merkle, Schirmer & Becker (2002) *Meth. Enzymol.* **347**, 382. **10.** Short, Steven, Griffin & Itzhaki (1981) *Brit. J. Cancer* **44**, 709. **11.** Rohozkova & Steven (1983) *Brit. J. Pharmacol.* **79**, 181. **12.** Takaishi, Morita, Kudo & Miyamoto (1984) *J. Allergy Clin. Immunol.* **74**, 296. **13.** Chan & Minta (1985) *Immunopharmacology* **10**, 61. **14.** Mallya & Van Wart (1987) *Biochem. Biophys. Res. Commun.* **144**, 101. **15.** Mallya & Van Wart (1989) *J. Biol. Chem.* **264**, 1594. **16.** Hardcastle, Hardcastle & Kelleher (1987) *J. Pharm. Pharmacol.* **39**, 572. **17.** Parente, Walsh, Girard, *et al.* (1989) *Mol. Pharmacol.* **35**, 26. **18.** Froscio, Murray & Hurst (1989) *Biochem. Pharmacol.* **38**, 2087. **19.** Mahoney, Hensey & Azzi (1989)

Biochem. Pharmacol. **38**, 3383. **20**. Shalit & Levi-Schaffer (1990) *Int. J. Immunopharmacol.* **12**, 403. **21**. Betts, Hurst, Murphy & Cleland (1990) *Biochem. Pharmacol.* **39**, 1233. **22**. Marki & Stanton (1992) *Arzneimittelforschung* **42**, 328. **23**. Lazarevic, Yan, Swedler, Rasenick & Skosey (1992) *Arthritis Rheum.* **35**, 857. **24**. Roberts & Shaw (1998) *Biochem. Pharmacol.* **55**, 1291. **25**. Jeon, Byun & Jue (2003) *Exp. Mol. Med.* **35**, 61. **26**. Chircorian & Barrios (2004) *Bioorg. Med. Chem. Lett.* **14**, 5113. **27**. Gunatilleke & Barrios (2006) *J. Med. Chem.* **49**, 3933. **28**. Omata, Folan, Shaw, *et al.* (2006) *Toxicol. In Vitro* **20**, 882.

Aureobasidin A

This cyclic depsipeptide antibiotic (FW$_\text{Aureosasidin-A}$ = 1101.43; CAS 127785-64-2) from *Aureobasidium pullulans* R106 is toxic (typically at 0.1-0.5 µg/mL) against *Saccharomyces cerevisiae, Schizosaccharo-myces pombe, Candida glabrata, Aspergillus nidulans,* and *A. niger.* Aureobasdin A inhibits inositolphosphoryl-ceramide synthase. Note that the antibiotic contains *N*-methyl-L-valine, L-alloisoleucine, *N*-methyl-L-phenyl-alanine, and *N*-methyl-3-hydroxy-L-valine. **Target(s):** inositolphosphoryl-ceramide synthase, *or* ceramide inositolphosphoryltransferase, inhibited by aureobasidin A (1,2). **1**. Fischl, Liu, Browdy & Cremesti (2000) *Meth. Enzymol.* **311**, 123. **2**. Lynch (2000) *Meth. Enzymol.* **311**, 130.

Auristatin, *See* Dolastatin 10; Monomethyl auristatin E

Aurothiomalate

This immunosuppressive (FW = 390.08 g/mol; CAS 12244-57-4; Symbol: AuTM), also known as sodium aurothiomalate, gold sodium thiomalate, Myocrisin®, and sodium 2-(auriosulfanyl)-3-carboxypropanoate, is an orally active nonsteroidal anti-inflammatory agent for treating rheumatism. **See also** Auranofin Even so, the demonstrated effectiveness of low-dose methotrexate (MTX) has led to a significant decline in using sodium aurothiomalate (1). **Mode of Action:** AuTM markedly inhibits DNA binding by the transcription factor AP-1, with less potent effects for AP-2, NF-1 and TFIID (2). In studies of the interactions of the progesterone receptor (*or* PR) with its DNA response element (*or* PRE), AuTM was found to regulate gene expression, again suggesting that inhibition of binding of a transcription factor to DNA is a likely mechanism (3). Electrophoretic mobility-shift assays showed that AP-1 DNA binding is inhibited by gold(I) thiolates and selenite, with IC$_{50}$ values occurring of 5 µM and 1 µM, respectively. Thiomalate was without effect in the absence of gold(I), and other metal ions inhibited at higher concentrations, in a rank order correlating with their thiol binding affinities. Cysteine-to-serine mutants demonstrated that these effects of Au(I) and selenite require Cys272 and Cys154 in the DNA-binding domains of Jun and Fos, respectively (3). Gold(I) thiolates and selenite did not inhibit nonspecific protein binding to the AP-1 site and were at least an order of magnitude less potent as inhibitors of sequence-specific binding to the AP-2, TFIID, or NF1 sites compared with the AP-1 site (3). *Note*: Given the chemical reactivity of AuTM, is is likely that this agent may inhibit other thiol-dependent processes within cells. **1**. Kean & Kean (2008) *Inflammopharmacol.* **16**, 112. **2**. Handel, Sivertsen, Watts, Day & Sutherland (1993) *Agents Actions Suppl.* **44**, 219. **3**. Handel, Watts, deFazio, Day & Sutherland (1995) *Proc. Natl. Acad. Sci. U.S.A.* **92**, 4497.

aUY11, *See* 5-(Perylen-3-yl)ethynyl-arabino-uridine

AUY922, *See* Luminespib

AV-412

This second-generation, oral dual kinase inhibitor (FW = 507.01 g/mol; CAS 451492-95-8), also known as *N*-(4-(3-chloro-4-fluorophenylamino)/-7-(3-methyl-3-(4-methylpiperazin-1-yl)but-1-ynyl)quinazolin-6-yl)acryl-amide, targets both EGFR and ErbB2 tyrosine kinases, including EGFR$^\text{L858R}$ and EGFR$^\text{T790M}$ mutants that are clinically resistant to the EGFR-specific kinase inhibitors, such as erlotinib and gefitinib. AV-412 inhibited EGFR variants and ErbB2 in the nM range, with over 100x selectivity over other kinases; however, ABL and FLT1 are both moderately sensitive. In cells, AV-412 inhibited autophosphorylation of EGFR and ErbB2 with respective IC$_{50}$ values of 43 and 282 nM. In animal studies using cancer xenograft models, AV-412 (30 mg/kg) demonstrated complete inhibition of tumor growth of the A431 and BT-474 cell lines, which overexpress EGFR and ErbB2, respectively. **1**. Suzuki, Fujii, Ohya, *et al.* (22007) *Cancer Sci.* **98**, 1977.

AV-951, *See* Tivozani

AVE1330A, *See* Avibactam

AVI-4658, *See* Eteplirsen

Avian Erythrocyte NAD⁺-dependent ADP-ribosyltransferase

This ribosyltransferase catalyzes the reaction of NAD$^+$ with an acceptor substrate to produce nicotinamide and the ADP-ribosylated acceptor. *Cholera* enterotoxin catalyzes the mono(ADP-ribosyl)ation of a number of GTP-binding proteins. The avian erythrocyte enzyme has been shown to modify both tubulin and MAP-2 and impair microtubule assembly. **1**. Purich & Scaife (1988) in *Enzyme Dynamics and Regulation* (Chock, Huang, Tsou & Wang, eds.), pp. 217-223, Springer-Verlag, New York. **2**. Scaife, Wilson & Purich (1992) *Biochemistry* **31**, 310. **3**. Raffaelli, Scaife & Purich (1992) *Biochem. Biophys. Res. Commun.* **184**, 414.

Avibactam

This non-β-lactam inhibitor of β-lactamase (FW = 261.27 g/mol; CAS 1192500-31-4), also known as NLX104 and AVE1330A, shows a spectrum of action against Classes A and C β-lactamase as well as selected Class D β-lactamase, enzymes that often confer resistance to β-lactam antibiotics. By forming a high-affinity complex with its target enzymes, avibactam enhances the antibacterial activity of certain β-lactam drugs, such as ceftaroline. With over 1000 known β-lactamase, the action of avibactam is apt to depend on the invidual active-site interactions. **Mode of Action:** Acylation and deacylation rates have been measured for the clinically important enzymes CTX-M-15, KPC-2, *E. cloacae* AmpC, *P. aeruginosa* AmpC, OXA-10, and OXA-48 (1). The efficiency of acylation (k_on/K_i) varies across the enzyme spectrum, from 1.1 x 10^1 M^{-1}s^{-1} for OXA-10 to 1.0 x 10^5 M^{-1}s^{-1} for CTX-M-15. Inhibition of OXA-10 was shown to follow a reversible covalent mechanism (see below), and the acylated OXA-10 displayed the longest residence time for deacylation, with a half-life of greater than 5 days. The inhibited enzyme forms are stable to hydrolysis for all enzymes with the exception of KPC-2, which displays slow hydrolytic route that involved fragmentation of the acyl-avibactam complex. In the case of TEM-1 lactamase, avibactam slowly covalently acylates its target, and the acylated enzyme subsequently undergoes slow deacylation (k_off = 0.045 min^{-1}) regenerating avibactam intact (2). **Use as a Combined Drug:** Combination of avibactim with extended-spectrum cephalosporins or aztreonam shows promise in inhibiting *Klebsiella pneumoniae* (KP) isolates harboring carbapenemases, *or* KPCs (3). Given abundant experience with ceftazidime and the significant improvement avibactam provides against contemporary β-lactamase-producing Gram-negative pathogens, this combination will likely play a role in the treatment of complicated urinary tract infections (as monotherapy) and complicated intra-abdominal infections (in combination with metronidazole) caused (or suspected to be caused) by otherwise resistant pathogens, such as extended spectrum β-lactamase-, AmpC-, *or Klebsiella pneumoniae* carbapenemase-producing *Enterobacteriaceae* and multidrug-resistant *P. aeruginosa* (4). **1**. Ehmann, Jahic, Ross, *et al*. (2013) *J. Biol. Chem.* **288**, 27960. **2**. Ehmann, Jahic, Ross, *et al*. (2012) *Proc. Natl. Acad. Sci.* **109**, 11663. **3**. Endimiani (2009) *Antimicrob. Agents Chemother.* **53**, 3599. **4**. Lagacé-Wiens, Walkty & Karlowsky (2014) *Core Evid.* **9**, 13.

Avidin

This bacteriostatic egg-white glycoprotein (MW = 54.7 kDa; CAS 1405-69-2; $Abs_{10mg/mL}$ = 15.5 at 282nm) contains four identical subunits, each containing a ultrahigh-affinity binding site ($K_d \approx 10^{-15}$ M) for biotin, a coenzyme essential in many carboxylation reactions. The avidin-biotin interaction is among the strongest known noncovalent interactions, making it the constant subject of biophysical inquiry. By binding ~15 μg biotin per mg, avidin (which makes up makes up ~0.05% of total egg protein, or around 1.8 mg/egg) deprives invading microorganisms of this vital coenzyme. Many microorganisms partly by-pass this effect by possessing enzymes that covalently link biotin to ε-NH_2 groups of active-site lysyl residues within biotin-dependent carboxylases. Upon cooking, biotin-free avidin is fully denatured, thus imparting no effect on those consuming cooked eggs. **Applications in Cytology:** The avidin-biotin interaction provides a simple and sensitive immunoenzymatic method to localize antigens (Ag) in formalin-fixed tissues, typically involving a biotin-labeled secondary antibody (Ab), followed by the addition of avidin-biotin-peroxidase complex, and affording superior results over the use of unlabeled antibodies (1). The availability of biotin-binding sites in the Ab-Ag complex is created by the incubation of a relative excess of avidin with biotin-labeled peroxidase. The oligomeric structure of avidin serves to bridge biotin-labeled peroxidase molecules to biotin-labeled peroxidase molecules. When compared on a mol-for-mol basis, the extreme affinity of avidin-biotin interactions rivals or exceeds that of nearly all known Ab-Ag interactions (***See also*** *Streptavidin*). **Biochemical Applications with Recombinant Fusion Proteins:** In *Escherichia coli*, the biotin carboxy carrier protein (BCCP) is biotinylated by BirA, a biotin ligase that covalently attaches a biotin to the amino group of a lysine residue present within the recognition sequence within BCCP. A minimal biotinylation sequence has been found from screens of combinatorial peptide libraries; this 13-residue peptide (*Sequence*: GLNDIFEAQKIEW), along with a 13-residue long variant (*Sequence*: GLNDIFEAQKIEWHE), termed the AviTag™, have been identified as effective *in vivo* and *in vitro* substrates for the BirA enzyme (2). When targets proteins are fused to the AviTag and co-expressed *in vivo* along with BirA, they can be biotinylated in bacteria, yeast, insect, or mammalian cells. Furthermore, when recombinant proteins are fused to the AviTag and incubated *in vitro* with purified BirA, they can be biotinylated efficiently on the central lysine residue in the AviTag. **1**. Hsu, Raine & Fanger (1981) *J. Histochem. Cytochem.* **29**, 577. **2**. Kay, Thai & Volgina (2009) *Meth. Mol. Biol.* **498**, 185.

AVL-301, *See Rociletinib*

AVN-944

This orally available antineoplastic agent (FW = 310.36 g/mol; CAS 297730-17-7), also known as VX-944 and (*R*)-1-cyanobutan-2-yl-(*S*)-1-(3-(3-(3-methoxy-4-(oxazol-5-yl)phenyl)ureido)phenyl)ethyl)carbamate, inhibits inosine 5'-monophosphate dehydrogenase (IMPDH), the enzyme catalyzing the conversion of IMP into XMP, leading to the biosynthesis of GMP, GDP, GTP, and ultimately dGTP. IMPDH is overexpressed in some cancer cells, particularly in hematological malignancies, and AVN944 appears to have a selective cytotoxic effect on cancer cells (1). Reduction of GTP in normal cells results in a temporary slowing of cell growth only.

AVN944 treatment results in dose-dependent cell growth inhibition in all studied prostate cancer cells, independently of their androgen sensitivity (1). AVN944 treatment arrests LNCaP cells in G_1-phase and and androgen-independent 22Rv1, DU145 and PC-3 cells in S-phase (1). AVN-944 treatment leads to the rapid disappearance of nucleostemin, a positive regulator of cell proliferation that is highly expressed in a variety of stem cells, tumors, and tumor cell lines (2). **1**. Floryk & Thompson (2008) *Int. J. Cancer* **123**, 2294. **2**. Huang, Itahana, Zhang & Mitchell (2009) *Cancer Res.* **69**, 3004.

Avosentan

This potent, selective and orally available endothelin A receptor blocker (FW = 479.51 g/mol; CAS 290815-26-8; Soluble in DMSO), also known by its code names SPP301 and Ro 67-0565, induces the dose-dependent reduction in the fractional renal excretion of sodium (up to 8.7% at 50 mg avosentan), with a paralleled dose-related increase in proximal sodium reabsorption. Avosentan reduces albuminuria, when added to standard treatment in patients with type 2 diabetes and overt nephropathy, but induces significant fluid overload and congestive heart failure (2). **1**. Smolander, Vogt, Maillard, *et al*. (2009) *Clin. Pharmacol. Ther.* **85**, 628. **2**. Mann, Green, Jamerson, *et al*. (2010) *J. Am. Soc. Nephrol.* **21**, 527.

Axitinib

This orally bioavailable anti-angiogenesis agent (FW = 386.47 g/mol; CAS 319460-85-0), also known by its code name AG013736, its trade name Inlyta®, and its systematic name *N*-methyl-2-[[3-[(*E*)-2-pyridin-2-ylethenyl]-1*H*-indazol-6-yl]sulfanyl]benzamide, is a tyrosine kinase inhibitor that targets vascular endothelial growth factor receptors VEGFR-1 (IC_{50} = 0.1 nM), VEGFR-2 (IC_{50} = 0.3 nM), and VEGFR-3 (IC_{50} = 0.1-0.3 nM), as well as platelet-derived growth factor receptor, *or* PDGFR (IC_{50} = 1.6 nM), and cKIT, *or* CD117 (IC_{50} = 1.7 nM). Axitinib is cytostatic in xenograft models for breast cancer, renal cell carcinoma, and other tumors, presumably by acting as a vasculature-disrupting agent, *or* VDA (1-7). After inhibition of VEGF signaling, some normal capillaries regress in a systematic sequence of events initiated by a cessation of blood flow and followed by apoptosis of endothelial cells, migration of pericytes away from regressing vessels, and formation of empty basement membrane sleeves that can facilitate capillary regrowth (1). A Phase-3 trial indicates its superiority in prolonging progression-free survival (PFS) in previously treated renal cell carcinoma (RCC) patients, with axitinib-treated patients exhibiting a median PFS of 6.7 months *versus* 4.7 months for sorafenib (8). Axitinib potently inhibits CYP1A2, K_i = 0.11 μM (9). **1**. Baffert, Le, Sennino, *et al*. (2006) *Am. J. Physiol. Heart Circ. Physiol.* **290**, H547. **2**. Hu-Lowe, Zou, Grazzini, *et al*. (2008) *Clin. Cancer Res.* **14**, 7272. **3**. Rossler, Monnet, Farace, *et al*. (2011) *Int. J. Cancer* **128**, 2748. **4**. Wilmes, Pallavicini, Fleming, *et al*. (2007) *Magn. Reson. Imaging* **25**, 319. **5**. Verbeek, Alves, de Groot, *et al*. (2011) *J. Clin. Endocrinol. Metab.* **96**, E991. **6**. Porta, Tortora, Linassier, *et al*. (2012) *Med. Oncol.* **29**, 1896. **7**. Rini, Escudier, Tomczak, *et al*. (2011) *Lancet* **378**, 1931. **8**. Mittal, Wood & Rini (2012) *Biol. Ther.* **2**, 5. **9**. Gu, Hibbs, Ong, Edwards & Murray (2014) *Biochem. Pharmacol.* **88**, 245.

AZ20

This novel potent inhibitor (FW = 412.51 g/mol; CAS 1233339-22-4; Solubility: 83 mg/mL DMSO; <1 mg/mL H$_2$O), also named AZ 20 and 4-[4-[(3R)-3-methyl-4-morpholinyl]-6-[1-(methylsulfonyl)cyclopropyl]-2-pyrimidinyl]-1H-indole, selectively targets ATR kinase, a serine/threonine-protein kinase also known as ataxia telangiectasia and Rad3-related protein (*or* ATR) kinase with IC$_{50}$ of 5 nM, 8-fold selectivity over mTOR. **1.** Foote, Blades, Cronin, *et al.* (2013) *J. Med. Chem.* **56**, 2125.

AZ-27

This RSV inhibitor (FW = 633.77 g/mol) targets Respiratory Syncytial Virus, inhibiting both mRNA transcription and genome replication in cell-based mini-genome assays, suggesting it targets a step common to both of these RNA synthesis processes. Analysis in an *in vitro* transcription run-on assay, containing RSV nucleocapsids, showed that AZ-27 inhibits (IC$_{50}$ ≈ 15 nM) the synthesis of transcripts from the 3′-end of the genome to a greater extent than those from the 5′-end, indicating that it inhibits transcription initiation. In experiments assaying polymerase activity on the promoter, AZ-27 inhibits transcription and replication initiation. The RSV polymerase also utilizes the promoter sequence to perform a back-priming reaction. Addition of AZ-27 has no effect on the addition of up to three nucleotides by back-priming, but inhibits further extension of the back-primed RNA. These findings suggest that the RSV polymerase is likely to adopt different conformations to execute its separate functions at the promoter. **1.** Tiong-Yip, Aschenbrenner, Johnson, *et al.* (2014) *Antimicrob. Agents Chemother.* **58**, 3867. **2.** Noton, Nagendra, Dunn, *et al.* (2015) *J. Virol.* **89**, 7786.

AZ191

This potent and selective DYRK inhibitor (FW = 429.53 g/mol), also systematically named N-(2-methoxy-4-(4-methylpiperazin-1-yl)phenyl)-4-(1-methyl-1H-pyrrolo[2,3-c]pyridin-3-yl)pyrimidin-2-amine, selectively targets DYRK1B, a <u>D</u>ual specificity t<u>y</u>rosine-phosphorylation-<u>r</u>egulated <u>k</u>inase, with an IC$_{50}$ value of 17 nM, or about 5x and 110x higher affinity than for DYRK1A and DYRK2. AZ191 is effective both *in vitro* and in cells. DYRK1B has also been proposed to promote the turnover of Cyclin D$_1$, a key regulator of mammalian G$_1$-S-phase transitions. Cyclin D$_1$ is phosphorylated on Thr-286 by GSK3β (*or* glycogen synthase kinase 3β), thereby promoting its degradation. **1.** Ashford, Oxley, Kettle, et al. (2014) *Biochem. J.* **457**, 43.

AZ3146

This selective inhibitor (FW = 452.55 g/mL; CAS 1124329-14-1; Solubility: 28 mg/mL DMSO; <1 mg/mL Water), also known as 9-cyclopentyl-2-(2-methoxy-4-(1-methylpiperidin-4-yloxy)phenylamino)-7-methyl-7H-purin-8(9H)-one, inhibits MPS1 protein kinases (IC$_{50}$ = 35 nM), which are found widely, but not ubiquitously, in eukaryotes and regulate kinetochore in both the chromosome attachment and the spindle checkpoints. MPS1 kinases also regulate centrosome function. **1.** Hewitt, Tighe, Santaguida, *et al.* (2010) *J. Cell Biol.* **190**, 25.

AZ3976

This small molecule PAI-1 inhibitor (FW = 317.35 g/mol) targets Plasminogen Activator Inhibitor type-1 (IC$_{50}$ = 26 μM, in enzyme assays; IC$_{50}$ = 16 μM, in a plasma clot lysis assay), not by binding to active PAI-1, but to latent PAI-1 (K_d = 0.29 μM at 35 °C). The observed binding stoichiometry (0.94 mol AZ3974 per mol PAI-1) was determined by isothermal calorimetry. The X-ray structure of AZ3976·PAI-1$_{latent}$ revealed that the inhibitor is lodged within the flexible-joint region, with the entrance to the cavity located between α-helix D and β-strand 2A. AZ3976 inhibits PAI-1 by enhancing the latency transition of active PAI-1, and, because AZ3976 only has measurable affinity for latent PAI-1, the likely mechanism of inhibition entails AZ3976 binding to the small fraction of PAI-1$_{latent}$ in equilibrium with PAI-1$_{active}$. These findings suggest new routes for designing small molecule drugs that target PAI-1. **1.** Fjellström, Deinum, Sjögren, *et al.* (2013) *J. Biol. Chem.*, **288**, 873.

AZ 6102

This moderately orally available TNKS2/TNKS1 inhibitor (FW = 428.53 g/mol; CAS 1645286-75-4; Solubility: 50 mM in DMSO), also named *rel*-2-[4-[6-[(3R,5S)-3,5-dimethyl-1-piperazinyl]-4-methyl-3-pyridinyl]phenyl]-3,7-dihydro-7-methyl-4H-pyrrolo[2,3-d]pyrimidin-4-one, targets tankyrase TNKS2 (IC$_{50}$ = 1 nM) and TNKS1 (IC$_{50}$ = 3 nM), exhibiting >100x greater selectivity for TNKS1/2 over PARP-1 (IC$_{50}$ = 2.0 μM), PARP-2 (IC$_{50}$ = 0.5 μM) and PARP-6 (IC$_{50}$ = 3.0 μM). Tankyrases are poly-ADP-ribosyltransferases that are involved in Wnt signaling pathway, telomere length, and vesicle trafficking. They activate the Wnt signaling pathway by poly-ADP-ribosylating AXIN1 and AXIN2, which are key components of the β-catenin destruction complex, such that poly-ADP-ribosylated target proteins are recognized by RNF146, the latter mediating their ubiquitination and later degradation. Germ-line mutations of several Wnt pathway components (*e.g.*, Axin, APC, and ß-catenin), are known to be oncogenic. Inhibition of the poly(ADP-ribose) polymerase (PARP) catalytic domain of the tankyrases (TNKS1 and TNKS2) is known to inhibit the Wnt pathway via increased stabilization of Axin. AZ-6102 inhibits Wnt signaling in DLD-1 cells (IC$_{50}$ = 5 nM). **1.** Johannes, Almeida, Barlaam, *et al.* (2015) *ACS Med. Chem. Lett.* **6**, 254.

AZ 10606120

This P2X$_7$ purine receptor antagonist (FW = 422.57 g/mol; CAS 607378-18-7; Solubility: 100 mM in DMSO), also named N-[2-[[2-[(2-hydroxyethyl)amino]ethyl]amino]-5-quinolinyl]-2-tricyclo[3.3.1.1^{3,7}]dec-1-ylacetamide, exhibits K_d values of 1.4 and 19 nM for human and rat receptors respectively. P2X$_7$ is a ligand-gated Ca^{2+}/Mg^{2+} cation channel that opens in response to the binding of extracellular ATP, resulting in cell depolarization. **1**. Michel, et al. (2007) Brit. J. Pharmacol. **151**, 103. **2**. Michel & Fonfria (2007) Brit. J. Pharmacol. **152**, 523. **3**. Michel, et al. (2008) Brit. J. Pharmacol. **153**, 737. **4**. Adinolfi, et al. (2012) Cancer Res. **72**, 2957.

8-Azaadenine

This bioavailable adenine analogue (FW = 136.12 g/mol) inhibits xanthine oxidase, K_i = 0.66 µM. When incorporated in DNA, 8-azaadenine base pairs with thymine (also with cytosine), and ab initio calculations suggest these base-pairs may be 7 kcal/mol more stable than Adenine::Thymine base pairs. **Target(s):** hypoxanthine(guanine) phosphoribosyl-transferase (1); xanthine oxidase (2); xanthine phosphoribosyltransferase (1); xanthine dehydrogenase (3,4). **1**. Naguib, Iltzsch, el Kouni, Panzica & el Kouni (1995) Biochem. Pharmacol. **50**, 1685. **2**. Sheu, Lin & Chiang (1997) Anticancer Res. **17**, 1043. **3**. Aretz, Kaspari & Klemme (1981) Z. Naturforsch. C **36**, 933. **4**. Pérez-Vicente, Alamillo, Cárdenas & Pineda (1992) Biochim. Biophys. Acta **1117**, 159.

25-Azacholesterol

This cholesterol analogue (FW$_{free-base}$ = 387.65 g/mol; CAS 1973-61-1) inhibits sterol 25-C-methyltransferase, or cycloartenol 24-C-methyltransferase, K_i = 10 nM (1,3-8) and cholestenol Δ-isomerase, or sterol Δ^{7,8}-isomerase (2). **1**. Oehlschlager, Angus, Pierce, Pierce & Srinivasan (1984) Biochemistry **23**, 3582. **2**. Nes, Zhou, Dennis, et al. (2002) Biochem. J. **367**, 587. **3**. Mangla & Nes (2000) Bioorg. Med. Chem. **8**, 925. **4**. Nes (2000) Biochim. Biophys. Acta **1529**, 63. **5**. Venkatramesh, Guo, Jia & Nes (1996) Biochim. Biophys. Acta **1299**, 313. **6**. Ator, Schmidt, Adams & Dolle (1989) Biochemistry **28**, 9633. **7**. Rahier, Génot, Schuber, Benveniste & Narula (1984) J. Biol. Chem. **259**, 15215. **8**. Nes, Guo & Zhou (1997) Arch. Biochem. Biophys. **342**, 68.

25-Azacycloartanol

This cycloartenol analogue (FW$_{free-base}$ = 429.73 g/mol), also known as 25-azacycloartenol, inhibits plant sterol biosynthesis and cycloartenol 24-C-methyltransferase (Prototheca wickerhamii K_i = 3 nM). **1**. Rahier, Bouvier, Cattel, Narula & Benveniste (1983) Biochem. Soc. Trans. **11**, 537. **2**. Zhao & Nes (2003) Arch. Biochem. Biophys. **420**, 18. **3**. Nes, Song, Dennis, et al. (2003) J. Biol. Chem. **278**, 34505. **4**. Mangla & Nes (2000) Bioorg. Med. Chem. **8**, 925. **5**. Nes (2000) Biochim. Biophys. Acta **1529**, 63. **6**. Rahier, Génot, Schuber, Benveniste & Narula (1984) J. Biol. Chem. **259**, 15215. **7**. Nes, Guo & Zhou (1997) Arch. Biochem. Biophys. **342**, 68.

5-Azacytidine

This nucleoside antibiotic (FW = 244.21 g/mol), also known as 1-β-D-ribofuranosyl-5-azacytosine, is a natural product of Streptoverticillius ladakanus that is effective against Gram-negative organisms, inhibiting the biosynthesis of DNA and RNA. Observations that 5-azacytidine was incorporated into DNA (1) and that, when present in DNA, inhibited DNA methylation, led to widespread use of 5-azacytidine and 5-aza-2'-deoxycytidine (Decitabine) to demonstrate the correlation between loss of methylation in specific gene regions and activation of the associated genes. Upon incorporation into DNA results in instability. Azacytidine resistance in myelodysplastic syndrome and acute myeloid leukemia appears to reverse azacytidine-mediated activation of apoptosis signaling and to the re-establish G$_2$/M checkpoint (2). Independently of its DNA methylation effects, 5-AzaC selectively and potently reduces expression of PCSK9, HMGCR, FASN, which are key genes in cholesterol and lipid metabolism, and this effect is observed in all tested cell lines and in vivo in mouse liver (3). Treatment with 5-AzaC perturb subcellular cholesterol homeostasis, impeding activation of sterol regulatory element-binding proteins (SREBPs). Through inhibition of UMP synthase, 5-AzaC also strongly induces expression of 1-acylglycerol-3-phosphate O-acyltransferase 9 (AGPAT9) and promoted triacylglycerol synthesis and cytosolic lipid droplet formation. Remarkably, complete reversal was obtained by co-addition of either UMP or cytidine (3). **Target(s):** protein biosynthesis (4,5); DNA biosynthesis (5); tRNA (cytosine-5) methyltransferase (7,8); DNA (cytosine-5)methyltransferase (9-11); adenosine deaminase (12); UTP synthase (3). **1**. Scott & Tomkins (1975) Meth. Enzymol. **40**, 273. **2**. Sripayap, Nagai, Uesawa, et al. (2013) Exp. Hematol. **42**, 294. **3**. Poirier, Samami, Mamarbachi, et al. (2014) J. Biol. Chem. **289**, 18736. **4**. Doskocil, Paces & Sorm (1967) Biochim. Biophys. Acta **145**, 771. **5**. Simson & Baserga (1971) Lab. Invest. **24**, 464. **6**. Reichman & Penman (1973) Biochim. Biophys. Acta **324**, 282. **7**. Lu & Randerath (1979) Cancer Res. **39**, 940. **8**. Lu & Randerath (1980) Cancer Res. **40**, 2701. **9**. Friedman (1981) Mol. Pharmacol. **19**, 314. **10**. Friedman & Cheong (1984) Biochem. Pharmacol. **33**, 2675. **11**. Christman (2002) Oncogene **21**, 5483. **12**. Pechan, Pechanova, Krizko, Starkova & Cihak (1983) Neoplasma **30**, 109.

(3R,4S)-1-[(8-Aza-9-deazaadenin-9-yl)methyl]-4-(benzylthiomethyl)-3-hydroxy-pyrrolidine

This immucillin nucleoside analogue (FW = 359.47 g/mol), which mimics a highly dissociative transition-state possessing ribooxacarbenium ion character, potently inhibits 5'-methylthioadenosine/S-adenosyl-homocysteine nucleosidase, K_i' = 400 fM (1) and S-methyl-5'-thioadenosine phosphorylase, K_i = 55 nM (2). When the ring nitrogen is protonated, this analogue mimics the charge of oxa-carbenium ion intermediate that forms transiently in many glycosidase reactions. This property results in tight binding. **1.** Singh, Evans, Lenz, *et al.* (2005) *J. Biol. Chem.* **280**, 18265. **2.** Evans, Furneaux, Lenz, *et al.* (2005) *J. Med. Chem.* **48**, 4679.

5-Aza-2'-Deoxycytidine

This 2'-deoxycytidine analogue (FW = 228.21 g/mol; CAS 2353-33-5), also known generically as Decitabine, by its trade name Dacogen® and systematically as 4-amino-1-(2-5'-deoxy-β-D-*erythro*-pentofuranosyl)-1,3,5-triazin-2(1*H*)-one, is a powerful DNA methylase inhibitor. Dacogen is specifically indicated for the treatment of multiple types of myelodysplastic syndromes, including previously treated and untreated, de novo and secondary myelodysplastic syndromes (MDS) of all French-American-British subtypes. **Primary Mode of Action:** Upon metabolic phosphorylation and direct incorporation into DNA, the modified DNA inhibits DNA methyltransferase, bringing about hypomethylation, cellular differentiation, and/or apoptosis. Decitabine also inhibits DNA methylation *in vitro*, which is achieved at concentrations that do not suppress DNA synthesis. Decitabine-induced hypomethylation in neoplastic cells may restore normal function to genes that are critical for the control of cellular differentiation and proliferation. In rapidly dividing cells, the cytotoxicity of decitabine may also be attributed to the formation of covalent adducts between DNA methyltransferase and decitabine incorporated into DNA. Non-proliferating cells are relatively insensitive to decitabine. **Effect on Inorganic Phosphorus-Induced Mineralization of Vascular Smooth Muscle Cells:** 5-aza-dC increase the expression and activity of alkaline phosphatase (ALP) and reduce DNA methylation in the ALP promoter region in the Human Aortic Smooth Muscle Cells, *or* HASMCs (5). Treatment with 5-aza-dC and downregulation of the DNMT1 expression both promotes P_i-induced mineralization of HASMCs. Moreover, both treatment with phosphonoformic acid (*or* PFA), a sodium-dependent phosphate transporter inhibitor, and suppression of the ALP expression inhibits the 5-aza-dC-promoted mineralization of HASMCs. **1.** Adams, Fulton & Kirk (1982) *Biochim. Biophys. Acta* **697**, 286. **2.** Paluska, Hrubá, Madar, *et al.* (1982) *Immunobiology* **162**, 288. **3.** Wilson, Jones & Momparler (1983) *Cancer Res.* **43**, 3493. **4.** Karahoca & Momparler (2013) *Clin. Epigenetics* **5**, 3. **5.** Azechi, Sato, Sudo & Wachi (2014) *J. Atheroscler. Thromb.* **21**, 463.

5-Aza-5,6-dihydro-2'-deoxycytidine

This pro-drug (FW = 115.12 g/mol), also known as KP1212, undergoes metabolic phosphorylation to the deoxynucleoside triphosphate, followed by its RNA-directed DNA polymerase-catalyzed incorporation into HIV DNA, causing lethal A-to-G and G-to-A mutations in the viral genome, both in tissue culture and in HIV positive patients on KP1212 monotherapy. KP1212 base pairs promiscuously with A and G, and variable-temperature NMR and 2-dimensional infrared spectroscopy shows that KP1212 exists as a broad ensemble of interconverting tautomers, which its enolic forms dominating. KP1212's mutagenic properties were assessed empirically by *in vitro* and *in vivo* replication of a single-stranded vector containing a single KP1212. It was found that KP1212 paired ambiguously with both A (10% of the time) and G (90% of the time), a mutation frequency that is sufficient for pushing a viral population over its error catastrophe limit. **1.** Li, Fedeles, Singh, *et al.* (2014) *Proc. Natl. Acad. Sci. USA* **111**, E3252.

25-Aza-24,25-dihydrozymosterol

This cholesterol analogue ($FW_{free-base}$ = 387.65 g/mol) inhibits sterol 24-*C*-methyltransferase (*Saccharomyces cerevisiae*, K_i = 5 nM). **1.** Pierce, Mueller, Unrau & Oehlschlager (1978) *Can. J. Biochem.* **56**, 794. **2.** Oehlschlager, Angus, Pierce, Pierce & Srinivasan (1984) *Biochemistry* **23**, 3582. **3.** Nes (2000) *Biochim. Biophys. Acta* **1529**, 63.

3-Azageranylgeranyl Diphosphate

This analogue of geranylgeranyl diphosphate (FW = 454.46 g/mol) has a K_i value of 15 nM for protein geranylgeranyltransferase, type I (1,2) and inhibits farnesyl*trans*transferase, *or* geranylgeranyl-diphosphate synthase (3-5). **1.** Yokoyama & Gelb (2001) *The Enzymes*, 3rd. ed., **21**, 105. **2.** Zhang, Moomaw & Casey (1994) *J. Biol. Chem.* **269**, 23465. **3.** Sagami, Korenaga, Ogura, Steiger, Pyun & Coates (1992) *Arch. Biochem. Biophys.* **297**, 314. **4.** Sagami, Morita & Ogura (1994) *J. Biol. Chem.* **269**, 20561. **5.** Szabo, Matsumura, Fukura, *et al.* (2002) *J. Med. Chem.* **45**, 2185.

8-Azaguanine

This toxic guanine analogue (FW = 152.12 g/mol; CAS 134-58-7; Soluble in dilute acid), also known as pathocidin and 5-amino-1,4-dihydro-7*H*-1,2,3-triazolo[4,5-*d*]pyrimidin-7-one, is readily incorporated into ribonucleic acids and acts as a purine antimetabolite. In most cases, 8-azaguanine inhibits translation and/or causes translation errors. Subsequent to its laboratory synthesis, 8-azaguanine was found to be a natural antibiotic produced by *Streptomyces spectabalis*. **Target(s):** xanthine oxidase (1-3); adenosine deaminase (3-5); methionine *S* adenosyltransferase (6,21); GTP cyclohydrolase I (7); xanthine dehydrogenase (8,24,25); translation ()9; pterin deaminase (10,18,19); rRNA maturation (11); ribonucleotide reductase (12); DNA biosynthesis (12); protein biosynthesis (13-15); uricase (16); sepiapterin deaminase (17); agmatine deaminase (20); tRNA guanine transglycosylase, also alternative substrate (21); hypoxanthine(guanine) phosphoribosyltransferase (22,23); xanthine phosphoribosyltransferase (22). **1.** Baker & Hendrickson (1967) *J. Pharmaceu. Sci.* **56**, 955. **2.** Baker (1967) *J. Pharmaceut. Sci.* **56**, 959. **3**. Webb (1966) *Enzyme and Metabolic Inhibitors*, vol. 2, Academic Press, New York. **4.** Zielke & Suelter (1971) *The Enzymes*, 3rd ed., **4**, 47. **5.** Feigelson & Davidson (1956) *J. Biol. Chem.* **223**, 65. **6.** Berger & Knodel (2003) *BMC Microbiol.* **3**, 12. **7.** Yoneyama, Wilson & Hatakeyama (2001) *Arch. Biochem. Biophys.* **388**, 67. **8.** Ribeiro (1988) *Biomed. Biochim. Acta.* **47**, 107. **9.** Rivest, Irwin & Mandel (1982) *Biochem. Pharmacol.* **31**, 2505. **10.** Takikawa, Kitayama-Yokokawa & Tsusue (1979) *J. Biochem.* **85**, 785. **11.** Weiss & Pitot (1974) *Arch. Biochem. Biophys.* **160**, 119. **12.** Brockman, Shaddix, Laster & Schabel (1970) *Cancer Res.* **30**, 2358. **13.** Zimmerman (1968) *Biochim. Biophys. Acta* **157**, 378. **14.** Mahadevan & Bhagwat (1969) *Indian J. Biochem.* **6**, 169. **15.** Zimmerman & Greenberg (1965) *Mol. Pharmacol.* **1**, 113. **16.** Norris & Roush (1950) *Arch. Biochem.* **28**, 465. **17.** Tsusue & Mazda (1977) *Experientia* **33**, 854. **18.** Tsusue, Takikawa & Yokokawa (1978) *Dev. Biochem.* **4**, 153. **19.** Rembold & Simmersbach (1969) *Biochim. Biophys. Acta* **184**, 589. **20.** Legaz, Iglesias & Vicente (1983) *Z. Pflanzenphysiol.* **110**, 53. **21.** Berger & Knodel (2003) *BMC Microbiol.* **3**, 12. **22.** Naguib,

Iltzsch, el Kouni, Panzica & el Kouni (1995) *Biochem. Pharmacol.* **50**, 1685. **23**. Schimandle, Mole & Sherman (1987) *Mol. Biochem. Parasitol.* **23**, 39. **24**. Aretz, Kaspari & Klemme (1981) *Z. Naturforsch. C* **36**, 933. **25**. Yen & Glassman (1967) *Biochim. Biophys. Acta* **146**, 35.

8-Aza-immucillin-H

This nucleotide analogue (FW = 266.24 g/mol), also known as (1*S*)-1-(7-hydroxy-1*H*-pyrazolo[4,3-*d*]pyrimidin-3-yl)-1,4-dideoxy-1,4-imino-D-ribitol, is a high-affinity transition-state mimic for purine nucleosidase. When the ring nitrogen is protonated, this analogue mimics the charge of oxa-carbenium ion intermediate that forms transiently in many glycosidase reactions. This property results in tight binding. **1**. Versées, Barlow & Steyaert (2006) *J. Mol. Biol.* **359**, 331.

1-Azakenpaullone

This cell-permeable protein kinase inhibitor (FW = 328.17 g/mol; CAS 676596-65-9: Solubility: 66 mg/mL DMSO; <1 mg/mL H_2O), also named 9-bromo-7,12-dihydro-pyrido[3,2':2,3]azepino[4,5-*b*]indol-6(5*H*)-one, selectively and potently targets glycogen synthase kinase 3β, *or* GSK3β (IC_{50} = 18 nM) with 100x less inhibitory action against cyclin-dependent kinases (CDKs) (1-3). 1-Azakenpaullone has become a gold standard for GSK3 inhibition (3). **1**. Zaharevitz, Gussio, Leost, *et al.* (1999) *Cancer Res.* **59**, 2566. **2**. Schultz, Link, Leost, *et al*, (1999) *J. Med. Chem.* **42**, 2909. **3**. Kunick, Lauenroth, Leost, Meijer & Lemcke (2004) *Bioorg. Med. Chem. Lett.* **14**, 413.

25-Azalanosterol

This steroid analogue (FW = 429.73 g/mol) inhibits sterol 24-*C*-methyltransferase (K_i = 7 nM). **1**. Mangla & Nes (2000) *Bioorg. Med. Chem.* **8**, 925. **2**. Venkatramesh, Guo, Jia & Nes (1996) *Biochim. Biophys. Acta* **1299**, 313. **3**. Nes, Jayasimha & Song (2008) *Arch. Biochem. Biophys.* **477**, 313. **4**. Zhou, Lepesheva, Waterman & Nes (2006) *J. Biol. Chem.* **281**, 6290. **5**. Nes (2005) *Biochem. Soc. Trans.* **33**, 1189. **6**. Nes, Jayasimha, Zhou, *et al.* (2004) *Biochemistry* **43**, 769. **7**. Nes, Guo & Zhou (1997) *Arch. Biochem. Biophys.* **342**, 68.

Azalanstat

This imidazole-containing agent (FW = 429.96 g/mol), also known as RS-21607, inhibits both inducible heme oxygenase-1 and constitutive heme oxygenase-2 (rat IC_{50} = 5.3 and 24.5 μM, respectively). The *m*-aminophenyl analogue is a slightly stronger inhibitor of heme oxygenase-1 (IC_{50} = 1 μM). Azalanstat is a stronger inhibitor of lanosterol 14α-demethylase (rat and human K_i = 2.5 and 0.79 nM, respectively). The other three stereoisomers are weaker inhibitors of the demethylase. **Target(s):** aromatase, *or* CYP19 (3,4); cholestanetriol 26-monooxygenase, *or* cholesterol 27-monooxygenase (4); cholesterol 7α-monooxygenase (CYP7)3,4; cholesterol monooxygenase, side-chain cleaving (4); CYP1A1 (4); CYP1A2 (4); CYP2C9 (4); CYP2D6 (4); heme oxygenase (1,2,5-7); progesterone 6β-monooxygenase, *or* CYP3A (3,4); progesterone 16α-monooxygenase (4); steroid 11β monooxygenase (3,4); steroid 17α-monooxygenase, *or* steroid 17α/20-lyase (3,4); steroid 21 monooxygenase (4); sterol 14-demethylase, *or* CYP5 (1); lanosterol 14α-demethylase (3,4,8). **1**. Roman, Riley, Vlahakis, *et al.* (2007) *Bioorg. Med. Chem.* **15**, 3225. **2**. Vlahakis, Kinobe, Bowers, *et al.* (2005) *Bioorg. Med. Chem. Lett.* **15**, 1457. **3**. Walker, Kertesz, Rotstein, *et al.* (1993) *J. Med. Chem.* **36**, 2235. **4**. Swinney, So, Watson, *et al.* (1994) *Biochemistry* **33**, 4702. **5**. Vlahakis, Kinobe, Bowers, *et al.* (2006) *J. Med. Chem.* **49**, 4437. **6**. Kinobe, Vlahakis, Vreman, *et al.* (2006) *Brit. J. Pharmacol.* **147**, 307. **7**. Kinobe, Dercho & Nakatsu (2008) *Can. J. Physiol. Pharmacol.* **86**, 577. **8**. Sloane, So, Leung, *et al.* (1995) *Gene* **161**, 243.

L-Azaserine

This substituted serine (FW = 173.13 g/mol; CAS 115-02-6), also known as O^3-(diazoacetyl)-L-serine and 3-[(diazoacetyl)oxy]alanine, is an L-glutamine analogue and inhibits many glutamine-dependent enzymes. Azaserine is also produced by species of *Streptomyces*. Solutions must always be freshly prepared. **Target(s):** phosphoribosyl-formylglycinamidine synthetase, *or* formylglycinamide ribonucleotide amidotransferase (1,8,10-13,16-20,24-27,29,32); amido-phosphribosyl-transferase (1,7,11,12, 15,16,31,54); γ-glutamyl transpeptidase, *or* γ-glutamyltransferase (2,3,14,21,23,55-57); NAD⁺ synthetase (glutamine-hydrolyzing) (4,5,11,15,16,51); carbamoyl-phosphate synthetase (6,11); glutamine:fructose-6-phosphate aminotransferase, isomerizing (7,52,53); anthranilate synthase (9,15,30,33); CTP synthetase (11); GMP synthetase (11,28,35,50); phosphoribosyl-formylglycinamide cyclo-ligase (12); glutaminase (15); glutamine uptake (22); glutamin-(asparagin-)ase, *or* glutaminase-asparaginase (34,36,37); glutamate synthase (38,39); γ-glutamyl peptide-hydrolyzing enzyme (40); γ-glutamyl hydrolase41; glutamate synthase (ferredoxin) (46-45); asparagine synthetase (46-49). **1**. Buchanan (1982) *Meth. Enzymol.* **87**, 76. **2**. Tate & Meister (1977) *Proc. Natl. Acad. Sci. U.S.A.* **74**, 931. **3**. Allison (1985) *Meth. Enzymol.* **113**, 419. **4**. Zalkin (1985) *Meth. Enzymol.* **113**, 297. **5**. Imsande, Preiss & Handler (1963) *Meth. Enzymol.* **6**, 345. **6**. Kaseman & Meister (1985) *Meth. Enzymol.* **113**, 305. **7**. Ghosh & Roseman (1962) *Meth. Enzymol.* **5**, 414. **8**. Flaks & Lukens (1963) *Meth. Enzymol.* **6**, 52. **9**. Tamir & Srinivasan (1970) *Meth. Enzymol.* **17A**, 401. **10**. Rando (1977) *Meth. Enzymol.* **46**, 28. **11**. Pinkus (1977) *Meth. Enzymol.* **46**, 414. **12**. Lewis & Hartman (1978) *Meth. Enzymol.* **51**, 171. **13**. Buchanan, Ohnoki & Hong (1978) *Meth. Enzymol.* **51**, 193. **14**. Meister, Tate & Griffith (1981) *Meth.*

Enzymol. **77**, 237. **15**. Meister (1962) *The Enzymes*, 2nd ed., **6**, 247. **16**. Shaw (1970) *The Enzymes*, 3rd ed., **1**, 91. **17**. French, Dawid & Buchanan (1963) *J. Biol. Chem.* **238**, 2186. **18**. Baker (1967) *Design of Active-Site-Directed Irreversible Enzyme Inhibitors*, Wiley, New York. **19**. Levenberg, Melnick & Buchanan (1957) *J. Biol. Chem.* **225**, 163. **20**. Webb (1966) *Enzyme and Metabolic Inhibitors*, vol. **2**, p. 333, Academic Press, New York. **21**. Repetto, Letelier, Aldunate & Morello (1987) *Comp. Biochem. Physiol. B* **87**, 73. **22**. Hsu, Marshall, McNamara & Segal (1980) *Biochem. J.* **192**, 119. **23**. Tate & Ross (1977) *J. Biol. Chem.* **252**, 6042. **24**. French, Dawid, Day & Buchanan (1963) *J. Biol. Chem.* **238**, 2171. **25**. Dawid, French & Buchanan (1963) *J. Biol. Chem.* **238**, 2178. **26**. Hartman, Levenberg & Buchanan (1955) *J. Amer. Chem. Soc.* **77**, 501. **27**. Schroeder, Allison & Buchanan (1969) *J. Biol. Chem.* **244**, 5856. **28**. Iarovaia, Mardashev & Debov (1967) *Vopr. Med. Khim.* **13**, 176. **29**. Mizobuchi & Buchanan (1968) *J. Biol. Chem.* **243**, 4853. **30**. Tamir & Srinivasan (1969) *J. Biol. Chem.* **244**, 6507. **31**. Wyngaarden (1972) *Curr. Top. Cell Regul.* **5**, 135. **32**. Chu & Henderson (1972) *Biochem. Pharmacol.* **21**, 401. **33**. Queener, Queener, Meeks & Gunsalus (1973) *J. Biol. Chem.* **248**, 151. **34**. Steckel, Roberts, Philips & Chou (1983) *Biochem. Pharmacol.* **32**, 971. **35**. Page, Bakay & Nyhan (1984) *Int. J. Biochem.* **16**, 117. **36**. Lebedeva, Kabanova & Berezov (1985) *Biull. Eksp. Biol. Med.* **100**, 696. **37**. Lebedeva, Kabanova & Berezov (1986) *Biochem. Int.* **12**, 413. **38**. Carlberg & Nordlund (1991) *Biochem. J.* **279**, 151. **39**. Marques, Florencio & Candau (1992) *Eur. J. Biochem.* **206**, 69. **40**. Mineyama, Mikami & Saito (1995) *Microbios* **82**, 7. **41**. Waltham, Li, Gritsman, Tong & Bertino (1997) *Mol. Pharmacol.* **51**, 825. **42**. Cullimore & Sims (1981) *Phytochemistry* **20**, 597. **43**. Galvan, Marquez & Vega (1984) *Planta* **162**, 180. **44**. Haeger, Danneberg & Bothe (1983) *FEMS Microbiol. Lett.* **17**, 179. **45**. Suzuki & Gadal (1982) *Plant Physiol.* **69**, 848. **46**. Lea & Fowden (1975) *Proc. R. Soc. Lond. B Biol. Sci.* **192**, 13. **47**. Reitzer & Magasanik (1982) *J. Bacteriol.* **151**, 1299. **48**. Stewart (1979) *Plant Sci. Lett.* **14**, 269. **49**. Patterson & Orr (1968) *J. Biol. Chem.* **243**, 376. **50**. Tesmer, Stemmler, Penner-Hahn, Davisson & Smith (1994) *Proteins* **18**, 394. **51**. Yu & Dietrich (1972) *J. Biol. Chem.* **247**, 4794. **52**. Chmara, Andruszkiewicz & Borowski (1985) *Biochim. Biophys. Acta* **870**, 357. **53**. Ghosh, Blumenthal, Davidson & Roseman (1960) *J. Biol. Chem.* **235**, 1265. **54**. King, Boounous & Holmes (1978) *J. Biol. Chem.* **253**, 3933. **55**. Suzuki, Kumagai & Tochikura (1986) *J. Bacteriol.* **168**, 1325. **56**. Takahashi, Zukin & Steinman (1981) *Arch. Biochem. Biophys.* **207**, 87. **57**. Takahashi, Steinman & Ball (1982) *Biochim. Biophys. Acta* **707**, 66.

Azathioprine

This mercaptopurine cancer agent, proliferation inhibitor, and immunosuppressant (FW = 277.27 g/mol; CAS 446-86-6 and 55774-33-9 (sodium salt); λ_{max} at 276 nm in 0.1 N HCl (ε = 18200 $M^{-1}cm^{-1}$) and 280 nm in 0.1 N NaOH (ε = 17300 $M^{-1}cm^{-1}$)), also named 6-[(1-methyl-4-nitro-1*H*-imidazol-5-yl)sulfanyl]-7*H*-purine, is a 6-mercaptopurine pro-drug. Prior to the advent of cyclosporin an FK506, azathioprine was often used in combination with prednisone as an immunosuppressant in kidney transplantation. **Mechanism of Inhibitory Action:** In the presence of glutathione, azathioprine is converted nonenzymatically to the active drug, 6-mercaptopurine (*See 6-Mercaptopurine*). Azathioprine has a superior therapeutic index relative to 6-mercaptopurine. Its major adverse effect is bone marrow suppression, one that can prove life-threatening in individuals with a deficiency in thiopurine *S*-methyltransferase (*Reaction: S*-Adenosyl-L-Methionine + 6-Mercaptopurine ⇌ 6-*S*-Methylpurine + *S*-Adenosyl-L-Homoserine). Azathioprine's metabolite *S*-methyl-6-thioinosine monophosphate also mimics the action of IMP in the allosteric feedback inhibition of the first step in purine biosynthesis, a reaction catalyzed by amidophosphoribosyl-transferase. **Pharmacokinetics:** Azothioprine bioavailability varies between 30-90%, owing to individual variability in its inactivation within the liver. Highest plasma concentrations are reached after one to two hours, with an average plasma half-life of 26-80 minutes for azathioprine and 200-300 minutes for drug plus metabolites. 6-

Mercaptopurine is converted to its nucleotide form by HGPRT and eventually gives rise to thioguanosine triphosphate and thio-deoxyguanosine triphosphate by metabolism of thioinosine monophosphate. **Key Pharmacokinetic Parameters:** *See* Appendix II in Goodman & Gilman's THE PHARMACOLOGICAL BASIS OF THERAPEUTICS, 12th Edition (Brunton, Chabner & Knollmann, eds.) McGraw-Hill Medical, New York. **Target(s):** methionine *S*-adenosyltransferase (1); phosphoribosylamino-imidazole-carboxamide formyltransferase (2). **1**. Berger & Knodel (2003) *BMC Microbiol.* **3**, 12. **2**. Ha, Morgan, Vaughn & Baggott (1990) *Biochem. J.* **272**, 339.

6-Azathymine

This thymine analogue (FW = 127.10 g/mol; CAS 932-53-6) has the following inhibitor targets: (*S*)-3-amino-2-methylpropionate:pyruvate aminotransferase, *or* L-3-aminoisobutyrate aminotransferase (1); DNA biosynthesis (2); 4-aminobutyrate aminotransferase (3,6); D-3-aminobutyrate aminotransferase (4); (*R*)-3-amino-2-methylpropionate: pyruvate aminotransferase, *or* D-3-aminoisobutyrate aminotransferase, weakly inhibited (5). **1**. Tamaki, Fujimoto, Sakata & Matsuda (2000) *Meth. Enzymol.* **324**, 376. **2**. Prusoff (1957) *J. Biol. Chem.* **226**, 901. **3**. Tamaki, Kubo, Aoyama & Funatsuka (1983) *J. Biochem.* **93**, 955. **4**. Tamaki, Kaneko, Mizota, Kikugawa & Fujimoto (1990) *Eur. J. Biochem.* **189**, 39. **5**. Kontani, Kaneko, Kikugawa, Fujimoto & Tamaki (1993) *Biochim. Biophys. Acta* **1156**, 161. **6**. Buu & van Gelder (1974) *Can. J. Physiol. Pharmacol.* **52**, 674.

6-Azauracil

This uracil analogue (FW = 113.08 g/mol; CAS 461-89-2), also known as 1,2,4-triazine-3,5-(2*H*,4*H*)-dione, inhibits 3-amino-isobutyrate and 4-aminobutyrate aminotransferases. **Target(s):** (*R*)-3-amino-2-methyl-propionate:pyruvate aminotransferase, *or* D-3-aminoisobutyrate aminotransferase (1,3,7); (*S*)-3-amino-2-methylpropionate:pyruvate amino-transferase, *or* L-3-aminoisobutyrate aminotransferase (1); 4-aminobutyrate amino-transferase (2,8); orotidine-5'-phosphate decarboxylase (4); β-alanine aminotransferase (5); uracil DNA glycohydrolase (6); alanine:glyoxylate aminotransferase (7); orotate phosphoribosyltransferase (9). **1**. Tamaki, Fujimoto Sakata & Matsuda (2000) *Meth. Enzymol.* **324**, 376. **2**. Tamaki, Kubo, Aoyama & Funatsuka (1983) *J. Biochem.* **93**, 955. **3**. Tamaki, Kaneko, Mizota, Kikugawa & Fujimoto (1990) *Eur. J. Biochem.* **189**, 39. **4**. Handschumacher & Pasternak (1958) *Biochim. Biophys. Acta* **30**, 451. **5**. Tamaki, Aoyama, Kubo, Ikeda & Hama (1982) *J. Biochem.* **92**, 1009. **6**. Bensen & Warner (1987) *Plant Physiol.* **83**, 149. **7**. Kontani, Kaneko, Kikugawa, Fujimoto & Tamaki (1993) *Biochim. Biophys. Acta* **1156**, 161. **8**. Buu & van Gelder (1974) *Can. J. Physiol. Pharmacol.* **52**, 674. **9**. Javaid, el Kouni & Iltzsch (1999) *Biochem. Pharmacol.* **58**, 1457.

6-Azauridine 5'-Monophosphate

This nucleotide analogue ($FW_{free-acid}$ = 325.17 g/mol) inhibits orotidylate decarboxylase and blocks pyrimidine biosynthesis. With protonation at N3, this analogue may resemble a putative zwitterionic intermediate preceding decarboxylation. **Target(s):** orotidine-5'-phosphate decarboxylase, *or* orotidylate decarboxylase, K_i = 7 x 10^{-7} M for yeast enzyme (1-5,7-22); orotate phosphoribosyltransferase (2); IMP dehydrogenase (6). **1**. Silva & Hatfield (1978) *Meth. Enzymol.* **51**, 143. **2**. Jones, Kavipurapu & Traut (1978) *Meth. Enzymol.* **51**, 155. **3**. Pasternak & Handschumacher (1959) *J. Biol. Chem.* **234**, 2992. **4**. Handschumacher (1960) *J. Biol. Chem.* **235**, 2917. **5**. Webb (1966) *Enzyme and Metabolic Inhibitors*, vol. 2, p. 472, Academic Press, New York. **6**. Exinger & Lacroute (1992) *Curr. Genet.* **22**, 9. **7**. Saenger, Suck, Knappenberg & Dirkx (1979) *Biopolymers* **18**, 2015. **8**. Levine, Brody & Westheimer (1980) *Biochemistry* **19**, 4993. **9**. Bell & Jones (1991) *J. Biol. Chem.* **266**, 12662. **10**. Levine, Brody & Westheimer (1980) *Biochemistry* **19**, 4993. **11**. Porter & Short (2000) *Biochemistry* **39**, 11788. **12**. Miller, Butterfoss, S. A. Short & R. Wolfenden (2001) *Biochemistry* **40**, 6227. **13**. Miller & Wolfenden (2002) *Ann. Rev. Biochem.* **71**, 847. **14**. Brody & Westheimer (1979) *J. Biol. Chem.* **254**, 4238. **15**. Shostak & Jones (1992) *Biochemistry* **31**, 12155. **16**. Krungkrai, Prapunwattana, Horii & Krungkrai (2004) *Biochem. Biophys. Res. Commun.* **318**, 1012. **17**. Wu, Christendat, Dharamsi & Pai (2000) *Acta Crystallogr. Sect. D* **56**, 912. **18**. Wu, Gillon & Pai (2002) *Biochemistry* **41**, 4002. **19**. Pragobpol, Gero, Lee & O'Sullivan (1984) *Arch. Biochem. Biophys.* **230**, 285. **20**. Shoaf & Jones (1973) *Biochemistry* **12**, 4039. **21**. Langdon & Jones (1987) *J. Biol. Chem.* **262**, 13359. **22**. Smiley & Jones (1992) *Biochemistry* **31**, 12162.

24-Azazymosterol

This steroid analogue (FW = 387.65 g/mol)) inhibits cycloartenol 24-*C*-methyltransferase, *or* SMT (K_i = 15 nM), or about four orders of magnitude more tightly the substrate cycloartenol (K_m = 28 μM), supporting the likely intermediacy of the carbenium ion reaction intermediate (1,2). Ergosterol (but not cholesterol or sitosterol) inhibits this methyltransferase (K_i = 80 μM). This combination of results suggests that the interrelationships of substrate functional groups within the active center of a $\Delta24^{25}$ to $\Delta25^{27}$-SMT could be approximated, thereby facilitating the rational design and testing of novel C-methylation inhibitors (2). In the earlier study (1), an isosteric analogue of the natural substrate zymosterol, in which the 26/27-gem-dimethyl groups were joined to form a cyclopropylidene function was found to be a potent irreversible mechanism-based inactivator of SMT enzyme activity that exhibits competitive-type inhibition (K_i = 80 μM; k_{inact} = 1.5 min^{-1}) **1** Mangla & Nes (2000) *Bioorg. Med. Chem.* **8**, 925. **2**. Nes, Guo & Zhou (1997) *Arch. Biochem. Biophys.* **342**, 68.

25-Azazymosterol, *Similar in action to 24-Azazymosterol*

AZD0530, *See* Saracatinib

AZ 628

This pan-Raf inhibitor (FW = 451.52 g/mol; CAS 878739-06-1; Solubility: 90 mg/mL DMSO, <1 mg/mL H_2O), also named 3-(2-cyanopropan-2-yl)-*N*-(4-methyl-3-(3-methyl-4-oxo-3,4-dihydroquinazolin-6-ylamino)-

phenyl)benzamide, targets the proto-oncogene B-Raf, *or* BRAF, *or*, more formally, serine/threonine-protein kinase B-Raf (IC_{50} = 105 nM), $BRAF^{V600E}$ (IC_{50} = 34 nM), and c-Raf-1 (IC_{50} = 29 nM), with off-target inhibition of VEGFR2, DDR2, Lyn, Flt1, FMS, and others (1). **Mechanism of Action:** Frequently mutated in human cancer, MAPK/ERK pathway consists of a RAS family GTP protein that is activated in response to extracellular signaling and recruits a RAF kinase family member to the cell membrane. Once activated, RAF signals through MAP/ERK kinase to activate ERK and its downstream effectors to regulate cell differentiation, proliferation, senescence, and survival. Mutations in BRAF (especially $BRAF^{V600E}$) are found in approximately 30% of all human cancers (melanoma, colorectal, and thyroid cancers), making the BRAF pathway a druggable target for therapy. **1**. Montagut, Sharma, Shioda, *et al.* (2008) *Cancer Res.* **68**, 4853.

AZ-960

This potent and selective JAK2 protein kinase inhibitor (FW = 354.36 g/mol; CAS 905586-69-8; Solubility = 70 mg/mL DMSO), also named 5-fluoro-2-[[(1*S*)-1-(4-fluorophenyl)ethyl]amino]-6-[(5-methyl-1*H*-pyrazol-3-yl)amino]-3-pyridinecarbonitrile, targets Janus-associated kinase-2 kinase. The $JAK2^{V617F}$ mutation is thought to play a critical role in the pathogenesis of polycythemia vera, essential thrombocythemia, and idiopathic myelofibrosis. AZ960 inhibits JAK2 kinase (K_i = = 0.45 nM, *in vitro*). Treatment of TEL-JAK2 driven Ba/F3 cells with AZ960 blocks STAT5 phosphorylation and potently inhibits cell proliferation (GI_{50} = 25 nM). AZ960 demonstrates selectivity for TEL-JAK2-driven STAT5 phosphorylation and cell proliferation when compared with cell lines driven by similar fusions of the other JAK kinase family members. JAK2 inhibition induces apoptosis by direct and indirect regulation of the anti-apoptotic protein BCL-xL. **1**. Gozgit, Bebernitz, Patil, *et al.* (2008) *J. Biol. Chem.* **283**, 32334.

AZD217, *See* Cediranib

AZD1080

This orally active, brain-permeable protein kinase inhibitor (FW = 334.37 g/mol; CAS 612487-72-6), also named 2-hydroxy-3-[5-(4-morpholinyl-methyl)-2-pyridinyl]-1*H*-indole-5-carbonitrile, selectively targets human GSK3α (K_i = 6.9 nM) and GSK3β (K_i = 31 nM), showing >14x lower inhibitory action against CDK2, CDK5, CDK1 and Erk2. AZD1080 inhibits human tau phosphorylation. The inhibitory pattern on tau phosphorylation reveals a prolonged pharmacodynamic effect predicting less frequent dosing in humans. Subchronic but not acute, administration with AZD1080 reverses MK-801-induced deficits, as measured by long-term potentiation in hippocampal slices and in a cognitive test in mice. Such findings suggest that reversal of synaptic plasticity deficits in dysfunctional systems requires longer term modifications of proteins downstream of GSK3β signaling. **1**. Georgievska, Sandin, Doherty, *et al.* (2013) *J. Neurochem.* **125**, 446.

AZD1152, See *Barasertib*

AZD1208

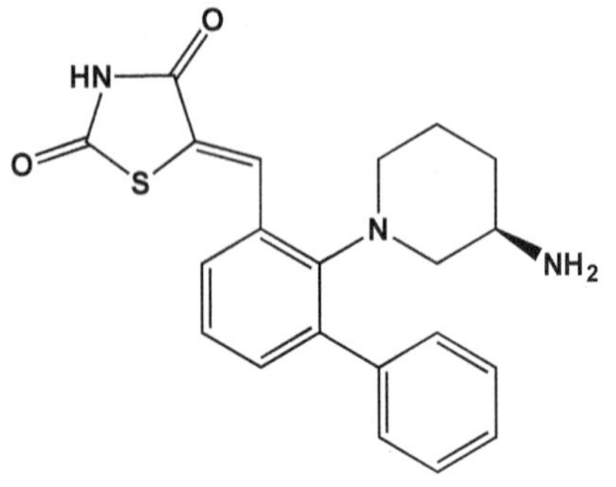

This orally available protein kinase inhibitor (FW = 379.48 g/mol; CAS 1204144-28-4; Soluble in DMSO, Insoluble in H_2O), also named 5-[[2-[(3R)-3-aminopiperidin-1-yl]biphenyl-3-yl]methylidene]-1,3-thiazolidine-2,4-dione, targets the proto-oncogene serine/threonine-protein kinases PIM-1, PIM-2 and PIM-3, often interrupting the G_1/S cell cycle transition and inducing apoptosis in PIM-overexpressing cells. Pim kinases are downstream effectors of ABL, JAK2, and Flt-3 oncogenes that are critical for driving tumorigenesis (1). PIM1, PIM2 and PIM3 were first recognized as pro-viral integration sites for the Moloney Murine Leukemia virus (2). Unlike other kinases, they possess a hinge region which creates a unique binding pocket for ATP. Absence of a regulatory domain makes these proteins are constitutively active once transcribed. Upregulation of Pim kinases is observed in several types of leukemias (especially Acute Myelogenous Leukemia (AML)) and lymphomas, with PIM-1, PIM-2 and PIM-3 promoting cancer cell proliferation and survival. AZD1208, demonstrates efficacy in preclinical models of acute myeloid leukemia (2). 1. Keeton, McEachern, Dillman, *et al.* (2013) *Blood* **123**, 905. 2. Swords, Kelly, Carew, *et al.* (2011) *Curr. Drug Targets* **12**, 2059.

AZD1480

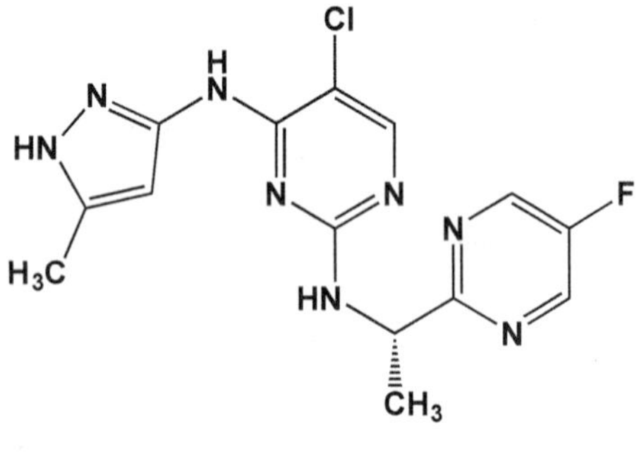

This ATP-competitive protein kinase inhibitor (FW = 348.77 g/mol; CAS 935666-88-9) targets Janus kinases JAK1 (IC_{50} = 1.3 nM) and JAK2 (IC_{50} = 0.26 nM), inducing arrest at the G_2/M junction and cell death by its inhibition of Aurora kinases. AZD1480 suppresses the growth of human solid tumor xenografts harboring persistent Stat3 activity, a downstream target of JAK2 (1). Through its inhibition of phosphorylation of JAK2, STAT3 and MAPK signaling proteins, AZD1480 (at low-μM concentrations) blocks proliferation and induces apoptosis in myeloma cell lines (2). AZD1480 likewise inhibits constitutive and stimulus-induced phosphorylation of JAK2 and STAT-3 in glioblastoma multiforme, a devastating and presently incurable disease (3). AZD1480 effectively inhibits tumor angiogenesis and metastasis, as mediated by STAT3 in stromal cells as well as tumor cells (4). Treatment of HCT116, HT29 and SW480 human colorectal cancer cell lines with AZD-1480 results in decreased phosphorylated JAK2 and phosphorylated STAT3, as well as the decreased expression of STAT3-targeted genes c-Myc, cyclin D2 and IL-6 (5). 1. Hedva, Huszar, Herrmann, *et al.* (2009) *Cancer Cell* **16**, 487. 2. Scuto, Krejci, Popplewell, *et al.* (2011) *Leukemia* **25**, 538. 3. McFarland, Ma, Langford, *et al.* (2011) *Mol. Cancer Ther.* **10**, 2384. 4. Xin, Herrmann, Reckamp, *et al.* (2011) *Cancer Res.* **71**, 6601. 5. Wang, Hu, Guo, *et al.* (2014) *Oncol. Rep.* **32**, 1991.

AZD-2281, See *Olaparib*

AZD2624

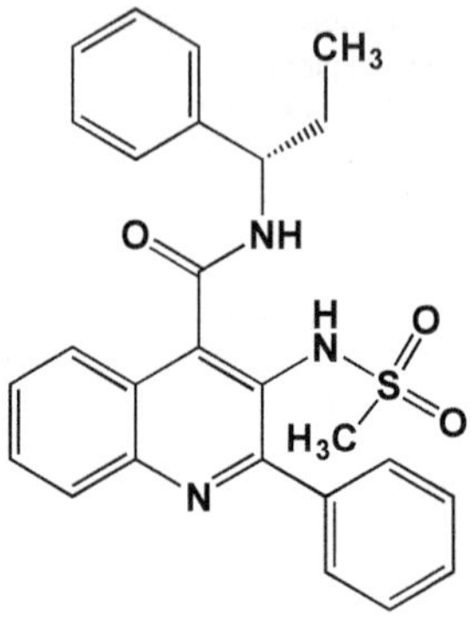

This neurokinin receptor antagonist (FW = 459.56 g/mol; CAS 941690-55-7) selectively targets Neurokinin 3, *or* NK_3, receptors, which are driven by tachykinins released from intrinsic primary afferent neurones (IPANs). AZD shows high selectivity for NK_3 receptors (K_i = 2 nM; calcium flux IC_{50} = 2.6 nM) with >100x selectivity over a 184 other receptors, enzymes, and ion channels, including NK_2R, NK_1R, and CCK_2R. The major human metabolite, AZ12592232, has only slightly weaker human NK3R antagonist potency (calcium flux IC_{50} = 9.0 nM), with >10x selectivity of toward NK_2R. CYP3A4 and CYP3A5 appear to be the primary enzymes forming the pharmacologically active ketone metabolite (M_1), whereas CYP3A4, CYP3A5, and CYP2C9 form the hydroxylated metabolite (M_2). Apparent K_m values are 1.5 and 6.3 μM for the formation of M_1 and M_2 in human liver microsomes, respectively. 1. Li, Zhou, Ferguson, *et al.* (2010) *Xenobiotica* **40**, 721.

AZD2858

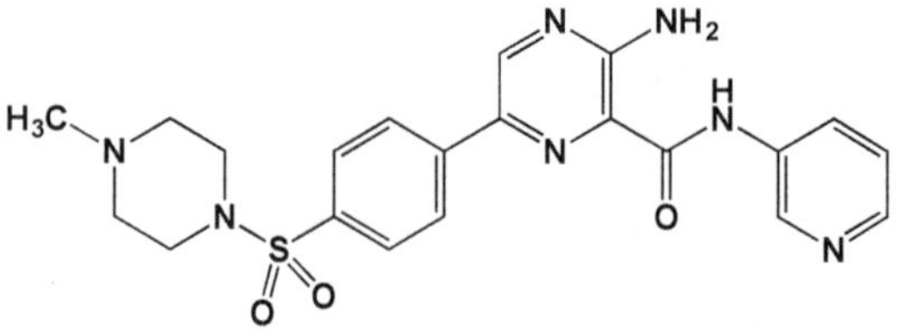

This selective protein kinase inhibitor (FW = 453.52 g/mol; CAS 486424-20-8; Solubility: 7 mg/mL DMSO), also named 3-amino-6-[4-[(4-methyl-1-piperazinyl)sulfonyl]phenyl]-N-3-pyridinyl-2-pyrazine carboxamide, targets glycogen synthase kinase-3 (*or* GSK-3), IC_{50} = 68 nM, thereby activating Wnt signal transduction pathway. Treatment of human osteoblast cells with 1 μM AZD2858 *in vitro* increased β-catenin levels (1). In rats, oral AZD2858 treatment caused a dose-dependent increase in trabecular bone mass compared to control after a two-week treatment (1). AZD2858 produced time-dependent changes in serum bone turnover biomarkers and increased bone mass over 28 days exposure in rats (2). AZD2858 likewise drives mesenchymal cells into the osteoblastic pathway, leading to direct bone repair in an otherwise unstable fracture milieu (3). 1. Marsell, Sisask, Nilsson, *et al.* (2012) *Bone* **50**, 619. 2. Gilmour, O'Shea, Fagura, *et al.* (2013) *Toxicol. Appl. Pharmacol.* **272**, 399. 3. Sisask, Marsell, Sundgren-Andersson, *et al.* (2013) *Bone* **54**, 126.

AZD3293

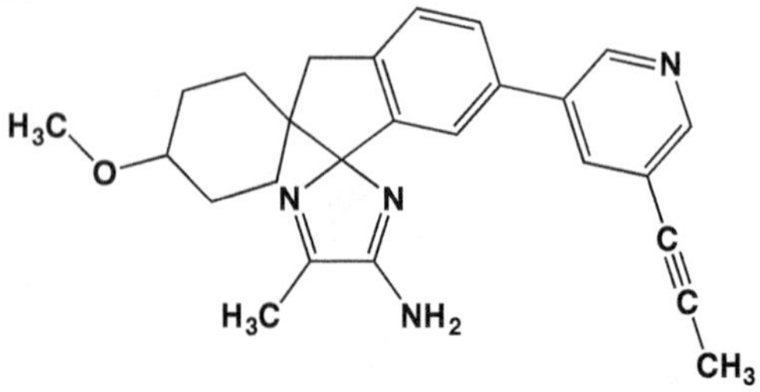

This orally active β-secretase "sheddase" inhibitor (FW = 412.54 g/mol), also named 4-methoxy-5"-methyl-6'-[5-(prop-1-yn-1-yl)pyridin-3-yl]-3'H-dispiro[cyclohexane-1,2'-indene-1',2"-imidazole]-4-amine and LY3314814, targets BACE1 (*or* β-site Amyloid Precursor Protein Cleaving Enzyme *or* Memapsin-2), an aspartic-acid protease important in the formation of myelin sheaths in peripheral nerve cells and in the pathogenesis of Alzheimer's disease. BACE1 cleaves the APP protein to release the C99 fragment, itself a substrate for subsequent γ-secretase cleavage to form Aβ peptide, reducing amyloid-related toxicity in Alzheimer's disease. AZD3293 is a blood-brain barrier (BBB) penetrating BACE1 inhibitor with unique slow off-rate kinetics (1). The in vitro potency of AZD3293 was demonstrated in several cellular models, including primary cortical neurons. In vivo in mice, guinea pigs, and dogs, AZD3293 displayed significant dose- and time-dependent reductions in plasma, cerebrospinal fluid, and brain concentrations of Aβ40, Aβ42, and sAβPPβ (1). The in vitro potency of AZD3293 in mouse and guinea pig primary cortical neuronal cells was correlated to the in vivo potency expressed as free AZD3293 concentrations in mouse and guinea pig brains (1). AZD3293 is the only BACE1 inhibitor for which prolonged suppression of plasma Aβ with a QW dosing schedule has been reported (2).   **1.** Eketjäll, Janson, Kaspersson, *et al.* (2016) *J. Alzheimers Dis.* **50**, 1109.   **2.** Cebers, Alexander, Haeberlein, *et al.* (2017) *J. Alzheimers Dis.* **55**, 1039.

AZD3463

This novel, orally bioavailable ALK inhibitor (FW = 448.95 g/mol; CAS 1356962-20-3; Solubility: 24 mg/mL DMSO; <1 mg/mL H$_2$O), also named *N*-[4-(4-amino-1-piperidinyl)-2-methoxyphenyl]-5-chloro-4-(1*H*-indol-3-yl)-2-pyrimidinamine, inhibits Anaplastic Lymphoma Kinase (ALK) with a K_i of 0.75 nM and overcomes multiple mechanisms of acquired crizotinib resistance. AZD3463 inhibits ALK in cells as demonstrated by its ability to decrease ALK autophosphorylation in tumor cell lines containing ALK fusions, including DEL (ALCL NPM-ALK), H3122 (NSCLC EML4-ALK) and H2228 (NSCLC EML4-ALK) mutant forms. Inhibition of ALK is associated with perturbations in downstream signaling including ERK, AKT and STAT3 pathways leading to preferential inhibition of proliferation in the ALK fusion containing cell lines in vitro. AZD3463 also demonstrates the ability to dose dependently inhibit pALK in xenograft tumors in vivo resulting in stasis or regression. **Other Target(s):** AZD3463 also inhibits additional receptor tyrosine kinases including insulin growth factor receptor (IGF1R) with equivalent potency. **1.** Drew, Cheng, Engelman, *et al.* (2013) *Cancer Res.* **73**, Suppl. 1, Abstr. 919.

AZD3965

This orally bioavailable MCT1 inhibitor (FW = 515.51 g/mol; CASs = 733809-45-5 and 1448671-31-5), also named (*S*)-5-(4-hydroxy-4-methylisoxazolidine-2-carbonyl)-1-isopropyl-3-methyl-6-((3-methyl-5-(trifluoromethyl)-1*H*-pyrazol-4-yl)methyl)-thieno[2,3-*d*]pyrimidine-2,4(1*H*, 3*H*)-dione, selectively targets the Monocarboxylate Transporter-1 (IC$_{50}$ = 1.6 nM) and is ~6x less active toward MCT2. AZD3965 does not inhibit MCT4, even at 10 μM. MCT1 and MCT4 are primarily involved in lactate transport, targeting highly glycolytic cancer cells, especially those that are hypoxic. The latter property suggests that the combined use with a vascular-disrupting anticancer agent (such as combrestatin) may improve the inhibitory action of AZD-3965 against cancer cells. **1.** Polański, Hodgkinson, Fusi, *et al.* (2014) *Clin. Cancer Res.* **20**, 926.

AZD4547

This potent and selective FGFR inhibitor (FW = 463.57 g/mol; CAS 1035270-39-3; Solubility (25°C) = 92 mg/mL DMSO; <1 mg/mL water), also known as *N*-(5-(3,5-dimethoxyphenethyl)-1*H*-pyrazol-3-yl)-4-((3*S*,5*R*)-3,5-dimethylpiperazin-1-yl)benzamide, is active against the receptor-coupled tyrosine kinase activity of both the wild-type and mutant forms of fibroblast growth factor receptors (1). AZD4547 exhibits weaker activity against FGFR4, and VEGFR2(KDR), with little activity observed against IGFR, CDK2, and p38. A 1.65 Å resolution structure of AZD4547 bound to the kinase domain of FGFR1 reveals extensive drug-protein interactions, an integral network of water molecules, and tight closure of the FGFR1 P-loop, the latter forming a long, narrow crevice into which AZD4547 binds (2). Such findings likely explain the sub-nM K_i for its action as a FGFR1 inhibitor. **Targets:** FGFR1, K_i = 0.2 nM; FGFR1, K_i = 2.5 nM; FGFR1, K_i = 1.8 nM (1). **1.** Gavine, *et al.* (2012) *Cancer Res.* **72**, 2045. **2.** Yosaatmadja, Patterson, Smaill & Squire (2015) *Acta Crystallogr. D Biol. Crystallogr.* **71**, 525.

AZD5438

This cyclin kinase-directed inhibitor (FW = 371.46 g/mol; CAS 602306-29-6; Solubility: 70 mg/mL DMSO; <1 mg/mL Water; Formulation: dissolve in hydroxy-propyl-methyl-cellulose), systematically named 4-(1-isopropyl-2-methyl-1*H*-imidazol-5-yl)-*N*-(4-(methylsulfonyl)-phenyl)-pyrimidin-2-amine, targets CDK1, CDK2 and CDK9 with IC_{50} values of 16 nM, 6 nM, and 20 nM, respectively, inducing cell cycle arrest in a range of tumor cell lines. AZD5438 also inhibits the kinase activity of p25-cdk5 and glycogen synthase kinase 3β with IC$_{50}$ of 14 nM and 17 nM, respectively. **Cyclin Target Selectivity:** Cdk1 (weak, if any), Cdk2 (weak, if any), Cdk3 (weak, if any), Cdk4 (weak, if any), Cdk5 (weak, if any), Cdk6 (weak, if any), Cdk7 (++), Cdk8 (weak, if any), Cdk9 (weak, if any), Cdk10 (weak, if any). **1.** Byth, *et al.* (2009) *Mol. Cancer Ther.* **8**, 1856.

AZD 5582

This potent dimeric Smac mimetic (FW = 1088.21 g/mol; Solubility: 100 mM in H$_2$O; 100 mM in DMSO), also named 3,3'-[2,4-hexadiyne-1,6-diylbis[oxy[(1*S*,2*R*)-2,3-dihydro-1*H*-indene-2,1-diyl]]]bis[*N*-methyl-L-alanyl-(2*S*)-2-cyclohexylglycyl-L-prolinamide, targets X-linked Inhibitor of

Apoptosis Protein (or XIAP) and cellular Inhibitor of Apoptosis Proteins 1 and 2 (cIAP1 and cIAP2), withe IC$_{50}$ values of 15, 15 and 21 nM for XIAP, cIAP1 and cIAP2, respectively. A protein that promotes cytochrome c-dependent activation by eliminating the inhibition via IAP, Smac is normally found in mitochondria, but escapes into the cytosol when a cell is primed for apoptosis during the final so-called execution step of caspase activation. When administered intravenously to MDA-MB-231 xenograft-bearing mice, AZD5582 results in cIAP1 degradation and caspase-3 cleavage within tumor cells, resulting in substantial tumor regression after two weekly doses of 3.0 mg/kg. Antiproliferative effects are observed with AZD 5582 in only a small subset of the over 200 cancer cell lines examined, consistent with other published IAP inhibitors. AZD 5582 binds to the BIR3 domain of XIAP to prevent interaction with caspase-9. Causes degradation of cIAP1 and cIAP2 and induces apoptosis in MDA-MB-231 breast cancer cells. **1**. Hennessy, Adam, Aquila, *et al.* (2013) *J. Med. Chem.* **56**, 9897.

AZD6140, See *Ticagrelor*

AZD6244, See *Selumetinib*

AZD6738

This orally active, and selective ATR kinase inhibitor (FW = 412.51 g/mol; CAS 1352226-88-0; Solubility: 80 mg/mL DMSO; < 1 mg/mL H$_2$O) targets Ataxia Telangiectasia/Rad3-related kinase (IC$_{50}$ = 1 nM), a serine/threonine protein kinase that activates checkpoint signaling upon genotoxic stresses (*e.g.*, ionizing radiation, ultraviolet light, or replication stalling), thereby serving as a DNA damage sensor. In a model consisting of primary human Chronic Myelogenous Leukemia (CL) cells with biallelic TP53 or ATM inactivation xenotransplanted into NOD/Shi-scid/IL-2Rγ mice, AZD6738 provides potent and specific inhibition of ATR signalling with compensatory activation of ATM/p53 pathway in cycling CLL cells in the presence of genotoxic stress (1). In p53 or ATM-defective cells, AZD6738 treatment results in replication fork stalls and accumulation of unrepaired DNA damage, as evidenced by γH2AX and 53BP1 foci formation, carried through into mitosis and resulting in cell death by mitotic catastrophe (1). AZD6738 displays selective cytotoxicity towards ATM- or p53-deficient CLL cells (1). AZD6738 also potentiates cisplatin and gemcitabine cytotoxicity in Non-Small-Cell Lung Carcinoma (NSCLC) cell lines with intact ATM kinase signaling. It also potently synergizes with cisplatin in ATM-deficient NSCLC cells (2). When used in combination, cisplatin and AZD6738 resolve ATM-deficient lung cancer xenografts. **1**. Kwok, Davies, Agathanggelou, *et al.* (2015) *Lancet* **385**, Suppl. 1, S58. **2**. Vendetti, Lau, Schamus, *et al.* (2015) *Oncotarget* **6**, 44289.

AZD7545

This [pyruvate dehydrogenase] kinase-2 inhibitor (FW = 478.88 g/mol) maintains the active dephospho-form of pyruvate dehydrogenase *in vivo*, thereby improving blood glucose control in Type II diabetes. **1**. Mayers, Butlin, Kilgour, *et al.* (2003) *Biochem. Soc. Trans.* **31**, 1165. **2**. Kato, Li, Chuang & Chuang (2007) *Structure* **15**, 992. **3**. Roche & Hiromasa (2007) *Cell. Mol. Life Sci.* **64**, 830. **4**. Mayers, Leighton & Kilgour (2005) *Biochem. Soc. Trans.* **33**, 367. **5**. Tuganova, Klyuyeva & Popov (2007) *Biochemistry* **46**, 8592.

AZD8931

This signal transduction kinase inhibitor (FW = 473.94 g/mol; CAS 848942-61-0), also named 2-(4-[4-(3-chloro-2-fluorophenylamino)-7-methoxyquinazolin-6-yloxy]piperidin-1-yl)-*N*-methylacetamide, reversibly inhibits EGFR (IC$_{50}$ = 4 nM), erbB2 (IC$_{50}$ = 3 nM), and erbB3 (IC$_{50}$ = 4 nM) phosphorylation in cells. In proliferation assays, AZD8931 is significantly more potent than gefitinib or lapatinib in specific squamous cell carcinoma of the head and neck and non-small cell lung carcinoma cell lines. *In vivo*, AZD8931 inhibits xenograft growth in a range of models, while significantly affecting EGFR, erbB2, and erbB3 phosphorylation and downstream signaling pathways, apoptosis, and proliferation. In EGF and HRG ligand-driven cell systems, AZD8931 is more potent than gefitinib or lapatinib. Metabolic disposition of AZD8931 (2). **1**. Hickinson, Klinowska, Speake, *et al.* (2010) *Clin. Cancer Res.* **16**, 1159. **2**. Ballard, Swaisland, Malone, *et al.* (2014) *Xenobiotica* **44**, 1083.

AZD9272

This orally bioavailable and brain-penetrant potent and selective mGlu$_5$ antagonist (FW = 284.22 g/mol; CAS 327056-26-8; Solubility: 50 mM in DMSO), also named 3-fluoro-5-[3-(5-fluoro-2-pyridinyl)-1,2,4-oxadiazol-5-yl]benzonitrile, with IC$_{50}$ values are 2.6 and 7.6 nM for rat and human receptors, respectively. AZD9272 exhibits >3900-fold selectivity for mGlu$_5$ over other mGlu receptors. AZD9272 does not share discriminative effects with drugs of abuse of the cocaine, benzodiazepine, or NMDA antagonist types. A partial (−)-Δ^9-THC–like effect was found. AZD9272 caused a partial but dose-dependent increase in (−)-Δ^9-THC–appropriate responding, with a maximal effect of 37.39%, and then a decrease with no dose-dependent effects on response rates. **1**. Raboisson, Breitholtz-Emanuelsson, Dahllöf, *et al.* (2012) *Bioorg. Med. Chem. Lett.* **22**, 6974.

AZD9291

This novel third-generation TKI (FW = 499.62 g/mol; CAS 1421373-65-0), also named osimertinib and Tagrisso$^®$, is an orally available antineoplastic agent that selectively and irreversibly inhibits both EGFRm$^+$-sensitizing and EGFR^{T790M} resistance-conferring mutants, while sparing wild-type EGFR (1,2). When studied *in vitro*, AZD9291 had IC$_{50}$ values of 1, 12, 5, and 184 nM toward EGFR$^{L858R/T790M}$, EGFR^{L858R}, EGFR^{L861Q}, and wild-type EGFR (1). AZD-9291 binds to the EGFR kinase, forming a covalent bond with Cysteine-797 in the ATP binding site (3). This mono-anilino–pyrimidine compound is structurally distinct from other third-generation EGFR TKIs and, in preclinical studies, likewise irreversibly inhibits signaling pathways and cellular growth in both EGFRm$^+$ and EGFRm$^+$/T790M$^+$ mutant cell lines *in vitro*, showing lower activity against wild-type EGFR lines. The result is profound and sustained tumor regression in *EGFR*-mutant tumor xenograft and transgenic models. In clinical trials, AZD 9291 was highly active in $EGFR^{T790M}$ lung cancer patients experiencing disease progression during prior therapy with other EGFR-directed TKIs (4). **Binding Characteristics:** X-ray structural models of AZD9291 bound to EGFR showed (a) that its pyrimidine core forms two hydrogen bonds to Met-793 in the hinge region, (b) that its indole group is favorably oriented relative to an adjacent gatekeeper residue, (c) that its amine moiety occupies the solvent channel, and (d) that its acrylamide moiety is situated near Cys-797, the nucleophile with which it reacts to form a covalent adduct (1). While sharing a number of structural features with the other third-generation TKIs, such as WZ4002 and CO-1686, (*e.g.*, location of the thiol-reactive electrophile, positioning of heteroatom-linked pyrimidine 4-substituents, and the presence of a pyrimidine 5-substituent), AZD9291 is architecturally unique (1). Indeed, its electrophile resides on the pyrimidine C-2 substituent ring, its pyrimidine 4-substituent is C-linked and heterocyclic, and the pyrimidine 5-position on AZD9291 is unsubstituted (1). **1.** Cross, Ashton, Ghiorghiu, *et al.* (2014) *Cancer Discov.* **4**, 1046. **2.** Finlay, Anderton, Ashton, *et al.* (2014) *J. Med. Chem.* **57**, 8249. **3.** Ward, Anderton, Ashton, *et al.* (2013) *J. Med. Chem.* **56**, 7025. **4.** Jänne, Yang, Kim, *et al.* (2015) *New Engl. J. Med.* **372**, 1689.

AzddTTP, *See* 3'-Azido-2,3'-dideoxy-5'-thymidine 5'-Triphosphate

Azelnidipine

This dihydropyridine–class calcium channel antagonist (FW = 582.64 g/mol; CAS 123524-52-7), also known as Calblock® and O^3-[1-[di(phenyl) methyl]azetidin-3-yl]-O^5-propan-2-yl-2-amino-6-methyl-4-(3-nitrophenyl)-1,4-dihydropyridine-3,5-dicarboxylate, selectively inhibits L-type calcium channels and is indicated for the treatment of patients with hypertension. **1.** Wellington & Scott (2003) *Drugs* **63**, 2613.

3-*O*-(2(*S*)-Azetidinylmethoxy)pyridine

This hygroscopic pyridine, also known as A-85380 (FW$_{free-base}$ = 164.21 g/mol), is a neuronal nicotinic acetylcholine receptor agonist (1,2). For the human $\alpha_4\beta_2$ neuronal acetylcholine receptor, K_i = 50 pM. **1.** Sullivan, Donnelly-Roberts, Briggs, *et al.* (1996) *Neuropharmacology* **35**, 725. **2.** Abreo, Lin, Garvey, *et al.* (1996) *J. Med. Chem.* **39**, 817.

6β-Azido-7α-acetoxyandrost-4-ene-3,17-dione

This steroid (FW = 385.46 g/mol) inhibits CYP19, *or* aromatase, K_i = 14 nM. Inhibitors of human placental aromatase (P-450^{arom}) may be useful in treating estrogen-dependent diseases, including breast cancer. **1.** Njar, Grun & Hartmann (1995) *J. Enzyme Inhib.* **9**, 195.

2-Azidoadenosine 5'-Diphosphate

This purine-2 position photoreactive ADP analogue (FW$_{free-acid}$ = 468.22 g/mol) will covalently modify adenine nucleotide-dependent reactions/processes. **Target(s):** CF$_1$ ATPase (1,3,5,11); F$_1$ ATPase (2,4,7-10); ADP/ATP transporter (6). **1.** Colman (1990) *The Enzymes*, 3rd ed., **19**, 283. **2.** Garin, Vincon, Gagnon & Vignais (1994) *Biochemistry* **33**, 3772. **3.** Czarnecki, Abbott & Selman (1982) *Proc. Natl. Acad. Sci. U.S.A.* **79**, 7744. **4.** Cross, Cunningham, Miller, *et al.* (1987) *Proc. Natl. Acad. Sci. U.S.A.* **84**, 5715. **5.** Melese, Xue, Stempel & Boyer (1988) *J. Biol. Chem.* **263**, 5833. **6.** Dalbon, Brandolin, Boulay, Hoppe & Vignais (1988) *Biochemistry* **27**, 5141. **7.** Milgrom & Boyer (1990) *Biochim. Biophys. Acta* **1020**, 43. **8.** Chernyak & Cross (1992) *Arch. Biochem. Biophys.* **295**, 247. **9.** Edel, Hartog & Berden (1992) *Biochim. Biophys. Acta* **1101**, 329. **10.** Martins & Penefsky (1994) *Eur. J. Biochem.* **224**, 1057. **11.** Possmayer, Hartog, Berden & Graber (2000) *Biochim. Biophys. Acta* **1456**, 77.

8-Azidoadenosine 5'-Diphosphate

This purine-8 position photoreactive ADP analogue (FW$_{free-acid}$ = 468.22 g/mol) will covalently modify adenine nucleotide-dependent reactions/processes. **Target(s):** F$_1$ ATPase, *or* ATP synthase (1,2); Ca^{2+}-transporting ATPase (3); apyrase (4). **1.** Garin, Vincon, Gagnon & Vignais (1994) *Biochemistry* **33**, 3772. **2.** Sloothaak, Berden, Herweijer & Kemp (1985) *Biochim. Biophys. Acta* **809**, 27. **3.** Lacapere, Garin, Trinnaman & Green (1993) *Biochemistry* **32**, 3414. **4.** Marti, Gomez de Aranda & Solsona (1997) *Brain Res. Bull.* **44**, 695.

8-Azidoadenosine 5'-Monophosphate

This purine-8 position photoreactive AMP analogue (FW$_{free-acid}$ = 388.24 g/mol) will covalently modify adenine nucleotide-dependent reactions/processes. **Target(s):** fructose-1,6-bisphosphatase, at AMP regulatory site (1,4); Amidophosphoribosyltransferase (2); nicotinamide nucleotide transhydrogenase (3); methotrexate transporter (5); ADPglucose synthetase (6,7); AMP nucleosidase (8). **1.** Colman (1990) *The Enzymes*, 3rd ed., **19**, 283. **2.** Zhou, Charbonneau, Colman & Zalkin (1993) *J. Biol. Chem.* **268**, 10471. **3.** Hu, Persson, Hoog, *et al.* (1992) *Biochim. Biophys. Acta* **1102**, 19. **4.** Marcus & Haley (1979) *J. Biol. Chem.* **254**, 259. **5.** Henderson, Zevely & Huennekens (1979) *J. Biol. Chem.* **254**, 9973. **6.** Larsen & Preiss (1986) *Biochemistry* **25**, 4371. **7.** Larsen, Lee & Preiss (1986) *J. Biol. Chem.* **261**, 15402. **8.** DeWolf, Fullin & Schramm (1979) *J. Biol. Chem.* **254**, 10868.

2-Azidoadenosine 5'-Triphosphate

This purine-2 photoreactive ATP analogue (FW$_{free-acid}$ = 548.20 g/mol) will covalently modify adenine nucleotide-dependent reactions/processes. **Target(s):** CF$_1$ ATPase (1,4); F$_1$ ATPase (2,3,6, but see ref. 8); 2-5A synthetase (5); dynein ATPase (7); Na$_+$/K$^+$-exchanging ATPase (9); creatine kinase (10); rubisco activase (11); V-ATPase, coated vesicle (12); H$^+$-ATPase (13). **1.** Colman (1990) *The Enzymes*, 3rd ed., **19**, 283. **2.** van Dongen, de Geus, Korver, Hartog & Berden (1986) *Biochim. Biophys. Acta* **850**, 359. **3.** Cross, Cunningham, Miller, *et al.* (1987) *Proc. Natl. Acad. Sci. U.S.A.* **84**, 5715. **4.** Melese, Xue, Stempel & Boyer (1988) *J. Biol. Chem.* **263**, 5833. **5.** Suhadolnik, Li, Sobol & Haley (1988) *Biochemistry* **27**, 8846. **6.** Bragg & Hou (1989) *Biochim. Biophys. Acta* **974**, 24. **7.** King, Haley & Witman (1989) *J. Biol. Chem.* **264**, 10210. **8.** Hartog, Edel, Lubbers & Berden (1992) *Biochim. Biophys. Acta* **1100**, 267. **9.** Tran, Huston & Farley (1994) *J. Biol. Chem.* **269**, 6558. **10.** Olcott, Bradley & Haley (1994) *Biochemistry* **33**, 11935. **11.** Salvucci, Chavan, Klein, Rajagopolan & Haley (1994) *Biochemistry* **33**, 14879. **12.** Zhang, Vasilyeva, Feng & Forgac (1995) *J. Biol. Chem.* **270**, 15494. **13.** Possmayer, Hartog, Berden & Graber (2001) *Biochim. Biophys. Acta* **1510**, 378.

8-Azidoadenosine 5'-Triphosphate

This purine-8 position photoreactive AMP analogue (FW$_{free-acid}$ = 548.20 g/mol) will covalently modify adenine nucleotide-dependent reactions/processes. (*See also 2-Azido-adenosine 5'-Triphosphate*) **Target(s):** Na$^+$/K$^+$-exchanging ATPase (1,19,44); DNA nucleotidylexotransferase, *or* terminal deoxyribonucleotidyl-transferase (2); 6-phosphofructo-2-kinase (3,5,13); recA protein (4,22); glutamine synthetase (6); DNA-directed RNA polymerase II, mammalian (7,16); F$_1$ ATPase, *or* ATP synthase (8,11,20,21,23,24,29); protein kinase A, *or* cAMP-dependent protein kinase (9); phosphorylase kinase (10); DNA-directed DNA polymerase (12,35); DNA polymerase I (12); dynein ATPase (14,25,27,34); DNA-directed RNA polymerase, *Escherichia coli* (15); asparagine synthetase (17); carbamoyl phosphate synthetase I (18); ATPase (19); apyrase (26,53); glucose-1-phosphate adenylyltransferase, *or* ADPglucose synthetase (28); hexokinase (30); kinesin ATPase (31); lysosomal H$^+$ pump (32); 2-5A synthetase (33); DNA polymerase α (35); DNA primase (35); RNA-directed RNA polymerase, rotavirus (36); T7 RNA polymerase (37); ecto-ATPase (38,39); phosphatidylinositol 4-kinase (40); selenophosphate synthetase (41); termination factor ρ (42); P-glycoprotein ATPase, *or* xenobiotic transporting ATPase, *or* multidrug-sensitizing protein (43); creatine kinase (45,56); integrase, HIV-1 (46); rubisco activase (47); Ca^{2+}-transporting ATPase (48); thymidine kinase (49); elongation factor-2 (50); pyruvate:orthophosphate dikinase (51); deoxyuridine triphosphatase (52); K$^+$ channel, ATP sensitive (54); succinyl-CoA synthetase (55); phosphofructokinase-1 (57); arginyl-tRNA synthetase (58,59); isoleucyl-tRNA synthetase (60); protein-disulfide oxidoreductase (61); guanylate cyclase (62); polyamine-transporting ATPase (63); dUTP diphosphatase (64); [isocitrate dehydrogenase (NADP$^+$)] kinase (65); selenide,water dikinase (66-68); polynucleotide adenylyltransferase, *or* poly(A) polymerase (69). **1.** Schoner & Scheiner-Bobis (1988) *Meth. Enzymol.* **156**, 312. **2.** Abraham, Haley & Modak (1983) *Biochemistry* **22**, 4197. **3.** Pilkis, Claus, Kountz & el-Maghrabi (1987) *The Enzymes*, 3rd ed., **18**, 3. **4.** Colman (1990) *The Enzymes*, 3rd ed., **19**, 283. **5.** el-Maghrabi, Pate, D'Angelo, *et al.* (1987) *J. Biol. Chem.* **262**, 11714. **6.** Tanaka & Kimura

(1991) *J. Biochem.* **110**, 780. **7.** Freund & McGuire (1986) *Biochemistry* **25**, 276. **8.** Scheurich, Schafer & Dose (1978) *Eur. J. Biochem.* **88**, 253. **9.** Hoppe & Freist (1979) *Eur. J. Biochem.* **93**, 141. **10.** King, Carlson & Haley (1982) *J. Biol. Chem.* **257**, 14058. **11.** Hollemans, Runswick, Fearnley & Walker (1983) *J. Biol. Chem.* **258**, 9307. **12.** Abraham & Modak (1984) *Biochemistry* **23**, 1176. **13.** Sakakibara, Kitajima & Uyeda (1984) *J. Biol. Chem.* **259**, 8366. **14.** Pfister, Haley & Witman (1984) *J. Biol. Chem.* **259**, 8499. **15.** Woody, Vader, Woody & Haley (1984) *Biochemistry* **23**, 2843. **16.** Freund & McGuire (1986) *J. Cell. Physiol.* **127**, 432. **17.** Larsen & Schuster (1992) *Arch. Biochem. Biophys.* **299**, 15. **18.** Powers-Lee & Corina (1987) *J. Biol. Chem.* **262**, 9052. **19.** Haley & Hoffman (1974) *Proc. Natl. Acad. Sci. U.S.A.* **71**, 3367. **20.** Eul, Risi, Schafer & Dose (1983) *Biochem. Int.* **6**, 723. **21.** Hollemans, Runswick, Fearnley & Walker (1983) *J. Biol. Chem.* **258**, 9307. **22.** Knight & McEntee (1985) *J. Biol. Chem.* **260**, 867. **23.** Sloothaak, Berden, Herweijer & Kemp (1985) *Biochim. Biophys. Acta* **809**, 27. **24.** Van der Bend, Duetz, Colen, Van Dam & Berden (1985) *Arch. Biochem. Biophys.* **241**, 461. **25.** Pfister, Haley & Witman (1985) *J. Biol. Chem.* **260**, 12844. **26.** LeBel & Beattie (1986) *Biochem. Cell. Biol.* **64**, 13. **27.** Pratt (1986) *J. Biol. Chem.* **261**, 956. **28.** Lee, Mukherjee & Preiss (1986) *Arch. Biochem. Biophys.* **244**, 585. **29.** van Dongen, de Geus, Korver, Hartog & Berden (1986) *Biochim. Biophys. Acta* **850**, 359. **30.** Nemat-Gorgani & Wilson (1986) *Arch. Biochem. Biophys.* **251**, 97. **31.** Porter, Scholey, Stemple, *et al.* (1987) *J. Biol. Chem.* **262**, 2794. **32.** Cuppoletti, Aures-Fischer & Sachs (1987) *Biochim. Biophys. Acta* **899**, 276. **33.** Suhadolnik, Li, Sobol & Haley (1988) *Biochemistry* **27**, 8846. **34.** King, Haley & Witman (1989) *J. Biol. Chem.* **264**, 10210. **35.** Bodner & Bambara (1990) *Cancer Biochem. Biophys.* **11**, 7. **36.** Valenzuela, Pizarro, Sandino, *et al.* (1991) *J. Virol.* **65**, 3964. **37.** Knoll, Woody & Woody (1992) *Biochim. Biophys. Acta* **1121**, 252. **38.** Hohmann, Kowalewski, Vogel & Zimmermann (1993) *Biochim. Biophys. Acta* **1152**, 146. **39.** Rodriguez-Pascual, Torres & Miras-Portugal (1993) *Arch. Biochem. Biophys.* **306**, 420. **40.** Nickels & Carman (1993) *J. Biol. Chem.* **268**, 24083. **41.** Kim, Veres & Stadtman (1993) *J. Biol. Chem.* **268**, 27020. **42.** Stitt & Stitt (1994) *J. Biol. Chem.* **269**, 5009. **43.** al-Shawi, Urbatsch & Senior (1994) *J. Biol. Chem.* **269**, 8986. **44.** Tran, Scheiner-Bobis, Schoner & Farley (1994) *Biochemistry* **33**, 4140. **45.** Olcott, Bradley & Haley (1994) *Biochemistry* **33**, 11935. **46.** Lipford, Worland & Farnet (1994) *J. Acquir. Immune. Defic. Syndr.* **7**, 1215. **47.** Salvucci, Chavan, Klein, Rajagopolan & Haley (1994) *Biochemistry* **33**, 14879. **48.** Webb & Dormer (1995) *Biochim. Biophys. Acta* **1233**, 1. **49.** Rechtin, Black, Mao, Lewis & Drake (1995) *J. Biol. Chem.* **270**, 7055. **50.** Guillot, Vard & Reboud (1996) *Eur. J. Biochem.* **236**, 149. **51.** McGuire, Carroll, Yankie, *et al.* (1996) *Biochemistry* **35**, 8544. **52.** Roseman, Evans, Mayer, Rossi & Slabaugh (1996) *J. Biol. Chem.* **271**, 23506. **53.** Marti, Gomez de Aranda & Solsona (1997) *Brain Res. Bull.* **44**, 695. **54.** Tanabe, Tucker, Matsuo, *et al.* (1999) *J. Biol. Chem.* **274**, 3931. **55.** Joyce, Fraser, Brownie, *et al.* (1999) *Biochemistry* **38**, 7273. **56.** David & Haley (1999) *Biochemistry* **38**, 8492. **57.** Knoche, Monnich, Schafer & Kopperschlager (2001) *Arch. Biochem. Biophys.* **385**, 301. **58.** Gerlo, Freist & Charlier (1982) *Hoppe-Seyler's Z. Physiol. Chem.* **363**, 365. **59.** Charlier & Gerlo (1979) *Biochemistry* **18**, 3171. **60.** Freist, von der Haar & Cramer (1981) *Eur. J. Biochem.* **119**, 151. **61.** Pedone, Ren, Ladenstein, Rossi & Bartolucci (2004) *Eur. J. Biochem.* **271**, 3437. **62.** Ruiz-Stewart, Tiyyagura, *et al.* (2004) *Proc. Natl. Acad. Sci. U.S.A.* **101**, 37. **63.** Kashiwagi, Endo, Kobayashi & Igarashi (1995) *J. Biol. Chem.* **270**, 25377. **64.** Roseman, Evans, Mayer, Rossi & Slabaugh (1996) *J. Biol. Chem.* **271**, 23506. **65.** Varela & Nimmo (1988) *FEBS Lett.* **231**, 361. **66.** Kim, Veres & Stadtman (1993) *J. Biol. Chem.* **268**, 27020. **67.** Veres, Kim, Scholz & Stadtman (1994) *J. Biol. Chem.* **269**, 10597. **68.** Low, Harney & Berry (1995) *J. Biol. Chem.* **270**, 21659. **69.** Chen & Sheppard (2004) *J. Biol. Chem.* **279**, 40405.

8-Azido-ADP-glucose

This photoaffinity labeling reagent and sugar-nucleotide derivative (FW$_{free-acid}$ = 630.36 g/mol) inhibits glucose-1-phosphate adenylyltransferase, *or* ADPglucose synthase. In the dark, the synthase utilizes 8-azido-ADPglucose as a substrate. When illuminated at 254 nm, however, this analogue specifically and covalently modifies the enzyme, with concomitant loss of catalytic activity. ADPglucose also protects the enzyme from covalent modification by the photoreagent. **1.** Colman (1990) *The Enzymes*, 3rd ed., **19**, 283. **2.** Lee, Mukherjee & Preiss (1986) *Arch. Biochem. Biophys.* **244**, 585.

γ-(*p*-Azidoanilide)adenosine 5'-Triphosphate

This photoaffinity labeling reagent and ATP analogue (FW$_{free-acid}$ = 623.31 g/mol) inhibits several ATP dependent enzymes. Not all aminoacyl-tRNA synthetases are inhibited by this amidate. **Target(s):** tryptophanyl-tRNA synthetase (1,2,5); phenylalanyl-tRNA synthetase (3,5); arginine kinase (4); creatine kinase4,6; arginyl-tRNA synthetase5; valyl-tRNA synthetase5; isoleucyl-tRNA synthetase (5); leucyl-tRNA synthetase (5,7); threonyl-tRNA synthetase (5); K$^+$ channel, ATP-sensitive (8). **1.** Kisselev, Favorova & Kolaleva (1979) *Meth. Enzymol.* **59**, 234. **2.** Akhveridan, Kiselev, Knorre, Lavrik & Nevinskii (1976) *Dokl. Akad. Nauk. SSSR* **226**, 698. **3.** Ankilova, Knorre, Kravchenko, Lavrik & Nevinsky (1975) *FEBS Lett.* **60**, 172. **4.** Vandest, Labbe & Kassab (1980) *Eur. J. Biochem.* **104**, 433. **5.** Bulychev, Lavrik & Nevinskii (1980) *Mol. Biol. (Moscow)* **14**, 558. **6.** Akopian, Gazariants, Mkrtchian, Nersova & Lavrik (1981) *Biokhimiia* **46**, 262. **7.** Krauspe & Lavrik (1983) *Eur. J. Biochem.* **132**, 545. **8.** Tanabe, Tucker, Ashcroft, *et al.* (2000) *Biochem. Biophys. Res. Commun.* **272**, 316.

4-Azidochalcone Oxide

This photoaffinity labeling reagent (FW = 265.27 g/mol) inhibits soluble epoxide hydrolase (IC$_{50}$ = 1.6 μM). The 4'-azido analogue also inhibits (IC$_{50}$ = 4.8 μM). **1.** Prestwich & Hammock (1985) *Proc. Natl. Acad. Sci. U.S.A.* **82**, 1663.

2'-Azido-2'-deoxycytidine 5'-Diphosphate

This nucleotide (FW$_{free-acid}$ = 428.19 g/mol) inhibits ribonucleotide reductase. **1.** Brush & Kozarich (1992) *The Enzymes*, 3rd ed., **20**, 317. **2.** Thelander, Hobbs & Eckstein (1977) *Meth. Enzymol.* **46**, 321. **3.** Akerblom & Reichard (1985) *J. Biol. Chem.* **260**, 9197. **4.** Bianchi, Borella, Calderazzo, *et al.* (1994) *Proc. Natl. Acad. Sci. U.S.A.* **91**, 8403. **5.** Behravan, Sen, Rova, *et al.* (1995) *Biochim. Biophys. Acta* **1264**, 323. **6.** Ator, Stubbe & Spector (1986) *J. Biol. Chem.* **261**, 3595. **7.** Salowe, Ator & Stubbe (1987) *Biochemistry* **26**, 3408. **8.** Engström, Eriksson, Thelander & Akerman (1979) *Biochemistry* **18**, 2941.

3'-Azido-3'-deoxythymidine

This pro-drug (FW = 267.24 g/mol; *Abbreviation*: AZT), also known as Zidovudine and 3'-azido-2',3'-dideoxythymidine, is metabolically transformed to its 5'-triphosphate, which then potently inhibits HIV replication. AZT was the first drug found to be effective in treating individuals with HIV-AIDS, and provided much-needed relief until other, more effective antivirals came on line. SLC28A1, the Na$^+$-dependent nucleoside transporter selective for pyrimidine nucleosides and adenosine, also transports Zidovudine. **Target(s):** HIV-1 reverse transcriptase, inhibited by the triphosphate (1-3,5,9); avian myeloblastosis virus reverse transcriptase, inhibited by the triphosphate (4); poly(ADP-ribose) polymerase (6); cytidine deaminase (7); 5'-nucleotidase (8); RNA-directed DNA polymerase (1-5,9); dTMP kinase, *or* thymidylate kinase (10-12); deoxynucleoside kinase (13); thymidine kinase (13,14); phosphorylase, weakly inhibited (15); uridine phosphorylase, rat K_i = 56 μM (15); glucuronosyl-.transferase, substrate for UGT2B7 (16). **Key Pharmacokinetic Parameters:** *See* Appendix II in Goodman & Gilman's *THE PHARMACOLOGICAL BASIS OF THERAPEUTICS*, 12th Edition (Brunton, Chabner & Knollmann, eds.) McGraw-Hill Medical, New York (2011). **1.** Le Grice, Cameron & Benkovic (1995) *Meth. Enzymol.* **262**, 130. **2.** Wilson, Porter & Reardon (1996) *Meth. Enzymol.* **275**, 398. **3.** Balzarini & De Clercq (1996) *Meth. Enzymol.* **275**, 472. **4.** Eriksson, Vrang, Bazin, Chattopadhyaya & Oberg (1987) *Antimicrob. Agents Chemother.* **31**, 600. **5.** Furman, Fyfe, St. Clair, *et al.* (1986) *Proc. Natl. Acad. Sci. U.S.A.* **83**, 8333. **6.** Pivazyan, Birks, Wood, Lin & Prusoff (1992) *Biochem. Pharmacol.* **44**, 947. **7.** Cacciamani, Vita, Cristalli, *et al.* (1991) *Arch. Biochem. Biophys.* **290**, 285. **8.** Garvey, Lowen & Almond (1998) *Biochemistry* **37**, 9043. **9.** El Safadi, Vivet-Boudou & Marquet (2007) *Appl. Microbiol. Biotechnol.* **75**, 723. **10.** Munier-Lehmann, Pochet, Dugué, *et al.* (2003) *Nucleosides Nucleotides Nucleic Acids* **22**, 801. **11.** Fioravanti, Adam, Munier-Lehmann & Bourgeois (2005) *Biochemistry* **44**, 130. **12.** Pochet, Dugué, Labesse, Delepierre & Munier-Lehmann (2003) *ChemBioChem* **4**, 742. **13.** Johansson, Van Rompay, Degrève, Balzarini & Karlsson (1999) *J. Biol. Chem.* **274**, 23814. **14.** Gustafson, Chillemi, Sage & Fingeroth (1998) *Antimicrob. Agents Chemother.* **42**, 2923. **15.** Veres, Neszmélyi, Szabolcs & Dénes (1988) *Eur. J. Biochem.* **178**, 173. **16.** Picard, Ratanasavanh, Premaud, Le Meur & Marquet (2005) *Drug Metab. Dispos.* **33**, 139.

3'-Azido-3'-deoxythymidine 5'-Diphosphate

This dideoxynucleotide (FW = 426.19 g/mol), often abbreviated AzddTDP and AZTDP and also called 3'-azido-2',3'-dideoxythymidine 5'-diphosphate, is a dTDP analogue and inhibits viral Rauscher murine leukemia virus reverse transcriptase, *or* RNA-directed DNA polymerase. **1.** Ono, Nakane, Herdewijn, Balzarini & De Clerq (1988) *Nucl. Acids Res.* **20**, 5.

3'-Azido-3'-deoxythymidine 5'-Monophosphate

This modified nucelotide (FW$_{\text{free-acid}}$ = 347.22 g/mol) inhibits DNA polymerase. **Target(s):** simian virus 40 origin-dependent replication (1); DNA polymerase-associated 3'→5' exonuclease activity (2); ribonuclease H (3); dTMP kinase, *or* thymidylate kinase (4-6). **1.** Bebenek, Thomas, Roberts, Eckstein & Kunkel (1993) *Mol. Pharmacol.* **43**, 57. **2.** Bridges, Faraj & Sommadossi (1993) *Biochem. Pharmacol.* **45**, 1571. **3.** Zhan, Tan, Scott, *et al.* (1994) *Biochemistry* **33**, 1366. **4.** Munier-Lehmann, Chaffotte, Pochet & Labesse (2001) *Protein Sci.* **10**, 1195. **5.** Fioravanti, Adam, Munier-Lehmann & Bourgeois (2005) *Biochemistry* **44**, 130. **6.** Haouz, Vanheusden, Munier-Lehmann, *et al.* (2003) *J. Biol. Chem.* **278**, 4963.

3'-Azido-2,3'-dideoxy-5'-thymidine 5'-Triphosphate

This dideoxynucleotide (FW$_{\text{free-acid}}$ = 507.18 g/mol), often abbreviated AzddTTP and AZTTP and also called 3'-azido 2',3'-dideoxythymidine 5'-triphosphate, is a dTTP analogue and inhibits viral reverse transcriptases. **Target(s):** HIV reverse transcriptase, IC$_{50}$ = 0.05 μM (1,3-5,10-16,19,20,23,25); DNA polymerase β, IC$_{50}$ = 31 μM (2,4,5); DNA polymerase γ (4,5,17); avian myeloblastosis virus reverse (6); telomerase (7-9); feline immunodeficiency virus reverse transcriptase (15,26); feline leukemia virus reverse transcriptase (26); DNA helicase (18); nucleoside-triphosphatase (18); mouse mammary tumor virus reverse transcriptase (21); Rauscher murine leukemia virus reverse transcriptase (22,24); RNA-directed DNA polymerase (1,3-6,10-16,19-26); DNA-directed DNA polymerase (2,4,5,17,27); integrase, HIV (27). **1.** Abbotts & Wilson (1992) *J. Enzyme Inhib.* **6**, 35. **2.** Brown & Wright (1995) *Meth. Enzymol.* **262**, 202. **3.** Wilson, Porter & Reardon (1996) *Meth. Enzymol.* **275**, 398. **4.** De Clercq (1992) *AIDS Res. Human Retrovir.* **8**, 119. **5.** White, Parker, Macy, *et al.* (1989) *Biochem. Biophys. Res. Commun.* **161**, 393. **6.** Eriksson, Vrang, Bazin, Chattopadhyaya & Oberg (1987) *Antimicrob. Agents Chemother.* **31**, 600. **7.** Strahl & Blackburn (1996) *Mol. Cell Biol.* **16**, 53. **8.** Strahl & Blackburn (1994) *Nucl. Acids Res.* **22**, 893. **9.** Yamaguchi, Takahashi, Jinmei, Takayama & Saneyoshi (2003) *Nucleosides Nucleotides Nucleic Acids* **22**, 1575. **10.** Furman, Fyfe, St. Clair, *et al.* (1986) *Proc. Natl. Acad. Sci. U.S.A.* **83**, 8333. **11.** Vrang, Bazin, Remaud, Chattopadhyaya & Oberg (1987) *Antiviral Res.* **7**, 139. **12.** Jaju, Beard & Wilson (1995) *J. Biol. Chem.* **270**, 9740. **13.** Carroll, Geib, Olsen, *et al.* (1994) *Biochemistry* **33**, 2113. **14.** Ono, Ogasawara, Iwata, *et al.* (1986) *Biochem. Biophys. Res. Commun.* **140**, 498. **15.** North, Cronn, Remington & Tandberg (1990) *Antimicrob. Agents Chemother.* **34**, 1505. **16.** Reardon & Miller (1990) *J. Biol. Chem.* **265**, 20302. **17.** Lewis, Simpson & Meyer (1994) *Circ. Res.* **74**, 344. **18.** Locatelli, Gosselin, Spadari & Maga (2001) *J. Mol. Biol.* **313**, 683. **19.** Shaw-Reid, Feuston, Munshi, *et al.* (2005) *Biochemistry* **44**, 1595. **20.** El Safadi, Vivet-Boudou & Marquet (2007) *Appl. Microbiol. Biotechnol.* **75**, 723. **21.** Taube, Loya, Avidan, Perach & Hizi (1998) *Biochem. J.* **329**, 579. **22.** Ono, Nakane, Herdewijn, Balzarini & De Clerq (1988) *Nucl. Acids Res.* **20**, 5. **23.** Nissley, Radzio, Ambrose, *et al.* (2007) *Biochem. J.* **404**, 151. **24.** Ono, Ogasawara, Iwata, *et al.* (1986) *Biochem. Biophys. Res. Commun.* **140**, 498. **25.** Quiñones-Mateu, Soriano, Domingo & Menéndez-Arias (1997) *Virology* **236**, 364. **26.** Operario, Reynolds & Kim (2005) *Virology* **335**, 106. **27.** Acel, Udashkin, Wainberg & Faust (1998) *J. Virol.* **72**, 2062.

8-Azidoguanosine 5'-Triphosphate

This photoaffinity labeling reagent and GTP analogue (FW$_{\text{free-acid}}$ = 564.19 g/mol) photolabels RNA polymerase II and elongation factor-2, while inhibiting cytosolic phosphoenolpyruvate carboxykinase (GTP) without photolabeling. **Target(s):** DNA-dependent RNA polymerase II (1,2); elongation factor-2 (3); transglutaminase (4); phosphoenolpyruvate carboxykinase, GTP-requiring (5); glutamate dehydrogenase (6,7); GTP cyclohydrolase IIa (8). **1.** Freund & McGuire (1986) *Biochemistry* **25**, 276. **2.** Freund & McGuire (1986) *J. Cell. Physiol.* **127**, 432. **3.** Guillot, Vard & Reboud (1996) *Eur. J. Biochem.* **236**, 149. **4.** Achyuthan & Greenberg (1987) *J. Biol. Chem.* **262**, 1901. **5.** Lewis, Haley & Carlson (1989) *Biochemistry* **28**, 9248. **6.** Shoemaker & Haley (1993) *Biochemistry* **32**, 1883. **7.** Cho, Ahn, Lee & Choi (1996) *Biochemistry* **35**, 13907. **8.** Graham, Xu & White (2002) *Biochemistry* **41**, 15074.

3-Azido-5-methoxy-2-methyl-6-(3,7-dimethyloctyl)-1,4-benzoquinone

This photoaffinity labeling reagent and quinone analogue (FW = 333.43 g/mol), also known as azido-Q and 3-azido-2-methyl-5-methoxy-6-(3,7-dimethyl[^{3}H]octyl)-1,4-benzoquinone, inhibits plastoquinol:plastocyanin reductase (*Reaction*: plastoquinol + 2 oxidized plastocyanin + 2 H$^+_{\text{[Side-1]}}$ ⇌ plastoquinone + 2 reduced plastocyanin + 2 H$^+_{\text{[Side-2]}}$). Maximum inactivation (~45%) is observed when 30 mol azido-Q is used per mol cytochrome f (*or* cytochrome b_6f complex). The extent of the decrease in activity upon illumination correlates with the amount of azido-Q incorporated into the protein. **1.** Doyle, Li, Yu & Yu (1989) *J. Biol. Chem.* **264**, 1387.

2A-Azido-NAD$^+$

This photoaffinity labeling reagent and (NAD$^+$) derivative (FW = 704.45 g/mol) photolabels NAD$^+$-dependent oxidoreductases. **Target(s):** NAD$^+$-dependent enzymes (1); 15-hydroxyprostaglandin dehydrogenase (2); glutamate dehydrogenase (3-5); NAD(P)H:quinone oxidoreductase (6); isocitrate dehydrogenase, NADP$^+$-dependent (7). (*See 2-Azidoadenosine for location of azido group*) **1.** Colman (1997) *Meth. Enzymol.* **280**, 186. **2.** Ensor & Tai (1997) *Meth. Enzymol.* **280**, 204. **3.** Kim & Haley (1990) *J. Biol. Chem.* **265**, 3636. **4.** Cho, Yoon, Ahn, Choi & Kim (1998) *J. Biol. Chem.* **273**, 31125. **5.** Yoon, Cho, Kwon, Choi & Cho (2002) *J. Biol. Chem.* **277**, 41448. **6.** Deng, Zhao, Iyanagi & Chen (1991) *Biochemistry* **30**, 6942. **7.** Sankaran, Chavan & Haley (1996) *Biochemistry* **35**, 13501.

3'-O-[3-(4-Azido-2-nitroanilino)propionyl]adenosine 5'-Triphosphate

This photoaffinity labeling reagent and substituted ATP (FW$_{free-acid}$ = 740.37 g/mol), also known as arylazido-β-alanine-adenosine 5'-triphosphate, inhibits myosin ATPase (1), F$_1$ ATPase (1,2), and oxidative phosphorylation (3). **1**. Guillory & Jeng (1977) *Meth. Enzymol.* **46**, 259. **2**. Schäfer (1986) *Meth. Enzymol.* **126**, 649. **3**. Schäfer, Lücken & Lübben (1986) *Meth. Enzymol.* **126**, 682.

Azilsartan & Azilsartan Medoxomil

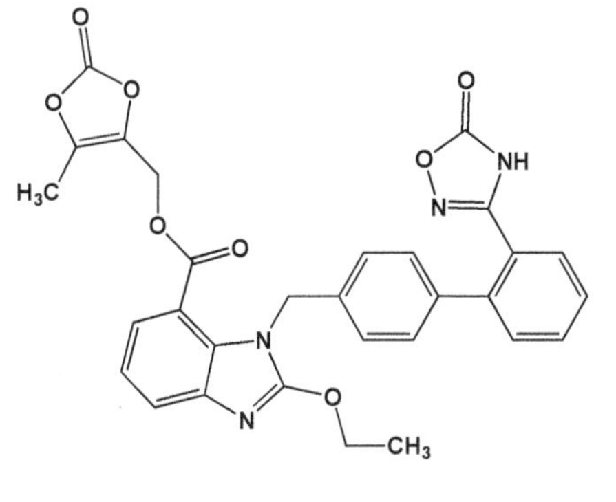

pro-drug

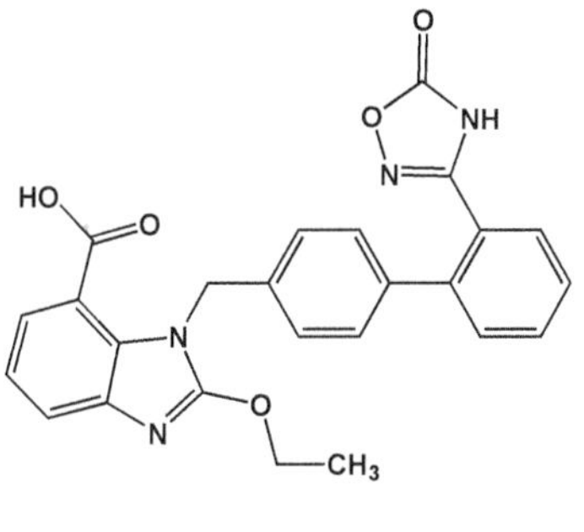

active drug

This angiotensin II receptor antagonist and its membrane-permeant pro-drug (FW$_{drug}$ = 456.45 g/mol; CAS 147403-03-0; FW$_{pro-drug}$ = 568.53 g/mol; CAS 863031-21-4), also named TAK-536 (drug) and TAK-491 (pro-drug), Edarbi™, and (5-methyl-2-oxo-1,3-dioxol-4-yl)methyl 2-ethoxy-1-([2'-(5-oxo-4,5-dihydro-1,2,4-oxadiazol-3-yl)biphenyl-4-yl]methyl)-1*H*-benzimid-azole-7-carboxylate, lowers blood pressure by blocking the binding of the vasopressor hormone, angiotensin II, to the angiotensin Type-1 receptor (*or* AT$_1$-receptor), IC$_{50}$ = 45 nM (1). Blocking of AT$_1$ receptors reduces blood pressure by promoting vasodilation, decreasing vasopressin secretion, and reducing aldosterone production/secretion. The pro-drug is hydrolyzed to the active moiety in the gastrointestinal (GI) tract during the absorption phase. The estimated absolute bioavailability of azilsartan is 60%. Absorption is unaffected by food, and peak plasma concentrations are reached within several hours before its eventual deactivation by cytochrome P450 (CYP2C9), biological $t_{1/2} \approx 11$ hours. The U.S. FDA approved Edarbi for the treatment of high blood pressure in adults in February, 2011. **1**. Ojima, Igata, Tanaka, *et al.* (2011) *J. Pharmacol. Exp* **336**, 801.

Azimilide

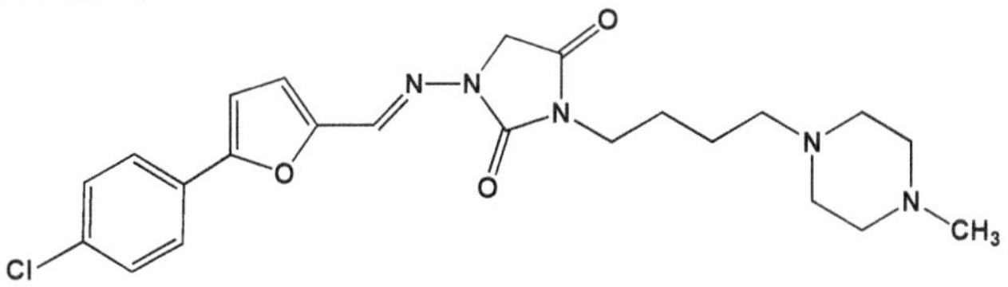

This class III antiarrhythmic drug and ion channel blocker (FW = 457.95 g/mol; CAS 149908-53-2), also known by the code name NE-10064 and systematic name, 1-({(*E*)-[5-(4-chlorophenyl)furan-2-yl]methylidene}amino)-3-[4-(4-methylpiperazin-1-yl)butyl]-imidazolidine-2,4-dione, blocks both the slow-activating I$_{Ks}$ and rapidly activating I$_{Kr}$ components of the delayed rectifier potassium current, distinguishing its action from conventional potassium channel blockers such as sotalol and dofetilide,

which block only I$_{Kr}$ (1). Azimilide prolongs the cardiac refractory period in a dose-dependent manner, as manifested by increases in action potential duration, QTc interval, and effective refractory period. **1**. Karam, Marcello, Brooks, Corey & Moore (1998) *Am. J. Cardiol.* **81**, 40D.

Aziridine, *See* Ethylenimine

Azithromycin

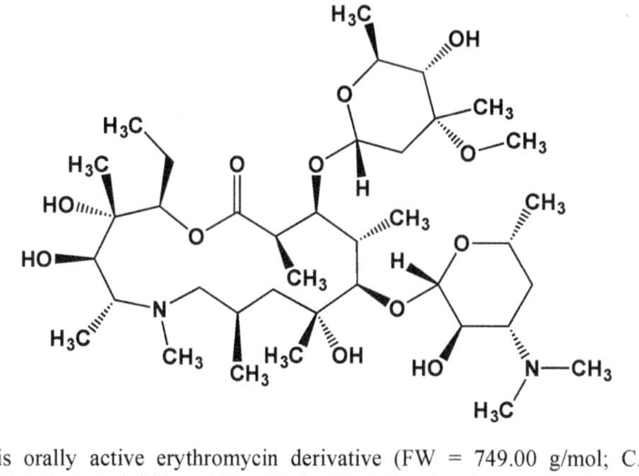

This orally active erythromycin derivative (FW = 749.00 g/mol; CAS 83905-01-5; Solubility: 100 mM in DMSO), also known by the trade names Zithromax®, Azyth®, Sumamed®, is a macrolide antibiotic that targets the assembly of 50S ribosomal subunits and inhibits the transpeptidation step in many Gram-positive and Gram-negative microorganisms. Azithromycin's large volume of distribution (23 L/kg) and peak serum level (0.4 μg/mL) indicate an extensive tissue distribution, allowing it to achieve high, bactericidal levels at sites of infection. It is also taken up by cells (even phagocytes), localizing mainly in cell granules and cytosol and allowing it to be effective against suchh intracellular pathogens as *Listeria monocytogenes*. Azithromycin is also effective against respiratory, urogenital, and dermal infections. **Microbial Targets:** Aerobic & Facultative Gram-positive Microorganisms: *Staphylococcus aureus* (including MRSA), *Streptococcus agalactiae*, *S. pneumoniae*; *S. pyogenes*; Aerobic & Facultative Gram-negative Microorganisms: *Haemophilus influenza*, *Neisseria gonorrhoeae*, *Bordetella pertussis*, *Legionella pneumophila*. **Key Pharmacokinetic Parameters:** *See* Appendix II in Goodman & Gilman's *THE PHARMACOLOGICAL BASIS OF THERAPEUTICS*, 12th Edition (Brunton, Chabner & Knollmann, eds.) McGraw-Hill Medical, New York. **1**. Girard, *et al.* (1987) *Antimicrob. Agents Chemother.* **31**, 1948. **2**. Retsema, *et al.* (1987) *Antimicrob. Agents Chemother.* 31, 1939. **3**. Champney & Burdine (1995) *Antimicrob. Agents Chemother.* 39 2141. **4**. Parnham, Erakovic, Giamarellos-Bourboulis, *et al.* (2014) *Pharmacol Ther.* **143**, 225. **IUPAC:** 13-[(2,6-dideoxy-3-*C*-methyl-3-*O*-methyl-α-L-ribo-hexopyranosyl)oxy]-2-ethyl-3,4,10-trihydroxy-3,5,6,8,10,12,14-hepta-methyl-11-[[3,4,6-trideoxy-3-(dimethylamino)-β-D-xylohexopyranosyl]oxy]-1-oxa-6-azacyclopentadecan-15-one.

Azoramide

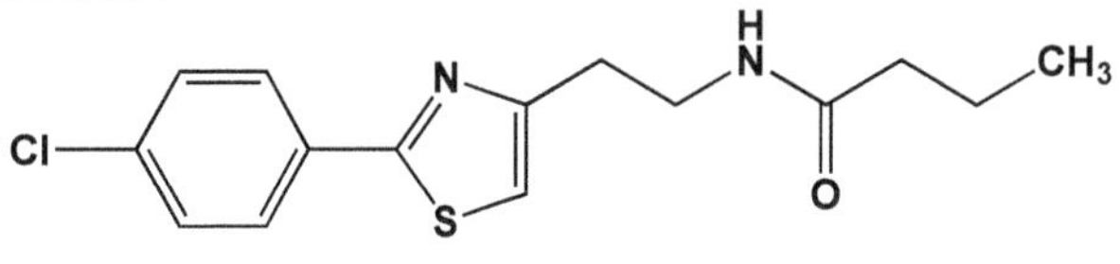

This first-in-class, dual-function UPR modulator (FW = 308.83 g/mol; CAS 932986-18-0; Soluble to 100 mM in DMSO), also named *N*-[2-[2-(4-chlorophenyl)-4-thiazolyl]ethyl]butanamide, is a orally bioavailable drug that displays the unique ability to modulate the unfolded protein response by acutely increasing the output of endoplasmic reticulum (ER) protein folding, while also inducing chaperone expression and chronically promoting ER homeostasis. Treatment of cells with azoramide is strongly protective against chemically induced (*e.g.*, by thapsigargin or tunica-mycin), hypoxia-induced, lipotoxicity-induced, and protein misfolding-induced ER stress. When used at 1-25 μM, azoramide stimulates the expression of multiple chaperone proteins, enhancing ER chaperone capacity and inducing phosphorylation of eukaryotic translation initiation factor 2α subunit (eIF2α), thereby reducing protein biosynthesis. At 150 mg/kg, azoramide exerts antidiabetic activity in stress are associated both *ob/ob* and diet-induced obese mice, improving insulin sensitivity and glucostasis, as well as protecting pancreatic β-cells against ER stress.

Azoramide thus improves insulin sensitivity, glucose tolerance, and β-cell function in multiple obese mouse preclinical models. ER plays a critical role in protein, lipid, and glucose metabolism as well as cellular calcium signaling and homeostasis. Perturbation of ER function and chronic ER with many pathologies ranging from diabetes and neurodegenerative diseases to cancer and inflammation. Such proof-of-principle data show that small-molecule modulators of adaptive UPR pathways can be identified with functional phenotypic screens targeting the ER and confer benefits in the treatment of ER stress-mediated pathologies. Azoramide was also effective in preventing the death of cells expressing a mutant form of rhodopsin-identified in autosomal dominant human retinitis pigmentosa and associated with protein misfolding and ER stress. 1. Fu, Yalcin, Lee, *et al.* (2015) *Sci. Transl. Med.* **7**, 292ra98.

AZT, *See* *3'-Azido-3'-deoxythymidine*

AZTTP, *See* *3'-Azido-2,3'-dideoxy-5'-thymidine 5'-Triphosphate*

– B –

B581

This cell-permeable CAAX peptidomimetic (FW = 470.70 g/mol; $FW_{trifluoroacetate-salt}$ = 583.71 g/mol; CAS 149759-96-6), named as N-[2(S)-(2(R)-2-amino-3-mercaptopropylamino)-3-methylbutyl]-L-phenylalanyl-L-methionine), binds to and prevents protein farnesyltransferase (IC_{50} = 21 nM) from interacting with C-terminal L-Cys-L-Val-L-Phe-L-Met substrate-recognition sequences that are present in its natural substrates. Otherwise hydrophilic proteins associate with membranes by means of enzymatic attachment of hydrophobic moieties to their C-termini. Whereas prenylation occurs in the cytosol, post-prenylation processing is accomplished on the cytoplasmic surface of the endoplasmic reticulum and Golgi apparatus. B581 inhibits prenylation and processing of H-ras and lamin A. B581 specifically blocks farnesylated, but not geranylgeranylated or myristylated, oncogenic ras signaling and transformation. **1.** Cox, Garcia, Westwick, *et al.* (1994) *J. Biol. Chem.* **269**, 19203. **2.** Garcia, Rowell, Ackermann, Kowalczyk & Lewis (1993) *J. Biol. Chem.* **268**, 18415.

B-995, *See* Daminozide

B827-33, *See* Etomoxir

BI6727, *See* Volasertib

Baccharin

This anti-leukemia agent (FW = 562.61 g/mol; CAS 61251-97-6), also named (2'S,3'R,4'S, 7'R,9R,10S)-7'-deoxo-2'-deoxy-2',3':9,10-bisoxy-9,10-dihydro-4'-hydroxy-7'-[(R)-1-hydroxyethyl]verrucarin A, from the toxic South American plant known as green propolis (*Baccharis megapotamica*), is a likely mycotoxin that was later modified by *Baccharis megapotamica* (1,2). Baccharin is also a potent competitive inhibitor (K_i = 56 nM) of human aldo-keto reductase (AKR) 1C3, also known as 17β-hydroxysteroid dehydrogenase (Type 5) and prostaglandin F synthase. AKR is regarded as a druggable target in the treatment of prostate and breast cancers. Baccharin shows no significant inhibition toward the AKR1C1, AKR1C2, and AKR1C4 isoforms (3). Moreover, non-prenyl analogues of baccharin as selective and potent inhibitors for AKR1C3 (4). **1.** Kupchan, Jarvis, Dailey *et al.* (1976) *J. Am. Chem. Soc.* **98**, 7092. **2.** Akao, Maruyama, Matsumoto *et al.* (2003) *Biol. Pharm. Bull.* **26**, 1057. **3.** Endo, Matsunaga, Kanamori, *et al.* (2012) *J. Nat. Prod.* **75**, 716. **4.** Endo, Hu, Matsunaga, *et al.* (2014) *Bioorg. Med. Chem.* **22**, 5220.

Bacillusin A

This novel antibiotic (FW = 1157.58 g/mol) from *Bacillus amyloliquefaciens* AP183, is a macrodiolide composed of dimeric 4-hydroxy-2-methoxy-6-alkenylbenzoic acid lactones with conjugated pentaene-hexahydroxy polyketide chains, Head-to-head comparisons of antibiotics against methicillin-resistant *Staphyllococcus aureus* (ATCC 33591) are: Bacillusin A (Minimal Inhibitory Concentration = 1.2 μg/mL); Vancomycin (MIC = 1.6 μg/mL); Ciprofloxacin (MIC = 0.4 μg/mL), and Methicillin (MIC > 100 μg/mL). Head-to-head comparisons of antibiotics against methicillin-resistant *Enterococcus faecium* (ATCC 700221) are: Bacillusin A (MIC = 0.6 μg/mL); Vancomycin (MIC > 100 μg/mL); Ciprofloxacin (MIC > 100 μg/mL), and Methicillin (MIC > 100 μg/mL). **1.** Ravu, Jacob, Chen, *et al.* (2015) *J. Nat. Prod.* **78**, 924.

Bacillus thuringiensis Crystal Proteins

These biopesticides are *B.* thuringiensis-derived protein toxins, also known as Bt Toxins, that are composed of parasporal crystalline inclusions containing so-called Crystal (or Cry) and Cytolytic (or Cyt, also known as δ-endotoxins) toxins, which are cytotoxic/cytostatic toward a wide range of insect orders, nematodes as well as human-cancer cells. Cry and Cyt toxins are biosynthesized at the onset of sporulation and during stationary growth phase, mainly forming parasporal crystalline inclusions. Upon ingestion by susceoptible insects, the protein crystals are solubilized in the midgut, followed by proteolytic activation by midgut proteases. They then bind to specific insect cell membrane receptors, resulting in cell disruption and insect death. More than 700 *cry* gene sequences have been identified, most often located on large plasmids. Cry proteins contain: Domain I, a seven α-helix cluster membrane-perforating domain; a central or middle Domain II, consisting of three antiparallel β-sheets that play an important role in toxin-receptor interactions; and a galactose-binding Domain III, consisting of a two-antiparallel β-sheet sandwich that is also involved in receptor binding and pore formation. The extended pro-region in Cry proteins comprises the less well defined Domains IV, V, VI, and VII. Cyt proteins constitute a smaller, distinct group of crystal proteins with insecticidal activity against several dipteran larvae, particularly mosquitoes and black flies. **1.** Höfte & Whiteley (1989) *Microbiol. Rev.* **53**, 242. **2.** Knowles & Dow (1993) *Bioessays* **15**, 469. **3.** Raymond, Johnston, Nielsen-LeRoux, Lereclus & Crickmore (2010) *Trends Microbiol.* **18**, 189. 4. Roh, Choi, Li, Jin & Je (2007) *J. Microbial Biotech.* **17**, 547. **5.** Schnepf, Crickmore, van Rie, *et al.* (1998) *Microbiol. Mol. Biol. Rev.* **62**, 775. **6.** Bravo, Gill & Soberón (2007) *Toxicon.* **49**, 423. **7.** Palma, Muñoz, Berry Murillo & Caballero (2014) *Toxins* (Basel) **6**, 3296.

Bacitracins

This group of oligopeptide antibiotics, produced by *Bacillus subtilis* and *B. lichenformis*, inhibits cell wall peptidoglycan synthesis (1,5,8,11) Most commercial sources of bacitracin contain at least nine distinct components, the most common being bacitracin A (FW = 1422.71 g/mol; containing both D- and L-aminoacyl residues; Very soluble in water, methanol, and ethanol). It is relatively stable in acidic conditions but is unstable above pH 9. Bacitracin A inhibits the dephosphorylation of undecapenyl diphosphate and dolichyl diphosphate, thereby inhibiting cell wall biosynthesis. Bacitracin A also inhibits many peptidases and proteinases, of which a few are listed below. It has also been reported to activate aminopeptidase B. **Target(s):** pyroglutamyl-peptidase, bacterial (2); insulysin (3,36-40); pitrilysin, *or* protease Pi (4,10,36,41); protein glycosylation (1,5,8.11); 4-hydroxy-benzoate polyprenyltransferase (6); dolichyl-diphosphatase, *or* dolichyl pyrophosphate phosphatase (7,8,24,28,32-34); dipeptidyl peptidase IV, *or* post-proline dipeptidyl aminopeptidase (9,21,46); dolichyl-phosphate *N* acetylglucosaminyltransferase (11); protein-disulfide isomerase (12,13,29, 30); thermitase (14,42); savinase (14); subtilisin (14,42); *Treponema denticola* oligopeptidase (15); *Neurospora crassa* insulinase (16); leucyl aminopeptidase (17,45); enkephalin aminopeptidase (18); membrane-bound alanyl aminopeptidase (19); dolichyl-monophosphate phosphatase (20); receptor-mediated endocytosis of α₂-macroglobulin-protease complexes (22); extracellular protease of *Bacillus* licheniformis (23); undecaprenyl-diphospho-muramoylpentapeptide β-*N*-acetylglucos-aminyl transferase, weakly inhibited (25); dolichol kinase (26); geranyl*trans*transferase, *or* farnesyl-diphosphate synthase (27); oligosaccharide-diphosphodolichol diphosphatase (31); undecaprenyl-diphosphatase (35); peptidase K, *or* proteinase K (42); prolyl oligopeptidase (43,44); tripeptide aminopeptidase (47); CDP-glycerol glycerophospho-transferase (48); *N*-acetylglucosaminyl-diphosphodolichol *N*-acetylglucosaminyl-transferase (49); cellulose synthase, UDP-forming (50); protein-disulfide reductase (glutathione) (51,52). **1**. Izaki, Matsuhashi & Strominger (1966) *Meth. Enzymol.* **8**, 487. **2**. Robert-Baudouy, Clauziat & Thierry (1998) in *Handb. Proteolytic Enzymes* (Barrett, Rawlings & Woessner, eds.), p. 791, Academic Press, San Diego. **3**. Roth (1998) in *Handb. Proteolytic Enzymes* (Barrett, Rawlings & Woessner, eds.), p. 1362, Academic Press, San Diego. **4**. *ibid.* p. 1367. **5**. Elbein (1983) *Meth. Enzymol.* **98**, 135. **6**. Gupta & Rudney (1985) *Meth. Enzymol.* **110**, 327. **7**. Scher & Waechter (1985) *Meth. Enzymol.* **111**, 547. **8**. Elbein (1987) *Meth. Enzymol.* **138**, 661. **9**. Emerson (1989) *Meth. Enzymol.* **168**, 365. **10**. Stepanov (1995) *Meth. Enzymol.* **248**, 675. **11**. Schwarz & Datema (1980) *Trends Biochem. Sci.* **5**, 65. **12**. Essex, Li, Miller & Feinman (2001) *Biochemistry* **40**, 6070. **13**. Tager, Kroning, Thiel & Ansorge (1997) *Exp. Hematol.* **25**, 601. **14**. Pfeffer-Hennig, Dauter, Hennig, *et al.* (1996) *Adv. Exp. Med. Biol.* **379**, 29. **15**. Makinen, Makinen, Loesche & Syed (1995) *Arch. Biochem. Biophys.* **316**, 689. **16**. Kole, Smith & Lenard (1992) *Arch. Biochem. Biophys.* **297**, 199. **17**. Mantle, Lauffart & Gibson (1991) *Clin. Chim. Acta* **197**, 35. **18**. de Souza, Bruno & Carvalho (1991) *Comp. Biochem. Physiol. C* **99**, 363. **19**. Jahreis & Aurich (1990) *Biomed. Biochim. Acta* **49**, 339. **20**. Steen, Van Dessel, de Wolf, Lagrou, Hilderson & Dierick (1987) *Int. J. Biochem.* **19**, 427. **21**. Browne & O'Cuinn (1983) *Eur. J. Biochem.* **137**, 75. **22**. Van Leuven, Marynen, Cassiman & Van den Berghe (1981) *FEBS Lett.* **134**, 83. **23**. Vitkovic & Sadoff (1977) *J. Bacteriol.* **131**, 891. **24**. Presper & Heath (1983) *The Enzymes*, 3rd ed. (Boyer, ed.), **16**, 449. **25**. Meadow, Anderson & Strominger (1964) *Biochem. Biophys. Res. Commun.* **14**, 382. **26**. Steen, Van Dessel, De Wolf, *et al.* (1987) *Int. J. Biochem.* **19**, 419. **27**. Takahashi & Ogura (1981) *J. Biochem.* **89**, 1581. **28**. Siewert & Strominger (1967) *Proc. Natl. Acad. Sci. U.S.A.* **57**, 767. **29**. Barbouche, Miquelis, Jones & Fenouillet (2003) *J. Biol. Chem.* **278**, 3131. **30**. Mizunaga, Katakura, Miura & Maruyama (1990) *J. Biochem.* **108**, 846. **31**. Belard, Cacan & Verbert (1988) *Biochem. J.* **255**, 235. **32**. Scher & C. Waechter (1984) *J. Biol. Chem.* **259**, 14580. **33**. Wedgwood & Strominger (1980) *J. Biol. Chem.* **255**, 1120. **34**. Applekvist, Chojnacki & Dallner (1981) *Biosci. Rep.* **1**, 619. **35**. Stone & Strominger (1971) *Proc. Natl. Acad. Sci. U.S.A.* **68**, 3223. **36**. Ding, Becker, Suzuki & Roth (1992) *J. Biol. Chem.* **267**, 2414. **37**. Kole, Smith & Lenard (1992) *Arch. Biochem. Biophys.* **297**, 199. **38**. Werlen, Offord & Rose (1994) *Biochem. J.* **302**, 907. **39**. Bennett, Duckworth & Hamel (2000) *J. Biol. Chem.* **275**, 36621. **40**. Garcia, Fenton & Rosner (1988) *Biochemistry* **27**, 1337. **41**. Anastasi, Knight & Barrett (1993) *Biochem. J.* **290**, 601. **42**. Betzel, Rachev, Dolashka & Genov (1993) *Biochim. Biophys. Acta* **1161**, 47. **43**. Besedin & Rudenskaya (2003) *Russ. J. Bioorg. Chem.* **29**, 1. **44**. Daly, Maskrey, Mantle & Pennington (1985) *Biochem. Soc. Trans.* **13**, 1161. **45**. Mantle, Lauffart & Gibson (1991) *Clin. Chim. Acta* **197**, 35. **46**. Brownlees, Williams, Brennan & Halton (1992) *Biol. Chem. Hoppe-Seyler* **373**, 911. **47**. Hiraoka & Harada (1993) *Mol. Cell. Biochem.* **129**, 87. **48**. Burger & Glaser (1964) *J. Biol. Chem.* **239**, 3168. **49**. Kaushal & Elbein (1986) *Plant Physiol.* **81**, 1086. **50**. Haass, Hackspacher & Franz (1985) *Plant Sci.* **41**, 1. **51**. Varandani (1989) in *Coenzymes and Cofactors, Glutathione, Chem. Biochem. Med. Aspects Pt. A* (Dolphin, Poulson & Avromonic, eds.) vol. **3**, p. 753, Wiley, New York. **52**. Roth (1981) *Biochem. Biophys. Res. Commun.* **96**, 431.

Baclofen

This γ-aminobutyric acid (GABA) derivative (FW = 213.66 g/mol; CAS 1134-47-0), also named (*RS*)-4-amino-3-(4-chlorophenyl)butanoic acid, is a GABA_B receptor agonist marketed under the trade names Kemstro®, Lioresal®, Liofen®, Gablofen®, Beklo® and Baclosan® for the treatment of spasticity and early research stages of alcoholism. Baclofen increases expression of GABA_B receptors has been detected in several human cancer cells and tissues, baclofen inhibits tumor growth in rat models (1). GABA_B receptor activation not only induces suppressing the proliferation and migration of various human tumor cells but also results in inactivation of cAMP-responsive element binding protein (CREB) and ERK in tumor cells (1). P-type Ca^{2+} channels in cerebellar Purkinje neurons were inhibited by GABA as well as baclofen (2). In cross-chopped rat cortical slice preparations, (–)-baclofen inhibits forskolin-stimulated adenylyl cyclase activity and augments noradrenaline-stimulated adenylyl cyclase activity (3). Baclofen inhibition of the hyperpolarization-activated cation current in rat substantia nigra zona compacta neurons appears to be secondary to potassium current activation (4). **Key Pharmacokinetic Parameters:** *See* Appendix II in Goodman & Gilman's THE PHARMACOLOGICAL BASIS OF THERAPEUTICS, 12ᵗʰ Edition (Brunton, Chabner & Knollmann, eds.) McGraw-Hill Medical, New York (2011). **1**. Jiang, Su, Zhang, *et al.* (2012) *J. Histochem. Cytochem.* **60**, 269. **2**. Mintz & Bean (1993) *Neuron* **10**, 889. **3**. Knight & Bowery (1996) *Neuropharmacology* **35**, 703. **4**. Watts, Williams, & Henderson (1996) *J. Neurophysiol.* **76**, 2262.

Bafilomycin A₁

This toxic macrolide antibiotic (FW = 622.84 g/mol; CAS 88899-55-2) from *Streptomyces griseus*, potently inhibits the proton-pumping vacuolar ATPase, *or* V-ATPase (IC₅₀ = 0.44 nM), thereby blocking acid influx (1). Bafilomycin not only dissipates the low endosomal pH but also blocks transport from early to late endosomes in HeLa cells. At doses of 0.1-1 μM, Bafilomycin A₁ completely inhibits lysosome acidification. Bafilomycin A₁ also prevents maturation of autophagic vacuoles by inhibiting fusion between autophagosomes and lysosomes in rat hepatoma cell line, H-4-II-E cells (2). Inhibition of the fusion was reversible, and the autophagosomes changed into autolysosomes after the removal of the inhibitor. Bafilomycin A₁ also prevented the appearance of endocytosed horse radish peroxidase in autophagic vacuoles. These results suggested that acidification of the lumenal space of autophagosomes or lysosomes by V-ATPase is important for the fusion between autophagosomes and lysosomes (2). **Target(s):** Na⁺/K⁺-exchanging ATPase, moderately inhibited (1); Ca²⁺-transporting ATPase, moderately inhibited (1); K⁺-transporting ATPase (1,4-6); vesicle-fusing ATPase (2); cadmium-exporting ATPase (7). **1**. Bowman, Siebers & Altendorf (1988) *Proc. Natl. Acad. Sci. U.S.A.* **85**, 7972. **2**. Yamamoto, Tagawa, Yoshimori, *et al.* (1998) *Cell Struct. Funct.* **23**, 33. **3**. Yoshimori, Yamagata, Yamamoto *et al.* (2000) *Mol. Biol.* **11**, 747. **4**. Hafer, Siebers & Bakker (1989) *Mol. Microbiol.* **3**, 487. **5**. Siebers & Altendorf (1989) *J. Biol. Chem.* **264**, 5831. **6**. Abee, Siebers, Altendorf & Konings (1992) *J. Bacteriol.* **174**, 6911. **7**. Tsai, Yoon & Lynn (1992) *J. Bacteriol.* **174**, 116.

Bafilomycin B₁, *Similar in action to* Bafilomycin B₁

Bafilomycin C₁, *Similar in action to* Bafilomycin B₁

BAL9141, See Ceftobiprole

Baicalein

This flavone (FW = 270.24 g/mol; CAS 491-67-8), also known as 5,6,7-trihydroxy-2-phenyl-chromen-4-one, is formed upon metabolic hydrolysis of baicalin, a flavone prodrug found in the Chinese medicinal herb *Huang-chin*

(*See Baicalin*). It inhibits leukotriene biosynthesis and is also a free radical scavenger. Alkaline solutions have a greenish-brown color, while strong acidic solutions are yellow with a green fluorescence. Baicalein inhibits CYP2C9 ($K_i \approx 2$ µM), a cytochrome P450 that metabolizes many drugs, including NSAIDs (Celecoxib, Ibuprofen, Naproxen), Sulfonylureas (Glipizide, Glibenclamide, Glimepiride), Angiotensin II receptor antagonists (Irbesartan and Losartan), Warfarin, Sildenafil, and Tamoxifen. **Target(s):** 12-lipoxygenase, *or* arachidonate 12-lipoxygenase, $K_i = 0.14$ µM (1,13,43-48); 17β-hydroxysteroid dehydrogenase (2); NADP+-dependent malic enzyme (3); phosphoenolpyruvate carboxylase (3); thyroid type I deiodinase (4); steroid 5α-reductase (5); aromatase, *or* CYP19 (6); RNA directed DNA polymerase (7,8); aldose reductase (9,12); 5-lipoxygenase, *or* arachidonate 5 lipoxygenase (10,21,40-42); aryl hydrocarbon hydroxylase, unspecific monooxygenase (11,25); sialidase, mouse liver (14); NAD(P)H dehydrogenase, *or* DT-diaphorase (15,18); DNA polymerase γ (16); DNA polymerase I, *Escherichia coli* (16); hyaluronidase (17); xanthine oxidase (19); UDP glucuronosyltransferase (20); α-glucosidase, *or* sucrase (22,29); Raf-1 protein kinase (23,27); integrase, HIV-1 (24); CYP3A4, *or* testosterone 6β-hydroxylation (25); CYP1A2, caffeine *N*-demethylation (25); VHR dual-specificity protein-tyrosine phosphatase (26); 15-lipoxygenase, *or* arachidonate 15-lipoxygenase; $Ki = 0.18$ µM (28,43-45); lipoxygenase, soybean (28,49); protein kinase CK2 (30); lactoylglutathione lyase, *or* glyoxalase I (31); membrane alanyl aminopeptidase, *or* aminopeptidase N (32); DNA-directed DNA polymerase (16,33); inositol-trisphosphate 3-kinase (34); glycogen phosphorylase (35); fatty-acid synthase (36); 1-alkylglycerophosphocholine *O*-acetyltransferase (37,38); tyrosinase, monophenol monooxygenase (39); 8-lipoxygenase, *or* arachidonate 8-lipoxygenase (40). **1.** Sekiya & H. Okuda (1982) *Biochem. Biophys. Res. Commun.* **105**, 1090. **2.** Le Lain, Nicholls, Smith & Maharlouie (2001) *J. Enzyme Inhib.* **16**, 35. **3.** Pairoba, Colombo & Andreo (1996) *Biosci. Biotechnol. Biochem.* **60**, 779. **4.** Ferreira, Lisboa, Oliveira, *et al.* (2002) *Food Chem. Toxicol.* **40**, 913. **5.** Hiipakka, Zhang, Dai, Dai & Liao (2002) *Biochem. Pharmacol.* **63**, 1165. **6.** Kao, Zhou, Sherman, Laughton & Chen (1998) *Environ. Health Perspect.* **106**, 85. **7.** Ono, Nakane, Fukushima, Chermann & Barre-Sinoussi (1990) *Eur. J. Biochem.* **190**, 469. **8.** Ono, Nakane, Fukushima, Chermann & Barre-Sinoussi (1989) *Biochem. Biophys. Res. Commun.* **160**, 982. **9.** Kohda, Tanaka, Yamaoka, *et al.* (1989) *Chem. Pharm. Bull. (Tokyo)* **37**, 3153. **10.** McMillan, D. J. Masters, Sterling & Bernstein (1985) in *Prostaglandins, Leukotrienes, and Lipoxins* (Bailey, ed.), p. 655, Plenum Press, New York. **11.** Friedman, West, Sugimura & Gelboin (1985) *Pharmacology* **31**, 203. **12.** Li, Mao, Du, *et al.* (1987) *Yan Ke Xue Bao* **3**, 93. **13.** Nadler, Natarajan & Stern (1987) *J. Clin. Invest.* **80**, 1763. **14.** Nagai, Yamada & Otsuka (1989) *Planta Med.* **55**, 27. **15.** Liu, Liu, Iyanagi, *et al.* (1990) *Mol. Pharmacol.* **37**, 911. **16.** Ono & Nakane (1990) *J. Biochem.* **108**, 609. **17.** Kakegawa, Matsumoto & Satoh (1992) *Chem. Pharm. Bull.* **40**, 1439. **18.** Chen, Hwang & Deng (1993) *Arch. Biochem. Biophys.* **302**, 72. **19.** Chang, Lee, Lu & Chiang (1993) *Anticancer Res.* **13**, 2165. **20.** Yokoi, Narita, Nagai, *et al.* (1995) *Jpn. J. Cancer Res.* **86**, 985. **21.** Ghosh & Myers (1997) *Biochem. Biophys. Res. Commun.* **235**, 418. **22.** Nishioka, Kawabata & Aoyama (1998) *J. Nat. Prod.* **61**, 1413. **23.** Nakahata, Kyo, Kutsuwa & Ohizumi (1999) *Nippon Yakurigaku Zasshi* **114** Suppl. 1, 215P. **24.** Ahn, Lee, Kim, *et al.* (2001) *Mol Cells* **12**, 127. **25.** Kim, Lee, Kim, *et al.* (2002) *J. Toxicol. Environ. Health A* **65**, 373. **26.** Lee, Oh, Kim, *et al.* (2002) *Planta Med.* **68**, 1063. **27.** Nakahata, Tsuchiya, Nakatani, Ohizumi & Ohkubo (2003) *Eur. J. Pharmacol.* **461**, 1. **28.** Sadik, Sies & Schewe (2003) *Biochem. Pharmacol.* **65**, 773. **29.** Gao, Nishioka, Kawabata & Kasai (2004) *Biosci. Biotechnol. Biochem.* **68**, 369. **30.** Lin, Liu, Chen, Chen & Liang (2004) *Ai Zheng* **23**, 874. **31.** Ito, Sadakane, Shiotsuki & Eto (1992) *Biosci. Biotechnol. Biochem.* **56**, 1461. **32.** Bauvois & Dauzonne (2006) *Med. Res. Rev.* **26**, 88. **33.** Spampinato, Pairoba, Colombo, Benediktsson & Andreo (1994) *Biosci. Biotechnol. Biochem.* **58**, 822. **34.** Mayr, Windhorst & Hillemeier (2005) *J. Biol. Chem.* **280**, 13229. **35.** Jakobs, Fridrich, Hofem, Pahlke & Eisenbrand (2006) *Mol. Nutr. Food Res.* **50**, 52. **36.** Li & Tian (2004) *J. Biochem.* **135**, 85. **37.** Hurst & Bazan (1997) *J. Ocul. Pharmacol. Ther.* **13**, 415. **38.** Yamazaki, Sugatani, Fujii, *et al.* (1994) *Biochem. Pharmacol.* **47**, 995. **39.** Gao, Nishida, Saito & Kawabata (2007) *Molecules* **12**, 86. **40.** Schweiger, Fürstenberger & Krieg (2007) *J. Lipid Res.* **48**, 553. **41.** Furakawa, Yoshimoto, Ochi & Yamamoto (1984) *Biochim. Biophys. Acta* **795**, 458. **42.** Schneider & Bucar (2005) *Phytother. Res.* **19**, 81. **43.** Vasquez-Martinez, Ohri, Kenyon, Holman & Sepúlveda-Boza (2007) *Bioorg. Med. Chem.* **15**, 7408. **44.** Deschamps, Kenyon & Holman (2006) *Bioorg. Med. Chem.* **14**, 4295. **45.** Cho, Ueda, Tamaoka, *et al.* (1991) *J. Med. Chem.* **34**, 1503. **46.** Chen, Wang, Li, *et al.* (2008) *Dig. Dis. Sci.* **53**, 181. **47.** Endsley,

Aggarwal, Isbell, *et al.* (2007) *Int. J. Cancer* **121**, 984. **48.** Higuchi, Tanii, Koriyama, Mixukami & Yoshimoto (2007) *Life Sci.* **80**, 1856. **49.** Ahmad, Iqbal, Nawaz, *et al.* (2006) *Chem. Biodiv.* **3**, 996.

Baicalin

This baicalein glucuronide (FW = 446.36 g/mol; CAS 21967-41-9), also known as 5,6-dihydroxy-4-oxygen-2-phenyl-4*H*-1-benzopyran-7-β-D-glucopyranose acid, is a flavone prodrug found in the Chinese medicinal herb Huang-chin (*Scutellaria baicalensis*) that is hydrolyzed to baicalein (*See Baicalein*). Baicalin is a slow, tight-binding inhibitor of Jack Bean urease, rapidly forming initial BA-urease complex ($K_i = 3.9 \times 10^{-3}$ M) that slowly isomerizes to the final complex (overall inhibition constant of $K_i^* = 0.15 \times 10$ µM). Inhibition can be reversed by dithiothreitol but not dilution of substrate. Baicalin also inhibits prolyl oligopeptidase (2), a cytosolic serine peptidase that hydrolyzes proline-containing peptides at the carboxy terminus of proline residues and is associated with schizophrenia, bipolar affective disorder, and related neuropsychiatric disorders. (*See Pramiracetam*) **1.** Tan, Su, Wu, *et al.* (2013) *Scientific World J.* **2013**, 879501. **2.** Tarragó, Kichik, Claasen, *et al.* (2008) *Bioorg. Med. Chem.* **16**, 7516.

BAL8557, *See* Isavuconazole & Isavuconazonium sulfate

Balofloxacin

This carboxyquinolone-class antibiotic (FW = 389.42 g/mol; CAS 127294-70-6; Symbol: BLFX), also named 1-cyclopropyl-6-fluoro-8-methoxy-7-(3-methylaminopiperidin-1-yl)-4-oxoquinoline-3-carboxylate, is active against *Staphylococcus aureus* ($MIC_{90} = 0.2$ µg/mL), methicillin-resistant *S. aureus* ($MIC_{90} = 6.25$ µg/mL), *Staphylococcus epidermidis* ($MIC_{90} = 0.2$ µg/mL), *Streptococcus pneumoniae* ($MIC_{90} = 0.4$ µg/mL), and *Streptococcus pyogenes* ($MIC_{90} = 0.4$ µg/mL). For 82 ciprofloxacin-resistant staphylococci ($MIC_{90} = 100$ µg/mL), Q-35 was the most active of the new quinolones tested ($MIC_{90} = 6.25$ µg/mL). The MIC_{90} values for *Escherichia coli*, *Enterobacter aerogenes*, and *Pseudomonas aeruginosa* were 0.2, 0.78, and 12.5 µg/mL, respectively. **1.** Ito, Otsuki & Nishino (1992) *Antimicrob. Agents Chemother.* **36**, 1708.

BAM15

This novel mitochondrial uncoupler (FW = 340.29 g/mol), also named (2-fluorophenyl)-{6-[(2-fluorophenyl)amino]-(1,2,5-oxadiazolo[3,4-*e*]pyrazin-5-yl)}amine, exhibits a broad effective range of H^+ gradient-dissipating action without affecting plasma membrane electrophysiology. With FCCP as an equipotent positive control, BAM15 was found to be fully able to increase mitochondrial respiration in the presence of oligomycin over a broader concentration range than FCCP in both myoblasts and hepatocytes. BAM15 and FCCP had similar effects on mitochondrial depolarization in L6 myoblasts treated with concentrations of each uncoupler at 1 and 10 µM, as measured by tetramethylrhodamine (TMRM) fluorescence. Neither FCCP nor BAM15 donates electrons to the Electron Transport Chain. Although some non-protonophoric uncouplers increase proton transport into the matrix via interaction with the mitochondrial inner membrane adenine nucleotide translocase (ANT), neither BAM15 nor FCCP requires the ANT to increase mitochondrial respiration. **1**. Kenwood, Weaver, Bajwa, *et al.* (2013) *Mol. Metab.* **3**, 114.

Bambuterol

Bambuterol

Terbutaline

This β_2-adrenergic agonist pro-drug and long-acting bronchodilator (FW = 367.44 g/mol; CAS 81732-46-9), also named (*RS*)-5-[2-(*tert*-butylamino)-1-hydroxyethyl]benzene-1,3-diyl bis(dimethylcarbamate), is used in the management of asthma and chronic obstructive pulmonary disease (1,2). (*See also Formoterol; Salmeterol*) Bambuterol is an extremely effective inhibitor (*i.e.,* weak alternative substrate) of cholinesterase, when assayed with butyrylthiocholine as the substrate (IC$_{50}$ = 1.7 x 10^{-8} M), but is 2400-fold less efficient in inhibiting cholinesterase (with acetylthiocholine as substrate, IC$_{50}$ = 4.1 x 10^{-5} M) (3). However, preincubation of blood with bambuterol in the absence of thiocholine ester substrate is essential for obtaining maximal inhibition. The inhibition exerted by bambuterol after such preincubation was reversible and noncompetitive (3). Because butyrylthiocholine is the preferred substrate for cholinesterase (EC 3.1.1.8) and acetylthiocholine for acetylcholinesterase (EC 3.1.1.7), such results indicate that bambuterol is a selective and potent inhibitor of cholinesterase. Such findings facilitated the discovery that bambuterol is the inactive *bis*-*N*,*N*-dimethylcarbamate prodrug of the active β_2-adrenoceptor agonist terbutaline (FW = 225.28 g/mol; CAS 23031-25-6; IUPAC: (*RS*)-5-[2-(*tert*-butylamino)-1-hydroxyethyl]-benzene-1,3-diol). The latter is formed through hydrolytic and/or oxidative pathways catalyzed by plasma cholinesterase (EC 3.1.1.8) and cytochrome P450 (CYP) enzymes. Terbutaline itself is a fast-acting bronchodilator that is often used as a short-term asthma treatment; it is also used as a tocolytic agent to delay premature labor. Note that the complex bambuterol-to-terbutaline biotransformation is likely to be a factor contributing to the long duration of action of bambuterol. Cholinesterases also share homologies with cell adhesion molecules, and *in vitro* neurite growth from various neuronal tissues of the chick embryo can be modified by the action of bambuterol (4). **1**. Pedersen, Laursen, Gnosspelius, Faurschou & Weeke (1985) *Eur. J. Clin. Pharmacol.* **29**, 425. **2**. Holstein-Rathlou, Laursen, Madsen, *et al.* (1986) *Eur. J. Clin. Pharmacol.* **30**, 7. **3**. Tunek & Svensson (1988) *Drug Metab. Dispos.* **16**, 759. **4**. Layer, Weikert & Alber (1993) *Cell Tissue Res.* **273**, 219.

b-AP15

This deubiquitinase inhibitor (FW = 421.31 g/mol; CAS 1009817-63-3), also named 3,5-bis[[(4-nitrophenyl)methylene]-1-(1-oxo-2-propen-1-yl)]-(3*E*,5*E*)-4-piperidinone, targets two proteasome-associated ubiquitin carboxyl-terminal hydrolase-14, *or* USP14, and Ubiquitin Carboxyl-terminal Hydrolase isozyme L5 UCHL5, IC$_{50}$ = 2.1 μM, resulting in a rapid accumulation of high-molecular-weight ubiquitin conjugates and functional shutdown of proteasome. Interestingly, b-AP15 displays several differences with respect to bortezomib including insensitivity to over-expression of the anti-apoptotic mediator Bcl-2 and anti-tumor activity in solid tumor models.

Inhibition of DUBs blocked the processing and release of interleukin IL-1β in both mouse and human macrophages. DUB activity was necessary for inflammasome association as DUB inhibition also impaired ASC oligomerization and caspase-1 activation without directly blocking caspase-1 activity. These data reveal the requirement for DUB activity in a key reaction of the innate immune response and highlight the therapeutic potential of DUB inhibitors for chronic auto-inflammatory diseases. (*See also Bortezomib, Eeeyarestatin I*) **1**. Lopez-Castejon, Luheshi, Compan, *et al.* (2013) *J. Biol. Chem.* **288**, 2721.

BAPN, *See β-Aminopropionitrile*

BAPTA

BAPTA

BAPTA-AM

This calcium ion chelator (FW$_{free-acid}$ = 476.44 g/mol), also known as 1,2-bis(*o*-aminophenoxy)-ethane-*N*,*N*,*N'*,*N'*-tetraacetic acid, has high affinity of Ca^{2+}, with a K_d of 110 nM in 0.1 M KCl (1) The K_d value for Mg^{2+} is 17 mM. BAPTA has a λ_{max} at 254 nm (ε = 16000 M^{-1}cm^{-1}), which, upon binding of Ca^{2+}, shifts to 274 nm (ε = 4200 M^{-1}cm^{-1}). BAPTA is much less affected to changes in pH than EGTA: two pK_a values are below 4 and the other two are 5.47 and 6.36. Note that, while one may suspect that BAPTA would be an effector with Ca^{2+}-dependent systems, it may also be a modulator in Ca^{2+}-independent systems: *e.g.*, it has been reported that BAPTA exhibits a potent Ca^{2+}-independent actin and microtubule depolymerizing activity in most cell types. The tetrakis(acetoxymethyl ester), often abbreviated BAPTA-AM (FW = 764.69 g/mol), is a commercially available, membrane permeable agent that is de-esterified to BAPTA within the cell. **Target(s):** phosphoinositide phospholipase C (2); tubulin polymerization, microtubule assembly (3); actin polymerization (3); RNA biosynthesis (4); mannosyl-oligosaccharide 1,2-α mannosidase, *or* mannosidase I (5); Ca^{2+}/calmodulin-dependent protein kinase II (6); phosphatidylinositol-4-phosphate 3-kinase (7); 1,3-β-glucan synthase (8). **1**. Tsien (1980) *Biochemistry* **19**, 2396. **2**. Hardie (2005) *Cell Calcium* **38**, 547. **3**. Saoudi, Rousseau, Doussiere, *et al.* (2004) *Eur. J. Biochem.* **271**, 3255. **4**. Shang & Lehrman (2004) *Biochemistry* **43**, 9576. **5**. Forsee, Palmer & Schutzbach (1989) *J. Biol. Chem.* **264**, 3869. **6**. Tsui, Inagaki & Schulman (2005) *J. Biol. Chem.* **280**, 9210. **7**. Crljen, Volinia & Banfic (2002) *Biochem. J.* **365**, 791. **8**. Kamat, Garg & Sharma (1992) *Arch. Biochem. Biophys.* **298**, 731.

Barbourin

This platelet aggregation activation inhibitor (MW = 7700.52 g/mol; CAS 135402-55-0; Accession Number = P22827; Sequence: EAGEECDCGSP ENPCCDAATCKLRPGAQCADGLCCDQCRFMKKGTVCRVA**KGD**WN DDTCTGQSADCPRNGLYG) was isolated from the venom of the Southeastern Pigmy Rattlesnake (*Sistrurus miliarius barbouri*), the only of 52 venoms specific for integrin GPIIb-IIIa *versus* other integrins. Barbourin is highly homologous to other peptides of the viper venom GPIIb-IIIa antagonist family, but contains a Lys-Gly-Asp (KGD) in place of the canonical Arg-Gly-Asp (RGD) sequence needed to inhibit receptor function. Barbourin represents a new structural model for designing potent and GPIIb-

IIIa-specific, platelet aggregation inhibitors (**See** *Eptifibatide for details on Mechanism of Action*). **1**. Scarborough, Rose, His, *et al.* (1991) *J. Biol. Chem.* **266**, 9359.

Barasertib

This highly selective Aurora kinase inhibitor (FW = 507.56 g/mol; CAS 722544-51-6; Solubility: 102 mg/mL DMSO; <1 mg/mL H_2O), alternatively named AZD1152-AQPA and systematically as 2-(5-(7-(3-(ethyl(2-hydroxyethyl)amino)propoxy) quinazolin-4-ylamino)-1*H*-pyrazol-3-yl)-*N*-(3-fluorophenyl)acetamide, targets Aurora B (IC_{50} = 0.37 nM) and inhibits proliferation of malignant hematopoietic cells (*e.g.*, HL-60, NB4, MOLM13, PALL-1, PALL-2, MV4-11, EOL-1, THP-1, and K562 cells, with IC_{50} values of 3-40 nM). Aurora kinases are serine/threonine protein kinases that play various roles in mitosis. Aurora B helps to regulate chromosome alignment, kinetochore-microtubule bi-orientation, spindle assembly checkpoint activation, and cytokinesis. Its higher expression levels and activity correlate with malignancy of solid tumors in prostate, colon, pancreas, breast, and thyroid. **Target(s):** AZD1152 shows 3000x lower activity against Aurora A (IC_{50} = 1.37 μM), with even less activity against fifty other serine-threonine and tyrosine kinases, including FLT3, JAK2, and Abl. **1**. Yang, Ikezoe, Nishioka, *et al.* (2007) *Blood* **110**, 2034. **2**. Wilkinson, Odedra, Heaton, *et al.* (2007) *Clin. Cancer Res.* **13**, 3682.

Bardoxolone Methyl

This oleanane triterpenoid (FW = 505.70 g/mol; CAS 218600-53-4; Solubility: 1 mg/mL DMSO; <1 mg/mL H_2O), also known as RTA 402, TP-155, NSC 713200, CDDO Methyl Ester, and 2-cyano-3,12-dioxooleana-1,9(11)-dien-28-oic acid methyl ester, targets <u>I</u>nhibitor of nuclear factor <u>K</u>appa-B <u>K</u>inase subunit <u>β</u>, *or* IKKβ, a key component of the cytokine-activated intracellular signaling pathway that triggers immune responses (1). Significantly, CDDO methyl ester inhibits proliferation of myeloid leukemia cells, inducing both differentiation and apoptosis (2). CDDO-Me induces loss of mitochondrial membrane potential, induction of caspase-3 cleavage, increase in annexin V binding, and DNA fragmentation – all suggesting induction of apoptosis (2). CDDO-Me induces pro-apoptotic Bax protein that preceded caspase activation. CDDO-Me also inhibits ERK1/2 activation, as indicated by inhibition of mitochondrial ERK1/2 phosphorylation and blocking of Bcl-2 phosphorylation, rendering the latter less anti-apoptotic (2). CDDO-Me inhibits NF-κB through inhibition of IκBα kinase, resulting in the suppression of expression of NF-κB-regulated gene products (*Readouts*: IAP2, cFLIP, TRAF1, survivin, and Bcl-2) as well as inhibition of proliferation (*Readouts*: cyclin d_1 and c-myc), and angiogenesis (*Readouts*: VEGF, Cox-2, and MMP-9) (3). CDDO-Me thus enhances apoptosis induced by TNF and chemotherapeutic agents (3). Bardoxolone methyl exhibits potent pro-apoptotic and anti-inflammatory activity, potently inhibiting interferon γ-induced nitric oxide synthesis in mouse macrophages, IC_{50} = 0.1 nM (4). Bardoxolone methyl pretreatment uniquely confers protection against lipopolysaccharide (LPS) challenge by modulating the *in vivo* immune response to LPS, indicating that it represents a novel oral agent for use in LPS-mediated inflammatory diseases. Significantly, a number of well-recognized naturally occurring or synthetic anti-inflammatory compounds possessing a Michael-type acceptor (*e.g.*, thymoquinone (TQ), the paracetamol metabolite NAPQI, the 5-LO inhibitor AA-861, and bardoxolone methyl) are all direct covalent 5-LO enzyme inhibitors that target the catalytically relevant Cysteines 416 and 418 (6). Their actiom as irreversible Michael acceptor moietied that interact with required cysteines of proteins/enzymes may expain why they are effective and sustained enzyme activity modulators. **1**. Honda, Rounds, Bore, *et al.* (2000) *J. Med. Chem.* **43**, 4233. **2**. Konopleva, Tsao, Ruvolo, *et al.* (2002) *Blood* **99**, 326. **3**. Shishodia, Sethi, Konopleva, Andreeff & Aggarwal (2006) *Clin. Cancer Res.* **12**, 1828. **4**. Liby, Royce, Williams, *et al.* (2007) *Cancer Res.* **67**, 2414. **5**. Auletta, Alabran, Kim, Meyer & Letterio (2010) *J. Interferon Cytokine Res.* **30**, 497. **6**. Maucher, Rühl, Kretschmer, *et al.* (2017) *Biochem. Pharmacol.* **125**, 55.

Baricitinib

This selective, ATP-competitive protein kinase inhibitor (FW = 371.42 g/mol; CAS 1187594-09-7; Solubility: 70 mg/mL DMSO, <1 mg/mL H_2O), also known by its code names LY3009104 and INCB028050 as well as its systematic name 1-(ethylsulfonyl)-3-[4-(7*H*-pyrrolo[2,3-*d*]-pyrimidin-4-yl)-1*H*-pyrazol-1-yl]-3-azetidineacetonitrile, targets JAK1 and JAK2, with respective IC_{50} values of 5.9 nM and 5.7 nM, showing little effect on JAK3 (IC_{50} = 560 nM) (1-3). In cell-based assays, INCB028050 proved to be a potent inhibitor of JAK signaling and function. **Primary Mode of Inhibitory Action:** Janus kinases JAK1, JAK2, JAK3, and the related enzyme TYK2 are critical components of signaling mechanisms used by many cytokines and growth factors, including several that are elevated in patients with RA. Cytokines such as interleukin-6, -12 and -23 and both type 1 and type 2 interferons signal through these pathways. JAK-dependent cytokines have been implicated in the pathogenesis of inflammatory and autoimmune diseases, suggesting that JAK inhibitors may be useful for the treatment of a broad range of inflammatory conditions. At concentrations <50 nM, baricitinib inhibits intracellular signaling of multiple proinflammatory cytokines including IL-6 and IL-23 (3). Significant efficacy, as assessed by improvements in clinical, histologic and radiographic signs of disease, was achieved in a rat adjuvant arthritis model with baricitinib providing partial and/or periodic inhibition of JAK1/JAK2 and no inhibition of JAK3. Such findings suggest that fractional inhibition of JAK1 and JAK2 may be sufficient for significant pharmacological activity in autoimmune diseases, such as rheumatoid arthritis. A related JAK inhibitor, Tofacitinib, is approved in the U.S. for the treatment of RA. **Target(s):** Baricitinib shows moderate (~10x) selectivity against Tyk2 (IC_{50} = 53 nM) and marked selectivity over the unrelated c-Met (IC_{50} > 10,000 nM) and Chk2 (IC_{50} > 1,000 nM) kinases as well as ~100x higher IC_{50} values against Abl, Akt1, AurA, AurB, CDC2, CDK2, CDK4, CHK2, c-kit, EGFR, EphB4, ERK1, ERK2, FLT-1, HER2, IGF1R, IKKα, IKKβ, JNK1, Lck, MEK1, p38α, p70S6K, PKA, PKCα, Src, and ZAP70. **1**. Keystone, Taylor, Genovese, (2012) Annals *Rheumatic Disease* **71** (*Supplement* 3), 152. **2**. Genovese, Keystone, Taylor, *et al.* (2012) *Arthritis & Rheumatism* **64**, S1049. **3**. Fridman, Scherle, Collins, *et al.* (2010) *J. Immunol.* **184**, 5298.

Barium Ions

This Group 2 stable alkali earth element (Atomic Weight = 137.33; Atomic Number = 56; Symbol: Ba; Atomic radius = 222 pm; Covalent radius = 215 pm; Van der Waals radius = 268 pm) forms divalent ions (Radius = 149.57 pm (compare to 114.21 pm for Ca^{2+})) and lies two rows beneath calcium in the Periodic Table, shares other properties with members of its group. Barium can be highly toxic to plants and animals upon acute and chronic exposure. Inhaled dust containing insoluble barium compounds can accumulate in the lungs, causing a benign condition called baritosis. Generated in many industrial processes, barium targets the potassium inward rectifier channels (IRCs) in the KCNJx gene family. Extracellular barium enters the potassium-conducting pore, blocking potassium conduction. As a competitive potassium channel antagonist, divalent barium ion blocks the passive efflux of intracellular potassium, resulting in a shift of extracellular

K^+ into intracellular compartments, resulting in a significant decrease of K^+ in the blood plasma. Although relatively insoluble in water, barium carbonate (Formula = $BaCO_3$; FW = 197.34 g; CAS 513-77-9) becomes toxic when hydrolyzed stomach acid, allowing further absorption of Ba^{2+} within the gastrointestinal tract. (Other insoluble barium compounds (notably barium sulfate) are inefficient sources of Ba^{2+} ion and are of little concern in humans.) Ingestion of $BaCO_3$ often results in gastrointestinal effects (vomiting, diarrhea), cardiovascular effects (*e.g.*, arrhythmias and hypertension), neuromuscular (*e.g.*, abnormal reflexes and paralysis), respiratory effects (*e.g.*, arrest and even failure), as well as metabolic effects (*e.g.*, hypokalemia). Treatment of barium poisoning uses sodium sulfate (administered orally) to form insoluble barium sulfate in the intestinal tract, with potassium supplementation.

Batimastat

This potent, broad-spectrum matrix metalloprotease, *or* MMP, inhibitor (FW = 477.64 g/mol; CAS 130370-60-4; Solubility: 96 mg/mL DMSO, <1 mg/mL H_2O), also known by its code name BB-94 and its systematic name $(2R,3S)$-N^4-hydroxy-N^1-[(1S)-2-(methylamino)-2-oxo-1-(phenylmethyl) ethyl]-2-(2-methylpropyl)-3-[(2-thienylthio)methyl]butanediamide, targets MMP-1 (IC_{50} = 3 nM), MMP-2 (IC_{50} = 4 nM), MMP-9 (IC_{50} = 4 nM), MMP-7 (IC_{50} = 6 nM) and MMP-3 (IC_{50} = 22 nM). Matrix metalloproteinases have been implicated in the growth and spread of metastatic tumors, and Batimastat not only prevents colonization of secondary organs by B16-BL6 cells, but limits the growth of solid tumors (1). Animals receiving BB-94 (30 mg/kg, i.p., once daily for 60 days, followed by 3 times weekly) show a reduction in the median primary tumor weight from 293 mg in the control group to 144 mg in the treated group (3). BB-94 treatment also reduces the incidence of local and regional invasion. Batimastat also reduces *in vivo* growth of experimental hemangiomas, most probably by blocking endothelial cell recruitment by the transformed cells or by interfering with cell organization in vascular structures (3). **Target(s):** *Trimeresurus mucrosquamatus* (Taiwan habu) venom metalloproteinases (4); matrix metalloproteinases (5); interstitial collagenase, *or* MMP1 (5,15); stromelysin, *or* MMP3 (5,10); matrilysin, *or* MMP7 (5,20); gelatinase A, *or or* MMP2 (5); gelatinase B, *or* MMP9 (5); neutrophil collagenase, *or* MMP8 (6,19); atrolysin C, *or* *Crotalus atrox* metalloendopeptidase c (10); membrane-type 1 matrix metalloproteinase, *or* MMP14 (11,12,20); membrane-type 3 matrix metalloproteinase, *or* MMP-16 (12,17); ADAM 17 endopeptidase, *or* tumor necrosis factor-α converting enzyme; TACE (13-15); α-secretase (15); ADAM TS-4 endopeptidase, *or* aggrecanase (16); macrophage elastase, *or* MMP12 (18). **1.** Chirivi, Garofalo, Crimmin, *et al.* (1994) *Int. J. Cancer* **58** 460. **2.** Wang, Fu, Brown, Crimmin & Hoffman (1994) *Cancer Res.* **54**, 4726. **3.** Taraboletti, Garofalo, Belotti, *et al.* (1995) *J. Natl. Cancer Inst.* **87**, 293. **4.** Huang, Chiou, Ko & Wang (2002) *Eur. J. Biochem.* **269**, 3047. **5.** Brown (1995) *Adv. Enzyme Regul.* **35**, 293. **6.** Grams, Crimmin, Hinnes, *et al.* (1995) *Biochemistry* **34**, 14012. **7.** Botos, Scapozza, Zhang, Liotta & Meyer (1996) *Proc. Natl. Acad. Sci. U.S.A.* **93**, 2749. **8.** Kinoshita, Sato, Takino, Itoh, Akizawa & Seiki (1996) *Cancer Res.* **56**, 2535. **9.** Yamamoto, Tsujishita, Hori, *et al.* (1998) *J. Med. Chem.* **41**, 1209. **10.** Epps, Poorman, Petzold, *et al.* (1998) *J. Protein Chem.* **17**, 699. **11.** Chesneau, Becherer, Zheng, *et al.* (2003) *J. Biol. Chem.* **278**, 22331. **12.** Lang, Braun, Sounni, *et al.* (2004) *J. Mol. Biol.* **336**, 213. **13.** Schlöndorff, Becherer & Blobel (2000) *Biochem. J.* **347**, 131. **14.** Becherer, Lambert & Andrews (2000) *Handbook of Experimental Pharmacology* (von der Helm, Korant & Cheronis, eds.) **140**, 235. **15.** Parkin, Trew, Christie, *et al.* (2002) *Biochemistry* **41**, 4972. **16.** Sugimoto, Takahashi, Yamamoto, Shimada & Tanzawa (1999) *J. Biochem.* **126**, 449. **17.** Shofuda, Hasenstab, Kenagy, *et al.* (2001) *FASEB J.* **15**, 2010. **18.** Lang, Kocourek, Braun, *et al.* (2001) *J. Mol. Biol.* **312**, 731. **19.** Gioia, Fasciglione, Marini, *et al.* (2002) *J. Biol. Chem.* **277**, 23123. **120.** Abramson, Conner, Nagase, Neuhaus & Woessner (1995) *J. Biol. Chem.* **270**, 16016.

Batrachotoxin

This lipid-soluble neurotoxin (FW = 526.67 g/mol; CAS 23509-16-2; LD_{50} = ~2 μg/kg mouse body weight.), isolated from the skin of the Columbian poison-dart frog *Phyllobates aurotaenis*, enhances Na^+ conductance, promoting opening of voltage-gated Na^+ channels, inducing depolarization of the resting membrane potential (1,2). It does so by binding to the sodium channel, keeping the membrane permeable to sodium ions in an all-or-none manner. This results in hyperexcitability of excitable tissues, followed by convulsions, paralysis, and death in animals. The most common use of batrachotoxin is in darts for bloguns used in hunting by the Noanamá Chocó and Emberá Chocó Indians of western Colombia. Batrachotoxin has also been identified in the feathers of the passerine birds of New Guinea *Pitohui dichrous*, *Pitohui kirhocephalus*, and *Ifrita kowaldi* (3). The less toxic batrachotoxin A is the deesterified steroid. **1.** Conn (1983) *Meth. Enzymol.* **103**, 401. **2.** Wang & Wang (2003) *Cell Signal* **15**, 151. **3.** Dumbacher, Spande & Daly (2000) *Proc. Natl. Acad. Sci. U.S.A.* **97**, 12970.

Batroxostatin

This cysteine-rich, 71-residue disintegrin (MW = 7602.46 g/mol; CAS 130357-67-4), from the snake *Bothrops atrox* (*Barba amarilla*) contains an RGD loop that facilitates high-affinity binding to integrins in a manner that inhibits ADP-induced platelet aggregation (IC_{50} = 133 nM), about 1000-times higher than that of Arg-Gly-Asp-Ser tetrapeptide. In addition, batroxostatin was about 400-times more potent than Arg-Gly-Asp-Ser at inhibiting melanoma cell adhesion to fibronectin. When covalently attached to plastic culture dishes, batroxostatin promoted adhesion of melanoma cells. **1.** Rucinski, Niewiarowski, Holt, Soszka & Knudsen (1990) *Biochim. Biophys. Acta* **1054**, 257.

Bauhinia ungulata Factor Xa Inhibitor

This secreted Kunitz-type inhibitor (MW = 19,238 g/mol, based on primary sequence), isolated from *Bauhinia ungulata* (orchid tree) seeds, is a potent inhibitor of plasma kallikrein (K_i = 6.9 nM) and coagulation factor Xa (K_i = 18.4 nM) (1,2). **Target(s):** chymotrypsin (2); coagulation factor Xa (1,2); coagulation factor XIIa (2); plasma kallikrein (1-3); plasmin (2); trypsin (2). **1.** Oliva, Andrade, Batista, *et al.* (1999) *Immunopharmacol.* **45**, 145. **2.** Oliva, Andrade, Juliano, *et al.* (2003) *Curr. Med. Chem.* **10**, 1085. **3.** Oliva, Santomauro-Vaz, Andrade, *et al.* (2001) *Biol. Chem.* **382**, 109.

Bax Inhibitor-1

This evolutionarily conserved integral membrane protein (MW = 26538 g/mol; Accession Number = P55061; Abbreviation: BI-1) is an anti-apoptotic factor (1), mediating a calcium leak, or safety valve, that provides for Ca^{2+} homeostasis by balancing active Ca^{2+} uptake by the endoplasmic reticulum (2). BI-1 represents the highly conserved and widely distributed Transmembrane Bax Inhibitor Motif (*or* TMBIM) proteins that share a canonical seven-transmembrane-helix fold featuring two triple-helix sandwiches wrapped around a central C-terminal helix (2). Structures show a pH-dependent switch that reversibly interconverts closed and open conformations, as governed by the pK_a of a perturbed pair of conserved aspartate residues and allowing pH to regulate Ca^{2+} influx in proteoliposomes (2). **1.** Xu & Reed (1998) *Molec. Cell* **1**, 337. **2.** Chang, Bruni, Kloss, *et al.* (344) *Science* **344**, 1131.

BAY 11-7082

This protein kinase inhibitor (FW = 207.31 g/mol; CAS 19542-67-7; λ_{max} = 251 nm), also named 3-[(4-methylphenyl)sulfonyl]-(2E)-propenenitrile, targets NF-κB activation, selectively and irreversibly blocking TNF-α-induced phosphorylation of IκB-α without affecting phosphorylation of constitutive IκB-α (1). **Mechanism of Inhibitory Action:** NF-κB transcription factor regulates expression of inflammatory cytokines, various chemokines and immunoreceptors, as well as cell adhesion molecules. When stationed within the cytoplasm, NF-κB is kept inactive through its binding to the inhibitory factor IκB; however, certain stimuli result in IκB phosphorylation and ubiquitin-mediated degradation, freeing NF-κB for translocation to the nucleus. In endothelial cells, IκB-α phosphorylation and degradation occur within 15 min of TNFα treatment, allowing NF-κB to translocate to the nucleus to activate gene expression. Treatment of humnan vascular endothelial cells (HUVEC) with TNFα results in rapid loss of IκB-α from the cytoplasm (1). BAY11-7082 stabilizes IκB-α in a dose-dependent manner (IC$_{50}$ ≈ 10 μM). There is a clear correlation between the concentration of drug that stabilizes IκB-α, the concentration that inhibits nuclear levels of NF-κB, and the concentration that inhibits adhesion molecule expression. More recent studies (2) demonstrate that BAY 11-7082 prevents ubiquitin conjugation to Ubc13 and UbcH7 by forming a covalent adduct with their cysteine residues via Michael addition at the C-3 atom of BAY 11-7082, followed by the release of 4-methylbenzenesulfinate. BAY 11-7082 stimulated Lys48-linked polyubiquitin chain formation in cells and protected Hypoxia-Inducible Factor-1α (HIF1α) from proteasomal degradation, suggesting it inhibits the proteasome. These results indicate that the anti-inflammatory effects of BAY 11-7082, its ability to induce B-cell lymphoma and leukemic T-cell death and to prevent the recruitment of proteins to sites of DNA damage are exerted via inhibition of components of the ubiquitin system– not by inhibiting NF-κB (2). BAY11-7082 also inhibits proliferation and promotes apoptosis in breast carcinoma MCF-7 cells by inhibiting phosphorylation of ATP citrate lyase (3). **1.** Pierce, Schoenleber, Jesmok, *et al.* (1997) *J. Biol. Chem.* **272**, 21096. **2.** Strickson, Campbell, Emmerich, *et al.* (2013) *Biochem. J.* **451**, 427. **3.** Huang, Su, Wei, *et al.* (2015) *Xi Bao Yu Fen Zi Mian Yi Xue Za Zhi.* **31**, 1458.

BAY 12-8039, *See* Moxifloxacin

BAY 12-9566, *See* Tanomastat

BAY 43-9006, *See* Sorafenib

BAY 57-1293

This prototypical primase-helicase inhibitor (FW = 402.49 g/mol; CAS 348086-71-5; IUPAC Name: *N*-[5-(aminosulfonyl)-4-methyl-1,3-thiazol-2-yl]-*N*-methyl-2-[4-(2-pyridinyl)phenyl]acetamide), also known as Pritelivir, represents a new class of antiviral compounds inhibits replication of herpes simplex virus (HSV) Types 1 and 2 in the nanomolar range *in vitro* by interfering with the enzymatic activity of the viral primase-helicase complex. In rodent models of HSV infection, the antiviral activity of BAY 57-1293 *in vivo* was found to be superior compared to all compounds currently used to treat HSV infections (1,2). BAY 57-1293 is well tolerated, significantly reduces time to healing, prevents rebound of disease after cessation of treatment and, most importantly, reduces the frequency and severity of recurrent disease. Indeed, Pritelivir reduces the rates of genital HSV shedding and days with lesions in a dose-dependent manner in otherwise healthy men and women with genital herpes (3). Variants resistant to BAY 57-1293 retained sensitivity to the nucleoside analogue, ACV. Moreover, single amino acid substitutions in the HSV-1 helicase protein that confer resistance to BAY 57-1293 are associated with increased or decreased virus growth characteristics in tissue culture (4). BAY 57-1293 also reduces the formation of Alzheimer's disease-like β-amyloid (Aβ) and abnormal tau (P-tau) in HSV-1 infected cells in culture (5). **1.** Kleymann, Fischer, Betz, *et al.* (2002) *Nature Med.* **8**, 392. **2.** Betz, Fischer, Kleymann, *et al.* (2002) *Antimicrob. Agents Chemother.* **46**, 1766. **3.** Wald, Corey, Timmler, *et al.* (2014) *New Engl. J. Med.* **370**, 201. 4. Biswas, Jennens & Field (2007) *Arch. Virol.* **152**, 1489. **5.** Wozniak, Frost & Itzhaki (2013) *Antiviral Res.* **99**, 401.

BAY 59-7939, *See* Rivaroxaban

BAY 60-7550

This selective cGMP/cAMP phosphodiesterase PDE2A inhibitor (FW = 476.60 g/mol; CAS 439083-90-6; Solubility: 10 mM DMSO), systematically named 2-[(3,4-dimethoxyphenyl)methyl]-7-[(1*R*)-1-hydroxyethyl]-4-phenylbutyl]-5-methyl-imidazo[5,1-f][1,2,4]triazin-4(1*H*)-one, targets Type 2 cyclic nucleotide phosphodiesterase (human, IC$_{50}$ = 4.7 nM; bovine, IC$_{50}$ = 2.0 nM), a dual-function enzyme that, when stimulated by 3',5'-cyclicGMP (cGMP), preferentially catalyzes hydrolysis of 3',5'-cyclicAMP (cAMP). BAY 60-7550 is 50x more selective for PDE2 than PDE1 and >100x selective relative to PDE5 PDE3B, PDE4B, PDE7B, PDE8A, PDE9A, PDE10A, and PDE11A (1). Use of BAY-60-7550 also indicates that PDE2 is responsible for the degradation of newly synthesized cGMP in cultured neurons and hippocampal slices (1). Inhibition of PDE2 enhanced long-term potentiation of synaptic transmission without altering basal synaptic transmission. Inhibition of PDE2 also improved the performance of rats in social and object recognition memory tasks, and reversed MK801-induced deficits in spontaneous alternation in mice in a T-maze (1). Treatment of mice with L-buthionine-(S,R)-sulfoximine (300 mg/kg) induces oxidative stress and also causes anxiety-like behavioral effects in elevated plus-maze, open-field, and hole-board tests, most likely through the NADPH oxidase pathway (2). These effects are antagonized by Bay 60-7550 (3 mg/kg), which decreases oxidative stress and expression of NADPH oxidase subunits in amygdala, hypothalamus, and cultured neurons. The resulting increase in cGMP also promotes phosphorylation of vasodilator-stimulated phosphoprotein (VASP) at Ser-239, suggesting a role of cGMP-Protein Kinase G signaling in the reduction of anxiety (2). PDE inhibitors also enhance object recognition, and the effects of PDE inhibition on cognitive functions may result from higher cerebrovascular function (3). In the spatial location task, PDE5 inhibition of cGMP hydrolysis by vardenafil only enhances early-phase consolidation, and PDE4 inhibition of cAMP hydrolysis by rolipram only enhances late-phase consolidation; on the other hand, PDE2 inhibition of cAMP and cGMP hydrolysis by Bay 60-7550, enhances both (3). These results underscore the specific effects of cAMP and cGMP on memory consolidation (object and spatial memory) and provide evidence that the underlying mechanisms of PDE inhibition on cognition are independent of cerebrovascular effects. Bay 60-7550 treatment also indicates that phosphodiesterase 2A is a major negative regulator of iNOS expression in lipopolysaccharide-treated mouse alveolar macrophages (4). BAY 60-7550 reverses functional impairments induced by brain ischemia by decreasing hippocampal neurodegeneration and enhancing hippocampal neuronal plasticity (5). **1.** Boess, Hendrix, van der Staay, *et al.* (2004) *Neuropharmacol.* **47**, 1081. **2.** Masood, Nadeem, Mustafa & O'Donnell (2008) *J. Pharmacol. Exp. Ther.* **326**, 369. **3.** Rutten, Van Donkelaar, Ferrington, *et al.* (2009) *Neuropsychopharmacol.* **34**, 1914. **4.** Rentsendorj, D'Alessio & Pearse (2014) *J. Leukoc. Biol.* **96**, 907. **5.** Soares, Meyer, Milani, *et al.* (2017) *Eur. J. Neurosci.* **45**, 510.

BAY 63-2521, *See* Riociguat

BAY 73-4506, *See* Regorafenib

BAY 85-3934, *See* Molidustat

BAY 87-2243

This Hif1α inhibitor (FW = 525.53 g/mol; CAS 1227158-85-1; Solubility: <1 mg/mL DMSO or H_2O) targets the transcription factor hypoxia-inducible factor-1 (HIF-1), which plays an essential role in tumor development, tumor progression, and resistance to chemo- and radiotherapy. BAY 87-2243 inhibits HIF-1α and HIF-2α accumulation under hypoxic conditions in the H460 Non-Small Cell Lung Cancer (NSCLC) cell line but is without effect on HIF-1α protein levels that are induced by such hypoxia mimetics asdesferrioxamine or cobalt chloride. BAY 87-2243 has no effect on HIF target gene expression levels in RCC4 cells lacking Von Hippel-Lindau (VHL) activity; nor does it affect the activity of HIF prolyl hydroxylase-2. Antitumor activity of BAY 87-2243, suppression of HIF-1α protein levels, and reduction of HIF-1 target gene expression *in vivo* have been demonstrated in a H460 xenograft model. BAY 87-2243 does not inhibit cell proliferation under standard conditions. Upon glucose depletion, a condition favoring mitochondrial ATP generation as energy source, BAY 87-2243 inhibits cell proliferation in the low-nM range. In a mouse model for BRAF-mutant melanoma, BAY 87-2243-mediated complex I inhibition induces melanoma cell death *in vitro* and reduces melanoma tumor growth in various mouse models *in vivo* (2). This effect is mediated through BAY 87-2243-induced stimulation of mitochondrial ROS production, leading to oxidative damage and subsequent cell death. BAY 87-2243 displays increased anti-tumor efficacy compared to single agent treatment, when combined with the BRAF inhibitor vemurafenib in nude mice with BRAF mutant melanoma xenografts. **1**. Ellinghaus, Heisler, Unterschemmann, *et al.* (2013) *Cancer Med.* **2**, 611. **2**. Schöckel, Glasauer, Basit, *et al.* (2015) *Cancer Metab.* **3**, 11.

BAY e 9736, *See* Nimodipine

Bayer-2502, *See* Nifurtimox

BAY U6751

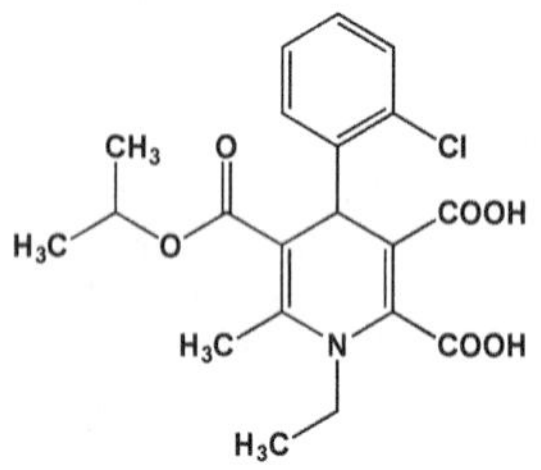

This dihydropyridine (FW = 407.85 g/mol), also known as (−)-(S)-3-isopropyl 4-(2-chlorophenyl)-1,4-dihydro-1-ethyl-2-methylpyridine-3,5,6-tricarboxylate and BAY W1807, inhibits glycogen phosphorylase (K_i = 1.6 nM). It is competitive with respect to AMP. Note that the term BAY U6751 refers to the racemic mixture, of which the active enantiomer is BAY W1807. **1**. Gregus & Németi (2007) *Toxicol. Sci.* **100**, 44. **2**. Zographos, Oikonomakos, Tsitsanou, *et al.* (1997) *Structure* **5**, 1413. **3**. Bergans, Stalmans, Goldmann, Vanstapel (2000) *Diabetes* **49**, 1419.

BAZ2-ICR

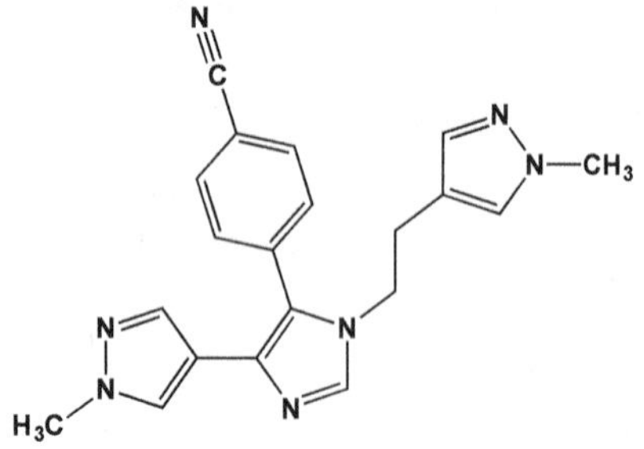

This selective bromodomain inhibitor (FW = 357.41 g/mol; Solubility: 100 mM in DMSO), named 4-[4-(1-methyl-1*H*-pyrazole-4-yl)-1-[2-(1-methyl-1*H*-pyrazol-4-yl)ethyl]-1*H*-imidazol-5-yl]benzonitrile, targets BAZ2A (K_d = 109 nM) and BAZ2B (K_d = 170 nM), showing 15x selectivity for the BAZ2 bromodomain over the CERC2 bromodomain as well as >100-fold selectivity over other bromodomain inhibitors. BAZ2A/B belong to a ubiquitously expressed bromodomain containing protein family characterized by a C-terminal bromodomain that lies adjacent to a PHD finger and a WACZ motif. BAZ2B is believed to regulate nucleosome mobilization by the ATP-dependent chromatin remodeling factor ISWI. **1**. Drouin, McGrath, Vidler, *et al.* (2015) *J. Med. Chem.* **58**, 2553.

Bazedoxifene

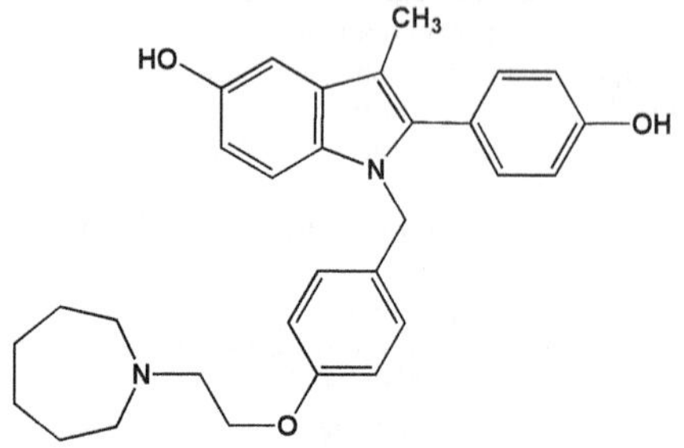

This potent and selective estrogen receptor modulator, *or* SERM (FW = 530.65 g/mol; CAS 198481-33-3; Solubility: 00 mM in DMSO), also named 1-[[4-[2-(hexahydro-1*H*-azepin-1-yl)ethoxy]phenyl]methyl]-2-(4-hydroxyphenyl)-3-methyl-1*H*-indol-5-ol, selectivity targets the extrogen receptor ERα (IC_{50} = 26 nM), with weaker action against ERβ (IC_{50} = 99 nM), inhibiting 17β-estradiol-induced proliferation of MCF-7 cells. Bazedoxifene represents a promising new treatment for osteoporosis, one with a potential for less uterine and vasomotor effects than selective estrogen receptor modulators now used clinically. **1**. Komm, Kharode, Bodine *et al.* (2005) *Endocrinology* **146**, 3999.

BC21, *See* β-Catenin/Tcf Inhibitor V

Bcl-xL Anti-apoptotic Factor
This Bcl-2 family member (MW = 26063 g), also known as B-cell lymphoma-extra large, is 233-residue pro-survival mitochondria transmembrane protein that prevents cytochrome *c* release and is often overexpressed in cancer cells. This *Bcl-xL* gene was first recognized through its involvement in the *t*(14;18) chromosomal translocations found in B-cell lymphomas, where it contributes to neoplastic cell expansion by preventing cell turnover due to programmed cell death. BCL-2 overexpression also occurs in many other types of human tumors (*e.g.*, prostate cancer, colon cancer, and lung cancer) and has been associated with chemoresistance and radioresistance in some malignant cell types. Its overexpression is an indicator of a poor prognosis for recovery.

BCX-1777, *See* Immucillin-H

Bedaquiline

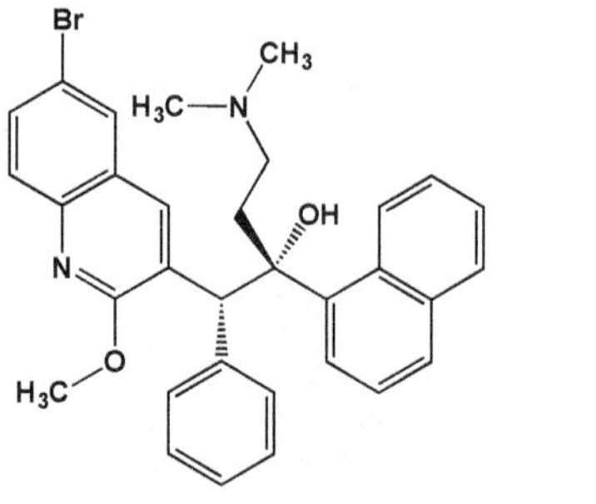

This diarylquinoline-class anti-tuberculosis drug (FW = 554.43 g/mol; CAS 843663-66-1), also known by the tradename Sirturo® and code names TMC207 and R207910, is the first new antibiotic against multi-drug-resistant tuberculosis (1). *Mycobacterium tuberculosis* represents a global threat, with >2 billion people harboring latent TB and >9 million new TB cases. Two million die of TB every year, and ~500,000 become multidrug-resistant. A single dose of bedaquiline inhibits mycobacterial growth for 1 week, and plasma levels associated with efficacy in mice are well tolerated in healthy human volunteers (1). Mutants selected *in vitro* suggest that the drug targets the proton pump of ATP synthase. Six distinct mutations (*e.g.*, Asp-28-Gly, Asp-28-Ala, Leu-59-Val, Glu-61-Asp, Ala-63-Pro, and Ile-66-Met) within ATP Synthase subunit *c* confer resistance against some diarylquinoline drugs (2). **1**. Andries, Verhasselt, Guillemont, *et al.* (2005) *Science* **307**, 223. **2**. Segala, Sougakoff, Nevejans-Chauffour, Jarlier & Petrella (2012) *Antimicrob. Agents Chemother.* **56**, 2326.

BeKm 1

RPTDIKCSESYQCFPVCKSRFGKTNGRCVNGFCDCF

This ERG channel inhibitor (FW = 4091.65 g/mol; CAS 524962-01-4; Soluble to 1 mg/mL in H_2O), from the Central Asian scorpion *Buthus eupeus*, targets human $K_V11.1$ Ether-a-go-go-Related Gene potassium ion currents (1,2). BeKm-1 inhibits hERG1 channels (IC_{50} = 3.3 nM), but has little or no effect, even at 100 nM, on hEAG, hSK1, rSK2, hIK, hBK, KCNQ1/KCNE1, KCNQ2/3, and KCNQ4 channels. BeKm-1 shares a molecular scaffold (consisting of a short α-helix and a triple-stranded antiparallel β-sheet) in common with other short scorpion toxins (3). Site-directed mutagenesis identified Tyr-11, Lys-18, Arg-20, and Lys-23 as key residues for binding to hERG channels. All four are located in the α-helix and its following loop, whereas the canonical functional site of other short scorpion toxins lies within the β-sheet (3). BeKm-1 significantly prolongs QTc intervals in isolated rabbit hearts (4.7 at 10 nM, 16.3% at 100 nM), concentrations that inhibit hERG1 channels (4). **1**. Fillipov, Kozlov, Pluzhnikov, Grishin & Brown (1996) *FEBS Lett*. **384**, 277. **2**. Korolkova, Kozlov, Lipkin, *et al.* (2001) *J. Biol. Chem*. **276**, 9868. **3**. Korolkova, Bocharov, Angelo, *et al.* (2002) *J. Biol. Chem*. **277**, 43104. **4**. Qu, Fang, Gao, Chui & Vargas (2011) *J. Pharmacol. Exp. Ther*. **337**, 2.

Belinostat

This synthetic epigenetic modulator (FW = 318.35 g/mol; CAS 866323-14-0), also known as PXD101 and the IUPAC name, (2*E*)-*N*-hydroxy-3-[3-(phenylsulfamoyl)phenyl]prop-2-enamide, inhibits histone deacetylase (IC_{50} = 27 nM) and induces a concentration-dependent increase (over the 0.2–5 μM range) in histone H4 acetylation and alters expression of genes located on DNA associated with its parent histone octamer (1). A simple and sensitive high-performance liquid chromatography ultraviolet method has been developed for the quantification of PXD101 in human plasma (2). In Rhesus monkeys, belinostat is cleared rapidly from plasma with a half-life of 1.0 h, a mean residence time of 0.47 h, and a clearance of 425 mL/min/m^2 (3). CSF drug exposure is <1% of plasma drug exposure and <10% of free (non-protein bound) plasma drug exposure. The DNA-methylation inhibitor decitabine and histone-deacetylase inhibitor belinostat increases the efficacy of chemotherapeutic agents in tumors that acquired drug resistance due to DNA methylation and gene silencing (4). **1**. Plumb, Finn, Williams, *et al.* (2003) *Mol. Cancer Ther*. **2**, 721. **2**. Zhang, Goh, Khoo, Yeo & Lee (2007) *Ther. Drug Monit*. **29**, 231. **3**. Warren, McCully, Dvinge. *et al.* (2008) *Cancer Chemother. Pharmacol*. **62**, 433. **4**. Steele, Finn, Brown & Plumb (2009) *Br. J. Cancer*. **100**, 758.

Benazepril

This ACE-directed pro-drug ($FW_{free-acid}$ = 424.50 g/mol; CAS 86541-75-5), also known as 3-((1-(ethoxycarbonyl)-3-phenylpropyl)amino)-2,3,4,5-tetrahydro-2-oxo-(*S*-(*R**,*R**))-1*H*-1-benzazepine-1-acetic acid, Lotensin and CGS 14824A, is rapidly metabolized to its diacid benazeprilat (CGS 14831; $FW_{free-diacid}$ = 396.44 g/mol), the latter a potent, long-acting inhibitor of peptidyl-dipeptidase A, *or* angiotensin I-converting enzyme, IC_{50} = 1.7 nM (1,2). Lotensin is used to treat hypertension, congestive heart failure, and chronic renal failure. The antihypertensive effects of benazepril begin as early as 30 min after a single dose, and those effects during consecutive dosing are also sustained for 24 h with a lesser diurnal variation in blood pressure (3). **1**. Watthey, Stanton, Desai, Babiarz & Finn (1985) *J. Med. Chem*. **28**, 1511. **2**. Van Dyck, Novakova, Van Schepdael & Hoogmartens (2003) *J. Chromatogr. A* **1013**, 149. **3**. Shionoiri, Ueda, Minamisawa, *et al.* (1992) *J. Cardiovasc. Pharmacol*. **20**, 348.

Benazeprilat, *See* Benazepril

Bendamustine

This nitrogen mustard and anticancer drug ($FW_{free-acid}$ = 358.26 g/mol; CAS 16506-27-7), also known by its code name SDX-105, its trade names Treanda™, Treakisym™, Ribomustin™, and Levact™, as well as by its systematic name 4-[5-[bis(2-chloroethyl)amino]-1-methylbenzimidazol-2-yl]butanoic acid, is a relatively nonspecific DNA alkylating agent that causes intra- and inter-strand cross-links. Bendamustine is used in the treatment of chronic lymphocytic leukemia (CLL), Hodgkin's disease, non-Hodgkin's lymphoma, multiple myeloma and lung cancer. **Pharmacokinetics:** After intravenous infusion, >95% of the drug becomes protein-bound, mainly to albumin; however, only free bendamustine is active. Bendamustine is metabolized by liver cytochrome p450, and elimination (renal) is biphasic, with an initial half-life of 6–10 minutes and a terminal half-life of approximately 30 minutes. **1**. Tageja, Nishant, Nagi & Jasdeepa (2010) *Cancer Chemother. Pharmacol*. **66**, 413.

Bengamide

This potent NF-κB activation inhibitor and anti-inflammatory and anti-tumor agent (FW = 598.81 g/mol; CAS 104947-69-5; Soluble in DMSO), also named (6*E*)-6,7,8,9-tetradeoxy-*N*-[(3*S*,6*S*)-hexahydro-1-methyl-2-oxo-6-[(1-oxotetradecyl)oxy]-1*H*-azepin-3-yl]-8-methyl-2-*O*-methyl-D-gulonon-6-enonamide, is a a novel sponge-derived marine natural product with broad spectrum antitumor activity, decreasing IκBα phosphorylation and attenuating LPS-induced nitric oxide production as well as expression of TNF-α, IL-6 and MCP. Bengamide also suppresses HeLa and HCT116 cell proliferation. **1**. Hu, Dang, Tenney, *et al.* (2007) *Chem. Biol*. **14**, 764.

Benidipine

This oral, once-daily antihypertensive agent (FW = 505.57 g/mol; CAS 105979-17-7), also known by its code name KW-3049, trade name Coniel®, and systematic name O^5-methyl,O^3-[(3*R*)-1-(phenylmethyl)piperidin-3-yl] 2,6-dimethyl-4-(3-nitrophenyl)-1,4-dihydropyridine-3,5-dicarboxylate, is a dihydropyridine-class calcium ion blocker that is effective against L-, N-, and T-channels, with potent and selective inhibitory action on cardiac slow calcium channels (1). Benidipine bound stereospecifically to nitrendipine binding sites of rat myocardium with high affinity (K_i = 0.13 nM) and to the rat brain $α_1$-adrenergic receptor (K_i = 1.2 μM). KW-3049 exhibited no remarkable binding affinity to $α_2$ adrenergic, β-adrenergic, D_2 dopamine, H_1 histamine, S_2 serotonin, A_1 adenosine, A_2 adenosine and muscarinic cholinergic receptors at 100 μM (2). **1**. Yoshitake, Kubo, Ikeda, *et al.* (1986) *Nihon. Yakurigaku Zasshi*. **88**, 179. **2**. Ishii, Nishida, Oka & Nakamizo (1988) *Arzneimittelforschung*. **38**, 1677.

Benocyclidine

This piperazine-class psychostimulant and dopamine reuptake inhibitor (FW = 620.82 g/mol) also known as BCP, BTCP, and benzothiophenyl-cyclohexylpiperidine, is highly selective, showing negligible affinity for the NMDA receptor and lacking anticonvulsant, anesthetic, hallucinogenic, or dissociative effects (1,2). Its high-affinity interactions with dopamine transporters commend it for use as a cellular localization reagent (3,4). **1.** Vignon, Pinet, Cerruti, Kamenka & Chicheportiche (1998) *Eur. J. Pharmacol.* **148**, 427. **2.** Chaudieu, Vignon, Chicheportiche, *et al.* (1989). Pharmacol. Biochem. *Behav.* **32**, 699. **3.** Filloux, Hunt, Wamsley (1989) *Neurosci. Lett.* **100**, 105. **4.** Maurice, Vignon, Kamenka & Chicheportiche (1989) *Neurosci Lett.* **101**, 234.

Benomyl

Benomyl

Carbendazim

S-Methyl-N-butylthiocarbamate sulfoxide

This benzimidazole-based fungicide (FW = 290.32 g/mol) inhibits tubulin polymerization and strongly suppresses microtubule dynamic instability *in vitro* (1-3) The active component is its methyl *N*-(benzimidazol-2-yl)carbamate (or, carbendazim; FW = 191.19 g/mol), which inhibits 1-3-polygalacturonase (4) and cellulase (4). Carbendazim interferes with initial events of MT polymerization, possibly GTP binding, with microtubule-associated proteins (MAPs) moderating this effect (5). Another downstream metabolite, *S*-methyl *N*-butylthiocarbamate sulfoxide (FW = 163.24 g/mol), inhibits aldehyde dehydrogenase (6,7). **1.** Friedman & Platzer (1978) *Biochim. Biophys. Acta* **544**, 605. **2.** Kilmartin (1981) *Biochemistry* **20**, 3629. **3.** Gupta, Bishop, Peck, *et al.* (2004) *Biochemistry* **43**, 6645. **4.** Ohazurike (1996) *Nahrung* **40**, 150. **5.** Winter, Straandgard & Miller (2001) *Toxicol. Sci.* **59**, 138 **6.** Staub, Quistad & Casida (1998) *Chem. Res. Toxicol.* **11**, 535. **7.** Staub, Quistad & Casida (1999) *Biochem. Pharmacol.* **58**, 1467.

Bentazon

This thiadiazinol (FW = 240.28 g/mol; CAS 25057-89-0), also known as Basagran® and 3-(1-methylethyl)-1*H*-2,1,3 benzothiadiazin-4(3*H*)-one 2,2-dioxide, is a selective post-emergence herbicide used to control the growth of cocklebur, Canada thistle, and yellow nutsedge. Bentazon is a contact herbicide, meaning that it injures only to the parts of the plant to which it is applied. **Target(s):** photosystem II, as primary mode of herbicidal action (1); acetyl-CoA carboxylase, weakly inhibited (2). **1.** Silverman, Petracek, Heiman, Ju, Fledderman & Warrior (2005) *J. Agric. Food Chem.* **53**, 9769. **2.** Rendina & Felts (1988) *Plant Physiol.* **86**, 983.

Benzalkonium Chloride

This cationic surface-active disinfectant and preservative (FW_{lauryl-homologue} = 340.00 g/mol; CAS 8001-54-5; Soluble in Ethanol and Acetone), also named alkyldimethylbenzyl-ammonium chloride, refers to several cationic detergents that are widely used in various pharmaceutical preparations, including eye, ear, and nasal drops as well as leave-on antiseptics. (*See also specific agent*) Benzalkonium chloride has also been used in the isolation and purification of proteins and as an antimicrobial agent. Commercial sources typically contain a mixture of alkyl homologues, predominantly with a dodecyl (lauryl) side-chain. Standard concentrates are manufactured as 50% and 80% w/w solutions, and sold under trade names such as BC50, BC80, BAC50, and BAC80. **Target(s):** acetylcholinesterase, esterase activity (1); butyrylcholinesterase, esterase activity (but activates the aryl acylamidase activity) (1); trypsin (2); chymotrypsin (2); histamine release (3); G proteins (4); serralysin (5); chitosanase (6). **1.** Jaganathan & Boopathy (2000) *Bioorg. Chem.* **28**, 242. **2.** Feldbau & Schwabe (1971) *Biochemistry* **10**, 2131. **3.** Read & Kiefer (1979) *J. Pharmacol. Exp. Ther.* **211**, 711. **4.** Patarca, Rosenzwei, Zuniga & Fletcher (2000) *Crit. Rev. Oncog.* **11**, 255. **5.** Nakajima, Mizusawa & Yoshida (1974) *Eur. J. Biochem.* **44**, 87. **6.** Cruz Camarillo, Sanchez Perez, Rojas Avelizapa, Gomez Ramirez & Rojas Avelizapa (2004) *Folia Microbiol.* **49**, 94.

Benzamidine Riboside

Benzamide Riboside

Benzamide Riboside 5'-Triphosphate

Benzamide Riboside Adenine Dinucleotide (BAD⁺)

This pseudonucleoside and pro-drug (FW = 253.25 g/mol; CAS 3804827-0; IUPAC Name: (2*R*,3*S*,4*R*,5*R*)-2-(hydroxymethyl)-5-[6-[[(*Z*)-4-hydroxy-2-methylbut-2-enyl]amino]purin-9-yl]oxolane-3,4-diol)), also known as norzeatin riboside, is a synthetic nicotinamide riboside analogue (1) that is taken up by human cells and metabolically activated to form benzamide riboside 5'-mono, di- and tri-phosphates as well as benzamide riboside-adenine dinucleotide (*or* BAD⁺). The latter is a potent inhibitor of IMP dehydrogenase, IC$_{50}$ = 2 μM (2). Phosphorylation to BR 5'-P (BRMP) is catalyzed by adenosine kinase, and conversion of the latter to benzamide riboside-adenine dinucleotide (*or* BAD⁺) is catalyzed by NMN adenylyltransferase. BAD⁺ is a more potent inhibitor of IMPDH than either BR and BRMP, the cytotoxicity of BR is thus more closely connected with the metabolism to BAD⁺. While a weak PARP inhibitor, BAD⁺ is an extremely potent inhibitor of cell proliferation. Survival-relevant genes (*e.g.,* cdc25A, akt, bcl-2, and transferrin receptor) are repressed by BR, whereas the expression level of the apoptosis enforcing gene c-myc persists (3). At high BR concentrations, DNA double-strand breaks occur hours before necrosis. There is also a dramatic decrease of intracellular ATP. Restoration of ATP by addition of adenosine or provision of sufficient glucose prevents BR-promoted necrosis and favors apoptosis instead (3). By inhibiting nicotinamide adenine dinucleotide kinase, BAD⁺ also reduces cellular pools of NADP⁺ and NADPH, the lack of which destabilizes dihydrofolate reductase (4). Likely effects of BAD⁺ on NAD⁺-dependent histone deacetylase have yet to be reported. BAD⁺ also inhibits cell growth by down-regulating DHFR protein. Supporting this mode of action is the finding that CCRF-CEM/R human T-cell lymphoblasic leukemia cells (which are MTX-resistant as a consequence of DHFR gene amplification and overexpression) are more resistant to BR than are parental cells. BAD⁺ reduces cellular levels of NADP⁺ and NADPH by inhibiting nicotinamide adenine dinucleotide kinase (NADK), depleting NADPH, and destabilizing DHFR. NADK inhibition therefore represents another approach for down-regulating DHFR and inhibiting cancer cell growth (5) **Target(s):** IMP dehydrogenase (2,5); malate dehydrogenase (3); dihydrofolate reductase (4); nicotinamide adenine dinucleotide kinase, *or* NADK (5). **1.** Krohn, Heins, Wielckens (1992) *J. Med. Chem.* **35**, 511. **2.** Jayaram, Gharehbaghi, Jayaram, *et al.* (1992) *Biochem. Biophys. Res. Commun.* **186**, 1600. **3.** Polgar, Gfatter, Uhl, *et al.* (2002) *Curr. Med. Chem.* **9**, 765. **4.** Roussel, Johnson-Farley, Kerrigan, *et al.* (2012) *Cancer Biol. Ther.* **13**, 1290. **5.** Roussel, Johnson-Farley, Kerrigan, *et al.* (2012) *Cancer Biol. Ther.* **13**, 1290.

3-(2-Benzamidoethyl)-5-(pyridin-3-ylmethyl)-1-benzenepropionate

This sulfonamide ($FW_{\text{free-acid}}$ = 374.46 g/mol) inhibits human thromboxane-A synthase (IC_{50} = 22 nM) and is also a thromboxane receptor antagonist. Oral administration at a dose of 5 mg/kg to conscious dogs produces long-lasting thromboxane synthase inhibition and thromboxane receptor blockade, as measured by inhibition of U46619-induced platelet aggregation *ex vivo*. **1.** Dickinson, Dack, Long & Steele (1997) *J. Med. Chem.* **40**, 3442.

Benzbromarone

This CaCC blocker (FW = 424.08 g/mol; CAS 3562-84-3; Solubility: 100 mM in DMSO), also named (3,5-dibromo-4-hydroxyphenyl)(2-ethyl-3-benzofuranyl)methanone, targets transmembrane proteins 16A and 16B (*or* TMEM16A and TMEM16B), which are activated by uncomplexed intracellular Ca^{2+} and by Ca^{2+}-mobilizing stimuli. Benzbromarone inhibits ATP-induced mucin secretion and metacholine-induced airway smooth muscle contraction *in vitro*. **1.** Huang, et al. (2012) *Proc. Natl. Acad. Sci. U.S.A.* **109**, 16354. **2.** Gallos, *et al.* (2013) *Am. J. Physiol. Lung Cell Mol. Physiol.* **305**, 625.

(R)-2-(Benzenesulfonyl)-1,2,3,4-tetrahydro-isoquinolin-3-carboxylic Acid Hydroxamate

This isoquinoline derivative (FW = 331.37 g/mol) inhibits human neutrophil collagenase, matrix metalloproteinase-8, IC_{50} = 20 nM. The hydroxamate moiety most likely forms a coordination complex with an active-site Zn^{2+}. A 1.7-Å resolution X-ray structure of the catalytic domain of MMP-8 was used to make highly predictive 3D-QSAR models of the steric, electrostatic, and hydrophobic complementarity of enzyme and inhibitor. It was possible to compensate the weaker zinc binding properties of carboxylates by introducing optimal fitting at P1' residues. The final QSAR model agrees with all experimental data for the binding topology and provides clear guidelines and accurate activity predictions for novel MMP-8 inhibitors. **1.** Matter, Schwab, Barbier, *et al.* (1999) *J. Med. Chem.* **42**, 1908.

N-1H-Benzimidazol-2-yl-N'-(3-bromo-5-chloro-6-ethoxybenzyl)propane-1,3-diamine

This substituted benzimidazole (FW = 437.77 g/mol) inhibits *Staphylococcus aureus* methionyl-tRNA synthetase, *or* MRS (IC_{50} = 3.8 nM), showing very good antibacterial activity against panels of antibiotic-resistant *Staphylococci* and *Enterococci*. Its design was guided by a Structure-Activity Relationship (SAR) study suggesting that the right-hand side pharmacophore for MRS inhibition has now been defined as an NH-C-NH functionality in the context of a bicyclic heteroaromatic system. **1.** Jarvest, Armstrong, Berge, *et al.* (2004) *Bioorg. Med. Chem. Lett.* **14**, 3937.

Benznidazole

This orally available 5-nitrofuran and anti-trypanosomal agent (FW = 260.25 g/mol; CAS 22994-85-0), known by its trade name Rochagan® and Radanil® as well as its systematic name *N*-benzyl-2-(2-nitro-1*H*-imidazol-1-yl)acetamide, is the front-line treatment used against American trypanosomiasis, a parasitic infection caused by *Trypanosoma cruzi* and the trypanosomal diseases, Chagas Disease and Sleeping Sickness. While benznidazole activation was once thought to generate reductive metabolites that damage DNA and deplete thiols, recent studies suggest benznidazole activation is likely to involve trypanosomal type I nitroreductase catalysis of a NADH-dependent, oxygen-insensitive reaction to form 4,5-dihydro-4,5-dihydroxyimidazole (1). Breakdown of this product releases glyoxal, a reactive dialdehyde that, in the presence of guanosine, generates guanosine-glyoxal adducts with attendant toxicity. **1.** Hall & Wilkinson (2012) *Antimicrob. Agents Chemother.* **56**, 115.

Benzocaine

This widely used local anesthetic (FW = 165.19 g/mol; CAS 94-09-7), known systematically as ethyl 4-aminobenzoate, is a common ingredient in topical pain-relieving formulations, including the gel used by dentists to numb gums prior to injection of nerve blocks as well as in cough drops and cough syrup. Benzocaine is a membrane-permeable NaV1.5-directed anesthetic that stops the propagation of an action potential. Its lower affinity (IC_{50} ~800 μM) is most likely the consequence of its neutral charge (pK_a = 3.5), making it unable to replicate π-cation binding interactions typical of lidocaine (pK_a = 7.7) and other tertiary amine-containing anesthetics (1). At 2 mM benzocaine, Q_{max} is 0.85 of control values, $V_{\frac{1}{2}}$ is shifted from −62 mV to −76 mV, and the slope factor *s* is flattened from −12 mV to −17 mV, suggesting that benzocaine acts similarly to lidocaine, albeit without use-dependence (2). Benzocaine-associated methemoglobinemia is reversed by methylene blue treatment, most often with clinical recovery (3). **1.** Becker & Reed (2006) *Anesth. Prog.* **53**, 98. **2.** Hanck, Nikitina, McNulty, *et al.* (2009) *Circ. Res.* **105**, 492. **3.** Taleb, Ashraf, Valavoor & Tinkel (2013) *Am. J. Cardiovasc. Drugs* **13**, 325.

Benzoate (Benzoic Acid)

This naturally occurring aromatic acid ($FW_{free-acid}$ = 122.12 g/mol; CAS 65-85-0; MP = 122.4°C, pK_a = 4.19) is metabolized by butyrate-CoA ligase, forming *S*-benzoyl-CoA, which is further metabolized to hippuric acid through the action of glycine *N*-acyltransferase. Because benzoate is metabolically conjugated to glycine (a nonessential amino acid) to form hippurate (*N*-benzoylglycine), it has been used for decades as a therapeutic regimen to reduce ammonia toxicity in patients with inborn defects in Urea Cycle enzymes. Benzoic acid is also an efficient peroxide scavenger, forming benzoyl peroxide, and nontoxic food additive that inhibits the growth of mold, yeast and some bacteria. Benzoic acid is, however, ineffective against common meat-spoilage bacteria, including *Brochothrix thermosphacta, Carnobacterium piscicola, Lactobacillus curvatus, Lactobacillus sake, Pseudomonas fluorescens,* and *Serratia liquefaciens.* Acceptable daily intake, as established by the World Health Organization, is 5 mg/kg in healthy human subjects. **Target(s):** D-amino-acid oxidase (1,2,7,8,11,15,20,23,26,27, 29,33,38,47); L-amino-acid oxidase (3,9,11,15, 23,39); UDP-glucuronosyl-transferase (4); amine oxidase (5); tyrosinase, *or* monophenol monooxygenase (6,23,25,36,71-76); carboxypeptidase A (10,13); diamine oxidase (12); xanthine oxidase (14); phenylalanine ammonia-lyase (16,54); creatine kinase (17); aspartate aminotransferase (18); salicylate monooxygenase (19); chymotrypsin, weakly inhibited (21,22,28,31,42,51); catechol oxidase (23,44,86,87); polyphenol oxidase (86,87); lactate dehydrogenase, weakly inhibited (23,24); glucose dehydrogenase, weakly inhibited (23,24); γ-butyrobetaine dioxygenase, γ-butyrobetaine hydroxylase, K_i = 2.6 mM (30); γ-glutamyl transpeptidase, transfer activity (32); acetoin dehydrogenase (34); mandelonitrile lyase (35,57,58); carbonyl reductase (34); esterase (37); peroxidase (40); pyrophosphatase.*or* inorganic diphosphatase (41); choline acetyltransferase (43); leucyl aminopeptidase (45); urease (46); pyruvate carboxylase (48); mandelate racemase (49,50); L-3-cyanoalanine synthase, *or* β-cyanoalanine synthase (52); serine-sulfate ammonia-lyase (53); hydroxymandelonitrile lyase (55,56); 4,5-dihydroxyphthalate decarboxylase (59); *o*-pyrocatechuate decarboxylase, weakly inhibited (60,61); haloacetate dehalogenase (62); nicotinamidase (63); peptidyl-dipeptidase A, angiotensin I-converting enzyme (64); 5'-nucleotidase (65); tyrosine-ester sulfotransferase (66); nicotinate phosphoribosyltransferase (67); nicotinate glucosyltransferase (68); nicotinate *N*-methyltransferase (69); 4-methoxybenzoate monooxygenase, *or* *O*-demethylating (70); catechol oxidase (74); polyphenol oxidase (75); 4 hydroxybenzoate 3-monooxygenase (77-80); salicylate 1-monooxygenase (81); procollagen-lysine 5-dioxygenase (82); procollagen-proline 3-dioxygenase (82); procollagen-proline 4-dioxygenase (82,83); 15 lipoxygenase, *or* arachidonate 15-lipoxygenase (84); catechol 2,3-dioxygenase (85). **1.** Burton (1955) *Meth. Enzymol.* **2**, 199. **2.** Yagi (1971) *Meth. Enzymol.* **17B**, 608. **3.** Ratner (1955) *Meth. Enzymol.* **2**, 204. **4.** Dutton & Storey (1962) *Meth. Enzymol.* **5**, 159. **5.** Yasunobu & Smith (1971) *Meth. Enzymol.* **17B**, 698. **6.** Lerch (1987) *Meth. Enzymol.* **142**, 165. **7.** Dixon & Kleppe (1965) *Biochim. Biophys. Acta* **96**, 383. **8.** Hellerman, Lindsay & Bovarnick (1946) *J. Biol. Chem.* **163**, 553. **9.** Zeller & Maritz (1945) *Helv. Chim. Acta* **28**, 365. **10.** Elkins-Kaufman, Neurath & De Maria (1949) *J. Biol. Chem.* **178**, 645. **11.** Krebs (1951) *The Enzymes*, 1st ed. (Sumner & Myrbäck, eds.), **2** (Part 1), 499. **12.** Zeller (1951) *The Enzymes*, 1st ed. (Sumner & Myrbäck, eds.), **2** (Part 1), 536. **13.** Smith (1951) *The Enzymes*, 1st ed. (Sumner & Myrbäck, eds.), **1** (Part 2), 793. **14.** Bray (1963) *The Enzymes*, 2nd ed. (Boyer, Lardy & Myrbäck, eds.), **7**, 533. **15.** Meister & Wellner (1963) *The Enzymes*, 2nd ed. (Boyer, Lardy & Myrbäck, eds.), **7**, 609. **16.** Hanson & Havir (1972) *The Enzymes*, 3rd ed. (Boyer, ed.), **7**, 75. **17.** Watts (1973) *The Enzymes*, 3rd ed. (Boyer, ed.), **8**, 383. **18.** Braunstein (1973) *The Enzymes*, 3rd ed. (Boyer, ed.), **9**, 379. **19.** Massey & Hemmerich (1975) *The Enzymes*, 3rd ed. (Boyer, ed.), **12**, 191. **20.** Bright & Porter (1975) *The Enzymes*, 3rd ed. (Boyer, ed.), **12**, 421. **21.** Baker (1967) *Design of Active-Site-Directed Irreversible Enzyme Inhibitors*, Wiley, New York. **22.** Wallace, A. N. Kurtz & C. Niemann (1963) *Biochemistry* **2**, 824. **23.** Webb (1966) *Enzyme and Metabolic Inhibitors*, vol. **2**, Academic Press, New York. **24.** von Euler (1942) *Ber.* **75**, 1876. **25.** Menon, Fleck, Yong & Strothkamp (1990) *Arch. Biochem. Biophys.* **280**, 27. **26.** Sarower, Matsui & Abe (2003) *J. Exp. Zoolog. Part A Comp. Exp. Biol.* **295**, 151. **27.** Wellner & Scannone (1964) *Biochemistry* **3**, 1746. **28.** Huang & C. Niemann (1952) *J. Amer. Chem. Soc.* **74**, 5963. **29.** Yagi, Osawa & Okada (1959) *Biochim. Biophys. Acta* **35**, 102. **30.** Ng, Hanauske-Abel & Englard (1991) *J. Biol. Chem.* **266**, 1526. **31.** Wildnauer & Canady (1966) *Biochemistry* **5**, 2885. **32.** Gardell & Tate (1983) *J. Biol. Chem.* **258**, 6198. **33.** Nishino, Massey & Williams (1980) *J. Biol. Chem.* **255**, 3610. **34.** Hara, Seiriki, Nakayama & Sawada (1985) *Prog. Clin. Biol. Res.* **174**, 291. **35.** Yemm & Poulton (1986) *Arch. Biochem. Biophys.* **247**, 440. **36.** Duckworth

& Coleman (1970) *J. Biol. Chem.* **245**, 1613. **37.** Weber & King (1935) *J. Biol. Chem.* **108**, 131. **38.** Klein & Kamin (1941) *J. Biol. Chem.* **138**, 507. **39.** Blanchard, Green, Nocito & Ratner (1944) *J. Biol. Chem.* **155**, 421. **40.** Lück (1958) *Enzymologia* **19**, 227. **41.** Naganna (1950) *J. Biol. Chem.* **183**, 693. **42.** Neurath & Gladner (1951) *J. Biol. Chem.* **188**, 407. **43.** Korey, de Braganza & Nachmansohn (1951) *J. Biol. Chem.* **189**, 705. **44.** Aerts & R. Vercauteren (1964) *Enzymologia* **28**, 1. **45.** Ludewig, Lasch, Kettmann, Frohne & Hanson (1971) *Enzymologia* **41**, 59. **46.** Onodera (1915) *Biochem. J.* **9**, 544. **47.** Klein & Austin (1953) *J. Biol. Chem.* **205**, 725. **48.** Griffith, Cyr, Egan & Tremblay (1989) *Arch. Biochem. Biophys.* **269**, 201. **49.** Maggio, Kenyon, Mildvan & Hegeman (1975) *Biochemistry* **14**, 1131. **50.** Hegeman, Rosenberg & Kenyon (1970) *Biochemistry* **9**, 4029. **51.** Shiao (1970) *Biochemistry* **9**, 1083. **52.** Macadam & Knowles (1984) *Biochim. Biophys. Acta* **786**, 123. **53.** Tudball & Thomas (1973) *Eur. J. Biochem.* **40**, 25. **54.** Pridham & Woodhead (1974) *Biochem. Soc. Trans.* **2**, 1070. **55.** Lauble, Miehlich, Foerster, Wajant & Effenberger (2002) *Biochemistry* **41**, 12043. **56.** Gruber & Kratky (2004) *J. Polym. Sci. A* **42**, 479. **57.** Jorns (1980) *Biochim. Biophys. Acta* **613**, 203. **58.** Wehtje, Adlercreutz & Mattiasson (1988) *Appl. Microbiol. Biotechnol.* **29**, 419. **59.** Nakazawa & Hayashi (1978) *Appl. Environ. Microbiol.* **36**, 264. **60.** Santha, Rao & Vaidyanathan (1996) *Biochim. Biophys. Acta* **1293**, 191. **61.** Kamath, Dasgupta & Vaidyanathan (1987) *Biochem. Biophys. Res. Commun.* **145**, 586. **62.** Goldman (1965) *J. Biol. Chem.* **240**, 3434. **63.** Albizati & Hedrick (1972) *Biochemistry* **11**, 1508. **64.** Oshima & Nagasawa (1977) *J. Biochem.* **81**, 57. **65.** Marseno, Hori & Miyazawa (1993) *J. Agric. Food Chem.* **41**, 1208. **66.** Duffel (1994) *Chem. Biol. Interact.* **92**, 3. **67.** Gaut & Solomon (1971) *Biochem. Pharmacol.* **20**, 2903. **68.** Taguchi, Sasatani, Nishitani & Okumura (1997) *Biosci. Biotechnol. Biochem.* **61**, 720. **69.** Taguchi, Nishitani, Okumura, Shimabayashi & Iwai (1989) *Agri. Biol. Chem.* **53**, 2867. **70.** Bernhardt, Erdin, Staudinger & Ullrich (1973) *Eur. J. Biochem.* **35**, 126. **71.** Lee (2002) *J. Agric. Food Chem.* **50**, 1400. **72.** Abdel-Halim, Marzouk, Mothana & Awadh (2008) *Pharmazie* **63**, 405. **73.** Shiino, Watanabe & Umezawa (2008) *J. Enzyme Inhib. Med. Chem.* **23**, 16. **74.** Ben-Shalom, Kahn, Harel & Mayer (1977) *Phytochemistry* **16**, 1153. **75.** Güllcin, Küfrevioglu & Oktay (2005) *J. Enzyme Inhib. Med. Chem.* **20**, 297. **76.** McIntyre & Vaughan (1975) *Biochem. J.* **149**, 447. **77.** Fernandez, Dimarco, Ornston & Harayama (1995) *J. Biochem.* **117**, 1261. **78.** Spector & Massey (1972) *J. Biol. Chem.* **247**, 4679. **79.** Sterjiades (1993) *Biotechnol. Appl. Biochem.* **17**, 77. **80.** Hosokawa & Stanier (1966) *J. Biol. Chem.* **241**, 2453. **81.** White-Stevens & Kamin (1972) *J. Biol. Chem.* **247**, 2358. **82.** Majamaa, Turpeenniemi-Hujanen, Latipää, *et al.* (1985) *Biochem. J.* **229**, 127. **83.** Majamaa, Hanauske-Abel, Günzler & Kivirikko (1984) *Eur. J. Biochem.* **138**, 239. **84.** Russell, Scobbie, Duthie & Chesson (2008) *Bioorg. Med. Chem.* **16**, 4589. **85.** Kobayashi, Ishida, Horiike, *et al.* (1995) *J. Biochem.* **117**, 614. **86.** Kanade, Suhas, Chandra & Gowda (2007) *FEBS J.* **274**, 4177. **87.** Güllçin, Küfrevioglu & Octay (2005) *J. Enzyme Inhib. Med. Chem.* **20**, 297.

N-[(1,3-Benzodioxol-5-yl)methyl]-1-[6-chloro-2-(1*H*-imidazol-1-yl)pyrimidin-4-yl]piperazine-2-acetamide

This synthetic pyrimidine (FW = 455.90 g/mol) is a potent inhibitor of nitric-oxide synthase activity (IC_{50} = 1.1 nM), blocking dimerization of the inducible enzyme. This agent displays a 1000-times greater selectivity for inhibiting iNOS versus endothelial NOS dimerization in a cell-based assay. The crystal structure of inhibitor bound to the monomeric iNOS oxygenase domain revealed the inhibitor-heme interaction, substantially perturbing the substrate binding site as well as the dimerization interface. These studies demonstrated that this small molecule acts by allosterically disrupting protein-protein interactions at the dimer interface. **1.** McMillan, Adler, Auld, *et al.* (2000) *Proc. Natl. Acad. Sci. U.S.A.* **97**, 1506.

2-(Benzofuran-2-yl)-N-(4-(cyclohexylaminocarbonyl)phenylsulfonyl)tryptamine

This modified tryptamine (FW = 541.66 g/mol) potently inhibits 15-lipoxygenase (IC$_{50}$ = 12 nM), demonstrating that highly potent 15-LO inhibitors may be generated by introducing aryl and heteroaryl substituents at the indole C-2 position of simple tryptamine and homotryptamine precursors. Many substituents are tolerated on the aryl-sulfonamide region of the molecule, providing a potential handle for modulating their binding properties. The unprecedented potency and lipoxygenase selectivity profiles of the tryptamine sulfonamides make them attractive lead moleculess for optimizing pharmacokinetic properties and assessing models of 15-LO-mediated diseases, including atherosclerosis. 1. Weinstein, Liu, Gu, *et al.* (2005) *Bioorg. Med. Chem. Lett.* **15**, 1435.

5-(Benzofuran-2-yl)-6-(6-morpholin-4-ylpyridin-3-ylethynyl)pyrimidin-4-ylamine

This pyrimidine (FW = 397.44 g/mol) inhibits cytosolic adenosine kinase (IC$_{50}$ = 1 nM). Because the use of adenosine receptor agonists is plagued by dose-limiting cardiovascular side effects, there is great interest in developing adenosine kinase inhibitors (AKIs) as a means for raising steady-state adenosine concentrations. 1. Matulenko, Paight, Frey, *et al.* (2007) *Bioorg. Med. Chem.* **15**, 1586.

(2S)-2-{[(3aS,6S,6aR)-4-(1,3-Benzothiazol-2-yl)-6-methyl-5-oxohexa-hydropyrrolo[3,2-b]pyrrol-1(2H)-yl]carbonyl}-N-(4-isopropylphenyl)pyrrolidine-1-carboxamide

This pyrrolidine-5,5'-*trans*-lactam (FW = 530.67 g/mol) inhibits human cytomegalovirus assemblin (K_i = 10 nM). The combination of high potency against HCMV Δ^{Ala} protease and high human plasma stability produced compounds with significant *in vitro* antiviral activity against human cytomegalovirus, with potency equivalent to that of ganciclovir. 1. Borthwick, Davies, Ertl, *et al.* (2003) *J. Med. Chem.* **46**, 4428.

4-((3E)-1-(1H-Benzotriazol-1-yl)-1-(4-(difluoro(phosphono)methyl)benzyl)-4-phenylbut-3-en-1-yl)benzoic acid Methyl Ester

This phosphonate (FW = 601.55 g/mol) inhibits protein-tyrosine-phosphatase (IC$_{50}$ = 39 nM) through a conformation-assisted inhibition mechanism that accounts for its considerable selectivity for protein-tyrosine phosphatase-1B over T-cell protein-tyrosine phosphatase. Its design exploited the conservative L119V substitution between the two enzymes to synthesize a PTP-1B inhibitor that is an order of magnitude more selective over TCPTP. 1. Asante-Appiah, Patel, Desponts, *et al.* (2006) *J. Biol. Chem.* **281**, 8010.

N-(Benzoylaminothiocarbonyl)-N-(4-nitrobenzyl)-β-alanine Hydroxamate

This hydroxamate (FW = 402.43 g/mol) inhibits interstitial collagenase (K_i = 32 nM), gelatinase A (K_i = 2 nM), neutrophil collagenase (K_i = 1.8 nM), gelatinase B (K_i = 1.6 nM), and microbial collagenase (K_i = 15 nM). Its 4-nitrobenzyl moiety is an efficient P$_2$'-anchoring moiety, and the β-alanyl scaffold is a promising replacement for the α-amino acyl to obtain potent MMP/ChC inhibitors. 1. Scozzafava, Ilies, Manole & Supuran (2000) *Eur. J. Pharm. Sci.* **11**, 69.

(E,E)-8-O-(4-Benzoylbenzyl)-3,7-dimethyl-2,6-octadien-1,8-diol Diphosphate

This substrate analogue (FW = 524.44 g/mol) targets protein farnesyltransferase (K_i = 45 nM). The 3-benzoylbenzyl analogue also inhibits (K_i = 49 nM). Crystallographic analysis of one of these analogues bound to PFTase reveals that the diphosphate moiety and the two isoprene units bind in the same positions occupied by the corresponding atoms in FPP when bound to PFTase. However, the benzophenone group protrudes into the acceptor protein binding site and prevents the binding of the second (protein) substrate. 1. Turek-Etienne, Strickland & Distefano (2003) *Biochemistry* **42**, 3716.

N-Benzoyl-N-(4-cyano-3-(naphthalen-1-yl)phen-1-yl)-N-(3-methyl-3H-imidazol-4-ylmethyl)amine

This imidazole-containing agent (FW = 442.52 g/mol) inhibits rat protein farnesyltransferase (IC_{50} = 14 nM). An X-ray crystal structure of the inhibitor bound to rat farnesyltransferase is also presented. 1. Curtin, Florjancic, Cohen, *et al.* (2003) *Bioorg. Med. Chem. Lett.* **13**, 1367.

6-Benzoylheteratisine

This benzoylated diterpene alkaloid ($FW_{free-base}$ = 495.62 g/mol), structurally related to aconitine, has been identified in different *Aconitum* species, which are utilized as analgesics in traditional Chinese folk medicine. 6-Benzoylheteratisine suppresses the spread of seizure activity and has an anticonvulsive potential. It also inhibits voltage-gated Na^+ channels in rat brain synaptosomes. Note that this alkaloid is photosensitive and should be stored in the dark. 6-benzoylheteratisine is structurally related to aconitine, and patch-clamp studies on cultured rat hippocampal pyramidal cells revealed an inhibitory action of 6-benzoylheteratisine on whole cell Na^+ currents, suggesting that 6-benzoylheteratisine is a naturally occurring antagonist of the Na^+ channel activator aconitine (2). **See Aconitine 1.** Gutser & Gleitz (1998) *Neuropharmacol.* **37**, 1139. **2.** Ameri & Simmet (1999) *Eur. J. Pharmacol.* 386, 187.

N^1-Benzoyl-N^4-[N-(4-methoxybenzoyl)-L-valyl-L-prolyl]-1,4(S)-diamino-2,2-difluoro-5-methyl-3-oxohexane

This orally bioavailable peptide analogue (FW = 613.68 g/mol) inhibits leukocyte elastase (K_i = 6.4 nM). Development of such compounds demonstrates that peptidyl trifluoromethyl ketone inhibitors can achieve high levels of oral activity and bioavailability, commending their use as therapeutic agents in the treatment of elastase-associated disorders. **1.** Veale, Bernstein, Bohnert, *et al.* (1997) *J. Med. Chem.* **40**, 3173.

(R)-2-(4-Benzoylphenylsulfonyl)-1,2,3,4-tetrahydroisoquinoline-3-carboxylic Hydroxamate

This isoquinoline (FW = 593.93 g/mol) inhibits human neutrophil collagenase, *or* matrix metalloproteinase-8, IC_{50} = 3 nM. (R)-2-(4-benzoylphenylsulfonyl)-1,2,3,4-tetrahydro-isoquinolin-3-carboxylic acid, by contrast, has an IC_{50} value of 300 nM. **1.** Matter, Schwab, Barbier, *et al.* (1999) *J. Med. Chem.* **42**, 1908.

N-{N-[3-(2(S)-Benzoylpyrrolidine-1-carbonyl)benzoyl]-L-prolyl}pyrrolidine

This dipeptide analogue (FW = 501.58 g/mol) inhibits pig prolyl oligopeptidase (IC_{50} = 18 nM). In the series of compounds where the *N*-acyl group was a Boc group, the 5(R)-tert-butyl group increased the potency strongly. A similar effect is not observed for the 5(S)-*tert*-butyl group. In the series of compounds, where the *N*-acyl group was a 4-phenylbutanoyl group, the 5(R)-tert-butyl, 5(R)-methyl and 5(S)-methyl groups was found to be ineffective. **1.** Wallén, Christiaans, Saarinen, *et al.* (2003) *Bioorg. Med. Chem.* **11**, 3611.

5-Benzylacyclouridine

This substituted uracil (FW = 276.29 g/mol), also known as 5-benzyl-1-(2'-hydroxy-ethoxy-methyl)uracil, inhibits uridine phosphorylase, K_i = 98 nM. **Target(s):** uridine phosphorylase (1-7); nucleoside transport (8); thymidine phosphorylase (9). **1.** Jimenez, Kranz, Lee, Gero & O'Sullivan (1989) *Biochem. Pharmacol.* **38**, 3785. **2.** Niedzwicki, Chu, el Kouni, Rowe & Cha (1982) *Biochem. Pharmacol.* **31**, 1857. **3.** Bu, Settembre, el Kouni & Ealick (2005) *Acta Crystallogr. D Biol. Crystallogr.* **61**, 863. **4.** Naguib, el Kouni, Chu & Cha (1987) *Biochem. Pharmacol.* **36**, 2195. **5.** Bu, Settembre, el Kouni & Ealick (2005) *Acta Crystallogr. Sect. D* **61**, 863. **6.** Temmink. de Bruin, Turksma, *et al.* (2007) *Int. J. Biochem. Cell Biol.* **39**, 565. **7.** el Kouni, el Kouni & Naguib (1993) *Cancer Res.* **53**, 3687. **8.** Lee, el Kouni, Chu & Cha (1984) *Cancer Res.* **44**, 3744. **9.** Liu, Cao, Russell, Handschumacher & Pizzorno (1998) *Cancer Res.* **58**, 5418.

α-Benzylaminobenzylphosphonic Acid

This phosphonic acid ($FW_{free-acid}$ = 277.26 g/mol) inhibits human prostatic acid phosphatase (IC_{50} = 4 nM). The enhanced potency is thought to be the consequence of four interaction sites, including the phosphate binding region, hydrophobic moieties in the benzylamino and phenylphosphonic acid, as well as rigidity produced by an internal salt bridge between the phosphonate and the α-amino group. **1.** Beers, Schwender, Loughney, *et al.* (1996) *Bioorg. Med. Chem.* **4**, 1693.

(E)-7-[4-[4-[(Benzylamino)carbonyl]-2-oxazolyl]phenyl]-7-(3-pyridyl)hept-6-enoate

This substituted oxazolecarboxamide ($FW_{\text{free-acid}}$ = 481.55 g/mol) inhibits human thromboxane-A synthase (IC_{50} = 25 nM) and is a thromboxane receptor antagonist (K_d = 135 nM). **1**. Takeuchi, Kohn, True, *et al.* (1998) *J. Med. Chem.* **41**, 5362.

2-Benzylamino-2-deoxy-D-glucosyl-α1,6-D-*myo*-inositol-1-phosphoryl-*sn*-1,2-dipalmitoylglycerol

This substrate analogue (FW = 1063.33 g/mol) inhibits *Trypanosoma brucei* *N*-acetylglucosaminylphosphatidyl-inositol deacetylase (IC_{50} = 8 nM), showing that (a) the de-*N*-acetylases show little specificity for the lipid moiety; (b) the 3'-OH group of the GlcNAc residue is needed for substrate recognition, whereas the 6'-OH group is dispensable and the 4'-OH, while not required for recognition, cannot be epimerized or substituted; (c) the parasite enzyme acts on analogues containing βGlcNAc or aromatic *N*-acyl groups, whereas the human enzyme cannot; (d) three GlcNR-PI analogues are de-*N*-acetylase inhibitors, one of which is a suicide inhibitor; (e) the suicide inhibitor most likely forms a carbamate or thiocarbamate ester to an active-site hydroxy-amino acid or Cys or residue such that inhibition is reversed by certain nucleophiles. **1**. Smith, Crossman, Borissow, *et al.* (2001) *EMBO J.* **20**, 3322.

6-(Benzylamino)-2-(3-hydroxypropyl)purine

This purine (FW = 283.34 g/mol) inhibits cytokinin 7β-glucosyltransferase. The most effective inhibitors are the cytokinin analogues 3-methyl-7-*n*-pentylaminopyrazolo[4,3-*d*]pyrimidine, which act competitively (K_i = 22 µM), and the diaminopurine, 6-benzylamino-2-(2-hydroxyethylamino)-9-methylpurine (K_i = 3.3 µM). However these compounds were ineffective as inhibitors of the cytokinin-alanine synthase which was inhibited competitively by IAA (K_i = 70 µM) and related compounds, especially 5,7-dichloro-IAA (K_i 0.4 µM). Certain urea derivatives were moderately effective inhibitors of the enzymes (K_i = 100µM). **1**. Parker, Entsch & Letham (1986) *Phytochemistry* **25**, 303.

2-[[[Benzylamino]methyl](hydroxyphosphinoyl)methyl]pentanedioic Acid

This phosphinic acid (FW = 329.29 g/mol) inhibits glutamate carboxypeptidase II (IC_{50} = 59 nM) and prevents neurodegeneration in a middle cerebral artery occlusion model of cerebral ischemia. In addition, in the chronic constrictive model of neuropathic pain, it significantly attenuates the hypersensitivity observed with saline-treated animals. **1**. Jackson, Tays, Maclin, *et al.* (2001) *J. Med. Chem.* **44**, 4170.

N-(*R*,*S*)-2-Benzyl-3[(*S*)-(2-amino-4-methylthiobutyldithio)-1-oxopropyl]-L-phenylalanine, *See RB 101*

5-(3-Benzylaminophenyl)-1*H*-1,2,3-triazole

This aryl-1,2,3-triazole (FW = 250.30 g/mol), also known as *N*-benzyl-3-(1*H*-1,2,3-triazol-5-yl)aniline, inhibits human methionyl aminopeptidase type 2 ($K_{i,\text{app}}$ = 29 nM for the Co(II)-substituted protein). **1**. Kallander, Lu, Chen, *et al.* (2005) *J. Med. Chem.* **48**, 5644.

*N*¹-Benzyl-*N*⁴-[*N*-(benzyloxycarbonyl)-L-valyl-L-prolyl]-1,4(*S*)-diamino-2,2-difluoro-5-methyl-3-oxohexane

This peptide analogue (FW = 599.70 g/mol) inhibits leukocyte elastase (K_i = 1.6 nM), a chymotrypsin-like serine proteinase that is secreted by neutrophils and macrophages during inflammation, thereby destroying both invasive bacteria and host tissue. **1**. Veale, Bernstein, Bohnert, *et al.* (1997) *J. Med. Chem.* **40**, 3173.

[(2*R*,4*R*,5*S*)-2-Benzyl-5-(*N*-(*t*-butoxycarbonyl)-amino)-4-hydroxy-6-phenyl-hexanoyl]-L-leucyl-L-phenylalaninamide

This tetrapeptide isostere (FW = 672.87 g/mol), also known as L-685458 and γ-secretase inhibitor X, inhibits amyloid β-protein precursor γ-secretase activity: IC_{50} = 17 nM. It is a 100-fold weaker inhibitor of cathepsin D. Note that the hydroxyethylene dipeptide isostere has the opposite stereochemistry as required for inhibition of the HIV-1 aspartyl protease. **1**. Shearman, Beher, Clarke, *et al.* (2000) *Biochemistry* **39**, 8698. **2**. Li, Lai, Xu, *et al.* (2000) *Proc. Natl. Acad. Sci. U.S.A.* **97**, 6138.

Benzyl (*S*)-1-((2*S*,3*R*)-4-((*R*)-4-(*tert*-butylcarbamyl)thiazolidin-3-yl)-3-hydroxy-1-phenylbutan-2-ylamino)-4-amino-1,4-dioxobutan-2-ylcarbamate

This peptide analogue (FW = 599.75 g/mol) inhibits HIV-1 retropepsin (K_i = 6 nM) and HIV-2 retropepsin (K_i = 25 nM). Although the sequences of the proteinases from HIV-1 and HIV-2 are less than 50% identical overall, their enzymatic properties are quite similar. There are only five active-site residues that are different between the two proteinases, and this report focuses on homo- and hetero-dimer effects on inhibitor binding. **1**. Griffiths, Tomchak, Mills, *et al.* (1994) *J. Biol. Chem.* **269**, 4787.

2-Benzyl-5-(6-(4-(2-chlorophenoxy)piperidin-1-yl)pyridazin-3-yl)1,3,4-oxadiazole

This potent, selective, orally bioavailable pyridazine derivative (FW = 447.93 g/mol) inhibits stearoyl-CoA 9-desaturase (IC_{50} = 12 nM, mouse). It and related SCD1 inhibitors are based on a common pyridazine carboxamide template. **1**. Liu, Lynch, Freeman, *et al.* (2007) *J. Med. Chem.* **50**, 3086.

N-Benzyl-N-(4-chlorophenylsulfonyl-aminocarbonyl)-glycine hydroxamate

This tight-binding hydroxamate-based protease inhibitor (FW = 397.84 g/mol) targets proteases, including: interstitial collagenase, matrix metallo-proteinase-1, *or* MMP-1 K_i = 143 nM; gelatinase A, *or* matrix metalloproteinase-2, K_i = 10 nM; gelatinase B, *or* matrix metalloproteinase-9, K_i = 15 nM; neutrophil collagenase, *or* matrix metalloproteinase-8 (K_i = 15 nM), and microbial collagenase (K_i = 15 nM). **1.** Ilies, Banciu, Scozzafava, *et al.* (2003) *Bioorg. Med. Chem.* **11**, 2227.

Benzyl Cyanide

This nitrile (FW = 117.15 g/mol; CAS 140-29-4), also known as phenylacetonitrile, occurs naturally in plants (*e.g.*, garden cress). It is an oily liquid (melting point = −23.8°C; boiling point = 233.5°C) that has a very low solubility in water. It is miscible with ethanol. Benzyl cyanide is a courtship-inhibition pheromone for mature gregarious male desert locusts, *Schistocerca gregaria*. Benzyl cyanide is also an anti-aphrodisiac, a chemical agent or pheromone passed from males to females of the butterfly *Pieris brassicae* during mating, rendering the recipient less attractive to conspecific males. Wings and legs, in particular the fore wings, have been identified as the main releasing sites. Given its utility in the chemical synthesis of phenobarbital, methylphenidate, and other amphetamines, benzyl cyanide appears on List I of the U.S. Drug Enforcement Agency. **Target(s):** monoamine oxidase (1); courtship (2,3); dopamine β-monooxygenase (4). **1.** Bright & Porter (1975) *The Enzymes*, 3rd ed. (Boyer, ed.), **12**, 421. **2.** Seidelmann, Weinert & Ferenz (2003) *J. Insect Physiol.* **49**, 1125. **3.** Fatouros, Huigens, van Loon, Dicke & Hilker (2005) *Nature* **433**, 704. **4.** Colombo, Rajashekhar, Giedroc & Villafranca (1984) *J. Biol. Chem.* **259**, 1593.

N-Benzyl-N-(4-cyano-3-(naphthalen-1-yl)phen-1-yl)-N-(3-methyl-3H-imidazol-4-ylmethyl)amine

This highly substituted imidazole-containing biphenyl (FW = 428.54 g/mol) inhibits rat protein farnesyltransferase (IC$_{50}$ = 1.5 nM). This and several of these analogues, are potent inhibitors of farnesyltransferase (K_i < 1 nM), showing >300-fold selectivity over geranylgeranyltransferase and cellular IC$_{50}$ values ≤ 80 nM. An X-ray crystal structure of the inhibitor bound to rat farnesyltransferase is also presented. **1.** Curtin, Florjancic, Cohen, *et al.* (2003) *Bioorg. Med. Chem. Lett.* **13**, 1367.

N-[(R)-2-Benzyl-5-cyano-4-oxopentanoyl]-L-phenylalanine

This acylated amino acid (FW$_{free-acid}$ = 378.43 g/mol) is a mechanism-based inactivator of human neutral endopeptidase 24.11, *or* neprilysin), with inactivation occurring via a ketenimine intermediate that is then intercepted by an active-site nucleophile to irreversibly generate an enzyme-inhibitor compound. Inactivation is likely to result from NEP-catalyzed formation of ketenimine intermediates, which are subsequently trapped by an active-site nucleophile. **1.** Levy, Taibi, Mobashery & Ghosh (1993) *J. Med. Chem.* **36**, 2408.

1-Benzylcyclopropylamine

This cyclopropylamine (FW$_{free-base}$ = 147.22 g/mol) is a mechanism-based inactivator pf momoamine oxidase, with one mole of the enzyme inactivated (at an active-site cysteine) for every 2.3 moles of the amine converted to product. Monoamine oxidase B is inactivated more rapidly than monoamine oxidase A. **1.** Brush & Kozarich (1992) *The Enzymes*, 3rd ed. (Sigman, ed.), **20**, 317. **2.** Silverman & Zieske (1985) *J. Med. Chem.* **28**, 1953. **3.** Silverman & Hiebert (1988) *Biochemistry* **27**, 8448.

9-Benzyl-9-deazaguanine

This pseudonucleoside analogue (FW = 240.26 g/mol) inhibits calf spleen purine-nucleoside phosphorylase (K_i = 12 nM). **1.** Bennett, Allan, Noker, *et al.* (1993) *J. Pharmacol. Exp. Ther.* **266**, 707. **2.** Parker, Allan, Niwas, Montgomery & Bennett (1994) *Cancer Res.* **54**, 1742. **3.** Kimble, Hadala, Ludewig, *et al.* (1995) *Inflamm. Res.* **44** Suppl. 2, S181. **4.** Castilho, Postigo, de Paula, *et al.* (2006) *Bioorg. Med. Chem.* **14**, 516. **5.** Bzowska, Kulikowska & Shugar (2000) *Pharmacol. Ther.* **88**, 349.

(2S,3R)-2-Benzyl-3,4-epoxybutanoic Acid

This epoxide (FW$_{free-acid}$ = 192.21 g/mol) irreversibly inhibits carboxypeptidase A, covalently modifying Glu270 in the process. The (2R,3S)-enantiomer acts likewise, but the other diastereoisomers are ineffective. **1.** Lee & Kim (2002) *Bioorg. Med. Chem.* **10**, 913. **2.** Kim & Kim (1991) *J. Amer. Chem. Soc.* **113**, 3200. **3.** Yun, Park, Kim, *et al.* (1992) *J. Amer. Chem. Soc.* **114**, 2281. **4.** Ryu, Choi & Kim (1997) *J. Amer. Chem. Soc.* **119**, 38. **5.** Cho, Kim, Lee & Choi (2001) *Biochemistry* **40**, 10197.

O^4-Benzylfolic acid

This modified folate (FW = 521.53 g/mol) inhibits methylated-DNA:[protein]-cysteine *S*-methyltransferase, *or* O^6-alkylguanine DNA alkyltransferase. The activity of O^4-benzylfolic acid is particularly noteworthy, because it is roughly 30-times more active than O^6-

benzylguanine against the wild-type alkyltransferase. It is also capable of inactivating the P140K mutant alkyltransferase that is resistant to inactivation by O^6-benzylguanine. **1**. McMurry (2007) *DNA Repair* **6**, 1161. **2**. Nelson, Loktionova, Pegg & Moschel (2004) *J. Med. Chem.* **47**, 3887.

N-Benzyl-β-D-galactosylamine

This synthetic *N*-glycoside (FW$_{\text{free-base}}$ = 269.30 g/mol) is a potent inhibitor of β-galactosidase (K_i = 9.5 nM). Acid catalysis of *O*-galactoside hydrolysis and tight binding of basic inhibitors is observed only when Mg^{2+} is bound to the enzyme. Formation of an ion-pair by proton transfer from the acidic group required for catalysis to the bound galactosylamine is proposed as the cause of the enhanced affinity. The magnitude of the observed effects indicates strong shielding of the active site from the aqueous environment. The pH-dependence of K_i for 1-d-galactosylpiperidine and d-galactosyl-benzene shows that no deprotonation from the active site occurs up to pH 10. The hydrophobic nature of the aglycon site inferred from earlier studies was confirmed by the high affinity of *N*-benzyl- (K_i = 9 × 10^{-9}M) and *N*-heptyl-β-d-galactosylamine (K_i = 0.6 × 10^{-9}M). Such results suggest a model for catalysis, wherein the charge of an essential carboxylate group is neutralized by an adjacent cationic group. **1**. Mooser (1992) *The Enzymes*, 3rd ed., **20**, 187. **2**. Legler & Herrchen (1983) *Carbohyd. Res.* **116**, 95.

O^6-Benzylguanine

This modified guanine (FW = 241.25 g/mol) inactivates human O^6-alkylguanine-DNA alkyltransferase via the formation of an *S*-benzyl-L-cysteinyl residue and stoichiometric guanine. **Target(s):** DNA helicase, weakly inhibited (1); methylated-DNA:[protein]-cysteine *S*-methyltransferase, IC$_{50}$ = 0.2 μM (2-11); nucleoside triphosphatase, weakly inhibited (1). **1**. Borowski, Niebuhr, Mueller, *et al.* (2001) *J. Virol.* **75**, 3220. **2**. Pegg, Boosalis, Samson, *et al.* (1993) *Biochemistry* **32**, 11998. **3**. Goodtzova, Crone & Pegg (1994) *Biochemistry* **33**, 8385. **4**. Elder, Margison & Rafferty (1994) *Biochem. J.* **298**, 231. **5**. McMurry (2007) *DNA Repair* **6**, 1161. **6**. Nelson, Loktionova, Pegg & Moschel (2004) *J. Med. Chem.* **47**, 3887. **7**. Walter, Sung, Intano & Walter (2001) *Mutat. Res.* **493**, 11. **8**. Terashima, Kawata, Sakumi, Sekiguchi & Kohda (1997) *Chem. Res. Toxicol.* **10**, 1234. **9**. Sabharwal & Middleton (2006) *Curr. Opin. Pharmacol.* **6**, 355. **10**. Verbeek, Southgate, Gilham & Margison (2008) *Brit. Med. Bull.* **85**, 17. **11**. Shibata, Glynn, McMurry, *et al.* (2006) *Nucl. Acids Res.* **34**, 1884.

2(*S*)-Benzyl-3-[hydroxy(1'(*R*)-aminoethyl)phosphinyl]propanoyl-L-phenylalanine

This phosphinate-containing peptidomimetic (FW = 416.41 g/mol) resemble a required tetrahedral catalytic intermediate, making it a potent transition-state inhibitor of membrane alanyl aminopeptidase, *or* aminopeptidase N (K_i = 0.59 nM). **1**. Noble, Luciani, Da Nacimento, *et al.* (2000) *FEBS Lett.* **467**, 81. **2**. Jardinaud, Banisadr, Noble, *et al.* (2004) *Biochimie* **86**, 105.

(*RS*)-2-Benzyl-3-mercaptopropionic Acid

This β-mercaptoacid (FW$_{\text{free-acid}}$ = 196.27 g/mol), also known as SQ 14603 and (*RS*)-2-(mercaptomethyl)-3-phenylpropionic acid, is a strong inhibitor of carboxypeptidase A, but more weakly inhibits carboxypeptidase B and angiotensin-converting enzyme. **Target(s):** carboxypeptidase A, K_i = 11 nM (1-4); carboxypeptidase B, K_i = 163 μM (1); β-lactamase (metallo-β-lactamase) (5); peptidyl-dipeptidase A, *or* angiotensin-converting enzyme, K_i = 58 μM (1). **1**. Ondetti, Condon, Reid, *et al.* (1979) *Biochemistry* **18**, 1427. **2**. Park & Kim (2002) *J. Med. Chem.* **45**, 911. **3**. Lee & Kim (2003) *Bioorg. Med. Chem.* **11**, 4685. **4**. Chung & Kim (2001) *Bioorg. Med. Chem.* **9**, 185. **5**. Jin, Arakawa, Yasuzawa, *et al.* (2004) *Biol. Pharm. Bull.* **27**, 851.

N^1-Benzyl-N^4-[*N*-(4-methoxybenzoyl)-L-valyl-L-prolyl]-1,4(*S*)-diamino-2,2-difluoro-5-methyl-3-oxohexane

This peptide analogue (FW = 600.71 g/mol) inhibits leukocyte elastase (K_i = 0.35 nM). This and related compounds demonstrate peptidyl trifluoromethyl ketone inhibitors can achieve high levels of oral bioavailability, indicating they may prove to be useful as agents for treating elastase-associated diseases. **1**. Veale, Bernstein, Bohnert, *et al.* (1997) *J. Med. Chem.* **40**, 3173.

3-(3-Benzyl-1,2,4-oxadiazol-5-yl)-6-(4-(2-fluorophenoxy)piperidin-1-yl)pyridazine

This pyridazine derivative (FW = 431.47 g/mol) inhibits stearoyl-CoA 9-desaturase (mouse IC$_{50}$ = 7 nM). Stearoyl-CoA desaturase 1 (SCD1) catalyzes the committed step in the biosynthesis of monounsaturated fatty acids from saturated, long-chain fatty acids. Studies with SCD1 knockout mice have established that these animals are lean and protected from leptin deficiency-induced and diet-induced obesity, with greater whole body insulin sensitivity than wild-type animals. **1**. Liu, Lynch, Freeman, *et al.* (2007) *J. Med. Chem.* **50**, 3086.

N^1-[2-(Benzyloxy)benzyl]-4-oxo-4-(1-piperidinyl)-1,3-(*S*)-butanediamine

This substituted piperidine (FW = 382.53 g/mol) is a strong inhibitor of dipeptidyl-peptidase II (IC$_{50}$ = 0.39 nM). The 4-(benzyloxy)benzyl analogue also inhibits (IC$_{50}$ = 0.41 nM; this analogue is also a weak inhibitor of dipeptidyl-peptidase IV: IC$_{50}$ = 192 μM). **1**. Senten, van der Veken, De Meester, *et al.* (2004) *J. Med. Chem.* **47**, 2906.

O-{[(1R)-[[N-(Benzyloxycarbonyl)-L-alanyl]amino]ethyl]hydroxy phosphinyl}-L-3-phenyllactate

This peptidomimetic phosphonate ester (FW = 402.34 g/mol) is a potent tetrahedral transition-state analogue and inhibitor of bovine carboxypeptidase A, or CPA (K_i = 3 pM). Additional enzyme-inhibitor interactions for tripeptide phosphonates secure a binding mode in which a Pi portion of the inhibitor is clearly bound by the corresponding Si binding subsite. The phosphinyl groups of these phosphonates coordinate to the active-site zinc in a manner that has been proposed as a characteristic feature of the general-base (Zn^{2+}-hydroxyl or Zn^{2+}-water) mechanism for the CPA-catalyzed reaction. 1. Kim & Lipscomb (1991) *Biochemistry* **30**, 8171.

1-(N-Benzyloxycarbonyl-L-alanyl)-5-(N-Benzyloxycarbonyl)-L-leucyl)carbohydrazide

This carbohydrazide (FW = 542.59 g/mol) strongly inhibits cathepsin K (K_i = 1 nM) and cathepsin S (K_i = 44 nM), but exerts little inhibition on cathepsins B and L. X-ray, MS, and NMR data indicate that different intermediates or transition states are being represented that are dependent on the conditions of measurement and the specific groups flanking the carbonyl in the inhibitor. The species observed crystallographically are most consistent with tetrahedral intermediates that may be close approximations of those that occur during substrate hydrolysis. Initial kinetic studies suggest the possibility of irreversible and reversible active-site modification. 1. Thompson, Halbert, Bossard, *et al.* (1997) *Proc. Natl. Acad. Sci. U.S.A.* **94**, 14249.

N-[1-[[[(Benzyloxycarbonyl)amino]phenyl]ethyl]-hydroxyphosphinyl]-L-leucyl-L-alanine

This tetrahedral reaction intermediate analogue (FW = 519.54 g/mol), also named N-(benzyloxycarbonyl)-phenylalanyl[PO₂NH]-L-leucyl-L-alanine, is a strong inhibitor of thermolysin (K_i = 68 pM). 1. Christianson & Lipscomb (1988) *J. Amer. Chem. Soc.* **110**, 5560. 2. Holden, Tronrud, Monzingo, Weaver & Matthews (1987) *Biochemistry* **26**, 8542. 3. Baltora-Rosset, Aboubeker, Dupradeau, *et al.* (1999) *J. Biomol. Struct. Dyn.* **16**, 1061.

N-{N-[N-(Benzyloxycarbonyl)-L-asparaginyl]-(3S-amino-2R-hydroxy-4-phenylbutyl)}piperidine-2(S)-carboxylic t-Butyl Amide

This pseudopeptide (FW = 595.74 g/mol), also known as Z-Asn-Phe[CH(OH)CH₂N]PIC-NhtBu, contains a tetrahedral mimic, making it a powerful inhibitor of HIV-1 protease (IC₅₀ = 18 nM). Antiviral activity was observed in the nM range in three different cell systems, as assessed by p24

antigen and syncytium formation. 1. Roberts, Martin, *et al.* (1990) *Science* **248**, 358.

N-(Benzyloxycarbonyl)-D-aspartic Acid

This modified amino acid (FW$_{free-acid}$ = 267.24 g/mol) inhibits aspartoacylase. Canavan disease, an autosomal recessive disorder, is characterized biochemically by N-acetylaspartic aciduria and aspartoacylase (N-acyl-L-aspartate amidohydrolase; EC 3.5.1.15) deficiency. Aspartoacylase is localized to white matter, whereas the N-acetylaspartic acid concentration is threefold higher in gray matter than in white matter. 1. Kaul, Casanova, Johnson, Tang & Matalon (1991) *J. Neurochem.* **56**, 129.

N$^\alpha$-(Benzyloxycarbonyl)-L-aspartyl-L-glutamyl-L-valyl-L-aspartate α-Semialdehyde

This potent tetrapeptide aldehyde (FW = 594.57 g/mol) is a potent inhibitor of caspase-8 (K_i = 2 nM). Aldehyde inhibitors can bind reversibly, forming adducts with the active-site residues. Caspase-8 initiates the Fas-mediated pathway, in which the downstream executioner caspase-3 is a physiological target. These cysteine proteases are specific for substrates possessing an aspartic acid residue at the P¹ position and a four-residue recognition motif lying N-terminal to the cleavage site. The aldehyde warhead of this inhibitor forms a tetrahedral thioacetal that mimics the catalytic transition state. 1. Blanchard, Donepudi, Tschopp, *et al.* (2000) *J. Mol. Biol.* **302**, 9.

N-(Benzyloxycarbonyl)glutathione

This glutathione derivative (FW$_{free-acid}$ = 441.46 g/mol) inhibits hydroxyacylglutathione hydrolase, or glyoxalase II or GLX2-2 (K_i = 5.6 μM). Analysis of site-directed mutants of GLX2-2 demonstrated that tight binding of these inhibitors is not due to interactions of the protecting groups with hydrophobic amino acids on the surface of the enzyme. Instead, MM2 calculations predict that the lowest energy structures of the unbound, doubly substituted inhibitors are similar to those of a bound inhibitor. 1. Yang, Sobieski, Carenbauer, *et al.* (2003) *Arch. Biochem. Biophys.* **414**, 271.

N-(Benzyloxycarbonyl)-L-isoleucyl-L-(glutamyl-O-t-butylester)-L-alanyl-L-leucinal

This tetrapeptide aldehyde (FW = 618.77 g/mol), also known as proteasome inhibitor I (PSI) and Z-IE(O-t-Bu)AL CHO, inhibits the chymotrypsin-like activity of the mammalian proteasome. Aldehyde inhibitors can bind reversibly, forming adducts with the active-site residues. It also prevents activation of NF-κB and stabilizes a newly phosphorylated form of I κB-α that is still bound to NF-κB. 1. Gonzalez-Flores, Guerra-Araiza, Cerbon, Camacho-Arroyo & Etgen (2004) *Endocrinology* **145**, 2328. 2. Figueiredo-Pereira, Berg & Wilk (1994) *J. Neurochem.* **63**, 1578. 3. Traenckner, Wilk & Baeuerle (1994) *EMBO J.* **13**, 5433. 4. Griscavage, Wilk & Ignarro (1996) *Proc. Natl. Acad. Sci. U.S.A.* **93**, 3308. 5. Tanaka, Sawada & Sawada (2000) *Comp. Biochem. Physiol. C* **125C**, 215. 6. Fernández Murray, Biscoglio & Passeron (2000) *Arch. Biochem. Biophys.* **375**, 211.

1-(*N*ᵅ-Benzyloxycarbonyl-L-leucylamino)-3-(phenoxyphenyl sulfonamido)-2-propanone

This substituted diaminopropanone (FW = 567.66 g/mol) is a strong reversible inhibitor of cathepsin K (apparent K_i = 1.8 nM), a key enzyme in osteoclast-mediated bone resorption. Its design is based on the poorly electrophilic 1,3-bis(acylamino)-2-propanone scaffold. **1**. Yamashita, Smith, Zhao, *et al.* (1997) *J. Amer. Chem. Soc.* **119**, 11351.

1-(*N*ᵅ-Benzyloxycarbonyl-L-leucyl-L-leucyl-L-leucinal

This cell-permeable tripeptide aldehyde (FW = 475.63 g/mol), also known as MG-132 and Cbz-Leu-Leu-Leu-CHO, is a strong inhibitor of the chymotrypsin-like activity of the 26S proteasome (K_i = 4 nM). It also inhibits cathepsin K (K_i = 1.4 nM), activates c-Jun N-terminal kinase and thus initiating apoptosis, and inhibits NF-κB activation (IC$_{50}$ = 3 μM). In addition, it inhibits both β- and γ-secretase. **Target(s):** proteasome endopeptidase complex (1,2,4,9-12); cathepsin K (3); β-secretase (5); γ secretases (5-7); DNA-directed DNA polymerase (8); DNA polymerase α (8); DNA polymerase β (8); DNA polymerase γ (8); DNA polymerase I (8); DNA nucleotidylexotransferase, weakly inhibited (8). **1**. Rock, Gramm, Rothstein, *et al.* (1994) *Cell* **78**, 761. **2**. Grisham, Palombella, Elliott, *et al.* (1999) *Meth. Enzymol.* **300**, 345. **3**. Votta, Levy, Badger, *et al.* (1997) *J. Bone Miner. Res.* **12**, 1396. **4**. Lee & Goldberg (1996) *J. Biol. Chem.* **271**, 27280. **5**. Steinhilb, Turner & Gaut (2001) *J. Biol. Chem.* **276**, 4476. **6**. Pinnix, Musunuru, Tun, *et al.* (2001) *J. Biol. Chem.* **276**, 481. **7**. Klafki, Abramowski, Swoboda, Paganetti & Staufenbiel (1996) *J. Biol. Chem.* **271**, 28655. **8**. Taguchi, Matsukage, Ito, Saito & Kawashima (1992) *Biochem. Biophys. Res. Commun.* **185**, 1133. **9**. Wojcikiewicz, Xu, Webster, Alzayady & Gao (2003) *J. Biol. Chem.* **278**, 940. **10**. Tanaka, Sawada & Sawada (2000) *Comp. Biochem. Physiol. C* **125C**, 215. **11**. Fernández Murray, Biscoglio & Passeron (2000) *Arch. Biochem. Biophys.* **375**, 211. **12**. Eleuteri & Angeletti (2003) *FEBS Lett.* **547**, 7.

1-(*N*ᵅ-Benzyloxycarbonyl-L-leucyl-L-phenylalanyl-(3-acetamido)-L-alaninal

This aldehyde-containing transition-state analogue (FW = 524.62 g/mol), also known as AG6084, inhibits picornain 3C (rhinovirus 3C protease), K_i = 6 nM. **1**. Patick & Potts (1998) *Clin. Microbiol. Rev.* **11**, 614. **2**. Webber, Okano, Little, *et al.* (1998) *J. Med. Chem.* **41**, 2786.

((*S*)-1-{[(3*R*,*S*)-((*S*)-2-Benzyloxycarbonyl(methyl)-amino-4-methylpentanoylamino)-4-oxopyrrolidin-1-yl]methanoyl}-3-methylbutyl)carbamic Acid Benzyl Ester

This acyclic ketone-based inhibitor (FW = 605.71 g/mol) targets cathepsin K ($K_{i,app}$ = 0.6 nM), consistent with reaction of the active-site thiol at the ketone to form a hemithioketal adduct. **1**. Marquis, Ru, Zeng, *et al.* (2001) *J. Med. Chem.* **44**, 725.

1-(*N*-Benzyloxycarbonyl-L-phenylalanine Chloromethyl Ketone

This halomethyl ketone (FW = 331.80 g/mol), often abbreviated zPCK, is a site-specific inhibitor of chymotrypsin, particularly bovine chymotrypsin Aγ. **Target(s):** chymotrypsin (1,8,13); Rt41A protease, *Thermus* strain (2); nicotinamidase (3); thermomycolin (4); carboxypeptidase C, *or* carboxypeptidase Y (5,6,11,16,17); *Thermomonospora* protease (7); subtilisin (9,10,13); nicotinamidase (12); thermitase (13); endopeptidase La (14); peptidyl glycinamidase, *or* carboxamidopeptidase (15). **1**. Shaw (1972) *Meth. Enzymol.* **25**, 655. **2**. Toogood & Daniel (1998) in *Handb. Proteolytic Enzymes* (Barrett, Rawlings & Woessner, eds.), p. 297, Academic Press, San Diego. **3**. Su & Chaykin (1971) *Meth. Enzymol.* **18B**, 185. **4**. Gaucher & Stevenson (1976) *Meth. Enzymol.* **45**, 415. **5**. Hayashi (1976) *Meth. Enzymol.* **45**, 568. **6**. Powers (1977) *Meth. Enzymol.* **46**, 197. **7**. Kristjansson & Kinsella (1990) *Int. J. Pept. Protein Res.* **36**, 201. **8**. Shaw (1970) *The Enzymes*, 3rd ed. (Boyer, ed.), **1**, 91. **9**. Shaw & Ruscica (1968) *J. Biol. Chem.* **243**, 6312. **10**. Cohen (1970) *The Enzymes*, 3rd ed. (Boyer, ed.), **1**, 147. **11**. Hayashi, Bai & Hata (1975) *J. Biol. Chem.* **250**, 5221. **12**. Gillam, Watson & Chaykin (1973) *Arch. Biochem. Biophys.* **157**, 268. **13**. Brömme & Fittkau (1985) *Biomed. Biochim. Acta* **44**, 1089. **14**. Waxman & Goldberg (1985) *J. Biol. Chem.* **260**, 12022. **15**. Simmons & Walter (1980) *Biochemistry* **19**, 39. **16**. Jung, Ueno & Hayashi (1999) *J. Biochem.* **126**, 1. **17**. Hayashi, Bai & Hata (1974) *J. Biochem.* **76**, 1355.

1-(*N*-Benzyloxycarbonyl-L-phenylalanyl-L-phenylalanine Diazomethyl Ketone

This diazomethyl ketone (*Sequence*: Cbz-Phe-PheCH-N$_2$ or Z-FF-CHN$_2$; FW = 470.53 g/mol), also known as L-1-diazo-3-[(*N*-carbobenzyloxy)-L-phenylalanyl]amino-4-phenyl-2-butanone, can be used to selectively inhibit cathepsin L (IC$_{50}$ = 50 nM) in the presence of cathepsin B. **Target(s):** cathepsin L (1-3,6,9,13,15,17-25); cathepsin B (3,15,17); papain (4,7,8); clostripain (5); cruzipain (10); cathepsin K (11); cathepsin S (12-15); cathepsin N (15); cathepsin H (16); dipeptidyl peptidase I, *or* cathepsin C (26). **1**. Kirschke (1998) in *Handb. Proteolytic Enzymes* (Barrett, Rawlings & Woessner, eds.), p. 617, Academic Press, San Diego. **2**. Barrett & Kirschke (1981) *Meth. Enzymol.* **80**, 535. **3**. Shaw & Green (1981) *Meth. Enzymol.* **80**, 820. **4**. Buttle (1994) *Meth. Enzymol.* **244**, 539. **5**. Green & Shaw (1981) *J. Biol. Chem.* **256**, 1923. **6**. Kirsche & Shaw (1981) *Biochem. Biophys. Res. Commun.* **101**, 454. **7**. Leary, Larsen, Watanabe & Shaw (1977) *Biochemistry* **16**, 5857. **8**. Buttle, Ritonja, Dando, *et al.* (1990) *FEBS Lett.* **262**, 58. **9**. Rifkin, Vernillo, Kleckner, *et al.* (1991) *Biochem. Biophys. Res. Commun.* **179**, 63. **10**. Cazzulo, Stoka & Turk (1997) *Biol. Chem.* **378**, 1. **11**. Brömme, Okamoto, Wang & Biroc (1996) *J. Biol. Chem.* **271**, 2126. **12**. Shaw, Mohanty, Colic, Stoka & Turk (1993) *FEBS Lett.* **334**, 340. **13**. Kirschke, Schmidt & Wiederanders (1986) *Biochem. J.* **240**, 455. **14**. Shi, Munger, Meara, Rich & Chapman (1992) *J. Biol. Chem.* **267**, 7258. **15**. Maciewicz & Etherington (1988) *Biochem. J.* **256**, 433. **16**. Schwartz & Barrett (1980) *Biochem. J.* **191**, 487. **17**. Crawford, Mason, Wikstrom & Shaw (1988) *Biochem. J.* **253**, 751. **18**. Mason, Green & Barrett (1985) *Biochem. J.* **226**, 233. **19**. Mason, Taylor & Etherington (1984) *Biochem. J.* **217**, 209. **20**. Marks & Berg (1987) *Arch. Biochem. Biophys.* **259**, 131. **21**. Recklies & Mort (1985) *Biochem. Biophys. Res. Commun.* **131**, 402. **22**. Day, Dalton, Clough, *et al.* (1995) *Biochem. Biophys. Res. Commun.* **217**, 1. **23**. Mason (1986) *Biochem. J.* **240**, 285. **24**. McDonald & Kadkhodayan (1988) *Biochem. Biophys. Res. Commun.* **151**, 827. **25**. Dufour, Obled,

Valin, *et al.* (1987) *Biochemistry* **26**, 5689.   **26**. Hola-Jamriska, Dalton, Aaskow & Brindley (1999) *Parasitology* **118**, 275.

1-(*N*-Benzyloxycarbonyl-D-phenylalanyl-L-prolyl-boro-(methoxy-propyl)-glycine-pinanediol

This peptide analogue (FW = 659.59 g/mol) inhibits thrombin (K_i = 8.9 nM) but is a much weaker inhibitor of both trypsin (K_i = 1.1 µM) and plasmin (K_i = 15.7 µM). **1**. Tapparelli, Metternich, Ehrhardt, *et al.* (1993) *J. Biol. Chem.* **268**, 4734.

N-[*N*-(Benzyloxycarbonyl)thioprolyl]thioprolinal

This peptide analogue (FW = 366.46 g/mol) inhibits prolyl oligopeptidase, K_i = 0.01 nM. Aldehyde warhead inhibitors can bind reversibly, forming adducts with the active-site residues. **1**. Yoshimoto & Ito (1998) in *Handb. Proteolytic Enzymes* (Barrett, Rawlings & Woessner, eds.), p. 372, Academic Press, San Diego. **2**. Polgár (1994) *Meth. Enzymol.* **244**, 188. **3**. Tsuru, Yoshimoto, Koriyama & Furukawa (1988) *J. Biochem.* **104**, 580. **4**. Besedin & Rudenskaya (2003) *Russ. J. Bioorg. Chem.* **29**, 1. **5**. Oyama, Aoki, Amano, *et al.* (1997) *J. Ferment. Bioeng.* **84**, 538.

N$^\alpha$-(Benzyloxycarbonyl)-L-valyl-L-alanyl-L-aspartate Fluoromethyl Ketone

This tripeptide halomethyl ketone ($FW_{free-acid}$ = 467.49 g/mol; Soluble to 9.35 mg/ml in DMSO; CAS RegistryNumber = 187389-52-2), also known as Z-VAD-FMK and caspase inhibitor VI, is a broad-spectrum caspase inhibitor that blocks caspase-mediated apoptosis. Z-VAD-FMK inhibits caspase processing (IC_{50} = 0.0015–5.8 mM, depending on enzyme and cell type). The aspartate methyl ester derivative (FW = 467.49 g/mol), also known as caspase inhibitor I, is far more cell-permeable. **Target(s):** caspases (1-7,11); caspase-1 (11); caspase-2 (11); caspase-3 (11); caspase-4 (11); caspase 5 (11); caspase-6 (11,14); caspase-7 (11); caspase-8 (11-13); caspase-9 (8,9,11); caspase-10 (10). **1**. Thornberry (1998) in *Handb. Proteolytic Enzymes* (Barrett, Rawlings & Woessner, eds.), p. 737, Academic Press, San Diego. **2**. Denmeade, Lin, Tombal & Isaacs (1999) *Prostate* **39**, 269. **3**. Gastman, Johnson, Whiteside & Rabinowich (1999) *Cancer Res.* **59**, 1422. **4**. Gregoli & Bondurant (1999) *J. Cell Physiol.* **178**, 133. **5**. Fearnhead, Dinsdale & Cohen (1995) *FEBS Lett.* **375**, 283. **6**. Zhu, Fearnhead & Cohen (1995) *FEBS Lett.* **374**, 303. **7**. Amstad, Yu, Johnson, *et al.* (2001) *Biotechniques* **31**, 608. **8**. Li, Nijhawan, Budihardjo, *et al.* (1997) *Cell* **91**, 479. **9**. Sadhukhan, Leone, Lull, *et al.* (2006) *Protein Expr. Purif.* **46**, 299. **10**. Kominami, Takagi, Kurata, *et al.* (2006) *Genes Cells* **11**, 701. **11**. Garcia-Calvo, Peterson, Leiting, *et al.* (1998) *J. Biol. Chem.* **273**, 32608. **12**. Yu, Alva, Su, *et al.* (2004) *Science* **304**, 1500. **13**. Dohrman, Kataoka, Cuenin, *et al.* (2005) *J. Immunol.* **174**, 5270. **14**. Nyormoi, Wang & Bar-Eli (2003) *Apoptosis* **8**, 371.

Benzyloxycarbonyl-L-valyl-L-phenylalaninal

This potent, cell-permeable, aldehyde-containing peptidomimetic (FW = 382.45 g/mol; CAS 88191-84-8; Sequence: z-Val-Phe-al; Solubility = 26 mg/mL DMSO,<<1 mg/mL H_2O), also known as MDL 28170, targets calpain I and II inhibitor, protects rat erythrocyte membrane-associated cytoskeletal proteins from proteolytic degradation (IC_{50} = 1 microM) which occurs when the cells are rendered permeable to Ca^{2+} (1). Calpain I and II is a Ca^{2+}-activated non-lysosomal protease. MDL 28170 reduces capsaicin-mediated cell death in cultured dorsal root ganglion neurons (2). Although calpain 1 and protein kinase C (PKC) is known to form a complex in rabbit skeletal muscle, MDL 28170 does not block the activation of the purified PKC by Ca^{2+} and phosphatidylserine (3). When tested in two models of experimental Parkinson Disorder induced by the neurotoxin 1-methyl-4-phenyl 1,2,3,6-tetrahydropyridine and the environmental toxin, rotenone, treatment with MDL-28170 prevents neuronal death and restore functions (4). **1**. Mehdi, Angelastro, Wiseman & Bey (1988) *Biochem. Biophys. Res. Commun.* **157**, 1117. (*Erratum* in: *Biochem. Biophys. Res. Commun.* **159**, 371.) **2**. Chard, Bleakman, Savidge & Miller (1995) *Neuroscience* **65**, 1099. **3**. Savart, Pallet, Letard, Bossuet & Ducastaing (1991) *Biochimie* **73**, 1409. **4**. Samantaray, Ray & Banik (2008) *CNS Neurol. Disord. Drug Targets* **7**, 305.

(*E*)-7-[4-[4-[[[2-(Benzyloxy)ethyl]amino]carbonyl]-2-oxazolyl]phenyl]-7-(3-pyridyl)-hept-6-enoate

This substituted oxazolecarboxamide ($FW_{free-acid}$ = 525.60 g/mol) inhibits human thromboxane-A synthase (IC_{50} = 48 nM) and is a thromboxane receptor antagonist (K_d = 156 nM). **1**. Takeuchi, Kohn, True, *et al.* (1998) *J. Med. Chem.* **41**, 5362.

5-(4-Benzyloxyphenyl)-4*S*-(7-phenylheptanoylamino)pentanoate

This sPLA$_2$ inhibitor ($FW_{free-acid}$ = 487.64 g/mol), also known as 5-(4-benzyloxyphenyl)-4*S*-(7-phenylheptanoyl-amino)pentanoic Acid, targets secretory phospholipase A$_2$: IC_{50} = 29 nM for the human nonpancreatic enzyme. **1**. Arumugam, Arnold, Proctor, *et al.* (2003) *Brit. J. Pharmacol.* **140**, 71. **2**. Hansford, Reid, Clark, *et al.* (2003) *Chem. Bio. Chem.* **4**, 101.

4-(Benzyloxypropoxy)-3-[(naphth-2-yl)methoxy]piperidine

This substituted piperidine (FW = 481.63 g/mol) is a strong inhibitor of human renin, with an IC_{50} value of 8 nM for the recombinant protein. **1.** Vieira, Binggeli, Breu, *et al.* (1999) *Bioorg. Med. Chem. Lett.* **9**, 1397.

N-(1-Benzyl-4-piperidinyl)-2,4-dichlorobenzamide

This anticonvulsant (FW = 363.28 g/mol) inhibits betaine/GABA transporter-1 (BGT-1), a recently recognized target for treating epilepsy. While a selective noncompetitive inhibitor of the human BGT-1, BPDBA exhibits no significant inhibitory activity against the three human γ-amino-butyrate transporter subtypes GAT-1, GAT-2 and GAT-3. GABA transporters are essential regulators of GABAergic neurons through the continuous uptake of the neurotransmitter from the synaptic cleft and extrasynaptic space. (**See** *Tiagabine*) **1.** Kragholm, Kvist, Madsen, *et al.* (2013) *Biochem. Pharmacol.* **86**, 521.

5'-(Benzylthio)-immucillin-A

This pseudonucleoside analogue (FW = 357.46 g/mol), systematically known as (1*S*)-5-(benzylthio)-1-(9-deazaadenin-9-yl)-1,4-dideoxy-1,4-imino-D-ribitol, inhibits 5'-methylthioadenosine/*S*-adenosylhomocysteine nucleosidase, $K_i' = 12$ pM (1) and *S*-methyl-5'-thioadenosine phosphorylase, $K_i = 26$ nM (2). **1.** Singh, Evans, Lenz, *et al.* (2005) *J. Biol. Chem.* **280**, 18265. **2.** Evans, Furneaux, Schramm, Singh & Tyler (2004) *J. Med. Chem.* **47**, 3275.

N-Benzyl-*N*-(2-toluenesulfonylaminocarbonyl)glycine Hydroxamate

This hydroxamate (FW = 376.41 g/mol) inhibits a number of metalloproteinases: interstitial collagenase ($K_i = 162$ nM), gelatinase A ($K_i = 18$ nM), neutrophil collagenase ($K_i = 25$ nM), gelatinase B ($K_i = 24$ nM), and microbial collagenase ($K_i = 17$ nM). **Target(s):** gelatinase A, *or* matrix metalloproteinase 2; gelatinase B, *or* matrix metalloproteinase 9; interstitial collagenase, *or* matrix metalloproteinase; microbial collagenase; neutrophil collagenase, *or* matrix metalloproteinase 8. **1.** Ilies, Banciu, Scozzafava, *et al.* (2003) *Bioorg. Med. Chem.* **11**, 2227.

Bepridil

This substituted pyrrolidine ($FW_{free-base} = 366.55$ g/mol) is a calcium channel blocker that inhibits the Na^+/Ca^{2+} exchange, exhibiting antiarrhythmic, antianginal, and vasodilatory properties. Bepridil inhibits the growth of certain brain tumors *in vitro*. Bepridil also binds to troponin C. **Target(s):** Ca^{2+} channels (1,2); ATP-sensitive K^+ channels (3); Na^+-activated K^+ channels (3). **1.** Galizzi, Borsotto, Barhanin, Fosset & Lazdunski (1986) *J.*

Biol. Chem. **261**, 1393. **2.** Yatani, Brown & Schwartz (1986) *J. Pharmacol. Exp. Ther.* **237**, 9. **3.** Li, Sato & Arita (1999) *J. Pharmacol. Exp. Ther.* **291**, 562.

Berbamine

This antihypertensive alkaloid (FW = 608.72 g/mol; CAS 478-61-5; *Symbol*: BBM) from *Berberis lyceum* and *Mahonia swaseyi* is a dihydropyridine-class calcium ion blocker (1). Berbamine also inhibits KM3 cell growth by inducing G_1 arrest as well as apoptosis, by blocking NF-κB signaling through up-regulation of A20, down-regulation of IKKα, p-IκBα, and then inhibition of p65 nuclear translocation, all with a resultant decrease in the expression of the downstream targets of NF-κB (2). BBM potently suppresses liver cancer cell proliferation and induces cancer cell death by targeting Ca^{2+}/calmodulin-dependent protein kinase II (CAMKII). BBM likewise inhibits *in vivo* tumorigenicity of liver cancer cells in NOD/SCID mice. These effects are recapitulated by short hairpin RNA-mediated CAMKII knockdown, whereas CAMKII overexpression promotes cancer cell proliferation and increases the resistance of liver cancer cells to BBM. CAMKII was hyperphosphorylated in liver tumors compared with the paired peri-tumor tissues, supporting a role of CAMKII in promoting human liver cancer progression and the clinical potential of BBM in liver cancer therapies (3). **1.** Khan, Qayum & Qureshi (1969) *Life Sci.* **8**, 993. **2.** Liang, Xu, Zhang & Zhao (2009) *Acta Pharmacol Sin.* **30**, 1659. **3.** Meng, Li, Ma, *et al.* (2013) *Mol. Cancer Ther.* **12**, 2067.

Beraprost

This orally active epoprostenol analogue (FW = 398.50 g/mol; CAS 88475-69-8; Soluble to 25 mM in DMSO), also named TRA-418 and 2,3,3a,8b-tetrahydro-2-hydroxy-1-(3-hydroxy-4-methyl-1-octen-6-yn-1-yl)-1*H*-cyclopenta[*b*]benzofuran-5-butanoic acid, is a potent agonist for the Prostacyclin Receptor, a member of the G-protein coupled receptor family. Prostacyclin, the major product of cyclooxygenase in macrovascular endothelium, elicits a potent vasodilation and inhibition of platelet aggregation through binding to this receptor. Beraprost potently inhibits ADP-induced platelet aggregation ($pIC_{50} = 8.26$) and P-selectin expression in vitro ($pIC_{50} = 8.56$). It also increases vasodilation and reduces pulmonary hypertension *in vivo*. TRA-418 inhibited platelet GPIIb/IIIa activation as well as induction of P-selectin expression by adenosine 5'-diphosphate, Thrombin Receptor Agonist Peptide 1-6 (Ser-Phe-Leu-Leu-Arg-Asn-NH₂), and U-46619 in the presence of epinephrine (2). TRA-418 also inhibits platelet aggregation induced by those platelet-stimulants in Ca^{2+}-chelating anticoagulant, citrate and in nonchelating anticoagulant, *d*-phenylalanyl-l-prolyl-l-arginyl-chloromethyl ketone (PPACK). The TP-receptor antagonist SQ-29548 inhibited only U-46619+epinephrine-induced GPIIb/IIIa activation, P-selectin expression, and platelet aggregation. The IP-receptor agonist beraprost sodium inhibited platelet activation. Beraprost also inhibited platelet aggregation induced by platelet stimulants we tested in citrate and in PPACK (2). The GPIIb/IIIa inhibitor abciximab blocked GPIIb/IIIa activation and platelet aggregation. However, abciximab showed slight inhibitory effects on P-selectin expression. TRA-418 is more advantageous as an antiplatelet agent than TP-receptor antagonists or IP-receptor agonists separately used. TRA-418 showed a different inhibitory profile from abciximab in the effects on P-selectin expression (2). **1**. Saito, Tatsumi, Kasahara, Tani & Kuriyama (1995) *Nihon. Kyobu. Shikkan. Gakkai. Zasshi.* **33**, 497. **2**. Miyamoto, Yamada, Ikezawa, *et al.* (2003) *Br. J. Pharmacol.* **140**, 889.

Berberine

This yellow isoquinoline-based alkaloid (FW$_{\text{free-base}}$ = 336.37 g/mol; FW$_{\text{chloride}}$ = 371.82 g/mol; CAS 2086-83-1) is an acetylcholinesterase inhibitor that also induces apoptosis in promyelocytic leukemia HL-60 cells, modulates the expression and function of P-glycoprotein 170 (multi-drug resistance transporter) in hepatoma cells, and inhibits platelet aggregation induced by ADP, arachidonate, and collagen in rats. It is a fluorescent stain for heparin in mast cells. Berberine inhibits β-catenin transcriptional activity and attenuates anchorage-independent growth, reducing cellular levels of active β-catenin and increasing the expression of E-cadherin (1). A screen of synthetic 13-arylalkyl berberine derivatives identified compounds exhibiting activities superior to those of the naturally occurring parent substance with more than 100-fold lower EC$_{50}$ values for Wnt-repression. Berberine is also a partial agonist of α$_2$ adrenoceptors. With rat cerebral cortex synaptosomes, berberine inhibits the release of glutamate evoked by the K$^+$ channel blocker 4-aminopyridine, an effect that can be prevented by the chelating extracellular Ca^{2+} ions or by the vesicular transporter inhibitor bafilomycin A$_1$ (2). Berberine inhibition is insensitive to the glutamate transporter inhibitor DL-*threo*-β-benzyl-oxyaspartate. The inhibitory effect of berberine on glutamate release is associated with a reduction in the depolarization-induced increase in cytosolic free Ca^{2+} (2). Berberine is a component of the Chinese herb medicine Huanglian (*Coptis chinensis*). *Note*: Tetrahydroberberine is canadine (*See Canadine*). **Target(s)**: lactate dehydrogenase (3); malate dehydrogenase (4); acetylcholinesterase (4,7); alcohol dehydrogenase (5,10); (*RS*)-norcoclaurine 6-*O*-methyltransferase (6); electron transport (8); Na$^+$/K$^+$-exchanging ATPase (9); diamine oxidase (11); microbial collagenase (12); elastase (13); K$^+$ channels (IK1, IK, and HERG channels) (14,16,18,24); HIV-1 reverse transcriptase (15); β-fructofuranosidase, *or* invertase (17); Ca^{2+} channels, L- and T-type (19); arylamine *N*-acetyltransferase (20,21,22); lipase, *or* triacylglycerol lipase (22); monoamine oxidase (23); telomerase (25); sortase (26); reticuline oxidase (27); mitogen-activated protein kinase kinase (28); (*S*)-tetrahydroprotoberberine *N*-methyltransferase (30); (*S*)-scoulerine 9-*O*-methyltransferase (31); 3'-hydroxy-N-methyl-(*S*)-coclaurine 4'-*O*-methyltransferase (32). **1**. Albring, Weidemüller, Mittag, *et al.* (2013) *Biofactors* **39**, 652. **2**. Lin, Lin, Lu, Huang & Wang (2013) *PLoS One* **8**, e67215. **3**. Kapp & Whiteley (1991) *J. Enzyme Inhib.* **4**, 233. **4**. Whiteley & Daya (1995) *J. Enzyme Inhib.* **9**, 285. **5**. Brändén, Jörnvall, Eklund & Furugren (1975) *The Enzymes*, 3rd ed. (Boyer, ed.), **11**, 103. **6**. Sato, Tsujita, Katagiri, Yoshida & Yamada (1994) *Eur. J. Biochem.* **225**, 125. **7**. Sobotka & Antopol (1937) *Enzymologia* **4**, 189. **8**. Schewe & Muller (1976) *Acta Biol. Med. Ger.* **35**, 1019. **9**. Lee, MacFarlane, Zee-Cheng & Cheng (1977) *J. Pharm. Sci.* **66**, 986. **10**. Kovar, Durrova & Skursky (1979) *Eur. J. Biochem.* **101**, 507. **11**. Vaidya, Rajagopalan, Kale & Levine (1980) *J. Postgrad. Med.* **26**, 28. **12**. Tanaka, Metori, Mineo, *et al.* (1991) *Yakugaku Zasshi* **111**, 538. **13**. Tanaka, Metori, Mineo, *et al.* (1993) *Planta Med.* **59**, 200. **14**. Hua & Wang (1994) *Yao Xue Xue Bao* **29**, 576. **15**. Gudima, Memelova, Borodulin, *et al.* (1994) *Mol. Biol. (Moscow)* **28**, 1308. **16**. Wang, Zheng & Zhou (1996) *Eur. J. Pharmacol.* **316**, 307. **17**. Rojo, Quiroga, Vattuone & Sampietro (1997) *Biochem. Mol. Biol. Int.* **43**, 1331. **18**. Wu, Yu, Jan, Li & Yu (1998) *Life Sci.* **62**, 2283. **19**. Xu, Zhang, Ren & Zhou (1997) *Zhongguo Yao Li Xue Bao* **18**, 51. **20**. Chung, Wu, Chu, *et al.* (1999) *Food Chem. Toxicol.* **37**, 319. **21**. Lin, Chung, Wu, *et al.* (1999) *Amer. J. Chin. Med.* **27**, 265. **22**. Grippa, Valla, Battinelli, *et al.* (1999) *Biosci. Biotechnol. Biochem.* **63**, 1557. **23**. Kong, Cheng & Tan (2001) *Planta Med.* **67**, 74. **24**. Li, Yang, Zhou, Xu & Li (2001) *Acta Pharmacol. Sin.* **22**, 125. **25**. Sriwilaijareon, Petmitr, Mutirangura, Ponglikitmongkol & Wilairat (2002) *Parasitol. Int.* **51**, 99. **26**. Kim, Shin, Oh, *et al.* (2004) *Biosci. Biotechnol. Biochem.* **68**, 421. **27**. Steffens, Nagakura & Zenk

(1985) *Phytochemistry* **24**, 2577. **28**. Liang, Ting, Yin, *et al.* (2006) *Biochem. Pharmacol.* **71**, 806. **29**. Wu, Tsou, Ho, *et al.* (2005) Curr. Microbiol. **51**, 255. **30**. O'Keefe & Beecher (1994) *Plant Physiol.* **105**, 395. **31**. Sato, Takeshita, Fitchen, Fujiwara & Yamada (1993) *Phytochemistry* **32**, 659. **32**. Frenzei & Zenk (1990) *Phytochemistry* **29**, 3505.

Berninamycin A

This fungal antibiotic (FW = 1146.12 g/mol; CAS 58798-97-3; Soluble in DMSO), also known as Antibiotic U27810, from *Streptomyces bernensis* inhibits protein biosynthesis by binding to a complex formed by ribosomal 23S and the L11 protein, thereby disrupting the function of the ribosome's A site and inhibiting protein synthesis in Gram-positive bacteria (1,2). In this respect, it has the same mode of action as thiostrepton. *Streptomyces bernensis* and *Streptomyces azureus*, which produce berninamycin and thiostrepton, respectively, possess similar ribosomal RNA methylases that mediate specific pentose-directed methylation of 23S ribosomal RNA, rendering the modified ribosomes resistant to these antibiotics (2). **1**. Reusser (1969) *Biochemistry* **8**, 3303. **2**. Thompson, Cundliffe & Stark (1982) *J. Gen. Microbiol.* **128**, 875.

Berterion

This sulforaphane analogue and antimetastatic agent (FW = 175.43 g/mol; CAS 4430-42-6; Physical State = Liquid; Soluble in ethanol, DMSO, and chloroform; Store at -20° C; M.P. = 155-156° C), also named 5-methylthiopentyl isothiocyanate, is found in cruciferous vegetables (including Chinese cabbage and rucola leaves) and mustard oil. Berteroin inhibits LPS-induced degradation of Inhibitor of κBα (*or* IκBα) as well as NF-κB p65 translocation into the nucleus and DNA binding. Berteroin also suppresses degradation of IL-1 Receptor Associated Kinase and phosphorylation of Transforming Growth Factor β-Activated Kinase-1. Berteroin also inhibits LPS-induced phosphorylation of p38 MAPK, ERK1/2, and AKT. **1**. Jung, Jung, Cho, *et al.* (2014) *Int. J. Mol. Sci.* **15**, 20686.

Beryllium & Beryllium Ions

Tetraaquo Be(II) Ion   **Be(II)F$_3$·H$_2$O**   **Metaphosphate Anion**

This Group 2 (*or* IIA) alkaline element (Symbol: Be; Atomic Number = 4; Atomic Weight = 9.012182; CAS CAS No. 7440-41-7), positioned directly above magnesium in the Periodic Table, has oxidation states of 0, 1, and 2, with ionic radii of 0.44 and 0.35 Å for Be$^+$ and Be^{2+}, respectively (1-3). Beryllium fluoride ion is frequently employed as an isostere of phosphoryl and metaphosphate intermediates that form by various phosphotransferases, phosphohydrolases, transporters, and molecular motors. **CAUTION:** Epidemiological studies indicate an increased risk of lung cancer in occupational groups exposed to beryllium or beryllium compounds, a finding that is consistent with the carcinogenicity of beryllium and beryllium compounds in humans. While sweet to the taste, is extremely toxic. **Target(s):** myosin ATPase, inhibited by BeF_x (4,17,54); calcium-exporting ATPase (6); β-galactosidase (7,29); peptidyltransferase (8,9); thymidine kinase (10,22); [glycogen synthase] kinase (11); nitrogenase (12); P-glycoprotein ATPase (13); gelsolin severing of F-actin (14); adenylate kinase (15); inositol-phosphate phosphatase, *or myo*-inositol monophosphatase (16,56,57); bromoperoxidase (18); cytochrome P450 (19); casein kinase-1 (20); phosphatase (21); Na^+/K^+-exchanging ATPase (23,24); phosphoglucomutase (25,26,34,50,51); triokinase (27); creatine kinase (28); alkaline phosphatase (30,32,36,40) (42,44,45,47,61); enolase (31); pancreatic ribonuclease, *or* ribonuclease A, by $BeCl_2$, (33); phosphatidate phosphatase, by $BeCl_2$ (35); phospholipase D, by $BeCl_2$ (36); lipase (38); pyrophosphatase, *or* inorganic diphosphatase (39); glucose-6-phosphate isomerase (43); hexokinase (46); phosphomannomutase (48); phosphoacetylglucosamine mutase, *or N*-acetylglucosamine-phosphate mutase (49); pyruvate decarboxylase (52); nitrogenase, by BeF_2 (53); Xaa-Pro Dipeptidase, *or* prolidase, *or* imidodipeptidase (55); histidinol-phosphatase (58); phosphatidate phosphatase (59); acid phosphatase (60). **1.** Vauquelin (1798) *Ann. Chim.* **26**, 155. **2.** Bussy (1828) *J. Chim. Médic.* **4**, 455. **3.** Wöhler (1828) *Ann. Phys.* **13**, 577 and (1828) *Ann. Chim.* **39**, 77. **4.** Maruta, Henry, Sykes & Ikebe (1993) *J. Biol. Chem.* **268**, 7093. **5.** Allen, Lahan, Way & Janmey (1996) *J. Biol. Chem.* **271**, 4665. **6.** Murphy & Coll (1993) *J. Biol. Chem.* **268**, 23307. **7.** Wallenfels (1962) *Meth. Enzymol.* **5**, 212. **8.** Monro (1970) *Meth. Enzymol.* **19**, 797. **9.** Monro (1971) *Meth. Enzymol.* **20**, 472. **10.** Bresnick (1978) *Meth. Enzymol.* **51**, 360. **11.** Ryves, Dajani, Pearl & Harwood (2002) *Biochem. Biophys. Res. Commun.* **290**, 967. **12.** Clarke, Yousafzai & Eady (1999) *Biochemistry* **38**, 9906. **13.** Sankaran, Bhagat & Senior (1997) *Biochemistry* **36**, 6847. **14.** Allen, Laham, Way & Janmey (1996) *J. Biol. Chem.* **271**, 4665. **15.** Garin & Vignais (1993) *Biochemistry* **32**, 6821. **16.** Faraci, Zorn, Bakker, Jackson & Pratt (1993) *Biochem. J.* **291**, 369. **17.** Phan & Reisler (1992) *Biochemistry* **31**, 4787. **18.** Tromp, Van de Wever (1991) *Biochim. Biophys. Acta* **1079**, 53. **19.** Teixeira, Yasaka, Silva, Oshiro & Oga (1990) *Toxicology* **61**, 293. **20.** Cummings, Kaser, Wiggins, Ord & Stocken (1982) *Biochem. J.* **208**, 141. **21.** Toda, Koide & Yoshitoshi (1971) *J. Biochem.* **69**, 73. **22.** Mainigi & Bresnick (1969) *Biochem. Pharmacol.* **18**, 2003. **23.** Toda (1968) *J. Biochem.* **64**, 457. **24.** Toda, Hashimoto, Asakura & Minakami (1967) *Biochim. Biophys. Acta* **135**, 570. **25.** Aldridge & Thomas (1966) *Biochem. J.* **98**, 100. **26.** Joshi (1982) *Meth. Enzymol.* **89**, 599. **27.** Bystrykh, de Koning & Harder (1990) *Meth. Enzymol.* **188**, 445. **28.** O'Sullivan & Morrison (1963) *Biochim. Biophys. Acta* **77**, 142. **29.** Wallenfels & Malhotra (1960) *The Enzymes*, 2nd ed. (Boyer, Lardy & Myrbäck, eds.), **4**, 409. **30.** Stadtman (1961) *The Enzymes*, 2nd ed. (Boyer, Lardy & Myrbäck, eds.), **5**, 55. **31.** Malmström (1961) *The Enzymes*, 2nd ed. (Boyer, Lardy & Myrbäck, eds.), **5**, 471. **32.** Fernley (1971) *The Enzymes*, 3rd ed. (Boyer, ed.), **4**, 417. **33.** Richards & Wyckoff (1971) *The Enzymes*, 3rd ed. (Boyer, ed.), **4**, 647. **34.** Ray & Peck (1972) *The Enzymes*, 3rd ed. (Boyer, ed.), **6**, 407. **35.** Esko & Raetz (1983) *The Enzymes*, 3rd ed. (Boyer, ed.), **16**, 207. **36.** Plocke & Vallee (1962) *Biochemistry* **1**, 1039. **37.** Kanfer & McCartney (1994) *FEBS Lett.* **337**, 251. **38.** Yang & Hsu (1944) *J. Biol. Chem.* **155**, 137. **39.** Naganna & Menon (1948) *J. Biol. Chem.* **174**, 501. **40.** Klemperer, Miller & Hill (1949) *J. Biol. Chem.* **180**, 281. **41.** Grier, Hood & Hoagland (1949) *J. Biol. Chem.* **180**, 289. **42.** Varma & Srinivasan (1954) *Enzymologia* **17**, 116. **43.** Alvarado (1963) *Enzymologia* **26**, 12. **44.** Spencer & Macrae (1972) *Enzymologia* **42**, 329. **45.** Lindenbaum, White, Schubert & Graves (1952) *J. Biol. Chem.* **196**, 273. **46.** Sols & Crane (1954) *J. Biol. Chem.* **206**, 925. **47.** Schubert & Limdenbaum (1954) *J. Biol. Chem.* **208**, 359. **48.** Murata (1976) *Plant Cell Physiol.* **17**, 1099. **49.** Fernandez-Sorensen & Carlson (1971) *J. Biol. Chem.* **246**, 3485. **50.** Takamiya & Fukui (1978) *J. Biochem.* **84**, 569. **51.** Daugherty, Kraemer & Joshi (1975) *Eur. J. Biochem.* **57**, 115. **52.** Leblova, M. Malik & M. Fojta (1989) *Biologia (Bratisl.)* **44**, 329. **53.** Igarashi & Seefeldt (2003) *Crit. Rev. Biochem. Mol. Biol.* **38**, 351. **54.** Park, Ajtai & Burghardt (1999) *Biochim. Biophys. Acta* **1430**, 127. **55.** Fujii, Nagaoka, Imamura & Shimizu (1996) *Biosci. Biotechnol. Biochem.* **60**, 1118. **56.** Creba, Carey & McCulloch (1988) *Biochem. Soc. Trans.* **16**, 557. **57.** Nigou, Dover & Besra (2002) *Biochemistry* **41**, 4392. **58.** Houston & Graham (1974) *Arch. Biochem. Biophys.* **162**, 513. **59.** Spitzer & Johnston (1978) *Biochim. Biophys. Acta* **531**, 275. **60.** Andrews & Pallavicini (1973) *Biochim. Biophys. Acta* **321**, 197. **61.** Ezawa, Kuwahara, Sakamoto, Yoshida & Saito (1999) *Mycologia* **91**, 636.

Beryllium Fluoride, *See Beryllium and Beryllium Ions; Fluorine and Fluoride Ions*

Besifloxacin

This fourth-generation, fluoroquinolone-class antibiotic (FW = 393.84 g/mol; CAS 141388-76-3), also known as Besivance®, BOL-303224-A, SS734, and 7-[(3*R*)-3-aminoazepam-1-yl]-8-chloro-1-cyclopropyl-6-fluoro-4-oxo-1,4-dihydroquinoline-3-carboxylic acid, displays activity in *in vitro* antimicrobial efficacy studies as well as efficacy in an *in vivo* murine infection model (1). BOL-303224-A demonstrates excellent ocular pharmacokinetics in rabbits, with ocular mean residence times >7 hours, reaching conjunctival concentrations that exceed the MIC_{90} for nonresistant ophthalmic isolates for >12 hours following a single dose (1). Besifloxacin acts as an anti-inflammatory agent in monocytes *in vitro*, a property that may enhance its efficacy in ocular infections (2). It exhibits broad-spectrum *in vitro* activity against aerobic and anaerobic bacteria, based on studies on 2,690 clinical isolates representing 40 species (3). **1.** Ward, Lepage & Driot (2007) *J. Ocul. Pharmacol. Ther.* **23**, 243. **2.** Zhang & Ward (2008) *J. Antimicrob. Chemother.* **61**, 111. **3.** Haas, Pillar, Zurenko, *et al.* (2009) *Antimicrob. Agents Chemother.* **53**, 3552.

Bestatin

This potent aminopeptidase inhibitor (FW = 308.38 g/mol; CASs = 8970-76-6 (free acid) and 65391-42-6 (HCl salt); pK_1 = 3.1; pK_2 = 8.1), also known by the nonproprietary name, ubenimex, and systematically as (2*S*,3*R*)-3-amino-2-hydroxy-4-phenylbutanoyl)-L-leucine, is usually effective in the 1–25 nM range, and, except for nardilysin, is without effect on endoproteinases. Bestatin inhibits the enzymatic degradation of oxytocin, vasopressin, and enkephalins. It also potently inhibits arginyl aminopeptidase (*or* aminopeptidase B), leukotriene A_4 hydrolase, alanyl aminopeptidase (*or* aminopeptidase M/N), leucyl/cystinyl aminopeptidase (*or* oxytocinase, vasopressinase), and membrane dipeptidase (*or* leukotriene D_4 hydrolase). Aqueous solutions are stable for about one day, but 1 mM stock solutions in methanol are stable for one month at –20°C. The D-leucine analogue is a slightly weaker inhibitor of leucyl aminopeptidase and aminopeptidase B. The (2*S*,3*S*)-3-amino-2-hydroxy-4-phenylbutanoyl-L-leucine and D-leucine stereoisomers also inhibit these two aminopeptidases. **Target(s):** membrane alanyl aminopeptidase, *or* aminopeptidase M, *or* aminopeptidase N (1,2,21,24,27,29,43,73,83,108-120); leucyl amino-peptidase (3,11,12,20,24,39,57,58,73,96,102, 104,121-137); bacterial pyroglutamyl-peptidase (4); cystinyl aminopeptidase, *or* oxytocinase (5,104-107); yeast aminopeptidase Ape2 (7); leukotriene-A4 hydrolase (8,23,28,138-144); aminopeptidase B, also inhibited by the 3*S*-amino-and-D-leucine stereoisomer (9,20,22,24,82-96); aminopeptidase Ey (10,63); aminopeptidase Y, *or* aminopeptidase Co (13,67,68); bacterial leucyl aminopeptidase, *or* *Vibrio* aminopeptidase (14,70-74); aminopeptidase T (15); tripeptide aminopeptidase, *or* tripeptidyl-peptidase (16,47,98-103); Xaa-Trp aminopeptidase, *or* aminopeptidase W (17,29,65,66); cytosol nonspecific dipeptidase (18,30,51,102); nardilysin (19,32-34); mouse ascites tumor dipeptidase (24,25); lysine-specific aminopeptidase (26); glutamyl aminopeptidase, *or* aminopeptidase A (29,81); biotinidase (31); mitochondrial intermediate peptidase (35); envelysin (36); cathepsin H (37); chymase (38); glutamate carboxypeptidase II, *or N*-acetylated-γ-linked-acidic dipeptidase (40); tripeptidyl-peptidase I, weakly inhibited (41); dipeptidyl-peptidase IV (42); dipeptidyl-peptidase III, weakly inhibited (44); dipeptidyl-peptidase II, weakly inhibited (45,46); β-Ala-His Dipeptidase, *or* carnosinase (48); membrane Dipeptidase, *or* renal Dipeptidase, *or* dehydropeptidase (49; however, see 50); Xaa-Pro Dipeptidase, *or* prolidase, *or* imidodipeptidase (52.77); Xaa-methyl-His

Dipeptidase, *or* anserinase, *or* *N*-acetylhistidine deacetylase, weakly inhibited (53); PepB aminopeptidase (peptidase B) (54); aminopeptidase (17,55,56,59-62); methionyl aminopeptidase (64); cytosol alanyl aminopeptidase (69); Xaa-Pro aminopeptidase, *or* aminopeptidase P (75-80); prolyl aminopeptidase, *or* *Aneurinibacillus* sp. (97); leucyl aminopeptidase/cystinyl aminopeptidase (104); *Salmonella enterica serovar Typhimurium* Peptidase B (137). **1**. Tate (1985) *Meth. Enzymol.* **113**, 471. **2**. Turner (1998) in *Handb. Proteolytic Enzymes* (Barrett, Rawlings & Woessner, eds.), p. 996, Academic Press, San Diego. **3**. Umezawa, Aoyagi, Suda, Hamada & Takeuchi (1976) *J. Antibiot.* **29**, 97. **4**. Robert-Baudouy, Clauziat & Thierry (1998) in *Handb. Proteolytic Enzymes* (Barrett, Rawlings & Woessner, eds.), p. 791, Academic Press, San Diego. **5**. Mizutani, Tsujimoto & Nakazato (1998) in *Handb. Proteolytic Enzymes* (Barrett, Rawlings & Woessner, eds.), p. 1008, Academic Press, San Diego. **6**. Dando & Barrett (1998) in *Handb. Proteolytic Enzymes* (Barrett, Rawlings & Woessner, eds.), p. 1013, Academic Press, San Diego. **7**. Caprioglio (1998) in *Handb. Proteolytic Enzymes* (Barrett, Rawlings & Woessner, eds.), p. 1016, Academic Press, San Diego. **8**. Haeggström (1998) in *Handb. Proteolytic Enzymes* (Barrett, Rawlings & Woessner, eds.), p. 1022, Academic Press, San Diego. **9**. Foulon, Cadel & Cohen (1998) in *Handb. Proteolytic Enzymes* (Barrett, Rawlings & Woessner, eds.), p. 1026, Academic Press, San Diego. **10**. Ichishima (1998) in *Handb. Proteolytic Enzymes* (Barrett, Rawlings & Woessner, eds.), p. 1030, Academic Press, San Diego. **11**. Sträter & Lipscomb (1998) in *Handb. Proteolytic Enzymes* (Barrett, Rawlings & Woessner, eds.), p. 1384, Academic Press, San Diego. **12**. Wood (1998) in *Handb. Proteolytic Enzymes* (Barrett, Rawlings & Woessner, eds.), p. 1389, Academic Press, San Diego. **13**. Yasuhara (1998) in *Handb. Proteolytic Enzymes* (Barrett, Rawlings & Woessner, eds.), p. 1429, Academic Press, San Diego. **14**. Chevier & D'Orchymont (1998) in *Handb. Proteolytic Enzymes* (Barrett, Rawlings & Woessner, eds.), p. 1433, Academic Press, San Diego. **15**. Motoshima & Kaminogawa (1998) in *Handb. Proteolytic Enzymes* (Barrett, Rawlings & Woessner, eds.), p. 1452, Academic Press, San Diego. **16**. Harada (1998) in *Handb. Proteolytic Enzymes* (Barrett, Rawlings & Woessner, eds.), p. 1510, Academic Press, San Diego. **17**. Hooper (1998) in *Handb. Proteolytic Enzymes* (Barrett, Rawlings & Woessner, eds.), p. 1513, Academic Press, San Diego. **18**. Bauer (1998) in *Handb. Proteolytic Enzymes* (Barrett, Rawlings & Woessner, eds.), p. 1520, Academic Press, San Diego. **19**. Cohen, Pierotti, Chesneau, Foulon & Prat (1995) *Meth. Enzymol.* **248**, 703. **20**. Mathe (1991) *Biomed. Pharmacother.* **45**, 49. **21**. Rich, Moon & Harbeson (1984) *J. Med. Chem.* **27**, 417. **22**. Kawata, Takayama, Ninomiya & Makisumi (1980) *J. Biochem.* **88**, 1601. **23**. Evans & Kargman (1992) *FEBS Lett.* **297**, 139. **24**. Taylor (1993) *FASEB J.* **7**, 290. **25**. Patterson (1989) *J. Biol. Chem.* **264**, 8004. **26**. Hui & Hui (2006) *Neurochem. Res.* **31**, 95. **27**. Bauvois & Dauzonne (2006) *Med. Res. Rev.* **26**, 88. **28**. Andberg, Wetterholm, Medina & Haeggström (2000) *Biochem. J.* **345**, 621. **29**. Tieku & Hooper (1992) *Biochem. Pharmacol.* **44**, 1725. **30**. Peppers & Lenney (1988) *Biol. Chem. Hoppe-Seyler* **369**, 1281. **31**. Oizumi & Hayakawa (1991) *Biochim. Biophys. Acta* **1074**, 433. **32**. Chesneau, Pierotti, Barre, *et al.* (1994) *J. Biol. Chem.* **269**, 2056. **33**. Foulon, Cadel, Prat, *et al.* (1997) *Ann. Endocrinol.* **58**, 357. **34**. Gluschankof, Gomez, Morel & Cohen (1987) *J. Biol. Chem.* **262**, 9615. **35**. Kalousek, Isaya & Rosenberg (1992) *EMBO J.* **11**, 2803. **36**. Fan & Katagiri (2001) *Eur. J. Biochem.* **268**, 4892. **37**. Raghav, Kamboj, Parnami & Singh (1995) *Indian J. Biochem. Biophys.* **32**, 279. **38**. Kido, Fukusen & Katunuma (1985) *Arch. Biochem. Biophys.* **239**, 436. **39**. Mantle, Lauffart & Gibson (1991) *Clin. Chim. Acta* **197**, 35. **40**. Robinson, Blakely, Couto & Coyle (1987) *J. Biol. Chem.* **262**, 14498. **41**. Vines & Warburton (1998) *Biochim. Biophys. Acta* **1384**, 233. **42**. Malík, Busek, Mares, *et al.* (2003) *Adv. Exp. Med. Biol.* **524**, 95. **43**. Vanha-Perttula (1988) *Clin. Chim. Acta* **177**, 179. **44**. Abramic, Schleuder, Dolovcak, *et al.* (2000) *Biol. Chem.* **381**, 1233. **45**. Sakai & Kojima (1987) *J. Chromatogr.* **416**, 131. **46**. Mentlein & Struckhoff (1989) *J. Neurochem.* **52**, 1284. **47**. Mantle (1991) *Clin. Chim. Acta* **196**, 135. **48**. Otani, Okumura, Hashida-Okumura & Nagai (2005) *J. Biochem.* **137**, 167. **49**. Campbell, Di Shih, Forrester & Zahler (1988) *Biochim. Biophys. Acta* **956**, 110. **50**. McIntyre & Curthoys (1982) *J. Biol. Chem.* **257**, 11915. **51**. Lenney (1990) *Biol. Chem. Hoppe-Seyler* **371**, 167 and 433. **52**. Morel, Frot-Coutaz, Aubel, Portalier & Atlan (1999) *Microbiology* **145**, 437. **53**. Yamada, Tanaka, Sameshima & Furuichi (1993) *Comp. Biochem. Physiol. B Comp. Biochem.* **106**, 309. **54**. Mathew, Knox & Miller (2000) *J. Bacteriol.* **182**, 3383. **55**. Tisljar & Wolf (1993) *FEBS Lett.* **322**, 191. **56**. Millership, Chappell, Okhuysen & Snowden (2002) *J. Parasitol.* **88**, 843. **57**. Taylor, Peltier, Jahngen, *et al.* (1992) *Biochemistry* **31**, 4141. **58**. Nankya-Kitaka, Curley, Gavigan, Bell & Dalton (1998) *Parasitol. Res.* **84**, 552. **59**. Nishiwaki & Hayashi (2001) *Biosci.* *Biotechnol. Biochem.* **65**, 424. **60**. Desimone, Kruger, Wessel, *et al.* (2000) *J. Chromatogr. B* **737**, 285. **61**. Cristofoletti & Terra (2000) *Biochim. Biophys. Acta* **1479**, 185. **62**. Noguchi, Nagata, Koganei, *et al.* (2002) *J. Agric. Food Chem.* **50**, 3540. **63**. Ichishima, Yamagata, Chiba, Sawaguchi & Tanaka (1989) *Agric. Biol. Chem.* **53**, 1867. **64**. Zhang, Huang, Cali, *et al.* (2005) *Folia Parasitol.* **52**, 182. **65**. Gee & Kenny (1987) *Biochem. J.* **246**, 97. **66**. Tieku & Hooper (1993) *Biochem. Soc. Trans.* **21**, 250S. **67**. Achstetter, Ehmann & Wolf (1982) *Biochem. Biophys. Res. Commun.* **109**, 341. **68**. Yasuhara, Nakai & Ohashi (1994) *J. Biol. Chem.* **269**, 13644. **69**. Yamamoto, Li, Huang, Ohkubo & Nishi (1998) *Biol. Chem.* **379**, 711. **70**. Izawa, Ishikawa, Tanokura, Ohta & Hayashi (1997) *J. Agric. Food Chem.* **45**, 4897. **71**. Bertin, Lozzi, Howell, *et al.* (2005) *Infect. Immun.* **73**, 2253. **72**. Deejing, Yoshimune, Lumyong & Moriguchi (2005) *J. Ind. Microbiol. Biotechnol.* **32**, 269. **73**. Wilkes & Prescott (1985) *J. Biol. Chem.* **260**, 13154. **74**. Dong, Cheng, Wang, *et al.* (2005) *Microbiology* **151**, 2017. **75**. McDonnell, Fitzgerald, Fhaoláin, Jennings & O'Cuinn (1997) *J. Dairy Res.* **64**, 399. **76**. Lasch, Koelsch, Steinmetzer, Neumann & Demuth (1988) *FEBS Lett.* **227**, 171. **77**. Harbeck & Mentlein (1991) *Eur. J. Biochem.* **198**, 451. **78**. Matsumoto, Erickson & Kim (1995) *J. Nutr. Biochem.* **6**, 104. **79**. Hooper, Hryszko & Turner (1990) *Biochem. J.* **267**, 509. **80**. Achstetter, Ehmann & Wolf (1983) *Arch. Biochem. Biophys.* **226**, 292. **81**. Iturrioz, Reaux-Le Goazigo & Llorens-Cortes (2004) in *Aminopeptidases in Biology and Disease* (Hooper & Lendaechel, eds.), vol. **2**, 229, Springer. **82**. Sharma, Padwal-Desai & Ninjoor (1989) *Biochem. Biophys. Res. Commun.* **159**, 464. **83**. Ishiura, Yamamoto, Yamamoto, Nojima, Aoyagi & Sugita (1987) *J. Biochem.* **102**, 1023. **84**. Tanioka, Hattori, Masuda, *et al.* (2003) *J. Biol. Chem.* **278**, 32275. **85**. Cadel, Piesse, Gouzy-Darmon, Cohen & Foulon (2004) in *Proteases in Biology and Disease* (Hooper & Lendeckel, eds.) vol. **2**, 113, Springer. **86**. Mercado-Flores, Noriega-Reyes, Ramirez-Zavala, Hernandez-Rodriguez & Villa-Tanaca (2004) *FEMS Microbiol. Lett.* **234**, 247. **87**. Millership, Chappell, Okhuysen & Snowden (2002) *J. Parasitol.* **88**, 843. **88**. Cadel, Pierotti, Foulon, *et al.* (1995) *Mol. Cell. Endocrinol.* **110**, 149. **89**. Goldstein, Nelson, Kordula, Mayo & Travis (2002) *Infect. Immun.* **70**, 836. **90**. Petrov, Fagard & Lijnen (2004) *Cardiovasc. Res.* **61**, 724. **91**. Kawata, Takayama, Ninomiya & Makisumi (1980) *J. Biochem.* **88**, 1601. **92**. Nagata, Mizutani, Nomura, *et al.* (1991) *Enzyme* **45**, 165. **93**. Belbacene, Mari, Rossi & Auberger (1993) *Eur. J. Immunol.* **23**, 1948. **94**. Mantle, Lauffart & Pennington (1984) *Biochem. Soc. Trans.* **12**, 826. **95**. Yamada, Sukenaga, Fujii, Abe & Takeuchi (1994) *FEBS Lett.* **342**, 53. **96**. Suda, Aoyagi, Takeuchi & Umezawa (1976) *Arch. Biochem. Biophys.* **177**, 196. **97**. Murai, Tsujimoto, Matsui & Watanabe (2004) *J. Appl. Microbiol.* **96**, 810. **98**. Morgan & Donlan (1985) *Eur. J. Biochem.* **146**, 429. **99**. Sanz, Mulholland & Toldra (1998) *J. Agric. Food Chem.* **46**, 349. **100**. Hayashi & Oshima (1980) *J. Biochem.* **87**, 1403. **101**. Hiraoka & Harada (1993) *Mol. Cell. Biochem.* **129**, 87. **102**. Frick & Wolfenden (1985) *Biochim. Biophys. Acta* **829**, 311. **103**. Sachs & Marks (1982) *Biochim. Biophys. Acta* **706**, 229. **104**. Matsumoto, Rogi, Yamashiro, *et al.* (2000) *Eur. J. Biochem.* **267**, 46. **105**. Itoh & Nagamatsu (1995) *Biochim. Biophys. Acta* **1243**, 203. **106**. Yamamoto, Ishiura & Sugita (1988) *Biomed. Res.* **9**, 11. **107**. Tisljar & Wolf (1993) *FEBS Lett.* **322**, 191. **108**. Xu & Li (2005) *Curr. Med. Chem. Anticancer Agents* **5**, 281. **109**. Bauvois & Dauzonne (2006) *Med. Res. Rev.* **26**, 88. **110**. Hua, Tsukamoto, Taguchi, *et al.* (1998) *Biochim. Biophys. Acta* **1383**, 301. **111**. Jamadar, Jamdar, Dandekar & Harikumar (2003) *J. Food Sci.* **68**, 438. **112**. Contreras-Rodriguez, Ramirez-Zavala, Contreras, *et al.* (2003) *Infect. Immun.* **71**, 5238. **113**. Yamauchi, Ejiri & Tanaka (2001) *Biosci. Biotechnol. Biochem.* **65**, 2802. **114**. Fetterer, Miska & Barfield (2005) *J. Parasitol.* **91**, 1280. **115**. Yoshimoto, Tamesa, Gushi, Murayama & Tsuru (1988) *Agric. Biol. Chem.* **52**, 217. **116**. Luciani, Marie-Claire, Ruffet, *et al.* (1998) *Biochemistry* **37**, 686. **117**. Huang, Takahara, Kinouchi, *et al.* (1997) *J. Biochem.* **122**, 779. **118**. Terashima & Bunnett (1995) *J. Gastroenterol.* **30**, 696. **119**. Thielitz, Bukowska, Wolke, *et al.* (2004) *Biochem. Biophys. Res. Commun.* **321**, 795. **120**. Gros, Giros & Schwartz (1985) *Biochemistry* **24**, 2179. **121**. Millership, Chappell, Okhuysen & Snowden (2002) *J. Parasitol.* **88**, 843. **122**. Acosta, Goni & Carmona (1998) *J. Parasitol.* **84**, 1. **123**. Ichishima, Yamagata, Chiba, Sawaguchi & T. Tanaka (1989) *Agric. Biol. Chem.* **53**, 1867. **124**. Lauffart, McDermott, Gibson & Mantle (1988) *Biochem. Soc. Trans.* **16**, 850. **125**. Kohno, Kanda & Kanno (1986) *J. Biol. Chem.* **261**, 10744. **126**. Morty & Morehead (2002) *J. Biol. Chem.* **277**, 26057. **127**. Gu, Holzer & Walling (1999) *Eur. J. Biochem.* **263**, 726. **128**. McCarthy, Stack, Donnelly, *et al.* (2004) *Int. J. Parasitol.* **34**, 703. **129**. Xu, Shawar & Dresden (1990) *Exp. Parasitol.* **70**, 124. **130**. Kumagai, Watanabe & Fujimoto (1991) *Biochem. Med. Metab. Biol.* **46**, 110. **131**. Gibson, Biggins, Lauffart, Mantle & McDermott (1991) *Neuropeptides* **19**, 163. **132**. Sträter & W. N. Lipscomb (2004) in *Handbook of*

Metalloproteins (Messerschmidt, Huber, Wieghardt & Poulos, eds.) **3**, 199, Wiley, New York. **133.** Mikkonen (1992) *Physiol. Plant* **84**, 393. **134.** Burley, David & Lipscomb (1991) *Proc. Natl. Acad. Sci. U.S.A.* **88**, 6916. **135.** Gardiner, Trenholme, Skinner-Adams, Stack & Dalton (2006) *J. Biol. Chem.* **281**, 1741. **136.** Jösch, Klotz & Sies (2003) *Biol. Chem.* **384**, 213. **137.** Mathew, Knox & Miller (2000) *J. Bacteriol.* **182**, 3383. **138.** Mueller, Samuelsson & Haeggström (1995) *Biochemistry* **34**, 3536. **139.** Haeggström, Kull, Rudberg, Tholander & Thunnissen (2002) *Prostaglandins* **68-69**, 495. **140.** Tsuge, Ago, Aoki, *et al.* (1994) *J. Mol. Biol.* **238**, 854. **141.** Kull, Ohlson & Haeggström (1999) *J. Biol. Chem.* **274**, 34683. **142.** Chen, Li, Wang, *et al.* (2003) *J. Natl. Cancer Inst.* **95**, 1053. **143.** Orning, Krivi & Fitzpatrick (1991) *J. Biol. Chem.* **266**, 1375. **144.** Clamagirand, Cadel, Barre & Cohen (1998) *FEBS Lett.* **433**, 68.

Betahistine

This vasodilator (FW$_{free-base}$ = 136.20 g/mol; CAS 5638-76-6) is a structural analogue of histamine, a weak histamine H$_1$ receptor agonist, and a more potent H$_3$ receptor antagonist (1). Betahistine improves vestibular compensation in animal models of unilateral vestibular dysfunction. Betahistine has been used to treat Meniere's disease, vertigo, and peripheral vascular arterial disease (2). **1.** Lacour & Sterkers (2001) *CNS Drugs* **15**, 853. **2.** Allison & Barnes (1971) *Tex. Med.* **67**, 85.

Betaxolol

This selective β$_1$-receptor blocker (FW = 307.43 g/mol; CAS 63659-18-7; IUPAC: (*RS*)-1-{4-[2-(cyclopropylmethoxy)ethyl]phenoxy}-3-(isopropyl amino)propan-2-ol), also known by the trade names Betoptic®, Betoptic S®, Lokren®, and Kerlone®, is used to treatment of hypertension and glaucoma (1). Given its selective, it has fewer systemic side-effects than non-selective β-blockers, for example, not causing bronchospasm (mediated by beta2 receptors) as timolol may. Betaxolol shows greater affinity for β-receptors than metoprolol. In addition to its effect on the heart, betaxolol reduces the intraocular pressure, an effect thought to be caused by reducing the production of the aqueous humor within the eye. Reduction in intraocular pressure reduces the risk of damage to the optic nerve and loss of vision in patients with elevated intraocular pressure due to glaucoma. **1.** Buckley, Goa & Clissold, (1990) *Drugs* **40**, 75.

Bevacizumab

This humanized monoclonal antibody (MW = 149.2 kDa; CAS 216974-75-3), known by the tradename Avastin®, is an angiogenesis inhibitor that targets vascular endothelial growth factor A (VEGF-A). Because VEGF is the key angiogenic factor in tumors, blocking VEGF signal transduction can lead to tumor growth arrest and inhibition of metastasis. The rationale for Avastin therapy is premised on findings that high VEGF expression correlates with (a) reduced overall survival, (b) disease progression, (c) greater risk of relapse, (d) lymph node involvement, and (e) malignant pleural effusion. By binding directly bind to VEGF-A, Avastin blocks its interaction with endothelial cell VEGF receptors, thereby inhibiting neovascularization and depriving cancer cells of vital nutrients and oxygen. Avastin is approved for: *metastatic colorectal cancer* (mCRC), when started with the first or second intravenous 5-FU–based chemotherapy for metastatic cancer; advanced-stage *nonsquamous, non–small cell lung cancer* (NSCLC), when administered in combination with carboplatin and paclitaxel in patients who have not received chemotherapy for their advanced disease; *metastatic renal cell cancer* (mRCC) when used with interferon-α; and *glioblastoma multiforme* (GBM) in adult patients whose cancer has progressed after prior treatment.

Bevantolol

This antihypertensive agent (FW = 345.43 g/mol; CAS 59170-23-9), also known by the code name NC-1400 and systematically as (*RS*)-[2-(3,4-dimethoxyphenyl)ethyl]-[2-hydroxy-3-(3-methylphenoxy)propyl]amine, is a dual-action β-blocker and calcium ion channel blocker. Bevantolol is a selective β$_1$-adrenoceptor blocker with some blocking activity of α$_1$-adrenoceptors (1). **1.** Gross, Buck, Warltier & Hardman (1979) *J. Cardiovasc. Pharmacol.* **1**, 139. **2.** Takayanagi, Kizawa, Iwasaki & Nakagoshi (1987) *Gen. Pharmacol.* **18**, 87.

Bevirimat

This first-in-class, natural product-based antiviral (FW = 584.83 g/mol; CAS 174022-42-5; *Abbreviation*: BVM), also known as 3-*O*-(3',3'-dimethylsuccinyl)betulinate (DSB), 3β-(3-carboxy-3-methyl-butanoyloxy)-lup-20(29)-en-28-oic acid, PA-457, and MPC-4326, is a novel HIV-1 maturation inhibitor that blocks proteolytic processing of the Gag capsid precursor (CA-SP1) into mature capsid (CA) protein. Bevirimat binds to the Gag polypeptide at the CA/SP1 cleavage site. The net result of BVM treatment is the release of immature, noninfectious viral particles. Despite its promising mode of action, about half of all patients have viruses containing genetic polymorphisms in the Gag SP1 (at positions 6 to 8) protein do not respond to BVM treatment. **1.** Lu, Salzwedel, Wang, *et al.* (2011) *Antimicrob. Agents Chemother.* **55**, 3324.

Bexarotene

This third-generation retinoid receptor X agonist and antineoplastic agent (FW = 348.48 g/mol; CAS 153559-49-0), also named LGD1069, Targretin® and 4-[1-(5,6,7,8-tetrahydro-3,5,5,8,8-pentamethyl-2-naphthalenyl)ethenyl] benzoate, is a first-line therapy for early-stage Cutaneous T-Cell Lymphomas (CTCLs), a group of relatively uncommon lymphoproliferative disorders characterized by the presence of a malignant population of T cells that is localized to the skin at presentation (1). Bexarotene is a retinoid X receptor (RXR) agonist (EC$_{50}$ = 24, 25 and 33 nM for RXRβ, RXRγ and RXRα, respectively) that blocks cell cycle progression, induces apoptosis and differentiation, prevents multidrug resistance, and inhibits angiogenesis and metastasis. In clinical trials, bexarotene gel demonstrated efficacy for the topical treatment of cutaneous lesions in Stage IA or IB CTCL patients with refractory or persistent disease following other therapies or who cannot tolerate other therapies, including: topical corticosteroids; topical chemotherapy with mechlorethamine or carmustine; local and total skin electron beam therapy; and UV-A or UV-B phototherapy (1,2). A promising development was the finding that bexarotene also blocks calcium-permeable ion channels formed by neurotoxic Alzheimer's β-amyloid peptides (3). **1.** Bischoff, Gottardis, Moon, *et al.* (1998) Cancer Res. 58, 4479. **2.** Martin (2003) *J. Drugs Dermatol.* **2**, 155. **3.** Fantini, Di Scala, Yahi, *et al.* (2014) *ACS Chem. Neurosci.* **5**, 216.

BEZ235

This ATP site-competitive, dual pan-PI3K/mTOR and antineoplastic agent (FW = 469.55; CASs = 915019-65-7 (free base) and 1028385-32-1 (4-methylbenzenesulfonate salt); Solubility (25°C): 1 mg/mL DMSO, <1 mg/mL Water), IUPAC: 2-methyl-2-(4-(3-methyl-2-oxo-8-(quinolin-3-yl)-2,3-dihydro-imidazo[4,5-c]-quinolin-1-yl)phenyl)-propane-nitrile, targets phosphoinositide 3-kinases in the PI3K/mTOR signaling pathway, inhibiting PI3K p110α (IC$_{50}$ = 4 nM), PI3K p110γ (IC$_{50}$ = 5 nM), PI3K p110δ (IC$_{50}$ = 7 nM), and PI3K p110β (IC$_{50}$ = 75 nM) (1). In cellular settings using human tumor cell lines, this inhibitor effectively and specifically blocks the dysfunctional activation of the PI3K pathway, inducing G$_1$ cell-cycle arrest in *in vivo* models of human cancer. BEZ235 also significantly reduces phosphorylation of the mTOR-activated kinase p70S6K (2,3). In vivo NVP-BEZ235 treatment of colonic tumors resulted in a 56% decrease in cellular proliferation, but the absence of an induction in apoptosis. Although induction of apoptosis by NVP-BEZ235 treatment has been reported in lung, breast, renal cell carcinoma, and ovarian models, the absence of apoptosis has been found after treatment of glioma and sarcoma models. Furthermore, *in vivo* NVP-BEZ235 treatment of colonic tumors resulted in a 75% decrease in microvessel density after NVP-BEZ235 treatment (2). use of NVP-BEZ235 and sorafenib has greater antitumor benefit compared to either drug alone and thus provides a strategy for treating metastatic renal cell carcinoma (3). **1**. Maira, Stauffer, Breuggen, *et al.* (2008) *Mol. Cancer Ther.* **7**, 1851. **2**. Roper, *et al.* (2011) *PLoS One* **6**, e25132. **3**. Roulin, *et al.* (2011) *Mol. Cancer* **10**, 90.

BFA, See *Brefeldin A*

BGB324, See *R428*

BGC 638

This nonpolyglutamatable folate (FW = 631.64 g/mol), also known as 6S-CB300638, inhibits thymidylate synthase (K_i = 0.24 nM) and is transported into α-folate receptor-overexpressing tumors (1,2). The low expression of the α-FR in normal tissues, particularly those sensitive to TS inhibitors, together with BGC638's low affinity for the reduced-folate carrier, suggests it may have potential as an antitumor agent and is likely to have a high therapeutic index (1). **1**. Theti, Bavetsias, Skelton, *et al.* (2003) *Cancer Res.* **63**, 3612. **2**. Gibbs, Theti, Wood, *et al.* (2005) *Cancer Res.* **65**, 11721.

BGC 945, See *ONX-0801*

BGT226

This novel Class I PI3K/mTOR inhibitor (FW$_{free-base}$ = g/mol; FW$_{maleate-salt}$ = 650.60 g/mol; CAS 1245537-68-1; Solubility: 30 mg/mL DMSO; <1 mg/mL H$_2$O), also known as NVP-BGT226 and 8-(6-methoxypyridin-3-yl)-3-methyl-1-(4-(piperazin-1-yl)-3-(trifluoromethyl)phenyl)-1H-imidazo[4,5-c]quinolin-2(3H)-one, targets PI3Kα (IC$_{50}$ = 4 nM), PI3Kβ (IC$_{50}$ = 63 nM), and PI3Kγ (IC$_{50}$ = 38 nM). BGT226 is active against all tested cancer cell lines, and cross-resistance is not observed in the cisplatin-resistant cell line (1). Activation of the AKT/mTOR signal cascade is suppressed by BGT226 in a concentration- and time-dependent manner, with cells accumulating in the G$_0$-G$_1$ phase, attended by concomitant loss in the S-phase. TUNEL assays and analysis of caspase 3/7 and PARP indicate that BGT226 induces cancer cell death through an apoptosis-independent pathway (1). BGT226 induces autophagy, as indicated by the aggregation and upregulation of the microtubule-associated protein light chain 3B-II, and p62 degradation. Gene silencing of Beclin1 or cotreatment of the autophagosome inhibitor, 3-methyladenine, inhibits the BGT226-induced autophagy and led to the retrieval of colony survival (1). In a xenografted animal model, BGT226 delays tumor growth in a dose-dependent manner, along with suppressed cytoplasmic expression of p-p70 S6 kinase and the presence of autophagosome formation (1). BGT226 inhibits growth in common myeloma cell lines and primary myeloma cells at nM-concentrations in a time-dependent and dose-dependent manner (2). BGT226 also has a potent cytotoxic effect on normoxic and hypoxic hepatocarcinoma cells, inactivating p-Akt and p-S6 at less than 10 nM (3). **1**. Chang, Tsai, Wu, *et al.* (2011) *Clin. Cancer Res.* **17**, 7116. **2**. Baumann, Schneider, Mandl-Weber, Oduncu & Schmidmaier (2012) *Anticancer Drugs* **23**, 131. **3**. Simioni, Cani, Martelli, *et al.* (2015) *Oncotarget* **6**, 6597.

BI397, See *Dalbavancin*

BI 2536

This protein kinase inhibitor (FW = 521.66 g/mol; CASs = 755038-02-9, 876126-71-5; Solubility: 100 mg/mL DMSO; <1 mg/mL H$_2$O), also named (R)-4-(8-cyclopentyl-7-ethyl-5-methyl-6-oxo-5,6,7,8-tetrahydropteridin-2-ylamino)-3-methoxy-N-(1-methylpiperidin-4-yl)benzamide), targets the Polo-like serine/threonine kinase Plk1 (IC$_{50}$ = 0.83 nM), bringing about G$_2$/M cell-cycle arrest and cell death. **1**. Steegmaier, *et al.* (2007) *Curr. Biol.* **17**, 316. **2**. Nappi, *et al.* (2009) *Cancer Res.* **69**, 1916.

BI-7273, Similar in action to *BI-9564*

BI-9564

This potent, selective and cell-permeable BRD9 and BRD7 inhibitor (FW = 353.41 g/mol; CAS 1883429-22-8; Solubility: 20 mM in DMSO; 10 mM in 1eq. HCl, with gentle warming), also named 4-[4-[(dimethylamino)methyl]-2,5-dimethoxyphenyl]-2-methyl-1,2-dihydro-2,7-naphthyridin-1-one, targets Bromodomain-containing proteins 9 and 7, with respective K_d values of 14 and 239 nM, and exhibiting higher selectivity for BRD9/7 *versus* 48 other bromodomains, 324 kinases, and 55 GPCRs. BRD9/7 are components of SWI/SNF (*or* switch/sucrose nonfermentable), a chromatin remodelling complex that also plays a role in oncogenesis. In acute leukemias, it was found that the SWI/SNF complex supports an oncogenic transcriptional program. In the absence of the SWI/SNF ATPase Brg1, leukemic cells arrest in G$_1$ and differentiate. Notably, BI-9654 shows efficacy in a xenograft model for acute myelogenous leukemia (AML). **1**. Karim & Schönbrunn (2016) *J. Med. Chem.* **59**, 4459. **2**. Martin, Koegl, Bader, *et al.* (2016) *J. Med. Chem.* **59**, 4462.

BI6727, See *Volasertib*

BI 224436

This novel non-catalytic-site HIV integrase inhibitor (MW = 441.51 g/mol) targets the 3'-processing step, blocking interaction of LTR DNA with the enzyme as well as the interaction between the enzyme and the transcriptional coactivator LEDGF, without inhibiting strand-transfer (1). When tested against different laboratory strains of HIV-1, antiviral EC_{50} values are below 15 nM. A 6000x higher BI 224436 concentration is required to evoke cytotoxicity. Antiviral potency is reduced by a factore of ~2.1 in the presence of 50% human serum. By virtue of its steep dose-response curve slope, however, BI 224436 exhibits serum-shifted EC_{95} values ranging from 22 to 75 nM. This latter property reflects the fact that binding of BI 224436 is highly cooperative. EC_{95} values are therefore only ~2.4x higher than their corresponding EC_{50} values (*i.e.*, % Inhibition = {$I_{max} \times [I]$}/{$[I]^n + (EC_{50})^n$},with a Hill coefficient *n* of 4). Passage of virus in the presence of inhibitor gives rise to resistance substitutions mapping to a conserved allosteric pocket near the integrase's catalytic core. BI 224436 also retains full antiviral activity against recombinant viruses encoding well-known INSTI resistance substitutions. BI 224436 exhibits drug-like ADME properties *in vitro*. **1.** Fenwick, Amad, Bailey, *et al.* (2014) *Antimicrob. Agents Chemother.* **58**, 3233.

Biapenem

This novel, broad-spectrum carbapenem antibiotic (FW = 350.39 g/mol; CAS 120410-24-4), also named LJC10,627 and (4*R*,5*S*,6*S*)-3-(6,7-dihydro-5*H*-pyrazolo[1,2-*a*][1,2,4]triazol-8-ium-6-ylsulfanyl)-6-(1-hydroxyethyl)-4-methyl-7-oxo-1-azabicyclo[3.2.0]hept-2-ene-2-carboxylate, exhibits MIC values ranging from 0.1–25 μg/mL for 90% of tested members of the family *Enterobacteriaceae*, including ceftazidime-resistant strains (1). Biapenem is more stable than imipenem, meropenem and panipenem to hydrolysis by human renal dihydropeptidase-I (DHP-I), and therefore does not require the coadministration of a DHP-I inhibitor (2). Upon intravenous administration, biapenem penetrates well into various tissues (*e.g.*, lung tissue) and body fluids (*e.g.*, sputum, pleural effusion, abdominal cavity fluid). **1.** Ubukata, Hikida, Yoshida M, *et al.* (1990) *Antimicrob. Agents Chemother.* **34**, 994. **2.** Perry & Ibbotson (2002) *Drugs* **62**, 2221.

BIBF 1120, See *Nintedanib*

BIBW2992, See *Afatinib*

Bicalutamide

This synthetic androgen (FW = 430.37 g/mol; CAS 90357-06-5; Solubility: 86 mg/mL DMSO, <1 mg/mL H_2O), also known by its trade names Casodex®, Cosudex®, Calutide®, Kalumid®, and its systematic name, *N*-[4-cyano-3-(trifluoromethyl)phenyl]-3-[(4-fluorophenyl)sulfonyl]-2-hydroxy-2-methylpropanamide, is an androgen receptor antagonist (IC_{50} = 0.16 μM), preventing binding of testosterone and other androgens, and has been used in the treatment of prostate cancer. Casodex does not cross the blood-brain barrier exists, excluding this antiandrogen from CNS sites of androgen negative feedback and accounting for its peripherally selective antihormonal profile. **Key Pharmacokinetic Parameters:** *See* Appendix II in Goodman & Gilman's THE PHARMACOLOGICAL BASIS OF THERAPEUTICS, 12th Edition (Brunton, Chabner & Knollmann, eds.) McGraw-Hill Medical, New York (2011). **1.** Furr, Valcaccia, Curry, *et al.* (1987) *J. Endocrinol.* **113**, R7. **2.** Freeman, Mainwaring & Furr (1989) *Brit. J. Cancer* **60**, 664.

Bicarbonate

This carbonic acid monoanion (FW_{NaHCO3} = 84.01 g/mol CAS 144-55-8 (NaHCO₃ Salt); FW_{KHCO3} = 100.12 g/mol, CAS 298-14-6 (KHCO₃ Salt); $FW_{NH4HCO3}$ = 79.06 g/mol, CAS 1066-33-7 (NH₄HCO₃ Salt); Apparent pK_{a1} of carbonic acid is 6.35 [Note that the true value is ~3.8.] and pK_{a2} is 10.33 at 25°C) is a key physiological buffer in all living organisms. (***See also*** *Carbon Dioxide; Carbonic Acid*) Aqueous solutions of sodium bicarbonate form carbon dioxide and sodium carbonate spontaneously ~20°C. **Target(s):** membrane dipeptidase (1); succinate dehydrogenase (2,6); glycosyl-phosphatidylinositol phospholipase D, *or* glycoprotein phospholipase D (3,28); glutamyl endopeptidase (4); glucose-6-phosphatase (5,17); *Ustilago sphaerogena* ribonuclease (7); ATP synthesis (8); glutamate dehydrogenase (9); aconitase (10,11); aspartate aminotransferase (12); carbonic anhydrase (13,18); *Clostridium pasteurianum* F_1F_o ATP synthase (14); creatine kinase (15); phosphoenolpyruvate carboxykinase (16); glucose-6-phosphate dehydrogenase (19,23); adenylo-succinate synthetase (20); acetyl-CoA synthetase, *or* acetate:CoA ligase (21); phosphoenolpyruvate mutase (22); threonine ammonia-lyase, *or* threonine dehydratase (24); oxaloacetate decarboxylase (25); fumarylacetoacetase (26); AMP deaminase (27); adenosine deaminase (27); phosphoprotein phosphatase (29); pyrophosphate-dependent phosphofructokinase, *or* diphosphate:fructose-6-phosphate 1-phosphotransferase (30); starch phosphorylase, weakly inhibited (31); glycine amidinotransferase (32); ferredoxin:$NADP^+$ reductase (33). **1.** Campbell (1970) *Meth. Enzymol.* **19**, 722. **2.** Hatefi (1978) *Meth. Enzymol.* **53**, 27. **3.** Deeg & Davitz (1995) *Meth. Enzymol.* **250**, 630. **4.** Birktoft & Breddam (1994) *Meth. Enzymol.* **244**, 114. **5.** Nordlie (1971) *The Enzymes*, 3rd ed. (Boyer, ed.), **4**, 543. **6.** Hatefi & Stiggall (1976) *The Enzymes*, 3rd ed. (Boyer, ed.), **13**, 175. **7.** Glitz & Dekker (1964) *Biochemistry* **3**, 1391. **8.** Lodeyro, Calcaterra & Roveri (2001) *Biochim. Biophys. Acta* **1506**, 236. **9.** Baverel & Lund (1979) *Biochem. J.* **184**, 599. **10.** Stepinski & Angielski (1976) *Acta Biochim. Pol.* **23**, 203. **11.** Stepinski & Angielski (1975) *Curr. Probl. Clin. Biochem.* **4**, 65. **12.** Lysiak, Pienkowska-Vogel, Szutowicz & Angielski (1975) *Acta Biochim. Pol.* **22**, 211. **13.** Vullo, Franchi, Gallori, *et al.* (2003) *J. Enzyme Inhib. Med. Chem.* **18**, 403. **14.** Das & Ljungdahl (2003) *J. Bacteriol.* **185**, 5527. **15.** Noda, Nihei & Morales (1960) *J. Biol. Chem.* **235**, 2830. **16.** Chang, Maruyama, Miller & Lane (1966) *J. Biol. Chem.* **241**, 2421. **17.** Dyson, Anderson & Nordlie (1969) *J. Biol. Chem.* **244**, 560. **18.** Pocker & Stone (1968) *Biochemistry* **7**, 2936. **19.** Anderson, Horne & Nordlie (1968) *Biochemistry* **7**, 3997. **20.** Markham & Reed (1977) *Arch. Biochem. Biophys.* **184**, 24. **21.** Frenkel & Kitchens (1981) *Meth. Enzymol.* **71**, 317. **22.** Seidel & Knowles (1994) *Biochemistry* **33**, 5641. **23.** Horne, Anderson & Nordlie (1970) *Biochemistry* **9**, 610. **24.** Pagani, Leoncini, Terzuoli, Guerranti & Marinello (1990) *Enzyme* **43**, 122. **25.** Labrou & Clonis (1999) *Arch. Biochem. Biophys.* **365**, 17. **26.** Braun & Schmidt (1973) *Biochemistry* **12**, 4878. **27.** Conway & Cooke (1939) *Biochem. J.* **33**, 479. **28.** Stieger, Diem, Jakob & Brodbeck (1991) *Eur. J. Biochem.* **197**, 67. **29.** Nakai & Thomas (1974) *J. Biol. Chem.* **249**, 6459. **30.** Mahajan & Singh (1989) *Plant Physiol.* **91**, 421. **31.** Kamogawa & Fukui (1973) *Biochim. Biophys. Acta* **302**, 158. **32.** Humm, Fritsche, Mann, Goehl & Huber (1997) *Biochem. J.* **322**, 771. **33.** Zanetti (1981) *Plant Sci. Lett.* **23**, 55.

Bicuculline

This neurotoxin and convulsant ($FW_{free-base}$ = 367.36 g/mol; CAS 485-49-4) is a $GABA_A$ receptor antagonist, IC_{50} = 3 μM (1-3). **Target(s):** acetylcholinesterase (4,5); nicotinic acetylcholine receptor (6). **1.** Curtis (1973) *Proc. Aust. Assoc. Neurol.* **9**, 145. **2.** Chebib & Johnston (1999) *Clin. Exp. Pharmacol. Physiol.* **26**, 937. **3.** Curtis, Duggan, Felix & Johnston (1970) *Nature* **226**, 1222. **4.** Svenneby & Roberts (1973) *J. Neurochem.* **21**, 1025. **5.** Breuker & Johnston (1975) *J. Neurochem.* **25**, 903. **6.** Demuro, Palma, Eusebi & Miledi (2001) *Neuropharmacology* **41**, 854.

BI-D1870

This cell-permeant, ATP-competitive inhibitor (FW = 391.42 g/mol; CAS 501437-28-1: Solubility = 78 mg/mL DMSO; <1 mg/mL H_2O), also known as 2-[(3,5-difluoro-4-hydroxyphenyl)amino]-7,8-dihydro-5,7-dimethyl-8-(3-methylbutyl)-6(5*H*)-pteridinone, targets the p90 Ribosomal protein S6 Kinase α-1, 2, 3, and 4, *or* RSK1, RSK2, RSK3, and RSK4, with IC_{50} values of 31 nM, 24 nM, 18 nM, and 15 nM, respectively (1). RSK isoforms are activated by Extracellular-signal-Regulated Kinases ERK1 and ERK2 in response to growth factors, phorbol esters and other agonists. BI-D1870 shows 10x to 100x selectivity for RSK over MST2, GSK-3β, MARK3, CK1 and Aurora B protein kinases. When tested *in vitro*, BI-D1870 exhibits IC_{50} values for inhibits RSK1, RSK2, RSK3 and RSK4 are in the 10–30 nM (1). Although assumed to be a remarkably specific inhibitor of RSK isoforms, with a >500-fold greater selectivity over nine other AGC kinases, later work assessing the ability of BI-D1870 to protect enzyme active sites from alkylation demonstrated that several other kinases interact with BI-D1870 (2). Subsequent direct activity assays confirmed that Slk, Lok and Mst1 protein kinases are inhibited by BI-D1870 and that phosphorylation of some of their substrates is blocked by BI-D1870 in living cells (2). Such results suggest that the specificity of BI-D1870 cannot be taken for certain. **1.** Gopal, Sapkota, Cummings & Newell (2007) *Biochem. J.* **401** (Pt 1), 29. **2.** Edgar, Trost, Watts & Zaru (2013) *Biosci. Rep.* 34, e00091.

BIIB021

This purine scaffold-based, orally available HSP90 inhibitor (FW = 318.76 g/mol; CAS 848695-25-0), known as 6-chloro-9-((4-methoxy-3,5-dimethyl-pyridin-2-yl)methyl)-9*H*-purin-2-amine and CNF2024, inhibits HSP90 with K_i and EC_{50} of 1.7 nM and 38 nM, respectively. **Primary Mode of Action:** BIIB021 binds within the ATP-binding pocket of Hsp90, interfering with its chaperone function and resulting in client protein degradation (including HER-2, Akt, and Raf-1), with consequential tumor growth inhibition. BIIB021 inhibits tumor cell proliferation with IC_{50} from 0.06-0.3 μM. BIIB021 up-regulates expression of the heat shock proteins Hsp70 and Hsp27. BIIB021 depletes NF-κB and sensitizes Hodgkin's lymphoma cells for natural killer cell-mediated cytotoxicity (2). BIIB021 also shows preferential cytotoxicity towards Kaposi's sarcoma-associated herpesvirus-associated primary effusion lymphomas (PEL) cells, as compared to non-PEL cells. BIIB021's cytotoxic effect against PEL is associated with induction of cell-cycle arrest and apoptosis (1). BIIB021 blocks the expression of a number of cellular proteins involved in cell cycle regulation and apoptosis. BIIB021 also blocks constitutive NF-κB activity present in PEL cells in part by blocking the interaction of vFLIP K13 with the IKK complex subunits. **1.** Lundgren, Zhang, Brekken, *et al.* (2009) *Mol. Cancer Ther.* 8, 921. **2.** Böll, Eltaib, Reiners, *et al.* (2009) *Clin. Cancer Res.* 15, 5108. **3.** Gopalakrishnan, Matta & Chaudhary (2013) *Clin. Cancer Res.* 19, 5016.

BIIB037, *See Aducanumab*

Bikinin

This ATP-competitive Arabidopsis GSK-3 inhibitor, (FW = 273.08 g/mol; CAS 188011-69-0), also named 4-(5-bromopyridin-2-ylamino)-4-oxobutanoic acid, is a strong activator of brassinosteroid (BR) signaling (1). Bikinin directly binds the GSK3 BIN2 and also inhibits the activity of six other Arabidopsis GSK3s. Genome-wide transcript analyses demonstrate that simultaneous inhibition of seven GSK3s is sufficient to activate BR responses (1). inhibited the SnRK2.3 kinase activity and its T180 phosphorylation in vivo by blocking BIN2 phosphorylation of SnRK2.3 on Thr-180 (2). **1.** De Rybel, Audenaert, Vert, *et al.* (2009) *Chem Biol.* 16, 594. **2.** Cai, Liu, Wang, *et al.* (2014) *Proc. Natl. Acad. Sci. U S A.* 111, 9651.

Bilirubin

This photosensitive tetrapyrrole metabolite ($FW_{free-acid}$ = 584.67 g/mol), also known as bilirubin IXα, is the primary bile pigment that is formed in heme degradation in the biliverdin reductase reaction. Billirubin is a free radical scavenger that protects membrane lipids and neurons from oxidatively generated radicals. **Target(s):** HIV-1 protease, K_i = 0.8 μM (1); protoporphyrinogen oxidase (2,28); electron transport, as an uncoupler (3,12); glutathione *S*-transferase (4,5,7,15,17,29,31-34); prostaglandin-D synthase (6,22); lactoylglutathione lyase, *or* glyoxalase I (8); alcohol dehydrogenase (9); glutamate dehydrogenase (10); isocitrate dehydrogenase, NAD^+ (11); malate dehydrogenase (13,16,25); protein kinase A (14); Na^+/K^+-exchanging ATPase (18); 5'-nucleotidase (19); phospholipase A_2 (20); nitric-oxide synthase (21); protein kinase C (23,26); coproporphyrinogen III oxidase (24); aspartate aminotransferase (25); glycerol-3-phosphate dehydrogenase, mildly inhibited (25); uroporphyrinogen I synthase (27); biliverdin reductase, mildly inhibited (30); glucuronosyltransferase, substrate for UGT1A1 (35-38); 5-aminolevulinate synthase (39). **1.** McPhee, Caldera, Bemis, *et al.* (1996) *Biochem. J.* **320**, 681. **2.** Ferreira & Dailey (1988) *Biochem. J.* **250**, 597. **3.** Cowger, Igo & Labbe (1965) *Biochemistry* **4**, 2763. **4.** Di Ilio, Aceto, Del Boccio, *et al.* (1988) *Eur. J. Biochem.* **171**, 491. **5.** Boyer & Vessey (1987) *Hepatology* **7**, 843. **6.** Urade, Fujimoto, Ujihara & Hayaishi (1987) *J. Biol. Chem.* **262**, 3820. **7.** Kaplowitz, Kuhlenkamp & Clifton (1975) *Proc. Soc. Exp. Biol. Med.* **149**, 234. **8.** Douglas & Sharif (1983) *Biochim. Biophys. Acta* **748**, 184. **9.** Flitman & Worth (1966) *J. Biol. Chem.* **241**, 669. **10.** Yamaguchi (1970) *J. Biochem.* **68**, 441. **11.** Ogasawara, Watanabe & Goto (1973) *Biochim. Biophys. Acta* **327**, 233. **12.** Noir, Boveris, Garaza Pereira & Stoppani (1972) *FEBS Lett.* **27**, 270. **13.** Kashiwamata, Niwa, Katoh & Higashida (1975) *J. Neurochem.* **24**, 189. **14.** Constantopoulos & Matsaniotis (1976) *Cytobios* **17**, 17. **15.** Ketley, Habig & Jakoby (1975) *J. Biol. Chem.* **250**, 8670. **16.** Kashiwamata, Niwa & Katoh (1975) *J. Neurochem.* **25**, 7. **17.** Kaplowitz, Kuhlenkamp & Clifton (1975) *Proc. Soc. Exp. Biol. Med.* **149**, 234. **18.** Kashiwamata, Goto, Semba & Suzuki (1979)

J. Biol. Chem. **254**, 4577. **19**. De la Morena Garcia & Gonzalez Isabel (1981) *Rev. Clin. Esp.* **161**, 233. **20**. Jameel, Frey, Frey, Gowda & Vishwanath (2005) *Mol. Cell Biochem.* **276**, 219. **21**. Tkachenko (2000) *Membr. Cell Biol.* **13**, 369. **22**. Beuckmann, Aoyagi, Okazaki, *et al.* (1999) *Biochemistry* **38**, 8006. **23**. Amit & Boneh (1993) *Clin. Chim. Acta* **223**, 103. **24**. Rossi, Attwood & Garcia-Webb (1992) *Biochim. Biophys. Acta* **1135**, 262. **25**. McLoughlin & Howell (1987) *Biochim. Biophys. Acta* **93**, 7. **26**. Sano, Nakamura & Matsuo (1985) *Pediatr. Res.* **19**, 587. **27**. Kohashi, Tse & Piper (1984) *Life Sci.* **34**, 193. **28**. Corrigall, Siziba, Maneli, *et al.* (1998) *Arch. Biochem. Biophys.* **358**, 251. **29**. Blanchette & Singh (1999) *Mar. Biotechnol. (NY)* **1**, 74. **30**. Tenhunen, Ross, Marver & Schmid (1970) *Biochemistry* **9**, 298. **31**. Liebau, Eckelt, Wildenburg, *et al.* (1997) *Biochem. J.* **324**, 659. **32**. Vander Jagt, Hunsaker, Garcia & Royer (1985) *J. Biol. Chem.* **260**, 11603. **33**. Vander Jagt, Wilson & Heidrich (1981) *FEBS Lett.* **136**, 319. **34**. Lee (1984) *Biochem. Soc. Trans.* **12**, 30. **35**. Watanabe, Nakajima, Ohashi, Kume & Yokoi (2003) *Drug Metab. Dispos.* **31**, 589. **36**. Matern, Matern, Schelzig & Gerok (1980) *FEBS Lett.* **118**, 251. **37**. Rao, Rao & Breuer (1976) *Biochim. Biophys. Acta* **452**, 89. **38**. Tachibana, Tanaka, Masubuchi & Horie (2005) *Drug Metab. Dispos.* **33**, 803. **39**. Scholnick, Hammaker & Marver (1972) *J. Biol. Chem.* **247**, 4132.

Bilobalide

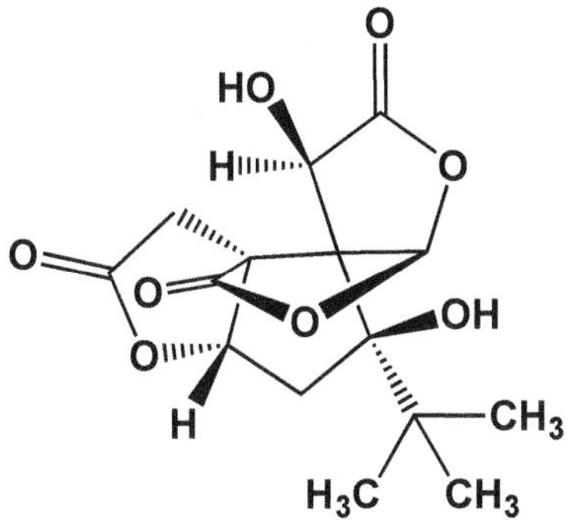

This plant natural product (FW = 326.30 g/mol; CAS 33570-04-6) from the gingko tree (*Ginkgo biloba*) is a terpene trilactone that exhibits anticonvulsant properties (1). Bilobalide has multiple mechanisms of action (*e.g.*, acting as a GABAA receptor antagonist (/1), preserving mitochondrial ATP synthesis, inhibiting staurosporine-induced apoptotic damage, suppressing hypoxia-induced membrane deterioration in the brain, and increasing the expression the mitochondrial DNA-encoded COX III subunit of cytochrome *c* oxidase and the ND1 subunit of NADH dehydrogenase). Bilobalide was later synthesized in E. J. Corey's laboratory in the late 1980s (2,3). **1**. Ivic, Sands, Fishkin, *et al.* (2003) *J. Biol. Chem.* **278**, 49279. **2**. Corey & Su (1987) *J. Amer. Chem. Soc.* **109**, 7534. **3**. Corey & Su (1988) *Tetrahedron Lett.* **29**, 3423.

BIMT-17, *See Flibanserin*

Binimetinib

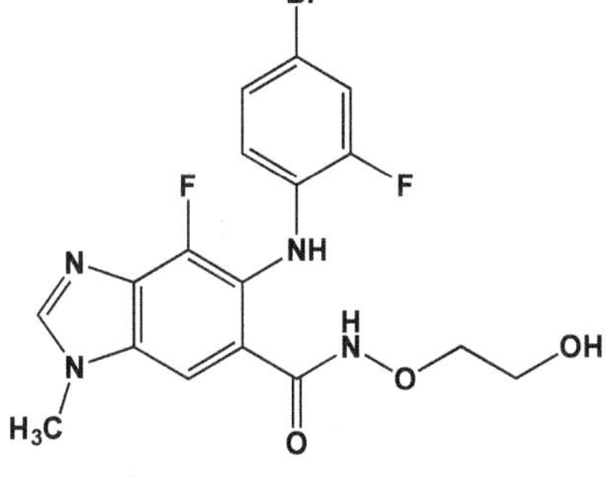

This MEK inhibitor (FW = 441.23 g/mol; CAS 606143-89-9; Solubility: 88 mg/mL DMSO, when warmed), also named MEK162, ARRY-162, ARRY-438162, and 5-((4-bromo-2-fluorophenyl)amino)-4-fluoro-*N*-(2-hydroxy-ethoxy)-1-methyl-1*H*-benzo[*d*]imidazole-6-carboxamide, targets Mitogen-Activated Protein Kinase Kinase (MAPKK), also known as MAP2K and MEK, which phosphorylates Mitogen-Activated Protein Kinase (MAPK). (IC$_{50}$ = 12 nM) and is the first targeted therapy to show activity in patients with NRAS -mutated melanoma. **1**. Ascierto, Schadendorf, Berking, *et al.* (2013) *Lancet Oncol.* **14**, 249.

BIO, *See* (2'Z,3'E)-6-Bromoindirubin-3'-oxime

BIO 5192

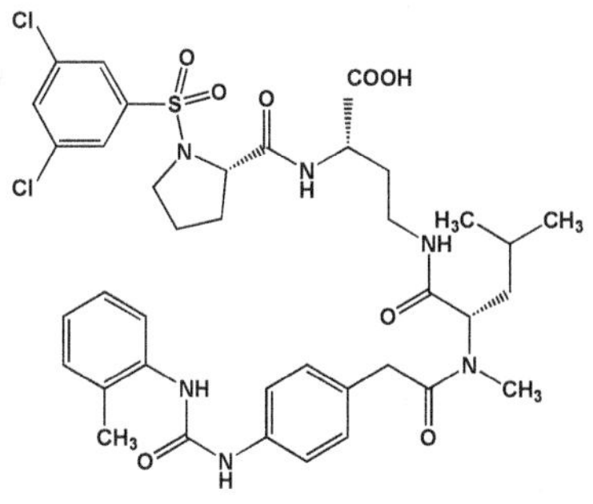

This selective integrin inhibitor (FW = 832.82 g/mol; CAS Number = 327613-57-0; Soluble to 100 mM in DMSO and to 50 mM in 1eq. NaOH) targets integrin α$_4$β$_1$, also known as Very Late Antigen-4 (VLA-4) with a K_d < 10 pM and IC$_{50}$ value of 1.8 nM. BIO 5192 shows high selectivity for α$_4$β$_1$ versus integrins α$_9$β$_1$ (IC$_{50}$ = 138 nM), α$_2$β$_1$ (IC$_{50}$ = 1053 nM), α$_4$β$_7$ (IC$_{50}$ > 500 nM), and α$_{11b}$β$_3$ (IC$_{50}$ > 10,000 nM). Integrin α$_4$β$_1$ plays an important role in inflammatory processes by regulating the migration of lymphocytes into inflamed tissues, suggesting that BIO 5192 offers a new way to treat human inflammatory diseases. Significantly, interruption of the Vascular Cell Adhesion Molecule-Very Late Antigen-4 (*or* VCAM-1/VLA-4) axis with BIO5192 results in a 30-fold increase in mobilization of murine hematopoietic stem and progenitors (HSPCs) over basal levels. A 3x enhancement of HSPC mobilization was observed when plerixafor (AMD3100), a small molecule inhibitor of the CXCR-4/SDF-1 axis, was combined with BIO5192. (**See** *Plerixafor*) Furthermore, the combination of granulocyte colony-stimulating factor (G-CSF), BIO5192, and plerixafor enhanced mobilization by 17x compared with G-CSF alone. **1**. Leone, Giza, Gill, *et al.* (2003) *J. Pharmacol. Exp. Ther.* **305**, 1150. **2**. Ramirez, Rettig, Uy, *et al.* (2009) *Blood* **114**, 1340. **IUPAC**: (2*S*)-2-[[[(2*S*)-1-[(3,5-dichloro-phenyl)sulfonyl]-2-pyrrolidinyl]carbonyl]-amino]-4-[[(2*S*)-4-methyl-2-[methyl[2-[4-[[[(2-methyl-phenyl)amino]carbonyl]-amino]phenyl]acetyl]amino]-1-oxopentyl]amino]butanoate

Bipinnatins

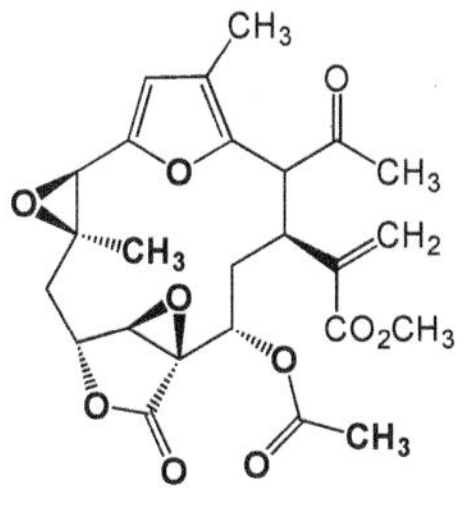

Bipinnatin A

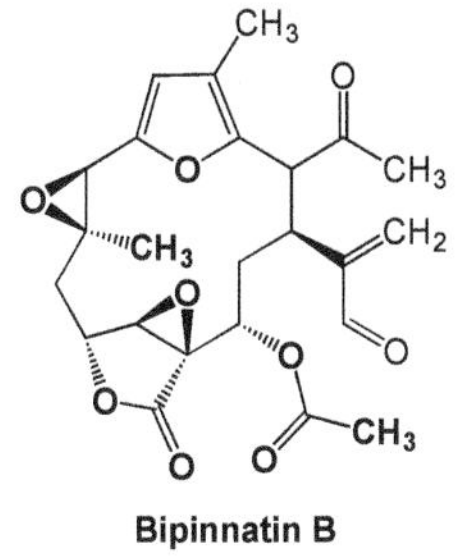

Bipinnatin B

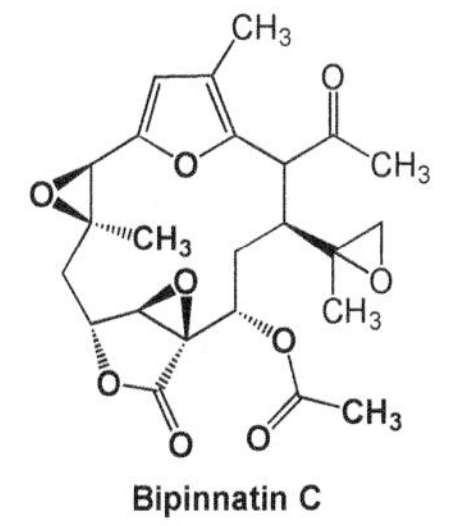

Bipinnatin C

These furanocembrenolide-class diterpenes (FW$_{\text{Bipinnatin-A}}$ = 488.49 g/mol; CAS (Bipinnatin A) = 99552-28-0; FW$_{\text{Bipinnatin-B}}$ = 458.46 g/mol; CAS (Bipinnatin B) = 99552-24-6; FW$_{\text{Bipinnatin-C}}$ = 460.48 g/mol; CAS (Bipinnatin C) = 123483-20-5), isolated from the bipinnate sea plume *Antillogorgia bipinnata*, a sea fan found in the eastern Caribbean Sea, are naturally occurring marine neurotoxins that irreversibly inhibit nicotinic acetylcholine receptors by forming a covalent bond with Tyrosine-190 in the α-subunit of the receptor (**See also** *Lophotoxin*). The parent species of the bipinnatins display little, if any, affinity for the nicotinic receptor. Preincubation of the toxins appeared to produce a single, relatively stable, active toxin species that irreversibly inhibited the two acetylcholine-binding sites on the nicotinic receptor with two distinguishable apparent pseudo first-order rates. The difference in the rates of irreversible inhibition of the two binding sites on the receptor was exploited to selectively inhibit one site for the pharmacological investigation of the other. The bipinnatins preferentially inhibited the binding site near the αδ-subunit interface that displays low affinity for metocurine and high affinity for acetylcholine. The bimolecular reaction constants for the interaction of the bipinnatins with the nicotinic receptor decreased in the order: Bipinnatin-B > Bipinnatin-A > Bipinnatin-C, for both acetylcholine-binding sites. The ratio of the bimolecular reaction constants for the two binding sites on the receptor was not the same for the three bipinnatins, suggesting that the reaction of the bipinnatins with the nicotinic receptor is sensitive to differences in the structure of the two acetylcholine-binding sites. **1**. Groebe, Dumm & Abramson (1994) *J. Biol. Chem.* **269**, 8885.

BI-RG-587, *See* Nevirapine

Birinapant

This rotationally symmetric SMAC mimetic antagonist and anticancer agent (FW = 806.94 g/mol; CAS 1260251-31-7) targets cellular Inhibitor of Apoptosis Protein (*or* cIAP1; K_d of <1 nM), arguably the most potent mammalian caspase inhibitor, showing single-agent efficacy due to its pan-IAP antagonism and causing rapid cIAP1 degradation, caspase activation, PARP cleavage, and NF-κB activation (1). When administered in combination with TNF-α, Birinapant is effective against a melanoma cell line with acquired resistance to BRAF inhibitors (2). **1**. Allensworth, Sauer, Lyerly, Morse & Devi (2013) *Breast Cancer Res. Treat.* **137**, 359. **2**. Krepler, *et al.* (2013) *Clin. Cancer Res.* **19**, 1784.

1,5-Bis(4-allyldimethylammoniumphenyl)pentan-3-one

This extremely hazardous quaternary amine (FW$_{\text{dibromide}}$ = 566.42 g/mol), also known as BW284c51, is a potent and selective inhibitor of acetylcholinesterase. It has also been used to ascertain the tissue localization of acetylcholinesterases in many organisms. **Target(s):** acetylcholine esterase (1-10); nicotinic acetylcholine receptors (11,12); cholinesterase, *or* butyrylcholinesterase (13-17); aryl-acylamidase (9). **1**. Di Patre, Mathes & Butcher (1993) *J. Histochem. Cytochem.* **41**, 129. **2**. Dupree & Bigbee (1994) *J. Neurosci. Res.* **39**, 567. **3**. Augustinsson (1960) *The Enzymes*, 2nd ed. (Boyer, Lardy & Myrbäck, eds.), **4**, 521. **4**. Adamson, Ayers, Deussen & Graham (1975) *Biochem. J.* **147**, 205. **5**. Ralston, Rush, Doctor & Wolfe (1985) *J. Biol. Chem.* **260**, 4312. **6**. Hussein, Grigg & Selkirk (1999) *Exp. Parasitol.* **91**, 144. **7**. Sanders, Mathews, Sutherland, *et al.* (1996) *Comp. Biochem. Physiol. B* **115**, 97. **8**. Hussein, Chacón, Smith, Tosado-Acevedo & Selkirk (1999) *J. Biol. Chem.* **274**, 9312. **9**. Costagli & Galli (1998)

Biochem. Pharmacol. **55**, 1733. **10**. Mikalsen, Andersen & Alexander (1986) *Comp. Biochem. Physiol. C* **83**, 447. **11**. Olivera-Bravo, Ivorra & Morales (2005) *Brit. J. Pharmacol.* **144**, 88. **12**. Fayuk & Yakel (2004) *Mol. Pharmacol.* **66**, 658. **13**. deVos & Dick (1992) *Comp. Biochem. Physiol. C* **103**, 129. **14**. Rodriguez-Fuentes & Gold-Bouchot (2004) *Mar. Environ. Res.* **58**, 505. **15**. McClellan, Coblentz, Sapp, *et al.* (1998) *Eur. J. Biochem.* **258**, 419. **16**. García-Ayllón, Sáez-Valero, Muñoz-Delgado & Vidal (2001) *Neuroscience* **107**, 199. **17**. Boeck, Schopfer & Lockridge (2002) *Biochem. Pharmacol.* **63**, 2101.

trans-1,4-Bis(2-chlorobenzylaminomethyl)-cyclohexane

This cholesterol biosynthesis inhibitor (FW$_{\text{free-base}}$ = 151.16 g/mol), often referred to as AY-9944, selectively inhibits 7-dehydrocholesterol reductase (1,2,4-6), but also inhibits sterol Δ^{24}-reductase (3) and cholestenol Δ isomerase, *or* sterol $\Delta^{7,8}$-isomerase (7). **1**. Dempsey (1969) *Meth. Enzymol.* **15**, 501. **2**. Dvornik, Kraml & Bagli (1966) *Biochemistry* **5**, 1060. **3**. Steinberg & Avigan (1969) *Meth. Enzymol.* **15**, 514. **4**. Dempsey (1967) in *Progress in Biochemical Pharmacology* (Kritchevsky, Paoletti & Steinberg, eds.), vol. 2, pp. 21-29, Karger, New York. **5**. Koroly & Dempsey (1981) *Lipids* **16**, 755. **6**. Kraml, Bagli & Dvornik (1964) *Biochem. Biophys. Res. Commun.* **15**, 455. **7**. Moebius, Reiter, Bermoser, *et al.* (1998) *Mol. Pharmacol.* **54**, 591.

bis(2-Chloroethyl)ethylamine

This strong vesicant and chemical warfare agent (FW$_{\text{free-base}}$ = 170.08 g/mol), also known as nitrogen mustard, HN-1, and (2,2'-dichloro)triethylamine, is a liquid with a melting point of –34°C. While the free base is only slightly soluble in water, the hydrochloride is very soluble. Bis(2-chloroethyl)ethylamine has a faint odor of fish and exhibits strong vesicant activity, albeit not as potent as sulfur mustard. HN-1 is reported to be a DNA and RNA alkylating agent, inhibit both transcription and translation. Although there are no reports that bis(2-chloroethyl)ethylamine has ever been used on the battlefield, several nations currently have stockpiles of sulfur and nitrogen mustards. **CAUTION:** Strong vesicant! Special training is required for safe handling. The free base of bis(2-chloroethyl)ethylamine should only be handled in a suitably vented chemical hood, and use of a gas mask is recommended. Fully protective clothing should be worn to prevent any possibility of contact with skin. **Target(s):** hexokinase (1,2); choline dehydrogenase (3,4). **1**. McDonald (1955) *Meth. Enzymol.* **1**, 269. **2**. Colowick (1951) *The Enzymes*, 1st ed. (Sumner & Myrbäck, eds.), **2** (Part 1), 114. **3**. Singer (1963) *The Enzymes*, 2nd ed. (Boyer, Lardy & Myrbäck, eds.), **7**, 345. **4**. Hatefi & Stiggall (1976) *The Enzymes*, 3rd ed. (Boyer, ed.), **13**, 175.

1,6-Bis(cyclohexyloximinocarbonylamino)hexane

This reagent (FW = 394.51 g/mol), also known as RHC-80267, RG 80267, and 1,6-di(*O*-(carbamoyl)cyclohexanone oxime)hexane, is a potent and selective inhibitor of diacylglycerol lipase activity in canine platelets (I$_{50}$ = 4 μM) and in a variety of mammalian cells (1.2). It also inhibits glucose- and carbachol-induced insulin release from intact islets (3) and is an inhibitor of angiotensin II- and ATP-induced synthesis of 6-keto-prostaglandin F1α. **Target(s):** diacylglycerol lipase (1-8); lipase, *or* triacylglycerol lipase (8); lipoprotein lipase (8). **1**. Shinoda, Suzuki, Oiso & Kozawa (1997) *Arterioscler. Thromb. Vasc. Biol.* **17**, 295. **2**. Shinoda, Kozawa, Suzuki, *et al.* (1997) *Eur. J. Endocrinol.* **136**, 207. **3**. Konrad, Major & Wolf (1994) *Biochemistry* **33**, 13284. **4**. Mason-Garcia, Clejan, Tou & Beckman (1992) *Amer. J. Physiol.* **262**, C1197. **5**. Sutherland & Amin (1982) *J. Biol. Chem.* **257**, 14006. **6**. Rindlisbacher, Reist & Zahler (1987) *Biochim. Biophys. Acta* **905**, 349. **7**. Franson & Rosenblum (1984) *Thromb. Res.* **36**, 32. **8**.

Galatioto & Zahler (1993) *J. Neurochem.* **60**, 32.   **9**. Carroll & Severson (1992) *Lipids* **27**, 30.

Bisdethiobis(methylthio)gliotoxin

This naturally occurring indole derivative (FW = 356.47 g/mol; CAS 74149-38-5), also known as bis(methylthio)gliotoxin and dimethylgliotoxin, is a platelet-acivating factor (PAF) antagonist that inhibits PAF-induced platelet aggregation (1). The parent compound, gliotoxin is a dual inhibitor of famesyltransferase and geranylgeranyl-transferase. Bisdethio-bis(methylthio)gliotoxin is a minor metabolite of *Gliocladium deliquescens*, Radiocarbon tracer experiments showed it is formed, apparently irreversibly, from gliotoxin (2). It has also been synthesized from gliotoxin by reduction and methylation. Bisdethio-bis(methylthio)gliotoxin also binds Cu^+ and Ag^+ ions. **1**. Okamoto, Yoshida, Uchida, *et al.* (1986) *Chem. Pharm. Bull.* **34**, 340. **2**. Kirby, Robins, Sefton & Talekar (1980) *J. Chem. Soc., Perkin Trans.* **1**, 119.

1-(2-[Bis(4-fluorophenyl)methoxy]ethyl)-4-(3-phenylpropyl) piperazine

This piperazine derivative ($FW_{free-base}$ = 450.57 g/mol; supplied as the dihydrochloride salt), also known as GBR 12909 and vanoxerine, inhibits dopamine transport (K_i = 9 nM). This high-affinity dopamine reuptake inhibitor was originally prepared in the late 1970s as a potential antidepressant, but was later considered for its use in treating cocaine addiction. **Target(s):** dopamine reuptake (1-3,6); TTX-sensitive Na^+ channels (4; CYP2D1 (5). **1**. Borowsky & Kuhn (1993) *Brain Res.* **613**, 251. **2**. Heikkila & Manzino (1984) *Eur. J. Pharmacol.* **103**, 241. **3**. Reith, Coffey, Xu & Chen (1994) *Eur. J. Pharmacol.* **253**, 175. **4**. Mike, Karoly, Vizi & Kiss (2003) *Neuroreport* **14**, 1945. **5**. Niznik, Tyndale, Sallee, *et al.* (1990) *Arch. Biochem. Biophys.* **276**, 424. **6**. Preti (2000) *Curr. Opin. Investig. Drugs* **1**, 241.

(2R,3R,4R,5R)-N^1,N^6-Bis[(1S,2R)-2-hydroxy-1-indanyl]-2,5-bis[(2E)-3-(3,4-methylenedioxyphenyl)-2-propenyloxy]-3,4-dihydroxyhexane-1,6-diamide

This peptide analogue (FW = 788.81 g/mol) inhibits hemoglobin-degrading aspartic proteases, plasmepsin I and plasmepsin II (K_i = 0.5 and 14 nM, respectively. It shows no measurable affinity to the human enzyme cathepsin D. **1**. Ersmark, Feierberg, Bjelic, *et al.*(2004) *J. Med. Chem.* **47**, 110.

(1α,2β,3α)-9-[2,3-Bis(hydroxy-methyl)-cyclobutyl]-guanine

This carbocyclic guanosine analogue (FW = 265.27 g/mol), also known as lobucavir, inhibits a number of viral DNA polymerases, after metabolism to form the triphosphate metabolite. It is also a carbocyclic analogue of oxetanocin G and has been used in the treatment of certain viral infections (*e.g.*, hepatitis B virus). **Target(s):** viral DNA polymerases (1-3); HSV-1 and HSV-2 replication (1,2); *Varicella* zoster virus replication (1,2); human cytomegalovirus replication (1,2). **1**. Prisbe & Chen (1996) *Meth. Enzymol.* **275**, 425. **2**. Field, Tuomari, McGeever-Rubin, *et al.* (1990) *Antiviral Res.* **13**, 41. **3**. Terry, Cianci & Hagen (1991) *Mol. Pharmacol.* **40**, 591.

9-[2,3-Bis(hydroxymethyl)cyclobutyl]guanine Triphosphate

This carbocyclic analogue ($FW_{free-acid}$ = 504.20 g/mol) of oxetanocin G triphosphate, also known as lobucavir triphosphate, is likewise an analogue of guanosine 5'-triphosphate and a nonobligate chain terminator in viral DNA biosynthesis. **Target(s):** telomerase (1); DNA polymerase, viral (2-4); HIV reverse transcriptase (2). **1**. Takahashi, Amano, Saneyoshi, Maruyama & Yamaguchi (2003) *Nucl. Acids Res. Suppl.* (3), 285. **2**. Izuta, Shimada, Kitagawa, *et al.* (1992) *J. Biochem.* **112**, 81. **3**. Seifer, Hamatake, Colonno & Standring (1998) *Antimicrob. Agents Chemother.* **42**, 3200. **4**. Terry, Cianci & Hagen (1991) *Mol. Pharmacol.* **40**, 591.

Bislysine

This symmetric lysine analogue (FW = 348.49 g/mol) inhibits dihydrodipicolinate synthase (DHDPS), which catalyzes the condensation of pyruvate and β-aspartate semialdehyde (ASA) to form a cyclic product that dehydrates to form dihydrodipicolinate, the latter required for bacterial peptidoglycan biosynthesis in *Campylobacter jejuni*. DHDPS is allosterically regulated by lysine. This novel inhibitor is a mixed partial inhibitor with respect to pyruvate (the first substrate) and a noncompetitive partial inhibitor with respect to ASA. Bislysine binds to all forms of the enzyme with a K_i near 200 nM, >300x more tightly than lysine. Hill plots show its inhibition is cooperative, indicating that the allosteric sites are not independent, despite their distal locations on opposite sides of the protein tetramer. DHDPS^{Y110F}, a mutant enzyme that is resistant to lysine inhibition, is strongly inhibited by this novel inhibitor, suggesting bislysine-based agents may be a promising strategy for antibiotic development. **1**. Skovpen, Conly, Sanders & Palmer (2016) *J. Am. Chem. Soc.* **138**, 2014.

4,5-Bis(4-methoxyphenyl)-1,2-selenazole

This isoselenazole inhibits cyclooxygenase-1 (IC_{50} = 6 nM), cyclooxygenase-2 (IC_{50} = 8 μM), and 5-lipoxygenase, or arachidonate 5-lipoxygenase (IC_{50} = 8 μM), portending their use as multi-target non-steroidal anti-inflammatory drugs (MTNSAIDs) which can intervene into the inflammatory processes via different mechanisms of action. **1**. Scholz, Ulbrich & Dannhardt (2008) *Eur. J. Med. Chem.* **43**, 1152.

1,2-Bis(methylsulfonyl)-1-(2-chloroethyl)-2-[(methylamino) carbonyl]hydrazine

This antitumor prodrug (FW = 272.33 g/mol), also known as VNP40101M, generates a chloroethylating agent that alkylates DNA at the O^6-position of guanine residues. It also produces a carbamoylating agent, methyl isocynate, which inhibits O^6-alkylguanine-DNA alkyltransferase. **1**. Baumann, Shyam, Penketh, *et al.* (2004) *Cancer Chemother. Pharmacol.* **53**, 288.

Bisoprolol

This oral cardiospecific β-blocker ($FW_{free-base}$ = 325.45 g/mol; CAS 66722-44-9; IUPAC: (*RS*)-1-{4-[(2-isopropoxyethoxy)methyl]phenoxy}-3-(isopropylamino)propan-2-ol), also named Zebeta® (oral tablets as fumarate salt), selectively targets $β_1$-adrenergic receptors, blocking epinephrine stimulation of $β_1$-adrenoreceptors that are concentrated in the heart muscle cells and heart conduction tissue as well as kidney juxtaglomerular cells. Bisoprolol is used to treat hypertension, coronary heart disease, arrhythmias, ischemic heart diseases, and myocardial infarction after the acute event. Given such versatility, bisoprolol is understandably included in the World Health Organization's List of Essential Medicines. **Pharmacokinetics:** Bioavailability > 90%; Hepatic metabolism (50%) by CYP2D6 and CYP3A4; Biological $t_{1/2}$ = 10-12 hours. **1**. Bühring, Sailer, Faro, *et al.* (1986) *J. Cardiovasc. Pharmacol.* **8** (*Supplement* 11), S21–8. **3**. Leopold (1986) *J. Cardiovasc. Pharmacol.* **8** (*Supplement* 11), S16–20. **2**. Horikiri, Suzuki & Mizobe (1998) *J. Pharmaceut. Sci.* **87**, 289.

Bis(oxalato)platinate(II), Potassium

This platinum(II) complex (FW = 449.32 g/mol; $FW_{dihydrate}$ = 485.35 g/mol; CAS 14244-64-5) slowly inactivates D-alanyl-D-alanine carboxypeptidase (*or* DD-peptidase) with a second-order rate constant of about 0.06 $M^{-1}s^{-1}$ at a single binding site near the active-site Zn^{2+}. 3-Mercaptopropionate (racemic) and 3-mercaptoisobutyrate (L-isomer) inhibit the enzyme competitively with K_i values ~5 nM, whereas classical β-lactam compounds are very weak inhibitors. **1**. Charlier, Dideberg, Jamoulle, *et al.* (1984) *Biochem. J.* **219**, 763.

2,3-Bis(phospho)-D-glycerate

This O_2 transport modulator ($FW_{free-acid}$ = 266.04 g/mol; CAS 138-81-8), also called D-2,3-bisphosphoglycerate (2,3-BPG), D-glycerate 2,3-bisphosphate, and formerly 2,3-diphosphoglycerate, is an important intermediate (cofactor) in the reaction catalyzed by phosphoglycerate mutase (*Reaction*: 1,3-BPG ⇌ 2,3-BPG), an enzyme unique to erythrocytes and placental cells. 2,3-BPG synthesis is enhanced at lower pH, a condition occurring during altitude-induced acidosis (hypoxia). The Bohr Effect, which defines the reciprocal binding of H^+ and O_2, also suggests how preferential binding of 2,3-BPG to deoxygenated human hemoglobin A, thereby reducing the latter's affinity for O_2. By dumping a greater fraction of transported O_2, 2,3-BPG helps to attenuate circulatory and cerebral edema associated with severe bouts of Mountain Sickness. Phosphoglycerate mutase possesses a phosphatase activity, which is likely to reduce 2,3-BPG content at lower altitude or normal pH. (Diamox treatment favors the lowering of blood pH and elevation in 2,3-BPG within red blood cells). *Note*: 2,3-BPG also activates 3-hydroxy-3 methylglutaryl-CoA reductase, the rate-controlling enzyme (NADH-dependent, EC 1.1.1.88; NADPH-dependent, EC 1.1.1.34) in the mevalonate pathway producing cholesterol and other essential isoprenoids. **Target(s):** 6-phosphofructokinase (1,13,18,35); acylphosphatase, *or* 1,3-diphosphoglycerate phosphatase (2,43,44); AMP deaminase (3,32,45); 3-phosphoglycerate kinase (4,59-61); casein kinase I (5); casein kinase II (6,7,21,24-26); hemoglobin oxygenation (8,9); inositol-polyphosphate 5-phosphatase, *or* inositol-1,4,5 trisphosphate 5-phosphatase (12,20,27,46-52); glucose-6-phosphate dehydrogenase (14); pyruvate kinase (15,19,30,66-68); phospholipase D (16); inositol 1,3,4-trisphosphate 5/6-kinase (17); phytase (22); 3 deoxy-7-phosphoheptulonate synthase, *or* 3-deoxy-D-*arabino*heptulosonate-7-phosphate synthetase (28); ribose-phosphate diphosphokinase, *or* phosphoribosyl-pyrophosphate synthetase (29,36,56-58); hexokinase (31,70,71); phosphoglucomutase (33,38-40); transaldolase (34); phosphopentomutase (37); aspartate ammonia-lyase, K_i = 1.1 mM (41,42); phosphoinositide 5 phosphatase (53); [β-adrenergic-receptor] kinase, weakly inhibited (54,55); inositol-trisphosphate 3 kinase (62); glucose-1,6-bisphosphate synthase (63-65); glycerate kinase (69); glycogen synthase (72); starch phosphorylase (73); 1-deoxy-D-xylulose-5-phosphate synthase (74). **1**. Kemp (1975) *Meth. Enzymol.* **42**, 67. **2**. Ramponi (1975) *Meth. Enzymol.* **42**, 409. **3**. Nathans, Chang & Deuel (1978) *Meth. Enzymol.* **51**, 497. **4**. Kuntz & Krietsch (1982) *Meth. Enzymol.* **90**, 103. **5**. Hathaway, Tuazon & Traugh (1983) *Meth. Enzymol.* **99**, 308. **6**. Hathaway & Traugh (1983) *Meth. Enzymol.* **99**, 317. **7**. Roach (1984) *Meth. Enzymol.* **107**, 81. **8**. Benesch & Benesch (1967) *Biochem. Biophys. Res. Commun.* **26**, 162. **9**. Chanutin & Curnish (1967) *Arch. Biochem. Biophys.* **121**, 96. **10**. Dische (1976) *Trends Biochem. Sci.* **1**, N269. **11**. Dische (1940) *Bull. Soc. Chim. Biol.* **23**, 1140. **12**. Rana, Sekar, Hokin & MacDonald (1986) *J. Biol. Chem.* **261**, 5237. **13**. Espinet, Bartrons & Carreras (1988) *Comp. Biochem. Physiol. B* **90**, 453. **14**. Ozer, Aksoy & Ogus (2001) *Int. J. Biochem. Cell. Biol.* **33**, 221. **15**. Kaloyianni, Cotoglou & Giagtzoglou (2000) *J. Comp. Physiol. [B]* **170**, 85. **16**. Kusner, Hall & Schlesinger (1996) *J. Exp. Med.* **184**, 585. **17**. Hughes, Kirk & Michell (1994) *Biochim. Biophys. Acta* **1223**, 57. **18**. Kaloyianni, Kotinis & Gounaris (1994) *Comp. Biochem. Physiol. Biochem. Mol. Biol.* **107**, 479. **19**. Podesta & Plaxton (1991) *Biochem. J.* **279**, 495. **20**. Fowler & Brännström (1990) *Biochem. J.* **271**, 735. **21**. Gonzatti & Traugh (1988) *Biochem. Biophys. Res. Commun.* **157**, 134. **22**. Martin & Luque (1985) *Comp. Biochem. Physiol. B* **80**, 557. **23**. Morin, Noble & Srikantaiah (1984) *Experientia* **40**, 953. **24**. Hathaway & Traugh (1984) *J. Biol. Chem.* **259**, 7011. **25**. Hathaway & Traugh (1984) *J. Biol. Chem.* **259**, 2850. **26**. Gonzatti-Haces & Traugh (1982) *J. Biol. Chem.* **257**, 6642. **27**. Downes, Mussat & Michell (1982) *Biochem. J.* **203**, 169. **28**. Simpson & Davidson (1976) *Eur. J. Biochem.* **70**, 501. **29**. Becker, Kostel & Meyer (1975) *J. Biol. Chem.* **250**, 6822. **30**. Ng & Hamilton (1975) *J. Bacteriol.* **122**, 1274. **31**. Rijksen & Staal (1977) *Biochim. Biophys. Acta* **485**, 75. **32**. Zielke & Suelter (1971) *The Enzymes*, 3rd ed. (Boyer, ed.), **4**, 47. **33**. Ray & Peck (1972) *The Enzymes*, 3rd ed. (Boyer, ed.), **6**, 407. **34**. Tsolas & Horecker (1972) *The Enzymes*, 3rd ed. (Boyer, ed.), 7, 259. **35**. Bloxham & Lardy (1973) *The Enzymes*, 3rd ed. (Boyer, ed.), **8**, 239. **36**. Switzer (1974) *The*

Enzymes, 3rd ed. (Boyer, ed.), **10**, 607. **37**. Webb (1966) *Enzyme and Metabolic Inhibitors*, vol. **2**, p. 413, Academic Press, New York. **38**. Chiba, Ueda & Hirose (1976) *Agric. Biol. Chem.* **40**, 2423. **39**. Fazi, Piacentini, Piatti & Accorsi (1990) *Prep. Biochem.* **20**, 219. **40**. Takamiya & Fukui (1978) *J. Biochem.* **84**, 569. **41**. Falzone, Karsten, Conley & Viola (1988) *Biochemistry* **27**, 9089. **42**. Viola (2000) *Adv. Enzymol. Relat. Areas Mol. Biol.* **74**, 295. **43**. Pazzagli, Ikram, Liguri, *et al.* (1993) *Ital. J. Biochem.* **42**, 233. **44**. Liguri, Camici, Manao, *et al.* (1986) *Biochemistry* **25**, 8089. **45**. Lian & Harkness (1978) *Biochim. Biophys. Acta* **341**, 27. **46**. Verjans, De Smedt, Lecocq, *et al.* (1994) *Biochem. J.* **300**, 85. **47**. Milani, Volpe & Pozzan (1988) *Biochem. J.* **254**, 525. **48**. Hodgkin, Craxton, Parry, *et al.* (1994) *Biochem. J.* **297**, 637. **49**. Communi, Lecocq & Erneux (1996) *J. Biol. Chem.* **271**, 11676. **50**. Erneux, Govaerts, Communi & Pesesse (1998) *Biochim. Biophys. Acta* **1436**, 185. **51**. Hansen, Johanson, Williamson & Williamson (1987) *J. Biol. Chem.* **262**, 17319. **52**. Hansbro, Foster, Hogan, Ozaki & Denborough (1994) *Arch. Biochem. Biophys.* **311**, 47. **53**. Palmer, Théolis, Cook & Byers (1994) *J. Biol. Chem.* **269**, 3403. **54**. Benovic (1991) *Meth. Enzymol.* **200**, 351. **55**. Benovic, Stone, Caron & Lefkowitz (1989) *J. Biol. Chem.* **264**, 6707. **56**. Fox & Kelley (1972) *J. Biol. Chem.* **247**, 2126. **57**. Nosal, Switzer & Becker (1993) *J. Biol. Chem.* **268**, 10168. **58**. Becker (2001) *Prog. Nucleic Acid Res. Mol. Biol.* **69**, 115. **59**. Nojima, Oshima & Noda (1979) *J. Biochem.* **85**, 1509. **60**. Szilágyi & Vas (1998) *Biochemistry* **37**, 8551. **61**. Ali & Brownstone (1976) *Biochim. Biophys. Acta* **445**, 89. **62**. Carrasco & Figueroa (1995) *Comp. Biochem. Physiol. B* **110**, 747. **63**. Maliekal, Sokolova, Vertommen, Veiga-da-Cunha & Van Schaftingen (2007) *J. Biol. Chem.* **282**, 31844. **64**. Rose, Warms & Wong (1977) *J. Biol. Chem.* **252**, 4262. **65**. Ueda, Hirose, Sasaki & Chiba (1978) *J. Biochem.* **83**, 1721. **66**. Wu & Turpin (1992) *J. Phycol.* **28**, 472. **67**. Calomenopoulou, Kaloyianni & Beis (1989) *Comp. Biochem. Physiol. B Comp. Biochem.* **93**, 697. **68**. Lin, Turpin & Plaxton (1989) *Arch. Biochem. Biophys.* **269**, 228. **69**. Yoshida, Fukuta, Mitsunaga, Yamada & Izumi (1992) *Eur. J. Biochem.* **210**, 849. **70**. de Cesar, Colepicolo, Rosa & Rosa (1997) *Comp. Biochem. Physiol. B* **118**, 395. **71**. Magnani, Stocchi, Serafini, *et al.* (1983) *Arch. Biochem. Biophys.* **226**, 377. **72**. Mied & Bueding (1979) *J. Parasitol.* **65**, 14. **73**. Lee & Braun (1973) *Arch. Biochem. Biophys.* **156**, 276. **74**. Eubanks & Poulter (2003) *Biochemistry* **42**, 1140.

Bisulfite Ion

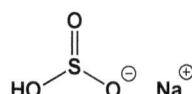

This anionic food preservative (FW = 104.06 g/mol (Sodium Salt); CAS 7631-90-5; Solubility = 0.42 g/mL) is a powerful antioxidant and food additive (E222 on the European Scale) that likewise inhibits many enzymes. Bisulfite ion has the same geometry as bicarbonate and metaphosphate. In winemaking and dried fruit preservation, sodium bisulfite releases sulfur dioxide gas to increase its bacterial and fungal killing action. (**See also** *Sulfur Dioxide; Metabisulfite*) In erythrocytes, anion transport of bisulfite and diffusion of its conjugate acid in the form of SO_2 make the rate of bisulfite influx through the anion exchange pathway at least 100-fold faster than sulfate ion. **Bisulfite Sequencing:** Providing the ability to analyze the methylation state of every cytosine accurately, bisulfite sequencing has become the gold standard for DNA methylation analysis. When coupled with next-generation sequencing technology, it allows for an unbiased genome-wide analysis of DNA methylation and various analyses can be performed on the altered sequence. Combining bisulfite treatment and high-throughput sequencing generates high-accuracy methylome, providing reference for differential DNA methylation analysis across large-scale samples. **Note:** Because bisulfite slowly oxidizes in air to form sulfate, reagent bottles should be tightly closed. Replace reagent after one year.
Target(s): mannitol-1-phosphate dehydrogenase (1); hydroxylamine reductase (2); ethanolamine oxidase (3); acetoacetate decarboxylase (4); histidine ammonia-lyase (5,17,45,46); L-2-aminoadipate-6-semialdehyde dehydrogenase (6,22); amine oxidase (7,11); mannosidostreptomycin hydrolase (8); arylformamidase (9,50-53); urease (12); L-lactate dehydrogenase (13,27); glucose oxidase (14); cytochrome *c* oxidase, complex IV (15); carbonic anhydrase (16,36,38,39); aminomalonate decarboxylase, *or* aspartate 4-decarboxylase (18); cystathionine γ-lyase, *or* cysteine desulfhydrase (19,43); *Clostridium pasteurianum* F_1F_o ATP synthase (20); Lactoperoxidase (21); lipase (23); pyruvate decarboxylase (10,24); Arylsulfatase (25); formate dehydrogenase (26); glycogen phosphorylase (28); starch phosphorylase (28); papain (29); DNA

polymerase I (30); phospholipase A, lysosomal (31); phospholipase C (31,34); Lactoperoxidase (32); lysozyme (33); xanthine oxidase (35); F_1F_o ATP synthase, *Clostridium pasteurianum* (37); homogentisate dioxygenase (40); kynureninase (41); bisphosphoglycerate mutase (42); phenylalanine ammonia-lyase (44); urocanate hydratase, *or* urocanase (47); sphinganine-1-phosphate aldolase (48); acetoacetate decarboxylase (49); adenosylhomocysteinase (54); bisphosphoglycerate phosphatase (55); leucine aminotransferase (56); catechol oxidase (57); tyrosinase, *or* monophenol monooxygenase (57); nitronate monooxygenase (58); laccase (59); catechol oxidase (60-62); polyphenol oxidase (60-62); nitrate reductase, cytochrome (63). **1**. Wolff & Kaplan (1955) *Meth. Enzymol.* **1**, 346. **2**. Zucker & Nason (1955) *Meth. Enzymol.* **2**, 416. **3**. Narrod & Jakoby (1966) *Meth. Enzymol.* **9**, 354. **4**. Westheimer (1969) *Meth. Enzymol.* **14**, 231. **5**. Rechler & H. Tabor (1971) *Meth. Enzymol.* **17B**, 63. **6**. Rodwell (1971) *Meth. Enzymol.* **17B**, 188. **7**. Yasunobu & Smith (1971) *Meth. Enzymol.* **17B**, 698. **8**. Inamine & Demain (1975) *Meth. Enzymol.* **43**, 637. **9**. Katz, Brown & Hitchcock (1987) *Meth. Enzymol.* **142**, 225. **10**. Vennesland (1951) *The Enzymes*, 1st ed. (Sumner & Myrbäck, eds.), **2** (part 1), 183. **11**. Zeller (1951) *The Enzymes*, 1st ed. (Sumner & Myrbäck, eds.), **2** (part 1), 536. **12**. Varner (1960) *The Enzymes*, 2nd ed. (Boyer, Lardy & Myrbäck, eds.), **4**, 247. **13**. Schwert & Winer (1963) *The Enzymes*, 2nd ed. (Boyer, Lardy & Myrbäck, eds.), **7**, 142. **14**. Bentley (1963) *The Enzymes*, 2nd ed. (Boyer, Lardy & Myrbäck, eds.), **7**, 567. **15**. Yonetani (1963) *The Enzymes*, 2nd ed. (Boyer, Lardy & Myrbäck, eds.), **8**, 41. **16**. Lindskog, Henderson, Kannan, *et al.* (1971) *The Enzymes*, 3rd. ed. (Boyer, ed.), **5**, 587. **17**. Hanson & Havir (1972) *The Enzymes*, 3rd ed. (Boyer, ed.), **7**, 75. **18**. Thanassi & Fruton (1962) *Biochemistry* **1**, 975. **19**. Lawrence & Smythe (1943) *Arch. Biochem.* **2**, 225. **20**. Das & Ljungdahl (2003) *J. Bacteriol.* **185**, 5527. **21**. Ohlsson (1984) *Eur. J. Biochem.* **142**, 233. **22**. Calvert & Rodwell (1966) *J. Biol. Chem.* **241**, 409. **23**. Weinstein & Wynne (1936) *J. Biol. Chem.* **112**, 649. **24**. Wallerstein & Stern (1945) *J. Biol. Chem.* **158**, 1. **25**. Dodgson (1959) *Enzymologia* **20**, 301. **26**. Kanamori & Suzuki (1968) *Enzymologia* **35**, 185. **27**. Ciaccio (1966) *J. Biol. Chem.* **241**, 1581. **28**. Kamogawa & Fukui (1973) *Biochim. Biophys. Acta* **302**, 158. **29**. Fujimoto, Nakagawa, Ishimitsu & Ohara (1983) *Chem. Pharm. Bull. (Tokyo)* **31**, 992. **30**. Mallon & Rossman (1983) *Chem. Biol. Interact.* **46**, 101. **31**. Eisen, Bartolf & Franson (1984) *Biochim. Biophys. Acta* **793**, 10. **32**. Ohlsson (1984) *Eur. J. Biochem.* **142**, 233. **33**. Nishida, Morita, Ono & Shimakawa (1987) *Chem. Pharm. Bull. (Tokyo)* **35**, 2519. **34**. Gamache, Fawzy & Franson (1988) *Biochim. Biophys. Acta* **958**, 116. **35**. Fish, Massey, Sands & Dunham (1990) *J. Biol. Chem.* **265**, 19665. **36**. Hakansson, Carlsson, Svensson & Liljas (1992) *J. Mol. Biol.* **227**, 1192. **37**. Das & Ljungdahl (2003) *J. Bacteriol.* **185**, 5527. **38**. Innocenti, Zimmerman, Ferry, Scozzafava & Supuran ((2004) *Bioorg. Med. Chem. Lett.* **14**, 3327. **39**. Pocker & Stone (1968) *Biochemistry* **7**, 2936. **40**. Schepartz & Boyle (1953) *J. Biol. Chem.* **205**, 185. **41**. Jakoby & Bonner (1953) *J. Biol. Chem.* **205**, 699. **42**. Sasaki, Ikura, Sugimoto & Chiba (1975) *Eur. J. Biochem.* **50**, 581. **43**. Matsuo & Greenberg (1958) *J. Biol. Chem.* **234**, 507. **44**. Kalghatgi & Subba Rao (1975) *Biochem. J.* **149**, 65. **45**. Hernandez, Phillips & Zon (1994) *Biochem. Mol. Biol. Int.* **32**, 189. **46**. Hernandez & Phillips (1994) *Biochem. Biophys. Res. Commun.* **201**, 1433. **47**. Hacking, Bell & Hassall (1978) *Biochem. J.* **171**, 41. **48**. Shimojo, Akino, Miura & Schroepfer (1976) *J. Biol. Chem.* **251**, 4448. **49**. Kimura, Yasuda, Tanigaki-Nagae, *et al.* (1986) *Agric. Biol. Chem.* **50**, 2509. **50**. Serrano & Nagayama (1991) *Comp. Biochem. Physiol. B Comp. Biochem.* **99**, 281. **51**. Shinohara & OIshiguro (1970) *Biochim. Biophys. Acta* **198**, 324. **52**. Bode & Birnbaum (1979) *Biochem. Physiol. Pflanz.* **174**, 26. **53**. Jakoby (1954) *J. Biol. Chem.* **207**, 657. **54**. Fujioka & Takata (1981) *J. Biol. Chem.* **256**, 1631. **55**. Sheibley & Hass (1976) *J. Biol. Chem.* **251**, 6699. **56**. Pathre, Singh, Viswanathan & Sane (1987) *Phytochemistry* **26**, 2913. **57**. Ben-Shalom, Kahn, Harel & Mayer (1977) *Phytochemistry* **16**, 1153. **58**. Kido, Soda, Suzuki & Asada (1976) *J. Biol. Chem.* **251**, 6994. **59**. Ishigami, Hirose & Yamada (1988) *J. Gen. Appl. Microbiol.* **34**, 401. **60**. Motoda (1979) *J. Ferment. Technol.* **57**, 79. **61**. Sellés-Marchart, Casado-Vela & Bru-Martínez (2006) *Arch. Biochem. Biophys.* **446**, 175. **62**. Wong, Luh & Whitaker (1971) *Plant Physiol.* **48**, 19. **63**. Sadana & McElroy (1957) *Arch. Biochem. Biophys.* **67**, 16.

Bivalirudin

This intravenously administered, direct thrombin inhibitor (FW = 2180.29 g/mol; CAS 128270-60-0; *Sequence*: FPRPGGGGNGDFEEIPEEYL, with N-terminal D-phenylalanine), also known by its trade name Angiomax™, is a hirudin analogue and potent anticoagulant, exhibiting high potency (K_i = 2 nM) by simultaneously occupying the fibrinogen-binding region of thrombin's active site and its anion-binding exosite. Angiomax is indicated as a thrombin-specific anticoagulant to be used in in patients taking aspirin with unstable angina undergoing percutaneous transluminal coronary angioplasty (PTCA). By inhibiting both thrombin's catalytic and exosite domains, bivalirudin enjoys antithrombotic potency that is a thousand times greater than achieved by inhibitors acting at either domain alone, (1). Bivalirudin is particularly at sites of deep arterial injury, and because it degraded proteolytically, its action ceases within 1 hour of its last dose. **1.** Kelly, Maraganore, Bourdon, Hanson & Harker (1992) *Proc. Natl. Acad. Sci. U.S.A.* **89**, 6040.

BIX01294

This histone methyl transferase inhibitor (FW$_{\text{free-base}}$ = 490.65 g/mol; FW$_{\text{tri-HCl}}$ = 600.00 g/mol; CAS 935693-62-2; IUPAC: 2-(hexahydro-4-methyl-1*H*-1,4-diazepin-1-yl)-6,7-dimethoxy-*N*-[1-(phenylmethyl)-4-piperidinyl]-4-quinazolinamine) selectively targets G9a HMTase, IC$_{50}$ = 1.7 μM (1), with lower inhibitory effectiveness against HMTase G9a-like protein (GLP; IC$_{50}$ = 38 μM). BIX01294 has little or no inhibitory effect on other known HMTases (1). Methylation of histone H3 at lysine 9 (H3K9) occurs in heterochromatin, which requires H3K9me$_3$, and in euchromatin, which requires H3K9me$_1$ and H3K9me2, formed mainly by G9a and GLP. H3K9me$_1$ and H3K9me$_2$ are the only silencing marks lost when tumor suppressor genes in colorectal and breast cancer cells are reactivated after treatment with the DNA demethylation drug, 5-aza-2'-deoxycytidine. G9a HMTase and GLP are thus considered to be important druggable targets. Moreover, when used along with the calcium channel activator BayK8644, BIX01294 also facilitates the *in vitro* generation of induced pluripotent stem cells from somatic cells (2). The structural basis for BIX-01294 inhibition of G9a-like protein lysine methyltransferase is well understood (3). **1.** Kubicek, O'Sullivan, August, *et al.* (2007) *Mol. Cell* **25**, 473. **2.** Shi, Desponts, Do, *et al.* (2008) *Cell Stem Cell* **3**, 568. **3.** Chang, Zhang, Horton, *et al.* (2009) *Nature Struct. Mol. Biol.* **16**, 312.

BIX 02189

This selective MEK/ERK pathway inhibitor (FW = 440.54 g/mol; CAS 1265916-41-3; Solubility: 100 mM in DMSO; 100 mM for monohydrochloride salt), also named (3*Z*)-3-[[[3-[(dimethylamino)methyl]phenyl]amino]phenylmethylene]-2,3-dihydro-*N,N*-dimethyl-2-oxo-1*H*-indole-6-carboxamide, targets Mitogen/Extracellular signal-regulated Kinase-5, *or* MEK5 (IC$_{50}$ = 1.5 nM), and Extracellular-signal-Regulated Kinase, *or* ERK5 (IC$_{50}$ = 59 nM), showing far weaker effects on MEK1 (IC$_{50}$ = 6200 nM), MEK2 (IC$_{50}$ = >6200 nM), ERK1 (IC$_{50}$ >6200 nM), JNK2 (IC$_{50}$ >6200 nM), TGFβR1 (IC$_{50}$ = 580 nM), EGFR (IC$_{50}$ >6300 nM), and STK16 (IC$_{50}$ >6300 nM) (1). **Pharmacology:** ERK5 is an atypical MAPK that is activated in the heart by pressure overload. ERK5 regulates cardiac hypertrophy and hypertrophy-induced apoptosis, and silencing ERK5 expression or inhibiting its kinase activity with BIX02189 reduces Myocyte Enhancer Factor-2 (MEF2) transcriptional activity and blunted hypertrophic responses in neonatal rat cardiomyocytes, *or* NRCMs (2). By applying BIX02189 to NRCMs, ERK5 phosphorylation is blocked without affecting other MAP kinases (2). BIX02189 induces apoptosis in acute

myeloid leukemia tumor cells without affecting T cells from healthy donors (3). **1.** Tatake, O'Neill, Kennedy, *et al.* (2008) *Biochem. Biophys. Res. Commun.* **377**, 120. **2.** Kimura, Jin, Zi, *et al.* (2010) *Circ. Res.* **106**, 961. **3.** Lopez-Royuela, Rathore, Allende-Vega, *et al.* (2014) *Int. J. Biochem. Cell Biol.* **53C**, 253.

BKM-120, *See Buparlisib*

BKT140

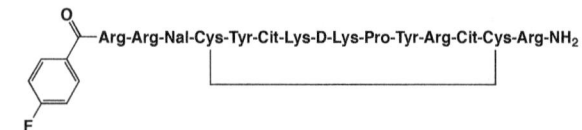

This orally bioavailable, disulfide cross-linked, polyphemusin-II-derived peptide antagonist (sequence shown above, with Nal = naphthylamine, Cit = citrulline, D-Lys = D-Lysine, and Cys = half-cystine), also known as 4-F-benzoyl-TN14003 and BL-8040, targets CXC Chemokine Receptor-4 (CXCR4), a G-protein-coupled receptor that is directly involved in tumor progression, tumor angiogenesis, metastasis, and cancer cell survival (1,2). **Primary Mode of Inhibitory Action:** CXCR4 is over-expressed in more than 70% of human cancers and its expression often correlates with disease severity. BKT140 exhibits high affinity (K_i = 0.8nM) and a low rate of dissociation from its receptor. BKT140 prevents the binding of stromal derived factor-1 (SDF-1 *or* CXCL12) to CXCR4 activation, resulting in decreased tumor cell proliferation and migration. In addition, inhibition of CXCR4 may mobilize hematopoietic cell egress from the bone marrow and into blood. BKT140 significantly and preferentially stimulates multiple myeloma apoptotic cell death, as indicated by induced morphological changes, phosphatidylserine externalization, decreased mitochondrial membrane potential, caspase-3 activation, sub-G$_1$ cell-cycle arrest, as well as DNA double-stranded breaks (2). **1.** Tamamura, Hiramatsu, Mizumoto, *et al.* (2003) *Org. Biomol. Chem.* **1**, 3663. **2.** Beider K, Begin M, Abraham, *et al.* (2011) *Exp. Hematol.* **39**, 282.

BL-4162A, *See Anagrelide*

BL-8040, *See BKT140*

Blasticidin S

This pyrimidine antibiotic (FW = 422.43 g/mol; CAS 2079-00-7) from *Streptomyces griseochromogenes* inhibits protein biosynthesis through its action as a mixed-type noncompetitive inhibitor of the peptidyltransferase (1-10). Blasticidin S has been used to suppress rice blast disease. Structures of the large ribosomal subunit of *Haloarcula marismortui* at 3.0Å resolution indicate the poses taken by bound anisomycin, chloramphenicol, sparsomycin, blasticidin S, and virginiamycin M (9). Two hydrophobic crevices, one at the peptidyl transferase center and the other at the entrance to the peptide exit tunnel play roles in antibiotic binding. Blasticidin S base pairs with the P-loop and thereby mimics C74 and C75 of a P-site bound tRNA. Most ribotoxic antibiotics bind to sites that overlap those of either peptidyl-tRNA or aminoacyl-tRNA, consistent with their functioning as competitive inhibitors of peptide bond formation (13). The cytidine moiety of blasticidin S makes a Watson-Crick base-pair with the corresponding guanine base at the P-site of the 50S ribosome. Blasticidin S is enzymatically deaminated to its uracil-containing form, which cannot bind to the P-site and likewise lacks antibiotic aactivity. **Target(s):** peptidyltransferase (1-13); DNA biosynthesis (14); arginase (15); TMV RNA biosynthesis (16); ribonuclease P (17); cytidine deaminase (18). **1.** Gottesman (1971) *Meth. Enzymol.* **20**, 490. **2.** Pestka (1974) *Meth. Enzymol.* **30**, 261. **3.** Carrasco, Battaner & Vazquez (1974) *Meth. Enzymol.* **30**, 282. **4.** Jiménez (1976) *Trends Biochem. Sci.* **1**, 28. **5.** Coutsogeorgopoulos (1967) *Biochemistry* **6**, 1704. **6.** Yamaguchi & Tanaka (1966) *J. Biochem.* **60**, 632. **7.** Yamaguchi, Yamamoto & Tanaka (1965) *J. Biochem.* **57**, 667. **8.** Sikorski, Cerna, Rychlik & Legocki (1977) *Biochim. Biophys. Acta* **475**, 123. **9.** Kalpaxis, Theocharis & Coutsogeorgopoulos

(1986) *Eur. J. Biochem.* **154**, 267. **10**. Dinos & Kalpaxis (1997) *Pharmazie* **52**, 875. **11**. Kalpaxis, Theocharis & Coutsogeorgopoulos (1986) *Eur. J. Biochem.* **154**, 267. **12**. Ioannou, Coutsogeorgopoulos & Drainas (1997) *Anal. Biochem.* **247**, 115. **13**. Hansen, Moore & Steitz (2003) *J. Mol. Biol.* **330**, 1061. **14**. Sullia & Griffin (1977) *Biochim. Biophys. Acta* **475**, 14. **15**. Shimotohno, Iida, Takizawa & Endo (1994) *Biosci. Biotechnol. Biochem.* **58**, 1045. **16**. Hirai, Wildman & Hirai (1968) *Virology* **36**, 646. **17**. Kalavrizioti, Vourekas, Tekos, *et al.* (2003) *Mol. Biol. Rep.* **30**, 9. **18**. Faivre-Nitschke, Grienenberger & Gualberto (1999) *Eur. J. Biochem.* **263**, 896.

(–)-Blebbistatin

This selective, cell-permeant, nonmuscle myosin II inhibitor (FW = 292.34 g/mol; CAS 674289-55-5) binds to myosin II, altering van der Waals contacts, destroying a critical salt bridge, and blocking the myosin's mechanochemical action. Blebbistatin inhibits contraction of the cleavage furrow without disrupting mitosis or contractile ring assembly (1). Blebbistatin therefore blocks cell blebbing rapidly and reversibly (1,2). Illumination of blebbistatin-treated bovine aortic endothelial cells at 365 and 450-490 nm, but not 510-560 or 590-650 nm, caused dose-dependent cell death (3). Illumination of blebbistatin alone at 365 and 450-490 nm changed its absorption and emission spectra, suggesting that the structure had been chaged. While nontoxic, the resultant photocompounds no longer disrupted myosin distribution in cells, indicating consequential loss of blebbistatin's pharmacological activity. Fluorescence microscopy also demonstrated that upon illumination, blebbistatin became bound to cells and to protein-coated glass, suggesting its toxicity may arise, at least in part, from light-induced reactions. Blebbistatin should be used in a manner that controls for these photochemical effects. **Target(s):** myosin II ATPase, IC_{50} = 2 μM (1); [myosin light-chain] kinase (2). **1**. Straight, Cheung, Limouze, *et al.* (2003) *Science* **299**, 1743. **2**. Lucero, Stack, Bresnick & Shuster (2006) *Mol. Biol. Cell* **17**, 4093. **3**. Kolega (2004) *Biochem Biophys Res Commun.* **320**, 1020.

Bleomycins

This set of glycopeptide antibiotics and anticancer agents (Bleomycin A_1: FW = 1415.55 g/mol; CAS 11056-06-7; R = –NH(CH₂)₃–SOCH₃); Bleomycin A_2, R = –NH(CH₂)₃–S⁺CH₃); Bleomycin B_1, R = –NH(CH₂)₄–NCH(=NH)NH₂); Bleomycin A_5, R = –NH(CH₂)₃–NH(CH₂)₄–NH₂); and Bleomycin A_6, R = –NH(CH₂)₃–NH(CH₂)₄–NH(CH₂)₃–NH₂), often abbreviated BLM and produced by *Streptomyces verticillus*, are DNA strand-breaking agents that exploit their interaction with Fe(II) ion. More than two-hundred bleomycins have been isolated from different strains of *Streptomyces*, with molecular weights ranging from 1000 to 10,000. Bleomycin's activity is associated with single-strand DNA cleavage through the generation of hydroxyl free radicals. Although the exact *in vivo* mechanism of bleomycin's action as an antineoplastic agent is unknown, available evidence indicates that the main mode of action is the inhibition of DNA synthesis with lesser effects arising from effects on RNA and protein synthesis. Bleomycin can stop the cell cycle at G_2 and is often used to induce synchrony in cell cultures. **Primary Mode of Action:** Bleomycin forms chelation complexes with Fe^{2+}, Co^{2+}, Cu^{2+}, Ni^{2+}, and Zn^{2+}. Upon binding of a bleomycin-Fe^{2+}-O_2 complex to DNA, the coplanar bithiazole moiety intercalates in the minor groove of DNA, whereupon the iron is oxidized to

Fe^{3+}, and the resulting complex of bleomycin acts on deoxyribose. Bleomycin interacts with either Fe(II) and O_2 or Fe(III) and H_2O_2 to form an activated complex that attacks DNA (1). Under aerobic conditions, both reactions yield similar quantities of free bases as well as products consisting of the base plus deoxyribose carbon-atoms 1 to 3. Under anaerobic conditions, activated bleomycin releases only free base. The yield of free base is the same under aerobic or anaerobic conditions, when DNA is in excess. When DNA concentration is limiting, more base is released under anaerobic than under aerobic conditions. Drug self-destruction proceeds as quickly and completely in the presence or absence of O_2. **Pharmacokinetics:** Bleomycin is rapidly absorbed following intramuscular (IM) or subcutaneous (SC), administration reaching peak plasma concentrations in 30 to 60 minutes and exhibiting a systemic bioavailability of 100% (IM) and 70% (SC). Upon IV bolus administration, bleomycin is widely distributed, with a mean volume of distribution of 17.5 L/m². Protein binding has not been studied. Bleomycin is inactivated by a cytosolic cysteine proteinase enzyme, bleomycin hydrolase, which is widely distributed in normal tissues (exception: skin and lungs, where toxicity is highest. **Target(s):** DNA polymerase (2,5,7,8,12); protein-glutamine γ-glutamyltransferase (transglutaminase), by bleomycin alone or copper-bleomycin (3); RNA polymerase (2,6,11); DNA ligase (4,7); dopamine β-monooxygenase (9); tyrosine monooxygenase (10). **1**. Burger, Peisach & Horwitz (1982) *J. Biol. Chem.* **257**, 3372. **2**. Müller (1980) in *Inhibitors of DNA and RNA Polymerases* (Sarin & Gallo, eds.) p. 207, Pergamon Press, Oxford. **3**. Griffin, Barnes, Wynne & Williams (1978) *Biochem. Pharmacol.* **27**, 1211. **4**. Miyaki, Ono & Umezawa (1971) *J. Antibiot.* **24**, 587. **5**. Müller, Yamazaki & Zahn (1972) *Biochem. Biophys. Res. Commun.* **46**, 1667. **6**. Shishido (1973) *J. Antibiot.* **26**, 501. **7**. Ono, Miyaki, Taguchi & Ohashi (1976) *Prog. Biochem. Pharmacol.* **11**, 48. **8**. DiCioccio & Srivastava (1976) *Cancer Res.* **36**, 1664. **9**. Matsui, Kato, Yamamoto, *et al.* (1980) *J. Antibiot.* **33**, 435. **10**. Oka, . Kato, Takita, *et al.* (1980) *J. Antibiot.* **33**, 1043. **11**. Capobianco, Musella, Pelella, Farina & Zarrilli (1981) *Boll. Soc. Ital. Biol. Sper.* **57**, 138. **12**. FitzGerald & Wick (1981) *Biochem. Biophys. Res. Commun.* **101**, 734.

BMN 673

This novel, orally bioavailable poly(ADP-ribose) polymerase, or PARP, inhibitor (FW = 380.35 g/mol; CAS 1207456-01-6), also known as (8S,9R)-5-fluoro-8-(4-fluorophenyl)-9-(1-methyl-1H-1,2,4-triazol-5-yl)-8,9-dihydro-2H-pyrido[4,3,2-de]phthalazin-3(7H)-one, targets PARP-mediated DNA repair (IC_{50} = 0.58 nM) of single-strand DNA breaks by the base-excision repair pathway. By enhancing the accumulation of DNA strand breaks, BMN 673 promotes genomic instability, eventually leading to apoptosis. BMN 673 exhibits selective anti-tumor cytotoxicity and elicits DNA repair biomarkers at much lower concentrations than earlier generation PARP1/2 inhibitors, including olaparib, rucaparib and veliparib. BMN 673 shows remarkable anti-tumor activity *in vivo*, with strong action against xenografted tumors carrying defects in DNA repair due to BRCA mutations or PTEN deficiency. Synergistic or additive anti-tumor effects are observed, when BMN 673 was combined with temozolomide, SN38 or platinum drugs. **1**. Shen, Rehma, Feng, *et al.* (2013) *Clin. Cancer Res.* **19**, 5003.

BMS-232632, *See* Atazanavir

BMS-265246

This cyclin -directed inhibitor (FW = 345.34 g/mol; CAS 582315-72-8; Solubility: 20 mg/mL DMSO; <1 mg/mL Water), also known systematically as (4-butoxy-1*H*-pyrazolo[3,4-*b*]pyridin-5-yl)(2,6-difluoro-4-methylphenyl) methanone, targets CDK1/cyclin B and CDK2/cyclin E with IC_{50} of 6 nM and 9 nM, respectively. **Cyclin Target Selectivity:** Cdk1 (+++), Cdk2 (+++), Cdk3 (weak, if any), Cdk4 (+), Cdk5 (weak, if any), Cdk6 (weak, if any), Cdk7 (weak, if any), Cdk8 (weak, if any), Cdk9 (weak, if any), Cdk10 (weak, if any). **1**. Misra, *et al.* (2003) Bioorg. *Med. Chem. Lett.* **13**, 2405. **2**. Sutherland, *et al.* (2011) *Mol. Cancer Ther.* **10**, 242.

BMS-354825, *See* *Dasatinib*

BMS-387032, *See* *SNS-032*

BMS-477118, *See* *Saxagliptin*

BMS 512148, *See* *Dapagliflozin*

BMS-540215, *See* *Brivanib*

BMS 566419

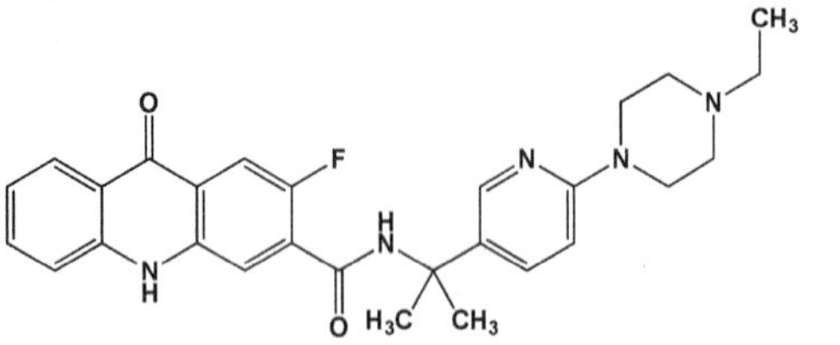

This IMPDH inhibitor (FW = 487.58 g/mol; CAS 566161; Solubility: 50 mM in H_2O (mono-HCl salt); 100 mM in DMSO (free base)), also named *N*-[1-[6-(4-ethyl-1-piperazinyl)-3-pyridinyl]-1-methylethyl]-2-fluoro-9,10-dihydro-9-oxo-3-acridinecarboxamide, targets inosine monophosphate dehydrogenase isoform-2, a key enzyme in *de novo* guanosine nucleotide synthesis and required for neuron survival. Steady-state kinetics showed that BMS 566419 is a reversible uncompetitive inhibitor relative to IMP (K_i = 25 nM). **1**. Watterson, Chen, Zhao, *et al.* (2007) *J. Med. Chem.* 50, 3730.

BMS-626529

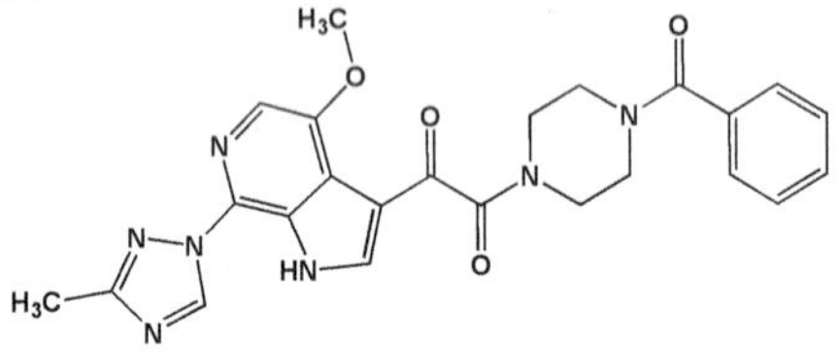

This novel small-molecule HIV-1 attachment inhibitor (FW = 473.49 g/mol; CAS 701213-36-7) is active against both CCR5- and CXCR4-tropic viruses and functions (1). HIV-1 attachment inhibitors represent a new class of drug candidates that prevent the initial interaction between virus and host cell by binding to the viral envelope protein gp120 and blocking attachment of the virus to the CD4 receptor on CD4+ T-cells. The efficacy of BMS-626529 is virus-dependent, due to heterogeneity within gp120 (1). Genotypic substitutions in HIV-1 gp120 (*e.g.*, mainly Met-426-Leu or Ser-375-Met, with Met-434-Ile and Met-475-Ile contributing to a lesser extent) confer phenotypic resistance to BMS-626529 (2). A prodrug (*See BMS-663068*) is currently in clinical development. Administration of BMS-663068 for 8 days with or without ritonavir resulted in substantial declines in plasma HIV-1 RNA levels and was generally well tolerated (2). **1**. Nowicka-Sans, Gong, McAuliffe, *et al.* (2012) *Antimicrob. Agents Chemother.* **56**, 3498. **2**. Zhou, Nowicka-Sans, McAuliffe, *et al.* (2013) *J. Antimicrob. Chemother.* **69**, 573.

BMS-663068, *See* *Fostemsavir*

BMS-707035

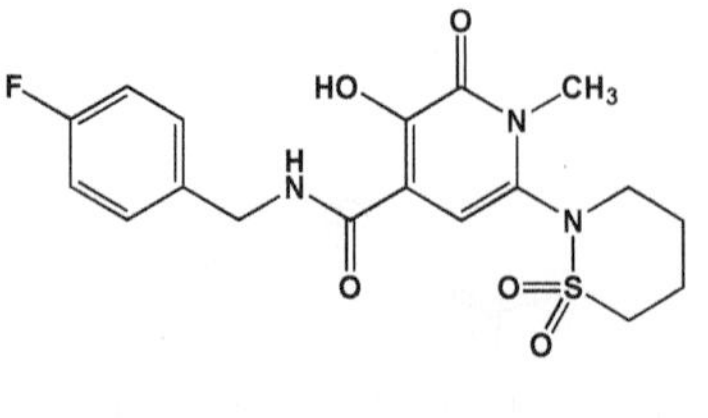

This specific integrase (IN) inhibitor (FW = 410.42 g/mol; CAS 729607-74-3; Solubility: 38 mg/mL DMSO; <1 mg/mL H_2O), also known as *N*-[(4-fluorophenyl)methyl]-1,6-dihydro-5-hydroxy-1-methyl-6-oxo-2-(tetrahydro-1,1-dioxido-2*H*-1,2-thiazine-4-pyrimidinecarboxamide, targets HIV-1 integrase, IC_{50} = 15 nM (1,2). **1**. Dicker, Samanta, Li, *et al.* (2007) *J. Biol. Chem.* **282**, 31186. **2**. Langley, Samanta, Lin, et al. (2008) *Biochemistry* **47**, 13481.

BMS-708163

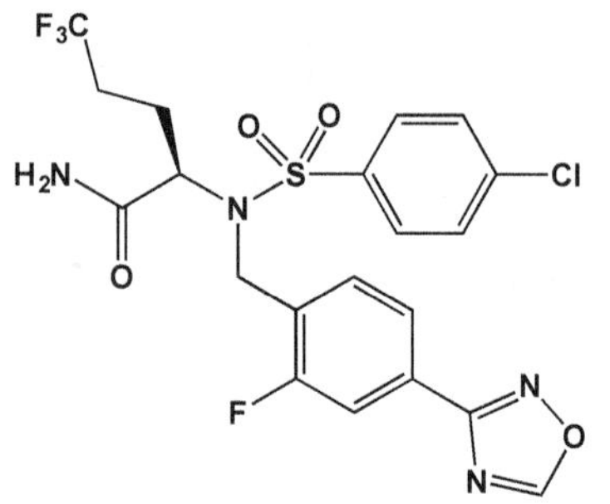

This orally bioavailable, selective protease inhibitor (FW = 520.88 g/mol; CAS 1146699-66-2; Solubility: 100 mg/mL DMSO; <1 mg/mL Water; Formulation for Animal Studies: 99% PEG-400, 1% Tween-80 (in rats) or 94% labrafil-1944, 5% ethanol, 1% tween-80 (in dogs)), known as Avagacestat and (2*R*)-2-(*N*-(2-fluoro-4-(1,2,4-oxadiazol-3-yl)benzyl)-4-chlorophenylsulfonamido)-5,5,5-trifluoropentanamide, targets γ-secretase (GS), thereby inhibiting the formation of Aβ40 and Aβ42, with IC_{50} values of 0.3 nM and 0.27 nM, respectively. **1**. Gillman, Starrett, Parker, *et al.* (2010) *Med. Chem. Lett.* **1**, 120.

BMS-790052, *See* *Daclatasvir*

BMS-791325

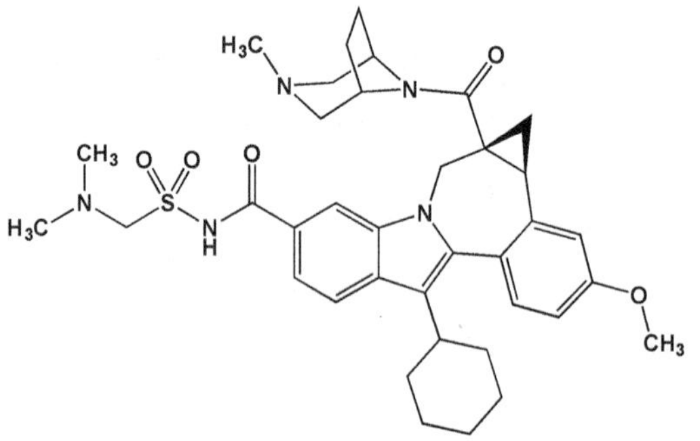

This potent allosteric inhibitor (FW = 673.87 g/mol) targets hepatitis C virus NS5B polymerase IC_{50} <0.5 nM (biochemical assay) and EC_{50} = 0.8-6 nM (replicon assay). Unlike prior cyclopropyl-fused indolobenzazepine inhibitors, BMS-791325 has improved solubility and membrane permeability properties as well as reduced off-target activities, directed most notably against human pregnane X receptor (hPXR) transactivation. **1**. Gentles, Ding, Bender, *et al.* (2014) *J. Med. Chem.* **57**, 1855.

BMS-936558, *See* *Nivolumab*

BMS-986001

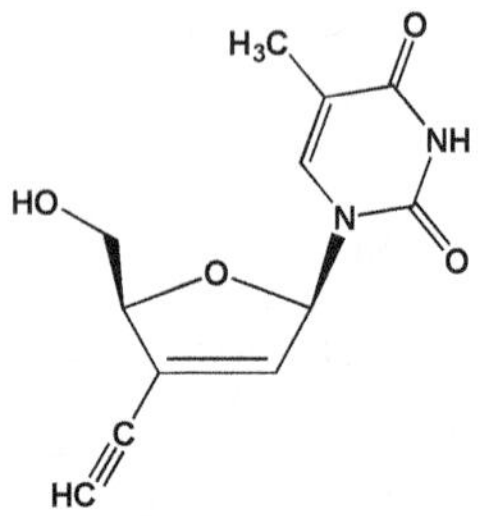

This nucleoside reverse transcriptase inhibitor, *or* NRTI (FW = 248.24 g/mol; CAS 1097733-37-3; *Symbol:* 4'-Ed4T), also known as 4'-ethynyl-d4T, OBP-601, festinavir, and 1-[(2*R*,5*R*)-5-ethynyl-5-(hydroxymethyl)-2*H*-furan-2-yl]-5-methylpyrimidine-2,4-dione, is a powerful anti-HIV-1 agent that is more active than the parent compound stavudine (d4T), while much less toxic to host cells (1). 4'-Ed4T is also to mitochondrial DNA synthesis. BMS-986001 is a better substrate for human thymidine kinase than d4T and is very much more resistant to catabolism by thymidine phosphorylase (1). Administration of BMS-986001 for 10 days resulted in substantial decreases in plasma HIV-1 RNA levels for all dose groups and was generally well tolerated. (2). **1**. Tanaka, Haraguchi, Kumamoto, Baba & Cheng (2005) *Antivir. Chem. Chemother.* **16**, 217. **2**. Cotte, Dellamonica, Raffi, *et al.* (2013) *J. Acquir. Immune Defic. Syndr.* **63**, 346.

Boceprevir

This oral HCV inhibitor (FW = 519.69 g/mol; CAS 394730-60-0), also named (1*R*,5*S*)-*N*-[3-amino-1-(cyclobutylmethyl)-2,3-dioxopropyl]-3-[2(*S*)-[[[(1,1-dimethylethylamino] carbonyl]amino]-3,3-dimethyl-1-oxobutyl]-6,6-dimethyl-3-azabicyclo[3.1.0]hexane-2(S)-carboxamide, SCH 503034, and Victrelis™, selectively targets hepatitis C virus non-structural protein 3 (NS3) serine protease (1). **Primary Mode of Action:** The NS3 protease is required for cleavage and processing of most nonstructural hepatitis C proteins. SCH-503034 is a nonpeptidomimetic drug that initially binds to NS3/4A protease to form a low-affinity, reversible E·I complex (K_i = 5 μM), followed by formation of a second higher-affinity (covalent) E-I complex (K_i = 20 nM). This increase in potency can be explained by its slow dissociation rate, forming complexes with a 5-hour dissociation half-life (2). Inhibitor potency correlated with changes in the buried surface area upon its binding to the active site, with greatest contributions to the binding energy arising from hydrophobic interactions of the P_1 and P_2 groups as they bind to the S_1 and S_2 pockets (3). **1**. Venkatraman, Bogen, Arasappan, *et al.* (2006) *J. Med. Chem.* **49**, 6074. **2**. Flores, Strawbridge, Ciaramella & Corbau (2009) *Biochim. Biophys. Acta* **1794**, 1441. **3**. Madison, Prongay, Guo, *et al.* (2008) *J. Synchrotron Radiat.* **15** (Part 3), 204.

BOL-303224-A, *See* Besifloxacin

Bongkrekic Acid

This toxic tricarboxylic acid (FW$_{free-acid}$ = 486.61 g/mol; CAS 11076-19-0), produced *Pseudomonas cocovenenans*, is a potent competitive inhibitor of the mitochondrial ATP-ADP translocator, effectively blocking nucleotide binding to the carrier (1-7). The name is derived from bongkrek, a moldy coconut product produced in Indonesia. The toxin accumulates when *P. cocovenenans* outgrows the mold. Klingenberg *et al.* (3) investigated bongkrekate binding to mitochondrial membrane to examine the reorienting site mechanism. The inferred mode of inhibition requires bongkrekate to bind to the single carrier site only from the inner face of the mitochondrial membrane (*i.e.*, the matrix side)). They confirmed the pH-dependent accumulation of [^{3}H]bongkrekate inside the mitochondria which superimposes onto the binding at carrier sites. By breaking the membrane with Lubrol or sonication, binding to the carrier sites could be titrated, and a K_d value of approximately 5 x 10^{-8} M was determined. The presence of ADP or ATP increases the amount of binding but does not alter the K_d. [^{35}S]Atractylate is displaced by [^{3}H]bongkrekate at a 1:1 molar ratio; this displacement is dependent on ADP concentration with the K_m = 0.5 x 10^{-6} M. *See also Atractyloside* The isomer known as isobongkrekic acid, which has a *cis*-double bond at the dicarboxylic acid end of the molecule, has similar biological activity. **Target(s):** Adenine nucleotide translocator, ADP/ATP carrier (1-5); ATPase (6); bromelain, stem (7); papain (7); ficain, *or* ficin (7). **1**. Vignais, Brandolin, Lauquin & Chabert (1979) *Meth.*

Enzymol. **55**, 518. **2**. Klingenberg, Appel, Babel & Aquila (1983) *Eur. J. Biochem.* **131**, 647. **3**. Henderson & Lardy (1970) *J. Biol. Chem.* **245**, 1319. **4**. Weidemann, Erdelt & Klingenberg (1970) *Biochem. Biophys. Res. Commun.* **39**, 363. **5**. Lauquin & Vignais (1976) *Biochemistry* **15**, 2316. **6**. Henderson (1972) *Biochem. J.* **127**, 321. **7**. Murachi, Kamei & Soedigdo (1982) *Toxicon* **20**, 1011.

Borate (Boric Acid)

Borate   Boric Acid

This conjugate base (FW = 61.83 g/mol; CAS 11129-12-7) of boric acid is the likely consequence of boric acid acting as a Lewis acid (*i.e.*, as an electron pair acceptor), as does boron trifluoride, to form B(OH)$_4^-$. Borate is soluble in cold water (1 g/18 mL) and inhibits many enzymes. The neutral species has been shown to inhibit urease by bridging the enzyme's pair of catalytic nickel ions. Boric acid also forms a complex with serine that inhibits γ-glutamyl transpeptidase. (*See also Serine-Borate Complex*) Sodium borate Na$_2$B$_4$O$_7$, also called sodium tetraborate, is often used as an antiseptic and detergent. The decahydrate is known as Borax. **Dionex BorateTrap™:** Because borate can be leached from borosilicate laboratory glassware, and because borate binds to vicinal hydroxyl groups in many carbohydrates, one cannot discount the possibility of borate effects on carbohydrate-utilizing enzymes. The Dionex BorateTrap is a 4 × 50 mm column placed between the eluent pump and an HPLC injection valve to remove borate contamination from eluents. Its placed prior the injection valve assures that the trap will have no effect on the efficiencies or retention times of the carbohydrate analytes. The trap itself is packed 20-μm high-capacity resin that has high selectivity for borate. The resin used in the BorateTrap is functionalized with a polyol capable of complexing borate in the presence of hydroxide. The BorateTrap eliminates peak tailing for mannose, fructose, and reduced monosaccharides, resulting from contamination of the eluent from borate. Therefore, the BorateTrap is highly recommended for optimal performance during carbohydrate analysis. **Target(s):** L-iditol dehydrogenase (1); deoxyribonuclease I (2); glycerol-2-phosphatase (3); 5'-nucleotidase (4,32,76); steryl-sulfatase (5); aspartate-semialdehyde dehydrogenase (6); γ-glutamyltransferase, plant (7); clostripain (8,11); L-gulonolactone oxidase (9); dehydro-L-gulonate decarboxylase (10); catechol *O*-methyltransferase (12); alcohol dehydrogenase (13,28,41,51); catechol oxidase, *or* o-diphenol oxidase (14); arginase (15,18,19,56-59); peptide-*N*^{4}-(*N*-acetyl-β glucosaminyl)asparagine amidase (16); glutamyl endopeptidase (17); glutaminase (20); alkaline phosphatase (21,25,31); glycogen synthase (22); xanthine oxidase (23); carboxypeptidase B, above pH 8 (24); acid phosphatase (26,77-79); glucose-6-phosphatase (27); urease (29,36-38,55); formimidoylglutamase, *or* formimino-glutamase (30); phosphodiesterase (31); arylsulfatase A, *or* cerebroside-sulfatase (33); arylsulfatase B, *or N*-acetylgalactosamine-4-sulfatase (33); cholinesterase (34); β-lactamase (35); tyrosinase (40); glyceraldehyde-3-phosphate dehydrogenase (43); venom exonuclease (44); kynureninase, weakly inhibited) (45); tetrahydrofolate synthetase (46,47); carboxy-*cis,cis*-muconate cyclase, under acidic conditions (48); cyclamate sulfohydrolase (49); nucleotide diphosphatase (50); cytidine deaminase (52); methylenediurea deaminase (53); allantoate deaminase (54); 4-methyleneglutaminase (60-62); pantothenase (63); bacterial leucyl aminopeptidase, weakly inhibited (64); aminopeptidase B (65); *N*-acetylgalactosamine-4-sulfatase, *or* arylsulfatase B (66,67); glycosulfatase (68); steryl-sulfatase, inhibited by metaborate and tetraborate (69-71); Arylsulfatase (72,73); 3-phosphoglycerate phosphatase (74); 3- or 4-phytase (75); alkaline phosphatase (80,81); arginine kinase (82); arginine:pyruvate transaminase (83); peptidyltransferase, weakly inhibited (84); protein-histidine *N*-methyltransferase, weakly inhibited (85); cyclopropane-fatty-acyl-phospholipid synthase (86); xanthine oxidase (87); 4 hydroxybenzoate 3-monooxygenase (88); lysine 2-monooxygenase (89); catechol 2,3-dioxygenase (90); catalase, weakly inhibited (91); catechol oxidase (92); polyphenol oxidase (92) **1**. Wolff (1955) *Meth. Enzymol.* **1**, 348. **2**. McDonald (1955) *Meth. Enzymol.* **2**, 437. **3**. Morton (1955) *Meth. Enzymol.* **2**, 533. **4**. Heppel & Hilmoe (1955) *Meth. Enzymol.* **2**, 546. **5**. Sandberg & Jenkins (1969) *Meth. Enzymol.* **15**, 684. **6**. Hegeman, Cohen & Morgan (1970) *Meth. Enzymol.* **17A**, 708. **7**. Thompson (1970) *Meth. Enzymol.* **17A**, 894. **8**. Ullmann & Bordusa (1998) in *Handb. Proteolytic Enzymes* (Barrett, Rawlings & Woessner, eds.), p. 759, Academic Press, San

Diego. **9**. Chatterjee (1970) *Meth. Enzymol.* **18A**, 28. **10**. Kagawa & Shimazono (1970) *Meth. Enzymol.* **18A**, 46. **11**. Mitchell & Harrington (1970) *Meth. Enzymol.* **19**, 635. **12**. Beattie & Weersink (1992) *J. Inorg. Biochem.* **46**, 153. **13**. Weser (1968) *Hoppe-Seyler's Z. Physiol. Chem.* **349**, 1479. **14**. Weser (1968) *Hoppe-Seyler's Z. Physiol. Chem.* **349**, 982. **15**. Reczkowski & Ash (1994) *Arch. Biochem. Biophys.* **312**, 31. **16**. Tarentino & Plummer (1994) *Meth. Enzymol.* **230**, 44. **17**. Birktoft & Breddam (1994) *Meth. Enzymol.* **244**, 114. **18**. Greenberg (1951) *The Enzymes*, 1st ed. (Sumner & Myrbäck, eds.), **1** (part 2), 893. **19**. Greenberg (1960) *The Enzymes*, 2nd ed. (Boyer, Lardy & Myrbäck, eds.), **4**, 257. **20**. Roberts (1960) *The Enzymes*, 2nd ed. (Boyer, Lardy & Myrbäck, eds.), **4**, 285. **21**. Stadtman (1961) *The Enzymes*, 2nd ed. (Boyer, Lardy & Myrbäck, eds.), **5**, 55. **22**. Leloir & Cardini (1962) *The Enzymes*, 2nd ed. (Boyer, Lardy & Myrbäck, eds.), **6**, 317. **23**. Bray (1963) *The Enzymes*, 2nd ed. (Boyer, Lardy & Myrbäck, eds.), **7**, 533. **24**. Folk (1971) *The Enzymes*, 3rd ed. (Boyer, ed.), **3**, 57. **25**. Fernley (1971) *The Enzymes*, 3rd ed. (Boyer, ed.), **4**, 417. **26**. Hollander (1971) *The Enzymes*, 3rd ed. (Boyer, ed.), **4**, 449. **27**. Nordlie (1971) *The Enzymes*, 3rd ed. (Boyer, ed.), **4**, 543. **28**. Li, Ulmer & Vallee (1963) *Biochemistry* **2**, 482. **29**. Todd & Hausinger (1989) *J. Biol. Chem.* **264**, 15835. **30**. Kaminskas, Kimhi & Magasanik (1970) *J. Biol. Chem.* **245**, 3536. **31**. Zittle (1947) *J. Biol. Chem.* **167**, 297. **32**. Heppel & Hilmoe (1951) *J. Biol. Chem.* **188**, 665. **33**. Bleszynski & Leznicki (1967) *Enzymologia* **33**, 373. **34**. Gayrel & Lagreu (1951) *Toulouse Med.* **52**, 778. **35**. Kiener & Waley (1978) *Biochem. J.* **169**, 197. **36**. Breitenbach & Hausinger (1988) *Biochem. J.* **250**, 917. **37**. Mazurkiewicz, Bingham, Runswick & Ang (1993) *Ann. Clin. Biochem.* **3**, 215. **38**. Benini, Rypniewski, Wilson, Mangani & Ciurli (2004) *J. Amer. Chem. Soc.* **126**, 3714. **39**. Lenti (1953) *Arch. Sci. Biol. (Bologna)* **37**, 84. **40**. Yasunobu & Norris (1957) *J. Biol. Chem.* **227**, 473. **41**. Roush & Gowdy (1961) *Biochim. Biophys. Acta* **52**, 200. **42**. Smith & Johnson (1976) *Biochemistry* **15**, 560. **43**. Wolny (1977) *Eur. J. Biochem.* **80**, 551. **44**. Richards & Laskowski, Jr. (1969) *Biochemistry* **8**, 1786. **45**. Jakoby & Bonner (1953) *J. Biol. Chem.* **205**, 699. **46**. Shane (1982) in *Peptide Antibiotics: Biosynthesis and Functions: Enzymatic Formation of Bioactive Peptides and Related Compounds* (Kleinkauf & von Döhren., eds.) W. de Gruyter, Berlin, pp. 353. **47**. Cichowicz, Foo & Shane (1981) *Mol. Cell. Biochem.* **39**, 209. **48**. Thatcher & Cain (1975) *Eur. J. Biochem.* **56**, 193. **49**. Nimura, Tokiedo & Yamaha (1974) *J. Biochem.* **75**, 407. **50**. Kornberg & Pricer (1950) *J. Biol. Chem.* **182**, 763. **51**. Still (1940) *Biochem. J.* **34**, 1177. **52**. Ashley & Bartlett (1984) *J. Biol. Chem.* **259**, 13615. **53**. Jahns, Schepp & Kaltwasser (1997) *Can. J. Microbiol.* **43**, 1111. **54**. Winkler, Polacco, Blevins & Randall (1985) *Plant Physiol.* **79**, 787. **55**. Pearson & Smith (1943) *Biochem. J.* **37**, 148. **56**. Han, Moore & Viola (2002) *Bioorg. Chem.* **30**, 81. **57**. Moreno-Vivian, Soler & Castillo (1992) *J. Biochem.* **204**, 531. **58**. Colleluori, Morris & Ash (2001) *Arch. Biochem. Biophys.* **389**, 135. **59**. Kaysen & Strecker (1973) *Biochem. J.* **133**, 779. **60**. Ibrahim, Lea & Fowden (1984) *Phytochemistry* **23**, 1545. **61**. Winter & Dekker (1991) *Plant Physiol.* **95**, 206. **62**. Powell & Dekker (1983) *J. Biol. Chem.* **258**, 8677. **63**. Airas (1983) *Anal. Biochem.* **134**, 122. **64**. Baker, Wilkes, Bayliss & J. M. Prescott (1983) *Biochemistry* **22**, 2098. **65**. Hopsu, Mäkinen & Glenner (1966) *Arch. Biochem. Biophys.* **114**, 567. **66**. Atsumi, Kawai, Seno & Anno (1972) *Biochem. J.* **128**, 983. **67**. Lloyd & Fielder (1968) *Biochem. J.* **109**, 14. **68**. Lloyd & Fielder (1967) *Biochem. J.* **105**, P33. **69**. Iwamori, Moser & Kishimoto (1976) *Arch. Biochem. Biophys.* **174**, 199. **70**. Roy (1954) *Biochim. Biophys. Acta* **15**, 300. **71**. Daniel (1985) *Isozymes, Curr. Top. Biol. Med. Res.* **12**, 189. **72**. Sakurai, Isobe & Shiota (1980) *Agric. Biol. Chem.* **44**, 1. **73**. Lien & Schreiner (1975) *Biochim. Biophys. Acta* **384**, 168. **74**. Randall & Tolbert (1971) *J. Biol. Chem.* **246**, 5510. **75**. Powar & Jagannathan (1982) *J. Bacteriol.* **151**, 1102. **76**. Ikura & Horikoshi (1989) *Agric. Biol. Chem.* **53**, 645. **77**. Rossi, Palma, Leone & Brigliador (1981) *Phytochemistry* **20**, 1823. **78**. Shimada, Shinmyo & Enatsu (1977) *Biochim. Biophys. Acta* **480**, 417. **79**. Harsanyi & Dorn (1972) *J. Bacteriol.* **110**, 246. **80**. Sakurai, Toda & Shiota (1981) *Agric. Biol. Chem.* **45**, 1959. **81**. Fernandez, Gascon & Schwenke (1981) *Curr. Microbiol.* **6**, 121. **82**. Brown & Grossman (2004) *Arch. Insect Biochem. Physiol.* **57**, 166. **83**. Yang & Lu (2007) *J. Bacteriol.* **189**, 3954. **84**. Cerná & Rychlík (1980) *FEBS Lett.* **119**, 342. **85**. Vijayasarathy & Rao (1987) *Biochim. Biophys. Acta* **923**, 156. **86**. Iwig, Uchida, Stromberg & Booker (2005) *J. Amer. Chem. Soc.* **127**, 11612. **87**. De Renzo (1956) *Adv. Enzymol. Relat. Subj. Biochem.* **17**, 293. **88**. Shoun, Arima & Beppu (1983) *J. Biochem.* **93**, 169. **89**. Flashner & Massey (1974) *J. Biol. Chem.* **249**, 2579. **90**. Kachhy & Modi (1976) *Indian J. Biochem. Biophys.* **13**, 234. **91**. Chatterjee & Sanwal (1993) *Mol. Cell. Biochem.* **126**, 125. **92**. Jang, Shin & Song (2005) *Food Sci. Biotechnol.* **14**, 117.

S-(2-Boronoethyl)-L-cysteine

This boronic acid-containing arginine analogue (FW = 190.00 g/mol; CAS 63107-40-4; *Abbreviation*: BEC) is a slow, tight-binding competitive inhibitor of arginase (*Reaction*: L-Arginine + H_2O $\rightleftharpoons$ Urea + L-Ornithine), the binuclear manganese metalloenzyme catalyzing the last step in the Urea Cycle. Upon binding, the boronic acid moiety undergoes nucleophilic attack by the enzyme's metal-bridging hydroxide ion, yielding a tetrahedral boronate anion that bridges the binuclear manganese cluster and mimics the tetrahedral intermediate (or a flanking transition state configuration) in arginase catalysis (1). BEC is also a potent inhibitor of human arginase II with K_i values of 0.31 μM and 30 nM at pH 7.5 and 9.5, respectively (2). Inhibition of arginase by BEC increases the circulating concentration of arginine, thereby increasing nitric oxide synthesis and significantly enhancing NO-dependent smooth muscle relaxation in human penile corpus cavernosum. Arginase is thus a potential druggable target for the treatment of erectile dysfunction. **1**. Kim, Cox, Baggio, *et al.* (2001) *Biochemistry* **40**, 2678. **2**. Colleluori & Ash (2001) *Biochemistry* **40**, 9356.

Bortezomib

This FDA-approved alkyl boronate drug for relapsed multiple myeloma (FW = 384.24 g/mol; CAS 179324-69-7; Solubility: 75 mg/mL DMSO; <1 mg/mL Water), also known as PS-341, Velcade, and [(1*R*)-3-methyl-1-({(2*S*)-3-phenyl-2-[(pyrazin-ylcarbonyl)amino]propanoyl}amino)butyl] boronic acid, is a potent proteasome inhibitor, K_i = 0.6 nM (1). **Mechanism of Action:** Bortezomib is an *N*-protected dipeptide and can be written as Pyz-Phe-boroLeu, where Pyz- is pyrazinoic acid, Phe is phenylalanine, and the last residue is a leucine analogue with a boronic acid in place of a carboxylic acid. It is the sp^3 orbital hybridization of the boronate that is thought to mimic a tetrahedral reaction intermediate, thereby conferring high-affinity binding (K_i = 0.6 nM). **Pharmacodynamics:** Bortezomib-mediated 26S proteasome inhibition causes cell-cycle arrest and induces apoptosis in CD-30^+ anaplastic large cell lymphoma (2). Bortezomib-treated cells form aggregates of ubiquitin-conjugated proteins ("aggresomes") *in vitro* and *in vivo*, resulting in apoptosis. Bortezomib reverses defective G-CSF-triggered granulocytic differentiation of $CD34^+$ cells from CN patients *in vitro*, an effect that was accompanied by restoration of Lymphoid Enhancer-binding Factor-1 (LEF-1) protein levels and LEF-1 mRNA autoregulation. This inhibitory effect helped to define a novel mechanism of LEF-1 down-regulation in CN patients via enhanced LEF-1 protein ubiquitination and degradation by hyperactivated Signal Transducer/Activator of Transcription-5 (STAT5) (3). Bortezomib is currently used to eliminate malignant plasma cells in multiple myeloma patients. It is also effective in depleting both alloreactive plasma cells in acute Ab-mediated transplant rejection and their autoreactive counterparts in animal models of lupus and myasthenia gravis (MG). Significantly, bortezomib (at 10 nM or higher) kills long-lived plasma cells in cultured thymus cells from nine early-onset MG patients, consistently halting their spontaneous production of autoantibodies against the acetylcholine receptor, but also of total IgG (4). Bortezomib appears to be a promising treatment for MG and possibly other Ab-mediated autoimmune or allergic disorders, especially when given in short courses at modest doses before the standard immunosuppressive drugs have taken effect (4). **1**. Adams & Kauffman (2004) *Cancer Invest.* **22**, 304. **2**. Bonvini, Zorzi, Basso & Rosolen (2007) *Leukemia* **21**, 838. **3**. Gupta, Kuznetsova, Klimenkova, *et al.* (2014) *Blood* **123**, 2550. **4**. Gomez, Willcox, Vrolix, *et al.* (2014) *J. Immunol.* **193**, 1055.

Bottromycin A₂

This major component (FW = 823.07 g/mol; CAS 15005-62-6) of a series of structurally related *Streptomyces* antibiotics inhibits protein biosynthesis by blocking the binding of aminoacyl-tRNA to the polyribosomes. **1**. Pestka (1974) *Meth. Enzymol.* **30**, 261. **2**. Carrasco, Battaner & Vazquez (1974) *Meth. Enzymol.* **30**, 282. **3**. Jiménez (1976) *Trends Biochem. Sci.* **1**, 28. **4**. Tanaka, Sashikata, Yamaguchi & Umezawa (1966) *J. Biochem.* **60**, 405. **5**. Lin, Kinoshita & Tanak (1968) *J. Antibiot.* **21**, 471. **6**. Otaka & Kaji (1976) *J. Biol. Chem.* **251**, 2299. **7**. Otaka & Kaji (1983) *FEBS Lett.* **153**, 53.

Botulinum Toxin

This bacterial toxin and neurotoxin (FW = 149.3 kDa; CAS 93384-43-1; abbreviated as BTX or BoNT) from Gram-positive *Clostridium botulinum* are ADP-ribosyltransferases and proteases that give rise to the clinical manifestations of botulism. **Classification:** BTX consists of seven antigenically and serologically distinct neurotoxins (types A, B, C₁, C₂, C₃, D, E, F, and G) that are otherwise functionally and structurally similar. Human botulism is caused mainly by types A, B, E, and (rarely) F. Types C and D cause toxicity only in animals. Each botulinum toxin possesses individual potencies, necessitating special care to avoid medication errors. Recent changes to the established drug names by the FDA were intended to reinforce these differences and prevent medication errors. They include the following: exotoxins, which inhibit the release of catecholamines from the adrenal medulla and block the release of acetylcholine at neuromuscular junctions. The seven distinct types of exotoxin are now labeled A through G, each consisting of two peptide chains linked by a disulfide bond. Chain I of botulinum toxin C₂ catalyzes the ADP-ribosylation of G-actin and prevents actin polymerization. **Neurotoxic Effects:** Following the attachment of the botulinum toxin's heavy chain to one or more proteins on the exterior surface of axon terminals, the toxin is taken by neurons via endocytosis. The light chain cleaves endocytotic vesicles and enters the cytoplasm, where it proteolytically degrades the SNAP-25 protein, a type of SNARE protein involved in vesicle trafficking. SNAP-25 is essential for vesicle fusion, a step that is required for release of neurotransmitters, particular acetylcholine, from the axon endings Botulinum toxin thereby prevents neurosecretory vesicles from docking/fusing with the nerve synapse plasma membrane and releasing their neurotransmitters. **Clinical Applications:** Injectable formulations of OnabotulinumtoxinA (INN generic name) are now used to treat blepharospasm or strabismus, to relieve neck pain, resulting from cervical dystonia, and for muscle stiffness in the elbow, wrist, and finger muscles in adult patients with upper limb spasticity. Botox is often employed to prevent chronic migraine and cosmetically to improve the appearance of deep facial lines or furrows between eyebrows and creases in the skin around the eyes. Patients with urinary incontinence and overactive bladder can also benefit from appropriate application. **Target(s):** catecholamine release (1,2); acetylcholine release (1,2); actin polymerization (3). **1**. Wictome & Shone (1998) *Symp. Ser. Soc. Appl. Microbiol.* **27**, 87S. **2**. Dressler, Saberi & Barbosa (2005) *Arq. Neuropsiquiatr.* **63**, 180. **3**. Wille, Just, Wegner & Aktories (1992) *J. Biol. Chem.* **267**, 50.

Botulinum Toxin C3 ADP-ribosyltransferase

This 211-residue bacterial toxin (MW = 23.5 kDa) catalyzes the ADP ribosylation of small GTPases, particularly Rho-like proteins that control actin polymerization, cell motility, and related aspects of cancer cell metastasis. Treatment of cells with a cell-permeable chimeric C3 toxin led to complete localization of modified Rho to the cytosolic fraction based on the complexation of ADP-ribosylated Rho with the Guanine-nucleotide Dissociation Inhibitor-1 (GDI-1) (3). The modified complex was resistant to phosphatidyl-inositol 4,5-bisphosphate- and GTPγS-induced release of Rho from GDI-1. ADP-ribosylation therefore leads to entrapment of Rho in the GDI-1 complex. Note also that covalent modification is not the only mode of C3 action. Indeed, C3 transferases from *Clostridium botulinum* and *C.*

limosum form a GDI-like complex with the Ras-like low molecular mass GTPase Ral, inhibiting its action *without* ADP-ribosylation (4). **1**. Aktories & Frevert (1987) *Biochem. J.* **247**, 363. **2**. Popoff, Boquet, Gill & Eklund (1990) *Nucleic Acids Res.* **18**, 1291. **3**. Genth, Gerhard, Maeda, *et al.* (2003) *J. Biol. Chem.* **278**, 28523. **4**. Vogelsgesang, Pautsch & Aktories (2007) *Naunyn. Schmiedebergs Arch. Pharmacol.* **374**, 347.

Bouvardin

This cyclic hexapeptide antitumor agent (FW = 772.86 g/mol; CAS 64755-14-2), from *Bouvardia ternifolia* (scarlet bourvardia or trumpetilla), also known as NSC 259968, inhibits protein synthesis in intact eukaryotic cells and cell-free systems. Bouvardin inhibits EF₁-dependent binding of aminoacyl-tRNA and EF₂-dependent translocation of peptidyl-tRNA but does not affect the nonenzymic translocation since this reaction does not require EF₂. Bouvardin contains L- and D-alanine, and *N*-methyl-L-tyrosine. **1**. Jolad, Hoffman, Torrance, *et al.* (1977) *J. Amer. Chem. Soc.* **99**, 8040. **2**. Chitnis, Alate & Menon (1981) *Chemotherapy* **27**, 126. **3**. Zaera, Santamaria, Zalacain & Jimenez (1982) *Rev. Esp. Oncol.* **29**, 587. **4**. Zalacain, Zaera, Vazquez & Jimenez (1982) *FEBS Lett.* **148**, 95.

Bovine Pancreatic Trypsin Inhibitor

This monomeric 58-residue polypeptide (MW = 6511.44 g/mol; CAS 9087-70-1; Isoelectric Point = pH 10.5), known as Trasylol, forms an extremely tight complex with trypsin, chymotrypsin, kallikrein, and plasmin, blocking substrate access to their active sites (1-4). BPTI is a member of the pancreatic trypsin inhibitor (Kunitz) family, which is a family of serine protease inhibitors. (**See** *Trypsin Inhibitors; Aprotinin*) These proteins usually have conserved cysteine residues that participate in forming disulfide bonds at Cys₅-Cys₅₅, Cys₁₄-Cys₃₈, and Cys₃₀-Cys₅₁. One of the most stable proteins known, BPTI is inert to urea-, thermal- and acid-denaturation, the latter exhibiting a mid-point at 80 °C at pH 2.1 Surprisingly, this cationic protein also blocks magnesium channels required for growth by *Saccharomyces cerevisiae* and the human pathogen *Candida albicans* (5). **Mechanism of Inhibition:** BPTI may be classified as a slow-binding, active site-directed inhibitor. Bond angles in about seven residues near the BPTI N-terminus are distorted in such a way that they mimic key geometric aspects of substrate binding in trypsin's catalytic transition state. Stopped-flow kinetic studies using the prflavin-displacement assay (6), indicate that, at neutral pH, BPTI binds to trypsin in a diffusion-controlled reaction, with an initial dissociation constant of 10⁻⁵ M, followed by slow isomerization to te final complex (*k* = 350 sec⁻¹). The pre-equilibrium step contributes about one-third of the overall Gibbs energy change for formation of the final complex. The pH-dependence for BPTI-trypsin binding is explained by assuming that binding requires the unprotonated histidine residue in trypsin's active site. **Target(s):** trypsin (1,2,4); kallikrein (3,4); plasmin (3,4); chymotrypsin (4). **1**. Kassell, Radicevic, Berlow, Peanasky & Laskowski (1963) *J. Biol. Chem.* **238**, 3274. **2**. Kassell, Radicevic, Ansfield & Laskowski (1965) *Biochem. Biophys. Res. Commun.* **18**, 255. **3**. Kraut & Bhargava (1967) *Hoppe-Seyler's Z. Physiol. Chem.* **348**, 1498. **4**. Ascenzi, Bocedi, Bolognesi, *et al.* (2003) *Curr. Protein Pept. Sci.* **4**, 231. **5**. Bleackley, Hayes, Parisi, *et al.* (2014) *Mol. Microbiol.* **92**, 1188. **6**. Quast, Engel, Heurman, Krause & Steffin (1974) *Biochemistry* **13**, 2512.

Bowman–Birk Protease Inhibitors

These plant serpins (MW = ~8 kDa; CAS 37330-34-0) belong to MEROPS serine proteinase inhibitor family I12, clan IF and mainly inhibit serine peptidases of the S1 family, but also inhibit S3 peptidases (1-3). BBI family members have a duplicated structure, consisting of two topologically distinct inhibitory sites. They interact with target enzymes via an exposed surface loop that adopts the canonical proteinase inhibitory conformation. The net result is formation of an inactive noncovalent (or a cleaved and extremely tightly bound) protease-inhibitor complex. The interaction loop is a particularly well-defined, disulfide-linked, short β-sheet region (2), and synthetic peptide mimics of the BBI have been prepared and investigated (4,5). **1**. Rawlings, Tolle & Barrett (2004) *Biochem. J.* **378**, 705. **2**. Brauer, Nievo, McBride, *et al.* (2003) *J. Biomol. Struct. Dyn.* **20**, 645. **3**. Rawlings, Waller, Barrett & Bateman (2014) MEROPS database of proteolytic enzymes, substrates and inhibitors, *Nucleic Acids Res.* **42**, D503. **4**. McBride & Leatherbarrow (2001) *Curr. Med. Chem.* **8**, 909. **5**. McBride, Watson, Brauer, Jaulent & Leatherbarrow (2002) *Biopolymers* **66**, 79.

Bozepinib

This potent antitumor agent (FW = 521.33 g/mol), sysyematically known as (*RS*)-2,6-dichloro-9-[1-(*p*-nitrobenzenesulfonyl)-1,2,3,5-tetrahydro-4,1-benzoxazepin-3-yl]-9*H*-purine, induces apoptosis in breast and colon cancer cells via inhibition (IC$_{50}$ = 0.3–2 μM) of the the double-stranded RNA-dependent protein kinase, PKR, but not p53. Moreover, interferon enhances cytotoxicity of bozepinib in colon and breast cancer cells by increasing apoptotic cell death. **1**. Marchal, Carrasco, Ramirez, *et al.* (2013) *Drug Des. Devel. Ther.* **7**, 1301.

BPTES

This synthetic allosteric inhibitor (FW = 524.68 g/mol; CAS 314045-39-1; Soluble to 20 mM in DMSO), also known as *N,N'*-[thiobis(2,1-ethanediyl-1,3,4-thiadiazole-5,2-diyl)]bisbenzeneacetamide, selectively targets glutaminase GLS1 (IC$_{50}$ = 3.3 μM), with weaker action against GLS2 and γ-glutamyl transpeptidase. The level of a splice variant of GLS1 (GAC) is elevated in certain cancers. BPTES induces cell death of P493 human lymphoma B cells *in vitro* and reduces tumor volume of P493 cell xenografts in mice. **1**. Shukla, Ferraris, Thomas, *et al.* (2012) *J. Med. Chem.* **55**, 10551. **2**. DeLaBarre, Gross, Fang, *et al.* (2011) *Biochemistry* **50**, 10764.

BR-A-657, *See* Fimasartan

Bradykinin Potentiating Peptide 5a
This snake venom pentapeptide (*Sequence: pyro*-EKWAP; FW = 611.69 g/mol) from *Bothrops jararaca* selectively potentiates the action of bradykinin and acts as a substrate inhibitor for angiotensin I-converting enzyme (*K*$_i$ = 0.09 μM for the human testicular enzyme). BPP$_{5a}$ is also a weak inhibitor (not substrate) of pyroglutamyl-peptidase II (*K*$_i$ = 1.05 mM). **Target(s):** peptidyl-dipeptidase A, *or* angiotensin I-converting enzyme (1-6); pyroglutamyl peptidase II (7). **1**. Farias, Sabatini, Sampaio, *et al.* (2006) *Biol. Chem.* **387**, 611. **2**. Soffer (1976) *Ann. Rev. Biochem.* **45**, 73. **3**. Lanzillo & Fanburg (1976) *Biochim. Biophys. Acta* **445**, 161. **4**. Das & Soffer (1975) *J. Biol. Chem.* **250**, 6762. **5**. Ondetti & Cushman (1982) *Ann. Rev. Biochem.* **51**, 283. **6**. Cheung & Cushman (1973) *Biochim. Biophys. Acta* **293**, 451. **7**. O'Connor & O'Cuinn (1985) *Eur. J. Biochem.* **150**, 47.

Bradykinin Potentiating Peptide 9a
This snake venom nonapeptide (FW = 1101.26 g/mol), also known as teprotide, V-6-1, and SQ-20881, from the venom of *Bothrops jararaca* potentiates the action of bradykinin, while inhibiting angiotensin I-converting enzyme (IC$_{50}$ = 0.01 μM for the human testicular enzyme). The corresponding octapeptide lacking one C-terminal proline residues also inhibits. **1**. Farias, Sabatini, Sampaio, *et al.* (2006) *Biol. Chem.* **387**, 611. **2**. Takada, Hiwada & Kokubu (1981) *J. Biochem.* **90**, 1309. **3**. Ward & Sheridan (1982) *Biochim. Biophys. Acta* **716**, 208. **4**. Ondetti, Williams, Sabo, *et al.* (1971) *Biochemistry* **10**, 4033. **5**. Dorer, Kahn, Lentz, Levine & Skeggs (1976) *Biochim. Biophys. Acta* **429**, 220. **6**. Das & Soffer (1975) *J. Biol. Chem.* **250**, 6762. **7**. Lanzillo & Fanburg (1977) *Biochim. Biophys. Acta* **491**, 339. **8**. Sakharov, Danilov & Sukhova (1987) *Anal. Biochem.* **166**, 14. **9**. Strittmatter, Thiele, Kapiloff & Snyder (1985) *J. Biol. Chem.* **260**, 9825. **10**. Cotton, Hayashi, Cuniasse, *et al.* (2002) *Biochemistry* **41**,

6065. **11**. Cheung & Cushman (1973) *Biochim. Biophys. Acta* **293**, 451. **12**. Oshima, Gecse & Erdös (1974) *Biochim. Biophys. Acta* **350**, 26. **13**. Nakajima, Oshima, Yeh, Igic & Erdös (1973) *Biochim. Biophys. Acta* **315**, 430.

Brassinazole

This triazole-type inhibitor (FW = 327.81; CAS 224047-41-0; Brz2001), systematically named 4-(4-chlorophenyl)-2-phenyl-3-[[1,2,4]triazol-1-yl]butan-2-ol, targets brassinosteroid biosynthesis. The latter are plant steroids and phytohormones that promote organ growth through their combined effect on cell expansion and division, plant architecture, vascular differentiation, fertility and flowering, photo-morphogenesis, tolerance to biotic and abiotic stresses, as well as senescence. Brassinazole's inhibitory action is reversed by brassinolide, which showed good recovery after addition brassinazole-treated plants. Although brassinazole is structurally similar to pacrobutrazol, a gibberellin biosynthesis inhibitor, it does not show any evidence of recovery upon addition of gibberellin. Brassinazole also induces morphological changes in dark-grown *Arabidopsis*. **1**. Asami, Min, Nagata *et al.* (2000) *Plant Physiol.* **123**, 93.

Brefeldin A

This fungal metabolite (FW = 280.36 g/mol; CAS 20350-15-6; Symbol: BFA; Solubility: 4 mg/mL (14 mM) DMSO; <1 mg/mL (<1 mM) H$_2$O), isolated from *Penicillium brefeldianum*, *P. decumbens*, and *P. cyaneum*, inhibits activation of ADP-ribosylation factor (Arf), thereby disrupting the structure and function of the Golgi apparatus and inhibiting protein secretory traffic. Its effects are not limited to the Golgi apparatus, but are evident throughout the central vacuolar system, mainly by inducing tubule formation (*or* "tubulation") of the endosomal membrane vesicle system, the trans-Golgi network, and lysosomes. While BFA only modestly affects endocytic membrane transport in cultured fibroblasts, it profoundly alters endocytic pathways within polarized cells, such as MDCK cells. **Mechanism of Action:** ADP-ribosylation factors are found ubiquitously in eukaryotic cells. Although Arfs are initially formed as soluble elements within the cytosol, they undergo N-terminal myristoylation, allowing them to become firmly associated with Golgi and vesicle membranes. Arfs function as regulators of vesicular traffic and actin remodeling. A canonical GDP/GTP phosphoryl switch undergoes conformational changes within Arf Switch-1 and Arf Switch-2 regions, which tightly bind to the γ-phosphoryl of GTP, but only weakly, or not at all, bind GDP. Arf·GTP undergoes hydrolysis to generate Arf·GDP·P$_i$, and the latter must dissociate and be replaced by GTP for the secretory interaction cycle to start anew. Unassisted Arf GDP-GTP exchange is very slow, however, requiring catalysis by a family of guanine nucleotide-exchange factors. Formation of transport vesicles carrying cargo proteins between the plasma membrane, the trans-Golgi network (TGN) and endosomes relies in part on vesicle coats containing clathrin and adaptor protein (AP) complexes. AP-1 is mainly found at the TGN, but also on endosomes and specialized granules, depending on the cell type. AP-3 is present on endosomal compartments and is involved in the trafficking of cargo proteins to lysosomes and lysosome-related organelles. Association of AP-1 and -3, but not AP-2, with the membrane of endosomes or the TGN is mediated by Arf1. Arfs contain a Sec7 domain that binds brefeldin A and disrupts trafficking and induces organelle disintegration. The net effect is that BFA alters membrane transport by inhibiting the association of vesicle coat proteins, thereby blocking the formation of transport vesicles. **Glucose Transporter Trafficking:** Brefeldin A also appears to regulate the trafficking of the GLUT4 glucose transporter, inducing its relocation from intracellular stores to the cell surface and inducing apoptosis in human cancer cells (operating on a p53-independent manner). GLUT4 redistribution is normally regulated by insulin-mediated signaling. BFA causes the activation of the insulin receptor, IRS-1, Akt-2, and AS160 components of the insulin pathway. This response is mediated through phosphoinositol-3-kinase (PI3K) and Akt kinase, and BFA-mediated activation of the insulin pathway results in Akt-mediated phosphorylation of the insulin-responsive transcription factor FoxO1. This event leads to exclusion of FoxO1 from the nucleus, decreasing transcription of the insulin-responsive gene SIRT-1. **Target(s):** guanine nucleotide-exchanging proteins (1-3); protein transport and processing (4,5); memapsin (6). **1.** Pacheco-Rodriguez, Moss & Vaughan (2001) *Meth. Enzymol.* **329**, 300. **2.** Pacheco-Rodriguez, Moss & Vaughan (2002) *Meth. Enzymol.* **345**, 397. **3.** Dinter & Berger (1998) *Histochem. Cell Biol.* **109**, 571. **4.** Misumi, Misumi, Miki, *et al.* (1986) *J. Biol. Chem.* **261**, 11398. **5.** Klausner, Donaldson & Lippincott-Schwartz (1992) *J. Cell Biol.* **116**, 1071. **6.** Fluhrer, Capell, Westmeyer, *et al.* (2002) *J. Neurochem.* **81**, 1011.

Brentuximab vedotin

This antibody-drug conjugate, *or* ADC (MW ≈ 151 kDa; CAS 914088-09-8; *Symbol*: cAC10-vcMMAE), also known by the trade name Adcetris®, is directed against CD30, a cell surface protein expressed in classical Hodgkin lymphoma (*or* HL) and systemic anaplastic large cell lymphoma (*or* sALCL). The chimeric monoclonal antibody cAC10 induces growth arrest of CD30+ cell lines *in vitro* and has pronounced antitumor activity in severe combined immunodeficiency (SCID) mouse xenograft models of Hodgkin disease. Adcetris comprises an anti-CD30 monoclonal antibody joined by a protease-cleavable linker to monomethyl auristatin E (*or* MMAE), a cytotoxic agent. (**See** *Monomethyl Auristatin E*) The linker is stable in the bloodstream but releases MMAE upon internalization into CD30-expressing tumor cells, resulting in death of targeted cells. **1.** Francisco, Cerveny, Meyer, *et al.* (2003) *Blood* **102**, 1458.

Brevetoxins

Brevetoxin A

Brevetoxin B

This collection of cyclic polyether natural products (CASs = 98225-48-0 and 98112-41-5; Symbol: Btx or PbTx-2) from the dinoflagellate *Karenia brevis* consists of neurotoxins that bind to voltage-gated sodium channels, giving rise to neurotoxic shellfish poisoning. Symptoms include nausea, vomiting, slurred speech. Brevetoxins bind at a common binding site (*i.e.*, receptor-site 5) on the α-subunit of this neuronal transmembrane protein. Once bound, the sodium channels remain permanently opened, at the resting membrane potential, which produces a continuous entry of sodium ions in most excitable cells. Such a sodium entry has various consequences on sodium-dependent physiological mechanisms, consisting in a membrane depolarization that, in turn, causes spontaneous and/or repetitive action potential discharges and thereby increases membrane excitability. These neuronal discharges may be transient or continuous depending on the preparation and the toxin tested (1-3). **Brevetoxin Types:** Btx-1 (*or* PbTx-1) = Btx A + R = –CH$_2$C(=CH$_2$)CHO, FW = 853.05 g/mol; Btx-2 (*or* PbTx-2) = Btx B + R = –CH$_2$C(=CH$_2$)CHO, FW = 881.06 g/mol; Btx-3 (*or* PbTx-3) = Btx B + R = –CH$_2$C(=CH$_2$)CH$_2$OH, FW = 852.98 g/mol; Btx-5 (*or* PbTx-5) = Btx B (acetylated at Position-38) + R = –CH$_2$C(=CH$_2$)CH$_2$OH, FW = 895.04 g/mol; Btx-6 (*or* PbTx-6) = Btx B (epoxide in place of double-bond at C27 & C28) + R = –CH$_2$C(=CH$_2$)CH$_2$OH, FW = 897.06 g/mol; Btx-7 (*or* PbTx-7) = Btx A + R = –CH$_2$C(=CH$_2$)CH$_2$OH, FW = 855.07 g/mol; Btx-8 (*or* PbTx-8) = Btx B + R = CH$_2$COCH$_2$Cl, FW = 903.50 g/mol; Btx-9 (*or* PbTx-9) = Btx B + R = –CH$_2$CH(CH$_3$)CH$_2$OH, FW = 885.09 g/mol; and Btx-10 (*or* PbTx-10) = Btx A + R = –CH$_2$CH(-CH$_3$)CH$_2$OH, FW = 857.08 g/mol. **1.** Mattei, Molgó, Legrand & Benoit (1999) *J. Soc. Biol.* **193**, 329. **2.** Benoit, Mattei, Ouanounou, *et al.* (2002) *Cell. Mol. Biol. Lett.* **7**, 317. **3.** Stommel & Watters (2004) *Curr. Treat. Options Neurol.* **6**, 105.

Brevilin A

This naturally occurring sesquiterpenoid (FW = 346.40 g/mol; CAS 16503-32-5) from the volatile oils of the plant *Centipeda minima* inhibits (IC$_{50}$ = 10 µg/mL) STAT3 phosphorylation in the Jak/STAT signal transduction pathway of A549R cells. Brevilin A showed high specificity on Janus Kinase activity and following STAT3 signaling without directly affecting some other signals, including p65, AKT and GSK-3β phosphorylation, as well as Src kinase activity. Brevilin A is also an antibiotic with activity against *Giardia* (IC$_{50}$ = 16 µM) and *Entamoeba histolitica* (IC$_{50}$ = 4–9 µM). **1.** Chen X, Du Y, Nan, *et al.* (2013) *PLoS One* **8**, e63697.

Brexpiprazole

This serotonin-dopamine activity modulator, *or* SDAM (FW = 433.60 g/mol; CAS 913611-97-9), also known as OPC-34712, Rexulti®, 7-{4-[4-(1-benzothiophen-4-yl)piperazin-1-yl]butoxy}quinolin-2(1*H*)-one, is an atypical antipsychotic drug that acts as a dopamine D$_{2L}$ (K_i = 1.1 nM) and D$_3$ (K_i = 0.3 nM) receptor partial agonist and a partial agonist of 5-HT$_{1A}$ receptors (K_i = 0.12 nM). Brexpiprazole is also an antagonist of the 5-HT$_{2A}$ (K_i = 0.47 nM), 5-HT$_{2B}$ (K_i = 1.9 nM), 5-HT$_7$ (K_i = nM), α$_{1B}$-adrenergic (K_i = 0.17 nM), α$_{2C}$-adrenergic (K_i = 0.59 nM), and histamine H$_1$ receptors (K_i = 19 nM). It has negligible affinity for the mACh receptors. Rexulti is an FDA-approved add-on medication for major depressive disorder in adults. **See** *Aripiprazole* **1.** Maeda, Sugino, Akazawa, *et al.* (2014) *J. Pharmacol. Exp. Ther.* **350**, 589.

Brilacidin

This membrane-lyzing defensin mimetic (FW = 932.98 g/mol; CAS 1224095-98-0; IUPAC: *N,N′*-bis[3-{[5-(carbamimidoylamino)pentanoyl]amino}-2-[(3*R*)-pyrrolidin-3-yloxy]-5-(trifluoromethyl)phenyl]pyrimidine-4,6-dicarboxamide)) is bactericidal (not simply bacteriostatic like most antibiotic) toward a broad spectrum of bacteria, but not mammalian cells. Brilacidin kills many Gram-positive bacteria including methicillin-resistant *Staphylococcus aureus* (MRSA) and *Enterococcus faecium* as well as Gram-negative bacteria, including *Eschericia coli* and *Klebsiella pneumonia* (NDM). This action of this drug commend it for use in controlling Acute Bacterial Skin and Skin Structure Infections (ABSSSI). A five-day treatment course is usually effective in the treatment of many infections, and, given its direct membrane-disrupting action, bacteria are highly unlikely to acquire resistance to brilacidin. **1.** Mensa, Howell, Scott & DeGrado (2014) *Antimicrob. Agents Chemother.* **58**, 5136.

Brincidofovir, See *Cidofovir*

Brivanib

This ATP-competitive VEGFR inhibitor (FW = 370.38 g/mol; CAS 649735-46-6; Solubility: 74 mg/mL DMSO, <1 mg/mL H_2O), also named (R)-1-(4-(4-fluoro-2-methyl-1H-indol-5-yloxy)-5-methylpyrrolo[1,2-f][1,2,4]triazin-6-yloxy)propan-2-ol and Brivanib®, targets human Vascular Endothelial Growth Factor-2 Receptor, *or* VEGFR-2 (IC_{50} = 25 nM; K_i = 26 nM), with VEGFR-1 (IC_{50} = 0.38 μM) and FGFR-1 (IC_{50} = 0.148 μM) (1). Because brivanib has a high oral dose (60 mg/kg) relative to its intravenous dose (10 mg/kg), its alanine ester is a more effective oral drug (**See Brivanib Alanate**) (2). BMS-540215 is only weakly active against PDGFRβ, EGFR, LCK, PKCα or JAK-3, all with IC_{50} values >1.9 μM. BMS-540215 selectively inhibits VEGF- and FGF-stimulated endothelial cell (HUVEC) proliferation *in vitro*, with respective IC_{50} values of 40 and 276 nM (1). Furthermore, brivanib-induced growth inhibition was associated with a decrease in phosphorylated VEGFR-2 at Tyr(1054/1059), increased apoptosis, reduced microvessel density, inhibition of cell proliferation, and down-regulation of cell cycle regulators. The levels of FGFR-1 and FGFR-2 expression in these xenograft lines were positively correlated with its sensitivity to brivanib-induced growth inhibition (2). It also shows broad-spectrum *in vivo* antitumor activity over multiple dose levels and induces stasis in large tumors, with potential for treating hepatocellular carcinoma (HCC). **1**. Bhide, Cai, Zhang, *et al.* (2006) *J. Med. Chem.* **49**, 2143. **2**. Huynh, Ngo, Fargnoli, *et al.* (2008) *Clin. Cancer Res.* **14**, 6146.

Brivanib Alanate

This dual VEGFR/FGFR inhibitor (FW = 441.46 g/mol; CAS 649735-63-7), also named (S)-(R)-1-((4-((4-fluoro-2-methyl-1H-indol-5-yl)oxy)-5-methylpyrrolo[2,1-f][1,2,4]triazin-6-yl)oxy)propan-2-yl 2-aminopropanoate, targets vascular endothelial growth factor receptor and fibroblast growth factor receptor tyrosine kinases, inducing growth inhibition in mouse models of human hepatocellular carcinoma (1). Brivanib-induced growth inhibition is associated with inactivation of VEGFR-2, increased apoptosis, a reduction in microvessel density, inhibition of cell proliferation, and down-regulation of cell cycle regulators, including cyclin D_1, Cdk-2, Cdk-4, cyclin B1, and phospho-*c*-Myc. **1**. Huynh, Ngo, Fargnoli, *et al.* (2008) *Clin. Cancer Res.* **14**, 6146.

Brodifacoum

This potent vitamin K antagonist (FW = 523.43 g/mol; CAS 56073-10-0), also named Bromfenacoum, Superwarfarin, WBA 8119, and its IUPAC name 3-[3-[4-(4-bromophenyl)phenyl]-1,2,3,4-tetrahydronaphthalen-1-yl]-2-hydroxychromen-4-one, is famously lethal 4-hydroxycoumarin anticoagulant and biologically active ingredient in rodenticides and rat baits, such as D-Con®, Prufe I® and Prufe II®, Ramik®, Talon-G®, Ratak®, and Contrac®. (**See** *Dicumarol*) Its toxic potency is about 200x greater than warfarin, with a half-life as much as 60x longer. Such lethal and powerful action as a blood clotting inhibitor requires care to protect house pets and unintended animal targets. Baiting with medium oatmeal or wheat containing brodifacoum (0.0005–0.002 % wt/wt) completely controls infestations of warfarin-resistant rats on farms, when maintained until no longer required (1). Acute LD_{50} values: rats (oral), 0.27 mg/kg body weight; mice (oral), 0.40 mg/kg; rabbits (oral), 0.30 mg/kg; cats (oral), 0.25 mg/kg; and dogs (oral), 0.25 mg/kg. Such potency led to its designation as extremely toxic. Moreover, its long biological half-life also qualifies brodifacoum as noxious environmental pollutant. In humans, brodifacoum is poorly responsive to the administration of vitamin K (2). **1**. Rennison & Dubock (1978) *J. Hygiene (London)* **80**, 77. **2**. Wallace, Worsnop, Paull & Mashford (1990) *Austral. New Zeal. J. Med.* **20**, 713.

Bromazepam

This lipophilic valium-like benzodiazepine (FW = 316.20 g/mol; CAS 1812-30-2; IUPAC Name: 7-bromo-5-(pyridin-2-yl)-1H-benzo[e][1,4]diazepin-2(3H)-one), is a long-lasting anxiolytic agent used to treat anxiety or panic states. Like diazepam, bromazepam is a positive allosteric modulator of $GABA_A$ receptors, promoting GABA binding, which in turn increases the total conduction of chloride ions across the neuronal cell membrane. Like other benzodiazepines, bromazepam also inhibits the acetylcholine receptor-operated potassium current. Bromazepam is mainly metabolized by oxidative pathways within the liver. **1**. Marketed as Lectopam®, Lexotan®, Lexilium®, Lexaurin®, Brazepam®, Rekotnil®, Bromaze®, Somalium® and Lexotanil®.

Bromelain Inhibitors

This set of eight isoforms (MW = 5,700-5,900 g/mol, based on SDS polyacrylamide gel electrophoresis), from pineapple stem, inhibits bromelain. These inhibitors consist of a light chain (10 or 11 residues) and a heavy chain (40 or 41 residues) that are cross-linked with disulfide bonds. There are two distinct domains, each of which is formed by a three-stranded anti-parallel β-sheet. They are structurally related to the Bowman-Birk trypsin/chymotrypsin inhibitor. The K_i value for bromelain inhibitor VI is 270 nM. **Target(s):** stem bromelain (1-3); papain (1); ficain (1). **1**. Heinrikson & Kézdy (1976) *Meth. Enzymol.* **45**, 740. **2**. Hatano, Sawano & Tanokura (2002) *Biol. Chem.* **383**, 1151. **3**. Reddy, Keim, Heinrikson & Kézdy (1975) *J. Biol. Chem.* **250**, 1741.

2-Bromoacetamide

This alkylating agent (FW = 137.96 g/mol; CAS 683-57-8; MP = 88–90°C; Water-soluble) is a potent irreversible inhibitor of enzymes containing reactive thiols, ε-amino groups, and/or imidazole side-chain groups. In certain instances, methionyl residues also undergo alkylation. The reaction is typically a bimolecular process showing no evidence of the rate-saturation kinetics characteristic of affinity labeling reagents. The reactivity of sulfhydryl groups is usually highest, increasing in proportion to the fraction of the more nucleophilic thiolate species (optimal between pH 6 and 8.5). However, solution pH in excess of 8.5 should be avoided to minimize side-reactions, particularly with deprotonated amino groups. Because bromoacetamide slowly decomposes in aqueous solutions, fresh stock solutions should be prepared. *Note:* This compound must not be confusing with CH_3–C(=O)N(H)Br, the common brominating agent N-bromoacetamide. **Target(s):** alcohol dehydrogenase(1-3); arylamine *N*-acetyltransferase (4); papain (5). **1.** Plapp (1982) *Meth. Enzymol.* **87**, 469. **2.** Shaw (1970) *The Enzymes*, 3rd ed. (Boyer, ed.), **1**, 91. **3.** Woronick (1961) *Acta Chem. Scand.* **15**, 2062. **4.** Wang, Vath, Gleason, Hanna & Wagner (2004) *Biochemistry* **43**, 8234. **5.** Glazer & Smith (1971) *The Enzymes*, 3rd ed. (Boyer, ed.), **3**, 501.

N^6-(*p*-Bromoacetamidobenzyl)adenosine 5'-Monophosphate

This haloacetylated nucleotide ($FW_{\text{free-acid}}$ = 573.30 g/mol) irreversibly inactivates dehydrogenases and kinases through its action at their respective NAD^+ and ATP binding. It also irreversibly labels the allosteric AMP activator site on muscle glycogen phosphorylase *b* with an attendant increase in enzyme activity. **1.** Colman (1990) *The Enzymes*, 3rd ed., **19**, 283. **2.** Suzuki, Eguchi & Imahori (1977) *J. Biochem.* **81**, 1147.

2(*R*)-[2-(4'-Bromobiphenyl-4-yl)ethyl]-4(*S*)-(*n*-butyl)-1,5-pentanedioic Acid 1-(α(*S*)-*tert*-Butylglycine Methylamide) Amide

This substituted peptidomimetic (FW = 531.58 g/mol), also known as 2(*R*)-[2-[4-(4-bromophenyl)phenyl]ethyl]-4(*S*) butylpentanedioate 1-((*S*)-*tert*-butylglycine methylamide) amide, inhibits gelatinase A (K_i = 17 nM) and stromelysin 1 (K_i = 11 nM). **1.** Esser, Bugianesi, Caldwell, *et al.* (1997) *J. Med. Chem.* **40**, 1026.

6-Bromo-2-(4-bromophenyl)-4-quinolinecarboxylic Acid

This substituted quinoline analogue ($FW_{\text{free-acid}}$ = 407.06 g/mol) is a strong inhibitor of *Candida albicans* prolyl-tRNA synthetase (IC_{50} = 25 nM, with high selectivity over the corresponding human enzyme. **1.** Yu, Hill, Yu, *et al.* (2001) *Bioorg. Med. Chem. Lett.* **11**, 541.

4*S*-[(6-Bromo-2*H*-chromene-3-carbonyl)amino]-5-(2-oxopyrrolidin-3*S*-yl)pent-2-enoic Acid Ethyl Ester

This Michael acceptor-containing ester (FW = 462.08 g/mol) irreversibly inhibits picornain 3C. No cytotoxicity is observed at the limits of the assay concentration. A crystal structure of one of the more potent inhibitors covalently bound to HRV-2 3CP is also provided. **1.** Johnson, Hua, Luu, *et al.* (2002) *J. Med. Chem.* **45**, 2016.

Bromocolchicine

This halogenated mitotic spindle poison (FW = 478.34 g/mol) is a highly specific affinity label for tubulin, blocking tubulin polymerization and microtubule assembly (*See also* Colchicine). **1.** Schmitt & Atlas (1976) *J. Mol. Biol.* **102**, 743. **2.** Atlas & Schmitt (1977) *Meth. Enzymol.* **46**, 567.

Bromocriptine

This nonspecific monoamine receptor agonist and lactation inhibitor (FW = 654.60 g/mol; CAS 25614-03-3), also known as (5'α)-2-bromo-12'-hydroxy-5'-(2-methylpropyl)-3',6',18-trioxo-2'-(propan-2-yl)-ergotaman, inhibits the release of glutamate, by reversing the glutamate GLT1 transporter (1). An ergoline derivative, bromocriptine dose-dependently inhibits d-[^{3}H]aspartate release elicited by chemical anoxia or high KCl, while no changes occurred in uptake. The inhibitory action of bromocriptine is unaffected by sulpiride, a dopamine D_2 receptor antagonist. On the other hand, bromocriptine is without effect on swelling-induced d-[^{3}H]aspartate release, which is mediated by volume-regulated anion channels. *In vivo*, bromocriptine suppresses the excessive elevation of glutamate levels in gerbils subjected to transient forebrain ischemia in a manner similar to DHK. Taken together, such results provide evidence that bromocriptine inhibits excitatory amino acid release via reversed operation of GLT-1 without altering forward transport (1). In Europe, use of bromocriptine to inhibit lactation is restricted to situations in which breastfeeding must be stopped for medical reasons. **Agonist Targets:** Dopamine Receptor Family: D_1 (K_i = 680 nM); D_5 (K_i = 500 nM); D_2 (K_i = 3 nM); D_3 (K_i = 5.4 nM); D_4 (K_i = 330 nM); Serotonin Receptor Family: 5-HT_{1A} (K_i = 13 nM); 5-HT_{1B} (K_i = 355 nM); 5-HT_{1D} (K_i = 11 nM); 5-HT_{2A} (K_i = 107 nM); 5-HT_{2B} (K_i = 56 nM); 5-HT_{2C} (K_i = 740 nM); 5-HT6 (K_i = 33 nM); Adrenergic α Receptor Family: $α_{1A}$ (K_i = 4 nM); $α_{1B}$ (K_i = 1.4 nM); $α_{1D}$ (K_i = 1.1 nM); $α_{2A}$ (K_i = 11 nM); $α_{2B}$ (K_i = 35 nM); $α_{2C}$ (K_i = 28 nM); Adrenergic β Receptor Family: $β_1$ (K_i = 590 nM); and $β_2$ (K_i = 740 nM). **1.** Shirasaki, Sugimura, Sato, *et al.* (2010). *Eur. J. Pharmacol.* **643**, 48.

5-Bromo-2'-deoxyuridine

This pyrimidine nucleoside (FW = 307.10 g/mol), abbreviated as 5BrdU and also known as broxuridine, is metabolized to 5-bromo-2'-deoxyuridine 5'-triphosphate (**See below**), which is enzymatically incorporated into DNA, replacing thymidine in the process and rendering the DNA product sensitive to visible light (1). 5BdUr also inhibits cell differentiation. **Indirect Detection of Intracellular DNA Synthesis:** 5-Bromodeoxyuridine substitutes for pyrimidine bases during DNA syntheesis, a property permitting the detection of cells in S-phase using BrdU-specific antibodies: **Step-1.** Incubate cells with 10 μM BrdU for 30-120 min. **Step-2.** Wash three times with 2-3 mL PBS (One liter containing 8.01 g/L sodium chloride, 0.2g KCl; 1.78g $Na_2HPO_4 \cdot 2H_2O$, 0.27g KH_2PO_4, adjusted to pH 7.4). **Step-3.** Fix cells in 2-4% paraformaldehyde for 5 min. **Step-4.** Wash again in PBS. **Step-5.** Denature with 2 M HCl for 30 min at 37°C. **Step-6.** Rinse away HCl. **Step-7.** Neutralize by washing three times with one mL borate buffer (0.1 M, pH 8.5, 5 min). **Step-8.** Wash three more times with PBS. **Step-9.** Wash again PBS, containing 0.1% (w/v) Tween-20 and 2% Serum, incubating for 30 min at 37°C. **Step-10.** Aspirate away the buffer, and incubate with anti-BrdU antibody (Pierce Monoclonal Antibody BU-1 Catalog Number MA3-071; Nominal titer = 1:1,000 to 1:2,000) for 1-2 hr at room temperature. **Step-11.** Remove antibody by three washes with PBS/Tween-20. **Step-12.** Incubate with secondary antibody, for 30-40 min at room temperature. **Step-13.** Wash three more times with PBS/Tween-20. **Step-14.** Stain with DAPI, wash, and mount. **Target(s):** viral DNA polymerase, via its 5'-triphosphate (2); dihydroorotase (3); thymidine kinase (4,5,12-14); DNA biosynthesis (6,7); deoxycytidine deaminase (8); cytidine deaminase (9); dTMP kinase, *or* thymidylate kinase (10,11); NAD^+ ADP-ribosyltransferase, *or* (poly(ADP-ribose) polymerase (15); thymidine phosphorylase (16). **See also 5-Ethynyluridine** 1. Scott & Tomkins (1975) *Meth. Enzymol.* **40**, 273. 2. Prisbe & Chen (1996) *Meth. Enzymol.* **275**, 425. 3. Bresnick & Blatchford (1964) *Biochim. Biophys. Acta* **81**, 150. 4. Baker & Neenan (1972) *J. Med. Chem.* **15**, 940. 5. Bresnick & Thompson (1965) *J. Biol. Chem.* **240**, 3967. 6. Beltz & Visser (1955) *J. Amer. Chem. Soc.* **77**, 736. 7. Bardos, Levin, Herr & Gordon (1955) *J. Amer. Chem. Soc.* **77**, 4279. 8. Le Floc`h & Guillot (1974) *Phytochemistry* **13**, 2503. 9. Hosono & Kuno (1973) *J. Biochem.* **74**, 797. 10. Munier-Lehmann, Pochet, Dugué, *et al.* (2003) *Nucleosides Nucleotides Nucleic Acids* **22**, 801. 11. Pochet, Dugué, Labesse, Delepierre & Munier-Lehmann (2003) *ChemBioChem* **4**, 742. 12. Swinton & Chiang (1979) *Mol. Gen. Genet.* **175**, 399. 13. Gustafson, Chillemi, Sage & Fingeroth (1998) *Antimicrob. Agents Chemother.* **42**, 2923. 14. Chraibi & Wright (1983) *J. Biochem.* **93**, 323. 15. Rankin, Jacobson, Benjamin, Moss & Jacobson (1989) *J. Biol. Chem.* **264**, 4312. 16. Avraham, Yashphe & Grossowicz (1988) *FEMS Microbiol. Lett.* **56**, 29.

5-Bromo-2'-deoxyuridine 5'-Triphosphate

This halogenated dUTP derivative ($FW_{free-acid}$ = 547.04 g/mol; Soluble in water), often abbreviated 5-Br-dUTP, is a useful probe of DNA biosynthesis (**See entry above**). Brominated nucleotides inhibit DNA replication by acting as poor alternative substrates, resulting in their incorporation into DNA. Modified DNA is produced and cell growth is inhibited. 5-Br-dUTP should be stored in a desiccator in the cold and protected from light. **Target(s):** dCMP deaminase (1); DNA polymerase (2); thymidine kinase (3). 1.

Sergott, Debeer & Bessman (1971) *J. Biol. Chem.* **246**, 7755. 2. Prisbe & Chen (1996) *Meth. Enzymol.* **275**, 425. 3. Okazaki & Kornberg (1964) *J. Biol. Chem.* **239**, 275.

S-(4-Bromo-2,3-dioxobutyl)-Coenzyme A

This haloketone-containing acyl-CoA analogue (FW = 884.47 g/mol) is an irreversible inhibitor for several enzymes that bind short-chain fatty-acyl-CoA substrates (1-8). Note that the fatty acid-like chain is joined to Coenzyme A by a thioether linkage, not a thiolester bond. A convenient chemical synthesis has been described (3). **Target(s):** acetyl-CoA carboxylase (1,3); fatty-acid synthase (2,3,5-7); citrate synthase (3,4); acetyl-CoA hydrolase (3); acetoacetyl-CoA thiolase (3); 3-hydroxy-3-methylglutaryl-CoA reductase (4); β-oxoacyl synthase (condensing) (5,6); β-oxoacyl reductase (5); β-hydroxyacyl dehydratase (5); enoyl reductase (5); carnitine *O*-palmitoyltransferase (8). 1. Ahmad & Ahmad (1981) *Meth. Enzymol.* **71**, 16. 2. Clements, Barden, Ahmad & Ahmad (1979) *Biochem. Biophys. Res. Commun.* **86**, 278. 3. Barden, Owens & Clements (1981) *Meth. Enzymol.* **72**, 580. 4. Colman (1990) *The Enzymes*, 3rd ed. (Sigman & Boyer, eds.), **19**, 283. 5. Clements, Barden, Ahmad, Chisner & Ahmad (1982) *Biochem. J.* **207**, 291. 6. Katiyar, Pan & Porter (1983) *Eur. J. Biochem.* **130**, 177. 7. Katiyar, Pan & Porter (1982) *Biochem. Biophys. Res. Commun.* **104**, 517. 8. Lund (1987) *Biochim. Biophys. Acta* **918**, 67.

2-(4-Bromo-2,3-dioxobutylthio)adenosine 2',5'-Bisphosphate

This synthetic nucleotide ($FW_{free-acid}$ = 622.24 g/mol) is an affinity label for isocitrate dehydrogenase ($NADP^+$). Biphasic inactivation kinetics are observed, with a fast initial phase resulting in partially active enzyme (6-7% residual activity), followed by a slower phase leading to total inactivation. The inactivation rate constant exhibits a nonlinear dependence on reagent concentration, consistent with the formation of a reversible complex with the enzyme prior to irreversible modification. The enzyme incorporates this reagent to a limited extent and is protected against inactivation by $NADP^+$ and NADPH. 1. Bailey & Colman (1987) *Biochemistry* **26**, 6858.

Bromoenol lactone

This alkylating agent (FW = 317.18 g/mol; CAS 88070-98-8; λ_{max} = 224 nm; Symbol = BEL) inactivates phospholipase A_2, *or* PLA_2, in a linear, time-dependent manner. Inactivation of the PLA_2 by 5,5'-dithiobis(2-nitrobenzoic acid) before treatment with [^{3}H]BEL results in the near complete lack of radiolabel incorporation, a finding that is fully consistent with covalent irreversible suicide inhibition of the enzyme. Labeling of the 80-kDa band rather than 40-kDa band distinguishes the macrophage Ca^{2+}-independent PLA_2, providing strong evidence that the 80-kDa protein is the catalytic subunit (1). BEL also inhibits cellular phosphatidic acid phosphohydrolase (PAP) activity in P388D1 macrophages (IC_{50} = 8 μM), blocking incorporation of exogenous arachidonate and palmitate into diacylglycerol and triacylglycerol (2). 1. Ackermann, Conde-Frieboes & Dennis (1995) *J. Biol. Chem.* **270**, 445. 2. Balsinde & Dennis (1996) *J. Biol Chem.* **271**, 31937.

4-(4-Bromo-2-fluorophenylamino)-3-fluoro-1-methyl-pyridin-2(¹H)-one 5-(O-cyclopropylmethyl)hydroxamate

This methylpyridinone (FW = 428.23 g/mol) inhibits mitogen-activated protein kinase kinase (IC_{50} = 6.8 nM). This inhibitor exhibits excellent cellular potency and good pharmacokinetic properties. It inhibits ERK phosphorylation in HT-29 tumors from mouse xenograft studies. **1.** Wallace, Lyssikatos, Blake, *et al.* (2006) *J. Med. Chem.* **49**, 441.

4-Bromo-3-hydroxybenzyloxyamine

This substituted hydroxylamine (FW = 218.05 g/mol), also known as *O*-(4-bromo-3 hydroxybenzyl)hydroxylamine, α-(aminooxy)-6-bromo-*m*-cresol, NSD 1055, Brocresine, and Bromcresine, inhibits 5-aminolevulinate aminotransferase (1), arginine decarboxylase (2,15,16), histidine decarboxylase (3,6,7,9), diamine oxidase (3), phosphatidylserine decarboxylase (4), aromatic-L-amino-acid decarboxylase (Dopa decarboxylase (5,7,14), *S*-adenosylmethionine decarboxylase (8,12,13), threonine ammonia-lyase, *or* threonine dehydratase (10), phenylalanine decarboxylase (11), and dihydroxyphenylalanine aminotransferase (17). **1.** Turner & Neuberger (1970) *Meth. Enzymol.* **17A**, 188. **2.** Smith (1983) *Meth. Enzymol.* **94**, 176. **3.** Kobayashi, Kupelian & Maudsley (1970) *Biochem. Pharmacol.* **19**, 1761. **4.** Satre & Kennedy (1978) *J. Biol. Chem.* **253**, 479. **5.** Sourkes (1987) *Meth. Enzymol.* **142**, 170. **6.** Zachariae, Brodthagen & Sondergaard (1969) *J. Invest. Dermatol.* **53**, 341. **7.** Ellenbogen, Kelly, Taylor & Stubbs (1973) *Biochem. Pharmacol.* **22**, 939. **8.** Pösö, Sinervirta, Himberg & Jänne (1975) *Acta Chem. Scand. B* **29**, 932. **9.** Levine (1966) *Science* **154**, 1017. **10.** Desai, Laub & Anita (1972) *Phytochemistry* **11**, 277. **11.** David, Dairman & Udenfriend (1974) *Arch. Biochem. Biophys.* **160**, 561. **12.** Pösö, Sinervirta & Jänne (1975) *Biochem. J.* **151**, 67. **13.** Manen & Russell (1974) *Biochemistry* **13**, 4729. **14.** Jung (1986) *Bioorg. Chem.* **14**, 429. **15.** Smith (1979) *Phytochemistry* **18**, 1447. **16.** Ramakrishna & Adiga (1975) *Eur. J. Biochem.* **59**, 377. **17.** Waterhouse, Chia & Lees (1979) *Mol. Pharmacol.* **15**, 108.

6-Bromoindirubin-3'-oxime

These protein kinase inhibitors (FW_{BIO} = 356.19 g/mol; CAS 667463-62-9; Solubility: 70 mg/mL DMSO, <1mg/mL H_2O), also known as BIO, 6BIO, and systematically as 6-bromo-3-[(3*E*)-1,3-dihydro-3-(hydroxyimino)-2*H*-indol-2-ylidene]-1,3-dihydro-(3*Z*)-2*H*-indol-2-one, targets glycogen synthase kinase-3 (*or* GSK-3), with an IC_{50} value of 5 nM for α and β forms. BIO also shows greater than 16x selectivity over CDK5 and is also a weaker pan-JAK inhibitor, with IC_{50} values of 30 nM, 1.5 μM, 8.0 μM, and 0.5 μM for TYK2, JAK1, JAK2 and JAK3 (1,2). (*See the structural related inhibitor 1-Azakenpaullone*) BIO, but not 1-methyl-BIO, closely mimicked Wnt signaling in *Xenopus* embryos (1,2). BIO-induced apoptosis of human melanoma cells is associated with reduced phosphorylation of JAKs and STAT3 in both dose- and time-dependent manners. Consistent with inhibition of STAT3 signaling, expression of the anti-apoptotic protein Mcl-1 was down-regulated. **Maintaining Embryonic Stem Cells Pluripotency:** Human and mouse embryonic stem cells (ESCs) self-renew indefinitely while maintaining the ability to generate all three germ-layer derivatives. Importantly, Wnt pathway activation by BIO is sufficient to maintain the undifferentiated phenotype in both types of ESCs and sustains expression of the pluripotent state-specific transcription factors Oct-3/4, Rex-1 and Nanog (3). Such results suggest that BIO and related GSK-3-specific inhibitors may off practical advantages in regenerative medicine. Bio can also participate in controlling the proliferative capability of the highly differentiated cardiomyocytes (4). Activation of Wnt/β-catenin and inhibition of Notch signaling pathways, as mediated by simultaneous co-application of BIO and the γ-secretase inhibitor *N*-[(3,5-difluorophenyl)acetyl]-L-alanyl-2-phenylglycine-1,1-dimethylethyl ester (*or* DAPT), efficiently induces intestinal differentiation of ESCs cultured on feeder cells. These findings that Wnt and Notch signaling function to pattern the anterior-posterior axis of the DE and control intestinal differentiation (5). **6-Bromoindirubin-3'-acetoxime:** This cell-permeable BIO analogue ($FW_{BIO-acetoxime}$ = 398.21 g/mol; CAS 667463-85-6) is active against herpes simplex virus-1 (HSV-1) infection in human oral epithelial cells, the latter representing a natural target cell type. BIO-acetoxime reduces viral yields and the expression of different classes of viral proteins. BIO-acetoxime may suppress viral gene expression and protect oral epithelial cells from HSV-1 infection. **Tyrian Purple:** BIO's indirubin nucleus is related to the famous Tyrian purple dye that the Phoenicians isolated from the gastropod mollusk *Hexaplex trunculus* and gained favor in antiquity, because it did not fade, actually becoming more brilliant upon weathering and exposure to sunlight. The inhibitory properties of BIO suggest that 6-bromoindirubin provides a new scaffold for the development of selective and potent pharmacological inhibitors of GSK-3 (1). **1.** Meijer, Skaltsounis, Magiatis, *et al.* (2003) *Chem. Biol.* **10**, 1255. **2.** Liu, Nam, Tian, *et al.* (2011) *Cancer Res.* **71**, 3972. **3.** Sato Meijer, Skaltsounis, Greengard & Brivanlou (2004) *Nature Med.* **10**, 55. **4.** Tseng, Engel & Keating (2006) *Chem. Biol.* **13**, 957. **5.** Ogaki, Shiraki, Kume & Kume (2013) *Stem Cells* **31**, 1086. **6.** Hsu & Hung (2013) *Arch. Virol.* **58**, 1287.

2-Bromolysergic Acid Diethylamide

This halogenated LSD derivative (FW = 402.33 g/mol), also known as bromolysergide and bromo-LSD, is a serotonin antagonist that lacking the powerful hallucinogenic effects exhibited by LSD. **Target(s):** histamine *N*-methyltransferase (1,2); serotonin receptor (3); aryl-acylamidase (4,5). **1.** Axelrod (1971) *Meth. Enzymol.* **17B**, 766. **2.** Brown, Tomchick & Axelrod (1959) *J. Biol. Chem.* **234**, 2948. **3.** Bennett & Snyder (1975) *Brain Res.* **94**, 523. **4.** Paul & Halaris (1976) *Biochem. Biophys. Res. Commun.* **70**, 207. **5.** Hsu, Paul, Halaris & Freedman (1977) *Life Sci.* **20**, 857.

2-Bromopalmitate

This α-halofatty acid (FW = 335.33 g/mol; CAS 18263-25-7; *Symbols*: 2-BP and Br-C16) inhibits palmitate oxidation by approximately 40% and 60%, when added at 50 and 100 μM (1). Pretreatment with both concentrations also inhibited lipolysis in washed cells (1). Br-C16 also inhibits human palmitoyl acyltransferases (2), a family of enzymes sharing a conserved DHHC (*or* Asp-His-His-Cys) motif (3). Actively inhibitory forms of 2-bromopalmitate are the CoA and carnitine esters, which are also extremely powerful and specific inhibitors of mitochondrial fatty acid oxidation (4). Indeed, 2-bromopalmitoyl-CoA, when added as such or formed from 2-bromopalmitate, inhibits the carnitine-dependent mitochondrial oxidation of palmitate or palmitoyl-CoA, but not the oxidation of palmitoylcarnitine. 2-Bromopalmitoylcarnitine inhibits the oxidation of palmitoylcarnitine as well as that of palmitate or palmitoyl-CoA, but has no effect on succinate oxidation. Significantly, 2-bromopalmitate also inhibits palmitoylation of

Phospholipid Scramblase-1 (or PLSCR1), a Ca^{2+}-binding, endofacial plasma membrane protein contributing to the trans-bilayer movement (or "scrambling") of phosphatidylserine and other phospholipids normally present on cytoplasmic face of membrane bilayers to the outer leaflet, upon influx of calcium ion (5). Accumulation of PS in outer leaflet is suppressed by ATP-dependent flippases, and loss of ATP formation in apoptotic cells allows PS to accumulate to levels that initiate phagocytosis and clearance by mononuclear macrophages. PLSCR1's cysteine-rich sequence (184-CCCPCC-189) is required for palmitoylation, and mutation of these five cysteines abrogates PLSCR1 trafficking to the plasma membrane, keeping nearly all of the expressed protein in the nucleus. **Other Target(s):** 2-Bromopalmitate also increases expression of Uncoupling Protein-2 (or UCP2), a mitochondrial protein involved in controlling energy expenditure (6). This α-halofatty acid also inhibits sex pheromone production, after topical treatment of pheromonal glands with 2-bromopalmitate (in DMSO), in the lepidopteran insects *Spodoptera littoralis*, *Thaumotopoea pityocampa*, and *Bombyx mori* (7). 2-Bromopalmitate reduces protein deacylation by inhibition of acyl-protein thioesterase enzymatic activities (8). Mass spectrometry analysis of enriched 2-bromopalmitate targets identified PAT enzymes, transporters, and many palmitoylated proteins, with no observed preference for CoA-dependent enzymes. Such findings raise uncertainty whether 2-bromopalmitate or 2-bromopalmitoyl-CoA blocks *S*-palmitoylation by inhibiting protein acyl transferases, or by blocking palmitate incorporation by direct covalent competition, highlighting the likely promiscuity of 2BP (9). **1**. Fong, Leu & Chai (1997) *Biochim Biophys Acta* **1344**, 65. **2**. Ducker, Griffel, Smith, *et al.* (2006) *Mol. Cancer Ther.* **5**, 1647. **3**. Jennings, Nadolski, Ling, *et al.* (2009) *J. Lipid Res.* **50**, 233. **4**. Chase & Tubbs (1972) *Biochem J.* **129**, 55. **5**. Wiedmer, Zhao, Nanjundan & Sims (2003) *Biochemistry.* 42, 1227. **6**. Viguerie-Bascands, Saulnier-Blache, Dandine, *et al.* (1999) *Biochem. Biophys. Res. Commun.* **256**, 138. **7**. Hernanz, Fabrias & Camps (1997) *J. Lipid Res.* **38**, 1988. **8**. Pedro, Vilcaes, Tomatis, *et al.* (2013) *PLoS One* **8**, e75232. **9**. Davda, El Azzouny, Tom, *et al.* (2013) *ACS Chem. Biol.* **8**, 1912.

4-[(3-Bromophenyl)amino]-6,7-dimethoxyquinazoline

This potent tyrosine protein kinase inhibitor (FW = 360.21 g/mol), also known as tyrphostin AG 1517, PD 153035, compound 32, and SU 5271, inhibits epidermal growth factor receptor (EGFR) protein-tyrosine kinase, K_i = 6 pM (1-5). **1**. Bridges, Zhou, Cody, *et al.* (1996) *J. Med. Chem.* **39**, 267. **2**. Fry, Kraker, McMichael, *et al.* (1994) *Science* **265**, 1093. **3**. Varkondi, Schäfer, Bökönyi, *et al.* (2005) *J. Recept. Signal Transduct. Res.* **25**, 45. **4**. MacKintosh & MacKintosh (1994) *Trends Biochem. Sci.* **19**, 444. **5**. Ren & Baumgarten (2005) *Amer. J. Physiol.* **288**, H2628.

N-[2-((3-(4-Bromophenyl)-2-propenyl)amino)ethyl]-5-isoquinolinesulfonamide

This substituted isoquinoline (FW$_{free-base}$ = 446.37 g/mol), commonly referred to as H-89 and *N*-[2-((*p*-bromocinnamyl)amino)ethyl]-5-isoquinolinesulfonamide, inhibits a number of protein kinases. H-89 also inhibits phosphatidylcholine biosynthesis in HeLa cells. **Target(s):** cAMP-dependent protein kinase, *or* protein kinase A (K_i = 0.048 μM) (1-10); mitogen- and stress activated protein kinase (12); p70 ribosomal protein S$_6$ kinase (2); Rho-dependent protein kinase II (2); checkpoint kinase (2); AMP-activated protein kinase (2); [mitogen-activated protein kinase]-activated protein kinase 2 (2); cGMP-dependent protein kinase, *or* protein kinase G (3); [myosin light-chain] kinase (K_i = 28.3 μM) (3); casein kinase I (K_i = 38.3 μM) (3); casein kinase II (K_i = 30 μM) (3); protein kinase C (K_i =

31.7 μM) (3); Ca^{2+}/calmodulin-dependent protein kinase II (3); *Renilla*-luciferin 2-monooxygenase (11). **1**. Hidaka, Watanabe & Kobayashi (1991) *Meth. Enzymol.* **201**, 328. **2**. Davies, Reddy, Caivano & Cohen (2000) *Biochem. J.* **351**, 95. **3**. Chijiwa, Mishima, Hagiwara, *et al.* (1990) *J. Biol. Chem.* **265**, 5267. **4**. Patel, Soulages, Wells & Arrese (2004) *Insect Biochem. Mol. Biol.* **34**, 1269. **5**. Su, Chen, Zhou & Vacquier (2005) *Dev. Biol.* **285**, 116. **6**. Lombardo, Armas, Weiner & Calcaterra (2007) *FEBS J.* **274**, 485. **7**. Rapacciuolo, Suvarna, Barki-Harrington, *et al.* (2003) *J. Biol. Chem.* **278**, 35403. **8**. D'Souza, Agarwal & Morin (2005) *J. Biol. Chem.* **280**, 26233. **9**. Gardner, Delos Santos, Matta, Whitt & Bahouth (2004) *J. Biol. Chem.* **279**, 21135. **10**. Liu, Large & Albert (2005) *J. Physiol.* **562**, 395. **11**. Herbst, Allen & Zhang (2009) *PloS One* **4**, e5642.

Bromophycolide A

This *Callophycus serratus* natural product (FW = 665.30 g/mol) and its natural and synthetic analogues target the crystallization of heme that is released during the life cycle of the human malaria parasite. **Mechanism of Inhibitory Action:** Plasmodium parasites produce and excrete plasmepsin, a pepsin-like enzyme that degrades hemoglobin to supply essential nutrients to the parasite, with concomitant release of α-hematin. The latter, also known as ferriprotoporphyrin IX, is a toxic pro-oxidant that catalyzes reactive oxygen species production. This substance is itself toxic to the parasite, which relies on heme crystallization to reduce its toxicity. Bromophycolide A binds to α-hematin (K_d = 1.8 x 10^{-5} M), thereby inhibiting the crystallization/detoxification process. **1**. Stout, Cervantes, Prudhomme, *et al.* (2011) *Chem. Med. Chem.* **6**, 1572. **2**. Teasdale, Prudhomme, Torres, *et al.* (2013) *ACS Med. Chem. Lett.* **4**, 989.

Bromopyruvate (Bromopyruvic Acid)

This α-halo-keto acid (FW$_{free-acid}$ = 166.96 g/mol; CAS 1113-59-3; Symbol: 3BrPA) is an active-site-directed irreversible enzyme inhibitor and lactate/pyruvate transport inhibitor. A classical example is *Escherichia coli* isocitrate lyase (*Reaction*: Isocitrate ⇌ Glyoxylate + Succinate), where, upon reaction with 3-bromopyruvate, Cys-195 is alkylated with full loss of catalytic activity. Inactivation by 3BrPA exhibits saturation kinetics, with protection against inactivation exerted by isocitrate and glyoxylate, but not succinate (1). Touted to be a molecular "Trojan horse" in cancer chemotherapy, 3-bromopyruvate is a lactate analogue that is thought to selectively enter those cancerous cells exhibiting the Warburg effect, whereupon it quickly dissipates glycolytic and mitochondrial ATP production, resulting in tumor destruction without apparent harm to the animals (2,3). Although hexokinase-II was once thought to be responsible for 3BrPA inhibition of glycolysis, pyruvylation of glyceraldehyde 3-phosphate dehydrogenase (GAPDH) by 3BrPA not only inhibits glycolysis, but also appears to be the primary intracellular target in 3BrPA-mediated cancer cell death (4). At higher concentrations, however, bromoyruvate is apt to be indiscriminant in its cytotoxicity, most likely the result of nonspecific enzyme alkylation. Given its large number of targets, the cellular action of bromopyruvate is far from straightforward. **Target(s):** ribonuclease A, *or* pancreatic ribonuclease (5,17,80); 3-phosphoglycerate dehydrogenase (6,26); 2-dehydro-3-deoxy-6-phosphogluconate aldolase, *or* 6-phospho-2-dehydro-3 deoxygluconate aldolase, *or* 2-keto-3-deoxy-6-phosphogluconate aldolase (7,8,12,14,16,20,31,35,38,40); pyruvate, orthophosphate dikinase (9,105); aspartate aminotransferase

(10,11,34,39,40); 6-phospho-2-dehydro-3-deoxygalactonate aldolase (12,14,45); isocitrate lyase (12,21,60,81); 3-dehydroquinate synthase (12); carbonic anhydrase I (12,18,63); *N*-acetylneuraminate lyase (12,32,94-96); pyruvate carboxylase (12); glutamate decarboxylase (12,42,47,50,100); malic enzyme (12,41,48,53); malonyl-CoA decarboxylase (9); alcohol dehydrogenase (15); 2-dehydro-3 deoxyphosphoheptonate aldolase, *or* 3-deoxy-D-*arabino*-heptolosonate 7-phosphate synthase (16); fructose-1,6-bisphosphate aldolase (19,86); L-lactate dehydrogenase (22,25); succinate dehydrogenase (23,36); glycerate dehydrogenase (24); pyruvate kinase (27,51,83); pyruvate dehydrogenase complex (28,30,37,46,57,58); phosphoenolpyruvate carboxylase (29,52,64); pyruvate decarboxylase (30); glutamate dehydrogenase (33); serine hydroxymethyltransferase, *or* glycine hydroxymethyltransferase (44,111); flavocytochrome b (29,54); choline transport (55); phosphoenolpyruvate carboxykinase (56); 3-phosphoshikimate 1-carboxyvinyltransferase, *or* 5-enolpyruvylshikimate-3-phosphate synthase (59,73,108); octopine dehydrogenase (9); tryptophanase, *or* tryptophan indole-lyase (62); dipeptide transporter (65); acetolactate synthase, *or* acetohydroxyacid synthase (66); phosphofructo-1-kinase (67); glutathione *S*-transferase (68); fatty acid synthase (69); lactate oxidase (70); pyruvate:ferredoxin oxidoreductase (71); 2-keto-4-hydroxyglutarate aldolase (72); arginase (74); 4-aminobutyrate aminotransferase (75); histidine aminotransferase (76); dihydrodipicolinate synthase (77); 4-oxalocrotonate tautomerase (78); 3-deoxy-D-*manno*-octulosonic acid 8-phosphate synthase (79); 2-methylisocitrate lyase (82,85); malate synthase (81); anthranilate synthase (87); GMP synthetase (88); dihydrodipicolinate synthase (89,90); methylisocitrate lyase (91); 4-hydroxy-2-oxoglutarate aldolase (92,93); 2-dehydro-3-deoxy-6-phosphogalactonate aldolase (95); threonine aldolase (96); malonyl-CoA decarboxylase (101,102); pantotheine hydrolase (103); *N*-acyl-D-amino-acid deacylase (104); phosphoenol-pyruvate:protein phosphotransferase (103); 3-deoxy-8 phosphooctulonate synthase (107); UDP-*N*-acetylglucosamine 1-carboxyvinyltransferase (109); acetolactate synthase (110). **1.** Ko & McFadden (1990) *Arch. Biochem. Biophys.* **278**, 373. **2.** Pedersen (2007) *J. Bioenerg. Biomembr.* **39**, 211. **3.** Ko, Pedersen & Geschwind (2001) *Cancer Lett.* **173**, 83. **4.** Ganapathy-Kanniappan, Geschwind, Kunjithapathan, *et al.* (2009) *Anticancer Res.* **29**, 4909. **5.** Plapp (1982) *Meth. Enzymol.* **87**, 469. **6.** Sallach (1966) *Meth. Enzymol.* **9**, 216. **7.** Meloche, Ingram & Wood (1966) *Meth. Enzymol.* **9**, 520. **8.** Hammerstedt, Möhler, Decker, Ersfeld & Wood (1975) *Meth. Enzymol.* **42**, 258. **9.** Milner, Michaels & Wood (1975) *Meth. Enzymol.* **42**, 199. **10.** Birchmeier & Christen (1977) *Meth. Enzymol.* **46**, 41. **11.** Sigman & Mooser (1975) *Ann. Rev. Biochem.* **44**, 889. **12.** Hartman (1977) *Meth. Enzymol.* **46**, 130. **13.** Kolattukudy, Poulose & Kim (1981) *Meth. Enzymol.* **71**, 150. **14.** Meloche (1981) *Trends Biochem. Sci.* **6**, 38. **15.** Rashed & Rabin (1968) *Eur. J. Biochem.* **5**, 147. **16.** Shaw (1970) *The Enzymes*, 3rd ed., **1**, 91. **17.** Richards & Wyckoff (1971) *The Enzymes*, 3rd ed. **4**, 647. **18.** Lindskog, Henderson, Kannan, *et al.* (1971) *The Enzymes*, 3rd. ed., **5**, 587. **19.** Horecker, Tsolas & Lai (1972) *The Enzymes*, 3rd ed., **7**, 213. **20.** Wood (1972) *The Enzymes*, 3rd ed., **7**, 281. **21.** Spector (1972) *The Enzymes*, 3rd ed., **7**, 357. **22.** Holbrook, Liljas, Steindel & Rossmann (1975) *The Enzymes*, 3rd ed. **11**, 191. **23.** Hatefi & Stiggall (1976) *The Enzymes*, 3rd ed. **13**, 175. **24.** Holzer & Holldorf (1957) *Biochem. Z.* **329**, 292. **25.** Busch & Nair (1957) *J. Biol. Chem.* **229**, 377. **26.** Walsh & Sallach (1965) *Biochemistry* **4**, 1076. **27.** Blumberg & Stubbe (1975) *Biochim. Biophys. Acta* **384**, 120. **28.** Korotchkina, & Patel (1999) *Arch. Biochem. Biophys.* **369**, 277. **29.** Gonzalez, Iglesias & Andreo (1986) *Arch. Biochem. Biophys.* **245**, 179. **30.** Lowe & Perham (1984) *Biochemistry* **23**, 91. **31.** Meloche (1967) *Biochemistry* **6**, 2273. **32.** Barnett & Kolisis (1974) *Biochem. J.* **143**, 487. **33.** Baker & Rabin (1969) *Eur. J. Biochem.* **11**, 154. **34.** Morino & Okamoto (1970) *Biochem. Biophys. Res. Commun.* **40**, 600. **35.** Meloche (1970) *Biochemistry* **9**, 5050. **36.** Sanborn, Felberg & Hollocher (1971) *Biochim. Biophys. Acta* **227**, 219. **37.** Maldonado, Oh & Frey (1972) *J. Biol. Chem.* **247**, 2711. **38.** Meloche, Luczak & Wurster (1972) *J. Biol. Chem.* **247**, 4186. **39.** Okamoto & Morino (1973) *J. Biol. Chem.* **248**, 82. **40.** Meloche (1973) *J. Biol. Chem.* **248**, 6945. **41.** Chang & Hsu (1973) *Biochem. Biophys. Res. Commun.* **55**, 580. **42.** Fonda & DeGrella (1974) *Biochem. Biophys. Res. Commun.* **56**, 451. **43.** Birchmeier & Christen (1974) *J. Biol. Chem.* **249**, 6311. **44.** Akhtar, El-Obeid & Jordan (1975) *Biochem. J.* **145**, 159. **45.** Meloche & Monti (1975) *Biochemistry* **14**, 3682. **46.** Grande, Bresters, de Abreu, de Kok & Veeger (1975) *Eur. J. Biochem.* **59**, 355. **47.** Fonda (1976) *J. Biol. Chem.* **251**, 229. **48.** Chang & Hsu (1977) *Biochemistry* **16**, 311. **49.** Mulet & Lederer (1977) *Eur. J. Biochem.* **73**, 443. **50.** Tunnicliff & Ngo (1978) *Int. J. Biochem.* **9**, 249. **51.** Yun & Suelter (1979) *J. Biol. Chem.* **254**, 1811. **52.** Kameshita, Tokushige, Izui & Katsuki (1979) *J. Biochem.* **86**, 1251. **53.** Pry & Hsu (1980) *Biochemistry* **19**, 951. **54.** Alliel, Mulet & Lederer (1980) *Eur. J. Biochem.* **105**, 343. **55.**

Jope & Jenden (1980) *J. Neurochem.* **35**, 318. **56.** Silverstein, Lin, Fanning & Hung (1980) *Biochim. Biophys. Acta* **614**, 534. **57.** Lowe & Perham (1984) *Biochemistry* **23**, 91. **58.** Apfel, Ikeda, Speckhard & Frey (1984) *J. Biol. Chem.* **259**, 2905. **59.** Steinrücken & Amrhein (1984) *Eur. J. Biochem.* **143**, 351. **60.** Jameel, El-Gul & McFadden (1985) *Arch. Biochem. Biophys.* **236**, 72. **61.** Thome, Pho & Olomucki (1985) *Biochimie* **67**, 249. **62.** Honda & Tokushige (1985) *J. Biochem.* **97**, 851. **63.** Sundaram, Rumbolo, Grubb, Strisciuglio & Sly (1986) *Amer. J. Hum. Genet.* **38**, 125. **64.** Gonzalez, Iglesias & Andreo (1986) *Arch. Biochem. Biophys.* **245**, 179. **65.** Miyamoto, Ganapathy & Leibach (1986) *J. Biol. Chem.* **261**, 16133. **66.** Silverman & Eoyang (1987) *J. Bacteriol.* **169**, 2494. **67.** Banas, Gontero, Drews, *et al.* (1988) *Biochim. Biophys. Acta* **957**, 178. **68.** Ricci, Del Boccio, Pennelli, *et al.* (1989) *J. Biol. Chem.* **264**, 5462. **69.** Tian, Yang & Hsu (1989) *Biochim. Biophys. Acta* **998**, 310. **70.** Duncan, Wallis & Azari (1989) *Biochem. Biophys. Res. Commun.* **164**, 919. **71.** Williams, Leadlay & Lowe (1990) *Biochem. J.* **268**, 69. **72.** Vlahos & Dekker (1990) *J. Biol. Chem.* **265**, 20384. **73.** Huynh (1991) *Arch. Biochem. Biophys.* **284**, 407. **74.** Turkoglu & Ozer (1992) *Int. J. Biochem.* **24**, 937. **75.** Blessinger & Tunnicliff (1992) *Biochem. Cell Biol.* **70**, 716. **76.** Roos, Mattern, Schrempf & Bormann (1992) *FEMS Microbiol. Lett.* **76**, 185. **77.** Laber, Gomis-Ruth, Romao & Huber (1992) *Biochem. J.* **288**, 691. **78.** Stivers, Abeygunawardana, Mildvan, *et al.* (1996) *Biochemistry* **35**, 803. **79.** Salleh, Patel & Woodard (1996) *Biochemistry* **35**, 8942. **80.** Wang, Wang & Zhao (1996) *Biochem. J.* **320**, 187. **81.** Sharma, Sharma, Höner zu Bentrup, *et al.* (2000) *Nature Struct. Biol.* **7**, 663. **82.** Brock, Darley, Textor & Buckel (2001) *Eur. J. Biochem.* **268**, 3577. **83.** Acan & Ozer (2001) *J. Enzyme Inhib.* **16**, 457. **84.** Smith, Huang, Miczak, *et al.* (2003) *J. Biol. Chem.* **278**, 1735. **85.** Grimm, Evers, Brock, *et al.* (2003) *J. Mol. Biol.* **328**, 609. **86.** Simizu & Bacila (1975) *An. Acad. Bras. Cienc.* **47**, 131. **87.** Zalkin & Kling (1968) *Biochemistry* **7**, 3566. **88.** Patel, Moyed & Kane (1977) *Arch. Biochem. Biophys.* **178**, 652. **89.** Laber, Gomis-Rueth, Romao & Huber (1992) *Biochem. J.* **288**, 691. **90.** Dereppe, Bold, Ghisalba, Ebert & Schaer (1991) *Plant Physiol.* **98**, 813. **91.** Brock, Darley, Textor & Buckel (2001) *Eur. J. Biochem.* **268**, 3577. **92.** Anderson, Scholtz & Schuster (1985) *Arch. Biochem. Biophys.* **236**, 82. **93.** Vlahos & Dekker (1990) *J. Biol. Chem.* **265**, 20384. **94.** Schauer, Sommer, Kruger, van Unen & Traving (1999) *Biosci. Rep.* **19**, 373. **95.** Barnett, Corina & Rasool (1971) *Biochem. J.* **125**, 275. **96.** DeVries & Binkley (1972) *Arch. Biochem. Biophys.* **151**, 234. **97.** Höner zu Bentrup, Miczak, Swenson & Russell (1999) *J. Bacteriol.* **181**, 7161. **98.** Meloche & Monti (1975) *Biochemistry* **14**, 3682. **99.** Akhtar & El-Obeid (1972) *Biochim. Biophys. Acta* **258**, 791. **100.** Tunnicliff (1990) *Int. J. Biochem.* **27**, 1235. **101.** Buckner, Kolattukudy & Poulose (1976) *Arch. Biochem. Biophys.* **177**, 539. **102.** Kim & Kolattukudy (1978) *Arch. Biochem. Biophys.* **190**, 585. **103.** Ricci, Nardini, Chiaraluce, Dupre & Cavallini (1986) *Biochim. Biophys. Acta* **870**, 82. **104.** Wakayama, Yada, Kanda, *et al.* (2000) *Biosci. Biotechnol. Biochem.* **64**, 1. **105.** Yoshida & Wood (1978) *J. Biol. Chem.* **253**, 7650. **106.** Saier, Schmidt & Lin (1980) *J. Biol. Chem.* **255**, 8579. **107.** Hedstrom & Abeles (1988) *Biochem. Biophys. Res. Commun.* **157**, 816. **108.** Sost, Chulz & Amrhein (1984) *FEBS Lett.* **173**, 238. **109.** Anwar & Vlaovic (1980) *Biochim. Biophys. Acta* **616**, 389. **110.** Lee, Choi & Yoon (2002) *Bull. Korean Chem. Soc.* **23**, 765. **111.** Akhtar & El-Obeid (1972) *Biochim. Biophys. Acta* **258**, 791.

N-Bromosuccinimide

This common protein-modifying reagent (FW = 178 g/mol; CAS 128-08-5; Abbreviatin: NBS; water solublility = 0.015 g/mL, M.P. = 173-175°C) reacts somewhat selectively under acidic conditions (pH ≤ 4, typically using acetate or formate buffers) with tryptophanyl residues in peptides/proteins. Indole-ring oxidation (detected as an absorbance loss at 280 nm or a gain at 261 nm; $\varepsilon = 10,300$ $M^{-1}cm^{-1}$ in the latter case) induces peptide-bond cleavage on its carboxyl side. (Note: No cleavage occurs at or above neutral pH.) NBS treatment often oxidizes the sulfur atom(s) in cysteinyl, cystinyl, and methionyl residues; histidyl and tyrosyl residues are also oxidized by NBS, albeit more slowly. Accordingly, the proximal cause of lost biological activity of an NBS-treated peptide/protein is often initially uncertain. **Note:** *N*-Bromosuccinimide is an eye and skin irritant, necessitating appropriate handling. Avoid inhalation of the easily dispersed dry powder.

O^6-(4-Bromothenyl)-8-aza-7-deazaguanine

This modified guanine (FW = 327.19 g/mol) inactivates methylated-DNA:[protein]-cysteine *S*-methyltransferase, *or* O^6-alkylguanine-DNA alkyltransferase (I_{50} = 6 nM). The methyl transferase active site also tolerates O^6-substituted guanines, where the side chain can be quite large. However, it does not tolerate those with an aromatic or heteroaromatic ring with an "ortho" substituent. **1.** McMurry (2007) *DNA Repair* **6**, 1161.

O^6-(4-Bromothenyl) guanine

This modified guanine (FW = 327.19 g/mol), also known as lomeguatrib, inactivates methylated-DNA:[protein]-cysteine *S*-methyltransferase, *or* O^6-alkylguanine-DNA alkyl-transferase (I_{50} = 3.4 n M). The 5-bromo analogue also inhibits (I_{50} = 4.5 nM). Note that the O^6-(4-bromothien-3-ylmethyl)-guanine isomer is a weaker inhibitor. **1.** McMurry (2007) *DNA Repair* **6**, 1161. **2.** Verbeek, Southgate, Gilham & Margison (2008) *Brit. Med. Bull.* **85**, 17. **3.** Shibata, Glynn, McMurry, *et al.* (2006) *Nucl. Acids Res.* **34**, 1884. **4.** Juillerat, Heinis, Sielaff, *et al.* (2005) *ChemBioChem* **6**, 1263.

5-(5-Bromothien-2-yl)-2'-deoxyuridine

This nucleoside (FW = 389.23 g/mol) is an antiherpetic agent that inhibits viral DNA polymerase. It also inhibits herpes simplex virus type 1 thymidine kinase (IC_{50} = 3.5 µM; the 3 bromothien-2-yl) analogue is a weaker inhibitor). **Target(s):** DNA polymerase, via its 5'-triphosphate (1,2); thymidine kinase (3). **1.** Prisbe & Chen (1996) *Meth. Enzymol.* **275**, 425. **2.** Wigerinck, Pannecouque, Snoeck, *et al.* (1991) *J. Med. Chem.* **34**, 2383. **3.** Creuven, Evrard, Olivier, *et al.* (1996) *Antiviral Res.* **30**, 63.

5-Bromotryptophan

This halogenated amino acid (FW = 283.12 g/mol; CAS 6548-09-0) is a relatively potent inhibitor of sickle hemoglobin polymerization (1,2), indoleamine 2,3-dioxygenase, also alternative substrate (3,4), and tyrosine 3-monooxygenase (5,6). **1.** Li, Lin & Johnson (1995) *Biochem. J.* **308**, 251. **2.** De Croos, Sangdee, Stockwell, *et al.* (1990) *Med. Chem.* **33**, 3138. **3.** Southan, Truscott, Jamie, *et al.* (1996) *Med. Chem. Res.* **6**, 343. **4.** Macchiarulo, Camaioni, Nuti & Pellicciari (2009) *Amino Acids* **37**, 219. **5.**

Moore & Dominic (1971) *Fed. Proc.* **30**, 859. **6.** McGeer, McGeer & Peters (1967) *Life Sci.* **6**, 2221.

5-Bromouracil

This halogenated thymine analogue (FW = 190.98 g/mol; CAS 51-20-7) can be incorporated into DNA and is a powerful mutagen for a number of organisms. **Target(s):** dihydropyrimidine dehydrogenase (1); glucose-6-phosphate dehydrogenase (2,3); orotate phosphoribosyltransferase (4); pancreatic ribonuclease (5); thymidine phosphorylase, rabbit liver, I_{50} = 180 µM (6-9,17,18); uracil-DNA glycosylase (10); uracil phosphoribosyl-transferase (16); uridine phosphorylase, K_i = 230 µM (11-13); (*R*)-3-amino-2-methylpropionate:pyruvate aminotransferase, weakly inhibited (14); NAD^+ ADP-ribosyltransferase, *or* poly(ADP-ribose) polymerase (15); thymine dioxygenase (19). **1.** Porter, Chestnut, Merrill & Spector (1992) *J. Biol. Chem.* **267**, 5236. **2.** Noltmann & Kuby (1963) *The Enzymes*, 2nd ed., **7**, 223. **3.** Hochster (1961) *Biochem. Biophys. Res. Commun.* **6**, 289. **4.** Jones, Kavipurapu & Traut (1978) *Meth. Enzymol.* **51**, 155. **5.** Richards & Wyckoff (1971) *The Enzymes*, 3rd ed., **4**, 647. **6.** Baker (1967) *Design of Active-Site-Directed Irreversible Enzyme Inhibitors*, Wiley, New York. **7.** Baker, Kawazu & McClure (1967) *J. Pharm. Sci.* **56**, 1081. **8.** Baker & Kawazu (1967) *J. Med. Chem.* **10**, 316. **9.** Baker & Kelley (1971) *J. Med. Chem.* **14**, 812. **10.** Seal, Arenaz & Sirover (1987) *Biochim. Biophys. Acta* **925**, 226. **11.** Jimenez, Kranz, Lee, Gero & O'Sullivan (1989) *Biochem. Pharmacol.* **38**, 3785. **12.** Baker & Kelley (1970) *J. Med. Chem.* **13**, 461. **13.** Jimenez, Kranz, Lee, Gero & O'Sullivan (1989) *Biochem. Pharmacol.* **38**, 3785. **14.** Kaneko, Kontani, Kikugawa & Tamaki (1992) *Biochim. Biophys. Acta* **1122**, 45. **15.** Banasik, Komura, Shimoyama & Ueda (1992) *J. Biol. Chem.* **267**, 1569. **16.** Dai, Lee & O'Sullivan (1995) *Int. J. Parasitol.* **25**, 207. **17.** Blank & Hoffee (1975) *Arch. Biochem. Biophys.* **168**, 259. **18.** Miszczak-Zaborska & Wozniak (1997) *Z. Naturforsch. C* **52**, 670. **19.** Bankel, Lindstedt & Lindstedt (1977) *Biochim. Biophys. Acta* **481**, 431.

5-Bromouridine

This halogenated pyrimidine nucleoside (FW = 323.10 g/mol; CAS 957-75-5) is a thymidine analogue that can be incorporated (via the 5'-triphosphate) into polynucleic acids. (**See also** *5-Bromo-2' deoxyuridine*) **Target(s):** pancreatic ribonuclease (1); dihydroorotase (2); pre-mRNA splicing (3); cytidine deaminase (4); NAD^+ ADP-ribosyltransferase, *or* poly(ADP-ribose) polymerase (5). **1.** Richards & Wyckoff (1971) *The Enzymes*, 3rd ed. (Boyer, ed.), **4**, 647. **2.** Bresnick & Blatchford (1964) *Biochim. Biophys. Acta* **81**, 150. **3.** Sierakowska, Shukla, Dominski & Kole (1989) *J. Biol. Chem.* **264**, 19185. **4.** Cacciamani, Vita, Cristalli, *et al.* (1991) *Arch. Biochem. Biophys.* **290**, 285. **5.** Banasik, Komura, Shimoyama & Ueda (1992) *J. Biol. Chem.* **267**, 1569.

5-Bromouridine 5'-Triphosphate

This halogenated UTP derivative (FW = 563.04 g/mol; CAS 93882-11-2), often abbreviated 5-Br-UTP, is often used in the study of RNA and DNA biosynthesis. Acting as an inhibitor of DNA replication, this brominated nucleotide is incorporated into RNA and DNA, producing mismatch DNA and inhibiting normal cell proliferation. **Target(s):** CTP synthetase, *Escherichia coli* (1,2); transcription termination (3); UMP kinase (4); α-1,3-mannosyl-glycoprotein 2-β-N-acetylglucosaminyl-transferase (5). **1.** Scheit & Linke (1982) *Eur. J. Biochem.* **126**, 57. **2.** Scheit & Linke (1981) *Nucl. Acids Res.* **9**, 229. **3.** Shuman & Moss (1989) *J. Biol. Chem.* **264**, 21356. **4.** Evrin, Straut, Slavova-Azmanova, *et al.* (2007) *J. Biol. Chem.* **282**, 7242. **5.** Nishikawa, Pegg, Paulsen & Schachter (1988) *J. Biol. Chem.* **263**, 8270.

(*E*)-5-(2-Bromovinyl)-2'-deoxyuridine

This halogenated nucleoside (FW = 333.14 g/mol), also known as brivudin and 5-[(*E*)-2-bromoethenyl]-2'-deoxyuridine, is an antiherpetic agent that inhibits viral DNA polymerase. BvdU is a highly potent and selective inhibitor of herpes simplex virus type 1 and varicella-zoster virus infections. **Target(s):** DNA polymerase, viral, via the 5'-triphosphate (1-4); HSV-1 replication (1-4); 5' nucleotidases (5); dTMP kinase, *or* thymidylate kinase (6-8); deoxynucleoside kinase, also alternative substrate (9,10); thymidine kinase (9,11). **1.** Prisbe & Chen (1996) *Meth. Enzymol.* **275**, 425. **2.** Wigerinck, Pannecouque, Snoeck, *et al.* (1991) *J. Med. Chem.* **34**, 2383. **3.** De Clercq (2005) *Med. Res. Rev.* **25**, 1. **4.** Allaudeen, Kozarich, Bertino & De Clercq (1981) *Proc. Natl. Acad. Sci. U.S.A.* **78**, 2698. **5.** Garvey, Lowen & Almond (1998) *Biochemistry* **37**, 9043. **6.** Haouz, Vanheusden, Munier-Lehmann, *et al.* (2003) *J. Biol. Chem.* **278**, 4963. **7.** Munier-Lehmann, Pochet, Dugué, *et al.* (2003) *Nucleosides Nucleotides Nucleic Acids* **22**, 801. **8.** Pochet, Dugué, Labesse, Delepierre & Munier-Lehmann (2003) *ChemBioChem* **4**, 742. **9.** Johansson, Van Rompay, Degrève, Balzarini & Karlsson (1999) *J. Biol. Chem.* **274**, 23814. **10.** Balzarini, Degrève, Hatse, *et al.* (2000) *Mol. Pharmacol.* **57**, 811. **11.** Rechtin, Black, Mao, Lewis & Drake (1995) *J. Biol. Chem.* **270**, 7055.

BS-181

This cyclin kinase inhibitor (FW$_{free-base}$ = 416.99 g/mol; CAS 1092443-52-1); Solubility: 83 mg/mL DMSO; 3 mg/mL H$_2$O; Formulation: Dissolve in 10% DMSO/50 mM HCl/ 5% Tween 20/85% saline), systematically as named N^5-(6-aminohexyl)-N^7-benzyl-3-isopropylpyrazolo[1,5-*a*]pyrimidine-5,7-diamine, targets CDK7 with IC$_{50}$ value of 21 nM. Testing of other CDKs as well as another 69 kinases showed that BS-181 only inhibited CDK2 at concentrations lower than 1 μM, with CDK2 being inhibited 35-fold less potently (*IC*$_{50}$ = 880 nM) than CDK7. **Cyclin Target Selectivity:** Cdk1 (+++), Cdk2 (++), Cdk3 (weak, if any), Cdk4 (weak, if any), Cdk5 (weak, if any), Cdk6 (weak, if any), Cdk7 (weak, if any), Cdk8 (weak, if any), Cdk9 (++), Cdk10 (weak, if any). **1.** Ali, *et al.* (2009) *Cancer Res.* **69**, 6208.

BSc2118

This potent aldehyde-containing tripeptide mimetic (FW = 533.31 g/mol; CAS 863924-64-5) targets the 20 S proteasome (with lowest IC$_{50}$ values of 58 nm (chymotrypsin-like activity), 53 nm (trypsin-like activity), and 100 nm (caspase-like activity)), forming covalent adducts to active-site threonines (1). Given its critical role in protein turnover with effect on vital cellular pathways, the ubiquitin-proteasome system advanced to a promising new target in cancer treatment. Cell cycle regulating peptides, pro- and anti-apoptotic proteins as well as transcription factors, all necessary for cell survival, are degraded proteasome-dependently. Proteasomal inhibition results in accumulation of myriad of regulatory proteins, leading to confounding signals toward cell cycle progression and induction of apoptosis. In OPM-2, RPMI-8226, U266 multiple myeloma (MM) cell lines and primary MM cells, BSc2118 causes dose-dependent growth inhibitory effects (2). After 48 hours, dose-dependent apoptosis occurs both in cell lines and primary myeloma cells. G$_2$-M cell cycle arrest occurs after 24 hours, also causing marked inhibition of intracellular proteasome activity, an increase in intracellular p21 levels, and an inhibition of NF-κB activation (2). **1.** Braun, Umbreen, Groll, *et al.* (2005) *J. Biol. Chem.* **280**, 28394. **2.** Sterz, Jakob, Kuckelkorn, *et al.* (2010) *Eur. J. Hematol.* **85**, 99.

BSF-208075, *See* Ambrisentan

BTB-1

This mitotic kinesin inhibitor (FW = 297.71 g/mol; CAS 86030-08-2; Solubility: 100 mM in DMSO), also known as 4-chloro-2-nitro-1-(phenylsulfonyl)-benzene, targets Kif18A inhibitor (IC$_{50}$ = 1.7 μM), a cytosleletal motor that is frequently overexpressed in solid tumors (1). Kif18A reduces the amplitude of preanaphase oscillations and slows poleward movement during anaphase (2). BTB-1 is ATP-competitive and microtubule- (*or* MT-) uncompetitive and is selective, showing much weaker action against Kif4, Eg5, MKLP-1, MKLP-2, MPP1, CENP-E and MCAK (1). Kif18A's *in vitro* MT-gliding activity is also blocked reversibly (1). BTB-1 only inhibits the ATPase activity of the MT-bound motor protein. **1.** Catarinella, Grüner, Strittmatter, Marx & Mayer (2009) *Angew. Chem. Int. Ed. Engl.* **48**, 9072. **2.** Braun, Möckel, Strittmatter, *et al.* (2015) *ACS Chem. Biol.* **10**, 554.

BTB06584

This (FW = 417.82 g/mol; CAS 219793-45-0; Solubility: 84 mg/mL DMSO; < 1 mg/mL H$_2$O), also named 2-nitro-5-(phenylsulfonyl)phenyl 4-chlorobenzoate, inhibits F$_1$Fo-ATPase activity, with no effect on mitochondrial membrane potential (ΔΨ$_m$) or O$_2$ consumption. ATP consumption was significantly slowed after inhibition of respiration, and ischemic cell death was reduced. BTB efficiency was increased by IF$_1$ overexpression, and reduced by silencing the protein. **1.** Ivanes, Faccenda, Gatliff (2014) *Brit. J. Pharmacol.* **171**, 4193.

Budesonide

This glucocorticosteroid (FW = 430.53 g/mol; CAS 51333-22-3), also known as Rhinocort™, Pulmicort™, Budacort™, Ucerus™, and 16,17-(butylidenebis(oxy))-11,21-dihydroxy-(11β,16α)-pregna-1,4-diene-3,20-dione, reduces inflammation in the respiratory and digestive tract, depending on its mode of administration (1). Compared to prednisolone, budesonide has been associated with less bone density loss, and has little influence on the hypothalamic-pituitary-adrenal axis, with lower incidence of systemic manifestations than similar medications. The [22R]-epimer was more susceptible to liver biotransformation than the [22S]-epimer (2). Interleukins IL-2 and IL-4 counteract budesonide inhibition of GM-CSF and IL-10, but not of IL-8, IL-12 or TNF-α production by human mononuclear blood cells (3). **1**. Thalén & Brattsand (1979) *Arzneimittel-forschung*. 29, 1687. **2**. Andersson, Edsbäcker, Ryrfeldt & von Bahr (1982) *J. Steroid Biochem*. 16, 787. **3**. Larsson, Löfdahl & Linden (1999) *Br. J. Pharmacol*. 127, 980.

Bufalin

This cardioactive C-24 steroid (FW = 386.52 g/mol; CAS 465-21-4; Solubility: 5 mg/mL chloroform, 25 mg/mL DMSO), also named 3,14-dihydroxybufa-20,22-dienolide;5β,20(22)-bufadienolide-3β,14-diol and 3β,14-dihydroxy-5β,20(22)-bufadienolide, is the major digoxin-like immunoreactive component of *Chansu*, a traditional Chinese medicine prepared from the skin and venom-containing parotid gland of a poisonous toad (1). Bufalin increases the doubling time of three prostate cancer cell lines, inducing apoptosis and the caspase-3 activity. Expression of other proapoptoptic factors, such as mitochondrial Bax and cytosolic cytochrome *c*, were also increased. Bufalin also reduces serum-induced invasiveness of human hepatocellular cancer SK-Hep1 cells, markedly inhibiting MMP-2 and MMP-9 activities (2). It also attenuates phosphoinisitide-3-kinase (PI3K) and AKT phosphorylation was associated with reduced levels of NF-κB. Bufalin also suppresses protein levels of FAK, Rho A, VEGF, MEKK3, MKK7, and uPA. Such observations indicate that bufalin is as an antiinvasive agent that inhibits MMP-2 and -9 and alters PI3K/AKT and NF-κB signaling pathways (2). **1**. Takai, Kira, Ishii, *et al*. (2012) *Asian Pacific J. Cancer Prev*. 13, 399. **2**. Chen, Lu, Hsu, *et al*. (2015) *Environ. Toxicol*. 30, 74.

α-Bungarotoxin

This 74-residue snake neurotoxin (MW = 7983 Da; Accession Number = 720920A; CAS 11032-79-4; Symbol: Bgt) from the venom of the many-banded krait *Bungarus multicinctus* is a nicotinic cholinergic receptor antagonist that prevents opening of post-synaptic nicotinic receptor ion channels. α-Bungarotoxin is far more selective for the α7 receptor (IC50 =1.6 nM) than α3β4 receptor (IC50 > 3 μM). Because fluorescently tagged α-bungarotoxin binds to the nAChR with high affinity and defined stoichiometry, it may be employed to determine the location and amount of nAChR on the surface of muscle cells. **1**. Berg, Kelly, Sargent, Williamson & Hall (1972) *Proc. Natl. Acad. Sci. U.S.A.* 69, 147. **2**. Harvey, ed. (1991) *Snake Toxins*, Pergamon.

β-Bungarotoxin

This snake neurotoxin (MW = 15 kDa; CAS 12778-32-4) from the venom of the many-banded krait (*Bungaros multicinctus*) is a presynaptic terminus-seeking phospholipase A2 (of which there are two forms, designated β1 and β2) catalyzing phospholipid hydrolysis, thereby impairing acetylcholine release. **1**. Chen & Lee (1970) *Virchows Arch B Cell Pathol*. **6**, 318. **2**. Howard (1975) *Biochem. Biophys. Res. Commun*. **67**, 58. **3**. Strong, Goerke, Oberg & Kelly (1976) *Proc. Natl. Acad. Sci. USA* **73**, 178.

Buparlisib

This orally available, pan-class 1 PI3K inhibitor (FW = 410.17; CAS 944396-07-0; Solubility: 82 mg/mL DMSO; <1 mg/mL H_2O) also known as NVP-BKM-120 and BKM-120, as well as 5-[2,6-di(4-morpholinyl)-4-pyrimidinyl]-4-(trifluoromethyl)-2-pyridinamine, targets all four class I PI3K isoforms in biochemical assays with at least 50-fold selectivity against other protein kinases (1). Buparlisib displays mixed-type inhibition, affecting both the V_{max} and the K_m for ATP. It is also active against the most common somatic PI3Kα mutations, but does not significantly inhibit the related class III (Vps34) and class IV (mTOR, DNA-PK) PI3K kinases (1). Consistent with this mechanism of action, NVP-BKM120 decreases the cellular levels of p-Akt in mechanistic models and relevant tumor cell lines, as well as downstream effectors in a concentration-dependent and pathway-specific manner. NVP-BKM120 demonstrated anti-proliferative activity in 11 human gastric cancer cell lines by decreasing mTOR downstream signaling (2). In studies nf human glioblastoma multiforme (GBM) cells *in vitro* and *in vivo*, Koul, *et al*. (3) observed a differential sensitivity pattern with respect to p53 status, with cells containing wild-type p53 proving to be more sensitive than those with mutated or deleted p53. NVP-BKM120 showed differential forms of cell death on the basis of p53 status of the cells with p53 wild-type cells undergoing apoptotic cell death and p53 mutant/deleted cells having a mitotic catastrophe cell death. NVP-BKM120 mediates mitotic catastrophe mainly through Aurora B kinase (3). Knockdown of p53 in p53 wild-type U87 glioma cells displayed microtubule misalignment, multiple centrosomes, and mitotic catastrophe cell death (3). **1**. Maira, Pecchi, Huang, *et al*. (2012) *Mol. Cancer Ther*. **11**, 317. **2**. Park, *et al*. (2012) *Int. J. Oncol*. **40**, 1259. **3**. Koul, *et al*. (2012) *Clin. Cancer Res*. **18**, 184.

Buprenorphine

Buprenorphine   **Norbuprenorphine**

This semi-synthetic thebaine-based prodrug (FW = 467.44 g/mol; CAS 52485-79-7) is metabolized to norbuprenorphine (FW = 415.55 g/mol; CAS 78715-23-8), the pharmacologically active form that acts as an μ-opioid receptor antagonist (*or* more accurately a partial agonist). Buprenorphine is a strong analgesic with marked narcotic antagonist activity, producing side effects similar to those observed with other morphine-like compounds, including respiratory depression (3). Buprenorphine is used (a) at lower dosages (~200 μg) to control moderate acute pain in non-opioid-tolerant individuals, (b) at dosages ranging from 20-70 μg/hour to control moderate chronic pain, and (c) at higher dosages (>2 mg) to treat opioid addiction.

Buprenorphine is also employed to maintain sobriety. Dependence liability of buprenorphine may be lower than that of other older morphine-like drugs. Due to its partial agonist properties, buprenorphine offers advantages over methadone in managing opioid addiction (*i.e.,* decreased respiratory depression, less sedation, less withdrawal symptoms, lower risk of toxicity, and decreased risk of diversion). **Targets:** Buprenorphine is a nonselective, mixed agonist/antagonist opioid receptor modulator, acting at the μ-Opioid Receptor (*or* MOR), K_i = 1.5 nM; Intrinsic Activity = 28.7%; the κ-Opioid Receptor (*or* KOR), K_i = 2.5 nM, Intrinsic Activity = 0%; δ-Opioid Receptor (*or* DOR): K_i = 6.1 nM, Intrinsic Activity = 0%; and Nociceptin Receptor (*or* ORL-1), K_i = 77.4 nM; Intrinsic Activity = 15.5%. **Key Pharmacokinetic Parameters:** *See* Appendix II in Goodman & Gilman's *THE PHARMACOLOGICAL BASIS OF THERAPEUTICS*, 12th Edition (Brunton, Chabner & Knollmann, eds.) McGraw-Hill Medical, New York (2011). **1.** Buprenorphine IUPAC Name: (2*S*)-2-[(5*R*,6*R*,7*R*,14*S*)-9α-cyclopropylmethyl-4,5-epoxy-6,14-ethano-3-hydroxy-6-methoxymorphinan-7-yl]-3,3-dimethyl butan-2-ol; **Norbuprenorphine IUPAC Name:** (5α,6β,14β,18*R*)-18-[(1*S*)-1-hydroxy-1,2,2-trimethylpropyl]-6-methoxy-18,19-dihydro-4,5-epoxy-6,14-ethenomorphinan-3-ol. **2.** Dum & Herz (1981) *Brit. J. Pharmacol.* **74**, 627. **3.** Heel, Brogden, Speight & Avery (1979) *Drugs* **17**, 81.

Bupropion

This β-ketoamphetamine-type antidepressant and smoking cessation aid (FW = 239.74 g/mol; CAS 34841-39-9), also known as Wellbutrin®, Zyban®, and (±)-2-(*tert*-butylamino)-1-(3-chlorophenyl)propan-1-one, is said to be a norepinephrine and dopamine reuptake inhibitor, although its weak effect on dopamine levels calls into question the latter action (1). Bupropion also acts noncompetitively as a neuronal acetylcholine receptor antagonist. Unlike many antidepressants, however, bupropion shows no serotonergic activity. It also lacks typical antidepressant side effects, *e.g.,* sexual dysfunction, weight gain, and sedation (2). **Pharmacokinetics:** Upon oral administration, bupropion is rapidly absorbed with first-order kinetics and subsequently eliminated via biphasic kinetics, with a redistribution $t_{1/2}$ of ~1 hour and an elimination $t_{1/2}$ of 11-14 hours. Widely distributed throughout the body, bupropion is extensively metabolized, both oxidatively and reductively, to form as many as six metabolites, of which some are pharmacologically active. While bupropion does not inhibit monoamine oxidase, it exerts no effect on serotonin uptake. It minimally alters the reuptake of norepinephrine at presynaptic sites. Importantly, bupropion does not act on postsynaptic β-adrenergic down-regulation or presynaptic dopamine uptake. Gene variants in CYP2C19 are associated with altered *in vivo* bupropion pharmacokinetics (3). **Key Pharmacokinetic Parameters:** *See* Appendix II in Goodman & Gilman's *THE PHARMACOLOGICAL BASIS OF THERAPEUTICS*, 12th Edition (Brunton, Chabner & Knollmann, eds.) McGraw-Hill Medical, New York (2011). **1.** Soroko, Mehta, Maxwell, Ferris & Schroeder (1977) *J. Pharm. Pharmacol.* **29**, 767. **2.** Stahl, Pradko, Haight, *et al.* (2004) *Prim. Care Companion J. Clin. Psychiatry.* **6**, 159. **3.** Zhu, Zhou, Cox, *et al.* (2014) *Drug Metab. Dispos.* **42**, 1971.

Buspirone

This non-benzodiazepine serotonin receptor partial agonist (FW = 385.50 g/mol; CAS 36505-84-7) targets 5-HT$_{1A}$ receptors, exerting both anxiolytic and antidepressant effects (1). Buspirone differs from benzodiazepines, inasmuch as it fails to stimulate or inhibit ³H-benzodiazepine binding and does not affect GABA binding/uptake (2). Buspirone lacks anticonvulsant activity, interacts minimally with CNS depressants, and does not cause muscle relaxation. In animal models, its tranquilizing activity is characterized by the ability to tame aggression, to block conditioned avoidance responses, to inhibit shock-induced fighting, and to attenuate shock-induced suppression of drinking (2). Because it lacks the anticonvulsant, sedative, and muscle-relaxant properties associated with other anxiolytics, buspirone has been termed "anxioselective" (3).

Moreover, in contrast to the benzodiazepines, buspirone operates on a matrix of neural targets (see below) to treat anxiety effectively, while preserving arousal and attentional processes (4). *In vitro*, buspirone is metabolized by CYP3A4. Moreover, its active metabolites show a mechanism of action different from that of the parent compound (5). Given that buspirone appears to be useful in various neurological and psychiatric disorders (*e.g.,* attenuating side-effects of Parkinson's disease therapy, ataxia, depression, social phobia, and behaviour disturbances following brain injury, Alzheimer's disease, dementia and attention deficit disorder), it is unfortunate that buspirone's mechanism(s) of action are incompletely defined (6). **Key Pharmacokinetic Parameters:** *See* Appendix II in Goodman & Gilman's *THE PHARMACOLOGICAL BASIS OF THERAPEUTICS*, 12th Edition (Brunton, Chabner & Knollmann, eds.) McGraw-Hill Medical, New York (2011). **Target(s):** 5-HT$_{1A}$ receptors, K_d = 29 nM, as an agonist; 5-HT$_{2A}$ receptors, K_d = 138 nM, as an agonist; 5-HT$_{2B}$ receptors, K_d = 214 nM, as an agonist; 5-HT$_{2C}$ receptors, K_d = 490 nM, as an agonist; D$_2$ receptors, moderate an antagonist; D$_3$ receptors, as an antagonist; D$_4$ receptors, as an antagonist; α$_1$ receptors, as an agonist; and α$_{1D}$ receptors, as an agonist. **1.** Goldberg & Finnerty (1979) *Am. J. Psychiatry* **136**, 1184. **2.** Riblet, Taylor, Eison & Stanton (1982) *J. Clin. Psychiatry* **43**, 11. **3.** Taylor, Eison, Riblet & Vandermaelen (1985) *Pharmacol. Biochem. Behav.* **23**, 687. **4.** Eison & Temple (1986) *Am. J. Med.* **80**, 1. **5.** Garattini (1985) *Clin. Pharmacokinet.* **10**, 216. **6.** Loane & Politis (2012) *Brain Res.* **21**, 1461.

Busulfan

This antineoplastic drug (FW = 246.30 g/mol; CAS 55-98-1), also known as Myleran®, Busulfex IV®, and 1,4-butanediol dimethanesulfonate, is a cell cycle non-specific DNA alkylating agent that penetrates cancer cells and induces apoptosis. (**See also the water-soluble analogue,** *Treosulfan*) Busulfan undergoes nucleophilic attack by the N^7-atom of guanine, forming intra-strand DNA-DNA crosslinks between guanine and nearby adenine bases as well as between guanine and nearby guanine bases. By preventing replication as well as the action of DNA repair enzymes, busulfan-treated cells undergo programmed cell death. Initially developed to treat chronic myelogenous leukemia in view of its suppressive effect on peripheral granulocyte counts, Busulfan was later found to exert broad myelosuppressive effects. Indeed, high doses of busulfan produce myeloablation, depleting bone marrow precursors upon repeated dosing. In combination with Cyclophosphamide, Busulfan is an alternative to total-body irradiation for allogeneic bone marrow transplantation. Given is low solubility, busulfan exhibits erratic bulsufan absorption, necessitating close monitoring to achieve predictable myelosuppressive levels. Busulfan is now the cornerstone for high-dose treatment of various hematologic malignancies and immunodeficiencies. **Key Pharmacokinetic Parameters:** *See* Appendix II in Goodman & Gilman's *THE PHARMACOLOGICAL BASIS OF THERAPEUTICS*, 12th Edition (Brunton, Chabner & Knollmann, eds.) McGraw-Hill Medical, New York (2011). **1.** Elson, Galton & Till (1958) *Brit. J. Hematol.* **4**, 355.

L-Buthionine Sulfoximine

Butathione Sulfoximine

Butathione Sulfoximine Phosphate

This transition-state analogue precursor (FW = 222.31 g/mol; CAS 5072-26-4; *Symbol*: BSO), also known as *S*-*n*-butyl-homocysteine sulfoximine, undergoes ATP-dependent phosphorylation to the transition-state analogue L-buthionine sulfoximine phosphate, the latter potently inhibiting γ-glutamylcysteine synthetase. This reagent is without effect on glutamine synthetase (**See instead** L-Methionine Sulfoximine and L-Methionine Sulfoximine Phosphate). **Depletion of Cellular Glutathione Content:** Treatment of many cell types of buthionine sulfoximine results in a depletion of intracellular glutathione. In fed male mice, liver and kidney GSH levels are depleted by BSO in a dose-dependent manner, with maximum effect (i.e., reduction to 35% of initial levels) occurring at intraperitoneal doses of 0.8-1.6 g/kg body weight. At such doses, maximum effects on γ-glutamylcysteine synthetase (γ-GCS) and GSH are observed 2-4 hours after BSO administration, with initial γ-GCS activity and GSH restored ~16 hours post-BSO. Because cellular glutathione plays essential protective roles against radiation-induced cell death, BSO also increases radiation sensitivity. **1**. Griffith (1982) J. Biol. Chem. **257**, 13704. **2**. Seelig & Meister (1985) Meth. Enzymol. **113**, 379. **3**. Griffith (1987) Meth. Enzymol. **143**, 286. **4**. Meister (1995) Meth. Enzymol. **252**, 26. **5**. Griffith & Meister (1979) J. Biol. Chem. **254**, 7558. **6**. Dethmers & Meister (1981) Proc. Natl. Acad. Sci. U.S.A. **78**, 7492. **7**. Huang, Moore & Meister (1988) Proc. Natl. Acad. Sci. U.S.A. **85**, 2464. **8**. Kelly, Antholine & Griffith (2002) J. Biol. Chem. **277**, 50. **9**. Misra & Griffith (1998) Protein Expr. Purif. **13**, 268. **10**. Janowiak & Griffith (2005) J. Biol. Chem. **280**, 11829.

Butoconazole

This fungicide (FW = 411.78 g/mol; CAS 64872-77-1), also known as Femstat™, Femstat-3™, Gynazole-1™, and systematically as 1-[4-(4-chlorophenyl)-2-[(2,6-dichlorophenyl)-sulfanyl]butyl]-1H-imidazole (* = chiral center), is believed to inhibit steroid synthesis, targeting the conversion of lanosterol to ergosterol and modifying membrane composition/function. At 80 μM, butoconazole was strictly fungistatic against early stationary-phase Candida albicans cells. During early log-phase growth, butoconazole were highly lethal at 20 μM (1). In 1995, butoconazole was approved for treatment of vaginal yeast infections. **1**. Beggs (1985) J. Antimicrob. Chemother. **16**, 397.

N-(N-(t-Butoxycarbonyl)-L-alanyl)-(R)-7-amino-2-(4-methoxy benzenesulfonyl)-1,2,3,4-tetrahydroisoquinoline-3-carboxylic acid hydroxamate

This isoquinoline derivative (FW = 548.62 g/mol) inhibits human neutrophil collagenase, or MMP-8 (IC_{50} = 2 nM), fitting well with catalytic site, based on steric, electrostatic, and hydrophobic complementarity. **1**. Matter, Schwab, Barbier, et al. (1999) J. Med. Chem. **42**, 1908.

N-[3-(t-Butoxycarbonylamino)-4-methoxyphenylsulfonyl]-N-(4-nitrobenzyl)-β-alanine hydroxamate

This hydroxamate-based protease inhibitor (FW = 507.54 = g.mol) inhibits interstitial collagenase (K_i = 24 nM), gelatinase A (K_i = 4 nM), neutrophil collagenase (K_i = 7 nM), gelatinase B (K_i = 6 nM), and microbial collagenase (K_i = 11 nM). **1**. Scozzafava, Ilies, Manole & Supuran (2000) Eur. J. Pharm. Sci. **11**, 69.

(2R,3S,11'S,8'S)-3-[t-Butoxycarbonylamino]-4-phenyl-1-[7',10'-dioxo-8'-(1-methylpropyl)-2'-oxa-6',9'-diazabicyclo[11.2.2] heptadeca-13',15',16'-trien-11'-yl]amino]butan-2-ol

This hydrolytically stable macrocyclic peptidomimetic agent (FW = 596.77 g/mol) is a potent inhibitor of HIV-1 protease: K_i = 4 nM. Such cyclic analogues are potent inhibitors of HIV protease, and the crystal structures show them to be structural mimics of acyclic peptides, binding in the active site of HIV protease via the same interactions. Each macrocycle is restrained to adopt a β-strand conformation which is preorganized for protease binding. An unusual feature of the binding of C-terminal macrocyclic inhibitors is the interaction between a positively charged secondary amine and a catalytic aspartate of HIV protease. A bicyclic inhibitor binds similarly through its secondary amine that lies between its component N-terminal and C-terminal macrocycles. **1**. Abbenante, March, Bergman, et al. (1995) J. Amer. Chem. Soc. **117**, 10220. **2**. Martin, Begun, Schindeler, et al. (1999) Biochemistry **38**, 7978.

N-[N-(t-Butoxycarbonyl)-L-phenylalanyl-L-histidyl]-(4-amino-5-cyclohexyl-2,2-difluoro-3-oxopentanoyl)-L-isoleucyl-2-pyridylmethylamide

This peptidomimetic ($FW_{free-base}$ = 836.98 g/mol) inhibits renin (IC_{50} = 2.5 nM). The readily hydrated fluoro ketone is thought to mimic the tetrahedral intermediate that forms during the enzyme-catalyzed hydrolysis of a peptidic bond. The fluoro ketone is shown to be a much more effective inhibitor than the corresponding nonfluorinated ketone, which acts as a pseudosubstrate. **1**. Thaisrivongs, Pals, Kati, et al. (1986) J. Med. Chem. **29**, 2080.

N-(t-Butoxycarbonyl)-D-phenylalanyl-L-prolyl-L-boroarginine

This boronyl-tripeptide ($FW_{chloride}$ = 554.88 g/mol) is a potent inhibitor of thrombin (K_i = 3.6 pM), but a much weaker inhibitor of plasma kallikrein (K_i = 0.4 nM), coagulation factor Xa (K_i = 0.94 nM), plasmin (K_i = 0.46 nM), and t-plasminogen activator (K_i = 5.7 nM). Peptides containing alpha-aminoboronic acids with neutral side chains are highly effective reaction intermediate analog inhibitors of the serine proteases leukocyte elastase, pancreatic elastase, and chymotrypsin. **1**. Kettner, Mersinger & Knabb (1990) J. Biol. Chem. **265**, 18289.

N-(*t*-Butoxycarbonyl)-L-valyl-L-methionyl-[(2*R*,4*S*,5*S*)-5-amino-4-hydroxy-2,7-dimethyl-octanoyl]-L-valine Benzylamide

This pentapeptide analogue (FW = 722.01 g/mol), also known as GT-1017, contains a nonhydrolyzable -(*S*)CH(OH)CH$_2$-, a hydroxyethylene isostere often denoted by Ψ, that replaces the peptide bond between the Leu-3 and Ala-4 residues. GT-1017 inhibits memapsin-2 (β-secretase; K_i = 2.5 nM) and memapsin-1 (K_i = 1.2 nM) (1-3). Note that the methionine sulfone analogue (GT-1026) also inhibits, with respective K_i values of 9.4 and 44.7 nM. **1**. Ghosh, Hong & Tang (2002) *Curr. Med. Chem.* **9**, 1135. **2**. Ghosh, Bilcer, Harwood, *et al.* (2001) *J. Med. Chem.* **44**, 2865. **3**. Turner, Loy, Nguyen, *et al.* (2002) *Biochemistry* **41**, 8742.

N^α-{4-[2-(*t*-Butylaminosulfonyl)phenyl]-2-methyl-benzoyl}-L-boroarginine *O*,*O*-(+)-pinanediol ester

This transition-state mimic (FW$_{hydrochloride}$ = 704.55 g/mol) potently inhibits thrombin and trypsin, with K_i values of 0.21 and 0.70 nM, respectively. Modified peptides possessing a boronate functional group in place of the carboxyl group are "sticky" for many proteaszes. **1**. Quan, Wityak, Dominguez, *et al.* (1997) *Bioorg. Med. Chem. Lett.* **7**, 1595.

Butylated hydroxytoluene

This radical scavenger and widely used antioxidant (FW = 220.36 g/mol; CAS 128-37-0; IUPAC Name: 2,6-bis(1,1-dimethylethyl)-4-methylphenol; Symbol: BHT; Soluble in most oils) inhibits food spoilage by terminating free radical reactions through the conversion of peroxy radicals (R–O$_2$˙) to hydroperoxides (R–OOH). This is accomplished by donating a hydrogen atom (*Reaction*: R–O$_2$˙ + BHT–OH → ROOH + BHT–O˙). where R is alkyl or aryl, and where BHT–OH indicates the hydroxyl group of butylated hydroxytoluene. The key to BHT's effectiveness is the formation of a highly stabilized oxygen free radical that is poorly reactive due to steric contraints.

t-Butyl-bicyclo[2.2.2]-phosphorothionate

This highly hindered, membrane-permeant phosphorothioate (FW = 222.25 g/mol) is a potent convulsant, GABA$_A$ receptor antagonist (1,2) and chloride channel blocker (3). While slightly soluble in water, solubility is increased in the presence of nonionic detergents. TBPS is also soluble in dimethyl sulfoxide. **CAUTION**: Avoid inhalation and contact with skin contact, especially when dissolved in DMSO). **1**. Casida, Palmer & Cole (1985) *Mol. Pharmacol.* **28**, 246. **2**. Squires, Casida, Richardson & Saederup (1983) *Mol. Pharmacol.* **23**, 326. **3**. Tehrani, Vaidyanathaswamy, Verkade & Barnes (1986) *J. Neurochem.* **46**, 1542.

n-Butylboronic Acid

This hygroscopic boronic acid (FW = 101.94 g/mol; CAS 69190-62-1; M.P.= 94-96°C), also known as butaneboronic acid and butylboron dihydroxide, inhibits a number of hydrolytic enzymes. The respective K_i values for cholesterol esterase and cutinase is 223 and 3.2 μM. **Target(s)**: bacterial leucyl aminopeptidase, *or Aeromonas* aminopeptidase (1,2,8-11); cholesterol esterase (3,4); chymotrypsin (5); lipoprotein lipase (4); methionyl aminopeptidase (6); lipase, *or* triacylglycerol lipase (7); cutinase (12). **1**. Prescott, Wagner, Holmquist & Vallee (1985) *Biochemistry* **24**, 5350. **2**. Baker & Prescott (1985) *Biochem. Biophys. Res. Commun.* **130**, 1154. **3**. Feaster & Quinn (1997) *Meth. Enzymol.* **286**, 231. **4**. Sutton, Stout, Hosie, Spencer & Quinn (1986) *Biochem. Biophys. Res. Commun.* **134**, 386. **5**. Antonov, Ivanina, Berezin-Martinek (1970) *FEBS Lett.* **7**, 23. **6**. Copik, Waterson, Swierczek, Bennett & Holz (2005) *Inorg. Chem.* **44**, 1160. **7**. Garner (1980) *J. Biol. Chem.* **255**, 5064. **8**. Baker, Wilkes, Bayliss & Prescott (1983) *Biochemistry* **22**, 2098. **9**. Schürer, Lanig & Clark (2004) *Biochemistry* **43**, 5414. **10**. Bennett (2002) *Curr. Top. Biophys.* **26**, 49. **11**. De Paola, Bennett, Holz, Ringe & Petsko (1999) *Biochemistry* **38**, 9048. **12**. Sebastian & Kolattukudy (1988) *Arch. Biochem. Biophys.* **263**, 77.

N-Butyl-1-deoxygalactonojirimycin

This sugar analogue (FW$_{free-base}$ = 219.28 g/mol) is a strong inhibitor of glycolipid biosynthesis but has no effect on the maturation of *N*-linked oligosaccharides or on lysosomal glucocerebrosidase. When the ring nitrogen is protonated, this analogue mimics the charge of oxa-carbenium ion intermediate that forms transiently in many glycosidase reactions, a property that results in tight binding. It has been effective in depleting glycosphingolipids in Type I Gaucher disease patients. **Target(s)**: ceramide glucosyltransferase (1); α-galactosidase A (2,3); β-galactosidase (4); lactase/phlorizin-hydrolaseglycosylceramidase (5,6). **1**. Platt, Neises, Karlsson, Dwek & Butters (1994) *J. Biol. Chem.* **269**, 27108. **2**. Fan, Ishii, Asano & Suzuki (1999) *Nat. Med.* **5**, 112. **3**. Asano, Ishii, Kizu, *et al.* (2000) *Eur. J. Biochem.* **267**, 4179. **4**. Tominaga, Ogawa, Taniguchi, *et al.* (2001) *Brain Dev.* **23**, 284. **5**. Wilkinson, Gee, Dupont, *et al.* (2003) *Xenobiotica* **33**, 255. **6**. Sesink, Arts, Faassen-Peters & Hollman (2003) *J. Nutr.* **133**, 773.

N-Butyl-1-deoxynojirimycin

This sugar analogue (FW = 219.28 g/mol), also known as Miglustat, Vevesca, and Zavesca, inhibits processing α-glucosidases as well as ceramide glucosyltransferase. Because the ring nitrogen is protonated, this analogue mimics the charge of oxa-carbenium ion intermediate that forms transiently in many glycosidase reactions, a property that very often results in tight binding. Its oral administration prevents lysosomal accumulation of glucocerebroside that occurs in patients with Type I Gaucher's disease. *N*-

Butyl-1- deoxynojirimycin also inhibits human immunodeficiency virus (HIV) replication (EC$_{50}$ = 56 μM) by inhibiting processing α-glucosidase activities. **Target(s):** ceramide glucosyltransferase (1,2,9-11); mannosyl-oligosaccharide glucosidase, *or* glucosidase I (2-5); glucan 1,3-α-glucosidase II, *or* mannosyl-oligosaccharide α-glucosidase II (2-4,6); acid α-glucosidase (4); α-1,6- glucosidase (4); glucosylceramidase (7,8).   **1.** Riboni, Viani & Tettamanti (2000) *Meth. Enzymol.* **311**, 656.   **2.** Platt, Neises, Dwek & Butters (1994) *J. Biol. Chem.* **269**, 8362.   **3.** Tian, Wilcockson, Perry, Rudd & Dwek (2004) *J. Biol. Chem.* **279**, 39303.   **4.** Andersson, Reinkensmeier, Butters, Dwek & Platt (2004) *Biochem. Pharmacol.* **67**, 697.   **5.** Schweden, Borgmann, Legler & Bause (1986) *Arch. Biochem. Biophys.* **248**, 335.   **6.** Gamberucci, Konta, Colucci, *et al.* (2006) *Biochem. Pharmacol.* **72**, 640.   **7.** Legler & Liedtke (1985) *Biol. Chem. Hoppe-Seyler* **366**, 1113.   **8.** Osiecki-Newman, Fabbro, Legler, Desnick & Grabowski (1987) *Biochim. Biophys. Acta* **915**, 87.   **9.** Boucheron, Desvergnes, Compain, *et al.* (2005) *Tetrahedron* **16**, 1747.   **10.** Faugeroux, Genisson, Andrieu-Abadie, *et al.* (2006) *Org. Biomol. Chem.* **4**, 4437.   **11.** Ichikawa & Hirabayashi (1998) *Trends Cell Biol.* **8**, 198.

Butyrate (Butyric Acid)

This HDAC inhibitor (FW$_{free-acid}$ = 88.11 g/mol; CAS 107-92-6; FW$_{Na-salt}$ = 110.09 g/mol; CAS 156-54-7), also known as butanoate (the conjugate base of butanoic acid), is a major colonic lumen metabolite formed by bacterial fermentation of dietary fiber. Butyrate also serves as a critical mediator of inflammatory response in the colon. Its action as a histone deacetylase inhibitor is believed to underlie its ability to suppress colonic inflammation by inhibition of IFN-γ/STAT1 signaling pathways. **Discovery of HDAC Inhibition:** The discovery of butyrate inhibition of histone deacetylases followed early reports that sodium butyrate halts DNA synthesis, arrests cell proliferation, alters cell morphology and increases or decreases gene expression (1). This finding was followed by the discovery that *n*-butyrate causes histone modification in HeLa and Friend erythroleukemic cells (2) and that sodium butyrate inhibits histone deacetylation in cultured cells (3). Later papers added evidence that butyrate suppresses histone deacetylation, both *in vivo* and *in vitro* (4,5), leading to accumulation of multi-acetylated forms of histones H3 and H4 and increased DNase I-sensitivity of the associated DNA sequences (6). HDAC8 action on a histone H4 peptide substrate, for example, is inhibited by sodium butyrate, as is deacetylation of nucleosomal core histones (7). **Effect on Fetal Hemoglobin Production:** Intermittent therapy in which butyrate is administered for 4 days, followed by 10-24 days without drug, induces fetal globin gene expression in 9 of every 11 patients, with the mean Hb F in this group increased 7-21 % over the course of 39 weeks (8). Butyrate is likewise a HDAC inhibition-mediated fetal globin gene inducer in patients with β-thalassemia (9). **Other Actions of Butyrate on Histone Acetylation:** Butyrate inhibition of HDAC activity affects the expression of ~2% of mammalian genes. Promoters of butyrate-responsive genes have butyrate response elements, and a model has been proposed in which inhibition of Sp1/Sp3-associated HDAC activity leads to histone hyperacetylation and transcriptional activation of the p21(Wafl/Cip1) gene, with the latter inhibiting cyclin-dependent kinase 2 activity and thereby arresting cell cycling (10).   **1.** Prasad & Sinha   (1976) *In Vitro* **12**, 125.   **2.** Riggs, Whittaker, Neumann & Ingram (1977) *Nature* **268**, 462.   **3.** Candido, Reeves & Davie (1978) *Cell* **14**, 105.   **4.** Sealy & Chalkley (1978) *Cell* **14**, 115.   **5.** Boffa, Vidali, Mann & Allfrey (1978) *J. Biol. Chem.* **253**, 3364.   **6.** Vidali, Boffa, Bradbury & Allfrey (1978) *Proc. Natl. Acad. Sci. U.S.A.* **75**, 2239.   **7.** Buggy, Sideris, Mak, *et al.* (2000) *Biochem. J.* **350** (Part 1), 199. **8.** Atweh, Sutton, Nassif, *et al.* (1999) *Blood* **93**,1790.   **9.** Perrine, Castaneda, Boosalis, *et al.* (2005) *Ann. N. Y. Acad. Sci.* **1054**, 257.   **10.** Davie (2003) *J. Nutr.* **133** (Supplement), 2485S.

Butyrolactone I

This natural product (FW = 424.45 g/mol), first identified in species of *Aspergillus* and named systematically as 2,5-dihydro-4-hydroxy 2-([4-hydroxy-3-(3-methyl-2-butenyl)phenyl]methyl)-3-(4-hydroxyphenyl)-5-oxo-2-furancarboxylic acid methyl ester, competes with ATP in cyclin-dependent kinase reactions. The IC$_{50}$ values for these kinases are typically 0.6-1.5 μM. Butyrolactone I also inhibits cell proliferation by the inhibiting pRb phosphorylation in IMR32 cells, causing both G$_1$ and G$_2$ arrest and stimulating apoptosis. **Target(s):** cyclin-dependent kinases 1 and 2 (1,2,4); cdk5 (1); protein kinase C (1); protein kinase A (1); MAP kinase (1); casein kinase II (1); [tau protein] kinase (3).   **1.** Meijer & Kim (1997) *Meth. Enzymol.* **283**, 113.   **2.** Kitagawa, Okabe, Ogino, *et al.* (1993) *Oncogene* **8**, 2425.   **3.** Alvarez, Toro, Caceres & Maccioni (1999) *FEBS Lett.* **459**, 421.   **4.** Woodard, Li, Kathcart, *et al.* (2003) *J. Med. Chem.* **46**, 3877.

BvdU, *See* *(E)-5-(2-Bromovinyl)-2'-deoxyuridine*

BW755C, *See* *3-Amino-1-(3-trifluoromethyl)phenyl-2-pyrazoline*

BW759U, *See* *Ganciclovir*

BW284c51, *See* *1,5-Bis(4-allyldimethylammoniumphenyl)pentan-3-one*

BX-795

This nonselective signal transduction inhibitor and antineoplastic agent (FW = 591.47 g/mol; CAS 702675-74-9; Solubility: 120 mg/mL DMSO; <1 mg/mL Water), systematically named *N*-(3-(5-iodo-4-(3-(thiophene-2-carboxamido)propylamino)pyrimidin-2-ylamino)phenyl)pyrrolidine-1-carboxamide, targets the phosphoinositide 3-kinase/3-phosphoinositide-dependent kinase 1 (PDK1)/Akt signaling pathway plays a key role in cancer cell growth, survival, and tumor angiogenesis and represents a promising target for anticancer drugs. BX-795 targets TBK1 (IC$_{50}$ = 6 nM), IKKε (IC$_{50}$ = 41 nM), PDK-1 (IC$_{50}$ = 111 nM), and ERK8 (IC$_{50}$ = 140 nM), but acts more broadly by inhibiting MARK1 (IC$_{50}$ = 55 nM), MARK2 (IC$_{50}$ = 53 nM), MARK4 (IC$_{50}$ = 19 nM), NUAK1 (IC$_{50}$ = 5 nM), VEGFR (IC$_{50}$ = 157 nM), MLK1 (IC$_{50}$ = 50 nM), MLK2 (IC$_{50}$ = 46 nM), MLK3 (IC$_{50}$ = 42 nM). BX795 also inhibits the phosphorylation of JNK1/2 and p38α MAPK in MEFs stimulated with IL-1α or TNFα. The effect of BX795 on the activation of JNK1/2 and p38α MAPK does not result from the inhibition of TBK1/IKKε. **1.** Feldman, *et al.* (2005) *J. Biol. Chem.* **280**, 19867.   **2.** Bain, *et al.* (2007) *Biochem J.* **408**, 297.   **3.** Clark, *et al.* (2009) *J. Biol. Chem.* **284**, 14136.

BY-1023, *See* *Pantoprazole*

BYL719, *See* *Alpelisib*

– C –

C, *See* Carbon; Cysteine; Cytidine; Cytosine

C1 Esterase Inhibitor (*or* C1 Inhibitor)

This serpin (CAS 80295-38-1; *Abbreviation*: C1 INH), known commercially as Cinryze®, is the only known plasma inhibitor of the activated homologous serine proteases C1s and C1r in the complement cascade. C1-INH is also a major inhibitor of plasma kallikrein and factor XIIa. Hereditary angioedema (HAE), a rare genetic deficiency in functional C1-esterase inhibitor, is characterized by recurrent episodes of angioedema in the absence of associated urticaria. Therapy includes use of plasma-derived and recombinant C1 inhibitors, kallikrein inhibitors, and bradykinin B_2-receptor antagonists. Prophylactic measures include use of plasma-derived C1 inhibitors, attenuated androgens, and antifibrinolytic agents, albeit with significant adverse events. **Primary Mode of Action:** C1-INH regulates the activation of the classical complement pathway, and genetic or acquired deficiencies of C1-INH result in recurrent episodes of angioedema (1-3). Like many other serpins, C1 INH is a slow-binding inhibitor (K_i = 0.03 nM) that target proteases by forming a stable covalent serpin-enzyme complex (*or* SEC), resulting in the cleavage of the P_1–$P_{1'}$ bond with the reactive center loop of the serpin, generating a COOH-terminal peptide (M_r = 4152). This large-scale structural rearrangement exposes a binding site for low-density lipoprotein receptor-related protein (LRP), a multifunctional endocytic receptor that mediates the cellular uptake of the C1 INH-C1s and C1 IHN-C1r complexes, as well as other SEC's and unrelated ligands, such as lipoproteins, lipoprotein lipase, proteinase- or methylamine-activated α_2-macroglobulin, bacterial toxins, and lactoferrin. Early blockade of complement activation by recombinant human C1-inhibitor prevents acute antibody-mediated rejection in pre-sensitized recipients (4). (*For a discussion of the likely mechanism of inhibitory action, See Serpins; also α_1-Antichymotrypsin*) **Berinert P™:** This purified, pasteurized, and lyophilized concentrate of human plasma-derived C1 esterase inhibitor is an FDA-approved intravenous treatment for acute abdominal or facial attacks of hereditary angioedema (HAE) in adult and adolescent patients (5,6). 1. Arlaud & Thielens (1993) *Meth. Enzymol.* **223**, 61. 2. Bock, Skriver, Nielsen, *et al.* (1986) *Biochemistry* **25**, 4292. 3. Nilsson, Sjoholm & Wiman (1983) *Biochem. J.* **213**, 617. 4. Tillou, Poirie & Le Bas-Bernardet (2010) *Kidney Int.* **78**, 152. 5. Craig, Levy, Wasserman, *et al.* (2009) *J. Allergy & Clin. Immunol.* **124**, 801. 6. De Serres, Gröner & Lindner (2003) *Transfusion & Apheresis Science* **29**, 247.

C-1

This PKC inhibitor and vasodilator (FW = 277.34 g/mol; CAS 84468-24-6; Solubility: 100 mM in DMSO), also named 1-(5-isoquinoline sulfonyl)piperazine, targets protein kinase C at µM concentrations, markedly inhibiting the release of superoxide anion from human PMN stimulated with phorbol myristate acetate or the synthetic diacylglycerol, 1-oleoyl-2-acetyl glycerol. This action agrees with previous findings that protein kinase C is the intracellular target for these agonists. In contrast, superoxide anion production (stimulated by the complement anaphylatoxin peptide C5a or the synthetic chemotaxin formyl-methionyl-leucyl-phenylalanine (f-Met-Leu-Phe)) were not inhibited by C-1. These observations suggest that the agonist-stimulated respiratory burst of human neutrophils employs parallel pathways, of which only one utilizes the calcium and phospholipid-dependent protein kinase. 1. Gerard, McPhail, Marfat, *et al.* (1986) *J. Clin. Invest.* **77**, 61.

C21

This selective PRMT inhibitor (FW = 2166.94 g/mol; CAS 1229236-78-5; Soluble: 2 mg/mL H_2O; Sequence: Acetyl-SGXGKGGKGLGKGG AKRHRKV, where X = N^5-(2-chloro-1-iminoethyl)]-ornithine)) targets Protein Arginine Methyl-Transferase 1, *or* PRMT1 (IC$_{50}$ = 1.8 µM), with 5x selectivity for PRMT1 over PRMT6 and >250x over PRMT3 and CARM1 (1). Protein arginine methyltransferases catalyze the *S*-adenosyl-methionine-dependent post-translational methylation of arginine residues, with PRMT1 the major isozyme that generates most of the asymmetrically dimethylated arginine (**See** Asymmetric Dimethylarginine) subsequently liberated *in vivo* by proteolysis. C21 is a chloroacetam'''idine-containing peptide irreversibly modifies an active-site and inactivates the enzyme selectively. Plots of the pseudo first order rate constants of inactivation versus C21 concentration are saturatable, consistent with a standard two-step inactivation process: *Reaction* 1: E + I $\rightleftharpoons$ E·I; *Reaction* 2: E·I $\rightarrow$ E–I, where E·I is a reversible complex, and E–I is the resulting covalent adduct, K_i = [E][I]/[E·I] $\leq$ 0.8 µM, k_{inact} = 3.1 min^{-1}, and k_{inact}/K_i = 4.6 × 10^6 M^{-1}min^{-1}. The fact that higher substrate concentrations protects against inactivation strongly suggests that enzyme inactivation involves modification of an active-site nucleophile. No inactivation is observed no inhibition was observed with H21, a control peptide lacking the chloro group. 1. Obianyo, Causey, Osborne, *et al.* (2010) *Chembiochem.* **11**, 1219.

C36

This thiazolium-based AGE breaker (FW = 310.19 g/mol), also known as 3-benzyloxy-carbonylmethyl-4-methylthiazol-3-ium bromide, selectively cleaves glucose-derived protein cross-links both *in vitro* and *in vivo*, when administered at 30-50 mg/kilogram. Formation of <u>A</u>dvanced <u>G</u>lycation <u>E</u>nd-products, or AGE, is a normal aging process, one that is accelerated in diabetes and likely to contribute to Alzheimer's Disease. AGE cross-links accumulate on long-lived proteins, especially collagen and elastin, often with harmful consequences for cardiovascular and renal systems. Proteins are mainly cross-linked as α-dicarbonyl adducts (also called Amadori diones), and, akin to the well-known coenzyme thiamin pyrophosphate, C36 acts catalytically to cleave these α-dicarbonyl cross-links, thus freeing the proteins. In so doing, C36 also blocks AGE fibrosis-associated gene expression and improves diabetes-associated cardiovascular dysfunction in rat models. (Given the high circulating concentrations of thiamin, one cannot discount the vitamin's role in mitigating AGE cross-link accumulation.) 1. Cheng, Wang, Long, *et al.* (2007) *Br. J. Pharmacol.* **152**, 1196.

C75

This fatty acyl lactone (FW$_{free-acid}$ = 254.33 g/mol; CAS 191282-48-1)), systematically named as tetrahydro-4-methylene-2*R*-octyl-5-oxo-3*S*-furancarboxylic acid, inhibits fatty-acid synthase (1-5). C75 is also cytotoxic to many human cancer cell lines, an effect believed to be mediated by the accumulation of malonyl-coenzyme A in cells with an upregulated FAS pathway. **See also** C75-CoA **Target(s):** enoyl-[acyl-carrier-protein] reductase (1); fatty-acid synthase (1-5); β-ketoacyl [acyl-carrier-protein] synthase I (1); thiolesterase (1). 1, Rendina & Cheng (2005) *Biochem. J.* **388**, 895. 2. Loftus, Jaworsky, Frehywot, *et al.* (2000) *Science* **288**, 237. 3. Schmid, Rippmann, Tadayyon & Hamilton (2005) *Biochem. Biophys. Res. Commun.* **328**, 1073. 4. Rohrbach, Han, Gan, *et al.* (2005) *Eur. J. Pharmacol.* **511**, 31. 5. Lupu & Menendez (2006) *Curr. Pharm. Biotechnol.* **7**, 483. 6. Pizer, Thupari, Han, *et al.* (2000) *Cancer Res.* **60**, 213.

C-1311, See Symadex

C646

This potent cell-permeable p300/CBP histone acetyltransferase (HAT) inhibitor (FW = 445.42 g/mol; CAS 328968-36-1; Solubility: 30 mM in DMSO), named systematically as (Z)-4-(4-((5-(4,5-dimethyl-2-nitrophenyl)furan-2-yl)methylene)-3-methyl-5-oxo-4,5-dihydro-1H-pyrazol-1-yl)benzoic acid, competes at the acetyl-CoA binding pocket (IC_{50} ~400 nM), selectively inducing cell-cycle arrest and apoptosis and blocking the growth of human melanoma, leukemia, lung, and prostate cancer cells in vitro (1-3). **Mechanism of Action:** Histone acetylation (Reaction: Histone + Acetyl-S-CoA $\rightleftharpoons$ N^ε-lysyl-Histone + CoASH) leads to charge neutralization of lysines residues, thereby reducing the affinity of acetylated histone tails for DNA by creating a more open state for transcription and forming binding sites for other effector molecules. C646 blocks dynamic acetylation of histone H3 Lys-4 trimethylation in mouse and fly cells, and locally across the promoter and start-site of inducible genes in the mouse, thereby disrupting RNA polymerase II association and the activation of these genes. The histone acetyltransferase p300 promotes self-renewal of acute myelogenous leukemia cells by acetylating AML1-ETO fusion protein (AE) and facilitating its downstream gene expression as a transcriptional co-activator, suggesting that p300 may be a potential therapeutic target for AE-positive AML (3). C646 inhibited cellular proliferation, reduced colony formation, evoked partial cell cycle arrest in G_1 phase, and induced apoptosis in AE-positive AML cell lines and primary blasts isolated from leukemic mice and AML patients (3). **1.** Bowers, Yan, Mukherjee, et al. (2010) Chem. Biol. **17**, 471. **2.** Crump, et al. (2011) Proc. Natl. Acad Sci. USA **108**, 7814. **3.** Gao, Lin, Ning, et al. (2013) PLoS One **8**, e55481.

CA-028, See N-(trans-Epoxysuccinyl)-L-isoleucyl-L-proline

CA-030, See N-(trans-Epoxysuccinyl)-L-isoleucyl-L-proline

CA-074, See N-[L-3-trans-(Propylcarbamoyl)oxirane-2-carbonyl]-L-isoleucyl-L-proline

Cabazitaxel

This semi-synthetic taxane (taxoid) (FW = 835.93 g/mol; CAS 183133-96-2), also known as XRP-6258 and Jevtana®, is a microtubule- (MT) stabilizing drug that, like Paclitaxel (Taxol), interrupts normal MT cytoskeletal dynamics and forms a deranged cytoskeleton. Cabazitaxel binds to and stabilizes MTs, thereby inhibiting cell division, inducing cell-cycle arrest at G_2/M, and inhibiting tumor cell proliferation. When used in combination with prednisone, cabazitaxel is highly effective in treating hormone-refractory prostate cancer. It is also effective against multidrug- and docetaxel-resistant cancer cell lines, demonstrating a survival benefit over mitoxantrone and prednisone in patients who failed docetaxel-based chemotherapy. Unlike paclitaxel, cabazitaxel is a poor substrate for the P-glycoprotein multidrug resistance transporter. **1.** Galsky, Dritselis, Kirkpatrick & Oh (2010) Nature Rev. Drug Discov. **9**, 677.

Cabergoline

This long-acting dopamine D_2 receptor agonist (FW = 451.60 g/mol; CAS 81409-90-7), also named 6aR,9R,10aR)-N-[3-(dimethylamino)propyl]-N-(ethylcarbamoyl)-7-prop-2-enyl-6,6a,8,9,10,10a-hexahydro-4H-indolo[4,3-fg]quinoline-9-carboxamide, is used to treat: hyperprolactinemia; adjunctive therapy of prolactin-producing pituitary gland tumors; early-phase Parkinson's disease (as a monotherapy) and progressive-phase Parkinson's disease (in a combination with levodopa and carbidopa. Prolactin-lowering effects occur rapidly and, after a single dose, were evident at the end of follow up (21 days) in puerperal women, and up to 14 days in patients with hyperprolactinaemia (1). **Targets:** Serotonin 5-HT$_{1A}$ Receptor, K_i = 20 nM (as agonist); Serotonin 5-HT$_{1B}$ Receptor, K_i = 480 nM (action unknown); Serotonin 5-HT$_{1D}$ Receptor, K_i = 8.7 nM (action unknown); Serotonin 5-HT$_{2A}$ Receptor, K_i = 6.2 nM (as agonist); Serotonin 5-HT$_{2B}$ Receptor, K_i = 1.2 nM (as agonist); Serotonin 5-HT$_{2C}$ Receptor, K_i = 690 nM (as agonist); Adrenergic α_{1A} Receptor, K_i = 288 nM (action unknown); Adrenergic α_{1B} Receptor, K_i = 60 nM (action unknown); Adrenergic α_{1D} Receptor, K_i = 166 nM (action unknown); Adrenergic α_{2A} Receptor, K_i = 12 nM (action unknown); Adrenergic α_{2B} Receptor, K_i = 70 nM (as antagonist); Adrenergic α_{2C} Receptor, K_i = 22 nM (action unknown); Adrenergic β_1 Receptor, K_i >10,000 nM (action unknown); Adrenergic β_2 Receptor, K_i >10,000 nM (action unknown); Dopamine D_1 Receptor, K_i = 214 nM (as agonist); Dopamine D_{2S} Receptor, K_i = 0.6 nM (as agonist); Dopamine D_{2L} Receptor, K_i = 0.9 nM (as agonist); Dopamine D_3 Receptor, K_i = 0.8 nM (as agonist); Dopamine D_4 Receptor, K_i = 56 nM (as agonist); Dopamine D_5 Receptor, K_i = 22.4 nM (action unknown). **1.** Rains, Bryson & Fitton (1995) Drugs **49**, 255.

Cabozantinib

This oral, pan-tyrosine kinase inhibitor and antiangiogenic agent (FW = 501.51 g/mol; CAS 849217-68-1; Soluble to 100 mM in DMSO), also known as XL184, Cometriq®, Cabometyx®, and N-(4-((6,7-dimethoxyquinolin-4-yl)oxy)phenyl)-N-(4-fluorophenyl)-cyclopropane-

1,1-dicarboxamide, targets angiogenesis and metastasis in cancers with dysregulated mitogenic and angiogenic signaling pathways, particularly within the VEGFR and MET transduction cascades. In 2012, cabozantinib was approved by FDA for the treatment of metastatic medullary thyroid cancer. Cabometyx is also approved by FDA for the treatment of patients with advanced renal cell carcinoma who have received prior antiangiogenic therapy. A simple, rapid and sensitive HPLC-MS/MS method for detecting and quantifying cabozantinib is available (3). **Target(s):** VEGFR (IC_{50} = 0.035 nM); c-Met (IC_{50} = 1.3 nM); KIT (IC_{50} = 4.6 nM); RET (IC_{50} = 5.2 nM); FLT4 (IC_{50} = 6 nM); Axl (IC_{50} = 7 nM); Flt3 (IC_{50} = 11.3 nM); Flt1 (IC_{50} = 12 nM); and Tie2 (IC_{50} = 14.3 nM). **1.** Torres, Zhu, Bill, *et al.* (2011) *Clin. Cancer Res.* **17**, 3943. **2.** Gild, Bullock, Robinson & Clifton-Bligh (2011) *Nat. Rev. Endocrinol.* **7**, 617. **3.** Su, Li, Ji, *et al.* (2015) *J. Chromatogr. B Analyt. Technol. Biomed. Life Sci.* **985**, 119.

Cacodylate

This organoarsenical ($FW_{Sodium-Salt}$ = 159.98 g/mol; CAS 124-65-2; MP = 195-196°C; Solubility = 0.2 g/mL; pK_a = 6.27 at 25°C), also known as dimethylarsinate, is the conjugate base of cacodylic acid and is a common pH buffer salt. Given its structural similarity to orthophosphate and phosphate esters, care should be exercised when phosphate is a substrate or product. Cacodylate is also redox-active at pH < 6.5, inhibiting many sulfhydryl-dependent enzymes (1). **Caution:** Avoid contact with skin. Cacodylate and cacodylic acid are toxic to most animals and plants. Combined in a 4:1 ratio (w/w), sodium cacodylate and cacodylic acid were widely employed by the U.S. military under the name Agent Blue to kill grasses and rice plants. The aim was defoliate supply routes and reduce foodstuffs used by the North Vietnamese. **Target(s):** adenosine-phosphate deaminase (2,9); aspartate aminotransferase (3); galactosyltransferase (4); nitrogenase (5); phenylalanyl-tRNA synthetase (6); *Ustilago sphaerogena* ribonuclease (7); galactarate dehydratase (8); acetylesterase (10). **1.** Allison & Purich (1979) *Meth. Enzymol.* **63**, 3. **2.** Zielke & Suelter (1971) *The Enzymes*, 3rd ed. (Boyer, ed.), **4**, 47. **3.** Braunstein (1973) *The Enzymes*, 3rd ed. (Boyer, ed.), **9**, 379. **4.** Boyle, Cook & Peters (1988) *Clin. Chim. Acta* **172**, 291. **5.** Bulen & LeComte (1972) *Meth. Enzymol.* **24**, 456. **6.** Kull & Jacobson (1971) *Meth. Enzymol.* **20**, 220. **7.** Glitz & Dekker (1964) *Biochemistry* **3**, 1391. **8.** Blumenthal & Jepson (1966) *Meth. Enzymol.* **9**, 665. **9.** Yates (1969) *Biochim. Biophys. Acta* **171**, 299. **10.** Zhu, Larsen, Basran, Bruce & Wilson (2003) *J. Biol. Chem.* **278**, 2008.

Cadaverine

This toxic diamine ($FW_{free-base}$ = 102.18 g/mol; CAS 462-94-2; pK_a values of 9.13 and 10.25), also known as pentamethylenediamine and 1,5-diaminopentane, is formed by lysine decarboxylase. It is an excellent substrate for amine oxidase. Cadaverine has a putrid odor, contributing to the odor of decaying meat. Indeed, deacaying carcasses produce putrescine and cadaverine, both of which are volatile chemotractants for vultures. These facts that remind us to keep this substance in a dessicator, preferably above a sulfuric acid trap. Cadaverine in its free base form readily absorbs atmospheric CO_2, and aqueous solutions must be freshly prepared and kept in tightly sealed vessels. **Target(s):** tRNA adenylyltransferase (1); spermidine synthase (putrescine aminopropyltransferase, also an alternative substrate (2,42-45); kynureninase (3,29); porins (4-7); lipoxygenase-18, *or* ornithine decarboxylase (9,10,27,28); histamine *N*-methyltransferase (11); protein kinase C (12); conversion of proacrosin to acrosin (13); agmatinase (14); arginine decarboxylase (15,26); T_4 polynucleotide ligase (16); ouabain-sensitive and K⁺-dependent *p*-nitrophenyl-phosphatase (17); amino-acid *N*-acetyltransferase, *or* *N*-acetylglutamate synthetase (18,49); pyridoxal kinase (19); RNA-directed RNA polymerase (20); putrescine oxidase (21); indolethylamine *N*-methyltransferase, *or* amine *N*-methyltransferase (22); lysyl-tRNA synthetase (23,24); diaminopimelate decarboxylase (25); agmatine deaminase (30-32); acetylspermidine

deacetylase (33); pyridoxal kinase (34); aminolevulinate aminotransferase, mildly inhibited (35); ornithine δ-aminotransferase (36,37); deoxyhypusine synthase, weakly inhibited (38); homospermidine synthase, spermidine specific (39); homospermidine synthase (40,41); protein-glutamine γ-glutamyltransferase, *or* transglutaminase; also as an alternative substrate (46,47); diamine *N*-acetyltransferase (48); putrescine *N*-methyltransferase (50,51); deoxyhypusine monooxygenase, weakly inhibited (52). **1.** Deutscher (1974) *Meth. Enzymol.* **29**, 706. **2.** Pegg (1983) *Meth. Enzymol.* **94**, 294. **3.** Jakoby & Bonner (1953) *J. Biol. Chem.* **205**, 709. **4.** Samartzidou, Mehrazin, Xu, Benedik & Delcour (2003) *J. Bacteriol.* **185**, 13. **5.** Samartzidou & Delcour (1999) *FEBS Lett.* **444**, 65. **6.** Iyer & Delcour (1997) *J. Biol. Chem.* **272**, 18595. **7.** delaVega & Delcour (1995) *EMBO J.* **14**, 6058. **8.** Maccarrone, Baroni & Finazzi-Agro (1998) *Arch. Biochem. Biophys.* **356**, 35. **9.** Solano, Penafiel, Solano & Lozano (1988) *Int. J. Biochem.* **20**, 463. **10.** Nechaeva, Fateeva, Iarygin & Brodskii (1984) *Biull. Eksp. Biol. Med.* **97**, 158. **11.** Tachibana, Taniguchi, Fujiwara & Imamura (1986) *Exp. Mol. Pathol.* **45**, 257. **12.** Schatzman, Mazzei, Turner, *et al.* (1983) *Biochem. J.* **213**, 281. **13.** Parrish, Goodpasture, Zaneveld & Polakoski (1979) *J. Reprod. Fertil.* **57**, 239. **14.** Khramov (1976) *Biokhimiia* **41**, 553. **15.** Rosenfeld & Roberts (1976) *J. Bacteriol.* **125**, 601. **16.** Raae, Kleppe & Kleppe (1975) *Eur. J. Biochem.* **60**, 437. **17.** Tashima & Hasegawa (1975) *Biochem. Biophys. Res. Commun.* **66**, 1344. **18.** Haas & Leisinger (1974) *Biochem. Biophys. Res. Commun.* **60**, 42. **19.** Gäng & von Collins (1973) *Int. J. Vitam. Nutr. Res.* **43**, 318. **20.** Lazarus & Itin (1973) *Arch. Biochem. Biophys.* **156**, 154. **21.** DeSa (1972) *J. Biol. Chem.* **247**, 5527. **22.** Porta, Camardella, Esposito & Della Pietra (1977) *Biochem. Biophys. Res. Commun.* **77**, 1196. **23.** Levengood, Ataide, Roy & Ibba (2004) *J. Biol. Chem.* **279**, 17707. **24.** Takita, Ohkubo, Shima, *et al.* (1996) *J. Biochem.* **119**, 680. **25.** Ray, Bonanno, Rajashankar, *et al.* (2002) *Structure* **10**, 1499. **26.** Das, Bhaduri, Bose & Ghosh (1996) *J. Plant Biochem. Biotechnol.* **5**, 123. **27.** Lee & Cho (2001) *J. Biochem. Mol. Biol.* **34**, 408. **28.** Pandit & Ghosh (1988) *Phytochemistry* **27**, 1609. **29.** Soda & Tanizawa (1979) *Adv. Enzymol. Relat. Areas Mol. Biol.* **49**, 1. **30.** Chaudhuri & Ghosh (1985) *Phytochemistry* **24**, 2433. **31.** Yanagisawa & Suzuki (1981) *Plant Physiol.* **67**, 697. **32.** Sindhu & Desai (1979) *Phytochemistry* **18**, 1937. **33.** Santacroce & Blankenship (1982) *Proc. West. Pharmacol. Soc.* **25**, 113. **34.** Gäng & von Collins (1973) *Int. J. Vitam. Nutr. Res.* **43**, 318. **35.** Neuberger & Turner (1963) *Biochim. Biophys. Acta* **67**, 342. **36.** Strecker (1965) *J. Biol. Chem.* **240**, 1225. **37.** Yasuda, Tanizawa, Misono, Toyama & Soda (1981) *J. Bacteriol.* **148**, 43. **38.** Jakus, Wolff, Park & Folk (1993) *J. Biol. Chem.* **268**, 13151. **39.** Böttcher, Adolph & Hartmann (1993) *Phytochemistry* **32**, 679. **40.** Srivenugopal & Adiga (1980) *Biochem. J.* **190**, 461. **41.** Tait (1979) *Biochem. Soc. Trans.* **7**, 199. **42.** Raina, Hyvönen, Eloranta, *et al.* (1984) *Biochem. J.* **219**, 991. **43.** Graser & Hartmann (2000) *Planta* **211**, 239. **44.** Samejima & Yamanoha (1982) *Arch. Biochem. Biophys.* **216**, 213. **45.** Samejima & Nakazawa (1980) *Arch. Biochem. Biophys.* **201**, 241. **46.** Wu, Lai & Tsai (2005) *Int. J. Biochem. Cell Biol.* **37**, 386. **47.** Ickeson & Apelbaum (1987) *Plant Physiol.* **84**, 972. **48.** Seiler & al-Therib (1974) *Biochim. Biophys. Acta* **354**, 206. **49.** Haas & Leisinger (1974) *Biochem. Biophys. Res. Commun.* **60**, 42. **50.** Stenzel, Teuber & Dräger (2006) *Planta* **223**, 200. **51.** Walton, Peerless, Robins, *et al.* (1994) *Planta* **193**, 9. **52.** Abbruzzese, Park, Beninati & Folk (1989) *Biochim. Biophys. Acta* **997**, 248.

Cadazolid

This fluoroquinolone-class antibiotic (FW = 585.55 g/mol; CAS 1025097-10-2), also named 1-cyclopropyl-6-fluoro-7-[4-({2-fluoro-4-[(5R)-5-(hydroxymethyl)-2-oxo-1,3-oxazolidin-3-yl]phenoxy}methyl)-4-hydroxy piperidin-1-yl]-4-oxo-1,4-dihydroquinolin-3-carboxylate, is a broad-spectrum systemic antibacterial agent that targets Type II DNA topoisomerases (gyrases), which are required for bacterial replication and transcription. Cadazolid is effective against *Clostridium difficile*, a major cause of drug resistant diarrhea in the elderly. (For the prototypical member of this antibiotic class, *See Ciprofloxacin*) 1. Rashid, Dalhoff, Weintraub & Nord (2014) *Anaerobe* 28, 216.

Caffeic Acid

trans-isomer   **cis-isomer**

This yellow-colored anti-oxidant (FW$_{free-acid}$ = 180.16 g/mol), also known as 3,4-dihydroxycinnamic acid, is typically found in plants in conjugated form. Caffeic acid is a key intermediate in the biosynthesis of lignin, a major contributor to plant biomass. Caffeic inhibits leukotriene biosynthesis and stimulates low-density lipoprotein oxidation. It also significantly reduces lipid peroxidation and decreases DNA damage in UVB-irradiated cells. **Target(s):** xanthine oxidase (1,6,12,30,53); phenylalanine ammonia-lyase (2,16,23,36-40); *N*-acetylindoxyl oxidase (3); indoleacetate oxidase (4); horseradish peroxidase (5); aromatic-amino-acid decarboxylase, *or* dopa decarboxylase (7,19,21); γ-butyrobetaine dioxygenase.*or* γ-butyrobetaine hydroxylase; K_i = 6 μM (8); α-amylase (9,27,42); β-amylase (9); 5-lipoxygenase, *or* arachidonate 5-lipoxygenase (10,11,59); glutathione *S*-transferase (13,14); DNA methyltransferase (15); starch synthase (17); thiamin transport (18); vitamin D$_3$ 1α-monooxygenase (20); vitamin D$_3$ 24-monooxygenase (20); lipase, pancreatic (22); arylamine *N*-acetyltransferase (24); phosphorylase kinase (25); protein kinase A (25); protein kinase C (25); neutrophil elastase (26); trypsin (27); gelatinase B (matrix metalloproteinase 9)28; sucrase, weakly inhibited (29); 4-coumaroyl-CoA synthetase, *or* 4-coumarate:CoA ligase (31,32); DNA topoisomerase I (33); chorismate mutase (34); phenylpyruvate tautomerase (35); peptidyl-dipeptidase A, *or* angiotensin I-converting enzyme, weakly inhibited (41); chlorogenate hydrolase (43); phenol sulfotransferase, *or* aryl sulfotransferase (44); shikimate kinase (45); 3-deoxy-7-phosphoheptulonate synthase (46); anthocyanin 5-*O*-glucoside 6'''-*O*-malonyltransferase (47); quinate *O*-hydroxycinnamoyltransferase (48); arylamine *N*-acetyltransferase (49-51); catechol *O*-methyltransferase (52); tyrosinase, *or* monophenol monooxygenase (54,55); phenylalanine 4-monooxygenase (56); hyoscamine (6*S*)-dioxygenase (57); procollagen-proline 4-dioxygenase (58); 15-lipoxygenase, *or* arachidonate 15-lipoxygenase (60); 12-lipoxygenase, *or* arachidonate 12-lipoxygenase (61); 3,4-dihydroxyphenylacetate 2,3-dioxygenase (62); lipoxygenase (63); protocatechuate 3,4-dioxygenase (64,65). 1. Chiang, Lo & Lu (1994) *J. Enzyme Inhib.* 8, 61. 2. Hanson & Havir (1972) *The Enzymes*, 3rd ed. (Boyer, ed.), 7, 75. 3. Beevers & French (1954) *Arch. Biochem. Biophys.* 50, 427. 4. Rabin & Klein (1957) *Arch. Biochem. Biophys.* 70, 11. 5. Gamborg, Wetter & Neish (1961) *Can. J. Biochem. Physiol.* 39, 1113. 6. Chang, Chang, Lu & Chiang (1994) *Anticancer Res.* 14, 501. 7. Webb (1966) *Enzyme and Metabolic Inhibitors*, vol. 2, p. 311, Academic Press, New York. 8. Ng, Hanauske-Abel & Englard (1991) *J. Biol. Chem.* 266, 1526. 9. Broda (1966) *Acta Pol. Pharm.* 23, 577. 10. Koshihara, Neichi, Murota, *et al.* (1984) *Biochim. Biophys. Acta* 792, 92. 11. Koshihara, Neichi, Murota, *et al.* (1983) *FEBS Lett.* 158, 41. 12. Chan, Wen & Chiang (1995) *Anticancer Res.* 15, 703. 13. Ploemen, van Ommen, de Haan, Schefferlie & van Bladeren (1993) *Food Chem. Toxicol.* 31, 475. 14. Das, Bickers & Mukhtar (1984) *Biochem. Biophys. Res. Commun.* 120, 427. 15. Lee & Zhu (2005) *Carcinogenesis* 27, 269. 16. Kalghatgi & Subba Rao (1975) *Biochem. J.* 149, 65. 17. Wieweg & de Fekete (1976) *Acta Physiol. Lat. Amer.* 26, 415. 18. Schaller & Holler (1976) *Int. J. Vitam. Nutr. Res.* 46, 143. 19. Barboni, Voltattorni, D'Erme, Fiori, Minelli & Rosei (1982) *Life Sci.* 31, 1519. 20. Crivello (1986) *Arch. Biochem. Biophys.* 248, 551. 21. Rosei (1987) *Pharmacol. Res. Commun.* 19, 663. 22. Karamac & Amarowicz (1996) *Z. Naturforsch. [C]* 51, 903. 23. Lim, Park & Lim (1997) *Mol. Cells* 7, 715. 24. Lo & Chung (1999) *Anticancer Res.* 19, 133. 25. Nardini, Scaccini, Packer & Virgili (2000) *Biochim. Biophys. Acta* 1474, 219. 26. Loser, Kruse, Melzig & Nahrstedt (2000) *Planta Med.* 66, 751. 27. Rohn, Rawel & Kroll (2002) *J. Agric. Food Chem.* 50, 3566. 28. Park, Kim & Kim (2005) *Toxicology* 207, 383. 29. Matsui, Ebuchi, Fukui, *et al.* (2004) *Biosci. Biotechnol. Biochem.* 68, 2239. 30. Nguyen, Awale, Tezuka, *et al.* (2006) *Planta Med.* 72, 46. 31. Harding, Leshkevich,

Chiang & Tsai (2002) *Plant Physiol.* 128, 428. 32. Feutry & Letouze (1984) *Phytochemistry* 23, 1557. 33. Stagos, Kazantzoglou, Magiatis, *et al.* (2005) *Int. J. Mol. Med.* 15, 1013. 34. Woodin & Nishioka (1973) *Biochim. Biophys. Acta* 309, 211 and 224. 35. Molnar & Garai (2005) *Int. Immunopharmacol.* 5, 849. 36. Dahiya (1993) *Indian J. Exp. Biol.* 31, 874. 37. Dubery & Smit (1994) *Biochim. Biophys. Acta* 1207, 24. 38. Campbell & Ellis (1992) *Plant Physiol.* 98, 62. 39. Sarma & Sharma (1999) *Phytochemistry* 50, 729. 40. Jorrin & Dixon (1990) *Plant Physiol.* 92, 447. 41. Actis-Goretta, Ottaviani & Fraga (2006) *J. Agric. Food Chem.* 54, 229. 42. Funke & Melzig (2005) *Pharmazie* 60, 796. 43. Schöbel & Pollmann (1980) *Z. Naturforsch. C* 35, 699. 44. Yeh & Yen (2003) *J. Agric. Food Chem.* 51, 1474. 45. Bowen & Kosuge (1979) *Plant Physiol.* 64, 382. 46. Rubin & Jensen (1985) *Plant Physiol.* 79, 711. 47. Suzuki, Nakayama, Yonekura-Sakakibara, *et al.* (2001) *J. Biol. Chem.* 276, 49013. 48. Rhodes, Wooltorton & Lourencq (1979) *Phytochemistry* 18, 1125. 49. Kukongviriyapan, Phromsopha, Tassaneeyakul, *et al.* (2006) *Xenobiotica* 36, 15. 50. Makarova (2008) *Curr. Drug Metab.* 9, 538. 51. Lo & Chung (1999) *Anticancer Res.* 19, 133. 52. Bonifácio, Palma, Almeida & Soares-da-Silva (2007) *CNS Drug Reviews* 13, 352. 53. Chang, Lee, Chen, *et al.* (2007) *Free Rad. Biol. Med.* 43, 1541. 54. Parvez, Kang, Chung & Bae (2007) *Phytother. Res.* 21, 805. 55. Lee (2002) *J. Agric. Food Chem.* 50, 1400. 56. Koizumi, Matsushima, Nagatsu, *et al.* (1984) *Biochim. Biophys. Acta* 789, 111. 57. Hashimoto & Yamada (1987) *Eur. J. Biochem.* 164, 277. 58. Majamaa, Günzler, Hanauske-Abel, Myllylä & Kivirikko (1986) *J. Biol. Chem.* 261, 7819. 59. Furakawa, Yoshimoto, Ochi & Yamamoto (1984) *Biochim. Biophys. Acta* 795, 458. 60. Russell, Scobbie, Duthie & Chesson (2008) *Bioorg. Med. Chem.* 16, 4589. 61. Flatman, Hurst, McDonald-Gibson, Jonas & Slater (1986) *Biochim. Biophys. Acta* 883, 7. 62. Que, Widom & Crawford (1981) *J. Biol. Chem.* 256, 10941. 63. Pinto, Tejeda, Duque & Macías (2007) *J. Agric. Food Chem.* 55, 5956. 64. Durham, Sterling, Ornston & Perry (1980) *Biochemistry* 19, 149. 65. Hou, Lillard & Schwartz (1976) *Biochemistry* 15, 582.

Caffeine

This naturally occurring CNS stimulant and diuretic (FW = 194.19 g/mol; CAS 58-08-2), also known as 1,3,7-trimethylxanthine, is a secondary metabolite in coffee (*Coffea arabica* and *Coffea canephora*) and tea (*Camellia sinensis*) plants. It competitively inhibits 3',5'-cyclic AMP phosphodiesterase, thereby sustaining the cAMP concentration, which stimulates Protein Kinase (PKA). In this way, caffeine maintains the high phosphorylation state of PKA's protein substrates. Elevated cAMP concentrations also activate popeye domain-containing proteins (*or* Popdc), cyclicAMP-gated ion channels as well as Exchange Proteins Activated by cAMP (*or* EPAC), including RAPGEF3. (*See Adenosine 3',5'-Cyclic Monophosphate*) **Other Actions of Caffeine:** Caffeine is also a nonselective low-affinity adenosine receptor antagonist. Adenosine generally inhibits adenyl cyclase (AC) through the action of high-affinity A$_1$ receptors and stimulates AC via low-affinity A$_2$ receptors. Caffeine binds to adenosine receptors without activating them, thereby acting as a competitive inhibitor and reversing the latter's effects on the release of acetylcholine, dopamine, norepinephrine, GABA, and serotonin. Caffeine is also known to inhibit DNA repair. First isolated from coffee beans in 1821, it was later synthesized by Nobelist Emil Fischer (see References. 10 & 11). **Caffeine Consumption:** Caffeine is widely used to ward off drowsiness, restore alertness, and elevate mood. While mildly habit-forming, moderate caffeine consumption has no direct link to any serious health risks. Caffeine is classified by the Food and Drug Administration as "Generally Recognized as Safe", with toxic doses >10 g for adults, much higher than typically consumed (300-500 mg per day). An 8-ounce cup of brewed coffee contains 90-200 mg caffeine, whereas a 12-ounce can of Coca Cola® contains 32 mg (DietCoke®, 42 mg). Generally regarded as safe when consumed by healthy adults, high consumption is ill-advised for cardiac patients and pregnant women. **Pharmacokinetics:** Caffeine is rapidly and nearly completely absorbed from caffeinated beverages, reacing peak blood levels within 30-60 min. The apparent first-order caffeine elimination rate constant (0.16 hour^{-1}, when administered as a 50-mg initial intravenous dose, and 0.10 hour^{-1} for 750 mg) decreases linearly with dose (1). The total body clearance rate of 0.98 mL/min/kg is dose-independent. The apparent volume of distribution is high and trends upwards with increasing dose (1). Primary metabolites are paraxanthine, theobromine and theophylline, dimethylxanthines that also contribute to caffeine's pharmacological effects (*See also Theophylline*). Caffeine activates the PI3K/Akt pathway and prevents apoptotic cell death in a Parkinson's disease model of SH-SY5Y cells (2). **Target(s):** 3',5'-cyclic-nucleotide phosphodiesterase (3-5,25-28); purine-nucleoside phosphorylase (6,11); D-amino-acid oxidase (7,10); photoalkylated DNA endonuclease, *or* DNA (apurinic or apyrimidinic site) lyase (8); DNA insertase (8); dATP(dGTP)-DNA purinetransferase (13,34); adenosine receptor (14); ATP diphosphatase, *or* ATP pyrophosphatase (15); adenosine receptor (16,17); acetylcholinesterase (18); guanylate cyclase (19); lactoylglutathione lyase, *or* glyoxalase I (20); adenosine deaminase (21); purine nucleosidase, weakly inhibited (22); adenine deaminase (22); hypoxanthine(guanine) phosphoribosyltransferase (22,38); 2',3'-cyclic-nucleotide 3'-phosphodiesterase (23); 3',5'-cyclic-GMP phosphodiesterase, *or* cGMP phospho-diesterase (24, 25, 28); alkaline phosphatase (29); hydroxyacylglutathione hydrolase, *or* glyoxalase II (20); phosphatidylinositol-4-phosphate 3-kinase (29); 1-phosphatidylinositol 4 kinase (33); thymidine kinase (33,34); NAD$^+$ ADP-ribosyltransferase, *or* poly(ADP-ribose) polymerase (35-37); xanthine phosphoribosyl-transferase (38); glycogen synthase (39); glycogen phosphorylase (40-46). **1.** Newton, Broughton, Lind, *et al.* (1981) *Eur. J. Clin. Pharmacol.* **21**, 45. **2.** Nakaso, Ito & Nakashima (2008) *Neurosci. Lett.* **432**, 146. **3.** Nair (1966) *Biochemistry* **5**, 150. **4.** Cheung (1967) *Biochemistry* **6**, 1079. **5.** Drummond & Yamamoto (1971) *The Enzymes*, 3rd ed. (Boyer, ed.), **4**, 355. **6.** Friedkin & Kalckar (1961) *The Enzymes*, 2nd ed. (Boyer, Lardy & Myrbäck, eds.), **5**, 237. **7.** Meister & Wellner (1963) *The Enzymes*, 2nd ed. (Boyer, Lardy & Myrbäck, eds.), **7**, 609. **8.** Friedberg, Bonura, Radany & Love (1981) *The Enzymes*, 3rd ed. (Boyer, ed.), **14**, 251. **9.** Livneh & Sperling (1981) *The Enzymes*, 3rd ed. (Boyer, ed.), **14**, 549. **10.** Burton (1951) *Biochem. J.* **48**, 458. **11.** Fischer (1882) *Ber.* **15**, 453. **12.** Koch (1958) in *Symposium on Information Theory in Biology* (Gatlinberg, Tennessee, 1956), pp. 136-147, Pergamon Press, New York. **13.** Deutsch & Linn (1979) *Proc. Natl. Acad. Sci. U.S.A.* **76**, 141. **14.** Evoniuk, Jacobson, Shamim, Daly & Wurtman (1987) *J. Pharmacol. Exp. Ther.* **242**, 882. **15.** Kawamura & Nagano (1975) *Biochim. Biophys. Acta* **397**, 207. **16.** Fredholm (1982) *Acta Physiol Scand.* **115**, 283. **17.** Ferre, Fredholm, Morelli, Popoli & Fuxe (1997) *Trends Neurosci.* **20**, 482. **18.** Nachmansohn & Schneemann (1945) *J. Biol. Chem.* **159**, 239. **19.** Sun, Shapiro & Rosen (1974) *Biochem. Biophys. Res. Commun.* **61**, 193. **20.** Oray & Norton (1980) *Biochem. Biophys. Res. Commun.* **95**, 624. **21.** Singh & Sharma (2000) *Mol. Cell. Biochem.* **204**, 127. **22.** Nolan & Kidder (1979) *Biochem. Biophys. Res. Commun.* **91**, 253. **23.** Sprinkle (1989) *CRC Crit. Rev. Clin. Neurobiol.* **4**, 235. **24.** Morishima (1975) *Biochim. Biophys. Acta* **410**, 310. **25.** Methven, Lemon & Bhoola (1980) *Biochem. J.* **186**, 491. **28.** Lim, Palanisamy & Ong (1986) *Arch. Microbiol.* **146**, 142. **29.** Lim, Woon, Tan & Ong (1989) *Int. J. Biochem.* **21**, 909. **30.** Pannbacker, Fleischman & Reed (1972) *Science* **175**, 757. **31.** Al-Saleh (2002) *J. Nat. Toxins* **11**, 357. **32.** Shepherd (2005) *Acta Physiol. Scand.* **183**, 3. **33.** Buckley (1977) *Biochim. Biophys. Acta* **498**, 1. **34.** Chraibi & Wright (1983) *J. Biochem.* **93**, 323. **35.** Sandlie & Kleppe (1980) *FEBS Lett.* **110**, 223. **36.** Livneh, Elad & Sperling (1979) *Proc. Natl. Acad. Sci. U.S.A.* **76**, 1089. **37.** Rankin, Jacobson, Benjamin, Moss & Jacobson (1989) *J. Biol. Chem.* **264**, 4312. **38.** Kofler, Wallraff, Herzog, Schneider, Auer & Schweiger (1993) *Biochem. J.* **293**, 275. **39.** Burtscher, Klocker, Schneider, *et al.* (1987) *Biochem. J.* **248**, 859. **40.** Naguib, Iltzsch, el Kouni, Panzica & el Kouni (1995) *Biochem. Pharmacol.* **50**, 1685. **41.** Moses, Bashan & Gutman (1972) *Eur. J. Biochem.* **30**, 205. **42.** Chen, Gong, Liu, *et al.* (2008) *Chem. Biodivers.* **5**, 1304. **43.** Gregus & Németi (2007) *Toxicol. Sci.* **100**, 44. **44.** Chen, Liu, Zhang, *et al.* (2006) *Bioorg. Med. Chem. Lett.* **16**, 2915. **45.** Li, Lu, Su, *et al.* (2008) *Planta Med.* **74**, 287. **46.** Tanabe, Kobayashi & Matsuda (1987) *Agric. Biol. Chem.* **51**, 2465. **47.** Tanabe, Kobayashi & Matsuda (1988) *Agric. Biol. Chem.* **52**, 757. **48.** Wen, Sun, Liu, *et al.* (2008) *J. Med. Chem.* **51**, 3540. **49.** Berndt & Rösen (1984) *Arch. Biochem. Biophys.* **228**, 143. **50.** Burkhardt & Wegener (1994) *J. Comp. Physiol. B* **164**, 261.

CAL-101, *See Idelalisib*

Calcineurin Autoinhibitory Peptide

This specific calcineurin inhibitor (FW = 2930.35 g/mol; CAS 148067-21-4; *Sequence*: ITSFEEAKGLDRINERMPPRRDAMP), corresponding to residues 457-482 of the calmodulin-binding domain of calcineurin (PP2B), inhibits Mn^{2+}-stimulated calcineurin activity (IC$_{50}$ = 10 μM, with ^{32}P-myosin light chain as a substrate). Calcineurin autoinhibitory peptide has no effect on Ni^{2+}-stimulated enzyme activity. It does not inhibit CaM kinase II, protein phosphatase 1 or 2A. **1.** Hashimoto, *et al.* (1990) *J. Biol. Chem.* **265**, 1924. **2.** Yokoyama & Wang (1994) *FEBS Lett.* **337**, 128. **3.** Perrino, *et al.* (1995) *J. Biol. Chem.* **270**, 340.

Calcipotriol

This vitamin D$_3$ analogue and and anti-psoriasis drug (FW = 412.61 g/mol; CAS 112828-00-9), also named (1*R*,3*S*,5*E*)-5-{2-[(1*R*,3a*S*,4*Z*,7a*R*)-1-[(2*R*,3*E*)-5-cyclopropyl-5-hydroxypent-3-en-2-yl]-7a-methyloctahydro-1*H*-inden-4-ylidene]ethylidene}-4-methylidene-cyclohexane-1,3-diol and calcitriene, is a vitamin D receptor ligand that inhibits epidermal cell proliferation and enhances cell differentiation. Psoriasis is characterized by hyperproliferation and abnormal differentiation of epidermal keratinocytes resulting from a disordered immune response. In patients with chronic plaque psoriasis, calcipotriol (applied as an ointment twice daily) is significantly more effective than betamethasone valerate and dithranol. Topically applied calcipotriol has low hypercalcemic potential (1). Enstilar® (calcipotriene and betamethasone dipropionate) is an FDA-approved aerosol foam for the topical treatment of plaque psoriasis in patients 18 years of age and older. **1.** Murdoch & Clissold (1992) *Drugs* **43**, 415.

Calcitonin

CGNLSTCMLGTYTQDFNKFHTFPQTAIGVGAP-NH$_2$

Human Calcitonin

CGNLSTCVLGKLSQELHKLQTYPRTNTGSGTP-NH$_2$

Salmon Calcitonin

This calcium-regulating hormone and antiresorptive drug (MW = 3548 g), also known as thyrocalcitonin, is FDA-approved, intravenously or intranasally administered drug (prepared synthetically, based on the salmon sequence) for inhibiting bone loss in osteoporosis. Calcitonin lowers blood calcium ion levels in four ways: by reducing Ca^{2+} absorption by the intestine, by inhibiting osteoclast activity in bones, by stimulating osteoblastic activity in bones, and by reducing renal tubular cell reabsorption of Ca^{2+}, followed by urinary excretion. Calcitonin has a short absorption half-life (10–15 minutes) and a somewhat longer elimination half-life (50–80 minutes). **Mechanism of Action:** Calcitonin reduces blood Ca^{2+} and opposes the action of parathyroid hormone (PTH). In humans, calcitonin is produced by thyroid parafollicular cells (C-cells). Calcitonin protects the skeleton during episodes of increased calcium demand for growth, pregnancy and lactation. Calcitonin exerts its bone-sparing effects by targeting osteoclasts, but not ondontoclasts, suiting it as an inhibitor bone loss without altering the dynamics of dental root surface remodeling.

activating the G-proteins often found on osteoclasts. **Calcitonin Receptor:** Found on osteoclasts, kidney and regions of the spine and brain, the calcitonin receptor is a G protein-coupled receptor, which is coupled by Gs (*i.e.*, cAMP) to adenylate cyclase and thereby to the generation of cAMP in target cells. It also binds to G_q-coupled (*i.e.*, phospholipase-linked) transduction pathways to regulate calcium homeostasis. **Salmon Calcitonin:** Sharing structural features with calcitonins, salmon calcitonin also has 32 residues, terminating in prolinamide and containing a cystine linkage between Cys-1 and Cys-7 (1). Although differing in their sequences, human, porcine, bovine and salmon calcitonins are homologous in nine of their thirty-two positions. Moreover, to exert its hypocalcemic effect, all 32 residues are required. The much higher biological potency of salmon calcitonin commends its use as a drug. Nasal salmon calcitonin exerted a significant hypocalcemic effect, but did not interfere with antigen- or mitogen-induced expansion of T-lymphocytes, suggesting it is unlikely that nasal salmon calcitonin affects cell-mediated immunity in healthy subjects (2). Salmon calcitonin is effective in relieving pain associated with post-menopausal osteoporosis, especially in resolving pain during the first month following a vertebral bone crush (3). **1.** Niall, Keutmann, Copp & Potts (1969) *Proc. Natl. Acad. Sci. U.S.A.* **64**, 771. **2.** Thamsborg, Møller, Kollerup & Sørensen (1993) *Bone Miner.* **20**, 245. **3.** Knopp-Sihota, Newburn-Cook, Homik, Cummings & Voaklander (2012) *Osteoporos Int.* **23**, 17.

Calcium Ion

This alkaline earth ion (Symbol, Ca; Atomic Number = 20; Atomic Weight = 40.08; Pauling Ionic Radius = 1.14 Å) is a group 2 (or group IIA) element, situated directly beneath magnesium in the Periodic Table. [*Note:* The Shannon-Prewitt Scale for effective ionic radius takes into account that the effective ionic radius r depends on the coordination number, such that for $n = 6$, $r = 1.00$ Å; for $n = 7$, $r = 1.06$ Å; for $n = 8$, $r = 1.12$ Å; for $n = 9$, $r = 1.18$ Å; for $n = 10$, $r = 1.23$ Å; ; for $n = 12$, $r = 1.34$ Å.] Calcium is an essential element in all living organisms (although some fungi may be exceptions). It is a major constituent of bones, teeth, shells, and other hard tissues. It is involved in the regulation of a number of enzymes and binds to quite a few proteins; *e.g.*, calmodulin, calcium-dependent ATPases, myosin, and calcium pumps. A significant number of proteins use calcium ions as a cofactor, including tryptophan dehydrogenase, protein-glutamine γ-glutamyltransferase, [myosin light-chain] kinase, calcium/ calmodulin-dependent protein kinase, 1-D-*myo*-inositol-trisphosphate 3-kinase, phospholipase A_2, calpain, leucolysin, envelysin, stromelysins, gelatinases, neutrophil collagenase, russellysin, macrophage elastase, apyrase, and 3-ketovalidoxylamine C-N-lyase. Calcium ions are messengers in nerve and muscle responses and they initiate chemotropic responses in pollen. The high degree of spatial and temporal precision of calcium ion control of cellular processes requires the highly coordinated interplay of calcium ion channels, stores, and oscillations. Accumulators of calcium include mammalian bone, many invertebrate shells, and some red algae. **Radioisotopes:** The nuclide with the longest half-life is ^{41}Ca: 1.03×10^5 years (via nuclear disintegration by electron capture: 0.421 MeV). Calcium-47 (half-life = 4.536 days; β-emitter, 1.988 MeV) is used in a number of metabolic studies, as is calcium-45, a β-emitter (0.257 MeV) with a half-life of 162.7 days. A special note of precaution should be presented here. **Caution:** Bone is the critical organ in terms of calcium dose, and the effective biological half-life of cacium ion in humans is nearly twenty years. The experimenter using radioactive calcium is at risk, and double-gloving and great caution is recommended for those working with radioactive calcium. **Inhibitory Targets:** Divalent calcium cation (Ionic Radius = 0.99 Å) combines readily with oxygen ligands (chiefly water, phosphates, polyphosphates, and carboxylates to reversibly form stable metal ion complexes. More than 300 enzymes are known to be inhibited by divalent calcium, far too many to list here. For most, the inhibitory effect occurs at calcium ion concentrations well above the physiologic range found in most cells.

Calcium Ion Channel Blockers

In search of desired pharmacologic outcomes over the last half century, researchers have uncovered the unique properties of a wide spectrum of calcium ion channel blockers. The following partial list tallies calcium ion blockers and modulators that are described in this reference. *See also* *Amlodipine; Amrinone; Anandamide; Anipamil; Aranidipine; Azelnidipine; Azimilide; Benidipine; Bepridil; Berbamine; Bevantolol; Canadine; Carboxyamidotriazole; Caroverine; Cilnidipine; Cinnarizine; Clevidipine; Conotoxins; Fantofarone; Fasudil; Flunarizine; Gabapentin; Magnesium ion; Manoalide; Mibefradil; Naftopidil;*

Nicardipine; Niguldipine; Nimodipine; Ochratoxin A; Oxodipine; Pregabalin; Prenylamine; Valproate; Valpromide; Verapamil.

Calmidazolium Chloride

This calmodulin antagonist (FW = 687.71 g/mol; CAS 57265-65-3), also known as compound R-24571, inhibits calmodulin-dependent cyclic nucleotide phosphodiesterase I ($IC_{50} = 0.01$ μM) and skeletal muscle sarcoplasmic reticulum Ca^{2+}-exporting ATPase ($K_i = 0.06$ μM). Stock solutions are typically prepared in dimethyl sulfoxide and are stable for several weeks. Wnen in aqueous solutions, calmidazolium chloride binds to borosilicate laboratory glassware. **Target(s):** 3',5'-cyclic-nucleotide phospho-diesterase (1,5,9,23); Ca^{2+}-transporting ATPase (1,4,9,11,22); calpains (2); adenylate cyclase (3,15,16,20); Na^+ channels (6); Ca^{2+}-dependent protein kinase (7,10.11); Ca^{2+} channels (8); NADH: semidehydroascorbate oxidoreductase (12); myosin light-chain kinase (13,19); phosphatidylinositol-specific phospholipase C (14); calmodulin-stimulated protein phosphatase (17); F_1 ATPase (18); inositol-1,4,5-trisphosphate 5-phosphatase (21); Ca^{2+}/calmodulin dependent protein kinase (24); NAD^+ kinase (25); calmodulin-lysine *N*-methyltransferase (26,27); heme oxygenase (28); nitric-oxide synthase (29-32). **1.** Gietzen, Sadorf & Bader (1982) *Biochem. J.* **207**, 541. **2.** Zhang & Johnson (1988) *J. Enzyme Inhib.* **2**, 163. **3.** Toscano & Gross (1991) *Meth. Enzymol.* **195**, 91. **4.** Anderson, Coll & Murphy (1984) *J. Biol. Chem.* **259**, 11487. **5.** Silver, Connell, Dillon, *et al.* (1989) *Cardiovasc. Drug Ther.* **3**, 675. **6.** Ichikawa, Urayama & Matsumoto (1991) *J. Membrane Biol.* **120**, 211. **7.** Yuasa & Muto (1992) *Arch. Biochem. Biophys.* **296**, 175. **8.** Nakazawa, Higo, Abe, *et al.* (1993) *Brit. J. Pharmacol.* **109**, 137. **9.** Gietzen, Wuthrich & Bader (1981) *Biochem. Biophys. Res. Commun.* **101**, 418. **10.** Mazzei, Schatzman, Turner, Vogler & Kuo (1984) *Biochem. Pharmacol.* **33**, 125. **11.** Lamers & Stinis (1983) *Cell Calcium* **4**, 281. **12.** Sun, Crane & Morre (1983) *Biochem. Biophys. Res. Commun.* **115**, 952. **13.** Zimmer & Hofmann (1984) *Eur. J. Biochem.* **142**, 393. **14.** Benedikter, Knopki & Renz (1985) *Z. Naturforsch. [C]* **40**, 68. **15.** Gross, Toscano & Toscano (1987) *J. Biol. Chem.* **262**, 8672. **16.** Ahlijanian & Cooper (1987) *J. Pharmacol. Exp. Ther.* **241**, 407. **17.** Manalan & Werth (1987) *Circ. Res.* **60**, 602. **18.** De Meis, Tuena de Gomez Puyou & Gomez Puyou (1988) *Eur. J. Biochem.* **171**, 343. **19.** Sobieszek (1989) *Biochem. J.* **262**, 215. **20.** Haunso, Simpson & Antoni (2003) *Mol. Pharmacol.* **63**, 624. **21.** Fowler & Eriksson (1992) *Cell Signal* **4**, 723. **22.** Wright & van Houten (1990) *Biochim. Biophys. Acta* **1029**, 241. **23.** Ryan & Toscano (1985) *Arch. Biochem. Biophys.* **241**, 403. **24.** Rodriguez-Mora, LaHair, Howe, McCubrey & Franklin (2005) *Exp. Opin. Ther. Targets* **9**, 791. **25.** Gallais, de Crescenzo & Laval-Martin (2001) *Aust. J. Plant Physiol.* **28**, 363. **26.** Pech & Nelson (1994) *Biochim. Biophys. Acta* **1199**, 183. **27.** Wright, Bertics & Siegel (1996) *J. Biol. Chem.* **271**, 12737. **28.** Boehning, Sedaghat, Sedlak & Snyder (2004) *J. Biol. Chem.* **279**, 30927. **29.** Schmidt & Murad (1991) *Biochem. Biophys. Res. Commun.* **181**, 1372. **30.** Lacza, Puscar, Figueroa, *et al.* (2001) *Free Radic. Biol. Med.* **31**, 1609. **31.** Bush, Gonzalez, Griscavage & Ignarro (1992) *Biochem. Biophys. Res. Commun.* **185**, 960. **32.** Sheng & Ignarro (1996) *Pharmacol. Res.* **33**, 29.

Calmodulin-Dependent Protein Kinase II, Fragment 281-309

This 29-residue fragment (MW = 3374.09 g; *Sequence:* MHRQET VDCLKKFNARRKLKGAILTTML; Isoelectric Point = 10.45) is a tight-binding inhibitor of Ca^{2+}/calmodulin-dependent protein kinase II ($IC_{50} = 80$

nM). The oligopeptide contains elements of the calmodulin-binding, inhibitory, and autophosphorylation domains of the protein kinase. It is also an alternative substrate of protein kinase C, undergoing phosphorylation at Thr-286. Replacement of this residue (*i.e.*, [Thr-286-Ala]-calmodulin-dependent protein kinase II fragment 281-309) results in a synthetic peptide that also inhibits the kinase (IC50 = 2 μM). **1.** Colbran, Fong, Schworer & Soderling (1988) *J. Biol. Chem.* **263**, 18145. **2.** Colbran, Smith, Schworer, Fong & Soderling (1989) *J. Biol. Chem.* **264**, 4800. **3.** Fukunaga, Stoppini, Miyamoto & Muller (1993) *J. Biol. Chem.* **268**, 7863.

Calmodulin-Dependent Protein Kinase II, Fragment 290-309

This peptide (MW = 2202.78 g; *Sequence*: LKKFNARRKLKGAILTTML; Isoelectric Point = 12.03) is a tight-binding inhibitor of Ca^{2+}/calmodulin-dependent protein kinase II (IC_{50} = 52 nM). The oligopeptide contains the calmodulin-binding, inhibitory, and autophosphorylation domains of the protein kinase. **Target(s):** Ca^{2+}/calmodulin-dependent protein kinase II (1-3); calmodulin-lysine *N*-methyltransferase (4). **1.** Payne, Fong, Ono, *et al.* (1988) *J. Biol. Chem.* **263**, 7190. **2.** Basavappa, Mangel, Scott & Liddle (1999) *Biochem. Biophys. Res. Commun.* **254**, 699. **3.** James, Vorherr & Carafoli (1995) *Trends Biochem. Sci.* **20**, 38. **4.** Wright, Bertics & Siegel (1996) *J. Biol. Chem.* **271**, 12737.

Calpastatin

This naturally occurring non-lysosomal protein (MW = 126 kDa; Accession: BAA03747) inhibits calpains, the calcium-activated proteases that control steps in cell fusion, mitosis and meiosis. Calpains are heterodimers of a small regulatory subunit and one of three large catalytic subunits, designated Calpain 1, Calpain 2 and Calpain p94. The three subdomains of this intrinsically disordered protein bind to wrap around calpain, making contact with three non-active-site regions and blocking substrate access. Seven isoforms of the human protein are produced by alternative splicing. The K_i value reported with μ-calpain (calpain I) is 32 nM. **Target(s):** calpain-1, *or* μ-calpain (1,3-12,14); calpain-2, *or* m-calpain (2-12); picornain 3C, weakly inhibited (13). **1.** Sorimachi & Suzuki (1998) in *Handb. Proteolytic Enzymes* (Barrett, Rawlings & Woessner, eds.), p. 643, Academic Press, San Diego. **2.** *Previous reference*, p. 649. **3.** Zhang & Johnson (1988) *J. Enzyme Inhib.* **2**, 163. **4.** Asada, Ishino, Shimada, *et al.* (1989) *J. Enzyme Inhib.* **3**, 49. **5.** Hamakubo, Ueda, Takano & Murachi (1990) *J. Enzyme Inhib.* **3**, 203. **6.** Kambayashi & Sakon (1989) *Meth. Enzymol.* **169**, 442. **7.** Murachi (1983) *Trends Biochem. Sci.* **8**, 167. **8.** Wendt, Thompson & Goll (2004) *Biol. Chem.* **385**, 465. **9.** Sazontova, Matskevich & Arkhipenko (1999) *Pathophysiology* **6**, 91. **10.** Murachi, Tanaka, Hatanaka & Murakami (1981) *Adv. Enzyme Regul.* **19**, 407. **11.** Yoshimura, Tsukahara & Murachi (1984) *Biochem. J.* **223**, 47. **12.** Betts, Weinsheimer, Blouse & Anagli (2003) *J. Biol. Chem.* **278**, 7800. **13.** Hata, Sato, Sorimachi, Ishiura & Suzuki (2000) *J. Virol. Methods* **84**, 117. **14.** Kiss, Bozoky, Kovács, *et al.* (2008) *FEBS Lett.* **582**, 2149.

Calpeptin

This potent dipeptide aldehyde inhibitor (FW = 362.47 g/mol; CAS 117591-20-5; Soluble to 72 mg/mL in DMSO), also known as N^{α}-(benzyloxycarbonyl)-L-leucyl-L-norleucinal and Z-LNle-CHO, targets calpain-1 (IC_{50} = 52 nM), calpain-2 (IC_{50} = 34 nM), and papain (IC_{50} = 138 nM). Peptide aldehyde hydrates most often resemble tetrahedral intermediates formed by H_2O attack on peptide substrates. Alternatively, these inhibitors may form adducts with an active-site nucleophile, again by resembling intermediates in a double-displacement mechanism. Calpeptin also modulates the processing of the β-amyloid precursor protein. Calpeptin is cell-permeable and soluble in dimethyl sulfoxide and dimethylformamide. It is also a weak inhibior of HIV protease (IC_{50} = 160

μM). **Target(s):** calpain-1, μ-calpain (1,2); calpain-2, *or* m-calpain (2,6,7); papain (1); protein-tyrosine phosphatase (3); peptide deformylase (*N*-formylmethionylaminoacyl-tRNA deformylase (4); HIV retropepsin, weakly inhibited (5); cathepsin V (8); cathepsin K (9); phosphatidylinositol-4-phosphate 3-kinase (10). **1.** Tsujinaka, Kajiwara, Kambayashi, *et al.* (1988) *Biochem. Biophys. Res. Commun.* **153**, 1201. **2.** Saito & Nixon (1993) *Neurochem. Res.* **18**, 231. **3.** Schoenwaelder & Burridge (1999) *J. Biol. Chem.* **274**, 14359. **4.** Durand, Green, O'Connell & Grant (1999) *Arch. Biochem. Biophys.* **367**, 297. **5.** Sarubbi, Seneci, Angelastro, *et al.* (1993) *FEBS Lett.* **319**, 253. **6.** Han, Weinman, Boldogh, Walker & Brasier (1999) *J. Biol. Chem.* **274**, 787. **7.** Azarian, Schlamp & Williams (1993) *J. Cell Sci.* **105**, 787. **8.** Brömme, Li, Barnes & Mehler (1999) *Biochemistry* **38**, 2377. **9.** Brömme, Okamoto, Wang & Biroc (1996) *J. Biol. Chem.* **271**, 2126. **10.** Crljen, Volinia & Banfic (2002) *Biochem. J.* **365**, 791.

Calphostin C

This red-colored naturally occurring polycyclic (FW = 790.78 g/mol; CAS 121263-19-2) from *Cladosporium cladosporioides*, named (2*R*)-1-[3,10-dihydroxy-12-[(2*R*)-2-(4-hydroxyphenoxy)carbonyloxypropyl]-2,6,7,11-tetramethoxy-4,9-dioxoperylen-1-yl]propan-2-yl] benzoate, selectively inhibits protein kinase C, IC_{50} = 50 nM. It is a weaker inhibitor of other protein kinases: *e.g.*, protein kinase A (IC_{50} > 50 μM), protein kinase G (IC_{50} > 25 μM), p60v-*src* (IC_{50} > 50 μM), and [myosin light-chain] kinase (IC_{50} > 5 μM). Calphostin C absorbs light strongly in the visible and ultraviolet range of the spectra. Inhibition by calphostin C is dependent on exposure to light: ordinary fluorescent light is sufficient for full activation of the inhibitor. Light-activated calphostin C inhibits protein kinase C isozymes by covalent modification of the lipid binding regulatory domain. **Target(s):** protein kinase C (1-7,11,12); cGMP-dependent protein kinase (1,2); [myosin-light chain] kinase (1,2); cAMP-dependent protein kinase (1,2); p60v-*aro* protein kinase (1,2); phospholipase D (8,9); L-type Ca^{2+} channels (10); inositol-trisphosphate 3-kinase (13). **1.** Kobayashi, Nakano, Morimoto & Tamaoki (1989) *Biochem. Biophys. Res. Commun.* **159**, 548. **2.** Tamaoki & Nakano (1990) *Biotechnology* **8**, 732. **3.** Gopalakrishna, Chen & Gundimeda (1995) *Meth. Enzymol.* **252**, 132. **4.** Tamaoki (1991) *Meth. Enzymol.* **201**, 340. **5.** Eriksson, Toivola, Sahlgren, Mikhailov & Härmälä-Braskén (1998) *Meth. Enzymol.* **298**, 542. **6.** Bruns, Miller, Merriman, *et al.* (1991) *Biochem. Biophys. Res. Commun.* **176**, 288. **7.** Gopalakrishna, Chen & Gundimeda (1992) *FEBS Lett.* **314**, 149. **8.** Sciorra, Hammond & Morris (2001) *Biochemistry* **40**, 2640. **9.** Dubyak & Kertesy (1997) *Arch. Biochem. Biophys.* **341**, 129. **10.** Hartzell & Rinderknecht (1996) *Amer. J. Physiol.* **270**, C1293. **11.** Bazán-Tejeda, Argüello-García, Bermúdez-Cruz, Robles-Flores & Ortega-Pierres (2007) *Arch. Microbiol.* **187**, 55. **12.** Nozawa, Nishihara, Akizawa, *et al.* (2004) *J. Pharmacol. Sci.* **94**, 233. **13.** Mayr, Windhorst & Hillemeier (2005) *J. Biol. Chem.* **280**, 13229.

Calycanthine

This highly toxic alkaloid (FW$_{free-base}$ = 346.48 g/mol) from *Calycanthus floridus* (Carolina allspice), *C. glaucus* (Eastern sweetshrub), *Chimonanthus praecox* (wintersweet), *Idiospermum australiense* (a primitive flowering plant), and *Psychotria colorata*, is a very powerful convulsant and poison. It has been proposed its convulsant action arises from the inhibition of γ-aminobutyrate release as a result of interactions with L-type Ca^{2+} channels and by inhibiting γ-aminobutyrate-mediated chloride currents at GABA$_A$ receptors. **1.** Chebib, Duke, Duke, *et al.* (2003) *Toxicol. Appl. Pharmacol.* **190**, 58.

Calyculin A

This marine toxin (FW$_{free-acid}$ = 1009.18 g/mol) from the sponge *Discodermia calyx* is a phosphorylated polyketide that has an IC$_{50}$ value of 1-2 nM for protein phosphatase 1 and < 1 nM for protein phosphatase 2A. Calyculin A is an antitumor agent and has been observed to prevent apoptosis in Burkitt's lymphoma cell line BM13674. **Target(s):** [myosin-light-chain] phosphatase (1); protein phosphatase type 1 (2-7,9, 10; protein phosphatase type 2A (2-4,6,10); phosphoprotein phosphatase (2-13); protein phosphatase type 4 (10,13); protein phosphatase type 5 (10). **1.** Parizi, Howard & Tomasek (2000) *Exp. Cell Res.* **254**, 210. **2.** Song & Lavin (1993) *Biochem. Biophys. Res. Commun.* **190**, 47. **3.** Eriksson, Toivola, Sahlgren, Mikhailov & Härmälä-Braskén (1998) *Meth. Enzymol.* **298**, 542. **4.** Yasumoto, Murata, Oshima, Matsumoto & Clardy (1984) in *Seafood Toxins, ACS Symp. Series* **262**, 207. **5.** Kita, Matsunaga, Takai, *et al.* (2002) *Structure* **10**, 715. **6.** Ishihara, Martin, Brautigan, *et al.* (1989) *Biochem. Biophys. Res. Commun.* **159**, 871. **7.** Stubbs, Tran, Atwell, *et al.* (2001) *Biochim. Biophys. Acta* **1550**, 52. **8.** Cheng, Wang, Gong, *et al.* (2001) *Neurochem. Res.* **26**, 425. **9.** Andrioli, Zaini, Viviani & da Silva (2003) *Biochem. J.* **373**, 703. **10.** Gallego & Virshup (2005) *Curr. Opin. Cell Biol.* **17**, 197. **11.** Solow, Young & Kennelly (1997) *J. Bacteriol.* **179**, 5072. **12.** Sayed, Whitehouse & Jones (1997) *J. Endocrinol.* **154**, 449. **13.** Cohen, Philp & Vázquez-Martin (2005) *FEBS Lett.* **579**, 3278.

Camphorquinone-10-sulfonate

1S-isomer   1R-isomer

This water-soluble reagent (FW$_{free-acid}$ = 246.28 g/mol), also known as camphorquinone-10-sulfonic acid, reacts reversibly with arginyl residues in peptides and proteins. **Target(s):** ferrochelatase (1,2); myosin light-chain kinase (3); spermidine/spermine N^1-acetyltransferase (4); choline acetyltransferase (5); lactoylglutathione lyase, *or* glyoxalase I (6). **1.** Dailey & Fleming (1986) *J. Biol. Chem.* **261**, 7902. **2.** Dailey, Fleming & Harbin (1986) *J. Bacteriol.* **165**, 1. **3.** Pearson & Kemp (1986) *Biochim. Biophys. Acta* **870**, 312. **4.** Della Ragione, Erwin & Pegg (1983) *Biochem. J.* **213**, 707. **5.** Mautner, Pakula & Merrill (1981) *Proc. Natl. Acad. Sci. U.S.A.* **78**, 7449. **6.** Schasteen & Reed (1983) *Biochim. Biophys. Acta* **742**, 419.

(*S*)-(+)-Camptothecin

This alkaloid (FW = 348.36 g/mol; CAS 7689-03-4; unstable when stored in frozen aqueous solutions) from the Chinese tree *Camptotheca acuminata* intercalates into DNA, rendering the latter unstable toward alkali. This alkaloid binds to a transient topoisomerase-DNA complex and inhibit the resealing of a single-strand nick that the enzyme creates to relieve superhelical tension in duplex DNA. The inactive complex is then ubiquinated and hydrolyzed by proteasomes, thus depleting the cell of DNA topoisomerase I (*See also CPT-11*). Camptothecin induces apoptosis in a variety of cells. **Target(s):** DNA topoisomerase I (1-6); DNA topoisomerase II (7,8); sphinganine kinase, *or* sphingosine kinase (9); PARP-1 (10; *Note:* Camptothecin inhibition of poly(ADP-ribose) polymerase-1 converts PARP1 into a dominant-negative molecule that poisons the ability of DNA repair machinery to participate in either PARP1-dependent or PARP1-independent repair of camptothecin-induced DNA damag.). **1.** Hertzberg, Caranfa & Hecht (1989) *Biochemistry* **28**, 4629. **2.** Scott & Tomkins (1975) *Meth. Enzymol.* **40**, 273. **3.** Hsiang, Hertzberg, Hecht & Liu (1985) *J. Biol. Chem.* **260**, 14873. **4.** Rahier, Eisenhauer, Gao, Thomas & Hecht (2005) *Bioorg. Med. Chem.* **13**, 1381. **5.** Argaman, Bendetz-Nezer, Matlis, Segal & Priel (2003) *Biochem. Biophys. Res. Commun.* **301**, 789. **6.** Heath-Pagliuso, Cole & Kmiec (1990) *Plant Physiol.* **94**, 599. **7.** Saijo, Enomoto, Hanaoka & Ui (1990) *Biochemistry* **29**, 583. **8.** Walker & Saravia (2004) *J. Parasitol.* **90**, 1155. **9.** Cuvillier (2007) *Anticancer Drugs* **18**, 105. **10.** Patel, Flatten, Schneider, *et al.* (2012) *J. Biol. Chem.* **287**, 4198.

Canagliflozin

This orally available Type-2 diabetes drug (FW = 444.52 g/mol; CAS 842133-18-0; Symbol: CNF), also known by the trade name: Invokana® and systematically as (1*S*)-1,5-anhydro-1-(3-{[5-(4-fluorophenyl)-2-thienyl]methyl}-4-methylphenyl)-D-glucitol, inhibits the sodium-glucose transporter, Subtype 2 (SGLT2), a carrier responsible for >90% of kidney glucose reabsorption (1-3). Canagliflozin lowers fasting plasma glucose and hemoglobin A$_{1c}$ levels in a dose-dependent manner. **Primary Mode of Inhibitory Action:** SGLT2 inhibitors block glucose reabsorption in the proximal tubule, thereby lowering the renal threshold for glucose, increasing urinary glucose excretion, and reducing serum glucose in hyperglycemic patients. Canagliflozin improves glycemic control in an insulin-independent fashion through inhibition of glucose reuptake in the kidney. This novel mechanism of action offers potential advantages over other antihyperglycemic agents, including a relatively low hypoglycemia risk and weight loss-promoting effects. **Pharmacokinetics:** Canagliflozin shows high oral bioavailability (~85%). Its half-life is 10.6 hours at a 100-mg oral dose and 13.1 hours with a 300-mg dose, attaining peak plasma concentrations in 1-2 hours and steady-state levels within 4-5 days. **Other Targets:** CNF inhibits all UDP-glucuronosyltransferase UGT1A subfamily enzymes, but the greatest inhibition is observed with UGT1A1, UGT1A9, and UGT1A10 ($IC_{50} \leq 10$ μM) (4). CNF also inhibits recombinant and human liver microsomal UGT1A9, with K_i values ranging from 1.4–3.0 μM, depending on the substrate (propofol/4-methylumbelliferone). K_i values for CNF inhibition of UGT1A1 were approximately 3-fold higher (4). Given such findings, CNF inhibition of UGT1A1 and UGT1A9 *in vivo* cannot be discounted as a source of potential drug-drug interactions (4).

Anion Gap Ketoacidosis: In some subjects, canagliflozin therapy can give rise to severe anion gap metabolic acidosis and euglycemic diabetic ketoacidosis, *or* DKA (5). This form of metabolic acidosis is characterized by a high anion gap (*i.e.*, a medical value based on the concentrations of ions in a patient's serum). An anion gap is usually considered to be high if the value exceeds 11 mEq/L. High anion gap metabolic acidosis is caused generally by the body producing too much acid or not producing enough bicarbonate. This is often due to an increase in lactic acid or ketoacids, or it may be a sign of kidney failure. DKA can develop when insulin levels are too low or during prolonged fasting. (*See also Dapagliflozin; Empagliflozin, Tofogliflozin, Luseogliflozin*) 1. Nomura, Sakamaki, Hongu, *et al.* (2010) *J. Med. Chem.* **53**, 6355. 2. Sha, Devineni, Ghosh, *et al.* (2011) *Diabetes Obes. Metab.* **13**, 669. 3. Nisly, Kolanczyk & Walton (2013) *Am. J. Health Syst. Pharm.* **70**, 311. 4. Pattanawongsa, Chau, Rowland & Miners (2015) *Drug Metab. Dispos.* **43**, 1468. 5

Canakinumab

This humanized anti-human IL-1β monoclonal IgGκ antibody (MW = 145.2 kDa; CAS 914613-48-2) is also known by its code name ACZ885 and trade name Ilaris®. **Primary Mode of Inhibitory Action:** Canakinumab binds to IL-1β (K_d = 30-50 pM), blocking its interaction with IL-1 receptor, without binding to IL-1a or IL-1 receptor antagonist (1). Interleukin-1 is a pro-inflammatory cytokine causing fever, anorexia, joint destruction, and tissue damage/remodeling, which are reversed/reduced by specific IL-1 blockers. Ilaris is an FDA-approved drug specifically indicated for treating cryopyrin-associated periodic syndrome (CAPS), a spectrum of autoinflammatory syndromes, including familial cold autoinflammatory syndrome (FCAS), the Muckle-Wells syndrome (MWS), and neonatal-onset multisystem inflammatory disease (NOMID). All are associated with mutations in *NLRP3*, the gene encoding cryopyrin, an adaptor protein that is essential for inflammasome activation in response to signalling pathways triggered by ATP, nigericin, maitotoxin, *Staphylococcus aureus*, or *Listeria monocytogenes*. **Clinical Applications:** Treatment with subcutaneous canakinumab once every eight weeks was associated with rapid remission of symptoms in most patients with CAPS (2). Ilaris is also approved for the treatment of active Systemic Junenile Idiopathic Arthritis (SJIA) in patients aged two and older. Biweekly administration offers advantages over treatment with the human IL-1 receptor antagonist Anakinra, which must be injected daily and is often poorly tolerated (3). Canakinumab is also effective in the prophylaxis and management of acute gout, reducing its pain and risk of new attacks. Its long *in vivo* half-life contributes to its prolonged anti-inflammatory effects (4). 1. Dhimolea (2010) *MAbs* **2**, 3. 2. Lachmann, Kone-Paut, Kuemmerle-Deschner, *et al.* (2009) *New Engl. J. Med.* **360**, 2416. 2. Church & McDermott (2009) *Curr. Opin. Mol. Ther.* **11**, 81. 3. Schlesinger (2012) *Expert Opin. Biol. Ther.* **12**, 1265.

L-Canaline

This ornithine analogue (FW = 134.14 g/mol; CAS 496-93-5), also known as L-2-amino-4-(aminooxy)butyric acid, is an inhibitor of many pyridoxal-dependent enzymes, reacting with the coenzyme to form a covalently-bound oxime that inactivates the enzyme. Hydroxylamines are known to possess enhanced nucleophilicity, especially toward aldehydes, acyl anhydrides, and acyl-phosphates. The D-enantiomer also inactivates ornithine and alanine transaminases, albeit more slowly. L-Canaline reductase catalyzes an NADPH-dependent reductive cleavage of L-canaline to L-homoserine and ammonia. L-Canaline is found in many legumes, including the jack bean (*Canavalia ensiformis*). In higher plants, canaline inhibits production of ethylene, a key metabolic signaling molecule. **Target(s):** ornithine carbamoyltransferase (1,7); tyrosine aminotransferase (1,11,16,31); ornithine δ-aminotransferase (1,4,5,8,14,15, 26-30); ornithine decarboxylase (1,6,24); aromatic-amino-acid decarboxylase (1); diamine oxidase (1); 1-aminocyclopropane-1-carboxylate synthase (2,21); alanine

aminotransferase (3); serine hydroxymethyl-transferase, *or* glycine hydroxymethyltransferase (9); γ-cystathionase, *or* cystathionine γ-lyase (10); γ-aminobutyrate aminotransferase, GABA aminotransferase (12); lysine transport, weakly inhibited (13); aspartate aminotransferase (17); methionine aminotransferase (18); arginase (20); cystathionine β-lyase (22); adenosyl-methionine decarboxylase, weakly inhibited (23); pyridoxal kinase (25). 1. Klosterman (1979) *Meth. Enzymol.* **62**, 483. 2. Adams & Yang (1981) *Trends Biochem. Sci.* **6**, 161. 3. Worthen, Ratliff, Rosenthal, Trifonov & Crooks (1996) *Chem. Res. Toxicol.* **9**, 1293. 4. Kito, Sanada & Katunuma (1978) *J. Biochem.* **83**, 201. 5. Seiler (2000) *Curr. Drug Targets* **1**, 119. 6. Rosenthal (1997) *Life Sci.* **60**, 1635. 7. Kekomaki, Rahiala & Räiha (1969) *Ann. Med. Exp. Biol. Fenn.* **47**, 33. 8. Rosenthal & Dahlman (1990) *J. Biol. Chem.* **265**, 868. 9. Baskaran, Prakash, Savithri, Radhakrishnan & Appaji Rao (1989) *Biochemistry* **28**, 9613. 10. Beeler & Churchich (1976) *J. Biol. Chem.* **251**, 5267. 11. Ohisalo, Andersson & Pispa (1977) *Biochem. J.* **163**, 411. 12. Watts & Atkins (1984) *Mol. Biochem. Parasitol.* **12**, 207. 13. Mokrasch (1987-1988) *Membr. Biochem.* **7**, 249. 14. Bolkenius, Knodgen & Seiler (1990) *Biochem. J.* **268**, 409. 15. Shah, Shen & Brunger (1997) *Structure* **5**, 1067. 16. Heilbronn, Wilson & Berger (1999) *J. Bacteriol.* **181**, 1739. 17. Berger, Wilson, Wood & Berger (2001) *J. Bacteriol.* **183**, 4421. 18. Berger, English, Chan & Knodel (2003) *J. Bacteriol.* **185**, 2418. 19. Boyar & Marsh (1982) *J. Amer. Chem. Soc.* **104**, 1995. 20. Kitagawa (1939) *J. Agr. Chem. Soc. Japan* **15**, 267. 21. Boller, Herner & Kende (1979) *Planta* **145**, 293. 22. Gentry-Weeks, Keith & Thompson (1993) *J. Biol. Chem.* **268**, 7298. 23. Pösö, Sinervirta & Jänne (1975) *Biochem. J.* **151**, 67. 24. Ono, Inoue, Suzuki & Takeda (1972) *Biochim. Biophys. Acta* **284**, 285. 25. McCormick & Snell (1961) *J. Biol. Chem.* **236**, 2085. 26. Seiler (2000) *Curr. Drug Targets* **1**, 119. 27. Shah, Shen & Brunger (1997) *Structure* **5**, 1067. 28. Shiono, Hayasaka & Mizuno (1981) *Exp. Eye Res.* **32**, 475. 29. Rosenthal & Dahlman (1990) *J. Biol. Chem.* **265**, 868. 30. Gafan, Wilson, Berger & Berger (2001) *Mol. Biochem. Parasitol.* **118**, 1. 31. Dietrich, Lorber & Kern (1991) *Eur. J. Biochem.* **201**, ·399.

L-Canavanine

This toxic L-arginine analogue (FW = 176.18 g/mol; CAS 543-38-4; pK_a = 2.35 for carboxyl; pK_a = 7.01 for hydroxyguanidine group; pK_a = 9.22 for the amino groups), also known as L-2-amino-4-(guanidinooxy)butyric acid, from the jack bean (*Canavalia ensiformis*). When incorporated into proteins, many of the latter are non-functional, because the hydroxyguanidine moiety has very different electrostatics. L-Canavanine also inhibits the growth of certain bacterial strains. L-Canavanine is an alternative substrate and competitive inhibitor to many arginine-utilizing enzymes, including nitric-oxide synthases. **Target(s):** *N*-acetylglutamate kinase (1,6,37); arginine decarboxylase, as alternative substrate (2,10,12,24-28); argininosuccinate lyase (3,4); ornithine aminotransferase (5,38,39); alcohol dehydrogenase, particularly the yeast enzyme (7); arginine deaminase, mechanism-based inhibitor (8,9,14,29,30); arginine transport, competitive (11,13); lysine permease, competitive (13); nitric-oxide synthase, K_i = 11.1 μM (15-18,20,21,40); group I self-splicing intron (19); arginyl-tRNA synthetase (23); Arginase (31-33); arginine kinase (34-36). 1. Dénes (1970) *Meth. Enzymol.* **17A**, 269. 2. Smith (1983) *Meth. Enzymol.* **94**, 176. 3. Ratner (1962) *The Enzymes*, 2nd ed., **6**, 495. 4. Ratner (1972) *The Enzymes*, 3rd ed., **7**, 167. 5. Braunstein (1973) *The Enzymes*, 3rd ed., **9**, 379. 6. Dénes (1973) *The Enzymes*, 3rd ed., **9**, 511. 7. Brändén, Jörnvall, Eklund & Furugren (1975) *The Enzymes*, 3rd ed., **11**, 103. 8. Webb (1966) *Enzyme and Metabolic Inhibitors*, vol. **2**, p. 353, Academic Press, New York. 9. Oginsky & Gehrig (1952) *J. Biol. Chem.* **198**, 799. 10. Blethen, Boeker & Snell (1968) *J. Biol. Chem.* **243**, 1671. 11. Boller, Durr & Wiemken (1975) *Eur. J. Biochem.* **54**, 81. 12. Ramakrishna & Adiga (1975) *Eur. J. Biochem.* **59**, 377. 13. Beckerich & Heslot (1978) *J. Bacteriol.* **133**, 492. 14. Smith, Ganaway & Fahrney (1978) *J. Biol. Chem.* **253**, 6016. 15. Schmidt, Nau, Wittfoht, *et al.* (1988) *Eur. J. Pharmacol.* **154**, 213. 16. McCall, Boughton-Smith, Palmer, Whittle & Moncada (1989) *Biochem. J.* **261**, 293. 17. Boje & Fung (1990) *J. Pharmacol. Exp. Ther.* **253**, 20. 18. Persson, Midtvedt, Leone & Gustafsson (1994) *Eur. J. Pharmacol.* **264**, 13. 19. Liu & Leibowitz (1995) *Nucl. Acids Res.* **23**, 1284. 20. Liaudet, Feihl, Rosselet, *et al.*

(1996) *Clin. Sci. (London)* **90**, 369. **21**. Babu, Frey & Griffith (1999) *J. Biol. Chem.* **274**, 25218. **22**. Boyar & Marsh (1982) *J. Amer. Chem. Soc.* **104**, 1995. **23**. Allende & Allende (1964) *J. Biol. Chem.* **239**, 1102. **24**. Smith (1979) *Phytochemistry* **18**, 1447. **25**. Das, Bhaduri, Bose & Ghosh (1996) *J. Plant Biochem. Biotechnol.* **5**, 123. **26**. Choudhuri & Ghosh (1982) *Agric. Biol. Chem.* **46**, 739. **27**. Rosenfeld & Roberts (1976) *J. Bacteriol.* **125**, 601. **28**. Balbo, Patel, Sell, *et al.* (2003) *Biochemistry* **42**, 15189. **29**. Knodler, Sekyere, Stewart, Schofield & Edwards (1998) *J. Biol. Chem.* **273**, 4470. **30**. Lu, Li, Feng, *et al.* (2005) *J. Amer. Chem. Soc.* **127**, 16412. **31**. Colleluori & Ash (2001) *Biochemistry* **40**, 9356. **32**. Patchett, Daniel & Morgan (1991) *Biochim. Biophys. Acta* **1077**, 291. **33**. Kaysen & Strecker (1973) *Biochem. J.* **133**, 779. **34**. Pereira, Alonso, Ivaldi, *et al.* (2003) *J. Eukaryot. Microbiol.* **50**, 132. **35**. Pereira, Alonso, Paveto, *et al.* (2000) *J. Biol. Chem.* **275**, 1495. **36**. Baker (1976) *Insect Biochem.* **6**, 449. **37**. Faragó & Dénes (1967) *Biochim. Biophys. Acta* **136**, 6. **38**. Strecker (1965) *J. Biol. Chem.* **240**, 1225. **39**. Kalita, Kerman & Strecker (1976) *Biochim. Biophys. Acta* **429**, 780. **40**. Knowles, Merrett, Salter & Moncada (1990) *Biochem. J.* **270**, 833.

Cancerous Inhibitor of Protein Phosphatase 2A

This 905-residue human protein (MW = 102,185 g; NCBI Reference Sequence = NP 065941.2; *Symbol* = CIP2A) targets Protein Phosphatase 2A (PP2A), an enzyme that dephosphorylates *c*-Myc, a constitutively expressed form of Myc transcription factor in many cancer cell types. CIP2A is overexpressed in most human cancers is associated with altered expression of epithelial-mesenchymal transition markers, increased lymph node metastasis, and poorer prognosis in pancreatic ductal adenocarcinoma (1). CIP2A strongly interacts with NIMA-related kinase 2 (NEK2 serine/threonine kinase) during the G_2/M phase, thereby enhancing NEK2 kinase activity and facilitate centrosome separation in a PP1 and PP2A-independent manner (2). stage. CIP2A depletion impairs cell-cycle progression, resulting in aberrant centrosome separation, altered mitotic spindle dynamics, and SAC activation. CIP2A plays a role in regulating centrosome separation through the activation of NEK2A. **1**. Wang, Gu, Ma, Zhang, Bian & Cao (2013) *Tumor Biol.* **34**, 2309. **2**. Jeong, Lee, Park, *et al.* (2014) *J. Biol. Chem.* **289**, 28.

Candesartan & Candesartan Cilexetil

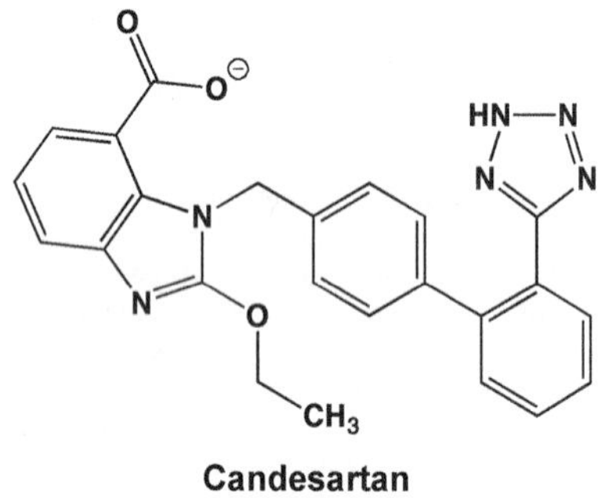

Candesartan

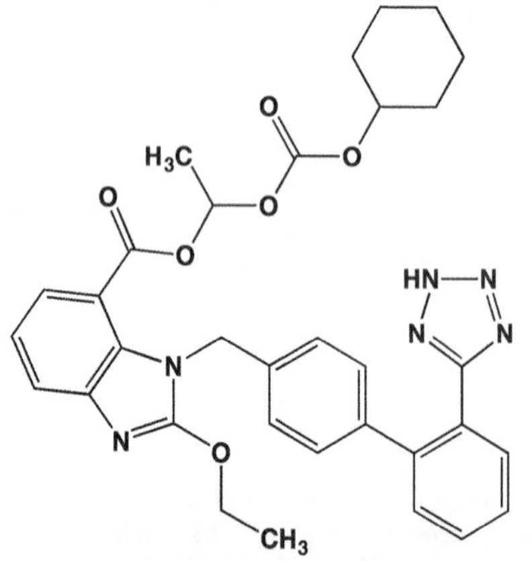

Candesartan Cilexetil

This angiotensin II receptor antagonist (FW = 440.45 g/mol; CAS 139481-59-7), also known as 2-ethoxy-1-({4-[2-(2*H*-1,2,3,4-tetrazol-5-yl)phenyl]phenyl}methyl)-1*H*-1,3-benzodiazole-7-carboxylic acid, is an antihyper-tensive agent. Its prodrug candesartan cilexetil (FW = 610.67 g/mol) is marketed under the trade names Blopress®, Atacand®, Amias®, and Ratacand®. Candesartan exhibits strong receptor binding affinity, with slow dissociation (1,2). Candesartan cilexetil is metabolized by esterases in the intestinal wall, attended by absorption of candesartan. The clinical benefits of candesartan may extend beyond its proven antihypertensive effects to a wider range of complications across the cardiovascular continuum, including diabetes, left ventricular hypertrophy, atherosclerosis and stroke (3). (***See also*** *Eprosartan; Irbesartan; Losartan; Olmesartan; Telmisartan; Valsartan*) **Pharmacological Parameters:** Bioavailability = 42 %; Food Effect? NO; Drug $t_{1/2}$ = 3.5–4 hours; Metabolite $t_{1/2}$ = 3–11 hours; Drug's Protein Binding = 99.5 %; Metabolite's Protein Binding = very low; Route of Elimination = 33 % Renal, 67 % Hepatic. **1**. Mizuno, Niimura, Tani, *et al.* (1992) *Life Sci.* **51**, PL183. **2**. Shibouta, Inada, Ojima, *et al.* (1993) *J. Pharmacol. Exp. Ther.* **266**, 114. **3**. Meredith (2007) *Curr. Med. Res. Opin.* **23**, 1693.

Cangrelor

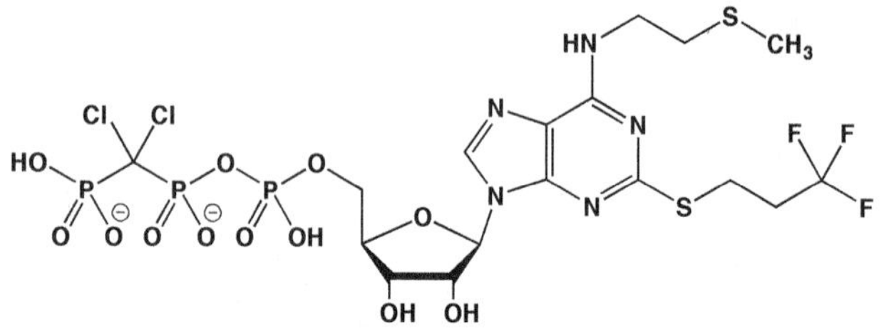

This antiplatelet drug (FW = 776.36 g/mol; CAS 163706-06-7), also kown as Kengreal®, potently inhibits $P2Y_{12}$, the G_i class of ADP-chemosensing G protein-coupled receptors that promote platelet aggregation (1). Although cangrelor is itself an ATP derivative, it does not require metabolic conversion (nor, as a dichloromethyldiphosphonate, can it be) to the ADP species for its pharmacodynamics properties. Cangrelor rapidly achieves steady-state concentrations, showing a clearance rate of 50 L/h and a $t_{1/2} \approx$ 3 minutes. Kengreal is indicated for reducing thrombotic cardiovascular events in patients with coronary artery disease and as a bridging therapy in patients with acute coronary syndrome, or with stents, who face an increased risk for thrombotic events (*e.g.*, stent thrombosis) when oral $P2Y_{12}$ therapy is interrupted because of surgery (2). IUPAC: also named [dichloro-[[[(2*R*,3*S*,4*R*,5*R*)-3,4-dihydroxy-5-[6-(2-methylsulfanylethyl amino)-2-(3,3,3-trifluoropropyl-sulfanyl)purin-9-yl]oxolan-2-yl]methoxy hydroxyphosphoryl]oxyhydroxy-phosphoryl]methyl]phosphonate **1**. Ingall, Dixon, Bailey, *et al.* (1999) *J. Med. Chem.* **42**, 213. **2**. Lhermusier, Baker & Waksman (2015) *Am. J. Cardiol.* **115**, 1154.

Cannabidiol

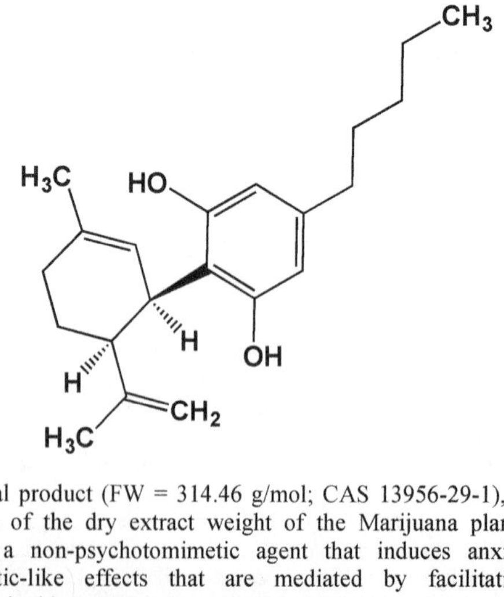

This natural product (FW = 314.46 g/mol; CAS 13956-29-1), comprising some 40% of the dry extract weight of the Marijuana plant *Cannabis sativa*, is a non-psychotomimetic agent that induces anxiolytic- and antipsychotic-like effects that are mediated by facilitation of the endocannabinoid system or by activation of 5-HT_{1A} (serotonin) receptors. Cannabidiol exhibits low affinity for cannabinoid CB_1 and CB_2 receptors (1), but can block anandamide reuptake (2) as well as its metabolism by fatty acid amide hydrolase (3). **1**. Thomas, Gilliam, Burch, Roche & Seltzman (1998) *J. Pharmacol. Exp. Ther.* **285**, 285. **2**. Bisogno, Hanus,

De Petrocellis *et al.* (2001) *Br. J. Pharmacol.* **134**, 845. **3**. Watanabe, Ogi, Nakamura, *et al.* (1998) *Life Sci.* 62, 1223.

Cantharidin

This terpene carboxylic anhydride and PP2A inhibitor (FW = 196.20 g/mol; CAS 56-25-7; Symbol: CA; IUPAC Name: 2,6-dimethyl-4,10-dioxatricyclo[5.2.1.0^{2,6}]decane-3,5-dione), from the Spanish fly (*Lytta vesicatoria*), is a poisonous blister agent used topically to remove warts and tattoos as well as to treat the small papules of *Molluscum contagiosum.* **Identification as Phosphoprotein Phosphatase-2A Inhibitor:** The toxic effects of cantharidin as well as those of its herbicidal analogue, endothall, are attributable to their highly affine and specific interactions with cantharidin-binding protein, *or* CBP (1). The latter is the protein phosphatase 2A (PP2A), an $\alpha\beta$ heterodimer (α-subunit, 61-kDa β-subunit, 39-kDa). CBP catalyzes the dephosphorylation of phosphorylase *a*, a reaction that is sensitive not only to okadaic acid, cantharidin and its analogues. Indeed, okadaic acid is a potent inhibitor of [^{3}H]cantharidin binding to CBP (1,2). The PP2A inhibitory properties of a series of anhydride modified cantharidin analogues indicates that only those analogues that can undergo facile ring opening of the anhydride moiety display significant inhibition (3). NMR experiments suggest that 7-oxobicyclo[2.2.1]heptane-2,3-dicarboxylic acid was the sole species under PP2A assay conditions (3). **Role in Insect Defense:** Cantharidin is secreted by the male blister beetle as a copulatory gift during mating. The female recipient then cover her eggs with cantharidin, which imparts a disagreeable odor to ward off predators, EC$_{50}$ ≈ 10 μM (4). **Mode of Vesicant Action:** Cantharidin is readily absorbed by epidermal cell membranes, activation the release of serine proteases that cleave disintegrate desmosomal plaques, resulting in tonofilament detachment and skin blistering that eventually heals without scarring. That said, the widespread presence of PP2A in animals and plants, and its inhibition (K_i = 80 nM in rat brain) by cantharidin is likely to underlie many of the toxic effects of cantharidin and its analogues. The vesicant action of cantharidin is thought to account for the notorious use of Spanish fly as an aphrodisiac. **Antineoplastic Properties:** Based on its use in traditional Chinese medicine, cantharidin was found to be a potent antitumor agent. Such action goes beyond its inhibition of PP2A. Indeed, cantharidin induces cancer cell death, with an IC$_{50}$ value of 4.2 μM by blocking Heat Shock Factor 1 (HSF1) binding to promoters and inhibiting expression of Heat Shock Protein 70, *or* HSP70, as well as Bcl-2-associated athanogene domain-3, *or* BAG3 (5). However, cantharidin does not inhibit NF-κB luciferase reporter activity, suggesting it is not a general transcription inhibitor. Moreover, PP2A siRNA or okadaic acid fails to block HSF1 activity, again suggesting cantharidin inhibits HSF1 in a PP2A-independent manner (5). **1**. Li & Casida (1992) *Proc. Natl. Acad. Sci. U.S.A.* **89**, 11867. **2**. Eldridge & Casida (1995) *Toxicol. Appl. Pharmacol.* **130**, 95. **3**. McCuskey, Keane, Mudgee, *et al.* (2000) *Eur. J. Med. Chem.* **35**, 957. **4**. Carrel & Eisner (1974) *Science* **183**, 755. **5**. Kim, Kim, Kwon & Han (2013) *J. Biol. Chem.* **288**, 28713.

Capillarisin & "Thiocapillarisin"

Capillarisin

"Thiocapillarisin"

This naturally occurring flavone from *Artemisia capillaris* (FW = 316.26 g/mol; CAS 56365-38-9; Symbol: CPS), also named 5,7-dihydroxy-2-(4-hydroxyphenoxy)-6-methoxychromen-4-one, exerts a relaxing effect on rabbit penile corpus cavernosum (PCC) by activating the NO-cGMP and adenylyl cAMP signaling pathways. Capillarisin may well become an alternative medicine for patients wishing to use natural products to improve erectile function or in those who do not respondd adequately to cGMP phosphodiesterase isozyme-5 inhibitors (1). Capillarisin also specifically inhibits both constitutive and inducible STAT3 activation at tyrosine-705, but not at serine-727, in human multiple myeloma cells (2). CPS also blocks STAT3 constitutive activity and nuclear translocation, as mediated through the inhibition of activation of upstream JAK1, JAK2, and c-Src kinases (2). Capillorisin also inhibits aldol reductase (3). Its thio analogue is [(4-hydroxyphenyl)thio]-7-isopropoxy-5,6-dimethoxy-4*H*-chromen-4-one (FW = 332.33 g/mol) (4). **1**. Kim, Choi, Bak, *et al.* (2011) *Phycotherapy Res.* **26**, 805. **2**. Lee, Chiang, Nam, *et al.* (2013) *Cancer Lett.* **345**, 130. **3**. Yamaguchi, Sato, Chin, *et al.* (1988) *Wakan Iyaku Gakkaishi* **5** , 374. **4**. Igarashi, Kumazawa, Ohshima, *et al.* (2005) *Chem. Pharm. Bull.* **53**, 1088.

Capecitabine

This orally bioavailable 5-fluorouracil pro-drug (FW$_{free\text{-}acid}$ = 359.35 g/mol; CAS 154361-50-9), systematically named pentyl [1-(3,4-dihydroxy-5-methyltetrahydrofuran-2-yl)-5-fluoro-2-oxo-1*H*-pyrimidin-4-yl]carbamate, is enzymatically transformed to 5-FU, whereupon it is then metabolized to form the potent thymidylate synthase inhibitor, 5-fluoro-dUMP. Capecitabine activation follows a pathway employing three enzymatic steps and two intermediates, 5'-deoxy-5-fluorocytidine (5'-dFCR) and 5'-deoxy-5-fluorouridine (5'-dFUR). The *N*^{4}-pentyloxycarbonyl group is hydrolytically removed by liver carboxylesterases to form 5'-deoxy-5-fluorocytidine, which suffers subsequent enzymatic deamination to 5'-deoxy-5-fluorouridine. Absent any susceptibility to 5'phosphorylation, the latter is instead converted by thymidine phosphorylase into 5-fluorouracil. (*See Fluorouracil*) **Key Pharmacokinetic Parameters:** *See* Appendix II in Goodman & Gilman's THE PHARMACOLOGICAL BASIS OF THERAPEUTICS, 12th Edition (Brunton, Chabner & Knollmann, eds.) McGraw-Hill Medical, New York (2011). **1**. Parker (2009) *Cmem. Rev.* **109**, 2880.

Capsaicin

This pungent secondary plant metabolite (FW = 305.42 g/mol; CAS 404-86-4), also known as 8-methyl-*N*-vanillyl-6-nonenamide, is the primary irritatant found in hot chilies and red peppers, giving them and their powdered forms (*e.g.*, paprika, and cayenne) a stinging sensation and spiciness. **Primary Mode of Action:** At sub-micromolar concentrations, capsaicin targets certain pain-sensing neurons, activating the mammalian Transient Receptor Potential Vanilloid-1 (TRPV1) receptor involved in nociception. At micro- to millimolar concentrations, capsaicin is used in clinical and *in vitro* studies, modulating the function of a large number of seemingly unrelated membrane proteins. Capsaicin also regulates voltage-dependent sodium channels by altering lipid bilayer elasticity, promoting channel inactivation, similar to other amphiphiles that decrease bilayer stiffness. By activating TRPV1, capsaicin allows more QX-314 (membrane-impermeant lidocaine derivative) to enter primary sensory nociceptor neurons, thereby synergizing lidocaine's anesthetic effects (*See Lidocaine*). Birds reportedly express a capsaicin-insensitive vanilloid receptor, relieving them capsaicin's sting and allowing them to serve as far-flying vectors for seed dispersal. **Capsaicin Patch:** Containing 8% capsaicin by weight, a transdermal patch marketed under the trade name Qutenza™ relieves pain by targeting neurons in the area of skin where pain is being experienced. In 2009, Qutenza received FDA approval for the use of capsaicin to relieve neuropathic pain associated with post-herpetic neuralgia. **Target(s):** voltage-dependent Na$^+$ channels (1); NADH dehydrogenase, *or* complex I (2-4,8,9); CYP2A2 (5); CYP3A1 (5); CYP2C1 (5); CYP2B1 (5); CYP2B (25); CYP2C6 (5); tyrosyl-tRNA synthetase (6); 5-lipoxygenase, *or* arachidonate 5-lipoxygenase (10,11); inhibits chloride secretion in colonic epithelial cells independently of TRPV1 (12); human platelet aggregation and thromboxane biosynthesis (13); blocks constitutive and interleukin-6-inducible STAT3 activation (14); inhibits protein synthesis in TRPV1-expressing HEK cells, IC$_{50}$ = 15.6 nM (15); stimulates microtubule depolymerization within 10 min (16); inhibits catecholamine secretion and synthesis by blocking Na$^+$ and Ca^{2+} influx through a vanilloid receptor-independent pathway in bovine adrenal medullary cells (17); inhibits VEGF-induced vessel sprouting in rat aortic ring and Matrigel plug assays (18); inhibits lipid peroxidation *via* the radical scavenging properties of capsaicin's C7-benzyl carbon (19); inhibits activation of voltage-gated sodium currents in capsaicin-sensitive trigeminal ganglion neurons (20); inhibits NADH-ubiquinone oxidoreductases (21); blocks Ca^{2+} channels in isolated rat trigeminal and hippocampal neurons (22); disrupts axoplasmic transport in sensory neurons by disorganizing microtubules and neurofilaments (23). **1.** Lundbaek, Birn, Tape, *et al.* (2005) *Mol. Pharmacol.* **68**, 680. **2.** Fang, Wang & Beattie (2001) *Eur. J. Biochem.* **268**, 3075. **3.** Bogachev, Murtazina & Skulachev (1996) *J. Bacteriol.* **178**, 6233. **4.** Yagi (1991) *J. Bioenerg. Biomembr.* **23**, 21. **5.** Zhang, Hamilton, Stewart, Strother & Teel (1993) *Anticancer Res.* **13**, 2341. **6.** Cochereau, Sanchez, Bourhaoui & Creppy (1996) *Toxicol. Appl. Pharmacol.* **141**, 133. **7.** Jordt & Julius (2002) *Cell* **108**, 421. **8.** Yagi (1990) *Arch. Biochem. Biophys.* **281**, 305. **9.** Shimomura, Kawada & Suzuki (1989) *Arch. Biochem. Biophys.* **270**, 573. **10.** Schneider & Bucar (2005) *Phytother. Res.* **19**, 81. **11.** Flynn, Rafferty & Boctor (1986) *Prostaglandins Leukot. Med.* **24**, 195. **12.** Bouyer, Tang, Weber, *et al.* (2013) *Am. J. Physiol. Gastrointest. Liver Physiol.* **304**, G142. **13.** Raghavendra & Naidu (2009) *Prostaglandins Leukot Essent Fatty Acids* **81**, 73; **14.** Bhutani, Pathak & Nair (2007) *Clin Cancer Res.* **13**, 3024; **15.** Han, McDonald, Bianchi, *et al.* (2007) *Biochem. Pharmacol.* **73**, 1635; **16.** Takahashi, Toyohira, Ueno, Tsutsui & Yanagihara (2006) *Naunyn. Schmiedebergs Arch. Pharmacol.* **374**, 107; **17.** Min, Han, Kim, *et al.* (2004) *Cancer Res.* **64**, 644; **18.** Kogure, Goto, Nishimura, *et al.* (2002) *Biochim. Biophys. Acta* **1573**, 84; **19.** Liu, Oortgiesen, Li & Simon (2001) *J. Neurophysiol.* **85**, 745; **20.** Satoh, Miyoshi, Sakamoto & Iwamura (1996) *Biochim. Biophys. Acta* **1273**, 21; **21.** Wilkinson, Kim, Cho, *et al.* (1996) *Arch. Biochem. Biophys.* **336**, 275. **22.** Kopanitsa, Panchenko, Magura, Lishko & Krishtal (1995) *Neuroreport.* **6**, 2338; **23.** Kawakami, Hikawa, Kusakabe, *et al.* (1993) *J. Neurobiol.* **24**, 545.

Capsazepine

This synthetic capsaicin analogue (FW = 376.91 g/mol; CAS 138977-28-3; IUPAC: *N*-[2-(4-chlorophenyl)ethyl]-1,3,4,5-tetrahydro-7,8-dihydroxy-2*H*-2-benzazepine-2-carbothioamide, is a vanilloid receptor antagonist, the first found to be competitive with capsaicin. **1.** Urban & Dray (1991) *Neurosci. Lett.* **134**, 9. **2.** Bevan, Hothi, Hughes, *et al.* (1992) *Brit. J. Pharmacol.* **107**, 544. **3.** Walpole, Bevan, Bovermann, *et al.* (1994) *J. Med. Chem.* **37**, 1942.

Captopril

This antihypertensive drug, (FW$_{free-acid}$ = 217.29 g/mol; CAS 62571-86-2) also called (2*S*)-1-(3-mercapto-2-methylpropionyl)-L-proline, D-2-methyl-3-mercaptopropanoyl-L-proline, and SQ-14225, is a reversible competitive

inhibitor of LTA$_4$ hydrolase (IC$_{50}$ = 11 mM) and angiotensin I-converting enzyme, *or* peptidyl-dipeptidase A; IC$_{50}$ = 23–35 nM. The first orally active ACE inhibitor, Captopril also inhibits angiogenesis and slows the growth of experimental tumors in rats. It also inhibits apoptosis in human lung epithelial cells (IC$_{50}$ = ~320 nM). Significantly, a study of 5492 patients (aged 25-66 years; with diastolic blood pressure ≥100 mm Hg measured on two occasions) randomly assigned Captopril or conventional antihypertensive treatment (*i.e.,* diuretics or β-blockers) showed that captopril did not differ from conventional treatment in efficacy in preventing cardiovascular morbidity and mortality (72). **Target(s):** peptidyl-dipeptidase A, *or* angiotensin I-converting enzyme (1,2,9-12,14,17,18,31,33,57); leukotriene-A$_4$ hydrolase (3,66-69); invertebrate peptidyl-dipeptidase A$_4$, *or* peptidyl-dipeptidase Dcp (dipeptidyl carboxypeptidase) (5,13); tripeptide aminopeptidase (6,65); peptidyl-dipeptidase B (7,32); *Streptomyces* peptidyl-dipeptidase (8); tripeptidyl carboxypeptidase activity of peptidyl dipeptidase A (14); Xaa-Pro aminopeptidase, *or* aminopeptidase P (19,60,61,63); β-lactamase, *or* metallo-β lactamase (20,21); bontoxilysin (22-26); tentoxilysin (24,25); saccharolysin (27); nephrosin (28); meprin A (29); lysine carboxypeptidase (30); carboxypeptidase A, weakly inhibited; K_i = 620 μM (31); Xaa-Trp aminopeptidase, *or* aminopeptidase W (58,59); Xaa-Pro dipeptidase (62); glutamyl aminopeptidase, *or* aminopeptidase A, K_i = 29 μM (64); tyrosinase, *or* monophenol monooxygenase (70); catechol oxidase (70); lactoperoxidase (71); peroxidase (71). **1.** Cushman, Cheung, Sabo & Ondetti (1977) *Biochemistry* **16**, 5484. **2.** Corvol & Williams (1998) in *Handb. Proteolytic Enzymes*, p. 1066, Academic Press, San Diego. **3.** Haeggström (1998) in *Handb. Proteolytic Enzymes*, p. 1022. **4.** Isaac & Coates (1998) in *Handb. Proteolytic Enzymes*, p. 1076. **5.** Henrich & Klein (1998) *op. cit.*, p. 1121. **6.** Harada (1998) *op. cit.*, p. 1510. **7.** Barrett (1998) *op. cit.*, p. 1529. **8.** Maruyama (1998) *op. cit.*, p. 1530. **9.** Baudin & Beneteau-Burnat (1999) *J. Enzyme Inhib.* **14**, 447. **10.** Stewart, Weare & Erdös (1981) *Meth. Enzymol.* **80**, 442. **11.** Ryan (1988) *Meth. Enzymol.* **163**, 194. **12.** Corvol, Williams & Soubrier (1995) *Meth. Enzymol.* **248**, 283. **13.** Conlin & Miller (1995) *Meth. Enzymol.* **248**, 567. **14.** van Sande, Inokuchi, Nagamatsu, Scharpe, Neels & Van Camp (1985) *Urol. Int.* **40**, 100. **17.** Lanzillo & Fanburg (1977) *Biochemistry* **16**, 5491. **18.** Ondetti, Rubin & Cushman (1977) *Science* **196**, 441. **19.** Hooper, Hryszko, Oppong & Turner (1992) *Hypertension* **19**, 281. **20.** Bounaga, Galleni, Laws & Page (2001) *Bioorg. Med. Chem.* **9**, 503. **21.** Garcia-Saez, Hopkins, Papamicael, *et al.* (2003) *J. Biol. Chem.* **278**, 23868. **22.** Schiavo, Malizio, Trimble, *et al.* (1994) *J. Biol. Chem.* **269**, 20213. **23.** Schiavo, Shone, Rossetto, Alexander & Montecucco (1993) *J. Biol. Chem.* **268**, 11516. **24.** Schiavo, Benfenati, Poulain, *et al.* (1992) *Nature* **359**, 832. **25.** Montecucco & Schiavo (1994) *Mol. Microbiol.* **13**, 1. **26.** Schiavo, Rossetto, Catsicas, *et al.* (1993) *J. Biol. Chem.* **268**, 23784. **27.** Achstetter, Ehmann & Wolf (1985) *J. Biol. Chem.* **260**, 4585. **28.** Hung, Huang, Huang, Huang & Chang (1997) *J. Biol. Chem.* **272**, 13772. **29.** Sterchi, Naim, Lentze, Hauri & Fransen (1988) *Arch. Biochem. Biophys.* **265**, 105. **30.** Skidgel, Weerasinghe & Erdös (1989) *Adv. Exp. Med. Biol.* **247A**, 325. **31.** Ondetti, Condon, Reid, *et al.* (1979) *Biochemistry* **18**, 1427. **32.** Harris & Wilson (1984) *Arch. Biochem. Biophys.* **233**, 667. **33.** Kase, Hazato, Shimamura, Kiuchi & Katayama (1985) *Arch. Biochem. Biophys.* **240**, 330. **34.** Takada, K. Hiwada & T. Kokubu (1981) *J. Biochem.* **90**, 1309. **35.** Kawamura, Oda & Muramatsu (2000) *Comp. Biochem. Physiol. B* **126**, 29. **36.** Yoshida & Nosaka (1990) *J. Neurochem.* **55**, 1861. **37.** Ward & Sheridan (1982) *Biochim. Biophys. Acta* **716**, 208. **38.** Velletri, Billingsley & Lovenberg (1985) *Biochim. Biophys. Acta* **839**, 71. **39.** Kawamura, Kikuno, Oda & Muramatsu (2000) *Biosci. Biotechnol. Biochem.* **64**, 2193. **40.** Deddish, Wang, Jackman, *et al.* (1996) *J. Pharmacol. Exp. Ther.* **279**, 1582. **41.** Polanco, Miguel, Agapito & Recio (1992) *J. Endocrinol.* **132**, 261. **42.** Van Dyck, Novakova, Van Schepdael & Hoogmartens (2003) *J. Chromatogr. A* **1013**, 149. **43.** Miska, Croseck & Schill (1988) *Biol. Chem. Hoppe-Seyler* **369**, 493. **44.** Miska, Croseck & Schill (1990) *Andrologia* **22** Suppl. 1, 178. **45.** Garats, Nikolskaya, Binevski, Pozdnev & Kost (2001) *Biochemistry (Moscow)* **66**, 429. **46.** Binevski, Sizova, Pozdnev & Kost (2003) *FEBS Lett.* **550**, 84. **47.** Wei, Alhenc-Gelas, Soubrier, *et al.* (1991) *J. Biol. Chem.* **266**, 5540. **48.** Ehlers, Maeder & Kirsch (1986) *Biochim. Biophys. Acta* **883**, 361. **49.** Lanzillo, Stevens, Dasarathy, Yotsumoto & Fanburg (1985) *J. Biol. Chem.* **260**, 14938. **50.** Wei, Clauser, Alhenc-Gelas & Corvol (1992) *J. Biol. Chem.* **267**, 13398. **51.** Araujo, Melo, *et al.* (2000) *Biochemistry* **39**, 8519. **52.** Cushman, Cheung, Sabo, Rubin & Ondetti (1979) *Fed. Proc.* **38**, 2778. **53.** Harris, Ohlsson & Wilson (1981) *Anal. Biochem.* **111**, 227. **54.** Yokosawa, Endo, Ohgaki, Maeyama & Ishii (1985) *J. Biochem.* **98**, 1293. **55.** Strittmatter, Thiele, Kapiloff & Snyder

(1985) *J. Biol. Chem.* **260**, 9825. **56.** Takeuchi, Shimizu, Ohishi, *et al.* (1989) *J. Biochem.* **106**, 442. **57.** Ondetti & Cushman (1982) *Ann. Rev. Biochem.* **51**, 283. **58.** Gee & Kenny (1987) *Biochem. J.* **246**, 97. **59.** Tieku & Hooper (1992) *Biochem. Pharmacol.* **44**, 1725. **60.** Ryan, Valido, Berryer, Chung & Ripka (1992) *Biochim. Biophys. Acta* **1119**, 140. **61.** Orawski & Simmons (1995) *Biochemistry* **34**, 11227. **62.** Harbeck & Mentlein (1991) *Eur. J. Biochem.* **198**, 451. **63.** Achstetter, Ehmann & Wolf (1983) *Arch. Biochem. Biophys.* **226**, 292. **64.** Iturrioz, Rozenfeld, Michaud, Corvol & Llorens-Cortes (2001) *Biochemistry* **40**, 14440. **65.** Sachs & Marks (1982) *Biochim. Biophys. Acta* **706**, 229. **66.** Mueller, Samuelsson & Haeggström (1995) *Biochemistry* **34**, 3536. **67.** Haeggström, Kull, Rudberg, Tholander & Thunnissen (2002) *Prostaglandins* **68 69**, 495. **68.** Orning, Krivi & Fitzpatrick (1991) *J. Biol. Chem.* **266**, 1375. **69.** Thunnissen, Andersson, Samuelsson, Wong & Haeggström (2002) *FASEB J.* **16**, 1648. **70.** Kim & Uyama (2005) *Cell. Mol. Life Sci.* **62**, 1707. **71.** Bhuyan & Mugesh (2008) *Inorg. Chem.* **47**, 6569. **72.** Hansson, Lindholm, Niskanen, *et al.* (1999) *Lancet* **363**, 611.

Carbamazepine

Carbamazepine

Oxcarbamazepine

This common antiepileptic drug (AED) and neuropathic pain drug (FW = 236.27 g/mol; CASs = 298-46-4 and 85756-57-6; IUPAC Name: 5*H*-dibenzo[*b,f*]azepine-5-carboxamide; Abbreviation: CBZ (*not* Cbz for carbobenzoxy), also known as Tegretol®), binds to and stabilizes the inactivated state of voltage-gated sodium channels, making fewer channels available for subsequent opening. Carbamazepine is also a GABA receptor agonist, potentiating those made up of α_1, β_2, and γ_2 subunits. *Note*: Dangerous and potentially carbamazepine hypersensitivity, presenting as Stevens–Johnson syndrome and toxic epidermal necrolysis, occurs in carbamazepine-treated individuals possessing the human leukocyte antigen allele, HLA-B*1502. **Drug Interactions:** Carbamazepine is one of the most commonly prescribed antiepileptic drugs and is also used in the treatment of trigeminal neuralgia and psychiatric disorders, particularly bipolar depression. Because of its widespread and long term use, carbamazepine is frequently prescribed in combination with other drugs, leading to the possibility of drug interactions (1). Carbamazepine is a potent inducer of CYP3A4 and other oxidative enzyme system in the liver, and it may also increase glucuronyltransferase activity. This results in the acceleration of the metabolism of concurrently prescribed anticonvulsants, particularly valproic acid, clonazepam, ethosuximide, lamotrigine, topiramate, tiagabine and remacemide (1). The metabolism of many other drugs such as tricyclic antidepressants, antipsychotics, steroid oral contraceptives, glucocorticoids, oral anticoagulants, cyclosporin, theophylline, chemotherapeutic agents and cardiovascular drugs can also be induced, leading to a number of clinically relevant drug interactions (1). **Comparison of Carbamazepine & Oxcarbazepine Pharmacodynamics:** Oxcarbazepine (FW = 252.27 g/mol; CAS 28721-07-5; IUPAC: 10,11-dihydro-10-oxo-5*H*-dibenz(*b,f*)azepine-5-carboxamide; Symbol: OXC, also known as Trileptal®) is a so-called modern antiepileptic drug (AED) that is used as both monotherapy and adjunctive therapy for the treatment of partial seizures with or without secondary generalization in adults and children older than 4 years (USA) or 6 years (Europe) (2). Although OXC is a structural variant of carbamazepine, there are significant differences between the two drugs. The mechanism of action of OXC involves mainly blockade of sodium currents, but differs from CBZ by modulating different types of calcium channels (2). In contrast to CBZ, which is oxidized by the cytochrome P-450 system, OXC undergoes reductive metabolism at its keto moiety to form the monohydroxy derivative (MHD), which is glucuronidated and excreted in the urine. The involvement of the hepatic cytochrome P-450-dependent enzymes in the metabolism of OXC is minimal, making why it might be more effectively combined with other AEDs that show drug interactions with CBZ. Such findings indicate that OXC and CBZ are distinctly different medications (2). **Target(s):** Carbamazepine strongly attenuates lipopolysaccharide-induced production of NO and iNOS protein at concentrations of 5, 10, and 20 µM (3). Carbamazepine also directly inhibits adipocyte differentiation through activation of the ERK 1/2 pathway (4). It inhibits agonist-stimulated inositol lipid metabolism in rat hippocampus (5). CBZ increases acetylation of histone H4 in the HepG$_2$ liver carcinoma cell line by inhibiting both HDAC 3 and HDAC 7, which respectively represent class I and II histone de-acetylases respectively (6). The IC$_{50}$ of ~2 mM is considerably lower than therapeutic plasma levels of 25–50 mM typically achieved in CBZ-treated patients. **1.** Spina, Pisani & Perucca (1996) *Clin. Pharmacokinet.* **31**, 198. **2.** Schmidt & Elger (2004) *Epilepsy Behav.* **5**, 627. **3.** Wang, Hsiao, Lin, *et al.* (2014) *Pharm. Biol.* **52**, 1451. **4.** Turpin, Muscat, Vatier, *et al.* (2013) *Brit. J. Pharmacol.* **168**, 139. **5.** McDermott & Logan (1989) *Brit. J. Pharmacol.* **98**, 581. **6.** Beutler, Li, Nicol & Walsh (2005) *Life Sci.* **76**, 3107.

N-Carbamoyl-L-Glutamate

This *N*-acetyl-glutamate isostere (FW = 190.15 g/mol; CAS 1188-38-1), also known as carglumic acid and Carbaglu®, is used in the treatment of hyperammonemia resulting from primary and secondary deficiencies in *N*-acetyl-L-glutamate synthase (*or* NAGS), the mammalian mitochondrial enzyme that catalyzes acetyl transfer from Acetyl-CoA to L-glutamate (1,2). (NAGS deficiency is, in fact, the rarest Urea Cycle disorder.) **Primary Mode of Action:** *N*-Acetyl-L-glutamate is required by liver cells to allosterically activate Carbamoyl-Phosphate Synthase I (CPS-I), the first committed step in the Urea Cycle. Any impairment in the synthesis of *N*-acetyl-glutamate results in lower CPS-I activity, attended by hyperammonemia. In contrast to *N*-acetyl-glutamate, carbaglu is taken up by liver cells and readily transported into the mitochondrial compartment (3), where it can activate CPS-I. At constant urea production, carbaglu decreases the concentration of ammonia; whereas, at constant ammonia concentrations, it increases urea production (4). Carbaglu should, in principle, also be effective for treating hyperammonemia that is incidental to mutations in CPS-I that lower its affinity for *N*-acetyl-glutamate. **Human Studies:** The efficacy of Carbaglu in the treatment of hyperammonemia due to NAGS deficiency was evaluated in a retrospective review of the clinical course of 23 NAGS deficiency patients who received Carbaglu treatment for a median of 7.9 years (Range: 0.6 – 20.8 years) (1). Short-term efficacy was evaluated using mean and median change in plasma ammonia levels from baseline to days 1 to 3. Persistence of efficacy was evaluated using long-term mean and median change in plasma ammonia level. **Likely Inhibitory Targets:** As a *N*-acetyl-glutamate isostere, carbaglu is likely to inhibit those enzymes known to be inhibited by *N*-acetyl-glutamate, including *N*-formylglutamate deformylase, 4-methyleneglutaminase, glutamate carboxypeptidase II, glutamate carboxypeptidase, and glutamate formimidoyltransferase (*See N-acetyl-glutamate*). **1.** CARBAGLU (carglumic-acid) NDA 022562 by ORPHAN EUROPE. **2.** Caldovic, Morizono & Daikhin (2004) *J. Pediatr.* **145**, 552. **3.** Rubio & Grisolia (1981) *Enzyme* (Basel) **26**, 233. **4.** Meijer, Lof, Ramos & Verhoeven (1985) *Eur. J. Biochem.* **148**, 189.

Carbamoyl Phosphate

This key Urea Cycle and Pyrimidine Nucleotide Biosynthesis intermediate (FW$_{free-acid}$ = 141.02 g/mol; CAS 90-55-6), also known as carbamyl phosphate, is an important precursor of arginine and urea biosynthesis as well as pyrimidine nucleotides. It was identified as the naturally occurring carbamoyl group donor by Nobelist Lipmann and coworkers Mary Ellen Jones and Leonard Spector. Carbamoyl phosphate is a product of the reactions catalyzed by carbamoyl phosphate synthetase-I, *or* CPSI (using mitochondrial ammonia in the first committed step of the Urea Cycle), carbamoyl-phosphate synthetase-II, *or* CPSII (cytoplasmic glutamine in the first committed step in the Pyrimidine Nucleotide Pathway). It is also formed by carbamate kinase. Carbamoyl-P is a substrate for aspartate carbamoyltransferase, ornithine carbamoyltransferase, oxamate carbamoyl-transferase, putrescine carbamoyl-transferase, 3-hydroxymethylcephem carbamoyltransferase, and lysine carbamoyltransferase. It is also an alternative substrate for a number of enzymes (*e.g.*, glucose-6-phosphatase and biotin carboxylase). **Stability:** Carbamoyl-P decomposes nonenzymatically in water, with the rate of orthophosphate release dependent on pH. In acidic conditions, the products are ammonium ions, orthophosphate, and carbon dioxide. In basic conditions the products are orthophosphate, water, and cyanate. pH-rate profile studies and use of isotopic probes indicate that different hydrolytic mechanisms operate for the monoanion and dianion forms, resulting in P–O and C–O bond cleavage, respectively (See Ref. 9). **Target(s):** glutamine synthetase (1,2,7,8,13,20-22,25); pyruvate kinase (3); acylphosphatase (4); nitrate reductase (5); threonine dehydratase (6,18); nitrogenase (10,16; however, see 17); formyltetrahydrofolate synthetase (11); hemoglobin S polymerization (12); carbonic anhydrase (14); alanine aminotransferase (18); glutamate dehydrogenase (19); asparate aminotransferase (18); threonine deaminase (18); asparagine synthetase (23); adenylosuccinate synthetase (24); ornithine decarboxylase, inhibited at elevated concentrations (26). **1.** Meister (1985) *Meth. Enzymol.* **113**, 185. **2.** Rhee, Chock & Stadtman (1985) *Meth. Enzymol.* **113**, 213. **3.** Tuominen & Bernlohr (1975) *Meth. Enzymol.* **42**, 157. **4.** Ramponi (1975) *Meth. Enzymol.* **42**, 409. **5.** Rigano & Aliotta (1975) *Biochim. Biophys. Acta* **384**, 37. **6.** Pagani, Ponticelli, Terzuoli, Leoncini & Marinello (1991) *Biochim. Biophys. Acta* **1077**, 233. **7.** Meister (1974) *The Enzymes*, 3rd ed. (Boyer, ed.), **10**, 699. **8.** Stadtman & Ginsburg (1974) *The Enzymes*, 3rd ed. (Boyer, ed.), **10**, 755. **9.** Allen & Jones (1964) *Biochemistry* **3**, 1238. **10.** Seto & Mortenson (1974) *J. Bacteriol.* **117**, 805. **11.** Buttlaire, Balfe, Wendland & Himes (1979) *Biochim. Biophys. Acta* **567**, 453. **12.** Kraus & Kraus (1971) *Biochem. Biophys. Res. Commun.* **44**, 1381. **13.** Dahlquist & Purich (1975) *Biochemistry* **14**, 1980. **14.** Rusconi, Innocenti, Vullo, *et al.* (2004) *Bioorg. Med. Chem. Lett.* **14**, 5763. **15.** Jones, Spector & Lipmann (1955) *J. Amer. Chem. Soc.* **77**, 819. **16.** Lawrie (1979) *J. Bacteriol.* **139**, 115. **17.** Gordon, Shah & Brill (1981) *J. Bacteriol.* **148**, 884. **18.** Pagani, Leoncini, Vannoni, *et al.* (1994) *Life Sci.* **54**, 775. **19.** Veronese, Piszkiewicz & Smith (1972) *J. Biol. Chem.* **247**, 754. **20.** Pahuja & Reid (1985) *Exp. Eye Res.* **40**, 75. **21.** Tate & Meister (1971) *Proc. Natl. Acad. Sci. U.S.A.* **68**, 781. **22.** Kapoor & Bray (1968) *Biochemistry* **7**, 3583. **23.** Hongo, Matsumoto & Sato (1978) *Biochim. Biophys. Acta* **522**, 258. **24.** Ogawa, Shiraki, Matsuda, Kakiuchi & Nakagawa (1977) *J. Biochem.* **81**, 859. **25.** Singh & Singh (1990) *Arch. Int. Physiol. Biochim.* **98**, 95. **26.** Yarlett, Goldberg, Moharrami & Bacchi (1993) *Biochem. J.* **293**, 487.

Carba-NAD⁺

This NAD⁺ analogue (FW$_{free-acid}$ = 662.47 g/mol), also known as carbanicotinamide adenine dinucleotide, has a D-2,3-dihydroxycyclopentyl moiety in placing of the D-ribosyl residue in NAD⁺ and inhibits NADase. Carba-NAD⁺ is reduced to Carba-NADH by yeast and horse liver alcohol dehydrogenase. **Target(s):** NAD⁺ Nucleosidase, *or* NADase, *or* NAD⁺ glycohydrolase (1-3,5); cholera toxin ADPribosyltransferase (3); poly(ADP-ribose) polymerase (4). **1.** Oppenheimer & Handlon (1992) *The Enzymes*, 3rd ed., **20**, 453. **2.** Slama & Simmons (1988) *Biochemistry* **27**, 183. **3.** Slama & Simmons (1989) *Biochemistry* **28**, 7688. **4.** Ruf, Rolli, de Murcia & Schulz (1998) *J. Mol. Biol.* **278**, 57. **5.** Muller-Steffner, Slama & Schuber (1996) *Biochem. Biophys. Res. Commun.* **228**, 128.

Carbenicillin

This semisynthetic penicillin-based antibiotic (FW = 378.41 g/mol; CAS 4697-36-3), also known as α-phenyl(carboxymethylpenicillin), inhibits peptidoglycan biosynthesis. While the free acid is sparingly soluble in water, sodium and potassium salts are very soluble. Although more resistant than ampicillin to degradation by β-lactamases, carboxypenicillins are still susceptible. **Target(s):** neprilysin, *or* enkephalinase (1,6); plasmin, weakly inhibited (2); urokinase (3); serine type D-Ala-D-Ala carboxypeptidase (4,5,10,11); β-lactamase, *or* cephalosporinase (7,8); muramoylpentapeptide carboxypeptidase (9). **1.** Livingston, Smith, Sewell & Ahmed (1992) *J. Enzyme Inhib.* **6**, 165. **2.** Higazi & Mayer (1989) *Biochem. J.* **260**, 609. **3.** Higazi &Mayer (1988) *Thromb. Haemost.* **60**, 305. **4.** Martin, Schilf & Maskos (1976) *Eur. J. Biochem.* **71**, 585. **5.** Frère, Ghuysen & Perkins (1975) *Eur. J. Biochem.* **57**, 353. **6.** Williams, Sewell, Smith & Gonzalez (1989) *J. Enzyme Inhib.* **3**, 91. **7.** Toda, Inoue & Mitsuhashi (1981) *J. Antibiot.* **34**, 1469. **8.** Power, Galleni, Ayala & Gutkind (2006) *Antimicrob. Agents Chemother.* **50**, 962. **9.** Diaz-Mauriño, Nieto & Perkins (1974) *Biochem. J.* **143**, 391. **10.** Coyette, Ghuysen & Fontana (1978) *Eur. J. Biochem.* **88**, 297. **11.** Marquet, Nieto & Diaz-Mauriño (1976) *Eur. J. Biochem.* **68**, 581.

Carbenoxolone

This anti-inflammatory glucocorticoid (FW$_{free-acid}$ = 570.77 g/mol; CASs 5697-56-3 (free acid), 74203-92-2 (dicholine salt), and 7421-40-1 (disodium salt)), also known as (3β,20β)-3-(3-carboxy-1-oxopropoxy)-11-oxoolean-12-en-29-oic acid and 18β-glycyrrhetic acid hydrogen succinate, is a water-soluble derivative of glycyrrhetinic acid. **Target(s):** 11β-hydroxysteroid dehydrogenase type (11,2,8,9,12,13,16,20); pepsin (3,4,7); cAMP phosphodiesterase (5); cGMP phosphodiesterase (5); 15-hydroxyprostaglandin dehydrogenase (6,7); Δ¹³-prostaglandin reductase (6); trypsin (7); chymotrypsin (7); elastase (7); Δ⁵-reductase (7); 3α-hydroxysteroid dihydro-genase (10); 3α– (or, 20β–) hydroxysteroid dehydrogenase (11,14,20); *cis*-retinol/androgen dehydrogenase (15,19); linoleate Desaturase (17); 7α-hydroxysteroid dehydrogenase (18); Ca²⁺ channels, voltage-gated (21); DNA polymerase β (22). **1.** Robinzon & Prough (2005) *Arch. Biochem. Biophys.* **442**, 33. **2.** Sandeep, Andrew, Homer, *et al.* (2005) *Diabetes* **54**, 872. **3.** Roberts & Taylor (1973) *Clin. Sci.* **44**, 6P. **4.** Roberts & Taylor (1973) *Clin. Sci. Mol. Med. Suppl.* **42**, 213. **5.** Vapaatalo, Linden, Metsa-Ketela, Kangasaho & Laustiola (1978) *Experientia* **34**, 384. **6.** Peskar (1980) *Scand. J. Gastroenterol. Suppl.* **65**, 109. **7.** Roberts & Taylor (1980) *Scand. J. Gastroenterol. Suppl.* **65**, 11. **8.** Monder, Stewart, Lakshmi, Valentino, Burt & Edwards (1989) *Endocrinology* **125**, 1046. **9.** Mercer & Krozowski (1992) *Endocrinology* **130**, 540. **10.** Akao, Akao, Hattori, Namba & Kobashi (1992) *Chem. Pharm. Bull. (Tokyo)* **40**, 1208. **11.** Ghosh, Erman, Pangborn, Duax & Baker (1992) *J. Steroid Biochem. Mol. Biol.* **42**, 849. **12.** Jellinck, Monder, McEwen & Sakai (1993) *J. Steroid Biochem. Mol. Biol.* **46**, 209. **13.** Yang & Yu (1994) *J. Steroid Biochem. Mol. Biol.* **49**, 245. **14.** Ghosh, Erman, Wawrzak, Duax & Pangborn (1994) *Structure* **2**, 973. **15.** Boerman & Napoli (1995) *Biochemistry* **34**, 7027. **16.** Reeves (1995) *Amer. J. Physiol.* **268**, C1467. **17.** Norman, Pillai & Baker (1995) *FEBS Lett.* **368**, 135. **18.**

Song, Chen, Dean, Redinger & Prough (1998) *J. Biol. Chem.* **273**, 16223. **19.** Su, Chai, Kahn & Napoli (1998) *J. Biol. Chem.* **273**, 17910. **20.** Duax, Ghosh & Pletnev (2000) *Vitam. Horm.* **58**, 121. **21.** Vessey, Lalonde, Mizan, *et al.* (2004) *J. Neurophysiol.* **92**, 1252. **22.** Hu, Horton, Gryk, *et al.* (2004) *J. Biol. Chem.* **279**, 39736.

Carbidopa

This nonmetabolizable L-DOPA isostere and Parkinson drug (FW = 225.22 g/mol; CAS 28860-95-9), also named (2S)-3-(3,4-dihydroxyphenyl)-2-hydrazino-2-methylpropanoic acid, inhibits aromatic-L-amino-acid decarboxylase (*or* DOPA decarboxylase), a required enzyme in both tryptophan to serotonin biosynthesis as well as in the conversion of L-DOPA to Dopamine (DA). Although L-DOPA can cross the blood brain barrier, dopamine cannot. By preventing peripheral DDC catalysis of L-DOPA conversion to dopamine, carbidopa can be used to treat Parkinson Disease (PD), a condition arising from dopaminergic neuron death located within the substantia nigra. It thus enhances entry of L-DOPA into dopamine-deficient regions of PD brains. Blood-brain barrier to carbidopa (MK-486) and Ro 4-4602, peripheral dopa decarboxylase inhibitors. **Key Pharmacokinetic Parameters:** *See* Appendix II in Goodman & Gilman's *THE PHARMACOLOGICAL BASIS OF THERAPEUTICS*, 12th Edition (Brunton, Chabner & Knollmann, eds.) McGraw-Hill Medical, New York (2011). **1.** Clark, Oldendorf & Dewherst (1973) *J. Pharm. Pharmacol.* **25**, 416. **2.** Lieberman, Derby, Feigenson *et al.* (1973) *Dis. Nerv. Syst.* **34**,167.

Carbinoxamine

This antihistamine (FW = 290.79 g/mol; CAS 486-16-8), also known by its trade names Clistin®, Palgic®, Rondec®, and Rhinopront®, and Rotoxamine® (the latter as the maleate salt of the (−)-enantioner), as well as systematically as 2-[(4-chlorophenyl)-pyridin-2-ylmethoxy]-*N,N*-dimethylethanamine, targets histidine H_1 receptors. **1.** Garat, Landa, Richeri & Tracchia (1956) *J. Allergy* **27**, 57. **2.** Barouh, Dall, Patel & Hite (1971) *J. Med. Chem.* **14**, 834. **3.** Oishi, Shishido, Yamori & Saeki (1994) *Naunyn. Schmiedebergs Arch. Pharmacol.* **349**, 140.

Carbon Monoxide

This colorless and odorless metabolic signaling gas (FW = 28.01 g/mol; CAS 630-08-0; M.P. = −199°C; B.P. = −191.5°C; bond length = 0.113 nm; shown above are contributing resonance forms) binds to and inhibits a wide range of metalloproteins, particularly heme proteins and enzymes, displaying very high affinity for hemoglobin (Hb), myoglobin, neuroglobin, cytochrome *c* oxidase, and CYP's. Carbon monoxide's linear diatomic structure allows its vacant *p*-orbitals to form an end-on coordinate bond with suitable metal ions, especially Fe(II) coordinated to heme.

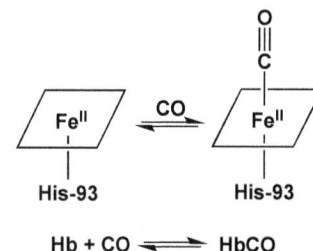

$$Hb + CO \rightleftharpoons HbCO$$

Carbon monoxide binding to heme proteins is also photolabile, and excitation of HbCO by a single 6-psec pulse of 530-nm light results in prompt photodissociation (first-order constant = 0.9×10^{11} sec^{-1}). That heme-bound carbon monoxide photodissociates has proven to be an important tool in the characterization of heme proteins and their involvement in electron transport chains that create the proton motive force driving both photophosphorylation and oxidative phosphorylation. **Binding to Hemoglobin:** With >250-300 times greater affinity than diatomic oxygen for adult hemoglobin (HbA), carbon monoxide can fully displace hemoglobin-bound O_2. At sub-saturating levels, however, carbon monoxide has the seemingly paradoxical effect of greatly increasing hemoglobin's apparent affinity for O_2. This property, which is anticipated in the Monod-Wyman-Changeux concerted transition model for allostery, explains Claude Bernard's observation of substantial levels of hemoglobin-bound O_2 in the veinous circulation of individuals exposed to subsaturating CO. In the concerted transition model, hemoglobin exists in an all-or-nothing dynamic equilibrium between ligand-free R_0 and T_0 states, as defined by the allosteric constant $\mathcal{L}$, equal to $[T_0]/[R_0] \approx 700$-800. O_2 binds preferentially to R_0, generating R_1 (*i.e.*, $(Hb(O_2)_1)$, which then binds a second, third and fourth O_2 molecule to form R_2 (*i.e.*, $(Hb(O_2)_2)$, then R_3 (*i.e.*, $(Hb(O_2)_3)$, and finally R_4 (*i.e.*, $(Hb(O_2)_4)$. Note further that the affinity of R_1, R_2, R_3, and R_1 for O_2 are all equivalent, and cooperative binding only occurs, because T_0 is far more stable than R_0. This property indicates that hemoglobin resists oxygenation, except at high O_2 pressure within the lung (where pO_2 is ~100 mm Hg) and that hemoglobin undergoes substantial deoxygenation in peripheral circulation (where pO_2 is ~30 mm Hg). This results in an S-shaped O_2 saturation curve, where little binding occurs at low O_2 pressure. The inflection point (or "tipping point") is defined O_2 partial pressure needed for $\sum[R_i]/\sum[T_i] = 1$. This condition is attained when the oxygen partial pressure high enough to fully form $Hb(O_2)_1$ and partially form $Hb(O_2)_2$. However, because CO binds with much higher affinity than O_2, binding of a single CO molecule drives hemoglobin into the R state, increasing the the concentration of ligand-free subunits that are available for O_2 binding. **Role in Metabolic Signaling:** CO also plays an significant role as a cellular messenger in mammals in a manner akin to nitric oxide, and CO is a likely alternative activator for guanylate cyclase. Heme oxygenase and 3,4-dihydroxyquinoline 2,4-dioxygenase catalyze reactions that produce CO as a second messenger for CO-activated guanylate cyclases, thereby abetting 3'5'-cyclicGMP synthesis. Enzymes utilizing CO as a substrate include carbon monoxide oxygenase (or, cytochrome *b*-561), carbon monoxide oxidase, and carbon monoxide dehydrogenase. Neurons contain neuroglobin, a myoglobin-like CO-binding protein that is thought to play a role in modulating CO-dependent processes. **Target(s):** tryptophan 2,3-dioxygenase (1,8,9,137); catechol oxidase (2); hydrogenase (3,20,28,29,33); steroid 21-monooxygenase (4,37); cytochrome *c* oxidase, *or* Complex IV (5,6,17,18,27, 32,36,45-47,51,165-172); steroid 11b-monooxygenase (7,37,55,83-86); corticosterone 18 monooxygenase (7); steroid 21-monooxygenase (7); sulfite reductase (10,35); sulfite reductase (NADPH) (11,35); dopamine β-monooxygenase (12,34,82); mitochondrial electron transport (5,27); photosynthetic electron transport (13); nitrogenase (14,15,38,58-64,66); heme oxygenase (16); cytochrome a_3 (19,32); cytochrome a_1 (19); cytochrome *o* (19); cytochrome *d* (19); cytochrome *c* (19); cytochrome *c'* (19); cytochrome P450 systems (19,34,37,40,41); ferredoxin:nitrite reductase (21); calcidiol 1-monooxygenase, *or* 25-hydroxyvitamin D_1 α-hydroxylase (22,123); leukotriene B_4 20-hydroxylase (24); cholesterol 7α-monooxygenase (25); hemoglobin (26,52); lipoxygenase (31); nitrite reductase (*e.g.*, from *Achromobacter fischeri*) (35); cholesterol monooxygenase, side-chain-cleaving, *or* CYP11A1 (37,56); steroid 17α-monooxygenase (37); cystathionine *b* synthase (39); flavin-containing monooxygenase, *or* dimethylaniline-*N*-oxide aldolase (40,41); lanosterol 14 α-demethylase (42); nitrogen fixation (43); nitric oxide synthase (44); catalase (48); carbonic anhydrase (49,50); formate hydrogen-lyase (53); aliphatic aldoxime dehydratase (54); cystathionine *b* synthase (57); coenzyme-B sulfoethylthiotransferase, *or* methyl-CoM reductase (65); taxadiene 5α-hydroxylase (67); linalool 8-monooxygenase (68); ecdysone 20-monooxygenase (69-72); progesterone 11a-monooxygenase (73); steroid 21-monooxygenase, *or* CYP21A1 (74,75); steroid 17a monooxygenase (76); heme oxygenase (77,78); (*S*)-canadine synthase (79); salutaridine synthase (80); tyrosinase, *or* monophenol monooxygenase (23,30,81); alkane 1-monooxygenase (87,88); abietadiene hydroxylase (89); abietadienol hydroxylase (89); 8-dimethylallyl-naringenin 2'-hydroxylase (90); psoralen synthase (91); (+)-abscisate 8'-hydroxylase (92); deoxysarpagine hydroxylase (93); flavonoid 3',5' hydroxylase (94); 2-hydroxyisoflavanone synthase (95); glyceollin synthase (96); ent-kaurene oxidase (97); vinorine hydroxylase (98); 7-deoxyloganin 7-hydroxylase (99); tabersonine 16-hydroxylase (100); sterol 14-demethylase, *or* CYP51 (101-103); dihydrochelirubine 12-monooxygenase (104); dihydrosanguinarine 10-monooxygenase (105); protopine 6-monooxygenase (106); 4'-methoxyisoflavone 2'-hydroxylase (107); isoflavone 3'-hydroxylase (107); (*S*)-limonene 3-monooxygenase (108); (*S*)-limonene 6-monooxygenase (108); (*S*)-limonene 7-monooxygenase (108); nitric-oxide synthase (109-111); methyltetrahydro-protoberberine 14-monooxygenase (112); leukotriene B$_4$ 20-monooxygenase (113 115); flavonoid 3'-monooxygenase (116-118); cholestanetriol 26-monooxygenase (119-122); calcidiol 1 monooxygenase (124,125); benzoate 4-monooxygenase (126,127); trans-cinnamate 4-mono-oxygenase (128); nitric-oxide dioxygenase (129,130); deacetoxyvindoline 4-hydroxylase (131); tryptophan 2' dioxygenase, weakly inhibited (132,133); indoleamine 2,3-dioxygenase (134,135); indole 2,3 dioxygenase (136); hydrogenase (acceptor) (138-147); 5,10-methenyl-tetrahydromethanopterin hydrogenase (148,149); coenzyme F420 hydrogenase (150); ferredoxin hydrogenase (142,151-156); cytochrome c_3 hydrogenase (157-160); hydrogen dehydrogenase (NADP$^+$) (161); L-ascorbate peroxidase (162); cytochrome *c* peroxidase (163); fatty-acid peroxidase (164).

1. Knox (1955) *Meth. Enzymol.* **2**, 242. **2.** Dawson & Magee (1955) *Meth. Enzymol.* **2**, 817. **3.** San Pietro (1955) *Meth. Enzymol.* **2**, 861. **4.** Hayano & Dorfman (1962) *Meth. Enzymol.* **5**, 503. **5.** Slater (1967) *Meth. Enzymol.* **10**, 48. **6.** Wharton & Tzagoloff (1967) *Meth. Enzymol.* **10**, 245. **7.** Rosenthal & Narasimhulu (1969) *Meth. Enzymol.* **15**, 596. **8.** Ishimura (1970) *Meth. Enzymol.* **17A**, 429. **9.** Yamamoto & Hayaishi (1970) *Meth. Enzymol.* **17A**, 434. **10.** Asada, Tamura & Bandurski (1971) *Meth. Enzymol.* **17B**, 528. **11.** Siegel & Kamin (1971) *Meth. Enzymol.* **17B**, 539. **12.** Kaufman (1971) *Meth. Enzymol.* **17B**, 754. **13.** Izawa & Good (1972) *Meth. Enzymol.* **24**, 355. **14.** Bulen & LeComte (1972) *Meth. Enzymol.* **24**, 456. **15.** Burns & Hardy (1972) *Meth. Enzymol.* **24**, 480. **16.** Schacter (1978) *Meth. Enzymol.* **52**, 367. **17.** Errede, Kamen & Hatefi (1978) *Meth. Enzymol.* **53**, 40. **18.** Wilson & Erecinska (1978) *Meth. Enzymol.* **53**, 191. **19.** Smith (1978) *Meth. Enzymol.* **53**, 202. **20.** Mortenson (1978) *Meth. Enzymol.* **53**, 286. **21.** Vega, Cárdenas & Losada (1980) *Meth. Enzymol.* **69**, 255. **22.** Lobaugh, Almond & Drezner (1986) *Meth. Enzymol.* **123**, 159. **23.** Lerch (1987) *Meth. Enzymol.* **142**, 165. **24.** Soberman & Okita (1988) *Meth. Enzymol.* **163**, 349. **25.** Chiang (1991) *Meth. Enzymol.* **206**, 483. **26.** Mathews & Olson (1994) *Meth. Enzymol.* **232**, 363. **27.** Rapoport & Schewe (1977) *Trends Biochem. Sci.* **2**, 186. **28.** Mayhew & O'Connor (1982) *Trends Biochem. Sci.* **7**, 18. **29.** Umbreit (1951) *The Enzymes*, 1st ed, **2** (Part 1), 329. **30.** Dawson & Tarpley (1951) *The Enzymes*, 1st ed., **2** (Part 1), 454. **31.** Holman & Bergström (1951) The Enzymes, 1st ed. **2** (Part 1), 559. **32.** Yonetani (1963) *The Enzymes*, 2nd ed., **8**, 41. **33.** Purec, Krasna & Rittenberg (1962) *Biochemistry* **1**, 270. **34.** Ullrich & Duppel (1975) *The Enzymes*, 3rd ed., **12**, 253. **35.** Hatefi & Stiggall (1976) *The Enzymes*, 3rd ed., **13**, 175. **36.** Caughey, Wallace, Volpe & Yoshikawa (1976) *The Enzymes*, 3rd ed., **13**, 299. **37.** Coon & Koop (1983) *The Enzymes*, 3rd ed., **16**, 645. **38.** Webb (1966) *Enzyme and Metabolic Inhibitors*, **2**, 292, Academic Press, New York. **39.** Taoka, West & Banerjee (1999) *Biochemistry* **38**, 2738. **40.** Ziegler & Pettit (1966) *Biochemistry* **5**, 2932. **41.** Machinist, Orme-Johnson & Ziegler (1966) *Biochemistry* **5**, 2939. **42.** Mitropoulos, Gibbons & Reeves (1976) *Steroids* **27**, 821. **43.** Lind & Wilson (1942) *Arch. Biochem.* **1**, 59. **44.** White & Marletta (1992) Biochemistry 31, 6627. **45.** Altschul, Abrams & Hogness (1939) *J. Biol. Chem.* **130**, 427. **46.** Haas (1943) *J. Biol. Chem.* **148**, 481. **47.** Wainio, Cooperstein, Kollen & Eichel (1948) *J. Biol. Chem.* **173**, 145. **48.** Malowan (1938) *Enzymologia* **5**, 89. **49.** van Goor (1948) *Enzymologia* **13**, 73. **50.** Kiese & Hastings (1940) *J. Biol. Chem.* **132**, 281. **51.** Ball, Strittmatter & Cooper (1951) *J. Biol. Chem.* **193**, 635. **52.** Hill (1913) *Biochem. J.* **7**, 471. **53.** Stephenson & Stickland (1932) *Biochem. J.* **26**, 712. **54.** Oinuma, Hashimoto, Konishi, *et al.* (2003) *J. Biol. Chem.* **278**, 29600. **55.** Wilson & Harding (1970) *Biochemistry* **9**, 1615. **56.** Wilson & Harding (1970) *Biochemistry* **9**, 1621. **57.** Taoka, West & Banerjee (1999) *Biochemistry* **38**, 2738. **58.** Seefeldt, Dance & Dean (2004) *Biochemistry* **43**, 1401. **59.** Igarashi & Seefeldt (2003) *Crit. Rev. Biochem. Mol. Biol.* **38**, 351. **60.** Mortenson & Thorneley (1979) *Ann. Rev. Biochem.* **48**, 387. **61.** Christiansen, Cash, Seefeldt & Dean (2000) *J. Biol. Chem.* **275**, 11459. **62.** Shah, Ugalde, Imperial & Brill (1984) *Ann. Rev. Biochem.* 53, 231. **63.** Fisher, Dilworth, Kim & Newton (2000) *Biochemistry* **39**, 10855. **64.** Kim, Newton & Dean (1995) *Biochemistry* **34**, 2798. **65.** Gunsalus & Wolfe (1980) *J. Biol. Chem.* **255**, 1891. **66.** Benton, Laryukhin, Mayer, *et al.* (2003) *Biochemistry* **42**, 9102. **67.** Hefner, Rubenstein, Ketchum, *et al.* (1996) *Chem. Biol.* **3**, 479. **68.** Ullah, Murray, Bhattacharyya, Wagner & Gunsalus (1990) *J. Biol. Chem.* **265**, 1345. **69.** Feyereisen & Durst (1978) *Eur. J. Biochem.* **88**, 37. **70.** Greenwood & Rees (1984) *Biochem. J.* **223**, 837. **71.** Johnson & Rees (1977) *Biochem. J.* **168**, 513. **72.** Grebenok, Galbraith, Benveniste & Feyereisen (1996) *Phytochemistry* **42**, 927. **73.** Jayanthi, Madyastha & Madyastha (1982) *Biochem. Biophys. Res. Commun.* **106**, 1262. **74.** Hiwatashi & Ichikawa (1981) *Biochim. Biophys. Acta* **664**, 33. **75.** Ryan & Engel (1957) *J. Biol. Chem.* **225**, 103. **76.** Nakajin, Shively, Yuan & Hall (1981) Biochemistry 20, 4027. **77.** Kutty & Maines (1982) *J. Biol. Chem.* **257**, 9944. **78.** Trakshel, Kutty & Maines (1986) *J. Biol. Chem.* **261**, 11131. **79.** Rueffer (1993) *Chem. Listy* **87**, 215. **80.** Gerardy & Zenk (1993) *Phytochemistry* **32**, 79. **81.** McIntyre & Vaughan (1975) *Biochem. J.* **149**, 447. **82.** Friedman & Kaufman (1965) *J. Biol. Chem.* **240**, 4763. **83.** Watanucki, Tilley & Hall (1978) *Biochemistry* **17**, 127. **84.** Watanucki, Tilley & Hall (1977) *Biochim. Biophys. Acta* **483**, 236. **85.** Sato, Ashida, Suhara, *et al.* (1978) *Arch. Biochem. Biophys.* 190, 307. **86.** Suzuki, Sanga, Chikaoka & Itagaki (1993) *Biochim. Biophys. Acta* **1203**, 215. **87.** Cardini & Jurtshul (1970) *J. Biol. Chem.* 245, 2789. **88.** Nakayama & Shoun (1994) *Biochem. Biophys. Res. Commun.* **202**, 586. **89.** Funk & Croteau (1994) Arch. Biochem. Biophys. 308, 258. **90.** Yamamoto, Yatou & Inoue (2001) *Phytochemistry* **58**, 651. **91.** Wendorff & Matern (1986) *Eur. J. Biochem.* **161**, 391. **92.** Krochko, Abrams, Loewen, Abrams & Cutler (1998) *Plant Physiol.* **118**, 849. **93.** Yu, Ruppert & Stöckigt (2002) *Bioorg. Med. Chem.* **10**, 2479. **94.** Menting, Scopes & Stevenson (1994) *Plant Physiol.* **106**, 633. **95.** Kochs & Grisebach (1986) *Eur. J. Biochem.* **115**, 311. **96.** Welle & Grisebach (1988) *Arch. Biochem. Biophys.* **263**, 191. **97.** Ashman, Mackenzie & Bramley (1990) *Biochim. Biophys. Acta* **1036**, 151. **98.** Falkenhagen & Stöckigt (1995) *Z. Naturforsch.* C 50, 45. **99.** Katano, Yamamoto, Iio & Inoue (2001) *Phytochemistry* 58, 53. **100.** St-Pierre & De Luca (1995) *Plant Physiol.* **109**, 131. **101.** Aoyama, Yoshida & Sato (1984) *J. Biol. Chem.* **259**, 1661. **102.** Aoyama, Horiuchi & Yoshida (1996) *J. Biochem.* **120**, 982. **103.** Kahn, Bak, Olsen, Svendsen & Møller (1996) *J. Biol. Chem.* **271**, 32944. **104.** Kammerer, De-Eknamkul & Zenk (1994) *Phytochemistry* **36**, 1409. **105.** De-Eknamkul, Tanahashi & Zenk (1992) *Phytochemistry* **31**, 2713. **106.** Tanahashi & Zenk (1990) *Phytochemistry* **29**, 1113. **107.** Clemens, Hinderer, Wittkampf & Barz (1993) *Phytochemistry* **32**, 653. **108.** Karp, Mihaliak, Harris & Croteau (1990) *Arch. Biochem. Biophys.* **276**, 219. **109.** Stuehr & Ikeda-Saito (1992) *J. Biol. Chem.* **267**, 20547. **110.** White & Marletta (1992) *Biochemistry* **31**, 6627. **111.** Rusche, Spiering & Marletta (1998) *Biochemistry* **37**, 15503. **112.** Rueffer & Zenk (1987) *Tetrahedron Lett.* **28**, 5307. **113.** Sumimoto, Takeshige & Minakami (1988) *Eur. J. Biochem.* **172**, 315. **114.** Kikuta, Kusunose, Sumimoto, *et al.* (1998) *Arch. Biochem. Biophys.* **355**, 201. **115.** Sumimoto, Kusunose, Gotoh, Kusunose & Minakami (1990) *J. Biochem.* **108**, 215. **116.** Larson & Bussard (1986) *Plant Physiol.* **80**, 483. **117.** Doostdar, Shapiro, Niedz, *et al.* (1995) *Plant Cell Physiol.* **36**, 69. **118.** Hagmann, Heller & Grisebach (1983) *Eur. J. Biochem.* **134**, 547. **119.** Okuda, Masumoto & Ohyama (1988) *J. Biol. Chem.* **263**, 18138. **120.** Sato, Atsuta, Imai, Taniguchi & Okuda (1977) *Proc. Natl. Acad. Sci. U.S.A.* **74**, 5477. **121.** Taniguchi, Hoshita & Okuda (1973) *Eur. J. Biochem.* **40**, 607. **122.** Okuda, Weber & Ullrich (1977) *Biochem. Biophys. Res. Commun.* **74**, 1071. **123.** Hagenfeldt, Pedersen & Björkhem (1988) *Biochem. J.* **250**, 521. **124.** Gray, Omdahl, Ghazarian & DeLuca (1972) *J. Biol. Chem.* **247**, 7528. **125.** Sakaki, Sawada, Takeyama, Kato & Inouye (1999) *Eur. J. Biochem.* **259**, 731. **126.** McNamee & Durham (1985) *Biochem. Biophys. Res. Commun.* **129**, 485. **127.** Fukuda, Nakamura, Sukita, Ogawa & Fujii (1996) *J. Biochem.* **119**, 314. **128.** Billett & Smith (1978) *Phytochemistry* **17**, 1511. **129.** Mowat, Gazur, Campbell & Chapman (2010) *Arch. Biochem. Biophys.* **493**, 37. **130.** Gardner (2005) *J. Inorg. Biochem.* **99**, 247. **131.** De Carolis & De Luca (1993*) J. Biol. Chem.* **268**, 5504. **132.** Takai & O. Hayaishi (1987) *Meth. Enzymol.* **142**, 195. **133.** Takai, Ushiro, Noda, *et al.* (1977) *J. Biol. Chem.* **252**, 2638. **134.** Sono (1989) *Biochemistry* **28**, 5400. **135.** Sono (1990) *Biochemistry* **29**, 1451. **136.** Kunapuli & Vaidyanathan (1983) *Plant Physiol.* **71**, 19. **137.**

Matsumura, Osada & Aiba (1984) *Biochim. Biophys. Acta* **786**, 9. **138.** Zadvorny, Zorin, Gogotov & Gorlenko (2004) *Biochemistry* (Moscow) **69**, 164. **139.** Lamle, Albracht & Armstrong (2005) *J. Amer. Chem. Soc.* **127**, 6595. **140.** Long, Liu, Chen, *et al.* (2007) *J. Biol. Inorg. Chem.* **12**, 62. **141.** George, Kurkin, Thorneley & Albracht (2004) *Biochemistry* **43**, 6808. **142.** Ghirardi, Posewitz, Maness, *et al.* (2007) *Ann. Rev. Plant Biol.* **58**, 71. **143.** Serebryakova, Medina, Zorin, Gogotov & Cammack (1996) *FEBS Lett.* **383**, 79. **144.** Adams (1987) *J. Biol. Chem.* **262**, 15054. **145.** Hatchikian, Forget, Fernandez, Williams & Cammack (1992) *Eur. J. Biochem.* **209**, 357. **146.** Shima, Pilak, Vogt, *et al.* (2008) *Science* **321**, 572. **147.** Meuer, Bartoschek, Koch, Künkel & Hedderich (1999) *Eur. J. Biochem.* **265**, 325. **148.** Lyon, Shima, Buurman, *et al.* (2004) *Eur. J. Biochem.* **271**, 195. **149.** Lyon, Shima, Boecher, *et al.* (2004) *J. Amer. Chem. Soc.* **126**, 14239. **150.** Livingston, Fox, Orme-Johnson & Walsh (1987) *Biochemistry* **26**, 4228. **151.** Adams & Mortenson (1984) *J. Biol. Chem.* **259**, 7045. **152.** Chen & Blanchard (1978) *Biochem. Biophys. Res. Commun.* **84**, 1144. **153.** Thauer, Käufer, Zähringer & Jungermann (1974) *Eur. J. Biochem.* **42**, 447. **154.** Erbes & Burris (1978) *Biochim. Biophys. Acta* **525**, 45. **155.** Pereira, Tavares, Moura, Moura & Huynh (2001) *J. Amer. Chem. Soc.* **123**, 2771. **156.** Kemner & Zelkus (1994) *Arch. Microbiol.* **161**, 47. **157.** Lalla-Maharajh, Hall, Cammack, Rao & Le Gall (1983) *Biochem. J.* **209**, 445. **158.** Yagi, Honya & Tamiya (1968) *Biochim. Biophys. Acta* **153**, 699. **159.** Odom & Peck (1984) *Ann. Rev. Microbiol.* **38**, 551. **160.** Yagi, Kimura, Daidoji, Sakai & Tamura (1976) *J. Biochem.* **79**, 661. **161.** van Haaster, Silva, Hagedoom, Jongejam & Hagen (2008) *J. Bacteriol.* **190**, 1584. **162.** Dalton, Hanus, Russell & Evans (1987) *Plant Physiol.* **83**, 789. **163.** Tano, Sakai, Sugio & Imai (1977) *Agric. Biol. Chem.* **41**, 323. **164.** Martin & Stumpf (1959) *J. Biol. Chem.* **234**, 2548. **165.** Wastyn, Achatz, Molitor & Peschek (1988) *Biochim. Biophys. Acta* **935**, 217. **166.** Nichols & Chance (1974) in *Mol. Mech. Oxygen Activ.* (Hayaishi, ed.), p. 479, Academic Press, New York. **167.** Anemüller & Schäfer (1990) *Eur. J. Biochem.* **191**, 297. **168.** Collman, Dey, Barile, Ghosh & Decréau (2009) *Inorg. Chem.* **48**, 10528. **169.** Cooper, Davies, Psychoulis, *et al.* (2003) *Biochim. Biophys. Acta* **1607**, 27. **170.** Rascati & Parsons (1979) *J. Biol. Chem.* **254**, 1586. **171.** Häfele, Scherer & Böger (1988) *Biochim. Biophys. Acta* **934**, 186. **172.** Brunori, Antonini, Malatesta, Sarti & Wilson (1987) *Adv. Inorg. Biochem.* **7**, 93.

Carboplatin

This platinum-based antineoplastic agent (FW = 371.25 g/mol; CAS 41575-94-4), also known by NSC-241-240 and its IUPAC name *cis*-diammine(cyclobutane-1,1-dicarboxylate-*O*,*O*′)platinum(II), is a cisplatin analogue that damages DNA. It is considerably less nephrotoxic, ototoxic, and neurotoxic than its parent compound, cisplatin. Its main drawback is its myelosuppressive effect, causing blood cell and platelet output of bone marrow to decrease to as low as 10% the usual output. **Primary Mode of Action:** Carboplatin primarily considered as a DNA-damaging anticancer drug, forming different types of bifunctional adducts with cellular DNA (1-4). While cisplatin and carboplatin are structurally distinct, they form similar adducts at the same sites of the DNA that are recognized by a number of cellular proteins. (*See Cisplatin; Oxaliplatin*) Mismatch repair proteins and some damage-recognition proteins may show selectivity, contributing to the differences in its cytotoxicity and tumor range relative to other platinum-based antineoplastic agents. (The last reference deals with metal complexes in medicine with a focus on enzyme inhibition.) **Carboplatin Cytotoxicity:** Toxicity arises from its interference with transcription and/or DNA replication mechanisms, damaging tumors *via* induction of apoptosis (possibly by means of generating Reactive Oxygen Species, *or* ROS), as mediated by the activation of various signal transduction pathways, including calcium signaling, death-receptor signaling, and mitochondrial apoptotic pathways. Its cytotoxicity and apoptosis are not exclusively induced in cancer cells; thus, carboplatin might also lead to diverse side-effects such as neuro- and/or renal-toxicity or bone marrow-suppression. **Mechanisms of Drug Resistance:** These include changes in cellular uptake, drug efflux, increased detoxification, inhibition of apoptosis and increased DNA repair. To minimize carboplatin

resistance, combinatorial therapies have proven more effective to defeat cancers. **Key Pharmacokinetic Parameters:** *See* Appendix II in Goodman & Gilman's THE PHARMACOLOGICAL BASIS OF THERAPEUTICS, 12th Edition (Brunton, Chabner & Knollmann, eds.) McGraw-Hill Medical, New York (2011). **1.** Frezza, Hindo, Chen, *et al.* (2010) *Pharm. Des.* **16**, 1813. **2.** Rosenberg (1980) in *Nucleic Acid-Metal Ion Interactions* (Spiro, ed.) John Wiley & Sons, Inc., New York. **1**, 1. **3.** Desoize & Madoulet, (2002) *Crit. Rev. Oncol. Hematol.* **42**, 317. **4.** Che, Siu, (2010) *Curr. Opin. Chem. Biol.* **14**, 255.

Carbovir

This carbocyclic 2′,3′-dideoxy-2′,3′-didehydroguanosine analogue (FW = 249.27 g/mol; CAS 118353-05-2; Symbol: CBV) is metabolized to its 5′-triphosphate at concentrations sufficient to inhibit HIV reverse transcriptase. There was no evidence of CBV degradation by purine nucleoside phosphorylase, and the $t_{1/2}$ of CBV-triphosphate is 2.5 h, similar to that of zidovudine 5′-triphosphate (AZT-triphosphate). Unlike the latter, however, CBV- triphosphate levels decline without evidence of a plateau. CBV did not affect the metabolism of AZT; nor does AZT affect CBV metabolism. The small fraction of CBV that is incorporated into DNA is increased by incubation with mycophenolic acid, an IMP dehydrogenase inhibitor. CBV specifically inhibits the incorporation of nucleic acid precursors into DNA, but is without effect on the incorporation of radiolabeled precursors into RNA or protein. **Mode of Action:** Carbovir-TP is a potent inhibitor of HIV-1 reverse transcriptase, with K_i values (0.09 μM with 16S rRNA template; 0.05 μM with gapped duplex DNA template) are similar to that observed by AZT-TP, ddGTP, and ddTTP. The kinetic constants for incorporation of these nucleotide analogues into DNA (K_m = 0.96 μM with 16S rRNA template; 0.74 μM with gapped duplex DNA template) by HIV-1 reverse transcriptase are similar to those seen for their respective natural nucleotides. In addition, the incorporation of either carbovir-TP or AZT-TP in the presence of dGTP or dTTP, respectively, indicates that the mechanism of inhibition by these two nucleotide analogs is due to their incorporation into the DNA resulting in chain termination. **Target(s):** Carbovir-TP is not a potent inhibitor of DNA polymerases α, β, or γ, or DNA primase. CBV does not decrease the level of TTP, dGTP, dCTP, or dATP, indicating that the cytotoxicity of CBV was due to inhibition of DNA synthesis. **1.** Parker, White, Shaddix, *et al.* (1991) *J. Biol. Chem.* **266**, 1754. **2.** Parker, Shaddix, Bowdon, *et al.* (1993) *Antimicrob. Agents Chemother.* **37**, 1004.

Carboxin

This well-established, globally registered Electron Transport System Complex-II inhibitor (FW = 235.30 g/mol; CAS 5234-68-4; IUPAC: 5,6-dihydro-2-methyl-1,4-oxathiine-3-carboxanilide), marketed under the trade names Vitavax®, Anchor®, Vitaflo® and Provax®, is a broad-spectrum succinate dehydrogenase inhibitor (*or* SDHI) fungicide. Carboxin is especially effective on bunts and rhizoctonia and is used on wheat, barley, oats, rye, triticale, rice, maize, cotton, sugar beets, peanuts, soybeans, peas, sunflower, linseed, flax, rapeseed and potatoes. Carboxin and its primary metabolite (carboxin sulfoxide) are absorbed by germinating seeds, traveling upward in the plant by transpiration system and moving into the lower stem and first leaves. Carboxin does not redistribute into new growth as the plant develops and is not found in the seeds or upper leaves of crops grown from treated seed. Animals excrete carboxin and its degradation products rapidly, and no accumulation in tissues has been reported. In mammals, carboxin shows low-level, acute oral, dermal and inhalational toxicity; without skin or eye irritation. It is, however, a skin sensitizer. Carboxin also inhibits dichlorophenol indophenol (DCIP) reductase activities by these membranes in a manner both qualitatively and quantitatively similar to the inhibition of oxidation of the various substrates. **1**. Tucker & Lillich (1974) *Antimicrob. Agents Chemother.* **6**, 572.

Carboxyamidotriazole

This indirect PRPP synthetase inhibitor and calcium channel blocker (FW = 424.67 g/mol; CAS 99519-84-3), also known by its code name L-651,582 and its systematic name 5-amino-1-[4-(4-chlorobenzoyl)-3,5-dichlorobenzyl]-1,2,3-triazole-4-carboxamide, is an antiproliferative and antiparasitic agent that inhibits mammalian nucleotide metabolism (1). It inhibits ^{3}H-hypoxanthine, ^{14}C-adenine, and ^{14}C-formate incorporation into nucleotide pools, suggesting depletion of phosphoribosyl pyrophosphate stores required for *de novo* nucleotide synthesis. L-651,582 also inhibits ^{3}H-hypoxanthine incorporation into nucleotide pools with glucose, uridine, or ribose as the carbon source, suggesting a block at PRPP synthetase. Drug treatment does not kill cells, but instead reduces the fraction of cells in S and G_2/M, while increasing the population in G_1 (1). L651582 also blocks muscarinic m_5 receptor-stimulated $^{45}Ca^{2+}$ influx and release of arachidonic acid at low micromolar concentrations, suggesting it is an effective calcium ion blocker (2). Attesting to carboxyamidotriazole's action as a calcium antagonist is its ability to block L-type calcium channels in a voltage-dependent manner, with the apparent dissociation constant for the high affinity state is 0.5 µM. The IC_{50} for blocking T-type calcium channels is 3.5 µM. The inhibition of cellular proliferation and the production of arachidonate metabolites by L-651,582 may be the result of its nearly equipotent blockade of receptor-operated and voltage-gated calcium channels (3). **1**. Hupe, Behrens & Boltz (1990) *J. Cell Physiol.* **144**, 457. **2**. Felder, Ma, Liotta & Kohn (1991) *J. Pharmacol. Exp. Ther.* **257**, 967. **3**. Hupe, Boltz & Cohen, *et al.* (1991) *J. Biol. Chem.* **266**, 10136.

2-Carboxyarabinitol 1-phosphate

This endogenous diurnal regulator (FW = 258.12 g/mol; CAS 106777-19-9; *Symbol*: CA 1-P), also known as (2-*C*-phosphohydroxymethyl)-D-ribonic acid, inhibits ribulose-bisphosphate carboxylase by binding tightly (K_d = 32 nM) to the carbamoylated form of the enzyme (1). This inhibitor was extracted from leaves and copurified with the Rbu-1,5-P_2 carboxylase of the leaves. Further purification by ion-exchange chromatography, adsorption to purified Rbu-1,5-P_2 carboxylase, barium precipitation, and HPLC separation yielded a phosphorylated compound that was a strong inhibitor of Rbu-1,5-P_2 carboxylase. GC/MS as well as ^{13}C- and H^1-NMR proved its structure. This compound differs from the Rbu-1,5-P(2) carboxylase transition-state analogue, 2-carboxyarabinitol 1,5-bisphosphate (K_d = 10 pM), only by its lack of the C-5 phosphate group. In *Phaseolus vulgaris*, dark/light regulation of Rubisco activity is principally achieved by synthesis/degradation of CA 1-P (2). **See also** *2-Carboxyarabinitol 1,5-bisphosphate* **1**. Berry, Lorimer, Pierce, *et al.* (1987) *Proc. Natl. Acad. Sci. U.S.A.* **84**, 734. **2**. Holbrook, Turner & Polans (1992) *Photosynth. Res.* **32**, 37.

2-Carboxyarabinitol 1,5-bisphosphate

This transition-state inhibitor (FW = 258.12 g/mol; *Symbol*: CA 1,5-P_2), targets ribulose-bisphosphate (Rbu-1,5-P_2) carboxylase by binding extremely tightly. Diurnal change in carboxylase activity is regulated by an endogenous inhibitor that binds tightly to the activated (carbamoylated) form of Rbu-1,5-P_2 carboxylase. When extracted from leaves of *Phaseolus vulgaris*, the inhibitor copurified with the carboxylase, and further purification yielded 2-carboxy-arabinitol 1-phosphate [(2-*C*-phosphohydroxymethyl)-D-ribonic acid]. Synthetic 2-carboxy-D-arabinitol 1-phosphate, but not 2-carboxy-D-arabinitol 5-phosphate, was kinetically identical to that of the isolated, naturally occurring compound. Note that the isolated compound differs from the transition-state analogue (2-carboxyarabinitol 1,5-bisphosphate) by lacking a C-5 phosphate group, resulting in lower affinity for the monophosphate (K_d = 32 nM) than the bisphosphate (K_d < 10 pM). The less tightly bound compound is active in a light-dependent, reversible regulation of Rbu-1,5-P_2 carboxylase *in vivo*. **See also** *2-Carboxyarabinitol 1-phosphate* **1**. Berry, Lorimer, Pierce, *et al.* (1987) *Proc. Natl. Acad. Sci. U.S.A.* **84**, 734.

Carboxyatractyloside

This toxic glycoside (FW$_{dipotassium\ salt}$ = 847.01 g/mol; CAS 33286-30-5; M.P. = 270-275°C; Soluble in water but insoluble in ethanol), also called gummiferin, is more effective (K_d = 10^{-7} M) than atractyloside in inhibiting the adenine nucleotide translocator (1-4), thereby inhibiting oxidative phosphorylation. Unlike atractyloside, carboxy-atractyloside inhibits the translocase noncompetitively. In addition, plant mitochondria are significantly more sensitive to carboxyatractyloside than atractyloside. Note that carboxyatractyloside can be converted to atractyloside by pyrolysis. **1**. Vignais, Brandolin, Lauquin & Chabert (1979) *Meth. Enzymol.* **55**, 518. **2**. Luciani (1971) *Life Sci.* **10**, 961. **3**. Vignais, Vignais & Defaye (1973) *Biochemistry* **12**, 1508. **4**. Silva Lima & Denslow (1979) *Arch. Biochem. Biophys.* **193**, 368.

S-(4-Carboxybutyl)-L-homocysteine

This dual-substrate analogue (FW = 235.30 g/mol) binds with high affinity to the active site of the human hepatic betaine:homocysteine *S*-methyltransferase (K_i = 6.5 µM), presumably due to the L-isomer only. The corresponding sulfoxide and sulfone are both weaker inhibitors. **1**. Skiba, Wells, Mangum & Awad (1987) *Meth. Enzymol.* **143**, 384. **2**. Awad, Whitney, Skiba, Mangum & Wells (1983) *J. Biol. Chem.* **258**, 12790. **3**. Castro, Gratson, Evans, *et al.* (2004) *Biochemistry* **43**, 5341. **4**. Jiracek, Collinsova, Rosenberg, *et al.* (2006) *J. Med. Chem.* **49**, 3982. **5**. Szegedi, Castro, Koutmos & Garrow (2008) *J. Biol. Chem.* **283**, 8939.

(*S*)-γ-Carboxyglutamate

This clotting enzyme-derived amino acid (FW = 191.14 g/mol; pK_a values = 1.7, 3.2, 4.75, and 9.9; Symbol: Gla) is obtained principally from the proteolytic degradation of posttranslationally carboxylated Ca^{2+}-binding enzymes and proteins, mainly those in the blood coagulation cascade (*e.g.*, prothrombin, and coagulation factors VII, IX, and X) (1-4). Carboxyglutamate is also found in the bone mineralization protein osteocalcin, in invertebrate ion channel blockers (*e.g.*, conotoxins) and in NMDA receptor antagonists (*e.g.*, conantokins). It is produced via vitamin K-dependent carboxylation of specific glutamyl residues in protein substrates. γ-Carboxyglutamate was first identified as a component of prothrombin in 1974 (3,4). The free amino acid is relatively stable in alkaline solutions, but decomposes in acidic solutions ($t_{1/2}$ = 10 minutes in 0.05 M HCl at 100° C). Note: Because γ-carboxyglutamate is a calcium ion chelator, one should test a putative target enzyme's activity in the presence of other calcium-ion chelators to exclude susceptibility to other metal ion chelator before inferring that carboxyglutamate binds to a inhibitor site on that enzyme. **Target(s):** group II metabotropic glutamate receptors1; group III metabotropic glutamate receptors1; glutamate dehydrogenase, K_i = 0.38 μM (2). **1.** Schoepp, Jane & Monn (1999) *Neuropharmacology* **38**, 1431. **2.** Federici, Ricci, Matarese, *et al.* (1979) *Arch. Biochem. Biophys.* **196**, 304. **3.** Stenflo, Fernlund, Egan & Roepstorff (1974) *Proc. Natl. Acad. Sci. U.S.A.* **71**, 2730. **4.** Nelsestuen, Zytkovicz & Howard (1974) *J. Biol. Chem.* **249**, 6347.

N-[(2R)-2-(Carboxymethyl)-4-methylpentanoyl]-L-tryptophan-(S)-methyl-benzylamide

This peptidomimetic ($FW_{free-acid}$ = 463.58 g/mol), also known as GM 1489, is a potent inhibitor of certain matrix metalloproteinases (MMPs), including MMP-1 (or interstitial collagenase, K_i = 0.2 nM), MMP-2 (or gelatinase A, K_i = 0.5 μM), MMP-3 (or stromelysin 1, K_i = 20 μM), MMP-8 (or neutrophil collagenase, K_i = 0.1 μM), and MMP-9 (or gelatinase B, K_i = 0.1 μM). **1.** Holleran, Galardy, Gao, *et al.* (1997) *Arch Dermatol Res.* **289**,138.

2-[[1-(Carboxymethyl)-2,3,4,5-tetrahydro-5-hydroxy-2-oxo-1H-1-benzazepine-3-yl]amino]-4-phenylbutanoate

This benzazepinone derivative ($FW_{free-acid}$ = 412.44 g/mol) is a potent inhibitor of human peptidyl-dipeptidase A, *or* angiotensin-I-converting enzyme, IC_{50} = 8 nM. **1.** Ksander, Erion, Yuan, *et al.* (1994) *J. Med. Chem.* **37**, 1823.

[4S-[4α,7α(R*),12bβ]]-7-[S-(1-Carboxy-3-phenylpropyl)amino]-1,2,3,4,6,7,8,12b-octahydro-6-oxopyrido[2,1-a]-[2]benzazepine-4-carboxylate

This tricyclic peptidomimetic ($FW_{free-acid}$ = 436.51 g/mol) of a substrate C-terminal dipeptide linked with a 4-phenylbutanoic acid fragment, also known as MDL 27,088, inhibits peptidyl-dipeptidase A, *or* angiotensin I converting enzyme with K_i ≈ 4 pM. The structurally related compound MDL 27,788 inhibits angiotensin I-converting enzyme with K_i ≈ 46 pM. **1.** Giroux, Beight, Dage & Flynn (1989) *J. Enzyme Inhib.* **2**, 269.

2-[[[2-Carboxy-3-phenylpropyl]hydroxyphosphinoyl] methyl]pentanedioate

This phosphinate-containing tricarboxylic acid (FW = 372.31 g/mol) inhibits glutamate carboxypeptidase II (IC_{50} = 2 nM), presumably by mimicking the tetrahedral adduct formed during catalysis. In this case, a methylene also serves as a mimic for the amino that wpould be present in a glutamate containing peptide. **1.** Jackson, Tays, Maclin, *et al.* (699Ä) *J. Med. Chem.* **44**, 4170.

(3R,4S,5R)-5-((R)-1-Carboxy-1-phosphonoethoxy)-4-hydroxy-3-phosphono-oxycyclohex-1-enecarboxylate

This tetrahedral-intermediate analogue (FW = 405.17 g/mol) targets phosphoshikimate 1 carboxyvinyltransferase, *or* 3-enolpyruvoyldhikimate-5 phosphate synthase (K_i = 16 nM). The (*S*)-1-carboxy-1-phosphonoethoxy analogue is a slightly weaker inhibitor (K_i = 75 nM). **1.** Priestman, Healy, Becker, *et al.* (2005) *Biochemistry* **44**, 3241. **2.** Alberg & Bartlett (1989) *J. Amer. Chem. Soc.* **111**, 2337.

2-C-Carboxy-D-ribitol 1,5-Bisphosphate

This modified ribitol ($FW_{free-acid}$ = 356.12 g/mol) inhibits ribulose-bisphosphate carboxylase (K_d = 1.5 μM). **1.** Calvin (1954) *Fed. Proc., Fed. Amer. Soc. Exp. Biol.* **13**, 697. **2.** Wishnick, Lane & Scrutton (1970) *J. Biol. Chem.* **245**, 4939. **3.** Siegel & Lane (1975) *Meth. Enzymol.* **42**, 472. **4.** Schloss, Phares, Long, *et al.* (1982) *Meth. Enzymol.* **90**, 522. **5.** Siegel, Wishnick & Lane (1972) *The Enzymes*, 3rd ed., **6**, 169. **6.** Pierce, Tolbert & Barker (1980) *Biochemistry* **19**, 934.

Cardamonin

This chalconoid and Wnt/β-catenin inhibitor (FW = 270.27 g/mol; CAS 19309-14-9; IUPAC Name: (*E*)-1-(2,4-dihydroxy-6-methoxyphenyl)-3-phenylprop-2-en-1-one) from the seeds of *Alpinia katsumadai hayata* inhibits RANKL-induced NF-κB activation, with its suppressive effects correlating with inhibition of IκBα kinase and inhibition of phosphorylation and degradation of IκBα, an inhibitor of NF-κB (1). Furthermore, cardamonin also down-regulated RANKL-induced phosphorylation of MAPK including ERK and p38 MAPK. Cardamonin suppresses the RANKL-induced differentiation of monocytes to osteoclasts in a dose-dependent and time-dependent manner, suggesting that cardamonin may prove to be a valuable agent in suppressing osteoporosis. Cardamonin also inhibits pigmentation in melanocytes by suppressing the Wnt/β-catenin, key elements in a signaling pathway that regulates stem cell pluripotency and fate decisions during development by integrating signals from other pathways, including retinoic acid, FGF, TGF-β, and BMP. Wnt ligand is a secreted glycoprotein that binds to Frizzled receptors, triggering displacement of the multifunctional kinase GSK-3β from a regulatory APC/Axin/GSK-3β-complex. Cardamonin also reduces COX and iNOS expression by inhibiting p65NF-κB nuclear translocation and I-κB phosphorylation in RAW 264.7 macrophage cells (3). **1.** Sung, Prasad, Yadav, *et al.* (2013) *PLoS One* **8**, e64118. **2.** Cho, Ryu, Jeong, *et al.* (2009) *Biochem. Biophys. Res. Commun.* **390**, 500. **3.** Israf, Khaizurin, Syahida, Lajis & Khozirah (2007) *Mol. Immunol.* **44**, 673.

Cardiotonic Steroids

These digitalis-like inhibitors, or CTS, exert their effects on cardiac cells by binding to specific binding sites on the extracellular loops connecting transmembrane helices TM$_1$ to TM$_2$, TM$_5$ to TM$_6$, and TM$_7$ to TM$_8$ of the α-subunit of the Na$^+$/K$^+$-ATPase. Originally considered to regulate renal sodium transport and arterial pressure, these substances are now known to regulate cell growth, differentiation, apoptosis, and fibrosis, the modulation of immunity and of carbohydrate metabolism as well as CNS function and human behavior. Remarkably, digitalis-like cardiotonic steroids were constituents of herbal remedies administered to cardiac patients in Roman times. Those in the bufadienolide class were routinely used in clinical practice more than 1000 years ago. **See also:** *Ouabain, Digitoxin, Digoxin, Convallatoxin, SC4453, Bufalin, Gitaloxin, Digoxigenin, Actodigin, Oleandrin, Digitoxigenin, Gitoxin, Strophanthidin, Gitoxigenin, Lanatosides A, B and C, ✔and ⚘ Acetyl-digoxin, ✔and ⚘Methyl-digoxin.*

Cardizem, See *Diltiazem*

Carfilzomib

This epoxyketone-containing tetrapeptide (FW = 719.91 g/mol; CASs = 868540-17-4, 1140908-84-4 (TFA salt), 1140908-85-5 (methanesulfonate); Solubility: 50 mg/mL DMSO; <1 mg/mL H$_2$O; Animal Studies Formulation: 10% sulfobutylether β-cyclodextrin in 10 mM citrate, pH 3.5), also known as PR-171, selectively and irreversibly inhibits the chymotrypsin-like activity of proteasomes (IC$_{50}$ < 5 nM). Inhibition of proteasome-mediated proteolysis results in a build-up of misfolded and polyubiquinated proteins, including consequentially dysfunctional chaperonins, which may cause cell cycle arrest, apoptosis, and inhibition of tumor growth. Carfilzomib inhibits proliferation in a variety of cell lines

and patient-derived neoplastic cells, including multiple myeloma, and induced intrinsic and extrinsic apoptotic signaling pathways and activation of c-Jun-N-terminal kinase (JNK). This second-generation proteasome inhibitor, based on the naturally occurring epoxomicin, is intended to be a potential treatment for patients with multiple myeloma and solid tumors. (**See also** *Bortezomib; Ixazomib; Oprozomib*) **IUPAC:** (αS)-α-[[2-(4-morpholinyl)acetyl]amino]benzenebutanoyl-L-leucyl-*N*-[(1*S*)-3-methyl-1-[[(2*R*)-2-methyl-2-oxiranyl]carbonyl]butyl]-L-phenylalaninamide **1.** Kuhn, *et al.* (2007) *Blood* **110**, 3281. **2.** Kuhn, *et al.* (2011) *Curr. Cancer Drug Targets* **11**, 285.

Cariporide

This orally bioavailable Na$^+$/H$^+$ exchange inhibitor (FW = 283.35 g/mol; CAS 159138-80-4; Soluble to 100 mM in DMSO), also named *N*-(aminoiminomethyl)-4-(1-methylethyl)-3-(methylsulfonyl)benzamide and HOE 642, selectively targets NHE1 (IC$_{50}$ = 0.05 μM), with much weaker action against NHE3 (IC$_{50}$ = 3 μM) and NHE2 (IC$_{50}$ = 1000 μM) (1). Subtype specifity was determined by ^{22}Na$^+$ uptake inhibition in a fibroblast cell line separately expressing subtype Isoforms 1, 2 or 3. Cariporide attenuates ischemia-induced cardiomyocyte apoptosis *in vitro* and reduces cardiac arrhythmia *in vivo* (1). Although mild, apotosis can be rapidly induced in isolated hearts by a relatively brief period of ischemia without reperfusion. Loss of mitochondrial membrane potential is a critical step of the death pathway, and cariporide prevents H$_2$O$_2$-induced loss of mitochondrial membrane potential (3). Significantly, cariporide sensitizes leukemic cells to tumor necrosis factor related apoptosis-inducing ligand by up-regulating Death Receptor-5 via the endoplasmic reticulum stress-CCAAT/enhancer binding protein homologous protein-dependent mechanism (4). **See** *Zoniporide* **1.** Scholz, Albus, Counillon, *et al.* (1995) *Cardiovasc. Res.* **29**, 260. **2.** Chakrabarti, Hoque & Karmazyn (1997) *J. Mol. Cell Cardiol.* **29**, 3169. **3.** Teshima, Akao, Jones & Marbán (2003) *Circulation* **108**, 2275. **4.** Li, Chang, Wang, *et al.* (2014) *Leuk. Lymphoma* **55**, 2135.

Cariprazine

This orally active, brain-penetrant dopamine D$_3$ receptor partial agonist (FW = 427.41 g/mol; CAS 839712-12-8), also known as *N'*-[*trans*-4-[2-[4-(2,3-dichlorophenyl)-1-piperazinyl]-ethyl]cyclohexyl]-*N*,*N*-dimethylurea, RGH-188, and Vraylar®, is specifically indicated for treatment of schizophrenia and acutely as a treatment for manic or mixed episodes associated with bipolar I disorder. For the former, Vraylar's efficacy was demonstrated at 1.5 to 9 mg/day, and, for the latter, efficacy was shown at 3 to 12 mg/day. Unlike many D$_2$ and 5-HT$_{2A}$ receptor-targeting antipsychotics, cariprazine is a D$_2$ and D$_3$ partial agonist, showing higher affinity for the latter (1). D$_2$ and D$_3$ receptors are regarded as important targets treating schizophrenia, inasmuch as overstimulation of dopamine receptors is believed to be a possible causative factor of schizophrenia. Acting as an antagonist, aripirazine inhibits overstimulated dopamine receptors, but, acting as an agonist, it stimulates the same receptors, when endogenous dopamine levels are low. In animal studies using acute phencyclidine (PCP) treatment to model the cognitive deficits of schizophrenia, cariprazine pretreatment significantly diminished PCP-triggered cognitive deficits, and studies on knockout mice show that dopamine D$_3$ receptors contribute to this effect (2). Activation of dopamine D$_2$ receptors (D$_2$R) modulates G-protein/cAMP-dependent signaling and engages Akt-GSK-3 signaling through D$_2$R/β-arrestin 2 scaffolding of Akt and PP2A (3). This G protein-independent pathway may be important in mediating the antimanic effects of mood stabilizers and antipsychotics. The mood stabilizer lithium influences behavior and Akt/GSK-3 signaling in mice and many antipsychotics have been shown to more potently antagonize the activity of the β-arrestin-2 pathway relative to the G protein-dependent pathway. Cariprazine was more potent than aripiprazole in inhibiting isoproterenol-induced cAMP although both compounds showed similar maximum efficacy. In assays of D_2R/β-arrestin 2-dependent interactions, cariprazine showed very weak partial agonist activity, unless the levels of receptor kinase were increased; as an antagonist it showed similar potency to haloperidol and around five times greater potency than aripiprazole (3). Although D_2 and D_3 receptor agonists (*e.g.*, dopamine, pramipexole, apomorphine, 7-OH-DPAT, quinpirole) fully activate $[^{35}S]$GTPγS binding in membrane preparation from rat striatum, HEK293-D_2, and CHO-D_3 cells, cariprazine and aripiprazole did not show G-protein activation at concentrations up to 10 μM (1). **Pharmodynamic Profile:** Serotonin 5-HT$_{1A}$ receptor, K_i = 3 nM (acting as a partial agonist); 5-HT$_{2A}$ receptor, K_i = 19 nM (acting as an inverse agonist); 5-HT$_{2B}$ receptor, K_i = 0.58 nM (acting as an inverseagonist); 5-HT$_{2C}$ receptor, K_i = 134 nM (acting as an inverse agonist); 5-HT$_7$ receptor, K_i = 111 nM (acting as an antagonist); Dopamine D$_{2L}$ receptor, K_i = 0.49 nM (acting as a partial agonist); D$_{2S}$ receptor, K_i = 0.69 nM (acting as a partial agonist); D$_3$ receptor, K_i = 0.09 nM (acting as a partial agonist); Histamine H$_1$ receptor, K_i = 23 nM (acting as an inverse agonist). **1.** Kiss, Horváth, Némethy, *et al.* (2010) *J. Pharmacol. Exp. Ther.* 333, 328. **2.** Zimnisky, Chang, Gyertyán *et al.* (2013) *Psychopharmacol.* (Berl) 226, 91. **3.** Gao, Peterson, Masri, *et al.* (2015) *Pharmacol Res Perspect.* 3, e00073.

Carmustine

This mustard gas-related β-chloronitrosourea anticancer drug (FW = 214.05 g/mol; CAS 154-93-8; *Abbreviation*: BCNU), also known as 1,3-bis(2-chloroethyl)-1-nitrosourea, is a cell cycle-independent DNA alkylating agent that forms interstrand crosslinks, thereby preventing DNA replication and transcription (1,2). Given its ability to cross the blood-brain barrier, carmustine is often used to treat brain tumors. A single BCNU dose (50 mg/kg i.p.) to mice significantly inhibited glutathione reductase activity in liver within 10 min, and in lung and heart within 30 min (3). Maximal inhibition was reached at 4 h in all tissues, and glutathione reductase activity remained significantly depressed for 48 h in liver and for 96 h in lung and heart. BCNU doses <25 mg/kg were without effect on reductase activity. As noted elsewhere (**See** *Nitrosoureas*), cell redox mechanisms greatly influence cell susceptibility of nitrosoureas. **1.** Ahmann, Hahn & Bisel (1972) *Cancer Chemother. Rep.* 56, 93. **2.** Carter, Schabel, Broder & Johnston (1972) *Adv. Cancer Res.* 16, 273. **3.** Kehrer (1983) *Toxicol. Lett.* 17, 63.

L-Carnitine

This hydrophilic quaternary amine and key fatty acid-conjugating metabolite (FW$_{inner-salt}$ = 161.20 g/mol; carboxyl group pK_a = 3.80 at 25°C), also known as (*R*)-(–)-carnitine is found in high abundance in skeletal muscle, liver, and yeast, where participates in the transport of acyl groups across the mitochondrial membrane. Carnitine facilitates the transport of long-chain fatty acids (as fatty acyl-carnitine) into the mitochondrial matrix, where they undergo β-oxidation to acetyl CoA and the citric acid cycle. Carnitine is not metabolized, is excreted in urine, and undergoes highly efficient renal reabsorption. Carnitine is itself a product of the reactions catalyzed by acylcarnitine hydrolase and carnitinamidase. Because L-carnitine can deplete cellular acetyl-CoA stores via carnitine acetyl transferase (*Reaction*: Carnitine + Acetyl-CoA ⇌ Acetyl-carnitine + CoA), it would be expected to reduce histone acetylation. On the contrary, L-carnitine actually increases histone acetylation and induces accumulation of acetylated histones both in normal thymocytes and cancer cells. This effect appears to result from its inhibition of histone deasacetylases HDAC-I and HDAC-II by binding to their active sites (1). **Target(s):** caspase-3 (2); caspase-7 (2); caspase-8 (2); D-carnitine dehydrogenase (3); phosphatidylcholine:sterol acyltransferase, *or* LCAT (4); succinate:3-hydroxy-3-methylglutarate CoA-transferase (5); glycerophosphocholine cholinephosphodiesterase (6); palmitoyl-CoA hydrolase, *or* acyl-CoA

hydrolase, activated at low concentrations (7); lysophospholipase (8); choline sulfotransferase (9); hydroxylysine kinase (10); glycerone-phosphate *O*-acyltransferase, *or* dihydroxyacetone-phosphate *O*-acyltransferase, inhibited by DL-carnitine (11). **1.** Huang, Liu, Guo, *et al.* (2012) *PLoS One* 7. e49062 **2.** Mutomba, Yuan, Konyavko, *et al.* (2000) *FEBS Lett.* 478, 19. **3.** Hanschmann & Kleber (1997) *Biochim. Biophys. Acta* 1337, 133. **4.** Albers, Chen & Lacko (1986) *Meth. Enzymol.* 129, 763. **5.** Francesconi, Donella-Deana, Furlanetto, *et al.* (1989) *Biochim. Biophys. Acta* 999, 163. **6.** Sok (1998) *Neurochem. Res.* 23, 1061. **7.** Berge & Farstad (1979) *Eur. J. Biochem.* 95, 89. **8.** Wright, Payne, Santangelo, *et al.* (2004) *Biochem. J.* 384, 377. **9.** Orsi & Spencer (1964) *J. Biochem.* 56, 81. **10.** Hiles & Henderson (1972) *J. Biol. Chem.* 247, 646. **10.** Hajra (1968) *J. Biol. Chem.* 243, 3458.

Caroverine

This N-type calcium channel blocker (FW = 365.47 g/mol; CAS Registry = 23465-76-1), also known as Spasmium®, Spadon®, and 1-[2-(diethylamino)ethyl]-3-(4-methoxybenzyl)quinoxalin-2(1*H*)-one, is a spasmolytic drug and an oto-neural protective agent that is also a competitive antagonist of AMPA receptors, a noncompetitive antagonist of NMDA receptors, and a potent antioxidant. **1.** Nohl, Bieberschulte, Dietrich, *et al.* (2003). *BioFactors* 19, 79.

Carvedilol

This potent receptor antagonist and cardiovascular drug (FW = 406.48 g/mol; CAS 72956-09-3; Solubility: 100 mM in DMSO), also known 1-(9*H*-carbazol-4-yloxy)-3-[[2-(2-methoxyphenoxy)ethyl]amino]-2-propanol and BM 14190, targets the β$_1$-adrenoceptor (K_i = 0.81 nM), β$_2$-adrenoceptor (K_i = 0.96 nM), and α$_1$-adrenoceptor (K_i = 2.2 nM), displaying both antihypertensive and peripheral vasodilatory activity. Carvedilol also blocks cardiac inward-rectifier K$^+$ (K$_{IR}$) channels and voltage-dependent Ca^{2+} channels. It also exhibits antioxidant properties at much higher concentrations. Carvedilol also produced a concentration-dependent inhibition of vascular smooth muscle cell migration induced by platelet-derived growth factor, with an IC$_{50}$ value of 3 μM. **Key Pharmacokinetic Parameters:** *See* Appendix II in Goodman & Gilman's *THE PHARMACOLOGICAL BASIS OF THERAPEUTICS*, 12th Edition (Brunton, Chabner & Knollmann, eds.) McGraw-Hill Medical, New York (2011). **1.** Ohlstein, Douglas, Sung, *et al.* (1993) *Proc. Natl. Acad. Sci. U.S.A.* 90, 6189. **2.** Pönicke, Heinroth-Hoffmann & Brodde (2002) *J. Pharmacol. Exp. Ther.* 301, 71.

αS1-Casomorphin

This κ-acting opioid agonist (*Sequence*: Tyr-Val-Pro-Phe-Pro; FW = 621.73 g/mol; CAS 51871-47-7) is a noncompetitive inhibitor of nitric oxide synthase (1). This pentapeptide was isolated from hydrolysis of human S1 casein (residues: 158-162). Both the pentapeptide and its amide inhibit the proliferation of Thr47Asp human breast cancer cells (2). **1.** Kampa, Hatzoglou, Notas, *et al.* (2001) *Cell Death Differ.* 8, 943. **2.** Kampa. Loukas, Hatzoglou, *et al.* (1996) *Biochem. J.* 319, 903.

Castanospermine

This indolizine alkaloid (FW = 189.21 g/mol; CAS 79831-76-8), also named (1S,6S,7R,8R,8aR)-1,2,3,5,6,7,8,8a-octahydroindolizine-1,6,7,8-tetrol, from the seeds of *Castanospermum austrate*, is a potent glucosidase inhibitor with antiviral activity. In liver, both neutral (pH 6.5) and acidic (pH 4.5) α-glucosidase activities are inhibited, but the former is more susceptible (1). Castanospermine is also a potent inhibitor of α-glucocerebrosidase in fibroblast extracts, whereas other lysosomal glycosidases (*i.e.*, α- or β-galactosidase, α- or β-mannosidase, β-*N*-acetylhexosaminidase, β-glucuronidase, α- or β-L-fucosidase) are insensitive to castanospermine (2). Castanospermine inhibits the processing of the oligosaccharide portion of the influenza viral hemagglutinin (3). Castanospermine is a competitive inhibitor of amyloglucosidase at both pH 4.5 and 6.0, when assayed with the *p*-nitrophenyl-α-D-glucoside (4). It is also a competitive inhibitor of almond emulsin β-glucosidase at pH 6.5; however, inhibition was of the mixed type at pH 4.5 to 5.0 (4). In all cases, castanospermine was a much better inhibitor at pH 6.0 to 6.5 than at lower pH values, and the pK_a for castanospermine is 6.09. The authors suggest that the alkaloid is probably more active in its unprotonated form, which comports with the finding that the *N*-oxide of castanospermine, while still a competitive inhibitor, is 50-100x less active than was castanospermine. The latter's activity is not markedly altered by pH. These results probably explain why castanospermine is a good inhibitor of the glycoprotein processing enzyme, glucosidase I, since which is a neutral hydrolase (4). It should also be noted that bicyclic iminosugars, such as castanospermine and Celgosivir®, inhibit Dengue virus production by inhibiting endoplasmic reticulum α-glucosidases, and not by reducing the activity of glycolipid-processing enzymes (5). *See also Celgosivir* **1.** Saul, Ghidoni, Molyneux & Elbein (1985) *Proc. Natl. Acad. Sci. U.S.A.* **82**, 93. **2.** Saul, Molyneux & Elbein (1984) *Arch. Biochem. Biophys.* **230**, 668. **3.** Pan, Hori, Saul, *et al.* (1983) *Biochemistry* **22**, 3975. **4.** Saul, Molyneux & Elbein (1984) *Arch. Biochem. Biophys.* **230**, 668. **5.** Sayce, Alonzi, Killingbeck, *et al.* (2016) *PLoS Negl. Trop. Dis.* **10**, e0004524.

β-Catenin/Tcf Inhibitor V

This cell-permeable, dimeric copper complex (FW = 378.60 g/mol; CAS 691005-38-6), also known as BC21, is a Wnt signaling inhibitor that binds to β-catenin, thereby competitively and reversibly blocking the β-catenin/Tcf4 interaction (IC$_{50}$ = 5.0 μM). Named for *Drosophila* **W**ingless-**T**ype proteins, Wnts are secreted glycolipoproteins that regulate embryonal development and tissue homeostasis. Wnt binds to certain receptors, resulting in the inactivation of the β-catenin destruction complex to initiate Wnt-responsive gene expression after associating with TCF (T-cell factor) in the nucleus and activating cell proliferation. β-Catenin over-activation/over-expression is often observed in tumorigenesis. By disrupting key regulatory steps, BC21 decreases the viability of β-catenin overexpressing HCT116 colon cancer cells that harbor the β-catenin mutation, and more significantly, it inhibits the clonogenic activity of these cells (1). **1.** Tian, Han, Yan, *et al.* (2012) *Biochemistry* **51**, 724.

Caudatin

This natural product and potential antitumor agent (FW = 490.63 g/mol; CAS 38395-02-7), also named (3β,12β,14β,17α)-12-[[(2E)-3,4-dimethyl-1-oxo-2-pentenyl]oxy]-3,8,14,17-tetrahydroxypregn-5-en-20-one, from the traditional Chinese herbal medicine *bai shou wu*, the root tuber of *Cynanchum auriculatum* Royle ex Wight, inhibits carcinomic human alveolar basal epithelial cell growth and angiogenesis through modulating GSK3β/β-catenin pathway (1). In HepG$_2$ cells treated with caudatin, Western blotting indicated that the levels of Bcl-2 are down-regulated after caudatin treatment, whereas the expression of Bax is up-regulated (2). Furthermore, caudatin-induced apoptosis is accompanied by activation of caspase-3, -9, and poly(ADP-Ribose) Polymerase (PARP) (2). Treatment with caudatin also induced phosphorylation of extracellular-signal regulating kinase (ERK) and c-Jun N-terminal kinase (JNK). Such results demonstrate caudatin inhibits cell proliferation via DNA synthesis reduction and induces caspase-dependent apoptosis in HepG$_2$ cells (2). **1.** Fei, Cui, Zhang, Zhao & Wang (2012) *J. Cell Biochem.* **113**, 3403. **2.** Fei, Chen, Xiao, Chen & Wang (2012) *Mol. Biol. Rep.* **39**, 131.

CB-839

This potent, orally bioavailable glutaminase inhibitor (FW = 571.58 g/mol) binds reversibly at an allosteric site on both KGA and GAC splice forms of the ubiquitously expressed mitochondrial enzyme glutaminase, *or* GLS (1). CB-839 does not inhibit the related enzyme glutaminase 2 (GLS2), whose expression is restricted primarily to liver. **Rationale:** Inhibition of mitochondrial glutaminase activity, as achieved by small molecule inhibitors or genetic knockdown, exerts anti-tumor activity across a variety of tumor types, including lymphoma, glioma, breast, pancreatic, non-small cell lung, and renal cancers. Development of CB-839 as a GLS inhibitor exploits knowledge that BPTES (*bis*-2-(5-phenylacetamido-1,2,4-thiadiazol-2-yl)ethyl sulfide) forms inactive glutaminase homotetramer through the binding of two inhibitor molecules at the interface between a pair of homodimers. CB-839 is most likely a slow, tight-binding inhibitor, and IC$_{50}$ values < 50 nM following ≥ 1-hour pre-incubation with recombinant human GAC-GLS, or ~13x lower than with BPTES. **1.** Gross, Demo, Dennison, *et al.* (2014) *Mol. Cancer Ther.* **13**, 890.

CB 300945, *See ONX-0801*

CBR 5884

This small-molecule anti-cancer agent (FW = 336.39 g/mol; CAS 681159-27-3; Soluble to 100 mM in DMSO), also named ethyl 5-[(2-furanylcarbonyl)amino]-3-methyl-4-thiocyanato-2-thiophenecarboxylate, targets 3-phosphoglycerate dehydrogenase (IC$_{50}$ = 33 μM), thereby reducing the *de novo* serine synthesis in cancer cells and inhibiting proliferation of cancer cell lines characterized by high PHGDH expression levels. CBR-5884 is a time-dependent noncompetitive inhibitor that disrupts enzyme oligomerization. CBR-5884 was identified by screening a library of 800,000 drug-like compounds. **1.** Mullarky, Lucki, Beheshti Zavareh, *et al.* (2016) *Proc. Natl. Acad. Sci. U.S.A.* **113**, 1778.

CC-223

This potent and orally bioavailable mTOR inhibitor (FW = 397.47 g/mol; CAS 1228013-30-6; Solubility: 80 mg/mL DMSO; <1 mg/mL H$_2$O), also named 7-(6-(2-hydroxypropan-2-yl)pyridin-3-yl)-1-((*trans*)-4-methoxycyclohexyl)-3,4-dihydropyrazino[2,3-*b*]pyrazin-2(1*H*)-one, selectively targets mTOR kinase (IC$_{50}$ = 16 nM), with >200-fold selectivity over the related PI3K-α (IC$_{50}$ = 4.0 μM). Of the PI3K-related kinases tested, CC-223 shows no significant inhibition of ATR or SMG1, but inhibits DNA-PK (IC$_{50}$ = 0.84 μM). When screened in a single-point assay against a commercially available panel of 246 kinases, only FLT4 (IC$_{50}$ = 0.65 μM) and cFMS (IC$_{50}$ = 28 nM) were substantially inhibited below 1 μM. The kinase selectivity of CC-223 was confirmed upon evaluation in cellular systems using ActivX KiNativ profiling. Other than mTOR kinase, no kinase target was identified when HCT 116 or A549 cells were treated for 1 hour with 1 μmol/L CC-223 and assayed for kinase activity. Like rapamycin, CC-223 inhibits the direct and indirect mTORC1 substrates p-p70S6K and pS6RP. However, unlike rapamycin, CC-223 more fully inhibits mTORC1 activity, demonstrating a concentration-dependent reduction in the direct marker p-4EBP1^{T46}. CC-223 also demonstrates inhibition of the mTORC2 complex, as assessed by reduction in phosphorylation of the direct substrate AKT^{S473} and the downstream markers PRAS40^{T246} and GSK3β^{S9}. Using a panel consisting of twenty-three diffused large B-cell lymphoma (DLBCL), four follicular lymphoma (FL), 5 acute myelogenous leukemia (AML), six mantle cell lymphoma (MCL), and two anaplastic large cell lymphoma (ALCL) lines, potent growth inhibitory effects were observed with GI$_{50}$ < 0.7 μM in 33 of 40 lines and GI$_{50}$ values between 1-5 μM in the remaining lines. **1.** Mortensen, Fultz, Xu, *et al.* (2015) *Mol. Cancer Ther.* **14**, 1295.

CC 0651

This allosteric Cdc34 inhibitor (FW = 442.29 g/mol; CAS 1319207-44-7; Solubility: 100 mM in 1eq. NaOH; 100 mM in DMSO), also named (2*R*,3*S*,4*S*)-5-[4-(3,5-dichlorophenyl)phenyl]-2,3-dihydroxy-4-(2-methoxy acetamido)pentanoate, targets human Cdc34-mediated ubiquitination of p27^{Kip1} (IC$_{50}$ = 1.72 μM), with selectivity over Uba1, Ube2G1, UbcH7, UbcH5, Ube2N (Ubc13), Ube2R2, SMURF2, SspH1 and Rnf168. Structure determination revealed that CC0651 inserts into a cryptic binding pocket on hCdc34 distant from the catalytic site, causing subtle but wholesale displacement of E2 secondary structural elements (1). CC0651 analoguees also inhibit proliferation of human cancer cell lines and caused accumulation of the SCF (*or* Skp2) substrate p27 (*or* Kip1). CC0651 does not affect hCdc34 interactions with E$_1$ or E$_3$ enzymes or the formation of the ubiquitin thioester but instead interferes with the discharge of ubiquitin to acceptor lysine residues. E$_2$ enzymes are thus susceptible to noncatalytic site inhibition and may represent a viable class of drug target in the UPS (1). CC-0651 traps a weak interaction between ubiquitin and the E$_2$ donor ubiquitin-binding site (2). A structure of the ternary CC0651-Cdc34A-ubiquitin complex revealed that the inhibitor engages a composite binding pocket formed from Cdc34A and ubiquitin. CC0651 also suppresses the spontaneous hydrolysis rate of the Cdc34A-ubiquitin thioester without decreasing the interaction between Cdc34A and the RING domain subunit of the E$_3$ enzyme (2). Stabilization of otherwise weak interactions between ubiquitin and UPS enzymes appears to be a strategy for selectively inhibiting different UPS activities (2). **1.** Ceccarelli, Tang, Pelletier, *et al.* (2011) *Cell* **145**, 1075. **2.** Huang, Ceccarelli, Orlicky, *et al.* (2014) *Nature Chem. Biol.* **10**, 156.

CC-10004, *See* **Aprimilast**

CCCP

This mitochondria/chloroplast-inhibiting protonophore, *or* small-molecule transmembrane carrier (FW = 204.62 g/mol; CAS 555-60-2), also known as carbonyl cyanide *m*-chlorophenyl hydrazone, is cytotoxic, dissipating the proton motive forces needed to drive oxidative phosphorylation and photophosphorylation, gradually causing cell destruction and death (1-3). CCCP inhibits chloroplast function as well as plant mitochondria by uncoupling of the proton chemiosmotic gradient established by the photon capture and the electron transport chain, thereby reducing or blocking ATP synthase. (**See** *FCCP*) **Intracellular Target:** Although oxidative phosphorylation is blocked by CCCP, the agent's actual target is the cytochrome *c* oxidase complex. CCCP binds to the cytochrome *aa*$_3$, K_d = 0.27 μM, with weaker binding to other sites in the cytochrome *c* oxidase complex (4). The electrostatic interaction between CCCP and cytochrome *c* oxidase involves an ionizable group (pK_a = 6.64) on cytochrome *c* oxidase. CCCP's interaction at low-affinity binding sites of cytochrome *c* oxidase induces the shift of the anion spectrum of CCCP toward the UV (4). **Mechanism of Protonophore Action:** After binding within a phospholipid membrane bilayer, each protonophore resides within that membrane throughout multiple rounds, shuttling protons one-by-one, forth or back across the membrane, depending on the transmembrane potential. This mechanism explains the efficiency of protonophore, which relies on rapid protonation/deprotonation rather than carrier desorption after each round of proton transport. In the following model, transmembrane proton transport is facilitated by intramembrane-resident weak acids like CCCP. In such a scheme, there are four relevant parameters: β$_A$ and β$_{HA}$, describing the proton binding to and release from the weak acid's anionic and neutral forms, respectively, and κ$_A$ and κ$_{HA}$, the rate constants for the movement of A$^-$ and HA within and across the membrane's interior.

In this diagram (*excerpted from* (5) *with permission of the authors and publisher, AEPress, Bratislava, Slovakia*), the circled positive and negative signs indicate an implied potential difference across the membrane. The rate constants k_R and k_D refer to the heterogenous reassociation and dissociation, respectively, of an aqueous proton with an anion carrier molecule adsorbed to the membrane. The rate constants k'_A and k''_A refer to the voltage-dependent translocation of the anion A$^-$ from the *i*-(memb/aq)-interface to the *ii*-(memb/aq)-interface. The rate constant k_{HA} refers to the movement of HA between the two interfaces. The parameters β$_A$, β$_{HA}$, κ$_A$, and κ$_{HA}$ were determined by equilibrium dialysis, electrophoretic mobility, membrane potential, membrane conductance, and spectrophotometric measurements (5). Based on these equilibrium and steady-state measurements on diphytanoyl phosphatidylcholine/ chlorodecane membranes, β$_A$ = β$_{HA}$ = 1.4 x 10^{-3} cm, κ$_A$ = 175 s^{-1} and κ$_{HA}$ = 12,000 sec^{-1}. The model predicts a single exponential decay of the current in a voltage-clamp experiment. The model also predicts that the decay in the voltage across the membrane following an intense current pulse of short duration (approximately 50 nsec) can be described by the sum of two exponentials. The magnitudes and time constants of the relaxations that we observed in both voltage-clamp and charge-pulse experiments agree well with the predictions of the model for all values of pH, voltage and CCCP concentrations. **CCCP Effects on Autophagy:** Treatment of HeLa cells with CCCP causes redistribution of mitochondrially targeted dyes (*e.g.,* DiOC6, TMRM, MTR, and MTG) from mitochondria to cytosol, and then to lysosomal compartments (6). Relocalization of mitochondrial dyes to lysosomal compartments was brought about by retargeting the dye, rather than mitochondrial component delivery to the lysosome. CCCP interfered with lysosomal function and autophagosomal degradation in both yeast and mammalian cells, inhibited starvation-induced mitophagy in mammalian cells, and blocked the induction of mitophagy in yeast cells. Such results demonstrate that CCCP inhibits autophagy at both the initiation and lysosomal degradation stages. In addition, our data demonstrated that caution should be taken when using organelle-specific dyes in conjunction with strategies affecting membrane potential. **Effect on Choramphenicol & Ciprofloxacin:** Although chloramphenicol and ciprofloxacin are highly effective against *Enterobacter aerogenes*, certain clinical isolates possess the highly active AcrAB drug efflux pump that reduces their intracellular concentrations to noncytostatic levels (). The MICs are reduced in drug-insensitive isolates, but not sensitive bacterial strains, when CCCP is added as an uncoupling agent for systems maintaining transmembrane ion gradients. The intracellular chloramphenicol concentration, for example, rises 2.3x in the presence of CCCP, suggesting that transport energization or the organization of the efflux channel is impaired (7,8). **1**. Hanstein (1976) *Biochim. Biophys. Acta* **456**, 129. **2**. Cunarro & Weiner (1975) *Biochim. Biophys. Acta* **387**, 234. **3**. Terada (1981) *Biochim. Biophys. Acta* **639**, 225. **4**. Bona, Antalík, Gazová, *et al.* (1993) *Gen. Physiol. Biophys.* **12**, 533. **5**. Kasianowicz, Benz & McLaughlin (1984) *J. Membr. Biol.* **82**, 179. **6**. Padman, Bach, Lucarelli, Prescott & Ramm (2013) *Autophagy* **9**, 1862. **7**. Mallea, Chevalier, Eyraud & Pages (2002) *Biochem. Biophys. Res. Commun.* **293**, 1370. **8**. Pakzad, Zayyen, Taherikalani, Boustanshenas & Lari (2013) *GMS Hyg. Infect. Control* **8**, Doc 15.

CCEP-32496

This potent BRAF/cRaf inhibitor (FW = 517.46 g/mol; CAS 1188910-76-0; Solubility: 9 mg/mL DMSO; <1 mg/mL H_2O), also named *N*-[3-[(6,7-dimethoxy-4-quinazolinyl)oxy]phenyl]-*N*-[5-(2,2,2-trifluoro-1,1-dimethylethyl)-3-isoxazolyl]urea, targets wild type BRAF (K_d = 36 nM), BRAF^{V600E} (K_d = 14 nM) and c-Raf (K_d = 39 nM), but is also against Abl-1, c-Kit, Ret, PDGFRβ and VEGFR2 (1,2). Affinity for MEK-1, MEK-2, ERK-1 and ERK-2 is insignificant. Notably, cellular expression of mutant BRAF^{V600E} results in constitutive activation of the MAPK pathway, which can lead to uncontrolled growth. Significant oral efficacy was observed in a 14-day BRAF^{V600E}-dependent human Colo-205 tumor xenograft mouse model, upon dosing at 30 and 100 mg/kg BID (1). CEP-32496 exhibits single oral dose (10-55 mg/kg) pharmacodynamic inhibition of both pMEK and pERK in BRAF^{V600E} colon carcinoma xenografts in nude mice (2). Moreover, sSustained tumor stasis and regressions are observed with oral administration (30-100 mg/kg twice daily) against BRAF^{V600E} melanoma and colon carcinoma xenografts, with no adverse effects (2). **1**. Rowbottom, Faraoni, Chao, *et al.* (2012) *J. Med. Chem.* **55**, 1082. **2**. James, Ruggeri, Armstrong, *et al.* (2012) *Mol. Cancer Ther.* **11**, 930.

CCI-779, *See Temsirolimus*

CCT128930

This potent, ATP-competitive inhibitor (F.Wt. = 341.84; CAS 885499-61-6; Solubility (at 25°C): 65 mg/mL DMSO; <1 mg/mL H2O), known systematically as 4-(4-chlorobenzyl)-1-(7*H*-pyrrolo[2,3-*d*]pyrimidin-4-yl)piperidin-4-amine, targets Akt2 (also known as Protein Kinase B, *or* PKB), PKA, and p70S6K, with IC$_{50}$ values of 6 nM, 168 nM, and 120 nM, respectively. CCT128930 exhibits marked antiproliferative activity and inhibits phosphorylation of AKT substrates in multiple tumor cell lines *in vitro*, consistent with its action as an AKT inhibitor. CCT128930 also causes a G_1 arrest in PTEN-null U87MG human glioblastoma cells,

consistent with AKT pathway blockade. CCT128930 also blocks phosphorylation of several downstream AKT biomarkers in U87MG tumor xenografts, indicating AKT inhibition *in vivo*. Antitumor activity is observed with CCT128930 in U87MG and HER2-positive, PIK3CA-mutant BT474 human breast cancer xenografts, behavior that is consistent with its pharmacokinetic and pharmacodynamic properties. **1**. Yap, Walter, Hunter *et al.* (2011) *Mol. Cancer Ther.* **10**, 360.

CCT 241533

This potent Chk inhibitor (FW = 515.41 g/mol; CAS 1262849-73-9; Soluble to 10 mM in H_2O (with gentle warming); 100 mM in DMSO), also known as (3*R*,4*S*)-4-[[2-(5-fluoro-2-hydroxyphenyl)-6,7-dimethoxy-4-quinazolinyl]amino]-α,α-dimethyl-3-pyrrolidinemethanol, targets the serine/threonine-protein checkpoint kinase Chk2 (IC$_{50}$ = 3 nM), showing >63-fold selectivity over Chk1. CCT 241533 inhibits Chk2 activation in response to etoposide-induced DNA damage in HT29 cells. It also blocks ionizing radiation-induced apoptosis of mouse thymocytes. **1**. Caldwell, Welsh, Matijssen, *et al.* (2011) *J. Med. Chem.* **54**, 580.

CDN 1163

This allosteric SERCA2 activator (FW = 320.38 g/mol; CAS 892711-75-0; Soluble to 100 mM in DMSO), also known as 4-(1-methylethoxy)-*N*-(2-methyl-8-quinolinyl)benzamide, targets the <u>Sarco</u>/<u>E</u>ndoplasmic <u>R</u>eticulum Ca^{2+}-<u>A</u>TPase, *or* SR Ca^{2+}-ATPase, a divalent cation pump that medates the ATP-dependent transfer of Ca^{2+} from the cytosol to the lumen of the SR during muscle relaxation. Dysregulation of ER Ca^{2+} homeostasis triggers ER stress, leading to insulin resistance in obesity and diabetes. Impaired SERCA function has emerged as a major contributor to ER stress. CDN-1163 increases Ca^{2+}-ATPase activity and Ca^{2+} uptake by ER microsomes from obese mice, rescuing HEK cells from ER stress-induced cell death.It also reduces fasting glucose levels and adipose tissue weight, and increases energy expenditure in *ob/ob* mice. CDN1163 also mproves glucose tolerance and reverses hepatic steatosis in *ob/ob* mice. **1**. Kang, Dahl, Hsieh, *et al.* (2016) *J. Biol. Chem.* **291**, 5185.

CDP-choline

This metabolite and anti-senility drug (FW$_{zwitterion}$ = 488.33 g/mol; CAS 987-78-0; Water-soluble; UV/Visible Spectrum is that of CDP), also called cytidine-5'-diphosphocholine, Citicoline, and Cognizin®, is required for the biosynthesis of phospholipids and sphingomyelins (1,2). CDP-choline is a ready source of the choline needed by cholinergic neurons, improving the cognitive and memory disturbances which are observed after a head trauma. CDP-choline activates the biosynthesis of structural phospholipids in the neuronal membranes, increases cerebral metabolism and acts on the levels of various neurotransmitters. Thus, it has been experimentally proven that CDP-choline increases noradrenaline and dopamine levels in the CNS. It also restores the activity of mitochondrial ATPase and of membranal Na+/K+ ATPase. It also accelerates the reabsorption of cerebral edema in various experimental models. CDP-choline also stimulates acetylcholinesterase and Na$^+$/K$^+$-exchanging ATPase reactions (3) It inhibits phospholipase A$_2$ activation (4). Heating CDP-choline in 1.0 M H$_2$SO$_4$ for eighteen minutes results in 50% conversion to CMP and phosphocholine. **Target(s):** ethanolamine phospho-transferase (1,7-10); choline kinase (2,12); glycogenin glucosyltransferase (5,14); FAD diphosphatase (6); nucleotide diphosphatase (6); choline-phosphate cytidylyl-transferase, product inhibition (11); 1,4-β-D-xylan synthase (13). **1.** Moore (1981) *Meth. Enzymol.* **71**, 596. **2.** Ishidate & Nakazawa (1992) *Meth. Enzymol.* **209**, 121. **3.** Plataras, Tsakiris & Angelogianni (2000) *Clin. Biochem.* **33**, 351. **4.** Arrigoni, Averet & Cohadon (1987) *Biochem. Pharmacol.* **36**, 3697. **5.** Manzella, Ananth, Oegema, *et al.* (1995) *Arch. Biochem. Biophys.* **320**, 361. **6.** Byrd, Fearney & Kim (1985) *J. Biol. Chem.* **260**, 7474. **7.** Coleman & Bell (1977) *J. Biol. Chem.* **252**, 3050. **8.** Sparace, Wagner & Moore (1981) *Plant Physiol.* **67**, 922. **9.** Yang, Moroney & Moore (2004) *Arch. Biochem. Biophys.* **430**, 198. **10.** Kanoh & Ohno (1976) *Eur. J. Biochem.* **66**, 201. **11.** Mages, Rey, Fonlupt & Pacheco (1988) *Eur. J. Biochem.* **178**, 367. **12.** Haubrich (1973) *J. Neurochem.* **21**, 315. **13.** Dalessandro & Northcote (1981) *Planta* **151**, 61. **14.** Meezan, Manzella & Roden (1995) *Trends Glycosci. Glycotechnol.* **7**, 303.

CDP-ethanolamine

This metabolite (FW$_{zwitterion}$ = 446.25 g/mol; CAS 3036-18-8; Soluble in water; UV/Visible Spectrum is that of CDP), also called cytidine-5'-diphosphoethanolamine, ihhibits diacylglycerol choline-phosphotransferase (1-4) and ethanolamine-phosphate cytidylyltransferase, product inhibition (5). **1.** Coleman & Bell (1977) *J. Biol. Chem.* **252**, 3050. **2.** Kanoh & Ohno (1981) *Meth. Enzymol.* **71**, 536. **3.** Kanoh & Ohno (1976) *Eur. J. Biochem.* **66**, 201. **4.** Sparace, Wagner & Moore (1981) *Plant Physiol.* **67**, 922. **5.** Sundler (1975) *J. Biol. Chem.* **250**, 8585.

CDP-D-glucose

This metabolite (FW$_{free-acid}$ = 565.32 g/mol; CAS 2906-23-2; Soluble in water; UV/Visible Spectrum is that of CDP), is an intermediate in lipopolysaccharide biosynthesis. CDP-D-glucose is a product of the reaction catalyzed by glucose-1-phosphate cytidylyltransferase and is a substrate for CDP-glucose 4,6-dehydratase. **Target(s):** UDP-glucuronate 4-epimerase (1,3,4); glycogenin glucosyltransferase (2,9); UDP sugar diphosphatase (5); [protein-P$_{II}$] uridylyltransferase (6); glucose-1-phosphate cytidylyltransferase, by product inhibition (7,8); β-1,4-mannosyl-glycoprotein 4-β-N-acetylglucosaminyltransferase (10); ceramide glucosyl-transferase, weakly inhibited (11). **1.** Gaunt, Ankel & Schutzbach (1972) *Meth. Enzymol.* **28**, 426. **2.** Manzella, Ananth, Oegema, *et al.* (1995) *Arch. Biochem. Biophys.* **320**, 361. **3.** Gaunt, Maitra & Ankel (1974) *J. Biol. Chem.* **249**, 2366. **4.** Munoz, Lopez, de Frutos & Garcia

(1999) *Mol. Microbiol.* **31**, 703. **5.** Glaser, Melo & Paul (1967) *J. Biol. Chem.* **242**, 1944. **6.** Engleman & Francis (1978) *Arch. Biochem. Biophys.* **191**, 602. **7.** Kimata & Suzuki (1966) *J. Biol. Chem.* **241**, 1099. **8.** Lindqvist, Kaiser, Reeves & Lindberg (1994) *J. Biol. Chem.* **269**, 122. **9.** Meezan, Manzella & Roden (1995) *Trends Glycosci. Glycotechnol.* **7**, 303. **10.** Ikeda, Koyota, Ihara, *et al.* (2000) *J. Biochem.* **128**, 609. **11.** Shukla & Radin (1990) *Arch. Biochem. Biophys.* **283**, 372.

CDP-glycerol

This nucleotide alcohol (FW$_{free-acid}$ = 477.26 g/mol), also known as cytidine 5'-diphosphate glycerol, is a component in lipid metabolism, the diacyl derivative being an intermediate in the biosynthesis of phosphatidylinositol. Note that this metabolite is unstable in acidic and alkaline conditions. Treatment for thirty minutes in 1 M HCl at 100°C produces a mixture of CMP, sn-glycerol 3-phosphate, and sn-glycerol 2-phosphate. Treatment with concentrated ammonia at 100°C for one hour forms CMP and glycerol 2,3-cyclic phosphate. **Target(s):** CDP-ribitol ribitolphosphotransferase, weakly inhibited (1); teichoic acid synthase (1); glycerol-3-phosphate cytidylyltransferase (2,3). **1.** Fiedler & Glaser (1974) *J. Biol. Chem.* **249**, 2684. **2.** Park, Sweitzer, Dixon & Kent (1993) *J. Biol. Chem.* **268**, 16648. **3.** Badurina, Zolli-Juran & Brown (2003) *Biochim. Biophys. Acta* **1646**, 196.

CE3F4

This potent cAMP-binding protein inhibitor (FW = 351.01 g/mol; CAS 143703-25-7; Soluble to 100 mM in DMSO), also named 5,7-dibromo-6-fluoro-2-methyl-1,2,3,4-tetrahydroquinoline-1-carbaldehyde, with positions numbered for any tetrahydroquinoline skeleton, targets Epac (or Exchange protein directly activated by 3',5'-cyclic-AMP). The latter has two isoforms, each consisting of a cAMP-binding regulatory region and a GTP Exchange Factor (or GEF) region that catalyzes GTP exchange for GDP bound to the Ras-like small GTPases Rap1 and Rap2. Epac proteins contain Dishevelled, Egl-10, and pleckstrin domains, followed by an evolutionally conserved cAMP-binding domain (similar to that found in PKA) and bacterial transcriptional factor cAMP-receptor protein. In the absence of cAMP, the regulatory region containing the cAMP-binding domain directly interacts with the catalytic region, blocking its GEF activity. CE3F4 blocks Epac1 guanine nucleotide exchange reaction on Rap1 in intact cells. Because CE3F4 did not exert its antagonism by directly competing with cAMP for binding to Epac1, it is difficult to predict whether it could also inhibit Epac2. CE3F4 proved useful in demonstrating that Epac1 promotes autophagy during cardiomyocyte hypertrophy (2). **1.** Courilleau, Bisserier & Jullian (2012) *J. Biol. Chem.* **287**, 44192. **2.** Laurent, et al. (2015) *Cardiovasc. Res.* **105**, 55.

Cediranib

This indole-ether quinazoline-based vascular-disrupting agent and pan-VEGF receptor tyrosine kinase inhibitor (FW = 450.51 g/mol; CAS 288383-20-0; Solubility: 90 mg/mL DMSO, <1 mg/mL H_2O), also known by its code names AZD2171 and NSC-732208, its trade name Recentin®, and by its IUPAC name 4-[(4-fluoro-2-methyl-1H-indol-5-yl)oxy]-6-methoxy-7-[3-(pyrrolidin-1-yl)propoxy]quinazoline, is a highly potent vascular endothelial growth factor receptor-2 tyrosine kinase (VEGFR2) inhibitor, IC_{50} = 0.5 nM (1). Once-daily oral administration of AZD2171 shows promise in limiting the progression of solid malignancies by suppressing tumor-induced angiogenesis. Cediranib inhibits Flt1 and Flt4, with IC_{50} values of 5 nM and ≤3 nM; however, when assayed with ELISA to measure phosphorylated products, 10 μM Cediranib showed little or no inhibition of KDR, c-Kit, PDGFRα, PDGFRβ, CSF-1R, FGFR1, Src, Abl, epidermal growth factor receptor (EGFR), ErbB2, Aurora A, or Aurora B activity in the presence of 100 μM ATP. Cediranib reverses ABCB1- and ABCC1-mediated MDR by directly inhibiting drug efflux (2). **1**. Wedge, Kendrew, Hennequin, *et al.* (2005) *Cancer Res.* **65**, 4389. **2**. Tao, Liang, Wang, *et al.* (2009) *Cancer Chemother. Pharmacol.* **64**, 961.

Ceefourin 1

This MRP4 inhibitor (FW = 293.42 g/mol; CAS 315702-40-0; Solubility: 100 mM in DMSO), also named 5-[(2-benzothiazolylthio)methyl]-2,4-dihydro-4-methyl-3H-1,2,4-triazole-3-thione, selectively targets Multidrug Resistance-associated Protein-4 (IC_{50} = 2.6 μM). It and the structurally related compound Ceefourin 2 were discovered using high-throughput screening and were found to be highly selective for MRP4 over other ABC transporters, including P-glycoprotein (P-gp), ABCG2 (Breast Cancer Resistance Protein; BCRP) and MRP1 (multidrug resistance protein 1; ABCC1). Both compounds are more potent MRP4 inhibitors in cellular assays than MK-571, the most widely used inhibitor, and they require lower concentrations for a comparable level of inhibition. Ceefourin 1 and Ceefourin 2 have low cellular toxicity, and high microsomal and acid stability. Sensitizes MRP4 over-expressing cells to the DNA topoisomerase-1 inhibitor SN 38, which is formed metabolically from the pro-drug CPT-11 (**See** *CPT-11*). **1**. Cheung, Flemming, Watt, *et al.* (2014) *Biochem. Pharmacol.* **91**, 97.

CEP-18770, *See Delanzomib*

Cefadroxil

This broad-spectrum cephalosporin-class antibiotic (FW = 363.39 g/mol; CAS 66592-87-8),1 also known as Duricef and (6R,7R)-7-{[(2R)-2-amino-2-(4-hydroxyphenyl)acetyl]amino}-3-methyl-8-oxo-5-thia-1-azabicyclo[4.2.0]oct-2-ene-2-carboxylic acid, is highly effective against Gram-positive and Gram-negative bacterial infections (*e.g.*, *Escherichia coli*, (MIC = 8 μg/mL), *Staphylococcus aureus*, (MIC = 1-2 μg/mL), and *Streptococcus pneumoniae* (MIC = 1-10 μg/mL). Oral cefadroxil in doses of 0-6-1-8 g per day given on twice or three times daily schedules was effective in the treatment of thirty-six patients with infections such as abscesses, carbuncles, cellulitis, furunculosis and impetigo (1). Cefadroxil is almost completely absorbed from the gastrointestinal tract. After doses of 500 mg and 1 g by mouth, peak plasma concentrations of about 16 and 30

micrograms/mL, respectively, are obtained after 1.5 to 2 hours. **1**. Cordero (1976) *J. Int. Med. Res.* **4**, 176.

Cefepime

This fourth-generation cephalosporin antibiotic (FW = 480.56 g/mol; CAS 88040-23-7; Route: Intravenous *or* Intramuscular) is highly effective against *Pseudomonas aeruginosa*, *Staphylococcus aureus*, *Streptococcus pneumoniae*, and certain *Enterobacteriaceae* (1,2). Unlike other cephalosporins, which are readily degraded by many plasmid- and chromosome-mediated β-lactamases, cefepime is stable. Upon intravenous infusion, cefepime reaches peak plasma concentrations of 193 μg/mL during the 30-min infusion (3). The mean plasma elimination $t_{1/2}$ is 2.1 hours. Penetration into inflammatory fluid is rapid, with mean peak levels of 91.5 μg/mL at 0.9 hour post-infusion. Urinary elimination accounted for 98.9% of the dose within 8 h. Therapeutic plasma levels (>2 μg/mL) are present for >8 hours after infusion, suggesting that 2 or 3 doses daily would be sufficient to treat infections due to susceptible organisms (3). **Key Pharmacokinetic Parameters:** *See* Appendix II in Goodman & Gilman's *THE PHARMACOLOGICAL BASIS OF THERAPEUTICS*, 12th Edition (Brunton, Chabner & Knollmann, eds.) McGraw-Hill Medical, New York (2011). **1**. Okamoto, Nakahiro, Chin & Bedikian (1993) *Clin. Pharmacokinet.* **25**, 88. **2**. Clynes, Scully & Neu (1989) *Diagn. Microbiol. Infect. Dis.* **12**, 257. **3**. Nye, Shi, Andrews & Wise (1989) *J. Antimicrob. Chemother.* **24**, 23.

Cefixime

This third-generation cephalosporin (FW = 455.46 g/mol; CAS 79350-37-1), also named Suprax and (6R,7R)-7-{[2-(2-amino-1,3-thiazol-4-yl)-2-(carboxymethoxyimino)acetyl]amino}-3-ethenyl-8-oxo-5-thia-1-azabicyclo[4.2.0]oct-2-ene-2-carboxylic acid, is an orally active antibiotic that is often used to treat bacterial infections of the ear, urinary tract, and upper respiratory tract. Its broad spectrum of antimicrobial action includes both gram-positive and -negative bacteria including many beta-lactamase-producing strains of streptococci, *Haemophilus influenzae*, *Neisseria gonorrhoeae*, and the majority of the Enterobacteriaceae (1). Activity of cefixime against *Staphylococcus aureus*, *Enterococci*, *Listeria monocytogenes*, and *Pseudomonas* spp. is poor. It is especially active against *Escherichia coli* (MIC = 15 ng/mL to 4 μg/mL), *Haemophilus influenzae* (MIC ≤4 ng/mL to 4 μg/mL) and *Proteus mirabilis* (MIC 8 to 60 ng/mL). **1**. Shimada K.(1987) *Japan J. Antibiot.* **40**, 1537.

Cefotaxime

This third-generation cephalosporin (FW = 455.47 g/mol; CAS 63527-52-6), also known by the trade name Claforan®, is a broad-spectrum antibiotic that is active against both Gram-positive and Gram-negative bacteria. It is slightly less active than cefazolin against *Staphylococcus aureus*, but 4x to 300x as active as carbenicillin against gram-negative organisms, including *Pseudomonas aeruginosa*, *Pseudomonas cepacia*, *Enterobacter cloacae*, and *Serratia marcescens* (2). Cefotaxime was the most active compound against *Enterobacteriaceae* members and 20x to 100x more active than cefoxitin against the indole-positive *Proteus* group. **1**. Braveny, Dickert & Machka (1979) *Infection* 7, 231. **2**. Masuyoshi, Arai, Miyamoto & Mitsuhashi (1980) *Antimicrob Agents Chemother.* **18**, 1.

Ceftaroline Fosamil

This advanced-generation cephalosporin antibiotic (FW = g/mol; CAS 866021-48-9 and 400827-46-5), also known by its code names PPI-0903 and TAK-599, its trade names Teflaro® in the U.S. and Zinforo® in the Europe, and its IUPAC name (1). is active against Methicillin-Resistant *Staphylococcus aureus* (MRSA) and broad-spectrum activity against Gram-positive bacteria (2-4). In 2010, ceftaroline is approved by the FDA for the treatment of community-acquired bacterial pneumonia and acute bacterial skin infections. **1**. (6R,7R)-7-[(2Z)-2-ethoxyimino-2-[5-(phosphonoamino)-1,2,4-thiadiazol-3-yl]acetyl]amino]-3-[4-(1-methylpyridin-1-ium-4-yl)-1,3-thiazol-2-yl]sulfanyl]-8-oxo-5-thia-1-azabicyclo[4.2.0]oct-2-ene-2-carboxylate. **2**. Andes & Craig (2006) *Antimicrob. Agents Chemother.* **50**, 1376. **3**. Jacqueline, Caillon, Le Mabecque, *et al.* (2007) *Antimicrob. Agents Chemother.* **51**, 3397. **4**. Stee & Rybak (2010) *Pharmacotherapy* **30**, 375.

Ceftazidime

This third-generation cephalosporin (FW = 546.58 g/mol; CAS 72558-82-8), also named Fortaz®, Tazicef®, GR-20263, and (6R,7R,Z)-7-(2-(2-aminothiazol-4-yl)-2-(2-carboxypropan-2-yloxyimino)acetamido)-8-oxo-3-(pyridinium-1-ylmethyl)-5-thia-1-azabicyclo[4.2.0]oct-2-ene-2-carboxylate, is bacteriocidal (MIC = 60-500 ng/mL) for *Enterobacteriaceae*, *Haemophilus influenzae*, *Neisseria gonorrhoeae*, and Lancefield group A β-hemolytic *Streptococci*, but acts far more weakly against *Pseudomonas aeruginosa* (MIC = 2 µg/mL), *Staphylococcus aureus* (MIC = 16 µg/mL), and *Bacteroides fragilis* (MIC > 128 µg/mL) (1). When administered i.m. or i.v., ceftazidime is widely distributed in body fluids and tissues, exhibiting relatively low binding to serum proteins (2). It is eliminated primarily through the urine and, depending upon the type of infection, can be administered every 8-12 hours. Because of its broad spectrum and its activity against penicillin-resistant and aminoglycoside-resistant *Pseudomonas*, ceftazidime can be used as a single agent in place of combination therapy in patients with cystic fibrosis (2). **Key Pharmacokinetic Parameters:** *See* Appendix II in Goodman & Gilman's *THE PHARMACOLOGICAL BASIS OF THERAPEUTICS*, 12th Edition (Brunton, Chabner & Knollmann, eds.) McGraw-Hill Medical, New York (2011). **1**. Wise, Andrews & Bedford (1980) *Antimicrob. Agents Chemother.* **17**, 884. **2**. Smith (1984) *Clin. Pharm.* **3**, 373.

Ceftobiprole

This so-called fifth-generation cephalosporin antibiotic (FW = 534.57 g/mol; CASs = 209467-52-7 and 252188-71-9), also known by its code name BAL9141 and by trade names Zeftera® and Zevtera®, is highly effective against methicillin-resistant *Staphylococcus aureus*, penicillin-resistant *Streptococcus pneumoniae*, *Pseudo-monas aeruginosa*, and certain enterococci. It also comparable to a vancomycin/ceftazidime combination for treating skin and soft tissue infections. **1**. Entenza, Hohl, Heinze-Krauss, Glauser & Moreillon (2002) *Antimicrob. Agents Chemother.* **46**, 171. **2**. Zbinden, Pünter & von Graevenitz (2002) *Antimicrob. Agents Chemother.* **46**, 871. **3**. Dauner, Nelson & Taketa (2010) *Am. J. Health Syst. Pharm.* **67**, 983.

Ceftriaxone

This third-generation cephalosporin (FW = 554.58 g/mol; CAS 73384-59-5), also known as Ro 13-9904 and Rocephin®, demonstrates broad-spectrum activity against Gram-positive bacteria as well as many Gram-negative bacteria. It is commonly used intravenously to treat sexually transmitted infections caused by *Neisseria gonorrhoeae*. Ceftriaxone inhibits bacterial cell wall synthesis by inhibiting the transpeptidation reaction in the synthesis of peptidoglycan elements of the bacterial cell wall. Many β-lactam antibiotics are transcriptional activators of the Excitatory Amino Acid Transporter-2 (*or* EAAT2), resulting in increased EAAT2 protein levels (2). Treatment of animals with ceftriaxone increases EAAT2 expression and glutamate transport activity in brain. CEF has neuroprotective effects in both in vitro and in vivo models based on its ability to inhibit neuronal cell death by preventing glutamate excitotoxicity. Although ceftriaxone has also been used as a treatment for attenuating attenuates cue-induced cocaine relapse (3), its cerebral penetration in surgery patients is quite low, amounting to <2% of serum concentrations (4). Clavulanic acid may be a better alternative (5). (*See Clavulanic Acid*) **Key Pharmacokinetic Parameters:** *See* Appendix II in Goodman & Gilman's *THE PHARMACOLOGICAL BASIS OF THERAPEUTICS*, 12th Edition (Brunton, Chabner & Knollmann, eds.) McGraw-Hill Medical, New York (2011). **1.** Neu, Meropol & Fu (1981) *Antimicrob. Agents Chemother.* **19**, 414. **3.** Kim, Lee, Kegelman (2011) *J. Cell Physiol.* **226**, 2484. **3.** Rao & Sari (2012) *Curr. Med. Chem.* **19**, 5148. **4.** Lucht, Dorche, Aubert (1990) *J. Antimicrob. Chemother.* **26**, 81. **5.** Schroeder, Tolman & McKenna (2014) *Drug Alcohol Depend.* **142**, 41.

Celecoxib

This cyclooxygenase inhibitor and non-steroidal anti-inflammatory drug (FW = 381.38 g/mol; CAS 169590-42-5; *Symbol*: CE), also known by its code name SC-58635, its trade names Celebrex™ or Celebra™ (for arthritis) and Onsenal™ (for polyps) and systematically as 4-[5-phenyl-3 (trifluoromethyl)-1*H*-pyrazol-1-yl]benzenesulfonamide, is a significantly better inhibitor of COX-2 (IC$_{50}$ = 32 nM) than COX-1 (IC$_{50}$ = 55 µM). In mice, low doses of celecoxib [(0.5-1 mg/kg), but *not* a higher dose (5 mg/kg)] attenuate inflammation-associated gut barrier failure, thereby reducing entry of pathogenic bacteria (6). Celecoxib is a highly permeable drug that is readily absorbed throughout the GI tract, such that dissolution of solid dosage forms may be a rate-limiting factor for absorption. *Note*: Replacement of the methyl group on the phenyl ring by a second trifluoromethyl group abrogates COX-2 inhibition, but enhances other anti-inflammatory properties (*See 4-Trifluoromethyl-Celecoxib*) **Other Targets:** cyclooxygenase (1,4,5); steroid sulfotransferase (*or* SULT2A1), celecoxib switches steroid sulfation from the 3-*O*-position to the 17β-*O*-site (2); thiopurine *S*-methyltransferase (3); 5-lipoxygenase, *or* arachidonate 5-lipoxygenase (5). In prostate cancer cells, CE also inhibits the protein kinase B/Akt pathway (7,8), a signal transduction route known to play a significant role in cancer cell growth and survival. CE, but not rofecoxib, also inhibits cell growth in human prostate cell lines that do not produce COX2 (9). Kulp et al. (2004) used CE and a COX2 inactive analogue to study their effect on prostate cancer cells. By inhibiting ER Ca²⁺-ATPases, CE increase intracellular calcium in prostate cancer cells and osteoblasts, unlike other NSAIDs (10,11). CE also inhibits adenyl cyclase (11). **Key Pharmacokinetic Parameters:** *See* Appendix II in Goodman & Gilman's *THE PHARMACOLOGICAL BASIS OF THERAPEUTICS*, 12th Edition (Brunton, Chabner & Knollmann, eds.) McGraw-Hill Medical, New York (2011). **1.** Penning, Talley, Bertenshaw, *et al.* (1997) *J. Med. Chem.* **40**, 1347. **2.** Cui, Booth-Genthe, Carlini, Carr & Schrag (2004) *Drug Metab. Dispos.* **32**, 1260. **3.** Oselin & Anier (2007) *Drug Metab. Dispos.* **35**, 1452. **4.** Scholz, Ulbrich & Dannhardt (2008) *Eur. J. Med. Chem.* **43**, 1152. **5.** Sud'ina, Pushkareva, Shephard & Klein (2008) *Prostaglandins Leukot. Essent. Fatty Acids* **78**, 99. **6.** Short, Wang, Castle, *et al.* (2013) *Lab Investig.* **93**, 1265. **7.** Hsu, Ching, Wang, *et al.* (2000) *J. Biol. Chem.* 275: 11397. **8.** Kulp, Yang, Hung, *et al.* (2004) *Cancer Res.* **64**, 1444. **9.** Patel, Subbaramaiah, Du, *et al.* (2005) *Clin. Cancer Res.* **11**, 1999. **10.** Johnson, Hsu, Lin, *et al.* (2002) *Biochem. J.* **366**, 831. **11.** Saini, Gessell-Lee & Peterson (2003) *Inflammation* **27**, 79.

Celgosivir

This castanospermine prodrug (FW = 259.30 g/mol; CAS 121104-96-9), also named 6-*O*-butanoylcastanospermine and (1*S*,6*S*,7*S*,8*R*,8*aR*)-1,7,8-trihydroxyoctahydro-6-indolizinyl butyrate, is most likely taken up by cells where it is hydrolyzed to castanospermine, a potent glucosidase inhibitor that inhibits Dengue virus production *in vitro* by inhibiting endoplasmic reticulum α-glucosidases, not by reducing the activity of glycolipid-processing enzymes (1). Dengue infection is the most common mosquito-borne viral disease worldwide. Although celgosivir is an antiviral agent in a lethal mouse model of antibody-enhanced DENV infection, results of a recent clinical trial in hunans suggest celgosivir does not significantly reduce viral load or fever burden in patients with dengue (2). **See also** *Castanospermine* **1.** Sayce, Alonzi, Killingbeck, *et al.* (2016) *PLoS Negl. Trop. Dis.* **10**, e0004524. **2.** Low, Sung, Wijaya, *et al.* (2014) *Lancet Infect. Dis.* **14**. 706.

CEM-101, *See Solithromycin*

Centimitor-1

This cytostatic anticancer agent (FW$_{\text{free-acid}}$ = 466.34 g/mol; Abbreviation: Centi-1), also known as *N'*-(3-bromo-4-hydroxybenzylidene)-2-(9-oxo-10(9*H*)-acridinyl)acetohydrazide, reduces microtubule (MT) dynamics and disrupts mitosis (1). Cent-1's actions phenocopy those of Rigosertib, a related inhibitor said to target PI3K and PLK signaling pathways (**See** *Rigosertib*). Depending on the target cell, Cent-1 induces a transient mitotic arrest, followed by abnormal exit from M-phase and/or cell death. Cent-1 also induces multipolarity and centrosome fragmentation. Notably, Cent-1 and rigosertib modulate microtubule dynamics and the behavior of proteins that bind to microtubule (+)-ends. Although Cent-1 and rigosertib do not show any obvious effects on the appearance of MTs at low concentrations, identical treatments cause discernable changes in the localization of MT end-binding protein EB1 in interphase cells, with EB1 comets appearing to be much shorter and fragmented. In A549 EGFP-α-tubulin cells, 6-hour exposure to 5 µM Cent-1 caused a similar phenotype as in HeLa cells (*e.g.,* chromosome misalignment, multipolarity, and prolonged mitotic delay). The same concentration only shows minor effects on interphase MTs. (These cells were much more sensitive to rigosertib treatment than HeLa cells, and treatment with 0.25 µM rigosertib completely abolished microtubules; at 0.55 µM, rigosertib did not eliminate MTs but was sufficient for mitotic arrest.) Both Cent-1 and rigosertib cause centrosome fragmentation and reduced the amount of centrosome-associated γ-tubulin. They also significantly retard MT dynamics in interphase cells, shorten spindle length in mitosis, and decrease tension across sister kinetochores in mitotic chromosomes. They also cause the delocalization of NuMA and

EB1 in mitotic cells. Taken together, these data suggest Cent-1 and rigosertib impair MT-mediated processes during M phase. Acentrosomal spindle poles, reduced interkinetochore tension, mislocalization of MT end-tracking proteins, and chromosome misalignment are consequences of treating cells with low doses of well-studied MT drugs, again supporting the inference that these compounds may modulate microtubule dynamics. Taken together, these data suggest Cent-1 and rigosertib impair MT-mediated processes during M phase. It is noteworthy that neither Cent-1 nor Rigoserib perturb MT self-assembly *in vitro*. Moreover, the current findings do not distinguish between direct and indirect drug-induced mechanisms for altering microtubule processes and thereby hindering mitosis. **1.** Mäki-Jouppila, Laine, Rehnberg, *et al.* (2014) *Mol. Cancer Ther.* **13**, 1054.

Centrinone B

This PLK4 inhibitor (FW = 631.67 g/mol; CAS 1798871-31-4; Solubility: 50 mM DMSO), also named 2-[[2-fluoro-4-[[(2-fluoro-3-nitrophenyl)methyl]sulfonyl]phenyl]thio]-5-methoxy-*N*-(5-methyl-1*H*-pyrazol-3-yl)-6-(1-piperidinyl)-4-pyrimidinamine, targets the Polo-like Kinase-4 (K_i = 0.6 nM), depleting centriole and centrosome levels *in vitro*. Centrinone induces cell cycle arrest in normal human cell lines in a p53-dependent manner. **1.** Wong, Anzola, Davis, *et al.* (2015) *Science* **348**, 1155.

CEP-32215

This potent, selective, and orally bioavailable inverse agonist (FW = 383.49 g/mol), also known as 3-(1'-cyclobutylspiro[4*H*-1,3-benzodioxine-2,4'-piperidin]-6-yl)-5,5-dimethyl-1,4-dihydro-pyridazin-6-one, targets the histamine H_3 receptor (*or* H_3R), a G-coupled histamine receptor that affects wakefulness. Histaminergic neurons are located exclusively in the tuberomammillary nucleus of the posterior hypothalamus and project to most regions of the brain. The wake-promoting effects of histamine are modulated via H_1 receptors, as indicated by the ability of brain-penetrant H_1 antagonists to induce sedation. Histamine also produces similar effects via H_3R activation, which results in the inhibition of histamine synthesis and release via the autoreceptor mechanism. CEP-32215 exhibits drug-like properties, with high affinity for human H_3R (K_i = 2.0 nM) and rat H_3R (K_i = 3.6 nM) (1,2). It is an antagonist (K_b = 0.3 nM) and inverse agonist (EC_{50} = 0.6 nM) in binding assays employing [^{35}S]guanosine 5'-*O*-(γ-thio)-triphosphate. Upon oral dosing, occupancy of H_3R by CEP-32215 can be estimated in *ex vivo* binding measurements on rat cortical slices (ED_{50} = 0.1 mg/kg, p.o.). Functional antagonism is also demonstrable in brain as inhibition of *R*-α-methylhistamine-induced drinking in the rat dipsogenia model (ED_{50} = 0.92 mg/kg). CEP-32215 also significantly increases wake duration in the rat EEG model at 3-30 mg/kg, p.o. (1). Increased motor activity, sleep rebound or undesirable events (such as spike wave or seizure activity) is not observed following doses up to 100 mg/kg p.o., indicating an acceptable therapeutic index. CEP-32215 may have potential utility in the treatment of a variety of sleep disorders (1). Wang, Przyuski, Roemmele, *et al.* (2) describe the synthesis of CEP-

32215 and provide details on the formation of its key spiroketal moiety as well as discovery of a novel Suzuki coupling approach for synthesis of the backbone of the molecule. **1.** Hudkins, Gruner, Radditz, *et al.* (2016) *Neuropharmacol.* **106**, 37. **2.** Haas & Panula (2016) *Neuropharmacol.* **106**, 1. **3.** Wang, Przyuski, Roemmele, *et al.* (2013) *Org. Proc. Res. Develop.* **17**, 846.

Cephalexin

This orally active semi-synthetic cephalosporin antibiotic (FW$_{free-acid}$ = 347.39 g/mol; CAS 15686-71-2), also known by the Lilly tradename Keflex™ and by the systematic name, (6*R*,7*R*)-7-{[(2*R*)-2-amino-2-phenylacetyl]amino}-3-methyl-8-oxo-5-thia-1-azabicyclo-[4.2.0]oct-2-ene-2-carboxylate, inhibits bacterial peptido-glycan biosynthesis. Oligopeptidic drugs, such as this β-lactam antibiotic and angiotensin-converting enzyme inhibitors, often rely on the same carriers to gain entry into human and animal cells, a property that can give rise to competitive pharmacokinetic interactions. **Key Pharmacokinetic Parameters:** *See* Appendix II in Goodman & Gilman's *THE PHARMACOLOGICAL BASIS OF THERAPEUTICS*, 12th Edition (Brunton, Chabner & Knollmann, eds.) McGraw-Hill Medical, New York (2011). **Target(s):** serine-type D-Ala-D-Ala carboxypeptidase, *or* penicillin-binding protein-5 (1,5,8); penicillin-binding protein-4 (2,6); penicillin-binding protein-3 (3,7); peptidoglycan biosynthesis (4,5); penicillin-binding protein-1 (7); penicillin-binding protein-2 (7). **1.** Wilkin (1998) in *Handb. Proteolytic Enzymes*, p. 418, Academic Press, San Diego. **2.** Wilkin (1998) in *Handb. Proteolytic Enzymes*, p. 435, Academic Press, San Diego. **3.** Wientjes & Nanninga (1991) *Res. Microbiol.* **142**, 333. **4.** Wickus & Strominger (1972) *Meth. Enzymol.* **28**, 687. **5.** Fuad, Frère, Ghuysen, Duez & Iwatsubo (1976) *Biochem. J.* **155**, 623. **6.** Miyamoto, Yamaguchi, Abu Sayed, *et al.* (1997) *Microbiol. Res.* **152**, 227. **7.** Barbour (1981) *Antimicrob. Agents Chemother.* **19**, 316. **8.** Coyette, Perkins, Polacheck, Shockman & Ghuysen (1974) *Eur. J. Biochem.* **44**, 459.

Cephalosporin C

This antibiotic (FW = 415.42 g/mol; CAS 28240-09-7 (potassium salt)), originally obtained from a species of *Cephalosporium*, but made synthetically (1,2), inhibits peptidoglycan and cell wall biosynthesis. Cephalosporin C has pK_a values of < 2.6, 3.1, and 9.8. **Target(s):** serine-type D-Ala-D-Ala carboxypeptidase, *or* penicillin-binding protein-5 (2,8); penicillin-binding protein-4 (4); β-lactamase, *or* penicillinase (5); peptide-glycan biosynthesis (6); glutaryl 7-aminocephalosporic-acid acylase, also a weak substrate (7); undecaprenyldiphospho-muramoylpentapeptide β-*N*-acetylglucosaminyl-transferase, *or* MurG transferase (9). **1.** Woodward, Heusler, Gosteli, *et al.* (1966) *J. Amer. Chem. Soc.* **88**, 852. **2.** Woodward (1966) *Science* **153**, 487. **3.** Wilkin (1998) in *Handb. Proteolytic Enzymes*, p. 418, Academic Press, San Diego. **4.** Wilkin (1998) in *Handb. Proteolytic Enzymes*, p. 435, Academic Press, San Diego. **5.** Abraham & Newton (1956) *Biochem. J.* **63**, 628. **6.** Izaki, Matsuhashi & Strominger (1968) *J. Biol. Chem.* **243**, 3180. **7.** Lee, Chang, Liu & Chu (1998) *Biotechnol. Appl. Biochem.* **28** (Pt. 2), 113. **8.** Kelly, Knox, Zhao, Frère & Ghuysen (1989) *J. Mol. Biol.* **209**, 281. **9.** Ravishankar, Kumar, Chandrakala, *et al* (2005) *Antimicrob. Agents Chemother.* **49**, 1410.

Cerebratulus Toxin A-III

This marine worm toxin (MW = 9.8 kDa), one of four homologous *Cerebratulus lacteus* protein toxins, consists of a 95-residue polypeptide, cross-linked by three disulfide bonds and mainly consisting of α-helix. Toxin A-III permeabilizes a variety of cells as well as liposomes made from a variety of phospholipids, apparently forming large pores permitting release of large proteins almost as rapidly as small organic molecules and inorganic ions. At sublytic concentrations, the toxin also inhibits protein kinase C and endogenous voltage-gated cation-selective sodium, potassium channels in nervous and cardiovascular systems. In view of its small size and helical structure, A-III remains a promising probe for investigating pore-forming protein toxin insertion into biological membranes and the ensuing steps in pore formation. **1.** Kem (1994) *Toxicol.* **87**, 189.

Ceritinib

This potent EML4-ALK signal transduction inhibitor and antineoplastic agent (FW = 558.14 g/mol; CAS 1032900-25-6; Solubility: 56 mg/mL DMSO; <1 mg/mL H_2O), also known as LDK378, Zykadia®, and 5-chloro-N^4-[2-[(1-methylethyl)sulfonyl]phenyl]-N^2-[5-methyl-2-(1-methylethoxy)-4-(4-piperidinyl)phenyl]-2,4-pyrimidinediamine, targets ALK, the Activin receptor-Like Kinase activity of Transforming Growth Factor-β (TGF-β) receptors, with IC_{50} of 0.2 nM, showing 40x and 35x selectivity against IGF-1R and InsR, respectively. Zykadia inhibits autophosphorylation of ALK, ALK-mediated phosphorylation of the downstream signaling protein STAT3, as well proliferation of ALK-dependent cancer cells in *in vitro* and *in vivo* assays. Non-small cell lung cancers often harbor anaplastic lymphoma kinase (ALK) mutations (*e.g.,* Leu-1196-Met, Gly-1269-Ala, Ile-1171-Thr, and Ser-1206-Tyr) that invariably lead to crizotinib resistance; however, ceritinib can overcome several crizotinib-resistant mutations and is potent against several *in vitro* and *in vivo* laboratory models of acquired resistance to crizotinib (3,4). Ceritinib was approved by the FDA Zykadia for the treatment of patients with anaplastic lymphoma kinase (ALK)-positive metastatic non-small cell lung cancer who have progressed on or are intolerant to crizotinib. Brain accumulation of ceritinib is restricted by P-glycoprotein (P-GP/ABCB1) and breast cancer resistance protein (BCRP/ABCG2), but ceritinib export can be specifically inhibited by zosuquidar (ABCB1 inhibitor) and Ko143 (ABCG2 inhibitor) (5). **1.** Marsilje, Pei, Chen, *et al.* (2013) *J. Med. Chem.* **56**, 5675. **2.** Chen, Jiang & Wang (2013) *J. Med. Chem.* **56**, 5673. **3.** Friboulet, Li, Katayama, *et al.* (2014) *Cancer Discov.* **4**, 662. **4.** Shaw, Kim, Mehra, *et al.* (2014) *New Engl. J. Med.* **370**, 1189. **5.** Kort, Sparidans, Wagenaar, *et al.* (2015) *Pharmacol Res.* **201**, 102.

Cerivastatin

This cholesterol-reducing statin ($FW_{free-acid}$ = 459.56 g/mol; CAS 145599-86-6; IUPAC Name: (3R,5S,6E)-7-[4-(4-fluorophenyl)-5-(methoxymethyl)-2,6-bis(propan-2-yl)pyridin-3-yl]-3,5-dihydroxyhept-6-enoic acid, potently inhibits 3-hydroxy-3-methylglutaryl-CoA reductase: K_i = 5.7 nM at 37°C. (Cerivastatin was withdrawn from the market in August, 2001 due to risk of serious rhabdomyolysis.) Cerivastatin inhibits cholesterol biosynthesis in the human cell lines with a similar IC_{50} value. At daily doses of 0.2 mg/day, low density lipoprotein-cholesterol, total cholesterol, and triacylglycerol levels were significantly reduced in individuals with type IIa hypercholesterolemia. In addition, high density lipoprotein-cholesterol levels increased. **Target(s):** 3-hydroxy-3-methylglutaryl-CoA reductase (1,2); CYP2C8 (3). **1.** Istvan & Deisenhofer (2001) *Science* **292**, 1160. **2.** Carbonell & Freire (2005) *Biochemistry* **44**, 11741. **3.** Tornio, Pasanen, Laitila, Neuvonen & Backman (2005) *Basic Clin. Pharmacol. Toxicol.* **97**, 104.

CEP-701, *See Lestaurtinib*

Certolizumab pegol

This anti-TNFα drug (MW = 91 kDa; CAS 428863-50-7; *Abbreviation*: CZP), also known as CDP870 and the trade name Cimzia™, is monoclonal antibody approved for the treatment of rheumatoid arthritis (RA) and Crohn's Disease (CD). Composed of a humanized antigen-binding fragment (Fab) conjugated to polyethylene glycol, CZP lacks an Fc fragment and is thus unlike other agents. Pegylation increases certolizumab's plasma half-life to ~14 days. In mice with collagen-induced arthritis, CZP penetrated inflamed joint tissue to a greater extent and for a longer time period than either adalimumab or infliximab and the degree of penetration correlated better with the level of inflammation within the tissue (1). Unlike other anti-TNFα agents (*e.g.,* infliximab, adalimumab, and etanercept), certolizumab pegol does not stimulate apoptosis in a variety of *in vitro* assays, a finding that suggests apoptosis is not essential for the efficacy of anti-TNFα agents in Crohn's Disease (2). **1.** Horton, Sudipto Das, Paul Emery (2011) *Int. J. Clin. Rheumatol.* **6**, 517. **2.** Nesbitt, Fossati & Bergin (2007) *Inflamm. Bowel Dis.* **13**, 1323.

Cerulenin

This antifungal antibiotic and irreversible protein acylation inhibitor (FW = 223.27 g/mol; CAS 17397-89-6), also named (2R,3S)-3-[(4E,7E)-nona-4,7-dienoyl]oxirane-2-carboxamide, inhibits fatty acid and steroid biosynthesis, targeting β-keto-acyl-ACP synthase (k = 88 $M^{-1}s^{-1}$ at 0 °C and pH 6.5, or ~90x faster than by iodoacetamide) in the Fatty Acid Synthase and blocking its interactions with malonyl-CoA (1,2). The enzyme was protected against inhibition by prior treatment with acetyl-CoA but not malonyl-CoA. Cerulenin has no effect on the malonyl-CoA decarboxylase activity of the iodoacetamide-treated enzyme (2). When the enzyme is first incubated with cerulenin, malonyl-CoA decarboxylase activity is not detectable, even after treatment of the enzyme with iodoacetamide, suggesting cerulenin reacts with the peripheral SH-groups, leading to inactivation (2). The structure of the complex between the antibiotic cerulenin and its target, β-ketoacyl-acyl carrier protein synthase has been published (3). Both cerulenin (~10 µg/ml) and TOFA (~1 µg/ml) are effective in blocking the incorporation of radiolabeled acetate into palmitate; however, TOFA reduces malonyl-CoA levels, rather than elevating them. Cerulenin inhibits growth of yeast-type fungi and several kinds of filamentous fungi, with *Candida stellatoidea* most sensitive among them. This antifungal activity is not reversed by amino acids, and purine and pyrimidine derivatives, but it is reversed by ergocalciferol, and to a lesser extent by retinol, thiamine, pantothenic acid, lauric acid and oleic acid (4). Cerulenin also suppresses expression of the p185(HER2) oncoprotein and tyrosine-kinase activity in breast and ovarian HER2 overexpressors (5). Cerulenin is also a potent inhibitor of antigen processing by antigen-presenting cells (6). **1.** Vance, Goldberg, Mitsuhashi & Bloch (1972) *Biochem. Biophys. Res. Commun.* 48, 649. **2.** Kawaguchi, Tomoda, Nozoe, Omura, Okuda (1982) *J. Biochem.* (Tokyo) **92**, 7. **3.** Moche, Schneider, Edwards, *et al.* (1999) *J. Biol. Chem.* **274**, 6031. **4.** Nomura, Horiuchi, Omura & Hata (1972) *J. Biochem.* (Tokyo) **71**, 783. **5.** Menendez, Vellon, Mehmi, *et al.* (2004) *Proc. Natl. Acad. Sci. U.S.A.* **101**,

10715. **6**. Falo, Benacerraf, Rothstein & Rock (1987) *J. Immunol.* **139**, 3918.

Cetirizine

This once-daily, orally bioavailable, second-generation antihistamine (FW = 366.69 g/mol; CAS 83881-51-0), also known as Zyrtec® and Reactine® and (±)-[2-[4-[(4-chlorophenyl)phenylmethyl]-1-piperazinyl]ethoxy]acetic acid, is selective H_1 receptor inverse agonist, with rapid onset of action. Cetirizine is a major human metabolite of hydroxyzine, another histamine H_1-antagonist. Unlike its parent compound, cetirizine only poorly penetrates the blood-brain barrier, thereby minimizing the sedative and anticholinergic effects observed with hydroxyzine (1). At a nominal dose of 5–20 mg in adult humans, cetirizine provides greater protection against histamine-induced bronchospasm than hydroxyzine, which is oxidized to cetirizine by the action of liver alcohol dehydrogenase (**See** *Hydroxyzine*) (1). Compared to placebo, cetirizine significantly decreased the eosinophil attraction at skin sites challenged with grass pollen and compound 48/80 (2). With the FDA-mandated withdrawal of both terfenadine and astemizole (which produce unexpected and significant prolongation of the cardiac QT interval, with other more serious cardiac effects) from the U.S. market, current widely available second-generation antihistamines include, loratadine, cetirizine, and fexofenadine. **Key Pharmacokinetic Parameters:** *See* Appendix II in Goodman & Gilman's *THE PHARMACOLOGICAL BASIS OF THERAPEUTICS*, 12th Edition (Brunton, Chabner & Knollmann, eds.) McGraw-Hill Medical, New York (2011). **1**. Brik, Tashkin, Gong, Dauphinee & Lee (1987) *J. Allergy Clin. Immunol.* **80**, 51. **2**. Fadel, Herpin-Richard, Rihoux & Henocq (1987) *Clin. Allergy* **17**, 373.

CFTR Modulators

Lumacaftor

Ivacaftor

These agents (marketed in combination as Orkambi®) target Phe^{506}-deleted Cystic Fibrosis Transmembrane Conductance Regulator (CFTR$^{\Delta F508}$), a chloride channel present on the epithelial cells of multiple organs. CFTR$^{\Delta F508}$ is misfolded, resulting in defective cellular processing/trafficking as well as proteasome-mediated degradation. Lumacaftor (FW = 452.41 g/mol; CAS 936727-05-8; also named VX-809 and 3-{6-{[1-(2,2-difluoro-1,3-benzodioxol-5-yl)cyclopropyl]carbonyl]amino}-3-methyl pyridin-2-yl}benzoate) increases the conformational stability of CFTR$^{\Delta F508}$, resulting in increased processing/trafficking of mature protein to the cell surface. Ivacaftor (FW = 392.49 g/mol; CAS 873054-44-5; also named *N*-(2,4-di-*tert*-butyl-5-hydroxyphenyl)-4-oxo-1,4-dihydroquinoline-3-carbox amide and VX-770) is a CFTR potentiator that facilitates increased chloride transport by increasing the channel-open probability of the CFTR protein, once incorporated into the cell surface. *In vitro* experiments showed that lumacaftor and ivacaftor act directly on the CFTR in primary human bronchial epithelial cultures, resulting in increased chloride ion transport. Orkambi received FDA's breakthrough therapy designation

because the sponsor demonstrated through preliminary clinical evidence that the drug may offer a substantial improvement over available therapies.

CGP-33101, *See* Rufinamide

CGP-37157

This widely employed benzothiazepine-based calcium ion blocker (FW = 342.22 g/mol; CAS 75450-34-9; *Symbol* = CGP), also known as 7-chloro-5-(2-chlorophenyl)-1,5-dihydro-4,1-benzothiazepin-2(3*H*)-one, selectively inhibits the Na$^+$-Ca^{2+} exchanger of cardiac mitochondria (IC$_{50}$ = 0.36 μM), without affecting the L-type voltage-dependent calcium channel, the Na$^+$-Ca^{2+} exchanger, or the Na$^+$-K$^+$-ATPase of the cardiac sarco-lemma, or the Ca^{2+}-ATPase of the cardiac sarcoplasmic reticulum (1,2). Mitochondrial Na$^+$-Ca^{2+} exchange activity is determined by monitoring intramitochondrial [Ca^{2+}]$_{free}$ in isolated heart mitochondria loaded with the Ca^{2+}-sensitive fluorophore fura-2. CGP-37157 had no effect on the calcium current, recorded by whole-cell voltage clamp in isolated neonatal ventricular myocytes. CGP-37157, at or below 10 μM, was without effect on the activities of the Na$^+$-Ca^{2+} exchanger and Na$^+$-K$^+$-ATPase in isolated cardiac sarcolemmal vesicles or on activity of the Ca^{2+}-ATPase in isolated cardiac sarcoplasmic reticulum vesicles. Such findings data suggest CGP-37157 is a potent, selective, and specific inhibitor of mitochondrial Na$^+$-Ca^{2+} exchange when used at or below 10 μM. CGP37157 also modulates mitochondrial Ca^{2+} homeostasis in cultured rat dorsal root ganglion neurons (3). CGP37157 was also useful in demonstrating that mitochondria accumulate Ca^{2+} following intense glutamate stimulation of cultured rat forebrain neurons. **Other Target(s):** CGP-37157 also directly modulates ryanodine receptor channels (RyRs) and/or sarco/endoplasmic reticulum Ca^{2+}-stimulated ATPase (SERCA) by CGP (5). In the presence of ruthenium red (an inhibitor of RyRs), CGP decreased SERCA-mediated Ca^{2+} uptake of cardiac and skeletal sarcoplasmic reticulum (SR) microsomes, with IC$_{50}$ values of 6.6 and 9.9 μM, respectively. The CGP effects on SERCA activity correlated with a decreased V_{max} of ATPase activity of SERCA-enriched skeletal SR fractions. CGP (≥ 5 μM) also increased RyR-mediated Ca^{2+} leak from skeletal SR microsomes. Planar bilayer studies confirmed that both cardiac and skeletal RyRs are directly activated by CGP-37157, with EC$_{50}$ values of 9.4 and 12.0 μM, respectively. Therefore, the reported action of CGP on cellular Ca^{2+} homeostasis of cardiac, skeletal muscle, and other nonmuscle systems requires further analysis to take into account the contribution of other CGP-sensitive Ca^{2+} transporters. **1**. Cox, Conforti, Sperelakis & Matlb (1993) *J. Cardiovasc. Pharmacol.* **21**, 595. **2**. Cox & Matlib (1993) *TiPS* **14**, 408. **3**. Baron & Thayer (1997) *Eur. J. Pharmacol.* **340**, 295. 4. White & Reynolds (1997) *J. Physiol.* **498**, 31. **5**. Neumann, Diaz-Sylvester, Fleischer & Copello (2011) *Mol. Pharmacol.* **79**, 141.

CGP 41251

This staurosporine derivative (FW = 556.64 g/mol), also known as PKC412 and 4'-N-benzoylstaurosporine, selectively inhibits *in vitro* protein kinase C (1). PKCα (IC$_{50}$ = 24 nM), PKCβI (IC$_{50}$ = 17 nM), PKCβII (IC$_{50}$ = 32 nM), PKCγ (IC$_{50}$ = 18 nM), PKCδ(IC$_{50}$ = 360 nM), PKCη (IC$_{50}$ = 60 nM), and PKCϵ (IC$_{50}$ = 4.5 μM). PKCζ is very weakly inhibited. **Target(s):** phosphatidylinositol 3-kinase (1); protein kinase C (2-4); vascular endothelial growth factor (VEGF) receptor kinase (4). **1**. Berggren, Gallegos, Dressler, Modest & Powis (1993) *Cancer Res.* **53**, 4297. **2**. Marte, Meyer, Stabel, *et al.* (1994) *Cell Growth Differ.* **5**, 239. **3**. Kessels, Krause & Verhoeven (1993) *Biochem. J.* **292**, 781. **4**. Fabbro, Buchdunger, Wood, *et a.* (1999) *Pharmacol Ther.* **82**, 293.

CGP 52622A

This hydroxylamine derivative (FW$_{\text{free-base}}$ = 130.19 g/mol) competitively inhibits *Plasmodium falciparum* ornithine decarboxylase (K_i = 20.4 nM). **1**. Birkholtz, Joubert, Neitz & Louw (2003) *Proteins* **50**, 464. **2**. Das Gupta, Krause-Ihle, Bergmann, *et al.* (2005) *Antimicrob. Agents Chemother.* **49**, 2857. **3**. Krause, Lüersen, Wrenger, *et al.* (2000) *Biochem. J.* **352**, 287.

CGP-57148, *See* Imatinib

CGP 74514A

This cell-permeable CDK/cyclin inhibitor (FW = 385.89 g/mol; CAS 1173021-98-1), also named 2-N-(2-aminocyclohexyl)-6-N-(3-chloro phenyl)-9-ethylpurine-2,6-diamine, potently and selectively targets CDK/cyclin B (IC$_{50}$ = 25 nM) and shows weaker action against PKCα (IC$_{50}$ = 6.1 μM), PKA (IC$_{50}$ = 125 μM) and EGFR (IC$_{50}$ > 10 μM). In human leukemia cells, CGP74514A induces complex changes in cell cycle-related proteins accompanied by extensive mitochondrial damage, caspase activation, and apoptosis (1). Exposure of U937 monocytic leukemia cells to minimally toxic concentrations of CGP74514A for 3 hours along with the PI3K inhibitor LY294002 markedly decreases Akt phosphorylation (2). **1**. Dai, Dent & Grant (2002) *Cell Cycle* **1**, 143. **2**. Yu, Rahmani, Dai, *et al. Cancer Res.* **63**, 1822.

CGS 9343B

This potent, selective and cell-permeable calmodulin antagonist (FW = 544.61 g/mol; CAS 109826-27-9; Soluble to 50 mM in DMSO), also known as Zaldaride® and 1,3-dihydro-1-[1-[(4-methyl-4H,6H-pyrrolo[1,2-a][4,1]benzoxazepin-4-yl)methyl]-4-piperidinyl]-2H-benzimidazol-2-one, targets calmodulin-stimulated cAMP phosphodiesterase activity (IC$_{50}$ = 3.3 μM). CGS 9343B is 3.8 times more potent than trifluoperazine (IC$_{50}$ = 12.7

μM) as an inhibitor of calmodulin activity (1). CGS 9343B does not inhibit protein kinase C activity at concentrations up to 100 μM, whereas trifluoperazine inhibits protein kinase C activity (IC$_{50}$ = 44 μM (1). CGS 9343B weakly displaces [^{3}H]spiperone from postsynaptic dopamine receptors with an IC$_{50}$ value of 4.8 μM, while IC$_{50}$ for trifluoperazine is 18 nM (1). CGS-9343B reversibly blocks voltage-activated Ca^{2+}, Na$^+$, and K$^+$ currents in differentiated rat pheochromocytoma (PC12) cells (2). It also inhibits nicotinic acetylcholine receptor (nAChR) channel currents, but not inward ion currents evoked by extracellular ATP. Depolarization-induced intracellular Ca^{2+} transients are almost completely inhibited in growth cones and cell bodies by CGS 9343B (2). CGS9343B also prevents estrogen-induced transcriptional activation by the estrogen receptor (ER), without altering basal transcription (3). The inhibition is dose-dependent and independent of the time of estrogen stimulation. **1**. Norman, Ansell, Stone, Wennogle & Wasley (1987) *Mol. Pharmacol.* **31**, 535. **2**. Neuhaus & Reber (1992) *Eur. J. Pharmacol.* **226**, 183. **3**. Li, Li & Sacks (2003) *J. Biol. Chem.* **278**, 1195.

CGS 13945, *See* Pentopril (and Pentoprilat)

CGS 14824A, *See* Benazepril

CGS27023A, *See* MMI270

CH5132799

This orally active PI3K inhibitor (FW = 377.42 g/mol; CAS 1007207-67-1; Solubility: 12 mg/mL DMSO) selectively targets PI3Kα (IC$_{50}$ = 14 nM), PI3Kβ (IC$_{50}$ = 120 nM), PI3Kδ (IC$_{50}$ = 500 nM), PI3Kγ (IC$_{50}$ = 36 nM), but shows less inhibition of class II PI3Ks, class III PI3k and mTOR and also no inhibitory activity (IC$_{50}$ > 10 μM) against twenty-six protein kinases. PI3K phosphorylates PtdIns(4,5)P$_2$, producing phosphoinositide PtdIns(3,4,5)P$_3$, which plays a central role in cell proliferation and human cancer survival. **1**. Ohwada, Ebiike, Kawada, *et al.* (2011) *Bioorg. Med. Chem. Lett.* **21**, 1767. **2**. Tanaka, Yoshida, Tanimura, *et al.* (2011) *Clin. Cancer Res.* **17**, 3272.

CH5164840

This anti-proliferative agent (FW = 385.49 g/mol) targets the molecular chaperone Hsp90 (Heat shock protein 90), inducing the degradation of cell-stabilizing and oncogenic client proteins, altering the proliferation and/or survival of treated cells. CH5164840 exhibits potent antitumor efficacy with regression in NCI-N87 and BT-474 tumor xenograft models and significantly enhances antitumor efficacy against gastric and breast cancer models, when used in combination with the HER2-directed agents, trastuzumab and lapatinib. CH5164840 binds tightly to Hsp90α (k_{on} = 5.6 x 10^5 M^{-1}s^{-1}; k_{off} = 2.9 x 10^{-4} s^{-1}; K_d = 0.52 nM) and Hsp90β (k_{on} = 6.6 x 10^5 M^{-1}s^{-1}; k_{off} = 9.2 x 10^{-4} s^{-1}; K_d = 1.4 nM). Hsp90 is considered an attractive target in anticancer therapy for several reasons; first, its role in the regulation of its client proteins, many of which are cancer-related proteins, such as kinases, transcription factors, and steroid receptors. Hsp90 inhibition induces degradation of these clients, leading to the inhibition of multiple signaling pathways that regulate tumor cell proliferation and survival. **1**. Ono, Yamazaki, Nakanishi, *et al.* (2012) *Cancer Sci.* **103**, 342.

CH5183284

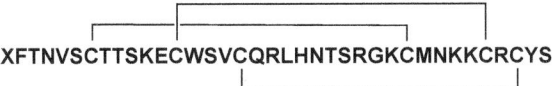

This selective and orally available mitogen-activated protein kinase kinase (MEK) inhibitor, *or* phosphoinositide 3-kinase (PI3K) inhibitor (FW = 356.38 g/mol; CAS 1265229-25-1; Solubility: 71 mg/mL DMSO; < 1 mg/mL H_2O), also named Debio-1347 and (5-amino-1-(2-methyl-3*H*-benzo[d]-imidazol-5-yl)-1*H*-pyrazol-4-yl)(1*H*-indol-2-yl)methanone, targets fibroblast growth factor receptors FGFR1 (IC_{50} = 9.3 nM), FGFR2 (IC_{50} = 7.6 nM), FGFR3 (IC_{50} = 22 nM), and FGFR4 (IC_{50} = 290 nM) (1). The FGFR family consists of FGFR1, FGFR2, FGFR3, and FGFR4, each of which is bound by a subset of 22 fibroblast growth factor (FGF) ligands. FGFRs are activated by ligand-dependent or ligand-independent dimerization that leads to intermolecular phosphorylation. FGFR substrate 2 (FRS2) is a key adaptor protein that is phosphorylated by FGFR. Phosphorylated FRS2 recruits several other adaptor proteins and activates the mitogen-associated protein kinase (MAPK) or PI3K/AKT pathways. In FGFR-altered cancer cells, <u>D</u>ual-<u>S</u>pecificity <u>P</u>hosphatase <u>6</u> (DUSP6) was more significantly suppressed by CH5183284 than by CH5126766. CH5126766 suppressed phospho-ERK, but not the downstream ERK effector DUSP6, suggesting that CH5126766 does not strongly suppress MAPK, resulting in the resistance of FGFR-altered cancer cells to MEK inhibition. It thus appears that CH5183284 exerts its inhibitory effects by suppressing ERK signaling without feedback activation of the MAPK pathway via FGFR (2). **1**. Nakanishi, Akiyama, Tsukaguchi, *et al.* (2014) *Mol. Cancer Ther.* **3**, 2547. **2**. Nakanishi, Mizuno, Sase, *et al.* (2015) *Mol. Cancer Ther.* **14**, 2831.

CH5424802, *See Alectinib*

Chaetomellic Acid A

This natural product ($FW_{free-acid}$ = 326.48 g/mol; CAS 148796-51-4) from *Chaetomella acutiseta* inhibits protein farnesyltransferase (IC_{50} = 55 nM), competitively with respect to farnesyl diphosphate and noncompetitive with respect to Ras). At higher concentrations, it will also inhibit protein geranyl-geranyltransferases I and II (IC_{50} values of 92 and 34 μM, respectively). **Target(s):** protein farnesyltransferase (1-3,5); protein geranylgeranyltransferase type I (2); protein geranylgeranyltransferase type II (2); rubber *cis*-polyprenylcistransferase (4); penicillin binding protein-1b (6); peptidoglycan glycosyltransferase (6). **1**. Gibbs (2001) *The Enzymes*, 3rd ed., **21**, 81. **2**. Gibbs, Pompliano, Mosser, *et al.* (1993) *J. Biol. Chem.* **268**, 7617. **3**. Lingham, Silverman, Bills, *et al.* (1993) *Appl. Microbiol. Biotechnol.* **40**, 370. **4**. Mau, Garneau, Scholte, *et al.* (2003) *Eur. J. Biochem.* **270**, 3939. **5**. Courdavault, Thiersault, Courtois, *et al.* (2005) *Plant Mol. Biol.* **57**, 855. **6**. Garneau, Qiao, Chen, Walker & Vederas (2004) *Bioorg. Med. Chem.* **12**, 6473.

Charybdotoxin

XFTNVSCTTSKECWSVCQRLHNTSRGKCMNKKCRCYS

This 37-residue neurotoxin (FW = 4295.95 g/mol; CAS 95751-30-7: Symbol: ChTX) from the venom of the Deathstalker scorpion (*Leiurus quinquestriatus hebraeus*) is a relatively specific inhibitor of the big conductance Ca^{2+}-activated K^+ channel (1). This channel is activated by depolarizing voltages and has a voltage-sensitivity that is modulated by cytoplasmic Ca^{2+} levels. Its conduction is also unusual, showing high ionic

selectivity and a very high unitary conductance (1). Purified ChTX potently and selectively inhibits the 220-pS Ca^{2+}-activated K^+ channel present in GH3 anterior pituitary cells and primary bovine aortic smooth muscle cells (2). It reversibly blocks channel activity by interacting at the external pore of the channel protein (K_d = 2.1 nM). The ChTX primary structure is similar to neurotoxins of diverse origin, suggesting ChTX is a member of a superfamily of ion-channel modifiers (2). ChTX-sensitive K_{Ca} channels were highly activated to regulate the myogenic tone in the resting state of carotid, femoral and mesenteric arteries from spontaneously hypertensive rats (SHR). The increased K_{Ca} channel functions in SHR arteries appeared to be secondary to the increased Ca^{2+} influx via L-type voltage-dependent Ca^{2+} channels in the resting state of these arteries (3). ChTX blocks homotetrameric, voltage-gated K^+ channels and its Shaker channel (*i.e.*, His replacing wild-type Phe-425) by binding near the outer entrance to the pore in one of four indistinguishable orientations (4). (**See also Paxilline**) **1**. Miller, Moczydlowski, Latorre & Phillips (1985) *Nature* **313**, 316. **2**. Gimenez-Gallego, Navia, Reuben, *et al.* (1988) *Proc. Natl. Acad. Sci. USA* **85**, 3329. **3**. Asano, Masuzawa-Ito & Matsuda (1993) *Brit. J. Pharmacol.* **108**, 214. **4**. Thompson & Begenisich (2000) *Biophys. J.* **78**, 2382.

Chemical Synthetic Lethality

This chemotherapeutic strategy has its roots in phenomenon known as Synthetic Lethality, which requires the experimentally or environmentally imposed co-occurrence of two genetic events (*e.g.*, gene fusions, mutations, and delections) for an organism or cell toi die. Two genes are said to be synthetic lethal, if a mutation in either is compatible with viability, but mutations in both genes results prove lethal. In genetics, synthetic lethal interactions provide insights into the functional relationships between/among genes that are relatively easy to score in screens. In the analogous phenomenon, known as Chemical Synthetic Lethality (CSL), death occurs when a particular gene fusion, deletion, or mutation(s), but not the corresponding wild-type allele, confers lethality in the presence of a particular drug. CSL experiments often provide (a) insights into the pathophysiologic role(s) of a particular gene product, (b) elucidation of mechanism(s) for a pharmacologic effect, and (c) new strategies for treating cancers by inducing mutations that confer drug lethality. The classical example is the action of imatinib on cancer cells expressing the Bcr-Abl fusion protein (**See Imatinib**). Many protein kinase-directed inhibitors also show enhanced cell killing in the presence of mutations within their ATP binding domains. Many cancers respond to chemotherapy by mutations placed under strongly selective conditions when cultured in the presence of a chemotherapeutic agent.

Chemotactic Factor Inactivator

This serum protein (often abbreviated CFI), first isolated from the α-globulin fraction of human serum, inhibits chemotactic factor C5a-directed neutrophil chemotaxis by binding to the C5a cochemotaxin Gc-globulin, a vitamin-D-binding protein, and inhibiting the capacity of Gc-globulin to enhance the chemotactic activity of C5a (1). Small amounts (10^{-10} mol) of purified human CFI suppress leukocytic infiltration, permeability changes, and hemorrhage associated with acute immune complex-induced injury in rats (2). *In vitro*, human CFI inhibits the chemotactic activity generated in complement-activated rat serum. The inhibitory effects of human CFI are not seen if it is first heat inactivated. The highly chemotactic synthetic peptide Met-Leu-Phe is completely hydrolyzed by CFI preparations, indicating that CFI is most likely an aminopeptidase that inhibits the activity of chemotactic substrates by catalyzing their proteolytic turnover (3). The observed heterogeneity in CFI activity released by phagocytosing neutrophils suggests that the predominant inactivator activity of the C5 chemotactic fragment is most likely elastase and/or cathepsin G (4). **1**. Maderazo, Woronick & Ward (1988) *Meth. Enzymol.* **162**, 223. **2**. Johnson, Anderson & Ward (1977) *J. Clin. Invest.* **59**, 951. **3**. Ward & Ozols (1976) *J. Clin. Investig.* **58**, 123. **4**. Brozna, Senior, Kreutzer & Ward (1977) *J. Clin. Investig.* **60**, 1280.

CHF-2819, *See Ganstigmine*

Chidamide

This synthetic benzamide-type HDAC inhibitor (FW = 390.40 g/mol; CAS 743420-02-2), also known by the code names CS055 and HBI-8000 and systematically as *N*-(2-amino-5-fluorophenyl)-4-[[[1-oxo-3-(3-pyridinyl)-2-propen-1-yl]amino]methyl]benzamide, mainly targets Class I HDACi, potently inhibiting histone deacetylases HDAC-1, -2 and -3, all at low nanomolar concentrations and inducing G_1 arrest (1).. The results showed that at low concentrations (<1 μM), CS055. At moderate concentrations (0.5–2 μM), CS055 induced differentiation of primary myeloid leukemia cells, as determined by the increased expression of the myeloid differentiation marker CD11b. At relatively high concentrations (2–4 μM), CS055 potently induced caspase-dependent apoptosis. Co-treatment with the ROS (reactive oxygen species) scavengers *N*-acetyl-L-cysteine or Tiron blocked CS055-induced cell differentiation and apoptosis, suggesting an essential role for ROS in these effects. Chidamide treatment leads to increased acetylation of histone H3 at Lys9/Lys18 and H4 at Lys8, resulting ultimately in activation of gene transcription (2). **1**. Gong, Xie, Yi & Li (2012) *Biochem. J.* **443**, 735. **2**. Ning, Li, Newman *et al.* (2012) *Cancer Chemother .Pharmacol.* **69**, 901.

Ching001

This tubulin polymerization inhibitor and spindle poison (FW = 438.42 g/mol; Solubility: dissolve in DMSO, then dilute 20x into buffer or growth medium) shows specific cytotoxic action against lung cancer cell lines, with little effect against normal lung cells (1). IC_{50} values for various cell lines were: CL1-0, 2 μM; CL1-5, 1.9 μM; A549, 1.2 μM; H1299, 2.6 μM; and MRC5, 8.3 μM for cells treated with Ching001 for 48 h. Aurora B, survivin and p-MPM2 remained high after Ching001 treatment compared with DMSO control, a result indicating M-phase arrest. Significantly, Ching001 inhibits *in vivo* xenograft growth by mitotic arrest and apoptosis, but without inducing apoptosis in host tissue surrounding of xenograft. Consistent with the general observation that prolong M-phase arrest leads to chromosome segregation failure and death, results with Ching001 are consistent with M-phase arrest followed by DNA damage and apoptotic cell death. *See also Podophyllotoxin; Colchicine; Vinbastine* **1**. Chen, Tang, Li, *et al.* (2013) *PLoS One* **8**, e62082.

CHIR-258

This benzimidazole (FW = 392.44 g/mol), also known as TKI258, inhibits fibroblast growth factor receptor-3 kinase, IC_{50} = 5 nM (1-3). **Target(s):** fibroblast growth factor kinase (2,3); platelet-derived growth factor receptor β kinase (3); polo kinase (1); receptor protein-tyrosine kinase (2); signal-regulated kinase (2); vascular endothelial growth factor receptor kinase (3). **1**. Johnson, Stewart, Woods, Giranda & Luo (2007) *Biochemistry* **46**, 9551. **2**. Trudel, Wei, Wiesmann, *et al.* (2005) *Blood* **105**, 2941. **3**. Lee, Lopes de Menezes, Vora, *et al.* (2005) *Clin. Cancer Res.* **11**, 363.

CHIR-99021

This GSK-3 inhibitor (F.Wt. = 465.34; CAS 252917-06-9; Solubility (25°C): 1 mg/mL DMSO, <1 mg/mL Water), also known as CT99021 and systematically named 6-(2-(4-(2,4-dichlorophenyl)-5-(4-methyl-1*H*-imidazol-2-yl)pyrimidin-2-ylamino)ethylamino)nicotinonitrile, targets glycogen synthase kinase 3α and 3β, with IC_{50} values of 10 nM and 7 nM, respectively. Glycogen synthase kinase-3 can negatively regulate several aspects of insulin signaling, and elevated levels of GSK-3 have been reported in skeletal muscle from diabetic rodents and humans. CHIR 99021 stabilizes free cytosolic β-catenin and inhibits adipogenesis by blocking Inducing CCAAT/Enhancer-binding Protein-α and Peroxisome Proliferator-activated Receptor-γ (1). Preadipocyte differentiation is inhibited when 3T3-L1 cells are exposed to CHIR 99021 for any 24-hour period during the first 3 days of adipogenesis. **1**. Ring, *et al.* (2003) *Diabetes*, 52, 588. **2**. Bennett, *et al.* (2002) *J. Biol. Chem.* 277, 30998. **3**. Mussmann R, *et al.* (2007) *J. Biol. Chem.* 282, 12030. **4**. Trowbridge, *et al.* (2006) *Nature Med.* 12, 89.

Chloramphenicol

This natural product (FW = 323.13 g/mol; CAS 56-75-7; M.P. = 150.5-151.5°C; Low solublity in water (2.5 mg/mL at 25°C), but more soluble in organic solvents; λ_{max} = 278 nm; ε = 9530 $M^{-1}cm^{-1}$), also known as Chlorocid, Chlorasol, and Leukomycin (not to be confused with leucomycin) and IUPAC name, 2,2-dichloro-*N*-[1,3-dihydroxy-1-(4-nitrophenyl)propan-2-yl]acetamide, is a broad spectrum antibiotic first isolated from *Streptomyces venezuelae* in 1948. It was introduced commercially in 1949 and was a breakthrough in the treatment of typhoid fever. Chloramphenicol inhibits the growth of most bacteria; however, some resistant microorganisms have emerged. Chloramphenicol binds to the 50S ribosomal subunit ($K_d \approx 2$ μM). There is also at least one low-affinity binding site on the 30S subunit. Chloramphenicol blocks protein biosynthesis by inhibiting the peptidyltransferase, competing with substrates binding to the A-site. It also inhibits protein biosynthesis in the mitochondria and chloroplast. While eukaryotic protein biosynthesis is not generally affected, chloramphenicol is hepatotoxic. **Target(s):** D-amino acid oxidase (1); peptidyltransferase (2-4,6,8,11,12,17-22); CYP2B1 and several other cytochrome P450 systems (5,10); urease (7); peptidyl-tRNA hydrolase, *or* aminoacyl tRNA hydrolase (9); NADH dehydrogenase (ubiquinone), *or* Complex I (13); actomyosin ATPase (14); protein-synthesizing GTPase, elongation factor (15); aryldialkylphosphatase, *or* paraoxonase (16); methane monooxygenase (23). **1**. Yagi (1971) *Meth. Enzymol.* **17B**, 608. **2**. Fernandez-Muñoz, R. E. Monro & D. Vazquez (1971) *Meth. Enzymol.* **20**, 481. **3**. Gottesman (1971) *Meth. Enzymol.* **20**, 490. **4**. Pestka (1974) *Meth. Enzymol.* **30**, 261. **5**. Halpert & Stevens (1991) *Meth. Enzymol.* **206**, 540. **6**. Jiménez (1976) *Trends Biochem. Sci.* **1**, 28. **7**. Reithel (1971) *The Enzymes*, 3rd ed., **4**, 1. **8**. Lucas-Lenard & Beres (1974) *The Enzymes*, 3rd ed., **10**, 53. **9**. Tate & Caskey (1974) *The Enzymes*, 3rd ed., **10**, 87. **10**. Ator & Ortiz de Montellano (1990) *The Enzymes*, 3rd ed., **19**, 213. **11**. Coutsogeorgopoulos (1967) *Biochemistry* 6, 1704. **12**. Sypherd, Strauss & Treffers (1962) *Biochem. Biophys. Res. Commun.* 7, 477. **13**. Freeman & Haldar (1967) *Biochem. Biophys. Res. Commun.* 28, 8. **14**. Mugikura, Miyazaki & Nagai (1956) *Enzymologia* 17, 321. **15**. Campuzano & Modolell (1981) *Eur. J. Biochem.* 117, 27. **16**. Sinan, Kockar, Gencer, Yildirim & Arslan (2006) *Biochemistry (Moscow)* 71, 46. **17**. Spirin & Asatryan (1976) *FEBS Lett.* 70, 101. **18**. Polacek & Mankin (2005) *Crit. Rev. Biochem. Mol. Biol.* 40, 285. **19**. Rychlik (1966) *Biochim. Biophys. Acta* 114, 425. **20**. Drainas, Kalpaxis & Coutsogeorgopoulos (1987) *Eur. J. Biochem.* 164, 53. **21**. Traut & Monro

(1964) *J. Mol. Biol.* **10**, 63. **22.** Michelinaki, Mamos, Coutsogeorgopoulos & Kalpaxis (1997) *Mol. Pharmacol.* **51**, 139. **23.** Jahng & Wood (1996) *Appl. Microbiol. Biotechnol.* **45**, 744.

Chlordiazepoxide

This first-in-class benzodiazepine anxiolytic (FW = 299.75 g/mol; CAS 58-25-3; IUPAC Name: 7-chloro-2-methylamino-5-phenyl-3*H*-1,4-benzodiazepine-4-oxide), known famously as Librium, is indicated as a short-term treatment (2–4 weeks) for severe anxiety that is apt to subject a patient to otherwise unacceptable distress. Like diazepam, chlordiazepoxide is a positive allosteric modulator of GABA$_A$ receptors, promoting GABA binding, which in turn increases the total conduction of chloride ions across the neuronal cell membrane. Although its half-life is 5-30 hours, the active metabolite, demethyldiazepam, has a much longer half-life (36-200 hours). Often used recreationally, chlordiazepoxide is a Schedule IV controlled drug under the Convention on Psychotropic Substances. **1.** López-Muñoz, Alamo & García-García (2011) *J. Anxiety Disord.* **25**, 554.

Chlordimeform

This acaricide (FW = 196.68 g/mol; CAS 6164-98-3), also named *N'*-(4-chloro-2-methylphenyl)-*N,N*-dimethylmethanimidamide, acts against motile forms of mites and ticks as well as the eggs and early instar larval forms of certain *Lepidoptera*. An interesting characteristic is that, in some cases, chlordimeform acts as an insect sedative, causing ineffective egg-laying behavior. Rather than a direct pesticide, it interferes with amine-mediated control of nervous and endocrine systems, mainly through a build-up of 5-hydroxytryptamine and, to a lesser extent, norepinephrine. Chlordimeform also inhibits monoamine oxidase from rat liver *in vitro*, and, in the American cockroach, it directly stimulates the heart and blocking octopamine-dependent stimulation of adenylate cyclase in the CNS. **1.** Matsumura & Beeman (1976) *Environ. Health Perspect.* **14**, 71.

Chlorfenapyr

This pro-insecticide (FW = 407.62 g/mol; CAS 122453-73-0), also known as 4-bromo-2-(4-chlorophenyl)-1-ethoxymethyl-5-trifluoromethyl-1*H*-pyr-role-3-carbonitrile, is taken up by target organisms and metabolized into CL 303268, a mitochondrial electron transport inhibitor. Chlorfenapyr is a broad spectrum insecticide/acarcide that is used to control whiteflies, thrips, caterpillars mites, leaf miners and aphids. Because this insecticide is toxic to birds, EPA restricts its use to non-food crops grown in greenhouses.

Chlorhexidine

This membrane-active disinfectant (FW$_{free-base}$ = 505.45 g/mol; CAS 55-56-1), also known as 1,1'-hexamethylenebis[5-(4-chlorophenyl)biguanide], is an antiseptic, used by surgeons ands in plaque control. The dihydrochloride has a solubility in water of about 0.06 g/100 mL at 20°C; however, it is less soluble under elevated pH conditions. Because chlorhexidine is protonated at multiple sites, it is apt to bind tightly to phospholipids. **Target(s):** ATPase, membrane (1,3); however, see Reference 6); cation transport (1); dextransucranase, weakly inhibited (2); β-lactamase (3); succinate dehydrogenase, weakly inhibited (3); penicillin-binding protein-7, weakly inhibited (3); histamine *N*-methyltransferase (4); protease, subgingival plaque (5); oligopeptidase, *Treponema denticola* (7); gelatinase A, *or* matrix metalloproteinase-2 (8); gelatinase B, *or* matrix metalloproteinase-9 (8); collagenase 2, *or* matrix metalloproteinase 8 (8); fructosyltramsferase (9); gingipain R, *or* Arg-gingipain (10); trypanothione reductase (11); urate oxidase (12); IgA-specific serine endopeptidase (13). **1.** Harold, Baarda, Baron & Abrams (1969) *Biochim. Biophys. Acta* **183**, 129. **2.** Christensen & Kilian (1977) *Acta Odontol. Scand.* **35**, 119. **3.** Chopra, Johnson & Bennett (1987) *J. Antimicrob. Chemother.* **19**, 743. **4.** Harle & Baldo (1988) *Biochem. Pharmacol.* **37**, 385. **5.** Radford, Homer, Naylor & Beighton (1992) *Arch. Oral Biol.* **37**, 245. **6.** Kuyyakanond & Quesnel (1992) *FEMS Microbiol. Lett.* **79**, 211. **7.** Makinen, Makinen, Loesche & Syed (1995) *Arch. Biochem. Biophys.* **316**, 689. **8.** Gendron, Grenier, Sorsa & Mayrand (1999) *Clin. Diagn. Lab. Immunol.* **6**, 437. **9.** Steinberg, Bachrach, Gedalia, Abu-Ata & Rozen (2002) *Eur. J. Oral Sci.* **110**, 374. **10.** Houle, Grenier, Plamondon & Nakayama (2003) *FEMS Microbiol. Lett.* **221**, 181. **11.** Meiering, Inhoff, Mies, *et al.* (2005) *J. Med. Chem.* **48**, 4793. **12.** Bentley & Truscoe (1969) *Enzymologia* **37**, 285. **13.** Bleeg, Reinholdt & Kilian (1985) *FEBS Lett.* **188**, 357.

Chlorimuron-ethyl

This sulfonylurea herbicide (FW = 414.83 g/mol; CAS 90982-32-4; soluble in common organic solvents), also known as DPX-F6025, inhibits acetolactate synthase, *or* acetohydroxy acid synthase (K_i = 3.3 nM). Chlorimuron-ethyl is often used to control weed growth in soybean fields. **1.** Ray (1986) *Trends Biochem. Sci.* **11**, 180. **2.** Pang, Guddat & Duggleby (2003) *J. Biol. Chem.* **278**, 7639. **3.** Pang, Guddat & Duggleby (2004) *Acta Crystallogr. D Biol. Crystallogr.* **60**, 153. **4.** McCourt, Pang, Guddat & Duggleby (2005) *Biochemistry* **44**, 2330. **5.** Hill, Pang & Duggleby (1997) *Biochem. J.* **327**, 891. **6.** Choi, Yu, Hahn, Choi & Yoon (2005) *FEBS Lett.* **579**, 4903.

N^δ-Chloroacetyl-L-Ornithine

This chloroacetylated amino acid (FW = 208.64 g/nol; Symbol: NCAO) targets ornithine decarboxylase (ODC), the enzyme that forms putrescine by decarboxylation of the urea cycle metabolite, ornithine. ODC is the first committed step in the synthesis of polyamines (*e.g.*, putrescine, spermidine and spermine). Despite the presence of a susceptible haloacetamido group, NCAO is a reversible competitive inhibitor (K_i = 59 μM). Its cytotoxic and antiproliferative effects are concentration- and time-dependent, with EC$_{50,72h}$ values of 15.8, 17.5 and 10.1 μM for HeLa, MCF-7 and HepG2 cells, respectively. At 500 μ, NCAO completely inhibits growth of all cancer cells after 48 h treatment, with almost no effect on normal cells. Putrescine reverses NCAO effects on MCF-7 and HeLa cells, indicating its antiproliferative effects are due to ODC inhibition. **1.** Medina-Enríquez, Alcántara-Farfán, Aguilar-Faisal, *et al.* (2014) *J. Enzyme Inhib. Med. Chem.* **18**, 1.

3-Chloroacetylpyridine Adenine Dinucleotide Phosphate

This NADP$^+$ analogue (FW = 726.42 g/mol; CAS 39938-03-9; Abbreviation: ac3PdADP$^+$), also known as 3-chloroacetyl-pyridinium adenine nucleotide phosphate, is an affinity label for NADP$^+$-dependent dehydrogenases. Active site-directed chloroketone-containing reagents bind to and react with sulfhydryl groups and less so with primary amines located within the coenzyme binding region. **Target(s):** Aspartate-β-semialdehyde dehydrogenase (irreversible inactivation with pseudo-first-order kinetics; inactivation prevented by NADP$^+$ and NADPH, but not the substrate; incorporation of 1 mol ac3PdADP$^+$/dimer totally inactivates the enzyme) (1); estradiol 17β-dehydrogenase (2); 6-phosphogluconate dehydrogenase (3). **1**. Biellmann, Eid & Hirth (1980) *Eur. J. Biochem.* **104**, 65. **2**. Biellmann, Goulas, Nicolas, Descomps & Crastes De Paulet (1979) *Eur. J. Biochem.* **99**, 81. **3**. Biellmann, Goulas & Dallocchio (1978) *Eur. J. Biochem.* **88**, 433.

4-(3-Chloroanilino)quinazoline

This substituted quinazoline (FW = 255.71 g/mol; CAS 88404-44-8), also known as *N*-(*m*-chlorophenyl)-4-aminoquinazoline, is a strong inhibitor of epidermal growth factor receptor protein-tyrosine kinase (apparent K_i = 16 nM). **1**. Wakeling, Barker, Davies, *et al.* (1996) *Breast Cancer Res. Treat.* **38**, 67. **2**. Ward, Cook, Slater, *et al.* (1994) *Biochem. Pharmacol.* **48**, 659.

Chlorocholine Chloride

This quaternary amine (FW = 158.07 g/mol; *Abbreviation*: CCC), also known as 2-chloroethyltrimethylammonium chloride, chlormequat chloride, and, is a synthetic plant growth retardant. It is also a gibberellin antagonist. **Target(s):** choline kinase (1); acetylcholinesterase (2,3); copalyl-diphosphate synthase (4); *ent* kaurene synthase (4,6); choline *O*-acetyltransferase (5); beyerene biosynthesis (6); sandaracopimaradiene biosynthesis (6); trachylobane biosynthesis (6); phenylalanine ammonia-lyase (7); choline sulfotransferase (8); choline monooxygenase (9). **1**. Ishidate & Nakazawa (1992) *Meth. Enzymol.* **209**, 121. **2**. Krupka (1965) *Biochemistry* **4**, 429. **3**. Friess & McCarville (1954) *J. Amer. Chem. Soc.* **76**, 2260. **4**. Rademacher (2000) *Annu. Rev. Plant Physiol. Plant Mol. Biol.* **51**, 501. **5**. Hersh & M. Peet (1977) *J. Biol. Chem.* **252**, 4796. **6**. Robinson & West (1970) *Biochemistry* **9**, 80. **7**. El-Shora (2002) *Plant Sci.* **162**, 1. **8**. Renosto & Segel (1977) *Arch. Biochem. Biophys.* **180**, 416. **9**. Burnet, Lafontaine & Hanson (1995) *Plant Physiol.* **108**, 581.

3-[(2S,4S)-4-(2-Chloro-4-cyanophenyl)amino-2-pyrrolidinylcarbonyl-1,3-thiazolidine

This substituted thiazolidine (FW = 336.85 g/mol) inhibits dipeptidyl-peptidase IV (IC$_{50}$ = 18 nM for the human enzyme). The 3-chloro-4-cyanophenyl analogue is a slightly stronger inhibitor (IC$_{50}$ = 11 nM). **1**. Sakashita, Akahoshi, Kitajima, Tsutsumiuchi & Hayashi (2006) *Bioorg. Med. Chem.* **14**, 3662.

4-[6-(Cyclobutylamino)-imidazo[1,2-b] pyridazin-3-yl]-2-fluoro-N-{[(2S,4R)-4-fluoropyrrolidin-2-yl] methyl}benzamide

This protein kinase inhibitor (FW = 427.47 g/mol), also known as Compound D, targets IκB kinaseβ (*or* IKKβ), which plays a critical role in nuclear factor-κB (NF-κB) activation and proinflammatory cytokine production in various inflammatory diseases including rheumatoid arthritis. **Pharmacodynamics:** Phosphorylation of inhibitor of κB (IκB) is catalyzed by the IκB kinase (IKK) complex, which consists of two catalytic subunits, IKKα and IKKβ, and a regulatory subunit, NF-κB essential modulator. Ample evidence indicates that IKKβ, but not IKKα, is required for NF-κB activation in response to inflammatory stimuli, suggesting that IKKβ is a key factor in the production of pro-inflammatory cytokines in inflammatory conditions. **Signaling Pathway Target(s):** NF-κB (IC$_{50}$ = 0.15 μM); Activator Protein-1, *or* AP-1 (IC$_{50}$ = >20 μM); Nuclear Factor of Activated T-cells, *or* NFAT (IC$_{50}$ > 20 μM); Serum Response Element, *or* SRE (IC$_{50}$ = 19 μM); cAMP Response Element, *or* CRE (IC$_{50}$ > 20 μM); Interferon-Stimulated Response Element, *or* ISRE (IC$_{50}$ = 16 μM). **1**. Tanaka, Toki, Yokoyama, *et al.* (2014) *Biol. Pharm. Bull.* **37**, 87.

5'-Chloro-5'-deoxyformycin A

This nonglycosidic or C-nucleoside analogue (FW = 284.68 g/mol; CAS 72453-27-1), also known as 5'-chloroformycin, inhibits methylthioadenosine nucleosidase (K_i = 322 nM). Its choro group prevents any possibility of metabolic 5'-phosphorylation within cells, making it an ideal nucleoside analogue. **Target(s):** adenosylhomocysteine nucleosidase (1-4); methylthiuoadenosine/adenosylhomocysteine nucleosidase (1-4); *S*-methyl-5'-thioadenosine phosphorylase (5,6). **1**. Kushad, Richardson & Ferro (1985) *Plant Physiol.* **79**, 525. **2**. Kushad, Orvos & Ferro (1992) *Physiol. Plant.* **86**, 532. **3**. Della Ragione, Porcelli, Carteni-Farina, Zappia & Pegg (1985) *Biochem. J.* **232**, 335. **4**. Dunn, Bryant & Kerr (1994) *Phytochem. Anal.* **5**, 286. **5**. Della Ragione, Takabayashi, Mastropietro, *et* *al.* (1996) *Biochem. Biophys. Res. Commun.* **223**, 514.   **6**. Savarese, Crabtree & Parks (1981) *Biochem. Pharmacol.* **30**, 189.

9-(6-Chloro-5,6-dideoxy-β-D-*ribo*-hex-5-ynofuranosyl)-9*H*-purin-6-amine

This halo-akynyne nucleoside analogue (FW = 295.69 g/mol) binds to and inactivates adenosylhomocysteinase. The enzyme mediates water addition at the 5'-position of compound to produce an α-halomethyl ketone intermediate, which is subsequently attacked by the proximal, active-site nucleophile (Lys318) to form an enzyme-inhibitor covalent adduct (lethal event). In a parallel pathway (nonlethal event), addition of water at the 6'-position produces an acyl halide, which is released into solution and chemically degrades into adenosine, halide ion, and sugar-derived products. **1**. Yang, Yin, Wnuk, Robins & Borchardt (2000) *Biochemistry* **39**, 15234.

4'-Chloro-3α-(diphenylmethoxy)tropane

This modified tropane (FW$_{free-base}$ = 341.88 g/mol), which inhibits dopamine transport (IC$_{50}$ = 15 nM), increases locomotor activity. While exhibiting lower efficacy than cocaine, it does not exhibit cocaine-like behavioral profiles. **1**. Newman, Allen, Izenwasser & Katz (1994) *J. Med. Chem.* **37**, 2258. **2**. Katz, Izenwasser, Kline, Allen & Newman (1999) *J. Pharmacol. Exp. Ther.* **288**, 302.

(2-Chloroethyl)₂-ethylamine, See *Bis(2-chloroethyl)ethylamine*

N-(2-Chloroethyl)-*N*-methyl-2-chloro-2-phenethylamine

This dichloro phenethylamine derivative (FW$_{free-base}$ = 232.15 g/mol) rearranges in aqueous solution to form the aziridinium ion, (*N*-(2-chloroethyl)-*N*-methyl-2-phenylaziridinum cation), from *N*-(2-chloroethyl)-*N*-methyl-2-chloro-2-phenethylamine, a quaternary amine that inhibits acetylcholinesterase. **1**. O'Brien (1969) *Biochem. J.* **113**, 713. **2**. Shaw (1970) *The Enzymes*, 3rd ed., **1**, 91. **3**. Froede & Wilson (1971) *The Enzymes*, 3rd ed., **5**, 87.

6-[4-(2-Chloro-5-fluorophenoxy)piperidin-1-yl]pyridazine-3-carboxamide

This potent, orally bioavailable pyridazine (FW = 350.78 g/mol) inhibits stearoyl-CoA 9-desaturase (SCD), with an IC$_{50}$ value of 1.0 nM in mice. This enzyme catalyzes the rate-limiting step in the cellular synthesis of monounsaturated fatty acids from saturated fatty acids. Because the ratio of saturated to monounsaturated fatty acids strongly infuences membrane fluidity, alterations in this ratio lead to various disorders, including cardiovascular disease, obesity, non-insulin-dependent diabetes mellitus, hypertension, neurological diseases, immune disorders, and cancer. Studies with SCD-1 knockout mice established that animals treated with this agent are lean and are protected from leptin deficiency-induced and diet-induced obesity, showing a greater whole-body insulin sensitivity than wild-type animals. **1**. Liu, Lynch, Freeman, *et al.* (2007) *J. Med. Chem.* **50**, 3086.

Chlorogenate (*or* Chlorgenic Acid)

This conjugated phenol (FW = 354.31 g/mol; CAS 327-97-9; Symbol: CGA), also named 3-Caffeoylquinate, 3-Caffeoylquinic (3-CQA), 3-*O*-Caffeoylquinic acid, Heriguard, and systematically (1*S*,3*R*,4*R*,5*R*)-3-{[(2*E*)-3-(3,4-dihydroxyphenyl)prop-2-enoyl]oxy}-1,4,5-trihydroxycyclo-hexane carboxylate, is a well-known tannin found in coffee, tea, and fruit. Acid hydrolysis produces caffeic acid. Some bacteria and fungi possess a chlorogenate hydrolase (EC 3.1.1.42) that catalyzes the reaction: Chlorogenate + H$_2$O ⇌ Caffeate + Quinate. The enzyme hydroxycinnamoyl-CoA:shikimate/quinate hydroxy-cinnamoyltransferase reported converts this chlorogenate to caffeoyl-CoA. Chlorogenic acid has a low solubility in water (approximately 4% at 25°C) with the solubility increasing with temperature. The reported pK_a value is 2.66 at 27°C. Chlorogenate also reduces ferrylmyoglobin. (*Note*: The term "chlorogenic" has nothing to do with chlorine (of which it has none) and is instead based on the Greek word χλωρός (green), because, upon oxidation, this compound generates a characteristic emission in that region of the visible light spectrum.) **Target(s):** indoleacetate oxidase (1,2); horseradish peroxidase (3); α-amylase (4,21); trypsin (4); lysozyme (4); arylamine *N*-acyltransferase (5,25); benzyloxyresorufin *O*-dealkylase, *or* cytochrome P450 (6); ethoxyresorufin *O*-deethylase, *or* cytochrome P450, *or* CYP1A1 (6,7); methoxyresorufin *O*-demethylase, *or* cytochrome P450, CYP1A2 (6,7); pentoxy-*O*-dealkylase, cytochrome P450, *or* CYP2B (7); glucose-6-phosphatase (8,22); glucose-6-phosphate translocase (10,22); Na$^+$-dependent D-glucose uptake (11); glutathione *S*-transferase (12); carbonyl reductase (13); arginase (14); DNA methyltransferase (15); chorismate mutase (16); phenylpyruvate tautomerase (17); phenylalanine ammonia-lyase, weakly inhibited (18); angiotensin I-converting enzyme, weakly inhibited (19); hyaluronoglucosaminidase, *or* hyaluronidase (20); phenol sulfotransferase, *or* aryl sulfotransferase (23); inositol-polyphosphate multikinase (24); tyrosinase, monophenol monooxygenase; weakly inhibited (26); *trans*-cinnamate 4-monooxygenase, weakly inhibited (27). **Implications for Treating Rheumatoid Arthritis:** Chorogenic acid is also a major component of *Caulis lonicera*, an herb commonly used in Chinese folk medicine to treat rheumatoid arthritis (RA), which is characterized by persistent inflammation of the synovial tissues of joints, resulting in loss of joint function. CGA inhibits the inflammatory proliferation of RSC-364 cells by inducing apoptosis, as mediated by IL-6. CGA can also suppress the expression levels of key molecules in the JAK/STAT and NF-κB signaling pathways, thereby inhibiting the activation of these signaling pathways in the inflammatory response through IL-6-mediated signaling (28). Such action results in the inhibition of the inflammatory proliferation of synoviocytes, indicating that CGA may afford a novel therapeutic agent to inhibit inflammatory hyperplasia of the synovium by inducing synoviocyte apoptosis in RA patients (28). **1**. Rabin & Klein (1957) *Arch. Biochem. Biophys.* **70**, 11. **2**. Sondheimer & Griffin (1960) *Science* **131**, 672. **3**. Gamborg, Wetter & Neish (1961) *Can. J. Biochem. Physiol.* **39**, 1113. **4**. Rohn, Rawel & Kroll (2002) *J. Agric. Food Chem.* **50**, 3566. **5**. Tsou, Hung, Lu, *et al.* (2000) *Microbios* **101**, 37. **6**. Teel & Huynh (1998)

Cancer Lett. **133**, 135. **7**. Baer-Dubowska, Szaefer & Krajka-Kuzniak (1998) *Xenobiotica* **28**, 735. **8**. Arion, Canfield, Ramos, *et al.* (1997) *Arch. Biochem. Biophys.* **339**, 315. **9**. Hoffmann, Maury, Martz, Geoffroy & Legrand (2003) *J. Biol. Chem.* **278**, 95. **10**. Hemmerle, Burger, Below, *et al.* (1997) *J. Med. Chem.* **40**, 137. **11**. Welsch, Lachance & Wasserman (1989) *J. Nutr.* **119**, 1698. **12**. Das, Bickers & Mukhtar (1984) *Biochem. Biophys. Res. Commun.* **120**, 427. **13**. Wermuth (1981) *J. Biol. Chem.* **256**, 1206. **14**. Reifer & Augustyniak (1968) *Bull. Acad. Pol. Sci. Biol.* **16**, 139. **15**. Lee & Zhu (2006) *Carcinogenesis* **27**, 269. **16**. Woodin & Nishioka (1973) *Biochim. Biophys. Acta* **309**, 211. **17**. Molnar & Garai (2005) *Int. Immunopharmacol.* **5**, 849. **18**. Jorrin & Dixon (1990) *Plant Physiol.* **92**, 447. **19**. Actis-Goretta, Ottaviani & Fraga (2006) *J. Agric. Food Chem.* **54**, 229. **20**. Girish & Kemparaju (2005) *Biochemistry (Moscow)* **70**, 948. **21**. Funke & Melzig (2005) *Pharmazie* **60**, 796. **22**. Van Schaftingen & Gerin (2002) *Biochem. J.* **362**, 513. **23**. Yeh & Yen (2003) *J. Agric. Food Chem.* **51**, 1474. **24**. Mayr, Windhorst & Hillemeier (2005) *J. Biol. Chem.* **280**, 13229. **25**. Lo & Chung (1999) *Anticancer Res.* **19**, 133. **26**. Karioti, Protopappa, Magoulas & Skaltsa (2007) *Bioorg. Med. Chem.* **15**, 2708. **27**. Billett & Smith (1978) *Phytochemistry* **17**, 1511. **28**. Lou, Zhou, Liu, *et al.* (2016) *Exp. Ther. Med.* **11**, 2054.

5-Chloro-6-[(2-iminopyrrolidin-1-yl)methyl]uracil

This substituted uracil (FW = 242.67 g/mol), often abbreviated TPI, is a potent inhibitor of thymidine phosphorylase (K_i = 20 nM). Thymidine phosphorylase (TP), which catalyzes the reversible conversion of thymidine to thymine, is identical to an angiogenic factor, platelet-derived endothelial cell growth factor. TP is expressed at higher levels in a wide variety of solid tumors than in the adjacent nonneoplastic tissues. Patients with TP-positive colon and esophageal tumors have a poorer prognosis than those with TP-negative tumors. That this inhibitor's positively charged ring system improves binding (1-10) is consident with the finding that structurally related bioreductive nitroimidazolyl pro-drugs rearrange to form potent aminoimidazolylmethyluracil inhibitors of thymidine thosphorylase (9). **1**. Reigan, Gbaj, Stratford, Bryce & Freeman (2008) *Eur. J. Med. Chem.* **43**, 1248. **2**. Yano, Tada, Kazuno, *et al.* (1996) *PCT Int. Appl.*, 108. **3**. Suzuki, Fukushima, Asao, Yamada & Akiyama (1997) *Proc. Amer. Assoc. Cancer Res.* **38**, 101. **4**. Akiyama, Furukawa, Sumizawa, *et al.* (2004) *Cancer Sci.* **95**, 851. **5**. Liekens, Bronckaers, Pérez-Pérez & Balzarini (2007) *Biochem. Pharmacol.* **74**, 1555. **6**. Nencka, Votruba, Hrebabecky, *et al.* (2007) *J. Med. Chem.* **50**, 6016. **7**. Gbaj, Edwards, Reigan, *et al.* (2006) *J. Enzyme Inhib. Med. Chem.* **21**, 69. **8**. McNally, Rajabi, Gbaj, *et al.* (2007) *J. Pharm. Pharmacol.* **59**, 537. **9**. Reigan, Edwards, Gbaj, *et al.* (2005) *J. Med. Chem.* **48**, 392. **10**. Norman, Barry, Bate, *et al.* (2008) *Bioorg. Med. Chem.* **16**, 3866.

p-Chloromercuribenzoate

This sulfhydryl-blocking reagent (FW$_{\text{free-acid}}$ = 357.16 g/mol; *Symbol*: PCMB) inhibits numerous enzymes. It is especially useful as the reaction with a thiol causes an increase in absorbance at 250 nm. Inhibition is often reversed by the addition of sulfhydryl-containing molecules, such as cysteine, 2-thioethanol, and dithiothreitol. PCMB also inhibits oxidative phosphorylation. It is highly toxic and extreme care should be exercised in its use. Importantly, PCMB undergoes hydrolysis, readily converting to *p*-hydroxymercuribenzoate (PHMB), when the sodium salt is prepared by dissolving PCMB in 200 mM NaOH, followed by adjustment of the solution to a lower pH. The reactive species in aqueous solutions is therefore the mercuric cation ($^-$OOC–C_6H_4–Hg$^+$). Since the chloride ion is simply a counterion, one would predict that any salt would be equally effective. Nevertheless, enzymes have been reported to be inhibited by both PCMB and PHMB, albeit with different rates (the difference in rates may

be due in part to the role of chloride ions as an additional effector). (*See also p-Hydroxymercuribenzoate*) Over 1000 enzymes show some degree of inhibition, making PCMB a notoriously highly unselective thiol-reacting reagent.

1-(5-Chloronaphthalene-1-sulfonyl)-1*H*-hexahydro-1,4-diazepine

This diazepine (FW$_{\text{free-base}}$ = 324.83 g/mol), commonly referred to as ML-9, is a strong inhibitor of [myosin light-chain] kinase (IC$_{50}$ = 3.8 μM). Other kinases are inhibited less effectively: *e.g.*, protein kinase A (IC$_{50}$ = 32 μM) and protein kinase C (IC$_{50}$ = 54 μM). It also inhibits Natural Killer cell activity and catecholamine secretion in intact and permeabilized chromaffin cells. **Target(s)**: [myosin light-chain] kinase (1-5); protein kinase A (1,4); protein kinase C (1,4); mitogen- and stress-activated protein kinase 1 (4); p70 ribosomal protein S6 kinase (4); Rho-dependent protein kinase II (4). **1**. Saitoh, Ishikawa, Matsushima, Naka & Hidaka (1987) *J. Biol. Chem.* **262**, 7796. **2**. Hidaka, Watanabe & Kobayashi (1991) *Meth. Enzymol.* **201**, 328. **3**. Eriksson, Toivola, Sahlgren, Mikhailov & Härmälä-Braskén (1998) *Meth. Enzymol.* **298**, 542. **4**. Bain, McLauchlan, Elliott & Cohen (2003) *Biochem. J.* **371**, 199. **5**. Connell & Helfman (2006) *J. Cell Sci.* **119**, 2269.

3-[4-(2-Chlorophenoxy)piperidin-1-yl]-6-(pyridin-3-yl)pyridazine

This pyridazine derivative (FW = 354.84 g/mol) reversibly inhibits stearoyl-CoA 9-desaturase (mouse IC$_{50}$ < 4 nM). This enzyme catalyzes the rate-limiting step in the cellular synthesis of monounsaturated fatty acids from saturated fatty acids. Because a proper ratio of saturated to monounsaturated fatty acids is required for optimal membrane fluidity, alterations in this ratio have been implicated in various disease states, including cardiovascular disease, obesity, non-insulin-dependent diabetes mellitus, hypertension, neurological diseases, immune disorders, and cancer. **1**. Liu, Lynch, Freeman, *et al.* (2007) *J. Med. Chem.* **50**, 3086.

p-Chloro-L-phenylalanine

This halogenated amino acid (FW = 199.64 g/mol; CAS 7424-00-2; Soluble to 100 mM in water and to 100 mM in DMSO) depletes tissue serotonin levels. **Target(s)**: tryptophan 5-monooxygenase (1,4,13-16); phenylalanine 4-monooxygenase (2,3,5,8,17,18); pyruvate kinase (6,9); 3-deoxy-7-phosphoheptulonate synthase, *or* 3-deoxy-D-*arabino*heptulosonate-7-phosphate synthetase (7); alkaline phosphatase, mildly inhibited by racemic mixture (10); phenylalanyl-tRNA synthetase, K_i = 6.8 mM (11); prephenate dehydratase (12). **1**. Ichiyama & Nakamura (1970) *Meth. Enzymol.* **17A**, 449. **2**. Kaufman (1978) *Meth. Enzymol.* **53**, 278. **3**. Kaufman (1987) *Meth. Enzymol.* **142**, 3. **4**. Fujisawa & H. Nakata (1987) *Meth. Enzymol.* **142**, 93. **5**. Guroff (1969) *Arch. Biochem. Biophys.* **134**, 610. **6**. Kayne (1973) *The Enzymes*, 3rd ed., **8**, 353. **7**. Simpson &

Davidson (1976) *Eur. J. Biochem.* **70**, 509. **8.** Miller, McClure & Shiman (1976) *J. Biol. Chem.* **251**, 3677. **9.** Schwark, Singhal & Ling (1970) *Life Sci. I 9*, 939. **10.** Fishman & Sie (1971) *Enzymologia* **41**, 141. **11.** Santi & Danenberg (1971) *Biochemistry* **10**, 4813. **12.** Dopheide, Crewther & Davidson (1972) *J. Biol. Chem.* **247**, 4447. **13.** Nakata & Fujisawa (1982) *Eur. J. Biochem.* **124**, 595. **14.** Izikki, Hanoum, Marcos, *et al.* (2007) *Amer. J. Physiol. Lung Cell Mol. Physiol.* **293**, L1045. **15.** Hamdan & Ribeiro (1999) *J. Biol. Chem.* **274**, 21746. **16.** Nagai, Hamada, Kai, Tanoue & Nagayama (1997) *Comp. Biochem. Physiol. B Biochem. Mol. Biol.* **116**, 161. **17.** Woo, Gillam & Woolf (1974) *Biochem. J.* **139**, 741. **18.** Ledley, Grenett & Woo (1987) *J. Biol. Chem.* **262**, 2228.

p-Chloro-L-phenylalanine Methyl Ester

This membrane-permeant chlorinated amino acid ($FW_{HCl\text{-}Salt}$ = 250.12 g/mol; CAS 63024-26-0) undergoes metabolic desterification to generate of *p*-chloro-L-phenylanine, a 5-HT-depleting agent. (*See p-Chloro-L-phenylalanine*) Injectable solutions should be prepared fresh daily by dissolving into sterile 0.9% saline (78 mg/mL *p*-CPME HCl-salt), adjusted to pH 6.0 with 0.1 M NaOH. When injected into rats at a dose of 200 mg/kg (calculated as the free base), this treatment depletes 94% of 5-HT. **1.** Lapiz-Bluhm, Soto-Piña, Hensler & Morilak (2009) *Psychopharmacol.* (Berlin) **202**, 329.

*N*¹-[4-Chloro-D-phenylalanyl-3-(naphth-1-yl)-L-alanyl]-1-amino-3-guanidinopropane

This peptidomimetic TK inhibitor ($FW_{free\text{-}base}$ = 494.02 g/mol) targets tissue kallikrein (K_i = 2.2 nM), showing a high degree of selectivity for TK over several other serine proteases. This inhibitor reduces eosinophilia in an allergic inflammation model, and the observed *in vivo* effects provide further evidence for the involvement of TK and kinins in the pathophysiology of allergic diseases such as asthma. 1. Evans, Jones, Pitt, *et al.* (1996) *Immunopharmacology* **32**, 117.

8-(4-Chlorophenylthio)guanosine 3',5'-Cyclic Monophosphorothioate, *Rp*-Isomer

This lipophilic guanine nucleotide (FW = 503.88 g/mol), also known as 8-*p*CPT-cGMPS, *Rp*-isomer, inhibits cGMP-dependent protein kinase (IC_{50} =

0.5 μM). **1.** Butt, Eigenthaler & Genieser (1994) *Eur. J. Pharmacol.* **269**, 265. **2.** West, Meno, Nguyen, *et al.* (2003) *J. Cardiovasc. Pharmacol.* **41**, 444. **3.** Andoh, Chiueh & Chock (2003) *J. Biol. Chem.* **278**, 885. **4.** Zhang, J. Gesmonde, S. Ramamoorthy & G. Rudnick (2007) *J. Neurosci.* **27**, 10878. **5.** Li, Zhang, Marjanovic, Ruan & Du (2004) *J. Biol. Chem.* **279**, 42469. **6.** Schlossmann & Hofmann (2005) *Drug Discov. Today* **10**, 627. **7.** MacPherson, Lohmann & Davies (2004) *J. Biol. Chem.* **279**, 40026.

5'-(4-Chlorophenylthio)-immucillin-A

This nucleoside analogue (FW = 365.87 g/mol), named systematically as (1*S*)-5-(4-chlorophenylthio)-1-(9-deazaadenin-9-yl)-1,4-dideoxy-1,4-imino-D-ribitol and (1*S*)-1-(9-deazaadenin-9-yl)-5-(4-chlorophenylthio)-1,4-dideoxy-1,4-imino-D-ribitol, targets 5'-methylthioadenosine/*S*-adenosylhomocysteine nucleosidase, K_i' = 2 pM (1). The *meta*-chloro analogue is a weaker inhibitor, K_i' = 20 pM. The 4-chlorophenyl analogue also inhibits human *S*-methyl-5'-thioadenosine phosphorylase, K_i = 0.166 nM (2). The 3-chlorophenyl analogue is a slightly weaker inhibitor. **1.** Singh, Evans, Lenz, *et al.* (2005) *J. Biol. Chem.* **280**, 18265. **2.** Evans, Furneaux, Schramm, Singh & Tyler (2004) *J. Med. Chem.* **47**, 3275.

(3*R*,4*S*)-4-(4-Chlorophenylthiomethyl)-1-[(9-deazaadenin-9-yl)methyl]-3-hydroxy-pyrrolidine

This nucleoside analogue (FW = 377.90 g/mol) potently inhibits 5'-methylthioadenosine/*S*-adenosylhomocysteine Nucleosidase, K_i' = 4.7 x 10^{-14} M (1,2) and *S*-methyl-5'-thioadenosine phosphorylase, K_i' = 10 pM (3,4). The *meta*-chloro analogue is a weaker inhibitor of the nucleosidase (K_i' = 0.75 pM) and the phosphorylase (K_i' = 0.27 nM). **1.** Singh, Evans, Lenz, *et al.* (2005) *J. Biol. Chem.* **280**, 18265. **2.** Singh, Lee, Nùñez, Howell & Schramm (2005) *Biochemistry* **44**, 11647. **3.** Singh, Shi, Evans, *et al.* (2004) *Biochemistry* **43**, 9. **4.** Evans, Furneaux, Lenz, *et al.* (2005) *J. Med. Chem.* **48**, 4679.

2-Chloro-3-phytyl-1,4-naphthoquinone

This anticoagulant (FW = 471.12 g/mol; CAS 30949-95-2), also known as chloro-K1, is a competitive vitamin K_1 (phylloquinone) antagonist. **Target(s):** vitamin K-dependent carboxylase, *or* protein-glutamate carboxylase (1-8); phylloquinone monooxygenase (2,3-epoxidizing) (9). **1.** O'Reilly (1976) *Ann. Rev. Med.* **27**, 245. **2.** Suttie (1980) *Trends Biochem. Sci.* **5**, 302. **3.** Lowenthal, MacFarlane & McDonald (1960) *Experentia* **16**, 428. **4.** Johnson (1980) *Meth. Enzymol.* **67**, 165. **5.** Suttie, Canfield & Shah (1980) *Meth. Enzymol.* **67**, 180. **6.** Brody & Suttie (1984) *Meth.*

Enzymol. **107**, 552. **7.** Cocchetto & Bjornsson (1986) *Haemostasis* **16**, 321. **8.** Esmon & Suttie (1976) *J. Biol. Chem.* **251**, 6238. **9.** Sadowski, Schnoes & Suttie (1977) *Biochemistry* **16**, 3856.

3-Chloropropham

This tubulin-directed toxic agent (FW = 213.66 g/mol; CAS 101-21-3; M.P. = 40°C; Water-insoluble; Soluble in most organic solvents), also known as chlorpropham, *N*-(3-chlorophenyl) *O*-isopropyl carbamate, and *O*-isopropyl *N*-(3-chlorophenyl) carbamate, is a potent herbicide and inhibitor of plant cell division. This plant growth regulator is used for preemergence control of grass weeds in alfalfa, lima beans, snap beans, blueberries, carrots, cranberries, clover, garlic, seed grass, onions, spinach, sugar beets, tomatoes, safflower, soybeans, gladioli and woody nursery stock. It is also used to inhibit potato sprouting and for sucker control in tobacco. Chlorpropham is available in emulsifiable concentrate and liquid formulations. 3-Chloropropham is also toxic to animals. The oral LD$_{50}$ for chlorpropham in rats ranges from 1,200 to 3,800 mg/kg, and the 4-hour inhalation LC$_{50}$ in rats is >32 mg/L (4). **Target(s):** microtubule assembly (tubulin polymerization) (1-3). **1.** Yokoyama, Kaji, Nishimura & Miyaji (1990) *J. Gen. Microbiol.* **136**, 1067. **2.** Walker (1982) *J. Gen. Microbiol.* **128**, 61. **3.** Brown & Bouck (1974) *J. Cell Biol.* **61**, 514.

Chloroquine

This antimalarial drug and autophagy innhibitor (FW$_{free-base}$ = 319.88 g/mol; CAS 54-05-7), also known as N^4-(7-chloro-4-quinolinyl)-N^1,N^1-diethyl-1,4-pentanediamine and SN7618, prevents the malarial parasite from disposing of toxic heme products formed upon proteolysis of hemoglobin in infected red blood cells. **Primary Mode of Antimalarial Action:** This and similarly acting drugs accumulate in the acid food vacuoles of the intraerythrocytic-stage malaria parasite. The primary function of the food vacuole is to harvest nutrients derived from proteolysis of ingested red cell haemoglobin, thus providing the growing parasite with essential amino acids. The breakdown of hemoglobin within the food vacuole releases heme, which can damage biological membranes and inhibit many enzymes. To detoxify heme, the parasite has an enzyme that inserts heme into an insoluble crystalline material known as haemozoin, a polymer of heme molecules, each joined to the next by a link between the central ferric ion of one heme and a side-group carboxylate of a neighboring heme. This heme polymerase is inhibited by quinoline-containing drugs such as chloroquine and quinine, providing a rationale for the highly stage-specific anti-malarial action of these drugs (1). Chloroquine may also inhibit ribonucleotide metabolism in the parasite. **Other Modes of Inhibitory Action:** Chloroquine also intercalates into double-stranded DNA, and this property may be exploited to increase transfection efficiency. CQ also induces acute hemolysis in glucose-6 phosphate dehydrogenase-deficient individuals. Chloroquine is also a weak base that is trapped in acidic organelles such as lysosomes, thereby increasing vacuolar pH. The change in pH also causes accumulation of autophagosomes in cells by inhibiting the fusion of autophagosomes with lysosomes. Chloroquine is a more effective inhibitor than ammonium chloride, thus requiring a 100-fold lower concentration. CQ also appears to trigger multiple death pathways. **Key Pharmacokinetic Parameters:** *See* Appendix II in Goodman & Gilman's THE *PHARMACOLOGICAL BASIS OF THERAPEUTICS*, 12th Edition (Brunton, Chabner & Knollmann, eds.) McGraw-Hill Medical, New York (2011). **Target(s):** cathepsin B (2,14,15); photophosphorylation uncoupler (3);

alcohol dehydrogenase (4,6); DNA-directed RNA polymerase (5); heme polymerase (1,7); D-amino-acid oxidase (8); lanosterol synthase (9); phosphoenolpyruvate carboxylase (10); bontoxilysin (11); cathepsin D (12); cathepsin F (13); sphingomyelin phosphodiesterase, *or* sphingomyelinase (16); acid phosphatase (17); palmitoyl-[protein] hydrolase (18); phospholipase A$_1$ (19); phospholipase A$_2$ (19); sterol esterase, *or* cholesterol esterase (20); acetylcholinesterase (21); Ca^{2+}/calmodulin-dependent protein kinase (22); DNA polymerase (23); duck hepatitis B virus reverse transcriptase (23); RNA-directed DNA polymerase (23); histamine *N*-methyltransferase, both enantiomers inhibit (24,25); diferric-transferrin reductase (26); quinine 3-monooxygenase (27); ribosyldihydro-nicotinamide dehydrogenase, quinone-requiring (28); synergistically potentiates anti-cancer property of artemisinin by promoting ROS dependent apoptosis (29). **1.** Slater & Cerami (1992) *Nature* **355**, 167. **2.** Mycek (1970) *Meth. Enzymol.* **19**, 285. **3.** Izawa & Good (1972) *Meth. Enzymol.* **24**, 355. **4.** Fiddick & Heath (1967) *Nature* **213**, 628. **5.** Chamberlin (1974) *The Enzymes*, 3rd ed., **10**, 333. **6.** Brändén, Jörnvall, Eklund & Furugren (1975) *The Enzymes*, 3rd ed., **11**, 103. **7.** Chou & Fitch (1993) *Biochem. Biophys. Res. Commun.* **195**, 422. **8.** Hellerman, Lindsay & Bovarnick (1946) *J. Biol. Chem.* **163**, 553. **9.** Chen & Leonard (1984) *J. Biol. Chem.* **259**, 8156. **10.** McDaniel & Siu (1972) *J. Bacteriol.* **109**, 385. **11.** Burnett, Schmidt, Stafford, *et al.* (2003) *Biochem. Biophys. Res. Commun.* **310**, 84. **12.** Woessner, Jr. (1973) *J. Biol. Chem.* **248**, 1634. **13.** Dingle, Blow, Barrett & Martin (1977) *Biochem. J.* **167**, 775. **14.** Banno, Yano & Nozawa (1983) *Eur. J. Biochem.* **132**, 563. **15.** Olstein & Liener (1983) *J. Biol. Chem.* **258**, 11049. **16.** Watanabe, Sakuragawa, Arima & Satoyoshi (1983) *J. Lipid Res.* **24**, 596. **17.** Leake, Heald & Peters (1982) *Eur. J. Biochem.* **128**, 557. **18.** Lu, Verkruyse & Hofmann (2002) *Biochim. Biophys. Acta* **1583**, 35. **19.** Löffler & Kunze (1987) *FEBS Lett.* **216**, 51. **20.** Slotte & Ekman (1986) *Biochim. Biophys. Acta* **879**, 221. **21.** Katewa & Katyare (2005) *Drug Chem. Toxicol.* **28**, 467. **22.** Boulton, Gregory & Cobb (1991) *Biochemistry* **30**, 278. **23.** Offensperger, Walter, Offensperger, *et al.* (1988) *Virology* **164**, 48. **24.** Donatelli, Marchi, Giuliani, Gustafsson & Pacifici (1994) *Eur. J. Clin. Pharmacol.* **47**, 345. **25.** Thithapandha & Cohn (1978) *Biochem. Pharmacol.* **27**, 263. **26.** Sun, Navas, Crane, Morré & Löw (1987) *J. Biol. Chem.* **262**, 15915. **27.** Zhao & Ishizaki (1997) *J. Pharmacol. Exp. Ther.* **283**, 1168. **28.** Graves, Kwiek, Fadden, *et al.* (2002) *Mol. Pharmacol.* **62**, 1364. **29.** Ganguli, Choudhury, Datta, *et al.* (2014) *Biochimie* (Part B) **107**, 338.

Chlorothricin

This membrane-interacting agent (FW$_{free-acid}$ = 955.49 g/mol; CAS 34707-92-1), isolated from a strain of *Streptomyces antibioticus*, is macrolide-type antibiotic that is active against Gram-positive bacteria. Chlorothricin also inhibits cholesterol biosynthesis by impeding the mevalonate pathway (1). **Target(s):** pyruvate carboxylase, noncompetitive inhibition (2,3,5); malate dehydrogenase (4). **1.** Kawashima, Nakamura, Ohta *et al.* (1992) *J. Antibiot.* **45**, 207. **2.** Schindler & Zahner (1973) *Eur. J. Biochem.* **39**, 591. **3.** Schindler & Zahner (1972) *Arch. Mikrobiol.* **82**, 66. **4.** Schindler (1975) *Eur. J. Biochem.* **51**, 579. **5.** Schindler & M. C. Scrutton (1975) *Eur. J. Biochem.* **55**, 543.

Chlorotoxin

This 36-residue marine toxin (FW = 4004.75 g/mol; CAS 163515-35-3; Sequence: MCMPCFTTDHQMARKCDDCCGGKGRGKCYGPQCLCR), from the venom of the deathstalker scorpion *Leiurus quinquestriatus*, blocks small-conductance chloride channels, when applied to the cytoplasmic surface (1). Chlorotoxin is also a reliable and specific histopathological marker for tumors of neuroectodermal origin, suggesting that chlorotoxin derivatives with cytolytic activity may have therapeutic potential for these cancers (2). **1.** DeBin, Maggio & Strichartz (1993) *Am. J. Physiol.* **264**, C361. **2.** Lyons, O'Neal & Sontheimer (2002) *Glia* **39**, 162.

Chlorphenamine

This first-generation alkylamine antihistamine (FW = 274.79 g/mol; CAS 113-92-8), also known as chlorpheniramine, 3-(4-chlorophenyl)-*N*,*N*-dimethyl-3-pyridin-2-ylpropan-1-amine as well as various trade names (*e.g.*, Antagonate®, Chlor-Trimeton®, Efidac 24®, Kloromin®, Phenetron®, and Pyridamal 100®), targets H₁ receptors and is also a Serotonin-Norepinephrine Reuptake Inhibitor (*or* SNRI). Chlorphenamine has a serum half-life of approximately 20 hours in adults, and elimination from the body is primarily by metabolism to mono-demethyl and di-demethyl forms (1). It is also combined with codeine and other cough-suppressing substances in the treatment of persistent coughing associated with the common cold. **Key Pharmacokinetic Parameters:** *See* Appendix II in Goodman & Gilman's THE PHARMACOLOGICAL BASIS OF THERAPEUTICS, 12ᵗʰ Edition (Brunton, Chabner & Knollmann, eds.) McGraw-Hill Medical, New York (2011). **1.** Rumore (1984) *Drug Intell. Clin. Pharm.* **18**, 701.

Chlorpromazine

This photosensitive antipsychotic drug and tranquilizer (FW_free-base = 318.87 g/mol; CAS 50-53-3; *Abbreviation* = Cpz), also known as Largactil™, Thorazine™ and systematically as 2-chloro-10-(3-dimethylaminopropyl)phenothiazine, acts on a variety of receptors within the central nervous system, producing anticholinergic, antidopaminergic, antihistaminic, and weak antiadrenergic effects. It was first prepared in the early 1950s in efforts to find better sedatives. It is a potent inhibitor of dopamine-stimulated adenylate cyclase and a mammalian multidrug resistance modulator. Chlorpromazine also inhibits oxidative phosphorylation and photophosphorylation, is a histamine H₁ receptor and dopamine D₂ receptor antagonist, and inhibits the calmodulin stimulation of phosphodiesterase, *or* cyclic-nucleotide phosphodiesterase I, IC₅₀ = 17 µM (10) and transglutaminase. Chlorpromazine also inhibits mouse brain nitric-oxide synthase and prevents lipopolysaccharide induction of the synthase in murine lung. **Chlorpromazine Inhibition of Clathrin-Mediated Endocytosis:** Chlorpromazine (30-50 µM for 20-30 min) blocks endocytosis of clatharin-coated endosomes by targeting clathrin's heavy chain, *or* chc (11,12), often inhibiting the uptake of vesicle cargo (*e.g.,* transferrin), some viruses (*e.g,* adenovirus) as well as endocytosis-mediated dye uptake. Indeed, FlAsH-FALI-mediated chc photoinactivation of chc or control treatment of synapses with chlorpromazine does not show a statistical difference in membrane uptake or the size of internalized membranous structures (13). That said, the promiscuity chlorpromazine inhibition (*see below*) suggests that one must avoid its indiscriminant use as an endocytosis inhibitor. Moreover, given chlorpromazine's light sensitivity, one cannot discount the likelihood that chorpromazine's potency and/or inhibitory action is itself photo-dependent. **Key Pharmacokinetic Parameters:** *See* Appendix II in Goodman & Gilman's THE PHARMACOLOGICAL BASIS OF THERAPEUTICS, 12ᵗʰ Edition (Brunton, Chabner & Knollmann, eds.) McGraw-Hill Medical, New York (2011). **Target(s):** NAD(P)H dehydrogenase, *or* DT diaphorase (1,23,33); malate dehydrogenase (acceptor) (2); D-amino acid oxidase (3); histamine *N*-methyl-transferase (4,34,76); mitochondrial and photophos-phorylation uncoupler (5); protein kinase C (6,10,27); calmodulin (7,10); sterol acyltransferase, *or* cholesterol acyltransferase (8); indoleamine *N*-methyltransferase (9); nitric-oxide synthase (25); 3',5'-cyclic-nucleotide phosphodiesterase (10,24,29,55,72); Ca²⁺-transporting ATPase (10,17,26); myosin light-chain kinase (10); [β-adrenergic-receptor]

kinase (11); dopamine-stimulated adenylate cyclase (15,16); multidrug resistance, as modulator of (15,16); Na⁺/K⁺-exchanging ATPase (17,22,35,48); actin polymerization (18); CYP2D6 (19); [pyruvate dehydrogenase] phosphatase (20); CYP2B1 (21); carbonyl reductase (28); phospholipase A₂ (30); lysosomal sphingomyelinase (31); histamine H₁ receptor (32); phosphatidylcholine:sterol acyltransferase, *or* LCAT (36); succinate-semialdehyde dehydrogenase (37); dopamine D₂ receptor (38,39); myosin ATPase (40); ATP synthase (F₀F₁ ATPase; H⁺-transporting two-sector ATPase (41,42); ATPase, *Escherichia coli* (32); TF₁ ATPase, stimulated at concentrations below 0.6 mM at 23°C (42); carnosine synthetase, inhibition at millimolar levels (43); phosphatidylinositol diacylglycerol-lyase, *or* phosphatidyl-inositol-specific phospholipase C, inhibits at pH 7, but stimulates at pH 5.5 (44); adenylate cyclase (45,46); 1-aminocyclopropane-1-carboxylate synthase (47); thiamin-triphosphatase (49-51); nucleoside-diphosphatase (52); Xaa-methyl-His dipeptidase, *or* anserinase,*or* *N*-acetylhistidine deacetylase (53); glycosylphosphatidylinositol phospholipase D (54); [pyruvate dehydrogenase (acetyl-transferring)]-phosphatase (55); phosphatidate phosphatase (55-61); phosphoserine phosphatase (62,63); phospholipase A₁ (64); diacylglycerol lipase (65); acylglycerol lipase, *or* monoacylglycerol lipase; weakly inhibited (65); sterol esterase, *or* cholesterol esterase (65); cholinesterase (butyrylcholinesterase; K₁ = 3.22 µM (67); phenol sulfotransferase, *or* aryl sulfotransferase (68); choline-phosphate cytidylyltransferase (69,70); diphosphoinositol pentakisphosphate kinase (74); NAD⁺ kinase (76); carnitine *O*-octanoyltransferase (77); carnitine *O*-palmitoyltransferase (75); flavin-containing mono-oxygenase (79); 5-lipoxygenase, *or* arachidonate 5-lipoxygenase (78); 12-lipoxygenase, *or* arachidonate 12-lipoxygenase (78); prostaglandin synthase (78); tyrpanothione-disulfide reductase (79,80). **1.** Ernster (1967) *Meth. Enzymol.* **10**, 309. **2.** Benziman (1969) *Meth. Enzymol.* **13**, 129. **3.** Yagi (1971) *Meth. Enzymol.* **17B**, 608. **4.** Axelrod (1971) *Meth. Enzymol.* **17B**, 766. **5.** Izawa & Good (1972) *Meth. Enzymol.* **24**, 355. **6.** Kikkawa, Minakuchi, Takai & Nishizuka (1983) *Meth. Enzymol.* **99**, 288. **7.** Weiss (1983) *Meth. Enzymol.* **102**, 171. **8.** Billheimer (1985) *Meth. Enzymol.* **111**, 286. **9.** Porta (1987) *Meth. Enzymol.* **142**, 668. **10.** Hidaka, Inagaki, Nishikawa & Tanaka (1988) *Meth. Enzymol.* **159**, 652. **11.** Wang, Rothberg & Anderson (1993) *J. Cell Biol.* **123**, 1107. **12.** Blanchard, Belouzard, Goueslain, *et al.* (2006) *J. Virol.* **80**, 6964. **13.** Kasprowicz, Kuenen, Miskiewicz, *et al.* (2008) *J. Cell Biol.* **182**, 1007. **14.** Benovic (1991) *Meth. Enzymol.* **200**, 351. **15.** Iversen (1975) *Science* **188**, 1084. **16.** Iversen (1976) *Trends Biochem. Sci.* **1**, 121. **17.** Bhattacharyya & Sen (1999) *Mol. Cell Biochem.* **198**, 179. **18.** Milzani, DalleDonne & Dalledonne (1999) *Arch. Biochem. Biophys.* **369**, 59. **19.** Shin, Soukhova & Flockhart (1999) *Drug Metab. Dispos.* **27**, 1078. **20.** Bak, Huh, Hong & Song (1999) *Biochem. Mol. Biol. Int.* **47**, 1029. **21.** Murray (1992) *Biochem. Pharmacol.* **44**, 1219. **22.** Hackenberg & Krieglstein (1972) *Naunyn Schmiedebergs Arch. Pharmacol.* **274**, 63. **23.** Martius (1963) *The Enzymes*, 2nd ed., **7**, 517. **24.** Drummond & M. Yamamoto (1971) *The Enzymes*, 3rd ed., **4**, 355. **25.** Palacios, Padron, Glaria, *et al.* (1993) *Biochem. Biophys. Res. Commun.* **196**, 280. **26/.** Hasselbach (1974) *The Enzymes*, 3rd ed., **10**, 431. **27.** Kikkawa & Nishizuka (1986) *The Enzymes*, 3rd ed., **17**, 167. **28.** Wermuth (1981) *J. Biol. Chem.* **256**, 1206. **29.** Marshak, Lukas & Watterson (1985) *Biochemistry* **24**, 144. **30.** Lindahl & Tagesson (1993) *Inflammation* **17**, 573. **31.** Maor, Mandel & Aviram (1995) *Arter`iosclero. Thromb. Vasc. Biol.* **15**, 1378. **32.** Hill, Ganellin, Timmerman, *et al.* (1997) *Pharmacol. Rev.* **49**, 253. **33.** Ernster, Ljunggren & Danielson (1960) *Biochem. Biophys. Res. Commun.* **2**, 48 and 88. **34.** Brown, Tomchick & Axelrod (1959) *J. Biol. Chem.* **234**, 2948. **35.** Van Dyke & Scharschmidt (1987) *Amer. J. Physiol.* **253**, G613. **36.** Albers, Chen & Lacko (1986) *Meth. Enzymol.* **129**, 763. **37.** Cash, Maitre, Rumigny, Weissman-Nanopoulos & Mandel (1982) *Prog. Clin. Biol. Res.* **114**, 379. **38.** Anden, Butcher, Corrodi, Fuxe & Ungerstedt (1970) *Eur. J. Pharmacol.* **11**, 303. **39.** Anden, Roos & Werdinius (1964) *Life Sci* **3**, 149. **40.** Grillo, Rinaudo & Vergani (1964) *Enzymologia* **27**, 41. **41.** Laikind, Goldenberg & Allison (1982) *Biochem. Biophys. Res. Commun.* **109**, 423. **42.** Bullough, Kwan, Laikind, Yoshida & Allison (1985) *Arch. Biochem. Biophys.* **236**, 567. **43.** Ng & Marshall (1976) *Experientia* **32**, 839. **44.** Allan & Michell (1974) *Biochem. J.* **142**, 591. **45.** Asbury, Cook & Wolff (1978) *J. Biol. Chem.* **253**, 5286. **46.** Carricarte, Bianchini, Muschietti, *et al.* (1988) *Biochem. J.* **249**, 807. **47.** Mattoo, Adams, Patterson & Lieberman (1983) *Plant Sci. Lett.* **28**, 173. **48.** Schuurman, Stekhoven & Bonting (1981) *Physiol. Rev.* **61**, 1. **49.** Iwata, Baba, Matsuda & Terashita (1974) *Thiamine* (*Proc. Pap. U.S.-Jpn. Semin.*, 2nd. Meeting), 213. **50.** Iwata, Baba, Matsuda & Terashita (1975) *J. Neurochem.* **24**, 1209. **51.** Iwata, Baba, Matsuda, Terashita & Ishii (1974) *Jpn. J. Pharmacol.* **24**, 825.

52. Sano, Matsuda & Nakagawa (1988) *Eur. J. Biochem.* **171**, 231. **53.** Lenney, Baslow & Sugiyama (1978) *Comp. Biochem. Physiol. B* **61**, 253. **54.** Malik & Low (1986) *Biochem. J.* **240**, 519. **55.** Ryan & W. A. Toscano (1985) *Arch. Biochem. Biophys.* **241**, 403. **56.** Bak, Huh, Hong & Song (1999) *Biochem. Mol. Biol. Int.* **47**, 1029. **57.** Fleming & Yeaman (1995) *Biochem. J.* **308**, 983. **58.** Kanoh, Imai, Yamada & Sakane (1992) *J. Biol. Chem.* **267**, 25309. **59.** Nanjundan & Possmayer (2000) *Exp. Lung Res.* **26**, 361. **60.** Bowley, Cooling, Burditt & Brindley (1977) *Biochem. J.* **165**, 447. **61.** Martin, Hales & Brindley (1987) *Biochem. J.* **245**, 347. **62.** Shetty & Shetty (1991) *Neurochem. Res.* **16**, 1203. **63.** Hawkinson, Acosta-Burruel, Ta & Wood (1997) *Eur. J. Pharmacol.* **337**, 315. **64.** Wainszelbaum, Isola, Wilkowsky, *et al.* (2001) *Biochem. J.* **355**, 765. **65.** Rindlisbacher, Reist & Zahler (1987) *Biochim. Biophys. Acta* **905**, 349. **66.** Sando & Rosenbaum (1985) *J. Biol. Chem.* **260**, 15186. **67.** Darvesh, McDonald, Penwell, *et al.* (2005) *Bioorg. Med. Chem.* **13**, 211. **68.** Fernando, Sakakibara, Nakatsu, *et al.* (1993) *Biochem. Mol. Biol. Int.* **30**, 433. **69.** Pelech, Jetha & Vance (1983) *FEBS Lett.* **158**, 89. **70.** Mansbach & Arnold (1986) *Biochim. Biophys. Acta* **875**, 516. **71.** Safrany (2004) *Mol. Pharmacol.* **66**, 1585. **72.** Mata, Gamboa, Macias, *et al.* (2003) *J. Agric. Food Chem.* **51**, 4559. **73.** Gallais, de Crescenzo & Laval-Martin (2001) *Aust. J. Plant Physiol.* **28**, 363. **74.** Puzkin & V. Raghuraman (1985) *J. Biol. Chem.* **260**, 16012. **75.** Pagot & Belin (1996) *Appl. Environ. Microbiol.* **62**, 3864. **76.** Thithapandha & Cohn (1978) *Biochem. Pharmacol.* **27**, 263. **77.** Demirdoegen & Adall (2005) *Cell Biochem. Funct.* **23**, 245. **78.** Hamasaki & Tai (1985) *Biochim. Biophys. Acta* **834**, 37. **79.** Iribarne, Paulino, Aguilera & Tapia (2009) *J. Mol. Graph. Model.* **28**, 371. **80.** Chan, Yin, Garforth, *et al.* (1998) *J. Med. Chem.* **41**, 148.

Chlorpropamide

This antidiabetic agent (FW = 276.74 g/mol; CAS 94-20-2; Soluble in water at pH 6 (2.2 mg/mL) but practically insoluble at pH 7.3), also known as chloropropamide and *N*-propyl-*N'*-(*p*-chlorobenzene sulfonyl)urea, is an inhibitor of cholesterol biosynthesis. **Mechanism of Action in Type II Diabetes:** As with other sulfonylureas, chlorpropamide acts by increasing the insulin secretion and is only effective in patients who have some pancreatic beta cell function. **Target(s):** cholesterol biosynthesis (1); lathosterol oxidase (2); rat renal organic anion transporter (3); aldehyde dehydrogenase (4,5,9); urease (6,7); cAMP phosphodiesterase (8). **1.** McDonald & Dalidowicz (1962) *Biochemistry* **1**, 1187. **2.** Dempsey (1969) *Meth. Enzymol.* **15**, 501. **3.** Uwai, Saito, Hashimoto & Inui (2000) *Eur. J. Pharmacol.* **398**, 193. **4.** Lee, Elberling & Nagasawa (1992) *J. Med. Chem.* **35**, 3641. **5.** Ting & Crabbe (1983) *Biochem. J.* **215**, 351. **6.** Garcia, Molina & Herrera (1978) *Rev. Esp. Fisiol.* **34**, 145. **7.** Herrera, Sabater & Molina (1977) *Rev. Esp. Fisiol.* **33**, 37. **8.** Brooker & Fichman (1971) *Biochem. Biophys. Res. Commun.* **42**, 824. **9.** Deitrich & Hellerman (1963) *J. Biol. Chem.* **238**, 1683.

Chlorprothixene

This antipsychotic drug (FW = 315.87 g/mol; CAS 113-59-7) is an inhibitor of dopamine-stimulated adenylate cyclase and a dopamine receptor antagonist (1) There are two distinct forms of this pharmaceutical: an α- or *cis*-isomer (*i.e.*, the *Z*-isomer) and a β- or *trans*-form. α-Chlorprothixene is the more active form, *K*i = 37 nM (2). **Target(s):** dopamine-stimulated adenylate cyclase (1,2); dopamine β-hydroxylase (3); adenosylmethionine decarboxylase (4); transport ATPase (5). **1.** Iversen (1975) *Science* **188**, 1084. **2.** Iversen (1976) *Trends Biochem. Sci.* **1**, 121. **3.** Palatini, Dabbeni-Sala & Finotti (1984) *Biochem. Int.* **9**, 675. **4.** Hietala, Lapinjoki & Pajunen (1984) *Biochem. Int.* **8**, 245. **5.** Ebadi & Carver (1970) *Eur. J. Pharmacol.* **9**, 190.

Chlorpyrifos

This organophosphothioate-based insecticide (FW = 350.59 g/mol; CAS 2921-88-2; IUPAC: *O,O*-diethyl-*O*-(3,5,6-trichloro-2-pyridyl)phosphorothionate) undergoes hydrolysis to form the potent acetylcholinesterase inhibitor, chlorpyrifos oxon (*See Chlorpyrifos Oxon*). **Target(s):** acetylcholinesterase (1,4,8); cutinase (2); palmitoyl-CoA hydrolase, *or* acyl-CoA hydrolase (3); CYP6D1 (5); ATPase, Ca²⁺-stimulated (6); kinesin (7); acylglycerol lipase, *or* monoacylglycerol lipase (8); fatty acid amide hydrolase, *or* anandamide amidohydrolase (8); carboxyl-esterase (8-10); neuropathy target esterase (8). **1.** Rigterink & Kenaga (1966) *J. Agric. Food Chem.* **14**, 304. **2.** Sebastian & Kolattukudy (1988) *Arch. Biochem. Biophys.* **263**, 77. **3.** Larson & Kolattukudy (1985) *Arch. Biochem. Biophys.* **237**, 27. **4.** Huff, Corcoran, Anderson & Abou-Donia (1994) *J. Pharmacol. Exp. Ther.* **269**, 329. **5.** Scott (1996) *Insect Biochem. Mol. Biol.* **26**, 645. **6.** Barber, Hunt & Ehrich (2001) *J. Toxicol. Environ. Health A* **63**, 101. **7.** Gearhart, Sickles, Buccafusco, Prendergast & Terry (2007) *Toxicol. Appl. Pharmacol.* **218**, 20. **8.** Quistad, Klintenberg, Caboni, Liang & Casida (2006) *Toxicol. Appl. Pharmacol.* **211**, 78. **9.** Zhang, Qiao & Lan (2004) *Environ. Toxicol.* **19**, 154. **10.** Valles, Oi & Strong (2001) *Insect Biochem. Mol. Biol.* **31**, 715.

Chlorsulfuron

This triazine derivative (FW = 357.78 g/mol; CAS 64902-72-3), also known as DPX-4189, and 2-chloro-*N*-[[(4-methoxy-6-methyl-1,3,5-triazin-2-yl)amino]carbonyl]benzenesulfonamide, is a selective pre- and post-emergence herbicide that is active on many grasses and broadleaf weeds (*e.g.*, Canada thistle, wild buckwheat, hemp nettle). Chlorsulfuron noncompetitively inhibits purified barley acetolactate synthase with respect to pyruvate, yielding an initial apparent inhibition constant of 68 nM (calculated from the initial phase of inhibition) and a final steady-state dissociation constant of 3 nM. Inasmuch as acetolactate is needed in the valine and leucine synthesis, these findings appear to explain why low-level application (8-30 liters per hectare) is sufficient for chlorsulfuron'n herbicidal action. **1.** Durner, Gailus & Boger (1994) *FEBS Lett.* **354**, 71. **2.** Hill, Pang & Duggleby (1997) *Biochem. J.* **327**, 891. **3.** Chang & Duggleby (1997) *Biochem. J.* **327**, 161. **4.** McCourt, Pang, Guddat & Duggleby (2005) *Biochemistry* **44**, 2330. **5.** Southan & Copeland (1996) *Physiol. Plant.* **98**, 824.

7-Chlortetracycline

This tetracycline-based antibiotic (FW = 478.89 g/mol; CAS 64-72-2), also named aureomycin, 7-chlorotetracycline and [4S-(4α,4aα,5aα,6β,12aα)]-7-chloro-4-(dimethylamino)-1,4,4a,5,5a,6-11,12a-octahydro-3,6,10,12,12a

-pentahydroxy-6-methyl-1,11-dioxo-2-naphthacenecarboxamide, inhibits protein biosynthesis by inhibiting the binding of molecules of aminoacyl-tRNA to the acceptor site (A-site) of ribosomes. **Target(s):** D-amino-acid oxidase (1,2); cystathionine γ-lyase, *or* cysteine desulfhydrase (3); guanosine-3',5'-bis(diphosphate) 3'-diphosphatase (4); HIV-1 reverse transcriptase (5); NADH:cytochrome oxidoreductase (6); NAD(P)H dehydrogenase (quinone) (7); polynucleotide phosphorylase, *or* polyribonucleotide nucleotidyltransferase (8); protein biosynthesis (9-12); RNA-directed DNA polymerase (5); peptidyltramsferase (12); anhydro-tetracycline monooxygenase (13). **1.** Yagi (1971) *Meth. Enzymol.* **17B**, 608. **2.** Meister & Wellner (1963) *The Enzymes*, 2nd ed., **7**, 609. **3.** Bernheim & Deturk (1953) *Enzymologia* **16**, 69. **4.** Richter (1980) *Arch. Microbiol.* **124**, 229. **5.** Wondrak, Lower & Kurth (1988) *J. Antimicrob. Chemother.* **21**, 151. **6.** Colaizzi, Knevel & Martin (1965) *J. Pharm. Sci.* **54**, 1425. **7.** Wosilait & Nason (1954) *J. Biol. Chem.* **208**, 785. **8.** Simuth, Zelinka & Polek (1975) *Biochim. Biophys. Acta* **379**, 397. **9.** Pestka (1974) *Meth. Enzymol.* **30**, 261. **10.** Laskin & Chan (1964) *Biochem. Biophys. Res. Commun.* **14**, 137. **11.** Lucas-Lenard & Beres (1974) *The Enzymes*, 3rd ed., **10**, 53. **12.** Traut & Monro (1964) *J. Mol. Biol.* **10**, 63. **13.** Behal, Neuzil & Hostalek (1983) *Biotechnol. Lett.* **5**, 537.

Chlorthalidone

This thiazide-like diuretic (FW = 338.76 g/mol; CAS 77-36-1; Symbol: CTDN), also known as chlortalidone and (*RS*)-2-chloro-5-(1-hydroxy-3-oxo-2,3-dihydro-1*H*-isoindol-1-yl)benzene-1-sulfonamide, as well as gigroton, novatalidon, and uridon, prevents sodium and chloride reabsorption by inhibiting the Na$^+$/Cl$^-$ symporter within the distal convoluted tubule. It also decreases the glomerular filtration rate, further reducing its efficacy in patients with renal insufficiency. By increasing sodium ion delivery to the distal renal tubule, chlorthalidone indirectly increases potassium excretion via the sodium-potassium exchange mechanism (*i.e.*, apical ROM K$^+$/Na$^+$ channels, coupled with basolateral Na$^+$/K$^+$ ATPases). The latter effect can result in hypokalemia (and hypochloremia) as well as a mild metabolic alkalosis. Nonetheless, the chlorthalidone's diuretic efficacy is unaffected by the patient's acid-base balance. CTDN reduces pulse wave velocity, predictor of cardiovascular events (CVEs) and a measure of central aortic stiffness associated with endothelial dysfunction (2). On the other hand, CTDN fosters hypokalemia, hyperglycemia, sympathetic discharge, and the renin-angiotensin-aldosterone system, but these potentially harmful effects do not appear to materially reduce CTDN's ability to prevent CVEs (2). **Key Pharmacokinetic Parameters:** *See* Appendix II in Goodman & Gilman's THE *PHARMACOLOGICAL BASIS OF THERAPEUTICS*, 12th Edition (Brunton, Chabner & Knollmann, eds.) McGraw-Hill Medical, New York (2011). **1.** Ford (1960) *Curr. Ther. Res. Clin. Exp.* **2**, 347. **2.** Roush, Buddharaju, Ernst & Holford (2013) *Curr. Hypertens. Rep.* **15**, 514.

Cholate

This bile acid (FW$_{free-acid}$ = 408.58 g/mol; CAS 81-25-4; pK_a = 6.4; Critical Micelle Concentration = 9-15 mM), also called 3α,7α,12α-trihydroxycholanic acid, is a cholesterol metabolite found in bile and is often conjugated to glycine (*as* glycocholate) or taurine (*as* taurocholate). In the liver of cholate-fed rats, excess levels of bile salts activate a GAPDH-mediated transnitrosylation cascade that provides feedback

inhibition of bile salt synthesis (1). Cholate is an FDA-approved drug (Cholbam®) that is specifically indicated for the treatment of bile acid synthesis disorders resulting from single enzyme defects and for the adjunctive treatment of peroxisomal disorders, including Zellweger spectrum disorders in patients exhibiting manifestations of liver disease, steatorrhea or complications from decreased fat soluble vitamin absorption. Cholate's inhibition of scores of enzymes is likely to be due mainly to its indiscriminant and nonphysiologic detersive actions. **Use in Membrane Solubilization:** Cholate is used widely to prepare solubilized membrane-associated enzymes/proteins. Enzyme activity can usually be differentially solubilized with 0.4 to 1.2% sodium cholate. The classical example is Ephraim Racker's groundbreaking preparation of the proton-translocating F$_O$F$_1$-ATPase from bovine heart mitochondria by extraction of submitochondrial particles with cholate, followed by fractionation with ammonium sulfate, and sucrose gradient centrifugation in the presence of methanol, deoxycholate, and lysolecithin (2). **1.** Rodríguez-Ortigosa, Celay, Olivas, *et al.* (2014) *Gastroenterology* 147, 1084. **2.** Serrano, Kanner & Racker (1976) *J. Biol. Chem.* **251**, 2453.

Cholera Toxin

This bacterial enterotoxin (MW = 84 kDa; CAS 9012-63-9), also known as choleragen, is a multimeric protein produced by *Vibrio cholerae*. The toxin uses its B-subunit to bind to a specific receptor on sensitive cells, whereupon the A-subunit combines with cellular ADP-ribosylating factor to form an ADP ribosyltransferase that covalently modifies and permanently activates the heterotrimeric GTPase or GTP-regulatory protein in the adenylate cyclase pathway. Once ADP-ribosylated, the GS·GTP complex cannot reconvert to its metabolic inactive form, and the adenylate cyclase remains active, overproducing cAMP, inhibiting sodium channels, and activating the chloride channel.

Cholesterol

This principal steroid (FW = 386.66 g/mol; CAS 57-88-5; M.P. (anhydrous) = 148.5°C; slowly oxidizes in air), also named cholest-5-en-3β-ol, is found in free and esterified forms in all vertebrate tissues, particularly abundant in the plasma membrane and certain vesicles. Required to make and maintain animal cell membranes; cholesterol modulates membrane fluidity over the range of physiological temperatures. Its hydroxyl group interacts with the polar groups of membrane phospholipids and sphingolipids, whereas its hydrocarbon rings are deeply embedded in the membrane, lying alongside the nonpolar fatty-acid chains of other lipids. Cholesterol is also the precursor for most natural steroids. Excess cholesterol is harmful to humans, and randomised trials have shown that interventions that lower LDL cholesterol concentrations can significantly reduce the incidence of coronary heart disease and other major vascular events in a wide range of individuals. **Target(s):** β-glucuronidase (1); glutamate dehydrogenase (1); sphingomyelin phosphodiesterase, *or* sphingomyelinase (2,4,10); F$_1$F$_o$ ATPase, weakly inhibited (3); steryl sulfatase (5,9); cholestenol Δ-isomerase, *or* sterol Δ7,8-isomerase (6); cholesterol-5,6-oxide hydrolase (7); steryl-β-glucosidase (8); retinyl-palmitate esterase (11); sterol esterase, *or* cholesterol esterase (12); ethanolaminephosphotransferase (13); bile-acid-CoA:amino acid *N*-acyltransferase (14); cycloartenol 24-*C*-methyltransferase (15); 27-hydroxycholesterol 7α-monooxygenase (16). **1.** Tappel & Dillard (1967) *J. Biol. Chem.* **242**, 2463. **2.** Brady (1983) *The Enzymes*, 3rd ed., **16**, 409. **3.** Zheng & Ramirez (1999) *Eur. J. Pharmacol.* **368**, 95. **4.** Maziere, Wolf, Maziere, *et al.* (1981) *Biochim. Biophys. Res. Commun.* **100**, 1299. **5.** Notation & Ungar (1969) *Biochemistry* **8**, 501. **6.** Nes, Zhou, Dennis, *et al.* (2002) *Biochem. J.* **367**, 587. **7.** Nashed, Michaud, Levin & Jerina (1985) *Arch. Biochem. Biophys.* **241**, 149. **8.** Kalinowska & Wojciechowski (1978) *Phytochemistry* **17**, 1533. **9.** Norkowska & Gniot-Szulzycka (2002) *J. Steroid Biochem. Mol. Biol.* **81**, 263. **10.** Liu, Nilsson & Duan (2002) *Lipids* **37**, 469. **11.** Blaner, Halperin, Stein, Stein & Goodman (1984) *Biochim. Biophys. Acta* **794**, 428. **12.** Wee & Grogan

(1993) *J. Biol. Chem.* **268**, 8158. **13**. Vecchini, Roberti, Freysz & Binaglia (1987) *Biochim. Biophys. Acta* **918**, 40. **14**. Kimura, Okuno, Inada, Ohyama & Kido (1983) *Hoppe-Seyler's Z. Physiol. Chem.* **364**, 637. **15**. Zhao & Nes (2003) *Arch. Biochem. Biophys.* **420**, 18. **16**. Souidi, Parquet, Dubrac, *et al.* (2000) *Biochim. Biophys. Acta* **1487**, 74.

Cholesteryl 3-Sulfate

This sulfated sterol (FW$_{free-acid}$ = 466.73 g/mol), also known as cholesterol 3-sulfate, inhibits pancreatic elastase. It has been identified in a variety of cells as well as in low-density lipoproteins. Cholesteryl 3-sulfate activates protein kinase C. **Target(s):** pancreatic elastase (1); deoxyribonuclease I (2); trypsin (2); plasmin (3); thrombin (3); cholesteryl ester transfer protein (4); cholesterol monooxygenase, side-chain cleaving, *or* CYP11A1 (5-7); 3-hydroxy-3-methylglutaryl-CoA reductase (8). **1**. Ito, Iwamori, Hanaoka & Iwamori (1998) *J. Biochem.* **123**, 107. **2**. Iwamori, Suzuki, Kimura & Iwamori (2000) *Biochim. Biophys. Acta* **1487**, 268. **3**. Iwamori, Iwamori & Ito (1999) *J. Biochem.* **125**, 594. **4**. Connolly, Krul, Heuvelman & Glenn (1996) *Biochim. Biophys. Acta* **1304**, 145. **5**. Lambeth & Xu (1989) *Endocr. Res.* **15**, 85. **6**. Xu & Lambeth (1989) *J. Biol. Chem.* **264**, 7222. **7**. Lambeth, Xu & Glover (1987) *J. Biol. Chem.* **262**, 9181. **8**. Williams, Hughes-Fulford & Elias (1985) *Biochim. Biophys. Acta* **845**, 349.

Choline

This nutritionally essential, water-soluble quaternary amine (FW$_{free-base}$ = 104.17 g/mol; CAS 62-49-7), also named trimethyl-β-hydroxyethylammonium, is a substituent of the cationic head-group of lecithin (phosphatidylcholine) as well as the neurotransmitter, acetylcholine. In liver, phosphatidylethanolamine *N*-methyltransferase catalyses the synthesis of PtdCho through the sequential methylation of phosphatidylethanolamine, using *S*-adenosylmethionine (AdoMet). It is also the major dietary source of methyl groups via synthesis of AdoMet. In brain, the high-affinity choline transporter (CHT1) is present on the presynaptic terminal of cholinergic neurons, taking up choline formed by the action of acetylcholinesterase. Na$^+$-dependent uptake of choline by CHT1 is the rate-limiting step for resynthesis of acetylcholine. In view of its wide-ranging roles in human metabolism, choline deficiency is now thought to exacerbate liver disease, atherosclerosis, and certain neurological disorders. For women, the adequate choline intake is 425 mg/day; for adult men, 550 mg/day. (*See also Acetyl-choline*) **Target(s):** phospholipase D (1); trimethylamine dehydrogenase (2); diamine oxidase (3); acetyl-cholinesterase (4,5,9,10); carnitine 3-dehydrogenase (6); adenylate cyclase (7); ethanolamine kinase (8,27-31); carnitine dehydratase (11); NTPase (12); *N*-acetylmuramoyl-L-alanine amidase (13); aryl-acylamidase (14); coagulation factor Xa15); choline sulfatase (16,17); glycerophosphocholine cholinephospho-diesterase (18-20); phosphoethanolamine/phosphocholine phosphatase, weakly inhibited (21); 4-nitrophenylphosphatase activity of Na$^+$/K$^+$-exchanging ATPase (22); acid phosphatase (23); sinapine esterase (24); cholinesterase, *or* butyrylcholinesterase (25); acetylcholinesterase (26); hydroxylysine kinase, weakly inhibited (32,33); sucrose:sucrose fructosyltransferase (34); carnitine *O*-acetyltransferase (35,36); betaine:homocysteine *S*-methyltrransferase (37). **1**. Kates & Sastry (1969) *Meth. Enzymol.* **14**, 197. **2**. McIntire (1990) *Meth. Enzymol.* **188**, 250. **3**. Zeller (1951) *The Enzymes*, 1st ed., **2** (part 1), 536. **4**. Wilson (1960) *The Enzymes*, 2nd ed., **4**, 501. **5**. Krupka (1965) *Biochemistry* **4**, 429. **6**. Schöpp, Sorger, Kleber & Aurich (1969) *Eur. J. Biochem.* **10**, 56. **7**. Steer & Wood (1981) *J. Biol. Chem.* **256**, 9990. **8**. Weinhold & Rethy (1972) *Biochim. Biophys. Acta* **276**, 143. **9**. Wilson (1952) *J. Biol. Chem.* **197**, 215. **10**. Wilson (1954) *J. Biol. Chem.* **208**, 123. **11**. Jung, Jung & Kleber (1989) *Biochim. Biophys. Acta* **1003**, 270. **12**. Plesner, Juul, Scriver & Aalkjaer (1991) *Biochim. Biophys. Acta* **1067**, 191. **13**. Romero, Lopez & Garcia (1990) *J. Virol.* **64**,

137. **14**. Oommen & Balasubramanian (1979) *Eur. J. Biochem.* **94**, 135. **15**. Monnaie, Arosio, Griffon, *et al.* (2000) *Biochemistry* **39**, 5349. **16**. Lucas, Burchiel & Segel (1972) *Arch. Biochem. Biophys.* **153**, 664. **17**. Takebe (1961) *J. Biochem.* **50**, 245. **18**. Sok (1998) *Neurochem. Res.* **23**, 1061. **19**. Lee, Kim, Kim, Myung & Sok (1997) *Neurochem. Res.* **22**, 1471. **20**. Baldwin & Cornatzer (1968) *Biochim. Biophys. Acta* **164**, 195. **21**. Massimelli, Beassoni, Forrellad, *et al.* (2005) *Curr. Microbiol.* **50**, 251. **22**. Mendonca, Masui, McNamara, Leone & Furriel (2007) *Comp. Biochem. Physiol. A* **146**, 534. **23**. Lisa, Garrido & Domenech (1984) *Mol. Cell. Biochem.* **63**, 113. **24**. Nurmann & Strack (1979) *Z. Naturforsch. C* **34**, 715. **25**. Cengiz, Cokugras & Tezcan (2002) *J. Protein Chem.* **21**, 145. **26**. Ciliv & Oezand (1972) *Biochim. Biophys. Acta* **284**, 136. **27**. Sung & Johnstone (1967) *Biochem. J.* **105**, 497. **28**. Ishidate, Furusawa & Nakazawa (1985) *Biochim. Biophys. Acta* **836**, 119. **29**. Brophy, Choy, Toone & Vance (1977) *Eur. J. Biochem.* **78**, 491. **30**. Infante & Kinsella (1976) *Lipids* **11**, 727. **31**. Weinhold & Rethy (1974) *Biochemistry* **13**, 5135. **32**. Hiles & Henderson (1972) *J. Biol. Chem.* **247**, 646. **33**. Chang (1977) *Enzyme* **22**, 230. **34**. Chevalier & Rupp (1993) *Plant Physiol.* **101**, 589. **35**. Fritz & Schultz (1965) *J. Biol. Chem.* **240**, 2188. **36**. White & Wu (1973) *Biochemistry* **12**, 841. **37**. Waditee & Incharoensakdi (2001) *Curr. Microbiol.* **43**, 107.

Chondramide

These *Chondromyces crocatus*-derived actin polymerization inhibitors (FW$_{Chondramide-A}$ = 646.78 g/mol, R$_1$ = –OCH$_3$, R$_2$ = –H; FW$_{Chondramide-B}$ = 681.23 g/mol, R$_1$ = –OCH$_3$, R$_2$ = –Cl; FW$_{Chondramide-C}$ = 616.76 g/mol, R$_1$ = –H, R$_2$ = –H; MW$_{Chondramide-D}$ = 651.20 g/mol, R$_1$ = –H, R$_2$ = –Cl) are novel depsipeptides that are structurally related to the marine natural product, jasplakinolide (*See Jasplakinolide*). These cyclic depsipeptides were stabilize actin filaments, showing similarities to the action of phalloidin (*See Phalloidin*). Their cytostatic effects give IC$_{50}$ values ranging from 3 to 85 nM, on the same order of magnitude as that for cytochalasin D and jasplakinolide (1). Unlike the latter, however, the chondramides can be produced in large amounts by fermentation. All four chondramides readily enter the cells and cause severe shape malformations when applied during growth of unicellular green alga *Micrasterias denticulate* (2). **1**. Kunze, Jansen, Sasse, Höfle & Reichenbach (1995) *J. Antibiot.* (Tokyo) **48**, 1262. **2**. Holzinger & Lütz-Meindl (2001) *Cell Motil. Cytoskeleton* **48**, 87.

Chorismate

This conjugate base of chorismic acid (FW$_{free-acid}$ = 226.19 g/mol), known systematically as (3*R*-4*R*)3-[(1-carboxyethenyl)oxy]-4-hydroxy-1,5-cyclohexadiene-1-carboxylic acid, is a branch-point intermediate in the biosynthesis of L-tryptophan, L-tyrosine, and L-phenylalanine *via* the shikimate pathway in microorganisms, fungi, and higher plants (1-3). Chorismate is also a precursor in the biosynthesis of folate, enterochelin, ubiquinone, and menaquinone. Care should always be exercised when working with chorismate since it slowly decomposes to prephenate and *p*-hydroxybenzoate (2). Commercial preparations may also have these substances present in small amounts. **Target(s):** prephenate dehydrogenase (5); glycogen synthase (6); 3-deoxy-7 phosphoheptulonate synthase, *or* 2-keto-3-deoxy-D-*arabino*-heptulosonate-7-phosphate synthase (7-9). **1**. Gibson, Gibson, Doy & Morgan (1962) *Nature* **195**, 1173. **2**. Gibson (1968) *Biochem. Prep.* **12**, 94. **3**. Gibson & Jackson (1963) *Nature* **198**, 388. **4**. Gibson (1999) *Trends Biochem. Sci.* **24**, 36. **5**. Fischer & Jensen (1987) *Meth. Enzymol.* **142**, 503. **6**. Rothman & Cabib (1967) *Biochemistry* **6**, 2107. **7**. Gosset, Bonner & Jensen (2001) *J. Bacteriol.* **183**, 4061. **8**. Whitaker, Fiske & Jensen (1982) *J. Biol. Chem.* **257**, 12789. **9**. Wu, Sheflyan & Woodard (2005) *Biochem. J.* **390**, 583.

Chromium Adenosine 5'-Triphosphate Complex

Λ-Isomer (*or* Lambda-Isomer) of Cr(III) Complex

Δ-Isomer (*or* Delta-Isomer) of Cr(III) Complex

These oxygen exchange-inert, tervalent metal ion-nucleotide complexes often inhibit many ATP-dependent enzymes by mimicking the geometries of one-to-one complexes of ATP with Mg^{2+}, Ca^{2+}, and Mn^{2+} complexes (1,2). Because favorable solution conditions can be chosen to prevent the release and rebinding of the tervalent metal ion from ATP, each of the two bidentate isomers can exist in pseudo-axial and pseudo-equatorial forms. Determination of which of these stable metal ion-ATP complexes is a dead-end competitive inhibitor relative to the phosphoryl-donor substrate $MgATP^{2-}$, $CaATP^{2-}$, or $MnATP^{2-}$ allows the experimenter to infer the likely geometry of the enzyme-bound $MgATP^{2-}$, $CaATP^{2-}$, or $MnATP^{2-}$ (1). Because Cr(III) is paramagnetic, distances can be measured by measuring the effects of Cr(III) on the NMR signals of nearby atoms when the Cr(III)-nucleotide complex binds to the surface of a macromolecule (2). (**See also** *Exchange-Inert Complexes; specific chromium nucleotide*) **Target(s):** pyruvate carboxylase (3); hexokinase (4,8,10,11,13); aminoacyl-tRNA synthetases (5); pyruvate, orthophosphate dikinase (5); F₁ ATP synthase5; creatine kinase7,9; acetate kinase (11); glycerol kinase (11,22); pyruvate kinase (11); 3-phosphoglycerate kinase (11, 14); Na⁺/K⁺-exchanging ATPase (12); glucokinase (13); formyltetrahydrofolate synthetase (15); ketohexokinase, *or* hepatic fructokinase (16); Ca^{2+}-ATPase (17); malyl-CoA synthetase, *or* malate:CoA ligase (18); apyrase (19); glucose-1-phosphate adenylyltransferase (20); sulfate adenylyltransferase (21). **1**. Cleland (1982) *Meth. Enzymol.* **87**, 159. **2**. Villafranca (1982) *Meth. Enzymol.* **87**, 180. **3**. Armbruster & F. B. Rudolph (1976) *J. Biol. Chem.* **251**, 320. **4**. Danenberg & Cleland (1975) *Biochemistry* **14**, 28. **5**. Santi, Webster & Cleland (1974) *Meth. Enzymol.* **29**, 620. **6**. Ciskanik & Dunaway-Mariano (1986) *J. Enzyme Inhib.* **1**, 113. **7**. Steinke, Bacon & Schuster (1987) *Arch. Biochem. Biophys.* **258**, 482. **8**. Purich, Fromm & Rudolph (1973) *Adv. Enzymol.* **39**, 249. **9**. Schimerlik & W. W. Cleland (1973) *J. Biol. Chem.* **248**, 8418. **10**. DePamphilis & Cleland (1973) *Biochemistry* **12**, 3714. **11**. Janson & Cleland (1974) *J. Biol. Chem.* **249**, 2572. **12**. Pauls, Bredenbrocker & Schoner (1980) *Eur. J. Biochem.* **109**, 523. **13**. Monasterio & Cárdenas (2003) *Biochem. J.* **371**, 29. **14**. Serpersu, Summitt & Gregory (1992) *J. Inorg. Biochem.* **48**, 203. **15**. Mejillano, Wendland, Everett, Rabinowitz & Himes (1986) *Biochemistry* **25**, 1067. **16**. Raushel & Cleland (1977) *Biochemistry* **16**, 2169. **17**. Moreira, Rios & Barrabin (2005) *Biochim. Biophys. Acta* **1708**, 411. **18**. Hersh (1974) *J. Biol. Chem.* **249**, 6264. **19**. Kettlun, Uribe, Calvo, *et al.* (1982) *Phytochemistry* **21**, 551. **20**. Preiss (1978) *Adv. Enzymol. Relat. Areas Mol. Biol.* **46**, 317. **21**. Farley, Cryns, Yang & Segel (1976) *J. Biol. Chem.* **251**, 4389. **22**. Janson & Cleland (1974) *J. Biol. Chem.* **249**, 2562.

Chromium Guanosine 5'-Triphosphate

Λ-Isomer (*or* Lambda-Isomer) of Cr(III) Complex

Δ-Isomer (*or* Delta-Isomer) of Cr(III) Complex

These oxygen exchange-inert, tervalent metal ion-nucleotide complexes inhibit many GTP-dependent enzymes by mimicking the geometries of one-to-one complexes of GTP with Mg^{2+}, Ca^{2+}, and Mn^{2+} complexes (1,2). Because favorable solution conditions can be chosen to prevent the release and rebinding of the tervalent metal ion from GTP, each of the two bidentate isomers can exist in pseudo-axial and pseudo-equatorial forms. Determination of which of these stable metal ion-GTP complexes is a dead-end competitive inhibitor relative to the phosphoryl-donor substrate $MgGTP^{2-}$, $CaGTP^{2-}$, or $MnGTP^{2-}$ allows the experimenter to infer the likely geometry of the enzyme-bound $MgGTP^{2-}$, $CaGTP^{2-}$, or $MnGTP^{2-}$ (1). Because Cr(III) is paramagnetic, distances can be measured by measuring the effects of Cr(III) on the NMR signals of nearby atoms when the Cr(III)-nucleotide complex binds to the surface of a macromolecule (2). (**See** *Exchange-Inert Complexes; specific chromium nucleotide*) **Target(s):** acetate kinase (1,2); creatine kinase (2); glycerokinase (2); pyruvate kinase (2); phosphoglycerate kinase (2); hexokinase (2); phosphoenolpyruvate carboxykinase (GTP) (3,4) **1**. Todhunter, Reichel & Purich (1976) *Arch. Biochem. Biophys.* **174**, 120. **2**. Janson & Cleland (1974) *J. Biol. Chem.* **249**, 2572. **3**. Kramer & Nowak (1988) *J. Inorg. Biochem.* **32**, 135. **4**. Hlavaty & Nowak (2000) *Biochemistry* **39**, 1373.

Chromium(IV) Toxicity

Chromium is a trace metal of clinical importance in humans and has been widely used (as the radionuclide ⁵¹Cr) to quantify blood cell pool sizes, intravascular kinetics, and red cell lifespan and turnover. Hexavalent chromium is also a widely occurring industrial agent and pollutant, occurs in the form of chromate ion. CrO_4^{2-} resembles sulfate (SO_4^{2-}), both in its chemical structure and electronic charge, and chromate is taken up by human cells through the action of sulfate transporters. In normal human leukocytes, $^{51}CrO_4^{2-}$ uptake is unidirectional over a 1 hr incubation with extracellular chromate concentrations up to 200 mumoles/liter (1). Chromate and phosphate also inhibit each other's uptake and translocation in arsenic hyperaccumulator *Pteris vittata* (2) In humans, Cr(III) is not transported into cells, but is generated from Cr(VI), with the latter first reduced Cr(V), itself a precursor of Cr(III). Incubation of Cr(VI) with ascorbate generates Cr(V), Cr(IV) and ascorbate-derived carbon-centered alkyl radicals, as well as formyl radicals (3). Hydrogen peroxide causes the generation of hydroxyl radicals (OH) and much higher levels of Cr(V), showing that OH˙ can be generated via a Cr(IV)-mediated Fenton-like reaction (*i.e.*, Cr(IV) + H_2O_2 → Cr(V) + OH˙ + OH⁻). Moreover, the presence of 1,10-phenanthroline and deferoxamine inhibit the formation of both OH˙ and Cr(V) from the reaction of Cr(VI) with ascorbate in the presence of H_2O_2. Electrophoresis also demonstrates that Cr(IV)/ascorbate-derived free radicals causes DNA double-strand breaks (3). Hydroxyl radicals generated by Cr(V)⁻ and Cr(IV)-mediated Fenton-like reactions likewise cause DNA double-strand breaks. Hydroxyl radicals generated by Cr(IV) and Cr(V) from H_2O_2 cause hydroxylation of 2'-deoxyguanine to form 8-hydroxy-2'-deoxyguanine (3). **1**. Lilien, Spivak & Goldman (1970) *J. Clin. Invest.* **49**, 1551. **2**. de Oliveira, Lessl, Gress, *et al.* (2015) *Environ. Pollut.* **197**, 240. **3**. Shi, Mao, Knapton, *et al.* (1994) *Carcinogenesis* **15**, 2475.

Chromomycin A₃

This antibiotic (FW = 1183.3 g/mol), also called toyomycin, from *Streptomyces griseus* and inhibits RNA biosynthesis in bacterial, fungi, and animal cells. It binds to the DNA minor groove (only in the presence of divalent cations like Mg^{2+}) and can serve as a fluorescent DNA stain. The ultraviolet/visible spectrum has λ_{max} values (in ethanol) of 230, 281, 304, 318, 330, and 412 nm (with corresponding molar extinction coefficients of 24500, 55500, 7080, 8320, 6920, and 11700 $M^{-1}cm^{-1}$, respectively). **Target(s):** DNA-directed RNA polymerase (1,5,8); DNA-directed DNA polymerase (3,4,6); DNA topoisomerase II, *or* DNA gyrase (2,7). **1**. Kajiro & Kamiyama (1967) *J. Biochem.* **62**, 424. **2**. Simon, Wittig & Zimmer (1994) *FEBS Lett.* **353**, 79. **3**. Honikel & Santo (1972) *Biochim. Biophys. Acta* **269**, 354. **4**. Honikel, Sippel & Hartmann (1968) *Z. physiol. Chem.* **349**, 957. **5**. Kersten, Kersten, Szybalski & Fiandt (1966) *Biochemistry* **5**, 236. **6**. Müller, Yamazaki & Zahn (1972) *Enzymologia* **43**, 1. **7**. Christmann-Franck, Bertrand, Goupil-Lamy, *et al.* (2004) *J. Med. Chem.* **47**, 6840. **8**. Sethi (1971) *Prog. Biophys. Mol. Biol.* **23**, 67.

Chromostatin-20

This eicosapeptide (*Sequence* SDEDSDGDRPQASPGLG-PGP; MW = 1953.95 g; Isoelectric Point = 3.71) is released via protease activity on chromogranin A, a ubiquitous 48-kDa secretory protein present in adrenal medulla, anterior pituitary, central and peripheral nervous system, endocrine gut, thyroid, parathyroid, and endocrine pancreas. Chromostatin-20 inhibits the release of catecholamines from chromaffin cells (by activating a protein phosphatase in these cells) and inhibits secretagogue-induced Ca^{2+} influx. **1**. Galindo, Rill, Bader & Aunis (1991) *Proc. Natl. Acad. Sci. U.S.A.* **88**, 1426.

Chymostatin A

This is a high-affinity protease inhibitor (FW = 607.71 g/mol; CAS 9076-44-2), also named *N*-(-*N*ᵅ-carbonyl-Cpd-X-phenylalaninal)-L-phenylalanine, where Cpd is capreomycidine, *i.e.*, [*S*,*S*]-α-(2-iminohexahydro-4 pyrimidyl)glycine, and X refers to L-leucyl (Chymostatin A), L-valyl (Chymostatin B), or L-isoleucyl (Chymostatin C) residues (1). Commmercial chymostatin is typically a mixture of all three forms, with Chymostatin A the principal component. The chymostatins are potent inhibitors of chymotrypsin-like proteinases: *e.g.*, chymases ($K_i \approx 20$–30 nM) and mast cell proteinase II ($K_i \approx 0.6$ µM). Chymostatin also

inhibits cysteine proteinases including members of the cathepsin family. One target, chymase, is a chymotrypsin-like protease that is abundant in secretory granules from mast cells. Chymase is a key enzyme in the local renin-angiotensin system (RAS) that generates angiotensin II (Ang II) in a manner that operates independently of angiotensin converting enzyme (ACE). **Properties:** Chymostatin is typically prepared as a 10 mM stock solution (in dimethylsulfoxide or glacial acetic acid), stored at –20° C (stable for months under these conditions), and used at 20-200 µM (final concentration). **Target(s):** chymotrypsin C, *or* caldecrin (1); chymase (2,20,95,96,98-101); stratum corneum chymotryptic enzyme (3); Rt41A protease, *Thermus* strain (4); bacillopeptidase F (5); cerevisin, *or* yeast proteinase B (6,35,37,97); archaelysin (7); proteinase I of *Sulfolobus solfataricus* (7); carboxypeptidase D (8); proteasome endopeptidase complex (9,23,31,40-46); cell-associated endopeptidase of Trichophyton (10); falcipain (11); *Fasciola* cysteine endopeptidases (12); cathepsin L (13,20,69,70); picornain 3C (14,26-28,66); picornain 2A (15,26,64,65); peptidyl-dipeptidase Dcp (16,112); protease IV (17,85); chymotrypsin (18,20,29,33,34,87,96,103,106, 107); carboxypeptidase C, *or* carboxypeptidase Y (18,109-111); cathepsin B (18,20,72-80); cathepsin D (18); brachyuran (19,38); cathepsin G (20,96,103); seminin (20); cathepsin H (20,68); signal peptidase II (21); *Coccidioides* endopeptidase (22); endopeptidase La (24); glycyl endopeptidase (25,67); elastase (29); leukocyte elastase, weakly inhibited (29); cysteine proteinase from *Trypanosoma cruzi* (30); sheddase (32); cathepsin A, weakly inhibited (36); pseudomonalisin, *or* pseudomonopepsin, $K_i = 45$ nM (39); pitrilysin, *or* protease P_i (47); serralysin, weakly inhibited (48); coccolysin (49); procollagen C-endopeptidase (50); meprin A (51,52); thimet oligopeptidase (53); envelysin (54-57); neprilysin, weakly inhibited (58); polyporopepsin (59); candidapepsin (60); cruzipain (61); cathepsin V (62); cathepsin K (63); clostripain (71); kallikrein-8, *or* neuropsin (81); stratum corneum chymotryptic enzyme, kallikrein-7 (82); xanthomonolisin, *or* xanthomonopepsin, weakly inhibited (83); hepacivirin, *or* hepatitis C virus NS3 serine proteinase (84); signal peptidase I, *or* *Escherichia coli* (85); oligopeptidase B (86); streptogrisin A (87-89); granzyme B (90); furin (91); pancreatic elastase II (92); thermitase (93); subtilisin Sendai (94); tryptase (96); chymotrypsin (98-100,103); tissue kallikrein, dog (102); trypsin, gastric porcine (104); trypsin, gypsy moth (105); peptidyl-glycinamidase, *or* carboxamidopeptidase (108); cytosol alanyl aminopeptidase (113); tripeptide aminopeptidase (114); membrane alanyl aminopeptidase, *or* aminopeptidase N (115); protein-glutamate methylesterase (116). **1**. Tomomura (1998) in *Handb. Proteolytic Enzymes*, p. 40, Academic Press, San Diego. **2**. Caughey (1998) in *Handb. Proteolytic Enzymes*, p. 66, Academic Press, San Diego. **3**. Egelrud (1998) in *Handb. Proteolytic Enzymes*, p. 87, Academic Press, San Diego. **4**. Toogood & Daniel (1998) in *Handb. Proteolytic Enzymes*, p. 297, Academic Press, San Diego. **5**. Hageman (1998) in *Handb. Proteolytic Enzymes*, p. 301, Academic Press, San Diego. **6**. Naik & Jones (1998) in *Handb. Proteolytic Enzymes*, p. 317, Academic Press, San Diego. **7**. Vanoni & Tortora (1998) in *Handb. Proteolytic Enzymes*, p. 334, Academic Press, San Diego. **8**. Remington (1998) in *Handb. Proteolytic Enzymes*, p. 398, Academic Press, San Diego. **9**. Seemüller, Dolenc & Lupas (1998) in *Handb. Proteolytic Enzymes*, p. 495, Academic Press, San Diego. **10**. Lambkin, Hamilton & Hay (1998) in *Handb. Proteolytic Enzymes*, p. 531, Academic Press, San Diego. **11**. Rosenthal (1998) in *Handb. Proteolytic Enzymes*, p. 585, Academic Press, San Diego. **12**. Wijffels (1998) in *Handb. Proteolytic Enzymes*, p. 606, Academic Press, San Diego. **13**. Kirschke (1998) in *Handb. Proteolytic Enzymes*, p. 617, Academic Press, San Diego. **14**. Skern (1998) in *Handb. Proteolytic Enzymes* , p. 705, Academic Press, San Diego. **15**. Skern (1998) in *Handb. Proteolytic Enzymes*, p. 713, Academic Press, San Diego. **16**. Henrich & Klein (1998) in *Handb. Proteolytic Enzymes*, p. 1121, Academic Press, San Diego. **17**. Miller (1998) in *Handb. Proteolytic Enzymes*, p. 1589, Academic Press, San Diego. **18**. Umezawa (1976) *Meth. Enzymol.* **45**, 678. **19**. Grant, Eisen & Bradshaw (1981) *Meth. Enzymol.* **80**, 722. **20**. Seglen (1983) *Meth. Enzymol.* **96**, 737. **21**. Sankaran & Wu (1995) *Meth. Enzymol.* **248**, 169. **22**. Rawlings & Barrett (1994) *Meth. Enzymol.* **244**, 19. **23**. Rivett, Savory & Djaballah (1994) *Meth. Enzymol.* **244**, 331. **24**. Goldberg, Moerschell, Chung & Maurizi (1994) *Meth. Enzymol.* 244, 350. **25**. Buttle (1994) *Meth. Enzymol.* **244**, 539. **26**. Skern & Liebig (1994) *Meth. Enzymol.* **244**, 583. **27**. Orr, Long, Kay, Dunn & Cameron (1989) *J. Gen. Virol.* **70**, 2931. **28**. Kay & Dunn (1990) *Biochim. Biophys. Acta* **1048**, 1. **29**. Feinstein, Malemud & Janoff (1976) *Biochim. Biophys. Acta* **429**, 925. **30**. Cazzulo, Couso, Raimondi, Wernstedt & Hellman (1989) *Mol. Biochem. Parasitol.* **33**, 33. **31**. Mason (1990) *Biochem. J.* **265**, 479. **32**. Lee, Alpaugh, Nguyen, *et al.* (2000) *Biochem. Biophys. Res. Commun.*

279, 116. **33**. Tatsuta, Mikami, Fujimoto, Umezawa & Umezawa (1973) *J. Antibiot.* **26**, 625. **34**. Umezawa, Aoyagi, Morishima, Kunimoto & Matsuzaki (1970) *J. Antibiot.* **23**, 425. **35**. Lenney (1975) *J. Bacteriol.* **122**, 1265. **36**. Matsuda & Misaka (1975) *J. Biochem.* **78**, 31. **37**. Fujishiro, Sanada, Tanaka & Katunuma (1980) *J. Biochem.* **87**, 1321. **38**. Grant & Eisen (1980) *Biochemistry* **19**, 6089. **39**. Wlodawer, Li, Gustchina, *et al.* (2001) *Biochemistry* **40**, 15602. **40**. Akaishi, Sawada & Yokosawa (1996) *Biochem. Mol. Biol. Int.* **39**, 1017. **41**. Tanaka, Sawada & Sawada (2000) *Comp. Biochem. Physiol. C* **125C**, 215. **42**. Dick, Moomaw, DeMartino & Slaughter (1991) *Biochemistry* **30**, 2725. **43**. Fernández Murray, Biscoglio & Passeron (2000) *Arch. Biochem. Biophys.* **375**, 211. **44**. Ishiura, Tsukahara & Sugita (1990) *Int. J. Biochem.* **22**, 1195. **45**. Ozaki, Fujinami, Tanaka, *et al.* (1992) *J. Biol. Chem.* **267**, 21678. **46**. Lomo, Coetzer & Lonsdale-Eccles (1997) *Immunopharmacology* **36**, 285. **47**. Fricke, Betz & Friebe (1995) *J. Basic Microbiol.* **35**, 21. **48**. Kim, Tamanoue, Jeohn, *et al.* (1997) *J. Biochem.* **121**, 82. **49**. Fernandez-Espla, Garault, Monnet & Rul (2000) *Appl. Environ. Microbiol.* **66**, 4772. **50**. Hojima, van der Rest & Prockop (1985) *J. Biol. Chem.* **260**, 15996. **51**. Yamaguchi, Fukase, Sugimoto, Kido & Chihara (1994) *Biol. Chem. Hoppe-Seyler* **375**, 821. **52**. Beynon, Shannon & Bond (1981) *Biochem. J.* **199**, 591. **53**. Akopyan, Couedel, Orlowski, Fournie-Zaluski & B. P. Roques (1994) *Biochem. Biophys. Res. Commun.* **198**, 787. **54**. Fan & Katagiri (2001) *Eur. J. Biochem.* **268**, 4892. **55**. Nomura, Tanaka, Kikkawa, Yamaguchi & Suzuki (1991) *Biochemistry* **30**, 6115. **56**. Post, Schuel & Schuel (1988) *Biochem. Cell. Biol.* **66**, 1200. **57**. Nakatsuka (1985) *Dev. Growth Differ.* **27**, 653. **58**. Beynon, Shannon & Bond (1981) *Biochem. J.* **199**, 591. **59**. Kobayashi, Kusakabe & Murakami (1985) *Agric. Biol. Chem.* **49**, 2393. **60**. Negi, Tsuboi, Matsui & Ogawa (1984) *J. Invest. Dermatol.* **83**, 32. **61**. Cazzulo, Stoka & Turk (1997) *Biol. Chem.* **378**, 1. **62**. Brömme, Li, Barnes & Mehler (1999) *Biochemistry* **38**, 2377. **63**. Brömme, Okamoto, Wang & Biroc (1996) *J. Biol. Chem.* **271**, 2126. **64**. Sommergruber, Ahorn, Zöphel, *et al.* (1992) *J. Biol. Chem.* **267**, 22639. **65**. Liebig, EZiegler, Yan, *et al.* (1993) *Biochemistry* **32**, 7581. **66**. Hata, Sato, Sorimachi, Ishiura & Suzuki (2000) *J. Virol. Methods* **84**, 117. **67**. Buttle, Ritonja, Dando, *et al.* (1990) *FEBS Lett.* **262**, 58. **68**. Schwartz & Barrett (1980) *Biochem. J.* **191**, 487. **67**. Aranishi, Ogata, Hara, Osatomi & Ishihara (1997) *Comp. Biochem. Physiol. B* **118B**, 531. **70**. Yamashita & Konagaya (1990) *Comp. Biochem. Physiol. B* **96B**, 247. **71**. Kembhavi, Buttle, Rauber & Barrett (1991) *FEBS Lett.* **283**, 277. **72**. Towatari, Kawabata & Katunuma (1979) *Eur. J. Biochem.* **102**, 279. **73**. Banno, Yano & Nozawa (1983) *Eur. J. Biochem.* **132**, 563. **74**. Etherington (1976) *Biochem. J.* **153**, 199. **75**. Aranishi, Hara, Osatomi & Ishihara (1997) *Comp. Biochem. Physiol. B* **117B**, 579. **76**. Suhar & Marks (1979) *Eur. J. Biochem.* **101**, 23. **77**. Wada & Tanabe (1988) *J. Biochem.* **104**, 472. **78**. Hirao, Hara & Takahashi (1984) *J. Biochem.* **95**, 871. **79**. Jiang, Lee & Chen (1999) *J. Agric. Food Chem.* **42**, 1073. **80**. Kawada, Hara, Morimoto, Hiruma & Ishibashi (1995) *Int. J. Biochem. Cell Biol.* **27**, 175. **81**. Shimizu, Yoshida, Shibata, *et al.* (1998) *J. Biol. Chem.* **273**, 11189. **82**. Lundstrom & Egelrud (1991) *Acta Derm. Venereol.* **71**, 471. **83**. Oda, Sugitani, Fukuhara & Murao (1987) *Biochim. Biophys. Acta* **923**, 463. **84**. Mori, Yamada, Kimura, *et al.* (1996) *FEBS Lett.* **378**, 37. **85**. Ichihara, Beppu & Mizushima (1984) *J. Biol. Chem.* **259**, 9853. **86**. Caldas, Cherqui, Pereira & Simoes (2002) *Appl. Environ. Microbiol.* **68**, 1297. **87**. Tomkinson, Galpin & Beynon (1992) *Biochem. J.* **286**, 475. **88**. Delbaere & Brayer (1985) *J. Mol. Biol.* **183**, 89. **89**. Delbaere & Brayer (1980) *J. Mol. Biol.* **139**, 45. **90**. Poe, Blake, Boulton, *et al.* (1991) *J. Biol. Chem.* **266**, 98. **91**. Cieplik, Klenk & Garten (1998) *Biol. Chem.* **379**, 1433. **92**. Azuma, Banshou & Suzuki (2001) *J. Protein Chem.* **20**, 577. **93**. Brömme & Kleine (1984) *Curr. Microbiol.* **11**, 317. **94**. Yamagata, Isshiki & Ichishima (1995) *Enzyme Microb. Technol.* **17**, 653. **95**. Kido, Fukusen & Katunuma (1985) *Arch. Biochem. Biophys.* **239**, 436. **96**. Powers, Tanaka, Harper, *et al.* (1985) *Biochemistry* **24**, 2048. **97**. Kominami, Hoffschulte & Holzer (1981) *Biochim. Biophys. Acta* **661**, 124. **98**. Akahoshi, Ashimori, Sakashita, *et al.* (2001) *J. Med. Chem.* **44**, 1286. **99**. Akahoshi, Ashimori, Sakashita, *et al.* (2001) *J. Med. Chem.* **44**, 1297. **100**. Akahoshi, Ashimori, Yoshimura, *et al.* (2001) *Bioorg. Med. Chem.* **9**, 301. **101**. Ferry, Gillet, Bruneau, *et al.* (2001) *Eur. J. Biochem.* **268**, 5885. **102**. Ohnishi, Ikekita, Atomi, *et al.* (1992) *Protein Seq. Data Anal.* **5**, 1. **103**. Stein & Strimpler (1987) *Biochemistry* **26**, 2611. **104**. Jeohn, Serizawa, Iwamatsu & Takahashi (1995) *J. Biol. Chem.* **270**, 14748. **105**. Valaitis (1995) *Insect Biochem. Mol. Biol.* **25**, 139. **106**. Asgeirsson & Bjarnason (1991) *Comp. Biochem. Physiol. B* **99B**, 327. **107**. Peterson, Fernando & Wells (1995) *Insect Biochem. Mol. Biol.* **25**, 765. **108**. Simmons & Walter (1980) *Biochemistry* **19**, 39. **109**. Liu, Tachibana, Taira, *et al.* (2004) *J. Ind. Microbiol. Biotechnol.* **31**, 572. **110**. Liu, Tachibana, Taira, Ishihara & Yasuda (2004) *J. Ind. Microbiol. Biotechnol.*

31, 23. **111**. Satoh, Kadota, Oheda, *et al.* (2004) *J. Antibiot.* **57**, 316. **112**. Henrich, Becker, Schroeder & Plapp (1993) *J. Bacteriol.* **175**, 7290. **113**. Yamamoto, Li, Huang, Ohkubo & Nishi (1998) *Biol. Chem.* **379**, 711. **114**. Sachs & Marks (1982) *Biochim. Biophys. Acta* **706**, 229. **115**. Tokioka-Terao, Hiwada & Kokubu (1984) *Enzyme* **32**, 65. **116**. Veeraragavan & Gagnon (1987) *Biochem. Biophys. Res. Commun.* **142**, 603.

CI-1033

CI-1033

Cysteinyl Covalent Adduct

This Michael adduct-forming PKI (FW = 485.94; CAS 267243-28-7; Solubility: 20 mM in H_2O; 100 mM in DMSO), also known as *N*-[4-[(3-chloro-4-fluorophenyl)amino]-7-[3-(4-morpholinyl)propoxy]-6-quinazol-inyl]-2-propenamide, PD 183805, and Canertinib (1), and its congeners target epidermal growth factor receptor (EGFR) kinase irreversibly (k_{inact}/K_i = 10^5-10^7 $M^{-1}s^{-1}$) with a low specific reactivity ($k_{inact} \leq 2.1 \times 10^{-3}$ s^{-1}) (2). The latter is compensated for by the inhibitors' high binding affinities ($K_i <$ 1 nM). (*See* Fig. 2 and Table 1 of Reference-2). Quinazoline-based covalent drugs (*e.g.*, dacomitinib, afatinib, and CI-1033) are potent inhibitors of wild-type EGFR autophosphorylation in A549 tumor cells (IC_{50} =2–12 nM). The reversible binding affinity of EGFR covalent inhibitors correlates highly with antitumor cell potency. Because these inhibitors target a cysteine residue, the effects of its oxidation on enzyme catalysis and inhibitor pharmacology must be considered. CI-1033 is also classified as a pan-ErbB inhibitor for EGFR and ErbB2 with IC_{50} values of 1.5 nM and 9 nM, respectively, and showing little or no inhibitory action on PDGFR, FGFR, InsR, PKC, or CDK1/2/4. Canertinib was the first irreversible protein kinase inhibitor to enter clinical trial. **1**. Smaill, Rewcastle, Loo, *et al.* (2000) *J. Med. Chem.* **43**, 1380. **2**. Schwartz, Kuzmic, Solowiej, *et al.* (2013) *Proc. Natl. Acad. Sci. U.S.A.* **111**, 173.

CI-1040, *See* PD 184352

Cialis, *See* Tadalafil

CID29950007

This first-in-class G-protein signal-transduction inhibitor and cytostatic agent (FW = 485.38 g/mol), also known as 4-(5-(4-methoxyphenyl)-3-phenyl-4,5-dihydro-1H-pyrazol-1-yl)benzene-sulfonam, antagonizes GTP binding to Cdc42 by binding at a location other than its active site and without affecting other GTPases, including the same-family small GTPases, Rac and Rho. In binding curves fitted to a hyperbolic one-site binding equation, the observed change in maximal fluorescence, B_{max}, indicated a noncompetitive, or allosteric, binding mode, wherein CID2950007 binds at a site that is topologically distinct from the GTP site. Moreover, when 3T3 fibroblasts were incubated with CID2950007 (10 μM) for 1 h prior to bradykinin treatment, both the number and average length of cdc42-induced filopodia decreased significantly. (*Note*: Because CID2950007 must be dissolved in DMSO and subsequently diluted into aqueous solution, the inhibitory effect of the compound is demonstrated by its ability to counteract the stimulatory effects of both DMSO and the bradykinin cue.) CID2950007 also inhibited actin-mediated cell migration in a dose-dependent manner. At 3 μM, CID2950007 reduced cell migration to less than 50% of the untreated control. CID2950007 likewise inhibits Cdc42-regulated and actin-mediated internalization of Sin Nombre virus. **1**. Hong, Kenney, Phillips, *et al.* (2013) *J. Biol. Chem.* **288**, 8531.

Cidofovir

Cidofovir

Brincidofovir

This CMV DNA polymerase inhibitor and intravenous antiviral (FW = 265.16 g/mol; CAS 113852-37-2; Symbol: CDV), also named Vistide® and ({[(*S*)-1-(4-amino-2-oxo-1,2-dihydropyrimidin-1-yl)-3-hydroxypropan-2-yl]oxy}methyl)phosphonic acid, is primarily used as an treatment for cytomegalovirus (CMV) retinitis, an AIDS-associated infection (1,2). The minimum concentration required to reduce CMV plaque formation by 50% is 0.07 μg/mL (2). CDV diphosphate, the putative antiviral metabolite of CDV, is a competitive inhibitor relative to dCTP and is an alternate substrate for human CMV DNA polymerase (3). The slower rate of incorporation of CDVpp is mainly due to the higher K_m value of CDVpp toward the enzyme-primer-template complexes (3). The pro-drug Brincidofovir (FW = 561.70 g/mol; CAS 444805-28-1; IUPAC: [(2*S*)-1-(4-amino-2-oxopyrimidin-1-yl)-3-hydroxypropan-2-yl]oxymethyl-(3-hexa decoxypropoxy)phosphinate) is active against adenoviruses, BK virus, herpes simplex viruses, and Ebola virus. Beyond its well-recognized antiviral activity, cidofovir exerts anticancer actions, both *in vitro* and with animal models, including T cell lymphoma (4). With human breast, colon, liver, lung, prostate, and thyroid carcinomas, CDV reduced expression of procaspase-3 and increased expression of PARP p85 fragment, indicating apoptosis can be inducted in a virus-independent manner (5). It also increases the pro-apoptotic proteins p53, cytochrome c and caspase-3 as well as decreases the survival protein Bcl-x. Evidence of nuclear DNA fragmentation clearly indicated that CDV promotes apoptotic cell death. **Key Pharmacokinetic Parameters:** *See* Appendix II in Goodman & Gilman's *THE PHARMACOLOGICAL BASIS OF THERAPEUTICS*, 12th Edition (Brunton, Chabner & Knollmann, eds.) McGraw-Hill Medical, New York (2011). **1**. Bronson, Ghazzouli, Hitchcock, Webb & Martin (1989) *J. Med. Chem.* **32**,

1457. **2**. Snoeck, Sakuma, De Clercq, Rosenberg & Holy (1988) *Antimicrob. Agents Chemother.* **32**, 1839. **3**. Xiong, Smith, Kim, Huang & Chen (1996) *Biochem. Pharmacol.* **51**, 1563. **4**. Chan, Tse, Kwong (2017) *Onco Targets Ther.* **10**, 347. **5**. Catalani, Palma, Battistelli, *et al.* (2017) *Toxicol. In Vitro* **41**,49.

Ciguatoxins

Ciguatoxin 1

These structurally comples marine toxins (Ciguatoxin 1: FW = 1112.22 g/mol; CAS 11050-21-8), while found in a variety of fish and is the toxin involved in ciguatera fish poisoning, originate from the dinoflagellate *Gambierdiscus toxicus*. They participate in sodium channel activation, thereby stimulating the release of neurotransmitters (1,2). The ciguatoxins bind strongly to site-5 of the voltage-sensitive sodium channel. Purified ciguatoxin (EC_{50} = 0.62 ng/mL) stimulates neurotransmitter release, lowering the net accumulation of γ-aminobutyric acid and dopamine by brain synaptosomes (3). This effect is completely inhibited by tetrodotoxin ($K_{0.5}$ = 4 nM). Electrophysiological studies on neuroblastoma cells indicate that ciguatoxin induces a membrane depolarization which is prevented by tetrodotoxin and which is due to an action that increases Na^+ permeability (3). **See also** *Maitotoxin, Scaritoxin; Palytoxin*. **Target(s):** K^+ channels (2). **1**. Bidard, Vijverberg, Frelin, *et al.* (1984) *J. Biol. Chem.* **259**, 8353. **2**. Birinyi-Strachan, Gunning, Lewis & Nicholson (2005) *Toxicol. Appl. Pharmacol.* **204**, 175. **3**. Bidard, Vijverberg, Frelin, *et al.* (1984) *J. Biol. Chem.* **259**, 8353.

CK0238273, *See Ispinesib*

Cilazapril & Cilazaprilat

cilazapril (pro-drug)   **cilazaprilat (active drug)**

This antihypertensive pro-drug ($FW_{pro-drug}$ = 417.31 g/mol; CAS 88768-40-5), also known by its tradenames Inhibace®, Vascace® and Dynorm®, and by its systematic name (4*S*,7*S*)-7-[[(2*S*)-1-ethoxy-1-oxo-4-phenylbutan-2-yl]amino]-6-oxo-1,2,3,4,7,8,9,10-octahydropyridazino[1,2-*a*]diazepine-4-carboxylate, is metabolically de-esterified to cilazaprilat, which targets Angiotensin-Converting Enzyme-1, *or* ACE1 ($IC_{50} \approx$ 2 nM)) and is used to treat hypertension and congestive heart failure. Its tolerability profile is similar to other frequently prescribed antihypertensive drugs, such as sustained-release propranolol, enalapril, captopril, atenolol and hydrochlorothiazide. In addition to its blood pressure-lowering abilities, cilazapril can moderate the proliferative response seen in vessels after vascular injury caused by techniques such as ballooning. **1**. Szucs (1991) *Drugs* **41**(Suppl 1), 18.

Cilengitide

This first-in-class, RGD-containing cyclic peptide (FW = 588.46 g/mol; CAS 188968-51-6; *Sequence*: cyclo[-RGDfV-]), also known as EMD 121974, is a selective α_v-integrin inhibitor and angiogenesis disruptor (1). **Primary Mode of Inhibitory Action:** During angiogenesis, α_v-integrin are overexpressed and in abundance on the endothelial cell surface, where they facilitate growth and survival of newly forming vessels. Accordingly, blocking α_v-integrin function by disrupting ligand binding should produce an anti-angiogenic effect. α_v-Integrins (*i.e.*, $\alpha_v\beta_3$, $\alpha_v\beta_5$, and $\alpha_5\beta_1$) bind with high affinity to argininyl-glycyl-aspartyl- (RGD-) registers present in the matrix proteins fibronectin, vitronectin, osteopontin, collagens, thrombospondin, fibrinogen, and von Willebrand factor. Cilengitide competitively inhibits (IC$_{50}$ = 0.3-5 nM, or 100-1000x more tightly than linear RGD-containing peptides) the binding of these proteins to $\alpha_v\beta_3$, $\alpha_v\beta_5$, and $\alpha_5\beta_1$ integrins, thereby acting as a vasculature-disrupting agent (2). Integrin binding to ligands in the ECM induces conformational changes in the integrin's structure and contributes to clustering of heterodimers into oligomers. These changes initiate intracellular signals through multiple activation of signaling proteins, a process known as "outside-in signaling" that controls cell polarity, cytoskeletal structure, gene expression and cell survival (2). Cilengitide treatment of Malignant Pleural Mesothelioma (MPM) cells results in detachment and subsequent death of anoikis-sensitive cells due to antagonism of $\alpha_v\beta_3$ and $\alpha_v\beta_5$. **1.** Burke, DeNardo, Miers, *et al.* (2002) *Cancer Res.* **62**, 4263. **2.** Mas-Moruno, Rechenmacher & Kessler (2010) *Anticancer Agents Med. Chem.* **10**, 753. **3.** Cheng, van Zandwijk & Reid (2014) *PLoS One* **9**, e90374.

Cilnidipine

This dihydropyridine-class, N-type calcium channel blocker (FW = 492.52 g/mol; CAS 132203-70-4), also known by the code name FRC-8653 and systematically as *O*3-(2-methoxyethyl)-*O*^{5}-[(*E*)-3-phenylprop-2-enyl]-2,6-dimethyl-4-(3-nitrophenyl)-1,4-dihydro-pyridine-3,5-dicarboxylate, is an antihypertensive agent with combined action as an angiotensin converting enzyme (ACE) inhibitor, a diuretic, and a β-adrenergic blocking agent (1). Cilnidipine is rapidly metabolized in human liver microsomes. Dehydrogenation of dihydropyridine ring by CYP3A (*or cytochrome P-4503A*) is crucial for cilnidipine elimination (2). **1.** Yoshimoto, Hashiguchi, Dohmoto, *et al.* (1992) *J. Pharmacobiodyn.* **15**, 25. **2.** Liu, Zhao, Li, Qian & Wang (2003) *Acta Pharmacol. Sin.* **24**, 263.

Cilostamide

This cyclic nucleotide phosphodiesterase inhibitor (FW = 342.43 g/mol; CAS 68550-75-4), also known by its code name OPC 3689, targets human cGMP Phosphodiesterase 3B (K_i= 5 nM) and human cGMP Phosphodiesterase 3A (K_i= 20 nM) (1-10). Indeed, specific concentration-dependent inhibition of the photolabelling of PDE III by ^{32}P-cGMP is observed with the following PDE inhibitors: trequinsin (IC$_{50}$ = 13 nM), lixazinone (IC$_{50}$ = 22 nM), milrinone (IC$_{50}$ = 56 nM), cilostamide (IC$_{50}$ = 70 nM), siguazodan (IC$_{50}$ = 117 nM) and 3-isobutyl 1-methylxanthine (IBMX) (IC$_{50}$ = 3950 nM). This approach demonstrated that one can evaluate of the inhibitory effects of compounds on the photolabelling of platelet PDE III, thus providing a simple quantitative means of investigating their actions at a molecular level that avoids the need to purify the enzyme. Cilostamide also blocks induction of germinal vesicle breakdown (GVBD) by Akt, suggesting that the activity of a PDE is required for Akt action (3). Cilostamidealso reduces DNA binding of p-Stat3, increases cAMP levels, and blocks the effects of insulin and IGF-1 on meiosis. Cilostamide activates K$_{ATP}$ channels, producing hyperpolarization in rat mesentery artery studies. In F/B cells, cilostamide inhibits the IGF-1 induced phosphorylation of the apoptotic protein, Bad. **Target(s):** cAMP-PDE activity in cytosolic extracts from human monocytes is also much more sensitive to inhibition by rolipram than by cilostamide (2); ADP-induced platelet aggregation, IC$_{50}$ = 16.8 μM; cGMP-inhibited phosphodiesterase, IC$_{50}$ = 70 nM **1**. Tang, Jang & Haslam (1994) *Eur. J. Pharmacol.* **268**, 105. **2**. Verghese, McConnel, Strickland *et al.* (1995) *J. Pharmacol. Exp. Ther.* **272**, 1313. **3**. Andersen, Roth & Conti (1998) *J. Biol. Chem.* **273**, 18705. **4**. Ahmad, Cong, Stenson, *et al.* (2000) *J. Immunol.* **164**: 4678. **5**. Zhao, *et al.* (2002) *Nat. Neurosci.* **5**, 727. **6**. Abi-Gerges, *et al.* (2009) *Circ. Res.* **105**, 784. **7**. Hanson, *et al.* (2008) *Am. J. Physiol. Heart Circ. Physiol.* **295**, H786. **8**. Park, *et al.* (2009) *Acta Anaesthesiol. Scand.* **53**, 1043. **9**. Kansui, *et al.* (2009) *Clin. Exp. Pharmacol. Physiol.* **36**: 729. **10**. Dimopoulos, *et al.* (2009) *Vascul. Pharmacol.* **50**, 78.

Cilostazol

This antithrombotic agent and vasodilator (FW = 369.47 g/mol; CAS 73963-72-1), also known by its code name OPC-13013, its trade name Pietal®, and its systematic name 6-[4-(1-cyclohexyl-1*H*-tetrazol-5-yl)butoxy]-3,4-dihydro-2(1*H*)-quinolinone, was first approved in the United States in 1999 to reduce symptoms of intermittent claudication. It is a platelet aggregation inhibitor that selectively inhibits cyclic nucleotide phosphodiesterase III, IC$_{50}$ = 0.19 μM (1-7). Pietal alleviates intermittent muscle ache associated with peripheral vascular disease. Cilostazol weakly inhibits type IV phosphodiesterase. Cilostazol also stimulates large-conductance, calcium-activated potassium channels. In N2a cells that express human APP (Swedish mutation N2aSwe) and endogenously overproducing Aβ-peptide, cilostazol (3-30 μM) treatment attenuates Aβ-amyloid-induced tauopathy via activation of CK2α/SIRT1 and suppresses tau acetylation (Ac-tau) and tau phosphorylation (P-tau) (8). Cilostazol also shows promise for treating microvascular complications associated with diabetes mellitus (9). **Target(s):** 3',5'-cyclic-nucleotide phosphodiesterase (1-7); adenosine uptake (4); 3',5'-cyclicGMP phosphodiesterase (6); phosphodiesterase 2 (6); phosphodiesterase 3A (6,7); phosphodiesterase 3B (6); phosphodiesterase 4 (6); phosphodiesterase 5 (6); phosphodiesterase 7 (6). **1**. Hidaka, Inagaki, Nishikawa & Tanaka (1988) *Meth. Enzymol.* **159**, 652. **2**. Tokuyama, Yokoyama, Arakawa, Morikawa & Kuroume (1994) *Pharmacology* **48**, 301. **3**. Kimura (1995) *Nippon Yakurigaku Zasshi* **106**, 205. **4**. Liu, Shakur, Yoshitake, & Kambayashi Ji (2001) *Cardiovasc. Drug Rev.* **19**, 369. **5**. Zhang, Ke & Colman (2002) *Mol. Pharmacol.* **62**, 514 **6**. Sudo, Tachibana, Toga, *et al.* (2000) *Biochem. Pharmacol.* **59**, 347. **7**. Hambleton, Krall, Tikishvili, *et al.* (2005) *J. Biol. Chem.* **280**, 39168. **8**. Lee, Shin, Park, *et al.* (2013) *J. Neurosci. Res.* **92**, 206.

Cimetidine

This substituted guanidine (FW = 252.34 g/mol; CAS 51481-61-9; pK_a = 6.8; Solubility: 0.11 g/mL H_2O, increasing upon addition of dilute HCl), commonly known by its proprietary name Tagamet™ and systematically referred to as N-cyano-N'-methyl-N''-[2-[[(5-methyl-1H-imidazol-4-yl)methyl]thio]ethyl]guanidine, is used in the treatment of ulcers and gastrointestinal bleeding. Originally classified as a histamine H_2-receptor antagonist, cimetidine has been reclassified as an inverse agonist. It inhibits gastric acid secretion and reduces pepsin release. In addition, cimetidine inhibits the expression of E-selectin on endothelial cell surfaces and thus acts to block tumor cell adhesion. For his work on antagonists, including cimetidine, Sir James Black shared the 1988 Nobel Prize in Physiology or Medicine. **Target(s):** aldehyde oxidase (1,2); diamine oxidase (3,16); CYP2C9 (4); CYP2C19 (4); CYP2D6 (4,5); CYP3A (4,12,13); CYP1A2 (5,11,14); CYP3A4/5 (4,5); alcohol dehydrogenase (6,10,18); CYP2C6 (7); CYP2C11 (7–9,14); CYP2B1/2 (14); CYP2A1/2 (14); CYP2E1 (15); alkaline phosphatase (16); phospholipase A_2 (17); histamine H_2 receptor (19,20). **1.** Lake, Ball, Kao, *et al.* (2002) *Xenobiotica* **32**, 835. **2.** Renwick, Ball, Tredger, *et al.* (2002) *Xenobiotica* **32**, 849. **3.** Wantke, Hemmer, Focke, *et al.* (2001) *Urol. Int.* **67**, 59. **4.** Furuta, Kamada, Suzuki, *et al.* (2001) *Xenobiotica* **31**, 1. **5.** Martinez, Albet, Agundez, *et al.* (1999) *Clin. Pharmacol. Ther.* **65**, 369. **6.** Dawidek-Pietryka, Szczepaniak, Dudka & Mazur (1998) *Arch. Toxicol.* **72**, 604. **7.** Levine, Law, Bandiera, Chang & Bellward (1998) *J. Pharmacol. Exp. Ther.* **284**, 493. **8.** Roos & Mahnke (1996) *Biochem. Pharmacol.* **52**, 73. **9.** Levine & Bellward (1995) *Drug Metab. Dispos.* **23**, 1407. **10.** Stone, Hurley, Peggs, *et al.* (1995) *Biochemistry* **34**, 4008. **11.** Spaldin, Madden, Pool, Woolf & Park (1994) *Brit. J. Clin. Pharmacol.* **38**, 15. **12.** Wrighton & Ring (1994) *Pharm. Res.* **11**, 921. **13.** Lewis, Lau, Duran, Wolf & Sikic (1992) *Cancer Res.* **52**, 4379. **14.** Chang, Levine & Bellward (1992) *J. Pharmacol. Exp. Ther.* **260**, 1450. **15.** Persson, Terelius & Ingelman-Sundberg (1990) *Xenobiotica* **20**, 887. **16.** Metaye, Mettey, Lehuede, Vierfond & Lalegerie (1988) *Biochem. Pharmacol.* **37**, 4263. **17.** Hirohara, Sugatani, Okumura, Sameshima & Saito (1987) *Biochim. Biophys. Acta* **919**, 231. **18.** Hernandez-Munoz, Caballeria, Baraona, *et al.* (1990) *Alcohol Clin. Exp. Res.* **14**, 946. **19.** Smit, Leurs, Alewijnse, *et al.* (1996) *Proc. Natl. Acad. Sci. U.S.A.* **93**, 6802. **20.** Brimblecombe, Duncan, Durant, *et al.* (1975) *Brit. J. Pharmacol.* **53**, 435.

Cinchonine

This photosensitive alkaloid (FW$_{free-base}$ = 294.40 g/mol; CAS 118-10-5) from the bark of a number of trees (*e.g.*, *Cinchona succiruba* and *C. micrantha*) is structurally related to quinine. Like quinine, it has been used as an antimalarial agent. Cinchonine also induces apoptosis. **Target(s):** amine oxidase (copper-containing) (1,2); choline transport (3,6); Ca^{2+} influx (4); multidrug resistance (MDR) efflux pump, stereospecifically and noncompetitively, thereby potentiating the action of drugs that are typically transported by this pump (5); strictosidine synthase (7). **1.** Pec & Frebort (1991) *J. Enzyme Inhib.* **4**, 327. **2.** Zajoncova, Frebort, Luhova, *et al.* (1999) *Biochem. Mol. Biol. Int.* **47**, 47. **3.** Ebel, Hollstein & Gunther (2002) *Biochim. Biophys. Acta* **1559**, 135. **4.** Shah, Nawaz, Virani, *et al.* (1998) *Biochem. Pharmacol.* **56**, 955. **5.** Wigler & Patterson (1993) *Biochim. Biophys. Acta* **1154**, 173. **6.** Pelassy & Aussel (1993) *Pharmacology* **47**, 28. **7.** Stevens, Giroud, Pennings & Verpoorte (1993) *Phytochemistry* **33**, 99.

Cinerins

Cinerin I

Cinerin II

These botanical insecticides are active constituents in the pyrethrum flowers *Chrysanthemum cinerariaefolium* or *Pyrethrum cinerariaefolium*. Cinerin I (FW = 316.44 g/mol) is a viscous liquid that is inactivated in air. It should be stored in the cold and in the dark. Cinerin II (FW = 360.45 g/mol) is likewise a viscous liquid that is inactivated by air. Both are practically insoluble in water. Note that the term cinerin also refers to a mixture of I and II. In addition, both cinerins are components of the pyrethrins. ***See also*** Pyrethrins

Cinnamate (Cinnamic Acid)

***trans*-Cinnamate**   ***cis*-Cinnamate**

This naturally occurring acid (FW$_{free-acid}$ = 148.16 g/mol; CAS 621-82-9; M.P. = 133°C; pK_a = 4.46 at 25°C), also known as (2E)-3-phenylpropenoic acid, is a product of the reactions catalyzed by phenylalanine ammonia-lyase and cinnamoyl-CoA:phenyllactate CoA-transferase. Cinnamic acid is a substrate for *trans*-cinnamate 4-monooxygenase, *trans*-cinnamate 2-monooxygenase, and cinnamate β-D-glucosyltransferase. **Target(s):** carboxy-peptidase C, *or* carboxypeptidase Y (1,12,13); melilotate 3-monooxygenase (2); tyrosinase, *or* monophenol monooxygenase (3,5,16-18); D-amino-acid oxidase, K_i = 0.56 mM (3,7); chymotrypsin (4); esterase (6); phenylpyruvate tautomerase (the *cis*-isomer is a stronger inhibitor than *trans*-cinnamate: K_i = 83 μM and 5.5 mM, respectively) (8,9); benzoylformate decarboxylase, inhibited by *trans*-cinnamate (10); carboxypeptidase A (11); diphosphomevalonate decarboxylase (14); phospho-mevalonate kinase (14); quinate O-hydroxy-cinnamoyltransferase (15); benzoate 4-monooxygenase, inhibited by *trans*-cinnamate (19); *trans*-cinnamate 4-monooxygenase, inhibited by *cis*-cinnamate (20); catechol oxidase (21); polyphenol oxidase (21). **1.** Hayashi (1976) *Meth. Enzymol.* **45**, 568. **2.** Husain, Schopfer & Massey (1978) *Meth. Enzymol.* **53**, 543. **3.** Webb (1966) *Enzyme and Metabolic Inhibitors*, vol. 2, Academic Press, New York. **4.** Gerig & Halley (1984) *Arch. Biochem. Biophys.* **232**, 467. **5.** Tan, Zhu & Lu (2002) *Chin. Med J (Engl.)* **115**, 185. **6.** Weber & King (1935) *J. Biol. Chem.* **108**, 131. **7.** Klein & Austin (1953) *J. Biol. Chem.* **205**, 725. **8.** Pirrung, Chen, Rowley & McPhall (1993) *J. Amer. Chem. Soc.* **115**, 7103. **9.** Molnar & Garai (2005) *Int. Immunopharmacol.* **5**, 849. **10.** Weiss, Garcia, Kenyon, Cleland & Cook (1988) *Biochemistry* **27**, 2197. **11.** Ludwig (1973) *Inorg. Biochem.* **1**, 438. **12.** Bai & Hayashi (1979) *J. Biol. Chem.* **254**, 8473. **13.** Bai, Hayashi & Hata (1975) *J. Biochem.* **77**, 81. **14.** Shama, Bhat & Ramasarma (1979) *Biochem. J.* **181**, 143. **15.** Rhodes, Wooltorton & Lourencq (1979) *Phytochemistry* **18**, 1125. **16.** Lee (2002) *J. Agric. Food Chem.* **50**, 1400. **17.** Anosike & Ayaebene (1982) *Phytochemistry* **21**, 1889. **18.** Orenes-Pinero, Garcia-Carmona & Sanchez-Ferrer (2007) *J. Mol. Catal. B* **47**, 143. **19.** McNamee & Durham (1985) *Biochem. Biophys. Res. Commun.* **129**, 485. **20.** Pfändler, Scheel, Sandermann & Grisebach (1977) *Arch. Biochem. Biophys.* **178**, 315. **21.** Mazzafera & Robinson (2000) *Phytochemistry* **55**, 285.

Cinnamycin

This lanthionine-containing peptide antibiotic (FW = 2041.29 g/mol; CAS 110655-58-8) from *Staphylococcus cinnamoneus* is a bacteriocin that is effective against *Bacilli*, *Clostyridia* and *Mycobacterium* (1,2). The sequence of this antibiotic is shown with single uppercase letters referring to the standard aminoacyl residues, Asp(OH) to hydroxyaspartate, Ala–S–Ala to lanthionine, Abu–S–Ala to β-methyl-lanthionine, and Ala–NH–K to lysinoalanine. **Primary Mode of Action:** Cinnamycin binds to phosphatidylethanolamine within cell membranes and induces cytolysis by means of transbilayer lipid movement, resulting in exposure of PE to the outer leaflet of the plasma membrane. Tight binding of cinnamycin to PE provides a means for detecting PE. Cinnamycin indirectly inhibits phospholipase A₂ (3). **1.** Fredenhagen, Märki, Fendich, *et al.* (1991) in *Nisin and Novel Lantibiotics* (Jung & Sahl, eds.), p. 131, Escom Publ., Leiden. **2.** Makino, Baba, Fujimoto, *et al.* (2003) *J. Biol. Chem.* **278**, 3204. **3.** Märki, Hanni, Fredenhagen & van Oostrum (1991) *Biochem. Pharmacol.* **42**, 2027.

Cinnarizine

This calcium channel blocker and antihistamine (FW = 368.51 g/mol; CAS Registry = 298-57-7), also known by the trade names Stugeron® and Stunarone® as well as systematically as (*E*)-1-(diphenylmethyl)-4-(3-phenylprop-2-enyl)piperazine, is an antihypertensive agent used to relieve nausea and vomiting due to motion sickness and chemotherapy as well as vertigo, tinnitus, and Meniere's syndrome. By virtue of its ability to promote cerebral blood flow, cinnarizine is used to treat cerebral apoplexy, post-trauma cerebral symptoms, and cerebral arteriosclerosis.

CINPA 1

This potent CAR antagonist (FW = 395.49 g/mol; CAS 102636-74-8), also named Ethyl [5-[(diethylamino)acetyl]-10,11-dihydro-5*H*-dibenz[*b,f*]azepin-3-yl]carbamate, targets the constitutive androstane receptor (IC₅₀ = 70 nM). Constitutive androstane receptor (CAR) and pregnane X receptor (PXR) are xenobiotic sensors that enhance the detoxification and elimination of xenobiotics and endobiotics by modulating the expression of genes encoding drug-metabolizing enzymes and transporters. Elevated levels of drug-metabolizing enzymes and efflux transporters (arising from CAR activation in various cancers) promote the elimination of chemotherapeutic agents, leading to reduced therapeutic effectiveness and acquired drug resistance. **Target(s):** CINPA-1 exhibits

>90-fold selectivity for CAR over pregnane X receptor (PXR). CINPA-1 binding to CAR is selective over a panel of other nuclear receptors including FXR, LXR, Peroxisome Proliferator-Activated Receptor-γ, *or* PPARγ and RXR. **1.** Cherian, Lin, Wu & Chen (2015) *Mol. Pharmacol.* **87**, 878.

Ciprofloxacin

This second-generation fluoroquinolone antibiotic (FW_free-base = 331.35 g/mol; CAS 85721-33-1), also known as Cipro and by its IUPAC name, 1-cyclopropyl-6-fluoro-4-oxo-7-(piperazin-1-yl)quinoline-3-carboxylic acid, achieves its bacteriocidal action by inhibiting DNA topoisomerase II, *or* DNA gyrase (1-4). **Mechanism of Action:** Genetic and biochemical studies have demonstrated that the A-subunit of DNA gyrase is the ciprofloxacin target, and inhibition of purified DNA gyrase correlates with antibacterial activity. When administered orally, ciprofloxacin exhibits 70% bioavailability, attaining peak serum levels (1.5–2.9 μg/mL) after a single 500-mg dose. Ciprofloxacin is considerably more active against the Gram-negative bacteria tested than other agents tested. In most cases, body fluid/tissue concentrations equal or exceed those in concurrent serum samples. **Ciprofloxacin Resistance:** While gentamicin-resistant bacteria (*Enterobacteriaceae*, *Pseudomonas aeruginosa*, and methicillin-resistant *Staphyloccus aureus*) are highly susceptible to Cipro, but resistance has also emerged. Mutations are selected first in DNA gyrase within gram-negative bacteria and topoisomerase IV within gram-positive bacteria (5). Additional resistance-conferring mutations occur in genes controlling Cipro accumulation, and further resistance is mediated by plasmids producing Qnr, a 218-residue protein belonging to the pentapeptide repeat family that shares sequence homology with McbG immunity protein and that protects DNA gyrase from the action of microcin B17, a 3.1-kDa bactericidal peptide. **Effect on Protein Tyrosine Kinase Inhibitors on Ciprofloxacin Phototoxicity:** TKI anticancer agents, such as gefitinib and imatinib, inhibit ABCG2 transporter-mediated drug efflux at the blood-retinal barrier, where it limits retinal exposure to phototoxic compounds, such as ciprofloxacin and other fluoroquinolone-class antibiotics (*e.g.*, amifloxacin, balofloxacin, difloxacin, gemifloxacin, levofloxacin, norfloxacin) (6). Concurrent administration of ABCG2 inhibitors with photoreactive fluoroquinolone antibiotics may result in retinal damage. **Key Pharmacokinetic Parameters:** *See* Appendix II in Goodman & Gilman's *THE PHARMACOLOGICAL BASIS OF THERAPEUTICS*, 12ᵗʰ Edition (Brunton, Chabner & Knollmann, eds.) McGraw-Hill Medical, New York (2011). **1.** Flamm, Vojtko, Chu, *et al.* (1995) *Antimicrob. Agents Chemother.* **39**, 964. **2.** Gruger, Nitiss, Maxwell, *et al.* (2004) *Antimicrob. Agents Chemother.* **48**, 4495. **3.** Sutcliffe, Gootz & Barrett (1989) *Antimicrob. Agents Chemother.* **33**, 2027. **4.** Inoue, Sato, Fujii, *et al.* (1987) *J. Bacteriol.* **169**, 2322. **5.** Jacoby (2005) *Clin. Infect. Disc.* **41** (*Supplement* 2), S-120. **6.** Mealey, Dassanayake & Burke (2014) *Oncology* **87**, 364.

Cisplatin

This yellow-colored anti-cancer drug (FW = 300.05 g/mol; CAS 15663-27-1; *Symbol*: *cis*-Pt(NH₃)₂Cl₂), also known as Platinol®, *cis*-diamminedichloroplatinum (II), and *cis*-dichloro-diammineplatinum(II), is a powerful, intravenously administered antineoplastic agent and mutagen. **Chemical Properties:** This electroneutral, four coordinate platinum complex has two chloride ion ligands situated adjacent to one another. As in organic chemistry, the prefix *cis* indicates that the chloro groups are next to each other, an apparent requirement for its anticancer action. Although no longer ionic, chloride ligands of Pt(NH₃)₂Cl₂ retain some of their electronic charge after forming a coordinate covalent bond with the central metal ion. Other similar antineoplastic have bromide, carbonate, oxalate or malonate as leaving groups. Such complexes are stable enough to reach their intended target, before nonenzymatic hydrolysis occurs:

$$[Pt(NH_3)_2Cl_2] + H_2O \rightleftharpoons [Pt(NH_3)_2Cl(H_2O)]^{1+} + Cl^-$$

$$[Pt(NH_3)_2Cl(H_2O)]^{1} + 2H_2O \rightleftharpoons [Pt(NH_3)_2(H_2O)_2]^{2+} + 2Cl^-$$

When administered in physiologic saline, the abundant free chloride ion rebinds to *cis*-monochlorodiammine platinum(II), preventing water molecules from occupying the metal ions inner coordination site. (Other platinum(II) complexes, such as *cis*-dinitratodiammineplatinum(II) hydrolyze more rapidly, with loss of nitrate groups and are ineffective as anticancer agents.) Reaction with DNA, the main biologic target, requires disubstitution, presumably as a sequence of two mono-substitution reactions. Most ligand substitution reactions occur by the Eigen-Tamm mechanism, where the anion leaves, creating a vacant site (or inner coordination sphere) for reaction with DNA. The anticancer action of cisplatin stems from the relative ease of chlorine ligand substitution with nucleophilic species, such as purine and pyrimidine bases of DNA. **History:** The structure of cisplatin was determined in 1893 by the Swiss chemist Alfred Werner, who formulated the basic rules of metal ion coordination chemistry. He predicted that ammonia bonds to a metal ion like Pt(II) by donating its lone pair of electrons in a dative bond (later termed a *coordinate covalent bond* by G. N. Lewis), with an overall bond-strength comparable to many covalent bonds. Werner's discovery of the square-planar geometry of cisplatin was abetted by the relatively poor reactivity of square planar Pt(II) compounds, unless situated between pairs of nearby nucleophiles. **Primary Mode of Action:** Cisplatin is primarily a DNA-damaging anticancer drug, forming different types of bifunctional adducts with cellular DNA, most often reacting with adjacent N^7-positions guanine and forming 1,2-intrastrand cross-links (1,2). The 1,2-intrastrand adducts produced by cisplatin causes DNA kinking and leads to thermal destabilization. Cisplatin stimulates the immune response by activating macrophages and other cells of the immune system. It increases cellular NF-kB content. Mismatch repair proteins and some damage-recognition proteins may show selectivity, contributing to the differences in its cytotoxicity and tumor range relative to other platinum-based antineoplastic agents. The ATP-hydrolysis motif in the mechanoenzyme RAD51D is required for resistance to DNA interstrand crosslinking agents and interaction with RAD51C (3). **Other Biological Effects:** Cisplatin induces structural lesions in DNA that sequester high-mobility group (HMG) proteins that are important inflammatory mediator, thereby triggering programmed cell death. Histone H2AX phosphorylation occurs in response to replication fork damage caused by cisplatin-induced DNA lesions, most likely interstrand crosslinks (4). Moreover, prior treatment with NH_4Cl increases the rate of cell apoptosis and activates caspase-3 (5). **Key Pharmacokinetic Parameters:** See Appendix II in Goodman & Gilman's THE PHARMACOLOGICAL BASIS OF THERAPEUTICS, 12th Edition (Brunton, Chabner & Knollmann, eds.) McGraw-Hill Medical, New York (2011). **Target(s):** acetylcholinesterase (6); thioredoxin reductase (4,7); glutaredoxin (8); thioredoxin (7); *Pst*I restriction endonuclease, type II site-specific deoxyribonuclease (9); dihydropteridine reductase (10); deoxyribonuclease (10); glutathione peroxidase (12); glyceraldehyde-3-phosphate dehydrogenase (13); Ca^{2+}-transporting ATPase (14); Na^+/K^+-exchanging ATPase (14); Mg^{2+}-ATPase (14); adenosylhomocysteinase (15); DNA-directed RNA polymerase (16); bacteriophage T7 RNA polymerase (16); RNA polymerase II (16); arylamine *N*-acetyltransferase (17); ribonucleoside diphosphate reductase (18); diferric-transferrin reductase (19). **1.** Poklar, Pilch, Lippard, et al. (1996) Proc. Natl. Acad. Sci. U.S.A. **93**, 7606. **2.** Ali-Osman, Rairkar & Young (1995) Cancer Biochem. Biophys. **14**, 231. **3.** Gruver, Miller, Rajesh, et al. (2005) Mutagenesis **20**, 433. **4.** Olive & Bánáth (2009) *Cytometry B Clin Cytom.* **76**, 79. **5.** Xu, Wang, Ding, *et al.* (2013) *Oncol Rep.* **30**, 1195. **6.** al-Jafari, Kamal & Duhaiman (1995) *J. Enzyme Inhib.* **8**, 281. **7.** Becker, Gromer, Schirmer & Muller (2000) *Eur. J. Biochem.* **267**, 6118. **8.** Arner, Nakamura, Sasada, *et al.* (2001) *Free Radic. Biol. Med.* **31**, 1170. **9.** Wells, Klein & Singleton (1981) *The Enzymes*, 3rd ed., **14**, 157. **10.** Armarego & Ohnishi (1987) *Eur. J. Biochem.* **164**, 403. **11.** Link & Tempel (1991) *J. Cancer Res. Clin. Oncol.* **117**, 549. **12.** Milano, Caldani, Khater, *et al.* (1988) *Biochem. Pharmacol.* **37**, 981. **13.** Aull, Allen, Bapat, *et al.* (1979) *Biochim. Biophys. Acta* **571**, 352. **14.** Nechay & Neldon (1984) *Cancer Treat Rep.* **68**, 1135. **15.** Impagnatiello, Franceschini, Oratore & Bozzi (1996) *Biochimie* **78**, 267. **16.** Tornaletti, Patrick, Turchi & Hanawalt (2003) *J. Biol. Chem.* **278**, 35791. **17.** Ragunathan, Dairou, Pluvinage, *et al.* (2008) *Mol. Pharmacol.* **73**, 1761. **18.** Smith & Douglas (1989) *Biochem. Biophys. Res. Commun.* **162**, 715. **19.** Sun, Navas, Crane, Morré & Löw (1987) *J. Biol. Chem.* **262**, 15915.

Citalopram

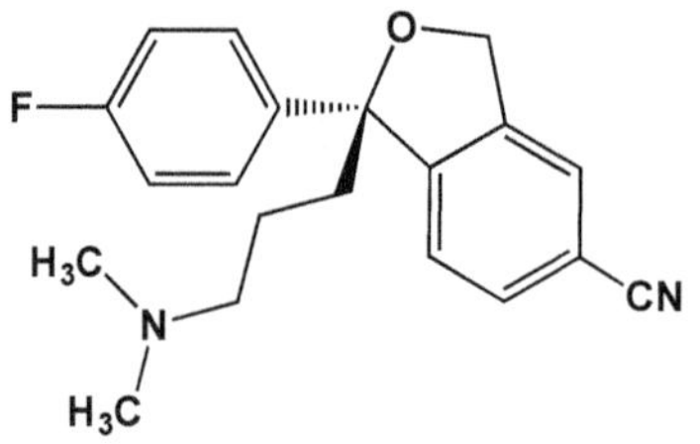

This orally available, single-daily-dose, selective serotonin reuptake inhibitor, *or* SSRI (FW = 324.39 g/mol; CAS 59729-33-8; Code Name = Lu 10-171), marketed under the trade names Celexa™, Cipramil™, Cipram™, Cital™, Citox™, Dalsan™, Emocal™, Recital™, Sepram™, and Seropram™, and named systematically as (*RS*)-1-[3-(dimethylamino)propyl]-1-(4-fluorophenyl)-1,3-dihydroisobenzofuran-5-carbonitrile, is a FDA-approved drug for treating major depression. Only the (*S*)-(+) enantiomer (shown above) has the desired clinical effect. **Mode of Action:** The primary pharmacological action of SSRIs is presynaptic inhibition of serotonin (5-HT) reuptake, thereby increasing the concentration of this amine in the synaptic cleft. Serotonin exerts its effects through interactions with any of seven distinct 5-HT receptor families, and interaction between serotonin and post-synaptic receptors mediates a wide range of its actions. Citalopram potentiates serotonergic transmission considerably, possibly by producing very strong inhibition of uptake without simultaneous blockade of the postsynaptic 5-HT receptors (1). Significantly, the antidepressant responses to selective serotonin reuptake inhibitors (citalopram or paroxetine) are abolished in mice unable to synthesize histamine due to either targeted disruption of histidine decarboxylase gene (*or* $HDC^{-/-}$) or injection of α-fluoromethylhistidine, a suicide inhibitor of this enzyme (2). Such findings demonstrate that SSRIs selectively require the integrity of the brain histamine system to exert their preclinical responses. (*See also Paroxetine; α-Fluoromethyl histidine*) **Metabolism:** Citalopram metabolites show weak 5-HT uptake-inhibiting properties and are devoid of noradrenaline uptake-inhibiting properties. In this respect, it clearly differs from the tricyclic antidepressants, which possess effects both on 5-HT and NA uptake (3). Inhibition of 5-HT uptake *in vitro* is competitive and not connected with an increased efflux of 5-HT. Lu10-171 and its metabolites only inhibit DA uptake in extremely high concentrations and in this respect they are even weaker than chlorimipramine and other tricyclic thymoleptics (3). **1.** Pawłowski, Ruczyńska & Górka (1981) *Psychopharmacology* (Berlin) **74**, 161. **2.** Munari, Provensi, Passani, *et al.* (2015) *Int. J. Neuropsychopharmacol.* **18**, pyv045. **3.** Hyttel (1977) *Psychopharmacology* (Berlin) **51**, 225.

Citrate (Citric Acid)

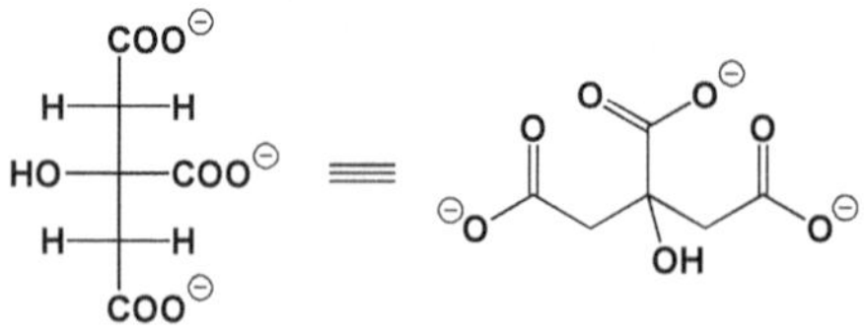

This tricarboxylic acid (FW$_{free-acid}$ = 192.13 g/mol; CAS 77-92-9; M.P. = 153°C), which is a key intermediate in the Tricarboxylic Acid and Glyoxylate Cycles, has pK_a values of 3.128, 4.761, and 6.396 at 25°C, with corresponding dpK_a/dT values of −0.0024, −0.0016, and ~0, respectively. Citrate is frequently used as a metal-ion chelator, binding magnesium, calcium, manganese, iron, copper, nickel, cobalt, and zinc ions. Formation constants (*i.e.*, $K = [Me^{2+} \cdot Citrate]/ [Me^{2+}][Citrate]$) for both Mg^{2+} and Ca^{2+} are ~4000 M^{-1}, and many Mg^{2+}- and Ca^{2+}-dependent processes as well as metalloenzymes are inhibited at elevated citrate levels. Depletion of plasma calcium ion blocks blood clotting, and citrate is widely used as an anticoagulant in blood collection. It enjoys the advantage that, upon re-infusion, its rapid metabolism fully restores normal clotting kinetics.

Citreoviridin

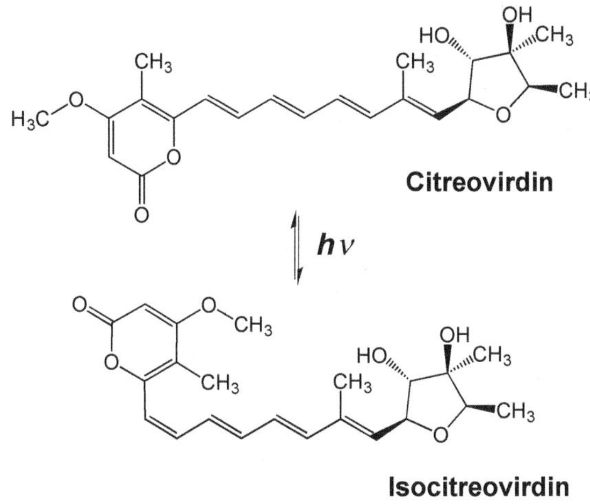

Citreovirdin

$$h\nu$$

Isocitreovirdin

This polyene neurotoxin (FW = 402.49 g/mol; CAS 25425-12-1), first isolated from moldy rice (a source of molds in the genera *Penicillium* and *Aspergillus*), is a strong inhibitor of F_1 ATPase (K_i = 4.5 μM). Note that citreoviridin and isocitreoviridin are interconverted by light. Ambient laboratory illumination is sufficient to generate substantial amounts of isocitreoviridin. Isocitreoviridin has without effect on F_1 ATPase. **Target(s):** ATP synthase (1-4); transketolase (5). **1.** Linnett & Beechey (1979) *Meth. Enzymol.* **55**, 472. **2.** Sayood, Suh, Wilcox & Schuster (1989) *Arch. Biochem. Biophys.* **270**, 714. **3.** Linnett, Mitchell, Osselton, Mulheirn & Beechey (1978) *Biochem. J.* **170**, 503. **4.** Gause, Buck & Douglas (1981) *J. Biol. Chem.* **256**, 557. **5.** Datta & Ghosh (1981) *Folia Microbiol. (Praha)* **26**, 408.

L-Citrulline

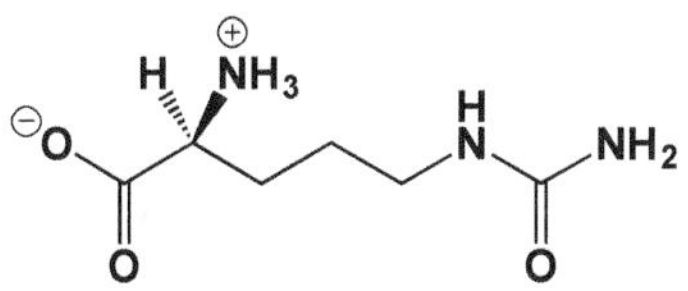

This Urea Cycle intermediate (FW = 175.19 g/mol; pK_a = 2.43 and 9.69; water-soluble, but insoluble in methanol, ethanol, and diethyl ether), also known as 2-amino-5-ureidovalerate and N^5-carbamoylornithine, was first identified in the watermelon *Citrullus vulgaris* (1). Citrulline is biosynthesized trough the action of ornithine transcarbamylase (Reaction: Ornithine + Carbamoyl-P $\rightleftharpoons$ L-Citrulline + P_i). It is also formed by the action of nitric oxide synthases (*Reaction*: 2 L-Arginine + 3 NADPH + 1 H^+ + 4 O_2 $\rightleftharpoons$ 2 L-Citrulline +2 Nitric Oxide + 4 H_2O + 3 $NADP^+$), and citrulline-specific antibodies may be used to detect sites of nitric oxide formation. While stable in acids, citrulline is unstable in bases, producing ornithine, ammonia, and carbon dioxide. **Target(s):** *N*-acetylglutamate kinase (2,6,14,15); arginine decarboxylase (3,11,12); arginase, weakly inhibited (4,13); saccharopine dehydrogenase (5,7); arginyl-tRNA synthetase (8); argininosuccinate lyase (9,10); amino-acid *N*-acetyltransferase (16). **1.** Koga & Odaka (1914) *Tokyo Kagakukai Shi* **35**, 579. **2.** Dénes (1970) *Meth. Enzymol.* **17A**, 269. **3.** Boeker & Snell (1971) *Meth. Enzymol.* **17B**, 657. **4.** Hunter & Downs (1945) *J. Biol. Chem.* **157**, 427. **5.** Ameen, Palmer & Oberholzer (1987) *Biochem. Int.* **14**, 589. **6.** Dénes (1973) *The Enzymes*, 3rd ed. (Boyer, ed.), **9**, 511. **7.** Fujioka & Nakatani (1972) *Eur. J. Biochem.* **25**, 301. **8.** Williams, Yem, McGinnis & Williams (1973) *J. Bacteriol.* **115**, 228. **9.** Raushel & Nygaard (1983) *Arch. Biochem. Biophys.* **221**, 143. **10.** Yu & Howell (2000) *Cell. Mol. Life Sci.* **57**, 1637. **11.** Blethen, Boeker & Snell (1968) *J. Biol. Chem.* **243**, 1671. **12.** Ramakrishna & Adiga (1975) *Eur. J. Biochem.* **59**, 377. **13.** Beruter, Colombo & Bachmann (1978) *Biochem. J.* **175**, 449. **14.** Haas & Leisinger (1975) *Eur. J. Biochem.* **52**, 377. **15.** Faragó & Dénes (1967) *Biochim. Biophys. Acta* **136**, 6. **16.** Marvil & Leisinger (1977) *J. Biol. Chem.* **252**, 3295.

CK666

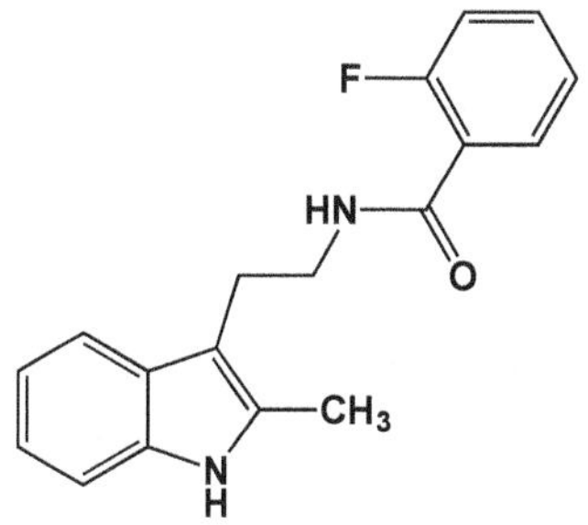

This cell-permeable actin cytoskeleton assembly inhibitor (FW = 296.34 g/mol; CAS 442633-00-3; Solubility: ≥25 mg/mL DMSO), also known as CK-0944666 and 2-fluoro-*N*-[2-(2-methyl-1*H*-indol-3-yl)ethyl] benzamide, targets the actin-related protein Arp3, IC_{50} = 5-17 μM, inserting into its hydrophobic core, stabilizing an inactive state, and preventing rearrangement of the Arp2/3 actin filament-initiation complex (1). During actin filament branch formation, Arp2 moves 31 Å relative to Arp3 from its position in inactive Arp2/3 complex, and the location of another inhibitor (CK-636) in the binding pocket between Arp2 and Arp3 suggests that both CK-636 and CK-666 lock Arp2/3 into an inactive conformation. CK666 prevents Arp3 from accessing its "short-pitch" conformation in the Arp2/3's active filament assembly-promoting state (2). CK666 reduces formation of actin filament rocket tails by *Listeria* in infected SKOV3 cells (1), suggesting that, at least in these cells, the latter process involves the Arp2/3 complex. CK666 also alters cortactin's interaction with Arp2/3 complex, which plays a role in regulating the activity of the Epithelial Na^+ Channel (ENaC), a cellular component that apparently interacts with the cortical cytoskeleton (3). CK666 potently disrupts the sub-cortical meiotic spindle position in treated MII oocytes, with the resulting spindles detached from the cortex, re-positioning themselves near the center of the oocytes (4). Such findings suggest that the asymmetric MII spindle position is dynamically maintained as a result of balanced forces governed in part by Arp2/3 effects on actin filament formation. **1.** Nolen, Tomasevic, Russell, *et al.* (2009) *Nature* **460**, 1031. **2.** Hetrick, Han, Helgeson & Nolen (2013) *Chem. Biol.* **20**, 701. **3.** Ilatovskaya, Pavlov, Levchenko, Negulyaev & Staruschenko (2011) *FASEB J.* **25**, 2688. **4.** Yi, Unruh, Deng, *et al.* (2011) *Nature Cell Biol.* **13**, 1252.

CK0106023

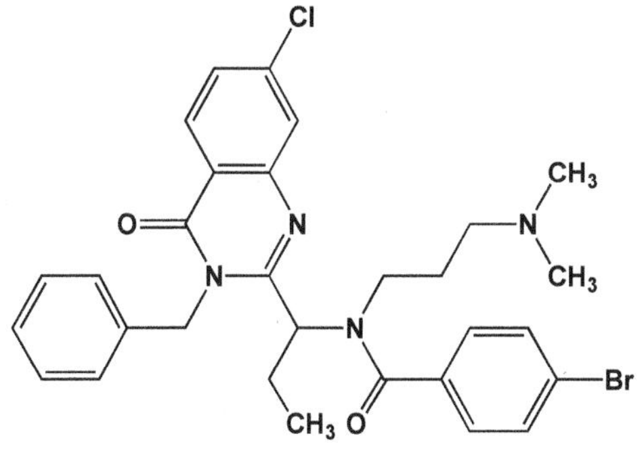

This novel, quinazolinone-based kinesin spindle inhibitor, *or* KSI (FW = 595.97 g/mol; CAS 336115-72-1; chiral center indicated by asterisk) targets the human Eg5, a kinesin that plays an essential role in forming bipolar mitotic spindles. Eg5 is required for cell cycle progression through mitosis, and addition of CK0106023 gives rise to mitotic arrest and growth inhibition in several human tumor cell lines. CK0106023 is an allosteric kinesin spindle protein inhibitor that arrests the motor domain ATPase, K_i = 12 nM. Among the five kinesins tested, CK0106023 showed clear specificity for Eg5. In tumor-bearing mice, CK0106023 exhibited antitumor activity comparable to or exceeding that of paclitaxel and caused the formation of monopolar mitotic figures identical to those produced in cultured cells. (***See also*** *Monastrol; Terpendole E; S-Trityl-L-Cysteine*) **1.** Sakowicz, Finer, Beraud, *et al.* (2004) *Cancer Res.* **64**, 3276.

Cladribine

This antineoplastic agent (FW = 285.69 g/mol; CAS 4291-63-8), also known as 2-chloro-2'-deoxyadenosine, is an with anti-leukemic drug with immunosuppressive properties. Cladribine 5'-triphosphate is enzymatically incorporated into DNA in place of dATP. The action of DNA polymerases with this modified DNA is significantly reduced and the modified DNA is also more susceptible to 3'→5' exonucleases. Cladribine is resistant to adenosine deaminase and inhibits DNA repair. Its triphosphate inhibits ribonucleotide reductase. Do not confuse cladribine with 2'-chloro-2'-deoxyadenosine. **Target(s):** DNA biosynthesis1; ribonucleotide reductase (2,3,5); adenosine deaminase, by the α- and β-isomers of cladribine, with K_i values of 60 and 21 μM, respectively (4); deoxynucleoside kinase (6); deoxycytidine kinase (6). **1.** Hentosh & Grippo (1994) *Mol. Pharmacol.* **45**, 955. **2.** Tsimberidou, Alvarado & Giles (2002) *Expert Rev. Anticancer Ther.* **2**, 437. **3.** Griffig, Koob & Blakley (1989) *Cancer Res.* **49**, 6923. **4.** Simon, Bauer, Tolman & Robins (1970) *Biochemistry* **9**, 573. **5.** Parker, Bapat, Shen, Townsend & Cheng (1988) *Mol. Pharmacol.* **34**, 485. **6.** Johansson, Van Rompay, Degrève, Balzarini & Karlsson (1999) *J. Biol. Chem.* **274**, 23814.

o-Cl-Amidine

This protein arginine deiminase (PAD) inhibitor (FW = 424.80 g/mol; CAS 913723-61-2; Soluble in methanol), systematically named as N^α-(2-carboxyl)benzoyl-N^δ-(2-chloro-1-iminoethyl)-L-ornithine amide, is a potent irreversible inhibitor that covalently modifies an active-site cysteine residue. Catalyzing posttranslational citrullination (*i.e.*, conversion of protein guanidinium groups of into ureido groups), PADs transform arginine residues into citrulline residues. By removing positively side-chains from proteins, these deiminases fundamentally alter the strength and/or specificity of regulatory interactions between target proteins with their intracellular binding partners. These enzymes (of which there are five in humans: PAD-1, PAD-2, PAD-3, PAD-4, and PAD-6) regulate gene transcription, cellular differentiation, as well as the innate immune response, and they are up-regulated in a number of human diseases, including rheumatoid arthritis, ulcerative colitis, and cancer. (1) Relevant inhibitor parameters are: PAD-1 (IC_{50} = 0.84 μM; k_{inact} = ND ; K_i = ND; k_{inact}/K_i = 106400 M^{-1}/min^{-1}); PAD-2 (IC_{50} = 6.2 μM; k_{inact} = ND; K_i = ND; k_{inact}/K_i = 14100 M^{-1}/min^{-1}); PAD-3 (IC_{50} = 0.69 μM; k_{inact} = 0.3 min⁻¹; K_i = 29 μM; k_{inact}/K_i = 10345 M^{-1}/min^{-1}); and PAD-4 (IC_{50} = 2.2 μM; k_{inact} = 0.5 min⁻¹; K_i = 13 μM; k_{inact}/K_i = 38000 M^{-1}/min^{-1}) (2). Cl-amidine is cytotoxic to HL-60, MCF7, and HT-29 cancer cell lines, with IC_{50} values of 0.25, 0.05 and 1 μM, respectively (3). (*See also o-F-Amidine*) **1.** Slade, Subramanian, Fuhrmann & Thompson (2013) *Biopolymers* **101**, 133. **2.** Causey, Jones, Slack, *et al.* (2011) *J. Med. Chem.* **54**, 6919. **3.** Slack, Causey & Thompson (2011) *Cell Mol. Life Sci.* **68**, 709.

(3*R*,5*R*)-Clavulanate (Clavulanic Acid)

This oxapenicillin analogue ($FW_{free-acid}$ = 199.16 g/mol; CAS RegistryNumber = 58001-44-8; Symbol: CA), also known as [2*R*-(2α,3*Z*,5α)]-3-(2-hydroxyethylidene)-7-oxo-4-oxa-1-azabicyclo[3.2.0]-heptane-2-carboxylic acid, from *Streptomyces clavuligerus*, is a weak antibiotic that greatly improves the effectiveness of β-lactam antibiotics by inhibiting antibiotic deactivation by Class A serine β-lactamases (penicillinases) (1-14). It was the first naturally-occurring, oxygen-containing, fused β-lactam identified. **Mechanism of Action:** Clavulanate is destroyed by β-lactamases and simultaneously inhibits it by producing two catalytically inactive forms (1). One of these is transiently stable and decomposes to free enzyme (k = 3.8 x 10⁻³ s⁻¹), while the other corresponds to an irreversibly inactivated form. The transient complex is formed from the Michaelis complex at a rate (k ≈ 3 x 10⁻² s⁻¹) which is some threefold faster than the rate of formation of the irreversibly inactivated complex. The transient complex is, therefore, the principle enzyme form present after short time periods. In the presence of excess clavulanate, however, all the enzyme accumulates into the irreversibly inactivated form. The number of clavulanate turnovers occurring prior to complete inactivation is 115 (1). Both the transiently inhibited and the irreversibly inactivated species show a marked increase in the absorbance at 281 nm that is proportional to the loss in enzyme activity (2). Hydroxylamine treatment of irreversibly inactivated enzyme restores about one-third of the catalytic activity, with a concomitant decrease in absorbance at 281 nm. Polyacrylamide isoelectric focusing of the irreversibly inactivated enzyme shows three bands of approximately equal intensity, different from native enzyme. Upon hydroxylamine treatment, one of the three bands disappears and now focuses identically with native enzyme. **Augmentin®:** This combination antibiotic therapy (CAS 74469-00-4), innovated by Beecham Pharmaceutics in the mid-1980s, uses amoxicillin and clavulanic acid for the treatment of a wide range of bacterial infections (15). Augmentin restores efficacy against amoxicillin-resistant bacteria, including *Klebsiella*. Augmentin is not effective against *Pseudomonas* infections. **Other Actions:** Although the β-lactam antibiotic ceftriaxone (CTX) reduces cocaine reinforcement and relapse by activating the glutamate transporter subtype 1 (GLT-1), its required IV administration and poor brain penetrability limit its therapeutic utility for indications related to CNS diseases. An a structural analogue of CTX that retains the β-lactam core required for GLT-1 activity, but displays enhanced brain penetrability and oral activity relative to CTX (16). GLT-1 transporter expression in the nucleus accumbens of mice treated with repeated CA (1-10 mg/kg) is enhanced relative to saline-treated mice. Repeated CA treatment reduces the reinforcing efficacy of cocaine in mice maintained on a progressive-ratio (PR) schedule of reinforcement, but did not affect acquisition of cocaine self-administration under fixed-ratio responding or acquisition or retention of learning. Such findings suggest CA activates the cellular glutamate reuptake system in the brain reward circuit and reduce cocaine's reinforcing efficacy at a 100-times lower dose than CTX (16). **1.** Fisher, Charnas & Knowles (1978) *Biochemistry* **17**, 2180. **2.** Charnas, Fisher& Knowles (1978) *Biochemistry* **17**, 2185. **3.** Reading & Cole (1977) *Antimicrob. Agents Chemother.* **11**, 852. **4.** Spratt, Jobanputra & Zimmerman (1977) *Antimicrob. Agents Chemother.* **12**, 406. **5.** Fisher, Charnas & Knowles (1978) *Biochemistry* **17**, 2180. **6.** Brown, Aplin & Schofield (1996) *Biochemistry* **35**, 12421. 7. Rizwi, Tan, Fink & Virden (1989) *Biochem J.* **258**, 205. **8.** Guo, Huynh, Dmitrienko, Viswanatha & Clarke (1999) *Biochim. Biophys. Acta* **1431**, 132. **9.** Voha, Docquier, Rossolini & Fosse (2006) *Antimicrob. Agents Chemother.* **50**, 2673. **10.** Hedberg, Lindqvist, Tuner & Nord (1992) *Eur. J. Clin. Microbiol. Infect. Dis.* **11**, 1100. **11.** Matthew (1978) *FEMS Microbiol. Lett.* **4**, 241. **12.** Matthew & Sykes (1977) *J. Bacteriol.* **132**, 341. **13.** Poirel, Brinas, Verlinde, Ide & Nordmann (2005) *Antimicrob. Agents Chemother.* **49**, 3743. **14.** Ogawara, Minagawa & Nishizaki (1978) *J. Antibiot.* **31**, 923. **15.** Leigh, Bradnock & Marriner (1981) *J. Antimicrob. Chemother.* **7**, 229. **16.** Kim, John, Langford, *et al.* (2015) *Amino Acids* **48**, 649.

Clemizole

This TRPC blocker (FW$_{free-base}$ = 325.84 g/mol; CAS 1163-36-6; Solubility: 20 mM in water; 50 mM in DMSO), also named 1-[(4-chlorophenyl)methyl]-2-(1-pyrrolidinylmethyl)-1H-benzimidazole, selectively targets the Transient Receptor Potential Cation-5 channel, *or* TRPC5 (IC$_{50}$ = 1.1 μM), with weaker action against TRPC4 (IC$_{50}$ = 6.4 μM), TRPC3 (IC$_{50}$ = 9.1 μM), TRPC6 (IC$_{50}$ = 11.3 μM), and TRPC7 (IC$_{50}$ = 26.5 μM) channels. Clemizole exhibits minimal effect on TRPM3, TRPM8 and TRPV1, TRPV2, TRPV3 and TRPV4 channels (1). Clemizole is also TRPC1:TRPC5 heteromers as well as native TRPC5-like currents in the U-87 glioblastoma cell line. It is also a histamine H$_1$ receptor antagonist, as tested in longitudinal colon strips (2). **1**. Richter, Schaefer & Hill (2014) *Mol. Pharmacol.* **86**, 514. **2**. Aguilar, Morales-Olivas & Rubio (1986) *Brit. J. Pharmacol.* **88**, 501.

Clevidipine

This intravenous ultrashort-acting antihypertensive (FW = 456.32 g/mol; CAS 166432-28-6; Trade name: Cleviprex™; IUPAC: O^3-(butanoyloxymethyl)-O^5-methyl-(4R)-4-(2,3-dichloro-phenyl)-2,6-dimethyl-1,4-dihydropyridine-3,5-dicarboxylate) is a dihydropyridine calcium channel antagonist that selectively relaxes (EC$_{50}$ = 4 μM) small artery smooth muscle lining (1). Clevidipine is rapidly hydrolyzed by circulating esterases, exhibiting *in vitro* half-lives in blood of 0.6 min (rat), 16 min (dog) and 5.8 min (humans) (2). It is specifically indicated when oral therapy is either infeasible or undesirable. **1**. Huraux, Makita, Szlam, Nordlander & Levy (1997) *Anesth. Analg.* **85**, 1000. **2**. Ericsson, Tholander & Regårdh (1999) *Eur. J Pharm Sci.* **8**, 29.

Clevudine

This (deoxy)thymidine analogue and nucleos(t)ide reverse transcriptase inhibitor, *or* NRTI (FW = 260.22 g/mol; CAS 69256-17-3; *Abbreviation*: CLV), also known by the trade names Levovir® and Revovir® and its systematic name 1-[(2S,3R,4S,5S)-3-fluoro-4-hydroxy-5-(hydroxymethyl) oxolan-2-yl]-5-methylpyrimidine-2,4-dione, is an antiviral prodrug that is indicated for the treatment of hepatitis B virus, *or* HBV (1). Clevudine is

differentiated from other antivirals by (a) its unusual phosphorylation pathway to the active triphosphate in uninfected and infected liver cells; (b) an inhibitory mechanism that interferes with multiple steps in the intracellular life cycle of hepatitis B virus; (c) a long *in vivo* half-life; and (d) significant reduction of covalently closed circular DNA (cccDNA), when tested in animal models (2). Upon cellular uptake, clevudine is phosphorylated to its mono-, di-, and tri-phosphate forms, with the latter presumed to be the active inhibitor and antiviral drug. The major CLV metabolite formed in human primary hepatocytes was CLV 5'-monophosphate, whereas in the hepatoma cell lines the major metabolite was 5'-triphosphate (3). The level of CLV 5'-triphosphate was similar in both cell types. In primary hepatocytes the conversion of CLV 5'-monophosphate to the corresponding 5'-diphosphate was the rate-limiting step in CLV phosphorylation; the level of CLV phosphorylation was dependent upon exogenous drug concentration and exposure time. CLV 5'-triphosphate accumulated rapidly with peak levels observed after approximately 8 h. Kinetic analyses using the HBV endogenous polymerase assay revealed that clevudine-5'-triphosphate noncompetitively inhibited DNA chain elongation, suggesting that it binds to and distorts the HP active site in a unique manner (4). High-dose treatment with clevudine induces mitochondrial defects associated with mtDNA depletion and impairs glucose-stimulated insulin secretion in insulin-releasing cells (5). Although there is no indication that clevudine is carcinogenic, several accounts of clevudine-associated myopathy have appeared. **1**. Dienstag (2008) New Engl. J. Med. **359**, 1486. **2**. Anderson (2009) *Drugs Today* **45**, 331. **3**. Niu, Murakami & Furman (2008) *Antivir. Ther.* **13**, 263. **4**. Jones, Murakami, Delaney, Furman & Hu (2013) *Antimicrob. Agents Chemother.* **57**, 4181. **5**. Jang, Quan, Das, *et al.* (2012) *BMC Gastroenterol.* **12**, 4.

Clinafloxacin

This fluoroquinolone-class antibiotic (FW = 365.79 g/mol; CAS 105956-99-8), also named 7-(3-aminopyrrolidin-1-yl)-8-chloro-1-cyclopropyl-6-fluoro-4-oxoquinoline-3-carboxylic acid, is a broad-spectrum systemic antibacterial agent that targets Type II DNA topoisomerases (gyrases), which are required for bacterial replication and transcription. It has been used for parenteral and oral administration in subjects with serious infections. For the prototypical member of this antibiotic class, *See Ciprofloxacin*

Clindamycin

This semisynthetic lincosamide antibiotic (FW = 424.99 g/mol; CAS 18323-44-9), also known as (7S)-7-chloro-7-deoxylincomycin, is effective against anaerobic microorganisms, particularly β-lactamase-producing strains of the *Bacteroides* species. Clindamycin inhibits protein synthesis by acting on the 50S ribosomal subunits of bacteria, with binding sites composed exclusively of segments of 23S ribosomal RNA at the peptidyl transferase cavity (1). Clindamycin does not with any ribosomal protein. **Key Pharmacokinetic Parameters:** *See* Appendix II in Goodman & Gilman's *THE PHARMACOLOGICAL BASIS OF THERAPEUTICS*, 12th Edition (Brunton, Chabner & Knollmann, eds.) McGraw-Hill Medical, New York (2011). **Target(s):** respiratory burst of neutrophils (2,5); protein biosynthesis (3,4); peptidyl-transferase (6,7). **1.** Schlünzen, Zarivach, Harms, *et al.* (2001) *Nature* **413**, 814. **2.** Perry, Hand, Edmondson & Lambeth (1992) *J. Immunol.* **149**, 2749. **3.** Spizek & Rezanka (2004) *Appl. Microbiol. Biotechnol.* **64**, 455. **4.** Heman-Ackah & Garrett (1972) *J. Med. Chem.* **15**, 152. **5.** Hand, Hand & King-Thompson (1990) *Antimicrob. Agents Chemother.* **34**, 863. **6.** Douthwaite (1992) *J. Bacteriol.* **174**, 1333. **7.** Polacek & Mankin (2005) *Crit. Rev. Biochem. Mol. Biol.* **40**, 285.

Clobenpropit

This substituted imidazole (FW$_{free-base}$ = 308.83 g/mol; CAS 145231-35-2) is a potent histamine H$_3$ receptor antagonist. Clobenpropit can cross the blood-brain barrier. **Target(s):** cholesterol 24-hydroxylase, K_d = 80 nM (1); histamine H$_3$ receptor (2-4). **1.** Mast, White, Bjorkhem, *et al.* (2008) *Proc. Natl. Acad. Sci. U.S.A.* **105**, 9546. **2.** Hill, Ganellin, Timmerman, *et al.* (1997) *Pharmacol. Rev.* **49**, 253. **3.** Barnes, Brown, Clarke, *et al.* (1993) *Eur. J. Pharmacol.* **250**, 147. **4.** Kathmann, Schlicker, Detzner & Timmerman (1993) *Naunyn Schmiedebergs Arch. Pharmacol.* **348**, 498.

Clofibrate

This lipid-lowering agent (FW = 242.70 g/mol; CAS 637-07-0), also known by its systematic name, ethyl 2-(4-chlorophenoxy)-2-methylpropanoate, and its tradename Atromid-S®, is a used to control the high circulating cholesterol and triacylglyceride levels by increasing lipoprotein lipase activity and promoting the conversion of VLDL to LDL. Clofibrate also inhibits pyruvate dehydrogenase from *Fusarium culmorum* (2). Clofibrate inhibition of 11-β-hydroxylation in bovine adrenal cortex mitochondria (3). **1.** Robillard, Fontaine, Chinetti, Fruchart & Staels (2005) *Handbook Expmtl. Pharmacol.* p. 389. **2.** Madhosingh & Orr (1985) *J. Environ. Sci. Health B* **20**, 201. **3.** McIntosh, Uzgiris & Salhanick (1970) *Endocrinology* **86**, 656.

Clonazepam

This intermediate-acting benzodiazepine (FW = 315.72 g/mol; CAS 1622-61-3; IUPAC: 5-(2-chlorophenyl)-7-nitro-2,3-dihydro-1,4-benzo iazepin-2-one), also known as Rivotril, Linotril, Clonotril, and Clonex, mainly facilitates GABAergic transmission within the brain by its direct effect on benzodiazepine receptors and exhibits anxiolytic, anticonvulsant, muscle relaxant, amnestic, sedative, and hypnotic properties. With once daily oral administration, clonazepam is rapidly absorbed and passes quickly from blood to brain. Its biological half-life of 22-32 hours allows it to reach and sustain its 5-50 ng/mL therapeutic serum concentration, with once daily dosing. **Key Pharmacokinetic Parameters:** *See* Appendix II in Goodman & Gilman's *THE PHARMACOLOGICAL BASIS OF THERAPEUTICS*, 12th Edition (Brunton, Chabner & Knollmann, eds.) McGraw-Hill Medical, New York (2011). **1.** Pinder, Brogden, Speight & Avery (1976) *Drugs* **12**, 321. **2.** Browne (1976) *Arch Neurol.* **33**, 326. **3.** Hvidberg & Dam (1976) *Clin. Pharmacokinet.* **1**, 161.

Clonidine

This α$_2$ adrenergic receptor agonist (FW$_{free-base}$ = 230.10 g/mol; CAS 4205-90-7), also named *N*-(2,6-dichlorophenyl)-4,5-dihydro-1*H*-imidazol-2-amine, is a nonopiate antihypertensive agent. Other uses include treatment of neuropathic pain, sleep hyperhidrosis, and as an anesthetic by veterinarians. It has been used to reduce the withdrawal symptoms during smoking, alcohol, or opiate cessation. Gold and coworkers (1-3) proposed that clonidine could be an effective nonopiate treatment for opiate withdrawal distress, based on their hypothesis that opiate withdrawal syndrome is the result of hyperactivity (or hyperexcitability) of noradrenergic neurons in the locus coeruleus. (*See also Lofexidine*) Clonidine-displacing substance, thought to be the endogenous ligand for imidazoline receptors, was recently identified as agmatine, *or* 1-amino-4-guanidinobutane (*See Agmatine*). **Key Pharmacokinetic Parameters:** *See* Appendix II in Goodman & Gilman's *THE PHARMACOLOGICAL BASIS OF THERAPEUTICS*, 12th Edition (Brunton, Chabner & Knollmann, eds.) McGraw-Hill Medical, New York (2011). **Target(s):** amine oxidase (copper-containing) (4,5); polyamine oxidase (5); trypsin (5); nitric oxide synthase (5). **1.** Gold, Redmond & Kleber (1978). *Lancet* I, 929; *Lancet* II, 599. **2.** Charney, Heninger & Kleber (1986) *Amer. J. Psych.* **143**, 817. **3.** Gold (2001) "Clonidine" in *Encyclopedia of Drugs, Alcohol, and Addictive Behavior*. **4.** Ercolini, Angelini, Federico, *et al.* (1998) *J. Enzyme Inhib.* **13**, 465. **5.** Federico, Leone, Botta, *et al.* (2001) *J. Enzyme Inhib.* **16**, 147.

α-Clopenthixol

This anti-psychotic drug (FW = 400.97 g/mol; CAS 982-24-1) is a potent inhibitor of dopamine-stimulated adenylate cyclase and is a dopamine receptor antagonist (1). Note that there are two distinct forms of this pharmaceutical: an α- or *cis*-isomer (*i.e.*, the *Z*-isomer) and a β- (*or trans*-) form. α-Clopenthixol is the more active form, K_i = 17 nM. Both stereoisomers enhance human neutrophil chemotaxis (2). **1.** Iversen (1975) *Science* **188**, 1084. **2.** Rechnitzer, Kristiansen & Kharazmi (1985) *Acta Pathol. Microbiol. Immunol. Scand. [C]* **93**, 199.

Clopidogrel

This antithrombotic agent (FW$_{free-base}$ = 321.83 g/mol; CAS 113665-84-2; Trade Name = Plavix™) is a selective thienopyridine-class inhibitor of P2Y12 ADP (*or* purinergic) receptors, and the subsequent ADP-mediated activation of the glycoprotein GPIIb/IIIa complex, thereby inhibiting platelet aggregation. Clopidogrel blocks the binding of ADP and its analogues to the receptor that suppresses platelet adenylate cyclase activity. The effectiveness of clopidogrel is dependent on its enzymatic conversion to an active metabolite by CYP2C19. Quantification of Vasodilator-activated serine phosphoprotein phosphorylation status (so-called "VASP assay") is probably the most specific assay for evaluating clopidogrel inhibition of the P2Y12 receptor. Oral bioavailability of clopidogrel is low (<50%), a consequence of its poor water solubility. When prepared as a capmul microemulsion, clopidogrel solubility can be enhanced by 80x, compared with distilled water at pH = 7.4 (2). Finally, VASP is the prototypical motor protein for the actoclampin molecular motors (3), and because actin-based motility underlies platelet "pancaking" during clot formation, one cannot discount the likelihood that clopidogrel-induced VASP phosphorylation lies at the heart of its potent hemodynamic properties. **Key Pharmacokinetic Parameters:** *See* Appendix II in Goodman & Gilman's THE PHARMACOLOGICAL BASIS OF THERAPEUTICS, 12th Edition (Brunton, Chabner & Knollmann, eds.) McGraw-Hill Medical, New York (2011). **1.** Mills, Puri, Hu, *et al.* (1992) *Arterioscler. Thromb.* **12**, 430. **2.** Patel, Kukadiya, Mashru, Surti & Mandal (2010) *Iran J. Pharm. Res.* **9**, 327. **3.** Purich (2016) *Biophys. Chem.* **209**, 41.

Clorazepate

Clorazepate   Demethyldiazepam

This benzodiazepine-class pro-drug (FW = 314.72 g/mol; CASs = 23887-31-2; 57109-90-7 (potassium salt)), also known as Tranxene®, Novo-Clopate®, and 7-chloro-2,3-dihydro-2-oxo-5-phenyl-1*H*-1,4-benzodiaze pine-3-carboxylic acid, is rapidly converted to the active metabolite demethyldiazepam (*or* nordazepam), which exhibits anxiolytic, anticonvulsant, sedative, hypnotic, and skeletal muscle relaxant properties. **Key Pharmacokinetic Parameters:** *See* Appendix II in Goodman & Gilman's THE PHARMACOLOGICAL BASIS OF THERAPEUTICS, 12th Edition (Brunton, Chabner & Knollmann, eds.) McGraw-Hill Medical, New York (2011). **1.** Nicholson, Stone, Clarke & Ferres (1976) *Brit. J. Clin. Pharmacol.* **3**, 429. **2.** Dixon, Brooks, Postma, *et al.* (1976) *Clin. Pharmacol. Ther.* **20**, 450.

Clostridium difficile Toxins A & B

These large clostridial cytotoxins, designated Toxin A (MW ≈ 308 kDa) and Toxin B (MW ≈ 270 kDa) are K$^+$-dependent UDP-glucose glucosyltransferases that inactivate certain GTPases by catalyzing the coupling of a sugar residue (utilizing UDP-D-glucose or UDP-D-*N*-acetylglucosamine) to a threonyl residue in the targeted GTP-binding protein. This glucosylation results in a variety of effects including dysregulation of the actin cytoskeleton, cell rounding, cytotoxicity, and altered cellular signaling (*C. difficile* is a leading cause of nosocomial diarrhea). Clinically relevant Toxin A-negative, toxin B-positive (A$^-$B$^+$) strains of *Clostridium difficile* are known to cause diarrhea and colitis in humans, a finding that indicates both toxins are not simultaneously required for toxin-associated pathogenesis. *Note*: These toxins also exhibit a UDP-glucose hydrolyzing activity. **Target(s):** Rho GTPase (1-3); Rac GTPase (1,2); Cdc-42 (1,2); Rap (1); Ral (1). **1.** Moos & von Eichel-Streiber (2000) *Meth. Enzymol.* **325**, 114. **2.** Ciesla & Bobak (1998) *J. Biol. Chem.* **273**, 16021. **3.** Aktories (1997) *J. Clin. Invest.* **99**, 827. **4.** Jank, Giesemann & Aktories (2007) *Glycobiology* **17**, 15R.

Clotrimazole

This antifungal agent (FW$_{free-base}$ = 344.84 g/mol; CAS 23593-75-1), sold under the brand name Canesten or Lotrimin™ and also known as 1-[(2-chlorophenyl)diphenylmethyl]-1*H*-imidazole, is a nonspecific cytochrome P450 inhibitor, a specific inhibitor of Ca^{2+}-activated K$^+$ channels, and also acts as a calmodulin antagonist. Note that clotrimazole hydrolyzes rapidly upon heating in aqueous acids. (*See TRAM-34*) **Target(s):** 1,25-dihydroxycholecalciferol metabolism, via cytochrome P450 (1); nitric oxide synthase (2,8); Ca^{2+}-transporting ATPase (3); L-type calcium currents (4); Ca^{2+}-activated K$^+$ channels (5); cytochrome P450, nonspecific inhibitor (6,7); cyclic-nucleotide phosphodiesterase, calmodulin-dependent (8); hemo-peroxidase (9); TRMP2 channels (10); 6-phosphofructokinase (11); taxadiene 5α-hydroxylase (12); steroid 11β-monooxygenase (13); abietadiene hydroxylase (14); abietadienol hydroxylase (14); (*R*)-limonene

6-monooxygenase (15); tabersonine 16-hydroxylase (16); sterol 14-demethylase, CYP51 (17,18); 27-hydroxy-cholesterol 7α-monooxygenase (19); (*S*)-limonene 3-monooxygenase (20); (*S*)-limonene 6-monooxygenase (20,21); (*S*)-limonene 7-monooxygenase (weakly inhibited (20); calcidiol 1-monooxygenase (22); nitric-oxide dioxygenase (23,24). **1.** Napoli, Martin & Horst (1991) *Meth. Enzymol.* **206**, 491. **2.** Griffith & Stuehr (1995) *Ann. Rev. Physiol.* **57**, 707. **3.** Snajdrova, Xu & Narayanan (1998) *J. Biol. Chem.* **273**, 28032. **4.** Thomas, Karmazyn, Zygmunt, Antzelevitch & Narayanan (1999) *Brit. J. Pharmacol.* **126**, 1531. **5.** Wu, Li, Jan & Shen (1999) *Neuropharmacology* **38**, 979. **6.** Yan, Rafferty, Caldwell & Masucci (2002) *Eur. J. Drug Metab. Pharmacokinet.* **27**, 281. **7.** Suzuki, Kurata, Nishimura, Yasuhara & Satoh (2000) *Eur. J. Drug Metab. Pharmacokinet.* **25**, 121. **8.** Wolff, Datto & Samatovicz (1993) *J. Biol. Chem.* **268**, 9430. **9.** Trivedi, Chand, Srivastava, *et al.* (2005) *J. Biol. Chem.* **280**, 41129. **10.** Hill, McNulty & Randall (2004) *Naunyn Schmiedebergs Arch. Pharmacol.* **370**, 227. **11.** Zancan, Rosas, Marcondes, Marinho-Carvalho & Sola-Penna (2007) *Biochem. Pharmacol.* **73**, 1520. **12.** Hefner, Rubenstein, Ketchum, Gibson, Williams & Croteau (1996) *Chem. Biol.* **3**, 479. **13.** Denner, Vogel, Schmalix, Doehmer & Bernhardt (1995) *Pharmacogenetics* **5**, 89. **14.** Funk & Croteau (1994) *Arch. Biochem. Biophys.* **308**, 258. **15.** Bouwmeester, Konings, Gershenzon, Karp & Croteau (1999) *Phytochemistry* **50**, 243. **16.** St-Pierre & De Luca (1995) *Plant Physiol.* **109**, 131. **17.** McLean, Warman, Seward, *et al.* (2006) *Biochemistry* **45**, 8427. **18.** Trösken, Adamska, Arand, *et al.* (2006) *Toxicology* **228**, 24. **19.** Martin, Bean, Rose, Habib & Seckl (2001) *Biochem. J.* **355**, 509. **20.** Karp, Mihaliak, Harris & Croteau (1990) *Arch. Biochem. Biophys.* **276**, 219. **21.** Croteau, Karp, Wagschal, *et al.* (1991) *Plant Physiol.* **96**, 744. **22.** Friedrich, Diesing, Cordes, *et al.* (2006) *Anticancer Res.* **26**, 2615. **23.** Gardner (2005) *J. Inorg. Biochem.* **99**, 247. **24.** Helmick, Fletcher, Gardner, *et al.* (2005) *Antimicrob. Agents Chemother.* **49**, 1837.

Cloxacillin

This semisynthetic penicillin-class antibiotic (FW = 435.88 g/mol; CAS 61-72-3), also named Cloxapen®, Cloxacap®, Tegopen®, and (2*S*,5*R*,6*R*)-6-{[3-(2-chlorophenyl)-5-methyloxazole-4-carbonyl]amino}-3,3-dimethyl-7-oxo-4-thia-1-azabicyclo[3.2.0]heptane-2-carboxylic acid, is often used against *Staphylococci* producing β-lactamase, because its large side-chain prevents binding to and deactivation by this hydrolase (*See also Floxacillin*).

Clozapine

This second-generation antipsychotic agent and atypical antipsychotic (FW$_{free-base}$ = 326.83 g/mol; CAS 5786-21-0), also named Clozaril® and 8-chloro-11-(4-methylpiperazin-1-yl)-5*H*-dibenzo[*b,e*][1,4]diazepine, is a relatively nonspecific antagonist for D$_4$-dopamine receptors as well as 5-HT2A, 5-HT2C, 5-HT3, 5-HT6, and 5-HT7 serotonin receptors. It also downregulates 5-HT2A and 5-HT6 receptors, but upregulates 5-HT7 receptors. Given its tendency to induce agranulocytosis, its FDA-approved use for treating schizophrenia is premised on frequent blood testing for absolute neutrophil counts. **Target(s):** D$_4$-dopamine receptor (1,6); 5-HT3 serotonin receptor (2); 5-HT1C serotonin receptor (3); 5-HT6 serotonin receptor (4,6); 5-HT7 serotonin receptor (4,6); 5-HT2A serotonin receptor (5,6); 5-HT2C serotonin receptor (5,6); α_1-adrenergic receptor (6); glucose transport (7). **Receptor Affinity for Clozapine:** Serotonin transporter (SERT), K_i = 1.6 μM; Norepinephrine transporter (NET or SLC6A2), K_i = 3.2 μM; Dopamine transporter (DAT), K_i = 10 μM; Serotonin Receptor 5-HT$_{1A}$, K_i = 124 nM; 5-HT$_{1B}$, K_i = 520 nM; 5-HT$_{1D}$, K_i = 1.36 μM; 5-HT$_{2A}$, K_i = 5 nM; 5-HT$_{2B}$, K_i = 8 nM; 5-HT$_{2C}$, K_i = 9 nM; 5-HT$_3$, K_i = 240 nM; 5-HT$_{5A}$, K_i = 3.9 μM; 5-HT$_6$, K_i = 13 nM; 5-HT$_7$, K_i = 18 nM; Dopamine receptor D$_1$, K_i = 270 nM; D$_2$, K_i = 160 nM; D$_3$, K_i = 270 nM; D$_4$, K_i = 26 nM; D$_5$, K_i = 0.25 μM; Adrenergic Receptor α_{1A}, K_i = 1.7 nM; α_{1B}, K_i = 7 nM; α_{2A}, K_i = 27 nM; α_{2B}, K_i = 37 nM; α_{2C}, K_i = 6 nM; β_1, K_i = 5 μM; β_2, K_i = 1.65 μM; Muscarinic Acetylcholine Receptor M$_1$, K_i = 6 nM; M$_2$, K_i = 37 nM; M$_3$, K_i = 19 nM; M$_4$, K_i = 15 nM; M$_5$, K_i = 15 nM; Histamine Receptor H$_1$, K_i = 1 nM; H$_2$, K_i = 153 nM; H$_3$, K_i = 10 μM; H$_4$, K_i = 660 nM. **Key Pharmacokinetic Parameters:** *See* Appendix II in Goodman & Gilman's THE PHARMACOLOGICAL BASIS OF THERAPEUTICS, 12th Edition (Brunton, Chabner & Knollmann, eds.) McGraw-Hill Medical, New York (2011). **1.** Van Tol, Bunzow, Guan, *et al.* (1991) *Nature* **350**, 610. **2.** Hermann, Wetzel, Pestel, *et al.* (1996) *Biochem. Biophys. Res. Commun.* **225**, 957. **3.** Kuoppamaki, Syvalahti & Hietala (1993) *Eur. J. Pharmacol.* **245**, 179. **4.** Roth, Craigo, Choudhary, *et al.* (1994) *J. Pharmacol. Exp. Ther.* **268**, 1403. **5.** Bergqvist, Dong & Blier (1999) *Psychopharmacology (Berlin)* **143**, 89. **6.** Meltzer (1994) *J. Clin. Psychiatry* **55** Suppl. B, 47. **7.** Ardizzone, Bradley, Freeman & Dwyer (2001) *Brain Res.* **923**, 82.

CMP-*N*-acetylneuraminate

This modified sialic acid (FW$_{free-acid}$ = 614.46 g/mol; CAS 3063-71-6), also known as CMP-sialate, is an unstable metabolite that functions as a sialate donor. The dry solid will decompose to CMP and *N*-acetylneuraminate at –20°C at a rate of about 5-10% per day. It is completely hydrolyzed in aqueous solutions at pH 4 and 37°C within one hour (83% hydrolyzed in five minutes at 23°C in 10 mM HCl). CMP-sialate is more stable at neutral or alkaline pH; however, hydrolysis still occurs. Hence, solutions have to be freshly prepared. **Target(s):** *N*-acylneuraminate cytidylyltransferase (1,2); UDP-*N*-acetylglucosamine 2-epimerase (3-7). **1.** Rodríguez-Aparicio, Luengo, Gonzalez-Clemente & Reglero (1992) *J. Biol. Chem.* **267**, 9257. **2.** Samuels, Gibson & Miller (1999) *Biochemistry* **38**, 6195. **3.** Sommar & Ellis (1972) *Biochim. Biophys. Acta* **268**, 581. **4.** Kikuchi & Tsuiki (1980) *Tohoku J. Exp. Med.* **131**, 209. **5.** Hinderlich, Stäsche, Zeitler & Reutter (1997) *J. Biol. Chem.* **272**, 24313. **6.** Kikuchi & Tsuiki (1973) *Biochim. Biophys. Acta* **327**, 193. **7.** Effertz, Hinderlich & Reutter (1999) *J. Biol. Chem.* **274**, 28771.

CMPD101

This potent and selective GRK2/3 inhibitor (FW = 466.46 g/mol; CAS 865608-11-3; Soluble to 100 mM in DMSO), also known as 3-[[[4-methyl-5-(4-pyridinyl)-4*H*-1,2,4-triazol-3-yl]methyl]amino]-*N*-[[2-(trifluoromethyl)phenyl]methyl]benzamide, targets the G protein-coupled receptor kinases GRK2 (IC$_{50}$ = 54 nM) and GRK3 (IC$_{50}$ = 32 nM). Five amino acid residues that surround the inhibitor binding site are likely to contribute to inhibitor selectivity (1). Notably, GRK2 Overexpression is strongly linked to heart failure, and GRK2 is a target for the treatment of cardiovascular disease. Cmpd101 inhibits the desensitization of the G-protein-activated inwardly-rectifying potassium current evoked by receptor-saturating concentrations of methionine-enkephalin (Met-Enk), [d-Ala(2), N-MePhe(4), Gly-ol(5)]-

enkephalin (DAMGO), endomorphin-2, and morphine in rat and mouse locus coeruleus (LC) neurons (2). In LC neurons from GRK3-knockout mice, Met-Enk-induced desensitization is unaffected, suggesting a role for GRK2 in MOPr desensitization. Quantitative analysis of the loss of functional MOPrs following acute agonist exposure revealed that Cmpd101 only partially reversed MOPr desensitization. Inhibition of extracellular signal-regulated kinase 1/2, protein kinase C, c-Jun N-terminal kinase, or GRK5 does not inhibit the Cmpd101-insensitive component of desensitization. In HEK 293 cells, Cmpd101 produces almost complete inhibition of DAMGO-induced MOPr phosphorylation at Ser(375), arrestin translocation, and MOPr internalization (2). **1.** Thal, Yeow, Schoenau, Huber & Tesmer (2011) *Mol. Pharmacol.* **80**, 294. **2.** Lowe, Sanderson, Cooke, *et al.* (2015) *Mol. Pharmacol.* **88**, 347.

CNF-2024

This orally available synthetic Hsp90 inhibitor (FW = 308.76 g/mol; CAS 848695-25-0), also known as BIIB021 and systematically named 6-chloro-9-(4-methoxy-3,5-dimethylpyridin-2-ylmethyl)-9*H*-purin-2-ylamine, targets Hsp90α (K_i = 9 nM), also displaying excellent antiproliferative activity against various tumor cell lines (IC$_{50}$ = 30 nM in MCF7 cells (1). Like other Hsp90 inhibitors, CNF-2024 interacts with the N-terminal ATP-binding site to prevent ATP binding and stop the chaperone cycle, leading to degradation of multiple client proteins involved in tumor progression. CNF-2024 induces degradation of HER-2 *in vitro* (EC$_{50}$ = 38 nM) in MCF-7 cells, and also inhibits growth and promotes cell death in a variety of human tumor cells. Metabolic profiling revealed that BIIB021 is extensively metabolized primarily *via* hydroxylation of the methyl group (at position-M^7), *O*-demethylation (at position-M^2), and to a lesser extent by glutathione conjugation (at position-M^8 and position-M^9) (2). Position-M^7 was further metabolized to form the carboxylic acid (at position-M^3) and glucuronide conjugate (at position-M^4) (2). **1.** Kasibhatla, Hong, Biamonte, *et al.* (2013) *J. Med. Chem.* **50**, 2767. **2.** Xu, Woodward, Dai & Prakash (2013) *Drug Metab. Dispos.* **41**, 2133.

CNI-1493, *See* Semapimod

CNTO 328, *See* Siltuximab

CNX-419, *See* Rociletinib

CNX-774

This orally available, irreversible enzyme inhibitor (FW = 499.51 g/mol; CAS 1202759-32-7; Solubility: 100 mg/mL DMSO; <1 mg/mL H$_2$O), also known as 4-(4-((4-((3-acrylamidophenyl)-amino)-5-fluoropyrimidin-2-yl)amino)phenoxy)-*N*-methylpicolinamide, selective targets Bruton's tyrosine kinase (Btk), a member of the Tec family of kinases that is involved in B-lymphocyte development, differentiation, and B cell receptor signaling. CNX-774 forms a site-directed covalent bond with Cys-481, a non-conserved amino acid within the enzyme's active site. CNX-774 has potent inhibitory activity towards the intended target, Btk, while achieving remarkable specificity in a variety of assays designed to assess off-target reactivity towards abundant cellular thiols and blood proteins. **1**. Labenski, Chaturvedi, Evans, *et al.* (2011) 17th North. Amer. Meet. Int. Soc. Study Xenobiot. *Abst.* P211.

CO-1686, *See Rociletinib*

Coagulation Factor Va Fragment 323-331

This acidic nonapeptide (*Sequence*: EYFIAAEEV; FW = 1070.16 g/mol; Isoelectric Point = 3.67), corresponding to residues 323-331 of coagulation factor Va, inhibits coagulation factor Xa (K_i = 5.7 µM). The addition of an N-terminal tryptophanyl residue (*i.e.*, fragment 322-331) is also a strong inhibitor. **1**. Kalafatis & Beck (2002) *Biochemistry* **41**, 12715.

Coβ-4-Ethylphenyl-cob(III)alamin

This vitamin B_{12} analogue ($FW_{free-acid}$ = 1434.53 g/mol; Symbol: EtPhCbl) is a synthetic organometallic that binds to transcobalamin (TC), a plasma protein facilitating cellular cobalamin (Cbl) uptake. When assayed *in vitro*, key enzymes do not convert EtPhCbl to the active coenzyme forms of cobalamin, suggesting that administration of EtPhCbl may cause cellular Cbl deficiency. Plasma total Cbl concentration was higher in animals treated with EtPhCbl (129 nmol/L) than in CNCbl treated animals (88 nmol/L). However, the organ levels of total Cbl were significantly lower in animals treated with EtPhCbl compared to CNCbl treated animals or controls, notably in the liver (157 pmol/g *versus* 604 pmol/g, and 443 pmol/g, respectively). **1**. Mutti, Ruetz, Birn, Kräutler & Nexo (2013) *PLoS One* **8**, e75312.

Cobalamin

This name refers to the active form (FW = 1355.70 g/mol; CAS Registry Number = 13408-78-1) of vitamin B_{12}, where R = 5'-deoxyadenosyl (5'-deoxyadenosyl-cobalamin (*or* AdoCbl)), methyl (*or* methylcobalamin (MeCbl)); CN (*or* vitamin B_{12} (CNCbl)), or H_2O (aquocobalamin (*or* H_2OCbl)). **See also** *Adeninylpentylcobalamin; [5-(Adenosin-5'-O-yl)pentyl]cobalamin; Coβ-4-Ethylphenyl-cob(III)alamin; Cyanocobalamin; Methylcobalamin*

Cobalt Ions

These Group 9 (or Group VIIIA) ions (Symbol: Co^{2+} and Co^{3+}; Atomic Number: 27; Atomic Weight: 58.93; Ionic Radii = 0.72 Å for Co^{2+} and 0.63 Å for Co^{3+}) are found in several enzymes. Required by a dozen vitally important enzymes, vitamin B_{12}, *or* cobalamin, contains a central cobalt ion, with four of its six coordination sites provided by the corrin ring, a fifth by a dimethylbenzimidazole group, and the sixth serving as a versatile reaction center that is occupied by a cyano group (*i.e.*, Co–CN) in cyanocobalamin, a hydroxyl group (*i.e.*, Co–OH) in hydroxocobalamin, a methyl group (*i.e.*, Co–CH3) in methylcobalamin, or a 5'-deoxyadenosyl group (*i.e.*, Co–C5' atom of the deoxyribose) in adenosylcobalamin. Essential cobalt ions are also found in nitrile hydratase, methionine aminopeptidase-2, benzyloxycarbonylglycine hydrolase, *myo*-inosose-2 dehydratase, 3-dehydroquinate synthase, and isobutyryl-CoA mutase. Co^{2+} often substitutes for Zn^{2+} in zinc-dependent enzymes, allowing its spectroscopic and magnetic properties to serve as useful probes in otherwise spectrally silent zinc. Cobalt-nucleotide complexes have also proven to be extremely useful enzyme inhibitors that can be used to deduce key aspects about an enzyme's substrate binding order and active-site stereochemistry. (**See** *specific cobalt nucleotide; Cobalamin*) **1**. Cleland (1982) *Meth. Enzymol.* **87**, 159. **2**. Villafranca (1982) *Meth. Enzymol.* **87**, 180.

Cobimetinib

This orally active MEK inhibitor (FW = 531.31 g/mol; CAS 934660-93-2; Solubility: 100 mg/mL DMSO; < 1 mg/mL H_2O), also known as GDC-0973, XL-518, and Cotellic®, as well as the IUPAC name [3,4-difluoro-2-(2-fluoro-4-iodoanilino)phenyl]{3-hydroxy-3-[(2S)-piperidin-2-yl]azetidin-1-yl}methanone, targets MAP/ERK kinase, *or* MEK (IC_{50} = 4.2 nM), exhibiting strong cellular potency against various tumor types, particularly in wild-type BRAF and $BRAF^{V600E}$ (EC_{50} < 2 µM) or KRAS mutant cancer cell lines (1). KRAS- and BRAF-activating mutations drive tumorigenesis through constitutive activation of the MAPK pathway. Cotellic is indicated for the treatment of patients with unresectable or metastatic melanoma with a $BRAF^{V600E}$ or $BRAF^{V600K}$ mutation, in combination with vemurafenib. **Pharmacokinetics:** Cobimetinib is well absorbed (F_a = 0.88) and is extensively metabolized, with intestinal first-pass metabolism contributing in its disposition (2). Subsequent to a single 20-mg oral dose of [^{14}C]-cobimetinib in healthy male subjects, the major circulating species were unchanged cobimetinib (20.5%) and M16, the glycine conjugate of hydrolyzed cobimetinib, (18.3%) (2). Other circulating metabolites were minor, accounting for less than 10% of drug-related material in plasma. **1**. Hoeflich, Merchant, Orr, *et al.* (2012) *Cancer Res.* **72**, 210. **2**. Hatzivassiliou, Haling, Chen, *et al.* (2013) *Nature* **501**, 232. (*Erratum: Nature* (2013) **502**, 258.)

α-Cobratoxin

This nAChR-directed neurotoxin (MW = 7842.12 g/mol; CAS 769933-79-1), from the Indo-Chinese spitting cobra (*Naja siamensis*) and the Chinese cobra (*Naja atra*), is a potent antagonist that targets nicotinic acetylcholine receptor, causing paralysis by preventing acetylcholine binding to nAChR. **See also** *α–Bungarotoxin*

Cobra Venom Factor

This complement-activating protein complex (MW = 149 kDa; Abbreviation: CVF), also known as cobra venom anti-complementary protein, from *Naja naja kaouthia*, is a nontoxic disulfide-linked, three-chain glycoprotein resembling Complement C3b, the latter being the activated form of complement component C3 formed upon proteolytic cleavage of an Arg-Ser bond in C3's α-chain. As part of the innate immune system, the complement cascade helps (hence "complements") antibodies and phagocytic cells to clear pathogens from an infected organism. Like C3b, CVF forms a C3/C5 convertase with factor B in the presence of factor D and Mg^{2+}. While exhibiting the activity of C3b, CVF resembles the structure of C3b's degradation product C3c, which cannot form the active C3/C5 convertase. CVF is often used to "decomplement" laboratory animals (1), thereby facilitating the study of complement roles in host defense, immune response, and pathogenesis. The CVF crystal structure reveals a CUB domain (composed of Complement C1r/C1s, Uegf, and Bmp1) positioned in identical position to that in C3b (2). The similarly positioned CUB and slightly displaced C345c domains of CVF could play a key role in forming C3-convertase by providing primary binding sites for factor B (2). Humanized CVF shows great promise for therapeutically depleting complement in humans (3). **1**. Ballow & Cochrane (1969) *J. Immunol.* **103**, 944. **2**. Krishnan, *et al.* (2009) *Structure* (London) **17**, 611. **3**. Vogel, Fritzinger, Hew, Thorne & Bammert (2004) *Mol. Immunol.* **41**, 191.

Cocaine

This tropane alkaloid (FW$_{free-base}$ = 303.36 g/mol; CAS 50-36-2), named systematically as [1*R*-(*exo,exo*)]-3-(benzoyloxy)-8-methyl-8-azabicyclo[3.2.1]octane-2-carboxylic acid methyl ester, is prespared from extracts of coca leaves (*Erythroxylon coca*). First used as a local anesthetic by indigenous South Americans, cocaine's highly addictive nature soon became evident. It readily crosses the blood-brain barrier, where it binds to neurotransmitter transporters, thereby blocking re-uptake of serotonin, dopamine, and norepinephrine. As such, cocaine is often called a "triple-uptake inhibitor". **Pharmacokinetics:** Depending on its purity, chemical form (free base *versus* HCl salt), as well as the mode of its administration (*e.g.*, intravenous, intranasal, oral route, or as a suppository), cocaine concentration(s) in affected organs may differ greatly. Its elimination half-life is 0.5–1.1 hour after intravenous administration and 0.9–1.5 hour after nasal or oral administration. After nasal inhalation, cocaine's bioavailability is about 60%. Hydrolysis of the benzoyl group to the ecgonine methyl ester is catalyzed by plasma cholinesterase. The methyl ester within cocaine also spontaneously hydrolyzes, forming benzoyl ecgonine. *N*-Demethylation is mediated by microsomal cytochrome P-450 (CYP3A), producing norcocaine, which possesses a pharmacological activity comparable to that of cocaine. Accumulation of the latter can be prevented by CYP3A inhibitors, such as triacetyloleandomycin, cannabidiol, or gestodene. **Mechanism of Cocaine Reward:** The strong relationship between dopamine transporter inhibition and cocaine's behavioral effects suggests that cocaine's primary target is the dopamine (DA) transporter (DAT). Support for the DA Hypothesis for Cocaine Dependence comes from the observation the that the primary cocaine "receptor" and the DA transporter are one and the same protein. Molecular genetic studies on cocaine reward provided important insights regarding the DA Hypothesis. (Cocaine reward refers to the altered behavior of humans or other animals seeking to maintain access to cocaine.) While cocaine reward is not eliminated in mice carrying a knock-out for the DAT, it is eliminated in mice carrying combined gene knock-outs for both DAT and the serotonin transporter (SERT), as assessed in the conditioned place preference (CPP) paradigm (1). In the absence of DAT alone, there is greater participation in cocaine reward by serotonin (SERT) and norepinephrine (NET) transporters. Taken together, these studies indicate important requirements for several monoaminergic system genes to fully explain cocaine reward, in particular those expressed by dopamine and serotonin systems (1). **Target(s):** β-fructofuranosidase, *or* invertase) (2); dopamine and biogenic amine reuptake (3,6); acetylcholine receptor (4); HERG-encoded K⁺ channel (5); acetylcholinesterase (6); esterase, nonspecific (7); sphingomyelin phosphodiesterase, *or* sphingomyelinase (8); phospholipase A₁ (8); amphetamine-activated Rho GTPase signaling that mediates dopamine transporter internalization (9). **1**. Sora, Hall, Andrews, *et al.* (2001) *Proc. Natl. Acad. Sci. U.S.A.* **98**, 5300. **2**. Neuberg & Mandl (1950) *The Enzymes*, 1st ed., **1** (Part 1), 527. **3**. Maldonado-Irizarry, Stellar & Kelley (1994) *Pharmacol. Biochem. Behav.* **48**, 915. **4**. Chen, Banerjee & Hess (2004) *Biochemistry* **43**, 10149. **5**. Ferreira, Crumb, Carlton & Clarkson (2001) *J. Pharmacol. Exp. Ther.* **299**, 220. **5**. Blakely & Bauman (2000) *Curr. Opin. Neurobiol.* **10**, 328. **7**. Nachmansohn & Schneemann (1945) *J. Biol. Chem.* **159**, 239. **8**. Nassogne, Lizarraga, N'Kuli, *et al.* (2004) *Toxicol. Appl. Pharmacol.* **194**, 101. **9**. Wheeler, Underhill, Stolz, *et al.* (2015) *Proc. Natl. Acad. Sci. U.S.A.* **112**, E7138.

Codeine

This narcotic pain-relieve and cough suppressant (FW$_{free-base}$ = 299.37 g/mol; CAS 76-57-3), first isolated from opium in 1832, is an important antitussive agent that is only slightly analgesic and mildly habit forming. Codeine is readily prepared by *N*-methylation of morphine. Codeine is a potent agonist, albeit not as powerful as morphine, of the G-protein coupled μ-, δ-, and κ-opioid receptors. Its modest analgesic action may also be due, in part, to its conversion to morphine. **Codeine Metabolism:** In humans, codeine is converted to the following six metabolites: codeine-6-glucuronidem, 81 %; norcodeine, 2%, morphine, 0.6 %, morphine-3-glucuronide, 2%, morphine-6-glucuronide, 0.8 %, and normorphine, 2.4 % (1). Individuals lacking sufficient CYP2D6 fail to *O*-dealkylate codeine and cannot form morphine. The *in vivo* half-life of codeine is 1.5 hours, that of codeine-6-glucuronide is 2.8 hours, and that of morphine-3-glucuronide 1.7 hours (1). The plasma AUC for codeine-6-glucuronide is approximately 10x higher than that of codeine. Protein binding of codeine and codeine-6-glucuronide *in vivo* was 56 and 34 %, respectively. *The in vitro* protein binding of norcodeine was 24 %, compared to 47% for morphine, 24 % for normorphine, 27% for morphine-3-glucuronide, and 37% for morphine-6-glucuronide (1). **PK Summary:** Onset of action, 15–30 minutes; Oral Bioavailability, ~90%; Biological half-life, 2.5–3 hours. Duration of action: 4–6 hours. **Target(s):** glucuronosyl-transferase (2,3); oxidative phosphorylation (4); alcohol dehydrogenase (5). **1** Vree &Verwey-van Wissen (1992) *Biopharm. Drug Dispos.* **13**, 445. **2**. Puig & Tephly (1986) *Mol. Pharmacol.* **30**, 558. **3**. Tephly (1990) *Chem. Res. Toxicol.* **3**, 509. **4**. Chistiakov & Gegenava (1976) *Biokhimiia* **41**, 1272. **5**. Roig, Bello, Burguillo, Cachaza & Kennedy (1991) *J. Pharm. Sci.* **80**, 267.

Coenzyme A

This heat-stable coenzyme (FW = 767.54 g/mol; CAS 85-61-0; Water-soluble; Stable in the pH 4–7 range; λ$_{max}$ = 259.5 nm, ε = 16800 M⁻¹cm⁻¹; pK$_a$ = 9.6 for thiol group), commonly abbreviated CoA or CoASH, serves as an acyl carrier in a number of enzyme-catalyzed processes, forming a thiol ester intermediate. Coenzyme A should be stored dry at −20°C. Low or high pH results in hydrolysis: 31% of coenzyme A is hydrolyzed in 24 hours at pH 8 and 25°C. Aqueous solutions oxidized in air to the disulfide, and freshly prepared solutions should always be used. When present at physiologic concentration (*i.e.*, < 0.1 mM), Coenzyme A inhibits enzymes only weakly.

Coformycin

This transition-state inhibitor (FW = 284.27 g/mol; CAS 11033-22-0) derives its powerful inhibitory action from the sp^3-hybridized carbon that mimics the tetrahedral intermediate in the adenosine deaminase reaction. **Target(s):** adenosine deaminase, slow, tight-binding inhibition, $K_i = 10^{-13}$ M (1-8); AMP deaminase (9-11); adenine deaminase (12); adenosine kinase (13). **1.** Agarwal, Spector & Parks, Jr. (1977) *Biochem. Pharmacol.* **26**, 359. **2.** Sawa, Fukagawa, Homma, Takeuchi & Umezama (1967) *J. Antibiot.* **204**, 227. **3.** Castro & Britt (2001) *J. Enzyme Inhib.* **16**, 217. **4.** Wolfenden (1977) *Meth. Enzymol.* **46**, 15. **5.** Agarwal & Parks (1978) *Meth. Enzymol.* **51**, 502. **6.** Nygaard (1978) *Meth. Enzymol.* **51**, 508. **7.** Morrison (1982) *Trends Biochem. Sci.* **7**, 102. **8.** Philips, Robbins & Coleman (1987) *Biochemistry* **26**, 2893. **9.** Agarwal & Parks (1977) *Biochem. Pharmacol.* **26**, 663. **10.** Thakkar, Janero, Sharif, Hreniuk & Yarwood (1994) *Biochem. J.* **300**, 359. **11.** Sollitti, Merkler, Estupinan & Schramm (1993) *J. Biol. Chem.* **268**, 4549. **12.** Kidder & Nolan (1979) *Proc. Natl. Acad. Sci. U.S.A.* **76**, 3670. **13.** de Jong & Kalkman (1973) *Biochim. Biophys. Acta* **320**, 388.

Coformycin 5'-Monophosphate

This inhibitor (FW = 363.24 g/mol) is a transition-state analogue in the reaction catalyzed by AMP deaminase, $K_i = 55$ pM (1,2). The transition state for the hydrolysis of AMP by AMP deaminase has been characterized through the determination of multiple kinetic isotope effects, and the experimentally established transition state show a bond order of 0.8 to the attacking water nucleophile, a full bond order to the exocyclic 6-amino group, re-hybridization of C-6 of the purine ring to sp^3, and protonation of N-1 by Glu633 (3). The inferred transition state is on the path to formation of an unstable tetrahedral intermediate, in which the exocyclic amine undergoes rapid protonation, followed by its departure. In this mechanism, the highest energetic barrier on the reaction coordinate is attack by the zinc-activated water. The electrostatic potential surface for the purine ring in the transition state has been determined by molecular orbital calculations and compared to that of (*R*)-coformycin 5'-monophosphate (4). The electrostatic nature of the late transition state with a protonated amine leaving group differs both from the transition state determined by kinetic isotope effects and from that of (*R*)-coformycin analogues, providing further evidence that the nature of the enzyme-stabilized transition state for adenine deamination involves an early transition state with a partially bonded hydroxyl group. The observed tight binding inhibition by (*R*)-coformycin analogues as transition-state inhibitors results from the similarity of the partial charges on the inhibitors to that of the enzymatic

transition state stabilized by AMP deaminase (4). **1.** Frieden, Gilbert, Miller & Miller (1979) *Biochem. Biophys. Res. Commun.* **91**, 278. **2.** Freiden, Kurtz & Gilbert (1980) *Biochemistry* **19**, 5303. **3.** Merkler, Kline, Weiss & Schramm (1993) *Biochemistry* **32**, 12993. **4.** Kline & Schramm (1994) *J. Biol. Chem.* **269**, 22385.

COH29

This novel ribonucleotide reductase (RNR) inhibitor (FW = 422.24 g/mol) targets both recombinant and cellular human RNR enzymes (IC$_{50}$ = 18 μM) and overcomes hydroxyurea and gemcitabine resistance in cancer cells. COH29 effectively inhibited proliferation of most cell lines in the NCI 60 human cancer cell panel, with notable cytostatic action against ovarian cancer and leukemia and little effect on normal fibroblasts or endothelial cells. COH29 treatment reduced tumor growth, as compared to vehicle, when tested in mouse xenograft models of human cancer. COH29 binds near the C-terminal tail of RNR, as proposed by computer modeling and verified by site-directed mutagenesis and NMR. **1.** Zhou, Su, Hu, *et al.* (2013) *Cancer Res.* **73**, 6484.

Colcemid

This microtubule-directed alkaloid (FW = 371.43 g/mol; CAS 477-30-5), also known as demecolcine, is from the meadow saffron (*Colchicum autumnale*) and binds rapidly and reversibly to tubulin (in contrast to colchicine). This agent inhibits α,β-tubulin dimer polymerization, induces microtubule disassembly (occasionally attended by the onset of apoptosis), and inhibits mitotic and meiotic spindle formation. **Colcemid Washout:** Whereas colchicine's interactions with tubulin are virtually irreversible, colcemid is only weakly bound. This property commends its use in experiments requiring reversible disruption of the microtubule cytoskeleton within living cells. When treated with 5-20 μM colcemid, the microtubule cytoskeleton of most eukaryotic cells depolymerizes within 10-20 minutes. Upon rinsing colcemid-treated cells three times with several mL of colcemid-free culture medium (at 10-min intervals), colcemid effuses from the cell, followed by microtubule reassembly at microtubule organizing centers (MTOCs), such as the centrosome. Like colchicine, colcemid is photolabile and should be stored in the dark. (**See** *Colchicine*) **1.** Scott & Tomkins (1975) *Meth. Enzymol.* **40**, 273. **2.** Mareel & De Brabander (1978) *J. Natl. Cancer Inst.* **61**, 787. **3.** Bershadsky & Gelfand (1981) *Proc. Natl. Acad. Sci. U.S.A.* **78**, 3610. **4.** Ray, Bhattacharyya & Biswas (1984) *Eur. J. Biochem.* **142**, 577.

Colchicine

This toxic alkaloid (FW = 399.44 g/mol; CAS 64-86-8; Symbol: Clc; Solubility: 45 mg/mL H_2O, >150 mg/mL DMSO; M.P. = 142-150°C; Extremely Photosensitive), from the meadow saffron *Colchicum autumnale*, blocks tubulin polymerization (microtubule self-assembly), impairing mitosis and meiosis (1-10). **General Properties:** At very low doses, cochicine interferes with mitotic spindle formation, but need not stop cell division entirely. This property has been exploited to stably increase ploidy in certain plants that are then propagated vegetatively. Higher ploidy fish have also been produced in this manner. However, exposure of most animal cells to 10^{-7} M colchicine for 6-8 hours arrests cell division in metaphase (often called "colchicine mitosis" or "C-mitosis"). Cochicine induces apoptosis, activates the JNK/SAPK signaling pathway, and disrupts neutophil migration/ phagocytosis, the latter property commending its centuries-old use in the treatment of the exquisite arthritic pain associated with Gout. Colchicine itself exhibits a notoriously low therapeutic index, requiring cautious dosage and management. A colchicine transporter, if it exists, has yet to be demonstrated. The high cellular content of tubulin most likely accounts for the consistent finding that colchicine accumulates within cells at levels many times higher than its extracellular concentration (*Consult* Table II in Reference 8). **Mode of Tubulin (Tb) Binding:** Colchicine is a slow, tight-binding ligand of the αβ-tubulin dimer. The time-course for colchicine-tubulin association is complex, consisting of a relatively fast step, which is responsible for most of the reaction (as detected by attendant fluorescence changes), followed by one or more slower isomerizations that, when complete, increase the binding amplitude by 5-10% (4). Analysis of the fast step showed that, under pseudo-first-order conditions, the association rate constant increases linearly with colchicine or tubulin concentration, but deviates from linearity at high concentrations. The simplest binding mechanism is a slow ligand-induced conformational change, beginning as a weak Tb·Clc complex (characterized by an initial dissociation constant of 0.15 mM) to conformationally mature tubulin-drug complexes, ultimately reaching a stable binding equilibrium defined by a sub-picomolar dissociation constant. (*Other Agents That Bind at the Colchicine Site:* Combretastatin, Podophyllotoxin, 3,4,5-Trimethoxychalcones, Indolephenstatins) **Substoichiometric Action on MT Assembly & MT-dependent Processes:** Colchicine inhibits MT self-assembly at concentrations far below the solution concentration of tubulin dimers (5-7). Unlike vinblastine, which can bind directly to MT ends (8), colchicine first binds to αβ-Tb dimers. Colchicine blocks microtubule-dependent processes through the limited incorporation of αβ-Tubulin·Colchicine complex onto MT (+)-ends. The latter entails the limited co-polymerization of α,β-Tubulin and αβ-Tubulin·Colchicine (9,10), attended by the impaired ability of colchicine-containing microtubules to fulfill their roles in chromosome and organelle translocation. Lower concentrations of colchicine (*e.g.*, 0.2 µM) or α,β-Tubulin·Colchicine complex (*e.g.*, 0.1 µM) inhibit the exchange of tubulin dimers at MT ends, often without appreciable net polymer loss (8). **Cellular Microtubules (MTs) as Colchicine Targets:** When added in excess, colchicine has the classic effects of inhibiting tubulin polymerization or causing MT depolymerization. At lower concentrations, however, the very same agent suppresses MT assembly/disassembly dynamics in subtle ways, often with no noticeable change the abundance of assembled microtubules, but blocking mitosis and meiosis (11). Tubulin·Colchicine complex (*e.g.*, 0.1 µM), for example, inhibits tubulin exchange at MT ends without appreciable polymer loss (8). Colchicine also suppresses the dynamics of MT assembly/disassembly, greatly reducing the rates of MT growth as well as increasing the duration of non-growing/non-shortening states. Within living cells, highly dynamic spindle microtubules are, in fact, among the most sensitive targets for tubulin-directed drugs, suggesting that managing tubulin dimer addition/release at microtubule ends is highly relevant to anticancer chemotherapy (11). Analysis of microtubule growth times also supports the view that catastrophic MT disassembly also depends on the

"age" of MTs (12). Colchicine and vinblastine accelerate aging in a manner that depends on the presence of MT end-binding proteins. Of course, cellular MT dynamics also depends on the effects of microtubule-associated proteins, such as MAP2 and Tau proteins, which bind to and stabilize MTs. **Colchicine Toxicity:** In humans, an acute dose of ~0.8 mg colchicine per kilogram is typcally fatal. Nausea, vomiting, abdominal pain, and diarrhea, with a massive loss of fluid and electrolytes are the first clinical symptoms of colchicine poisoning. Stomach lavage and rapid gastric decontamination using activated charcoal are crucial in relieving toxicity. **Colchicine Metabolism:** Colchicine disposition involves both active biliary and renal excretion. In mammals, a substantial fraction undergoes hepatic demethylation prior to excretion. Tracer studies with [^{3}H]colchicine in human liver samples show that the formation of its main metabolites, 3-demethylcolchicine (3DMC) and 2-demethylcolchicine (2DMC), is linear with incubation time, cytochrome (P450) content, and substrate concentration (13). Upon incubation of colchicine (5 nM) with microsomes in the presence of a NADPH-generating system for 60 min, 9.8% and 5.5% of colchicine was 3DMC and 2DMC, respectively, the formation of which correlates with nifedipine oxidase activity, a marker of CYP3A4 activity, but not with the metabolic markers of CYP2A6, CYP2C19, CYP2C9, CYP2D6, and CYP2E1 activities. Chemical inhibition of CYP3A4 by preincubation with gestodene (40 µM) or troleandomycin (40 µM) reduced the formation of 3DMC and 2DMC by 70 and 80%, respectively, whereas quinidine, diethyldithiocarbamate, and sulfaphenazole had no inhibitory effect. Such findings suggest that co-administration of colchicine with CYP3A4 inhibitors/substrates may increase colchicine toxicity. **See also** *Colcemid, Combretastatin, Podophyllotoxin, Paclitaxel (Taxol), Vinblastine, Vincristine.* **Colchicine as a Treatment for Gouty Arthritis:** For centuries, oral colchicine has been a mainstay alternative for treating gout attacks, particularly for those unable to take anti-inflammatory painkillers. This alkaloid reduces the number of motile white blood cells responding to chemotactic cues emitted by pain sites. Colchicine is effective in breaking the cycle of cytokine-driven inflammasome response, thereby reducing the swelling and pain of a gout attack. Low-dose colchicine (1.8 mg, taken over 1 hour) should be at the ready for gout patients seeking to reduce pain. It is well tolerated in patients with acute gout. **Target(s):** *myo*-inositol 2-dehydrogenase (14); xanthine oxidase (15; urate-ribonucleotide phosphorylase (16); P-glycoprotein (multidrug-resistance protein), *or* xenobiotic transporting ATPase (17); thiamin-triphosphatase, weakly inhibited (18); α-tubulin *N*-acetyltransferase, inhibition due to MT depolymerization (19). **1.** Dustin (1984) *Microtubules*, 2nd. ed., pp. 482, Springer-Verlag, Berlin. **2.** Taylor (1965) *J. Cell Biol.* **25**, 125. **3.** Borisy & Taylor (1967) *J. Cell Biol.* **34**, 525. **4.** Garland (1978) *Biochemistry* **17**, 4266. **5.** Olmsted & Borisy (1973) *Biochemistry* **12**, 4282, **6.** Margolis & Wilson (1977) *J. Biol. Chem.* **252**, 7006. **7.** Margolis, Rauch & Wilson (1980) *Biochemistry* **19**, 5550. **8.** Jordan & Wilson (1998) *Meth. Enzymol.* **298**, 252. **9.** Sternlicht & Ringel (1979) *J. Biol. Chem.* **254**, 10540. **10.** Sternlicht, Ringel & Szasz (1979) *Biophys. J.* **42**, 255. **11.** Jordan & Wilson (2004) *Nature Rev. Cancer* **4**, 253. **12.** Mohan, Katrukha, Doodhi, *et al.* (2013) *Proc. Natl. Acad. Sci. U.S.A.* **110**, 8900. **13.** Tateishi, Soucek, Caraco, Guengerich & Wood (1997) **53**, 111. **14.** Larner (1962) *Meth. Enzymol.* **5**, 326. **15.** Roussos (1967) *Meth. Enzymol.* **12A**, 5. **16.** Laster & Blair (1963) *J. Biol. Chem.* **238**, 3348. **17.** Shapiro & Ling (1997) *Eur. J. Biochem.* **250**, 130. **18.** Iwata, Baba, Matsuda, Terashita & Ishii (1974) *Jpn. J. Pharmacol.* **24**, 825. **19.** Maruta, Greer & Rosenbaum (1986) *J. Cell Biol.* **103**, 571.

Colistins

These heterodetic ring-containing and cationic decapeptide antibiotics (FW = 1155.45 g/mol; CAS 1264-72-8; Source: *Bacillus colistinus*), also known as polymyxin E, characteristically contain L-2,4-diaminobutyrate (L-Dab) residues that play a role in their antibiotic activity. Their N-terminal residue is acylated with a 6-methyloctanoyl or 6-methylheptanoyl group. Colistins A and B are identical to polymyxins E_1 and E_2. **Mechanism of Antibiotic Action:** Colistins are polycationic surface-active detergents and mainly interacts electrostatically with the outer membrane of Gram-negative bacteria, displacing Mg^{2+} and Ca^{2+} from the membrane lipids, disrupting the outer membrane, and releasing essential lipopolysaccharides. Resultant changes in bacterial membrane permeability leads to leakage of cell contents as well as cell lysis and death. **Spectrum of Antibiotic Action:** These antibiotics display a relatively narrow antibacterial spectrum of activity, exhibiting significant activity against multi-resistant *Pseudomonas aeruginosa*, *Acinetobacter baumannii*, and *Klebsiella pneumoniae*. Colistins interact electrostatically with the outer membrane of Gram-negative bacteria and displace divalent cations. Colistin and polymyxin B are increasingly used as the last-line therapy for treating multidrug-resistant Gram-negative bacterial infections. *See also Polymyxins* **1**. Li, Nation, Milne, Turnidge & Coulthard (2005) *Int. J. Antimicrob. Agents* **25**, 11. **2**. Lim, Ly, Anderson, *et al.* (2010) *Pharmacotherapy* **30**, 1279.

Colesevelam

This oral cholesterol-reducing medication (FW = indefinite polymer; CAS 182815-44-7) consists of a random poly-allylamine that contains numerous cationic side-chains (consisting of 1-chloro-2,3-epoxypropane, [6-(allylamino)hexyl]trimethylammonium chloride and *N*-allyldecylamine) that trap bile acids and block their reabsorption. Colesevelam is safe and efficacious alone or in combination with HMG-CoA reductase inhibitors (statins) in reducing low-density lipoprotein cholesterol (LDL-C) levels. It may also be beneficial for patients who cannot tolerate other lipid lowering therapies, such as organ transplant recipients, cholestatic liver disesase, and end-stage renal disease. **1**. Steinmetz & Schonder (2005) *Cardiovasc. Drug Rev.* **23**, 15.

Combretastatin A$_4$

This tubulin-directed vascular disrupting agent, *or* VDA (FW = 316.34 g/mol; *Abbreviation*: CA4) and its phosphorylated pro-drug CA4P (FW = 396.32 g/mol; CAS 117048-59-6; Source: the arid shrub *Combretum caffrum*), systematically named 2-methoxy-5-[(Z)-2-(3,4,5-trimethoxy phenyl)vinyl]phenol *or* 3,4,5-trimethoxy-3'-hydroxy-4'-methoxy-stilbene, binds reversibly to the β-subunit of the tubulin αβ-heterodimer (IC_{50} = 2-3 μM), thereby preventing microtubule self-assembly with consequential vasculature disruption, particularly the abnormal vasculature associated with cancer and macular degeneration (1-4). Such action has been attributed to the ability of these agents to selectively destroy the central regions of tumors, areas widely believed to contain cell populations

resistant to cytotoxic therapies (2,3). The presence of its trimethoxyphenyl nucleus, a feature also found in colchicine and podophyllotoxin, first suggested that combretatstin may bind to the colchicine/podophyllotoxin binding site, albeit with reduced affinity. A virtue of combretastatin A$_4$ pharmacodynamics is its reversibility of inhibition and rapid *in vivo* clearance, the combined actions of which greatly limit neurotoxicity. The water-soluble pro-drug is combretastatin A$_4$ phosphate or fosbretabulin. Recent work on the more potent VDA, Combretistatin A$_1$ *cis* (*or* Oxi5703) indicates that its greater toxity arises from formation of highly reactive ortho-quinone, attended by free radical generation and ensuing apotosis (4). The additional phenolic hydroxyl group in CA$_1$ *versus* CA$_4$ is believed to make the former a moreeffective free radical generator. **Chemosensitizing Acute Myeloid Leukemia (AML) cells by Targeting Bone Marrow Endothelial Cells:** When AML cells are co-cultured with bone marrow endothelial cells (BMECs), a greater proportion of leukemic cells accumulate in G_0/G_1, suggesting a strategy of using combretastatins to force BMECs to lose their flat phenotype, to degrade their cytoskeleton, to cease growth, and to impair migration (5). Combretastatins also induce down-regulation of BMEC adhesion molecules known to tether AML cells, including VCAM-1 and VE-cadherin. When AML-BMEC co-cultures were treated with combretastatins, a significantly greater proportion of AML cells dislodged from BMECs and entered G_2/M cell cycle, suggesting enhanced susceptibility to cell cycle agents. Indeed, the combination of combretastatin and cytotoxic chemotherapy showed enhanced additive AML cell death (5). *In vivo* mice xenograft studies confirmed this finding by showing complete AML regression after treatment with combretastatins and cytotoxic chemotherapies. Beyond highlighting the pathologic role of BMECs in the leukemia microenvironment as a protective reservoir of disease, these results support a new strategy for using vascular-targeting combretastatins in combination with cytotoxic chemotherapy to treat AML (5). **1**. Pettit, Singh, Hamel, *et al.* (1989) *Experientia* **45**, 209. *Erratum*: *Experientia* **45**, 680. **2**. Dark, Hill, Prise, *et al.* (1997) Cancer Res. **57**, 1829. **3**. Chaplin, Horsman & Siemann (2006) *Curr. Opin. Investig. Drugs* **7**, 522. **4**. Rice, Campo, Lepler, Rojiana & Siemann (2011) *Microvasc. Res.* **81**, 44. **5**. Bosse, Wasserstrom, Meacham, *et al.* (2016) *Exp. Hematol.* **44**, 363.

Compound 666-15

This potent and selective CREB inhibitor (FW = 620.22 g/mol; CAS 1433286-70-4; Soluble to 100 mM in DMSO), also named 666-15 and 3-(3-aminopropoxy)-*N*-[2-[[3-[[(4-chloro-2-hydroxyphenyl)amino]carbonyl]-2-naphthalenyl]oxy]ethyl]-2-naphthalenecarboxamide hydrochloride, targets Cyclic adenosine monophosphate Response Element Binding protein-mediated gene transcription and cancer cell growth (IC_{50} = 81 nM). 666-15 also potently inhibits cancer cell growth without harming normal cells. In an *in vivo* MDA-MB-468 xenograft model, 666-15 completely suppresses the tumor growth without overt toxicity. **1**. Xie, Li, Kassenbrock, *et al.* (2015) *J. Med. Chem.* **58**, 5075.

Conantokins

These peptide toxins from marine gastropods (genus *Conus*) typically contain seventeen to twenty-seven amino acyl residues with a number of γ-carboxyglutamyl residues and no disulfide bonds. They bind to *N*-methyl-D-aspartate-sensitive glutamate channels. Conantokin G from *Conus geographus* induces a sleep-like state in mice, upon intracerebral injection. It is selective for NMDA receptors containing the NR2B subunit and has the sequence GEZZLQZNQZLIRZKSN–NH$_2$ where Z indicates γ-carboxyglutamyl residues. Related toxins: Conantokin-T: GEZZYQKM LZNLRZAEVKKNA–NH$_2$; Conantokin L: (GEZZVAKMAAZLARZDA VN–NH$_2$, and Conantokin R: GEZZVAKMAAZLARZNIAKGCKVN CYP). **1**. McIntosh, Olivera, Cruz & Gray (1984) *J. Biol. Chem.* **259**, 14343. **2**. Rivier, Galyean, Simon, *et al.* (1987) *Biochemistry* **26**, 8508. **3**.

Haack, Rivier, Parks, *et al.* (1990) *J. Biol. Chem.* **265**, 6025. **4**. Layer, Wagstaff & White (2004) *Curr. Med. Chem.* **11**, 3073. **5**. Jimenez, Donevan, Walker, *et al.* (2002) *Epilepsy Res.* **51**, 73. **6**. White, McCabe, Armstrong, *et al.* (2000) *J. Pharmacol. Exp. Ther.* **292**, 425. **7**. Teichert, Jimenez, Twede, *et al.* (2007) *J. Biol. Chem.* **282**, 36905.

Conantokin G

This *Conus* genus toxin and NMDA antagonist (FW = 2264.21 g/mol; CAS 93438-65-4; *Sequence*: GEXXLQXNQXLIRXKSN-NH$_2$, where X = γ-carboxyglutamyl residue; Solubility: 1 mg/mL H$_2$O; Symbol: Con G) a potent and selective antagonist of *N*-methyl-D-aspartate-evoked currents in murine cortical neurons (IC$_{50}$ = 480 nM). The slow onset of Con G block could be prevented by coapplication with high concentrations of NMDA or (*RS*)-3-(2-carboxypiperazine-4-yl)-propyl-1-phosphonate, the latter a competitive antagonist (1). In oocytes expressing NR1a/NR2B receptors, Con G produced a rightward shift in the concentration-response curve for NMDA, providing support for a competitive interaction with the NMDA-binding site (1). Con G blocks NMDA-evoked current in mouse cortical neurons (IC$_{50}$ = 480 nM). Con G is also a neuroprotective agent, with an excellent therapeutic window for the potential intervention against ischemic/excitotoxic brain injury (2). Single and multiple administrations of Con G preferentially attenuates the methamphetamine-induced increases in tissue levels of these neuropeptides in the substantia nigra (3). **1**. Donevan & McCabe (2000) *Mol. Pharmacol.* **58**, 614. **2**. Williams, Dave, Phillips, *et al.* (2000) *J. Pharmacol. Exp. Ther.* **294**, 378. **3**. Bush, McCabe & Hanson (2000) *Eur. J. Pharmacol.* **387**, 55.

Conantokin R

GEXXVAKMAAXLARXNIAKGCKVNCYP

This NMDA antagonist from the fish-hunting snail *Conus radiatus* (FW = 3098.40 g/mol; CAS 202925-60-8; Solubility: 1 mg/mL H$_2$O; Symbol: Con R; Structure: X = γ-carboxyglutamyl residue) is a potent, noncompetitive *N*-methyl-D-aspartate receptor antagonist (IC$_{50}$ = 93 nM) that reportedly shows NR2 subunit selectivity (1). Con R inhibits inward currents evoked by NMDA in central nervous system neurons (IC$_{50}$ = 350 nM) and exhibits broad anticonvulsant and antiparkinsonian activity *in vivo* at doses devoid of behavioral toxicity (2). The amino acid residue at sequence position-5 in the conantokin peptides partially governs subunit-selective antagonism of recombinant NMDA receptors (3). **1**. Blandl, Warder, Prorok & Castellino (2000) *FEBS Lett.* **470**, 139. **2**. White, *et al* (2000) *J. Pharmacol. Exp. Ther.* **292**, 425. **3**. Klein, *et al* (2001) *J. Biol. Chem.* **276**, 26860.

Conantokin T

This NMDA antagonist from the fish-hunting snail *Conus radiatus* (FW = 2683.8 g/mol; CAS 127476-26-0; Sequence: GEXXYQKMLXNL RXAEVKKNA, where X = γ-carboxyglutamyl residue Symbol: Con T;) is a noncompetitive *N*-methyl-D-aspartate receptor antagonist (IC$_{50}$ = 0.4 μM). Conantokin T inhibits Ca^{2+} influx and glutamate-induced toxicity in central nervous system neurons. It exhibits age-dependent physiological effects, inducing a sleep-like state in young mice, but hyperactivity in older mice. **1**. Haack, *et al* (1990) *J. Biol. Chem.* **265**, 6025. **2**. Klein, *et al* (1999) *Neuropharmacol.* **38**, 1819. **3**. Warder, *et al* (2001) *J. Neurochem.* **77**, 812.

Concanamycins

These cytotoxic macrolides (FW$_{Concanamycin-A}$ = 866.10 g/mol; R = C(=O)NH$_2$; CAS 98932-70-8; FW$_{Concanamycin-B}$ = 852.06 g/mol; R = H; CAS 81552-33-2; FW$_{Concanamycin-C}$ = 823.06 g/mol; CAS 81552-34-3) from *Streptomyces diastatochromogenes* inhibit vacuolar type H$^+$-ATPases and will induce apoptosis. Concanamycins A, B and C inhibit the acidification of rat liver lysosomes at 10^{-11}-10^{-9} M. **1**. Dröse, Bindseil, Bowman, *et al.* (1993) *Biochemistry* **32**, 3902. **2**. Nishihara, Akifusa, Koseki, *et al.* (1995) *Biochem. Biophys. Res. Commun.* **212**, 255. **3**. Fendler, Dröse, Altendorf & Bamberg (1996) *Biochemistry* **35**, 8009.

Concanavalin A

This homotetrameric plant lectin (MW = 110 kDa; CAS 11028-71-0) from jack bean has four metal ion-binding site (Mn^{2+} or Ca^{2+}) and four sites for α-D-mannopyranoses or α-D-glucopyranoses with unmodified hydroxyl groups at positions C-1, C-4, and C-6. Concanavalin A agglutinates erythrocytes and precipitates polysaccharides. Concanavalin A also binds to many glycoproteins (*e.g.*, immunoglobulin G). **Target(s):** procollagen C-endopeptidase (1,12); procollagen N-endopeptidase (2,13,14); thrombin (3); superoxide dismutase (4); γ-glutamyl hydrolase, weakly inhibited (7); 5'-nucleotidase (8 10,19-28); gametolysin (11); u-plasminogen activator, *or* urokinase (15); peptidyl-glycinamidase, *or* carboxamidopeptidase (16); peptidyl-dipeptidase A, *or* angiotensin I-converting enzyme (17); β-fructofuranosidase, *or* invertase (18); chlorophyllase (29); [β-adrenergic-receptor] kinase (30); procollagen galactosyltransferase (31); procollagen-proline 3-dioxygenase (32); procollagen-proline 4-dioxygenase (33). **1**. Kessler (1998) in *Handb. Proteolytic Enzymes*, p. 1236, Academic Press, San Diego. **2**. Bruckner-Tuderman (1998) in *Handb. Proteolytic Enzymes*, p. 1244, Academic Press, San Diego. **3**. Lundblad, Kingdon & Mann (1976) *Meth. Enzymol.* **45**, 156. **4**. Munkres (1990) *Meth. Enzymol.* **186**, 249. **5**. Jones & Johns (1916) *J. Biol. Chem.* **28**, 59. **6**. Sumner (1919) *J. Biol. Chem.* **37**, 137. **7**. Silink, Reddel, Bethel & Rowe (1975) *J. Biol. Chem.* **250**, 5982. **8**. Riordan & Slavik (1974) *Biochim. Biophys. Acta* **373**, 356. **9**. Stefanovic, Mandel & Rosenberg (1975) *J. Biol. Chem.* **250**, 7081. **10**. Williamson, Morre & Shen-Miller (1976) *Cell Tissue Res.* **170**, 477. **11**. Jaenicke, Kuhne, Spessert, Wahle & Waffenschmidt (1987) *Eur. J. Biochem.* **170**, 485. **12**. Hojima, van der Rest & Prockop (1985) *J. Biol. Chem.* **260**, 15996. **13**. Hojima, McKenzie, van der Rest & Prockop (1980) *J. Biol. Chem.* **264**, 11336. **14**. Hojima, Morgelin, Engel, *et al.* (1994) *J. Biol. Chem.* **269**, 11381. **15**. Takahashi, Kwaan, Koh & Tanabe (1992) *Biochem. Biophys. Res. Commun.* **182**, 1473. **16**. Simmons & Walter (1980) *Biochemistry* **19**, 39. **17**. Lanzillo & Fanburg (1976) *Biochim. Biophys. Acta* **445**, 161. **18**. Quiroga, Vattuone & Sampietro (1995) *Biochim. Biophys. Acta* **1251**, 75. **19**. Grondal & Zimmermann (1987) *Biochem. J.* **245**, 805. **20**. Camici, Fini & Ipata (1985) *Biochim. Biophys. Acta* **840**, 6. **21**. García-Ayllón, Campoy, Vidal & Muñoz-Delgado (2001) *J. Neurosci. Res.* **66**, 656. **22**. Turnay, Olmo, Navarro, Gavilanes & Lizarbe (1992) *Mol. Cell. Biol.* **117**, 23. **23**. Zekri, Harb, Bernard & Meflah (1988) *Eur. J. Biochem.* **172**, 93. **24**. Harb, Meflah, Duflos & Bernard (1983) *Eur. J. Biochem.* **137**, 131. **25**. Lamers, Heyliger, Panagia & Dhalla (1983) *Biochim. Biophys. Acta* **742**, 568. **26**. Dornand, Bonnafous & Mani (1978) *Eur. J. Biochem.* **87**, 459. **27**. Willadsen, Nielsen & Riding (1989) *Biochem. J.* **258**, 79. **28**. Picher, Burch, Hirsh, Spychala & Boucher (2003) *J. Biol. Chem.* **278**, 13468. **29**. Terpstra (1982) *Biochim. Biophys. Acta* **681**, 233. **30**. Wang & Liu (2003) *J. Biol. Chem.* **278**, 15809. **31**. Risteli (1978) *Biochem. J.* **169**, 189. **32**. Tryggvason, Majamaa, Risteli & Kivirikko (1979) *Biochem. J.* **183**, 303. **33**. Guzman, Berg & Prockop (1976) *Biochem. Biophys. Res. Commun.* **73**, 279.

(*S*)-Coniine

This alkaloid and peripheral neurotoxin (FW = 127.23 g/mol; CAS 3238-60-6 (*R/S*-mixture), 5985-99-9 (*R*-enantiomer), and 458-88-8 (*S*-enantiomer); D$^{0°}$ = 0.8626 and D$^{19°}$ = 0.8438, refractive index, $n^{23°}_D$ = 1.4505, dextrorotatory, with [α]$^{19°}_D$ = +15.7°), also named (2*S*)-2-propylpiperidine, is the chief toxic constituent in Poison Hemlock (*Conium maculatum*). A dose of <100 mg is usually fatal in humans, and similar per-kilogram body-weight toxicities obtain for livestock. Respiratory paralysis usually occurs within 20-30 minutes, followed by death in 2-3 hours. Coniine blocks the nicotinic receptor on the post-synaptic membrane of the neuromuscular junction, but there is some CNS involvement at higher doses. In 399 BC, Socrates is said to have been executed with a drink containing hemlock extract. **1.** Vetter (1998) *Toxicon.* **36**, 13.

Conoidin A

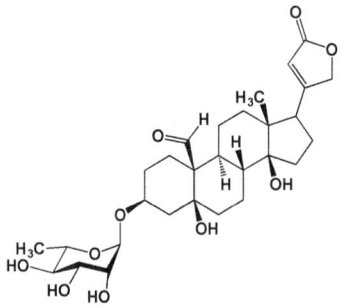

This potent antimalarial (FW = 347.99 g/mol; CAS 18080-67-6; IUPAC Name: 2,3-bis(bromo-methyl)-1,4-dioxidequinoxaline; Symbol: BBMQ) is an inhibitor of host cell invasion by the protozoan parasite *Toxoplasma gondii* and exerts potent cytocidal action against cultured *Plasmodium falciparum*. Conoidin A irreversibly inhibits peroxiredoxins, a widely conserved and important family of enzymes that function in antioxidant defense and participate in redox signaling pathways. Peroxiredoxins use a reactive, peroxidatic cysteine residue to reduce ROS such as H_2O_2, resulting in the formation of a cysteine sulfenic acid (Cys-SOH). Changes in human PrxII expression are associated with cancer, cardiovascular dysfunction and neurodegeneration, suggesting these redox enzymes are attractive druggable targets. Conoidin A inhibits (IC_{50} = 23 μM) *Toxoplasma gondii* peroxiredoxin II (*or* TgPrxII) through binding to and covalent modification of its peroxidatic Cysteine-47 (1). At 5 μM, conoidin A can also inhibit the glucose oxidase-mediated hyperoxidation of mammalian peroxiredoxin I and II, but not peroxiredoxin III. The compound also inhibits the hyperoxidation of mammalian PrxII in response to oxidative stress. **1.** Haraldsen, Liu, Botting, *et al.* (2009) *Org. Biomol. Chem.* **7**, 3040.

α-Conotoxin AuIB

GCCSYPPCFATNPDC-NH₂

This oligopeptides toxin (FW = 1572.76 g/mol; CAS 216299-21-7) from the venom of the "court cone," *Conus aulicus*, selectively blocks $\alpha_3\beta_4$ nAChRs expressed in *Xenopus* oocytes with an IC_{50} of 0.75 μM, a k_{on} of 1.4 x 10^6 min⁻¹M⁻¹, a k_{off} of 0.48 min⁻¹, and a K_d of 0.5 μM. Furthermore, α-conotoxin AuIB blocks the $\alpha_3\beta_4$ receptor with >100-fold higher potency than other receptor subunit combinations, including $\alpha_2\beta_2$, $\alpha_2\beta_4$, $\alpha_3\beta_2$, $\alpha_4\beta_2$, $\alpha_4\beta_4$, and $\alpha_1\beta_1\gamma\delta$. **1.** Luo, *et al.* (1998) *J. Neurosci.* **18**, 8571. **2.** Nai, *et al.* (2003) *Mol. Pharmacol.* **63**, 311. **3.** Park, *et al.* (2006) *Pflugers Arch.* **452**, 775.

α-Conotoxin EI

RDXCCYHPTCNMSNPQIC-NH₂

This *Conus* peptide (FW = 2093.40 g/mol; CAS 170663-33-9) is a selective antagonist of neuromuscular $\alpha_1\beta_1\gamma\delta$ nicotinic receptors (1). α-Conotoxin EI displays selectivity for α/δ sites over α/γ sites in *Torpedo*. α-Conotoxin EI has the same disulfide framework as α4/7 conotoxins targeting neuronal nicotinic acetylcholine receptors, but antagonizes the neuromuscular receptor as do the α3/5 and α A conotoxins. **1.** Martinez, *et al.* (1995) *Biochemistry* **34**, 14519. **2.** Park, *et al.* (2001) *J. Biol. Chem.* **276**, 49028.

ω-Conotoxin GVIA

CKSXGSSCSXTSYNCCRSCNXYTKRCY-NH₂

This *Conus* peptide (FW = 3037 g/mol; CAS 106375-28-4) targets N-type calcium channels (IC_{50} = 0.15 nM). **1.** Sato, et al (1993) *Biochem. Biophys. Res. Comm.* **194**, 1292. **2.** Wright & Angu (1997) *J. Cardiovasc. Pharmacol.* **30**, 392.

Ω-Conotoxin MVIIA

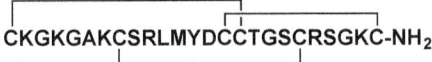

CKGKGAKCSRLMYDCCTGSCRSGKC-NH₂

This 25-residue, disulfide-bridged marine toxin (FW = 2639.13 g/mol; CAS 107452-89-1) from the venom of the sea snail *Conus geographus* that

binds to neuronal N-type calcium channels (1-3). This conotoxin forms a compact folded structure, presenting a loop between Cys-8 and Cys-15 that contains a set of residues critical for channel binding (4). The loop does not have a unique defined structure, nor is it intrinsically flexible. Indeed, multi-dimensional NMR is consistent with a short triple-stranded antiparallel β-sheet (5). The overall β-sheet topology is the same as that reported for Ω-conotoxin GVIA, another N-type calcium channel blocker. The orientation of β-stranded structure is similar to each other, suggesting that the conserved disulfide bond combination is essential for the molecular folding. **Mode of Inhibitory Action:** Ω-Conotoxin produces persistent block of L- and N-type Ca^{2+} channels in dorsal root ganglion neurons but only transiently inhibits T-type channels. Its actions are neuron-specific, blocking high-threshold Ca^{2+} channels in sensory, sympathetic, and hippocampal neurons of vertebrates, but not in cardiac, skeletal, or smooth muscle cells (1-3). The toxin interacts with an external site near the Ca^{2+} channel. **Clinical Aspects:** The potency and selectivity of ω-conotoxin on voltage-gated Ca^{2+} channels commend its use in the treatment of neuropathic pain states, especially those in which Ca^{2+} channel activity is characteristically aberrant (6). While ω-conotoxins show analgesic efficacy, the means of drug delivery and adverse effects limit its use. Synthetic Ω-conotoxin (Ziconotide®) is an FDA-approved peptidic drug indicated for the treatment of severe chronic pain, when administered intrathecally. **1.** Feldman, Olivera & Yoshikami (1987) *FEBS Lett.* **214**, 295. **2.** Olivera, Cruz, de Santos, *et al.* (1987) *Biochemistry* **26**, 2086. **3.** McCleskey, Fox, Feldman *et al.* (1987) *Proc Natl Acad Sci U.S.A.* **84**, 4327. **4.** Atkinson, Kieffer, Dejaegere, Sirockin & Lefèvre (2000) *Biochemistry* **39**, 3908. **5.** Kohno, Kim, Kobayashi, *et al.* (1995) *Biochemistry* **34**, 10256. **6.** Hannon & Atchison (2113) *Marine Drugs* **11**, 680.

μ-Conotoxin PIIIA

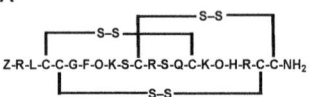

Z-R-L-C-C-G-F-O-K-S-C-R-S-Q-C-K-O-H-R-C-C-NH₂

This marine predatory toxin (FW = 2204 g/mol; *Symbol*: μ-PIIIA), also known as μ-Conotoxin P3.7, from *Conus pupurascens* is a sodium channel blocker that reversibly blocks mammalian skeletal muscle voltage-gated Na^+ channels, *or* $Na_v1.4$ (IC_{50} = 44 nM), Type II ($Na_v1.2$) and PN1 ($Na_v1.7$) neuronal voltage-gated Na^+ channels (IC_{50} = 640 nM) by binding to Site-I domain of the channel. is of exceptional interest because, unlike the previously characterized μ-conotoxin GIIIA, μ-PIIIA irreversibly blocks amphibian muscle Na^+ channels, providing a useful tool for probing synaptic electrophysiology and for discriminating among different sodium channel subtypes. *Note:* The Z and O in the above structure represent N-terminal pyroglutamate and 4-*trans*-hydroxyproline, respectively. **1.** Shon, Olivera, Watkins, *et al.* (1998) *J. Neurochem.* **18**, 4473.

Conotoxin Vc1.1, *See* ACV 1

Conus Toxins

This collection of ~100,000 neuroactive compounds, all derive from 500-700 different cone snail species, represents one of the most versatile and structurally diverse groups of marine natural products and metabolic inhibitors. They allow the relatively immotile *Conus* members to paralyze and prey upon faster-moving fish and arthropods. (*See also Conantokins; μ Conotoxin PIIIA; Ω-Conotoxin MVIIA*). **1.** Han, Teichert, Olivera & Bulaj (2008) *Curr. Pharm. Design* **14**, 2462.

Convallatoxin

This cardiac glycoside (FW = 550.65 g/mol) from flowers of Lily-of-the-Valley (*Convallaria majalis*) is structurally similar to digitoxin.

Convallatoxin has also been reported in *Ornithogalum umbellatum* (sleepydick, also known as star-of-Bethlehem) and *Antiaris toxicaria* (ipoh tree). Convalllatoxin consists of L-rhamnose and strophanthidin. **Target(s):** acetylcholinesterase, weakly inhibited (1); Na$^+$/K$^+$-exchanging ATPase (2). **1**. Hanke, Nelson & Baskin (1991) *J. Appl. Toxicol.* **11**, 119. **2**. Ozaki, Nagase & Urakawa (1985) *Eur. J. Biochem.* **152**, 475.

Copanlisib

This intravenous pan-Class-I PI3K inhibitor (FW = 480.52 g/mol; CAS 1032568-63-0), also known as BAY 80-6946 and 2-amino-*N*-[2,3-dihydro-7-methoxy-8-[3-(4-morpholinyl)propoxy]imidazo[1,2-*c*]quinazolin-5-yl]-5-pyrimidinecarboxamide, targets PI3Kα (IC$_{50}$ = 0.5 nM), PI3Kβ (IC$_{50}$ = 3.7 nM), PI3Kγ (IC$_{50}$ = 6.4 nM), and PI3Kδ (IC$_{50}$ = 3.7 nM) in cell-free assays. BAY 80-6946 inhibits AKT phosphorylation by PI3Kα in cells, with 10x selectivity relative to PI3Kβ (1). BAY 80-6946 shows superior antitumor activity (>40x) in PIK3CA mutant and/or HER2 overexpression, as compared with HER2-negative and wild-type PIK3CA breast cancer cell lines. In addition, BAY 80-6946 induces apoptosis in a subset of tumor cells with aberrant activation of PI3K as a single agent (1). *In vivo*, single IV administration of BAY-80-6946 exhibits higher exposure and prolonged inhibition of pAKT levels in tumors versus plasma. BAY 80-6946 is efficacious in tumors with activated PI3K. when dosed either continuously or intermittently. **1**. Liu, Rowley, Bull, *et al.* (2013) *Mol. Cancer Ther.* **12**, 2319. **2**. Göckeritz, Kerwien, Baumann, *et al.* (2015) *Int. J. Cancer* **137**, 2234.

Copper Ions

These Group 11 (or Group IB) metal ions Cu^{1+} and Cu^{2+} (Atomic Number = 29; Atomic Weight = 63.54) have valences of 1 and 2, with ionic radii of 0.96 and 0.72 Å. The radioisotope with the longest half-life is ^{67}Cu: 2.580 days. Modest levels of ionic copper have been found to be essential for living organisms (1). Ligation schemes have been classified (2): **Type-I Copper Centers** – These proteins have a single copper atom coordinated by two histidine residues and a cysteine residue in a trigonal planar structure, with a fourth variable ligand occupying an axial site. (*Examples:* Amicyanin, Plastocyanin, and Pseudoazurin). **Type-II Copper Centers** – These proteins have a square planar coordination scheme with ligands having N or O atoms. **Type-III Copper Centers** – These proteins have a pair of copper centers, each coordinated by three histidine residues. (*Examples:* Hemocyanin and Tyrosinase). **Binuclear Copper A Centers** – These proteins havehe two copper atoms are coordinated by two histidines, one methionine, a protein backbone carbonyl oxygen, and two bridging cysteine residues. (*Examples:* Cytochrome *c* oxidase and Nitrous-oxide reductase). **Copper B Centers** – These proteins have a copper atom coordinated to three histidines in trigonal pyramidal geometry. (*Example:* cytochrome *c* oxidase). **Tetranuclear Copper Z Centers** – These proteins have four copper ions are coordinated by as many as seven histidine residues and bridged by a sulfur atom. (*Example:* Nitrous-oxide reductase). Copper ions also play catalytically essential roles in the following enzymes: Amine oxidase (copper-containing), L-Ascorbate oxidase, Bilirubin oxidase, Catechol oxidase, Dopamine β-monooxygenase, 2-Furoyl-CoA dehydrogenase, Galactose oxidase, Hexose oxidase, Indole 2,3-dioxygenase, Laccase, Lysyl oxidase, Nitrite reductase (NO-forming), Nitrous-oxide reductase, Peptidylglycine monooxygenase, Polyphenol oxidase, Quercetin 2,3-dioxygenase, Cu,Zn-Superoxide dismutase, and Tyrosinase (3,4). Even so, numerous other enzymes (>480 in the author's literature survey) are inhibited by cupric or cuprous ions. Wilson's disease (also known as Hepatolenticular Degeneration) is an autosomal recessive disorder in the ATP7B gene, resulting in uncontrolled copper accumulation in various tissues and manifested by liver failure as well as neurological or psychiatric symptoms. Recent work shows that Cu^{2+} potently blocks

steady-state GABA currents mediated by extrasynaptic δ subunit-containing GABA$_A$ receptors with an IC$_{50}$ of 65 nM (compare with IC$_{50}$ = 85 μM for synaptic γ subunit-containing γ-GABA$_A$ receptors (5). **1**. DiSilvestro & Cousins (1983) *Annu. Rev. Nutrition* **3**, 261. **2**. Holm, Kennepohl & Solomon (1996), Chemical Reviews **96**, 2239. **3**. Purich & Allison (2000) *The Enzyme Reference*, Academic Press, New York. **4**. Purich (2010) *Enzyme Kinetics: Catalysis & Control*, Academic Press, New York. **5**. McGee, Houston & Brickley (2013) *J. Neurosci.* **33**, 13431.

Coprine

This disulfiram-like constituent (FW = 202.21 g/mol; CAS 58919-61-2) from the mushroom *Coprinus atramentarius*, also named *N*-(1-hydroxycyclopropyl)-L-glutamine, inhibits the low-K_m form of rat liver aldehyde dehydrogenase *in vivo*, increasing the acetaldehyde level in blood during ethanol metabolism. Coprine *per se* does not inhibit the low-K_m enzyme *in vitro*; however, its hydrolytic product, 1-aminocyclo-propanol, is a potent inhibitor both *in vitro* and *in vivo*. A rapid onset of inhibition is observed after administration of coprine, and the inhibition is long-lasting. Such findings suggest that 1-aminocyclopropanol is responsible for the inhibition caused by coprine *in vivo*. **1**. Tottmar & Lindberg (1977) *Acta Pharmacol. Toxicol. (Copenhagen)* **40**, 476. **2**. Pettersson & Tottmar (1982) *J. Neurochem.* **39**, 628. **3**. Wiseman & Abeles (1979) *Biochemistry* **18**, 427. **4**. Lindberg, Bergman & Wickberg (1977) *J. Chem. Soc. [Perkin 1]* **1977** (6), 684.

Coproporphyrin I

The protoporphyrin precursor (FW$_{free-acid}$ = 654.72 g/mol) is present at elevated in individuals with certain porphyrias; note that trace amount are observed in normal human adults. **Target(s):** glutathione *S*-transferase (1); heme oxygenase, inhibited by the iron complex (2); uroporphyrinogen decarboxylase (3); uroporphyrinogen-III synthase (4). **1**. Smith, Nuiry & Awasthi (1985) *Biochem. J.* **229**, 823. **2**. Frydman, Tomaro, Buldain, *et al.* (1981) *Biochemistry* **20**, 5177. **3**. Jones & Jordan (1993) *Biochem. J.* **293**, 703. **4**. Levin (1971) *Biochemistry* **10**, 4669.

Coproporphyrin III

The heme precursor (FW$_{free-acid}$ = 654.72 g/mol) is elevated in patients with certain porphyrias. **Target(s):** coproporphyrinogen-III oxidase (1); lactoylglutathione lyase, *or* glyoxalase I (2); glutathione *S*-transferase (3); ferrochelatase (4); uroporphyrinogen decarboxylase (5); uroporphyrinogen III synthase (6); heme oxygenase, inhibited by the iron complex (7). **1**.

Labbe (1997) *Meth. Enzymol.* **281**, 367. **2**. Douglas & Sharif (1983) *Biochim. Biophys. Acta* **748**, 184. **3**. Smith, Nuiry & Awasthi (1985) *Biochem. J.* **229**, 823. **4**. Dailey, Jones & Karr (1989) *Biochim. Biophys. Acta* **999**, 7. **5**. Jones & Jordan (1993) *Biochem. J.* **293**, 703. **6**. Levin (1971) *Biochemistry* **10**, 4669. **7**. Frydman, Tomaro, Buldain, *et al.* (1981) *Biochemistry* **20**, 5177.

Coptisine

This cationic alkaloid (FW$_{chloride}$ = 355.78 g/mol; CAS 3486-66-6), first isolated from the roots of *Coptis japonica*, is a component of a number of traditional medicines. **Target(s):** alcohol dehydrogenase (1); microbial collagenase, *Clostridium histolyticum* (2); NADH dehydrogenase, Complex I (3); monoamine oxidase A, K_i = 3.3 μM (4); H$^+$/K$^+$-ATPase (5); elastase (6). **1**. Kovar, Stejskal & Matyska (1985) *J. Enzyme Inhib.* **1**, 35. **2**. Tanaka, Metori, Mineo, *et al.* (1991) *Yakugaku Zasshi* **111**, 538. **3**. Barreto, Pinto, Arrabaca & Pavao (2003) *Toxicol. Lett.* **146**, 37. **4**. Ro, Lee, Lee & Lee (2001) *Life Sci.* **70**, 639. **5**. Satoh, Nagai, Seto & Yamauchi (2001) *Yakugaku Zasshi* **121**, 173. **6**. Tanaka, Metori, Mineo, *et al.* (1993) *Planta Med.* **59**, 200.

Coralyne

This toxic alkaloid and topoisomerase I inhibitor (FW$_{chloride}$ = 399.87 g/mol; CAS 6872-73-7 (free base)) is a DNA-intercalating agent, preferentially interacting with A:T base pairs and thereby interfering with the action of DNA-directed enzymes. Coralyne also binds preferentially to single-stranded poly(A), $K_{d,apparent}$ = 0.55 × 10^{-6} M, and a maximal binding stoichiometry b_{max} of one coralyne per four bases. (1). Binding to poly(A) is enthalpically driven, inducing a secondary structure with a melting temperature of 60 °C. **Target(s):** reverse transcriptase (2,3); DNA polymerase (2,3); RNA polymerase, moderately inhibited (2,3); polyadenylate polymerase, moderately inhibited (1); catechol *O*-methyltransferase (4); tRNA methyltransferase (4); DNA topoisomerase I, main target (5-7); DNA topoisomerase II, mildly inhibited (6,7). **1**. Xing, Song, Ren, *et al.* (2005) *FEBS Lett.* **589**, 5035. **2**. Sethi (1976) *Cancer Res.* **36**, 2390. **3**. Sethi (1977) *Ann. N. Y. Acad. Sci.* **284**, 508. **4**. Lee, MacFarlane, Zee-Cheng & Cheng (1977) *J. Pharm. Sci.* **66**, 986. **5**. Wang, Rogers & Hecht (1996) *Chem. Res. Toxicol.* **9**, 75. **6**. Gatto, Sanders, Yu, *et al.* (1996) *Cancer Res.* **56**, 2795. **7**. Meng, Liao & Pommier (2003) *Curr. Top. Med. Chem.* **3**, 305.

Cordycepin

This uncommon deoxynucleoside (FW = 251.25g/mol; M.P. = 225-226°C; λ$_{max}$ (in ethanol) at 260 nm, ε = 14600 M^{-1}cm^{-1}), also known as 3'-deoxyadenosine, from *Cordyceps militaris* is metabolically converted to its 5'-triphosphate derivative (***See also*** *Cordycepin 5'-Triphosphate*), which inhibits the synthesis of mRNA polyA tails, thereby blocking mRNA processing and protein biosynthesis. Cordycepin has also been reported to

inhibit purine, pyrimidine, and DNA biosynthesis. **Target(s):** adenosine kinase (2,13,14); nucleic acid methylation (3); polynucleotide adenylyltransferase, inhibited with excess cordycepin (4); adenosylhomocysteinase (5); RNA polymerase, via the triphosphate (6); rhodopsin kinase (7); GMP synthetase (8); tyrosyl-tRNA synthetase, weakly inhibited by the free nucleoside (9); adenosine deaminase (10); adenosine Nucleosidase (11); deoxycytidine kinase (12). **1**. Scott & Tomkins (1975) *Meth. Enzymol.* **40**, 273. **2**. Drabikowska, Halec & Shugar (1985) *Z. Naturforsch. [C]* **40**, 34. **3**. Kredich (1980) *J. Biol. Chem.* **255**, 7380. **4**. Maale, Stein & Mans (1975) *Nature* **255**, 80. **5**. Fabianowska-Majewska, Duley & Simmonds (1994) *Biochem. Pharmacol.* **48**, 897. **6**. Shigeura, Boxer, Meloni & Sampson (1966) *Biochemistry* **5**, 994. **7**. Palczewski, McDowell & Hargrave (1988) *J. Biol. Chem.* **263**, 14067. **8**. Spector & Beeham (1975) *J. Biol. Chem.* **250**, 3101. **9**. Santi & Peña (1973) *J. Med. Chem.* **16**, 273. **10**. Daddona, Wiesmann, Lambros, Kelley & Webster (1984) *J. Biol. Chem.* **259**, 1472. **11**. Guranowski & Schneider (1977) *Biochim. Biophys. Acta* **482**, 145. **12**. Krenitsky, Tuttle, Koszalka, *et al.* (1976) *J. Biol. Chem.* **251**, 4055. **13**. Palella, Andres & Fox (1980) *J. Biol. Chem.* **255**, 5264. **14**. Datta, Das, Sen, *et al.* (2006) *Biochem. J.* **394**, 35.

Cordycepin 5'-Triphosphate

This nucleotide (FW$_{free-acid}$ = 488.16 g/mol; Soluble in water and ethanol), also called 3'-deoxyadenosine 5'-triphosphate, terminates elongation of mRNA polyadenylate tails requir, owing to the absence of the 3'-hydroxyl group. **Target(s):** polynucleotide adenylyltransferase, *or* poly(A) polymerase (1,3,11,13,14,22-28); DNA primase (2); protein kinase (4); DNA-directed RNA polymerase (5,9-12,16,29); translocation (6); isoleucyl-tRNA synthetase (7); picornavirus RNA polymerase (8); RNA polymerase I and II (9); 2',5'-oligoadenylate synthetase (15); DNA ligase (ATP), *Vaccinia* virus (17); phenylalanyl-tRNA synthetase (18,22); arginyl-tRNA synthetase (19,20,22); seryl-tRNA synthetase (21); lysyl-tRNA synthetase (19,22); isoleucyl-tRNA synthetase (22,23); threonyl-tRNA synthetase (19); DNA nucleotidylexotransferase, *or* terminal deoxyribonucleotidyl-transferase (21); influenza virus RNA polymerase (26); 1-phosphatidylinositol 4-kinase (30). **1**. Edmonds (1990) *Meth. Enzymol.* **181**, 161. **2**. Izuta, Kohsaka-Ichikawa, Yamaguchi & Saneyoshi (1996) *J. Biochem.* **119**, 1038. **3**. Shuman & Moss (1988) *J. Biol. Chem.* **263**, 8405. **4**. Berman (1988) *Amer. J. Trop. Med. Hyg.* **38**, 298. **5**. Nolan & Fehr (1987) *Antimicrob. Agents Chemother.* **31**, 1734. **6**. Dietrich, Teissere, Job & Job (1985) *Nucl. Acids Res.* **13**, 6155. **7**. Freist, F. von der Haar & Cramer (1981) *Eur. J. Biochem.* **119**, 151. **8**. Panicali & Nair (1978) *J. Virol.* **25**, 124. **9**. Desrosiers, Rottman, Boezi & Towle (1976) *Nucl. Acids Res.* **3**, 325. **10**. Chamberlin (1974) *The Enzymes*, 3rd ed., **10**, 333. **11**. Maale, G. Stein & R. Mans (1975) *Nature* **255**, 80. **12**. Shigeura & Boxer (1964) *Biochem. Biophys. Res. Commun.* **17**, 758. **13**. Edmonds (1982) *The Enzymes*, 3rd ed. (Boyer, ed.), **15**, 217. **14**. Blakesley & Boezi (1975) *Biochim. Biophys. Acta* **414**, 133. **15**. Ball (1982) *The Enzymes*, 3rd ed., **15**, 281. **16**. Shigeura, Boxer, Meloni & Sampson (1966) *Biochemistry* **5**, 994. **17**. Shuman (1995) *Biochemistry* **34**, 16138. **18**. Gabius, Freist & Cramer (1982) *Hoppe-Seyler's Z. Physiol. Chem.* **363**, 1241. **19**. Freist, Sternbach, von der Haar & Cramer (1978) *Eur. J. Biochem.* **84**, 499. **20**. Freist, Sternbach & Cramer (1981) *Eur. J. Biochem.* **119**, 477. **21**. Freist, von der Haar, Sprinzl, Cramer (1976) *Eur. J. Biochem.* **64**, 389. **22**. Freist & Cramer (1980) *Eur. J. Biochem.* **107**, 47. **23**. Freist, von der Haar & Cramer (1981) *Eur. J. Biochem.* **119**, 151. **21**. Müller, Zahn & Arendes (1978) *FEBS Lett.* **94**, 47. **22**. Kurl, Holmes, Verney & Sidransky (1988) *Biochemistry* **27**, 8974. **23**. Koch & Niessing (1978) *FEBS Lett.* **96**, 354. **24**. Pellicer, Salas & Salas (1978) *Biochim. Biophys. Acta* **519**, 149. **25**. Das Gupta, Li, Thomson & Hunt (1995) *Plant Sci.* **110**, 215. **26**. Hooker, Strong, Adams, *et al.* (2001) *Nucl. Acids Res.* **29**, 2691. **27**. Rose & Jacob (1976) *Eur. J. Biochem.* **67**, 11. **28**. Tsiapalis, Dorson & Bollum (1975) *J. Biol. Chem.* **250**, 4486. **29**. Sethi (1971) *Prog. Biophys. Mol. Biol.* **23**, 67. **30**. Vogel & Hoppe (1986) *Eur. J. Biochem.* **154**, 253.

Corm-3

This ruthenium-based carbonyl (FW = 294.61 g/mol; CAS 475473-26-8; Soluble to 100 mM in H$_2$O), also known as CO-Releasing Molecule-3 and tricarbonylchloro(glycinato)ruthenium, readily undergoes nonenzymic hydrolysis, liberating several equivalents of carbon monoxide. RuII(CO)$_2$–protein complexes are formed upon the hydrolytic decomposition products of CORM-3 with histidine residues exposed on the surface of proteins, whereupon CO is released in aqueous solution, cells, and mice (1). CO release can be detected by mass spectrometry (MS) or confocal microscopy using a CO-responsive turn-on fluorescent probe (1). Such findings demonstrate that plasma proteins can act as CO carriers after *in vivo* administration of CORM-3 (1). Although CORM-3 is ionic and cannot penetrate cell membranes, CO does so at diffusion-controlled rates. At 75 µM, CORM-3 shows no sign of cytotoxicity (2). CORM-3 relieves acute myocardial ischemia/reperfusion (I/R) injury in hyperglycemic streptozotocin-treated rats (STZ rats) (3). Given carbon monoxide's role as a gaseous signaling molecule, CORM-3 is likely to enjoy broad use as a CO source for cellular studies. Carbon monoxide-releasing molecules are emerging as a new class of pharmacological agents that regulate important cellular function by liberating CO in biological systems. *Note*: Inasmuch as CO binds extremely tightly to heme-containing proteins and enzymes, CORM-3 is likely to interfere with some heme-dependent processes. **1**. Chaves-Ferreira, Albuquerque, Matak-Vinkovic, *et al.* (2015) *Angew. Chem. Int. Ed.* **54**, 1172. **2**. Bani-Hani, Greenstein, Mann, Green & Motterlini (2006) *J. Pharmacol. Exp. Ther.* **318**, 1315. **3**. Filippo *et al* (2012) *Naunyn. Schmiedebergs. Arch. Pharmacol.* **385**, 137.

Coronaridine

This alkaloid (FW = 338.44 g/mol; CAS 467-77-6; also named 18-carbomethoxyıbogamıne and ıbogamine-18-carboxylic acid methyl ester), isolated from the rainforest perennial *Tabernanthe iboga* (1) binds to µ-opioid receptors (K_i = 2.0 µM), δ-opioid receptors (K_i = 8.1 µM), and κ-opioid receptors (K_i = 4.3 µM), NMDA receptor (K_i = 6.2 µM, as an antagonist) (2,3), σ$_2$ sites (K_i = 201 nM) and low affinity for σ$_2$ sites (K_i = 8.5 µM) (4) and nicotinic AChRs, as an antagonist. Coronaridine is also a acetylcholinesterase inhibitor (5), a voltage-gated sodium channel blocker, and displays estrogenic activity in rodents. Coronaridine Coronaridine inhibits the Wnt signaling pathway by decreasing β-catenin mRNA expression (6). **1**. Pawelka & Stöckigt (1983) *Plant Cell Rep.* **2**, 105. **2**. Popik, Layer, Fossom, *et al.* (1996) *J. Pharmacol. Exp. Ther.* **275**, 753. **3**. Layer, Skolnick, Bertha, *et al.* (1996) *Eur. J. Pharmacol.* **309**, 159. **4**. Bowen, Vilner, Williams, *et al.* (1995) *Eur. J. Pharmacol.* **279**, R1. **5**. Andrade, Lima, Pinto, *et al.* (2005) *Bioorg. Med Chem.* **13**, 4092. **6**. Ohishi, Toume, Arai, *et al.* (2015) *Bioorg. Med. Chem. Lett.* **25**, 3937.

3α- and 3β-Corosolic Acid

α-isomer   β-isomer

These natural triterpenes (FW$_{free-acid}$ = 472.71 g/mol) inhibit rat DNA polymerases α and β and human DNA topoisomerase II (IC$_{50}$ = 43.8, 38.8, and 43 µM for the 3β-isomer, respectively, and 60.2, 80, and 83 µM for the 3α-epimer). The term corosolic acid typically refers to the 3β-isomer, whereas 3α-triterpene is also known as 3-epicorosolic acid. **Target(s)**: sterol *O*-acyltransferase (acyl-CoA:cholesterol *O*-acyltransferase; ACAT; inhibited by 3β-corosolate (1); diacylglycerol *O*-acyltransferase, IC$_{50}$ = 44.3 µM (2); DNA-directed DNA polymerase (3); DNA polymerase α (3); DNA polymerase β (3); DNA topoisomerase II (3); glycogen phosphorylase (4,7); protein kinase C (5); protein-tyrosine-phosphatase 1B (6). **1**. Kim, Han, Chung, *et al.* (2005)*Arch. Pharm. Res.* **28**, 550. **2**. Dat, Cai, Rho, *et al.* (2005) *Arch. Pharm. Res.* **28**, 164. **3**. Mizushina, Ikuta, Endoh, *et al.* (2003) *Biochem. Biophys. Res. Commun.* **305**, 365. **4**. Wen, Sun, Liu, *et al.* (2005) *Bioorg. Med. Chem. Lett.* **15**, 4944. **5**. Ahn, Hahm, Park, Lee & Kim (1998) *Planta Med.* **64**, 468. **6**. Na, Yang, He, *et al.* (2006) *Planta Med.* **72**, 261. **7**. Wen, Sun, Liu, *et al.* (2008) *J. Med. Chem.* **51**, 3540.

Corticosterone

This glucocorticoid (FW = 346.47 g/mol; Insoluble in water but soluble in most organic solvents; λ_{max} = 240 nm), also known as 4-pregnene-11β,21-diol-3,20-dione, is a prominent adrenal cortex steroid. Corticosterone participates in the regulation of protein, carbohydrate, and lipid metabolism, modulating cell growth and gene expression as well as serving as an aldosterone precursor. Corticosterone activates both mineralocorticoid and glucocorticoid receptors. **Target(s):** xanthine oxidase (1); NADH dehydrogenase (ubiquinone) (2); F$_1$F$_O$ ATPase (3); glucose 6-phosphate dehydrogenase, mildly inhibited (4); glucose transport, weakly inhibited (5); glutathione *S*-transferase (6); aldehyde dehydrogenase (7); adenosine deaminase (8); cortisol sulfotransferase (9,10); steroid sulfotransferase (11); steroid 11β-monooxygenase (12). **1**. Roussos (1967) *Meth. Enzymol.* **12A**, 5. **2**. Jensen (1959) *Nature* **184**, 451. **3**. Zheng & Ramirez (1999) *Eur. J. Pharmacol.* **368**, 95. **4**. Criss & McKerns (1969) *Biochim. Biophys. Acta* **184**, 486. **5**. Horner, Packan & Sapolsky (1990) *Neuroendocrinology* **52**, 57. **6**. Maruyama, Niitsu, Takahashi, *et al.* (1985) *Nippon Naibunpi Gakkai Zasshi* **61**, 893. **7**. Douville & Warren (1968) *Biochemistry* **7**, 4052. **8**. Singh & R. Sharma (2000) *Mol. Cell. Biochem.* **204**, 127. **9**. Singer (1979) *Arch. Biochem. Biophys.* **196**, 340. **10**. Singer & Bruns (1980) *Can. J. Biochem.* **58**, 660. **11**. Singer, Gebhart & Hess (1978) *Can. J. Biochem.* **56**, 1028. **12**. Delorme, Piffeteau, Viger & Marquet (1995) *Eur. J. Biochem.* **232**, 247.

Cortisol

This stress-activated glucocorticoid (FW = 362.47 g/mol; CAS 50-23-7), also known as hydrocortisone and 4-pregnene-11β,17α,21 triol-3,20-dione, is the major human glucocorticoid and is secreted by the adrenal cortex. Synthesized in the *zona fasciculata* of the adrenal cortex, cortisol helps to control protein, carbohydrate, and lipid metabolism. In the early fasting state, cortisol stimulates gluconeogenesis and also activates anti-stress and anti-inflammatory pathways. Cortisol also plays an indirect, role in liver and muscle glycogenolysis. Cortisol exerts profound anti-inflammatory and antiallergy actions. At a cellular level, it and other glucocorticoids (*e.g.*, prednisone, methylprednisolone, and dexamethasone) inhibit the production and action of inflammatory cytokines. At high doses, glucocorticoids impair immune function, thereby reducing cell-mediated immune reactions and reducing antibody production and action. Such properties can be used by transplant surgeoins to prevent organ rejection and by rheumatologists to treat certain allergic or autoimmune diseases (*e.g.*, rheumatoid arthritis and disseminated lupus erythematosus). Even so, their beneficial effects are frequently offset by the occurrence of serious side effects at high doses, especially when administered over a long period of time. The Hypothalmic-Pituitary-Adrenal Axis relies on circulating cortisol action in a negative feedback loop. Low blood cortisol levels leads to release of Corticotrope Releasing Hormone (or CRH), thereby activating the anterior pituitary and provoking the release of Adrenocorticotropic Hormone (ACTH), which in turn stimulates the synthesis of more cortisol within the adrenal gland to make more cortisol. In addition to cortisol, the adrenal gland also releases androgen, leading to hyperandrogenism, giving rise to the symptoms commonly associated with Cortisone Reductase Deficiency (*See Cortisone*). **Target(s):** xanthine oxidase (1); 21-hydroxysteroid dehydrogenase (NAD$^+$) (2,3,6); NADH dehydrogenase (ubiquinone) (4); glutamine:D-fructose-6-phosphate aminotransferase (5); steroid sulfatase (7); 3',5'-cyclic-nucleotide phosphodiesterase (8); steroid sulfotransferase (9). **1.** Roussos (1967) *Meth. Enzymol.* **12A**, 5. **2.** Monder & Furfine (1969) *Meth. Enzymol.* **15**, 667. **3.** Talalay (1963) *The Enzymes*, 2nd ed., **7**, 177. **4.** Jensen (1959) *Nature* **184**, 451. **5.** Chatterjee & Stefanovich (1976) *Arzneimittelforschung* **26**, 502. **6.** Monder & White (1962) *Biochem. Biophys. Res. Commun.* **8**, 383. **7.** Notation & Ungar (1969) *Biochemistry* **8**, 501. **8.** Orellana, Jedlicki, Allende & Allende (1984) *Arch. Biochem. Biophys.* **231**, 345. **9.** Singer, Gebhart & Hess (1978) *Can. J. Biochem.* **56**, 1028.

Cortisone

This cortisol-generating steroid (FW = 360.45 g/mol) exhibits approximately twice the anti-inflammatory as corticosterone, with less Na$^+$ retention potency (1-3). Cortisone is reversibly converted *in vivo* to cortisol through the action of 11β-hydroxysteroid dehydrogenase Type 1 (*or* 11β-HSD1), also known as cortisone reductase. (*See Cortisol for details of its actions*) Under certain conditions, cortisone has antiglucocorticoid activity (3). **Target(s):** xanthine oxidase (1); NADH dehydrogenase (ubiquinone) (2); glucose-6-phosphate dehydrogenase, weakly inhibited (4); histidine decarboxylase (5); thymidine kinase (6); GABA$_A$ receptor (7); protein biosynthesis (8); steroid sulfatase, weakly inhibited (9); cortisol sulfotransferase (10); steroid sulfotransferase (11). **1.** Roussos (1967) *Meth. Enzymol.* **12A**, 5. **2.** Jensen (1959) *Nature* **184**, 451. **3.** Scott & Tomkins (1975) *Meth. Enzymol.* **40**, 273. **4.** Gaylor, Chang, Nightingale, Recio &. Ying (1965) *Biochemistry* **4**, 1144. **5.** Parrot & Laborde (1961) *J. Physiol. (Paris)* **53**, 441. **6.** Kaneko & LePage (1970) *Proc. Soc. Exp. Biol. Med.* **133**, 229. **7.** Ong, Kerr, Capper & Johnston (1990) *J. Pharm. Pharmacol.* **42**, 662. **8.** Clark (1953) *J. Biol. Chem.* **200**, 69. **9.** Notation & Ungar (1969) *Biochemistry* **8**, 501. **10.** Singer & Bruns (1980) *Can. J. Biochem.* **58**, 660. **11.** Singer, Gebhart & Hess (1978) *Can. J. Biochem.* **56**, 1028.

Cortistatin-8

Pro-Cys-Phe-D-Trp-Lys-Thr-Cys-Lys-NH$_2$

This synthetic cortistatin analogue (FW = 1009.25; CAS 485803-62-1; Solubility: 2 mg/mL in acetonitrile) lacks the somatostatin-binding capacity of full-length cortistatin, yet retains its GHS-R-binding capacity. CST8 was found to stimulate growth hormone (GH) release only at low doses (10^{-15} M), but does not reduce GH secretion stimulated by GHRH, ghrelin, or low-dose, full-length cortistatin. Later work, however, suggests that, when administered either as bolus or as continuous infusion, CST-8 fails to modify spontaneous, ghrelin- or hexarelin-stimulated secretion of GH, PRL, ACTH and cortisol, suggesting that CST-8 is devoid of any modulatory action on either spontaneous or ghrelin-stimulated somatotroph, lactotroph and corticotroph secretion in humans *in vivo* (2). **1.** Luque, Peinado, Gracia-Navarro, *et al.* (2006) *J. Mol. Endocrinol.* **36**, 547. **2.** Prodam, Benso, Gramaglia, *et al.* (2008) *Neuropeptides* **42**, 89.

Cotinine

This substituted pyridine (FW$_{free-base}$ = 176.22 g/mol; CAS Eegistry Number = 486-56-6) is the major nicotine metabolite. With an *in vivo* half-life of 20 hours, continin is detectable in blood, saliva, and urine for 4-5 days after tobacco use. **Target(s):** 7-ketosteroid reductase (1); aromatase, *or* CYP19 (2); steroid 11β monooxygenase (3,5); 3α-hydroxysteroid dehydrogenase (4); CYP2E1, weakly inhibited (6). **1.** Yeh, Barbieri & Friedman (1989) *J. Steroid Biochem.* **33**, 627. **2.** Barbieri, Gochberg & Ryan (1986) *J. Clin. Invest.* **77**, 1727. **3.** Barbieri, York, Cherry & Ryan (1987) *J. Steroid Biochem.* **28**, 25. **4.** Meikle, Liu, Taylor & Stringham (1988) *Life Sci.* **43**, 1845. **5.** Barbieri, Friedman & Osathanondh (1989) *J. Clin. Endocrinol. Metab.* **69**, 1221. **6.** Van Vleet, Bombick & Coulombe, Jr. (2001) *Toxicol. Sci.* **64**, 185.

Coumermycins

Coumermycin A$_1$

These antibiotics (FW = 1110.10 g/mol), which are structurally related to novomycin and chloromycin, inhibit DNA gyrase and DNA biosynthesis. **Target(s):** DNA gyrase (1,2,8); DNA topoisomerase I (3,11,12); DNA topoisomerase II (4,5,9,13,18); *Molluscum contagiosum* virus topoisomerase (6); human immunodeficiency virus type 1 (HIV 1) integrase (7); DNA topoisomerase I, yeast, weakly inhibited (5); DNA ligase (ATP), Vaccinia virus (10); DNA-directed DNA polymerase (19). **1.** Gellert (1981) *The Enzymes*, 3rd ed., **14**, 345. **2.** Gellert, O'Dea, Itoh & Tomizawa (1976) *Proc. Natl. Acad. Sci. U.S.A.* **73**, 4474. **3.** Foglesong & Bauer (1984) *J. Virol.* **49**, 1. **4.** Goto & Wang (1982) *J. Biol. Chem.* **257**, 5866. **5.** Goto, Laipis & Wang (1984) *J. Biol. Chem.* **259**, 10422. **6.** Hwang, Wang & Bushman (1998) *J. Virol.* **72**, 3401. **7.** Mazumder, Neamati, Sommadossi, *et al.* (1996) *Mol. Pharmacol.* **49**, 621. **8.** Cozzarelli (1980) *Science* **207**, 953. **9.** Assairi (1995) *Lett. Pept. Sci.* **2**, 169. **10.** Shuman (1995) *Biochemistry* **34**, 16138. **11.** Riou, Douc-Rasy & Kayser (1986) *Biochem. Soc. Trans.* **14**, 496. **12.** Shaffer & Traktman (1987) *J. Biol. Chem.* **262**, 9309. **13.** Gellert (1981) *Ann. Rev. Biochem.* **50**, 879. **14.** Melendy & Ray (1989) *J. Biol. Chem.* **264**, 1870. **15.** Vosberg (1985) *Curr. Top. Microbiol. Immunol.* **114**, 19. **16.** Saijo, Enomoto, Hanaoka & Ui (1990) *Biochemistry* **29**, 583. **17.** Assairi (1994) *Biochim. Biophys. Acta* **1219**, 107. **18.** Riou, Douc-Rasy & Kayser (1986) *Biochem. Soc. Trans.* **14**, 496. **19.** Acel, Udashkin, Wainberg & Faust (1998) *J. Virol.* **72**, 2062.

CP-358774, *See Erlotinib*

CP 55,940

This endocannabinoid receptor probe (FW = 376.58 g/mol; CAS 83002-04-4; 100 mM in DMSO), also named 2-[(1R,2R,5R)-5-hydroxy-2-(3-hydroxy-propyl) cyclohexyl]-5-(2-methyloctan-2-yl)phenol, a full agonist at both CB_1 (K_i = 0.58 nM) and CB_2 (K_i = 0.68 nM) receptors, but acts as an antagonist at GPR55, the putative CB_3 receptor. Displacement of [3H]-labeled CP 55,940 has proven to be a useful tool for evaluating the binding properties of putative CB_1 and CB_2 receptor ligands. **1**. Avdesh, Cornelisse, & Martin-Iverson (2012) *Psychopharmacol.* (Berl) **220**, 405. **2**. Griffin, *et al.* (1998) *J. Pharmacol. Exp. Ther.* **285**, 553. **3**. Thomas, *et al.* (1998) *J. Pharmacol. Exp. Ther.* **285**, 285. **4**. Wiley, *et al.* (1995) *Neuropharmacol.* **34**, 669.

CP-66,248, *See Tenidap*

CP-945598

This substituted adenine and cannabinoid receptor antagonist (FW$_{HCl-Salt}$ = 287.33 g/mol; CAS 446859-33-2; Solubility: 1 mg/mL DMSO (with warming); <1 mg/mL H_2O), also known as 1-[8-(2-chlorophenyl)-9-(4-chlorophenyl)-9H-purin-6-yl]-4-(ethylamino)-piperidine carboxamide, potently and selectively targets human CB_1 receptor with greater than 10,000x selectivity for CB_1 (binding K_i = 0.7 nM; functional assay K_i = 0.12 nM). When tested *in vivo*, CP-945598 reverses cannabinoid agonist-mediated responses, reducing food intake, and increasing energy expenditure and fat oxidation in rodents (1). *In vivo*, CP-945,598 reverses four cannabinoid agonist-mediated CNS-driven responses (*e.g.*, hypo-locomotion, hypothermia, analgesia, and catalepsy) to a synthetic cannabinoid receptor agonist (2). CP-945,598 exhibits dose and concentration-dependent anorectic activity in two models of acute food intake in rodents, namely fast-induced re-feeding and spontaneous, nocturnal feeding. **1**. Griffith, Hadcock, Black, *et al.* (2009) *J. Med. Chem.* **52**, 234. **2**. Hadcock, Griffith, Iredale, et al. (2010) *Biochem. Biophys. Res. Commun.* **394**, 366.

CP-99,219, *See Trovafloxacin*

CP-724,714

This potent signal transduction inhibitor (FW = 469.53 g/mol; CAS 537705-08-1; Solubility: 94 mg/mL DMSO; (E)-2-methoxy-N-(3-(4-(3-methyl-4-(6-methylpyridin-3-yloxy)phenyl-amino)quinazolin-6-yl)allyl) acetamide) targets HER2/ErbB2 (IC_{50} = 10 nM), with >640x selectivity against EGFR (IC_{50} = 6.4 µM) and >1,000-fold less potent for IR, IGF-1R, PDGFRβ, VGFR2, abl Src, c-Met c-jun NH_2-terminal kinase-2 (*or* JNK-2), JNK-3, ZAP-70, cyclin-dependent kinase (CDK)-2, and CDK-5 (1). CP-724,714 induces G_1 cell-cycle block in erbB2-overexpressing human breast carcinoma cells and inhibits erbB2 autophosphorylation in xenografts, when administered p.o. to athymic mice. It also induces a marked reduction of extracellular signal-regulated kinase and Akt phosphorylation, tumor cell apoptosis, and release of caspase-3. CP-724,714 was discontinued from further clinical development due to unexpected hepatotoxicity via both hepatocellular injury and hepatobiliary cholestatic mechanisms (2). CP-724,714 decreases mitochondrial membrane potential in primary cultured human hepatocytes. It also demonstrates an uptake clearance ~4x greater than its efflux clearance into bile. It inhibits MDR1 and BSEP transporters which contribute to CP-724,714 and bile acid elimination. **1**. Jani, Finn, Campbell, *et al.* (2007) *Cancer Res.* **67**, 9887. **2**. Feng, Xu, Bi, *et al.* (2009) *Toxicol. Sci.* **108**, 492.

CP-690550, *See Tofacitinib*

CPAF Inhibitory Peptide

This inhibitory peptide (FW = 3055.39 g/mol; *Sequence*: SEFY SPMVPHFWAELRNHYATSGLKS) targets the chlamydial protease/proteasome-like activity factor (CPAF) that degrades host molecules to facilitate infection by the intracellular pathogen *Chlamydia* (1). Derived from the self-inhibitory fragment of the CPAF zymogen, the peptide blocks substrate access to the active site (2). Treating *Chlamydia trachomatis*-infected HeLa cells with this nanomolar-affinity inhibitor prevents CPAF-catalyzed cleavage of the intermediate filament vimentin. Brown & Grieshaber (2013) also found that, while chladmydia-infected lysates actively degraded cyclin B_1 and the spindle assembly checkpoint protein securin, such proteolysis is inhibited by the CP AF inhibitory peptide. A sequence-scrambled peptide (*Sequence*: WNHSSTMGYFLLPAEVEKY SFSPHRA) containing the same amino acids did not block proteolysis of either cyclin B_1 or securin. **1**. Bednar, Jorgensen, Valdivia & McCafferty (2011) *Biochemistry* **50**, 7441. **2**. Huang, Feng, Chen, *et al.* (2008) *Cell Host Microbe* **4**, 529. **3**. Brown & Grieshaber (2013) *Journal*, **vol**, page.

CPI-613

This potent lipoic acid derivative (FW = 388.59 g/mol; CAS 95809-78-2; Solubility: 80 mg/mL DMSO, <1 mg/mL Water), also known systematically as 6,8-bis(benzylthio)octanoic acid or 6,8-dibenzyllipoic acid, targets the mitochondrial enzymes pyruvate dehydrogenase and α-ketoglutarate dehydrogenase, with as EC_{50} = 120 µM in human lung cancer cells and human sarcoma cells. **1**. Zachar, *et al.* (2011) *J. Mol. Med.* *(Berlin)* **89**, 1137.

CPP-115, See *Vigabatrin*

CPT-11

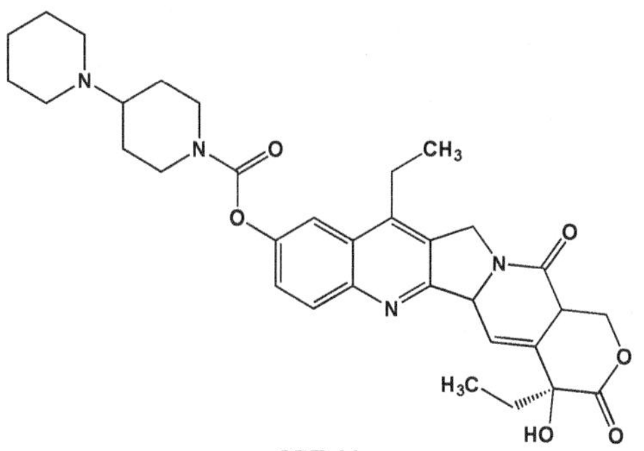

CPT-11

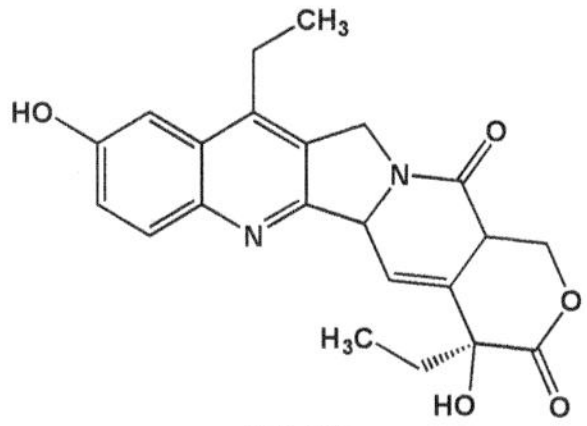

SN-38

This DNA Topoisomerase-I pro-drug (FW = 623.14 g/mol; CAS 100286-90-6; Solubility: 5 mM in H_2O with gentle warming; 100 mM in DMSO; also named Irinotecan, Camptothecin-11, and (4S)-4,11-diethyl-3,4,12,14-tetrahydro-4-hydroxy-3,14-dioxo-1*H*-pyrano[3',4':6,7]-indolizino[1,2*b*]quinolin-9-yl[1,4'-bipiperidine]-1'-carboxylic acid ester) is hydrolyzed metabolically to form SN-38, the pharmacologically active metabolite SN 38 (FW = 392.40 g/mol; CAS 86639-52-3; IUPAC: 7-ethyl-10-hydroxy camptothecin and (4S)-4,11-diethyl-4,9-dihydroxy-1*H*-pyrano[3',4':6,7] indolizino[1,2-*b*]quinoline-3,14(4*H*,12*H*)-dione) (1-3). SN-38 then inhibits DNA topoisomerase I (IC_{50} values are 0.74 and 1.9 µM in P388 and Ehrlich cells, respectively). CPT-11 inhibits both DNA synthesis (IC_{50} = 77 nM) and RNA synthesis (IC_{50} = 1.3 µM respectively) but does not affect protein synthesis. CPT-11 displays potent antitumor activity against a range of human tumor cells: HCT-116 (IC_{50} = 3.3 nM), BEL-7402 (IC_{50} = 13 nM), HL60 (IC_{50} = 19 nM), and HeLa (IC_{50} = 22 nM). Both SN-38 and its parent molecule, camptothecin (**See** *Camptothecin*), show strong, time-dependent inhibition against DNA synthesis of P388 cells. By alkaline and neutral elution assays, it was demonstrated that SN-38 caused much more frequent DNA single-strand breaks in P388 cells than did CPT-11. The same content of SN-38 and a similar frequency of single-strand breaks were detected in the cells treated with SN-38 at 0.1 microM or with CPT-11 at 100 microM (3). (*Note*: Reference-2 also established that plasma carboxylesterase-deficient SCID mice, bearing human tumor xenografts, may represent a more accurate model for antitumor studies with this drug and other agents metabolized by carboxylesterases. **1**. Vassal, *et al.* (1996) *Br. J. Cancer.* **74**, 537. **2**. Morton , *et al* (2005) *Can. Chemother. Pharmacol.* **56**, 629. Kawato, Aonuma, Hirota, Kuga & Sato (1991) *Cancer Res.* **51**, 4187.

CPYPP

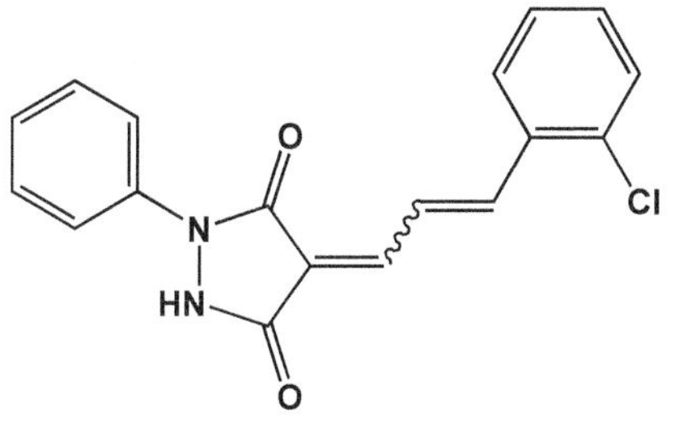

This DOCK2-Rac1 interaction inhibitor (FW = 324.76 g/mol; CAS 310460-39-0; Soluble to 50 mM in DMSO), also named 4-[3-(2-chlorophenyl)-2-propen-1-ylidene]-1-phenyl-3,5-pyrazolidinedione, both reduces Dedicator of Cytokinesis-2 (DOCK1) -induced Rac1 activation by interfering with DOCK1 and DOCK5 binding (1,2). CPYPP binds to DOCK2 DHR-2 domain in a reversible manner and inhibits its catalytic activity *in vitro* (1). When lymphocytes are treated with CPYPP, both chemokine receptor- and antigen receptor-mediated Rac activation are blocked, resulting in marked reduction of chemotactic response and T cell activation (1). CPYPP also exhibits selectivity over DOCK9. CPYPP reduces chemotactic responses *in vitro* and *in vivo*, and activates T cells *in vitro* and attenuates HER2-mediated breast cancer cell migration *in vitro* (2). **1**. Nishikimi, Uruno, Duan, *et al.* (2012) *Chem. Biol.* **19**, 488. **2**. Laurin, Fradet, Blangy, *et al.* (2008) *Proc. Natl. Acad. Sci U.S.A.* **105**, 5446.

CPZEN-45

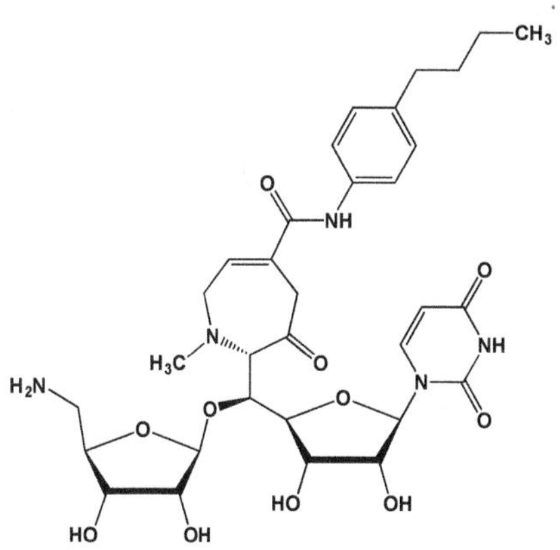

This caprazene butylanilide (FW = 673.72 g/mol) targets the *Mycobacterium tuberculosis* GlcNAc-1-phosphate transferase WecA, which catalyzes the first step in synthesis of the mycobacterial cell wall core. Unlike caprazamycins, CPZEN-45 strongly inhibits glycerol incorporation in growing cultures, showing antibacterial activity against caprazamycin-resistant strains, including a strain overexpressing translocase-I (MraY), a caprazamycin target involved in peptidoglycan biosynthesis. By contrast, CPZEN-45 is ineffective against a strain overexpressing undecaprenyl-phosphate-GlcNAc-1-phosphate transferase (TagO) that involved in the teichoic acid biosynthesis. **1**. Ishizaki, Hayashi, Inoue, *et al.* (2013) *J. Biol. Chem.* **288**, 30309.

CRA-024781, See *PCI-24781*

CRD-401, See *Diltiazem*

Creatine

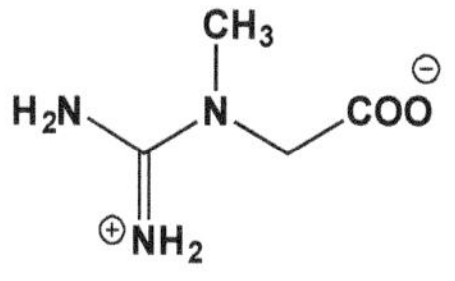

This guanidinium group-containing metabolite (FW = 131.13 g/mol; CAS 57-00-1), also known as *N*-amidinosaccosine and *N*-(aminoiminomethyl)-*N*-methylglycine, is frequently found mainly in skeletal and cardiac muscle, where it undergoes reversible transphosphorylation by creatine kinase to form phosphocreatine. The carboxyl group has a pK_a of 2.63 at 25°C, and the pK_a of the guanidinium group is ~14. Creatine is a product of the reactions catalyzed by guanidinoacetate *N*-methyltransferase, creatininase, and phosphoamidase. Phosphocreatine is readily converted nonenzymatically, particularly under acidic conditions, to creatinine. **Target(s):** pyrophosphatase, inorganic diphosphatase (1); acetylcholinesterase (2); amylase (3); acid phosphatase (4); arginine kinase, weakly inhibited (5); guanidinoacetate kinase, weakly inhibited (6). **1**. Naganna (1950) *J. Biol. Chem.* **183**, 693. **2**. Bergmann, Wilson & Nachmansohn (1950) *J. Biol. Chem.* **186**, 693. **3**. Mystkowski (1932) *Biochem. J.* **26**, 910. **4**. Gonzales & Meizel (1973) *Biochim. Biophys. Acta* **320**, 166. **5**. Pereira, Alonso, Ivaldi, *et al.* (2003) *J. Eukaryot. Microbiol.* **50**, 132. **6**. Shirokane, Nakajima & Mizusawa (1991) *Agric. Biol. Chem.* **55**, 2235.

Creatine Phosphate, See *Phosphocreatine*

Creatinine

This metabolite ($FW_{free-base}$ = 113.12 g/mol; CAS 60-27-5), is readily produced from creatine or, perhaps more likely, from phosphocreatine, the latter proceeding by intramolecular *N*-to-*O* phosphoryl migration and subsequent amino group attack on the transient acyl-P intermediate. Creatinine formation is virtually linear with muscle mass, resulting in higher amounts in males. The so-called creatine clearance rate (~25 mg/kg body weight) serves as a useful index of renal function. **Target(s):** histidine ammonia-lyase (1); arginine deaminase (2); D-amino-acid oxidase (3); acetylcholine esterase (4). **1**. Leuthardt (1951) *The Enzymes*, 1st ed., **1** (part 2), 1156. **2**. Petrack, Sullivan & Ratner (1957) *Arch. Biochem. Biophys.* **69**, 186. **3**. Nohara, Suzuki, Kinoshita & Watanabe (2002) *Nephron* **91**, 281. **4**. Bergmann, Wilson & Nachmansohn (1950) *J. Biol. Chem.* **186**, 693.

Crenolanib

This orally active and potent protein tyrosine kinase inhibitor (FW = 441.58 g/mol; CAS 670220-88-9), also known by its code name CP-868596 and ARO-002 as well as by its systematic name 1-(2-(5-((3-methyloxetan-3-yl)methoxy)-1*H*-benzo[*d*]imidazol-1-yl)quinolin-8-yl)piperidin-4-amine, selectively targets platelet-derived growth factor receptor PDGFRα and PDGFRβ, with K_d values of 2.1 nM and 3.2 nM. Crenolanib also potently inhibits PDGFRα's Aspartate-842-Valine mutation, but not Valine-561-Aspartate mutation, with >100x higher selective for PDGFR over c-Kit, VEGFR-2, TIE-2, FGFR-2, EGFR, erbB2, and Src protein kinases (1). PDGFRα mutations have been detected in a small population of melanoma patients, suggesting melanoma patients harboring certain mutations may benefit from crenolanib treatment (2). Mutations of the type III receptor tyrosine kinase (RTK) FLT3 occur in approximately 30% of acute myeloid leukemia (AML) patients and lead to constitutive activation, and crenolanib inhibits both FLT3/ITD and resistance-conferring FLT3/D835 mutants *in vivo* (3). Key attributes of highly successful TKI drugs are (a) kinase selectivity and (b) invulnerability to secondary resistance-conferring kinase domain (KD) mutations. Notably, crenolanib not only displays a high degree of selectivity for FLT3 and other class III RTKs, but it also appears to be invulnerable to resistance-conferring KD mutations, when probed *in vitro* (4). Whereas several type II FLT3 TKIs (quizartinib, sorafenib, and ponatinib) target the related class III RTK KIT, crenolanib elicits cytotoxicity in FLT3–mutant AML, but largely spares KIT inhibition. With its limited activity against other kinases, crenolanib may prove to be less toxicity than other type II FLT3 TKIs (4). **1**. Heinrich, Griffith, McKinley, *et al.* (2012) *Clin. Cancer Res.* **18**, 4375. **2**. Dai, Kong, Si, *et al.* (2013) *Clin. Cancer Res.* **19**, 6935. **3**. Galanis, Ma, Rajkhow, *et al.* (2013) *Blood* **123**, 94. **4**. Smith, Lasater, Lin, *et al.* (2014) *Proc. Natl. Acad. Sci. U.S.A.* **111**, 5319.

p-Cresol

This common organic reagent (FW = 108.14 g/mol; CAS 106-44-5; pK_a = 10.17; M.P. = 35.5°C B/P. = 201.8°C; Solubility: 2.5 g/100 mL H_2O at 50°C), also known as 4-methylphenol, is a mechanism-based inactivator of dopamine β-monooxygenase (k_{inact} = 2.0 min^{-1} at pH 5.0), resulting in the covalent modification a tyrosyl residue. **Target(s):** dopamine β-

monooxygenase (1,3-6,8); L-ascorbate oxidase (2); ascorbate peroxidase (7); RCK1 (*or* Kv1.1) K$^+$ channels (9); oxidative phosphorylation (10); arylsulfatase (11); glycogen phosphorylase (12); polyphenol oxidase (13,14); tyrosinase, *or* monophenol monooxygenase (13,14). **1**. Ator & Ortiz de Montellano (1990) *The Enzymes*, 3rd ed. (D. S. Sigman & Boyer, eds.), **19**, 213. **2**. Gaspard, Monzani, Casella, *et al.* (1997) *Biochemistry* **36**, 4852. **3**. Kruse, DeWolf, Chambers & Goodhart (1986) *Biochemistry* **25**, 7271. **4**. Goodhart, DeWolf, Jr., & Kruse (1987) *Biochemistry* **26**, 2576. **5**. DeWolf, Carr, Varrichio, *et al.* (1988) *Biochemistry* **27**, 9093. **6**. Southan, DeWolf & Kruse (1990) *Biochim. Biophys. Acta* **1037**, 256. **7**. Chen & Asada (1990) *J. Biol. Chem.* **265**, 2775. **8**. Kim & Klinman (1991) *Biochemistry* **30**, 8138. **9**. Elliott & Elliott (1997) *Mol. Pharmacol.* **51**, 475. **10**. Kitagawa (2001) *Drug Chem. Toxicol.* **24**, 39. **11**. Beil, Kehrli, James, *et al.* (1995) *Eur. J. Biochem.* **229**, 385. **12**. Soman & Philip (1974) *Biochim. Biophys. Acta* **358**, 359. **13**. Anosile & Ayaebene (1981) *Phytochemistry* **20**, 2625. **14**. Anosike & Ayaebene (1982) *Phytochemistry* **21**, 1889.

Crizotinib

This orally bioavailable, ATP-competitive, small molecule inhibitor and antineoplastic drug (FW = 450.34 g/mol; CAS 877399-52-5; Solubility: 25 mg/mL DMSO, with warming, but very poorly soluble in H_2O (maximum ~10-20 μM)), also known by the code name PF-02341066, the trade name Xalkori$^®$, and the systematic name, 3-[(1*R*)-1-(2,6-dichloro-3-fluorophenyl)ethoxy]-5-(1-piperidin-4-ylpyrazol-4-yl)pyridin-2-amine, targets the anaplastic lymphoma kinase ALK (IC_{50} = 11 nM) and c-ros oncogene-1 (ROS1) receptor tyrosine kinase (IC_{50} = 0.62 nM) and is used to treat non-small cell lung carcinoma (NSCLC). **Mode of Inhibitory Action:** The t(2;5) chromosomal translocation leads to the expression of an oncogenic kinase fusion protein known as nucleophosmin-anaplastic lymphoma kinase, *or* NPM-ALK, which has been implicated in the pathogenesis of anaplastic large-cell lymphoma *or* ALCL (1-2). Upon treatment with PF-02341066, ALCL cells undergo G_1-S phase cell cycle arrest and apoptosis. Crizotinib potently inhibits c-Met-dependent proliferation, migration, or invasion of human tumor cells *in vitro* (IC_{50} = 5-20 nM) as well as HGF-stimulated endothelial cell survival or invasion and serum-stimulated tubulogenesis *in vitro* (1). Crizotinib-resistant patient samples reveal a variety of point mutations in the kinase domain of ALK, including the L1196M gatekeeper mutation (3). **1**. Christensen, Zou, Arango, *et al.* (2007) *Mol. Cancer Ther.* **6**, 3314. **2**. Zou, Li, Lee, *et al.* (2007) *Cancer Res.* **67**, 4408. **3**. Heuckmann, Hölzel, Sos, *et al.* (2011) *Clin. Cancer Res.* **17**, 7394.

CRL-40028, *See* Adrafinil

Crotonaldehyde

This toxic aldehyde (FW = 70.091 g/mol; CAS 123-73-9; Solubility: 18.1 g/100 mL H_2O at 20°C), also known as 2-butenal is a strong lacrimator and cigarette smoke constituent. Commercial crotonaldehyde is mainly the *trans* isomer (M.P. = –76.5°C; B.P. = 104°C). **Target(s):** aldehyde dehydrogenase (1,2); glutathione *S*-transferase (3-5); glutathione-disulfide reductase (6); cytochrome *c* reductase, mildly inhibited (7); xanthine oxidase (8); xanthine oxidase, inhibited by *trans*-isomer (9). **1**. Seegmiller (1955) *Meth. Enzymol.* **1**, 511. **2**. Mitchell & Petersen (1993) *Drug Metab. Dispos.* **21**, 396. **3**. Fujita & Hossain (2003) *Plant Cell Physiol.* **44**, 481. **4**.

Iersel, Ploemen, Struik, *et al.* (1996) *Chem. Biol. Interact.* **102**, 117. **5**. van Iersel, Ploemen, Lo Bello, Federici & van Bladeren (1997) *Chem. Biol. Interact.* **108**, 67. **6**. Vander Jagt, Hunsaker, Vander Jagt, *et al.* (1997) *Biochem. Pharmacol.* **53**, 1133. **7**. Cooper, Witmer & Witz (1987) *Biochem. Pharmacol.* **36**, 627. **8**. Villela (1963) *Enzymologia* **25**, 261. **9**. Bounds & Winston (1991) *Free Radic. Biol. Med.* **11**, 447.

Crotonate

This α,β-unsaturated acid (FW$_{free-acid}$ = 86.09 g/mol; CAS 3724-65-0: pK_a = 4.70 at 25°C; M.P. = 71.6°C; B.P. = 185°C), also known as (*E*)-2-butenoic acid. Inhibits fumarase (1,2), D-amino-acid oxidase (1), glutathione *S*-transferase (3), urocanate hydratase, *or* urocanase) (4), guanidinobutyrase (5); prostaglandin-endoperoxide synthase, *or* cyclooxygenase (6). **1**. Webb (1966) *Enzyme and Metabolic Inhibitors*, vol. **2**, Academic Press, New York. **2**. Jacobsohn (1953) *Enzymologia* **16**, 113. **3**. Dierickx (1984) *Res. Commun. Chem. Pathol. Pharmacol.* **45**, 471. **4**. Lane, Scheuer, Thill & Dyll (1976) *Biochem. Biophys. Res. Commun.* **71**, 400. **5**. Yorifuji, Sugai, Matsumoto & Tabuchi (1982) *Agric. Biol. Chem.* **46**, 1361. **6**. Ryan & Davis (1988) *Biochem. Soc. Trans.* **16**, 398.

Crotonoyl-CoA

This acyl-CoA derivative (FW = 853.41 g/mol; CAS 102680-35-3), also known as (*E*)-2-butenoyl-CoA, is an intermediate in the β-oxidation pathway for fatty acid degradation. **Target(s):** acetyl-CoA synthetase, *or* acetate:CoA ligase (1); acyl-CoA dehydrogenase (2,3); butyrate:acetoacetate CoA-transferase (4); fatty-acid synthase (5); 3-hydroxyacyl-[acyl-carrier protein] dehydrase (5). **1**. Preston, Wall & Emerich (1990) *Biochem. J.* **267**, 179. **2**. Beinert (1963) *The Enzymes*, 2nd ed., **7**, 447. **3**. Hauge (1956) *J. Amer. Chem. Soc.* **78**, 5266. **4**. Sramek & Freman (1975) *Arch. Biochem. Biophys.* **171**, 14. **5**. Katiyar, Pan & Porter (1983) *Eur. J. Biochem.* **130**, 177.

Crufomate

This insecticide (FW = 291.71 g/mol; CAS 299-86-5), also named Ruelene and 4-*t*-butyl-2-chlorophenyl *N*-methyl-*O*-methylphosphoramidate, inhibits insect cholinesterase and is mainly used for controling cattle grubs, lice, and horn flies. Ruelene is practically insoluble in water and is relatively stable as an aqueous dispersion at pH 7 and below. In rat intestine (2), ^{14}C-labeled cruformate is metabolized to form: 4-*tert*-butyl-2-chlorophenyl methyl phosphoramidate (25%), 2-chloro-4(2-hydroxy-1,1-dimethylethyl) phenyl methyl methyl-phosphoramidate (19%), 2-[3-chloro-4-[[(methoxy) (methyl-amino)phosphoinyl]oxy]phenyl]-2-methylpropionic acid (2%), 4-*tert*-butyl-2-chlorophenol (0.8% free form, and 6% as its glucuronide), and the aromatic glucuronide of 2-chloro-4(2-hydroxy-1,1-dimethylethyl) phenol (1%). **1**. Neely, Unger, Blair & Nyquist (1964) *Biochemistry* **3**, 1477. **2**. Pekas, Bakke, Giles & Price (1977) *J. Environ. Sci. Health B.* **12**, 261.

Cryptotanshinone

This natural product (FW = 296.36 g/mol; CAS 35825-57-1; Solubility: 10 mM in DMSO; red-colored; IUPAC: 1,2,6,7,8,9-hexahydro-1,6,6-trimethyl[1,2-*b*]furan-10,11-dione), a major tanshinone metabolite in the

Chinese medicinal root *Danshen* (*Salvia miltiorrhiza*) that exhibits antitumor activity by inhibiting phosphorylation at Tyr705 of Signal Transducer and Activator of Transcription-3, *or* STAT3, (IC$_{50}$ = 4.6 µM) in a JAK2-independent mechanism and blocking STAT3 dimerization in DU145 prostate cancer cells (1). Cryptotanshinone also shows antibacterial and anti-inflammatory activity. It acts as an antidiabetes and antiobesity agent via its ability to activate AMP-activated protein kinase, *or* AMPK (2). It improves cognitive impairment in Alzheimer's disease transgenic mice by inhibiting acetylcholinesterase (IC$_{50}$ = 4.1 µM) and reducing Aβ-peptide formation (3). Cryptotanshinone also ameliorates scopolamine-induced amnesia in Morris water maze task (4). **1**. Shin, *et al.* (2009) *Cancer Res.* **69**, 193. **2**. Kim, *et al.* (2007) *Mol. Pharmacol.* **72**, 62. **3**. Zhang, *et al.* (2009) *Neurosci. Lett.* **452** 90. **3**. Wong, *et al.* (2009) *Planta Med.* **76**, 288.

CS055, See *Chidamide*

CS-600, See *Loxoprofen*

CS-706, See *Apricoxib*

CSG452, See *Tofogliflozin*

CT99021, See *CHIR-99021*

CTA095

This novel dual tyrosine protein kinase inhibitor (FW = 555.68 g/mol), also named 2-(dibenzo[*b,d*]furan-2-yl)-7-phenyl-1-(3-(piperidin-1-yl)propyl)-1*H*-imidazo[4,5-*g*]-quinoxalin-6(5*H*)-one, targets EtK (IC$_{50}$ = 60 nM) and Src (IC$_{50}$ = 60 nM), inducing autophagy and apoptosis, as well as exerting synergistic effects with autophagy modulators (*e.g.*, 10 µM chloroquine) in prostate cancer cells (1). By contrast, the IC$_{50}$ values for Abl, Axl, Btk, EGFR, Itk, Mer, and Yes exceeded 10 µM. A member of the Btk nonreceptor tyrosine kinase family, Etk (also known as Bmx) has a modular structure that include an N-terminal pleckstrin homology (PH) domain, a Tec homology domain, a SH2 domain, a SH3 domains, an autophosphorylation site within the SH3 domain, and a C-terminal kinase domain. Etk interacts with and inactivates p53, and overexpression of Etk in prostate cancer cells confers resistance to androgen deprivation and photodynamic therapy. Src is a proto-oncogene tyrosine-protein kinase (also known as cellular-Src, or simply c-Src) is likewise a modular non-receptor protein tyrosine kinase that contains an SH2 domain, an SH3 domain, and a tyrosine kinase domain. CTA095 blocks the ATP binding pocket in Src kinase, whereas it inhibits ATP binding in Etk by inducing a conformational change via the so-called "back-pocket". Treatment of PC3 cells with CTA095 decreased Etk and Src autophoshorylation as well as phosphorylation of the downstream signal transducing kinases Stat3 and Akt. CTA095 has little effect on the survival of the immortalized normal prostate cell RWPE1, suggesting growth and survival are more dependent on Etk and Src. **1**. Guo, Liu, Bhardwaj, *et al.* (2013) *PLoS One* **8**, e70910.

CTS-1027, See *RS-130830*

Cucurbitacins

These bitter-tasting plant steroids ($FW_{cucurbita-5-ene}$ = 412.75 g/mol), derived from unsaturated variant cucurbita-5-ene, or 19-(10→9β)-abeo-10α-lanost-5-ene (shown above) are glucocorticoid-like cytotoxic triterpene cytotoxins found in the family *Cucurbitaceae*, including the common pumpkins and gourds, where they serve to discourage consumption by herbivores.

Curare, See Tubocurarine

Curcumin

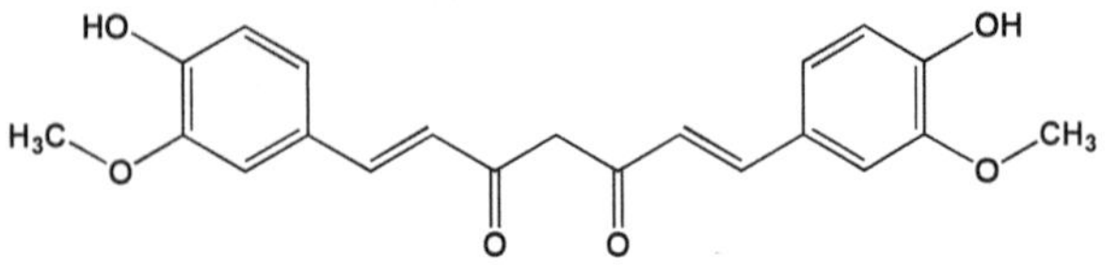

This symmetrical orange-yellow dione (FW = 368.39 g/mol; CAS 458-37-7), also known as tumeric yellow, and diferuloylmethane and named systematically as 1,7-bis(4-hydroxy-3-methoxyphenyl)-1,6-heptadiene 3,5-dione, is obtained from the roots of *Curcuma longa* and is the major active component of turmeric, the primary spice in Indian curries. Curcumin is an anti-tumor promoter, inducing apoptosis in prostate cancer cells, as well as an anti-inflammatory agent that inhibits both cyclooxygenase (IC_{50} = 52 μM) and 5-lipoxygenase (IC_{50} = 8 μM). It also inhibits induction of nitric-oxide synthase biosynthesis in activated macrophages and is an anti-oxidant that enhances wound healing. Although its effects on only a few of its known targets were assessed, Phase I clinical trials indicate curcumin is safe in healthy subjects, even at high doses (12 g/day). **Target(s):** DNA topoisomerase II (1); cyclooxygenase (2,42,44); 5-lipoxygenase (2,8,44); epidermal growth factor receptor (EGFR) protein-tyrosine kinase (3,25); protein farnesyltransferase (4); glucuronosyl transferase, *or* UDPglucuronosyltransferase (5,56); COP9 signalosome (CSN) protein kinase (6,7,15); DNA-directed DNA polymerase (9); DNA polymerase λ (9); membrane alanyl aminopeptidase, *or* aminopeptidase N (10,47,48); phenol sulfotransferase (11,29); cell survival signal protein kinase B/Akt (12); c-jun/AP 1 (13,18); protein kinase C (14,23,28,37,51); inositol 1,3,4-trisphosphate 5/6-kinase (15); inositol 1,4,5 trisphosphate-sensitive Ca^{2+} channel, *or* inositol 1,4,5-trisphosphate receptor (16); Ca^{2+}-transporting ATPase (17,19); CYP1A1 (20); CYP1A2 (20); CYP2B1 (20); soybean lipoxygenase (21); β-carotene 15,15' monooxygenase (22); F_OF_1 ATPase (24); p185*neu* protein-tyrosine kinase (26); c-Jun N-terminal (JNK) signaling pathway (27); protein kinase A (28); plant Ca^{2+}-dependent protein kinase (28); Δ^5 desaturase (30,38,39); Δ^6-desaturase (30,38); HIV-1 integrase (31,32); xanthine dehydrogenase and xanthine oxidase (33); phosphorylase kinase (34); tyrosinase (35); HIV-1 retropepsin (36); HIV-2 retropepsin (36); polyunsaturated fatty acid chain elongation (39); glutathione *S*-transferase (40); nitric-oxide synthase, inducible (43); phenylpyruvate tautomerase (45); NAD(P)H dehydrogenase (46); hyaluronoglucosaminidase, *or* hyaluronidase (49); glucan 1,4-α-glucosidase, *or* glucoamylase (50); IκB kinase, weakly inhibited (52); protein-tyrosine kinase (26,53); Syk protein-tyrosine kinase (53); inositol 1,3,4-trisphosphate 5/6-kinase (54); glutathione *S*-transferase (55); histone acetyltransferase (57,58); arylamine *N*-acetyltransferase (59,60). **1**. Martin-Cordero, Lopez-Lazaro, Galvez & Ayuso (2003) *J. Enzyme Inhib. Med. Chem.* **18**, 505. **2**. Flynn, Rafferty & Boctor (1986) *Prostagland. Leuk. Med.* **22**, 357. **3**. Korutla & Kumar (1994) *Biochim. Biophys. Acta* **1224**, 597. **4**. Kang, Son, Yang, et al. (2004) *Nat. Prod. Res.* **18**, 295. **5**. Basu, Kole, Kubota & Owens (2004) *Drug Metab. Dispos.* **32**, 768. **6**. Uhle, Medalia, Waldron, et al.(2003) *EMBO J.* **22**, 1302. **7**. Bech-Otschir, Kraft, Huang, et al. (2001) *EMBO J.* **20**, 1630. **8**. Prasad, Raghavendra, Lokesh & Naidu (2004) *Prostaglandins Leukot. Essent. Fatty Acids* **70**, 521. **9**. Mizushina, Hirota, Murakami, et al. (2003) *Biochem. Pharmacol.* **66**, 1935. **10**. Shim, Kim, Cho, et al. (2003) *Chem. Biol.* **10**, 695. **11**. Vietri, Pietrabissa, Mosca, Spisni & Pacifici (2003) *Xenobiotica* **33**, 357. **12**. Chaudhary & Hruska (2003) *J. Cell Biochem.* **89**, 1. **13**. Takeshita, Chen, Watanabe, Kitano & Hanazawa (1995) *J. Immunol.* **155**, 419. **14**. Gopalakrishna & Gundimeda (2002) *J. Nutr.* **132**, 3819S. **15**. Sun, Wilson & Majerus (2002) *J. Biol. Chem.* **277**, 45759. **16**. Dyer, Khan, Bilmen, et al. (2002) *Cell Calcium* (2002) **31**, 45. **17**. Bilmen, Khan, Javed & Michelangeli (2001) *Eur. J. Biochem.* **268**, 6318. **18**. Shih & Claffey (2001) *Growth Factors* **19**, 19. **19**. Logan-Smith, Lockyer, East & Lee (2001) *J. Biol. Chem.* **276**, 46905. **20**. Thapliyal & Maru (2001) *Food Chem. Toxicol.* **39**, 541. **21**. Skrzypczak-Jankun, McCabe, Selman & Jankun (2000) *Int. J. Mol. Med.* **6**, 521. **22**. Nagao, Maeda, Lim,

Kobayashi & Terao (2000) *J. Nutr. Biochem.* **11**, 348. **23**. Gopalakrishna & Jaken (2000) *Free Radic. Biol. Med.* **28**, 1349. **24**. Zheng & Ramirez (2000) *Brit. J. Pharmacol.* **130**, 1115. **25**. Dorai, Gehani & Katz (2000) *Mol. Urol.* **4**, 1. **26**. Hong, Spohn & Hung (1999) *Clin. Cancer* Kawashima, Akimoto & Yamada (1992) *Lipids* **27**, 509. **39**. Fujiyama-Fujiwara, Umeda & Igarashi (1992) *J. Nutr. Sci. Vitaminol.* (1992) **38**, 353. **40**. van Iersel, Ploemen, Lo Bello, Federici & van Bladeren (1997) *Chem. Biol. Interact.* **108**, 67. **41**. Iersel, Ploemen, Struik, et al. (1996) *Chem. Biol. Interact.* **102**, 117. **42**. Zhang, Altorki, Mestre, Subbaramaiah & Dannenberg (1999) *Carcinogenesis* **20**, 445. **43**. Rao, Kawamori, Hamid & Reddy (1999) *Carcinogenesis* **20**, 641. **44**. Huang, Newmark & Frenkel (1997) *J. Cell. Biochem. Suppl.* **27**, 26. **45**. Molnar & Garai (2005) *Int. Immunopharmacol.* **5**, 849. **46**. Tsvetkov, Asher, Reiss, et al. (2005) *Proc. Natl. Acad. Sci. U.S.A.* **102**, 5535. **47**. Xu & Li (2005) *Curr. Med. Chem. Anticancer Agents* **5**, 281. **48**. Bauvois & D. Dauzonne (2006) *Med. Res. Rev.* **26**, 88. **49**. Girish & Kemparaju (2005) *Biochemistry (Moscow)* **70**, 708 and 948. **50**. Vijayakumar & Divakar (2005) *Biotechnol. Lett.* **27**, 1411. **51**. Mahmmoud (2007) *Brit. J. Pharmacol.* **150**, 200. **52**. Burke (2003) *Curr. Opin. Drug Discov. Devel.* **6**, 720. **53**. Gururajan, Dasu, Shahidain, et al. (2007) *J. Immunol.* **178**, 111. **54**. Sun, Wilson & Majerus (2002) *J. Biol. Chem.* **277**, 45759. **55**. Hayeshi, Mutingwende, Mavengere, Masiyanise & Mukanganyama (2007) *Food Chem. Toxicol.* **45**, 286. **56**. Basu, Ciotti, Hwang, et al. (2004) *J. Biol. Chem.* **279**, 1429. **57**. Costi, Di Santo, Artico, et al. (2007) *J. Med. Chem.* **50**, 1973. **58**. Morimoto, Sunagawa, Kawamura, et al. (2008) *J. Clin. Invest.* **118**, 868. **59**. Kukongviriyapan, Phromsopha, Tassaneeyakul, et al. (2006) *Xenobiotica* **36**, 15. **60**. Makarova (2008) *Curr. Drug Metab.* **9**, 538.

CVS-1123

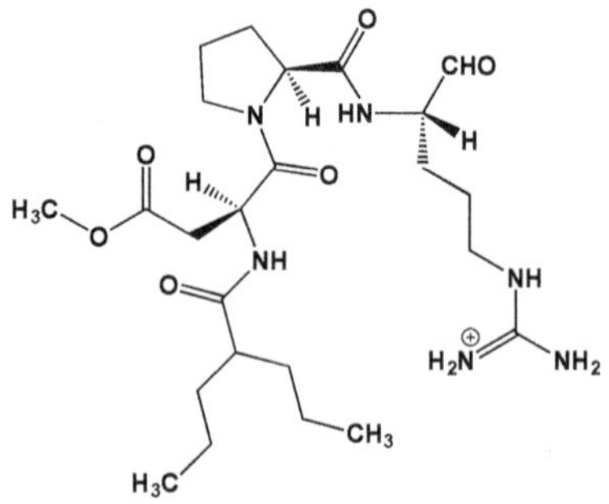

This L-arginal-containing peptidomimetic ($FW_{free-base}$ = 510.63 g/mol) is a potent inhibitor of thrombin that prevents coronary artery thrombosis in the conscious dog. CVS-1123 is a slow-binding competitive inhibitor of thrombin (IC_{50} = 1.13 nM) as well as a potent anticoagulant in plasma *in vitro*. It is a weaker inhibitor of coagulation factor Xa and plasmin, with respective IC_{50} values of 257 and 315 nM. **1**. Cousins, Friedrichs, Sudo, et al. (1996) *Circulation* **94**, 1705.

CW069

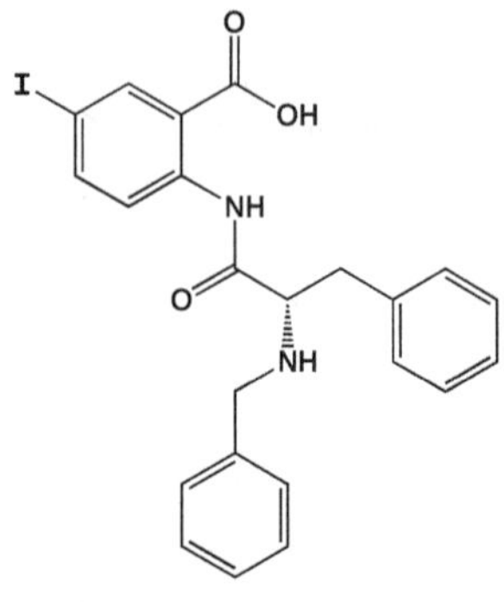

This inhibitor (FW = 500.34 g/mol) is an allosteric inhibitor (IC_{50} = 75 μM) of the (–)-end-directed, kinesin-related microtubule motor HSET, inducing multipolar mitosis in cancer cells with extra centrosomes and causing apoptosis by means of catastrophic aneuploidy. Notably, CW069 does not disrupt mitosis in normal human fibroblast cells. The increased incidence of CW069-induced multipolar mitoses in N1E-115 cells recapitulates the phenotype obtained by siRNA-mediated HSET depletion. CW069 also reduces centrosome clustering in BT549 and MDA-MB-231 breast cancer cells. Unlike KSP or CENP-E inhibitors, CW069 does not cause mitotic delay in HeLa cells with normal centrosome numbers, nor does it decrease the clonogenic capacity of primary adult human bone marrow cells. Thus, CW069 is less likely to not cause neutropenia toxicity in normal cells. Selective inhibition of HSET is consistent with computational modeling that suggests that dynamic conformational selection with CW069 cannot be achieved by the closely related KSP. The motor antagonism between plus-end-directed KSP and minus-end-directed HSET is understood to be responsible for establishing a proper bipolar spindle during mitosis (2), and KSP inhibition has been shown to result in monopolar spindle formation as well as mitotic delay (3,4). **1.** Watts, Richards, Bender, *et al.* (2013) *Chem. Biol.* **20**, 1399. **2.** Mountain, Simerly, Howard, *et al.* (1999) *J. Cell Biol.* **147**, 351. **3.** Gatlin and Bloom (2010) *Seminar Cell Dev. Biol.* **21**, 248. **4.** Mayer, Kapoor, Haggarty, *et al.* (1999) *Science* **286**, 971.

CX-4945

This orally bioavailable ATP-competitive protein kinase inhibitor and anti-angiogenic agent (FW = 349.77 g/mol; CAS 1009820-21-6), also named 5-(3-chlorophenyl-amino)benzo[*c*][2,6]naphthyridine-8-carboxylic acid, targets casein kinase CK2α, with a K_i of 0.38 and an intracellular IC$_{50}$ of 1 nM, exhibiting broad-spectrum anti-proliferative activity in multiple cancer cell lines (1,2). CX-4945 attenuates PI3K/Akt signaling, as evidenced by dephosphorylation of Akt at its CK2-specific site (S129) as well as other canonical phosphorylation sites (S473 and T308). CX-4945 also inhibits proliferation of human umbilical vein endothelial cells (HUVECs), resulting in cell-cycle arrest and apoptosis. In models of angiogenesis, CX-4945 inhibits HUVEC migration, capillary tube formation, and transcription of CK2-dependent Hypoxia-Induced Factor-11α (*HIF-1α*) gene in cancer cells. In this respect, may also prove to be a useful vascular-disrupting agent (VDA) that starves solid tumors of required nutrients. In A549 cells, CX-4945 inhibits the TGF-β1-induced cadherin switch and the activation of key signaling molecules involved in Smad (Smad2/3, Twist and Snail), non-Smad (Akt and Erk), Wnt (β-catenin) and focal adhesion signaling pathways (FAK, Src and paxillin) that cooperatively regulate the overall process of EMT (3). As a result, CX-4945 inhibits the migration and invasion of A549 cells accompanied with the downregulation of MMP-2 and 9 (3). CX-4945 also blocks TGF-β1-induced epithelial-to-mesenchymal transition in A549 human lung adenocarcinoma cells (4). **1.** Siddiqui-Jain, Drygin, Streiner *et al.* (2010) *Cancer Res.* **70**, 10288. **2.** Pierre, Chua, O'Brien, *et al.* (2011) *J. Med. Chem.* **54**, 635. **3.** Ku, Park, Ryu *et al.* (2013) *Bioorg. Med. Chem. Lett.* **23**, 5609. **4.** Kim, Hwan & Kim (2013) *PLoS One* **8**, e74342.

CXI-benzo-84

This tubulin-directed anticancer agent (FW = 402.52 g/mol), first identified by screening cells treated with a series of benzimidazoles, inhibits cell-cycle progression in metaphase, accumulating spindle assembly checkpoint proteins Mad2 and BubR1 on kinetochores and subsequently activating apoptosis. CXI-benzo-84 also depolymerizes both interphase and mitotic microtubules (MTs), perturbs MT binding of EB1 protein, and inhibits MT self-assembly. It likewise inhibits tubulin GTPase activity *in vitro* and

binds to tubulin at a single binding site (K_d = 1.2 µM), most likely by docking within the colchicine binding pocket. **1.** Rai, Gupta, Kini, *et al.* (2013) *Biochem. Pharmacol.* **86**, 378.

Cyanate

$$N\equiv C—O^{\ominus}$$

This conjugate base of cyanic acid N≡C–OH (FW$_{Na-Salt}$ = 65.01 g/mol; CAS 917-61-3 (sodium salt); FW$_{K-Salt}$ = 81.12 g/mol; CAS 590-28-3 (potassium salt); FW$_{NH3-Salt}$ = 59.05 g/mol; CAS 22981-32-4 (ammonium salt)) is an electrophile that reacts with amino, sulfhydryl, tyrosyl, imidazole, and carboxyl groups in proteins. Only the reaction with amino groups generates a stable product, R–NHC(=O)NH$_2$. Because an unprotonated amine is required for reaction, α-amino groups are more reactive than ϵ-amino groups, due to their lower pK_a values. This property permits preferential modification of α-amino groups. *Note*: Solutions of cyanate salts must be prepared freshly, as cyanate decomposes in aqueous environments. **Target(s):** ribonuclease (1,13); papain (2,9); trypsin (3); chymotrypsin (3); subtilisin (3,9,10); nitrate reductase (4); tubulin polymerization (6); formate dehydrogenase (6); hemoglobin S polymerization (7,8,23); alkaline phosphatase (11,27); pyrophosphatase (12); carbonic anhydrase (14,15,18,20, 28,32-36); glutamate dehydrogenase (16,26); phospho-ribosylformylglycinamidine synthetase (19,30);glycogen phosphorylase (17,22); carbamoyl-phosphate synthetase, glutamine (21,29); complement component C3/C5 (2)4; complement factor, *or* complement component C3b inactivator (24); glucose-6-phosphate dehydrogenase (25); fumarase (31); catalase (31); amidase (37); mucorpepsin (38); glycosylphosphatidylinositol phospholipase D (39); glutathione dehydrogenase (ascorbate), mildly inhibited (40). **1.** Stark, Stein & Moore (1960) *J. Biol. Chem.* **235**, 3177. **2.** Sluyterman (1967) *Biochim. Biophys. Acta* **139**, 439. **3.** Shaw, Stein & Moore (1964) *J. Biol. Chem.* **239**, PC671. **4.** Rigano & . Aliotta (1975) *Biochim. Biophys. Acta* **384**, 37. **5.** Mellado, Slebe & Maccioni (1982) *Biochem. J.* **203**, 675. **6.** Jollie & Lipscomb (1990) *Meth. Enzymol.* **188**, 331. **7.** Cerami & Manning (1971) *Proc. Natl. Acad. Sci. U.S.A.* **68**, 1180. **8.** Harkness (1976) *Trends Biochem. Sci.* **1**, 73. **9.** Cohen (1970) *The Enzymes*, 3rd ed., **1**, 147. **10.** Markland & Smith (1971) *The Enzymes*, 3rd ed., **3**, 561. **11.** Fernley (1971) *The Enzymes*, 3rd ed., **4**, 417. **12.** Josse & Wong (1971) *The Enzymes*, 3rd ed., **4**, 499. **13.** Richards & Wyckoff (1971) *The Enzymes*, 3rd ed., **4**, 647. **14.** Lindskog, Henderson, Kannan, *et al.* (1971) *The Enzymes*, 3rd. ed., **5**, 587. **15.** Thorslund & Lindskog (1967) *Eur. J. Biochem.* **3**, 117. **16.** Smith, Austen, Blumenthal & Nyc (1975) *The Enzymes*, 3rd ed., **11**, 293. **17.** Huang & Madsen (1966) *Biochemistry* **5**, 116. **18.** Lindskog (1966) *Biochemistry* **5**, 2641. **19.** Schroeder, Allison & Buchanan (1969) *J. Biol. Chem.* **244**, 5856. **20.** Tu & Silverman (1985) *Biochemistry* **24**, 5881. **21.** Anderson & Carlson (1975) *Biochemistry* **14**, 3688. **22.** Avramovic & Madsen (1968) *J. Biol. Chem.* **243**, 1656. **23.** Manning & Acharya (1984) *Amer. J. Pediatr. Hematol. Oncol.* **6**, 51. **24.** Schultz & Arnold (1975) *J. Immunol.* **115**, 1558. **25.** Glader & Conrad (1972) *Nature* **237**, 336. **26.** Veronese, Piszkiewicz & Smith (1972) *J. Biol. Chem.* **247**, 754. **27.** Carey & Butterworth (1969) *Biochem. J.* **111**, 745. **28.** Pocker & Stone (1968) *Biochemistry* **7**, 2936 and 3021. **29.** Anderson, Carlson, Rosenthal & Meister (1973) *Biochem. Biophys. Res. Commun.* **55**, 246. **30.** Buchanan, Ohnoki & Hong (1978) *Meth. Enzymol.* **51**, 193. **31.** Shyadehi & Harding (2002) *Biochim. Biophys. Acta* **1587**, 31. **32.** Elleby, Chirica, Tu, Zeppezauer & Lindskog (2001) *Eur. J. Biochem.* **268**, 1613. **33.** Yachandra, Powers & Spiro (1983) *J. Amer. Chem. Soc.* **105**, 6596. **34.** Engberg, Millqvist, Pohl & Lindskog (1985) *Arch. Biochem. Biophys.* **241**, 628. **35.** Sanyal, Pessah & Maren (1981) *Biochim. Biophys. Acta* **657**, 128. **36.** Hurt, Tu, Laipis & Silverman (1997) *J. Biol. Chem.* **272**, 13512. **37.** Gregoriou & Brown (1979) *Eur. J. Biochem.* **96**, 101. **38.** Rickert (1972) *Biochim. Biophys. Acta* **271**, 93. **39.** Stieger, Diem, Jakob & Brodbeck (1991) *Eur. J. Biochem.* **197**, 67. **40.** Amako, Ushimaru, Ishikawa, *et al.* (2006) *J. Nutr. Sci. Vitaminol.* **52**, 89.

Cyanide

$$\left[^{\ominus}:C\equiv N: \longleftrightarrow :C\equiv \ddot{N}:^{\ominus} \longleftrightarrow {}_{\ominus}^{..}\ddot{C}\equiv N:^{\oplus}\right]$$
Principal Resonance Form

This extremely poisonous ion and potent metalloenzyme inhibitor (*Symbol*: ⁻CN), commonly supplied as the potassium (FW = 65.12 g/mol; CAS 151-50-8) or sodium salt (FW = 49.01 g/mol; CAS 143-33-9), emits a characteristic odor of bitter almonds. Its protonated form, hydrogen cyanide

(*Symbol*: HCN) is rapid-acting asphyxiant. The main target of ⁻CN and HCN is cytochrome oxidase, an essential respiratory chain enzyme (see below). **Chemical Properties:** Hydrogen cyanide is colorless and odorless gas (*See also* Hydrogen Cyanide; Zyklon-B). When hydrated, HCN dissociates (pK_a = 9.31 at 25°C), and at physiologic pH the major species is HCN rather than CN⁻. CN⁻ can still be the active species even at pH values where its concentration is significantly lower than that of HCN. In the cyanide anion, the combined presence of a lone pair and negative charge makes the carbon atom the reactive nucleophilic species in associative S_N2 substitutions reactions. **Primary Mode of Inhibitor Action:** Cyanide ion combines rapidly with heme iron in cytochrome a_3, a component of mitochrondrial cytochrome oxidase, preventing intracellular oxygen utilization and instead forcing cells to rely on anaerobic metabolism, to produce excess lactic acid, and to experience metabolic acidosis. Extended X-ray absorption fine structure (EXAFS) revealed that cyanide binds to purified cytochrome oxidase in a multiphasic manner, suggesting that the resting enzyme is a mixture of several forms. Upon reduction and reoxidation (*i.e.*, conversion to the "oxygenated" form), however, the kinetics for cyanide binding become monophasic, and all preparations of the oxygenated form combine with cyanide at the same rate (1). Note that cyanide also exhibits high affinity for the ferric iron of methemoglobin. **Toxicity:** Whether ingested as free CN⁻ or inhaled as HCN, cyanide toxicity is usually evident within seconds At high dose, death often ensues within 6-8 minutes. For gaseous hydrogen cyanide, the LC_{50} is 100-300 ppm (or 1-3 mg/m³), resulting in death within 10-60 minutes. Inhalation of 2000 ppm HCN causes death within one minute. The LD_{50} for ingested cyanide is 50-200 mg, or 2-6 mg sodium cyanide per kg body weight. Upon contact with skin, the LD_{50} for hydrogen cyanide gas is 100 mg/kg body weight. Modest amounts of cyanide are also inevitably present in certain foods, and rhodanese (*Reaction*: CN⁻ + SSO_3^{2-} ⇌ SCN^{1-} + SO_3^{2-}), a liver enzyme, can convert trace quantities of cyanide at a rate of ~0.5 fmol/kg-min, forming the far less toxic thiocyanate which is excreted in urine. Cyanide toxicity also results from improperly prepared cassava, a cyanide-rich food (containing the cyanogenic glucosides, linamarin and lotaustralin) that is a staple in the diets of over 500 million in the developing world. In severe cases, cassava toxicity results in ataxic neuropathy and vision failure. Another source of cyanide is smoke from burning wood at campsites, fireplaces, and especially house and forest fires. The considerable affinity of cyanide for ionic cobalt commends the use of hydroxocobalamin as a detoxifying agent. Hydroxocobalamin is safe when administered as a 5-gram intravenous dose, effectively counteracting toxicity at low to moderate whole blood cyanide levels. **CAUTION:** Extreme care must be exercised when handling cyanide. All reagent bottles and stock solutions must be sealed tightly and protected from light. One must never combine potassium or sodium cyanide (in solid *or* liquid form) with strong acid, because gaseous HCN forms instantly. (Even the addition of potassium or sodium cyanide to a physiologic buffer (pH 7.4) can result in the formation of hydrogen cyanide gas.) All work must be carried out in a recently certified chemical fume hood, with the window properly positioned to assure adequate suction. Laboratory workers must follow all recommended institutional protocols for safe handling and disposal of cyanide. **Target(s):** Anaerobic fermentation (2); ascorbate oxidase (3); catalase (1,4); nitrate reductase, from spinach (*Spinacea oleracea*) (5); cytochrome oxidase, bimolecular rate constant = 10^6 M⁻¹s⁻¹ and K_d = 50 nM, or less (6); cytochrome P450-mediated drug oxidation in rat liver microsomes (7); glutathione peroxidase, ovine erythrocyte (8); heme oxygenase (9); 4-hydroxybenzoyl-CoA reductase (10); lignin peroxidase, from *Phanerochaete chrysosporium* (11); phloem translocation (12); stearyl-coenzyme A desaturase, reversible inhibition (13); superoxide dismutase, cochlear (14); vitamin K-dependent carboxylase (15). **1.** Naqui, Kumar, Ching, Powers, Chance (1984) *Biochemistry* **23**, 6222. **2.** Meyerhof & Kaplan (1952) *Arch. Biochem. Biophys.* **37**, 375. **3.** Strothkamp & Dawson (1977) *Biochemistry* **16**, 1926. **4.** Wolfe, Beers & Sizer (1957) *Arch. Biochem. Biophys.* **72**, 353. **5.** Notton & Hewitt (1971) *FEBS Lett.* **18**, 19. **6.** Jones, Bickar, Wilson, *et al.* (1984) *Biochem. J.* **220**, 57. **7.** Kitada, Chiba, Kamataki & Kitagawa (1977) *Japanese J. Pharmacol.* **27**, 601. **8.** Kraus, Prohaska & Ganther (1980) *Biochim Biophys Acta.* **615**, 19. **9.** Li, Syvitski, Auclair, de Montellano & La Mar (2004) *J. Biol. Chem.* **279**, 10195. **10.** Johannes, Unciuleac, Friedrich *et al.* (2008) *Biochemistry* **47**, 4964. **11.** de Ropp, La Mar, Wariishi & Gold (1991) *J. Biol. Chem.* **266**, 15001. **12.** Giaquinta (1977) *Plant Physiol.* **59**, 178. **13.** Hiwatashi, Ichickawa & Yamano (1975) *Biochim Biophys Acta.* **388**, 397. **14.** Pierson & Gray (1982) *Hearing Res.* **6**, 141. **15.** Dowd & Hamm (1991) *Proc. Natl. Acad. Sci. U.S.A.* **88**, 10583. **See** *Zyklon-B*

Cyanidin

This flavylium cation (FW$_{chloride}$ = 322.70 g/mol; brownish-red solid; λ_{max} = 535 nm in methanol and HCl), also known as the 3,5,7,3',4'-pentahydroxyflavylium cation, is the aglycon of many the plant pigments anthocyanins, mekocyanin, chrysanthemin, and cyanin). **Target(s):** aflatoxin biosynthesis (1); glucose uptake (2); CYP2A1 (3); CYP3A2 (3); epidermal growth-factor receptor protein-tyrosine kinase (4); DNA topoisomerase I (5); DNA topoisomerase II (5); glycogen phosphorylase (6). **1.** Norton (1999) *J. Agric. Food Chem.* **47**, 1230. **2.** Park (1999) *Biochem. Biophys. Res. Commun.* **260**, 568. **3.** Dai, Jacobson, Robinson & Friedman (1997) *Life Sci.* **61**, PL75. **4.** Meiers, Kemeny, Weyand, Gastpar, von Angerer & Marko (2001) *J. Agric. Food Chem.* **49**, 958. **5.** Habermeyer, Fritz, Barthelmes, *et al.* (2005) *Chem. Res. Toxicol.* **18**, 1395. **6.** Jakobs, Fridrich, Hofem, Pahlke & Eisenbrand (2006) *Mol. Nutr. Food Res.* **50**, 52.

N-(*N*-Cyanoacetyl-L-phenylalanyl)-L-phenylalanine

This modified dipeptide (FW$_{free-acid}$ = 379.42 g/mol) is a mechanism-based inactivator of human neprilysin (neutral endopeptidase 24.11) and angiotensin-converting enzyme (or, peptidyl dipeptidase A), by intercepting an enzyme-bound ketenimine intermediate. It was the first mechanism-base inhibitor of angiotensin-converting enzyme. Note that the enzyme catalyzes the hydrolysis of the inhibitor, yielding the partition ratio (*i.e.*, k_{cat}/k_{inact}) of 8300. **1.** Levy, Taibi, Mobashery & Ghosh (1993) *J. Med. Chem.* **36**, 2408. **2.** Ghosh, Said-Nejad, Roestamadji & Mobashery (1992) *J. Med. Chem.* **35**, 4175.

β-Cyano-L-alanine

This toxic amino acid (FW = 114.10 g/mol; CAS 6232-19-5), found in the legume *Vicia sativa* (common vetch), β-cyano-L-Alanine (BCA) reversibly inhibits cystathionine γ-lyase, *or* CSE (IC_{50} = 6.5 μM), a major H_2S-producing enzyme that is primarily present in both smooth muscle and endothelium as well as in periadventitial adipose tissues (1–3). CSE is a versatile pyridoxal-P-dependent catalyst for the conversion of (a) cystathionine into cysteine, α-ketobutyrate, and NH₃; (b) L-homoserine into H₂O, NH₃ and 2-oxobutanoate; (c) L-cystine into thiocysteine, pyruvate and NH₃, and (d) L-cysteine into pyruvate, NH₃ and H₂S. Regulation of H₂S production from CSE is controlled by a complex integration of transcriptional, posttranscriptional, and posttranslational mechanism. β-Cyano-L-alanine blocks H₂S synthesis in rat liver preparations. **Target(s):** tryptophanase (4); aspartate 4-decarboxylase, K_i = 0.27 mM for the *Alcaligenes faecalis* enzyme (5–7); cystathionine β-lyase (8–12); alanine aminotransferase (13,14); D-amino-acid oxidase (15); UDP-*N*-acetylmuramate: L-alanine ligase, *or* UDP-*N* acetylmuramoyl-L-alanine synthetase, *or* L-alanine-adding enzyme (16,17,18); asparaginyl-tRNA synthetase (19); serine-sulfate ammonia-lyase (20); asparaginase, as slow alternative substrate (21,22); *O*-phosphoserine sulfhydrylase (23). **1.** Nagasawa, Kanzaki & Yamada (1987) *Meth. Enzymol.* **143**, 486. **2.** Pfeffer & Ressler (1967) *Biochem. Pharmacol.* **16**, 2299. **3.** Nagasawa, Kanzaki & Yamada (1984) *J. Biol. Chem.* **259**, 10393. **4.** Morino & Snell (1970) *Meth. Enzymol.* **17A**, 439. **5.** Tate, Novogrodsky, Soda, Miles & Meister (1970) *Meth. Enzymol.* **17A**, 681. **6.** Boeker & Snell (1972) *The Enzymes*, 3ʳᵈ ed., **6**, 217. **7.** Tate & Meister (1969) *Biochemistry* **8**, 1660. **8.** Giovanelli (1987) *Meth. Enzymol.* **143**, 443. **9.** Giovanelli & Mudd (1971) *Biochim. Biophys. Acta* **227**, 654. **10.** Dwivedi, Ragin & J. R. Uren (1982) *Biochemistry* **21**, 3064. **11.** Alting, Engels, van Schalkwijk & Exterkate (1995) *Appl. Environ. Microbiol.* **61**, 4037. **12.** Giovanelli, Owens & Mudd (1971) *Biochim. Biophys. Acta* **227**, 671. **13.** Ator & Ortiz de Montellano (1990) *The Enzymes*, 3ʳᵈ ed., **19**, 213. **14.** Alston, Porter, Mela & Bright (1980) *Biochem. Biophys. Res. Commun.* **92**, 299. **15.** Miura, Shiga, Miyake, Watari & Yamano (1980) *J. Biochem.* **87**, 1469. **16.** Liger, Blanot & van Heijenoort (1991) *FEMS Microbiol. Lett.* **64**, 111. **6. 17.** Liger, Masson, Blanot, van Heijenoort & Parquet (1995) *Eur. J. Biochem.* **230**, 80. **18.** Ishiguro (1982) *Can. J. Microbiol.* **28**, 654. **19.** Lea & Fowden (1973) *Phytochemistry* **12**, 1903. **20.** Tudball & Thomas (1973) *Eur. J. Biochem.* **40**, 25. **21.** Lea, Fowden & Miflin (1978) *Phytochemistry* **17**, 217. **22.** Chang & Farnden (1981) *Arch. Biochem. Biophys.* **208**, 49. **23.** Mino & Ishikawa (2003) *FEBS Lett.* **551**, 133.

N-(3-Cyanobenzyl)-*N*-(4-cyano-3-(2-tolyl)phen-1-yl)-*N*-(3-methyl-3*H*-imidazol-4-ylmethyl)amine

This imidazole-containing biphenyl (FW= 417.52 g/mol) inhibits rat protein farnesyltransferase, *or* FTase (IC₅₀ = 0.97 nM). The 4-cyanobenzyl analogue is a slightly weaker inhibitor (IC₅₀ = 10 nM). This inhibitor is cellularly active (≤80 nM) and shows >300-fold selectivity for FTase *versus* geranylgeranyltransferase (GGTase) selective. An X-ray crystal structure of inhibitor bound to rat farnesyltransferase is also presented. **1.** Curtin, Florjancic, Cohen, *et al.* (2003) *Bioorg. Med. Chem. Lett.* **13**, 1367.

4-(4-Cyanobenzoyl)-2,5-dimethyl-1-(2(*R*)-hydroxy-2-methyl-3,3,3-trifluoro-propionyl)piperazine

This substituted piperazine (FW = 383.37 g/mol), also known as Nov3r, inhibits pyruvate dehydrogenase (acetyl transferring)] kinase (IC₅₀ = 6.7 μM). **1.** Mann, Dragland, Vinluan, *et al.* (2000) *Biochim. Biophys. Acta* **1480**, 283. **2.** Aicher, Anderson, Bebernitz, *et al.* (1999) *J. Med. Chem.* **42**, 2741. **3.** Aicher, Anderson, Gao, *et al.* (2000) *J. Med. Chem.* **43**, 236. **4.** Roche & Hiromasa (2007) *Cell. Mol. Life Sci.* **64**, 830. **5.** Knoechel, Tucker, Robinson, *et al.* (2006) *Biochemistry* **45**, 402. **6.** Mayers, Leighton & Kilgour (2005) *Biochem. Soc. Trans.* **33**, 367.

(5*E*)-6-(3-(2-Cyano-3-benzylguanidino)phenyl)-6-(3-pyridyl)hex-5-enoate

This guanidine derivative (FW_free-acid = 439.52 g/mol) inhibits human thromboxane-A synthase (IC₅₀ = 11 nM) and is a thromboxane receptor antagonist (IC₅₀ = 32 nM). It also inhibits the collagen-induced platelet aggregation in human platelet-rich plasma and whole blood. **1.** Soyka, Guth, Weisenberger, Luger & Müller (1999) *J. Med. Chem.* **42**, 1235.

2(*S*)-{2-[(2(*S*)-{2-[3-(4-Cyanobenzyl)-3*H*-imidazol-4-yl]acetylamino}-3(*S*)-methyl-pentyl)naphthalen-1-ylmethylamino]acetylamino}-4-(methylsulfanyl)butyric Acid

This compound (FW = 668.86 g/mol), also known as Merck 21, is a potent inhibitor of Ras farnesyltransferase (IC₅₀ = 0.15 nM). The IC₅₀ is 67 nM for protein geranylgeranyl-transferase I. **1.** Anthony, Gomez, Schaber, *et al.* (1999) *J. Med. Chem.* **42**, 3356. **2.** Gibbs (2001) *The Enzymes*, 3rd ed., **21**, 81.

Cyanoborohydride

This anionic reducing agent (FW_Na-Salt = 62.842 g/mol; Solubility = 2 g/mL H₂O) is a powerful reducing reagent that is more stable and more selective than borohydride. Cyano-borohydride reduces aldehydes and ketones to their corresponding alcohols. It will not reduce carboxyl groups, amides, ethers, lactones, or epoxides. It is frequently used for reductive amination of aldehydes, ketones, imines, as well as reductive alkylation of amines and hydrazines. **Applications:** Given its relative stability in aqueous solutions and preference as an imine reductant, cyanoborohydride has found many applications in protein and enzyme chemistry. Indeed, this agent is very useful for trapping covalent intermediates formed during enzyme catalysis (1-3). Antibodies to carbohydrates: preparation of antigens by coupling carbohydrates to proteins by reductive amination with cyanoborohydride (4); Use in the preparation of biologically active nitroxides (5); Reduction of nicotinamide adenine dinucleotides by sodium cyanoborohydride (6); Factors affecting cyanoborohydride reduction of aromatic Schiff's bases in proteins (7); Protein labeling by reductive methylation with sodium cyanoborohydride: effect of cyanide and metal ions on the reaction (8); Labeling of proteins by reductive methylation using sodium cyanoborohydride (9); Reversibility of the ketoamine linkages of aldoses with proteins (10); Detection and determination of protein-bound formaldehyde: Improved recovery of formaldehyde by reduction with sodium cyanoborohydride (11); Cyanoborohydride reduction of rhodopsin (12); Identification of bound pyruvate essential for the activity of phosphatidylserine decarboxylase of *Escherichia coli* (13); Chemical trapping of complexes of dihydroxyacetone phosphate with muscle fructose-1,6-bisphosphate aldolase (14); Use of *p*-aminobenzoic acid and tritiated cyanoborohydride for the detection of pyruvoyl residues in proteins (15); Inactivation of 5-enolpyruvylshikimate 3-phosphate synthase by its substrate analogue pyruvate in the presence of sodium cyanoborohydride (16); Identification of lysine-153 as a functionally important residue in UDP-galactose 4-epimerase from *Escherichia coli* (17); Proteins containing reductively aminated disaccharides: chemical and immunochemical characterization (18); Characterization of the products of reduction of the collagen cross-links in skin, tendon and bone with sodium cyanoborohydride (19); Lectin purification on affinity columns containing reductively aminated disaccharides (20). **1.** Allison & Purich (2002) *Meth. Enzymol.* **354**, 455. **2.** Frey & Hegestad *Enzymatic Reaction Mechanisms* (2007) Oxford University Press, New York. **3.** Purich (2010) *Enzyme Kinetics: Catalysis & Control*, Academic Press, NY. **4.** Gray (1978) *Meth. Enzymol.* **50**, 155. **5.** Rosen (1974) *J. Med. Chem.* **17**, 358. **6.** Avigad (1979) *Biochim. Biophys. Acta.* **571**, 171. **7.** Chauffe & Friedman (1977) *Adv. Exp. Med. Biol.* **86A**, 415. **8.** Jentoft & Dearborn (1980) *Anal. Biochem.* **106**, 186. **9.** Jentoft & Dearborn (1979) *J. Biol. Chem.* **254**, 4359. **10.** Acharya & Sussman (1984) *J. Biol. Chem.* **259**, 4372. **11.** Brunn & Klostermeyer (1983) *Z. Lebensm. Unters. Forsch.* **176**, 367. **12.** Fager (1982) *Meth. Enzymol.* **81**, 288. **13.** Satre & Kennedy (1978) *J. Biol. Chem.* **253**, 479. **14.** Kuo & Rose (1985) Biochemistry **24**, 3947. **15.** van Poelje & Snell (1987) *Anal. Biochem.* **161**, 420. **16.** Huynh (1992) *Biochem. Biophys. Res. Commun.* **185**, 317. **17.** Swanson & Frey (1993) Biochemistry 32, 13231. **18.** Gray, Schwartz & Kamicker (1978) *Prog. Clin. Biol. Res.* **23**, 583. **19.** Robins & Bailey (1977) *Biochem J.* **163**, 339. **20.** Baues & Gray (1977) *J. Biol. Chem.* **252**, 57.

Cyanocobalamin

This vitamin B_{12} analogue (FW = 1355.37 g/mol; CAS 68-19-9), is a precursor to the physiologic cobalamins, because its cyanide group is readily replaced by other substituents. Aqueous solutions are photolabile and will generate hydroxocobalamin. Strongly acidic or alkaline conditions will slowly hydrolyze the amide groups, even at room temperature. It will react with cyanide at pH 10 to produce dicyanocobalamin. **Target(s):** ethanolamine ammonia-lyase (1-3,5,6); propanediol dehydratase (2,8); glycerol dehydratase (2,7); HIV-1 integrase (4); ribonucleoside-triphosphate reductase, K_i = 42.6 μM (9). **1.** Kaplan & Stadtman (1971) *Meth. Enzymol.* **17B**, 818. **2.** Abeles (1971) *The Enzymes*, 3rd ed. (Boyer, ed.), **5**, 481. **3.** Stadtman (1972) *The Enzymes*, 3rd ed. (Boyer, ed.), **6**, 539. **4.** Weinberg, Shugars, Sherman, Sauls & Fyfe (1998) *Biochem. Biophys. Res. Commun.* **246**, 393. **5.** Bradbeer (1965) *J. Biol. Chem.* **240**, 4675. **6.** Blackwell & Turner (1978) *Biochem. J.* **175**, 555. **7.** Smiley & Sobolov (1962) *Arch. Biochem. Biophys.* **97**, 538. **8.** Sauvageot, Pichereau, Louarme, *et al.* (2002) *Eur. J. Biochem.* **269**, 5731. **9.** Suto, Poppe, Rétey & Finke (1999) *Bioorg. Chem.* **27**, 451.

(2S)-2-Cyano-1-[(2S,4S)-4-(5-cyano-2-pyridyl)amino-2-pyrrolidinylcarbonyl]-pyrrolidine

This substituted (FW_{free-base} = 310.36 g/mol) inhibits dipeptidyl-peptidase IV (IC_{50} = 0.25 nM for the human enzyme). Based on pharmacokinetic experiments in rats, the representative compound 11, which displayed high oral bioavailability (BA=83.9%) and long half-life in plasma (t(1/2)=5.27 h), was found to have an excellent pharmacokinetic profile. **1.** Sakashita, Akahoshi, Kitajima, Tsutsumiuchi & Hayashi (2006) *Bioorg. Med. Chem.* **14**, 3662.

3-Cyano-6-{[2-([2-((2S)-2-cyanopyrrolidin-1-yl)-2-oxoethyl]-amino)-ethyl]-amino}-pyridine

This substituted pyrrolidone (FW = 298.35 g/mol), also known as NVP-DDP728, is a strong inhibitor of dipeptidyl peptidase IV (IC_{50} = 1.4 nM for the human enzyme). **Target(s):** cytosol nonspecific Dipeptidase, *or* prolyl Dipeptidase (1); dipeptidyl-peptidase II, weakly inhibited (1); dipeptidyl-peptidase IV (1,2). **1.** Jiaang, Chen, Hsu, *et al.* (2005) *Bioorg. Med. Chem. Lett.* **15**, 687. **2.** Sakashita, Akahoshi, Kitajima, Tsutsumiuchi & Hayashi (2006) *Bioorg. Med. Chem.* **14**, 3662.

(5E)-6-(3-(2-Cyano-3-cyclohexylguanidino)phenyl)-6-(3-pyridyl)hex-5-enoate

This guanidine derivative (FW_{free-acid} = 431.54 g/mol) inhibits human thromboxane-A synthase (IC_{50} = 3 nM) and is a thromboxane receptor antagonist (IC_{50} = 47 nM). Note that the structural isomer having the guanidino group located in the *para*-position of the phenyl moiety is a significantly weaker inhibitor (IC_{50} = 0.18 and 1.3 μM for the synthase and receptor, respectively). **1.** Soyka, Guth, Weisenberger, Luger & Müller (1999) *J. Med. Chem.* **42**, 1235.

(S)-N-(1-Cyanocyclopropyl)-3-(methylsulfonyl)-2-(((S)-2,2,2-trifluoro-1-(4-fluoro-phenyl)ethyl)amino)propanoate

This orally bioavailable protease inhibitor (FW = 407.38 g/mol), also kown as Compound 6, selectively targets cathepsin S, *or* CTSS (IC$_{50}$ = 2.9 µM), with weaker action against cathepsins K, V, L and B. Cathepsin S is a lysosomal protease that is expressed under inflammatory conditions, including autoimmune disorders, cardiovascular disease, and cancer. Compound 6 blocks both murine and human cell invasion and disrupts endothelial tube formation, underscoring its anti-angiogenic properties. Selectivity of Compound 6 towards CTSS was confirmed using an MC38 knock-down model, whereby treatment of CTSS-depleted cells with Compound 6 produced no further reduction in invasiveness. **1**. Wilkinson, Young, Burden, Williams & Scott (2016) *Mol. Cancer* **15**, 29.

α-Cyano-4-hydroxycinnamate

This surface matrix molecule and pyruvate transport inhibitor (FW = 189.17 g/mol; CAS 28166-41-8; IUPAC Name: (*E*)-2-cyano-3-(4-hydroxyphenyl)prop-2-enoate), which is often used to desorb and mobilize peptides and nucleotides into the ion analyzer stream in MALDI mass spectrometers, also strongly inhibits the transport of pyruvate, but not acetate or butyrate, into liver mitochondria and erythrocytes (1). In the red cells, lactate uptake is also inhibited. α-Cyano-4-hydroxycinnamate is a non-competitive inhibitor (K_i = 6.3 µM), whereas phenyl-pyruvate is a competitive inhibitor (K_i = 1.8 mM). At 100 µM, α-cyano-4-hydroxycinnamate rapidly and almost totally inhibited O$_2$ uptake by rat heart mitochondria oxidizing pyruvate (1). Inhibition could be detected at concentrations of inhibitor as low as 1 µM, although inhibition requires more time to develop at this concentration (1). Importantly, inhibition is reversed by diluting out the inhibitor. In Ehrlich ascites tumor cells, lactate transport is inhibited competitively by (a) a variety of other substituted monocarboxylic acids (*e.g.*, pyruvate, K_i = 6.3 mM), which were themselves transported; (b) the non-transportable analogues α-cyano-4-hydroxycinnamate (K_i = 0.5 mM), α-cyano-3-hydroxycinnamate (K_i = 2mM) and DL-*p*-hydroxyphenyl-lactate (K_i = 3.6 mM); and (c) the thiol-group reagent mersalyl (K_i = 125 µM) (2). (**See also** UK5099, another *mitochondrial pyruvate transport inhibitor*) **1**. Halestrap (1975) *Biochem. J.* **148**, 85. **2**. Spencer & Lehninger (1976) *Biochem. J.* **154**, 405.

2α-Cyano-17β-hydroxy-4,4,17α-trimethylandrost-5-en-3-one

This semisynthetic steroid (FW = 355.52 g/mol), commonly referred to as cyanoketone, is a strong inhibitor of 3β-hydroxysteroid dehydrogenase/steroid Δ^5-isomerase. **Target(s):** 3β-hydroxysteroid dehydrogenase/steroid Δ^5-isomerase (1-6,10); cholesterol monooxygenase, side-chain-cleaving, *or* CYP11A1 (7,8); cytochrome *c* oxidase (7); steroid Δ isomerase (9). **1**. Edwards, O'Conner, Bransome & Braselton (1976) *J. Biol. Chem.* **251**, 1632. **2**. Cooke (1996) *J. Steroid Biochem. Mol. Biol.* **58**, 95. **3**. Takahashi, Luu-The & Labrie (1990) *J. Steroid Biochem. Mol. Biol.* **37**, 231. **4**. Sharp & Penning (1988) *Steroids* **51**, 441. **5**. Yu, Ku, Chang & Hsu (1987) *Cell Tissue Res.* **250**, 585. **6**. van der Vusse, Kalkman & van der Molen (1974) *Biochim. Biophys. Acta* **348**, 404. **7**. Graves, Uzgiris, Querner, *et al.* (1978) *Endocrinology* **102**, 1077. **8**. Farese (1970) *Steroids* **15**, 245. **9**. Neville & Engel (1968) *Endocrinology* **83**, 873. **10**. Ohno, Matsumoto, Watanabe & Nakajin (2004) *J. Steroid Biochem. Mol. Biol.* **88**, 175.

4-*N*-(4-Cyanophenyl)amino-5-(4-ethoxyphenyl)-7-(β-D-*erythro*-furanosyl)-pyrrolo[2,3-*d*]pyrimidine

This tubercidin analogue (FW = 457.49 g/mol) inhibits human adenosine kinase (IC$_{50}$ = 2 nM). Because the use of adenosine receptor agonists is plagued by dose-limiting cardiovascular side effects, there is great interest in developing adenosine kinase inhibitors (AKIs) as a means for raising steady-state adenosine concentrations. **1**. Boyer, Ugarkar, Solbach, *et al.* (2005) *J. Med. Chem.* **48**, 6430.

2α-Cyanoprogesterone

This steroid (FW = 339.48 g/mol) is an irreversible active-site-directed inhibitor of *Pseudomonas testosteroni* steroid Δ-isomerase (1). It is also a potent inhibitor of the bovine corpus luteum β-hydroxysteroid dehydrogenase, K_i = 15 nM (2,3). **1**. Penning (1985) *Biochem. J.* **226**, 469. **2**. Sharp, Senior & Penning (1985) *Biochem. J.* **230**, 587. **3**. Sharp & Penning (1988) *Steroids* **51**, 441.

2α-Cyano-4,4,17α-trimethyl-Δ^5-androsten-17β-ol-3-one

This synthetic steroid (FW = 355.52 g/mol), also known as cyanoketone, inhibits many steroidogenic processes (1). **Target(s):** cytochrome *c* oxidase (2); cholesterol monooxygenase, side-chain cleaving (2,3,7); Δ^5-3β-hydroxysteroid dehydrogenase (4-7); steroid Δ-isomerase (1,6); 20α-hydroxysteroid dehydrogenase, moderately inhibited (7); estradiol receptor (8). **1**. Talalay & Benson (1972) *The Enzymes*, 3rd ed. (Boyer, ed.), **6**, 591. **2**. Graves, Uzgiris, Querner, *et al.* (1978) *Endocrinology* **102**, 1077. **3**. Farese (1970) *Steroids* **15**, 245. **4**. Josso (1970) *Biol. Reprod.* **2**, 85. **5**. van der Vusse, Kalkman & van der Molen (1974) *Biochim. Biophys. Acta* **348**, 404. **6**. Edwards, O'Conner, Bransome & Braselton (1976) *J. Biol. Chem.* **251**, 1632. **7**. Rabe, Kiesel, Kellermann, *et al.* (1983) *Fertil. Steril.* **39**, 829. **8**. Wolfson, Richards & Rotenstein (1983) *J. Steroid Biochem.* **19**, 1817.

Cyanuric Acid

This toxic triazine (FW = 129.08 g/mol), also known as 5-azabarbituric acid, is a disinfectant and a reagent for organic synthesis. It is also a precusor for cyanic acid. Cyanuric acid is slightly soluble in water (0.2% at 25°C) and has pK_a values of 6.88, 11.40, and 13.5. The keto form, also known as isocyanuric acid, is the main tautomer in acidic solutions whereas the enol form is more stable in basic solutions. **Target(s):** barbiturase (1); orotate phosphoribosyl-transferase (2); urate oxidase, or uricase, K_i = 30 μM (3,4); xanthine oxidase (5). **1**. Soong, Ogawa, Sakuradani & Shimizu (2002) *J. Biol. Chem.* **277**, 7051. **2**. Javaid, el Kouni & Iltzsch (1999) *Biochem. Pharmacol.* **58**, 1457. **3**. Fridovich (1965) *J. Biol. Chem.* **240**, 2491. **4**. Truscoe (1968) *Enzymologia* **34**, 337. **5**. Fridovich (1965) *Biochemistry* **4**, 1098.

Cyc-116

This potent mitotic protein kinase inhibitor (FW = 368.46 g/mol; CAS 693228-63-6; Solubility = 24 mg/mL DMSO (with warming)), also named 4-(2-amino-4-methylthiazol-5-yl)-*N*-(4-morpholinophenyl)pyrimidin-2-amine, targets Aurora A (K_i = 8.0 nM) and Aurora B (K_i = 9.2 nM). **Other Targets:** VEGFR2 (K_i = 44 nM); Flt3 (K_i = 44 nM); and 50x greater potency toward Aurora A and B than other CDKs. Cyc-116 is inactive against PKA, Akt/PKB, PKC, no effect on GSK-3α/β, CK2, Plk1 and SAPK2A. **1**. Wang, Midgley, Scaërou, *et al.* (2010) *J. Med. Chem.* **53**, 4367.

CYC202, *See* Roscovitine

cyclic-AMP, *See* Adenosine 3',5'-Cyclic Monophosphate

cyclic Adenosine 3',5'-Monophosphate, *See* Adenosine 3',5'-Cyclic Monophosphate

cyclic-GMP, *See* Guanosine 3',5'-Cyclic Monophosphate

cyclic Guanosine 3',5'-Monophosphate, *See* Guanosine 3',5'-Cyclic Monophosphate

Cyclin-Dependent Kinase Inhibitor p21

This potent, high-affinity Cdk-binding protein (MW = 21 kDa), also known as p21^{CIP1} and 21-kDa Cdk-interacting protein-1 (Cip1), inhibits the phosphorylation of retinoblastoma tumor suppressor protein (Rb) by cyclin A-Cdk2, cyclin E-Cdk2, cyclin D1-Cdk4, and cyclin D2-Cdk4 complexes (1). Cyclin-dependent kinase Cdk2 associates with cyclins A, D, and E and has been implicated in controlling the G₁-to-S cell cycle transition in mammals. 1α,25-Dihydroxyvitamin D₃ inhibits pancreatic cancer cell proliferation by up-regulating the expression of cyclin-dependent kinase inhibitor p21 (2). **1**. Harper, Adami, Wei, Keyomarsi & Elledge (1993) *Cell* **75**, 805. **2**. Kanemaru, Maehara & Chijiiwa (2013) *Hepatogastro-enterol.* **60**, 1199.

Cycloartenol

This widely occurring phytosteroid precursor (FW = 426.73 g/mol; CAS 469-38-5), itself a substrate for cycloartenol 24-C-methyltransferase, inhibits trypsin and chymotrypsin, with respective K_i values of 25 and 420 μM, respectively). **Target(s):** sterol acyltransferase, or cholesterol acyltransferase (1); cycloeucalenol cyclo-isomerase, weakly inhibited (2); chymotrypsin (3); trypsin (3). **1**. Billheimer (1985) *Meth. Enzymol.* **111**, 286. **2**. Rahier, Taton & Beneviste (1989) *Eur. J. Biochem.* **181**, 615. **3**. Rajic, Akihisa, Ukiya, *et al.* (2001) *Planta Med.* **67**, 599.

Cyclobenzaprine

This widely used oral muscle relaxant (FW$_{free-base}$ = 275.39 g/mol; FW$_{HCl}$ = 311.85 g/mol; CAS 303-53-7), also named 3-(5*H*-dibenzo[*a,d*]cyclohepten-5-ylidene)- *N,N*-dimethyl-1-propanamine, is a centrally acting drug that most likely activates locus coeruleus neurons, leading to an increased release of noradrenaline in the ventral horn of the cord and the subsequent inhibitory action of noradrenaline on alpha motoneurons (1). Other studies indicate cyclobenzaprine is a 5-HT₂ receptor antagonist and that its muscle relaxant effect is due to inhibition of serotonergic, not noradrenergic, descending systems in the spinal cord (2). **1**. Commissing, Karoum, Reiffenstein & Neff (1981) *Can. J. Physiol. Pharmacol.* **59**, 37. **2**. Kobayashi, Hasegawa & Ono (1996) *Eur. J. Pharmacol.* **311**, 29.

(3*R*,4*S*)-4-(Cyclobutylthiomethyl)-1-[(9-deazaadenin-9-yl)methyl]-3-hydroxy-pyrrolidine

This transition-state mimic (FW$_{free-base}$ = 319.43 g/mol) inhibits *Escherichia coli* 5'-methylthioadenosine/*S*-adenosylhomocysteine nucleosidase, or MTAN (K_i' = 2.2 pM), an enzyme that hydrolyzes its substrates to form adenine and 5-methylthioribose (MTR) or *S*-ribosylhomocysteine (SRH). 5'-Methylthioadenosine (MTA) is a by-product of polyamine synthesis and SRH is a precursor to the biosynthesis of one or more quorum sensing autoinducer molecules. MTAN is therefore involved in quorum sensing, recycling MTA from the polyamine pathway via adenine phosphoribosyltransferase and recycling MTR to methionine. Hydrolysis of MTA by *E. coli* MTAN involves a highly dissociative transition state with ribooxacarbenium ion character. This iminoribitol mimics the transition state of MTAN, a property surmised from its extraordinary affinity as an inhibitor. **1**. Singh, Evans, Lenz, *et al.* (2005) *J. Biol. Chem.* **280**, 18265.

Cycloheximide

This antibiotic and rodenticide (FW = 281.35 g/mol; CAS 66-81-9; Solubility: 2.1 g/100 mL H₂O at 2°C), also known as 3-[2(3,5-dimethyl-2-oxocyclohexyl)-2-hydroxyethyl]glutarimide, from *Streptomyces griseus* inhibits eukaryotic protein biosynthesis by interferring with chain elongation and translocation, *i.e.*, inhibition of the peptidyltransferase activity on the 60S subunit of the ribosome (1-4). Cycloheximide is without effect on prokaryotic and mitochondrial protein biosynthesis. Displaying significant toxicity (*e.g.,* DNA damage, teratogenesis, and birth defects), cycloheximide is unsuitable as a human medicine. **Use in Estimating *In Vivo* Half-life of Proteins:** Because most proteins undergo continual turnover, the steady-state concentration depends on their rate constants for synthesis (k_{syn}) and degradation (k_{deg}), the latter often defined as a first-order process, with $t_{1/2.} = 0.693/k_{deg}$. In the presence of cycloheximide, one may use SDS gel electrophoresis and Western blotting (with a protein-spacific antibody) to determine $t_{1/2.}$ (4). The *in vivo* turnover of the rat liver tyrosine transaminase, for example, was measured by a label and chase procedure under conditions where the amount of enzyme undergoes no change. The half-life of the ^{14}C-labeled enzyme in this basal condition was found to be 1.5 hours. Inhibitors of protein synthesis (cycloheximide or puromycin) do not appreciably influence the basal enzyme level over a 5-hour period, although these drugs will block hormonal induction of this enzyme. In pulse-labeling experiments, cycloheximide blocked transaminase synthesis almost completely (4). Such findings showed that enzyme degradation kinetics can be determined when rat liver protein synthesis is blocked by cycloheximide. **Target(s):** protein biosynthesis, eukaryotic (1-4,8); peptidyltransferase (5); alcohol dehydrogenase, yeast (6); DNA-directed RNA polymerase (7); agmatine deaminase (9); ferric-chelate reductase (10); stearoyl-CoA 9-desaturase (11). **1.** Pestka (1974) *Meth. Enzymol.* **30**, 261. **2.** Carrasco, Battaner & Vazquez (1974) *Meth. Enzymol.* **30**, 282. **3.** Scott & Tomkins (1975) *Meth. Enzymol.* **40**, 273. **4.** Kenney (1967) *Science* **156**, 525. **5.** Dresios, Panopoulos, Frantziou & Synetos (2001) *Biochemistry* **40**, 8101. **6.** Sund & Theorell (1963) *The Enzymes*, 2nd ed., **7**, 25. **7.** Chambon (1974) *The Enzymes*, 3rd ed., **10**, 261. **8.** Trakatellis, Montjar & Axelrod (1965) *Biochemistry* **4**, 2065. **9.** Legaz, Iglesias & Vicente (1983) *Z. Pflanzenphysiol.* **110**, 53. **10.** Schmidt & Bartels (1998) *Protoplasma* **203**, 186. **11.** Jeffcoat (1977) *Biochem. Soc. Trans.* **5**, 811.

2-[5-[(Cyclohexylsulfamoyl)amino]-2-(4-fluorophenyl)-6-oxo-1,6-dihydro-1-pyrimidinyl]-*N*-[1-isopropyl-2-oxo-3,3,3-trifluoropropyl]acetamide

This substituted 5-aminopyridin-6-one (FW = 575.58 g/mol) inhibits human leukocyte elastase (K_i = 1.6 nM), relying on its trifluoromethylketone as the chemically active warhead to react with the active-site Ser-195. Such molecules are more bioavailable than boronate and haloketone inhibitors. **1.** Veale, Bernstein, Bryant, *et al.* (1995) *J. Med. Chem.* **38**, 98.

3-(Cyclohexylthio)-1,1,1-trifluoropropan-2-one

This affinity-directed halomethyl ketone (FW = 225.25 g/mol), also known as 3-cyclohexylsulfanyl-1,1,1-trifluoro-2-propanone, inhibits juvenile-hormone esterase (IC$_{50}$ = 5.2 μM) and acetylcholinetserase (IC$_{50}$ = 0.71 μM). This study indicates that, by varying the substituent on the sulfide moiety, potent "transition-state" inhibitors can be developed for a wide variety of esterases and proteases. The most powerful inhibitor tested was 3-octylthio-1,1,1-trifluoro-2-propanone, with an IC$_{50}$ of 2.3×10^{-9} M on JH esterase. **1.** Hammock, Abdel-Aal, Mullin, Hanzlik & Roe (1984) *Pest. Biochem. Physiol.* **22**, 209.

Cyclo(L-histidyl-L-prolyl)

This naturally occurring diketopiperazine (FW$_{free-base}$ = 234.26 g/mol), produced in the degradation of thyrotropin-releasing factor, inhibits 2-deoxy-D-glucose-stimulated pancreatic secretion of prolactin. **Target(s):** prolactin secretion (1,2); chitinase (3). **1.** Fragner, Presset, Bernad, *et al.* (1997) *Amer. J. Physiol.* **273**, E1127. **2.** Prasad (1998) *Thyroid* **8**, 969. **3.** Houston, Synstad, Eijsink, *et al.* (2004) *J. Med. Chem.* **47**, 5713.

Cycloleucine

This highly toxic leucine/valine analogue (FW = 129.16 g/mol; CAS 52-52-8), also known as 1-aminocyclopentanecarboxylic acid, is an *N*-methyl-D-aspartate (NMDA) receptor antagonist (acting at the glycine site), an inhibitor of *S*-adenosyl-L-methionine synthesis (and therefore all SAM-dependent methyltransferases), and amino acid transport. When administered as a single dose intraperitoneally (2 mg/g body weight) to 21-day-old young mice and 6-10-week adults aged, brain SAM concentrations are reduced, with levels of methionine greatly elevated (1). The immature developing central nervous system is much more vulnerable than the fully myelinated adult brain and spinal cord. **Target(s):** flagellar motor (2); methionine *S*-adenosyltransferase (3,6,7,9-13); NMDA receptor (4,5); lysine transport (8); cystine transport (8). **1.** Lee, Surtees & Duchen (1992) *Brain* **115**, 935. **2.** Macnab (1986) *Meth. Enzymol.* **125**, 563. **3.** Mudd (1973) *The Enzymes*, 3rd ed. (Boyer, ed.), **8**, 121. **4.** Snell & Johnson (1988) *Eur. J. Pharmacol.* **151**, 165. **5.** Hershkowitz & Rogawski (1989) *Brit. J. Pharmacol.* **98**, 1005. **6.** Slapeta, Stejskal & Keithly (2003) *FEMS Microbiol. Lett.* **225**, 271. **7.** Caboche & Hatzfeld (1978) *J. Cell Physiol.* **97**, 361. **8.** Craan & Bergeron (1975) *Can. J. Physiol. Pharmacol.* **53**, 1027. **9.** Chou & Talalay (1972) *Biochemistry* **11**, 1065. **10.** Berger & Knodel (2003) *BMC Microbiol.* **3**, 12. **11.** Porcelli, Cacciapuoti, Carteni-Farina & Gambacorta (1988) *Eur. J. Biochem.* **177**, 273. **12.** Hote, Sahoo, Jani, *et al.* (2007) *J. Nutr. Biochem.* **19**, 384. **13.** Yarlett, Garofalo, Goldberg, *et al.* (1993) *Biochim. Biophys. Acta* **1181**, 68.

Cyclooctatin

This natural product (FW = 322.49 g/mol), produced by a strain of *Streptomyces melanosporofaciens* Strepto MI614-43F2, competitively inhibits lysophospholipase, K_i = 4.8 μM. **1.** Aoyagi, Aoyama, Kojima, *et al.* (1992) *J. Antibiot.* **45**, 1587.

Cyclopamine

This teratogenic *Veratrum* alkaloid (FW$_{free\text{-}base}$ = 411.63 g/mol), also known as 11-deoxyjervine, inhibits the sonic hedgehog signaling pathway by direct binding to the heptahelical bundle of Smoothened (*i.e.*, it is a sonic-hedgehog signaling antagonist). The name (derived from *Cyclops*, the one-eyed giant in Homer's *Odyssey*) refers to one-eyed lamb that born to ewes who had consumed the leaves of *Veratrum californicum*, the corn lily plant that produces cyclopamine. **Target(s):** sonic hedgehog signalling pathway (1,2,3); P-glycoprotein-mediated drug transport (4). **1.** Incardona, Gaffield, Kapur & Roelink (1998) *Development* **125**, 3553. **2.** Incardona, Gaffield, Lange, *et al.* (2000) *Dev. Biol.* **224**, 440. **3.** Chen, Taipale, Cooper & Beachy (2002) *Genes Dev.* **16**, 2743. **4.** Lavie, Harel-Orbital, Gaffield & Liscovitch (2001) *Anticancer Res.* **21**, 1189.

1,2-Cyclopentanedicarboxylic Acid

This succinic acid analogue (FW$_{free\text{-}acid}$ = 158.15 g/mol; CAS 1461-97-8) inhibits Xaa-Pro dipeptidase (K_i = 0.5 μM) and succinate dehydrogenase. **Target(s):** Xaa-Pro Dipeptidase, *or* prolidase, inhibited by *trans*-isomer (1-3); succinate dehydrogenase, inhibited by the *trans*-isomer (4,5) however, another report lists both geometric isomers as inhibitors (6). **1.** Mock (1998) in *Handb. Proteolytic Enzymes*, p. 1403, Academic Press, San Diego. **2.** Mock & Green (1990) *J. Biol. Chem.* **265**, 19606. **3.** Radzicka & Wolfenden (1990) *J. Amer. Chem. Soc.* **112**, 1248. **4.** Webb (1966) *Enzyme and Metabolic Inhibitors*, vol. **2**, pp. 37 & 241, Academic Press, New York. **5.** Seaman & Houlihan (1950) *Arch. Biochem.* **26**, 436. **6.** Tietze & Klotz (1952) *Arch. Biochem. Biophys.* **35**, 355.

2-(Cyclopentylcarbonylamino)-6-{3R-[(2-naphthylmethyl)amino]cyclopentyl-1S-carboxamido}benzothiazole

This benzothiazole (FW = 512.68 inhibits *Candida albicans* N-myristoyltransferase (IC$_{50}$ = 15 nM). The 3S,1R-isomer also inhibits (IC$_{50}$ = 11 nM). **1.** Yamazaki, Kaneko, Suwa, *et al.* (2005) *Bioorg. Med. Chem.* **13**, 2509.

8-Cyclopentyl-6-ethyl-5-methyl-2-[4-(piperazin-1-yl)phenylamino]-8H-pyrido[2,3-d]pyrimidin-7-one

This pyridopyrimidinone (FW = 431.56 g/mol) inhibits cyclin-dependent kinase 4, *or* Cdk4, IC$_{50}$ = 25 nM). The pyrido[2,3-*d*]pyrimidin-7-one template had been identified previously as a privileged structure for inhibition of ATP-dependent kinases, and introduction of a methyl substituent at the C-5 position of the pyrido[2,3-d]pyrimidin-7-one conferred excellent selectivity for Cdk4 versus other Cdks and representative tyrosine kinases. Further optimization identified highly potent and Cdk4-selective inhibitor showing potent antiproliferative activity against human tumor cells *in vitro*. **1.** VanderWel, Harvey, McNamara, *et al.* (2005) *J. Med. Chem.* **48**, 2371.

8-Cyclopentyl-N³-[3-(4-(fluorosulfonyl)benzoyloxy)-propyl-N¹-propylxanthine

This substituted xanthine (FW = 506.56 g/mol) is an irreversible A$_1$ adenosine receptor antagonist. **1.** Srinivas, Shryock, Scammells, *et al.* (1996) *Mol. Pharmacol.* **50**, 196. **2.** Scammells, Baker, Belardinelli & Olsson (1994) *J. Med. Chem.* **37**, 2704. **3.** van Muijlwijk-Koezen, Timmerman, van der Sluis, *et al.* (2001) *Bioorg. Med. Chem. Lett.* **11**, 815. **4.** Beauglehole, Baker & Scammells (2000) *J. Med. Chem.* **43**, 4973.

2-[(4-Cyclopentylmethoxy-3-diethylsulfamoyl-benzoyl)amino]benzoic Acid

This aromatic sulfonamide (FW$_{free\text{-}acid}$ = 474.58 g/mol) inhibits *Enterococcus faecalis* β-ketoacyl-[acyl-carrier-protein] synthase III (IC$_{50}$ = 2.1 μM). **1.** Nie, Perretta, Lu, *et al.* (2005) *J. Med. Chem.* **48**, 1596. **2.** Ashek, San Juan & Cho (2007) *J. Enzyme Inhib. Med. Chem.* **22**, 7.

Cyclophellitol

This epoxide-containing cyclitol (FW = 176.17 g/mol; CAS 126661-83-4), from the mushroom *Phellinus* sp. and named as (1S,2R,3S,4R,5R,6R)-5-hydroxymethyl-7-oxabicyclo[41.0]heptane-2,3,4-triol inhibits β-glucosidase (1-4), glucosyl ceramidase (1), and β-xylosidase, the latter weakly (3). **1.** Atsumi, Nosaka, Iinuma & Umezawa (1992) *Arch. Biochem. Biophys.* **297**, 362. **2.** Atsumi, Umezawa, Iinuma, *et al.* (1990) *J. Antibiot.* **43**, 49. **3.** Atsumi, Iinuma, Nosaka & Umezawa (1990) *J. Antibiot.* **43**, 1579. **4.** Gloster, Madsen & Davies (2007) *Org. Biomol. Chem.* **5**, 444.

Cyclophosphamide

Cyclophosphamide   **4-Hydroxycyclophosphamide**

This anticancer drug (FW = 261.09 g/mol), also known as Endoxan™ and Cytoxan™, is an alkylating agent that crosslinks DNA, causing strand breakage, and inducing lethal mutations. Cyclophosphamide can also alkylate protein sulfhydryl and amino groups. Microsomal mixed function hydroxylases act on cyclophosphamide to produce 4-hydroxycyclophosphamide (and its tautomer, aldophosphamide), which is the reactive metabolite. **Target(s):** acetylcholinesterase (1,5,6); DNA polymerase (2); cholinesterase, *or* butyrylcholinesterase (3,4,6). **Key Pharmacokinetic Parameters:** *See* Appendix II in Goodman & Gilman's *THE PHARMACOLOGICAL BASIS OF THERAPEUTICS*, 12th Edition (Brunton, Chabner & Knollmann, eds.) McGraw-Hill Medical, New York (2011). **1.** al-Jafari, Kamal & Alhomida (1997) *J. Enzyme Inhib.* **11**, 275. **2.** Helge, Oberdisse & Engels (1968) *Naunyn Schmiedebergs Arch. Exp. Pathol. Pharmakol.* **260**, 139. **3.** Wolff (1965) *Klin. Wochenschr.* **43**, 819. **4.** Lalka & Bardos (1975) *Biochem. Pharmacol.* **24**, 455. **5.** al-Jafari, Duhaiman & Kamal (1995) *Toxicology* **96**, 1. **6.** Fujii, Ohnoshi, Namba & Kimura (1982) *Gan To Kagaku Ryoho* **9**, 831.

Cyclopiazonate

β-cyclopiazonate   **α-cyclopiazonate**

This toxic metabolite (FW = 336.39 g/mol; CAS 18172-33-3) from certain strains of *Penicillium cyclopium* and *Aspergillus flavus* depletes intracellular Ca^{2+} stores. Cyclopiazonate is a product of the reaction catalyzed by β-cyclopiazonate dehydrogenase. **1.** Suzuki, Muraki, Imaizumi & Watanabe (1992) *Brit. J. Pharmacol.* **107**, 134. **2.** Seidler, Jona, Vegh & Martonosi (1989) *J. Biol. Chem.* **264**, 17816. **3.** Goeger, Riley, Dorner & Cole (1988) *Biochem. Pharmacol.* **37**, 978. **4.** Plenge-Tellechea, Soler & Fernandez-Belda (1997) *J. Biol. Chem.* **272**, 279. **5.** Sorin, Rosas & Rao (1997) *J. Biol. Chem.* **272**, 9895. **6.** Goeger & Riley (1989) *Biochem Pharmacol.* **38**, 3995.

Cyclopropanecarboxylate

This hypoglycemic agent ($FW_{free-acid}$ = 86.09 g/mol; M.P. = 17-19°C; B.P. = 182-184°C) inhibits metabolism of pyruvate, branched-chain α-keto acids, acetoacetate, and hydroxybutyrate. Its methyl ester (FW = 100.12 g/mol; CAS 2868-37-3) is cell-permeant and undergoes enzymatic hydrolysis to the free acid. **Target(s):** monocarboxylic acid transporter (1); β-ureidopropionase, IC$_{50}$ = 8 μM (2); glucuronosyltransferase, IC$_{50}$ = 1 μM (3). **1.** Buxton, Bahl, Bressler & Olson (1983) *Metabolism* **32**, 736. **2.** Walsh, Green, Larrinua & Schmitzer (2001) *Plant Physiol.* **125**, 1001. **3.** Bichlmaier, Kurkela, Joshi, *et al.* (2007) *J. Med. Chem.* **50**, 2655.

Cyclopropanone & Hydrate

This reactive ketone (FW = 56.064 g/mol; $FW_{hydrate}$ = 74.079 g/mol; CAS 5009-27-8), which exists primarily as the *gem*-diol or hydrate in aqueous solutions, inhibits several oxidases and oxidoreductases., In the case of horseradish peroxidase, the *gem*-diol undergoes oxidative ring opening to form the primary free radical of propionic acid which alkylates the porphyrin ring (1,2). The newly inserted new propionic acid side chain is substituted for one of the methine protons of the heme. **Target(s):** horseradish peroxidase, inhibited by *gem*-diol (1,2); alcohol oxidase (3,4); aldehyde dehydrogenase (NAD$^+$), inhibited by the *gem* diol (5-7); methanol dehydrogenase, inhibited by *gem*-diol (8,9); alcohol dehydrogenase (quinoprotein), inhibited by *gem*-diol (10). **1.** Schmid, Hittmair, Schmidhammer & Jasani (1989) *J. Histochem. Cytochem.* **37**, 473. **2.** Wiseman, Nichols & Kolpak (1982) *J. Biol. Chem.* **257**, 6328. **3.** van der Klei, Bystrykh & Harder (1990) *Meth. Enzymol.* **188**, 420. **4.** Cromartie (1982) *Biochem. Biophys. Res. Commun.* **105**, 785. **5.** Wiseman & Abeles (1979) *Biochemistry* **18**, 427. **6.** Weiner, Freytag, Fox & Hu (1982) *Prog. Clin. Biol. Res.* **114**, 91. **7.** Nagasawa, Elberling & DeMaster (1984) *J. Med. Chem.* **27**, 1335. **8.** Frank & Duine (1990) *Meth. Enzymol.* **188**, 202. **9.** Ator & Ortiz de Montellano (1990) *The Enzymes*, 3rd ed., **19**, 213. **10.** Dijkstra, Frank, Jongejan & Duine (1984) *Eur. J. Biochem.* **140**, 369.

(±)-*N*-Cyclopropyl-α-methylbenzylamine

This substituted benzylamine ($FW_{free-base}$ = 161.25 g/mol) inactivates both monoamine oxidases A and B, with a preference for the later (1-3). Inactivation occurs via attachment to a cysteinyl residue as well as the flavin cofactor. Trimethylamine dehydrogenase is also inhibited by this agent (4). **1.** Brush & Kozarich (1992) *The Enzymes*, 3rd ed., **20**, 317. **2.** Silverman (1984) *Biochemistry* **23**, 5206. **3.** Silverman & Hiebert (1988) *Biochemistry* **27**, 8448. **4.** Mitchell, Nikolic, Jang, *et al.* (2001) *Biochemistry* **40**, 8523.

Cyclosarin

This extremely toxic chemical warfare agent (FW = 180.16 g/mol; M.P. = –12°C; B.P. = 239°C), also called Agent GF and methyl cyclohexylfluorophosphonate, is a potent acetylcholinesterase inhibitor. It is a liquid at room temperature is readily absorbed through the skin (the LD$_{50}$ is 0.35 g/person). **Target(s):** acetylcholinesterase (1-4); cholinesterase (butyrylcholinesterase) (1,3). **1.** Worek, Eyer & Szinicz (1998) *Arch. Toxicol.* **72**, 580. **2.** Worek, Widmann, Knopff & Szinicz (1998) *Arch. Toxicol.* **72**, 237. **3.** Schopfer, Voelker, Bartels, Thompson & Lockridge (2005) *Chem. Res. Toxicol.* **18**, 747. **4.** Worek, Thiermann, Szinicz & Eyer (2004) *Biochem. Pharmacol.* **68**, 2237.

D-Cycloserine

This isoxazolidinone ($FW_{free-base}$ = 109.09 g/mol; CAS 68-41-7; Symbol: DCS), also known as D-4-amino-3-isoxazolidinone and D-oxamycin, has been isolated from *Streptomyces garyphalus* sive *orchidaceus*. Note that neutral or acidic solutions are unstable. Solutions buffered at pH 10 can be stored at 4°C for one week. It inhibits the biosynthesis of bacterial cell walls in Gram-negative microorganisms As a cyclic analogue of D-alanine, cycloserine targets alanine racemase (ALR) and D-alanine:D-alanine ligase (DDL), two crucial enzymes important in the cytosolic stages of peptidoglycan synthesis. Cycloserine is also a partial agonist at the glycine modulatory site of the NMDA (*N*-methyl-D-aspartate) receptor. **Cinical Application:** Although less potent than first-line antituberculosis drugs such as isoniazid (INH) and streptomycin, D-cycloserine is a highly effective second-line drug against *Mycobacterium tuberculosis*, the causative agent of tuberculosis. Less drug-resistant *M. tuberculosis* is reported for DCS than INH and rifampin. **Target(s):** D-amino-acid aminotransferase, *or* D-alanine aminotransferase (1,3,12,21,26,55,56); D-alanyl-D-alanine synthetase, *or* D-alanine:D-alanine ligase (2,13-16,19,31,32,34-36); D-alanine Hydroxymethyltransferase (4,11); amino acid racemase (5); D-3-aminoisobutyrate:pyruvate aminotransferase (6); β-alanine:α-ketoglutarate aminotransferase (6); cystathionine γ-lyase (7,23); ω-amino-acid:pyruvate aminotransferase, *or* β-alanine:pyruvate aminotransferase (8); serine *C*-palmitoyltransferase (9,27); alanine racemase (10,19,25,31, 32,35,36,41-43); 4-aminobutyrate aminotransferase (13,57-59); serine hydroxymethyltransferase (glycine hydroxy-methyltransferase) (17,18,28,65-70); glutamate decarboxylase (20); DOPA decarboxylase (20); kynureninase (22); 2,2-dialkylglycine decarboxylase (pyruvate) (α,α-dialkylamino acid aminotransferase), with the L-enantiomer is a stronger inhibitor (24,30); asparagine synthetase, weakly inhibited (28); amine oxidase, pea (33); isopenicillin-N epimerase (37); 2-aminohexano-6-lactam racemase (38); asparagine:oxo-acid aminotransferase (60); aspartate racemase (39); amino-acid racemase (40); 3-chloro-D-alanine dehydrochlorinase (44); cystathionine γ-lyase (45); diaminopropionate ammonia-lyase (46); kynureninase (47,48); arylacetonitrilase, weakly inhibited (49); adenosylmethionine:8-amino-7-oxononanoate transaminase (50); taurine:2-oxoglutarate aminotransferase, weakly inhibited (51,52); serine:glyoxylate aminotransferase (53); 2-aminoethylphosphonate: pyruvate aminotransferase (54); alanine aminotransferase (61); cystathionine γ-synthase (62); cysteine synthase (63); 8-amino-7-oxononanoate synthase (64). **1.** Jones, Soper, Ueno & Manning (1985) *Meth. Enzymol.* **113**, 108. **2.** Ito, Nathenson, Dietzler, Anderson & Strominger (1966) *Meth. Enzymol.* **8**, 324. **3.** Martinez-Carrion & Jenkins (1970) *Meth. Enzymol.* **17A**, 167. **4.** Miles (1971) *Meth. Enzymol.* **17B**, 341. **5.** Soda & Osumi (1971) *Meth. Enzymol.* **17B**, 629. **6.** Klosterman (1979) *Meth. Enzymol.* **62**, 483. **7.** Nagasawa, Kanzaki & Yamada (1987) *Meth. Enzymol.* **143**, 486. **8.** Yonaha, Toyama & Soda (1987) *Meth. Enzymol.* **143**, 500.**9.** Merrill & Wang (1992) *Meth. Enzymol.* **209**, 427. **10.** Adams (1972) *The Enzymes*, 3rd ed., **6**, 479. **11.** Rader & Huennekens (1973) *The Enzymes*, 3rd ed., **9**, 197. **12.** Braunstein (1973) *The Enzymes*, 3rd ed., **9**, 379. **13.** Webb (1966) *Enzyme and Metabolic Inhibitors*, vol. **2**, p. 359, Academic Press, New York. **14.** Neuhaus & Lynch (1964) *Biochemistry* **3**, 471. **15.** Lacoste, Poulsen, Cassaigne & Neuzil (1979) *Curr. Microbiol.* **2**, 113. **16.** Neuhaus & J. L. Lynch (1962) *Biochem. Biophys. Res. Commun.* **8**, 377. **17.** Ramesh & N. Appaji Rao (1980) *Biochem. J.* **187**, 623. **18.** Miyazaki, Toki, Izumi & Yamada (1987) *Eur. J. Biochem.* **162**, 533. **19.** Noda, Kawahara, Ichikawa, *et al.* (2004) *J. Biol. Chem.* **279**, 46143. **20.** Dengler, Rauchs & Rummel (1962) *Ned. Milit. Geneeskd. Tijdschr.* **243**, 366. **21.** Martinez-Carrion & Jenkins (1965) *J. Biol. Chem.* **240**, 3547. **22.** Illiano, Picciafuoco, Soscia, Rogliani & Della Pietra (1967) *Life Sci.* **6**, 1759. **23.** Brown, Hudgins & Roszell (1969) *J. Biol. Chem.* **244**, 2809. **24.** Lamartiniere, Itoh & Dempsey (1971) *Biochemistry* **10**, 4783. **25.** Lambert & Neuhaus (1972) *J. Bacteriol.* **110**, 978. **26.** Soper & Manning (1981) *J. Biol. Chem.* **256**, 4263. **27.** Sundaram & Lev (1984) *J. Neurochem.* **42**, 577. **28.** Manohar, Rao &. Rao (1984) *Biochemistry* **23**, 4116. **29.** Maul & Schuster (1986) *Arch. Biochem. Biophys.* **251**, 585. **30.** Malashkevich, Strop, Keller, Jansonius & Toney (1999) *J. Mol. Biol.* **294**, 193. **31.** Reitz, Slade, & Neuhaus (1967) *Biochemistry* **6**, 2561. **32.** Strominger, Ito & Threnn (1960) *J. Amer. Chem. Soc.* **82**, 998. **33.** Suzuki & Yamasaki (1970) *Enzymologia* **39**, 57. **34.** Zawadzke, Bugg & Walsh (1991) *Biochemistry* **30**, 1673. **35.** Vicario, Green & Katzen (1987) *J. Antibiot.* **15**, 209. **36.** Lugtenberg (1972) *J. Bacteriol.* **110**, 26. **37.** Usui & Yu (1989) *Biochim. Biophys. Acta* **999**, 78. **38.** Ahmed, Esaki, Tanaka & Soda (1983) *Agric. Biol. Chem.* **47**, 1887. **39.** Shibata, Watanabe, Yoshikawa, *et al.* (2003) *Comp. Biochem. Physiol. B* **134**, 307. **40.** Asano & Endo (1988) *Appl. Microbiol. Biotechnol.* **29**, 523. **41.** Shibata, Shirasuna, Motegi, *et al.* (2000) *Comp. Biochem. Physiol. B* **126**, 599. **42.** Fenn, Stamper, Morollo & Ringe (2003) *Biochemistry* **42**, 5775. **43.** Strych, Penland, Jimenez, Krause & Benedik (2001) *FEMS Microbiol. Lett.* **196**, 93. **44.** Nagasawa, Ohkishi, Kavakami, *et al.* (1982) *J. Biol. Chem.* **257**, 13749. **45.** Nagasawa, Kanzaki & Yamada (1984) *J. Biol. Chem.* **259**, 10393. **46.**

Nagasawa, Tanizawa, Satoda & Yamada (1988) *J. Biol. Chem.* **263**, 958. **47.** Moriguchi, Yamamoto & Soda (1973) *Biochemistry* **12**, 2969. **48.** Tanizawa & Soda (1979) *J. Biochem.* **85**, 901. **49.** Nagasawa, Mauger & Yamada (1990) *Eur. J. Biochem.* **194**, 765. **50.** Izumi, Sato, Tani & Ogata (1975) *Agric. Biol. Chem.* **39**, 175. **51.** Toyama, Misono & Soda (1978) *Biochim. Biophys. Acta* **523**, 75. **52.** Yonaha, Toyama & Soda (1985) *Meth. Enzymol.* **113**, 102. **53.** Izumi, Yoshida & Yamada (1990) *Eur. J. Biochem.* **190**, 285. **54.** Dumora, Lacoste & Cassaigne (1983) *Eur. J. Biochem.* **133**, 119. **55.** Yonaha, Misono, Yamamoto & Soda (1975) *J. Biol. Chem.* **250**, 6983. **56.** Ogawa & Fukuda (1973) *Biochem. Biophys. Res. Commun.* **52**, 998. **57.** Tamaki, Kubo, Aoyama & Funatsuka (1983) *J. Biochem.* **93**, 955. **58.** Yonaha & Toyama (1980) *Arch. Biochem. Biophys.* **200**, 156. **59.** Yonaha, Suzuki & Toyama (1985) *Eur. J. Biochem.* **146**, 101. **60.** Maul & Schuster (1986) *Arch. Biochem. Biophys.* **251**, 585. **61.** Orzechowski, Socha-Hanc & Paszkowski (1999) *Acta Biochim. Pol.* **46**, 447. **62.** Kanzaki, Kobayashi, Nagasawa & Yamada (1987) *Eur. J. Biochem.* **163**, 105. **63.** Nakamura, Iwahashi & Eguchi (1984) *J. Bacteriol.* **158**, 1122. **64.** Izumi, Morita, Tani & Ogata (1973) *Agric. Biol. Chem.* **37**, 1327. **65.** Appalji Rao, Talwar & Savithri (2000) *Int. J. Biochem. Cell Biol.* **32**, 405. **66.** Rao & Rao (1982) *Plant Physiol.* **69**, 11. **67.** Miyazaki, Toki, Izumi & Yamada (1987) *Agric. Biol. Chem.* **51**, 2587. **68.** Manohar, Ramesh & Rao (1982) *J. Biosci.* **4**, 31. **69.** Schirch (1982) *Adv. Enzymol. Relat. Areas Mol. Biol.* **53**, 83. **70.** Manohar, Rao & Rao (1984) *Biochemistry* **23**, 4116.

L-Cycloserine

This isoxazolidinone (FW$_{\text{free-base}}$ = 109.09 g/mol), also known as L-4-amino-3-isoxazolidinone and L-oxamycin, inhibits several pyridoxal-phosphate-dependent enzymes. It also inhibits sphingosine biosynthesis. Note that neutral or acidic solutions are unstable. Solutions buffered at pH 10 can be stored at 4°C for one week. **Inhibition of γ-Aminobutyric Acid Aminotransferase:** The inactivation of pig brain GABA-AT by L-cycloserine is time-dependent (1). Treatment of the enzyme with [^{14}C]-L-cycloserine, followed by rapid gel filtration, gives a stoichiometry of 1.1 mol radiolabeled inhibitor per mol enzyme. Inactivation of [^{3}H]pyridoxal 5'-phosphate-reconstituted GABA-AT with L-cycloserine followed by dialysis or denaturation also leads to the release of radioactivity from the enzyme. Electrospray ionization tandem mass spectrometry of the isolated cycloserine-coenzyme adduct suggests it is a tautomeric form of the Schiff base between pyridoxamine 5'-phosphate and oxidized cycloserine. *See also* D-Cycloserine. **Target(s):** taurine:2-oxoglutarate aminotransferase (2,31); L-alanine aminotransferase (3,11,12); ornithine aminotransferase (4,32); D-alanine hydroxymethyl-transferase, *or* 2-methylserine hydroxymethyltransferase (5,10); methionine γ-lyase (6,21,27); serine *C*-palmitoyltransferase (7-9,18, 20,35-40); glutamine:pyruvate aminotransferase (11); glutamate decarboxylase (12); asparagine aminotransferase (12); alanine racemase (13,25,26); aspartate aminotransferase (14,17,23,34); cystathionine γ-lyase (15); 2,2-dialkylglycine decarboxylase (pyruvate), *or* α,α-dialkylamino acid aminotransferase (16,22,29); aspartate racemase (24); homocysteine desulfhydrase (28); D-amino-acid aminotransferase (31); kynurenine: oxoglutarate aminotransferase (33). **1.** Silverman & Olson (1995) *Bioorg. Med. Chem.* **3**, 11. **2.** Yonaha, Toyama & Soda (1985) *Meth. Enzymol.* **113**, 102. **3.** Jenkins & Saier (1970) *Meth. Enzymol.* **17A**, 159. **4.** Jenkins & Tsai (1970) *Meth. Enzymol.* **17A**, 281. **5.** Miles (1971) *Meth. Enzymol.* **17B**, 341. **6.** Esaki & Soda (1987) *Meth. Enzymol.* **143**, 459. **7.** Merrill & Wang (1992) *Meth. Enzymol.* **209**, 427. **8.** Dickson, Lester & Nagiec (2000) *Meth. Enzymol.* **311**, 3. **9.** Lynch (2000) *Meth. Enzymol.* **311**, 130. **10.** Rader & Huennekens (1973) *The Enzymes*, 3rd ed., **9**, 197. **11.** Braunstein (1973) *The Enzymes*, 3rd ed., **9**, 379. **12.** Webb (1966) *Enzyme and Metabolic Inhibitors*, vol. **2**, p. 360, Academic Press, New York. **13.** Wang & Walsh (1978) *Biochemistry* **17**, 1313. **14.** Karpeiskii & Breusov (1965) *Biokhimiia* **30**, 153. **15.** Brown, Hudgins & Roszell (1969) *J. Biol. Chem.* **244**, 2809. **16.** Lamartiniere, Itoh & Dempsey (1971) *Biochemistry* **10**, 47 **17.** Janski & Cornell (1981) *Biochem. J.* **194**, 1027. **18.** Sundaram & Lev (1984) *J. Neurochem.* **42**, 577. **19.** Sundaram & Lev (1984) *Biochem. Biophys. Res. Commun.* **119**, 814. **20.** Sundaram & Lev (1984)

Antimicrob. Agents Chemother. **26**, 211. **21**. Lockwood & Coombs (1991) *Biochem. J.* **279**, 675. **22**. Malashkevich, Strop, Keller, Jansonius & Toney (1999) *J. Mol. Biol.* **294**, 193. **23**. Baptist (1967) *Enzymologia* **33**, 161. **24**. Shibata, Watanabe, Yoshikawa, *et al.* (2003) *Comp. Biochem. Physiol. B* **134**, 307. **25**. Shibata, Shirasuna, Motegi, *et al.* (2000) *Comp. Biochem. Physiol. B* **126**, 599. **26**. Fenn, Stamper, Morollo & Ringe (2003) *Biochemistry* **42**, 5775. **27**. Tanaka, Esaki & Soda (1977) *Biochemistry* **16**, 100. **28**. Thong & Coombs (1985) *IRCS Med. Sci. Libr. Compend.* **13**, 493. **29**. Esaki, Watanabe, Kurihara & Soda (1994) *Arch. Microbiol.* **161**, 110. **30**. Toyama, Misono & Soda (1978) *Biochim. Biophys. Acta* **523**, 75. **31**. Martinez-Carrion & W. T. Jenkins (1965) *J. Biol. Chem.* **240**, 3547. **32**. Yasuda, Tanizawa, Misono, Toyama & Soda (1981) *J. Bacteriol.* **148**, 43. **33**. Asada, Sawa, Tanizawa & Soda (1986) *J. Biochem.* **99**, 1101. **34**. Ryan, Bodley & Fottrell (1972) *Phytochemistry* **11**, 957. **35**. Hanada, Nishijima, Fujita & Kobayashi (2000) *Biochem. Pharmacol.* **59**, 1211. **36**. Lynch & Fairfield (1993) *Plant Physiol.* **103**, 1421. **37**. Holleran, Williams, Gao & Elias (1990) *J. Lipid Res.* **31**, 1655. **38**. Weiss & Stoffel (1997) *Eur. J. Biochem.* **249**, 239. **39**. Perry (2002) *Biochim. Biophys. Acta* **1585**, 146. **40**. Ikushiro, Hayashi & Kagamiyama (2004) *Biochemistry* **43**, 1082.

Cyclosporin A

This cyclic undecapeptide antibiotic and immunosuppressant (FW = 1202.63 g/mol; CAS 59865-13-3; *Symbol*: CyA; Solubility: 50 mg/mL DMSO) targets cyclophilin, and the resulting complex in turn inhibits calcineurin, an essential phosphoprotein phosphatase controlling immune cell signal transduction. Cyclosporin A was initially isolated in 1969 from *Tolypocladium inflatum* (*Beauveria nivea*), a fungus found in a soil sample, by Sandoz biologist Hans Frey. Later investigations led by Hartmann Stähelin, also at Sandoz, demonstrated its powerful immunosuppressive action, a property commending its use to prevent graft-versus-host disease in bone-marrow transplantation and organ rejection in kidney, heart, and liver transplant patients. Cyclosporin is also approved in the U.S. for the treatment of rheumatoid arthritis and psoriasis. **Primary Mode of Action:** Cyclophylin is a ubiquitously expressed low-molecular-weight basic protein. Its peptidyl prolyl *cis-trans* isomerase (PPIase) activity regulates protein folding and trafficking. High cyclophilin expression correlates with poor outcome for patients with inflammatory diseases, including rheumatoid arthritis, asthma and sepsis. Cyclophilin comprises ~0.1–0.6% of total cytosolic protein by weight and is highly abundant in thymus, the organ from which it was first isolated. The Cyclophylin·Cyclosporin complex binds to and inhibits calcineurin (*or* Phosphoprotein Phosphatase 2B), K_i = 2.3 nM. Calcineurin inhibition results in a complete block in the translocation of the cytosolic component of the Nuclear Factor of Activated T cells (NF-AT), resulting in a failure to activate the genes regulated by the NF-AT transcription factor. These genes include those required for B-cell help (such as interleukin (IL-4) and CD40 ligand) as well as those necessary for T-cell proliferation (such as IL-2). It also inhibits nitric oxide synthesis induced by interleukin-1α, lipopolysaccharides, and TNF-α. **Clinical Applications:** Cyclosporin A is a powerful immunosuppressive agent that lacks myelotoxicity, making it unique among nonsteroidal immunosuppressing drugs and commending it for preventing organ rejection in recipients of kidney, liver, bone marrow and pancreas transplants. CyA may also be used to treat autoimmune disorders. Its utility in preventing kidney rejection was demonstrated by Roy Calne and

colleagues at Cambridge University for kidney transplants and for liver transplants by Thomas Starzl at University Presbyterian Hospital in Pittsburgh. **Key Pharmacokinetic Parameters:** *See* Appendix II in Goodman & Gilman's THE PHARMACOLOGICAL BASIS OF THERAPEUTICS, 12th Edition (Brunton, Chabner & Knollmann, eds.) McGraw-Hill Medical, New York (2011). **Structural Features:** Unusual residues present within cyclosporin A include: L-2-aminobutyrate, *N*-methylglycine (sarcosine), L-*N*-methyl-leucine, D-alanine, L-*N*-methylvaline, and (4*R*)-4-([(*E*)-2-butenyl]-4,*N*-dimethyl-L-threonine. Cyclosporin H is identical to cuclosporin A, except for a D-*N*-methylvaline at position eleven instead of the L-isomer. CyH is devoid of most biological effects. The substituted Cyclosporin A analogue NIM811 binds to cyclophilin, but the resulting complex cannot inhibit calcineurin, and thus lacks Cyclosporin A's immunosuppressive activity (*See also* N-Methyl-4-isoleucine Cyclosporin). For these reasons, CyH and NIM811 are frequently used as controls in experiments on CyA. Given the presence of so many unusual amino acid residues in cyclosporin, it should not be surprising that it is formed in fungi by nonribosomal peptide synthetic machinery. **Target(s):** calcineurin (protein phosphatase 2B), as the Cyclosporin A·Cyclophilin complex (1-4,28-31); peptidylprolyl isomerase, K_i = 1.6 nM for the human isoform-1 (5-19); P-glycoprotein (20,22-25); xenobiotic-transporting ATPase, multidrug-resistance protein (20-25); phospholipid-translocating ATPase, flippase (25); protein phosphatase 2A, indirectly as the CyA·Cyclophilin complex (1-4,26-31); 27-hydroxycholesterol 7α-monooxygenase (32); tomato bushy stunt tombusvirus (TBSV) replication, acting as a cell-originated virus retriction factor (33).

1. Eriksson, Toivola, Sahlgren, Mikhailov & Härmälä-Braskén (1998) *Meth. Enzymol.* **298**, 542. **2**. Liu, Farmer, Lane, *et al.* (1991) *Cell* **66**, 807. **3**. Fruman, Klee, Bierer & Burakoff (1992) *Proc. Natl. Acad. Sci. U.S.A.* **89**, 3686. **4**. Swanson, Born, Zydowsky, *et al.* (1992) *Proc. Natl. Acad. Sci. U.S.A.* **89**, 3741. **5**. Wang, Kim, Bakhtiar & Germanas (2001) *J. Med. Chem.* **44**, 259. **6**. Gallo, Rossi, Saviano, *et al.* (1998) *J. Biochem.* **124**, 880. **7**. Gasser, Gunning, Budelier & Brown (1990) *Proc. Natl. Acad. Sci. U.S.A.* **87**, 9519. **8**. Dornan, Page, Taylor, *et al.* (1999) *J. Biol. Chem.* **274**, 34877. **9**. Picken, Eschenlauer, Taylor, Page & Walkinshaw (2002) *J. Mol. Biol.* **322**, 15. **10**. Compton, Davis, MacDonald & Bächinger (1992) *Eur. J. Biochem.* **206**, 927. **11**. Zeng, MacDonald, Bann, *et al.* (1998) *Biochem. J.* **330**, 109. **12**. Valle, Troiani, Lazzaretti, *et al.* (2005) *Parasitol. Res.* **96**, 199. **13**. Schmidt, Tradler, Rahfeld, *et al.* (1996) *Mol. Microbiol.* **21**, 1147. **14**. Holzman, Egan, Edalji, Simmer, Helfrich, Taylor & Burres (1991) *J. Biol. Chem.* **266**, 2474. **15**. Bergsma, Eder, Gross, *et al.* (1991) *J. Biol. Chem.* **266**, 23204. **16**. Kofron, Kuzmic, Kishore, *et al.* (1992) *J. Amer. Chem. Soc.* **114**, 2670. **17**. Breiman, Fawcett, Ghirardi & Mattoo (1992) *J. Biol. Chem.* **267**, 21293. **18**. Takahashi, Hayano & Suzuki (1989) *Nature* **337**, 473. **19**. Fischer, Wittmann-Liebold, Lang, Kiefhaber & Schmid (1989) *Nature* **337**, 476. **20**. Sharom, Liu, Romsicki & Lu (1999) *Biochim. Biophys. Acta* **1461**, 327. **21**. Paul, Breuninger & Kruh (1996) *Biochemistry* **35**, 14003. **22**. Lu, Liu & Sharom (2001) *Eur. J. Biochem.* **268**, 1687. **23**. Garrigues, Escargueil & Orlowski (2002) *Proc. Natl. Acad. Sci. U.S.A.* **99**, 10347. **24**. Wang, Casciano, Clement & Johnson (2001) *Drug Metab. Dispos.* **29**, 1080. **25**. Romsicki & Sharom (2001) *Biochemistry* **40**, 6937. **26**. Hall, Feekes, Don, *et al.* (2006) *Biochemistry* **45**, 3448. **27**. Cohen (2004) in *Topics in Current Genetics* (Arino & Alexander, eds.) Springer **5**, 1. **28**. Hallaway & O'Kane (2000) *Meth. Enzymol.* **305**, 391. **29**. Roberts, Sternberg & Chappell (1997) *Parasitology* **114**, 279. **30**. Dobson, May, Berriman, Del Vecchio, *et al.* (1999) *Mol. Biochem. Parasitol.* **99**, 167. **31**. Haddy & Rusnak (1994) *Biochem. Biophys. Res. Commun.* **200**, 1221. **32**. Souidi, Parquet, Dubrac, *et al.* (2000) *Biochim. Biophys. Acta* **1487**, 74. **33**. Kovalev & Nagy (2013) *J. Virol.* **87**, 13330.

Cyclothiazide

This substituted benzothiadiazine (FW = 389.88 g/mol; CAS 2259-96-3) enhances glutamatergic transmission and is widely used as a blocker for α-amino-3-hydroxy-5-methyl-4-isoxazolepropionate-type glutamate receptor desensitization (*i.e.*, it is a positive AMPA receptor modulator). **Target(s):** GABA$_A$ receptor (1); α-amino-3-hydroxy-5-methyl-4-isoxazolepropionic acid type glutamate receptor desensitization (2,3). **1.** Deng & Chen (2003) *Proc. Natl. Acad. Sci. U.S.A.* **100**, 13025. **2.** Yamada & Tang (1993) *J. Neurosci.* **13**, 3904. **3.** Patneau, Vyklicky & Mayer (1993) *J. Neurosci.* **13**, 3496.

CYP Inhibition

Drug-drug interactions (DDIs) are associated with severe adverse effects that may lead to the patient requiring alternative therapeutics and could ultimately lead to drug withdrawal from the market if they are severe. Cytochrome P$_{450}$ enzymes, *or* CYPs within human liver endoplasmic reticulum *in vivo* (microsomes *in vitro*) modify drugs and xenobiotic compounds, most often targeting them for elimination, but, in some instances, paradoxically activate agents to form more bioreactive and/or cytotoxic metabolites. When two drugs (or nutriceuticals) are alternative substrates for the same CYP, the two can give rise to drug-drug interactions, meaning that they (or metabolically derived forms thereof) modify each other's pharmacokinetics. It is therefore common practice to screen for inhibition of drug turnover when another drug (or nutriceutical) is added to the test system as well as inhibition due to metabolites formed upon preincubation of an agent with liver microsomal preparations. Well-designed experiments can minimize artifacts that arise from protein binding, metabolic instability of the test agents, using reference (or marker) substrates. These substrates and their corresponding reaction products are readyl assayed by standardized procedures. In all, there are 60 CYP genes in humans. Common variations (*or* polymorphisms) in cytochrome P$_{450}$ genes can affect the function of the enzymes. The effects of polymorphisms are most prominently seen in the breakdown of medications. Depending on the gene and the polymorphism, drugs can be metabolized quickly or slowly. Typical cytochrome P$_{450}$ enzymes and their well-characterized marker substrate and modification include: CYP1A2 monooxygenase (Phenacetin, *O*-Dealkylation); CYP2A6 mixed-function oxidase (Coumarin, 7-Hydroxylation); CYP2B6 monooxygenase (Efavirenz, 8-Hydroxylation); CYP2C8 mixed-function oxidase; (Amodiaquine, *N*-Dealkylation); CYP2C8 mixed-function oxidase (Paclitaxel, 6-Hydroxylation); CYP2C9 oxidase and epoxidase (Diclofenac, 4'-Hydroxylation); CYP2C19 mixed-function oxidase (*S*-Mephenytoin, 4'-Hydroxylation); CYP2D6 mixed-function oxidase (Dextromethorphan, *O*-Demethylation); CYP2E1 mixed-function oxidase (Chlorzoxazone, 6-Hydroxylation); CYP3A4/5 hydroxylase (Testosterone, 6β-Hydroxylation); CYP3A4/5 mixed-function oxidase (Nifedipine, Oxidation); CYP3A4/5 hydroxylase (Midazolam, 1'-Hydroxylation); CYP3A4/5 hydroxylase (Atorvastatin, *ortho*-Hydroxylation); and CYP4A11 hydroxylase (Lauric acid, 12-Hydroxylation).

Cypermethrin

This synthetic insecticide (FW = 416.30 g/mol; CAS 52315-07-8), also known as α-cyano-3-phenoxybenzyl 3-(2,2-dichlorovinyl) 2,2-dimethylcyclopropanecarboxylate, is a strong inhibitor of calcineurin, *or* protein phosphatase 2B (IC$_{50}$ = 40 pM) (1-4). **Target(s):** glutathione *S* transferase (5); glutamate uptake (6); acetylcholinesterase (7); monoamine oxidase A (8); protein phosphatase 2A (9); phosphoprotein phosphatase (1-4,9); carboxylesterase (10). **1.** Enan & F. Matsumura (1992) *Biochem. Pharmacol.* **43**, 1777. **2.** Wang & Stelzer (1994) *NeuroReport* **5**, 2377. **3.** Fakata, Swanson, Vorce & Stemmer (1998) *Biochem. Pharmacol.* **55**, 2017. **4.** Enz & Pombo-Villar (1997) *Biochem. Pharmacol.* **54**, 321. **5.** da Silva Vaz, Torino Lermen, Michelon, *et al.* (2004) *Vet. Parasitol.* **119**, 237. **6.** Wu, Xia, Shi & Liu (1999) *Wei Sheng Yan Jiu* **28**, 261. **7.** Rao & Rao (1995) *J. Neurochem.* **65**, 2259. **8.** Rao & Rao (1993) *Mol. Cell Biochem.* **124**, 107. **9.** Hall, Feekes, Don, *et al.* (2006) *Biochemistry* **45**, 3448. **10.** Valles, Oi & Strong (2001) *Insect Biochem. Mol. Biol.* **31**, 715.

Cyprodime

This photosensitive morphinanone (FW$_{free-base}$ = 356.25 g/mol; CAS 118111-51-6 (bromide salt)) is a selective μ opioid receptor antagonist, with a K_i of 8.1 nM for the rat brain receptor and 26.6 nM for the guinea pig brain receptor (1,2). The EC$_{50}$ value for morphine-stimulated [^{35}S]GTPγS binding is increased about 500x in the presence of 10 μM cyprodime. **1.** Marki, Monory, Otvos, *et al.* (1999) *Eur. J. Pharmacol.* **383**, 209. **2.** Schmidhammer, Burkard, Eggstein-Aeppli & Smith (1989) *J. Med. Chem.* **32**, 418.

Cyproterone Acetate

This synthetic steroid ester (FW = 416.94 g/mol; CAS 427-51-0), also known as 6-chloro-1β,2β-dihydro-17-hydroxy-3'*H*-cyclopropa(1,2)-pregna-1,4,6-triene-3,20-dione acetate, exhibits anti-androgen activity and is a progestogen. Cyproterone also exhibits partial agonistic and antagonistic actions and displays partial progestational and glucocorticoid action. Its action is not pure antiandrogenic. Cyproterone inhibits leukocyte migration through endothelial cell monolayers and is used in the treatment of prostate cancer, acne, and hirsutismus. **Target(s):** androgen receptor (2,3); 17β-hydroxysteroid dehydrogenase (4); steroid Δ isomerase (5). **1.** Scott & Tomkins (1975) *Meth. Enzymol.* **40**, 273. **2.** Simental, Sar & Wilson (1992) *J. Steroid Biochem. Mol. Biol.* **43**, 37. **3.** Berrevoets, Umar & Brinkmann (2002) *Mol. Cell. Endocrinol.* **198**, 97. **4.** Jarabak & Sack (1969) *Biochemistry* **8**, 2203. **5.** Weintraub, Vincent, Baulieu & Alfsen (1977) *Biochemistry* **16**, 5045.

Cys-AAX tetrapeptides

These peptides KKSKTKCVIM (FW = 1165.52 g/mol; isoelectric point = 10.2), TKCVIM (FW = 693.92 g/mol; isoelectric point = 7.9), and CVIM (FW = 464.64 g/mol; isoelectric point = 5.5), corresponding to the COOH-terminal 10-, 6-, and 4-residues in human p21^{ras}, target farnesyl:protein transferase, inhibiting the transfer of the farnesyl moiety from farnesyl pyrophosphate to the cysteine-acceptor residue in p21^{ras} proteins, thereby blocking membrane localization of ras and disrupting ras-mediated signal transduction. The sequence-scrambled eptides CNFDNPVSQKTT and TKVCIM serve as suitable controls. **1.** Reiss, Goldstein, Seabra, Casey & Brown (1990) *Cell* **62**, 81.

Cystatin

This family of cysteine protease inhibitors, originally isolated from chicken egg white (alternative name ovocystatin) is divided into three families: (a) Stefin family proteins (MW = 11–12 kDa) have no disulfide bonds and no covalently linked carbohydrates; (b) Cystatin family members are slightly larger proteins (MW 13–14 kDa) contain two disulfide bonds near the C-terminus; Kininogens family members include higher molecular weight kininogens (approximately 120 kDa) and lower molecular weight proteins (68 kDa). The kininogens are glycoproteins containing three cystatin-like domains, each with at least two disulfide bonds. Notably, some cystatins either lost or never acquired an inhibitory capacity.

Cysteamine

$$H_2N\diagup\diagdown\diagup^{SH}$$

This mercaptan (FW$_{hydrochloride}$ = 113.61 g/mol; CAS 60-23-1; Soluble in water and ethanol; pK_a values of 8.35 (thiol) and 10.81 (amine)), also called thioethanolamine and 2-mercaptoethylamine, is formed by the decarboxylation of cysteine. Cysteamine is readily oxidized to its disulfide form, requiring cysteamine solutions to be prepared freshly. Cysteamine is a disulfide-group reducing agent, and is expected to inhibit enzymes requiring disulfide linkages for catalytic activity. It also affords protection against oxidative stress associated with radiation damage. **Target(s):** 5-aminolevulinate aminotransferase (1,47); threonine ammonia-lyase, *or* threonine dehydratase (2,17); cysteine dioxygenase (3,58,59); histidine ammonia-lyase (4); urease (5,10); HIV-1 replication (6); amine oxidase (7); lactate dehydrogenase (8); alkaline phosphatase (9,21,27); spermidine synthase (11); aldehyde dehydrogenase (12); dopamine β-hydroxylase (13,14); cytochrome P450 isotype, *or* thiobenzamide *S*-monooxygenase (15); NADH: cytochrome *c* reductase (16); NADPH:cytochrome *c* reductase (16); steroid 11β-monooxygenase (18,19); *Agkistrodon halys* venom proteinases (20); β-galactosidase (22); aspartate 4-decarboxylase (23); carnitine palmitoyltransferase (24); dihydroorotase (25); nitrile hydratase (26,36); cysteinyl-tRNA synthetase (28,29); prolyl/cysteinyl-tRNA synthetase, *Methanocaldococcus jannaschii* (30); alanyl tRNA synthetase (31); phenylalanine racemase, ATP-hydrolyzing (32); 3-chloro-D-alanine dehydrochlorinase (33); cystathionine γ-lyase (34); diaminopropionate ammonia-lyase (35); adenylylsulfatase (37); isopenicillin-N synthase (38); guanidinobutyrase39; leishmanolysin (40); leucyl aminopeptidase (41); leukotriene-A$_4$ hydrolase, IC$_{50}$ = 27 μM (42); thioglucosidase, myrosinase (43); glycerophosphocholine cholinephosphodiesterase (44); 5'-nucleotidase (45); 3-mercaptopyruvate sulfurtransferase, weakly inhibited (46); spermidine synthase (48,49); protein-glutamine γ-glutamyltransferase, *or* transglutaminase (50-52); glutaminyl-peptide cyclotransferase (53-55); acetylserotonin *O*-methyltransferase, *or* hydroxyindole *O*-methyltransferase (56); deoxyhypusine monooxygenase (57). **1**. Turner & Neuberger (1970) *Meth. Enzymol.* **17A**, 188. **2**. Leoncini, Pagani, Marinello & Keleti (1989) *Biochim. Biophys. Acta* **994**, 52. **3**. Yamaguchi & Hosokawa (1987) *Meth. Enzymol.* **143**, 395. **4**. Hanson & Havir (1972) *The Enzymes*, 3rd ed. (Boyer, ed.), 7, 75. **5**. Metelitsa, Tarun, Puchkaev & Losev (2001) *Prikl. Biokhim. Mikrobiol.* **37**, 190. **6**. Bergamini, Ventura, Mancino, *et al.* (1996) *J. Infect. Dis.* **174**, 214. **7**. Fontana, Costa, Pensa & Cavallini (1993) *Physiol. Chem. Phys. Med. NMR* **25**, 121. **8**. Sharma & Schwille (1992) *Biochem. Int.* **27**, 431. **9**. Japundzic, Rakic-Stojiljkovic & Levi (1991) *Scand. J. Gastroenterol.* **26**, 523. **10**. Todd & Hausinger (1989) *J. Biol. Chem.* **264**, 15835. **11**. Hibasami, Kawase, Tsukada, *et al.* (1988) *FEBS Lett.* **229**, 243. **12**. Watanabe, Hobara & Nagashima (1986) *Ann. Nutr. Metab.* **30**, 54. **13**. Terry & Craig (1985) *Neuroendocrinology* **41**, 467. **14**. Smythe, Bradshaw, Gleeson & Nicholson (1985) *Life Sci.* **37**, 841. **15**. Duffel & Gillespie (1984) *J. Neurochem.* **42**, 1350. **16**. Mull, Hinkelbein, Gertz & Flemming (1977) *Biochim. Biophys. Acta* **481**, 407. **17**. Lusini, Di Stefano, Di Bello, Pagani & Marinello (1975) *Boll. Soc. Ital. Biol. Sper.* **51**, 1565. **18**. Flemming & Seydewitz (1974) *Experientia* **30**, 989. **19**. Flemming, Geierhass & Seydewitz (1973) *Biochem. Pharmacol.* **22**, 1241. **20**. Iwanaga, Oshima & Suzuki (1976) *Meth. Enzymol.* **45**, 459. **21**. Agus, Cox & Griffin (1966) *Biochim. Biophys. Acta* **118**, 363. **22**. Wallenfels & Weil (1972) *The Enzymes*, 3rd ed. (Boyer, ed.), 7, 617. **23**. Thanassi & Fruton (1962) *Biochemistry* **1**, 975. **24**. Bressler (1970) in *Comprehensive Biochemistry* (Florkin & Stotz, eds.), vol. **18**, p. 331, Elsevier, Amsterdam. **25**. Christopherson & Jones (1980) *J. Biol. Chem.* **255**, 3358. **26**. Nagasawa, Nanba, Ryuno, Takeuchi & Yamada (1987) *Eur. J. Biochem.* **162**, 691. **27**. Japundzic & Levi (1987) *Biochem. Pharmacol.* **36**, 2489. **28**. Pan, Yu, Duh & Lee (1976) *J. Chin. Biochem. Soc.* **5**, 45. **29**. Burnell & Whatley (1977) *Biochim. Biophys. Acta* **481**, 266. **30**. Ambrogelly, Ahel, Polycarpo, *et al.* (2002) *J. Biol. Chem.* **277**, 34749. **31**. Webster (1961) *Biochim. Biophys. Acta* **49**, 141. **32**. Schroeter-Kermani, von Döhren & Kleinkauf (1986) *Biochim. Biophys. Acta* **883**, 345. **33**. Nagasawa, Ohkishi, Kavakami, *et al.* (1982) *J. Biol. Chem.* **257**, 13749. **34**. Braunstein & Goryachenkova (1984) *Adv. Enzymol. Relat. Areas Mol. Biol.* **56**, 1. **35**. Nagasawa, Tanizawa, Satoda & Yamada (1988) *J. Biol. Chem.* **263**, 958. **36**. Nagasawa & Yamada (1987) *Biochem. Biophys. Res. Commun.* **147**, 701. **37**. Li & Schiff (1991) *Biochem. J.* **274**, 355. **38**. Palissa, von Dohren, Kleinkauf, Ting & Baldwin (1989) *J. Bacteriol.* **171**, 5720. **39**. Yorifuji, Shimizu, Hirata, *et al.* (1992) *Biosci.* *Biotechnol. Biochem.* **56**, 773. **40**. Chaudhuri, Chaudhuri, Pan & Chang (1989) *J. Biol. Chem.* **264**, 7483. **41**. Xu, Shawar & Dresden (1990) *Exp. Parasitol.* **70**, 124. **42**. Ollmann, Hogg, Muñoz, *et al.* (1995) *Bioorg. Med. Chem.* **3**, 969. **43**. Uda, Kurata & Arakawa (1986) *Agric. Biol. Chem.* **50**, 2741. **44**. Sok (1998) *Neurochem. Res.* **23**, 1061. **45**. Liu & Sok (2000) *Neurochem. Res.* **25**, 1475. **46**. Vachek & Wood (1972) *Biochim. Biophys. Acta* **258**, 133. **47**. Neuberger & Turner (1963) *Biochim. Biophys. Acta* **67**, 342. **48**. Haider, Eschbach, de Soudas Dias, *et al.* (2005) *Mol. Biochem. Parasitol.* **142**, 224. **49**. Hibasami, Kawase, Tsukada, *et al.* (1988) *FEBS Lett.* **229**, 243. **50**. Jeitner, Delikatny, Ahlqvist, Capper & Cooper (2005) *Biochem. Pharmacol.* **69**, 961. **51**. Siegel & Khosla (2007) *Pharmacol. Ther.* **115**, 232. **52**. Jeon, Lee, Jang, *et al.* (2004) *Exp. Mol. Med.* **36**, 576. **53**. Cynis, Rahfeld, Stephan, *et al.* (2008) *J. Mol. Biol.* **379**, 966. **54**. Schilling, Cynis, von Bohlen, *et al.* (2005) *Biochemistry* **44**, 13415. **55**. Schilling, Lindner, Koch, *et al.* (2007) *Biochemistry* **46**, 10921. **56**. Sugden & Klein (1987) *J. Biol. Chem.* **262**, 6489. **57**. Abbruzzese, Park & Folk (1986) *J. Biol. Chem.* **261**, 3085. **58**. Yamaguchi, Hosokawa, Kohashi, *et al.* (1978) *J. Biochem.* **83**, 479. **59**. Dominy, Simmons, Karplus, Gehring & Stipanuck (2006) *J. Bacteriol.* **188**, 5561.

Cysteic Acid

This amino acid sulfonate (FW$_{free-acid}$ = 169.16 g/mol; CAS 498-40-8; Water-soluble; pK_a values of 1.3 (for R–SO$_3$H), 1.89, and 8.7 at 25°C), derived from L-cysteine and also known as L-cysteate, 3-sulfo-L-alanine and L-cysteinesulfonate, is a aspartate isostere that stimulates the activity of methionyl-tRNA synthetase. **Target(s):** aspartate 1-decarboxylase (1,12); kynurenine:oxoglutarate aminotransferase (3); sulfinoalanine decarboxylase, *or* cysteinesulfinate decarboxylase, also alternative substrate (4,5,9-11); pantothenate synthetase, *or* pantoate:β-alanine ligase (6); cysteinyl-tRNA synthetase (7); serine-sulfate ammonia-lyase, weakly inhibited (8); N^4-(β-N-acetylglucosaminyl)-L-asparaginase (13); glutamate carboxypeptidase II, N-acetylated-α-linked-acidic dipeptidase (14); aspartate aminotransferase (15); serine *O*-acetyltransferase (16); acetylserotonin *O*-methyltransferase, *or* hydroxyindole *O*-methyltransferase (17); homogentisate 1,2-dioxygenase, weakly inhibited (18). **1**. Williamson (1985) *Meth. Enzymol.* **113**, 589. **2**. Weinstein & Griffith (1988) *J. Biol. Chem.* **263**, 3735. **3**. Kocki, Luchowski, Luchowska, *et al.* (2003) *Neurosci. Lett.* **346**, 97. **4**. Do & Tappaz (1996) *Neurochem. Int.* **28**, 363. **5**. Jacobsen, Thomas & Smith (1964) *Biochim. Biophys. Acta* **85**, 103. **6**. Miyatake, Nakano & Kitaoka (1978) *J. Nutr. Sci. Vitaminol.* **24**, 243. **7**. Burnell & Whatley (1977) *Biochim. Biophys. Acta* **481**, 266. **8**. Tudball & Thomas (1973) *Eur. J. Biochem.* **40**, 25. **9**. Jin & Jones (1996) *J. Biochem. Mol. Biol.* **29**, 335. **10**. Tang, Hsu, Sun, *et al.* (1996) *J. Biomed. Sci.* **3**, 442. **11**. Weinstein & Griffith (1987) *J. Biol. Chem.* **262**, 7254. **12**. Williamson & Brown (1979) *J. Biol. Chem.* **254**, 8074. **13**. Risley, Huang, Kaylor, Malik & Xia (2001) *J. Enzyme Inhib.* **16**, 269. **14**. Robinson, Blakely, Couto & Coyle (1987) *J. Biol. Chem.* **262**, 14498. **15**. Martins, Mourato & de Varennes (2001) *J. Enzyme Inhib.* **16**, 251. **16**. Kredich & Tomkins (1966) *J. Biol. Chem.* **241**, 4955. **17**. Karahasanoglu & Ozand (1972) *J. Neurochem.* **19**, 411. **18**. Knox & Edwards (1955) *J. Biol. Chem.* **216**, 479.

L-Cysteine

This thiol-containing amino acid (FW = 121.16 g/mol; CAS 52-90-4; Water-soluble; pK_a values of 1.92, 8.37 (SH), and 10.70), also called (*R*)-2-amino-3-mercaptopropanoate and symbolized by Cys or C, is one of the twenty proteinogenic amino acids. There are two codons for L-cysteine: UGU and UGC. Cysteinyl residues in proteins are among the most conserved; these residues are interchanged in homologous proteins most frequently with seryl residues. Cysteinyl residues are roughly evenly distributed between the surface and the interior of folded proteins. They are frequently found in β-pleated sheets and are often at the N-termini of α-helices. The sulfhydryl groups of cysteinyl residues are quite reactive and are at the active sites of many enzymes. The pK_a of this group is about 8.37. However, that pKa may very greatly with the microenvironment. Two cysteinyl residues can form a disulfide bond in a folded protein, provided that the protein conformation permits this oxidation. Thiol groups can also interact strongly with metal ions. In addition, they can undergo post-translational modification (*e.g.*, phosphorylation, palmitoylation, farnesylation, and geranylgeranylation).

Cystine

This amino acid disulfide (FW = 240.30 g/mol; CAS 923-32-0; pK_a values are < 1, 2.1, 8.02, and 8.71 at 35°C; Low Solubility = 10 mg/100 mL H_2O at 25°C), derived from two cysteine molecules, is the predominant circulating form of L-cysteine in humans. Because L-cystine is readily produced from L-cysteine under physiological conditions, elevated levels of L-cysteine in the diet, blood plasma, and urine can lead to formation of cystine calculi. Enzymes with reactive active-site thiol groups may form an Enz-S–S-Cys upon exposure to cystine (*Reaction*: Enz-SH + Cys-S–S-Cys ⇌ Enz-S–S-Cys + Cys-SH). **Target(s):** homogentisate 1,2-dioxygenase (1,44); diaminopimelate decarboxylase (2,29); inosamine-phosphate amidinotransferase (3,4); diaminopimelate decarboxylase (5); 3α-hydroxysteroid dehydrogenase (5); phosphoenolpyruvate carboxykinase (GTP), by both D- and L-cystine (7); oxytocin aminopeptidase (8); glutathione *S*-transferase (9); guanylate cyclase (10,11,20); phosphorylase phosphatase (11,18); tyrosine aminotransferase (11); cysteine dioxygenase, by L-cystine (12,43); succinate dehydrogenase (13); 6-phosphogluconate dehydrogenase (14); catalase (15); phosphatidate phosphatase, weakly inhibited (16); γ-glutamyl cyclotransferase, weakly inhibited (17); calpains (19); gluconate dehydratase (21); peroxidase (22); pyrophosphatase, *or* inorganic diphosphatase (23); choline oxidase (24); glutamate decarboxylase (25); succinyl-CoA synthetase (26); glutathione synthetase, moderately inhibited (27); phenylalanine racemase, ATP-hydrolyzing, inhibited by DL-cystine (28); sulfur reductase (29); isopenicillin-N synthase (31); pantetheine hydrolase (32); cystinyl aminopeptidase, *or* oxytocinase (33); phosphodiesterase I, *or* 5'-exonuclease. weakly inhibited (34); [phosphorylase] phosphatase (35); creatine kinase (36); pyruvate kinase (37,38); phosphoribulokinase (39); serine *O*-acetyltransferase (40); thiol *S*-methyltransferase (41); betaine:homocysteine *S*-methyltransferase, inhibited by both D- and L-cystine (42). 1. Edwards & Knox (1955) *Meth. Enzymol.* 2, 292. 2. Work (1962) *Meth. Enzymol.* 5, 864. 3. Walker & Walker (1970) *Meth. Enzymol.* 17A, 1012. 4. Walker (1975) *Meth. Enzymol.* 43, 451. 5. White (1971) *Meth. Enzymol.* 17B, 140. 6. Terada, Nanjo, Shinagawa, *et al.* (1993) *J. Enzyme Inhib.* 7, 33. 7. Ballard & Hopgood (1976) *Biochem. J.* 154, 717. 8. Itoh & Nagamatsu (1995) *Biochim. Biophys. Acta* 1243, 203. 9. Nishihara, Maeda, Okamoto, *et al.* (1991) *Biochem. Biophys. Res. Commun.* 174, 580. 10. Kamisaki, Waldman & Murad (1986) *Arch. Biochem. Biophys.* 251, 709. 11. Gilbert (1984) *Meth. Enzymol.* 107, 330. 12. Yamaguchi & Hosokawa (1987) *Meth. Enzymol.* 143, 395. 13. Schlenk (1951) *The Enzymes*, 1st ed., 2 (Part 1), 316. 14. Noltmann & Kuby (1963) *The Enzymes*, 2nd ed., 7, 223. 15. Pihl, Lange & Evang (1961) *Acta Chem. Scand.* 15, 1271. 16. Coleman & Hübscher (1962) *Biochim. Biophys. Acta* 56, 479. 17. Taniguchi & Meister (1978) *J. Biol. Chem.* 253, 1799. 18. Shimazu, Tokutake & Usami (1978) *J. Biol. Chem.* 253, 7376. 19. Di Cola & Sacchetta (1987) *FEBS Lett.* 210, 81. 20. Brandwein, Lewicki & Murad (1981) *J. Biol. Chem.* 256, 2958. 21. Bender & Gottschalk (1973) *Eur. J. Biochem.* 40, 309. 22. Guilbault, Brignac & Zimmer (1968) *Anal. Chem.* 40, 190. 23. Naganna (1950) *J. Biol. Chem.* 183, 693. 24. Eadie & Bernheim (1950) *J. Biol. Chem.* 185,

731. 25. Roberts & Frankel (1951) *J. Biol. Chem.* 190, 505. 26. Wider & Tigier (1971) *Enzymologia* 41, 217. 27. Gupta, Srivastava & Banu (2005) *Exp. Parasitol.* 111, 137. 28. Schroeter-Kermani, von Döhren & Kleinkauf (1986) *Biochim. Biophys. Acta* 883, 345. 29. Asada, Tanizawa, Kawabata, Misuno & Soda (1981) *Agric. Biol. Chem.* 45, 1513. 30. Zöphel, Kennedy, Beinert & Kroneck (1988) *Arch. Microbiol.* 150, 72. 31. Palissa, von Dohren, Kleinkauf, Ting & Baldwin (1989) *J. Bacteriol.* 171, 5720. 32. Ricci, Nardini, Chiaraluce, Dupre & Cavallini (1986) *Biochim. Biophys. Acta* 870, 82. 33. Krishna & Kanagasabapathy (1989) *J. Endocrinol.* 121, 537. 34. Sakura, Nagashima, Nakashima & Maeda (1998) *Thromb. Res.* 91, 83. 35. Gratecos, Detwiler, Hurd & Fischer (1977) *Biochemistry* 16, 4812. 36. Pereira Oliveira, Rodrigues-Junior, Rech & Wannmacher (2007) *Arch. Med. Res.* 38, 164. 37. Feksa, Cornelio, Dutra-Filho, *et al.* (2005) *Int. J. Dev. Neurosci.* 23, 509. 38. Feksa, Cornelio, Dutra-Filho, *et al.* (2004) *Brain Res.* 1012, 93. 39. Lebreton, Graciet & Gontero (2003) *J. Biol. Chem.* 278, 12078. 40. Nozaki, Shigeta, Saito-Nakano, Imada & Kruger (2001) *J. Biol. Chem.* 276, 6516. 41. Attieh, Sparace & Saini (2000) *Arch. Biochem. Biophys.* 380, 257. 42. Finkelstein, Harris & Kyle (1972) *Arch. Biochem. Biophys.* 153, 320. 43. Sakakibara, Yamaguchi, Hosokawa, Kohashi & Ueda (1976) *Biochim. Biophys. Acta* 422, 273. 44. Knox & Edwards (1955) *J. Biol. Chem.* 216, 479.

CYT997

This novel orally active tubulin polymerization inhibitor (FW = 434.54 g/mol), also known as Lexibulin®, exhibits potent both cytotoxic and vascular disrupting activities *in vitro* and *in vivo* (1,2). CYT997 blocks the cell cycle at the G_2/M boundary, and Western blot analysis indicates an increase in phosphorylated Bcl-2, along with increased expression of cyclin B_1 (2). CYT997 prevents *in vitro* tubulin polymerization (IC_{50} ~3 μM) and reversibly disrupts the microtubule network in cells, as visualized using fluorescence microscopy. CYT997 is active against DU145 human prostate carcinoma cells, IC_{50} = 73 nM; A549 human lung carcinoma cells, IC_{50} = 21 nM; Ramos human Burkitt's lymphoma cells, IC_{50} = 80 nM; KHOS/NP human osteosarcoma cells, IC_{50} = 101 nM; A375 human melanoma cells, IC_{50} = 49 nM; HCT-15 human colon carcinoma cells, IC_{50} = 52 nM; HT1376 human bladder carcinoma cells, IC_{50} = 93 nM; BT-20 human breast carcinoma cells, IC_{50} = 58 nM; A431 human epithelial carcinoma cells, IC_{50} = 51 nM; PA-1 human ovarian teratocarcinoma, IC_{50} = 48 nM; U937 human leukemic monocyte lymphoma, IC_{50} = 21 nM; HepG2 human hepatocellular liver carcinoma cells, IC_{50} = 9 nM; TF-1 human erythroleukemia cells, IC_{50} = 32 nM; Baf3/TelJAK2 constitutively active JAK2 cells, IC_{50} = 23 nM; PC3 human prostate carcinoma; cells, IC_{50} = 27 nM; and K562 human chronic myelogenous leukemia cells, IC_{50} = 20 nM (2). Treatment of acute myeloid leukemia cells with CYT997 resulted in G_2/M phase cell cycle arrest, and induced apoptosis through the activation of extrinsic and intrinsic apoptotic pathways (3). 1. Burns, Harte, Bu, *et al.* (2009) *Bioorg. Med. Chem. Lett.* 19, 4639. 2. Li, Yeh, Song, *et al.* (2013) *Anticancer Drugs* 24, 1047. 3. Chen, Yang, Xu, *et al.* (2013) *Exp. Ther. Med.* 6, 299.

Cytarabine

This nucleoside (FW = 243.22 g/mol; CAS 147-94-4; *Abbreviation: ara*-C or Ara-C), also known as 1-β-D-arabinofuranosylcytosine and *ara*-cytidine, is an antiviral agent and S phase-acting antineoplastic agent. Ara-C is highly effective in treating acute myelogenous leukemia (AML). This agent is incorporated into leukemic cell DNA, and the extent of this incorporation correlates with loss of clonogenic survival. When metabolically converted to cytarabine 5'-triphosphate (*ara*-CTP), the latter is a potent inhibitor of DNA synthesis, but not RNA biosynthesis. (A nominal inhibitory concentration is 10 µM.) The incorporated Ara-C behaves as a DNA chain terminator, and the extent of (Ara-C)DNA formation correlates with inhibition of DNA synthesis. (**See** *Cytarabine 5'-Triphosphate*) **Key Pharmacokinetic Parameters:** *See* Appendix II in Goodman & Gilman's *THE PHARMACOLOGICAL BASIS OF THERAPEUTICS*, 12th Edition (Brunton, Chabner & Knollmann, eds.) McGraw-Hill Medical, New York (2011). **Target(s):** deoxycytidine kinase, also alternative substrate (1-7); dCMP deaminase (8); deoxynucleoside kinase (2); deoxyguanosine kinase (9). **1.** Kessel (1968) *J. Biol. Chem.* **243**, 4739. **2.** Johansson, Van Rompay, Degrève, Balzarini & Karlsson (1999) *J. Biol. Chem.* **274**, 23814. **3.** Someya, Shaddix, Tiwari, Secrist & Parker (2003) *J. Pharmacol. Exp. Ther.* **304**, 1314. **4.** Yamada, Goto & Ogasawara (1983) *Biochim. Biophys. Acta* **761**, 34. **5.** Durham & Ives (1970) *J. Biol. Chem.* **245**, 2276. **6.** Momparler & Fischer (1968) *J. Biol. Chem.* **243**, 4298. **7.** Coleman, Stoller, Drake & Chabner (1975) *Blood* **46**, 791. **8.** Riva, Barra, Cano, *et al.* (1990) *J. Cell. Pharmacol.* **1**, 79. **9.** Yamada, Goto & Ogasawara (1982) *Biochim. Biophys. Acta* **709**, 265.

Cytarabine 5'-Triphosphate

This nucleotide analogue (FW = 482.15 g/mol), also known as cytosine 1-β-D-arabinofuranoside 5'-triphosphate and *ara* CTP, is a potent inhibitor of DNA biosynthesis, but *not* RNA biosynthesis. **Target(s):** deoxyadenosine kinase (1,5); DNA polymerase (10,11); DNA polymerase I (2); DNA polymerase α (12); DNA-directed DNA polymerase (2-4,6-12); DNA primase (12); deoxycytidine kinase (13,14). **1.** Anderson (1973) *The Enzymes*, 3rd ed., **9**, 49. **2.** Kornberg & Kornberg (1974) *The Enzymes*, 3rd ed., **10**, 119. **3.** Loeb (1974) *The Enzymes*, 3rd ed., **10**, 173 **4.** Brown & Wright (1995) *Meth. Enzymol.* **262**, 202. **5.** Krygier & Momparler (1971) *J. Biol. Chem.* **246**, 2752. **6.** Kimball & Wilson (1968) *Proc. Soc. Exp. Biol. Med.* **127**, 429. **7.** Graham & Whitmore (1970) *Cancer Res.* **30**, 2636. **8.** Momparler (1969) *Biochem. Biophys. Res. Commun.* **34**, 464. **9.** Scott & Tomkins (1975) *Meth. Enzymol.* **40**, 273. **10.** Gass & Cozzarelli (1974) *Meth. Enzymol.* **29**, 27. **11.** Burrows & Goward (1992) *Biochem. J.* **287**, 971. **12.** Kuchta, Ilsley, Kravig, Schubert & Harris (1992) *Biochemistry* **31**, 4720. **13.** Datta, Shewach, Mitchell & Fox (1989) *J. Biol. Chem.* **264**, 9359. **14.** Wang, Kucera & Capizzi (1993) *Biochim. Biophys. Acta* **1202**, 309.

Cytidine

This pyrimidine nucleoside (FW = 243.22 g/mol; CAS 65-46-3) is a precursor for the biosynthesis of cytidine 5'-mono-, di-, and tri-phosphates. See cytidine 5'-diphosphate for spectral data. **Target(s):** aspartate carbamoyltransferase (1,4-8,13); cytosine deaminase (2,12); pancreatic ribonuclease, *or* ribonuclease A (3); 5'-nucleotidase (5,9); arginase (10); dihydroorotase (11); lactoylglutathione lyase, *or* glyoxalase I (14); aspartate ammonia-lyase (15); uridine nucleosidase (16); purine nucleosidase (17,18); ribonuclease T2(19); CMP-*N* acylneuraminate phosphodiesterase (20,21); acid phosphatase (22); hydroxyacylglutathione hydrolase, *or* glyoxalase II (14); *N*-acylneuraminate cytidylyltransferase (23); adenosine kinase (24). **1.** Jones (1962) *Meth. Enzymol.* **5**, 903. **2.** Ipata & Cercignani (1978) *Meth. Enzymol.* **51**, 394. **3.** Richards &

Wyckoff (1971) *The Enzymes*, 3rd ed., **4**, 647. **4.** Jacobson & Stark (1973) *The Enzymes*, 3rd ed., **9**, 225. **5.** Webb (1966) *Enzyme and Metabolic Inhibitors*, vol. **2**, Academic Press, New York. **6.** Smith, Jr., & Sullivan (1960) *Biochim. Biophys. Acta* **39**, 554. **7.** Gerhart & Pardee (1962) *J. Biol. Chem.* **237**, 891. **8.** Bresnick (1963) *Biochim. Biophys. Acta* **67**, 425. **9.** Segal & Brenner (1960) *J. Biol. Chem.* **235**, 471. **10.** Rosenfeld, Dutta, Chheda & Tritsch (1975) *Biochim. Biophys. Acta* **410**, 164. **11.** Bresnick & Blatchford (1964) *Biochim. Biophys. Acta* **81**, 150. **12.** Ipata, Marmocchi, Magni, Felicioli & Polidoro (1971) *Biochemistry* **10**, 4270. **13.** Yates & Pardee (1956) *J. Biol. Chem.* **221**, 757. **14.** Oray & Norton (1980) *Biochem. Biophys. Res. Commun.* **95**, 624. **15.** Noh, Kwon, Kim, Lee & Yoon (2000) *J. Biochem. Mol. Biol.* **33**, 366. **16.** Kurtz, Exinger, Erbs & Jund (2002) *Curr. Genet.* **41**, 132. **17.** Ogawa, Takeda, Xie, *et al.* (2001) *Appl. Environ. Microbiol.* **67**, 1783. **18.** Koszalka & Krenitsky (1979) *J. Biol. Chem.* **254**, 8185. **19.** Irie & Ohgi (1976) *J. Biochem.* **80**, 39. **20.** van Dijk, Maier & van den Eijnden (1976) *Biochim. Biophys. Acta* **444**, 816. **21.** Kean & Bighouse (1974) *J. Biol. Chem.* **249**, 7813. **22.** Uerkvitz (1988) *J. Biol. Chem.* **263**, 15823. **23.** Bravo, Barrallo, Ferrero, *et al.* (2001) *Biochem. J.* **358**, 585. **24.** Long & Parker (2006) *Biochem. Pharmacol.* **71**, 1671.

Cytidine 5'-Diphosphate

This pyrimidine nucleotide (FW$_{free-acid}$ = 403.18 g/mol) is a precursor for phospholipids and sphingomyelins. It is also a component of certain nucleotide coenzymes. Alkali salts of CDP are very soluble in water, whereas the free acid is less soluble. The UV spectrum of CDP at pH 7 (or pH 11) has a λ_{max} at 271 nm (ε = 9100 M^{-1}cm^{-1}). At pH 2, the λ_{max} is 280 nm (ε = 12800 M^{-1}cm^{-1}). **Target(s):** pyruvate carboxylase (1,15); aspartate carbamoyltransferase (2,11,12,78-80); cytosine deaminase (3,20,29); cytidine deaminase (3); β-galactoside α-2,3-sialyltransferase (4,58,59); α-*N*-acetyl-galactosaminide α-2,6-sialyltransferase (4); [3-methyl-2-oxobutanoate dehydrogenase, *or* 2-methylpropanoyl-transferring)] phosphatase (5,8); membrane-bound sialyltransferase of *Escherichia coli* (6); carbamoyl-phosphate synthetase (7); isocitrate dehydrogenase (9); glutamine synthetase (10); 5'-nucleotidase (13,33); dihydroorotase (14); glycogenin glucosyltransferase, self-glucosylation (16); α-*N*-acetylneuraminate α-2,8-sialyltransferase (17); dolichol kinase (18); polyribonucleotide nucleotidyltransferase (19); phosphoribosyl-pyrophosphate synthetase (21); glucose-6-phosphate dehydrogenase (22); adenylate cyclase (23); lactoylglutathione lyase, *or* glyoxalase I (24); UDP-glucuronate decarboxylase (25); CDP-glycerol diphosphatase (26); nucleotide diphosphatase (27); dCTP deaminase (28); CMP-*N*-acylneuraminate phospho-diesterase (30,31); fructose-1,6-bisphosphatase (32); hydroxyacylglutathione hydrolase, *or* glyoxalase II (24); [branched-chain α-keto-acid dehydrogenase] kinase (34); CDP-diacylglycerol: Inositol-3 phophatidyltransferase (35,36); serine-phosphoethanolamine synthase (37); diacylglycerol cholinephosphotransferase (38,39); ethanolaminephosphtransferase (38,39); [protein-PII] uridylyltransferase (40); *N*-acylneuraminate cytidylyl-transferase (41-44); phosphatidate cytidylyltransferase (45); glucose-1-phosphate cytidylyltransferase (46); ribose phosphate diphosphokinase, *or* phosphoribosyl-pyrophosphate synthetase (47); cytidylate kinase, product inhibition (48); diacylglycerol kinase (49,50); deoxycytidine kinase (51); 1-phosphatidylinositol 4-kinase (52,53); fucokinase (54); glucuronokinase (55); lactosylceramide α-2,3-sialyltransferase (56,57); α *N*-acetylneuraminate α-2,8-sialyltransferase (57); α-*N*-acetylgalactosaminide α-2,6 sialyltransferase (59); β-galactoside α-2,6-sialyltransferase (60-62); protein xylosyltransferase (63); xanthine phosphoribosyltransferase (64); nicotinate-nucleotide diphosphorylase (65,66); hypoxanthine(guanine) phospho-ribosyl-transferase, *or* HGPRT (67); glycoprotein 3-α-L-fucosyltransferase (68,69); glycogenin glucosyltransferase (70); β-1,4-mannosyl-glycoprotein 4-β-*N*-acetylglucos-aminyltransferase (71); protein *N*-acetylglucos-aminyltransferase (72); globoside α-*N*-acetylgalactos-aminyltransferase (73); glycoprotein 6-α-L-fucosyl-transferase (74); procollagen galactosyltransferase (75); polypeptide *N*-acetylgalactosaminyltransferase (76); glycol-protein fucosyl galactoside α-*N*-acetylgalactosaminyl-transferase (77). **1**. Scrutton, Olmsted & Utter (1969) *Meth. Enzymol.* **13**, 235. **2**. Adair & Jones (1978) *Meth. Enzymol.* **51**, 51. **3**. Ipata & Cercignani (1978) *Meth. Enzymol.* **51**, 394. **4**. Sadler, Beyer, Oppenheimer, *et al.* (1982) *Meth. Enzymol.* **83**, 458. **5**. Damuni & Reed (1988) *Meth. Enzymol.* **166**, 321. **6**. Ortiz, Reglero, Rodriguez-Aparicio & Luengo (1989) *Eur. J. Biochem.* **178**, 741. **7**. Aoki & Oya (1987) *Comp. Biochem. Physiol. B* **87**, 655. **8**. Damuni, Merryfield, Humphreys & Reed (1984) *Proc. Natl. Acad. Sci. U.S.A.* **81**, 4335. **9**. Friga & Farkas (1981) *Arch. Microbiol.* **129**, 331. **10**. Bhandari & Nicholas (1981) *Aust. J. Biol. Sci.* **34**, 527. **11**. Webb (1966) *Enzyme and Metabolic Inhibitors*, vol. 2, p. 468, Academic Press, New York. **12**. Gerhart & Pardee (1962) *J. Biol. Chem.* **237**, 891. **13**. Madrid-Marina & Fox (1986) *J. Biol. Chem.* **261**, 444. **14**. Bresnick & Blatchford (1964) *Biochim. Biophys. Acta* **81**, 150. **15**. Scrutton & Utter (1965) *J. Biol. Chem.* **240**, 3714. **16**. Manzella, Ananth, Oegema, *et al.* (1995) *Arch. Biochem. Biophys.* **320**, 361. **17**. Eppler, Morre & Keenan (1980) *Biochim. Biophys. Acta* **619**, 332. **18**. Burton, Scher & Waechter (1979) *J. Biol. Chem.* **254**, 7129. **19**. Singer (1963) *J. Biol. Chem.* **238**, 336. **20**. Ipata, Marmocchi, Magni, Felicioli & Polidoro (1971) *Biochemistry* **10**, 4270. **21**. Wong & Murray (1969) *Biochemistry* **8**, 1608. **22**. Horne, Anderson & Nordlie (1970) *Biochemistry* **9**, 610. **23**. Yang & Epstein (1983) *J. Biol. Chem.* **258**, 3750. **24**. Oray & Norton (1980) *Biochem. Biophys. Res. Commun.* **95**, 624. **25**. Suzuki, Watanabe, Masumura & Kitamura (2004) *Arch. Biochem. Biophys.* **431**, 169. **26**. Glaser (1965) *Biochim. Biophys. Acta* **101**, 6. **27**. Wise, Anderson & Anderson (1997) *Vet. Microbiol.* **58**, 261. **28**. Beck, Eisenhardt & Neuhard (1975) *J. Biol. Chem.* **250**, 609. **29**. Balestreri, Felicioli & Ipata (1973) *Biochim. Biophys. Acta* **293**, 443. **30**. van Dijk, Maier & van den Eijnden (1976) *Biochim. Biophys. Acta* **444**, 816. **31**. Kean & Bighouse (1974) *J. Biol. Chem.* **249**, 7813. **32**. Fujita & Freese (1979) *J. Biol. Chem.* **254**, 5340. **33**. Newby, Luzio & Hales (1975) *Biochem. J.* **146**, 625. **34**. Reed, Damuni & Merryfield (1985) *Curr. Top. Cell. Regul.* **27**, 41. **35**. Antonsson & L. S. Klig (1996) *Yeast* **12**, 449. **36**. Antonsson (1994) *Biochem. J.* **297**, 517. **37**. Allen & Rosenberg (1968) *Biochim. Biophys. Acta* **151**, 504. **38**. Coleman & Bell (1977) *J. Biol. Chem.* **252**, 3050. **39**. Percy, Carson, Moore & Waechter (1984) *Arch. Biochem. Biophys.* **230**, 69. **40**. Engleman & Francis (1978) *Arch. Biochem. Biophys.* **191**, 602. **41**. Rodríguez-Aparicio, Luengo, Gonzalez-Clemente & A. Reglero (1992) *J. Biol. Chem.* **267**, 9257. **42**. Schmelter, Ivanov, Wember, *et al.* (1993) *Biol. Chem. Hoppe Seyler* **374**, 337. **43**. Corfield, Schauer & Wember (1979) *Biochem. J.* **177**, 1. **44**. Bravo, Barrallo, Ferrero, *et al.* (2001) *Biochem. J.* **358**, 585. **45**. Hanenberg, Heim, Wissing & Wagner (1993) *Plant Sci.* **88**, 13. **46**. Kimata & Suzuki (1966) *J. Biol. Chem.* **241**, 1099. **47**. Fox & Kelley (1972) *J. Biol. Chem.* **247**, 2126. **48**. Ruffner & Anderson (1969) *J. Biol. Chem.* **244**, 5994. **49**. Kato & Takenawa (1990) *J. Biol. Chem.* **265**, 794. **50**. Wissing & Wagner (1992) *Plant Physiol.* **98**, 1148. **51**. Kozai, Sonoda, Kobayashi & Sugino (1972) *J. Biochem.* **71**, 485. **52**. Steinert, Wissing & Wagner (1994) *Plant Sci.* **101**, 105. **53**. Yamakawa & Takenawa (1988) *J. Biol. Chem.* **263**, 17555. **54**. Kilker, Shuey & Serif (1979) *Biochim. Biophys. Acta* **570**, 271. **55**. Gillard & Dickinson (1978) *Plant Physiol.* **62**, 706. **56**. Melkerson-Watson & Sweeley (1981) *J. Biol. Chem.* **266**, 4448. **57**. Eppler, Morré & Keenan (1980) *Biochim. Biophys. Acta* **619**, 332. **58**. Chandrasekaran, Xue, Xia, *et al.* (2008) *Biochemistry* **47**, 320. **59**. Sadler, Rearick, Paulson & Hill (1979) *J. Biol. Chem.* **254**, 4434. **60**. Scudder & Chantler (1981) *Biochim. Biophys. Acta* **660**, 136. **61**. Datta & Paulson (1995) *J. Biol. Chem.* **270**, 1497. **62**. Weinstein, de Souza-e-Silva & Paulson (1982) *J. Biol. Chem.* **257**, 13845. **63**. Casanova, Kuhn, Kleesiek & Götting (2008) *Biochem. Biophys. Res. Commun.* **365**, 678. **64**. Miller, Adamczyk, Fyfe & Elion (1974) *Arch. Biochem. Biophys.* **165**, 349. **65**. Shibata & Iwai (1980) *Agric. Biol. Chem.* **44**, 119. **66**. Taguchi & Iwai (1976) *Agric. Biol. Chem.* **40**, 385. **67**. Nagy & Ribet (1977) *Eur. J. Biochem.* **77**, 77. **68**. De Vries, Storm, Rotteveel, *et al.* (2001) *Glycobiology* **11**, 711. **69**. Shinoda, Morishita, Sasaki, *et al.* (1997) *J. Biol. Chem.* **272**, 31992. **70**. Meezan, Manzella & Roden (1995) *Trends Glycosci. Glycotechnol.* **7**, 303. **71**. Ikeda, Koyota, Ihara, *et al.* (2000) *J. Biochem.* **128**, 609. **72**. Haltiwanger, Blomberg & Hart (1992) *J. Biol. Chem.* **267**, 9005. **73**. Ishibashi, Ohkubo & Makita (1977) *Biochim. Biophys. Acta* **484**, 24. **74**. Ihara, Ikeda & Taniguchi (2006) *Glycobiology* **16**, 333. **75**. Risteli (1978) *Biochem. J.* **169**, 189. **76**. Takeuchi, Yoshikawa, Sasaki & Chiba (1985) *Agric. Biol. Chem.* **49**, 1059. **77**. Navaratnam, Findlay, Keen & Watkins (1990) *Biochem. J.* **271**, 93. **78**. Burns, Mendz & Hazell (1997) *Arch. Biochem. Biophys.* **347**, 119. **79**. Santiago & West (2003) *J. Basic Microbiol.* **43**, 75. **80**. Adair & Jones (1972) *J. Biol. Chem.* **247**, 2308.

Cytidine 5'-Triphosphate

This pyrimidine nucleotide (FW$_{free-acid}$ = 483.16 g/mol; λ$_{max}$ = 271 nm; ε = 9000 M^{-1}cm^{-1}) plays a role in RNA and DNA biosynthesis as well as in the formation of many complex lipids, synthesis of CDP-glucose, and in the phosphorylation of dolichol. CTP is also a powerful allosteric feedback inhibitor of *Escherichia coli* aspartate carbamoyltransferase (1,2). Sodium and potassium salts are very soluble in water. CTP forms tight complexes with many metal ions. Metal ion-free CTP is likely to be inactive in CTP-dependent enzyme reactions. The stability constants are reported to be similar to those for MgATP^{2-}, CaATP^{2-}, and MnATP^{2-}, which are 73000, 35000, and 100,000 M^{-1}, respectively. **Target(s):** aspartate carbamoyltransferase (1,2,23,25,26, 34,39,40, 41,43,61,63,185-199); glutamine synthetase (3,35,52,58); *myo*-inositol oxidase (4,200); lactose synthase (5); pyruvate carboxylase (6,49); [glutamine-synthetase] adenylyltransferase (7,111); *N*-acetylglucosamine kinase (8); pyruvate, orthophosphate dikinase (9,10); arylsulfate sulfotransferase (11); carbamoyl-phosphate synthetase (12,45,67,68); GMP synthetase (13); uridine kinase (14-16,33,140-147); cytosine deaminase (17,59); cytidine deaminase (17); thiamin-phosphate pyrophosphorylase (18); β-galactoside α-2,3-sialyltransferase (19); α-*N*-acetyl-galactosaminide α-2,6-sialyltransferase (19); [3-methyl-2 oxobutanoate dehydrogenase (2-methylpropanoyl-transferring)] phosphatase ([branched-chain α-keto-acid dehydrogenase] phosphatase) (20,28); polynucleotide adenylyltransferase (21,37); *Pyrococcus furiosus* carbamate kinase (22); 5'→3' exoribonuclease (24); membrane-bound sialyltransferase of *Escherichia coli* (27); 5'-nucleotidase (30,47,60,62,94-97); 3',5'-cyclic-nucleotide phospho-diesterase (31,46); fumarase (32); cytidylate kinase (33,125); tRNA adenylyltransferase (36); phosphoprotein phosphatase types 1 and 2A (38); membrane Dipeptidase, *or* renal Dipeptidase (42); guanylate cyclase (44); dihydroorotase (48); glycogenin glucosyltransferase, self-glucosylation (50); NAD(P)H dehydrogenase (quinone) (51); α-*N*-acetylneuraminate α-2,8-sialyltransferase (53); deoxyadenosine kinase (54); diacylglycerol kinase (55); membrane dipeptidase (56); glutathione peroxidase (57); amidophosphoribosyl-transferase (64); phosphoribosyl-pyrophosphate synthetase (65); asparagine synthetase (66); phosphoglucomutase (69); guanylate cyclase (70,71); adenylate cyclase (72); glucose-6-phosphate dehydrogenase (73); lactoylglutathione lyase, *or* glyoxalase I (74); aspartate ammonia-lyase, moderate (75); UDP-glucuronate decarboxylase, weakly inhibited (76); phosphoenolpyruvate carboxylase (77); adenylylsulfatase (78); NAD$^+$ diphosphatase (79); CDP-glycerol diphosphatase (80); exopolyphosphatase (81); nucleotide diphosphatase (82); 5-hydroxypentanoate CoA-transferase (83); GTP cyclo-hydrolase I (84); dCTP deaminase (85); arginine deaminase (86); NMN nucleosidase (87); uridine nucleosidase (88); β-fructofuranosidase, *or* invertase (89); CMP-*N*-acylneuraminate phosphodiesterase (90,91); phospho-diesterase I, *or* 5'-exonuclease (92); multiple inositol poly-phosphate phosphatase, inositol-1,3,4,5-tetrakisphosphate 3-phosphatase (93); nucleotidases (94); phosphoprotein phosphatase (95); fructose-1,6-bisphosphatase (96); 3' nucleotidases (94); phosphatidate phosphatase (98); hydroxyacylglutathione hydrolase, *or* glyoxalase II (74); pyruvate,orthophosphate dikinase (99); CDP-glycerol glycerophospho-transferase (100); CDP diacylglycerol:inositol 3-phosphatidyltransferase (101,102); CDP-diacylglycerol:serine *O*-phosphatidyltransferase (103); diacylglycerol cholinephospho-transferase (104); ethanolamine-phosphotransferase (104,105); 2-*C*-methyl-D-erythritol-4-phosphate cytidylyltransferase (106); [protein-P$_{II}$] uridylyltransferase (107); *N*-acylneuraminate cytidylyl-transferase (108-110); tRNA adenylyltransferase (112-118); polynucleotide adenylyltransferase, *or* poly(A) polymerase (119-121); ribose-phosphate diphosphokinase, *or* phosphoribosyl-pyrophosphate synthetase (122-124); carbamate kinase (126); deoxyguanosine kinase (127); diacylglycerol kinase (128,129);

ethanolamine kinase (130); polynucleotide 5'-hydroxyl kinase (131,132); deoxycytidine kinase (133,134); 1-phosphatidylinositol 4-kinase (135,136); *N*-acetylglucosamine kinase (137); 1-phosphofructokinase (138); fucokinase (139); pyruvate kinase (148); 6-phosphofructokinase (149,150); glucokinase, as weak alternative substrate (151,152); methionine *S*-adenosyltransferase (153-155); thiamin-phosphate diphosphorylase (156); lactosylceramide α-2,3-sialyltransferase (157,158); α-*N*-acetylneuraminate α-2,8-sialyltransferase (158); β-galactoside α-2,3 sialyltransferase (159-162); α-*N*-acetylgalactosaminide α-2,6-sialyltransferase (161,163); β-galactoside α-2,6-sialyltransferase (164-166); protein xylosyltransferase (167); xanthine phosphoribosyltransferase (168); nicotinate-nucleotide diphosphorylase, carboxylating (169,170); uracil phosphoribosyltransferase (171-173); adenine phosphoribosyltransferase (174); [Skp1-protein] hydroxyproline *N*-acetylglucosaminyltransferase, weakly inhibited (175); polypeptide *N*-acetylgalactos-aminyltransferase (176); 1,3-β-glucan synthase (177); glucomannan 4-β mannosyltransferase, weakly inhibited (178); sucrose-phosphate synthase (179,180); ATP:citrate lyase (ATP:citrate synthase)181; citrate (*si*)-synthase, weakly inhibited (182); protein-glutamine γ-glutamyltransferase, *or* transglutaminase (183); ornithine carbamoyltransferase (184); 3-hydroxyanthranilate oxidase, weakly inhibited (201). **1.** Yang, Kirschner & Schachman (1978) *Meth. Enzymol.* **51**, 35. **2.** Adair & Jones (1978) *Meth. Enzymol.* **51**, 51. **3.** Rhee, Chock & Stadtman (1985) *Meth. Enzymol.* **113**, 213. **4.** Charalampous (1962) *Meth. Enzymol.* **5**, 329. **5.** Babad & Hassid (1966) *Meth. Enzymol.* **8**, 346. **6.** Scrutton, Olmsted & Utter (1969) *Meth. Enzymol.* **13**, 235. **7.** Shapiro & Stadtman (1970) *Meth. Enzymol.* **17A**, 910. **8.** Datta (1975) *Meth. Enzymol.* **42**, 58. **9.** South & Reeves (1975) *Meth. Enzymol.* **42**, 187. **10.** Milner, Michaels & Wood (1975) *Meth. Enzymol.* **42**, 199. **11.** Konishi-Imamura, Kim, Koizumi & Kobashi (1995) *J. Enzyme Inhib.* **8**, 233. **12.** Mori & Tatibana (1978) *Meth. Enzymol.* **51**, 111. **13.** Spector (1978) *Meth. Enzymol.* **51**, 219. **14.** Orengo & Kobayashi (1978) *Meth. Enzymol.* **51**, 299. **15.** Valentin-Hansen (1978) *Meth. Enzymol.* **51**, 308. **16.** Anderson (1978) *Meth. Enzymol.* **51**, 314. **17.** Ipata & Cercignani (1978) *Meth. Enzymol.* **51**, 394. **18.** Penttinen (1979) *Meth. Enzymol.* **62**, 68. **19.** Sadler, Beyer, Oppenheimer, *et al.* (1982) *Meth. Enzymol.* **83**, 458. **20.** Damuni & Reed (1988) *Meth. Enzymol.* **166**, 321. **21.** Edmonds (1990) *Meth. Enzymol.* **181**, 161. **22.** Uriarte, Marina, Ramón-Maiques, *et al.* (2001) *Meth. Enzymol.* **331**, 236. **23.** Purcarea (2001) *Meth. Enzymol.* **331**, 248. **24.** Slobin (2001) *Meth. Enzymol.* **342**, 282. **25.** Kantrowitz, Pastra-Landis & Lipscomb (1980) *Trends Biochem. Sci.* **5**, 124. **26.** Kantrowitz, Pastra-Landis & Lipscomb (1980) *Trends Biochem. Sci.* **5**, 150. **27.** Ortiz, Reglero, Rodriguez-Aparicio & Luengo (1989) *Eur. J. Biochem.* **178**, 741. **28.** Damuni, Merryfield, Humphreys & Reed (1984) *Proc. Natl. Acad. Sci. U.S.A.* **81**, 4335. **29.** O'Sullivan & Smithers (1979) *Meth. Enzymol.* **63**, 294. **30.** Drummond & Yamamoto (1971) *The Enzymes*, 3rd ed., **4**, 337. **31.** Drummond & Yamamoto (1971) *The Enzymes*, 3rd ed., **4**, 355. **32.** Hill & Teipel (1971) *The Enzymes*, 3rd. ed., **5**, 539. **33.** Anderson (1973) *The Enzymes*, 3rd ed., **9**, 49. **34.** Jacobson & Stark (1973) *The Enzymes*, 3rd ed., **9**, 225. **35.** Stadtman & Ginsburg (1974) *The Enzymes*, 3rd ed., **10**, 755. **36.** Deutscher (1982) *The Enzymes*, 3rd ed., **15**, 183. **37.** Edmonds (1982) *The Enzymes*, 3rd ed., **15**, 217. **38.** Ballou & Fischer (1986) *The Enzymes*, 3rd ed., **17**, 311. **39.** Webb (1966) *Enzyme and Metabolic Inhibitors*, vol. 2, p. 468, Academic Press, New York. **40.** Gerhart & Pardee (1962) *J. Biol. Chem.* **237**, 891. **41.** Bresnick (1963) *Biochim. Biophys. Acta* **67**, 425. **42.** Harper, Rene & Campbell (1971) *Biochim. Biophys. Acta* **242**, 446. **43.** Gerhart & Schachman (1965) *Biochemistry* **4**, 1054. **44.** Hardman & Sutherland (1969) *J. Biol. Chem.* **244**, 6363. **45.** O'Neal & Naylor (1969) *Biochem. J.* **113**, 271. **46.** Cheung (1967) *Biochemistry* **6**, 1079. **47.** Madrid-Marina & Fox (1986) *J. Biol. Chem.* **261**, 444. **48.** Bresnick & Blatchford (1964) *Biochim. Biophys. Acta* **81**, 150. **49.** Scrutton & Utter (1965) *J. Biol. Chem.* **240**, 3714. **50.** Manzella, Ananth, Oegema, *et al.* (1995) *Arch. Biochem. Biophys.* **320**, 361. **51.** Koli, Yearby, Scott & Donaldson (1969) *J. Biol. Chem.* **244**, 621. **52.** Dahlquist & Purich (1975) *Biochemistry* **14**, 1980. **53.** Eppler, Morre & Keenan (1980) *Biochim. Biophys. Acta* **619**, 332. **54.** Krygier & Momparler (1971) *J. Biol. Chem.* **246**, 2752. **55.** Daleo, Piras & Piras (1976) *Eur. J. Biochem.* **68**, 339. **56.** Harper, Rene & Campbell (1971) *Biochim. Biophys. Acta* **242**, 446. **57.** Little, Olinescu, Reid & O'Brien (1970) *J. Biol. Chem.* **245**, 3632. **58.** Kapoor & Bray (1968) *Biochemistry* **7**, 3583. **59.** Ipata, Marmocchi, Magni, Felicioli & Polidoro (1971) *Biochemistry* **10**, 4270. **60.** Cercignani, Serra, Fini, *et al.* (1974) *Biochemistry* **13**, 3628. **61.** Kleppe & Spaeren (1967) *Biochemistry* **6**, 3497. **62.** Ipata (1968) *Biochemistry* **7**, 507. **63.** Changeux, Gerhart & Schachman (1968) *Biochemistry* **7**, 531. **64.** Hill & L. L. Bennett, Jr.

(1969) *Biochemistry* **8**, 122. **65.** Wong & Murray (1969) *Biochemistry* **8**, 1608. **66.** Hongo, Matsumoto & Sato (1978) *Biochim. Biophys. Acta* **522**, 258. **67.** Durbecq, C. Legrain, A. Roovers, A. Piérard & N. Glansdorff (1997) *Proc. Natl. Acad. Sci. U.S.A.* **94**, 12803. **68.** Purcarea, G. Hervé, R. Cunin & D. R. Evans (2001) *Extremophiles* **5**, 229. **69.** Maino & Young (1974) *J. Biol. Chem.* **249**, 5176. **70.** Zwiller, Basset & Mandel (1981) *Biochim. Biophys. Acta* **658**, 64. **71.** Tsai, Manganiello & Vaughan (1978) *J. Biol. Chem.* **253**, 8452. **72.** Ide (1971) *Arch. Biochem. Biophys.* **144**, 262. **73.** Horne, Anderson & Nordlie (1970) *Biochemistry* **9**, 610. **74.** Oray & Norton (1980) *Biochem. Biophys. Res. Commun.* **95**, 624. **75.** Noh, Kwon, Kim, Lee & Yoon (2000) *J. Biochem. Mol. Biol.* **33**, 366. **76.** Suzuki, Watanabe, Masumura & Kitamura (2004) *Arch. Biochem. Biophys.* **431**, 169. **77.** Singal & R. Singh (1986) *Plant Physiol.* **80**, 369. **78.** Bayley-Wood, Dodgson & Rose (1969) *Biochem. J.* **112**, 257. **79.** Nakajima, Fukunaga, Sasaki & Usami (1973) *Biochim. Biophys. Acta* **293**, 242. **80.** Glaser (1965) *Biochim. Biophys. Acta* **101**, 6. **81.** Proudfoot, Kuznetsova., Brown, *et al.* (2004) *J. Biol. Chem.* **279**, 54687. **82.** Krishnan & Rao (1972) *Arch. Biochem. Biophys.* **149**, 336. **83.** Eikmanns & Buckel (1990) *Biol. Chem. Hoppe-Seyler* **371**, 1077. **84.** Yoo, Han, Ko & Bang (1998) *Arch. Pharm. Res.* **21**, 692. **85.** Beck, Eisenhardt & Neuhard (1975) *J. Biol. Chem.* **250**, 609. **86.** Manca de Nadra, Pesce de Ruiz Holgado & Oliver (1984) *J. Appl. Biochem.* **6**, 184. **87.** Imai (1987) *J. Biochem.* **101**, 163. **88.** Raggi-Ranieri & Ipata (1971) *Ital. J. Biochem.* **20**, 27. **89.** Lee & Sturm (1996) *Plant Physiol.* **112**, 1513. **90.** van Dijk, Maier & van den Eijnden (1976) *Biochim. Biophys. Acta* **444**, 816. **91.** Kean & Bighouse (1974) *J. Biol. Chem.* **249**, 7813. **92.** Picher & Boucher (2000) *Amer. J. Respir. Cell Mol. Biol.* **23**, 255. **93.** Höer, Höer & Oberdisse (1990) *Biochem. J.* **270**, 715. **94.** Proudfoot, Kuznetsova, Brown, *et al.* (2004) *J. Biol. Chem.* **279**, 54687. **95.** Polya & Haritou (1988) *Biochem. J.* **251**, 357. **96.** Fujita & Freese (1979) *J. Biol. Chem.* **254**, 5340. **97.** Newby, Luzio & Hales (1975) *Biochem. J.* **146**, 625. **98.** Berglund, Björkhem, Angelin & Einarsson (1989) *Biochim. Biophys. Acta* **1002**, 382. **99.** Reeves (1971) *Biochem. J.* **125**, 531. **100.** Burger & Glaser (1964) *J. Biol. Chem.* **239**, 3168. **101.** Antonsson & Klig (1996) *Yeast* **12**, 449. **102.** Antonsson (1994) *Biochem. J.* **297**, 517. **103.** Yamashita & Nikawa (1997) *Biochim. Biophys. Acta* **1348**, 228. **104.** Percy, Carson, Moore & Waechter (1984) *Arch. Biochem. Biophys.* **230**, 69. **105.** Coleman & Bell (1977) *J. Biol. Chem.* **252**, 3050. **106.** Richard, Lillo, Tetzlaff, *et al.* (2004) *Biochemistry* **43**, 12189. **107.** Engleman & Francis (1978) *Arch. Biochem. Biophys.* **191**, 602. **108.** Rodríguez-Aparicio, Luengo, Gonzalez-Clemente & Reglero (1992) *J. Biol. Chem.* **267**, 9257. **109.** Bravo, Barrallo, Ferrero, *et al.* (2001) *Biochem. J.* **358**, 585. **110.** Potvin, Raju & Stanley (1995) *J. Biol. Chem.* **270**, 30415. **111.** Ebner, Wolf, Gancedo, Elsässer & Holzer (1970) *Eur. J. Biochem.* **14**, 535. **112.** McGann & Deutscher (1980) *Eur. J. Biochem.* **106**, 321. **113.** Rether, Bonnet & Ebel (1974) *Eur. J. Biochem.* **50**, 281. **114.** Dullin, Fabisz-Kijowska & Walerych (1975) *Acta Biochim. Pol.* **22**, 279. **115.** Best & Novelli (1971) *Arch. Biochem. Biophys.* **142**, 527. **117.** Deutscher (1973) *Biochem. Biophys. Res. Commun.* **52**, 216. **118.** Dullin, Fabisz-Kijowska & Walerych (1975) *Acta Biochim. Pol.* **22**, 279. **119.** Sarkar, Cao & Sarkar (1997) *Biochem. Mol. Biol. Int.* **41**, 1045. **120.** Pellicer, Salas, M. L. Salas (1978) *Biochim. Biophys. Acta* **519**, 149. **121.** Roggen & Slegers (1985) *Eur. J. Biochem.* **147**, 225. **122.** Fox & Kelley (1972) *J. Biol. Chem.* **247**, 2126. **123.** Tatibana, Kita, Taira, *et al.* (1995) *Adv. Enzyme Regul.* **35**, 229. **124.** Switzer & Sogin (1973) *J. Biol. Chem.* **248**, 1063. **125.** Liou, Dutschman, Lam, Jiang & Cheng (2002) *Cancer Res.* **62**, 1624. **126.** Manca de Nadra, Nadra Chaud, Pesce de Ruiz Holgado & Oliver (1986) *Biotechnol. Appl. Biochem.* **8**, 46. **127.** Yamada, Goto & Ogasawara (1982) *Biochim. Biophys. Acta* **709**, 265. **128.** Kato & Takenawa (1990) *J. Biol. Chem.* **265**, 794. **129.** Wissing & Wagner (1992) *Plant Physiol.* **98**, 1148. **130.** Brophy, Choy, Toone & Vance (1977) *Eur. J. Biochem.* **78**, 491. **131.** Austin, Sirakoff, Roop & Moyer (1978) *Biochim. Biophys. Acta* **522**, 412. **132.** Levin & Zimmerman (1976) *J. Biol. Chem.* **251**, 1767. **133.** Kozai, Sonoda, Kobayashi & Sugino (1972) *J. Biochem.* **71**, 485. **134.** Datta, Shewach, Mitchell & Fox (1989) *J. Biol. Chem.* **264**, 9359. **135.** Steinert, Wissing & Wagner (1994) *Plant Sci.* **101**, 105. **136.** Yamakawa & Takenawa (1988) *J. Biol. Chem.* **263**, 17555. **137.** Datta (1970) *Biochim. Biophys. Acta* **220**, 51. **138.** Van Hugo & Gottschalk (1974) *Eur. J. Biochem.* **48**, 455. **139.** Kilker, Shuey & Serif (1979) *Biochim. Biophys. Acta* **570**, 271. **140.** Cihak (1976) *Hoppe-Seyler's Z. Physiol. Chem.* **357**, 345. **141.** Vidair & Rubin (2005) *Proc. Natl. Acad. Sci. U.S.A.* **102**, 662. **142.** Orengo & Saunders (1972) *Biochemistry* **11**, 1761. **143.** Cihak (1975) *FEBS Lett.* **51**, 133. **144.** Appleby, Larson, Cheney, *et al.* (2005) *Acta Crystallogr. Sect. D* **61**, 278. **145.** Suzuki, Koizumi, Fukushima, Matsuda & Inagaki (2004) *Structure* **12**, 751. **146.** Payne, Cheng & Traut (1985) *J. Biol. Chem.* **260**, 10242. **147.**

Orengo (1969) *J. Biol. Chem.* **244**, 2204. **148.** Evans & Ratledge (1985) *Can. J. Microbiol.* **31**, 479. **149.** Isaac & Rhodes (1986) *Phytochemistry* **25**, 339. **150.** Sapico & Anderson (1969) *J. Biol. Chem.* **244**, 6280. **151.** Porter, Chassy & Holmlund (1982) *Biochim. Biophys. Acta* **709**, 178. **152.** Doelle (1982) *Eur. J. Appl. Microbiol. Biotechnol.* **14**, 241. **153.** Schröder, Eichel, Breinig & Schröder (1997) *Plant Mol. Biol.* **33**, 211. **154.** Chou & Talalay (1973) *Biochim. Biophys. Acta* **321**, 467. **155.** Porcelli, Cacciapuoti, Carteni-Farina & Gambacorta (1988) *Eur. J. Biochem.* **177**, 273. **156.** Kayama & Kawasaki (1973) *Arch. Biochem. Biophys.* **158**, 242. **157.** Melkerson-Watson & Sweeley (1981) *J. Biol. Chem.* **266**, 4448. **158.** Eppler, Morré & Keenan (1980) *Biochim. Biophys. Acta* **619**, 332. **159.** Kurosawa, Hamamoto, Inoue & Tsuji (1995) *Biochim. Biophys. Acta* **1244**, 216. **160.** Rearick, Sadler, Paulson & R. L. Hill (1979) *J. Biol. Chem.* **254**, 4444. **161.** Sadler, Rearick, Paulson & Hill (1979) *J. Biol. Chem.* **254**, 4434. **162.** Westcott, Wolf & Hill (1985) *J. Biol. Chem.* **260**, 13109. **163.** Sadler, Rearick & Hill (1979) *J. Biol. Chem.* **254**, 5934. **164.** Scudder & Chantler (1981) *Biochim. Biophys. Acta* **660**, 136. **165.** Weinstein, de Souza-e-Silva & Paulson (1982) *J. Biol. Chem.* **257**, 13845. **166.** Nagpurkar, Hunt & Mookerjea (1996) *Int. J. Biochem. Cell Biol.* **28**, 1337. **167.** Casanova, Kuhn, Kleesiek & Götting (2008) *Biochem. Biophys. Res. Commun.* **365**, 678. **168.** Miller, Adamczyk, Fyfe & Elion (1974) *Arch. Biochem. Biophys.* **165**, 349. **169.** Shibata & Iwai (1980) *Agric. Biol. Chem.* **44**, 119. **170.** Taguchi & Iwai (1976) *Agric. Biol. Chem.* **40**, 385. **171.** Linde & Jensen (1996) *Biochim. Biophys. Acta* **1296**, 16. **172.** Arent, Harris, Jensen & Larsen (2005) *Biochemistry* **44**, 883. **173.** Jensen, Arent, Larsen & Schack (2005) *FEBS J.* **272**, 1440. **174.** Nagy & Ribet (1977) *Eur. J. Biochem.* **77**, 77. **175.** Teng-umnuay, van der Wel & West (1999) *J. Biol. Chem.* **274**, 36392. **176.** Takeuchi, Yoshikawa, Sasaki & Chiba (1985) *Agric. Biol. Chem.* **49**, 1059. **177.** Kamat, Garg & Sharma (1992) *Arch. Biochem. Biophys.* **298**, 731. **178.** Piro, Zuppa, Dalessandro & Northcote (1993) *Planta* **190**, 206. **179.** Sinha, Pathre & Sane (1997) *Phytochemistry* **46**, 441. **180.** Salerno & Pontis (1978) *Planta* **142**, 41. **181.** Antranikian, Herzberg & Gottschalk (1982) *J. Bacteriol.* **152**, 1284. **182.** Okabayashi & E. Nakano (1979) *J. Biochem.* **85**, 1061. **183.** Kawashima (1991) *Experientia* **47**, 709. **184.** Ruepp, Müller, Lottspeich & Soppa (1995) *J. Bacteriol.* **177**, 1129. **185.** Burns, Mendz & Hazell (1997) *Arch. Biochem. Biophys.* **347**, 119. **186.** Santiago & West (2003) *J. Basic Microbiol.* **43**, 75. **187.** De Vos, Xu, Hulpiau, Vergauwen & Van Beeumen (2007) *J. Mol. Biol.* **365**, 379. **188.** Vickrey, Herve & Evans (2002) *J. Biol. Chem.* **277**, 24490. **189.** Adair & Jones (1972) *J. Biol. Chem.* **247**, 2308. **190.** Durbecq, Thia-Toong, Charlier, *et al.* (1999) *Eur. J. Biochem.* **264**, 233. **191.** Allewell (1989) *Ann. Rev. Biophys. Biophys. Chem.* **18**, 71. **192.** Kantrowitz & Lipscomb (1990) *Trends Biochem. Sci.* **15**, 53. **193.** Mort & Chan (1975) *J. Biol. Chem.* **250**, 653. **194.** Rabinowitz, Hsiao, Gryncel, *et al.* (2008) *Biochemistry* **47**, 5881. **195.** Xu, Zhang, Liang, *et al.* (1998) *Microbiology* **144**, 1435. **196.** England & Herve (1992) *Biochemistry* **31**, 9725. **197.** Zhang & Kantrowitz (1991) *J. Biol. Chem.* **266**, 22154. **198.** Purcarea, Erauso, Prieur & Herve (1994) *Microbiology* **140**, 1967. **199.** Van Boxstael, Maes & Cunin (2005) *FEBS J.* **272**, 2670. **200.** Charalampous (1959) *J. Biol. Chem.* **234**, 220. **201.** Nair (1972) *Arch. Biochem. Biophys.* **153**, 139.

O-(Cytidin-5‘-yl) 2-(5-Acetamido-2,6-anhydro-3,5-dideoxy-D-erythro-L-gluco-nonanoate-2-yl)ethylphosphonate

This CMP-sialic acid analogue (FW$_{free-acid}$ = 571.51 g/mol) inhibits rat *N*-acetyllactosaminide α-2,3-sialyltransferase (IC$_{50}$ = 47 nM) and α-*N*-acetylgalactosaminide α-2,6-sialyltransferase (IC$_{50}$ = 0.34 μM). **1.** Izumi, Wada, Yuasa & Hashimoto (2005) *J. Org. Chem.* **70**, 8817.

Cytisine

This toxic alkaloid (FW = 190.24 g/mol; CASs = 485-35-8; IUPAC Name: (1*R*,5*S*)-1,2,3,4,5,6-hexahydro-1,5-methano-8*H*-pyrido[1,2*a*][1,5]diazocin-8-one), also known as baptitoxine and sophorine, is a partial acetylcholine agonist, with high affinity for nicotinic receptors. Cytisine was a lead molecule in the development of the smoking-cessation drug Chantix (*See Varenicline*). **1.** Pérez, Méndez-Gálvez & Cassels (2012) *Nat Prod Rep.* **29**, 555.

Cytochalasin B

This cell-permeable fungal metabolite (FW = 479.62 g/mol) from the plant pathogen *Drechslera dematioideum* (formerly *Heiminthosporium dematioideum*) is a mycotoxin that binds to the barbed end of actin filaments, reversibly inhibiting their elongation and shortening (1). (*See Cytochalasin D for details on Mechanism of Action*). It exhibits cytostatic activity, inhibiting glucose transport, cytokinesis, phagocytosis, pinocytosis, cell adhesion, and cell division. Cytochalasin B also inhibits platelet aggregation and induces nuclear extrusion. **Target(s):** actin polymerization (2,4,5,7,8,13,14); glucose transport (6,9-12,15); cell division; Ca^{2+} transport (3); phagocytosis (16,17). **1.** Sampath & Pollard (1991) *Biochemistry* **30**, 1973. **2.** Rosenshine, Ruschkowski & Finlay (1994) *Meth. Enzymol.* **236**, 467. **3.** Jande & Liskova-Kiar (1981) *Calcif. Tissue Int.* **33**, 143. **4.** Flanagan & Lin (1980) *J. Biol. Chem.* **255**, 835. **5.** Lin, Tobin, Grumet & Lin (1980) *J. Cell Biol.* **84**, 455. **6.** Lin, Lin & Flanagan (1978) *Proc. Natl. Acad. Sci. U.S.A.* **75**, 329. **7.** Yahara, Harada, Sekita, Yoshihira & Natori (1982) *J. Cell Biol.* **92**, 69. **8.** Sampath & Pollard (1991) *Biochemistry* **30**, 1973. **9.** Kletzien, J. F. Perdue & Λ. Springer (1972) *J. Biol. Chem.* **247**, 2964. **10.** Mizel & Wilson (1972) *J. Biol. Chem.* **247**, 4102. **11.** Estensen & Plagemann (1972) *Proc. Natl. Acad. Sci. U.S.A.* **69**, 1430. **12.** Lachaal, Rampal, Lee, Shi & Jung (1996) *J. Biol. Chem.* **271**, 5225. **13.** Urbanik & Ware (1989) *Arch. Biochem. Biophys.* **269**, 181. **14.** Wodnicka, Pierzchalska, Bereiter-Hahn & Kajstura (1992) *Folia Histochem. Cytobiol.* **30**, 107. **15.** Jung & Rampal (1977) *J. Biol. Chem.* **252**, 5456. **16.** Davies, Fox, Polyzonis, Allison & Harwell (1973) *Lab. Invest.* **28**, 16. **17.** Davies & Allison (1978) *Front. Biol.* **46**, 143.

Cytochalasin D

This fungal metabolite (FW = 507.63 g/mol) from *Metarrhizium anisopliae* inhibits actin polymerization and actin-based motility (1-4). When present μM concentrations, cytochalasin D (CD) (as well as cytochalasin B (CB) inhibits elongation at both ends of the filament, ~95% at the barbed (*or* plus) end and ~50% at the pointed (*or* minus) end, so that the two ends contribute about equally to the rate of growth (2). Half-maximal inhibition of elongation at the barbed end is at 0.1 μM CB and 0.02 μM CD for ATP·actin and at 0.1 μM CD for ADP·actin (2). At the pointed end, CD inhibits elongation by ATP·actin and ADP·actin about equally (2). At 2 μM, the cytochalasins reduce the association and dissociation rate constants in parallel for both ADP·actin and ATP·actin, reducing their effects on the critical concentrations (2). The dependence of the elongation rate on the concentration of both cytochalasin and actin can be explained quantitatively by a mechanism that includes the effects of cytochalasin binding to actin monomers (3) and a partial cap of the barbed end of the filament by the complex of ADP·actin and cytochalasin. *Note*: *In vivo* actin-based motility relies on actoclampin-type, filament end-tracking motors (5,6), thereby casting great doubt on the physiologic relevance of the *in vitro* studies of CD effects on free actin filaments. *Caution*: Cytochalasin D must be stored in the dark, because double bonds in the macrocycle slowly undergo *trans*-to-*cis* transition in the presence of light. **1**. Bonder & Mooseker (1986) *J. Cell Biol.* **102**, 282. **2**. Sampath & Pollard (1991) *Biochemistry* **30**, 1973. **3**. Godette & Frieden (1986) *J. Biol. Chem.* **261**, 5974. **4**. Coluccio & Tilney (1984) *J. Cell Biol.* 99, 529. **5**. Dickinson & Purich (2002) *Biophys. J.* **82**, 605. **6**. Dickinson, Caro & Purich (2004) *Biophys. J.* **87**, 2838.

cytosine

This naturally occurring pyrimidine (FW = 111.10 g/mol; CAS 71-30-7Solubility: 0.77 g/100 mL H_2O at 25°C; λ_{max} = 267 nm at pH 7 (ε = 6100

$M^{-1}cm^{-1}$; pK_a values of 4.60 and 12.16), also known as 4-amino-2-hydroxypyrimidine, is a component of RNA and DNA as well as certain antibiotics and complex lipids. **Target(s):** D-amino-acid oxidase, weakly inhibited (1); arginase (2); deoxycytidine kinase, K_i = 1.5 mM (3); β-glucosidase (4); 3-hydroxyanthranilate oxidase (5); hypoxanthine(guanine) phosphoribosyl-transferase (6); xanthine phosphoribosyl-transferase (6). **1**. Walaas & Walaas (1956) *Acta Chem. Scand.* **10**, 122. **2**. Rosenfeld, Dutta, Chheda & Tritsch (1975) *Biochim. Biophys. Acta* **410**, 164. **3**. Krenitsky, Tuttle, Koszalka, *et al.* (1976) *J. Biol. Chem.* **251**, 4055. **4**. Dale, Ensley, Kern, Sastry & Byers (1985) *Biochemistry* **24**, 3530. **5**. Nair (1972) *Arch. Biochem. Biophys.* **153**, 139. **6**. Naguib, Iltzsch, el Kouni, Panzica & el Kouni (1995) *Biochem. Pharmacol.* **50**, 1685.

Cytosporone B

This naturally occurring NR4A1 (*or* Nur77) agonist (FW = 322.41 g/mol; CAS 321661-62-5; Soluble to 100 mM in DMSO), also named Csn-B and ethyl 3,5-dihydroxy-2-(1-oxooctyl)-benzeneacetate, targets Nerve Growth Factor IB, *or* NGFIB (K_d = 0.85 μM), also known as Nur77 or NR4A1 (Nuclear Receptor subfamily 4 group A member 1). NR4A1 is an endogenous inhibitor of transforming growth factor-β (TGF-β) signaling. **1**. Zhan, Du, Chen, *et al.* (2008) *Nature Chem. Biol.* **4**, 548. **2**. Palumbo-Zerr, Zerr, Distler, *et al.* (2015) *Nature Chem. Biol.* **8**, 576.

– D –

D, See *Aspartate and Aspartic Acid; Dihydrouridine*

D15

This endocytosis inhibitor (FW = 1566.78 g/mol; CAS 251939-41-0; Sequence: PPPQVPSRPNRAPPG; Solubility: 1 mg/mL H_2O) blocks the interaction of dynamin with amphiphysins 1 and 2. D-15 impairs α-amino-3-hydroxy-5-methyl-4-isoxazolepropionate (AMPA) receptor cycling and blocks 3,5-dihydroxyphenylglycine- (DHPG-) induced long-term depression (LTD). **1**. Wigge & McMahon (1998) *Trends Neursci.* **21**, 339. **2**. Lüscher, *et al.* (1999) *Neuron* **24**, 649. **3**. Xiao, *et al.* (2001) *Neuropharmacol.* **41**, 664.

D145, See *Memantine*

D609

This antiviral, antioxidant and antitumor agent (FW = 266.46 g/mol; CAS 83373-60-8; Soluble to 100 mM in water), also named *O*-(octahydro-4,7-methano-1*H*-inden-5-yl)carbonopotassium dithioate and tricyclodecan-9-yl-xanthogenate, selectively and competitively inhibits phosphatidyl choline-specific phospholipase C, *or* PC-PLC (K_i = 6.4 μM) (1-3). D609 suppresses LPS- and IFNγ-induced NO production (IC_{50} = 20 mg/ml) and blocks oxidative glutamate toxicity in nerve cells *in vivo* (2,3). D609 inhibits ionizing radiation-induced oxidative damage by acting as a potent antioxidant (4). **1**. Tschiakowsky, *et al.* (1994) *Brit. J. Pharmacol.* **113**, 664. **2**. Amtmann (1996) *Drugs Exp. Clin. Res.* **22**, 287. **3**. Li, *et al.* (1998) *Proc. Natl. Acad. Sci. U.S.A.* **95**, 7748. **4**. Zhou, *et al.* (2001). *J. Pharmacol. Exp. Ther.* **298**, 103.

D 4476

This selective casein kinase inhibitor (FW = 398.41 g/mol; CAS 301836-43-1; Soluble to 100 mM in DMSO and to 50 mM in ethanol), also named 4-[4-(2,3-dihydro-1,4-benzodioxin-6-yl)-5-(2-pyridinyl)-1*H*-imidazol-2-yl]benzamide, targets CK1δ some 20–30-fold more potently than PKD1 or p38α MAPK. No other protein kinases in the panel were inhibited to a significant extent. D4476 is recommended for inhibiting CK1 isoforms in cell-based assays (EC_{50} = 50–100 μM). **1**. Bain, Plater, Elliott, *et al.* (2007) *Biochem J.* **408**, 297.

D-64131

This tubulin polymerization inhibitor (FW = 251.28 g/mol; CAS 74588-78-6; Soluble to 100 mM in DMSO), also named (5-methoxy-1*H*-indol-2-yl)phenylmethanone, is cytotoxic, blocking tumor cell proliferation *in vitro* (IC_{50} = 74 nM) and preventing growth of mouse tumor models after oral administration. In contrast to colchicine, vincristine, nocodazole, or taxol, D-64131 does not significantly affect tubulin's GTPase activity (1). No cross-resistance toward cell lines with multidrug resistance/multidrug resistance protein independent resistance phenotypes were evident (2). In animal studies, no signs of systemic toxicity were observed after p.o. dosages of up to 400 mg/kg of D-64131. In xenograft experiments with the human amelanoic melanoma (MEXF 989), D-64131 is highly active, with treatment resulting in tumor growth delay of 23 days at 400 mg/kg (2). **1**. Mahboobi, Pongratz, Hufsky, *et al.* (2001) *J. Med. Chem.* **44**, 4535. **2**. Beckers, Reissmann, Schmidt, *et al.* (2002) *Cancer Res.* **62**, 3113.

dA, See *2'-Deoxyadenosine*

DA-1229, See *Evogliptin*

Dabigatran

Intracellular Drug: Dabigatran

Double Prodrug: Dabigatran Etexilate

This non-vitamin K antagonist oral anticoagulant, *or* NOAC (FW = 627.73 g/mol; CAS 211915-06-9), known as dabigatran and dabigatran etexilate as well as by the trade names Pradaxa® (U.S.A., Canada, Europe, and Australia) and Prazaxa® (Japan), is a direct inhibitor of thrombin, *or* Clotting Factor IIa (1). Dabigatran holds the distinction of being the first oral anticoagulant available in the U.S. since approval of the indirect inhibitor, warfarin. With a predictable anticoagulation response, dabigatran does not usually require routine anticoagulation monitoring (**See comment below**). Dabigatran was found to be noninferior to warfarin in preventing stroke and systemic embolism in patients with atrial fibrillation, a finding that motivated FDA approval. Dabigatran exhibits efficacy in preventing venous thromboembolism in patients undergoing total hip or knee replacement surgery. Drug excretion is slowed by strong p-glycoprotein pump inhibitors (*e.g.*, quinidine, verapamil, clarithromycin (Zitromax®) and amiodarone (Cordarone®)), increasing plasma dabigatran levels. Dosing should be reduced when p-glycoprotein pump inhibitors are used. **Mode of Inhibitor Action:** Based on the crystal structure of the peptide-like thrombin inhibitor NAPAP complexed with bovine thrombin, a new class of nonpeptidic inhibitors was developed, using a 1,2,5-trisubstituted benzimidazole as the central scaffold and optimized to achieve thrombin inhibition in the lower nanomolar range, mainly from nonpolar, hydrophobic interactions (1). For improved *in vivo* potency, overall hydrophilicity was increased by introducing carboxylate groups. The resultiung polar compound (BIBR 953) exhibited a favorable activity profile *in vivo*, and this zwitterionic molecule was converted into the double-prodrug (BIBR 1048), which showed strong oral activity in different animal species (1). **Monitoring:** Although superior to vitamin K antagonists and heparinoids in several respects, NOACs retain the ability to cause haemorrhage and, despite claims to the contrary, may need monitoring (2). Circulating levels of dabigatran are increased in affected patients, because the drug is mainly eliminated by renal excretion. Six analytical methods for determining plasma, serum, and urine dabigatran have been compared (3). A sensitive monoclonal antibody-based competitive ELISA has been developed to analyze free dabigatran in serum ultrafiltrate for therapeutic drug monitoring and pharmacokinetic studies (4). **Reversal of Dabigatran Anticoagulation:** Although dabigatran therapy is associated with an increased risk of bleeding, Idarucizumab (MW = 47.8 kDa; CAS 1362509-93-0; Trade Name: Praxbind®; Symbol: aDabi-Fab) is an antigen-binding fragment (*or* Fab) of a monoclonal antibody that binds free and thrombin-bound dabigatran stoichiometrically (5). X-ray crystal structure data of dabigatran in complex with aDabi-Fab reveals many structural similarities of dabigatran recognition to those in thrombin. A network of higher-affinity interactions, however, allows idarucizumab to bind ~350 times more tightly to dabigatran than to thrombin (5). Moreover, clinical studies demonstrated that idarucizumab completely neutralizes dabigatran's anticoagulative properties within minutes, when administered at massive 5-gram doses (6). **1**. Hauel, Nar, Priepke, *et al.* (2002) *J. Med. Chem.* 45, 1757. **2**. Blann (2014) *Brit. J. Biomed. Sci.* **71**, 158. **3**. Du, Weiss, Christina, *et al.* (2015) *Clin. Chem. Lab Med.* **53**, 1237. **4**. Oiso, Morinaga, Goroku, *et al.* (2015) *Ther. Drug Monit.* **37**, 594. **5**. Schiele, van Ryn, Canada, *et al.* (2013) *Blood* **121**, 3554. **6**. Pollack, Reilly, Eikelboom, *et al.* (2015) *New Engl. J. Med.* **373**, 511.

Dabrafenib

This orally active and selective, ATP-competitive, protein kinase inhibitor (FW = 519.56 g/mol; CAS 1195765-45-7; Solubility: 30 mg/mL DMSO; <1 mg/mL H_2O), also known as GSK2118436 and Tafinlar®, targets B-Raf, B-Raf^{V600E} and c-Raf, with IC_{50} values of 3.2, 0.8 and 5.0 nM, respectively (1,2). After one or two 28-day courses of Tafinlar treatrment in thirteen patients with tumors bearing a BRAF mutation, there was no detectable impact on serum cytokine levels, peripheral blood cell counts, leukocyte subset frequencies, and memory CD4$^+$ and CD8$^+$ T-cell recall responses (3). Such findings suggest that combining this BRAF^{V600E} inhibitor with immunotherapy should not impair immune responses (3). Other similarly acting drugs are: Ipilimumab, Vemurafenib, and Trametinib. **1**. Laquerre, Arnone, Moss, *et al.* (2009) *EORTC International Conference.* Abstract B88. **2**. Greger, Eastman, Zhang, *et al.* (2012) *Mol. Cancer Ther.* **11**, 909. **3**. Hong, Vence, Falchook, *et al.* (2012) *Clin. Cancer Res.* **18**, 2326. (*Erratum: Clin. Cancer Res.* **18**, 3715.)

Daclatasvir

This pan-genotypic hepatitis C virus NS5A replication inhibitor (FW = 738.89 g/mol; CAS 1009119-64-5; Solubility = 150 mg/mL DMSO, <1 mg/mL H_2O), also known by the code name BMS-790052, the trade name Daklinza®, and systematically as *N,N'*-[[1,1'-biphenyl]-4,4'-diylbis[1*H*-imidazole-5,2-diyl-(2*S*)-2,1-pyrrolidinediyl[(1*S*)-1-(1-methylethyl)-2-oxo-2,1-ethane-diyl]]]biscarbamic acid *C,C'*-dimethyl ester, targets HCV NS5A (EC_{50} = 9-50 pM), a zinc-binding and proline-rich hydrophilic phosphoprotein associated with replicating RNA in a cytoplasmic replication complex. It is active against a broad range of HCV replicon genotypes and the JFH-1 genotype 2a infectious virus. Administration of a single 100-mg dose in two patients infected with genotype 1b virus resulted in a 2000x reduction in mean viral load (1). BMS-790052 also alters the subcellular localization of the NS5A non-structural viral protein. HCV replicates via only RNA-based intermediates with no incorporation of the viral genome into the host cell chromosome, as occurs with HIV-1. HCV is therefore curable, with cure defined as virus-free plasma 24 weeks after completion of therapy. In 2013, the investigational all-oral 3DAA Regimen (*or* Daclatasvir/Asunaprevir/BMS-791325) received Breakthrough Therapy Designation, and, as of early 2014, Daclatasvir has been favorably evaluated in more than 5500 patients (5). It was approved in Europe in August, 2014. **1**. Gao, Nettles, Belema, *et al.* (2010) *Nature* **465**, 96. **2**. Lee, *et al.* (2011) *Virology* **414**, 10. **3**. Wang, Jia, Huang, *et al.* (2012) *Antimicrob. Agents Chemother.* **56**, 1588. **4**. O'Boyle, Nower, Lemm, *et al.* (2005) *Antimicrob. Agents Chemother.* **49**, 1346. **5**. Belema & Meanwell (2014) *J. Med. Chem.* **57**, 5057.

Dacomitib

This orally available, second-generation pan-HER inhibitor (FW = 469.95 g/mol; CAS 1110813-31-4; Solubility = 19 mg/mL DMSO, <1 mg/mL H_2O), also known by its code name PF-00299804 and its systematic name (*E*)-*N*-[4-(3-chloro-4-fluoroanilino)-7-methoxy-quinazolin-6-yl]-4-(1-piperidyl)but-2-enamide, irreversibly inactivates epidermal growth factor receptor, IC_{50} = 6 nM, and is also effective against gefitinib-resistant nonsmall cell lung cancers (NSCLCs) with ERBB2 mutations as well as those harboring the EGFR Thr-790-Met mutation. HER2 amplification occurs in 20-25% of patients with breast cancer and is associated with a poor prognosis and is a validated target for therapy. Dacomitinib overcomes the acquired resistance to trastuzumab and also maintains considerable antiproliferative activity in the cell lines acquiring resistance to lapatinib. **Other Target(s):** ErbB1, IC_{50} = 6 nM; ErbB2, IC_{50} = 46 nM; ErbB4, IC_{50} = 74 nM. **1**. Engelman, Zejnullahu, Gale, *et al.* (2007) *Cancer Res.* **67**, 11924. **2**. Gonzales, Hook, Althaus, *et al.* (2008) *Mol. Cancer Ther.* **7**, 1880. **3**. Nam, Ching, Kan, *et al.* (2012) *Mol. Cancer Ther.* **11**, 439. **4**. Kalous, Conklin, Desai, *et al.* (2012) *Mol. Cancer Ther.* **11**, 1978.

Dactolisib

This orally available, dual ATP-competitive PI3K/mTOR protein kinase inhibitor (FW = 469.55 g/mol; CAS 915019-65-7: Solubility: 1 mg/mL DMSO, <1 mg/mL H_2O), also named BEZ235, NVP-BEZ235, and 2-methyl-2-(4-(3-methyl-2-oxo-8-(quinolin-3-yl)-2,3-dihydroimidazo[4,5-*c*]quinolin-1-yl)phenyl)propane nitrile, targets p110α (IC_{50} = 4 nM), p110γ (IC_{50} = 5 nM), p110δ (IC_{50} = 7 nM), p110β (IC_{50} = 75 nM), mTOR, or p70S6K (IC_{50} = 6 nM), and ATR (IC_{50} = 21 nM), with little inhibition of Akt and PDK1 (1). *In vitro* treatment of colorectal cancer (CRC) cell lines with NVP-BEZ235 results in transient PI3K blockade, sustained decreases in mTORC1 and mTORC2 signaling, and a corresponding decrease in cell viability, with a median IC_{50} = 9.0-14.3 nM (2). Treatment of 786-0 and Caki-1 cells with NVP-BEZ235 or sorafenib reduces tumor cell proliferation and increases tumor cell apoptosis *in vitro*, and the combined inhibitory action of NVP-BEZ235 and sorafenib is more effective than either compound alone (3). *Ex vivo* pharmacokinetic and pharmacodynamic analysis of tumor tissues show a time-dependent correlation between compound concentration and PI3K/Akt pathway inhibition (4). Dual inhibition of PI3K/mTOR with NVP-BEZ235 induces growth arrest in metastatic renal cell carcinoma (RCC) cell lines both *in vitro* and *in vivo* more effectively than inhibition of TORC1 alone (5). EGF-stimulated IL-8 production, phosphorylation of Akt and Erk, and human non-small cell lung cancer (NSCLC) SPC-A1 cell proliferation and movement is inhibited by EGFR inhibitor (Erlotinib), PI3K inhibitor (GDC-0941 BEZ-235 and SHBM1009), and ERK1/2 inhibitor (PD98059) (6). The mutational statuses of PTEN and K-Ras appear to be useful predictors of sensitivity to NVP-BEZ235 in certain endometrial carcinomas (7). NVP-BEZ235 is cytotoxic to T-cell acute lymphoblastic leukemia (T-ALL) patient lymphoblasts displaying pathway activation, where the drug dephosphorylated eukaryotic initiation factor 4E-binding protein 1, at variance with rapamycin (8). **Other Target(s):** BEZ235 is also very potent against Ataxia-Telangiectasia Mutated (ATM), Rad3-related (ATR), and the catalytic subunit of DNA-dependent protein kinase, *or* DNA-PKcs (9). NVP-BEZ235 also potently inhibited both DNA-PKcs and ATM kinases and attenuated the repair of IR-induced DNA damage in tumors (10). **1**. Maira, Stauffer, Brueggen, *et al.* (2008) *Mol. Cancer Ther.* **7**, 1851. **2**. Roper, Richardson, Wang, *et al.* (2011) *PLoS One* **6**, e25132. **3**. Roulin, Waselle, Dormond-Meuwly, *et al.* (2011) *Mol. Cancer* **10**, 90. **4**. Maira, Stauffer, Brueggen *et al.* (2008) *Mol. Cancer Ther.* **7**, 1851. **5**. Cho, Cohen, Panka, *et al.* (2010) *Clin. Cancer Res.* **16**, 3628. **6**. Zhang, Wang, Zhang, *et al.* (2012) *J. Cell Physiol.* **227**, 35. **7**. Shoji, Oda, Kashiyama, *et al.* (2012) *PLoS One* **7**, e37431. **8**. Chiarini, Grimaldi, Ricci, *et al.* (2010) *Cancer Res.* **70**, 8097. **9**. Toledo, Murga, Zur, *et al.* (2011) *Nature Struct. Mol. Biol.* **18**, 721. **10**. Gil Del Alcazar, Gillam, Mukherjee, *et al.* (2013) *Clin. Cancer Res.* **20**, 1235.

β-DADF

This multisubstrate analogue ($FW_{free-acid}$ = 815.65 g/mol), encompassing both the AICAR and folate moieties within a single covalently linked entity, inhibits phosphoribosyl-aminoimidazolecarboxamide formyltransferase (K_i = 20 nM) (1). β-DADF is intimately bound at the dimer interface of the transformylase domains, with the majority of AICAR interactions occurring within one subunit, whereas the primary interactions to the folate occur in the opposing subunit (2). **1.** Wall, Shim & Benkovic (1999) *J. Med. Chem.* **42**, 3421. **2.** Wolan, Greasley, Wall, Benkovic & Wilson (2003) *Biochemistry* **42**, 10904.

Daidzein

This isoflavone and phytoestrogen (FW = 254.24 g/mol; CAS 486-66-8; Pale yellow solid; Soluble in ethanol and diethyl ether), the aglycon of daidzin as found in legumes, is an analogue of genistein, but, unlike genistein, does not inhibit protein-tyrosine kinases. Daidzein is thus a useful control in experiments designed to evaluate the effects of genistein. Daidzein also inhibits casein kinase II and blocks Swiss 3T3 cells in the G_1 cell cycle phase. *Note*: Daidzein is occasionally referred to as diadzein. **Target(s):** aldehyde dehydrogenase (1); Ca^{2+} channels (2); casein kinase II (3); 3',5'-cyclic nucleotide phosphodiesterase (4); cyclooxygenase, weakly inhibited (5); CYP1A1 (6); DNA-directed DNA polymerase (7); DNA polymerase (7); DNA topoisomerase II, weakly inhibited (7); estrone sulfotransferase (8,9); fatty-acid synthase (10); F_oF_1 ATP synthase, mitochondrial (11); $GABA_A$ receptor (12); β-galactosidase (13); α-glucosidase (14); GLUT1, glucose transporter (15); glutathione *S*-transferase (16); 3-hydroxy-3-methylglutaryl-CoA reductase (17,18); 3β-hydroxysteroid dehydrogenase (19,20); inositol-trisphosphate 3-kinase (21); 5-lipoxygenase, *or* arachidonate 5-lipoxygenase (22); 15-lipoxygenase, *or* arachidonate 15-lipoxygenase (23); lipoxygenase, soybean (22,23); NF-κB and AP-1 and suppressed secretion of urokinase-type plasminogen activator from breast cancer cells (24); phenol sulfotransferase, *or* aryl sulfotransferase (9,25-27); phenylalanine ammonia lyase, weakly inhibited (28); phosphodiesterase (34); steroid Δ-isomerase (20); steroid 5α-reductase (29); steroid sulfotransferase (8). **1.** Keung, Klyosov & Vallee (1997) *Proc. Natl. Acad. Sci. U.S.A.* **94**, 1675. **2.** Dobrydneva, Williams, Morris & Blackmore (2002) *J. Cardiovasc. Pharmacol.* **40**, 399. **3.** Higashi & Ogawara (1994) *Biochim. Biophys. Acta* **1221**, 29. **4.** Ko, Shih, Lai, Chen & Huang (2004) *Biochem. Pharmacol.* **68**, 2087. **5.** You, Jong & Kim (1999) *Arch. Pharm. Res.* **22**, 18. **6.** Shertzer, Puga, Chang, *et al.* (1999) *Chem. Biol. Interact.* **123**, 31. **7.** Sun, Woo, Cassady & Snapka (1998) *J. Nat. Prod.* **61**, 362. **8.** Ohkimoto, Liu, Suiko, Sakakibara & Liu (2004) *Chem. Biol. Interact.* **147**, 1. **9.** Harris, Wood, Bottomley, *et al.* (2004) *J. Clin. Endocrinol. Metab.* **89**, 1779. **10.** Li, Ma, Wang & Tian (2005) *J. Biochem.* **138**, 679. **11.** Zheng & Ramirez (2000) *Brit. J. Pharmacol.* **130**, 1115. **12.** Huang, Fang & Dillon (1999) *Brain Res. Mol. Brain Res.* **67**, 177. **13.** Hazato, Naganawa, Kumagai, Aoyagi & Umezawa (1979) *J. Antibiot.* **32**, 217. **14.** Kim, Kwon & Son (2000) *Biosci. Biotechnol. Biochem.* **64**, 2458. **15.** Martin, Kornmann & Fuhrmann (2003) *Chem. Biol. Interact.* **146**, 225. **16.** Hayeshi, Mutingwende, Mavengere, Masiyanise & Mukanganyama (2007) *Food Chem. Toxicol.* **45**, 286. **17.** Sung, Choi, Lee, Park & Moon (2004) *Biosci. Biotechnol. Biochem.* **68**, 1051. **18.** Sung, Lee, Park & Moon (2004) *Biosci. Biotechnol. Biochem.* **68**, 428. **19.** Ohno, Matsumoto, Watanabe & Nakajin (2004) *J. Steroid Biochem. Mol. Biol.* **88**, 175. **20.** Wong & Keung (1999) *J. Steroid Biochem. Mol. Biol.* **71**, 191. **21.** Mayr, Windhorst & Hillemeier (2005) *J. Biol. Chem.* **280**, 13229. **22.** Mahesha, Singh & Rao (2007) *Arch. Biochem. Biophys.* **461**, 176. **23.** Sadik, Sies & Schewe (2003) *Biochem. Pharmacol.* **65**, 773. **24.**

Valachovicova, Slivova, Bergman, Shuherk & Sliva (2004) *Int. J. Oncol.* **25**, 1389. **25.** Mesia-Vela & Kauffman (2003) *Xenobiotica* **33**, 1211. **26.** Ebmeier & Anderson (2004) *J. Clin. Endocrinol. Metab.* **89**, 5597. **27.** Meng, Shankavaram, Chen, *et al.* (2006) *Cancer Res.* **66**, 9656. **28.** Jorrin & Dixon (1990) *Plant Physiol.* **92**, 447. **29.** Hiipakka, Zhang, Dai, Dai & Liao (2002) *Biochem. Pharmacol.* **63**, 1165.

Daidzin

This glucoside (FW = 416.38 g/mol; CAS 552-66-9), also known as daidzein 7-glucoside, from soybean meal (*Soja hispida*) is the major active principle of an ancient Chinese herbal treatment for alcohol addiction (*i.e.*, utilizing extracts of the root and flower of kudzu, *Pueraria lobata*). **Target(s):** aldehyde dehydrogenase (1-4); CYP1A1 (5); DNA topoisomerase II (6); lipoxygenase, soybean (7). **1.** Gao, Li & Keung (2001) *J. Med. Chem.* **44**, 3320. **2.** Keung, Klyosov & Vallee (1997) *Proc. Natl. Acad. Sci. U.S.A.* **94**, 1675. **3.** Keung, Lazo, Kunze & Vallee (1995) *Proc. Natl. Acad. Sci. U.S.A.* **92**, 8990. **4.** Keung & Vallee (1993) *Proc. Natl. Acad. Sci. U.S.A.* **90**, 1247. **5.** Shertzer, Puga, Chang, *et al.* (1999) *Chem. Biol. Interact.* **123**, 31. **6.** Martin-Cordero, Lopez-Lazaro, Pinero, Ortiz, Cortes & Ayuso (2000) *J. Enzyme Inhib.* **15**, 455. **7.** Mahesha, Singh & Rao (2007) *Arch. Biochem. Biophys.* **461**, 176.

Dalazatide

pY-AEEA-RSCIDTIPKSRCTAFQCKHSMKYRLSFCRKTCGTC-NH₂

This first-in-class potassium channel blocker (MW = 4442.10 g/mol; CAS 1081110-69-1), also known as SHK-186, is a modified form of sea anemone (*Stichodactyla helianthus*) SHK toxin that selectively targets the $K_V1.3$ voltage-gated potassium ion channel, a druggable target for treatment of autoimmune disorders (*e.g.*, multiple sclerosis, lupus, psoriasis, ulcerative colitis, uveitis, and asthma). SHK-186 blocks the activation of autoreactive $K_V1.3$-overexpressing, effector memory T cells (1-4). Dalazatide's improved stability arises from amidation of its C-terminus and incorporation of a selectivity-enhancing phosphotyrosine substituent at its N-terminus. This peptide exhibits >100x greater selectivity for $K_V1.3$ versus $K_V1.1$, and >1000x versus $K_V1.6$. ShK-192 (a related analogue with a nonhydrolyzable phosphotyrosine surrogate, norleucine in place methionine, and a C-terminal amide) docks to Kv1.3, with the N-terminal *p*-phosphonophenylalanine group interacting at the junction of two channel monomers by forming a salt-bridge with Lys[411] within the channel. After a single subcutaneous injection of 100 μg/kg, approximately 100–200 pM concentrations of active peptide is detectable in the blood of Lewis rats 24, 48, and 72 hour after the injection. Other non-hydrolysable analogues of the N-terminal pTyr, such as *p*- phosphonphenylalanine (Ppa), prepared with and without substitutions of Met21 (ShK-190) with the nonoxidizable Nle residue (ShK-192), show improved stability profiles with selectivity profiles that are similar to ShK-186. [EWSS]ShK, an N-terminally extended ShK containing Glu-Trp-Ser-Ser, inhibits $K_V1.3$ (IC_{50} = 34 pm), with 158x selectivity over $K_V1.1$ and >2900x affinity for $K_V1.3$ over $K_V1.2$ and $K_{Ca}1.3$ channels (4). Another highly $K_V1.3$-selective ShK analogue, [EWSS]ShK, is based entirely on protein amino acids and can be produced by recombinant expression, making it a valuable addition to the complement of therapeutic candidates for the treatment of autoimmune diseases. *Note*: Mouse models are unsuitable models for evaluating $K_V1.3$ blockers in immune functions, inasmuch as: (a) $K_V1.3$ does not set the membrane potential in mouse T cells; (b) beyond the presence of $K_V1.3$, mouse T cells also possess $K_V1.1$, $K_V1.2$, $K_V1.6$ and $K_V3.1$; (c) mouse T cell proliferation and cytokine secretion are unaffected by $K_V1.3$ blockers; and (d) $K_V1.3$ knockout mouse has no immune T cell phenotype. **Targets:**

$K_V1.1$ (K_d = 7 nM); $K_V1.2$ (K_d = 48 nM); **$K_V1.3$ (K_d = 69 pM)**; $K_V1.4$ (K_d = 0.1 μM); $K_V1.5$ (K_d = 0.1 μM); $K_V1.6$ (K_d = 0.02 μM); $K_V1.7$ (K_d = 0.1 μM); $K_V2.1$ (K_d = 0.1 μM); $K_V3.1$ (K_d = 0.1 μM); K_V 3.2 (K_d = 0.02 μM); hERG (K_d = 0.1 μM); $K_{Ca}1.1$ (K_d = 0.1 μM); $K_{Ca}2.2$ (K_d = 0.1 μM); $K_{Ca}2.3$ (K_d = 0.1 μM); $K_{Ca}3.1$ (K_d = 0.12 μM); $K_{ir}2.1$ (K_d = 0.1 μM); $Na_V1.2$ (K_d = 0.1 μM); $Na_V1.4$ (K_d = 0.1 μM); $Ca_V1.2$ (K_d = 0.1 μM). **1**. Beeton, Pennington, Wulff, *et al.* (2005) *Mol. Pharmacol.* **67**, 1369. **2**. Pennington, Chang, Chauhan, *et al.* (2015) *Mar. Drugs* **13**, 529. **3**. Pennington, Beeton, Galea, *et al.* (2009) *Mol. Pharmacol.* **75**, 762. **4**. Chang, Huq, Chhabra, *et al.* (2015) *FEBS J.* **282**, 2247.

Dalbavancin

This once-weekly, semisynthetic lipoglycopeptide antibiotic (FW = 1799.72 g/mol; CAS 171500-79-1), also known by its code name BI397 as well as it trade name Zeven®, possesses excellent *in vitro* activity versus a variety of Gram-positive pathogens, including *Enterococci* and *Clostridia* spp., *Streptococcus pyogenes*, methicillin-resistant *Staphylococcus aureus* (MRSA) and methicillin-resistant *Staphylococcus epidermidis* (MRSE). Upon infusion, plasma dalbavancin concentrations reach their maximum, with an averaged $t_{1/2}$ of 181 hours, a mean volume of distribution of 9.75 L, and a clearance of 0.0473 L/hour. **1**. Leighton, Gottlieb, Dorr, *et al.* (2004) *Antimicrob. Agents Chemother.* **48**, 940. **2**. Chen, Zervos & Vazquez (2007) *Int. J. Clin. Pract.* **61**, 853.

Dalcetrapib

This CETP inhibitor (FW = 389.59 g/mol; CAS 211513-37-0), also known by its code name JTT-705 and its systematic name *S*-[2-({[1-(2-ethylbutyl)cyclohexyl]carbonyl}amino)phenyl] 2-methylpropanethioate, targets cholesterylester transfer protein, which promotes the transfer of cholesteryl esters from antiatherogenic HDLs to pro-atherogenic apolipoprotein B (apoB)-containing lipoproteins, including VLDLs, VLDL remnants, IDLs, and LDLs. A deficiency of CETP is associated with increased HDL levels and decreased LDL levels, a profile that is typically antiatherogenic. Therefore, CETP inhibition should result in higher circulating levels of HDL particles (1-4). Inhibition of CETP activity by JTT-705 not only increased the quantity of HDL, including HDL-C levels and charge-based HDL subfractions, but also favorably affected the size distribution of HDL subpopulations and the apolipoprotein and enzyme composition of HDL in rabbits (4). Despite encouraging animal studies supporting this hypothesis, subsequent studies indicated that the pharmacological manipulation of HDL-C by this CETP inhibitor does not improve the cardiovascular outcomes and may have unacceptable side-effects (5). This agent nonetheless remains a useful tool in dissecting the molecular and cellular mechanisms that manage cholesterol levels in animals. The IC_{50} in human serum is approximately 2 μM. Other CETP inhibitors include Anacetrapib, Evacetrapib, and Torcetrapib. **1**. Okamoto,

Yonemori, Wakitani, *et al.* (2000) *Nature* **406**, 203. **2**. Kobayashi, Okamoto, Otabe, Bujo & Saito (2002) *Atherosclerosis* **162**, 131. **3**. de Grooth, Kuivenhoven, Stalenhoef *et al.* (2002) *Circulation* 105, 2159. **4**. Zhang, Fan & Shimoji, *et al.* (2004) *Arterioscler. Thromb. Vasc Biol.* **24**, 1910. **3**. Wright (2013) *Curr. Opin. Cardiol.* **28**, 389.

Dalfopristin

This Streptogramin-like antibiotic (FW = 690.85 g/mol; CAS 112362-50-2), also named Synercid® (when combined with quinupristin) and (3*R*,4*R*,5*E*,10*E*,12*E*,14*S*,26*R*,26a*S*)-26-[[2-(diethylamino)ethyl]sulfonyl]-8, 9,14,15,24,25,26,26a-octahydro-14-hydroxy-3-isopropyl-4,12-dimethyl-3*H* -21,18-nitrilo-1*H*,22*H*-pyrrolo-[2,1-*c*][1,8,4,19]-dioxadiazacyclotetracosine -1,7,16,22(4*H*,17*H*)-tetrone, binds to sites located on the ribosome's 50S subunit. Dalfopristin binding results in a conformational change of the ribosome, increasing affinity for quinupristin. The resultant stable two-drug-ribosome complex inhibits protein synthesis by preventing peptide-chain formation and blocking the extrusion of newly formed peptide chains. Resistance is mediated by (a) enzymatic drug inactivation, (b) drug efflux or active transport, and (c) most commonly, conformational alterations in ribosomal target-binding sites. (**See** *Quinupristin*) **1**. Allington & Rivey (2001) *Clin. Ther.* **23**, 24.

Daminozide

This plant growth regulator ($FW_{free-acid}$ = 160.17 g/mol; CAS 1596-84-5; M.P. = 154-155°C; Water-soluble), also known as Alar®, B-995, dimethazide, and SADH (*i.e.*, succinic acid mono(2,2-dimethylhydrazide), is a potent and selective inhibitor of the KDM2/7 family of human JmjC histone demethylases, which catalyze the *N*-demethylation of $N^ε$-methyl lysine residues in histones. Kinetic and crystallographic studies reveal that daminozide chelates the active site metal via its hydrazide carbonyl and dimethylamino groups. Daminozide's traditional action is that of a synthetic gibberellin antagonist that inhibits indole-3-acetate (IAA) synthesis and increases the number of blossoms on fruit trees. It has been used mainly in apple orchards to control flower development, to reduce premature fruit fall, and to improve color development, size, and storage. Daminozide is a suspected carcinogen in laboratory animals and is known to decompose to *N*',*N*'-dimethylhydrazine, another potential carcinogen, when apple homogenates are heat treated. **Target(s):** amine oxidase, pea (1); gibberellin activity and biosynthesis (2,3); gibberellin 2β dioxygenase (4); JmjC histone demethylases (5). **1**. Suzuki & Yamasaki (1970) *Enzymologia* **39**, 57. **2**. Pullman, Mein, Johnson & Zhang (2005) *Plant Cell Rep.* **23**, 596. **3**. Rademacher (2000) *Ann. Rev. Plant Physiol. Plant Mol. Biol.* **51**, 501. **4**. Smith & MacMillan (1984) *J. Plant Growth Regul.* **2**, 251. **5**. Rose, Woon & Tumber (2012) *J. Med. Chem.* **26**, 6639.

Damnacanthal

This orange-yellow anthraquinone (FW = 282.25 g/mol; CAS 477-84-9), also known as 3-hydroxy-1-methoxyanthraquinone-2 aldehyde, from the stem bark and roots of *Morinda lucidais* and *M. citrafolia*, is a strong inhibitor of p56*lck* protein-tyrosine kinase: IC$_{50}$ = 620 nM) with respect to an exogenous peptide phosphorylation); and (IC$_{50}$ = 17 nM) for p56*lck* autophosphorylation. It is a weaker inhibitor of protein kinase A, protein kinase C, epidermal growth factor receptor protein-tyrosine kinase, insulin receptor protein-tyrosine kinase, platelet-derived growth factor receptor protein-tyrosine kinase, p60*src* protein-tyrosine kinase, and p59*fyn* protein-tyrosine kinase. There is also a modest (7-20x), but highly statistically significant, selectivity for p56lck over the homologous enzymes p60src and p59fyn. Damnacanthal is competitive for peptide binding site, but mixed noncompetitive with respect to ATP (1). **Target(s):** DNA topoisomerase II (1); protein-tyrosine kinase, p56*lck* and p60*src*, non specific protein-tyrosine kinase (2-5); LIM-kinases-1, which plays a crucial role in various cell activities, including migration, division and morphogenesis, by phosphorylating and inactivating cofilin (6). **1.** Tosa, Iinuma, Asai, *et al.* (1998) *Biol. Pharm. Bull.* **21**, 641. **2.** Faltynek, Schroeder, Mauvais, *et al.* (1995) *Biochemistry* **34**, 12404. **3.** Kusaka, Kimura, Kusaka, *et al.* (2003) *J. Neurosurg.* **99**, 383. **4.** Hawkins & Marcy (2001) *Protein Expr. Purif.* **22**, 211. **5.** Hollósy & Kéri (2004) *Curr. Med. Chem. Anticancer Agents* **4**, 173. **6.** Ohashi, Sampei, Nakagawa, *et al.* (2014) *Mol. Biol. Cell* **25**, 828.

Danazol

This synthetic steroid and anterior pituitary suppressant (FW = 337.46 g/mol; CAS 17230-88-5), also known as 17α-pregna-2,4-dien-20-ynol[2,3-*d*]isoxazol-17β-ol and Danocrine, exhibits anti-estrogenic and anti-progestogenic activity, while displaying weak androgenic properties. Danazol also inhibits gonadotropins and suppresses gametogenesis and steroidogenesis. **Target(s):** aromatase, *or* CYP19 (1); cytochrome P450-dependent activities, *or* testosterone monooxygenase, *or* aminopyrine dealkylase, *or* 7-ethoxycoumarin dealkylase (2); estrone sulfatase (3-6); 3β-hydroxysteroid dehydrogenase (7); 17β-hydroxysteroid dehydrogenase (8); steroid 17,20 lyase (7); steroid 11β-monooxygenase (9); steroid 17α-monooxygenase (7); steroid 21-monooxygenase (9); steroid sulfotransferase (10); steryl sulfatase (3-6,11,12). **1.** Kitawaki, Noguchi, Amatsu, *et al.* (1997) *Biol. Reprod.* **57**, 514. **2.** Konishi, Morita, Ono & Shimakawa (1990) *Yakugaku Zasshi* **110**, 49. **3.** Pasqualini, Chetrite & Nestour (1996) *J. Endocrinol.* **150** Suppl., S99. **4.** Selcer & Li (1995) *J. Steroid Biochem. Mol. Biol.* **52**, 281. **5.** Nguyen, Ferme, Chetrite & Pasqualini (1993) *J. Steroid Biochem. Mol. Biol.* **46**, 17. **6.** Evans, Rowlands, Jarman & Coombes (1991) *J. Steroid Biochem. Mol. Biol.* **39**, 493. **7.** Arakawa, Mitsuma, Iyo, *et al.* (1989) *Endocrinol. Jpn.* **36**, 387. **8.** Blomquist, Lindemann & Hakanson (1984) *Steroids* **43**, 571. **9.** Barbieri, Osathanondh, Canick, Stillman & Ryan (1980) *Steroids* **35**, 251. **10.** Bamforth, Dalgliesh & Coughtrie (1992) *Eur. J. Pharmacol.* **228**, 15. **11.** Selcer, Kabler, Sarap, Xiao & Li (2002) *Steroids* **67**, 821. **12.** Carlstrom, Doberl, Pousette, Rannevik & Wilking (1984) *Acta Obstet. Gynecol. Scand. Suppl.* **123**, 107.

Danofloxacin

This fluoroquinolone-class antibiotic (FW = 357.37 g/mol; CAS 112398-08-0), also named Advocin® (injectable) and 1-cyclopropyl-6-fluoro-7-[(1S,4S)-3-methyl-3,6-diazabicyclo[2.2.1]heptan-6-yl]-4-oxoquinoline-3-carboxylic acid, is a broad-spectrum systemic antibacterial agent that targets Type II DNA topoisomerases (gyrases), which are required for bacterial replication and transcription. It has been used to combat respiratory infections in calves, sheep and poultry. (For the prototypical member of this antibiotic class, **See** *Ciprofloxacin*)

Danoprevir

This macrocyclic peptidomimetic (FW = 717.85 g/mol; CASs = 850876-88-9, 916881-67-9, 1001913-18-3, 1225266-12-5; Solubility: 140 mg/mL DMSO; <1 mg/mL Water), also known as RG7227, ITMN-191, RO5190591, and systematically as 4-fluoro-1,3-dihydro-(2R,6S,13aS,14aR,16aS)-14a-[[(cyclopropylsulfonyl)amino]carbonyl]-6-[[(1,1-dimethylethoxy)-carbonyl]amino]-2H-isoindole-2-carboxylic acid, targets the NS3/4A protease of hepatitis C virus (HCV) with IC$_{50}$ = 0.2-3.5 nM (1-4). The inhibitor's time-dependent effects suggest a two-step binding mechanism with a low overall dissociation rate. Progress-curve analysis indicates the formation of an initial collision complex (EI) that isomerizes to a highly stable complex (EI*), from which danoprevir dissociates very slowly (5). The dissociation constant (K_i) for EI is 100 nM, and the rate constant for EI-to-EI* conversion is 6.2 x 10^{-2} s^{-1}. The rate constant for dissociation from the EI* complex is 3.8 x 10^{-5} s^{-1}, with a complex half-life calculated to be ~5 hours and a true biochemical potency (K_i^*) of 62 pM (5). Surface plasmon resonance studies as well as assessments of enzyme reactivation after dilution of EI* confirm slow dissociation, with a considerably longer dissociation half-life (5). **1.** Seiwert, *et al.* (2008) *Antimicrob. Agents Chemother.* **52**, 4432. **2.** Bartels, *et al.* (2008) *J. Infect. Dis.* **198**, 800. **4.** Imhof, *et al.* (2011) *Hepatology* **53**, 1090. **5.** Rajagopalan, Misialek, Stevens *et al.* (2009) *Biochemistry* **48**, 2559.

N$^{\alpha}$-Dansyl-L-arginine 4-ethylpiperidineamide

This substituted arginine (FW$_{hydrochloride}$ = 539.14 g/mol), also known as N$^{\alpha}$-dansyl-L-arginine N,N-(3-ethyl-1,5 pentanediyl)amide and DAPA, is a potent inhibitor of thrombin (K_i = 37 nM), and becomes fluorescent when bound. It is a weak inhibitor of trypsin, plasmin, coagulation factor Xa, and u-plasminogen activator. **Target(s):** acetylcholinesterase, weakly inhibited (1); adenylate cyclase (2); butyrylcholinesterase (1); cholinesterase (1,3,4); thrombin (4-6). **1.** Brimijoin, Mintz & Prendergast (1983) *Biochem. Pharmacol.* **32**, 699. **2.** Wilks, Campbell & Hui (1986) *Biol. Reprod.* **35**, 877. **3.** Hijikata-Okunomiya, Okamoto, Tamao & Kikumoto (1988) *J. Biol. Chem.* **263**, 11269. **4.** Okamoto & Hijikata-Okunomiya (1993) *Meth. Enzymol.* **222**, 328. **5.** Hibbard, Nesheim & Mann (1982) *Biochemistry* **21**, 2285. **6.** Nesheim, Prendergast & Mann (1979) *Biochemistry* **18**, 996.

N-Dansylcadaverine

This monodansylated diamines (FW$_{\text{free-base}}$ = 335.47 g/mol), also known as N-dansyl-1,5-pentanediamine, is an alternative substrate for many transglutaminases and also inhibits clathrin mediated receptor endocytosis. **Target(s):** [β-adrenergic-receptor] kinase (1); 17-ketosteroid reductase (2); protein-glutamine γ-glutamyltransferase, or transglutaminase (3-9). **1.** Wang & Liu (2003) *J. Biol. Chem.* **278**, 15809. **2.** Moger (1982) *Can. J. Physiol. Pharmacol.* **60**, 858. **3.** Siegel & Khosla (2007) *Pharmacol. Ther.* **115**, 232. **4.** Wu, Lai & Tsai (2005) *Int. J. Biochem. Cell Biol.* **37**, 386. **5.** Carvajal-Vallejos, Campos, Fuentes-Prior, *et al.* (2007) *Biotechnol. Lett.* **29**, 1255. **6.** Brobey & Soong (2006) *Exp. Parasitol.* **114**, 94. **7.** Korner, Schneider, Purdon & Bjornsson (1989) *Biochem. J.* **262**, 633. **8.** Mottahedeh & Marsh (1998) *J. Biol. Chem.* **273**, 29888. **9.** Porta, De Santis, Esposito, *et al.* (1986) *Biochem. Biophys. Res. Commun.* **138**, 596.

N-Dansyl-N-(4-nitrobenzyl)-β-alanine Hydroxamate

This hydroxamate (FW = 472.52 g/mol) inhibits interstitial collagenase (K_i = 89 nM), gelatinase A, or matrix metalloproteinase-2, (K_i = 54 nM), neutrophil collagenase, or matrix metalloproteinase-8 (K_i = 60 nM), gelatinase B, or matrix metalloproteinase-9 (K_i = 67 nM), and microbial collagenase (K_i = 18 nM). **1.** Scozzafava, MIlies, Manole & Supuran (2000) *Eur. J. Pharm. Sci.* **11**, 69.

Dantrolene

This hydantoin (FW = 314.26 g/mol; CAS 85008-71-5), also named systematically as 1-([5-(p-nitrophenyl)furfurylidene]amino)hydantoin, is a skeletal muscle relaxant that inhibits Ca^{2+} release from sarcoplasmic reticulum. Dantrolene is used in the treatment of malignant hyperthermia, a disorder associated with increased intracellular free Ca^{2+}. With mounting evidence that high intracellular calcium ion is a contributing factor in bacterial sepsis, dantrolene may be useful in managing bloodstream-borne infections. **Target(s):** Ca^{2+} release from sarcoplasmic reticulum (1,2); carbonic anhydrase (3); glucose-6 phosphate dehydrogenase (4). **1.** Harrison (1988) *Brit. J. Anaesth.* **60**, 279. **2.** Song, Karl, Ackerman & Hotchkiss (1993) *Proc. Natl. Acad. Sci. U.S.A.* **90**, 3933. **3.** Gulcin, Beydemir & Buyukokuroglu (2004) *Biol. Pharm. Bull.* **27**, 613. **4.** Beydemir, Gulcin, Kufrevioglu & Ciftci (2003) *Pol. J. Pharmacol.* **55**, 787.

Dapagliflozin

This Type-2 diabetes drug (FW = 408.87 g/mol; CAS 461432-26-8; Symbol:), also known as BMS 512148, BMS-512148, the trade name Farxiga™, and systematic name (2S,3R,4R,5S,6R)-2-[4-chloro-3-(4-ethoxybenzyl)phenyl]-6-(hydroxymethyl)-tetrahydro-2H-pyran-3,4,5-triol) inhibits (IC$_{50}$ = 60 nM) the sodium-glucose transporter (SGLT2), Subtype 2, a carrier responsible for >90% of kidney glucose reabsorption. The IC$_{50}$ for SGLT2 is 1.1 nM, or >1000-times lower than that for SGLT1 (IC$_{50}$ = 1.4

μM), such that the drug does not interfere with intestinal glucose absorption (2). **Mechanism of Action:** The kidney plays an important role in glucose homeostasis, primarily by the reabsorption of filtered glucose, a process known as 'glucuresis'. Located in the proximal convoluted tubule, SGLT2 is responsible for the majority of glucose reabsorption by the kidney. SGLT2 inhibitors offer a novel approach to treat T2D and reduce hyperglycemia by increasing urinary excretion of glucose. In humans, SGLT2 inhibition with Farxiga (10 mg) resulted in excretion of approximately 70 grams of glucose in the urine per day at Week-12. Farxiga increases the risk of genital mycotic infections. Other SGLT2 inhibitors include canagliflozin, empagliflozin, tofogliflozin, and luseogliflozin. In 2014, dapagliflozin was approved by the Food & Drug Administration to improve glycemic control, when used along with dieting and exercise in adult patients afflicted with T2DM. **Other Targets:** Dapaglitflozin also increases hepatic glucose production through the induction of glucagon secretion in islet α-cells (2). DPF inhibits UDP-glucuronosyltransferases UGT1A1, UGT1A9, and UGT1A10, with IC$_{50}$ values ranging from 39 to 66 μM (3). The K_i for DPF inhibition of UGT1A1 was 81 μM, whereas the K_i values for inhibition of UGT1A9 ranged from 12 to 15 μM. Based on the *in vitro* K_i values and plasma concentrations reported in the literature, DPF may be excluded as a perpetrator of drug-drug interactions arising from inhibition of UGT enzymes (3). (***See also*** *Canagliflozin; Empagliflozin, Tofogliflozin, Luseogliflozin*) **1.** Meng, Ellsworth, Nirschl, *et al.* (2008) *J. Med. Chem.* **51**, 1145. **2.** Bonner, Kerr-Conte, Gmyr, *et al.* (2015) *Nature Med.* **21**, 512. **3.** Pattanawongsa, Chau, Rowland & Miners (2015) *Drug Metab. Dispos.* **43**, 1468.

Dapoxetine

This selective serotonin reuptake inhibitor, *or* SSRI (FW = 305.41 g/mol; CASs = 119356-77-3 (free base) and 129938-20-1 (HCl salt)), also known as LY 210448, Priligy®, Westoxetin®, and (S)-N,N-dimethyl-3-(naphthalen-1-yloxy)-1-phenylpropan-1-amine, was the first agent specifcially developed for the treatment of premature ejaculation in men 18–64 years old. By increasing serotonin's action at the post synaptic cleft, dapoxetine delays ejaculation. Dapoxetine is structurally related to fluoxetine and was initially pursued as an antidepressant. **1.** Dresser, Desai, Gidwani, *et al.* (2006) *Internat. J. Impotence Res.* **18**, 104.

Dapsone

This antibacterial agent (FW = 248.31 g/mol; CAS 80-08-0; Soluble in ethanol and methanol; Low solubility in H_2O), also known as 4,4'-diaminodiphenyl sulfone (*i.e.*, DDS) and 4,4'-sulfonylbisbenzeneamine, is an effective treatment for against noninfectious inflammatory diseases. Dapsone is frequently used against the causative agent of leprosy (*Mycobacterium leprae*), as well as an adjunct in treating malaria. Dapsone acts as a folate synthetase inhibitor by via competition with *para*-aminobenzoate for the active site of dihydropteroate synthetase and reacting with the substrate (7,8-dihydro-6-hydroxymethylpterinopyrophosphate) to form tightly bound 7,8-dihydropteroic acid analogue (1). Dapsone has a dramatic beneficial effect in dermatitis herpetiformis, processes in which the polymorphonuclear leukocytes (PMNL) and immune complexes appear to be important for skin lesions to develop (2). Pruritus disappears and inflammatory eruptions clear within a few days of dapsone therapy. At dapsone concentrations (1-30 μg/mL) comparable to therapeutic doses, it interferes primarily with the myeloperoxidase (MPO)-H_2O_2-halide-mediated cytotoxic system in PMNL (2). No effect was observed on locomotion, chemotaxis, phagocytic ingestion, oxidative metabolism, or the release of lysosomal enzymes (2). Radioligand binding studies using human neutrophils exposed to 10-100 μM dapsone indicate that it antagonizes association of leukotriene B$_4$ (LTB$_4$) with its specific receptor sites, reducing biologic response in LTB$_4$-stimulated neutrophil chemotaxis (3). A physiologic model of LTB$_4$-dependent inflammation in mice is antagonized by systemic administration of dapsone (3). **Key Pharmacokinetic Parameters:**

See Appendix II in Goodman & Gilman's THE PHARMACOLOGICAL BASIS OF THERAPEUTICS, 12th Edition (Brunton, Chabner & Knollmann, eds.) McGraw-Hill Medical, New York (2011). **Target(s):** cyclooxygenase (4); dihydropteroate synthase (5-17); hydroxymethyl-dihydropterin pyrophosphokinase-dihydropteroate synthase (11); lactoperoxidase (18); lipoxygenase (19); 5-lipoxygenase (20); myeloperoxidase (21-24); peroxidase, eosinophil (21,22); tubulin polymerization, *or* microtubule self-assembly (25). **1.** Seydel, Richter & Wempe (1980) *Int. J. Lepr. Other Mycobact. Dis.* **48**, 18. **2.** Stendahl, Molin & Dahlgren (1978) *J. Clin. Invest.* **62**, 214. **3.** Maloff, Fox, Bruin & Di Meo (1988) *Eur. J. Pharmacol.* **158**, 85. **4.** Ruzicka, Wasserman, Soter & Printz (1983) *J. Allergy Clin. Immunol.* **72**, 365. **5.** Ho (1980) *Meth. Enzymol.* **66**, 553. **6.** Zhang & Meshnick (1991) *Antimicrob. Agents Chemother.* **35**, 267. **7.** Merali, Zhang, Sloan & Meshnick (1990) *Antimicrob. Agents Chemother.* **34**, 1075. **8.** Allegra, Boarman, Kovacs, *et al.* (1990) *J. Clin. Invest.* **85**, 371. **9.** McCullough & Maren (1974) *Mol. Pharmacol.* **10**, 140. **10.** McCullough & Maren (1973) *Antimicrob. Agents Chemother.* **3**, 665. **11.** Kasekarn, Sirawaraporn, Chahomchuen, Cowman & Sirawaraporn (2004) *Mol. Biochem. Parasitol.* **137**, 43. **12.** Nopponpunth, Sirawaraporn, Greene & Santi (1999) *J. Bacteriol.* **181**, 6814. **13.** Williams, Spring, Harris, Roche & Gillis (2000) *Antimicrob. Agents Chemother.* **44**, 1530.m **14.** Triglia, Menting, Wilson & Cowman (1997) *Proc. Natl. Acad. Sci. U.S.A.* **94**, 13944. **15.** Berglez, Iliades, Sirawaraporn, Coloe & Macreadie (2004) *Int. J. Parasitol.* **34**, 95. **16.** Iliades, Meshnick & Macreadie (2005) *Antimicrob. Agents Chemother.* **49**, 741. **17.** Fernley, Iliades & Macreadie (2007) *Anal. Biochem.* **360**, 227. **18.** Shin, Tomita & Lonnerdal (2000) *J. Nutr. Biochem.* **11**, 94. **19.** Klegeris & McGeer (2003) *J. Leukoc. Biol.* **73**, 369. **20.** Wozel & Lehmann (1995) *Skin Pharmacol.* **8**, 196. **21.** Bozeman, Learn & Thomas (1992) *Biochem. Pharmacol.* **44**, 553. **22.** Bozeman, Learn & Thomas (1990) *J. Immunol. Methods* **126**, 125. **23.** Kettle & Winterbourn (1991) *Biochem. Pharmacol.* **41**, 1485. **24.** Stendahl, Molin & Lindroth (1983) *Int. Arch. Allergy Appl. Immunol.* **70**, 277. **25.** Rajagopalan & Gurnani (1983) *Biochem. Biophys. Res. Commun.* **116**, 128.

DAPT

This orally bioavailable γ-secretase Notch inhibitor (FW = 432.46 g/mol; CAS 208255-80-5; Solubility: 85 mg/mL DMSO; <1 mg/mL H$_2$O), known as GSI-IX and systematically as (*S*)-*tert*-butyl 2-((*S*)-2-(2-(3,5-difluorophenyl)acetamido)propanamido)-2-phenylacetate, targets γ-secretase (GS), with IC$_{50}$ = 20 nM, thereby reducing the amount of secreted Aβ peptides into the culture medium in HEK293 cells (1-3). DAPT also induces caspase-dependent and caspase-independent apoptosis in lung squamous cell carcinoma cells by inhibiting Notch receptor signaling pathway. DAPT impairs recovery from lipopolysaccharide-induced inflammation in rat brain (2). When treated with DAPT, paclitaxel-insensitive breast cancer cells show restored sensitivity to the microtubule-directed drug, suggesting the Notch pathway affects paclitaxel sensitivity (4). **1.** Cao, *et al.* (2012) *APMIS* **120**, 441. **2.** Nasoohi, Hemmati, Moradi & Ahmadiani Nasoohi (2012) *Neuroscience* **210**, 99. **3.** Dovey, John, Anderson, *et al.* (2001) *J. Neurochem.* **76**, 173. **4.** Zhao, Ma, Gu & Fu (2014) *Chin. Med. J.* (Engl). **127**, 442.

Daptomycin

This intravenous lipopeptide antibiotic (FW = 1619.71 g/mol; CAS 103060-53-3), also named LY-146032, Cubicin®, and N-decanoyl-L-W-L-N-L-D-L-T-G-L-Orn-L-D-D-A-L-D-G-D-S-*threo*-3-CH$_3$-L-E-3-anthranil-oyl-L-A-[egr]$_1$-lactone, from the soil saprotroph *Streptomyces roseosporus*, exerts its antibiotic effect by inserting itself into the cell membrane in a phosphatidylglycerol-dependent fashion, whereupon it aggregates (1-3). The latter alters membrane curvature, creating pores that leak ions, causing persistent depolarization and inhibiting protein, DNA, and RNA synthesis. Daptomycin is bactericidal against Gram-positive bacteria, including *Enterococci* (also glycopeptide-resistant *Enterococci*, *or* GRE), *Staphylococci* (also methicillin-resistant *S. aureus*, *or* MRSA), viridans group *Streptococci*, and *Corynebacteria*. *Listeria monocytogenes* is markedly less susceptible to LY-146032 (MIC$_{90}$ = 16 µg/mL) than vancomycin (MIC$_{90}$ = 1 µg/mL). Vancomycin and daptomycin are the first-line antibiotic choices for MRSA bacteremia (3). **Key Pharmacokinetic Parameters:** *See* Appendix II in Goodman & Gilman's THE PHARMACOLOGICAL BASIS OF THERAPEUTICS, 12th Edition (Brunton, Chabner & Knollmann, eds.) McGraw-Hill Medical, New York (2011). **1.** Eliopoulos, Willey, Reiszner, *et al.* (1986) *Antimicrob. Agents Chemother.* **30**, 532. **2.** Fass & Helsel (1986) *Antimicrob. Agents Chemother.* **30**, 781. **3.** Holland, Arnold & Fowler (2014) *JAMA* **312**, 1330.

Darapladib

This investigational atherosclerosis drug (FW = 665.79 g/mol; CAS 356057-34-6) produces sustained inhibition of plasma lipoprotein-associated phospholipase A$_2$ (or Lp-PLA$_2$) activity in patients receiving intensive atorvastatin therapy (1). Lp-PLA$_2$ is abundant in the necrotic core of coronary lesions, and its pro-inflammatory reaction products are likely to contribute to cell death and increase the risk of plaque rupture and ensuing blockage of vital blood vessels. Changes in IL-6 and hs-CRP after 12-week course on darapladib (160 mg) suggest a possible reduction in inflammatory burden (1). Lp-PLA$_2$ inhibition by darapladib prevents necrotic core expansion, a key determinant of plaque vulnerability (2). Darapladib exhibits favorable pharmacokinetics, minimal predicted drug-drug interactions, sustained blood levels with once-daily oral dosing and limited inhibition of other PLA$_2$ isozymes (3). In a double-blind trial with 15,828 stable coronary heart disease patients randomly assigned to receive either once-daily darapladib (at a dose of 160 mg) or placebo, however, darapladib did not significantly reduce the risk of the primary composite end point of cardiovascular death, myocardial infarction, or stroke darapladib versus placebo, when added to standard of care (3). **1.** Mohler, Ballantyne, Davidson, *et al.* (2008) *J. Am. Coll. Cardiol.* **51**, 1632. **2.** Serruys, García-García, Buszman, *et al.* (2008) Circulation **118**, 1172. **3.** Riley & Corson (2009) *IDrugs* **12**, 648. **3.** White, Held, Stewart, *et al.* (2014) *New Engl. J. Med.* **370**, 1702.

Darexaban

This orally bioavailable direct Factor Xa inhibitor (FW = 474.56 g/mol; CASs = 365462-23-3 and 365462-24-4 (maleate salt)), also known by the code name YM150 and systematically named *N*-(3-hydroxy-2-{[4-(4-methyl-1,4-diazepan-1-yl)benzoyl]amino}phenyl)-4-methoxybenzamide, is a potent anticoagulant and antithrombotic agent used for the prevention of post-surgical venous thromboembolism, stroke in patients with atrial fibrillation, as well as ischemic events in acute coronary syndrome. The hydroxy moiety of YM-150 is rapidly transformed into its glucuronide conjugate, YM-222714 (1). Both darexaban and its glucuronide selectively inhibit human factor Xa with K_i values of 0.03 and 0.02 µM (2). They show anticoagulant activity in human plasma, with concentrations of darexaban and darexaban glucuronide needed for doubling prothrombin time of 1.2 and 0.95 µM, respectively (2). **1.** Hirayama, Koshio, Ishihara, *et al.* (2011) *J. Med. Chem.* **54**, 8051. **2.** Iwatsuki, Sato, Moritani, *et al.* (2011) *Eur. J. Pharmacol.* **673**, 49.

Darunavir

This second-generation nonpeptidic HIV protease inhibitor (FW = 5477.67 g/mol; CAS 206361-99-1; IUPAC: [(1*R*,5*S*,6*R*)-2,8-dioxabicyclo[3.3.0]oct-6-yl] *N*-[(2*S*,3*R*)-4-[(4-aminophenyl)sulfonyl-(2-methylpropyl)amino]-3-hydroxy-1-phenylbutan-2-yl]carbamate), also known as TMC114 and Prezista®, is used to treat Human Immunodeficiency Virus infection by targeting post-translational fragmentation of the single HIV polyprotein into its component structural proteins and replicative enzymes. Darunavir interacts with the catalytic aspartates and the backbone within the active site (making hydrogen bonds to residues Asp-25, Asp-25', Asp-29, Asp-30, Asp-30', and Gly-27), allowing Darunavir to competitively inhibit (K_d = 4.5 x 10^{-12} M) polyprotein substrate access to the active site (1,2). Even with the reduction in binding affinity to the multidrug-resistant HIV protease, TMC114 still binds with an affinity that is more than 1.5 orders of magnitude tighter than the first-generation inhibitors (3). Darunivir is best paired with ritonavir, a potent CYP3A and P-glycoprotein inhibitor, greatly increasing Darunivir's efficacy. **Key Pharmacokinetic Parameters:** *See* Appendix II in Goodman & Gilman's THE PHARMACOLOGICAL BASIS OF THERAPEUTICS, 12th Edition (Brunton, Chabner & Knollmann, eds.) McGraw-Hill Medical, New York (2011). **1.** Koh, Nakata, Maeda, *et al.* (2003) *Antimicrob. Agents Chemother.* **47**, 3123. **2.** Tie, Boross, Wang, *et al.* (2004) *J. Mol. Biol.* **338**, 341. **3.** King, Prabu-Jeyabalan, Nalivaika, *et al.* (2004) *J. Virol.* **78**, 12012.

Darvon, See Propoxyphene

DAS645

This experimental herbicide (FW = 489.81 g/mol), also known as [1-*tert*-butyl- 3- (2,4- dichlorophenyl)- 5- hydroxy- 1*H*- pyrazol- 4- yl][2- chloro- 4-(methanesulfonyl)phenyl]-methanone, inhibits plant 4-hydroxy-phenylpyruvate dioxygenase (IC$_{50}$ = 12 nM). DAS645 exhibits a remarkable (>1600x) selectivity toward the *Arabidopsis thaliana* enzyme compared to its mammalian counterpart. **1.** Yang, Pflugrath, Camper, *et al.* (2004) *Biochemistry* **43**, 10414.

Dasatinib

This multitargeting Bcr-Abl tyrosine protein kinase inhibitor (FW = 485.05 g/mol; CAS 302962-49-8), also known as BMS-354825 and Sprycel, targets Abl (IC$_{50}$ < 1 nM), Src (IC$_{50}$ = 0.5 nM) and c-Kit (IC$_{50}$ = 79 nM) (1-5). Dasatinib induces DNA damage and activates DNA repair pathways, leading to senescence only in non-small cell lung cancer (NSCLC) that does not harbor targetable kinase mutations or translocations. NSCLC cells harboring kinase-inactivating BRAF mutations (KIBRAF) undergo senescence, when treated with dasatinib. Moreover, dasatinib-induced senescence is dependent on Chk1 and p21 proteins, both of which are known to mediate DNA damage-induced senescence (6). Dasatinib treatment also leads to a marked decrease in TAZ but not YAP protein levels. Over-expression of TAZ inhibits dasatinib-induced senescence (6). **Target(s):** Bcr-Abl protein-tyrosine kinase (1-5); Btk protein-tyrosine kinase (5); protein tyrosine kinase (1-5); ribosyl-dihydronicotinamide dehydrogenase, quinone-requiring (7); Src protein-tyrosine kinase (3,5); Tec protein-tyrosine kinase (5). **1.** Tauchi & Ohyashiki (2006) *Int. J. Hematol.* **83**, 294. **2.** Shah, Tran, Lee, *et al.* (2004) *Science* **305**, 399. **3.** Lombardo, Lee, Chen, *et al.* (2004) *J. Med. Chem.* **47**, 6658. **4.** Cortes, Jabbour, Kantarjian, *et al.* (2007) *Blood* **110**, 4005. **5.** Hantschel, Rix, Schmidt, *et al.* (2007) *Proc. Natl. Acad. Sci. U.S.A.* **104**, 13283. **6.** Peng, Sen, Mazumdar, *et al.* (2016) *Oncotarget* **7**, 565. **7.** Winger, Hantschel, Superti-Furga & Kuriyan (2009) *BMC Struct. Biol.* **9**, 7.

Daunorubicin

This antibiotic and antitumor agent (FW$_{free-base}$ = 527.53 g/mol; CAS 20830-81-3), also known as daunomycin, from *Streptomyces peucetius* readily permeates cells, intercalating into double-stranded DNA, with the aminosugar component docking into the minor groove and thereby inhibiting replication (1). Aqueous solutions are pink under acidic conditions and blue with basic pH. The amino sugar can by removed by treatment with acid. It should be stored in a desiccator in the dark and at 4°C. Stock solutions should be prepared just prior to use. **Primary Mode of Action:** Daunorubicin consists of a tetracyclic structure called daunomycinone that is covalently linked to an amino sugar (daunosamine).Daunorubicin's antimitotic and cytotoxic effects arise by a number of mechanisms. It complexes with DNA by intercalation, inhibiting topoisomerase II activity by stabilizing the DNA-topoisomerase II complex, and preventing the religation phase of the ligation-religation reaction cycle catalyzed by topoisomerase II catalyzes. Single-strand and double-strand DNA breaks result. Daunorubicin may also inhibit DNA polymerase activity, impair regulation of gene expression, and promote/reinforce free radical damage to DNA. Daunorubicin also inhibits RNA transcription. Daunorubicin also exhibits antitumor effects against a wide spectrum of animal tumors, either grafted or spontaneous. Daunorubicin is frequently used in the treatment of acute leukemias. **Redox Cycling in Mitochondria:** Complex I (most likely the NADH dehydrogenase flavin) is the mitochondrial site of anthracycline reduction. During forward electron transport, the anthracyclines doxorubicin (Adriamycin) and daunorubicin act as one-electron acceptors for BH-SMP (*i.e.* they are reduced to semiquinone radical species), but only when NADH was used as substrate (27). Succinate and ascorbate are without effect. Inhibitor experiments (with rotenone, amytal, piericidin A) indicate that the anthracycline reduction site lies on the substrate side of ubiquinone. (*See also* WP631) **Target(s):** carbonyl reductase (2); DNA-directed DNA polymerase (3-5); DNA-directed RNA polymerase (3,6-12); DNA helicase (13); DNA ligase (ATP) (14); DNA topoisomerase II (4,15-18); inositol trisphosphate 3-kinase (19); nitric-oxide synthase (20); nucleoside-triphosphatase (13); P glycoprotein (21); procollagen-proline 4-dioxygenase (22-24); reverse transcriptase (25); transcription (1); Xaa-Pro Dipeptidase, *or* prolidase, *or* imidodipeptidase (26); xenobiotic-transporting ATPase, *or* multidrug-resistance protein (21).

1. Scott & Tomkins (1975) *Meth. Enzymol.* **40**, 273. **2**. Imamura, Koga, Higuchi, *et al.* (1997) *J. Enzyme Inhib.* **11**, 285. **3**. Zunina, Gambetta & Di Marco (1975) *Biochem. Pharmacol.* **24**, 309. **4**. Spadari, Pedrali-Noy, Focher, *et al.* (1986) *Anticancer Res.* **6**, 935. **5**. Müller, Yamazaki & Zahn (1972) *Enzymologia* **43**, 1. **6**. Kersten, Kersten, Szybalski & Fiandt (1966) *Biochemistry* **5**, 236. **7**. Portugal, Martin, Vaquero, *et al.* (2001) *Curr. Med. Chem.* **8**, 1. **8**. Wilson, Grier, Reimer, *et al.* (1976) *J. Med. Chem.* **19**, 381. **9**. Gabbay, Grier, Fingerle, *et al.* (1976) *Biochemistry* **15**, 2062. **10**. Gabbay (1976) in *Inter. J. Quantum Chem., Quantum Biology Symposium No. 3*, p. 217. **11**. Yen, Wilson, Pearce & Gabbay (1984) *J. Pharmaceut. Sci.* **73**, 1575. **12**. Sethi (1971) *Prog. Biophys. Mol. Biol.* **23**, 67. **13**. Borowski, Niebuhr, Schmitz, *et al.* (2002) *Acta Biochim. Pol.* **49**, 597. **14**. Ciarrocchi, Lestingi, Fontana, Spadari & Montecucco (1991) *Biochem. J.* **279**, 141. **15**. Zunino & Capranico (1990) *Anticancer Drug Des.* **5**, 307. **16**. Bodley, Liu, Israel, *et al.* (1989) *Cancer Res.* **49**, 5969. **17**. Insaf, Danks & Witiak (1996) *Curr. Med. Chem.* **3**, 437. **18**. Bergerat, Gadelle & Forterre (1994) *J. Biol. Chem.* **269**, 27663. **19**. Mayr, Windhorst & Hillemeier (2005) *J. Biol. Chem.* **280**, 13229. **20**. Luo & Vincent (1994) *Biochem. Pharmacol.* **47**, 2111. **21**. Shapiro & Ling (1997) *Eur. J. Biochem.* **250**, 130. **22**. Günzler, Hanauske-Abel, Myllylä, *et al.* (1988) *Biochem. J.* **251**, 365. **23**. Merriweather, Guenzler, Brenner & Unnasch (2001) *Mol. Biochem. Parasitol.* **116**, 185. **24**. Kivirikko, Myllylä & Pihlajaniemi (1989) *FASEB J.* **3**, 1609. **25**. Dhananjaya & Antony (1987) *Indian J. Biochem. Biophys.* **24**, 265. **26**. Muszynska, Wolczynski & Palka (2001) *Eur. J. Pharmacol.* **411**, 17. **27**. Davies & Doroshow (1986) *J. Biol. Chem.* **261**, 3060.

Dauricine

This toxic alkaloid (FW = 624.77 g/mol; CAS 524-17-4), also named as 4-((1,2,3,4-tetrahydro-6,7-dimethoxy-2-methyl-1-isoquinolinyl)methyl)-2-(4-((1,2,3,4-tetrahydro-6,7-dimethoxy-2-methyl-1-isoquinolinyl)methyl)phen-oxy)phenol, from the North American moonseed vine *Menispermum canadense* is a highly potent calcium channel blocker. Its antiarrhythmic effects are well documented in experimental arrhythmic models and in arrhythmic cardiac patients. Dauricine blocks cardiac transmembrane Na^+, K^+ and Ca^{2+} ion currents, but differs from quinidine and sotalol, which exhibit reverse use-dependent effects. Dauricine prolongs action potential duration (APD) in a normal use-dependent manner. **1**. Qian (2002) *Acta Pharmacol. Sin.* **23**, 1086.

Daurinol

This novel arylnaphthalene lignan (FW = 350.33 g/mol) from the ethnopharmacological plant *Haplophyllum dauricum*, resembles etoposide structurally and likewise inhibits (10-100 μM) the topoisomerase IIα in an ATP concentration-dependent manner and suppressed the ATP hydrolysis activity of the enzyme. Daurinol inhibits human topoisomerase IIα without the formation of the DNA cleavable complex, suggesting it is a catalytic inhibitor that interferes with at least one step of the topoisomerase catalytic cycle. (Daurinol does not inhibit topoisomerase I, which lacks an ATP binding domain.) However, daurinol treatment does not cause DNA damage or nuclear enlargement *in vitro*. **1**. Kang, Oh, Yun, *et al.* (2011) *Neoplasia* **13**, 1043.

DB-02

This antiviral (FW = 370.51), also named 6-(cyclohexylmethyl)-5-ethyl-2-((2-oxo-2-phenylethyl)thio)pyrimidin-4(3*H*)-one, targets HIV-1 reverse transcriptase (HIV1RT) and displays potent anti-HIV-1 activity against laboratory adapted strains and primary isolated strains including different subtypes and tropism strains (EC_{50} values ranging from 2.4 to 42 nM). DB-02 shows very low cytotoxicity ($CC_{50} > 1$mM) to tested cell lines and peripheral blood mononuclear cells (PBMCs). Genotypic resistance profiles revealed V106A was the major resistance contributor for the compound. Molecular docking analysis showed that DB-02 located in the hydrophobic pocket of HIV1RT, interacting with Lys101, Val106, Leu234, His235. DB-02 also showed non-antagonistic effects to four approved antiretroviral drugs. All studies indicated that DB-02 would be a potential NNRTI with low cytotoxicity and improved activity. **1**. Zhang, Lu, Wang, *et al.* (2013) *PLoS One* **8**, e81489.

DBeQ

This potent and ATP-competitive protein kinase inhibitor (FW = 364.40 g/mol; CAS 1159824-67-5; Solubility: 68 mg/mL DMSO) selectively and reversibly targets p97 (IC_{50} = 2.4 μM), the AAA ATPase that is also known as Transitional Endoplasmic Reticulum ATPase (*or* TER ATPase) also known as Valosin-Containing Protein (*or* VCP). ((*Note*: Unlike DBeQ, epoxomicin is irreversible proteasome inhibitor.)) P97 is a ubiquitin segregase that remodels multimeric protein complexes by extracting polyubiquitinated proteins for recycling or degradation by the proteasome, playing vital roles in mitochondrial quality control, autophagy, vesicle transport/fusion, 26S proteasome function, and peroxisome assembly. DBeQ blocks multiple processes shown by RNAi to depend on p97, including degradation of ubiquitin fusion degradation and endoplasmic reticulum-associated degradation pathway reporters, as well as autophagosome maturation. DBeQ also potently inhibits cancer cell growth and is more rapid than a proteasome inhibitor at mobilizing the executioner caspase-3 and caspase-7 (1). A distinguishing feature between DBeQ and conventional proteasome inhibitors is that DBeQ is a 'separation of function' inhibitor that affects both the UPS and autophagic protein degradation pathways and rapidly activates cell death (2). **1**. Chou, Brown, Minond, *et al.* (2011) *Proc. Natl. Acad. Sci. USA* **108**, 4834. **2**. Chou & Deshaies (2011) *Autophagy* **7**, 1091. **3**. Hauler, Mallery, McEwan, Bidgood & James (2012) *Proc. Natl. Acad. Sci. USA* **109**, 19733.

DBL-583

This fat-soluble, piperazine-class psychostimulant pro-drug (FW = 620.82 g/mol), also known as GBR-decanoate, undergoes slow ester-bond hydrolysis, thereby releasing decanoate along with the active dopamine reuptake inhibitor GBR-12935 on a time-scale suitable for once-monthly dosing. GBR-decanoate elevates basal synaptic dopamine levels and blocks methamphetamine-evoked dopamine release in a persistent manner, without significantly perturbing dopamine receptor function. **1**. Baumann, Phillips, Ayestas, *et al.* (2002) *Ann. New York Acad. Sci.* **965**, 92. **2**. Berger, Janowsky, Vocci *et al.* (1985) *Eur. J. Pharmacol.* **107**, 289. **3**. Janowsky, Vocci, Berger, *et al.* (1987) *J. Neurochemi.* **49**, 617. **4**. Zhu, Green, Bardo & Dwoskin (2004) *Behav. Brain Res.* **148**, 107.

DCCD, *See* *N,N'-Dicyclohexylcarbodiimide*

DCCI, *See* *N,N'-Dicyclohexylcarbodiimide*

107498

This synthetic, single-dose, transmission-blocking and chemoprotective antimalarial (FW = 448.23 g/mol) targets the *Plasmodium falciparum* elongation factor-2 (PfeEF-2), the GTP hydrolysis-dependent mechanoenzyme required for ribosome translocation along mRNA during protein synthesis. DD107498 exhibits an EC_{50} of 1 nM against *P. falciparum* (Strain 3D7), with >20,000 times lower toxicity to human MRC5 and Hep-G2 cell lines *in vitro*. In malaria-infected SCID mice, this antimalarial's effects are observable within one replication cycle (48 hours), inhibiting development and resulting in trophozoites with an uncharacteristically condensed cytoplasm. A single dose reduces the parasite burden by 90%. DD107498 shows promising drug-like properties ($\log P$ = 3.2; adequate H_2O solubility = 0.22 mM; FW < 500; limited hydrogen bonding; long half-life in circulation (*e.g.*, 16-19 hours in mice), commending its development as an orally bioavailable antimalarial. **1**. Baragaña, Hallyburton, Lee, *et al.* (2015) *Nature* **522**, 315.

DDE, *See* *1,1-Dichloro-2,2-Bis(p-chlorophenyl)-ethylene*

DDR1-IN-1

This protein tyrosine kinase inhibitor (FW_{di-HCl} = 625.51 g/mol; CAS 1449685-96-4; Solubility: 100 mM in H_2O; 50 mM in DMSO), also named *N*-[3-[(2,3-dihydro-2-oxo-1*H*-indol-5-yl)oxy]-4-methylphenyl]-4-[(4-ethyl-1-piperazinyl)methyl]-3-(trifluoromethyl)benzamide, selectively targets the collagen-binding Discoidin Domain Receptor-1 (*or* DDR1) tyrosine kinase (IC_{50} = 105 nM), with weaker action against DDR2 (IC_{50} = 413 nM), inhibiting integrin-induced DDR1 autophosphorylation in an osteosarcoma cell line. **Pharmacology:** The DDR1 receptor tyrosine kinase is activated by matrix collagens and has been implicated in numerous cellular functions, including adhesion, proliferation, differentiation, migration, and invasion. The G707A mutation within the hinge region of DDR1 confers >20x resistance to DDR1-IN-1 and can be used to establish whether cellular process is DDR1-dependent. Ponatinib is more potent, inhibiting DDR1 and DDR2 with an IC_{50} of 9nM (2). Whereas imatinib and ponatinib bind potently to both the DDR and ABL kinases, the hydrophobic interactions of the ABL P-loop appear poorly satisfied by DDR1-IN-1 suggesting a structural basis for its DDR1 selectivity (2). **1**. Kim, Tan, Weisberg, *et al.* (2013) *ACS Chem Biol.* **8**, 2145 (*Erratum: ACS Chem. Biol.* (2014) **9**, 840.). **2**. Canning, Tan, Chu, *et al.* (2014) *J. Mol. Biol.* **426**, 2457.

DDT, *See* *1,1-Trichloro-2,2-bis(p-chlorophenyl)-ethane*

DDVP, *See* *O-2,2-Dichlorovinyl-O,O-dimethyl Phosphate*

(1*S*)-1-(9-Deazaadenin-9-yl)-1,4-imino-1,4,5,6,7-pentadeoxy-D-*ribo*-heptitol, *See* *5'-Deoxy-5'-ethyl-immucillin-A*

(1*S*)-1-(9-Deazaadenin-9-yl)-1,4-imino-1,4,5-trideoxy-D-ribitol, *See* *5'-Deoxy-immucillin-A*

(3*R*,4*S*)-1-[(9-Deazaadenin-9-yl)methyl]-3-hydroxy-4-(propylthiomethyl)pyrrolidine

This putative transition-state analogue ($FW_{free-base}$ = 309.44 g/mol) inhibits 5'-methylthioadenosine/*S*-adenosyl-homocysteine nucleosidase, K_i' = 0.58 nM (1) and *S*-methyl-5'-thioadenosine phosphorylase, K_i' = 0.12 nM (2). **1**. Singh, Evans, Lenz, *et al.* (2005) *J. Biol. Chem.* **280**, 18265. **2**. Evans, Furneaux, Lenz, *et al.* (2005) *J. Med. Chem.* **48**, 4679.

3-Deazaadenosine

This adenosine analogue (FW = 266.26 g/mol; CAS 6736-58-9; Symbol: DZA) inhibits *S*-adenosyl-homocysteinase as well as leukocyte adhesion to tumor necrosis factor-stimulated human endothelial cells. DZA may activate intrinsic apoptosis by stimulating BAX activation, thereby releasing cytochrome *c*. Because the use of adenosine receptor agonists is plagued by dose-limiting cardiovascular side effects, there is great interest in developing adenosine kinase inhibitors (AKIs) as a way for to raise steady-state adenosine concentrations. **Target(s):** adenosine deaminase (1); adenosine kinase, also weak alternative substrate (2,3); adenosylhomocysteinase, *or* *S*-adenosyl-homocysteinase (4-10); *S*-adenosylmethionine decarboxylase (11); glycine *N*-methyltransferase (12); histone-arginine *N*-methyltransferase (13); phosphatidyl-ethanolamine methyltransferase (14); purine nucleosidase, *or* purine-specific nucleoside *N*-ribohydrolase (15,16). **1**. Lupidi, Riva, Cristalli & Grifantini (1982) *Ital. J. Biochem.* **31**, 396. **2**. Long & Parker (2006) *Biochem. Pharmacol.* **71**, 1671. **3**. Long, Allan, Luo, *et al.* (2007) *J. Antimicrob. Chemother.* **59**, 118. **4**. Chiang (1987) *Meth. Enzymol.* **143**, 377. **5**. Ishikura, Nakamura, Sugawara, *et al.* (1983) *Nucl. Acids Symp. Ser.* (12), 119. **6**. Bader, Brown, Chiang & Cantoni (1978) *Virology* **89**, 494. **7**. Kim, Zhang, Chiang &

Cantoni (1983) *Arch. Biochem. Biophys.* **226**, 65. **8**. Impagnatiello, Franceschini, Oratore & Bozzi (1996) *Biochimie* **78**, 267. **9**. Bozzi, Parisi & Martini (1993) *J. Enzyme Inhib.* **7**, 159. **10**. Shea, Ashline, Ortiz, Milhalik & Rogers (2004) *NeuroMol. Med.* **5**, 171. **11**. Gordon, Brown & Chiang (1983) *Biochem. Biophys. Res. Commun.* **114**, 505. **12**. Kloor, Karnahl & Kömpf (2004) *Biochem. Cell Biol.* **82**, 369. **13**. Gupta, Jensen, Kim & Paik (1982) *J. Biol. Chem.* **257**, 9677. **14**. Esko & Raetz (1983) *The Enzymes*, 3rd ed., **16**, 207. **15**. Parkin (1996) *J. Biol. Chem.* **271**, 21713. **16**. Versées, Decanniere, Pellé, *et al.* (2001) *J. Mol. Biol.* **307**, 1363.

7-Deazaadenosine, *See* Tubercidin

S-(3-Deazaadenosyl)-L-homocysteine

This *S*-adenosylhomocysteine analogue (FW = 383.43 g/mol) inhibits numerous methyltransferases most likely by mimicking the strong product inhibition of methylases by *S*-adenosyl-homocysteine. **Target(s):** *S*-adenosylmethionine decarboxylase (1); 1-aminocyclopropane-1-carboxylate synthase (2,3); guanidinoacetate *N*-methyltransferase, K_i = 39 μM (4); mRNA (guanine-N^7-)methyltransferase (5,6); mRNA (nucleoside-2'-*O*-)methyltransferase (5); RNA methyltransferase (7); thiol *S*-methyltransferase (8). **1**. Gordon, Brown & Chiang (1983) *Biochem. Biophys. Res. Commun.* **114**, 505. **2**. Miura & Chiang (1985) *Anal. Biochem.* **147**, 217. **3**. Jakubowicz (2002) *Acta Biochim. Pol.* **49**, 757. **4**. Im, Chiang & Cantoni (1979) *J. Biol. Chem.* **254**, 11047. **5**. Pugh & Borchardt (1982) *Biochemistry* **21**, 1535. **6**. Pugh, Borchardt & Stone (1977) *Biochemistry* **16**, 3928. **7**. Backlund, Carotti & Cantoni (1986) *Eur. J. Biochem.* **160**, 245. **8**. Borchardt & Cheng (1978) *Biochim. Biophys. Acta* **522**, 340.

9-Deaza-9-[4(*R*)-hydroxy-2(*S*)-(hydroxymethyl)pyrrolidin-1-ylmethyl]adenine

This nucleoside analogue (FW = 263.30 g/mol), also known as 4'-deaza-1'-aza-2'-deoxy-1'-(9-methylene) immucillin-A and DADMe-Immucillin-A, inhibits purine nucleoside phosphorylase, K_i = 30 pM. When the ring nitrogen is protonated, this analogue mimics the charge on the oxa-carbenium ion intermediate that is believed to form transiently in many glycosidase reactions. Transition-state mimmicry thus results in tight binding. **1**. Rinaldo-Matthis, Wing, Ghanem, *et al.* (2007) *Biochemistry* **46**, 659.

3-Deazaneplanocin A

This carbocyclic adenosine analogue (FW = 298.73 g/mol; CAS 120964-45-6; *Symbol* = DZNep; Solubility: 10 mM in H₂O), also named (1*S*,2*R*,5*R*)-5-(4-amino-1*H*-imidazo[4,5-*c*]pyridin-1-yl)-3-(hydroxymethyl)-3-cyclopentene-1,2-diol, inhibits histone methyltransferase, decreasing global DNA methylation. **Mechanism of Action:** ZNep inhibits *S*-adenosylhomocysteine (AdoHcy) hydrolase, the enzyme responsible for reversible hydrolysis of AdoHcy to adenosine and homocysteine (1,2),

thereby indirectly inhibiting of various *S*-adenosyl-methionine-dependent methylation reactions (3). DZNep can thus be considered as an inhibitor of SAM-dependent methyltransferase. DZNep has recently been reported to decrease protein levels of PRC2 (4). As DZNep can induce efficient apoptotic cell death in cancer cells but not in normal cells (4), it has been widely examined as a possible epigenetic therapeutic agent for the treatment of various cancers, including lung cancer (5), gastric cancer (6), myeloma (7), acute myeloid leukemia (8), and lymphoma (9). **EZH2 Degradation:** An inhibitor of *S*-adenosyl-methionine-dependent methyltransferase, DZNep targets the degradation of EZH2, a core component of Polycomb Repressive Complex-2 (PRC2) that plays a role in transcriptional repression through H3K27 trimethylation (10). DZNep preferentially induces apoptosis in various hematological malignancies, suggesting that EZH2 may be a new target for epigenetic treatment. As PRC2 participates in epigenetic silencing of a subset of GATA-1 target genes during erythroid differentiation, inhibition of EZH2 may influence erythropoiesis. DZNep treatment significantly induces erythroid differentiation of K562 cells, as assessed by benzidine staining and quantitative RT-PCR analysis for representative erythroid-related genes including globins (10). **1**. Glazer, Hartman, Knode, *et al.* (1986) *Biochem. Biophys. Res. Commun.* **135**, 688. **2**. Glazer, Knode, Tseng, Haines & Marquez, (1986) *Biochem. Pharmacol.* **35**, 4523. **3**. Chiang (1998) *Pharmacol. Ther.* **77**, 115. **4**. Tan, Yang, Zhuang, *et al.* (2007) *Genes Dev.* **21**, 1050. **5**. Kikuchi, Takashina, Kinoshita, *et al.* (2012) *Lung Cancer* **78**, 138. **6**. Cheng, Itahana, Lei, *et al.* (2012) *Clin. Cancer Res.* **18**, 4201. **7**. Xie, Bi, Cheong, *et al.* (2011) *PLoS One* **6**, e21583. **28**. Fiskus, Wang, Sreekumar, *et al.* (2009) *Blood* **114**, 2733. **29**. Fiskus, Rao, Balusu, *et al.* (2012) *Clin. Cancer Res.* **18**, 6227. **10**. Fujiwara, Saitoh & Inoue, *et al.* (2014) *J. Biol. Chem.* **289**, 2131.

5-Deazatetrahydrofolate

This tetrahydrofolate analogue (FW = 444.45 g/mol) inhibits deoxycytidylate 5-hydroxymethyltransferase (1,2) and phosphoribosylglycinamide formyltransferase, K_i = 5.1 nM (3,4). The inability of 5-deazatetrahydrofolate to stimulate enzyme-catalyzed tritium exchange from [5-(^{3}H)]nucleotides into solvent suggests that N^5-atom of tetrahydrofolate is the base that deprotonates the nucleotide. **1**. Butler, Graves & Hardy (1994) *Biochemistry* **33**, 10521. **2**. Hardy, Graves & Nalivaika (1995) *Biochemistry* **34**, 8422. **3**. Baldwin, Tse, Gossett, *et al.* (1991) *Biochemistry* **30**, 1997. **4**. Sanghani & Moran (1997) *Biochemistry* **36**, 10506.

3-Deazathiamin Pyrophosphate

This noncleavable thiamin pyrophosphate analogue (FW_free-acid = 423.32 g/mol) inhibits transketolase, K_i = 12 nM (1,2). Removal of the pyrophosphoryl group greatly reduces its inhibitory effectiveness (K_i = 18 μM). An efficient ten-step synthesis of 3-deazathiamin from commercially available α-acetyl-γ-butyrolactone (3). **1**. Thomas, De Meese, Le Huerou, *et al.* (2008) *Bioorg. Med. Chem. Lett.* **18**, 509. **2**. Mann, Perez-Melero, Hawksley & Leeper (2004) *Org. Biomol. Chem.* **2**, 1732. **3**. Hawksley, Griffin & Leeper (2001) *J. Chem. Soc., Perkin Trans.* **1**, 144.

9-Deaza-9-(3-thienylmethyl)guanine

This nucleoside analogue (FW = 246.29 g/mol), also known as PD 141955, inhibits purine nucleoside phosphorylase (IC$_{50}$ = 25 nM). The 2-thienyl analogue is also a potent inhibitor (IC$_{50}$ = 21 nM). The 7-deaza-9-(2-thienylmethyl)guanine isomer is significantly weaker (IC$_{50}$ = 674 nM). (**See also** *Immucillin-H and Immucillin-G*) **1**. Gilbertsen, Josyula, Sircar, *et al.* (1992) *Biochem. Pharmacol.* **44**, 996. **2**. Castilho, Postigo, de Paula, *et al.* (2006) *Bioorg. Med. Chem.* **14**, 516. **3**. Bzowska, Kulikowska & Shugar (2000) *Pharmacol. Ther.* **88**, 349.

3-Deazauracil

This pyrimidine analogue (FW = 111.10 g/mol), also known as 2,4-dihydroxypyridine and 2,4-pyridinediol, inhibits dihydrouracil dehydrogenase. *See also 3-Deazauridine 5′ Triphosphate* **Target(s):** dihydrouracil dehydrogenase (1); 2,5-dihydroxypyridine 5,6-dioxygenase (2); orotate phosphoribosyltransferase (3). **1**. Naguib, el Kouni & Cha (1989) *Biochem. Pharmacol.* **38**, 1471. **2**. Cain, Houghton & Wright (1974) *Biochem. J.* **140**, 293. **3**. Javaid, el Kouni & Iltzsch (1999) *Biochem. Pharmacol.* **58**, 1457.

3-Deazauridine 5'-Triphosphate

This nucleotide (FW$_{free-acid}$ = 470.13 g/mol), often abbreviated 3-deazaUTP, inhibits bovine calf liver CTP synthetase. The prodrug is 3-deazauridine (*i.e.*, 2,4-dihydroxypyridine riboside). **1**. Weinfeld, Savage & McPartland (1978) *Meth. Enzymol.* **51**, 84. **2**. McPartland, Wang, Bloch & Weinfeld (1974) *Cancer Res.* **34**, 3107.

Debio-1347, See CH5183284

S-(2,3-Decadienoyl)-N-acetylcysteamine

This acylated acyl-carrier protein analogue (FW = 269.41 g/mol), also known as *N*-acetylcysteamine 2,3-decadienoate thioester, inhibits 3-hydroxydecanoyl-[acyl-carrier protein] dehydratase (IC$_{50}$ = 53 nM) of the fatty acid synthase complex. It forms nonenzymatically from *S*-(3-decynoyl)-*N*-acetylcysteamine. **1**. Bloch (1971) *The Enzymes*, 3rd ed., **5**, 441. **2**. Morisaki & Bloch (1972) *Biochemistry* **11**, 309.

Decamethonium Bromide

This skeletal muscle relaxant (FW = 418.30 g/mol), also known as *N,N,N,N′,N′,N′*-hexamethyl-1,10 decane-diaminium dibromide and syncurine, is a nicotinic acetylcholine receptor partial agonist and neuromuscular blocking agent. Decamethonium halide salts are readily soluble in water; however, unlike the slowly hydrolyzable analogue succinylcholine dibromide, decamethonium halide solutions are stable. **Target(s):** acetylcholinesterase (1-10); acid phosphatase (11); aryl-acylamidase (10); butyrylcholinesterase (12). **1**. Desire, Blanchet, Definod & Arnaud (1975) *Biochimie* **57**, 1359. **2**. Agbaji, Gerassimidis & Hider (1984) *Comp. Biochem. Physiol. C* **78**, 211. **3**. Morel & Massoulie (1997) *Biochem. J.* **328**, 121. **4**. Saxena, Redman, Jiang, *et al.* (1997) *Biochim. Biophys. Acta* **1339**, 253. **6**. Eichler, Anselment, Sussman, Massoulie & Silman (1994) *Mol. Pharmacol.* **45**, 335. **7**. Munoz-Delgado & Vidal (1986) *Biochem. Int.* **13**, 625. **8**. Danilov (1967) *Farmakol. Toksikol.* **30**, 664. **9**. Bergman, Wilson & Nachmansohn (1950) *Biochim. Biophys. Acta* **6**, 217. **10**. Costagli & Galli (1998) *Biochem. Pharmacol.* **55**, 1733. **11**. Lisa, Garrido & Domenech (1984) *Mol. Cell. Biochem.* **63**, 113. **12**. Seto & Shinohara (1988) *Arch. Toxicol.* **62**, 37.

Decanoyl-CoA

This short-chain fatty acyl Coenzyme A thioester (FW = 921.79g/mol), a common intermediate in the β-oxidation of fatty acids, inhibits prolyl endopeptidase acetyl-CoA *C*-acyltransferase (1); aralkyl-amine *N*-acetyltransferase (2); butyrate:acetoacetate CoA-transferase (3); glucose-6-phosphatase (4); glycylpeptide *N* tetradecanoyltransferase, *or* protein *N*-myristoyltransferase (5); 3-hydroxy-3-methylglutaryl-CoA synthase (6); (*S*)-methylmalonyl-CoA hydrolase (7); prolyl endopeptidase, *K*$_i$ = 9 μM (8); [pyruvate dehydrogenase (acetyl-transferring)] kinase (9); and stearoyl-CoA 9-desaturase (10-12). **1**. Olowe & Schulz (1980) *Eur. J. Biochem.* **109**, 425. **2**. Ferry, Loynel, Kucharczyk, *et al.* (2000) *J. Biol. Chem.* **275**, 8794. **3**. Sramek & Frerman (1975) *Arch. Biochem. Biophys.* **171**, 14. **4**. Methieux & Zitoun (1996) *Eur. J. Biochem.* **235**, 799. **5**. Glover, Goddard & Felsted (1988) *Biochem. J.* **250**, 485. **6**. Middleton (1972) *Biochem. J.* **126**, 35. **7**. Kovachy, Copley & Allen (1983) *J. Biol. Chem.* **258**, 11415. **8**. Yamakawa, Shimeno, Soeda & Nagamatsu (1990) *Biochim. Biophys. Acta* **1037**, 302. **9**. Rahmatullah & Roche (1985) *J. Biol. Chem.* **260**, 10146. **10**. Joshi, Prasad & Sreekrishna (1981) *Meth. Enzymol.* **71**, 252. **11**. Sreekrishna & Joshi (1980) *Biochim. Biophys. Acta* **619**, 267. **12**. Prasad & Joshi (1979) *J. Biol. Chem.* **254**, 6362.

Decatransin

This novel hydrophobic decadepsipeptide (MW = 1184.62 g/mol; with non-proteinogenic pipecolate and homoleucine residues at positions 2, 6, 9, and 3, 4, 7, respectively, indicating its likely synthesis by a non-ribosomal peptide synthetase) from the saprophyte fungus *Chaetosphaeria tulasneorum* inhibits co- and post-translational translocation across the Sec61/SecYEG translocon. Its name alludes to the cotransins, a group of heptadepsipeptide compounds without detectable molecular similarity that were previously found to be mammalian translocon inhibitors. **1**. Junne, Wong, Studer, *et al.* (2015) *J. Cell Sci.* **128**, 1217.

Decernotinib

This potent and selective JAK inhibitor (FW = 392.38 g/mol; CAS 944842-54-0; Solubility: 200 mM in DMSO, <1 mM H$_2$O), also code-named VX-509 and systematically named 2-methyl-2-[[2-(1H-pyrrolo[2,3-b]pyridin-3-yl)-4-pyrimidinyl]amino]-N-(2,2,2-trifluoroethyl)-(2R)-butanamide, targets Janus Kinase isoform-3, or JAK3 (K_i = 2.5 nM), showing >4x selectivity over JAK1, JAK2, and the nonreceptor tyrosine-protein kinase TYK2, respectively (1,2). JAK3 is of particular interest due to its importance in immune function and its expression, which is largely confined to lymphocytes, thus limiting the potential impact of JAK3 inhibition on nonimmune physiology. VX-509 treatment resulted in dose-dependent reduction in ankle swelling and paw weight and improved paw histopathology scores in the rat collagen-induced arthritis model (2). In a mouse model of oxazolone-induced delayed-type hypersensitivity, VX-509 reduced the T-cell-mediated inflammatory response in skin. These findings demonstrate that VX-509 is a selective and potent inhibitor of JAK3 in vitro and modulates proinflammatory response in models of immune-mediated diseases, such as collagen-induced arthritis and delayed-type hypersensitivity (2). **1.** Farmer, Ledeboer, Hoock, *et al.* (2015) *J. Med. Chem.* **58**, 7195. **2.** Mahajan, Hogan, Shlyakhter, *et al.* (2015) *J. Pharmacol. Exp. Ther.* **353**, 405.

Decitabine

This DNMT-inhibiting cytosine nucleoside analogue and hypomethylating agent (FW = 228.21 g/mol; CAS 2353-33-5; Solubility = 45 mg/mL DMSO, 10 mg/mL H$_2$O; *Symbol* = 5aza-CdR), also known by its code name NSC127716 its systematically as 4-amino-1-((2R,4S,5R)-4-hydroxy-5-(hydroxymethyl)-tetrahydrofuran-2-yl)-1,3,5-triazin-2(1H)-one, targets KG1a (IC$_{50}$ = 4.3 nM) and HL-60 (IC$_{50}$ = 0.43 μM), two DNA methyltransferases (DNMTs). DNA methylation is an epigenetic mechanism for establishing long-term gene silencing during development and cell commitment. Aberrant DNA methylation is found at gene promoters in many cancers, often attended by silencing of tumor suppressor genes. Decitabine is metabolically phosphorylated, whereupon it is incorporated into DNA. Decitabine targets DNA-linked DNMTs and unbound DNMT1 for proteasomal degradation, leading to DNA demethylation. DNMT inhibitors show no interference with cell-cycle progression whereas cytarabine treatment results in S-phase arrest. At high doses, the DNA is not able to recover and cell death occurs. At lower doses, the formed adducts are degraded by the proteosome, after which the DNA is restored. DNA synthesis is then resumed in the absence of DNMT-1. As a consequence the aberrant DNA methylation pattern can no longer be reproduced toward the daughter strands. In this way, a low dose of decitabine is able to induce re-expression of previously silenced genes. Reactivation of cell cycle-regulating genes that were initially silenced due to hypermethylation may induce cell differentiation, reduce proliferation, and/or increase apoptosis of the daughter cells. Upon removal of decitabine, tumor suppressor genes (TSGs) are resilenced, albeit at varying rates. 5-aza-CdR exhibits high antineoplastic activity against anaplastic large cell lymphoma (ALCL), a rare CD30 positive non-Hodgkin lymphoma of T-cell origin (2). **1.** Shaker, Bernstein, Momparler & Momparler (2003) *Leuk. Res.* **27**, 437. **2.** Hassler, Klisaroska, Kollmann, *et al.* (2012) *Biochimie* **94**, 2297.

Decoyinine

This *Streptomyces hygroscopicus*-derived nucleoside antibiotic (FW = 279.26 g/mol; CAS 2004-04-8; Solubility: 20 mg/mL warm DMSO; 1 mg/mL water), also known as U-7984, angustmycin A, decoyinin, and 5,6-didehydropsico-furanosyladenine, inhibits GMP synthetase, thereby depleting GTP stores in treated organisms and inhibiting GTP-dependent processes (1-7). The rate of wall synthesis in *Bacillus subtilis* is decreased by 50% within 5 min after decoyinine addition (1) this decrease was prevented by the presence of guanosine. By comparing the effects of other inhibitors of cell wall synthesis, it became clear that decoyinine also inhibits the final portion of the cell wall biosynthetic pathway (*i.e.*, after the steps inhibited by bacitracin or vancomycin) (6) such effects are reversed upon addition of sufficient guanosine. Decoyinine is highly effective against mycobacteria. **Target(s):** GMP synthetase (1-5); bacterial cell wall formation (6); deoxycytidine kinase, K_i = 9.3 mM (7); GMP synthetase (2-7); 5'-methylthioadenosine nucleosidase (8,9); NAD$^+$ synthetase (10,11); ribose-phosphate diphosphokinase (4); depletion of guanine nucleotides, resulting in sporulation of *Bacillus subtilis* (12); ppGpp formation in *Bacillus subtilis* (13); melanoma cell invasion *in vitro* and tumorigenicity in immunocompromised mice (14). **1.** Nakamura & Lou (1995) *J. Biol. Chem.* **270**, 7347. **2.** Slechta (1960) *Biochem. Biophys. Res. Commun.* **3**, 596. **3.** Bloch & Nichol (1964) *Biochem. Biophys. Res. Commun.* **16**, 400. **4.** Spector & Beecham (1975) *J. Biol. Chem.* **250**, 3101. **5.** Uratani, Lopez & Freese (1983) *J. Bacteriol.* **154**, 261. **6.** Krenitsky, Tuttle, Koszalka, *et al.* (1976) *J. Biol. Chem.* **251**, 4055. **7.** Lou, Nakamura, Tsing, *et al.* (1995) *Protein Expr. Purif.* **6**, 487. **8.** Guranowski, Chiang & Cantoni (1983) *Meth. Enzymol.* **94**, 365. **9.** Guranowski, Chiang & Cantoni (1981) *Eur. J. Biochem.* **114**, 293. **10.** Zalkin (1985) *Meth. Enzymol.* **113**, 297. **11.** Spencer & Preiss (1967) *J. Biol. Chem.* **242**, 385. **12.** Lopez, Marks & Freese (1979) *Bichim. Biophys. Acta* **587**, 238. **13.** Ikehara, Okamoto & Sugae (1982) *J. Biochem.* **91**, 1089. **14.** Bianchi-Smiraglia, Wawrzyniak, Bagati, *et al.* (2015) *Cell Death Differ.* **22**, 1858.

S-Decylglutathione

This glutathione conjugate (FW = 434.58 g/mol) is a potent inhibitor of both the peroxidase and transferase activities associated with glutathione S-transferase. It is also a strong inhibitor of glyoxalase I. **Target(s):** glutathione S-transferase (1,2); lactoylglutathione lyase, *or* glyoxalase I (3); xenobiotic-transporting ATPase, multidrug-resistance protein (4,5). **1.** Chen & Juchau (1998) *Biochem. J.* **336**, 223. **2.** Burgess, Yang, Chang, *et al.* (1987) *Biochem. Biophys. Res. Commun.* **142**, 441. **3.** Regoli, Saccucci & Principato (1996) *Comp. Biochem. Physiol. C, Pharmacol. Toxicol. Endocrin.* **113**, 313. **4.** Loe, Stewart, Massey, Deeley & Cole (1997) *Mol. Pharm.* **51**, 1034. **5.** Mao, Deeley & Cole (2000) *J. Biol. Chem.* **275**, 34166.

Defensins

These anti-infective oligopeptides (MW = 3934.57 g/mol; Human Neutrophil Peptide-1, *or* Defensin HNP-1 Sequence: ACYCRIPAC IAGERRYGTCIYQGRLWAFCC, with disulfides at Cys2-S–S-Cys30, Cys4-S–S-Cys19, and Cys9-S–S-Cys29), permeabilize membranes by first binding electrostatically and then forming multimeric pores. The defensins are cationic and typically contain 29-36 aminoacyl residues joined together by three disulfide bonds, and have a predominantly β-sheet structure (1,2). There are three types of defensins in primates. α-Defensins are found in neutrophils and Paneth cells of the small intestine while β-defensins protect the skin and the mucous membranes of the respiratory, genitourinary, and gastrointestinal tracts. θ-Defensins, expressed only in Old World monkeys, lesser apes, and orangutans, are lectins with broad-spectrum antiviral efficacy. Human neutrophil peptide-2 (Defensin HNP-2) but lacks the N-terminal alanyl residue. Except for an N-terminal aspartyl residue instead of alanyl, Human Defensin HNP-3 is identical to HNP-1, whereas HNP-4 has the 33-residue sequence VCSCRLVFCRRTELRVGNCLIGGVSFTYCC RV. Larger defensin oligopeptides are found in plants and insects. Human β-Defensin-1 has the sequence DHYNCVSSGGQCLYSACPIFTKIQGTC YRGKAKCCK, joined by disulfide bonds at Cys5-S–S-Cys34, Cys12-S–S-Cys27, and Cys17-S–S-Cys35. **1.** White, Wimley & Selsted (1995) *Curr. Opin. Struct. Biol.* **5**, 521. **2.** Ganz & Lehrer (1995) *Pharmacol. Ther.* **66**, 191.

Deferiprone

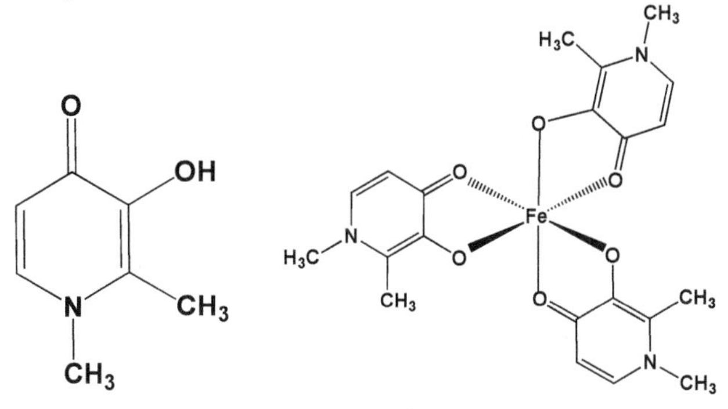

Deferiprone   (Deferiprone)₃Fe(III) Complex

This bacterial siderophore and orally active iron-binding drug (FW = 139.15 g/mol; CAS 30652-11-0; pK_a values, 3.3 and 9.7; Forms red-colored complexes with iron; *Abbreviation*: DMHP), also known by its trade name Ferriprox® and its IUPAC name 3-hydroxy-1,2-dimethylpyridin-4(1*H*)-one, isolated from *Streptomyces pilosus*, is used in the treatment of thalassemia major, an autosomal recessive blood disorder anemia that releases iron. Deferiprone reduces complicating effects of iron overload. Biochemical studies showed that deferiprone is more efficient than desferrioxamine at releasing iron from ferritin, but less so in the removal of iron from the other two polynuclear iron forms (*i.e.*, hemosiderin and FeCl₃ precipitate), when each was present at the same iron content (1). When delivered to rabbits at 10 and 50 mg/kg/day, deferiprone also significantly decreases levels of Aβ40 and Aβ42 as well as BACE1, the enzyme that initiates cleavage of amyloid-β protein precursor to yield Aβ (2). (**See also** *Deferoxamine*) **Target(s):** acid phosphatase (3); catechol *O*-methyltransferase (4); cyclooxygenase (5); deoxyhypusine monooxygenase (6,7); HIV-1 replication (8); lipoxygenase (5); 5-lipoxygenase (9,10); procollagen-proline 4-dioxygenase (7); purple acid phosphatase (3); ribonucleotide reductase (9); tryptophan 5-monooxygenase (4); tyrosine 3-monooxygenase (4). **1.** Kontoghiorghes, Chambers, Hoffbrand (1987) *Biochem. J.* **241**, 87. **2.** Prasanthi, Schrag, Dasari, *et al.* (2012) *J. Alzheimers Dis.* **30**, 167. **3.** Hayman & Cox (1994) *J. Biol. Chem.* **269**, 1294. **4.** Waldmeier, Buchle & Steulet (1993) *Biochem. Pharmacol.* **45**, 2417. **5.** Barradas, Jeremy, Kontoghiorghes, *et al.* (1989) *FEBS Lett.* **245**, 105. **6.** Andrus, Szabo, Grady, *et al.* (1998) *Biochem. Pharmacol.* **55**, 1807. **7.** Clement, Hanauske-Abel, Wolff, Kleinman & Park (2002) *Int. J. Cancer* **100**, 491. **8.** Georgiou, van der Bruggen, Oudshoorn, *et al.* (2000) *J. Infect. Dis.* **181**, 484. **9.** Kayyali, Porter, Liu, *et al.* (2001) *J. Biol. Chem.* **276**, 48814. **10.** Liu, Kayyali, Hider, Porter & Theobald (2002) *J. Med. Chem.* **45**, 631.

Deferoxamine, See *Desferrioxamine B*

Dehydroepiandrosterone

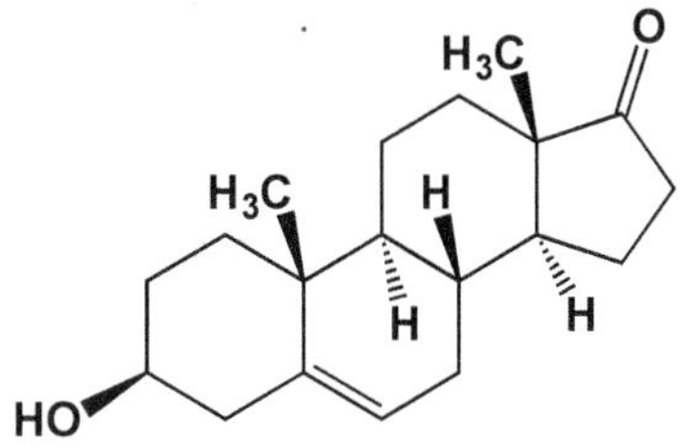

This steroid (FW = 288.43 g/mol; CAS 53-43-0), also known as prasterone, *trans*-dehydroandrosterone, dehydroisoandrosterone, 5-androsten-3β-hydroxy-17-one, and 3β-hydroxyandrost-5-en-17-one and abbreviated DHEA and DHAS, is produced by the adrenal gland and is, with its sulfate conjugate (prasterone 3-sulfate), the most abundant steroid hormone in the circulation. Dehydro-epiandrosterone is an intermediate in androgen and estrogen biosynthesis and is the immediate precursor to androstenedione. **Target(s):** androst-4-ene-3,17-dione monooxygenase (1); cortisol sulfotransferase (2-4); estrone sulfotransferase (5); F₁F₀ ATPase (6); glucose-6-phosphate dehydrogenase (7-12); glutamate dehydrogenase (11); 27-hydroxycholesterol 7α-monooxygenase, also a substrate (13,14); 17β-hydroxysteroid dehydrogenase (15); 21-hydroxysteroid dehydrogenase, NAD⁺-requiring (16); steroid 21-monooxygenase (17); steroid sulfotransferase, also alternative substrate for the enzyme from many

sources (1,18,19); sterol *O*-acyltransferase, *or* cholesterol *O*-acyltransferase, *or* ACAT (20); steryl sulfatase (21,22). **1.** Itagaki (1986) *J. Biochem.* **99**, 825. **2.** Singer, Gebhart & Hess (1978) *Can. J. Biochem.* **56**, 1028. **3.** Singer (1979) *Arch. Biochem. Biophys.* **196**, 340. **4.** Singer & Bruns (1980) *Can. J. Biochem.* **58**, 660. **5.** Rozhin, Huo, Zemlicka & Brooks (1977) *J. Biol. Chem.* **252**, 7214. **6.** Zheng & Ramirez (1999) *Eur. J. Pharmacol.* **368**, 95. **7.** Noltmann & Kuby (1963) *The Enzymes*, 2nd ed., 7, 223. **8.** Levy (1963) *J. Biol. Chem.* **238**, 775. **9.** Criss & McKerns (1969) *Biochim. Biophys. Acta* **184**, 486. **10.** McKerns & Kaleita (1960) *Biochem. Biophys. Res. Commun.* **2**, 344. **11.** Douville & Warren (1968) *Biochemistry* **7**, 4052. **12.** Raineri & Levy (1970) *Biochemistry* **9**, 2233. **13.** Pettersson, Holmberg, Axelson & Norlin (2008) *FEBS J.* **275**, 1778. **14.** Norlin & Wikvall (1998) *Biochim. Biophys. Acta* **1390**, 269. **15.** Bonney, Reed & James (1983) *J. Steroid Biochem.* **18**, 59. **16.** Monder & Furfine (1969) *Meth. Enzymol.* **15**, 667. **17.** Sharma (1964) *Biochemistry* **3**, 1093. **18.** Liu, Apak, Lehmler, Robertson & Duffel (2006) *Chem. Res. Toxicol.* **19**, 1420. **19.** Chang, Shi, Rehse & Lin (2004) *J. Biol. Chem.* **279**, 2689. **20.** Simpson & Burkhart (1980) *Arch. Biochem. Biophys.* **200**, 79. **21.** Notation & Ungar (1969) *Biochemistry* **8**, 501. **22.** Dibbelt & Kuss (1983) *Hoppe-Seyler's Z. Physiol. Chem.* **364**, 187.

Dehydroascorbate

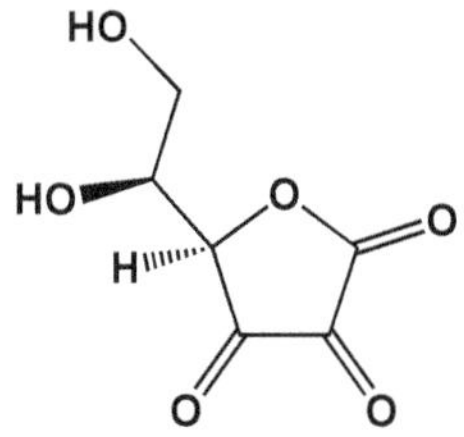

This oxidized vitamin C metabolite and ROS-producing substance (FW = 174.11 g/mol; CAS 490-83-5; Symbol: DHA; Soluble to 10 mM in water, with gentle warming, and to 100 mM in DMSO), also named L-*threo*-2,3-hexodiulosonic acid γ-lactone, is a redox-active compound having both beneficial and potentially harmful metabolic consequences. DHA undergoes nonenzymatic hydration to the following species:

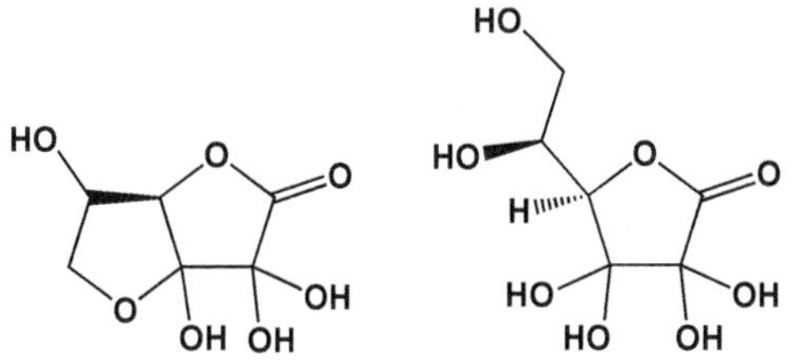

These hydrated forms may alter the distribution and kinetics of DHA. Although ascorbate itself cannot penetrate the blood-brain barrier, DHA enters cells possessing the GLUT1 glucose transporter, which is widely distributed in fetal tissues, with highest levels in erythrocytes and endothelial cells of adults. DHA thereby accumulates in the cytoplasm, endoplasmic reticulum, and mitochondria. (Note: Dehydroascorbic acid is often added to cell and tissue culture medium to promote the uptake of vitamin C in cell types that do not contain ascorbic acid transporters.) DHA also induces oxidative stress, as it is reduced to vitamin C, an action that can deplete intracellular stores of reduced glutathione and generate reactive oxygen species (ROS). The latter can inactivate glyceraldehyde 3-phosphate dehydrogenase (GAPDH) and induce death of glycolytically active KRAS or BRAF mutant cells, but not KRAS and BRAF wild-type cells. High-dose vitamin C impairs tumor growth in Apc/Kras(G12D) mutant mice. Vitamin C antagonizes the cytotoxic effects of antineoplastic drugs in human hematopoietic cancers by preserving mitochondrial membrane potential (2). Indeed, vitamin C antagonizes fas ligand-mediated mitochondrial membrane depolarization (3). Previous studies have also shown that cisplatin, doxorubicin, etoposide and several other agents result in mitochondrial membrane depolarization (4, 5). **See** *Ascorbate* **1.** Yun, Mullarky, Lu, *et al.* (2015) *Science* **350**, 1391. **2.** Heaney, Gardner, Karasavvas, *et al.* (2008) *Cancer Res.* 68, 8031. **3.** Perez-Cruz, Carcamo & Golde (2003) *Blood* **102**, 336. **4.** Decaudin, Geley, Hirsch, *et al.* (1997) *Cancer Res.* 57, 62. **5.** Costantini, Jacotot, Decaudin & Kroemer (2000) *J. Natl. Cancer Inst.* **92**, 1042.

3,4-Dehydro-L-proline

This proline derivative (FW = 113.12 g/mol; CAS 4043-88-3), also known as (*S*)-3-pyrroline-2-carboxylic acid and 3,4-didehydro-L-proline, inhibits prolyl-tRNA synthetase. It also blocks cell wall assembly and cell division in tobacco protoplasts. *See also 3,4-Dehydro-DL-proline* **Target(s):** γ-glutamyl kinase, *or* glutamate 5-kinase (1); peptidyl-prolyl hydroxylase (2); procollagen-proline monooxygenase (3,4); proline oxidase (5); prolyl-tRNA synthetase (6-8). **1.** Krishna & Leisinger (1979) *Biochem. J.* **181**, 215. **2.** Cooper, Heuser & Varner (1994) *Plant Physiol.* **104**, 747. **3.** Kerwar & Felix (1976) *J. Biol. Chem.* **251**, 503. **4.** Salvador, Tsai, Marcel, Felix & Kerwar (1976) *Arch. Biochem. Biophys.* **174**, 381. **5.** Dashman, Lewinson, Felix & Schwartz (1979) *Res. Commun. Chem. Pathol. Pharmacol.* **24**, 143. **6.** Norton (1964) *Arch. Biochem. Biophys.* **106**, 147. **7.** Kerwar (1979) *Arch. Biol. Med. Exp. (Santiago)* **12**, 359. **8.** Norris & Fowden (1972) *Phytochemistry* **11**, 2921.

Dehydrosinefungin

This *S*-adenosylhomocysteine analogue and anti-trypanosomal antibiotic (FW = 379.38 g/mol), also known as A9415C, inhibits numerous SAM-dependent methyltransferases. **Target(s):** catechol *O*-methyl-transferase (1); cyclopropane-fatty-acyl-phospholipid synthase (2,3); dimethyl macrocin *O*-methyltransferase (4); histamine methyltransferase (1); histone arginine *N*-methyltransferase (5); histone-lysine *N*-methyltransferase (6,7); macrocin *O*-methyltransferase (8); mRNA-(guanine-N^7-)-methyltransferase (7,9,10); mRNA (nucleoside-2'-*O*-) methyltransferase (7,9,10); norepinephrine *N*-methyltransferase (1,7); phenylethanolamine *N*-methyltransferase (1); protein-arginine methyltransferase (6,7); protein-glutamate *O*-methyltransferase (7,11,13,14; protein-*S*-isoprenylcysteine *O*-methyltransferse (7); spermidine synthase (12); spermine synthase (12). **1.** Fuller & Nagarajan (1978) *Biochem. Pharmacol.* **27**, 1981. **2.** Wang, Grogan & Cronan (1992) *Biochemistry* **31**, 11020. **3.** Smith & Norton (1980) *Biochem. Biophys. Res. Commun.* **94**, 1458. **4.** Kreuzman, Turner & Yeh (1988) *J. Biol. Chem.* **263**, 15626. **5.** Gupta, Jensen, Kim & Paik (1982) *J. Biol. Chem.* **257**, 9677. **6.** Vedel, Lawrence, Robert-Gero & Lederer (1978) *Biochem. Biophys. Res. Commun.* **85**, 371. **7.** Robert-Gero, Pierre, Vedel, *et al.* (1980) in *Enzyme Inhibitors* (Brodbeck, ed.), p. 61, Verlag Chemie, Weinheim. **8.** Bauer, Kreuzman, Dotzlaf & Yeh (1988) *J. Biol. Chem.* **263**, 15619. **9.** Pugh, Borchardt & Stone (1978) *J. Biol. Chem.* **253**, 4075. **10.** Pugh & Borchardt (1982) *Biochemistry* **21**, 1535. **11.** Pierre & Robert-Gero (1980) *FEBS Lett.* **113**, 115. **12.** Hibasami, Borchardt, Chen, Coward & Pegg (1980) *Biochem. J.* **187**, 419. **13.** Kim (1984) *Meth. Enzymol.* **106**, 295. **14.** Rollins & Dahlquist (1980) *Biochemistry* **19**, 4627.

Dehydroxymethylepoxyquinomicin

This first-in-class, selective NF-κB inhibitor (FW = 261.06 g/mol, CAS 287194-40-5), also known as DHMEQ (–)-DHMEQ, (–)-DHM₂EQ, and 2-hydroxy-*N*-((1*S*,2*S*,6*S*)-2-hydroxy-5-oxo-7-oxabicyclo[4.1.0]hept-3-en-3-yl)benzamide, inhibits p65 and p50 NF-κB subunit translocation into the nucleus, reduces DNA-binding activity of the RelA/p65, suppresses NF-κB-dependent expression of the Inhibitor of Apoptosis (IAP)-family proteins (*e.g.*, cIAP-1, cIAP-2, and XIAP), and inhibits *de novo* synthesis of inhibitor of NF-κB α (1). The transcriptional factor NF-κB and the Rel/NF-κB signaling pathway regulate cellular proliferation, apoptotic response, and oncogenesis, with elevation of basal NF-κB levels playing a critical role in carcinogenesis and making it an attractive target for molecular therapy. Over the 0.1–5 μg/mL range, DHMEQ induces a caspase-mediated apoptotic response that is blocked by the c-Jun N-terminal kinase inhibitor SP600125, but not by inhibitors of mitogen-activated protein/extracellular signal-regulated kinase kinase or p38 (1). While normal human thyrocytes resist DHMEQ-induced apoptosis, higher doses of DHMEQ result in necrotic-like killing of both normal and malignant thyrocytes. In nude mice, DHMEQ inhibits tumor growth without observable side effects, showing greater numbers of apoptotic cells in histologic sections (1). DHMEQ also synergizes with chemotherapy agents in classical Hodgkin lymphoma (2). **1.** Starenki, Namba, Saenko, *et al.* (2004) *Clin. Cancer Res.* **10**, 6821. **2.** Celegato, Borghese, Umezawa, *et al.* (2014) *Cancer Lett.* **349**, 26.

DEL 22379

This ERK inhibitor (FW = 444.53 g/mol; CAS 181223-80-3; Solubility: 50 mM in DMSO; Store at –20°C), also named *N*-[2,3-dihydro-3-[(5-methoxy-1*H*-indol-3-yl)methylene]-2-oxo-1*H*-indol-5-yl]-1-piperidine propanamide, targets the dimerization of Extracellular signal–Regulated Kinase, *or* Mitogen-Activated Protein Kinase (IC₅₀ ~ 0.5 μM, without effecting phosphorylation or its intrinsic protein kinase activity. Del-22379 suppresses the growth of tumor cells expressing RAS-ERK oncogenes *in vitro*, inducing apoptosis and preventing tumor progression *in vivo*. In nearly 50% of human malignancies, unregulated RAS-ERK signaling occurs, making ERK a target for chemotherapeutic intervention. Upon activation, ERK must dimerize for ERK's extranuclear (but not for nuclear) signaling cascade to operate. By blocking sub-localization-specific signalling, rather than all ERK-mediated signalling, Del-22379 impedes oncogenic RAS-ERK signaling. Moreover, by targeting regulatory protein-protein interactions, rather than catalytic activities, Del22379 is a highly effective antitumor agent. **1.** Herrero, Pinto, Colón-Bolea *et al.* (2015) *Cancer Cell* **28**, 170. **2.** Ryan, Der, Wang-Gillam & Cox (2015) *Trends Cancer* **1**, 183.

Delafloxacin

This fluoroquinolone-class antibiotic (FW = 440.76 g/mol; CAS 189279-58-1), also named RX-3341 and 1-(6-amino-3,5-difluoro-2-pyridyl)-8-chloro-6-fluoro-7-(3-hydroxyazetidin-1-yl)-4-oxo-quinoline-3-carboxylic acid, is a broad-spectrum systemic antibacterial agent that targets Type II DNA topoisomerases (gyrases), which are required for bacterial replication and transcription. (For the prototypical member of this antibiotic class, *See Ciprofloxacin*) Delafloxacin has a lower MIC₉₀ than other quinolones against Gram-positive bacteria, such as methicillin-resistant *Staphylococcus aureus* (MRSA). Unlike other fluoroquinolones, which are inherently zwitterionic, delafloxacin is anionic, resulting in its 10x greater accumulation in both bacteria and cells at acidic pH. This property confers an advantage in delafloxacin's eradication of MRSA in acidic environments. **1.** Lemaire, Tulkens & Van Bambeke (2011) *Antimicrob. Agents Chemother.* **55**, 649.

Delamanid

This multi-drug-resistant TB drug (FW = 534.49 g/mol; CAS 681492-22-8), codenamed OPC-67683, blocks the synthesis of mycolic acids in *Mycobacterium tuberculosis*, thereby destabilising its cell wall (1). Delamanid is frequently used in combination with standard treatments, such ethambutol, isoniazid, pyrazinamide, rifampicin, aminoglycoside antibiotics, and quinolones. free of mutagenicity and to possess highly potent activity against TB, including MDR-TB, as shown by its exceptionally low minimum inhibitory concentration (MIC) range of 6-24 ng/mL *in vitro* and is highly effective therapeutic activity at low doses *in vivo* (1). In a randomized, placebo-controlled, multinational clinical trial, 481 patients (nearly all of whom were negative for the human immunodeficiency virus) with pulmonary multidrug-resistant tuberculosis were assigned to receive delamanid (dose of 100 mg twice daily in 161 patients or 200 mg twice daily in 160 patients), or placebo (160 patients) for 2 months. The primary efficacy end-point was the proportion of patients with sputum-culture conversion in liquid broth medium at 2 months. Delamanid is associated with an increase in sputum-culture conversion at two months among patients with multidrug-resistant tuberculosis, suggesting that delamanid may enhance treatment options for multidrug-resistant tuberculosis (2). The European Medicines Agency (EMA) has conditionally issued marketing authorization for adults with multidrug-resistant pulmonary tuberculosis. 1. Matsumoto, Hashizume, Tomishige, *et al.* (2006) *PLoS Med.* 3, e466. 2. Gler, Skripconoka, Sanchez-Garavito, *et al.* (2012) *New Engl. J. Med.* 366, 2151.

Delanzomib

This orally bioavailable proteasome inhibitor (FW = 413.28 g/mol; CAS 847499-27-8; Solubility: 83 mg/mL DMSO), code named CEP-18770 and known as (*R*)-1-((2*S*,3*R*)-3-hydroxy-2-(2-phenylpicolinamido)butanamido)-3-methylbutan-2-ylboronic acid, is a boronate peptidomimetic that targets the chymotrypsin-like activity at the β5 proteolytic site of the 20S proteasome (IC$_{50}$ = 3.8 nM), with only marginal inhibition of the tryptic and peptidylglutamyl activities of the proteosome. Delanzomib inhibits the chymotrypsin-like activity of the proteasome that down-modulates the nuclear factor-κB (NF-κB) activity and the expression of several NF-κB downstream effectors. Delanzomib induces apoptotic cell death in multiple myeloma (MM) cell lines and in primary purified CD138-positive explant cultures from untreated and bortezomib-treated MM patients. *In vitro*, it has a strong antiangiogenic activity and potently represses RANKL-induced osteoclastogenesis (1). Importantly, delanzomib exhibits a favorable cytotoxicity profile toward normal human epithelial cells, bone marrow progenitors, and bone marrow-derived stromal cells. Intravenous and oral administration resulted in a more sustained pharmacodynamic inhibition of proteasome activity in tumors relative to normal tissues, complete tumor regression of MM xenografts and improved overall median survival in a systemic model of human MM (1). When combined with melphalan or bortezomib, delanzomib induces synergistic inhibition of MM cell viability *in vitro* (2). In multiple myeloma xenograft models, the addition of delanzomib (IV route) to melphalan completely prevented the growth of both melphalan-sensitive and melphalan-resistant tumors. Combination of Delanzomib (IV) and bortezomib induced complete regression of bortezomib-sensitive tumors and markedly delayed progression of bortezomib-resistant tumors compared to treatment with either agent alone

(2). Single agent delanzomib (PO) also showed marked anti-MM effects in these xenograft models (2). 1. Piva, Ruggeri, Williams *et al.* (2008) *Blood* 111, 2765. 2. Sanchez, Li, Steinberg, *et al.* (2010) *Brit. J. Haematol.* 148, 569.

Delavirdine

This orally active antiviral (FW = 456.57 g/mol; CAS 147221-93-0; Soluble to 75 mM in DMSO), also named 1-[3-[(1-methylethyl)amino]-2-pyridinyl]-4-[[5-[(methylsulfonyl)amino]-1*H*-indol-2-yl]carbonyl]piperazine, also known as BHAP and U-90152, inhibits HIV-1 reverse transcriptase (RNA-dependent DNA polymerase), showing selectivity over other cellular polymerases: HIV-1 reverse transcriptase (IC$_{50}$ = 0.26 μM), DNA polymerase α (IC$_{50}$ = 440 μM) and DNA polymerase δ (IC$_{50}$ > 550 μM). Delavirdine prevents HIV-1 spread *in vivo* and blocks viral replication in peripheral blood lymphocytes. **Target(s):** CYP3A4 (1-3); CYP2C9 (3); CYP2C19 (3); CYP2D6 (3); reverse transcriptase (*e.g.*, HIV-1, HIV-2, and SIV) (4-10); RNA-directed DNA polymerase (4-10). 1. Zhou, Yung Chan, Cher Goh, *et al.* (2005) *Clin. Pharmacokinet.* 44, 279. 2. Zhou, Chan, Lim, *et al.* (2004) *Curr. Drug Metab.* 5, 415. 3. Voorman, Payne, Wienkers, Hauer & Sanders (2001) *Drug Metab. Dispos.* 29, 41. 4. Tucker, Lumma & Culberson (1996) *Meth. Enzymol.* 275, 440. 5. Balzarini & De Clercq (1996) *Meth. Enzymol.* 275, 472. 6. Althaus, Chou, Gonzales, *et al.* (1994) *Biochem. Pharmacol.* 47, 2017. 7. Freimuth (1996) *Adv. Exp. Med. Biol.* 394, 279. 8. Witvrouw, Pannecouque, Van Laethem, *et al.* (1999) *AIDS* 13, 1477. 9. El Safadi, Vivet-Boudou & Marquet (2007) *Appl. Microbiol. Biotechnol.* 75, 723. 10. Nissley, Radzio, Ambrose, *et al.* (2007) *Biochem. J.* 404, 151.

Deltamethrin

This synthetic insecticide (FW = 505.21 g/mol), also known as (*S*)-α-cyano-3-phenoxybenzyl (1*R*)-*cis*-3-(2,2 dibromo-vinyl)-2,2-dimethylcyclopropane carboxylate, is a potent inhibitor of protein phosphatase 2B (calcineurin; IC$_{50}$ = 0.1 nM). Deltamethrin also keeps sodium channels open for unusually long periods of time, causing a prolonged sodium current, which, in turn, leads to hyperexcitation of the nervous system. **Target(s):** adenylate cyclase (1); ATPase (2); chloride channe (l), voltage-dependent (3); glutathione *S*-transferase (4-6); protein phosphtase 2B, *or* calcineurin (7, however, see ref's. 8 and 9). 1. Sahib, Prasada Rao & Desaiah (1987) *J. Appl. Toxicol.* 7, 75. 2. Rao, Chetty & Desaiah (1984) *J. Toxicol. Environ. Health* 14, 257. 3. Forshaw, Lister & Ray (1993) *Neuropharmacology* 32, 105. 4. da Silva Vaz, Torino Lermen, Michelon, *et al.* (2004) *Vet. Parasitol.* 119, 237. 5. Wongtrakul, Sramala, Prapanthadara & Ketterman (2005) *Insect Biochem. Mol. Biol.* 35, 197. 6. Yamamoto, Fujii, Aso, Banno & Koga (2007) *Biosci. Biotechnol. Biochem.* 71, 553. 7. Enan & Matsumura (1992) *Biochem. Pharmacol.* 43, 1777. 8. Fakata, Swanson, Vorce & Stemmer (1998) *Biochem. Pharmacol.* 55, 2017 9. Enz & Pombo-Villar (1997) *Biochem. Pharmacol.* 54, 321.

Delta-Sleep-Inducing Peptide, See *L-Tryptophanyl-L-alanylglycylglycyl-L-aspartyl-L-alanyl-L-serylglycyl-L-glutamate*

α-Dendrotoxins

These oligopeptides from the venom of the Eastern green mamba (*Dendroaspis angusticeps*) block Kv1.1, Kv1.2, and Kv1.6 channels, with IC$_{50}$ values in the nanomolar range. They induce repetitive firing in neurons, thereby facilitating transmitter release. Although homologous to Kunitz-type serine protease inhibitors, the denrotoxins have little or no anti-protease activity. One example is α-Dendrotoxin-1 (MW = 7072 g/mol; *Sequence*: ZPRRKLCILHRNPGRCYDKIPAFYYNQKKKQCERFDWSG CGGNSNRFKTIEECRRTCIG), where Z designates a 5-oxoprolyl residue, and Cys$_7$-S–S-Cys$_{57}$, Cys$_{16}$-S–S-Cys$_{40}$, and Cys$_{32}$-S–S-Cys$_{53}$ are disulfide bonds. α-Dendrotoxin-2 is identical, except for an aspartyl residue in place of an asparagine at Residue-12. (*See also Mast Cell Degranualting Peptide*) **1.** Gasparini, Danse, Lecoq, *et al.* (1998) *J. Biol. Chem.* **273**, 25393. **2.** Harvey (1997) *Gen. Pharmacol.* **28**, 7. **3.** Harvey & Robertson (2004) *Curr. Med. Chem.* **11**, 3065. **4.** Benishin, Sorensen, Brown, Krueger & Blaustein (1988) *Mol. Pharmacol.* **34**, 152.

Dendrotoxin-1, *Similar in mode of action to* α-Dendrotoxin

Denibulin

This vascular-disrupting agent, *or* VDA (FW$_{HCl\text{-}salt}$ = 421.90 g/mol; FW$_{free\text{-}base}$ = 385.44 g/mol; CAS 284019-34-7; 779356-64-8, also named MN-029 and methyl-[5-[[4-[[(2S)-2-aminopropanoyl]amino]phenyl] sulfanyl]-1*H*-benzimidazol-2-yl]arbamate monohydrochloride, reversibly binds to the colchicine binding site on tubulin and inhibits microtubule assembly, resulting in disruption of the cytoskeleton of tumor vascular endothelial cells. Preclinical studies demonstrated MN-029 could evoke rapid vascular shutdown in solid tumors, dose-dependent secondary tumor cell killing, and effective enhancement of the antitumor effects of radiation and cisplatin chemotherapy. **1.** Shi, Wenyin & Siemann (2005) *Anticancer Res.* **25**, 3899. **2.** Lee & Gewirtz (2008) *Drug Devel. Res.* **69**, 352. **3.** Traynor, Gordon, Alberti, *et al.* (2010) *Investig. New Drugs* **28**, 509.

Denosumab

This intravenous drug (MW = 144.7 kDa; CAS 615258-40-7), also known by its code name AMG 162 and trade names Xgeva® and Prolia®, fully humanized monoclonal antibody that binds to and inhibits the action of RANKL (*or* Receptor Activator of NFκB Ligand), a member of the tumor necrosis factor (TNF) superfamily and a major mediator of bone loss via activation of osteoclastogenesis. RANKL overproduction occurs in degenerative bone diseases such as rheumatoid and psoriatic arthritis (1). Denosumab can be used to treat osteoporosis, treatment-induced bone loss, bone metastases, rheumatoid arthritis, and multiple myeloma. It is specifically indicated for the treatment of adults and skeletally mature adolescents with giant cell tumors of the bone that are unresectable or likely would be expected to result in severe morbidity upon surgical resection (2). **1.** Bekker, Holloway, Rasmussen, *et al.* (2004) *J. Bone Miner. Res.* **19**, 1059. **2.** Branstetter, Nelson, Manivel, *et al.* (2012) *Clin. Cancer Res.* **18**, 4415.

2'-Deoxyadenosine

This purine nucleoside (FW = 251.25 g/mol; CAS 958-09-8; λ$_{max}$ at 260 nm (pH 7); Molar Absorptivity (ε$_M$) = 15,200 M^{-1}cm^{-1}; Symbol: dA is typically obtained from the hydrolysis of DNA. 2'-dA is phosphorylated to dATP, and, when present in excess, the latter inhibits ribonucleotide reductase and suppresses cell proliferstion. **Target(s):** acid phosphatase (1); adenosine kinase, as a substrate inhibitor (2-5); adenosine nucleosidase (6,7); adenosylhomocysteinase, *or* S-adenosyl-homocysteinase (8-18); adenylate cyclase (19-21); cob(I)yrinic acid a,c-diamide adenosyltransferase, weakly

inhibited (22); cytidine deaminase (23-26); deoxycytidine kinase (27); deoxyguanosine kinase, also alternative substrate (28-30); dihydroorotase (31); glycine *N*-methyltransferase, weakly inhibited (32); GMP synthetase (33); histone-arginine *N*-methyltransferase (34); NADase (*or* NAD$^+$ glycohydrolase) (35); 5'-nucleotidase (36,37); purine nucleosidase (38); ribonuclease T2 (39; ribonuclease U2 (40); thymidine phosphorylase (41). **1.** Uerkvitz (1988) *J. Biol. Chem.* **263**, 15823. **2.** Palella, Andres & Fox (1980) *J. Biol. Chem.* **255**, 5264. **3.** Datta, Das, Sen, *et al.* (2006) *Biochem. J.* **394**, 35. **4.** Kidder (1982) *Biochem. Biophys. Res. Commun.* **107**, 381. **5.** Datta, Bhaumik & Chatterjee (1987) *J. Biol. Chem.* **262**, 5515. **6.** Guranowski & Schneider (1977) *Biochim. Biophys. Acta* **482**, 145. **7.** Chen & Kristopeit (1981) *Plant Physiol.* **68**, 1020. **8.** Chiang (1987) *Meth. Enzymol.* **143**, 377. **9.** Ator & Ortiz de Montellano (1990) *The Enzymes*, 3rd ed., **19**, 213. **10.** Hohman, Guitton & Veron (1984) *Arch. Biochem. Biophys.* **233**, 785. **11.** Fabianowska-Majewska, Duley & Simmonds (1994) *Biochem. Pharmacol.* **48**, 897. **12.** Magnuson, Perryman, Decker & Magnuson (1984) *Int. J. Biochem.* **16**, 1163. **13.** Shimizu, Shiozaki, Ohshiro & Yamada (1984) *Eur. J. Biochem.* **141**, 385. **14.** Impagnatiello, Franceschini, Oratore & Bozzi (1996) *Biochimie* **78**, 267. **15.** Bozzi, Parisi & Martini (1993) *J. Enzyme Inhib.* **7**, 159. **16.** Kloor, Kurz, Fuchs, Faust & Osswald (1996) *Kidney Blood Press. Res.* **19**, 100. **17.** Ichikawa, Sato & Tomita (1985) *J. Biochem.* **97**, 189. **18.** Knudsen & Yall (1972) *J. Bacteriol.* **112**, 569. **19.** Dessauer (2002) *Meth. Enzymol.* **345**, 112. **20.** Jayaswal, Bressan & Handa (1985) *FEMS Microbiol. Lett.* **27**, 313. **21.** Dessauer & Gilman (1997) *J. Biol. Chem.* **272**, 27787. **22.** Johnson, M. L. Buszko & T. A. Bobik (2004) J. Bacteriol. **186**, 7881. **23.** Vita, Amici, Cacciamani, Lanciotti & Magni (1985) *Biochemistry* **24**, 6020. **24.** Hosono & Kuno (1973) *J. Biochem.* **74**, 797. **25.** Cacciamani, Vita, Cristalli, *et al.* (1991) *Arch. Biochem. Biophys.* **290**, 285. **26.** Vita, A. Amici, M. Lanciotti, T. Cacciamani & G. Magni (1986) *Ital. J. Biochem.* **35**, 145A. **27.** Wang, Kucera & Capizzi (1993) *Biochim. Biophys. Acta* **1202**, 309. **28.** Yamada, Goto & Ogasawara (1982) *Biochim. Biophys. Acta* **709**, 265. **29.** Green & Lewis (1979) Biochem. J. **183**, 547. **30.** Barker & Lewis (1981) *Biochim. Biophys. Acta* **658**, 111. **31.** Bresnick & Blatchford (1964) *Biochim. Biophys. Acta* **81**, 150. **32.** Kloor, Karnahl & Kömpf (2004) *Biochem. Cell Biol.* **82**, 369. **33.** Spector & Beecham (1975) *J. Biol. Chem.* **250**, 3101. **34.** Gupta, Jensen, Kim & Paik (1982) *J. Biol. Chem.* **257**, 9677. **35.** Schuber, Pascal & Travo (1978) *Eur. J. Biochem.* **83**, 205. **36.** Garvey, Lowen & Almond (1998) *Biochemistry* **37**, 9043. **37.** Olson & Fraser (1974) *Biochim. Biophys. Acta* **334**, 156. **38.** Koszalka & Krenitsky (1979) *J. Biol. Chem.* **254**, 8185. **39.** Irie & K. Ohgi (1976) *J. Biochem.* **80**, 39. **40.** Yasuda & Inoue (1982) *Biochemistry* **21**, 364. **41.** Blank & Hoffee (1975) *Arch. Biochem. Biophys.* **168**, 259.

5'-Deoxyadenosine

This deoxynucleoside (FW = 251.25 g/mol; λ$_{max}$ at 260 nm (pH 7); Molar Absorptivity (ε$_M$) = 15200 M^{-1}cm^{-1}) is a component of *S*-adenosyl-L-methionine and 5'-deoxyadenosylcobalamin (vitamin B$_{12}$). **Target(s):** adenosine kinase (1-3); adenosylhomocysteinase (4); biotin synthase (5); deoxycytidine kinase, K$_i$ = 2 mM (6); GMP synthetase (7-9); *S*-methyl-5'-thioadenosine phosphorylase (10); rhodopsin kinase, K$_i$ = 10 μM (11). **1.** Palella, Andres & Fox (1980) *J. Biol. Chem.* **255**, 5264. **2.** Miller, Adamczyk, Miller, *et al.* (1979) *J. Biol. Chem.* **254**, 2346. **3.** Galazka, Striepen & Ullman (2006) *Mol. Biochem. Parasitol.* **149**, 223. **4.** Shimizu, Shiozaki, Ohshiro & Yamada (1984) *Eur. J. Biochem.* **141**, 385. **5.** Ollagnier-de-Choudens, Mulliez & Fontecave (2002) *FEBS Lett.* **532**, 465. **6.** Krenitsky, Tuttle, Koszalka, *et al.* (1976) *J. Biol. Chem.* **251**, 4055. **7.** Spector & Beecham (1975) *J. Biol. Chem.* **250**, 3101. **8.** Nakamura & Lou (1995) *J. Biol. Chem.* **270**, 7347. **9.** Spector, Jones, Krenitsky & Harvey (1976) *Biochim. Biophys. Acta* **452**, 597. **10.** Fabianowska-Majewska, Duley, Fairbanks, Simmonds & Wasiak (1994) *Acta Biochim. Pol.* **41**, 391. **11.** Palczewski, McDowell & Hargrave (1988) *J. Biol. Chem.* **263**, 14067.

2'-Deoxyadenosine 3',5'-Cyclic Monophosphate

This unnatural cyclic nucleotide (FW = 251.25 g/mol; λ_{max} at 260 nm (pH 7); Molar Absorptivity (ε_M) = 15200 $M^{-1}cm^{-1}$) is more active than adenosine 3',5'-cyclic monophosphate as a substrate for mammalian cAMP phosphodiesterase. It is also an effective activator of the glycogen phosphorylase cascade (1). **Target(s):** poly(ADP-ribose) glycohydrolase (2)   **1.** Braun, Hechter & Baer (1969) *Proc. Soc. Exp. Biol. Med.* **132**, 233.   **2.** Tavassoli, Tavassoli & Shall (1983) *Eur. J. Biochem.* **135**, 449.

2'-Deoxyadenosine 5'-Diphosphate

This deoxynucleotide (FW = 411.20 g/mol; λ_{max} = 259 nm; Molar Absorptivity (ε_M) = 15400 $M^{-1}cm^{-1}$; Abbreviation: dADP) is an intermediate in the dATP biosynthesis. In eukaryotes, ADP is converted to dADP by ribonucleotide reductase (RNR). In prokaryotes, dADP is formed from dATP, the latter the product of bacterial RNR action on ATP. **Target(s):** amidophosphoribosyltransferase (1); *Aspergillus* nuclease S (12); cytidylate kinase (2); deoxya kinase (4,5); deoxycytidine kinase6; deoxyguanosine kinase (7); DNA directed DNA polymerase (8); dTMP kinase, *or* thymidylate kinase (9-12); gluconokinase, weakly inhibited (13); glucose-1-phosphate cytidylyltransferase, weakly inhibited (14); 5'-nucleotidase (15); polynucleotide phosphorylase, *or* polyribonucleotide nucleotidyltransferase (16-20); ribose phosphate diphosphokinase, *or* phosphoribosyl-pyrophosphate synthetase (21-23).   **1.** Hill & Bennett (1969) *Biochemistry* **8**, 122.   **2.** Oleson & Hoganson (1981) *Arch. Biochem. Biophys.* **211**, 478.   **3.** Ruffner & Anderson (1969) *J. Biol. Chem.* **244**, 5994.   **4.** Anderson (1973) *The Enzymes*, 3rd ed., **9**, 49.   **5.** Krygier & Momparler (1968) *Biochim. Biophys. Acta* **161**, 578.   **6.** Kozai, Sonoda, Kobayashi & Sugino (1972) *J. Biochem.* **71**, 485.   **7.** Yamada, Goto & Ogasawara (1982) *Biochim. Biophys. Acta* **709**, 265.   **8.** Bollum (1968) *Meth. Enzymol.* **12B**, 591. 9. Jong & Campbell (1984) *J. Biol. Chem.* **259**, 14394.   **10.** Kielley (1970) *J. Biol. Chem.* **245**, 4204.   **11.** Lee & Cheng (1977) *J. Biol. Chem.* **252**, 5686.   **12.** Tamiya, Yusa, Yamaguchi, *et al.* (1989) *Biochim. Biophys. Acta* **995**, 28.   **13.** Coffee & Hu (1972) *Arch. Biochem. Biophys.* **149**, 549.   **14.** Kimata & Suzuki (1966) *J. Biol. Chem.* **241**, 1099.   **15.** Madrid-Marina & Fox (1986) *J. Biol. Chem.* **261**, 444.   **16.** Littauer & Soreq (1982) *The Enzymes*, 3rd ed., **15**, 517.   **17.** Bon, T. Godefroy & M. Grunberg-Manago (1970) *Eur. J. Biochem.* **16**, 363.   **18.** Godefroy-Colburn & Stetondji (1972) *Biochim. Biophys. Acta* **272**, 417.   **19.** Chou & Singer (1971) *J. Biol. Chem.* **246**, 7497.   **20.** Hunt & Cowles (1977) *J. Gen. Microbiol.* **102**, 403.   **21.** Roth, White & Deuel (1978) *Meth. Enzymol.* **51**, 12.   **22.** Sonoda, Kita, Ishijima, *et al.* (1997) *J. Biochem.* **122**, 635.   **23.** Switzer & Sogin (1973) *J. Biol. Chem.* **248**, 1063.

2'-Deoxyadenosine 5'-Monophosphate

This deoxynucleotide (FW = 331.23 g/mol; λ_{max} = 259 nm; Molar Absorptivity (ε_M) = 15400 $M^{-1}cm^{-1}$; *Symbol*: dAMP is released from DNA upon treatment with certain diesterases and is formed in the deoxyadenosine kinase reaction. **Target(s):** adenine phosphoribosyltransferase (1); adenylosuccinate lyase (2,3); adenylosuccinate synthetase (4-7); alcohol dehydrogenase (8); Amidophosphoribosyltransferase (5,10); aspartate carbamoyl-transferase (5,10); *Aspergillus* nuclease S1 (11-13); dCMP deaminase, *or* deoxycytidylate deaminase (5,14-23); deoxyadenosine kinase (24); deoxycytidine kinase (25); deoxyribonuclease (26); dihydroorotase (27); DNA-directed DNA polymerase (28); φ29 DNA polymerase (29); fructose-1,6-bisphosphatase (30-33); IMP cyclohydrolase (34); micrococcal nuclease (35); NAD$^+$ nucleosidase, *or* NAD$^+$ glycohydrolase, *or* NADase (36); 3'-nucleotidase (not a substrate for the *Leishmania donovani* enzyme (37,38); nucleotide diphosphatase (39); phosphodiesterase I, *or* 5'-exonuclease (40); ribonuclease H (41); ribonuclease T2 (42); ribonuclease U2 (43).   **1.** Nagy & Ribet (1977) *Eur. J. Biochem.* **77**, 77.   **2.** Spector, Jones & Elion (1979) *J. Biol. Chem.* **254**, 8422.   **3.** Spector (1977) *Biochim. Biophys. Acta* **481**, 741.   **4.** Baker (1967) *Design of Active-Site-Directed Irreversible Enzyme Inhibitors*, Wiley, New York.   **5.** Webb (1966) *Enzyme and Metabolic Inhibitors*, vol. **2**, Academic Press, New York.   **6.** Wyngaarden & Greenland (1963) *J. Biol. Chem.* **238**, 1054.   **7.** Van der Weyden & Kelly (1974) *J. Biol. Chem.* **249**, 7282.   **8.** Li & Vallee (1964) *J. Biol. Chem.* **239**, 792.   **9.** Hill & Bennett (1969) *Biochemistry* **8**, 122.   **10.** Bresnick (1962) *Cancer Res.* **22**, 1246.   **11.** Shishido & Ando (1982) *Cold Spring Harbor Monogr. Ser.* **14**, 155.   **12.** Oleson & Hoganson (1981) *Arch. Biochem. Biophys.* **211**, 478.   **13.** Wiegand, Godson & Radding (1975) *J. Biol. Chem.* **250**, 8848.   **14.** Scarano, Bonaduce & de Petrocellis (1960) *J. Biol. Chem.* **235**, 3556.   **15.** Maley & Maley (1959) *J. Biol. Chem.* **234**, 2975.   **16.** Geraci, Rossi & Scarano (1967) *Biochemistry* **6**, 183.   **17.** Mastrantonio, Nucci, Vaccaro, Rossi & Whitehead (1983) *Eur. J. Biochem.* **137**, 421.   **18.** Rossi, Geraci & Scarano (1967) *Biochemistry* **6**, 3640.   **19.** Scarano, Geraci & Rossi (1967) *Biochemistry* **6**, 3645.   **20.** Maley, Lobo & Maley (1993) *Biochim. Biophys. Acta* **1162**, 161.   **21.** Raia, Nucci, Vaccaro, Sepe & Rella (1982) *J. Mol. Biol.* **157**, 557.   **22.** Nucci, Raia, Vaccaro, *et al.* (1978) *J. Mol. Biol.* **124**, 133.   **23.** Nucci, Raia, Vaccaro, Rossi & Whitehead (1991) *Arch. Biochem. Biophys.* **289**, 19.   **24.** Anderson (1973) *The Enzymes*, 3rd ed. (Boyer, ed.), **9**, 49.   **25.** Datta, Shewach, Mitchell & Fox (1989) *J. Biol. Chem.* **264**, 9359.   **26.** Chen & Grossman (1985) *J. Biol. Chem.* **260**, 5073.   **27.** Bresnick & Blatchford (1964) *Biochim. Biophys. Acta* **81**, 150.   **28.** Bollum (1968) *Meth. Enzymol.* **12B**, 591.   **29.** Lázaro, Blanco & Salas (1995) *Meth. Enzymol.* **262**, 42.   **30.** Pontremoli (1966) *Meth. Enzymol.* **9**, 625.   **31.** Fujita & Freese (1979) *J. Biol. Chem.* **254**, 5340.   **32.** Rao, Rosen & Rosen (1969) *Biochemistry* **8**, 4904.   **33.** Funayama, Molano & Gancedo (1979) *Arch. Biochem. Biophys.* **197**, 170.   **34.** Szabados, Hindmarsh, Phillips, Duggleby & Christopherson (1994) *Biochemistry* **33**, 14237.   **35.** Cuatrecasas, Fuchs & Anfensen (1967) *J. Biol. Chem.* **242**, 1541, 3063, and 4759.   **36.** Schuber, Pascal & Travo (1978) *Eur. J. Biochem.* **83**, 205.   **37.** Gbenle & Dwyer (1992) *Biochem. J.* **285**, 41.   **38.** Gbenle (1993) *Biochim. Biophys. Acta* **1203**, 162.   **39.** Bachorik & Dietrich (1972) *J. Biol. Chem.* **247**, 5071.   **40.** Razzell (1961) *J. Biol. Chem.* **236**, 3031.   **41.** Lee, Hong, Kim, Rho & Jung (1997) *Biochem. Biophys. Res. Commun.* **233**, 401.   **42.** Irie & Ohgi (1976) *J. Biochem.* **80**, 39.   **43.** Yasuda & Inoue (1982) *Biochemistry* **21**, 364.

2'-Deoxyadenosine 5'-Triphosphate (*or* dATP)

This key nucleotide (FW$_{free-acid}$ = 491.18 g/mol; λ_{max} at 259 nm at pH 7 (ε_M = 15400 $M^{-1}cm^{-1}$) and 258 nm at pH 2 (ε_M = 14800 $M^{-1}cm^{-1}$) is one of four DNA polymerases substrates. **Target(s):** adenine phosphoribosyltransferase (1); amidophosphoribosyl transferase (2); AMP deaminase (3); *Aspergillus* nuclease S$_1$ (4-6); ATP:citrate lyase (*or* ATP:citrate synthase) (7,8); *Bacillus subtilis* ribonuclease (9-11); CDP reductase (12); dCTP deaminase (13); deoxyadenosine kinase (14-19); deoxycytidine kinase (20-22); deoxyguanosine kinase (23-25); DNA helicase (26); dTMP kinase, *or* thymidylate kinase, weakly inhibited (27); exopolyphosphatase (28); glucose-1-phosphate cytidylyl-transferase (29); glucuronokinase (30); guanylate cyclase (31,32); 3-hydroxyanthranilate oxidase (68); [hydroxymethylglutaryl-CoA reductase (NADPH)] kinase (33); leucyl-tRNA synthetase (34); lysyl-tRNA synthetase (35); nucleoside-triphosphatase, alternative substrate (26); nucleotidases (28); 3'-nucleotidase (28); 5'-nucleotidase (28,36); 2',5'-oligoadenylate synthetase (37); 1-phosphatidylinositol 4-kinase (38); polynucleotide adenylyl-transferase (*or,* poly(A) polymerase) (39-43); polynucleotide 5'-hydroxyl-kinase (44,45); pyruvate, orthophosphate dikinase (46); ribonucleoside diphosphate reductase (12,47-50,56-67); ribose-phosphate diphospho-kinase (*or* phosphoribosyl pyrophosphate synthetase) (51); RNA-directed RNA polymerase (52); threonyl-tRNA synthetase (35); tRNA cytidylyltransferase (53); tRNA nucleotidyltransferase (53); tryptophanyl-tRNA synthetase (54); tyrosyl-tRNA synthetase (55). **1**. Nagy & Ribet (1977) *Eur. J. Biochem.* **77**, 77. **2**. Hill & Bennett (1969) *Biochemistry* **8**, 122. **3**. Yabuki & Ashihara (1992) *Phytochemistry* **31**, 1905. **4**. Shishido & Ando (1982) *Cold Spring Harbor Monogr.* Ser. 14, 155. **5**. Oleson & Hoganson (1981) *Arch. Biochem. Biophys.* **211**, 478. **6**. Wiegand, Godson & Radding (1975) *J. Biol. Chem.* **250**, 8848. **7**. Antranikian, Herzberg & Gottschalk (1982) *J. Bacteriol.* **152**, 1284. **8**. Ishii, Igarashi & Kodama (1989) *J. Bacteriol.* **171**, 1788. **9**. Yamasaki & Arima (1969) *Biochem. Biophys. Res. Commun.* **37**, 430. **10**. Yamasaki & Arima (1967) *Biochim. Biophys. Acta* **139**, 202. **11**. Yamasaki & Arima (1970) *Biochim. Biophys. Acta* **209**, 475. **12**. Langelier, Dechamps & Buttin (1978) *J. Virol.* **26**, 547. **13**. Beck, Eisenhardt & Neuhard (1975) *J. Biol. Chem.* **250**, 609. **14**. Anderson (1973) *The Enzymes*, 3rd ed., **9**, 49. **15**. Krygier & Momparler (1971) *J. Biol. Chem.* **246**, 2752. **16**. Krygier & Momparler (1968) *Biochim. Biophys. Acta* **161**, 578. **17**. Ma, Hong & Ives (1995) *J. Biol. Chem.* **270**, 6595. **18**. Ikeda, Ma & Ives (1994) *Biochemistry* **33**, 5328. **19**. Welin, Wang, Eriksson & Eklund (2007) *J. Mol. Biol.* **366**, 1615. **20**. Yamada, Goto & Ogasawara (1983) *Biochim. Biophys. Acta* **761**, 34. **21**. Kozai, Sonoda, Kobayashi & Sugino (1972) *J. Biochem.* **71**, 485. **22**. Datta, Shewach, Mitchell & Fox (1989) *J. Biol. Chem.* **264**, 9359. **23**. Yamada, Goto & Ogasawara (1982) *Biochim. Biophys. Acta* **709**, 265. **24**. Green & Lewis (1979) *Biochem. J.* **183**, 547. **25**. Eriksson, Munch-Petersen, Johansson & Eklund (2002) *Cell. Mol. Life Sci.* **59**, 1327. **26**. Locatelli, Gosselin, Spadari & Maga (2001) *J. Mol. Biol.* **313**, 683. **27**. Smith & Eakin (1975) *Arch. Biochem. Biophys.* **167**, 61. **28**. Proudfoot, Kuznetsova, Brown, *et al.* (2004) *J. Biol. Chem.* **279**, 54687. **29**. Kimata & Suzuki (1966) *J. Biol. Chem.* **241**, 1099. **30**. Gillard & Dickinson (1978) *Plant Physiol.* **62**, 706. **31**. Garbers & Gray (1974) *Meth. Enzymol.* **38**, 196. **32**. Durham (1976) *Eur. J. Biochem.* **61**, 535. **33**. Henin, Vincent & Van den Berghe (1996) *Biochim. Biophys. Acta* **1290**, 197. **34**. Maruztky, Flossdorf & Kula (1976) *Nucl. Acids Res.* **3**, 2067. **35**. Freist, Sternbach, von der Haar & Cramer (1978) *Eur. J. Biochem.* **84**, 499. **36**. Madrid-Marina & Fox (1986) *J. Biol. Chem.* **261**, 444. **37**. Ball (1982) *The Enzymes*, 3rd ed., **15**, 281. **38**. Vogel & Hoppe (1986) *Eur. J. Biochem.* **154**, 253. **39**. Edmonds (1982) *The Enzymes*, 3rd ed., **15**, 217. **40**. Sarkar, Cao & Sarkar (1997) *Biochem. Mol. Biol. Int.* **41**, 1045. **41**. Koch & Niessing (1978) *FEBS Lett.* **96**, 354. **42**. Hunt, Meeks, Forbes, Das Gupta & Mogen (2000) *Biochem. Biophys. Res. Commun.* **272**, 174. **43**. Tsiapalis, Dorson & Bollum (1975) *J. Biol. Chem.* **250**, 4486. **44**. Austin, Sirakoff, Roop & Moyer (1978) *Biochim. Biophys. Acta* **522**, 412. **45**. Levin & Zimmerman (1976) J. Biol. Chem. 251, 1767. **46**. Reeves (1971) *Biochem. J.* **125**, 531. **47**. Moore (1967) *Meth. Enzymol.* **12A**, 155. **48**. Thelander, Sjöberg & Eriksson (1978) *Meth. Enzymol.* **51**, 227. **49**. Chang & Cheng (1980) *Cancer Res.* **40**, 3555. **50**. Hofmann, Feller, Pries & Follmann (1985) *Biochim. Biophys. Acta* **832**, 98. **51**. Sonoda, Kita, Ishijima, *et al.* (1997) *J. Biochem.* **122**, 635. **52**. Hung, Gibbs & Tsiang (2002) *Antiviral Res.* **56**, 99. **53**. Hill & Nazario (1973) *Biochemistry* **12**, 482. **54**. Jakubowski & Pawelkiewicz (1975) *Eur. J. Biochem.* **52**, 301. **55**. Santi & Peña (1973) *J. Med. Chem.* **16**, 273. **56**. Cory, Sato & Brown (1986) *Adv. Enzyme Regul.* **25**, 3. **57**. Holmgren (1981) *Curr. Top. Cell. Regul.* **19**, 47. **58**. Averett, Lubbers, Elion & Spector (1983) *J. Biol. Chem.* **258**, 9831. **59**. Huszar & Bacchetti (1981) *J. Virol.* **37**, 580. **60**. Kucera & Paulus (1982) *Arch. Biochem. Biophys.* **214**, 114. **61**. Stubbe (1990) *J. Biol. Chem.* **265**, 5329. **62**. Berglund (1972) *J. Biol. Chem.* **247**, 7276. **63**. Feller & Follmann (1976) *Biochem. Biophys. Res. Commun.* **70**, 752. **64**. Larsson & Reichard (1966) *J. Biol. Chem.* **241**, 2533. **65**. Rofougaran, Vodnala & Hofer (2006) *J. Biol. Chem.* **281**, 27705. **66**. Cory, Rey, Carter & Bacon (1985) *J. Biol. Chem.* **260**, 12001. **67**. Moore & Hurlbert (1966) *J. Biol. Chem.* **241**, 4802. **68**. Nair (1972) *Arch. Biochem. Biophys.* **153**, 139.

2'-Deoxy-*S*-adenosyl-L-homocysteine

This product analogue (FW$_{zwitterion}$ = 368.42 g/mol; λ_{max} at 260 nm (pH 7); ε_M of 15200 $M^{-1}cm^{-1}$) weakly inhibits mRNA (guanine-N^7-)-methyltransferase, mRNA (nucleoside 2'-*O*-)-methyltransferase, and DNA (cytosine-5)-methyltransferase. **Target(s):** DNA (cytosine-5-)-methyltransferase (1); *Hae*III DNA methyltransferase, K_i = 69 µM (1); *Hha*I DNA methyltransferase, K_i = 44 µM (1); mRNA (guanine-N^7-)-methyltransferase (2); mRNA (nucleoside 2'-*O*-)-methyltransferase, weakly inhibited (2). **1**. Cohen, Griffiths, Tawfik & Loakes (2005) *Org. Biomol. Chem.* **3**, 152. **2**. Pugh & Borchardt (1982) *Biochemistry* **21**, 1535.

Deoxycholate (Deoxycholic Acid)

This fat-emulsifying secondary bile acid (FW$_{free-acid}$ = 392.58 g/mol; CAS 83-44-3; Symbol: DCA; Critical Micelle Concentration (CMC) = 2.4-4 mM, depending on other solutes present), also known as (3α,5β,12α,20R)-3,12-dihydroxycholan-24-oic acid, is a mild detergent that is often employed in the isolation of membrane-associated proteins (1). Indeed, the proton-translocating ATP synthase of bovine heart mitochondria was first purified by extraction of submitochondrial particles with cholate, fractionation with ammonium sulfate, and sucrose gradient centrifugation in the presence of methanol, deoxycholate, and lysolecithin (2). Sodium deoxycholate is also used often to lyze cells. DCA (0.05-0.3 mmol/l) also inhibits colon cell proliferation by up to 92%, increasing the apoptosis rate by up to 4.5 times (3). DCA induces an increase in only G$_1$ fraction, with a concomitant drop in the S-phase fraction, when compared with the untreated cells. DCA increases intracellular reactive oxygen species, genomic DNA breakage, as well as activation of ERK1/2, caspase-3 and PARP. In warm-blooded animals, deoxycholic acid is used in the emulsification of fats for the absorption in the intestine. **Remodeling Submental Fat Deposits:** As many individuals age, fat accumulates beneath the chin in the submental fat pad, resulting in an unsightly "turkey gobbler" jawline. Significantly, the facial fat within the superficial subcutaneous layer is not one confluent layer, but is instead compartmentalized. Each border of these compartment is formed by fascial septae traveling from the deep fascia or periosteum and insert into the dermis. This anatomical feature commends the use of biochemical strategies for remodeling submental fat without the need for liposuction or surgical intervention. ATX-101 is a proprietary formulation (Tradename: Kybella®) of purified synthetic deoxycholic acid that, when injected appropriately, emulsifies and mobilizes submental fat, selectively destroying fat-storing cells (3). It is a FDA-approved agent for cosmetically recontouring the fat pad just below the chin, often with little effect on nearby tissues. **1**. Philippot (1971) *Biochim. Biophys. Acta* **225**, 201. **2**. Serrano, Kanner & Racker (1976) *J. Biol. Chem.* **251**, 2453. **3**. Zeng, Claycombe & Reindl (2015) *J. Nutr. Biochem.* **26**, 1022. **4**. Müller, Misund, Holien, *et al.* (2013) *PLoS One* **8**, e70430.

2'-Deoxycoformycin

This nucleoside (FW = 268.27 g/mol; CAS 53910-25-1; λ_{max} = 282 nm (at pH 7), Molar Absorptivity (ε_M) = 8000 $M^{-1}cm^{-1}$), also known as pentostatin and covidarabine, from *Streptomyces antibioticus* is a tetrahedral transition-state analogue of a likely intermediate in the reaction catalyzed by adenosine deaminase (K_i = 0.22 nM). Its kinetic behavior is consistent with its classification as a slow-, tight-binding inhibitor ($t_{1/2}$ = 25–30 hr). (**See** *Coformycin for comments on likely mechanism of action*) **Target(s):** adenine deaminase (1); adenosine deaminase (2-17); adenosine kinase (18,19) adenosylhomocysteinase (20); *S*-adenosylhomocysteine deaminase (21); AMP deaminase (5); DNA ligase (22). **1.** Kidder & Nolan (1979) *Proc. Natl. Acad. Sci. U.S.A.* **76**, 3670. **2.** Agarwal & Parks (1978) *Meth. Enzymol.* **51**, 502. **3.** Frieden, Kurtz & Gilbert (1980) *Biochemistry* **19**, 5303. **4.** Cha (1976) *Biochem. Pharmacol.* **25**, 2695. **5.** Agarwal & Parks (1977) *Biochem. Parmacol.* **26**, 663. **6.** Rogler-Brown, Agarwal & Parks (1978) *Biochem. Pharmacol.* **27**, 2289. **7.** Lupidi, Marmocchi & Cristalli (1998) *Biochem. Mol. Biol. Int.* **46**, 1071. **8.** Wang & Quiocho (1998) *Biochemistry* **37**, 8314. **9.** Marrone, Straatsma, Briggs, *et al.* (1996) *J. Med. Chem.* **39**, 277. **10.** Kati, Acheson & Wolfenden (1992) *Biochemistry* **31**, 7356. **11.** Wolfenden, Wentworth & Mitchell (1977) *Biochemistry* **16**, 5071. **12.** Philips, Robbins & Coleman (1987) *Biochemistry* **26**, 2893. **13.** Iwaki-Egawa & Watanabe (2002) *Comp. Biochem. Physiol. B* **133**, 173. **14.** Iwaki-Egawa, Namiki & Watanabe (2004) *Comp. Biochem. Physiol. B* **137**, 247. **15.** Daddona, Wiesmann, Lambros, Kelley & Webster (1984) *J. Biol. Chem.* **259**, 1472. **16.** Lupidi, Marmocchi & Cristalli (1998) *Biochem. Mol. Biol. Int.* **46**, 1071. **17.** Wiginton, Coleman & Hutton (1981) *Biochem. J.* **195**, 389. **18.** Long & Parker (2006) *Biochem. Pharmacol.* **71**, 1671. **19.** Kidder (1982) *Biochem. Biophys. Res. Commun.* **107**, 381. **20.** Refaie (1988) *J. Egypt. Med. Assoc.* **71**, 235. **21.** Zulty & M. K. Speedie (1989) *J. Bacteriol.* **171**, 6840. **22.** Lamballe, Le Prise, Le Gall & David (1989) *Leukemia* **3**, 97.

11-Deoxycorticosterone

This mineralocorticoid steroid (FW = 330.47 g/mol; CAS 64-85-7; M.P. = 141-142°C), also known as 4-pregnen-21-ol-3,20-dione, DOC, cortexone, Kendall's desoxy compound B, and Reichstein's compound Q, was first isolated from the adrenal cortex and is a steroid that is biosynthesized from progesterone. Although a potent mineralocorticoid, DOC is virtually devoid of glucocorticoid action. Itt also exhibits anticonvulsant properties in animals, where it can be converted to the neurosteroid $5\alpha,3\alpha$-tetrahydrodeoxycorticosterone. This steroid was first isolated in the laboratories of Kendall and Reichstein, who shared the 1950 Nobel Prize in Physiology or Medicine. **Target(s):** aldehyde dehydrogenase (1); D-amino-acid oxidase (2); ascorbate oxidase (3); cortisol sulfotransferase (4,5); 21-hydroxysteroid dehydrogenase, NAD^+-requiring (6-8); NADH dehydrogenase, ubiquinone-requiring, *or* NADH:cytochrome *c* reductase (9,10); steroid Δ-isomerase (11); steroid sulfotransferase (12); sterol *O*-acyltransferase, *or* cholesterol *O*-acyltransferase; ACAT (13,14); tryptophan 2,3-dioxygenase (15). **1.** Douville & Warren (1968) *Biochemistry* **7**, 4052. **2.** Hayano, Dorfman & Yamada (1950) *J. Biol. Chem.* **186**, 603. **3.** Stark & Dawson (1963) *The Enzymes*, 2nd ed., **8**, 297. **4.** Singer (1979) *Arch. Biochem. Biophys.* **196**, 340. **5.** Singer & Bruns (1980) *Can. J. Biochem.* **58**, 660. **6.** Monder & Furfine (1969) *Meth. Enzymol.* **15**, 667. **7.** Talalay (1963) *The Enzymes*, 2nd ed., **7**, 177. **8.**

Monder & White (1962) *Biochem. Biophys. Res. Commun.* **8**, 383. **9.** Jensen (1959) *Nature* **184**, 451. **10.** Vernon, Mahler & Sarkar (1952) *J. Biol. Chem.* **199**, 599. **11.** Weintraub, Vincent, Baulieu & Alfsen (1977) *Biochemistry* **16**, 5045. **12.** Singer, Gebhart & Hess (1978) *Can. J. Biochem.* **56**, 1028. **13.** Billheimer (1985) *Meth. Enzymol.* **111**, 286. **14.** Simpson & Burkhart (1980) *Arch. Biochem. Biophys.* **200**, 79. **15.** Braidman & Rose (1970) *Biochem. J.* **118**, 7P.

2'-Deoxycytidine 5'-Diphosphate (*or*, dCDP)

This ribonucleotide reductase substrate ($FW_{free-acid}$ = 389.19 g/mol; $FW_{trisodium-salt}$ = 453.12; CAS 151151-32-5; λ_{max} = 271 nm (pH 7 and 12; with Molar Absorptivities (ε_M) = 9300 and 9000 $M^{-1}cm^{-1}$, respectively) and 280 nm at pH 2 (ε_M = 13500 $M^{-1}cm^{-1}$) inhibits DNA polymerase. In eukaryotes, CDP is converted to dCDP by ribonucleotide reductase (RNR). In prokaryotes, dCDP is formed from dCTP, the latter the product of bacterial RNR action on ATP. **Target(s):** adenylosuccinate synthetase, weakly inhibited (1); cytidylate kinase (2,3); dCTP diphosphatase, also acts as an alternative substrate (4); deoxyadenosine kinase (2,5,6); deoxycytidine kinase (2,7 11); deoxyguanosine kinase (12); DNA-directed DNA polymerase (13); dolichol kinase, alternative product inhibition (14); 5'-nucleotidase (15); polynucleotide 5'-hydroxyl-kinase (16); polynucleotide phosphorylase (17-19). **1.** Van der Weyden & Kelly (1974) *J. Biol. Chem.* **249**, 7282. **2.** Anderson (1973) *The Enzymes*, 3rd ed. (Boyer, ed.), **9**, 49. **3.** Ruffner & Anderson (1969) *J. Biol. Chem.* **244**, 5994. **4.** Zimmerman & Kornberg (1961) *J. Biol. Chem.* **236**, 1380. **5.** Krygier & Momparler (1971) *J. Biol. Chem.* **246**, 2752. **6.** Krygier & Momparler (1968) *Biochim. Biophys. Acta* **161**, 578. **7.** Kozai, Sonoda, Kobayashi & Sugino (1972) *J. Biochem.* **71**, 485. **8.** Durham & Ives (1970) *J. Biol. Chem.* **245**, 2276. **9.** Datta, Shewach, Mitchell & Fox (1989) *J. Biol. Chem.* **264**, 9359. **10.** Kessel (1968) *J. Biol. Chem.* **243**, 4739. **11.** Momparler & Fischer (1968) *J. Biol. Chem.* **243**, 4298. **12.** Yamada, Goto & Ogasawara (1982) *Biochim. Biophys. Acta* **709**, 265. **13.** Bollum (1968) *Meth. Enzymol.* **12B**, 591. **14.** Sumbilla & Waechter (1985) *Arch. Biochem. Biophys.* **238**, 75. **15.** Madrid-Marina & Fox (1986) *J. Biol. Chem.* **261**, 444. **16.** Levin & Zimmerman (1976) *J. Biol. Chem.* **251**, 1767. **17.** Maley (1958) *Fed. Proc.* **17**, 267. **18.** Maley & Ochoa (1958) *J. Biol. Chem.* **233**, 1538. **19.** Webb (1966) *Enzyme and Metabolic Inhibitors*, vol. **2**, p. 474, Academic Press, New York.

2'-Deoxycytidine 5'-Monophosphate (*or*, dCMP)

This deoxynucleotide (FW = 306.19 g/mol; λ_{max} = 271 nm (pH 7 and 12; ε_M = 9300 and 9000 $M^{-1}cm^{-1}$, respectively) and 280 nm at pH 2 (ε_M = 13500 $M^{-1}cm^{-1}$) is released from DNA upon treatment with certain phosphodiesterases (*e.g.*, *Escherichia coli* phosphodiesterase). **Target(s):** adenylosuccinate synthetase, weakly inhibited (1); amidophospho-ribosyltransferase, weakly inhibited (2); D-amino-acid oxidase (3); aspartate carbamoyltransferase (4-6); cytidine deaminase (7-9); cytidylate kinase, as alternative substrate and inhibitor (10); deoxyadenosine kinase (11,12); deoxycytidine kinase, by product inhibition (11,13-18); deoxyguanosine kinase, weakly inhibited (19); dihydroorotase (20); DNA-directed DNA polymerase (21); φ29 DNA polymerase (22); fructose-1,6-bisphosphatase (23); NMN nucleosidase (24); phosphodiesterase I, *or* 5'-exonuclease (25);

polynucleotide 5'-hydroxyl-kinase (26); [protein-P$_{II}$] uridylyltransferase (27); pyruvate carboxylase (28); ribonuclease T$_2$ (29); uracil phosphoribosyltransferase (30). **1**. Van der Weyden & Kelly (1974) *J. Biol. Chem.* **249**, 7282. **2**. Holmes, McDonald, McCord, Wyngaarden & Kelley (1973) *J. Biol. Chem.* **248**, 144. **3**. McCormick, Chassy & Tsibris (1964) *Biochim. Biophys. Acta* **89**, 447. **4**. Webb (1966) *Enzyme and Metabolic Inhibitors*, vol. 2, p. 468, Academic Press, New York. **5**. Gerhart & Pardee (1962) *J. Biol. Chem.* **237**, 891. **6**. Bresnick (1963) *Biochim. Biophys. Acta* **67**, 425. **7**. Vita, Cacciamani, Natalini, *et al.* (1989) *Comp. Biochem. Physiol. B* **93**, 591. **8**. Cacciamani, Vita, Cristalli, *et al.* (1991) *Arch. Biochem. Biophys.* **290**, 285. **9**. Faivre-Nitschke, Grienenberger & Gualberto (1999) *Eur. J. Biochem.* **263**, 896. **10**. Kohno, Kumagai & Tochikura (1983) *Agric. Biol. Chem.* **47**, 19. **11**. Anderson (1973) *The Enzymes*, 3rd ed., **9**, 49. **12**. Krygier & Momparler (1968) *Biochim. Biophys. Acta* **161**, 578. **13**. Kozai, Sonoda, Kobayashi & Sugino (1972) *J. Biochem.* **71**, 485. **14**. Durham & Ives (1970) *J. Biol. Chem.* **245**, 2276. **15**. Datta, Shewach, Mitchell & Fox (1989) *J. Biol. Chem.* **264**, 9359. **16**. Kessel (1968) *J. Biol. Chem.* **243**, 4739. **17**. Momparler & Fischer (1968) *J. Biol. Chem.* **243**, 4298. **18**. Wang, Kucera & Capizzi (1993) *Biochim. Biophys. Acta* **1202**, 309. **19**. Green & Lewis (1979) *Biochem. J.* **183**, 547. **20**. Bresnick & Blatchford (1964) *Biochim. Biophys. Acta* **81**, 150. **21**. Bollum (1968) *Meth. Enzymol.* **12B**, 591. **22**. Lázaro, Blanco & Salas (1995) *Meth. Enzymol.* **262**, 42. **23**. Fujita & Freese (1979) *J. Biol. Chem.* **254**, 5340. **24**. Imai (1979) *J. Biochem.* **85**, 887. **25**. Sabatini & Hotchkiss (1969) *Biochemistry* **8**, 4831. **26**. Levin & Zimmerman (1976) *J. Biol. Chem.* **251**, 1767. **27**. Engelman & Francis (1978) *Arch. Biochem. Biophys.* **191**, 602. **28**. Scrutton & Utter (1965) *J. Biol. Chem.* **240**, 3714. **29**. Irie & Ohgi (1976) *J. Biochem.* **80**, 39. **30**. Natalini, Ruggieri, Santarelli, Vita & Magni (1979) *J. Biol. Chem.* **254**, 1558.

2'-Deoxycytidine 5'-Triphosphate (*or*, dCTP)

This deoxynucleotide (FW$_{free-acid}$ = 467.16 g/mol; λ_{max} = 271 nm (pH 7 and 12; with Molar Absorptivities (ε_M) = 9300 and 9000 M^{-1}cm^{-1}, respectively) and 280 nm at pH 2 (ε_M = 13500 M^{-1}cm^{-1}), is one of four substrates for DNA polymerases. **Target(s):** *N*-acylneuraminate cytidylyl-transferase (1); amidophosphoribosyl-transferase (2); aspartate carbamoyltransferase (3,4); CDP-glycerol diphosphatase (5); CTP synthetase, weakly inhibited (6); cytidylate kinase (7,8); deoxyadenosine kinase (7,9-12); deoxycytidine kinase (7,13-28); deoxyguanosine kinase (29,30); deoxynucleoside kinase (31); dTMP kinase, *or* thymidylate kinase (32); dUTP diphosphatase (33); exopolyphosphatase (34); 3-hydroxyanthranilate oxidase (56); nucleotidases (34); 3'-nucleotidase (34); 5'-nucleotidase (34,35); phosphatidate cytidylyltransferase (36-38); phosphoribosyl-pyrophosphate synthetase, *or* ribose-5-phosphate pyrophosphokinase (39); polynucleotide 5'-hydroxyl-kinase (40,41); pyruvate carboxylase (42); ribonucleoside-diphosphate reductase (43); RNA-directed RNA polymerase (44,45); thymidine kinase (46-52); thymidine triphosphatase (53); tRNA cytidylyltransferase (54); tRNA nucleotidyltransferase (54); uracil phospho-ribosyltransferase (55). **1**. Schmelter, Ivanov, Wember, *et al.* (1993) *Biol. Chem. Hoppe Seyler* **374**, 337. **2**. Hill & Bennent (1969) *Biochemistry* **8**, 122. **3**. Webb (1966) *Enzyme and Metabolic Inhibitors*, vol. 2, p. 468, Academic Press, New York. **4**. Gerhart & Pardee (1962) *J. Biol. Chem.* **237**, 891. **5**. Glaser (1965) *Biochim. Biophys. Acta* **101**, 6. **6**. Wadskov-Hansen, Willemoës, Martinussen, *et al.* (2001) *J. Biol. Chem.* **276**, 38002. **7**. Anderson (1973) *The Enzymes*, 3rd ed., **9**, 49. **8**. Liou, Dutschman, Lam, Jiang & Cheng (2002) *Cancer Res.* **62**, 1624. **9**. Krygier & Momparler (1971) *J. Biol. Chem.* **246**, 2752. **10**. Krygier & Momparler (1968) *Biochim. Biophys. Acta* **161**, 578. **11**. Ikeda, Ma & Ives (1994) *Biochemistry* **33**, 5328. **12**. Welin, Wang, Eriksson & Eklund (2007) *J. Mol. Biol.* **366**, 1615. **13**. Ives & Wang (1978) *Meth. Enzymol.* **51**, 337. **14**. Bouffard, Laliberte & Momparler (1993) *Biochem. Pharmacol.* **45**, 1857. **15**. Yamada, Goto & Ogasawara

(1983) *Biochim. Biophys. Acta* **761**, 34. **16**. Kozai, Sonoda, Kobayashi & Sugino (1972) *J. Biochem.* **71**, 485. **17**. Durham & Ives (1970) *J. Biol. Chem.* **245**, 2276. **18**. Kim & Ives (1989) *Biochemistry* **28**, 9043. **19**. Datta, Shewach, Mitchell & Fox (1989) *J. Biol. Chem.* **264**, 9359. **20**. Kessel (1968) *J. Biol. Chem.* **243**, 4739. **21**. Momparler & Fischer (1968) *J. Biol. Chem.* **243**, 4298. **22**. Wang, Kucera & Capizzi (1993) *Biochim. Biophys. Acta* **1202**, 309. **23**. Coleman, Stoller, Drake & Chabner (1975) *Blood* **46**, 791. **24**. Ives & Durham (1970) *J. Biol. Chem.* **245**, 2285. **25**. Hughes, Hahn, Reynolds & Shewach (1997) *Biochemistry* **36**, 7540. **26**. Johansson & Karlsson (1995) *Biochem. Pharmacol.* **50**, 163. **27**. Mani, Usova, Eriksson & Cass (2004) *Nucleosides Nucleotides Nucleic Acids* **23**, 1343. **28**. Eriksson, Munch-Petersen, Johansson & Eklund (2002) *Cell. Mol. Life Sci.* **59**, 1327. **29**. Yamada, Goto & Ogasawara (1982) *Biochim. Biophys. Acta* **709**, 265. **30**. Green & Lewis (1979) *Biochem. J.* **183**, 547. **31**. Knecht, Petersen, Munch-Petersen & Piskur (2002) *J. Mol. Biol.* **315**, 529. **32**. Nelson & Carter (1969) *J. Biol. Chem.* **244**, 5254. **33**. Bergman, Nyman & Larsson (1998) *FEBS Lett.* **441**, 327. **34**. Proudfoot, Kuznetsova, Brown, *et al.* (2004) *J. Biol. Chem.* **279**, 54687. **35**. Madrid-Marina & Fox (1986) *J. Biol. Chem.* **261**, 444. **36**. Belendiuk, Mangnall, Tung, Westley & Getz (1978) *J. Biol. Chem.* **253**, 4555. **37**. Carman & Kelley (1992) *Meth. Enzymol.* **209**, 242. **38**. Kelley & Carman (1987) *J. Biol. Chem.* **262**, 14563. **39**. Wong & Murray (1969) *Biochemistry* **8**, 1608. **40**. Austin, Sirakoff, Roop & Moyer (1978) *Biochim. Biophys. Acta* **522**, 412. **41**. Levin & Zimmerman (1976) *J. Biol. Chem.* **251**, 1767. **42**. Scrutton & Utter (1965) *J. Biol. Chem.* **240**, 3714. **43**. Kucera & Paulus (1982) *Arch. Biochem. Biophys.* **214**, 114. **44**. Alaoui-Lsmaili, Hamel, L'Heureux, *et al.* (2000) *J. Hum. Virol.* **3**, 306. **45**. Hung, Gibbs & Tsiang (2002) *Antiviral Res.* **56**, 99. **46**. Cheng (1978) *Meth. Enzymol.* **51**, 365. **47**. Bresnick, Thompson, Morris & Liebelt (1964) *Biochem. Biophys. Res. Commun.* **16**, 278. **48**. Ellims & Van der Weyden (1981) *Biochim. Biophys. Acta* **660**, 238. **49**. Barroso, Carvalho & Flatmark (2005) *Biochemistry* **44**, 4886. **50**. Wang, Saada & Eriksson (2003) *J. Biol. Chem.* **278**, 6963. **51**. Kit, Leung & Trkula (1973) *Arch. Biochem. Biophys.* **158**, 503. **52**. Taylor, Stafford & Jones (1972) *J. Biol. Chem.* **247**, 1930. **53**. Dahlmann (1984) *Hoppe-Seyler's Z. Physiol. Chem.* **365**, 1263. **54**. Hill & Nazario (1973) *Biochemistry* **12**, 482. **55**. Natalini, Ruggieri, Santarelli, Vita & Magni (1979) *J. Biol. Chem.* **254**, 1558. **56**. Nair (1972) *Arch. Biochem. Biophys.* **153**, 139.

2-Deoxy-2,3-didehydro-*N*-acetylneuraminate

This derivative of neuraminic acid (FW$_{free-acid}$ = 291.26 g/mol), also known as 2,3-dehydro-2-deoxysialic acid and abbreviated as Neu5Ac2en, was the first substrate-based inhibitor of neuraminidase. While a strong inhibitor of influenza neuraminidase, it has proven to be not selective for the influenza enzyme and thus not suitable as an anti-influenza drug. **Target(s):** *N*-acylneuraminate cytidylyltransferase (1); exo-α-sialidase, *or* neuraminidase, *or* sialidase (2-28). **1**. Corfield, Schauer & Wember (1979) *Biochem. J.* **177**, 1. **2**. Meindl & Tuppy (1969) *Hoppe-Seyler's Z. Physiol. Chem.* **350**, 1088. **3**. Miller, Wand & Flashner (1978) *Biochem. Biophys. Res. Commun.* **83**, 1479. **4**. Gottschalk & Bhargava (1971) *The Enzymes*, 3rd ed., **5**, 321. **5**. Meindl, Bodo, Palese, Schulman & Tuppy (1974) *Virology* **58**, 457. **6**. Holzer, von Itzstein, Jin, *et al.* (1993) *Glycoconj. J.* **10**, 40. **7**. Iriyama, Takeuchi, Shiraishi, *et al.* (2000) *Comp. Biochem. Physiol. B* **126**, 561. **8**. Wang, Tong, Grant & Cihlar (2001) *J. Virol. Methods* **98**, 53. **9**. Byers, Tarelli, Homer & Beighton (2000) *J. Med. Microbiol.* **49**, 235. **10**. Nok & Rivera (2003) *Parasitol. Res.* **89**, 302. **11**. Teufel, Roggentin & Schauer (1989) *Biol. Chem. Hoppe-Seyler* **370**, 435. **12**. Schauer & Wember (1989) *Biol. Chem. Hoppe-Seyler* **370**, 183. **13**. von Nicolai, Hammann, Werner & Zilliken (1983) *FEMS Microbiol. Lett.* **17**, 217. **14**. Warner, Chang, Ferrari, *et al.* (1993) *Glycobiology* **3**, 455. **15**. Michalski, Corfield & Schauer (1986) *Biol. Chem. Hoppe-Seyler* **367**, 715. **16**. Waters, Corfield, Eisenthal & Pennok (1994) *Biochem. J.* **301**, 777. **17**. Arora & L. F. Gabriel (1986) *Biochim. Biophys. Acta* **884**, 73. **18**. Sastre, Cobaleda, Cabezas & Villar (1991) *Biol. Chem. Hoppe-Seyler* **372**, 923. **19**. Miyagi, Hata, Hasegawa & Aoyagi (1993) *Glycoconjugate J.* **10**, 45. **20**. Hoyer, Roggentin, Schauer & Vimr (1991) *J. Biochem.* **110**, 462. **21**. Kobayashi, Ito, Ikeda, Tannaka & Saito (2000) *J. Biochem.* **127**, 569. **22**. Amino, Porto, Chammas, Egami &

Schenkman (1998) *J. Biol. Chem.* **273**, 24575. **23.** Engstler, Reuter & Schauer (1992) *Mol. Biochem. Parasitol.* **54**, 21. **24.** Nok, Nzelibe & Yako (2003) *Z. Naturforsch. C* **58**, 594. **25.** Burg & Muthing (2001) *Carbohydr. Res.* **330**, 335. **26.** Buratai, Nok, Ibrahim, Umar & Esievo (2005) *Cell Biochem. Funct.* **24**, 71. **27.** Inoue, Lin, Inoue, *et al.* (2001) *Biochem. Biophys. Res. Commun.* **280**, 104. **28.** Schauer, Wember & Tschesche (1984) *Hoppe-Seyler's Z. Physiol. Chem.* **365**, 419.

2-Deoxy-2,3-didehydro-*N*-trifluoroacetylneuraminate

This neuraminic acid derivative (FW$_{free-acid}$ = 345.23 g/mol) is a substrate-based inhibitor of neuraminidase. While a strong inhibitor of influenza neuraminidase (K_i = 0.8 μM), it has proven to be ineffective against influenza infection. **1.** Palese, Schulman, Bodo & Meindl (1974) *Virology* **59**, 490.

5'-Deoxy-*N*⁴,5-diphenyltubercidin

This deoxytubercidin derivative (FW = 402.45 g/mol) inhibits human adenosine kinase (IC$_{50}$ = 0.5 nM). Adenosine is an endogenous neuromodulator that is produced in the central and the peripheral nervous systems and exerts anticonvulsant, anti-inflammatory, and analgesic properties. Efforts to use adenosine receptor agonists are plagued by dose-limiting cardiovascular side effects. As an alternative, the use of adenosine kinase inhibitors (AKIs) as potential antiseizure agents was demonstrated, showing an adenosine receptor mediated therapeutic effect in the absence of overt cardiovascular side effects. 1. Boyer, Ugarkar, Solbach, *et al.* (2005) *J. Med. Chem.* **48**, 6430.

5'-Deoxy-5'-ethylimmucillin-A

This transition-state, pseudonucleoside analogue (FW$_{free-base}$ = 265.32 g/mol) inhibits 5'-methylthioadenosine/*S*-adenosyl-homocysteine nucleosidase, K_i' = 38 pM (1) and is a tight, slow-binding inhibitor of human *S*-methyl-5' thioadenosine phosphorylase, K_i = 44 nM (2). When the ring nitrogen is protonated, this analogue mimics the charge of oxa-carbenium ion intermediate that forms transiently in many glycosidase reactions, and this property results in tight binding. 1. Singh, Evans, Lenz, *et al.* (2005) *J. Biol. Chem.* **280**, 18265. 2. Evans, Furneaux, Schramm, Singh & Tyler (2004) *J. Med. Chem.* **47**, 3275.

5'-Deoxy-5'-(ethylthio)adenosine

This nucleoside (FW = 311.36 g/mol; λ$_{max}$ at 260 nm (pH 7) with a molar absorptivity of 15200 M^{-1}cm^{-1}), also known as 5'-ethylthioadenosine, is an alternative substrate and competitive inhibitor of *Escherichia coli* 5'-methylthioadenosine/*S*-adenosylhomocysteine nucleosidase (K_i = 0.13 μM). **Target(s):** adenosylhomocysteine nucleosidase (1,2); *S*-adenosyl-methionine cyclotransferase (3); deoxycytidine kinase, K_i = 0.7 mM (4); 5'-methylthioadenosine/*S* adenosylhomocysteine nucleosidase (1,2,5,6); *S*-methyl-5'-thioadenosine phosphorylase (7); spermidine synthase, *or* putrescine aminopropyltransferase (8-12); spermine synthase, *or* spermidine aminopropyltransferase (9,10); tRNA methyltransferases (13); tRNA (uracil-5-)-methyltransferase (14). **1.** Cornell, Swarts, Barry & Riscoe (1996) *Biochem. Biophys. Res. Commun.* **228**, 724. **2.** Ferro, Barrett & Shapiro (1976) *Biochim. Biophys. Acta* **438**, 487. **3.** Mudd (1959) *J. Biol. Chem.* **234**, 87. **4.** Krenitsky, Tuttle, Koszalka, *et al.* (1976) *J. Biol. Chem.* **251**, 4055. **5.** Kushad, Richardson & Ferro (1985) *Plant Physiol.* **79**, 525. **6.** Kushad, Orvos & Ferro (1992) *Physiol. Plant.* **86**, 532. **7.** Ferro, Wrobel & Nicolette (1979) *Biochim. Biophys. Acta* **570**, 65. **8.** Raina, Hyvönen, Eloranta, *et al.* (1984) *Biochem. J.* **219**, 991. **9.** Pegg (1983) *Meth. Enzymol.* **94**, 294. **10.** Hibasami, Borchardt, Chen, Coward & Pegg (1980) *Biochem. J.* **187**, 419. **11.** Hibasami, Kawase, Tsukada, *et al.* (1988) *FEBS Lett.* **229**, 243. **12.** Bowman, Tabor & Tabor (1973) *J. Biol. Chem.* **248**, 2480. **13.** Kerr & Borek (1973) *The Enzymes*, 3rd ed. (Boyer, ed.), **9**, 167. **14.** Tscheme & Wainfan (1978) *Nucl. Acids Res.* **5**, 451.

L-1-Deoxyfucomycin

This presumptive transition-state mimic (FW = 147.17 g/mol), also called 5-amino-1,5-dideoxy-L-fucopyranose and 1-deoxy-L-*fuco* nojirimycin, inhibits α-L-fucosidase: K_i = 10 nM for the human liver enzyme. When the ring nitrogen is protonated, this analogue mimics the charge of oxa-carbenium ion intermediate that forms transiently in many glycosidase reactions, a property that often leads to very tight binding. **Target(s):** α-L-fucosidase (1-5); glycoprotein 3-α-L-fucosyltransferase (1,3 fucosyltransferase V, weakly inhibited, but inhibition increases synergistically in the presence of GDP (6). **1.** Mooser (1992) *The Enzymes*, 3rd ed., **20**, 187. **2.** Winchester, Barker, Baines, *et al.* (1990) *Biochem. J.* **265**, 277. **3.** Gramer, Schaffer, Sliwkowski & Goochee (1994) *Glycobiology* **4**, 611. **4.** Berteau, McCort, Goasdoue, Tissot & Daniel (2002) *Glycobiology* **12**, 273. **5.** Dumas, Kajimoto, Liu, C.-H. *et al.* (1992) *Bioorg. Med. Chem. Lett.* **2**, 33. **6.** Murray, Takayama, Schultz & Wong (1996) *Biochemistry* **35**, 11183.

1-Deoxygalactonojirimycin

This presumptive transition-state mimic (FW = 163.17 g/mol), also known as 5-amino-1,5-dideoxy-D-*galacto*-pyranose and 1-deoxy D-*galacto*-nojirimycin, competitively inhibits galactosidases (1). When the ring nitrogen is protonated, this analogue mimics the charge of oxa-carbenium ion intermediate that forms transiently in many glycosidase reactions, a property that results in tight binding. 1-Deoxygalactomycin has been used in the treatment of patients with Fabry disease, a disorder of glycosphingolipid metabolism caused by deficiency of lysosomal α-galactosidase A (2). This agent binds to and stabilizes the mutant enzyme ($IC_{50} = 0.04$ μM), allowing it to be processed and targeted for lysosomal disposal. **Target(s):** galactinol-sucrose galactosyltransferase, or raffinose synthase (1); α-galactosidase (2-5); β-galactosidase (2,6); α-L-fucosidase (7); lactase/phlorizin hydrolase, or glycosylceramidase (8); sucrose α-glucosidase, or sucrase (9). **1.** Peterbauer, Mach, Mucha & Richter (2002) *Planta* **215**, 839. **2.** Mooser (1992) *The Enzymes*, 3rd ed., **20**, 187. **3.** Fan, Ishii, Asano & Suzuki (1999) *Nat. Med.* **5**, 112. **4.** Asano, Ishii, Kizu, *et al.* (2000) *Eur. J. Biochem.* **267**, 4179. **5.** Ishii, Suzuki & Fan (2000) *Arch. Biochem. Biophys.* **377**, 228. **6.** Tominaga, Ogawa, Taniguchi, *et al.* (2001) *Brain Dev.* **23**, 284. **7.** Berteau, McCort, Goasdoue, Tissot & Daniel (2002) *Glycobiology* **12**, 273. **8.** Wacker, Keller, Falchetto, Legler & Semenza (1992) *J. Biol. Chem.* **267**, 18744. **9.** Karley, Ashford, Minto, Pritchard & Douglas (2005) *J. Insect Physiol.* **51**, 1313.

2-Deoxy-D-galactose

β-anomer   α-anomer

This deoxy sugar (FW = 164.16 g/mol; CAS 1949-89-9), also known as 2-deoxy-D-talose, inhibits brain glycoprotein fucosylation. It also induces amnesia, but does not affect the K^+-stimulated release of noradrenaline from rat striatal slices or that of dopamine from striatal synaptosomes. Long-term memory in chicks is also blocked by this fucose analogue. 2-Deoxy-D-galactose is also an alternative substrate for galactokinase. **Target(s):** aldose 1-epimerase (1); galactokinase, also as an alternative substrate (2); galactose oxidase, also as an alternative substrate (3); β-galactosidase (4); glycoprotein fucosylation (5). **1.** Bailey, Fishman, Kusiak, Mulhern & Pentchev (1975) *Meth. Enzymol.* **41**, 471. **2.** Mathai & Beutler (1967) *Enzymologia* **33**, 224. **3.** Yip & Dain (1968) *Enzymologia* **35**, 368. **4.** Huber, Roth & Bahl (1990) *J. Protein Chem.* **15**, 621. **5.** Jork, Grecksch & Matthies (1986) *Pharmacol. Biochem. Behav.* **25**, 1137.

2-Deoxy-D-glucose

β-anomer   α-anomer

This deoxy sugar (FW = 164.16 g/mol; CAS 154-17-6), also known as 2-deoxy-D-mannose and 2-deoxy-D-*arabino*-hexose, is a glucose antimetabolite that has also been used as an antiviral agent. This deoxy sugar appears to potentiate the action of various anticancer treatments (including etoposide, arsenic trioxide, and radiation), suggesting that impairment of glycolysis increases cell susceptibility selective cancer cell toxicity. **Hexokinase Inhibition:** 2DG is a well-known nonmetabolizable inhibitor of glycolysis – not directly, but only after hexokinase-catalyzed phosphorylation to form 2-deoxy-D-glucose 6-phosphate. The latter is a potent product and allosteric inhibitor of mammalian hexokinase ($K_i \approx 20$-50 μM), but is a much weak inhibitor of yeast hexokinase ($K_i \approx 5$-10 mM) (1). **Hexokinase-Independent Pathway:** 2DG's anticancer effects cannot be fully explained by hexokinase inhibition alone. It induces a transient expression of p21 and continuous expression of p53 in colorectal cancer cells (SW620). Treatment also causes cell-cycle arrest at G_0/G_1 phase, inducing apoptosis through the mitochondrial pathway. The effects of 2DG on p21 and p53 protein levels are totally independent of its inhibitory effect on either hexokinase or ATP levels (2). **Target(s):** aldose 1-epimerase (3); α-amylase (4); concanavalin A (5); dolichyl-phosphate β-glucosyltransferase (6); endo-1,4-β-xylanase (7); glucan 1,3-α-glucosidase, or glucosidase II, or mannosyl oligosaccharide glucosidase II (8); glucan-1,3-β-glucosidase, weakly inhibited (9); glucokinase, also alternative substrate (10); glucose oxidase (11-13); glucose-6-phosphate isomerase (14); glycolysis (15); hexokinase, also alternative substrate that generates nonmetabolizable 2-deoxyglucose 6-P (1,10,16-18); *myo*-inositol-1-phosphate synthase (19); lactose synthase (20); laminaribiose phosphorylase (21); mannokinase, also weak alternative substrate (22); protein

glycosylation (6,23-25); UDP-*N*-acetylglucosamine:dolichyl-phosphate *N*-acetylglucosamine phosphotransferase (23). **1.** Purich, Rudolph & Fromm (1973) *Adv. Enzymol.* **39**, 259. **2.** Muley, Olinger & Tummala (2015) *Nutr. Cancer* **67**, 514. **3.** Bailey, Fishman, Kusiak, Mulhern & Pentchev (1975) *Meth. Enzymol.* **41**, 471. **4.** Oosthuizen, Naude, Oelofsen, Muramoto & Kamiya (1994) *Int. J. Biochem.* **26**, 1313. **5.** Goldstein, Hollerman & Smith (1965) *Biochemistry* **4**, 876. **6.** Schwarz & Datema (1982) *Meth. Enzymol.* **83**, 432. **7.** Khanna & Gauri (1993) *Enzyme Microb. Technol.* **15**, 990. **8.** Presper & Heath (1983) *The Enzymes*, 3rd ed., **16**, 449. **9.** Fleet & Phaff (1975) *Biochim. Biophys. Acta* **401**, 318. **10.** Monasterio & Cardenas (2003) *Biochem. J.* **371**, 29. **11.** Keilin & Hartree (1946) *Nature* **157**, 801. **12.** Keilin & Hartree (1948) *Biochem. J.* **42**, 221 and 230. **13.** Bentley (1963) *The Enzymes*, 2nd ed., **7**, 567. **14.** Wick, Drury, Nakada & Wolfe (1957) *J. Biol. Chem.* **224**, 963. **15.** Webb (1966) *Enzyme and Metabolic Inhibitors*, vol. 2, p. 386, Academic Press, New York. **16.** Bertoni (1981) *J. Neurochem.* **37**, 1523. **17.** Bertoni & Weintraub (1984) *J. Neurochem.* **42**, 513. **18.** Petit, Blázquez & Gancedo (1996) *FEBS Lett.* **378**, 185. **19.** Charalampous & Chen (1966) *Meth. Enzymol.* **9**, 698. **20.** Ebner (1973) *The Enzymes*, 3rd ed., **9**, 363. **21.** Goldemberg, Maréchal & De Souza (1966) *J. Biol. Chem.* **241**, 45. **22.** Sabater, Sebastián & Asensio (1972) *Biochim. Biophys. Acta* **284**, 406. **23.** Elbein (1983) *Meth. Enzymol.* **98**, 135. **24.** Elbein (1987) *Meth. Enzymol.* **138**, 661. **25.** Schwarz & Datema (1980) *Trends Biochem. Sci.* **5**, 65.

2-Deoxy-D-glucose 6-Phosphate

This phosphorylated deoxy sugar ($FW_{free-acid} = 244.14$ g/mol; CAS 33068-19-8) inhibits a number of glucose-dependent enzymes. Although a relatively weak inhibitor of brain hexokinase, 2-deoxyglucose 6-P is effective in mimicking the release of hexokinase from mitochondria (**See also** 2-Deoxy-D-glucose). **Target(s):** glucose-6-phosphatase, weakly inhibited (1); glucose-6-phosphate dehydrogenase (1); glucose-6-phosphate isomerase (1,2); glycogen synthase (1); inositol-phosphate phosphatase, or 1-L-*myo*-inositol-1-phosphatase (3); inositol-3-phosphate synthase, or *myo*-inositol-1 phosphate synthase (3-9); phosphomannomutase (10). **1.** Gauthier, Denis-Pouxviel & Murat (1990) *Internat. J. Biochem.* **22**, 419. **2.** Noltmann (1972) *The Enzymes*, 3rd ed., **6**, 271. **3.** Charalampous & Chen (1966) *Meth. Enzymol.* **9**, 698. **4.** Wong, Mauck & Sherman (1982) *Meth. Enzymol.* **90**, 309. **5.** Loewus, Bedgar & Loewus (1984) *J. Biol. Chem.* **259**, 7644. **6.** Barnett, Rasheed & Corina (1973) *Biochem. Soc. Trans.* **1**, 1267. **7.** Wong & Sherman (1985) *J. Biol. Chem.* **260**, 11083. **8.** Gumber, Loewus & Loewus (1984) *Arch. Biochem. Biophys.* **231**, 372. **9.** Donahue & Henry (1981) *J. Biol. Chem.* **256**, 7077. **10.** Guha & Rose (1985) *Arch. Biochem. Biophys.* **243**, 168.

2'-Deoxyguanosine (or, dG)

This purine nucleoside (FW = 267.24 g/mol; CAS 207121-55-9; λ_{max} at 253 nm, with ε = 13000 $M^{-1}cm^{-1}$; Water-soluble) is an important DNA building block. **Target(s):** adenosine kinase (1); cytidine deaminase, weakly inhibited (2); cytosine deaminase (3); dihydroorotase (4); *S*-methyl-5'-deoxyadenosine phosphorylase (5); 5'-nucleotidase (6); purine nucleosidase (7); ribonuclease T_2 (8); ribonucleotide reductase (9). **1.** Kidder (1982) *Biochem. Biophys. Res. Commun.* **107**, 381. **2.** Hosono & Kuno (1973) *J. Biochem.* **74**, 797. **3.** Yu, Sakai & Omata (1976) *Agric. Biol. Chem.* **40**, 543 and 551. **4.** Bresnick & Blatchford (1964) *Biochim. Biophys. Acta* **81**, 150. **5.** Koszalka & Krenitsky (1986) *Adv. Exp. Med. Biol.* **195B**, 559. **6.** Garvey, Lowen & Almond (1998) *Biochemistry* **37**, 9043. **7.** Koszalka &

Krenitsky (1979) *J. Biol. Chem.* **254**, 8185. **8**. Irie & Ohgi (1976) *J. Biochem.* **80**, 39. **9**. Carter & Cory (1989) *Adv. Enzyme Regul.* **29**, 123.

2'-Deoxyguanosine 5'-Diphosphate

This deoxynucleotide (FW$_{free\text{-}acid}$ = 427.20; λ_{max} at 253 nm, with ε = 13000 $M^{-1}cm^{-1}$), often abbreviated dGDP, inhibits DNA polymerase, deoxyguanosine kinase, deoxyadenosine kinase, and 5'-nucleotidase. In eukaryotes, GDP is converted to dGDP by ribonucleotide reductase (RNR). In prokaryotes, dGDP is formed from dGTP, the latter produced by bacterial RNR. **Target(s):** adenylosuccinate synthetase (1); Amidophosphoribosyltransferase (2); dCMP deaminase, slightly inhibited (3); deoxyadenosine kinase (4,5); deoxycytidine kinase (6); deoxyguanosine kinase (7-10); DNA-directed DNA polymerase (11); dTMP kinase, *or* thymidylate kinase (12,13); galactoside 2-α-L-fucosyltransferase (14); glycoprotein 6-α-L-fucosyltransferase (15); GTP diphosphokinase (16); 5'-nucleotidase (17); 8-oxo-2'-deoxyguanosine 5'-triphosphate pyrophosphohydrolase (18); thymidine kinase (19). **1**. Van der Weyden & Kelly (1974) *J. Biol. Chem.* **249**, 7282. **2**. Hill & Bennett (1969) *Biochemistry* **8**, 122. **3**. Ellims, Kao & Chabner (1981) *J. Biol. Chem.* **256**, 6335. **4**. Anderson (1973) *The Enzymes*, 3rd ed. (Boyer, ed.), **9**, 49. **5**. Krygier & Momparler (1968) *Biochim. Biophys. Acta* **161**, 578. **6**. Kozai, Sonoda, Kobayashi & Sugino (1972) *J. Biochem.* **71**, 485. **7**. Yamada, Goto & Ogasawara (1983) *FEBS Lett.* **157**, 51. **8**. Yamada, Goto & Ogasawara (1982) *Biochim. Biophys. Acta* **709**, 265. **9**. Barker & Lewis (1981) *Biochim. Biophys. Acta* **658**, 111. **10**. Gower, Carr & Ives (1979) *J. Biol. Chem.* **254**, 2180. **11**. Bollum (1968) *Meth. Enzymol.* **12B**, 591. **12**. Kielley (1970) *J. Biol. Chem.* **245**, 4204. **13**. Tamiya, Yusa, Yamaguchi, *et al.* (1989) *Biochim. Biophys. Acta* **995**, 28. **14**. Bella & Kim (1971) *Biochem. J.* **125**, 1157. **15**. Ihara, Ikeda & Taniguchi (2006) *Glycobiology* **16**, 333. **16**. Sy & Akers (1976) *Biochemistry* **15**, 4399. **17**. Madrid-Marina & Fox (1986) *J. Biol. Chem.* **261**, 444. **18**. Bialkowski & Kasprzak (2003) *Free Radic. Biol. Med.* **35**, 595. **19**. Chraibi & Wright (1983) *J. Biochem.* **93**, 323.

2'-Deoxyguanosine 5'-Monophosphate

This deoxynucleotide (FW$_{free\text{-}acid}$ = 348.23 g/mol; λ_{max} at 253 nm, with ε = 13000 $M^{-1}cm^{-1}$), often abbreviated dGMP, is an important component in DNA and can be released from DNA via the actions of certain phosphodiesterases (*e.g.*, *Escherichia coli* phosphodiesterase). dGMP is readily hydrolyzed by dilute mineral acids. **Target(s):** adenylosuccinate synthetase (1-5); amidophosphoribosyl-transferase (6); *Basidobolus haptosporus* nuclease Bh1 (7); cytosine deaminase (8,9); dCMP deaminase. *Or* deoxycytidylate deaminase (1,10-17); deoxyadenosine kinase (18,19); deoxyguanosine kinase, product inhibition (20-24); deoxynucleotide 3'-phosphatase (25); dihydro-orotase (26); DNA-directed DNA polymerase (27); φ29 DNA polymerase (28); dTMP kinase, *or* thymidylate kinase (29,30); fructose-1,6 bisphosphatase (31); NMN nucleosidase (32); ribonuclease T2 (33); thymidylate 5'-phosphatase (34). **1**. Webb (1966) *Enzyme and Metabolic Inhibitors*, vol. **2**, Academic Press, New York. **2**. Wyngaarden & Greenland (1963) *J. Biol. Chem.* **238**, 1054. **3**. Spector, Jones & Elion (1979) *J. Biol. Chem.* **254**, 8422. **4**. Van der Weyden & Kelly (1974) *J. Biol. Chem.* **249**, 7282. **5**. Spector & Miller (1976) *Biochim. Biophys. Acta* **445**, 509. **6**. Hill & Bennett (1969) *Biochemistry* **8**, 122. **7**. Desai & Shankar (2000) *Eur. J. Biochem.* **267**, 5123. **8**. Yu, Sakai & Omata (1976) *Agric. Biol. Chem.* **40**, 543. **9**. West (1988) *Experientia*

41, 1563. **10**. Maley (1967) *Meth. Enzymol.* **12A**, 170. **11**. Scarano, Bonaduce & de Petrocellis (1960) *J. Biol. Chem.* **235**, 3556. **12**. Geraci, Rossi & Scarano (1967) *Biochemistry* **6**, 183. **13**. Rossi, Geraci & Scarano (1967) *Biochemistry* **6**, 3640. **14**. Scarano, Geraci & Rossi (1967) *Biochemistry* **6**, 3645. **15**. Maley, Lobo & Maley (1993) *Biochim. Biophys. Acta* **1162**, 161. **16**. Raia, Nucci, Vaccaro, Sepe & Rella (1982) *J. Mol. Biol.* **157**, 557. **17**. Ellims, Kao & Chabner (1981) *J. Biol. Chem.* **256**, 6335. **18**. Anderson (1973) *The Enzymes*, 3rd ed., **9**, 49. **19**. Krygier & Momparler (1968) *Biochim. Biophys. Acta* **161**, 578. **20**. Yamada, Goto & Ogasawara (1982) *Biochim. Biophys. Acta* **709**, 265. **21**. Green & Lewis (1979) *Biochem. J.* **183**, 547. **22**. Barker & Lewis (1981) *Biochim. Biophys. Acta* **658**, 111. **23**. Gower, Carr & Ives (1979) *J. Biol. Chem.* **254**, 2180. **24**. Andersen & Neuhard (2001) *J. Biol. Chem.* **276**, 5518. **25**. Magnusson (1971) *Eur. J. Biochem.* **20**, 225. **26**. Bresnick & Blatchford (1964) *Biochim. Biophys. Acta* **81**, 150. **27**. Bollum (1968) *Meth. Enzymol.* **12B**, 591. **28**. Lázaro, Blanco & Salas (1995) *Meth. Enzymol.* **262**, 42. **29**. Bello, M. J. van Bibber & M. J. Bessman (1961) *Biochim. Biophys. Acta* **53**, 194. **30**. Pochet, Dugué, Labesse, Delepierre & Munier-Lehmann (2003) *ChemBioChem* **4**, 742. **31**. Fujita & Freese (1979) *J. Biol. Chem.* **254**, 5340. **32**. Imai (1979) *J. Biochem.* **85**, 887. **33**. Irie & Ohgi (1976) *J. Biochem.* **80**, 39. **34**. Price & Fogt (1973) *J. Biol. Chem.* **248**, 1372.

2'-Deoxyguanosine 5'-Triphosphate

This nucleotide (FW$_{free\text{-}acid}$ = 507.18 g/mol; FW$_{trisodium\text{-}salt}$ = 573.13 g/mol; CAS 93919-41-6; λ_{max} at 253 nm, with ε = 13000 $M^{-1}cm^{-1}$) is a DNA polymerase substrate. Although dGTP is not a substrate in any other metabolic process, a curious finding is that dGTP is lodged with the nonexchangeable nucleotide site of αβ-tubulin dimers within microtubules within neurons treated with nerve growth factor (1). The likely explanation is that intracellular stores of GDP (and consequentially GTP) are depleted by ribonucleotide reductase in NGF-stimulated cells, such that newly synthesized tubulin must use dGTP in place of GTP as the α and β-tubulin monomers assemble into αβ-tubulin dimers. **Target(s):** amidophospho-ribosyltransferase (2); deoxyadenosine kinase (3-6); deoxycytidine kinase (7); deoxyguanosine kinase (8-15); DNA helicase (16); exopolyphosphatase (17); glucose-1 phosphate cytidylyltransferase (18); GTP cyclohydrolase I (19,20); GTP diphosphokinase (21); guanylate cyclase (22,23); guanylate kinase (3); 3-hydroxyanthranilate oxidase (40); methionine *S*-adenosyltransferase (24); nucleoside-triphosphatase, alternative substrate (16); nucleotidases (17); 3' nucleotidases (17); 5'-nucleotidase (17,25); phosphoribosyl-pyrophosphate synthetase, *or* ribose-5 phosphate pyrophosphokinase (26); polynucleotide 5'-hydroxyl-kinase (27,28); ribonucleoside diphosphate reductase (29-36); RNA-directed RNA polymerase (37,38); thymidine kinase (39). **1**. Angelastro & Purich (1992) *J. Biol. Chem.* **267**, 25685. **2**. Hill & Bennett, Jr. (1969) *Biochemistry* **8**, 122. **3**. Anderson (1973) *The Enzymes*, 3rd ed., **9**, 49. **4**. Krygier & Momparler (1971) *J. Biol. Chem.* **246**, 2752. **5**. Krygier & Momparler (1968) *Biochim. Biophys. Acta* **161**, 578. **6**. Ikeda, Ma & Ives (1994) *Biochemistry* **33**, 5328. **7**. Kozai, Sonoda, Kobayashi & Sugino (1972) *J. Biochem.* **71**, 485. **8**. Yamada, Goto & Ogasawara (1983) *FEBS Lett.* **157**, 51. **9**. Gower, Carr & Ives (1979) *J. Biol. Chem.* **254**, 2180. **10**. Yamada, Goto & Ogasawara (1982) *Biochim. Biophys. Acta* **709**, 265. **11**. Green & Lewis (1979) *Biochem. J.* **183**, 547. **12**. Eriksson, Munch-Petersen, Johansson & Eklund (2002) *Cell. Mol. Life Sci.* **59**, 1327. **13**. Barker & Lewis (1981) *Biochim. Biophys. Acta* **658**, 111. **14**. Park & Ives (2002) *J. Biochem. Mol. Biol.* **35**, 244. **15**. Ma, Hong & Ives (1995) *J. Biol. Chem.* **270**, 6595. **16**. Locatelli, Gosselin, Spadari & Maga (2001) *J. Mol. Biol.* **313**, 683. **17**. Proudfoot, Kuznetsova, Brown, *et al.* (2004) *J. Biol. Chem.* **279**, 54687. **18**. Kimata & Suzuki (1966) *J. Biol. Chem.* **241**, 1099. **19**. Blau & Niederwieser (1984) *Biochem. Clin. Aspects Pteridines* **3**, 77. **20**. Yim & Brown (1976) *J. Biol. Chem.* **251**, 5087. **21**. Sy & Akers (1976) *Biochemistry* **15**, 4399. **22**. Hardman & Sutherland (1969) *J. Biol. Chem.* **244**, 6363. **23**. Durham (1976) *Eur. J. Biochem.* **61**, 535. **24**. Chou & Talalay (1973) *Biochim. Biophys. Acta* **321**, 467. **25**. Madrid-Marina & I. H. Fox (1986) *J. Biol. Chem.* **261**, 444. **26**. Wong & Murray (1969)

Biochemistry **8**, 1608. **27**. Austin, Sirakoff, Roop & Moyer (1978) *Biochim. Biophys. Acta* **522**, 412. **28**. Levin & Zimmerman (1976) *J. Biol. Chem.* **251**, 1767. **29**. Moore (1967) *Meth. Enzymol.* **12A**, 155. **20**. Holmgren (1981) *Curr. Top. Cell. Regul.* **19**, 47. **31**. Averett, Lubbers, Elion & Spector (1983) *J. Biol. Chem.* **258**, 9831. **32**. Kucera & Paulus (1982) *Arch. Biochem. Biophys.* **214**, 114. **33**. Berglund (1972) *J. Biol. Chem.* **247**, 7276. **34**. Rofougaran, Vodnala & Hofer (2006) *J. Biol. Chem.* **281**, 27705. **35**. Cory, Rey, Carter & Bacon (1985) *J. Biol. Chem.* **260**, 12001. **36**. Moore & Hurlbert (1966) *J. Biol. Chem.* **241**, 4802. **37**. Alaoui-Lsmaili, Hamel, L'Heureux, *et al.* (2000) *J. Hum. Virol.* **3**, 306. **38**. Hung, Gibbs & Tsiang (2002) *Antiviral Res.* **56**, 99. **39**. Gröbner (1979) *J. Biochem.* **86**, 1607. **40**. Nair (1972) *Arch. Biochem. Biophys.* **153**, 139.

5'-Deoxy-5'-[(3-hydrazinopropyl)methylamino]adenosine

This synthetic nucleoside (FW$_{free-base}$ = 352.40 g/mol; λ_{max} at 260 nm (pH 7) with a molar absorptivity of 15200 M^{-1}cm^{-1}) inactivates human adenosylmethionine decarboxylase by forming a hydrazone derivative at the pyruvate prosthetic group. **1**. White, Arnett, Secrist & Shannon (1994) *Virus Res.* **31**, 255. **2**. Shantz, Stanley, Secrist & Pegg (1992) *Biochemistry* **31**, 6848.

5'-Deoxyimmucillin-A

This transition-state analogue (FW$_{free-base}$ = 236.25 g/mol), also known as (1*S*)-1-(9-deazaadenin-9-yl)-1,4,5-trideoxy-1,4 imino-D-ribitol, inhibits 5'-methylthioadenosine/*S*-adenosyl-homocysteine nucleosidase, K_i' = 13 nM (1). It is also a tight, slow-binding inhibitor of human *S*-methyl-5'-thioadenosine phosphorylase, K_i = 720 nM (2). When the ring nitrogen is protonated, this analogue mimics the charge of oxa-carbenium ion intermediate that forms transiently in many glycosidase reactions, and this property results in tight binding. **1**. Singh, Evans, Lenz, *et al.* (2005) *J. Biol. Chem.* **280**, 18265. **2**. Evans, Furneaux, Schramm, Singh & Tyler (2004) *J. Med. Chem.* **47**, 3275.

5'-Deoxy-5'-(isobutylthio)adenosine

This *S*-adenosylhomocysteine analogue (FW = 339.42 g/mol; λ_{max} at 260 nm (pH 7) with a molar absorptivity of 15200 M^{-1}cm^{-1}), also known as 5'-isobutylthioadenosine, inhibits polyamine biosynthesis as well as many methyltransferases. It also inhibits *Escherichia coli* 5'-methylthioadenosine/*S* adenosyl-homocysteine nucleosidase (IC$_{50}$ = 0.74 μM). **Target(s):** *S*-adenosylhomocysteinase (1,2); adenosylhomocysteine nucleosidase (3); 1-aminocyclopropane-1-carboxylate synthase (4); amino-propyltransferases, *or* spermidine/s permine synthase (5); cAMP phosphodiesterase (6); cholinephosphate cytidylyltransferase, *or* CTP:choline phosphate cytidylyltransferase (7); 5'-deoxy-5'-methylthioadenosine phosphorylase (8); glycine *N* methyltransferase (3);

histone-lysine *N*-methyltransferase (9); 5'-methylthioadenosine/*S*-adenosylhomocysteine nucleosidase (10); methylthio-adenosine nucleosidase (11); nucleoside transport (12); 5'-nucleotidase (13,14); protein-arginine methyltransferase (9,15); protein-*S* isoprenyl-cysteine *O*-methyltransferase (15); spermidine synthase (5); spermine synthase (5); sugar transport (12); tRNA (adenine-*N*^{1}-)-methyltransferase (16); tRNA (guanine-*N*^{2}-)-methyltransferase (17); tRNA methyltransferases (18). **1**. Pierre, Richou, Lawrence, Robert-Gero & Vigier (1977) *Biochem. Biophys. Res. Commun.* **76**, 813. **2**. Della Ragione & Pegg (1983) *Biochem. J.* **210**, 429. **3**. Kloor, Karnahl & Kömpf (2004) *Biochem. Cell Biol.* **82**, 369. **4**. Icekson & Apelbaum (1983) *Biochem. Biophys. Res. Commun.* **113**, 586. **5**. Hibasami, Tanaka & Nagai (1982) *Biochem. Pharmacol.* **31**, 1649. **6**. Zimmerman, Schmitges, Wolberg, Deeprose & Duncan (1981) *Life Sci.* **28**, 647. **7**. de Blas, Adler, Shih, *et al.* (1984) *Proc. Natl. Acad. Sci. U.S.A.* **81**, 4353. **8**. Abbruzzese, Della Pietra & Porta (1983) *J. Neurochem.* **40**, 487. **9**. Vedel, Lawrence, Robert-Gero & Lederer (1978) *Biochem. Biophys. Res. Commun.* **85**, 371. **10**. Cornell, Swarts, Barry & Riscoe (1996) *Biochem. Biophys. Res. Commun.* **228**, 724. **11**. Kushad, Richardson & Ferro (1985) *Plant Physiol.* **79**, 525. **12**. Pierre & Robert-Gero (1979) *FEBS Lett.* **101**, 233. **13**. Skladanowski, Sala & Newby (1989) *Biochem. J.* **262**, 203. **14**. Skladanowski & Newby (1990) *Biochem. J.* **268**, 117. **15**. Robert-Gero, Pierre, Vedel, *et al.* (1980) in *Enzyme Inhibitors* (Brodbeck, ed.), p. 61, Verlag Chemie, Weinheim. **16**. Brahmachari & Ramakrisnan (1984) *Arch. Microbiol.* **140**, 91. **17**. Pierré, Berneman, Vedel, Robert-Géro & Vigier (1978) *Biochem. Biophys. Res. Commun.* **81**, 315. **18**. Vedel, Robert-Géro, Legraverend, Lawrence & Lederer (1978) *Nucl. Acids Res.* **5**, 2979.

5'-Deoxy-5'-(isopropylthio)adenosine

This nucleoside (FW = 324.39 g/mol; λ_{max} at 260 nm (pH 7) with a molar absorptivity of 15200 M^{-1}cm^{-1}), also known as 5'-isopropyl-thioadenosine, inhibits *Escherichia coli* 5' methylthioadenosine/*S*-adenosylhomocysteine nucleosidase (IC$_{50}$ = 1.1 μM). 5'-(*p*-Nitrophenyl)thioadenosine was most potent, with a K_i of 20 nM. **1**. Cornell, Swarts, Barry & Riscoe (1996) *Biochem. Biophys. Res. Commun.* **228**, 724.

D-1-Deoxymannojirimycin

This presumptive transition-state mimic (FW$_{free-base}$ = 163.17 g/mol), also called 5-amino-1,5-dideoxy-D-mannopyranose and 1-deoxy-D-*manno*-nojirimycin, inhibits glycoprotein processing. When the ring nitrogen is protonated, this analogue mimics the charge distribution of a presumptive oxa-carbenium ion intermediate likely to form transiently in many glycosidase reactions. This property often results in very tight binding. **Target(s):** α-L-fucosidase (1-4); α-glucosidase, moderately inhibited (2); glycoprotein processing (5-10); β-D-hexosaminidase, moderately inhibited (2); α-mannosidase (2,9-17); β-mannosidase (11); mannosyl-oligosaccharide glucosidase, *or* glucosidase I, weakly inhibited (18); mannosyl-oligosaccharide 1,2-α-mannosidase, *or* mannosidase I (5-11,19-31); mannosyl oligosaccharide 1,3-1,6-α-mannosidase, *or* mannosidase II (2,32-34); trehalose (35). **1**. Winchester, Barker, Baines, *et al.* (1990) *Biochem. J.* **265**, 277. **2**. Siriwardena, Strachan, El-Daher, *et al.* (2005) *ChemBioChem* **6**, 845. **3**. Berteau, McCort, Goasdoué, Tissot & Daniel (2002) *Glycobiology* **12**, 273. **4**. Dumas, Kajimoto, Liu, Wong, Berkowitz & Danishefsky (1992) *Bioorg. Med. Chem. Lett.* **2**, 33. **5**. Elbein (1987) *Meth. Enzymol.* **138**, 661. **6**. Tulsiani & Touster (1989) *Meth. Enzymol.* **179**, 446. **7**. Kaushal & Elbein (1989) *Meth. Enzymol.* **179**, 452. **8**. Kaushal & Elbein (1994) *Meth. Enzymol.* **230**, 316. **9**. Elbein (1987) *Ann. Rev. Biochem.* **56**, 497. **10**. Szumilo, Kaushal, Hori & Elbein (1986) *Plant Physiol.* **81**, 383. **11**. Mooser (1992) *The Enzymes*, 3rd ed., **20**, 187. **12**. Woo, Miyazaki, Hara, Kimura & Kimura (2004) *Biosci. Biotechnol.*

Biochem. **68**, 2547.   **13.** Vázquez-Reyna, Balcázar-Orozco & Flores-Carreón (1993) *FEMS Microbiol. Lett.* **106**, 321.   **14.** Vázquez-Reyna, Ponce-Noyola, Calvo-Méndez, López-Romero & Flores-Carreón (1999) *Glycobiology* **9**, 533.   **15.** Yamashiro, Itoh, Yamagishi, *et al.* (1997) *J. Biochem.* **122**, 1174.   **16.** Porwoll, Fuchs & Tauber (1999) *FEBS Lett.* **449**, 175.   **17.** Schneikert & Herscovics (1994) *Glycobiology* **4**, 445.   **18.** Schweden, Borgmann, Legler & Bause (1986) *Arch. Biochem. Biophys.* **248**, 335.   **19.** Fuhrmann, Bause, Legler & Ploegh (1984) *Nature* **307**, 755.   **20.** Slusarewicz & Warren (1995) *Glycobiology* **5**, 154.   **21.** Schweden, Legler & Bause (1986) *Eur. J. Biochem.* **157**, 563.   **22.** Hamagashira, Oku, Mega & Hase (1996) *J. Biochem.* **119**, 998.   **23.** Gonzalez, Karaveg, Vandersall-Nairn, Lal & Moremen (1999) *J. Biol. Chem.* **274**, 21375.   **24.** Forsee, Palmer & Schutzbach (1989) *J. Biol. Chem.* **254**, 3869.   **25.** Schutzbach & Forsee (1990) *J. Biol. Chem.* **265**, 2546.   **26.** Kimura, Yamaguchi, Suehisa & Tagaki (1991) *Biochim. Biophys. Acta* **1075**, 6.   **27.** Ren, Bretthauer & Castellino (1995) *Biochemistry* **34**, 2489.   **28.** Schweden & Bause (1989) *Biochem. J.* **264**, 347.   **29.** Bause, Breuer, Schweden, Roeser & Geyer (1992) *Eur. J. Biochem.* **208**, 451.   **30.** Tempel, Karaveg, Liu, *et al.* (2004) *J. Biol. Chem.* **279**, 29774.   **31.** Desmet, Nerinckx, Stals, *et al.* (2002) *Anal. Biochem.* **307**, 361.   **32.** Shah, Kuntz & Rose (2003) *Biochemistry* **42**, 13812.   **33.** Ren, Castellino & Bretthauer (1997) *Biochem. J.* **324**, 951.   **34.** van den Elsen, Kuntz & Rose (2001) *EMBO J.* **20**, 3008.   **35.** Asano, Kato & Matsui (1996) *Eur. J. Biochem.* **240**, 692.

5'-Deoxy-5'-(methylthio)tubercidin

This nucleoside (FW = 297.09 g/mol), also known as 5'-(methylthio)tubercidin and 5'-methylthio-7-deazaadenosine, inhibits 5'-methylthioadenosine/S-adenosylhomocysteine nucleosidase (IC$_{50}$ = 0.75 μM) an polyamine biosynthesis. **Target(s):** adenosylhomocysteine nucleosidase (1-3); 5'-methylthioadenosine/S adenosylhomocysteine nucleosidase (1-3); S-methyl-5'-thioadenosine phosphorylase (4-9); polyamine aminopropyltransferase (10); spermidine synthase, *or* putrescine aminopropyltransferase (11-14); spermine synthase, *or* spermidine aminopropyltransferase (11,12).   **1.** Cornell, Swarts, Barry & Riscoe (1996) *Biochem. Biophys. Res. Commun.* **228**, 724.   **2.** Della Ragione, Porcelli, Cartenì-Farina, Zappia & Pegg (1985) *Biochem. J.* **232**, 335.   **3.** Lee, Cornell, Riscoe & Howell (2003) *J. Biol. Chem.* **278**, 8761.   **4.** White, Riscoe & Ferro (1983) *Biochim. Biophys. Acta* **762**, 405.   **5.** Zappia, Oliva, Cacciapuoti, *et al.* (1978) *Biochem. J.* **175**, 1043.   **6.** Pankaskie & Lakin (1987) *Biochem. Pharmacol.* **36**, 2063.   **7.** Della Ragione, Takabayashi, Mastropietro, *et al.* (1996) *Biochem. Biophys. Res. Commun.* **223**, 514.   **8.** White, Vandenbark, Barney & Ferro (1982) *Biochem. Pharmacol.* **31**, 503.   **9.** Ferro, Wrobel & Nicolette (1979) *Biochim. Biophys. Acta* **570**, 65.   **10.** Cacciapuoti, Porcelli, Carteni-Farina, Gambacorta & Zappia (1986) *Eur. J. Biochem.* **161**, 263.   **11.** Pegg (1983) *Meth. Enzymol.* **94**, 294.   **12.** Hibasami, Borchardt, Chen, Coward & Pegg (1980) *Biochem. J.* **187**, 419.   **13.** Samejima & Yamanoha (1982) *Arch. Biochem. Biophys.* **216**, 213.   **14.** Sindhu & Cohen (1984) *Plant Physiol.* **74**, 645.

D-1-Deoxynojirimycin

This presumptive transition-state mimic (FW$_{free-base}$ = 163.17 g/mol; CAS 19130-96-2), also named 5-amino-1,5-dideoxy-D-glucopyranose and molanoline, is produced by reduction of nojirimycin with a platinum catalyst or NaBH$_4$. 1-Deoxynojirimycin is more stable than nojirimycin. When the ring nitrogen is protonated, this analogue mimics the charge of oxa-carbenium ion intermediate that forms transiently in many glycosidase reactions, a property that often results in tight binding. It has also been identified in the roots a nd leaves of mulberry trees as well as in strains of

Bacillus and *Streptomyces*. While an excellent inhibitor of α-glucosidase activities under *in vitro* conditions, 1-deoxynojirimycin has often shown to be only a poor or moderate inhibitor *in vivo*. 1-Deoxynojirimycin also inhibits human immunodeficiency virus (HIV) replication (EC$_{50}$ = 560 μM) by inhibiting processing α-glucosidase activities. **Target(s):** α-amylase (1); amylo-α-1,6-glucosidase/4-α-glucanotransferase, *or* glycogen debranching enzyme (2,3)/; coniferin β-glucosidase (4); cyclomaltodextrin glucanotransferase (5,6); exo-(1→4)-α-D-glucan lyase (K_i = 0.13 μM)7,8; α-L fucosidase (9); glucan 1,3-α-glucosidase, *or* glucosidase II (1,10-17); glucan 1,3-β-glucosidase (18); glucan 1,4-α-glucosidase, *or* glucoamylase, *or* amyloglucosidase (19-21); α-glucosidase (1,11,22-26); β-glucosidase (23,27-29); glycoprotein processing (1,10,22); glucosylceramidase (30-34); lactase/phlorizin hydrolase, *or* glycosylceramidase (35); mannosyl-oligosaccharide glucosidase, *or* glucosidase I (1,10-12,22,36-47); oligosaccharide 4-α-D-glucosyl-transferase (48); phosphatidyl-choline:sterol *O*-acyl-transferase, *or* LCAT (49); sucrose α-glucosidase, *or* sucrase (22,50,51); thioglucosidase, *or* myrosinase, *or* sinigrinase (52); trehalose (53); α,α-trehalose phosphorylase (54); α,α-trehalose phosphorylase, configuration-retaining (55,56).   **1.** Elbein (1987) *Meth. Enzymol.* **138**, 661.   **2.** Bollen & Stalmans (1989) *Eur. J. Biochem.* **181**, 775.   **3.** Liu, Madsen, Braun & Withers (1991) *Biochemistry* **30**, 1419.   **4.** Watt, Ono & Hayashi (1998) *Biochim. Biophys. Acta* **1385**, 78.   **5.** Bovetto, Villette, Fontaine, Sicard & Bouquelet (1992) *Biotechnol. Appl. Biochem.* **15**, 59.   **6.** Kanai, Haga, Yamane & Harata (2001) *J. Biochem.* **129**, 593.   **7.** Yu, Christensen, Kragh, Bojsen & Marcussen (1997) *Biochim. Biophys. Acta* **1339**, 311.   **8.** Lee, Yu & Withers (2003) *Biochemistry* **42**, 13081   **9.** Berteau, McCort, Goasdoue, Tissot & Daniel (2002) *Glycobiology* **12**, 273.   **10.** Kaushal & Elbein (1989) *Meth. Enzymol.* **179**, 452.   **11.** Kaushal & Elbein (1994) *Meth. Enzymol.* **230**, 316.   **12.** Presper & Heath (1983) *The Enzymes*, 3rd ed. (Boyer, ed.), **16**, 449.   **13.** Takeuchi, Kamata, Yoshida, Kameda & Matsui (1990) *J. Biochem.* **108**, 42.   **14.** Hirano, Ziak, Kamoshita, *et al.* (2000) *Glycobiology* **10**, 1283.   **15.** Hentges & Bause (1997) *Biol. Chem.* **378**, 1031.   **16.** Kaushal, Pastuszak, Hatanka & Elbein (1990) *J. Biol. Chem.* **265**, 16271.   **17.** Pelletier, Marcil, Sevigny, *et al.* (2000) *Glycobiology* **10**, 815.   **18.** Kruse & Cole (1992) *Infect. Immun.* **60**, 4350.   **19.** De Mot & Verachtert (1987) *Eur. J. Biochem.* **164**, 643.   **20.** Fierobe, Clarke, Tull & Svensson (1998) *Biochemistry* **37**, 3753.   **21.** Sauer, Sigurskjold, Christensen, *et al.* (2000) *Biochim. Biophys. Acta* **1543**, 275.   **22.** Schwarz & Datema (1984) *Trends Biochem. Sci.* **9**, 32.   **23.** Mooser (1992) *The Enzymes*, 3rd ed., **20**, 187.   **24.** Kato, Kato, Kano, *et al.* (2005) *J. Med. Chem.* **48**, 2036.   **25.** Bravo-Torres, Calvo-Méndez, Flores-Carreón & López-Romero (2003) *Antonie Leeuwenhoek* **84**, 169.   **26.** Bravo-Torres, Villagómez-Castro, Calvo-Méndez, Flores-Carreón & López Romero (2004) *Int. J. Parasitol.* **34**, 455.   **27.** Odoux, Chauwin & Brillouet (2003) *J. Agric. Food Chem.* **51**, 3168.   **28.** Dale, Ensley, Kern, Sastry & Byers (1985) *Biochemistry* **24**, 3530.   **29.** Langston, Sheehy & Xu (2006) *Biochim. Biophys. Acta* **1764**, 972.   **30.** Legler & Liedtke (1985) *Biol. Chem. Hoppe-Seyler* **366**, 1113.   **31.** Legler (1988) *NATO ASI Ser. A, Life Sci.* **150**, 63.   **32.** Osiecki-Newman, Fabbro, Legler, Desnick & Grabowski (1987) *Biochim. Biophys. Acta* **915**, 87.   **33.** Maret, Salvayre, Potier, *et al.* (1988) *NATO ASI Ser. A, Life Sci.* **150**, 57.   **34.** Grace, Graves, Smith & Grabowski (1990) *J. Biol. Chem.* **265**, 6827.   **35.** Wacker, Keller, Falchetto, Legler & Semenza (1992) *J. Biol. Chem.* **267**, 18744.   **36.** Zeng & Elbein (1998) *Arch. Biochem. Biophys.* **355**, 26.   **37.** Schweden, Borgmann, Legler & Bause (1986) *Arch. Biochem. Biophys.* **248**, 335.   **38.** Bause, Erkens, Schweden & Jaenicke (1986) *FEBS Lett.* **206**, 208.   **39.** Hettkamp, Legler & Bause (1984) *Eur. J. Biochem.* **142**, 85.   **40.** Shailubhai, Pratta & Vijay (1987) *Biochem. J.* **247**, 555.   **41.** Bause, Schweden, Gross & Orthen (1989) *Eur. J. Biochem.* **183**, 661.   **42.** Hettkamp, Bause & Legler (1982) *Biosci. Rep.* **2**, 899.   **43.** Zamarripa-Morales, Villagómez-Castro, Calvo-Méndez, Flores-Carreón & López Romero (1999) *Exp. Parasitol.* **93**, 109.   **44.** Herscovics (1999) *Biochim. Biophys. Acta* **1473**, 96.   **45.** Kalz-Fuller, Bieberich & Bause (1995) *Eur. J. Biochem.* **231**, 344.   **46.** Herscovics (1999) *Biochim. Biophys. Acta* **1426**, 275.   **47.** Szumilo, Kaushal & Elbein (1986) *Arch. Biochem. Biophys.* **247**, 261.   **48.** Nebinger (1986) *Biol. Chem. Hoppe-Seyler* **367**, 169.   **49.** Collet & Fielding (1991) *Biochemistry* **30**, 3228.   **50.** Hanozet, Pircher, Vanni, Oesch & Semenza (1981) *J. Biol. Chem.* **256**, 3703.   **51.** Karley, Ashford, Minto, Pritchard & Douglas (2005) *J. Insect Physiol.* **51**, 1313.   **52.** Scofield, Rossiter, Witham, *et al.* (1990) *Phytochemistry* **29**, 107.   **53.** Kyosseva, Kyossev & Elbein (1995) *Arch. Biochem. Biophys.* **316**, 821.   **54.** Kizawa, Miyagawa & Sugiyama (1995) *Biosci. Biotechnol. Biochem.* **59**, 1908.   **55.** Wannet, Op den Camp, Wisselink, *et al.* (1998) *Biochim. Biophys. Acta* **1425**, 177.   **56.** Nidetzky & Eis (2001) *Biochem. J.* **360**, 727.

L-1-Deoxynojirimycin

This enantiomer ($FW_{free-base}$ = 163.17 g/mol) of the potent, competitive α-glucosidase inhibitor and presumptive transition-state mimic, D-1 deoxynojirimycin, is a noncompetitive inhibitor with a K_i value of 5.6 μM for the rice enzyme. When the ring nitrogen is protonated, this analogue mimics the charge of oxa-carbenium ion intermediate that forms transiently in many glycosidase reactions, a property that results in tight binding. This agent is a useful nonbinding control for its D-enantiomer. *See D-1-Deoxynojirimycin* 1. Kato, Kato, Kano, *et al.* (2005) *J. Med. Chem.* **48**, 2036.

3',4'-Deoxynorlaudanosolinecarboxylate

This dopamine-derived tetrahydroisoquinoline (FW = 299.33 g/mol) is a novel inhibitor of human liver dihydropteridine reductase. It is produced by condensation of phenylpyruvate with dopamine and has been detected in the urine of children with PKU children and in the urine and brain of rats with experimentally induced hyperphenyl-alaninemia. **Target(s):** dihydro-pteridine reductase (1); dopamine β-monooxygenase (2); tyrosine 3-monooxygenase (3). 1. Shen, Smith, Davis, Brubaker & Abell (1982) *J. Biol. Chem.* **257**, 7294. 2. Lasala & Coscia (1979) *Science* **203**, 283. 3. Coscia, Burke, Galloway, *et al.* (1980) *J. Pharmacol. Exp. Ther.* **212**, 91.

5'-Deoxy-5'-(propylthio)adenosine

This nucleoside (FW = 349.42 g/mol), also known as 5'-propylthioadenosine, is an alternative substrate and competitive inhibitor of *Escherichia coli* 5'-methylthioadenosine/*S*-adenosylhomocysteine nucleo-sidase (K_i = 0.05 μM). **Target(s):** adenosylhomocysteine nucleosidase (1,2); 5'-methylthioadenosine/*S*-adenosylhomocysteine nucleosidase (1,2); *S*-methyl-5'-thioadenosine phosphorylase (3). 1. Cornell, Swarts, Barry & Riscoe (1996) *Biochem. Biophys. Res. Commun.* **228**, 724. 2. Ferro, Barrett & Shapiro (1976) *Biochim. Biophys. Acta* **438**, 487. 3. Ferro, Wrobel & Nicolette (1979) *Biochim. Biophys. Acta* **570**, 65.

2'-Deoxyribavirin 5'-Triphosphate

This nucleotide analogue ($FW_{free-acid}$ = 452.15 g/mol) inhibits mRNA (nucleoside-2'-*O*-)-methyltransferase. It is formed metabolically from ribavirin, one of the few nucleoside analogues widely used to treat RNA

virus infections. Ribavirin 5'-triphosphate inhibits dengue virus 2'-*O*-methyltransferase NS5 domain ($NS5MTase_{DV}$), competing with GTP to bind to $NS5MTase_{DV}$. A 2.6-Å crystal structure of a ternary complex consisting of $NS5MTase_{DV}$, ribavirin 5'-triphosphate, and *S*-adenosyl-L-homocysteine afforded a detailed atomic of the specificity-conferring properties. 1. Benarroch, Egloff, Mulard, *et al.* (2004) *J. Biol. Chem.* **279**, 35638.

2-Deoxyribose 5-Phosphate

α-anomer   β-anomer

This phosphorylated deoxyribose ($FW_{free-acid}$ = 214.11 g/mol; CAS 102916-66-5) inhibits ribose-5-phosphate adenylyltransferase and the 1-deoxy derivative is a component of DNA. Note that phosphopentomutase can catalyzed the interconversion of 2-deoxy-α-D-ribose 1-phosphate and 2-deoxy-α-D-ribose 5-phosphate utilizing 2-deoxy-α-D-ribose 1,5-bisphosphate as a cofactor. **Target(s):** fructose-1,6-bisphosphatase (1); orotidine-5'-phosphate decarboxylase (2); ribose 5-phosphate adenylyltransferase (3,4); ribose-phosphate diphosphokinase, *or* phosphoribosyl pyrophosphate synthetase (5); ribulose-phosphate 3-epimerase (6). 1. Ganson & Fromm (1984) *Curr. Top. Cell. Regul.* **24**, 197. 2. Miller, Butterfoss, Short & Wolfenden (2001) *Biochemistry* **40**, 6227. 3. Evans (1971) *Meth. Enzymol.* **23**, 566. 4. Evans & Pietro (1966) *Arch. Biochem. Biophys.* **113**, 236. 5. Fox & Kelley (1972) *J. Biol. Chem.* **247**, 2126. 6. Wood (1979) *Biochim. Biophys. Acta* **570**, 352.

3'-Deoxythymidine (*or*, 3'-dT)

This dideoxynucleoside (FW = 226.23 g/mol; CAS 3416-05-5), also known as 1-(β-D-2',3' dideoxyribofuranosyl)thymine, inhibits thymidine kinase and thymidine phosphorylase. (*See also 3'-Deoxythymidine 5'-Triphosphate*) **Target(s):** deoxynucleoside kinase (1); 5'-nucleotidase (2); poly(ADP-ribose)polymerase (3); thymidine kinase (1,4); thymidine phosphorylase (4,5). 1. Johansson, Van Rompay, Degrève, Balzarini & Karlsson (1999) *J. Biol. Chem.* **274**, 23814. 2. Garvey, Lowen & Almond (1998) *Biochemistry* **37**, 9043. 3. Pivazyan, Birks, Wood, Lin & Prusoff (1992) *Biochem. Pharmacol.* **44**, 947. 4. Baker (1967) *Design of Active-Site-Directed Irreversible Enzyme Inhibitors*, Wiley, New York. 5. Panova, Alexeev, Kuzmichov, *et al.* (2007) *Biochemistry (Moscow)* **72**, 21.

5'-Deoxythymidine (5'-dT)

This dideoxy nucleoside (FW = 226.23 g/mol; CAS 3458-14-8), also known as 1-(β-D-2',5'-dideoxyribofuranosyl)thymine, inhibits thymidine kinase and dTMP kinase. Note that the pentose portion of this molecule is actually 2',5'-dideoxyribose. **Target(s):** 5'-nucleotidase (1); thymidine kinase (2-5); thymidylate kinase, *or* dTMP kinase (6,7). 1. Garvey, Lowen & Almond (1998) *Biochemistry* **37**, 9043. 2. Baker (1967) *Design of Active-Site-Directed Irreversible Enzyme Inhibitors*, Wiley, New York. 3. Baker,

Schwan & Santi (1966) *J. Med. Chem.* **9**, 66.   **4**. Baker & Neenan (1972) *J. Med. Chem.* **15**, 940.   **5**. Cheng & Prusoff (1974) *Biochemistry* **13**, 1179.   **6**. Kara & Duschinsky (1969) *Biochim. Biophys. Acta* **186**, 223.   **7**. Munier-Lehmann, Pochet, Dugue, *et al.* (2003) *Nucleosides Nucleotides Nucleic Acids* **22**, 801.

2′,3′-Deoxythymidine 5′-Triphosphate

This dideoxy nucleotide (FW$_\text{free-acid}$ = 466.17 g/mol), also known as 2′,3′-dideoxythymidine-5′-triphosphate, inhibits the action of DNA polymerases via chain termination. It is used in the dideoxy-method of Nobelist Fred Sanger in sequencing DNA (**See** *2′,3′-Dideoxythymidine 5′-Triphosphate*) **Target(s):** adenovirus DNA polymerase (1); dCMP deaminase, weakly inhibited (2); DNA-directed DNA polymerase (3-22); DNA nucleotidylexotransferase, *or* terminal deoxynucleotidyltransferase (13,14,23); DNA polymerase I (14); DNA polymerase α (8,10); DNA polymerase β (14); DNA polymerase γ (4,6); DNA polymerase δ (19); DNA polymerase λ (13); DNA polymerase POL4 (9); duck hepitits B virus DNA-directed DNA polymerase (12); HIV-1 reverse transcriptase (5,24-29); Rauscher murine leukemia virus reverse transcriptase (30); RNA-directed DNA polymerase (5,12,24-31); RNA-directed RNA polymerase (32); simian immunodeficiency virus (SIV) reverse transcriptase (27); visna virus reverse transcriptase (31).   **1**. Mentel, Kurek, Wegner, *et al.* (2000) *Med. Microbiol. Immunol.* **189**, 91.   **2**. Sergott, Debeer & Bessman (1971) *J. Biol. Chem.* **246**, 7755.   **3**. Makioka, Stavros, Ellis & Johnson (1993) *Parasitology* **107**, 135.   **4**. Izuta, Saneyoshi, Sakurai, *et al.* (1991) *Biochem. Biophys. Res. Commun.* **179**, 776.   **5**. Parker, White, Shaddix, *et al.* (1991) *J. Biol. Chem.* **266**, 1754.   **6**. Izuta, Saneyoshi, Sakurai, *et al.* (1991) *Nucl. Acids Symp. Ser.* (25), 79.   **7**. Sanger, Nicklen & Coulson (1977) *Proc. Natl. Acad. Sci. U.S.A.* **74**, 5463.   **8**. Ward & Weissbach (1986) *Meth. Enzymol.* **118**, 97.   **9**. Budd & Campbell (1995) *Meth. Enzymol.* **262**, 108.   **10**. Brown & Wright (1995) *Meth. Enzymol.* **262**, 202.   **11**. Atkinson, Deutscher, Kornberg, Russell & Moffatt (1969) *Biochemistry* **8**, 4897.   **12**. Lofgren, Nordenfelt & Oberg (1989) *Antiviral Res.* **12**, 301.   **13**. Di Santo & Maga (2006) *Curr. Med. Chem.* **13**, 2353.   **14**. Matthes, Lehmann, Drescher, Büttner & Langen (1985) *Biomed. Biochim. Acta* **44**, K63.   **15**. Meißner, Heinhorst, Cannon & Börner (1993) *Nucl. Acids Res.* **21**, 4893.   **16**. Matsuda, K. Takami, A. Sono & K. Sakaguchi (1993) *Chromosoma* **102**, 631.   **17**. Spampinato, Pairoba, Colombo, Benediktsson & Andreo (1994) *Biosci. Biotechnol. Biochem.* **58**, 822.   **18**. Burrows & C. R. Goward (1992) *Biochem. J.* **287**, 971.   **19**. Peck, Germer & Cress (1992) *Nucl. Acids Res.* **20**, 5779.   **20**. Weissbach (1979) *Arch. Biochem. Biophys.* **198**, 386.   **21**. Gray & Wong (1992) *J. Biol. Chem.* **267**, 5835.   **22**. Sen, Mukhopadhyay, Wetzel & Biswas (1994) *Acta Biochim. Pol.* **41**, 79.   **23**. Ono (1990) *Biochim. Biophys. Acta* **1049**, 15.   **24**. De Clercq (1992) *AIDS Res. Human Retrovir.* **8**, 119.   **25**. White, Parker, Macy, *et al.* (1989) *Biochem. Biophys. Res. Commun.* **161**, 393.   **26**. Le Grice, Cameron & Benkovic (1995) *Meth. Enzymol.* **262**, 130.   **27**. Wu, Chernow, Boehme, *et al.* (1988) *Antimicrob. Agents Chemother.* **32**, 1887.   **28**. Nissley, Radzio, Ambrose, *et al.* (2007) *Biochem. J.* **404**, 151.   **29**. Quinones-Mateu, Soriano, Domingo & Menendez-Arias (1997) *Virology* **236**, 364.   **30**. Ono, Ogasawara, Iwata, *et al.* (1986) *Biochem. Biophys. Res. Commun.* **140**, 498.   **31**. Frank, McKernan, Smith & Smee (1987) *Antimicrob. Agents Chemother.* **31**, 1369.   **32**. Hung, Gibbs & Tsiang (2002) *Antiviral Res.* **56**, 99.

2-Deoxy-2-ureido-D-glucosyl-α1,6-(2-O-octyl)-D-*myo*-inositol-1 phosphoryl-*sn*-1,2-dipalmitoylglycerol

This analogue (FW = 1123.41 g/mol) is a mechanism-based inhibitor of *Trypanosomal* *N*-acetylglucosaminyl-phosphatidyl-inositol deacetylase

(IC$_{50}$ = 8 nM). This suicide inhibitor is likely to form a carbamate (or thiocarbamate) ester to an active-site Ser, Thr or Cys residue, such that inhibition is reversed by certain nucleophiles.   **1**. Smith, Crossman, Borissow, *et al.* (2001) *EMBO J.* **20**, 3322.

2′-Deoxyuridine

This deoxynucleoside (FW = 228.20 g/mol; λ$_\text{max}$ = 262 nm at pH 1 and 7, both with ε values of 10200 M^{-1} cm^{-1}) inhibits dihydroorotase and thymidine kinase. **Target(s):** acid phosphatase (1); cytidine deaminase (2-5); dihydroorotase (6); dTMP kinase, *or* thymidylate kinase (7,8); dUTP diphosphatase (9); nucleotidases (10); 5′-nucleotidase (11); ribonuclease T2 (12); thymidine kinase (13-15); thymidylate synthase (16).   **1**. Uerkvitz (1988) *J. Biol. Chem.* **263**, 15823.   **2**. Vita, Amici, Cacciamani, Lanciotti & Magni (1985) *Biochemistry* **24**, 6020.   **3**. Hosono & S. Kuno (1973) *J. Biochem.* **74**, 797.   **4**. Cacciamani, Vita, Cristalli, *et al.* (1991) *Arch. Biochem. Biophys.* **290**, 285.   **5**. Vita, Amici, Lanciotti, Cacciamani & Magni (1986) *Ital. J. Biochem.* **35**, 145A.   **6**. Bresnick & Blatchford (1964) *Biochim. Biophys. Acta* **81**, 150.   **7**. Munier-Lehmann, Pochet, Dugué, *et al.* (2003) *Nucleosides Nucleotides Nucleic Acids* **22**, 801.   **8**. Pochet, Dugué, Labesse, Delepierre & Munier-Lehmann (2003) *ChemBioChem* **4**, 742.   **9**. Nord, Larsson, Kvassman, Rosengren & Nyman (1997) *FEBS Lett.* **414**, 271.   **10**. Hoglund & Reichard (1990) *J. Biol. Chem.* **265**, 6589.   **11**. Garvey, Lowen & Almond (1998) *Biochemistry* **37**, 9043.   **12**. Irie & Ohgi (1976) *J. Biochem.* **80**, 39.   **13**. Bresnick & Thompson (1965) *J. Biol. Chem.* **240**, 3967.   **14**. Baker & Neenan (1972) *J. Med. Chem.* **15**, 940.   **15**. Swinton & Chiang (1979) *Mol. Gen. Genet.* **175**, 399.   **16**. Nakata, Tsukamoto, Miyoshi & Kojo (1987) *Biochim. Biophys. Acta* **924**, 297.

2′-Deoxyuridine 5′-Diphosphate

This nucleotide (FW$_\text{free-acid}$ = 388.16 g/mol; λ$_\text{max}$ = 262 nm at pH 1 and 7, both with ε values of 10200 M^{-1} cm^{-1}; *Symbol*: dUDP, often abbreviated dUDP, inhibits carbamoyl-phosphate synthetase II and 5′ nucleotidase. **Target(s):** carbamoyl-phosphate synthetase, glutamine-hydrolyzing (1); deoxycytidine kinase (2,3); dTMP kinase, *or* thymidylate kinase (4,5); dUTP diphosphatase (6-11); α-1,3-mannosyl glycoprotein 2-β-*N*-acetylglucosaminyl-transferase, weakly inhibited (12); 5′-nucleotidase (13); thymidine kinase (14).   **1**. Mori & Tatibana (1978) *Meth. Enzymol.* **51**, 111.   **2**. Kozai, Sonoda, Kobayashi & Sugino (1972) *J. Biochem.* **71**, 485.   **3**. Wang, Kucera & Capizzi (1993) *Biochim. Biophys. Acta* **1202**, 309.   **4**. Jong & Campbell (1984) *J. Biol. Chem.* **259**, 14394.   **5**. Tamiya, Yusa, Yamaguchi, *et al.* (1989) *Biochim. Biophys. Acta* **995**, 28.   **6**. Nord, Larsson, Kvassman, Rosengren & Nyman (1997) *FEBS Lett.* **414**, 271.   **7**. Bergman, Nyman & Larsson (1998) *FEBS Lett.* **441**, 327.   **8**. Hidalgo-Zarco, Camacho, Bernier-Villamor, *et al.* (2001) *Protein Sci.* **10**, 1426.   **9**. Camacho, Hilgado-Zarco, Bernier-Villamor, Ruiz-Pérez & González-Pacanowska (2000) *Biochem. J.* **346**, 163.   **10**. Persson, Harkiolaki, McGeehan & Wilson (2001) *Acta Crystallogr. Sect. D* **57**, 876.   **11**. Kovari, Barabas, Takacs, *et al.* (2004) *J. Biol. Chem.* **279**, 17932.   **12**. Nishikawa, Pegg, Paulsen & Schachter (1988) *J. Biol. Chem.* **263**, 8270.   **13**. Madrid-Marina & Fox (1986) *J. Biol. Chem.* **261**, 444.   **14**. Swinton & Chiang (1979) *Mol. Gen. Genet.* **175**, 399.

2'-Deoxyuridine 5'-Monophosphate

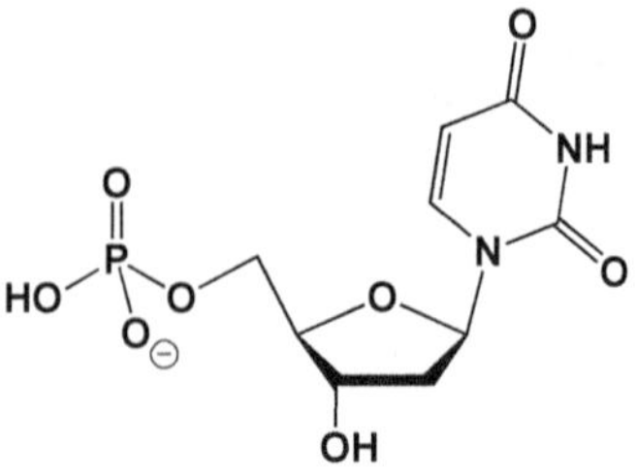

This deoxynucleotide (FW$_{free-acid}$ = 306.19 g/mol; λ_{max} = 262 nm at pH 1 and 7, both with ε values of 10200 M^{-1}cm^{-1}; *Symbol*: dUMP) is a product of the reaction catalyzed by dCMP deaminase. **Target(s):** dCMP deaminase, deoxycytidylate deaminase (1-3); deoxycytidine kinase (4); dihydroorotase (5); dUTP diphosphatase, as product inhibition (6-8); orotate phosphoribosyltransferase (9); orotidine-5'-phosphate decarboxylase (9,10); phospho-diesterase I, *or* 5'-exonuclease (11); [protein-P$_{II}$] uridylyltransferase (12); thymidylate kinase, *or* dTMP kinase (13-15); thymidylate 5'-phosphatase (16); undecaprenyl phosphate galactose phosphotransferase (17); uracil phosphoribosyltransferase (18). **1.** Maley (1967) *Meth. Enzymol.* **12A**, 170. **2.** Geraci, Rossi & Scarano (1967) *Biochemistry* **6**, 183. **3.** Rossi, Geraci & Scarano (1967) *Biochemistry* **6**, 3640. **4.** Datta, Shewach, Mitchell & Fox (1989) *J. Biol. Chem.* **264**, 9359. **5.** Bresnick & Blatchford (1964) *Biochim. Biophys. Acta* **81**, 150. **6.** Nord, Larsson, Kvassman, Rosengren & Nyman (1997) *FEBS Lett.* **414**, 271. **7.** Hidalgo-Zarco, Camacho, Bernier-Villamor, *et al.* (2001) *Protein Sci.* **10**, 1426. **8.** Camacho, Hilgado-Zarco, Bernier-Villamor, Ruiz-Perez & Gonzalez-Pacanowska (2000) *Biochem. J.* **346**, 163. **9.** Jones, Kavipurapu & Traut (1978) *Meth. Enzymol.* **51**, 155. **10.** Miller, Butterfoss, Short & Wolfenden (2001) *Biochemistry* **40**, 6227. **11.** Sabatini & Hotchkiss (1969) *Biochemistry* **8**, 4831. **12.** Engleman & Francis (1978) *Arch. Biochem. Biophys.* **191**, 602. **13.** Cheng & Prusoff (1973) *Biochemistry* **12**, 2612. **14** Jong & Campbell (1984) *J. Biol. Chem.* **259**, 14394. **15.** Nelson & Carter (1969) *J. Biol. Chem.* **244**, 5254. **16.** Price & S. Fogt (1973) *J. Biol. Chem.* **248**, 1372. **17.** Osborn & Tze-Yuen (1968) *J. Biol. Chem.* **243**, 5145. **18.** Natalini, Ruggieri, Santarelli, Vita & Magni (1979) *J. Biol. Chem.* **254**, 1558.

2'-Deoxyuridine 5'-Triphosphate

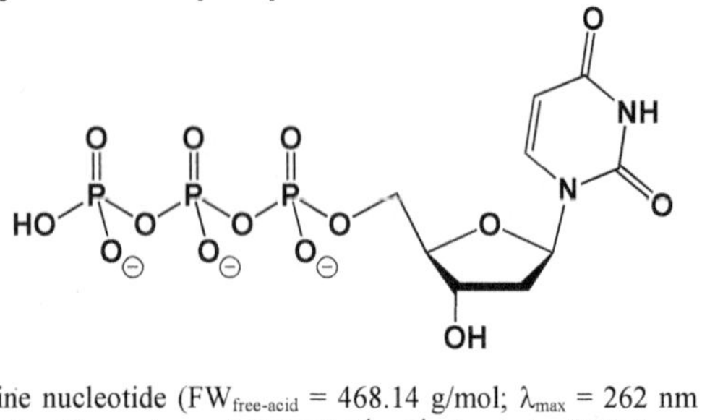

This purine nucleotide (FW$_{free-acid}$ = 468.14 g/mol; λ_{max} = 262 nm at pH 1 and 7, both with ε values of 10200 M^{-1}cm^{-1}; Symbol: dUTP is a mutation-causing alternative substrate for DNA polymerase. **Target(s):** amidophosphoribosyl-transferase (1); carbamoyl-P synthetase, glutamine-hydrolyzing (2); cytidine deaminase (3); dCMP deaminase (4-7); dCTP deaminase (8-10); dCTP deaminase, dUMP-forming (11); deoxycytidine kinase (12); deoxyguanosine kinase (13); glucose-1 phosphate thymidylyltransferase (14); 5'-nucleotidase (15); ribonucleoside-diphosphate reductase (16); thymidine kinase (17,18); UMP kinase, weakly inhibited (19); uracil phosphoribosyltransferase (20). **1.** Hill & Bennent (1969) *Biochemistry* **8**, 122. **2.** Mori & Tatibana (1978) *Meth. Enzymol.* **51**, 111. **3.** Hosono & Kuno (1973) *J. Biochem.* **74**, 797. **4.** Maley (1967) *Meth. Enzymol.* **12A**, 170. **5.** Sergott, Debeer & Bessman (1971) *J. Biol. Chem.* **246**, 7755. **6.** Ellims, Kao & Chabner (1981) *J. Biol. Chem.* **256**, 6335. **7.** Moore, Ciesla, Changchien, Maley & Maley (1994) *Biochemistry* **33**, 2104. **8.** Neuhard (1978) *Meth. Enzymol.* **51**, 418. **9.** Beck, Eisenhardt & Neuhard (1975) *J. Biol. Chem.* **250**, 609. **10.** Price (1974) *J. Virol.* **14**, 1314. **11.** Li, Xu, Graham & White (2003) *J. Biol. Chem.* **278**, 11100. **12.** Datta, Shewach, Mitchell & Fox (1989) *J. Biol. Chem.* **264**, 9359. **13.** Yamada, Goto & Ogasawara (1982) *Biochim. Biophys. Acta* **709**, 265. **14.** Melo & Glaser (1961) *J. Biol. Chem.* **240**, 398. **15.** Madrid-Marina & Fox (1986) *J. Biol. Chem.* **261**, 444. **16.** Moore & Hurlbert (1966) *J. Biol. Chem.* **241**, 4802. **17.** Swinton & Chiang (1979) *Mol. Gen. Genet.* **175**,

399. **18.** Chraibi & Wright (1983) *J. Biochem.* **93**, 323. **19.** Evrin, Straut, Slavova-Azmanova, *et al.* (2007) *J. Biol. Chem.* **282**, 7242. **20.** Andersen, Smith & Mygind (1992) *Eur. J. Biochem.* **204**, 51.

Dequalinium Chloride

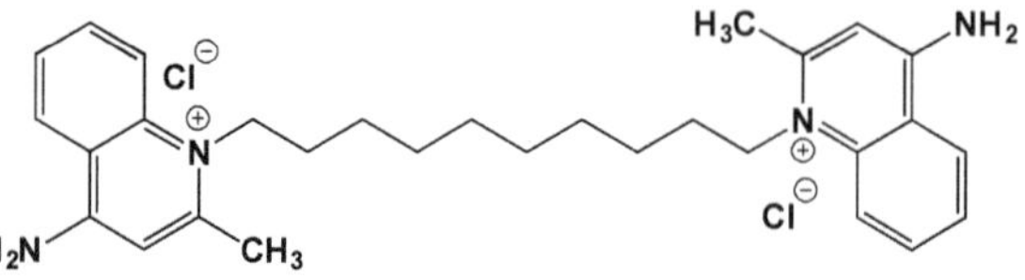

This antimicrobial agent (FW = 527.58 g/mol; CAS 522-51-0), also known as 1,1'-decamethylenebis-4-aminoquinaldinium dichloride, blocks small conductance Ca^{2+}-activated K$^+$ channels. It also inhibits protein kinase C translocation by inhibiting the formation of the PKC:RACK-1 (receptor for activated C kinase-1) complex. **Target(s):** ATPase, Ca2+-dependent, of CF$_1$ (1); electron transport (2); F$_1$ ATPase and TF$_1$ ATPase (1,3-7); H$^+$-ATPase, vacuolar (8); K$^+$ channels, apamin-sensitive and small conductance Ca^{2+}-activated (9-12); NADH:ubiquinone reductase (NADH dehydrogenase (2); protein kinase C (13-15). **1.** Ren & Allison (1997) *J. Biol. Chem.* **272**, 32294. **2.** Anderson, Patheja, Delinck, *et al.* (1989) *Biochem. Int.* **19**, 673. **3.** Paik, Jault & Allison (1994) *Biochemistry* **33**, 126. **4.** Zhuo, Paik, Register & Allison (1993) *Biochemistry* **32**, 2219. **5.** Bullough, Ceccarelli, Roise & Allison (1989) *Biochim. Biophys. Acta* **975**, 77. **6.** Zhuo & Allison (1988) *Biochem. Biophys. Res. Commun.* **152**, 968. **7.** Zhuo, Paik, Register & Allison (1993) *Biochemistry* **32**, 2219. **8.** Moriyama, Patel & Futai (1995) *FEBS Lett.* **359**, 69. **9.** Dunn (1994) *Eur. J. Pharmacol.* **252**, 189. **10.** Castle, Haylett, Morgan & Jenkinson (1993) *Eur. J. Pharmacol.* **236**, 201. **11.** Rosenbaum, Gordon-Shaag, Islas, *et al.* (2004) *J. Gen. Physiol.* **123**, 295. **12.** Rosenbaum, Islas, Carlson & Gordon (2003) *J. Gen. Physiol.* **121**, 37. **13.** Rotenberg & Sun (1998) *J. Biol. Chem.* **273**, 2390. **14.** Rotenberg, Smiley, Ueffing, *et al.* (1990) *Cancer Res.* **50**, 677. **15.** Sullivan, Stone, Marshall, Uberall & Rotenberg (2000) *Mol. Pharmacol.* **58**, 729.

Desferrioxamine

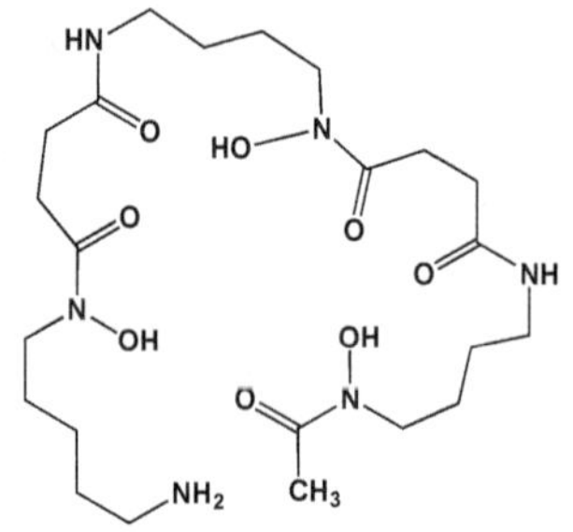

Metal-free Siderophore

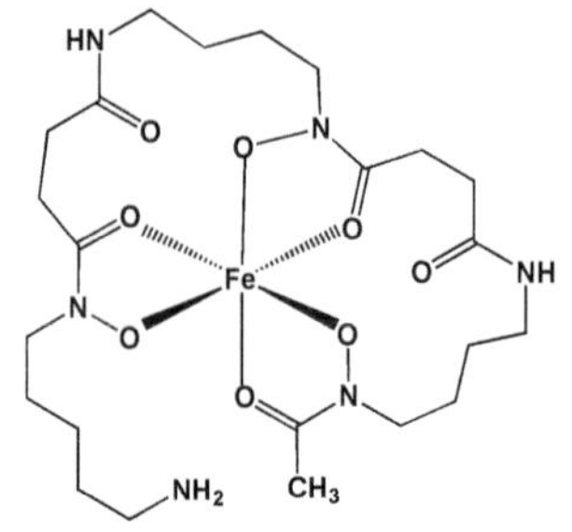

Iron-bound Siderophore

This metal ion chelator (FW$_{free-base}$ = 560.69 g/mol; CAS 70-51-9; Symbol: DFX; also known as deferoxamine mesylate, *or* desferral (FW = 656.79 g/mol; CAS 138-14-7; Soluble to 100 mM in water)), a siderophore produced by the Actinobacterium *Streptomyces pilosus*, binds Fe^{3+} (complex shown above) and other trivalent cations extremely tightly, thereby inhibiting many iron-dependent enzymes, including ferrochelatase, which catalyzes the last step in heme biosynthesis. **Chemical Properties:** Desferrioxamine forms a red-colored, one-to-one complex. Formation constants: (K_F = [Complex]/[Cation]$_{free}$ × [Ligand]$_{free}$): Mg^{2+}, K_F = 10^4 M^{-1};

Ca^{2+}, $K_F = 10^2$ M^{-1}; Sr^{2+}, $K_F = 10$ M^{-1}; Fe^{3+}, $K_F = 10^{31}$ M^{-1}; Co^{2+}, $K_F = 10^{11}$ M^{-1}; Ni^{2+}, $K_F = 10^{10}$ M^{-1}; Cd^{2+}, $K_F = 10^8$ M^{-1}; Zn^{2+}, $K_F = 10^{11}$ M^{-1}. **As Therapy for Iron Overload:** The Fe^{3+}-complex, as shown above, is referred to as ferrioxamine. Desferrioxamine has been used to treat chronic iron overload, also known as primary and secondary hemochromatosis (**See also** *Ferrioxamines*). In humans, normal total iron is 4-5 g, but rises to as much as 20 g in overload. Because iron is so well retained, regularly transfused patients often have excess iron. One-hundred mg desferrioxamine binds 9.35 mg iron. Desferrioxamine also induces apoptosis in HL-60 cells and can arrest cells in the G_1 phase. **Effects on Immune System:** HIF-1α protein levels, HIF-1 DNA-binding activity, and HIF-1 transcriptional activity can be increased by exposure of cells to desferrioxamine (1). By contrast, the hypoxia mimetic $CoCl_2$ induces of HIF-1α accumulation with no attendant increase in COX-2 expression. DFX-induced increases in COX-2 expression and HIF-1 α protein level are attenuated by addition of ferric citrate, suggesting an iron-chelating role in of DFX induction of COX-2 and HIF-1α protein. Treatment witrh PD98059 (a non-ATP competitive MEK inhibitor (IC_{50} = 2 μM) in a cell-free assay) significantly inhibits the induction of COX-2 protein and accumulation of HIF-1α, suggesting that DFX-induced increase of HIF-1α and COX-2 protein was mediated, at least in part, through the ERK signaling pathway. In addition, pretreatment with NS-398 (IC_{50} = 1.77 μM for human COX-2 *versus* IC_{50} = 75 μM for human COX-1) to inhibit COX-2 activity also effectively suppresses DFX-induced HIF-1α accumulation in human colon cancer cells, providing the evidence that COX-2 plays as a regulator of HIF-1;α accumulation in DFX-treated colon cancer cells. Such findings suggest iron metabolism regulate HIF-1α protein by modulating cyclooxygenase-2 signaling pathway, acting, in part, through inhibition of prolyl hydroxylases. **Target(s):** β-carotene 15,15'-monooxygenase (2); cyclooxygenase (3); deoxyhypusine monooxygenase (4-6); dihydroceramide desaturase (7); laccase (8); lipoxygenase (3); phenylalanine 4-monooxygenase (9); procollagen-proline 4-dioxygenase (6,10); ribonucleoside-diphosphate reductase (11,12); tryptophan 5-monooxygenase (13,14). **1.** Woo, Lee, Park & Kwon (2006) *Biochem. Biophys. Res. Commun.* **343**, 8. **2.** During, Smith, Piper & Smith (2001) *J. Nutr. Biochem.* **12**, 640. **3.** Barradas, Jeremy, Kontoghiorghes, *et al.* (1989) *FEBS Lett.* **245**, 105. **4.** Abbruzzese, Park & Folk (1986) *J. Biol. Chem.* **261**, 3085. **5.** Park, Cooper & Folk (1982) *J. Biol. Chem.* **257**, 7217. **6.** Clement, Hanauske-Abel, Wolff, Kleinman & Park (2002) *Int. J. Cancer* **100**, 491. **7.** Schulze, Michel & van Echten-Deckert (2000) *Meth. Enzymol.* **311**, 22. **8.** El-Shora, Youssef & Khalaf (2008) *Biotechnology* **7**, 35. **9.** Goreish, Bednar, Jones, Mitchell & Steventon (1993) *Drug Metabol. Drug Interact.* **20**, 159. **10.** Myllyharju (2008) *Ann. Med.* **40**, 402. **11.** Lammers & Follmann (1983) *Struct. Bonding* **54**, 27. **12.** Shao, Zhou, Zhu, *et al.* (2005) *Biochem. Pharmacol.* **69**, 627. **13.** Cash, Vayer, Mandel & Maitre (1985) *Eur. J. Biochem.* **149**, 239. **14.** D'Sa, Arthur, States & Kuhn (1996) *J. Neurochem.* **67**, 900.

N-Desmethylclozapine

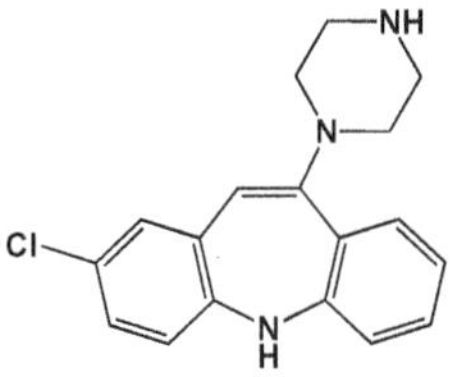

This potent 5-HT receptor antagonist ($FW_{free-base}$ = 312.80 g/mol; CAS 6104-71-8) is also an allosteric at muscarinic-1 receptor agonist and potentiates *N*-methyl-D-aspartate receptor activity. It may contribute to clozapine's clinical activity in schizophrenics through modulation of both muscarinic and glutamatergic neurotransmission (1). **Target(s):** $GABA_A$ receptors (2); glucose transport (3); histamine H_3 receptor (4); serotonin 5-HT_2, 5-HT_6, and 5-HT_7 receptors (5,6). **1.** Mallorga, Wittmann, Jacobson, *et al.* (2003) *Proc. Natl. Acad. Sci. U.S.A.* **100**, 13674. **2.** Wong, Kuoppamaki, Hietala, *et al.* (1996) *Eur. J. Pharmacol.* **314**, 319. **3.** Ardizzone, Bradley, Freeman & Dwyer (2001) *Brain Res.* **923**, 82. **4.** Schlicker & Marr (1996) *Naunyn Schmiedebergs Arch. Pharmacol.* **353**, 290. **5.** Kuoppamaki, Syvalahti & Hietala (1993) *Eur. J. Pharmacol.* **245**, 179. **6.** Roth, Craigo, Choudhary, *et al.* (1994) *J. Pharmacol. Exp. Ther.* **268**, 1403.

Desmolaris

This novel anticoagulant (MW = 21.5 kDa) from the salivary gland of the vampire bat (*Desmodus rotundus*) is a naturally occurring tissue factor pathway inhibitor (TFPI) that is a slow, tight binding inhibitor of FXIa,

showing noncompetitive behavior (pM-range K_d) that is modulated by heparin. Arg-32 in the Kunitz-1 domain is critical for protease inhibition. Desmolaris also inhibits FXa with lower affinity. It also binds kallikrein and reduces bradykinin generation in plasma activated with kaolin. Kunitz-2 and the C-terminus domains mediate interaction of Desmolaris with heparin and are required for optimal inhibition of FXIa and FXa. **1.** Ma, Mizurini, Assumpção, *et al.* (2013) *Blood* **122**, 4094.

Destruxins

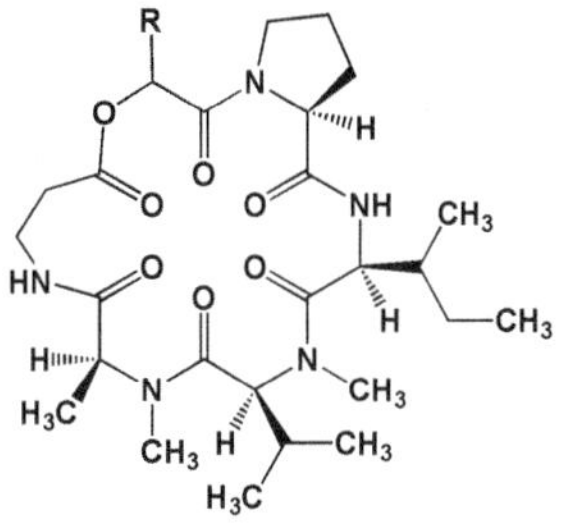

Destruxin A:
　R = –CH₂CH=CH₂
Destruxin B:
　R = –CH₂CH(CH₃)₂
Destruxin C:
　R = –CH₂CH(CH₃)CH₂OH
Destruxin E:
　R = –CH₂CHCH₂ (epoxide O)

These lactone antibiotics and naturally occurring insecticides (Destruxin A, FW = 577.72 g/mol, CAS 6686-70-0; Destruxin B, FW = 578.73 g/mol; Destruxin C, FW = 609.76 g/mol; Destruxin E, FW = 593.34 g/mol; CAS 2503-26-6) from the entomopathogenic fungus *Metarhizium anisopliae* and related fungi, paralyzes and kills insects. The fungus is not infectious or toxic to mammals; however, inhalation of spores can induce allergic reactions. At least thirty-five structurally related destruxins have been isolated from these fungi. In addition to being insecticidal, the destruxins also exhibit phytotoxic, antiviral, nematocidal, and immunomodulating activities. They exhibit some effects on cellular phosphorylation and calcium balance. In addition, destruxins have been reported to induce erythropoietin production in mammalian cells. Note that the destruxin can bind calcium ions and have ionophoric properties. **Target(s):** Na^+-transporting two-sector ATPase, blocked by destruxin B (1-4); vacuolar type H^+-ATPase, inhibited by destruxins B and E (5,6). **1.** Murata, Takase, Yamato, Igarashi & Kakinuma (1997) *J. Biol. Chem.* **272**, 24885. **2.** Kakinuma, Yamato & Murata (1999) *J. Bioenerg. Biomembr.* **31**, 7. **3.** Murata, Kawano, Igarashi, Yamato & Kakinuma (2001) *Biochim. Biophys. Acta* **1505**, 75. **4.** Murata, Igarashi, Kakinuma & Yamato (2000) *J. Biol. Chem.* **275**, 13415. **5.** Togashi, Kataoka & Nagai (1997) *Immunol. Lett.* **55**, 139. **6.** Muroi, Shiragami & Takatsuki (1994) *Biochem. Biophys. Res. Commun.* **205**, 1358.

Desulfo-Coenzyme A

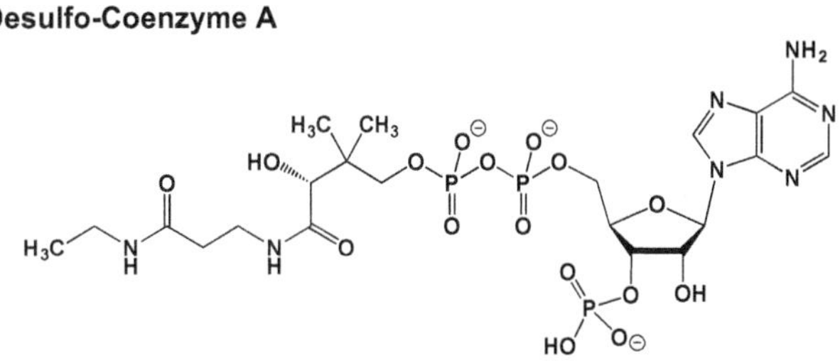

This coenzyme A analogue (FW = 735.46 g/mol; λ_{max} = 259 nm; ε = 15400 $M^{-1}cm^{-1}$) inhibits many coenzyme A-dependent enzymes and has been used for affinity chromatography, such as in the purification of phosphate acetyltransferase (1). **Target(s):** acetyl-CoA *C*-acetyltransferase (2); *N*-acetylneuraminate 4-*O*-acetyltransferase (3); aminoglycoside $N^{3'}$-acetyltransferase (4); aralkylamine *N*-acetyltransferase (5); citrate (*si*)-synthase (6); CO-methylating acetyl-CoA synthase, *or* carbon monoxide dehydrogenase (7,8); heparan-α glucosaminide *N*-acetyltransferase (9); histone acetyltransferase (10,11); homocitrate synthase (12); 3 hydroxy-3-methylglutaryl-CoA reductase (13); 3-hydroxy-3-methylglutaryl-CoA synthase (14); 2 isopropylmalate synthase (15); 3-oxoacid CoA-transferase, *or* 3-ketoacid CoA-transferase; succinyl CoA:3-oxoacid CoA-transferase (16); phosphate acetyltransferase (17-23); serine acetyltransferase (24); succinyl-CoA synthetase (25); xenobiotic acetyltransferase from *Pseudomonas aeruginosa* (26). **10.** Tanner, Langer, Kim & Denu (2000) *J. Biol. Chem.* **275**, 22048. **11.** Lau, Courtney, Vassilev, *et al.* (2000) *J. Biol. Chem.* **275**, 21953. **12.** Andi, West & Cook (2004) *Biochemistry* **43**, 11790. **13.** Roitelman & Shechter (1989) *J. Lipid Res.* **30**, 97. **14.** Middleton (1972) *Biochem. J.* **126**, 35. **15.** Tracy & Kohlhaw (1977) *J. Biol. Chem.* **252**, 4085. **16.** Whitty, Fierke & Jencks (1995) *Biochemistry* **34**, 11678.

17. Shimizu (1970) *Meth. Enzymol.* **18A**, 322. **18.** Iyer & Ferry (2001) *J. Bacteriol.* **183**, 4244. **19.** Henkin & Abeles (1976) *Biochemistry* **15**, 3472. **20.** Lawrence & Ferry (2006) *J. Bacteriol.* **188**, 1155. **21.** Duhr, Owens & Barden (1983) *Biochim. Biophys. Acta* **749**, 84. **22.** Shimizu, Suzuki, Kameda & Abiko (1969) *Biochim. Biophys. Acta* **191**, 550. **23.** Suzuki (1969) *Biochim. Biophys. Acta* **191**, 559. **24.** Johnson, Huang, Roderick & Cook (2004) *Arch. Biochem. Biophys.* **429**, 115. **25.** Collier & Nishimura (1978) *J. Biol. Chem.* **253**, 4938. **26.** Beaman, Sugantino & Roderick (1998) *Biochemistry* **37**, 6689.

Desvenlafaxine

Desvenlafaxine   **Venlafaxine**

This serotonin/norepinephrine reuptake inhibitor, *or* SNRI (FW = 263.38 g/mol; CAS 93413-62-8), also named Pristiq®, Desfax®, *O*-desmethylvenlafaxine, and 4-[2-dimethylamino-1-(1-hydroxycyclohexyl)ethyl]phenol, targets the serotonin transporter, *or* SERT (K_i = 60 nM) and the norepinephrine (*or* noradrenaline) transporter, *or* NET (K_i = 2.9 µM). Desvenlafaxine is itself the major active metabolite of another SNRI known as Venlafaxine (FW = 277.40 g/mol; CAS 93413-69-5; Trade Names: Effexor®, Effexor XR®, Trevilor®, Lanvexin®). SNRIs block or delay the reuptake of serotonin and norepinephrine by the presynaptic nerves, thereby increasing the levels of these neurotransmitters in the synapse and elevating mood. Both desvenlafaxin and venlafaxine have proved to be effective antidepressants for treating Major Depressive Disorder (MDD), showing favorable safety and tolerability profiles. The need for suitable alternatives in treating MDD is indicated by the fact that a majority of individuals with MDD do not achieve and sustain a recovered state (1). **1.** Kornstein, McIntyre, Thase & Boucher (2014) *Expert Opin. Pharmacother.* **15**, 1449.

Dexamethasone

This synthetic glucocorticoid (FW = 392.47 g/mol; CAS 50-02-2), also known as DEX and 9α-fluoro-11β,17α,21-trihydroxy-16α-methylpregna-1,4-diene-3,20-dione, is an anti-inflammatory agent that regulates T cell survival, growth, and differentiation. It also induces liver tyrosine aminotransferase and inhibits the induction of nitric-oxide synthase. Dexamethasone also inhibits spontaneous apoptosis in primary cultures of human and rat hepatocytes via Bcl-2 and Bcl-xL induction. Dexamethasone is also the most commonly used type of steroid for prostate cancer (**See also** *Leuprolide*). *Note*: Dexamethasone is the 16-epimer of betamethasone. **Target(s):** choline *O*-acetyltransferase (1); cortisol sulfotransferase (2-4); hyaluronoglucosaminidase, *or* hyaluronidase (5); 3α-hydroxysteroid dehydrogenase (6); phospholipase A₂ (7). **1.** Wattanathorn, Kotchabhakdi, Casalotti, Baldwin & Govitrapong (1996) *Eur. J. Pharmacol.* **313**, 69. **2.** Singer, Gebhart & Hess (1978) *Can. J. Biochem.* **56**, 1028. **3.** Singer (1979) *Arch. Biochem. Biophys.* **196**, 340. **4.** Singer & Bruns (1980) *Can. J. Biochem.* **58**, 660. **5.** Girish & Kemparaju (2005) *Biochemistry (Moscow)* **70**, 948. **6.** Penning, I. Mukharji, S. Barrows & P. Talalay (1984) *Biochem. J.* **222**, 601. **7.** Blackwell, Flower, Nijkamp & Vane (1978) *Brit. J. Pharmacol.* **62**, 79.

Dexniguldipine

This photosensitive diester (FW$_{free-base}$ = 609.72 g/mol), also known as *R*(–)-niguldipine, is the enantiomer of a potent L-type Ca²⁺ channel blocker. It will also act as an L-type Ca²⁺ channel blocker and α1A-adrenergic receptor antagonist, albeit not as effectively. Note that this diester adsorbs strongly to plastic surfaces, requiring one to take measures to passivate plasticware used in an assay. **Target(s):** α₁A-adrenoceptor (1); Ca²⁺ channels, L-type (1); DNA topoisomerase (2); drug transport by P-glycoprotein (3); protein kinase C (4,5). **1.** Boer, Grassegger, Schudt & Glossmann (1989) *Eur. J. Pharmacol.* **172**, 131. **2.** Straub, Boesenberg, Gekeler & Boege (1997) *Biochemistry* **36**, 10777. **3.** Hollt, Kouba, Dietel & Vogt (1992) *Biochem. Pharmacol.* **43**, 2601. **4.** Patterson, Beckman, Klotz, Mallia & Jeter (1996) *J. Cancer Res. Clin. Oncol.* **122**, 465. **5.** Schuller, Orloff & Reznik (1994) *J. Cancer Res. Clin. Oncol.* **120**, 354.

Dexon

This photosensitive fungicide (FW = 219.18 g/mol; CAS 140-56-7), also known as *p*-(*N,N*-dimethylamino)benzenediazosulfonate sodium salt and Fenaminosulf, Phenaminosulf, Deksona, Dexoxon, Lesan, Gold Orange MP, is metabolized to *N,N*-dimethyl-*p*-phenylenediamine by *Pseudomonas fragi*. **Caution:** Fenaminosulf is readily absorbed through the skin. Note: Dexon is also a proprietary name for polyglycolic acid, which is frequently used absorbable suture material. **Target(s):** electron transport (1-4); NADH dehydrogenase (1-4). **1.** Rapoport & Schewe (1977) *Trends Biochem. Sci.* **2**, 186. **2.** Schewe, Hiebsch & Halangk (1975) *Acta Biol. Med. Ger.* **34**, 1767. **3.** Schewe & Hiebsch (1977) *Acta Biol. Med. Ger.* **36**, 961 **4.** Muller & Schewe (1977) *Acta Biol. Med. Ger.* **36**, 967.

Dexrazoxane

This cyclic methyl-EDTA derivative (FW = 268.27 g/mol; CAS 24584-09-6), also known as Cardioxane®, Zinecard®, and 4-[(2*S*)-2-(3,5-dioxopiperazin-1-yl)propyl]piperazine-2,6-dione, is a prodrug, site-specific cardioprotective agent that protects against anthracycline-induced cardiac toxicity (1). Dexrazoxane undergoes metabolic hydrolysis to methyl-EDTA, a powerful metal ion-chelator (**See** *EDTA*). It is approved in the US and Europe as a cardioprotectant in women receiving doxorubicin to treat advanced and/or metastatic breast cancer. Dexrazoxane is also approved for patients with advanced cancer receiving anthracyclines, significantly reducing the incidence of anthracycline-induced congestive heart failure and adverse cardiac events in women with advanced breast cancer or adults with soft tissue sarcomas or small-cell lung cancer, but does not improve progression-free and overall patient survival (1). Dexrazoxane also induces topoisomerase IIα (TOP2A)-dependent cell death, γ-H2AX accumulation and increased tail moment in neutral comet assays (2). Dexrazoxane induces DNA damage responses, as indicated by enhanced levels of γ-H2AX/53BP1 foci, ATM (ataxia telangiectasia mutated), ATR (ATM/Rad3-related), Chk1 and Chk2 phosphorylation, and by accumulation of p53 (2). Dexrazoxane-induced γ-H2AX accumulation was dependent on ATM. ATF3 protein was induced by dexrazoxane in a concentration- and time-dependent manner, which was abolished in TOP2A-depleted cells and in cells pre-incubated with ATM inhibitor. Knockdown of ATF3 gene expression by siRNA triggered apoptosis in control cells and diminished the p53 protein level in both control and dexrazoxane-treated cells (2). These effects is accompanied by increased γ-H2AX accumulation and delayed repair of dexrazoxane-induced DNA double-strand breaks (2). Earlier studies suggested that Dexrazoxane may prevent doxorubicin-induced DNA damage via depleting both topoisomerase II isoforms (3). **1**. Cvetković & Scott (2005) *Drugs* **65**, 1005. **2**. Deng, Yan, Nikolova, *et al.* (2015) *Brit. J. Pharmacol.* **172**, 2246. **3**. Deng, Yan, Jendrny, *et al.* (2014) *BMC Cancer* **14**, 842.

Dextroamphetamine, *See Amphetamine*

Dextromethorphan

This over-the-counter morphinan-class sedative (FW = 271.40 g/mol; CAS 125-71-3), frequently found in combination (with doxylamine and acetaminophen) in cough syrups and cold remedies, inhibits dopamine reuptake, activates σ-receptors, and blocks open NMDA (*N*-methyl-D-aspartate) channels. It is metabolized by cytochrome P450 systems (particularly CYP2D6 and CYP3A4). Because roughly 5-10% of the Caucasian population is CYP2D6-deficient, such individuals metabolize dextromethorphan poorly. Unlike its stereoisomer, levomethorphan, dextromethorphan has no analgesic properties and is nonaddictive. Levomethorphan and the racemic mixture racemethorphan are controlled substances. **Target(s):** Ca^{2+} channels, voltage-dependent (1); $\alpha_3\beta_4$-nicotinic acetylcholine receptor (2); NMDA-controlled ion channels (3). **1**. Carpenter, Marks, Watson & Greenberg (1988) *Brain Res.* **439**, 372. **2**. Jozwiak, Hernandez, Kellar & Wainer (2003) *J. Chromatogr. B Analyt. Technol. Biomed. Life Sci.* **797**, 373. **3**. Wong, Coulter, Choi & Prince (1988) *Neurosci. Lett.* **85**, 261.

DHMEQ, *See Dehydroxymethylepoxyquinomicin*

N,N'-Diacetylchitobiose

This reducing disaccharide (FW = 424.41 g/mol), also known as 4-*O*-(2-acetamido-2-deoxy-β-D-glucopyranosyl)-2-acetamido-2-deoxy-D-glucose, is the repeating unit of chitin and the limit hydrolase product for many chitinases. **Target(s):** β-*N*-acetylglucosaminidase, also as an alternative substrate (1,2); β-*N* acetylhexosaminidase (1-5); *Bandeiraea simplicifolia* lectin II (6); chitinase (7-9); gorse (*Ulex europeus*) phytohemaagglutinin II (10); lysozyme (11,12); peptide-*N*⁴-(*N*-acetyl-β-glucosaminyl)-asparagine amidase (13); *Wistaria floribunda* phytomitogen (14). **1**. Cohen (1986) *Plant Sci.* **43**, 93. **2**. Mommsen (1980) *Biochim. Biophys. Acta* **612**, 361. **3**. Ohtakara (1988) *Meth. Enzymol.* **161**, 462. **4**. Keyhani & Roseman (1996) *J. Biol. Chem.* **271**, 33425. **5**. Ohtakara, Yoshida, Murakami & Izumi (1981) *Agric. Biol. Chem.* **45**, 239. **6**. Ebisu & Goldstein (1978) *Meth. Enzymol.* **50**, 350. **7**. Sakai, A. Yokota, Kurokawa, Wakayama & Moriguchi (1998) *Appl. Environ. Microbiol.* **64**, 3397. **8**. Bhushan (2000) *J. Appl. Microbiol.* **88**, 800. **9**. Molano, Polacheck, Duran & Cabib (1979) *J. Biol. Chem.* **254**, 4901. **10**. Osawa & Matsumoto (1972) *Meth. Enzymol.* **28**, 323. **11**. Imoto, Johnson, North, Phillips & Rupley (1972) *The Enzymes*, 3rd ed., **7**, 665. **12**. Bernard, Canioni, Cozzone, Berthou & Jolles

(1990) *Int. J. Protein Res.* **36**, 46. **13**. Suzuki, Kitajima, Inoue & Inoue (1995) *J. Biol. Chem.* **270**, 15181. **14**. Osawa & Toyoshima (1972) *Meth. Enzymol.* **28**, 328.

Diacetylmorphine, *See Heroin*

P^1,P^5-Di(adenosine-5') pentaphosphate

This naturally occurring multisubstrate, geometrical analogue ($FW_{free-acid}$ = 916.36 g/mol; CAS 75522-97-3 (free-acid); $FW_{Penta-Lithium-Salt}$ = 946.03 g/mol; CAS 94108-02-8 (penta-lithium salt), often abbreviated ApppppA or Ap₅A, is stored in secretory granules of thrombocytes, chroaffin, and neuronal cells. Ap₅A was the second diadenosine polyphosphate found to inhibit adenylate kinase, *or* myokinase (*See P^1,P^4-di(adenosine-5') tetraphosphate*). This property is itself especially useful in view of the fact that adenylate kinase is a frequent contaminant in enzyme preparatoions. Ap₅A is also an antagonist at Ca^{2+}-coupled P2Y1 receptors, acting as a stress mediator in the cardiovascular system. **Target(s):** acetyl-CoA synthetase, *or* acetate:CoA ligase (1); adenosine kinase (2-4); adenylate kinase, $K_i = 10^{-5}$ M (5-25); angiogenin (26); Ca^{2+}-coupled P2Y1 receptors (27); carbamate kinase (28,29); carbamoyl phosphate synthetase (30,31); creatine kinase, weakly inhibited (32); cytidylate kinase (ATP:NMP phosphotransferase, UMP/CMP kinase) (33,34); terminal deoxyribonucleotidyltransferase (35-37); dynein ATPase, weakly inhibited (38); ecto-adenylate kinase (39); F_1 ATPase (40); firefly luciferase, *or Photinus*-luciferin monooxygenase (41); K^+-channel, ATP-sensitive (42,43); nucleoside-triphosphate:adenylate kinase (8,44); 5'-nucleotidase (45); protein kinase C (46); ribonuclease A (26); ribonucleotide reductase (47); tetrahydrofoalte synthetase (48). **1**. Oberlies, Fuchs & Thauer (1980) *Arch. Microbiol.* **128**, 248. **2**. Delaney, Blackburn & Geiger (1997) *Eur. J. Pharmacol.* **332**, 35. **3**. Bone, Cheng & Wolfenden (1986) *J. Biol. Chem.* **261**, 16410. **4**. Rotllan & Miras Portugal (1985) *Eur. J. Biochem.* **151**, 365. **5**. Lienhard & Secemski (1973) *J. Biol. Chem.* **248**, 1121. **6**. Wolfenden (1977) *Meth. Enzymol.* **46**, 15. **7**. Egner, Tomasselli & Schulz (1987) *J. Mol. Biol.* **195**, 649. **8**. Ulschmid, Rahlfs, Schirmer & Becker (2004) *Mol. Biochem. Parasitol.* **136**, 211. **9**. Hamada & Kuby (1978) *Arch. Biochem. Biophys.* **190**, 772. **10**. Font & Gautheron (1980) *Biochim. Biophys. Acta* **611**, 299. **11**. Dinbergs & Lindmark (1989) *Exp. Parasitol.* **69**, 150. **12**. Munier-Lehmann, Burlacu-Miron, Craescu, Mantsch & Schultz (1999) *Proteins Struct. Funct. Genet.* **36**, 238. **13**. Bhaskara, Dupré, Lengsfeld, *et al.* (2007) *Mol. Cell* **25**, 647. **14**. Sheng, Li & Pan (1999) *J. Biol. Chem.* **274**, 22238. **15**. Neufang, Müller & Knobloch (1983) *Arch. Microbiol.* **134**, 153. **16**. Tomasselli & Noda (1980) *Eur. J. Biochem.* **103**, 481. **17**. Kuby, Hamada, Gerber, *et al.* (1978) *Arch. Biochem. Biophys.* **187**, 34. **18**. Kinukawa, Nomura & Vacquier (2007) *J. Biol. Chem.* **282**, 2947. **19**. Sperlagh & Vizi (2007) *Neurochem. Res.* **32**, 1978. **20**. Van der Ljin, Barrio & Leonard (1979) *Biochemistry* **18**, 5557. **21**. Villa, Pérez-Pertejo, García-Estrada, *et al.* (2003) *Eur. J. Biochem.* **270**, 4339. **22**. Müller & Schulz (1992) *J. Mol. Biol.* **224**, 159. **23**. Kurebayashi, Kodama & Ogawa (1980) *J. Biochem.* **88**, 871. **24**. Lacher & Schäfer (1993) *Arch. Biochem. Biophys.* **302**, 391. **25**. Kleczkowski, Randall & Zahler (1990) *Z. Naturforsch. C* **45**, 607. **26**. Kumar, Jenkins, Jardine & Shapiro (2003) *Biochem. Biophys. Res. Commun.* **300**, 81. **27**. Vigne, Breittmayer & Frelin (2000) *Brit. J. Pharmacol.* **129**, 1506. **28**. Uriarte, Marina, Ramón-Maiques, *et al.* (2001) *Meth. Enzymol.* **331**, 236. **29**. Marina, Uriarte, Barcelona, *et al.* (1998) *Eur. J. Biochem.* **253**, 280. **30**. Powers, Griffith & Meister (1977) *J. Biol. Chem.* **252**, 3558. **31**. Durbecq, Legrain, Roovers, Piérard & Glansdorff (1997) *Proc. Natl. Acad. Sci. U.S.A.* **94**, 12803. **32**. Szasz, Gerhardt & Gruber (1977) *Clin. Chem.* **23**, 1888. **33**. Wiesmuller, Scheffzek, Kliche, *et al.* (1995) *FEBS Lett.* **363**, 22. **34**. Zhou, Lacroute & Thornburg (1998) *Plant Physiol.* **117**, 245. **35**. Pandey, Amrute, Satav & Modak (1987) *FEBS Lett.* **213**, 204. **36**. Pandey & Modak (1987) *Biochemistry* **26**, 2033. **37**. Di Santo & Maga (2006) *Curr. Med. Chem.* **13**, 2353. **38**. Shimizu (1987) *J. Biochem.* **102**, 1159. **39**. Picher & Boucher (2003) *J. Biol. Chem.* **278**, 11256. **40**. Vogel & Cross (1991) *J. Biol. Chem.* **266**, 6101. **41**. Pojoga, Moose & Hilderman (2004) *Biochem. Biophys. Res. Commun.* **315**, 756. **42**. Jovanovic, Alekseev & Terzic (1997) *Biochem. Pharmacol.* **54**, 219. **43**. Jovanovic, Alekseev & Terzic (1996) *Naunyn Schmiedebergs Arch. Pharmacol.* **353**, 241. **44**. Yang, Yu, Wu, *et al.*

(2005) *Parasitol. Res.* **95**, 406. **45**. Minelli, Moroni & Mezzasoma (1997) *Biochem. Mol. Med.* **61**, 95. **46**. Leventhal & Bertics (1991) *Biochemistry* **30**, 1385. **47**. Wasternack, Sillero & Sillero (1991) *Biochem. Int.* **23**, 151. **48**. Cichowicz & B. Shane (1987) *Biochemistry* **26**, 513.

P¹,P⁴-Di(adenosine-5') tetraphosphate

This naturally occurring multisubstrate, geometric analogue (FW = 924.31 g/mol (tetrasodium salt); CAS 5542-28-9), often abbreviated Ap₄A, corresponding to the fusion of two ADP nucleotides linked by their terminal phosphate groups, is stored in secretory granules of thrombocytes, chromaffin, and neuronal cells. Ap₄A was the second diadenosine polyphosphate found to inhibit adenylate kinase, *or* myokinase (1). (***See the more potent inhibitor*** *P¹,P⁴-Di(adenosine-5') pentaphosphate*). Ap₄A also inhibits adenosine kinase and carbamoyl-phosphate synthetase. It is also an antagonist at Ca²⁺-coupled P2Y1 receptors and act as a stress mediator in the cardiovascular system. Ap₄A stimulates phospholipase D activity and induces nitric oxide release in endothelial cells. **Target(s):** adenylate kinase, *or myokinase* (1-4); adenosine kinase (5-8); ATP adenylyltransferase (9,10); Ca²⁺-coupled P2Y1 receptors (11); carbamoyl-phosphate syntheatase, weakly inhibited (12); casein kinase II (13); creatine kinase, weakly inhibited (14); deoxycytidine kinase (15); dephospho-[reductase kinase] kinase (16); DNA nucleotidyl-exotransferase, *or* terminal deoxyribo-nucleotidyltransferase (17-19); endoribonuclease VI (20); *Escherichia coli* DnaK protein 5'-nucleotidase activity (21); F₁ ATPase (22); firefly luciferase, *or Photinus*-luciferin monooxygenase (23); [3-hydroxy-3-methylglutaryl-CoA reductase (NADPH)] kinase (16); K⁺-channel, ATP-sensitive (24); poly(ADP-ribose) glycohydrolase (25); pp60src protein kinase (26); protein kinase C (27); ribonucleotide reductase (28); RNA ligase (ATP) (29). **1**. Purich & Fromm (1972) *Biochim. Biophys. Acta* **276**, 563. **2**. Lienhard & Secemski (1973) *J. Biol. Chem.* **248**, 1121. **3**. Bhaskara, Dupré, Lengsfeld, *et al.* (2007) *Mol. Cell* **25**, 647. **4**. Van der Ljin, Barrio & Leonard (1979) *Biochemistry* **18**, 5557. **5**. Bone, Cheng & Wolfenden (1986) *J. Biol. Chem.* **261**, 5731. **6**. Delaney, Blackburn & Geiger (1997) *Eur. J. Pharmacol.* **332**, 35. **7**. Bone, Cheng & Wolfenden (1986) *J Biol Chem.* **261**, 16410. **8**. Rotllan & Miras Portugal (1985) *Eur. J. Biochem.* **151**, 365. **9**. Booth & Guidotti (1995) *J. Biol. Chem.* **270**, 19377. **10**. Guranowski & Blanquet (1986) *J. Biol. Chem.* **261**, 5943. **11**. Vigne, Breittmayer & Frelin (2000) *Brit. J. Pharmacol.* **129**, 1506. **12**. Powers, Griffith & Meister (1977) *J. Biol. Chem.* **252**, 3558. **13**. Pype & Slegers (1993) *Enzyme Protein* **47**, 14. **14**. Marina, Uriarte, Barcelona, *et al.* (1998) *Eur. J. Biochem.* **253**, 280. **15**. Datta, Shewach, Mitchell & Fox (1989) *J. Biol. Chem.* **264**, 9359. **16**. Weekes, Hawley, Corton, Shugar & Hardie (1994) *Eur. J. Biochem.* **219**, 751. **17**. Pandey, Amrute, Satav & Modak (1987) *FEBS Lett.* **213**, 204. **18**. Ono, Iwata, Nakamura & Matsukage (1980) *Nucl. Acids Symp. Ser.* (8), s187. **19**. Ono, Iwata, Nakamura & Matsukage (1980) *Biochem. Biophys. Res. Commun.* **95**, 34. **20**. Grau & Heredia (1988) *FEBS Lett.* **236**, 291. **21**. Bochner, Zylicz & Georgopoulos (1986) *J. Bacteriol.* **168**, 931. **22**. Vogel & Cross (1991) *J. Biol. Chem.* **266**, 6101. **23**. Pojoga, Moose & Hilderman (2004) *Biochem. Biophys. Res. Commun.* **315**, 756. **24**. Jovanovic, Alekseev & Terzic (1997) *Biochem. Pharmacol.* **54**, 219. **25**. Tanuma, Kawashima & Endo (1986) *J. Biol. Chem.* **261**, 965. **26**. Levy, Sorge, Drum & Maness (1983) *Mol. Cell. Biol.* **3**, 718. **27**. Leventhal & Bertics (1991) *Biochemistry* **30**, 1385. **28**. Wasternack, Sillero & Sillero (1991) *Biochem. Int.* **23**, 151. **29**. Juodka & Labeikyte (1991) *Nucleosides Nucleotides* **10**, 367.

Diallyl disulfide

This garlic oil constituent (FW = 146.27 g/mol; CAS 2179-57-9) increases cellular production of glutathione *S*-transferase (GST), an enzyme that binds and deactivates many electrophilic toxins. A concentration dependent inhibition of was found When incubated with microsomal preparations, diallyl disulfide inhibits 3-hydroxy-3-methylglutaryl CoA (HMG CoA) reductase in a concentration-dependent manner that cannot by reversed by

eith dithiothreitol or reduced glutathione (1). The authors suggest diallyl disulfide form an internal protein disulfide linkage that is inaccessible for reduction by thiol agents. Diallyl disulfide also arrests unsynchronized human colon tumor cells (HCT-15) at the G₂/M stage of the cell cycle by inducing ERK phosphorylation and altering gene expression profiles (2). Diallyl sulfide also inhibits diethylstilbestrol-induced DNA damage in human breast epithelial MCF-10A cells (3). *Note:* Diallyl disulfide is a skin irritant and allergen that readily penetrates most latex gloves. Its median LD₅₀ for oral intake in rats is 260 mg per kg of body weight. **1**. Kumar, Banerji, Kurup & Ramasarma (1991) *Biochim. Biophys. Acta.* **1078**, 219. **2**. Knowles & Milner (2003) *J. Nutr.* **133**, 2901. **3**. McCaskill, Rogan & Thomas (2014) *Steroids* **92**, 96.

L-2,4-Diaminobutyrate

This amino acid (FW$_{neutral}$ = 118.14 g/mol; pK_a values of 1.85 (α-COOH), 8.24 (α-NH₂), and 10.44 (γ-NH₂) at 20°C), found in species of vetch (*Lathyrus* sp.), is occasionally found in place of the *meso*-diaminopimelate residue in bacterial cell wall peptidoglycans and is also a component of polymyxins. It has also been observed in certain higher plants. **Target(s):** 4-aminobutyrate aminotransferase (1); γ-aminobutyrate uptake (2); 5-aminovalerate aminotransferase (3); arginine racemase (4); carnitine uptake (5); diaminopimelate decarboxylase (6,7); diaminopropionate ammonia-lyase (8); hydroxylysine kinase (9); D-lysine 5,6-aminomutase (10); D-ornithine 4,5-aminomutase, inhibited by the racemic mixture (11); ornithine carbamoyltransferase (12-17); ornithine decarboxylase (18). **1**. Tunnicliff, Ngo, Rojo-Ortega & Barbeau (1977) *Can. J. Biochem.* **55**, 479. **2**. Taberner & Roberts (1978) *Eur. J. Pharmacol.* **52**, 281. **3**. Der Garabedian (1986) *Biochemistry* **25**, 5507. **4**. Yorifuji (1971) *J. Biol. Chem.* **246**, 5093. **5**. Virmani, Conti, Spadoni, Rossi & Arrigoni-Martelli (1995) *Pharmacol. Res.* **31**, 211. **6**. Asada, Tanizawa, Kawabata, Misuno & Soda (1981) *Agric. Biol. Chem.* **45**, 1513. **7**. Rosner (1975) *J. Bacteriol.* **121**, 20. **8**. Nagasawa, Tanizawa, Satoda & Yamada (1988) *J. Biol. Chem.* **263**, 958. **9**. Hiles & Henderson (1972) *J. Biol. Chem.* **247**, 646. **10**. Morley & Stadtman (1970) *Biochemistry* **9**, 4890. **11**. Chen, Hsui, Lin, Ren & Wu (2004) *Eur. J. Biochem.* **271**, 4293. **12**. Nakamura & Jones (1970) *Meth. Enzymol.* **17A**, 286. **13**. O'Neal, Chen, Reynolds, Meghal & Koeppe (1968) *Biochem. J.* **106**, 699. **14**. Legrain & Stalon (1976) *Eur. J. Biochem.* **63**, 289. **15**. Templeton, Reinhardt, Collyer, Mitchell & W. W. Cleland (2005) *Biochemistry* **44**, 4408. **16**. Marshall & Cohen (1972) *J. Biol. Chem.* **247**, 1654. **17**. Pierson, Cox & Gilbert (1977) *J. Biol. Chem.* **252**, 6464. **18**. Guirard & Snell (1980) *J. Biol. Chem.* **255**, 5960.

4,6-Diamino-1-(3',4'-dichlorophenyl)-1,2-dihydro-2,2-dimethyl-*s*-triazine

This dihydrotriazine (FW = 286.16 g/mol), also known as NSC-3077, reversibly inhibits dihydrofolate reductases from pigeon liver, *Escherichia coli*, T₂ phage, mouse leukemia L1210, and Walker 256 rat tumor (IC₅₀ ≈ 15, 160, 19, 6.4, and 2.9 nM respectively). **1**. Baker (1967) *Design of Active-Site-Directed Irreversible Enzyme Inhibitors*, Wiley, New York. **2**. Baker (1967) *J. Med. Chem.* **10**, 912. **3**. Baker, Janson & Vermeulen (1969) *J. Med. Chem.* **12**, 898. **4**. Baker (1968) *J. Med. Chem.* **11**, 483. **5**. Baker, Vermeulen, Ashton & Ryan (1970) *J. Med. Chem.* **13**, 1130. **6**. Baker (1971) *Ann. N. Y. Acad. Sci.* **186**, 214.

2,6-Diaminopurine

This folate antagonist (FW = 150.14 g/mol; CAS 1904-98-9; λ_{max} at 241 and 282 nm at pH 1.9 (ε = 9550 and 10000 $M^{-1}cm^{-1}$, respectively), also known as 2-aminoadenine, is a reversible inhibitor of RNA biosynthesis. 2,6-Diaminopurine is metabolically converted to ribonucleotide derivatives. Discovery of its antifolate properties (1) stimulated interest in antifolate cancer chemotherapy. **Target(s):** adenine deaminase (2,3); adenine phosphoribosyltransferase (*Plasmodium falciparum* but not human) (4); adenosine deaminase (2,5-7); dihydrofolate reductase, K_i = 2.1 μM (1,8-11); hypoxanthine(guanine) phosphoribosyltransferase, *or* HGPRT (12); methionine *S*-adenosyltransferase (13); xanthine oxidase, also alternative substrate (14); and xanthine phosphoribosyltransferase (12). **1**. Hitchings, Elion, VanderWerff & Falco (1948) *J. Biol. Chem.* **174**, 765. **2**. Webb (1966) *Enzyme and Metabolic Inhibitors*, vol. 2, p. 466, Academic Press, New York. **3**. Remy (1961) *J. Biol. Chem.* **236**, 2999. **4**. Queen, Vander Jagt & Reyes (1989) *Biochim. Biophys. Acta* **996**, 160. **5**. Agarwal & Parks (1978) *Meth. Enzymol.* **51**, 502. **6**. Feigelson & Davidson (1956) *J. Biol. Chem.* **223**, 65. **7**. Daddona, Wiesmann, Lambros, Kelley & Webster (1984) *J. Biol. Chem.* **259**, 1472. **8**. Zakrzewski (1963) *J. Biol. Chem.* **238**, 1485. **9**. Webb (1966) *Enzyme and Metabolic Inhibitors*, vol. 2, p. 583, Academic Press, New York. **10**. Bertino, Perkins & Johns (1965) *Biochemistry* **4**, 839. **11**. Baker & Santi (1967) *J. Heterocyclic Chem.* **4**, 216. **12**. Naguib, Iltzsch, el Kouni, Panzica & el Kouni (1995) *Biochem. Pharmacol.* **50**, 1685. **13**. Berger & Knodel (2003) *BMC Microbiol.* **3**, 12. **14**. Wyngaarden (1957) *J. Biol. Chem.* **224**, 453.

3,5-Diamino-1*H*-1,2,4-triazole

This free radical scavenger and antitumor agent (FW = 99.095 g/mol), also known as guanazole and 3,5-diamino-1,2,4-triazole, inhibits ribonucleotide reductase, *or* RNR (1-8) and glutaminyl-peptide cyclotransferase (9). RNR is known to engage multiple catalytically essential free radicals, of which one or more is (are) intercepted by guanazole. Inhibition of deoxyribonucleotide biosynthesis blocks DNA synthesis and tumor cell proliferation. Rapid intravenous injection of guanazole produces a sharp drop in arterial blood pressure, an increase in right ventricular pressure, a slight decrease in heart rate, and a rise in central and portal venous pressures. These changes are transient, lasting 5–10 min, and can be partially ameliorated by prior treatment with antihistaminic, antiserotonin, and adrenergic blocking drugs (10). **1**. Spector & Jones (1985) *J. Biol. Chem.* **260**, 8694. **2**. Moore & Hurlbert (1985) *Pharmacol. Ther.* **27**, 167. **3**. Brockman, Shaddix, Laster & Schabel (1970) *Cancer Res.* **30**, 235. **4**. Lammers & Follmann (1983) *Struct. Bonding* **54**, 27. **5**. Cory, Sato & Lasater (1981) *Adv. Enzyme Regul.* **19**, 139. **6**. Holmgren (1981) *Curr. Top. Cell. Regul.* **19**, 47. **7**. Averett, Lubbers, Elion & Spector (1983) *J. Biol. Chem.* **258**, 9831. **8**. FitzGerald, Rosowsky & Wick (1984) *Biochem. Biophys. Res. Commun.* **120**, 1008. **9**. Huang, Liu & Wang (2005) *Protein Expr. Purif.* **43**, 65. **10**. Vick & Herman (1970) *Toxicol. Applied Pharmacol.* **16**, 108.

Dianemycin

This open-chain macrotetralide (FW$_{free-acid}$ = 888.51 g/mol; CASs = 35865-33-9 and 65101-87-3 (sodium salt)), with structural features similar to nigericin and monensin, is an ionophore and an uncoupler of oxidative phosphorylation. Dianemycin was first isolated from *Streptomyces hydroscopicus*. Dianemycin inhibits cation transport in mitochondria (4). This agent also exhibits antibiotic properties against coccidial infections in chickens (5). **1**. Izawa & Good (1972) *Meth. Enzymol.* **24**, 355. **2**. Reed (1979) *Meth. Enzymol.* **55**, 435. **3**. Thore, Keister, Shavit & San Pietro

(1968) *Biochemistry* **7**, 3499. **4**. Lardy, Johnson & McMurray (1958) *Arch. Biochem. Biophys.* **78**, 587. **5**. Shumard & Callender (1967) *Antimicr. Agents & Chemoth.* **1967**, 369.

Diamox, *See* Acetazolamide

Dianicline

This orally bioavailable and selective nAChR partial agonist (FW$_{di-HCl}$ = 289.20 g/mol; CAS 292634-27-6; Solubility: 100 mM in H_2O ; 20 mM in DMSO), also named SSR-591813 and (5*aS*,8*S*,10*aR*)-5*a*,6,9,10-tetrahydro-7*H*,11*H*-8,10*a*-methanopyrido[2′,3′:5,6]pyrano[2,3-*d*]azepine, selectively targets the $\alpha_4\beta_2$ nicotinic acetylcholine receptor (IC$_{50}$ = 105 nM), exhibiting 20-fold greater action toward $\alpha_4\beta_2$ over other subtypes (1). By activating $\alpha_4\beta_2$ nAChRs and reducing the reinforcing effects of nicotine upon smoking by competing with nicotine, high-affinity $\alpha_4\beta_2$ nAChR partial agonists are desired for relieving nicotine craving and withdrawal symptoms during quit-smoking attempts. Dianicline also increases dopamine turnover in the nucleus accumbens of the rat *in vivo*. ***See also*** Varenicline **1**. Rollema, Shrikhande, Ward, *et al.* (2010) *Brit. J. Pharmacol.* **160**, 334. **2**. *Drug Trial Case Report:* https://clinicaltrials.gov/ct2/show/NCT00356967

Diazaborine

This experimental drug (FW = 313.14 g/mol; CAS 22959-81-5) inhibits maturation of rRNAs for the large ribosomal subunit, K_d = 0.12 mM(a). **Primary Mode of Inhibitory Action:** In eukaryotic cells, ribosomes are assembled through the coordinated action of >200 transacting factors, making ribosome biogenesis a target for antimicrobial and anti-tumor chemotherapy. In this study, we investigated the mechanism of action of the first specific inhibitor. Diazaborine's direct target is the AAA-ATPase Drg1, a key pre-60S ribosome maturation factor in yeast, with the drug blocking Rlp24 release from pre-60S particles, halting any further maturation. Diazaborine binds within the second AAA domain of Drg1, where it inhibits ATP hydrolysis in this site, even though the drug does not act as a competitive inhibitor for ATP. ATP binding is a prerequisite for diazaborine to recognize Drg1. **Other Target(s):** Nucleotide binding is also required for inhibitor interaction with the enoyl-ACP reductase (FabI), a diazaborine target in Gram-negative bacteria. bacteria. For FabI, NAD$^+$ cofactor must likewise first bind to the enzyme to create the site for inhibitor binding (2,3), with diazaborine 's boron atom forming a covalent bond with the 2'OH of the NMN moiety of the coenzyme (26,27). **1**. Loibl, Klein, Prattes, *et al.* (2013) *J. Biol. Chem.* **289**, 3913. **2**. Bergler, Wallner, Ebeling, *et al.* (1994) *J. Biol. Chem.* **269**, 5493. **3**. Kater, Koningstein, Nijkamp, H. J., and Stuitje, A. R. (1994) *Plant Mol. Biol.* **25**, 771. **4**. Baldock, Rafferty, Sedelnikova, *et al.* (1996) *Acta Crystallogr. D Biol. Crystallogr.* **52**, 1181. **5**. Levy, Baldock, Wallace, *et al.* (2001) *J. Mol. Biol.* **309**, 171.

Diazepam

This powerful and widely used tranquilizer (FW = 284.74 g/mol; CAS 439-14-5), also known by its trade name Valium® and IUPAC name 7-chloro-1,3-dihydro-1-methyl-5-phenyl-2*H*-1,4-benzodiazepin-2-one, anti-anxiety agent and muscle relaxant. **Mode of Action:** Diazepam enhances the effect of the GABA by binding to the benzodiazepine site on the GABA$_A$ receptor, resulting in CNS depression. It also retards adenosine reuptake, with a consequential rise in extracellular adenosine. Overuse often leads to habituation or addiction. Diazepam was synthesized in 1959 by Sternbach and Reeder at Hofmann-La Roche (1). **Pharmacokinetics:** After a lag time (~12 min) between ingestion and absorption, first-order absorption proceed with a mean half-life of 19 min (*range*: 0–96 min), reaching peak plasma concentrations within 0.9 hour after dosage (*range*: 0.25–2.5 hr). Age and sex were without effect on absorption kinetics. Peak plasma concentrations averaged 150-160 ng/mL, somewhat higher in women than men owing to differences in body weight. Dominant features of diazepam pharmacokinetics are its relatively low rate of elimination and the formation of the biologically active metabolic products, chiefly desmethyldiazepam and oxazepam. Diazepam elimination rates depend on the extent of plasma protein binding, amount of drug and duration of administration, as well as the age and liver function of the patient. The normal half-life is 1–2 days, but can be as long as 3–4 in subjects age 60 and over. With sub-chronic administration, diazepam elimination rates drop to 20–70% that observed in healthy subjects. The major metabolite desmethyldiazepam has a $t_{1/2}$ of 51 h and a clearance rate of 11 mL/min, but accumulates after multiple doses of diazepam, because its elimination is much slower than diazepam. **Key Pharmacokinetic Parameters:** *See* Appendix II in Goodman & Gilman's THE PHARMACOLOGICAL BASIS OF THERAPEUTICS, 12th Edition (Brunton, Chabner & Knollmann, eds.) McGraw-Hill Medical, New York (2011). **Target(s):** acetylcholinesterase (2,3); acyl-CoA:cholesterol acyltransferase, *oe* ACAT (4); adenylate cyclase (5,6); apyrase (2); Ca^{2+}/calmodulin-dependent protein kinase (7,8); cAMP-specific phosphodiesterase (9,10); cholinesterase, *or* butyrylcholinesterase (3); cytochrome P450, *or* CYP2B (11); glucuronosyltransferase, *or* UDP-glucuronyltransferase (12,13); nitric-oxide synthase (14); quinine 3-monooxygenase (15); succinate semialdehyde dehydrogenase (16); xanthine oxidase (17). **1.** Sternbach & Reeder (1961) *J. Org. Chem.* **26**, 4936. **2.** Schetinger, Porto, Moretto, *et al.* (2000) *Neurochem. Res.* **25**, 949. **3.** Chiou, Lai, Tsai, *et al.* (2005) *J. Chin. Chem. Soc.* **52**, 843. **4.** Bell (1985) *Lipids* **20**, 75. **5.** Niles & Wang (1999) *Pharmacol. Toxicol.* **85**, 153. **6.** Niles & Hashemi (1990) *Biochem. Pharmacol.* **40**, 2701. **7.** Babcock-Atkinson, Norenberg, Norenberg & Neary (1989) *Brain Res.* **484**, 399. **8.** DeLorenzo, Burdette & Holderness (1981) *Science* **213**, 546. **9.** Cherry, Thompson & Pho (2001) *Biochim. Biophys. Acta* **1518**, 27. **10.** Margiotta, Borasio, Ardizzone & Lippe (1989) *Gen. Pharmacol.* **20**, 341. **11.** Nims, Prough, Jones, *et al.* (1997) *Drug Metab. Dispos.* **25**, 750. **12.** del Villar, Sanchez, Letelier & Vega (1981) *Res. Commun. Chem. Pathol. Pharmacol.* **33**, 433. **13.** Hara, Nakajima, Miyamoto & Yokoi (2007) *Drug Metab. Pharmacokinet.* **22**, 103. **14.** Fernandez-Cancio, Fernandez-Vitos, Imperial & Centelles (2001) *Brain Res. Mol. Brain Res.* **96**, 87. **15.** Zhao & Ishizaki (1997) *J. Pharmacol. Exp. Ther.* **283**, 1168. **16.** Sawaya, Horton & Meldrum (1975) *Epilepsia* **16**, 649. **17.** Korotkina, Papin, Voronina & Karelin (1988) *Biull. Eksp. Biol. Med.* **106**, 565.

Diazinon

This membrane-penetrating pro-pesticide (FW = 304.35 g/mol; CAS 333-41-5; colorless to amber-colored liquid), also known as phosphorothioic acid *O,O*-diethyl-*O*-[6-methyl-2-(1-methylethyl)-4-pyrimidinyl] ester, inhibits cholinesterases. Upon entry into the body, it is hydrolyzed and oxidatively metabolized to diaxozon, an organophosphate compound that is a far more effective acetylcholine esterase inhibitor (1-3). **Target(s):** fatty acid amide hydrolase (4); glutathione *S* transferase (5); kynurenine formamidase (6). **1.** Kamal & Al-Jafari (2000) *J. Enzyme Inhib.* **15**, 201. **2.** Gasser (1953) *Zeit. Naturforsch.* **8B**, 225. **3.** Dybing & Sognen (1958) *Acta Pharmacol. Toxicol. (Copenhagen)* **14**, 23. **4.** Quistad, Sparks & Casida (2001) *Toxicol. Appl. Pharmacol.* **173**, 48. **5.** da Silva Vaz, Torino

Lermen, Michelon, *et al.* (2004) *Vet. Parasitol.* **119**, 237. **6.** Seifert & Pewnim (1992) *Biochem. Pharmacol.* **44**, 2243.

N-Diazoacetyl-DL-norleucine Methyl Ester

This affinity labeling reagent (FW = 213.24 g/mol; CAS 7013-09-4; *Abbreviation*: DAN) inhibits a number of peptidases (particularly aspartic proteinases like pepsin and cathepsins), often in the presence of cupric ions (*i.e.*, Cu^{2+} and Ag$^+$ often facilitate the inhibition). Attack on the diazomethyl carbon by a carboxyl group on the target enzyme yields a protein-bound ester that will hydrolyze at pH > 9.5. **Mode of Pepsin Inactivation:** A copper-complexed carbene is most likely the reactive species, an idea that is supported by the finding that dimethylsulfonium phenacylide, which is known to form such a complex, rapidly inactivates pepsin in the presence of Cu(II). Reaction of a carbene is thought to occur at a protonated carboxyl, with specificity determined by the proximity of an ionized carboxyl group that serves to orient the positively charged inhibitor molecule. The proposed mechanism of inactivation is compatible with the presence at or near the active site of pepsin of two carboxyl groups with markedly different pK values. The attacking group is thus a protonated carboxyl on the enzyme, the action of which generates a positively charged adduct that is stabilized/oriented by a nearby ionized carboxyl group. (*See Means & Feeney (1971) Chemical Modification of Proteins, Holden-Day, San Francisco*) This property probably explains why aspartic proteinases are especially susceptible to DAN. **Target(s):** acrocylindropepsin (1,2); aspergillopepsin I (2-5); aspergillopepsin II (3,6,7); bovine leukemia virus retropepsin (8); candidapepsin (9; other references report no inhibition); canditropsin (10); cardosin A (11); cardosin B (12); cathepsin D (13-18); cathepsin E (19); chymosin, *or* rennin (20,21); *Fasciola hepatica* acid protease22; gastricsin, *or* pepsin C (23-27); lysozyme (28); mouse submandibular renin (25,29); mucorpepsin (2,20,30-32); nepenthesin (33-35); *Penicillium duponti* acid protease, closely related to penicillopepsin (36); penicillopepsin (25,36-42); pepsin A (2,24,25,43-57); physaropepsin (58-62); phytepsin (63); poly(3-hydroxybutyrate) depolymerase (64-66); polyporopepsin (67,68); pro-opiomelanocortin converting enzyme (yapsin A) (69,70); renin (2,71); rhizopuspepsin (2,20,38,72-75); rhodotorulapepsin (2,76-79); saccharopepsin (80,81); thermopsin (82-84); todarepsin (17); Wai21.A1 protease (85,86); Wp22.A1 protease (85); yapsin 3 (87). **1.** Takahashi (1998) in *Handb. Proteolytic Enzymes*, p. 867, Academic Press, San Diego. **2.** Takahashi & W. J. Chang (1976) *J. Biochem.* **80**, 497. **3.** Chang, Horiuchi, Takahashi, Yamasaki & Yamada (1976) *J. Biochem.* **80**, 975. **4.** Yagi, Fan, Tadera & Kobayashi (1986) *Agric. Biol. Chem.* **50**, 1029. **5.** Iio & Yamasaki (1976) *Biochim. Biophys. Acta* **429**, 912. **6.** Takahashi (1998) in *Handb. Proteolytic Enzymes*, p. 971, Academic Press, San Diego. **7.** Iio & Yamasaki (1976) *Biochim. Biophys. Acta* **429**, 912. **8.** Ménard & Guillemain (1998) in *Handb. Proteolytic Enzymes*, p. 940, Academic Press, San Diego. **9.** Nelson & Young (1987) *J. Gen. Microbiol.* **133**, 1461. **10.** Foundling (1998) in *Handb. Proteolytic Enzymes*, p. 902, Academic Press, San Diego. **11.** Pires (1998) in *Handb. Proteolytic Enzymes*, p. 843, Academic Press, San Diego. **12.** Pires (1998) in *Handb. Proteolytic Enzymes*, p. 844, Academic Press, San Diego. **13.** Keilova (1970) *FEBS Lett.* **6**, 312. **14.** Woessner, Jr. (1977) *Adv. Exp. Med. Biol.* **95**, 313. **15.** von Clausbruch & Tschesche (1988) *Biol. Chem. Hoppe-Seyler* **369**, 683. **16.** Cunningham & J. Tang (1976) *J. Biol. Chem.* **251**, 4528. **17.** Komai, Kawabata, Amano, Lee & Ichishima (2004) *Comp. Biochem. Physiol. B* **137**, 373. **18.** Pohl, Bures & Slavik (1981) *Collect. Czech. Chem. Commun.* **46**, 3302. **19.** Kageyama (1995) *Meth. Enzymol.* **248**, 120. **20.** Takahshi, Mizobe & Chang (1972) *J. Biochem.* **71**, 161. **21.** Chang & Takahashi (1974) *J. Biochem.* **76**, 467. **22.** Rupova & Keilova (1979) *Z. Parasitenkd.* **61**, 83. **23.** Tang (1998) in *Handb. Proteolytic Enzymes*, p. 823, Academic Press, San Diego. **24.** Ryle (1970) *Meth. Enzymol.* **19**, 316. **25.** Sigman & Mooser (1975) *Ann. Rev. Biochem.* **44**, 889. **26.** Martin, Trieu-Cuot, Collin & Ribadeau Dumas (1982) *Eur. J. Biochem.* **122**, 31. **27.** Kageyama & Takahashi (1976) *J. Biochem.* **80**, 983. **28.** Brecher, Rasmusson, Riley & VanDenBerghe (1999) *Fundam. Clin. Pharmacol.* **13**, 107. **29.** Suzuki,

Murakami, Nakamura & Inagami (1998) in *Handb. Proteolytic Enzymes*, p. 856, Academic Press, San Diego. **30**. Beppu & Nishiyama (1998) in *Handb. Proteolytic Enzymes*, p. 890, Academic Press, San Diego. **31**. Rickert & McBride-Warren (1977) *Biochim. Biophys. Acta* **480**, 262. **32**. Takahshi, Chang & Arima (1976) *J. Biochem.* **80**, 61. **33**. Woessner (1998) in *Handb. Proteolytic Enzymes*, p. 846, Academic Press, San Diego. **34**. Athauda, Matsumoto, Rajapakshe, *et al.* (2004) *Biochem. J.* **381**, 295. **35**. Takahashi, Chang & Ko (1974) *J. Biochem.* **76**, 897. **36**. Emi, Myers & Iacobucci (1976) *Biochemistry* **15**, 842. **37**. Hofmann (1998) in *Handb. Proteolytic Enzymes*, p. 878, Academic Press, San Diego. **38**. Sodek & Hofmann (1970) *Meth. Enzymol.* **19**, 372. **39**. Hofmann (1976) *Meth. Enzymol.* **45**, 434. **40**. Matsubara & Feder (1971) *The Enzymes*, 3rd. ed., **3**, 721. **41**. Mains & Hofmann (1974) *Can. J. Biochem.* **52**, 1018. **42**. Gripon (1976) *Biochimie* **58**, 747. **43**. Tang (1998) in *Handb. Proteolytic Enzymes*, p. 805, Academic Press, San Diego. **44**. Kassell & Meitner (1970) *Meth. Enzymol.* **19**, 337. **45**. Rajagopalan, Stein & Moore (1966) *J. Biol. Chem.* **241**, 4295. **46**. Lundblad & Stein (1969) *J. Biol. Chem.* **244**, 154. **47**. Llewellin & Green (1975) *Biochem. J.* **151**, 319. **48**. Tang, Kageyama & Takahashi (1988) *Eur. J. Biochem.* **177**, 251. **49**. Iliadis, Zundel & Brzezinski (1994) *FEBS Lett.* **352**, 315. **50**. Shaw (1970) *The Enzymes*, 3rd ed., 91. **51**. Fruton (1971) *The Enzymes*, 3rd ed., **3**, 119. **52**. Pichova & Kostka (1990) *Comp. Biochem. Physiol. B* **97**, 89. **53**. Kageyama, Moriyama & Takahashi (1983) *J. Biochem.* **94**, 1557. **54**. Athauda, Tanji, Kageyama & Takahashi (1989) *J. Biochem.* **106**, 920. **55**. Kageyama & Takahashi (1980) *J. Biochem.* **88**, 635. **56**. Kageyama & Takahashi (1976) *J. Biochem.* **79**, 455. **57**. Tanji, Kageyama & Takahashi (1988) *Eur. J. Biochem.* **177**, 251. **58**. Murakami-Murofushi (1998) in *Handb. Proteolytic Enzymes*, p. 870, Academic Press, San Diego. **59**. Nishii, Ueki, Miyashita, *et al.* (2003) *Biochem. Biophys. Res. Commun.* **301**, 1023. **60**. Murakami-Murofushi, Takahashi, Minowa, *et al.* (1990) *J. Biol. Chem.* **32**, 19898. **61**. Murakami-Murofushi, Takahashi, Murofushi & Takahashi (1995) *Adv. Exp. Med. Biol.* **362**, 565. **62**. Nishii, Ueki, Miyashita, *et al.* (2003) *Biochem. Biophys. Res. Commun.* **301**, 1023. **63**. Kervinen, Törmäkangas, Runeberg-Roos, *et al.* (1995) *Adv. Exp. Med. Biol.* **362**, 241. **64**. Brucato & Wong (1991) *Arch. Biochem. Biophys.* **290**, 497. **65**. Sadacco, Nocerino, Dubini-Paglia, Seves & Elegir (1997) *J. Environ. Polym. Degrad.* **5**, 57. **66**. Han & Kim (2002) *J. Microbiol.* **40**, 20. **67**. Kobayashi (1998) in *Handb. Proteolytic Enzymes*, p. 893, Academic Press, San Diego. **68**. Kobayashi, Kusakabe & Murakami (1983) *Agric. Biol. Chem.* **47**, 1921. **69**. Loh & Cawley (1995) *Meth. Enzymol.* **248**, 136. **70**. Loh (1986) *J. Biol. Chem.* **261**, 11949. **71**. Suzuki, Murakami, Nakamura & Inagami (1998) in *Handb. Proteolytic Enzymes*, p. 851, Academic Press, San Diego. **72**. Nakamura & Takahashi (1978) *J. Biochem.* **84**, 1593. **73**. Graham, Sodek & Hofmann (1973) *Can. J. Biochem.* **51**, 789. **74**. Mizobe, Takahashi & Ando (1973) *J. Biochem.* **73**, 61. **75**. Ichishima, Ojima, Yamagata, Hanzawa & Nakamura (1995) *Phytochemistry* **38**, 27. **76**. Murao (1998) in *Handb. Proteolytic Enzymes*, p. 866, Academic Press, San Diego. **77**. Oda, Funakoshi & Murao (1973) *Agric. Biol. Chem.* **37**, 1723. **78**. Kanazawa (1977) *J. Biochem.* **81**, 1739. **79**. Liu, Ohtsuki & Hatano (1973) *J. Biochem.* **73**, 671. **80**. Meussendoerffer, Tortora & Holzer (1980) *J. Biol. Chem.* **255**, 12087. **81**. Nowak & Tsai (1989) *Can. J. Microbiol.* **35**, 295. **82**. Tang & Lin (1998) in *Handb. Proteolytic Enzymes*, p. 980, Academic Press, San Diego. **83**. Lin & Tang (1995) *Meth. Enzymol.* **248**, 156. **84**. Fusek, Lin & Tang (1990) *J. Biol. Chem.* **265**, 1496. **85**. Toogood & Daniel (1998) in *Handb. Proteolytic Enzymes*, p. 978, Academic Press, San Diego. **86**. Prescott, Peek & Daniel (1995) *Int. J. Biochem. Cell Biol.* **27**, 729. **87**. Loh (1998) in *Handb. Proteolytic Enzymes*, p. 910, Academic Press, San Diego.

N-Diazoacetyl-L-phenylalanine Methyl Ester

This reagent (FW = 247.25 g/mol) inhibits a number of peptidases, particularly aspartic proteinases, most often in the presence of Cu^{2+} and Ag^+ (*See* Means & Feeney (1971) *Chemical Modification of Proteins*, Holden-Day, San Francisco). Attack on the diazomethyl carbon by a carboxyl group on the

target enzyme yields a protein-bound ester that hydrolyzes at pH above 9.5. **Target(s):** penicillopepsin, with Cu^{2+} present (1); pepsin A (2-7). **1**. Sodek & Hofmann (1970) *Meth. Enzymol.* **19**, 372. **2**. Wilcox (1972) *Meth. Enzymol.* **25**, 596. **3**. Sigman & Mooser (1975) *Ann. Rev. Biochem.* **44**, 889. **4**. Delpierre & Fruton (1966) *Proc. Natl. Acad. Sci. U.S.A.* **56**, 1817. **5**. Iliadis, Zundel & Brzezinski (1994) *FEBS Lett.* **352**, 315. **6**. Fruton (1971) *The Enzymes*, 3rd ed. (Boyer, ed.), **3**, 119. **7**. Irvine & Elmore (1979) *Biochem. J.* **183**, 389.

6-Diazo-5-oxo-L-norleucine

This toxic glutamine analogue (FW = 171.16 g/mol; *Abbreviation*: DON) inhibits many enzymes in amino acid and nucleotide biosynthesis, inactivating glutamine-hydrolyzing synthetases. Such enzymes generate nascent ammonia by attack of an active-site SH on the γ-amide, rendering these enzyme especially susceptible to DON activation by reaction with the same thiol. DON is also produced by a *Streptomyces* species. Prepare freshly and keep cold at pH 4.5-6.5. **Target(s):** D-alanine γ-glutamyltransferase (1); ω-amidase (2); amidophospho-ribosyltransferase (3-15; 4-amino-3-deoxychorismate synthase (16); anthranilate synthase (6,8,17-19); asparagine synthetase (6,20-24); carbamoyl-phosphate synthetase II, glutamine-hydrolyzing (6,25-29); CTP synthetase (6,30-33); glutamate synthase (6,34); glutaminase (6,9,35-40); glutamin-(asparagin-)ase (41-45); glutamine:fructose-6-P transaminase (5,6,8,37,46-59); [glutamine synthetase] adenylyltransferase (60,93); glutamine transport (61,62); glutaminyl-tRNA synthetase, inhibiting the glutamine dependent reaction but not the ammonia-dependent activity (63); γ-glutamyl peptide-hydrolyzing enzyme (64); γ-glutamyl transpeptidase, *or* γ-glutamyltransferase (38,65-77); GMP synthetase (6,78-83); 4-methyleneglutaminase, weakly inhibited (84); NAD^+ synthetase, glutamine-hydrolyzing (6,85,86); nitrogenase (87); phosphoribosylformyl-glycinamidine synthetase (4,6,9,37,88-92). **1**. Kawasaki, Ogawa & Sasaoka (1982) *Biochim. Biophys. Acta* **716**, 194. **2**. Calderon, Morett & Mora (1985) *J. Bacteriol.* **161**, 807. **3**. Zalkin (1985) *Meth. Enzymol.* **113**, 264. **4**. Flaks & Lukens (1963) *Meth. Enzymol.* **6**, 52. **5**. Aronson (1998) in *Handb. Proteolytic Enzymes*, p. 482, Academic Press, San Diego. **6**. Pinkus (1977) *Meth. Enzymol.* **46**, 414. **7**. Lewis & Hartman (1978) *Meth. Enzymol.* **51**, 171. **8**. Meister (1962) *The Enzymes*, 2nd ed., **6**, 247. **9**. Shaw (1970) *The Enzymes*, 3rd ed., **1**, 91. **10**. Hartman (1963) *J. Biol. Chem.* **238**, 3036. **11**. Tso, Hermodson & Zalkin (1982) *J. Biol. Chem.* **257**, 3532. **12**. Messenger & Zalkin (1979) *J. Biol. Chem.* **254**, 3382. **13**. Hill & Bennent (1969) *Biochemistry* **8**, 122. **14**. Kim, Wolle, Haridas, *et al.* (1995) *J. Biol. Chem.* **270**, 17394. **15**. King, Boounous & Holmes (1978) *J. Biol. Chem.* **253**, 3933. **16**. Viswanathan, Green & Nichols (1995) *J. Bacteriol.* **177**, 5918. **17**. Tamir & Srinivasan (1970) *Meth. Enzymol.* **17A**, 401. **18**. Queener, Queener, Meeks & Gunsalus (1973) *J. Biol. Chem.* **248**, 151. **19**. Zalkin & Hwang (1971) *J. Biol. Chem.* **246**, 6899. **20**. Jayaram, Cooney, Milman, Homan & Rosenbluth (1976) *Biochem. Pharmacol.* **25**, 1571. **21**. Larsen & S. Schuster (1992) *Arch. Biochem. Biophys.* **299**, 18. **22**. Mehlhaff & Schuster (1991) *Arch. Biochem. Biophys.* **284**, 143. **23**. Milman & Cooney (1979) *Biochem. J.* **181**, 51. **24**. Reitzer & Magasanik (1982) *J. Bacteriol.* **151**, 1299. **25**. Kaseman & Meister (1985) *Meth. Enzymol.* **113**, 305. **26**. Levenberg (1970) *Meth. Enzymol.* **17A**, 244. **27**. Casey & Anderson (1983) *J. Biol. Chem.* **258**, 8723. **28**. Khedouri, Anderson & Meister (1966) *Biochemistry* **5**, 3552. **29**. Yip & Knox (1970) *J. Biol. Chem.* **245**, 2199. **30**. Zalkin (1985) *Meth. Enzymol.* **113**, 282. **31**. Long & Koshland (1978) *Meth. Enzymol.* **51**, 79. **32**. Koshland, Jr., & A. Levitzki (1974) *The Enzymes*, 3rd ed., **10**, 539. **33**. Hofer, Steverding, Chabes, Brun & Thelander (2001) *Proc. Natl. Acad. Sci. U.S.A.* **98**, 6412. **34**. Marques, Florencio & Candau (1992) *Eur. J. Biochem.* **206**, 69. **35**. Holcenberg (1985) *Meth. Enzymol.* **113**, 257. **36**. Hartman (1971) *The Enzymes*, 3rd ed., **4**, 79. **37**. Webb (1966) *Enzyme and Metabolic Inhibitors*, vol. 2, Academic Press, New York. **38**. Willis & Seegmiller (1977) *J. Cell Physiol.* **93**, 375. **39**. Ardawi & Newsholme (1984) *Biochem. J.* **217**, 289. **40**. Morehouse & Curthoys (1981) *Biochem. J.* **193**, 709. **41**. Steckel, Roberts, Philips & Chou (1983) *Biochem. Pharmacol.* **32**, 971. **42**. Lebedeva, Kabanova & Berezov (1986) *Biochem. Int.* **12**, 413. **43**. Lebedeva, Kabanova & Berezov (1985) *Biull.*

Eksp. Biol. Med. **100**, 696. **44**. Ortlund, Lacount, Lewinski & Lebioda (2000) *Biochemistry* **39**, 1199. **45**. Roberts, Holcenberg & Dolowy (1972) *J. Biol. Chem.* **247**, 84. **46**. Zalkin (1985) *Meth. Enzymol.* **113**, 278. **47**. Pogell (1962) *Meth. Enzymol.* **5**, 408. **48**. Ghosh & Roseman (1962) *Meth. Enzymol.* **5**, 414. **49**. Ghosh, Blumenthal, Davidson & Roseman (1960) *J. Biol. Chem.* **235**, 1265. **50**. Badet, Vermoote, Haumont, Lederer & LeGoffic (1987) *Biochemistry* **26**, 1940. **51**. Chmara, Andruszkiewicz & Borowski (1985) *Biochim. Biophys. Acta* **870**, 357. **52**. Wojciechowski, Milewski, Mazerski & Borowski (2005) *Acta Biochim. Pol.* **52**, 647. **53**. Huynh, Gulve & Dian (2000) *Arch. Biochem. Biophys.* **379**, 307. **54**. Ellis & Sommar (1972) *Biochim. Biophys. Acta* **267**, 105. **55**. Trujillo & Gan (1974) *Int. J. Biochem.* **5**, 515. **56**. Badet, Vermoote & Le Goffic (1988) *Biochemistry* **27**, 2282. **57**. Badet, Vermoote, Haumont, Lederer & Le Goffic (1987) *Biochemistry* **26**, 1940. **58**. Mouilleron & Golinelli-Pimpaneau (2007) *Protein Sci.* **16**, 485. **59**. Mouilleron, Badet-Denisot & Golinelli-Pimpaneau (2006) *J. Biol. Chem.* **281**, 4404. **60**. Ebner, Wolf, Gancedo, Elsässer & Holzer (1970) *Eur. J. Biochem.* **14**, 535. **61**. Goldstein (1975) *Amer. J. Physiol.* **229**, 1027. **62**. Hundal, Mackenzie, Taylor & Rennie (1990) *Biochem. Soc. Trans.* **18**, 944. **63**. Jahn, Kim, Ishino, Chen & Soll (1990) *J. Biol. Chem.* **265**, 8059. **64**. Mineyama, Mikami & Saito (1995) *Microbios* **82**, 7. **65**. Tate & Meister (1977) *Proc. Natl. Acad. Sci. U.S.A.* **74**, 931. **66**. Allison (1985) *Meth. Enzymol.* **113**, 419. **67**. Meister, Tate & Griffith (1981) *Meth. Enzymol.* **77**, 237. **68**. Repetto, Letelier, Aldunate & Morello (1987) *Comp. Biochem. Physiol. B* **87**, 73. **69**. Horiuchi, Inoue & Morino (1980) *Eur. J. Biochem.* **105**, 93. **70**. Tate & Ross (1977) *J. Biol. Chem.* **252**, 6042. **71**. Takahashi, Steinman & Ball (1982) *Biochim. Biophys. Acta* **707**, 66. **72**. Suzuki, Kumagai & Tochikura (1986) *J. Bacteriol.* **168**, 1325. **73**. Takahashi, Zukin & Steinman (1981) *Arch. Biochem. Biophys.* **207**, 87. **74**. Moallic, Dabonne, Colas & Sine (2006) *Protein J.* **25**, 391. **75**. Minami, Suzuki & Kumagai (2003) *Enzyme Microb. Technol.* **32**, 431. **76**. Nakayama, Kumagai & Tochikura (1984) *J. Bacteriol.* **160**, 341. **77**. Nakano, Okawa, Yamauchi, Koizumi & Sekiya (2006) *Biosci. Biotechnol. Biochem.* **70**, 369. **78**. Zalkin (1985) *Meth. Enzymol.* **113**, 273. **79**. Sakamoto (1978) *Meth. Enzymol.* **51**, 213. **80**. Patel, Moyed & Kane (1975) *J. Biol. Chem.* **250**, 2609. **81**. Zalkin & Truitt (1977) *J. Biol. Chem.* **252**, 5431. **82**. Chittur, Klem, Shafer & Davisson (2001) *Biochemistry* **40**, 876. **83**. Patel, Moyed & Kane (1977) *Arch. Biochem. Biophys.* **178**, 652. **84**. Winter & Dekker (1991) *Plant Physiol.* **95**, 206. **85**. Zalkin (1985) *Meth. Enzymol.* **113**, 297. **86**. Yu & Dietrich (1972) *J. Biol. Chem.* **247**, 4794. **87**. Yoch & Gotto (1982) *J. Bacteriol.* **151**, 800. **88**. Buchanan (1982) *Meth. Enzymol.* **87**, 76. **89**. Rando (1977) *Meth. Enzymol.* **46**, 28. **90**. Buchanan, Ohnoki & Hong (1978) *Meth. Enzymol.* **51**, 193. **91**. Baker (1967) *Design of Active-Site-Directed Irreversible Enzyme Inhibitors*, Wiley, New York. **92**. Levenberg, Melnick & Buchanan (1957) *J. Biol. Chem.* **225**, 163. **93**. Ebner, Wolf, Gancedo, Elsässer & Holzer (1970) *Eur. J. Biochem.* **14**, 535.

6-Diazo-5-oxo-L-norleucylglycine

This dipeptide analogue (FW = 228.21 g/mol; CAS 78081-74-0) irreversibly inactivates γ-glutamyltranspeptidase, while the latter remains bound to renal brush border membrane vesicles. Notably, this same reagent does not inactivate the purified enzyme. The presence of L-cysteinylglycine *S*-acetyldextran polymer (Mr ≈ 500,000), which does not permeate membranes, protected the membrane-bound transferase from inactivation by 6-diazo-5-oxo-L-norleucyglycine. These findings suggest that the norleucylglycine derivative was hydrolyzed by peptidase(s) bound to the outer surface of the brush border membranes and that the 6-diazo-5-oxo-L-norleucine thus released acts as an affinity-labeling reagent for the membrane-bound transferase. Upon hydrolysis, this dipeptide liberates 6-diazo-5-oxo-L-norleucine, which then inhibits the membrane-bound transpeptidase. **1**. Inoue & Morino (1981) *Proc. Natl. Acad. Sci. U.S.A.* **78**, 46.

5-Diazo-4-oxo-L-norvaline

This asparagine analogue (FW = 157.13 g/mol; *Abbreviation*: DONV) inhibits asparagine-dependent enzymes. **Target(s)**: N^4-[β-*N*-acetylglucosaminyl]-L-asparaginase, *or* glycosylasparaginase (1-7); asparaginase (8-17); asparagine synthetase (18,19); asparagine transport (20); glutamin-(asparagin-)ase, *or* glutaminase/asparaginase (21-23). **1**. Aronson (1998) in *Handb. Proteolytic Enzymes*, p. 482, Academic Press, San Diego. **2**. Tarentino & Maley (1972) *Meth. Enzymol.* **28**, 782. **3**. Tarentino & Plummer (1993) *Biochem. Biophys. Res. Commun.* **197**, 179. **4**. Kaartinen, Williams, Tomich, *et al.* (1991) *J. Biol. Chem.* **266**, 5860. **5**. Tollersrud, Hofmann & Aronson (1988) *Biochim. Biophys. Acta* **953**, 353. **6**. Tarentino & Maley (1969) *Arch. Biochem. Biophys.* **130**, 295. **7**. Noronkoski, Stoineva, Petkov & Mononen (1997) *FEBS Lett.* **412**, 149. **8**. Wriston (1970) *Meth. Enzymol.* **17A**, 732. **9**. Pinkus (1977) *Meth. Enzymol.* **46**, 414. **10**. Handschumacher (1977) *Meth. Enzymol.* **46**, 432. **11**. Wriston & Yellin (1973) *Adv. Enzymol. Relat. Areas Mol. Biol.* **39**, 185. **12**. Shaw (1970) *The Enzymes*, 3rd ed., **1**, 91. **13**. Wriston (1971) *The Enzymes*, 3rd ed., **4**, 101. **14**. Handschumacher, Bates, Chang, Andrews & Fischer (1968) *Science* **161**, 62. **15**. Manna, Sinha, Sadhukhan & Chakrabarty (1995) *Curr. Microbiol.* **30**, 291. **16**. Mesas, Gil & Martin (1990) *J. Gen. Microbiol.* **136**, 515. **17**. Chang & Farnden (1981) *Arch. Biochem. Biophys.* **208**, 49. **18**. Jayaram, Cooney, Milman, Homan & Rosenbluth (1976) *Biochem. Pharmacol.* **25**, 1571. **19**. Milman & Cooney (1979) *Biochem. J.* **181**, 51. **20**. Willis & Woolfolk (1975) *J. Bacteriol.* **123**, 937. **21**. Ortlund, Lacount, Lewinski & Lebioda (2000) *Biochemistry* **39**, 1199. **22**. Roberts (1976) *J. Biol. Chem.* **251**, 2119. **23**. Roberts, Holcenberg & Dolowy (1972) *J. Biol. Chem.* **247**, 84.

(*E,E*)-8-(2-Diazo-3,3,3-trifluoropropionyloxy)geranyl Pyrophosphate

This farnesyl pyrophosphate analogue (FW$_{\text{free-acid}}$ = 466.24 g/mol; *Abbreviation*: DATFP-GPP) also known as (*E,E*)-(2-diazo-3 trifluoropropionyloxy)geranyl diphosphate, is a photolabile reagent: irradiation with ultraviolet light in the presence of DATFP-GPP, Mg^{2+}, isopentenyl pyrophosphate, and undecaprenyl pyrophosphate synthase results in inactivation of the enzyme, K_i = 0.22 μM (1-3). Tritiated derivatives photolabel undecaprenyl pyrophosphate synthase (3) and protein farnesyltransferase (4). **Target(s)**: di-*trans*,poly-*cis*-decaprenyl-*cis*transferase, *or* undecaprenyl pyrophosphate synthase; *cis*-prenyl-transferase, *or* C55-pyrophosphate synthetase (1-3); protein farnesyl-transferase (4,5). **1**. Baba & Allen (1984) *Biochemistry* **23**, 1312. **2**. Allen & Baba (1985) *Meth. Enzymol.* **110**, 117. **3**. Baba, Muth, C. M. Allen (1985) *J. Biol. Chem.* **260**, 10467. **4**. Omer, Kral, Diehl, *et al.* (1993) *Biochemistry* **32**, 5167. **5**. Das & Allen (1991) *Biochem. Biophys. Res. Commun.* **181**, 729.

Diazoxide

This antihypertensive agent (FW = 230.67 g/mol; CAS 364-98-7), also known as 7-chloro-3-methyl-2*H*-1,2,4-benzathiadiazine 1,1 dioxide, is a selective ATP-sensitive K^+ channel activator. Diazoxide is also a muscle relaxant, acting by increasing membrane permeability to potassium ions, which switches off voltage-gated calcium ion channels that inhibits the generation of an action potential. **Target(s)**: 3',5'-cyclic-nucleotide phosphodiesterase (1); glycerol-3-phosphate dehydrogenase, islet mitochondria (2); succinate dehydrogenase (3). **1**. Drummond & Yamamoto (1971) *The Enzymes*, 3rd ed., **4**, 355. **2**. MacDonald (1981) *J. Biol. Chem.* **256**, 8287. **3**. Schafer, Portenhauser & Trolp (1971) *Biochem. Pharmacol.* **20**, 1271.

Diazoxon

This toxic triester organophosphate (FW = 288.28 g/mol; CAS 962-58-3), also known as *O,O*-diethyl-*O*-(2-isopropyl-4-methylpyrimid-6-yl)phosphate, is a cholinesterase inhibitor. Diazoxon is an alternative substrate for paraoxonase. **Target(s):** acetylcholinesterase (1); acylaminoacyl-peptidase (2); acylglycerol lipase, *or* monoacylglycerol lipase (3); 1-alkyl-2-acetylglycero-phosphocholine esterase, *or* platelet-activating factor deacetylase (4); arylformamidase, *or* kynurenine formamidase (5); cholinesterase, *or* butyrylcholinestaerase (1,6); fatty acid amide hydrolase, *or* anandamide amidohydrolase (3). **1.** Schopfer, Voelker, Bartels, Thompson & Lockridge (2005) *Chem. Res. Toxicol.* **18**, 747. **2.** Richards, Johnson & Ray (2000) *Mol. Pharmacol.* **58**, 577. **3.** Quistad, Klintenberg, Caboni, Liang & Casida (2006) *Toxicol. Appl. Pharmacol.* **211**, 78. **4.** Quistad, Fisher, Owen, Klintenberg & Casida (2005) *Toxicol. Appl. Pharmacol.* **205**, 149. **5.** Seifert & Casida (1979) *Pestic. Biochem. Physiol.* **12**, 273. **6.** Yuknavage, Fenske, Kalman, Keifer & Furlong (1997) *J. Toxicol. Environ. Health* **51**, 35.

Dibekacin

This semisynthetic antibiotic (FW$_{\text{free-base}}$ = 451.52 g/mol; CAS 34493-98-6 (free base) and 58580-55-5 (sulfate salt)) inhibits initiation of DNA replication. It is an analogue of kanamycin and is effective against kanamycin-resistant microorganisms. Dibekacin is usually formulated in combination with sulbenicillin, a penicillin derivative **Target(s):** phospholipase A (1,2); phospholipase A$_1$ (3); phospholipase A$_2$ (3); phospholipase C (1,2). **1.** Hostetler & Hall (1982) *Biochim. Biophys. Acta* **710**, 506. **2.** Hostetler & Hall (1982) *Proc. Natl. Acad. Sci. U.S.A.* **79**, 1663. **3.** Carlier, Laurent & Tulkens (1984) *Arch. Toxicol. Suppl.* **7**, 282.

1,2:5,6-Dibenzanthracene

This polycyclic DNA intercalater and cancer-causing agent (FW = 278.35 g/mol; CAS 53-70-3), also known as dibenz[*a,h*]anthracene, is isolated from coal tar and is carcinogenic. Dibenzanthracene was isolated and identified by Kennaway and coworkers in 1932 (1-3) and represents the first purified carcinogen, along with its 3-methyl derivative. Yamagiwa and Ichikawa (4) succeeded in evoking cancer by the application of coal tar to the skin of a rabbit's ear in 1914-1915. Kennaway and coworkers later showed that polycyclic hydrocarbons could cause cancer (5) and they soon obtained benzo[*a*]pyrene. Thus, while the early work provided the foundation for the field of chemical carcinogenesis, later work demonstrated the need for initial bioactivation as aryl epoxides. Such polycyclic hydrocarbons (PAHs) often contain a "bay region" that are especially susceptible to enzymatic epolxidation (*e.g.*, benzo[*a*]pyrene is converted to benzo[a]pyrene-7,8-diol-9,10-epoxide). There is a strong coincidence of

Guanine-to-Thymine transversion hotspots in lung cancers and sites of preferential formation of PAH adducts along the p53 gene. Endogenously methylated CpG dinucleotides are the preferred sites for G-to-T transversions, accounting for more than 50% of such mutations in lung tumors (6). The same dinucleotide, when present within CpG-methylated mutational reporter genes, is the target of G to T transversion hotspots in cells exposed to the model PAH compound benzo[a]pyrene-7,8-diol-9,10-epoxide (6). **1.** Cook, Hieger, Kennaway & Mayneord (1932) *Proc. R. Soc. (B)* **111**, 455. **2.** Haddow (1974) *Perspect. Biol. Med.* **17**, 543. **3.** Rubin (2001) *Carcinogenesis* **22**, 1903. **4.** Yamagiwa & Ichikawa (1918) *J. Cancer Res.* **3**, 1. **5.** Kennaway & Hieger (1930) *Brit. Med. J.* **1**, 1044. **6.** Pfeifer, Denissenko, Olivier, *et al.* (2002) *Oncogene* **21**, 7435.

Dibenzazepine

This nitrogen heterocycle (FW = 193.25 g/mol; CAS 256-96-2; *Symbol*: DBZ), also known as 5*H*-dibenzo[*b,f*]azepine, is a Notch/γ-secretase inhibitor. **Likely Mode of Inhibitor Action:** The idea that γ-secretase inhibitors originally developed for treating Alzheimer's disease might redirect cell fate follows from the fact that Alzheimer precursor protein and Notch are both γ-secretase substrates. Small intestine epithelium self-renewal requires orderly emergence of crypt and villus compartments, and proliferative crypt cell conversion into post-mitotic goblet cells requires removal/down-regulation of the common Notch pathway transcription factor CSL/RBP-J. This phenotype is obtained by blocking the Notch cascade with γ-secretase inhibitors like dibenzazepine (1,2). Moreover, these same inhibitors also induce goblet cell differentiation in adenomas in mice carrying a mutation of the Apc tumor suppressor gene (2). Dibenzazepine likewise attenuates angiotensin II-induced abdominal aortic aneurysm in ApoE knockout mice by multiple mechanisms (3). **Note:** Given the fact that dibenzaazepine is a nucleus in many tricyclic antidepressants (*e.g.*, clomipramine, desipramine, doxepin, imipramine, imipraminoxide, lofepramine, metapramine, opipramol, quinupramine, and trimipramine), the specificity of DBZ action as a Notch/γ-secretase inhibitor is questionable. **1.** Milano, McKay, Dagenais, *et al.* (2004) *Toxicol. Sci.* **82**, 341. **2.** van Es, van Gijn, Riccio, *et al.* (2005) *Nature* **435**, 959. **2.** Zheng, Li, Tian, *et al.* (2013) *PLoS One* **8**, e83310.

Dibucaine

This hygroscopic quinolone (FW$_{\text{free-base}}$ = 343.47 g/mol; CAS 85-79-0; Very soluble in water and ethanol), also known as 2-butoxy-*N*-[2-(diethylamino)-ethyl]-4-quinolinecarboxamide, cinchocaine, and Percaine, is a long-acting local anesthetic. Given its toxicity, dibucaine's current use is restricted to spinal and topical anesthesia. **Target(s):** acetylcholinesterase (1,2); 1-acylglycerophosphocholine *O*-acyltransferase, *or* lysophosphatidylcholine *O*-acyltransferase (3); Ca^{2+}-dependent cyclic-dinucleotide phospho-diesterase (4); Ca^{2+}-transporting ATPase (5-7); choline-phosphate cytidylyltransferase (8); cholinesterase, *or* butyrylcholinesterase (1,9-11); cytochrome *c* oxidase (12-15); F$_1$ ATPase (16); guanylate cyclase, calmodulin-dependent (17); histamine *N*-methyltransferase (18); D-3-hydroxybutyrate dehydrogenase (19); lecithin:cholesterol acyltransferase, *or* LCAT (20); 5-lipoxygenase, *or* arachidonate 5-lipoxygenase; ID$_{50}$ = 66 μM (21); Lysophospholipase (22); monoamine oxidase (23); NADH dehydrogenase (13); NADPH:ferrihemoprotein reductase (24); Na$^+$/K$^+$-exchanging ATPase (2,6,25); phosphatidyl-inositol diacylglycerol-lyase, *or* phosphatidylinositol-specific phospholipase C, inhibits at pH 7, but stimulates at pH 5.5 (26); phospholipase A$_2$ (22,27,28); protein kinase C29-31; stearoyl-CoA desaturase (24); sterol *O*-acyltransferase, *or* cholesterol *O*-acyltransferase, *or* ACAT (32,33); sterol esterase, *or* cholesterol esterase (34); succinate dehydrogenase (13); ubiquinol:cytochrome *c* oxidoreductase (13). **1.** Schnurr (1967) *Arzneimittelforschung* **17**, 1577. **2.** Sidek, Nyquist-Battie & Vanderkooi (1984) *Biochim. Biophys. Acta* **801**, 26. **3.** Sanjanwala, Sun & MacQuarrie (1989) *Arch. Biochem. Biophys.* **271**, 407.

4. Hidaka, Inagaki, Nishikawa & Tanaka (1988) *Meth. Enzymol.* **159**, 652. 5. Black, Jarett & McDonald (1980) *Biochim. Biophys. Acta* **596**, 359. **6.** Roed & Brodal (1981) *Acta Pharmacol. Toxicol. (Copenhagen)* **48**, 65. **7.** Vincenzi, Adunyah, Niggli & Carafoli (1982) *Cell Calcium* **3**, 545. **8.** Pelech, Jetha & Vance (1983) *FEBS Lett.* **158**, 89. **9.** Augustinsson (1950) *The Enzymes*, 1st. ed., **1** (part 1), 443. **10.** Elamin (2003) *J. Biochem. Mol. Biol.* **36**, 149. **11.** Primo-Parmo, Lightstone & La Du (1997) *Pharmacogenetics* **7**, 27. **12.** Singer (1982) *Biochem. Pharmacol.* **31**, 527. **13.** Chazotte & Vanderkooi (1981) *Biochim. Biophys. Acta* **636**, 153. **14.** Vanderkooi & Chazotte (1982) *Proc. Natl. Acad. Sci. U.S.A.* **79**, 3749. **15.** Stringer & Harmon (1990) *Biochem. Pharmacol.* **40**, 1077. **16.** Chazotte, Vanderkooi & Chignell (1982) *Biochim. Biophys. Acta* **680**, 310. **17.** Muto, Kudo & Nozawa (1983) *Biochem. Pharmacol.* **32**, 3559. **18.** Thithapandha & Cohn (1978) *Biochem. Pharmacol.* **27**, 263. **19.** Gotterer (1969) *Biochemistry* **8**, 641. **20.** Bell & Hubert (1980) *Lipids* **15**, 811. **21.** Hamasaki & Tai (1985) *Biochim. Biophys. Acta* **834**, 37. **22.** Kunze, Nahas, Traynor & Wurl (1976) *Biochim. Biophys. Acta* **441**, 93. **23.** Yasuhara, Wada, Sakamoto & Kamijo (1982) *Jpn. J. Pharmacol.* **32**, 213. **24.** Umeki & Nozawa (1986) *Biol. Chem. Hoppe-Seyler* **367**, 61. **25.** Hudgins & Bond (1984) *Biochem. Pharmacol.* **33**, 1789. **26.** Allan & Michell (1974) *Biochem. J.* **142**, 591. **27.** Scherphof & Westenberg (1975) *Biochim. Biophys. Acta* **398**, 442. **28.** Wilschut, Regts, Westenberg & Scherphof (1976) *Biochim. Biophys. Acta* **433**, 20. **29.** Kikkawa, Minakuchi, Takai & Nishizuka (1983) *Meth. Enzymol.* **99**, 288. **30.** Kikkawa & Nishizuka (1986) *The Enzymes*, 3rd ed. (Boyer & E. G. Krebs, eds.), **17**, 167. **31.** Mori, Takai, Minakuchi, Yu & Nishizuka (1980) *J. Biol. Chem.* **255**, 8378. **32.** Bell & Hubert (1980) *Biochim. Biophys. Acta* **619**, 302. **33.** Bell (1981) *Biochim. Biophys. Acta* **666**, 58. **34.** Traynor & Kunze (1975) *Biochim. Biophys. Acta* **409**, 68.

Dobutamine

This sympathomimetic drug (FW = 301.38 g/mol; CAS 34368-04-2), also named (*RS*)-4-(2-{[4-(4-hydroxyphenyl)butan-2-yl]amino}ethyl)benzene-1,2-diol, is used to treat heart failure and cardiogenic shock by a primary mechanism involving direct stimulation of β_1 receptors of the sympathetic nervous system. Dobutamine stimulates β_1-adrenoceptors of the heart, increasing contractility and cardiac output. Showing no effect on the release of norepinephrine, dobutamine is less prone to induce hypertension than dopamine. It does not exhibit the chronotropic or vasodepressor actions of isoproterenol, the vasopressor actions of norepinephrine or dopamine, or the arrhythmogenic potential of any of these three agents. Yet, like isoproterenol and norepinephrine, dobutamine acts directly on the cardiac β_1 receptors and so has inotropic efficacy that is substantially greater than that of the digitalis glycosides. Like other catecholamines, dobutamine has an immediate onset but a brief duration of action that makes possible titration of the desired amount of inotropic effect by controlling the rate of intravenous infusion. **1.** Tuttle & Mills (1975) *J. Circ. Res.* **36**, 185.

3,5-Di(*t*-butyl)-4-hydroxybenzylidenemalononitrile

This respiratory chain uncoupler (FW = 282.39 g/mol),[1] also known as SF 6847, is an effective protonophore at concentrations of 30 nM, corresponding to less that 0.2 molecules uncoupler per repiratory chain, suggesting a lack of a stoichiometric relationship between uncoupler molecules and cytochrome *c* oxidase or between the former and phosphorylation assemblies. **1.** Muraoka & Terada (1972) *Biochim. Biophys. Acta* **275**, 271. **2.** Terada & van Dam (1975) *Biochim. Biophys. Acta* **387**, 507. **3.** Heytler (1979) *Meth. Enzymol.* **55**, 462.

$N^6,O^{2'}$-Dibutyryladenosine 3',5'-Cyclic Monophosphate

This cell-permeant cyclic AMP analogue ($FW_{free-acid}$ = 469.39 g/mol; CAS 362-74-3; *Abbreviation*: dibutyryl-cAMP, alson known as Bucladesine, activates cAMP-dependent protein kinases, including the glycogen phosphorylase system. It is less susceptible to hydrolysis by phosphodiesterases. It is used to sustain the cAMP activation in treated cells. In HeLa cells, the active form is N^6-monobutyryladenosine 3',5'-monophosphate (1) The dibutyryl form is also converted nonenzymatically to the monobutyryl form in slightly alkaline conditions (*e.g.*, pH 8.5). **Target(s):** acetyl-CoA carboxylase (2); 3',5'-cyclic-GMP phosphodiesterase, *or* cGMP phosphodiesterase (3,4); 3',5'-cyclic-nucleotide phosphodiesterase, *or* cAMP phosphodiesterase (4-6); poly(ADP-ribose) glycohydrolase (7); rhodopsin kinase, weakly inhibited (8). **1.** Kaukel, Mundhenk & Hilz (1972) *Eur. J. Biochem.* **27**, 197. **2.** Allred & Roehrig (1973) *J. Biol. Chem.* **248**, 4131. **3.** Morishima (1975) *Biochim. Biophys. Acta* **410**, 310. **4.** Pannbacker, Fleischman & Reed (1972) *Science* **175**, 757. **5.** Cohen (1979) *Phytochemistry* **18**, 943. **6.** Lim, Palanisamy & Ong (1986) *Arch. Microbiol.* **146**, 142. **7.** Miwa, Tanaka, Matsushima & Sugimura (1974) *J. Biol. Chem.* **249**, 3475. **8.** Weller, Virmaux & Mandel (1975) *Proc. Natl. Acad. Sci. U.S.A.* **72**, 381.

N^2,2'-*O*-Dibutyrylguanosine 3',5'-Cyclic Monophosphate

This cell-permeant cGMP analogue ($FW_{free-acid}$ = 485.39 g/mol; λ_{max} = 260 nm at pH 7.0; ε = 16700 $M^{-1}cm^{-1}$) activates protein kinase G. It is slightly more resistant to the action of specific phosphodiesterases than cGMP. The lipolytic activity of dibutyryl-cGMP is considerably lower than that of dibutyryl-cAMP (1). This synthetic agent has also been shown to increase intracellular calcium levels (2,3) and inhibit thrombin-induced arachidonate release in platelets (4) Dibutyryl cGMP. Note that dibutyryl-cGMP is slowly converted to N^2-monobutyryl-cGMP upon prolonged standing in aqueous solutions. **1.** Braun, Hechter & Baer (1969) *Proc. Soc. Exp. Biol. Med.* **132**, 233. **2.** Rooney, Joseph, Queen & Thomas (1996) *J. Biol. Chem.* **271**, 1981. **3.** Sato & Kawatani (1998) *Brain Res.* **813**, 203. **4.** Sane, Bielawska, Greenberg & Hannun (1989) *Biochem. Biophys. Res. Commun.* **165**, 708.

Dichloroacetate

This α-dihalo acid (FW = 128.94 g/mol; CAS 79-43-6; B.P. = 193-194°C; *Abbreviation*: DCA) is a caustic liquid that is miscible with water and ethanol. Dichloroacetic acid has been used as an oral antidiabetic agent that reduces blood glucose and lipids without stimulating insulin secretion. DCA inhibits the phosphorylation of pyruvate dehydrogenase as well as gluconeogenesis. It is an irreversible inhibitor of the glutathione *S*-transferase ζ reaction, wherein glutathione first displaces chloride from dichloroacetate, and an active-site cysteinyl residue then displaces the second chlorine atom, resulting in a covalently modified and inactivated enzyme. DCA promotes local conformational changes in [pyruvate dehydrogenase] kinase isozyme-1 that are communicated to both nucleotide-binding and lipoyl-binding sub-sites, leading to the inactivation of enzyme activity. **Target(s):** [branched-chain α-keto-acid dihydrogenase] kinase, *or* [3-methyl-2 oxobutanoate dehydrogenase (acetyl-transferring)] kinase (2-8); dichloromethane dehalogenase (9); glutathione *S*-transferase (10-13); haloalcohol dehalogenase (14); 3-hydroxy-3-methylglutaryl-CoA reductase (15,16); pantothenate synthetase (17); [pyruvate dehydrogenase (acetyl-transferring)] kinase (1,18-28). **1.** Kato, Li, Chuang & Chuang (2007) *Structure* **15**, 992. **2.** Paxton (1988) *Meth. Enzymol.* **166**, 313. **3.** Popov, Shimomura, Hawes & Harris (2000) *Meth. Enzymol.* **324**, 162. **4.** Randle, Patston & Espinal (1987) *The Enzymes*, 3rd ed., **18**, 97. **5.** Paxton & Harris (1984) *Arch. Biochem. Biophys.* **231**, 58. **6.** Harris, Paxton & DePaoli-Roach (1982) *J. Biol. Chem.* **257**, 13915. **7.** Shimomura, Nanaumi, Suzuki, Popov & Harris (1990) *Arch. Biochem. Biophys.* **283**, 293. **8.** Paxton & Harris (1982) *J. Biol. Chem.* **257**, 14433. **9.** Kohler-Staub & Leisinger (1985) *J. Bacteriol.* **162**, 676. **10.** Keys, Schultz, Mahle & Fisher (2004) *Toxicol. Sci.* **82**, 381. **11.** Anderson, Liebler, Board & Anders (2002) *Chem. Res. Toxicol.* **15**, 1387. **12.** Anderson, Board, Gargano & Anders (1999) *Chem. Res. Toxicol.* **12**, 1144. **13.** Dierick (1984) *Res. Commun. Chem. Pathol. Pharmacol.* **44**, 327. **14.** Assis, Sallis, Bull & Hardman (1998) *Enzyme Microb. Technol.* **22**, 568. **15.** Harwood, Schneider & Stacpoole (1984) *J. Lipid Res.* **25**, 967. **16.** Stacpoole, Harwood & Varnado (1983) *J. Clin. Invest.* **72**, 1575. **17.** Van Oorschot & Hilton (1963) *Arch. Biochem. Biophys.* **100**, 289. **18.** Pettit, Yeaman & Reed (1983) *Meth. Enzymol.* **99**, 331. **19.** Roach (1984) *Meth. Enzymol.* **107**, 81. **20.** Harris, Kuntz & Simpson (1988) *Meth. Enzymol.* **166**, 114. **21.** Kerbey, Randle, Cooper, *et al.* (1976) *Biochem. J.* **154**, 327. **22.** Pratt & Roche (1979) *J. Biol. Chem.* **254**, 7191. **23.** Mann, Dragland, Vinluan, *et al.* (2000) *Biochim. Biophys. Acta* **1480**, 283. **24.** Baker, Yan, Peng, Kasten & Roche (2000) *J. Biol. Chem.* **275**, 15773. **25.** Hiromasa, Hu & Roche (2006) *J. Biol. Chem.* **281**, 12568. **26.** Mayers, Leighton & Kilgour (2005) *Biochem. Soc. Trans.* **33**, 367. **27.** Klyuyeva, Tuganova & Popov (2007) *FEBS Lett.* **581**, 2988. **28.** Popov, Kedishvili, Zhao, *et al.* (1994) *J. Biol. Chem.* **268**, 26602.

5,6-Dichlorobenzimidazole 1-β-D-Ribofuranoside

This synthetic nucleoside (FW = 319.14 g/mol), also known as 5,6-dichlorobenzimidazole riboside and 5,6-dichloro-1-β-D-ribofuranosyl-benzimidazole, is metabolically transformed to its mono-, di-, and triphosphate forms, with the latter inhibiting RNA biosynthesis and causing premature termination of transcription. **Target(s):** casein kinase-2, $K_i = 7$ μM (1-4); cdc2 kinase (2,5); cyclin-dependent kinase-9, *or* carboxyl-terminal domain protein kinase (6-9); DNA replication (10); [glycogen synthase] kinase 3 (GSK3), *or* [tau protein] kinase (2); helicase (11,12); insulin-stimulated nuclear and cytosolic p70 S6 kinase (13); non-specific serine/threonine protein kinase (1-4); rhodopsin kinase ($K_i = 4$ μM (14); RNA polymerase, *or* RNA polymerase II (11,15-21); [RNA-polymerase]-subunit kinase, *or* CTD kinase (8,22 24); transcription factor IIH-associated protein kinase (7). **1.** Zandomeni, Carrera Zandomeni, Shugar & Weinmann (1986) *J. Biol. Chem.* **261**, 3414. **2.** Moreno, Munoz-Montano & Avila (1996) *Mol. Cell. Biochem.* **165**, 47. **3.** Zandomeni (1989) *Biochem. J.* **262**, 469. **4.** Mottet, Ruys, Demazy, Raes & Michiels (2005) *Int. J. Cancer* **117**, 764. **5.** Villa-Moruzzi (1992) *FEBS Lett.* **304**, 211. **6.** Shan, Zhuo, Chin, *et al.* (2005) *J. Biol. Chem.* **280**, 1103. **7.** Yankulov, Yamashita, Roy, Egly & Bentley (1995) *J. Biol. Chem.* **270**, 23922. **8.** Cisek & Corden (1991) *Meth. Enzymol.* **200**, 301. **9.** Shima, Yugami, Tatsuno, *et al.* (2003) *Genes Cells* **8**, 215. **10.** Hand & Tamm (1977) *Exp.*

Cell Res. **10**, 343. **11.** Seghal, Darnell & Tamm (1976) *Cell* **9**, 473. **12.** Borowski, Deinert, Schalinski, *et al.* (2003) *Eur. J. Biochem.* **270**, 1645. **13.** Kim & Kahn (1997) *Biochem. Biophys. Res. Commun.* **234**, 681. **14.** Palczewski, Kahn & Hargrave (1990) *Biochemistry* **29**, 6276. **15.** Tamm, Hand & Caliguiri (1976) *J. Cell Biol.* **69**, 229. **16.** Dreyer & Hausen (1978) *Nucl. Acids Res.* **5**, 3325. **17.** Zandomeni, Mittleman, Bunick, Ackerman & Weinmann (1982) *Proc. Natl. Acad. Sci. U.S.A.* **79**, 3167. **18.** Hensold, Barth & Stratton (1996) *J. Cell Physiol.* **168**, 105. **19.** Zandomeni, Bunick, Ackerman, Mittleman & Weinmann (1983) *J. Mol. Biol.* **167**, 561. **20.** Mittleman, Zandomeni & Weinmann (1983) *J. Mol. Biol.* **165**, 461. **21.** Giardina & Lis (1993) *J. Biol. Chem.* **268**, 23806. **22.** Medlin, Uguen, Taylor, Bentley & Murphy (2003) *EMBO J.* **22**, 925. **23.** Jacobs, Ogiwara & Weiner (2004) *Mol. Cell. Biol.* **24**, 846. **24.** Stevens & Maupin (1989) *Biochem. Biophys. Res. Commun.* **159**, 508.

3-{[(4,7-Dichlorobenzoxazol-2-yl)methyl]amino}-5-ethyl-6-methylpyridin-2(1H)-one

This pyridin-2-one (FW = 352.22 g/mol), also known as L-697,661, inhibits HIV-1 reverse transcriptase, IC_{50} = 19 nM (1-5). P236L-HIV-1 RT shows that substitution at amino acid 236 changes the shape of the binding pocket in a way that confers resistance to bisheteroarylpiperazines (BHAPs) but sensitizes the mutant enzyme to L-697,661 (6). **1.** Tucker, Lumma & Culberson (1996) *Meth. Enzymol.* **275**, 440. **2.** Balzarini & De Clercq (1996) *Meth. Enzymol.* **275**, 472. **3.** Le Grice, Cameron & Benkovic (1995) *Meth. Enzymol.* **262**, 130. **4.** De Clercq (1992) *AIDS Res. Human Retrovir.* **8**, 119. **5.** Saari, Wai, Fisher, *et al.* (1992) *J. Med. Chem.* **35**, 3792. **6.** Fan, Evans, Rank, *et al.* (1995) *FEBS Lett.* **359**, 233.

Dichloro(2-chlorovinyl)arsine

This liquid vesicant (FW = 207.32 g/mol; M.P. = 0.1°C; B.P. [decomposes] = 190°C; Insoluble in H2O; Soluble in Most Organic Solvents), known as Lewisite, is a potent respiratory poison and chemical warfare agent. Lewisite reacts with protein-SH groups. **Caution:** Extreme care must be exercised in its handling British Lewisite. As little as 0.5 mL applied to the skin can cause severe symptoms, with 2 mL causing death. Inhalation of high concentrations can cause death with as ten minutes. The antidote is 2,3-dimercapto-1-propanol, better known as British Anti-Lewisite (BAL). **Target(s):** acid phosphatase (1); alanine aminotransferase (4); alcohol dehydrogenase, yeast (2,3); arginase (4); choline oxidase (4); cholinesterase (1,3,4); α-glycerophosphate dehydrogenase (5); hexokinase (6,7); homogentisate 1,2-dioxygenase (8.9); lactate dehydrogenase (5); pyruvate oxidase, *or* pyruvate dehydrogenase (6,10,11); succinate dehydrogenase (5,6); succinate oxidase (1,5); sulfhydryl dependent enzymes (12). **1.** Barron, Miller, Bartlett, Meyer & Singer (1947) *Biochem. J.* **41**, 69. **2.** Sund & Theorell (1963) *The Enzymes*, 2nd ed., **7**, 25. **3.** Mounter & Whittaker (1953) *Biochem. J.* **53**, 167. **4.** Thompson (1947) *J. Physiol. (London)* **105**, 370. **5.** Peters, Sinclair & Thompson (1946) *Biochem. J.* **40**, 516. **6.** Dixon & Needham (1946) *Nature* **158**, 432. **7.** Bailey & Webb (1948) *Biochem. J.* **42**, 60. **8.** Flamm & Crandall (1963) *J. Biol. Chem.* **238**, 389. **9.** Nozaki (1974) in *Mol. Mech. Oxygen Activ.* (Hayaishi, ed.), p. 135, Academic Press, New York. **10.** Peters (1949) *Symposia Soc. Exptl. Biol.* **3**, 36. **11.** Peters, Stocken & Thompson (1945) *Nature* **156**, 616. **12.** Goldman & Dacre (1989) *Rev. Environ. Contam. Toxicol.* **110**, 75.

5,7-Dichlorokynurenate

This kynurenic acid derivative ($FW_{free-acid}$ = 258.06 g/mol; CAS 131123-76-7; Symbol: 5,7-DCKA) is an NMDA (*N*-methyl-D-aspartate) receptor antagonist (K_i = 79 nM) and is associated with the strychnine-insensitive glycine-binding site of the receptor. At 10 μM, 5,7-DCKA antagonizes the ability of NMDA to stimulate the binding of the radiolabeled ion channel blocker *N*-[^{3}H][1-(2-thienyl)cyclohexyl]-piperidine ([^{3}H]TCP). Glycine overcomes this effect, and in the presence of 5,7-DCKA enhanced [^{3}H]TCP binding to antagonist-free levels. 5,7-DCKA completely and noncompetitively antagonized several NMDA receptor-mediated biochemical and electrophysiological responses. **1.** Baron, Harrison, Miller, *et al.* (1990) *Mol. Pharmacol.* **38**, 554. **2.** Baron, Siegel, Slone, *et al.* (1991) *Eur. J. Pharmacol.* **206**, 149. **3.** McNamara, Smith, Calligaro, *et al.* (1990) *Neurosci. Lett.* **120**, 17.

Dichloromethylenediphosphonate

This collagenase inhibitor ($FW_{free-acid}$ = 244.89 g/mol), also known as dichloromethylene bisphosphonic acid and clodronic acid, inhibits interstitial and neutrophil and blocks osteoclast-mediated bone resorption. This bisphosphonate is also used to treat bone metastases and the management of skeletal complications associated with bone metastases in patients with breast cancer. **Target(s):** adenosine kinase (1); collagenase-3, *or or* MMP-13 (2); enamelysin, *or* MMP-20 (2); gelatinase A, *or* MMP-2 (2); gelatinase B, *or* MMP-9 (2); interstitial collagenase, *or* MMP-1, IC_{50} = 150 μM (2,3); macrophage elastase, *or* matrix MMP-12 (2); MMP-14, *or* membrane-type MMP-1 (4); neutophil collagenase, *or* MMP-8 (2,5); pyrophosphatase (6-8); pyrophosphate-dependent phosphofructokinase, *or* diphosphate:fructose-6-phosphate 1-phosphotransferase (9); ribokinase (10); stromelysin-1, matrix metalloproteinase-3 (2); vacuolar H⁺-transporting pyrophosphatase (8). **1.** Park, Singh & Gupta (2006) *Mol. Cell. Biochem.* **283**, 11. **2.** Heikkila, Teronen, Moilanen, *et al.* (2002) *Anticancer Drugs* **13**, 245. **3.** Teronen, Konttinen, Lindqvist, *et al.* (1997) *Calcif. Tissue Int.* **61**, 59. **4.** Heikkila, Teronen, Hirn, *et al.* (2003) *J. Surg. Res.* **111**, 45. **5.** Teronen, Konttinen, Lindqvist, *et al.* (1997) *J. Dent. Res.* **76**, 1529. **6.** Granstrom (1983) *Arch. Oral Biol.* **28**, 453. **7.** Felix & Fleisch (1975) *Biochem. J.* **147**, 111. **8.** McIntosh & Vaidya (2002) *Int. J. Parasitol.* **32**, 1. **9.** Bruchhaus, Jacobs, Denart & Tannich (1996) *Biochem. J.* **316**, 57. **10.** Park, van Koeverden, Singh & Gupta (2007) *FEBS Lett.* **581**, 3211.

(2,4-Dichlorophenoxy)acetate

This herbicide ($FW_{free-acid}$ = 221.04 g/mol; CAS 94-75-7; *Abbreviation*: 2,4-D; pK_a = 2.64; Free acid is relatively insoluble (0.06 g/100 g H₂O;); Sodium salts is soluble (3.5 g/100 g), is a synthetic auxin. The herbicidal properties of 2,4-D were discovered in the early 1940s in both the U.S. and U.K. When sprayed on plants at the correct concentrations, broad-leaf organisms are killed, but not narrow-leaf plants such as grasses. Earlier so-called inorganic herbicides remain in the soil much longer than 2,4-D. Given its ease of synthesis, 2,4-D and related compounds dominated the herbicide market until the arrival of more effective agents in te late 1960s. **CAUTION:** Care should be exercised to avoid direct contact with skin or inhalation of stock solutions, as overdoses can cause dermatitis and convulsions. **Target(s):** benzoyl-CoA synthetase, *or* benzoate:CoA ligase (1); glutathione *S*-transferase (2,3); glycerol-3-phosphate dehydrogenase (4);

indole-3-acetate β-glucosyltransferase (5); Na⁺/K⁺-exchanging ATPase (6); palmitoyl-CoA hydrolase, *or* acyl-CoA hydrolase (7); pectin lyase (8); phosphatidate phosphatase (9); zeatin 9-aminocarboxyethyltransferase (10). **1.** Gregus, Halaszi & Klaassen (1999) *Xenobiotica* **29**, 547. **2.** Singh & Awasthi (1985) *Toxicol. Appl. Pharmacol.* **81**, 328. **3.** Vessey & Boyer (1984) *Toxicol. Appl. Pharmacol.* **73**, 492. **4.** Berrada, Naya, Iddar & Bourhim (2002) *Mol. Cell. Biochem.* **231**, 117. **5.** Leznicki & R. S. Bandurski (1988) *Plant Physiol.* **88**, 1481. **6.** Cascorbi & Foret (1991) *Ecotoxicol. Environ. Saf.* **21**, 38. **7.** Dixon, Osterloh & Becker (1990) *J. Pharm. Sci.* **79**, 103. **8.** Albersheim (1966) *Meth. Enzymol.* **8**, 628. **9.** Scherer & Morre (1978) *Biochem. Biophys. Res. Commun.* **84**, 238. **10.** Parker, Entsch & Letham (1986) *Phytochemistry* **25**, 303.

6-[4-(2,5-Dichlorophenoxy)piperidin-1-yl]-*N*-methylpyridazine-3-carboxamide

This pyridazine derivative (FW = 381.26 g/mol) inhibits stearoyl-CoA 9-desaturase, IC_{50} = 7.0 nM. Stearoyl-CoA desaturase 1 (SCD1) catalyzes the committed step in the biosynthesis of monounsaturated fatty acids from saturated, long-chain fatty acids. Studies with SCD1 knockout mice have established that these animals are lean and protected from leptin deficiency-induced and diet-induced obesity, with greater whole body insulin sensitivity than wild-type animals. Such findings point to SCD1 as an attractive druggable target in efforts to develop anti-obesity drugs. **1.** Liu, Lynch, Freeman, *et al.* (2007) *J. Med. Chem.* **50**, 3086.

S-1,2-Dichlorovinyl-L-cysteine

This cysteine analogue (FW = 216.09 g/mol) reacts with and inactivates cysteine *S*-conjugate β-lyase as well as a number of transaminases. **Target(s):** L-alanine:glyoxalate aminotransferase (1); aspartate aminotransferase (2,3); branched-chain-amino-acid aminotransferase (4); cysteine *S*-conjugate β-lyase, alternative substrate and suicide inhibitor (5); cysteine conjugate *S*-oxidase (6); glutathione-disulfide reductase (7); kynurenine:oxoglutarate aminotransferase (8,9); lipoyl dehydrogenase (7). **1.** Cooper, Krasnikov, Okuno & Jeitner (2003) *Biochem. J.* **376**, 169. **2.** Kato, Asano & Cooper (1996) *Dev. Neurosci.* **18**, 505. **3.** Cooper, Bruschi, Iriarte & Martinez-Carrion (2002) *Biochem. J.* **368**, 253. **4.** Cooper, Bruschi, Conway & Hutson (2003) *Biochem. Pharmacol.* **65**, 181. **5.** Stevens & Jakoby (1983) *Mol. Pharmacol.* **23**, 761. **6.** Sausen & Elfarra (1990) *J. Biol. Chem.* **265**, 6139. **7.** Lock & Schnellmann (1990) *Toxicol. Appl. Pharmacol.* **104**, 180. **8.** Guidetti, Okuno & Schwarcz (1997) *J. Neurosci. Res.* **50**, 457. **9.** Okuno, Nishikawa & Nakamura (1996) *Adv. Exp. Med. Biol.* **398**, 455.

O-2,2-Dichlorovinyl-*O*,*O*-dimethyl Phosphate

This toxic organophosphate (FW = 220.98 g/mol; *Abbreviation*: DDVP), also known as *O*,*O*-dimethyl-*O*-2,2-dichlorovinyl phosphate, and dichlorvos, is a serine proteinase/esterase inhibitor that has been used as a pesticide to control mushroom flies, aphids, spider mites, caterpillars, thrips, and white flies in crops as well as parasitic worm infections in mammals. **Target(s):** acetylcholinesterase (1-7); acylaminoacyl-peptidase (8); acylglycerol lipase, *or* monoacylglycerol lipase (9); carboxylesterase (10); cholinesterase, *or* butyrylcholinesterase (2,3,11-16); chymotrypsin

(11); fatty acid amide hydrolase, *or* anandamide amidohydrolase (9); glutaminase (17); juvenile-hormone esterase (18); lipase, *or* triacylglycerol lipase (11); monoamine oxidase (19); trypsin (11); urethanase (20). **1.** Aharoni & O'Brien (1968) *Biochemistry* **7**, 1538. **2.** Kovarik, Ciban, Radic, Simeon-Rudolf & Taylor (2006) *Biochem. Biophys. Res. Commun.* **342**, 973. **3.** Schopfer, Voelker, Bartels, Thompson & Lockridge (2005) *Chem. Res. Toxicol.* **18**, 747. **4.** Shi, Boyd, Radic & Taylor (2001) *J. Biol. Chem.* **276**, 42196. **5.** Walsh, Dolden, Moores, *et al.* (2001) *Biochem. J.* **359**, 175. **6.** Bentley, Jones & Agnew (2005) *Mol. Biochem. Parasitol.* **141**, 119. **7.** Boublik, Saint-Aguet, Lougarre, *et al.* (2002) *Protein Eng.* **15**, 43. **8.** Richards, Johnson & Ray (2000) *Mol. Pharmacol.* **58**, 577. **9.** Quistad, Klintenberg, Caboni, Liang & Casida (2006) *Toxicol. Appl. Pharmacol.* **211**, 78. **10.** Siddalinga Murthy & Veerabhadrappa (1996) *Insect Biochem. Mol. Biol.* **26**, 287. **11.** Mounter, Tuck, Alexander & Dien (1957) *J. Biol. Chem.* **226**, 873. **12.** Reiff, Lambert & Natoff (1971) *Arch. Int. Pharmacodyn. Ther.* **192**, 48. **13.** Ward & Glicksberg (1971) *J. Amer. Vet. Med. Assoc.* **158**, 457. **14.** Bueding, Liu & Rogers (1972) *Brit. J. Pharmacol.* **46**, 480. **15.** Ecobichon & Comeau (1973) *Toxicol. Appl. Pharmacol.* **24**, 92. **16.** Sanchez-Hernandez & Sanchez (2002) *Environ. Toxicol. Chem.* **21**, 2319. **17.** Nag (1992) *Indian J. Exp. Biol.* **30**, 543. **18.** Pratt (1975) *Insect Biochem.* **5**, 595. **19.** Nag & Nandy (2001) *Indian J. Exp. Biol.* **39**, 802. **20.** Kobashi, Takebe & Sakai (1990) *Chem. Pharm. Bull.* **38**, 1326.

Diclofop

This phenoxypropionic acid ($FW_{free-acid}$ = 327.16 g/mol; CAS 40843-25-2) is a post-emergence herbicide that inhibits acetyl-CoA carboxylase. **1.** Harwood (1988) *Trends Biochem. Sci.* **13**, 330. **2.** Rendina, Felts, Beaudoin, *et al.* (1988) *Arch. Biochem. Biophys.* **265**, 219. **3.** Alban, Baldet & Douce (1994) *Biochem. J.* **300**, 557. **4.** Rendina, Craig-Kennard, Beaudoin & Breen (1990) *J. Agric. Food Chem.* **38**, 1282. **5.** Evenson, Gronwald & Wyse (1994) *Plant Physiol.* **105**, 671. **6.** Motel, Gunther, Clauss, *et al.* (1993) *Z. Naturforsch.* **48**, 294. **7.** Focke, Gieringer, Schwan, *et al.* (2003) *Plant Physiol.* **133**, 875.

Diclofenac

This nonsteroidal anti-inflammatory and analgesic (FW = 296.15 g/mol; CAS 15307-86-5) is used to treat rheumatoid arthritis, degenerative joint disease, ankylosing spondylitis as well as pain resulting from minor surgery, trauma and menstruation. Although its mode of therapeutic action remains unclear, Diclofenac inhibits the conversion of arachidonic acid to prostaglandin G_2 by the cyclooxygenase (COX-2) activity of prostaglandin endoperoxide synthases (1). The IC_{50} values for time-dependent inhibition of wild type COX-2 (K_i = 77 nM), Arg-120-Ala COX-2 (K_i = 257 nM), and Tyr-355-Phe COX-2 (K_i = 137 nM) were roughly comparable, whereas the Ser-530-Ala enzyme is completely resistant to diclofenac inhibition (1). Diclofenac also suppresses the fast tetrodotoxin-sensitive (TTX-S) sodium current (K_d = 14 µM) and the slow tetrodotoxin-resistant (TTX-R) sodium current (K_d = 97 µM) in rat dorsal root ganglion neurons at a holding potential of -80 mV (2). Diclofenac has no effect on the kinetic parameters of the activation process in either sodium current, but produces shifts of the steady-state inactivation curves in the hyperpolarizing direction in both types of sodium currents in a dose-dependent manner (2). Diclofenac induces apoptosis in oral cavity cancer cells by increasing the expression of the nonsteroidal anti-inflammatory drug-activated gene NAG-1 (3). However, it suppresses apoptosis induced by endoplasmic reticulum stresses by inhibiting caspase signaling (4). Toxic effects of Diclofenac on hepatocytes *in vitro* is likely to be caused by drug-induced mitochondrial impairment, together with a futile consumption of NADPH (5). **Key**

Pharmacokinetic Parameters: *See* Appendix II in Goodman & Gilman's *THE PHARMACOLOGICAL BASIS OF THERAPEUTICS*, 12th Edition (Brunton, Chabner & Knollmann, eds.) McGraw-Hill Medical, New York (2011). **1.** Rowlinson, Kiefer, Prusakiewicz, *et al.* (2003) *J. Biol. Chem.* **278**, 45763. **2.** Lee, Kim, Shin, *et al.* (2003) *Brain Res.* **992**, 120. **3.** Kim, Yoon, Kim, *et al.* (2004) *Biochem. Biophys. Res. Commun.* **325**, 1298. **4.** Yamazaki, Muramoto, Oe, *et al.* (2006) *Neuropharmacol.* **50**, 558. **5.** Bort, Ponsoda, Jover, *et al.* (1999) *J. Pharmacol. Exp. Ther.* **288**, 65.

Dicrotophos

This cell-permeant organophosphate insecticide (FW = 237.19 g/mol; CAS 141-66-2; M.P. < 25°C), also known as Bidrin® and dimethyl-2-dimethylcarbamoyl 1-methylvinyl phosphate, inhibits acetylcholinesterase (1) and arylformamidase (2). The *E*-isomer is moreactive as an insecticidally than the *Z*-isomer. Dicrotophos is classified as a restricted-use pesticide. **1.** Aharoni & O'Brien (1968) *Biochemistry* **7**, 1538. **2.** Seifert & Casida (1979) *Pestic. Biochem. Physiol.* **12**, 273.

Dictyostatin

This tubulin polymerization-promoting and microtubule-stabilizing agent (FW = 518.73 g/mol), also named (3Z,5E,7S,8R,10R,11Z,13R,14S,15R,17R,20S,22S)-22-[(2R,3Z)-3,5-hexadien-2-yl]-8,10,14,20-tetrahydroxy-7,13,15,17-tetramethyloxacyclodocosa-3,5,11-trien-2-one, from a *Lithistida* sponge arrested cells in the G_2/M junction of the cell cycle (1). Staining of with tubulin-specific antibodies revealed multiple aster formation and microtubule bundling patterns similar to that observed with paclitaxel (1). Dictyostatin crosses the blood-brain barrier in mice and exhibits extended brain retention (2). A single dose of dictyostatin to mice causes prolonged microtubule stabilization in brain, whereas discodermolide, a structurally related MT-stabilizer, shows significantly less brain exposure. The total synthesis of dictyostatin has been reported (3). **1.** Isbrucker, Cummins, Pomponi, Longley & Wright (2003) *Biochem. Pharmacol.* **66**, 75. **2.** Brunden, Gardner, James, *et al.* (2013) ACS Med. Chem. Lett. **4**, 886. **3.** O'Neil & Phillips (2006) *J. Am. Chem. Soc.* **128**, 5340.

Dicoumarol (*or* Dicumarol)

Dicumarol

Vitamin K₁

This substituted coumarin and vitamin K mimic (FW = 336.30 g/mol; CAS 66-76-2), also known as 3,3'-methylene-bis(4-hydroxycoumarin), targets Vitamin K epoxide reductase (EC 1.1.4.1), which reduces vitamin K after it has been oxidized during enzymatic carboxylation of glutamic acid residues in blood coagulation enzymes. Dicoumarol thus interferes with vitamin K-dependent formation of γ-carboxyglutamyl (Gla) residues in coagulation enzymes, thereby inhibiting/preventing blood clotting. It is also a protonophoric uncoupler of oxidative phosphorylation (EC_{50} = 30-70 μM) and photophosphorylation (1-3). It also inhibits NADH dehydrogenase (4), or DT diaphorase (K_i = 10 nM), binding to the oxidized form of the enzyme (i.e., competitively versus NADH). **History:** Used as cattle feed, vast quantities of sweet clover spoiled during several wet growing seasons in the 1920s, attended by a concomitant rise in cattle "bleeding disease". Link's group at the Uviversity of Wisconsin isolated the hemorrhagic agent (5), determining the structure and devising a synthesis (6) for this dihydroxylated derivative of coumarol, the latter responsible for the fragrance of freshly mowed grass. Dicoumarol quickly found use in controlling blood clotting and led to the synthesis of brodifacoum, a highly potent anticoagulant and widely used rodenticide (**See** Brodifacoum) (7). **Target(s):** aldehyde oxidase (8); ATP phospho-ribosyltransferase, Escherichia coli, K_d of 50 μM (9); azoreductase (10,11); brefeldin-A-dependent ADP-ribosylation (12,13); carbonyl reductase (14); γ-carboxylation of glutamyl residues (15); cellulose biosynthesis (53); cholestanetriol 26-monooxygenase, weakly inhibited (54); cytochrome P450 (17); dihydrolipoamide dehydrogenase (18); electron transport (2); ferricyanide reductase (19); gap junctions (20); glutathione peroxidase (21); glutathione S-transferase (16,21,22); hyaluronidase (23); 15-ketoprostaglandin Δ^{13}-reductase (24); L-lactate oxidase (25); microtubule depolymerization (26); NADH dehydrogenase (4,27); NADH dehydrogenase, quinone-requiring (20,28,29); NADH-hexacyanoferrate(III) reductae (30) ; NADH-linked and NAD(P)H-linked nitro-reductases (31); NADPH dehydrogenase (32); NAD(P)H dehydrogenase, or DT diaphorase (20,27,29,32-39); NADPH-linked α,β-ketoalkene double-bond reductase (40); nitrofuran reductase (41); 1-nitropyrene nitroreductase (42); nitroreductase (10,43); nuclear factor κB (or, NF-κB) (44); oxidative and photophosphorylation uncoupler (1-3,45); phospho-diesterase (46); quinone reductase (39); ribosyldihydro-nicotinamide dehydrogenase, quinone-requiring (55-58); rubredoxin:NAD$^+$ reductase (47); salicylhydroxamate reductase (48); steroid 11β-monooxygenase (49,50); thyroxine 5'-deiodinase (51,52). **1.** Slater (1967) Meth. Enzymol. **10**, 48. **2.** Izawa & Good (1972) Meth. Enzymol. **24**, 355. **3.** Heytler (1979) Meth. Enzymol. **55**, 462. **4.** Hatefi & Rieske (1967) Meth. Enzymol. **10**, 235. **5.** Campbell & Link (1941) J. Biol. Chem. **138**, 21. **6.** Stahmann, Huebner & Link (1941) J. Biol. Chem. **138**, 513. **7.** Ikawa, Stahmann & Link (1944) J. Amer. Chem. Soc. **66**, 902. **8.** Rajagopalan, Fridovich & Handler (1962) J. Biol. Chem. **237**, 922. **9.** Dall-Larsen, Kryvi & Klungsoyr (1976) Eur. J. Biochem. **66**, 443. **10.** Douch (1975) Xenobiotica **5**, 657. **11.** Huang, Miwa, Cronheim & Lu (1979) J. Biol. Chem. **254**, 11223. **12.** Mironov, Colanzi, Polishchuk, et al. (2004) Eur. J. Cell Biol. **83**, 263. **13.** Colanzi, Mironov, Weigert, et al. (1997) Adv. Exp. Med. Biol. **419**, 331. **14.** Wermuth (1981) J. Biol. Chem. **256**, 1206. **15.** Stenflo (1978) Adv. Enzymol. Relat. Areas Mol. Biol. **46**, 1. **16.** Depeille, Cuq, Mary, et al. (2004) Mol. Pharmacol. **65**, 897. **17.** Adams & Notides (1987) Toxicol. Appl. Pharmacol. **88**, 113. **18.** Okamoto (1971) Meth. Enzymol. **18B**, 67. **19.** Baroja-Mazo, Del Valle, Rua, et al. (2004) J. Bioenerg. Biomembr. **36**, 481. **20.** Abdelmohsen, Stuhlmann, Daubrawa & Klotz (2005) Arch. Biochem. Biophys. **434**, 241. **21.** Mays & Benson (1992) Biochem. Pharmacol. **44**, 921. **22.** Rauch & Nauen (2004) Insect Biochem. Mol. Biol. **34**, 321. **23.** Beiler & Martin (1947) J. Biol. Chem. **171**, 507. **24.** Kitamura, Katsura & Tatsumi (1993) Biochem. Mol. Biol. Int. **30**, 839. **25.** Akimenko & Medentsev (1976) Biokhimiia **41**, 665. **26.** Madari, Panda, Wilson & Jacobs (2003) Cancer Res. **63**, 1214. **27.** Aloj & Giuditta (1967) J. Neurochem. **14**, 955. **28.** Duhaiman & Rabbani (1996) J. Enzyme Inhib. **11**, 13. **29.** McGuire, Pesche & Fanning (1963) Nature **200**, 71. **30.** Berczi, Fredlund & Moller (1995) Arch. Biochem. Biophys. **320**, 65. **31.** Kitamura, Narai & Tatsumi (1983) J. Pharmacobiodyn. **6**, 18. **32.** Koli, Yearby, Scott & Donaldson (1969) J. Biol. Chem. **244**, 621. **33.** Ernster (1967) Meth. Enzymol. **10**, 309. **34.** Lind, Cadenas, Hochstein & Ernster (1990) Meth. Enzymol. **186**, 287. **35.** Martius (1963) The Enzymes, 2nd ed., 7, 517. **36.** Ernster, Ljunggren & Danielson (1960) Biochem. Biophys. Res. Commun. **2**, 48 and 88. **37.** Preusch, Siegel, Gibson & Ross (1991) Free Radic. Biol. Med. **11**, 77. **38.** Raw, Nogueira & Filho (1961) Enzymologia **23**, 123. **39.** Wosilait & Nason (1954) J. Biol. Chem. **208**, 785. **40.** Kitamura & Tatsumi (1990) Arch. Biochem. Biophys. **282**, 183. **41.** Tatsumi, Doi, Yoshimura, Koga & Horiuchi (1982) J. Pharmacobiodyn. **5**, 423. **42.** Kinouchi & Ohnishi (1983) Appl. Environ. Microbiol. **46**, 596. **43.** Koder & Miller (1998) Biochim. Biophys. Acta **1387**, 395. **44.** Jing, Yi, Chen, Q. S. Hu, G. Y. Shi, H. Li & X. M. Tang (2004) Acta Biochim. Biophys. Sin. (Shanghai) **36**, 235. **45.** Laruelle & Godfroid (1975) J. Med. Chem. **18**, 85. **46.** Ochoa de Aspuru & Zaton (1996) J. Enzyme Inhib. **10**, 135. **47.** Petitdemange, Marczak, Blusson & Gay (1979) Biochem. Biophys. Res. Commun. **91**,

1258. **48.** Katsura, Kitamura & Tatsumi (1993) Arch. Biochem. Biophys. **302**, 356. **49.** Williamson & O'Donnell (1967) Can. J. Biochem. **45**, 340. **50.** Williamson & O'Donnell (1969) Biochemistry **8**, 1300. **51.** Goswami & Rosenberg (1983) Endocrinology **112**, 1180. **52.** Goswami, Leonard & Rosenberg (1982) Biochem. Biophys. Res. Commun. **104**, 1231. **53.** Hopp, Romero & Pont Lezica (1978) FEBS Lett. **86**, 259. **54.** Taniguchi, Hoshita & Okuda (1973) Eur. J. Biochem. **40**, 607. **55.** Mailliet, Ferry, Vella, et al. (2005) Biochem. Pharmacol. **71**, 74. **56.** Wu, Knox, Sun, et al. (1997) Arch. Biochem. Biophys. **347**, 221. **57.** Nosjean, Ferro, Cogé, et al. (2000) J. Biol. Chem. **275**, 31311. **58.** Zhao, Yang, Holtzclaw & Talalay (1997) Proc. Natl. Acad. Sci. U.S.A. **94**, 1669.

5,10-Dideazatetrahydrofolate

This folic acid analogue (FW = 442.45 g/mol), also known as lometrexol and 5,10-dideazatetrahydrofolic acid, inhibits purine biosynthesis and phosphoribosylglycinamide formyltransferase (K_i = 2-9 nM). The 6S-isomer also inhibits and is an alternative substrate for folylpolyglutamate synthetase. The di-, tri-, tetra-, penta- and hexaglutamate derivatives are stronger inhibitors of phosphoribosylglycinamide formyltransferase (e.g., K_i = 0.1-0.3 nM for (6R)-5,10-dideazatetrahydrofolate hexaglutamate). **Target(s):** 10-formyltetrahydrofolate dehydrogenase (1); phospho-ribosylaminoimidazole carboxamide formyltransferase (2-4); phosphoribosyl-glycinamide formyltransferase, or GAR transformylase (4-13); tetrahydrofolate synthetase (14). **1.** Cook & Wagner (1995) Arch. Biochem. Biophys. **321**, 336. **2.** Vergis, Bulock, Fleming & Beardsley (2001) J. Biol. Chem. **276**, 7727. **3.** Bulock, Beardsley & Anderson (2002) J. Biol. Chem. **277**, 22168. **4.** Cheng, Chong, Hwang, et al. (2005) Bioorg. Med. Chem. **13**, 3577. **5.** Baldwin, Tse, Gossett, et al. (1991) Biochemistry **30**, 1997. **6.** Beardsley, Moroson, Taylor & Moran (1989) J. Biol. Chem. **264**, 328. **7.** Sanghani & Moran (1997) Biochemistry **36**, 10506. **8.** Habeck, Leitner, Shackelford, et al. (1994) Cancer Res. **54**, 1021. **9.** DeMartino, Hwang, Connelly, Wilson & Boger (2008) J. Med. Chem. **51**, 5441. **10.** Antle, Donat, Hua, et al. (1999) Arch. Biochem. Biophys. **370**, 231. **11.** Chen, Schulze-Gahmen, Stura, Inglese, et al. (1992) J. Mol. Biol. **227**, 283. **12.** Almassy, Janson, Kann & Hostomska (1992) Proc. Natl. Acad. Sci. U.S.A. **89**, 6114. **13.** Deng, Wang, Cherian, et al. (2008) J. Med. Chem. **51**, 5052. **14.** Liani, Rothem, Bunni, et al. (2003) Int. J. Cancer **103**, 587.

(Z)-4',5'-Didehydro-5'-deoxy-5'-fluoroadenosine

This adenine nucleoside derivative (FW = 267.22 g/mol), also known as MDL-28842, is an irreversible inhibitor of S-adenosylhomocysteinase, an enzyme that converts S-adenosylhomocysteine, a potent endogenous inhibitor of S-adenosylmethionine-mediated methyltransferase reactions, to adenosine and l-homocysteine. MDL 28,842 suppressed the proliferation of all cells in a dose-dependent manner, and significantly increased keratinocyte differentiation at a concentration of 1 μm. Following incubation with MDL 28,842, the methylation indices (ratio of S-adenosyl-methionine/S-adenosylhomocysteine) of undifferentiated keratinocytes and squamous cell carcinoma lines were significantly decreased. Such findings demonstrate that the inhibitory effect of MDL 28842 on squamous carcinoma cells and keratinocyte proliferation may result from inhibition of S-adenosylhomocysteine hydrolase activity. The antiproliferative activity of MDL 28,842 against squamous carcinoma cells and keratinocytes suggests a potential role for MDL 28,842 as a novel therapeutic agent for neoplastic and hyperproliferative disorders of the skin. The (E)-isomer is also an inhibitor, albeit less effective. **1.** Mehdi, Jarvi, Koehl, McCarthy & Bey (1990) J. Enzyme Inhib. **4**, 1. **2.** Paller, Arnsmeier, Clark & Mirkin (1993) Cancer Res. **53**, 6058. **3.** Yuan, Yeh, Liu & Borchardt (1993) J. Biol.

Chem. **268**, 17030. **4**. McCarthy, Jarvi, Matthews, *et al.* (1989) *J. Amer. Chem. Soc.* **111**, 1127. **5**. Bitonti, Baumann, Jarvi, McCarthy & McCann (1990) *Biochem. Pharmacol.* **40**, 601. **6**. Jarvi, McCarthy, Mehdi, *et al.* (1991) *J. Med. Chem.* **34**, 647.

Didemnins

Didemnin A, R = H–

Didemnin B, R = CH₃CHOHCO–N

Didemnin C, R = CH₃CHOHCO–

These depsipeptides from a species of tunicate *Trididemnum* exhibit potent cytostatic and immunosuppressive effects commending them as antineoplastic agent. They inhibit protein and DNA biosynthesis and are uncompetitive inhibitors of palmitoyl protein thioesterase (1). Protein biosynthesis is inhibited at the elongation stage. The main agents are Didemnin A (FW = 943.19 g/mol), Didemnin B (FW = 1112.37 g/mol), and Didemnin C (FW = 1015.26 g/mol). All three have low solubility in water but are soluble in ethanol and chloroform. **Target(s):** DNA biosynthesis (1); palmitoyl-[protein] hydrolase (2,3); protein biosynthesis (1,4,5). **1**. Li, Timmins, Wallace, *et al.* (1984) *Cancer Lett.* **23**, 279. **2**. Linder (2001) *The Enzymes*, 3rd ed., **21**, 215. **3**. Meng, Sin & Crews (1998) *Biochemistry* **37**, 10488. **4**. Ahuja, Geiger, Ramanjulu, *et al.* (2000) *J. Med. Chem.* **43**, 4212. **5**. Ahuja, Vera, SirDeshpande, *et al.* (2000) *Biochemistry* **39**, 4339.

2',3'-Dideoxyadenosine

This dideoxynucleoside (FW = 235.25 g/mol; CAS 4097-22-7; *Symbol*: ddA) is converted metabolically to 2',3'-dideoxyadenosine 5'-triphosphate, and the latter exhibits antiviral activity and inhibits bacterial DNA synthesis, both by terminating polydeoxynucleotide elongation (1) (***See also** 2',3'-Dideoxyadenosine 5'-Triphosphate*) The 5'-triphosphate is also a potent inhibitor of viral reverse transcriptases. **Target(s):** adenosyl-homocysteinase, *or S*-adenosylhomocysteinase, weakly inhibited (2,3); deoxycytidine kinase (4); duck hepadnavirus replication (5); HIV-1 replication (6); HIV-2 replication (7); human T-lymphotropic virus type III, *or* HTLV-III)/lymphadenopathy-associated virus (6); 5' nucleotidases (8); rhodopsin kinase (9); RNA-directed DNA polymerase (10); simian acquired immunodeficiency syndrome-associated type D retrovirus reverse transcriptase (10). **1**. Toji & Cohen (1969) *Proc. Natl. Acad. Sci. U.S.A.* **63**, 871. **2**. Kim, Zhang, Cantoni, Montgomery & Chiang (1985) *Biochim. Biophys. Acta* **829**, 150. **3**. Fabianowska-Majewska, Duley & Simmonds (1994) *Biochem. Pharmacol.* **48**, :897. **4**. Datta, Shewach, Mitchell & Fox (1989) *J. Biol. Chem.* **264**, 9359. **5**. Lee, Luo, Suzuki, Robins & Tyrrell (1989) *Antimicrob. Agents Chemother.* **33**, 336. **6**. Mitsuya & Broder (1986) *Proc. Natl. Acad. Sci. U.S.A.* **83**, 1911. **7**. Mitsuya & Broder (1988) *AIDS Res. Hum. Retroviruses* **4**, 107. **8**. Garvey, Lowen & Almond (1998) *Biochemistry* **37**, 9043. **9**. Palczewski, McDowell & Hargrave (1988) *J. Biol. Chem.* **263**, 14067. **10**. Tsai, Follis & Benveniste (1988) *AIDS Res. Hum. Retroviruses* **4**, 359.

2',3'-Dideoxyadenosine 5'-Triphosphate

This dideoxynucleotide (FW = 475.19 g/mol; *Symbol*: ddATP) is a strong inhibitor of viral reverse transcriptase (*e.g.*, HIV-1 reverse transcriptase has a K_i value of 0.22 μM and an IC_{50} value of 0.083 μM. This dATP analogue is also used in the dideoxy DNA sequencing method innovated by Nobel Laureate Frederick Sanger. **Target(s):** DNA-directed DNA polymerase (1-5); DNA nucleotidylexotransferase, *or* terminal deoxynucleotidyl-transferase (6,7); DNA polymerase α, weakly inhibited (5); DNA polymerase (13); DNA primase (5); feline immunodeficiency virus reverse transcriptase (8,9); feline leukemia virus reverse transcriptase (9); HIV-1 reverse transcriptase (10-12); phenylalanyl-tRNA synthetase (13); poly(A) polymerase (14); simian immunodeficiency virus reverse transcriptase (15); ribonucleoside diphosphate reductase (16); RNA-directed DNA polymerase (8-12,15,17); RNA-directed RNA polymerase, weakly inhibited (18); visna virus reverse transciptase (17). **1**. Sanger, Nicklen & Coulson (1977) *Proc. Natl. Acad. Sci. U.S.A.* **74**, 5463. **2**. Brown & Wright (1995) *Meth. Enzymol.* **262**, 202. **3**. Toji & Cohen (1969) *Proc. Natl. Acad. Sci. U.S.A.* **63**, 871. **4**. Mentel, Kurek, Wegner, *et al.* (2000) *Med. Microbiol. Immunol.* **189**, 91. **5**. Kuchta, Ilsley, Kravig, Schubert & Harris (1992) *Biochemistry* **31**, 4720. **6**. Ono (1990) *Biochim. Biophys. Acta* **1049**, 15. **7**. Di Santo & Maga (2006) *Curr. Med. Chem.* **13**, 2353. **8**. Cronn, Remington, Preston & North (1992) *Biochem. Pharmacol.* **44**, 1375. **9**. Operario, Reynolds & Kim (2005) *Virology* **335**, 106. **10**. Le Grice, Cameron & Benkovic (1995) *Meth. Enzymol.* **262**, 130. **11**. De Clercq (1992) *AIDS Res. Human Retrovir.* **8**, 119. **12**. Perno, Yarchoan, Cooney, *et al.* (1988) *J. Exp. Med.* **168**, 1111. **13**. Gabius, Freist & Cramer (1982) *Hoppe-Seyler's Z. Physiol. Chem.* **363**, 1241. **14**. Shuman & Moss (1988) *J. Biol. Chem.* **263**, 8405. **15**. Kraus, Behr, Baier, König & Kurth (1990) *Eur. J. Biochem.* **192**, 207. **16**. Cory, Sato & Brown (1986) *Adv. Enzyme Regul.* **25**, 3. **17**. Frank, McKernan, Smith & Smee (1987) *Antimicrob. Agents Chemother.* **31**, 1369. **18**. Hung, Gibbs & Tsiang (2002) *Antiviral Res.* **56**, 99.

2',3'-Dideoxycytidine

This dideoxynucleoside (FW = 211.22 g/mol; CAS 7481-89-2; *Symbol*: ddC), also known as Zalcitabine, is converted metabolically to 2',3'-dideoxyadenosine 5'-triphosphate, and the latter inhibits viral reverse transcriptases. ddCTP will also induce apoptosis in glioblastoma cells. SLC28A1, the Na⁺-dependent nucleoside transporter selective for pyrimidine nucleosides and adenosine, also transports zalcitabine. **Target(s):** deoxycytidine kinase (1-3); deoxynucleoside kinase (1); DNA polymerases, viral, via the 5'-triphosphate (4); HIV-1 reverse transcriptase, via the 5'-triphosphate (4-8); human T lymphotropic virus type III, *or* HTLV-III)/lymphadenopathy-associated virus (6); 5'-nucleotidase (9); RNA-directed DNA polymerase (4-8,10); simian acquired immuno-deficiency syndrome-associated type D retrovirus reverse transcriptase (10); thymidine kinase (1). **1**. Johansson, Van Rompay, Degrève, Balzarini & Karlsson (1999) *J. Biol. Chem.* **274**, 23814. **2**. Datta, Shewach, Hurley, Mitchell & Fox (1989) *Biochemistry* **28**, 114. **3**. Datta, Shewach, Mitchell & Fox (1989) *J. Biol. Chem.* **264**, 9359. **4**. Wilson, Porter & Reardon (1996) *Meth. Enzymol.* **275**, 398. **5**. De Clercq (1992) *AIDS Res. Human Retrovir.* **8**, 119. **6**. Balzarini & De Clercq (1996) *Meth. Enzymol.* **275**, 472. **7**. Mitsuya & Broder (1986) *Proc. Natl. Acad. Sci. U.S.A.* **83**, 1911. **8**. El Safadi, Vivet-Boudou & Marquet (2007) *Appl. Microbiol. Biotechnol.* **75**, 723. **9**. Garvey, Lowen & Almond (1998) *Biochemistry* **37**, 9043. **10**. Tsai, Follis & Benveniste (1988) *AIDS Res. Hum. Retroviruses* **4**, 359.

2',3'-Dideoxycytidine 5'-Triphosphate

This deoxynucleotide analogue (FW$_{free-acid}$ = 451.16 g/mol; *Symbol*: ddCTP), which is formed by the transport and metabolic phosphorylation of Zalcitabine (*See 2',3'-Dideoxycytidine*), inhibits viral reverse transcriptases. This nucleotide is used in the dideoxy DNA sequencing method innovated by Nobel Laureate Frederick Sanger. **Target(s):** DNA-directed DNA polymerase (1-4); DNA helicase (5); DNA nucleotidylexotransferase, *or* terminal deoxy-nucleotidyltransferase (6,7); DNA polymerase α (4); DNA primase (4); duck hepatitis B virus reverse transcriptase (8); HIV-1 reverse transcriptase (9-11); nucleoside-triphosphatase, alternative substrate (5); RNA-directed DNA polymerase (8-12); RNA directed RNA polymerase (13,14); Visna virus reverse transcriptase (12). **1.** Sanger, Nicklen & Coulson (1977) *Proc. Natl. Acad. Sci. U.S.A.* **74**, 5463. **2.** Brown & Wright (1995) *Meth. Enzymol.* **262**, 202. **3.** Mentel, Kurek, Wegner, *et al.* (2000) *Med. Microbiol. Immunol.* **189**, 91. **4.** Kuchta, Ilsley, Kravig, Schubert & Harris (1992) *Biochemistry* **31**, 4720. **5.** Locatelli, Gosselin, Spadari & Maga (2001) *J. Mol. Biol.* **313**, 683. **6.** Ono (1990) *Biochim. Biophys. Acta* **1049**, 15. **7.** Di Santo & Maga (2006) *Curr. Med. Chem.* **13**, 2353. **8.** Kassianides, Hoofnagle, Miller, *et al.* (1989) *Gastroenterology* **97**, 1275. **9.** Le Grice, Cameron & Benkovic (1995) *Meth. Enzymol.* **262**, 130. **10.** Wilson, Porter & Reardon (1996) *Meth. Enzymol.* **275**, 398. **11.** De Clercq (1992) *AIDS Res. Human Retrovir.* **8**, 119. **12.** Frank, McKernan, Smith & Smee (1987) *Antimicrob. Agents Chemother.* **31**, 1369. **13.** Alaoui-Lsmaili, Hamel, L'Heureux, *et al.* (2000) *J. Hum. Virol.* **3**, 306. **14.** Hung, Gibbs & Tsiang (2002) *Antiviral Res.* **56**, 99.

β-D-2',3'-Dideoxy-3'-fluoroguanosine

This deoxynucleoside analogue (FW = 269.24 g/mol) is converted metabolically to 2',3'-dideoxy-3'-fluorognosine 5'-triphosphate, and the latter inhibits HIV-1 reverse transcriptase via its 5'-triphosphate, with profound antival action when aministered in its orally bioavailable prodrug form, MIV-210. **Target(s):** adenovirus DNA polymerase (1); hepatitis B virus DNA polymerase (2-4); HIV-1 reverse transcriptase (5); RNA-directed DNA polymerase (1-5). **1.** Mentel, Kurek, Wegner, *et al.* (2000) *Med. Microbiol. Immunol.* **189**, 91. **2.** Matthes, Reimer, von Janta-Lipinski, Meisel & Lehmann (1991) *Antimicrob. Agents Chemother.* **35**, 1254. **3.** Hafkemeyer, Keppler-Hafkemeyer, al Haya, *et al.* (1996) *Antimicrob. Agents Chemother.* **40**, 792. **4.** Schröder, Holmgren, Oberg & Löfgren (1998) *Antiviral Res.* **37**, 57. **5.** El Safadi, Vivet-Boudou & Marquet (2007) *Appl. Microbiol. Biotechnol.* **75**, 723.

2',3'-Dideoxyguanosine

This dideoxynucleoside (FW = 251.24 g/mol; CAS 85326-06-3; *Symbol*: ddG) is converted metabolically to 2',3'-dideoxyguanosine 5'-triphosphate, and the latter inhibits hepatitis B virus replication (1,2) and has anti-HIV activity. (*See also 2',3'-Dideoxyguanosine 5'-Triphosphate*) **Target(s):** human T-lymphotropic virus type III (HTLV-III)/lymphadenopathy associated virus (3); 5'-nucleotidase4; ribozyme self-splicing, *Tetrahymena* (5). **1.** Aoki-Sei, O'Brien, Ford, *et al.* (1991) *J. Infect. Dis.* **164**, 843. **2.** Korba,

Xie, Wright, *et al.* (1996) *Hepatology* **23**, 958. **3.** Mitsuya & Broder (1986) *Proc. Natl. Acad. Sci. U.S.A.* **83**, 1911. **4.** Garvey, Lowen & Almond (1998) *Biochemistry* **37**, 9043. **5.** Bass & Cech (1986) *Biochemistry* **25**, 4473.

2',3'-Dideoxyguanosine 5'-Triphosphate

This dideoxy nucleotide (FW$_{free-acid}$ = 491.18 g/moll *Symbol*: ddGTP) terminates DNA biosynthesis and inhibits reverse transcriptases. This dGTP analogue is also used in the dideoxy DNA chain-termination sequencing method innovated by Nobel Laureate Frederick Sanger. **Target(s):** DNA-directed DNA polymerase (1-5); DNA nucleotidylexotransferase, *or* terminal deoxynucleotidyltransferase (6,7); DNA polymerase, via chain termination (1,8); DNA polymerase α (4); DNA polymerase β (2); DNA polymerase γ (2); duck hepatitis B virus reverse transcriptase (3,9); HIV-1 reverse transcriptase (2,10-14); mouse mammary tumor virus reverse transcriptase (15); RNA directed DNA polymerase (2,3,9-17); RNA-directed RNA polymerase (18,19); simian immunodeficiency virus (SIV) reverse transcriptase (12); telomerase (20-22); Ty1 reverse transcriptase (17); Visna virus reverse transcriptase (16). **1.** Sanger, Nicklen & Coulson (1977) *Proc. Natl. Acad. Sci. U.S.A.* **74**, 5463. **2.** White, Parker, Macy, *et al.* (1989) *Biochem. Biophys. Res. Commun.* **161**, 393. **3.** Lofgren, Nordenfelt & Oberg (1989) *Antiviral Res.* **12**, 301. **4.** Ono & Nakane (1990) *Biomed. Pharmacother.* **44**, 115. **5.** Zhu & Ito (1994) *Biochim. Biophys. Acta* **1219**, 267. **6.** Ono (1990) *Biochim. Biophys. Acta* **1049**, 15. **7.** Di Santo & Maga (2006) *Curr. Med. Chem.* **13**, 2353. **8.** Brown & Wright (1995) *Meth. Enzymol.* **262**, 202. **9.** Howe, Robins, JWilson & Tyrrell (1996) *Hepatology* **23**, 87. **10.** Le Grice, Cameron & Benkovic (1995) *Meth. Enzymol.* **262**, 130. **11.** De Clercq (1992) *AIDS Res. Human Retrovir.* **8**, 119. **12.** Wu, Chernow, Boehme, *et al.* (1988) *Antimicrob. Agents Chemother.* **32**, 1887. **13.** Debyser, Vandamme, Pauwels, *et al.* (1992) *J. Biol. Chem.* **267**, 11769. **14.** Parker, White, Shaddix, *et al.* (1991) *J. Biol. Chem.* **266**, 1754. **15.** Taube, Loya, Avidan, M. Perach & A. Hizi (1998) *Biochem. J.* **329**, 579. **16.** Frank, McKernan, Smith & Smee (1987) *Antimicrob. Agents Chemother.* **31**, 1369. **17.** Wilhelm, Boutabout & Wilhelm (2000) *Biochem. J.* **348**, 337. **18.** Alaoui-Lsmaili, Hamel, L'Heureux, *et al.* (2000) *J. Hum. Virol.* **3**, 306. **19.** Hung, Gibbs & Tsiang (2002) *Antiviral Res.* **56**, 99. **20.** Strahl & Blackburn (1996) *Mol. Cell Biol.* **16**, 53. **21.** Tendian & Parker (2000) *Mol. Pharmacol.* **57**, 695. **22.** Strahl & Blackburn (1994) *Nucl. Acids Res.* **22**, 89.

1,4-Dideoxy-1,4-imino-D-arabinitol

This fructose analogue (FW = 133.15 g/mol), also known as 2-hydroxymethyl-3,4-dihydroxypyrrolidine, first isolated from the fruits of *Angylocalyx boutiquenus*, is a strong inhibitor of liver glycogen phosphorylase ($K_i \approx 0.4$ μM) and hence glycogenolysis. **Target(s):** α-glucosidase (1,2); glycogen phosphorylase (3-6); isomaltase (1); mannosyl oligosaccharide 1,2-α-mannosidase, *or* α-mannosidase I (1,2); mannosyl-oligosaccharide 1,3-1,6-α mannosidase, *or* α-mannosidase II (1,2); processing α-glucosidase II, mannosyl-oligosaccharide glucosidase II (1,2); thioglucosidase, *or* myrosinase, *or* sinigrinase (7); trehalose (1,8). **1.** Asano, Oseki, Kizu & Matsui (1994) *J. Med. Chem.* **37**, 3701. **2.** Asano, Kizu, Oseki, *et al.* (1995) *J. Med. Chem.* **38**, 2349. **3.** Fosgerau, Westergaard, Quistorff, *et al.* (2000) *Arch. Biochem. Biophys.* **380**, 274. **4.** Andersen, Rassov, Westergaard & Lundgren (1999) *Biochem. J.* **342**, 545. **5.** Latsis, Andersen & Agius (2002) *Biochem. J.* **368**, 309. **6.** Gregus & Németi (2007) *Toxicol. Sci.* **100**, 44. **7.** Scofield, Rossiter, Witham, *et al.* (1990) *Phytochemistry* **29**, 107. **8.** Asano, A. Kato & K. Matsui (1996) *Eur. J. Biochem.* **240**, 692.

2,6-Dideoxy-2,6-imino-7-*O*-(β-D-glucopyranosyl)-D-*glycero*-L-*gulo*-heptitol

This sugar alcohol (FW = 355.34 g/mol), also known as 7-*O*-β-D-glucopyranosyl-α-homonojirimycin and MDL 25637, from *Lobelia sessilifolia* inhibits α-glucosidases. **Target(s):** α-amylase, weakly inhibited (1); glucan 1,3-α-glucosidase, *or* glucosidase II, *or* mannosyl-oligosaccharide glucosidase II (2-4); glucoamylase (1); α-glucosidase (1,5); isomaltase (1); lactase, weakly inhibited (1); maltase (1); mannosyl-oligosaccharide glucosidase I, weakly inhibited (4); sucrase (1); trehalose (1). **1.** Rhinehart, Robinson, Liu, *et al.* (1987) *J. Pharmacol. Exp. Ther.* **241**, 915. **2.** Kaushal & Elbein (1989) *Meth. Enzymol.* **179**, 452. **3.** Kaushal, Pastuszak, Hatanka & Elbein (1990) *J. Biol. Chem.* **265**, 16271. **4.** Kaushal, Pan, Tropea, *et al.* (1988) *J. Biol. Chem.* **263**, 17278. **5.** Ikeda, Takahashi, Nishida, *et al.* (2000) *Carbohydr. Res.* **323**, 73.

1,4-Dideoxy-1,4-imino-D-mannitol

This sugar alcohol (FW$_{free-base}$ = 163.17 g/mol; CAS 100937-52-8) inhibits various α-mannosidases. **Target(s):** glycoprotein processing (1-6); α-mannosidase (1,3,4,7,8); mannosyl-oligosaccharide 1,2-α-mannosidase, mannosidase I (4); mannosyl-oligosaccharide 1,3-1,6-α-mannosidase, mannosidase II (2-7,9). **1.** Elbein (1987) *Meth. Enzymol.* **138**, 661. **2.** Kaushal & Elbein (1989) *Meth. Enzymol.* **179**, 452. **3.** Kaushal & Elbein (1994) *Meth. Enzymol.* **230**, 316. **4.** Palamarczyk, Mitchell, Smith, Fleet & Elbein (1985) *Arch. Biochem. Biophys.* **243**, 35. **5.** Moremen (2002) *Biochim. Biophys. Acta* **1573**, 225. **6.** Elbein (1991) *FASEB J.* **5**, 3055. **7.** Winchester, al Daher, Carpenter, *et al.* (1993) *Biochem. J.* **290**, 743. **8.** Cenci di Bello, Fleet, Namgoong, Tadano & Winchester (1989) *Biochem. J.* **259**, 855. **9.** Kaushal, Szumilo, Pastuszak & Elbein (1990) *Biochemistry* **29**, 2168.

2',3'-Dideoxyinosine

This dideoxy nucleoside analogue (FW = 236.23 g/mol; CAS 69655-05-6; *Symbol*: ddI; [2',3'-³H]-ddI; CAS 124516-24-1), also known as didanosine and sold under the trade names Videx and Videx EC, is metabolized to ddATP (*Likely Pathway*: ddI → ddIMP → ddAMP-Succinate → ddAMP → ddADP → ddATP), whereupon it inhibits viral reverse transcriptase by competing with natural dATP and acts as a chain terminator upon incorporation into DNA by viral reverse transcriptases. Approved for the treatment of HIV-AIDS by the Food and Drug Administration in 1991, didanosine became the first generic anti-HIV drug marketed in the U.S. **Target(s):** DNA polymerases, viral, by the 5'-triphosphate (1); HIV-1 reverse transcriptase, by the 5'-triphosphate (1-5); 5'-nucleotidase (6); RNA-directed DNA polymerase (1-5). **1.** Wilson, Porter & Reardon (1996) *Meth. Enzymol.* **275**, 398. **2.** Balzarini & De Clercq (1996) *Meth. Enzymol.* **275**, 472. **3.** Mitsuya & Broder (1986) *Proc. Natl. Acad. Sci. U.S.A.* **83**, 1911. **4.** Russell & Klunk (1989) *Biochem. Pharmacol.* **38**, 1385. **5.** El Safadi, Vivet-Boudou & Marquet (2007) *Appl. Microbiol. Biotechnol.* **75**, 723. **6.** Garvey, Lowen & Almond (1998) *Biochemistry* **37**, 9043.

2',3'-Dideoxyinosine 5'-Triphosphate

This dideoxy nucleotide (FW$_{free-acid}$ = 476.17 g/mol; FW$_{TetraNaSalt}$ = 564.09 g/mol; Symbol: ddITP), also known as didanosine 5'-triphosphate, inhibits viral reverse transcriptases. **Target(s):** HIV-1 reverse transcriptase (1-3); RNA-directed DNA polymerase (1-3). **1.** Russell & Klunk (1989) *Biochem. Pharmacol.* **38**, 1385. **2.** Wilson, Porter & Reardon (1996) *Meth. Enzymol.* **275**, 398. **3.** El Safadi, Vivet-Boudou & Marquet (2007) *Appl. Microbiol. Biotechnol.* **75**, 723.

2',3'-Dideoxyuridine 5'-Triphosphate

This dideoxynucleotide (FW = 452.14 g/mol; *Symbol*: ddUTP) is a strong inhibitor of the HIV reverse transcriptase (K_i = 0.05 μM) and the avian myeloblastosis virus enzyme (K_i = 1.0 μM). However, the bacterial DNA polymerase I, mammalian DNA polymerase α, terminal deoxyribonucleotidyltransferase, and Moloney murine leukemia virus reverse transcriptase were resistant to ddUTP. **1.** Hao, Cooney, Farquhar, *et al.* (1990) *Mol. Pharmacol.* **37**, 157. **2.** Le Grice, C. E. Cameron & S. J. Benkovic (1995) *Meth. Enzymol.* **262**, 130. **3.** De Clercq (1992) *AIDS Res. Human Retrovir.* **8**, 119.

Dieldrin

This highly toxic pesticide (FW = 380.91 g/mol), once a widely used against a range of insects, targets the GABA$_A$ receptor- chloride ion channel complex. The initial transient component of the chloride current was blocked more than the late sustained component, resulting in unbalanced sodium accumulation. Action on GABA-mediated processes is thought to result in hyperexcitation of the nervous system, a finding that fits with the well-known convulsant action of this agent. From 1950 to 1974, dieldrin was widely used to control insects threatening cotton, corn and citrus crops. It was also used to control locusts and mosquitoes as well as to preserve wood, especially as a long-lasting agent for termite control. Dieldrin's high fat solubility allows it to become concentrated in the food chain. Its remarkable chemical stability accounts for its extremely low rate of biological turnover, a property that poses a challenge for bioremediation. Owing to its harmful effects on humans, fish, and wildlife, most uses of dieldrin were banned in 1987 U.S., where it is no longer produced. The EPA now classifies dieldrin as a persistent, bioaccumulative, and toxic pollutant. **Target(s):** ATPase (1,2); Ca^{2+}-transporting ATPase (3,4); chitin synthase (5); Cl$^-$ uptake, γ-aminobutyrate-dependent (6-10); F$_1$F$_o$ ATPase (2); lactate dehydrogenase (11). **1.** Desaiah & Koch (1975) *Biochem. Biophys. Res. Commun.* **64**, 13. **2.** Mehrotra, Bansal & Desaiah (1982) *J. Appl. Toxicol.* **2**, 278. **3.** Mehrotra, Moorthy, Reddy & Desaiah (1989) *Toxicology* **54**, 17. **4.** Janik & Wolf (1992) *J. Appl. Toxicol.* **12**, 351. **5.** Leighton, Marks & Leighton (1981) *Science* **213**, 905. **6.** Bloomquist & Soderlund (1985) *Biochem. Biophys. Res. Commun.* **133**, 37. **7.** Narahashi, Frey, Ginsburg & Roy (1992) *Toxicol. Lett.* **64-65** Spec. No., 429. **8.** Nagata & Narahashi (1994) *J. Pharmacol. Exp. Ther.* **269**, 164. **9.** Pomes, Rodriguez-Farre & Sunol (1994) *J. Pharmacol. Exp. Ther.* **271**, 1616. **10.** Narahashi, Ginsburg, Nagata, Song & Tatebayashi (1998) *Neurotoxicology* **19**, 581. **11.** Hendrickson & Bowden (1976) *J. Agric. Food Chem.* **24**, 757.

Diethanolamine

This aminoalkyldiol (FW$_{free-base}$ = 105.14 g/mol; CAS 111-42-2; M.P. = 28°C; B.P. = 268.8°C), also known as 2,2'-iminobisethanol, is a common buffer salt, with a pK_a of 8.88 at 25°C (dpK_a/dT = –0.025). **Target(s):** acetylenecarboxylate hydratase (1); amylo-α-1,6-glucosidase/4-α glucano-transferase, *or* glycogen debranching enzyme, K_i = 16 mM (2); choline sulfotransferase (3); choline uptake (4); formiminotetrahydrofolate cyclodeaminase (5); α-glucosidase (6); β-glucuronidase (6); β-lysine 5,6-aminomutase (7); lysine 2-monooxygenase (8); oligo-1,6-glucosidase, *or* isomaltase (9); sucrose α-glucosidase, *or* sucrase-isomaltase (9). **1.** Yamada & Jakoby (1959) *J. Biol. Chem.* **234**, 941. **2.** Nelson, Kolb & Larner (1969) *Biochemistry* **8**, 1419. **3.** Orsi & Spencer (1964) *J. Biochem.* **56**, 81. **4.** Lehman-McKeeman & Gamsky (1999) *Biochem. Biophys. Res. Commun.* **262**, 600. **5.** Uyeda & Rabinowitz (1963) *Meth. Enzymol.* **6**, 380. **6.** Balbaa, Abdel-Hady, el-Rashidy, *et al.* (1999) *Carbohydr. Res.* **317**, 100. **7.** Baker, van der Drift & Stadtman (1973) *Biochemistry* **12**, 1054. **8.** Flashner & Massey (1974) *J. Biol. Chem.* **249**, 2579. **9.** Kano, Usami, Adachi, Tatematsu & Hirano (1996) *Biol. Pharm. Bull.* **19**, 341.

3β-(2-Diethylaminoethoxy)androst-5-en-17-one

This synthetic steroid (FW$_{free-base}$ = 387.61 g/mol), also known as U-18666A, inhibits cholesterol biosynthesis. Subcutaneous injections occasionally result in the development of irreversible nuclear cataracts. For more information on the cholesterol synthesis inhibiting properties of U18666A and the role of sterol metabolism and trafficking in numerous pathophysiological processes, see (1). **Target(s):** cholesterol biosynthesis (1-5); cholestenol Δ-isomerase, *or* sterol Δ^{7,8}-isomerase (6,7); cholesterol transport (8,9); cycloartenol synthase (10); 2,3-oxidosqualene cyclase (5,11); sterol 24-C methyltransferase (K_i = 14.8 μM (12); sterol Δ^{24}-reductase (13). **1.** Cenedella (2009) *Lipids* **44**, 477. **2.** Cenedella (1980) *Biochem. Pharmacol.* **29**, 2751. **3.** Cenedella & Bierkamper (1979) *Exp. Eye Res.* **28**, 673. **4.** Bierkamper & Cenedella (1978) *Brain Res.* **150**, 343. **5.** Sexton, Panini, Azran & Rudney (1983) *Biochemistry* **22**, 5687. **6.** Moebius, Reiter, Bermoser, *et al.* (1998) *Mol. Pharmacol.* **54**, 591. **7.** Bae, Seong & Paik (2001) *Biochem. J.* **353**, 689. **8.** Harmala, Porn, Mattjus & Slotte (1994) *Biochim. Biophys. Acta* **1211**, 317. **9.** Sparrow, Carter, Ridgway, Cook & Byers (1999) *Neurochem. Res.* **24**, 69. **10.** Fenner & Raphiou (1995) *Lipids* **30**, 253. **11.** Duriatti, Bouvier-Nave, Benveniste, *et al.* (1985) *Biochem. Pharmacol.* **34**, 2765. **12.** Rahier, Génot, Schuber, Benveniste & Narula (1984) *J. Biol. Chem.* **259**, 15215. **13.** Steinberg & Avigan (1969) *Meth. Enzymol.* **15**, 514.

Diethylcarbamazine

This substituted piperazine (FW$_{free-base}$ = 199.30 g/mol), also known as *N,N*-diethyl-4-methyl-1-piperazinecarboxamide, is frequently used to treat individuals infected by filarial parasites. It reportedly interferes with arachidonate metabolism of filarial parasites and host. **Target(s):** acetylcholinesterase (1,2); catalase (3); Ca^{2+}-transporting ATPase (4); glutathione peroxidase (3); leukotriene-A$_4$ synthase, *or* 5-lipoxygenase (5,6); leukotriene-C$_4$ synthase (7,8); Na$^+$/K$^+$ exchanging ATPase (4); tubulin polymerization, *or* microtubule assembly (9). **1.** Fujimaki, Sakamoto, Shimada, Kimura & Aoki (1989) *Southeast Asian J. Trop. Med. Public Health* **20**, 179. **2.** Bhattacharya, Singh, Misra & Rathaur (1997) *Trop. Med. Int. Health* **2**, 686. **3.** Batra, Chatterjee & Srivastava (1990) *Biochem. Pharmacol.* **40**, 2363. **4.** Agarwal, Tekwani, Shukla & Ghatak (1990)

Indian J. Exp. Biol. **28**, 245. **5.** Pace-Asciak & Smith (1983) *The Enzymes*, 3rd ed. (Boyer, ed.), **16**, 543. **6.** Lonigro, Sprague, Stephenson & Dahms (1988) *J. Appl. Physiol.* **64**, 2538. **7.** Bach & Brashler (1986) *Biochem. Pharmacol.* **35**, 425. **8.** Nicholson, Ali, Klemba, *et al.* (1992) *J. Biol. Chem.* **267**, 17849. **9.** Fujimaki, Ehara, Kimura, Shimada & Aoki (1990) *Biochem. Pharmacol.* **39**, 851.

1,1-Diethyl-2-hydroxy-2-nitrosohydrazine

This substituted hydrazine (FW = 133.15 g/mol), also known as the diethylamine-nitric oxide complex, is a synthetic nitric oxide donor that readily nitrosylates a cysteinyl residue of rat ornithine decarboxylase under aerobic conditions. **Target(s):** adenovirus proteinase (1); DNA ligase (2); dopamine β-monooxygenase (3); epidermal growth factor receptor protein-tyrosine kinase (4); formamidopyrimidine-DNA glycolyase (5); insulin receptor kinase (6); K$^+$ current, glutamate-induced (7); ornithine decarboxylase (8). **1.** Cao, Baniecki, McGrath, *et al.* (2003) *FASEB J.* **17**, 2345. **2.** Graziewicz, Wink & Laval (1996) *Carcinogenesis* **17**, 2501. **3.** Zhou, Espey, Chen, *et al.* (2000) *J. Biol. Chem.* **275**, 21241. **4.** Estrada, Gomez, Martin-Nieto, *et al.* (1997) *Biochem. J.* **326**, 369. **5.** Wink & Laval (1994) *Carcinogenesis* **15**, 2125. **6.** Schmid, Hotz-Wagenblatt & Droge (1999) *Antioxid. Redox Signal.* **1**, 45. **7.** Sawada, Ichinose & Anraku (2000) *J. Neurosci. Res.* **60**, 642. **8.** Bauer, Buga, Fukuto, Pegg & Ignarro (2001) *J. Biol. Chem.* **276**, 34458.

N,N-Diethyl-3-methylbenzamide

This widely used insect repellant (FW = 191.27 g/mol; CAS 134-62-3; Physical State: slightly yellow oil, with a density of 0.998 g/mL), also known as *N,N*-diethyl-*m*-toluamide (the name giving rise to its famous abbreviation, "DEET") is directly sensed in mosquitoes by DEET-detecting olfactory receptor neurons, *or* ORNs (1). It is also highly effective against chiggers, fleas, leeches, ticks, and other biting insects. Earlier work had suggested DEET modulates the physiological response of lactic acid-sensitive ORNs in the antennae of the yellow fever mosquito (*Aedes aegypti*), leading to the idea that DEET interferes with and/or inhibits olfactory responses to an attractive, naturally occurring chemical signal (2). Jamming the olfactory system had been supported by behavioral observations indicating that lactic acid is a mosquito attractant and that DEET inhibits attraction to lactic acid (4, 5). DEET's reported attenuation of mosquito response to 1-octen-3-ol, a volatile substance present in human sweat and breath, also supported the earlier conclusion (6). The work of Syed & Leal (1), however, demonstrated that DEET itself is detected by specific ORNs situated on antennae of *Cx quinquefasciatus*, that DEET induces avoidance in sugar-seeking male and female mosquitoes, and that DEET also causes reduced landing of females in the vicinity of an attractive, warm, and black surface. Such findings support the conclusion that mosquitoes smell and avoid DEET (1). Intriguingly, DEET was originally isolated from the female Pink Bollworm Moth (*Pectinophora gossypiella*), suggesting a role in warding off suitors and/or predators. *Note*: DEET content in commercial repellants varies from 4–100% (*weight/weight*). While higher concentrations are no more effective, they remain effective over a longer duration. Depending upon the period of outdoor activity, products having lower DEET content may require reapplication. The U.S. Centers for Disease Control and Prevention recommends using products with 30-50% DEET to prevent the spread of pathogens carried by DEET-sensitive insects. **1.** Syed & Leal (2008) *Proc. Natl. Acad. Sci. U.S.A.* **105**, 13598. **2.** 2. Davis & Sokolove (1976) *J. Comp. Physiol.* **105**, 43. **3.** Davis (1985) *J. Med. Entomol.* **22**, 237. **4.** Boeckh, *et al.* (1996) *Pestic. Sci.* **48**, 359. **5.** Dogan, Ayres & Rossignol (1999) *Med. Vet. Entomol.* **13**, 97. **6.** Ditzen, Pellegrino & Vosshall (2008) *Science* **319**, 1838. **7.** Jones & Jacobson (1968) *Science* **159**, 99.

O,O-Diethyl O-(p-Nitrophenyl) Phosphate

This toxic phosphotriester (FW = 275.20 g/mol; CAS 311-45-5; Oily liquid; Soluble in organic solvents; Moderate water solubility (2.4 mg/mL at 25°C), also called Paraoxon and E600, is an acetylcholinesterase inhibitor that has found wide use as an insecticide. This enzyme is inhibited via phosphorylation of an active-site seryl residue. Paraoxon is a substrate for several lactonases, named paraoxonases (PON's), for lack of information on their natural substrates. PON-1 is synthesized in the liver and transported to the plasma, where it prevents LDL oxidation; PON-2 is expressed ubiquitously and protect cells against cytosolic oxidative damage; and PON3 is similar to PON-1, but is not regulated by inflammation or levels of oxidized LDL. **Target(s):** acetylcholinesterase (1-14); acetylsalicylate deacetylase (15); acylaminoacyl peptidase (16); acylglycerol lipase, *or* monoacylglycerol lipase (17-20); alkylamidase (21); amidase (22); arylformamidase (23-25); bis(2-ethylhexyl)phthalate esterase (26); butyrylesterase (27); carboxylesterase (1,19,28-43); cholinesterase, *or* butyrylcholinesterase (1,3,44-49); chymotrypsin (1,50,51; however, see also ref. 52); cutinase (53-55); diacylglycerol lipase (20); fatty acid amide hydrolase, *or* anandamide amidohydrolase (18); β-glucosidase (56); hormone-sensitive lipase (57,58); juvenile hormone esterase (59-64); lipase, *or* triacylglycerol lipase (20,65-77,93); lipoprotein lipase (67,68,78); phosphatidylcholine acyltransferase (79); phospholipase A₂ (80,81); phospholipase B, *Penicillium notatum* (82); sialate O-acetylesterase (83-88); sinapine esterase (89); sterol esterase, *or* cholesterol esterase (90,91); tropinesterase (92); trypsin (1). **1.** Cohen, Oosterbaan & Berends (1967) *Meth. Enzymol.* **11**, 686. **2.** Kousba, Sultatos, Poet & Timchalk (2004) *Toxicol. Sci.* **80**, 239. **3.** Schopfer, Voelker, Bartels, Thompson & Lockridge (2005) *Chem. Res. Toxicol.* **18**, 747. **4.** Worek, Thiermann, Szinicz & Eyer (2004) *Biochem. Pharmacol.* **68**, 2237. **5.** Talesa, Romani, Antognelli, Giovannini & Rosi (2001) *Chem. Biol. Interact.* **134**, 151. **6.** Geyer, Muralidharan, Cherni, *et al.* (2005) *Chem. Biol. Interact.* **157-158**, 331. **7.** Frasco, Fournier, Carvalho & Guilhermino (2006) *Aquat. Toxicol.* **77**, 412. **8.** Ma, He & Zhu (2004) *Pestic. Biochem. Physiol.* **78**, 67. **9.** Boublik, Saint-Aguet, Lougarre, *et al.* (2002) *Protein Eng.* **15**, 43. **10.** Cheng, Wang, Ding & Zhao (2004) *J. Appl. Entomol.* **128**, 292. **11.** Rosenfeld & Sultatos (2006) *Toxicol. Sci.* **90**, 460. **12.** Kardos & Sultatos (2000) *Toxicol. Sci.* **58**, 118. **13.** Pralavorio & Fournier (1991) *Biochem. Genet.* **30**, 77. **14.** Rosenfeld, Kousba & Sultatos (2001) *Toxicol. Sci.* **63**, 208. **15.** Kim, Yang & Jakoby (1990) *Biochem. Pharmacol.* **40**, 481. **16.** Richards, Johnson & Ray (2000) *Mol. Pharmacol.* **58**, 577. **17.** Somma-Delpéro, Valette, Lepetit-Thévenin, *et al.* (1995) *Biochem. J.* **312**, 519. **18.** Quistad, Klintenberg, Caboni, Liang & Casida (2006) *Toxicol. Appl. Pharmacol.* **211**, 78. **19.** Keough, de Jersey & Zerner (1985) *Biochim. Biophys. Acta* **829**, 164. **20.** Stam, Broekhoven-Schokker & Hülsmann (1986) *Biochim. Biophys. Acta* **875**, 76. **21.** Chen & Dauterman (1971) *Biochim. Biophys. Acta* **250**, 216. **22.** Alt, Heymann & Krisch (1975) *Eur. J. Biochem.* **53**, 357. **23.** Bailey & Wagner (1974) *J. Biol. Chem.* **249**, 4439. **24.** Menge (1979) *Hoppe-Seyler's Z. Physiol. Chem.* **360**, 185. **25.** Arndt, Junge, Michelsen & Krisch (1973) *Hoppe-Seyler's Z. Physiol. Chem.* **354**, 1583. **26.** Krell & Sandermann (1984) *Eur. J. Biochem.* **143**, 57. **27.** Hojring & Svensmark (1988) *Arch. Biochem. Biophys.* **260**, 351. **28.** Scott & Zerner (1975) *Meth. Enzymol.* **35**, 208. **29.** Heymann & Mentlein (1981) *Meth. Enzymol.* **77**, 333. **30.** Krisch (1971) *The Enzymes*, 3rd ed. (Boyer, ed.), **5**, 43. **31.** Højring & Svensmark (1977) *Biochim. Biophys. Acta* **481**, 500. **32.** Højring & Svensmark (1988) *Arch. Biochem. Biophys.* **260**, 351. **33.** Junge & Heymann (1979) *Eur. J. Biochem.* **95**, 519. **34.** McGhee (1987) *Biochemistry* **26**, 4101. **35.** Ketterman, Jayawardena & Hemingway (1991) *Biochem. Soc. Trans.* **19**, 305S. **36.** Zhang, Qiao & Lan (2004) *Environ. Toxicol.* **19**, 154. **37.** Valles, Oi & Strong (2001) *Insect Biochem. Mol. Biol.* **31**, 715. **38.** Saboori & Newcombe (1990) *J. Biol. Chem.* **265**, 19792. **39.** Siddalinga Murthy & Veerabhadrappa (1996) *Insect Biochem. Mol. Biol.* **26**, 287. **40.** Ketterman, Jayawardena & Hemingway (1992) *Biochem. J.* **287**, 355. **41.** Canaan, Maurin, Chahinian, *et al.* (2004) *Eur. J. Biochem.* **271**, 3953. **42.** Manco, di Gennaro, de Rosa & Rossi (1994) *Eur. J. Biochem.* **221**, 965. **43.** Cummins & Edwards (2004) *Plant J.* **39**, 894. **44.** Davison (1953) *Biochem. J.* **54**, 583. **45.** Riov & Jaffe (1973) *Plant Physiol.* **51**, 520. **46.** Mehrani (2004) *Proc. Biochem.* **39**, 877. **47.**

Sanchez-Hernandez & Sanchez (2002) *Environ. Toxicol. Chem.* **21**, 2319. **48.** Chemnitius, Chemnitius, Haselmeyer, Kreuzer & Zech (1992) *Biochem. Pharmacol.* **43**, 823. **49.** Ashani, Segev & Balan (2004) *Toxicol. Appl. Pharmacol.* **194**, 90. **50.** Desnuelle (1960) *The Enzymes*, 2nd ed., **4**, 93. **51.** Cunningham (1954) *J. Biol. Chem.* **207**, 443. **52.** Jansen, Curl & Balls (1951) *J. Biol. Chem.* **190**, 557. **53.** Purdy & Kolattukudy (1975) *Biochemistry* **14**, 2832. **54.** Martinez, Nicolas, van Tilbeurgh, *et al.* (1994) *Biochemistry* **33**, 83. **55.** Soliday & Kolattukudy (1976) *Arch. Biochem. Biophys.* **176**, 334. **56.** Park, Bae, Sung, Lee & Kim (2001) *Biosci. Biotechnol. Biochem.* **65**, 1163. **57.** Ben Ali, Chahinian, Petry, *et al.* (2004) *Biochemistry* **43**, 9298. **58.** Remaury, Laurell, Grober, *et al.* (1995) *Biochem. Biophys. Res. Commun.* **207**, 175. **59.** Stauffer, T. Shiotsuki, W. Chan & B. D. Hammock (1997) *Arch. Insect Biochem. Physiol.* **34**, 203. **60.** Abdel-Aal & Hammock (1986) *Science* **233**, 1073. **61.** Pratt (1975) *Insect Biochem.* **5**, 595. **62.** Sparks & Rose (1983) *Insect Biochem.* **13**, 633. **63.** McCaleb, Reddy & Kumaran (1980) *Insect Biochem.* **10**, 273. **64.** Sparks & Hammock (1979) *Insect Biochem.* **9**, 411. **65.** Khoo & Steinberg (1975) *Meth. Enzymol.* **35**, 181. **66.** Sigman & Mooser (1975) *Ann. Rev. Biochem.* **44**, 889. **67.** Ransac, Gargouri, Marguet, *et al.* (1997) *Meth. Enzymol.* **286**, 190. **68.** Hofstee (1960) *The Enzymes*, 2nd ed., **4**, 485. **69.** Desnuelle (1972) *The Enzymes*, 3rd ed., **7**, 575. **70.** Derewenda, Brzozowski, Lawson & Derewenda (1992) *Biochemistry* **31**, 1532. **71.** Eastmond (2004) *J. Biol. Chem.* **279**, 45540. **72.** Saxena, Davidson, Sheoran & Giri (2003) *Process Biochem.* **39**, 239. **73.** Garner (1980) *J. Biol. Chem.* **255**, 5064. **74.** Deb, Daniel, Sirakova, *et al.* (2006) *J. Biol. Chem.* **281**, 3866. **75.** Maylié, Charles & Desnuelle (1972) *Biochim. Biophys. Acta* **276**, 162. **76.** Derewenda, Brzozowski, Lawson & Derewenda (1992) *Biochemistry* **31**, 1532. **77.** Jaeger, Adrian, Meyer, Hancock & Winkler (1992) *Biochim. Biophys. Acta* **1120**, 315. **78.** Olivecrona & Lookene (1997) *Meth. Enzymol.* **286**, 102. **79.** Glomset (1969) *Meth. Enzymol.* **15**, 543. **80.** Akiba, Dodia, Chen & Fisher (1998) *Comp. Biochem. Physiol. B* **120**, 393. **81.** Rice, Southan, Boyd, *et al.* (1998) *Biochem. J.* **330**, 1309. **82.** Saito, Sugatani & Okumura (1991) *Meth. Enzymol.* **197**, 446. **83.** Schauer (1987) *Meth. Enzymol.* **138**, 611. **84.** Higa, Manzi, Diaz & Varki (1989) *Meth. Enzymol.* **179**, 409. **85.** Higa, Manzi & Varki (1989) *J. Biol. Chem.* **264**, 19435. **86.** Schauer, Reuter & Stoll (1988) *Biochimie* **70**, 1511. **87.** Schauer, Reuter, Stoll, *et al.* (1988) *Biol. Chem. Hoppe-Seyler* **369**, 1121. **88.** Guimarães, Bazan, Castagnola, Diaz, Copeland, Gilbert, Jenkins, Varki & Zlotnik (1996) *J. Biol. Chem.* **271**, 13697. **89.** Nurmann & Strack (1979) *Z. Naturforsch. C* **34**, 715. **90.** Feaster & Quinn (1997) *Meth. Enzymol.* **286**, 231. **91.** Nègre, Salvayre, Rogalle, Dang & Douste-Blazy (1987) *Biochim. Biophys. Acta* **918**, 76. **92.** Moog & Krisch (1974) *Hoppe-Seyler's Z. Physiol. Chem.* **355**, 529. **93.** Krysan & Guss (1971) *Biochim. Biophys. Acta* **239**, 349.

Diethyl Pyrocarbonate

This highly reactive reagent (FW = 162.14 g/mol; CAS 1609-47-8), also known as ethoxyformic anhydride and diethyl dicarbonate, is frequently used to ethoxyformylate proteins. DEPC is an especially effective inactivator of enzymes requiring a catalytic histidyl residue. The classical example is ribonuclease, with which it reacts to block catalysis totally. **Chemical Stability:** Aqueous solutions of have a short half-life ($t_{1/2}$ = 7 min at pH 7 (25°C); $t_{1/2}$ = 24 min at pH 6). The presence of Tris buffer accelerates decomposition ($t_{1/2}$ = 1.25 min at pH 7.5 in the presence of Tris; $t_{1/2}$ decreases at higher pH). Therefore, solutions should always be freshly prepared from a tightly sealed reagent bottle. **Histidine Modification & Reactivation:** The *N*-carboethoxyhistidyl product absorbs ultraviolet light at 240 nm (Δε = 3200 M⁻¹cm⁻¹) and is moderately unstable ($t_{1/2}$ = 55 hours at pH 7; shorter under acidic or alkaline conditions). Note that the diethyl pyrocarbonate is hydrolyzed in aqueous solutions to ethanol and carbonic acid with a half-life of about 25 minutes at neutral pH. Modification of histidyl residues is preferred at acid pH values. DEPC-inactivated enzymes often undergo reactivation upon treatment with neutral hydroxylamine. **Reactivity with Other Residues:** Diethyl pyrocarbonate also reacts with tyrosyl residues, amino groups, and occasionally sulfhydryl groups. Reversal of amino and sulfhydryl group modification does not occur with hydroxylamine. Reaction with tyrosyl residues results in a decrease in absorbance at 278 nm. Because this reagent is slowly hydrolyzed in water, solutions should be freshly prepared. **DEPC-Treated Water/Buffers:** Water or buffer (limited to buffer salts that are both heat-stable solutes and that do not react wirh DEPC) is treated with by addition of DEPC to yield a final concentration of 0.1% (vol/vol) for at least 2 hours at 37 °C, before the solution is autoclaved (> 15 min) to inactivate any traces of DEPC that remain. Although the resulting solution is now RNase-free, it is easily re-contaminated (by experimenter lacking gloves and face masks), especially by microbe-labeled aerosols, or tiny droplets of spittle, or by using untreated plastic and glass labware has not been treated with DEPC. **Procedure for Minimizing RNA Degradation:** Given its remarkable stability and high isoelectric point (pI = 12.8), ribonuclease is stable, sticky and positively charged; such properties that allow it to bind firmly to borosilicate glassware that it can degrade RNA during isolation from various cells/tissues. For this reason, glassware must be treated with 0.1% (wt/vol) DEPC, after which the glassware should be extensively washed with ribonuclease-free water to remove all traces of DEPC. **CAUTION:** Care should be exercised in its handling since concentrated solutions can be irritating to skin, eyes, and mucous membranes. Note that DEPC will decompose to carbon dioxide and ethanol; cases of spontaneous hazardous chemical explosions of unopened bottles of diethyl pyrocarbonate have been reported. **1.** Scofield, Sun & Pettinger (1992) *Biotechniques* **12**, 820. **2.** Rosén & Fedorsák (1966) *Biochim. Biophys. Acta* **130**, 401. **3.** Melchior & Fahrney (1970) *Biochemistry* **9**, 251. **4.** Mendelsohn & Young (1978) *Biochim. Biophys. Acta* **519**, 461. **5.** Irie, Ohgi & Iwama (1977) *J. Biochem.* **82**, 1701. **6.** Jones (1976) *Biochem. Biophys. Res. Commun.* **69**, 469. **7.** Wolf, Lesnaw & Reichmann (1970) *Eur. J. Biochem.* **13**, 519. **8.** Fedorcsák & Ehrenberg (1966) *Acta Chem. Scand.* **20**, 107. **9.** Safarian, Moosavi-Movahedi, Hosseinkhani, *et al.* (2003) *J. Protein Chem.* **22**, 643.

Diethylstilbestrol

This synthetic estrogen and endocrine-disrupting agent (FW = 268.36 g/mol; CAS 6898-97-1; Insoluble in water; Soluble in ethanol; M.P. = 169-172°C), also known as stilbestrol and 4,4'-[(1*E*)-1,2-diethyl-1,2 ethenediyl]bisphenol, and abbreviated DES, is a nonsteroidal agent that was once widely used used in estrogen therapy as well as an antineoplastic hormone. Even so, the Food and Drug Administration in the U.S. published warnings about the use of diethylstilbestrol by pregnant women in 1971, recognizing it as a perinatal carcinogen that led to clear cell adenocarcinoma in women and experimental animals. Diethylstilbestrol was also used as a growth promoter in livestock. Exposure during critical periods of differentiation permanently alters the programming of estrogen target tissues, resulting in benign and malignant abnormalities in the reproductive tract later in life. **Note:** Although the term "stilbestrol" is frequently used as a synonym for diethylstilbestrol, stilbestrol technically refers to 4,4'-dihydroxystilbene. **Target(s):** A_1 ATPase (1,2); aldehyde dehydrogenase (3-5); aldose 1-epimerase, *or* mutarotase (6-8); anion-activated ATPase (9); apyrase (10); arylamine *N*-acetyltransferase (11); ATP-ADPase (12); ATPases (*e.g.*, H$^+$-transporting ATPase) (9,13-25); Ca^{2+} channels (26); Ca^{2+}-transporting ATPase (27); cortisol sulfotransferase (28-30); DNA topoisomerase I (31); dTMP kinase (32); electron transport (33); estrogen-2/4-hydroxylase (34); estrogen 2-monooxygenase (35); estrone sulfatase (36); estrone sulfotransferase (37); F$_o$F$_1$ ATPase, *or* ATP synthase, *or* H$^+$-transporting two-sector ATPase (38-40); glutamate dehydrogenase (41-45); 3α-hydroxysteroid dehydrogenase (46); 3(*or* 17)β-hydroxysteroid dehydrogenase (46-51); 17β-hydroxysteroid dehydrogenase (50,52); kainate-induced currents (53); Na$^+$/K$^+$-exchanging ATPase (39,54); Na$^+$-transporting two-sector ATPase (55); oxaloacetate decarboxylase, Na$^+$-pumping (56); pyruvate kinase (57); RNA polymerase II (58); steroid *N*-acetylglucosaminyltransferase (59); steroid Δ-isomerase (60); testosterone 17β-dehydrogenase (46); tubulin polymerization, *or* microtubule assembly (61-65). **1.** Lemker, Gruber, Schmid & Muller (2003) *FEBS Lett.* **544**, 206. **2.** Lemker, Ruppert, Stoger, Wimmers & Muller (2001) *Eur. J. Biochem.* **268**, 3744. **3.** Kitson, Kitson & Moore (2001) *Chem. Biol. Interact.* **130-132**, 57. **4.** Julian & Duncan (1977) *Biochem. J.* **161**, 123. **5.** Koivula & Koivusalo (1975) *Biochim. Biophys. Acta* **410**, 1. **6.** Bailey, Fishman, Kusiak, Mulhern & Pentchev (1975) *Meth. Enzymol.* **41**, 471. **7.** Fishman, Pentchev & Bailey (1973) *Biochemistry* **12**, 2490. **8.** Mulhern, Fishman, Kusiak & Bailey (1973) *J. Biol. Chem.* **248**, 4163. **9.** Tognoli (1985) *Eur. J. Biochem.* **146**, 581. **10.** Kettlun, Alvarez, Quintar, *et al.* (1994) *Int. J. Biochem.* **26**, 437. **11.** Kawamura, Westwood, Wakefield, *et al.* (2008)

Biochem. Pharmacol. **75**, 1550. **12.** Tognoli & Marre (1981) *Biochim. Biophys. Acta* **642**, 1. **13.** Dufour, Amory & Goffeau (1988) *Meth. Enzymol.* **157**, 513. **14.** Serrano (1988) *Meth. Enzymol.* **157**, 533. **15.** Uchida, Ohsumi & Anraku (1988) *Meth. Enzymol.* **157**, 544. **16.** Sekler & Pick (1993) *Plant Physiol.* **101**, 1055. **17.** Kotyk & Dvorakova (1992) *Biochim. Biophys. Acta* **1104**, 293. **18.** Sekler, Glaser & Pick (1991) *J. Membr. Biol.* **121**, 51. **19.** Gronberg & Flatmark (1988) *FEBS Lett.* **229**, 40. **20.** Linker & Wilson (1985) *J. Bacteriol.* **163**, 1258. **21.** Tu & Sliwinski (1985) *Arch. Biochem. Biophys.* **241**, 348. **22.** Uchida, Ohsumi & Anraku (1985) *J. Biol. Chem.* **260**, 1090. **23.** Eilam, Lavi & Grossowicz (1984) *Microbios.* **41**, 177. **24.** Vara & Serrano (1982) *J. Biol. Chem.* **257**, 12826. **25.** Serrano (1980) *Eur. J. Biochem.* **105**, 419. **26.** Dobrydneva, Williams, Katzenellenbogen, Ratz & Blackmore (2003) *Thromb. Res.* **110**, 23. **27.** Martinez-Azorin, Teruel, Fernandez-Belda & Gomez-Fernandez (1992) *J. Biol. Chem.* **267**, 11923. **28.** Singer, Gebhart & Hess (1978) *Can. J. Biochem.* **56**, 1028. **29.** Singer (1979) *Arch. Biochem. Biophys.* **196**, 340. **30.** Singer & Bruns (1980) *Can. J. Biochem.* **58**, 660. **31.** Oda, Sato, Kodama & Kaneko (1993) *Biol. Pharm. Bull.* **16**, 708. **32.** Nelson & Carter (1967) *Mol. Pharmacol.* **3**, 341. **33.** Schulz, Link, Chaudhuri & Fittler (1990) *Cancer Res.* **50**, 5008. **34.** Chakraborty, Davis & Dey (1988) *J. Steroid Biochem.* **31**, 231. **35.** Purba, Back & Breckenridge (1986) *J. Steroid Biochem.* **24**, 1091. **36.** Hobkirk, Cardy, Nilsen & Saidi (1982) *J. Steroid Biochem.* **17**, 71. **37.** Rozhin, Huo, Zemlicka & Brooks (1977) *J. Biol. Chem.* **252**, 7214. **38.** Kipp & Ramirez (2001) *Endocrine* **15**, 165. **39.** Zheng & Ramirez (1999) *Eur. J. Pharmacol.* **368**, 95. **40.** McEnery & Pedersen (1986) *J. Biol. Chem.* **261**, 1745. **41.** Tappel & Dillard (1967) *J. Biol. Chem.* **242**, 2463. **42.** Frieden (1963) *The Enzymes*, 2nd ed., **7**, 3. **43.** Smith, Austen, Blumenthal & Nyc (1975) *The Enzymes*, 3rd ed., **11**, 293. **44.** Colon, Plaitakis, Perakis, Berl & Clarke (1986) *J. Neurochem.* **46**, 1811. **45.** Pons, Michel, Descomps & Crastes de Paulet (1978) *Eur. J. Biochem.* **84**, 257. **46.** Hasebe, Hara, Nakayama, *et al.* (1987) *Enzyme* **37**, 109. **47.** Talalay (1962) *Meth. Enzymol.* **5**, 512. **48.** Talalay (1963) *The Enzymes*, 2nd ed., **7**, 177. **49.** Webb (1966) *Enzyme and Metabolic Inhibitors*, vol. 2, p. 449, Academic Press, New York. **50.** Blomquist, Lindemann & Hakanson (1984) *Steroids* **43**, 571. **51.** Talalay & Dobson (1953) *J. Biol. Chem.* **205**, 823. **52.** Jarabek & Sack (1969) *Biochemistry* **8**, 2203. **53.** Ishibashi, Okuya, Shimada & Takahama (2000) *Jpn. J. Pharmacol.* **84**, 225. **54.** Robinson (1970) *Biochem. Pharmacol.* **19**, 1852. **55.** Müller, Aufurth & Rahlfs (2001) *Biochim. Biophys. Acta* **1505**, 108. **56.** Dimroth & Thomer (1992) *FEBS Lett.* **300**, 67. **57.** Kayne (1973) *The Enzymes*, 3rd ed., **8**, 353. **58.** Oda, Sato, Kitajima & Yasukochi (1991) *Chem. Pharm. Bull. (Tokyo)* **39**, 2627. **59.** Collins, Jirku & Layne (1968) *J. Biol. Chem.* **243**, 2928. **60.** Weintraub, Vincent, Baulieu & Alfsen (1977) *Biochemistry* **16**, 5045. **61.** Albertini, Friederich, Holderegger & Wurgler (1988) *Mutat. Res.* **201**, 283. **62.** Chaudoreille, Peyrot, Braguer & Crevat (1987) *Mol. Pharmacol.* **32**, 731. **63.** Sato, Murai, Oda, *et al.* (1987) *J. Biochem.* **101**, 1247. **64.** Sharp & Parry (1985) *Carcinogenesis* **6**, 865. **65.** Chaudoreille, Peyrot, Braguer, Codaccioni & Crevat (1991) *Biochem. Pharmacol.* **41**, 685.

DIF-3

This photosensitive ketone (FW = 272.73 g/mol), also known as 1-(3-chloro-2,6-dihydroxy-4-methoxyphenyl)-1-hexanone and differentiation-inducing factor 3, was first observed in the slime mold *Dictyostelium discoideum* and inhibits the cell cycle by inducing arrest at the G_0/G_1 transition. DIF-3 suppresses cyclin D_1 expression and activates [glycogen synthase] kinase-3β. **1.** Masento, Morris, Taylor, *et al.* (1988) *Biochem. J.* **256**, 23. **2.** Morris, Masento, Taylor, Jermyn & Kay (1988) *Biochem. J.* **249**, 903.

Difloxacin

This oral fluoroquinolone-class antibiotic ($FW_{free-base}$ = 399.40 g/mol; CAS 98106-17-3 (free base) and 91296-86-5 (HCl salt)), also named 6-fluoro-1-(4-fluorophenyl)-7-(4-methylpiperazin-1-yl)- 4-oxoquinoline-3-carboxylate, is a broad-spectrum agent that targets DNA Topoisomerase Type II (gyrase), an enzyme required for bacterial replication and transcription. **Pharmacokinetics:** Difloxacin is rapidly and nearly fully absorbed after oral administration, with ~50% is bound to circulating plasma proteins. It is metabolized primarily by the liver through glucuronidation and secreted in the bile. Mean value of the peak plasma concentration, C_{max} = 1.8 µg/mL; Time to reach peak plasma concentration, t_{max} = 2.8 hours; elimination half-life, $t_{1/2}$ = 9.3 hours; Area under the curve =14.5 µg·hr/mL; Total body clearance rate = 375 mL/kg/hour; Steady-state volume of Distribution = 3.8 L/kg. **Antibiotic Spectrum:** Most intestinal pathogens are sensitive (1). *Acinetobacter, Pseudomonas aeruginosa,* and other *Pseudomonas* species (except *Pseudomonas maltophilia*) are usually sensitive. (Ciprofloxacin is more active against these organisms.) Difloxacin displays antistaphylococcal activity. Against streptococci, including enterococci and pneumococci, the its activity is moderate or poor. *Haemophilus influenzae* and *Branhamella catarrhalis* are very sensitive, as are gonococci and meningococci. Activity against *Chlamydia trachomatis* and mycoplasmas is borderline. Organisms associated with nonspecific vaginal infection are not very sensitive. (For the prototypical member of this antibiotic class, **See** *Ciprofloxacin*) 1. Phillips & King (1988) *Rev. Infect. Dis.* **10** (*Supplement* 1), S70.

Diflunisal

This potent anti-inflammatory analgesic agent ($FW_{free-acid}$ = 250.20 g/mol; CAS 22494-42-4), also known as 5-(2',4'-difluorophenyl)salicylic acid, inhibits prostaglandin biosynthesis. **Target(s):** catechol sulfotransferase (1); cathepsin G (2); Ca^{2+}-transporting ATPase (3); GABA$_A$ receptor (4,5); lyso-platelet activating factor:acetyl-CoA acetyltransferase (6); medium-chain fatty-acyl-CoA synthetase, *or* butyryl-CoA synthetase, *or* butyrate:CoA ligase (7-9); oxidative phosphorylation, as an uncoupler (10); phenol sulfotransferase (1); prostaglandin synthase, *or* cyclooxygenase-2 and -1 (11-14); UDP-glucuronosyltransferase (15,16). 1. Vietri, De Santi, Pietrabissa, Mosca & Pacifici (2000) *Eur. J. Clin. Pharmacol.* **56**, 81. **2.** Lentini, Ternai & Ghosh (1987) *Biochem. Int.* **15**, 1069. **3.** Holguin (1988) *Biochem. Pharmacol.* **37**, 4035. **4.** Woodward, Polenzani & Miledi (1994) *J. Pharmacol. Exp. Ther.* **268**, 806. **5.** Smith, Oxley, Malpas, Pillai & Simpson (2004) *J. Pharmacol. Exp. Ther.* **311**, 601. **6.** White & Faison (1988) *Prostaglandins* **35**, 939. **7.** Kasuya, Yamaoka, Igarashi & Fukui (1998) *Biochem. Pharmacol.* **55**, 1769. **8.** Kasuya, Igarashi & Fukui (1999) *Chem. Biol. Interact.* **118**, 233. **9.** Kasuya, Hiasa, Kawai, Igarashi & Fukui (2001) *Biochem. Pharmacol.* **62**, 363. **10.** McDougall, Markham, Cameron & Sweetman (1983) *Biochem. Pharmacol.* **32**, 2595. **11.** Majerus & Stanford (1977) *Brit. J. Clin. Pharmacol.* **4** Suppl. 1, 15S. **12.** Humes, Winter, Sadowski & Kuehl (1981) *Proc. Natl. Acad. Sci. U.S.A.* **78**, 2053. **13.** Shen (1983) *Pharmacotherapy* **3**, 3S. **14.** Young, Panah, Satchawatcharaphong & Cheung (1996) *Inflamm. Res.* **45**, 246. **15.** Vietri, Pietrabissa, Mosca & Pacifici (2000) *Eur. J. Clin. Pharmacol.* **56**, 659. **16.** Mano, Usui & Kamimura (2005) *Biopharm. Drug Dispos.* **26**, 35.

3-(2,6-Difluorobenzamido)-5-(4-ethoxyphenyl) thiophene-2-carboxylic acid

This synthetic probe (FW = 403.39 g/mol), first identified by high throughput screening, inhibits *Escherichia coli* UT189 bacterial capsule biogenesis, with an IC_{50} value of 4.5 µM and a greater than 10x selectivity index, demonstrating its worthiness as a lead for developing highly effective therapeutic agents against community-acquired infections caused by uropathogenic *E. coli* (UPEC) (1). This probe should promote further studies on K capsule formation. 1. Noah, Ananthan, Evans, *et al.* (2012) *Probe Reports from the NIH Molecular Libraries Program,* National Center for Biotechnology Information (U.S.), PMID: 23762949.

2,2-Difluoro-5-hexyne-1,4-diamine

This mechanism-based inhibitor ($FW_{free-base}$ = 148.16 g/mol), also known as 1,4-diamino-2,2-difluoro-5-hexyne and 3,6-diamino-5,5-difluoro 1-hexyne, inactivates mammalian ornithine decarboxylase, K_i = 10 µM and k_{inact} = 0.29 min^{-1} (1). In rats, it produces a rapid, long-lasting, and dose-dependent decrease of ornithine decarboxylase activity in the ventral prostate, testis, and thymus. In contrast with 5-hexyne-1,4-diamine (2), 2,2-difluoro-5-hexyne-1,4-diamine is not a substrate of mitochondrial monoamine oxidase. 1. Kendrick, Danzin & Kolb (1989) *J. Med. Chem.* **32**, 170. **2.** Danzin, Casara, Claverie & Grove (1983) *Biochem. Pharmacol.* **32**, 941.

[(3S,4S)-1,1-Difluoro-3-(hexadecanoylamino)-4-hydroxy-4-phenylbutyl]phosphonate

This sphingomyelin analogue ($FW_{free-acid}$ = 518.60 g/mol) inhibits both neutral and acidic sphingomyelinase (IC_{50} = 3.3 µM for the bovine neutral enzyme). Magnesium-dependent neutral sphingomyelinase (N-SMase) present in plasma membranes is an enzyme that can be activated by stress in the form of inflammatory cytokines, serum deprivation, and hypoxia. As key component of these signaling pathways, N-SMase inhibition is an important strategy to prevent neuron death from ischemia. 1. Soeda, Tsuji, Ochiai, *et al.* (2004) *Neurochem. Int.* **45**, 619.

1-(3,5-Difluoro-4-hydroxybenzyl)-1H-imidazole-2-thiol

This potent multisubstrate geometric inhibitor (FW = 206.27 g/mol) targets dopamine β-monooxygenase (K_i = 39 nM). By incorporating a phenethylamine substrate mimic and an oxygen mimic into a single molecule, this inhibitor exhibit both the kinetic properties and the potency expected for multisubstrate inhibitors. Steady-state kinetic experiments with such inhibitors supports the proposed pH-dependent changes in substrate binding order (2) and a mechanism whereby the inhibitor binds specifically to the reduced Cu^+ form of enzyme at both the phenethylamine substrate site and the active-site copper atom(s). Yonetani-Theorell double-inhibition analysis indicates mutually exclusive binding of the inhibitor substructures p-cresol and 1-methylimidazole-2-thiol, suggesting an extremely short inter-site distance between the phenethylamine binding site and the active-site copper ion(s) (1). 1. Kruse, DeWolf, Chambers & Googhart (1986) *Biochemistry* **25**, 7271. **2.** Ahn & Klinman (1983) *Biochemistry* **22**, 3106.

cis-(1,1-Difluoro-2-[2-(hypoxanthin-9-ylmethyl)-tetrahydropyranos-3-yl]ethyl-phosphonate

This nucleotide analogue (FW = 378.27 g/mol), also known as Yokomatsu compound, inhibits purine-nucleoside phosphorylase (K_i = 2.7 nM). **1.** Iwanow, Magnowska, Yokomatsu, Shibuya & Bzowska (2003) *Nucleosides Nucleotides Nucleic Acids* **22**, 1567. **2.** Glavas-Obrovac, Suver, Hikishima, Yokomatsu & Bzowska (2007) *Nucleosides Nucleotides Nucleic Acids* **26**, 989.

α-(Difluoromethyl)arginine

This arginine derivative (FW = 224.21 g/mol) irreversibly inactivates arginine decarboxylase. **1.** Bey, Casara, Vevert & Metcalf (1983) *Meth. Enzymol.* **94**, 199. **2.** Bitonti, Casara, McCann & Bey (1987) *Biochem. J.* **242**, 69. **3.** Kallio, McCann & Bey (1981) *Biochemistry* **20**, 3163. **4.** Kierszenbaum, Wirth, McCann & Sjoerdsma (1987) *Proc. Natl. Acad. Sci. U.S.A.* **84**, 4278. **5.** Birecka, Bitonti & McCann (1985) *Plant Physiol.* **79**, 515. **6.** Feirer, Mignon & Litvay (1984) *Science* **223**, 1433. **7.** Hernandez & Schwarcz de Tarlovsky (1999) *Cell. Mol. Biol.* **45**, 383.

α-(Difluoromethyl)ornithine

This anticancer agent and antiparasitic drug (FW = 182.17 g/mol; CAS 70052-12-9; Symbol: DFMO), also known as Eflornithine, acts as a specific and mechanism-based irreversible inhibitor of ornithine decarboxylase, the rate-limiting enzyme in polyamine biosynthesis. The L-enantiomer has the lower K_d value of 1.3 μM *versus* 28.3 μM for the D-isomer; however, the k_{inact} values are roughly identical. It inhibits B16 melanoma-induced angiogenesis *in ovo* and proliferation of vascular endothelial cells *in vitro*. It is also effective against African trypanosomiasis. **Target(s):** lysine decarboxylase (1,2); ornithine decarboxylase (2-15); protein-glutamine γ-glutamyl-transferase, *or* transglutaminase (16). **1.** Takatsuka, Onoda, Sugiyama, *et al.* (1999) *Biosci. Biotechnol. Biochem.* **63**, 1063. **2.** Lee & Cho (2001) *Biochem. J.* **360**, 657. **3.** Steven, Williams, Warne & Tucker (1989) *J. Enzyme Inhib.* **3**, 133. **4.** Tyagi, Tabor & Tabor (1983) *Meth. Enzymol.* **94**, 135. **5.** Bey, Casara, Vevert Metcalf (1983) *Meth. Enzymol.* **94**, 199. **6.** Seely, Pösö & Pegg (1983) *Meth. Enzymol.* **94**, 206. **7.** McCann, Bacchi, Hanson, Nathan, Hutner & Sjoerdsma (1983) *Meth. Enzymol.* **94**, 209. **8.** Fozard & Part (1983) *Meth. Enzymol.* **94**, 213. **9.** Qu, Ignatenko, Yamauchi, *et al.* (2003) *Biochem. J.* **375**, 465. **10.** Birecka, Bitonti & McCann (1985) *Plant Physiol.* **79**, 515. **11.** Qu, Ignatenko, Yamauchi, *et al.* (2003) *Biochem. J.* **375**, 465. **12.** Niemann, von Besser & Walter (1996) *Biochem. J.* **317**, 135. **13.** Schaeffer & Donatelli (1990) *Biochem. J.* **270**, 599. **14.** Poulin, Lu, Ackermann, Bey & Pegg (1992) *J. Biol. Chem.* **267**, 150. **15.** Yarlett, Goldberg, Moharrami & Bacchi (1993) *Biochem. J.* **293**, 487. **16.** Delcros, Roch & Quash (1984) *FEBS Lett.* **171**, 221.

N-[N-(3,5-Difluorophenacetyl)]-L-alanyl]-3-(S)-amino-1-methyl-5-phenyl-1,3-dihydro-benzo[e](1,4)-diazepin-2-one

This peptidyl benzodiazepinone (FW = 490.51 g/mol), also known as γ-Secretase Inhibitor XVIII and Compound E, is a strong noncompetitive inhibitor of γ-secretase (IC$_{50}$ for Aβ$_{total}$ = 300 pM). Because this compound possesses a hydroxyethylene dipeptide isostere of aspartyl protease transition state analogues, potent inhibition undicates that γ-secretase may be an aspartyl protease. **1.** Seiffert, Bradley, Rominger, *et al.* (2000) *J. Biol. Chem.* **275**, 34086. **2.** Tian, Sobotka-Briner, Zysk, *et al.* (2002) *J. Biol. Chem.* **277**, 31499.

N-[N-3,5-Difluorophenacetyl]-L-alanyl-(S)-phenyl-glycine Methyl Ester

This cell-permeable dipeptide, also known as γ-secretase inhibitor XVI and DAPM (FW = 390.39 g/mol), inhibits γ-secretase (IC$_{50}$ Aβ ~ 10 nM in 7PA2 cells). It is also reported to block the formation of amyloid-β dimers and trimers. Such naturally secreted oligomers of amyloid beta protein potently inhibit hippocampal long-term potentiation *in vivo*. Importantly, treatment of cells with γ-secretase inhibitors prevents oligomer formation at doses that allow appreciable monomer production, and such medium no longer disrupts LTP, indicating synaptotoxic Aβ oligomers can be targeted therapeutically. **1.** Walsh, Klyubin, Fadeeva, *et al.* (2002) *Nature* **416**, 535.

N-{(S)-2-[[5-(2,4-Difluorophenyl)-2-[(phosphono-methyl)amino]pent-4-ynoyl]}-L-leucyl-L-alanine

This arylacetylene-containing agent (FW = 491.43 g/mol) is a strong inhibitor of endothelin-converting enzyme 1, *or* ECE-1 (IC$_{50}$ = 8 nM) and a weak inhibitor of neprilysin (IC$_{50}$ = 5.8 μM). The final step in the posttranslational processing of Endothelin-1 (ET-1), the most potent peptidic vasoconstrictor to date, is the conversion of its precursor by ECE-1, a metalloprotease displaying high sequence identity with neutral endopeptidase 24.11 (NEP), especially at the catalytic center. This inhibitor potently blocks ET-1 production *in vivo*, as demonstrated by the big ET-1-induced pressor response in rats. **1.** Wallace, Moliterni, Moskal, *et al.* (1998) *J. Med. Chem.* **41**, 1513.

N-({4-[Difluoro(phosphono)methyl]phenyl}acetyl)-L-aspartyl-4-[difluoro-(phosphono)-methyl]-L-phenylalaninamide

This cell-impermeable phosphonate (FW = 658.41 g/mol) is a strong inhibitor of human protein-tyrosine phosphatase 1B (K_i = 2.4 nM). A cell-permeable analogue is reported in which the terminal amide is linked to –CH$_2$CH$_2$–S–S–CH$_2$CH$_2$–NH–(D-Arg)$_8$–NH$_2$. The disulfide is then reductively cleaved upon entry into the cytoplasm. **1**. Liang, Lee, J. Liang, Lawrence & Zhang (2005) *J. Biol. Chem.* **280**, 24857. **2**. Zhang (2003) *Acc. Chem. Res.* **36**, 385. **3**. Lee, Xie, Luo, *et al.* (2006) *Biochemistry* **45**, 234.

Difluoropine

This tropane-derived dopamine reuptake inhibitor (FW = 401.45 g/mol; CAS 156774-35-5), also known as O-620 and systematically named methyl (1*S*,2*S*,3*S*,5*R*)-3-[bis(4-fluorophenyl)methoxy]-8-methyl-8-aza-bicyclo[3.2.1]octane-2-carboxylate, (IC$_{50}$ = 11 nM) and selective ligand for the dopamine transporter. **1**. Meltzer, Liang & Madras (1994) *J. Med. Chem.* **37**, 2001.

Difluorotoluene β-D-ribose C-glycoside

This nonpolar thymidine isostere (FW = 244.24 g/mol; Symbol: F) is an efficient alternative substrate for many DNA polymerases, replacing thymine and coding specifically for A:T base-pairing (1,2). Its triphosphate is thus a thymidine triphosphate shape analogue lacking Watson–Crick pairing ability but is nonetheless replicated with high sequence selectivity (1-3). The apparent efficiency of insertion of A opposite F is only ~4x lower than that opposite T. In addition, the selectivity for insertion of A rather than C, T, or G is surprisingly similar. With dT in the template, A is preferred by 3.0–4.5 log units versus the other three bases, and with F in the template, A is preferred by 2.9–4.2 log units. These data indicate that the energetic barrier leading to the newly synthesized A–F base pair is within 0.9 kcal of that for the A–T pair. Such observations suggest that conventional hydrogen bonds may not be necessary for the high efficiency and fidelity of DNA synthesis and that shape complementarity can play a more significant role in fidelity than previously appreciated. **1**. Liu, Moran & Kool (1997) *Chem. Biol.* **4**, 919. **2**. Guckian, Krugh &, Kool (1998)

Nature Struct. Biol. **5**, 954. **3**. Kool & Sintim (2006) *Chem. Commun.* (Cambridge) **35**, 3665.

cyclic-di-GMP

This rotationally symmetrical cyclic 5'-GMP diester (FW$_{di-Na}$ = 734.38 g/mol; CAS 61093-23-0; Soluble to 20 mM in water), also known as bis-(3'-5')-cyclic dimeric guanosine monophosphate, is an intracellular and intercellular signaling molecule in prokaryotes (1). In mammals, c-di-GMP is recognized by the direct innate immune sensor known as S̲timulator of I̲nterferon G̲enes (*or* STING), thereby initiating production of type I interferons through the TBK1/IRF3 axis (1,2). Upon infection with intracellular pathogens, STING induces type I interferon production, protecting infected cells and nearby cells from local infection by binding to the same cell that secretes it (*via* autocrine signaling) as well as nearby cells (*via* paracrine signaling.) The helicase DDX41 may play a role in the recognition of c-di-GMP upstream of STING (3). Intranasal c-di-GMP-adjuvanted plant-derived H5 influenza vaccine induces multifunctional Th1 CD4$^+$ cells and strong mucosal and systemic antibody responses in mice. **1**. Jin, *et al.* (2011) *J. Immunol.* **187**, 2595. **2**. Burdette, *et al.* (2011) *Nature* **478**, 515. **3**. Parvatiyar, *et al.* (2012) *Nature Immunol.* 13, 1155. **4**. Madhun, *et al.* (2011) *Vaccine* **29**, 4973.

Digitonin

This toxic steroid saponin (FW = 1229.33 g/mol; CAS 11024-24-1) from the foxglove *Digitalis purpurea* prefentially binds 3β-sterol, commending its use in (a) cholesterol determinations, (b) receptor solubilization, and (c) peripheral membrane permeabilization. For the latter, one typically be by working in the 55-75 μM range. Note that the term digitonin has also been used to refer to a mixture of four steroid saponins of *Digitalis purpurea* of which the above steroid glycoside is the major component, necessitating care when purchasing digitonin. **Target(s):** α-*N*-acetylneuraminate α-2,8-sialyltransferase, activated at lower concentrations (1); [β-adrenergic-receptor] kinase (2,3); carnitine *O*-palmitoyltransferase (4); diacylglycerol:sterol *O*-acyltransferase (5); flavonoid 3',5'-hydroxylase (26); galactolipase (6); 1,3-β glucan synthase (7); glucomannan 4-β-mannosyltransferase, weakly inhibited (8); β-glucosidase (9); glucuronosyltransferase (10); glyceryl-ether monooxygenase (11); 3-hydroxy-3-methylglutaryl-CoA reductase12; lipase, *or* triacylglycerol lipase (13); mRNA (guanine-*N*^{7}-)-methyltransferase (14); NADH dehydrogenase (15; Na$^+$/K$^+$-exchanging ATPase (16,17); 2,3-oxidosqualene-lanosterol cyclase (18); P glycoprotein (19); prenylated-protein carboxyl methyltransferase (20); rhodopsin kinase (21,22); squalene epoxidase (18); sterol *O*-acyltransferase, *or* cholesterol *O*-acyltransferase, *or* ACAT (23-25); xenobiotic transporting ATPase, multidrug-resistance protein (19). **1**. Eppler, Morré & Keenan (1980) *Biochim. Biophys. Acta* **619**, 318. **2**. Benovic (1991) *Meth. Enzymol.* **200**, 351. **3**. Benovic, Mayor, Staniczewski, Lefkowitz & Caron (1987) *J. Biol. Chem.* **262**, 9026. **4**. Murthy & Pande (1987) *Biochem. J.* **248**, 727. **5**. Garcia & Mudd (1981) *Meth. Enzymol.* **71**, 768. **6**. O'Sullivan, Warwick & Dalling (1987) *J. Plant Physiol.* **131**, 393. **7**. Beaulieu, Tang, Yan, *et al.* (1994) *Antimicrob. Agents Chemother.* **38**, 937. **8**. Piro, Zuppa, Dalessandro & Northcote (1993)

Planta **190**, 206.   **9**. Glew, Peters & Christopher (1976) *Biochim. Biophys. Acta* **422**, 179.   **10**. Matern, Matern, Schelzig & Gerok (1980) *FEBS Lett.* **118**, 251.   **11**. Taguchi & Armarego (1998) *Med. Res. Rev.* **18**, 43.   **12**. Kawachi & Rudney (1970) *Biochemistry* **9**, 1700.   **13**. Bhardwaj, Raju & Rajasekharan (2001) *Plant Physiol.* **127**, 1728.   **14**. Ramadevi, Burroughs, Mertens, Jones & Roy (1998) *Proc. Natl. Acad. Sci. U.S.A.* **95**, 13537.   **15**. Nason, Garrett, Nair, Vasington & Detwiler (1964) *Biochem. Biophys. Res. Commun.* **14**, 220.   **16**. Shcheglova, Raikhman, Lopina & Boldyrev (1981) *Biokhimiia* **46**, 62.   **17**. Robinson (1980) *Biochim. Biophys. Acta* **598**, 543.   **18**. Eilenberg, Klinger, Przedecki & Shechter (1989) *J. Lipid Res.* **30**, 1127.   **19**. Garrigues, Escargueil & Orlowski (2002) *Proc. Natl. Acad. Sci. U.S.A.* **99**, 10347.   **20**. Philips & Pillinger (1995) *Meth. Enzymol.* **256**, 49.   **21**. Shichi, Somers & Yamamoto (1983) *Meth. Enzymol.* **99**, 362.   **22**. Shichi & Somers (1978) *J. Biol. Chem.* **253**, 7040.   **23**. Billheimer (1985) *Meth. Enzymol.* **111**, 286.   **24**. Erickson, Shrewsbury, Brooks & Meyer (1980) *J. Lipid Res.* **21**, 930.   **25**. Balasubramaniam, Venkatesan, Mitropoulos & Peters (1978) *Biochem. J.* **174**, 863.   **26**. Menting, Scopes & Stevenson (1994) *Plant Physiol.* **106**, 633.

Digitoxin

This cardiotonic glycoside (FW = 764.95 g/mol; CAS 71-63-6), a secondary glycoside from the leaves of *Digitalis purpurea* and *D. lanata*, inhibits H^+/K^+-exchanging ATPase (1) and the Na^+/K^+-exchanging ATPase (2-9). **1**. Modyanov, Pestov, Adams, *et al.* (2003) *Ann. N. Y. Acad. Sci.* **986**, 183.   **2**. Forbush & Hoffman (1979) *Biochim. Biophys. Acta* **555**, 299.   **3**. Contreras, Flores-Maldonado, Lazaro, *et al.* (2004) *J. Membr. Biol.* **198**, 147.   **4**. Albers, Koval & Siegel (1968) *Mol. Pharmacol.* **4**, 324.   **5**. Sastry & Phillis (1977) *Can. J. Physiol. Pharmacol.* **55**, 170.   **6**. Gubitz, Akera & Brody (1977) *Biochim. Biophys. Acta* **459**, 263.   **7**. Erdmann (1978) *Arzneimittelforschung* **28**, 531.   **8**. Brown & Erdmann (1984) *Arzneimittelforschung* **34**, 1314.   **9**. Balzan, D'Urso, Ghione, Martinelli & Montali (2000) *Life Sci.* **67**, 1921.

Digoxigenin

This less-soluble digoxin aglycon (FW = 390.52 g/mol; CAS 1672-46-4) inhibits the Na^+/K^+-exchanging ATPase. Because anti-digoxigenin antibodies exhibit both high affinity and high specificity, digoxigenin has found wide use as a tag for immunolocalization and ELISA assays. **1**. Balzan, D'Urso, Ghione, Martinelli & Montali (2000) *Life Sci.* **67**, 1921.   **2**. Senn, Lelievre, Braquet & Garay (1988) *Brit. J. Pharmacol.* **93**, 803.   **3**. Ahmed, Rohrer, Fullerton, *et al.* (1983) *J. Biol. Chem.* **258**, 8092.

Digoxin

This cardiotonic glycoside (FW = 780.95 g/mol; CAS 20830-75-5), a secondary glycoside from the leaves of foxglove (*Digitalis purpurea*), inhibits H^+/K^+ exchanging ATPase (1) and Na^+/K^+-exchanging ATPase (2-7).   **Mechanism of Action:** Digoxin inhibits Na^+,K^+-ATPase, increasing intracellular sodium ion concentration and thereby stimulating $Na^+ \rightleftarrows Ca^{2+}$ exchange and increasing the intracellular Ca^{2+} concentration. Digoxin's beneficial pharmacologic effects relate to its direct actions on cardiac muscle and its indirect actions on the cardiovascular system via effects on the autonomic nervous system. The pharmacologic consequences of these direct and indirect effects are: (a) its *positive inotropic effect* – increased force and velocity of myocardial systolic contractions, (b) its *neurohormonal deactivating effect* – decreased activation of the sympathetic nervous system and renin-angiotensin system, and (c) its *vagomimetic effect* – reduced heart rate and an decreased conduction velocity through the AV node. Digoxin's beneficial effects in treating heart failure are mediated by its positive inotropic and neurohormonal deactivating effects, whereas its effects in atrial arrhythmias are linked to its vagomimetic actions. **Action as HIF-1α Protein Inhibitor:** Digoxin was identified as an inhibitor of Hypoxia-Inducible Factor-1α (HIF-1α) protein translation and HIF-2α mRNA expression in a cell-based screen of a library of drugs (at 100 nM in the case of A) in clinical trials (8). Notably, digoxin prolongs tumor latency and inhibits tumor xenograft growth in mice, when treatment is initiated before subcutaneous implantation of P493-Myc, P493-Myc-Luc, PC3, and Hep3B cells. It also arrests tumor growth, when treatment is initiated after the establishment of PC3 and P493-Myc tumor xenografts. Digoxin inhibits the expression of HIF-1α protein and the HIF-1 target genes *VEGF*, *GLUT1*, *HK1*, and *HK2*; and (iv) does not effectively inhibit the growth of xenografts derived from transfected cells with enforced HIF-1α expression. Therapeutic plasma concentrations of digoxin in cardiac patients (10–30 nM) are somewhat lower than the concentration required for maximal inhibition of HIF-1α expression in hypoxic cancer cells after 24 h of drug treatment; however, effective inhibition at lower concentrations might be observed after more prolonged exposure of cells to the drug.   (**See also** *Proscillaridan A*; *Oabain*).   **Key Pharmacokinetic Parameters:** *See* Appendix II in Goodman & Gilman's THE PHARMACOLOGICAL BASIS OF THERAPEUTICS, 12th Edition (Brunton, Chabner & Knollmann, eds.) McGraw-Hill Medical, New York (2011). **1**. Modyanov, Pestov, Adams, *et al.* (2003) *Ann. N. Y. Acad. Sci.* **986**, 183.   **2**. Ruoho & Kyte (1977) *Meth. Enzymol.* **46**, 523.   **3**. Contreras, Flores-Maldonado, Lazaro, *et al.* (2004) *J. Membr. Biol.* **198**, 147.   **4**. Erdmann (1978) *Arzneimittelforschung* **28**, 531.   **5**. Matsui & Schwartz (1968) *Biochim. Biophys. Acta* **151**, 655.   **6**. Krstic, Krinulovic, Spasojevic-Tisma, *et al.* (2004) *J. Enzyme Inhib. Med. Chem.* **19**, 409.   **7**. Balzan, D'Urso, Ghione, Martinelli & Montali (2000) *Life Sci.* **67**, 1921.   **8**. Zhang, Qian, Tan, *et al.* (2008) *Proc. Natl. Acad. Sci. USA* **105**, 19579.

Dihydroasparagusic Acid

Dihydroasparagusic Acid   **Asparagusic Acid**

This sulfur-containing odorant (FW = 152.22 g/mol) is the reduced form of asparagusic acid. It is a radical-scavenging antioxidant that also competitively inhibits tyrosinase, a copper-containing enzyme involved in melanogenesis and food browning (1). When added to asparagus mitochondria, asparagusic and dihydroasparagusic acids strongly stimulate3 pyruvate oxidation, whereas α-lipoic acid was only slightly activating (2). Such results suggest that asparagusic acids may play essential roles in the pyruvate oxidation of asparagus. **1**. Venditti, Mandrone, Serrilli, *et al.* (2013) *J. Agric. Food Chem.* **61**, 6848. **2**. Yanagawa, Kato & Kitahara (1973) *Plant Cell Physiol.* **14**, 1213.

7,8-Dihydrobiopterin

This oxidized folate (FW = 239.23 g/mol; CAS 6779-87-9) inhibits dihydroneopterin aldolase (1); GTP cyclohydrolase I (2-4); tryptophan 5-monooxygenase (5), and tyrosine 3-monooxygenase (6). **1**. Deng, Callender & Dale (2000) *J. Biol. Chem.* **275**, 30139. **2**. Shen, Alam & Zhang (1988) *Biochim. Biophys. Acta* **965**, 9. **3**. Blau & Niederwieser (1986) *Biochim. Biophys. Acta* **880**, 26. **4**. Maita, Hatakeyama, Okada & Hakoshima (2004) *J. Biol. Chem.* **279**, 51534. **5**. McKinney, Teigen, Frøystein, *et al.* (2001) *Biochemistry* **40**, 15591. **6**. Goodwell, Sabatier & Stevens (1998) *Biochemistry* **39**, 13437.

Dihydrofolate

This folic acid derivative (FW$_{\text{free-acid}}$ = 443.42 g/mol; CAS 4033-27-6; Slowly oxidized in air, particularly in alkaline pH; 0.3 M 2-mercaptoethanol protects against decomposition), also known as 7,8-dihydropteroyl-L-glutamate, inhibits folate-dependent reactions. **Target(s):** amidophospho-ribosyltransferase, inhibited by dihydrofolate polyglutamates (1); arylamine *N*-acetyltransferase (2); deoxycytidylate 5-hydroxy-methyltransferase (3); dihydroneopterin aldolase (4); dihydropteroate synthase (5,6); folylpolyglutamate synthetase (7); glutathione synthetase (8); GTP cyclohydrolase I (9; however reference 10 indicates no such inhibition); methenyl-tetrahydrofolate cyclohydrolase (11); N^5,N^{10} methenyl-tetrahydrofolate synthetase, K_i = 50 μM (12,13); methylene-tetrahydrofolate reductase, NADPH (14); phosphoribosyl-aminoimidazolecarboxamide formyltransferase , *or* AICAR transformase, weakly inhibited (15); serine hydroxymethyl-transferase, *or* glycine hydroxymethyltransferase; partial inhibitor (16). **1**. Schoettle, Crisp, Szabados & Christopherson (1997) *Biochemistry* **36**, 6377. **2**. Wang, Vath, Kawamura, *et al.* (2005) *Protein J.* **24**, 65. **3**. Lee, Gautam-Basak, Wooley & Sander (1988) *Biochemistry* **27**, 1367. **4**. Mathis & Brown (1970) *J. Biol. Chem.* **245**, 3015. **5**. Mouillon, Ravanel, Douce & Rebeille (2002) *Biochem. J.* **363**, 313. **6**. Vinnicombe & Derrick (1999) *Biochem. Biophys. Res. Commun.* **258**, 752. **7**. Masurekar & Brown (1980) *Meth. Enzymol.* **66**, 648. **8**. Kato, Chihara, Nishioka, *et al.* (1987) *J. Biochem.* **101**, 207. **9**. Shen, Alam & Zhang (1988) *Biochim. Biophys. Acta* **965**, 9. **10**. De Saizieu, Vankan & van Loon (1995) *Biochem. J.* **306**, 371. **11**. Kirk, Chen, Imeson & Cossins (1995) *Phytochemistry* **39**, 1309. **12**. Jolivet (1997) *Meth. Enzymol.* **281**, 162. **13**. Bertrand, MacKenzie & Jolivet (1987) *Biochim. Biophys. Acta* **911**, 154. **14**. Matthews (1986) *Meth. Enzymol.* **122**, 372. **15**. Allegra, Drake, Jolivet & Chabner (1985) *Proc. Natl. Acad. Sci. U.S.A.* **82**, 4881. **16**. Ramesh & Appaji Rao (1980) *Biochem. J.* **187**, 623.

Dihydrofolate Hexaglutamate, *See* Dihydropteroyl Hexaglutamate

Dihydrofolate Pentaglutamate, *See* Dihydropteroyl Pentaglutamate

Dihydrokainate

This glutamic acid analogue (FW = 215.25 g/mol) inhibits high-affinity [³H]-L-glutamate transport into synaptosomes and blocks the binding of radioligands to the NMDA (N-methyl-D-aspartate), KA (kainate), and QA (quisqualate) glutamate neurotransmitter receptor sites. **1**. Bridges, Stanley, Anderson, Cotman & Chamberlin (1991) *J. Med. Chem.* **34**, 71. 2. Pocock, Murphie &Nicholls (1988) *J. Neurochem.* **50**, 745. **3**. Koch, Kavanaugh, Esslinger, *et al.* (1999) *Mol. Pharmacol.* **56**, 1095.

α-Dihydrolipoate

This dithiol form of the acyl-transfer (FW = 208.35 g/mol; pK_a of 4.76 at 25°C (COOH group); Redox potential = –0.29 V (pH 7)), also called 6,8-thioctic acid (reduced), is a redox coenzyme α-(*R*)-lipoic acid that is found covalently attached to the ε-amino group of lysyl residues of transacetylases (subunits of α-ketoacid dehydrogenase complexes), thereby permitting processive acyl transfer as a sging arm akin to that of carboxybiocytin. *R*-stereoisomer is the naturally occurring form. **Target(s):** prostaglandin-endoperoxide synthase, *or* cyclooxygenase (1); [pyruvate dehydrogenase (acetyl-transferring)] kinase (2); stearoyl-CoA 9-desaturase (3); thiosulfate sulfurtransferase, rhodanese (4,5). **1**. Ohki, Ogino, Yamamoto, *et al.* (1977) *Proc. Natl. Acad. Sci. U.S.A.* **74**, 144. **2**. Korotchkina, Sidhu & Patel (2004) *Free Radic. Res.* **38**, 1083. **3**. Fulco & Bloch (1964) *J. Biol. Chem.* **239**, 993. **4**. Pagani, Bonomi & Cerletti (1983) *Biochim. Biophys. Acta* **742**, 116. **5**. Turkowsky, Blotevogel & Fischer (1991) *FEMS Microbiol. Lett.* **81**, 251.

1,4-Dihydro-3-nitrophenanthrolin-4-one-8-carboxylate

This phenanthrolinone (FW$_{\text{free-acid}}$ = 273.20 g/mol) is an Fe(II/III) chelator that inhibits chick procollagen-proline 4-dioxygenase, *or* prolyl 4-hydroxylase (IC$_{50}$ = 38 nM), thereby commending its use to control excessive collagen deposition in pathological fibrosis by continuously blocking collagen hydroxylation and allowing intracellular proteolysis to reduce excess collagen. **1**. Franklin, Morris, Edwards, Large & Stephenson (2001) *Biochem. J.* **353**, 333.

(S)-Dihydroorotate

This pyrimidine nucleotide biosynthesis intermediate (FW$_{\text{free-acid}}$ = 158.11 g/mol) inhibits barbiturase (1), carbamoyl-phosphate synthetase (2), dihydropyrimidinase, *or* hydantoinase (3), orotate phosphoribosyl-transferase (4), and 5-oxoprolinase (5). **1**. Soong, Ogawa, Sakuradani & Shimizu (2002) *J. Biol. Chem.* **277**, 7051. **2**. Potvin & Gooder (1975) *Biochem. Genet.* **13**, 125. **3**. Lee, Cowling, Sander & Pettigrew (1986) *Arch. Biochem. Biophys.* **248**, 368. **4**. Javaid, el Kouni & Iltzsch (1999) *Biochem. Pharmacol.* **58**, 1457. **5**. Van Der Werf, Griffith & Meister (1975) *J. Biol. Chem.* **250**, 6686.

20α-Dihydroprogesterone

This naturally occurring progestagen (FW = 316.48 g/mol), also known as 20α-hydroxypregn-4-en-3-one, is a functionally important progesterone metabolite, exhibiting 0.2-0.3 the potency of progesterone. **Target(s):** cholesterol acyltransferase, *or* ACAT (1); 17β-hydroxysteroid dihydrogenase (2); progesterone 5α-reductase (3); steroid 17α-monooxygenase, *or* steroid C17/20-lyase, *or* CYP17A1 (4,5). **1**. Chang & Doolittle (1983) *The Enzymes*, 3rd ed., **16**, 523. **2**. Blomquist, Lindemann & Hakanson (1984) *Steroids* **43**, 571. **3**. Kinoshita (1981) *Endocrinol. Jpn.* **28**, 499. **4**. Huang, Kominami & Takemori (1993) *Sci. China B* **36**, 411. **5**. Betz & Tsai (1979) *J. Steroid Biochem.* **10**, 39.

Dihydropteroyl Hexaglutamate

This polyglutamylated dihydrofolate (FW$_{free-acid}$ = 1089.00 g/mol) inhibits methylenetetrahydrofolate reductase (NADPH), K_i = 13 nM (1-3) and thymidylate synthase (4). **1**. Matthews (1986) *Meth. Enzymol.* **122**, 372. **2**. Matthews & Daubner (1982) *Adv. Enzyme Regul.* **20**, 123. **3**. Matthews & Baugh (1980) *Biochemistry* **19**, 2040. **4**. Kisliuk, Gaumont & Baugh (1974) *J. Biol. Chem.* **249**, 4100.

Dihydropteroyl Pentaglutamate

This polyglutamylated 7,8-dihydrofolate analogue (FW$_{free-acid}$ = 959.88 g/mol) targets amidophosphoribosyltransferase (1), N^5,N^{10}-methenyltetrahydrofolate synthetase, *or* 5-formyltetrahydrofolate cyclo-ligase; K_i = 3.8 μM (2,3), and phosphoribosylamino-imidazole carboxamide formyl-

transferase, *or* AICAR transformase (4-6). **1**. Schoettle, Crisp, Szabados & Christopherson (1997) *Biochemistry* **36**, 6377. **2**. Jolivet (1997) *Meth. Enzymol.* **281**, 162. **3**. Bertrand, MacKenzie & Jolivet (1987) *Biochim. Biophys. Acta* **911**, 154. **4**. Baggott, Vaughn & Hudson (1986) *Biochem. J.* **236**, 193. **5**. Allegra, Drake, Jolivet & Chabner (1985) *Proc. Natl. Acad. Sci. U.S.A.* **82**, 4881. **6**. Sugita, Aya, Ueno, Ishizuka & Kawashima (1997) *J. Biochem.* **122**, 309.

Dihydropyridine Ca²⁺ Channel Blockers

These antihypertensive agents, which all contain a dihydropyridine nucleus, reduce systemic vascular resistance and arterial pressure in the treatment of angina and tachycardia (1). While the vasodilatory action of Ca²⁺ entry blockers is due primarily to slow Ca²⁺ channel inhibition, some DHP-class blockers have additional actions contributing to vasodilation. For example, nifedipine and four related blockers selectively inhibit peak I phosphodiesterase activity (IC$_{50}$ = 2-3 μM), but are weak inhibitors of peak II phosphodiesterase (IC$_{50}$ ≥ 100 μM) (2). DHP Calcium Entry Blockers include: Amlodipine (Norvasc®); Aranidipine (Sapresta®); Azelnidipine (Calblock®); Barnidipine (HypoCa®); Benidipine (Coniel®); Cilnidipine (Atelec®, Cinalong®, and Siscard®); Clevidipine (Cleviprex®); Isradipine (DynaCirc® and Prescal®); Efonidipine (Landel®); Felodipine (Plendil®); Lacidipine (Motens® and Lacipil®); Lercanidipine (Zanidip®); Manidipine (Calslot® and Madipine®); Nicardipine (Cardene® and Carden SR®); Nifedipine (Procardia® and Adalat®); Nilvadipine (Nivadil®); Nimodipine (Nimotop®); Nisoldipine (Baymycard®, Sular®, and Syscor®); Nitrendipine (Cardif®, Nitrepin®, and Baylotensin®), Pranidipine (Acalas®). **1**. Takahashi & Ogura (1983) *FEBS Lett.* **152**, 191. **2**. Norman, Ansell & Phillips (1983) *Eur. J. Pharmacol.* **93**, 107.

DL-*threo*-Dihydrosphingosine

The D-enantiomer of this long-chain amino-diol (FW$_{free-base}$ = 301.51 g/mol) targets sphingosine kinase. The L-enantiomer (safingol) potentiates the effect of doxorubicin, inhibits protein kinase C (IC$_{50}$ = 37.5 μM), and inhibits the release of leukotriene B₄. **Target(s):** [myosin light-chain] kinase (1); protein kinase C (2-6); sphinganine kinase, *or* sphingosine kinase (7-16). **1**. Jinsart, Ternai & Polya (1991) *Plant Sci.* **78**, 165. **2**. Hoffmann, Leenen, Hafner, *et al.* (2002) *Anticancer Drugs* **13**, 93. **3**. Jarvis & Grant (1999) *Invest. New Drugs* **17**, 227. **4**. Carfagna, Young & Susick (1996) *Toxicol. Appl. Pharmacol.* **137**, 173. **5**. Schwartz, Jiang, Kelsen & Albino (1993) *J. Natl. Cancer Inst.* **85**, 402. **6**. Sachs, Safa, Harrison & Fine (1995) *J. Biol. Chem.* **270**, 26639. **7**. Tokuda, Kozawa, Harada & Uematsu (1999) *J. Cell Biochem.* **72**, 262. **8**. Lanterman & Saba (1998) *Biochem. J.* **332**, 525. **9**. zu Heringdorf, Lass, Alemany, *et al.* (1998) *EMBO J.* **17**, 2830. **10**. Pitson, D'andrea, Vandeleur, *et al.* (2000) *Biochem. J.* **350**, 429. **11**. Taha, Hannun & Obeid (2006) *J. Biochem. Mol. Biol.* **39**, 113. **12**. Buehrer & Bell (1992) *J. Biol. Chem.* **267**, 3154. **13**. Leiber, Banno & Tanfin (2007) *Amer. J. Physiol.* **292**, C240. **14**. Coursol, Le Stunff, Lynch, *et al.* (2005) *Plant Physiol.* **137**, 724. **15**. Choi, Kim & Kinet (1996) *Nature* **380**, 634. **16**. Kusner, Thompson, Melrose, *et al.* (2007) *J. Biol. Chem.* **282**, 23147.

Dihydrostreptomycin

This semisynthetic antibiotic (FW$_{free-base}$ = 583.60 g/mol) inhibits protein biosynthesis. **Target(s):** alcohol dehydrogenase, yeast (1); cysteine desulfhydrase, *or* cystathionine γ-lyase (2); diamine oxidase (3,4); F₁

ATPase (5); mechano-electric transduction channe (l6); oxidative phosphorylation (5,7); protein biosynthesis (8-10); protein-synthesizing GTPase, elongation factor (10); proton conduction by F_o (5); streptomycin 6-kinase (11); threonine deaminase (12). **1**. Sund & Theorell (1963) *The Enzymes*, 2nd ed., **7**, 25. **2**. Artman, Markenson & Olitzki (1955) *Proc. Soc. Exp. Biol. Med.* **90**, 584. **3**. Zeller (1951) *The Enzymes*, 1st ed., **2** (Part 1), 536. **4**. Zeller, Owen & Karlson (1951) *J. Biol. Chem.* **188**, 623. **5**. Guerrieri, Micelli, Massagli, Gallucci & Papa (1984) *Biochem. Pharmacol.* **33**, 2505. **6**. Kimitsuki & Ohmori (1993) *Brain Res.* **624**, 143. **7**. Bragg & Polglase (1963) *J. Bacteriol.* **86**, 1236. **8**. Pestka (1974) *Meth. Enzymol.* **30**, 261. **9**. Weisblum & Davies (1968) *Bacteriol. Rev.* **32**, 493. **10**. Campuzano & Modolell (1981) *Eur. J. Biochem.* **117**, 27. **11**. Walker & Walker (1967) *Biochim. Biophys. Acta* **148**, 335. **12**. Artman & Markenson (1958) *Enzymologia* **19**, 9.

5,6-Dihydrouridine

This reduced form of uridine (FW = 246.22 g/mol; λ_{max} = 235 nm; ε = 10100 $M^{-1}cm^{-1}$)), also known as 1-β-D-ribofuranosyl-5,6-dihydrouracil and symbolized by D, is a component in the dihydrouridine stem-loop region witin transfe RNA (tRNA). **Target(s):** cytidine deaminase (1-8); UDP-*N*-acetylglucosamine diphosphorylase, *or* *N*-acetylglucosamine-1-phosphate uridylyltransferase (9). **1**. Frick, Yang, Martinez & Wolfenden (1989) *Biochemistry* **28**, 9423. **2**. Wentworth & Wolfenden (1978) *Meth. Enzymol.* **51**, 401. **3**. Evans, Mitchell & Wolfenden (1975) *Biochemistry* **14**, 621. **4**. Vita. Amici, Cacciamani, Lanciotti & Magni (1985) *Biochemistry* **24**, 6020. **5**. Cacciamani, Vita, Cristalli, *et al.* (1991) *Arch. Biochem. Biophys.* **290**, 285. **6**. Vincenzetti, Cambi, Neuhard, *et al.* (1999) *Protein Expr. Purif.* **15**, 8. **7**. Cohen & Wolfenden (1971) *J. Biol. Chem.* **246**, 7561. **8**. Vita, Amici, Lanciotti, Cacciamani & Magni (1986) *Ital. J. Biochem.* **35**, 145A. **9**. Yamamoto, Moriguchi, Kawai & Tochikura (1980) *Biochim. Biophys. Acta* **614**, 367.

1,3-Dihydroxyacetone

This three-carbon metabolite (FW = 90.079 g/mol; Abbreviation: DHA), also known as glycerone and 1,3-dihydroxy-2-propanone, is the simplest ketose. It is frequently supplied as the solid dimer, 2,5 dihydroxydioxane-2,5-dimethanol (FW = 180.16 g/mol; melting point = 75-80°C), which is rapidly converted to dihydroxyacetone in aqueous solutions. **Target(s):** *N*-acetylneuraminate lyase (1); glutamine:fructose-6-phosphate aminotransferase (isomerizing), weakly inhibited (2;) ribose-5-phosphate isomerase, *or* phosphopentoisomerase (3,4). **1**. Aisaka, Igarashi, Yamaguchi & Uwajima (1991) *Biochem. J.* **276**, 541. **2**. Kikuchi, Ikeda & Tsuiki (1972) *Biochim. Biophys. Acta* **289**, 303. **3**. Webb (1966) *Enzyme and Metabolic Inhibitors*, vol. **2**, p. 411, Academic Press, New York. **4**. Agosin & Aravena (1960) *Enzymologia* **22**, 281.

Dihydroxyacetone Phosphate

This triose-phosphate metabolite ($FW_{free-acid}$ = 170.06 g/mol; pK_a values of 1.77 and 6.45; Abbreviated: DHAP), also known as glycerone phosphate, is an intermediate in glycolysis, gluconeogenesis, and photosynthetic CO_2 fixation. DHAP is unstable in alkaline conditions: complete hydrolysis occurs in twenty minutes in 1 M alkali at room temperature while 50% hydrolysis occurs in eight minutes in 1 M HCl at 100°C. Solutions should

be freshly prepared; however, solutions can be frozen for a few days at pH 4.5. **Target(s):** carboxypeptidase A (1); formate *C*-acetyltransferase, *or* pyruvate formate-lyase (2 5); glucose-6-phosphate isomerase (6); glutamine:fructose-6-phosphate aminotransferase (isomerizing), weakly inhibited (7); glycerol-3-phosphate *O*-acyltransferase (8-10); hexokinase (11); inositol-3-phosphate synthase, inositol-1-phosphate synthase, K_i = 700 µM at pH 7.2 (12); α-ketoaldehyde dehydrogenase (NADP⁺) (13); L-lactate dehydrogenase (14); phosphoenolpyruvate carboxykinase (ATP) (15); 6-phosphofructo-2-kinase/fructose-2,6-bisphosphatase (16-20); ribose-5 phosphate isomerase, *or* phosphopento-isomerase (21,22); starch (bacterial glycogen) synthase, weakly inhibited (23). **1**. Adelman & Lacko (1968) *Biochem. Biophys. Res. Commun.* **33**, 596. **2**. Takahashi, Abbe & Yamada (1982) *J. Bacteriol.* **149**, 1034. **3**. Asanuma & Hino (2000) *Appl. Environ. Microbiol.* **66**, 3773. **4**. Asanuma & Hino (2002) *Appl. Environ. Microbiol.* **68**, 3352. **5**. Melchiorsen, Jokumsen, Villadsen, Israelsen & Arnau (2002) *Appl. Microbiol. Biotechnol.* **58**, 338. **6**. Backhausen, Jöstingmeyer & Scheibe (1997) *Plant Sci.* **130**, 121. **7**. Kikuchi, Ikeda & Tsuiki (1972) *Biochim. Biophys. Acta* **289**, 303. **8**. Lewin, Schwerbrock, Lee & Coleman (2004) *J. Biol. Chem.* **279**, 13488. **9**. Kume, Shimizu & Seyama (1987) *J. Biochem.* **101**, 653. **10**. Bramley & Grigor (1982) *Biochem. Int.* **5**, 199. **11**. Magnani, Stocchi, Serafini, *et al.* (1983) *Arch. Biochem. Biophys.* **226**, 377. **12**. Migaud & Frost (1996) *J. Amer. Chem. Soc.* **118**, 495. **13**. Ray & Ray (1982) *J. Biol. Chem.* **257**, 10571. **14**. Gordon & Doelle (1976) *Eur. J. Biochem.* **67**, 543. **15**. Lea, Chen, Leegood & Walker (2001) *Amino Acids* **20**, 225. **16**. Cseke, Balogh, Wong, *et al.* (1984) *Trends Biochem. Sci.* **9**, 533. **17**. Villadsen & Nielsen (2001) *Biochem. J.* **359**, 591. **18**. Markham & Kruger (2002) *Eur. J. Biochem.* **269**, 1267. **19**. Stitt, Cseke & Buchanan (1984) *Eur. J. Biochem.* **143**, 89. **20**. Larondelle, Mertens, Van Schaftingen & Hers (1986) *Eur. J. Biochem.* **161**, 351. **21**. Webb (1966) *Enzyme and Metabolic Inhibitors*, vol. **2**, p. 411, Academic Press, New York. **22**. Agosin & Aravena (1960) *Enzymologia* **22**, 281. **23**. Ghosh & Preiss (1965) *Biochemistry* **4**, 1354.

6',7'-Dihydroxybergamottin

This psoralin (FW = 372.42 g/mol; CAS 145414-76-2), a component of grapefruit juice, inhibits aryl sulfotransferase, *or* phenol sulfotransferase (1); CYP1A2 (2), CYP3A4 (2-6), CYP1B1 (5), CYP2C9 (6), and CYP2D6 (6). Irreversible inhibition of CYP3A4 by 6',7'-dihydroxybergamottin and bergamottin accounts for the low turnover of some HMG-CoA inhibitors. **1**. Nishimuta, Ohtani, Tsujimoto, *et al.* (2007) *Biopharm. Drug Dispos.* **28**, 491. **2**. Tassaneeyakul, Guo, Fukuda, Ohta & Yamazoe (2000) *Arch. Biochem. Biophys.* **378**, 356. **3**. Edwards, Bellevue & Woster (1996) *Drug Metab. Dispos.* **24**, 1287. **4**. Edwards, Fitzsimmons, Schuetz, *et al.* (1999) *Clin. Pharmacol. Ther.* **65**, 237. **5**. Girennavar, Poulose, Jayaprakasha, Bhat & Patil (2006) *Bioorg. Med. Chem.* **14**, 2606. **6**. Girennavar, Jayaprakasha & Patil (2007) *J. Food Sci.* **72**, C417.

(2S,3R)-Dihydroxybutyramide 4-Phosphate

This deoxyxylulose 5-phosphate (DXP) analogue ($FW_{free-acid}$ = 215.10 g/mol) inhibits *Synechocystis* 1-deoxy-D-xylulose-5-phosphate reducto-isomerase, K_i = 90 µM., acting as a relatively weak competitive inhibitorswhen compared to fosmidomycin. **1**. Phaosiri & Proteau (2004) *Bioorg. Med. Chem. Lett.* **14**, 5309.

5-(3α,12α-Dihydroxy-5β-cholanamido)-1,3,4-thiadiazole-2-sulfonamide

This sulfonamide (FW = 544.78 g/mol) inhibits rainbow trout (*Oncorhynchus mykiss*) erythrocyte carbonic anhydrase, IC_{50} = 0.83 μM. Human carbonic anhydrase II is inhibited with IC_{50} = 93 nM. **1**. Bülbül, Hisar, Beydemir, Çiftçi & Küfrevioglu (2003) *J. Enzyme Inhib. Med. Chem.* **18**, 371. **2**. Bülbül, Saraçoglu, Küfrevioglu & Çiftçi (2002) *Bioorg. Med. Chem.* **10**, 2561.

Dihydrodeoxymorphine

This fast-acting opioid (FW = 271.36 g/mol; CAS 427-00-9; IUPAC: 4,5-α-epoxy-17-methylmorphinan-3-ol; also known as Desomorphine, Permonid, and by its street name, "Krokodil") is a easily synthesized morphine derivative possessing 8-10x more sedative and analgesic potency. There is no accepted medical use for desomorphine in the U.S., where it has been classified as a controlled substance since 1936. Krokodil is a crude (and hence dangerous) preparation used recreationally. **1**. "Desomorphine" (October, 2013) Office of Diversion Control Drug & Chemical Evaluation Section, U.S. Drug Enforcement Administration.

7,8-Dihydroxyflavone

This redox-active flavone (FW = 254.24 g/mol; CAS 38183-03-8; M.P. = 243-246°C), inhibits DT diaphorase, aromatase, and NADH dehydrogenase. **Target(s):** amine sulfotransferase, weakly inhibited (1); aromatase, *or* CYP19 (2,3); NADH dehydrogenase, *or* Complex I (4); NAD(P)H dehydrogenase (quinone), *or* DT diaphorase (5,6); phenol sulfotransferase (1); ribosyldihydronicotinamide dehydrogenase (quinone), weakly inhibited (7). **1**. Harris, Wood, Bottomley, *et al.* (2004) *J. Clin. Endocrinol. Metab.* **89**, 1779. **2**. Kao, Zhou, Sherman, Laughton & Chen (1998) *Environ. Health Perspect.* **106**, 85. **3**. Chen, Kao & Laughton (1997) *J. Steroid Biochem. Mol. Biol.* **61**, 107. **4**. Hodnick, Duval & Pardini (1994) *Biochem. Pharmacol.* **47**, 573. **5**. Lee, Westphal, de Haan, Aarts, Rietjens & van Berkel (2005) *Free Radic. Biol. Med.* **39**, 257. **6**. Chen, Wu, Zhang, *et al.* (1999) *Mol. Pharmacol.* **56**, 272. **7**. Wu, Knox, Sun, *et al.* (1997) *Arch. Biochem. Biophys.* **347**, 221.

Dihydroxyfumarate

This fumarate derivative (FW_{free-acid} = 148.07 g/mol; CAS 133-38-0; M.P. = 156°C) slowly autoxidizes at pH 6 and this oxidation is accelerated in the presence of metal ions such as Mn^{2+} and Cu^{2+}. Hydrogen peroxide, superoxide, and hydroxyl radicals are generated during the oxidation of dihydroxyfumarate by peroxidase (1,2). Dihydroxyfumarate should be stored in a desiccator in the cold (3). **Target(s):** glycerol dehydrogenase (4); lactoylglutathione lyase (glyoxalase I), however, the structure provided in the reference is for dihydroxymaleic acid (5); L-tartrate dehydrogenase/D-malate dehydrogenase (decarboxylating) (6). **1**. Pellmar & Lepinski (1992) *Brain Res.* **569**, 189. **2**. Halliwell (1977) *Biochem. J.* **163**, 441. **3**. Hartree (1953) *Biochem. Prep.* **3**, 56. **4**. Johnson, Levine & Lin (1985) *J. Bacteriol.* **164**, 479. **5**. Douglas & Nadvi (1979) *FEBS Lett.* **106**, 393. **6**. Giffhorn & Kuhn (1983) *J. Bacteriol.* **155**, 281.

3,4-Dihydroxyhydrocinnamate

This acid (FW_{free-acid} = 182.18 g/mol), also known as 3-(3,4-dihydroxyphenyl)propionic acid and dihydrocaffeic acid, inhibits: α-amylase, weakly inhibited (1); aromatic-amino-acid decarboxylase, *or* dopa decarboxylase (2); γ-butyrobetaine dioxygenase, *or* γ-butyrobetaine hydroxylase, K_i = 20 μM (3); 3,4 dihydroxyphenylacetate 2,3-dioxygenase (4); hyoscyamine (6*S*)-dioxygenase (5); phenylalanine ammonia-lyase (6,7); phenylalanine 4-monooxygenase, IC_{50} = 0.24 mM (8); procollagen-proline 4-dioxygenase (9,10); protocatechuate 3,4-dioxygenase (11); protocatechuate 4,5-dioxygenase (12). **1**. Funke & Melzig (2005) *Pharmazie* **60**, 796. **2**. Webb (1966) *Enzyme and Metabolic Inhibitors*, vol. **2**, p. 311, Academic Press, New York. **3**. Ng, Hanauske-Abel & Englard (1991) *J. Biol. Chem.* **266**, 1526. **4**. Que, Widom & Crawford (1981) *J. Biol. Chem.* **256**, 10941. **5**. Hashimoto & Yamada (1987) *Eur. J. Biochem.* **164**, 277. **6**. Dahiya (1993) *Indian J. Exp. Biol.* **31**, 874. **7**. Kalghatgi & Subba Rao (1975) *Biochem. J.* **149**, 65. **8**. Koizumi, Matsushima, Nagatsu, *et al.* (1984) *Biochim. Biophys. Acta* **789**, 111. **9**. Majamaa, Günzler, Hanauske-Abel, Myllylä & Kivirikko (1986) *J. Biol. Chem.* **261**, 7819. **10**. Kaska, Myllylä, Günzler, Gibor & Kivirikko (1988) *Biochem. J.* **256**, 257. **11**. Que, Lipscomb, Münck & Wood (1977) *Biochim. Biophys. Acta* **485**, 60. **12**. Ono, Nozaki & Hayaishi (1970) *Biochim. Biophys. Acta* **220**, 224.

3,4-DL-Dihydroxymandelate

This metabolite (FW_{free-acid} = 184.15 g/mol) is an intermediate in the formation of 4-hydroxy-1-methoxymandelate from norepinephrine. **Target(s):** γ-butyrobetaine dioxygenase, *or* γ-butyrobetaine hydroxylase, K_i = 28 μM (1); L-4-hydroxymandelate oxidase, inhibited by DL-mixture: K_i = 0.18 mM (2); hyoscyamine (6*S*) dioxygenase (3); imidazoleacetate 4-monooxygenase (4); phenol sulfotransferase, *or* aryl sulfotransferase (5); procollagen-proline 4-dioxygenase (6); protocatechuate 4,5-dioxygenase (7); tyrosine aminotransferase (8). **1**. Ng, Hanauske-Abel & Englard (1991) *J. Biol. Chem.* **266**, 1526. **2**. Bhat & Vaidyanathan (1976) *Eur. J. Biochem.* **68**, 323. **3**. Hashimoto & Yamada (1987) *Eur. J. Biochem.* **164**, 277. **4**. Watanabe, Kambe, Imamura, *et al.* (1983) *Anal. Biochem.* **130**, 321. **5**. Pennings, Vrielink, Wolters & van Kempen (1976) *J. Neurochem.* **27**, 915. **6**. Majamaa, Günzler, Hanauske-Abel, Myllylä & Kivirikko (1986) *J. Biol. Chem.* **261**, 7819. **7**. Ono, Nozaki & Hayaishi (1970) *Biochim. Biophys. Acta* **220**, 224. **8**. Jacoby & La Du (1964) *J. Biol. Chem.* **239**, 419.

2,5-Di(hydroxymethyl)-3*S*,4*R*-dihydroxypyrrolidine

This pyrrolidine alkaloid (FW_{free-base} = 163.17 g/mol), first isolated from the plant *Lonchocarpus sericeus*, inhibits glycoprotein processing. **Target(s):** α-glucosidase (1,2); β-glucosidase (1,3); glycoprotein processing (1,4,5); mannosyl oligosaccharide glucosidase, *or* glucosidase I (2,4-6); sucrose synthase (7); thioglucosidase, *or* myrosinase, *or* sinigrinase (8); trehalose

(1). **1.** Elbein (1987) *Meth. Enzymol.* **138**, 661. **2.** Kaushal & Elbein (1994) *Meth. Enzymol.* **230**, 316. **3.** Chinchetru, Cabezas & Calvo (1989) *Int. J. Biochem.* **21**, 469. **4.** Kaushal & Elbein (1989) *Meth. Enzymol.* **179**, 452. **5.** Elbein, Mitchell, Sanford, Fellows & Evans (1984) *J. Biol. Chem.* **259**, 12409. **6.** Szumilo, Kaushal & Elbein (1986) *Arch. Biochem. Biophys.* **247**, 261. **7.** Römer, Schrader, Günther, *et al.* (2004) *J. Biotechnol.* **107**, 135. **8.** Scofield, Rossiter, Witham, *et al.* (1990) *Phytochemistry* **29**, 107.

Dehydroxymethylepoxyquinomicin

This synthetic NF-κB nuclear translocation inhibitor and anti-inflammatory agent (FW = 261.23 g/mol; CAS 287194-40-5; *Abbreviations*: DHMEQ and (–)-DHMEQ) inhibits tumor necrosis factor-α (TNF-α)- and 12-*O*-tetradecanoylphorbol-13-acetate-induced transcriptional activity of NF-κB in human T cell leukemia Jurkat cells (1). It also inhibits TNF-α-induced DNA binding of nuclear NF-κB, but not phosphorylation and degradation of I-κB (1). DHMEQ likewise inhibits the TNF-α-induced nuclear accumulation of p65, a component of NF-κB (1). In terms of its ability to promote tumor cell apoptosis and cell-cycle arrest, DHMEQ inhibited the steady-state transcriptional activity of NF-κB in all hepatoma cells (2). DHMEQ blocked the constitutive DNA-binding activity and TNF-α-mediated nuclear translocation of NF-κB in Huh-7 cells. DHMEQ (5-20 μg/mL) dose-dependently reduced the viable cell number of all hepatoma cells. DHMEQ (20 μg/mL) induced apoptosis in all hepatoma cells, especially in Hep3B cells, and cell-cycle arrest in Huh-7 and HepG2 cells. These effects were accompanied by downregulation of proteins involved in anti-apoptosis (*e.g.*, Bcl-xL, XIAP or c-IAP2) and cell-cycle progression (*e.g.*, cyclin D₁), and induction of proteins involved in pro-apoptosis (*e.g.*, Bax) and cell-cycle retardation (*e.g.*, p21Wafl/Cip1). **Chemical Considerations:** DHMEQ's target protein is p65 (and other Rel homology proteins), making a covalent bond with a thiol of a p65 cysteinyl residue by an unusual *N*→*O* acyl shift (3,4). Although the hydroxyl group at the 2-position of the benzamide moiety is essential for the inhibitory activity, etherification of this group did not diminish the activity completely, suggesting it would be an allowable linker or biotin site (5). **Target(s):** enhances antitumor activity of taxanes in anaplastic thyroid cancer cells (6); suppresses growth and Type-I collagen accumulation in keloid fibroblasts (7); inhibition of active HIV-1 replication (8). **1.** Ariga, Namekawa, Matsumoto, Inoue & Umezawa (2002) *J. Biol. Chem.* **277**, 24625. **2.** Nishimura, Ishikawa & Matsumoto, *et al.* (2006) *Int. J. Oncol.* **29**, 713. **3.** Yamamoto, Horie, Takeiri, Kozawa & Umezawa (2008) *J. Med. Chem.* **51**, 5780. **4.** Kozawa, Kato, Teruya, Suenaga & Umezawa (2009) *Bioorg. Med. Chem. Lett.* **19**, 5380. **5.** Chaicharoenpong, Kato & Umezawa (2002) *Bioorg. Med. Chem.* **10**, 3933. **6.** Meng, Mitsutake, Nakashima, *et al.* (2008) *Endocrinology* **149**, 5357. **7.** Makino, Mitsutake, Nakashima, *et al.* (2008) *J. Dermatol. Sci.* **51**, 171. **8.** Miyake, Ishida, Yamagishi, *et al.* (2010) *Microbes Infect.* **12**, 400.

3,4-Dihydroxyphenylacetate

This catechol (FW_free-acid = 168.15 g/mol), also known as homoprotocatechuic acid and often abbreviated dopac, is a dopamine metabolite. **Target(s):** aldehyde oxidase, weakly inhibited (1); aromatic-amino-acid decarboxylase, *or* dopa decarboxylase (2); γ-butyrobetaine dioxygenase, *or* γ-butyrobetaine hydroxylase, K_i = 13 μM (3); dimethylaniline monooxygenase (4); glutamate decarboxylase (5); homogentisate 1,2 dioxygenase (6); 4-hydroxyphenylacetate 1-monooxygenase, K_i = 43 μM (4); 4-hydroxyphenylpyruvate dioxygenase (7); hyoscyamine (6*S*)-dioxygenase (8); imidazoleacetate 4 monooxygenase (9); indoleamine *N*-methyltransferase (10); phenol sulfotransferase, *or* aryl sulfotransferase (11); procollagen-proline 4-dioxygenase (12,13); protocatechuate 3,4-dioxygenase (14-16); protocatechuate 4,5-dioxygenase (17); tyrosine aminotransferase (18,19). **1.** Panoutsopoulos & Beedham

(2004) *Acta Biochim. Pol.* **51**, 649. **2.** Webb (1966) *Enzyme and Metabolic Inhibitors*, vol. **2**, p. 312, Academic Press, New York. **3.** Ng, Hanauske-Abel & Englard (1991) *J. Biol. Chem.* **266**, 1526. **4.** Hareland, Crawford, Chapman & Dagley (1975) *J. Bacteriol.* **121**, 272. **5.** Blindermann, Maitre, Ossola & Mandel (1978) *Eur. J. Biochem.* **86**, 143. **6.** Fernández-Cañón & Peñalva (1997) *Anal. Biochem.* **245**, 218. **7.** Lindstedt & Rundgren (1982) *J. Biol. Chem.* **257**, 11922. **8.** Hashimoto & Yamada (1987) *Eur. J. Biochem.* **164**, 277. **9.** Watanabe, Kambe, Imamura, *et al.* (1983) *Anal. Biochem.* **130**, 321. **10.** Porta (1987) *Meth. Enzymol.* **142**, 668. **11.** Pennings, Vrielink, Wolters & van Kempen (1976) *J. Neurochem.* **27**, 915. **12.** Majamaa, Günzler, Hanauske-Abel, Myllylä & Kivirikko (1986) *J. Biol. Chem.* **261**, 7819. **13.** Kaska, Myllylä, Günzler, Gibor & Kivirikko (1988) *Biochem. J.* **256**, 257. **14.** Durham, Sterling, Ornston & Perry (1980) *Biochemistry* **19**, 149. **15.** Fujisawa & Hayaishi (1968) *J. Biol. Chem.* **243**, 2673. **16.** Hou, Lillard & Schwartz (1976) *Biochemistry* **15**, 582. **17.** Ono, Nozaki & Hayaishi (1970) *Biochim. Biophys. Acta* **220**, 224. **18.** Granner & Tomkins (1970) *Meth. Enzymol.* **17A**, 633. **19.** Jacoby & La Du (1964) *J. Biol. Chem.* **239**, 419.

2,6-Dihydroxytetrahydropyran-2,6-dicarboxylate

This tetrahydrodipicolinate dicarboxylic acid (FW_free-acid = 206.15 g/mol) is a competitive inhibitor of 2,3,4,5-tetrahydropyridine-2,6-dicarboxylate *N*-succinyltransferase. The monohydroxy derivative has a K_{is} value of 58 nM. Based on the observed inhibition kinetics, the authors suggest a stereochemical model for the succinylation of tetrahydrodipicolinate dicarboxylic acid, *or* THDPA: a, the succinylase binds THDPA (in the L-configuration); b, imine hydration follows to yield 2-hydroxypiperidine-2,6-dicarboxylic acid in which the two carboxyl groups are trans; c, succinylation then occurs, attended by ring opening to give the acyclic product. 2-Hydroxytetrahydropyran-2,6-dicarboxylic acid is viewed as a transition state analogue that resembles the hydrated intermediate. **1.** Berges, DeWolf, Dunn, *et al.* (1986) *J. Biol. Chem.* **261**, 6160.

3,3'-Diindolylmethane

This secondary metabolite (FW = 248.31 g/mol; CAS 1968-05-4; *Symbol*: DIM), also known as 3-(1*H*-Indol-3-ylmethyl)-1*H*-indole and 3,3'-methylene-bis-1*H*-indole, is formed during the digestion of indole-3-carbinol, a metabolite in cruciferous vegetables (*Brassica* genus), including broccoli, Brussels sprouts, and cabbage. DIM suppresses lipopolysaccharide induced expression of inducible nitric oxide synthase and cyclooxygenase-2 in BV-2 microglia (1). DIM also attenuates the DNA-binding activity of the inflammatory transcription factor NF-κB, Such results suggest that DIM may have beneficial potential against brain inflammation and neurodegenerative diseases through the negative regulation of the NF-κB signal pathway in microglia (1). DIM also exerts a potent cytostatic effect in cultured human Ishikawa endometrial cancer cells by inducing the level of TGF-α transcripts by approximately 4x within 24 h of treatment (2). **Target(s):** trout cytochrome P450 (CYP) 1A-dependent ethoxyresorufin *O*-deethylase, potent noncompetitive inhibitor with low-micromolar K_i (3); down-regulation of pro-survival pathway in hormone independent prostate cancer (4); pure androgen receptor antagonist in humans (5). **1.** Kim, Kim, Kim, *et al.* (2013) *Toxicol Sci.* **137**, 158. **2.** Leong, Firestone & Bjeldanes (2001) *Carcinogenesis* **22**, 1809. **3.** Stresser, Bjeldanes, Bailey & Williams (1995) *J. Biochem. Toxicol.* **10**, 191. **4.** Garikapaty, Ashok, Tadi, Mittelman & Tiwari (2006) *Biochem. Biophys. Res. Commun.* **340**, 718. **5.** Le, Schaldach, Firestone & Bjeldanes (2003) *J. Biol. Chem.* **278**, 21136.

3,5-Diiodo-L-thyronine

This metabolite (FW = 525.08 g/mol; *Abbreviation*: T₂), an intermediate in the biosynthesis of the thyroid hormones, stimulates oxygen consumption and mitochondrial respiration. **Target(s):** aspartate aminotransferase, mildly inhibited (1); cAMP-dependent protein kinase (2); GABA$_A$ receptors (3); protein-disulfide isomerase (4); thyroxine 5-deiodinase (5); thyroxine 5' deiodinase (5,7). **1.** Smejkal & Smejkalová (1967) *Enzymologia* **33**, 320. **2.** Friedman, Lang & Burke (1978) *Endocr. Res. Commun.* **5**, 109. **3.** Martin, Padron, Newman, *et al.* (2004) *Brain Res.* **1004**, 98. **4.** Guthapfel, Gueguen & Quemeneur (1996) *Eur. J. Biochem.* **242**, 315. **5.** Galton & Hiebert (1987) *Endocrinology* **121**, 42. **6.** Wynn & Gibbs (1963) *J. Biol. Chem.* **238**, 3490. **7.** Koehrle, Auf'mkolk, Rokos, Hesch & Cody (1986) *J. Biol. Chem.* **261**, 11613.

3,5-Diiodo-L-tyrosine

This halogenated tyrosine (FW = 432.98 g/mol), also known as iodogorgoic acid and often abbreviated by DIT, is a product of enzyme-catalyzed post-translational modification of thyroglobulin in the pathway for formation of tri- and tetra-iodothyronines (1-4). It has also been reported in proteins isolated from corals, sponges, and other marine organisms, perhaps as a way to detoxify iodide present in oceans and waterways. DIT has pK_a values of 2.12, 6.48, and 7.82 at 25°C (another source lists 2.12, 5.32, and 9.48). The L-stereoisomer forms needles, when recrystallized from water or 70% ethanol. (Racemic DIT recrystallizes as double wedges from 70% ethanol and rectangular plates from water). The L-enantiomer also decomposes at 213°C, while 3,5-iodo-DL-tyrosine has a decomposition temperature of about 200°C. The specific optical rotation is +2.89° at 20°C of the D-line for 0.246 g of the L-isomer in 5 g of 4% HCl (+2.27° for 0.227 g in 5 g of 25% NH3). The λ_{max} is 310 nm (ε = 5920 M^{-1}cm^{-1}) in 40 mM KOH and 285 nm (ε = 2730 M^{-1}cm^{-1}) in 40 mM HCl. The solubility of the L-isomer in water is 0.617 g/L at 25°C (1.862 g/L at 50°C) whereas, for the racemic mixture the values are 0.340 g/L at 25°C and 0.773 g/L at 50°C. The L isomer forms a gelatinous precipitate upon prolonged boiling in dilute ethanol. Such a precipitate is not observed with the racemic mixture. Note: This amino acid is unstable in light (with loss of iodine) and should always be stored in the dark. Deiodination also occurs upon protein hydrolysis. 3,5-Diodotyrosine acts as an alternative substrate for a number of enzymes. These include thyroid peroxidase, lactoperoxidase, thyroid-hormone aminotransferase, and NADPH:ferredoxin reductase. **Target(s):** 3'5'-cyclicAMP phosphodiesterase (5); iodide transport (6); protein-disulfide isomerase (7); tyrosine 3-monooxygenase (8); tyrosine transport (9). **1.** Drechsel (1896) *Z. Biol.* **33**, 96. **2.** Wheeler & Jamieson (1905) *Amer. Chem. J.* **33**, 365. **3.** Harington & Randall (1929) *J. Soc. Chem. Ind.* **48**, 296. **4.** Foster (1929) *J. Biol. Chem.* **83**, 345. **5.** Law & Henkin (1984) *Res. Commun. Chem. Pathol. Pharmacol.* **43**, 449. **6.** Nasu & Sugawara (1994) *Eur. J. Endocrinol.* **130**, 601. **7.** Guthapfel, Gueguen & Quemeneur (1996) *Eur. J. Biochem.* **242**, 315. **8.** Moore & Dominic (1971) *Fed. Proc.* **30**, 859. **9.** Jara, Martinez-Liarte & Solano (1991) *Melanoma Res.* **1**, 15.

Dilantin, *See 5,5-Diphenylhydantoin*

Diltiazem

This slow calcium channel blocker (FW = 414.52 g/mol; CAS 42399-41-7), also known by the code name CRD-401, trade names Cardizem® and Tiazac®, and IUPAC Name *cis*-(+)-[2-(2-dimethylaminoethyl)-5-(4-methoxyphenyl)-3-oxo-6-thia-2-azabicyclo[5.4.0]undeca-7,9,11-trien-4-yl]ethanoate, is a non-dihydropyridine (*or* non-DHP) vasodilator that is indicated for the treatment of hypertension, angina pectoris, and some arrhythmias (1,2). Diltiazem relaxes the smooth muscles in the walls of arteries, dilating them and thereby promoting blood flow and lowering blood pressure. It also acts on the heart itself, reducing the rate, strength, and conduction speed of each contraction. In these respects, diltiazem exerts negative inotropic, chronotropic, as well as dromotropic effects. **Mechanism of Action:** When infused continuously into the canine renal artery, CRD-401 increased the urine flow and sodium excretion as well as renal blood flow in all the conditions of fluid loading tested (1). The glomerular filtration rate was enhanced under saline loading but not under water diuresis (1). At 1 μg/mL, diltiazem lowers the level of action potential plateau and shortened the duration in both ventricular and Purkinje fibers without change in maximum rate of rise (V$_{max}$) or resting potential (3). Contractile tension of ventricular muscle is also markedly decreased with shortening of plateau. Diltiazem causes a dose-dependent inhibiton of contractions as well as Ca^{2+} influx stimulated by α-adrenoceptor activation and high-K$^+$ depolarization (4). Diltiazem is roughly equipotent in inhibiting contractions induced by high-K$^+$ and a low concentration (10 nM) of norepinephrine (NE). The contractions induced by high NE concentrations (10^{-6}-10^{-5} M) Are more resistant to diltiazem inhibition (4). There is also a close relationship between diltiazem inhibition of Ca^{2+} influx and inhibition of contraction when either 40 mM K$^+$ or 10 nM NE was applied, but not when 1 μM NE was used. Diltiazem produces a noncompetitive inhibition of Ca^{2+}-induced contractions of depolarized rabbit aorta. There is also a lack of parallelism between the smooth muscle effects of removal of [Ca^{2+}]$_{ex}$ and of addition of diltiazem. Diltiazem appears to inhibit stimulated Ca^{2+} influx by interacting with the Ca^{2+} pathway involved in excitation rather than competing with Ca^{2+} for the entry (4). **Drug Interactions:** Diltiazem is slowly metabolized by the CYP3A4 (acting as an alternative substrate inhibitor), which can cause it to interact with a variety of other medications. By inhibiting hepatic cytochromes CYP3A4, CYP2C9 and CYP2D6, diltiazem has 800 known drug interactions. **1.** Yamaguchi, Ikezawa, Takada & Kiyomoto (1974) *Japan. J. Pharmacol.* **24**, 511. **2.** Himori, Ono & Taira (1975) *Japan. J. Pharmacol.* **25**, 350. **3.** Saikawa, Nagamoto & Arita (1977) *Jpn. Heart J.* **18**, 235. **4.** van Breemen, Hwang & Meisheri (1981) *J. Pharmacol. Exp. Ther.* **218**, 459.

Diisopropyl Fluorophosphate

This toxic substance (FW = 184.15 g/mol; CAS 55-91-4; M.P. = –82°C; B.P. = 183°C), also known as diisopropylphosphofluoridate and abbreviated DFP or DIFP, is a potent irreversible inactivator of many serine proteinases (*e.g.*, trypsin, chymotrypsin, and blood-clooting proteases) and serine esterases (most especially acetylcholinesterase). The action of DFP on the cholinesterases was arguably the first example of an inhibitor forming a covalent bond within an enzyme active site. *Note*: Aqueous solutions have a half-life of only about one hour at pH 7.5 and 25°C. **CAUTION:** All forms of DFP are dangerous, particularly the vapor. Inhibition of AChE increases the amount of acetylcholine (ACh) at central and peripheral sites of the nervous system, attended by excessive stimulation of muscarinic and nicotinic receptors. Signs of poisoning include hypersecretion, respiratory distress, tremor, seizures, convulsions, coma, and death. All work with pure DFP solutions must be carried out in an efficient chemical fume hood, using thick, chemically-inert rubber gloves (*never* latex surgical gloves), preferably while using a certified gas mask. A physician or nurse should be alerted of any plan to work with reagent-grade DFP, thus assuring access to antidotes (atropine sulfate and 2-pyridinealdoxime methiodide (*or* 2-PAM)) in pre-loaded autoinjectors, preferably unwrapped and available for immediate use. That said, neurotoxicity from inhalation is remarkably attenuated by using commercial DFP that is already dissolved in dry, water-miscible solvents, most often 2-propanol. The latter (typically made at 0.1-0.5 M DFP in dry 2-propanol) is stable for several months, when kept at –70°C. **1.** No. TE5075000, *Registry of Toxic Effects of Chemical Substances*, U.S. Government Printing Office, Washington, D.C. **2.** Saunders & Stacey (1948) *J. Chem. Soc.*, 695. **3.** Gould & Liener (1965) *Biochemistry* **4**, 90. **4.** Bouma, Miles, Beretta & Griffin *Biochemistry* **19**, 1151.

4,4'-Diisothiocyanatostilbene-2,2'-disulfonate

This photosensitive, amine/thiol-reactive crosslinker and transport inhibitor (FW$_{disodium-salt}$ = 498.49 g/mol; CAS 207233-90-7; *Abbreviation*: DIDS) reduces anion permeability by targeting volume-sensitive Cl$^-$ channels and the Cl$^-$/HCO$_3^-$ exchanger, often eliminating or reducing apoptotic hallmarks (*e.g.,* apoptotic volume decrease (AVD), caspase-3 activity and DNA fragmentation). As an isothiocyanate, DIDS also reacts with primary amines at pH ≥ 9-10, and, as such, is also employed in protein cross-linking experiments. This negatively charged amino/sulfhydryl-reactive reagent has been widely employed as a tool to study anion transport (1-4) and to identify possible coupling between the flux of anions and the transport of cations such as H$^+$ (5-9) or Ca^{2+} (10-13). DIDS also interacts with other enzymes not involved in anion translocation (14-16). Inner mitochondrial membrane permeabilization by Ca^{2+} ions in the presence of DIDS is mediated by the latter's reaction with membrane protein thiols (17). **Target(s):** *N*-acetylneuraminate 4-*O*-acetyltransferase (18,19); *N*-acetylneuraminate 7-*O*(or 9-*O*)-acetyltransferase (20) anion transport (21,22); apyrase (23); ATP synthase, *or* F$_o$F$_1$ ATPase (24); ATP transporter (25); cadmium-transporting ATPase (26); Ca^{2+}-transporting ATPase (27); channel-conductance-controlling ATPase, *or* cystic-fibrosis membrane conductance-regulating protein (28); Cl$^-$/HCO$_3^-$ exchanger (29); flavonol 7-*O*-β-glucosyltransferase (30); 5-formyltetrahydrofolate transport (31); glucosaminylgalactosylglucosylceramide β-galactosyltransferase (32); glucose-6-phosphatase (33,34); glucose-6-phosphate translocase (35,36); glutathione-transporting ATPase (26); H$^+$-exporting ATPase (37); H$^+$/K$^+$-exchanging ATPase (38); inositol 1,3,4-trisphosphate 5/6-kinase (39); Na$^+$/K$^+$-exchanging ATPase (38); nucleotide diphosphatase, ectonucleotide diphosphohydrolase (40-42); oligopeptide-transporting ATPase, oligopeptide permease (43); phenol sulfotransferase (44); protein-disulfide isomerase (45); quinate *O*-hydroxycinnamoyltransferase (46); shikimate *O*-hydroxycinnamoyltransferase (46); sphingosine *N*-acyltransferase (46); succinate dehydrogenase (24); tyrosyl sulfotransferase (44). **1.** Cabantchik, Knauf & Rothstein (1978) *Biochim. Biophys. Acta* **515**, 239. **2.** Zaki, Fasold, Schuhmann & Passow (1975) *J. Cell. Physiol.* **86**, 471. **3.** Leptke, Fasold, Pring & Passow (1976) *J. Membr. Biol.* **29**, 147. **4.** Kasai & Taguchi (1981) *Biochim. Biophys. Acta* **643**, 213. **5.** Lin (1981) *Plant Physiol.* **68**, 435. **6.** Bennett & Spanswick (1983) *J. Membr. Biol.* **71**, 95. **7.** Churchill & Sze (1983) *Plant Physiol.* **71**, 610. **8.** Xiao-Song, Stone & Racker (1983) *J. Biol. Chem.* **258**, 14834. **9.** Stone, Xiao-Song & Racker (1984) *J. Biol. Chem.* **259**, 2701. **10.** Campbell & McLennan (1980) *Ann. N. Y. Acad. Sci.* **358**, 328. **11.** Waisman, Gimble, Goodman & Rasmussen (1981) *J. Biol. Chem.* **256**, 415. **12.** Waisman, Smallwood, Lafreniere & Rasmussen H. (1982) *FEBS Lett.* **145**, 337. **13.** Romero & Ortiz (1988) *J. Membr. Biol.* **101**, 237. **14.** Pedemonte & Kaplan (1988) *Biochemistry* **27**, 7966. **15.** Pedemonte & Kaplan (1990) *Am. J. Physiol.* **258**, C1-C23. **16.** Guilherme, Meyer-Fernandes & Vieyra (1991) *Biochemistry* **30**, 5700. **17.** Bernardes, Meyer-Fernandes, Basseres, *et al.* (1994) *Biochim. Biophys. Acta* **1188**, 93. **18.** Tiralongo, Schmid, Thun, Iwersen & Schauer (2001) *Glycoconjugate J.* **17**, 849. **19.** Iwersen, Dora, Kohla, Gasa & Schauer (2003) *Biol. Chem.* **384**, 1035. **20.** Shen, Tiralongo, Kohla & Schauer (2004) *Biol. Chem.* **385**, 145. **21.** Horie, Yano & Watanabe (1993) *Res. Commun. Chem. Pathol. Pharmacol.* **79**, 117. **22.** Beavis & Davatol-Hag (1996) *J. Bioenerg. Biomembr.* **28**, 207. **23.** Alves-Ferreira, Dutra, Lopes, *et al.* (2003) *Curr. Microbiol.* **47**, 265. **24.** Bernardes, Meyer-Fernandes, Martins & Vercesi (1997) *Z. Naturforsch. [C]* **52**, 799. **25.** Mayinger & Meyer (1993) *EMBO J.* **12**, 659. **26.** Rebbeor, Connolly, Dumont & Ballatori (1998) *J. Biol. Chem.* **273**, 33449. **27.** Carafoli & Zurini (1982) *Biochim. Biophys. Acta* **683**, 279. **28.** van Kuijck, van Aubel, Busch, *et al.* (1996) *Proc. Natl. Acad. Sci. U.S.A.* **93**, 5401. **29.** Isayenkova, Wray, Nimtz, Strack & Vogt (2006) *Phytochemistry* **67**, 1598. **30.** Cai & Horne (2003) *Arch. Biochem. Biophys.* **410**, 161. **31.** Basu, Dastgheib, Ghosh, *et al.* (1998) *Acta Biochim. Pol.* **45**, 451. **32.** Speth & Schulze (1991) *Biochim. Biophys. Acta* **1068**, 217. **33.** Van Schaftingen & Gerin (2002) *Biochem. J.* **362**, 513. **34.** Zoccoli, Hoopes & Karnovsky (1982) *J. Biol. Chem.* **257**, 11296. **35.** Schneider & Chin (1988) *Meth. Enzymol.* **157**, 591.

36. Vega, Cabero & Mardh (1988) *Acta Physiol. Scand.* **134**, 543. **37.** Hughes, Kirk & Michell (1993) *Biochem. Soc. Trans.* **21**, 365S. **38.** dos Passos Lemos, de Sa Pinheiro, de Berredo-Pinho, *et al.* (2002) *Parasitol. Res.* **88**, 905. **39.** Barros, De Menezes, Pinheiro, *et al.* (2000) *Arch. Biochem. Biophys.* **375**, 304. **40.** Grobben, Claes, Roymans, *et al.* (2000) *Brit. J. Pharmacol.* **130**, 139. **41.** Hopfe & Henrich (2004) *J. Bacteriol.* **186**, 1021. **42.** Vargas, Tuong & Schwartz (1986) *J. Enzyme Inhib.* **1**, 105. **43.** Lambert & Freedman (1984) *Biochem. Soc. Trans.* **12**, 1042. **44.** Lotfy (1995) *Plant Physiol. Biochem.* **33**, 423. **45.** Mandon, Ehses, Rother, van Echten & Sandhoff (1992) *J. Biol. Chem.* **267**, 11144. **46.** Matsuda, Tonomura, Baba & Iwata (1991) *Int. J. Biochem.* **23**, 1111.

Diketene

This amine-reactive reagent (FW = 84.07 g/mol; CAS 674-82-8; M.P. = 34°C), also named 4-methylene-2-oxetanone, is frequently used to acetoacetylate proteins. Full catalytic activity is often recovered upon treatment of the modified protein with an equal volume of 2 Molar neutral hydroxylamine. (The latter is freshly prepared by combining equal volumes of 4 Molar NaOH and 4 Molar hydroxylamine hydrochloride). In the chemical laboratory, diketene reacts with alcohols and amines to form the corresponding acetoacetic acid derivatives, in a process likewise called acetoacetylation. (*See also Ketene*) Diketene has been reported to react with disulfide groups in DNA photo-lyase (1). **Target(s):** lysozyme (2,3); microbial collagenase (4); neutrophil collagenase (5); phosphopanto-thenoylcysteine decarboxylase (6); ribonuclease (2,7). **1.** Kim, Denson & Werbin (1977) *Photochem. Photobiol.* **26**, 417. **2.** Marzotto, Pajetta, Galzigna & Scoffone (1968) *Biochim. Biophys. Acta* **154**, 450. **3.** Imoto, Johnson, North, Phillips & Rupley (1972) *The Enzymes*, 3rd ed., **7**, 665. **4.** Bond, Steinbrink & Van Wart (1981) *Biochem. Biophys. Res. Commun.* **102**, 243. **5.** Mookhtiar, Wang & Van Wart (1986) *Arch. Biochem. Biophys.* **246**, 645. **6.** Scandurra, Consalvi, Politi & Gallina (1988) *FEBS Lett.* **231**, 192. **7.** Richards & Wyckoff (1971) *The Enzymes*, 3rd ed., **4**, 647.

4,6-Diketoheptanoate

This γ,ε-diketo acid (FW$_{free-acid}$ = 158.15 g/mol), also known as 4,6-dioxoheptanoic acid and succinyl acetone, inhibits heme biosynthesis by targeting porphobilinogen synthase, *or* δ-aminolevulinate dehydratase. 4,6-Diketoheptanoate is elevated in the urine of individuals with hereditary tyrosinemia. **1.** Ebert, Hess, Frykholm & Tschudy (1979) *Biochem. Biophys. Res. Commun.* **88**, 1382. **2.** Tschudy, Hess & Frykholm (1981) *J. Biol. Chem.* **256**, 9915. **3.** Cheung, Spencer, Timko & Shoolingin-Jordan (1997) *Biochemistry* **36**, 1148. **4.** Dhanasekaran, Chandra, Chandrasekhar Sagar, Rangarajan & Padmanaban (2004) *J. Biol. Chem.* **279**, 6934.

Dimenhydrinate

This motion-sickness drug (FW = 469.97 g/mol; CAS 523-87-5), also known as Dramamine, was originally compounded as a binary salt of the antihistamine diphenhydramine and the cAMP phosphodiesterse inhibitor 8-chlorotheophylline in attempts to balance the drowsiness induced by the former with the stimulation evoked by the latter (1). It is an H$_1$ histamine receptor antagonist, but also interacts, either directly or indirectly, with other neurotransmitter systems, including those using acetylcholine, serotonin, norepinephrine, dopamine, opioids, or adenosine. Sulfotransferase (2,3). **1.** Halpert, Olmstead & Beninger (2002) *Neurosci. Biobehav. Rev.* **26**, 61. **2.** Matsui, Takahashi, Miwa, Motoyoshi & Homma

(1995) *Biochem. Pharmacol.* **49**, 739.   **3**. Bamforth, Dalgliesh & Coughtrie (1992) *Eur. J. Pharmacol.* **228**, 15.

2,3-Dimercapto-1-propanesulfonate

This British anti-Lewisite analogue ($FW_{free-acid}$ = 188.29 g/mol) is a metal ion chelator often used an antidote for poisoning by heavy metal, particularly lead, mercury, cadmium, gold, silver, and arsenic. It has also been used to treat patients with Wilson's disease, a genetic disorder in which the body tends to retain copper. **Target(s):** D-Ala-D-Ala dipeptidase (1); organic anion transporter (2,3); peptide deformylase (4); porphobilinogen synthase, *or* δ-aminolevulinate dehydratase (5,6).   **1**. Wu & Walsh (1996) *J. Amer. Chem. Soc.* **118**, 1785.   **2**. Islinger, Gekle & Wright (2001) *J. Pharmacol. Exp. Ther.* **299**, 741.   **3**. Lungkaphin, Chatsudthipong, Evans, *et al.* (2004) *Amer. J. Physiol. Renal. Physiol.* **286**, F68.   **4**. Rajagopalan, Datta & Pei (1997) *Biochemistry* **36**, 13910.   **5**. Nogueira, Santos, Soares & Rocha (2004) *Environ. Res.* **94**, 254.   **6**. Nogueira, Soares, Nascimento, Muller & Rocha (2003) *Toxicology* **184**, 85.

2,3-Dimercapto-1-propanol

This powerful disulfide reducing agent (FW = 124.23 g/mol) is known famously as British anti-Lewisite (BAL), an antidote for Lewisite (*i.e.*, dichloro(2-chlorovinyl)arsine, *or* Cl–CH=CHAsCl₂), the latter a lethal agent developed by the U.S. but never used in chemical warfare. BAL is also a chelating agent and has been used in trapping redox-active heavy metal poisons, especially arsenic. Beginning in 1951, it has been used in the treatment of Wilson's disease. BAL is slowly oxidized to its disulfide in air (especially at alkaline pH) and must be kept in an air-tight container. **Target(s):** acetylcholinesterase, weakly inhibited (2,3); adenylylsulfatase (4); D-Ala-D-Ala dipeptidase (5); alcohol dehydrogenase, yeast (6); aldehyde oxidase (7); altronate dehydratase (8); aminopeptidase I (9); *o*-aminophenol oxidase (10,61); anthranilate 3-monooxygenase (11); ascorbate oxidase, *or* ascorbase (12); carbonic anhydrase (13); catechol oxidase (13-15); Ca²⁺ transport (16); cytochrome *c* oxidase (17); 2,3-dihydroxyindole 2,3-dioxygenase (60); electron transport, chloroplast (18); electron transport, mitochondrial (19-22); endothelin-converting enzyme (23); ethanolamine kinase (24); glutamate dehydrogenase (25); glutamate uptake (26); glycerol oxidase (27); glycerol-3-phosphate dehydrogenase (28); indoleacetaldoxime dehydratase (29,30); indole-3-acetic acid oxidase (31); laccase (62); lactaldehyde reductase (NADPH) (32); lysine carboxypeptidase, *or* lysine(arginine) carboxypeptidase, *or* carboxypeptidase N (33); methane monooxygenase (34,59); microbial collagenase (35,36); monophenol monooxygenase (15); neprilysin, weakly inhibited (37); nitrogenase (38); nucleotide diphosphatase (39,40); peptide deformylase (41); peptidyl-dipeptidase A, *or* angiotensin I-converting enzyme (42); peroxidase (13,31); phenylalanine 4-monooxygenase (43,44); phosphatidylethanolamine methyltransferase (45); 3'-phospho-adenylylsulfatase (46); phosphodiesterase I, *or* 5'-exonuclease (47); porphobilinogen synthase, *or* δ-aminolevulinate dehydratase (48,49); prostaglandin-endoperoxide synthase, *or* cyclooxygenase (50); *o*-pyrocatechuate decarboxylase (51); pyroglutamyl-peptidase II (52); L-ribulose-5-phosphate 4-epimerase (53); D-serine dehydratase, *or* D-serine ammonia-lyase (54); starch phosphorylase (55); thiopurine *S*-methyltransferase (56); tyrosinase, *or* monophenol monooxygenase (15,57); ubiquinol:cytochrome *c* reductase, *or* Complex III (58).   **1**. Peters, Stocken & Thompson (1945) *Nature* **156**, 616.   **2**. Augustinsson (1950) *The Enzymes*, 1st ed., **1** (Part 1), 443.   **3**. Barron, Miller & Meyer (1947) *Biochem. J.* **41**, 78.   **4**. Li & Schiff (1991) *Biochem. J.* **274**, 355.   **5**. Wu & Walsh (1996) *J. Amer. Chem. Soc.* **118**, 1785.   **6**. Sund & Theorell (1963) *The Enzymes*, 2nd ed.,**7**, 25   **7**. Hurwitz (1955) *J. Biol. Chem.* **212**, 757.   **8**. Ashwell (1962) *Meth. Enzymol.* **5**, 190.   **9**. Röhm (1985) *Eur. J. Biochem.* **146**, 633.   **10**. Subba Rao & Vaidyanathan (1970) *Meth. Enzymol.* **17A**, 554.   **11**. Nair & Vaidyanathan (1965) *Biochim. Biophys. Acta* **110**, 521.   **12**. Kostir, Jindra & Hrabetova (1955) *Cesk. Farm.* **4**, 17.   **13**. Dixon & Needham (1946) *Nature* **158**, 432.   **14**. Dawson & Magee (1955) *Meth. Enzymol.* **2**, 817.   **15**. Sharma & Ali (1980) *Phytochemistry* **19**, 1597.   **16**. Quinhones, Souza & Rocha (2001) *Neurochem. Res.* **26**, 251.   **17**.

Cooperstein (1963) *J. Biol. Chem.* **238**, 3606.   **18**. Shahak, Hind & Padan (1987) *Eur. J. Biochem.* **164**, 453.   **19**. Ksenzenko, Konstantinov, Khomutov, Tikhonov & Ruuge (1983) *FEBS Lett.* **155**, 19.   **20**. Kirschbaum & Wainio (1965) *J. Biol. Chem.* **240**, 462.   **21**. Slater (1949) *Biochem.* **45**, 14.   **22**. Slater (1950) *Biochem. J.* **46**, 484.   **23**. Ashizawa, Okumura, Kobayashi, *et al.* (1994) *Biol. Pharm. Bull.* **17**, 212.   **24**. Sung & R. M. Johnstone (1967) *Biochem. J.* **105**, 497.   **25**. Frieden (1963) *The Enzymes*, 2nd ed., **7**, 3.   **26**. Nogueira, Rotta, Tavares, Souza & Rocha (2001) *Neuroreport* **12**, 511.   **27**. Uwajima & Terada (1982) *Meth. Enzymol.* **89**, 243.   **28**. Baranowski (1963) *The Enzymes*, 2nd ed., **7**, 85.   **29**. Shulka & Mahadevan (1968) *Arch. Biochem. Biophys.* **125**, 873.   **30**. Shulka & Mahadevan (1970) *Arch. Biochem. Biophys.* **137**, 166.   **31**. Gibson & Liu (1978) *Arch. Biochem. Biophys.* **186**, 317.   **32**. Robinson (1966) *Meth. Enzymol.* **9**, 332.   **33**. Schweisfurth (1984) *Dtsch. Med. Wochenschr.* **109**, 1254.   **34**. Tonge, Harrison & Higgins (1977) *Biochem. J.* **161**, 333.   **35**. Seifter & Harper (1970) *Meth. Enzymol.* **19**, 613.   **36**. Seifter & Harper (1971) *The Enzymes*, 3rd ed., **3**, 649.   **37**. Fulcher & Kenny (1983) *Biochem. J.* **211**, 743.   **38**. Burns & Hardy (1972) *Meth. Enzymol.* **24**, 480.   **39**. Yano, Funakoshi & Yamashina (1985) *J. Biochem.* **98**, 1097.   **40**. Byrd, Fearney & Kim (1985) *J. Biol. Chem.* **260**, 7474.   **41**. Rajagopalan, Datta & Pei (1997) *Biochemistry* **36**, 13910.   **42**. Takada, Hiwada & Kokubu (1981) *J. Biochem.* **90**, 1309.   **43**. Kaufman (1962) *Meth. Enzymol.* **5**, 809.   **44**. Kaufman (1970) *Meth. Enzymol.* **17A**, 603.   **45**. Bremer (1969) *Meth. Enzymol.* **14**, 125.   **46**. Roy (1971) *The Enzymes*, 3rd ed., **5**, 1.   **47**. Yano, Funakoshi & Yamashina (1985) *J. Biochem.* **98**, 1097.   **48**. Emanuelli, Rocha, Pereira, Nascimento, Souza & Beber (1998) *Pharmacol. Toxicol.* **83**, 95.   **49**. Nogueira, Santos, Soares & Rocha (2004) *Environ. Res.* **94**, 254.   **50**. Ohki, Ogino, Yamamoto, *et al.* (1977) *Proc. Natl. Acad. Sci. U.S.A.* **74**, 144.   **51**. Subba Rao, Moore & Towers (1967) *Arch. Biochem. Biophys.* **122**, 466.   **52**. Bauer & Nowak (1979) *Eur. J. Biochem.* **99**, 239.   **53**. Deupree & Wood (1972) *J. Biol. Chem.* **247**, 3093.   **54**. Grillo, Fossa & Coghe (1965) *Enzymologia* **28**, 377.   **55**. Whelan (1955) *Meth. Enzymol.* **1**, 192.   **56**. Remy (1963) *J. Biol. Chem.* **238**, 1078.   **57**. Lerner, Fitzpatrick, Calkins & Summerson (1950) *J. Biol. Chem.* **187**, 793.   **58**. von Jagow & Link (1986) *Meth. Enzymol.* **126**, 253.   **59**. Colby & Dalton (1976) *Biochem. J.* **157**, 495.   **60**. Fujioka & Wada (1968) *Biochim. Biophys. Acta* **158**, 70.   **61**. Subba Rao & Vaidyanathan (1967) *Arch. Biochem. Biophys.* **118**, 388.   **62**. Walker (1968) *Phytochemistry* **7**, 1231.

3-[(1,4-Dimethoxynaphth-2-yl)methoxy]-4-[4-(2-methoxybenzyloxypropoxy)-phenyl]piperidine

This substituted piperidine ($FW_{free-base}$ = 571.71 g/mol) is a potent inhibitor of human renin (IC_{50} = 60 pM for the recombinant enzyme and 12 nM for human plasma renin), displaying potent and long-lasting blood pressure-lowering effects after oral administration to sodium-depleted conscious marmosets. **1**. Vieira, Binggeli, Breu, *et al.* (1999) *Bioorg. Med. Chem. Lett.* **9**, 1403.

3-[(3,4-Dimethoxyphenyl)sulfonyl]-1-(3,4-dimethylphenyl)imidazolidine-2,4-dione

This phenylimidazolidine-dione (FW = 404.44 g/mol), also known as SD 906, inhibits human chymase (IC_{50} = 22 nM). **1**. Ferry, Gillet, Bruneau, *et al.* (2001) *Eur. J. Biochem.* **268**, 5885.   **2**. Niwata, Fukami, Sumida, *et al.* (1997) *J. Med. Chem.* **40**, 2156.

Dimethylallyl Pyrophosphate

This isoprenoid metabolite (FW$_{free-acid}$ = 246.09 g/mol), also known as dimethylallyl diphosphate and 3-methylbut-2-enyl 1-diphosphate, is an intermediate in sterol and terpene biosynthesis. It is reasonably stable in neutral and alkaline solutions, but is unstable below pH 5.2. **Target(s):** geranyl-diphosphate cyclase, *or* bornyl-diphosphate synthase (1); geranyl*trans*transferase, *or* farnesyl-diphosphate synthase (2); limonene synthase (3); mevalonate kinase (4); pinene synthase (1,3). **1**. Wheeler & Croteau (1987) *J. Biol. Chem.* **262**, 8213. **2**. Glickman & Schmid (2007) *Assay Drug Dev. Technol.* **5**, 205. **3**. Cori & Rojas (1985) *Meth. Enzymol.* **110**, 406. **4**. Hinson, Chambliss, Toth, Tanaka & Gibson (1997) *J. Lipid Res.* **38**, 2216.

(Z)-3-[4-(Dimethylamino)benzylidenyl]indolin-2-one

This orange-red indolinone (FW = 264.33 g/mol; *Abbreviation*: DMBI) is a strong inhibitor of platelet-drived growth factor β-receptor (*or* β-PDGFR) protein-tyrosine kinase, IC$_{50}$ = 4 μM (1) and fibroblast growth factor receptor-1 (*or* FGFR1) protein-tyrosine kinase (IC$_{50}$ = 5 μM (1,2). **1**. Sun, Tran, Tang, *et al.* (1998) *J. Med. Chem.* **41**, 2588. **2**. Zaman, Vink, van den Doelen, Veeneman & Theunissen (1999) *Biochem. Pharmacol.* **57**, 57.

2-(Dimethylamino)ethyl Diphosphate

This transition-state analogue (FW = 250.10 g/mol), also known as 2-(dimethylamino)ethyl pyrophosphate, inhibits isopentenyl-diphosphate Δ-isomerase. The positively charged ammonium group is believed to mimic a tertiary carbocationic species formed during catalysis. The analogue forms a noncovalent complex with the enzyme with a K_d value less than 0.12 nM. **1**. Reardon & Abeles (1986) *Biochemistry* **25**, 5609. **2**. Muehlbacher & Poulter (1988) *Biochemistry* **27**, 7315. **3**. Muehlbacher & Poulter (1985) *J. Amer. Chem. Soc.* **107**, 8307. **4**. Reardon & Abeles (1985) *J. Amer. Chem. Soc.* **107**, 4078. **5**. Ramos-Valdivia, van der Heijden, Verpoorte & Camara (1997) *Eur. J. Biochem.* **249**, 161. **6**. Wouters, Oudjama, Ghosh, *et al.* (2003) *J. Amer. Chem. Soc.* **125**, 3198.

2-{[(2-{[2-(Dimethylamino)ethyl]sulfanyl}phenyl)-sulfonyl] amino}-5,6,7,8-tetrahydro-1-naphthalenecarboxylate

This substituted anthranilic acid sulfonamide (FW$_{free-acid}$ = 434.58 g/mol) inhibits methionyl aminopeptidase (IC$_{50}$ = 30 nM). **1**. Sheppard, Wang, Kawai, *et al.* (2006) *J. Med. Chem.* **49**, 3832.

(R)-2-(4-(4-Dimethylaminophenoxy)phenylsulfonyl)-1,2,3,4-tetrahydroisoquinolin-3-carboxylic Acid and Hydroxamate

These isoquinoline derivatives (FW$_{free-acid}$ = 467.55 g/mol; FW$_{hydroxamate}$ = 455.54 g/mol) inhibit human neutrophil collagenase, *or* Matrix Metalloproteinase-8, with respective IC$_{50}$ values of 30nM and 2 nM for the carboxylic acid and hydroxamate forms. **1**. Matter, Schwab, Barbier, *et al.* (1999) *J. Med. Chem.* **42**, 1908.

5-(4-Dimethylaminophenyl)-6-(6-morpholin-4-ylpyridin-3-ylethynyl)pyrimidin-4-ylamine

This pyrimidine (FW = 400.49 g/mol) inhibits cytosolic adenosine kinase (IC$_{50}$ = 2 nM). Unlike the substrate, which binds to the closed conformation of the enzyme, this compound binds tightly to a unique open conformation of AK. **1**. Matulenko, Paight, Frey, *et al.* (2007) *Bioorg. Med. Chem.* **15**, 1586.

(E)-3-[2-(4-(Dimethylamino)phenyl)vinyl]benzoate

This (*N,N*-dimethylamino)stilbene carboxylate (FW$_{free-acid}$ = 267.33 g/mol) is a noncompetitive inhibitor (K_{is} = 130 μM) and weak alternative substrate for mammalian flavin-containing monooxygenase. Note that the estimated K_m value is above the maximum solubility (~ 0.1 mM). The stilbene also inhibits bacterial cyclohexanone monooxygenase (K_{is} = 200 μM). In addition, the *o*-carboxy analogue (*i.e.*, (*E*)-2-[2-(4-(dimethylamino)-phenyl)vinyl]benzoic acid, is a slightly weaker inhibitor of flavin-containing monooxygenase (K_{is} = 150 μM) and cyclohexanone monooxygenase (K_{is} = 630 μM). **1**. Clement, Weide & Ziegler (1996) *Chem. Res. Toxicol.* **9**, 599.

N-(3-Dimethylaminopropyl)-N'-ethylcarbodiimide

This water-soluble carbodiimide (FW$_{hydrochloride}$ = 191.70 g/mol), also known as 1-ethyl-3-(3 dimethylaminopropyl)-carbodiimide and often abbreviated EDAC or EDC, is a hydrophilic analogue of *N,N'*-dicyclohexyl-carbodiimide that reacts with carboxyl groups, typically in the pH range 4.0-6.0 (Note: Amine- or carboxyl-containing buffers must obviously be avoided.). If an amine is present, an amide will be produced. Enzymes having catalytically essential carboxyl groups are often inhibited by this reagent when the condensation reaction is carried out in the presence of an amine such as glycine methyl ester). EDAC can also react with phosphate groups. This reagent can also be used to cross-link two proteins: *e.g.*, heavy meromyosin can be linked to smooth muscle myosin light-chain kinase. **Target(s):** acid phosphatase (1); *N*-acyl-D-amino-acid deacylase (2);

acylaminoacyl peptidase (3,4); adrenodoxin reductase (5); β-agarase6; aminopeptidase I (7); α-amylase/pullulanase (8); arginase (9); arginine esterase E-1 (10); argininosuccinate lyase (11); ascorbate 2-kinase (12); *Aspergillus* nuclease S1 (13); ATP:citrate lyase, *or* ATP:citrate synthase (14); bacterial leucyl aminopeptidase (15); cAMP-dependent protein kinase (16); catechol oxidase (78); ceramidase (17); chitinase (18); cholinesterase, *or* butyrylcholinesterase (19); cyclomaltodextrinase (20,21); electron transport, photosynthetic (22); endo 1,3(4)-β-glucanase (23); exo-(1→4)-α-D-glucan lyase (24,25); ferredoxin:NADP$^+$ reductase (26,27); F$_o$F$_1$ ATP synthase (28,29); β-D-fucosidase (30); α-galactosidase, in the presence of glycine ethyl ester (31); β-galactosidase (32); glucan 1,6-α-isomaltosidase (33); α-glucosidase (34); β-glucosidase (35); glutathione *S*-transferase (36); glycogen synthase I (37); (*S*)-2-haloacid dehalogenase (38); 2-haloacid dehalogenase, configuration-inverting (38,39); histidine decarboxylase (40); H$^+$/organic cation antiport (41); 3-hydroxyanthranilate 3,4-dioxygenase (76); 3-hydroxy-3-methylglutaryl-CoA synthase (42); laccase (77); lactose synthase, in the presence of glycinamide (43); lactoylglutathione lyase, *or* glyoxalase I (44,45); leader peptidase (46); mannosyl-oligosaccharide glucosidase, *or* glucosidase I (47); α-methylacyl-CoA epimerase (48); *N*-methyl-D-aspartate receptor (49); midgut microvillar aminopeptidase, *Tenebrio molitor* (50); NADPH:ferri-hemoprotein reductase (51); Na$^+$/K$^+$-exchanging ATPase (52); ornithine decarboxylase (53); palmitoyl-CoA hydrolase, *or* acyl-CoA hydrolase (54); papain (55); phosphate transport (56); phospholipase C, in the presence of *N*-(2,4-dinitrophenyl)ethylene diamines (57); polyamine transporter (58); polygalacturonidase, *or* pectinase (59); polyphenol oxidase (78); prepilin peptidase (60); protein-*S*-isoprenylcysteine *O*-methyltransferase (61); pullulanase (62); pyroglutamyl peptidase I (63); pyrophosphatase, inorganic diphosphatase (64); *Rhizopus stolonifer* nuclease, *or* nuclease Rsn (65); signal peptidase I (66); sulfite oxidase (79); thrombin (67,68); thymidylate synthase (69); transketolase (70); ubiquinone:cytochrome *c* oxidoreductase, *or* Complex III (71); vacuolar H$^+$ transporting ATPase (72,73); vacuolar H$^+$-transporting pyrophosphatase (74); xylan 1,4-β xylosidase (75). **1.** Nosaka (1990) *Biochim. Biophys. Acta* **1037**, 147. **2.** Wakayama, Yada, Kanda, *et al.* (2000) *Biosci. Biotechnol. Biochem.* **64**, 1. **3.** Kobayashi & Smith (1987) *J. Biol. Chem.* **262**, 11435. **4.** Scaloni, Jones, Barra, *et al.* (1992) *J. Biol. Chem.* **267**, 3811. **5.** Geren, O'Brien, Stonehuerner & Millett (1984) *J. Biol. Chem.* **259**, 2155. **6.** Ohta, Nogi, Miyazaki, *et al.* (2004) *Biosci. Biotechnol. Biochem.* **68**, 1073. **7.** Cristofoletti & Terra (2000) *Biochim. Biophys. Acta* **1479**, 185. **8.** Kim & Kim (1995) *Eur. J. Biochem.* **227**, 687. **9.** Turkoglu & Ozer (1991) *Int. J. Biochem.* **23**, 147. **10.** Gravett, Viljoen & Oosthuizen (1991) *Int. J. Biochem.* **23**, 1101. **11.** Garrard, Bui, Nygaard & Raushel (1985) *J. Biol. Chem.* **260**, 5548. **12.** Ahn, Moon, Kang & Cho (2004) *Biochimie* **86**, 151. **13.** Gite & Shankar (1992) *Eur. J. Biochem.* **210**, 437. **14.** Shashi, Bachhawat & Joseph (1990) *Biochim. Biophys. Acta* **1033**, 23. **15.** Mäkinen, Mäkinen, Wilkes, Bayliss & Prescott (1982) *Eur. J. Biochem.* **128**, 257. **16.** Buechler, Toner-Webb & Taylor (1991) *Meth. Enzymol.* **200**, 487. **17.** Kita, Okino & Ito (2000) *Biochim. Biophys. Acta* **1485**, 111. **18.** Milewski, O'Donnell & Gooday (1992) *J. Gen. Microbiol.* **138**, 2545. **19.** Ozer & Ozer (1987) *Arch. Biochem. Biophys.* **255**, 89. **20.** Kaulpiboon & Pongsawasdi (2005) *Enzyme Microb. Technol.* **36**, 168. **21.** Kaulpiboon & Pongsawasdi (2004) *J. Biochem. Mol. Biol.* **37**, 408. **22.** Trebst (1980) *Meth. Enzymol.* **69**, 675. **23.** Genta, Terra & Ferreira (2003) *Insect Biochem. Mol. Biol.* **33**, 1085. **24.** Yoshinaga, Fujisue, Abe, *et al.* (1999) *Biochim. Biophys. Acta* **1472**, 447. **25.** Nyvall, Pedersen, Kenne & Gacesa (2000) *Phytochemistry* **54**, 139. **26.** Medina, Peleato, Mendez & Gomez-Moreno (1992) *Eur. J. Biochem.* **203**, 373. **27.** Zanetti, Aliverti & Curti (1984) *J. Biol. Chem.* **259**, 6153. **28.** Satre, Lunardi, Dianoux, *et al.* (1986) *Meth. Enzymol.* **126**, 712. **29.** Steffens, Schneider, Herkenhoff, Schmid & Altendorf (1984) *Eur. J. Biochem.* **138**, 617. **30.** Zeng, Li, Gu & Zhang (1992) *Arch. Biochem. Biophys.* **298**, 226. **31.** Grossmann & Terra (2001) *Comp. Biochem. Physiol. B* **128**, 109. **32.** Gaunt & Huber (1987) *Int. J. Biochem.* **19**, 47. **33.** Takayanagi (2002) *J. Appl. Glycosci.* **49**, 57. **34.** Yang, Ge, Zeng & Zhang (1985) *Biochim. Biophys. Acta* **828**, 236. **35.** Marana, Jacobs-Lorena, Terra & Ferreira (2001) *Biochim. Biophys. Acta* **1545**, 41. **36.** Xia, Meyer, Chen, *et al.* (1993) *Biochem. J.* **293**, 357. **37.** Solov'eva & Sisse (1985) *Biokhimiia* **50**, 211. **38.** Liu, Kurihara, Hasan, *et al.* (1994) *Appl. Environ. Microbiol.* **60**, 2389. **39.** Nardi-Dei, Kurihara, Park, Esaki & Soda (1997) *J. Bacteriol.* **179**, 4232. **40.** Huynh & Snell (1986) *J. Biol. Chem.* **261**, 4389. **41.** Kim, Kim, Jung, Jung & Lee (1993) *J. Pharmacol. Exp. Ther.* **266**, 500. **42.** Chun, Vinarov & Miziorko (2000) *Biochemistry* **39**, 14670. **43.** Lin (1970) *Biochemistry* **9**, 984. **44.** Baskaran & Balasubramanian (1987) *Biochim. Biophys. Acta* **913**, 377. **45.** Lupidi, Bollettini, Venardi, Marmocchi & Rotilio (2001) *Prep. Biochem.*

Biotechnol. **31**, 317. **46.** Kim, Muramatsu & Takahashi (1995) *J. Biochem.* **117**, 535. **47.** Dhanawansa, Faridmoayer, van der Merwe, Li & Scaman (2002) *Glycobiology* **12**, 229. **48.** Schmitz, Fingerhut & Conzelmann (1994) *Eur. J. Biochem.* **222**, 313. **49.** Chazot, Fotherby & Stephenson (1993) *Biochem. Pharmacol.* **45**, 605. **50.** Cristofoletti & Terra (2000) *Biochim. Biophys. Acta* **1479**, 185. **51.** Inano & Tamaoki (1985) *J. Enzyme Inhib.* **1**, 47. **52.** Pedemonte & Kaplan (1986) *J. Biol. Chem.* **261**, 3632 and 16660. **53.** Lee & Cho (2001) *J. Biochem. Mol. Biol.* **34**, 408. **54.** Cheesbrough & Kolattukudy (1985) *Arch. Biochem. Biophys.* **237**, 208. **55.** Perfetti, Anderson & Hall (1976) *Biochemistry* **15**, 1735. **56.** Craik & Reithmeier (1984) *Biochim. Biophys. Acta* **778**, 429. **57.** Dennis (1983) *The Enzymes*, 3rd ed., **16**, 307. **58.** Torossian, Audette & Poulin (1996) *Biochem. J.* **319**, 21. **59.** Niture, Pant & Kumar (2001) *Eur. J. Biochem.* **268**, 832. **60.** LaPointe & Taylor (2000) *J. Biol. Chem.* **275**, 1502. **61.** Bolvin, Lin & Béliveau (1997) *Biochem. Cell Biol.* **75**, 63. **62.** Plant, Clemens, Morgan & Daniel (1987) *Biochem. J.* **246**, 537. **63.** Gonzales & Robert-Baudouy (1994) *J. Bacteriol.* **176**, 2569. **64.** Davis, Moses, Ndubuka & Ortiz (1987) *J. Gen. Microbiol.* **133**, 1453. **65.** Rangarajan & Shankar (2001) *J. Biochem. Mol. Biol. Biophys.* **5**, 309. **66.** Kim, Muramatsu & Takahashi (1995) *J. Biochem.* **117**, 535. **67.** Borders, Chan, Miner & Weerasuriya (1989) *FEBS Lett.* **255**, 365. **68.** Chan, Jorgensen & Borders (1988) *Biochem. Biophys. Res. Commun.* **151**, 709. **69.** Chen, Daron & Aull (1992) *J. Enzyme Inhib.* **5**, 259. **70.** Meshalkina, Kuimov, Kabakov, Tsorina & Kochetov (1984) *Biochem. Int.* **9**, 9. **71.** Gutweniger, Grassi & Bisson (1983) *Biochem. Biophys. Res. Commun.* **116**, 272. **72.** Uchida, Ohsumi & Anraku (1985) *J. Biol. Chem.* **260**, 1090. **73.** Uchida, Ohsumi & Anraku (1988) *Meth. Enzymol.* **157**, 544. **74.** Gordon-Weeks, Steele & Leigh (1996) *Plant Physiol.* **111**, 195. **75.** Kiss, Erdei & Kiss (2002) *Arch. Biochem. Biophys.* **399**, 188. **76.** Muraki, Taki, Hasegawa, Iwaki & Lau (2003) *Appl. Environ. Microbiol.* **69**, 1564. **77.** Salony, Garg, Baranwal, *et al.* (2008) *Biochim. Biophys. Acta* **1784**, 259. **78.** Kanade, Paul, Rao & Gowda (2006) *Biochem. J.* **395**, 551. **79.** Ritzmann & Bosshard (1986) *Eur. J. Biochem.* **159**, 493.

3-(1-(3-(Dimethylamino)propyl)-2-methyl-1*H*-indol-3-yl)-4-(2-methyl-1*H*-indol-3-yl)-1*H*-pyrrole-2,5-dione

This bisindolylmaleimide (FW = 440.55 g/mol) potently inhibits protein kinase CβII (IC$_{50}$ = 0.6 nM). The bound inhibitor adopts a nonplanar conformation in the ATP-binding site, with the kinase domain taking on an intermediate, open conformation. This PKCβII-inhibitor complex represents the first structural description of any conventional PKC kinase domain. **1.** Grodsky, Li, Bouzida, *et al.* (2006) *Biochemistry* **45**, 13970.

3-(Dimethylamino)-1-propyne

This flammable liquid, (CH$_3$)$_2$N–CH$_2$–C≡C–H (FW = 83.133 g/mol; B.P. = 79-83°C), irreversibly inactivates bovine liver mitochondrial monoamine oxidase, forming a covalent adduct with the enzyme-bound flavin (*i.e.*, an N-5 substituted dihydroflavin). **1.** Walsh, Cromartie, Marcotte & Spencer (1978) *Meth. Enzymol.* **53**, 437. **2.** Brush & Kozarich (1992) *The Enzymes*, 3rd ed. (D. S. Sigman, ed.), **20**, 317. **3.** Maycock, Abeles, Salach & Singer (1976) *Biochemistry* **15**, 114.

6-(Dimethylamino)purine

This modified purine (FW = 163.18 g/mol; λ_{max} = 275 nm (pH 7); ε = 17800 $M^{-1}cm^{-1}$), also known as N^6,N^6-dimethyladenine, inhibits cyclin-dependent kinases, and inhibiting mitosis. It absorbs ultraviolet light with). **Target(s):** adenine deaminase (1); 3',5'-cyclic-nucleotide phosphodiesterase (2); cyclin dependent kinases, or cdc2 (3-5); dual-specificity kinase (5); H1-histone kinase (6,7); methionine S-adenosyltransferase, weakly inhibited (8); protein kinase C (5); protein-tyrosine kinase (5); receptor protein-tyrosine kinase (5).   **1.** Jun & Sakai (1979) *J. Ferment. Technol.* **57**, 294. **2.** Kemp & Huang (1974) *Meth. Enzymol.* **38**, 240. **3.** Meijer & Kim (1997) *Meth. Enzymol.* **283**, 113. **4.** Westmark, Ghose & Huber (1998) *Nucl. Acids Res.* **26**, 4758. **5.** MacKintosh & MacKintosh (1994) *Trends Biochem. Sci.* **19**, 444. **6.** Yoo, Choe & Rho (2003) *Reprod. Domest. Anim.* **38**, 444. **7.** Szollosi, Kubiak, Debey, *et al.* (1993) *J. Cell Sci.* **104**, 861. **8.** Berger & Knobel (2003) *BMC Microbiol.* **3**, 12.

2-Dimethylamino-4,5,6,7-tetrabromobenzimidazole

This cell-permeable CK2 inhibitor (FW = 476.79 g/mol; PubChem CID: 5326976; Abbreviation: DMAT), also named systematically as 4,5,6,7-tetrabromo-N,N-dimethyl-1H-benzo[d]imidazol-2-amine, targets Casein Kinase-2 (K_i = 40 nM), a highly pleiotropic protein kinase that is implicated in cell cycle control, DNA repair, circadian rhythm and other cellular processes (1). CK2's high constitutive activity is believed to enhance tumor phenotype and the propagation of infectious diseases. Inhibition of CK2 with DMAT results in caspase-mediated killing of human breast cancer cells with acquired resistance to antiestrogens, while DMAT fails to kill parental MCF-7 cells (2). In studies with platelets, DMAT inhibits platelet aggregation and also suppresses activation of GPIIb/IIIa (3). DMAT inhibition is used to infer that CK2 promotes cancer cell viability by up-regulating cyclooxygenase-2 expression and enhanced prostaglandin E_2 production (4). DMAT treatment stimulates the purinosome assembly in HeLa cells (5). Direct comparison of purinosome-rich (cultured in purine-depleted medium) and normal cells shows a 3-fold increase in IMP concentration in purinosome-rich cells and similar levels of AMP, GMP, and ratios of AMP/GMP and ATP/ADP for both. In addition, a higher level of IMP is also observed in HeLa cells treated with DMAT. Furthermore, treatment increases in the de novo IMP/AMP/GMP biosynthetic flux rate under purine-depleted condition. **Other Targets:** When tested in a panel of 80 protein kinases DMAT and its parent compound TBI (or TBBz; 4,5,6,7-tetrabromo-1H-benzimidazole) are potent inhibitors of PIM (provirus integration site for Moloney murine leukemia virus)1, PIM2, PIM3, PKD1 (protein kinase D1), HIPK2 (homeodomain-interacting protein kinase 2) and DYRK1a (dual-specificity tyrosine-phosphorylated and -regulated kinase 1a) (6). Rescue experiments utilizing inhibitor-resistant CK2 mutants were unable to rescue the apoptosis associated with TBBz and DMAT treatment, suggesting the inhibitors had off-target effects (7). Chemoproteomics revealed that the detoxification enzyme quinone reductase 2 (QR2 or NQO2) is a likely off-CK2, or adventitious, target (7). Although kinetic analysis suggested that DMAT inhibits NQO2 by binding with similar affinities to the oxidized and reduced forms, crystal structure analysis showed that DMAT binds reduced NQO2 in a different way than in the oxidized state (8). In oxidized NQO2, TBBz and DMAT are deeply buried in the active site and make direct hydrogen and halogen bonds to the enzyme. In reduced NQO2, DMAT occupies a more peripheral region and hydrogen and halogen bonds with the enzyme are mediated through three water molecules. The active site of NQO2 is fundamentally different from the ATP binding site of CK2 (8). **1.** Pagano, Meggio, Ruzzene, *et al.* (2004) *Biochem. Biophys. Res. Commun.* **321**, 1040. **2.** Yde, Frogne, Lykkesfeldt *et al.* (2007) *Cancer Lett.* **256**, 229. **3.** Nakanishi, Toyoda, Tanaka, *et al.* (2010) *Thromb. Res.* **126**, 511. **4.** Yefi, Ponce, Niechi, *et al.* (2011) *J. Cell Biochem.* **112**, 3167. **5.** Zhao, Chiaro, Zhang, *et al* (2015) *J. Biol. Chem.* **290**, 6705. **6.** Pagano, Bain, Kazimierczuk, *et al.* (2008) *Biochem. J.* **415**, 353. **7.** Duncan, Gyenis, Lenehan, *et al.* (2008) *Mol. Cell Proteomics* **7**, 1077. **8.** Leung & Shilton (2015) *Biochemistry* **54**, 47.

Dimethylarginine, *See* Asymmetric Dimethyl Arginine

9,10-Dimethyl-1,2-benzanthracene

This so-called bay-area carcinogenic DNA intercalator (FW = 256.35 g/mol; CAS 57-97-6; M.P.= 122-123°C; *Abbreviation*: DMBA), also known as 7,12-dimethylbenz[*a*]anthracene, inhibits ribosyldihydro-nicotinamide dehydrogenase (quinone) (1). For best results, it agent should first be dissolved in ethanol and then diluted 20x into the reaction mixture. After the topical application of tritium-labeled DMBA to mouse skin, DMBA-bound DNA was fractionated into satellite and main-band components, demonstrating extensive binding of [^{3}H]-DMBA to mouse satellite DNA (2). **1.** Liao, Dulaney & Williams-Ashman (1962) *J. Biol. Chem.* **237**, 2981. **2.** Zeiger, Salomon, Kinoshita & Peacock (1972) *Cancer Res.* **32**, 643.

N-(1-(α,α-Dimethylbenzyl)pyrrolidin-3(R)-yl-(N-ethyl)amino sulfonyl)-5'-(4-methoxyphenyl)-2'-(2-thienyl)-1'H-histamine

This modified histamine (FW = 593.82 g/mol) inhibits 15-lipoxygenase, or arachidonate 15-lipoxygenase, IC_{50} = 10 nM. Non-symmetrical sulfamides are suitable replacements for the earlier arylsulfonamide-containing members of this series, and this agent is also a potent inhibitor of human 15-LO in a cell-based assay. **1.** Weinstein, Liu, Ngu, *et al.* (2007) *Bioorg. Med. Chem. Lett.* **17**, 5115.

3,3-Dimethylbutyrate

This betaine analogue (FW = 116.16 g/mol; CAS 1070-83-3), also known as *t*-butylacetic acid, inhibits betaine:homocysteine S-methyltransferase (1,2), dimethylglycine N-methyltransferase (3), and sarcosine/dimethylglycine N-methyltransferase (3). **1.** Skiba, Wells, Mangum & Awad (1987) *Meth. Enzymol.* **143**, 384. **2.** Skiba, Taylor, Wells, Mangum &Awad (1982) *J. Biol. Chem.* **257**, 14944. **3.** Waditee, Tanaka, Aoki, *et al.* (2003) *J. Biol. Chem.* **278**, 4932.

2,9-Dimethyl-9H-β-carbolinium Ion

This cation ($FW_{chloride}$ = 232.71 g/mol) is a strong inhibitor of mitochondrial respiration and is also a nigrostriatal neurotoxin, the latter a likely consequence of its structural resemblance to the neurotoxin 1-methyl-4-phenyl-1,2,3,6-tetrahydropyridine (MPTP). Binding studies with [^{3}H]-2,9-Dimethyl-9H-β-carbolinium ion showed it binds with high affinity to the chaperone member glucose regulated protein 78, carboxylesterase, cytochrome P450 2E1, monoamine oxidase B, and a Rho family G-protein. DC also binds to and inhibits triose-phosphate isomerase. **1.** Bonnet, Pavlovic, Lehmann & Rommelspacher (2004) *Neuroscience* **127**, 443.

N,N-Dimethyl-N-dodecylamine N-Oxide

This nonionic detergent (FW = 229.41g/mol), also known as lauryldimethylamine oxide (or, LDAO), Ammonyx LO, and Bardox-12, is an activator of F_1 ATPase. It is also used to solubilize membrane proteins and isolate channels. **Target(s):** Ca^{2+}-dependent ATPase of CF_1 (1); ceramide kinase (2); cycloartenol synthase (3); cytochrome-*c* oxidase (4); glutaconyl-CoA decarboxylase (5); guanylate cyclase (6); phosphatidylserine decarboxylase (7); phosphoinositide 5-phosphatase (8); sphinganine-1-phosphate aldolase (9); squalene:hopene cyclase (10); squalene-tetrahymanol cyclase (11). **1.** Ren & Allison (1997) *J. Biol. Chem.* **272**, 32294. **2.** Van Overloop, Gijsbers & Van Veldhoven (2006) *J. Lipid Res.* **47**, 268. **3.** Schmitt, Gonzales, Benveniste, Cerutti & Cattel (1987) *Phytochemistry* **26**, 2709. **4.** King & Drews (1976) *Eur. J. Biochem.* **68**, 5. **5.** Buckel (1986) *Meth. Enzymol.* **125**, 547. **6.** Fleischman & Denisevich (1979) *Biochemistry* **18**, 5060. **7.** Dowhan, Wickner & Kennedy (1974) *J. Biol. Chem.* **249**, 3079. **8.** Roach & Palmer (1981) *Biochim. Biophys. Acta* **661**, 323. **9.** Van Veldhoven (2000) *Meth. Enzymol.* **311**, 244. **10.** Ochs, Tappe, Gartner, Kellner & Poralla (1990) *Eur. J. Biochem.* **194**, 75. **11.** Saar, Kader, Poralla & Ourisson (1991) *Biochim. Biophys. Acta* **1075**, 93.

Dimethylenastron

This kinesin inhibitor (FW = 302.39 g/mol; CAS 863774-58-7; Solubility: 100 mM in DMSO; 50 mM in ethanol) targets mitotic motor kinesin Eg5 (IC_{50} = 200 nM) (1,2), thereby arresting mitosis and triggering apoptosis, and up-regulating Hsp-70 expression in multiple human myeloma cell lines (3). Hsp70 induction occurs at the transcriptional level via a *cis*-regulatory DNA element in Hsp70 promoter and is mediated by the phosphatidylinositol 3-kinase/Akt pathway. Eg5 inhibitor-mediated Hsp70 up-regulation is cytoprotective, because blocking Hsp70 induction directly by antisense or small interfering RNA or indirectly by inhibiting the phosphatidylinositol 3-kinase/Akt pathway significantly increases Eg5 inhibitor-induced apoptosis. Dimethylenastron also inhibits the migration and invasion of PANC1 pancreatic cancer cells, independent of suppressing the cell proliferation (3). **1.** Kaan, Ulaganathan, Rath, *et al.* (2010) *J. Med Chem.* **53**, 5676. **2.** Sun, Shi, Sun, *et al.* (2011) *Acta Pharmacol. Sin.* **32**, 1543. **3.** Liu, Aneja, Liu, *et al.* (2006) *J. Biol. Chem.* **281**, 18090.

N,N-Dimethylformamide

This organic solvent (FW = 73.09 g/mol; CAS 68-12-2; *Abbreviation*: DMF) is a combustible liquid (M.P. = –60.4°C; B.P. = 153.0°C) frequently used in organic synthesis, extractions, chromatography, spectrophotometry, and toxicology investigations. With a dipole moment of 3.86 D at 25°C (compare to a value of 1.87 for water) and a high dielectric constant of 36.71 at 25°C (compare to a value of 80.1 for water), DMF is miscible in water as well as many polar organic solvents. **Target(s):** *N*-acetyllactosamine synthase (1); acylaminoacyl-peptidase (2); adenylate cyclase, adenyl cyclase (3); alcohol dehydrogenase (4); aminoacylase, *or N*-acylamino acid amidohydrolase (5); arylformamidase (6); carboxylesterase (7); carboxypeptidase C, *or* carboxypeptidase Y (8,9); coagulation factor

IXa (10); cytosol alanyl aminopeptidase (11); endopeptidase La (12); hemoglobin S polymerization (13); lactose synthase (1); lanosterol synthase (14); lipase, *or* triacylglycerol lipase (15); membrane alanyl aminopeptidase, *or* aminopeptidase N (16); *N*-nitrosodimethylamine demethylase, *or* cytochrome P450 (17); pancreatic ribonuclease denaturation (18); papain (19); pepsin (20); prolyl oligopeptidase (21); soluble epoxide hydrolase (22). **1.** Pisvejcova, Rossi, Husakova, *et al.* (2006) *J. Mol. Catal. B* **39**, 98. **2.** Farries, Harris, Auffret & Aitken (1991) *Eur. J. Biochem.* **196**, 679. **3.** Huang, Smith & Zahler (1982) *J. Cyclic Nucleotide Res.* **8**, 385. **4.** Miroliaei & Nemat-Gorgani (2002) *Int. J. Biochem. Cell Biol.* **34**, 169. **5.** Boross, Kosary, Stefanovits-Banyai, Sisak & Szajani (1998) *J. Biotechnol.* **66**, 69. **6.** Santti (1969) *Hoppe-Seyler's Z. Physiol. Chem.* **350**, 1279. **7.** Sehgal, Tompson, Cavanagh & Kelly (2002) *Biotechnol. Bioeng.* **80**, 784. **8.** Lewis & Schuster (1991) *J. Biol. Chem.* **266**, 20818. **9.** Hayashi, Bai & Hata (1975) *J. Biochem.* **77**, 69. **10.** Stürzebecher, Kopetzki, Bode & Hopfner (1997) *FEBS Lett.* **412**, 295. **11.** Garner & Behal (1977) *Arch. Biochem. Biophys.* **182**, 667. **12.** Waxman & Goldberg (1985) *J. Biol. Chem.* **260**, 12022. **13.** Waterman, Yamaoka, Dahm, Taylor & Cottam (1974) *Proc. Natl. Acad. Sci. U.S.A.* **71**, 2222. **14.** Oliaro-Bosso, Schulz-Gasch, Balliano & Viola (2005) *ChemBioChem* **6**, 2221. **15.** Quyen, Le, Nguyen, *et al.* (2005) *Protein Expr. Purif.* **39**, 97. **16.** Wachsmuth, Fritze & Pfleiderer (1966) *Biochemistry* **5**, 175. **17.** Yoo, Cheung, Patten, Wade & Yang (1987) *Cancer Res.* **47**, 3378. **18.** Richards & Wyckoff (1971) *The Enzymes*, 3rd ed. (Boyer, ed.), **4**, 647. **19.** Szabelski, Stachowiak & Wiczk (2001) *Acta Biochim. Pol.* **48**, 1197. **20.** Fruton (1971) *The Enzymes*, 3rd ed., **3**, 119. **21.** Polgár (1994) *Meth. Enzymol.* **244**, 188. **22.** Schladt, Hartman, Wörner, Thomas & Oesch (1988) *Eur. J. Biochem.* **176**, 31.

Dimethylfumarate

This cell-permeant thiol-reactive multiple sclerosis drug (FW = 144.13 g/mol; CAS 624-49-7; *Abbreviation*: DMF), also known as BG-12 and its trade name Tecfidera®, reacts with and depletes intracellular glutathione (GSH) by covalent bond formation in a reaction that may be mediated by GSH-*S*-transferase. In patients with relapsing-remitting multiple sclerosis, dimethylfumarate significantly reduces the proportion of patients who had a relapse, lowers the annualized relapse rate, lessens the rate of disability progression, reduces the number of lesions on MRI (1). Although the mode of action is unclear, BG-12 is known to reduce the levels of certain pro-inflammatory cytokines that cause inflammation during an MS attack, while increasing the levels of anti-inflammatory cytokines (2-5). DMF is also a very effective hypoxic cell radiosensitizer, with an enhancement ratio (ER) of about 3 obtained by a 5-min exposure of cells to 5 mM DMF, without significant toxicity (6). **1.** Gold, Kappos, Arnold, *et al.* (2012) *New Engl. J Med.* **367**, 1098. **2.** Ockenfels, Schultewolter, Ockenfels, *et al.* (1998) *Br. J. Dermatol.* **139**, 390. **3.** de Jong, Bezemer, Zomerdijk, *et al.* (1996) *Eur. J. Immunol.* **26**, 2067. **4.** Lee, Linker & Gold (2008) *Int. MS J.* **15**, 12. **5.** Kappos, Gold & Miller (2008) Lancet 372, 1463. **6.** Held, Epp, Clark & Biaglow (1988) *Radiat. Res.* **115**, 495.

3,3-Dimethylglutarate

This dicarboxylic acid ($FW_{free-acid}$ = 160.17 g/mol; CAS 4839-46-7; M.P. = 83-85°C) is frequently employed as a buffer (pK_a values of 3.70 and 6.34 at 25°C, with respective dpK_a/dT values of +0.0076 and +0.0060). Do not confuse this dicarboxylic acid with dimethyl glutarate, the double ester of glutaric acid **Target(s):** adenosine-phosphate deaminase (1); 2-aminoadipate aminotransferase (2); γ-glutamyl hydrolase, weakly inhibited (3); iduronate-2-sulfatase (4); kynurenine:oxoglutarate aminotransferase (5,6); phospho-lipase D (7); pyridoxamine:oxaloacetate aminotransferase, weakly inhibited at pH 8 (8). **1.** Yates (1969) *Biochim. Biophys. Acta* **171**, 299. **2.** Deshmukh & Mungre (1989) *Biochem. J.* **261**, 761. **3.** Silink,

Reddel, Bethel & Rowe (1975) *J. Biol. Chem.* **250**, 5982.   **4.** Bielicki, Freeman, Clements & Hopwood (1990) *Biochem. J.* **271**, 75.   **5.** Mason (1959) *J. Biol. Chem.* **234**, 2770.   **6.** Mawal, Mukhopadhyay & Deshmukh (1991) *Biochem. J.* **279**, 595.   **7.** Chalifa, Mohn & Liscovitch (1990) *J. Biol. Chem.* **265**, 17512.   **8.** Wu & Mason (1964) *J. Biol. Chem.* **239**, 1492.

3,3-Dimethylglutaric Acid 2(*S*)-(cyclopentanecarbonyl) pyrrolidone *N*-(L-Prolyl)-2(*S*)-(hydroxyacetyl)pyrrolidine Diamide

This substrate analogue (FW = 510.61 g/mol) potently inhibits pig brain prolyl oligopeptidase (IC$_{50}$ = 0.32 nM). This agent has a significantly highter octanol:water partition constant, suiting it for targeting to the brain. **1.** Wallén, Christiaans, Jarho, *et al.* (2003) *J. Med. Chem.* **46**, 4543.

N,N-Dimethylglycine

This hygroscopic amino acid (FW = 103.12 g/mol) is an intermediate in the conversion of choline to glycine. **Target(s):** beteine:homocysteine methyltransferase, by product inhibition (1-4); choline oxidase (5); choline transport (6); sarcosine oxidase (7). **1.** Webb (1966) *Enzyme and Metabolic Inhibitors*, vol. **2**, p. 356, Academic Press, New York. **2.** Waditee & Incharoensakdi (2001) *Curr. Microbiol.* **43**, 107. **3.** Skiba, Taylor, Wells, Mangum & Awad (1982) *J. Biol. Chem.* **257**, 14944. **4.** Szegedi, Castro, Koutmos & Garrow (2008) *J. Biol. Chem.* **283**, 8939. **5.** Gadda, Powell & Menon (2004) *Arch. Biochem. Biophys.* **430**, 264. **6.** Lucchesi, Pallotti, Lisa & Domenech (1998) *FEMS Microbiol. Lett.* **162**, 123. **7.** Wagner, Trickey, Chen, Mathews & Jorns (2000) *Biochemistry* **39**, 8813.

Dimethyl(2-hydroxy-5-nitrobenzyl)sulfonium Halide

This water-soluble salt (FW$_{bromide}$ = 294.17 g/mol), also known as Koshland's Reagent I Water-Soluble, *or* K-IWS, modifies tryptophanyl and cysteinyl residues in peptides and proteins (*e.g.*, chymotrypsin and carboxypeptidase A$_5$) and can be utilized as an environmentally sensitive probe. Solutions are relatively stable at pH 3; however, aqueous solutions are rapidly hydrolyzed at neutral or alkaline pH (1). **Target(s):** chorismate mutase (2); creatine kinase (3); glyceraldehyde-3-phosphate dehydrogenase (4); lactose synthase (5); lysozyme (6); nucleoside transporter (7); pepsin (8); prephenate dehydratase (2); ribulose 1,5-bisphosphate carboxylase (9); taka-amylase A (10). **1.** Horton & Tucker (1970) *J. Biol. Chem.* **245**, 3397. **2.** Davidson (1987) *Meth. Enzymol.* **142**, 432. **3.** Zhou & Tsou (1985) *Biochim. Biophys. Acta* **830**, 59. **4.** Heilmann & Pfleiderer (1975) *Biochim. Biophys. Acta* **384**, 331. **5.** Hill & Brew (1975) *Adv. Enzymol. Relat. Areas Mol. Biol.* **43**, 411. **6.** Jorkasky, Pearson & Borders (1973) *Biochem. Biophys. Res. Commun.* **52**, 987. **7.** Thorne & Lowe (1991) *Biochem. Soc. Trans.* **19**, 418S. **8.** Llewellin & Green (1975) *Biochem. J.* **151**, 319.   **9.**

Purohit & Bhagwat (1990) *Indian J. Biochem. Biophys.* **27**, 81.   **10.** Kita, Fukazawa, Nitta & Watanabe (1982) *J. Biochem.* **92**, 653.

N^4-(2,2-Dimethyl-1(*S*)-methylcarbamoylpropyl)-*N*^{1}-hydroxy-2(*R*)-hydroxymethyl-3(*S*)-(4-methoxyphenyl)succinamide

This antiinflammatory agent (FW = 395.46 g/mol), also known as SDZ 242-484, inhibits matrix metalloproteinases as well as tumor necrosis factor-α activity. **Target(s):** ADAM 17 endopeptidase (tumor necrosis factor-α converting enzyme, *or* TACE, K_i = 0.6 nM (1,2); gelatinase A, matrix metalloproteinase-2, K_i = 0.5 nM; interstitial collagenase, *or* matrix metalloproteinase-1, K_i = 1.1 nM; stromelysin 1, *or* matrix metalloproteinase-3; K_i = 0.9 nM (1,2). **1.** Koch, Kottirsch, Wietfeld & Kusters (2002) *Org. Proc. Res. Dev.* **6**, 652. **2.** Kottirsch, Koch, Feifel & Neumann (2002) *J. Med. Chem.* **45**, 2289.

Dimethyloxalylglycine

This cell-permeable inhibitor and α-ketoglutarate mimic (FW = 175.14 g/mol; CAS 89464-63-1; Symbol: DMOG) targets HIFα prolyl hydroxylase (HIF-PH) competitively, leading to the stabilization of HIF and subsequent angiogenesis and glucose metabolism, when administered at concentrations between 0.1 and 1 mM. Hypoxia Inducible Factor (HIF), which is comprised of α and β subunits, regulates responses to hypoxia. Upon cellular exposure to hypoxic conditions, the HIF αβ is stabilized, binding to DNA transcriptionally activating genes linked to the cellular processes of angiogenesis and glucose metabolism (1). Under normal conditions, the HIF-a subunit is hydroxylated by HIF-α prolyl hydroxylase (HIF-PH), leading to ubiquitylation of HIF-alpha and its subsequent destruction (2). DMOG is a cell permeable competitive inhibitor of HIF-α prolyl hydroxylase (HIF-PH) leading to the stabilization of HIF and subsequent angiogenesis and glucose metabolism at 0.11 mM (3,4). **See also** *Desferrioxamine* **1.** Ivan, Kondo, *et al.* (2001) *Science* **292**, 464. **2.** Jaakkola, Mole, *et al.* (2001) *Science* **292**, 468. **3.** Cummins, Seeballuck, *et al.* (2008). *Gastroenterology* **134**, 156. **4.** Glassford, Yue, *et al.* (2007). *Am. J. Physiol. Endocrinol Metab.* **293**, E1590.

N,N-Dimethyl-2-phenylaziridinium Cation

This alkylating reagent (FW = 148.23 g/mol)), which forms in aqueous solution by intramolecular rearrangement of *N,N*-dimethyl-2-chloro-2 phenylethylamine, is an irreversible inhibitor of various acetylcholinesterases, including those from *Torpedo califorica* and cobra venom (*Naja naja oxiana*). An active-site tryptophanyl residue is modified. **Target(s):** acetylcholinesterase (1-7); antheraxanthin κ-cyclase (8); β-cyclohexenyl carotenoid epoxidase (9); lycopene β-cyclase (8). **1.** Purdie & Heggie (1970) *Can. J. Biochem.* **48**, 244. **2.** McIvor (1970) *Biochim. Biophys. Acta* **198**, 143. **3.** Purdie (1969) *Biochim. Biophys. Acta* **185**, 122.

4. Palumaa, Kiaembre & Iarv (1983) *Bioorg. Khim.* **9**, 1348. **5.** Palumaa & Jarv (1984) *Biochim. Biophys. Acta* **784**, 35. **6.** Soomets, Palumaa & Jarv (1987) *Bioorg. Khim.* **13**, 198. **7.** Kreienkamp, Weise, Raba, Aaviksaar & Hucho (1991) *Proc. Natl. Acad. Sci. U.S.A.* **88**, 6117. **8.** Bouvier, d'Harlingue & Camara (1997) *Arch. Biochem. Biophys.* **346**, 53. **9.** Bouvier, d'Harlingue, Hugueney, *et al.* (1996) *J. Biol. Chem.* **271**, 28861.

1,3-Dimethyl-6-(2-propoxy-5-methanesulfonylamidophenyl)pyrazol[3,4*d*]-pyrimidin-4-(5*H*)-one

This purine analogue (FW = 391.45 g/mol; *Abbreviation*: DMPPO) inhibits cGMP phosphodiesterase, isozyme-5 (K_i = 3 nM), with much weaker action on the other isozymes. **1.** Ni, Safai, Gardner & Humphreys (2001) *Kidney Int.* **59**, 1264. **2.** Coste & Grondin (1995) *Biochem. Pharmacol.* **50**, 1577. **3.** Delpy & le Monnier de Gouville (1996) *Brit. J. Pharmacol.* **118**, 1377.

N,N-Dimethylsphingosine

This D-*erythro*-sphingosine derivative (FW = 327.55 g/mol; CAS 119567-63-4; Symbol: DMS), also called *N,N*-dimethyl-D-*erythro*-4-*trans* sphinganine and *N,N*-dimethyl-(2*S*,3*R*,4*E*)-2-amino-4-octadecene-1,3-diol, is formed from sphingosine by the action of mouse brain *N*-methyltransferase, presumably with *S*-adenosylmethionine as the methyl donor (1). Significantly, labeling of sphingosine in human epidermoid carcinoma A431 cells occurs by culturing these cells in a medium containing [^{3}H]Serine, followed by extraction and isolation of sphingolipids by standard procedures. Bands corresponding to *N,N*-dimethylsphingosine and ceramide are observed, whereas sphingosine was virtually absent (2). Such findings clearly documented that DMS is a natural product. **Role in Pain:** Untargeted metabolomics showed that sphingomyelin-ceramide metabolism is altered in the dorsal horn of rats with neuropathic pain, attended by upregulated synthesis of *N,N*-dimethylsphingosine and induction of mechanically-induced pain hypersensitivity *in vivo* (3). DMS inhibits protein kinase C (IC$_{50}$ = 12 μM) and sphingosine kinase (K_i ≈ 0.01-0.3 mM; IC$_{100}$ ≈ 1-2 mM), the latter controlling the balance between proapoptotic ceramide and anti-apoptotic effects of the ceramide catabolite, sphingosine-1-phosphate (S1P). Although DMS blocks S1P production (4), the physiological concentrations of DMS are substantially lower than its K_i. Dimethylsphingosine also stimulates *src* kinase and sphingosine-dependent protein kinase and induces apoptosis in human leukemia HL-60 cells. DMS is likewise a weak activator of phospholipase Cδ. **Target(s):** ceramide kinase (5); DNA primase (6); phosphatidate phosphatase, *or* ecto-phosphatidic acid phosphohydrolase (7); protein kinase C (8-13); sphinganine kinase, *or* sphingosine kinase (4,12-21). Notably, phosphorylation of sphingosine to form sphingosine-1-P is inhibited by *N,N*-dimethylsphingosine (DMS) in a dose-dependent manner, but not by other structurally related sphingosine derivatives, including ceramide (4). The inhibition of sphingosine-1-P formation by DMS can be reproduced using a cell-free system (sphingosine kinase obtained from platelet cytosolic fractions) and much stronger than that by DL-*threo*-dihydrosphingosine, which had been considered to be the strongest inhibitor of sphingosine kinase. Administration of DMS to intact platelets results in a decrease in sphingosine-1-P and an increase in sphingosine (4). The autophosphorylation of EGF receptor in the absence of detergent is also strongly enhanced by *N,N*-dimethyl-D-*erythro*-sphingenine; this effect was even obvious in the absence of EGF and synergistic in its presence. Similar effects were not produced by *N,N*-dimethyl-L-*erythro*-sphingenine, D- and L-*erythro*-sphingenine, *N*-monomethyl-D-*erythro*-sphingenine, *N*-acetyl-D-

erythro-sphingenine, or the five *lyso*-glycosphingolipids tested (23). **1.** Igarashi, Hakomori (1989) *Biochem. Biophys. Res. Commun.* **164**, 1411. **2.** Igarashi, Kitamura, Toyokuni, *et al.* (1990) *J. Biol. Chem.* **265**, 5385. **3.** Patti, Yanes, Shriver, *et al.* (2012) *Nature Chem. Biol.* **8**, 232. **4.** Yatomi, Ruan, Megidish, *et al.* (1996) *Biochemistry* **35**, 626. **5.** Sugiura, Kono, Liu, *et al.* (2002) *J. Biol. Chem.* **277**, 23294. **6.** Tamiya-Koizumi, Murate, S uzuki, *et al.* (1997) *Biochem. Mol. Biol. Int.* **41**, 1179. **7.** English, Martin, Harvey, *et al.* (1997) *Biochem. J.* **324**, 941. **8.** Smith, Merrill, Obeid & Hannun (2000) *Meth. Enzymol.* **312**, 361. **9.** Kimura, Kawa, Ruan, *et al.* (1992) *Biochem. Pharmacol.* **44**, 1585. **10.** Igarashi, Hakomori, Toyokuni, *et al.* (1989) *Biochemistry* **28**, 6796. **11.** Khan, Dobrowsky, el Touny & Hannun (1990) *Biochem. Biophys. Res. Commun.* **172**, 683. **12.** Edsall, Van Brocklyn, Cuvillier, Kleuser & Spiegel (1998) *Biochemistry* **37**, 12892. **13.** Kim, Kim, Inagaki, *et al.* (2005) *Bioorg. Med. Chem.* **13**, 3475. **14.** Olivera, Kohama, Tu, Milstien & Spiegel (1998) *J. Biol. Chem.* **273**, 12576. **15.** Kono, Nishiuma, Nishimura, *et al.* (2007) *Amer. J. Respir. Cell Mol. Biol.* **37**, 395. **16.** Cuvillier (2007) *Anticancer Drugs* **18**, 105. **17.** Pitson, D'andrea, Vandeleur, *et al.* (2000) *Biochem. J.* **350**, 429. **18.** Taha, Hannun & Obeid (2006) *J. Biochem. Mol. Biol.* **39**, 113. **19.** Melendez, Carlos-Dias, Gosink, Allen & Takacs (2000) *Gene* **251**, 19. **20.** Leiber, Banno & Tanfin (2007) *Amer. J. Physiol.* **292**, C240. **21.** Coursol, Le Stunff, Lynch, *et al.* (2005) *Plant Physiol.* **137**, 724. **22.** Kee, Vit & Melendez (2005) *Clin. Exp. Pharmacol. Physiol.* **32**, 153. **23.** Liu, Sugiura, Nava, *et al.* (2000) *J. Biol. Chem.* **275**, 19513.

2,2-Dimethylsuccinate

This dicarboxylic acid (FW$_{free-acid}$ = 146.14 g/mol), also known as 2,2-dimethylsuccinic acid and *gem*-dimethylsuccinic acid, is a strong inhibitor of carboxypeptidase A and B, with K_i values of 1.7 and 28 μM, respectively (1). **Target(s):** glutamate decarboxylase, K_i = 24 mM (2); kynurenine aminotransferase, weakly inhibited (3); maleate hydratase, K_i = 25 μM (4). **1.** Asante-Appiah, Seetharaman, Sicheri, Yang & Chan (1997) *Biochemistry* **36**, 8710. **2.** Fonda (1972) *Biochemistry* **11**, 1304. **3.** Mason (1959) *J. Biol. Chem.* **234**, 2770. **4.** J. Van der Werf, Van der Tweel & Hartmans (1993) *Appl. Environ. Microbiol.* **59**, 2823.

Dimethyl Sulfate

This membrane-penetrating sulfate ester (FW = 126.13 g/mol; CAS 77-78-1; colorless liquid; M.P. = –27°C; decomposes at about 188°C; Soluble in water (2.8 g/100 mL at 18°C), diethyl ether, dioxane, and acetone) readily alkylates DNA and RNA and inhibits protein biosynthesis. Guanine residues are methylated primarilly at N^7, whereas adenine residues are modified at N^3 (1). Small amounts of O^6-methylguanine, N^3-methylcytosine, and N^1- and N^7-methyladenine are also formed. Methylation is a step in the Maxam-Gilbert method for sequencing DNA, one that is based on chemical modification of DNA and subsequent cleavage at specific sites (2). Although published two years after the Sanger's & Coulson's ground-breaking plus-minus sequencing method, the Maxam-Gilbert technique proved more popular, chiefly because purified DNA could be used directly, whereas the earliest version of the Sanger method required cloning of each read start for production of single-stranded DNA. That said, with the improvement of the chain-termination method, Maxam-Gilbert sequencing fell out of favor owing to its technical complexity, its use of a hazardous chemical, and difficulties in scale-up. **CAUTION:** Not surprisingly, DMS is also both carcinogenic and mutagenic. A strong pulmonary vesicant, DMS was a member of the chemical weapons arsenal. In humans, DMS is highly toxic and is attended by severe inflammation in the eye, respiratory epithelium, and skin, starting within minutes or hours. A brief, 10-min exposure to 500 mg DMS/m^3 is often fatal, with little initial discomfort, but proceeding to severe systemic effects on nerves, heart, liver, and kidneys. **Target(s):** carbamoyl-phosphate synthetase (3); protein biosynthesis (4). **1.** Lawley & Brookes (1963) *Biochem. J.* **89**, 12. **2.** Maxam & Gilbert (1977) *Proc. Natl. Acad. Sci.* **74**, 560. **3.** Kaseman (1985) Ph.D. thesis, Dept. Biochemistry, Cornell University Medical College, New York, New York. **2.** Chen, Kung & Bates (1976) *Chem. Biol. Interact.* **14**, 101.

Dimethyl Sulfoxide, *or* DMSO

This common laboratory solvent (FW = 78.13 g/mol; CAS 67-68-5; Density = 1.100 g/cm³; FP = 18.5 °C, *or* 65.3 °F; BP = 189 °C, *or* 372 °F) is an aprotic solvent that dissolves both polar and nonpolar compounds. Its significant dipole (3.9 D) and dielectric constant (46.68) make it miscible with water, but also soluble in many organic solvents (*e.g.*, ethanol, acetone, benzene, and chloroform). Its considerable skin-penetrating (percutaneous absorptive) properties also commend its use as a vehicle for rapid drug delivery. **DMSO Effects on Proteins:** Dimethyl sulfoxide stabilizes many proteins, especially multi-subunit proteins and supramolecular structures, the latter including microtubules, mitotic apparatus, *etc.* (1a). This effect is thought to result from the ability of particular DMSO:H2O solutions to alter the strength of hydrogen and hydrophobic bonding interactions. Preferential interactions can be expressed in terms of preferential binding of the cosolvent or its preferential exclusion (the latter also known as *preferential hydration*). The driving force is the perturbation by the protein of the chemical potential of the cosolvent. The measured change of the amount of water in contact with protein during the course of the reaction modulated by an osmolyte is a change in preferential hydration that is strictly a measure of the cosolvent chemical potential perturbation by the protein in the ternary water-protein-cosolvent system. Because DMSO also contains methyl groups, it also influences the strength of hydrophobic interactions. At a concentration of 20% or less, for example, DMSO drives glutamate dehydrogenase into the inactive monomer, and effect that is fully reversible by the allosteric activator ADP (1b). Higher DMSO levels result in irreversible inactivation. The predominant effect on β-glucuronidase is irreversible inactivation by 20% or more DMSO at 37 °C, but this enzyme is activated in 20% DMSO in the presence of high substrate levels. (1). DMSO also inhibits the clotting of fibrinogen by purified thrombin, but the major effect seems competition between thrombin and DMSO for fibrinogen binding sites. Such effects appear to be largely due to hydrophobic interactions between DMSO fibrinogen, even though DMSO also appears to reduce fibrin monomer aggregation mainly through effects on hydrophilic groups. These results suggest that reversible alterations in protein structure are the major effect of exposure of subunit proteins to low DMSO levels at low temperatures, whereas irreversible denaturation of subunit proteins may be appreciable a higher temperatures and higher DMSO concentrations (1). In the presence of 10% DMSO (vol/vol), phosphocellulose-purified tubulin assembles into microtubules (2). The reaction is rapid and strongly dependent on protein concentration. **DMSO-Mediated Drug Delivery:** The *stratum corneum*, corresponding to superficial-most skin layer, typically limits percutaneous absorption of drugs, especially those possessing multiple hydrogen bond-formin moeities. Factors altering solute transport across the skin include: presence of water, transport *decreases*; lipid solubility, transport *increases*; keratolytics, transport *increases*; surfactants, transport *increases*; molecular mass (transport *increases*, up to a critical point, then transport *decreases*); local hyperemia, transport *increases*; temperature, transport *increases*; dermatoses, transport *increases*; particle size, particle size (*i.e.*, the smaller the molecule, the faster its transport; greater effect than molecular weight, which shows a variable relation with absorption); viscosity, transport *decreases*; electrical impedance, transport *decreases*; regional variations for palm, forehead, *etc.* (3). DMSO's polar, strongly hygroscopic properties result in super-hydration of the stratum corneum with subsequent increase in permeability (3). While having no known effect on plasma membranes, DMSO can cause changes in keratin filaments. Its *in vivo* effects are short-lived (3). Dimethyl sulfoxide is a safe and effective mechanism for facilitating the transdermal delivery of both hydrophilic and lipophilic medications to provide localized drug delivery (4). The clinical use of pharmaceutical-grade DMSO as a penetration enhancer is supported by the robust data that have accumulated over the past three decades demonstrating the favorable safety and tolerability profile (4). DMSO is rapidly absorbed through the skin, carrying with it many dissolved substances. This property commends DMSO as a vehicle for delivering drugs by membrane penetration. **Utility for Dissolving Water-Insoluble Drugs & Metabolites:** DMSO is a convenient solvent for dissolving various water-insoluble drugs (*e.g.*, steroids, ionophorous antibiotics, antiviral nucleosides, taxol, colchicine, *etc.*) for biochemical experiments. The usual practice is to prepare a drug solution in neat DMSO at 50-100 times its final desired concentration, followed by the corresponding 50-100 times dilution into culture media or a drug target-containing solution. That said, DMSO is not inert. Although the exact mechanism remains unclear, DMSO lowers the critical micelle concentration (*c.m.c.*) for many lipids. It is also a highly effective hydroxyl radical scavenger. DMSO (>5% vol/vol) stabilizes microtubules and reduces the dynamnicity. For these reasons, the final DMSO concentration in experimental samples should be kept <1-2% (vol/vol) to minimize side-effects. **Inhibition of Inflammasome Activation:** DMSO has long been known to exert inflammatory properties that commend its use in treating pain associated with arthritis and cystitis. DMSO attenuates interleukin IL-1β maturation, Casp1 activity, and ASC pyroptosome formation *via* NLRP3 inflammasome activators (5). NLRC4 and AIM2 inflammasome activity are unaffected, suggesting that DMSO is a selective inhibitor of the NLRP3 inflammasomes. The anti-inflammatory effect of DMSO has been confirmed in animal, LPS-endotoxin sepsis, and inflammatory bowel disease models. DMSO also inhibits LPS-mediating IL-1s transcription (5). **DMSO Reductase:** This molybdenum-containing enzyme catalyzes the reduction of DMSO to dimethyl sulfide (DMS). The reaction is: $DMSO + 2H^+ + 2e \rightarrow DMS + H_2O$, where the electron source is unspecified. **Target(s):** acetylcholinesterase (11-14); *N*-acetyllactosamine synthase (15); adenosine deaminase (11,12); alcohol dehydrogenase (13,14); arylformamidase (15); asparaginase (16); aspartate ammonia lyase (17); bis(2-ethylhexyl)phthalate esterase (18); carboxylesterase (19); carboxypeptidase Y, *or* carboxypeptidase C (20); cathepsin B (21); cathepsin L (22); Ca^{2+}-transporting ATPase (23-26); chitinase (27); cholesterol-5,6-oxide hydrolase (28); cytochrome c_3 hydrogenase (69); dihydrofolate reductase, slightly inhibited (29); DNA-directed DNA polymerase (30-32); DNA polymerase ε (25,26); endopeptidase La (33); F_1 ATPase, *or* ATP synthase (34); glucuronosyltransferase (35,36); L-glutamine:D-fructose-6-phosphate aminotransferase (37); hemoglobin S polymerization (38); histone acetyltransferase (39); inositol-phosphate phosphatase, weakly inhibited (40); laccase (71,72); lactose synthase (10); lipase, *or* triacylglycerol lipase (41); matrilysin, *or* matrix metalloproteinase-7 (42); membrane alanyl aminopeptidase (43); methionine *S*-adenosyltransferase (44-49); Mg^{2+}-importing ATPase (26); Na^+/K^+-exchanging ATPase (50-52); *N*-nitrosodimethylamine demethylase, *or* cytochrome P450 (53); pancreatic ribonuclease denaturation (54); papain (50); penicillin-binding protein 1b (56); peptidoglycan glycosyltransferase (56); peroxidase (70); poly(ADP-ribose) synthetase (57); polyphenol oxidase (58); protein-*S* isoprenylcysteine *O*-methyltransferase (50); ribonuclease H (32); soluble epoxide hydrolase (60-62); streptopain (63); tannase (64); telomerase (65); thrombin (66); tyrosinase, *or* monophenol monooxygenase (58); urease (67); Xaa-Pro aminopeptidase, *or* aminopeptidase P (68).

1a. Timasheff (2002) *Proc. Natl. Acad. Sci. U.S.A.* **99**, 9721. **1b.** Henderson & Henderson (1975) *Ann. N. Y. Acad. Sci.* **243**, 38. **2.** Himes, Burton, Kersey & Pierson (1976) *Proc. Natl. Acad. Sci. U.S.A.* **73**, 4397. **3.** Brisson (1974) *Can. Med. Assoc. J.* **110**, 1182. **4.** Marren (2011) *Phys. Sports Med.* **39**, 75. **5.** Ahn, Kim, Jeung & Lee (2013) *Immunobiol.* **219**, 315. **6.** Jagota (1992) *Indian J. Med. Res.* **96**, 275. **7.** Watts & Hoogmoed (1984) *Biochem. Pharmacol.* **33**, 365. **8.** Plummer, Greenberg, Lehman & Watts (1983) *Biochem. Pharmacol.* **32**, 151. **9.** Shlafer, Matheny & Karow (1976) *Arch. Int. Pharmacodyn. Ther.* **221**, 21. **10.** Pisvejcova, Rossi, Husakova, *et al.* (2006) *J. Mol. Catal. B* **39**, 98. **11.** Centelles, Franco & Bozal (1988) *J. Neurosci. Res.* **19**, 258. **12.** Centelles, Franco & Bozal (1987) *Comp. Biochem. Physiol. B* **86**, 95. **13.** Kozhemiakin, Zelenin, Bonitenko, *et al.* (1990) *Vopr. Med. Khim.* **36**, 67. **14.** Perlman & Wolff (1968) *Science* **160**, 317. **15.** Santti (1969) *Hoppe-Seyler's Z. Physiol. Chem.* **350**, 1279. **16.** Wriston (1971) *The Enzymes*, 3rd ed., **4**, 101. **17.** Karsten & Viola (1991) *Arch. Biochem. Biophys.* **287**, 60. **18.** Krell & Sandermann (1984) *Eur. J. Biochem.* **143**, 57. **19.** Saboori & Newcombe (1990) *J. Biol. Chem.* **265**, 19792. **20.** Lewis & Schuster (1991) *J. Biol. Chem.* **266**, 20818. **21.** Kamboj, Pal & Singh (1990) *J. Biosci.* **15**, 397. **22.** Kirschke, Kembhavi, Bohley & Barrett (1982) *Biochem. J.* **201**, 367. **23.** Kosterin, Bratkova, Slinchenko & Zimina (1998) *Biofizika* **43**, 1037. **24.** McConnell, Wagoner, Keenan & Raess (1999) *Biochem. Pharmacol.* **57**, 39. **25.** Romero (1992) *Cell Calcium* **13**, 659. **26.** Taffet & Tate (1992) *Arch. Biochem. Biophys.* **299**, 287. **27.** Andronopoulou & Vorgias (2003) *Extremophiles* **7**, 43. **28.** Levin, Michaud, Thomas & Jerina (1983) *Arch. Biochem. Biophys.* **220**, 485. **29.** Aboge, Jia, Terkawi, *et al.* (2008) *Antimicrob. Agents Chemother.* **52**, 4072. **30.** Niranjanakumari & Gopinathan (1993) *J. Biol. Chem.* **268**, 15557. **31.** Bambara & Jessee (1991) *Biochim. Biophys. Acta* **1088**, 11. **32.** Auer, Landre & Myers (1995) *Biochemistry* **34**, 4994. **33.** Waxman & Goldberg

(1985) *J. Biol. Chem.* **260**, 12022. **34**. al-Shawi & Senior (1992) *Biochemistry* **31**, 886. **35**. Uchaipichat, Mackenzie, Guo, *et al.* (2004) *Drug Metab. Dispos.* **32**, 413. **36**. Kaji & Kume (2005) *Drug Metab. Dispos.* **33**, 403. **37**. Chatterjee & Stefanovich (1976) *Arzneimittelforschung* **26**, 502. **38**. Waterman, Yamaoka, Dahm, Taylor & Cottam (1974) *Proc. Natl. Acad. Sci. U.S.A.* **71**, 2222. **39**. Eberharter, Lechner, Goralik-Schramel & Loidl (1996) *FEBS Lett.* **386**, 75. **40**. Vincendon, Corti, Guindani, *et al.* (1996) *J. Antibiot.* **49**, 710. **41**. Lee, Kim, Lee, *et al.* (2001) *Enzyme Microb. Technol.* **29**, 363. **42**. Oneda & Inouye (2000) *J. Biochem.* **128**, 785. **43**. Wachsmuth, Fritze & Pfleiderer (1966) *Biochemistry* **5**, 175. **44**. Geller, Kotb, Jernigan & Kredich (1986) *Exp. Eye Res.* **43**, 997. **45**. Kotb & Kredich (1985) *J. Biol. Chem.* **260**, 3923. **46**. Yarlett, Garofalo, Goldberg, *et al.* (1993) *Biochim. Biophys. Acta* **1181**, 68. **47**. Suman, Shimizu & Tsukada (1986) *J. Biochem.* **100**, 67. **48**. Okada, Teraoka & Tsukada (1981) *Biochemistry* **20**, 934. **49**. Sullivan & Hoffman (1983) *Biochemistry* **22**, 1636. **50**. Chiou & Vesely (1995) *Life Sci.* **57**, 945. **51**. Robinson (1989) *Biochim. Biophys. Acta* **994**, 95. **52**. Swann (1983) *J. Biol. Chem.* **258**, 11780. **53**. Yoo, Cheung, Patten, Wade & Yang (1987) *Cancer Res.* **47**, 3378. **54**. Richards & Wyckoff (1971) *The Enzymes*, 3rd ed., **4**, 647. **55**. Szabelski, Stachowiak & Wiczk (2001) *Acta Biochim. Pol.* **48**, 995. **56**. Nakagawa, Tamaki, Tomioka & Matsuhashi (1984) *J. Biol. Chem.* **259**, 13937. **57**. Banasik & Ueda (1999) *J. Enzyme Inhib.* **14**, 239. **58**. Wissemann & Montgomery (1985) *Plant Physiol.* **78**, 256. **59**. De Busser, Van Dessel & Lagrou (2000) *Int. J. Biochem. Cell Biol.* **32**, 1007. **60**. Morisseau, Archelas, Guitton, *et al.* (1999) *Eur. J. Biochem.* **263**, 386. **61**. Meijer & Depierre (1985) *Eur. J. Biochem.* **150**, 7. **62**. Bellevik, Zhang & Meijer (2002) *Eur. J. Biochem.* **269**, 5295. **63**. Kortt & Liu (1973) *Biochemistry* **12**, 338. **64**. Kar, Banerjee & Bhattacharyya (2003) *Proc. Biochem.* **38**, 1285. **65**. Sharma, Raymond, Soda, *et al.* (1998) *Leuk. Res.* **22**, 663. **66**. Pal & Gertler (1983) *Thromb. Res.* **29**, 175. **67**. Reithel (1971) *The Enzymes*, 3rd ed., **4**, 1. **68**. Orawski & Simmons (1995) *Biochemistry* **34**, 11227. **69**. Grande, van Berkel-Arts, Bregh, K. van Dijk & C. Veeger (1983) *Eur. J. Biochem.* **131**, 81. **70**. Patel, Singh, Moir & Jagannadham (2008) *J. Agric. Food Chem.* **56**, 9236. **71**. Junghanns, Pecyna, Böhm, *et al.* (2009) *Appl. Microbiol. Biotechnol.* **84**, 1095. **72**. Park & Park (2008) *J. Microbiol. Biotechnol.* **18**, 670.

N,*N*-Dimethyltryptamine

This strong hallucinogen (FW$_{free-base}$ = 188.27 g/mol; CAS 61-50-7; M.P. = 44.6-46.8°C; B.P.= 60-80°C; pK_a = 8.68), also named 2-(1*H*-indol-3-yl)-*N*,*N*-dimethylethanamine, found in a number of plants, especially the leaves of the cohoba tree *Piptadenia peregrine* and *Banisteriopsis caapi* (also known as ayahuasca, caapi or yajé) is inactivated, when given orally, due to degradation by visceral monoamine oxidase (1). The psychedelic brew made in the Amazon is also rich in harmine and harmaline. A rabbit lung enzyme *N*-methylates serotonin and tryptamine to form the psychotomimetic metabolites, bufotenine and *N*,*N*-dimethyltryptamine (2). **Target(s):** amine *N*-methyltransferase (3); aromatic-L-amino-acid decarboxylase (4); aryl acylamidase, weakly inhibited (5); indoleamine *N*-methyltransferase (6). **1**. Barker, Monti & Christian (1981) *Int. Rev. Neurobiol.* **22**, 83. **2**. Axelrod (1961) *Science* **134**, 343. **3**. Ansher & Jakoby (1986) *J. Biol. Chem.* **261**, 3996. **4**. Baxter & Slaytor (1972) *Phytochemistry* **11**, 2763. **5**. Paul & Halaris (1976) *Biochem. Biophys. Res. Commun.* **70**, 207. **6**. Porta (1987) *Meth. Enzymol.* **142**, 668.

Diminazine Aceturate

This antiprotozoal agent (FW = 515.53 g/mol; CAS 908-54-3), also known as Berenil®, is frequently used in the treatment of trypanosomiasis in domestic livestock. **Target(s):** *S*-adenosylmethionine decarboxylase (1-7); Ca^{2+}-dependent ATPase (8); diamine *N*-acetyltransferase, *or*

spermidine/spermine *N*-acetyltransferase (9,10); diamine oxidase (2,11); DNA polymerase (12); DNA topoisomerase I (13-23); DNA topoisomerase II (16,24,25); polyamine oxidase, weakly inhibited (9); protein biosynthesis (26); reverse transcriptase (27); ribosome (26); RNA-directed DNA polymerase (27); trypanosomal oligopeptidase (28); trypsin (29). **1**. Karvonen, Kauppinen, Partanen & Pösö (1985) *Biochem. J.* **231**, 165. **2**. Balana-Fouce, Garzon Pulido, Ordonez-Escudero & Garrido-Pertierra (1986) *Biochem. Pharmacol.* **35**, 1597. **3**. Bitonti, Dumont & McCann (1986) *Biochem. J.* **237**, 685. **4**. Gupta, Shukla & Walter (1987) *Mol. Biochem. Parasitol.* **23**, 247. **5**. Da'dara, Mett & Walter (1998) *Mol. Biochem. Parasitol.* **97**, 13. **6**. Hugo & Byers (1993) *Biochem. J.* **295**, 203. **7**. Paulin, Brander & Pösö (1987) *Experientia* **43**, 174. **8**. Ariyibi, Odunuga & Olorunsogo (2001) *J. Vet. Pharmacol. Ther.* **24**, 233. **9**. Libby & Porter (1992) *Biochem. Pharmacol.* **44**, 830. **10**. Wittich & Walter (1990) *Mol. Biochem. Parasitol.* **38**, 13. **11**. Karvonen (1987) *Biochem. Pharmacol.* **36**, 2863. **12**. Marcus, Kopelman, Koll & Bacchi (1982) *Mol. Biochem. Parasitol.* **5**, 231. **13**. Fairfield, Bauer & Simpson (1985) *Biochim. Biophys. Acta* **824**, 45. **14**. Shaffer & Traktman (1987) *J. Biol. Chem.* **262**, 9309. **15**. Chen, Yu, Gatto & Liu (1993) *Proc. Natl. Acad. Sci. U.S.A.* **90**, 8131. **16**. Goto, Laipis & Wang (1984) *J. Biol. Chem.* **259**, 10422. **17**. Riou, Douc-Rasy & Kayser (1986) *Biochem. Soc. Trans.* **14**, 496. **18**. Vosberg (1985) *Curr. Top. Microbiol. Immunol.* **114**, 19. **19**. Shaffer & Traktman (1987) *J. Biol. Chem.* **262**, 9309. **20**. Melendy & Ray (1987) *Mol. Biochem. Parasitol.* **24**, 215. **21**. Yoshida, Nakata & Ichishima (1991) *Phytochemistry* **30**, 3885. **22**. Riou, Gabillot, Douc-Rasy, Kayser & Barrois (1983) *Eur. J. Biochem.* **134**, 479. **23**. Heath-Pagliuso, Cole & Kmiec (1990) *Plant Physiol.* **94**, 599. **24**. Portugal (1994) *FEBS Lett.* **344**, 136. **25**. Melendy & Ray (1989) *J. Biol. Chem.* **264**, 1870. **26**. Sinharay, Ali & Burma (1977) *Nucl. Acids Res.* **4**, 3829. **27**. Juca & Aoyama (1995) *J. Enzyme Inhib.* **9**, 171. **28**. Morty, Troeberg, Pike, *et al.* (1998) *FEBS Lett.* **433**, 251. **29**. Junqueira, Silva & Mares-Guia (1992) *Braz. J. Med. Biol. Res.* **25**, 873.

Dinaciclib

This cyclin kinase-directed inhibitor (FW = 396.49 g/mol; CAS 779353-01-4; Solubility: 80 mg/mL DMSO; <1 mg/mL Water; Formulation: 20% hydroxypropyl-β-cyclodextran), also known as SCH727965 and (*S*)-3-(((3-ethyl-5-(2-(2-hydroxyethyl)piperidin-1-yl)pyrazolo[1,5-*a*]pyrimidin-7-yl)-amino)methyl)pyridine 1-oxide, interacts with the acetyl-lysine recognition site within bromodomains of cyclin-dependent kinase (1,2). Checkpoint kinases are essential serine/threonine protein kinases that respond to DNA damage and stalled DNA replication. *In vitro*, SCH 727965 inhibits CDK2 (IC$_{50}$ = 1 nM), CDK5 (IC$_{50}$ = 1 nM), CDK1 (IC$_{50}$ = 3 nM), and CDK9 (IC$_{50}$ = 4 nM), blocking thymidine incorporation into DNA (1). SCH 727965 induces regression of established solid tumors in a range of mouse models following intermittent scheduling of doses below the maximally tolerated level (1). In human tumor cell lines, dinaciclib exposure leads to the accumulation of double-strand DNA breaks and cell death. **1**. Parry, *et al.* (2010) *Mol. Cancer Ther.* **9**, 2344. **2**. Guzi, *et al.* (2011) *Mol. Cancer Ther.* **10**, 591. **3**. Martin, Olesen, Georg & Schönbrunn (2013) *ACS Chem. Biol.* **8**, 2360.

Dinactin

This macrotetralide *Streptomyces* antibiotic (FW = 764.99 g/mol; CAS 20261-85-2; Soluble in ethyl acetate, DMSO, 100% ethanol (moderate), and methanol (moderate)), also known as Antibiotic AKD-1C and Antibiotic S-3466A, is a hydrophobic ionophore and nonactin homologue. Dinactin binds and transports monovalent cations with the following selectivity: $NH_4^+ > K^+ \approx Rb^+ > Cs^+ > Na^+$ (1-3). Dinactin blocks cytokine production (IC_{50} = 10 ng/mL) through a post-transcriptional mechanism, inhibiting T-cell proliferation induced by IL-2 as well as cytokine production at nanomolar levels for IL-2, IL-4, and IL-5 (4). Dinactin also reduces pulmonary eosinophilia when administered within 1 day of airway antigen challenge (4). (Note the structural similarity of dinactin to synthetic crown ethers used by Nobelist Donald Cram as models for metal ion Host:Guest interactions. In Cram's nomenclature, dinactin is an ester analogue of an 32:8 crown ether.) **1.** Pressman (1976) *Ann. Rev. Biochem.* **45**, 501. **2.** Haynes & Pressman (1974) *J. Membr. Biol.* **18**, 1. **3.** Graven, Lardy, Johnson & Rutter (1966) *Biochemistry* **5**, 1729. **4.** Umland, Shah, Jakway, *et al.* (1999) *Am. J. Respir Cell Mol. Biol.* **20**, 481.

3,5-Dinitro-o-cresol

This yellow herbicide and insecticide (FW = 198.14 g/mol; CAS 497-56-3), also known as dinitrocresol, 4,6-dinitro-*o*-cresol and 3,5-dinitro-2-hydroxytoluene, is a strong protonophore that acts as a classical uncoupler of oxidative phosphorylation. {Note: Care should be exercised in literature searches, because the name 3,5-dinitro-*o*-cresol also occasionally refers to 2-methyl-3,5-dinitrophenol.] **Target(s):** creatine kinase (1); oxidative phosphorylation, as uncoupler (2); pyruvate decarboxylase (3). **1.** Kuby & Noltmann (1962) *The Enzymes*, 2nd ed., **6**, 515. **2.** Castilho, Vicente, Kowaltowski & Vercesi (1997) *Int. J. Biochem. Cell Biol.* **29**, 1005. **3.** Vandendriessche (1941) *Enzymologia* **10**, 69.

2,4-Dinitrophenol

This protonophore and uncoupling agent (FW = 184.11 g/mol; CAS 51-28-5; M.P. = 112-114°C CAS 51-28-5; Pale yellow solid; pK_a value of 4.1–4.5, 2,4-dinitrophenolate anion is intensely yellow; colorless < pH 2.6 and yellow > pH 4.4) strongly inhibits mitochondrial oxidative phosphorylation (1-3) and mildly uncouples chloroplast photophosphorylation by increasing the transmembrane proton leak (3-8). Because mitochondria-derived ATP fuels cytosolic glucose phosphorylation, DNP inhibition of oxidative phosphorylation also reduces glycolytic flux. **Mode of Action:** With a relatively low octanol:water partition coefficient (*e.g.*, {(Solubility$_{n\text{-}Octanol}$)/(Solubility$_{Water}$)} = 1.5), DNP would be expected to have limited penetration of biomembranes. Nonetheless, DNP binds protons on one side of a membrane and shuttles them to the opposite side. The exact structure of the active protonophore remains unclear. Finkelstein (8) has proposed that the primary charge carrier in the membrane is neither H^+ or OH^-; rather, it is likely to be a dimer formed between the undissociated and dissociated form of the weak acid. He showed that all of the data on the action of these weak acids on thin lipid membranes is consistent with this idea. Although such associations and dissociations occur at random, the probability of proton binding is far greater on the side having the highest proton concentration. When sufficient DNP is present, maintenance of a proton gradient becomes highly unlikely. DNP also binds to the bilayer surface, modifying its Zeta Potential (*i.e.*, its charge density). (**See also** Uncouplers (Gradient-Dissipating Proton Carriers); 3,5-Dinitro-o-cresol; Dinoseb; FCCP) Finally, despite a slightly higher synthesis of cytosolic ATP in the presence of DNP, there is net mitochondrial uptake of ATP via the adenine nucleotide translocase, exchanging cytosolic ATP for mitochondrial ADP, the latter rising as oxidative phosphorylation fails in DNP's presence. **Hazzards of DNP Use for Weight Loss:** When taken regularly, DNP can increase the average metabolic rate in humans by ~11% for every 100 mg administered. Although once touted as a highly effective diet aid, DNP overdose often induces fatal hyperthermia as a direct consequence of uncoupler-induced thermogenesis. Other symptoms include tachycardia, diaphoresis, and rapid breathing. DNP is highly toxic, with unremitting high fever the greatest risk to poisoned individuals. Acute administration of 20–50 mg/kg in humans can be lethal, with formation of cataracts a common side-effect at lower dosage. To quote the famous mitochondrial researcher Efraim Racker (p. 155 of *A New Look at Mechanisms in Bioenergetics*, Academic Press, 1976), "...treatment eliminated not only the fat, but also the patients,... This discouraged physicians for awhile..." The majority of the damage results from unbridled thermogenesis, wherein heat is produced in place of ATP. While unappreciated by unwary dieters and bodybuilders intent on ridding their bodies of fat, the greatest threat of DNP consumption is that biological individuality makes its toxicity unpredictable and thus hazardous. Responding to a number of DNP-associated deaths in the 1930s, the U.S. Food & Drug Administration classified DNP as extremely dangerous and unfit for human consumption. **Controlled Release DNP:** A novel DNP formulation with a sustained-release coating results in lower peak plasma concentrations (1-5 µM) and sustained-release pharmacokinetics, thereby increasing the therapeutic index (*i.e.*, ratio of toxic to effective dose) by 25-fold over liver-targeted DNP and 1250-fold over unaltered DNP (9). Metabolic studies in rats receiving a 5-day continuous, low-dose intragastric infusion of DNP support the potential of using mitochondrial protonophores and other mitochondrial uncoupling agents to treat nonalcoholic fatty liver disease (NAFLD), metabolic syndrome, as well as Type 2 diabetes. **Target(s):** acylphosphatase (10); aldehyde oxidase (11); D-amino-acid oxidase (12); apyrase (13); arginine transport (14); ATP phosphoribosyl-transferase (15); calcidiol 1-monooxygenase, *or* 25-hydroxyvitamin D, *or* α-hydroxylase (29,59-61); catechol oxidase (63); cellulase (15); choline sulfotransferase (16); chymotrypsin (17); *trans*-cinnamate 4-monooxygenase (61); deoxycytidine kinase, weakly inhibited (19); dihydroorotate dehydrogenase (20); dimethylpropiothetin dethiomethylase (20); glucose-6-phosphate isomerase (22); glutaminase (23); glutamine synthetase (24); glycogen phosphorylase (25); hexokinase (26,27); 4-hydroxybenzoate nonaprenyl-transferase (27); isocitrate dehydrogenase (NAD^+) (30); leukocyte phagocytosis and clumping (31,32); lysine transport (14); malate dehydrogenase (33,34); NADH dehydrogenase (quinone) (35,36); NADH oxidase (36); NAD(P)H dehydrogenase (quinone), *or* diaphorase (35,37-39); 2-nitrophenol 2-monooxygenase (5,10); oxidative and photo-phosphorylation (1-9,41-44); phosphatidylglycero-phosphatase (45); phosphatidylserine decarboxylase (46); 3-phytase (47,48); polyphenol oxidase (63); pyruvate decarboxylase (49-51); rubredoxin:NAD^+ reductase (52); signal peptidase I, *or* leader peptidase (53-55); trypsin (18,56); very long chain fatty acid α-hydroxylase (57); xylose isomerase (58). **1.** Slater (1967) *Meth. Enzymol.* **10**, 48. **2.** Loomis & Lipmann (1948) *J. Biol. Chem.* **173**, 807. **3.** Heytler (1979) *Meth. Enzymol.* **55**, 462. **4.** Izawa & Good (1972) *Meth. Enzymol.* **24**, 355. **5.** Karlish & Avron (1968) *FEBS Lett.* **1**, 21. **6.** Neumann & Jagendorf (1964) *Biochem. Biophys. Res. Commun.* **16**, 562. **7.** Siow & Unrau (1968) *Biochemistry* **7**, 3507. **8.** Finkelstein (1970) *Biochim. Biophys. Acta.* **205**, 1. **9.** Terry, Zhang, Zhang, Boyer & Shulman (2015) *Science* **347**, 1253. **10.** Raijman, Grisolia & Edelhoch (1960) *J. Biol. Chem.* **235**, 2340. **11.** Rajagopalan, Fridovich & Handler (1962) *J. Biol. Chem.* **237**, 922. **12.** Yagi, Osawa & Okada (1959) *Biochim. Biophys. Acta* **35**, 102. **13.** Curdova, Jechova & Hostalek (1982) *Folia Microbiol.* **27**, 159. **14.** Maretzki & Thom (1970) *Biochemistry* **9**, 2731. **15.** Dall-Larsen, Kryvi & Klungsoyr (1976) *Eur. J. Biochem.* **66**, 443. **16.** Olutiola (1982) *Experientia* **38**, 1332. **17.** Orsi & Spencer (1964) *J. Biochem.* **56**, 81. **18.** Bonewell & Rossini (1969) *Ital. J. Biochem.* **18**, 457. **19.** Kessel (1968) *J. Biol. Chem.* **243**, 4739. **20.** Chen & Jones (1976) *Arch. Biochem. Biophys.* **176**, 82. **21.** Yoch, Ansede & Rabinowitz (1997) *Appl. Environ. Microbiol.* **63**, 3182. **22.** Alvarado (1963) *Enzymologia* **26**, 12. **23.** Goldstein, Richterich-Van Baerle & Dearborn (1957) *Enzymologia* **18**, 261 and 355. **24.** Richterich-Van Baerle, Goldstein & Dearborn (1957) *Enzymologia* **18**, 327. **25.** Soman & Philip (1974) *Biochim. Biophys. Acta* **358**, 359. **26.** Grillo & Cafiero (1964) *Biochim. Biophys. Acta* **82**, 92. **27.** Saltman (1953) *J. Biol. Chem.* **200**, 145. **28.** Uchida, Koizumi, Kawaji, Kawahara & Aida (1991) *Agric. Biol. Chem.* **55**, 2299. **29.** Lobaugh, Almond & Drezner (1986) *Meth. Enzymol.* **123**, 159. **30.** Stein, Kirkman & Stein (1967) *Biochemistry* **6**, 3197. **31.** Allison, Lancaster & Crosthwaite (1963) *Amer. J. Path.* **43**, 775. **32.** Allison & Lancaster (1964) *Ann. N. Y. Acad. Sci.* **116**, 936. **33.** Wedding, Hansch & Fukuto (1967) *Arch. Biochem. Biophys.* **121**, 9. **34.** Wuff & Ionesco (1947) *Compt. rend.* **225**, 263. **35.** Koli, Yearby, Scott & Donaldson (1969) *J. Biol. Chem.* **244**, 621. **36.** Suzuki (1966) *Enzymologia* **30**, 215. **37.** Wosilait & Nason (1955) *Meth. Enzymol.* **2**, 725. **38.** Martius (1963) *The Enzymes*, 2nd ed., **7**, 517. **39.** Raw, Nogueira & Filho (1961) *Enzymologia* **23**, 123. **40.** Zeyer & Kocher (1988) *J.*

Bacteriol. **170**, 1789. **41**. Hanstein (1976) *Trends Biochem. Sci.* **1**, 65. **41**. Terner (1951) *J. Biochem.* **49**, ii. **42**. Parker (1956) *Nature* **178**, 261. **43**. Middlebrook (1957) *Enzymologia* **18**, 337. **44**. Icho & Raetz (1983) *J. Bacteriol.* **153**, 722. **45**. Hovius, Faber, Brigot, Nicolay & de Kruijff (1992) *J. Biol. Chem.* **267**, 16790. **46**. Youssef, Ghareib & Nour el Dein (1987) *Zentralbl. Mikrobiol.* **142**, 397. **47**. Ghareib, Youssef & Nour el Dein (1988) *Zentralbl. Mikrobiol.* **143**, 397. **48**. Vandendriessche (1941) *Enzymologia* **10**, 69. **49**. Massart & Vandendriessche (1942) *Enzymologia* **10**, 244. **50**. Massart & Dufait (1941) *Enzymologia* **9**, 320. **51**. Petitdemange, Marczak, Blusson & Gay (1979) *Biochem. Biophys. Res. Commun.* **91**, 1258. **52**. Wolfe, Zwizinski & Wickner (1983) *Meth. Enzymol.* **97**, 40. **53**. Tschantz & Dalbey (1994) *Meth. Enzymol.* **244**, 285. **54**. Packer, Andre & Howe (1995) *Plant Mol. Biol.* **27**, 199. **55**. Rossini (1963) *Hoppe-Seyler's Z. Physiol. Chem.* **333**, 1. **56**. Kishimoto (1978) *Meth. Enzymol.* **52**, 310. **57**. Slein (1962) *Meth. Enzymol.* **5**, 347. **58**. Gray, Omdahl, Ghazarian & DeLuca (1972) *J. Biol. Chem.* **247**, 7528. **59**. Henry & Norman (1974) *J. Biol. Chem.* **249**, 7529. **60**. Paulson & DeLuca (1985) *J. Biol. Chem.* **260**, 11488. **61**. Billett & Smith (1978) *Phytochemistry* **17**, 1511. **62**. Xu, Zheng, Meguro & Kawachi (2004) *J. Wood Sci.* **50**, 260.

2,4-Dinitrophenyl 2-Deoxy-2-fluoro-β-D-glucopyranoside

This glucoside (FW = 348.24 g/mol) irreversibly inhibits certain β-glucosidases by trapping the enzyme as a covalent intermediate during catalysis (1). The intermediate is stable, but undergoes turnover in the presence of cellobiose, reactivating the enzyme by transglycosylation. Use of tritium-labeled inactivator facilitated isolation and sequencing of a radiolabeled peptide from this enzyme, confirming that the active site nucleophile is glutamate-274. Notably, this active-site glutamate and nearby amino acid residues are absolutely conserved in the homologous family F of cellulases. **Target(s):** glucan 1,3-β-glucosidase, *or* exo-β-(1,3)-glucanase (1,2); β-glucosidase (3,4); phlorizin hydrolase, *or* glycosylceramidase (5-8); lactase, *or* lactase/phlorizin hydrolase, *or* glycosyl-ceramidase (5-7). **1**. Tull, Withers, Gilkes, *et al.* (1991) *J. Biol. Chem.* **266**, 15621. **2**. Cutfield, Davies, Murshudov, *et al.* (1999) *J. Mol. Biol.* **294**, 771. **3**. Withers, Street, Bird & Dolphin (1987) *J. Amer. Chem. Soc.* **109**, 7530. **4**. Berrin, McLauchlan, Needs, *et al.* (2002) *Eur. J. Biochem.* **269**, 249. **5**. Mackey, Henderson & Gregory (2002) *J. Biol. Chem.* **277**, 26858. **6**. Day, Cañada, Diaz, *et al.* (2000) *FEBS Lett.* **468**, 166. **7**. Arribas, Herrero, Martin-Lomas, *et al.* (2000) *Eur. J. Biochem.* **267**, 6996. **8**. Paal, Ito & Withers (2004) *Biochem. J.* **378**, 141.

1,2-Dioctanoyl-*sn*-glycero-3-pyrophosphate

This phosphatidate analogue (FW$_\text{free-acid}$ = 517.47 g/mol) is an antagonist for lysophosphaidate receptors and was the first selective Edg-7 (LPA3) receptor antagonist (K_i = 106 nM). The K_i value for Edg-3 (LPA1) is 6.6 μM. **1**. Fischer, Nusser, Virag, *et al.* (2001) *Mol. Pharmacol.* **60**, 776.

Dioctyl Sulfosuccinate, Sodium Salt

This anionic detergent (FW = 444.57 g/mol), also known as Aerosol OT, AOT, docusate sodium, and bis(2 ethylhexyl) sodium sulfosuccinate, is

stable in acidic or neutral solution but will hydrolyze in alkaline solutions. Note that is it is bactericidal against Gram-positive microorganisms at neutral pH but is bactericidal against both Gram positive and negative organisms at pH 4. **Target(s):** acid phosphatase (1); α-amylase (2); cAMP phosphodiesterase (3); cutinase (4,5); lipase, *or* triacylglycerol lipase (6); phosphatidylinositol diacylglycerol-lyase, *or* phosphatidyl-inositol-specific phospholipase C (7). **1**. Lalitha & Mulimani (1997) *Biochem. Mol. Biol. Int.* **41**, 797. **2**. Gajjar, Dubey & Srivastava (1994) *Appl. Biochem. Biotechnol.* **49**, 101. **3**. Simon & Kather (1980) *Eur. J. Clin. Invest.* **10**, 231. **4**. Ternstrom, Svendsen, Akke & Adlercreutz (2005) *Biochim. Biophys. Acta* **1748**, 74. **5**. Goncalves, Serro, Aires-Barros & Cabral (2000) *Biochim. Biophys. Acta* **1480**, 92. **6**. Huang, Locy & Weete (2004) *Lipids* **39**, 251. **7**. Vizitiu, Kriste, Campbell & Thatcher (1996) *J. Mol. Recognit.* **9**, 197.

1,2-Dioleoyl-*sn*-glycerol

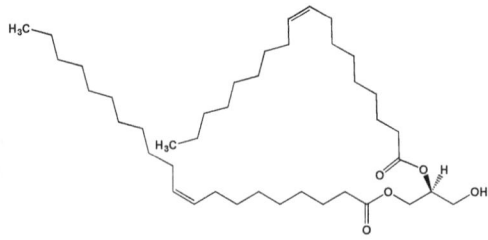

This diacylglycerol (FW = 621.00 g/mol), also known as 1,2-diolein, inhibits phospholipase C and CDP-diacylglycerol:serine *O*-phosphatidyl-transferase. It also activates protein kinase C. Note: Acyl group migration can occur during storage, especially under acidic conditions. **Target(s):** CDP-diacylglycerol:serine *O*-phosphatidyl-transferase (1,2); phosphatidate phosphatase (3); phospho-lipase C (4); sphingomyelin phosphodiesterase (5). **1**. Carman & Bae-Lee (1992) *Meth. Enzymol.* **209**, 298. **2**. Bae-Lee & Carmen (1990) *J. Biol. Chem.* **265**, 7221. **3**. Kanoh, Imai, Yamada & Sakane (1992) *J. Biol. Chem.* **267**, 25309. **4**. Haeffner & Wittmann (1992) *J. Lipid Mediat.* **5**, 237. **5**. Liu, Nilsson & Duan (2002) *Lipids* **37**, 469.

1,2-Dioleoyl-*sn*-glycerol 3-Phosphate

This phosphatide (FW$_\text{free-acid}$ = 700.98 g/mol), also known as dioleoyl phosphatidate, inhibits AMP deaminase and the Ca^{2+}-transporting ATPase. (**See also** *Phosphatidate and Phosphatidic Acid*) **Target(s):** AMP deaminase (1); Ca^{2+}-transporting ATPase (2); chymase (3); glycosyl-phosphatidylinositol phospholipase D (4); glyoxalase II, *or* hydroxyacylglutathione hydrolase (5); phosphoprotein phosphatase (6); protein phosphatase 1 (6). **1**. Tanfani, Kulawiak, Kossowska, *et al.* (1998) *Mol. Genet. Metab.* **65**, 51. **2**. Dalton, East, Mall, *et al.* (1998) *Biochem. J.* **329**, 637. **3**. Kido, Fukusen & Katunuma (1985) *Arch. Biochem. Biophys.* **239**, 436. **4**. Lee, Lee, Kim, Myung & Sok (1999) *Neurochem. Res.* **24**, 1577. **5**. Scirè, Tanfani, Saccucci, Bertoli & Principato (2000) *Proteins* **41**, 33. **6**. Plummer, Perreault, Holmes & Posse de Chaves (2005) *Biochem. J.* **385**, 685.

(2,3*S*,18*R*,19*R*)-Dioxidosqualene

This diepoxide (FW = 442.73 g/mol) inhibits (apparently reversibly) pig liver lanosterol synthase *or* 2,3-oxidosqualene-lanosterol cyclase (EC 5.4.99.7), with a K_i value of 21 nM. OSLC is the key enzyme of the ring cyclization reaction suffered by (3*S*)-2,3-oxidosqualene to give lanosterol as the final steroid precursor in fungi and mammalian systems. While the mechanism of this cyclization is still debated, most favor a stepwise process, that proceeds through a serie of conformationally rigid, partially cyclized carbocationic intermediates. **1.** Abad, Guardiola, Casas, Sánchez-Baeza & Messeguer (1996) *J. Org. Chem.* **61**, 7603.

2,4-Dioxopentanoate

This α,γ-diketo acid (FW$_{free-acid}$ = 130.10 g/mol), also known as 2,4-dioxopentanoic acid and 2,4-diketopentanoic acid, is an acetoacetate analogue that reversibly inhibits acetoacetate decarboylase ($K_i \approx 10^{-7}$ M). It reacts with an active-site lysyl residue to form a stable chromophore ($\lambda_{max} \approx$ 325 nm; $\varepsilon \approx$ 14,000 cm^{-1}M^{-1}), with the bound inhibitor (likely a *cis*-enamine) exhibiting a ten-minute dissociation half-life. **Target(s):** acetoacetate decarboxylase (1-7); 4-hydroxyphenylpyruvate dioxygenase (8); lactate dehydrogenase (9); phosphoenolpyruvate carboxykinase (GTP), weakly inhibited (10,11); ureidoglycolate lyase (12). **1.** Tagaki, Guthrie & Westheimer (1968) *Biochemistry* **7**, 905. **2.** Davies (1943) *Biochem. J.* **37**, 230. **3.** Seeley (1955) *Meth. Enzymol.* **1**, 624. **4.** Westheimer (1969) *Meth. Enzymol.* **14**, 231. **5.** Utter (1961) *The Enzymes*, 2nd ed., **5**, 319. **6.** Fridovich (1972) *The Enzymes*, 3rd ed., **6**, 255. **7.** Colman (1962) Ph.D. Thesis, Radcliffe College, Cambridge, Mass. **8.** Lindstedt & Rundgren (1982) *J. Biol. Chem.* **257**, 11922. **9.** Meister (1950) *J. Biol. Chem.* **184**, 117. **10.** Guidinger & Nowak (1990) *Arch. Biochem. Biophys.* **278**, 131. **11.** Silverstein, Lin, Fanning & Hung (1980) *Biochim. Biophys. Acta* **614**, 534. **12.** McIninch, McIninch & May (2003) *J. Biol. Chem.* **278**, 50091.

Dipeptidyl Peptidase IV Inhibitors

Glucagon-like peptide-1 (GLP-1), an incretin that displays a strong insulin secretion effect on pancreatic β-cells, contains a His-Ala cleavage site at the N-terminus of its active form. This site is targeted and cleaved by Dipeptidyl Peptidase IV (DPP-4), and inhibition of incretin proteolysis results in sustained insulin release. This protease also uses its S$_1$, S$_2$, S$_{1'}$, S$_{2'}$ and S$_2$ subsites to interacts with inhibitors: (a) Vildagliptin and Saxagliptin bind to the S$_1$ and S$_2$ subsites, (b) Alogliptin, Linagliptin and Trelagliptin bind to the S$_{1'}$ and/or S$_{2'}$ subsites in addition to the S$_1$ and S$_2$ subsites, and (c) Sitagliptin, Anagliptin, Teneligliptin and Omarigliptin bind to the S$_1$, S$_2$ and S$_2$ extensive subsites. The affinity of each DPP-4 inhibitor appears to reflect the binding energy available from interactions at these subsites as well as nearby steric interactions, but the efficacy of each is strongly strongly affected by its own ADME properties. ***See specific inhibitor***

Diphenhydramine

This tertiary amine (FW$_{free-base}$ = 255.36 g/mol; CAS 58-73-1), widely known as Benadryl and systematically as 2-diphenylmethoxy-*N,N*-dimethylethanamine, is a histamine H$_1$ receptor antagonist. Diphenhydramine is most often supplied as the hydrochloride and is readily absorbed. **Target(s):** amine oxidase (1); CYP2D6 (2); histamine H$_1$ receptor (3,4); histamine *N*-methyltransferase (5); Na$^+$ current (6). **1.** Tabor, Tabor & Rosenthal (1955) *Meth. Enzymol.* **2**, 390. **2.** Lessard, Yessine, Hamelin, *et al.* (2001) *J. Clin. Psychopharmacol.* **21**, 175. **3.** Hill, Ganellin, Timmerman, *et al.* (1997) *Pharmacol. Rev.* **49**, 253. **4.** Saitou, Kaneko, Sugimoto, Chen & Kamei (1999) *Biol. Pharm. Bull.* **22**, 1079. **5.** Horton, Sawada, Nishibori & Cheng (2005) *J. Mol. Biol.* **353**, 334. **6.** Kuo, Huang & Lou (2000) *Mol. Pharmacol.* **57**, 135.

Diphenylchloroarsine

This organoarsenical (FW = 264.59 g/mol; Referred to as CLARK I; Chemical warfare symbol = DA) is a riot control and vomit-inducing agent that is fast-acting rapid. Diphenylchloroarsine was first used in combat in 1917 by Germany in World War I. It has also been. **Target(s):** aconitase (1-3); arginine kinase (4); cholinesterase (5); creatine kinase (6,7); isocitrate dehydrogenase (8-10); phospholipase B (11); urease (12). **1.** Dickman (1961) *The Enzymes*, 2nd ed., **5**, 495. **2.** Glusker (1971) *The Enzymes*, 3rd ed., **5**, 413. **3.** Peters (1955) *Bull. Johns Hopkins Hosp.* **97**, 1. **4.** Morrison, Griffiths & Ennor (1957) *Biochem. J.* **65**, 143. **5.** Mounter & Whittaker (1953) *Biochem. J.* **53**, 167. **6.** Kuby & Noltmann (1962) *The Enzymes*, 2nd ed., **6**, 515. **7.** Rosenberg & Ennor (1955) *Biochim. Biophys. Acta* **17**, 261. **8.** Ochoa (1955) *Meth. Enzymol.* **1**, 699. **9.** Plaut (1963) *The Enzymes*, 2nd ed., **7**, 105. **10.** Lotspeich & Peters (1951) *Biochem. J.* **49**, 704. **11.** Dawson (1956) *Biochem. J.* **64**, 192. **12.** Rona & György (1920) *Biochem. Z.* **111**, 115.

Diphenyleneiodonium Chloride

This aromatic iodonium salt (FW$_{choride}$ = 314.55 g/mol; CAS 4673-26-1; *Abbreviation*: DPI) inhibits respiratory enzyme NADH dehydrogenase. Addition of 10 μM DPI abolished the reduction of both the FAD and the cytochrome *b* components of the NADPH oxidase. DPI also catalyses an exchange of Cl$^-$ for OH$^-$ across membranes and induces apoptosis. Another DPI target, aldehyde dehydrogenase II, catalyzes the vascular bioactivation of the antianginal drug nitroglycerin to yield nitric oxide, which in turn activates soluble guanylate cyclase, resulting in cGMP-mediated vasodilation (1). Note that the main pathway of ALDH II-catalyzed denitration yields 1,2-glycerol dinitrate and inorganic nitrite, but a minor reaction results in direct formation of NO (2). DPI is a competitive inhibitor (K_i = 2 μM) relative to nitroglycerin in the redox and nitrate esterase reactions catalyzed by ALDH II. Importantly, DPI inhibits the vasodilatory action of nitroglycerin in the same concentration range, providing strong evidence that ALDH II is responsible for transforming nitroglycerin into nitric oxide (or a related species). **Effect on Iodide Transport:** At concentrations used to those which inhibit the generation of O$_2$ derivatives, DPI activates iodide efflux but not of its analogues, monovalent technetium or rhodium in the PCCl3 rat thyroid cell line and in COS cell lines expressing the iodide transporter NIS. Therefore, the inhibitory effects of DPI, especially in thyroid cells, must be interpreted with caution (3). **Key Pharmacokinetic Parameters:** *See* Appendix II in Goodman & Gilman's *THE PHARMACOLOGICAL BASIS OF THERAPEUTICS*, 12th Edition (Brunton, Chabner & Knollmann, eds.) McGraw-Hill Medical, New York (2011). **Target(s):** Ca^{2+} currents (4); cytokinin dehydrogenase (5); hydrogenase (5); K$^+$ currents (4); NADH dehydrogenase, *or* Complex I (6,7); NADPH:cytochrome P450 reductase (8); NADPH oxidase (9,10); nitric-oxide dioxygenase (11,12); nitric-oxide synthase (13,14); oxidative phosphorylation (15); photophosphorylation (16); protoporphyrinogen oxidase (17); xanthine dehydrogenase (18,19); xanthine oxidase (20-22); zeaxanthin epoxidase (23). **1.** Neubauer, Neubauer, Wolkart, *et al.* (2013) *Mol. Pharmacol.* **84**, 407. *and* Wenzl, Beretta, Griesberger, *et al.* (2011) *Mol. Pharmacol.* **80**, 258. **2.** Massart, Giusti, Beauwens, *et al.* (2013) *FEBS Open Bio.* **4**, 55. **3.** Weir, Wyatt, Reeve, *et al.* (1994) *J. Appl. Physiol.* **76**, 2611. **4.** Galuszka, Frebort, Sebela, *et al.* (2001) *Eur. J. Biochem.* **268**, 450. **5.** Magnani, Doussiere & Lissolo (2000) *Biochim. Biophys. Acta* **1459**, 169. **6.** Ragan & Bloxham (1977) *Biochem. J.* **163**, 605. **7.** Hayes, Byrne, Shoubridge, Morgan-Hughes & Clark (1985) *Biochem. J.* **229**, 109. **8.** McGuire, Anderson, McDonald, Narayanasami & Bennett (1998) *Biochem. Pharmacol.* **56**, 881. **9.** Ellis, Mayer & Jones (1988) *Biochem. J.* **251**, 887. **10.** Doussiere & Vignais (1992) *Eur. J. Biochem.* **208**, 61. **11.** Igamberdiev, Seregélyes, Manac'h & Hill (2004) *Planta* **219**, 95. **12.** Schmidt & Mayer (2004) *FEBS Lett.* **577**, 199. **13.** Stuehr, Fasehun, Kwon, *et al.* (1991) *FASEB J.* **5**, 98. **14.** Wever, van Dam, van Rijn, de Groot & Rabelink (1997) *Biochem. Biophys. Res. Commun.* **237**, 340. **15.** Holland & Sherrat (1971) *Biochem. J.* **121**, 42. **16.** Watling-Payne & Selwyn (1974) *Biochem. J.* **142**, 65. **17.** Arnould, Berthon, Hubert, *et al.* (1997) *Biochemistry* **36**, 10178. **18.** Atmani, Baghiani, Harrison & Benboubetra (2005) *Int. Dairy J.* **15**, 1113. **19.** Yesbergenova, Yang, Oron, *et al.* (2005) *Plant J.* **42**, 862. **20.** Miller (2004) *FEBS Lett.* **562**, 129. **21.** Li, Samouilov, Liu & Zweier (2003) *Biochemistry* **42**, 1150. **22.** Li,

Samouilov, Liu & Zweier (2004) *J. Biol. Chem.* **279**, 16939. **23**. Büch, Stransky & Hager (1995) *FEBS Lett.* **376**, 45.

5,5-Diphenylhydantoin

This substituted imidazolidinedione (FW = 252.27 g/mol; CAS 57-41-0), widely known as Dilantin, stabilizes excitable membranes and is an anticonvulsant drug often used to treat neuropathic pain. Although first synthesized in 1908, its anticonvulsant properties were reported in 1938 (1,2), making it the first anti-epileptic drug not to act as a sedative at typical concentrations. Dilantin's discovery was not simply a landmark in the history of anti-convulsants; it also spurred research into discovering the differences between the numerous convulsive disorders. **Target(s):** acetoin dehydrogenase (3); alcohol dehydrogenase (NADP$^+$), *or* aldehyde reductase (3-5); 4-aminobutyrate aminotransferase (6); Ca^{2+} currents (7); Ca^{2+}-dependent protein kinase (8); carbonyl reductase (3); glycerol dehydrogenase (9); K$^+$ channels, delayed rectifier (10); Na$^+$ currents (11,12); procollagen-proline 4-dioxygenase (13); succinate-semialdehyde dehydrogenase (6). **1**. Merritt & Putnam (1938) *Arch. Neurol. Psychiatry* **39**, 1003. **2**. Merritt & Putnam (1938) *J. Amer. Med. Asso.* **111**, 1068. **3**. Hara, Seiriki, Nakayama & Sawada (1985) *Prog. Clin. Biol. Res.* **174**, 291. **4**. Erwin & Deitrich (1973) *Biochem. Pharmacol.* **22**, 2615. **5**. Davidson & Murphy (1985) *Prog. Clin. Biol. Res.* **174**, 251. **6**. Whittle & Turner (1978) *J. Neurochem.* **31**, 1453. **7**. Todorovic & Lingle (1998) *J. Neurophysiol.* **79**, 240. **8**. Delorenzo & Glaser (1976) *Brain Res.* **105**, 381. **9**. Flynn & Cromlish (1982) *Meth. Enzymol.* **89**, 237. **10**. Nobile & Lagostena (1998) *Brit. J. Pharmacol.* **124**, 1698. **11**. Rush & Elliott (1997) *Neurosci. Lett.* **226**, 95. **12**. Tomaselli, Marban & Yellen (1989) *J. Clin. Invest.* **83**, 1724. **13**. Kivirikko & Myllylä (1980) *Enzymol. Post-transl. Modif. Proteins* (Freedman & Hawkins, eds.) **1**, 53, Academic Press, New York.

Diphtheria Toxin

This toxic ADP-ribosyltransferase (MW$_{active-form}$ = 21 kDa; CAS 92092-36-9) from *Coryne-bacterium diphtheriae* inhibits protein synthesis by catalyzing the NAD$^+$-dependent ADP-ribosylation of Elongation Factor-2 (EF-2) of eukaryotes and archaebacteria, forming diphthamide, *or* 2-[3-carboxyamido-3-(trimethylammonio)propyl]histidine (1-12). The toxin begins as a 567-residue zymogen that, upon gaining entry into a target cell, undergoes cleavage of one disulfide bond and one peptide bond, thereby generating the active ADP-ribosyltransferase. The latter then catalyzes covalent modification of EF-2, preventing the transfer to the tRNA-bound amino acid to a growing polypeptide chain. In 1888, Roux and Yersin demonstrated that diphtheria culture filtrates (*i.e.*, absent intact bacilli) contained a toxin that, when injected into an animal, produced all the symptoms of diphtheria. Diphtheria was thus the first infectious disease demonstrated to be solely caused by a toxin, and it was also the first bacterial toxin to be purified into crystalline form (13). In humans, the lethal dose is ~0.1 μg of toxin per kg bodyweight. **1**. Passador & Iglewski (1994) *Meth. Enzymol.* **235**, 617. **2**. Oppenheimer & Handlon (1992) *The Enzymes*, 3rd ed., **20**, 453. **3**. Neville & Hudson (1986) *Ann. Rev. Biochem.* **55**, 195. **4**. Lai (1986) *Adv. Enzymol.* **58**, 99. **5**. Ueda & Hayaishi (1985) *Ann. Rev. Biochem.* **54**, 73. **6**. Pappenheimer (1977) *Ann. Rev. Biochem.* **46**, 69. **7**. Hayaishi & Ueda (1977) *Ann. Rev. Biochem.* **46**, 95. **8**. Pestka (1974) *Meth. Enzymol.* **30**, 261. **9**. Carrasco, Battaner & Vazquez (1974) *Meth. Enzymol.* **30**, 282. **10**. Carroll & Collier (1988) *Meth. Enzymol.* **165**, 218. **11**. Carroll, Barbieri & Collier (1988) *Meth. Enzymol.* **165**, 68. **12**. Jiménez (1976) *Trends Biochem. Sci.* **1**, 28. **13**. Pappenheimer, Jr. (1937) *J. Biol. Chem.* **120**, 543.

Diphyllin

This naturally occurring naphthofuran-containing lignin (FW = 380.35 g/mol; CAS 22055-22-7), named systematically as 9-(1,3-benzodioxol-5-yl)-4-hydroxy-6,7-dimethoxynaphtho[2,3-*c*]furan-1(3*H*)-one;4-Hydroxy-6,7-dimethoxy-9-(1,3-benzodioxole-5-yl)-1,3-dihydronaphtho[2,3-*c*] furan-1-one, from *Cleistanthus collinus*, potently inhibits acid influx (IC$_{50}$ = 0.6 nM) and directly inhibits vacuolar ATPase, *or* V-ATPase, (IC$_{50}$ = 17 nM), slightly more weakly than bafilomycin A$_1$ (IC$_{50}$ = 4 nM) (1). Remarkably, diphyllin is also cytostatic, inhibiting human SGC7901 gastric cancer cell proliferation, inducing apoptosis. It decreases LRP6 phosphorylation, but not the level of LRP6 (2), a diphyllin effect that was traced to altered Wnt/β-catenin signaling, which reduces expression of β-catenin, *c*-myc and cyclin D$_1$. Diphyllin also alters cellular susceptibility to influenza viruses through the inhibition of endosomal acidification, thus interfering with downstream virus replication (3). **1**. Sørensen, Henriksen, Neutzsky-Wulff, Dziegiel & Karsdal (2007) *J. Bone Miner. Res.* **22**, 1640. **2**. Shen, Zoua, Chen, *et al.* (2011) *Eur. J. Pharmacol.* **667**, 330. **3**. Chen, Cheng, Liu, *et al.* (2013) *Antiviral Res.* **143**, 32.

Dipropylcyclopentylxanthine

This A$_1$ receptor antagonist (FW = 304.39 g/mol; CAS 102146-07-6), also kown as DPCPX, PD-116,948, 8-cyclopentyl-1,3-dipropyl-7*H*-purine-2,6-dione and 8-cyclopentyl-*N*^{1}, *N*^{3}-dipropyl-7*H*-purine-2,6-dione, selectively targets A$_1$ adenosine receptors (1,2), but also potently inhibits 3',5'-cyclic-AMP phosphodiesterase-4, *or* PDE4 (3). **1**. Alonso, Schoijet, Torres & Flawiá (2006) *Mol. Biochem. Parasitol.* **145**, 40. **2**. Bader, Korthölt, Snippe & van Haastert (2006) *J. Biol. Chem.* **281**, 20018. **3**. Rascón, Soderling, Schaefer & Beavo (2002) *Proc. Natl. Acad. Sci. USA* **99**, 4714.

Dipyridamole

This vasodilator (FW = 504.63 g/mol; CAS 58-32-2) is a potent inhibitor of cyclic-nucleotide phosphodiesterase-5 (IC$_{50}$ = 0.9 μM). It also blocks nucleoside transport into mammalian cells and reduces platelet aggregation. **Target(s):** adenosine deaminase (1); adenosine kinase, inhibited at low substrate concentrations (2); adenosine uptake (3,4); 3',5'-cyclic-GMP phospho-diesterase (5-11); 3',5'-cyclic-nucleotide phosphodiesterase (12-20); cyclic-nucleotide phosphodiesterase isozyme-5 (5-7,9,10); cyclic-nucleotide phosphodiesterase isozyme-10 (20); cyclic-nucleotide phosphodiesterase isozyme-11 (14-17); *ecto*-nucleotidase (21); electron transport (22); glucose-6-phosphate dehydrogenase (23); guanylate cyclase, retinal (24,25). **1**. Sopena, Vina, Calderon, *et al.* (1970) *Rev. Esp. Fisiol.* **26**, 303. **2**. De Jong (1977) *Arch. Int. Physiol. Biochim.* **85**, 557. **3**. Kolassa, Pfleger & Rummel (1970) *Eur. J. Pharmacol.* **9**, 265. **4**. Thorn & Jarvis (1996) *Gen. Pharmacol.* **27**, 61. **5**. Clarke, Uezono, Chambers & Doepfner (1994) *Pulm. Pharmacol.* **7**, 81. **6**. McElroy & Philip (1975) *Life Sci.* **17**, 1479. **7**. Sudo, Tachibana, Toga, *et al.* (2000) *Biochem. Pharmacol.* **59**, 347. **8**. Thomas, Francis & Corbin (1990) *J. Biol. Chem.* **265**, 14964. **9**. Wang, Wu, Myers, *et al.* (2001) *Life Sci.* **68**, 1977. **10**. Su & Vacquier (2006) *Mol. Biol. Cell* **17**, 114. **11**. Yuasa, Mi-Ichi, Kobayashi, *et al.* (2005) *Biochem. J.* **392**, 221. **12**. Houslay, Pyne & Cooper (1988) *Meth. Enzymol.* **159**, 751. **13**. Grant & Colman (1984) *Biochemistry* **23**,

1801. **14.** Yuasa, Ohgaru, Asahina & Omori (2001) *Eur. J. Biochem.* **268**, 4440. **15.** Weeks, Zoraghi, Beasley, *et al.* (2005) *Int. J. Imp. Res.* **17**, 5. **16.** Hetman, Robas, Baxendale, *et al.* (2000) *Proc. Natl. Acad. Sci. U.S.A.* **97**, 12891. **17.** D'Andrea, Qiu, Haynes-Johnson, *et al.* (2005) *J. Histochem. Cytochem.* **53**, 895. **18.** D'Angelo, Garzia, Andre, *et al.* (2004) *Cancer Cell* **5**, 137. **19.** Kunz, Oberholzer & Seebeck (2005) *FEBS J.* **272**, 6412. **20.** Fujishige, Kotera & Omori (1999) *Eur. J. Biochem.* **266**, 1118. **21.** Connolly & Duley (2000) *Eur. J. Pharmacol.* **397**, 271. **22.** Sordahl, Allen, Schwartz & Pharo (1970) *Arch. Biochem. Biophys.* **138**, 44. **23.** Sopena, Vina, Pallardo & Cabo (1971) *Rev. Esp. Fisiol.* **27**, 317. **24.** Gorczyca (2000) *Meth. Enzymol.* **315**, 689. **25.** Gorczyca, Van Hooser & Palczewski (1994) *Biochemistry* **33**, 3217.

Discodermolide

This cell-penetrating microtubule-directed agent (FW = 491.83 g/mol; CAS 127943-53-7) causes cell-cycle arrest at the metaphase-anaphase transition, presumably due to its remarkable stabilizing effects on assembled microtubules (1). Discodermolide binds more tightly than paclitaxel to microtubules, displaying a stoichiometry of one molecule per tubulin dimer (1). Binding of discodermolide and taxol are mutually exclusive, suggesting some overlap in binding sites or binding conformations. Like taxol, discodermolide induces the polymerization of purified tubulin with and without microtubule-associated proteins (*or* MAPs), and the polymers formed are stable to cold- and calcium ion-induced depolymerization (2). Collectively, these properties represent hypernucleation, with nuclei forming spontaneously and efficiently, lowering the apparent tubulin critical concentration, and producing shorter polymers with discodermolide than with paclitaxel, when probed under a variety of reaction conditions (2). Discodermolide competitively inhibits (K_i = 0.4 μM) the binding of [^{3}H]paclitaxel to polymerized tubulin. Multidrug-resistant human colon and ovarian carcinoma cells overexpressing P-glycoprotein, which are 900x and 2800x resistant to paclitaxel, respectively, relative to the parental lines, retain significant sensitivity to discodermolide (*e.g.*, 25x and 89x more resistant relative to the parental lines). Ovarian carcinoma cells that are 20-30x more resistant to paclitaxel than the parental line on the basis of expression of altered β-tubulin polypeptides retained nearly complete sensitivity to discodermolide (2). (**See also** *Paclitaxel, Pelorusside, Laulimalide, Dictyostatin, Epothione, Sarcodictyin*) **1.** Hung, Chen & Schreiber (1996) *Chem. Biol.* **3**, 287. **2.** Kowalski, Giannakakou, Gunasekera, *et al.* (1997) *Mol. Pharmacol.* **52**, 613.

Disodium Tetraborate, *See* Borax

Distamycin A

This antibiotic (FW$_{free-base}$ = 481.51 g/mol; FW$_{hydrochloride}$ = 517.98 g/mol; CAS 6576-51-8), also called stallimycin, from *Streptomyces distallicus* has antiviral and oncolytic properties. Distamycin A binds to the minor groove of native DNA as well to denatured DNA. Both DNA and RNA polymerases are inhibited, with the latter inhibited more strongly in the presence of native DNA (1). Distamycin A is reported to target the initiation of RNA synthesis (2). **Target(s):** *Bal*I restriction endonuclease, type II site-specific deoxyribonuclease (3); DNA ligase (ATP) (4); DNA methyltransferase (5); DNA polymerase (6-8); DNA topoisomerase I (9-12); *Eco*RI restriction endonuclease, type II site-specific deoxyribonuclease (3,13); *Eco*RII restriction endonuclease, type II site-specific deoxyribonuclease (13); *Eco*RV restriction endonuclease, type II site-specific deoxyribonuclease (3); helicase (14,15); *Hin*DII restriction endonuclease, type II site specific deoxyribonuclease (13); *Hin*DIII restriction endonuclease, type II site-specific deoxyribonuclease (13); *Hpa*I

restriction endonuclease, type II site-specific deoxyribonuclease (13); *Hpa*II restriction endonuclease, type II site-specific deoxyribonuclease (13); *Nru*I restriction endonuclease, type II site-specific deoxyribonuclease (3); reverse transcriptase (16,17); RNA-directed DNA polymerase (16,17); RNA polymerase (1,2,18-20); *Vaccinia* virus DNA ligase (21). **1.** Zimmer, Puschendorf, Grunicke, Chandra & Venner (1971) *Eur. J. Biochem.* **21**, 269. **2.** Puschendorf, Peters, Wolf, Werchau & Grunicke (1971) *Biochem. Biophys. Res. Commun.* **43**, 617. **3.** Forrow, Lee, Souhami & Hartley (1995) *Chem. Biol. Interact.* **96**, 125. **4.** Montecucco, Lestingi, Rossignol, Elder & Ciarrocchi (1993) *Biochem. Pharmacol.* **45**, 1536. **5.** Tanaka, Hibasami, Nagai & Ikeda (1982) *Aust. J. Exp. Biol. Med. Sci.* **60**, 223. **6.** Puschendorf & Grunicke (1969) *FEBS Lett.* **4**, 355. **7.** Puschendorf & Grunicke (1969) *Hoppe-Seyler's Z. Physiol. Chem.* **350**, 1163. **8.** Müller, Yamazaki & Zahn (1972) *Enzymologia* **43**, 1. **9.** Thielmann, Popanda, Gersbach & Gilberg (1993) *Carcinogenesis* **14**, 2341. **10.** Mortensen, Stevnsner, Krogh, *et al.* (1990) *Nucl. Acids Res.* **18**, 1983. **11.** McHugh, Woynarowski, Sigmund & Beerman (1989) *Biochem. Pharmacol.* **38**, 2323. **12.** Bugreev, Vasyutina, Ryabinin, *et al.* (2001) *Antisense Nucleic Acid Drug Dev.* **11**, 137. **13.** Wells, Klein & Singleton (1981) *The Enzymes*, 3rd ed., **14**, 157. **14.** Brosh, Karow, White, *et al.* (2000) *Nucl. Acids Res.* **28**, 2420. **15.** Bachur, Johnson, Yu, *et al.* (1993) *Mol. Pharmacol.* **44**, 1064. **16.** Juca & Aoyama (1995) *J. Enzyme Inhib.* **9**, 171. **17.** Kotler & Becker (1971) *Nature New Biology* **243**, 212. **18.** Chandra, Zimmer & Thrum (1970) *FEBS Lett.* **7**, 90. **19.** Orfeo, Chen, Huang, Ward & Bateman (1999) *Biochim. Biophys. Acta* **1446**, 273. **20.** Piestrzeniewicz, Studzian, Wilmanska, Plucienniczak & Gniazdowski (1998) *Acta Biochim. Pol.* **45**, 127. **21.** Shuman (1995) *Biochemistry* **34**, 16138.

Disulfiram

This metabolic inhibitor and alcohol abuse therapeutic (FW = 296.56 g/mol; CAS 97-77-8), known also as bis(diethylthiocarbamoyl) disulfide, and Antabuse™, targets mammalian acetaldehyde dehydrogenase (Reaction: $CH_3CHO + NAD^+ \rightleftharpoons$ Acetate + NADH), leading to the accumulation of toxic levels of acetaldehyde and creating an unpleasant sensation upon alcohol consumption (1). **Primary Mode of Action:** Of the four human isozymes, only ALDH1 and ALDH2 are strongly inhibited. Based on its disulfiram inhibition characteristics, the F1 isozyme of the well-characterized horse liver ALDH is likely to be primarily responsible for oxidizing acetyldehyde produced during *in vivo* ethanol oxidation (2). Importantly, [^{14}C]-Disulfiram reacts with human liver aldehyde dehydrogenase E1 with attendant loss of catalytic activity and no incorporation of radiolabel, suggesting that disulfiram's action on ALDH does not involve sulfhydryl-disulfide interchange (1). An early claim that disulfiram and ethanol react to produce a quaternary ammonium compound that is responsible for disufuram's unpleasant effects is in error (3). **The "Disulfiram Reaction":** Alcohol ingestion by disulfiram-treated patients results in "The Disulfuram Reaction", referring to the strongly unpleasant combination of flushing, headache, nausea, head and neck throbbing, vomiting, sweating, thirst, chest pain, palpitations, dyspnea, hyperventilation, tachycardia, confusion, arrhythmias, and convulsions. In rare instances, disulfiram treatment can lead to other, far more serious effects, including acute hepatitis, fatigue, anorexia, nausea, abdominal pain, and jaundice. Nervous system side effects include malaise, lethargy, confusion, personality changes, disorientation, and memory impairment. **Target(s):** acetaldehyde dehydrogenase (1-2); alcohol dehydrogenase, K_i = 4 μM (4); 5-lipoxygenase, K_i = 0.5 μM, comparable to that of diphenyldisulfide (IC$_{50}$ = 0.2-0.4 μM), suggesting inhibition is the result of a thiol-disulfide exchange reaction (5); selective inhibition of alternative respiratory pathway in plant mitochondria, with onset of inhibition taking several minutes and fully prevented by dithiothreitol (6); DNA cleavage by hydroxyl radical-producing agents, such as 1,2-dimethylhydrazine, ascorbate, and cysteine (7); rat ovarian carbonyl reductase (8). **1.** Vallari & Pietruszko (1982) *Science* **216**, 637. **2.** Eckfeldt, Mope, Takio & Yonetani (1976) *J. Biol. Chem.* **251**, 236. **3.** Kitson (1977) *J. Stud. Alcohol.* **38**, 1771. **4.** Carper, Dorey & Beber (1987) *Clin. Chem.* **33**, 1906. **5.** Choo & Riendeau (1987) *Can. J. Physiol. Pharmacol.* **65**, 2503. **6.** Grover & Laties (1981) *Plant Physiol.* **68**, 393. **7.** Kuhnlein (1980) *Biochim. Biophys. Acta.* **609**, 75. **8.** Iwata, Inazu & Satoh (1992) *Japanese J. Pharmacol.* **58**, 167.

Dithiaden & Diathiaden S-Oxide

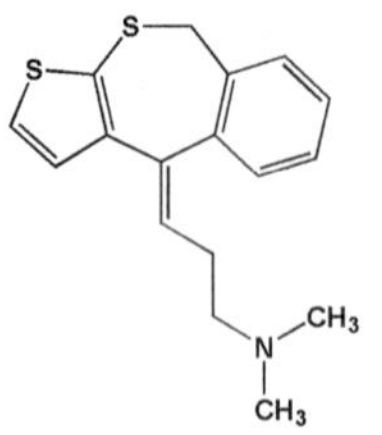

Dithiaden 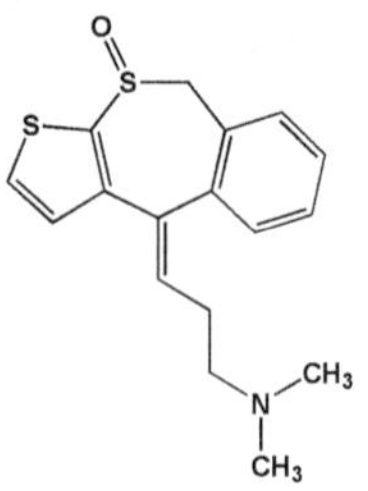   Dithiaden S-Oxide

This histamine receptor antagonist (FW = 301.47 g/mol; CAS 5802-61-9), also named Bisulepine and (3E)-N,N-dimethyl-3-(5H-thieno[2,3-c][2]benzothiepin-10-ylidene)propan-1-amine, and its oxidation metabolite, diathiden S-oxide (FW = 317.46 g/mol; CAS 11119-54-3), target H_1 receptors, both in pharmacological experiments and *in vivo*. In contrast to dithiaden, dithiaden S-oxide exerts a demonstrable inhibitory action as much as the subtoxic or sublethal dosage in experiments in mice and rats. Other experimental findings demonstrate higher selectivity of pharmacological effects of dithiadenoxide. ***See also*** *Loratadine* 1. Blehová & Metys (1992) *Cesk Farm.* **41**, 185.

Dithionite

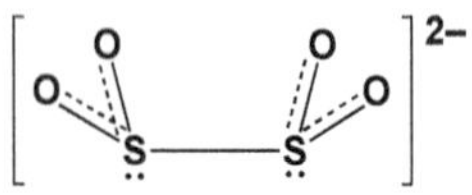

This powerful inorganic reducing agent (FW = 174.11 g/mol for disodium salt; CAS 14844-07-6; unusual 2.4-Å disulfide bond-length; Redox Potential = – 0.66 Volts relative to the standard hydrogen electrode), is of immense utility in biochemistry, with a redox potential that is sufficient to convert uncomplexed and porphyrin-bound Fe(III) to Fe(II), with concomitant generation of two molecules of SO_2 gas. Dithionite also reacts with heme-bound oxygen to produce deoxyhemoglobin. Treatment of intact red blood cells with dithionite can induce sickling in cells containing hemoglobin S. Dithionite also reduces NAD^+ to NADH. *Note*: Sodium dithionite oxidizes readily when damp or in solution. Solutions should always be freshly prepared in buffered conditions, preferably above pH 7.5. **Target(s):** acid phosphatase (1-9); ADP-ribosyl-[dinitrogen reductase] hydrolase (10); arylformamidase (11,12); aspartate aminotransferase (13); catalase (14,44); catechol 1,2 dioxygenase (43); CoB-CoM heterodisulfide reductase (45); CO-methylating acetyl-CoA synthase, *or* carbon monoxide dehydrogenase (15,16); cystathionine β-synthase (17); ferredoxin:NADP$^+$ reductase (18); hydrogenase (19); 4-hydroxybenzoate decarboxylase (20); 3-hydroxy-3-methylglutaryl CoA reductase, partially inhibited (21); 4-hydroxyquinoline 3-monooxygenase (22); indoleacetylglucose:inositol O-acyltransferase (23); indole 2,3-dioxygenase (24,25); ligninase H8, *or* lignin peroxidase H8 (26); methane monooxygenase (27); 2-(methylthio)ethanesulfonic acid reductase (28); nitrate reductase (29,30); nitrous oxide reductase (31); oxalate decarboxylase (32,33); phenol 2 monooxygenase (34); purple acid phosphatase (1,3,6); quercetin 2,3-dioxygenase (35); sulfur reductase (36); tetrachloroethene reductive dehalogenase (37); thiosulfate dehydrogenase, weakly inhibited (47); thiosulfate dehydrogenase (quinone) (46); urocanate hydratase, urocanase (38); vitamin K-dependent carboxylase, *or* protein-glutamate carboxylase (39); xanthine dehydrogenase (40-42). **1.** Fukushima, Bekker & Gay (1991) *Amer. J. Anat.* **191**, 228. **2.** Allen, Nuttleman, Ketcham & Roberts (1989) *J. Bone Miner. Res.* **4**, 47. **3.** Anderson & Toverud (1986) *Arch. Biochem. Biophys.* **247**, 131. **4.** Hammarström, Anderson, Marks & Toverud (1983) *J. Histochem. Cytochem.* **31**, 1167. **5.** Chambers, Peters, Glew, *et al.* (1978) *Metabolism* **27**, 801. **6.** Schlosnagle, Sander, Bazer & Roberts (1976) *J. Biol. Chem.* **251**, 4680. **7.** Schlosnagle, Bazer, Tsibris & Roberts (1974) *J. Biol. Chem.* **249**, 7574. **8.** Ketcham, Baumbach, Bazer & Roberts (1985) *J. Biol. Chem.* **260**, 5768. **9.** Hayman, Warburton, Pringle, Coles & Chambers (1989) *Biochem. J.* **261**, 601. **10.** Nielsen, Bao, Roberts & Ludden (1994) *Biochem. J.* **302**, 801. **11.** Katz, Brown & Hitchcock (1987) *Meth. Enzymol.* **142**, 225. **12.** Shinohara & Ishiguro (1970) *Biochim. Biophys. Acta* **198**, 324. **13.** Scandurra, Polidoro, Di Cola, Politi & Riordan (1975) *Biochemistry* **14**, 3701. **14.** Terzenbach & Blaut (1998) *Arch. Microbiol.* **169**, 503. **15.** Ragsdale & Wood (1985) *J. Biol. Chem.* **260**, 3970. **16.** Tan, Sewell & Lindahl (2002) *J. Amer. Chem. Soc.* **124**, 6277. **17.** Banerjee, Evande, Kabil, Ojha & Taoka (2003) *Biochim. Biophys. Acta* **1647**, 30. **18.**

Carrillo & Ceccarelli (2003) *Eur. J. Biochem.* **270**, 1900. **19.** Bennett, Lemon & Peters (2000) *Biochemistry* **39**, 7455. **20.** Gallert & Winter (1992) *Appl. Microbiol. Biotechnol.* **37**, 119. **21.** Dotan & Shechter (1982) *Biochim. Biophys. Acta* **713**, 427. **22.** Block & Lingens (1992) *Biol. Chem. Hoppe-Seyler* **373**, 249. **23.** Kesy & Bandurski (1990) *Plant Physiol.* **94**, 1598. **24.** Pundir, Garg & Rathore (1984) *Phytochemistry* **23**, 2423. **25.** Kunapuli & Vaidyanathan (1982) *Plant Sci. Lett.* **24**, 183. **26.** Tien & Kirk (1988) *Meth. Enzymol.* **161**, 238. **27.** Shaofeng, Shuben, Jiayin, *et al.* (2007) *Biosci. Biotechnol. Biochem.* **71**, 122. **28.** Olson, McMahon & Wolfe (1991) *Proc. Natl. Acad. Sci. U.S.A.* **88**, 4099. **29.** Kennedy, Rigaud & Trinchant (1975) *Biochim. Biophys. Acta* **397**, 24. **30.** Rigano & Aliotta (1975) *Biochim. Biophys. Acta* **384**, 37. **31.** Coyle, Zumft, Kroneck, Korner & Jakob (1985) *Eur. J. Biochem.* **153**, 459. **32.** Emiliani & Riera (1968) *Biochim. Biophys. Acta* **167**, 414. **33.** Tanner, Bowater, Fairhurst & Bornemann (2001) *J. Biol. Chem.* **276**, 43627. **34.** Neujahr & Gaal (1973) *Eur. J. Biochem.* **35**, 386. **35.** Oka, Simpson & Krishnamurty (1972) *Can. J. Microbiol.* **18**, 493. **36.** Sugio, Oda, Matsumoto, *et al.* (1998) *Biosci. Biotechnol. Biochem.* **62**, 705. **37.** Magnuson, Stern, Gossett, Zinder & Burris (1998) *Appl. Environ. Microbiol.* **64**, 1270. **38.** Gerlinger & Retey (1987) *Z. Naturforsch. C* **42**, 349. **39.** Johnson (1980) *Meth. Enzymol.* **67**, 165. **40.** Ziang & Edmondson (1996) *Biochemistry* **35**, 5441. **41.** Schräder, Rienhöfer & Andreesen (1999) *Eur. J. Biochem.* **264**, 862. **42.** Pérez-Vicente, Alamillo, Cárdenas & Pineda (1992) *Biochim. Biophys. Acta* **1117**, 159. **43.** Cha (2006) *J. Microbiol. Biotechnol.* **16**, 778. **44.** Tejera García, Iribame, Palma & Lluch (2007) *Plant Physiol. Biochem.* **45**, 535. **45.** Iwasaki, Kounosu, Aoshima, *et al.* (2002) *J. Biol. Chem.* **277**, 39642. **46.** Müller, Bandeiras, Ulrich, *et al.* (2004) *Mol. Microbiol.* **53**, 1147. **47.** Schook & Berk (1979) *J. Bacteriol.* **140**, 306.

Diumycin

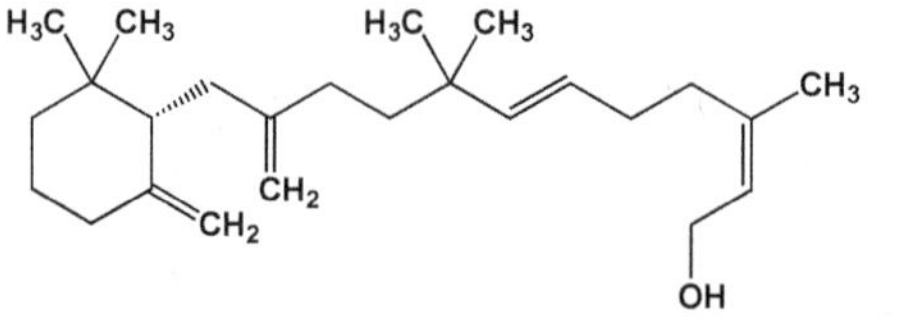

This phosphoglycolipid antibiotic (FW = 358.61 g/mol; CAS 11141-18-7) from *Streptomyces umbrinus* inhibits cell wall biosynthesis. Diumycin A inhibits *Staphylococcus aureus* cell wall synthesis, resulting in the accumulation of UDP-N-acetyl-muramyl-pentapeptide. Diumycin also inhibits *in vitro* peptidoglycan synthesis by particulate preparations of *Bacillus stearothermophilus* and *Escherichia coli* by blocking utilization of N-acetyl-glucosamine-N-acetyl-muramyl-pentapeptide It does not inhibit particulate D-alanine carboxypeptidase. Although details of diumycin's structure remain unclear, diumycin contains diumycinol, an unusual lipid alcohol (shown above). **Target(s):** N-acetyl-glucosaminyl-diphospho-dolichol N-acetyl-glucosaminyl-transferase (1,2); cell wall biosynthesis (3,4); dolichyl-phosphate β-glucosyltransferase (5); dolichyl-phosphate mannosyl-transferase (6-8); protein glycosylation (5,6,9,10); trehalose-phosphatase (11-13); α,α-trehalose phosphate synthase, UDP-forming (14); UDP-N-acetylglucosamine:dolichyl-phosphate N-acetylglucosamine phosphotransferase (5,10,15-17). **1.** Kean & Niu (1998) *Glycoconjugate J.* **15**, 11. **2.** Kaushal & Elbein (1986) *Plant Physiol.* **81**, 1086. **3.** Brown, Seinerova, Chan, *et al.* (1974) *Ann. N. Y. Acad. Sci.* **235**, 399. **4.** Lugtenberg, Hellings & van de Berg (1972) *Antimicrob. Agents Chemother.* **2**, 485. **5.** Schwarz & Datema (1982) *Meth. Enzymol.* **83**, 432. **6.** Elbein (1987) *Meth. Enzymol.* **138**, 661. **7.** Comley, Jaffe & Chrin (1982) *Mol. Biochem. Parasitol.* **5**, 19. **8.** Babczinski (1980) *Eur. J. Biochem.* **112**, 53. **9.** Elbein (1983) *Meth. Enzymol.* **98**, 135. **10.** Schwarz & Datema (1980) *Trends Biochem. Sci.* **5**, 65. **11.** Edavana, Pastuszak, Carroll, *et al.* (2004) *Arch. Biochem. Biophys.* **426**, 250. **12.** Klutts, Pastuszak, Edavana, *et al.* (2003) *J. Biol. Chem.* **278**, 2093. **13.** Pan & Elbein (1996) *Arch. Biochem. Biophys.* **335**, 258. **14.** Pan & Elbein (1996) *Arch. Biochem. Biophys.* **335**, 258. **15.** Kaushal & Elbein (1986) *Plant Physiol.* **82**, 748. **16.** Shailubhai, Dong-Yu, Saxena & Vijay (1988) *J. Biol. Chem.* **263**, 15964. **17.** Villemez & Carlo (1980) *J. Biol. Chem.* **255**, 8174.

Diuron

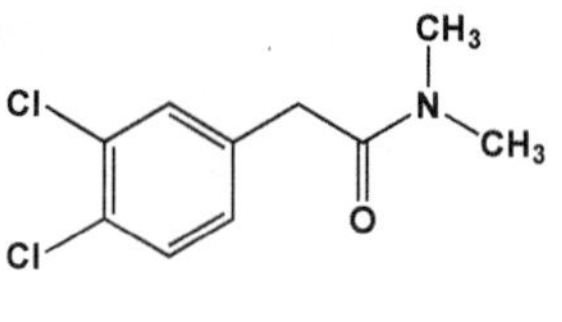

This urea derivative (FW = 233.10 g/mol; CAS 330-54-1), also known as Karmex and *N'*-(3,4-dichlorophenyl)-*N*,*N*-dimethylurea, is a herbicide used primarily to treat cotton crops. Diuron inhibits photosynthetic electron transport and the Hill reaction. There are two sites for diuron on photosystem II, located on the reducing and oxidizing sides of that protein complex. **Target(s):** electron transport, photosynthetic (1-5); plastoquinol:plastocyanin reductase (6); Ubiquinol; cytochrome-*c* reductase, *or* Complex III (7-9); unspecific monooxygenase (10). **1.** Dodge (1977) *Spec. Publ., Chem. Soc.* **29**, 7. **2.** Izawa & Good (1972) *Meth. Enzymol.* **24**, 355. **3.** Trebst (1980) *Meth. Enzymol.* **69**, 675. **4.** Duysens (1972) *Biophys. J.* **12**, 858. **5.** Hsu, Lee & Pan (1986) *Biochem. Biophys. Res. Commun.* **141**, 682. **6.** Krinner, Hauska, Hurt & Lockau (1982) *Biochim. Biophys. Acta* **681**, 110. **7.** Convent & Briquet (1978) *Eur. J. Biochem.* **82**, 473. **8.** Brasseur (1988) *J. Biol. Chem.* **263**, 12571. **9.** di Rago & Colson (1988) *J. Biol. Chem.* **263**, 12564. **10.** Gorinova, Nedkovska & Atanassov (2005) *Biotechnol. Biotechnol. Equip.* **19**, 105.

17-DMAG, *See* Alvespimycin

DMP266, *See* Efavirenz

DMXBA, *See* GTS-21

DNQX

This selective receptor antagonist (FW_diNa-Salt = 252.14 g/mol; CAS 2379-57-9; Solubility: 100 mM in DMSO), also named 6,7-dinitroquinoxaline-2,3-dione, potently and competitively target non-NMDA glutamate receptors, permitting the exploration of structure-activity relations for quisqualate and kainate receptors and the role of such receptors in synaptic transmission in the mammalian brain. **1.** Honoré, Davies, Drejer, *et al.* (1988) *Science* **241**, 701. **2.** Watkins, Krogsgaard-Larsen, Honoré (1990) *Trends Pharmacol Sci.* **11**, 25.

Docetaxel

This second-generation paclitaxel analogue (FW = 807.87 g/mol; CAS 114977-28-5; Solubility: 160 mg/mL DMSO; <1 mg/mL H$_2$O), also known as Taxotere, binds to assembled microtubules, stabilizing them against depolymerization and thereby reducing the dynamicity (*See Paclitaxel*). Docetaxel is especially cytotoxic toward proliferating cells, for which mitosis is either hampered or arrested as a consequence of microtubule bundling, followed by apoptosis or mitotic block. Docetaxel inhibits the survival of Hs746T human stomach cancer cells (IC$_{50}$ = 1 nM), AGS stomach cancer cells (IC$_{50}$ = 1 nM), HeLa human cervix cancer cells (IC$_{50}$ = 0.3 nM), CaSki human cervix cancer cells (IC$_{50}$ = 0.3 nM), BxPC3 human pancreatic cancer cells (IC$_{50}$ = 0.3 nM), and Capan-1 human pancreatic cancer cells (IC$_{50}$ = 0.3 nM). Side-effects include dose limiting neutropenia and peripheral neuropathy. Like paclitaxel, docetaxel is metabolized by cytochrome P450 (CYP3A4). Several docetaxel analogues are highly active in the NCI-60 cancer cell panel with broad efficacy and potency, exhibiting GI$_{50}$ values below 5 nM (7). **Key Pharmacokinetic Parameters:** *See* Appendix II in Goodman & Gilman's THE PHARMACOLOGICAL BASIS OF THERAPEUTICS, 12th Edition (Brunton, Chabner & Knollmann, eds.) McGraw-Hill Medical, New York (2011). **1.** Buey (2004) *Chem. Biol.* **11**, 225. **2.** Riou, *et al.* (1992) *Biochem. Biophys. Res. Commun.* **187**, 164. **3.** Balcer-Kubiczek, *et al.* (2006) *Chemotherapy* **52**, 231. **4.** Silvestrini, *et al.* (1993) *Stem Cells* **11**, 528. **5.** Tanaka, *et al.* (1996) *Eur. J. Cancer.* **32A**, 226. **6.** Hotchkiss, *et al.* (2002) *Mol. Cancer Ther.* **1**, 1191. **7.** Nicolaou & Valiulin (2013) *Org. Biomol. Chem.* **11**, 4154.

Dolasetron

This serotonin 5-HT$_3$ receptor antagonist and anti-emetic (FW = 324.37 g/mol; CAS 115956-12-2), also known by its trade name Anzemet® and systematic name (3*R*)-10-oxo-8-azatricyclo[5.3.1.0^{3,8}]undec-5-yl-1*H*-indole-3-carboxylate, is used to treat chemotherapy-induced nausea/vomiting. Dolasetron affects both peripheral and central nervous systems, reducing vagus nerve activity and thereby deactivating the 5-HT$_3$ receptor-dense vomiting center located in the medulla oblongata. (*Note*: Unlike other serotonin (5-HT) receptors, which are G-protein coupled receptors, 5-HT$_3$ is a ligand-gated ion channel.) Importantly, olasetron is without effect on dopamine receptors or muscarinic receptors. Dolasetron is actually a pro-drug requiring reduction of its prochiral carbonyl group to yield the pharmacologically active chiral secondary alcohol known as reduced dolasetron. Fortuitously, the (+)-(*R*)-enantiomer is preferentially formed by the endogenous carbonyl reductase, as it is more active as an anti-emetic (1). Other 5-HT$_3$ antagonists with similar effects include: tropisetron, ondansetron, palonosetron, and granisetron. The effectiveness of each depends on particular variants of 5-HT$_3$ receptors expressed by the patient, including changes in promoters for the receptor genes. **1.** Dow & Berg (1995) *Chirality* **7**, 342.

Dolastatin 10

This highly cytotoxic antimitotic peptide and antitumor agent (FW = 790.16 g/mol; CAS 110417-88-4; *Sequence*: Dolavaline-Valine-Dolaisoleucine-Dolaproine-Dolaphenine, with the last three amino acids unique to *D. auricularia*, terminating in the unusual primary amine, Dolaphenine) from the sea hare *Dolabella auricularia* and the marine cyanobacterium *Symploca* (sp. VP642), inhibits the growth of murine leukemia cells in culture (IC$_{50}$ = 0.5 nM), with a rise in the mitotic index (1-4). Dolastatin 10 binds at a distinct site for peptide antimitotic agents near the exchangeable nucleotide and vinca alkaloid binding sites (4). For comparison, the cytostatic effects of other drugs are: Maytansine (IC$_{50}$ = 0.5 nM), Rhizoxin (IC$_{50}$ = 1 nM), Vinblastine (IC$_{50}$ = 20 nM), and Phomopsin A (IC$_{50}$ = 7 μM). Inhibitory concentrations, using purified tubulin polymerization in glutamate buffer: Dolastatin 10 (IC$_{50}$ = 1.2 μM), Phomopsin A (IC$_{50}$ = 1.4 μM), Vinblastine (IC$_{50}$ = 1.5 μM), Maytansine (IC$_{50}$ = 3.5 μM), and Rhizoxin (IC$_{50}$ = 6.8 μM). **See also** *Monomethyl Auristatin E* **1.** Pettit, Kamano, Herald, *et al.* (1987) *J. Am. Chem. Soc.* **109**, 6883. **2.** Bai, Pettit & Hamel (1990) *Biochem Pharmacol.* **40**, 1859. **3.** Bai, Pettit, Hamel (1990) *Biochem. Pharmacol.* **39**, 1941. **4.** Bai, Pettit & Hamel (1990) *J. Biol. Chem.* **265**, 1714.

Dolichyl Phosphate

This phosphorylated 2,3-dihydropolyprenol (FW = Indefinite Polymer) functions as a carrier of mono- and oligosaccharides in glycoprotein and glycolipid production. It can also induce apoptosis in rat glioma C6 cells. Note that the term has been used for 2,3-dihydroprenols with as few as four isoprene subunits. Both *cis* and *trans* double bonds are present: frequently the two isoprene units (sometimes three) adjacent to the dihydro unit are *trans,* while the remaining are *cis.* The phospho monoester is stable in mild acid or alkaline treatment. **Target(s):** dolichylphosphate-mannose phophodiesterase (1); indole-3-acetate β-glucosyltransferase, weakly inhibited (2); mevalonate kinase (3). **1**. Tomita & Motokawa (1987) *Eur. J. Biochem.* **170**, 363. **2**. Leznicki & Bandurski (1988) *Plant Physiol.* **88**, 1481. **3**. Hinson, Chambliss, Toth, Tanaka & Gibson (1997) *J. Lipid Res.* **38**, 2216.

Dolutegravir

This integrase inhibitor, *or* INI, and once daily HIV/AIDS drug (FW = 419.38 g/mol; CAS 1051375-16-6; *Symbol*: DTG), also known by the code name S/GSK1349572 and systematic name (4R,12aS)-N-(2,4-difluorobenzyl)-7-hydroxy-4-methyl-6,8-dioxo-3,4,6,8,12,12a-hexahydro-2H-pyrido[1',2':4,5]-pyrazino[2,1-b][1,3]oxazine-9-carboxamide, potently inhibits HIV integrase, showing significant antiviral action in raltegravir-resistant patients. **Mode of Action:** HIV-1 integrase catalysis is a biphasic process: *Step*-1, integrase binds to viral cDNA, with attendant cleavage of two nucleotides, producing the enzyme-bound DNA "pre-integration" complex; and *Step*-2, DNA strand-transfer, a host nuclear process that dolutegravir inhibits by chelating two active-site Mg^{2+} ions and causing the viral integrase to disengage from the deoxyadenosine at the 3'-end of the viral DNA (1). Like other integrase strand-transfer inhibitors, DTG efficiently blocks viral replication *in vitro* (IC_{50} = 2-3 nM) and likewise suppresses viremia. Notably, DTG dissociates slowly from a wild-type integrase-DNA complex with an off-rate of 2.7×10^{-6} s^{-1} (*or* a dissociative $t_{1/2}$ of 71 h), significantly longer than the respective 8.8- and 2.7-hour half-lives for raltegravir and elvitegravir (2). Site-directed single and double mutations at sites affecting resistance suggest that DTG offers a significantly higher genetic barrier to resistance, as comapared to raltegravir and elvitegravir. Such findings suggest that the INI off-rate may be an important component of the mechanism of integrase inhibitor resistance (2). Site-directed single and double mutations at sites affecting and resistance suggest that DTG has a significantly higher genetic barrier to resistance seen with raltegravir and elvitegravir, providing evidence that the INI off-rate may be an important component of the mechanism of integrase inhibitor resistance (2). **Metabolism/Excretion:** S/GSK1349572 is primarily metabolized by glucuronidation by UGT1A1 (*or* Bilirubin UDP-glucuronosyltransferase), an enzyme that transforms small lipophilic molecules (including steroids, bilirubin, and a few hormones) into their water-soluble, excretable glucuronides. Drug interactions also occur with acetaminophen, atazanavir, atorvaststin, etoposide, fluvastatin, lovastatin, morphine, and simvastatin. These UGT1A1 inhibitors or alternative substrates result in dolutegravir retention by reducing its glucuronidation and excretion. **1**. Hare, Smith, Métifiot, *et al.* (2011) *Mol. Pharmacol.* **80**, 565 **2**. Hightower, Wang, Deanda, *et al.* (2011) *Antimicrob. Agents Chemother.* 55, 4552.

Dominant Negative Mutants

The term dominant-negative describes a gene or protein that exerts a dominant effect, similar to that described, when one copy of the gene gives a mutant phenotypic effect that prevents (or negatively impacts) a biological process such as a vital signal transduction pathway. Such an effect is possible when the active form of a protein is a homo- or hetero-dimer (or oligomer). If part of that protein complex is formed from a mutant protein that is defective in some functional aspect, then the complex may be able to self-assemble, but the defective component will dominantly inhibit other wild type components (subunits) of the complex, thus preventing the complex from carrying out its normal function. In this case the gene coding for the mutant component of complex is called a *dominant-negative gene,* and the protein it forms is called a *dominant-negative protein.* One example is the heterotrimeric 5'-AMP-activated protein kinase (AMPK), a metabolic switch that is only active, when assembled from fully functional α-, β-, and γ-subunits (1). When cellular AMP/ATP ratio is high, AMPK is phosphorylated by its upstream kinases, thereby generating catalytically active AMPK. The latter phosphorylates (and thus activates) target proteins/enzymes, serving to restore cellular ATP and GTP by enhancing ATP-regenerating processes (*e.g.,* glucose uptake, fatty acid β-oxidation, mitochondrial biogenesis), while simultaneously inhibiting ATP-consuming pathways (*e.g.,* nucleotide biosynthesis and ribosome function). There are two isoforms of the α-subunit, which is the subunit that catalyzes phosphoryl transfer by the fully assembled AMPK holoenzyme. The dominant-negative AMPKα1 recombinant adenovirus can be made by cellular expression of human α1 subunit having a D159A or D159N mutation in the ATP binding domain. The mutant AMPKα1 competes with wild type AMPKα1 for the binding with β- and γ-subunits; however, because mutant AMPKα1 lacks the capacity to bind ATP, the assembled AMPK is a dominant negative form (1). In this case, expression of α1(D159N) markedly inhibited both basal and stimulated activity of endogenous AMPK, but was without effect on the transcription of glucose-activated genes. These findings suggest that AMPK is involved in the inhibition of glucose-activated gene expression but not in the induction pathway. As a tool for introducing new genes to a target tissue, recombinant adenoviral vectors are especially effective, yielding a transduction efficiency of >90% for a gene-of-interest in many types of mammalian cells. Although adenovirus is currently used for a variety of purposes, including gene transfer *in vitro*, vaccination production, and gene therapy, another application is the construction of vectors containing recombinantly inactivated (*or* dominantly negative) subunits that, when expressed at sufficient concentrations, out-compete wild type subunits in the assembly of holoenzyme. Although construction of recombinant plasmids is one of the most time-consuming steps in generating recombinant adenovirus, premade recombinant high-titer adenoviruses is now commercially available for convenient use. Such vectors may also be made to express recombinant proteins that are tagged (with hemagglutinin, *or* HA (sequence = YPYDVPDYA), c-myc (sequence = EQKLISEEDL), or enhanced GFP) to mark successfully transduced cells in a potentially heterogeneous culture. **1**. Woods, Azzout-Marniche, Foretz, *et al.* (2000) *Mol. Cell Biol.* 20, 6704.

Domoate (Domoic Acid)

This highly toxic kainate analogue (FW = 311.33 g/mol; CAS 14277-97-5; pK_a values of 2.10, 3.72, 4.93, and 9.82), also named 3-(carboxymethyl)-4-(2-carboxy-1-methyl-1,3-hexadienyl)proline, is an excitatory amino acid isolated from the red alga *Chondria armata* (known in Japanese as domoi). It is a potent agonist at gutamate and kainate receptors and displays very high affinity for the AMPA/kainate receptor, damaging the hippocampus and amygdaloid nucleus. Treatment of cells with domoic acid leads to depletion of energy stores. Domoate also inhibits choline *O*-acetyltransferase (1). Domoic acid gives rise to the disorder known as Amnesic Shellfish Poisoning (*Symptoms*: nausea; vomiting; diarrhea; abdominal cramps; headache; dizziness; confusion; disorientation; short-term memory loss; motor weakness; seizures; profuse respiratory secretions; cardiac arrhythmias; coma and possibly even death). **1**. Loureiro-Dos-Santos, Reis, Kubruslyet al. (2001) *J. Neurochem.* 77, 1136.

Donepezil

This centrally acting ACHE inhibitor and palliative AD drug (FW = 379.49 g/mol; CAS 120014-06-4; IUPAC Name: (*RS*)-2-[(1-benzyl-4-piperidyl)methyl]-5,6-dimethoxy-2,3-dihydroinden-1-one), known widely by the trade names Aricept® and Donep®, targets acetylcholinesterase (IC_{50}

= 5.7 nM) and maintains the level of acetylcholine accumulating in the synaptic cleft, thereby sustaining the duration of the channel-open conformation of the acetylcholine receptor. Aricept helps to maintain memory, attention, social interactions, and speech in patients with declining mental function as a consequence of mild to moderate Alzheimer's Disease or vascular dementia. Lineweaver-Burk analysis of the inhibition data suggests that donepezil is a mixed-type (*i.e.*, $K_i \neq K_{ii}$) noncompetitive inhibitor of brain ACHE (1). It likewise elevates extracellular ACh, an action attended by stimulated catecholamine release (2). Donepezil is absorbed slowly, but completely, from the gut, reaching peak plasma levels in 3-4 hours and, with daily dosing, attains a steady-state level in 15-21 days (3). Within a relatively narrow range, there is a linear relationship between dose and pharmacodynamic effects, as measured by red blood cell acetylcholinesterase inhibition and clinical efficacy. Donepezil is principally excreted unchanged in the urine, but there is also hepatic metabolism; some of its metabolites may be active (3). **Key Pharmacokinetic Parameters:** *See* Appendix II in Goodman & Gilman's THE PHARMACOLOGICAL BASIS OF THERAPEUTICS, 12th Edition (Brunton, Chabner & Knollmann, eds.) McGraw-Hill Medical, New York (2011). **Other Target(s):** Donepezil also induces caspase-dependent apoptosis in human promyelocytic leukemia HL-60 cells (4). It also enhances Purkinje cell survival and alleviates motor dysfunction by inhibiting cholesterol synthesis in a murine model of Niemann Pick disease type C (5). **1.** Nochi, Asakawa & Sato (1995) *Biol. Pharm. Bull.* **18**, 1145. **2.** Giacobini, Zhu, Williams & Sherman (1996) *Neuropharmacol.* **35**, 205. **3.** Seltzer (2005) *Expert . Drug Metab. Toxicol.* **1**, 527. **4.** Ki, Park, Lee, *et al.* (2010) *Biol. Pharm. Bull.* **33**, 1054. **5.** Seo, Shin, Kim, *et al.* (2014) *J. Neuropathol. Expmtl. Neurol.* **73**, 234.

L-Dopa

This neurotransmitter (FW = 197.19 g/mol; CAS 59-92-7), also known as 3,4-dihydroxy-phenylalanine and 3-hydroxytyrosine, is an important intermediate in aromatic amino acid catabolism as well as in the synthesis of epinephrine and melanin. The L-isomer is also called levodopa and is used in the treatment of Parkinson Disease. **Target(s):** *N*-acetylindoxyl oxidase (1); aldehyde oxidase (2); catechol *O*-methyltransferase (3); 3-deoxy-7-phosphoheptulonate synthase (4); histidine decarboxylase (5); 3-hydroxyanthranilate oxidase (26); 4-hydroxyphenylpyruvate dioxygenase (25); kynureninase, inhibited by DL-mixture (6,7); lipoxygenase (8); monoamine oxidase (9); nitronate monooxygenase (10); phenylalanine ammonia lyase (11); phenylalanine 4-monooxygenase (12-14); prephenate dehydrogenase (15); pyridoxal kinase (16,17); ribonucleoside-diphosphate reductase (18); tryptophan 5-monooxygenase, inhibited by both D- and L-dopa (19,20); tubulin:tyrosine ligase, as weak alternative substrate (21,22); tyrosine aminotransferase, also as weak alternative substrate (23); tyrosine-phosphatase, by L-Dopa (24). **1.** Beevers & French (1954) *Arch. Biochem. Biophys.* **50**, 427. **2.** Panoutsopoulos & Beedham (2004) *Acta Biochim. Pol.* **51**, 649. **3.** Veser (1987) *J. Bacteriol.* **169**, 3696. **4.** McCandliss, Poling & Herrmann (1978) *J. Biol. Chem.* **253**, 4259. **5.** Webb (1966) *Enzyme and Metabolic Inhibitors*, vol. 2, p. 352, Academic Press, New York. **6.** Jakoby & Bonner (1953) *J. Biol. Chem.* **205**, 709. **7.** Soda & Tanizawa (1979) *Adv. Enzymol. Relat. Areas Mol. Biol.* **49**, 1. **8.** Holman & Bergström (1951) *The Enzymes*, 1st ed., **2** (part 1), 559. **9.** Naoi & Nagatsu (1987) *Life Sci.* **40**, 321. **10.** Kido, Soda & Asada (1978) *J. Biol. Chem.* **253**, 226. **11.** Koukol & Conn (1961) *J. Biol. Chem.* **236**, 2692. **12.** Kaufman (1978) *Meth. Enzymol.* **53**, 278. **13.** Kaufman (1987) *Meth. Enzymol.* **142**, 3. **14.** Letendre, Dickens & Guroff (1975) *J. Biol. Chem.* **250**, 6672. **15.** Fischer & Jensen (1987) *Meth. Enzymol.* **142**, 503. **16.** Neary, Meneely, Grever & Diven (1972) *Arch. Biochem. Biophys.* **151**, 42. **17.** Lainé-Cessac, Cailleux & Allain (1997) *Biochem. Pharmacol.* **54**, 863. **18.** FitzGerald, Rosowsky & Wick (1984) *Biochem. Biophys. Res. Commun.* **120**, 1008. **19.** Nakata & Fujisawa (1982) *Eur. J. Biochem.* **124**, 595. **20.** Naoi, Maruyama, Takahashi, Ota & Parvez (1994) *Biochem. Pharmacol.* **48**, 207. **21.** Deans, Allison & Purich (1992) *Biochem. J.* **286**, 243. **22.** Raybin & Flavin (1977) *Biochemistry* **16**, 2189. **23.** Jacoby & La Du (1964) *J. Biol. Chem.* **239**, 419. **24.** Fukami & Lipmann (1982) *Proc. Natl. Acad. Sci. U.S.A.* **79**, 4275. **25.** Lindstedt & Rundgren (1982) *J. Biol. Chem.* **257**, 11922. **26.** Morgan, Weimorts & Aubert (1965) *Biochim. Biophys. Acta* **100**, 393.

Dopamine

This catecholamine neurotransmitter (FW$_{free-base}$ = 153.18 g/mol; CAS 51-61-6; pK_a (hydroxyl) = 8.9; pK_a (amino) = 10.6; *Abbreviation*: L-Dopa; *Symbol*: DA), also known as 3-hydroxytyramine and 3,4-dihydroxyphenethylamine, is produced from L-dihydroxyphenylalanine and is a precursor to norepinephrine and epinephrine. Dopamine is photosensitive and readily oxidizes in aqueous solutions in alkaline pH. Within the brain, dopamine affects executive functions, motor control, motivation, arousal, reinforcement, and reward, as well as lower-level actions including lactation, sexual gratification, and nausea. **The Dopamine Cycle:** Dopamine is synthesized by 3,4-dihydroxyphenylalanine decarboxylase (*Reaction*: 3,4-dihydroxyphenylalanine $\rightleftharpoons$ L-Dopa + H$_2$O) in the cytoplasm, from which it is transported into synaptic vesicles by VMAT2 (Vesicular Monoamine Transporter-2), where it remains until vesicle mobilization is triggered by a stimulating action potential. (*See* **VMAT2 inhibitor Quinlobelane**) Upon migration to the peripheral membrane, these vesicles fuse with the latter in an exocytotic event that dumps dopamine into the synaptic cleft. Once released, dopamine binds to and activates postsynaptic dopamine receptors, therewith propagating a signal to a nearby postsynaptic neuron. Dopamine also binds to presynaptic autoreceptors, the occupancy of which inhibits dopamine synthesis and release. Presynaptic reuptake is mediated by the high-affinity dopamine transporter (DAT) and the low-affinity plasma membrane monoamine transporter (PMAT). Once returned to the cytosol, dopamine is again reloaded into vesicles by VMAT2. **Parkinson Disease:** This progressively debilitating motor system disorder is characterized by loss of dopamine-producing cells in the brainn resulting in (a) trembling of hands, arms, legs, jaw, and face, (b) stiffness of limbs and trunk; slowness of movement (*or* bradykinesia), and (c) impairment of postural balance and coordination. Parkinson Disease correlates with the progressive and marked loss of dopamine-containing cells in the *substantia nigra* as well as reduction in DAT and VMAT2 transporters within the remaining pigmented neurons. Side effects of dopamine therapy include: *Cardiovascular* – ventricular arrhythmia (at very high doses), ectopic beats, tachycardia, anginal pain, palpitation, conduction abnormalities, widened QRS complex, bradycardia, hypotension or hypertension, vasoconstriction; *Respiratory* – dyspnea; *Gastrointestinal*: nausea, vomiting; *Metabolic/Nutritional* – azotemia; *CNS* – headache and/or anxiety. **Circulating Dopamine:** Dopamine does not readily cross the blood–brain barrier, and its synthesis and functions in peripheral regions are independent of its synthesis and functions within the brain. Dopamine is also found in plasma at levels comparable to those of epinephrine, with most monosulfated at one of its two phenolic groups. Formation of 3-*O*-DA and 4-*O*-DA sulfate may represent a way of detoxifying dietary biogenic amines, thereby limiting their autocrine and/or paracrine effects as well as that of any endogenously generated DA. **Key Pharmacokinetic Parameters:** *See* Appendix II in Goodman & Gilman's THE PHARMACOLOGICAL BASIS OF THERAPEUTICS, 12th Edition (Brunton, Chabner & Knollmann, eds.) McGraw-Hill Medical, New York (2011). **Target(s):** aldehyde oxidase (1); alkaline phosphatase (2); aromatic-L-amino-acid decarboxylase, *or* dopa decarboxylase (3,4); arylsulfatase (5,6); dihydropteridine reductase (7-9); glutamate decarboxylase (10); γ-glutamylhistamine synthetase (11); Na$^+$/K$^+$-exchanging ATPase (12); phenylalanine 4-monooxygenase (13-18); procollagen-lysine 5-dioxygenase (19); pyridoxal kinase (20,21); tryptophan 5 monooxygenase (22-25); tyrosinase, *or* monophenol monooxygenase (26); tyrosine aminotransferase (27,28); tyrosine 3-monooxygenase (29-39); tyrosyl-tRNA synthetase (K_i = 2.5 mM for the *Escherichia coli* enzyme) (40); xanthine oxidase (1). **1.** Panoutsopoulos & Beedham (2004) *Acta Biochim. Pol.* **51**, 649. **2.** Pinoni & Lopez Mananes (2004) *J. Exp. Mar. Biol. Ecol.* **307**, 35. **3.** Nakazawa, Kumagai & Yamada (1987) *Agric. Biol. Chem.* **51**, 2531. **4.** Bender & Coulson (1977) *Biochem. Soc. Trans.* **5**, 1353. **5.** Okamura, Yamada, Murooka & Harada (1976) *Agric. Biol. Chem.* **40**, 2071. **6.** Murooka, Yim & Harada (1980) *Appl. Environ. Microbiol.* **39**, 812. **7.** Shen (1985) *J. Enzyme Inhib.* **1**, 61. **8.** Shen, Smith, Davis, Brubaker & Abell (1982) *J. Biol. Chem.* **257**, 7294. **9.** Shen (1983) *Biochim. Biophys. Acta* **743**, 129. **10.** Blindermann, Maitre, Ossola & Mandel (1978) *Eur. J. Biochem.* **86**, 143. **11.** Stein & Weinreich (1982) *J. Neurochem.* **38**, 204. **12.** Bertorello & Aperia (1987) *Acta Physiol. Scand.* **130**, 571. **13.** Kaufman (1978) *Meth. Enzymol.* **53**, 278. **14.** Kaufman (1987) *Meth. Enzymol.* **142**, 3. **15.**

Pember, Villafranca & Benkovic (1987) *Meth. Enzymol.* **142**, 50. **16.** Koizumi, Matsushima, Nagatsu, Iinuma, Takeuchi & Umezawa (1984) *Biochim. Biophys. Acta* **789**, 111. **17.** Letendre, Dickens & Guroff (1975) *J. Biol. Chem.* **250**, 6672. **18.** Bloom, Benkovic & Gaffney (1986) *Biochemistry* **25**, 4204. **19.** Murray, Cassell & Pinnell (1977) *Biochim. Biophys. Acta* **481**, 63. **20.** Neary, Meneely, Grever & Diven (1972) *Arch. Biochem. Biophys.* **151**, 42. **21.** Lainé-Cessac, Cailleux & Allain (1997) *Biochem. Pharmacol.* **54**, 863. **22.** Nakata & H. Fujisawa (1982) *Eur. J. Biochem.* **124**, 595. **23.** Hamdan & Ribeiro (1999) *J. Biol. Chem.* **274**, 21746. **24.** Naoi, Maruyama, Takahashi, Ota & Parvez (1994) *Biochem. Pharmacol.* **48**, 207. **25.** Winge, McKinney, Knappskog & Haavik (2007) *J. Neurochem.* **100**, 1648. **26.** Tudela, Garcia-Canovas, Varon, *et al.* (1987) *J. Enzyme Inhib.* **2**, 47. **27.** Granner & Tomkins (1970) *Meth. Enzymol.* **17A**, 633. **28.** Jacoby & La Du (1964) *J. Biol. Chem.* **239**, 419. **29.** Fujisawa & Okuno (1987) *Meth. Enzymol.* **142**, 63. **30.** Thöny, Calvo, Scherer, *et al.* (2008) *J. Neurochem.* **106**, 672. **31.** Nagatsu, Levitt & Udenfriend (1964) *J. Biol. Chem.* **239**, 2910. **32.** Chaube & Joy (2003) *J. Neuroendocrinol.* **15**, 273. **33.** Fitzpatrick (1988) *J. Biol. Chem.* **263**, 16058. **34.** Neckameyer, Holt & Paradowski (2005) *Biochem. Genet.* **43**, 425. **35.** Scholz, Toska, Luborzewski, *et al.* (2008) *FEBS J.* **275**, 2109. **36.** Gordon, Quinsey, Dunkley & Dickson (2008) *J. Neurochem.* **106**, 1614. **37.** Martínez, Haavik, Flatmark, Arrondo & Muga (1996) *J. Biol. Chem.* **271**, 19737. **38.** Hamdan & P. Ribeiro (1998) *J. Neurochem.* **71**, ˣ1369. **39.** Wallace (2007) *Synapse* **61**, 715. **40.** Santi & Peña (1973) *J. Med. Chem.* **16**, 273.

Doramapimod

This cell-permeable, high-affinity p38 kinase inhibitor (FW = 527.66 g/mol; CAS 285983-48-4; Soluble to 100 mM in DMSO), also named BIRB 796, 1-(5-*tert*-butyl-2-*p*-tolyl-2*H*-pyrazol-3-yl)-3-[4-(2-morpholin-4-ylethoxy) naphthalen-1-yl]urea, and *N*-[3-(1,1-dimethylethyl)-1-(4-methylphenyl)-1*H*-pyrazol-5-yl]-*N'*-[4-[2-(4-morpholinyl)ethoxy]-1-naphthalenyl]urea, selectively targets *all* p38 mitogen-activated protein kinases (with K_d values ranging from 50 to 100 pM), the mitogen-activated protein kinases that are responsive tosuch stress stimuli as cytokines, UV irradiation, heat shock, and osmotic shock, and are involved in cell differentiation, apoptosis and autophagy (1-4). BIRB796 blocks the stress-induced phosphorylation of the scaffold protein SAP97, further establishing that this is a physiological substrate of SAPK3/p38γ. BIRB-796 Exhibits no significant inhibition on a panel of related kinases. **Other Targets:** Doramapimod exhibits no significant inhibition on a panel of related protein kinases, but does inhibit JNK2α2 (IC_{50} = 98 nM) and c-Raf-1 (IC_{50} = 1.4 μM). It also inhibits LPS-induced TNFα production in human Peripheral Blood Mononuclear Cells *or* PBMCs (IC_{50} = 21 nM) and whole blood (IC_{50} = 960 nM). **1.** Regan, Capolino, Cirillo, *et al.* (2003) *J. Med. Chem.* **46**, 4676. **2.** Kuma, Sabio, Bain, *et al.* (2005) *J. Biol. Chem.* **280**, 19472. **3.** Goldstein, Kuglstatter, Lou & Soth (2010) *J. Med. Chem.* **53**, 2345. **4.** Laufer, Hauser, Domeyer, Kinkel & Liedtke (2008) *J. Med. Chem.* **51**, 4122.

Doravirine

This non-nucleoside reverse transcriptase inhibitor, *or* NNRTI (FW = 425.75 g/mol; CAS 1338225-97-0), also known as MK-1439 and 3-chloro-5-({1-[(4-methyl-5-oxo-4,5-dihydro-1*H*-1,2,4-triazol-3-yl)methyl] -2-oxo-4-(trifluoromethyl)-1,2-dihydro-3-pyridinyl}oxy)benzonitrile, exhibits IC_{50} values of 12, 9.7, and 9.7 nM, respectively, against wild-type, K103N and Y181C reverse transcriptase mutants in biochemical assays, while showing minimum off-target activities (1). The mutant profile of MK-1439 was superior overall to that of efavirenz (EFV) and comparable to that of etravirine (ETR) and rilpivirine (RPV). Furthermore, E138K, Y181C, and K101E mutant viruses that are associated with ETR and RPV were susceptible to MK-1439 with fold-change < 3 (1). **1.** Lai, Feng, Falgueyret, *et al.* (2013) *Antimicrob. Agents Chemother.* **58**, 1652.

Doripenem

This carbapenem-class antibiotic (FW = 420.50 g/mol; CAS 148016-81-3), also known as Finibax®, is a β-lactam antibiotic, exhibiting a broad action against both Gram-positive and Gram-negative bacteria. *In vivo*, doripenem inhibits cell wall synthesis by binding to penicillin-binding proteins (PBPs). Doripenem is inactive against Methicillin-Resistant *Staphylococcus aureus* (MRSA). Doripenem is also more active against *Pseudomonas aeruginosa* then other carbapenems. It is stable against β-lactamases, but is susceptible to hydrolysis by carbapenemases. **See also** *Carbapenem-Resistant Enterobacteriaceae; Ertapenem; Imipenem (Cilastatin); Meropenem.*

Dorsomorphin

This potent and reversible protein kinase inhibitor (FW = 472.41 g/mol; CAS 1219168-18-9; Solubility: >20 mg/mL DMSO; dihydrochloride is water-soluble), also named 6-[4-[2-(1-piperidinyl)ethoxy]phenyl]-3-(4-pyridinyl)-pyrazolo[1,5-a]pyrimidine, selectively targets 5'-AMP-activated protein kinase, *or* AMPK (K_i = 109 nM), a signal transduction enzyme that plays a central role in stimulating hepatic fatty acid oxidation (especially in skeletal muscle) as well as ketogenesis, and in inhibiting cholesterol synthesis, lipogenesis, triglyceride synthesis, adipocyte lipolysis and lipogenesis. AMPK also stimulates muscle glucose uptake and modulates insulin secretion by pancreatic β-cells. Dorsomorphin exhibits little or no inhibition of structurally related protein kinases, including ZAPK, SYK, PKθ, PKA, and JAK3. **1.** Zhou, Myers, Li, *et al.* (2001) *J. Clin. Invest.* **108**, 1167. **2.** Meley, Bauvy, Houben-Weerts, *et al.* (2006) *J. Biol. Chem.* **281**, 34870. **3.** Yu, Hong, Sachidanandan, *et al.* (2008) *Nature Chem. Biol.* **4**, 33. **4.** Kim, Kim, Kim, *et al.* (2011) *Atherosclerosis* **219**, 57.

Dorzolamide

This antiglaucoma drug (FW = 324.44 g/mol; CASs = 130693-82-2 and 120279-96-1), also known by the code names MK-507 and L-671,152, the trade name Trusopt® and the IUPAC name (4*S*,6*S*)-2-ethylamino-4-methyl-

5,5-dioxo-5λ^6,7-dithiabicyclo[4.3.0]nona-8,10-diene-8-sulfonamide, is a potent membrane-penetrant carbonic anhydrase inhibitor that decreases the production of aqueous humor, when administered as a 2% (wt/vol) topical solution (1,2). Dorzolamide targets CA-II (IC$_{50}$ = 0.18 nM, in vitro), and is far less active against CA-I (IC$_{50}$ = 600 nM). Dorzolamide penetrates the ciliary body, inhibits CA, and avoids side-effects of oral CA inhibitors. It reduces intra-ocular pressure (IOP), accelerates the retinal arteriovenous passage time, and improves vision in normotensive patients. **1**. Maren (1995) *J. Glaucoma* **4**, 49. **2**. Maren (2000) in *The Carbonic Anhydrases* (Chegwidder, Carter & Edwards, eds.) Birkhäuser Verlag, Basel, Switzerland.

Doxazosin

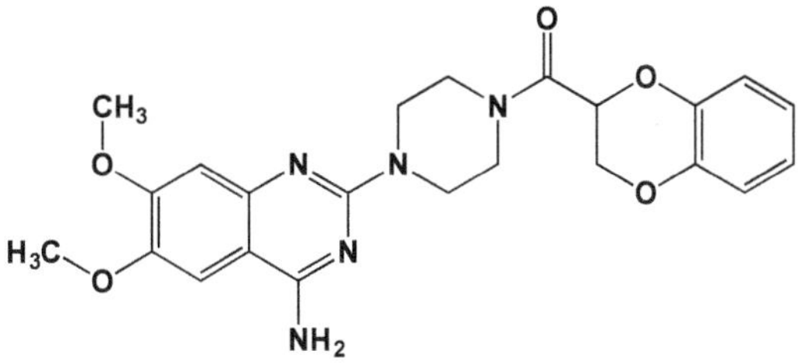

This quinazoline-based antihypertensive and α_1-selective adrenergic receptor antagonist (FW$_{free-base}$ = 451.58 g/mol; CAS 74191-85-8), also known by its code name UK33274, its trade names Cardura™ and Carduran™, and its systematic name (*RS*)-2-{4-[(2,3-dihydro-1,4-benzodioxin-2-yl)carbonyl]-piperazin-1-yl}-6,7-dimethoxyquinazolin-4-amine, relaxes vascular smooth muscle tone (vasodilation) and decreases peripheral vascular resistance. **Pharmacokinetics:** Upon oral administration of therapeutic doses, doxazosin reaches peak plasma levels in 2-3 hours, with a bioavailability of ~65%, reflecting its first-pass metabolism of by the liver. The mean elimination half-life of doxazosin is 11 hours. When administered with food, the mean maximum plasma concentration by 18% and the area under the concentration (AUC) versus time curve by 12%. **Key Pharmacokinetic Parameters:** *See* Appendix II in Goodman & Gilman's *THE PHARMACOLOGICAL BASIS OF THERAPEUTICS*, 12th Edition (Brunton, Chabner & Knollmann, eds.) McGraw-Hill Medical, New York (2011). **1**. Ali, Kossen, Timmermans & Van Zwieten (1980) *Brit. J. Pharmacol.* **68**, 113P. **2**. de Leeuw, Ligthart, Smout & Birkenhäger (1982) *Eur. J. Clin. Pharmacol.* **23**, 397.

Doxepin

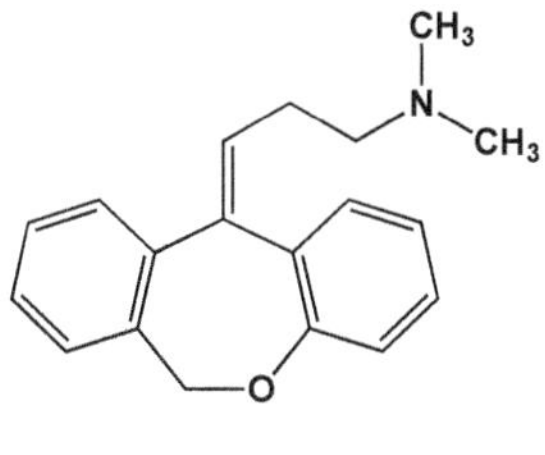

This antidepressant (FW$_{free-base}$ = 279.38 g/mol) is a histamine H$_1$ receptor antagonist as well as an antagonist at muscarinic cholinergic and α-adrenergic receptors. Commercial sources typically supply the hydrochloride as a mixture of the *E*-isomer and *Z*-isomer (shown above). **Key Pharmacokinetic Parameters:** *See* Appendix II in Goodman & Gilman's *THE PHARMACOLOGICAL BASIS OF THERAPEUTICS*, 12th Edition (Brunton, Chabner & Knollmann, eds.) McGraw-Hill Medical, New York (2011). **Target(s):** α-adrenergic receptor (1); glutathione *S*-transferase (2,3); histamine H$_1$ receptor (2,4,5); Mg^{2+}-ATPase (6); monoamine oxidase (7); muscarinic cholinergic receptor (1); Na$^+$/K$^+$-exchanging ATPase (6,8); norepinephrine uptake (9); protein kinase C (10); serotonin uptake (11). **1**. Richelson & Nelson (1984) *J. Pharmacol. Exp. Ther.* **230**, 94. **2**. Baranczyk-Kuzma, Kuzma, Gutowicz, Kazmierczak & Sawicki (2004) *Acta Biochim. Pol.* **51**, 207. **3**. Baranczyk-Kuzma, Sawicki, Kuzma & Jagiello (2001) *Pol. Merkuriusz Lek.* **11**, 472. **4**. Hill, Ganellin, Timmerman, *et al.* (1997) *Pharmacol. Rev.* **49**, 253. **5**. Kanba & Richelson (1983) *Eur. J. Pharmacol.* **94**, 313. **6**. Saha, Sengupta, Sirkar & Sengupta (1989) *Indian J. Med. Res.* **90**, 27. **7**. Roth (1975) *Life Sci.* **16**, 1309. **8**. Carfagna & Muhoberac (1993) *Mol. Pharmacol.* **44**, 129. **9**. Barth, Manns & Muscholl (1975) *Naunyn Schmiedebergs Arch. Pharmacol.* **288**, 215. **10**. Vaitla, Roshani, Holian, Cook & Kumar (1997) *Skin Pharmacol.* **10**, 191. **11**. Lingjaerde (1976) *Psychopharmacologia* **47**, 183.

Doxorubicin

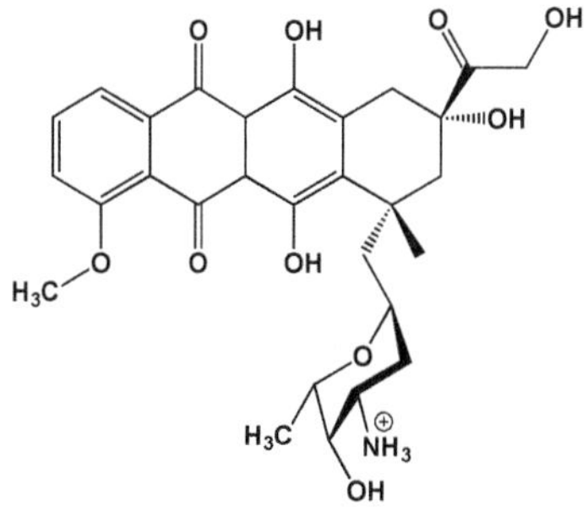

This photosensitive RNA/DNA-binding antibiotic (FW = 558.23 g/mol; CAS 23214-92-8), also called Adriamycin (referring to the hydrochloride salt) from *Streptomyces peucetius* intercalates into double-stranded DNA and inhibits replication. Doxorubicin impairs the function of DNA topoisomerase II, induces the formation of protein-associated double-strand breaks, and produces intrastrand cross-links. It also inhibits transcription. This antibody consists of a tetracyclic structure called adriamycinone covalently linked to an amino sugar (daunosamine) that can be removed by treatment with acid. Note that it tightly chelates Fe^{3+}. Aqueous solutions of doxorubicin are yellow-orange under acidic conditions, orange-red at neutral pH, and violet-blue with basic pH values (*i.e.*, above 9): the amino pK_a is 8.2. It should be stored in a desiccator in the dark and at 4°C. Stock solutions should be prepared just prior to use. Note: Doxorubicin also undergoes one-electron reduction. In the presence of dioxygen, the resulting semiquinone produces superoxide, leading to the formation of hydrogen peroxide and the reactive hydroxyl radical. The latter can damage DNA and membrane lipids. **Redox Cycling in Mitochondria:** Complex I (most likely the NADH dehydrogenase flavin) is the mitochondrial site of anthracycline reduction. During forward electron transport, the anthracyclines doxorubicin (Adriamycin) and daunorubicin act as one-electron acceptors for BH-SMP (*i.e.* they are reduced to semiquinone radical species), but only when NADH was used as substrate (38). Succinate and ascorbate are without effect. Inhibitor experiments (with rotenone, amytal, piericidin A) indicate that the anthracycline reduction site lies on the substrate side of ubiquinone. **Key Pharmacokinetic Parameters:** *See* Appendix II in Goodman & Gilman's *THE PHARMACOLOGICAL BASIS OF THERAPEUTICS*, 12th Edition (Brunton, Chabner & Knollmann, eds.) McGraw-Hill Medical, New York (2011). **Target(s):** *Ava*II restriction endonuclease (1); diferric-transferrin reductase (2); DNA helicase (3); DNA ligase, ATP-requiring (4,5); DNA ligase, NAD$^+$-requiring (4); DNA polymerase (6); DNA topoisomerase II (7-12); *Eco*RI restriction endonuclease (1); *Hae*III restriction endonuclease (1); *Hha*I restriction endonuclease (1); *Hpa*II restriction endonuclease (1); mitochondrial signal peptidase I, inner membrane protease (13); nitric-oxide synthase (14); nucleoside-triphosphatase (3); P-glycoprotein (15); phosphatidyl-*N*-methyl-ethanolamine *N*-methyltransferase (16); phospholipase A$_2$ (17); procollagen proline 4-dioxygenase (18-20); protein kinase C (21-23); RNA-directed DNA polymerase, *or* reverse transcriptase (24-28); RNA polymerase (6,29-33); *Sma*I restriction endonuclease (1); thioredoxin reductase, possible inhibition (34,35); Xaa-Pro Dipeptidase, *or* prolidase, *or* imidodipeptidase (36); xenobiotic-transporting ATPase, *or* multidrug-resistance protein (15). **1**. Corneo, Pogliani, Biassoni & Tripputi (1988) *Ric. Clin. Lab.* **18**, 19. **2**. Sun, Navas, Crane, Morré & Löw (1987) *J. Biol. Chem.* **262**, 15915. **3**. Borowski, Niebuhr, Schmitz, *et al.* (2002) *Acta Biochim. Pol.* **49**, 597. **4**. Ciarrocchi, MacPhee, Deady & Tilley (1999) *Antimicrob. Agents Chemother.* **43**, 2766. **5**. Ciarrocchi, Lestingi, Fontana, Spadari & Montecucco (1991) *Biochem. J.* **279**, 141. **6**. Zunina, Gambetta & Di Marco (1975) *Biochem. Pharmacol.* **24**, 309. **7**. Spadari, Pedrali-Noy, Focher, *et al.* (1986) *Anticancer Res.* **6**, 935. **8**. Chabner & Myers (1993) in *Cancer: Principles and Practice of Oncology* (DeVita, Hellman & Rosenberg, eds.), pp. 376-381, Lippincott, Philadelphia. **9**. Andersen, Bendixen & Westergaard (1996) *DNA Replication in Eucaryotic Cells, Cold Spring Harbor Laboratory Press*, p. 587. **10**. Insaf, Danks & Witiak (1996) *Curr. Med. Chem.* **3**, 437. **11**. Bergerat, Gadelle & Forterre (1994) *J. Biol. Chem.* **269**, 27663. **12**. Zhou, Guan & Kleinerman (2005) *Mol. Cancer Res.* **3**, 271. **13**. Howe & Floyd (2002) *The Enzymes*, 3rd ed., **22**, 101. **14**. Luo & Vincent (1994) *Biochem. Pharmacol.* **47**, 2111. **15**. Shapiro & Ling (1997) *Eur. J. Biochem.* **250**, 130. **16**. Iliskovic, Panagia, Slezák, *et al.* (1997) *Mol. Cell Biochem.* **176**, 235. **17**. Grataroli, Leonardi, Chautan, Lafont & Nalbone (1993) *Biochem. Pharmacol.* **46**, 349. **18**. Günzler, Hanauske-Abel, Myllylä, *et al.* (1988) *Biochem. J.* **251**, 365. **19**. Merriweather, Guenzler, Brenner & Unnasch (2001) *Mol. Biochem.*

Parasitol. **116**, 185. **20.** Kivirikko, Myllylä & Pihlajaniemi (1989) *FASEB J.* **3**, 1609. **21.** Wise, Glass, Chou, *et al.* (1982) *J. Biol. Chem.* **257**, 8489. **22.** Hannun, Foglesong & Bell (1989) *J. Biol. Chem.* **264**, 9960. **23.** Kikkawa & Nishizuka (1986) *The Enzymes*, 3rd ed., **17**, 167. **24.** Dhananjaya & Antony (1987) *Indian J. Biochem. Biophys.* **24**, 265. **25.** Nakashima, Yamamoto, Inouye & Nakamura (1987) *J. Antibiot.* **40**, 396. **26.** Bogdany & Csanyi (1982) *Neoplasma* **29**, 37. **27.** Matson, Fay & Bambara (1980) *Biochemistry* **19**, 2089. **28.** Tomita & Kuwata (1976) *Cancer Res.* **36**, 3016. **29.** Studzian, Wasowska, Piestrzeniewicz, *et al.* (2001) *Neoplasma* **48**, 412. **30.** Chuang, Nooteboom, Israel & Chuang (1984) *Biochem. Biophys. Res. Commun.* **120**, 946. **31.** Chuang & Chuang (1979) *Biochemistry* **18**, 2069. **32.** Gabbay, Grier, Fingerle, *et al.* (1976) *Biochemistry* **15**, 2062. **33.** Gabbay (1976) in *Inter. J. Quantum Chem., Quantum Biology Symposium No. 3*, p. 217. **34.** Mau & Powis (1992) *Biochem. Pharmacol.* **43**, 1613. **35.** Gromer, Merkle, Schirmer & Becker (2002) *Meth. Enzymol.* **347**, 382. **36.** Muszynska, Wolczynski & Palka (2001) *Eur. J. Pharmacol.* **411**, 17. **37.** Davies & Doroshow (1986) *J. Biol. Chem.* **261**, 3060.

Doxylamine

This first-generation antihistamine (FW = 270.38 g/mol; CAS 469-21-6 and 562-10-7 (succinate salt), also named (*RS*)-*N*,*N*-dimethyl-2-(1-phenyl-1-pyridin-2-ylethoxy)ethanamine, has short-term sedative properties, commending its use in combination with other drugs (typically acetaminophen and dextromethorphan) to provide night-time allergy and cold relief. Doxylamine is also combined with acetaminophen and codeine as an analgesic/calmative preparation that is often prescribed along with vitamin B6 (pyridoxine) to prevent morning sickness. **See also** *Acetaminophen, Dextromethorphan*

DPC 423

This oral anticoagulant (FW = 532.52 g/mol), named systematically as 1-[3-(aminomethyl)phenyl]-*N*-3-fluoro-[2'-(methylsulfonyl)[1,1'-biphenyl]-4-yl]-3-(trifluoromethyl)-1*H*-pyrazole-5-carboxamide, inhibits coagulation factor Xa (K_i = 0.15 nM, human). It is a much weaker inhibitor of thrombin, trypsin, plasma kallikrein, complement component I, activated protein C, and coagulation factor IXa, with K_i values of 6, 0.06, 0.061, 44 (IC50), 1.8, and 2.2 μM, respectively. **1.** Pinto, Orwat, Wang, *et al.* (2001) *J. Med. Chem.* **44**, 566.

DPCPX, See *Dipropylcyclopentylxanthine*

DPI-1

This dual prenylation inhibitor (FW_free-base = 471.60 g/mol), hence its intials, inhibits both protein farnesyltransferase and protein geranylgeranyl-transferase type II, with respective IC50 values of 0.05 and 3.6 nM. DPI blocks Ki-Ras prenylation and induces markedly higher levels of apoptosis relative to the separate actions of farnesyltransferase or geranylgeranyl-transferase inhibitors. **1.** Lobell, Omer, Abrams, *et al.* (2001) *Cancer Res.* **61**, 8758. **2.** Guida, Hamilton, Crotty & Sebti (2006) *J. Comp.-Aided Mol. Des.* **19**, 871.

DPX-4189, See *Chlorsulfuron*

DPX-5648, See *Sulfometuron-methyl*

DPX-F6025, See *Chlorimuron-ethyl*

DrKIn-II

This potent plasmin inhibitor and (FW = 6946.70 g/mol; Sequence: HDRPTFCNLAPESGRCRAHLRRIYYNLESNKCEVFFYGGCGGNDNN FSTWDECRHTCVGK) is a member of the venom trypsin inhibitor family that decreases the enzyme's amidolytic activity in a dose-dependent manner. In keeping with its action as a slow-binding, tight inhibitor, initial rapid binding (K_i = 6.7 nM) of DrKIn-II induces a slow enzyme conformational isomerization ($k_{forward}$ = 0.043 s^{-1} and $k_{reverse}$ = 0.0022 s^{-1},), thereby forming a much tighter enzyme-inhibitor complex, with a corrected overall inhibition constant K_i* of 0.2 nM. DrKIn-II also demonstrates anti-fibrinolytic activity in fibrin plate assay and significantly prolongs lysis of euglobulin clots. Screening of DrKIn-II against a panel of serine proteases indicates that plasmin is the preferential DrKIn-II target. Notably, while 1 μM DrKIn-II prolongs the activated partial thromboplastin time by 32 seconds, DrKIn-II fails to prolong the prothrombin time at all the concentrations tested, implying that, even at high concentrations, DrKIn-II has no inhibitory effect on serine proteases in the extrinsic coagulation pathway. DrKIn-II also prevents an increase of Fibrin/fibrinogen Degradation Product (FDP) in coagulation-stimulated mice and significantly reduces bleeding times in a murine tail bleeding model. **1.** Cheng & Tsai (2014) *Biochim. Biophys. Acta* **1840**, 153.

Dronedarone

This amiodarone-like antiarrhythmic agent (FW = 556.76 g/mol; CAS 141626-36-0), also known by the code name SR33589, the trade name Multaq™, and its IUPAC name *N*-(2-butyl-3-(*p*-(3-(dibutylamino)propoxy) benzoyl)-5-benzofuranyl)methanesulfonamide, is indicated for the treatment of atrial fibrillation and atrial flutter. Dronedarone inhibits (IC50 = 63 nM) the acetylcholine-activated potassium current in single cells isolated from sinoatrial node, either by disrupting the G-protein-mediated activation or by a direct inhibitory interaction with the channel protein (2). Dronedarone is a 10x less effective inhibitor of Na$^+$/Ca^{2+} exchange current in cardiac ventricular myocytes than amiodarone (3). **1.** Hodeige, Heyndrickx, Chatelain & Manning (1995) *Eur. J. Pharmacol.* **279**, 25. **2.** Altomare, Barbuti, Viscomi, Baruscotti & DiFrancesco (2000) *Br. J. Pharmacol.* **130**, 1315. **3.** Watanabe & Kimura (2008) *Naunyn Schmiedebergs Arch Pharmacol.* **377**, 371.

Droperidol

This potent butyrophenone-class receptor antagonist (FW = 379.44 g/mol; CAS 548-73-2), also named 1-[1-[4-(4-fluorophenyl)-4-oxobutyl]-1,2,3,6-

tetrahydro-4-pyridinyl]-1,3-dihydro-2*H*-benzimidazol-2-one, targets the dopamine D_2 receptor, with some histamine and serotonin antagonist activity (1). Droperidol has a central antiemetic action and effectively prevents postoperative nausea and vomiting in adults. It increases action potential duration (APD) of rabbit Purkinje fibers in a dose-dependent manner over the 10-300 μM range, when stimulated at 60 pulses/min (2). Droperidol exerts a dual effect on repolarization, prolongation with low concentrations with development of early after-depolarizations (*or* EADs) and subsequent triggered activity. Droperidol is commonly used as an antiemetic and antipsychotic in the United States since its approval in 1970 by the FDA. **1**. Parker, Brunton, Goodman, *et. al.* (2006). Goodman & Gilman's *The Pharmacological Basis of Therapeutics* (11 ed). New York, McGraw-Hill. **2**. Adamantidis, Kerram, Caron & Dupuis (1993) *J. Pharmacol. Exp. Ther.* **266**, 884.

Drug-Drug Interactions, See *CYP Inhibition*

DU-176b, See *Edoxaban*

DU-6859a, See *Sitafloxacin*

Duloxetine

This serotonin-norepinephrine reuptake inhibitor, *or* SNRI (FW = 297.42 g/mol; CAS 116539-59-4 (free base) and 136434-34-9 (hydrochloride)), also known by its code name LY248686, its trade names Cymbalta®, Ariclaim®, Xeristar®, Yentreve®, Duzela®, and Dulane®, as well as its systematic name (+)-(*S*)-*N*-methyl-3-(naphthalen-1-yloxy)-3-(thiophen-2-yl)propan-1-amine, is indicated for major depressive disorder (MDD), generalized anxiety disorder as well as osteoarthiritis and musculoskeletal pain (1,2). duloxetine has a low affinity for most 5-HT subtypes and for muscarinic, histamine H_1, α_1-adrenergic, α_2-adrenergic and dopamine D_2 receptors. Duloxetine inhibits 5-hydroxytryptamine (5-HT) and norepinephrine (NE) uptake in hippocampus slices of control rats, with IC_{50} values of 28 and 46 nM, respectively (3). Oral administration of duloxetine (3-12 mg/kg) produced a dose-dependent increase in 5-HT and NE output from the rat frontal cortex throughout the 4-hour observation period (4). **Key Pharmacokinetic Parameters:** *See* Appendix II in Goodman & Gilman's *THE PHARMACOLOGICAL BASIS OF THERAPEUTICS*, 12th Edition (Brunton, Chabner & Knollmann, eds.) McGraw-Hill Medical, New York (2011). **1**. Wong, Bymaster, Mayle, *et al.* (1993) *Neuropsychopharmacol.* 8, 23. **2**. Anttila & Leinonen (2002) Curr Opin Investig Drugs **3**, 1217. **3**. Kasamo, Blier & De Montigny (1996) *J. Pharmacol. Exp. Ther.* **277**, 278. **4**. Kihara & Ikeda (1995) *J. Pharmacol. Exp. Ther.* **272**, 177.

DuP-697

This irreversible and selective diaryl heterocycle-type COX inhibitor (FW = 411.30 g/mol; CAS 88149-94-4), also known as 5-bromo-2-(4-

fluorophenyl)-3-(4-(methylsulfonyl)phenyl)-thiophene, targets COX-2 (1). (**See** *structurally related Celecoxib; Rofecoxib*) When tested on isolated recombinant enzymes, DuP-697 is >50x more potent in the inhibition of COX-2 than COX-1 (2). The IC_{50} for human recombinant COX-2 decreases over time, with values of 80 and 40 nM at 5 and 10 minutes, respectively (3). The IC_{50} for the inhibition of human recombinant COX-1 over the same period is 9 μM (3). DuP-697 also attenuates the COX-1 inhibitory activity of nonselective COX inhibitors such as indomethacin (4). DuP-697 exerts antiproliferative, antiangiogenic and apoptotic effects on HT29 colorectal cancer cells, $IC_{50} = 4.28 \times 10^{-8}$ mol/L (5). **1**. Kargman, Wong, Greig, *et al.* (1996) *Biochem. Pharmacol.* **52**, 1113. **2**. Seibert, Masferrer, Needleman, *et al.* (1996) *Brit. J. Pharmacol.* **117**, 1016. **3**. Gierse, Hauser, Creely, *et al.* (1995) *Biochem. J.* **305**, 479. **4**. Rosenstock, Dannon, Rimon, *et al.* (1999) *Biochim Biophys. Acta* **1440**, 127. **5**. Altun, Turgut & Kaya (2014) *Asian Pac. J. Cancer Prev.* **15**, 3113.

DuP714, See *N-Acetyl-D-phenylalanyl-L-prolyl-L-boroarginine*

Dupilumab

This fully human whole monoclonal antibody (MW = 146,900 g/mol; CAS 1190264-60-8), directed against the α-subunit of IL-4 receptors, blocks IL-4 and IL-13 cytokine signaling pathways in type 2 helper T (*or* Th2) cell-mediated pathways. When administered as monotherapy or with topical corticosteroids, dupilumab results in rapid and significant improvement of Atopic Dermatitis (AD), a disorder that is characterized by Th2 cell-driven inflammation (1). In patients with persistent, moderate-to-severe asthma and elevated eosinophil levels, who used inhaled glucocorticoids and long-acting ß-agonists (LABAs), dupilumab therapy is associated with fewer asthma exacerbations, when LABAs and inhaled glucocorticoids are withdrawn, resulting in improved lung function and reduced levels of Th2-associated inflammatory markers (2). **1**. Hamilton, Ungar & Guttman-Yassky (2015) *Immunother.* 7, 1043. **2**. Wenzel, Ford, Pearlman, *et al.* (2013) New Engl. J. Med. **368**, 2455.

Durvalumab

This humanized anti-PD-L1 monoclonal antibody and antagonist (MW = 143.6 kDa; CAS 1428935-60-7), also named MEDI-4736, targets the Programmed Death Ligand-1, a 40-kDa type-1 transmembrane protein (also known as Cluster of Differentiation 274, *or* CD274, and B7 Homologue-1, *or* B7-H1) that plays a role in suppressing the immune system during pregnancy, tissue allografts, autoimmune disorders, cancer and hepatitis. To modulate activation/inhibition, PD-L1 binds relatively weakly to PD-1 (K_d = 0.7 μM), its ligand on activated T cells, B cells, and myeloid cells. By bindng to PD-L1, Durvalumab disrupts the PD-L1·PD-1 complex, thereby limiting immuosupresion.

Dutasteride

This 5-α reductase inhibitor, *or* 5α-RI (FW = 528.53 g/mol; CAS 164656-23-9), also known as Avodart® and (5α, 17β)-*N*-{2,5-bis(trifluoromethyl)phenyl}-3-oxo-4-azaandrost-1-ene-17-carboxamide, is used to treat benign prostate hyperplasia by inhibiting the conversion of testosterone to dihydrotestosterone, *or* DHT, by 3-oxo-5α-steroid 4-dehydrogenase, *or* 5-α reductase (*Reaction:* 3-oxo-5α-steroid + Acceptor ⇌ 3-oxo-Δ⁴-steroid + Reduced Acceptor) (1,2). 5α-Reductase thus catalyzes the rate-limiting step in the conversion of testosterone (T) to 5α-dihydroprogesterone (5α-DHP), progesterone to 5α-DHP, deoxycorticosterone (DOC) to 5β-dihydrodeoxycorticosterone, cortisol to

5α-dihydrocortisol, and aldosterone to 5α-dihydro-aldosterone. (*See also* Finasteride) Because 5-α reductase is also active in other organs, 5-α reductase inhibitors, such as dutasteride and fenasteride, often have side-effects, including sexual dysfunction, increased risk of high-grade prostate cancer, as well as delayed diagnosis and early treatment of prostate cancer. **Key Pharmacokinetic Parameters:** *See* Appendix II in Goodman & Gilman's *THE PHARMACOLOGICAL BASIS OF THERAPEUTICS*, 12th Edition (Brunton, Chabner & Knollmann, eds.) McGraw-Hill Medical, New York (2011). **1.** Bramson, Hermann, Batchelor, *et al.* (1997) *J. Pharmacol. Exp. Ther.* **282**, 1496. **2.** Frye, Bramson, Hermann, *et al.* (1998) *Pharm. Biotechnol.* **11**, 393.

Dvl-PDZ Domain Inhibitor, Peptide Pen-N₃

This penetratin-containing PDZ Domain peptidomimetic (FW = 3512.20 g/mol; Sequence: Ac-RQIKIWFQNRRMKWKKGGGEIVLWSDIP-NH₂), also known as Wnt Pathway Inhibitor III, consists of a 22-amino acid PDZ domain-binding element in Dvl (dishevelled) family proteins that is fused at the N-terminus with a 16-residue Antennapedia penetratin sequence to facilitate its entry into cells. (*Note*: "PDZ" is an acronym combining the first letters of three proteins (Postsynaptic density protein, *or* PSD95, Drosophila disc large tumor suppressor, *or* Dlg1, and Zonula occludens-1 protein, *or* zo-1) were first found to share this domain.) This inhibitor selectively inhibits up-regulation of cellular β-catenin accumulation and transcription activity upon Wnt3a stimulation (IC₅₀ = 11 μM) or Dvl1/2/3 overexpression in HEK29350 cultures, but not the Wnt3a- or Dvl1/2/3-independent, constitutively active Wnt pathway signaling in the APC-mutant HCT-15 colon cancer cells. Unlike Box5, the PDZ-interacting sequence in Pen-N3, originally identified via a random phage display screening, is not derived from Wnt or any known cellular protein. **1.** Zhang, Appleton, Weisman, *et al.* (2009) *Nature Chem. Biol.* **5**, 208.

DW-224a See *Zabofloxacin*

DX-88, See *Ecallantide*

DX-9065a

This novel anticoagulant (FW = 445.54 g/mol; CAS 155204-81-2), systematically named as (2*S*)-2-[4-[[(3*S*)-1-acetimidoyl-3-pyrrolidinyl]oxy]phenyl]-3-(7-amidino-2-naphthyl)propanoic acid, targets coagulation factor Xa (K_i = 41 nM). It is a significantly weaker inhibitor of thrombin (K_i > 2 mM), trypsin (Ki = 0.62 μM), chymotrypsin (K_i > 2 mM), plasmin (Ki = 23 μM), t-plasminogen activator (K_i = 21 μM), plasma kallikrein (Ki = 2.3 μM), and tissue kallikrein (K_i = 1 mM). It also inhibits proinflammatory events induced by gingipains and coagulation factor Xa. It exhibits strong anticoagulant actions *in vitro* and *in vivo* and inhibits proliferation of vascular smooth muscle cells. The crystal structure of DX-9065a with factor Xa was the first such structure reported for that serine proteinase. A zwitterion with high water solubility and low lipophilicity, human oral bioavailability of DX-9065a was low, amounting to only 2-3% in clinical trials. Compounding with polystyrene nanospheres coated with cationic poly(vinylamine) and cholestyramine greatly improved its oral bioavailablity. **Target(s):** coagulation factor Xa (1-6); plasma kallikrein (1); plasmin (1); t-plasminogen activator (1); trypsin (1). **1.** Hara, Yokoyama, Ishihara, *et al.* (1994) *Thromb. Haemost.* **71**, 314. **2.** Brandstetter, Kühne, Bode, *et al.* (1996) *J. Biol. Chem.* **271**, 29988. **3.** Kamata, Kawamoto, Honma, Iwama & Kim (1998) *Proc. Natl. Acad. Sci. U.S.A.* **95**, 6630. **4.** Ambrosini, Plescia, Chu, High & Altieri (1997) *J. Biol. Chem.* **272**, 8340. **4.** Leadley (2001) *Curr. Top. Med. Chem.* **1**, 151. **5.** Tanaka, Arai, Liu, *et al.* (2005) *Kidney Int.* **67**, 2123. **6.** Ieko, Tarumi, Takeda, *et al.* (2004) *J. Thromb. Haemost.* **2**, 612.

Dy 268

This FXR receptor antagonist (FW = 560.66 g/mol; CAS 1609564-75-1; Soluble to 100 mM in DMSO), also named 1-[(3-methoxyphenyl)methyl]-*N*-[4-methyl-3-(4-morpholinylsulfonyl)phenyl]-3-(4-methyl-phenyl)-1*H*-pyrazole-4-carboxamide, targets (IC₅₀ = 7.5 nM) the Farnesoid X Receptor (FXR), *or* Bile Acid Receptor (BAR), *or* NR1H4 (nuclear receptor subfamily 1, group H, member 4). In humans, this receptor is encoded by the *NR1H4* gene. Dy-268 exhibits no detectable FXR agonist activity and is not cytotoxic. Notably, it is the most potent FXR antagonist identified to date and possesses a promising *in vitro* profile that commends it as a tool for elucidating FXR's biological function. Known FXR agonists are Cafestol, Chenodeoxycholate, Obeticholate, and Fexaramine. **1.** Yu, Lin, Forman & Chen (2014) *Bioorg. Med. Chem.* **22**, 2919.

Dynasore

This cell-permeable dynamin inhibitor (FW_anhydrous = 322.31; CASs 1202867-00-2 (anhydrous) and 304448-55-3 for indefinite hydrate, $C_{18}H_{14}N_2O_4·xH_2O$; Solubility: > 10 mg/mL DMSO), named systematically as 3-hydroxynaphthalene-2-carboxylic acid (3,4-dihydroxybenzylidene)-hydrazide, blocks coated vesicle formation within seconds of addition. **Primary Mode of Action:** Dynamin is a large oligomerizing GTPase (M_r = 96K) that forms tube-like stuctures that are involved in clathrin-mediated endocytosis and other vesicular trafficking processes. As a mechanoenzyme, dynamin most likely uses the forces generated by GTP hydrolysis-driven conformational changes to facilitate release of nascent vesicles from parent membranes by first forming a helical structure around the neck of a nascent vesicle, whereupon GTP hydrolysis results in the lengthwise extension of this helix, elongating and then breaking the vesicle neck (1). Two types of coated pit intermediates accumulate during dynasore treatment: those that are U-shaped, half formed pits and those that are O-shaped, fully formed pits. Such structures uggesting that the active pinching-off process is interrupted by the drug (2). **Cellular Impact:** Dynasore inhibits dynamin-regulated receptor endocytosis, cholesterol trafficking, and the uptake of low-density lipoprotein (LDL) in HeLa cells (strongly inhibited) and human macrophages (moderately inhibited) (3). Such treatment leads to the abnormal accumulation of LDL and free cholesterol (FC) within the endolysosomal network. The delivery of regulatory cholesterol to the endoplasmic reticulum becomes deficient, thereby inhibiting transcriptional control of the three major sterol-sensitive genes: (a) sterol-regulatory element binding protein 2 (SREBP-2), (b) 3-hydroxy-3-methyl-coenzymeA reductase (HMGCoAR), and (c) low-density lipoprotein receptor (LDLR). Sequestration of cholesterol in the endolysosomal compartment also impairs both the active and passive cholesterol efflux in HMDM, emphasizing the importance of membrane trafficking in cholesterol homeostasis. **Effective Dosing:** Dynasore is effective at 80 μM (0.2% DMSO, final concentration (vol/vol)), when delivered in the presence of 1% lipoprotein-deficient serum (LPDS) medium to avoid dynasore binding/inactivation by serum proteins. Dynasore does not inhibit other functionally unrelated GTP-regulatory proteins. For other details on the inhibitory effects of dynasore on dynamin *in vitro* and advice on using this inhibitor to study the effects of dynasore on endocytosis in cells, see (4). Recent studies indicate that dynasore binds stoichiometrically to detergents typically used for *in vitro* drug screening, drastically reducing its potency (IC_{50} rises to 479 µM) and diminishing its utility as a research tool (5). The recently developed Dyngo Compound 4a, a dynasore derivative, is the most potent compound, exhibiting a 37x improvement in potency over dynasore for liposome-stimulated helical dynamin activity (5). **Target(s):** Dynamin-1 (2-9), Dynamin-2 (2), and the mitochondrial dynamin-related protein Drp1 (2); compensatory synaptic vesicle endocytosis (6); Atlastin-1, the dynamin-like GTPase responsible for spastic paraplegia SPG3A (7); *Toxoplasma gondii* invasion (8); hyperoxia-mediated NADPH oxidase activation and reactive oxygen species production in caveolin-enriched microdomains of the endothelium (9); polystyrene nanoparticle trafficking across rat alveolar epithelial cell monolayers via non-endocytic transcellular pathways (10); blocks botulinum neurotoxin type A endocytosis in neurons and delays botulism (11); RNase and penetratin sequence-containing protein uptake (12); CD4- and dynamin-dependent endocytosis of HIV-1 into plasmacytoid dendritic cells (13). **1.** Praefcke & McMahon, (2004) *Nat. Rev. Molec. Cell Biol.* **5**, 133. **2.** Macia, Ehrlich, Massol, *et al.* (2006) *Dev Cell.* **10**, 839. **3.** Girard, Paul, Fournier *et al.* (2011) *PLoS One* **6**, e29042. **4.** Kirchhausen, Macia & Pelish (2008) *Meth. Enzymol.* **438**, 77. **5.** McCluskey, Daniel, Hadzic, *et al.* (2013) *Traffic* **14**, 1272. **6.** Newton, Kirchhausen & Murthy (2006) *Proc. Natl. Acad. Sci. U.S.A.* **103**, 17955. **7.** Muriel, Dauphin, Namekawa, *et al.* (2009) *J. Neurochem.* **110**, 1607. **8.** Caldas, Attias & de Souza (2009) *FEMS Microbiol Lett.* **301**, 103. **9.** Singleton, Pendyala, Gorshkova, *et al.* (2009) *J. Biol. Chem.* **284**, 34964. **10.** Fazlollahi, Angelow, Yacobi, *et al.* (2011) *Nanomedicine* **7**, 588. **11.** Harper, Martin, Nguyen, *et al.* (2011) *J. Biol. Chem.* **286**, 35966. **12.** Chao & Raynes (2011) *Biochemistry* **50**, 8374. **13.** Pritschet, Donhauser, Schuster, *et al.* (2012) *Virology* **423**, 152.

Dyospirin

This DNA gyrase inhibitor (FW = 374.48 g/mol; CAS 28164-57-0), like others in the bis-naphthoquinone class, targets the ATP-hydrolyzing region within the N-terminal domain of GyrB, but acts allosterically (noncompetitively) with respect to ATP (1). The IC_{50} value of ~15 µM for diospyrin against *Mycobacterium tuberculosis* gyrase compares favorably with its minimal inhibitory concentration (MIC ~20 µM, or 8 µg/mL), indicating that its inhibitory action on gyrase B accounts for its antibiotic action. For comparison, gyrase B exhibits the following values for related inhibitors: 7-methyljuglone, IC_{50} = 30 µM; neodiospyrin, IC_{50} = 50 µM; isodiospyrin, IC_{50} = 100 µM; menadione, IC_{50} > 200 µM; shinanolone, IC_{50} > 200 µM; ciprofloxacin, IC_{50} = 10 µM; and novobiocin, IC_{50} = 1 µM. Dyosporin inhibits the growth of *Leishmania donovani* promastigotes, inducing topoisomerase I mediated DNA cleavage *in vitro* (2). **1.** Karkare, Chung, Collin, *et al.* (2013) *J. Biol. Chem.* **288**, 5149. **2.** Ray, Hazra, Mittra, Das & Majumder (1998) *Mol. Pharmacol.* **54**, 994.

E, *See Glutamate and Glutamic Acid*

E₁, *See Estrone*

E₂, *See 3,17-β-Estradiol*

E-64c, *See N-[(2S,3S)-trans-Epoxysuccinyl]-L-leucylamido-3-methylbutane*

E-64d

This cell- and lysosome-permeable thiol protease inhibitor (FW = 342.44 g/mol; CAS 88321-09-9; 20 mM in DMSO), also named (2S,3S)-3-[[[(1S)-3-methyl-1-[[(3-methylbutyl)amino]carbonyl]butyl]amino]carbonyl] -2-oxiranecarboxylic acid ethyl ester and Aloxistatin, targets cathepsins B and L as well as calpain. E64d also inhibits lysosomal proteases and interferes with autolysosomal digestion when used in combination with pepstatin A. **Alzheimer's Disease:** α-Secretase processing of amyloid precursor protein (APP) is an important pathway for decreasing secretion of neurotoxic amyloid β (Aβ). E-64d likewise ameliorates amyloid-β-induced reduction of sAPPα secretion by reversing ceramide-induced protein kinase C down-regulation in SH-SY5Y neuroblastoma cells (4). Such findings suggest ceramide may play an important role in sAPPα processing by modulating PKC activity. **1.** Wilcox & Mason (1992) *Biochem. J.* **285**, 495. **2.** Tanada, *et al.* (2005) *Autophagy* **1**, 84. **3.** Sato, *et al.* (2007) *Cancer Res.* **67**, 9677. 4. Kim, *et al.* (2008) *Autophagy* **4**, 659. **4.** Tanabe, Nakajima & Ito (2013) *Biochem. Biophys. Res. Commun.* **441**, 256.

E-3810

This novel dual VEGF/FGF receptor protein tyrosine kinase inhibitor (FW = 443.49 g/mnol; CAS 1058137-23-7) targets kinases associated with Vascular Endothelial Growth Factor Receptor family members (VEGFR-1, IC_{50} = 7 nM; VEGFR-2, IC_{50} = 25 nM; VEGFR-3, IC_{50} = 10 nM), Fibroblast Growth Factor Receptor family members (FGFR-1, IC_{50} = 17 nM; FGFR-2, IC_{50} = 83 nM; FGFR-3, IC_{50} = 238 nM; FGFR-4, IC_{50} = >1000 nM), Platelet-Derived Growth Factor Receptor family members (c-Kit, IC_{50} = 495 nM; PDGFRα, IC_{50} = 175 nM; PDGFRβ, IC_{50} = 525 nM), as well as Colony-Stimulating Factor 1 receptor, *or* CSF-1R, IC_{50} = 5 nM (1). When tested in a variety of tumor xenograft models, including early/late-stage subcutaneous and orthotopic models, E-3810 exhibited striking antitumor properties at well-tolerated oral doses administered daily (1). E-3810 remained active in tumors rendered nonresponsive to the general kinase inhibitor sunitinib as a consequence of a previous cycle of sunitinib treatment (1). FGFR inhibition results in enhanced E-3810 cytotoxic activity in cell lines with FGF-driven growth (2). Likewise, antitumor activity appears somewhat higher in tumor xenografts expressing high FGFR-1 and/or FGF ligand (3), suggesting a better clinical response

might be achieved in patients with deregulated FGF/FGFR signaling pathway. E-3810 also inhibits FGF-induced angiogenesis and reduces blood vessel density. **Other Targets:** A quantitative chemical genomics study (4) indicates that the following kinases are likely secondary targets Discoidin Domain Receptor tyrosine *kinase* 2, (DDR2), proto-oncogene tyrosine-protein kinase (YES), Lck/Yes-related Novel protein tyrosine kinase (LYN), Caspase-Recruiting Domain-containing Interleukin 1β-converting enzyme (ICE) Associated Kinase (CARDIAK), Ephrin type-A receptor 2 (EPHA2), and Cytokine-Suppressive mitogen-activating protein kinase (CSBP). **1.** Bello, Colella, Scarlato, *et al.* (2011) *Cancer Res.* **71**, 1396. **2.** Colella, *et al.* (2010) 22nd EORTC-NCI-AACR, Berlin, Germany, Abs. No. 191. **3.** Colella, *et al.*, 102nd AACR, Orlando, FL, USA, April 2011; Abs # 595. **4.** Colzani, Noberini, Romanenghi, *et al.* (2014) *Mol. Cell Proteomics* **13**, 1495.

E-5555, *See Atopaxar*

E5700

This quinuclidine derivative (FW = 433.55 g/mol) inhibits *Trypanosoma cruzi* and human squalene synthase (IC_{50} = 0.84 and 1.5 nM, respectively) (1). E5700 interferes with cellular proliferation and induce ultrastructural and lipid profile alterations in a *Candida tropicalis* strain that is resistant to fluconazole, itraconazole, and amphotericin B (2). **1.** Sealey-Cardona, Cammerer, Jones, *et al.* (2007) *Antimicrob. Agents Chemother.* **51**, 2123. **2.** Ishida, Visbal, Rodrigues, *et al.* (2011) *J. Infect. Chemother.* **17**, 563.

E7080, *See Lenvatinib*

E7107

This spliceosome inhibitor (FW = 704.95; CAS 630100-90-2) shows anti-proliferative activity of against various cancer cell lines by targeting spliceosome assembly by preventing tight binding of U2 snRNP to pre-mRNA. Although E7107 has no apparent effect on U2 snRNP integrity, it potently prevents an ATP-dependent conformational change exposing the branch point-binding region in U2 snRNP. (*See also* FR901464; *Spliceomycin A; Meayanmycin; Pladienolide B*) **1.** Kotake, Sagane, Owa, *et al.* (2007) *Nature Chem. Biol.* **3**, 570. **2.** Folco, Kaitlyn Coil & Reed (2011) *Genes Develop.* **25**, 440. **IUPAC Name:** (3R,6R,7S,8E,10S,11S,12E, 14E,16R,18R,19R,20R, 21S)-7-[(4-cycloheptylpiperazin-1-yl)carbonyl]oxy-3,6,16,21-tetrahydroxy-6,10,12,16,20-pentamethyl-18,19-epoyxtrichosa-8,12,14-trien-11-olide

E7389, *See Eribulin*

E7449

392

This oral and potent PARP inhibitor (FW = 317. 34 g/mol; CAS 1140964-99-3; Solubility: < 6.5 mg/mL DMSO) targets poly(ADP-ribose) polymerase PARP1 and PARP2, trapping PARP1 onto damaged DNA and potentiating the cytotoxic effects of both radiotherapy and chemotherapy. E7449 also inhibits PARP5a/5b, otherwise known as tankyrase1 and 2 (TNKS1 and 2), important regulators of canonical Wnt/β-catenin signaling. E7449 stabilizes axin and TNKS proteins resulting in β-catenin de-stabilization and significantly altered expression of Wnt target genes. **1.** McGonigle, Chen, Wu, *et al.* (2015) *Oncotarget* **6**, 41307.

E-616452, *See RepSox*

Early Mitotic Inhibitor 1

This regulatory protein (MW = 50146.38 g; *Symbol*: EMI1) plays an essential role in coordinating DNA synthesis and mitosis by deploying its 143-residue C-terminal domain to inhibit anaphase-promoting complex APC/C^{CDH1} functions, thus interfering with ubiquitin-mediated proteolysis required for cell division. **Mode of Action:** Ubiquitin action requires the E$_1$, E$_2$, E$_3$ enzymes, wherein E$_1$ catalyzes the formation of E$_2$-Ub thioester, and E$_3$s promote Ub ligation to specific protein substrates. Among the 600 predicted members of the largest human E$_3$ family is APC/C, which controls cell division by promoting timely Ub-mediated proteolysis of key regulatory proteins. APC/C^{CDH1} utilizes a two-step mechanism to mediate substrate poly-ubiquitination, first activating Ub transfer from an E$_2$ (often UBCH10 in human cells) to the substrate, and then allowing another E$_2$ (UBE2S in humans) to transfer Ub to the substrate-ligated form. EMI1's intrinsically disordered D-box, linker and tail elements, together with a structured zinc-binding domain, bind distinct regions of APC/C^{CDH1} to synergistically both block the substrate-binding site and inhibit ubiquitin-chain elongation. **1.** Frye, Brown, Petzold, *et al.* (2013) *Nature Struct. Molec. Biol.* **20**, 827.

Ebastine

This peripherally acting H$_1$ antihistamine (FW = 469.66 g/mol; CAS 90729-43-4; poor water solubility, necessitating use as micronized particles), also known systematically as 4-(4-benzhydryloxy-1-piperidyl)-1-(4-tert-butylphenyl)butan-1-one and by its trade names Evastin®, Kestine®, Ebastel®, and Aleva®, does not penetrate the blood–brain barrier to a significant amount and effective blocks the H$_1$ receptors in peripheral tissue, seldom causing drowsiness or sedation. **1.** Van Cauwenberge, De Belder & Sys (2004) *Expert Opin. Pharmacother.* **5**, 1807.

Ebelactone A

This unusual β-lactone (FW = 338.49 g/mol; CAS 76808-16-7) was first isolated in 1980 from a cultured strain of soil actinomycetes related to *Streptomyces aburaviensis*, Ebeleactone A is unstable and should be stored in the cold in the absence of light and moisture. Stock solutions in methanol, ethanol, or chloroform are stable for one week at 4°C. **Target(s):** acylaminoacyl-peptidase, irreversibly with covalent bond formation (1); esterase (2-4); *N*-formylmethionine aminopeptidase (3,4); lipase, *or* triacylglycerol lipase (3-5); retinyl-palmitate esterase (6). **1.** Scaloni, Jones, Barra, *et al.* (1992) *J. Biol. Chem.* **267**, 3811. **2.** Hoskins, Katz, Frieden, Ordorica & Young (1990) *Amer. J. Obstet. Gynecol.* **163**, 1944. **3.** Mandal (2002) *Org. Lett.* **4**, 2043. **4.** Uotani, Naganawa, Aoyagi & Umezawa (1982) *J. Antibiot.* **35**, 1670. **5.** Nonaka, Ohtaki, Ohtsuka, *et al.*

(1996) *J. Enzyme Inhib.* **10**, 57. **6.** Gollapalli & Rando (2003) *Biochemistry* **42**, 5809.

Ebelactone B

This unusual β-lactone (FW = 352.51 g/mol; CAS 76808-15-6) was first isolated in 1980 from a cultured strain of soil actinomycetes related to *Streptomyces aburaviensis*, Ebeleactone A is unstable and should be stored in the cold in the absence of light and moisture. Stock solutions in methanol, ethanol, or chloroform are stable for one week at 4°C. **Target(s):** carboxypeptidase C, *or* carboxypeptidase Y (1,2); esterase (3-6); *N*-formylmethionine aminopeptidase (5,6); lipase, *or* triacylglycerol lipase (5-7); protein C-terminal prenyl-cysteine methylesterase (4); retinyl-palmitate esterase (8). **1.** Majima, Ikeda, Kuribayashi, *et al.* (1995) *Eur. J. Pharmacol.* **284**, 1. **2.** Satoh, Kadota, Oheda, *et al.* (2004) *J. Antibiot.* **57**, 316. **3.** Hoskins, Katz, Frieden, Ordorica & Young (1990) *Amer. J. Obstet. Gynecol.* **163**, 1944. **4.** Dunten, Wait & Backlund (1995) *Biochem. Biophys. Res. Commun.* **208**, 174. **5.** Mandal (2002) *Org. Lett.* **4**, 2043. **6.** Uotani, Naganawa, Aoyagi & Umezawa (1982) *J. Antibiot.* **35**, 1670. **7.** Nonaka, Ohtaki, Ohtsuka, *et al.* (1996) *J. Enzyme Inhib.* **10**, 57. **8.** Gollapalli & Rando (2003) *Biochemistry* **42**, 5809.

Ebselen

This organoselenium antioxidant and anti-inflammatory (FW = 274.18 g/mol; CAS 60940-34-3), also known as 2-phenyl-1,2-benzisoselenazol-3(2*H*)-one, exhibits a glutathione peroxidase-like activity (*i.e.*, acting as an enzyme mimic) and also acts as a neuroprotectant, especially against acute focal ischemic injury (1-4). Ebselen binds covalently to thiol groups in proteins. Ebselen is a nonselective inhibitor of both cyclooxygenases 1 and 2 (COX-1 and COX-2). It is also an excellent peroxynitrite scavenger ($k \approx 10^6$ M^{-1}s^{-1}), yielding 2-phenyl-1,2-benzisoselenazol-3(2*H*)-one 1-oxide, the selenoxide of the parent molecule, as the sole selenium-containing product (5). A stoichiometry of 1 mol ebselen yielded 1 mol selenoxide per mole of peroxynitrite. It inhibits radiation-induced apoptosis in thymocytes, LPS-induced NF-κB nuclear translocation, and LPS-induced phosphorylation of JNK. Ebselen significantly reduces the glucose-induced increase in osmotic fragility and inhibited HbA1c formation. **Target(s):** cyclooxygenase (6,7,26); cytochrome b_5 reductase (8,9); cytochrome P450 reductase (9,10); electron transport (8,9); glutathione *S*-transferase (11); H$^+$ ATPase, plasma membrane (12); H$^+$/K$^+$ ATPase (13); 5-lipoxygenase, *or* arachidonate 5-lipoxygenase (7,26); 12-lipoxygenase, *or* arachidonate 12-lipoxygenase (6); 15-lipoxygenase, *or* arachidonate 15-lipoxygenase, IC$_{50}$ = 0.06 μM (7,14); NADPH oxidase (14-16); nitric-oxide synthase (17-20); papain (11); porphobilinogen synthase, *or* δ-aminolevulinate dehydratase (21); protein kinase C (15); thioredoxin reductase, as an alternative substrate (22,23); tumor necrosis factor-α-induced c-Jun *N*-terminal kinase activation (24); 1-alkylglycerophosphocholine *O*-acetyltransferase (25); HbA1c formation (26). **1.** Sies (1994) *Meth. Enzymol.* **234**, 476. **2.** Sies (1995) *Meth. Enzymol.* **252**, 341. **3.** Parnham & Kindt (1984) *Biochem. Pharmacol.* **33**, 3247. **4.** Maiorino, Roveri, Coassin & Ursini (1988) *Biochem. Pharmacol.* **37**, 2267. **5.** Masumoto & Sies (1996) *Chem. Res. Toxicol.* **9**, 262. **6.** Kuhl, Borbe, Fischer, Romer & Safayhi (1986) *Prostaglandins* **31**, 1029. **7.** Schewe, Schewe & Wendel (1994) *Biochem. Pharmacol.* **48**, 65. **8.** Laguna, Nagi, Cook & Cinti (1989) *Arch. Biochem. Biophys.* **269**, 272. **9.** Nagi, Laguna, Cook & Cinti (1989) *Arch. Biochem. Biophys.* **269**, 264. **10.** Wendel, Otter & Tiegs (1986) *Biochem. Pharmacol.* **35**, 2995. **11.** Nikawa, Schuch, Wagner & Sies (1994) *Biochem. Pharmacol.* **47**, 1007. **12.** Soteropoulos, Vaz, Santangelo, *et al.* (2000) *Antimicrob. Agents Chemother.* **44**, 2349. **13.** Tabuchi, Ogasawara & Furuhama (1994) *Arzneimittelforschung* **44**, 51. **14.** Walther, Holzhütter, Kuban, *et al.*

(1999) *Mol. Pharmacol.* **56**, 196. **15**. Cotgreave, Duddy, Kass, Thompson & Moldeus (1989) *Biochem. Pharmacol.* **38**, 649. **16**. Wakamura, Ohtsuka, Okamura, Ishibashi & Masayasu (1990) *J. Pharmacobiodyn.* **13**, 421. **17**. Wang, Komarov, Sies & de Groot (1992) *Hepatology* **15**, 1112. **18**. Hattori, Yui, Shinoda, *et al.* (1996) *Jpn. J. Pharmacol.* **72**, 191. **19**. Hattori, Inoue, Sase, *et al.* (1994) *Eur. J. Pharmacol.* **267**, R1. **20**. Zembowicz, Hatchett, Radziszewski & Gryglewski (1993) *J. Pharmacol. Exp. Ther.* **267**, 1112. **21**. Nogueira, Borges, Zeni & Rocha (2003) *Toxicology* **191**, 169. **22**. Engman, Cotgreave, Angulo, *et al.* (1997) *Anticancer Res.* **17**, 4599. **23**. Zhao, Masayasu & Holmgren (2002) *Proc. Natl. Acad. Sci. U.S.A.* **99**, 8579. **24**. Yoshizumi, Fujita, Izawa, *et al.* (2004) *Exp. Cell Res.* **292**, 1. **24**. Hurst & Bazan (1997) *J. Ocul. Pharmacol. Ther.* **13**, 415. **25**. Scholz, Ulbrich & Dannhardt (2008) *Eur. J. Med. Chem.* **43**, 1152. **26**. Soares, Folmer, Da Rocha & Nogueira (2014) *Cell Biol. Int.* **38**, 625.

Ecallantide

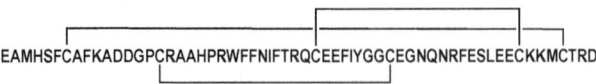

EAMHSFCAFKADDGPCRAAHPRWFFNIFTRQCEEFIYGGCEGNQNRFESLEECKKMCTRD

This recombinant polypeptide inhibitor (MW = 7053.7 g; CAS 460738-38-9), also known by the codename DX-88 and its tradename Kalbitor®, targets plasma kallikrein ($K_i \approx 25$ pM) and is an FDA-approved treatment for hereditary angioedema (*or* HAE) and for reducing blood loss during cardiothoracic surgery. Ecallantide's sequence is based on the first Kunitz domain (residues 20-79) within human tissue factor pathway inhibitor, with the following substitutions: Glu at position-20, Ala at position-21, Arg at position-36, Ala at position-38, His at position-39, Pro at position-40, and Trp at position-42. Plasma kallikrein plays a major role in the kallikrein-kinin contact cascade that forms bradykinin, a vasodilator that increases vascular permeability, activates inflammation, and produces pain. An autosomal dominant disorder, hereditary angioedema presents clinically as recurrent episodes of swelling resulting either from deficient production C1 inhibitor or mutations rendering the latter less effective. Acquired angioedema is associated with lymphoproliferative or autoimmune disease (2). Defective or missing C1-inhibitor allows activation of kallikrein, a protease responsible for bradykinin formation from its precursor, kininogen. **1**. Lehmann (2008) *Expert Opin. Biol. Ther.* **8**, 1187. **2**. Lock & Gompels (2007) *Curr. Allergy Asthma Rep.* **7**, 264. **3**.

Ecdysterone

This polyhydroxylated steroid (FW = 480.64 g/mol; CAS 5289-74-7), also named β-ecdysone, 20-hydroxyecdysone, and (2β,3β,5β,20*R*,22*R*)-2,3,14,20,22,25-hexahydroxycholest-7-en-6-one, is a hormone in many insects and crustaceans and has also been foun in certain plants. Ecdysterone, the most widely occurring ecdysteroid, participates in a number of physiological activities in these organisms. For example, it has a key role in insect development, ecdysis, metamorphosis, embryogenesis, ovulation, cell proliferation, gene regulation, and apoptosis. Ecdysterone binds to the ecdysone receptor resulting in the activation of primary response genes. The production of transcription factors results in the induction secondary response target genes. It is produced from ecdysone via ecdysone 20-monooxygenase and is a product inhibitor. **Target(s):** ecdysone *O*-acyltransferase (1); inositol-trisphosphate 3-kinase (2). **1**. Slinger & Isaac (1988) *Insect Biochem.* **18**, 779. **2**. Mayr, Windhorst & Hillemeier (2005) *J. Biol. Chem.* **280**, 13229.

Echinocandins

These cyclopeptide antibiotics (Echinicandin B: $R_1 = R_2 = OH$, FW = 797.81 g/mol, CAS 79411-15-7; Echinicandin C: $R_1 = OH$, $R_2 = H$, FW = 781.82 g/mol; Echinicandin D: $R_1 = H$, FW = 765.81 g/mol, CAS 71018-13-8) from *Aspergillus nidulans* and *A. ruglosis* inhibit cell wall biosynthesis and 1,3-β-D-glucan synthase, decreasing V_{max} without affecting the K_m value. **Target(s):** ATPase (1); 1,3-β-glucan synthase, inhibited by echinocandin B (1-7); 5'-nucleotidase (1). **1**. Surarit & Shepherd (1987) *J. Med. Vet. Mycol.* **25**, 403. **2**. Taft, Zugel & Selitrennikoff (1991) *J. Enzyme Inhib.* **5**, 41. **3**. Sawistowska-Schroder, Kerridge & Perry (1984) *FEBS Lett.* **173**, 134. **4**. Debono & Gordee (1994) *Annu. Rev. Microbiol.* **48**, 471. **5**. Kelly, Register, Hsu, Kurtz & Nielsen (1996) *J. Bacteriol.* **178**, 4381. **6**. Mazur, Morin, Baginsky, *et al.* (1995) *Mol. Cell Biol.* **15**, 5671. **7**. Kauss & Jeblick (1986) *Plant Physiol.* **80**, 7.

Echinomycin

This lactone antibiotic (FW = 1101.27 g/mol; CAS 512-64-1), also called quinomycin A, from *Streptomyces echinatus* is effective against Gram-positive organisms. It binds to DNA, but not RNA, and exhibits a preference for GpC sites. It will intercalate into DNA, altering the polynucleotide structure and inhibiting enzymes such as DNA topoisomerase II. **Target(s):** deoxyribonuclease I (1,2); DNA-directed DNA polymerase (3); DNA-directed RNA polymerase (4-7); DNA helicase (8); *Hpa*I restriction endonuclease (10). **1**. Low, Fox & Waring (1986) *Anticancer Drug Des.* **1**, 149. **2**. Bailly, Gentle, Hamy, Purcell & Waring (1994) *Biochem. J.* **300**, 165. **3**. May, Madine & Waring (2004) *Nucl. Acids Res.* **32**, 65. **4**. Waring & Wakelin (1974) *Nature* **252**, 653. **5**. Gause, Loshkareva & Zbarsky (1968) *Biochim. Biophys. Acta* **166**, 752. **6**. Takusagawa (1985) *J. Antibiot.* **38**, 1596. **7**. Sethi (1971) *Prog. Biophys. Mol. Biol.* **23**, 67. **8**. Bachur, Johnson, Yu, *et al.* (1993) *Mol. Pharmacol.* **44**, 1064. **9**. Yoshinari, Okada, Yamada, *et al.* (1994) *Jpn. J. Cancer Res.* **85**, 550. **10**. Malcolm, Moffatt, Fox & Waring (1982) *Biochim. Biophys. Acta* **699**, 211.

Echistatin α₁

This cysteine-rich 49-residue disintegrin (FW = 5417.05 g/mol; CAS 129038-42-2, from the venom of the saw-scaled viper *Echis carinatus*, inhibits platelet aggregation, IC_{50} = 30 nM. *Sequence:* ECESGPCCR NCKFLKEGTICKRARGDDMDDYCNGKTCDCPRNPHKGPAT, with 4 disulfide bridges (Cys2-Cys11, Cys7-Cys32, Cys8-Cys37, and Cys20-Cys39) and a canonical integrin-binding, **RGD**XXDD sequence, where X is any aminoacyl residue. **1**. Gan, Gould, Jacobs, Friedman & Polokoff (1988) *J. Biol. Chem.* **263**, 19827. **2**. Garsky, Lumma, Freidinger, *et al.* (1989) *Proc. Natl. Acad. Sci. U.S.A.* **86**, 4022.

Econazole

This substituted imidazole (FW$_{free-base}$ = 381.69 g/mol; CAS 27220-47-9 frequently supplied as the nitrate salt) is a broad-spectrum antifungal agent that inhibits ergosterol biosynthesis. When applied as an emulsified cream (1% weight/volume), Econazole is highly effective against mant dermatophytes (*Epidermophyton floccosum*, *Microsporum audouini*, *M. canis*, *M. gypseum*, *Trichophyton mentagrophytes*, *T. rubrum*, and *T. tonsurans*) as well as a few yeasts (*Candida albicans* and *Malassezia furfur*). **Target(s):** aromatase, CYP19 (1-6); Ca^{2+}-transporting ATPase (7); cholesterol monooxygenase, side-chain-cleaving, *or* CYP11A1 (8); CYP3A4 (9,10); CYP2E1 (11); CYP121 (12); 16-ene C19-steroid synthesizing enzyme (13); K$^+$ channels, Ca^{2+}-dependent (14); Na$^+$/I$^-$-cotransporter (15); nitric-oxide dioxygenase (16-18); nitric-oxide synthase (19); steroid 17,20-lyase (20); steroid 17α monooxygenase (20); sterol 14-demethylase, *or* CYP51 (12,21-24); TRPM2 channel, a Ca^{2+}-permeable non-selective cation channel (25). **1**. Doody, Murry & Mason (1990) *J. Enzyme Inhib.* **4**, 153. **2**. Corbin, Graham-Lorence, McPhaul, *et al.* (1988) *Proc. Natl. Acad. Sci. U.S.A.* **85**, 8948. **3**. Ayub & Levell (1988) *J. Steroid Biochem.* **31**, 65. **4**. France, Mason, Magness, Murry & Rosenfeld (1987) *J. Steroid Biochem.* **28**, 155. **5**. Mason, Carr & Murry (1987) *Steroids* **50**, 179. **6**. Ayub & Levell (1990) *Biochem. Pharmacol.* **40**, 1569. 7. Mason, Mayer & Hymel (1993) *Amer. J. Physiol.* **264**, C654. **8**. Kurokohchi, Nishioka & Ichikawa (1992) *J. Steroid Biochem. Mol. Biol.* **42**, 287. **9**. Fowler, Riley, Pritchard, *et al.* (2000) *Biochemistry* **39**, 4406. **10**. Monostory, Hazai & Vereczkey (2004) *Chem. Biol. Interact.* **147**, 331. **11**. Tassaneeyakul, Birkett & Miners (1998) *Xenobiotica* **28**, 293. **12**. McLean, Marshall, Richmond, *et al.* (2002) *Microbiology* **148**, 2937. **13**. Nakajin, Takahashi & Shinoda (1991) *J. Steroid Biochem. Mol. Biol.* **38**, 95. **14**. Alvarez, Montero & Garcia-Sancho (1992) *J. Biol. Chem.* **267**, 11789. **15**. Vroye, Beauwens, Van Sande, *et al.* (1998) *Pflugers Arch.* **435**, 259. **16**. Gardner (2005) *J. Inorg. Biochem.* **99**, 247. **17**. Helmick, Fletcher, Gardner, *et al.* (2005) *Antimicrob. Agents Chemother.* **49**, 1837. **18**. Hallstrom, Gardner & Gardner (2004) *Free Radic. Biol. Med.* **37**, 216. **19**. Bogle & Vallance (1996) *Brit. J. Pharmacol.* **117**, 1053. **20**. Ayub & Levell (1987) *J. Steroid Biochem.* **28**, 521. **21**. Jackson, Lamb, Kelly & Kelly (2000) *FEMS Microbiol. Lett.* **192**, 159. **22**. Vanden Bossche, Engelen & Rochette (2003) *J. Vet. Pharmacol. Ther.* **26**, 5. **23**. McLean, Warman, Seward, *et al.* (2006) *Biochemistry* **45**, 8427. **24**. Pietila, Vohra, Sanyal, *et al.* (2006) *Amer. J. Respir. Cell Mol. Biol.* **35**, 236. **25**. Hill, McNulty & Randall (2004) *Naunyn Schmiedebergs Arch. Pharmacol.* **370**, 227.

Ecotin

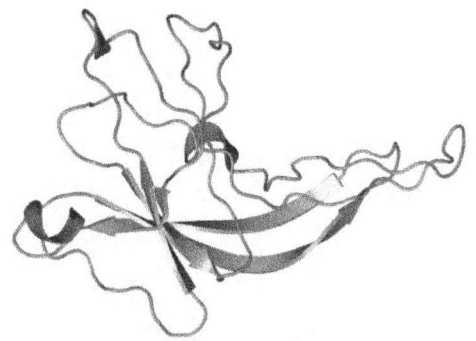

This serine proteinase inhibitor (MW = 32 kDa) is a homodimer of a 142-residue *Escherichia coli* protein (monomer show above), and targets coagulation factor XIIa (*K* = 90 nM) and human leukocyte elastase (*K* = 55 nM). **Target(s):** brachyurin (fiddler crab) collagenolytic serine protease-1 (1-3); cathepsin G, K_i = 15 pM (4); chymase (5); chymotrypsin (2,5-7); coagulation factor Xa (7,8); coagulation factor XIIa (2,7,9,10); elastase (2,5,6,10); granzyme B (2,11); kallikrein (7,10); trypsin, also inhibited by other ecotin variants (2,5-7,12-14); u-plasminogen activator, *or* urokinase;

also inhibited by other ecotin variants (5,13,14). Significantly, ecotin is also necessary for the early stages of *Burkholderia pseudomallei* colonization of murine cells, allowing replication following cell entry (15). *B. pseudomallei* is a Gram-negative soil saprophyte that causes the disease melioidosis where clinical symptoms can vary from localised infection to pneumonia and septic shock. An ecotin orthologue in *B. pseudomallei* inhibits elastase. **1**. Tsu, Perona, Schellenberger, Turck & Craik (1994) *J. Biol. Chem.* **269**, 19565. **2**. McGrath, Gillmor & Fletterick (1995) *Protein Sci.* **4**, 141. **3**. Perona, Tsu, Craik & Fletterick (1997) *Biochemistry* **36**, 5381. **4**. Sambrano, Huang, Faruqi, *et al.* (2000) *J. Biol. Chem.* **275**, 6819. **5**. Chung, Ives, Almeda & Goldberg (1983) *J. Biol. Chem.* **258**, 11032. **6**. McGrath, W. M. Hines, J. A. Sakanari, R. J. Fletterick & C. J. Craik (1991) *J. Biol. Chem.* **266**, 6620. **7**. Seymour, Lindquist, Dennis, *et al.* (1994) *Biochemistry* **33**, 3949. **8**. Wang, Hur, Sousa, *et al.* (2003) *Biochemistry* **42**, 7959. **9**. Ratnoff (1998) in *Handb. Proteolytic Enzymes*, p. 144, Academic Press, San Diego. **10**. Ulmer, Lindquist, Dennis & Lazarus (1995) *FEBS Lett.* **365**, 159. **11**. Waugh, Harris, Fletterick & Craik (2000) *Nature Struct. Biol.* **7**, 762. **12**. Halfon & Craik (1998) in *Handb. Proteolytic Enzymes*, p. 12, Academic Press, San Diego. **13**. Yang & Craik (1998) *J. Mol. Biol.* **279**, 1001. **14**. Wang, Yang & Craik (1995) *J. Biol. Chem.* **270**, 12250. **15**. Ireland, Marshall, Norville & Sarkar-Tyson (2014) *Microb. Pathog.* **68**, 55.

Ecteinascidin 743

This stereochemically complex antitumor agent (FW = 747.86 g/mol; CAS 114899-77-3), also known as Et-743, NSC 648766 and Trabectedin, from the Caribbean tunicate *Ecteinascidia turbinate* binds to the minor groove of DNA, alkylating guanine residues in a sequence-selective manner and inhibiting DNA topoisomerase I. Its unique mechanism of action triggers a cascade of events that interfere with several transcription factors, DNA binding proteins and DNA repair pathways, resulting in G$_2$-M cell cycle arrest and ultimately apoptosis. At therapeutic concentrations, Ecteinascidin 743 has selective anti-inflammatory and immunomodulatory properties on tumor microenvironment, decreasing the production of proinflammatory mediators (*e.g.*, such as IL-6 and CCL2) favoring tumor growth, progression, angiogenesis, and metastasization. Ecteinascidin 743 is marketed under the trade name Yondelis®. **1**. Meng, Liao & Pommier (2003) *Curr. Top. Med. Chem.* **3**, 305.

Eculizumab

This first-in-class humanized monoclonal antibody (MW = 148.3 kDa; CAS 219685-50-4), also known by the trade name Soliris®, targets Complement 5a, an important late-stage cascade component that, when once activated, stimulates host cells, attracts pro-inflammatory immune cells, and destroys cells by forming transmembrane pores. Eculizumab is a IgG2/4κ recombinant monoclonal antibody that is composed of two 448-residue heavy chains and two 214-residue light chains. Eculizumab contains constant regions from human IgG2 sequences and human IgG4 sequences and murine complementarity-determining regions grafted onto the human framework light-and heavy-chain variable regions. Eculizumab is produced in murine myeloma cell cultures, purified by standard methods, and delivered to patients as an IV infusion. Eculizumab is FDA-approved for the treatment of paroxysmal nocturnal hemoglobinuria (PNH) and atypical hemolytic uremic syndrome (aHUS). The former is a progressive and sometimes life-threatening disease associated with excessive hemolysis. The latter causes abnormal blood clots to form in small blood vessels, leading to kidney failure and damage to other vital organs. **1**. Kaplan (2002) *Curr. Opin. Investig. Drugs* **3**, 1017. **2**. Hillmen, Hall, Marsh, *et al.* (2004) *New Engl. J. Med.* **350**, 552. **3**. Rathbone, Kaltenthaler, Richards, *et al.* (2013) *BMJ Open* **3**, e003573.

ED-749, See *Edotecarin*

Edeines

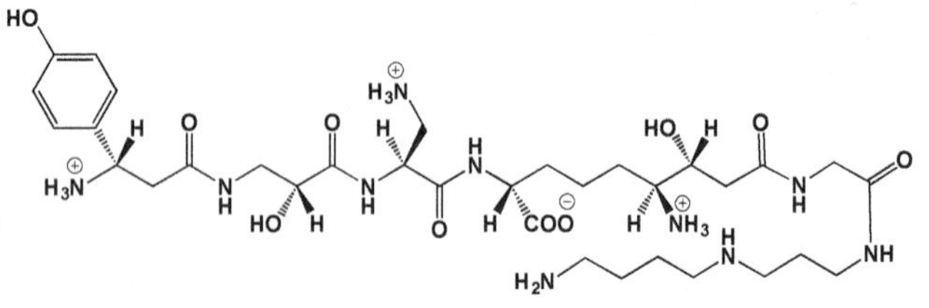

Edeine A₁

These basic antibiotics ($FW_{Edeine-A1}$ = 738.89 g/mol; CASs: 52452-77-4 (Edeine A), 52452-78-5 (Edeine B), 71750-64-6 (Edeine F)) from a strain of *Bacillus brevis* consist of linear oligopeptides linked to spermidine or guanylspermidine at their C-termini. Edeines specifically inhibit binding of aminoacyl-tRNAs to the P site of both 30S subunits and 70S ribosomes, as well as to eukaryotic 40S subunits. The amino acid components for the major edeines are β-tyrosine, isoserine, α,β-diaminopropionic acid, 2,6-diamino-7-hydroxyazelaic acid, and glycine. **Target(s):** arginase, inhibited by edeine B₁ (1); protein biosynthesis initiation (2-6). **1**. Shimotohno, Iida, Takizawa & Endo (1994) *Biosci. Biotechnol. Biochem.* **58**, 1045. **2**. Pestka (1974) *Meth. Enzymol.* **30**, 261. **3**. Carrasco, Battaner & Vázquez (1974) *Meth. Enzymol.* **30**, 282. **4**. Fresno & Vázquez (1979) *Meth. Enzymol.* **60**, 566. **5**. Ochoa & Mazumder (1974) *The Enzymes*, 3rd ed., **10**, 1. **6**. Dinos, Wilson, Teraoka, *et al.* (2004) *Mol. Cell* **13**, 113.

Edotecarin

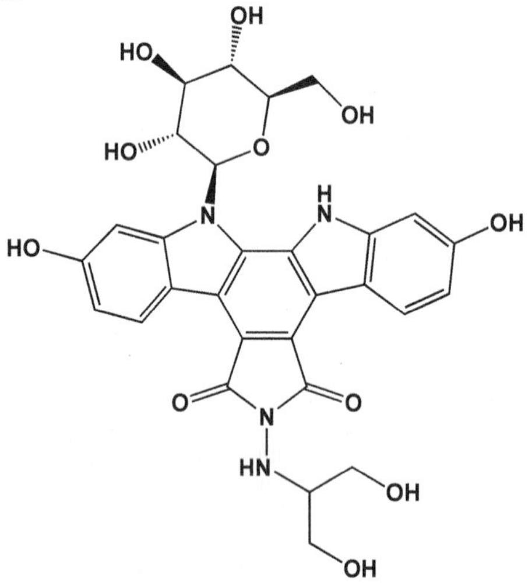

This topoisomerase inhibitor (FW = 608.55 g/mol; CAS 174402-32-5), also known by the code names J-107088 and ED-749 as well as its systematic name*, is a NB-506 derivative (**See** *NB-506*) that targets topoisomerase I by stabilizing the DNA-enzyme complex and enhancing single-strand DNA cleavage, thereby inhibiting DNA replication and decreasing tumor cell proliferation. This novel topoisomerase I inhibitor induces single-strand DNA cleavage more effectively than NB-506 or camptothecin (**See** *Camptothecin*) and at different DNA sequences, but the DNA-topoisomerase I complexes induced by edotecarin are more stable than those occurring after exposure to camptothecin or NB-506. The antitumor activity of edotecarin is less cell-cycle dependent than other topoisomerase I inhibitors. It is structurally related to staurosporine, but does not inhibit protein kinase. The antitumor activity of edotecarin has been tested *in vitro* and *in vivo*, and inhibition of tumor growth has been observed in breast, cervix, pharynx, lung, prostate, colon, gastric, and hepatic cancer models. Edotecarin is effective on cells that have acquired resistance related to P-glycoprotein multi-drug resistance. Drug synergy is observed *in vitro*, when edotecarin was tested in combination with cisplatin, 5-fluorouracil, etoposide, paclitaxel, doxorubicin, vincristine, camptothecin, and gemcitabine. Edotecarin is not a substrate for *in vitro* P450-mediated metabolism. **1**. Saif & Diasio (2005) *Clin. Colorectal Cancer* **5**, 27. ***IUPAC:** 6-((1,3-dihydroxypropan-2-yl)amino)-2,10-dihydroxy-12-((2R,3R, 4S,5S,6R)-3,4,5-trihydroxy-6-(hydroxymethyl)tetrahydro-2H-pyran-2-yl)-12,13-dihydro-5H-indolo[2,3-a]pyrrolo[3,4-c]carbazole-5,7(6H)-dione

Edoxaban

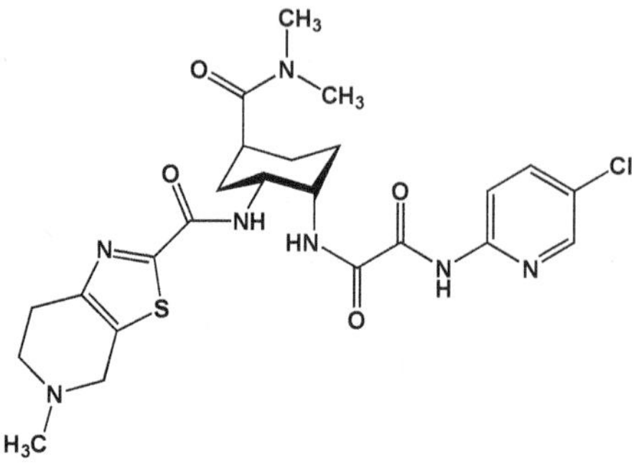

This oral anticoagulant (FW = 548.06 g/mol; CAS 912273-65-5), also known as DU-176b, Lixiana®, and *N'*-(5-chloropyridin-2-yl)-*N*-[(1S,2R,4S)-4-(dimethylcarbamoyl)-2-[(5-methyl-6,7-dihydro-4H-[1,3]thiazolo[5,4-c]pyridine-2-carbonyl)amino]cyclohexyl]oxamide, is a direct factor Xa inhibitor, with K_i values of 0.561 nM for free FXa and 2.98 nM for prothrombinase, exhibiting >10,000x selectivity toward factor Xa (1). DU-176b did not impair platelet aggregation by ADP, collagen or U46619, and, when administered *in vivo*, DU-176b dose-dependently inhibits thrombus formation in rat and rabbit thrombosis models, although bleeding time in rats is not significantly prolonged at an antithrombotic dose (1). Pharmacokinetic profiles are consistent across dose with rapid absorption, biphasic elimination, and terminal elimination half-life of 5.8 to 10.7 hours (2). It has been approved in Japan (2011) for prevention of venous thromboembolisms (VTE) following lower-limb orthopedic surgery and by the FDA (2015) for prevention of stroke and non–CNS systemic embolism. ***See other direct Factor Xa inhibitors:*** *Rivaroxaban, or Xarelto®); Apixaban, or Eliquis®; Betrixaban; Darexaban (discontinued); and Otamixaban injectable (discontinued).* **1**. Furugohri, Isobe, Honda, *et al.* (2008) *J. Thromb. Hemost.* **6**, 1542. **2**. Ogata, Mendell-Harary, Tachibana, *et al.* (2010) *J. Clin. Pharmacol.* 50, 743.

EDTA (*or Ethylenediamine tetraacetic acid*)

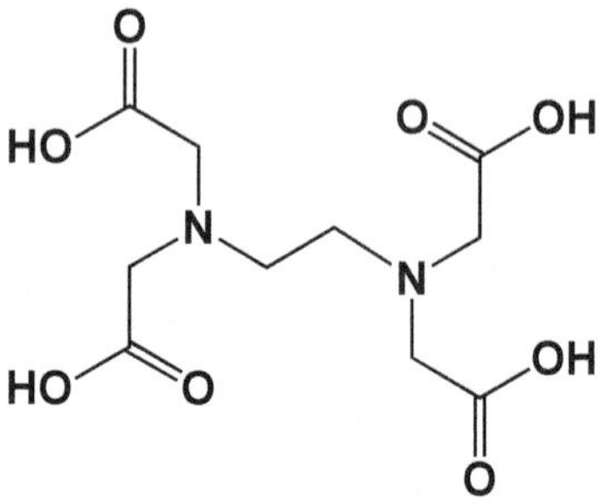

This widely used metal ion chelator ($FW_{free-acid}$ = 292.25 g/mol; CAS 60-00-4), also known as ethylenediamine-*N,N,N',N'*-tetraacetic acid and versene, is an important multidentate ligand and metal ion buffer. EDTA forms stable 1:1 complexes with di- or trivalent cations (structure shown above right).

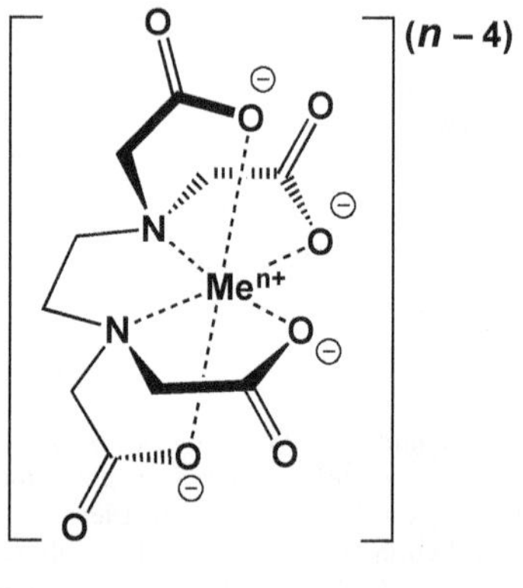

Its interactions with most monovalent ions are weak and are generally inconsequential. The pK_a values of EDTA (at an ionic strength of 0.1 and a

temperature of 20°C) are 2.0, 2.67, 6.16, and 10.26: thus, over the range from pH 4 to 5, most of the EDTA is present as the divalent anion, whereas between pH 7 and 9, the trivalent anion is more prevalent. Formation of a metal ion-EDTA complex will result in the displacement of the remaining protons from EDTA. Thus, $Me^{n+} + H_2(EDTA)^{2-} \rightleftharpoons M(EDTA)^{(n-4)+} + 2H^+$ between pH 4 and 5 and $M^{n+} + H(EDTA)^{3-} \rightleftharpoons M(EDTA)^{(n-4)+} + H^+$ between pH 7 and 9. The tetravalent anion of EDTA is the species of greatest importance in equilibrium calculations. The strength metal ion complexation is typically measured by its overall formation constant for the reaction, $M^n + EDTA^{4-} \rightleftharpoons M(EDTA)^{(n-4)}$, $K_{M(EDTA)} = [M(EDTA)^{(n-4)}]/([M^n][EDTA^{4-}])$. At 25°C, pH = 7, and ionic strength = 0.1 M, $\log K_{M(EDTA)}$ values are: Ca^{2+}, 10.6; Cd^{2+}, 16.6; Co^{2+}, 16.5; Cu^{2+}, 18.9; Fe^{3+}, 25.1; Mg^{2+}, 8.7; Mn^{2+}, 24.8; Ni^{2+}, 18.7; and Zn^{2+}, 16.7. Use of the commercially available K^+,Mg^{2+}:EDTA salt offers the convenience of avoiding wide swings in solution pH, when one seeks to buffer magnesium ion concentrations in many biochemical experiments. **A Word of Caution:** When working with enzymes exhibiting extreme affinity for metal ions, one should not overlook the possibility that commercial chelators can be an inadvertent source of metal ions. For example, a 10 mM solution of so-called reagent grade EDTA is likely to contain the following total metal ion concentrations: aluminum, ~0.6 μM; arsenic, ~0.03 μM; barium, ~0.1 μM; bismuth, ~5 μM; calcium ~0.7 μM; cadmium, ~0.1 μM; cobalt, ~0.2 μM; chromium, ~0.3 μM; copper, ~0.3 μM; iron, ~0.3 μM; lead, ~0.1 μM; lithium, ~2 μM; manganese, ~0.3 μM; molybdenum, ~0.6 μM; nickel, ~5 μM; potassium, ~10 μM; sodium, ~0.5 μM; strontium, ~0.1 μM. Of course, their respective uncomplexed (or "free") metal ion concentrations will be much lower. **See also EGTA** **1.** Perrin & Dempsey (1974) *Buffers for pH and Metal Ion Control*, Chapman & Hall, London.

Eeyarestatin I

This potent ERAD inhibitor (FW = 630.44 g/mol; CAS 412960-54-4; Symbol: ES_I) modulates endoplasmic reticulum- (*or* ER-) associated degradation (*or* ERAD), resulting in the formation of mislocalized polypeptides that are ubiquitinated and subsequently degraded as part of the Unfolded Protein Response (UPR). **Mechanism of Inhibitory Action:** For many substrates, retrotranslocation requires the action of ubiquitinating enzymes, which poly-ubiquitinate substrates emerging from the ER lumen, and of the p97-Ufd1-Npl4 ATPase complex, which hydrolyzes ATP to dislocate polyubiquitinated substrates into the cytosol (1,2). Polypeptides extracted by p97 are eventually transferred to the proteasome for destruction. Eeeyarestatin I specifically targets a component of the Sec61 complex that forms the membrane pore of the ER translocon. Eeeyarestatin I's cytotoxicity characteristics are caused by its activation of NOXA, a Bcl-2 homology3-only pro-apoptotic protein (3). ES_I acts by preventing the transfer of the nascent polypeptide from the co-translational targeting machinery to the Sec61 complex. These results suggest that ES_I can be a drug that modulates canonical protein transport from the cytosol into the mammalian ER both *in vitro* and *in vivo*. EerI and another ERAD inhibitor (bortezomib) elicit an integrated stress response program at the ER to activate the CREB/ATF transcription factors ATF3 and ATF4. The latter form a complex capable of binding to the NOXA promoter, a step that is required for NOXA activation. Eeeyarestatin I and bortezomib also block ubiquitination of histone H2A to relieve its inhibition on NOXA transcription (3). **1.** Wang, Li & Ye (2008) *J. Biol. Chem.* **283**, 7445. **2.** Cross, McKibbin, Callan, *et al.* (2009) *J. Cell. Sci.* **122**, 4393. **3.** Wang, Mora-Jensen, Weniger, *et al.* (2009) *Proc. Natl. Acad. Sci. U.S.A.* **106**, 2200.

Efavirenz

This benzoxazinone (FW = 315.68 g/mol; CAS 154598-52-4; *Symbol*: EFV), also known as L-743,726, DMP266, and Sustiva®, is a second-generation non-nucleoside reverse transcriptase inhibitor (NNRTI) with a resistance profile that overlaps with nevirapine and delavirdine. Unlike nucleoside RTIs, which bind at the enzyme's active site, NNRTIs act allosterically by binding to a distinct site known as the NNRTI pocket. Efavirenz has been in use clinically since 1996 in anti-retroviral therapy. An efavirenz metabolite (possibly 8-hydroxy-efavirenz) reportedly damages to neurons, resulting in impaired neurocognitive performance. In mice, EFV crosses the blood-brain barrier and inhibits creatine kinase in the cerebellum, hippocampus, striatum, and cortex (6). **Efavirenz-Induced Inhibition of Mitochondrial Complex I:** Exposure of cultured mouse hepatocytes to EFV results in rapid-onset cell injury, showing a narrow toxicity range, *i.e.*, no observable effect at 30μM and sub-maximal effects at 50μM (7). EFV also causes a dose-dependent decrease in cellular ATP levels, increased peroxynitrite formation, the latter with oxidation of protein thiols, including cyclophilin D (CypD). Notably, the superoxide scavenger Fe-TCP and the peroxynitrite decomposition catalyst Fe-TMPyP prevent such effects. Both ferroporphyrins completely protect from EFV injury, suggesting peroxynitrite plays a significant role (7). EFV also increases the $NADH/NAD^+$ ratio, inhibits Sirt3 activity, and leads to hyperacetylated lysine residues, including those in CypD. Hepatocytes isolated from Sirt3-null mice are protected against 40μM EFV, as compared to their wild-type controls. Such findings suggest that chemical inhibition of complex I activates multiple pathways leading to cell injury (7). **Key Pharmacokinetic Parameters:** *See* Appendix II in Goodman & Gilman's *THE PHARMACOLOGICAL BASIS OF THERAPEUTICS*, 12th Edition (Brunton, Chabner & Knollmann, eds.) McGraw-Hill Medical, New York (2011). **1.** Tucker, Lumma & Culberson (1996) *Meth. Enzymol.* **275**, 440. **2.** Young, Britcher, Tran, *et al.* (1995) *Antimicrob. Agents Chemother.* **39**, 2602. **3.** El Safadi, Vivet-Boudou & Marquet (2007) *Appl. Microbiol. Biotechnol.* **75**, 723. **4.** Nissley, Radzio, Ambrose, *et al.* (2007) *Biochem. J.* **404**, 151. **5.** Hang, Li, Yang, *et al.* (2007) *Biochem. Biophys. Res. Commun.* **352**, 341. **6.** Streck, Scaini, Rezin, *et al.* (2008) *Metab Brain Dis.* **23**, 485. **7.** Imaizumi, Kwang, Zhang & Boelsterli (2015) *Redox Biol.* **4**, 279.

Efegatran

This substituted tripeptide aldehyde (FW = 416.52 g/mol; CAS 105806-65-3), also known as LY 294468, GYKI-14766, and N^α-methyl-D-phenylalanyl-L-prolyl-L-arginal (D-MePhe-Pro-Arg-H), inhibits thrombin and is a powerful anticoagulant. Efegatran has been developed as an anticoagulant for the treatment of acute coronary syndromes. **1.** Nilsson, Sjoling-Ericksson & Deinum (1998) *J. Enzyme Inhib.* **13**, 11. **2.** Bajusz, Szell, Bagdy, *et al.* (1990) *J. Med. Chem.* **33**, 1729. **3.** Weitz (2003) *Thromb. Res.* **109** Suppl. 1, S17.

Efinaconazole

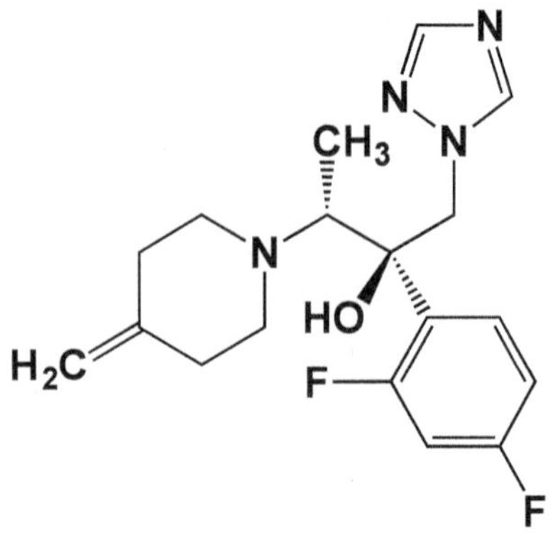

This triazole-based fungicide (FW = 348.39 g/mol; CAS 164650-44-6), also named (2R,3R)-2-(2,4-Difluorophenyl)-3-(4-methylene-1-piperidinyl)-1-(1H-1,2,4-triazol-1-yl)-2-butanol and marketed under the tradename Jublia®, dose-dependently decreases ergosterol production, with accumulation of 4,4-dimethylsterols and 4α-methylsterols. Its primary mode of action of is a reduction in ergosterol biosynthesis via inhibition of sterol 14α-demethylase. Efinaconazole induces morphological and ultrastructural changes in the hyphae in *Trichophyton mentagrophytes*, becoming more prominent at higher drug concentrations. It is also effective against *Candida albicans*. It also exhibits a lower minimum inhibitory concentration than observed with terbinafine, ciclopirox, itraconazole and amorolfine, when tested against *T. rubrum*, *T. mentagrophytes*, and *Candida albicans* (2). Efinaconazole is effective in the treatment of onychomycosis, a fungal infection of the matrix, bed, and/or plate of the toenail or fingernail. 1. Tatsumia, Nagashimaa, Shibanushib, *et al.* (2013) *Antimicrob. Agents Chemother.* **57**, 2405. 2. Gupta & Simpson (2014) *Expert Rev. Anti. Infect. Ther.* **12**, 743.

Eflornithine

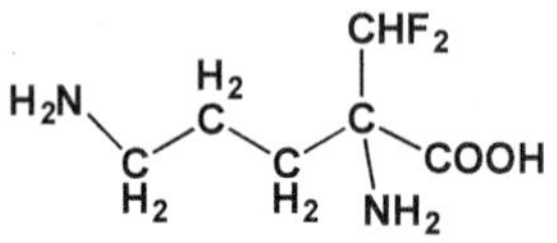

This antiprotozoal agent (FW = 182.17 g/mol; CAS 70050-56-5), known systematically as (RS)-2,5-diamino-2-(difluoromethyl)pentanoic acid, is used to treat Human African trypanosomiasis, *or* African Sleeping Sickness. Eflornithine is commonly known as the 'resurrection drug' in recognition of its often-spectacular capacity to revive and cure comatose Sleeping Sickness patients. It is also a specific, enzyme-activated irreversible inhibitor of the trypanosomatid spermidine biosynthetic enzyme, *S*-adenosylmethionine decarboxylase (1,2). Eflornithine is also a specific inhibitor of ornithine decarboxylase following growth stimulation, thereby blocking Ehrlich ascites tumor cell proliferation in culture (3). 1. Bacchi, Nathan, Hutner, McCann & Sjoerdsma (1980) *Science* **210**, 332. 2. Heby, Persson & Rentala (2007) *Amino Acids* **33**, 359. 3. Oredsson, Anehus & Heby (1980) *Biochem. Biophys. Res. Commun.* **94**, 151.

Eflucimibe

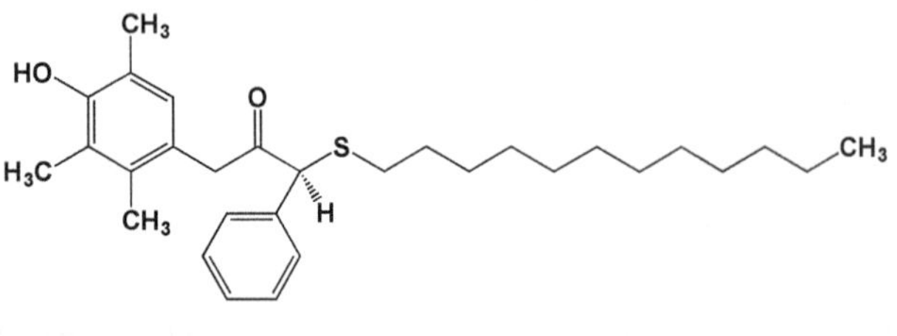

This acetanilide (FW = 453.74 g/mol; CAS 204192-08-5), also named F12511 and (S)-2',3',5'-trimethyl-4'-hydroxy-α-dodecylthio-α-phenyl-acetanilide, inhibits sterol O-acyltransferase (IC₅₀ = 39 nM for ACAT-I) and reduces plasma cholesterol levels. 1. Zamorano-León, Fernández-Sánchez, López Farré, *et al.* (2006) *J. Cardiovasc. Pharmacol.* **48**, 128. 2. Junquero, Oms, Carilla-Durand, *et al.* (2001) *Biochem. Pharmacol.* **61**, 97. 3. López-Farré, Sacristán, Zamorano-León, San-Martín & Macaya (2008) *Cardiovasc. Ther.* **26**, 65.

Efrapeptins

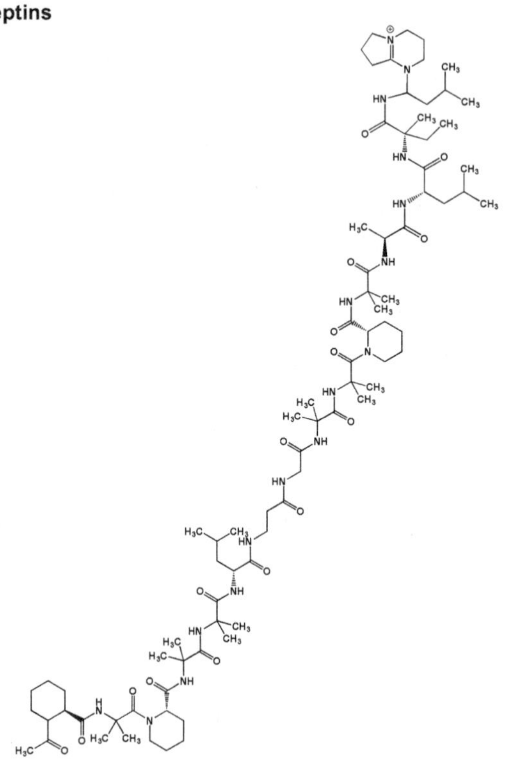

These fungal antibiotics and insecticides (Efrapeptin F (shown above): FW = 1635.10; CAS 131353-66-7), also known as efrastatins and A23871, from *Tolypocladium niveum* compete with ATP for binding to the F₁ fragment of mitochondrial ATP synthase, interacting directly with an active-site arginine that is required for nucleotide binding (1-13). Some mutant ATP synthases are efrapeptin-resistant. Each 15-residue peptides is acetylated at the N-terminus and are blocked at the C-terminus by 1-isobutyl-2-(2,3,4,6,7,8-hexahydropyrrolo[1,2-a]pyrimidin-1-io)ethylamine. (Efrapeptin C sequence: Ac-Pip-Aib-Pip-Aib-Aib-Leu-β-Ala-Gly-Aib-Aib-Pip-Aib-Gly-Leu-Aib-X, where Ac refers to acetyl, Pip is L-pipecolic acid, Aib is α-aminoisobutyric acid, β-Ala is β-alanine, and X refers to the bicyclic blocking group shown above. Efrapeptin D is identical, albeit with Iva (*i.e.*, *S*-isovaline) in place of Aib15, whereas Efrapeptin E contains Iva4 and Iva15. Efrapeptin F has Aib4, Ala13, and Iva15 and efrapeptin G has Iva4, Ala13, and Iva15. Its binding is associated with small structural changes in side chains of F1-ATPase around the binding pocket. Efrapeptin makes hydrophobic contacts with the α-helical structure in the γ-subunit, which traverses the cavity, and with subunit beta E and the two adjacent α-subunits (14). Two intermolecular hydrogen bonds could also form, helping to stabilize efrapeptin's two domains (residues 1-6 and 9-15, respectively), which are connected by a flexible region (β Ala-7 and Gly-8). Efrapeptin appears to inhibit F1-ATPase by blocking the conversion of subunit beta E to a nucleotide binding conformation, as would be required by an enzyme mechanism involving cyclic interconversion of catalytic sites (14). First noted for its action on chromatophores of purple bacterium *Rhodospirillum rubrum*, efrapeptins are potent inhibitors for mitochondrial and bacterial ATP synthases, with milder action against chloroplast ATP synthase. **Target(s):** ATP synthase, *or* F₁ ATPase (2-9); bacterial ATPase (2); CF₁ ATP synthase, *or* chloroplast ATP synthase (2,10; H⁺-transporting ATPase, lysosomal (11); RrF₁ ATPase (12); TF₁ ATPase (13). 1. Brinen & Clardy (1992) *J. Org. Chem.* **57**, 2306. 2. Linnett & Beechey (1979) *Meth. Enzymol.* **55**, 472. 3. Satre, Lunardi, Dianoux, *et al.* (1986) *Meth. Enzymol.* **126**, 712. 4. Nalin & Cross (1982) *J. Biol. Chem.* **257**, 8055. 5. Lardy, Reed & Lin (1975) *Fed. Proc.* **34**, 1707. 6. Cross & Kohlbrenner (1978) *J. Biol. Chem.* **253**, 4865. 7. Kasho, Allison & Boyer (1993) *Arch. Biochem. Biophys.* **300**, 293. 8. Reed & Lardy (1975) *J. Biol. Chem.* **250**, 3704. 9. Criddle, Johnston & Stack (1979) *Curr. Top. Bioenerg.* **9**, 89. 10. Lucero, Ravizzine & Vallejos (1976) *FEBS Lett.* **68**, 141. 11. Jonas, Smith, Allison, *et al.* (1983) *J. Biol. Chem.* **258**, 11727. 12. Weiss, McCarty & Gromet-Elhanan (1994) *J. Bioenerg. Biomembr.* **26**, 573. 13. Saishu, Kagawa & Shimizu (1983) *Biochem. Biophys. Res. Commun.* **112**, 822. 14. Andrews, Carroll, Ding, *et al.* (2013) *Proc. Natl. Acad. Sci.* **110**, 18934.

EGA

This host-membrane trafficking inhibitor (FW = 346.23 g/mol), also known as 4-bromobenzaldehyde-N-(2,6-dimethylphenyl)semi-carbazone, blocks (IC$_{50}$ = 1.7 µM) the entry of acid-dependent bacterial toxins (including *Bacillus anthracis* lethal toxin) as well as viruses into mammalian cells. EGA also delays lysosomal targeting and degradation of the EGF receptor, indicating that it targets host-membrane trafficking. However, EGA does not block endosomal recycling of transferrin, retrograde trafficking of ricin, phagolysosomal trafficking, or phagosome permeabilization by *Franciscella tularensis*. Because EGA does not neutralize acidic organelles, its mechanism of action is distinct from pH-raising agents, such as ammonium chloride and bafilomycin A$_1$. EGA represents a promising tool for investigating membrane trafficking as well as for treating infectious diseases. **1**. Gillespie, Ho, Balaji, *et al.* (2013) *Proc. Natl. Acad. Sci. U.S.A.* **110**, E4904.

EGTA

This tetracarboxylic acid (FW$_{free-acid}$ = 380.35 g/mol; CAS 67-42-5), also known as ethylene glycol bis(β-aminoethylether)-*N,N,N',N'*-tetraacetic acid, is a metal ion chelator that shows binding preference for Ca^{2+}. Unlike EDTA, which only forms hexa-coordinate complexes (*See EDTA*), EGTA contains two ether oxygens, allowing it to form both hexa- and octa-coordinate complexes (1). An example of the utility EGTA as a buffer of the free calcium ion concentration is the demonstration that calmodulin requires four bound calcium ions to fully activate 3',5'-cyclicAMP phosphodiesterase (2). **Equilibria:** Formation of a metal ion-EGTA (*or* Me·EGTA) complex results in the displacement of the remaining protons from EGTA. Thus, Me^n + H$_2$EGTA^{2-} ⇌ Me·EGTA$^{(n-4)}$ + 2H$^+$ between pH 4 and 5 and Me^n + HEGTA^{3-} ⇌ Me·EGTA$^{(n-4)}$ + H$^+$ between pH 7 and 9 (1). Hence, the tetravalent anion of EGTA is the species of greatest importance in equilibrium calculations. Metal ion complexation is measured by its formation constant, *or* stability constant ($K_{M(EDTA)}$ = [Me·EGTA$^{(n-4)}$]/([Me^n]×[EGTA^{4-}]) for the complexation reaction: Me^n + EGTA^{4-} ⇌ Me·EGTA$^{(n-4)}$, The following are values for log$K_{Me·EGTA}$ at 25°C, pH = 7, ionic strength = 0.1 Me·EGTA: 5.3 for Mg^{2+} ion 10.9 for Ca^{2+} ion, 12.1 for Mn^{2+} ion, 12.6 for Zn^{2+} ion, and 16.5 for Cd^{2+} ion. **1**. Perrin & Dempsey (1974) *Buffers for pH and Metal Ion Control*, Chapman & Hall, London. **2**. Huang, Chau, Chock, Wang & Sharma (1981) *Proc. Natl. Acad. Sci. U.S.A.* **78**, 871.

EHNA, *See* erythro-9-(2-Hydroxy-3-nonyl)adenine

EHop-016

This signal transduction inhibitor (FW = 430.56 g/mol), based on the structure of the Rac/Rac GEF inhibitor NSC23766, is cytostatic in metastatic cancer cells overexpressing the actin cytoskeletal modulator Rac, with an IC$_{50}$ value of 1.1 µM, or ~100-times lower than the IC$_{50}$ for NSC23766 (*See also NSC23766*). At concentrations ≤ 5 µM, EHop-016 is specific for Rac1 and Rac3. At higher concentrations, EHop-016 inhibits

Cdc42. In MDA-MB-435 cells that have high active levels of the Rac GEF Vav2, EHop-016 inhibits the association of Vav2 with a nucleotide-free Rac1(Gly-15-Ala mutant) having high affinity for activated GEFs. EHop-016 also inhibits the Rac activity of MDA-MB-231 metastatic breast cancer cells and reduces Rac-directed lamellipodia formation in both cell lines. EHop-016 also decreases Rac's downstream effects of p21-activated kinase 1 (or PAK1) as well as the directed migration of metastatic cells. Significantly, when present at effective concentrations below 5 µM, EHop-016 does not affect the viability of transformed mammary epithelial cells (MCF-10A) and reduces viability of MDA-MB-435 cells by only 20%. EHop-016 holds promise as a targeted therapeutic agent for the treatment of metastatic cancers characterized by high Rac activity. **1**. Montalvo-Ortiz, Castillo-Pichardo, Hernández, *et al.* (2012) *J. Biol. Chem.* **287**, 13228.

Ehrlich 594, *See* Acetarsone

EHT-0202, *See* Etazolate

EHT 1864

This Rac inhibitor (FW$_{di-HCl}$ = 581.47 g/mol; CAS 754240-09-0; FW$_{free-base}$ = 508.56 g/mol; Solubility: 100 mM in H$_2$O, 75 mM in DMSO), also named 5-(5-(7-(trifluoromethyl)quinolin-4-ylthio)pentyloxy)-2-(morpholino-methyl)-4H-pyran-4-one, blocks Rac signaling by direct binding to Rac1 (K_D = 40 nM), Rac1b (K_D = 50 nM), Rac2 (K_D = 60 nM), and Rac3 (K_D = 250 nM). EHT 1864 blocks Rac-, Ras-, and Tiam-induced growth transformation of NIH-3T3 fibroblasts (2). RAC1 inhibition by EHT 1864 likewise targets amyloid precursor protein processing by γ-secretase and decreases Aβ production, both *in vitro* and *in vivo* (3). EHT 1864 also markedly attenuates glucose-stimulated insulin secretion in the INS-1 832/13 pancreatic β-cell line, significantly reducing glucose-induced Rac1 activation and membrane targeting (4). It also suppresses glucose-induced activation of ERK1/2 and p53, but not Akt (4). **1**. Shutes, Onesto, Picard, *et al.* (2007) *J. Biol. Chem.* **282**, 35666. **2**. Onesto, Shutes, Picard, Schweighoffer & Der (2008) *Meth. Enzymol.* **439**, 111. **3**. Désiré, Bourdin, Loiseau, *et al.* (2005) *J. Biol. Chem.* **280**, 37516. **4**. Sidarala, Veluthakal, Syeda & Kowluru (2015) *Cell Signal.* **19**, 1691.

5,8,11,14-Eicosatetraynoate

This unsaturated fatty acid (FW$_{free-acid}$ = 296.41 g/mol; *Symbol*: ETYA) is the alkyne analogue of arachidonic acid, and, as such, inhibits a number of arachidonate-dependent enzymes. This substance can also be incorporated in cell lipids instead of arachidonate. **Target(s):** acyl-CoA synthetase (1,2); cyclooxygenase, *or* prostaglandin-endoperoxide synthase (2-4,8-11); cytochrome P450 (1); glutathione *S*-transferase (12); K$^+$ current, fast inactivating (13); linoleate diol synthase (14); linoleate 11-lipoxygenase, *or* manganese lipoxygenase (15,16); lipoxygenase (17); 5-lipoxygenase, *or* arachidonate 5-lipoxygenase (1,3,4,22); 8-lipoxygenase, *or* arachidonate 8-lipoxygenase (21); 12-lipoxygenase, *or* arachidonate 12-lipoxygenase (4,5,25-30); 15 lipoxygenase, *or* arachidonate 15-lipoxygenase (3,6,7,23,24); phospholipase A$_2$ (1,18-20). **1**. Tobias & Hamilton (1979) *Lipids* **14**, 181. **2**. Taylor, Morrison & Russell (1985) *Prostaglandins* **29**, 449. **3**. Pace-Asciak & Smith (1983) *The Enzymes*, 3rd ed., **16**, 543. **4**. Cerletti, Livio, Doni & De Gaetano (1983) *Biochim. Biophys. Acta* **759**, 125. **5**. Nugteren (1982) *Meth. Enzymol.* **86**, 49. **6**. Ator & Ortiz de Montellano (1990) *The Enzymes*, 3rd ed., **19**, 213. **7**. Narumiya & Salmon (1982) *Meth. Enzymol.* **86**, 45. **8**. Nugteren & Christ-Hazelhof (1987) *Prostaglandins* **33**, 403. **9**. Friedman, Lang & Burke (1975) *Biochim. Biophys. Acta* **397**, 331. **10**. Sheve, Belosludtsev, Demin, *et al.* (1993) *Bioorg. Khim.* **19**, 548. **11**. Miller, Munster, Wasvary, *et al.* (1994)

Biochem. Biophys. Res. Commun. **201**, 356.   **12**. Datta & Kulkarni (1994) *Toxicol. Lett.* **73**, 157.   **13**. Kehl (2001) *Can. J. Physiol. Pharmacol.* **79**, 338.   **14**. Brodowsky, Hamberg & Oliw (1994) *Eur. J. Pharmacol.* **254**, 43.   **15**. Su & Oliw (1998) *J. Biol. Chem.* **273**, 13072.   **16**. Cristea & Oliw (2006) *J. Biol. Chem.* **281**, 17612.   **17**. Kühn, Schewe & Rapoport (1984) *Biomed. Biochim. Acta* **43**, S358.   **18**. Lanni & Becker (1985) *Int. Arch. Allergy Appl. Immunol.* **76**, 214.   **19**. Holk, Rietz, Zahn, Quader & Scherer (2002) *Plant Physiol.* **130**, 90.   **20**. Jung & Kim (2000) *Plant Physiol.* **123**, 1057.   **21**. Fürstenberger, Hagedorn, Jacobi, *et al.* (1991) *J. Biol. Chem.* **266**, 15738.   **22**. Furakawa, Yoshimoto, Ochi & Yamamoto (1984) *Biochim. Biophys. Acta* **795**, 458.   **23**. Kühn, Barnett, Grunberger, *et al.* (1993) *Biochim. Biophys. Acta* **1169**, 80.   **24**. Bryant, Bailey, Schewe & Rapaport (1982) *J. Biol. Chem.* **257**, 6050.   **25**. Moody & Marnett (2002) *Biochemistry* **41**, 10297.   **26**. Yokoyama, Mizuno, Mitachi, *er al.* (1983) *Biochim. Biophys. Acta* **750**, 237.   **27**. Richards & Marnett (1997) *Biochemistry* **36**, 6692.   **28**. Yoshimoto, Miyamoto, Ochi & Yamamoto (1982) *Biochim. Biophys. Acta* **713**, 638.   **29**. Flatman, Hurst, McDonald-Gibson, Jonas & Slater (1986) *Biochim. Biophys. Acta* **883**, 7.   **30**. German, Bruckner & Kinsella (1986) *Biochim. Biophys. Acta* **875**, 12.

8Z,11Z,14Z-Eicosatrienoate

This unsaturated fatty acid ($FW_{free-acid}$ = 306.49 g/mol), also known as dihomo-(6,9,12)-linolenic acid and dihomo-γ-linolenic acid, is a member of the linoleic family of fatty acids. It is an intermediate in the biosynthesis of arachidonate from linoleate and is a precursor in the formation of the PG1 series of prostaglandins. **Target(s):** arachidonyl-CoA synthetase, *or* arachidonate:CoA ligase, also as an alternative substrtate (1-3); cyclooxygenase (4); DNA polymerase (5).   **1**. Wilson, Prescott & Majerus (1982) *J. Biol. Chem.* **257**, 3510.   **2**. Taylor, Sprecher & Russel (1985) *Biochim. Biophys. Acta* **833**, 229.   **3**. Neufeld, Sprecher, Evans & Majerus (1984) *J. Lipid Res.* **25**, 288.   **4**. Henry, Momin, Nair & Dewitt (2002) *J. Agric. Food Chem.* **50**, 2231.   **5**. Mizushina, Tanaka, Yagi, *et al.* (1996) *Biochim. Biophys. Acta* **1308**, 256.

Elacridar

This very potent drug efflux inhibitor (MW = 563.64 g/mol; CAS 143664-11-3; Solubility: 10 mM in DMSO), also known as GF120918 and named systematically as *N*-(4-(2-(6,7-dimethoxy-3,4-dihydroisoquinolin-2(1*H*)-yl)ethyl)phenyl)-5-methoxy-9-oxo-9,10-dihydroacridine-4-carboxamide, targets ABC transporters (P-glycoprotein) and Breast Cancer Resistance Protein (BCRP), reversing multidrug resistance when used at 0.02 to 0.1 μM *in vitro*. By blocking P-gp- and BCRP-mediated efflux, GF120918 increases the intracellular concentration of a co-administered drug, often potentiating its cytotoxicity and effectiveness as an anti-tumor agent (1). GF120918A has been used both *in vitro* and *in vivo* as a tool inhibitor of ABC transporters (or P-glycoprotein) to investigate the role of transporters in the disposition of test drug and to guide preclinical doses of candidate drugs (2).   **1**. Hyafil, Vergely, Du Vignaud & Grand-Perret (1993) *Cancer Res.* **53**, 4595.   **2**. Ward & Azzarano (2004) *J. Pharmacol. Exp. Ther.* **310**, 703.

Elafin

This human protein (MW = 12.6 kDa), also known as skin-derived antileukoproteinase (*or* SKALP) and elastase-specific inhibitor (ESI), inhibits elastase (1-5) and leukocyte elastase (5-9). The unprocessed human protein contains WAP-type core (WFDC) domain consisting of 117 residues and four disulfide bonds. Most WFDC gene members are localized to chromosome 20q12-q13 in two clusters: centromeric and telomeric. This gene belongs to the centromeric cluster. Expression of this gene is upgulated by bacterial lipopolysaccharides and cytokines. It is synthesized and secreted locally at sites of injury and is produced in response to primary cytokines.   **Target(s):** myeloblastin, *or* proteinase 3 (1,6-8,10-12); pancreatic elastase (4,7,13,14); stratum corneum chymotryptic enzyme, *or* kallikrein 7 (15).   **1**. Wiedow, Luademann & Utecht (1991) *Biochem. Biophys. Res. Commun.* **174**, 6.   **2**. Sallenave (2000) *Respir. Res.* **1**, 87.   **3**. Molhuizen & Schalkwijk (1995) *Biol. Chem. Hoppe-Seyler* **376**, 1.   **4**. Bieth (1998) in *Handb. Proteolytic Enzymes*, p. 42, Academic Press, San Diego.   **5**. Bieth (1998) in *Handb. Proteolytic Enzymes*, p. 54, Academic Press, San Diego.   **6**. Wiesner, Litwiller, Hummel, *et al.* (2005) *FEBS Lett.* **579**, 5305.   **7**. Zani, Nobar, Lacour, *et al.* (2004) *Eur. J. Biochem.* **271**, 2370.   **8**. Nobar, Zani, Boudier, Moreau & Bieth (2005) *FEBS J.* **272**, 5883.   **9**. Ying & Simon (1993) *Biochemistry* **32**, 1866.   **10**. Hoidal (1998) in *Handb. Proteolytic Enzymes*, p. 62, Academic Press, San Diego.   **11**. Hoidal, Rao & Gray (1994) *Meth. Enzymol.* **244**, 61.   **12**. Dolman, van de Wiel, Kam, *et al.* (1993) *Adv. Exp. Med. Biol.* **336**, 55.   **13**. Tsunemi, Matsuura, Sakakibara & Katsube (1996) *Biochemistry* **35**, 11570.   **14**. Tsunemi, Matsuura, Sakakibara & Katsube (1993) *J. Mol. Biol.* **232**, 310.   **15**. Franzke, Baici, Bartels, Christophers & Wiedow (1996) *J. Biol. Chem.* **271**, 21886.

Elaidic Acid

This unsaturated fatty acid ($FW_{free-acid}$ = 282.46 g/mol; CAS 112-79-8; M.P. = 44.5°C) is the *trans*-isomer of oleic acid, produced from the latter by treatment with heat or selenium (1).   **Target(s):** chondroitin AC lyase (2); chondroitin B lyase (2); chymase (3,4); Δ^5-desaturase (5); DNA-directed DNA polymerase (6); DNA polymerase α (6); glucokinase (7); hexokinase (8); hyaluronate lyase (2); linoleate isomerase (9); phospholipid-hydroperoxide glutathione peroxidase (10).   **1**. Swern & Scanlan (1953) *Biochem. Prep.* **3**, 118.   **2**. Suzuki, Terasaki & Uyeda (2002) *J. Enzyme Inhib. Med. Chem.* **17**, 183.   **3**. Kido, Fukusen & Katunuma (1985) *Arch. Biochem. Biophys.* **239**, 436.   **4**. Kido, Fukusen & Katunuma (1984) *Arch. Biochem. Biophys.* **230**, 610.   **5**. Rosenthal & Doloresco (1984) *Lipids* **19**, 869.   **6**. Mizushina, Tanaka, Yagi, *et al.* (1996) *Biochim. Biophys. Acta* **1308**, 256.   **7**. Lea & Weber (1968) *J. Biol. Chem.* **243**, 1096.   **8**. Stewart & Blakely (2000) *Biochim. Biophys. Acta* **1484**, 278.   **9**. Kepler, Tucker & Tove (1970) *J. Biol. Chem.* **245**, 3612.   **10**. Maiorino, Roveri, Gregolin & Ursini (1986) *Arch. Biochem. Biophys.* **251**, 600.

Elastatinal

This unusual peptide aldehyde and serine protease inhibitor (FW = 512.57 g/mol; CAS 51798-45-9; pK_a values are 3.7 and >11) from *Actinomycetes*, also known as *N*-(-*N*^α-carbonyl-Cpd-L-glutaminyl-L-alaninal)-L-leucine, where Cpd stands for capreomycidine, *or* (*S*,*S*)-α-(2-iminohexahydro-4-pyrimidyl)glycine, rapidly forms a hemicetal adduct between its aldehyde group and the active-site seryl residue of elastase (K_i ≈ 50 μM) and plasmin. Elastatinal is without effect on chymotrypsin and trypsin. Elastatinal is a water-soluble reagent that can be stored as a 10-mM stock solution (~20 mg/mL) and used at 0.5–1.0 mM final concentrations. Stock solutions are stable for one week at 4°C. **Target(s):** Bacillopeptidase F (1); *Bombyx mori* protease (2); brachyuran, *or* serine collagenolytic protease from greenshore crag (*Carcinus maenas*) (3); carboxypeptidase C (4); cathepsin B (5,6); envelysin (7); legumain (8)(; leucyl aminopeptidase (9); leukocyte elastase (10-12); α_2-macroglobulin protease complexes (13); multicatalytic proteinase (*Halocynthia roretzi* eggs) (14); pancreatic elastase (10,15-20); pancreatic elastase II (21); peptidyl-glycinamidase (22); picornain 2A (23-28); picornain 3C (24); post-proline cleavage enzyme, *or* prolyl oligopeptidase (29,30); protease IV (31,32); pyroglutamyl-peptidase I (9); signal peptidase I (*Escherichia coli*) (32); subtilisin Sendai, weakly inhibited (33); thermitase (34). **1.** Hageman (1998) in *Handb. Proteolytic Enzymes*, p. 301, Academic Press, San Diego. **2.** Maki & Yamashita (1997) *Insect Biochem. Mol. Biol.* **27**, 721. **3.** Roy, Colas & Durand (1996) *Comp. Biochem. Physiol. B Biochem. Mol. Biol.* **115**, 87. **4.** Liu, Tachibana, Taira, *et al.* (2004) *J. Ind. Microbiol. Biotechnol.* **31**, 572. **5.** Banno, Yano & Nozawa (1983) *Eur. J. Biochem.* **132**, 563. **6.** Olstein & Liener (1983) *J. Biol. Chem.* **258**, 11049. **7.** Nakatsuka (1985) *Dev. Growth Differ.* **27**, 653. **8.** Abe, Shirane, Yokosawa, *et al.* (1993) *J. Biol. Chem.* **268**, 3525. **9.** Mantle, Lauffart & Gibson (1991) *Clin. Chim. Acta* **197**, 35. **10.** Bartlett (1981) *Meth. Enzymol.* **80**, 581. **11.** Bieth (1998) in *Handb. Proteolytic Enzymes*, p. 54, Academic Press, San Diego. **12.** Feinstein, Malemud & Janoff (1976) *Biochim. Biophys. Acta* **429**, 925. **13.** Menendez-Arias, Risco & Oroszlan (1992) *J. Biol. Chem.* **267**, 11392. **14.** Saitoh, Yokosawa, Takahashi & Ishii (1989) *J. Biochem.* **105**, 254. **15.** Bieth (1998) in *Handb. Proteolytic Enzymes*, p. 42, Academic Press, San Diego. **16.** Umezawa (1976) *Meth. Enzymol.* **45**, 678. **17.** Okura, Morishima, Takita, *et al.* (1975) *J. Antibiot.* **28**, 337. **18.** Umezawa, Aoyagi, Okura, Morishima & Takeuchi (1973) *J. Antibiot.* **26**, 787. **19.** Smine & Gal (1995) *Mol. Mar. Biol. Biotechnol.* **4**, 295. **20.** Gildberg & Overbo (1990) *Comp. Biochem. Physiol. B* **97**, 775. **21.** Azuma, Banshou & Suzuki (2001) *J. Protein Chem.* **20**, 577. **22.** Simmons & Walter (1980) *Biochemistry* **19**, 39. **23.** Skern (1998) in *Handb. Proteolytic Enzymes*, p. 713, Academic Press, San Diego. **24.** Skern & Liebig (1994) *Meth. Enzymol.* **244**, 583. **25.** Molla, Hellen & Wimmer (1993) *J. Virol.* **67**, 4688. **26.** Sommergruber, Ahorn, Zöphel, *et al.* (1992) *J. Biol. Chem.* **267**, 22639. **27.** Liebig, Ziegler, Yan, *et al.* (1993) *Biochemistry* **32**, 7581. **28.** Wang, Johnson, Sommergruber & Shepherd (1998) *Arch. Biochem. Biophys.* **356**, 12. **29.** Moriyama & Sasaki (1983) *J. Biochem.* **94**, 1387. **30.** Sharma & Ortwerth (1994) *Exp. Eye Res.* **59**, 107. **31.** Miller (1998) in *Handb. Proteolytic Enzymes*, p. 1589, Academic Press, San Diego. **32.** Ichihara, Beppu & Mizushima (1984) *J. Biol. Chem.* **259**, 9853. **33.** Yamagata, Isshiki & Ichishima (1995) *Enzyme Microb. Technol.* **17**, 653. **34.** Brömme & Kleine (1984) *Curr. Microbiol.* **11**, 317.

Elbasvir

This highly potent and selective HCV antiviral (FW = 882.02 g/mol; CAS 1370468-36-2), also known by the code name MK-8742 and as dimethyl *N,N'*-([(6S)-6H-indolo[1,2-c][1,3]benzoxazine-3,10-diyl]bis{1H-imidazole-5,2-diyl-(2S)-pyrrolidine-2,1-diyl[(2S)-1-oxo-3-methylbutane-1,2-diyl]})-biscarbamate, targets the NS5a replication complex, a protein that plays a critical role in the lifecycle of Hepatitis C virus. Elbasvir is used in combination with grazoprevir, the resulting formulation, known as Zepatier®, is a highly effective, FDA-approved treatment of chronic HCV genotypes 1 or 4. **1.** Coburn, Meinke, Chang, *et al.* (2013) *ChemMedChem.* **8**, 1930.

Eliglustat

This orphan drug (FW = 404.54 g/mol; CAS 491833-29-5), also known as Genz-112638, Cerdelga® and *N*-[(1R,2R)-1-(2,3-dihydro-1,4-benzodioxin-6-yl)-1-hydroxy-3-(1-pyrrolidinyl)-2-propanyl]octanamide, is an oral inhibitor targeting glucosylceramide synthase (*Reaction*: UDP-Glucose + *N*-Acylsphingosine = UDP + D-Glucosyl-*N*-acylsphingosine), a Golgi enzyme that catalyzes the first step in the sequential addition of carbohydrate moieties in ganglioside biosynthesis. Eliglustat is used to treat Gaucher's disease, a genetic disorder leading to glucocerebroside (*or* glucosylceramide) over-accumulation in certain cells (white blood cells, especially mononuclear leukocytes) and organs (spleen, liver, kidneys, lungs, brain, and bone marrow), resulting in neurological complications, swelling of lymph nodes, anemia, and yellow fatty deposits on the white of the eye. Eliglustat improves bone mineral density, correcting an abnormal bone marrow MRI signal, as well as normalization of glucocerebroside and ganglioside GM3 levels (2). **1.** Lukina, Watman, Arreguin, *et al.* (2010) *Blood* **116**, 893. **2.** Cox (2010) *Curr. Opin. Investig. Drugs* **11**, 1169.

Ellagic Acid

This plant antioxidant and free radical scavenger (FW = 302.20 g/mol; CAS 476-66-4; Solubility: < 1 mg/mL DMSO or H_2O), also known as 4,4',5,5',6,6'-hexahydroxydiphenic acid 2,6,2',6'-dilactone, exhibits anti-mutagenic and anti-carcinogenic properties. Based on flow linear dichroism measurements, the neutral acid (pH 5.5) binds to double-stranded DNA (1). Ellagic acid protects DNA from chemical modification ny alkylating agents and inhibits the mutagenicity of the carcinogenic metabolite formed from benz[*a*]pyrene. It also inhibits γ-radiation (and thus hydroxyl radical) - induced lipid peroxidation in rat liver microsomes in a dose- and concentration-dependent manner and has been found to be a good scavenger of peroxynitrite. **Target(s):** aldose reductase (2-4); arylamine *N*-acetyltransferase (5,6,36); carbonic anhydrase, weakly inhibited (7); casein kinase 2, IC_{50} = 0.04 µM (8,9), compare to IC_{50} values of 2.9, 3.5, 4.3 and 9.4 µM for Lyn, PKA, Syk and FGR protein kinases, respectively); chitin synthase (10); CYP2A2 (11); CYP3A1 (11); CYP2B1 (11); CYP2B2 (11); CYP2C6 (11); CYP2C11 (11); CYP2E1 (12); cystathionine β-synthase (13); DNA gyrase (14,15); DNA topoisomerases, with IC_{50} values of 0.6 and 0.7 µg/mL for topo I and topo II, respectively (16, 17); 3-galactosyl-*N*-acetylglucosaminide 4-α-L-fucosyltransferase (18); glutathione-disulfide reductase (19); glutathione *S*-transferase (19-23); glycogen phosphorylase (24); glycoprotein 3-α-L-fucosyltransferase (18); H⁺/K⁺ ATPase (25); inositol-polyphosphate multikinase (17); inositol-trisphosphate 3-kinase (17); non-specific serine/threonine protein kinase (8,9); nucleoside diphosphate kinase activity (26); peptidyl-dipeptidase A, *or* angiotensin I-converting enzyme (27); phenol sulfotransferase (28); 3-phosphoglycerate kinase (29); phospholipase A_2 (30,31); p60*src* protein-tyrosine kinase (32); prostaglandin-endoperoxide synthase, *or* cyclooxygenase (37); protein kinase A (33); protein kinase C (34); thyroxine 5-deiodinase, *or* iodothyronine deiodinase (34,35); and the tautomerase activity of human macrophage migration inhibitory factor, *or* MIF (K_i = 1.97 µM, noncompetitively) (38). **1.** Thulstrup, Thormann, Spanget-Larsen & Bisgaard (1999) *Biochem. Biophys. Res. Commun.* **265**, 416. **2.** Shimizu, Horie, Terashima, *et al.* (1989) *Chem. Pharm. Bull.* **37**, 2531. **3.** Terashima, Shimizu, Horie & Morita (1991) *Chem. Pharm. Bull.* **39**, 3346.

4. Ueda, Tachibana, Moriyasu, Kawanishi & Alves (2001) *Phytomedicine* **8**, 377. **5.** Lo, Hsieh, Tsai & Chung (1999) *Drug Chem. Toxicol.* **22**, 555. **6.** Lin, Hung, Tyan, *et al.* (2001) *Urol. Res.* **29**, 371. **7.** Satomi, Umemura, Ueno, *et al.* (1993) *Biol. Pharm. Bull.* **16**, 787. **8.** Pagano, Cesaro, Meggio & Pinna (2006) *Biochem. Soc. Trans.* **34**, 1303. **9.** Duncan & Litchfield (2008) *Biochim. Biophys. Acta* **1784**, 33. **10.** Hwang, Ahn, Lee, *et al.* (2001) *Planta Med.* **67**, 501. **11.** Zhang, Hamilton, Stewart, Strother & Teel (1993) *Anticancer Res.* **13**, 2341. **12.** Wilson, Lewis, Cha & Gold (1992) *Cancer Lett.* **61**, 129. **13.** Walker & Barrett (1992) *Exp. Parasitol.* **74**, 205. **14.** Weinder-Wells, Altom, Fernandez, *et al.* (1998) *Bioorg. Med. Chem. Lett.* **8**, 97. **15.** Hilliard, Krause, Bernstein, *et al.* (1995) *Adv. Exp. Med. Biol.* **390**, 59. **16.** Constantinou, Stoner, Mehta, *et al.* (1995) *Nutr. Cancer* **23**, 121. **17.** Mayr, Windhorst & Hillemeier (2005) *J. Biol. Chem.* **280**, 13229. **18.** Niu, Fan, Sun, *et al.* (2004) *Arch. Biochem. Biophys.* **425**, 51. **19.** Kurata, Suzuki & Takeda (1992) *Comp. Biochem. Physiol. B* **103**, 863. **20.** Das, Bickers & Mukhtar (1984) *Biochem. Biophys. Res. Commun.* **120**, 427. **21.** Das, Singh, Mukhtar & Awasthi (1986) *Biochem. Biophys. Res. Commun.* **141**, 1170. **22.** Hayeshi, Mutingwende, Mavengere, Masiyanise & Mukanganyama (2007) *Food Chem. Toxicol.* **45**, 286. **23.** Liebau, Eckelt, Wildenburg, *et al.* (1997) *Biochem. J.* **324**, 659. **24.** Jakobs, Fridrich, Hofem, Pahlke & Eisenbrand (2006) *Mol. Nutr. Food Res.* **50**, 52. **25.** Murakami, Isobe, Kijima, *et al.* (1991) *Planta Med.* **57**, 305. **26.** Malmquist, Anzinger, Hirzel & Buxton (2001) *Proc. West Pharmacol. Soc.* **44**, 57. **27.** Kiss, Kowalski & Melzig (2004) *Planta Med.* **70**, 919. **28.** Eaton, Walle, Lewis, *et al.* (1996) *Drug Metab. Dispos.* **24**, 232. **29.** Hickey, Coutts, Tsang-Tan & Pogson (1995) *Biochem. Soc. Trans.* **23**, 607s. **30.** Marshall, Murphy, Marinari & Chang (1992) *Agents Actions* **37**, 60. **31.** Glaser, Sung, Hartman, *et al.* (1995) *Skin Pharmacol.* **8**, 300. **32.** Dow, Chou, Bechle, Goddard & Larson (1994) *J. Med. Chem.* **37**, 2224. **33.** Wang, Lu & Polya (1998) *Planta Med.* **64**, 195. **34.** Auf'mkolk, Kohrle, Gumbinger, Winterhoff & Hesch (1984) *Horm. Metab. Res.* **16**, 188. **35.** Auf'mkolk, Koehrle, Hesch & Cody (1986) *J. Biol. Chem.* **261**, 11623. **36.** Makarova (2008) *Curr. Drug Metab.* **9**, 538. **37.** Saeed, Butt & McDonald-Gibson (1981) *Biochem. Soc. Trans.* **8**, 443. **38.** Sarkar, Siddiqui, Mazumder, *et al.* (2015) *J. Agric. Food Chem.* **63**, 4988.

Ellipticine

This yellow alkaloid and antitumor agent (FW = 246.31 g/mol; CAS 519-23-3), also known as 5,11-dimethyl-6*H*-pyrido[4,3-*b*]carbazole, first isolated from the elliptic yellowwood or kopsia (*Ochrosia elliptica*), a flowering tree native of Australia and New Caledonia, is a DNA intercalating agent that causes double-strand breaks. Ellipticine also blocks p53 protein phosphorylation, inducing apotosis in many cell types. Ellipticine is also a strong inhibitor of CYP1A1 and DNA topoisomerase II. Its structure and chemical synthesis were reported by R. B. Woodward's laboratory (1). **Target(s):** bacteriophage T₄ DNA topoisomerase (2); cholesterol-5,6-oxide hydrolase (3); CYP1A1 (4-6); CYP1A2 (4); CYP2A6, *or* coumarin 7-monooxygenase (7); CYP3A27 (6); cytochrome P450 (4-9); DNA topoisomerase I (10,11); DNA topoisomerase II (2,10,12-18); (*S*)-limonene 3 monooxygenase (19); (*S*)-limonene 6-monooxygenase (19); (*S*)-limonene 7-monooxygenase (19); NADPH:cytochrome P450 reductase (20); poly(ADP-ribose) glycohydrolase (21); RNA polymerase (22); soluble epoxide hydrolase (23); UDPglucuronosyltransferase (8). **1.** Woodward, Iacobucci & Hochstein (1959) *J. Amer. Chem. Soc.* **81**, 4434. **2.** Huff & Kreuzer (1990) *J. Biol. Chem.* **265**, 20496. **3.** Palakodety, Vaz & Griffin (1987) *Biochem. Pharmacol.* **36**, 2424. **4.** Tassaneeyakul, Birkett, Veronese, *et al.* (1993) *J. Pharmacol. Exp. Ther.* **265**, 401. **5.** Beresford, Taylor, Ashcroft, *et al.* (1996) *Xenobiotica* **26**, 1013. **6.** Miranda, Henderson & Buhler (1998) *Toxicol. Appl. Pharmacol.* **148**, 237. **7.** Draper, Madan & Parkinson (1997) *Arch. Biochem. Biophys.* **341**, 47. **8.** Bostrom, Becedas & DePierre (2000) *Chem. Biol. Interact.* **124**, 103. **9.** Lesca, Rafidinarivo, Lecointe & Mansuy (1979) *Chem. Biol. Interact.* **24**, 189. **10.** Riou, Gabillot, Philippe, Schrevel & Riou (1986) *Biochemistry* **25**, 1471. **11.** Riou, Douc-Rasy & Kayser (1986) *Biochem. Soc. Trans.* **14**, 496. **12.** Evans, Ricanati, Horng & Mencl (1989) *Mutat. Res.* **217**, 53. **13.** Bergerat, Gadelle & Forterre (1994) *J. Biol. Chem.* **269**, 27663. **14.** Melendy & Ray (1989) *J. Biol. Chem.* **264**, 1870. **15.** Vosberg (1985) *Curr. Top. Microbiol. Immunol.* **114**, 19. **16.** Insaf, Danks & Witiak (1996) *Curr. Med. Chem.* **3**, 437. **17.** Bergerat, Gadelle & Forterre (1994) *J. Biol. Chem.* **269**, 27663. **18.** Froehlich-Ammon, Patchan, Osheroff & Thompson (1995) *J. Biol. Chem.* **270**, 14998. **19.** Karp, Mihaliak, Harris & Croteau (1990) *Arch. Biochem. Biophys.* **276**, 219. **20.** Guenthner, Kahl & Nebert (1980) *Biochem. Pharmacol.* **29**, 89. **21.** Tavassoli, Tavassoli & Shall (1985) *Biochim. Biophys. Acta* **827**, 228. **22.** Sethi (1981) *Biochem. Pharmacol.* **30**, 2026. **23.** Meijer & Depierre (1985) *Eur. J. Biochem.* **150**, 7.

Eluxadoline

This oral, peripherally active μ-opioid agonist (FW = 569.65 g/mol; CAS 864821-90-9), also named Viberzi® and 5-({[(2*S*)-2-amino-3-(4-carbamoyl-2,6-dimethylphenyl)propanoyl][(1*S*)-1-(4-phenyl-1*H*-imidazol-2-yl)ethyl]amino}methyl)-2-methoxybenzoate, is a FDA-approved drug for the treatment of diarrhea and abdominal pain in individuals with diarrhea-predominant irritable bowel syndrome, *or* IBS-D (1-3). Enteric neurons expressing μ opioid receptors contribute to opioid-induced constipation, and local activation by this drug of μ opioid receptor (*K* = 1.7 nM) reduces both secretion and motility, Eluxadoline is also a κ-opioid receptor agonist (*K* = 55 nM) and δ-opioid receptor antagonist (*K*ᵢ = 1.3 nM) that acts locally within the enteric nervous system (1). Pharmacokinetics in rats, mice and primates show low systemic exposure after oral administration, a finding that is consistent with Eluxadoline's local site action on the myenteric plexus and smooth muscle (1). As a new therapeutic agent for reducing symptoms of IBS with diarrhea in men and women, eluxadoline shows sustained efficacy over 6 months in patients receiving a twice-daily 100-mg dose (3). **1.** Wade, Palmer, McKenney, *et al.* (2012) *Br. J. Pharmacol.* **167**, 1111. **2.** Davenport, Covington, Bonifacio, McIntyre & Venitz (2015). *J. Clin. Pharmacol.* **55**, 534. **3.** Lembo, Lacy, Zuckerman, *et al.*. (2016) *New Engl. J. Med.* **374**, 242.

Elvitegravir

This viral integrase inhibitor (FW = 447.88 g/mol; CAS 697761-98-1; Solubility: 90 mg/mL DMSO; <1 mg/mL H₂O; Symbol: EVG), also known as JTK-303, GS-9137, and (*S*)-6-(3-chloro-2-fluorobenzyl)-1-(1-hydroxy-3-methylbutan-2-yl)-7-methoxy-4-oxo-1,4-dihydroquinoline-3-carboxylic acid, targets HIV-1 IIIB (IC₅₀ = 0.7 nM), HIV-2 EHO (IC₅₀ = 2.8 nM), and HIV-2 ROD (IC₅₀ = 1.4 nM). EVG inhibits HIV-1 replication (including various subtypes and multiple-drug-resistant clinical isolates) and HIV-2 replication, with an EC₅₀ values in the subnanomolar to nanomolar range. It can also inhibit the replication of several retroviruses and lentiviruses. Even so, elvitegravir is efficiently metabolized by hepatic and intestinal CYP3A and UDP-glucuronosyltransferases UGT1A1 and UGT1A3. The action of this drug may be combined with ritonavir, a strong CYP3A inhibitor, thereby prolonging its elimination half-life. T66I and E92Q substitutions within the active site contribute to EVG resistance. **1**. Shimura, Kodama, Sakagami, *et al*. (2008) *J. Virol*. **82**, 764.

EMATE, See *Estrone 3-O-Sulfamate*

EMD-1214063

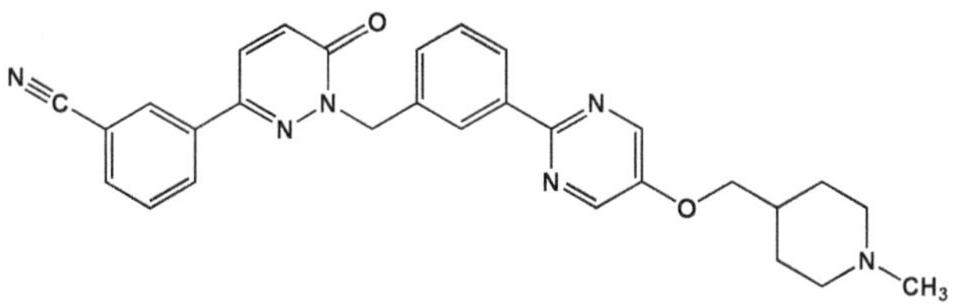

This MET protein kinase inhibitor (FW = 492.57 g/mol; CAS 1100598-32-0; Soluble to 20 mg/mL in DMSO), also known as 3-(1-(3-(5-((1-methylpiperidin-4-yl)methoxy)pyrimidin-2-yl)benzyl)-1,6-dihydro-6-oxopyridazin-3-yl)benzonitrile and HY-14721, is an antineoplastic agent that selectively disrupts MET signal transduction pathways, inducing apoptosis in tumor cells that overexpress this receptor. The tyrosine kinase MET, also known as <u>H</u>epatocyte <u>G</u>rowth <u>F</u>actor <u>R</u>eceptor (*or* HGFR) is the product of the proto-oncogene c-Met. Because MET plays key roles in cell growth and survival, invasion and metastasis, and angiogenesis, its overexpression or mutation results in the robust proliferation of many tumor cell types. **1**. Bladt, Faden, Friese-Hamim, *et al*. (2013) *Clin. Cancer Res*. **19**, 2941.

EMD 68843, See *Vilazodone*

EMD 121974, See *Cilengitide*

Emetin

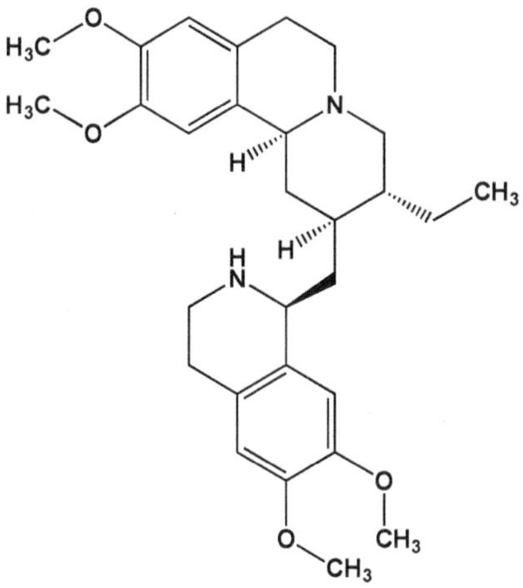

This toxic alkaloid (FW$_{\text{free-base}}$ = 480.65 g/mol; CAS 483-18-1; M.P. = 74°C; pK_a values of 6.77 and 6.64) from the roots of *Uragoga ipecacuanha* (ipecac) is a potent emetic, hence the name. Used clinically as an expectorant and an antiamoebic agent, emetine also interferes with translocation by stabilizing polyribosomes (1-4). It also induces apoptosis (5). Because emetine is a white powder that yellows upon exposure to light or heat, it should be stored in the dark and kept cold. While the free base is only slightly soluble in water, the dihydrochloride is very soluble (0.14 g/mL H$_2$O for the hydrated dihydrochloride). **Target(s):** deacetylipecoside synthase, weakly inhibited; IC$_{50}$ ≈ 1 mM (6); P glycoprotein (7); protein biosynthesis, elongation and translocation (1-4). **1**. Pestka (1974) *Meth. Enzymol*. **30**, 261. **2**. Carrasco, Battaner & Vazquez (1974) *Meth. Enzymol*. **30**, 282. **3**. Scott & Tomkins (1975) *Meth. Enzymol*. **40**, 273. **4**. Jiménez (1976) *Trends Biochem. Sci*. **1**, 28. **5**. Meijerman, Blom, de Bont, Mulder & Nagelkerke (1999) *Toxicol. Appl. Pharmacol*. **156**, 46. **6**. De-Eknamkul, Suttipanta & Kutchan (2000) *Phytochemistry* **55**, 177. **7**. Wang, Lew, Barecki, *et al*. (2001) *Chem. Res. Toxicol*. **14**, 1596.

EML4-ALK, See *Ceritinib*

EML 425

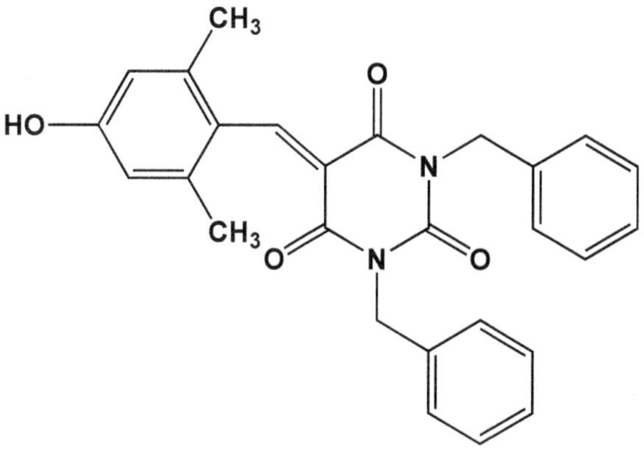

This cell-permeable non-competitive CBP/p300 inhibitor (FW = 440.39 g/mol; CAS 1675821-32-5; Soluble to 100 mM in DMSO), also named 5-[(4-hydroxy-2,6-dimethylphenyl)methylene]-1,3-bis(phenyl-methyl)-2,4,6 (1*H*,3*H*,5*H*)-pyrimidinetrione, targets CBP (IC$_{50}$ = 1.1 µM) and p300 (IC$_{50}$ = 2.9 µM), two paraloguous KAT3 histone acetyltransferases. **1**. Milite, Feoli, Sasaki, *et al*. (2015) *J. Med. Chem*. **58**, 2779.

Emodin

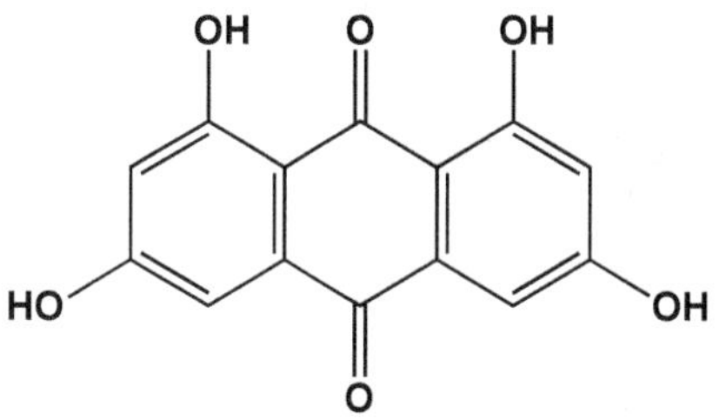

This orange anthraquinone derivative (FW = 270.24 g/mol; CAS 518-82-1), also known as 1,3,8-trihydroxy-6-methyl-9,10-anthracenedione, is found as a glycoside in rhubarb root, alder blackthorn, and other plants. It inhibits NK-κB activation and adhesion molecule expression. Emodin also inhibits p56lck protein-tyrosine kinase (IC$_{50}$ = 18.5 µM) and suppresses HER2/*neu* protein tyrosine kinase activity. Note that emodin disrupts model membranes, exhibiting a relatively high affinity for phospholipid membranes. It weakens hydrophobic interactions between hydrocarbon chains in phospholipid bilayers. **Target(s):** arylamine *N*-acetyltransferase (1); casein kinase (1), weakly inhibited (2); casein kinase 2 (2-6); cyclin-dependent kinase 2, weakly inhibited (2); electron transport (7); glucose 6-phosphate dehydrogenase (8); 11β-hydroxysteroid dehydrogenase (9); inositol-trisphosphate 3-kinase (10); Lyk protein-tyrosine kinase (11); monoamine oxidase B (12); [myosin light-chain] kinase (13); neuraminidase, IC$_{50}$ = 2.8 µM (14); non-specific serine/threonine protein kinase (2-6); oxidative phosphorylation, uncoupled (15); phosphatidylinositol 3-kinase (16); protein kinase A, weakly inhibited (2); protein kinase C, weakly inhibited (2); protein-tyrosine kinase, non-specific protein tyrosine kinase (11,17-19); tyrosinase, *or* monophenol monooxygenase (20). **1**. Chung, Wang, Wu, Chang & Chang (1997) *Food Chem. Toxicol*. **35**, 1001. **2**. Yim, Lee, Lee & Lee (1999) *Planta Med*. **65**, 9. **3**. Sarno, Moro, Meggio, *et al*. (2002) *Pharmacol. Ther*. **93**, 159. **4**. Battistutta, De Moliner, Sarno, Zanotti & Pinna (2001) *Protein Sci*. **10**, 2200. **5**. Duncan & Litchfield (2008) *Biochim. Biophys. Acta* **1784**, 33. **6**. Ahn, Min, Bae & Min (2006) *Exp. Mol. Med*. **38**, 55. **7**. Ubbink-Kok, Anderson & Konings (1986) *Antimicrob. Agents Chemother*. **30**, 147. **8**. Rychener & Steiger (1989) *Pharm. Acta Helv*. **64**, 8. **9**. Zhang & Wang (1997) *Zhongguo Yao Li Xue Bao* **18**, 240. **10**. Mayr, Windhorst & Hillemeier (2005) *J. Biol. Chem*. **280**, 13229. **11**. Mahabeleshwar & Kundu (2003) *J. Biol. Chem*. **278**, 52598. **12**. Kong, Cheng & Tan (2004) *J. Ethnopharmacol*. **91**, 351. **13**. Jinsart, Ternai & Polya (1992) *Biol. Chem. Hoppe Seyler* **373**, 903. **14**. Lee, Kim, Lee, *et al*. (2003) *Arch. Pharm. Res*. **26**, 367. **15**. Kawai, Kato, Mori, Kitamura & Nozawa (1984) *Toxicol. Lett*. **20**, 155. **16**. Frew, Powis, Berggren, *et al*. (1994) *Anticancer Res*. **14**, 2425. **17**. Jayasuriya, Koonchanok, Geahlen, McLaughlin & Chang (1992) *J. Nat. Prod*. **55**, 696. **18**. Zhang, Lau, Xi, *et al*. (2004) *Curr. Med. Chem. Anticancer Agents* **4**, 173. **20**. Leu, Hwang, Hu & Fang (2008) *Phytother. Res*. **22**, 552.

Empagliflozin

This oral Type-2 diabetes drug (FW = 450.91 g/mol; CAS 864070-44-0) also known as Jardiance® and (2S,3R,4R,5S,6R)-2-[4-chloro-3-[[4-[(3S)-oxolan-3-yl]oxyphenyl]-methyl]phenyl]-6-(hydroxymethyl)oxane-3,4,5-triol) is a potent and highly selective inhibitor of the sodium-glucose transporter (SGLT2), Subtype-2, a carrier responsible for >90% glucose reabsorption in the kidney. SGLT2 inhibitors target the kidney to remove excess glucose from the body, thus offering new options for managing Type-2 Diabetes. Other SGLT2 inhibitors include: canagliflozin, dapagliflozin, tofogliflozin, and luseogliflozin. **Pharmacokinetics:** Empagliflozin is rapidly absorbed, plasma concentrations peaking after 1.3-3.0 h, followed by a biphasic decline (3). In single rising-dose studies, mean terminal $t_{1/2}$ values range from 5.6-13.1 h; in multiple-dose studies, mean terminal $t_{1/2}$ values range from 10.3-18.8 h in. Following multiple oral doses, increases in exposure are dose-proportional, with trough concentrations remaining constant after treatment day-6, indicating attainment of steady state conditions. Oral clearance at steady state was similar to corresponding single-dose values, suggesting linear pharmacokinetics with respect to time (3). **1**. Grempler, Thomas, Eckhardt, *et al.* (2012) *Diabetes Obes. Metab.* **14**, 83. **2**. Luippold, Klein, Mark & Grempler (2012) *Diabetes Obes. Metab.* **14**, 601. **3**. Scheen (2014) *Clin. Pharmacokinet.* **53**, 213.

Emricasan

This first-in-class, cell-permeable protease inhibitor (FW = 569.50 g/mol; CAS 254750-02-2), also named IDN-6556, PF 03491390, and (S)-3-((2-(*tert*-butyl)phenyl)((S)-1-oxo-1-(2-oxoacetamido)propan-2-yl)amino)-4-oxo-5-(2,3,5,6-tetrafluorophenoxy)pentanoic acid, blocks caspase-mediated apoptosis by targeting caspase-1 (IC$_{50}$ = 0.4 nM), caspase-2 (IC$_{50}$ = 20 nM), caspase-3 (IC$_{50}$ = 2 nM), caspase-4 (IC$_{50}$ = 0.06 nM), caspase-5 (IC$_{50}$ = 0.01 nM), caspase-6 (IC$_{50}$ = 4 nM), caspase-7 (IC$_{50}$ = 0.2 nM), caspase-8 (IC$_{50}$ = 0.2 nM), caspase-9 (IC$_{50}$ = 0.3 nM), and caspase-10 (IC$_{50}$ = 1.4 nM). Hepatocyte apoptosis, the hallmark of Non-Alcoholic SteatoHepatitis (NASH), contributes to liver injury and fibrosis, processes sharing a final caspase-initiated apoptotic mechanism. To delay the progression to cirrhosis and end-stage liver disease, the FDA granted orphan drug status to emricasan for treating liver transplant patients with re-established fibrosis. **1**. Natori, Higuchi, Contreras & Gores (2003) *Liver Transpl.* **9**, 278. **2**. Canbay, Feldstein, Baskin-Bey, Bronk & Gores (2004) *J. Pharmacol. Exp. Ther.* **308**, 1191. **3**. Hoglen, Chen, Fisher, *et al.* (2004) *J. Pharmacol. Exp. Ther.* **309**, 634.

Emtricitabine

This 2',3'-dideoxycytidine isostere (FW = 247.25 g/mol; CAS 143491-57-0; Symbol: FTC), also named 4-amino-5-fluoro-1-[(2R,5S)-2-(hydroxymethyl)-1,3-oxathiolan-5-yl]-1,2-dihydropyrimidin-2-one and Emtriva (formerly Coviracil), is a nucleoside reverse transcriptase inhibitor, *or* NRTI, used to reduce HIV viral load and to increase the number of CD4+ T-cells. Inhibition requires phosphorylation of FTC to the corresponding triphosphate. Emtricitabine is commercially available and is approved by the FDA for treating HIV, but not HVB, infection. When employed as a single drug regimen, drug resistance occurs. The latter is less of a problem, when used as a fixed-dose combination (Truvada®) with tenofovir. The (–)-isomer is a potent inhibitor of viral replication (IC$_{50}$ = 10 nM), while the (+)-isomer is considerably less active. Both isomers showed minimal toxicity to HepG2 cells (IC$_{50}$ > 200 μM), showing minimal toxicity in the human bone marrow progenitor cell assay (1). (–)-FTC is a poor substrate for cytidine deaminase, which, like other catabolic enzymes, prefer the 1-β-D configuration of the (+)-isomer. In male CD rats, plasma clearance of 10 mg/kg body weight is a double-exponential ($t_{1/2}$ = 4.7 min in the first-phase; $t_{1/2}$ = 44 min in the second-phase), and total body clearance of FTC s 1.8 liters/hour/kg, with the oral bioavailability of 90% (2). The steady-state volume of distribution is 1.5 liters/kg. Increasing the dose to 100 mg/kg slows clearance and reduces oral bioavailability in the brains of rats (2). Emtricitabine is similar to lamivudine, and cross-resistance between the two drugs is near-universal (**See also** *Lamivudine*) **1**. Furman, Davis, Liotta, *et al.* (1992) *Antimicrob. Agents Chemother.* **36**, 2686. **2**. Frick, St John, Taylor, *et al.* (1993) *Antimicrob. Agents Chemother.* **37**, 2285.

EN-2234A, *See Nalbuphine*

Enalapril

This orally bioavailable ACE prodrug (FW = 376.45 g/mol; CAS 75847-73-3; Solubility: 100 mg/mL DMSO; <1 mg/mL Water), also known as Vasotec®, Glioten®, Renitec®, and named systematically as (2S)-1-[(2S)-2-{[(2S)-1-ethoxy-1-oxo-4-phenylbutan-2-yl]amino}propanoyl]-pyrrolidine-2-carboxylic acid, is metabolically de-esterified to the active drug enaprilat, which targets angiotensin-converting enzyme by competing with angiotensin I and blocking its conversion to angiotensin II. (**See** *Enalaprilat*) Inhibition of ACE results in decreased plasma angiotensin II. Enalaprilat may also acts on kininase II, an enzyme identical to ACE that degrades the vasodilator bradykinin.

Enalaprilat

This dicarboxylic acid (FW$_{free-acid}$ = 348.40 g/mol; CAS 76420-72-9), also known as N-[1(S)-carboxy-3-phenylpropyl]-L-alanyl-L-proline and MK-422, is the active form of Entinostatapril and is a potent inhibitor of angiotensin I-converting enzyme (IC$_{50}$ = 4.5 nM). (**See also** *Caprilat*) **Key**

Pharmacokinetic Parameters: *See* Appendix II in Goodman & Gilman's *THE PHARMACOLOGICAL BASIS OF THERAPEUTICS*, 12th Edition (Brunton, Chabner & Knollmann, eds.) McGraw-Hill Medical, New York (2011). **Target(s):** peptidyl-dipeptidase A, *or* angiotensin I-converting enzyme (1-14); Xaa-Pro aminopeptidase aminopeptidase P (15-17). **1.** Ryan (1988) *Meth. Enzymol.* **163**, 194. **2.** Corvol, Williams & Soubrier (1995) *Meth. Enzymol.* **248**, 283. **3.** Borek, Charlap & Frishman (1987) *Pharmacotherapy* **7**, 133. **4.** Patchett, Harris, Tristram, *et al.* (1980) *Nature* **288**, 280. **5.** Tocco, deLuna, Duncan, Vassil & Ulm (1982) *Drug Metab. Dispos.* **10**, 15. **6.** Reynolds (1984) *Biochem. Pharmacol.* **33**, 1273. **7.** Nagamori, Fujishima & Okada (1990) *Agric. Biol. Chem.* **54**, 999. **8.** Wei, Alhenc-Gelas, Soubrier, *et al.* (1991) *J. Biol. Chem.* **266**, 5540. **9.** Wei, Clauser, Alhenc-Gelas & Corvol (1992) *J. Biol. Chem.* **267**, 13398. **10.** Strittmatter, Thiele, Kapiloff & Snyder (1985) *J. Biol. Chem.* **260**, 9825. **11.** Hooper & Turner (1987) *Biochem. J.* **241**, 625. **12.** Sweet (1983) *Fed. Proc.* **42**, 167. **13.** Ondetti & Cushman (1982) *Ann. Rev. Biochem.* **51**, 283. **14.** Patchett & Cordes (1985) *Adv. Enzymol. Relat. Areas Mol. Biol.* **57**, 1. **15.** Hooper, Hryszko, Oppong & Turner (1992) *Hypertension* **19**, 281. **16.** Orawski & Simmons (1995) *Biochemistry* **34**, 11227. **17.** Lloyd, Hryszko, Hooper & Turner (1996) *Biochem. Pharmacol.* **52**, 229.

Enavatuzumab

This first-in-class humanized IgG1 monoclonal antibody (MW = 141.2 kDa; CAS 912628-39-8), also known by the code name PDR-192, targets TweakR, a receptor that binds TWEAK, *or* cytokine TNF-like weak inducer of apoptosis (1). TWEAK is a cell surface-associated, type-II transmembrane protein (2). Its smaller, yet still biologically active, form is shed into the extracellular milieu. Because TweakR is overexpressed in many solid tumor types, Enavatuzumab promises to be an effective growth-inhibiting anticancer drug. Treatment of sensitive cell lines with enavatuzumab, both *in vitro* and *in vivo* (in a xenograft model), activates both classical (*e.g.,* p50 and p65) as well as non-classical (*e.g.,* p52, RelB) NFκB pathways. Treatment with Enavatuzumab results in NFκB-dependent reduction in cell division, the likely consequence of activation of the cell-cycle inhibitor p21. **1.** Purcell, Kim, Tanlimco, *et al.* (2014) *Front Immunol.* **4**. 505. **2.** Wiley & Winkles (2003) *Cytokine Growth Factor Rev.* **14**, 241.

Endothall

This herbicide and defoliant ($FW_{free-acid}$ = 186.16 g/mol; CAS 145-73-3) is a potent inhibitor of protein phosphatase-2A (IC_{50} = 90 nM). Elevated concentrations are reported to inhibit protein phosphatase-1. Endothall also causes dose- and time-dependent cytostasis at the G_2/M junction in hepatocellular carcinoma cell lines. **1.** Li & Casida (1992) *Proc. Natl. Acad. Sci. U.S.A.* **89**, 11867. **2.** Li, Mackintosh & Casida (1993) *Biochem. Pharmacol.* **46**, 1435. **3.** Szöör, Fehér, Bakó, *et al.* (1995) *Comp. Biochem. Physiol. B* **112**, 515. **4.** Zabrocki, Swiatek, Sugajska, *et al.* (2002) *Eur. J. Biochem.* **269**, 3372.

Enduracidins

Enduracidin A (R - 10-methylundeca-2(Z),4(E)-dienoyl–)
Enduracidin B (R - 10-methyldodeca-2(Z),4(E)-dienoyl–)

These cyclic lipodepsipeptide antibiotics ($FW_{Enduracidin-A}$ = 2355.32 g/mol; CAS 34438-27-2), also known as enramycins, are produced by *Streptomyces fungicidicus* and are active against Gram-positive bacteria and inhibit cell wall biosynthesis, preferentially inhibiting the transglycosylation step of peptidoglycan biosynthesis, as compared with the MurG step. Enduracidins and ramplamin share a similar 17-residue peptide core, but differ mainly in the length of the acyl chain and the presence of the two D-mannose moieties found in ramoplanin. Enduracidins are often used as additives in livestock feed. **Target(s):** avian myeloblastosis virus reverse transcriptase (1); murein biosynthesis (2); penicillin-binding protein 1b (3); peptidoglycan glycosyltransferase (3); prolyl endopeptidase (4); RNA directed DNA polymerase (1). **1.** Inouye, Take, Nakamura, *et al.* (1987) *J. Antibiot.* **40**, 100. **2.** Tamura, Suzuki, Nishimura, Mizoguchi & Hirota (1980) *Proc. Natl. Acad. Sci. U.S.A.* **77**, 4499. **3.** Nakagawa, Tamaki, Tomioka & Matsuhashi (1984) *J. Biol. Chem.* **259**, 13937. **4.** Kimura, Kanou, Yamashita, Yoshimoto & Yoshihama (1997) *Biosci. Biotechnol. Biochem.* **61**, 1754.

Englerin A

This guaiane sesquiterpene (FW = 442.55 g/mol; CAS 1094250-15-3; IUPAC Name: (1*R*,3a*R*,4*S*,5*R*,7*R*,8*S*,8a*R*)-5-(glycoloyloxy)-7-isopropyl-1, 4-dimethyldecahydro-4,7-epoxyazulen-8-yl (2*E*)-3-phenylacrylate; Symbol: EA) from the poisonous stem bark of the Tanzanian plant *Phyllanthus engleri* (1), binds to and activates protein kinase Cθ (PKCθ), inducing an insulin-resistant phenotype, most likely by limiting access of kidney tumor cells to glucose. PKCθ phosphorylates Insulin Receptor Substrate 1, *or* IRS-1, which plays a key role in transmitting signals from the insulin and Insulin-like Growth Factor-1 (IGF-1) receptors to the intracellular PI3K/Akt signal transduction pathway and extracellular ERK/MAP kinase signal transduction pathways. The net effect is inhibition of the insulin pathway, limiting glucose uptake by renal cancer cells. EA also stimulates PKCθ-mediated phosphorylation and consequential activation of Heat Shock Factor 1, a key transcription factor and inducer of glucose dependence (2). By stimulating these opposing pathways, EA is synthetically lethal for glycolytic tumor cells expressing PKCθ and HSF1 (2). Significantly, Englerin A induces necrotic cell death in renal cancer cells, but not in normal kidney cells (3). EA-treated cells accumulated in G_2, indicating a block in G_2/M transition (4). Englerin A also inhibits activation of both AKT and ERK, which are activated in cancer and implicated in unrestricted cell proliferation and induction of autophagy. The phosphorylation status of the cellular energy sensor, AMPK, appeared unaffected by EA (4). **1.** Ratnayake, Covell, Ransom, Gustafson & Beutler (2009) *Org. Lett.* **11**, 57. **2.** Sourbier, Scroggins, Ratnayake, *et al.* (2013) *Cancer Cell* **23**, 228. **3.** Sulzmaier, Li, Nakashige, *et al.* (2012) *PLoS One* **7**, e48032. **4.** Williams, Yu, Diccianni, Theodorakis & Batova (2013) *J. Exp. Clin. Cancer Res.* **32**, 57.

ENMD-1198

2-Methoxyestradiol   ENMD-1198

This orally bioavailable microtubule-targeted 2-methoxyestradiol (*or* 2ME2) analogue (FW = 311.42 g/mol) exhibits a greater duration of action owing to the elimination of sites for metabolic conjugation (positions 3 and 17 in 2ME2) and oxidation (position 17 in 2ME2), with >65% remaining after 2-h incubation with hepatocytes. ENMD-1198 and its congeners (ENMD-1200 and ENMD-1237) bind the colchicine-binding site on tubulin, inducing G_2-M cell cycle arrest and apoptosis and also reducing hypoxia-inducible factor-1α (Hif-1α) levels. ENMD-1198 is also a potent vascular-disrupting agent (VDA) that leads to solid tumor necrosis by limiting tumor vascularization (1). Video microscopy confirms that ENMD-1198 completely disrupts preformed vascular structures within 2 hours, with extensive depolymerization of the microtubule network and accumulation of actin stress fibers and large focal adhesions in vascular endothelial cells (2). ENMD-1198 inhibited the phosphorylation of MAPK/Erk, PI-3K/Akt and FAK (3). Notably, ABT-737 sensitizes androgen-dependent castration-resistant prostate cancer PC3 cells to ENMD-1198-mediated, caspase-dependent apoptosis (4). **1**. LaVallee, Burke, Swartz, *et al.* (2008) *Mol. Cancer Ther.* **7**, 1472. **2**. Pasquier, Sinnappan, Munoz & Kavallaris (2010) *Mol. Cancer Ther.* **9**, 1408. **3**. Moser, Lang, Mori, *et al.* (2008) *BMC Cancer* **8**, 206. **4**. Parrondo, de Las Pozas, Reiner & Perez-Stable (2013) *Peer J.* **1**, e144.

Enniatins

These cyclodepsipeptide ionophores and antibiotics (Enniatin A, R = CH(CH$_3$)CH$_2$CH$_3$; Enniatin B, R = CH(CH$_3$)$_2$; Enniatin C, R = CH$_2$CH(CH$_3$)$_2$; FW$_{Enniatin-A}$ = 2643.44 g/mol; CAS 144470-22-4) from species of *Fusarium* inhibit acyl-CoA:cholesterol acyltransferase. Enniatin A is (cyclo(D-2-hydroxyisovaleryl-*N*-methyl-L-isoleucyl)$_3$. Enniatin A is cyclo(D-2-hydroxyisovaleryl-*N*-methyl-L-valyl)$_3$. Enniatin C is an artificial ionophore with an isobutyl side-chain. All bind metals ion, showing greatest selectivity for K$^+$, and and the resulting complex allows that ion to cross biomembranes. **Target(s):** calmodulin, inhibited by enniatin B (1); calmodulin-dependent cyclic nucleotide phosphodiesterase, inhibited by enniatin B (1); oxidative phosphorylation, inhibited by enniatin A (2); Pdr5p, an ABC transporters in *Saccharomyces cerevisiae* (3); sterol *O*-acyltransferase, *or* acyl-CoA:cholesterol acyltransferase; ACAT, inhibited by enniatins D, E, and F (4). **1**. Mereish, Solow, Bunner & Fajer (1990) *Pept. Res.* **3**, 233. **2**. Reed (1979) *Meth. Enzymol.* **55**, 435. **3**. Hiraga, Yamamoto, Fukuda, Hamanaka & Oda (2005) *Biochem. Biophys. Res. Commun.* **328**, 1119. **4**. Tomoda, Nishida, Huang, Masuma, Kim & Omura (1992) *J. Antibiot.* **45**, 1207.

Enoxacin

This broad-spectrum fluoroquinolone antibacterial and osteoclastogenesis inhibitor (FW = 320.32 g/mol; CAS 74011-58-8), also known by the trade names Almitil™, Bactidan™, Bactidron™, Comprecin™, Enoksetin™, Enoxen™, Enroxil™, Enoxin™, Enoxor™, Flumark™, Penetrex™, Gyramid™ and Vinone™, as well as its IUPAC name, 1-ethyl-6-fluoro-4-oxo-7-(piperazin-1-yl)-1,4-dihydro-1,8-naphthyridine-3-carboxylic acid, is widely used to treat urinary tract infections. A single 400-mg oral dose produces ≥ 95% bacteriological cure rates with *Neisseria gonorrhoeae*. Structurally related to nalidixic acid, enoxacin exhibits a broad spectrum of activity against Gram-negative and Gram-positive microorganisms, but is less active against Gram-positive organisms. Patients with complicated or uncomplicated urinary tract infections also respond to other antibacterial agents, including amoxicillin, cefuroxime axetil, cotrimoxazole (trimethoprim-sulfamethoxazole) or trimethoprim (1). Enoxacin binds to gyrase·DNA complexes, thereby inhibiting gyrase-mediated negative supercoiling of DNA (2-5). Visible light irradiation of DNA·Enoxacin complex can also result in single-strand breaks. Enoxacin also binds to the profilin-like actin-binding surface of the V-ATPase B$_2$-subunit, competitively inhibiting the B$_2$-subunit's interaction with actin and

disrupting osteoclastic bone resorption *in vitro*, without affecting osteoblast formation or bone mineralization (6). Enoxacin directly inhibits osteoclast formation without affecting cell viability, acting by a novel mechanism involving changes in posttranslational processing and trafficking of proteins (*e.g.*, ADAM, ADAM12, DC-STAMP, and TRAP5b) with known roles in osteoclast function. Such effects are downstream to enoxacin's action in blocking a binding interaction between A$_3$-containing V-ATPases and microfilaments (7). **Target(s):** DNA gyrase (2-5); Vacuolar ATPase and downstream effects on posttranslational processing in osteoclasts (6,7); CYP1A2, caffeine 3-demethylase, cytochrome P450 (8,9); DNA topoisomerase II (10,11); DNA topoisomerase IV (4); medium-chain-fatty-acyl-CoA synthetase, *or* butyryl-CoA synthetase, *or* butyrate:CoA ligase (12). **1**. Patel & Spencer (1996) *Drugs* **51**, 137. **2**. Traub (1984) *Chemotherapy* **30**, 379. **3**. Domagala, Hanna, Heifetz, *et al.* (1986) *J. Med. Chem.* **29**, 394. **4**. Yoshida, Nakamura, Bogaki, *et al.* (1993) *Antimicrob. Agents Chemother.* **37**, 839. **5**. Takahashi, Hayakawa & Akimoto (2003) *Yakushigaku Zasshi* **38**, 161. **6**. Toro, Ostrov, Wronski & Holliday (2012) *Curr. Prot. Pept. Sci* **13**, 180. **7**. Toro, Zuo, Ostrov, *et al.* (2012) *J. Biol. Chem.* **287**, 17894. **8**. Fuhr, Wolff, Harder, Schymanski & Staib (1990) *Drug Metab. Dispos.* **18**, 1005. **9**. Fuhr, Anders, Mahr, Sorgel & Staib (1992) *Antimicrob. Agents Chemother.* **36**, 942. **10**. Akahane, Hoshino, Sato, *et al.* (1991) *Chemotherapy* **37**, 224. **11**. Snyder & Cooper (1999) *Photochem. Photobiol.* **69**, 288. **12**. Kasuya, Hiasa, Kawai, Igarashi & Fukui (2001) *Biochem. Pharmacol.* **62**, 363.

Enrofloxacin

This fluoroquinolone-class antibiotic (FW = 359.40 g/mol; CAS 93106-60-6), also named Baytril® and 1-cyclopropyl-7-(4-ethylpiperazin-1-yl)-6-fluoro-4-oxo-1,4-dihydroquinoline-3-carboxylate, is a broad-spectrum systemic antibacterial agent that targets Type II DNA topoisomerases (gyrases), which are required for bacterial replication and transcription. It exerts broad-spectrum activity against Gram-positive and Gram-negative bacteria as well as against *Mycoplasma* spp. in various animals, including *Pseudomonas aeruginosa*, *Klebsiella*, *Escherichia coli*, *Enterobacter*, *Campylobacter*, *Shigella*, *Salmonella*, *Aeromonas*, *Hemophilus*, *Proteus*, *Yersinia*, *Serratia*, *Vibrio*, *Brucella*, *Chlamydia trachomatis*, *Staphylococcus* (including penicillinase-producing and methicillin-resistant strains), *Mycoplasma*, and *Mycobacterium*. Enrofloxacin is an FDA-approved drug for treating pets and domestic animals. For the prototypical member of this antibiotic class, *See Ciprofloxacin*

Entacapone

This pharmaceutical (FW = 305.29 g/mol; CAS 130929-57-6), also known as OR-611, is used in the treatment of Parkinson Disease and inhibits catechol *O*-methyltransferase (K_i = 14 nM for the rat liver enzyme), thereby preventing the L-DOPA conversion to 3-methoxy-4-hydroxy-L-phenylalanine. When given along with and carbidopa (the latter an aromatic amino acid decarboxylase inhibitor), plasma L-dopa levels are higher and remain so for a longer period than with administration of only L-dopa and a decarboxylase inhibitor. Entacapone is likely to influence the pharmacokinetics of drugs metabolized by COMT (*e.g.*, isoproterenol, epinephrine, norepinephrine, dopamine, dobutamine, α-methyldopa, apomorphine, isoetherine, and bitolterol). Note also that entacapone is structurally related to the tyrphostins. **Key Pharmacokinetic Parameters:** *See* Appendix II in Goodman & Gilman's *THE PHARMACOLOGICAL BASIS OF THERAPEUTICS*, 12th Edition (Brunton, Chabner & Knollmann, eds.) McGraw-Hill Medical, New York (2011). **1**. Nissinen, Linden, Schultz &

Pohto (1992) *Naunyn Schmiedebergs Arch. Pharmacol.* **346**, 262. **2**. Mannisto, Tuomainen & Tuominen (1992) *Brit. J. Pharmacol.* **105**, 569. **3**. Bonifácio, Palma, Almeida & Soares-da-Silva (2007) *CNS Drug Reviews* **13**, 352.

Entecavir

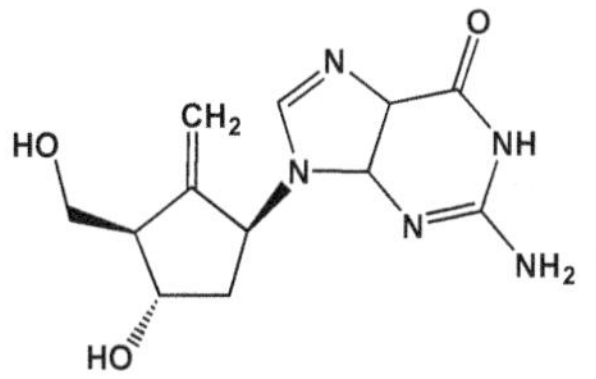

This orally available cyclopentyl guanosine analogue (FW = 277.28 g/mol; CAS 142217-69-4), known also as SQ-34676, Baraclude™ and 2-amino-9-[(1S,3R,4S)-4-hydroxy-3-(hydroxymethyl)-2-methylidenecyclopentyl]-6,9-dihydro-3H-purin-6-one, inhibits DNA polymerase and is extremely potent against hepatitis B virus (HBV) (ED$_{50}$ = 3 nM, compared with 200 nM for lamivudine) with relatively low toxicity, as indicated by its 30,000-nM cytostatic concentration (CC$_{50}$). (Entecavir is ineffective against HIV.) Although entecavir's use in the clinic is characterized by very low resistance rates, with potent and durable viral suppression, selection of certain combinations of amino acid substitutions (particularly L180M-S202G-M204V-V207I) in the hepatitis B virus (HBV) reverse transcriptase complicates treatment of chronic HBV infection and may affect the overlapping surface coding region (2). Clinical use of entecavir is indicated for treating chronic HBV infection in adults showing evidence of active viral replication and either persistently elevated serum aminotransferases or histologically discernable disease. (*See also Adefovir & Adefovir Dipivoxil; Lamivudine; Telbivudine; Tenofovir; Emtricitabine*) **1**. Julander, Colonno, Sidwell & Morrey (2003) *Antiviral Res.* **59**, 155. **2**. Rodriguez-Frías F, Tabernero D, Quer (2012) *PLoS One* **7**, e37874.

Enterostatin

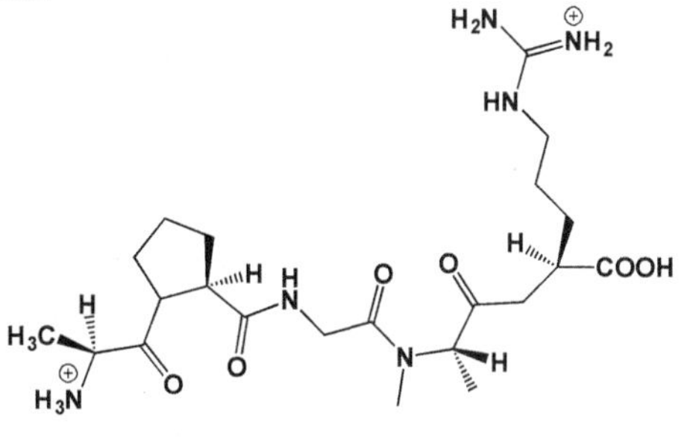

human and rabbit enterostatin

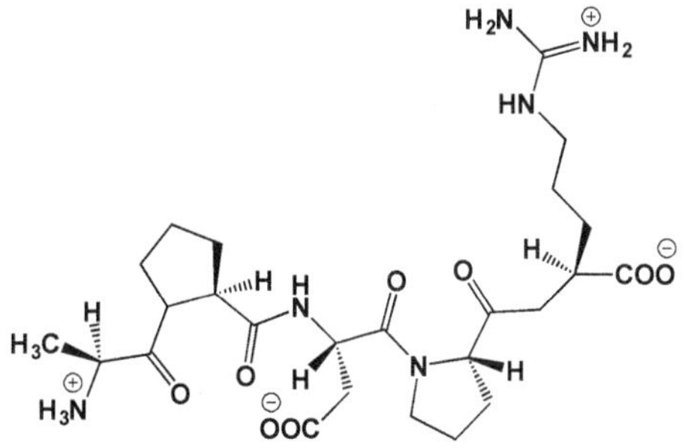

dog and pig enterostatin

This pentapeptide (FW = 305.29 g/mol; CAS 130929-57-6) selectively inhibits fat intake during normal feeding (1,2). The mechanism of action of this peptide is likely to be the inhibition of a μ-opioid-mediated pathway or an interaction the β-subunit of F$_1$F$_o$ ATP synthase (2,3). The human, rabbit, chicken, and rat sequence is Ala-Pro Gly-Pro-Arg (APGPR) while the dog, pig, ox, and horse sequence is Val-Pro-Asp-Pro-Arg (VPDPR). Eterstatin also inhibits insulin release (4,5). **1**. Erlanson-Albertsson & York (1997) *Obes. Res.* **5**, 360. **2**. Berger, Winzell, Mei & Erlanson-Albertsson (2004) *Physiol. Behav.* **83**, 623. **3**. Berger, Sivars, Winzell, *et al.* (2002) *Nutr. Neurosci.* **5**, 201. **4**. Ookuma & York (1998) *Int. J. Obes. Relat. Metab. Disord.* **22**, 800. **5**. Erlanson-Albertsson, Hering, Bretzel & Federlin (1994) *Acta Diabetol.* **31**, 160.

Entinostat

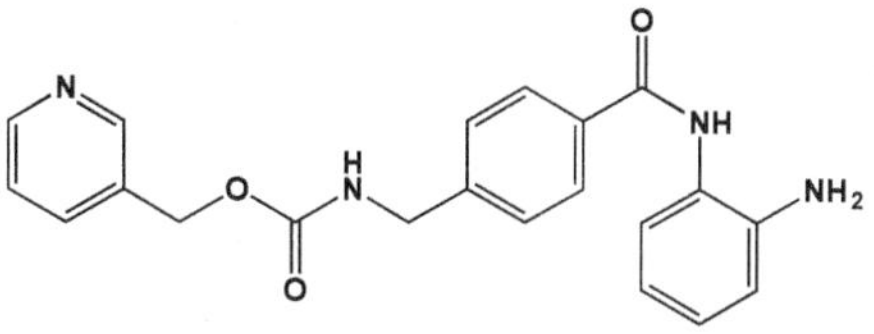

This Class-I histone deacetylase inhibitor (FW = 376.41 g/mol; CASs = 209783-80-2, 209784-79-2 (Maleic acid), 209784-80-5 (HCl); Solubility: 75 mg/mL DMSO; <1 mg/mL Water), also known as MS-275, SNDX-275, and named systematically as pyridin-3-ylmethyl 4-((2-aminophenyl) carbamoyl)benzylcarbamate, targets HDAC1 and HDAC3 with IC$_{50}$ values of 0.5 nM and 1.7 nM, respectively, resulting in a marked hyperacetylation of histone H4 (1). MS-275 is currently in Phase I/II clinical trials in recurrent advanced non-small cell lung cancer, combining with 5-azacytidine. Transient alteration of histone acetylation by entinostat alters the the fate of embryonic stem cells (5). **1**. Saito, *et al.* (1999) *Proc. Natl. Acad. Sci. USA* **96**, 4592. **2**. Rosato, *et al.* (2003) *Cancer Res.* **63**, 3637. **3**. Kato, *et al.* (2007) *Clin. Cancer Res.* **13**, 4538. **4**. Wegener, *et al.* (2003) *Chem Biol.* **10**, 61. **5**. Franci, Casalino, Petraglia, *et al.* (2013) *Biol. Open* **22**, 1070.

Entospletinib

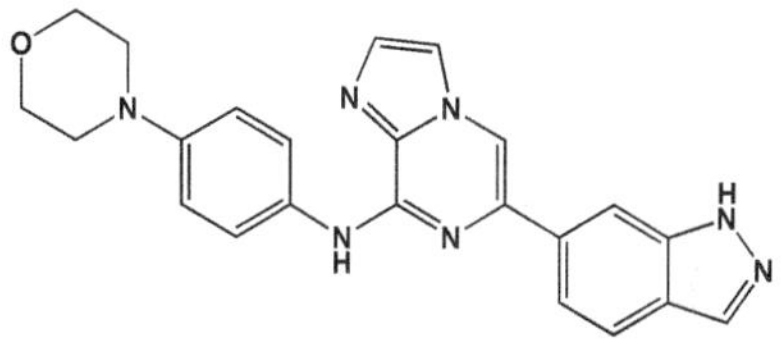

This orally bioavailable PKI (FW = 411.46 g/mol; CAS 1229208-44-9; Solubility: 82 mg/mL DMSO, with heating), also named 6-(1H-indazol-6-yl)-N-(4-morpholinophenyl)imidazo[1,2-a]pyrazin-8-amine and GS-9973, selectively targets Spleen Tyrosine Kinase, *or* Syk (IC$_{50}$ = 7.7 nM), a druggable enzyme with autoimmune, inflammatory, and cancer indications (1). PI3Kd and Syk inhibitors reduce Chronic Lymphocytic Leukemic (CLL) cell survival, and, when used in combination, induce synergistic growth inhibition and disrupt chemokine signaling at nM concentrations (2). Entospletinib demonstrates clinical activity in subjects with relapsed or refractory CLL with acceptable toxicity (3). **1**. Currie, Kropf, Lee, *et al.* (2014) *J. Med. Chem.* **57**, 3856. **2**. Burke, Meadows, Loriau, *et al.* (2014) *Oncotarget* **5**, 908. **3**. Sharman, Hawkins, Kolibaba, *et al.* (2015) *Blood* **125**, 2336.

Enzastaurin

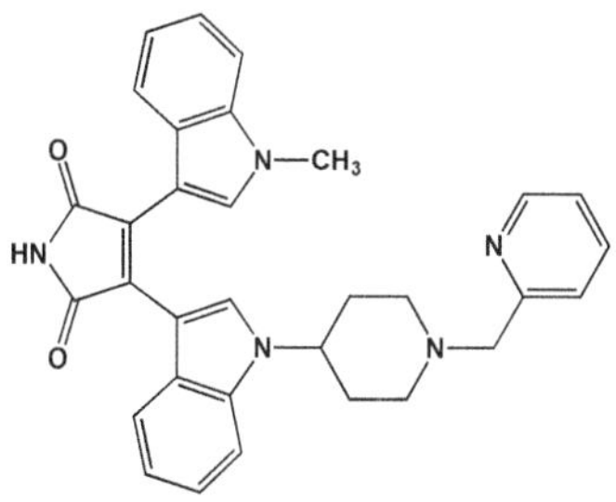

This synthetic bis-indolylmaleimide and potent ATP site-competitive PKI (FW = 515.62 g/mol; CAS 170364-57-5; Solubility: 20 mM in DMSO), also known as LY317615 and. 3-(1-methyl-1H-indol-3-yl)-4-[1-[1-(2-pyridinylmethyl)-4-piperidinyl]-1H-indol-3-yl]-1H-pyrrole-2,5-dione, selectively targets Protein Kinase-β, *or* CPKCβ (IC$_{50}$ = 6 nM), the activation of which has been implicated in tumor growth and invasiveness. Enzastaurin exhibits weaker action against PKCα (IC$_{50}$ = 39 nM), PKCγ (IC$_{50}$ = 83 nM) and PKCε (IC$_{50}$ = 110 nM). Its selective action against PKCβ also suppresses vascular endothelial growth factor (VEGF)-stimulated neo-angiogenesis (1). Treatment of cultured human tumor cells with enzastaurin induces apoptosis and suppresses cell proliferation. Enzastaurin also reduces the phosphorylation of ribosomal protein S6 (at Serine residues 240 and 244), and AKT (at Threonine-308) (2). Oral dosing with enzastaurin that yields plasma concentrations similar to those achieved in clinical trials was found to significantly suppress the growth of human glioblastoma and colon carcinoma xenografts. As in cultured tumor cells, enzastaurin treatment suppresses the phosphorylation of GSK3β in these xenograft tumor tissues (2). Enzastaurin also induces apoptosis in multiple myeloma cell lines in a caspase-independent manner, and enzastaurin's antimyeloma effect occurs by inhibiting signaling through the AKT pathway (3). Moreover, enzastaurin-induced cell death was not affected by interleukin-6 or ZVAD-fmk. In 2013, Lilly & Company announced that Phase-III enzastaurin trial did not meet primary endpoint as a monotherapy in the prevention of relapse in patients with diffuse large B-cell lymphoma (DLBCL). That study failed to show a statistically significant increase compared to placebo in disease-free survival in patients at high risk of relapse after rituximab-based chemotherapy. In contrast to two other PKC inhibitors, however, the combined use of enzastaurin with the immunotoxin SS1P produces synergistic cell death via apoptosis, thus displaying promise as a potential anticancer agent (4). **1.** Graff, McNulty, Hanna, *et al.* (2005) *Cancer Res.* **65**, 7462. **2.** Graff, McNulty, Hanna, *et al.* (2005) *Cancer Res.* **65**, 7462. **3.** Rizvi, Ghias, Davies, *et al.* (2006) *Mol. Cancer Ther.* **5**, 1783. **4.** Mattoo, Pastan & Fitzgerald (2013) *PLoS One* **8**, e75576.

Ep-475, See *N-[(2S,3S)-trans-Epoxysuccinyl]-L-leucylamido-3-methylbutane*

EP42675

This first-in-class parenteral anticoagulant (FW = 2355.54 g/mol) combines the action of a direct thrombin inhibitor, *or* DTI (the peptidomimetic moiety (an α-*N*-(2-naphthalenesulfonyl)glycyl-D-4-aminophenylalanylpiperidine, *or* α-NAPAP, analogue) and an indirect factor Xa, *or* FXa, inhibitor (an antithrombin-binding pentasaccharide moiety, *or* fondaparinux analogue) into a single molecular entity. EP42675 pharmacokinetics are dose-proportional and characterized by a low clearance, a small volume of distribution, a long terminal half-life. EP42675 pharmacodynamics are characterized by a long-lasting, dose-dependent increase in activated clotting time, ecarin clotting time, thrombin time, anti-FXa activity, activated partial thromboplastin time, prothrombin time, and a decrease in endogenous thrombin potential, when measured by a thrombin generation test. **1.** Gueret, Krezel, Fuseau, *et al.* (2014) *J. Thromb. Hemost.* **12**, 24.

Epalrestat

This adose reductase inhibitor (FW = 319.39 g/mol; CAS 82159-09-9; Symbol: EPS), also known by the code name ONO-2235, the trade names Eparel 50®, Aldonil®, Alrista®, and EPLISTAT 150 SR®, and by the systematic name 2-[(5Z)-5-[(E)-3-phenyl-2-methylprop-2-enylidene]-4-oxo-2-thioxo-3-thiazolidinyl]acetate, significantly improves motor nerve conduction velocity (MNCV) in diabetic rats at a minimal dose of 10 mg/kg/day by preventing aldose reductase-mediated accumulation of nontransported polyols (mainly sorbitol) that accumulate in diabetic hyperglycemia and osmotically damage neurons (1-4). Sorbitol content of the sciatic nerve and red blood cells after 2 weeks was concomitantly reduced in epalrestat-treated rats (sciatic nerve: 120 ± 13 *versus* 595 ± 146 nmol/g wet weight; red blood cell: 91 ± 21 *versus* 165 ± 39 nmol/g hemoglobin ($p < 0.01$) in 20 mg/kg/day-treated *versus* untreated animals

(1). Such findings indicate that, in diabetic hyperglycemia, epalrestat reduces intracellular sorbitol accumulation responsible for nerve and retina pathogenesis of late-onset complications of diabetes mellitus. EPS also increases the mRNA levels of γ-glutamylcysteine synthetase (γ-GCS), the enzyme catalyzing the first and rate-limiting step in *de novo* GSH synthesis, by activating <u>N</u>uclear <u>e</u>rythroid 2-related <u>f</u>actor-<u>2</u> (Nrf2), a key transcription factor that plays a central role in regulating the expression of γ-GCS (3). Knockdown of Nrf2 by siRNA suppressed the EPS-induced GSH biosynthesis, and pretreatment with EPS reduced the cytotoxicity induced by H_2O_2, *tert*-butylhydroperoxide, 2,2'-azo-*bis*(2-amidinopropane), and menadione, indicating EPS protects against oxidative stress. In view of EPS-induced biosynthesis of reduced glutathione (GSH) and the prior finding that paired helical filaments (PHFs) form upon disulfide coupling of oxidized glutathione (GSSG) to Tau protein (4), EPS may find therapeutic value in reducing the accumulation of neuropathic PHFs in Alzheimer's Disease. **1.** Kikkawa, Hatanaka, Yasuda, *et al.* (1983) *Diabetologia* **24**, 290. **2.** Kikkawa, Hatanaka, Yasuda, Kobayashi & Shigeta (1984) *Metabolism* **33**, 212. **3.** Sato, Yama, Murao, Tatsunami & Tampo (2013) *Redox Biol.* **2**, 15. **4.** DiNoto, Deture & Purich (2005) *Microsc. Res. Tech.* **67**, 156.

Ephedrine

This bronchodilator and vasoconstrictor ($FW_{free-base}$ = 165.24 g/mol; CAS 299-42-3; pK_a value of ephedrine is 9.54 at 25°C (dpK_a/dT = –0.022)), also known as (1*R*,2*S*)-2-methylamino-1 phenylpropan-1-ol, is an α- and β-adrenergic agonist from the stem and leaves of *Ephedra equisetina* and mentioned in the *Huangdi Neijing*, an ancient Chinese text on internal medicine. It is commonly used as a stimulant, appetite suppressant, concentration aid, decongestant, and to treat hypotension associated with anesthesia. Athough ephedrine efficacious in the treatment of numerous ailments, it has a long history of misuse. Ephedrine is a structural analogue of epinephrine. Ephedrine can be assayed enzymatically with EC 1.5.1.18, a bacterial ephedrine oxidoreductase. **Target(s):** acetylcholinesterase (1); diamine oxidase (2); β-glucosidase (3); monoamine oxidase, *or* amine oxidase (2,4-6); procollagen-lysine 5-dioxygenase, weakly inhibited (7). **1.** Ngiam & Go (1989) *Chem. Pharm. Bull. (Tokyo)* **37**, 2423. **2.** Zeller (1951) *The Enzymes*, 1st ed., **2** (part 1), 536. **3.** Dale, Ensley, Kern, Sastry & Byers (1985) *Biochemistry* **24**, 3530. **4.** Blaschko (1963) *The Enzymes*, 2nd ed., **8**, 337. **5.** Ulus, Maher & Wurtman (2000) *Biochem. Pharmacol.* **59**, 1611. **6.** Tabor, Tabor & Rosenthal (1954) *J. Biol. Chem.* **208**, 645. **7.** Murray, Cassell & Pinnell (1977) *Biochim. Biophys. Acta* **481**, 63.

Epicatechin

(–)-epicatechin   (+)-epicatechin

These flavanols and catechin epimers (FW = 290.27 g/mol; CAS 35323-91-2) are natural antioxidants and free radical scavengers. They also chelate iron and activate survival genes and cell signaling pathways. Note that the term (+)-epicatechin has also been incorrectly used to refer to the (2*S*,3*R*)-isomer which, while being a diastereoisomer, is not an epimer. **Target(s):** catechol *O*-methyltransferase (1-4); fatty-acid synthase (5); H^+/K^+-dependent ATPase (6); lipoxygenase, avocado (22); 5-lipoxygenase, *or* arachidonate 5-lipoxygenase (7); 15 lipoxygenase, arachidonate 15-lipoxygenase (8,9); Moloney murine leukemia virus reverse transcriptase (10); [myosin light-chain] kinase (11); pectate lyase (12-14); peptidyl-dipeptidase A, angiotensin I-converting enzyme, weakly inhibited (15,16); phenol sulfotransferase, *ot* aryl sulfotransferase; both (2*S*,3*S*)-(+)- and (2*R*,3*R*)-(–)-epicatechin inhibit weakly (17); prolyl oligopeptidase, inhibited by the (–)-isomer (18,19); ribonucleoside-diphosphate reductase, inhibited by (–)-epicatechin (20); RNA-directed DNA polymerase (10); xanthine oxidase (21). **1.** Chen, Wang, Lambert, *et al.* (2005) *Biochem. Pharmacol.* **69**, 1523. **2.** Nagai, Conney & Zhu (2004) *Drug Metab. Dispos.* **32**, 497. **3.** Rutherford, Le Trong, Stenkamp & Parson (2008) *J. Mol. Biol.* **380**, 120. **4.**

van Duursen, Sanderson, de Jong, Kraaij & van den Berg (2004) *Toxicol. Sci.* **81**, 316. **5**. Wang, Song, Guo & Tian (2003) *Biochem. Pharmacol.* **66**, 2039. **6**. Murakami, Muramatsu & Otomo (1992) *J. Pharm. Pharmacol.* **44**, 926. **7**. Schneider & Bucar (2005) *Phytother. Res.* **19**, 81. **8**. Schewe, Sadik, Klotz, *et al.* (2001) *Biol. Chem.* **382**, 1687. **9**. Sadik, Sies & Schewe (2003) *Biochem. Pharmacol.* **65**, 773. **10**. Chu, Hsieh & Lin (1992) *J. Nat. Prod.* **55**, 179. **11**. Jinsart, Ternai & Polya (1991) *Biol. Chem. Hoppe-Seyler* **372**, 819. **12**. Wattad, Dinoor & Prusky (1994) *Mol. Plant Microbe Interact.* **7**, 293. **13**. Tardy, Nasser, Robert-Baudouy & Hugouvieux-Cotte-Pattat (1997) *J. Bacteriol.* **179**, 2503. **14**. Pissavin, Robert-Baudouy & Hugouvieux-Cotte-Pattat (1998) *Biochim. Biophys. Acta* **1383**, 188. **15**. Actis-Goretta, Ottaviani & Fraga (2006) *J. Agric. Food Chem.* **54**, 229. **16**. Actis-Goretta, Ottaviani, Keen & Fraga (2003) *FEBS Lett.* **555**, 597. **17**. Harris, Wood, Bottomley, *et al.* (2004) *J. Clin. Endocrinol. Metab.* **89**, 1779. **18**. Fan, Tezuka & Kadota (2000) *Chem. Pharm. Bull.* **48**, 1055. **19**. Kobayashi, Miyase, Sano, *et al.* (2002) *Biol. Pharm. Bull.* **25**, 1049. **20**. Schroeder, Voevodskaya, Klotz, *et al.* (2005) *Biochem. Biophys. Res. Commun.* **326**, 614. **21**. Aucamp, Gaspar, Hara & Apostolides (1997) *Anticancer Res.* **17**, 4381. **22**. Marcus, Prusky & Jacoby (1988) *Phytochemistry* **27**, 323.

24(*R,S*)-25-Epiminolanosterol

This sterol (FW = 441.74 g/mol) inhibits sterol 24-*C*-methyltransferase, *or* cycloartenol 24-*C*-methyltransferase (*Helianthus annus* enzyme, K_i = 3 nM). **1**. Nes, Jansen, Norton, *et al.* (1991) *Biochem. Biophys. Res. Commun.* **177**, 566. **2**. Urbina, Vivas, Lazardi, *et al.* (1996) *Chemotherapy* **42**, 294. **3**. Urbina, Vivas, Visbal & Contreras (1995) *Mol. Biochem. Parasitol.* **73**, 199. **4**. Mangla & Nes (2000) *Bioorg. Med. Chem.* **8**, 925. **5**. Nes (2000) *Biochim. Biophys. Acta* **1529**, 63. **6**. Venkatramesh, Guo, Jia & Nes (1996) *Biochim. Biophys. Acta* **1299**, 313. **7**. Nes, Jayasimha & Song (2008) *Arch. Biochem. Biophys.* **477**, 313. **8**. Zhou, Lepesheva, Waterman & Nes (2006) *J. Biol. Chem.* **281**, 6290. **9**. Nes, Guo & Zhou (1997) *Arch. Biochem. Biophys.* **342**, 68.

(–)-Epinephrine

This catecholamine neurotransmitter and pharmacologic agent (FW$_{free-base}$ = 183.21 g/mol; CAS 51-43-4; White powder; M.P. = 211-212°C; discolors slowly upon exposure to air and light), once called adrenaline and systematically named as (–)-1-(3,4-dihydroxyphenyl)-2-(methylamino) ethanol, is produced from norepinephrine by the action of norepinephrine *N*-methyltransferase (or, phenylethanolamine *N*-methyltransferase). The pK_a of the secondary amine is 8.66 and the value for ionization of the first aromatic hydroxyl is 9.95. The active principle producing the pressor effects (1-4) was isolated by Takamine more than a century ago (2). Epinephrine binds to a variety of adrenergic receptors and is a nonselective agonist of α_1-, α_2-, β_1-, β_2- and β_3-adrenergic receptors, triggering physiologic effects (*e.g.*, increases heart rate, increases heart rate, and strengthens muscle contraction) as well as metabolic changes (*e.g.*, inhibition of insulin secretion, stimulation of glycogenolysis in the liver and muscle, and stimulation of glycolysis in muscle). β-Adrenergic receptor binding triggers glucagon secretion in the pancreas and increases adrenocorticotropic hormone (ACTH) secretion by the pituitary gland. With well-characterized vasoconstrictive effects, adrenaline is the drug of choice for treating anaphylaxis. **Target(s):** acylphosphatase (5); aromatic-L-amino-acid decarboxylase, *or* dopa decarboxylase (6,7); complement component C'3, guinea pig (8); CYP2C9 (9); dihydropteridine reductase (10,11); 4-hydroxyphenylpyruvate dioxygenase (33); lipoxygenase (12); 5-lipoxygenase, *or* arachidonate 5-lipoxygenase (13,14); 12-lipoxygenase, *or* arachidonate 12-lipoxygenase (13,15); 15-lipoxygenase, arachidonate 15-lipoxygenase (13); nitronate monooxygenase (16,17); phenylalanine 4-monooxygenase (18); procollagen-lysine 5-dioxygenase (19,20); procollagen-proline 4-dioxygenase (21); stearoyl-CoA 9-desaturase (22); steroid 11β-monooxygenase (23); tyrosine aminotransferase (24); tyrosine 3-monooxygenase (25-31); urease (32). **1**. Abel (1899) *Zeitschr. Physiol. Chem.* **28**, 318. **2**. Takamine (1901) *J. Soc. Chem. Ind.* **20**, 746. **3**. Aldrich (1901) *Amer. J. Physiol.* **5**, 457. **4**. Abel & R. deM. Taveau (1905) *J. Biol. Chem.* **1**, 1. **5**. Raijman, Grisolia & Edelhoch (1960) *J. Biol. Chem.* **235**, 2340. **6**. Fellman (1959) *Enzymologia* **20**, 366. **7**. Nakazawa, Kumagai & Yamada (1987) *Agric. Biol. Chem.* **51**, 2531. **8**. Shin & Mayer (1968) *Biochemistry* **7**, 3003. **9**. Gervasini, Martinez, Agundez, Garcia-Gamito & Benitez (2001) *Pharmacogenetics* **11**, 29. **10**. Shen (1985) *J. Enzyme Inhib.* **1**, 61. **11**. Shen (1983) *Biochim. Biophys. Acta* **743**, 129. **12**. Holman & Bergström (1951) *The Enzymes*, 1st ed., **2** (part 1), 559. **13**. Pace-Asciak & Smith (1983) *The Enzymes*, 3rd ed., **16**, 543. **14**. Furakawa, Yoshimoto, Ochi & Yamamoto (1984) *Biochim. Biophys. Acta* **795**, 458. **15**. Nugteren (1982) *Meth. Enzymol.* **86**, 49. **16**. Kido, Soda & Asada (1978) *J. Biol. Chem.* **253**, 226. **17**. Kido & Soda (1984) *Arch. Biochem. Biophys.* **234**, 468. **18**. Bloom, Benkovic & Gaffney (1986) *Biochemistry* **25**, 4204. **19**. Murray, Cassell & Pinnell (1977) *Biochim. Biophys. Acta* **481**, 63. **20**. Puistola, Turpeenniemi-Hujanen, Myllylä & Kivirikko (1980) *Biochem. Biophys. Acta* **611**, 51. **21**. Tuderman, Myllylä & Kivirikko (1977) *Eur. J. Biochem.* **80**, 341. **22**. Sreekrishna & Joshi (1980) *Biochim. Biophys. Acta* **619**, 267. **23**. Rosenthal & Narasimhulu (1969) *Meth. Enzymol.* **15**, 596. **24**. Webb (1966) *Enzyme and Metabolic Inhibitors*, vol. **2**, p. 305, Academic Press, New York. **25**. Fujisawa & Okuno (1987) *Meth. Enzymol.* **142**, 63. **26**. Moore & Dominic (1971) *Fed. Proc.* **30**, 859. **27**. Nagatsu, Levitt & Udenfriend (1964) *J. Biol. Chem.* **239**, 2910. **28**. Chaube & Joy (2003) *J. Neuroendocrinol.* **15**, 273. **29**. Fitzpatrick (1988) *J. Biol. Chem.* **263**, 16058. **30**. Neckameyer, Holt & Paradowski (2005) *Biochem. Genet.* **43**, 425. **31**. Gordon, Quinsey, Dunkley & Dickson (2008) *J. Neurochem.* **106**, 1614. **32**. Quastel (1933) *Biochem. J.* **27**, 1116. **33**. Lindstedt & Rundgren (1982) *J. Biol. Chem.* **257**, 11922.

Epirubicin

This photosensitive, semisynthetic antibiotic (FW$_{free-base}$ = 544.53 g/mol; CAS 56420-45-2) is a structural analogue of doxorubicin, with the sugar component the epimer of the component found in doxorubicin. **Target(s):** *Ava*II restriction endonuclease (1); DNA ligase (ATP) (2); DNA polymerase (3); DNA topoisomerase II (4); *Eco*RI restriction endonuclease (1); *Hae*III restriction endonuclease (1); *Hha*I restriction endonuclease (1); *Hpa*II restriction endonuclease (1); phospholipase A_2 (5); *Sma*I restriction endonuclease (1); type II site-specific DNase (1). **1**. Corneo, Pogliani, Biassoni & Tripputi (1988) *Ric. Clin. Lab.* **18**, 19. **2**. Ciarrocchi, Lestingi, Fontana, Spadari & Montecucco (1991) *Biochem. J.* **279**, 141. **3**. Tanaka, Yoshida & Kimura (1983) *Gann* **74**, 829. **4**. Spadari, Pedrali-Noy, Focher, *et al.* (1986) *Anticancer Res.* **6**, 935. **5**. Grataroli, Leonardi, Chautan, Lafont & Nalbone (1993) *Biochem. Pharmacol.* **46**, 349.

Epitestosterone

This steroid (FW = 288.43 g/mol; CAS 481-30-1), also called Δ^4-androsten-17α-ol-3-one and *cis*-testosterone, is the inactive isomer of testosterone, differing from the latter only with respect to the chirality of the hydroxyl group. Because exogenous administration of testosterone has no effect on epitestosterone levels, the measured ratio of testosterone to epitestosterone in urine has been used to detect drug doping in sports. Co-administration of testosterone and epitestosterone can mask doping. **Target(s):** aldehyde dehydrogenase (1); 11β-hydroxysteroid dehydrogenase (2); 17β-hydroxysteroid dehydrogenase (3); 21-hydroxysteroid dehydrogenase, NAD$^+$-requiring (4); steroid C17,20 lyase, *or* cholesterol monooxygenase, *or* side-chain cleaving (5); steroid 17α-monooxygenase (5); UDPglucuronosyltransferase (6). **1.** Douville & Warren (1968) *Biochemistry* **7**, 4052. **2.** Bicikova, Hill, Hampl & Starka (1997) *Horm. Metab. Res.* **29**, 465. **3.** Bicikova, Klak, Hill & Starka (2000) *Horm. Metab. Res.* **32**, 125. **4.** Monder & Furfine (1969) *Meth. Enzymol.* **15**, 667. **5.** Bicíková, Hampl, Hill & Stárka (1993) *J. Steroid Biochem. Mol. Biol.* **46**, 515. **6.** Falany, Green, Swain & Tephly (1986) *Biochem. J.* **238**, 65.

Eplivanserin

This potent and selective serotonin antagonist (FW$_{Free-Base}$ = 327.40 g/mol; FW$_{Hemi-Fumarate}$ = 386.42 g/mol; CAS 130580-02-8; Solubility: 50 mM in DMSO), also named SR-46349, SR-46349B, and (1*Z*,2*E*)-1-(2-2luorophenyl)-3-(4-hydroxyphenyl)-2-propen-1-one *O*-[2-(dimethyl-amino) ethyl]oxime hemifumarate, targets 5-HT$_{2A}$ (IC$_{50}$ = 5.8 nM), 5-HT$_{2B}$ (IC$_{50}$ = 120 nM), and 5-HT$_{2C}$ (IC$_{50}$ >100 nM), attenuating cocaine-induced hyperactivity. Eplivanserin increases dopamine (DA) release in rat medial prefrontal cortex (mPFC) and potentiates haloperidol-induced DA release in rat medial prefrontal cortex and nucleus accumbens. **1.** Bonaccorso, *et al.* (2002) *Neuropsychopharmacol.* **27**, 430. **2.** Filip, *et al.* (2004) *J. Pharmacol. Exp. Ther* **310**, 1246. **3.** Yadav, *et al.* (2011) *J. Pharmacol. Exp. Ther.* **339**, 99.

Epothilone A & B

Epothilone A   Epothilone B

These antitumor agents (FW$_{Epothilone-A}$ = 493.66; FW$_{Epothilone-B}$ = 507.69 g/mol; CAS 152044-53-6) from the myxobacterium *Sporangium cellulosum* inhibit microtubule depolymerization. Both have properties similar to the microtubule-stabilizing and cytoskeletal-disrupting drug paclitaxel (taxol). Epothilones A and B enjoy the advantage of being more water-soluble and can be produced in larger quantities than paclitaxel. The epothilones induce microtubule formation at low temperatures in the absence of GTP and/or microtubule-associated proteins (MAPs). They are also competitive inhibitors with respect to taxol binding sites on microtubules. These agents also inhibit the growth of cells overexpressing the P-glycoprotein efflux pump (multidrug resistance factor) and may eventually be useful for the treatment of multidrug-resistant tumors. *See* Colchicine; Maytansine; Paclitaxel (Taxol); Podophyllotoxin; Rhazinilam; Taxotere; Vinblastine **1.** Kowalski, Giannakakou & Hamel (1997) *J. Biol. Chem.* **272**, 2534. **2.** Bollag, McQueney, Zhu, *et al.* (1995) *Cancer Res.* **55**, 2325.

Epoxomicin

This epoxide-containing peptide (FW = 554.72 g/mol; CAS 134381-21-8; Solubility: 100 mg/mL DMSO), systematically named *N*-acetyl-*N*-methyl-L-isoleucyl-L-isoleucyl-*N*-[(1*S*)-3-methyl-1-[[(2*R*)-2-methyl-2-oxiranyl] carbonyl]butyl]-L-threoninamide, from *Actinomycetes* is a strong, irreversible inhibitor of the chymotrypsin-like, trypsin-like, and peptidylglutamyl peptide-hydrolyzing activities of the proteasome. Epoxomicin covalently binds to the LMP7, X, MECL1, and Z catalytic subunits of the proteasome. In HUVECs treated with 100 nM epoxomicin, there is 30x increase in the levels of p53 protein, a known proteasome target. **1.** Meng, Mohan, Kwok, *et al.* (1999) *Proc. Natl. Acad. Sci. U.S.A.* **96**, 10403. **2.** Sin, Kim, Elofsson, *et al.* (1999) *Bioorg. Med. Chem. Lett.* **9**, 2283. **3.** Vabulas & Hartl (2005) *Science* **310**, 1960. **4.** Wojcikiewicz, Webster, Alzayady & Gao (2003) *J. Biol. Chem.* **278**, 940.

2',3'-Epoxypropyl β-D-Glucopyranoside

This glycoside (FW = 236.22 g/mol) irreversibly inhibits yeast hexokinase, modifying an active-site cysteinyl residue in the process. **Target(s):** glucose transporter, *Escherichia coli* (1) hexokinase (2,3); licheninase (4). **1.** Garcia-Alles, Navdaeva, Haenni & Erni (2002) *Eur. J. Biochem.* **269**, 4969. **2.** Bessell (1973) *Chem. Biol. Interact.* **7**, 343. **3.** Bessell, Thomas & Westwood (1973) *Chem. Biol. Interact.* **7**, 327. **4.** Høj, Rodriguez, Stick & Stone (1989) *J. Biol. Chem.* **264**, 4939.

N-[(2*S*,3*S*)-*trans*-Epoxysuccinyl]-L-leucylamido-3-methylbutane

This cysteine proteinase inhibitor (FW$_{free-acid}$ = 314.38 g/mol), also known as *N*-[*N*-(L-3-transcarboxyoxirane-2-carbonyl)-L-leucyl]-3-methylbutylamine, Ep-475 and E-64c, targets cathepsins B, H, and L as well as calpain. With its low water solubility, this agent is typically prepared in dimethyl sulfoxide stock solutions and then diluted to a final 5% DMS value in enzyme assays. **Target(s):** calpain (1,2); calpain-2, *or* m-calpain (3,4); cathepsin B (5-10); cathepsin L (5-7,9,11-14); cathepsin N (9); cathepsin S (9); lactosylceramide α-2,3-sialyltransferase (15); papain (16-18); prohormone thiol protease (19); pyroglutamyl-peptidase I (20). **1.** Ishiura, Nonaka & Sugita (1981) *J. Biochem.* **90**, 283. **2.** Banik, Hogan, Jenkins, *J. Neurochem. Res.* **8**, 1389. **3.** Inomata, M. Nomoto, Hayashi, *et al.* (1984) *J. Biochem.* **95**, 1661. **4.** Azarian, Schlamp & Williams (1993) *J. Cell Sci.* **105**, 787. **5.** Barrett & Kirschke (1981) *Meth. Enzymol.* **80**, 535. **6.** Seglen (1983) *Meth. Enzymol.* **96**, 737. **7.** Barrett, Kembhavi, Brown, *et al.* (1982) *Biochem. J.* **201**, 189. **8.** Ohshita, Nikawa, Towatari & Katunuma (1992) *Eur. J. Biochem.* **209**, 223. **9.** Maciewicz & Etherington (1988) *Biochem. J.* **256**, 433. **10.** Yamamoto, Tomoo, Matsugi, *et al.* (2002) *Biochim. Biophys. Acta* **1597**, 244. **11.** McDonald & Kadkhodayan (1988) *Biochem. Biophys. Res. Commun.* **151**, 827. **12.** Mason, Green & Barrett (1985) *Biochem. J.* **226**, 233. **13.** Mason, Taylor & Etherington (1984) *Biochem. J.* **217**, 209..

14. McDonald & Kadkhodayan (1988) *Biochem. Biophys. Res. Commun.* **151**, 827. 15. Melkerson-Watson & Sweeley (1991) *Biochem. Biophys. Res. Commun.* **175**, 325. 16. Matsumoto, Yamamoto, Ohishi, *et al.* (1989) *FEBS Lett.* **245**, 177. 17. Yamamoto, Ishida & Inoue (1990) *Biochem. Biophys. Res. Commun.* **171**, 711. 18. Matsumoto, Sumiya, Kitamura & Ishida (1994) *Biochim. Biophys. Acta* **1208**, 268. 19. Azaryan & Hook (1994) *Arch. Biochem. Biophys.* **314**, 171. 20. Tsuru, Sakabe, Yoshimoto & Fujiwara (1982) *J. Pharmacobio-Dyn.* **5**, 859.

EPPTB

This TA$_1$ antagonist/inverse agonist (FW = 378.39 g/mol; CAS 1110781-88-8; Solubility: 100 mM in DMSO), also named *N*-(3-ethoxyphenyl)-4-(1-pyrrolidinyl)-3-(trifluoromethyl)benzamide, targets Trace Amine-associated Receptor-1, with respective IC$_{50}$ values are 27.5 (mouse), 4539 (rat) and 7487 nM (human), blocking TA$_1$ receptor-mediated activation of an inwardly rectifying K$^+$ current. The latter increases the firing frequency of dopamine neurons in the ventral tegmental area. EPPTB also increases dopamine potency at the D$_2$ receptor and displays inverse agonism, reducing basal cAMP levels *in vitro* (IC$_{50}$ = 19 nM). 1. Bradaia, Trube, Stalder, *et al.* (2009) *Proc. Natl. Acad. Sci. U.S.A.* **106**, 20081.

Eprosartan

This angiotensin II receptor antagonist (FW = 520.63 g/mol; CAS 133040-01-4), also known by the trade names Abbvie® and Eprozar® as well as the systematic name 4-({2-butyl-5-[2-carboxy-2-(thiophen-2-ylmethyl)eth-1-en-1-yl]-1*H*-imidazol-1-yl}methyl)benzoate, is an antihypertensive drug. When paired with hydrochlorothiazide, it is marketed as Teveten HCT® and Teveten Plus®. (*See also* Candesartan; Irbesartan; Losartan; Olmesartan; Telmisartan; Valsartan) **Key Pharmacological Parameters:** Bioavailability = 15 %; Food Effect? NO; Drug $t_{1/2}$ = 5–7 hours; Metabolite $t_{1/2}$ = Not Determined; Drug's Protein Binding = 98 %; Metabolite's Protein Binding = <1 %; Route of Elimination = 7 % Renal and 90 % Hepatic. 1. Weinstock, Keenan, Samanen, *et al.* (1991) *J. Med. Chem.* **34**, 1514.

Epsilon Toxin

This aerolysin-like, pore-forming toxin (MW = 28.7 kDa; Symbol: ETX; LD$_{50}$ = 100 ng/kg in mice) from *Clostridium perfringens* Types B and D strains, causes enterotoxemia in ruminants, particularly sheep (LD$_{50}$ = ~70 ng/kg). Such potency makes it a potential bioterrorism agent, now classified as a Category B biological agent by the Centers for Disease Control and Prevention. The 32.9-kDa protoxin is converted to an active toxin through proteolytic cleavage by specific proteases. ETX induces pore formation in eukaryotic cell membranes via detergent-resistant, cholesterol-rich lipid rafts that promote aggregation of ETX monomers into homo-heptamers. Sialidases (*or* neuraminidases) produced by *C. perfringens* enhance binding of the bacterium and ETX to cultured cells. 1. Stiles, Barth, Barth & Popoff (2013) *Toxins* (Basel) **5**, 2138.

Eptastigmine

This carbamate ester (FW = 359.51 g/mol; CAS 101246-68-8), also known as heptylphyso-stigmine, which inhibits acetylcholinesterase (1-8) and butyrylcholinesterase (4-8), with IC$_{50}$ values of 22 and 5 nM, respectively), has been used to treat Alzheimer's disease. 1. Brufani, Marta & Pomponi (1986) *Eur. J. Biochem.* **157**, 115. 2. Ogane, Giacobini & Messamore (1992) *Neurochem. Res.* **17**, 489. 3. Moriearty & Becker (1992) *Meth. Find. Exp. Clin. Pharmacol.* **14**, 615. 4. Iijima, Greig, Garofalo, *et al.* (1992) *Neurosci. Lett.* **144**, 79. 5. Liston, Nielsen, Villalobos, *et al.* (2004) *Eur. J. Pharmacol.* **486**, 9. 6. Costagli & Galli (1998) *Biochem. Pharmacol.* **55**, 1733. 7. Luo, Yu, Zhan, *et al.* (2005) *J. Med. Chem.* **48**, 986. 8. Giacobini (2003) *Neurochem. Res.* **28**, 515.

Eptifibatide

This glycoprotein IIb/IIIa (GPIIb/IIIa) antiplatelet drug (FW = 831.96 g/mol; CAS 188627-80-7), also known as Integrilin, is a cyclic heptapeptide from a GP IIb-IIIa inhibitory protein (*See Barbourin*). Eptifibatide is a so-called argininyl-glycyl-aspartyl (RGD) peptidomimetic that binds reversibly to platelet integrins, IC$_{50}$ = 0.28-0.34 µg/mL (1). In hirudinized blood, IC$_{50}$ values for eptifibatide were 1.5x to 3x higher than those obtained with citrated plasma. Concentration-dependent effect of glycoprotein IIb/IIIa antagonists on shear-induced platelet adhesion showed marked differences in potencies: IC$_{50}$ values of 34, 35, 91, 438, and 606 nM for DPC802, roxifiban, sibrafiban, lotrafiban, and orbofiban, respectively, and IC$_{50}$ values of 43, 430, and 5781 nM for abciximab, tirofiban, and eptifibatide, respectively (2). **Mode of Inhibitory Action:** In unstable angina, acute myocardial infarction, and ischemia after coronary angioplasty, one hallmark is thrombogenic platelet aggregation, the latter resulting when plaque rupture exposes subendothelial components (such as collagen and fibronectin). This event allows platelets to bind to fibrinogen via the glycoprotein IIb/IIIa integrin receptor, initating platelet activation and aggregation as well as coagulation. Glycoprotein IIb-IIIa shows a narrow tissue distribution in healthy subjects, where it is only found on platelets and platelet progenitors. Inhibition of glycoprotein IIb/IIIa integrin receptor function has thus emerged as a promising druggable target for managing acute ischemic coronary syndromes and acute ischemic complications of percutaneous coronary interventions. The venom of the southeastern pigmy rattlesnake (*Sistrurus miliarius barbouri*) was the only of 52 venoms specific for GPIIb-IIIa *versus* other integrins (3). Barbourin is highly homologous to other peptides of the viper venom GPIIb-IIIa antagonist family, but contains a KGD in place of the canonical RGD sequence required to block receptor function. 1. Harder, Klinkhardt, Graff, *et al.* (2000) *Thromb. Res.* **102**, 39. 2. Wang, Dorsam, Lauver, *et al.* (2000) *J. Pharmacol. Expmtl. Therap.* **303**, 38513. 3. Scarborough, Rose, His, *et al.* (1991) *J. Biol. Chem.* **266**, 9359.

EPZ-5676

This is an *S*-adenosyl methionine analogue (FW$_{\text{free-base}}$ = 529.73 g/mol; CAS 1380288-87-8; Solubility: 100 mg/mL DMSO; <1 mg/mL H$_2$O), also known as 9-[5-deoxy-5-[[cis-3-[2-[6-(1,1-dimethylethyl)-1*H*-benzimidazol-2-yl]ethyl]cyclobutyl](1-methylethyl)amino]-β-D-ribofuranosyl]-9H-Purin-6-amine, competitively inhibits the protein methyltransferase DOT1L (K_i = 80 pM), showing a greater than 37,000x selectivity versus all other PMTs tested, inhibits H3K79 methylation in tumors.

EPZ015866, See GSK591

Equanil, See Meprobamate

ER-30306, See Ravuconazole

Erastin

This cytotoxic ferroptosis activator (FW = 547.05 g/mol; CAS Rgistry Number = 571203-78-6; Solubility: 19 mg/mL DMSO), also named 2-[1-[4-[2-(4-chlorophenoxy)acetyl]-1-piperazinyl]-ethyl]-3-(2-ethoxyphenyl)-4(3*H*)-quinazolinone, induces a non-apoptotic death pathway in Ras-overexpressing cells, as evidenced by the absence of fragmentation or margination of chromatin (1,2). Ferroptosis depends upon intracellular iron, but not other metals, and is morphologically, biochemically, and genetically distinct from apoptosis, necrosis, and autophagy. Erastin-treated cells do not show DNA laddering, a hallmark of apoptotis that occurs when DNA is cleaved at regular intervals. Annexin V staining also shows no evidence of apotosis-associated accumulation of phosphatidylserine on the outer leaflet of the peripheral membrane. Most telling is the finding that caspase 3, a final executioner of apoptosis, remains in its inactivated, pro-caspase form, and the pan-caspase inhibitor BOC-D-fmk does not block erastin-induced cell death. Such findings indicate that erastin is a Ras-selective lethal compound operating by a novel mechanism for killing cells (2). Later work (3) showed that erastin acts through mitochondrial voltage-dependent anion channels (VDACs), thus representing a novel target for anti-cancer drug action. Erastin treatment of cells harboring oncogenic RAS causes cell death through an oxidative, non-apoptotic mechanism. RNA-interference-mediated knockdown of VDAC2 or VDAC3 brought on resistance to erastin, implicating these VDAC isoforms in the mechanism of erastin-mediated cell death (3). A radiolabelled erastin analogue also binds directly to VDAC2 (3). Erastin enhances cisplatin-based killing of lung cancer cells earing wild-type EGFR (4). Note that ferrostatin prevents erastin-induced ferroptosis in cancer cells, as well as glutamate-induced cell death in postnatal rat brain slices (*See Ferrostatin*). **1**. Dolma, Lessnick, Hahn & Stockwell (2003) *Cancer Cell* **3**, 285. **2**. Dixon, Lemberg, Lamprecht, *et al.* (2012) *Cell* **149**, 1060. **3**. Yagoda, von Rechenberg, Zaganjor, *et al.* (2007) *Nature* **447**, 864. **4**. Yamaguchi, Hsu, Chen, *et al.* (2013) *Clin. Cancer Res.* **19**, 845.

Eravacycline

This broad-spectrum intravenous and oral antibiotic (FW = 542.56 g/mol), also known as TP-434 and systematically as 7-fluoro-9-pyrrolidin-oacetamido-6-demethyl-6-deoxytetracycline, is a first-line empiric monotherapy for the treatment of multi-drug resistant (MDR) infections, including MDR Gram-negative bacteria. Eravacycline is a novel, fully synthetic tetracycline antibiotic exhibiting potent antibacterial activity against a broad spectrum of susceptible and multi-drug resistant bacteria, including Gram-negative, Gram-positive, atypical and anaerobic bacteria, with potential to treat the majority of patients as a first-line empiric monotherapy with convenient dosing, as well as the potential for intravenous-to-oral step-down therapy. In *in vitro* studies, eravacycline is highly effective against *Acinetobacter baumannii*, various well as clinically important *Enterobacteriaceae* species (including those resistant to the carbapenem class of antibiotics), and pathogenic anaerobes. **1**. Sutcliffe, O'Brien, Fyfe & Grossman (2013) *Antimicrob. Agents Chemother.* **57**, 5548.

Ercalciol

This photosensitive vitamin (FW = 396.66 g/mol; CAS 50-14-6), also known as ergocalciferol, calciferol, and vitamin D$_2$, is a secosteroid with vitamin D activity. It is metabolized to ercalcidiol and ercalcitriol, the latter the main biologically active form of the vitamin. Ercalciol is produced from ergosterol upon UV irradiation, a common practice in fortifying the vitamin D$_2$ content of cow's milk. **Target(s):** arylsulfatase A, *or* cerebroside-sulfatase (1); arylsulfatase B, *or*N-acetylgalactosamine-4-sulfatase (1); DNA-directed DNA polymerase (2); DNA polymerase α (2). **1**. Bleszynski & Leznicki (1967) *Enzymologia* **33**, 373. **2**. Mizushina, Xu, Murakami, *et al.* (2003) *J. Pharmacol. Sci.* **92**, 283.

Ergosterol

This mycosterol (FW = 396.66 g/mol; CAS 57-87-4; IUPAC: ergosta-5,7,22-trien-3β-ol), first isolated from rye in 1889, is converted to vitamin D_2 (ercalciol), via lumisterol and tachysterol, upon exposure to bright ultraviolet light. Because fungi require ergosterol to survive, its biosynthesis is a druggable target. Miconazole, itraconazole, and clotrimazole, for example, inhibit ergosterol formation from lanosterol, whereas the antifungal drug amphotericin B binds to ergosterol within fungal membranes, creating a polar pore that allows K^+ and H^+ to leak out. Large-scale cultivation of fungi yields ergosterol for photoconversion to Vitamin D used in dietary supplements. **Target(s):** alcohol O-acetyltransferase (1); cyclooxygenase (2,3); 3-hydroxy-3-methylglutaryl CoA reductase (4); sterol acyltransferase, or cholesterol acyltransferase (5); sterol 24-C methyltransferase, or cycloartenol 24-C-methyltransferase, K_i = 80 μM (6-8); sterol Δ^{24}-reductase (10). **1.** Yoshioka & Hashimoto (1981) *Agric. Biol. Chem.* **45**, 2183. **2.** Zhang, Mills & Nair (2002) *J. Agric. Food Chem.* **50**, 7581. **3.** Zhang, Mills & Nair (2003) *Phytomedicine* **10**, 386. **4.** Bard & Downing (1981) *J. Gen. Microbiol.* **125**, 415. **5.** T. Billheimer (1985) *Meth. Enzymol.* **111**, 286. **6.** Mangla & Nes (2000) *Bioorg. Med. Chem.* **8**, 925. **7.** Nes, Jayasimha & Song (2008) *Arch. Biochem. Biophys.* **477**, 313. **8.** Zhou, Lepesheva, Waterman & Nes (2006) *J. Biol. Chem.* **281**, 6290. **9.** Nes (2005) *Biochem. Soc. Trans.* **33**, 1189. **10.** Fernandez, Suarez, Ferruelo, Gomez-Coronado & Lasuncion (2002) *Biochem. J.* **366**, 109.

Ergotamine

This vasoconstricting indole alkaloid (FW = 581.67 g/mol) from ergot, a fungal parasite (*Claviceps purpurea*) of rye plants, is a serotonin 5-HT$_{1A}$, 5-HT$_{1B}$, 5-HT$_{1D}$, and 5-HT$_{1F}$ receptor agonist as well as act at dopamine D_2 receptors. It is also an α-adrenergic receptor antagonist, while acting as an agonist at low concentrations. Poisoning occurs from the consumption of rye bread contaminated with *C. purpurea*. The worst recorded case occurred in the Rhine Valley in 857 AD, when thousands perished. Due to its remarkable uterotonic and vasoconstrictor effects, ergotamine has also been used to induce childbirth and to control post-partum hemorrhage. Ergotamine hydrolysis produces lysergic acid. **Target(s):** acetylcholinesterase (1); α-adrenergic receptor (2); calmodulin-stimulated cyclic-nucleotide phosphodiesterase (2). **1.** Riedel, Kyriakopoulos & Nundel (1981) *Arzneimittelforschung* **31**, 1387. **2.** Earl, Prozialeck & Weiss (1984) *Life Sci.* **35**, 525.

Ergothioneine

L-Ergothioneine

This histidine derivative (FW = 229.30 g/mol; λ_{max} = 258 nm; ε = 16000 M^{-1}cm^{-1} at pH = 2–9; M.P. (decomposes) = 256-257°C; Solubility: 0.04 g/mL H_2O; Soluble in hot methanol, hot ethanol, and in acetone; tightly binds Zn^{2+} and Cu^{2+}) is a more effective reducing reagent than cysteine or glutathione, providing protection against UV damage and free radical damage. Many Gram-positive bacteria, including *Mycobacterium tuberculosis*, lack glutathione and instead produce mycothiol, bacillithiol and ergothioneine (ESH) as their major low-molecular-weight thiols. Humans do not synthesize ergothioneine, but possess the cation transporter (OCTN1) that shows high specificity for ESH uptake from dietary sources. **Target(s):** glutathione S-transferase P1 (1-2); polyphenol oxidase (3). **1.** van den Broeke & Beyersbergen van Henegouwen (1993) *J. Photochem. Photobiol. B* **17**, 279. **2.** Hayeshi, Mukanganyama, Hazra, Abegaz & Hasler (2004) *Phytother. Res.* **18**, 877. **3.** Hanlon (1971) *J. Med. Chem.* **14**, 1084.

Eribulin

This synthetic halichondrin B analogue and microtubule-directed anticancer agent (FW = 729.91 g/mol; CAS 253128-41-5), also known* by its code name E7389 and trade name Halaven®, is an FDA-approved treatment for patients with metastatic breast cancer who previously received at least two chemotherapeutic regimens for metastatic disease. (**See** *Halichondrin B*) that binds to tubulin with a sub-nM dissociation constant, suppressing microtubule (MT) dynamics, both *in vitro* and in cells, and operating by a unique mechanism that suppresses the growth phase of MT dynamic instability without suppressing the shortening phase (1). Suppression of MT dynamics in interphase cells occurs at eribulin concentrations that arrest mitosis and lead to apoptosis (2-4). Eribulin, paclitaxel, and Vinca alkaloids bind to tubulin and/or MTs in characteristically different ways. Paclitaxel binds along the inner surface of the microtubule, with maximal binding of one paclitaxel molecule per tubulin heterodimer. Vinblastine and eribulin bind with high affinity to a very small number of saturable sites on tubulin positioned at MT ends. The *in vitro* effects of these drugs also vary. Binding of 1-2 molecules of vinblastine to a microtubule end is sufficient to suppress MT treadmilling dynamics by 50%, whereas paclitaxel occupancy must be much higher to suppress microtubule dynamic instability to a similar. Yet, despite their chemical structural differences and their different binding modes on MTs, low concentrations of these drugs arrest mitosis by suppressing microtubule dynamics (5). Differences in the MT binding of eribulin, *Vinca* alkaloids, and paclitaxel also suggest that it is unlikely that the drugs all sterically block interaction with the kinetochore. Yet, they all suppress the stretching and relaxation movements between sister centromeres at concentrations that arrest mitosis, indicating that suppression of spindle MT dynamics on the same crucially important step in mitosis (5). Intravital microscopy with fluorescently labeled eribulin demonstrates that MDR1 drives eribulin resistance, even upon I.V. administration of HM30181, a third-generation MDR1 inhibitor (**See** *HM30181*) (6). When encapsulated within a nanoparticle delivery system, HM30181 reverses the multidrug-resistant phenotype, potentiating eribulin effects, both *in vitro* and *in vivo* in mice (6). (**See also**, *Colchicine, Paclitaxel, Vinblastine, Vincristine, Nocodazole*) **1.** Towle, Salvato, Budrow (2001) *Cancer Res.* **61**, 1013. **2.** Jordan, Kamath, Manna, *et al.* (2005) *Mol. Cancer Ther.* **4**, 1086. **3.** Dabydeen, Burnett, Bai, *et al.* (2006) *Mol. Pharmacol.* **70**, 1866. **4.** Kuznetsov, Towle, Cheng, *et al.* (2004) *Cancer Res.* **64**, 5760. **5.** Okouneva, Azarenko, Wilson, *et al.* (2008) *Mol. Cancer Ther.* **7**, 2003. **6.** Laughney, Kim, Sprachman, *et al.* (2014) Sci. Transl. Med. **6**, 261.

Eriodictyol

This flavanone (FW = 288.26 g/mol; CAS 552-58-9), also named (2S)-2-(3,4-dihydroxyphenyl)-5,7-dihydroxy-4-chromanone, from lemons and the native American plant Yerba Santa (*Eriodictyon californicum*) inhibits: Ca^{2+}-transporting ATPase (1), 3',5'-cyclic-nucleotide phosphodiesterase 3 (2), β-glucurinidase (3), glutathione S-transferase P (1-4), hyaluronidase (3), leucoanthocyanidin reductase (5), lysozyme (3); naringenin-chalcone synthase (6-8), tyrosinase (3), and xanthine oxidase (9-11). **1.** Thiyagarajah, Kuttan, Lim, Teo & Das (1991) *Biochem. Pharmacol.* **41**, 669. **2.** Ko, Shih, Lai, Chen & Huang (2004) *Biochem. Pharmacol.* **68**, 2087. **3.** Rodney, Swanson, Wheeler, Smith & Worrel (1950) *J. Biol. Chem.* **183**, 739. **4.** van Zanden, Geraets, Wortelboer, *et al.* (2004) *Biochem. Pharmacol.* **67**, 1607. **5.** Tanner, Francki, Abrahams, *et al.* (2003) *J. Biol. Chem.* **278**, 31647. **6.** Hinderer & Seitz (1985) *Arch. Biochem. Biophys.* **240**, 265. **7.** Peters, Schneider-Poetsch, Schwarz & Weissenböck (1988) *J. Plant Physiol.* **133**, 178. **8.** Saleh, Fritsch, Kreuzaler & Grisebach (1978) *Phytochemistry* **17**, 183. **9.** Nguyen, Awale, Tezuka, *et al.* (2004) *Biol. Pharm. Bull.* **27**, 1414. **10.** Nguyen, Awale, Tezuka, *et al.* (2006) *Planta Med.* **72**, 46. **11.** Dew, Day & Morgan (2005) *J. Agric. Food Chem.* **53**, 6510.

D-Eritadenine

This substituted purine ($FW_{free-acid}$ = 253.22 g/mol), also known as 2(*R*),3(*R*)-dihydroxy-4-(9-adenyl)butyric acid and (α*R*,β*R*)-6-amino-α,β-dihydroxy-9*H*-purine-9-butanoic acid, from the shiitake mushroom (*Lentinus edodes*) is hypocholesterolemic agent with wide-ranging effects, causing an increase in the liver phosphatidylethanolamine concentrations and a decrease in liver Δ^6-desaturase activity. **Target(s):** adenosylhomocysteinase, *or* S-adenosyl-homocysteine hydrolase (1,2); phosphatase (3). **1.** Huang, Komoto, Takata, *et al.* (2002) *J. Biol. Chem.* **277**, 7477. **2.** Schanche, Schanche, Ueland, Holy & Votruba (1984) *Mol. Pharmacol.* **26**, 553. **3.** Nemec & K. Slama (1989) *Experientia* **45**, 148.

Erlotinib

This HER1/EGFR inhibitor ($FW_{hydrochloride}$ = 429.90 g/mol; CASs = 183321-74-6 (free base), 183319-69-9 (hydrochloride), 248594-19-6 (methane sulfonate salt); Solubility (25°C): 3 mg/mL DMSO, <1 mg/mL H_2O), also known as CP-358774, OSI-774, NSC 718781, Tarceva®, and by systematically as *N*-(3-ethynylphenyl)-6,7-bis(2-methoxyethoxy) quinazolin-4-amine, blocks EGFR tyrosine kinase (IC_{50} = 2 nM) and reduces EGFR autophosphorylation in intact tumor cells (IC_{50} = 20 nM). This inhibition is selective for EGFR tyrosine kinase relative to other tyrosine kinases, both in assays of isolated kinases and whole cells. EGFR is frequently amplified and/or mutated in human tumors, and abnormal signaling from this receptor is believed to contribute to the malignant phenotype. At doses of 100 mg/kg, Erlotinib completely prevents EGF-induced autophosphorylation of EGFR in human HN5 tumors growing as xenografts in athymic mice and of the hepatic EGFR of the treated mice (1). Rapamycin synergizes with the epidermal growth factor receptor inhibitor erlotinib in non-small-cell lung, pancreatic, colon, and breast tumors (2). Erlotinib resistance typically occurs 8–12 months from commencing treatment, with over 50% of the cases caused by a mutation in the ATP binding pocket of the EGFR kinase domain. This Thr-790-Met mutation increases ATP binding affinity, thereby reducing the inhibitory effect of erlotinib (3). Such findiungs suggest that the Thr-790-Met mutation is a "generic" resistance mutation that would be expeted to reduce the potency of any ATP-competitive kinase inhibitor and that irreversible inhibitors overcome this resistance by covalent binding, not as a result of an alternative binding mode (3). The Hsp90 inhibitor CH5164840 further enhances the cytotoxic effects of erlotinib against non-small cell lung cancer (NSCLC) tumors, even those with EGFR overexpression and mutations (4). **Key Pharmacokinetic Parameters:** *See* Appendix II in Goodman & Gilman's THE PHARMACOLOGICAL BASIS OF THERAPEUTICS, 12th Edition (Brunton, Chabner & Knollmann, eds.) McGraw-Hill Medical, New York (2011). **1.** Moyer, Barbacci, Iwata, *et al.* (1997) **57**, 4838. **2.** Buck, Eyzaguirre, Brown, *et al.* (2006) **5**, 2676. **3.** Yun, Mengwasser, Toms, *et al.* (2008) *Proc. Natl. Acad. Sci. U.S.A.* **105**, 2070. **4.** Ono, Yamazaki, Tsukaguchi, *et al.* (2013) *Cancer Sci.* **104**, 1346.

Erythritol

D-isomer   **L-isomer**

This naturally occurring sugar alcohol (FW = 122.12 g/mol; CAS 149-32-6), also called *meso*-1,2,3,4-tetrahydroxybutane, has nearly twice the sweetness of sucrose. With pentaerythritol tetranitrate one of the most powerful explosives known, industrial-scale production uses *Torula corallina*, an organism that offers the highest erythritol yield. **Target(s):** aldose 1-epimerase (1); α-amylase (2); amylo-α-1,6-glucosidase/4-α glucanotransferase, *or* glycogen debranching enzyme (3); L-arabinose isomerase (4); β-galactosidase (5); glucan 1,3-β-glucosidase (6); α-glucosidase (5,7-10); β-glucosidase (5); glycerol kinase (11). **1.** Bailey, Fishman, Kusiak, Mulhern & Pentchev (1975) *Meth. Enzymol.* **41**, 471. **2.** Ali & Abdel-Moneim (1989) *Zentralbl. Mikrobiol.* **144**, 615. **3.** Gillard & Nelson (1977) *Biochemistry* **16**, 3978. **4.** Kim & Oh (2005) *J. Biotechnol.* **120**, 162. **5.** Kelemen & Whelan (1966) *Arch. Biochem. Biophys.* **117**, 423. **6.** Talbot & Vacquier (1982) *J. Biol. Chem.* **257**, 742. **7.** Jorgensen & Jorgensen (1967) *Biochim. Biophys. Acta* **146**, 167. **8.** Silva & Terra (1995) *Insect Biochem. Mol. Biol.* **25**, 487. **9.** Kanaya, Chiba, Shimomura & Nishi (1976) *Agric. Biol. Chem.* **40**, 1929. **10.** Takahashi, Shimomura & Chiba (1971) *Agric. Biol. Chem.* **35**, 2015. **11.** Eisenthal, Harrison & Lloyd (1974) *Biochem. J.* **141**, 305.

Erythritol 4-Phosphate

D-isomer   **L-isomer**

This phosphorylated sugar alcohol ($FW_{free-acid}$ = 202.1 g/mol; CAS 7183-41-7), also known as L-erythritol 1-phosphate, inhibits glucose-6-phosphate isomerase (1,2). D-Erythritol 4-P is a product of the reaction catalyzed by erythritol kinase. *Brucella abortus* erythritol kinase produces the L-isomer, also known as D-erythritol 1-phosphate (3). **Target(s):** glucose-6-phosphate isomerase (1,2); 2-*C*-methyl-D-erythritol-4-phosphate cytidylyltransferase, weakly inhibited by L-erythritol 4-phosphate (3). **1.** Noltmann (1972) *The Enzymes*, 3rd ed., **6**, 271. **2.** Chirgwin, Parsons & Noltmann (1975) *J. Biol. Chem.* **250**, 7277. **3.** Lillo, Tetzlaff, Sangari & Cane (2003) *Bioorg. Med. Chem. Lett.* **13**, 737.

Erythromycin

Erythromycin A

This class of related macrolide antibiotics, first obtained from *Streptomyces erythreus* (now *Saccharopolyspora erythraea*) by J. M. McGuire in 1949 at Eli Lilly Labs, are fermentation products: designated A, B, and C. Erythromycin A is the predominant form. All three inhibit prokaryotic protein biosynthesis by binding the 50S ribosomal subunit and inhibiting translocation (1-3). Erythromycin binds to ribosomal protein L15 in solution, and L16 appears to participate in erythromycin binding to ribosomes. They reportedly stimulate the dissociation of peptidyl-tRNA from the ribosome. Erythromycin A (FW = 733.94 g/mol) is soluble in water (approximately 2 mg/mL) consists of a large lactone glycosidically linked to two sugars, β-D-desosamine and α-L-cladinose (or, α-L-3-*O*-methylmycarose). Erythromycin B is virtually identical, with one hydroxyl group on the lactone ring and erythromycin C is identical to A, and mycarose replaces cladinose. Erythromycin A loses its biological activity at pH < 5. It absorbs light slightly in the ultraviolet: λ_{max} = 280 nm at pH 6.3 (ε = 50 $M^{-1}cm^{-1}$). The asymmetric synthesis of erythromycin A was accomplished in R. B. Woodward's laboratory (4). **Mode of Antibiotic Binding:** Erythromycin binds to *Escherichia coli* ribosomes in a two-step process (5). The first binding step (K_1 = 83 nM) is established rapidly, occurring at a low-affinity binding site situated at the entrance of the exit tunnel in the large ribosomal subunit, where macrolides bind mainly through hydrophobic portions. In the second step, a slow conformational change occurs as the antibiotic hydrophilic portion burrows deeper into the tunnel, attaining its higher-affinity state (k_{on} = 0.39 min^{-1}; k_{off} = 0.054 min^{-1}; k_{on}/k_{off} = 7.3; $K_d{}^*$ = $K_1 \times k_{on}/k_{off} \approx$ 10 nM). Polyamines attenuate erythromycin affinity for the ribosome at both steps in the sequential binding mechanism (5). **Key Pharmacokinetic Parameters:** *See* Appendix II in Goodman & Gilman's *THE PHARMACOLOGICAL BASIS OF THERAPEUTICS*, 12th Edition (Brunton, Chabner & Knollmann, eds.) McGraw-Hill Medical, New York (2011). **Target(s):** CYP3A4 (6); peptidyltransferase (7); protein biosynthesis (1-5,7-9). **1**. Pestka (1974) *Meth. Enzymol.* **30**, 261. **2**. Jiménez (1976) *Trends Biochem. Sci.* **1**, 28. **3**. Wilhelm & Corcoran (1967) *Biochemistry* **6**, 2578. **4**. Woodward, Logusch, Nambiar, *et al.* (1981) *J. Amer. Chem. Soc.* **103**, 3215. **5**. Petropoulos, Kouvela, Dinos & Kalpaxis (2008) *J. Biol. Chem.* **283**, 4756. **6**. Yamazaki & Shimada (1998) *Drug Metab. Dispos.* **26**, 1053. **7**. Rychlik (1966) *Biochim. Biophys. Acta* **114**, 425. **8**. Lucas-Lenard & Beres (1974) *The Enzymes*, 3rd ed., **10**, 53. **9**. Brock & Brock (1959) *Biochim. Biophys. Acta* **33**, 274.

D-Erythrose 4-Phosphate

$$
\begin{array}{c}
\text{CHO} \\
|\\
\text{HC—OH} \\
|\\
\text{HC—OH} \\
|\\
\text{CH}_2\text{OPO}_3\text{H}^-
\end{array}
$$

This phosphorylated tetrose (FW_{free-acid} = 200.08 g/mol; CAS 103302-15-4) is an intermediate in carbohydrate metabolism and is found in small quantities in mammalian muscle. It is more stable in acidic solutions than at neutral pH: neutral solutions should be kept cold and used immediately. It is an alternative substrate for ribulose-phosphate 3-epimerase. **Target(s):** acetyl-CoA synthetase (1); aspartate aminotransferase (2); fructose-1,6-bisphosphate aldolase (3); glucose-6-phosphate isomerase (4-15); glyceraldehyde-3 phosphate dehydrogenase (16); glycogen synthase, microsomal (17); *myo*-inositol-1-phosphate synthase (18); mannose-6-phosphate isomerase (5,12,13,19,20); phosphorylase phosphatase (17); orotate phosphoribosyltransferase, weakly inhibited (21); 6-phosphogluconate dehydrogenase (22); phosphoketolase (23); ribose-5-phosphate isomerase (24,25); triose-phosphate isomerase (26). **1**. O'Sullivan & Ettlinger (1976) *Biochim. Biophys. Acta* **450**, 410. **2**. Kopelovich, Sweetman & Nisselbaum (1972) *J. Biol. Chem.* **247**, 3262. **3**. Bais, James, Rofe & Conyers (1985) *Biochem. J.* **230**, 53. **4**. Gracy & Tilley (1975) *Meth. Enzymol.* **41**, 392. **5**. Noltmann (1972) *The Enzymes*, 3rd ed., **6**, 271. **6**. Webb (1966) *Enzyme and Metabolic Inhibitors*, vol. 2, p. 407, Academic Press, New York. **7**. Grazi, de Flora & Pontremoli (1960) *Biochem. Biophys. Res. Commun.* **2**, 121. **8**. Howell & Schray (1981) *Mol. Cell Biochem.* **37**, 101. **9**. Pradhan & Nadkarni (1980) *Biochim. Biophys. Acta* **615**, 474. **10**. Chirgwin, Parsons & Noltmann (1975) *J. Biol. Chem.* **250**, 7277. **11**. Backhausen, Jöstingmeyer & Scheibe (1997) *Plant Sci.* **130**, 121. **12**. Lee & Matheson (1984) *Phytochemistry* **23**, 983. **13**. Hansen, Wendorff & Schonheit (2004) *J. Biol. Chem.* **279**, 2262. **14**. Hansen, Schlichting, Felgendreher & Schonheit (2005) *J. Bacteriol.* **187**, 1621. **15**. Thomas (1981) *J. Gen. Microbiol.* **124**, 403. **16**. Iglesias, Serrano, Guerrero &

Losada (1987) *Biochim. Biophys. Acta* **925**, 1. **17**. Newman & Curnow (1985) *Mol. Cell. Biochem.* **66**, 151. **18**. Charalampous & Chen (1966) *Meth. Enzymol.* **9**, 698. **19**. Proudfoot, Payton & Wells (1994) *J. Protein Chem.* **13**, 619. **20**. Yu & Leary (2005) *Anal. Chem.* **77**, 5596. **21**. Victor, Greenberg & Sloan (1979) *J. Biol. Chem.* **254**, 2647. **22**. Moritz, Striegel, De Graaf & Sahm (2000) *Eur. J. Biochem.* **267**, 3442. **23**. Goldberg, Fessenden & Racker (1966) *Meth. Enzymol.* **9**, 515. **24**. Skrukrud, Gordon, Dorwin, *et al.* (1991) *Plant Physiol.* **97**, 730. **25**. Woodruff & Wolfenden (1979) *J. Biol. Chem.* **254**, 5866. **26**. Krietsch (1975) *Meth. Enzymol.* **41**, 438.

Erythrosine

This dye and color additive (FW_{disodium-salt} = 879.86 g/mol; CAS 49746-10-3), also called 2',4',5',7'-tetraiodofluorescein and erythrosin B, inhibits a number of antiport systems. A brown powder that produces a cherry-red solution in water, erythrosine has a λ_{max} = 524 nm. Frequently used as a stain for nerve cells and soil bacteria, erythrosine is also fluorescent. **Target(s):** acetylcholinesterase (1); AMP deaminase (2); anion channel, mitochondrial inner membrane (3); apyrase (4); Ca²⁺-transporting ATPase (5-9); cholinesterase (1); chloride transport (10); dopamine sulfotransferase (11); fumarase (12); galactosylceramide sulfotransferase (13); H⁺-ATPase (14,15); Na⁺/K⁺ exchanging ATPase (16); respiration, mitochondrial (17). **1**. Osman, Sharaf, el-Rehim & el-Sharkawi (2002) *Brit. J. Biomed. Sci.* **59**, 212. **2**. Yoshino & Kawamura (1978) *Biochim. Biophys. Acta* **526**, 640. **3**. Beavis & Davatol-Hag (1996) *J. Bioenerg. Biomembr.* **28**, 207. **4**. Komoszynski (1996) *Comp. Biochem. Physiol.* **B 113**, 581. **5**. Mignaco, Barrabin & Scofano (1997) *Biochim. Biophys. Acta* **1321**, 252. **6**. Mignaco, Lupi, Santos, Barrabin & Scofano (1996) *Biochemistry* **35**, 3886. **7**. Heffron, O'Callaghan & Duggan (1984) *Biochem. Int.* **9**, 557. **8**. Mugica, Rega & Garrahan (1984) *Acta Physiol. Pharmacol. Latinoam.* **34**, 16. **9**. Moreau, Daoud, Masse, Simoneau & Lafond (2003) *Mol. Reprod. Dev.* **65**, 283. **10**. Knauf, Strong, Penikas, Wheeler, Jr., & S. Q. Liu (1993) *Amer. J. Physiol.* **264**, C1144. **11**. Bamforth, Jones, Roberts & Coughtrie (1993) *Biochem. Pharmacol.* **46**, 1713. **12**. Quastel (1931) *Biochem. J.* **25**, 898. **13**. Zaruba, Hilt & Tennekoon (1985) *Biochem. Biophys. Res. Commun.* **129**, 522. **14**. Supply, Wach & Goffeau (1993) *J. Biol. Chem.* **268**, 19753. **15**. Wach & Graber (1991) *Eur. J. Biochem.* **201**, 91. **16**. Swann (1982) *Biochem. Pharmacol.* **31**, 2185. **17**. Reyes, Valim & Vercesi (1996) *Food Addit. Contam.* **13**, 5.

Escitalopram

This highly selective serotonin reuptake inhibitor, *or* SSRI (FW = 324.39 g/mol; CAS 128196-01-0), also known by the code name Lu-10-171, by its tradenames Lexapro® and Cipralex®, and as (*S*)-1-[3-(dimethylamino)propyl]-1-(4-fluorophenyl)-1,3-dihydro-isobenzofuran-5-carbonitrile, is an FDA-approved antidepressant for treating major depressive disorder and generalized anxiety disorder in adults. In humans, escitalopram reaches peak concentrations within 2-4 hours after the daily dose, and is eliminated slowly ($t_{1/2}$ = 30 hours). Long-term treatment with escitalopram does not induce changes in neurotransmitter receptors as seen with most tricyclic as well as newer atypical antidepressants. **Other Target(s):** Citalopram is an extremely potent inhibitor of neuronal serotonin (5-HT) uptake, but has no effect on the uptake of noradrenaline (NA) and dopamine (1). Citalopram has no antagonistic activity towards dopamine, NA, 5-HT, histamine, GABA, acetylcholine, and morphine receptors. In this way, it clearly deviates from many old and new antidepressant drugs exerting antagonistic effects towards some of these transmitters. Escitalopram blocks Human Ether-a-go-go-Related Gene (hERG) currents (IC_{50} = 2.6 µM for escitalopram; IC_{50} = 3.2 µM for citalopram) at supratherapeutic concentrations by preferentially binding to both the open and the inactivated states of the channels and by inhibiting the trafficking of hERG channel protein to the plasma membrane. The blocking of hERG by escitalopram was voltage-dependent, with a steep increase across the voltage range of channel activation. The blocking by escitalopram was frequency dependent. Rapid application of escitalopram induced a rapid and reversible blocking of the tail current of hERG. The extent of the blocking by escitalopram during the depolarizing pulse was less than that during the repolarizing pulse, suggesting that escitalopram has a high affinity for the open state of the hERG channel, with a relatively lower affinity for the inactivated state. Other selective serotonin reuptake inhibitors/antidepressants include fluoxetine, fluvoxamine, paroxetine, and sertraline. **Key Pharmacokinetic Parameters:** *See* Appendix II in Goodman & Gilman's *THE PHARMACOLOGICAL BASIS OF THERAPEUTICS*, 12th Edition (Brunton, Chabner & Knollmann, eds.) McGraw-Hill Medical, New York (2011). **1**. Hyttel (1982) *Prog. Neuropsychopharmacol. Biol. Psych.* **6**, 277. **2**. Chae, Jeon, Lee, *et al.* (2014) *Naunyn. Schmiedebergs. Arch. Pharmacol.* **387**, 23.

Eserine

This naturally occurring acetylcholinesterase inhibitor ($FW_{free-base}$ = 275.35 g/mol; CAS 57-47-6), also called physostigmine, from calabar beans (*Physostigma venenosum*) crosses the blood-brain barrier and binds at the anionic site of the cholinesterase (K_d = 0.3 µM), forming a carbamoyl ester with the hydroxyl group of a seryl residue. The resulting complex slowly degrades, threby allowing the enzyme to reactivate. Although eserine is toxic, these properties commend its use in assisting AD patients in long-term memory. Eserine is photosensitive and should be stored in the dark. Solutions should be used within a week of preparation. Storage should be in the cold, dark, in alkali-free glass containers, and free of metal ions. **Target(s):** acetylcholinesterase (1-23); acetylsalicylate deacetylase (24); amidase (25); arylacetamide deacetylase (26). **1**. Augustinsson (1950) *The Enzymes*, 1st. ed. (Sumner & Myrbäck, eds.), **1** (part 1), 443. **2**. Wilson (1960) *The Enzymes*, 2nd ed., **4**, 501. **3**. Main & Hastings (1966) *Science* **154**, 400. **4**. Winteringham & Fowler (1966) *Biochem. J.* **101**, 127. **5**. Wilson (1966) *Ann. N. Y. Acad. Sci.* **135**, 177. **6**. Kaur & Zhang (2000) *Curr. Med. Chem.* **7**, 273. **7**. Johansson & Nordberg (1993) *Acta Neurol. Scand.* Suppl. **149**, 22. **8**. Singh & Spassova (1998) *Comp. Biochem. Physiol. C Pharmacol. Toxicol. Endocrinol.* **119**, 97. **9**. Darvesh, Walsh, Kumar, *et al.* (2003) *Alzheimer Dis. Assoc. Disord.* **17**, 117. **10**. Joshi & Singh (2000) *Indian J. Biochem. Biophys.* **37**, 192. **11**. Nunes, Carvalho & Guilhermino (2005) *J. Enzyme Inhib. Med. Chem.* **20**, 369. **12**. Frasco, Fournier, Carvalho & Guilhermino (2006) *Aquat. Toxicol.* **77**, 412. **13**. Kasturi & Vasantharajan (1976) *Phytochemistry* **15**, 1345. **14**. Martín-Valmaseda, Sánchez-Yagüe, Cabezas & Llanillo (1995) *Comp. Biochem. Physiol. B Biochem. Mol. Biol.* **110**, 91. **15**. Hsiao, Lai, Liao & Feng (2004) *J. Agric. Food Chem.* **52**, 5340. **16**. Ma, He & Zhu (2004) *Pestic. Biochem. Physiol.* **78**, 67. **17**. Bentley, Jones & Agnew (2005) *Mol. Biochem. Parasitol.* **141**, 119. **18**. Talesa, Romani, Rosi & Giovanni (1996) *Eur. J. Biochem.* **238**, 538. **19**. Talesa, Romani, Calvitti, Rosi & Giovannini (1998) *Neurochem. Int.* **33**, 131. **20**. Li & Han (2202) *Arch. Insect Biochem. Physiol.* **51**, 37. **21**. Hannesson & DeVries (1990) *J. Neurosci. Res.* **27**, 84. **22**. Stojan & Zorko (1997) *Biochim. Biophys. Acta* **1337**, 75. **23**. Duhaiman, Alhomida, Rabbani, Kamal & al-Jafari (1996) *Biochimie* **77**, 46. **24**. Gresner, Dolník, Waczulíková, *et al.* (2006) *Biochim. Biophys. Acta* **1760**, 207. **25**. Alt, Heymann & Krisch (1975) *Eur. J. Biochem.* **53**, 357. **26**. Preuss & Svensson (1996) *Biochem. Pharmacol.* **51**, 1661.

ESI-05, *See* *4-Methylphenyl- 2, 4, 6- trimethylphenylsulfone*

Eslicarbazepine acetate

Eslicarbazepine acetate   Eslicarbazepin

This novel, once daily dose, voltage-gated Na^+ channel blocker and anticonvulsant prodrug (FW = 296.32 g/mol; CAS 236395-14-5), also known as Aptiom®, Zebinix®, Exalief® and (*S*)-10-acetoxy-10,11-dihydro-5*H*-dibenz[*b,f*]azepine-5-carboxamide, is converted to eslicarbazepine (FW = 254.28 g/mol; CAS 29331-92-8; also named (*S*)-(+)-Licarbazepine), a voltage-gated sodium channel blocker with anticonvulsant and mood-stabilizing effects. Eslicarbazepine acetate inhibits the release of glutamate evoked by 4-aminopyridine (4-AP) or veratridine in a concentration-dependent manner (2). Using conditions of stimulation (30 mM KCl), where Na^+ channels are inactivated, the eslicarbazepine does not inhibit either the Ca^{2+}-dependent or Ca^{2+}-independent release of glutamate (2). Eslicarbazepine acetate is also an alternative treatment in associated therapy in patients with partial epilepsy, who do not respond adequately to treatment in monotherapy. **1**. Ambrósio, Silva, Araújo, *et al.* (2000) *Eur. J. Pharmacol.* **406**, 191. **2**. Ambrósio, Silva, Malva, *et al.* (2001) *Biochem. Pharmacol.* **61**, 1271. **3**. Mauri-Llerda (2012) *Rev. Neurol.* **54**, 551.

ESP55016, *See* *ETC-1002*

Estazolam

This intermediate-acting oral triazolobenzodiazepine hypnotic agent (FW = 294.71 g/mol; CAS 29975-16-4; 8-chloro-6-phenyl-4*H*-1,2,4-triazolo(4,3-*a*)-1,4-benzodiazepine), also known by the trade names ProSom® and Eurodin®, is a Schedule IV benzodiazepine receptor agonist that is often used to treat insomnia, albeit at the risk of developing tolerance. Estazolam stabilizes the continuity of slow wave deep sleep (SWDS) with higher amplitude than comparable benzodiazepine receptor ligands. Animal studies indicate estazolam induces a drowsy pattern of spontaneous EEG, including high voltage slow waves and spindle bursts increase in the cortex and amygdala, with desynchronization of hippocampal θ-rhythms. It also exerts anxiolytic, anticonvulsant, sedative and skeletal muscle relaxant properties. Estazolam reaches peak plasma levels within 1–6 hours, and is excreted mainly as 4-hydroxyestazolam (formed by CYP3A4) with a half-life of 8–31 hours (1). Estazolam inhibits platelet-activating factor- (PAF-) induced aggregation in a dose-dependent manner, IC_{50} = 3.8 µM (2). Estazolam decreases histamine turnover in the brain, most likely via the benzodiazepine-GABA receptor complex (3). **1**. Mancinelli, Guiso, Garattini, Urso & Caccia (1985) *Xenobiotica* **15**, 257. **2**. Mikashima, Takehara, Muramoto, *et al.* (1987) *Jpn. J. Pharmacol.* **44**, 387. **3**. Oishi, Nishibori, Itoh & Saeki (1986) *Eur. J. Pharmacol.* **124**, 337.

Esterase Inhibitor C-1

This single-chain serpin (MW = 55.6 kDa), also known as complement C1 esterase inhibitor and C1 inhibitor, contains 478 aminoacyl residues and is heavily glycosylated (~30% carbohydrate). This protein controls complement activation and regulates all major kinin generating cascade systems. It also inhibits a number of serine proteinases; however, trypsin and chymotrypsin catalyze the degradation of the C1 inhibitor). (*For a discussion of the likely mechanism of inhibitory action, **See** Serpins; **also** α₁-Antichymotrypsin*) **Target(s):** coagulation factor Xia (1-6); coagulation factor XIIa (2,4,7-11); complement subcomponent C1r (2,4,12-18); complement subcomponent C1s (2,4,13-15,17-22); granzyme A, mouse (23); mannan-binding lectin-associated serine protease-2 (24-30); mouse nerve growth factor γ-subunit protease, possibly γ-renin (31); plasma kallikrein (2,4,13,16,32-35); plasmin (2,4,13,16,22,36-38); t plasminogen activator (39); u-plasminogen activator, *or* urokinase (4). **1**. Walsh (1998) in *Handb. Proteolytic Enzymes*, p. 153, Academic Press, San Diego. **2**. Sim & Reboul (1981) *Meth. Enzymol.* **80**, 43. **3**. Kurachi & Davie (1981) *Meth. Enzymol.* **80**, 211. **4**. Davis, Aulak, Zahedi, Bissler & Harrison (1993) *Meth. Enzymol.* **223**, 97. **5**. Wuillemin, Bleeker, Agterberg, *et al.* (1996) *Brit. J. Haematol.* **93**, 950. **6**. Meijers, Vlooswijk & Bouma (1988) *Biochemistry* **27**, 959. **7**. Ratnoff (1998) in *Handb. Proteolytic Enzymes*, p. 144, Academic Press, San Diego. **8**. Griffin & Cochrane (1976) *Meth. Enzymol.* **45**, 56. **9**. Pixley & Colman (1993) *Meth. Enzymol.* **222**, 51. **10**. Silverberg & Kaplan (1988) *Meth. Enzymol.* **163**, 68. **11**. Davie, Fujikawa, Kurachi & Kisiel (1979) *Adv. Enzymol.* **48**, 277. **12**. Arlaud & Thielens (1998) in *Handb. Proteolytic Enzymes*, p. 122, Academic Press, San Diego. **13**. Harpel (1976) *Meth. Enzymol.* **45**, 751. **14**. Sim (1981) *Meth. Enzymol.* **80**, 26. **15**. Arlaud & Thielens (1993) *Meth. Enzymol.* **223**, 61. **16**. Ratnoff, Pensky, Ogston & Naff (1969) *J. Exp. Med.* **129**, 315. **17**. Sim, Arlaud & Colomb (1980) *Biochim. Biophys. Acta* **612**, 433. **18**. Nilsson & Wiman (1983) *Eur. J. Biochem.* **129**, 663. **19**. Arlaud & Thielens (1998) in *Handb. Proteolytic Enzymes*, p. 125, Academic Press, San Diego. **20**. Salvesen, Catanese, Kress & Travis (1985) *J. Biol. Chem.* **260**, 2432. **21**. Kirschfink & Nurnberger (1999) *Mol. Immunol.* **36**, 225. **22**. Harpel & Cooper (1975) *J. Clin. Invest.* **55**, 593. **23**. Simon & Kramer (1994) *Meth. Enzymol.* **244**, 68. **24**. Ambrus, Gal, Kojima, *et al.* (2003) *J. Immunol.* **170**, 1374. **25**. Presanis, Hajela, Ambrus, Gal & Sim (2004) *Mol. Immunol.* **40**, 921. **26**. Rossi, Cseh, Bally, *et al.* (2001) *J. Biol. Chem.* **276**, 40880. **27**. Matsushita, Thiel, Jensenius, Terai & Fujita (2000) *J. Immunol.* **165**, 2637. **28**. Wong, Kojima, Dobo, Ambrus & Sim (1999) *Mol. Immunol.* **36**, 853. **29**. Gal, Harmat, Kocsis, *et al.* (2005) *J. Biol. Chem.* **280**, 33435. **30**. Sorensen, Thiel & Jensenius (2005) *Semin. Immunopathol.* **27**, 299. **31**. Faulmann, Young & Boyle (1987) *J. Immunol.* **138**, 4336. **32**. Colman & Bagdasarian (1976) *Meth. Enzymol.* **45**, 303. **33**. Silverberg & Kaplan (1988) *Meth. Enzymol.* **163**, 85. **34**. Schapira, de Agostini & Colman (1988) *Meth. Enzymol.* **163**, 179. **35**. Nilsson (1983) *Thromb. Haemost.* **49**, 193. **36**. Haselager, Goote & Vreeken (1976) *Thromb. Haemost.* **35**, 643. **37**. Highsmith, Burnett & Weirich (1979) *J. Lab. Clin. Med.* **93**, 459. **38**. Brown, Ravindran & Patston (2002) *Blood Coagul. Fibrinolysis* **13**, 711. **39**. Saksela (1985) *Biochim. Biophys. Acta* **823**, 35.

Esterastin

This immunomodulator (FW = 506.68 g/mol; CAS 67655-93-0), produced by actinomycetes, is structurally related to lipstatin. **Target(s):** acid lipase, lysosomal, $K_i \approx 80$ nM (competitive) (1); esterases (1-3); lipase, pancreatic, *or* triacylglycerol lipase (1). **1**. Imanaka, Moriyama, Ecsedi, *et al.* (1983) *J. Biochem.* **94**, 1017. **2**. Umezawa, Aoyagi, Hazato, *et al.* (1978) *J. Antibiot.* **31**, 639. **3**. Kondo, Uotani, Miyamoto, *et al.* (1978) *J. Antibiot.* **31**, 797.

3,17α-Estradiol

This membrane-permeant sterol (FW = 272.39 g/mol; CAS 50-28-2; λ_{max} = 280 nm; ε = 2,040 M⁻¹cm⁻¹; Insoluble in water), also known as 1,3,5-estratriene-3,17α-diol and oestradiol-17α, exhibits anti-estrogen activity (1). While a minor metabolite in humans, α-estradiol is more prevalent in other organisms. It has a) and. **Target(s):** F_oF_1 ATPase (2); glucuronosyltransferase (3); glutamate dehydrogenase (4); 3(or 17)β-hydroxysteroid dehydrogenase, K_i = 4.5 μM (5-7); 21-hydroxysteroid dehydrogenase, NAD⁺-dependent (8); steroid Δ-isomerase (9); xanthine oxidase (10). **1**. Scott & Tomkins (1975) *Meth. Enzymol.* **40**, 273. **2**. Zheng & Ramirez (1999) *Eur. J. Pharmacol.* **368**, 95. **3**. Rao, Rao & Breuer (1976) *Biochim. Biophys. Acta* **452**, 89. **4**. Douville & Warren (1968) *Biochemistry* **7**, 4052. **5**. Talalay (1963) *The Enzymes*, 2nd ed., 7, 177. **6**. Talalay & Dobson (1953) *J. Biol. Chem.* **205**, 823. **7**. Jarabak & Sack (1969) *Biochemistry* **8**, 2203. **8**. Monder & Furfine (1969) *Meth. Enzymol.* **15**, 667. **9**. Weintraub, Vincent, Baulieu & Alfsen (1977) *Biochemistry* **16**, 5045. **10**. Roussos (1967) *Meth. Enzymol.* **12A**, 5.

3,17β-Estradiol (*or* 17β-Estradiol)

This potent membrane-permeant estrogen (FW = 272.39 g/mol; CAS 50-28-2; Symbol = E2; λ_{max} = 280 nm; ε = 1900 M⁻¹cm⁻¹; water-insoluble), also known as 1,3,5-estratriene-3,17β-diol, 17β-estradiol, and oestradiol-17β, is synthesized in the ovary, placenta, adrenal cortex, and testis. **E2 Mode of Action:** β-Estradiol is mainly produced in ovarian follicles, but is also formed in fat, liver, adrenal, breast, and neural tissues. β-Estradiol induces the formation of gonadotropins, inducing the formation of female characteristics and modifying sexual behavior. In the classical E2 pathway, estradiol enters the cytoplasm, where it binds to heat shock protein (HSP). Two HSP·Estradiol complexes self-associate to form a homodimer, which enters the nucleus, binds to specific domains on estrogen response element (ERE), and induces gene transcription over the ensuing hours or days. Estradiol is also an agonist of the intracellular nonnuclear estrogen receptor GPER, which mediates a variety of rapid, non-genomic effects. 17β-Estradiol is a substrate for estrone sulfotransferase. Although the primary female sex hormone, 17β-estradiol is also found in males. **Mode of Inhibition of ATP Synthase:** 17β-Estradiol modulates ATP synthase by direct binding to the oligomycin sensitive-conferring component F_O, increasing the depolarization amplitude associated with ADP addition to energized mitochondria. This effect occurs when β-estradiol promotes uncoupling or "slipping" of the mechanochemical linkage of the synthase's F_O and F_1 components (1). β-estradiol depresses the respiratory control ratio (*i.e.,* the ratio of slopes of O₂ versus time curves, after and before ADP addition), reduces the P/O Ratio (*i.e.,* the ratio of the amount of ATP produced from the movement of two electrons through the ETS chain, by reduction of an oxygen atom) and decreases State-3 respiration (*i.e.,* respiration observed when ADP is added to intact, unpoisoned mitochondria in the presence of excess substrate), whereas it increases State-4 respiration (*i.e.,* oxygen consumption by isolated mitochondria on a particular substrate, in the absence of ADP or any metabolic poisons or inhibitors). By contrast, the uncoupler VFCCP remains unaltered. 17β-estradiol therefore appears to allow an energy-disappating back-leak of protons through ATP synthase's F_O component (1). Note also that estrogen receptors and hormone-responsive elements have been identified in mitochondria. **Target(s):** aldehyde oxidase (2,3); aromatase, *or* CYP19 (4); arylamine *N*-acetyltransferase (52); Ca^{2+} currents, voltage-dependent L-type (5); cortisol sulfotransferase (6,7); estrone sulfotransferase, also alternative substrate (8); ethoxyresorufin *O*-deethylation (9); glucose-6-phosphate dehydrogenase, weakly inhibited (10-15); glucuronosyltransferase, substrate for UGT1A1 (16-20); glutamate dehydrogenase (21,22); hydroxyindole *O*-methyltransferase (23); 17α-hydroxy-progesterone aldolase (24); 3(*or* 17)β-hydroxysteroid dehydrogenase (25,26); 21-hydroxysteroid dehydrogenase (NAD^+) (27-29); isocitrate dehydrogenase (10,11,13); 16-ketosteroid reductase (30); malate dehydrogenase (10,11); 6 phosphogluconate dehydrogenase (10,11,13); progesterone 5α-reductase (31); protein-disulfide isomerase (32); ribosyldihydro-nicotinamide dehydrogenase, weakly inhibited (56); steroid Δ-isomerase (33-39); steroid 16α-monooxygenase (40); steroid 17α-monooxygenase (41,53); steroid 21-monooxygenase (42); steroid sulfotransferase, also alternative substrate with substrate inhibition (43,44); sterol *O*-acyltransferase, *or* cholesteol *O*-acyltransferase, *or* ACAT (45); sterol esterase, *or* cholesterol esterase (46); steryl-sulfatase (47-49); tyrosinase (50); tyrosine 3-monooxygenase, low concentrations of estradiol stimulate activity (54,55); xanthine oxidase (56). **1.** Moreno, Moreira, Custódio & Santos (2013) *J. Bioenerg. Biomembr.* **45**, 261. **2.** Rajagopalan & Handler (1966) *Meth. Enzymol.* **9**, 364. **3.** Obach, Huynh, Allen & Beedham (2004) *J. Clin. Pharmacol.* **44**, 7. **4.** Shimizu, Yarborough & Osawa (1993) *J. Steroid Biochem. Mol. Biol.* **44**, 651. **5.** Nakajima, Kitazawa, Hamada, *et al.* (1995) *Eur. J. Pharmacol.* **294**, 625. **6.** Singer (1979) *Arch. Biochem. Biophys.* **196**, 340. **7.** Singer & Bruns (1980) *Can. J. Biochem.* **58**, 660. **8.** Rozhin, Huo, Zemlicka & Brooks (1977) *J. Biol. Chem.* **252**, 7214. **9.** Klinger, Lupp, Karge, *et al.* (2002) *Toxicol. Lett.* **128**, 129. **10.** McKerns (1962) *Biochim. Biophys. Acta* **63**, 552. **11.** McKerns (1963) *Biochim. Biophys. Acta* **71**, 710. **12.** McKerns (1963) *Biochim. Biophys. Acta* **69**, 425. **13.** McKerns (1963) *Biochim. Biophys. Acta* **73**, 507. **14.** McKerns & Kaleita (1960) *Biochem. Biophys. Res. Commun.* **2**, 344. **15.** Criss & McKerns (1969) *Biochim. Biophys. Acta* **184**, 486. **16.** Falany, Green, Swain & Tephly (1986) *Biochem. J.* **238**, 65. **17.** Zhang, Chando, Everett, *et al.* (2005) *Drug Metab. Dispos.* **33**, 1729. **18.** Watanabe, Nakajima, Ohashi, Kume & Yokoi (2003) *Drug Metab. Dispos.* **31**, 589. **19.** Matern, Matern, Schelzig & Gerok (1980) *FEBS Lett.* **118**, 251. **20.** Rao, Rao & Breuer (1976) *Biochim. Biophys. Acta* **452**, 89. **21.** Frieden (1963) *The Enzymes*, 2nd ed., **7**, 3. **22.** Douville & Warren (1968) *Biochemistry* **7**, 4052. **23.** Morton & Forbes (1989) *J. Pineal Res.* **6**, 259. **24.** Dalla Valle, Ramina, Vianello, Belvedere & Colombo (1996) *J. Steroid Biochem. Mol. Biol.* **58**, 577. **25.** Talalay (1962) *Meth. Enzymol.* **5**, 512. **26.** Raimondi, Olivier, Patrito & Flury (1989) *J. Steroid Biochem.* **32**, 413. **27.** Monder & Furfine (1969) *Meth. Enzymol.* **15**, 667. **28.** Talalay (1963) *The Enzymes*, 2nd ed., **7**, 177. **29.** Monder & White (1962) *Biochem. Biophys. Res. Commun.* **8**, 383. **30.** Murono & Payne (1976) *Biochim. Biophys. Acta* **450**, 89. **31.** Cheng & Karavolas (1975) *J. Biol. Chem.* **250**, 7997. **32.** Tsibris, Hunt, Ballejo, *et al.* (1989) *J. Biol. Chem.* **264**, 13967. **33.** Jarabak, Colvin, Moolgavkar & Talalay (1969) *Meth. Enzymol.* **15**, 642. **34.** Kawahara (1962) *Meth. Enzymol.* **5**, 527. **35.** Talalay & Benson (1972) *The Enzymes*, 3rd ed., **6**, 591. **36.** Zhao, Mildvan & Talalay (1995) *Biochemistry* **34**, 426. **37.** Wong & Keung (1999) *J. Steroid Biochem. Mol. Biol.* **71**, 191. **38.** Kawahara, Wang & Talalay (1962) *J. Biol. Chem.* **237**, 1500. **39.** Weintraub, Vincent, Baulieu & Alfsen (1977) *Biochemistry* **16**, 5045. **40.** Sano, Shibusawa, Yoshida, *et al.* (1980) *Acta Obstet. Gynecol. Scand.* **59**, 245. **41.** Coon & Koop (1983) *The Enzymes*, 3rd ed., **16**, 645. **42.** Yoshida, Sekiba, Yanaihara, *et al.* (1978) *Endocrinol. Jpn.* **25**, 349. **43.** Singer, Gebhart & Hess (1978) *Can. J. Biochem.* **56**, 1028. **44.** Sugiyama, Stolz, Sugimoto, *et al.* (1984) *Biochem. J.* **224**, 947. **45.** Simpson & Burkhart (1980) *Arch. Biochem. Biophys.* **200**, 79. **46.** Smith (1959) *J. Bacteriol.* **77**, 682. **47.** Notation & Ungar (1969) *Biochemistry* **8**, 501. **48.** Iwamori, Moser & Kishimoto (1976) *Arch. Biochem. Biophys.* **174**, 199. **49.** Payne (1972) *Biochim. Biophys. Acta* **258**, 473. **50.** Kline, Smith, Carland & Blackmon (1988) *J. Cell Physiol.* **134**, 497. **51.** Roussos (1967) *Meth. Enzymol.* **12A**, 5. **52.** Kawamura, Westwood, Wakefield, *et al.* (2008) *Biochem. Pharmacol.* **75**, 1550. **53.** Dalla Valle, Ramina, Vianello, Belvedere & Colombo (1996) *J. Steroid Biochem. Mol. Biol.* **58**, 577. **54.** Joy & Chaube (2004) *Fish Physiol.*

Biochem. **28**, 33. **55.** Chaube & Joy (2005) *Gen. Comp. Endocrinol.* **141**, 116. **56.** Wu, Knox, Sun, *et al.* (1997) *Arch. Biochem. Biophys.* **347**, 221.

Estramustine

This membrane-permeant antineoplastic agent (FW = 440.41 g/mol) is an ester of 3,17β-estradiol that inhibits tubulin assembly. **Target(s):** extracellular signal-regulated protein kinase 2, *or* ERK2 (1); tubulin polymerization, *or* microtubule self-assembly (2-6). **1.** Liu, Wang, Kreis, Budman & Adams (2001) *Brit. J. Cancer* **85**, 1403. **2.** Jordan (2002) *Curr. Med. Chem. Anti-Canc. Agents* **2**, 1. **3.** Stearns, Wang, Tew & Binder (1988) *J. Cell Biol.* **107**, 2647. **4.** Stearns & Tew (1988) *J. Cell Sci.* **89**, 331. **5.** Dahllof, Billstrom, Cabral & Hartley-Asp (1993) *Cancer Res.* **53**, 4573. **6.** Laing, Dahllof, Hartley-Asp, Ranganathan & Tew (1997) *Biochemistry* **36**, 871.

Estriol

This membrane-permeant sterol (FW = 288.39 g/mol; λ_{max} = 280 nm; Sparingly soluble in water; Soluble in chloroform, and dioxane), also called oestriol, 1,3,5-estratriene-3,16α,17β-triol, 1,3,5-oestratriene 3,16α,17β-triol, trihydroxyestrin, trihydroxyoestrin, and theelol, is the primary estrogen found in urine. It is a metabolite of 17β-estradiol and exhibits less biological activity (*i.e.,* it is a weak estrogen). **Target(s):** aldose 1-epimerase, *or* mutarotase (1); aromatase, *or* CYP19 (2); dihydropteridine reductase (3); estrone sulfotransferase (4); ethoxyresorufin *O*-deethylation (5); F_OF_1 ATPase (6); glucose-6-phosphate dehydrogenase (7-10); glucuronosyltransferase (11); 3-β-hydroxysteroid dehydrogenase (12); isocitrate dehydrogenase, weakly inhibited (7-9); 16-ketosteroid reductase (13); malate dehydrogenase, weakly inhibited (7,8); 6-phosphogluconate dehydrogenase (9); steroid Δ-isomerase (14); sterol 14-demethylase, *or* CYP51, K_i = 100 μM (15,16); tyrosinase (17); xanthine oxidase (18). **1.** Mulhern, Fishman, Kusiak & Bailey (1973) *J. Biol. Chem.* **248**, 4163. **2.** Shimizu, Yarborough & Osawa (1993) *J. Steroid Biochem. Mol. Biol.* **44**, 651. **3.** Shen & Abell (1983) *J. Neurosci. Res.* **10**, 251. **4.** Rozhin, Huo, Zemlicka & Brooks (1977) *J. Biol. Chem.* **252**, 7214. **5.** Klinger, Lupp, Karge, *et al.* (2002) *Toxicol. Lett.* **128**, 129. **6.** Zheng & Ramirez (1999) *Eur. J. Pharmacol.* **368**, 95. **7.** McKerns (1962) *Biochim. Biophys. Acta* **63**, 552. **8.** McKerns (1963) *Biochim. Biophys. Acta* **71**, 710. **9.** McKerns (1963) *Biochim. Biophys. Acta* **73**, 507. **10.** Criss & McKerns (1969) *Biochim. Biophys. Acta* **184**, 486. **11.** Rao, Rao & Breuer (1976) *Biochim. Biophys. Acta* **452**, 89. **12.** Marcus & Talalay (1955) *Proc. Roy. Soc.* **B144**, 116. **13.** Murono & Payne (1976) *Biochim. Biophys. Acta* **450**, 89. **14.** Weintraub, Vincent, Baulieu & Alfsen (1977) *Biochemistry* **16**, 5045. **15.** Eddine, von Kries, Podust, *et al.* (2008) *J. Biol. Chem.* 283, 15152. **16.** Podust, von Kries, Eddine, *et al.* (2007) *Antimicrob. Agents Chemother.* **51**, 3915. **17.** Kline, Smith, Carland & Blackmon (1988) *J. Cell Physiol.* **134**, 497. **18.** Roussos (1967) *Meth. Enzymol.* **12A**, 5.

Estrone 3-(Bromoacetate)

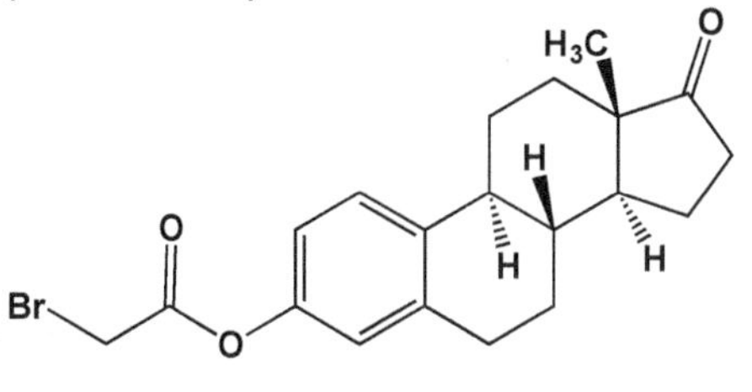

This membrane-permeant steroid haloester (FW = 391.30 g/mol), also known as 3-bromoacetoxyestrone, covalently modifies the active-site histidyl residue of human placental 17β,20α-hydroxysteroid dehydrogenase, or estradiol 17β-dehydrogenase, thereby inactivating the enzyme. **1**. Thomas, LaRochelle, Asibey-Berko & Strickler (1985) *Biochemistry* **24**, 5361. **2**. Thomas, Asibey-Berko & Strickler (1986) *J. Steroid Biochem.* **25**, 103. **3**. Murdock & Warren (1982) *Steroids* **39**, 165. **4**. Groman, Schultz & Engel (1977) *Meth. Enzymol.* **46**, 54. **5**. Groman, Schultz & Engel (1975) *J. Biol. Chem.* **250**, 5450. **6**. Murdock, Chin, Offord, Bradshaw & Warren (1983) *J. Biol. Chem.* **258**, 11460.

Estrone 3-O-Sulfamate

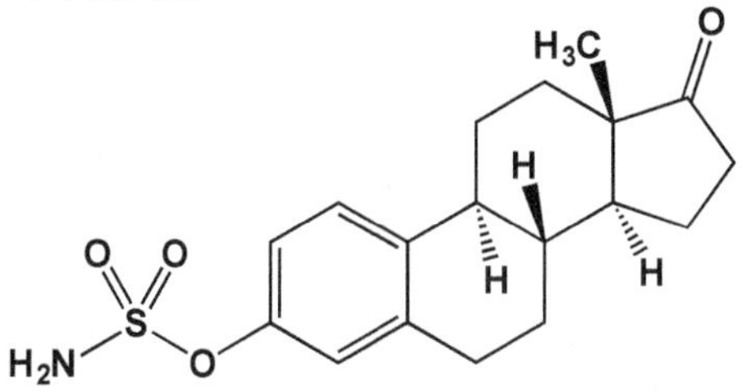

This steroid derivative (FW = 349.45 g/mol), often referred to as EMATE, inactivates steryl sulfatase (1-9) and carbonic anhydrase (10). **1**. Purohit, Williams, Howarth, Potter & Reed (1995) *Biochemistry* **34**, 11508. **2**. Horvath, Nussbaumer, Wolff & Billich (2004) *J. Med. Chem.* **47**, 4268. **3**. Woo, Howarth, Purohit, *et al.* (1998) *J. Med. Chem.* **41**, 1068. **4**. Walter, Liebl & von Angerer (2004) *J. Steroid Biochem. Mol. Biol.* **88**, 409. **5**. Fischer, Chander, Woo, *et al.* (2003) *J. Steroid Biochem. Mol. Biol.* **84**, 343. **6**. Shields-Botella, Bonnet, Duc, *et al.* (2003) *J. Steroid Biochem. Mol. Biol.* **84**, 327. **7**. Purohit, Reed, Morris, Williams & Potter (1996) *Ann. N.Y. Acad. Sci.* **784**, 40. **8**. Purohit, Potter, Parker & Reed (1998) *Chem. Biol. Interact.* **109**, 183. **9**. Bonser, Walker, Purohit, *et al.* (2000) *J. Endocrinol.* **167**, 465. **10**. Winum, Vullo, Casini, *et al.* (2003) *J. Med. Chem.* **46**, 2197.

ET-18-OCH₃, See *1-O-Octadecyl-2-O-methyl-sn-glycero-3-phosphocholine*

Et-743, See *Ecteinascidin 743*

Etamicastat

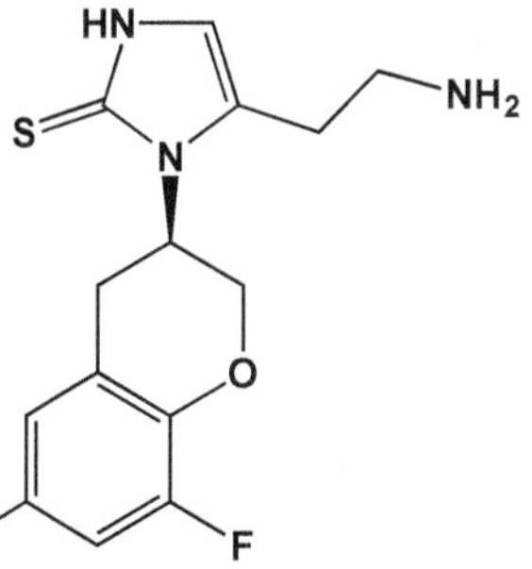

This novel dopamine-β-hydroxylase inhibitor (FW_free-base = 311.09 g/mol; CAS 760173-05-5), also named (*R*)-5-(2-aminoethyl)-1-(6,8-difluoro-chroman-3-yl)-1*H*-imidazole-2(3*H*)-thione hydrochloride, reduces both systolic and diastolic blood pressure (whether used alone or in combination with other antihypertensives) (1,2), decreases urinary excretion of noradrenaline in the spontaneously hypertensive rat with no change in heart rate (1-3), and lowers blood pressure in hypertensive patients (4). As a dopamine-β-hydroxylase inhibitor, etamicastat decreases norepinephrine levels in sympathetically innervated tissues. **1**. Igreja, Pires, Wright & Soares-da-Silva (2008) *Hypertension* **52**, E62. **2**. Igreja, Pires Wright, &

Soares-da-Silva (2011) *Hypertension* **58**, E161. **3**. Beliaev, Learmonth & Soares-da-Silva (2006) *J. Med. Chem.* **49**, 1191. **4**. Almeida, Nunes, Costa, *et al.* (2013) *Clin. Ther.* **35**, 1983.

Etamycin

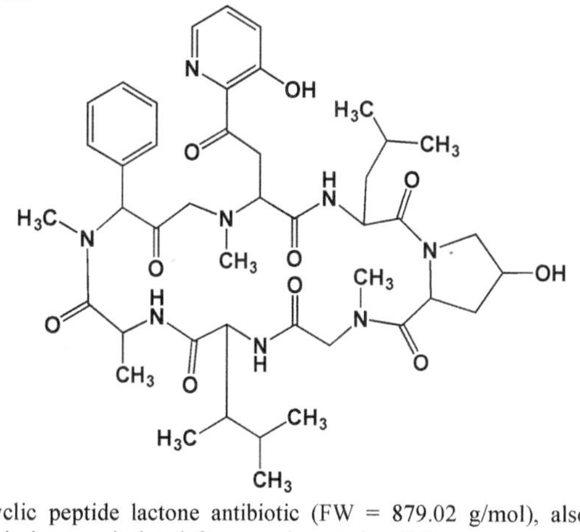

This cyclic peptide lactone antibiotic (FW = 879.02 g/mol), also called viridogrisein, was isolated from marine strain of *Streptomyces griseus* on the coast of Fiji. It is active against Gram-positive bacteria, including *Mycobacterium tuberculosis*, and some fungi. Etamycin exerts it action by arresting protein biosynthesis. Recent studies indicate that etamycin displays favorable time-kill kinetics compared with the first-line MRSA antibiotic (vancomycin) and conferred significant protection from mortality in a murine model of systemic lethal MRSA infection (2). **1**. Pestka (1974) *Meth. Enzymol.* **30**, 261. **2**. Haste, Perera, Maloney, *et al.* (2010) J. Antibiot. (Tokyo) **63**, 219.

Etazolate

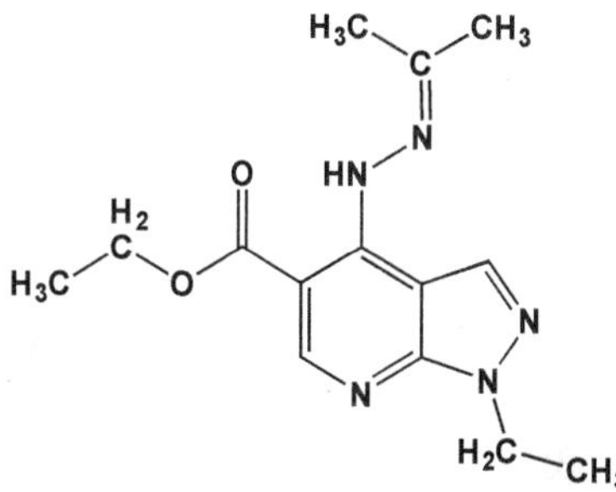

This anxiolytic drug (FW = 289.34 g/mol; CAS 51022-77-6), also known by the code names SQ-20009, EHT-0202, and systematically named ethyl 1-ethyl-4-[2-(propan-2-ylidene)hydrazinyl]-1*H*-pyrazolo[3,4-*b*]pyridine-5-carboxylate, is a positive allosteric modulator of the GABA_A receptor at the barbiturate binding site (1-3), an adenosine A₁ and A₂ receptor antagonist (4), and a PDE Isozyme-4-selective phosphodiesterase inhibitor (5,6). **1**. Zezula, Slany & Sieghart (1996) *Eur. J. Pharmacol.* **301**, 207. **2**. Remington & Baskys (1996) *Brain mechanisms and psychotropic drugs.* CRC Press, Boca Raton. **3**. Mishina, Kurachi, Yoshihisa (2000) *Pharmacology of ionic channel function: activators and inhibitors.* Springer, Berlin. **4**. Williams & Jarvis (1988) *Pharmacol. Biochem. Behav.* **29**, 433. **5**. Chasin, Harris, Phillips & Hess (1972) *Biochem. Pharmacol.* **21**, 2443. **6**. Wang, Myers, Wu, *et al.* (1997) *Biochem. Biophys. Res. Commun.* **234**, 320.

ETC-1002

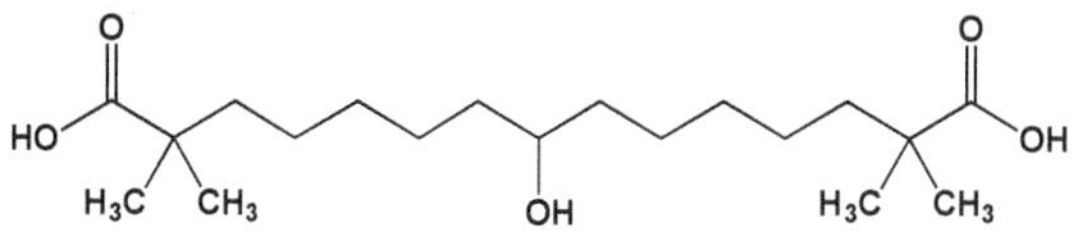

ETC-1002

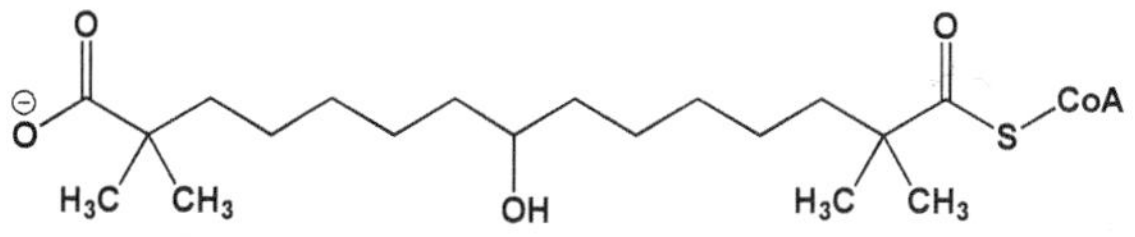

ETC-1002 Coenzyme A Thioester

This novel investigational drug (FW$_{free-acid}$ = 344.49 g/mol; CAS 738606-46-7; FW$_{CoA-ester}$ = 1110.00 g/mol), also known as 8-hydroxy-2,2,14,14-tetramethylpentadecanedioate and ESP55016, exhibits hypolipidemic, anti-atherosclerotic, anti-obesity, and glucose-lowering properties in preclinical disease models. ETC-1002 is a dual inhibitor of sterol and fatty acid synthesis and enhanced mitochondrial long-chain fatty acid β-oxidation. ETC-1002 works through two distinct mechanisms: (a) directly by ETC-1002 activation of AMP kinase pathway and (b) indirectly by ETC-1002 Coenzyme A thioester inhibition of hepatic ATP:citrate lyase (ACL). Pharmacological inhibition of ACL by (−)-hydroxycitrate is known to limit sterol and fatty acid synthesis, to upregulate LDL receptor activity, and to reduce plasma triglycerides and cholesterol levels. In HepG2 cells, ETC-1002 activates the AMPK pathway independent of energy depletion and calcium signaling. *In vivo*, ETC-1002$_{free-acid}$ is >100x more abundant than its CoA thioester in rat liver and is associated with AMPK activation. Rapid formation of ETC-1002-CoA and subsequent transient increases in citrate with concomitant reductions in acetyl-CoA, malonyl-CoA, and HMG-CoA in rat hepatocytes is consistent with ACL inhibition as well as dual inhibition of fatty acid and sterol synthesis. ETC-1002-CoA inhibits recombinant human ACL directly in a cell-free system. In seven Phase 1 and Phase 2a clinical studies, once daily dosing of ETC-1002 for 2-12 weeks has lowered LDL-C and reduced high-sensitivity C-reactive protein by up to 40%, with neutral to positive effects on glucose levels, blood pressure, and body weight (2). Importantly, use of ETC-1002 in statin-intolerant patients has shown statin-like lowering of LDL-C without the muscle pain and weakness responsible for discontinuation of statin use by many patients (2). **1.** Pinkosky, Filippov, Srivastava, *et al.* (2013) *J. Lipid Res.* **54**, 134. **2.** Filippov, Pinkosky & Newton (2014) *Curr. Opin. Lipidol.* **25**, 309.

Eteplirsen

Sequence: C-T-C-C-A-A-C-A-T-C-A-A-G-G-A-A-G-A-T-G-C-A-T-T-T-C-T-A-G

This intravenously administered, morpholino phosphorodiamidate antisense oligomer (FW = 10305.74 g/mol; CAS 1173755-55-9), also named AVI-4658, EXONDYS-51®, and (*P*-deoxy-*P*-(dimethylamino)][2',3'-dideoxy-2',3'-imino-2',3'-*seco*)(2'a→5')(C-m⁵U-C-C-A-A-C-A-m⁵U-C-A-A-G-G-A-A-G-A-m⁵U-G-G-C-A-m⁵U-m⁵U-m⁵U-C-m⁵U-A-G),5'-(P-(4-((2-(2-(2-hydroxy-ethoxy)ethoxy)ethoxy)carbonyl)-1-piperazinyl)-*N*,*N*-dimethyl-phosphonamide), is a systemic exon-skipping RNA drug for treating Duchenne Muscular Dystrophy (DMD). Eteplirsen triggers Exon-51 excision during pre-mRNA splicing of dystrophin RNA transcript, altering the downstream reading-frame of dystrophin. Administration of eteplirsen to a healthy person results in dystrophin mRNA production that does not lead to functional dystrophin protein. In DMD patients (characterized by the presence of certain frame-shifting mutations), however, eteplirsen restores the reading frame of the dystrophin mRNA, resulting in production of a novel internally deleted and therefore shorter dystrophin, attended by restoration of dystrophin-associated complex in a manner preserving functionality of the newly translated dystrophin (1). Eteplirsen-induced skipping of Exon-51 and restoration of dystrophin protein expression occurs in a dose-dependent, but variable, manner in DMD patients. In patients with Exon 51-amenable deletions (~13% of DMD patients), eteplirsen treatment produces a truncated in-frame dystrophin protein, similar to those found in Becker muscular dystrophy. a less severe dystrophinopathic condition (2). Exon skipping therefore appears to be a promising disease-modifying approach for DMD, restoring the open-reading frame and enabling production of functional dystrophin. Anthony *et al.* (3) described a standardized and reproducible method for assessing patient response to complement protein studies in preclinical and clinical exon skipping-based gene therapy studies for DMD. **Pharmacokinetics:** In male pediatric DMD patients, peak plasma concentration (C$_{max}$) occurs at the end of single or multiple IV infusion, after which there is multi-phasic decay with most of the drug eliminated within 24 hours (4). Approximate dose-proportionality is observed for 0.5–50 mg/kg/week, where the latter equals 1.7x the recommended therapeutic dose. No significant accumulation occurs after

weekly dosing, and the inter-subject variability in C$_{max}$ and AUC range from 20 to 55% (4). **1.** Cirak, Feng, Anthony, *et al.* (2012) *Mol. Ther.* **20**, 462. **2.** Mendell, Rodino-Klapac, Sahenk, *et al.* (2013) *Ann. Neurol.* **74**, 637. **3.** Anthony, Feng, Arechavala-Gomeza, *et al.* (2012) *Human Gene Ther. Meth.* **23**, 336. **4.** FDA Drug Profile on EXONDYS 51® Circular (Reference ID: 3987286), based on information supplied by Sarepta Therapeutics, Inc., Cambridge, MA.

Ethacrynic Acid

This diuretic agent (FW$_{free-acid}$ = 303.14 g/mol; pK_a' = 3.50), also known as [2,3-dichloro-4-(2-methylene-1-oxobutyl)phenoxy]acetic acid, reacts with thiol groups and inhibits many cysteine-dependent enzymes. Note that aqueous solutions at pH 7 are only stable for short periods of time; this stability decreases in alkaline pH as well as at higher temperatures. Ethacrynic acid is also a slow alternative substrate for the drug-inactivating enzyme, glutathione *S*-transferase, with an IC$_{50}$ in the low µM range. **Target(s):** acetoin dehydrogenase (1); adenylate cyclase (2); aldehyde reductase (3); aldo-keto reductase AKR7A5 (4); alkaline phosphatase (5); apyrase (6); asparagine synthetase (7); carbonyl reductase (8); Ca^{2+}-transporting ATPase (9); chloride-transporting ATPase (10-17); 3',5'-cyclic-nucleotide phosphodiesterase (18); DDT-dehydrochlorinase (19); glutathione *S*-transferase (19-27; with (27) suggesting ethacrynic acid is an allosteric inhibitor); glyceraldehyde 3-phosphate dehydrogenase (28); haloperidol reductase (29); hemoglobin S polymerization (30); histamine *N*-methyltransferase, moderately inhibited (59); histamine release (31); 11β-hydroxysteroid dehydrogenase (32); 15-hydroxyprostaglandin dehydrogenase (33-37); lactate dehydrogenase (38); Na$^+$-dependent ATPase, ouabain-insensitive (39,40); Na$^+$/K$^+$-exchanging ATPase (41,42); 4-nitrophenyl-phosphatase activity of Na$^+$/K$^+$-exchanging ATPase (47); oxidative phosphorylation (48,50); phosphate transport (51); proline dehydrogenase, *or* proline oxidase (52); prostaglandin E$_2$ 9-reductase (34,36,37,53); protein biosynthesis (54); thiol *S*-methyltransferase (55); thiopurine *S*-methyltransferase (56); tubulin polymerization (microtubule self-assembly (57,58). **1.** Hara, Seiriki, Nakayama & Sawada (1985) *Prog. Clin. Biol. Res.* **174**, 291. **2.** Ferrendelli, Johnson, Chang & Needleman (1973) *Biochem. Pharmacol.* **22**, 3133. **3.** Morpeth & Dickinson (1981) *Biochem. J.* **193**, 485. **4.** Hinshelwood, McGarvie & Ellis (2003) *Chem. Biol. Interact.* **143-144**, 263. **5.** Price (1980) *Clin. Chim. Acta* **101**, 313. **6.** LeBel, Poirier, Phaneuf, *et al.* (1980) *J. Biol. Chem.* **255**, 1227. **7.** Jayaram, Cooney, Milman, *et al.* (1975) *Biochem. Pharmacol.* **24**, 1787. **8.** Wermuth (1981) *J. Biol. Chem.* **256**, 1206. **9.** Vincenzi (1968) *Proc. West Pharmacol. Soc.* **11**, 58. **10.** Recasens, Pin & Bockaert (1987) *Neurosci. Lett.* **74**, 211. **11.** Simchowitz & De Weer (1986) *J. Gen. Physiol.* **88**, 167. **12.** Gerencser (1996) *CRC Crit. Rev. Biochem.* **31**, 303. **13.** Zeng, Higashida, Hara, *et al.* (1998) *Neurosci. Lett.* **258**, 85. **14.** Inagaki, Hara & Zeng (1996) *J. Exp. Zool.* **275**, 262. **15.** Zeng, Hara & Inagaki (1994) *Brain Res.* **641**, 167. **16.** Inoue, Hara, Zeng, *et al.* (1991) *Neurosci. Lett.* **134**, 75. **17.** Shiroya, Fukunaga, Akashi, *et al.* (1989) *J. Biol. Chem.* **264**, 17416. **18.** Drummond & Yamamoto (1971) *The Enzymes*, 3rd ed., **4**, 355. **19.** Prapanthadara, Promtet, Koottathep, Somboon & A. J. Ketterman (2000) *Insect Biochem. Mol. Biol.* **30**, 395. **20.** van Iersel, Ploemen, Lo Bello, Federici & van Bladeren (1997) *Chem. Biol. Interact.* **108**, 67. **21.** Iersel, Ploemen, Struik, *et al.* (1996) *Chem. Biol. Interact.* **102**, 117. **22.** Fujita & Hossain (2003) *Plant Cell Physiol.* **44**, 481. **23.** Tang & Chang (1995) *Biochem. J.* **309**, 347. **24.** Ahokas, Davies, Ravenscroft & Emmerson (1984) *Biochem. Pharmacol.* **33**, 1929. **25.** Leung (1986) *Biochem. Biophys. Res. Commun.* **137**, 195. **26.** Phillips & Mantle (1993) *Biochem. J.* **294**, 57. **27.** Musdal, Hegazy, Aksoy & Mannervik (2013) *Chem. Biol. Interact.* **205**, 53. **28.** Birkett (1973) *Mol. Pharmacol.* **9**, 209. **29.** Inaba & Kovacs (1989) *Drug Metab. Dispos.* **17**, 330. **30.** Orringer, Blythe, Whitney, Brockenbrough & Abraham (1992) *Amer. J. Hematol.* **39**, 39. **31.** Johansen (1983) *Eur. J. Pharmacol.* **92**, 181. **32.** Zhang, Lorenzo & Reidenberg (1994) *J. Steroid Biochem. Mol. Biol.* **49**, 81. **33.** Jarabak (1982) *Meth. Enzymol.* **86**, 126. **34.** Pace-Asciak & Smith (1983) *The Enzymes*, 3rd ed., **16**, 543. **35.** Tai & Hollander (1976) *Adv. Prostaglandin*

Thromboxane Res. **1**, 171. **36**. Hassid & Levine (1977) *Prostaglandins* **13**, 503. **37**. Korff & Jarabak (1980) *Prostaglandins* **20**, 111. **38**. Gunther & Ahlers (1976) *Arzneimittelforschung* **26**, 13. **39**. Orsenigo, Tosco, Esposito & Faelli (1988) *Int. J. Biochem.* **20**, 1411. **40**. Tosco, Orsenigo, Esposito & Faelli (1988) *Cell. Biochem. Funct.* **6**, 155. **41**. Sastry & Phillis (1977) *Can. J. Physiol. Pharmacol.* **55**, 170. **42**. Davis (1970) *Biochem. Pharmacol.* **19**, 1983. **43**. Charnock, Potter & McKee (1970) *Biochem. Pharmacol.* **19**, 1637. **44**. Banerjee, Khanna & Sen (1971) *Biochem. Pharmacol.* **20**, 1649. **45**. Furriel, McNamara & Leone (2001) *Comp. Biochem. Physiol. A* **130**, 665. **46**. Schuurman, Stekhoven & Bonting (1981) *Physiol. Rev.* **61**, 1. **47**. Mendonca, Masui, McNamara, Leone & Furriel (2007) *Comp. Biochem. Physiol. A* **146**, 534. **48**. Goldschmidt, Gaudemer & Gautheron (1976) *Biochimie* **58**, 713. **49**. Eknoyan, Sawa, Hyde, *et al.* (1975) *J. Pharmacol. Exp. Ther.* **194**, 614. **50**. Gaudemer & Foucher (1967) *Biochim. Biophys. Acta* **131**, 255. **51**. Le Quoc, Le Quoc & Gaudemer (1977) *Biochim. Biophys. Acta* **462**, 131. **52**. Karmar & Fitscha (1970) *Enzymologia* **39**, 101. **53**. Stone & Hart (1976) *Prostaglandins* **12**, 197. **54**. Buss, Kauten & Piatt (1985) *J. Antimicrob. Chemother.* **15**, 105. **55**. Weinshilboum, Sladek & Klumpp (1979) *Clin. Chim. Acta* **97**, 59. **56**. Woodson & Weinshilboum (1983) *Biochem. Pharmacol.* **32**, 819. **57**. Xu, Roychowdhury, Gaskin & Epstein (1992) *Arch. Biochem. Biophys.* **296**, 462. **58**. Luduena, Roach & Epstein (1994) *Biochem. Pharmacol.* **47**, 1677. **59**. Thithapandha & Cohn (1978) *Biochem. Pharmacol.* **27**, 263.

Ethambutol

This antitubercular pharmaceutical (FW$_{\text{free-base}}$ = 204.31 g/mol; CAS 74-55-5), also known as 2,2'-(1,2-ethanediyldiimino)bis-1-butanol and myambutal, inhibits arabinosyltransferases participating in arabinogalactan biosynthesis. It also exhibits ocular toxicity, with the optical nerve atrophy. **Target(s):** adenosylmethionine decarboxylase (1); arabinosyl-transferase and arabinogalactan biosynthesis (2–6); glucose-6-P dehydrogenase (7); glycerophosphate dehydrogenase (7); spermidine synthase (8,9). **Key Pharmacokinetic Parameters:** *See* Appendix II in Goodman & Gilman's *THE PHARMACOLOGICAL BASIS OF THERAPEUTICS*, 12th Edition (Brunton, Chabner & Knollmann, eds.) McGraw-Hill Medical, New York (2011). **1**. Paulin, Brander & Pösö (1987) *Experientia* **43**, 174. **2**. Rastogi, Goh & David (1990) *Antimicrob. Agents Chemother.* **34**, 759. **3**. Takayama & Kilburn (1989) *Antimicrob. Agents Chemother.* **33**, 1493. **4**. Wolucka, McNeil, de Hoffmann, Chojnacki & Brennan (1994) *J. Biol. Chem.* **269**, 23328. **5**. Mikusova, Slayden, Besra & Brennan (1995) *Antimicrob. Agents Chemother.* **39**, 2484. **6**. Lee, Brennan & Besra (1997) *Glycobiology* **7**, 1121. **7**. Storozhuk, Korochanskaia, Kolesnikov, Gontmakher & Movchan (1989) *Vrach Delo.* (4), 81. **8**. Paulin, Brander & Poso (1985) *Antimicrob. Agents Chemother.* **28**, 157. **9**. Poso, Paulin & Brander (1983) *Lancet* **2**, 1418.

Ethanol

This simple two-carbon alcohol CH$_3$CH$_2$OH (FW = 46.07 g/mol; CAS 64-17-5; Density: 0.789 g/cm^3 (at 25°C)) has numerous targets, including, but not limited to: the GABA$_A$ Receptor by acting as allosteric modulator, the Glycine Receptor by acting as positive and negative allosteric modulator; the NMDA Receptor by acting as negative allosteric modulator; the AMP$_A$ Receptor by acting as allosteric activato; the Kainate Receptor by acting as negative allosteric modulator; nicotinic Acetyl-Choline Receptors by acting both as positive and negative allosteric modulators; the Serotonin 5-HT$_3$ Receptor by acting as allosteric activator; the Glycine Reuptake System by acting as an inhibitor; the Adenosine Reuptake System by acting as an inhibitor; the L-type calcium channel by acting as blocker; and the G-protein-activated, inwardly rectifying K$^+$ (*or* GIRK) channel by acting as a channel opener. In humans, ethanol is rapidly absorbed through the stomach wall, such that ethanol consumption and the combined avtion of alcohol dehydrogenase (*Reaction*: CH$_3$CH$_2$OH + NAD$^+$ ⇌ CH$_3$CHO + NADH) and acetaldehyde dehydrogenase (*Reaction*: CH$_3$CHO + NAD$^+$ ⇌ CH$_3$COO$^-$ + NADH) bring about an almost immediate drop in the [NAD$^+$]/[NADH] ratio, both in the cytoplasm and mitochondria, that can persist for several

hours. The free cytoplasmic [NADP$^+$]/[NADPH] ratio in liver also decreases immediately after ethanol consumption, but eventually returns to control values. **A Word to Caution:** Ethanol inhibits too many enzymes to enumerate here, not to mention its indirect effects on those enzymes and metabolic processes sensitive to its immediate product, acetaldehyde. Ethanol also alters the [NAD+]/[NADH] ratio, thereby perturbing virtually the entirety of cellular energy metabolism. In view of the wide spectrum of ethanol targets, it is foolhearty to use ethanol as a vehicle for delivering inhibitors that are sparingly soluble in water. A better vehicle is 5-10% DMSO (vol/vol), with a drug-free sample serving as a control. Lest an unwary researcher overlook ethanol as a confounding variable in metabolic and pharmacologic experiments, ethanol use is discouraged, and solubilities of drugs in ethanol are rarely listed in this book.

Ethanolamine

This naturally occurring β-amino-alcohol (FW$_{\text{free-base}}$ = 61.084 g/mol; pK_a = 9.50 at 25°C; M.P. = 10.3°C; B.P. = 170.8°C), also known as 2-aminoethanol, is a key building block of several phospholipids. Ethanolamine is a product of the reactions catalyzed by alkylglycerol-phospho/ethanolamine phosphodiesterase and N-(long-chain-acyl)ethanolamine deacylase and is a substrate for ethanolamine oxidase, ethanolamine kinase, and ethanolamine ammonia-lyase. It is an alternative substrate for choline kinase. **Target(s):** acetylcholinesterase, very weakly inhibited (1,2); acetylenecarboxylate hydratase (3); alcohol dehydrogenase (4); alternansucrase (5); δ-aminolevulinate synthase (6,7); amylo-α-1,6-glucosidase/4-α-glucanotransferase, *or* glycogen debranching enzyme, weakly inhibited (8); carnosine synthetase (9); choline kinase (10–17); diaminopropionate ammonia-lyase (18); α-L-fucosidase (19); β-galactosidase (20); α-glucosidase (21); β glucuronidase (21); glutamate:ethylamine ligase, *or* theanine synthetase (22); glutaminyl-peptide cyclotransferase (23); glycerophosphocholine phosphodiesterase (24); glycine/sarcosine N-methyltransferase (25); hydroxylysine kinase, weakly inhibited (26,27); mannosyl-oligosaccharide glucosidase, *or* glucosidase I (28); oligo-1,6-glucosidase, *or* isomaltase (29); phospholipase D (30); protein glutamine γ-glutamyltransferase, *or* transglutaminase; K_i = 2.3 mM (31); D-serine ammonia-lyase, *or* D-serine dehydratase (32); L-serine ammonia-lyase, *or* L-serine dehydratase (33,34); sucrose α-glucosidase, *or* sucrase-isomaltase (29). **1**. Wilson (1952) *J. Biol. Chem.* **197**, 215. **2**. Wilson (1954) *J. Biol. Chem.* **208**, 123. **3**. Yamada & Jakoby (1959) *J. Biol. Chem.* **234**, 941. **4**. Ostrovskii & Artsukevich (1989) *Biokhimiia* **54**, 1888. **5**. Côté & Robyt (1982) *Carbohydr. Res.* **101**, 57. **6**. Jordan & Shemin (1972) *The Enzymes*, 3rd ed. (Boyer, ed.), **7**, 339. **7**. Varadharajan, Dhanasekaran, Bonday, Rangarajan & Padmanaban (2002) *Biochem. J.* **367**, 321. **8**. Nelson, Kolb & Larner (1969) *Biochemistry* **8**, 1419. **9**. Seely & Marshall (1982) in *Peptide Antibiotics: Biosynthesis and Functions: Enzymatic Formation of Bioactive Peptides and Related Compounds* (Kleinkauf & von Döhren, eds.) W. de Gruyter, Berlin, pp. 347. **10**. Ishidate & Nakazawa (1992) *Meth. Enzymol.* **209**, 121. **11**. Porter & Kent (1992) *Meth. Enzymol.* **209**, 134. **12**. Brophy, Choy, Toone & Vance (1977) *Eur. J. Biochem.* **78**, 491. **13**. Infante & Kinsella (1976) *Lipids* **11**, 727. **14**. Ulane, Stephenson & Farrell (1978) *Biochim. Biophys. Acta* **531**, 295. **15**. Choubey, Guha, Maity, *et al.* (2006) *Biochim. Biophys. Acta* **1760**, 1027. **16**. Porter & Kent (1990) *J. Biol. Chem.* **265**, 414. **17**. Ishidate, Furusawa & Nakazawa (1985) *Biochim. Biophys. Acta* **836**, 119. **18**. Nagasawa, Tanizawa, Satoda & Yamada (1988) *J. Biol. Chem.* **263**, 958. **19**. Alam & Balasubramanian (1978) *Biochim. Biophys. Acta* **524**, 373. **20**. Wallenfels & Weil (1972) *The Enzymes*, 3rd ed. (Boyer, ed.), **7**, 617. **21**. Balbaa, Abdel-Hady, el-Rashidy, *et al.* (1999) *Carbohydr. Res.* **317**, 100. **22**. Sasaoka, Kito & Onishi (1965) *Agric. Biol. Chem.* **29**, 984. **23**. Schilling, Cynis, von Bohlen, *et al.* (2005) *Biochemistry* **44**, 13415. **24**. Baldwin & Cornatzer (1968) *Biochim. Biophys. Acta* **164**, 195. **25**. Waditee, Tanaka, Aoki, *et al.* (2003) *J. Biol. Chem.* **278**, 4932. **26**. Hiles & Henderson (1972) *J. Biol. Chem.* **247**, 646. **27**. Chang (1977) *Enzyme* **22**, 230. **28**. Szumilo, Kaushal & Elbein (1986) *Arch. Biochem. Biophys.* **247**, 261. **29**. Kano, Usami, Adachi, Tatematsu & Hirano (1996) *Biol. Pharm. Bull.* **19**, 341. **30**. Kates & Sastry (1969) *Meth. Enzymol.* **14**, 197. **31**. Jeitner, Delikatny, Ahlqvist, Capper & Cooper (2005) *Biochem. Pharmacol.* **69**, 961. **32**. Federiuk & Shafer (1981) *J. Biol. Chem.* **256**, 7416. **33**. Newman & Kapoor (1980) *Can. J. Biochem.* **58**, 1292. **34**. Grabowski & Buckel (1991) *Eur. J. Biochem.* **199**, 89.

Ethanolamine O-Sulfate

This ethanolamine derivative (FW$_{free-acid}$ = 141.15 g/mol) resembles γ-aminobutyrate and inhibits 4-aminobutyrate aminotransferase. **Target(s):** ω-amino-acid:pyruvate aminotransferase, *or* β-alanine:pyruvate aminotransferase (1); 4-aminobutyrate aminotransferase (2-6). **1.** Yonaha, Toyama & Soda (1987) *Meth. Enzymol.* **143**, 500. **2.** Vasil'ev & Krylova (1976) *Biokhimiia* **41**, 1044. **3.** Fowler (1972) *Biochem. J.* **130**, 80P. **4.** Fowler & John (1972) *Biochem. J.* **130**, 569. **5.** Fowler & John (1981) *Biochem. J.* **197**, 149. **6.** Baxter, Fowler, Miller & Walker (1973) *Brit. J. Pharmacol.* **47**, 681P.

1,N⁶-Ethenoadenosine 5'-Triphosphate

This ATP analogue (FW = 539.18 g/mol), often abbreviated εATP, is a fluorescent probe of ATP binding. **Target(s):** adenylate kinase, also alternative substrate (1); glutamate dehydrogenase (2); GMP synthetase (3); phosphofructokinase (4,5); RNA ligase (ATP) (6). For details of fluorescent properties, consult (7). **1.** Hamada & Kuby (1978) *Arch. Biochem. Biophys.* **190**, 772. **2.** Colman (1990) *The Enzymes*, 3rd ed., **19**, 283. **3.** Spector (1978) *Meth. Enzymol.* **51**, 219. **4.** Roberts & Kellett (1980) *Biochem. J.* **189**, 561. **5.** Roberts & Kellett (1979) *Biochem. J.* **183**, 349. **6.** Iuodka, Labeikite & Sasnauskene (1993) *Biokhimiia* **58**, 857. **7.** Leonard (1984) *CRC Crit. Rev. Biochem.* **15**, 125.

Ethephon

This haloalkylphosphonic acid (FW$_{free-acid}$ = 144.49 g/mol; CAS 16672-87-0), also known as 2-chloroethyl)phosphonic acid, Ethrel, and Cepha, is a synthetic plant growth regulator. Ethephon is taken up by crops, whereupon it slowly releases ethylene, the latter stimulating fruit ripening, growth, *etc.* It is currently registered in the U.S. for use on apples, barley, blackberries, bromeliads, cantaloupes, cherries, coffee, cotton, cucumbers, grapes, guava, macadamia nuts, ornamentals, peppers, pineapples, rye, squash, sugarcane, tobacco, tomatoes, walnuts, wheat, etc. In animals, ethephon inhibits the activity of plasma butyrylcholinesterase in humans, dogs, rats, and mice, phosphorylating the esteratic site (1-4). Ethephon inhibits human butyrylcholinesterase (BChE) by making a covalent adduct on the active site Ser-198. Recent studies suggest that, in aqueous media at neutral to slightly alkaline pH, ~3% of the ethephon is converted ($t_{1/2}$ = 9.9 h at pH 7.0) into a cyclic oxaphosphetane, which is the actual BChE inhibitor that forms a 2-hydroxyethylphosphonate adduct on BChE at Ser-198. The other 97% of the ethephon breaks down ($t_{1/2}$ = 11-48 h at pH 7.0) to form ethylene, which is responsible for plant growth regulation (4). Note: Aqueous solutions are stable below pH 3.5; hence, formulations are typically acidic. Hydrolysis and ethylene generation occur at higher pH above 4.1 (5,6). **Target(s):** butyrylcholinesterase, *or* cholinesterase (1-4); aminopeptidase B (7). **1.** Haux, Lockridge & Casida (2002) *Chem. Res. Toxicol.* **15**, 1527. **2.** Haux, Quistad & Casida (2000) *Chem. Res. Toxicol.* **13**, 646. **3.** Hennighausen, Tiefenbach & Lohs (1977) *Pharmazie* **32**, 181. **4.** Liyasova, Schopfer, Kodani, *et al.* (2013) *Chem. Res. Toxicol.* **26**, 422. **5.** *The Agrochemicals Handbook* (1983) Royal Society of Chemistry, The University of Nottingham, UK. **6.** Worthing, ed. (1983) *The Pesticide Manual: A World Compendium.* 7th ed., Published by The British Crop Protection Council. **7.** Sharma, Padwal-Desai & Ninjoor (1989) *Biochem. Biophys. Res. Commun.* **159**, 464.

Ethionamide

Ethionamide   Ethionamide S-oxide   Postulated Intermediate

This antibiotic (FW = 166.24 g/mol; CAS 536-33-4; Symbol: ETA), also named 2-ethylpyridine-4-carbothioamide, Trecator®, and Trecator SC®, is active against *Mycobacter tuberculosis*, *M. bovis* and *M. segmatis*. ETA is often used as part of a five-drug regimen to treat Multidrug-Resistant Tuberculosis, *or* MDR-TB, as well as Extensively Drug-Resistant Tuberculosis, *or* XDR-TB (**See also** *Isoniazid*). Ethionamide is a prodrug that is oxidized by EtaA, a FAD-containing enzyme, to the corresponding S-oxide. The latter is further oxidized by the same enzyme to 2-ethyl-4-amidopyridine, presumably via the unstable sulfinic acid intermediate. In this way, ETA inhibits the synthesis of mycolic acids, long-chain fatty acids required within the cell walls of *M. tuberculosis*, *M. bovis* and *M. segmatis*. EtaA also oxidatively activates other thioamide antitubercular drugs (*e.g.*, thiacetazone, thiobenzamide, and isothionicotinamide) most likely accounting for TB's known crossover resistance toward structurally related drugs. **1.** Vannelli, Dykman & Ortiz de Montellano (2002) *J. Biol. Chem.* **277**, 12824.

Ethionine

This toxic and carcinogenic methionine analogue (FW = 163.24 g/mol), also known as S-ethylhomocysteine, is an alternative substrate for methionine S-adenosyltransferase. Ethionine toxicity is thought to arise by any or all of the following effects: (a) dead-end trapping of adenine nucleotide by ATP-dependent synthesis and over-accumulation of S-adenosyl-ethionine, which depresses synthesis of NAD+, proteins, and ucleic acids; (b) ethylation of RNA, DNA, or otherwise methylated low-molecular-weight metabolites, which, if ethylated, behave abnormally or compete with less abundant methylated metabolites; (c) ribosome-catalyzed incorporation of ethionine in place of methionine or isoleucince, leading to the formation of faulty proteins and/or enzymes; and (d) failure of cells to make cysteine from S-adenosyl-methionine. **Target(s):** O-acetylhomoserine aminocarboxypropyltransferase, *or* O-acetylhomoserine sulfhydrylase; inhibited by DL-ethionine (1,2); alkaline phosphatase, mildly inhibited by DL-ethionine (3); choline oxidase (4); homoserine O-acetyltransferase (5,6); methionine S adenosyltransferase, also alternative substrate (7-10); methionine S-methyltransferase, weakly inhibited (11); methionyl-tRNA synthetase, also alternative substrate (12); protein biosynthesis (13-16); RNA biosynthesis (17); sarcosine oxidase (4); tRNA (adenine-N¹-)-methyltransferase (18); tRNA (uracil-5-)-methyltransferase (19); tryptophanase, *or* tryptophan indole-lyase (20-23). **1.** Yamagata (1987) *Meth. Enzymol.* **143**, 465. **2.** Yamagata (1984) *J. Biochem.* **96**, 1511. **3.** Fishman & Sie (1971) *Enzymologia* **41**, 141. **4.** Swendseid, Swanson & Bethell (1953) *J. Biol. Chem.* **201**, 803. **5.** Shiio & Ozaki (1981) *J. Biochem.* **89**, 1493. **6.** Wyman & Paulus (1975) *J. Biol. Chem.* **250**, 3897. **7.** Berger & Knodel (2003) *BMC Microbiol.* **3**, 12. **8.** Schröder, Eichel, Breinig & Schröder (1997) *Plant Mol. Biol.* **33**, 211. **9.** Lu & Markham (2002) *J. Biol. Chem.* **277**, 16624. **10.** Yarlett, Garofalo, Goldberg, *et al.* (1993) *Biochim. Biophys. Acta* **1181**, 68. **11.** James, Nolte & Hanson (1995) *J. Biol. Chem.* **270**, 22344. **12.** Burnell (1981) *Plant Physiol.* **67**, 325. **13.** Farber & Corban (1958) *J. Biol. Chem.* **233**, 625. **14.** Villa-Trevino, Shull & Farber (1963) *J. Biol. Chem.* **238**, 1757. **15.** Hyams,

Sabesin, Greenberger & Isselbacher (1966) *Biochim. Biophys. Acta* **125**, 166. **16**. Simpson, Farber & Tarver (1950) *J. Biol. Chem.* **182**, 81. **17**. Villa-Trevino, Shull & Farber (1966) *J. Biol. Chem.* **241**, 4670. **18**. Brahmachari & Ramakrisnan (1984) *Arch. Microbiol.* **140**, 91. **19**. Tscheme & Wainfan (1978) *Nucl. Acids Res.* **5**, 451. **20**. Behbahani-Nejad, Dye & Suelter (1987) *Meth. Enzymol.* **142**, 414. **21**. Behbahani-Nejad, Suelter & Dye (1984) *Cur. Top. Cell. Reg.* **24**, 219. **22**. Suelter & Snell (1977) *J. Biol. Chem.* **252**, 1852. **23**. Snell (1975) *Adv. Enzymol. Relat. Areas Mol. Biol.* **42**, 287.

Ethopropazine

This anticholinergic and antiparkinsonian agent ($FW_{free-base}$ = 312.48 g/mol), also known as 10-(2 diethylaminopropyl)phenothiazine, inhibits butyrylcholinesterases (human K_i = 0.17 μM), but not usually acetylcholinesterases. **1**. Wheeler, Coleman & Finean (1972) *Biochim. Biophys. Acta* **255**, 917. **2**. Martin (1981) *J. Comp. Neurol.* **197**, 153. **3**. Singh, Singh & Agarwal (1982) *J. Physiol.* **78**, 467. **4**. Radic, Pickering, Vellom, Camp & Taylor (1993) *Biochemistry* **32**, 12074. **5**. Augustinsson (1960) *The Enzymes*, 2nd ed., **4**, 521. **6**. Riov & M. J. Jaffe (1973) *Plant Physiol.* **51**, 520. **7**. Darvesh, McDonald, Penwell, *et al.* (2005) *Bioorg. Med. Chem.* **13**, 211. **8**. Golicnik, Sinko, Simeon-Rudolf, Grubic & Stojan (2002) *Arch. Biochem. Biophys.* **398**, 23. **9**. Treskatis, Ebert & Layer (1992) *J. Neurochem.* **58**, 2236.

Ethosuximide

This anticonvulsant agent (FW = 141.17 g/mol; CAS 77-67-8), also known as 3-ethyl-3-methyl-2,5-pyrrolidinedione), is used in the treatment of absence seizures, the brief epileptic seizures experienced in childhood and adolescence and characterized by sudden loss of awareness and a characteristic electroencephalogram. Ethosuximide inhibits NADPH-linked aldehyde reductase (1,2) needed to form γ-hydroxybutyrate, which has been associated with the induction of absence seizures. Ethosuximide also blocks T-type Ca^{2+} channels (3,4). **1**. Roig, Bello, Burguillo, Cachaza & Kennedy (1991) *J. Pharm. Sci.* **80**, 267. **2**. Erwin & Deitrich (1973) *Biochem. Pharmacol.* **22**, 2615. **3**. Czapinski, Blaszczyk & Czuczwar (2005) *Curr. Top. Med. Chem.* **5**, 3. **4**. Lacinova (2004) *Curr. Drug Targets CNS Neurol. Disord.* **3**, 105.

(R)-2-(4-Ethoxycarbonylaminobenzenesulfonyl)-1,2,3,4-tetrahydroisoquinoline-3-carboxylic Acid Hydroxamate

This isoquinoline (FW = 419.45 g/mol) inhibits human neutrophil collagenase, *or* matrix metalloproteinase-8, IC_{50} = 30 nM. The 3-ethoxycarbonylamino analogue has an IC_{50} value of 300 nM. **1**. Matter, Schwab, Barbier, *et al.* (1999) *J. Med. Chem.* **42**, 1908.

2-Ethoxy-1-ethoxycarbonyl-1,2-dihydroquinoline

This *O*-substituted dihydroquinoline carbamate (FW = 247.29 g/mol; CAS 16357-59-8; Abbreviation: EEDQ) is a coupling reagent that reacts with carboxyl groups and is likewise an irreversible inhibitor of enzymes possessing catalytically essential carboxyl groups. When used as a carboxyl group-activating reagent in peptide synthesis, little or no racemization occurs. In its action as an enzyme-modifying reagent, transient formation of a mixed carbonic anhydride is indicated. The bulky dihydroquinoline moiety is apt to strongly influence the binding EEDQ to many proteins and enzymes. **Target(s):** acetylcholine receptor ionic channel (1); α_2-adrenergic receptor (2); aminoacylase (3); arylformamidase (4); bacterial leucyl aminopeptidase (5); Ca^{2+} channel, voltage-sensitive (6); Ca^{2+}-transporting ATPase (7); choline uptake (8); coccolysin (9); cytochrome bc_1 complex proton pump, mitochondrial (10); D_1 and D_2 dopamine receptors (11,12); F_1 ATPase (13-18); glucose transporter (19); β-glucosidase (20); H^+-exporting ATPase (21); histamine receptor (22); H^+/K^+-exchanging ATPase (23-25); D-β-hydroxybutyrate dehydrogenase (26); lysolecithin:lysolecithin acyltransferase, hydrolysis inhibited at pH 5 (27); membrane dipeptidase (renal dipeptidase; dehydropeptidase I (28); Na^+ channels (29); Na^+/H^+-antiporter (30); nicotinamide-nucleotide transhydrogenase (31-33); 5'-nucleotidase (34); penicillin acylase (35); prolyl aminopeptidase (36); serotonin receptor (37-39); TF_1 ATPase (40); thiamin-binding protein (41); tuna (*Thunnus albacares*) pyloric caeca aminopeptidase (42); UDP-glucose pyrophosphorylase, *or* glucose-1-phosphate uridylyltransferase (43). **1**. Eldefrawi, Eldefrawi, Mansour, *et al.* (1978) *Biochemistry* **17**, 5474. **2**. Adler, Meller & Goldstein (1985) *Eur. J. Pharmacol.* **116**, 175. **3**. Loffler & Schneider (1987) *Biol. Chem. Hoppe-Seyler* **368**, 481. **4**. Seifert & Casida (1979) *Pestic. Biochem. Physiol.* **12**, 273. **5**. Mäkinen, Mäkinen, Wilkes, Bayliss & Prescott (1982) *Eur. J. Biochem.* **128**, 257. **6**. Gopalakrishnan & Triggle (1990) *Biochem. Pharmacol.* **40**, 327. **7**. Argaman & Shoshan-Barmatz (1988) *J. Biol. Chem.* **263**, 6315. **8**. Vickroy & Malphurs (1994) *Biochem. Pharmacol.* **48**, 1281. **9**. Mäkinen, Clewell, An & Mäkinen (1989) *J. Biol. Chem.* **264**, 3325. **10**. Cocco, Di Paola, Papa & Lorusso (1999) *FEBS Lett.* **456**, 37. **11**. Hamblin & Creese (1983) *Life Sci.* **32**, 2247. **12**. Hess, Norman & Creese (1988) *J. Neurosci.* **8**, 2361. **13**. Satre, Lunardi, Dianoux, *et al.* (1986) *Meth. Enzymol.* **126**, 712. **14**. Allison, Bullough & Andrews (1986) *Meth. Enzymol.* **126**, 741. **15**. Khananshvili & Gromet-Elhanan (1983) *J. Biol. Chem.* **258**, 3720. **16**. Pougeois (1983) *FEBS Lett.* **154**, 47. **17**. Matsuno-Yagi & Hatefi (1984) *Biochemistry* **23**, 3508. **18**. Pougeois, Satre & Vignais (1978) *Biochemistry* **17**, 3018. **19**. Turner (1986) *J. Biol. Chem.* **261**, 1041. **20**. de la Mata, Castillon, Dominguez, Macarron & Acebal (1993) *J. Biochem.* **114**, 754. **21**. Serrano (1988) *Biochim. Biophys. Acta* **947**, 1. **22**. Detzner, Kathmann & Schlicker (1994) *Agents Actions* **41** Spec. No., C66. **23**. Sachs, Berglindh, Rabon, *et al.* (1980) *Ann. N. Y. Acad. Sci.* **358**, 118. **24**. Saccomani, Barcellona & Sachs (1981) *J. Biol. Chem.* **256**, 12405. **25**. De Pont & Bonting (1981) *New Compr. Biochem.* **2**, 209. **26**. Prasad & Hatefi (1986) *Biochemistry* **25**, 2459. **27**. Perez-Gil, Martin, Acebal & Arche (1990) *Arch. Biochem. Biophys.* **277**, 80. **28**. Adachi, Katayama, Nakazato & Tsujimoto (1993) *Biochim. Biophys. Acta* **1163**, 42. **29**. Park & Fanestil (1983) *Amer. J. Physiol.* **245**, F716. **30**. Niwano, Murakami, Okazawa & Konishi (1991) *Biochem. Int.* **25**, 173. **31**. Phelps & Hatefi (1984) *Biochemistry* **23**, 6340. **32**. Phelps & Hatefi (1985) *Arch. Biochem. Biophys.* **243**, 298. **33**. Yamaguchi & Hatefi (1993) *J. Biol. Chem.* **268**, 17871. **34**. Harb, Meflah, di Pietro, Bernard & Gautheron (1986) *Biochim. Biophys. Acta* **870**, 320. **35**. Martin, Mancheno & Arche (1993) *Biochem. J.* **291**, 907. **36**. Mäkinen, Syed, Mäkinen & Loesche (1987) *Curr. Microbiol.* **14**, 341. **37**. Biessen, Norder, Horn & Robillard (1988) *Biochem. Pharmacol.* **37**, 3959. **38**. Adham, Ellerbrock, Hartig, Weinshank & Branchek (1993) *Mol. Pharmacol.* **43**, 427. **39**. Ni, Camacho & Miledi (1997) *Proc. Natl. Acad. Sci. U.S.A.* **94**, 2715. **40**. Yoshida & Allison (1986) *J. Biol. Chem.* **261**, 5714. **41**. Nishimura, Sempuku, Nosaka & Iwashima (1984) *J. Biochem.* **96**, 1289. **42**. Hajjou & Le Gal (1995) *Biochim. Biophys. Acta* **1251**, 139. **43**. Signorini, Ferrari, Mariotti, Dallocchio & Bergamini (1989) *Biochem. J.* **264**, 799.

4-N-(4-Ethoxyphenyl)amino-5-phenyl-7-(β-D-erythro-furanosyl)pyrrolo[2,3-d]pyrimidine

This tubercidin analogue (FW = 417.49 g/mol) inhibits human adenosine kinase (IC$_{50}$ = 3 nM). Because the use of adenosine receptor agonists is plagued by dose-limiting cardiovascular side effects, there is great interest in developing adenosine kinase inhibitors (AKIs) as a means for raising steady-state adenosine concentrations. The 3-ethoxyphenyl analogue also inhibits (IC$_{50}$ = 2 nM). **1**. Boyer, Ugarkar, Solbach, et al. (2005) J. Med. Chem. **48**, 6430.

Ethoxzolamide

This diuretic (FW = 258.32 g/mol; CAS 452-35-7), also named 6-ethoxy-1,3-benzothiazole-2-sulfonamide, inhibits human carbonic anhydrases I and II, both with K_i values near 0.3 nM. Ethoxzolamide also inhibits carbonic anhydrase activity in kidney proximal tubules, decreasing reabsorption of water, sodium, potassium, and bicarbonate. It also decreases carbonic anhydrase in the CNS, increasing the seizure threshold in epilepsy. Ethoxzolamide is used to treat glaucoma and duodenal ulcers. **Target(s):** Ca^{2+} channels, high-voltage activated voltage-sensitive, encoded by the human α$_{1E}$ subunit (1); carbonic anhydrase (2-18). **1**. McNaughton, Davies & Randall (2004) J. Pharmacol. Sci. **95**, 240.   **2**. Chen & Kernohan (1967) J. Biol. Chem. **242**, 5813.   **3**. Gordon (1958) Amer. J. Ophthalmol. **46**, 41.   **4**. Garg (1974) Biochem. Pharmacol. **23**, 3153.   **5**. Schoenwald, Eller, Dixson & Barfknecht (1984) J. Med. Chem. **27**, 810.   **6**. Conroy & Maren (1995) Mol. Pharmacol. **48**, 486.   **7**. Maren, Parcell & Malik (1960) J. Pharmacol. Exp. Ther. **130**, 389.   **8**. Maren (1963) J. Pharmacol. Exp. Ther. **139**, 129 and 140.   **9**. Winum, Vullo, Casini, et al. (2003) J. Med. Chem. **46**, 2197.   **10**. Rumeau, Cuine, Fina, et al. (1996) Planta **199**, 79.   **11**. Brundell, Falkbring & Nyman (1972) Biochim. Biophys. Acta **284**, 311.   **12**. Elleby, Chirica, Tu, Zeppezauer & Lindskog (2001) Eur. J. Biochem. **268**, 1613.   **13**. Sanyal, Pessah & Maren (1981) Biochim. Biophys. Acta **657**, 128.   **14**. Wingo, Tu, Laipis & Silverman (2001) Biochim. Biophys. Res. Commun. **288**, 666.   **15**. Heck, Tanhauser, Manda, et al. (1994) J. Biol. Chem. **269**, 24742.   **16**. Earnhardt, Qian, Tu, et al. (1998) Biochemistry **37**, 10837.   **17**. Siffert & Gros (1984) Biochem. J. **217**, 727.   **18**. Hurt, Tu, Laipis & Silverman (1997) J. Biol. Chem. **272**, 13512.

N^ω-Ethyl-L-arginine

This arginine derivative (FW = 202.26 g/mol) inhibits nitric oxide synthase, with IC$_{50}$ values of 17.8, 42.8, and 12.2 μM for neuronal, inducible, and endothelial enzymes (1-3) and amino acid transporter y$^+$ (4). **1**. Griffith &

Kilbourn (1996) Meth. Enzymol. **268**, 375.   **2**. Griffith & Stuehr (1995) Ann. Rev. Physiol. **57**, 707.   **3**. Babu, Frey & Griffith (1999) J. Biol. Chem. **274**, 25218.   **4**. McDonald, Rouhani, Handlogten, et al. (1997) Biochim. Biophys. Acta **1324**, 133.

2-Ethylcaproic Acid

This carboxylic acid (FW$_{free-acid}$ = 144.21 g/mol), also known as 2-ethylhexanoic acid, is an industrial chemical and a toxic biotransformation product of the widely-used plasticizer, di(2-ethyl-hexyl)-phthalate. 2-Ethylcaproate inhibits acetyl-CoA carboxylase (1), aldehyde reductase (NADPH) (2), kynurenine aminotransferase (3), and UDP-glucuronosyltransferase (4). **1**. Sumida, Kaneko & Yoshitake (1996) Chemosphere **33**, 2201.   **2**. De Jongh, Schofield & Edwards (1987) Biochem. J. **242**, 143.   **3**. Mason (1959) J. Biol. Chem. **234**, 2770.   **4**. Dutton & Storey (1962) Meth. Enzymol. **5**, 159.

Ethylene

This plant ripening gas (FW = 28.054 g/mol; M.P. = –181 °C; B.P. = –102.4 °C at 700 mm Hg), also known as ethene and fruit-ripening hormone, occurs widely in plants, where it initiates flower opening, accelerates fruit maturation, and promotes leave abscission. It is produced from essentially all parts of higher plants, including leaves, stems, roots, flowers, fruits, tubers, and seeds. Ethylene also inhibits the biosynthesis of some auxins. The earliest report on the effects of ethylene was in 1901, when it was observed that the gas caused a loss in geotropism and inhibited epicotyl elongation in dark-grown pea seedlings (1). In 1943, it was shown that plants synthesized ethylene and used it to initiate ripening (2). Ethylene is a product of the reaction catalyzed by aminocyclopropanecarboxylate oxidase. Such details underscore the cynicism of plant scientists about the erroneous impression that nitric oxide was the first metabolic signaling gas. Synthetic ethylene-liberating agents (**See Ethephon**) induce fruit ripening, flowering, root initiation, cell expansion, seed germination, control of stress responses, leaf abscission, etc. **Target(s):** cytochrome P450 (3); polar auxin transport (4). **1**. Neljubow (1901) Beih. Bot. Zentralbl. **10**, 128.   **2**. Gane (1934) Nature **134**, 1008.   **3**. Ator & Ortiz de Montellano (1990) The Enzymes, 3rd ed., **19**, 213.   **4**. Burg & Burg (1967) Plant Physiol. **42**, 1224.

Ethylene Dibromide

This toxic dihaloalkane (FW = 187.86 g/mol; CAS 106-93-4; Symbol: EDB), also known as 1,2-dibromoethane, is a common intermediate in organic synthesis and industrial chemical. Once used as an anti-knocking additive for internal combustion engines in automotive and jets, EDB is still employed in the third world as an insecticide and fumigant in wheat, flour, bran, middlings, and bread. **CAUTION**: Human LD$_{50}$ = 90 mg/kg. Dibromoethane is also a demonstrated mutagen and carcinogen (1), meriting caution when used in the laboratory. Glutathione (GSH) reacts with EDB, and the resulting S-ethylbromide derivative participates in (GSH)-dependent base-substitution mutations (especially GC-to-AT transitions) in a variety of bacterial and eukaryotic systems. Known DNA adducts include: S-[2-(N^7-guanyl)ethyl]GSH, S-[2-(N^2-guanyl)ethyl]GSH, and I-[2-(O^6-guanyl)ethyl]GSH. These modifications resist DNA repair (2), thereby hampering the action of DNA polymerases (3). **1**. Fishbein (1976) Environ. Health Perspect. **14**, 39.   **2**. Nachtomi & Sarma (1977) Biochem. Pharmacol. **26**, 1941.   **3**. Kim & Guengerich (1998) Chem. Res. Toxicol. **11**, 311.

4-N-(3,4-Ethylenedioxyphenyl)amino-5-(2-methylphenyl)-7-(β-D-erythro-furanosyl)-pyrrolo[2,3-d]pyrimidine

This tubercidin analogue (FW = 460.49 g/mol) inhibits human adenosine kinase (IC$_{50}$ = 1 nM). Because the use of adenosine receptor agonists is plagued by dose-limiting cardiovascular side effects, there is great interest in developing adenosine kinase inhibitors (AKIs) as a means for raising steady-state levels of adenosine. **1.** Boyer, Ugarkar, Solbach, *et al.* (2005) *J. Med. Chem.* **48**, 6430.

Ethylene Glycol

This poisonous dihydroxyalkane HO–CH$_2$CH$_2$–OH (FW = 62.07 g/mol; CAS 107-21-1; density = 1.11 g/cm^3; miscible with H$_2$O; soluble in polar organic solvents), also named ethane-1,2-diol, is used in the synthesis of polyester fibers and polyethylene terephthalate, and a common radiator antifreeze additive. Ethylene glycol uses water or glycerol channels to gain entry into many cell types. **Toxicity:** Ethylene glycol's sweet taste puts children and unwary animals at risk, because it is readily oxidized to oxalate, a toxic metabolite (oral LD$_{50}$ = 0.8 g/kg in humans). **See** *Oxalate (or Oxalic Acid)*

Ethylenimine

This poisonous cyclic amine (FW = 43.07 g/mol; Polymerizes easily), also known as aziridine, reacts with sulfhydryl group(s) in proteins under slightly alkaline conditions, producing *S*-(2-aminoethyl)-cysteinyl residue(s). The latter is a structural analogue of a lysyl residue, in which the γ-methylene group has been replaced with sulfur (1). This derivatized cysteinyl residue is also susceptible to trypsin cleavage. If the cysteinyl residue was critical for catalysis, modification by ethylenimine will result in enzyme inactivation. Under acidic conditions, methionyl residues react slowly with ethyleneimine to form the corresponding sulfonium ions. **Target(s):** choline *O*-acetyltransferase (2); histidinol-phosphatase (3); lysozyme (4). **1.** Raftery & Cole (1963) *Biochem. Biophys. Res. Comm.* **10**, 467. **2.** Roskoski (1974) *J. Biol. Chem.* **249**, 2156. **3.** Houston & Millay (1974) *Biochim. Biophys. Acta* **370**, 216. **4.** Yamada, Imoto & Noshita (1982) *Biochemistry* **21**, 2187.

Ethylguanidine

This hygroscopic alkyl guanidine (FW$_{hydrochloride}$ = 123.59 g/mol) inhibits clostripain (anenzyme from *Clostridium histolyticum* that cleaves proteins on the carboxyl peptide bond of arginine) and other arginine-dependent enzymes. **Target(s):** acetylcholine-induced currents (1,2); arginine deaminase (3); arginine kinase (4); clostripain (5,6); 5-guanidino-2-oxopentanoate decarboxylase (7); lipase (8); trypsin (6,9). **1.** Farley & Narahashi (1983) *J. Physiol.* **337**, 753. **2.** Farley, Vogel & Narahashi (1986) *Pflugers Arch.* **406**, 629. **3.** Smith, Ganaway & Fahrney (1978) *J. Biol. Chem.* **253**, 6016. **4.** Pereira, Alonso, Paveto, *et al.* (2000) *J. Biol.*

Chem. **275**, 1495. **5.** Mitchell & Harrington (1971) *The Enzymes*, 3rd. ed., **3**, 699. **6.** Cole, Murakami & Inagami (1971) *Biochemistry* **10**, 4246. **7.** Vanderbilt, Gaby, Rodwell & Bahler (1975) *J. Biol. Chem.* **250**, 5322. **8.** Holmquist, Norin & Hult (1993) *Lipids* **28**, 721. **9.** Inagami & York (1968) *Biochemistry* **7**, 4045.

2-Ethylhexyl Phthalate

This monoester (FW$_{free-acid}$ = 278.35 g/mol), also known as mono(2-ethylhexyl) phthalate, is a common metabolite of bis(2-ethylhexyl) phthalate, a widely-used plasticizer in the plastic industry. Blood products stored in polyvinyl chloride bags often contain this diester, from which the monoester can be produced via action of a plasma enzyme. **Target(s):** aminocarboxymuconate-semialdehyde decarboxylase (1); phospholipase A$_2$ (2). **1.** Fukuwatari, Ohsaki, Fukuoka, Sasaki & Shibata (2004) *Toxicol. Sci.* **81**, 302. **2.** Labow, Meek, Adams & Rock (1988) *Environ. Health Perspect.* **78**, 179.

2-([3-[(7-Ethyl-5-iodo-1,2,3,4-tetrahydroquinolin-4-yl)amino]propyl]amino)-thieno[3,2-d]pyrimidin-4(1H)-one

This substituted thienopyrimidinone (FW = 523.44 g/mol) inhibits methionyl-tRNA synthetase, with IC$_{50}$ < 3 nM for the *Staphylococcus aureus* enzyme. **1.** Jarvest, Armstrong, Berge, *et al.* (2004) *Bioorg. Med. Chem. Lett.* **14**, 3937.

S-Ethylisothiouronium diethylphosphate

This antihypotensive drug (FW = 258.28 g/mol; CAS 21704-46-1), also known by the trade name Difetur®, is a specific inhibitor of inducible NO synthase that can be used to treat acute hypotension resulting from shock or hemorrhage. *S*-Ethylisothiouronium diethylphosphate is also used for the treatment of headaches, mainly migraines.

Ethylnitrosourea

This powerful mutagen (FW = 117.11 g/mol; CAS 759-73-9; *Abbreviation:* ENU), also known as 1-ethyl-1-nitrosourea, is a cell cycle-independent DNA alkylating agent that interferes with replication and transcription. (*See*

Nitrosourea) In a landmark study on chemically induced mutagenesis, Russell et al. (1) demonstrated that fractionating a single ethylnitrosourea dose from 100 mg/kg body weight into a series doses of 10 mg/kg injected intraperitoneally given at weekly intervals greatly reduces the mutation frequency. Moreover, because there is independent evidence that the doses of 10 and 100 mg/kg reach the germ cells in amounts directly proportional to the injected dose, the lower mutational response with the fractionated dose is attributed to repair. The mutation frequency was twelve times greater than that induced by high-dose X-radiation and thirty-six times that of procarbazine, another power mutagen. For a given gene in mice, ENU induces one new mutation in every seven hundred loci. Given such a high level of mutagenicity, ethylnitrosourea is not used as an anticancer drug. That said, ENU remains a versatile mechanistic probe in mouse molecular genetics (2,3). **1.** Russell, Hunsicker, Carpenter, Cornett & Guinn (1982) *Proc. Natl. Acad. Sci. U.S.A.* **79**, 3592. **2.** Gondo Fukumura Murata & Makino (2010) *Exp. Anim.* **59**, 5373. **3.** Oliver & Davies (2012) *Hum. Molec. Genet.* **21**, R72.

Ethyl Protocatechuate

This plant antioxidant (FW = 182.17 g/mol; CAS 3943-89-3), also named ethyl 3,4-dihydroxybenzoate (EDHB), inhibits procollagen-proline dioxygenase, *or* 4-prolyl hydroxylase (*Reaction*: Procollagen-L-proline + 2-Oxoglutarate (*or* α-Ketoglutarate) + O_2 → Procollagen-(2S,4R)-4-hydroxyproline + Succinate + CO_2), the collagen-processing enzyme that requires vitamin C as an anti-scurvy factor. As a structural analogue of α-ketoglutarate and ascorbate, EDHB markedly inhibits (K_i = 0.4 mM) the synthesis of 4-hydroxyproline in normal cell cultures, markedly reducing the synthesis and secretion of Type I and Type III procollagens (1). Control experiments indicate that the test compound does not affect the viability, proliferation, or plating efficiency of the cells, and it has little, if any, effect on the synthesis of noncollagenous proteins. Although a weaker Fe(III) chelator than desferrioxamine (*See Deferiprone; Desferrioxamine B; Ferrioxamines*), EDHB promotes iron deficiency when applied to cultured cells (2). EDHB triggers the activation of IRP1, a bifunctional iron-responsive element (IRE)-binding protein that controls iron metabolism by binding mRNA and repressing translation or degradation. EDHB modulates the coordinate expression of downstream IRP1 targets, such as the transferrin receptor and ferritin. Notably, EDHB has little, if any, effect on the hydroxylation of lysyl residues, and it does not affect total protein synthesis or DNA replication in these cells (3). EDHB also protects the myocardium by activating NOS and generating mitochondrial reactive oxygen species, *or* ROS (4). EDHB increases ROS generation by 50-75% in isolated rabbit cardiomyocytes. That this effect is mediated by NO is indicated by the inhibition by the NO synthase inhibitor N^ω-nitro-L-arginine methyl ester (*or* L-NAME), the guanylyl cyclase antagonist ODQ; and the Protein Kinase G blocker Rp-8-bromoguanosine-3',5'-cyclic phosphorothioate (4). Use of flow cytometry analysis demonstrated that ethyl-3,4-dihydroxybenzoate induces cell accumulation in S-phase, loss in mitochondrial membrane permeabilization, as well as caspase-dependent apoptosis (5). An expression profile analysis identified forty-six up-regulated and nine down-regulated genes in EDHB-treated esophageal cancer KYSE 170 cells, and these differentially expressed genes operate in several signaling pathways associated with cell cycle regulation and cellular metabolism (5). EDHB treatment also increases the protein levels of the hypoxia-induced protein (Hif-1α), the BCL2/adenovirus protein-interacting protein-3 (BNIP3), mammalian ortholog of *C. elegans* BEC-1 (Beclin), and N-myc Down-Regulated Gene-1 protein (NDRG1). **Other Target(s):** Because prolyl hydroxylase acts on proteins containing Xaa-Pro-Gly tripeptide motifs, it hydrolyxates prolyl residues in other proteins (*e.g.*, C1q, elastins, prion protein, argonaute-2, and several conotoxins). For this reason, in the absence of additional evidence, one cannot unambiguously define an effect of EDHB as the consequence of its inhibitory effects on

collagen processing. **1.** Sasaki, Majamaa & Uitto (1987) *J. Biol. Chem.* **262**, 9397. **2.** Wang, Buss, Chen, Ponka & Pantopoulos (2002) FEBS Lett. **529**, 309. **3.** Majamaa, Sasaki & Uitto (1987) *J. Invest. Dermatol.* **89**, 405. **4.** Philipp, Cui, Ludolph, *et al.* (2006) *Am. J. Physiol. Heart Circ. Physiol.* **290**, H450. **5.** Han, Li, Sun, *et al.* (2014) *PLoS One* **9**, e107204.

S-Ethyl-N-(4-[trifluoromethyl]phenyl)isothiourea

This reagent ($FW_{free-base}$ = 248.27 g/mol), also called ETPI and TFPI, inhibits endothelial nitric oxide synthase (IC_{50} = 9.4 μM), brain NOS (IC_{50} = 0.32 μM), and inducible NOS (IC_{50} = 37 μM (1,2). **1.** Shearer, Lee, Oplinger, *et al.* (1997) *J. Med. Chem.* **40**, 1901. **2.** Cooper, Mialkowski & Wolff (2000) *Arch. Biochem. Biophys.*

3'-Ethynyl-Cytosine

This novel synthetic cytosine nucleoside (FW = 267.24 g/mol), also known as TAS-106, Ecyd, and 1-(3-C-ethynyl-β-D-ribo-pento-furanosyl)-cytosine, inhibits RNA polymerases I, II and III. When tested *in vivo*, TAS-106 alone showed significant antitumor activity that was enhanced, when used in combination with cisplatin to inhibit the growth of several tumor xenografts (1,2). In a panel of six nasopharyngeal cancer (NPC) cell lines, ECyd effectively inhibited cellular proliferation, with IC_{50} values of 13 to 44 nM (3). (Cisplatin-resistant NPC cells were highly sensitive to ECyd, again in the low-nM range.). ECyd-mediated tumor growth inhibition is associated with G_2/M cell cycle arrest, PARP cleavage (a hallmark of apoptosis) and Bcl-2 downregulation. ECyd also significantly downregulates fructose-2,6-bisphosphatase (EC 3.1.3.46) (also known as p53 Tumor-suppressor protein- (TP53-) Induced Glycolysis and Apoptosis Regulator (TIGAR)), thereby lowering fructose-2,6-bisphosphate in ECyd-treated cells. Notably, ECyd-downregulation of TIGAR was also accompanied by marked depletion of NADPH, the major reducing agent that is critically required for cell proliferation and survival (2). Cell lines showing a high degree of cross-resistance to ECyd and EUrd have deletions in the gene for uridine/cytosine kinase-2, *or* UCK2, suggesting that the latter is responsible for the phosphorylation and activation of 3'-ethynyl nucleoside pro-drugs (3). **1.** Kazuno, Fujioka, Fukushima, *et al.* (2009) *Int. J. Oncol.* **34**, 1373. **2.** Lui, Lau, Cheung, *et al.* (2010) *Biochem. Pharmacol.* **79**, 1772. **3.** Murata, Endo, Obata, *et al.* (2004) *Drug Metab. Dispos.* **32**, 1178.

5-Ethynyl-2'-deoxyuridine

This thymidine analogue (FW = 252.23 g/mol; CAS 61135-33-9; IUPAC Name: 5-ethynyl-1-[(2R,4S,5R)-4-hydroxy-5-(hydroxymethyl)oxolan-2-yl]pyrimidine-2,4-dione; Symbol: EdU), which is readily taken up by living cells, is subsequently converted to EdUTP (5-ethynyl-2'-deoxyuridine 5'-triphosphate). The latter is then incorporated into cellular DNA (and mitochondrial DNA) during replication. Incorporated EdU can be visualized via "click" chemistry using fluorescent azides in a Cu(I)-catalyzed [3 + 2] cycloaddition (1-3).

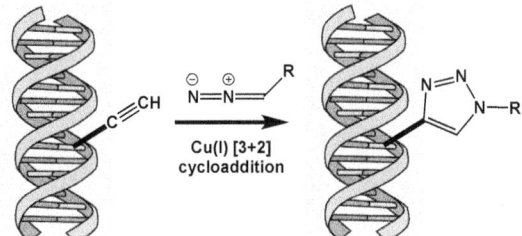

The resulting triazole ring is formed by stable covalent coupling to an Alexa Fluor®-conjugated azide group, affording a means for localizing subcellular sites of EdU incorporation. This detection method, which is highly sensitive and much faster than BrdU detection, enjoys the advantage that the required low-molecular-weight reagents have a much higher diffusivity and far greater penetration of tissue specimens, allowing for rapid, whole-mount staining of tissue and organ fragments. Moreover, because reaction with fluorescent azide probes does not require denaturation of the specimen, this method affords structural preservation that is suitable for high-resolution microscopic analysis *in vivo* of DNA repair, replication, recombination, and sister chromatid cohesion (1). Although EdU induces cell-cycle arrest at the G2/M interface in several breast cancer cell lines, and although it also specifically causes necrotic cell death, when used as a short pulse EdU at concentration that are only 1/200th the standard concentration of BrdU, such effects can largely be avoided (4). While EdU is employed as a replication marker more frequently than 5-ethynyl-2'-deoxycytidine (EdC), its cytotoxicity is commonly much higher than the toxicity of EdC (5). DNA analysis EdC-treated human cell lines showed that, while of all the cell lines contained EdU, none contained a detectable EdC in their DNA. The extent of EdC-related cytotoxicity was instead directly proportional to the content of EdU (6). Such effects can be explained by findings (a) EdC is a substrate for cytidine deaminase, thereby forming EdU, (b) EdC deamination *per se* does not prevent the conversion of EdC to EdCTP, the DNA polymerase substrate, and (c) EdCTP is not effectively recognized by the replication complex as a substrate for synthesis of nuclear DNA (5). **See** *Thymidine; 5-Bromo-2'-deoxyuridine (BrdU), 5-Ethynyluracil* **1.** Salic & Mitchison (2008) *Proc. Natl. Acad. Sci. U.S.A.* **105**, 2415. **2.** Buck, Bradford, Gee, *et al.* (2008) *Biotechniques* **44**, 927. **3.** Cappella, Gasparri, Pulici & Moll (2008) *Cytometry A.* **73**, 626. **4.** Diermeier-Daucher, Clarke, Hill, *et al.* (2009) *Cytometry A.* **75**, 535. **5.** Qu, Wang, Wang Z, *et al.* (2011) *Anal. Biochem.* **417**, 112. **5.** Ligasová, Liboska, Friedecký, *et al.* (2016) *Open Biology* **6**, 150172.

17α-Ethynylestradiol

This synthetic estrogen (FW = 296.41 g/mol), also known as 17α-ethinylestradiol and 19-nor-17α-pregna 1,3,5(10)-trien-20-yne-3,17β-diol, is a component of oral contraceptives. **Target(s):** aldehyde oxidase (1); CYP3A4 (2,3); CYP3A27 (4); CYP2B1 (5,6); CYP2C8 (7); 3β-hydroxysteroid dehydrogenase (8); 17β-hydroxysteroid dehydrogenase (9); kynurenine aminotransferase (10); Na$^+$/K$^+$-exchanging ATPase (11); steroid Δ-isomerase (12); steroid 16α-monooxygenase (13,14); thymidine 5'-monophosphate kinase (15); tryptophan 2,3-dioxygenase (16). **1.** Obach, Huynh, Allen & Beedham (2004) *J. Clin. Pharmacol.* **44**, 7. **2.** Jang, Wrighton & Benet (1996) *Biochem. Pharmacol.* **52**, 753. **3.** Lin, Kent & Hollenberg (2002) *J. Pharmacol. Exp. Ther.* **301**, 160. **4.** Miranda, Henderson & Buhler (1998) *Toxicol. Appl. Pharmacol.* **148**, 237. **5.** Von

Weymarn, Sridar & Hollenberg (2004) *J. Pharmacol. Exp. Ther.* **311**, 71. **6.** Lin, Kent, Zhang, Waskell & Hollenberg (2004) *J. Pharmacol. Exp. Ther.* **311**, 855. **7.** Walsky, Gaman & Obach (2005) *J. Clin. Pharmacol.* **45**, 68. **8.** Yoshida (1979) *Steroids* **33**, 9. **9.** Jarabak & Sack (1969) *Biochemistry* **8**, 2203. **10.** El-Zoghby, El-Sewedy, Saad, *et al.* (1976) *Biochem. Pharmacol.* **25**, 2411. **11.** Schwarz, H. E. Bostwick & M. S. Medow (1988) *Amer. J. Physiol.* **254**, G687. **12.** Weintraub, Vincent, Baulieu & Alfsen (1977) *Biochemistry* **16**, 5045. **13.** Sano, Shibusawa, Yoshida, *et al.* (1980) *Acta Obstet. Gynecol. Scand.* **59**, 245. **14.** Numazawa & Satoh (1989) *J. Steroid Biochem.* **32**, 85. **15.** Nelson & Carter (1967) *Mol. Pharmacol.* **3**, 341. **16.** Braidman & Rose (1970) *Biochem. J.* **118**, 7P.

5-Ethynyluracil

This substituted pyrimidine (FW = 136.11 g/mol) inactivates dihydropyrimidine dehydrogenase, initally forming a reversible complex (K_i = 1.6 μM), followed by inactivation (k_{inact} = 20 min^{-1}). It will also inactivate thymine dioxygenase (K_i = 22 μM; k_{inact} = 2.6 min^{-1}). Cell lines showing a high degree of cross-resistance to ECyd and EUrd have deletions in the gene for uridine/cytosine kinase-2, *or* UCK2, suggesting that the latter is responsible for the phosphorylation and activation of 3'-ethynyl nucleoside pro-drugs (1). **Target(s):** dihydropyrimidine dehydrogenase (2-4); thymine dioxygenase, *or* thymine 7-hydroxylase (5,6); xanthine oxidase (7). **1.** Murata, Endo, Obata, *et al.* (2004) *Drug Metab. Dispos.* **32**, 1178. **2.** Spector, Harrington & Porter (1993) *Biochem. Pharmacol.* **46**, 2243. **3.** Porter, Harrington, Almond, *et al.* (1994) *Biochem. Pharmacol.* **47**, 1165. **4.** Porter, Chestnut, Merrill & Spector (1992) *J. Biol. Chem.* **267**, 5236. **5.** Thornburg & Stubbe (1993) *Biochemistry* **32**, 14034. **6.** Lai, Wu & Stubbe (1995) *J. Amer. Chem. Soc.* **117**, 5023. **7.** Porter (1994) *J. Biol. Chem.* **269**, 27932.

Etidronate (Etidronic Acid)

This bone-seeking pyrophosphate analogue (FW = 204.01 g/mol; Very soluble in water; pK_a values of 1.35, 2.87, 7.03, and 11.3), also known as 1,1-bis(phosphono)ethanol, ethane-1-hydroxy-1,1-diphosphonate, and (1-hydroxyethylidene)bisphosphonate, inhibits bone resorption in tissue culture and in experimental animals and chelates metal ions. Etidronate is an oral and intravenously active drug that is used in the management of resorptive bone disease, treatment of Paget's disease, the prevention of heterotopic ossification, and in postmenopausal osteoporosis. By binding to bone surfaces, etidronate is readily absorbed by osteoclasts, with consequential inhibition of geranyl*trans*transferase and vacuolar H$^+$-transporting ATPase (*or* V-ATPase) as well as retraction of bone-resorbing osteoclast podosomes. **Target(s):** adenosine kinase (1); exopoly-phosphatase (2); geranyl*trans*transferase, *or* farnesyl diphosphate synthase (3,4); H$^+$-transporting ATPase, vacuolar, weakly inhibited (5); H$^+$-transporting pyrophosphatase, vacuolar (6); 3-phosphoglycerate kinase (7); pyrophosphatase (6,8-10); pyrophosphate-dependent phosphofructokinase (11,12); pyruvate,orthophosphate dikinase (13); ribokinase (14). **1.** Park, Singh & Gupta (2006) *Mol. Cell. Biochem.* **283**, 11. **2.** Leyhausen, Lorenz, Zhu, *et al.* (1998) *J. Bone Miner. Res.* **13**, 803. **3.** Montalvetti, Bailey, Martin, *et al.* (2001) *J. Biol. Chem.* **276**, 33930. **4.** Bergstrom, Bostedor, Masarachia, Reszka & Rodan (2000) *Arch. Biochem. Biophys.* **373**, 231. **5.** David, Nguyen, Barbier & Baron (1996) *J. Bone Miner. Res.* **11**, 1498. **6.** McIntosh & Vaidya (2002) *Int. J. Parasitol.* **32**, 1. **7.** Boyle, Fairbrother &

Williams (1989) *Eur. J. Biochem.* **184**, 535. **8**. Granstrom (1983) *Arch. Oral Biol.* **28**, 453. **9**. Cooperman (1982) *Meth. Enzymol.* **87**, 526. **10**. Felix & Fleisch (1975) *Biochem. J.* **147**, 111. **11**. Jimenez Cardoso, Cuevas Rosas, Jimenez Cardoso & Ortiz (1992) *Rev. Latinoam. Microbiol.* **34**, 275. **12**. Bruchhaus, Jacobs, Denart & Tannich (1996) *Biochem. J.* **316**, 57. **13**. Saavedra-Lira, Ramirez-Silva & Perez-Montfort (1998) *Biochim. Biophys. Acta* **1382**, 47. **14**. Park, van Koeverden, Singh & Gupta (2007) *FEBS Lett.* **581**, 3211.

Etodolac

This nonsteroidal anti-inflammatory drug, *or* NSAID (FW = 287.35 g/mol; CAS 41340-25-4), also named (*RS*)-2-(1,8-diethyl-4,9-dihydro-3*H*-pyrano[3,4-*b*]indol-1-yl)acetic acid, preferentially inhibits cyclooxygenase-2 and is widely used in the management of pain in patients with inflammatory arthritis. The apparent elimination half-life is 7 hours in man. Etodolac was extensively bound to serum proteins. Liver microsomal cytochrome P-450 levels were unaltered in rats given etodolac daily for 1 week. **1**. Cayen, Kraml, Ferdinandi, Greselin & Dvornik (1981) *Drug Metab. Rev.* **12**, 339.

Etomoxir

This photosensitive epoxide (FW$_{free-acid}$ = 298.77 g/mol; CAS 1049724-28-8), also known as B827-33, is an irreversible inhibitor of carnitine *O*-octanoyltransferase (1) and carnitine *O*-palmitoyltransferase (2-4). Inhibition of fatty acid β-oxidation by etomoxir impairs NADPH production and increases reactive oxygen species resulting in ATP depletion and cell death in human glioblastoma cells (5). In human glioblastoma SF188 cells, Etomoxir inhibition of fatty acid β-oxidation markedly reduces cellular ATP levels and viability (5). Etomoxir nhibition of fatty acid oxidation reduces the NADPH level. In the presence of reactive oxygen species scavenger tiron, however, ATP depletion is prevented without restoring fatty acid oxidation. Such findings suggest oxidative stress may lead to bioenergetic failure and cell death (5). **1**. Hegardt, Bach, Asins, *et al.* (2001) *Biochem. Soc. Trans.* **29**, 316. **2**. Gerondaes, Alberti & Agius (1988) *Biochem. J.* **253**, 169. **3**. Kuhajda & Ronnett (2007) *Curr. Opin. Invest. Drugs* **8**, 312. **4**. Ratheiser, Schneeweiss, Waldhaeusl, Fasching & Korn (1991) *Metabolism* **40**, 1185. **5**. Pike, Smift, Croteau, Ferrick & Wu (2011) *Biochim. Biophys. Acta.* **1807**, 726.

Etomoxiryl-CoA

This reactive acyl-coenzyme thiolester (FW = 1036.28g/mol), also known as 2-[6-(4-chlorophenoxy)hexyl]oxirane-2-acyl-CoA and B827 33-CoA, inhibits carnitine *O*-octanoyltransferase, *or* medium-chain carnitine acyltransferase (1-3) and carnitine *O*-palmitoyltransferase, *or* long-chain carnitine acyltransferase (1,4-6). **1**. Lilly, Chung, Kerner, VanRenterghem & Bieber (1992) *Biochem. Pharmacol.* **43**, 353. **2**. Chung & Bieber (1993) *J. Biol. Chem.* **268**, 4519. **3**. Lilly, Bugaisky, Umeda & Bieber (1990) *Arch. Biochem. Biophys.* **280**, 167. **4**. Bentebibel, Sebastian, Herrero, *et al.* (2006) *Biochemistry* **45**, 4339. **5**. Woeltje, Esser, Weis, *et al.* (1990) *J. Biol. Chem.* **265**, 10714. **6**. Murthy & Pande (1990) *Biochem. J.* **268**, 599.

Etoposide

This podophyllotoxin derivative (FW = 588.57 g/mol; CAS 33419-42-0), also known as VP-16-213, is an antineoplastic agent that acts as a DNA topoisomerase II inhibitor (IC$_{50}$ = 60 μM). It has major activity against a number of tumors, including germ cell neoplasms, small cell lung cancer, and malignant lymphoma. It induces apoptosis in mouse thymocytes and in HL-60 human leukemia cells. Etoposide forms a complex with DNA topoisomerases II and enhances strand cleavage and inhibits religation. **Target(s):** DNA topoisomerase II (1-15); P-glycoprotein (16); quinine 3-monooxygenase (17); xenobiotic-transporting ATPase, multidrug-resistance protein (16). **1**. DeLange, Carpenter, Choy & Newsway (1995) *J. Virol.* **69**, 2082. **2**. Onishi, Azuma, Sato, *et al.* (1993) *Biochim. Biophys. Acta* **1175**, 147. **3**. Hande (1998) *Eur. J. Cancer* **34**, 1514. **4**. McDonald, Eldredge, Barrows & Ireland (1994) *J. Med. Chem.* **37**, 3819. **5**. Lindsey, Bender & Osheroff (2005) *Chem. Res. Toxicol.* **18**, 761. **6**. Wada, Iida & Tanaka (2001) *Planta Med.* **67**, 659. **7**. Low, Orton & Friedman (2003) *Eur. J. Biochem.* **270**, 4173. **8**. Sullivan, Latham, Rowe & Ross (1989) *Biochemistry* **28**, 5680. **9**. Melendy & Ray (1989) *J. Biol. Chem.* **264**, 1870. **10**. Vosberg (1985) *Curr. Top. Microbiol. Immunol.* **114**, 19. **11**. Hammonds, Maxwell & Jenkins (1998) *Antimicrob. Agents Chemother.* **42**, 889. **12**. Zhou, Guan & Kleinerman (2005) *Mol. Cancer Res.* **3**, 271. **13**. Gruger, Nitiss, Maxwell, *et al.* (2004) *Antimicrob. Agents Chemother.* **48**, 4495. **14**. Dickey, Choi, Van Etten & Osheroff (2005) *Biochemistry* **44**, 3899. **15**. Dickey & Osheroff (2005) *Biochemistry* **44**, 11546. **16**. Shapiro & Ling (1997) *Eur. J. Biochem.* **250**, 130. **17**. Zhao, Kawashiro & Ishizaki (1998) *Drug Metab. Dispos.* **26**, 188.

ETP-46464

This potent and selective inhibitor (FW = 470.52 g/mol; CAS 1345675-02-6; Solubility = 6 mg/mL DMSO), also known as α,α-dimethyl-4-[2-oxo-9-(3-quinolinyl)-2*H*-[1,3]oxazino[5,4-c]quinolin-1(4*H*)-yl]benzeneaceto-nitrile and NVP-BEZ235, selectively targets the ataxia telangiectasia and Rad3-related protein, *or* ATR (IC$_{50}$ = 25 nM), a serine/threonine-specific protein kinase involved in sensing DNA damage and activating the DNA-damage checkpoint, resulting to cell cycle arrest. **1**. Toledo, Murga, Zur, *et al.* (2011) *Nature Struct. Mol. Biol.* **18**, 721.

Etravirin

This non-nucleoside reverse transcriptase inhibitor, *or* NNRTI (FW = 435.28 g/mol; CAS 269055-15-4; Solubility: 42 mg/mL DMSO; <1 mg/mL H_2O), also known by its code name TMC125 and its systematic chemical name 4-[[6-amino-5-bromo-2-[(4-cyanophenyl)amino]-4-pyrimidinyl]oxy]-3,5-dimethylbenzonitrile, is used in the treatment of HIV-AIDS. **Mode of Inhibitory Action:** Structural studies showed that etravirine can bind in at least two conformationally distinct modes, allowing torsional flexibility, *or* "wiggling", as well as significant repositioning and translation and rotation reorientation within the pocket, *or* "jiggling" (1). Such binding site adaptations appear to be critical for its potency against wild-type reverse transcriptase and a wide range of its drug-resistant mutant HIV-1 RTs (1). Etravirine is highly active against wild type HIV-1 (EC_{50} = 1.4 to 4.8 nM) and showed some activity against HIV-2 (EC_{50} = 3.5 μM) (2). It also inhibited a series of HIV-1 Group-M subtypes and circulating recombinant forms and a Group-O virus. Incubation of etravirine with human liver microsomal fractions suggested good metabolic stability, with only a 15% decrease in drug concentration and a 7% decrease in antiviral activity after 2 hours (2). **1**. Das, Clark, Lewi, *et al.* (2004) *J. Med. Chem.* **47**, 2550. **2**. Andries, Azijn, Thielemans, *et al.* (2004) *Antimicrob. Agents Chemother.* **48**, 4680.

Everlimus

This orally available 40-*O*-(2-hydroxyethyl)-rapamycin derivative (FW = 958.22 g/mol; CAS 159351-69-6), marketed under the code name RAD001 and tradenames Zortress®, Certican® and Afinitor®, is an allosteric mTORC1-specific inhibitor that and is an immunosuppressant used to prevent rejection of organ transplants and to treat renal cell cancer and other tumors. It is also an approved treatment for estrogen receptor-positive (ER⁺) breast cancer. The mTOR pathway regulates translation, and mTOR inhibitors impair synthesis of proteins encoded by mRNAs that contain a 5'-terminal oligopyrimidine tract (5'-TOP). Everolimus is a potent mTOR inhibitor of FKBP12 (IC_{50} = 1.6-2.4 nM), with action that is comparable to that of rapamycin (***See** Rapamycin and FK506*). It suppresses hypoxia-induced increases in Hypoxia Inducible Factor-1α (HIF-1α), inhibiting VEGF stimulation of vascular endothelial cells, thereby blocking tumor-induced neovascularization. At a molecular level, everolimus binds to FKB12, and the resulting complex blocks the activation of the TOR complex 1 complex, a cycle-specific kinase that activates p70 ribosomal S6 kinase (*or* p70S6k). Inhibition of the mTOR pathway prevents progression of the cell cycle from G_1 into the S phase, thereby suppressing interleukin-driven T-cell differentiation. Inactivation of the p70S6k in lymphocytes selectively inhibits ribosomal protein synthesis, thereby deactivating the immune response (4). mTOR plays an important role in several physiological processes, such that inhibition by everolimus also leads to various downstream consequences via an effect on nonimmune cells such as vascular smooth muscle cells, tubular epithelial cells, and fibroblasts (4). Everolimus also strongly inhibits proliferation of T cell lymphoma (TCL) cell lines, blocking phosphorylation of ribosomal S6, a raptor/mTORC1 target, without a compensatory activation of the rictor/mTORC2 target Akt (S475) (5). Mounting evidence suggests S6K1 signaling may be important for proliferation of ER-positive breast cancer cells. S6K1 controls proliferation by direct phosphorylation of ERα at Ser167, increasing ERα transcriptional activity as well as ER-dependent breast cancer cell proliferation. P-S6K1 is a marker of S6K1 activation that correlates with a poorer prognosis in patients with ER-positive, but not ER-negative, tumors. In a Phase II trial, 16 patients with relapsed TCL received everolimus (10 mg PO daily). The median progression-free survival was 4.1 months, and the median overall survival was 10.2 months. The median duration of response for the 7 responders was 8.5 months (5). These findings provide proof-of-concept evidence that mTORC1 pathway targeting is clinically relevant in TCL. (Note that other mTOR inhibitors differ in their actions: BEZ235 and GSK2126458 are ATP competitive mTOR inhibitors that target both PI3K and mTORC1/2; AZD8055, AZD2014 and KU-0063794 are ATP competitive mTOR inhibitors that target both mTORC1 and mTORC2; and GDC-0941 is a pan-PI3K inhibitor.) **1**. Sedrani, *et al.* (1998) *Transplant Proc.* **30**, 2192. **2**. Lane, *et al.* (2009) *Clin. Cancer Res.* **15**, 1612. **3**. Houghton (2010) *Clin. Cancer Res.* **16**, 1368. **4**. Ganschow, Pollok, Jankofsky & Junge (2014) *Clin. Exp. Gastroenterol.* **7**, 329. **5**. Witzig, Reeder, Han, *et al.* (2015) *Blood* **126**, 328.

Evogliptin

This novel, potent, and selective dipeptidyl peptidase IV inhibitor and investigative Type 2 diabetes drug (FW = 401.42 g/mol; CAS 1222102-29-5 (free-base); 1246960-27-9 (HCl salt); 1222102 -51-3 (tartrate salt)), also named as DA-1229 and (*R*)-4-[(*R*)-3-amino-4-(2,4,5-trifluorophenyl) butanoyl]-3-(*t*-butoxy-methyl)piperazin-2-one, controls glucose levels by preventing the breakdown of the incretin hormones glucose-dependent insulinotropic polypeptide (GIP) and glucagon-like peptide-1 (GLP-1), which stimulate insulin secretion in response to the increased levels of glucose in the period following meals. ***See also** Alogliptin; Linagliptin; Saxagliptin; Sitagliptin; Vildagliptin* **1**. Chae, Kim, Kim, *et al.* (2015) *PLoS One* **10**, e0144064.

Evolocumab

This injectable PCSK9 inhibitor (MW = 141.8 kDa; CAS 1256937-27-5), developed as AMG-145 by Amgen, is a fully humanized monoclonal antibody that targets Proprotein Convertase Subtilisin/Kexin type-9 (PCSK9), the enzyme that binds Epidermal Growth Factor-like repeat A (EGF-A) domain of the Low-Density Lipoprotein Receptor (LDLR) and promotes its degradation in the endosomal/lysosomal pathway. The rationale for monoclonal anti-PCSK9 antibody-based therapy is that, despite statin treatment, many patients do not reach recommended low-density lipoprotein cholesterol (LDL-c) targets. Indeed, a gain-of-function mutation in PCSK9 result in a Familial Hypercholesterolemia (FH) phenotype, again by promoting degradation of the LDL-R. AMG 145 prevents PCSK9/LDL-R interaction, thereby restoring LDL-R recycling and reducing plasma cholesterol levels. Evolocumab is administered intravenously every two to four weeks. In Phase-1 studies, AMG 145 significantly reduced serum LDL-c in healthy and hypercholesterolemic statin-treated subjects, including those with heterozygous familial hypercholesterolemia or taking the highest doses of atorvastatin or rosuvastatin, with an overall an adverse event profile similar to placebo (1). When administered every fourth week, evolocumab yielded rapid and substantial reductions in LDL-c in heterozygous familial hypercholesterolemia patients, despite intensive statin use, with or without ezetimibe, with minimal adverse events and good tolerability (2). In patients with heterozygous FH, evolocumab (140 mg IV every 2 weeks or 420 mg monthly) is well-tolerated, yielding similar and rapid 60% reductions in LDL cholesterol compared with placebo (3). (**See also** *Alirocumab*) **1**. Dias, Shaywitz, Wasserman, *et al.* (2012) *J. Am. Coll. Cardiol.* **60**, 1888. **2**. Raal, Scott, Somaratne, *et al.* (2012) *Circulation* **126**, 2408. **3**. Raal, Stein, Dufour, *et al.* (2015) *Lancet* **385**, 331.

EVP4593, *See* QNZ

EXEL-2880, *See* Foretinib

EXP 3174

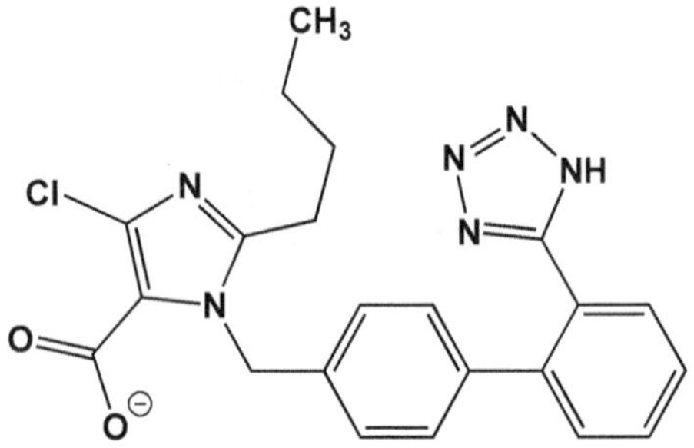

This noncompetitive AT$_1$ antagonist (FW = 436.89 g/mol; CAS 124750-92-1; Solubility: 100 mM DMSO), also named 2-butyl-4-chloro-1-[[2'-(2*H*-tetrazol-5-yl)[1,1'-biphenyl]-4-yl]methyl]-1*H*-imidazole-5-carboxylic acid, targets the Angiotensin-1 receptor (IC$_{50}$ = 37 nM), a member of the angiotensin group of G-protein-coupled receptors located primarily in the liver, kidney, adrenal gland and lung. EXP-3174 displaces [^{3}H]-angiotensin II from its specific binding sites in rat adrenal cortical membranes with an IC$_{50}$ of 3.7 x 10^{-8} M. EXP3174 does not alter blood pressure, but inhibits the pressor response to AII. In conscious renal artery-ligated rats, EXP3174 decreases blood pressure with an i.v. ED$_{30}$ of 0.038 mg/kg and a p.o. ED$_{30}$ of 0.66 mg/kg. **1**. Wong, Price, Chiu, *et al.* (1990) *J. Pharmacol. Exp. Ther.* **255**, 211.

Ezatiostat

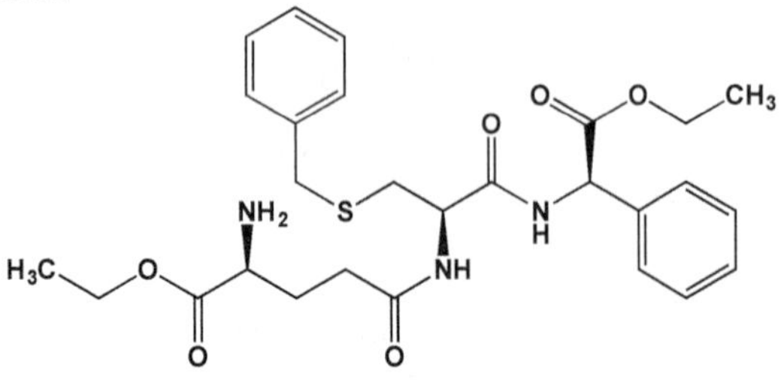

Ezatiostat

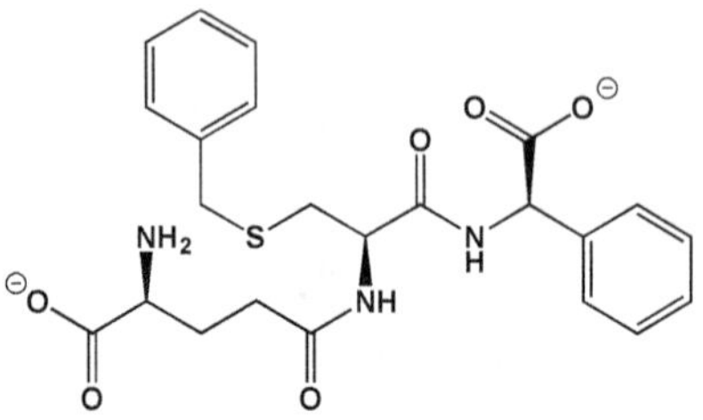

De-esterified Ezatiostat

De-esterified Ezatiostat

This glutathione analogue and prodrug (FW = 529.65 g/mol; CAS 168682-53-9), also known as Terrapin 199, TLK199, and (2*R*)-L-γ-glutamyl-*S*-(phenylmethyl)-L-cysteinyl-2-phenyl-glycine-1,3-diethyl ester, enters cells and undergoes enzymatic de-esterification, whereupon the carboxylate form selectively inhibits glutathione *S*-transferase GSTP1-1 (K_i = 400 nM), with much weaker inhibition of GSTα and GSTμ families, the latter with K_i values of 20-75 μM (1). Ezatiostat also inhibits multidrug resistance-associated protein1 (MRP1)-mediated drug resistance. **1**. O'Brien, Vulevic, Freer, *et al.* (1999) *J. Pharmacol. Exp. Ther.* **291**, 1348.

Ezetimibe

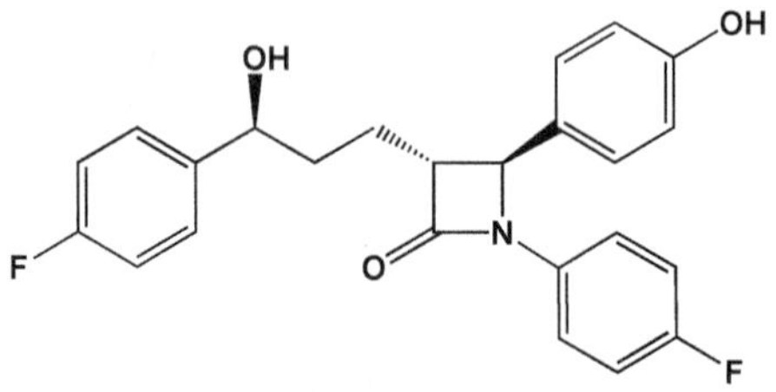

This rapidly absorbed, first-in-class cholesterol absorption inhibitor (FW = 409.40 g/mol; CAS 163222-33-1), also known by its code name SCH58235 and trade name Zetia®, and its active glucuronide metabolite (SCH60663) both lower LDL cholesterol (and increase HDL cholesterol) by impairing intestinal reabsorption of both dietary and hepatically excreted biliary cholesterol (1,2). The parent compound and its conjugated metabolite undergo enterohepatic recirculation, resulting in multiple peaks in the plasma concentration-time profile (3). Ezetimibe reduces plasma cholesterol levels from 964 to 374 mg/dL, from 726 to 231 mg/dL, and from 516 to 178 mg/dL in the Western, lowfat, and cholesterol-free diet groups, respectively (4). Ezetimibe markedly attenuates hepatic cholesterol accumulation and improves liver function in the lysosomal acid lipase-deficient mouse, a model for cholesteryl ester storage disease (5). **Primary Mechanism of Inhibitor Action:** Niemann-Pick C1-Like-1 (Accession No: AY437865.1) is the protein that mediates cholesterol absorption and serves as the primary ezetimibe target, as demonstrated (a) in binding assays, showing that labeled ezetimibe glucuronide binds specifically to a single site in brush border membranes and to human embryonic kidney 293 cells expressing NPC1L1 (K_D = 220 nM for human NPC1L1); (b) in quantitative binding experiments, confirming that affinities of ezetimibe to recombinant NPC1L1 are virtually identical to those observed for native enterocyte membranes; and (c) in NPC1L1-knockout mice, where ezetimibe no longer binds to membranes (6). **Key Pharmacokinetic Parameters:** *See* Appendix II in Goodman & Gilman's THE PHARMACOLOGICAL BASIS OF THERAPEUTICS, 12th Edition (Brunton, Chabner & Knollmann, eds.) McGraw-Hill Medical, New York (2011). **1**. Reiss, Burnett & Zaks (1999) *Bioorg Med Chem.* **7**, 2199. **2**. van Heek, Farley, Compton, *et al.* (2000) *Br. J. Pharmacol.* **129**, 1748. **3**. Ezzet, Krishna, Wexler, *et al.* (2001) *Clin. Ther.* **23**, 871. **4**. Davis, Compton, Hoos & Tetzloff (2001) *Arterioscler. Thromb. Vasc. Biol.* **21**, 2032. **5**. Chuang, Lopez, Posey & Turley (2013) *Biochem. Biophys. Res. Commun.* **443**, 1073. **6**. Garcia-Calvo, Lisnock, Bull, *et al.* (2005) *Proc. Natl. Acad. Sci. U.S.A.* **102**, 8132.

— F —

F, See *Fluorine (Fluoride Ions); Folate (Folic Acid); Phenylalanine*

F-244

This polyketide (FW = 323.41 g/mol), also known as (*E,E*)-11-(3(*R*)-hydroxymethyl-4-oxo-2(*R*)-oxetanyl) 3,5,7(*R*)-trimethyl-2,4-undecadienoic acid, L-659,699, and Hymeglusin, from *Fusarium* sp. and *Scopulariopsis* sp. is a potent inhibitor of hydroxymethyl-glutaryl-CoA synthase, or HMG-CoA synthase (IC_{50} = 54 nM for human enzyme). The lactone reacts with an enzyme thiol to form a thiolester. **1**. Pojer, Ferrer, Richard, *et al.* (2006) *Proc. Natl. Acad. Sci. U.S.A.* **103**, 11491. **2**. Greenspan, Yudkovitz, Lo, *et al.* (1987) *Proc. Natl. Acad. Sci. U.S.A.* **84**, 7488. **3**. Greenspan, Bull, Yudkovitz, Hanf & Alberts (1993) *Biochem. J.* **289**, 889. **4**. Nagegowda, Bach & Chye (2004) *Biochem. J.* **383**, 517. **5**. Rokosz, Boulton, Butkiewicz, *et al.* (1994) *Arch. Biochem. Biophys.* **312**, 1. **6**. Tomoda, Ohbayashi, Morikawa, Kumagai & Omura (2004) *Biochim. Biophys. Acta* **1636**, 22. **7**. Mayer, Louis-Flamberg, Elliott, Fisher & Leber (1990) *Biochem. Biophys. Res. Commun.* **169**, 610.

Factor VIII (FVIII) Inhibitors

These naturally occurring, high-affinity polyclonal anti-FVIII alloantibodies (MW = 150 kDa for the four chains of IgG4) neutralize the coagulative properties of blood clotting Factor VIII, but do not fix complement (1-3). **Factor VIII Therapy:** Hemophilia A is an X-linked recessive Factor VIII deficiency, which in males has a 1-in-5000 frequency showing no ethnic predominance. Factor VIII participates in blood coagulation as a cofactor for Factor IXa, which, in the presence of Ca^{2+} and phospholipids, forms a complex that converts factor X to its activated form Xa. The latter cleaves prothrombin at Arg-Thr and Arg-Ile bonds to yield the active thrombin, which, in turn, converts fibrinogen into fibrin. The mainstay for treating Hemophilia A is FVIII replacement, with hemostasis achieved through the use of either plasma or recombinant (exogenous) FVIII concentrates. FVIII replacement therapy is highly effective, until a patient produces inhibitory anti-FVIII alloantibodies. These antibodies bind to and interfere with the action of infused Factor VIII, necessitating the use of costly and much less effective alternatives, including activated Prothrombin Complex Concentrate, or aPCC, and recombinant factor VIIa. **Key Features of FVIII Inhibitors:** Although otherwise considered to be a non-pathologic subclass with anti-inflammatory antibodies, IgG4 correlates with the presence of high-titre FVIII inhibitors in hemophilia A. These IgG4 alloantibodies possess unique structural and functional properties, making them functionally monovalent as a result of undergoing so-called 'half-antibody exchange' *in vivo*. The recombined antibodies have two different antigen binding specificities. IgG4 antibody production is partly driven by T helper 2-derived cytokines that mediate allergic responses and IgE production. Formation of these FVIII inhibitors is a T-cell dependent event that involves antigen-presenting cells, B- and T-helper lymphocytes (4). Factor VIII contains three A-domains (designated A_1, A_2, A_3), one B-domain, and two C-domains (C_1 and C_2). Inhibitory antibodies are primarily directed against the A_2, A_3 and C_2 domains (5,6). Antibody binding at these domains blocks functional epitopes within FVIII (7), including those for interacting with Factor IX (FIX), phospholipid, and von Willebrand factor. Antibodies in inhibitor-producing patients target multiple FVIII epitopes, and these targets can change over time (8). **1**. Fulcher, de Graaf-Mahoney & Zimmerman (1987) *Blood* **69**, 1475. **2**. Gilles, Arnout, Vermylen & Saint-Remy (1993) *Blood* **82**, 2452. **3**. Lollar, (2004) *J. Thromb. Hemost.* **2**, 1082. **4**. Astermark (2006) *Hemophilia* **12** (*Suppl.* 3), 52. **5**. Fulcher, de Graaf Mahoney, Roberts, Kasper & Zimmerman (1985) *Proc. Natl. Acad. Sci. USA* **82**, 7728. **6**. Scandella, Mattingly, de Graaf, & Fulcher (1989) *Blood* **74**, 1618. **7**. Saint-Remy, Lacroix-Desmazes & Oldenburg (2004) *Hemophilia* **10** (*Suppl.* 4), 146. **8**. Fulcher, Lechner & de Graaf Mahoney (1988) *Blood* **72**, 1348.

Factor XIII Inhibitors

These autoantibodies target Factor XIII (FXIII), a blood-clotting enzyme that cross-links fibrin monomers, stabilizing clots (hence its initial name: Fibrin Stabilizing Factor) and wound healing. Acquired FXIII deficiency ia a rare condition that results in bleeding, despite normal coagulation test results. The mean age of onset is 60, regardless of sex, and is characterized by intramuscular, subcutaneous, external, or surgically induced bleeding. Clinical improvement occurs in patients receiving FXIII concentrate, cryoprecipitate, or plasma. Inhibitor reduction was observed in patients receiving rituximab, plasma exchange, intravenous immunoglobulin, steroids, and cyclophosphamide. **1**. Tone, James, Fergusson, *et al.* (2016) *Transfus. Med. Rev.* **30**, 123.

Factor XIIIa Fragment 72-97

This 26-residue oligopeptide (FW = 3226.65 g/mol; *Sequence*: NKLIVR RGQSFYVQIDFSRPYDPRRD) inhibits coagulation factor XIIIa (*or* transglutaminase, *or* protein-glutamine γ-glutamyltransferase), which catalyzes the formation of cross-link between fibrin and fibronectin or *N,N*'-dimethylcasein. Its inhibitory action demonstrated that regions outside the active site are important for substrate recognition in factor XIIIa and tissue transglutaminase. **1**. Achyuthan, Slaughter, Santiago, Enghild & Greenberg (1993) *J. Biol. Chem.* **268**, 21284.

FAD

This redox coenzyme (FW = 785.55 g/mol; E_0' = − 0.219 V at pH 7 and 30°C; λ_{max} at 263, 375, and 450 nm (ε = 38000, 9300, and 11300 $M^{-1}cm^{-1}$, respectively), known as flavin adenine dinucleotide, is the yellow, easily photobleached coenzyme (or prosthetic group for flavoenzymes). The reduced form is $FADH_2$, which has a molar extinction coefficient of only 980 $M^{-1}cm^{-1}$ at 450 nm. $FADH_2$ has a fluorescence maximum at 525-530 nm. FAD is reduced nonenzymatically to $FADH_2$ in the presence of dithionite or by H_2 in the presence of an appropriate catalyst. **Target(s):** acetolactate synthase (1); adenylyl sulfate kinase (2); alanine racemase (3); carbonyl reductase (4); 3',5'-cyclic-nucleotide phosphodiesterase (5); deoxyhypusine synthase (6); FAD synthetase, product inhibition (7-9); fructose-1,6-bisphosphatase (10); glucose-1-phosphate adenylyl-transferase (11); glutamate racemase (12,13); glutathione-disulfide reductase (14); glycerone kinase (15); glycogen phosphorylase *b* (16); isopenicillin-*N* isomerase (18); L-lactate oxidase (19); lysolecithin oxidase (20); NAD^+ diphosphatase (21); NADH:cytochrome *c* reductase, *or* NADH dehydrogenase (22); NAD(P)H dehydrogenase, *or* DT diaphorase (23-26); nicotinate-nucleotide pyrophosphorylase, carboxylating (27); nucleotide diphosphatase, also alternative substrate (28,29); palmitoyl-CoA hydrolase, *or* acyl-CoA hydrolase, weakly inhibited (30); phosphodiesterase I, *or* 5'-exonuclease (31); riboflavinase (32); ribose-phosphate diphosphokinase, *or* phosphoribosyl pyrophosphate synthetase (33); succinate oxidase (34); acyl-[acyl-carrier-protein] desaturase (36,37); very-long-chain fatty acid α-hydroxylase (35); *trans*-cinnamate 4-monooxygenase (38); *myo*-inositol oxidase (17,39); sulfur oxygenase/reductase (40); sulfur dioxygenase (41,42). **1**. Kaushal, Pabbi & Sharma (2003) *World J. Microbiol. Biotechnol.* **19**, 487. **2**. Schriek & Schwenn (1986) *Arch. Microbiol.* **145**, 32. **3**. Fujita, Okuma & Abe (1997) *Fish. Sci.* **63**, 440. **4**. Wermuth (1981) *J. Biol. Chem.* **256**, 1206. **5**. Lim, Palanisamy & Ong (1986) *Arch. Microbiol.* **146**, 142. **6**. Abid, Sasaki, Titani & Miyazaki (1997) *J. Biochem.* **121**, 769. **7**. Bowers-Komro, Yamada & McCormick (1989) *Biochemistry* **28**, 8439. **8**. Yamada, Merrill & McCormick (1990) *Arch. Biochem. Biophys.* **278**, 125. **9**. McCormick, Oka, Bowers-Komro, Yamada & Hartman (1997) *Meth. Enzymol.* **280**, 407. **10**. Fujita & Freese (1979) *J. Biol. Chem.* **254**, 5340. **11**. Preiss, Crawford, Downey, Lammel & Greenberg (1976) *J. Bacteriol.* **127**, 193. **12**. Adams (1972) *The Enzymes*, 3rd ed., **6**, 479. **13**. Diven (1969) *Biochim. Biophys. Acta* **191**, 702. **14**. Moroff & Kosow (1978) *Biochim. Biophys. Acta* **527**, 327. **15**. Cabezas, Costas, Pinto, Couto & Cameselle (2005) *Biochem. Biophys. Res. Commun.* **338**, 1682. **16**. Klinov, Kurganov, Pekel & Berezovskii (1986) *Biochem. Int.* **13**, 139. **17**. Charalampous (1962) *Meth. Enzymol.* **5**, 329. **18**. Usui & Yu (1989) *Biochim. Biophys. Acta* **999**, 78. **19**. Cousins (1956) *Biochem. J.* **64**, 297. **20**. Hoshino (1959) *J. Biochem.* **46**, 825. **21**. Nakajima, Fukunaga,

Sasaki & Usami (1973) *Biochim. Biophys. Acta* **293**, 242. **22**. Vernon, Mahler & Sarkar (1952) *J. Biol. Chem.* **199**, 599. **23**. Ernster (1967) *Meth. Enzymol.* **10**, 309. **24**. Martius (1963) *The Enzymes*, 2nd ed., 7, 517. **25**. Ernster, Ljunggren & Danielson (1960) *Biochem. Biophys. Res. Commun.* **2**, 48 and 88. **26**. Raw, Nogueira & Filho (1961) *Enzymologia* **23**, 123. **27**. Nishizuka & Nakamura (1970) *Meth. Enzymol.* **17A**, 491. **28**. Byrd, Fearney & Kim (1985) *J. Biol. Chem.* **260**, 7474. **29**. Krishnan & Rao (1972) *Arch. Biochem. Biophys.* **149**, 336. **30**. Berge & Dossland (1979) *Biochem. J.* **181**, 119. **31**. Udvardy, Marre & Farkas (1970) *Biochim. Biophys. Acta* **206**, 392. **32**. Kumar & Vaidyanathan (1964) *Biochem. Biophys. Acta* **89**, 127. **33**. Fox & Kelley (1972) *J. Biol. Chem.* **247**, 2126. **34**. Wadkins & Mills (1956) *Fed. Proc.* **15**, 377. **35**. Kishimoto (1978) *Meth. Enzymol.* **52**, 310. **36**. Jaworski & Stumpf (1974) *Arch. Biochem. Biophys.* **162**, 158. **37**. Nagai & Bloch (1968) *J. Biol. Chem.* **243**, 4626. **38**. Billett & Smith (1978) *Phytochemistry* **17**, 1511. **39**. Charalampous (1959) *J. Biol. Chem.* **234**, 220. **40**. Kletzin (1994) *Syst. Appl. Microbiol.* **16**, 534. **41**. Kletzin (1989) *J. Bacteriol.* **171**, 1638. **42**. Suzuki & Silver (1966) *Biochim. Biophys. Acta* **122**, 22.

Fadrozole

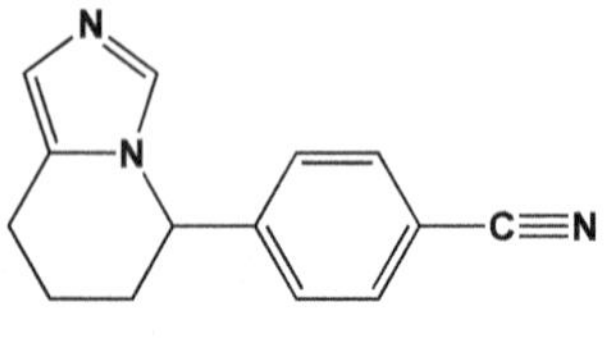

This antineoplastic agent (FW$_{free-base}$ = 223.28 g/mol; M.P. = 117-118°C), also called 4-(5,6,7,8-tetrahydroimidazo[1,5-*a*]pyridin-5-yl)benzonitrile, is a potent aromatase inhibitor (IC$_{50}$ = 52 nM). The hydrochloride is often referred to as CGS-16949A. **Target(s):** aromatase, *or* CYP19 (1-8); estrogen 2-hydroxylase (9); steroid 11β-monooxygenase (6-8,10); steroid 17α-monooxygenase, weakly inhibited (6-8); sterol 14-demethylase, *or* CYP51, weakly inhibited (11). **1**. Santen, Langecker, Santner, Sikka, Rajfer & Swerdloff (1990) *Endocr. Res.* **16**, 77. **2**. Moslemi & Seralini (1997) *J. Enzyme Inhib.* **12**, 241. **3**. Van Roey, Bullion, Osawa, Browne, Bowman & Braun (1991) *J. Enzyme Inhib.* **5**, 119. **4**. Steele, Mellor, Sawyer, Wasvary & Browne (1987) *Steroids* **50**, 147. **5**. Recanatini, Bisi, Cavalli, *et al.* (2001) *J. Med. Chem.* **44**, 672. **6**. Ulmschneider, Müller-Vieira, Mitrenga, *et al.* (2005) *J. Med. Chem.* **48**, 1796. **7**. Voets, Antes, Scherer, *et al.* (2006) *J. Med. Chem.* **49**, 2222. **8**. Ulmschneider, Müller-Vieira, Klein, *et al.* (2005) *J. Med. Chem.* **48**, 1563. **9**. Purba, King, Richert & Bhatnagar (1994) *J. Steroid Biochem. Mol. Biol.* **48**, 215. **10**. Müller-Vieira, M. Angotti & R. W. Hartmann (2005) *J. Steroid Biochem. Mol. Biol.* **96**, 259. **11**. Trösken, Adamska, Arand, *et al.* (2006) *Toxicology* **228**, 24.

Fagaronine

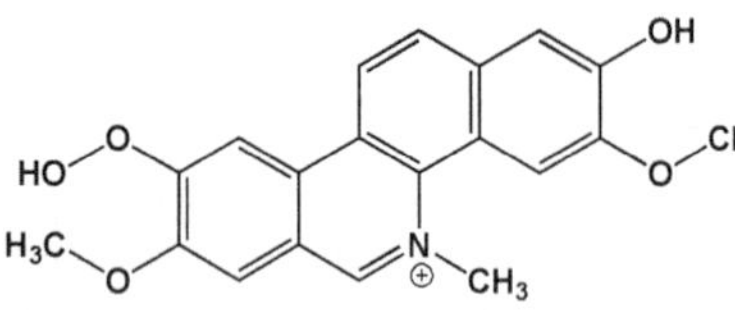

This benzo[*c*]phenanthridinium alkaloid (FW$_{chloride}$ = 385.85 g/mol), from trees of the genus *Zanthoxylum*, inhibits a number of DNA-dependent enzymes. **Target(s):** amine oxidase (copper-containing) (1); dipeptidyl-peptidase IV (2,3); DNA-directed DNA polymerase (4,5); DNA ligase (ATP), human (IC$_{50}$ = 27 μM) (6); DNA topoisomerase I (7-9); DNA topoisomerase II (7,9,10); polyadenylic acid polymerase (5); RNA-directed DNA polymerase, *or* reverse transcriptase (4,5,11,12); RNA polymerase (4,5). **1**. Luhova, Frebort, Ulrichova, *et al.* (1995) *J. Enzyme Inhib.* **9**, 295. **2**. Sedo, Malik, Vicar, Simanek & Ulrichova (2003) *Physiol. Res.* **52**, 367. **3**. Sedo, Vlasicova, Bartak, *et al.* (2002) *Phytother. Res.* **16**, 84. **4**. Sethi (1977) *Ann. N. Y. Acad. Sci.* **284**, 508. **5**. Sethi (1976) *Cancer Res.* **36**, 2390. **6**. Tan, Lee, Lee, *et al.* (1996) *Biochem. J.* **314**, 993. **7**. Prado, Michel, Tillequin, *et al.* (2004) *Bioorg. Med. Chem.* **12**, 3943. **8**. Fleury, Sukhanova, Ianoul, *et al.* (2000) *J. Biol. Chem.* **275**, 3501. **9**. Wang, Johnson & Hecht (1993) *Chem. Res. Toxicol.* **6**, 813. **10**. Larsen, Grondard, Couprie, *et al.* (1993) *Biochem. Pharmacol.* **46**, 1403. **11**. Sethi & Sethi (1975) *Biochem. Biophys. Res. Commun.* **63**, 1070. **12**. Sethi (1979) *J. Nat. Prod. (or, Lloydia)* **42**, 187.

Famciclovir

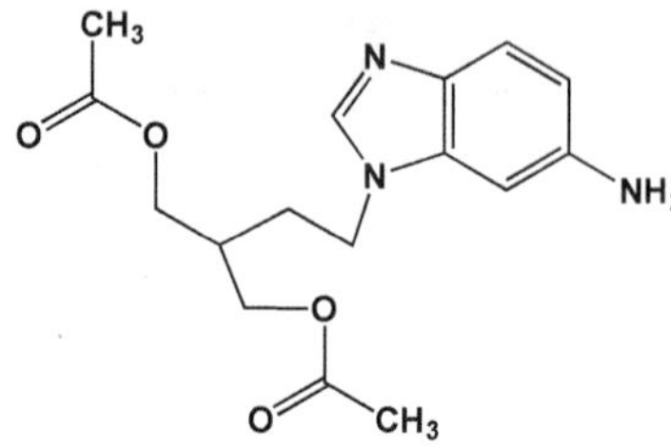

This acyclic guanosine analogue (FW = 321.34 g/mol; CAS = 104227-87-4; shown above in its familiar nucleoside-like form), also known by the trade name Famvir® and systematically as 2-[(acetyloxy)methyl]-4-(2-amino-9*H*-purin-9-yl)butyl acetate, is a penciclovir prodrug used to treat herpes virus infections, most often herpes zoster (shingles). Upon intracellular ester hydrolysis and subsequent metabolic phosphorylation, this prodrug forms penciclovir 5'-triphosphate, a potent inhibitor of viral DNA polymerases. (*See Penciclovir*) The widely used FDA-approved one-day treatment for cold sores consists of two 1000-mg doses (twelve hours apart) of famciclovir, initiated by the patient at first sensation of the tingling due to an otherwise soon-to-be erupting cold sore. There is a narrow window of opportunity at the start of an outbreak of herpes or cold sores, because the virus replicates most actively in those first hours. Interrupting that process appears to shorten the outbreak and to reduce the severity of infection. **Target(s):** Epstein-Barr virus replication; HSV-1 and HSV-2 replication; varicella zoster virus replication; viral DNA polymerases (via its phosphorylated form of penciclovir). **1**. Prisbe & Chen (1996) *Meth. Enzymol.* **275**, 425.

Famotidine

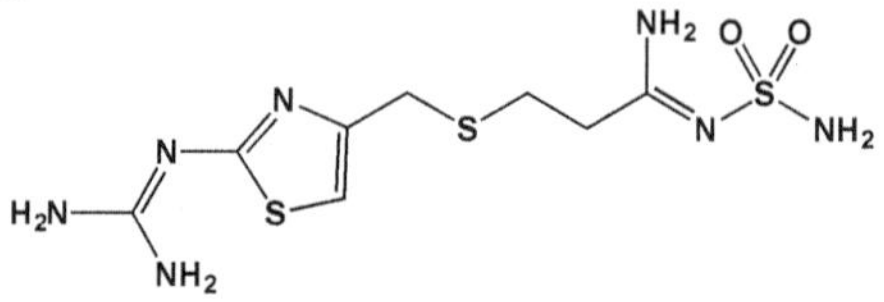

This well-known antacid medication (FW = 377.45 g/mol; CAS = 76824-35-6), also known by its trade names Pepcid®, Pepcidine®, and Gaster®, as well as by its systematic name 3-([2-(diaminomethyleneamino)thiazol-4-yl]methylthio)-*N'*-sulfamoylpropanimidamide, is a histamine H$_2$-receptor antagonist that inhibits stomach acid production and is often used to treatment peptic ulcer disease (PUD) and gastroesophageal reflux disease (GERD). Famotidine is about nine times more potent than ranitidine, and about thirty-two times more potent than cimetidine. Unlike cimetidine, however, famotidine is without effect on cytochrome P450 metabolism (3). Moreover, coadministration of famotidine had no effect on the pharmacokinetics of theophylline, diazepam, desmethyldiazepam, and phenytoin and no effect on the half-life of antipyrine and aminopyrine or on the prothrombin time ratios associated with warfarin therapy. **Key Pharmacokinetic Parameters:** *See* Appendix II in Goodman & Gilman's *THE PHARMACOLOGICAL BASIS OF THERAPEUTICS*, 12th Edition (Brunton, Chabner & Knollmann, eds.) McGraw-Hill Medical, New York (2011). **1**. Takeda, Takagi, Yashima & Maeno (1982) *Arzneimittelforschung.* **32**, 734. **2**. Takagi, Takeda & Maeno (1982) *Arch. Int. Pharmacodyn. Ther.* **256**, 49. **3**. Humphries (1987) *Scand. J. Gastroenterol.* **134** (Suppl.), 55.

Fananserin

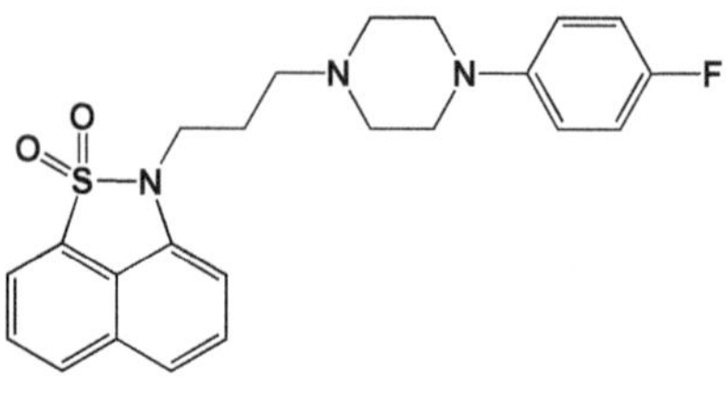

This serotonin receptor antagonist (FW = 425.52 g/mol; CAS = 127625-29-0; Soluble to 100 mM in DMSO), also named RP 62203 and 2-[3-[4-(4-fluorophenyl)-1-piperazinyl]propyl]-2*H*-naphthyl[1,8-*cd*]isothiazole-1,1-dioxide, selectively targets 5-HT$_{2A}$ receptors (K_i = 0.26 nM), but also binds to D$_4$ receptors (K_i = 2.9 nM), H$_1$ receptors (K_i = 25 nM), α$_1$ receptors (K_i = 38 nM), 5-HT$_{1A}$ receptors (K_i = 70 nM) and D$_2$ receptors (K_i = 726 nM) (1,2). In cortical slices from the neonatal rat, RP 62203 potently inhibited inositol phosphate formation (IC$_{50}$ = 7.76 nM) evoked by 5-HT. **1**.

Malleron, Comte, Gueremy, *et al.* (1991) *J. Med. Chem.* **34**, 2477. **2.** Doble, Girdlestone, Piot, *et al.* (1992) *Brit. J. Pharmacol.* **105**, 27.

Fantofarone

This antihypertensive agent (FW$_{free-base}$ = 550.72 g/mol; CAS = 114432-13-2), also known by its code name SR 33557 and as *N*-[2-(3,4-dimethoxy-phenyl)ethyl]-3-[4-[(2-isopropyl-1-indolizinyl)sulfonyl]phenoxy]-*N*-methyl propan-1-amine, is a calcium channel blocker. Fantofarone's blocking action on the different types of voltage-activated Ca^{2+} currents was investigated with the whole-cell patch-clamp method in chick dorsal root ganglion neurons (for T-, L- and N-type currents) and in rat cerebellar Purkinje neurons (for P-type current). Neuronal L-type Ca^{2+} channels are blocked totally by fantofarone in the microM range of concentration, as in skeletal muscle and cardiac cells at a holding membrane potential of –80 mV (IC$_{50}$ = 0.35 μM). N- and P-type channels are not very sensitive (IC$_{50}$ ≈ 5 microM). The T-type channel is unaffected. **Target(s):** Ca^{2+} channels (1); sphingomyelin phosphodiesterase, *or* acid sphingomyelinase (2,3). **1.** Schmid, Romey, Barhanin & Lazdunski (1989) *Mol. Pharmacol.* **35**, 766. **2.** Won, Im, Khan, Singh & Singh (2004) *J. Neurochem.* **88**, 583. **3.** Jaffrézou, Chen, Durán, *et al.* (1995) *Biochim. Biophys. Acta* **1266**, 1.

Farnesol

This branched-chain unsaturated alcohol (FW = 222.37 g/mol) is found in many essential oils (*e.g.*, oils of rose, musk, lemon grass, citronella, tuberose, *etc.*). Although there are four geometric isomers, the all-*trans* isomer is the one found in the natural oils. Occasionally, *cis,trans* farnesol is observed (*e.g.*, in petitgrain oil). This liquid serves as a pheromone in a number of organisms. Farnesol can also induce apoptosis. Note that farnesol is easily oxidized and can form cyclic products. **Target(s):** Ca^{2+} channels, N-type (1,2); *Candida albicans* biofilm formation (3); Ca^{2+} uptake (4,5); choline kinase (6); diclofenac 4-hydroxylase, *or* cytochrome P450 (7); dolichyldiphosphatase, weakly inhibited (8); ethoxycoumarin deethylase, cytochrome P450 (7); 3-hydroxy-3 methylglutaryl-CoA reductase (7); mevalonate kinase (10); prenylcysteine oxidase (11); squalene:hopene cyclase (12). **1.** Beedle & Zamponi (2000) *Biophys. J.* **79**, 260. **2.** Roullet, Spaetgens, Burlingame & Zamponi (1999) *J. Biol. Chem.* **274**, 25439. **3.** Ramage, Saville, Wickes & Lopez-Ribot (2002) *Appl. Environ. Microbiol.* **68**, 5459. **4.** Luft, Bychkov, Gollasch, *et al.* (1999) *Arterioscler. Thromb. Vasc. Biol.* **19**, 959. **5.** Roullet, Le Quan Sang, Luft, *et al.* (1997) *J. Hypertens.* **15**, 1723. **6.** Miquel, Pradines, Terce, Selmi & Favre (1998) *J. Biol. Chem.* **273**, 26179. **7.** Raner, A. Q. Muir, C. W. Lowry & B. A. Davis (2002) *Biochem. Biophys. Res. Commun.* **293**, 1. **8.** Scher & Waechter (1984) *J. Biol. Chem.* **259**, 14580. **9.** Rao, Newmark & Reddy (2002) *Cancer Detect. Prev.* **26**, 419. **10.** Hinson, Chambliss, Toth, Tanaka & Gibson (1997) *J. Lipid. Res.* **38**, 2216. **11.** Digits, Pyun, Coates & Casey (2002) *J. Biol. Chem.* **277**, 41086. **12.** Ochs, Tappe, Gartner, Kellner & Poralla (1990) *Eur. J. Biochem.* **194**, 75.

Farnesyl Diphosphate (*or*, Farnesyl Pyrophosphate)

This sesquiterpenyl diphosphate (FW = 382.32 g/mol) is a key intermediate in the biosynthesis of steroids and many terpenes. It is also the farnesyl donor in protein farnesyltransferases. Farnesyl diphosphate is unstable in acidic conditions: complete hydrolysis occurs in less than six hours at pH 1 and 22°C while hydrolysis at pH 4 takes about five days at that same temperature. Only 2% is hydrolyzed in three months at pH 8 and –20°C while 15% is hydrolyzed in one hour in 0.1 M KOH at 100°C. **Target(s):** dolichyl-diphosphatase (1); geranylgeranyl-diphosphatase (2); isopentenyl diphosphate Δ-isomerase (3); mevalonate kinase (4-12); rubber *cis*-polyprenyl*cis*transferase (13). **1.** Scher & Waechter (1984) *J. Biol. Chem.*

259, 14580. **2.** Nah, Song & Back (2001) *Plant Cell Physiol.* **42**, 864. **3.** Bruenger, Chayet & Rilling (1986) *Arch. Biochem. Biophys.* **248**, 620. **4.** Porter (1985) *Meth. Enzymol.* **110**, 71. **5.** Hinson, Chambliss, Toth, Tanaka & Gibson (1997) *J. Lipid Res.* **38**, 2216. **6.** Dorsey & Porter (1968) *J. Biol. Chem.* **243**, 4667. **7.** Voynova, Rios & Miziorko (2004) *J. Bacteriol.* **186**, 61. **8.** Schulte, van der Heijden & Verpoorte (2000) *Arch. Biochem. Biophys.* **378**, 287. **9.** Gray & Kekwick (1973) *Biochem. J.* **133**, 335. **10.** Gray & Kekwick (1972) *Biochim. Biophys. Acta* **279**, 290. **11.** Huang, Scott & Bennett (1999) *Protein Expr. Purif.* **17**, 33. **12.** Chu & Li (2003) *Protein Expr. Purif.* **32**, 75. **13.** Archer & Cockbain (1969) *Meth. Enzymol.* **15**, 476.

S-trans,trans-Farnesyl Thiosalicylic Acid Amide

This K-Ras 4B-targeting, farnesyl-cysteine mimic (FW = 357.55 g/mol; CAS = 1092521-74-8; Symbol: FTS) is taken up be cells, where it dislodges farnesylated Ras proteins from cell membranes by competing for their membrane binding sites, thus facilitating their degradation (1). **Mode of Inhibitor Action:** To regulate cell growth, differentiation and apoptosis, Ras proteins must remain anchored to the cytoplasmic surface of the plasma membrane by means of their common carboxy-terminal *S*-farnesylcysteine and a stretch of lysine residues (as in K-Ras 4B) or *S*-palmitoyl moieties (as in H-Ras, N-Ras and K-Ras 4A). FTS is not a farnesyltransferase inhibitor. Instead, it inhibits growth of NIH3T3 cells transformed by the non-palmitoylated K-Ras 4B(12V) or by its farnesylated (but unmethylated) K-Ras 4B(12) CVYM mutant. FTS does not alter the rate of degradation of the K-Ras 4B, SVIM mutant, which is not modified post-translationally, suggesting that only farnesylated Ras isoforms are substrates destined for facilitated degradation. The putative Ras-recognition sites (lying within domains in the cell membrane) appear to tolerate both C(15) and C(20) *S*-prenyl moeities, since geranylgeranyl thiosalicylic acid mimicked the growth-inhibitory effects of FTS in K-Ras 4B(12V)-transformed cells and FTS inhibited the growth of cells transformed by the geranylgeranylated K-Ras 4B(12V) CVIL isoform. **1.** Elad, Paz, Haklai, *et al.* (1999) *Biochim. Biophys. Acta* **1452**, 228.

S-all-trans-Farnesylthioacetic Acid

This *S*-farnesylated cysteine analogue (FW$_{free-acid}$ = 296.47 g/mol) inhibits protein-*S* isoprenylcysteine *O*-methyltransferase. **1.** Volker, Pillinger, Philips & Stock (1995) *Meth. Enzymol.* **250**, 216. **2.** Ma, Gilbert & Rando (1995) *Meth. Enzymol.* **250**, 226. **3.** Volker & Stock (1995) *Meth. Enzymol.* **255**, 65. **4.** Baron & Casey (2004) *BMC Biochem.* **5**, 19. **5.** De Busser, Van Dessel & Lagrou (2000) *Int. J. Biochem. Cell. Biol.* **32**, 1007. **6.** Shi & Rando (1992) *J. Biol. Chem.* **267**, 9547. **7.** Tan, Pérez-Sala, Canada & Rando (1991) *J. Biol. Chem.* **266**, 10719. **8.** Pérez-Sala, Gilbert, Tan & Rando (1992) *Biochem. J.* **284**, 835. **9.** Klein, Ben-Baruch, Marciano, *et al* (1994) *Biochim. Biophys. Acta* **1226**, 330.

S-Farnesylthiosalicylate

This farnesylated salicylate analogue (FW$_{free-acid}$ = 358.55 g/mol) is a nontoxic Ras antagonist that dislodges Ras from its membrane anchoring sites. It inhibits the growth of human Ha-Ras transformed Rat1 cells *in vitro* (EC$_{50}$ = 7.5 μM). It also suppresses the growth of prostate cancer *in vitro*. **Target(s):** mitogen-activated protein kinase kinase kinase, *or* MAPKKK (1); protein-*S*-isoprenylcysteine *O*-methyltransferase (2,3); Ras membrane association (4,5). **1.** Calipel, Lefevre, Pouponnot, *et al.* (2003) *J. Biol. Chem.* **278**, 42409. **2.** Hasne & Lawrence (1999) *Biochem. J.* **342**, 513. **3.** Marciano, Aharonson, Varsano, Haklai & Kloog (1997) *Bioorg. Med. Chem. Lett.* **7**, 1709. **4.** Marciano, Ben-Baruch, Marom, *et al.* (1995) *J.*

Med. Chem. **38**, 1267. **5.** Marom, Haklai, Ben-Baruch, *ewt al.* (1995) *J. Biol. Chem.* **270**, 22263.

Farnesyltransferase Inhibitor I, *See (E,E)-2-[(Dihydroxyphosphinyl) methyl]-3-oxo-3-[(3,7,11-trimethyl-2,6,10-dodecatrienyl)-amino]propanoate; N-[2(S)-[2(R)-amino-3-mercaptopropylamino]-3-methylbutyl]-L-phenylalanyl-L-methionine*

Farnesyltransferase Inhibitor II, *See (E,E)-2-[2-Oxo-2-[[(3,7,11-trimethyl-2,6,10-dodecatrienyl)oxy]amino]ethyl]-phosphonate; L-Cysteinyl-(4-aminobenzoyl)-L-methionine*

Farnesyltransferase Inhibitor III, *See (E,E)-[2-Oxo-2-[[(3,7,11-trimethyl-2,6,10-dodecatrienyl)-oxy]amino]ethyl]-phosphonic Acid (2,2-Dimethyl-1-oxopropoxy)methyl Ester; L-Cysteinyl-L-valyl-L-(2-naphthylalanyl)-L-methionine*

Farnesyltransferase Inhibitor 276, *See FTI-276 and FTI-277*

Farnesyltransferase Inhibitor 277, *See FTI-276 and FTI-277*

Fascaplysin

This red-colored marine CDK inhibitor (FW = 271.30 g/mol; CAS = 114719-57-2), first isolated from the sponge *Fascaplysinopsis Bergquist* (1), selectively targets cyclin-dependent kinase CDK4 (IC$_{50}$~0.4 μM), relative to CDK2 (IC$_{50}$ ~ 500 μM) and CDK6 (IC$_{50}$ ~ 3.4 μM) (1-3). Not surprisingly, fascaplysin is also an effective DNA intercalator and is likely to interfere with both replication and transcription (4). An excellent fit was obtained with two formation constants ($K_1 = 2.5 \times 10^6$ M^{-1}; $K_2 = 7.5 \times 10^4$ M^{-1}), putting the numbers of binding sites per base pair at $n_1 = 0.30$ and $n_2 = 0.33$), in good agreement with those obtained for cryptolepine, ellipticines, and acridines. **1**, Roll, *et al.* (1988) *J. Org. Chem.* **53**, 3276 (1988). **2.** Kirsch., *et al.* (2000). *J. Natur. Prod.* **63**, 825. **3.** Soni, *et al.* (2000) *Biochem. Biophys. Res. Communs.* **275**, 877. **4.** Hörmann, *et al.*, (2001) *Bioorg. Med. Chem.* **9**, 917.

Fasciculin 1 & 2

These snake venom toxins (MW$_{Fasciculin-1}$ = 6807 g/mol, *Sequence*: TMCY SHTTTSRAILTNCGENSCYRKSRRHPPKMVLGRGCGCPPGDDYLEV KCCTSPDKCNY; MW$_{Fasciculin-2}$ = 6758 g/mol, *Sequence*: TMCYSHTTT SRAILTNCGENSCYRKSRRHPPKMVLGRGCGCPPGDDYLEKCCTSP DKCNY) from the green mamba (*Dendroaspis angusticeps*) inhibit mammalian acetylcholinesterase with K_i values in the picomolar range. Both have disulfide bonds between Cys3-Cys22, Cys17-Cys39, Cys41-Cys52, and Cys53-Cys59, and differ only at Position-47. **Primary Mode of Action:** The acetylcholinesterase active site has a serine residue lying at the base of 20 Å-deep gorge lined with aromatic residues. Residues near the entrance to the gorge comprise a peripheral binding site for the fasciculins, and occupancy of that site by the snake neurotoxin reduces k_{cat} by a factor of 200, with a corresponding decrease in k_{cat}/K_m (1). **Target(s):** acetylcholinesterase, mammalian (1-10); cholinesterase, *or* butyryl-cholinesterase, weakly inhibited (10,11). **1.** Eastman, Wilson, Cerveñansky & Rosenberry (1995) *J. Biol. Chem.* **270**, 19694. **2.** Duran, Cervenansky, Dajas & Tipton (1994) *Biochim. Biophys. Acta* **1201**, 381. **3.** Karlsson, Mbugua & Rodriguez-Ithurralde (1984) *J. Physiol. (Paris)* **79**, 232. **4.** Radic & Taylor (2001) *J. Biol. Chem.* **276**, 4622. **5.** Cohen, Kronman, Chitlaru, *et al.* (2001) *Biochem. J.* **357**, 795. **6.** Shi, Boyd, Radic & Taylor (2001) *J. Biol. Chem.* **276**, 42196. **7.** Shi, Tai, McCammon, Taylor & Johnson (2003) *J. Biol. Chem.* **278**, 30905. **8.** Boyd, Marnett, Wong & Taylor (2000) *J. Biol. Chem.* **275**, 22401. **9.** Radic, Duran, Vellom, *et al.* (1994) *J. Biol. Chem.* **269**, 11233. **10.** Duran, Cervenansky & Karlsson (1996) *Toxicon* **34**, 959.

Fasentin

This selective glucose transport inhibitor (FW = 279.64 g/mol; CAS = 392721-37-8), also named *N*-[4-chloro-3-(trifluoromethyl)phenyl]-3-oxobutanamide, was first identified as a chemical sensitizer to the death receptor stimuli FAS and tumor necrosis factor apoptosis-inducing ligand.

Fasentin was later shown to alter expression of genes associated with nutrient and glucose deprivation by directly inhibiting GLUT1 (IC$_{50}$ = 10–15 μM). The latter is a glucose uniporter widely distributed in fetal tissues, with highest levels in erythrocytes and endothelial cells of adults. Fasentin interacts with a unique site in the intracellular channel of this transporter. Fasentin also blocks sorafenib-induced glucose uptake, potentiating its cytotoxic activity in SKBR3 cells (2). Persistent activation of AMPK by sorafenib impairment of glucose metabolism in both MCF-7 and SKBR3 cells as well as in the highly glycolytic MDA-MB-231 cells, resulting in cell death. This previously unrecognized long-term effect of sorafenib is mediated by AMPK-dependent inhibition of the mTORC1 pathway. Suppression of mTORC1 activity is sufficient for sorafenib to hinder glucose utilization in breast cancer cells, as demonstrated by the observation that the mTORC1 inhibitor rapamycin induced a comparable down-regulation of GLUT-1 expression and glucose uptake (2). **See also** *Sorafenib* **1**. Wood, Dalili, Simpson, *et al.* (2008) *Mol. Cancer Ther.* **7**, 3546. **2.** Fumarola, Caffarra, La Monica,*et al.* (2013) *Breast Cancer Res. Treat.* **141**, 67.

Fasidotril & Fasidotrilat

Fasidotrilat (active drug)

Fasidotril (prodrug)

This dual NEP/ACE inhibitor (FW$_{free-acid}$ = 311.36 g/mol) and its prodrug (FW = 443.52 g/mol) represent the first rationally designed prodrug and inhibitor for angiotensin-converting enzyme.This antihypertensive drug was at one time referred to as alatrioprilat, **Target(s):** neprilysin, K_i = 5.1 nM; peptidyl-dipeptidase A (angiotensin-converting enzyme, K_i = 9.8 nM. **1.** Gros, Noël, Souque, *et al.* (1991) *Proc. Natl. Acad. Sci. U.S.A.* **88**, 4210.

Fasciola hepatica Kunitz-type Inhibitor

This atypical protease inhibitor (MW = 7 kDa; Symbol: FhKT1), while possessing a canonical Kunitz-type serine protease inhibitory domain, exhibits no inhibitory activity towards serine proteases, but potently inhibits cysteine proteases, including the major secreted cathepsin L of *F. hepatica*, FhCL1 (K_i = 0.4 nM) and FhCL2 (K_i = 10 nM), as well asf human cathepsins L (K_i = 1.6 nM) and K (K_i = 5 nM). Disulfide reduction and alkylation alters FhKT1's structure and likewise abolishes its inhibitory properties. FhKT1 also prevents the auto-catalytic activation of FhCL1 and FhCL2 and forms stable complexes with the mature enzymes. Substitution of the unusual P1 Leu15 within the exposed reactive loop of FhKT1 for the more commonly found Arg (FhKT1Leu15/Arg15) has modest adverse effects on the cysteine protease inhibition, yet confers potent activity against trypsin (K_i = 1.5 nM). The arrangement of functional amino acids provides new insight into protease inhibition by KT inhibitors. **1.** Smith, Tikhonova, Jewhurst, *et al.* (2016) *J. Biol. Chem.* **291**, 19220.

Fasudil

Fasudil   **Ripasudil**

This modestly selective Rho kinase inhibitor and vasodilator ($FW_{free-base}$ = 291.37 g/mol; FW = 327.83 g/mol for the monohydrochloride; CAS = 103745-39-7 (free base) and 105628-07-7 (HCl Salt); Solubility: 5 mg/mL DMSO, 60 mg/mL Water), also known as HA-1077 and 1-(5-isoquinolinesulfonyl)homopiperazine, is an intracellular calcium ion antagonist that competitively inhibits the Ca^{2+}-induced contraction of depolarized rabbit aorta (1). (**See also** SC82510) Fasudil is a protein kinase inhibitor (2-9), with strongest action on protein kinase A (IC_{50} = 1.6 μM), protein kinase G (IC_{50} = 1.6 μM), Rho-dependent protein kinase (IC_{50} = 10.7 μM), and [myosin light-chain] kinase (IC_{50} = 36 μM) (2-5). Other targets include cGMP-dependent protein kinase (2), [mitogen-activated-protein-kinase]-activated protein kinase (4), mitogen- and stress-activated protein kinase (4), [myosin light-chain] kinase (2,6,7), phosphorylase kinase (4), p70 ribosomal protein S6 kinase (4), protein-kinase-C-related protein kinase (4). Fasudil selectively targets RhoA/Rho kinase (IC_{50} = 10.7 μM), an enzyme (abbreviated ROCK) that plays an important role in mediating vasoconstriction and vascular remodeling in the pathogenesis of pulmonary hypertension. ROCK also increases the expression and activity of angiotensinogen-converting enzyme (ACE). There are thus three likely consequences of ROCK inhibition: (a) an increase in eNOS expression, achieved by stabilizing eNOS mRNA; (b) reduction in circulating ACE and Ang-II, leading to a decrease in pulmonary vascular pressure; and (c) inhibition of extracellular signal-regulated kinase, or ERK (9-11). HA-1077 also prevents Rho-mediated myosin phosphatase inhibition in smooth muscle cells (12). Fasudil is used to treat cerebral vasospasm due to subarachnoid hemorrhage and also retards cognitive decline in stroke victims and Alzheimer's Disease. The fasudil derivative Ripasudil (FW = g/mol; CAS = 223645-67-8; Trade Name: Glanatec®; IUPAC: 4-fluoro-5-{[(2S)-2-methyl-1,4-diazepan-1-yl]sulfonyl}isoquinoline) is also a ROCK inhibitor that is used to treat glaucoma and ocular hypertension. **1.** Asano, Ikegaki, Satoh, et al. (1987) J. Pharmacol. Exp. Ther. **241**, 1033. **2.** Asano, Suzuki, Tsuchiya, et al. (1989) Brit. J. Pharmacol. **98**, 1091. **3.** Breitenlechner, Gassel, Hidaka, et al. (2003) Structure **11**, 1595. **4.** Davies, Reddy, Caivano & Cohen (2000) Biochem. J. **351**, 95. **5.** Patel, Soulages, Wells & Arrese (2004) Insect Biochem. Mol. Biol. **34**, 1269. **6.** Sasaki & Sasaki (1990) Biochem. Biophys. Res. Commun. **171**, 1182. **7.** Amano, Fukata, Shimokawa & Kaibuchi (2000) Meth. Enzymol. **325**, 149. **8.** Seto, Sasaki, Hidaka & Sasaki (1991) Eur. J. Pharmacol. **195**, 267. **9.** Kitaoka, Kitaoka, Kumai, et al. (2004) Brain Res. **1018**, 111. **9.** Ocaranze, Prvera, Novoa, et al. (2011) J. Hyperten. **29**, 706. **10.** Takemoto, Sun, Hiroko, Shimokawa & Liao (2002) J. Amer. Heart Assoc. **106**, 57. **11.** Liu, Ling, Wang, et al. (2011) Chin. Med. J. **124**, 3098. **12.** Nagumo, Sasaki, Ono, et al. (2000) Am. J. Physiol. **278**, C57.

Favipiravir

This broadly active small-molecule RNA-dependent RNA polymerase inhibitor (FW = 157.10 g/mol; CAS = 259793-96-9), also named 5-fluoro-2-oxo-1H-pyrazine-3-carboxamide, is active against influenza viruses, West Nile virus, yellow fever virus, foot-and-mouth disease virus as well as many other flaviviruses, arenaviruses, bunyaviruses and α-viruses, in some cases been effective, when initiated up to 5–7 days after virus infection, after the animals had already showed signs of illness (1). Significantly, Favipiravir does not inhibit host DNA and RNA synthesis or inosine 5'-monophosphate dehydrogenase (IMPDH) activity. **1.** Furutaa, Takahashia, Shiraki, et al. (2009) Antiviral Res. **82**, 95.

Favipiravir-β-ribosyltriphosphate

Favipiravir-β-ribosyltriphosphate

This substituted pyrazine (FW = 157.10 g/mol; CAS = 259793-96-9), also known as T-705 and 6-fluoro-3-hydroxy-2-pyrazinecarboxamide, inhibits viruses (e.g., influenza virus, arenaviruses, bunyaviruses, West Nile virus, yellow fever virus, and foot-and-mouth disease virus) utilizing viral RNA-dependent RNA polymerase (or RdRP) for genomic replication by induces a high rate of self-lethal mutations that generate a nonviable viral phenotype (1,2). **Mode of Inhibitory Action:** Favipiravir is a prodrug that is metabolized to its favipiravir-β-ribosyltriphosphate, which acts in a dose-dependent manner. Incorporation of T-705-RTP into viral RNA results in a 9x increase in A/Brisbane/59/2007 and a 5x increase in A/New Jersey/15/2007 in G→A transitions. In cells inoculated with a low virus dose, one observes virus-specific increases in RNA C→T transition after serial passage in treated, but not mock-treated, cells. G→A transitions are also increased at high infectious doses, but not significantly, reflecting dose-related rapid virus extinction. A→G mutations within the coding viral genome cause the production of functionally impaired mutant proteins. T-705 exerts its antiviral effects without significant toxicity to the host. **Inhibitory Dose:** IC_{50} values of favipiravir are 6.6 μM for seasonal influenza A/Brisbane/59/2007 virus; 5.3 μM for A/New Jersey/15/2007 virus; and 3.0 μM for both pandemic A (or H_1N_1) viruses, namely A/Denmark/524/2009 and A/Denmark/528/2009. T-705 was also effective against all four viruses in MDCK cells inoculated at a low dose, with EC_{50} values of 17 and 15 μM for the seasonal influenza viruses and 15 and 11 μM for the pandemic viruses (2). **1.** Furuta, Takahashi, Kuno-Maekawa, et al. (2005) Antimicrob. Agents Chemother. **49**, 981. **2.** Baranovich, Wong, Armstrong, et al. (2013) J. Virol. **87**, 3741.

Febuxostat

This gout-mitigating drug (FW = 316.34 g/mol; CAS = 144060-53-7), also known as TEI-6720 and systematically named as 2-(3-cyano-4-isobutoxyphenyl)-4-methyl-1,3-thiazole-5-carboxylic acid, is a strong noncompetitive inhibitor of xanthine oxidase (XO). **Mode of Action:** Although a non-purine, febuxostat inhibits xanthine oxidase by blocking the enzyme's molybdenum pterin center (1-4). Steady state kinetics measurements exhibit mixed type inhibition with K_i and K_i' values of 0.12 nM and 0.9 nM, respectively (2). Fluorescence titration showed that TEI-6720 bound very tightly to both the active and the inactive desulfo-form of the enzyme. The dissociation constant determined for the desulfo-form was 2 nM, too low to allow accurate measurement. The crystal structure of the active sulfo-form of milk xanthine dehydrogenase complexed with TEI-6720 revealed that inhibitor molecule binds in a narrow channel leading to the molybdenum-pterin active site of the enzyme, filling the channel as well as the immediate environment of the cofactor. Febuxistat inhibits the enzyme by preventing substrate binding (2). (**See Allopurinol**) Febuxostat is marketed under the trade name Uloric® in the U.S. and as Adenuric® and Febutaz® in the E.U. and India. Significantly, xanthine oxidase is also a major source of reactive oxygen species (ROS) and has been linked to the pathogenesis of atherosclerosis. Enhanced XO expression has been demonstrated in macrophages within atherosclerotic plaque and in aortic endothelial cells of ApoE(-/-) mice. Febuxostat suppresses plaque formation, reduces arterial ROS levels, and improves endothelial dysfunction in ApoE(-/-) mice without affecting plasma cholesterol levels (5). In vitro, febuxostat inhibits cholesterol crystal-induced ROS formation and inflammatory cytokine release in murine macrophages (5). **Target(s):** uridine transport (1); xanthine dehydrogenase (2,3); xanthine oxidase (2-4). **1.** Yamamoto, Moriwaki, Fujimura, et al. (2000) Pharmacology **60**, 34. **2.** Okamoto, Eger, Nishino, et al. (2003) J. Biol. Chem. **278**, 1848. **3.** Takano, Hase-Aoki, Horiuchi, et al. (2005) Life Sci. **76**, 1835. **4.** Osada, Tsuchimoto, Fukushima, et al. (1993) Eur. J. Pharmacol. **241**, 183. **5.** Nomura, Busso, Ives, et al. (2014) Sci. Rep. **4**, 4554.

Fedratinib

This potent and selective ATP-competitive JAK inhibitor (FW = 524.68 g/mol; CAS = 936091-26-8), also named SAR302503, TG101348, and *N*-(2-methyl-2-propyl)-3-{[5-methyl-2-({4-[2-(1-pyrrolidinyl)ethoxy]phenyl} amino)-4-pyrimidinyl]amino}benzenesulfonamide, targets Janus Kinase-2, *or* JAK2 (IC_{50} = 3 nM), showing efficacy in a murine model of myeloproliferative disease induced by the JAK2^{V617F} mutation (1). Fedratinib also inhibits the carrier-mediated uptake by hTHTR2 and transcellular flux of thiamine, such that oral absorption of dietary thiamine is significantly compromised by fedratinib dosing. The latter action appears to be a unique to fedratinib and is not shared by currently marketed JAK2 inhibitors, suggesting a molecular basis for the development of Wernicke's encephalopathy upon fedratinib treatment (2). On November 18, 2013, Sanofi announced announced that, following a thorough risk/benefit analysis, it would halt clinical trials and cancel plans for regulatory filings on fedratinib. Prior to that announcement, fedratinib was regarded as a promising investigational JAK2 inhibitor for the treatment of primary myelofibrosis – including those previously treated with ruxolitinib, polycythemia vera; and essential thrombocythemia. **1**. Wernig, Kharas, Okabe, *et al.* (2008) *Cancer Cell* **13**, 311. **2**. Zhang, Zhang, Diamond, *et al.* (2014) *Drug Metab. Dispos.* **42**, 1656.

Felbamate

This anticonvulsant agent (FW = 238.24 g/mol; CAS = 25451-15-4; Anatomical Therapeutic Chemical Classification: N03AX), also named (3-carbamoyloxy-2-phenylpropyl) carbamate, is an antagonist that binds at the NR2$_B$ subunit of the NMDA (*N*-methyl-D-aspartate) receptor, a heteromeric ligand-gated ion channel that interacts with multiple intracellular proteins using different subunits. NMDA receptors are composed of seven subunits: NR1, NR2$_A$, NR2$_B$, NR2$_C$, NR2$_D$, NR3$_A$, and NR3$_B$. Many studies have demonstrated NR2$_B$'s importance in synaptic signaling that modulate learning, memory processing, pain perception, and feeding behaviors. Felbamate is also a GABA$_A$ receptor agonist. Felbamate was launched in the U.S. as a new anti-epileptic pharmaceutical in 1993. **Target(s):** Ca^{2+} channels, voltage-sensitive (1); CYP2C19 (2,3); Na$^+$ channels, voltage sensitive (4,5); NMDA receptor (6-8). **1**. Stefani, Calabresi, Pisani, *et al.* (1996) *J. Pharmacol. Exp. Ther.* **277**, 121. **2**. Benedetti (2000) *Fundam. Clin. Pharmacol.* **14**, 301. **3**. Levy (1995) *Epilepsia* **36** Suppl. 5, S8. **4**. Srinivasan, Richens & Davies (1996) *Eur. J. Pharmacol.* **315**, 285. **5**. Taglialatela, Ongini, Brown, Renzo & Annunziato (1996) *Eur. J. Pharmacol.* **316**, 373. **6**. Domenici, Marinelli & Sagratella (1996) *Life Sci.* **58**, PL391. **7**. Kleckner, Glazewski, Chen & Moscrip (1999) *J. Pharmacol. Exp. Ther.* **289**, 886. **8**. Harty & Rogawski (2000) *Epilepsy Res.* **39**, 47.

Felbinac

This anti-inflammatory agent (FW$_{free-acid}$ = 212.25 g/mol), also known as 4-biphenylacetic acid and (1,1'-biphenyl)-4-acetic acid, is a fenbufen metabolite that inhibits prostaglandin biosynthesis. Ciprofloxacin and felbinac synergistically inhibit GABA$_A$ receptors. **Target(s):** GABA$_A$

receptor (1); medium-chain-fatty-acyl-CoA synthetase, *or* butyryl-CoA synthetase, *or* butyrate:CoA ligase, also, as an alternative substrate (2); prostaglandin biosynthesis (3-5). **1**. Green & Halliwell (1997) *Brit. J. Pharmacol.* **122**, 584. **2**. Kasuya, Hiasa, Kawai, Igarashi & Fukui (2001) *Biochem. Pharmacol.* **62**, 363. **3**. Child, Osterberg, Sloboda & Tomcufcik (1980) *Arzneimittelforschung* **30**, 695. **4**. Sloboda, Tolman, Osterberg & Panagides (1980) *Arzneimittelforschung* **30**, 716. **5**. Kohler, Tolman, Wooding & Ellenbogen (1980) *Thromb. Res.* **20**, 123.

Felodipine

This anti-hypertensive agent (FW = 284.25 g/mol; CAS = 72509-76-3; Solubility: 100 mM in DMSO, 100 mM in Ethanol), also named 4-(2,3-dichlorophenyl)-1,4-dihydro-2,6-dimethyl-3,5-pyridinedicarboxylic acid ethyl methyl ester, targets L-type Ca^{2+} channels, showing selective over N-, R-, P/Q- and T-type Ca^{2+} channels. Felodipine lowers arterial blood pressure without altering cardiac contractility. It is also used sometimes to treat Raynaud's syndrome and congestive heart failure. **Key Pharmacokinetic Parameters:** *See* Appendix II in Goodman & Gilman's *THE PHARMACOLOGICAL BASIS OF THERAPEUTICS*, 12th Edition (Brunton, Chabner & Knollmann, eds.) McGraw-Hill Medical, New York (2011). **1**. Ljung (1990) *J. Cardiovasc. Pharmacol.* **15**, S11. **2**. Furukawa et al (1999) *J. Pharmacol. Exp. Ther.* **291**, 464. **3**. Furukawa et al (2005) *J. Cardiovasc. Pharmacol.* **45**, 241.

Fenbendazole

This broad-spectrum antihelminthic (FW = 299.35 g/mol; CAS = 43210-67-9; *Abbreviation*: FZ), also known by the trade names Panacur™ and Safe-Guard™, and Intervet Panacur™ as well as by its systematic name, methyl *N*-(6-phenylsulfanyl-1*H*-benzoimidazol-2-yl)carbamate, is used to treat roundworms, hookworms, whipworms, and *Taenia* species tapeworms. Fenbendazole is a microtubule-directed drug that disrupts mitosis. The selectivity of these drugs for the parasite is explained on the basis of irreversible blockade in uptake of glucose in parasite, which leads to depletion of glycogen storage and degeneration of endoplasmic reticulum in the germinal layers, resulting in cell death (1). Recent studies demonstrated that exhibits a potent growth-inhibitory activity against cancer cell lines but not normal cells. IC_{50} values were ~1.2 and 1.6 µM, respectively, for H460 and A549 cell lines. One clue as to this mode of action is the unexpected finding that, upon fenbendazole treatment, cells accumulate a number of apoptosis regulatory proteins (*e.g.*, cyclins, p53, and IκBα) that are normally degraded by the ubiquitin-proteasome pathway (2). Proteasome inhibition results in antibody-detected accumulation of ubiquitylated forms of numerous proteins. Fenbendazole treatment of human non-small cell lung cancer cells also induces endoplasmic reticulum stress, production of reactive oxygen species, decreased mitochondrial membrane potential, as well as cytochrome *c* often associated with cancer cell death (2). **1**. Marriner, Morris Dickson & Bogan (1986) *Eur. J. Clin. Pharmacol.* **30**, 705. **2**. Dogra & Mukhopadhyay (2012) *J. Biol. Chem.* **287**, 30625.

Fenitrothion

This phosphorothioate insecticide (FW = 277.24 g/mol; CAS = 122-14-5), systematically named as *O,O*-dimethyl *O*-(3-methyl-4-nitrophenyl) phosphorothioate, inhibits cholinesterases. Resistant insects often have a symbiotic gut microbe that can metabolize and detoxify fenitrothion. This insecticide was employed for years to eradicate malaria-vectors. **Target(s):** acetylcholinesterase (1-3); carboxylesterase (4); cholinesterase (5,6); glutathione *S*-transferase (7); Na⁺/K⁺-exchanging ATPase (8). **1.** Hamilton, Hunter & Ruthven (1981) *Bull. EnviroContam. Toxicol.* **27**, 856. **2.** Kobayashi, Yuyama, Kudo & Matsusaka (1983) *Toxicology* **28**, 219. **3.** Ali, Eldeen & Hikal (2005) *Pestic. Biochem. Physiol.* **83**, 58. **4.** Tsujita, Okuda & Yamasaki (1982) *Biochim. Biophys. Acta* **715**, 181. **5.** Klaverkamp, Hobden & Harrison (1975) *Proc. West Pharmacol. Soc.* **18**, 358. **6.** Kovac, Markova & Kralik (1974) *J. Chromatogr.* **100**, 171. **7.** Yamamoto, Fujii, Aso, Banno & Koga (2007) *Biosci. Biotechnol. Biochem.* **71**, 553. **8.** Sancho, Ferrando & Andreu (1997) *Ecotoxicol. Environ. Saf.* **38**, 132

Fenofibrate

This prodrug ester (FW = 360.84 g/mol; M.P. = 80-81°C), also named propan-2-yl 2-{4-[(4-chlorophenyl)carbonyl]phenoxy}-2-methylpropanoate, inhibits membrane-bound hepatic 3-hydroxy-3-methylglutaryl-CoA reductase by altering membrane fluidity. Fenofibrate is mainly used to treat primary hypercholesterolemia or mixed dyslipidemia. The drug exhibits a hypocholesterolemic action due to decreased cholesterol biosynthesis via inhibition of 3-hydroxy-3-methylglutaryl-CoA reductase and increased low-density lipoprotein clearance via modulation of hepatic LDL receptors. It also activates 3α-hydroxysteroid dehydrogenase. Peroxisome proliferator-activated receptors (PPARs) are involved in lipid and glucose metabolic pathways. By activating PPARα, fenofibrate effectively improves the atherogenic lipid profile associated with type 2 diabetes mellitus and metabolic syndrome (2). Fenofibrate-related PPARα activation may enhance the expression of genes promoting hepatic FA β-oxidation. Fenofibrate also reduces hepatic insulin resistance. It also inhibits the expression of inflammatory mediators involved in non-alcoholic steatohepatitis pathogenesis. These include tumor necrosis factor-α, intercellular cell adhesion molecule-1, vascular cell adhesion molecule-1 and monocyte chemoattractant protein-1 (2). Its anti-inflammatory effects, when taken together with anti-oxidant actions, may prevent from Non-alcoholic steatohepatitis (or NASH) -related necroinflammation, apoptosis and fibrosis. These are attributed to inhibited expression of inflammatory mediators, including TNF-α, MCP-1, VCAM-1 and ICAM-1, together with reduced lipid peroxidation and reactive oxygen species formation. Note that the pharmacologically active agent is deesterified to fenofibric acid. **Key Pharmacokinetic Parameters:** *See* Appendix II in Goodman & Gilman's *THE PHARMACOLOGICAL BASIS OF THERAPEUTICS*, 12th Edition (Brunton, Chabner & Knollmann, eds.) McGraw-Hill Medical, New York (2011). **1.** Wülfert, Boissard, Legendre & Baron (1981) *Artery* **9**, 120. **2.** Kostapanos, Kei & Elisaf (2013) *World J. Hepatol.* **5**, 470.

Fenoprofen

This nonsteroidal anti-inflammatory agent (FW_free-acid = 242.27 g/mol), also named 2-(3-phenoxyphenyl)propanoic acid, targets prostaglandin synthase, *or* cyclooxygenase (1-3). The *S* enantiomer is more active than the *R*-isomer). Fenoprofen is used most often to relieve pain, tenderness, swelling, and stiffness caused by osteoarthritis and rheumatoid arthritis. It is also used to treat ankylosing spondylitis and gouty arthritis. Major urinary metabolites are fenoprofen glucuronide and 4'-hydroxyfenoprofen glucuronide. **Target(s):** aromatic amino-acid decarboxylase (4); hyaluronidase (5); long-chain-fatty-acyl CoA synthetase, *or* long-chain-fatty-acid:CoA ligase, by the *R*-isomer (6); palmitoyl-CoA synthetase (7). **1.** Jell & Sweatman (1976) *Can. J. Physiol. Pharmacol.* **54**, 161. **2.** Ali & McDonald (1978) *Thromb. Res.* **13**, 1057. **3.** Rubin, Knadler, Ho, Bechtol

& Wolen (1985) *J. Pharm. Sci.* **74**, 82. **4.** Pribova, Gregorova & Drsata (1992) *Pharmacol. Res.* **25**, 271. **5.** Joyce, Mack, Anderson & Zaneveld (1986) *Biol. Reprod.* **35**, 336. **6.** Knights & Jones (1992) *Biochem. Pharmacol.* **43**, 1465. **7.** Roberts & Knights (1992) *Biochem. Pharmacol.* **44**, 261.

Fenpropimorph

This synthetic fungicide (FW = 303.49 g/mol; CAS = 67564-91-4), also named *cis*-2,6-dimethyl-4-{2-methyl-3-[4-(2-methyl-2-propanyl)phenyl]-propyl}morpholine or (2R,6S)-4-[3-(4-tert-butylphenyl)-2-methyl-propyl]-2,6-dimethylmorpholine, inhibits phytosterol and cholesterol biosynthesis. Ergosterol biosynthesis is inhibited in the fungi, leading to the arrest of cell proliferation in the unbudded G₁ phase of the cell cycle. Cholesterol biosynthesis appears to be inhibited at a lanosterol demethylation step. **Target(s):** cholestenol Δ-isomerase, *or* sterol Δ^{8,7}-sterol isomerase (1-3); cholesterol oxidase (4); Δ^{14}-sterol reductase (1); Δ^{5,7}-sterol-Δ^7-reductase (5). **1.** Marcireau, Guilloton & Karst (1990) *Antimicrob. Agents Chemother.* **34**, 989. **2.** Moebius, Bermoser, Reiter, Hanner & Glossmann (1996) *Biochemistry* **35**, 16871. **3.** Moebius, Reiter, Bermoser, *et al.* (1998) *Mol. Pharmacol.* **54**, 591. **4.** Hesselink, Kerkenaar & Witholt (1990) *J. Steroid Biochem.* **35**, 107. **5.** Taton & Rahier (1991) *Biochem. Biophys. Res. Commun.* **181**, 465.

Fentanyl & Acetyl-fentanyl

Fentanyl

Acetylfentanyl

This much-abused synthetic, μ-opioid receptor agonist (FW_fentanyl = 336.48 g/mol; CAS = 437-38-7), also known by the trade names Sublimaze®, Actiq®, Durogesic®, Duragesic®, Fentora®, Matrifen®, Haldid®, Onsolis®, Instanyl®, Abstral®, and Lazanda®, is a general anesthetic is used to treat breakthrough pain (*i.e.*, pain that comes on suddenly and persists for a short period). It inhibits 3',5'-cyclic-AMP formation (IC₅₀ = 27 nM), but not K⁺-evoked release of noradrenaline. Possessing an analgesic potency that is 100x higher than morphine, intravenous fentanyl is best managed in the clinical setting, where it is used for anesthesia and analgesia, most often in operating rooms and intensive care units (1). **Fentanyl Pharmacokinetic Parameters:** Fentanyl displays a large apparent volume of distribution, short plasma half life and extensive biotransformation. With an *n*-Octanol/H₂O partition coefficient >700, fentanyl can be administered by various routes. Upon intravenous administration, it quickly passes to CNS (transfer t₀.₅ = 5–6 min), binding to μ-opioid G-protein-coupled receptors, thereby inhibiting pain neurotransmitter release by decreasing intracellular Ca²⁺ levels. Compared to morphine, fentanyl enjoys a shorter duration of action, the lack of hyperglycemic response to surgery, a decrease in catecholamine levels, and higher lipid solubility. Fentanyl is also delivered with a transdermal patch or an Actiq 'lollipop' for absorption across the buccal mucosa. Breakthrough Pain (*i.e.*, transitory pain reaching maximum severity in ~15 minutes and lasting ~60 minutes) is currently managed by administering fentanyl by transmucosal and intranasal routes. For additional information on fentanyl's pharmacokinetics, *See* Appendix II in Goodman & Gilman's *THE PHARMACOLOGICAL BASIS OF THERAPEUTICS*, 12th Edition (Brunton, Chabner & Knollmann, eds.) McGraw-Hill Medical, New York (2011). Fentanyl inhibits gastric cell growth and proliferation, with cell cycle arrest at the G₂/M junction (2). Fentanyl and acetylfentanyl (below) are easy to make and widely available designer drugs that have replacing heroin in their ilicit use. **Acetyl-fentanyl:** This recreational drug (FW = 322.44 g/mol; CAS: 3258-84-2; IUPAC Name: *N*-(1-phenethylpiperidin-4-yl)-*N*-phenylacetamide) is similar in its actions to fentanyl and possesses opioid-like *in vitro* binding affinity to μ-opioid receptors as well as μ-opioid receptor agonist effects. Acetyl fentanyl has not been approved for medical use in the United States, and there are no published studies on safety in human use. The potency of acetylfentanyl was about one-third that of fentanyl. As a μ-opioid receptor agonist, acetyl fentanyl often serves as a direct substitute for heroin or other μ-opioid receptor agonist substances in opioid dependent individuals. Acetyl-fentanyl has been detected in tablets that mimic pharmaceutical opiate products, in powder form and spiked on blotter papers. **1.** Heitz, Witkowski & Viscusi (2009) *Curr. Opin. Anaesthesiol.* **22**, 608. **2.** Kasai & Ikeda (2011) *Pharmacogenomics* **12**, 1305.

Fenticonazole

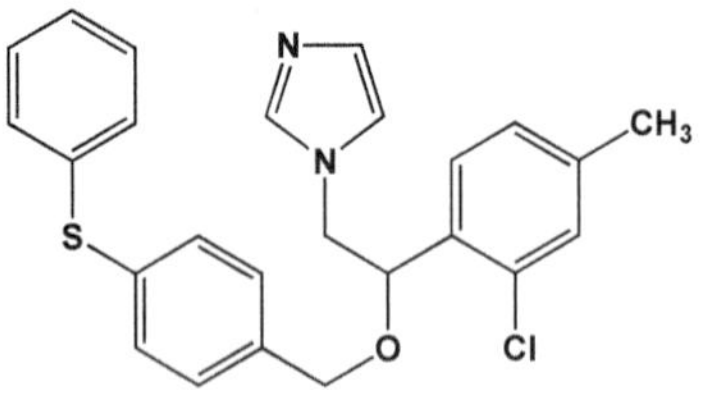

This azole-type fungicide (FW = 455.40 g/mol; CAS = 72479-26-6), also named 1-[2-(2,4-dichlorophenyl)-2-{[4-(phenylsulfanyl)phenyl]methoxy} ethyl]-1*H*-imidazole, resembles fluconazole by inhibiting 14α-demethylase, the cytochrome P450 that converts lanosterol to ergosterol. The latter is an essential component of the fungal cytoplasmic membrane. The antifungal activity spectrum of fenticonazole is very broad: dermatophytes, yeasts and dimorphic fungi were the most susceptible organisms (1). In the treatment of vulvovaginal candidiasis, fenticonazole is inserted as a 200-mg ovule into the vagina at bedtime on three successive nights (1). In an *in vitro* study, fenticonazole also exhibited excellent activity against gram-positive microorganisms. (**See also** *Fluconazole*) 1. Veronese, Salvaterra, Barzaghi (1981) *Arzneimittel-forschung.* **31**, 2133. **2.** Wiest & Ruffmann (1987) *J. Int. Med. Res.* **15**, 319.

Fenvalerate

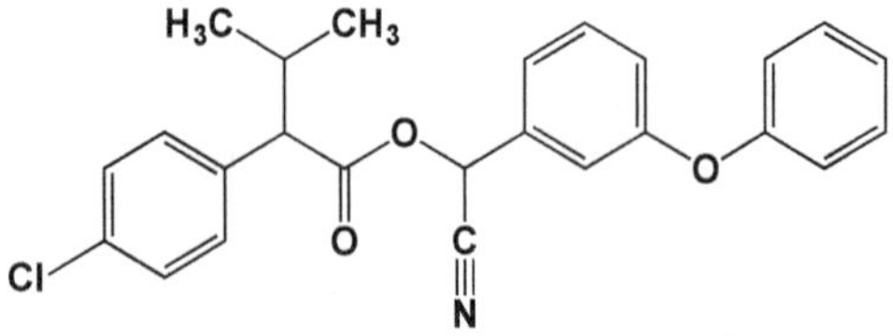

This synthetic insecticide (FW = 419.91 g/mol), used to control a wide range of pests, inhibits calcineurin (1-3) and induces depolarization by keeping Na$^+$ channels in an open state (4). Fenvalerate also decreases the phase-transition temperature of model membranes and localizes deep within the acyl chain region of the lipid, presumably weakening the acyl chain packing. Note: There are four stereoisomers of fenvalerate, a fact that is likely to explain differences in the literature concerning effects on calcineurin. **1.** Enan & Matsumura (1992) *Biochem.Pharmacol.* **43**, 1777. **2.** Fakata, Swanson, Vorce & Stemmer (1998) *Biochem. Pharmacol.* **55**, 2017. **3.** Enz & Pombo-Villar (1997) *Biochem. Pharmacol.* **54**, 321. **4.** Salgado & T. Narahashi (1993) *Mol. Pharmacol.* **43**, 626.

Ferrate Ion, *or* FeO$_4^{2-}$

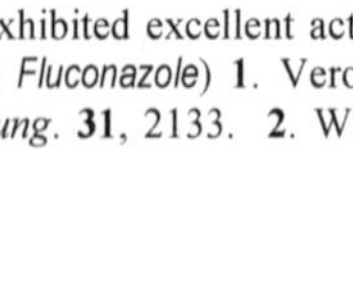

This powerful Fe(IV)-state oxidant and orthophosphate analogue, often supplied as K$_2$FeO$_4$ (FW = 198.04 g/mol) irreversibly inactivates glycogen phosphorylase *b* and adenylate kinase, in both cases reacting at tyrosyl residues. Ferrate ion also inactivates ribonuclease A by reacting at a histidyl residue. A tryptophanyl residue in triose-phosphate isomerase is also modified by ferrate. Although unstable at neutral and acidic pH, potassium ferrate can be kept at alkaline pH in the absence of reducing agents.

Chemical Properties: With a reduction potential varying from +2.2 V to +0.7 V in acidic and basic solutions, Fe(IV) is a powerful oxidizing agent throughout the entire pH range, respectively (1). In strong acids, H$_3$Fe(VI)O$_4^+$ and H$_2$Fe(VI)O$_4$ ions reduces, both rapidly and exothermally, to Fe(III) and oxygen. At pH 7, 98% of the agent exists in the HFe(VI)O$_4^{1-}$ species. At at pH 10, the oxygen ligands of Fe(VI) exchange very slowly with water. For a diagram detailing the complete pH-dependent speciation of ferrate ion, see Fig. 1 of the first reference. **Target(s):** acid phosphatase (2); adenylate kinase (3); alcohol dehydrogenase (2); alkaline phosphatase (2); arylsulfatase A (4,5); DNA-directed DNA polymerase (6); DNA polymerase I (6); fructose-1,6-bisphosphatase (2); fumarase (2); glycogen phosphorylase *b*, inhibited by potassium ferrate (7,8); lipoxygenase, soybean, weakly inhibited (2); murine leukemia virus reverse transcriptase (9); nitrate reductase (10); pancreatic ribonuclease, ribonuclease A (11,12); phosphoglucomutase (2); phosphoglycolate phosphatase (2); pyrophosphatase (2); RNA-directed DNA polymerase (9); triose-phosphate isomerase (13); venom 5'-nucleotidase (2). **1.** Sharma (2002) *Adv. Envir. Res.* **6**, 143. **2.** Rajababu & Axelrod (1978) *Arch. Biochem. Biophys.* **188**, 31. **3.** Crivellone, Hermodson & Axelrod (1985) *J. Biol. Chem.* **260**, 2657. **4.** Laidler (1991) *Folia Med. Cracov* **32**, 149. **5.** Laidler & Steczko (1986) *Acta Biochim. Pol.* **33**, 101. **6.** Basu, Williams & Modak (1987) *J. Biol. Chem.* **262**, 9601. **7.** Lee & Benisek (1976) *J. Biol. Chem.* **251**, 1553. **8.** Lee & Benisek (1978) *J. Biol. Chem.* **253**, 5460. **9.** Reddy, Nanduri, Basu & Modak (1991) *Biochemistry* **30**, 8195. **10.** Ramadoss, Steczko & Axelrod (1985) *Acta Biochim. Pol.* **32**, 179. **11.** Blackburn & Moore (1982) *The Enzymes*, 3rd ed., **15**, 317. **12.** Steczko, Walker, Hermodson & Axelrod (1979) *J. Biol. Chem.* **254**, 3254. **113.** Steczko, Hermodson, Axelrod & Dziembor-Kentzer (1983) *J. Biol. Chem.* **258**, 13148.

Ferrichrome & Ferrichrome A

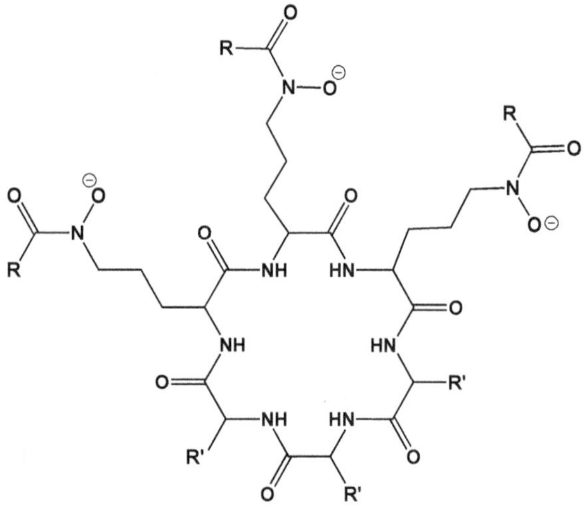

These ferric ion-chelating siderophores (FW$_{Ferrichrome}$ = 740.52 g/mol, R = CH$_3$, and R' = H; FW$_{Ferrichrome-A}$ = 1053.22 g/mol, R = *trans*-HOOCCH$_2$(CH$_3$)C=CH–, and R' = –CH$_2$OH; CAS = 15258-80-7) from *Ustilago sphaerogena* are cyclic hexapeptide, growth-promoting factors that competitively inhibit the antibiotic action of sideromycins. Six oxygen atoms from three hydroxamate groups bind Fe(III) in nearly perfect octahedral coordination. First isolated from the fungus *Ustilago sphaerogena*, the ferrichromes have been identified in other fungi. Note: *p*-Azidobenzoyloxyferricrocin is a photoactivatable analogue of ferrichrome (1). Albomycin, a derivative of ferrichrome with a bound thioribosyl-pyrimidine moiety, inhibits seryl-t-RNA synthetase. **Target(s):** gelatinase A (matrix metalloproteinase-2 (2). **1.** Salah el Din, Braun, Abdallah (1999) *Biometals* **12**, 151. **2.** Gendron, Grenier, Sorsa, Uitto & Mayrand (1999) *J. Periodontal Res.* **34**, 50.

Ferricyanide, *or* Fe(CN)$_6^{3-}$

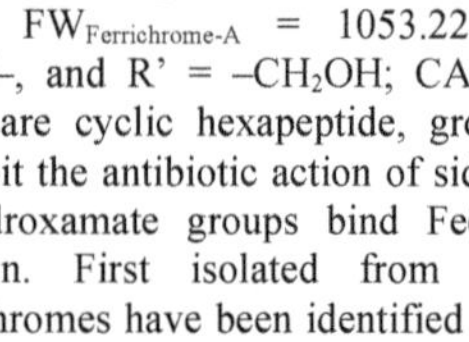

This hexacoordinate Fe(III)-cyanide complex (commonly supplied as $K_3Fe(CN)_6$, FW = 329.24 g/mol; CAS = 13746-66-2), also known as hexacyanoferrate(III), is frequently used as an electron acceptor in many enzyme-catalyzed reactions as well as an oxidant of sulfhydryl groups, resulting in the formation of ferrocyanide, $Fe(CN)_6^{4-}$. (**See Ferrocyanide for X-Gal Localization of β-Galactosidase**) The geometry of the metal ion complex does not change when switching between Fe(III) and Fe(II) oxidation states. **CAUTION:** Although a cyanide-containing substance, ferricyanide is considerably less toxic than potassium cyanide, a fact reflecting the overall stability of the octahedral complex. Ferricyanide will, however, react with mineral acids to release the extremely toxic gas, hydrogen cyanide.

Target(s): acid phosphatase (1); aconitase (2); adenylyl sulfate kinase (3); alkaline phosphatase (1); α-amylase (4); β-amylase (5); L-arabinose dehydrogenase (6,7); arylsulfatase (8); asclepain (9); asparaginase (10); aspartate aminotransferase (11); bromelain (9); carbonic anhydrase (12); carboxylesterase (13); chlorophyllase (14); cholinesterase (15); cytochrome c oxidase (16); deoxyribodipyrimidine photo-lyase (17,18); DNase (19); dextranase (20); diisopropylfluorophosphatase (21); ferredoxin:NADP⁺ reductase (60); formate acetyltransferase (22); β-fructofuranosidase (23,24); fructose 1,6-bisphosphate aldolase (25,26); galactose dehydrogenase (7,27); gentisate 1,2-dioxygenase (28); glucan endo-1,3-β-D-glucosidase (29); glycine acetyltransferase (30); homogentisate 1,2-dioxygenase (31); inositol oxygenase (61); β-lactamase, or penicillinase (32); lactate dehydrogenase (33); lipase (34); luciferase (35,36); mannose isomerase (37); mitochondrial processing peptidase (38); NADH dehydrogenase (39); NADH oxidase (40); papain (41,42); penicillopepsin (43); pepsin (44); phosphofructokinase45; 3-phosphoglycerate kinase (46); phospholipase C (47); phosphoprotein phosphatase (48); protein-phosphohistidine:sugar phosphotransferase (49); pyruvate decarboxylase (26); pyruvate oxidase (50); succinate dehydrogenase, or succinate oxidase (51-54); sulfite oxidase (55,56); transaldolase (26); transketolase (26); tryptophan 2,3 dioxygenase (57); tryptophan 2-monooxygenase (62); tryptophan synthase (58); β-tyrosinase, or tyrosine phenol-lyase (59). **1.** Sizer (1942) *J. Biol. Chem.* **145**, 405. **2.** Rahatekar & Rao (1963) *Enzymologia* **25**, 292. **3.** Schriek & Schwenn (1986) *Arch. Microbiol.* **145**, 32. **4.** Di Carlo & Redfern (1947) *Arch. Biochem.* **15**, 343. **5.** Weill & Caldwell (1945) *J. Amer. Chem. Soc.* **67**, 214. **6.** Doudoroff (1962) *Meth. Enzymol.* **5**, 342. **7.** Doudoroff, Contopoulou & Burns (1958) *Proc. Intern. Symp. Enzyme Chem., Tokyo Kyoto 1957*, p. 313, Maruzen, Tokyo. **8.** Takahashi (1960) *J. Biochem.* **47**, 230. **9.** Greenberg & Winnick (1940) *J. Biol. Chem.* **135**, 761. **10.** Gaffar & Shetna (1977) *Appl. Environ. Microbiol.* **33**, 508. **11.** Polyanovskii (1962) *Biochemistry (U.S.S.R.)* [Engl. transl.] **27**, 623. **12.** Walk & Metzner (1975) *Hoppe-Seyler's Z. Physiol. Chem.* **356**, 1733. **13.** Krisch (1963) *Biochem. Z.* **337**, 546. **14.** Klein & Vishniac (1961) *J. Biol. Chem.* **236**, 2544. **15.** Hargreaves (1955) *Arch. Biochem. Biophys.* **57**, 41. **16.** Kreke, Schaefer, Seibert & Cook (1950) *J. Biol. Chem.* **185**, 469. **17.** Saito & Werbin (1970) *Biochemistry* **9**, 2610. **18.** Werbin (1977) *Photochem. Photobiol.* **26**, 675. **19.** Watanabe & Yamafuji (1961) *Enzymologia* **23**, 353. **20.** Madhu & Prabhu (1985) *Enzyme Microb. Technol.* **7**, 279. **21.** Mounter, Floyd & Chanutin (1953) *J. Biol. Chem.* **204**, 221. **22.** McCormick, Ordal & Whiteley (1962) *J. Bacteriol.* **83**, 887. **23.** Sizer (1942) *J. Gen. Physiol.* **25**, 399. **24.** Myrbäck (1957) *Festschr. Arthur Stoll*, p. 551. **25.** Birkenhäger (1960) *Biochim. Biophys. Acta* **40**, 182. **26.** Christen (1977) *Meth. Enzymol.* **46**, 48. **27.** Doudoroff (1962) *Meth. Enzymol.* **5**, 339. **28.** Harpel & Lipscomb (1990) *Meth. Enzymol.* **188**, 101. **29.** Nagasaki, Nishioka, Mori & Yamamato (1976) *Agric. Biol. Chem.* **40**, 1059. **30.** Urata & Granick (1963) *J. Biol. Chem.* **238**, 811. **31.** Tokuyama (1959) *J. Biochem.* **46**, 1453. **32.** Smith (1963) *Nature* **197**, 900. **33.** Keyhani & Sattarahmady (2002) *Mol. Biol. Rep.* **29**, 163. **34.** Singer & Hofstee (1948) *Arch. Biochem.* **18**, 229. **35.** Green & McElroy (1955) *Meth. Enzymol.* **2**, 857. **36.** Shimomura, Johnson & Saiga (1963) *J. Cellular Comp. Physiol.* **61**, 275. **37.** Palleroni & Doudoroff (1955) *J. Biol. Chem.* **218**, 535. **38.** Deng, Zhang, Kachurin, et al. (1998) *J. Biol. Chem.* **273**, 20752. **39.** Frimmer (1960) *Biochem. Z.* **332**, 522. **40.** Dolin & Wood (1960) *J. Biol. Chem.* **235**, 1809. **41.** Smith (1951) *The Enzymes*, 1st ed., **1** (part 2), 793. **42.** Hellerman & Perkins (1934) *J. Biol. Chem.* **107**, 241. **43.** Mabrouk, Amr & Abdel-Fattah (1976) *Agric. Biol. Chem.* **40**, 419. **44.** Bohak (1970) *Meth. Enzymol.* **19**, 347. **45.** Bloxham & Lardy (1973) *The Enzymes*, 3rd ed., ed.), **8**, 239. **46.** Joao & Williams (1993) *Eur. J. Biochem.* **216**, 1. **47.** Takahashi, Sugahara & Ohsaka (1981) *Meth. Enzymol.* **71**, 710. **48.** Bargoni, Fossa & Sisini (1963) *Enzymologia* **26**, 65. **49.** Grenier, Waygood & Saier (1985) *Biochemistry* **24**, 47 and 4872. **50.** Kuratomi (1959) *J. Biochem.* **46**, 453. **51.** von Euler & Hellström (1939) *Arkiv Kemi Mineral. Geol.* **13B**, 1. **53.** Potter & K. P. DuBois (1943) *J.*

Gen. Physiol. **26**, 391. **53.** Vercauteren (1957) *Enzymologia* **18**, 97. **54.** Barron & Singer (1945) *J. Biol. Chem.* **157**, 221. **55.** Helton & Kirk (1999) *Inorg. Chem.* **38**, 4384. **56.** Kessler & Rajagopalan (1974) *Biochim. Biophys. Acta* **370**, 399. **57.** Ishimura (1970) *Meth. Enzymol.* **17A**, 429. **58.** Wada, Yoshimatsu, Koizumi, et al. (1958) *Proc. Intern. Symp. Enzyme Chem., Tokyo Kyoto 1957*, p. 148, Maruzen, Tokyo. **59.** Ichihara, Yashimatsu & Sakamoto (1956) *J. Biochem.* **43**, 803. **60.** Milani, Balconi, Aliverti, et al. (2007) *J. Mol. Biol.* **367**, 501. **61.** Reddy, Pierzchala & Hamilton (1981) *J. Biol. Chem.* **256**, 8519. **62.** Kosuge, Heskett & Wilson (1966) *J. Biol. Chem.* **241**, 3738.

Ferrocyanide

This anion (commonly supplied as the potassium salt, $K_4Fe(CN)_6$, $FW_{anhydrous} = 368.35$ g/mol; CAS = 13943-58-3; $FW_{trihydrate} = 422.388$ g/mol; CAS = 14459-95-1) inhibits many enzymes, particularly those requiring heme prosthetic groups. Potassium ferroxyanide trihydrate and sodium ferrocyanide decahydrate salts are eflorescent and slowly dehydrate at 60°C. (A related compound is sodium nitroprusside ($Na_2[Fe(CN)_5NO]$), a NO-releasing drug.). **X-Gal Localization of β-Galactosidase:** A combination of potassium ferrocyanide and potassium ferricyanide in phosphate-buffered solution (PBS) serves as a redox buffer for X-Gal, which is used to cleave β-galactosidase, resulting in a bright blue visualization method for marking the intracellular location of any protein, enzyme, or antibody conjugated to β-gal (1). *Step*-1. Aspirate media from dish of cultured cells, wash cells twice with PBS. *Step*-2. Fix cells in PBS/0.05% glutaraldehyde. *Step*-3. Incubate at room temperature for 5 min. *Step*-4. Remove PBS/0.05% Glutaraldehyde; then wash twice with PBS. *Step*-5. Stain with PBS containing 1 mM $MgCl_2$, 5 mM $K_4[Fe(II)(CN)_6]$, 5 mM $K_3[Fe(III)(CN)_6]$, plus 1 mg/ml X-gal. (*Note:* X-gal staining solution should be warmed to 37°C for 5-10 min before use.) *Step*-6. Incubate at room temperature or 37°C in dark until desired level of staining is observed (typically several hours to overnight). *Step*-7. Stop further development of staining by washing several times with PBS. *Step*-8. Examine cells under a bright-field microscope. (**Reagents:** *Washing Buffer:* 3.74 g $NaH_2PO_4·H_2O$ and 10.35 g Na_2HPO_4 (anhydrous) in 1 L H_2O; adjust to pH 7.3. *Fixing Buffer for X-gal Staining:* 0.1M phosphate buffer (pH 7.3) containing 5 mM EGTA, 2 mM $MgCl_2$ and 0.2% glutaraldehyde (Stored at 4°C for up to 4 months). *Wash Buffer for X-gal Staining:* 0.1M phosphate buffer (pH 7.3), containing 2 mM $MgCl_2$ (Stored at 4°C indefinitely). *X-gal Staining Buffer:* 0.1M phosphate buffer (pH 7.3), containing 2 mM $MgCl_2$, 5mM $K_4Fe(II)(CN)_6$ and 5mM $K_3Fe(III)(CN)_6$ (Store in the dark at 4°C) Prior to use, add 5-bromo-4-chloro-3-indolyl-β-D-galactoside (X-gal) to final concentration of 1 mg/ml and filter to remove any crystals.) **CAUTION:** Although a cyanide-containing substance, ferrocyanide is considerably less toxic than potassium cyanide, a fact reflecting the overall stability of the octahedral complex. Ferrocyanide will, however, react with mineral acids to release the extremely toxic gas, hydrogen cyanide. **Target(s):** aminopeptidase I (2); arylsulfatase A (3); arylsulfatase B, N-acetylgalactosamine-4-sulfatase (3); carbonic anhydrase (4); cytochrome c oxidase (5); dihydropteridine reductase (6); dopamine β-monooxygenase, also alternative substrate (7); ferredoxin:NADP(H) reductase, inhibited by Zn-ferrocyanide (8); inositol oxygenase (9); NADP⁺ reductase (10); pectin lyase (11); peptidylglycine monooxygenase (12); peroxidase (13); 3-phosphoglycerate kinase (14); urate oxidase, or uricase, weakly inhibited (15). **1.** MacGregor, Mogg, Burke & Casket (1987) *Somat Cell Mol Genet.* **13**, 253. **2.** Byun, Tang, Sloma, et al. (2001) *J. Biol. Chem.* **276**, 17902. **3.** Bleszynski & Leznicki (1967) *Enzymologia* **33**, 373. **4.** Walk & Metzner (1975) *Hoppe-Seyler's Z. Physiol. Chem.* **356**, 1733. **5.** Yu & Yu (1976) *Biochem. Biophys. Res. Commun.* **70**, 1115. **6.** Armarego & Ohnishi (1987) *Eur. J. Biochem.* **164**, 403. **7.** Rosenberg, Gimble & Lovenberg (1980) *Biochim. Biophys. Acta* **613**, 62. **8.** Dupuy, Rial & Ceccarelli (2004) *Eur. J. Biochem.* **271**, 4582. **9.** Reddy, Pierzchala & Hamilton (1981) *J. Biol. Chem.* **256**, 8519. **10.** Ricard, Nari & Diamantidis (1980) *Eur. J. Biochem.* **108**, 55. **11.** Afifi, Fawzi & Foaad (2002) *Ann. Microbiol.* **52**, 287. **12.** Evans, Blackburn & Klinman (2006) *Biochemistry* **45**, 15419. **13.** Barcelo, Pomar, Ferrer, et al. (2002) *Physiol. Plant.* **114**, 33. **14.** Joao & Williams (1993) *Eur. J. Biochem.* **216**, 1. **15.** Davidson (1942) *Biochem. J.* **36**, 252.

Ferrostatin-1

This potent and selective ferroptosis inhibitor (FW = 262.35 g/mol; CAS = 347174-05-4; Solubility: 52 mg/mL DMSO or Ethanol; <1 mg/mL H_2O; Symbol: Fer-1), also named 3-amino-4-(cyclohexylamino)benzoic acid ethyl ester, prevents erastin-induced ferroptosis in cancer cells, as well as glutamate-induced cell death in postnatal rat brain slices (EC_{50} = 60 nM). (*See Erastin*) Ferroptosis is dependent upon intracellular iron, but not other metals, and is morphologically, biochemically, and genetically distinct from apoptosis, necrosis, and autophagy. Given its ortho-diaminobenzene structure and the likelihood of intracellular ester hydrolysis, ferrostatin is likely to be an iron chelator. **1.** Dixon, Lemberg, Lamprecht, *et al.* (2012) *Cell* **149**, 1060.

Ferulate (Ferulic Acid)

trans-ferulate   *cis*-ferulate

These substituted cinnamic acids ($FW_{free-acid}$ = 194.19 g/mol), also known as 4-hydroxy-3-methoxycinnamic acid, itself a component of many plant esters, polyesters, and glycosides, also occurs in small amounts as the free acids in many plants. **Target(s):** α-amylase, weakly inhibited (1); arylamine *N*-acetyltransferase (2,3); chorismate mutase (4); 4-coumaroyl-CoA synthetase, *or* 4-coumarate:CoA ligase, as alternative substrate (5-7); diphosphomevalonate decarboxylase (8,9); ethoxyresorufin *O*-deethylase, *or* cytochrome P450 (10); feruloyl esterase (11); glutathione *S*-transferase (12,13); lipase, pancreatic, *or* triacylglycerol lipase (14); 15-lipoxygenase, soybean (15); methoxyresorufin *O*-demethylase, cytochrome P450 (10); peroxidase, horseradish (16); phenylalanine ammonia-lyase (17-19); phenylpyruvate tautomerase (20); quinate *O*-hydroxy-cinnamoyltransferase (21); thromboxane A_2 synthase (22); tyrosinase, *or* monophenol monooxygenase (23-25); xanthine oxidase (26). **1.** Funke & Melzig (2005) *Pharmazie* **60**, 796. **2.** Makarova (2008) *Curr. Drug Metab.* **9**, 538. **3.** Lo & Chung (1999) *Anticancer Res.* **19**, 133. **4.** Woodin & Nishioka (1973) *Biochim. Biophys. Acta* **309**, 211. **5.** Harding, Leshkevich, Chiang & Tsai (2002) *Plant Physiol.* **128**, 428. **6.** Feutry & Letouze (1984) *Phytochemistry* **23**, 1557. **7.** Gross & Zenk (1974) *Eur. J. Biochem.* **42**, 453. **8.** Lalitha, George & Ramasarma (1985) *Phytochemistry* **24**, 2569. **9.** Shama Bhat & T. Ramasarma (1979) *Biochem. J.* **181**, 143. **10.** Teel & Huynh (1998) *Cancer Lett.* **133**, 135. **11.** Garcia, Ball, Rodriguez, *et al.* (1998) *Appl. Microbiol. Biotechnol.* **50**, 213. **12.** Das, Bickers & Mukhtar (1984) *Biochem. Biophys. Res. Commun.* **120**, 427. **13.** Hayeshi, Mutingwende, Mavengere, Masiyanise & Mukanganyama (2007) *Food Chem. Toxicol.* **45**, 286. **14.** Karamac & Amarowicz (1996) *Z. Naturforsch. [C]* **51**, 903. **15.** Malterud & Rydland (2000) *J. Agric. Food Chem.* **48**, 5576. **16.** Gamborg, Wetter & Neish (1961) *Can. J. Biochem. Physiol.* **39**, 1113. **17.** Hanson & Havir (1972) *The Enzymes*, 3rd ed., **7**, 75. **18.** Sarma & Sharma (1999) *Phytochemistry* **50**, 729. **19.** Jorrin & Dixon (1990) *Plant Physiol.* **92**, 447. **20.** Molnar & Garai (2005) *Int. Immunopharmacol.* **5**, 849. **21.** Rhodes, Wooltorton & Lourencq (1979) *Phytochemistry* **18**, 1125. **22.** Wang, Gao, Huang & Zhu (1988) *Zhongguo Yao Li Xue Bao* **9**, 430. **23.** Lee (2002) *J. Agric. Food Chem.* **50**, 1400. **24.** Shirota, Miyazaki, Aiyama, Ichioka & Yokokura (1994) *Biol. Pharm. Bull.* **17**, 266. **25.** Karioti, Protopappa, Magoulas & Skaltsa (2007) *Bioorg. Med. Chem.* **15**, 2708. **26.** Chang, Lee, Chen, *et al.* (2007) *Free Rad. Biol. Med.* **43**, 1541.

Fesoterodine

Fesoterodine   5-Hydroxymethyl Tolterodine

This antimuscarinic agent (FW = 411.28 g/mol; CAS 286930-03-8; IUPAC: [2-[(1*R*)-3-(di(propan-2-yl)amino)-1-phenylpropyl]-4-(hydroxymethyl) phenyl]2-methylpropanoate), is a prodrug that is often delivered as its fumarate salt under the brand name Toviaz® to improve the symptoms of overactive bladder syndrome. Fesoterodine is converted by plasma esterases to its active metabolite, 5-hydroxymethyl tolterodine (*Symbol*: 5-HMT *or* SPM-7605). When tested *in vitro*, 5-HMT potently inhibits radioligand binding at all five human muscarinic receptor subtypes with equal affinity. Fesoterodine has a similar balanced selectivity profile but is less potent than 5-HMT (1). Both substances are competitive antagonists of cholinergic agonist-stimulated responses in human M_1-M_5 muscarinic receptor subtype-containing cell lines and has a similar potency and selectivity profile to the radioligand-binding studies (1). In the presence of the esterase inhibitor neostigmine, the concentration-response curve of fesoterodine is shifted to the right, suggesting that part of the activity is caused by metabolism to SPM 7605 by tissue enzymes (1). Fesoterodine is primarily eliminated as inactive metabolites, with significant renal excretion of 5-HMT. Significantly, fesoterodine and 5-HMT bind competitively and reversibly to muscarinic receptors with greater affinity in the human bladder mucosa and detrusor muscle than in the parotid gland (2). **1.** Ney, Pandita, Newgreen, *et al.* (2008) *BJU Internat.* **101**, 1036. **2.** Yoshida, Fuchihata, Kuraoka (2013) *Urology* **81**, 920.

Fexofenadine

This orally active, second-generation peripheral antihistamine (FW = 501.68 g/mol; CAS = 83799-24-0), also named Allegra®, Axodin®, and (±)-4-[1-hydroxy-4-[4-(hydroxydiphenyl-methyl)-1-piperidinyl]butyl]-α,α-dimethylbenzeneacetate, is a once daily, blockbuster H_1 receptor antagonist that does not cause sedation, because, as a zwitterionic species, it is less apt to pass the blood-brain barrier. As an active metabolite of the second-generation histamine H_1 receptor antagonist terfenadine, fexofenadine does not suffer the disadvantage of QT prolongation (1,2). In addition, unlike first-generation antihistamines, it is associated with few CNS adverse effects. By preventing the activation of the H_1 receptors by histamine, fexofenadine reduces the symptoms associated with allergies from occurring. It does not readily cross the blood–brain barrier, accounting for its failure to induce drowsiness or to show anticholinergic effects. **1.** Chen (2007) *Drugs* **8**, 301. **2.** Lappin, Shishikura, Jochemsen, *et al.* (2010) *Eur. J. Pharm. Sci.* 40, 125.

Fibronectin Fragment 1377-1388, *Similar in action to* Fibronectin Fragment 1371-1382

Fidarestat

This aldose reductase inhibitor (FW = 279.22 g/mol; CAS = 136087-85-9), also known by its code name SNK-860, its trade name Aldos®, and its systematic name (2*S*,4*S*)-6-fluoro-2',5'-dioxospiro[chroman-4,4'-imidazolidine]-2-carboxamide, is used to treat diabetic retinopathy and caractogenesis (1-3). (*See also Alrestatin*) Fifarestat has proven to be highly selective for aldose reductase (AR or AKR1B1) over aldehyde reductase (AKR1A1), a property that most likely explains its modest side effects *in vivo*. The structural basis for fifarestat selectivity has been reported (3,4). With 1-mg daily dosing of Type 2 diabetic patients, fidarestat had no effect on glycemic control (based on plasma glucose and HbA_{1c} levels), but normalized the sorbitol content of erythrocytes under fasting and postprandial conditions (5). Because glucose reduction increases intracellular osmotic pressure by the accumulation of nondiffusable sorbitol and because AKR1B1 action contributes to iron- and transferrin-related oxidative stress associated with cerebral ischemic injury, inhibition of AKR1B1 may have therapeutic potential against ischemic cerebral injury (6). **Other Targets:** 3-Deoxyglucosone and heparin-binding epidermal growth factor-like growth factor mRNA expression (7); phosphorylation of JNK and c-Jun occurs in fast-conduction mechanoceptors (8); ischemia-reperfusion-induced inflammatory response in rat retina (9). **1**. Yamaguchi, Miura, Usui, *et al.* (1994) *Arzneimittelforschung* **44**, 344.   **2**. Oka, Matsumoto, Sugiyama, Tsuruta & Matsushima (2000) *J. Med. Chem.* **43**, 2479.   **3**. El-Kabbani, Carbone, Darmanin, et al. (2005) *J. Med. Chem.* **48**, 5536.   **4**. Ruiz, Cousido-Siah, Mitschler, et al. (2013) *Chem. Biol. Interact.* **202**, 178.   **5**. Asano, Saito, Kawakami & Yamada (2002) *J. Diabetes & Its Complic.* 16, 133.   **6**. Lo, Cheung, Hung, *et al.,* (2007) *J. Cereb. Blood. Flow Metab.* **27**, 1496.   **7**. Li, Hamada, Nakashima, *et al.* (2004) *Biochem. Biophys. Res. Commun.* **314**, 370.   **8**. Middlemas, Agthong & Tomlinson (2006) *Diabetologia* **49**, 580.   **9**. Agardh, Agardh, Obrosova & Smith (2009) *Pharmacology* **84**, 257.

Fidaxomicin

This narrow-spectrum macrocyclic antibiotic (FW = 1058.04 g/mol; CAS = 873857-62-6), marketed under the trade names Dificid® and Dificlir®, and known previously by the code names OPT-80 and PAR-101, is an FDA-approved drug for the treatment of *Clostridium difficile*-associated diarrhea (1,2). The minimum inhibitory concentration for 90% of organisms for fidaxomicin against *C. difficile* ranged from 0.0078 to 2 µg/mL in *in vitro* studies. **Primary Mode of Action:** *C. difficile* is a Gram-positive, spore-forming, toxin-producing, anaerobic bacillus found ubiquitously and easily transmitted by the fecal-to-oral route. Fidaxomicin blocks initiation of transcription, but only when present before the formation of the "open promoter complex," in which the template DNA strands have separated and RNA synthesis has not yet begun. Its metabolite OP-1118 is virtually equipotent in its antibiotic action against *C. difficile*. **Target(s):** toxin production in *C. difficile*, an *in vitro* property that is not observed in vancomycin inhibition of *C. difficile* growth (4); spore production in *C. difficile* (5).   **1**. Tannock, Munro, Taylor, *et al.* (2010) *Microbiology* **156**, 3354.   **2**. Louie, Miller, Mullane (2011) *New Engl. J. Med.* **364**, 422.   **3**. Artsimovitch, Seddon & Sears (2012) *Clin. Infect. Dis.* **55**, (*Supplement* 2), S127.   **4**. Babakhani, Bouillaut, Sears, *et al.* (2013) *J. Antimicrob. Chemother.* **68**, 515.   **5**. Babakhani, Bouillaut, Gomez, *et al.* (2012) *Clin. Infect. Dis.* **55**, (*Supplement* 2), S162.

Fiduxosin

This α-receptor antagonist (FW$_{free-base}$ = 555.66 g/mol; FW$_{hydrochloride}$ = 592.11 g/mol; Insoluble in H$_2$O; 10 mg/mL DMSO), also known as 3-[4-((3aR,9bR)-cis-9-methoxy-1,2,3,3a,4,9b-hexahydro[1]benzopyrano[3,4-c]pyrrol-2-yl)butyl]-8-phenylpyrazino[2',3':4,5]-thieno[3,2-d]pyrimidine-2,4(1H,3H)-dione and ABT-980, targets α$_{1A}$- and α$_{1D}$-adrenoceptors (1-3), blocking both phenylephrine (PE)-induced intraurethral pressure (IUP) and

mean arterial pressure (MAP) responses after single oral dosing in conscious male beagles, with respective IC$_{50}$ values of 261 ng/mL and 1900 ng/mL (3). The high values reflect strong plasma protein binding (>99.8%) of the drug. **1**. Brune, Katwala, Milicic, *et al.* (2002) *J. Pharmacol. Exp. Ther.* **300**, 487.   **2**. Hancock, Buckner, Brune, *et al.* (2002) *J. Pharmacol. Exp. Ther.* **300**, 478.   **3**. Witte, Brune, Katwala, *et al.* (2002) *J. Pharmacol. Exp. Therap.* **300**, 495.

Figitumumab

This potent and fully humanized monoclonal anti-insulin-like growth factor-1 receptor antibody, *or* IGF1R, antibody (MW = 146.0 kDa; CAS = 943453-46-1), also known by its code name CP-751871 and others (1), is an investigational drug for treating adrenocortical carcinoma and non-small cell lung cancer (NSCLC). Pharmacokinetic analysis shw a dose-dependent increase in CP-751,871 and ~2-fold accumulation on repeated dosing in 21-day cycles (2). Figitumumab has an effective half-life of approximately 20 days, and it has been well tolerated in clinical studies when given alone or in combination with chemotherapy and targeted agents (3). **1**. Alternate Names = Immunoglobulin G$_2$, anti-(human insulin-like growth factor 1 receptor (EC.2.7.10.1 or CD221 antigen)); human monocle onal CP-751871 clone 2.13.2 γ$_2$-heavy chain (139-214')-disulfide with human monoclonal CP-751871 clone 2.13.2 κ light chain, dimer (227-227":228-228":231-231":234-234")-tetrakisdisulfide).   **2**. Haluska, Shaw, Batzel, *et al.* (2007) *Clin. Cancer Res.* **13**, 5834.   **3**. Gualberto & Karp (2009) *Clin. Lung Cancer* **10**, 273.

Filgotinib

This JAK1-selective inhibitor (FW = 425.51g/mol; CAS = 1206161-97-8; Solubility: 85 mg/mL DMSO; < 1 mg/mL H$_2$O), also named *N*-[5-[4-[(1,1-dioxido-4-thiomorpholinyl)methyl]phenyl][1,2,4] triazolo[1,5-*a*]pyridin-2-yl]cyclopropanecarboxamide and GLPG0634, targets Janus Kinases: JAK1 (IC$_{50}$ = 10 nM), JAK2 (IC$_{50}$ = 28 nM), JAK3 (IC$_{50}$ = 810 nM), and TYK2 (IC$_{50}$ = 116 nM). **Pharmacodynamics:** JAK1 plays a critical role in signal transduction in many Type I and Type II inflammatory cytokinee receptors, such that JAK1 inhibitors offer great promise fo treating immune-inflammatory diseases. After oral administration, exposure to filgotinib is dose-proportional, with an elimination half-life of 6 hours (2). Its major metabolite, formed by de-esterification, showed JAK1 selectivity, with higher exposure and lower potency than filgotinib. The relatively long duration of JAK1 inhibition following filgotinib dosing suggests that the maximum response is reached at a daily filgotinib dose of 200 mg.   **1**. Van Rompaey, Galien, van der Aar, *et al.* (2013) *J. Immunol.* **191**, 3568.   **2**. Namour, Diderichsen, Cox, *et al.* (2015) *Clin. Pharmacokinet.* **54**, 859.

Filibuvir

This non-nucleoside HCV inhibitor (FW = 503.64 g/mol; CAS = 877130-28-4), also known by the code name PF-868554, targets hepatitis C virus RNA-dependent RNA polymerase by binding noncovalently within the "Thumb-2" pocket, IC$_{50}$ of 19 nM *in vitro* (1). It is active against >95 % of genotype 1 replicons (EC$_{50}$ = 59 nM), with similar potencies against replicons containing patient-derived subtype 1a and subtype 1b NS5B sequences (2). In two Phase-1b clinical studies with HCV genotype 1-infected patients, filibuvir potently inhibited viral replication, with doses of ≥450 mg twice daily, resulting in >2.0-log-unit reductions in HCV RNA. Filibuvir is well tolerated, with no discontinuations, serious adverse events, or deaths reported. Mutations at Met-423 in NS5B were observed in 29 of 38 patients receiving filibuvir at ≥300 mg BID (4). M423I, M423T, and M423V mutations were all observed, both individually and in genotypic mixtures with the wild type (1). **1**. Love, Parge, Yu, *et al.* (2003) *J. Virol.* **77**, 7575. **2**. Shi, Herlihy & Graham (2009) *Antimicrob. Agents Chemother.* **53**, 2544. **3**. Wagner, Thompson, Kantaridis, *et al.* (2011) **54**, 50. **4**. Trokea, Lewisa, Simpson, *et al.* (2012) *Antimicrob. Agents Chemother.* **56**, 1331.

Filipin

Filipin III

This yellow-colored polyene antibiotic (FW = 654.83 g/mol; CAS = 11078-21-0; IUPAC: (3R,4S,6S,8S,10R,12R,14R,16S,17E,19E,21E,23E,25E,27S,28R)-4,6,8,10,12,14,16,27-Octahydroxy-3-[(1R)-1-hydroxyhexyl]-17,28-dimethyloxacyclooctacosa-17,19,21,23,25-pentaen-2-one), first isolated from *Streptomyces filipinensis* forms a complex with cholesterol or ergosterol and permeabilizes cells (note that it can form complexes with other 3β sterols). Because filipin is active mainly against cholesterol/ergosterol-containing membranes, it is effective at ~3 µg/mL as an inhibitor of the raft/caveolae endocytosis pathway in mamallian cells. There are at least four different filipins that vary in the number of hydroxyl groups present; filipin III (FW = 654.84 g/mol) is the major component. It is soluble in ethanol and methanol and very soluble in dimethylformamide. Because filipin is a mixture of filipin I (4%), II (25%), III (53%), and IV (18%), its should be referred to as the filipin complex. *Note*: Except dor amphotericin B and nystatin A1, filipin-like agents have proved to be too toxic for therapeutic applications. **Target(s)**: chitin synthase, weakly inhibited (1); H⁺-driven ATPase (2); photosystem I (3); sterol *O*-acyltransferase, *or* cholesterol *O*-acyltransferase, *or* ACAT (4,5); sterol 24-*C*-methyltransferase, slightly inhibited, and activated at high concentrations (6). **1**. Rast & Bartnicki-Garcia (1981) *Proc. Natl. Acad. Sci. U.S.A.* **78**, 1233. **2**. Kinne-Saffran & Kinne (1986) *Pflugers Arch.* **407** Suppl. 2, S180. **3**. Trebst (1980) *Meth. Enzymol.* **69**, 675. **4**. Billheimer (1985) *Meth. Enzymol.* **111**, 286. **5**. Erickson, Shrewsbury, Brooks & Meyer (1980) *J. Lipid Res.* **21**, 930. **6**. Mukhtar, Hakkou & Bonaly (1994) *Mycopathologia* **126**, 75.

Fimasartan

This non-peptide angiotensin II receptor antagonist (FW = 501.65 g/mol; CAS = 247257-48-3), also known as 2-(2-butyl-4-methyl-6-oxo-1-{[2'-(1*H*-tetrazol-5-yl)-4-biphenylyl]methyl}-1,6-dihydro-5-pyrimidinyl)-*N*,*N*-dimethylethanethioamide, its code name BR-A-657, and by the trade name Kanarb®, inhibits vasoconstriction by blocking AT₁ and by preventing aldosterone formation, thereby increasing water and salt excretion by the kidneys and decreasing overall blood volume. Fimasartan displaces

[¹²⁵I][Sar₁-Ile₈]-angiotensin II from its specific binding sites to AT₁ subtype receptors in membrane fractions of HEK-293 cells (IC₅₀ = 0.16 nM). **1**. Shin, Kim, Paik, *et al.* (2011) *Biomed. Chromatogr.* **25**, 1208. **2**. Yi, Kim, Yoon, *et al.* (2011) *J. Cardiovasc. Pharmacol.* **57**, 682.

Finasteride

This 5-α reductase inhibitor, *or* 5α-RI (FW = 372.55 g/mol; CAS = 98319-26-7), also known as MK-906, Proscar®, Propecia®, and *N*-(1,1-dimethylethyl)-3-oxo-(5α,17β)-4-azaandrost-1-ene-17-carboxamide, potently and competitively inhibits steroid 5α-reductase (IC₅₀ = 33 nM), thus blocking the formation of the strong androgen 5α-dihydrotestosterone. 5α-Reductase catalyzes the rate-limiting step in the conversion of testosterone (T) to 5α-dihydroprogesterone (5α-DHP), progesterone to 5α-DHP, deoxycorticosterone (DOC) to 5β-dihydrodeoxycorticosterone, cortisol to 5α-dihydrocortisol, and aldosterone to 5α-dihydro-aldosterone. (***See also*** *Dutasteride*) Finasteride was the first 5α-reductase inhibitor to receive clinical approval for the treatment of human benign prostatic hyperplasia and androgenetic alopecia (male pattern hair loss). Because 5-α reductase is also active in other organs, 5-α reductase inhibitors, such as finasteride and dutasteride, often have side-effects, including depression, sexual dysfunction, increased risk of high-grade prostate cancer, as well as delayed diagnosis and early treatment of prostate cancer. **Key Pharmacokinetic Parameters:** *See* Appendix II in Goodman & Gilman's *THE PHARMACOLOGICAL BASIS OF THERAPEUTICS*, 12ᵗʰ Edition (Brunton, Chabner & Knollmann, eds.) McGraw-Hill Medical, New York (2011). **1**. Rasmusson, Reynolds, Steinberg, *et al.* (1986) *J. Med. Chem.* **29**, 2298. **2**. Stoner (1990) *J. Steroid Biochem. Mol. Biol.* **37**, 375. **3**. Njar, Kato, Nnane, *et al.* (1998) *J. Med. Chem.* **41**, 902.

Fingolimod

This first-in-class myriocin derivative and immune modulator (FW_hydrochloride = 343.94 g/mol; CAS = 162359-56-0), also known as FTY720, Gilenya®, and 2-amino-2-[2-(4-octylphenyl)ethyl]-1,3-propanediol, is a sphingosine-1-phosphate (S1P) receptor-modulating drug that has been approved for the treatment of relapsing-remitting multiple sclerosis. (***See also*** *Myriocin*) Fingolimod prolongs allograft transplant survival in numerous models by inhibiting lymphocyte emigration from lymphoid organs (1). When metabolically phosphorylated by sphingosine kinase *in vivo*, FTY720 becomes a potent agonist for S1P₁, S1P₃, S1P₄, and S1P₅ sphingosine-1-phosphate receptors (2). FTY720 also stimulates multidrug transporter-dependent and cysteinyl leukotriene-dependent T cell chemotaxis to lymph nodes (3). Moreover, FTY720-mediated down-regulation of S1P₁ receptors on T and B lymphocytes impairs their egress from spleen, lymph nodes, and Peyer's patch (4). Fingolimod also enhances the activity of the sphingosine transporter Abcb1 (*or* Mdr1) and the leukotriene C₄ transporter Abcc1 (*or* Mrp1) and inhibits cytosolic phospholipase A₂ activity (4,5). FTY720 decreased the number of circulating blood lymphocytes and induced sequestration of circulating lymphocytes within secondary lymphoid tissues, decreasing lymphocyte infiltration into target organs (6). Moreover, rats pretreated with FTY720 show significantly lower arthritis incidence, delayed onset of arthritis, and suppressed disease activities, suggesting FTY720 is effective in preventing the occurrence and development of collagen-induced arthritis (CIA) in preclinical RA model (7). **1**. Brinkmann, Pinschewer, Feng, *et al.* (2001) *Transplantation* **72**, 764. **2**. Brinkmann, Davis, Heise, *et al.* (2002) *J. Biol. Chem.* **277**, 21453. **3**. Honig, Fu, Mao, *et al.* (2003) *J. Clin. Invest.* **111**, 627. **4**. Matloubian, Lo, Cinamon, *et al.* (2004) *Nature* **427** 355. **5**. Payne, Oskeritizian, Griffiths, *et al.* (2007)

Blood **109**, 1077. **6**. Chiba (2005) *Pharmacol. Ther.* **108**, 308. **7**. Matsuura, Imayoshi & Okumoto (2000) *Int. J. Immunopharmacol.* **22**, 323.

Fisetin

This naturally occurring flavonoid (FW = 286.24 g/mol; CAS = 528-48-3; IUPAC: 2-(3,4-dihydroxyphenyl)-3,7-dihydroxychromen-4-one) has anti-proliferative properties, often inducing apoptosis. Fisetin, for example, reduces the phosphorylation of p38 MAPK, but not that of ERK1/2, JNK1/2, or AKT in cervical cancer cells (1). Fisetin is also a potent sirtuin activator. **Target(s):** transcription of naked DNA by purified RNA polymerase II (2); oxidative burst of neutrophils, $IC_{50} \approx 100$ μM (3); Moloney murine leukemia virus reverse transcriptase (4); platelet aggregation, IC_{50} = 22 μM (5); protein kinase C (PKC), IC_{100} = 100 μM, competitive inhibitor with respect to ATP binding and noncompetitive with respect to protein substrate (6); osteoporosis by repression of NF-κB and MKP-1-dependent signaling pathways in osteoclasts (7). **1**. Chou, Hsieh, Yu, *et al.* (2013) *PLoS One* **8**, e71983. **2**. Nose (1984) *Biochem. Pharmacol.* **33**, 3823. **3**. Pagonis, Tauber, Pavlotsky & Simons (1986) *Biochem. Pharmacol.* **35**, 237. **4**. Chu, Hsieh & Lin (1992) *J. Nat. Prod.* **55**, 179. **5**. Tzeng, Ko, Ko & Teng (1991) *Thromb. Res.* **64**, 91. **6**. Ferriola, Cody & Middleton (1989) *Biochem. Pharmacol.* **38**, 1617. **7**. Léotoing, Wauquier, Guicheux, *et al.* (2013) *PLoS One* **8**, e68388.

FK228, See *Romidepsin*

FK463, See *Micafungin*

FK520, See *Ascomycin*

FK-506

This powerful immunosuppressant (FW = 804.03 g/mol; CAS = 109581-93-3), also known as tacrolimus, is one of two first-line agents (the other, cyclosporin A) responsible for the resounding success of organ transplantation in recent decades. A macrolide from *Streptomyces tsukubaensis*, FK-506 exhibits potent antifungal activity and is a CYP3A substrate (3), but its hallmark action is the inhibition of T-cell activation and, most significantly, the prevention of allograft rejection. **Primary Mode of Action:** Organ rejection is a multistep process that begins with alloantigen recognition, proceeds through lymphocyte activation and clonal expansion, and ends with graft inflammation, the latter attended by full-scale immune attack. Alloantigen recognition normally triggers a signal transduction pathway involving calcineurin, a Ca^{2+}- and calmodulin-dependent serine/threonine phosphoprotein phosphatase. The latter dephosphorylates the transcription factor NFATc (Nuclear Factor of Activated T-cells). Once dephosphorylated, cytoplasmic NFATc is translocated into the nucleus, where it up-regulates IL-2 expression, then stimulating killer T cell differentiation and clonal expansion. To prevent allograft destruction, one employs FK-506, which binds to FK-binding protein (*or* FKBP5), a highly promiscuous peptidylprolyl *cis/trans* isomerase that interacts with mature corticoid receptor hetero-complexes (*e.g.*, progesterone-, glucocorticoid-, mineralcorticoid-receptor complexes) as well as the 90-kDa heat shock

protein and P23 protein (*or* prostaglandin E synthase-3). It is the formation of the FK-506:FKBP5 complex, however, that noncompetitively inhibits calcineurin, thereby preventing NFATc dephosphorylation/ translocation and the sequelae of biochemical events culminating in allograft rejection. **Key Pharmacokinetic Parameters:** *See* Appendix II in Goodman & Gilman's THE PHARMACOLOGICAL BASIS OF THERAPEUTICS, 12th Edition (Brunton, Chabner & Knollmann, eds.) McGraw-Hill Medical, New York (2011). **Target(s):** calcineurin; FK-binding protein (1,2); glucuronosyltransferase (8); peptidylprolyl isomerase (9-12); membrane translocation of *Clostridium botulinum* C3 toxin in mammalian cells (13); interleukin-1β-induced angiopoietin-1, Tie-2 receptor, and vascular endothelial growth factor through down-regulation of JNK and p38 pathway (14); gliostatin/thymidine phosphorylase production, as induced by tumor necrosis factor-α in rheumatoid fibroblast-like synoviocytes (15). **1**. Handschumacher, Harding, Rice, Drugge & Speicher (1984) *Science* **226**, 544. **2**. Harding, Galat, Uehling & Schreiber (1989) *Nature* **341**, 758. **3**. Hebert, Blough, Townsend, *et al.* (2005) *J. Clin. Pharmacol.* **45**, 1018. **4**. Liu, Farmer, Lane, *et al.* (1991) *Cell* **66**, 807. **5**. Liu, Albers, Wandless, *et al.* (1992) *Biochemistry* **31**, 3896. **6**. Kissinger, Parge, Knighton, *et al.* (1995) *Nature* **378**, 641. **7**. Haddy & Rusnak (1994) *Biochem. Biophys. Res. Commun.* **200**, 1221. **8**. Hara, Nakajima, Miyamoto & Yokoi (2007) *Drug Metab. Pharmacokinet.* **22**, 103. **9**. Zeng, MacDonald, Bann, *et al.* (1998) *Biochem. J.* **330**, 109. **10**. Furutani, Iida, Yamano, K. Kamino & T. Maruyama (1998) *J. Bacteriol.* **180**, 388. **11**. Monaghan & Bell (2005) *Mol. Biochem. Parasitol.* **139**, 185. **12**. Hueros, Rahfeld, Salamini & Thompson (1998) *Planta* **205**, 121. **13**. Kaiser, Böhm, Ernst, *et al.* (2012) *Cell. Microbiol.* **14**, 1193. **14**. Choe, Lee, Park & Kim (2012) *Joint Bone Spine* **79**, 137. **15**. Yamagami, Waguri-Nagaya, Ikuta, *et al.* (2011) *Rheumatol Int.* **31**, 903.

FK-866

This antineoplastic agent (FW = 390.51 g/mol; CAS = 658084-64-1; *In Vitro* Solubility: <1 mg/mL H_2O, <1 mg/mL DMSO; *In Vivo*: Dissolve in 45% propylene glycol, then bring to 5% Tween 80), also known as Daporinad, APO866, and (*E*)-*N*-(4-(1-benzoylpiperidin-4-yl)butyl)-3-(pyridin-3-yl)acrylamide, is a potent noncompetitive inhibitor (K_i = 0.09 nM) of nicotinamide phosphoribosyltransferase, also known as NAMPRTase as well as pre-B-cell colony-enhancing factor 1 (PBEF1) and visfatin. This enzyme catalyzes the condensation of nicotinamide with PRPP to yield NMN, an intermediate step in NAD^+ biosynthesis, resulting in gradual NAD^+ depletion and cell death (1-5). FK-866 induces dose-dependent cytotoxicity in forty-one malignant hematologic cell types, including acute myeloid leukemia, acute lymphoblastic leukemia, mantle cell lymphoma, chronic lymphocytic leukemia, and T-cell lymphoma. In Hep-G_2 human liver carcinoma cells, FK-866-induced NAD^+ depletion results in apoptosis, $IC_{50} \approx 1$ nM (2). In healthy human smooth muscle cells, FK-866 induces premature senescence, an effect likely linked to decreased activity of the NAD^+-dependent enzyme SIRT1 (2). FK-866 also induces autophagy in SH-SY5Y neuroblastoma cells, as indicated by the formation of LC3-positive vesicles (3). NAMPT inhibition by FK-866 leads to the attenuation of glycolysis at the glyceraldehyde 3-phosphate dehydrogenase step due to the reduced availability of NAD^+ (6). Attenuation of glycolysis results in the accumulation of glycolytic intermediates before and at the glyceraldehyde 3-phosphate dehydrogenase step, promoting carbon overflow into the pentose phosphate pathway, as evidenced by the increased intermediate levels. Attenuation of glycolysis also causes decreased glycolytic intermediates after the glyceraldehyde 3-phosphate dehydrogenase step, thereby reducing carbon flow into serine biosynthesis and the TCA cycle (6). **1**. Hasmann & Schemainda (2003) *Cancer Res.* **63**, 7436. **2**. van der Veer, Ho, O'Neil, *et al.* (2007) *J. Biol. Chem.* **282**, 10841. **3**. Billington, Genazzani, Travelli, *et al.* (2008) *Autophagy* **4**, 385. **4**. Kim, Lee, Kim, *et al.* (2006) *J. Mol. Biol.* **362**, 66. **5**. Khan, Tao & Tong (2006) *Nat. Struct. Mol. Biol.* **13**, 582. **6**. Tan, Young, Lu, *et al.* (2013) *J. Biol. Chem.* **288**, 3500.

Flagyl, See *Metronidazole*

Flavopiridol

This intravenously administered, cyclin kinase-directed inhibitor (FW = 401.84 g/mol; CASs = 146426-40-6, 131740-09-5(HCl), 358739-41-0 (TFA); Solubility: 15 mg/mL DMSO; <1 mg/mL H_2O; Formulation: Dissolve in 50:50 mixture of Cremophor/ethanol; then diluted in water), also known as HMR-1275, L868275, Alvocidib, and systematically named 2-(2-chlorophenyl)-5,7-dihydroxy-8-((3S,4R)-3-hydroxy-1-methylpiperidin-4-yl)-4H-chromen-4-one, is a flavonoid from an indigenous plant in India, with potent and specific in vitro inhibition of all cdks, blocking cell cycle progression at the G_1/S and G_2/M boundaries. Alvocidib targets CDK1, CDK2, CDK4, CDK5, CDK6 and CDK9 with IC_{50} values of 30 nM, 170 nM, 100 nM, 170 nM, 80 nM and 20 nM, respectively (1). Flavopiridol also potently inhibits the activity of glycogen synthase kinase-3, or GSK-3 (IC_{50} = 280 nm) (2). 1. Kim, et al. (2000) J. Med. Chem. 43, 4126. 2. Lu, et al. (2005) J. Med. Chem. 48, 737. 3. Montagnoli, et al. (2008) Nat. Chem. Biol. 4, 357. 4. Carlson, et al. (1996) Cancer Res. 56, 2973. 5. Kim, et al. (2002) J. Med. Chem. 45, 3905. 6. Sekine, et al. (2008) J. Immunol. 180, 1954. 7. Motwani, et al. (2001) Clin. Cancer Res. 7, 4209.

Flavostatin

This 68-residue cysteine-rich inhibitor of platelet aggregation activation and zinc-metalloproteinase (MW = 56.3 kDa; Acession Number: P14530), is a disintegrin from the venom of the viper Trimeresurus flavoviridis that contains an RGD loop, thereby facilitating tight binding to integrins in a manner that inhibits ADP-, collagen-, and thrombin receptor agonist peptide-induced platelet aggregation in human platelet-rich plasma (IC_{50} = 60-100 nM). See also Flavoridin 1. Maruyama, Kawasaki, Sakai, et al. (1997) Peptides 18, 73. 2. Kawasaki, Sakai, Taniuchi, et al. (1996) Biochimie 78, 245.

Fleroxacin

This fluoroquinolone-class antibiotic (FW = 369.34 g/mol; CAS = 79660-72-3), also named 6,8-difluoro-1-(2-fluoroethyl)-7-(4-methylpiperazin-1-yl)-4-oxoquinoline-3-carboxylate and marketed as Quinodis® and Megalocin®, is a broad-spectrum systemic antibacterial that targets Type II DNA topoisomerases (gyrases), which are required for bacterial replication and transcription. Fleroxacin is active against many Gram-positive and Gram-negative bacteria, including Shigella, Salmonella, Escherichia coli, Branhamella catarrhalis, Haemophilus influenzae, Neisseria gonorrhoeae, Yersinia enterocolitica, Staphylococcus aureus, and Pseudomonas aeruginosa. For the prototypical member of this antibiotic class, See Ciprofloxacin

FLI-06

This first-in-class Notch signaling pathway inhibitor (FW = 438.52 g/mol; CAS = 313967-18-9; Solubility: <1 mg/mL DMSO or H_2O), also named cyclohexyl-2,7,7-trimethyl-4-(4-nitrophenyl)-5-oxo-1,4,5,6,7,8-hexahydro-quinoline-3-carboxylate, disrupts the Golgi apparatus in a manner distinct from that of brefeldin A and golgicide A. FLI-06 inhibits general secretion at a step before exit from the endoplasmic reticulum, which was accompanied by a tubule-to-sheet morphological transition of the ER, making FLI-06 the first small molecule to block secretory traffic at such an early stage. 1. Krämer, Mentrup, Kleizen, et al. (2013) Nature Chem Biol. 9, 731.

Flibanserin

This orally bioavailable serotonin agonist/antagonist (FW = 390.41 g/mol; CAS = 167933-07-5), known by its code name BIMT-17, its proposed tradename Addyi®, its systematic name, 1-(2-{4-[3-(trifluoromethyl) phenyl]piperazin-1-yl}ethyl)-1,3-dihydro-2H-benzimidazol-2-one, is a multifunctional post-synaptic 5-HT$_{1A}$ receptor agonist and 5-HT$_{2A}$ receptor antagonist that resets the poise between these neurotransmitter systems in a manner that is an effective as an antidepressant. Flibanserin shows moderate affinity for cerebral cortical 5-HT$_{1A}$ (pK_i = 7.72) and 5-HT$_{2A}$ (pK_i = 6.90) receptors, with no appreciable affinity for the other 5-HT receptor subtypes (1). It reduces forskolin-stimulated cAMP accumulation in the cerebral cortex (pEC50 = 6.09) and in the hippocampus (pEC_{50} = 6.50), and antagonizes 5-HT-induced phosphatidylinositol turnover (pK_i = 6.96) in the cerebral cortex (1). Its effect on cAMP accumulation was blocked by the 5-HT$_{1A}$ receptor antagonist tertatolol (1). Flibanserin also directly activates postsynaptic serotonin inhibitory responses in the rat cerebral cortex (2,3). **Treatment of Hypoactive Sexual Desire Disorder:** The lack of sexual desire in women causes distress, impacting negatively on personal relationships, quality of life, and health status. HSDD is characterized by reduced sexual fantasies and desire for sexual activity that causes marked distress or interpersonal difficulty and is not accounted for by coexisting conditions, use of medications, or relationship problems. Affecting nearly 10% of all women, HSDD is defined as the absence of sexual interest, desire and/or fantasy, leading to a significant emotional distress, interpersonal difficulties, and a drop in sexual satisfaction. Dubbed the "Female Viagra", flibanserin is also a treatment for female Hypoactive Sexual Desire Disorder (HSDD). Flibanserin appears to be an effective non-hormonal drug for treating HSDD in pre- and postmenopausal women, acting to increase dopamine and norepinephrine (both responsible for sexual excitement) and to decrease serotonin (responsible for sexual inhibition) (4). Flibanserin increases the frequency of satisfying sexual events as well as the intensity of sexual desire (4). Such properties commend its use to enhance sexual desire in pre-menopausal HSDD. Indeed, early clinical trials found that filbanserin increased the number of monthly "satisfying sexual encounters" by one, as compared to placebo. Flibanserin's mechanism of action in the treatment of HSDD is unknown. In 2015, an advisory committee voted to recommend FDA approval of flibanserin as a treatment of HSDD in pre-menopausal women. Although the antidepressant buproprion has shown some positive results in treating HSDD, flibanserin acts (at clinically relevant doses) as a 5-HT$_{1A}$ receptor agonist and secondarily as a 5-HT$_{2A}$ receptor antagonist, increasing dopamine and norepinephrine. Adverse side-effects include dizziness, nausea, fatigue and somnolence, which increase when the drug is taken during the day, especially when using any of the numerous moderate or strong cytochrome P-450 3A4 (CYP3A4) inhibitors such as some of the antiretroviral drugs, antihypertensive drugs, antibiotics, and alcohol. **Pharmacokinetics & Pharmacodynamics:** According to FDA filings, flibanserin exhibits linear and dose-proportional pharmacokinetics after single oral doses of 0.5 mg to 150 mg and after multiple doses of orally administered flibanserin with total daily doses ranging from 60–300 mg. The extent of exposure is increased 1.4x during once-daily administration of 100 mg flibanserin, as compared to a single dose. Flibanserin is rapidly absorbed, with 90% of the dose reaching the systemic circulation as flibanserin or metabolites. After oral administration, maximum observed plasma concentrations (C_{max}) are usually achieved in 45–60 minutes. Absolute bioavailability of flibanserin following oral dosing is 33%. CYP3A4 and to a minor extent CYP2D6 are the main cytochrome P450 isoenzymes in the oxidative metabolism of flibanserin. Flibanserin is excreted predominately as conjugated metabolites via the bile (~51% of the dose) and the kidney (~44% of the dose). Steady state is achieved after approximately three days, whereupon the mean terminal half-life of orally administered drug is ~12 hours. **1**. Borsini, Giraldo, Monferini, *et al.* (1995) *Naunyn. Schmiedebergs Arch. Pharmacol.* **352**, 276. **2**. Borsini, Ceci, Bietti & Donetti (1995) *Naunyn. Schmiedebergs Arch. Pharmacol.* **352**, 283. **3**. Borsini, Brambilla, Grippa & Pitsikas (1999) *Pharmacol. Biochem. Behav.* **64**, 137. **4**. Reviriego (2014) *Drugs Today* (Barc) **50**, 549.

Florfenicol

This potent bacterial protein biosynthesis inhibitor (FW = 358.22 g/mol; CAS = 73231-34-2), also known as Sch 25298 and 2,2-dichloro-*N*-[(1*R*,2*S*)-3-fluoro-1-hydroxy-1-(4-methanesulfonylphenyl)propan-2-yl]acetamide, is a fluorinated analogue of thiamphenicol and structural analogue of chloramphenicol (1,2). Significantly, the *Staphylococcus* and *Escherichia coli* Cfr resistance factor is an RNA methyltransferase that targets nucleotide A2503 and inhibits ribose methylation at nucleotide C2498, thereby causing resistance to chloramphenicol, florfenicol and clindamycin by interfering with the ribosome's peptidyltransferase activity. Florfenicol is solely used in veterinary medicine and was introduced into clinical use in the mid-1990s. **1**. Cannon, Harford & Davies (1990) *J. Antimicrob. Chemother.* **26**, 307. **2**. Schwarz, Kehrenberg, Doublet & Cloeckaert (2004) *FEMS Microbiol. Rev.* **28**, 519. **3**. Kehrenberg, Schwarz, Jacobsen, Hansen & Vester (2005) *Molec. Microbiol.* **57**, 1064.

Floxacillin

This semisynthetic isoxazole penicillin-class antibiotic (FW = 453.87 g/mol; CAS = 5250-39-5), also known as Flucloxacillin, Flopen®, Staphylex®, Softapen®, and (2*S*,5*R*,6*R*)-6-({[3-(2-chloro-5-fluorophenyl)-5-methyl-isoxazol-4-yl]carbonyl}-amino)-3,3-dimethyl-7-oxo-4-thia-1-azabicyclo[3.2.0]heptane-2-carboxylic acid, is a narrow-spectrum, penicillin-class β-lactam typically used to treat infections caused by susceptible Gram-positive bacteria, including β-lactamase-producing *Staphylococcus aureus*. Even so, floxacillin is ineffective against MRSA. Its binding to serum protein is similar to oxacillin and cloxacillin and less than that of dicloxacillin. Oral flucloxacillin reaches higher total and free serum levels than oxacillin and cloxacillin. Similarly, after intramuscular injection the free serum levels of flucloxacillin were higher than those of oxacillin, cloxacillin, and dicloxacillin (*See also Cloxacillin; Oxacillin*). Floxacillin inhibits cross-linking of the linear peptidoglycan polymer chains comprising the major peptidoglycan component of the cell wall of Gram-positive bacteria. **1**. Sutherland, Croydon & Rolinson (1970) *Brit. Med. J.* **4**, 455.

Fluazifop

Fluazifop

Fluazifop-butyl

This grass-selective herbicide (FW$_{\text{free-acid}}$ = 327.26 g/mol) inhibits acetyl-CoA carboxylase. The active metabolite is 2-[4-[[5-(trifluoromethyl)-2-pyridinyl]oxy]phenoxy]propionic acid. The *n*-butyl ester (FW = 383.37 g/mol) is a frequently used membrane-penetrating pro-drug. Note that the term fluazifop-P refers to the (*R*)-stereoisomer. **1**. Walker, Ridley, Lewis & Harwood (1988) *Biochem. J.* **254**, 307. **2**. Harwood (1988) *Trends Biochem. Sci.* **13**, 330. **3**. Herbert, Price, Alban, *et al.* (1996) *Biochem J.* **318**, 997. **4**. Price, Herbert, Moss, Cole & Harwood (2003) *Biochem. J.* **375**, 415.

Fluconazole

This azole (FW = 306.28 g/mol; CAS = 86386-73-4), named systematically as 2-(2,4-difluorophenyl)-1,3-bis(1*H*-1,2,4-triazol-1-yl)propan-2-ol and known by its trade name Diflucan®, is a orally active antifungal agent. **Primary Mode of Action:** Fluconazole inhibits the fungal cytochrome P450 enzyme 14α-demethylase, thus preventing conversion of lanosterol to ergosterol, the latter an essential component of the fungal cytoplasmic membrane (1-10). Subsequent accumulation of 14α-methyl sterols is also toxic to fungi. Note that the mammalian demethylase is much less sensitive to fluconazole. **Susceptibility:** Fluconazole is active against *Blastomyces dermatitidis*, *Candida* spp. (but not *C. krusei* and *C. glabrata*), *Coccidioides immitis*, *Cryptococcus neoformans Epidermophyton* spp., *Histoplasma capsulatum*, *Microsporum* spp., and *Trichophyton* spp. **Fluconazole Resistance:** In *Candida albicans*, resistance occurs as a consequence of mutations in the gene coding for 14α-demethylase. Such mutations prevent fluconazole from binding, while allowing the enzyme to bind the natural substrate, lanosterol. **Key Pharmacokinetic Parameters:** *See* Appendix II in Goodman & Gilman's THE PHARMACOLOGICAL BASIS OF THERAPEUTICS, 12th Edition (Brunton, Chabner & Knollmann, eds.) McGraw-Hill Medical, New York (2011). **Target(s):** CYP3A4 and CYP2C9 (11-13), glucuronosyltransferase (14), sterol 14-demethylase, *or* CYP51, K_d = 30 μM and sterol Δ^{22}-desaturase (15). **1**. Nitahara, Aoyama, Horiuchi, Noshiro & Yoshida (1999) *J. Biochem.* **126**, 927. **2**. Korosec, Acimovic, Seliskar, *et al.* (2008) *Bioorg. Med. Chem.* **16**, 209. **3**. McLean, Warman, Seward, *et al.* (2006) *Biochemistry* **45**, 8427. **4**. Eddine, von Kries, Podust, *et al.* (2008) *J. Biol. Chem.* **283**, 15152. **5**. Podust, von Kries, Eddine, *et al* (2007) *Antimicrob. Agents Chemother.* **51**, 3915. **6**. Matsuura, Yoshioka, Tosha, *et al.* (2005) *J. Biol. Chem.* **280**, 9088. **7**. Kudo, Ohi, Aoyama, *et al.* (2005) *J. Biochem.* **137**, 625. **8**. Mellado, Garcia-Effron, Buitrago, *et al.* (2005) *Antimicrob. Agents Chemother.* **49**, 2536. **9**. Trösken, Adamska, Arand, *et al.* (2006) *Toxicology* **228**, 24. **10**. Pietila, Vohra, Sanyal, *et al.* (2006) *Amer. J. Respir. Cell Mol. Biol.* **35**, 236. **11**. Houston, Humphrey, Matthew & Tarbit (1988) *Biochem. Pharmacol.* **37**, 401. **12**. La Delfa, Zhu, Mo & Blaschke (1989) *Drug Metab Dispos.* **17**, 49. **13**. Miners & Birkett (1998) *Brit. J. Clin. Pharmacol.* **45**, 525. **14**. Bowalgaha, Elliot, Mackenzie, *et al.* (2005) *Brit. J. Clin. Pharmacol.* **60**, 423. **15**. Lamb, Maspahy, Kelly, *et al.* (1999) *Antimicrob. Agents Chemother.* **43**, 1725.

Fludarabine

This adenosine analogue (FW = 285.23 g/mol; Symbol: F-ara-A), also known as 9-β-D-arabinofuranosyl-2-fluoroadenine, is resistant to adenosine deaminase and is phosphorylated by deoxycytidine kinase. Fludarabine is an antineoplastic agent for treating haematological malignancies, particularly B-cell chronic lymphocytic leukemia. **Mechanism of Action:** The free nucleoside F-ara-A enters cells and accumulates mainly as the 5'-

triphosphate, F-ara-ATP. The latter has multiple mechanisms of action, mainly affecting DNA synthesis, although incorporation into RNA and inhibition of transcription has been demonstrated. When used as a single agent or in combination with DNA-damaging agents to inhibit DNA repair processes, fludarabine is effective against indolent leukemia and lymphomas. **Key Pharmacokinetic Parameters:** *See* Appendix II in Goodman & Gilman's *THE PHARMACOLOGICAL BASIS OF THERAPEUTICS*, 12th Edition (Brunton, Chabner & Knollmann, eds.) McGraw-Hill Medical, New York (2011). **Target(s):** adenosine kinase (1); *S*-adenosylhomocysteinase (2-4); DNA (cytosine-5) methyltransferase (2); DNA biosynthesis (5,6); 5'-nucleotidase (7); RNA biosynthesis (8,9). **1.** Miller, Adamczyk, Miller, *et al.* (1979) *J. Biol. Chem.* **254**, 2346. **2.** Wyczechowska & Fabianowska-Majewska (2003) *Biochem. Pharmacol.* **65**, 219. **3.** Chiang (1987) *Meth. Enzymol.* **143**, 377. **4.** White, Shaddix, Brockman & Bennett (1982) *Cancer Res.* **42**, 2260. **5.** Brockman, Cheng, Schabel & Montgomery (1980) *Cancer Res.* **40**, 3610. **6.** Brockman, Schabel & Montgomery (1977) *Biochem. Pharmacol.* **26**, 2193. **7.** Hunsucker, Spychala & Mitchell (2001) *J. Biol. Chem.* **276**, 10498. **8.** Huang & Plunkett (1991) *Mol. Pharmacol.* **39**, 449. **9.** Huang, Sandoval, Van Den Neste, Keating & Plunkett (2000) *Leukemia* **14**, 1405.

Fludarabine 5'-Triphosphate

This nucleotide (FW$_{\text{free-acid}}$ = 522.15 g/mol) is an ATP analogue that is incorporated into DNA in a self-limiting manner at the 3'-termini. This incorporation into DNA prevents further elongation of the primers by DNA primase (primer RNA chain termination). The 3'→5' exonuclease of DNA polymerase ε is inhibited by fludarabine-terminated DNA. Note that it is also incorporated into RNA and RNA metabolism is inhibited. **Target(s):** DNA-directed DNA polymerase (1-4); DNA ligase, ATP-requiring (5); DNA polymerase α (1-4); ribonucleotide reductase (1-3). **1.** Tseng, Derse, Cheng, Brockman & Bennett (1982) *Mol. Pharmacol.* **21**, 474. **2.** White, Shaddix, Brockman & Bennett (1982) *Cancer Res.* **42**, 2260. **3.** Plunkett, Huang & Gandhi (1990) *Semin. Oncol.* **17**, Suppl. 8, 3. **4.** Huang, Chubb & Plunkett (1990) *J. Biol. Chem.* **265**, 16617. **5.** Yang, Huang, Plunkett, Becker & Chan (1992) *J. Biol. Chem.* **267**, 2345.

Flunarizine

This piperazine derivative (FW$_{\text{free-base}}$ = 404.50 g/mol), also known as Sibelium and (*E*)-1-[bis(4 fluorophenyl)methyl]-4-(3-phenyl-2-propenyl) piperazine, is a calcium channel blocker. It is often used as an anti-migraine agent. **Target(s):** Ca^{2+} channel (1); Ca2+-dependent ATPase (2,3); calmodulin binding (2,4); calmodulin dependent cyclic-nucleotide phosphodiesterase (5(; dopamine D$_2$ receptor (6); Na$^+$ channels, voltage dependent (7,8); NADH dehydrogenase, *or* Complex I (9); Na$^+$/K$^+$-exchanging ATPase (3,10); succinate dehydrogenase, *or* Complex II (9); vacuolar H$^+$-ATPase, uncoupling (11). **1.** Godfraind (1981) *Fed. Proc.* **40**, 2866. **2.** Santos, Lopes & Carvalho (1994) *Eur. J. Pharmacol.* **267**, 307. **3.** Chiang & Wei (1987) *Gen. Pharmacol.* **18**, 563. **4.** Lugnier, Follenius, Gerard & Stoclet (1984) *Eur. J. Pharmacol.* **98**, 157. **5.** Kubo, Matsuda, Kase & Yamada (1984) *Biochem. Biophys. Res. Commun.* **124**, 315. **6.** Ambrosio & Stefanini (1991) *Eur. J. Pharmacol.* **197**, 221. **7.** Fischer, Kittner, Regenthal & De Sarro (2004) *Basic Clin. Pharmacol. Toxicol.* **94**, 79. **8.** Abdul & Hoosein (2002) *Anticancer Res.* **22**, 1727. **9.** Veitch & Hue (1994) *Mol. Pharmacol.* **45**, 158. **10.** Dzurba, Breier, Slezak, *et al.* (1991) *Bratisl Lek Listy* **92**, 155. **11.** Terland & Flatmark (1999) *Neuropharmacology* **38**, 879.

Flunitrazepam

This benzodiazepine (FW = 313.29 g/mol), also known as Rohypnol and colloquially as "Roofie", is a hypnotic and anxiolytic agent that is often prescribed to individuals with insomnia. It has also become popular among alcohol and drug abusers, and its use has been documentd in many instances of date rape. The use of flunitrazepam is illegal in some nations. **Target(s):** glucurnosyltransferase UDPglucuronosyltransferase (1,2). **1.** Cheng, Rios, King, *et al.* (1998) *Toxicol. Sci.* **45**, 52. **2.** Tephly (1990) *Chem. Res. Toxicol.* **3**, 509.

N-(9-Fluorenylmethoxycarbonyl)alanylpyrrolidine-2-nitrile

This peptide analogue (FW = 389.45 g/mol), also known as Fmoc-alanyl-pyrrolidine-2-nitrile, can cross the blood-brain barrier and inhibit prolyl oligopeptidase: K_i = 5 nM. It also inhibits human lysosomal Pro-Xaa carboxypeptidase (IC$_{50}$ = 50 nM). **Target(s):** lysosomal Pro-Xaa carboxypeptidase (1); prolyl oligopeptidase (2). **1.** Shariat-Madar, Mahdi & Schmaier (2002) *J. Biol. Chem.* **277**, 17962. **2.** Li, Wilk & Wilk (1996) *J. Neurochem.* **66**, 2105.

S-[9-Fluorenylmethoxycarbonyl]glutathione

This glutathione derivative (FW = 528.56 g/mol) is a strong inhibitor of calf liver glyoxalase II, *or* hydroxyacylglutathione hydrolase, K_i = 2.1 μM. It also inhibits glyoxalase I (*or* lactoylglutathione lyase) and glutathione *S*-transferase, K_i = 17 and 25 μM, respectively. **1.** Chyan, Elia, Principato, *et al.* (1994-1995) *Enzyme Protein* **48**, 164. **2.** Yang, Sobieski, Carenbauer, *et al.* (2003) *Arch. Biochem. Biophys.* **414**, 271. **3.** Norton, Talesa, Yuan & Principato (1990) *Biochem. Int.* **22**, 411. **4.** Norton, Elia, Chyan, *et al.* (1993) *Biochem. Soc. Trans.* **21**, 545.

N-(9-Fluorenylmethoxycarbonyl)-*N*-methyl-L-phenylglycyl-L-leucyl-L-aspartyl-L-cyclohexylalanyl-L-aspartyl-L-phenylalanine

This hexapeptide (FW = 1028.15 g/mol) inhibits eukaryotic ribonucleotide reductase (RR), which catalyzes nucleoside diphosphate conversion to deoxynucleoside 5'-diphosphates needed for DNA biosynthesis, making RR a target for cancer therapy (1,2). RR catalysis requires formation of a complex between subunits R1 and R2, in which the R2 C-terminal peptide binds to R1. Crystal structures of heterocomplexes containing mammalian R2 C-terminal heptapeptide, P7 (Ac-1FTLDADF7) and its peptidomimetic P6 (1Fmoc(Me)PhgLDChaDF7) bound to *Saccharomyces cerevisiae* R1 (ScR1). P7 and P6, both of which inhibit ScRR, each bind at two contiguous sites containing residues that are highly conserved among eukaryotes. Such binding is quite distinct from that reported for prokaryotes. The Fmoc group in P6 makes several hydrophobic interactions that contribute to its enhanced potency in binding to ScR1. **1**. Cooperman, Gao, Tan, Kashlan & Kaur (2005) *Adv. Enzym. Regul.* **45**, 112. **2**. Xu, Fairman, Wijerathna, *et al.* (2008) *J. Med. Chem.* **51**, 4653.

Fluoride, *or* F⁻

This halide anion (typically supplied as NaF (FW = 41.99 g/mol; CAS 7681-49-4) or KF (FW = 58.10 g/mol; CAS 7789-23-3); Ionic radius = 1.33 Å), which is probably essential in trace amounts for mammals, accumulates in mammalian bone and teeth. The strength of skeleton and teeth, particularly the hydroxyapatite matrix, is greatly affected by the presence of F⁻. Noting as much in the 1920s, the nutritional biochemist Elmer McCollum proposed the now widespread practice of adding fluoride to city water supplies to prevent dental caries. (Contrary to silly claims that fluoridation of drinking water is an evil attempt to control the unruly masses, there simply is no evidence that fluoride has any mind- or mood-altering action.) The agents typically added to city water are sodium fluoride (NaF), fluorosilic acid (H_2SiF_6), or sodium fluorsilicate (Na_2SiF_6), typically at levels of 1-2 ppm. Many enzymes are inhibited by fluoride. The classical example is enolase, for which fluoride serves as a potent uncompetitive inhibitor (1-3). Another example is pyruvate kinase, which actually phosphorylates F⁻ in the presence of carbon dioxide (3). (Inhibition of enolase and pyruvate kinase most likely account for the bactericidal action of fluoride in drinking water, toothpaste, and mouthwash.) Acid phosphatases are likewise inhibited by fluoride (2). Fluoride suppresses DNA and protein biosynthesis in mammalian cells and in cell-free systems (4). Many other enzymes are inhibited at high fluoride concentrations; however, little inhibition is observed at physiologic levels. Treatment of erythrocytes with 5 mM NaF for 1–24 h causes progressive accumulation of cytosolic Ca^{2+} and phosphatidylserine (PS) on the outer membrane surface (5). After 1 h, these processes are suppressed by PKC inhibitors (staurosporine, GF 109203X and chelerythrine), but increased by PKC activator (PMA). Following 24 h, NaF-induced Ca^{2+} uptake and PS externalization were partly prevented by PMA or staurosporine, but not by GF 109293X and chelerythrine. **1**. Reiner (1947) *J. Gen. Physiol.* **30**, 367. **2**. Smith, Armstrong & Singer (1959) *Proc Soc Exp Biol Med.* **102**, 170. **3**. Rose (1969) in *Comprehensive Biochemistry* (Florkin & Stokes, eds.), vol. 17, p. 126, Elsevier, Amsterdam. **4**. Holland (1979) *Acta Pharmacol. Toxicol.* **45**, 96. **5**. Agalakova & Gusev (2013) *Toxicol. In Vitro* **27**, 2335.

Fluoroacetate

This α-halo acid ($FW_{free-acid}$ = 78.04 g/mol), found naturally in the leaves of the poisonous South African plant *Dichapetalum cymosum*, is transformed metabolically into fluorocitrate, a potent inactivator of the TCA Cycle enzyme aconitase (1,2). The sodium salt is used as a rodenticide (2). (*See 2R,3R-Fluorocitrate*) **CAUTION**: Fluoroacetate (human toxic dose = 0.5-2 mg/kg) is hazardous, if ingested. **Target(s)**: aconitase, aconitate hydratase (1,2); carbonate anhydrase (3); malate synthase (4-7); sarcosine dehydrogenase (8); tartronate semialdehyde reductase, 2-hydroxy-3-oxopropionate reductase (9); β-ureidopropionase (10). **1**. Morrison & Peters (1954) *Biochem. J.* **58**, 473. **2**. Treton & Peters (1978) *Agric. Biol. Chem.* **42**, 1201. **3**. Bertini, Luchinat & Scozzafava (1982) *Structure and Bonding* **48**, 45. **4**. Dixon & Kornberg (1962) *Meth. Enzymol.* **5**, 633. **5**. Higgins, Kornblatt & Rudney (1972) *The Enzymes*, 3rd ed., **7**, 407. **6**. Dixon, Kornberg & Lund (1960) *Biochim. Biophys. Acta* **41**, 217. **7**. Beeckmans, Khan, Kanarek & Van Driessche (1994) *Biochem. J.* **303**, 413. **8**. Mackenzie & Hoskins (1962) *Meth. Enzymol.* **5**, 738. **9**. Kornberg & Gotto (1966) *Meth. Enzymol.* **9**, 240. **10**. Wasternack, Lippmann & Reinbotte (1979) *Biochim. Biophys. Acta* **570**, 341.

3-Fluoro-*N*-acetylneuraminic Acid

This sialic acid derivative ($FW_{free-acid}$ = 327.26 g/mol; CAS = 921-40-4), also called 3-fluorosialate, is a reducing sugar that will inhibit *N*-acetyl-neuraminate lyase, *or N*-acetylneuraminate aldolase (1,2) and bacterial and viral sialidases (3). Diethylaminosulfur trifluoride (DAST) can be used to generate this inhibitor (4). **1**. Gantt, Millner & Binkley (1964) *Biochemistry* **3**, 1952. **2**. Sirbasku & Binkley (1970) *Biochim. Biophys. Acta* **198**, 479. **3**. Hagiwara, Kijima-Suda, Ido, Ohrui & Tomita (1994) *Carbohydr. Res.* **263**, 167. **4**. Hader & Watts (2013) *Carbohydr Res.* **374**, 23.

β-Fluoro-D-alanine

This modified amino acid (FW = 107.08 g/mol), also known as fludalanine and 3-fluoro-D-alanine is an antibacterial agent that inhibits alanine racemase. **Target(s)**: alanine racemase (1-3); amino-acid racemase (4); 1-aminocyclopropane-1 carboxylate deaminase (5); D-glutamate:D-amino acid aminotransferase, alternative substrate (6); serine Hydroxymethyl-transferase, *or* glycine Hydroxymethyltransferase (7). **1**. Walsh (1983) *Trends Biochem. Sci.* **8**, 254. **2**. Kollonitsch, Barash, Kahan & Kropp (1973) *Nature* **243**, 346. **3**. Esaki & Walsh (1986) *Biochemistry* **25**, 3261. **4**. Roise, Soda, Yagi & Walsh (1984) *Biochemistry* **23**, 5195. **5**. Walsh, Pascal, Johnston, *et al.* (1981) *Biochemistry* **20**, 7509. **6**. Soper & Manning (1981) *J. Biol. Chem.* **256**, 4263. **7**. Schirch (1982) *Adv. Enzymol. Relat. Areas Mol. Biol.* **53**, 83.

o-Fluoro-Amidine

This protein arginine deaminase inhibitor (FW = 408.40 g/mol; CAS = 877617-46-4; Soluble in methanol also known as *o*-F-amidine and systematically as N^α-(2-carboxyl)benzoyl-N^5-(2-fluoro-1-iminoethyl)-l-ornithine amide, is an active site-directed irreversible inhibitor. is a potent irreversible inhibitor that covalently modifies an active-site cysteine residue. Catalyzing posttranslational citrullination (*i.e.*, conversion of protein guanidinium groups of into ureido groups), PADs transform arginine residues into citrulline residues (1). By removing positively side-chains from proteins, these deiminases fundamentally alter the strength and/or specificity of regulatory interactions between target proteins with their intracellular binding partners. These enzymes (of which there are five in humans: PAD-1, PAD-2, PAD-3, PAD-4, and PAD-6) regulate gene transcription, cellular differentiation, as well as the innate immune response, and they are up-regulated in a number of human diseases, including rheumatoid arthritis, ulcerative colitis, and cancer. Relevant inhibitor parameters are: PAD-1 (IC$_{50}$ = 1.4 µM; k_{inact} = 1.7 min⁻¹; K_i = 9.4

μM; k_{inact}/K_i = 180900 M^{-1}/min^{-1}); PAD-2 (IC$_{50}$ = >50 μM; k_{inact} = ND; K_i = ND; k_{inact}/K_i = 7500 M^{-1}/min^{-1}); PAD-3 (IC$_{50}$ = 34 μM; k_{inact} = ND; K_i = ND; k_{inact}/K_i = 6700 M^{-1}/min^{-1}); and PAD-4 (IC$_{50}$ = 1.9 μM; k_{inact} = 0.5 min^{-1}; K_i = 16 μM; k_{inact}/K_i = 32500 M^{-1}/min^{-1}) (2). F-amidine is cytotoxic to HL-60, MCF7, and HT-29 cancer cell lines, with IC$_{50}$ values of 0.5, 0.5 and 1 μM, respectively (3). (**See also** o-Cl-Amidine) 1. Slade, Subramanian, Fuhrmann & Thompson (2013) *Biopolymers* **101**, 133. 2. Causey, Jones, Slack, *et al.* (2011) *J. Med. Chem.* **54**, 6919. 3. Slack, Causey & Thompson (2011) *Cell Mol. Life Sci.* **68**, 709.

(2R,3R)-Fluorocitrate

This highly toxic antimetabolite (FW$_{free-acid}$ = 210.12 g/mol) is a potent inhibitor of aconitase (IC$_{50}$ = 8 μM), blocking tricarboxylic acid (*or* Krebs) cycle in the process (1-6). The (–)-*erythro* diastereomer acts as a mechanism-based inhibitor, which is first converted to fluoro-*cis*-aconitate by the enzyme, followed by addition of hydroxide and loss of fluoride to form 4-hydroxy-*trans*-aconitate, an agent that binds extremely tightly, but reversibly, to the enzyme (7). Given sufficient time, submicromolar concentrations of (–)-*erythro*-fluorocitrate are sufficient to irreversibly inhibit citrate uptake by isolated brain mitochondria. (**See also** 3-Deoxy-3-fluorocitrate and 3-Deoxy-3-fluorocitric Acid) **Target(s):** aconitase, *or* aconitate hydratase (1-19); aconitate decarboxylase (20); ATP:citrate lyase, *or* ATP:citrate synthase (21); citrate (pro-*S*)-lyase (22); citrate transport (11,23); glutathione-citryl thioester formation (24); malic enzyme (25); 2-methylcitrate dehydratase, inhibited by DL-fluorocitrate (26); oxaloacetate tautomerase (27,28). 1. Glusker (1971) *The Enzymes*, 3rd ed., **5**, 413. 2. Dickman (1961) *The Enzymes*, 2nd ed., **5**, 495. 3. Peters, Wakelin, Butta & Thomas (1953) *Proc. Roy. Soc. London B* **140**, 497. 4. Morrison & Peters (1954) *Biochem. J.* **58**, 473. 5. Peters (1958) *Biochem. Pharmacol.* **1**, 101. 6. Uhrigshardt, Walden, John & Anemuller (2001) *Eur. J. Biochem.* **268**, 1760. 7. Lauble, Kennedy, Emptage, Beinert & Stout (1996) *Proc. Natl. Acad. Sci. U.S.A.* **93**, 13699. 8. Kent, Emptage, Merkle, *et al.* (1985) *J. Biol. Chem.* **260**, 6871. 9. Eanes & Kun (1974) *Mol. Pharmacol.* **10**, 130. 10. Villafranca & Platus (1973) *Biochem. Biophys. Res. Commun.* **55**, 1197. 11. Brand, Evans, Mendes-Mourao & Chappell (1973) *Biochem. J.* **134**, 217. 12. Carrell, Glusker, Villafranca, *et al.* (1970) *Science* **170**, 1412. 13. Guarriera-Bobyleva & Buffa (1969) *Biochem. J.* **113**, 853. 14. Treble, Lamport & Peters (1962) *Biochem. J.* **85**, 113. 15. Morrison & Peters (1954) *Process Biochem.* **56** (325th Meeting), xxxvi. 16. Peters (1961) *Biochem. J.* **79**, 261. 17. Brouquisse, M. Nishimura, J. Gaillard & R. Douce (1987) *Plant Physiol.* **84**, 1402. 18. De Bellis, Tsugeki, Alpi & Nishimura (1993) *Physiol. Plant* **99**, 485. 19. Uhrigshardt, Walden, John & Anemuller (2001) *Eur. J. Biochem.* **268**, 1760. 20. Bentley & Thiessen (1957) *J. Biol. Chem.* **226**, 703. 21. Antranikian, Herzberg & Gottschalk (1982) *J. Bacteriol.* **152**, 1284. 22. Rokita & Walsh (1983) *Biochemistry* **22**, 2821. 23. Kirsten, Sharma & Kun (1978) *Mol. Pharmacol.* **14**, 172. 24. Kun, Kirsten & Sharma (1977) *Proc. Natl. Acad. Sci. U.S.A.* **74**, 4942. 25. Krebs & Davies (1955) *Arch. Sci. Biol. (Bologna)* **39**, 533. 26. Aoki & Tabuchi (1981) *Agric. Biol. Chem.* **45**, 2831. 27. Belikova, Kotlyar & Vinogradov (1989) *FEBS Lett.* **246**, 17. 28. Vinogradov, Kotlyar, Burov & Belikova (1989) *Adv. Enzyme Regul.* **28**, 271.

5-Fluorocytosine

This pyrimidine (FW = 129.09 g/mol), also known as flucytosine, is an antifungal agent synthesized in 1957. Although it has no intrinsic antifungal capacity *per se*, it is taken up by cells and converted to 5-fluorouracil, which is then converted to nucleotide metabolites that inhibit fungal RNA and DNA synthesis. In addition, incorporation of 5-fluorocytosine into

RNA or DNA results in inhibition of the corresponding cytosine methyltransferases (*e.g.*, HhaI DNA (cytosine-5-)-methyltransferase-1). 5-Fluorocytosine is soluble in water (1.5 g/100 mL at 25°C) and has a λ$_{max}$ at 285 nm in 0.1 M HCl (ε = 8900 M^{-1}cm^{-1}). **Target(s):** DNA (cytosine-5-)-methyltransferase (1,2); hypoxanthine(guanine) phosphoribosyltransferase (3); xanthine phosphoribosyltransferase (3). 1. Osterman, DePillis, Wu, A. Matsuda & Santi (1988) *Biochemistry* **27**, 5204. 2. Santi, Garrett & Barr (1983) *Cell* **33**, 9. 3. Naguib, Iltzsch, el Kouni, Panzica & el Kouni (1995) *Biochem. Pharmacol.* **50**, 1685.

5-Fluoro-2'-deoxycytidine

This modified nucleoside (FW = 245.21 g/mol), a strong inhibitor of cell growth, is metabolized to 5-fluoro-2'-deoxyuridine 5'-monophosphate, which inhibits thymidylate synthase. In addition, incorporation into DNA results in inhibition of the corresponding DNA methyltransferases, *e.g.*, HaeIII DNA (cytosine-5-)-methyltransferase (1). **Target(s):** aspartate carbamoyltransferase (2); deoxycytidine kinase (3). 1. Chen, MacMillan, Chang, *et al.* (1991) *Biochemistry* **30**, 11018. 2. Bresnick (1962) *Cancer Res.* **22**, 1246. 3. Deibler, Reznik & Ives (1977) *J. Biol. Chem.* **252**, 8240.

5-Fluoro-2'-deoxycytidine 5'-Monophosphate

This modified deoxynucleotide (FW = 325.19 g/mol) inhibits deoxycytidylate 5-hydroxymethyltransferase competitively and is a tight-binding inhibitor of thymidylate synthase (1). (**See** 5-Fluoro-2'-deoxyuridine 5'-monophosphate) **Target(s):** deoxycytidylate 5-hydroxymethyltransferase (2); thymidylate synthase (1,3). 1. Liu & Santi (1994) *Biochim. Biophys. Acta* **1209**, 89. 2. Subramaniam, Wang, Mathews & Santi (1989) *Arch. Biochem. Biophys.* **275**, 11. 3. Hartmann & Heidelberger (1961) *J. Biol. Chem.* **236**, 3006

[^{18}F]-2-Fluoro-2-deoxy-D-glucose

This glucose analogue and PET tracer (FW = 181.15 g/mol; CAS = 63503-12-8; Symbol: FDG) is a marker for the tissue uptake of glucose enumerous tissues and organs (1,2). After [^{18}F]-FDG is injected into a patient, a PET scanner captures a two-dimensional or three-dimensional images of its distribution within the body. The kinetics of transport across the blood-brain barrier and metabolism in brain (hemisphere) of [^{14}C]2-FDG, [^{3}H]-2-deoxy-D-glucose (DG) and D-glucose in the pentobarbital-anesthetized adult rat show saturation kinetics when measured with the brain uptake index (BUI) method. The BUI for FDG was 54.3. Nonlinear regression analysis gave a K_m of 6.9 mM and V_{max} of 1.70 μmol/min/g (3). The K_i for glucose inhibition of FDG transport was 10.7 mM. The kinetic constants of influx (k_1) and efflux (k_2) for FDG were calculated from the K_m, V_{max}, and glucose concentrations of the hemisphere and plasma (2.3 μmol/g and 9.9 mM, respectively). The transport coefficient (k_1 for FDG divided by k_1 for glucose) was 1.67 and the phosphorylation constant was 0.55 (3). 1. Pacak, Tocik & Cerny (1969). *J. Chem. Soc. CHEM. COMM.* 77–77. 2. Ido, Wan,

Casella, *et al.* (1978) *J. Lab. Comp. Radiopharm.* **24**, 174. **3**. Crane, Pardridge, Braun & Oldendorf (1983) *J. Neurochem.* **40**, 160.

5-Fluoro-2'-deoxyuridine

This modified nucleoside (FW = 246.20 g/mol; Symbol: FUdR), also known as 2'-deoxy-5-fluorouridine and floxuridine, is a mutagen that is metabolically converted to the monophosphate, which in turn potently inhibits thymidylate synthase. The resulting imbalance in the composition of deoxynucleotide pools strongly inhibits DNA biosynthesis (*See 5-Fluoro-2'-deoxyuridine 5'-Monophosphate*). **Use in Synchronizing Aging Populations of Nematodes:** in 5-FUdR is commonly used to sterilize *Caenorhabditis elegans* to maintain a synchronized aging population of nematodes, without contamination by their progeny, in lifespan experiments. Although all somatic cells in the adult nematode are post-mitotic and do not require nuclear DNA synthesis, mitochondrial DNA (mtDNA) replicates independently of the cell cycle. Inhibition of mtDNA synthesis can lead to mtDNA depletion, which is linked to a number of diseases in humans. Even so, the 8-13-day half-life of mtDNA in adult *C. elegans* somatic cells greatly attenuates FUdR effects on various endpoints (*e.g.,* DNA copy numbers, DNA damage, steady-state ATP levels, nematode size, mitochondrial morphology, and lifespan). **1**. Rooney, Luz, González-Hunt, *et al.* (2014) *Exp Gerontol.* **56**, 69.

5-Fluoro-2'-deoxyuridine 5'-Monophosphate

This anticancer drug (FW = 326.18 g/mol), also known as fluorodeoxyuridylate and abbreviated 5-FdUMP, is a mechanism-based substrate that forms a covalent bond with FdUMP when an Enz-SH adds at the 6-position of the pyrimidine ring. This reaction requires the presence of 5,10-methylenetetrahydrofolate. By inhibiting thymidylate synthase, this deoxyuridine analogue inhibits DNA biosynthesis. **Target(s):** dCMP deaminase (1); deoxycytidylate 5-hydroxymethyltransferase (2-4); thymidylate synthase (5-8); thymidylate synthase (FAD) (9,10). **1**. Maley & Maley (1964) *J. Biol. Chem.* **239**, 1168. **2**. Butler, Graves & Hardy (1994) *Biochemistry* **33**, 10521. **3**. Graves & Hardy (1994) *Biochemistry* **33**, 13049. **4**. Hardy, Graves & Nalivaika (1995) *Biochemistry* **34**, 8422. **5**. Hartmann & C. Heidelberger (1961) *J. Biol. Chem.* **236**, 3006. **6**. Wataya & Santi (1977) *Meth. Enzymol.* **46**, 307. **7**. Carreras & Santi (1995) *Annu. Rev. Biochem.* **64**, 721. **8**. James, Pogolotti, Ivanetich, *et al.* (1976) *Biochem. Biophys. Res. Commun.* **72**, 404. **9**. Hunter, Gujjar, Pang & Rathod (2008) *PloS One* **3**, e2237. **10**. Graziani, Xia, Gurnon, *et al.* (2004) *J. Biol. Chem.* **279**, 54340.

2-Fluoro-1,3,5(10)-estratrien-3-ol-17-one

This fluorinated steroid (FW = 288.36 g/mol), also known as 2-fluoroestrone and 3-hydroxy-2-fluoro-1,3,5(10)-estratrien-17-one, noncompetitively inhibits *Pseudomonas testosteroni* steroid Δ-isomerase (K_i = 35 µM). **1**. Weintraub, Vincent, Baulieu & Alfsen (1977) *Biochemistry* **16**, 5045.

5'-(2-Fluoroethylthio)immucillin-A

This nucleoside analogue (FW$_{free-base}$ = 326.38 g/mol), also known as (1*S*)-1-(9-deazaadenin-9-yl)-5-(2-fluoroethylthio)-1,4-dideoxy-1,4-imino-D-ribitol, inhibits 5'-methylthio-adenosine/*S*-adenosyl-homocysteine nucleosidase, K_i' = 30 pM (1). It is a slightly weaker, slow-binding inhibitor of human *S*-methyl-5' thioadenosine phosphorylase, K_i = 20 nM (2). When the ring nitrogen is protonated, this analogue mimics the charge of oxa-carbenium ion intermediate that forms transiently in many glycosidase reactions, and this property results in tight binding. **1**. Singh, Evans, Lenz, *et al.* (2005) *J. Biol. Chem.* **280**, 18265. **2**. Evans, Furneaux, Schramm, Singh & Tyler (2004) *J. Med. Chem.* **47**, 3275.

2-Fluorofucose Triacetate & 2-Fluorofucose

These fucose analogues (FW$_{2-FF-Triacetate}$ = 294.23 g/mol; FW$_{2-FF}$ = 166.15 g/mol) are metabolically active, the former penetrating cells and the latter undergoing GTP-dependent conversion to GDP-2-fluoro-fucose, which inhibits fucosyl- or sialyltransferases (1). Because the latter cannot undergo fucosyl-transfer, its accumulates in cells and shuts down *de novo* synthesis of GDP-Fucose or CMP-NeuAc via a feedback loop, further decreasing fucose or sialic acid expression on the cell surface. Exposure of cells to this and other similar inhibitors remodels the cell surface and impairs selectin-mediated cell adhesion. For example, 2-FF also blocks the formation and tethering of sialyl-Lewis^x tetrasaccharides and structural variants on leukocytes and red blood cells to P- and E-selectins on activated endothelial cell surfaces (2). P- and E-selectins are known to be essential for vaso-occlusion in sickle cell disease (SCD), and 2FF inhibits vaso-occlusion in SCD mice. 2FF treatment of SCD mice also significantly reduces leukocyte rolling and adhesion along the vessel walls as well as the static adhesion of neutrophils and sickle red blood cells from 2FF-treated SCD mice to resting and activated endothelial cells (2). Such findings suggest 2FF may be beneficial for preventing or treating vaso-occlusive crises in SCD patients. Other novel monosaccharides that become incorporated and/or alter protein glycan structures 5-alkynylfucose (3,4), 5-thio-*N*-acetylglucosamine (5), 5- or 6-fluoro-galactofuranose (6), 3-fluoro-*N*-acetyl-neuraminic acid (7), azidosugars, such as *N*-azidoacetylgalactosamine or *N*-azidoacetylglucosamine (8,9), peracetylated 4-fluoro-glucosamine (10), and 5-thiofucose (11), **1**. Okeley, Alley, Anderson, *et al.* (2013) *Proc. Natl. Acad. Sci. U.S.A.* **110**, 5404. **2**. Belcher, Chen, Nguyen, *et al.* (2015) *PLoS One* **10**, e0117772. **3**. Hsu, *et al.* (2007) *Proc. Natl. Acad. Sci. U.S.A.* **104**, 2614. **4**. Sawa, *et al.* (2006) *Proc. Natl. Acad. Sci. U.S.A.* **103**, 1237. **5**. Gloster, *et al.* (2011) *Nature Chem. Biol.* **7**, 174. **6**. Brown, Rusek & Kiessling (2012) *J. Am. Chem. Soc.* **134**, 6552. **7**. Rillahan, *et al.* (2012) *Nature Chem. Biol.* **8**, 661. **8**. Agard & Bertozz (2009) *Acc. Chem Res.* **42**, 788. **9**. Chaubard, Krishnamurthy, Yi, Smith & Hsieh-Wilson (2012) *J. Am. Chem. Soc.* **134**, 4489. **10**. Barthel, et al. (2011) *J. Biol. Chem.* **286**, 21717. **11**. Zandberg, Kumarasamy, Pinto & Vocadlo (2012) *J. Biol. Chem.* **287**, 40021.

4-Fluoro-5-hydroxypentane-2,3-dione

This synthetic inhibitor of *Vibrio harveyi* quorum-sensing (FW = 134.11 g/mol), is a fluorinated derivative of (*S*)-4,5-dihydroxypentane-2,3-dione (shown on the right), an essential precursor of the naturally occurring autoinducer signaling molecule known as AI-2. The latter and its borate adduct act as a universal signaling molecule present in >70 bacterial species (1). 4-Fluoro-DPD completely inhibits bioluminescence and bacterial growth of *V. harveyi* BB170 strain at 12.5 μM and 100 μM, respectively. **1.** Kadirvel, Fanimarvasti, Forbes, *et al.* (2014) *Chem. Commun.* (Cambridge) **50**, 5000.

5-Fluoro-2-indolyldeschlorohalopemide

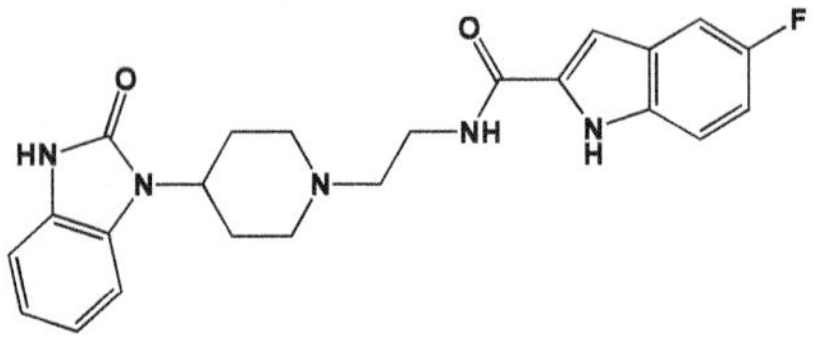

This halopemide derivative (FW = 421.50 g/mol; CAS = 939055-18-2; *Abbreviation*: FIPI), also known as *N*-[2-[4-(2,3-dihydro-2-oxo-1*H*-benzimidazol-1-yl)-1-piperidinyl]ethyl]-5-fluoro-1*H*-indole-2-carboxamide, targets phospholipase D isoforms PLD₁ (IC₅₀ = 25 nM) and PLD₂ (IC₅₀ = 20 nM), which catalyze phospholipid hydrolysis and release phosphatidic acid (PA) head groups. PA is itself a second messenger that promotes Ras activation, cell spreading, stress fiber formation, chemotaxis, and membrane vesicle trafficking. In rats, FIPI has a >5-hour half-life, a C_{max} that, at 363 nM, and bioavailability of 18%. (*See also* *N*-[2-(4-oxo-1-phenyl-1,3,8-triazaspiro[4,5]dec-8-yl)ethyl]-2-naphthalene-carboxamide) **1.** Monovich, Mugrage, Quadros, *et al.* (2007) *Bioorg. Med. Chem. Lett.* **17**, 2310. **2.** Su, Yeku, Olepu, *et al.* (2009) *Mol. Pharmacol.* **75**, 437.

(2-[4-[[2-(3-Fluoro-4-methoxyphenyl)-7,8-dihydro-6*H*-thiopyrano[3,2-*d*]pyrimidin-4-yl]amino]phenyl]acetate

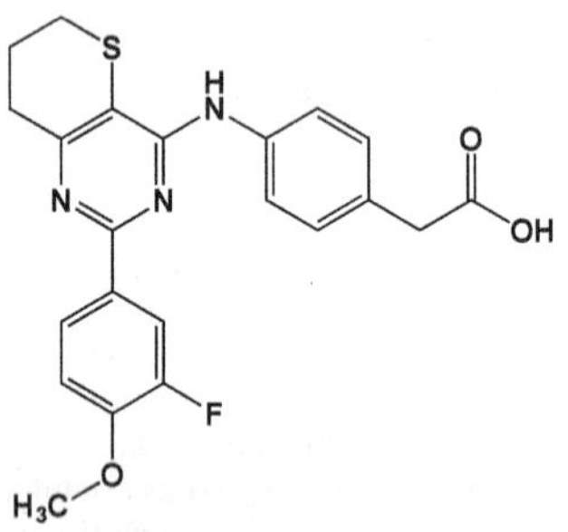

This orally active 3',5'-cyclicAMP phosphodiesterase isozyme 4B-selective inhibitor (FW = 425.28 g/mol) inhibits the activity of phosphodiesterases PDE4B and PDE4D in a concentration-dependent manner, both in humans (IC₅₀ values = 5.5 nM and 440 nM, with a selectivity factor of 80) and mice (IC₅₀ values = 26 nM and 760 nM, with a selectivity factor of 29), or about 10-15x more weakly than Roflumilast. By contrast, oral administration of this compound and Roflumilast to mice before LPS injection dose-dependently inhibits elevation of the plasma concentration of TNF-α, with respective IC₅₀ values of 29 mg/kg and 23 mg/kg. The therapeutic indices TI^TNF and TI^NEU for LPS-induced TNFα production and pulmonary neutrophilia are calculated by dividing the ID₅₀ value in gastric emptying by the ID₅₀ value in LPS injection–induced TNFα elevation and LPS inhalation–induced neutrophilia, respectively. The TI^TNF of this compound and Roflumilast are 9.0 and 0.2, respectively, indicating that this compound shows a reduced gastric adverse effect. The more important parameter TI^NEU for Compound A and Roflumilast are almost identical (*i.e.*, 1.0 and 0.5, respectively). **1.** Suzuki, Mizukami, Etori, *et al.* (2013) *J. Pharmacol. Sci.* **123**, 219.

(*S*)-α-(Fluoromethyl)histidine

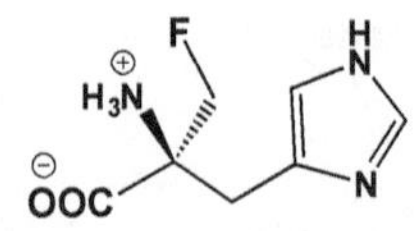

This histidine derivative (FW = 187.17 g/mol) inhibits histamine biosynthesis by acting as an irreversible, mechanism-based inhibitor of histidine decarboxylase (1-11). In the presence of the enzyme-activated

inhibitor α-FMH, decarboxylation of the analogue occurs, but the decarboxylated intermediate(s) bind covalently to the enzyme backbone to inhibit the enzyme irreversibly.

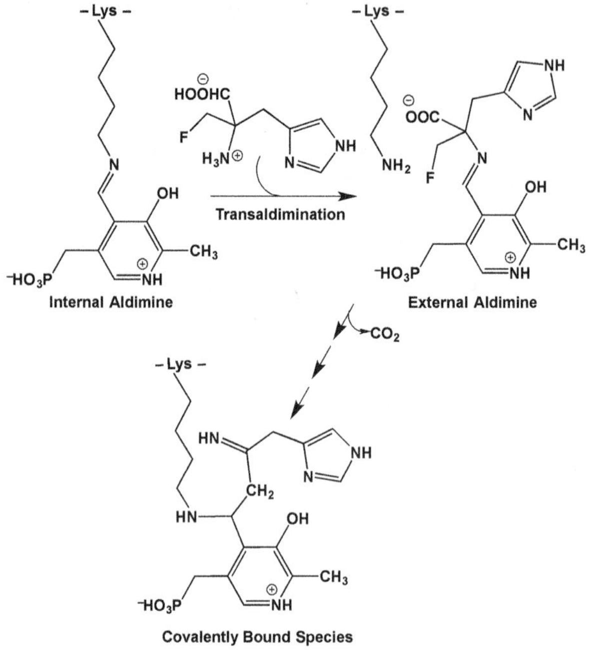

Fluoromethylhistidine has been used to deplete histamine concentrations in pharmacological studies. Indeed, it was also used to demonstrate that two SSRIs (namely Citalopram and Paroxetine) require the integrity of the brain histamine system to exert their preclinical responses (12). **1.** Kubota, Hayashi, Watanabe, Taguchi & Wada (1984) *Biochem. Pharmacol.* **33**, 983. **2.** Snell & Guirard (1986) *Meth. Enzymol.* **122**, 139. **3.** Garbarg, Barbin, Rodergas & Schwartz (1980) *J. Neurochem.* **35**, 1045. **4.** Tanase, Guirard & Snell (1985) *J. Biol. Chem.* **260**, 6738. **5.** Watanabe, Yamatodani, Maeyama & Wada (1990) *Trends Pharmacol. Sci.* **11**, 363. **6.** Yamakami, Sakurai, Kuramasu, *et al.* (2000) *Inflamm. Res.* **49**, 231. **7.** Tran & Snyder (1981) *J. Biol. Chem.* **256**, 680. **8.** Guirard & Snell (1987) *J. Bacteriol.* **169**, 3963. **9.** Mamune-Sato, Tanno, Maeyama, *et al.* (1990) *Biochem. Pharmacol.* **40**, 1125. **10.** Chudomelka, Ramaley & Murrin (1990) *Neurochem. Res.* **15**, 17. **11.** Skratt, Hough, Nalwalk & Barke (1994) *Biochem. Pharmacol.* **47**, 397. **12.** Munari, Provensi, Passani, *et al.* (2015) *Int. J. Neuropsychopharmacol.* **18**, pyv045.

3-Fluoromethyl-7-trifluoromethyl-1,2,3,4-tetrahydroisoquinoline

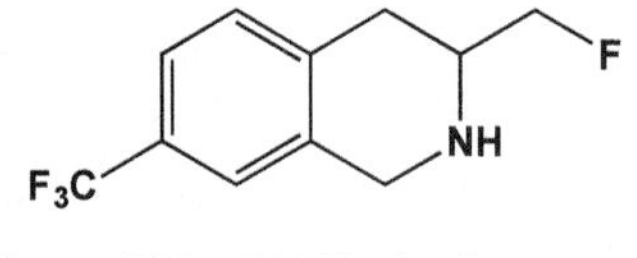

This substrate analogue (FW = 236.23 g/mol) targets phenylethanolamine *N*-methyltransferase (K_i = 30 nM) and is highly selective in its inhibition due to low affinity for the α₂-adrenoceptor. This report represents the first successful use of the β-fluorination of aliphatic amines to impart selectivity to a pharmacological agent, while still maintaining potency at the site of interest. **1.** Grunewald, Seim, Lu, Makboul & Criscione (12006) *J. Med. Chem.* **49**, 2939. **2.** Grunewald, Caldwell, Li, *et al.* (1999) *J. Med. Chem.* **42**, 3588.

6-Fluoromevalonate 5-Diphosphate

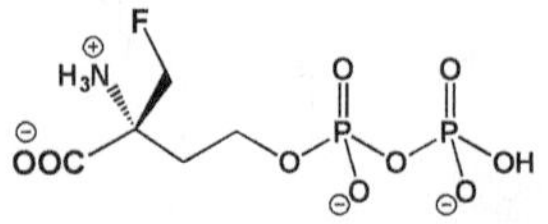

This substrate analogue (FW_free-acid = 326.11 g/mol) is a potent cholesterol biosynthesis inhibitor, targeting diphosphomevalonate decarboxylase (K_i = 37 nM), *or* mevalonate pyrophosphate decarboxylase. In a multienzyme assay for cholesterol biosynthesis, mevalonate 5-phosphate and mevalonate

5-pyrophosphate accumulate in the presence of 5 μM 6-fluoromevalonate. 6-Fluoromevalonate 5-pyrophosphate is more effective in inhibiting cholesterol biosynthesis. **1**. Nave, d'Orchymont, Ducep, Piriou & Jung (1985) *Biochem. J.* **227**, 247.

2-Fluoropalmitic Acid

This α-halo fatty acid (FW$_{free-acid}$ = 274.42 g/mol; CAS = 16518-94-8), also systematically named as 2-fluorohexadecanoic acid, inhibits incorporation of palmitate into sphingosine (IC$_{50}$ ≈ 0.05-0.2 mM) and accumulates in membrane lipids of cultured mammalian cells. Fluoropalmitate also competitively inhibits the cellular uptake of [^{3}H]-palmitate, consequently reducing the labeling of palmitoyl-CoA, glycerolipids and myelin proteolipid protein (1). The fact that fluoropalmitate competes so effectively with palmitate suggests that their uptake may be carrier-mediated. Furthermore, because the attachment of palmitate to protein cysteine residues is a common post-translational modification of both integral and peripheral membrane proteins, fluoropalmitate is likely to alter the intracellular distribution of peripheral membrane proteins by regulating membrane-cytosol exchange and/or by modifying the flux of the proteins through vesicular transport systems. **Target(s):** glycylpeptide N-tetradecanoyl-transferase, *or* protein N-myristoyltransferase (2); long-chain acyl-CoA synthetase (3). **1**. De Jesus & Bizzizero (2002) *Neurochem. Res.* **27**, 1669. **2**. Paige, Zheng, DeFrees, Cassady & Geahlen (1990) *Biochemistry* **29**, 10566. **3**. Soltysiak, Matsuura, Bloomer & Sweeley (1984) *Biochim. Biophys. Acta* **792**, 214.

m-Fluorophenylalanine

This substituted phenylalanine (FW = 183.18 g/mol) inhibits chorismate mutase, 3-deoxy-7-phosphoheptulonate synthase, prephenate dehydratase, and phenylalanine ammonia-lyase. **Target(s):** alkaline phosphatase, mildly inhibited by racemic mixture (1); chorismate mutase (2,3); 3-deoxy-7-phospho-heptulonate synthase, *or* 3-deoxy-D-*arabino*heptulosonate-7 phosphate synthetase (4); nonenzymatic hydroxylaminolysis of amino acid esters at −18°C, by racemic mixture (5); phenylalanine ammonia-lyase, inhibited by L-stereoisomer (6); prephenate dehydratase (7-9). **1**. Fishman & Sie (1971) *Enzymologia* **41**, 141. **2**. Sugimoto & Shiio (1980) *J. Biochem.* **88**, 167. **3**. Hertel, Hieke & Gröger (1991) *Acta Biotechnol.* **11**, 39. **4**. Simpson & Davidson (1976) *Eur. J. Biochem.* **70**, 509. **5**. Grant & Alburn (1965) *Biochemistry* **4**, 1913. **6**. Subba Rao & Towers (1970) *Meth. Enzymol.* **17A**, 581. **7**. Dopheide, Crewther & Davidson (1972) *J. Biol. Chem.* **247**, 4447. **8**. Ahmad, Wilson & Jensen (1988) *Eur. J. Biochem.* **176**, 69. **9**. Friedrich, Friedrich & Schlegel (1976) *J. Bacteriol.* **126**, 723.

p-Fluoro-L-phenylalanine

This substituted phenylalanine (FW = 183.18 g/mol; CAS = 1132-68-9) inhibits a number of enzymes, including 3-deoxy-7-phosphoheptulonate synthase. It will also substitute for phenylalanine in protein biosynthesis and inhibit mitosis. HeLa cells are arrested in the G$_2$ phase. **Target(s):** alkaline phosphatase, mildly inhibited by racemic mixture (1); arogenate dehydratase (2,3); 3-deoxy-7-phosphoheptulonate synthase, *or* 3-deoxy-D-*arabino*heptulosonate-7 phosphate synthetase (4,5); phenylalanine ammonia-lyase (6); phenylalanine 4-monooxygenase, weak alternative substrate (7,8); prephenate dehydratase (9-11); tryptophan aminotransferase (12); tyrosine 3-monooxygenase (13). **1**. Fishman & Sie (1971) *Enzymologia* **41**, 141. **2**. Fischer & Jensen (1987) *Meth. Enzymol.* **142**, 495. **3**. Zamir, Tiberio, Fiske, Berry & Jensen (1985) *Biochemistry* **24**, 1607. **4**. Simpson & Davidson (1976) *Eur. J. Biochem.* **70**, 509. **5**. Previc & Binkley (1964) *Biochem. Biophys. Res. Commun.* **16**, 162. **6**. Camm & Towers (1973) *Phytochemistry* **12**, 961. **7**. Kaufman (1978) *Meth. Enzymol.* **53**, 278. **8**. Woo, Gillam & Woolf (1974) *Biochem. J.* **139**, 741. **9**. Dopheide,

Crewther & Davidson (1972) *J. Biol. Chem.* **247**, 4447. **10**. Ahmad, Wilson & Jensen (1988) *Eur. J. Biochem.* **176**, 69. **11**. Friedrich, Friedrich & Schlegel (1976) *J. Bacteriol.* **126**, 723. **12**. George & Gabay (1968) *Biochim. Biophys. Acta* **167**, 555. **13**. Nagatsu, Levitt & Udenfriend (1964) *J. Biol. Chem.* **239**, 2910.

Fluorophosphate

This anion (available as disodium fluorophosphate Na$_2$FPO$_3$, FW = 143.95 g/mol; CAS = 7631-97-2) can be formed by the action of pyruvate kinase on fluoride ion and ATP (**See also** *Fluoride*). It is also an alternative substrate for sulfate adenylyltransferase. **Target(s):** enolase, *or* phosphopyruvate hydratase (1-4); lipase, *or* triacylglycerol lipase (5); phosphonoacetaldehyde hydrolase (6); phosphorylase phosphatase (7); pyruvate kinase, as an alternative product inhibitor (8); starch phosphorylase (9,10); succinate dehydrogenase (11). **1**. Verma & Dutta (1986) *Biochem. Int.* **13**, 555. **2**. Peters, Shorthous & Murray (1964) *Nature* **202**, 1331. **3**. Kaufmann & Bartholmes (1992) *Caries Res.* **26**, 110. **4**. Huether, Psarros & Duschner (1990) *Infect. Immun.* **58**, 1043. **5**. Bier (1955) *Meth. Enzymol.* **1**, 627. **6**. Olsen, Hepburn, Lee, Martin, Mariano &Dunaway-Mariano (1992) *Arch. Biochem. Biophys.* **296**, 144. **7**. Parrish & Graves (1976) *Biochem. Biophys. Res. Commun.* **70**, 1290. **8**. Mildvan, Leigh & Cohn (1967) *Biochemistry* **6**, 1805. **9**. Webb (1966) *Enzyme and Metabolic Inhibitors*, vol. 2, p. 406, Academic Press, New York. **10**. Rapp & Sliwinski (1956) *Arch. Biochem. Biophys.* **60**, 379. **11**. Bonner (1955) *Meth. Enzymol.* **1**, 722.

3-Fluorophosphoenolpyruvic Acid

E-isomer   *Z*-isomer

This fluorinated metabolite (FW$_{free-acid}$ = 186.03 g/mol) has two isomeric forms. The (*Z*)-isomer is an alternative substrate of phosphoenolpyruvate carboxykinase, whereas the (*E*)-isomer is a competitive inhibitor. The (*Z*)-isomer is a potent inhibitor of 3-phosphoshikimate 1 carboxyvinyl-transferase and UDP-*N*-acetylglucosamine 1-carboxy-vinyltransferase. **Target(s):** 3-deoxy-7-phosphoheptulonate synthase (1); phosphoenol-pyruvate carboxykinase, by the (*E*)-isomer (2); 3-phosphoshikimate 1-carboxyvinyltransferase, *or* 5-enolpyruvylshikimate-3-phosphate synthase, by the (*Z*)-isomer (3,4); UDP-*N* acetylglucosamine 1-carboxyvinyl-transferase, by both stereoisomers (5-7). **1**. Hu & Sprinson (1977) *J. Bacteriol.* **129**, 177. **2**. Duffy & Nowak (1984) *Biochemistry* **23**, 661. **3**. Walker, Ream, Simmons, *et al.* (1991) *BioMed. Chem. Lett.* **1**, 683. **4**. Gruys, Walker & Sikorski (1992) *Biochemistry* **31**, 5534. **5**. Kim, Lees, Haley & Walsh (1995) *J. Amer. Chem. Soc.* **117**, 1494. **6**. Kim, Lees & Walsh (1995) *J. Amer. Chem. Soc.* **117**, 6380. **7**. Kim, Lees, Kempsell, *et al.* (1996) *Biochemistry* **35**, 4923.

(E)-6-Fluoro-3-[2-(3-pyridyl)vinyl]-1H-indole

This indole (FW = 238.27 g/mol), also known as 680C91, is a strong and selective inhibitor of tryptophan 2,3-dioxygenase (K$_i$ = 51 nM), but does not inhibit indoleamine 2,3-dioxygenase, monoamine oxidases A and B, or serotonin uptake. **1**. Salter, Hazelwood, Pogson, Iyer & Madge (1995) *Biochem. Pharmacol.* **49**, 1435.

Fluoroquinolone Antibiotics

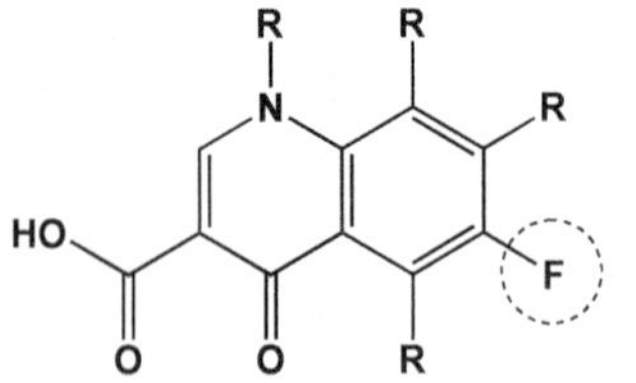

This family of synthetic broad-spectrum antibiotics share a common pharmacaphore (shown above), allowing them to exert their antibacterial action by preventing the unwinding and replication of bacterial DNA. Their slinically observed adverse effects on the CNS may be mediated, at least in part, through interactions with the benzodiazepine (BZD)-GABA$_A$-receptor complex. **See** *Alatrofloxacin; Amifloxacin; Balofloxacin; Besifloxacin; Cadazolid;* **Ciprofloxacin**; *Clinafloxacin; Danofloxacin; Delafloxacin; Difloxacin; Enoxacin; Enrofloxacin; Fleroxacin; Gatifloxacin; Gemifloxacin; Grepafloxacin; Levofloxacin; Lomefloxacin; Marbofloxacin; Moxifloxacin; Nadifloxacin; Norfloxacin; Ofloxacin; Orbifloxacin; Pazufloxacin; Pefloxacin; Prulifloxacin; Rufloxacin; Sitafloxacin; Sparfloxacin; Temafloxacin; Trovafloxacin; Zabofloxacin*

5'-(p-Fluorosulfonylbenzoyl)adenosine

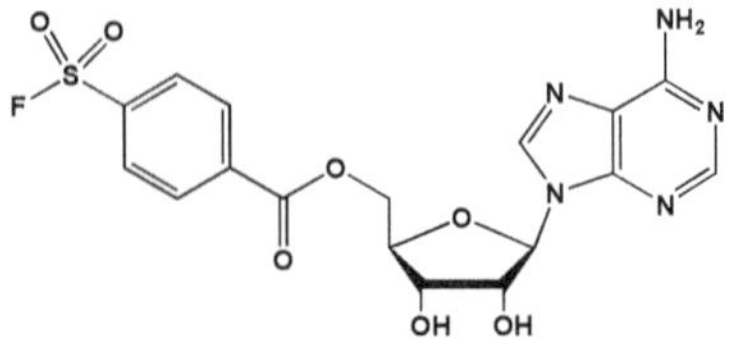

This substituted adenosine, often abbreviated 5'-FSBA (FW = 453.41 g/mol; CAS = 78859-42-4), has found widespread utility as an *in vitro* affinity label that permanently occupies active-site and/or allosteric regulatory sites for nucleotides or coenzymes. Given its lack of selectivity, experiments with 5'-FSBA are best conducted with purified enzymes. **Target(s):** acetyl-CoA carboxylase (1); actin (1); ˙*S*-adenosylhomocysteinase (1,2); adenylate cyclase (3); aldehyde reductase, *or* alcohol dehydrogenase (NADP$^+$) (1,4); Amidophosphoribosyl-transferase (5); anion-translocating ATPase (6); apyrase (7-12); asparagine synthetase (13,14); ATPase (1,15); Ca2+-dependent ATPase (107); calmodulin-dependent protein kinase (16); cAMP-dependent protein kinase, *or* protein kinase A (1,17-22); carbamoyl-phosphate synthetase (1,23-27); casein kinase (1,28,29); CF$_1$ ATPase (1,30); cGMP-dependent protein kinase, *or* protein kinase G (1,31); deoxyguanosine/ deoxyadenosine kinase (32); DNA-directed DNA polymerase (1); DNA-directed RNA polymerase, bacteriophage T$_7$ (33,34); DNA helicase (35); DNA polymerase I (1); ecto-ATPase (36,37); endopeptidase Clp (38); estradiol 17β-dehydrogenase (1,3,39,40); F$_1$ ATPase (1,41-56,106); glucokinase, polyphosphate- and ATP-dependent (57); glutamate dehydrogenase, binding at NADH inhibitory site (1,58-61); glutamine synthetase (1,62,63); glycerol kinase (1); glycine *N*-methyltransferase (1,108,109); histone H$_1$ kinase (64); *p*-hydroxybenzoate monooxygenase (44); [3 hydroxy-3-methylglutaryl-CoA reductase] kinase (1,65,66); 3α,20β-hydroxysteroid dehydrogenase (1,67); 3β-hydroxysteroid dehydrogenase (1,68-70); 20α-hydroxysteroid dehydrogenase (40); inositol 1,3,4-trisphosphate 5/6-kinase (71); [isocitrate dehydrogenase] kinase/phosphatase (72); malate dehydrogenase (1); [MAP kinase] kinase, *or* [mitogen-activated protein kinase] kinase (73); myosin (1); [myosin light-chain] kinase (74,75); NAD$^+$-dependent enzymes (1,76); NADH:cytochrome *b*$_5$ reductase (1); NADH dehydrogenase (77); NAD(P)H dehydrogenase, *or* DT diaphorase (1,78,79); Na$^+$/K$^+$-exchanging ATPase (1,80,81); nicotinamide-nucleotide transhydrogenase (1,82,83); nucleoside-triphosphatase (35); 5'-nucleotidase (84); oligopeptide transporting ATPase, *or* oligopeptide permease (85); 5-oxoprolinase (1,86); phosphate-transporting ATPase (87); phosphatidylinositol 3-kinase (88,89); phosphatidylinositol 4-kinase (90); phosphofructokinase (1); 6-phosphofructo-2-kinase (91); 6-phosphofructo-2-kinase/ fructose-2,6 bisphosphatase (1); 3-phosphoglycerate kinase (92); phosphoribosyl-aminoimidazolesuccinocarboxamide synthetase, *or* 4-[(*N*-succinylamino)carbonyl] 5-aminoimidazole ribonucleotide synthetase, *or* SAICAR synthetase (93); PRPP (or phosphoribosyl diphosphate) synthetase (1); phosphoribosyl-formylglycinamidine cyclo-ligase, *or* aminoimidazole ribonucleotide synthetase, *or* phosphoribosyl-aminoimidazole synthetase (94); phosphorylase kinase (95); luciferase (1); protein-tyrosine kinase, *or* non-specific protein-tyrosine kinase (96); protein-tyrosine kinase (*e.g.*,

pp60v-*src* and EGF receptor) (1,97-100); pyruvate kinase (1,58,101,102); *rec*A protein (1,103); rhodopsin kinase (104); ribulose-5-phosphate kinase (1); sphingosine kinase (105); Src protein-tyrosine kinase (96); steroid Δ-ene-isomerase (68-70); steroid 5α-reductase (77); xanthine dehydrogenase (1). **1**. Colman (1990) *The Enzymes*, 3rd ed. (D. S. Sigman & Boyer, eds.), **19**, 283. **2**. Takata & Fujioka (1984) *Biochemistry* **23**, 4357. **3**. Skurat, Yurkova, Khropov, Bulargina & Severin (1985) *Biochem. Int.* **10**, 743. **4**. Cronin & Flynn (1985) *Prog. Clin. Biol. Res.* **174**, 279. **5**. Zhou, Charbonneau, Colman & Zalkin (1993) *J. Biol. Chem.* **268**, 10471. **6**. Ching, Kaur, Karkaria, Steiner & Rosen (1991) *J. Biol. Chem.* **266**, 2327. **7**. Faudry, Lozzi, Santana, *et al.* (2004) *J. Biol. Chem.* **279**, 19607. **8**. Flores-Herrera, Uribe, Garcia-Perez, Milan & Martinez (2002) *Biochim. Biophys. Acta* **1585**, 11. **9**. Torres, Vasconcelos, Ferreira & Verjovski-Almeida (1998) *Eur. J. Biochem.* **51**, 516. **10**. Marti, Gomez de Aranda & Solsona (1996) *Biochim. Biophys. Acta* **1282**, 17. **11**. Cote, Ouellet & Beaudoin (1992) *Biochim. Biophys. Acta* **1160**, 246. **12**. Sevigny, Robson, Waelkens, *et al.* (2000) *J. Biol. Chem.* **275**, 5640. **13**. Boehlein, Walworth & Schuster (1997) *Biochemistry* **36**, 10168. **14**. Parr, Boehlein, Dribben, Schuster & Richards (1996) *J. Med. Chem.* **39**, 2367. **15**. Knowles & Leng (1984) *J. Biol. Chem.* **259**, 10919. **16**. King, D. J. Shell & A. P. Kwiatkowski (1988) *Arch. Biochem. Biophys.* **267**, 467. **17**. Taylor, Kerlavage & Zoller (1983) *Meth. Enzymol.* **99**, 140. **18**. Beebe & Corbin (1986) *The Enzymes*, 3rd ed., **17**, 43. **19**. Shaffer & Adams (1999) *Biochemistry* **38**, 12072. **20**. Hagiwara, Inagaki & Hidaka (1987) *Mol. Pharmacol.* **31**, 523. **21**. Bhatnagar, Hartl, Roskoski, Lessor & Leonard (1984) *Biochemistry* **23**, 4350. **22**. Zoller, Nelson & Taylor (1981) *J. Biol. Chem.* **256**, 10837. **23**. Boettcher & Meister (1980) *J. Biol. Chem.* **255**, 7129. **24**. Powers, Muller & Kafka (1983) *J. Biol. Chem.* **258**, 7545. **25**. Potter & Powers-Lee (1992) *J. Biol. Chem.* **267**, 2023. **26**. Marshall & Fahien (1985) *Arch. Biochem. Biophys.* **241**, 200. **27**. Corvi, Soltys & Berthiaume (2001) *J. Biol. Chem.* **276**, 45704. **28**. Feige, Cochet, Pirollet & Chambaz (1983) *Biochemistry* **22**, 1452.m **29**. Hathaway, Zoller & Traugh (1981) *J. Biol. Chem.* **256**, 11442. **30**. DeBenedetti & Jagendorf (1979) *Biochem. Biophys. Res. Commun.* **86**, 440. **31**. Hixson & Krebs (1981) *J. Biol. Chem.* **256**, 1122. **32**. Chakravarty, Ikeda & Ives (1984) *Biochemistry* **23**, 6235. **33**. Tunitskaya, Akbarov, Luchin, *et al.* (1990) *Eur. J. Biochem.* **191**, 99. **34**. Tunitskaia, Mishin, Tiurkin, Liakhov & Rechinskii (1988) *Mol. Biol. (Moscow)* **22**, 1642. **35**. Borowski, Niebuhr, Schmitz, *et al.* (2002) *Acta Biochim. Pol.* **49**, 597. **36**. Cerbon & Olguin (1997) *Microbios* **92**, 157. **37**. Filippini, Taffs, Agui & Sitkovsky (1990) *J. Biol. Chem.* **265**, 334. **38**. Maurizi, Thompson, Singh & Kim (1994) *Meth. Enzymol.* **244**, 314. **39**. Inano & Tamaoki (1985) *J. Steroid Biochem.* **22**, 681. **40**. Tobias & Strickler (1981) *Biochemistry* **20**, 5546. **41**. Allison, Bullough & Andrews (1986) *Meth. Enzymol.* **126**, 741. **42**. van der Zwet-de Graaff, Hartog & Berden (1997) *Biochim. Biophys. Acta* **1318**, 123. **43**. Hartog, Edel, Braham, Muijsers & Berden (1997) *Biochim. Biophys. Acta* **1318**, 107. **44**. van Berkel, Muller, Jekel, *et al.* (1988) *Eur. J. Biochem.* **176**, 449. **45**. Bullough, Kwan, Laikind, Yoshida & Allison (1985) *Arch. Biochem. Biophys.* **236**, 567. **46**. Matsuno-Yagi & Hatefi (1984) *Biochemistry* **23**, 3508. **47**. Wise, Duncan, Latchney, Cox & Senior (1983) *Biochem. J.* **215**, 343. **48**. Galante, Wong & Hatefi (1982) *Biochemistry* **21**, 680. **49**. Di Pietro, Godinot & Gautheron (1981) *Biochemistry* **20**, 6312. **50**. Esch & Allison (1978) *J. Biol. Chem.* **253**, 6100. **51**. Bullough & Allison (1986) *J. Biol. Chem.* **261**, 5722. **52**. Bullough, Verburg, Yoshida & Allison (1987) *J. Biol. Chem.* **262**, 11675. **53**. Jault & Allison (1993) *J. Biol. Chem.* **268**, 1558. **54**. Jault, Paik, Grodsky & Allison (1994) *Biochemistry* **33**, 14979. **55**. Esch & Allison (1979) *Anal. Biochem.* **95**, 39. **56**. Bullough, Yoshida & Allison (1986) *Arch. Biochem. Biophys.* **244**, 865. **57**. Phillips, Horn & Wood (1993) *Arch. Biochem. Biophys.* **300**, 309. **58**. Colman, Pal & Wyatt (1977) *Meth. Enzymol.* **46**, 240. **59**. Schmidt & Colman (1984) *J. Biol. Chem.* **259**, 14515. **60**. Saradambal, Bednar & Colman (1981) *J. Biol. Chem.* **256**, 11866. **61**. Pal, Wechter & Colman (1975) *J. Biol. Chem.* **250**, 8140. **62**. Tanaka & Kimura (1991) *J. Biochem.* **110**, 780. **63**. Nakano, Itho, Tanaka & Kimura (1990) *J. Biochem.* **107**, 180. **64**. Woodford & Pardee (1986) *J. Biol. Chem.* **261**, 4669. **65**. Ball, Dale, Weekes & Hardie (1994) *Eur. J. Biochem.* **219**, 743. **66**. Carling, Clarke & Hardie (1991) *Meth. Enzymol.* **200**, 362. **67**. Sweet & Samant (1981) *Biochemistry* **20**, 5170. **68**. Thomas, Myers & Strickler (1991) *J. Steroid Biochem. Mol. Biol.* **39**, 471. **69**. Rutherfurd, Chen & Shively (1991) *Biochemistry* **30**, 8116. **70**. Ishii-Ohba, Inano & Tamaoki (1986) *J. Steroid Biochem.* **25**, 555. **71**. Hughes, Kirk & Michell (1993) *Biochem. Soc. Trans.* **21**, 365S. **72**. Oudot, Jault, Jaquinod, *et al.* (1998) *Eur. J. Biochem.* **258**, 579. **73**. Adams & Parker (1992) *J. Biol. Chem.* **267**, 13135. **74**. Kennelly, Colburn, Lorenzen, *et al.* (1991) *FEBS Lett.* **286**, 217. **75**. Saitoh, Ishikawa, Matsushima, Naka & Hidaka (1987) *J. Biol. Chem.* **262**, 7796. **76**. Colman (1997) *Meth.*

Enzymol. **280**, 186. **77.** Golf & Graef (1982) *Steroids* **40**, 1. **78.** Chen & Liu (1992) *Mol. Pharmacol.* **42**, 545. **79.** Liu, Yuan, Haniu, *et al.* (1989) *Mol. Pharmacol.* **35**, 818. **80.** Johnson, Cooper & Winter (1986) *Biochim. Biophys. Acta* **860**, 549. **81.** Cooper & Winter (1980) *J. Supramol. Struct.* **13**, 165. **82.** Wakabayashi & Hatefi (1987) *Biochem. Int.* **15**, 915. **83.** Phelps & Hatefi (1985) *Biochemistry* **24**, 3503. **84.** Stochaj & Mannherz (1990) *Biochem. J.* **266**, 447. **85.** Hopfe & Henrich (2004) *J. Bacteriol.* **186**, 1021. **86.** Williamson & Meister (1982) *J. Biol. Chem.* **257**, 9161. **87.** Sarin, Aggarwal, Chaba, Varshney & Chakraborti (2001) *J. Biol. Chem.* **276**, 44590. **88.** Wymann, Bulgarelli-Leva, Zvelebil, *et al.* (1996) *Mol. Cell. Biol.* **16**, 1722. **89.** Ruiz-Larrea, Vicendo, Yaish, *et al.* (1993) *Biochem. J.* **290**, 609. **90.** Vereb, Balla, Gergely, *et al.* (2001) *Int. J. Biochem. Cell Biol.* **33**, 249. **91.** Pilkis, Claus, Kountz & El-Maghrabi (1987) *The Enzymes*, 3rd ed., **18**, 3. **92.** Jindal & Vishwanatha (1990) *J. Biol. Chem.* **265**, 6540. **93.** Firestine & Davisson (1994) *Biochemistry* **33**, 11917. **94.** Mueller, Oh, Kavalerchik, *et al.* (1999) *Biochemistry* **38**, 9831. **95.** Pickett-Gies & Walsh (1986) *The Enzymes*, 3rd ed., **17**, 395. **96.** Kamps, Taylor & Sefton (1984) *Nature* **310**, 589. **97.** Hunter & Cooper (1986) *The Enzymes*, 3rd ed., **17**, 191. **98.** Scoggins, Summerfield, Stein, Guyer & Staros (1996) *Biochemistry* **35**, 9197. **99.** Srivastava & Chiasson (1989) *Biochim. Biophys. Acta* **996**, 13. **100.** Buhrow, Cohen, Garbers & J. V. Staros (1983) *J. Biol. Chem.* **258**, 7824. **101.** Likos, Hess & Colman (1980) *J. Biol. Chem.* **255**, 9388. **102.** Annamalai & Colman (1981) *J. Biol. Chem.* **256**, 10276. **103.** Knight & McEntee (1985) *J. Biol. Chem.* **260**, 10177. **104.** Palczewski, McDowell & Hargrave (1988) *Biochemistry* **27**, 2306. **105.** Pitson, Moretti, Zebol, *et al.* (2002) *J. Biol. Chem.* **277**, 49545. **106.** Bullough, Brown, Saario & Allison (1988) *J. Biol. Chem.* **263**, 14053. **107.** Martin & Senior (1980) *Biochim. Biophys. Acta* **602**, 401. **108.** Fujioka & Ishiguro (1986) *J. Biol. Chem.* **261**, 6346. **109.** Fujioka & Takata, Konishi & Ogawa (1987) *Biochemistry* **26**, 5696.

5'-(p-Fluorosulfonylbenzoyl)-8-azidoadenosine

This bifunctional affinity label (FW = 494.42 g/mol), often abbreviated 8-N3-FSBA, which reacts chemically and/or photochemically, inactivates bovine heart mitochondrial F_1 ATPase with an apparent K_d of 0.47 mM at pH 8.0 and 23 °C in the absence of light. Irradiation of dark-inactivated enzyme with long-wavelength ultraviolet light produced cross-linked dimers consisting of α and β subunits. **Target(s):** F_1 ATPase (1); glutamate dehydrogenase (2). **1.** Zhuo, Garrod, Miller & Allison (1992) *J. Biol. Chem.* **267**, 12916. **2.** Colman (1990) *The Enzymes*, 3rd ed. (Sigman & Boyer, eds.), **19**, 283.

5'-(p-Fluorosulfonylbenzoyl)-1,N^6-ethenoadenosine

This fluorescent nucleotide analogue (FW = 477.43 g/mol), often abbreviated 5'-FSBεA, reacts irreversibly with bovine liver glutamate dehydrogenase, modifying one of its inhibitory GTP binding sites. The inactivation of rabbit muscle pyruvate kinase at pH 7.8 is biphasic at pH 7.8; the first phase proceeds rapidly to yield a partially active enzyme, followed by a slower rate that leads to total inactivation. **Target(s):** F_1 ATPase (1,2); fructose-1,6-bisphosphate aldolase (3,4); glutamate dehydrogenase (1,5,6); phosphofructokinase (3); phosphoglycerate kinase (3,7); pyruvate kinase (3,8). **1.** Verburg & Allison (1990) *J. Biol. Chem.* **265**, 8065. **2.** Grodsky, Dou & Allison (1998) *Biochemistry* **37**, 1007. **3.** Colman (1990) *The Enzymes*, 3rd ed., **19**, 283. **4.** Palczewski, Hargrave, Folta & Kochman (1985) *Eur. J. Biochem.* **146**, 309. **5.** Jacobson & Colman (1984) *Biochemistry* **23**, 6377. **6.** Jacobson & Colman (1983) *Biochemistry* **22**, 4247. **7.** Wiksell & Larsson-Raznikiewicz (1987) *J. Biol.*

Chem. **262**, 14472. **8.** Tomich & Colman (1985) *Biochim. Biophys. Acta* **827**, 344.

5'-(p-Fluorosulfonylbenzoyl)guanosine

This nucleotide analogue (FW = 469.41 g/mol) is an affinity label that reacts with bovine liver glutamate dehydrogenase to desensitize it irreversibly to inhibition by GTP, without affecting its intrinsic catalytic activity. **Target(s):** elongation factor EF-2 GTPase (protein-synthesizing GTPase) (1); glutamate dehydrogenase (2,3); phosphoenolpyruvate carboxykinase (2,4); pyruvate kinase (2,5,6); small monomeric GTPase (7); tubulin, after polymerization, no depolymerization occurred (2,8). **1.** Nilsson & Nygard (1984) *Biochim. Biophys. Acta* **782**, 49. **2.** Colman (1990) *The Enzymes*, 3rd ed., **19**, 283. **3.** Pal, Reischer, Wechter & Colman (1978) *J. Biol. Chem.* **253**, 6644. **4.** Jadus, Hanson & Colman (1981) B*iochem. Biophys. Res. Commun.* **101**, 884. **5.** Tomich, Marti & Colman (1981) *Biochemistry* **20**, 6711. **6.** Annamalai, Tomich, Mas & Colman (1982) *Arch. Biochem. Biophys.* **219**, 47. **7.** Sauvage, Rumigny & Maitre (1991) *Mol. Cell. Biochem.* **107**, 65. **8.** Steiner (1984) *Biochem. Biophys. Res. Commun.* **123**, 92.

5-Fluorotryptophan

This modified amino acid (FW = 222.22 g/mol; CAS = 154-08-5) inhibits a number of tryptophan-requiring enzymes. 5-Fluorotryptophan is an alternative substrate of tryptophanyl-tRNA synthetase and can be incorporated by ribosomes into proteins, allowing the use ^{19}F-NMR. It is also an activator of chorismate mutase. **Target(s):** alkaline phosphatase (1); anthranilate synthase (2,3); chorismate mutase (4); 3-deoxy 7-phosphoheptulonate synthase (5); indoleamine 2,3-dioxygenase, also as an alternative substrate (6); prephenate dehydratase (7); tryptophan 2,3-dioxygenase, also as a weak alternative substrate (8-11); tryptophan 5-monooxygenase (12,13); tryptophanyl-tRNA synthetase, as an alternative substrate (14,15); tyrosine 3-monooxygenase, moderately inhibited (16). **1.** Fishman & Sie (1971) *Enzymologia* **41**, 141. **2.** Matsukawa, Ishihara & Iwamura (2002) *Z. Naturforsch. [C]* **57**, 121. **3.** Widholm (1972) *Biochim. Biophys. Acta* **279**, 48. **4.** Hertel, Hieke & Gröger (1991) *Acta Biotechnol.* **11**, 39. **5.** Görisch & Lingens (1971) *Biochim. Biophys. Acta* **242**, 630. **6.** Southan, Truscott, Jamie, *et al.* (1996) *Med. Chem. Res.* **6**, 343. **7.** Bode, Melo & Birnbaum (1985) *J. Basic Microbiol.* **25**, 291. **8.** Webb (1966) *Enzyme and Metabolic Inhibitors*, vol. **2**, p. 325, Academic Press, New York. **9.** Brady & Feigelson (1975) *J. Biol. Chem.* **250**, 5041. **10.** Hitchcock & Katz (1988) *Arch. Biochem. Biophys.* **261**, 148. **11.** Feigelson & Brady (1974) in *Mol. Mech. Oxygen Activ.* (Hayaishi, ed.) p. 87, Academic Press, New York. **12.** Fujisawa & Nakata (1987) *Meth. Enzymol.* **142**, 93. **13.** Nakata & Fujisawa (1982) *Eur. J. Biochem.* **124**, 595. **14.** Xu, Love, Ma, *et al.* (1989) *J. Biol. Chem.* **264**, 4304. **15.** Favorova & Lavrik (1975) *Biokhimiia* **40**, 368. **16.** McGeer, McGeer & Peters (1967) *Life Sci.* **6**, 2221.

6-Fluorotryptophan

This modified amino acid (FW = 222.22 g/mol; CAS = 7730-20-3) inhibits a number of tryptophan-requiring enzymes. 5-Fluorotryptophan is an alternative substrate of tryptophanyl-tRNA synthetase and can be incorporated by ribosomes into proteins, allowing the use ^{19}F-NMR. It is also an activator of chorismate mutase. **Target(s):** anthranilate phosphoribosyltransferase (1); anthranilate synthase (2,3); indoleamine 2,3-dioxygenase, also as an alternative substrate (4,5); tryptophan 2,3-dioxygenase, weakly inhibited (1); tryptophan 5-monooxygenase (6-10); tryptophanyl-tRNA synthetase, as an alternative substrate (11,12). **1.** Webb (1966) *Enzyme and Metabolic Inhibitors*, vol. **2**, Academic Press, New York. **2.** Matsukawa, Ishihara & Iwamura (2002) *Z. Naturforsch. [C]* **57**, 121. **3.** Widholm (1972) *Biochim. Biophys. Acta* **279**, 48. **4.** Southan, Truscott, Jamie, *et al.* (1996) *Med. Chem. Res.* **6**, 343. **5.** Macchiarulo, Camaioni, Nuti & Pellicciari (2009) *Amino Acids* **37**, 219. **6.** Fujisawa & Nakata (1987) *Meth. Enzymol.* **142**, 93. **7.** Nicholson & Wright (1981) *Neuropharmacology* **20**, 335. **8.** Sugden & Fletcher (1981) *Psychopharmacology* **74**, 369. **9.** Nakata & Fujisawa (1982) *Eur. J. Biochem.* **124**, 595. **10.** Nukiwa, Toyama, Okita, Kataoka & Ichiyama (1974) *Biochem. Biophys. Res. Commun.* **60**, 1029. **11.** Xu, Love, Ma, Blum, *et al.* (1989) *J. Biol. Chem.* **264**, 4304. **12.** Favorova & Lavrik (1975) *Biokhimiia* **40**, 368.

5-Fluorouracil

This antineoplastic agent (FW = 130.08 g/mol; CAS 51-21-8; Decomposes at 282-283°C; λ_{max} = 265-266 nm; ε = 7070 $M^{-1}cm^{-1}$; Slightly soluble in H_2O; more so in acidic solutions), is also known as FU, and 5FU, is marketed under the tradename Efudex. **Metabolic Activation:** R5-Fluorouracil is readily absorbed by humans, whereupon it is converted to three main active metabolites: fluorodeoxyuridine monophosphate (FdUMP), fluorodeoxy-uridine triphosphate (FdUTP), and fluorouridine triphosphate (FUTP). The main mechanism of drug activation is by conversion of 5FU to fluorouridine monophosphate (FUMP), either directly by orotate phosphoribosyltransferase (OPRT), using 5′-phosphoribosylpyrophosphate (PRPP) as the cofactor, or indirectly via fluorouridine (FUR), a pathway requiring the sequential action of uridine phosphorylase and uridine kinase. FUMP is then phosphorylated to fluorouridine diphosphate (FUDP), which can be either further phosphorylated to the active metabolite fluorouridine triphosphate (FUTP), or converted to fluorodeoxyuridine diphosphate (FdUDP) by the action of ribonucleotide reductase (RNR). FdUDP is then be phosphorylated or dephosphorylated, generate the active metabolites FdUTP and FdUMP, respectively. Alternatively, FU is mobilized by thymidine phosphorylase, converting 5-FU to FUDR, with the latter then converted to FdUMP by thymidine kinase. FdUDP is then phosphorylated to FdUTP, a substrate for DNA polymerase. DNA(uracil) deglycosylase recognizes and removes fluorouracil, relying on subsequent action of repair enzymes to inser5 a thymine-containing monomer. At high (F)dUTP/dTTP ratios, the action of(uracil) deglycosylase only results in further incorporation of the false nucleotide. Such futile cycling (i.e., misincorporation, excision and repair) eventually results in DNA strand breaks and cell death. DNA damage due to dUTP misincorporation is highly dependent on the levels of the dUTPase, a pyrophosphatase that limits intracellular accumulation of dUTP. 5-FU is also incorporated into mRNA, resulting in translation errors. It is as FdUMP, however, that FU has its most potently killing effect. Dihydropyrimidine dehydrogenase converts 5-FU to dihydrofluorouracil (DHFU), the rate-limiting step of 5-FU catabolism in normal and tumor cells. Up to 80% of administered 5-FU is broken down by DPD in the liver. **Mechanism of Thymidylate Synthase Inhibition:** Thymidylate synthase normally catalyzes the conversion of dUMP to dTMP, with 5,10-methylene tetrahydrofolate (CH_2THF) serving as the methyl donor. UC San Francisco researcher Daniel Santi and coworkers demonstrated, however, that, in the presence of enzyme-bound $N^{5,10}$-methylenetetrahydrofolate, thymidylate synthase binds FdUMP, with the latter acting as a mechanism-based (or "suicide") inhibitor that recapitulates early steps in catalysis (*e.g.*, attack by the active-site thiol of Cys-198 at C^6 of FdUMP).

Suicide Inhibition

Failing to expel fluoride, expulsion of the C^5 proton of dUMP cannot proceed, and, by mimicking the electronic charge and geometry of an early reaction-cycle transition state, the highly stable covalent Enz-FdUMP adduct persists. FdUMP inhibits thymidylate synthase, preventing the latter from supplying replicating cells with dTMP, and hence dTTP, needed for DNA biosynthesis. Given its consideravble cytotoxicity, 5-fluorouracil is often administered as a pro-drug. Patients deficient in dihydropyrimidine dehydrogenase (DPD) experience profound systemic toxicity in response to 5-FU, owing to prolonged exposure to 5-FU from decreased drug catabolism. (**See** *Capecitabine*) **Key Pharmacokinetic Parameters:** *See* Appendix II in Goodman & Gilman's *THE PHARMACOLOGICAL BASIS OF THERAPEUTICS*, 12th Edition (Brunton, Chabner & Knollmann, eds.) McGraw-Hill Medical, New York (2011). **Target(s):** 4-aminobutyrate aminotransferase, K_i = 1.9 mM (1,2); (*R*)-3-amino-2-methylpropionate:pyruvate aminotransferase (1-4); (*S*)-3-amino-2-methylpropionate:pyruvate transaminase (3); orotate phosphoribosyltransferase, K_i = 0.27 mM (5,6); thymidine phosphorylase (7,8); thymidylate synthase, slow, tight-binding inhibition, K_i = 2 nM (9-13); thymine dioxygenase (14); uracil phosphoribosyltransferase (15); uridine phosphorylase (16,17). **1.** Kaneko, Kontani, Kikugawa & Tamaki (1992) *Biochim. Biophys. Acta* **1122**, 45. **2.** Tamaki, Kubo, Aoyama & Funatsuka (1983) *J. Biochem.* **93**, 955. **3.** Tamaki, Fujimoto, Sakata & Matsuda (2000) *Meth. Enzymol.* **324**, 376. **4.** Kontani, Kaneko, Kikugawa, Fujimoto & Tamaki (1993) *Biochim. Biophys. Acta* **1156**, 161. **5.** Javaid, el Kouni & Iltzsch (1999) *Biochem. Pharmacol.* **58**, 1457. **6.** Krungkrai, Aoki, Palacpac, *et al.* (2004) *Mol. Biochem. Parasitol.* **134**, 245. **7.** Baker (1967) *Design of Active-Site-Directed Irreversible Enzyme Inhibitors*, Wiley, New York. **8.** Miszczak-Zaborska & Wozniak (1997) *Z. Naturforsch. C* **52**, 670. **9.** Sigman & Mooser (1975) *Ann. Rev. Biochem.* **44**, 889. **10.** Heidelberger (1970) *Cancer Res.* **30**, 1549. **11.** Chalabi & Gutteridge (1977) *Biochim. Biophys. Acta* **481**, 71. **12.** Bijnsdorp, Comijn, Padron, Gmeiner & Peters (2007) *Oncol. Rep.* **18**, 287. **13.** Carpenter (1974) *J. Insect Physiol.* **20**, 1389. **14.** Holme & Lindstedt (1982) *Biochim. Biophys. Acta* **704**, 278. **15.** Carter, Donald, Roos & Ullman (1997) *Mol. Biochem. Parasitol.* **87**, 137. **16.** Jimenez, Kranz, Lee, Gero & O'Sullivan (1989) *Biochem. Pharmacol.* **38**, 3785. **17.** Baker & Kelley (1970) *J. Med. Chem.* **13**, 461.

5-Fluorouridine

This halogenated nucleoside and antitumor agent (FW = 266.18 g/mol; CAS = 316-46-1) is converted to the corresponding deoxynucleotide, the latter a potent inhibitor of thymidylate synthase. DNA biosynthesis is inhibited. In addition, 5-Fluoro-UTP can be incorporated into RNA, resulting in translation errors and in the inhibition of mRNA biosynthesis and rRNA processing (1-3). **Target(s):** (*R*)-3-amino-2-methylpropionate:pyruvate aminotransferase (weakly inhibited) (4); cytidine deaminase (5,6); dihydroorotase (7); RNA biosynthesis (1-3,8); thymidylate synthase (8); uridine nucleosidase (10). **1.** Wilkinson & Pitot (1973) *J. Biol. Chem.* **248**, 63. **2.** Wilkinson, Tlsty & Hanas (1975) *Cancer Res.* **35**, 3014. **3.** Lonn (1978) *Biochim. Biophys. Acta* **517**, 265. **4.** Kaneko, Kontani, Kikugawa & Tamaki (1992) *Biochim. Biophys. Acta* **1122**, 45. **5.** Cacciamani, Vita, Cristalli, *et al.* (1991) *Arch. Biochem. Biophys.* **290**, 285. **6.** Vincenzetti, Cambi, Neuhard, *et al.* (1999) *Protein Expr. Purif.* **15**, 8. **7.** Bresnick & Blatchford (1964) *Biochim. Biophys. Acta* **81**, 150. **8.** Wilkinson & Crumley (1976) *Cancer Res.* **36**, 4032. **9.** Danenberg (1977) *Biochim. Biophys. Acta* **473**, 73. **10.** Kurtz, Exinger, Erbs & Jund (2002) *Curr. Genet.* **41**, 132.

Fluoxetine

This widely used antidepressant (FW$_{free-base}$ = 309.33 g/mol; CAS = 54910-89-3), also named N-methyl-3-phenyl-3-[4-(trifluoromethyl)phenoxy] propan-1-amine and commonly known as Prozac™, is a selective serotonin reuptake inhibitor (SSRI) and was one of the earliest such inhibitors identified (1-4). **Primary Mode of Action:** The primary pharmacological action of SSRIs is presynaptic inhibition of serotonin (5-HT) reuptake, thereby increasing the concentration of this amine in the synaptic cleft. Serotonin exerts its effects through interactions with any of seven distinct 5-HT receptor families, and interaction between serotonin and post-synaptic receptors mediates a wide range of its actions. Fluoxetine is also an antagonist at the serotonin 5HT$_{2C}$ receptor. This drug also inhibits voltage-activated potassium and sodium channels (5). Indeed, when administered at clinically relevant concentrations, fluoxetine blocks *Shaker*-type K1 channel, *i.e.,* K$_V$1.3 (6). **Antiviral Properties:** Surprisingly, recent studies demonstrated the potent inhibition of coxsackievirus replication b fluoxetine (4). While it did not interfere with either viral entry, fluoxetine and its metabolite norfluoxetine markedly reduced the synthesis of viral RNA and protein. Fluoxetine's pharmacokinetics and safety commend it as a potential antiviral agent for enterovirus infections. **Key Pharmacokinetic Parameters:** *See* Appendix II in Goodman & Gilman's *THE PHARMACOLOGICAL BASIS OF THERAPEUTICS*, 12th Edition (Brunton, Chabner & Knollmann, eds.) McGraw-Hill Medical, New York (2011). **Target(s):** 5HT$_{2C}$ receptors (1); serotonin uptake (1-3); steroid 21-monooxygenase (7). **1.** Ni & Miledi (1997) *Proc. Natl. Acad. Sci. U.S.A.* **94**, 2036. **2.** Wong, Bymaster & Engleman (1995) *Life Sci.* **57**, 411. **3.** Wong, Horng, Bymaster, Hauser & Molloy (1974) *Life Sci.* **15**, 471. **4.** Zuo, Quinn, Kye, *et al.* (2012) *Antimicrob Agents Chemother.* **56**, 4838. **5.** Rae, Rich, Zamudio & Candia (1995) *Am. J. Physiol* **269**, C250. **6.** Choi, Hahn, Rhee, *et al.* (1999) *J. Pharmacol. Exp. Ther.* **291**, 2. **7.** Kishimoto, Hiroi, Sharaishi, *et al.* (2004) *Endocrinology* **145**, 699.

Flupenthixol

α-flupenthixol   β-flupenthixol

This anti-psychotic drug and neuroleptic agent (FW = 434.53 g/mol; CAS = 2709-56-0), also called flupenthixol and LC-44, is a potent inhibitor of dopamine-stimulated adenylate cyclase as well as a dopamine receptor antagonist (1,2). Note that there are two distinct forms of this pharmaceutical: an α- (or *cis*-) isomer (*i.e.*, the Z-isomer) and a β- (or *trans*-) form. α-Flupenthixol is the more active form with a nanomolar K_i value. Both isomers inhibit drug transport and reverse drug resistance mediated by the human multidrug transporter P-glycoprotein (3). α-Flupenthixol reportedly reverses the substrate inhibition observed with brain tyrosine monooxygenase (4) **Target(s):** dopamine receptor (1,2); dopamine-stimulated adenylate cyclase (1); HIV-1 replication (5); multidrug transporter (3). **1.** Iversen (1975) *Science* **188**, 1084. **2.** Post, Kennard & Horn (1975) *Nature* **256**, 342. **3.** Dey, Hafkemeyer, Pastan & Gottesman (1999) *Biochemistry* **38**, 6630. **4.** Mineeva, Kudrin, Kuznetsova & Raevskii (1982) *Biull. Eksp. Biol. Med.* **94**, 58. **5.** Kristiansen & Hansen (2000) *Int. J. Antimicrob. Agents* **14**, 209.

Flupirtine

This centrally acting, non-opioid, non-NSAID, non-steroidal analgesic (FW$_{free-base}$ = 304.32 g/mol; CAS = 56995-20-1; FW$_{maleate-salt}$ = 420.39 g/mol; CAS = 75507-68-5; Solubility (maleate salt): 84 mg/mL DMSO at 25 °C; <1 mg/mL H$_2$O at 25 °C), also known by the trade names Katadolon®, Awegal®, Efiret®, and Metanor® and its IUPAC name, ethyl {2-amino-6-[(4-fluorobenzyl)amino]pyridin-3-yl}carbamate, selectively targets G-protein-activated inwardly rectifying K$^+$ channels (GIRK), the opening of which leads to a stabilization of the resting membrane potential of neuronal cells and thus causes an indirect inhibition of the glutamatergic N-methyl-D-

aspartate (NMDA) receptor (1). Activation of GIRK channels leads to hyperpolarization of neuronal membrane, making the neuron becomes less excitable by stabilizing resting neuronal membrane. A drug that activates channels in this manner is called a $\underline{S}$elective $\underline{N}$euronal $\underline{P}$otassium $\underline{C}$hannel $\underline{O}$pener (SNEPCO). Flupirtine was the first of this new type of analgesic, muscle-relaxant and neuroprotective agent. That it exerts cerebral or spinal analgesia was demonstrated in rats by the finding that intracerebroventricular and intrathecal administration of showed dose-dependent analgesia, but not when applied systemically (3). Changes in prostaglandin synthesis are of minor importance for flupirtine's analgesic action *in vivo* (2). See (4) for a fuller discussion of the clinical pharmacology of flupirtine's properties as an analgesic and muscle relaxant. **1.** Kornhuber, Maler, Wiltfang J, *et al.* (1999) *Fortschr. Neurol. Psychiatr.* **67**, 466. **2.** Nickel, Herz, Jakovlev & Tibes (1985) *Arzneimittelforschung.* **35**, 1402. **3.** Darius & Schrör (1985) *Arzneimittelforschung.* **35**, 55. **4.** Harish, Bhuvana, Bengalorkar & Kumar (2012) *J. Anesthesiol. Clin. Pharmacol.* **28**, 172.

Fluphenazine

This antipsychotic drug (FW$_{free-base}$ = 437.53 g/mol) is a potent inhibitor of dopamine stimulated adenylate cyclase and is a D$_2$ (also D$_1$; albeit weaker) receptor antagonist. **Target(s):** adenylate cyclase (1,2); ATPase (3); Ca^{2+} channel, Type 1, inositol 1,4,5 trisphosphate-sensitive (4); Ca^{2+}-exporting ATPase (5); CYP1A2 (6); CYP2D6 (6); dopamine receptor, preference for D$_2$ (7); glutathione-disulfide reductase (18); hydroxyindole O-methyltransferase (8); K$^+$ channel (9); 5-lipoxygenase, *or* arachidonate 5-lipoxygenase; ID$_{50}$ = 21.6 μM (10); 12-lipoxygenase, *or* arachidonate 12-lipoxygenase, ID$_{50}$ = 27.5 μM (10); Na$^+$/K$^+$-exchanging ATPase (11); P glycoprotein (12); PI-specific phospholipase C (1)3; phosphodiesterase (14); prostaglandin synthase, ID$_{50}$ = 283 μM (10); protein kinase C (15); [pyruvate dehydrogenase (acetyl-transferring)]-phosphatase (16); tyrosyl sulfotransferase (17). **Key Pharmacokinetic Parameters:** *See* Appendix II in Goodman & Gilman's *THE PHARMACOLOGICAL BASIS OF THERAPEUTICS*, 12th Edition (Brunton, Chabner & Knollmann, eds.) McGraw-Hill Medical, New York (2011). **1.** Iversen (1975) *Science* **188**, 1084. **2.** Carricarte, Bianchini, Muschietti, *et al.* (1988) *Biochem. J.* **249**, 807. **3.** Palatini (1982) *Mol. Pharmacol.* **21**, 415. **4.** Khan, Dyer & Michelangeli (2001) *Cell Signal.* **13**, 57. **5.** de Meis (1991) *J. Biol. Chem.* **266**, 5736. **6.** Shin, Soukhova & Flockhart (1999) *Drug Metab. Dispos.* **27**, 1078. **7.** Christensen, Arnt & Svendsen (1985) *Psychopharmacol. Suppl.* **2**, 182. **8.** Hartley, Padwick & Smith (1972) *J. Pharm. Pharmacol.* **24**, Suppl, 100P. **9.** Muller, De Weille & Lazdunski (1991) *Eur. J. Pharmacol.* **198**, 101. **10.** Hamasaki & Tai (1985) *Biochim. Biophys. Acta* **834**, 37. **11.** Davis & Brody (1966) *Biochem. Pharmacol.* **15**, 703. **12.** Wang, Lew, Barecki, *et al.* (2001) *Chem. Res. Toxicol.* **14**, 1596. **13.** Benedikter, Knopki & Renz (1985) *Z. Naturforsch. [C]* **40**, 68. **14.** Janiec, Pytlik & Piekarska (1980) *Pol. J. Pharmacol. Pharm.* **32**, 297. **15.** Kikkawa & Nishizuka (1986) *The Enzymes*, 3rd ed., **17**, 167. **16.** Bak, Huh, Hong & Song (1999) *Biochem. Mol. Biol. Int.* **47**, 1029. **17.** Vargas, Tuong & Schwartz (1986) *J. Enzyme Inhib.* **1**, 105. **18.** Iribarne, Paulino, Aguilera & Tapia (2009) *J. Mol. Graph. Model.* **28**, 371. **19.** Chan, Yin, Garforth, *et al.* (1998) *J. Med. Chem.* **41**, 148.

Fluphenazine-*N*-mustard

This fluphenazine derivative (FW$_{free-base}$ = 455.97 g/mol; CAS = 83016-35-7; soluble in ethanol and water; forms fluphenazine by hydrolysis ($t_{1/2}$ = 24 hours) to), also known as fluphenazine-N-2-chloroethane, is an irreversible alkylator of calmodulin that prevents calmodulin-mediated activation of cyclic-nucleotide phosphodiesterase-1. **1.** Hait, Glazer, Kaiser, Cross & Kennedy (1987) *Mol. Pharmacol.* **32**, 404.

Flurbiprofen

This anti-inflammatory agent (FW$_{free-acid}$ = 244.27 g/mol; CAS = 5104-49-4) is a potent inhibitor of cyclooxygenase-1, or COX-1 (IC$_{50}$ = 0.8 nM). Commercial flurbiprofen is often a mixture of both enantiomers. **Target(s):** fatty acid amidohydrolase (1); prostaglandin-endoperoxide synthase (cyclooxygenase) (2-8); thiopurine S-methyltransferase (9). **1.** Fowler, Holt & Tiger (2003) *J. Enzyme Inhib. Med. Chem.* **18**, 55. **2.** Pace-Asciak & Smith (1983) *The Enzymes*, 3rd ed., **16**, 543. **3.** Range, Pang, Holland & Knox (2000) *Eur. Respir. J.* **15**, 751. **4.** Crook & Collins (1975) *Prostaglandins* **9**, 857. **5.** Rome & Lands (1975) *Prostaglandins* **10**, 813. **6.** Meade, Smith & DeWitt (1993) *J. Biol. Chem.* **268**, 6610. **7.** Barnett, Chow, Ives, *et al.* (1994) *Biochim. Biophys. Acta* **1209**, 130. **8.** Patrignani, Panara, Sciulli, *et al.* (1997) *J. Physiol. Pharmacol.* **48**, 623. **9.** Oselin & Anier (2007) *Drug Metab. Dispos.* **35**, 1452.

Fluticasone

Fluticasone   **Fluticasone Propionate**

This anti-inflammatory drug (FW = 444. 52 g/mol; CAS 90566-53-3; FW$_{propionate}$ = 500.57 g/mol; CAS = 80474-14-2 (propionate)), also named *S*-Fluoromethyl (6*S*,8*S*,9*R*,10*S*,11*S*,13*S*,14*S*, 16*R*,17*R*)-6,9-difluoro-11,17-dihydroxy-10,13,16-trimethyl-3-oxo-6,7,8,11,12,14,15,16-octahydrocyclopenta[*a*]phenanthrene-17-carbothioate, is a synthetic glucocorticoid tat optimizes therapeutic efficacy trough its lipophilicity and high one-pass metabolism, increasing lung retention and minimizing its systemic burden. Fluticasone is administered topically or inhaled in various formulations, often using its propionate ester. Its long retention time in the lung contributes to delayed absorption and metabolism, thereby accounting for its prolonged efficacy (1). Prolonged treatment with this topical corticosteroid inhibits allergen-induced early and late nasal responses and the associated tissue eosinophilia by inhibiting cells expressing IL-4 mRNA (2). *In vitro*, fluticasone propionate (PA) potently inhibits T-lymphocyte proliferation, cytokine generation, TNFα-induced adhesion molecule expression, interleukin-5-induced eosinophilia, mucosal edema and toluene 2,4-diisocyanate-induced mast cell proliferation, while promoting secretory leucocyte protease inhibitor production and eosinophil apoptosis (3). In human studies, FP demonstrates marked vasoconstrictor potency in normal subjects and inhibits antigen-induced mucosal platelet activating factor/eicosanoid production, T-lymphocytes and CD25$^+$ cells in patients with rhinitis (3). In mild asthmatics, FP reduces CD3$^+$, CD4$^+$, CD8$^+$ and CD25$^+$ cells, with an accompanying reduction in eosinophil and mast cell markers (3). **1.** Burton, Shaw, Schentag & Evans (2006) *Applied Pharmacokinetics & Pharmacodynamics: Principles of Therapeutic Drug Monitoring*, p. 266, Lippincott, Williams & Wilkins, Baltimore. **2.** Masuyama, Jacobson, Rak, *et al.* (1994) *Immunol.* **82**, 192. **3.** Fuller, Johnson & Bye (1995) Respir Med. **89** (Suppl. A) 3.

Fluvastatin

This statin (FW$_{free-acid}$ = 411.47 g/mol; CAS = 93957-54-1), also known by the trade names Lescol™, Canef™, and Vastin™ as well as its systematic name (3*R*,5*S*,6*E*)-7-[3-(4-fluorophenyl)-1-(propan-2-yl)-1*H*-indol-2-yl]-3,5-

dihydroxyhept-6-enoic acid, was the first wholly synthetic 3-hydroxy-3-methylglutaryl-CoA reductase inhibitor (IC$_{50}$ = 28 nM) to reach the market. Fluvastatin is effective in reducing both total and LDL cholesterol. Like other statins, fluvastatin is a substrate for Polypeptide 1B1, the organic anion uptake transporter expressed in hepatocytes. Intracellular drug concentrations are also determined by the efflux transporter Multidrug Resistance-associated Protein-2 (designated MRP2 in humans and Mrp2 in rat), which transport of numerous endogenous metabolites (*e.g.*, taurocholate) and xenobiotics (*e.g.*, methotrexate). **Target(s):** induction of inducible nitric oxide synthase, an inflammatory biomarker, in hepatocytes (4); FLT3 glycosylation in human and murine cells (5); high glucose-induced nuclear factor kappa B activation in renal tubular epithelial cells (6); carbonic anhydrase, fifteen human isoforms, hCA I-XIV (7); suppresses geranylgeranylation of small G-proteins of the Rho family, most likely by reducing mevalonate-derived intracellular stores of geranylgeranyl-pyrophosphate (8); mast cell degranulation without changing the cytoplasmic Ca^{2+} level (9 deposition of advanced glycation end products in aortic wall of cholesterol and fructose-fed rabbits (10); hepatitis C replication in humans, albeit with variable results (11). **1.** Istvan & Deisenhofer (2001) *Science* **292**, 1160. **2.** Parker, Clark, Sit, *et al.* (1990) *J. Lipid Res.* **31**, 1271. **3.** Corsini (2000) *J. Cardiovasc. Pharmacol. Ther.* **5**, 161. **4.** Tokuhara, Habara, Oishi, *et al.* (2012) *Hepatol Res.* **43**, 775. **5.** Williams, Li, Nguyen, *et al.* (2012) *Blood* **120**, 3069. **6.** Gao, Wu, Shui, *et al.* (2013) *J. Nephrol.* **26**, 289. **7.** Parkkila, Vullo, Maresca, *et al.* (2012) *Chem. Commun.* (Cambridge). **48**, 3551. **8.** Dunoyer-Geindre, Fish & Kruithof (2011) *Thromb. Haemost.* **105**, 461. **9.** Fujimoto, Oka, Murata, Hori & Ozaki (2009) *Eur. J. Pharmacol.* **602**, 432. **10.** Akira, Amano, Okajima, Hashimoto & Oikawa (2006) *Biol. Pharm. Bull.* **29**, 75. Erratum in: *Biol. Pharm. Bull.* (2007) **30**, 2412. **11.** Bader, Fazili, Madhoun, *et al.* (2008) *Amer. J. Gastroenterol.* **103**, 1383

FMN

This flavin coenzyme (FW = 514.36 g/mol; CAS = 6184-17-4), also known as riboflavin 5'-phosphate, is the yellow/orange coenzyme and/or prosthetic group for many redox enzymes (*i.e.*, flavoenzymes). The phosphate ester is hydrolyzed in acidic conditions, and the dimethylisoalloxazine ribitol bond is unstable in alkaline conditions: FMN is most stable at pH 6. Ammonium salts of FMN are very unstable. **Redox and Spectral Properties:** $E°_o$ = – 0.219 V at pH 7 and 30°C. FMN has λ$_{max}$ values at 266, 373, and 445 nm, with corresponding ε values of 31800, 10400, and 12500 M^{-1}cm^{-1}) in 0.1 M phosphate buffer at pH 7. The reduced form FMNH$_2$ has a molar extinction coefficient of only 870 M^{-1}cm^{-1} at 450 nm, with a fluorescence maximum at 525-530 nm. FMN is reduced to FMNH$_2$ in the presence of dithionite or by H$_2$ in the presence of a suitable catalyst. It is photolabile and solutions must be stored in the dark. **Target(s):** acetyl-CoA synthetase, acetate:CoA ligase (1); acyl-[acyl-carrier-protein] desaturase (29,30); AMP nucleosidase (4); γ-butyrobetaine dioxygenase (32); carbonyl reductase (5); cholest-5-ene-3β,7α-diol 3β-dehydrogenase (6); *trans*-cinnamate 4-monooxygenase (31); cytochrome b_5 reductase (7); deoxyhypusine synthase (8); ferredoxin:NADP$^+$ reductase (9); glutamate racemase (10-12); glutathione-disulfide reductase13; glycogen phosphorylase *b* (14); human cytomegalovirus UL80 protease (15); 2-hydroxypyridine 5-monooxygenase (16); *myo* inositol oxygenase (17,33); intron, group I (18); L-lactate oxidase (19); NADH:cytochrome *c* reductase, or NADH dehydrogenase (20); NAD(P)H dehydrogenase, or DT diaphorase, weakly inhibited (21-23); palmitoyl-CoA hydrolase, or acyl-CoA hydrolase (24); prostaglandin E$_2$ 9-reductase (25); riboflavin transglucosidase (26); succinate oxidase (27); sulfur dioxygenase (34); very-long-chain fatty acid α-hydroxylase (28). **1.** Preston, Wall & Emerich (1990) *Biochem. J.* **267**, 179. **2.** Walaas & Walaas (1956) *Acta Chem.*

Scand. **10**, 122. **3**. Crandall (1959) *Fed. Proc.* **18**, 208. **4**. Yoshino, Murakami & Tsushima (1976) *J. Biochem.* **80**, 839. **5**. Hara, Deyashiki, Nakagawa, Nakayama & Sawada (1982) *J. Biochem.* **92**, 1753. **6**. Wikvall (1981) *J. Biol. Chem.* **256**, 3376. **7**. Strittmatter (1959) *J. Biol. Chem.* **234**, 2665. **8**. Abid, Sasaki, Titani & Miyazaki (1997) *J. Biochem.* **121**, 769. **9**. San Pietro (1963) *Meth. Enzymol.* **6**, 439. **10**. Adams (1972) *The Enzymes*, 3rd ed., **6**, 479. **11**. Tanaka, Kato & S. Kinoshita (1961) *Biochem. Biophys. Res. Commun.* **4**, 114. **12**. Diven (1969) *Biochim. Biophys. Acta* **191**, 702. **13**. Moroff & Kosow (1978) *Biochim. Biophys. Acta* **527**, 327. **14**. Klinov, Kurganov, Pekel & Berezovskii (1986) *Biochem. Int.* **13**, 139. **15**. Baum, Ding, Siegel, *et al.* (1996) *Biochemistry* **35**, 5847. **16**. Sharma, Kaul & Shukla (1984) *Biol. Mem.* **9**, 43. **17**. Charalampous (1962) *Meth. Enzymol.* **5**, 329. **18**. Kim & Park (2000) *Biochim. Biophys. Acta* **1475**, 61. **19**. Cousins (1956) *Biochem. J.* **64**, 297. **20**. Vernon, H. R. Mahler & N. K. Sarkar (1952) *J. Biol. Chem.* **199**, 599. **21**. Ernster (1967) *Meth. Enzymol.* **10**, 309. **22**. Ernster, Ljunggren & Danielson (1960) *Biochem. Biophys. Res. Commun.* **2**, 48 and 88. **23**. Raw, Nogueira & Filho (1961) *Enzymologia* **23**, 123. **24**. Berge & Dossland (1979) *Biochem. J.* **181**, 119. **25**. Pace-Asciak & Smith (1983) *The Enzymes*, 3rd ed., **16**, 543. **26**. Tachibana, Katagiri & Yamada (1958) *Proc. Intern. Symp. Enzyme Chem., Tokyo Kyoto 1957*, p. 154, Maruzen, Tokyo. **27**. Eichel (1956) *Biochim. Biophys. Acta* **22**, 571. **28**. Kishimoto (1978) *Meth. Enzymol.* **52**, 310. **29**. Jaworski & Stumpf (1974) *Arch. Biochem. Biophys.* **162**, 158. **30**. Nagai & Bloch (1968) *J. Biol. Chem.* **243**, 4626. **31**. Billett & Smith (1978) *Phytochemistry* **17**, 1511. **32**. Lindstedt (1967) *Biochemistry* **6**, 1271. **33**. Charalampous (1959) *J. Biol. Chem.* **234**, 220. **34**. Suzuki & Silver (1966) *Biochim. Biophys. Acta* **122**, 22.

Folate (Folic Acid)

This photosensitive vitamin (FW = 441.40 g/mol; CAS = 59-30-3; Solubility: 0.0015 mg/mL H_2O at 25°C), also known as pteroylglutamate, is a precursor to the tetrahydrofolate coenzymes. Note also that the terms "folate" and "folic acid" are often used generically in reference to other pteroyl conjugates, having more than one glutamyl residue: *i.e.*, typically two to seven additional γ-glutamyl residues, linked together via isopeptide bonds). Folic acid has λ_{max} values of 282 and 350 nm at pH 7.0 (ε = 27000 and 7000 $M^{-1}cm^{-1}$, respectively). **Target(s):** adenosine deaminase (1); alcohol dehydrogenase (2); arylamine *N* acetyltransferase (3,4); dihydrofolate reductase, K_i = 0.11 μM (5,6); dihydropteroate synthase (7); formaldehyde dehydrogenase, rat (8); formimidoyl-tetrahydrofolate cyclodeaminase (9,10); *S*-formylglutathione hydrolase (11); glucose-6-P dehydrogenase (12); glutamate carboxypeptidase II (13); glutamate dehydrogenase (12,14); glutamate formimidoyltransferase (15-17); γ-glutamyl hydrolase, *or* pteroylpolyglutamate hydrolase (18); glycogen phosphorylase *b* (19); GTP cyclohydrolase I, weakly inhibited (20); *S*-(hydroxymethyl)glutathione dehydrogenase (8); N^5-lactate dehydrogenase (12); malate dehydrogenase (12); methenyl-tetrahydrofolate cyclohydrolase (21,22); methyltetrahydrofolate: homocysteine methyltransferase (23); N^5,N^{10}-methenyltetrahydro-folate synthetase (24); NAD(P)H dehydrogenase (25); NADPH dehydrogenase (quinone) (25); ribonuclease A (26,27); phenylpyruvate tautomerase (28); purine-nucleoside phosphorylase (29,30); thymidylate synthase (31-34); tRNA-guanine transglycosylase (35); xanthine oxidase (36-39**). 1**. Harbison & Fisher (1973) *Arch. Biochem. Biophys.* **154**, 84. **2**. Brändén, Jörnvall, Eklund & Furugren (1975) *The Enzymes*, 3rd ed., **11**, 103. **3**. Ward, Summers & Sim (1995) *Biochem. Pharmacol.* **49**, 1759. **4**. Wang, Vath, Kawamura, *et al.* (2005) *Protein J.* **24**, 65. **5**. Baker (1967) *Design of Active-Site-Directed Irreversible Enzyme Inhibitors*, Wiley, New York. **6**. Bertino, Perkins & Johns (1965) *Biochemistry* **4**, 839. **7**. Vinnicombe & Derrick (1999) *Biochem. Biophys. Res. Commun.* **258**, 752. **8**. Goodman & Tephly (1971) *Biochim. Biophys. Acta* **252**, 489. **9**. MacKenzie, Aldridge & Paquin (1980) *J. Biol. Chem.* **255**, 9474. **10**. Drury & MacKenzie (1977) *Can. J. Biochem.* **55**, 919. **11**. Uotila & Koivusalo (1974) *J. Biol. Chem.* **249**, 7664. **12**. Vogel, Snyder & Schulman (1963) *Biochem. Biophys. Res. Commun.* **10**, 97. **13**. Bzdega, Turi, Wroblewska, *et al.* (1997) *J. Neurochem.* **69**, 2270. **14**. White,

Yielding & Krumdieck (1976) *Biochim. Biophys. Acta* **429**, 689. **15**. MacKenzie (1980) *Meth. Enzymol.* **66**, 626. **16**. Miller & Waelsch (1957) *J. Biol. Chem.* **228**, 397. **17**. Beaudet & Mackenzie (1975) *Biochim. Biophys. Acta* **410**, 252. **18**. Chandler, Wang & Halsted (1986) *J. Biol. Chem.* **261**, 928. **19**. Klinov, Chebotareva, Sheiman, Birinberg & Kurganov (1987) *Bioorg. Khim.* **13**, 908. **20**. Shen, Alam & Zhang (1988) *Biochim. Biophys. Acta* **965**, 9. **21**. Kirk, Chen, Imeson & Cossins (1995) *Phytochemistry* **39**, 1309. **22**. Cohen & MacKenzie (1979) *Biochim. Biophys. Acta* **522**, 311. **23**. Whitfield, Steers & Weissbach (1970) *J. Biol. Chem.* **245**, 390. **24**. Jolivet (1997) *Meth. Enzymol.* **281**, 162. **25**. Koli, Yearby, Scott & Donaldson (1969) *J. Biol. Chem.* **244**, 621. **26**. Scheraga & Rupley (1962) *Adv. Enzymol.* **24**, 161. **27**. Blackburn & Moore (1982) *The Enzymes*, 3rd ed., **15**, 317. **28**. Molnar & Garai (2005) *Int. Immunopharmacol.* **5**, 849. **29**. Lewis (1978) *Arch. Biochem. Biophys.* **190**, 662. **30**. Barsacchi, Cappiello, Tozzi, *et al.* (1992) *Biochim. Biophys. Acta* **1160**, 163. **31**. Danenberg (1977) *Biochim. Biophys. Acta* **473**, 73. **32**. Kisliuk, Gaumont & Baugh (1974) *J. Biol. Chem.* **249**, 4100. **33**. Dolnick & Cheng (1978) *J. Biol. Chem.* **253**, 3563. **34**. Radparvar, Houghton & Houghton (1988) *Arch. Biochem. Biophys.* **260**, 342. **35**. Farkas, Jacobson & Katze (1984) *Biochim. Biophys. Acta* **781**, 64. **36**. Nishino & Tsushima (1986) *J. Biol. Chem.* **261**, 11242. **37**. Kalckar & Klenow (1948) *J. Biol. Chem.* **172**, 34. **38**. Hofstee (1949) *J. Biol. Chem.* **179**, 633. **39**. Williams (1950) *J. Biol. Chem.* **187**, 47.

Fomitellic Acids A, B, C, and D

These triterpenoids (Fomitellic Acid A: R_1 = –H, R_2 = –OH, R_3 = –OH; FW = 502.69 g/mol; Fomitellic Acid B: R_1 = –H, R_2 = –OH, R_3 = –H; FW = 486.68 g/mol; Fomitellic Acid C: R_1 = –H, R_2 = –OH, R_3 = –OCH$_3$; FW = 500.70 g/mol; Fomitellic Acid D: R_1 = –OH, R_2 = –H, R_3 = –H; FW = 486.68 g/mol), from a basidiomycete *Fomitella fraxinea* are potent inhibitors of eukaryotic DNA polymerases. Minimum inhibitory concentrations (MIC) ranges fro 35-130 μM. **Target(s):** DNA-directed DNA polymerase (1-4); DNA polymerase α (1,2,4); DNA polymerase β (1-4); DNA polymerase II, plant (1); DNA topoisomerase I (1); DNA topoisomerase II (1). **1**. Mizushina, Iida, Ohta, Sugawara & Sakaguchi (2000) *Biochem. J.* **350**, 757. **2**. Tanaka, Kitamura, Mizushina, Sugawara & Sakaguchi (1998) *J. Nat. Prod.* **61**, 193 and 1180. **3**. Mizushina, Tanaka, Kitamura, *et al.* (1998) *Biochem. J.* **330**, 1325. **4**. Obara, Nakahata, Mizushina, *et al.* (2000) *Life Sci.* **67**, 1659.

Foretinib

This orally available, ATP-competitive multikinase inhibitor and experimental anticancer drug (FW = 632.65 g/mol; CAS = 849217-64-7; Solubility: 125 mg/mL DMSO, <1 mg/mL H_2O), variously known as EXEL-2880, XL880, GSK1363089 and $N^{1'}$-[3-fluoro-4-[[6-methoxy-7-(3-morpholinopropoxy)-4-quinolyl]oxy]phenyl]-N^1-(4-fluorophenyl)cyclo propane-1,1-dicarboxamide, targets protein tyrosine kinase activities of human growth factor receptor (HGFR) and vascular endothelial growth factor receptor-2 (VEGFR-2), mainly c-Met (IC_{50} = 0.4 nM) and KDR (IC_{50} = 0.9 nM). Binding to c-Met and KDR is characterized by a very slow off-rate, consistent with crystallographic evidence that the inhibitor is deeply bound in the active-site cleft. Foretinib prevents anchorage-independent proliferation of tumor cells under both normoxic and hypoxic conditions. *In vivo*, these effects produce significant dose-dependent inhibition of tumor burden in an experimental model of lung metastasis (1). It also inhibits growth of gastric cancer cell lines by blocking inter-receptor tyrosine kinase networks (2). Foretinib is a more potent inhibitor than crizotinib (PF-02341066) of oncogenic ROS1 fusion protein, a druggable target in malignant transformation in lung adenocarcinoma, cholangiocarcinoma, and glioblastoma (4). **1**. Qian, Engst, Yamaguchi, *et al.* (2009) *Cancer Res.* 2009, **69**, 8009. **2**. Kataoka, Mukohara, Tomioka, *et al.* (2011) *Invest. New Drugs* **30**, 1352. **3**. Davare, Saborowski, Eide, *et al.* (2013) *Proc. Natl. Acad. Sci. U.S.A.* **110**, 19519.

Formaldehyde

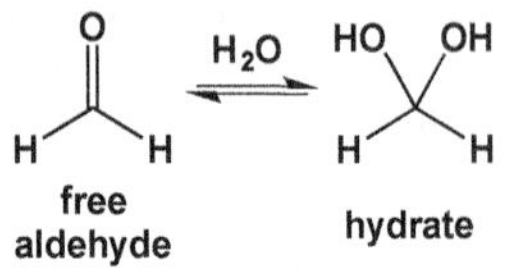

This one-carbon aldehyde (FW = 30.03 g/mol; CAS = 50-00-0; clear and colorless liquid; B.P. = –19 °C; M.P. = –92 °C; pungent odor, detectable at 1 ppm; Vapor Pressure = 10 mm Hg at –88 °C) is common industrial chemical and metabolite that is substantially hydrated to the *gem*-diol in body fluids (K_{eq} = [*gem*-diol]/[free aldehyde] > 1000 at pH 7). Upon condensation, gaseous formaldehyde converts to the cyclic trimer metaformaldehyde (*or* 1,3,5-trioxane, *or* (CH₂O)₃) and the linear polymer known as paraformaldehyde.

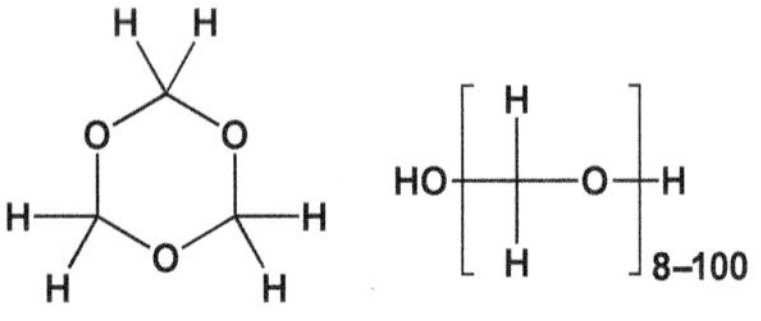

1,3,5-Trioxane   Paraformaldehyde

Formaldehyde reacts readily with ammonia, hydroxylamine, and hydrazine, as well as with alkyl- and aryl-amines. The tissue fixative and embalming agent known as formalin consists of a 37% (vol/vol) aqueous solution that usually contains 10–15% methanol. **Formol Titration:** This technique, invented by Sørensen, refers to the reaction of an amino acid with formaldehyde under alkaline conditions:

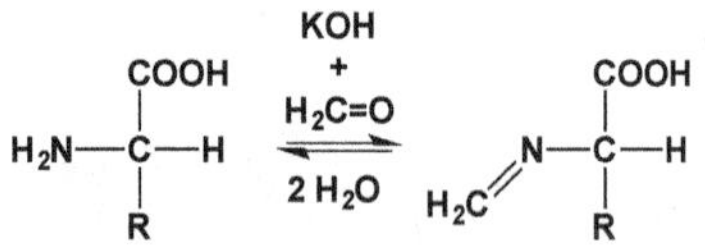

This reaction was used by French and Edsall to demonstrate the preponderance of amino acid zwitterion over neutral amino acid at neutral pH. Prior to the development of more sensitive methods, this technique was also used to determine protein concentration. **Health Risks & Carcinogenicity:** Formaldehyde is a ubiquitous environmental contaminant, with human exposures in industrial processes such as the manufacture of resins and adhesives, particle board, plywood, leather goods, paper, and pharmaceuticals. Formaldehyde is slowly released as a vapor over the lifetimes of these products. Exposure to formaldehyde also results in the formation of DNA-protein cross-links (DPCs) as a primary genotoxic effect. Detailed structural examination by NMR and mass spectrometry established that the cross-links between amino acids and single nucleosides involve formaldehyde-derived methylene bridges. Formaldehyde is also produced in the human body and is elevated in neurodegenerative conditions. Although the toxic potential of excess formaldehyde has been studied, little is known about the molecular mechanisms underlying its toxicity. The mechanisms by which formaldehyde causes cancer are not completely understood; however, formaldehyde clearly causes genetic damage in the nasal sinus of animals. Less is known about how it causes myeloid leukemia. **CAUTION:** Formaldehyde is regulated in the U.S. as a carcinogen (OSHA Standard 1910.1048) and is listed in Internation Agency for Research on Cancer as an IARC Group 2A "probable human carcinogen". This substance is classified as a "select carcinogen" under the criteria of the OSHA Laboratory Standard. It is moderately toxic by skin contact and inhalation. Exposure to gaseous formaldehyde irritates the eyes and respiratory tract, causing cough, dry throat, tightening of the chest, headache, a sensation of pressure in the head, as well as palpitations of the heart. Exposure to 0.1–5 ppm irritates eyes, nose, and throat; above 10 ppm causes severe lacrimation, burning in the nose and throat, and belabored

breathing. Acute exposure to concentrations at >25 ppm can cause serious injury, including fatal pulmonary edema. Formaldehyde has low acute toxicity via the oral route. Ingestion can cause irritation of the mouth, throat, and stomach, nausea, vomiting, convulsions, and coma. Lethal Dose oral (rat) = 500 mg/kg; Lethal Dose skin (rabbit) = 270 mg/kg; Lethal Inhaled Concentration (rat) = 203 mg/min for 2 hours. **1**. Epidemiology of chronic occupational exposure to formaldehyde: report of the Ad Hoc panel on health aspects of formaldehyde. (1988) *Toxicol Ind. Health.* **4**, 77. **2**. Goldmacher & Thilly (1983) Formaldehyde is mutagenic for cultured human cells. *Mutat. Res.* **116**, 417. **3**. Heck, Casanova-Schmitz, Dodd, *et al.* (1985) Formaldehyde (CH₂O) concentrations in the blood of humans and Fischer-344 rats exposed to CH₂O under controlled conditions. *Am Ind Hyg Assoc J.* **46**,1. **4**. Huennekens & Osborne (1959) Folic acid coenzymes and one-carbon metabolism. *Adv Enzymol.* **21**, 369. **5**. Natarajan, Darroudi, Bussman & van Kesteren-van Leeuwen (1983) Evaluation of the mutagenicity of formaldehyde in mammalian cytogenetic assays *in vivo* and *in vitro*. *Mutat. Res.* **122**, 355. **6**. Til, Woutersen, Feron, *et al.* (1989) Two-year drinking-water study of formaldehyde in rats. *Food Chem Toxicol.* **27**, 77. ***See also** entry for formaldehyde in NIOSH Pocket Guide to Chemical Hazards,* http://www.cdc.gov/niosh/npg/npgd0293.html

Formoterol

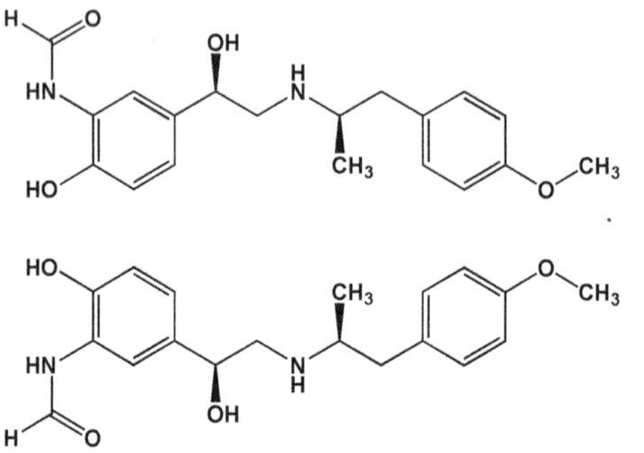

This bronchodilator (FW = 344.41 g/mol; CAS = 73573-87-2), also named *rac*-(*R,R*)-*N*-[2-hydroxy-5-[1-hydroxy-2-[1-(4-methoxyphenyl)propan-2-ylamino]ethyl]phenyl]formamide, is a long-acting β₂-adrenergic agonist that is used to manage asthma and chronic obstructive pulmonary disease, *or* COPD (1-3). Marketed under various trade names (including Foradil®, Foradile®, Oxeze®, Oxis®, Symbicort®, Atock®, Atimos®, and Perforomist®), formoterol is marketed as a dry-powder inhaler, a metered-dose inhaler, an oral tablet, and an inhalation solution. Formoterol inhibit both allergic (anti-IgE) and non-allergic histamine release (calcium ionophore and human complement component C5a) from human basophils (4). With its ability to cross the blood brain barrier, formoterol also improves cognitive function and promotes dendritic complexity in a mouse model of Down's Syndrome (5). (***See also** Salmeterol; Bambuterol*) **1**. Takeda & Takagi (1980) *Nihon Yakurigaku Zasshi* **76**, 185. **2**. Ida (1980) *Nihon Yakurigaku Zasshi* **76**, 633. **3**. Yamakido, Inamizu, Ikuta, *et al.* (1985) *Int. J. Clin. Pharmacol. Ther. Toxicol.* **23**, 461. **4**. Subramanian (1986) *Arzneimittel-forschung.* **36**, 502. **5**. Dang, Medina, Das, *et al.* (2013) *Biol. Pyschiat.* **75**, 179.

Formycin A

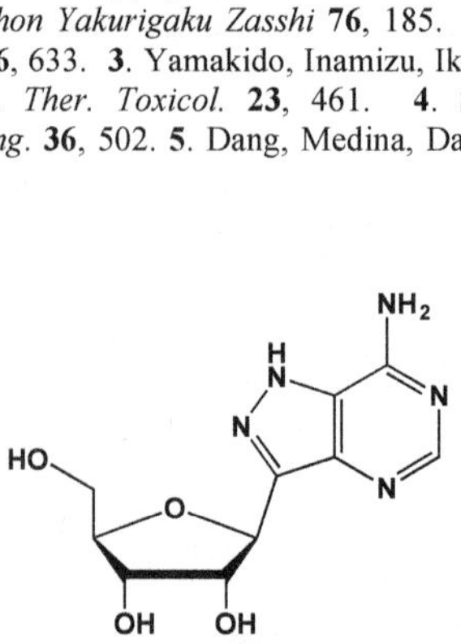

This highly fluorescent nucleoside antibiotic (FW = 267.24 g/mol; CAS = 6742-12-7), also named 7-amino-3-(β-D-ribofuranosyl)-1*H*-pyrazolo[4,3-*d*]pyrimidine and 8-aza-9-deazaadenosine, is a nonglycosidic adenosine isostere produced by *Nocardia interforma*. Formycin A is also metabolically phosphorylated to its mono-, di-, and tri-pohosphates, which inhibit purine biosynthesis and can also be incorporated into RNA. It likewise alters tRNA processing by increasing the degradation rate of abnormal precursors. Formycin A inhibits *Escherichia coli* and *Trichomonas vaginalis* purine-nucleoside phosphorylase with little effect on the mammalian enzyme. The purine-like base undergoes tautomerization when bound. One must not confuse this agent with formicin (*i.e.*, *N*-(hydroxymethyl)acetamide). **Target(s):** adenosine kinase, also as an alternative substrate (1,2); adenosyl-homocysteinase (3,4); adenosyl-homocysteine nucleosidase (5-7); AMP nucleosidase (8); diphosphate:purine-nucleoside kinase (9); GMP synthetase (10); 5'-methylthioadenosine/*S*-adenosylhomocysteine nucleosidase (5-7); 1 phosphatidyl-inositol 4-kinase (11); purine-nucleoside phosphorylase (9,12-23). **1.** Long & Parker (2006) *Biochem. Pharmacol.* **71**, 1671. **2.** Bhaumik & Datta (1989) *J. Biol. Chem.* **264**, 4356. **3.** Guranowski, Montgomery, Cantoni & Chiang (1981) *Biochemistry* **20**, 110. **4.** Shimizu, Shiozaki, Ohshiro & Yamada (1984) *Eur. J. Biochem.* **141**, 385. **5.** Singh, Evans, Lenz, *et al*, (2005) *J. Biol. Chem.* **280**, 18265. **6.** Cornell, Swarts, Barry & Riscoe (1996) *Biochem. Biophys. Res. Commun.* **228**, 724. **7.** Lee, Cornell, Riscoe & Howell (2003) *J. Biol. Chem.* **278**, 8761. **8.** de Wolf, Fullin & Schramm (1979) *J. Biol. Chem.* **254**, 10868. **9.** Munagala & Wang (2003) *Mol. Biochem. Parasitol.* **127**, 143. **10.** Spector & Beacham III (1975) *J. Biol. Chem.* **250**, 3101. **11.** Nishioka, Sawa, Hamada, *et al,* (1990) *J. Antibiot.* **43**, 1586. **12.** Wlodarczyk, Galitonov & Kierdaszuk (2004) *Eur. Biophys. J.* **33**, 377. **13.** Munagala & Wang (2002) *Biochemistry* **41**, 10382. **14.** Bzowska, Kulikowska & Shugar (1992) *Biochim. Biophys. Acta* **1120**, 239. **15.** Koellner, Bzowska, Wielgus-Kutrowska, *et al.* (2002) *J. Mol. Biol.* **315**, 351. **16.** Kierdaszuk, Modrak-Wojcik, Wierzchowski & Shugar (2000) *Biochim. Biophys. Acta* **1476**, 109. **17.** Bzowska, Kulikowska & Shugar (1990) *Z. Naturforsch. C* **45c**, 59. **18.** Bzowska, Kulikowska & Shugar (2000) *Pharmacol. Ther.* **88**, 349. **19.** Wielgus-Kutrowska, Tebbe, Schroder, *et al.* (1998) *Adv. Exp. Med. Biol.* **431**, 259. **20.** Koellner, Bzowska, Wielgus-Kutrowska, *et al.* (2002) *J. Mol. Biol.* **315**, 351. **21.** Bzowska, Kulikowska, Poopeiko & Shugar (1996) *Eur. J. Biochem.* **239**, 229. **22.** Maynes, Yam, Jenuth, *et al.* (1999) *Biochem. J.* **344**, 585. **23.** Munagala & Wang (2002) *Biochemistry* **41**, 10382.

Formycin A 5'-Monophosphate

This AMP analogue (FW = 347.22 g/mol) potently inhibits AMP nucleosidase. **Target(s):** adenine phosphoribosyl-transferase (1); AMP nucleosidase (2-12); rhodopsin kinase (13). **1.** Bashor, Denu, Brennan & Ullman (2002) *Biochemistry* **41**, 4020. **3.** Parkin, Mentch, Banks, Horenstein & Schramm (1991) *Biochemistry* **30**, 4586. **4.** Zhang, Cottet & Ealick (2004) *Structure* **12**, 1383. **5.** Giranda, Berman & Schramm (1989) *J. Biol. Chem.* **264**, 15674. **6.** Giranda, Berman & Schramm (1988) *Biochemistry* **27**, 5813. **7.** Giranda, Berman & Schramm (1986) *J. Biol. Chem.* **261**, 15307. **8.** DeWolf, Emig & Schramm (1986) *J. Biol. Chem.* **25**, 4132. **9.** Schramm & Reed (1980) *J. Biol. Chem.* **255**, 5795. **10.** Ehrlich & Schramm (1994) *Biochemistry* **33**, 8890. **11.** Leung & Schramm (1981) *J. Biol. Chem.* **256**, 12823. **12.** Leung & Schramm (1980) *J. Biol. Chem.* **255**, 10867. **13.** Palczewski, Kahn & Hargrave (1990) *Biochemistry* **29**, 6276.

Formycin B

This nucleoside analogue and antibiotic (FW = 268.23 g/mol; CAS = 13877-76-4), also known as 7-oxo-3-(β-D-ribofuranosyl)-1*H*-pyrazolo[4,3-*d*]pyrimidine and 8-aza-9-deazainosine 5'-phosphate, is produced by *Nocardia interforma*, *Streptomyces lavendilae*, and *S. roseochromogenes*. **Target(s):** adenosylhomocysteinase (1); GMP synthetase, weakly inhibited (2); purine nucleosidase, purine-specific nucleoside *N*-

ribohydrolase (3); purine-nucleoside phosphorylase (4-17); poly(ADP-ribose) polymerase (18,19). **1.** Shimizu, Shiozaki, Ohshiro & Yamada (1984) *Eur. J. Biochem.* **141**, 385. **2.** Spector & Beacham (1975) *J. Biol. Chem.* **250**, 3101. **3.** Parkin (1996) *J. Biol. Chem.* **271**, 21713. **4.** Stoeckler, Agarwal, Agarwal & Parks (1978) *Meth. Enzymol.* **51**, 530. **5.** Parks & Agarwal (1972) *The Enzymes*, 3rd ed., **7**, 483. **6.** Sheen, Kim & Parks, Jr. (1968) *Mol. Pharmacol.* **4**, 293. **7.** Schmidt, Walter & Konigk (1975) *Tropenmed. Parasitol.* **26**, 19. **8.** Willemot, Martineau, DesRosiers, *et al.* (1979) *Life Sci.* **25**, 1215. **9.** Kierdaszuk, Modrak-Wojcik, Wierzchowski & Shugar (2000) *Biochim. Biophys. Acta* **1476**, 109. **10.** Bzowska, Kulikowska & Shugar (1992) *Biochim. Biophys. Acta* **1120**, 239. **11.** Bzowska, Kulikowska & Shugar (1990) *Z. Naturforsch. C* **45c**, 59. **12.** Bzowska, Kulikowska & Shugar (2000) *Pharmacol. Ther.* **88**, 349. **13.** Wielgus-Kutrowska, Tebbe, Schroder, *et al.* (1998) *Adv. Exp. Med. Biol.* **431**, 259. **14.** Koellner, Bzowska, Wielgus-Kutrowska, *et al.* (2002) *J. Mol. Biol.* **315**, 351. **15.** Daddona, Wiesmann, Milhouse, *et al.* (1986) *J. Biol. Chem.* **261**, 11667. **16.** Wierzchowski, Iwanska, Bzowska & Shugar (2007) *Nucleosides Nucleotides Nucleic Acids* **26**, 849. **17.** Schimandle, Tanigoshi, Mole & Sherman (1985) *J. Biol. Chem.* **260**, 4455. **18.** Muller, Rohde, Steffen, *et al.* (1975) *Cancer Res.* **35**, 3673. **19.** Muller & Zahn (1975) *Experientia* **31**, 1014.

S-Formycinyl-L-homocysteine

This *S*-adenosylhomocysteine analogue (FW = 384.42 g/mol) inhibits 5'-methylthioadenosine/*S*-adenosyl-homocysteine nucleosidase, K_i = 9.7 nM. **Target(s):** adenosylhomocysteine nucleosidase (1); 5'-methylthio-adenosine/*S*-adenosylhomocysteine nucleosidase (1). **1.** Della Ragione, Porcelli, Carteni-Farina, Zappia & Pegg (1985) *Biochem. J.* **232**, 335.

*N*¹⁰-Formylfolate (*N*¹⁰-Formylfolic Acid)

This substituted folic acid (FW$_{free-acid}$ = 469.41 g/mol; λ_{max} = 256 nm; ε = 46000 M^{-1}cm^{-1}, in 0.01 M NaOH), also known as 10-formyl-pteroylglutamate, is a strong inhibitor of dihydrofolate reductase, K_i = 6.1 nM (1-4), methionyl-tRNA formyltransferase (5), phosphoribosyl-aminoimidazole carboxamide formyltransferase (6); thymidylate synthase (7). **1.** Baker (1967) *Design of Active-Site-Directed Irreversible Enzyme Inhibitors*, Wiley, New York. **2.** Bertino, Perkins & Johns (1965) *Biochemistry* **4**, 839. **3.** D'Urso-Scott, Uhoch & Bertino (1974) *Proc. Natl. Acad. Sci. U.S.A.* **71**, 2736. **4.** Friedkin, Plante, Crawford & Crumm (1975) *J. Biol. Chem.* **250**, 5614. **5.** Gambini, Crosti & Bianchetti (1980) *Biochim. Biophys. Acta* **613**, 73. **6.** Baggott, Vaughn & Hudson (1986) *Biochem. J.* **236**, 193. **7.** Banerjee, Bennett, Brockman, Sani & Temple, Jr. (1982) *Anal. Biochem.* **121**, 275.

N-Formyl-L-methionine

This formylated amino acid (FW$_{free-acid}$ = 177.22 g/mol), Symbol = fMet, is a residue at the N-terminus of a growing polypeptide chain in protein biosynthesis in prokaryotes, bacteriophages, mitochondria, and chloroplasts. The initiation codon AUG codes for *N*-formyl-L-methionine (occasionally by GUG or UUG, when these are the first codon). Note that free fMet is not formed in this pathway: fMet-tRNAfMet is produced by formylation of

Met-tRNAfMet *via* methionyl-tRNA formyltransferase. However, *N*-formyl-L-methionine is produced by the action of many aminopeptidases as well as by *N*-formylmethionyl-peptidase. **Target(s):** homoserine *O*-acetyltransferase (1); methionine *S*-methyltransferase, weakly inhibited (2); peptide deformylase, fMet is not a substrate for the *Escherichia coli* enzyme (3); pyrroline-2-carboxylate reductase (4). **1.** Shiio & Ozaki (1981) *J. Biochem.* **89**, 1493. **2.** James, Nolte & Hanson (1995) *J. Biol. Chem.* **270**, 22344. **3.** Meinnel, Blanquet & Dardel (1996) *J. Mol. Biol.* **262**, 375. **4.** Petrakis & Greenberg (1965) *Biochim. Biophys. Acta* **99**, 78.

cis-2-(2-Formylpyrrolidine-1-carbonyl)-1-cyclohexanecarboxylic Acid Benzyl Ester

This ester (FW = 343.42 g/mol) is a slow-binding inhibitor of prolyl oligopeptidase, *or* prolyl oligopeptidase, K_i = 3 nM. This non-peptide inhibitor binds at the same site as peptide inhibitors requiring L-configuration at the P_1 and P_2 binding pockets. **1.** Kolosko, Bakker, Faraci & Nagel (1994) *Drug Des. Discov.* **11**, 61.

N^5-Formyltetrahydrofolate (*or* N^5-Formyl-THF)

This tetrahydrofolate metabolite (FW$_{disodium-salt}$ = 517.41 g/mol; λ_{max} = 286 nm at pH 7 (ε = 31800 M^{-1}cm^{-1}) and 282 nm at pH 13 (ε = 32600 M^{-1}cm^{-1})), also known as leucovorin, folinic acid, and *N*-formyltetrahydropteroylglutamate, is a growth factor for *Leuconostoc citrovorum*. N^5-Formyl-THF is a substrate for glutamate formimidoyltransferase and 5-formyltetrahydrofolate cyclo-ligase. N^5-formyl-THF's ability to inhibit dihydrofolate reductase weakly stands as the basis for the so-called "methotrexate/leucovorin rescue" protocol for treating various neoplasias. While stable in neutral and alkaline solutions, this tetrahydrofolate derivative converts to N^5,N^{10}-methenyl-THF in acidic solutions. **Role of SHMT & Glycine in N^5-Formyl-THF Synthesis:** Serine hydroxymethyl-transferase (SHMT) is a pyridoxal-phosphate dependent enzyme that catalyzes the reversible interconversion of serine and tetrahydrofolate (THF) to glycine and N^5,N^{10}-methylene-THF, a reaction that generates one carbon unit for methionine, thymidylate and purine synthesis in the cytoplasm. SHMT is present in both the mitochondria (MSHMT) and the cytoplasm (CSHMT) of mammalian cells. SHMT also catalyzes a second reaction, the irreversible conversion of N^5,N^{10}-methenyl-THF to N^5-formyl-THF (1). Once synthesized, N^5-formyl-THF remains bound to the enzyme and acts as an effective slow-binding inhibitor of SHMT catalyzed reactions *in vitro* and *in vivo* (2,3). However, 5-formylTHF does not serve as a cofactor in the cell, and its role in cellular metabolism is not known. In the presence of excess glycine, the enzymatic activities of CSHMT and MTHFS constitute a futile cycle that is capable of buffering 5-formylTHF levels in the cytoplasm. Whereas the metabolic role of the futile cycle is not known, Girgis *et al.* (3) demonstrated that decreased levels of 5-formylTHF can result in impaired homocysteine remethylation in cultured human neuroblastoma. Elevations in 5-formylTHF levels result in decreased purine biosynthesis in MCF-7 cells (4). **Target(s):** glucose-6-phosphate dehydrogenase (5); glycine *N*-methyltransferase (6); glycogen phosphorylase *b* (7); methionyl-tRNA formyltransferase (8-10); serine hydroxy-methyltransferase, *or* glycine Hydroxymethyltransferase, also inhibited by the polyglutamylated form (11-16); thymidylate synthase (17). **1.** Stover & Schirch (1993) *Trends Biochem. Sci.* **18**, 102. **2.** Stover & Schirch (1991) *J. Biol. Chem.*, **266**, 1543. **3.** Girgis, Suh, Jolivet & Stover (1997) *J. Biol. Chem.* **272**, 4729. **4.** Bertrand & Jolivet (1989) *J. Biol. Chem.* **264**, 8843. **5.** Vogel, Snyder & Schulman (1963) *Biochem. Biophys. Res. Commun.* **10**, 97. **6.** Yeo & Wagner (1992) *J. Biol. Chem.* **267**, 24669. **7.** Klinov, Chebotareva, Sheiman, Birinberg & Kurganov (1987) *Bioorg.*

Khim. **13**, 908. **8.** Rader & Huennekens (1973) *The Enzymes*, 3rd ed. (Boyer, ed.), **9**, 197. **9.** Dickerman & Smith (1970) *Biochemistry* **9**, 1247. **10.** Gambini, Crosti & Bianchetti (1980) *Biochim. Biophys. Acta* **613**, 73. **11.** Schirch & Ropp (1967) *Biochemistry* **6**, 253. **12.** Girgis, Suh, Jolivet & Stover (1997) *J. Biol. Chem.* **272**, 4729. **13.** Stover & Schirch (1991) *J. Biol. Chem.* **266**, 1543. **14.** Ramesh & Appaji Rao (1980) *Biochem. J.* **187**, 623. **15.** Goyer, Collakova, Diaz de la Garza, *et al.* (2005) *J. Biol. Chem.* **280**, 26137. **16.** Chang, Tsai, Chen, Huang & Fu (2007) *Drug Metab. Dispos.* **35**, 2127. **17.** Friedkin, Plante, Crawford & Crumm (1975) *J. Biol. Chem.* **250**, 5614.

N^{10}-Formyltetrahydrofolate

This one-carbon folate (FW$_{disodium-salt}$ = 517.41 g/mol; λ_{max} = 258 nm between pH 7 to 9; ε = 22000 M^{-1}cm^{-1}), also known as N^{10}-formyltetrahydropteroylglutamate, is rapidly oxidized in air. At pH < 6, N^{10}-THF rearranges to form N^5,N^{10}-methenyltetrahydrofolate. In neutral or alkaline solutions, it isomerizes to produce N^5-formyltetrahydrofolate. The formyl group is lost in 0.1 M base. L-(+)-N^{10}-THF inhibits phosphoribosylglycinamide formyltransferase competitively (K_i = 0.75 μM). **Target(s):** N^5,N^{10}-methylenetetrahydrofolate dehydrogenase (1); N^5,N^{10}-methenyltetrahydrofolate cyclohydrolase (1); phosphoribosyl-glycinamide formyltransferase, inhibited by (6*S*)-stereoisomer (2,3); thymidylate synthase (4,5). **1.** Dev & Harvey (1978) *J. Biol. Chem.* **253**, 4245. **2.** Smith, Benkovic & Benkovic (1981) *Biochemistry* **20**, 4034. **3.** Inglese, Johnson, Shiau, Smith & Benkovic (1990) *Biochemistry* **29**, 1436. **4.** Balinska, Rhee, Whiteley, Priest & Galivan (1991) *Arch. Biochem. Biophys.* **284**, 219. **5.** Dolnick & Cheng (1978) *J. Biol. Chem.* **253**, 3563.

N^5-Formyltetrahydrohomofolate

This folic acid analogue inhibits 5-formyltetrahydrofolate cyclo-ligase, also known as N^5,N^{10}-methenyltetrahydrofolate synthetase; K_i = 1.4 μM for the MCF-7 human breast cancer cell enzyme. **1.** Schirch (1997) *Meth. Enzymol.* **281**, 146. **2.** Jolivet (1997) *Meth. Enzymol.* **281**, 162. **3.** Bertrand & Jolivet (1989) *J. Biol. Chem.* **264**, 8843. **4.** Huang & Schirch (1995) *J. Biol. Chem.* **270**, 22296.

Forskolin

This cell-permeable diterpenoid (FW = 408.54 g/mol), also known as colforsin, from the roots of *Coleus forskohlii*, activates adenyl cyclase (also known as adenylate cyclase), increasing cellular cAMP concentrations, thereby stimulating Protein Kinase A (cAMP-activated protein kinase). Treatment of rat renal mesangial cells with forskolin blunts activation of MAP kinase induced by platelet-derived growth factor and epidermal growth factor. The total synthesis of (±)-forskolin was accomplished independently (1,2). **Target(s):** ecdysone 20-monooxygenase (3); MAP kinase, *or* mitogen-activated protein kinase (4,5). **1.** Hashimoto, Sakata,

Sonegawa & Ikegami (1988) *J. Amer. Chem. Soc.* **110**, 3670. **2**. Corey, da Silva Jardine & Rohloff (1988) *J. Amer. Chem. Soc.* **110**, 3672. **3**. Keogh, Mitchell, Crooks & Smith (1991) *Experientia* **48**, 39. **4**. Li, Zarinetchi, Schrier & Nemenoff (1995) *Amer. J. Physiol.* **269**, C986. **5**. Siddhanti, Hartle & Quarles (1995) *Endocrinology* **136**, 4834.

Fosaprepitant

This antiemetic and aprepitant prodrug (FW = 614.40 g/mol; CAS = 172673-20-0), *or* [3-{[(2*R*,3*S*)-2-[(1*R*)-1-[3,5-bis(trifluoromethyl)phenyl] ethoxy]-3-(4-fluorophenyl)morpholin-4-yl]methyl}-5-oxo-2*H*-1,2,4-triazol-1-yl]phosphonic acid, is metabolically generates aprepitant, a selective high-affinity Substance P antagonist (*or* SPA) that blocks Substance P binding to the Neurokinin 1 (NK1) receptor (**See** *Aprepitant*).

Foscarnet

This intravenously administered antiviral drug (FW = 126.01 g/mol; CAS = 63585-09-1), also known as Foscavir and phosphonoformic acid, is a pyrophosphate mimic that binds to and inhibits many viral DNA polymerases (IC$_{50}$ = 10-100 μM), with little effect on human DNA polymerase. This drug is also active against RNA-dependent DNA polymerases. Foscarnet is used to treat drug-resistant cytomegalovirus (CMV), varicella-zoster virus, Epstein-Barr virus, as well as herpes simplex virus types 1 and 2. Phosphonoformic acid inhibited both chain elongation and exonuclease activities of HSV-1 DNA polymerase (uncompetitively with respect to activated DNA), K_i = 2.4 μM (1). Phosphonoacetate (PAA), which more closely resembles pyrophosphate, is of comparable effectiveness. Closely positioned mutations (*e.g.*, Leu-802-Met and Lys-805-Gln) in the conserved α-helix P in the nucleotide binding site of human cytomegalovirus DNA polymerase can respectively decrease and increase susceptibility to foscarnet, suggesting a unifying mechanism for CMV drug susceptibility (2). Indeed, recombinant viruses containing mutations Lys-805-Gln, Thr-821-Ile, and Lys-805-Gln/Thr-821-Ile had 0.8x, 5.3x and 4.8x increases in ganciclovir IC$_{50}$ values and 0.3-fold, 23.3-fold and 15.6-fold increases in foscarnet IC$_{50}$ values, respectively, compared with the wild-type virus (3). Phosphonopyruvate hydrolase, a novel bacterial carbon-phosphorus bond cleavage enzyme, is also inhibited by phosphonoformic acid (3). Foscarnet inhibits sodium-phosphate cotransporter, which mediates lithium reabsorption in the rat kidney (5). **Key Pharmacokinetic Parameters:** *See* Appendix II in Goodman & Gilman's *THE PHARMACOLOGICAL BASIS OF THERAPEUTICS*, 12th Edition (Brunton, Chabner & Knollmann, eds.) McGraw-Hill Medical, New York (2011). **1**. Derse & Cheng (1981) *J. Biol. Chem.* **256**, 8525. **2**. Tchesnokov, Gilbert, Boivin & Götte (2006) *J. Virol.* **80**, 1440. **3**. Martin, Arezki Azzi, Li & Boivin (2010) *Antiviral Ther.* **15**, 579. **4**. Kulakova, Wisdom, Kulakov & Quinn (2003) *J. Biol. Chem.* **278**, 23426. **5**. Uwai, Arima, Takatsu, *et al.* (2014) *Pharmacol Res.* **87**, 94.

Fosdevirine

This antiviral (FW = 413.80 g/mol; CAS = 1018450-26-4), also named IDX-899, GSK2248761A and carbamoyl-5-chloro-1*H*-indol-3-yl)-[3-(*E*)-2-

cyanovinyl)-5-methylphenyl)]-(*R*)-phosphinic acid methyl ester, is an HIV non-nucleoside reverse transcriptase inhibitor (1). In a study on fosdevirine's likely CNS toxicity, however, LC-MS and matrix-assisted laser desorption/ionization imaging confirmed the drug's gray matter disposition and metabolism (2). Moreover, Phase IIb studies to select a once-daily dose of fosdevirine were halted when 5 of 35 treatment-experienced subjects developed new-onset seizures after 4-weeks or longer exposure (3). The delayed onset of seizures and its persistence after discontinuation of fosdevirine is without precedent in antiretroviral drug development, underscoring the need for careful subject monitoring (3). Zhou, Pietropaolo, Damphousse *et al.* (2009) *Antimicrob. Agents Chemother.* **53**, 1739. **3**. Castellino, Groseclose, Sigafoos, *et al.* (2013) *Chem. Res. Toxicol.* **26**, 241. **4**. Margolis, Eron, Dejesus, *et al.* (2013) *Antivir. Ther.* **19**, 69.

Fosfomycin

This antibiotic (FW = 138.06 g/mol; CAS = 23155-02-4), also known as [(2*R*,3*S*)-3-methyloxiran-2-yl]phosphonic acid, phosphonomycin and phosphomycin, is produced by strains of *Streptomyces* and *Pseudomonas* and is an analogue of phosphoenolpyruvate. Fosfomycin inactivates UDP-*N*-acetylglucosamine 1-carboxyvinyltransferase, which catalyzes the first committed step in bacterial cell wall biosynthesis, attaching itself covalently with an active-site cysteinyl residue. Time-dependent inactivation is accelerated by the presence of the cosubstrate UDP-*N* acetylglucosamine, and the inactivation rate is further enhanced in the presence of the unreactive analogue 3-deoxy-UDP-*N*-acetylglucosamine (1). Fosfomycin activated maize leaf PEP carboxylase, but its Mg^{2+} complex inhibits (2). **Key Pharmacokinetic Parameters:** *See* Appendix II in Goodman & Gilman's *THE PHARMACOLOGICAL BASIS OF THERAPEUTICS*, 12th Edition (Brunton, Chabner & Knollmann, eds.) McGraw-Hill Medical, New York (2011). **Target(s):** acid phosphatase, inhibited weakly (3); 3-deoxy-7-phosphoheptulonate synthase (4); NF-κB activation (5); phosphoenolpyruvate carboxylase (2); UDP-*N*-acetylglucosamine 1-carboxyvinyltransferase, *or* MurA transferase (1,6-22). **1**. Marquardt, Brown, Lane, *et al.* (1994) *Biochemistry* **33**, 10646. **2**. Mujica-Jimenez, Castellanos-Martinez & Munoz-Clares (1998) *Biochim. Biophys. Acta* **1386**, 132. **3**. Haas, Redl, Leitner & Stöffler (1991) *Biochim. Biophys. Acta* **1074**, 392. **4**. Hu & Sprinson (1977) *J. Bacteriol.* **129**, 177. **5**. Yoneshima, Ichiyama, Ayukawa, Matsubara & Furukawa (2003) *Int. J. Antimicrob. Agents* **21**, 589. **6**. Leon, Garcia-Lobo & Ortiz (1983) *Antimicrob. Agents Chemother.* **24**, 276. **7**. Kahan, Kahan, Cassidy & Kropp H. (1974) *Ann. N.Y. Acad. Sci.* **235**, 364. **8**. Schönbrunn, Sack, Eschenburg, *et al.* (1996) *Structure* **4**, 1065. **9**. Kim, Lees, Kempsell, *et al.* (1996) *Biochemistry* **35**, 4923. **10**. Dunsmore, Miller, Blake, *et al.* (2008) *Bioorg. Med. Chem. Lett.* **18**, 1730. **11**. Anwar & Vlaovic (1980) *Biochim. Biophys. Acta* **616**, 389. **12**. Kedar, Brown-Driver, Reyes, *et al.* (2008) *Antimicrob. Agents Chemother.* **52**, 2009. **13**. Krekel, Oecking, Amrhein & Macheroux (1999) *Biochemistry* **38**, 8864. **14**. Samland, Amrhein & Macheroux (1999) *Biochemistry* **38**, 13162. **15**. Eschenburg, Priestman & Schönbrunn (2005) *J. Biol. Chem.* **280**, 3757. **16**. Krekel, Samland, Macheroux, Amrhein & Evans (2000) *Biochemistry* **39**, 12671. **17**. Thomas, Ginj, Jelesarov, Amrhein & Macheroux (2004) *Eur. J. Biochem.* **271**, 2682. **18**. Klein & Bachelier (2006) *J. Comput.-Aided Mol. Des.* **20**, 621. **19**. Dai, Parker & Bao (2002) *J. Chromatogr. B* **766**, 123. **20**. Skarzynski, Mistry, Wonacott, *et al.* (1996) *Structure* **4**, 1465. **21**. Venkateswaran, Lugtenberg & Wu (1973) *Biochim. Biophys. Acta* **293**, 570. **22**. Marquardt, Siegele, Kolter & Walsh (1992) *J. Bacteriol.* **174**, 5748.

Fosinopril & Fosinoprilat

Fosinopril

Fosinoprilat

This antihypertensive pro-drug (FW$_{fosinopril}$ = 566.63 g/mol; CAS = 98048-97-6), also named (2S,4S)-4-cyclohexyl-1-[2-[hydroxy(4-phenylbutyl)phosphoryl]acetyl]pyrrolidine-2-carboxylic acid and Monopril®, is metabolized to its pharmacologically active metabolite (FW$_{fosinoprilat}$ = 435.39 g/mol; CAS = 95399-71-6), the latter targeting peptidyl-dipeptidase A, *or* angiotensin-converting enzyme inhibitor (IC$_{50}$ = 1 nM). Unlike other ACE inhibitors that are primarily excreted by the kidneys, fosinopril is eliminated by both renal and hepatic pathways. **1.** Shionoiri, Naruse, Minamisawa, *et al.* (1997) *Clin. Pharmacokinet.* **32**, 460. **2.** Krapcho, Turk, Cushman, *et al.* (1988) *J. Med. Chem.* **31**, 1148.

Fosmidomycin

This antimalarial/antimicrobial agent (FW$_{free-acid}$ = 183.10 g/mol), also known as 3-(N-formyl-N-hydroxyamino)propylphosphonate, inhibits bacterial and plant 1-deoxy-D-xylulose-5-phosphate reductoisomerase (IC$_{50}$ = 8.2 nM) and suppresses the *in vitro* growth of multidrug-resistant *P. falciparum* strains. **1.** Jomaa, Wiesner, Sanderbrand, *et al.* (1999) *Science* **285**, 1573. **2.** Koppisch, Fox, Blagg & Poulter (2002) *Biochemistry* **41**, 236. **3.** Wiesner, Henschker, Hutchinson, Beck & Jomaa (2002) *Antimicrob. Agents Chemother.* **46**, 2889. **4.** Rodriguez-Concepcion, Ahumada, Diez-Juez, *et al.* (2001) *Plant J.* **27**, 213. **5.** Mueller, Schwender, Zeidler & Lichtenthaler (2000) *Biochem. Soc. Trans.* **28**, 792. **6.** Schwender, Muller, Zeidler & Lichtenthaler (1999) *FEBS Lett.* **455**, 140. **7.** Kuzuyama, Shimizu, Takahashi & H. Seto (1998) *Tetrahedron Lett.* **39**, 7913. **8.** Yajima, Hara, Sanders, *et al.* (2004) *J. Amer. Chem. Soc.* **126**, 10824. **9.** Kuzuyama, Takahashi, Takagi & Seto (2000) *J. Biol. Chem.* **275**, 19928. **10.** Mac Sweeney, Lange, Fernandes, *et al.* (2005) *J. Mol. Biol.* **345**, 115. **11.** Altincicek, Hintz, Sanderbrand, *et al.* (2000) *FEMS Microbiol. Lett.* **190**, 329. **12.** Grolle, Bringer-Meyer & Sahm (2000) *FEMS Microbiol. Lett.* **191**, 131.

Fostemsavir

This HIV prodrug (FW$_{free-acid}$ = 582.49 g/mol; FW = 704.63 g/mol (trishydroxymethylamine salt); CAS = 864953-29-7), also known as and {3-[(4-benzoyl-1-piperazinyl)-(oxo)acetyl]-4-methoxy-7-(3-methyl-1H-1,2,4-triazol-1-yl)-1H-pyrrolo[2,3-c]pyridin-1-yl}methylphosphate and BMS-663068, is a derivative of BMS-626529, a novel small-molecule HIV-1 attachment inhibitor (*See BMS-626529*). Because fostemsavir targets a different step of the viral life-cycle, it offers promise for individuals infected with virus that has achieved resistance to other HIV drugs. Administration of BMS-663068 for 8 days with or without ritonavir resulted in substantial declines in plasma HIV-1 RNA levels and was generally well tolerated. **1.** Nettles, Schürmann, Zhu, *et al.* (2012) *J. Infect. Dis.* **206**, 1002.

Fostriecin

This naturally-occurring phosphate monoester (FW$_{free-acid}$ = 404.33 g/mol; CAS = 87860-39-7), also known as phosphotrienin, from *Streptomyces pulveraceous* exhibits potent antitumor activity. It is a strong inhibitor of protein phosphatase 2A (IC$_{50}$ = 3.2 nM) and also inhibits protein phosphatase (1), albeit only at elevated concentrations (IC$_{50}$ = 131 μM). No effect has been observed on calcineurin (protein phosphatase 2B). Fostriecin also inhibits DNA topoisomerase II and will arrest cell growth at the G$_2$/M phase of the cell cycle. **Target(s):** DNA topoisomerase II (1-3); phosphoprotein phosphatase (4-13); protein phosphatase 1 (4,5); protein phosphatase 2A (4,6-10); protein phosphatase 4 (8,11,12). **1.** Boritzki, Wolfard, Besserer, Jackson & Fry (1988) *Biochem. Pharmacol.* **37**, 4063. **2.** de Jong, Mulder, Uges, *et al.* (1999) *Brit. J. Cancer* **79**, 882. **3.** Insaf, Danks & Witiak (1996) *Curr. Med. Chem.* **3**, 437. **4.** Walsh, Cheng & Honkanen (1997) *FEBS Lett.* **416**, 230. **5.** Stubbs, Tran, Atwell, *et al.* (2001) *Biochim. Biophys. Acta* **1550**, 52. **6.** Cheng, Balczon, Zuo, *et al.* (1998) *Cancer Res.* **58**, 3611. **7.** Hall, Feekes, Don, *et al.* (2006) *Biochemistry* **45**, 3448. **8.** Gallego & Virshup (2005) *Curr. Opin. Cell Biol.* **17**, 197. **9.** Kray, Carter, Pennington, *et al.* (2005) *J. Biol. Chem.* **280**, 35974. **10.** Zhang, Wu, Lei, Fang & Willis (2005) *Brain Res. Mol. Brain Res.* **138**, 264. **11.** Hastie & Cohen (1998) *FEBS Lett.* **431**, 357. **12.** Cohen, Philp & Vázquez-Martin (2005) *FEBS Lett.* **579**, 3278. **13.** Douglas, Moorhead, Ye & Lees-Miller (2001) *J. Biol. Chem.* **276**, 18992.

Fotemustine

This antineoplastic agent (FW = 315.69 g/mol; CAS = 92118-27-9) is a chloroethylnitrosourea that has been proposed to degrade in aqueous solutions to form a short-lived intermediate, 2-chloroethyl-diazohydroxide, which rapidly generates O^6-guanine adducts responsible for the drug's initial activity as well as a long-lived iminol tautomer responsible for the remaining O^6-guanine alkylation (1). **Target(s):** glutathione-disulfide reductase (2); ribonucleotide reductase (2,3); thioredoxin reductase (2). **1.** Hayes, Bartley, Parsons, Eaglesham & Prakash (1997) *Biochemistry* **36**, 10646. **2.** Schallreuter, Gleason & Wood (1990) *Biochim. Biophys. Acta* **1054**, 14. **3.** Schallreuter, Elgren, Nelson, *et al.* (1992) *Melanoma Res.* **2**, 393.

FR167653

This p38 kinase-selective inhibitor (FW$_{free-base}$ = 427.44 g/mol), also named 1-[7-(4-fluorophenyl)-1,2,3,4-tetrahydro-8-(4-pyridyl)pyrazolo[5,1-c][1,2,4]triazin-2-yl]-2-phenylethanedione, is a dual inhibitor, acting both *in vitro* and *in vivo* to reduce production of interleukin-1α (IC$_{50}$ = 0.84 μM), interleukin-1β (IC$_{50}$ = 0.09 μM) and tumor necrosis factor-α (TNFα) in human monocytes stimulated with lipopolysaccharide (IC$_{50}$ = 1.1 μM) and in human lymphocytes stimulated with phytohemagglutinin-M (IC$_{50}$ = 0.072 μM), preventing inflammation and suppressing the severity of adjuvant arthritis in a dose-dependent manner with little effect on body weight (1,2). FR167653 ameliorates endotoxin shock in rabbits and intravascular coagulation in rats (1,3) and also ameliorates cardiac dysfunction caused by chronic infusion of LPS in rats (4). Furthermore, FR167653 protects lung, liver, and heart against ischemia-reperfusion injury in dogs (5-7). FR167653 inhibits endogenous p38 kinase activity without affecting its activation, as observed in p38 kinase inhibitors, such as

SB203580 (8). This is due to the binding property of these p38 kinase inhibitors, i.e., this class of compounds competes with ATP at the ATP-binding site of p38 kinase, but does not bind to its phosphorylation sites (9,10). **1.** Yamamoto, Sakai, Yamazaki, Nakahara & Okuhara (1996) *Eur. J. Pharmacol.* **314**, 137. **2.** Shinozaki, Takagishi, Tsutsumi, *et al.* (2001) *Mod. Rheumatol.* **11**, 300. **3.** Yamamoto, Sakai, Yamazaki H, *et al.* (1997) *Eur. J. Pharmacol.* **327**, 169. **4.** Gardiner, Kemp, March & Bennett (1999) *J. Cardiovasc. Pharmacol.* **34**, 64. **5.** Kamoshita, Takeyoshi, Ohwada, Iino & Morishita (1997) *J. Heart Lung Transplant.* **16**, 1062. **6.** Karnitz & Abraham (1995) *Curr. Opin. Immunol.* **7**, 320. **7.** Kawano, Ogushi, Tani, *et al.* (1999) *J. Leukoc. Biol.* **65**, 80. **8.** Cuenda, Rouse, Doza, *et al.* (1995) *FEBS Lett.* **364**, 229. **9.** Frantz, Klatt & Pang (1998) *Biochemistry* **37**, 13846. **10.** Gum, McLaughlin, Kumar *et al.* (1998) *J. Biol. Chem.* **273**, 15605.

FR901228, *See Romidepsin*

FR901464

This natural product (FW = 491.62.65 g/mol), isolated from the fermentation broth of the bacterium *Pseudomonas* sp., is an anticancer compound that enhances the transcriptional activity of the SV40 promoter (1) and causes cell cycle arrest at the G_1 and G_2/M phases (2). FR901464 targets the spliceosome, a multicomponent complex that assembles on the newly synthesized pre-mRNA and catalyzes the removal of intervening sequences from transcripts (3). Spliceostatin A inhibits *in vitro* splicing and promotes pre-mRNA accumulation by binding to SF3b, a subcomplex of the U2 small nuclear ribonucleoprotein in the spliceosome. Treatment of cells with this agent results in leakage of (unspliced) pre-mRNA to the cytoplasm, where it is translated. Knockdown of SF3b by small interfering RNA induced phenotypes similar to those seen with spliceostatin A treatment. SF3b appears to have two functions, first in the splicing pre-mRNA and in retaining it. Albert *et al.* (4) describes the total syntheses, fragmentation studies, and antitumor/antiproliferative activities of FR901464 and its low picomolar analogue. They also showed that FR901464 was not very stable in phosphate buffer at pH 7.4 ($t_{1/2}$ = 4 hours at 37 °C). (*See also E7107; Meayamycin; Pladienolide B; Spliceostatin A*) **1.** Nakajima, Sato, Fujita, *et al.* (1996) *J. Antibiot.* (Tokyo) **49**, 1196. **2.** Nakajima, Hori, Terano H, *et al.* (1996) *J. Antibiot.* (Tokyo) **49**, 1204. **3.** Kaida, Motoyoshi, Tashiro, *et al.* (2007) *Nature Chem. Biol.* **3**, 576. **4.** Albert, Sivaramakrishnan, Naka, Czaicki & Koide (2007) *J Am Chem Soc.* 2007;129:2648 **IUPAC Name:** [(Z,2S)-4-[[(2R,3R,5S,6S)-6-[(2E,4E)-5-[(2R,3R,4S,6S)-4-(chloromethyl)-3,4,6-trihydroxy-6-methyloxan-2-yl]-3-methylpenta-2,4-dienyl]-2,5-dimethyloxan-3-yl]carbamoyl]but-3-en-2-yl] acetate

FRAX 486

This selective PAK1 inhibitor (FW = 513.39 g/mol; CAS = 1232030-35-1; Solubility: 100 mM in DMSO), also known as 6-(2,4-dichlorophenyl)-8-ethyl-2-[[3-fluoro-4-(1-piperazinyl)phenyl]amino]pyrido[2,3-*d*]pyrimidin-7(8*H*)-one, targets p21-activated kinase PAK1 (IC_{50} = 14 nM), with weaker action on PAK2 (IC_{50} = 33 nM), PAK3 (IC_{50} = 39 nM), and PAK4 (IC_{50} = 500 nM). **Pharmacologic Action:** Fragile X syndrome (FXS), the most common inherited form of autism and intellectual disability, is caused by silencing of the Fragile X mental retardation-1 (*Fmr1*) gene. *Fmr1* KO mice display phenotypes similar to symptoms observed in FXS, including hyperactivity, repetitive behaviors, and seizures, as well as abnormal dendritic spine density. Given the realization that defects in actin cytoskeletal dynamics alter dendritic spines in ways that impair synaptogenesis and/or synapse stability, attention focused on PAK1, a serine/threonine-protein kinase in signal transduction pathways linking Rho

GTPases to actin cytoskeleton reorganization and nuclear signaling in dendritic spines. FRAX486 was identified by high-throughput screening of a 12,000 kinase–focused small molecule library, with p21-activated kinase as the creening target. Treatment of adult *Fmr1* knock-out mice with FRAX486 reverses the spine density phenotype and likewise rescues every human-like, brain-related phenotypes observed in *Fmr1* knock-out mice, including abnormally high densities of dendritic spines, audiogenic seizures, and related attention deficit and hyperactivity. **1.** Dolan, Duron, Campbell, *et al.* (2013) *Proc. Natl. Acad. Sci. U.S.A.* **110**, 5671.

FRC-8653, *See Cilnidipine*

Frovatriptan

This orally bioavailable, tryptan-based 5HT receptor agonist (FW = 243.30 g/mol; CAS = 158930-17-7), also known by the trade name Frova® and systematically as (+)-(*R*)-3-methylamino-6-carboxamido-1,2,3,4-tetra-hydrocarbazole, is used in the treatment of migraine headaches and for the short-term prevention of menstrual migraine. Frovatriptan displays high agonist activity at mainly the serotonin 5-HT1B and 5-HT1D receptor subtypes, resulting in inhibition of the release of vasoactive neuropeptides by trigeminal nerves and inhibition of nociceptive neurotransmission. Its plasma half-life is 26 to 30h, suitable for once-daily dosing. **1.** Easthope & Goa (2001) *CNS Drugs* **15**, 969. **2.** Tepper, Rapoport & Sheftell (2002) *Arch. Neurol.* **59**, 1084.

D-Fructose

This hygroscopic D-ketohexose (FW = 180.16 g/mol; CAS = 57-48-7), also called fruit sugar and levulose, is the sweetest of all naturally occurring monosaccharides. As a reducing sugar, fructose supports nonenzymatic protein glycation. At 30°C in D_2O, D-fructose exists as: 2% α-pyranose, 70% β-pyranose, 5% α-furanose, and 23% β-furanose. **Target(s):** β-*N*-acetylhexosaminidase (1); acrosin (2); aldose 1-epimerase (3,4); catalase (5); catechol oxidase (45); choline acetyltransferase (6); concanavalin A (7); β-fructofuranosidase, *or* invertase (both anomers inhibit the yeast enzyme) (8-19); fructose-1,6-bisphosphate aldolase (20,21); fumarase (fumarate hydratase) (5;) α-galactosidase (22,23); β-galactosidase, weakly inhibited (24); glucan 1,4-α glucosidase, *or* glucoamylase; weakly inhibited (25); glucose dehydrogenase (NAD⁺) (26); glucose oxidase (27); α-glucosidase (maltase) (28,29); β-glucosidase (30,31); laminaribiose phosphorylase (44); mannokinase, also as an alternative substrate (32); naringinase (33); peroxidase (45,46); phosphoenolpyruvate carboxylase, weakly inhibited (34); α,α-phosphotrehalase, *or* trehalose-6-phosphate hydrolase (35); phytoene synthase (36); polyphenol oxidase (45,46); raucaffricine β-glucosidase (37); riboflavin phosphotransferase (38); ribokinase (39); superoxide dismutase, Cu/Zn (40); thioglucosidase, *or* myrosinase (41,42); vomilenine glucosyltransferase (43). **1.** Sakai, Narihara, Kasama, Wakayama & Moriguchi (1994) *Appl. Environ. Microbiol.* **60**, 2911. **2.** Anderson, Beyler, Mack & Zaneveld (1981) *Biochem. J.* **199**, 307. **3.** Bentley (1962) *Meth. Enzymol.* **5**, 219. **4.** Fishman, Pentchev & Bailey (1975) *Meth. Enzymol.* **41**, 484. **5.** Shyadehi & Harding (2002) *Biochim. Biophys. Acta* **1587**, 31. **6.** Augustinsson (1952) *The Enzymes*, 1st ed., **2** (part 2), 906. **7.** Goldstein, Hollerman & Smith (1965) *Biochemistry* **4**, 876. **8.** Neuberg & Mandl (1950) *The Enzymes*, 1st ed. (J. B. Sumner & K. Myrbäck, eds.), **1** (part 1), 527. **9.** Michaelis & Pechstein (1914) *Biochem. Z.* **60**, 79. **10.** Webb (1966) *Enzyme and Metabolic Inhibitors*, vol. **2**, p. 421, Academic Press, New York. **11.** Sayago, Vattuone, Sampietro & Isla

(2002) *J. Enzyme Inhib. Med. Chem.* **17**, 123. **12.** Ishimoto & Nakamura (1997) *Biosci. Biotechnol. Biochem.* **61**, 599. **13.** Isla, Vattuone, Gutierrez & Sampietro (1988) *Phytochemistry* **27**, 1993. **14.** Rojo, Quiroga, Vattuone & Sampietro (1998) *Phytochemistry* **49**, 965. **15.** Prado, Vattuone, Fleischmacher & Sampietro (1985) *J. Biol. Chem.* **260**, 4952. **16.** Lopez, Vattuone & Sampietro (1988) *Phytochemistry* **27**, 3077. **17.** Masuda, Takahashi & Sugawara (1987) *Agric. Biol. Chem.* **51**, 2309. **18.** Fotopoulos (2005) *J. Biol. Res.* **4**, 127. **19.** Quiroga, M. A. Vattuone & A. R. Sampietro (1995) *Biochim. Biophys. Acta* **1251**, 75. **20.** Taylor (1955) *Meth. Enzymol.* **1**, 310. **21.** Herbert, Gordon, Subrahmanyan & Green (1940) *Biochem. J.* **34**, 1108. **22.** Zapater, Ullah & Wodzinsky (1990) *Prep. Biochem.* **20**, 263. **23.** Schuler, Mudgett & Mahoney (1985) *Enzyme Microb. Technol.* **7**, 207. **24.** Itoh, Suzuki & Adachi (1982) *Agric. Biol. Chem.* **46**, 899. **25.** Buettner, Bode & Birnbaum (1987) *J. Basic Microbiol.* **27**, 299. **26.** Metzger, Wilcox & Wick (1964) *J. Biol. Chem.* **239**, 1769. **27.** Schepartz & Subers (1964) *Biochim. Biophys. Acta* **85**, 228. **28.** Gottschalk (1950) *The Enzymes*, 1st ed., **1** (Part 1), 551. **29.** Peruffo, Renosto & Pallavicini (1978) *Planta* **142**, 195. **30.** Dale, Ensley, Kern, Sastry & Byers (1985) *Biochemistry* **24**, 3530. **31.** Barbagallo, Spagna, Palmeri, Restuccia & Giudici (2004) *Enzyme Microb. Technol.* **35**, 58. **32.** Sebastian & Asensio (1972) *Arch. Biochem. Biophys.* **151**, 227. **33.** Nomura (1965) *Enzymologia* **29**, 272. **34.** Chen, Omiya, Hata & Izui (2002) *Plant Cell Physiol.* **43**, 159. **35.** Helfert, Gotsche & Dahl (1995) *Mol. Microbiol.* **16**, 111. **36.** Schledz, al-Babili, Von Lintig, *et al.* (1996) *Plant J.* **10**, 781. **37.** Schuebel, Stoeckigt, Feicht & Simon (1986) *Helv. Chim. Acta* **69**, 538. **38.** Katagiri, Yamada & Imai (1959) *J. Biochem.* **46**, 1119. **39.** Ogbunude, Lamour & Barrett (2007) *Acta Biochim. Biophys. Sin. (Shanghai)* **39**, 462. **40.** Ukeda, Hasegawa, Ishi & Sawamura (1997) *Biosci. Biotechnol. Biochem.* **61**, 2039. **41.** Tani, Ohtsuru & Hata (1974) *Agric. Biol. Chem.* **38**, 1623. **42.** Tsuruo & Hata (1968) *Agric. Biol. Chem.* **32**, 1420. **43.** Schuebel, Stoeckigt, Feicht & Simon (1986) *Helv. Chim. Acta* **69**, 538. **44.** Goldemberg, Maréchal & De Souza (1966) *J. Biol. Chem.* **241**, 45. **45.** Chisari, Barbagallo & Spagna (2008) *J. Agric. Food Chem.* **56**, 132. **46.** Chisari, Barbagallo & Spagna (2007) *J. Agric. Food Chem.* **55**, 3469.

D-Fructose 1,6-Bisphosphate

This key central pathway intermediate (FW_{free-acid} = 340.12 g/mol; CAS = 488-69-7; *Symbol*: FBP), formerly called D-fructose 1,6-diphosphate (FDP), is produced by the action of 6-phosphofructokinase, 1-phosphofructokinase, as well as ADP-specific phosphofructokinase. Calcium and barium salts of FBP are only slightly soluble in water. In addition, the solubility decreases at higher temperature. The β-anomer predominates in aqueous media: 77-80% at 30°C (1). **Target(s):** *N*-acetylglucosamine-6-phosphate deacetylase (2); acylphosphatase (3); adenylosuccinate synthetase (4-6); cAMP-dependent protein kinase phosphorylation of pyruvate kinase (7); 3′,5′-cyclic-GMP phosphodiesterase, *or* cGMP phosphodiesterase (8); 3-deoxy-7 phsophoheptulonate synthase, *or* 2-keto-3-deoxy-D-*arabino*-heptonate-7-phosphate synthase (9,10); β-fructofuranosidase, *or* invertase (11-13); fructose-2,6-bisphosphate 2-phosphatase (14); glucosamine 6-phosphate deaminase (15); glucose-1,6-bisphosphate synthase (16-18); glucose dehydrogenase (9,19); glucose-6-phosphate dehydrogenase (20); glucose-6-phosphate isomerase (21-24); glutamine:fructose 6-phosphate aminotransferase (25); glycerol kinase (26-36); glycerol 3-phosphate dehydrogenase (NAD$^+$) (37); glycogen phosphorylase (84); glycogen synthase (83); hexokinase (38,39); inositol-polyphosphate 5-phosphatase (40-43); isocitrate lyase (44-47); malate dehydrogenase, cytosolic (48); malate synthase (85); methylglyoxal synthase (49,50); nicotinate-nucleotide diphosphorylase (51); peptidase K (52); phosphoenol-pyruvate carboxykinase (ATP) (53); phosphoenolpyruvate carboxylase (54); phosphoenolpyruvate phosphatase (55); 6-phospho-β-galactosidase (56); phosphoglucomutase (57-62); β-phosphoglucomutase (63); 6-phosphogluconate dehydrogenase (decarboxylating) (64-68); 6-phospho β-glucosidase (69); phosphoprotein phosphatase (70); [protein-P_{II}] uridylyltransferase (71); pyrophosphatase (72); pyruvate kinase (73-77); ribose-phosphate diphosphokinase, *or* phosphoribosyl-pyrophosphate

synthetase, weakly inhibited (78); sedoheptulose-bisphosphatase (79); sucrose-phosphate synthase (80); thymidylate synthase, moderately inhibited (86); triose-phosphate isomerase (81); UDP-glucose 4-epimerase (82) **1.** Benkovic (1979) *Meth. Enzymol.* **63**, 370. **2.** Weidanz, Campbell, Moore, *et al.* (1996) *Brit. J. Haematol.* **95**, 645. **3.** Ramponi (1975) *Meth. Enzymol.* **42**, 409. **4.** Fischer, Muirhead & Bishop (1978) *Meth. Enzymol.* **51**, 207. **5.** Stayton, Rudolph & Fromm (1983) *Curr. Top. Cell. Regul.* **22**, 103. **6.** Borza, Iancu, Pike, Honzatko & Fromm (2003) *J. Biol. Chem.* **278**, 6673. **7.** El-Maghrabi, Haston, Flockhart, Claus & Pilkis (1980) *J. Biol. Chem.* **255**, 668. **8.** Hwang, Clark & Bernlohr (1974) *Biochem. Biophys. Res. Commun.* **58**, 707. **9.** Webb (1966) *Enzyme and Metabolic Inhibitors*, vol. 2, Academic Press, New York. **10.** Srinivasan & Sprinson (1954) *J. Biol. Chem.* **234**, 716. **11.** Liu, Huang, Chang & Sung (2006) *Food Chem.* **96**, 62. **12.** Prado, Vattuone, Fleischmacher & Sampietro (1985) *J. Biol. Chem.* **260**, 4952. **13.** Lee & Sturm (1996) *Plant Physiol.* **112**, 1513. **14.** Villadsen & Nielsen (2001) *Biochim. J.* **359**, 591. **15.** Weidanz, Campbell, DeLucas, *et al.* (1995) *Brit. J. Haematol.* **91**, 72. **16.** Rose, Warms & Kaklij (1975) *J. Biol. Chem.* **250**, 3466. **17.** Rose, Warms & Wong (1977) *J. Biol. Chem.* **252**, 4262. **18.** Ueda, Hirose, Sasaki & Chiba (1978) *J. Biochem.* **83**, 1721. **19.** Strecker (1955) *Meth. Enzymol.* **1**, 335. **20.** Sokolov & Trotsenko (1990) *Meth. Enzymol.* **188**, 339. **21.** Takama & Nosoh (1982) *Biochim. Biophys. Acta* **705**, 127. **22.** Howell & Schray (1981) *Mol. Cell Biochem.* **37**, 101. **23.** Noltmann (1972) *The Enzymes*, 3rd ed., **6**, 271. **24.** Jeong, Fushinobu, Ito, *et al.* (2003) *FEBS Lett.* **535**, 200. **25.** Kikuchi, Ikeda & Tsuiki (1972) *Biochim. Biophys. Acta* **289**, 303. **26.** Thorner (1975) *Meth. Enzymol.* **42**, 148. **27.** Ormo, Bystrom & Remington (1998) *Biochemistry* **37**, 16565. **28.** Liu, Faber, Feese, Remington & Pettigrew (1994) *Biochemistry* **33**, 10120. **29.** Thorner & Paulus (1973) *The Enzymes*, 3rd ed., **8**, 487. **30.** Nilsson, Thomson & Adler (1989) *Biochim. Biophys. Acta* **991**, 296. **31.** Yeh, Charrier, Paulo, *et al.* (2004) *Biochemistry* **43**, 362. **32.** Comer, Bruton & Atkinson (1979) *J. Appl. Biochem.* **1**, 259. **33.** Hayashi & Lin (1967) *J. Biol. Chem.* **242**, 1030. **34.** Thorner & Paulus (1973) *J. Biol. Chem.* **248**, 3922. **35.** Pawlyk & Pettigrew (2001) *Protein Expr. Purif.* **22**, 52. **36.** Kasinathan & Khuller (1984) *Lipids* **19**, 289. **37.** Beisenherz, Boltze, Bücher, *et al.* (1953) *Z. Naturforsch.* **8B**, 555. **38.** de Cesar, Colepicolo, Rosa & Rosa (1997) *Comp. Biochem. Physiol. B* **118**, 395. **39.** Magnani, Stocchi, Serafini, *et al.* (1983) *Arch. Biochem. Biophys.* **226**, 377. **40.** Rana, Sekar, Hokin & MacDonald (1986) *J. Biol. Chem.* **261**, 5237. **41.** Milani, Volpe & Pozzan (1988) *Biochem. J.* **254**, 525. **42.** Fowler & Brännström (1990) *Biochem. J.* **271**, 735. **43.** Hansen, Johanson, Williamson & Williamson (1987) *J. Biol. Chem.* **262**, 17319. **44.** McFadden (1969) *Meth. Enzymol.* **13**, 163. **45.** Tanaka, Nabeshima, Tokuda & Fukui (1977) *Agric. Biol. Chem.* **41**, 795. **46.** Johanson, Hill & McFadden (1974) *Biochim. Biophys. Acta* **364**, 327. **47.** DeLucas, Amor, Diaz, Turner & Laborda (1997) *Mycol. Res.* **101**, 410. **48.** Banaszak & Bradshaw (1975) *The Enzymes*, 3rd ed., **11**, 369. **49.** Murata, Fukuda, Watanabe, *et al.* (1985) *Biochem. Biophys. Res. Commun.* **131**, 190. **50.** Huang, Rudolph & Bennett (1999) *Appl. Environ. Microbiol.* **65**, 3244. **51.** Bhatia & Calvo (1996) *Arch. Biochem. Biophys.* **325**, 270. **52.** Orstan & Gafni (1991) *Biochem. Int.* **25**, 657. **53.** Lea, Chen, Leegood & Walker (2001) *Amino Acids* **20**, 225. **54.** Nakamura, Minoguchi & Izui (1996) *J. Biochem.* **120**, 518. **55.** Duff, Lefebvre & Plaxton (1989) *Plant Physiol.* **90**, 734. **56.** Calmes & Brown (1979) *Infect. Immun.* **23**, 68. **57.** Bartrons, Carreras, Climent & Carreras (1985) *Biochim. Biophys. Acta* **842**, 52. **58.** Hirose, Ueda & Chiba (1976) *Agric. Biol. Chem.* **40**, 2433. **59.** Chiba, Ueda & Hirose (1976) *Agric. Biol. Chem.* **40**, 2423. **60.** Fazi, Piacentini, Piatti & Accorsi (1990) *Prep. Biochem.* **20**, 219. **61.** Takamiya & Fukui (1978) *J. Biochem.* **84**, 569. **62.** Maino & Young (1974) *J. Biol. Chem.* **249**, 5176. **63.** Nakamura, Shirokane & Suzuki (1998) *J. Ferment. Bioeng.* **85**, 350. **64.** Silverberg & Dalziel (1975) *Meth. Enzymol.* **41**, 214. **65.** Bridges & Wittenberger (1975) *Meth. Enzymol.* **41**, 232. **66.** Adachi & Ameyama (1982) *Meth. Enzymol.* **89**, 291. **67.** Medina Puerta, Gallego-Iniesta & Garrido-Pertierra (1988) *Biochem. Int.* **17**, 479. **68.** Moritz, Striegel, De Graaf & Sahm (2000) *Eur. J. Biochem.* **267**, 3442. **69.** Wilson & Fox (1974) *J. Biol. Chem.* **249**, 5586. **70.** Titanji (1977) *Biochim. Biophys. Acta* **481**, 140. **71.** Engleman & Francis (1978) *Arch. Biochem. Biophys.* **191**, 602. **72.** Khandelwal & Hamilton (1981) *Can. J. Biochem.* **60**, 452. **73.** Lin, Turpin & Plaxton (1989) *Arch. Biochem. Biophys.* **269**, 228. **74.** Knowles, Smith, Smith & Plaxton (2001) *J. Biol. Chem.* **276**, 20966. **75.** Turner, Knowles & Plaxton (2005) *Planta* **222**, 1051. **76.** De Médicis, Laliberté & Vass-Marengo (1982) *Biochim. Biophys. Acta* **708**, 57. **77.** Gupta & Singh (1989) *Plant Physiol. Biochem.* **27**, 703. **78.** Fox & Kelley (1972) *J. Biol. Chem.* **247**, 2126. **79.** Traniello, Calcagno & Pontremoli (1971) *Arch. Biochem. Biophys.* **146**, 603. **80.** Harbron, Foyer & Walker (1981) *Arch. Biochem. Biophys.* **212**, 237. **81.** Tomlinson & Turner (1979)

Phytochemistry **18**, 1959. **82**. Lee, Kimura & Tochikura (1978) *Agric. Biol. Chem.* **42**, 731. **83**. Mied & Bueding (1979) *J. Parasitol.* **65**, 14. **84**. Tanabe, Kobayashi & Matsuda (1988) *Agric. Biol. Chem.* **52**, 757. **85**. Munir, Hattori & Shimada (2002) *Biosci. Biotechnol. Biochem.* **66**, 576. **86**. Lovelace, Gibson & Lebioda (2007) *Biochemistry* **46**, 2823.

D-Fructose 2,6-Bisphosphate

This central pathway metabolite ($FW_{\text{free-acid}}$ = 340.12 g/mol; CAS = 77164-51-3) regulates glycolysis and gluconeogenesis reciprocally by stimulating phosphofructokinase, while inhibiting fructose-1,6-bisphosphatase. **Target(s):** ATP:citrate lyase, *or* ATP:citrate synthase; weakly inhibited (1); fructose-1,6 bisphosphatase (2-26); inositol-polyphosphate 5-phosphatase, *or* inositol-1,4,5-trisphosphate 5 phosphomonoesterase, by the α-anomer (27-30); phosphoenolpyruvate carboxylase, mildly inhibited (31); 6-phosphofructokinase (32); phosphoglucomutase (33); sedoheptulose bisphosphatase (34); α,α-trehalose phosphorylase (35). **1**. Adams, Dack & Ratledge (2002) *Biochim. Biophys. Acta* **1597**, 36. **2**. Ganson & Fromm (1982) *Biochem. Biophys. Res. Commun.* **108**, 233. **3**. Hers, Hue & Van Schaftingen (1982) *Trends Biochem. Sci.* **7**, 329. **4**. Cseke, Balogh, Wong, *et al.* (1984) *Trends Biochem. Sci.* **9**, 533. **5**. Pilkis, Claus, Kountz & el-Maghrabi (1987) *The Enzymes*, 3rd ed., **18**, 3. **6**. Pilkis, el-Maghrabi, Pilkis & Claus (1981) *J. Biol. Chem.* **256**, 3619. **7**. Van Schaftingen & Hers (1981) *Proc. Natl. Acad. Sci. U.S.A.* **78**, 2861. **8**. Ganson & Fromm (1984) *Curr. Top. Cell. Regul.* **24**, 197. **9**. Kruger & Beevers (1984) *Plant Physiol.* **76**, 49. **10**. Kelley-Loughnane, Biolsi, Gibson, *et al.* (2002) *Biochim. Biophys. Acta* **1594**, 6. **11**. Iancu, Mukund, Fromm & Honzatko (2005) *J. Biol. Chem.* **280**, 19737. **12**. Nelson, Honzatko & Fromm (2004) *J. Biol. Chem.* **279**, 18481. **13**. Rittmann, Schaffer, Wendisch & Sahm (2003) *Arch. Microbiol.* **180**, 285. **14**. Rashid, Imanaka, Kanai, *et al.* (2002) *J. Biol. Chem.* **277**, 30649. **15**. Toyoda & Sy (1984) *J. Biol. Chem.* **259**, 8718. **16**. Van Tonder, Naude & Oelofsen (1991) *Int. J. Biochem.* **23**, 991. **17**. Ladror, Latshaw & Marcus (1990) *Eur. J. Biochem.* **189**, 89. **18**. Marcus, Rittenhouse, Moberly, *et al.* (1988) *J. Biol. Chem.* **263**, 6058. **19**. Noda, Hoffschulte & Holzer (1984) *J. Biol. Chem.* **259**, 7191. **20**. Mizunuma & Tashima (1986) *J. Biochem.* **99**, 1781. **21**. Dziewulska-Szwajkowska & Dzugaj (1999) *Comp. Biochem. Physiol. B* **122**, 241. **22**. Skalecki, Rakus, Wisniewski, Kolodziej & Dzugaj (1999) *Arch. Biochem. Biophys.* **365**, 1. **23**. Skalecki, Mularczyk & Dzugaj (1995) *Biochem. J.* **310**, 1029. **24**. Liu & Fromm (1988) *Arch. Biochem. Biophys.* **260**, 609. **25**. Rakus, Skalecki & Dzugaj (2000) *Comp. Biochem. Physiol. B* **127**, 123. **26**. Nelson, Honzatko & Fromm (2002) *J. Biol. Chem.* **277**, 15539. **27**. Rana, Sekar, Hokin & MacDonald (1986) *J. Biol. Chem.* **261**, 5237. **28**. Milani, Volpe & Pozzan (1988) *Biochem. J.* **254**, 525. **29**. Fowler & Brännström (1990) *Biochem. J.* **271**, 735. **30**. Hansen, Johanson, Williamson & Williamson (1987) *J. Biol. Chem.* **262**, 17319. **31**. Chen, Omiya, Hata & Izui (2002) *Plant Cell Physiol.* **43**, 159. **32**. Pelech, Cohen, Fisher, *et al.* (1984) *Eur. J. Biochem.* **145**, 39. **33**. Galloway, Dugger & Böack (1988) *Plant Physiol.* **88**, 980. **34**. Cadet & Meunier (1988) *Biochem. J.* **253**, 249. **35**. Miyatake, Kuramoto & Kitaoka (1984) *Biochem. Biophys. Res. Commun.* **122**, 906.

FTI-276 & FTI-277

FTI-276　　　　FTI-277

This protein farnesyltransferase inhibitor FTI-276 (FW = 433.60 g/mol, CAS = 170006-72-1) and its methyl ester prodrug FTI-277 (FW = 461.63 g/mol, CAS = 170006-73-2) are peptidomimetic analogues of L-cysteinyl-L-valyl-L-isoleucyl-L-methionine, in whuch the two central aminoacyl residues have been replaced with 2-phenyl-4 aminobenzoic acid. The reported K_i values for FTI-276 are 0.6 nM for protein farnesyltransferase and 50 nM for protein geranylgeranyltransferase type I. The IC50 value of FTI-277 for protein farnesyltransferase is 50 nM. FT-276 also antagonizes H and K-Ras oncogenic signaling. **Target(s):** protein farnesyltransferase

(1-5); protein geranylgeranyltransferase type I (1,2). **1**. Sebti & Hamilton (2000) *Meth. Enzymol.* **325**, 381. **2**. B. Gibbs (2001) *The Enzymes*, 3rd ed. (Tamanoi & Sigman, eds.), **21**, 81. **3**. Guida, Hamilton, Crotty & Sebti (2006) *J. Comp.-Aided Mol. Des.* **19**, 871. **4**. Yokoyama, Trobridge, Buckner, *et al.* (1998) *J. Biol. Chem.* **273**, 26497. **5**. Chakrabarti, Da Silva, Barger, *et al.* (2002) *J. Biol. Chem.* **277**, 42066.

FTI-2148

This protein farnesyltransferase inhibitor ($FW_{\text{free-acid}}$ = 452.57 g/mol), an L-Cys-L-Val-L-Ile-L-Met analogue with the two central aminoacyl residues replaced by 2-(2'-methylphenyl)-4-aminobenzoic acid, has a K_i value of 1.4 nM. The cell-permeant pro-drug FTI-2153 is its methyl ester. Because Ras malignant transformation requires posttranslational modification by farnesyltransferase (FTase), these inhibitors are antineoplastic agents. **1**. Sebti & Hamilton (2000) *Meth. Enzymol.* **325**, 381. **2**. Sun, Blaskovich, Knowles, *et al.* (1999) *Cancer Res.* **59**, 4919. **3**. Guida, Hamilton, Crotty & Sebti (2006) *J. Comp.-Aided Mol. Des.* **19**, 871.

FTY720, *See* Fingolimod

5FU, *See* 5-Fluorouracil

2'-Fucosyllactose

This principal milk oligosaccharide, *or* HMO (FW = 488. 44 g/mol; CAS = 41263-94-9), also known as α-L-fucosyl-(1→2)- β-D-galactosyl-(1→4)-D-glucose, comprising ~30% of all of HMO, inhibits the adhesion of *Pseudomonas aeruginosa* and enteric pathogens (*e.g.*, *Campylobacter jejuni*, enteropathogenic *Escherichia coli*, *Salmonella enterica*, serovar *fyris*) to human intestinal and respiratory cells. Human milk contains ~200 soluble oligosaccharides, of which many are unmetabolizable by infants, favoring monocultivation of *Bifidobacteria longum*, biovar *infantis*, in their intestinal tract, to the exclusion of pathogens. **1**. Weichert, Jennewein, Hüfner, *et al.* (2013) *Nutr. Res.* **33**, 831.

Fucoxanthin

This olive-green xanthophyll and brown algae accessory pigment (FW = 658.91 g/mol; CAS = 3351-86-8; λ_{max} = 510 nm), also named [(1S,3R)-3-hydroxy-4-[(3E,5E,7E,9E,11E,13E,15E)-18-[(1S,4S,6R)-4-hydroxy-2,2,6-trimethyl-7-oxabicyclo[4.1.0]heptan-1-yl]-3,7,12,16-tetramethyl-17-oxo-octadeca-1,3,5,7,9,11,13,15-octaenylidene]-3,5,5-trimethylcyclo-hexyl] ester, inhibits CYP1A1, CYP1A2 and CYP3A4. Fucoxanthin decreases rifampin-induced CYP3A4 and MDR1 expression by attenuating PXR-mediated CYP3A4 promoter activation and interaction between PXR and co-activator. Fucoxanthin markedly suppressed cell migration in wound healing assay and inhibited actin filament assembly (3). **1**. Satomi & Nishino (2013) *Oncol. Lett.* **6**, 860. **2**. Liu, Lim & Hu (2012) *Mar. Drugs* **10**, 242. **3**. Chung, Choi & Lee (2013) *Biochem. Biophys. Res. Commun.* **439**, 580.

FUD Peptide

This fibronectin fibrillogenesis inhibitor (FW = 4621.95 g/mol; Sequence: MGGQSESVEFTKDTQTGMSGQTTPQIETEDTKEPGVLMGGQSES), also named pUR4, mimics the cell-surface binding sites consisting of five N-terminal Type I modules within fibronectin. The Functional Upstream Domain (FUD) peptide contains 43 residues of the upstream domain and the first 6 residues from the repeat domain of *Streptococcus pyogenes* adhesin F_1, a protein that binds the N-terminal 70-kDa region of fibronectin with high affinity. FUD exhibits monophasic inhibition with an IC_{50} of ~10 nM. The 49-residue inhibitor blocks incorporation of both endogenous cellular fibronectin and exogenous plasma fibronectin into extracellular matrix and inhibits binding of 70-kDa fragment to fibronectin-null cells in a fibronectin-free system. Inhibition of matrix assembly has no effect on cell adhesion to substratum, cell growth, formation of focal contacts, or formation of stress fibers. As a disulfide-linked dimer of subunits composed of Type I, II, and III repeats, fibronectin binds to the endothelial cell surface by displaying its dominant cell-adhesive domain, a C-terminal heparin-binding domain, and a 70-kDa N-terminal domain that are recognized by integrins and syndecans (2). Inhibition of fibronectin fibrillogenesis on endothelial cells results in failure to achieve myosin-associated traction forces (2). *Note*: The control peptide Del29 (also called Δ^{29} and Control III-11C) lacks an aspartyl residue at position-29 of FUD and is without effect on fibronectin fibrillogenesis.   **1.** Tomasini-Johansson, Kaufman, Ensenberger, *et al.* (2001) *J. Biol. Chem.* **276**, 23430.   **2.** Zhou, Rowe, Hiraoka, *et al.* (2008) *Genes Dev.* **22**, 1231.

Fulvestrant

This first-in-class complete estrogen receptor antagonist, ER down-regulator, and second-line breast cancer therapy (FW = 606.77 g/mol; CAS = 129453-61-8), also known by the trade name Faslodex®, code names ZD9238 and ICI 182,780, and systematic name (7α,17β)-7-{9-[(4,4,5,5,5-pentafluoropentyl)sulfinyl]nonyl}estra-1,3,5(10)-triene-3,17-diol, is a synthetic ligand used to treat hormone receptor-positive metastatic breast cancer in postmenopausal women experiencing disease progression after anti-estrogen therapy. Estrogen receptor (ER) expression in breast cancer is a phenotype that can change during the natural history of the disease or during endocrine therapy. Fulvestrant also accelerates proteasomal degradation of estrogen receptors and reduces cell turnover index (CTI), a composite measurement of both proliferation and apoptosis (2). **1.** Robertson (2001) *J. Ster. Biochem. Mol. Biol.* **79**, 209.   **2.** Bundred, Anderson, Nicholson, *et al.* (2002) *Anticancer Res.* **22**, 2317.

Fumagillin

This antibiotic (FW$_{\text{free-acid}}$ = 458.55 g/mol; Photosensitive; Store in the dark), also named (2*E*,4*E*,6*E*,8*E*)-10-{[(3*R*,4*S*,5*S*,6*R*)-5-methoxy-4-[(2*R*)-2-methyl-3-(3-methylbut-2-enyl)oxiran-2-yl]-1-oxaspiro[2.5]octan-6-yl]oxy}-10-oxodeca-2,4,6,8-tetraenoic acid, from *Aspergillus fumigatus* suppresses the formation of new blood vessels and inhibits endothelial cell proliferation and angiogenesis. Its epoxide group reacts with an active-site histidyl residue in methionyl aminopeptidase type II (1-7). **1.** Liu, Widom, Kemp, Crews & Clardy (1998) *Science* **282**, 1324.   **2.** Luo, Li, Liu, *et al.* (2003) *J. Med. Chem.* **46**, 2631.   **3.** Schiffmann, Neugebauer & Klein (2006) *J. Med. Chem.* **49**, 511.   **4.** Son, Kwon, Jeong, *et al.* (2002) *Bioorg. Med. Chem.* **10**, 185.   **5.** Lowther, McMillen, Orville & Matthews (1998) *Proc. Natl. Acad. Sci. U.S.A.* **95**, 12153.   **6.** Klein, Schiffmann, Folkers, Piana & Rothlisberger (2003) *J. Biol. Chem.* **278**, 47862.   **7.** Brdlik & Crews (2004) *J. Biol. Chem.* **279**, 9475.

Fumarate

This unsaturated dicarboxylic acid (FW$_{\text{free-acid}}$ = 116.07 g/mol; Melting Point = 287°C; slightly soluble in water (0.63 g/100 mL at 20°C); pK_a values of 3.03 and 4.54 at 25°C), also known as (*E*)- or *trans*-butenedioate, is a key intermediate in the tricarboxylic acid cycle (*Fumarase Reaction*: Fumarate + H$_2$O $\rightleftharpoons$ Malate). It is also formed reactions catalyzed by argininosuccinate lyase (*Reaction*: Argininosuccinate $\rightleftharpoons$ Arginine + Fumarate) and AMP-succinase (*Reactions*: *N*-Succinyl-carboxyaminoimidazole ribonucleotide (SCAIR) $\rightleftharpoons$ Aminoimidazole-carboxamide ribonucleotide (AICAR) + Fumarate; *and* Adenylosuccinate $\rightleftharpoons$ AMP + Fumarate). (Maleate is fumarate's *cis* isomer.) **Target(s):** 2-(acetamidomethylene)succinate hydrolase (1); aconitase, *or* aconitate hydratase (2); adenylosuccinate synthetase (3,4); alanine aminotransferase (5); AMP deaminase (6); aspartate aminotransferase (7-9); aspartate ammonia-lyase, as product inhibitor (10); aspartate 4 decarboxylase (11); D-aspartate oxidase (12); γ-butyrobetaine:2-oxoglutarate dioxygenase, mildly inhibited (13); carbamoyl-phosphate synthetase (14); carboxy-*cis*,*cis*-muconate cyclase (15); [citrate lyase] deacetylase, weakly inhibited (16); cyanate hydratase, *or* cyanase (17); erythrose reductase (18); exo-α-sialidase, *or* neuraminidase, *or* sialidase (19); glutamate decarboxylase, K_i = 18 mM (20,21); glutamate dehydrogenase (22-24); glutamate formimidoyltransferase (55); glycogen synthase (25); 4-hydroxybenzoate 3-monooxygenase, weakly inhibited (56); hyoscyamine (6*S*)-dioxygenase (57); isocitrate lyase (26,27); leucine aminotransferase (28); malate dehydrogenase (29); malic enzyme, *or* malate dehydrogenase (decarboxylating) (30-32); nicotinate-nucleotide diphosphorylase (carboxylating) (33,34); oxaloacetate decarboxylase (35,36); palmitoyl-CoA hydrolase (37); pantothenase (38); peroxidase (39); phosphoenolpyruvate carboxykinase (40); phosphoenolpyruvate carboxylase (41-45); procollagen-proline 4-dioxygenase (58,59); pyrophosphatase (46); pyruvate kinase (47-49); succinyl-CoA synthetase (50); thiosulfate sulfotransferase, *or* rhodanese (51); urocanate hydratase, *or* urocanase (52-54).   **1.** Huynh & Snell (1985) *J. Biol. Chem.* **260**, 2379.   **2.** Eprintsev, Semenova & Popov (2002) *Biochemistry (Moscow)* **67**, 795.   **3.** Markham & Reed (1977) *Arch. Biochem. Biophys.* **184**, 24.   **4.** Gorrell, Wang, Underbakke, *et al.* (2002) *J. Biol. Chem.* **277**, 8817.   **5.** Agarwal (1985) *Indian J. Biochem. Biophys.* **22**, 102.   **6.** Chetty, Naidu, Moorthy & Swami (1982) *Arch. Int. Physiol. Biochim.* **90**, 293.   **7.** Tanaka, Tokuda, Tachibana, Taniguchi & Oi (1990) *Agric. Biol. Chem.* **54**, 625.   **8.** Martins, Mourato & de Varennes (2001) *J. Enzyme Inhib.* **16**, 251.   **9.** Rakhmanova & Popova (2006) *Biochemistry (Moscow)* **71**, 211.   **10.** Webb (1966) *Enzyme and Metabolic Inhibitors*, vol. 2, p. 355, Academic Press, New York.   **11.** Shibatani, Kakimoto, Kato, Nishimura & Chibata (1974) *J. Ferment. Technol.* **52**, 886.   **12.** Dixon & Kenworthy (1967) *Biochim. Biophys. Acta* **146**, 54.   **13.** Lindstedt (1967) *Biochemistry* **6**, 1271.   **14.** Jones (1962) *Meth. Enzymol.* **5**, 903.   **15.** Thatcher & Cain (1975) *Eur. J. Biochem.* **56**, 193.   **16.** Giffhorn & Gottschalk (1975) *J. Bacteriol.* **124**, 1052.   **17.** Anderson, Johnson, Endrizzi, Little & Korte (1987) *Biochemistry* **26**, 3938.   **18.** Lee, Koo & Kim (2002) *Appl. Environ. Microbiol.* **68**, 4534.   **19.** Engstler, Reuter & Schauer (1992) *Mol. Biochem. Parasitol.* **54**, 21.   **20.** Fonda (1972) *Biochemistry* **11**, 1304.   **21.** Gerig & Kwock (1979) *FEBS Lett.* **105**, 155.   **22.** Smith, Austen, Blumenthal & Nyc (1975) *The Enzymes*, 3rd ed., **11**, 293.   **23.** Bonete, Perez-Pomares, Ferrer & Camacho (1996) *Biochim. Biophys. Acta* **1289**, 14.   **24.** Hammer & Johnson (1988) *Arch. Microbiol.* **150**, 460.   **25.** Rothman & Cabib (1967) *Biochemistry* **6**, 2098.   **26.** McFadden & Howes (1963) *J. Biol. Chem.* **238**, 1737.   **27.** Johanson, Hill & McFadden (1974) *Biochim. Biophys. Acta* **364**, 327.   **28.** Pathre, Singh, Viswanathan & Sane (1987) *Phytochemistry* **26**, 2913.   **29.** Ochoa (1955) *Meth. Enzymol.* **1**, 735.   **30.** Kun (1963) *The Enzymes*, 2nd ed., **7**, 149.   **31.** van Heyningen & Pirie (1953) *Biochem. J.* **53**, 436.   **32.** Schimerlik & Cleland (1977) *Biochemistry* **16**, 565.   **33.** Shibata & Iwai (1980) *Biochim. Biophys. Acta* **611**, 280.   **34.** Iwai, Shibata & Taguchi (1979) *Agric. Biol. Chem.* **43**, 351.   **35.** Ochoa & Weisz-Tabori (1948) *J. Biol. Chem.* **174**, 123.   **36.** Sender, Martin, Peiru & Magni (2004) *FEBS Lett.* **570**, 217.   **37.** Berge & Dossland (1979) *Biochem. J.* **181**, 119.   **38.** Airas (1976) *Biochim. Biophys. Acta* **452**, 201.   **39.** Lück (1958) *Enzymologia* **19**, 227.   **40.** Wood, O'Brien & Michaels (1977) *Adv. Enzymol. Relat. Areas Mol. Biol.* **45**, 85.   **41.** Utter & Kolenbrander (1972) *The Enzymes*, 3rd ed., **6**, 117.   **42.** Cannata & Stoppani (1963) *J. Biol. Chem.* **238**, 1919.   **43.** Schwitzguebel & Ettlinger (1979) *Arch. Microbiol.*

122, 109. **44**. Gold & Smith (1974) *Arch. Biochem. Biophys.* **164**, 447. **45**. Chen, Omiya, Hata & Izui (2002) *Plant Cell Physiol.* **43**, 159. **46**. Naganna (1950) *J. Biol. Chem.* **183**, 693. **47**. Singh, Malhotra & Singh (2000) *Indian J. Biochem. Biophys.* **37**, 51. **48**. Singh, Malhotra & Singh (1998) *Indian J. Biochem. Biophys.* **35**, 346. **49**. Gibriel & Doelle (1975) *Microbios* **12**, 179. **50**. Leitzmann, Wu & Boyer (1970) *Biochemistry* **9**, 2338. **51**. Oi (1975) *J. Biochem.* **78**, 825. **52**. Phillips & George (1971) *Meth. Enzymol.* **17B**, 73. **53**. Lane, Scheuer, Thill & Dyll (1976) *Biochem. Biophys. Res. Commun.* **71**, 400. **54**. George & Phillips (1970) *J. Biol. Chem.* **245**, 528. **55**. Miller & Waelsch (1957) *J. Biol. Chem.* **228**, 397. **56**. Shoun, Arima & Beppu (1983) *J. Biochem.* **93**, 169. **57**. Hashimoto & Yamada (1987) *Eur. J. Biochem.* **164**, 277. **58**. Tuderman, Myllylä & Kivirikko (1977) *Eur. J. Biochem.* **80**, 341. **59**. Myllyharju (2008) *Ann. Med.* **40**, 402.

Fumonisins

These mycotoxins (Fumonisin B_1: MW_{B1} = 721.23 g/mol; CAS = 116355-83-0 (R_1 = OH, R_2 = OH); Fumonisin B_2: MW_{B2} = 705.83 g/mol; CAS = 116355-84-1 (R_1 = H, R_2 = OH); Fumonisin B_3: FW_{B3} = 705.83 g/mol; CAS = (R_1 = OH, R_2 = H); Fumonisin B_4: CAS = 136379-60-7 (R_1 = H, R_2 = H)) from *Fusarium moniliforme* and *F. proliferatum*, are sphinganine analogues inhibit ceramide synthase and block the biosynthesis of complex sphingolipids, promoting the accumulation of sphinganine and sphinganine 1-phosphate. More than eleven fumonisins have been identified; the most common are fumonisins B_1, B_2, and B_3. **Target(s):** H^+-ATPase (1); inositol-phosphoryl-ceramide synthase, *or* ceramide inositol-phosphoryltransferase (2); MAP kinase (3); protein-serine/threonine phosphatase (4); sphingosine *N*-acyltransferase, *or* ceramide synthase, *or* sphinganine *N*-acyltransferase (5-11). **1**. Gutiérrez-Nájera, Muñoz-Clares, Palacios-Bahena, *et al.* (2005) *Planta* **221**, 589. **2**. Wu, McDonough, Nickels, *et al.* (1995) *J. Biol. Chem.* **270**, 13171. **3**. Wattenberg, Badria & Shier (1996) *Biochem. Biophys. Res. Commun.* **227**, 622. **4**. Fukuda, Shima, Vesonder, *et al.* (1996) *Biochem. Biophys. Res. Commun.* **220**, 160. **5**. Merrill & Wang (1992) *Meth. Enzymol.* **209**, 427. **6**. Wang & Merrill (2000) *Meth. Enzymol.* **311**, 15. **7**. Lynch (2000) *Meth. Enzymol.* **311**, 130. **8**. Meredith (2000) *Meth. Enzymol.* **311**, 361. **9**. Wang, Norred, Bacon, Riley & Merrill (1991) *J. Biol. Chem.* **266**, 14486. **10**. Merrill, van Echten, Wang & Sandhoff (1993) *J. Biol. Chem.* **268**, 27299. **11**. Voss, Norred, Meredith, Riley & Saunders (2006) *J. Toxicol. Environ. Health A* **69**, 1387.

β-Funaltrexamine

This long-lasting morphine analogue ($FW_{free-base}$ = 454.52 g/mol; CAS = 72786-10-8; Soluble to 20 mM in H_2O), also known as (*E*)-4-[[5α,6β)-17-cyclopropylmethyl)-4,5-epoxy-3,14-dihydroxymorphinan-6-yl]amino]-4-oxo-2-butenoic acid methyl ester hydrochloride, is a μ-opioid antagonist and alkylating derivative of naltrexone that irreversibly alkylates the receptor. It is also a κ opioid receptor agonist. **1**. Rothman, Bykov, Mahboubi, *et al.* (1991) *Synapse* **8**, 86. **2**. Ward, Fries, Larson, Portoghese & Takemori (1985) *Eur. J. Pharmacol.* **107**, 323.

Furafylline

This synthetic xanthine analogue (FW = 260.25 g/mol), also known as 1,8-dimethyl-3-(2'-furfuryl)methylxanthine, is a potent, mechanism-based inhibitor of CYP1A2, K_i = 23 μM and k_{inact} = 0.87 min^{-1}. Furafylline has been used as a long-acting replacement for theophylline in the treatment of asthma. **1**. Sesardic, Boobis, Murray, *et al.* (1990) *Brit. J. Clin. Pharmacol.* **29**, 651. **2**. Tassaneeyakul, Birkett, Veronese, *et al.* (1994) *Pharmacogenetics* **4**, 281. **3**. Clarke, Ayrton & Chenery (1994) *Xenobiotica* **24**, 517. **4**. Kunze & Trager (1993) *Chem. Res Toxicol.* **6**, 649. **5**. Racha, Rettie & Kunze (1998) *Biochemistry* **37**, 7407.

Fusidic Acid

This bile salt-like antibiotic ($FW_{free-acid}$ = 516.72 g/mol; CAS = 6990-06-3; pK_a = 5.35) is from *Fusidium coccineum* is effective against Gram-positive microorganisms, readily inhibiting *Staphylococcus aureus*, but is without effect on Gram-negative bacteria, like *E. coli* (1-10). Fusidic acid inhibits the elongation factors as well as translocation of peptidyl-tRNA. Fusidic acid also inhibits aminoacyl-tRNA binding to ribosomes. **Mechanism of Inhibitory Action:** FA is a strong elongation inhibitor ($K_{0.5}$ ≈ 1 μM). Quench-flow and stopped flow experiments (11) that take advantage of the disparate time-scales for the inhibited (10 sec) and uninhibited (100 msec) elongation cycle gave rise to a detailed kinetic model showing that FA targets elongation factor G (EF-G) at an early Stage-I in the translocation process, then proceeding unhindered by FA to a Stage-II, whereupon the ribosome stalls. Stalling may also occur at a third stage of translocation (Stage-III), just before EF-G release from the post-translocation ribosome. **Target(s):** acid phosphatase (12); ATP-dependent transport of multi-drug resistance protein-2 substrates (13); bile salt transport pump (13,14); chloramphenicol acetyltransferase (15,16); protein biosynthesis (elongation) (1-10); protein-synthesizing GTPase (elongation factor) (16); vacuolar H^+ ATPase (17); *Veillonella alcalescens* ATPase (18). **1**. Pestka (1974) *Meth. Enzymol.* **30**, 261. **2**. Carrasco, Battaner & Vazquez (1974) *Meth. Enzymol.* **30**, 282. **3**. Scott & Tomkins (1975) *Meth. Enzymol.* **40**, 273. **4**. Jiménez (1976) *Trends Biochem. Sci.* **1**, 28. **5**. Lucas-Lenard & Beres (1974) *The Enzymes*, 3rd ed., **10**, 53. **6**. Harvey, Knight & Sih (1966) *Biochemistry* **5**, 3320. **7**. De Vendittis, De Paola, Gogliettino, *et al.* (2002) *Biochemistry* **41**, 14879. **8**. Malkin & Lipmann (1969) *Science* **164**, 71. **9**. Yamaki (1965) *J. Antibiot.* **18**, 228. **10**. Uritani & Miyazaki (1988) *J. Biochem.* **103**, 522. **11**. Borg, Holm, Shiroyama, *et al.* (2014) *J. Biol Chem.* **290**, 3440. **12**. Dassa, Cahu, Desjoyaux-Cherel & Boquet (1982) *J. Biol. Chem.* **257**, 6669. **13**. Bode, Donner, Leier & Keppler (2002) *Biochem. Pharmacol.* **64**, 151. **14**. Anwer & Hegner (1978) *Naunyn Schmiedebergs Arch. Pharmacol.* **302**, 329. **15**. Murray, Cann, Day, *et al.* (1995) *J. Mol. Biol.* **254**, 993. **16**. Bennett & Shaw (1983) *Biochem. J.* **215**, 29. **17**. Moriyama & Nelson (1988) *FEBS Lett.* **234**, 383. **18**. Yoshimura (1978) *J. Biochem.* **83**, 1231.

Fx-1006A, See **Tafamidis**

FXV673

This novel factor Xa inhibitor (FW$_{HCl-Salt}$ = 482.90 g/mol), also named 2-(R)-(3-carbamimidoylbenzyl)-3-(R)-[4-(1-oxypyridin-4-yl)benzoylamino]

butyric acid methyl ester, is a reversible, highly potent (K_i = 0.5 nM for fXa) and selective (K_i for thrombin = 3956 nM, trypsin = 301 nM, APC = 18491 nM, plasmin = 656 nM and t-PA = 8681 nM) inhibitor (1). FXV673 has a short biological half-life in rats, dogs and monkeys (0.27–0.33 h), and it is an effective antithrombotic agent after intravenous infusion in rat and canine carotid artery thrombosis models Compared to heparin and a GPIIb/IIIa receptor antagonist, RPR109891, infusion of FXV673 in dogs reperfused 50 and 38% more arteries with a trend for faster reperfusion (2). The incidence of reocclusion was significantly reduced by 75% with FXV673 infusion. **1.** Chu, Brown, Colussi, *et al.* (2000) *Blood,* **96** *Supplement* A236. **2.** Rebello, Bentley, Morgan, *et al.* (2001) *Br. J. Pharmacol.* **133**, 1190.

– G –

G, See Glucose; Glycine; Guanine; Guanosine

G-1

This potent and selective GPCR receptor agonist (FW = 412.28 g/mol; CAS 881639-98-1; Soluble to 100 mM in DMSO), also named (±)-1-[(3aR^*,4S^*,9bS^*)-4-(6-Bromo-1,3-benzodioxol-5-yl)-3a,4,5,9b-tetrahydro-3H-cyclopenta[c]quinolin-8-yl]ethanone, targets the G Protein-coupled Receptor-30, or GPR30, (K_i = 11 nM, EC_{50} = 2 nM), while displaying no activity at estrogen receptors ERα and ERβ, even 10 µM. This orphan receptor mediates the non-genomic signaling of 17β-estradiol in estrogen-sensitive cancer cells through the Epidermal Growth Factor Receptor (or EGFR) pathway. G-1 increases cytosolic Ca^{2+} and inhibits in vitro migration of SKBr3 cells (IC_{50} = 0.7 nM) and MCF-7 cells (IC_{50} = 1.6 nM) in response to chemoattractants. Blocks MCF-1 cell cycle progression at the G_1 phase. G-1 also displays therapeutic effects in the mouse EAE model of multiple sclerosis. (See also G-15; G-36) 1. Albanito, et al. (2007) G protein-coupled receptor 30 (GPR30) mediates gene expression changes and growth response to 17β-estradiol and selective GPR30 ligand G-1 in ovarian cancer cells. Cancer Res. 67, 1859. 2. Blasko, et al. (2009) J. Neuroimmunol. 214, 67. 3. Ariazi, et al. (2010) Cancer Res. 70, 1184.

G007-LK

This tankyrase inhibitor (FW = 529.96 g/mol; Solubility: 15 mg/mL DMSO; IUPAC: 4-{5-[(E)-2-{4-(2-chlorophenyl)-5-[5-(methylsulfonyl)pyridin-2-yl]-4H-1,2,4-triazol-3-yl]-ethenyl]-1,3,4-oxadiazol-2-yl} benzo nitrile) is an efficient and specific inhibitor of the canonical Wnt signaling, both in vitro and in vivo. JW74 rapidly reduces active β-catenin, attended subsequently by downregulation of the Wnt target genes AXIN2, SP5, and NKD1 (1). AXIN2 protein levels were strongly increased after compound exposure. G007-LK displays high selectivity toward Tankyrases 1 and 2, with biochemical IC_{50} values of 46 nM and 25 nM, respectively, and an intracellular IC_{50} value of 50 nM combined with an excellent pharmacokinetic profile in mice. (See also JW74) 1. Voronkov, Holsworth, Waaler, et al. (2013) J. Med. Chem. 56, 3012.

G-15

This high-affinity, selective GPCR antagonist (FW = 370.25 g/mol; CAS = 1161002-05-6; Soluble to 50 mM in DMSO), also named (3aS^*,4R^*,9bR^*)-4-(6-bromo-1,3-benzodioxol-5-yl)-3a,4,5,9b-3H-cyclopenta[c]quinoline,

targets GPR30 ($IC_{50} \approx$ 20 nM), a G protein-coupled, seven-helix transmembrane estrogen receptor that plays a role in the protective effects of estrogen in neurons (1,2). G-15 shows no significant binding to the classical estrogen receptors, ERα or ERβ, even at concentrations in the 1-10 µM range (1). Estrogen-dependent signaling proceeds by a pertussis toxin-sensitive pathway, indicating the involvement of $G_{i/o}$ heterotrimeric G proteins, attended by transactivation of EGFRs through the release of cell-surface heparin-bound EGF (3). A second phase of GPR30-dependent signaling via adenylyl cyclase leads to time-dependent attenuation of Erk activation (4). The GPR30- selective agonist G-1 partially attenuates the TNF-induced up-regulation of pro-inflammatory proteins (e.g., ICAM-1 and VCAM-1), and such effects are blocked completely by G-15 (5), a structural analogue of G-1 (1,2). Endothelial GPR30 represents a novel inflammatory response regulator and potential druggable target to prevent atherosclerosis and other inflammatory diseases (5). 1. Dennis, Burai, Ramesh, et al. (2009) Nature Chem. Biol. 5, 421. 2. Gingerich, Kim, Chalmers, et al. (2010) Neurosci. 170, 54. 3. Filardo, Quinn, Bland & Frackelton (2000) Mol. Endocrinol. 14, 1649. 4. Filardo, Quinn, Frackelton & Bland (2002) Mol. Endocrinol. 16, 70. 5. Chakrabarti & Davidge (2012) PLoS One 7, e52357.

G-36

This GPR30 antagonist (FW = 412.33 g/mol; CAS = 1392487-51-2; Soluble to 100 mM in DMSO), also named (±)-(3aR^*,4S^*,9bS^*)-4-(6-bromo-1,3-benzodioxol-5-yl)-3a,4,5,9b-tetrahydro-8-(1-methylethyl)-3H-cyclopenta[c]quinoline, selectively inhibits estrogen-mediated activation of phosphatidylinositide 3-kinase (PI3-K) by G protein-coupled receptor 30 (GPR30), but not by the estrogen receptor ERα. GPR30 is an integral membrane protein with high affinity for estradiol, but not other endogenous estrogens, such as estrone, estriol, progesterone, testosterone, and cortisol. It also inhibits estrogen- and G-1-mediated calcium mobilization (See G-1) as well as ERK1/2 activation, with no effect on EGF-mediated ERK1/2 activation. G36 inhibits estrogen- and G-1-stimulated proliferation of uterine epithelial cells in vivo. G-36 also inhibits estrogen-mediated calcium mobilization (IC_{50} = 112 nM). 1. Dennis, Field, Burai, et al. (2011) J. Steroid Biochem. Mol. Biol. 127, 358.

G-749

This FLT3 protein kinase inhibitor (FW = 521.42 g/mol) displays a unique kinase inhibition profile is highly potent in its action against FLT3 kinase, providing sustained inhibition of FLT3 phosphorylation and downstream effectors in FLT3-ITD expressing cell lines. G-749 is likewise highly potent against clinically known FLT3 mutants including gatekeeper and TKD that confer resistance to PKC412 and AC220. In comparison with PKC412 and AC220, G-749 shows several desirable characteristics to overcome other known drug resistances conferred by patient plasma, FLT3 ligand (FL) surge, and protection by stromal cells. Oral dosing of G-749 leads to complete tumor regression without relapse in the mouse xenograft model and increases survival in a bone marrow xenograft model. G-749 also shows potent anti-leukemic activity in patient blasts harboring FLT3-ITD, FLT3-TKD and FLT3-ITD/TKD mutations through inhibition of p-FLT3 and p-ERK1/2, including those with little or only minor response to AC220 and/or PKC412. **Target(s):** G-749 displays a unique inhibition pattern, with high potency against FLT3, FLT3-D835Y, and Mer (IC$_{50}$ ≈ 1 nM), receptor tyrosine kinases Ret, FLT1, Axl, Fms, FGFR1, and FGFR3 (IC$_{50}$ = 9-30 nM) and serine/threonine kinases Aurora B and C (IC$_{50}$ = 6-24 nM). Other kinases including c-KIT, PDGFRs, and EGFR were less potently inhibited (IC$_{50}$ > 300 nM); however, their mutants were significantly inhibited. G-749 is thus a novel and potent FLT3 inhibitor with a unique kinase inhibitory profile. **1.** Lee, Kim, Lee, *et al.* (2014) *Blood* **123**, 2209.

Gabaculine

This conformationally constrained GABA analogue (FW= 139.15 g/mol; CAS = 59556-17-1), also known as 3-amino-2,3-dihydrobenzoic acid, is a neurotoxin (produced by *Streptomyces toyocaenis*) that irreversibly inactivates ($t_{1/2}$ = 9 min) bacterial γ-aminobutyrate transaminase, when present at 3 × 10^{-7} M. Gabaculine is a mechanism-based inhibitor (K_i = 2.86 μM) for this pyridoxal phosphate-dependent transaminase. When transaminated on the enzyme, gabaculine is converted to a cyclohexatrienyl species with an exo double bond. Spontaneous aromatization transforms this unstable intermediate into *m*-carboxyphenylpyridoxamine phosphate, resulting in covalent and irreversible modification of the cofactor. (**See** *2-Aminobutyrate*) **Target(s):** alanine aminotransferase (1); D-amino acid aminotransferase, *or* ω-amino-acid:pyruvate aminotransferase (3,35); 4-amino-butyrate aminotransferase, *or* GABA aminotransferase (1,4-13,30-34); D-3-aminoisobutyrate aminotransferase, *or* (*R*)-3-amino-2 methylpropionate:pyruvate aminotransferase (14-16); L-3-aminoisobutyrate aminotransferase, *or* (*S*)-3-amino-2-methylpropionate:pyruvate aminotransferase (14); aminolevulinate aminotransferase (17); aspartate aminotransferase (1); dTDP-4-amino-4,6-dideoxy-D-glucose aminotransferase (18); glutamate-1-semialdehyde 2,1-aminomutase, *or* glutamate-1-semialdehyde aminotransferase (9,19-26); ornithine δ-aminotransferase (27-29,36-39); acetylornithine aminotransferase (40). **1.** Wood, Kurylo & Tsui (1979) *Neurosci. Lett.* **14**, 327. **2.** Jones, Soper, Ueno & Manning (1985) *Meth. Enzymol.* **113**, 108. **3.** Yonaha, Toyama & Soda (1987) *Meth. Enzymol.* **143**, 500. **4.** Rando & Bangerter (1976) *J. Amer. Chem. Soc.* **98**, 6762. **5.** Rando (1977) *Meth. Enzymol.* **46**, 28. **6.** Patel, Rothman, Cline & Behar (2001) *Brain Res.* **919**, 207. **7.** Pierard, Peres, Satabin, Guezennec & Lagarde (1999) *Exp. Brain Res.* **127**, 321. **8.** Behar & Boehm (1994) *Magn. Reson. Med.* **31**, 660. **9.** Rieble & Beale (1991) *Arch. Biochem. Biophys.* **289**, 289. **10.** Rando (1977) *Biochemistry* **16**, 4604. **11.** Rando & Bangerter (1977) *Biochem. Biophys. Res. Commun.* **76**, 1276. **12.** Fu & Silverman (1999) *Bioorg. Med. Chem.* **7**, 1581. **13.** Ator & Ortiz de Montellano (1990) *The Enzymes*, 3rd ed., **19**, 213. **14.** Tamaki, Fujimoto Sakata & Matsuda (2000) *Meth. Enzymol.* **324**, 376. **15.** Kaneko, Fujimoto, Kikugawa & Tamaki (1990) *FEBS Lett.* **276**, 115. **16.** Tamaki, Kaneko, Mizota, Kikugawa & Fujimoto (1990) *Eur. J. Biochem.* **189**, 39. **17.** McKinney & Ades (1991) *Int. J. Biochem.* **23**, 803. **18.** Hwang, Lee, Yang, Joo & Kim (2004) *Chem. Biol.* **11**, 915. **19.** Kannangara, Gough, Bruyant, *et al.* (1988) *Trends Biochem. Sci.* **13**, 139. **20.** Smith, Grimm, Kannangara & von Wettstein (1991) *Proc. Natl. Acad. Sci. U.S.A.* **88**, 9775. **21.** Hoober, Kahn, Ash, Gough & Kannangara (1988) *Carlsberg Res. Commun.* **53**, 11. **22.** Allison, Gough, Rogers & Smith (1997) *Mol. Gen. Genet.* **255**, 392. **23.** Bull, Breu, Kannangara, Rogers & Smith (1990) *Arch. Microbiol.* **154**, 56. **24.** Jahn, Chen & Söll (1991) *J. Biol. Chem.* **266**, 161. **25.** Kannangara & Schouboe (1985) *Carlsberg Res. Commun.* **50**, 179. **26.** Smith & Grimm (1992) *Biochemistry* **31**, 4122. **27.** Takechi, Kanda, Hori, Kurotsu & Saito (1994) *J. Biochem.* **116**, 955. **28.** Aniento, Garcia-Espana, Portoles, Alonso & Cabo (1988) *Mol. Cell Biochem.* **79**, 107. **29.** Jung & Seiler (1978) *J. Biol. Chem.* **253**, 7431. **30.** Jeffery, Rutherford, Witzman & Lunt (1988) *Biochem. J.* **249**, 795. **31.** Tamaki, Kubo, Aoyama & Funatsuka (1983) *J. Biochem.* **93**, 955. **32.** Yonaha & Toyama (1980) *Arch. Biochem. Biophys.* **200**, 156. **33.** Van Cauwenberghe & Shelp (1999) *Phytochemistry* **52**, 575. **34.** Yonaha, Suzuki & Toyama (1985) *Eur. J. Biochem.* **146**, 101. **35.** Yun, Lim, Cho & Kim (2004) *Appl. Environ. Microbiol.* **70**, 2529. **36.** Hervieu, Le Dily, Saos, Billard & Huault (1993) *Phytochemistry* **34**, 1231. **37.** Yasuda, Toyama, Rando, *et al.* (1980) *Agric. Biol. Chem.* **44**, 3005. **38.** Shah, Shen & Brunger (1997) *Structure* **5**, 1067. **39.** Yang & Kao (1999)

Plant Growth Regul. **27**, 189. **40.** Rajaram, Prasad, Ratna Prasuna, *et al.* (2006) *Acta Crystallogr. Sect. F* **62**, 980.

Gabapentin

This GABA analogue (FW = 171.24 g/mol; CAS = 60142-96-3; pK_a values of 3.68 and 10.70 at 25°C), also named 1-(aminomethyl)cyclohexaneacetate and Neurontin®, is an anticonvulsant that used to treat epilepsy, and is also effective in moderating neuropathic pain, hot flashes, and restless leg syndrome. It crosses the blood-brain barrier (by means of the System L amino acid transporter), increasing γ-aminobutyrate (GABA) concentrations in the central nervous system. Gabapentin binds with high affinity to α$_2$δ subunit of voltage-dependent calcium channels (K_d = 31 nM). The latter action inhibits the release of excitatory neurotransmitters. When tested *in vitro*, gabapentin modulates the action of the GABA synthetic enzyme, glutamic acid decarboxylase (GAD) and the glutamate synthesizing enzyme, branched-chain amino acid transaminase. MRI on human and rat brain indicates gabapentin increases GABA synthesis. It also increases non-synaptic GABA responses from neuronal tissues and reduces the release of several mono-amine neurotransmitters *in vitro*. **Key Pharmacokinetic Parameters:** *See* Appendix II in Goodman & Gilman's *THE PHARMACOLOGICAL BASIS OF THERAPEUTICS*, 12th Edition (Brunton, Chabner & Knollmann, eds.) McGraw-Hill Medical, New York (2011). **Target(s):** γ-aminobutyrate aminotransferase, very weakly inhibited, K_i = 17-20 mM (1); branched-chain-amino-acid aminotransferase, K_i = 0.8-1.4 mM (1-4); calcium channels, voltage dependent (5,6); dopamine release (7). Gabapentin has no action at brain GABA$_A$ or GABA$_B$ receptors. Gabapentin also activates glutamate dehydrogenase, serves as a non-NMDA receptor antagonist, interacts with NMDA receptors, binds to protein kinase C, interacts with inflammatory cytokines. **1.** Goldlust, Su, Welty, Taylor & Oxender (1995) *Epilepsy Res.* **22**, 1. **2.** Hutson, Berkich, Drown, *et al.* (1998) *J. Neurochem.* **71**, 863. **3.** Goto, Miyahara, Hirotsu, *et al.* (2005) *J. Biol. Chem.* **280**, 37246. **4.** Brosnan & Brosnan (2006) *J. Nutr.* **136**, 207S. **5.** Brown & Gee (1998) *J. Biol. Chem.* **273**, 25458. **6.** Gee, Brown, Dissanayake, *et al.* (1996) *J. Biol. Chem.* **271**, 5768. **7.** Reimann (1983) *Eur. J. Pharmacol.* **94**, 341.

Gabexate Mesylate

This nonpeptide protease inhibitor (FW$_{mesylate-salt}$ = 417.48 g/mol; CAS = 56974-61-9; Symbol: ε-GCA-CEP), also known as *p*-carbethoxyphenyl ε-guanidinocaproate methanesulfonate, and FOY, targets trypsin-like enzymes. Gabexate also binds divalent cations, especially Zn(II) and Cu(II). **Target(s):** amine oxidase, copper-containing (1-4); ancrod (5); arginine transport (6); coagulation factor Xa (3,5,7,8); crotalase5; kallikrein (3,5,8-11); mite protease, *or* Df protease, *or* *Dermatophagoides farinae* protease (12); nitric-oxide synthase (2,3,6); peptidyl-dipeptidase A, angiotensin-converting enzyme, mildly inhibited (13); phospholipase A$_1$ (14); phospholipase A$_2$ (15-18); plasmin (3,5,8,9,11); polyamine oxidase (2); progelatinase A$_1$ (9); thrombin (3,5,7,8,11); t-plasminogen activator (20); trypsin (2,3,5,8,11); tryptase, K_i = 95.1 nM (21,22); u-plasminogen activator, *or* urokinase (3,5,8); venombin A (5). **1.** Ercolini, Angelini, Federico, *et al.* (1998) *J. Enzyme Inhib.* **13**, 465. **2.** Federico, Leone, Botta, *et al.* (2001) *J. Enzyme Inhib.* **16**, 147. **3.** Cortesi, Ascenzi, Colasanti, *et al.* (1998) *J. Pharm. Sci.* **87**, 1335. **4.** Federico, Angelini, Ercolini, *et al.* (1997) *Biochem. Biophys. Res. Commun.* **240**, 150. **5.** Menegatti, Guaneri, Bolognesi, Ascenzi & Amiconi (1989) *J. Enzyme Inhib.* **2**, 249. **6.** Leoncini, Pascale & Signorello (2002) *Biochem. Pharmacol.* **64**, 277. **7.** Ohno, Kosaki, Kambayashi, Imaoka & Hirata (1980) *Thromb. Res.* **19**, 579. **8.** Menegatti, Bolognesi, Scalia, *et al.* (1986) *J. Pharm. Sci.* **75**, 1171. **9.**

Tamura, Ishimaru, Mori, Hirado & Fujii (1979) *Tokushima J. Exp. Med.* **26**, 81. **10**. Nakahara (1983) *Arzneimittelforschung* **33**, 969. **11**. Muramatu & Fujii (1972) *Biochim. Biophys. Acta* **268**, 221. **12**. Matsushima, Kodera, Ozawa, *et al.* (1992) *Biochem. Int.* **28**, 717. **13**. Bonner, Frericks, Buchsler & Kaufmann (1986) *Biol. Chem. Hoppe-Seyler* **367**, 963. **14**. Kunze & Bohn (1983) *Pharmacol. Res. Commun.* **15**, 451. **15**. Freise, Wittenberg & Magerstedt (1989) *Klin. Wochenschr.* **67**, 149. **16**. Schadlich, Buchler & Beger (1989) *Klin. Wochenschr.* **67**, 160. **17**. Hesse, Lankisch & Kunze (1984) *Pharmacol. Res. Commun.* **16**, 637. **18**. Freise, Magerstedt & Schmid (1983) *Enzyme* **30**, 209. **19**. Yoon, Jung, Kim, *et al.* (2004) *Clin. Cancer Res.* **10**, 4517. **20**. Itagaki, Yasuda, Morinaga, Mitsuda & Higashio (1991) *Agric. Biol. Chem.* **55**, 1225. **21**. Erba, Fiorucci, Pascarella, *et al.* (2001) *Biochem. Pharmacol.* **61**, 271. **22**. Mori, Itoh, Shinohata, *et al.* (2003) *J. Pharmacol. Sci.* **92**, 420.

GAE-654, See *Alaproclate*

D-Galactal

This galactose derivative (FW = 146.14 g/mol; CAS = 21193-75-9), also known as 1,2-dideoxy-D-*lyxo*-hex-1-enopyranose, inhibits β-galactosidases. The double bond imposes a half-chair conformation that mimics the geometry of *oxa*-carbenium ion intermediates often formed during glycosidase catalysis. Inhibition of bacterial β-galactosidase by D-galactal was found to be reversible and time-dependent (1,2). Rate constants observed for binding (2.7×10^2 M^{-1}s^{-1}) and release (4.6×10^{-3} s^{-1}) of galactal were found to be consistent with an apparent K_i of 1.4×10 μM for D-galactal as a competitive inhibitor. Efforts to trap and analyze the galactalenzyme complex were unsuccessful. **Target(s):** α-galactosidase (1); β-galactosidase (1-8); lactase/ phlorizin hydrolase, *or* glycosylceramidase (9). **1**. Wentworth & Wolfenden (1974) *Biochemistry* **13**, 4715. **2**. Wolfenden (1977) *Meth. Enzymol.* **46**, 15. **3**. Distler & G. W. Jourdian (1978) *Meth. Enzymol.* **50**, 514. **4**. Mooser (1992) *The Enzymes*, 3rd ed., **20**, 187. **5**. Lee (1969) *Biochem. Biophys. Res. Commun.* **35**, 161. **6**. Sekimata, Ogura, Tsumuraya, Hashimoto & Yamamoto (1989) *Plant Physiol.* **90**, 567. **7**. Coker, Sheridan, Loveland-Curtze, *et al.* (2003) *J. Bacteriol.* **185**, 5473. **8**. Frank & Somkuti (1979) *Appl. Environ. Microbiol.* **38**, 554. **9**. Arribas, Herrero, Martin-Lomas, *et al.* (2000) *Eur. J. Biochem.* **267**, 6996.

Galactarate

This dicarboxylic galactose derivative (FW$_{free-acid}$ = 210.14 g/mol; CAS = 526-99-8; pK_a values of 3.08 and 3.63; *meso* structure; No optical activity), also called galactaric acid mucic acid, is often used as a counter ion in formulations of cationic drugs. Galactarate is a product of the reaction catalyzed by uronate dehydrogenase and is a substrate for galactarate *O*-hydroxycinnamoyl-transferase and galactarate dehydratase. The free acid has a low solubility in water (0.33 g per 100 mL at 14°C) but the salt is very soluble. Galactaric acid does not react with Fehling's solution; however, it will reduce ammoniacal silver nitrate. **Target(s):** 2-dehydro-3-deoxyglucarate aldolase (1); glucarate dehydratase (2); β-glucuronidase (3-8); kynurenine aminotransferase, weakly inhibited (9). **1**. Fish & Blumenthal (1966) *Meth. Enzymol.* **9**, 529. **2**. Blumenthal (1966) *Meth. Enzymol.* **9**, 660. **3**. Levvy (1952) *Biochem. J.* **52**, 464. **4**. Webb (1966) *Enzyme and Metabolic Inhibitors*, vol. **2**, pp. 424, 427, Academic Press, New York. **5**. Spencer & Williams (1951) *Biochem. J.* **48**, 538. **6**. Mills, Paul & Smith (1953) *Biochem. J.* **53**, 232. **7**. Levvy, Hay & Marsh (1957) *Biochem. J.* **65**, 203. **8**. Marsh & Levvy (1958) *Biochem. J.* **68**, 610. **9**. Mason (1959) *J. Biol. Chem.* **234**, 2770.

Galactodeoxynojirimycin Aziridine

This unusual aminosugar analogue (FW = 145.16 g/mol), also called 1-deoxy-D-galactonojirimycin aziridine and *galacto*deoxy-nojirimycin aziridine, inhibits α-galactosidases. When the ring nitrogen is protonated, this analogue mimics the charge of oxa-carbenium ion intermediate believed to form transiently in many glycosidase reactions. Aziridines often react with protein thiols, thiol-containing purines and pyrimidines, as well as some oxygen-centered nucleophiles. **1**. Mooser (1992) *The Enzymes*, 3rd ed., **20**, 187.

Galactomannans

These polysaccharides (MW = indefinite; CAS = 11078-30-1), which consist of a poly(1-4)-linked β-D-mannopyranose backbone, with branchpoints at their 6-positions that are linked to α-D-galactopyranose, include fenugreek gum galactomannan, ~1 mannose per galactose; guar gum galactomannan, mannose:galactose, ~2 mannose per galactose; tara gum galactomannan, mannose:galactose, ~3 mannose per galactose; and locust bean gum galactomannan, ~4 mannose per galactose. Galactomannan and zymosan block epinephrine-induced particle transport in tracheal epithelial cells, which rely on ciliary beating to continuously purge pathogens from the lower airways. **1**. Weiterer, Kohlen, Veit, *et al.* (2015) *PLoS One* **10**, e0143163.

D-Galactono-1,4-lactone

This γ-lactone (FW = 178.14 g/mol; CAS = 2782-07-2; M.P. (hydrate) = 66°C), which is readily formed from D-galactonic acid, has an ester carbonyl that imposes a half-chair conformation that mimics the geometry of an *oxa*-carbenium ion or similar transition-state intermediates formed during glycosidase catalysis. *See also D-Galactono 1,5-lactone* **Target(s):** aldose 1-epimerase, *or* mutarotase (1); aryl-β-hexosidase (2); β-D-fucosidase (3,4); α-L-fucosidase, weakly inhibited (5); β-galactofuranosidase (6); α-galactosidase (7); β-galactosidase (8-20); galactosyl-ceramidase (21,22); ganglioside GM$_1$ β-galactosidase (14,23); β-glucosidase (24,25); lactase/phlorizin hydrolase, *or* glycosylceramidase (26,27); neutral glycosidase (28). **1**. Hucho & K. Wallenfels (1971) *Eur. J. Biochem.* **23**, 489. **2**. Distler & Jourdian (1978) *Meth. Enzymol.* **50**, 524. **3**. Wiederschain & Prokopenkov (1973) *Arch. Biochem. Biophys.* **158**, 539. **4**. Levvy & McAllan (1963) *Biochem. J.* **87**, 206. **5**. Reglero & Cabezas (1976) *Eur. J. Biochem.* **66**, 379. **6**. Wallis, Hemming & Peberdy (2001) *Biochim. Biophys. Acta* **1525**, 19. **7**. Grossmann & Terra (2001) *Comp. Biochem. Physiol. B* **128**, 109. **8**. Levvy & Conchie (1966) *Meth. Enzymol.* **8**, 571. **9**. Gatt (1969) *Meth. Enzymol.* **14**, 156. **10**. Meisler (1972) *Meth. Enzymol.* **28**, 820. **11**. Distler & Jourdian (1978) *Meth. Enzymol.* 50, 514. **12**. Wallenfels & Weil (1972) *The Enzymes*, 3rd ed., **7**, 617. **13**. Conchie & Hay (1959) *Biochem. J.* **73**, 327. **14**. Callahan & Gerrie (1975) *Biochim. Biophys. Acta* **391**, 141. **15**. Kuo & Wells (1978) *J. Biol. Chem.* **253**, 3550. **16**. Huber, Roth & Bahl (1990) *J. Protein Chem.* **15**, 621. **17**. Sekimata,

Ogura, Tsumuraya, Hashimoto & Yamamoto (1989) *Plant Physiol.* **90**, 567. **18**. Konno, Yamasaki & Katoh (1986) *Plant Sci.* **44**, 97. **19**. Li, Han, Chen & Chen (2001) *Phytochemistry* **57**, 349. **20**. Ferreira, Terra & Ferreira (2003) *Insect Biochem. Mol. Biol.* **33**, 253. **22**. Radin (1972) *Meth. Enzymol.* **28**, 834. **22**. Radin (1972) *Meth. Enzymol.* **28**, 844. **23**. Sloan (1972) *Meth. Enzymol.* **28**, 868. **24**. Heyworth & Walker (1962) *Biochem. J.* **83**, 331. **25**. Ferreira & Terra (1983) *Biochem. J.* **213**, 43. **26**. Malathi & Crane (1969) *Biochim. Biophys. Acta* **173**, 245. **27**. Kraml, Kolínská, Ellederová & Hirsová (1972) *Biochim. Biophys. Acta* **258**, 520. **28**. Cygan, Neufeld-Kaiser, Jara & Daniel (1997) *Comp. Biochem. Physiol. B Biochem. Mol. Biol.* **116**, 437.

D-Galactosamine

This amino sugar (FW$_{\text{free-base}}$ = 179.17 g/mol; pK_a = 7.70; CAS = 7535-00-4), also known as 2-amino-2-deoxy-D-galactose and chondrosamine, found in mucopolysaccharides, glycoproteins, and mucoproteins. 7-Galactosamine inhibits hepatic RNA biosynthesis. **Target(s):** β-*N*-acetylglucosaminidase (1,2); β-*N*-acetylhexosaminidase (1-6); L-arabinose dehydrogenase (7); chitinase (8(; galactose dehydrogenase (9); galactose oxidase, also an alternative substrate (10); β-galactosidase (11,12); glucose-1-phosphate uridylyltransferase (13); β-glucosidase, K_i = 790 mM for the sweet almond enzyme (14). **1**. Frohwein & Gatt (1967) *Biochemistry* **6**, 2775. **2**. Khar & Anand (1977) *Biochim. Biophys. Acta* **483**, 141. **3**. Frohwein & Gatt (1969) *Meth. Enzymol.* **14**, 161. **4**. Johnson, Mook & Brady (1972) *Meth. Enzymol.* **28**, 857. **5**. Potier, Teitelbaum, Melancon & Dallaire (1979) *Biochim. Biophys. Acta* **566**, 80. **6**. Gers-Barlag, Bartz & Ruediger (1988) *Phytochemistry* **27**, 3739. **7**. Doudoroff (1962) *Meth. Enzymol.* **5**, 342. **8**. Bhushan (2000) *J. Appl. Microbiol.* **88**, 800. **9**. Doudoroff (1962) *Meth. Enzymol.* **5**, 339. **10**. Yip & Dain (1968) *Enzymologia* **35**, 368. **11**. Huber & Gaunt (1982) *Can. J. Biochem.* **60**, 608. **12**. Nakao, Harada, Kodama, *et al.* (1994) *Appl. Microbiol. Biotechnol.* **40**, 657. **13**. Turnquist & Hansen (1973) *The Enzymes*, 3rd ed., **8**, 51. **14**. Dale, Ensley, Kern, Sastry & Byers (1985) *Biochemistry* **24**, 3530.

D-Galactose

α-D-galactopyranose   Fischer Projection   β-D-galactopyranose

This widely occurring D-aldohexose (FW = 180.16 g/mol; CAS = 59-23-4), found in free form and as a component of glycoproteins, mucopolysaccharides, lactose, galactans, gangliosides, and cerebrosides. Note that the enantiomer, L-galactose, also occurs naturally. β-D-galactopyranose is most prevalent (30% α-pyranose, 64% β-pyranose, 2.5% α-furanose, 3.5% β-furanose, and 0.02% straight chain in D$_2$O at 31°C). **Target(s):** α-*N*-acetylgalactosaminidase (1-3); β-*N*-acetyl-hexosaminidase, weakly inhibited (4,5); D-*threo*-aldose dehydrogenase (6); aldose 1-epimerase, *or* mutarotase (7-11); β-amylase (10,12); arabinogalactan endo-1,4-β-galactosidase (13); aralin (14); discoidins II (15); β-fructofuranosidase (invertase; inhibited by both anomers) (16,17); fructokinase (18); β- D-fucosidase (19-23); α-L-fucosidase (24,25); α-galactosidase (10,26-42); β-galactosidase (10,43-64); galactosylceramidase (65); galactosylgalactosylglucosyl-ceramidase, weakly inhibited (66,67); glucan 1,4-α-glucosidase, *or* glucoamylase (68); glucokinase (69); α-glucosidase (10,70,71); β-glucosidase (72-75); mannosyl-glycoprotein endo-β-*N*-acetylglucosaminidase (76); mucinaminylserine mucinaminidase (77); polypeptide *N*-acetylgalactosaminyltransferase (82); rRNA *N*-glycosylase (14); thioglucosidase, myrosinase (78,79); α,α-trehalose phosphorylase, configuration retaining (80); UDP-glucuronate 4-epimerase (81). **1**. Uda, Li & Li (1977) *J. Biol. Chem.* **252**, 5194. **2**. Chien (1986) *J.*

Chin. Biochem. Soc. **15**, 86. **3**. Itoh & Uda (1984) *J. Biochem.* **95**, 959. **4**. Gers-Barlag, Bartz & Ruediger (1988) *Phytochemistry* **27**, 3739. **5**. Sakai, Narihara, Kasama, Wakayama & Moriguchi (1994) *Appl. Environ. Microbiol.* **60**, 2911. **6**. Sasajima & Sinskey (1979) *Biochim. Biophys. Acta* **571**, 120. **7**. Bentley (1962) *Meth. Enzymol.* **5**, 219. **8**. Bailey, Fishman, Kusiak, Mulhern & Pentchev (1975) *Meth. Enzymol.* **41**, 471. **9**. Fishman, Pentchev & Bailey (1975) *Meth. Enzymol.* **41**, 484. **10**. Webb (1966) *Enzyme and Metabolic Inhibitors*, vol. **2**, Academic Press, New York. **11**. Bailey & Pentchev (1964) *Biochem. Biophys. Res. Commun.* **14**, 161. **12**. Wohl & Glimm (1910) *Biochem. Z.* **27**, 349. **13**. Nakano, Takenishi & Watanabe (1985) *Agric. Biol. Chem.* **49**, 3445. **14**. Tomatsu, Kondo, Yoshikawa, *et al.* (2004) *Biol. Chem.* **385**, 819. **15**. Barondes, Rosen, Frazier, Simpson & Haywood (1978) *Meth. Enzymol.* **50**, 306. **16**. Neuberg & Mandl (1950) *The Enzymes*, 1st ed. (J. B. Sumner & K. Myrbäck, eds.), **1** (part 1), 527. **17**. Isla, Vattuone, Gutierrez & Sampietro (1988) *Phytochemistry* **27**, 1993. **18**. Bueding & MacKinnon (1955) *J. Biol. Chem.* **215**, 495. **19**. Giordani & Lafon (1993) *Phytochemistry* **33**, 1327. **20**. Chinchetru, Cabezas & Calvo (1983) *Comp. Biochem. Physiol.* **75**, 719. **21**. Rodriguez, Cabezas & Calvo (1982) *Int. J. Biochem.* **14**, 695. **22**. Melgar, Cabezas & Calvo (1985) *Comp. Biochem. Physiol.* **80**, 149. **23**. Wiederschain & Prokopenkov (1973) *Arch. Biochem. Biophys.* **158**, 539. **24**. Grove & Serif (1981) *Biochim. Biophys. Acta* **662**, 246. **25**. Reglero & Cabezas (1976) *Eur. J. Biochem.* **66**, 379. **26**. Grossmann & Terra (2001) *Comp. Biochem. Physiol. B* **128**, 109. **27**. Bryant & Rao (2001) *J. Food Biochem.* **25**, 139. **28**. Zapater, Ullah & Wodzinsky (1990) *Prep. Biochem.* **20**, 263. **29**. King, White, Blaschek, *et al.* (2002) *J. Agric. Food Chem.* **50**, 5676. **30**. Gaudreault & Webb (1983) *Plant Physiol.* **71**, 662. **31**. Schuler, Mudgett & Mahoney (1985) *Enzyme Microb. Technol.* **7**, 207. **32**. Guimaraes, Tavares de Rezende, Moreira, Goncalves de Barros & Felix (2001) *Phytochemistry* **58**, 67. **33**. Oishi & Aida (1975) *Agric. Biol. Chem.* **39**, 2129. **34**. Lazo, Ochoa & Gascon (1978) *Arch. Biochem. Biophys.* **191**, 316. **35**. Soh, Ali & Lazan (2006) *Phytochemistry* **67**, 242. **36**. Sripuan, Aoki, Yamamoto, Tongkao & Kumagai (2003) *Biosci. Biotechnol. Biochem.* **67**, 1485. **37**. Dey & Kaus (1981) *Phytochemistry* **20**, 45. **38**. Mujer, Ramirez & Mendoza (1984) *Phytochemistry* **23**, 1251. **39**. Corchete & Guerra (1987) *Phytochemistry* **26**, 927. **40**. Ueno, Ikami, Yamauchi & Kato (1980) *Agric. Biol. Chem.* **44**, 2623. **41**. Chinen, Nakamura & Fukuda (1981) *J. Biochem.* **90**, 1453. **42**. Ohtakara, Mitsutomi & Uchida (1984) *Agric. Biol. Chem.* **48**, 1319. **43**. Distler & Jourdian (1978) *Meth. Enzymol.* **50**, 514. **44**. Nguyen, Splechtna, Steinböck, *et al.* (2006) *J. Agric. Food Chem.* **54**, 4989. **45**. Kuo & Wells (1978) *J. Biol. Chem.* **253**, 3550. **46**. Huber, Roth & Bahl (1990) *J. Protein Chem.* **15**, 621. **47**. Tanaka, Kagamiishi, Kiuchi & Horiuchi (1975) *J. Biochem.* **77**, 241. **48**. Miyazaki (1988) *Agric. Biol. Chem.* **52**, 625. **49**. Chakraborti, Sani, Banerjee & Sobti (2000) *J. Ind. Microbiol. Biotechnol.* **24**, 58. **50**. Sekimata, Ogura, Tsumuraya, Hashimoto & Yamamoto (1989) *Plant Physiol.* **90**, 567. **51**. Nakao, Harada, Kodama, *et al.* (1994) *Appl. Microbiol. Biotechnol.* **40**, 657. **52**. Ikura & Horikoshi (1979) *Agric. Biol. Chem.* **43**, 1359. **53**. Konno, Yamasaki & Katoh (1986) *Plant Sci.* **44**, 97. **54**. Li, Han, Chen & Chen (2001) *Phytochemistry* **57**, 349. **55**. Choi, Kim, Lee & Lee (1995) *Biotechnol. Appl. Biochem.* **22**, 191. **56**. Itoh, Suzuki & Adachi (1982) *Agric. Biol. Chem.* **46**, 899. **57**. Goodman & Pederson (1976) *Can. J. Microbiol.* **22**, 817. **58**. Cowan, Daniel, Martin & Morgan (1984) *Biotechnol. Bioeng.* **26**, 1141. **59**. Hung & Lee (2002) *Appl. Microbiol. Biotechnol.* **58**, 439. **60**. Nadder de Macias, Manca de Nadra, Strasser de Saad, Pesce de Ruiz Holgado & Oliver (1983) *J. Appl. Biochem.* **5**, 275. **61**. Levin & Mahoney (1981) *Antonie Leeuwenhoek* **47**, 53. **62**. Coker, Sheridan, Loveland-Curtze, *et al.* (2003) *J. Bacteriol.* **185**, 5473. **63**. Frank & Somkuti (1979) *Appl. Environ. Microbiol.* **38**, 554. **64**. Shaikh, Khire & Khan (1999) *Biochim. Biophys. Acta* **1472**, 314. **65**. Rushton & Dawson (1975) *Biochim. Biophys. Acta* **388**, 92. **66**. Poulos & Beckman (1978) *Clin. Chim. Acta* **89**, 35. **67**. Rietra, Tager & Borst (1972) *Biochim. Biophys. Acta* **279**, 436. **68**. Wong, Batt, Lee, Wagschal & Robertson (2005) *Protein J.* **24**, 455. **69**. Walker & Parry (1966) *Meth. Enzymol.* **9**, 381. **70**. Michaelis & Rona (1914) *Biochem. Z.* **60**, 62. **71**. Bravo-Torres, Villagómez-Castro, Calvo-Méndez, Flores-Carreón & López Romero (2004) *Int. J. Parasitol.* **34**, 455. **72**. Heyworth & Walker (1962) *Biochem. J.* **83**, 331. **73**. Dale, Ensley, Kern, Sastry & Byers (1985) *Biochemistry* **24**, 3530. **74**. Seidle, Marten, Shoseyov & Huber (2004) *Protein J.* **23**, 11. **75**. Harada, Tanaka, Fukuda, Hashimoto & Murata (2005) *Arch. Microbiol.* **184**, 215. **76**. Koide & Muramatsu (1975) *Biochem. Biophys. Res. Commun.* **66**, 411. **77**. Umemoto, Bhavanandan & Davidson (1977) *J. Biol. Chem.* **252**, 8609. **78**. Tani, Ohtsuru & Hata (1974) *Agric. Biol. Chem.* **38**, 1623. **70**. Tsuruo & Hata (1968) *Agric. Biol. Chem.* **32**, 1420. **80**. Eis & Nidetzky (2002) *Biochem. J.* **363**, 335. **81**. Munoz, Lopez, de Frutos &

Garcia (1999) *Mol. Microbiol.* **31**, 703. **82**. Wandall, Irazoqui, Tarp, *et al.* (2007) *Glycobiology* **17**, 374.

α-D-Galactose 1-Phosphate

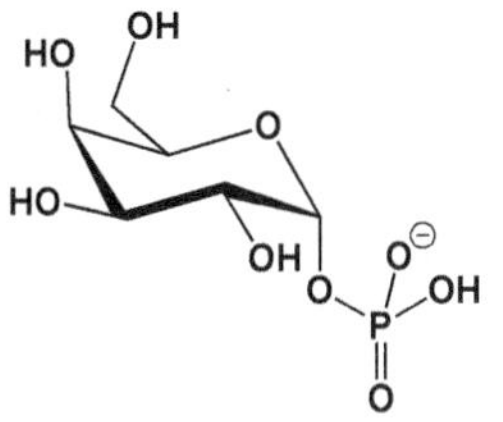

This phosphosugar (FW = 260.14 g/mol; CAS = 2255-14-3; pK_1 = 1.00 and pK_2 = 6.17) is a product galactokinase reaction and a substrate for UTP:hexose-1-phosphate uridylyltransferase, UDP-glucose:hexose-1-phosphate uridylyltransferase, galactose-1-phosphate thymidylyltransferase, and galactose-1 phosphatase. **Target(s):** β-phosphoglucomutase, K_i = 30 μM (1); plant polysaccharide biosynthesis (2); α,α-trehalose phosphorylase, configuration-retaining (3); UTP:glucose-1-phosphate uridylyltransferase, *or* glucose-1-phosphate uridylyltransferase (4-7). **1**. Tremblay, Zhang, Dai, Dunaway-Mariano & Allen (2005) *J. Amer. Chem. Soc.* **127**, 5298. **2**. Goring & Reckin (1968) *Naturwissenschaften* **55**, 40. **3**. Eis & Nidetzky (2002) *Biochem. J.* **363**, 335. **4**. Turnquist & Hansen (1973) *The Enzymes*, 3rd ed., **8**, 51. **5**. Oliver (1961) *Biochim. Biophys. Acta* **52**, 75. **6**. Lee, Kimura & Tochikura (1979) *J. Biochem.* **86**, 923. **7**. Turnquist, Gillett & Hansen (1974) *J. Biol. Chem.* **249**, 7695.

D-Galactose 6-Phosphate

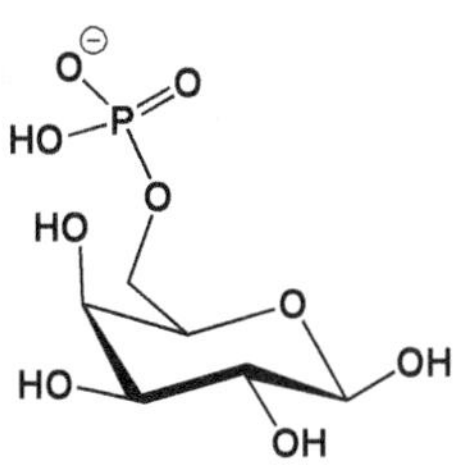

This phosphorylated hexose (FW = 260.14 g/mol; CAS = 6665-00-5) is the product of reactions catalyzed by 6-phospho-β-galactosidase and is a substrate for galactose 6-phosphate isomerase. It is also an alternative substrate of glucose-6-phosphate dehydrogenase. **Target(s):** β-galactosidase (1,2); glucose-6-phosphate isomerase (3); inositol-3-phosphate synthase, *or* inositol-1-phosphate synthase (4,5); mannose-6-phosphate isomerase, weakly inhibited (3,6); 6-phospho-β-galactosidase (7-11); phosphopentomutase (12); UTP:glucose-1-phosphate uridylyltransferase (13). **1**. Weinland (1956) *Hoppe-Seyler's Z. Physiol. Chem.* **306**, 66. **2**. Nadder de Macias, Manca de Nadra, Strasser de Saad, Pesce de Ruiz Holgado & Oliver (1983) *J. Appl. Biochem.* **5**, 275. **3**. Noltmann (1972) *The Enzymes*, 3rd ed., **6**, 271. **4**. RayChaudhuri, Hait, das Gupta, Bhaduri, R. Deb & Majumder (1997) *Plant Physiol.* **115**, 727. **5**. Naccarato, Ray & Wells (1974) *Arch. Biochem. Biophys.* **164**, 194. **6**. Gracey & Noltmann (1968) *J. Biol. Chem.* **243**, 5410. **7**. Calmes & Brown (1979) *Infect. Immun.* **23**, 68. **8**. Thompson (2002) *FEMS Microbiol. Lett.* **214**, 183. **9**. Hall (1979) *J. Bacteriol.* **138**, 691. **10**. Witt, Frank & Hengstenberg (1993) *Protein Eng.* **6**, 913. **11**. Johnson & McDonald (1974) *J. Bacteriol.* **117**, 667. **12**. Webb (1966) *Enzyme and Metabolic Inhibitors*, vol. 2, p. 413, Academic Press, New York. **13**. Lee, Kimura & Tochikura (1979) *J. Biochem.* **86**, 923.

Galactosylceramide

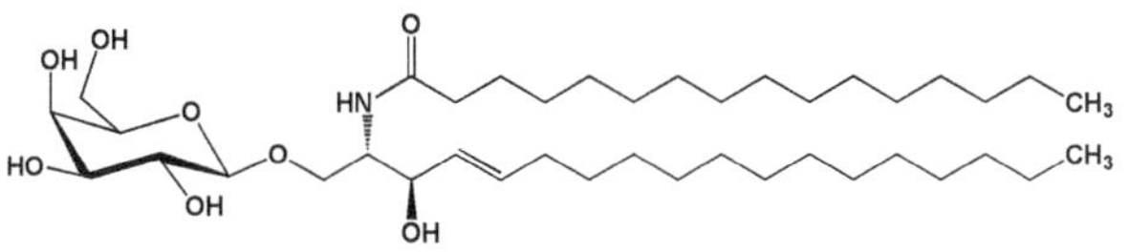

β-D-galactosyl-*N*-palmitoylsphingosine

This term refers to any D-galactosyl-*N*-acylsphingosine (FW = 700.06 g/mol for β-D-galactosyl-*N*-palmitoylsphingosine, as shown above), major lipids found in adult brain, with levels elevated in individuals with Krabbe's

disease and reduced levels in several demyelinating diseases. Galactosylceramide (GalCer) is an alternative receptor allowing human immunodeficiency virus (HIV)-1 entry into CD4-negative cells of neural and colonic origin. A soluble GalCer analogue, known as CA52(N15), is a new class of anti-HIV-1 agents that neutralize HIV-1 infection by masking of the V3 loop of the HIV-1 surface envelope glycoprotein gp120 (1). **Target(s):** *N*-acetyl-β-hexosaminidase (2); lactase/phlorizin hydrolase, *or* glycosylceramidase, as an alternative substrate (3). **1**. Fantini, Hamache, Delézay, *et al.* (1997) *J. Biol. Chem.* **272**, 7245. **2**. Frohwein Frohwein & Gatt (1967) *Biochemistry* **6**, 2783. **3**. Leese & Semenza (1973) *J. Biol. Chem.* **248**, 8170.

D-Galacturonate

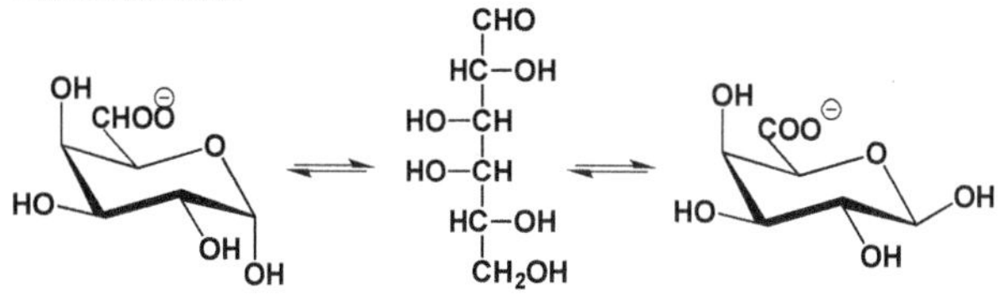

α-anomer	Fischer Projection	β-anomer

This uronic acid (FW = 194.14 g/mol; CAS = 14982-50-4; Soluble in water; Slightly soluble in hot ethanol; Insoluble in diethyl ether; M.P. (with decomposition) of 159°C and 166°C for the α- and β-anomers), also known as D-galacturonic acid, D-galactopyranuronic acid, is an abundant component in pectins, plant gums, galacturonides, and bacterial cell walls (1). **Target(s):** β-*N*-acetylhexosaminidase (2); aldose 1-epimerase (3); L-arabinose dehydrogenase (4); galactose dehydrogenase (5); galacturan 1,4-α-galacturonidase, *or* exopolygalacturonase (6,7); α-glucuronidase (8); β-glucuronidase (8-11); pectinesterase, weakly inhibited (12); polygalacturonidase, *or* pectinase (8,13,14). **1**. Link & Sell (1953) *Biochem. Prep.* **3**, 74. **2**. Johnson, Mook & Brady (1972) *Meth. Enzymol.* **28**, 857. **3**. Bailey, Fishman, Kusiak, Mulhern & Pentchev (1975) *Meth. Enzymol.* **41**, 471. **4**. Doudoroff (1962) *Meth. Enzymol.* **5**, 342. **5**. Doudoroff (1962) *Meth. Enzymol.* **5**, 339. **6**. Kester, Kusters-van Someren, Müller & Visser (1996) *Eur. J. Biochem.* **240**, 738. **7**. Kester, Benen & Visser (1999) *Biotechnol. Appl. Biochem.* **30**, 53. **8**. Webb (1966) *Enzyme and Metabolic Inhibitors*, vol. 2, pp. 424-427, Academic Press, New York. **9**. Levvy & Marsh (1960) *The Enzymes*, 2nd ed., **4**, 397. **10**. Levvy (1952) *Biochem. J.* **52**, 464. **11**. Marsh & Levvy (1958) *Biochem. J.* **68**, 610. **12**. Pitkänen, Heikinheimo & Pakkanen (1992) *Enzyme Microb. Technol.* **14**, 832. **13**. Sakai, Okushima & Sawada (1982) *Agric. Biol. Chem.* **46**, 2223. **14**. Blanco, Sieiro, Diaz & Villa (1994) *Can. J. Microbiol.* **40**, 974.

Galanal B

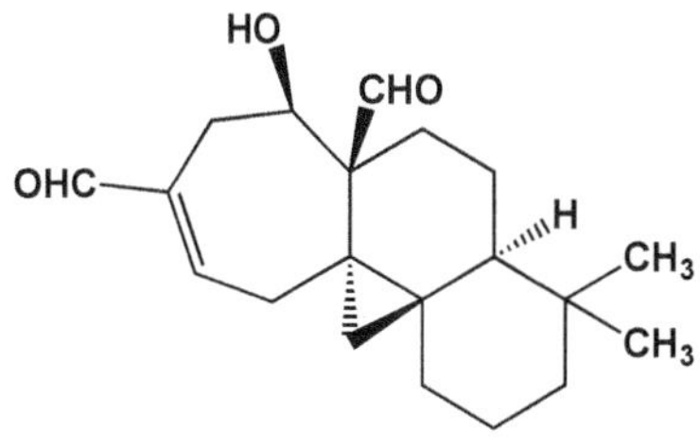

This putative immunomodulator from the Myoga flower (FW = 318.45 g/mol), also named (4aS,6aS,7R,11aR,11bS)-7-hydroxy-4,4,11b-trimethyl-1,2,3,4,4a,5,6,7,8,11,11a-decahydrocyclo-hepta[*a*]naphthalene-6a,9-dicarb-aldehyde and AC1MJ6B7, inhibits indoleamine 2,3-dioxygenase 1 (IDO1), which catalyzes the first (and rate-limiting) step in L-tryptophan degradation. IDO1 activity increases in various diseases, including tumors, autoimmune diseases, and different kinds of inflammatory disorders. Notably, IDO1 is induced by the interferon-γ-mediated effects of STAT1-α and IRF-1. Induction of IDO1 also occurs through an IFN-γ-independent mechanism and by lipopolysaccharide (LPS),the latter regulated both by the p38 mitogen-activated protein kinase (MAPK) pathway and nuclear factor-κB (NF-κB). Galanal proved to be a competitive inhibitor (IC$_{50}$ = 7.7 μM), attenuating L-kynurenine formation with an *in vitro* assay using recombinant human IDO1, but more potently (IC$_{50}$ = 45 nM) in cell-based assays. Noting that the stronger inhibitory effect of galanal than that of 1-methyl tryptophan, the authors posit that galanal may have potential for treating various immune-related diseases. **1**. Yamamoto, Yamamoto, Imai, *et al.* (2014) *PLoS One* **9**, e88789.

Galangin

This yellow flavone (FW = 270.24 g/mol; CAS = 548-83-4), derived from galanga root (*Alpininia officinarum*) and pines, chelates copper and inhibits various copper-dependent enzymes. **Target(s):** alcohol dehydrogenase (1); aromatase, *or* CYP19 (2); carbonyl reductase (3); cyclooxygenase (4); CYP1A1 (5,6); CYP1A2 (5,7); DNA topoisomerase II (8); estrone sulfotransferase (9); fatty-acid synthase, weakly inhibited (22,23); glutathione-disulfide reductase (10); glutathione *S*-transferase (11); inositol-trisphosphate 3-kinase (12); β-lactamase, *or* metallo-β-lactamase (13); 5-lipoxygenase, *or* arachidonate 5-lipoxygenase (4); 15-lipoxygenase, *or* arachidonate 15-lipoxygenase (25); lipoxygenase, soybean (25); [myosin light-chain] kinase (14); NADH oxidase, weakly inhibited (15); phenol sulfotransferase, *or* aryl sulfotransferase (9,16); phenylpyruvate tautomerase (17); protein tyrosine kinase, *or* non-specific protein-tyrosine kinase (18); ribosyldihydronicotinamide dehydrogenase (quinone), IC$_{50}$ = 3.0 μM (26,27); tyrosinase, *or* monophenol monooxygenase (19-21,24); suppresseion of β-catenin response transcription (CRT), which is aberrantly up-regulated in colorectal and liver cancer (28). **1.** Keung (1993) *Alcohol Clin. Exp. Res.* **17**, 1254. **2.** Kao, Zhou, Sherman, Laughton & Chen (1998) *Environ. Health Perspect.* **106**, 85. **3.** Imamura, Migita, Uriu, Otagiri & Okawara (2000) *J. Biochem.* **127**, 653. **4.** Hoult, Moroney & Payá (1994) *Meth. Enzymol.* **234**, 443. **5.** Zhai, Dai, Friedman & Vestal (1998) *Drug Metab. Dispos.* **26**, 989. **6.** Ciolino & Yeh (1999) *Brit. J. Cancer* **79**, 1340. **7.** Zhai, Dai, Wei, Friedman & Vestal (1998) *Life Sci.* **63**, PL119. **8.** Snyder & Gillies (2002) *Environ. Mol. Mutagen.* **40**, 266. **9.** Harris, Wood, Bottomley, *et al.* (2004) *J. Clin. Endocrinol. Metab.* **89**, 1779. **10.** Cipak, Berczeliova & Paulikova (2003) *Neoplasma* **50**, 443. **11.** van Zanden, Geraets, Wortelboer, *et al.* (2004) *Biochem. Pharmacol.* **67**, 1607. **12.** Mayr, Windhorst & Hillemeier (2005) *J. Biol. Chem.* **280**, 13229. **13.** Denny, Lambert & West (2002) *FEMS Microbiol. Lett.* **208**, 21. **14.** Jinsart, Ternai & Polya (1991) *Biol. Chem. Hoppe-Seyler* **372**, 819. **15.** Bohmont, Aaronson, Mann & Pardini (1987) *J. Nat. Prod.* **50**, 427. **16.** Eaton, Walle, Lewis, *et al.* (1996) *Drug Metab. Dispos.* **24**, 232. **17.** Molnar & J. Garai (2005) *Int. Immunopharmacol.* **5**, 849. **18.** Hollósy & Kéri (2004) *Curr. Med. Chem. Anticancer Agents* **4**, 173. **19.** Xie, Chen, Huang, Wang & Zhang (2003) *Biochemistry (Moscow)* **68**, 487. **20.** Kubo, Kinst-Hori, Chaudhuri, *et al.* (2000) *Bioorg. Med. Chem.* **8**, 1749. **21.** Kubo & Kinst-Hori (1999) *J. Agric. Food Chem.* **47**, 4121. **22.** Li & Tian (2004) *J. Biochem.* **135**, 85. **23.** Li & Tian (2003) *J. Enzyme Inhib. Med. Chem.* **18**, 349. **24.** Lu, Lin-Tao, Wang, Wei & Xiang (2007) *J. Enzyme Inhib. Med. Chem.* **22**, 433. **25.** Sadik, Sies & Schewe (2003) *Biochem. Pharmacol.* **65**, 773. **26.** Wu, Knox, Sun, Joseph, *et al.* (1997) *Arch. Biochem. Biophys.* **347**, 221. **27.** Boutin, Chatelain-Egger, Vella, Delagrange & Ferry (2005) *Chem.-Biol. Interact.* **151**, 213. **28.** Gwak, Oh, Cho, *et al.* (2011) *Molec. Pharmacol.* **79**, 1014.

Galanin

This oligopeptide (MW = 3210.52 g/mol; CAS = 88813-36-9), itself widely distributed in central and peripheral nervous system as well as the endocrine system, participates in the regulation of processes such as nociception, cognition, feeding behavior, and insulin secretion. Galinin inhibits acetylcholine release, increases the levels of growth hormone, prolactin and luteinizing hormone, inhibits glucose-induced insulin release, and affects gastrointestinal motility. The human peptide contains thirty aminoacyl residues: GWTLNSAGYLLGPHAVGNHRSFSDKNGLTS. The rat peptide contains 29 residues, with Ile-16, Asp-17, His-26, and Thr(NH$_2$) at position-29. The porcine peptide also contains 29 residues with Ile-12, Asp-17, His-23, Tyr-26, and Thr(NH$_2$) at position-29. **Target(s):** acetylcholine release (1-3); Ca^{2+}-channels, voltage-activated (1,2,4); glutamate release (1-3); insulin release (3,5,6). **1.** Vrontakis (2002) *Curr. Drug Targets CNS Neurol. Disord.* **1**, 531. **2.** Iismaa & Shine (1999) *Results Probl. Cell Differ.* **26**, 257. **3.** Bartfai, Hokfelt & Langel (1993) *Crit. Rev. Neurobiol.* **7**, 229. **4.** Palazzi, Felinska, Zambelli, *et al.* (1991) *J. Neurochem.* **56**, 739. **5.** Ahren, Arkhammar, Berggren & Nilsson (1986) *Biochem. Biophys. Res.*

Commun. **140**, 1059. **6.** McDonald, Dupre, Tatemoto, *et al.* (1985) *Diabetes* **34**, 192.

Galanin Fragment (1-13)-Neuropeptide Y Fragment (25-36) Amide

This neuropeptide fragment (FW = 2963.37 g/mol; Sequence: GWTLN-SAGYLLGPRHYINLITRQRY-NH$_2$), also known as M32, is a potent galanin receptor antagonist (IC$_{50}$ = 0.1 nM in rat hypothalamus and 0.01 nM in rat spinal cord). *See Galanin* **1.** Arvidsson, Land, Langel, Bartfai & Ehrenberg (1993) *Biochemistry* **32**, 7787. **2.** Kahl, Langel, Bartfai & Grundemar (1994) *Brit. J. Pharmacol.* **111**, 1129.

Galanin Fragment (1-13)-Spantide I

This chimeric neuropeptide and galanin receptor antagonist (MW = 2829.30 g/mol; Sequence: GWTLNSAGYLLGPRPKPQQWFWLL-NH$_2$) is a galanin receptor antagonist (IC$_{50}$ = 0.2 nM), significantly inhibiting galanin-induced consumption of a palatable wet cookie mash, when microinjected intraventricularly to satiated rats. It consists of the first thirteen residues of galangin linked to spantide I, a substance P analogue that is a potent NK-1 tachykinin receptor antagonist. *See Galanin* **1.** Crawley, Robinson, Langel & Bartfai (1993) *Brain Res.* **600**, 268.

Galanin Fragment (1-13)-Substance P Fragment (5-11) Amide

This chimeric neuropeptide fragment (MW = 2223.48 g/mol; Sequence: GWTLNSAGWLLGPQQFFGLM-NH$_2$), also known as M-15 and galantide, is a galanin receptor antagonist (IC$_{50}$ ≈ 0.1 nM) that blocks the galanin-mediated inhibition of glucose-induced insulin secretion. (*See Galanin*) **1.** Bartfai, Bedecs, Land, *et al.* (1991) *Proc. Natl. Acad. Sci. U.S.A.* **88**, 10961. **2.** Lindskog, Ahren, Land, Langel & Bartfai (1992) *Eur. J. Pharmacol.* **210**, 183.

Galanthamine

This water-soluble alkaloid and AChE inhibitor (FW$_{free-base}$ = 287.36 g/mol; CAS = 357-70-0; Soluble to 50 mM in water; IUPAC: (4a*S*,6*R*,8a*S*)-5,6,9,10,11,12-hexahydro-3-methoxy-11-methyl-4a*H*-[1]benzofuro[3a,3,2-*ef*][2]benzazepin-6-ol), also known as galantamine, from the Caucasian snowdrop (*Galanthus woronowii*) and several *Amaryllidaceae* plants, is a brain-selective acetylcholinesterase inhibitor that has been used in the treatment of individuals having mild to moderate Alzheimer's disease. Galantamine is a weaker inhibitor of butyrylcholinesterase. Because both esterases play a role in Aβ-aggregation during the early stages of senile plaque formation, they are regarded as critical targets for the effective management of AD by delaying the hydrolysis of acetylcholine in affected brain regions, while decreasing Aβ-peptide deposition. Galanthamine reverses neuromuscular paralysis by tubocurarine-like muscle relaxants and stimulates pre- and postsynaptic nicotinic receptors. Galanthamine partially reverses the effects of scopolamine-induced amnesia in rats. Galantamine is readily absorbed after oral administration, with a t_{max} of 52 min and a plasma elimination $t_{\frac{1}{2}}$ of 5.7 hours. **Key Pharmacokinetic Parameters:** See Appendix II in Goodman & Gilman's THE PHARMACOLOGICAL BASIS OF THERAPEUTICS, 12th Edition (Brunton, Chabner & Knollmann, eds.) McGraw-Hill Medical, New York (2011). **Target(s):** acetyl-cholinesterase (1-11); cholinesterase, *or* butyrylcholinesterase, mildly inhibited (1-5,11). **1.** Darvesh, Walsh, Kumar, *et al.* (2003) *Alzheimer Dis. Assoc. Disord.* **17**, 117. **2.** Harvey (1995) *Pharmacol. Ther.* **68**, 113. **3.** Sramek, Frackiewicz & Cutler (2000) *Expert Opin. Investig. Drugs* **9**, 2393. **4.** Luo, Yu, Zhan, *et al.* (2005) *J. Med. Chem.* **48**, 986. **5.** Giacobini (2003) *Neurochem. Res.* **28**, 515. **6.** Irwin & Smith (1960) *Biochem. Pharmacol.* **3**, 147. **7.** Pilger, Bartolucci, Lamba, Tropsha & Fels (2001) *J. Mol. Graph. Model.* **19**, 288. **8.** Bartolucci, Perola, Pilger, Fels & Lamba (2001) *Proteins* **42**, 182. **9.** Pietsch & M. Gütschow (2005) *J. Med. Chem.* **48**, 8270. **10.** Khalid, Azim, Parveen, Atta-ur-Rahman & Choudhary (2005) *Biochem. Biophys. Res. Commun.* **331**, 1528. **11.** Ahmad, Iqbal, Nawaz, *et al.* (2006) *Chem. Biodiv.* **3**, 996.

Galarmin

This neurophysin II C-terminal glycopeptide fragment (FW = 1453.57 g/mol; *Sequence*: AGAPEPAEPAQPGVY), also known as Proline-Rich Polypeptide-1, *or* PRP-1, from the neurosecretory granules of bovine neurohypophysis, is an immunomodulating cytokine that is produced by hypothalamic neurosecretory cells. Galarmin exerts an antiproliferative effect on the tumor cells of mesenchymal origin by inhibiting mTOR kinase and repressing cell cycle progression (1,2). Galarmin also inhibits Cu(II)-catalyzed decomposition of H_2O_2, thus preventing formation of HO$^{\bullet}$ and HOO$^{\bullet}$ radicals, with an E_0 value of 0.795 Volts versus Ag/Ag$^+_{aq}$ (3). **1**. Galoian, Temple & Galoyan (2011) *Tumor Biol.* **32**, 745. **2**. Galoian, Temple & Galoyan (2011) *Neurochem. Res.* **36**, 812. **3**. Tavadyan, Galoian, Harutunyan, Tonikyan & Galoyan (2010) *Neurochem. Res.* **35**, 947.

Galegine

This isoprenoid guanidine derivative (FW$_{sulfate-salt}$ = 225.27 g/mol; CAS = 20284-78-0), also called pentenylguanidine, isoamyleneguanidine, and (3-methyl-2-butenyl)guanidine, has been isolated from seeds of *Galega officinalis* (goat's rue or professor weed). Galegine is also the toxic principle of the Western Australian sedge *Schoenus asperocarpus*. Galegine (1-300 μM) also reduced isoprenaline-mediated lipolysis in adipocytes inhibited acetyl-CoA carboxylase activity in adipocytes and myotubes (1). At 500 μM, galegine down-regulated genes concerned with fatty acid synthesis, including fatty acid synthase and its upstream regulator SREBP. By inhibiting acetyl-CoA carboxylase, galegine reduces fatty acid synthesis and stimulates fatty acid oxidation, actions that may contribute to galegine's effect on reducing body weight. **Target(s):** arginine 2-monooxygenase (2); electron transport and oxidative phosphorylation (3-5); NADH dehydrogenase (3). **1**. Mooney, Fogarty, Stevenson, *et al.* (2008) *Brit. J. Pharmacol.* **153**, 1669. **2**. Olomucki, Dangba, Nguyen Van Thoai (1964) *Biochim. Biophys. Acta* **85**, 480. **3**. Slater (1967) *Meth. Enzymol.* **10**, 48. **4**. Alberti, Woods & Whalley (1973) *Eur. J. Clin. Invest.* **3**, 208. **5**. Desvages & Olomucki (1969) *B. Soc. Chim. Fr.* (9), 3229.

Galiellalactone

This cytostatic ascomycete hexaketide (FW = 194.23 g/mol; CAS = 133613-71-5; Soluble in 100% ethanol, methanol, DMF, and DMSO; *Symbol*: GL) targets IL-6-induced expression of Secreted Alkaline Phosphatase (SEAP) in castration-resistant prostate cancer by blocking the DNA binding of activated Signal Transducer and Activator of Transcription-3, *or* STAT3, a transcription factor that is constitutively active in several malignancies (1,2). Galiellalactone selectively reduces growth of STAT3-expressing prostate cancer cells, both *in vitro* and *in vivo*, and induces apoptosis in prostate cancer stem-like cells that express pSTAT3. Galiellalactone inhibits STAT3 binding to DNA in DU145 cell lysates without affecting phosphorylation status of STAT3 (3). Mass spectrometry of galiellalactone-treated STAT3 demonstrated covalent modification of Cys-367, Cys-468 and Cys-542 (1). Galiellalactone inhibits Stat3 without affecting tyrosine and serine phosphorylation of the Stat3 (2). GL inhibits vesicular stomatitis virus-recombinant HIV-1 infection and the NF-κB-dependent transcriptional activity of the HIV-LTR promoter by preventing the binding of NF-κB to DNA. GL neither affects the phosphorylation and degradation of the NF-κB inhibitory protein, IκBα, nor the phosphorylation and acetylation of the NF-κB p65 subunit. Instead, it prevents p65's association with importin-α₃, thereby impairing nuclear translocation of this transcription factor. GL also binds to p65, but not to importin-α₃. GL is thus a dual NF-κB/STAT3 inhibitor that will serve as a lead compound for developing novel drugs against viral infection, cancer,

and inflammatory diseases. **1**. Hellsten, Johansson, Dahlman, *et al.* (2008) *Prostate* **68**, 269. **2**. Hellsten, Johansson, Dahlman, Sterner & Bjartell (2011) *PLoS One* **6**, e22118. **3**. Don-Doncow, Escobar, Johansson, *et al.* (2014) *J. Biol. Chem.* **289**, 15969. **4**. Pérez, Soler-Torronteras, Collado, *et al.* (2014) *Chem. Biol. Interact.* **214**, 69.

Gallamine

This curare-like agent (FW$_{tri-iodide-salt}$ = 891.52; CAS = 65-29-2), named systematically as 2,2',2''-[benzene-1,2,3-triyltris(oxy)]tris(*N,N,N*-triethylethanaminium) triiodide, is a muscarinic receptor antagonist, exhibiting the following effectiveness (Muscarinic Receptor M$_2$ > Muscarinic Receptor M$_1$ = Muscarinic Receptor M$_4$ > Muscarinic Receptor M$_3$ = Muscarinic Receptor M$_5$). It also inhibits acetylcholinesterase and activates succinate:cytochrome *c* reductase. **Target(s):** acetylcholinesterase (1-6); acid phosphatase (7); muscarinic acetylcholine receptor (8-13). **1**. Munoz, Aldunate & Inestrosa (1999) *Neuroreport* **10**, 3621. **2**. Hasson & Liepin (1963) *Biochim. Biophys. Acta* **75**, 397. **3**. Kato (1972) *Mol. Pharmacol.* **8**, 575. **4**. Al-Jafari (1997) *Toxicol. Lett.* **90**, 45. **5**. Costagli & A. Galli (1998) *Biochem. Pharmacol.* **55**, 1733. **6**. Hussein, Chacón, Smith, Tosado-Acevedo & Selkirk (1999) *J. Biol. Chem.* **274**, 9312. **7**. Lisa, Garrido & Domenech (1984) *Mol. Cell. Biochem.* **63**, 113. **8**. Leppik, Miller, Eck & Paquet (1994) *Mol. Pharmacol.* **45**, 983. **9**. Trankle, Weyand, Schroter & Mohr (1999) *Mol. Pharmacol.* **56**, 962. **10**. Gnagey, Seidenberg & Ellis (1999) *Mol. Pharmacol.* **56**, 1245. **11**. Colquhoun & Sheridan (1981) *Proc. R. Soc. Lond. B Biol. Sci.* **211**, 181. **12**. Clark & Mitchelson (1976) *Brit. J. Pharmacol.* **58**, 323. **13**. Michel, Delmendo, Lopez & Whiting (1990) *Eur. J. Pharmacol.* **182**, 335.

Gallein

This G-protein signal transduction inhibitor (FW = 364.31 g/mol; CAS 2103-64-2; Solubility: 75 mM in DMSO), also named 3',4',5',6'-tetrahydroxyspiro[isobenzofuran-1(3*H*),9'-(9*H*)xanthen]-3-one, targets phosphoinositide 3-kinase, *or* PI3-kinase, and Rac1 activation in HL60 cells and chemotaxis in differentiated HL60 cells. Based on the SPR analysis, gallein bound to G$_{β1}$γ$_2$ with a K_d value of ~400 nM, in relatively close agreement with the IC$_{50}$ of 200 nM observed in a competition ELISA assay (1). Gallein inhibition has also demonstrated that PI3K signaling participates in mediating inhibitory odorant input to mammalian olfactory receptor neurons (2). **1**. Lehmann, Seneviratne & Smrcka (2008) *Mol. Pharmacol.* **73**, 410. **2**. Ukhanov, Brunert, Corey & Ache (2011) *J. Neurosci.* **31**, 273.

Galloflavin

This lactate dehydrogenase inhibitor (FW$_{free-acid}$ = 278.46 g/mol; FW$_{potassium-salt}$ = 316.26 g/mol; CAS = 568-80-9; Solubility: 5 mg/mL DMSO) targets both enzyme isoforms, LDH-A (K_i = 5.5 μM) and LDH–B (K_i = 15 μM).

By inhibiting lactate production and decreasing ATP synthesis, galloflavin induces oxidative stress and apoptosis, particularly in breast tumor cell lines. **1**. Farabegoli, *et al.* (2012) *Eur. J. Pharm. Sci.* **47**, 729. **2**. Manerba, *et al.* (2012) *ChemMedChem* **7**, 311. **3**. Fiume, *et al.* (2013) *Biochem. Biophys. Res. Commun.* **430**, 466.

3',3'-cyclic-GAMP

This STING agonist (FW$_{diNa-Salt}$ = 718.37 g/mol; Soluble to 10 mM in H$_2$O), also known as 3',3'-cGAMP and 3',3'-cyclic guanosine monophosphate-adenosine monophosphate, targets <u>St</u>imulator of <u>in</u>terferon <u>g</u>enes (STING), a receptor also known as transmembrane protein 173 (TMEM173), reducing B cell proliferation and inducing apoptosis of malignant B cells *in vitro*. 3',3'-cGAMP suppresses 5TGM1 multiple myeloma xenograft growth in immunodeficient mice and induces leukemic regression in Eµ-TCL1 mice. Notably, cyclic GMP-AMP synthase (*or* cGAS) is a DNA sensor that catalyzes the synthesis of cyclic GMP-AMP (cGAMP) from ATP and GTP in the presence of DNA (2). **1**. Tang, Zundell, Ranatunga, *et al.* (2016) *Cancer Res.* **76**, 2137. **2**. Wu, Sun, Chen, *et al.* (2013) *Science* **339**, 826.

Ganciclovir

This acyclic guanosine/deoxyguanosine analogue (FW = 255.23 g/mol; CAS = 82410-32-0), also known as 9-[(1,3-dihydroxy-2-propoxy)methyl] guanine, BW759U, and 9-[[2-hydroxy-1-(hydroxymethyl)ethoxy]methyl] guanine, is taken up by most human cells and metabolically phosphorylated to ganciclovir 5'-triphosphate, the latter a potent inhibitor of viral DNA polymerases, with relatively little host cell toxicity. 5'-Mono-, 3'5'-bis(mono-), 3',5'-cyclic monophosphate and 5'-triphosphate forms of DHPG inhibited cytomegalovirus plaque formation at 1-15 µM (1). The cyclic phosphate and homophosphonate were more active than the other compounds against murine cytomegalovirus (MCMV) *in vitro*. Phosphorylated ganciclovir is incorporated into the DNA of replicating eukaryotic cells, and the latter undergo arrest at the G$_2$-M checkpoint. **Key Pharmacokinetic Parameters:** *See* Appendix II in Goodman & Gilman's *THE PHARMACOLOGICAL BASIS OF THERAPEUTICS*, 12th Edition (Brunton, Chabner & Knollmann, eds.) McGraw-Hill Medical, New York (2011). **Target(s):** cytomegalovirus replication (2,3); DNA polymerases, viral (2-4); HSV replication (2-4); purine-nucleoside phosphorylase (5,6). **1**. Duke, Smee, Cernow, Boehme & Matthews (1986) *Antiviral Res.* **6**, 299. **2**. Prisbe & Chen (1996) *Meth. Enzymol.* **275**, 425. **3**. Cheng, Huang, Lin, *et al.* (1983) *Proc. Natl. Acad. Sci. U.S.A.* **80**, 2767. **4**. St. Clair, Miller, Miller, Lambe & Furman (1984) *Antimicrob. Agents Chemother.* **25**, 191. **5**. Kulikowska, Bzowska, Wierzchowski & Shugar (1986) *Biochim. Biophys. Acta* **874**, 355. **6**. Bzowska, Kulikowska & Shugar (2000) *Pharmacol. Ther.* **88**, 349.

Ganciclovir Diphosphate, *See* Ganciclovir

Ganciclovir Triphosphate

This acyclic GTP/dGTP analogue (FW = 463.19 g/mol; CAS = 86761-38-8) is a potent inhibitor of viral DNA polymerases. This nonelongating dGTP analogue is incorporated into the DNA of replicating eukaryotic cells and the cell cycle is stopped at the G$_2$-M checkpoint. **Target(s):** DNA polymerases, viral (1-4); purine nucleoside phosphorylase (5). **1**. Wilson, Porter & Reardon (1996) *Meth. Enzymol.* **275**, 398. **2**. Prisbe & Chen (1996) *Meth. Enzymol.* **275**, 425. **3**. St. Clair, Miller, Miller, Lambe & Furman (1984) *Antimicrob. Agents Chemother.* **25**, 191. **4**. Germershausen, Bostedor, Field, *et al.* (1983) *Biochem. Biophys. Res. Commun.* **116**, 360. **5**. Stein, Stoeckler, Li, *et al.* (1987) *Biochem. Pharmacol.* **36**, 1237.

Ganetespib

Ganetespib   **STA-1474**

This ATP site-directed enzyme inhibitor (FW = 364.40 g/mol; CAS = 888216-25-9; Solubility = 40 mg/mL DMSO and <1 mg/mL H$_2$O), also known as 5-(2,4-dihydroxy-5-isopropylphenyl)-4-(1-methyl-1*H*-indol-5-yl)-2*H*-1,2,4-triazol-3(4*H*)-one, inhibits Hsp90 (IC$_{50}$ = 4 nM), inducing apoptosis in OSA-8 cells, but with no effect on normal osteoblasts (1-4). Ganetespib is the active inhibitory metabolite of the water-soluble prodrug STA-1474 (FW$_{free-acid}$ = 444.38 g/mol; CASs = 1118915-78-8 (free acid) and 1118915-79-9 (sodium salt)) Ganetespib blocks Hsp90's "foldase" activity, resulting in the degradation of its 200-300 client proteins, including key signaling molecules (*e.g.*, HER2/neu, mutated EGFR, Akt, c-Kit, IGF-1R, PDGFRα, Jak1, Jak2, STAT3, STAT5, HIF-1α, CDC2, c-Met, and Wilms tumor-1 proteins) that markedly influence the course of malignant transformation (2). At low-nM concentrations, STA-9090 arrests cell proliferation and induces apoptosis in a wide variety of human cancer cell lines, including many that are receptor tyrosine kinase inhibitor- and tanespimycin-resistant. Moreover, STA-9090 administration leads to significant tumor shrinkage in several tumor xenograft models in mice and appears to be less toxic. STA-9090 shows better tumor penetration compared with tanespimycin. (*See also Alvespimycin; Ansamycin; Deguelin; Derrubone; Geldanamycin; Herbimycin; Macbecin; Radicicol; Tanespimycin; and the nonansamycin Hsp90 inhibitor KW-2478*) **1**. McCleese, Bear, Fossey, *et al.* (2009) *Int. J. Cancer* **125**, 2792. **2**. Wang, Trepel, Neckers & Giaccone (2010) *Curr. Opin. Investig. Drugs* **11**, 1466. **3**. Ying, Du, Sun, *et al.* (2012) *Mol. Cancer Ther.* **11**, 475. **4**. Lin, Bear, Du, *et al.* (2008) *Exp. Hematol.* **36**, 1266.

Ganglioside GD1a

This di-sialoganglioside (FW$_{free-acid}$ = 1838.08 g/mol; CAS = 12707-58-3), also known as disialoganglioside GD1a, consists of NeuAc(α2-3)Gal(β1-3)GalNAc(β1-4)(NeuAc(α2-3))Gal(β1-4)Glcβ1Cer, where Cer refers to ceramide. GD1a is the major ganglioside component in human brain and an inhibitor of protein kinase C. It can be converted to ganglioside GM$_1$ via the action of sialidases. GD1a is also inhibits Con A-stimulated mitogenesis in murine T-cells. **Target(s):** (N-acetylneuraminyl)-galactosylglucosyl-ceramide N-acetyl-galactosaminyltransferase (1); α-N-acetylneuraminate α-2,8-sialyltransferase (2); DNA-directed DNA polymerase (3); DNA polymerase-α (3); ganglioside galactosyltransferase (4); NAD$^+$ nucleosidase, or ecto-NAD$^+$ glycohydrolase, mildly inhibited (5,6); protein kinase C (7). **1**. Senn, Cooper, Warnke, Wagner & Decker (1981) *Eur. J. Biochem.* **120**, 59. **2**. Yusuf, Schwarzmann, Pohlentz & Sandhoff (1987) *Biol. Chem. Hoppe-Seyler* **368**, 455. **3**. Simbulan, Taki, Tamiya-Koizumi, et al. (1994) *Biochim. Biophys. Acta* **1205**, 68. **4**. Yip & Dain (1970) *Biochem. J.* **118**, 247. **5**. Hara-Yokoyama, Nagatsuka, Katsumata, et al. (2001) *Biochemistry* **40**, 888. **6**. Hara-Yokoyama, Kukimoto, Nishina, et al. (1996) *J. Biol. Chem.* **271**, 12951. **7**. Kreutter, Kim, Goldenring, et al. (1987) *J. Biol. Chem.* **262**, 1633.

Ganglionic Blocking Agents

These agents, also known as ganglioplegic agents, are nicotinic receptor antagonists, inhibiting synaptic transmission in autonomic ganglia, thereby reducing or suppressing sympathetic and the parasympathetic effects on autonomically innervated organs. Ganglionic blocking agents inhibit nicotinic postsynaptic receptors without preliminary stimulation. Such behavior contrasts with that of nicotine, which stimulates nicotinic postsynaptic receptors in moderate dose and inhibits them in very high dose. When ganglionic blocking agents are administered in sufficient dose, stimulation of sympathetic and parasympathetic preganglionic fibers is inactive, whereas stimulation of postganglionic fibers can still induce a response. Ganglionic blocking agents include: tetraethylammonium bromide, hexamethonium bromide, pentolinium bromide, mecamylamine, trimetaphan, pempidine, benzohexonium bromide, chlorisondamine and pentamine

Ganglioside GD1b

This disialoganglioside (FW$_{free-acid}$ = 1836.10 g/mol (based on sphingosine C18:1 and stearic acid); CAS = 19553-76-5), also known as disialoganglioside GD1b, consists of Gal(β1-3)GalNAc(β1-4)(NeuAc(α2-8)NeuAc(α2-3))Gal(β1-4)Glcβ1Cer where Cer refers to ceramide. GD1b is found in high concentrations in gliomas and astrocytoma and it inhibits protein kinase C. **Target(s):** NAD$^+$ nucleosidase, or ecto-NAD$^+$ glycohydrolase (1,2); protein kinase C (3). **1**. Hara-Yokoyama, Nagatsuka, Katsumata, et al. (2001) *Biochemistry* **40**, 888. **2**. Hara-Yokoyama, Kukimoto, Nishina, et al. (1996) *J. Biol. Chem.* **271**, 12951. **3**. Kreutter, Kim, Goldenring, et al. (1987) *J. Biol. Chem.* **262**, 1633.

Ganglioside GM₁

This monosialoganglioside (FW$_{free-acid}$ = 1545.82 g/mol (calculated on sphingosine C18:1 and stearic acid); CAS = 37758-47-7; Structure: Gal(β1-3)GalNAc(β1-4)(NeuAc(α2-3))Gal(β1 4)Glcβ1Cer, where Cer = ceramide), also known as monosialoganglioside GM₁, inhibits protein kinase C and can act as a receptor for cholera toxin. Ganglioside GM₁ accumulates in individuals with generalized gangliosidosis. **Target(s):** (N-acetyl-neuraminyl)galactosylglucosylceramide N-acetylgalactos-aminyl-transferase (1); DNA-directed DNA polymerase (2); DNA polymerase α (2); ganglioside galactosyltransferase (3); lactosylceramide α-2,3-sialyltransferase, also as an alternative substrate (4); phospholipase A₂ (5); protein kinase C (6). **1**. Senn, Cooper, Warnke, Wagner & Decker (1981) *Eur. J. Biochem.* **120**, 59. **2**. Simbulan, Taki, Tamiya-Koizumi, et al. (1994) *Biochim. Biophys. Acta* **1205**, 68. **3**. Yip & Dain (1970) *Biochem. J.* **118**, 247. **4**. Iber, van Echten & Sandhoff (1991) *Eur. J. Biochem.* **195**, 115. **5**. Yang, Farooqui & Horrocks (1996) *Adv. Exp. Med. Biol.* **416**, 309. **6**. Kreutter, Kim, Goldenring, et al. (1987) *J. Biol. Chem.* **262**, 1633.

Ganglioside GM₂

This monosialoganglioside (FW$_{free-acid}$ = 1383.70 (calculated for sphingosine C18:1 and stearic acid substituents); CAS = 19600-01-2) is a minor component of cell membranes and accumulates in individuals with Tay-Sachs disease and Sandhoff-Jatzkewitz disease. Ganglioside GM2, also known as monosialoganglioside GM2, consists of GalNAc(β1-4)(NeuAc(α2-3))Gal(β1 4)Glcβ1Cer where Cer refers to ceramide. GM2 is a weak substrate, albeit strong inhibitor, of N acetyl-β-hexosaminidase. **Target(s):** N-acetyl-β-hexosaminidase (1); (N-acetylneuraminyl)-galactosylglucosylceramide N-acetylgalactosaminyl-transferase (2); galactose oxidase (3). **1**. Frohwein & Gatt (1967) *Biochemistry* **6**, 2783. **2**. Senn, Cooper, Warnke, Wagner & Decker (1981) *Eur. J. Biochem.* **120**, 59. **3**. Yip & Dain (1968) *Enzymologia* **35**, 368.

Ganglioside GM₃

N-stearoyl-ganglioside G$_{M3}$

This monosialoganglioside (FW$_{free-acid}$ = 1180.50 (calculated on sphingosine C18:1 and stearic acid); CAS = 54827-14-4), also known as hematoside, is the major ganglioside in non-neuronal cell membranes. It inhibits a number of protein kinases (note that lysoganglioside GM₃ and de-N-acetyl ganglioside GM₃ are also inhibitors of protein kinase C). Ganglioside GM₃, also known as monosialoganglioside GM₃, consists of NeuAc(α2-3)Gal(β1-4)Glcβ1Cer where Cer refers to ceramide. **Target(s):** DNA-directed DNA polymerase (1); DNA polymerase α (1); DNA polymerase β (2); EGF receptor protein-tyrosine kinase (2-5); globotriaosyl-ceramide 3-β-N-acetylgalactos-aminyltransferase (6); lactosylceramide α-2,3-sialyl-transferase, product inhibition (7); monosialoganglioside sialyltransferase, also as an alternative substrate (8); phospholipase A₂ (9); protein kinase C (3,10); receptor protein-tyrosine kinase (2-5); sphingomyelin phosphodiesterase, or neutral sphingomyelinase (11). **1**. Simbulan, Taki, Tamiya-Koizumi, et al. (1994) *Biochim. Biophys. Acta* **1205**, 68. **2**. Song, Welti, Hafner-Strauss & Rintoul (1993) *Biochemistry* **32**, 8602. **3**. Igarashi, Nojiri, Hanai & Hakomori (1989) *Meth. Enzymol.* **179**, 521. **4**. Bremer, Schlessinger & Hakomori (1986) *J. Biol. Chem.* **261**, 2434. **5**. Yoon, Nakayama, Hikita, Handa & Hakomori (2006) *Proc. Natl. Acad. Sci. U.S.A.* **103**, 18987. **6**. Lockney & Sweely (1982) *Biochim. Biophys. Acta* **712**, 234. **7**. Burczak, Fairley & Sweeley (1984) *Biochim. Biophys. Acta* **804**, 442. **8**. Iber, van Echten & Sandhoff (1991) *Eur. J. Biochem.* **195**, 115. **9**. Yang, Farooqui & Horrocks (1996) *Adv. Exp. Med. Biol.* **416**, 309. **10**. Kreutter, Kim, Goldenring, et al. (1987) *J. Biol. Chem.* **262**, 1633. **11**. Lister, Crawford-Redick & Loomis (1993) *Biochim. Biophys. Acta* **1165**, 314.

Ganglioside GT1b

This trisialoganglioside (FW$_{free-acid}$ = 2126.30 (calculated on sphingosine C18:1 and stearic acid); CAS = 59247-13-1) modulates cellular differentiation, inhibits mitogenesis stimulated by lectins, and inhibits protein kinase C. Ganglioside GT1b, also known as trisialoganglioside GT1b, consists of NeuAc(α2-3)Gal(β1-3)GalNAc(β1-4)(NeuAc(α2-8)NeuAc(α2-3))Gal(β1 4)Glcβ1Cer where Cer refers to ceramide. It is converted to ganglioside GD1b via the action of sialidases. **Target(s):** α-N-acetylneuraminate α-2,8-sialyltransferase (1); (N acetylneuraminyl) galactosylglucosylceramide N-acetylgalactos-aminyltransferase (2); ganglioside galactosyltransferase (3); NAD$^+$ nucleosidase, or ecto-NAD+ glycohydrolase (4,5); protein kinase C (6,7). **1**. Yusuf, Schwarzmann, Pohlentz & Sandhoff (1987) *Biol. Chem. Hoppe-Seyler* **368**, 455. **2**. Senn, Cooper, Warnke, Wagner & Decker (1981) *Eur. J. Biochem.* **120**, 59. **3**. Yip & Dain (1970) *Biochem. J.* **118**, 247. **4**. Hara-Yokoyama, Nagatsuka, Katsumata, et al. (2001) *Biochemistry* **40**, 888. **5**. Hara-Yokoyama, Kukimoto, Nishina, et al. (1996) *J. Biol. Chem.* **271**, 12951. **6**. Igarashi, Nojiri, Hanai & Hakomori (1989) *Meth. Enzymol.* **179**, 521. **7**. Kreutter, Kim, Goldenring, et al. (1987) *J. Biol. Chem.* **262**, 1633.

Ganstigmine

This orally active physostigmine derivative (FW = 417.94 g/mol; CAS = 412044-92-9, for hydrochloride), also known as CHF-2819 and 2'-ethylphenylgeneserine N-oxide, inhibits acetylcholine esterase, IC$_{50}$ = 0.13 μM (1,2) and butyrylcholinesterase, IC$_{50}$ = 1.7 μM (1). Ganstigmine and its enantiomer CHF3360 over the 0.1 to 3 μM range significantly decrease neurodegeneration achieved by the addition of β-amyloid(25-35) by approximately 50%. Dose response curves of both substances were identical concerning effect size and concentration; however, because CHF3360 lacks acetylcholine inhibitor activity in the applied dose range, it appears that Ganstigmine provides significant neuroprotection independent from its cholinergic activity (3). **1**. Luo, Yu, Zhan, *et al*. (2005) *J. Med. Chem.* **48**, 986. **2**. Trabace, Cassano, Steardo, *et al*. (2000) *J. Pharmacol. Exp. Ther.* **294**, 187. **3**. Windisch, Hutter-Paier, Jerkovic, Imbimbo & Villetti (2003) *Neurosci. Lett.* **341**, 181.

GANT61

This Hedgehog inhibitor (FW = 429.61 g/mol; CAS = 500579-04-4; Solubility: <1 mg/mL DMSO, H$_2$O), also known as NSC136476 and 2,2'-[[dihydro-2-(4-pyridinyl)-1,3(2H,4H)-pyrimidinediyl]bis(methylene)]bis$N$, N-dimethylbenzenamine, targets GLI1- and GLI2-induced transcription (IC$_{50}$ = 5 μM), which constitutes the final step in the hedgehog signal transduction pathway (1). GANT61 displays selectivity over other pathways, among them TNF- and glucocorticoid-receptor gene transactivation. Profiling of cognate Hh pathway members revealed reduced expression of the key Hh signaling effectors, Smoothened (SMOH) and GLI, in chronic lymphocytic leukemia (CLL), whereas transcription levels of other pathway components resembled that observed with normal B-lymphocytes (2). Specific RNA interference knockdown experiments in a CLL-derived cell line confirmed the autonomous role of GLI in malignant cell survival. Analysis of the molecular mechanisms of GANT61-induced cytotoxicity in HT29 cells showed increased Fas expression and decreased expression of PDGFRα, which also regulates Fas (3). Furthermore, DR5 expression was increased whereas Bcl-2 (direct target of Gli2) was down-regulated following GANT61 treatment. Unlike cyclopamine, GANT61 induced transient cellular accumulation at G(1)-S (24 hours) and in early S-phase (32 hours), with elevated p21(Cip1), cyclin E, and cyclin A in HT29 cells (4). GANT61 induced DNA damage within 24 hours, with the appearance of p-ATM and p-Chk2. Pharmacologic inhibition of Gli1 and Gli2 by GANT61 or genetic inhibition by transient transfection of the Gli3 repressor (Gli3R) down-regulated Gli1 and Gli2 expression and induced γH2AX, PARP cleavage, caspase-3 activation, and cell death. GANT61 and rapamycin are synergistic in their inhibitory effects on Gli activiation in acute myeloid leukemic cells (5). Using compounds blocking SMO (cyclopamine and SANT1) or GLI1/GLI2 (GANT61) activity revealed that inhibition of Hh signaling at the level of GLI was most effective in reducing neuroblastoma growth (6). **1**. Lauth, Bergström, Shimokawa & Toftgård (2007) *Proc. Natl. Acad. Sci. U.S.A.* **104**, 8455. **2**. Desch, Asslaber, Kern, *et al*. (2010) *Oncogene* **29**, 4885. **3**. Mazumdar, DeVecchio, Shi, *et al*. (2011) *Cancer Res.* **71**, 1092. **4**. Mazumdar, Devecchio, Agyeman, Shi & Houghton (2011) *Cancer Res.* **71**, 5904. **5**. Pan, Li, Li, *et al*. (2012) *Leuk. Res.* **36**, 742. **6**. Wickström, Dyberg, Shimokawa, *et al*. (2013) *Int. J. Cancer* **132**, 1516.

Garcinol

This PCAF and HAT inhibitor (FW = 602.81 g/mol; CAS = 78824-30-3; Solubility: 100 mM in DMSO), also named Camboginol*, is a polyisoprenylated benzophenone derivative from the rind of *Garcinia indica* fruit that targets p300 (IC$_{50}$ ~ 7 μM) and PCAF (IC$_{50}$ ~ 5 μM) histone acetyltransferases *in vitro* (1). Also known as E1A binding protein, p300 regulates transcription factor-mediated cell growth and division. The p53-associated trancriptional coactivator PCAF plays a direct role in transcriptional regulation. Garcinol inhibits p300-mediated acetylation of p53, promoting apoptosis via caspase-3 activation. Additionally, this compound predominantly down-regulates global gene expression in HeLa cells. Garcinol also exhibits dose-dependent cancer cell-specific growth inhibition in ER-positive MCF-7 and ER-negative MDA-MB-231 breast cancer cells. Its anti-carcinogenic and antiinflammatory properties are also mediated through selective inhibition of of prostaglandin E$_2$ synthesis and 5-lipoxygenase. Garcinol regulates EMT and Wnt signaling pathways *in vitro* and *in vivo*, leading to anticancer activity against breast cancer cells (2). It is also effective in promoting *ex vivo* expansion of human hematopoietic stem cells (3). **1**. Balasubramanyam, *et al*. (2004) *J. Biol. Chem.* **279**, 33716. **2**. Ahmad, *et al*. (2012) *Mol. Cancer. Ther.* 11 2193. **3**. Nishino et al (2011) *PLoS One* **6**, e24298. ***IUPAC Name:** (1R,5R,7R)-3-(3,4-Dihydroxybenzoyl)-4-hydroxy-8,8-dimethyl-1,7-bis(3-methyl-2-buten-1-yl)-5-[(2S)-5-methyl-2-(1-methylethenyl)-4-hexen-1-yl]bicyclo[3.3.1]non-3-ene-2,9-dione

Gatifloxacin

This fluoroquinolone-based antibiotic (FW = 375.40 g/mol; CAS = 112811-59-3), also named AM-1155 and 1-cyclopropyl-6-fluoro-1,4-dihydro-8-methoxy-7-(3-methyl-1-piperazinyl)-4-oxo-3-quinolinecarboxylic acid, targets *Escherichia coli* type II topoisomerase, *or* DNA gyrase (IC$_{50}$ = 0.11 μg/mL) and *Staphylococcus aureus* topoisomerase IV (IC$_{50}$ = 13.8 μg/mL). Gatifloxacin exhibits potent activity against both Gram-positive and -negative bacteria. It is effective against sequentially acquired quinolone-resistant mutants and the *norA* transformant of *Staphylococcus aureus* (1). Gatifloxacin shows some inhibition of mammalian type II topoisomerases (2). It also stimulates short-term self-renewal in both human and mouse embryonic stem cells *in vitro* (3). Tubular secretion and reabsorption are major determinants of gatifloxacin's half-life in plasma, its efficacy, and its drug-drug interactions. **1**. Fukuda, *et al*. (1998) *Antimicrob. Agents Chemother.* **42**, 1917. **2**. Takei, *et al*. (1998) *Antimicrob. Agents Chemother.* **42**, 2678. **3**. Desbordes, *et al*. (2008) *Cell Stem Cell* **2**, 602.

GBR 12909, *See* 1-(2-[Bis(4-fluorophenyl)methoxy]ethyl)-4-(3-phenylpropyl)piperazine

GBR-12935

This dopamine reuptake inhibitor (FW$_{free-base}$ = 414.59 g/mol; CAS = 76778-22-8), named systematically as 1-(2-(diphenylmethoxy)ethyl)-4-(3-phenylpropyl)piperazine, is a high-affinity ligand suitable for specific labeling the dopamine transport complex (1). Specific binding of [^{3}H]GBR-12935 to striatal membranes is saturable and of high affinity. It is also sodium-dependent, with optimal binding observed at a sodium concentration of 120 mM. IC$_{50}$ values for the same drugs in inhibiting [^{3}H]dopamine uptake in a crude synaptosomal fraction from striatum are as follows: GBR-12921, IC$_{50}$ = 3.5 nM; GBR-12909, IC$_{50}$ = 3.5 nM;

mazindol, IC_{50} = 37 nM; nomifensine, IC_{50} = 49 nM; threo-(+)-methylphenidate, IC_{50} = 240 nM; cocaine, IC_{50} = 370 nM; erythro-(±)-methylphenidate, IC_{50} = 22,500 nM. In calculating the IC_{50} values, 4-7 concentrations of each drug were used in the presence of 30-50 nM [^{3}H]dopamine or 0.5-0.6 nM [^{3}H]GBR-12935. Early evidence that the dopamine transporter is the principal cocaine target was provided in experiments showing that a cocaine analogue (*i.e.*, [^{125}I]DEEP) and a GBR analogue labeled the same rat striatal membranes protein, based on molecular weight, similar pharmacological profile, similar sensitivity to neuraminidase (2). Moreover, GBR-12935 binds to the outward-facing conformational state of the transporter, where external DA binds to initiate its uptake (3). **Other Details:** For use of [^{3}H]GBR-12935 as in the quantitative autoradiographic localization of the dopamine transporters in the rat brain, *See Reference 4*. For slow-release form of hydroxy-GBR-12935 form, *See DBL-583*. **1**. Berger, Janowsky, Vocci, *et al.* (1985) *Eur. J. Pharmacol.* **107**, 289. **2**. Patel, Boja, Leve, *et al.* (1992) *Brain Res.* **576**, 173. **3**. Chen, Rickey, Berfield & Reith (2004) *J. Biol. Chem.* **279**, 5508. **4**. Dawson, Gehlert & Wamsley (1986) *Eur. J. Pharmacol.* **126**, 171.

GCK 1026, See *Scriptaid*

GDC-0032, See *Taselisib*

GDC-0152

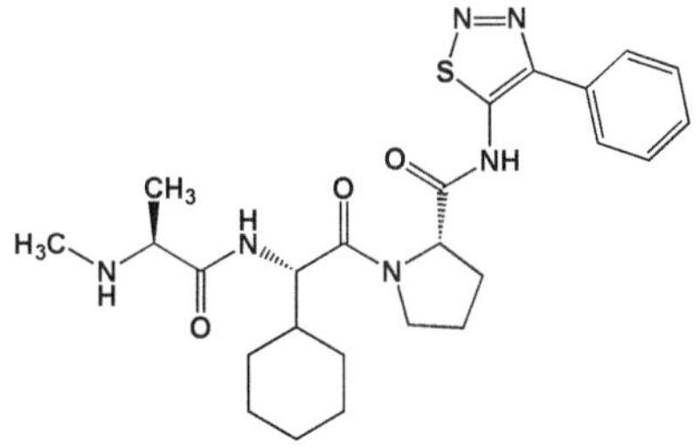

This potent IAP antagonist (FW = 498.64 g/mol; CAS = 873652-48-3; Solubility = 100 mg/mL DMSO or Ethanol, 3 mg/mL H_2O), also named *N*-methyl-L-alanyl-(2*S*)-2-cyclohexylglycyl-*N*-(4-phenyl-1,2,3-thiadiazol-5-yl)-L-prolinamide, targets the X-linked Inhibitor of Apoptosis Protein-Baculovirus Inhibitor of apoptosis protein Repeat proteins: XIAP-BIR3 (K_i = 28 nM), ML-IAP-BIR3 (K_i = 14 nM), cIAP1-BIR3 (K_i = 17 nM) and cIAP2-BIR3 (K_i = 43 nM), with much less affinity displayed toward cIAP1-BIR2 and cIAP2-BIR2. IAP family members, including X chromosome-linked IAP (XIAP), cellular IAP 1 (cIAP1), cellular IAP 2 (cIAP2), and melanoma IAP (ML-IAP), are often over-expressed in cancer cells, where they protect against a variety of pro-apoptotic stimuli. IAP proteins also been regulate signal transduction pathways associated with malignancy, targeting TNFα-mediated NF-κB activation via their C-terminal RING ubiquitin E$_3$-ligase domains, which have been shown to ubiquitinate receptor interacting protein (RIP)-1 and NF-κB inducing kinase, NIK. GDC-0152 can block protein-protein interactions that involve IAP proteins and pro-apoptotic molecules. GDC-0152 promotes degradation of cIAP1, induces activation of caspase-3/7, and leads to decreased viability of breast cancer cells without affecting normal mammary epithelial cells. GDC-0152 **Pharmacokinetics:** C_{max} = 53.7 μM; AUC = 203.5 hour·μM. **1**. Flygare, Beresini, Budha, *et al.* (2012) *J. Med. Chem.* **55**, 4101.

GDC-0199, See *Venetoclax*

GDC-0349

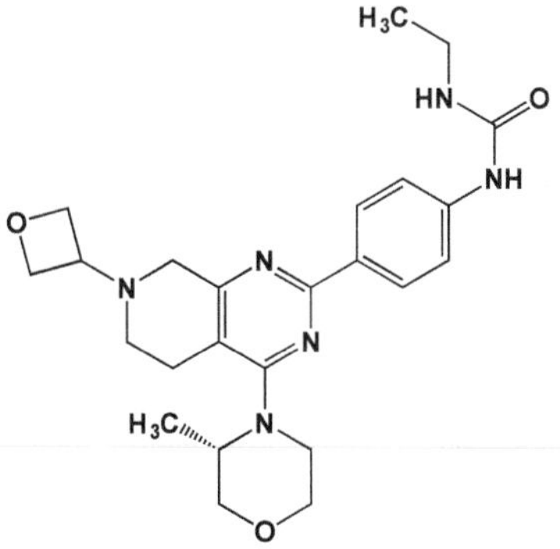

This potent and selective, pyrimidoaminotropane-based ATP-competitive inhibitor (FW = 452.55 g/mol; CAS = 1207360-89-1), also named RG7603

and *N*-ethyl-*N'*-[4-[5,6,7,8-tetrahydro-4-[(3*S*)-3-methyl-4-morpholinyl]-7-(3-oxetanyl)pyrido[3,4-*d*]pyrimidin-2-yl]phenyl]-urea, targets mTOR (K_i = 3.8 nM), with 790-fold inhibitory effect against PI3Kα and other 266 kinases. GDC-0349 inhibits downstream markers of mTOR, including phospho-4EBP1 and phospho-Akt(S473). **1**. Estrada, Shore, Blackwood, *et al.* (2013) *J. Med. Chem.* **56**, 3090.

GDC-0449, See *Vismodegib*

GDC-0941

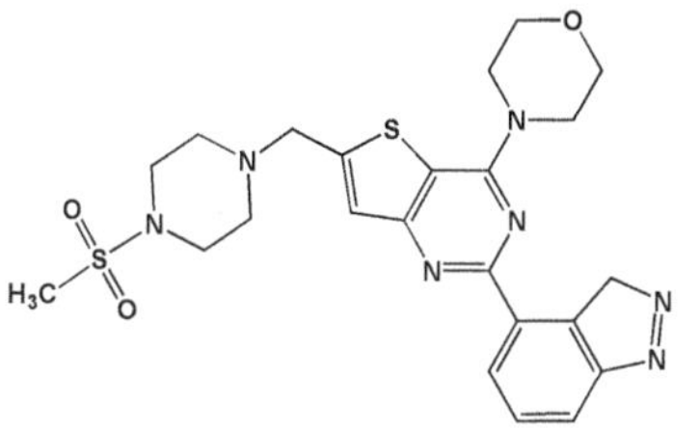

This potent phosphoinositide-3-kinase-directed signal-transduction pathway inhibitor (F.Wt. = 513.64; CAS = 154447-36-6 and 934389-88-5; Solubility (25°C): 44 mg/mL DMSO, <1 mg/mL H_2O), also known as Pictlsilib and 2-(1*H*-indazol-4-yl)-6-((4-(methylsulfonyl)piperazin-1-yl)methyl)-4-morpholinothieno[3,2-*d*]pyrimidine, inhibits phospho-inositide 3-kinases PI3Kα, PI3Kβ, PI3Kδ and PI3Kγ, with IC_{50} values of 3 nM, 33 nM, 3 nM and 75 nM, respectively. **1**. Folkes, et al. (2008) *J. Med. Chem.* 2008, 51, 5522. **2**. Junttila, et al. (2009) *Cancer Cell* 15, 429. **3**. Haagensen, et al. (2012) *Brit. J. Cancer* 106, 1386.

GDC-0994

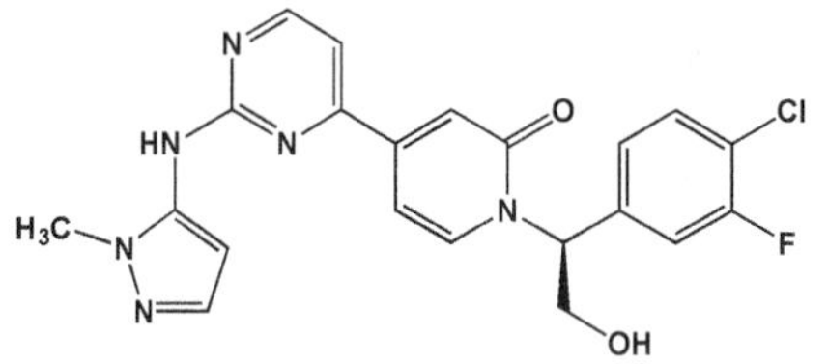

This potent ERK inhibitor (FW = 440.84 g/mol; CAS = 1453848-26-4; Solubility (25°C): 88 mg/mL DMSO, <1 mg/mL H_2O), also named (*S*)-1-(1-(4-chloro-3-fluorophenyl)-2-hydroxyethyl)-4-(2-((1-methyl-1*H*-pyrazol-5-yl)amino)pyrimidin-4-yl)pyridin-2(1*H*)-one, targets the Extracellular signal-Regulated Kinases ERK1 (K_i = 1.2 nM) and ERK2 (K_i = 0.3 nM), showing significant single-agent activity in multiple *in vivo* cancer models, including KRAS-mutant and BRAF-mutant human tumors in xenograft mice models. Orally administered GDC-0994 inhibits both ERK phosphorylation and activation of ERK-mediated signal transduction pathways, and subsequently prevents ERK-dependent tumor cell proliferation and survival. **1**. Robarge, Schwarz, Blake, *et al.* (2014) Discovery of GDC-0994, a potent and selective ERK1/2 inhibitor in early clinical development, Presentation Number DDT02-03, *Annual Meeting of Amer. Assoc. Cancer Res.*, San Diego.

GDP-(1-[4-[(biphenyl-4-ylmethyl)carbamoyl]butyl]-1*H*-[1,2,3]triazol-4-ylmethyl)

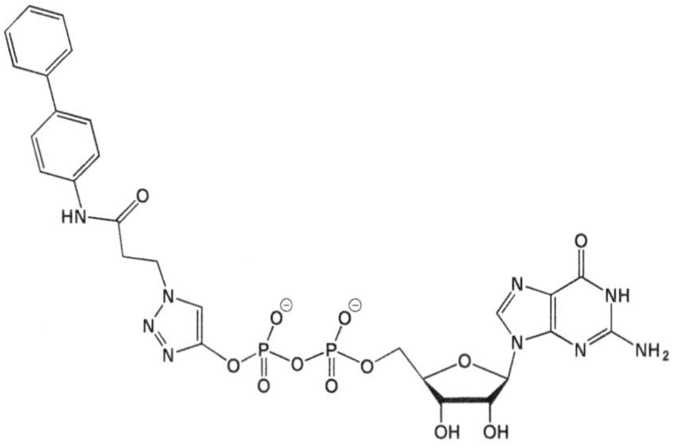

This GDP analogue (FW = 772.63 g/mol), also known as GDP-*N*-(biphenyl-4-ylmethyl)-5-(1*H*-1,2,3-triazol-1-yl)-pentanamide, inhibits glycoprotein 3-α-L-fucosyltransferase (K_i = 62 nM, human). **1**. Bryan, Lee & Wong (2004) *Bioorg. Med. Chem. Lett.* **14**, 3185. **2**. Sun, Lin, Ko, *et al.* (2007) *J. Biol. Chem.* **282**, 9973.

GDP-L-fucose

This sugar nucleotide (FW$_{\text{free-acid}}$ = 633.31 g/mol; CAS = 15839-70-0), also known as guanosine 5'-diphospho-β-L-fucose, is a product of the reactions catalyzed by GDP-L-fucose synthase and fucose-1-phosphate guanylyltransferase. It is a substrate for 3-galactosyl-*N*-acetylglucosaminide 4α-L-fucosyl-transferase, glycoprotein 6α-L-fucosyltransferase, galactoside 2α-L-fucosyltransferase, 4-galactosyl-*N*-acetylglucosaminide 3α-L-fucosyl-transferase, glycoprotein-3α-L-fucosyltransferase, and peptide-*O*-fucosyl-transferase.   **Target(s):**   α-*N*-acetyl-galactosaminide   α-2,6-sialyl-transferase (1); fucokinase (2,3); β-galactoside α-2,3-sialyltransferase1; GDP-mannose 4,6-dehydratase (4-13); GDP-mannose 3,5-epimerase (7); glucomannan   4-β-mannosyltransferase   (14);   mannose-1-phosphate guanylyl-transferase (4).   1. Sadler, Beyer, Oppenheimer, *et al.* (1982) *Meth. Enzymol.* **83**, 458.   2. Kilker, Shuey & Serif (1979) *Biochim. Biophys. Acta* **570**, 271.   3. Park, Pastuszak, Drake & Elbein (1998) *J. Biol. Chem.* **273**, 5685.   4. Glaser & Zarkowsky (1971) *The Enzymes*, 3rd ed., **5**, 465.   5. Somoza, Menon, Schmidt, *et al.* (2000) *Structure Fold Des.* **8**, 123.   6. Sullivan, Kumar, Kriz, *et al.* (1998) *J. Biol. Chem.* **273**, 8193.   7. Wolucka & Van Montagu (2003) *J. Biol. Chem.* **278**, 47483.   8. Sturla, Bisso, Zanardi, *et al.* (1997) *FEBS Lett.* **412**, 126.   9. Bisso, Sturla, Zanardi, De Flora & Tonetti (1999) *FEBS Lett.* **456**, 370.   10. Broschat, Chang & Serif (1985) *Eur. J. Biochem.* **153**, 397.   11. Somoza, Menon, Schmidt, *et al.* (2000) *Structure Fold. Des.* **8**, 123.   12. Sullivan, Kumar, Kriz, *et al.* (1998) *J. Biol. Chem.* **273**, 8193.   13. Yamamoto, Katayama, Onoda, *et al.* (1993) *Arch. Biochem. Biophys.* **300**, 694.   14. Piro, Zuppa, Dalessandro & Northcote (1993) *Planta* **190**, 206.

GDP-D-glucose

This sugar nucleotide (FW = 627.32 g/mol; CAS = 103301-72-0), also known as guanosine 5'-diphospho-α-D-glucopyranose, is a product of the reaction catalyzed by glucose-1-phosphate guanylyltransferase and is a substrate for cellulose synthase (GDP-forming), α,α-trehalose-phosphate synthase (GDP-forming), and GDP-glucosidase. **Target(s):** chitobiosyl-diphosphodolichol β-mannosyltransferase (1,2); CMP-*N* acylneuraminate phosphodiesterase (3); dolichyl-phosphate β-D-mannosyltransferase (4); GDP mannose 4,6-dehydratase (5-8); GDP-mannose 6-dehydrogenase (9); GDP-mannose 3,5-epimerase (10); glucomannan 4-β-mannosyl-transferase (11-14); glycogen phosphorylase (15-17); glycolipid 3-α-mannosyl-transferase (18,19); glycoprotein 3-α-L-fucosyl-transferase (20); glycoprotein 6-α-L-fucosyltransferase (21); maltose synthase (22); mannose-1 phosphate guanylyltransferase, *or* GDP-mannose pyrophos-phorylase (23,24); β-1,4-mannosyl glycoprotein 4-β-*N*-acetylglucosaminyl-transferase (25); sterol 3β-glucosyltransferase (26); α,α-trehalose-phosphate synthase (UDP-forming) (27); UDP-glucuronate 4-epimerase (28); UDP-sugar diphosphatase (29); xyloglucan 4-glucosyltransferase, weakly inhibited (30); xyloglucan 6 xylosyltransferase, weakly inhibited (31).   1. Kaushal & Elbein (1986) *Arch. Biochem. Biophys.* **250**, 38.   2. Kaushal & Elbein (1987) *Biochemistry* **26**, 7953.   3. Kean & Bighouse (1974) *J. Biol. Chem.* **249**, 7813.   4. Carlo & Villemez (1979) *Arch. Biochem. Biophys.* **198**, 117.   5. Liao & Barber (1972) *Biochim. Biophys. Acta* **276**, 85.   6. Sturla, Bisso, Zanardi, *et al.* (1997) *FEBS Lett.* **412**, 126.   7. Bisso, Sturla, Zanardi, De Flora & Tonetti (1999) *FEBS Lett.* **456**, 370.   8. Broschat, Chang & Serif (1985) *Eur. J. Biochem.* **153**, 397.   9. Roychoudhury, May, Gill, *et al.* (1989) *J. Biol. Chem.* **264**, 9380.   10. Wolucka & Van Montagu (2003) *J. Biol. Chem.* **278**, 47483.   11. Piro, Zuppa, Dalessandro & Northcote (1993) *Planta* **190**, 206.   12. Smith, Axelos & C. Péaud-Lenoël (1976) *Biochimie* **58**, 1195.   13. Elbein (1969) *J.*

Biol. Chem. **244**, 1608.   14. Dalessandro, Piro & Northcote (1986) *Planta* **169**, 564.   15. Thomas & Wright (1976) *J. Biol. Chem.* **251**, 1253.   16. Nader & Becker (1979) *Eur. J. Biochem.* **102**, 345.   17. Robson & Morris (1974) *Biochem. J.* **144**, 513.   18. Jensen & Schutzbach (1982) *J. Biol. Chem.* **257**, 9025.   19. Jensen & Schutzbach (1981) *J. Biol. Chem.* **256**, 12899.   20. Murray, Takayama, Schultz & Wong (1996) *Biochemistry* **35**, 11183.   21. Ihara, Ikeda & Taniguchi (2006) *Glycobiology* **16**, 333.   22. Schilling (1982) *Planta* **154**, 87.   23. Ning & Elbein (1999) *Arch. Biochem. Biophys.* **362**, 339.   24. Szumilo, Drake, York & Elbein (1993) *J. Biol. Chem.* **268**, 17943.   25. Ikeda, Koyota, Ihara, *et al.* (2000) *J. Biochem.* **128**, 609.   26. Staver, Glick & Baisted (1978) *Biochem. J.* **169**, 297.   27. Lapp, Patterson & Elbein (1971) *J. Biol. Chem.* **246**, 4567.   28. Munoz, Lopez, de Frutos & Garcia (1999) *Mol. Microbiol.* **31**, 703.   29. Glaser, Melo & Paul (1967) *J. Biol. Chem.* **242**, 1944.   30. Hayashi & Matsuda (1981) *J. Biol. Chem.* **256**, 11117.   31. Hayashi & Matsuda (1981) *Plant Cell Physiol.* **22**, 1571.

GDP-D-mannose

This sugar nucleotide (FW = 605.34 g/mol; CAS = 3123-67-9), also known as guanosine diphosphomannose and abbreviated GDPM, is required in the biosynthesis of mannans and is a precursor in the incorporation of mannosyl and fucosyl residues in glycoproteins and glycolipids. **Target(s):** CMP-*N*-acylneuraminate phosphodiesterase (1); FAD diphosphatase, also an alternative substrate (2); galactoside 2-α-L-fucosyltransferase (3); 3-galactosyl-*N*-acetylglucosaminide 4-α-L-fucosyltransferase (4); glycol-protein 3-α-L-fucosyltransferase (5); glycoprotein 6-α-L-fucosyltransferase (6); mannose-1-phosphate guanylyltransferase, product inhibition (7-9); nucleotide diphosphatase, as alternative substrate (10); UDP-*N*-acetylglucosamine:dolichyl   phosphate   *N*-acetylglucosaminephospho-transferase (11,12); UDP-glucuronate decarboxylase, mildly inhibited (13).   1. Kean & Bighouse (1974) *J. Biol. Chem.* **249**, 7813.   2. Byrd, Fearney & Kim (1985) *J. Biol. Chem.* **260**, 7474.   3. Bella & Kim (1971) *Biochem. J.* **125**, 1157.   4. Wong, Dumas, Ichikawa, *et al.* (1992) *J. Amer. Chem. Soc.* **114**, 7321.   5. Murray, Takayama, Schultz & Wong (1996) *Biochemistry* **35**, 11183.   6. Ihara, Ikeda & Taniguchi (2006) *Glycobiology* **16**, 333.   7. Fey, Elling & Kragl (1998) *Carbohydr. Res.* **305**, 475.   8. Szumilo, Drake, York & Elbein (1993) *J. Biol. Chem.* **268**, 17943.   9. Elling, Ritter & Verseck (1996) *Glycobiology* **6**, 591.   10. Haroz, Twu & Bretthauer (1972) *J. Biol. Chem.* **247**, 1452.   11. Kaushal & Elbein (1986) *Plant Physiol.* **82**, 748.   12. Zeng & Elbein (1995) *Eur. J. Biochem.* **233**, 458.   13. Suzuki, Watanabe, Masumura & Kitamura (2004) *Arch. Biochem. Biophys.* **431**, 169.

Gefitinib

This orally active, EGFR-directed protein kinase inhibitor (FWt = 446.90; CASs = 248594-19-6 (methanesulfonate); Solubility (25°C): 90 mg/mL DMSO, 1 mg/mL H$_2$O), also known as Iressa®, ZD-1839, and by its systematic name, *N*-(3-chloro-4-fluorophenyl)-7-methoxy-6-(3-morpholino-propoxy)quinazolin-4-amine, blocks phosphorylation of Tyr-1173, Tyr-992, Tyr-1173 and Tyr-992 sites in the NR6wtEGFR- and NR6W-cells with IC$_{50}$ values of 37 nM, 37nM, 26 nM and 57 nM, respectively (1-4). EGFR is frequently amplified and/or mutated in human tumors, and abnormal signaling from this receptor is believed to contribute to the malignant phenotype. Significantly, gefitinib was the first EGF receptor protein-tyrosine kinase inhibitor showing effects on solid tumors. Gefitinib (1-2 μM) significantly decreases EGFRvIII phosphotyrosine load, EGFRvIII-mediated proliferation as well as anchorage-independent growth (5). It likewise inhibits HER2-driven signaling and suppresses the growth of HER2-over-expressing tumor cells (6). While gefitinib treatment is effective against aberrant EGFR activity, chronic gefitinib treatment promotes ROS and mitochondrial dysfunction in lung cancer cells (7). Antioxidants alleviate ROS-mediated resistance. **1**. Woodburn, Morris, Kelly & Laight (1998) *Cell. Mol. Biol. Lett.* **3**, 348. **2**. Barker, Gibson, Grundy, *et al.* (2001) *Bioorg. Med. Chem. Lett.* **11**, 1911. **3**. Wakeling, Guy, Woodburn, *et al.* (2002) *Cancer Res.* **62**, 5749. **4**. Sirotnak, Zakowski, Miller, Scher & Kris (2000) *Clin. Cancer Res.* **6**, 4885. **5**. Pedersen, Pedersen, Ottesen & Poulsen (2005) *Br. J. Cancer* **93**, 915. **6**. Moasser, Basso, Averbuch & Rosen (2001) *Cancer Res.* 61, 7184. **7**. Okon, Coughlan, Zhang, Wang & Zou (2015) *J. Biol. Chem.* **290**, 9101.

Geldanamycin

This potent antitumor drug (FW = 559.27 g/mol; CAS = 30562-34-6; Symbols: GA and GDN), a benzoquinone ansamycin-class antibiotic first isolated from *Streptomyces hydroscopicus*, binds to Hsp90 heat shock proteins, inhibiting this chaperone's intrinsic ATP-hydrolyzing activity and stimulating client protein release. In view of its ability to bind to and inhibit Hsp90, treatment of cells with geldanamycin activates a heat shock response, even in the absence of heat shock. Moreover, pretreatment of cells with the Hsp90 inhibitors also delayed the rate of restoration of normal protein synthesis following a brief heat shock. (***See also*** *Alvespimycin; Ansamycin; Deguelin; Derrubone; Ganetespib; Herbimycin; Macbecin; Radicicol; Tanespimycin; and the nonansamycin Hsp90 inhibitor KW-2478*) **Mechanism of Action:** Geldanamycin blocks Hsp90-assisted folding of some 250-300 currently known client proteins. For example, when geldanamycin binds to Hsp90 (K_i = 10-30 nM), the HSP90's multimolecular interaction with Raf-1 is disrupted, destabilizing the latter. Geldanamycin also inhibits the oncogene product p185erbB-2 protein-tyrosine kinase and induces polyubiquitination and proteasomal degradation. *Note*: 17-DMAG is a water-soluble derivative of geldanamycin that potently inhibits Hsp90 (IC_{50} = 24 nM) and has excellent bioavailability and tissue distribution in animals (1-3). Geldanamycin also induces oxidative stress with concomitant glutathione (GSH) depletion, and *N*-acetylcysteine prevents the geldanamycin cytotoxicity by forming geldanamycin-N-acetylcysteine adduct (4). Renaturation of thermally denatured firefly luciferase in rabbit reticulocyte lysate (RRL) requires hsp90, hsc70, and other as yet unidentified RRL components (5). The effects of geldanamycin on the kinetics of the luciferase renaturation in RRL show that chaperone-mediated luciferase renaturation that obeys Michaelis-Menten kinetics. GA inhibits luciferase renaturation uncompetitively with respect to ATP and noncompetitively with respect to luciferase, indicating that it binds after ATP binds and that geldanamycin binds to both the hsp90 chaperone machine/ATP complex and the hsp90/ATP/luciferase complex. GA markedly decreased the K_{app} of the hsp90 chaperone machine for ATP, suggesting that GA increases the binding affinity of the hsp90 chaperone machinery for ATP or it slows the rate of ATP hydrolysis (6). Consistent with the notion that GA specifically binds hsp90 and inhibits its function, addition of hsp90, but not hsc70, p60, or p23, reversed GA-induced inhibition of luciferase renaturation in RRL. GA increases the steady-state levels of luciferase associated with hsp90/hsp70 chaperone machine complexes that contain p60 and blocks the association of the hsp90 cohort p23 with chaperone-bound luciferase (6). **Hsp90 Targets:** Humans have five Hsp90 isoforms: Hsp90-α_1 (the inducible cytoplasmic isoform (98.2-kDa) encoded by HSP90AA1), Hsp90-α_2 (a second inducible cytoplasmic isoform (84.7-kDA) encoded by HSP90AA2), Hsp90-β (the constitutive cytoplasmic isoform encoded by HSP90AB), Endoplasmin/GRP-94 (the ER protein (also known as gp96, grp94 and ERp99) encoded by HSP90B1), and

TNF Receptor-Associated Protein 1 (the mitochondrial isoform encoded by TRAP1). Hsp90's are homodimers, with each subunit containing a highly conserved 25-kDa N-terminal domain (NTD) with an ATP binding motif, a "charged linker" region, a 40-kDa middle domain, and a 12-kDa C-terminal domain (CTD). Geldanamycin and other benzoquinone ansamycins display an unusual capacity to revert tyrosine kinase-induced oncogenic transformation by preventing Hsp90 stabilization of oncogenic kinases against the same proteasome-mediated turnover. Geldanamycin inhibition of Hsp90 also induces apoptosis through inhibition of the PI3K/AKT signaling pathway and growth factor signaling. **Other Targets:** DNA replication (7); HtpG (8); nitric-oxide synthase (9); oncogene product p185/erbB-2 protein-tyrosine kinase (8,9); src protein-tyrosine kinase (10,12); terminal deoxyribonucleotidyl-transferase (13). **1**. Rowlands, Newbatt, Prodromou, *et al.* (2004) *Anal. Biochem.* **327**, 176. **2**. Chadli, Ladjimi, Baulieu & Catelli (1999) *J. Biol. Chem.* **274**, 4133. **3**. Panaretou, Prodromou, Roe, *et al.* (1998) *EMBO J.* **17**, 4829. **4**. Mlejnek & Dolezel (2014) *Chem. Biol. Interact.* **220**, 248. **5**. Schumacher, *et al.* (1994) *J. Biol. Chem.* **269**, 9493. **6**. Thulasiraman & Matts (1996) *Biochemistry* **35**, 13443. **7**. Yamaki, Iguchi-Ariga & Ariga (1989) *J. Antibiot.* **42**, 604. **8**. Joly, Ayres & Kilbourn (1997) *FEBS Lett.* **403**, 40. **9**. Schnur, Corman, Gallaschun, *et al.* (1995) *J. Med. Chem.* **38**, 3806. **10**. Mimnaugh, Chavany & Neckers (1996) *J. Biol. Chem.* **271**, 22796. **11**. Yamaki, Nakajima, Seimiya, Saya, Sugita & Tsuruo (1995) *J. Antibiot.* **48**, 1021. **12**. Hall, Schaeublin & Missbach (1994) *Biochem. Biophys. Res. Commun.* **199**, 1237. **13**. Srivastava, DiCioccio, Rinehart & Li (1978) *Mol. Pharmacol.* **14**, 442.

Gemcitabine

This synthetic cytidine analogue (FW = 263.20 g/mol; CAS = 95058-81-4), also known as 2'-deoxy-2',2'-difluorocytidine, is an antineoplastic agent that has been used in the treatment of early and advanced stage non-small cell lung cancer. After transport, gemcitabine undergoes metabolic phosphorylation steps to form gemcitabine 5'-triphosphate. (There are twelve known nucleoside transporters (NTs), and the susceptibility of any particular cell type to gemcitabine depends on the cell's NT subtype. NT deficiencies are also known that confer significant gemcitabine resistance.) Subsequent incorporation into DNA leads to cell death. Gemcitabine also induces JNK and p38 MAPK activation mediates increased apoptosis. **Key Pharmacokinetic Parameters:** *See* Appendix II in Goodman & Gilman's *THE PHARMACOLOGICAL BASIS OF THERAPEUTICS*, 12th Edition (Brunton, Chabner & Knollmann, eds.) McGraw-Hill Medical, New York (2011). **Target(s):** DNA biosynthesis (1,2); DNA polymerase (2); Na$^+$-dependent nucleoside transporter (3); ribonucleoside-diphosphate reductase (4,5); ribonucleoside triphosphate reductase (5,6); RNA biosynthesis (1). **1**. Ruiz van Haperen, Veerman, Vermorken & Peters (1993) *Biochem. Pharmacol.* **46**, 762. **2**. Huang, Chubb, Hertel, Grindey & Plunkett (1991) *Cancer Res.* **51**, 6110. **3**. Burke, Lee, Ferguson & Hammond (1998) *J. Pharmacol. Exp. Ther.* **286**, 1333. **4**. Cerqueira, Fernandes & Ramos (2007) *Recent Patents Anticancer Drug Discov.* **2**, 11. **5**. Sigmond, Kamphuis, Laan, *et al.* (2007) *Biochem. Pharmacol.* **73**, 1548. **6**. Cerqueira, Fernandes & Ramos (2007) *Chemistry* **13**, 8507.

Gemfibrozil

This fibric acid derivative (FW$_{free-acid}$ = 250.34 g/mol; CAS = 25812-30-0), also known as Lopid®, is a serum lipid-regulating agent that lowers very-low-density lipoprotein (VLDL) concentrations, while raising HDL levels. Gemfibrozil also decreases serum triacylglycerol levels and inhibits expression of inducible nitric-oxide synthase. **Key Pharmacokinetic Parameters:** *See* Appendix II in Goodman & Gilman's *THE*

PHARMACOLOGICAL BASIS OF THERAPEUTICS, 12th Edition (Brunton, Chabner & Knollmann, eds.) McGraw-Hill Medical, New York (2011). **Target(s):** acetyl-CoA carboxylase, possibly indirectly inhibited (1,2); CYP1A2, mildly inhibited (3); CYP2C8 (4); CYP2C9 (3); CYP2C19, mildly inhibited (3); diacylglycerol avyltransferase (5); fatty acid elongation (6); glucose-6-phosphate dehydrogenase (1); palmitoyl-CoA hydrolase, *or* acyl CoA hydrolase (7). **1.** Sánchez, Vazquez, Alegret, *et al.* (1993) *Life Sci.* **52**, 213. **2.** Munday & Hemingway (1999) *Adv. Enzyme Regul.* **39**, 205. **3.** Wen, Wang, Backman, Kivisto & Neuvonen (2001) *Drug Metab. Dispos.* **29**, 1359. **4.** Wang, Neuvonen, Wen, Backman & Neuvonen (2002) *Drug Metab. Dispos.* **30**, 1352. **5.** Zhu, Ganji, Kamanna & Kashyap (2002) *Atherosclerosis* **164**, 221. **6.** Sánchez, Vinals, Alegret, *et al.* (1992) *FEBS Lett.* **300**, 89. **7.** Sánchez, Alegret, Adzet, Merlos & Laguna (1992) *Biochem. Pharmacol.* **43**, 639.

Geminin

This highly conserved DNA replication inhibitor (MW_{human} = 23.5 kDa; Accession Number = NP 056979; Symbol: Gem and GMNN) is a nuclear protein with homologues in most eukaryotic organisms. To maintain genome integrity, DNA replication must occur only once during each cell cycle, and every step of pre-replication complex (*or* pRC) assembly process must be tightly regulated (1). Geminin was identified in *Xenopus* as a replication inhibitor that prevents binding of MCM proteins to chromatin (2). In synchronized HeLa cells, it is absent during G_1 phase, accumulates during S, G_2, and M phases, but disappears at the time of the metaphase-anaphase transition (3). Geminin forms a complex with Cdt1 and inhibits its interaction with MCM complexes (3-8). Geminin was also shown to inhibit DNA binding of Cdt1 *in vitro* (6). Encoded by a gene located on the short arm of Human Chromosome-16, this coiled-coil motif-containing protein inhibits DNA replication by binding to the DNA replication factor Cdt1, preventing incorporation of minichromosome maintenance proteins into the pRC. Geminin is expressed during S and G_2 phases of the cell cycle and is degraded later by the anaphase-promoting complex during the metaphase-anaphase transition. Among its putative functions are control of cell cycle, cellular proliferation, cell lineage commitment, and neural differentiation. Its increased expression may play a role in colon, rectal and breast cancer. **Mode of Inhibitory Action:** Geminin inhibits the initiation of eukaryotic DNA replication, a process that involves the assembly of pre-replicative complexes (pre-RCs) at origins of replication during the G_1 phase of the cell cycle (9). Geminin inhibits the association of HsCdt1 with DNA or with human origin recognition complex HsORC-HsCdc6-DNA complexes, but does not inhibit recruitment of HsMCM2-7 to DNA to form complexes containing all of the pre-RC proteins (9). Geminin is itself a component of these complexes and interacts directly with the HsMcm3 and MsMcm5 subunits of HsMCM2-7, as well as with HsCdt1 (9). Geminin is also required to restrain mesendodermal fate acquisition of early embryonic cells and is associated with both decreased Wnt signaling and enhanced Polycomb repressor complex retention at mesendodermal genes (10). **1.** Arias & Walter (2007) *Genes. Dev.* **21**, 497. **2.** McGarry & Kirschner (1998) *Cell* **93**, 1043. **3.** Wohlschlegel, Dwyer, Dhar, *et al.* (2000) *Science* **290**, 2309. **4.** Tada, Li, Maiorano, Méchali & Blow (2001) *Nat. Cell Biol.* **3**, 107. **5.** Mihaylov, Kondo, Jones, *et al.* (2002) *Mol. Cell. Biol.* **22**, 1868. **6.** Yanagi, Mizuno, You & Hanaoka (2002) *J. Biol. Chem.* **277**, 40871. **7.** Lee, Hong, Choi, *et al.* (2004) *Nature* **430**, 913. **8.** Yanagi, Mizuno, Tsuyama, *et al.* (2005) *J. Biol. Chem.* **280**, 19689. **9.** Wu, Lu, Santos, Frattini & Kelly (2014) *J. Biol. Chem.* **289**, 30810. **10.** Caronna, Patterson, Hummert & Kroll (2013) *Stem Cells* **31**, 1477.

Geneticin

This antibiotic ($FW_{free-base}$ = 496.56 g/mol; frequently supplied as the disulfate salt, FW = 692.71; CAS = 108321-42-2), also known as G418, is a gentamycin-related aminoglycoside that contains three monosaccharide units modified by hydroxyl, ammonium, and methyl groups. Genticin binds to both prokaryotic and eukaryotic ribosome A-sites, inhibiting peptide elongation. It is often used to maintain cells stably transfected with neomycin resistance genes (***See also*** *Kanamycin; Gentamicin; Neomycin*). **Target(s):** ornithine decarboxylase (1); phospholipase D (2); protein biosynthesis, elongation (3-5). **1.** Jackson, Goldsmith & Phillips (2003) *J. Biol. Chem.* **278**, 22037. **2.** Liscovitch, Chalifa, Danin & Eli (1991) *Biochem. J.* **279**, 319. **3.** Vicens & Westhof (2003) *J. Mol. Biol.* **326**, 1175. **4.** Bar-Nun, Shneyour & Beckmann (1983) *Biochim. Biophys. Acta* **741**, 123. **5.** Ursic, Kemp & Helgeson (1981) *Biochem. Biophys. Res. Commun.* **101**, 1031.

Genistein

This photosensitive isoflavone antioxidant (FW = 270.24 g/mol; CAS = 446-72-0), also known as 5,7-dihydroxy-3-(4-hydroxyphenyl)-4*H*-1-benzopyran-1-one, is a phytoestrogen found in a number of plants, particularly soy products. It is the aglycon of genistin and sophoricoside. This natural product is an inhibitor of protein-tyrosine kinases, including the autophosphorylation of epidermal growth factor receptor kinase (IC_{50} = 2.6 mM). The inhibition is competitive with respect to ATP and noncompetitive with respect to the phosphate acceptor. It also inhibits p60v-src protein-tyrosine kinase (IC_{50} = 25 μM) and certain other protein kinases (*e.g.*, protein-histidine kinase). It has very weak effects on the activity of protein kinase A, protein kinase C, and phosphorylase kinase. Genistein also inhibits tumor cell proliferation, induces tumor cell differentiation, inducing cell cycle arrest and apoptosis in Jurkat T-leukemia cells. **Target(s):** arylamine *N*-acetyltransferase (60,61); Ca^{2+} channels (1); cAMP-dependent protein kinase, *or* protein kinase A (2); cAMP phosphodiesterase (3); β-carotene (15), 15'-monooxygenase (62); catechol *O*-methyltransferase (4); chalcone isomerase (5); 3',5'-cyclic-GMP phosphodiesterase, cGMP phospho-diesterase (6); 3',5'-cyclic-nucleotide phosphodiesterase (6); CYP1A1 (7); diphosphoinositol-pentakisphosphate kinase (8); DNA topoisomerase I (9); DNA topoisomerase II (9-13); dual-specificity kinase (14-16); epidermal-growth-factor receptor protein-tyrosine kinase (17,18); estrone sulfotransferase (19,20); fatty-acid synthase (59); F_0F_1 ATP synthase, mitochondrial (21); focal adhesion kinase (22); GABAA receptor (23); β-galactosidase (24); α-glucosidase (25); GLUT1 (glucose transport) (26,27); glutathione S-transferase (58); histidine decarboxylase (4); 3-hydroxy-3-methylglutaryl-CoA reductase (28,29); 3β-hydroxysteroid dehydrogenase (30,31); inositol trisphosphate 3-kinase (32); iodide peroxidase, *or* thyroid peroxidase (65); 5-lipoxygenase, *or* arachidonate 5-lipoxygenase, K_i = 60 mM (63); 15-lipoxygenase, *or* arachidonate 15-lipoxygenase (64); lipoxygenase, soybean, IC_{50} = 107 mM (63,64); NADH oxidase (33); P-glycoprotein (34); phenol sulfotransferase, *or* aryl sulfotransferase (20,35-37); phenylalanine ammonia-lyase, weakly inhibited (38); phosphatidylinositol 3-kinase (39); phosphodiesterase (16); phosphodiesterase (26); phosphodiesterase (36); phosphodiesterase (46); phosphodiesterase (56); P_1-purinergic (adenosine) receptor (40); protein-disulfide isomerase (41); protein-histidine kinase (42); protein-histidine pros kinase (42); protein-tyrosine kinase (15,22,43-51); receptor protein-tyrosine kinase (15,17,18,52,53); S_6 kinase (54); src protein-tyrosine kinase (22,47-49); steroid D-isomerase (31,55); steroid 5a-reductase (56); steroid sulfotransferase (19); xenobiotic-transporting ATPase, multidrug-resistance protein (34,57); xanthine oxidase (66). **1.** Dobrydneva, Williams, Morris & Blackmore (2002) *J. Cardiovasc. Pharmacol.* **40**, 399. **2.** Patel, Soulages, Wells & Arrese (2004) *Insect Biochem. Mol. Biol.* **34**, 1269. **3.** Nichols & Morimoto (1999) *Arch. Biochem. Biophys.* **366**, 224. **4.** Umezawa, Tobe, Shibamoto, Nakamura & Nakamura (1975) *J. Antibiot.* **28**, 947. **5.** Dixon, Dey & Whitehead (1982) *Biochim. Biophys. Acta* **715**, 25. **6.** Ko, Shih, Lai, Chen & Huang (2004) *Biochem. Pharmacol.* **68**, 2087. **7.** Shertzer, Puga, Chang, *et al.* (1999) *Chem. Biol. Interact.* **123**, 31. **8.** Safrany (2004) *Mol. Pharmacol.* **66**, 1585. **9.** Okura, Arakawa, Oka, Yoshinari & Monden (1988) *Biochem. Biophys. Res. Commun.* **157**, 183. **10.** Markovits, Linassier, Fosse, *et al.* (1989) *Cancer Res.* **49**, 5111. **11.** McCabe &

Orrenius (1993) *Biochem. Biophys. Res. Commun.* **194**, 944. **12**. Andersen, Bendixen & Westergaard (1996) *DNA Replication in Eucaryotic Cells*, Cold Spring Harbor Laboratory Press, p. 587. **13**. Insaf, Danks & Witiak (1996) *Curr. Med. Chem.* **3**, 437. **14**. Rudrabhatla & Rajasekharan (2004) *Biochemistry* **43**, 12123. **15**. MacKintosh & MacKintosh (1994) *Trends Biochem. Sci.* **19**, 444. **16**. Lerner-Marmarosh, Shen, Torno, *et al.* (2005) *Proc. Natl. Acad. Sci. U.S.A.* **102**, 7109. **17**. Gargala, Baishanbo, Favennec, *et al.* (2005) *Antimicrob. Agents Chemother.* **49**, 4628. **18**. Benter, Juggi, Khan, *et al.* (2005) Mol. Cell. Biochem. **268**, 175. **19**. Ohkimoto, Liu, Suiko, Sakakibara & Liu (2004) *Chem. Biol. Interact.* **147**, 1. **20**. Harris, Wood, Bottomley, *et al.* (2004) *J. Clin. Endocrinol. Metab.* **89**, 1779. **21**. Zheng & Ramirez (2000) *Brit. J. Pharmacol.* **130**, 1115. **22**. Browe & Baumgarten (2003) *J. Gen. Physiol.* **122**, 689. **23**. Huang, Fang & Dillon (1999) *Brain Res. Mol. Brain Res.* **67**, 177. **24**. Hazato, Naganawa, Kumagai, Aoyagi & Umezawa (1979) *J. Antibiot.* **32**, 217. **25**. Wang, Ma, Pang, *et al.* (2004) *Bioorg. Med. Chem. Lett.* **14**, 2947. **26**. Vera, Reyes, Carcamo, *et al.* (1996) *J. Biol. Chem.* **271**, 8719. **27**. Martin, Kornmann & Fuhrmann (2003) *Chem. Biol. Interact.* **146**, 225. **28**. Sung, Choi, Lee, Park & Moon (2004) *Biosci. Biotechnol. Biochem.* **68**, 1051. **29**. Sung, Lee, Park & Moon (2004) *Biosci. Biotechnol. Biochem.* **68**, 428. **30**. Ohno, Matsumoto, Watanabe & Nakajin (2004) *J. Steroid Biochem. Mol. Biol.* **88**, 175. **31**. Wong & Keung (1999) *J. Steroid Biochem. Mol. Biol.* **71**, 191. **32**. Mayr, Windhorst & Hillemeier (2005) *J. Biol. Chem.* **280**, 13229. **33**. Bohmont, Aaronson, Mann & Pardini (1987) *J. Nat. Prod.* **50**, 427. **34**. Sharom, Liu, Romsicki & Lu (1999) *Biochim. Biophys. Acta* **1461**, 327. **35**. Mesia-Vela & Kauffman (2003) Xenobiotica 33, 1211. **36**. Ebmeier & Anderson (2004) *J. Clin. Endocrinol. Metab.* **89**, 5597. **37**. Meng, Shankavaram, Chen, *et al.* (2006) *Cancer Res.* **66**, 9656. **38**. Jorrin & Dixon (1990) *Plant Physiol.* **92**, 447. **39**. Berggren, Gallegos, Dressler, Modest & Powis (1993) *Cancer Res.* **53**, 4297. **40**. Okajima, Akbar, Abdul Majid, *et al.* (1994) *Biochem. Biophys. Res. Commun.* **203**, 1488. **41**. Winter, Klappa, Freedman, Lilie & Rudolph (2002) *J. Biol. Chem.* **277**, 310. **42**. Huang, Nasr, Kim & Matthews (1992) *J. Biol. Chem.* **267**, 15511. **43**. Rosenshine, Ruschkowski & Finlay (1994) *Meth. Enzymol.* **236**, 467. **44**. Akiyama & Ogawara (1991) *Meth. Enzymol.* **201**, 362. **45**. Eriksson, Toivola, Sahlgren, Mikhailov & Härmälä-Braskén (1998) *Meth. Enzymol.* **298**, 542. **46**. Akiyama, Ishida, Nakagawa, *et al.* (1987) *J. Biol. Chem.* **262**, 5592. **47**. Øvrevik, Låg, Schwarze & Refsnes (2004) *Toxicol. Sci.* **81**, 480. **48**. Ren & Baumgarten (2005) *Amer. J. Physiol.* **288**, H2628. **49**. Gao, Lau, Wong & Li (2004) *Cell. Signal.* **16**, 333. **50**. Du, Gao, Lau, *et al.* (2004) *J. Gen. Physiol.* **123**, 427. **51**. Burlando, Magnelli, Panfoli, Berti & Viarengo (2003) *Cell Physiol. Biochem.* **13**, 147. **52**. Mergler, Dannowski, Bednarz, *et al.* (2003) *Exp. Eye Res.* 77, 485. **53**. Ren & Baumgarten (2005) *Amer. J. Physiol.* **288**, H2628. **54**. Linassier, Pierre, Pecq & Pierre (1990) *Biochem. Pharmacol.* **39**, 187. **55**. Weintraub, F. Vincent, E. E. Baulieu & A. Alfsen (1977) *Biochemistry* **16**, 5045. **56**. Hiipakka, Zhang, Dai, Dai & Liao (2002) *Biochem. Pharmacol.* **63**, 1165. **57**. Paul, Breuninger & Kruh (1996) *Biochemistry* **35**, 14003. **58**. Hayeshi, Mutingwende, Mavengere, Masiyanise & Mukanganyama (2007) *Food Chem. Toxicol.* **45**, 286. **59**. Li, Ma, Wang & Tian (2005) *J. Biochem.* **138**, 679. **60**. Kukongviriyapan, Phromsopha, Tassaneeyakul, *et al.* (2006) Xenobiotica 36, 15. **61**. Kawamura, Westwood, Wakefield, *et al.* (2008) *Biochem. Pharmacol.* **75**, 1550. **62**. Nagao, Maeda, Lim, Kobayashi & Terao (2000) *Nutr. Biochem.* **11**, 348. **63**. Mahesha, Singh & Rao (2007) *Arch. Biochem. Biophys.* **461**, 176. **64**. Sadik, Sies & Schewe (2003) *Biochem. Pharmacol.* **65**, 773. **65**. Schmutzler, Bacinski, Gotthardt, *et al.* (2007) *Endocrinology* **148**, 2835. **66**. Lin, Zhang, Pan & Gong (2015) *J. Photochem. Photobiol. B* **153**, 463.

Genistin

This yellow glucoside (FW = 432.38 g/mol; CAS = 529-59-9), also known as genistein 7-*O*-β-D-glucopyranoside, is slightly soluble in hot water. Note that, while the aglycon genistein inhibits a number of protein-tyrosine kinases, genistin is inactive and has been used as a negative control. **Target(s):** aldehyde dehydrogenase (1); CYP1A1 (2); DNA nucleotidylexotransferase, terminal deoxyribonucleotidyltransferase (3); DNA topoisomerase II (4); lipoxygenase, soybean, IC_{50} = 10 μM (5);

phenylalanine ammonia-lyase, weakly inhibited (6). **1**. Keung & Vallee (1993) *Proc. Natl. Acad. Sci. U.S.A.* **90**, 1247. **2**. Shertzer, Puga, Chang, *et al.* (1999) *Chem. Biol. Interact.* **123**, 31. **3**. Uchiyama, Tagami, Kamisuki, *et al.* (2005) *Biochim. Biophys. Acta* **1725**, 298. **4**. Martin-Cordero, Lopez-Lazaro, Pinero, *et al.* (2000) *J. Enzyme Inhib.* **15**, 455. **5**. Mahesha, Singh & Rao (2007) *Arch. Biochem. Biophys.* **461**, 176. **6**. Jorrin & Dixon (1990) *Plant Physiol.* **92**, 447.

Gentamycins

Gentamycin C

Gentamycin A

These aminoglycosidic antibiotics ($FW_{Gentamycin-A}$ = 477.60 g/mol; CAS = 1403-66-3), often spelled gentamicin, inhibit protein biosynthesis (1-3) and are especially effective in the treatment of *Pseudomonas aeruginosa* infections. They are obtained by fermentation of *Micromonospora purpurea* or *M. echinospora*. Gentamycins C_1, C_2, and C_{1a} are structurally similar (for C_1: $R_1 = R_2 = CH_3$; for C_2: $R_1 = CH_3$ and R_2 = H; and C_{1a}: $R_1 = R_2$ =H) and contain garosamine, 2 deoxystreptamine, and a purpursamine residue that may be methylated. Gentamycin A, systematically referred to as *O*-2-amino-2-deoxy-α-D-glucopyranosyl-(1→4)-*O*-[3-deoxy-3-(methylamino)-α-D-xylopyranosyl-(1→6)]-2-deoxy-streptamine, is structurally similar to kanamycin C. **Key Pharmacokinetic Parameters:** *See* Appendix II in Goodman & Gilman's THE PHARMACOLOGICAL BASIS OF THERAPEUTICS, 12th Edition (Brunton, Chabner & Knollmann, eds.) McGraw-Hill Medical, New York (2011). **Target(s):** aminoglycoside N^6-acetyltransferase, *or* kanamycin 6'-acetyltransferase, inhibited by gentamycins A and C_1 (4-7); calreticulin, chaperone activity inhibited (8); glucose-6 phosphate dehydrogenase (9); kanamycin kinase, inhibited by gentamycin C_{1a} (4); leukocyte elastase, slightly inhibited (10); NADPH oxidase (11); ornithine decarboxylase (K_i = 1.6 mM (12); phospholipase A_1, inhibition due to substrate depletion (13); protein biosynthesis (1-3). **1**. Pestka (1974) *Meth. Enzymol.* **30**, 261. **2**. Jiménez (1976) *Trends Biochem. Sci.* **1**, 28. **3**. Milanesi & Ciferri (1966) *Biochemistry* **5**, 3926. **4**. Haas & Dowding (1975) *Meth. Enzymol.* **43**, 611. **5**. Martel, Masson, Moreau & Le Goffic (1983) *Eur. J. Biochem.* **133**, 515. **6**. Benveniste & Davies (1971) *Biochemistry* **10**, 1787. **7**. Le Goffic & Martel (1974) *Biochimie* **56**, 893. **8**. Horibe, Matsui, Tanaka, *et al.* (2004) *Biochem. Biophys. Res. Commun.* **323**, 281. **9**. Beydemir, Ciftci & Kufrevioglu (2002) *J. Enzyme Inhib. Med. Chem.* **17**, 271. **10**. Jones, Elphick, Pettitt, Everard & Evans (2002) *Eur. Respir. J.* **19**, 1136. **11**. Umeki (1995) *Comp. Biochem. Physiol. B Biochem. Mol. Biol.* **110**, 817. **12**. Henley, Mahran & Schacht (1988) *Biochem. Pharmacol.* **37**, 1679. **13**. Uchiyama, Miyazaki, Amakasu, *et al.* (1999) *J. Biochem.* **125**, 1001.

Gentiobiose

This disaccharide (FW = 342.30 g/mol; CAS = 5996-00-9), also known as 6-*O*-β-D-glucopyranosyl-D-glucose, is a component of many glycosides (*e.g.*, amygdalin) and oligosaccharides. The two anomeric forms reportedly have widely different melting points (*e.g.*, 190-195 °C *versus* 86 °C for the β- and α-anomers, respectively). **Target(s):** glucan 1,4-α-glucosidase, *or* glucoamylase (1); licheninase (2); membrane oligosaccharide glycerophosphotransferase (3); phosphatidyl-glycerol:membrane-oligosaccharide glycerophospho-transferase (4); α,α-trehalase, weakly inhibited (5). **1.** Fogarty & Benson (1983) *Eur. J. Appl. Microbiol. Biotechnol.* **18**, 271. **2.** Pigman (1951) *The Enzymes*, 1st ed., **1** (Part 2), 725. **3.** Goldberg, Rumley & Kennedy (1981) *Proc. Natl. Acad. Sci. U.S.A.* **78**, 5513. **4.** Jackson & Kennedy (1983) *J. Biol. Chem.* **258**, 2394. **5.** Lúcio-Eterovic, Jorge, Polizeli & Terenzi (2005) *Biochim. Biophys. Acta* **1723**, 201.

Gentisic Acid

This salicylate metabolite (FW$_{\text{free-acid}}$ = 154.12 g/mol; CAS = 490-79-9; pK_a = 2.93 at 25°C), also known as 2,5-dihydroxybenzoic acid, is a product of the 3-hydroxybenzoate 6-monooxygenase reaction and a substrate of gentisate 1,2-dioxygenase and gentisate decarboxylase. **Target(s):** γ-butyrobetaine dioxygenase, *or* γ-butyrobetaine hydroxylase (1); complement component C$_3$', guinea pig (2); 4,5-dihydroxybenzoate decarboxylase, weakly inhibited (3); diiodophenylpyruvate reductase (4); 3-galactosyl-*N*-acetylglucosaminide 4-α-L fucosyltransferase (5); glycolprotein 3-α-L-fucosyltransferase (5); 4-hydroxybenzoate 3-monooxygenase, NAD(P)H (6); 4-hydroxyphenylpyruvate dioxygenase (7); 15-lipoxygenase, *or* arachidonate 15-lipoxygenase (8); medium-chain fatty-acyl-CoA synthetase, *or* butyryl-CoA synthetase (9); orsellinate decarboxylase (10); procollagen-proline 4-dioxygenase (11); protocatechuate 3,4-dioxygenase (12); *o*-pyrocatechuatre decarboxylase (13); tannase (14); tyrosinase (15); tyrosine 3-monooxygenase (16). **1.** Ng, Hanauske-Abel & Englard (1991) *J. Biol. Chem.* **266**, 1526. **2.** Shin & Mayer (1968) *Biochemistry* **7**, 3003. **3.** Nakazawa & Hayashi (1978) *Appl. Environ. Microbiol.* **36**, 264. **4.** Zannoni (1970) *Meth. Enzymol.* **17A**, 665. **5.** Niu, Fan, Sun, *et al.* (2004) *Arch. Biochem. Biophys.* **425**, 51. **6.** Fujii & Kaneda (1985) *Eur. J. Biochem.* **147**, 97. **7.** Lindstedt & Rundgren (1982) *J. Biol. Chem.* **257**, 11922. **8.** Russell, Scobbie, Duthie & Chesson (2008) *Bioorg. Med. Chem.* **16**, 4589. **9.** Londesborough & Webster (1974) *The Enzymes*, 3rd ed., **10**, 469. **10.** Petterson (1965) *Acta Chem. Scand.* **19**, 2013. **11.** Majamaa, Günzler, Hanauske-Abel, Myllylä & Kivirikko (1986) *J. Biol. Chem.* **261**, 7819. **12.** Hou, Lillard & Schwartz (1976) *Biochemistry* **15**, 582. **13.** Santha, Rao & Vaidyanathan (1996) *Biochim. Biophys. Acta* **1293**, 191. **14.** Iibuchi, Minoda & Yamada (1972) *Agric. Biol. Chem.* **36**, 1553. **15.** Schved & Kahn (1992) *Pigment Cell Res.* **5**, 58. **16.** Nagatsu, Levitt & Udenfriend (1964) *J. Biol. Chem.* **239**, 2910.

Genz-112638, *See Eliglustat*

Geranial

This monoterpene aldehyde (FW = 152.24 g/mol; CAS = 5392-40-5; Light oil; Strong lemony fragrance; Practically insoluble in water; Miscible with ethanol and diethyl ether), also known as *trans*-citral, is a geometric isomer of neral and is found in many natural oils, including lemons and oranges. Geranial has also been observed in a number of insect pheromone mixtures. **Target(s):** aldehyde dehydrogenase (1-3); CYP1A1, inhibited by citral (4); CYP1A2, inhibited by citral (5); CYP2B4, inhibited by citral (5); retinal dehydrogenase, inhibited by citral (6,7); retinol dehydrogenase, inhibited by citral (8,9); tyrosinase, *or* monophenol monooxygenase, IC$_{50}$ = 1.6 mM (10). **1.** Kikonyogo, Abriola, Dryjanski & Pietruszko (1999) *Eur. J. Biochem.* **262**, 704. **2.** Pietruszko, Abriola, Izaguirre, *et al.* (1999) *Adv.*

Exp. Med. Biol. **463**, 79. **3.** Boyer & Petersen (1991) *Drug Metab. Dispos.* **19**, 81. **4.** Tomita, Okuyama, Ohnishi & Ichikawa (1996) *Biochim. Biophys. Acta* **1290**, 273. **5.** Raner, Vaz & Coon (1996) *Mol. Pharmacol.* **49**, 515. **6.** Gagnon, Duester & Bhat (2002) *Biochim. Biophys. Acta* **1596**, 156. **7.** Penzes, Wang & Napoli (1997) *Biochim. Biophys. Acta* **1342**, 175. **8.** Chen, Namkung & Juchau (1995) *Biochem. Pharmacol.* **50**, 1257. **9.** Connor & Smit (1987) *Biochem. J.* **244**, 489. **10.** Masuda, Odaka, Ogawa, Nakamoto & Kuninaga (2008) *J. Agric. Food Chem.* **56**, 597.

Geranyl Diphosphate (*or*, Geranyl Pyrophosphate)

This pyrophosphate ester, (FW$_{\text{free-acid}}$ = 314.21 g/mol; CAS = 763-10-0), often abbreviated GPP, is an intermediate in terpene biosynthesis. It is a product of the reaction catalyzed by dimethylallyl*trans*transferase and a substrate for geranyl*trans*transferase (farnesyl-diphosphate synthase), (4*S*)-limonene synthase, and geranyl-diphosphate cyclase. Note that geranyl diphosphate is very labile in acidic solutions and is relatively stable in neutral and alkaline solutions. Neryl pyrophosphate is the *cis* isomer. **Target(s):** diphospho-mevalonate decarboxylase, IC$_{50}$ = 65 μM for the rat enzyme (1); farnesyl-diphosphatase (2,3); isopentenyl-diphosphate Δ-isomerase (4-8); mevalonate kinase (1,9-16); squalene synthase (17). **1.** Qiu & Li (2006) *Org. Lett.* **8**, 1013. **2.** Nah, Song & Back (2001) *Plant Cell Physiol.* **42**, 864. **3.** Ha, Lee, Lee, Song & Back (2003) *Biol. Plant.* **47**, 477. **4.** Jones & Porter (1985) *Meth. Enzymol.* **110**, 209. **5.** Banthorpe, Doonan & Gutowski (1977) *Arch. Biochem. Biophys.* **184**, 381. **6.** Bruenger, Chayet & Rilling (1986) *Arch. Biochem. Biophys.* **248**, 620. **7.** Ramos-Valdivia, van der Heijden, Verpoorte & Camara (1997) *Eur. J. Biochem.* **249**, 161. **8.** Spurgeon, Sathymoorthy & Porter (1984) *Arch. Biochem. Biophys.* **230**, 446. **9.** Porter (1985) *Meth. Enzymol.* **110**, 71. **10.** Imblum & Rodwell (1974) *J. Lipid Res.* **15**, 211. **11.** Hinson, Chambliss, Toth, Tanaka & Gibson (1997) *J. Lipid Res.* **38**, 2216. **12.** Dorsey & Porter (1968) *J. Biol. Chem.* **243**, 4667. **13.** Gray & Kekwick (1973) *Biochem. J.* **133**, 335. **14.** Gray & Kekwick (1972) *Biochim. Biophys. Acta* **279**, 290. **15.** Huang, Scott & Bennett (1999) *Protein Expr. Purif.* **17**, 33. **16.** Chu & Li (2003) *Protein Expr. Purif.* **32**, 75. **17.** de Montellano, Wei, Castillo, Hsu & Boparai (1977) *J. Med. Chem.* **20**, 243.

Geranylgeranylacetone

This cytoprotective antioxidant and antiulcer drug (FW = 330.56 g/mol; CAS = 6809-52-5; often supplied as mixture of (5*E*,9*E*,13*E*) and (5*Z*,9*E*,13*E*) isomers; *Symbol*: GGA), known systematically as 6,10,14,18-tetramethyl-5,9,13,17-nonadecatetraen-2-one, induces expression of HSP70, HSPB8, and HSPB1 and protects against development of inflammatory bowel disease, hypoxic/ischemic brain injury, and spinal and bulbar muscular atrophy. Oral administration prevents development of gastric ulcer induced by a single, or repeated, oral aspirin administration (5 consecutive days). Effects were more potent than those of gefarnate. Intraduodenal administration, but not the intragastric administration, also inhibited the ulceration induced by aspirin in pylorus-ligated rats, while the intraduodenal administration of gefarnate did not (1). GGA may induce transcriptional activation of HSP genes, suggesting novel mechanism of gastric mucosal defense against stress (2). GGA-induced activation of Hsp90/AMPK significantly increases NO-mediated vasodilation in healthy subjects, as well as in smokers (3). **1.** Murakami, Oketani, Fujisaki, *et al.* (1982) *Jpn. J. Pharmacol.* **32**, 299. **2.** Hirakawa, Rokutan, Nikawa & Kishi (1995) *Gastroenterology* **111**, 345. **3.** Fujimura, Jitsuiki, Maruhashi, *et al.* (2012) *Arterioscler. Thromb. Vasc .Biol.* **32**, 153.

2*E*,6*E*,10*E*-Geranylgeranyl Diphosphate

This pyrophosphate ester (FW$_{free-acid}$ = 450.45 g/mol; CAS = 6699-20-3), also known as (2E,6E,10E)-3,7,11,15-tetramethyl-2,6,10,14-hexadeca-tetraen-1-yl diphosphate and *all-trans*-geranylgeranyl pyrophosphate, is a precursor in the biosynthesis of di- and tetraterpenes and polyprenols. It is also a product inhibitor of farnesyl-*trans*transferase (*i.e.*, geranylgeranyl-diphosphate synthase). **Target(s):** abietadiene synthase (1-3); farnesyl-diphosphatase (4,5); mevalonate kinase (6-9); *trans*-octaprenyl-*trans*transferase (10); protein farnesyltransferase (11-13). **1.** Peters & Croteau (2002) *Biochemistry* **41**, 1836. **2.** Peters, Ravn, Coates & Croteau (2001) *J. Amer. Chem. Soc.* **123**, 8974. **3.** Peters, Flory, Jetter, *et al.* (2000) *Biochemistry* **39**, 15592. **4.** Nah, Song & Back (2001) *Plant Cell Physiol.* **42**, 864. **5.** Ha, Lee, Lee, Song & Back (2003) *Biol. Plant.* **47**, 477. **6.** Hinson, Chambliss, Toth, Tanaka & Gibson (1997) *J. Lipid Res.* **38**, 2216. **7.** Gray & Kekwick (1973) *Biochem. J.* **133**, 335. **8.** Gray & Kekwick (1972) *Biochim. Biophys. Acta* **279**, 290. **9.** Chu & Li (2003) *Protein Expr. Purif.* **32**, 75. **10.** Teclebrhan, Olsson, Swiezewska & Dallner (1993) *J. Biol. Chem.* **268**, 23081. **11.** Reiss (1995) *Meth. Enzymol.* **250**, 21. **12.** Spence & Casey (2001) *The Enzymes*, 3rd ed., **21**, 1. **13.** Das & Allen (1991) *Biochem. Biophys. Res. Commun.* **181**, 729.

2Z,6E,10E-Geranylgeranyl Diphosphate

This pyrophosphate ester (FW = 450.45 g/mol), also named (2Z,6E,10E)-3,7,11,15-tetramethyl-2,6,10,14-hexadecatetraen-1-yl diphosphate and *cis,trans,trans*-geranylgeranyl pyrophosphate, is a precursor in the biosynthesis of di- and tetraterpenes and polyprenols. **Target(s):** farnesyl-*trans*transferase (geranylgeranyl-diphosphate synthase) (1,2); protein farnesyltransferase (3). **1.** Sagami, Morita & Ogura (1994) *J. Biol. Chem.* **269**, 20561. **2.** Sagami, Korenaga, Ogura, Steiger, Pyun & Coates (1992) *Arch. Biochem. Biophys.* **297**, 314. **3.** Das & Allen (1991) *Biochem. Biophys. Res. Commun.* **181**, 729.

Gevokizumab

This humanized monoclonal antibody (MW = 145.2 kDa; CAS = 1129435-60-4), known also be the code name XOMA-542, binds strongly to interleukin-1β (*or* IL-1β), a pro-inflammatory cytokine involved in non-infectious uveitis, including Behçet's uveitis, cardiovascular disease, and other auto-inflammatory diseases (1,2). Through its binding to IL-1β, Gevokizumab inhibits IL-1 receptor activation, thereby modulating the cellular signaling events producing inflammation. Based on SPR measurements, the association rate constant (k_{on}) is 1.19 x 10^7 M^{-1}s^{-1}, and the dissociation rate constant (k_{off}) is 4.6 x 10^{-3} s^{-1}, giving an equilibrium dissociation constant K_D value of 0.39 nM (3). Gevokizumab significantly reduces the affinity of IL-1 binding to sIL-1RI (4.8 nM) by reducing the observed association rate for IL-1 addition to sIL-1RI (1.21 x 10^6 M^{-1}s^{-1}), with no detectable impact on the dissociation rate constant (3). **1.** Owyang, Maedler, Gross, *et al.* (2010) *Endocrinology* **151**, 2515. **2.** Roell, Issafras, Bauer, *et al.* (2010) *J. Biol. Chem.* **285**, 20607. **3.** Issafras, Corbin, Goldfine & Roell (2013) *J. Pharmacol. Exp. Ther.* **348**, 202.

GF-15

This synthetic griseofulvin analogue (FW = 422.91 g/mol; CAS = 126-07-8), systematically named as (2S,6'R)-(7-chloro-4,6-dimethoxybenzofuran-3-one)-2-spiro-1'-(2'-hexoxy-6'-methyl-cyclohex-2'-en-4'-one), is a potent, first-in-class inhibitor of centrosomal clustering in malignant cells (1). Earlier SAR studies of griseofulvin analogues demonstrated that their antifungal activity did not correlate with their effects on centrosomal clustering (2,3). Significantly, at concentrations, where GF-15 had little or no impact on tubulin polymerization, mitotic spindle tension was markedly reduced upon exposure of dividing cells to GF-15 (1). Cells with conditional centrosome amplification are more sensitive to GF-15 than griseofulvin (1). In a wide array of tumor cell lines, the observed IC$_{50}$ was 1–5 µM for inhibiting centrosomal clustering and promoting apoptotic cell death. Treatment of mouse xenograft models of human colon cancer and multiple myeloma resulted in tumor growth inhibition and significantly prolonged survival. (**See** *Griseofulvin*) **1.** Rønnest, Rebacz, Markworth, *et al.* (2009) *J. Med. Chem.* **52**, 3342. **2.** Rebacz, Larsen, Clausen, *et al.* (2007) *Cancer Res.* **67**, 6342. **3.** Rønnest, Raab, Anderhub, *et al.* (2012) *J. Med. Chem.* **55**, 652.

GF109203X

This potent PKC inhibitor (FW = 412.49 g/mol; CAS = 133052-90-1; Solubility: 80 mg/mL DMSO, <1 mg/mL H$_2$O), also known as bisindolylmaleimide I and 3-[1-[3-(dimethylamino)propyl]-1H-indol-3-yl]-4-(1H-indol-3-yl)-1H-pyrrole-2,5-dione, targets PKCα (IC$_{50}$ = 20 nM), PKCβI (IC$_{50}$ = 17 nM), PKCβII (IC$_{50}$ = 16 nM), and PKCγ (IC$_{50}$ = 20 nM), with more than 3000x selectivity over EGFR, PDGFR and insulin receptor. GF 109203X efficiently prevents PKC-mediated phosphorylation of an Mr = 47,000 protein in platelets and of an Mr = 80,000 protein in Swiss 3T3 cells, but failed to prevent PKC-independent phosphorylation (1). GF 109203X inhibited collagen- and α-thrombin-induced platelet aggregation as well as collagen-triggered ATP secretion; however, ADP-dependent reversible aggregation was not modified (1). Intriguingly, GF 109203X probably influences multidrug resistance (MDR) via direct binding to P-gp (2,3). Co-treatment of the cells with AG1478 (an EGFR inhibitor) and GF109203X produced an additive effect on carbachol-stimulated ERK1/2 activation, suggesting that the EGFR and PKC pathways act in parallel (4). See also (5) for a discussion of the nonspecific actions of PKC inhibitors on cardiovascular ion channels in addition to their PKC-inhibiting functions. In RANKL-treated primary mouse bone marrow cells, GF109203X inhibits processes driven by Receptor Activator of NFκB Ligand (*or* RANKL), reducing expression of key osteoclastic genes (*e.g.*, cathepsin K, calcitonin receptor, tartrate resistant acid phosphatase (TRAP), and the proton pump subunit V-ATPase-d2) (6). Expression of these proteins depends on RANKL-induced NF-κB and NFAT transcription factor actions; both are reduced in osteoclast progenitor populations upon treatment with GF109203X (6). GF109203X inhibits RANKL-induced calcium oscillation (6). (***Other PKC inhibitors in clinical development include:*** *LY333531; ISIS 3521, or CGP 64128A; Bryostatin 1; GF109203x; Ro 32-0432; Ro 31-8220; Go 6976; Go 7611; CPR 1006; and Balanol, or SPC 100840*) **1.** Toullec, Pianetti, Coste, *et al.* (1991) *J. Biol. Chem.* **266**, 15771. **2.** Gekeler, Boer, Uberall, *et al.* (1996) *Brit. J. Cancer* **74**, 897. **3.** Gekeler, Boer, Ise, *et al.* (1995) *Biochem. Biophys. Res. Commun.* **206**, 119. **4.** Park & Cho (2012) *Mol. Cell Biochem.* **370**, 191. **5.** Son, Hong, Kim, Firth & Park (2011) *BMB Rep.* **44**, 559. **6.** Yao, Li, Zhou, *et al.* (2015) *J. Cell Physiol.* **230**, 1235.

GF120918, See *Elacridar*

GGTI-287, See *N-4-[2(R)-Amino-3-mercaptopropyl]amino-2-phenylbenzoyl-L-leucine*

GGTI-297, See *N-4-[2(R)-Amino-3-mercaptopropyl]amino-2-naphthylbenzoyl-L-leucine*

Ghrelin

This 28-residue orexigenic peptide (MW = 3611.23 g; CAS 258279-04-8; Sequence: GS(O-n-CH₃(CH₂)₆C(=O)O)SFLSPEHQRVQQRKESKKPPAK LQPR) displays strong Growth Hormone (GH)-releasing activity, which is mediated by the activation of the so-called GH secretagogue receptor type 1a (GHS-R1a), which is concentrated in the hypothalamus–pituitary unit, but is also distributed in other central and peripheral tissues. Ghrelin decreases the firing rate and burst frequency of GnRH neurons in metestrous, but not in proestrous mice. Ghrelin also decreases the firing rate of GnRH neurons in males. These effects are prevented by the ghrelin receptor antagonist JMV2959. (*See* JMV2959) **1**. Farkas, Vastagh, Sárvári & Liposits (2013) *PLoS One* **8**, e78178.

Gibberellic Acid

This diterpenoid acid and phytohormone (FW$_{free-acid}$ = 346.38 g/mol; CAS = 77-06-5;; Symbol: GA₃), also known as gibberellin A₃, the first pure gibberellin to be identified in *Gibberella fujikuroi* (1), is a powerful plant growth stimulator and also induces amylase production, stimulates flower production, and modulates sub-apical cell division. The total chemical synthesis of gibberellic acid was reported by Corey and coworkers (2). **Target(s):** DNA-cytosine methyltransferase (3); gibberellin 2β-dioxygenase (4-6); indole-3 acetate β-glucosyltransferase, weakly inhibited (7); indole-3-ethanol oxidase (8). **1**. Cross (1954) *J. Chem. Soc.* **1954**, 4670. **2**. Corey, Danheiser, Chandrasekaran, *et al.* (1978) *J. Amer. Chem. Soc.* **100**, 8034. **3**. Vlasova, Demidenko, Kirnos & Vanyushin (1995) *Gene* **157**, 279. **4**. King, Junttila, Mander & Beck (2004) *Physiol. Plant.* **120**, 287. **5**. Park, Nakajima, Sakane, *et al.* (2005) *Biosci. Biotechnol. Biochem.* **69**, 1498. **6**. Park, Nakajima, Hasegawa & Yamaguchi (2005) *Biosci. Biotechnol. Biochem.* **69**, 1508. **7**. Leznicki & Bandurski (1988) *Plant Physiol.* **88**, 1481. **8**. Zhu & Scott (1995) *Biochem. Mol. Biol. Int.* **35**, 423.

Gibberellins

These plant hormones play important roles controlling in seed germination by altering the production of enzyme needed to mobilize food production used for growth of new cells. Some gibberellins also affect the length of stems, thereby affecting the growth of fruit clusters, as in grapes. Such effects are mediated by modulating chromosomal transcription in the cell layer (*i.e.*, aleurone layer) that wraps around the endosperm tissue in the seeds of rice, wheat, corn, *etc.* Show above is gibberellin A₁ (FW$_{free-acid}$ = 348.40 g/mol; Symbol: GA₃) is a bioactive phytohormone, first identified in the fungus *Gibberella fujikuroi* (1) and later purified from runner bean seeds (*Phaseolus coccineus*). GA₃ was the first gibberellin identified in higher plants (2). **Target(s):** gibberellin 2β-dioxygenase (IC₅₀ = 0.6 µM (3-5). **1**. Takahashi, Kitamura, Kawarada, *et al.* (1955) *Bull. Agric. Chem. Soc.* **19**, 267. **2**. MacMillan & Suter (1958) *Naturwissenschaften* **45**, 46. **3**. King, Junttila, Mander & Beck (2004) *Physiol. Plant.* **120**, 287. **4**. Smith & MacMillan (1986) *Planta* **167**, 9. **5**. Park, Nakajima, Hasegawa & Yamaguchi (2005) *Biosci. Biotechnol. Biochem.* **69**, 1508.

Gimeracil

This uracil analogue and antitumor agent (FW = 145.54 g/mol; CAS = 103766-25-2; Solubility: 25 mg/mL DMSO, < 1 mg/mL H₂O), also named 5-chloro-2,4-dihydroxypyridine, targets dihydropyrimidine dehydrogenase

(DPD), an enzyme that degrades 5-fluorouracil in blood. Gimeracil also exhibits radiosensitizing effects by partially inhibiting homologous recombination in the repair of DNA double-strand breaks. **1**. Shirasaka, *et al.* (1996) *Cancer Res.* **56**, 2602. **2**. Patt, *et al.* (2004) *J. Clin. Oncol.* **22**, 271. **3**. Shirao, et al. (2004) *Cancer* **100**, 2355. **4**. Kelly, et al. (2005) *J. Clin. Oncol.* **23**, 4553.

Ginkgotoxin

This weakly neurotoxic B₆ antivitamin (FW = 183.21 g/mol), also named 4'-O-methylpyridoxine and 2-methyl-4-(methoxymethyl)-5-(hydroxy-methyl)pyridin-3-ol, inhibits pyridoxine kinase (*Reaction*: Pyridoxine + MgATP²⁻ ⇌ PLP + MgADP), resulting in a deficiency of pyridoxal 5-phosphate (PLP), the cofactor required by >160 essential PLP-dependent enzymes in primary and secondary metabolism (1). Gingkotoxin is found in the seeds and leaves of *Ginkgo biloba* and is present in Ginkgo medications, homoeopathic preparations, as well as boiled Japanese Ginkgo foods, albeit in amounts too low to exert detrimental effects (2). Its phosphorylated form, 4'-O-methylpyridoxine-5'-phosphate, inhibits the 65-kDa form of human brain glutamate decarboxylase, but only at nonphysiologically high concentrations (3). Structural studies show ginkgotoxin and theophylline bound at the kinase phosphoryl-acceptor substrate site and involve similar protein-ligand interactions as the natural substrate (1). **1**. Gandhi, Desai, Ghatge, *et al.* (2012) *PLoS One.* **7**, e40954. **2**. Arenz, Klein, Fiehe, *et al.* (1996) *Planta Med.* **62**, 548. **3**. Buss, Drewke, Lohmann, Piwonska & Leistner (2001) *J. Med. Chem.* **44**, 3166.

Ginsenoside RD

This steroid glycoside and triterpene saponin (FW = 963.15 g/mol; CAS = 52705-93-8) from the traditional Chinese medicinal herb ginseng (*Panax*) attenuates myocardial ischemia/reperfusion injury via Akt/GSK-3β signaling and inhibition of the mitochondria-dependent apoptotic pathway. Ginsenoside RD also inhibits Ca²⁺-influx via receptor and store-operated Ca²⁺ channels in vascular smooth muscle cells (2). **1**. Wang, Li, Wang, *et al.* (2013) *PLoS One* **8**, e70956. **2**. Guan, Zhou, Zhang, *et al.* (2006) *Eur. J. Pharmacol.* **548**, 129.

Gliclazide

This oral hypoglycemic (FW = 323.42 g/mol; CAS 21187-98-4; N-(hexahydrocyclopenta[c]pyrrol-2(1H)-ylcarbamoyl)-4-methylbenzene sulfonamide) stimulates insulin secretion by binding to the sulfonylurea receptor and closing the K⁺ channels in pancreatic β-cells, leading to the depolarization and causing voltage-dependent opening of Ca²⁺ channels and Ca²⁺ influx. Gliclazide is widely used in the treatment of non-insulin-dependent diabetes mellitus (Type-2 diabetes). **Target(s):** ATPase, Ca²⁺-dependent (1); K⁺ channels, ATP-dependent (2,3); protein phosphatase (4). **1**. Gronda, Rossi & Gagliardino (1988) *Biochim. Biophys. Acta* **943**, 183. **2**.

Gribble & Ashcroft (1999) *Diabetologia* **42**, 845. **3**. Pulido, Casla, Suarez, *et al.* (1996) *Diabetologia* **39**, 22. **4**. Gagliardino, Rossi & Garcia (1997) *Acta Diabetol.* **34**, 6.

Givinostat

This potent, orally available histone deacetylase, or HDAC, inhibitor (FW = 475.95 g/mol; CAS = 732302-99-7; Solubility: 95 mg/mL DMSO, <1 mg/mL H_2O) targets HDAC2, HDAC1B and HDAC1A with IC_{50} values of 10 nM, 7.5 nM and 16 nM, respectively (1-3). ITF2357 is neuroprotective, improving functional recovery and inducing glial apoptosis after experimental traumatic brain injury (4). In insulin-producing β-cells, givinostat did not upregulate expression of the anti-inflammatory genes Socs1-3 or sirtuin-1 but reduced levels of IL-1β + IFN-γ-induced proinflammatory Il1a, Il1b, Tnfα, Fas, Cxcl2, and reduced cytokine-induced ERK phosphorylation (5). **Pharmacokinetics:** After administration of a 50-mg dose, the mean maximal plasma concentrations reached 104 nmol/L at 2 hours, with a half-life of 6.9 hours (6). After administration of a 100-mg dose, the maximal concentration reached 199 nmol/L at 2.1 hours with a half-life of 6.0 hours. Repeat doses for 7 consecutive days of 50, 100 or 200 mg resulted in nearly the same kinetics. **1**. Leoni, Fossati, Lewis, *et al.* (2005) *Mol. Med.* **11**, 1. **2**. Golay, Cuppini, Leoni, *et al.* (2007) *Leukemia* **21**, 1892. **3**. Lewis, Blaabjerg, Størling, *et al.* (2011) *Mol Med.* **17**, 369. **4**. Shein, Grigoriadis, Alexandrovich, *et al.* (2009) *FASEB J.* **23**, 4266. **5**. Christensen, Gysemans, Lundh, *et al.* (2014) *Proc. Natl. Acad. Sci. U.S.A.* **111**, 1055. **6**. Furlan, Monzani, Reznikov (2011) *Mol. Med.* **17**, 353.

Gliotoxin

This sulfur-containing mycotoxin, immunosuppressant,and apoptosis-promoting factor (FW = 326.40 g/mol; CAS = 67-99-2; IUPAC Name: (3*R*,5a*S*,6*S*,10a*R*)-2,3,5a,6-tetrahydro-6-hydroxy-3-(hydroxymethyl)-2-methyl-10*H*-3a,10a-epidithiopyrazinol[1,2a]-indole-1,4-dione, is produced by a number of fungi and exhibits antiviral properties and immunomodulating activity. Gliotoxin reportedly acts via covalent interaction with proteins through mixed disulfide bond formation; hence, gliotoxin has been shown to inhibit a number of thiol-requiring enzymes. It was first synthesized by Nobelist R. B. Woodward. Gliotoxin is also a potent NOTCH2 transactivation inhibitor, completely blocking formation of DNA-bound NOTCH2 complexes in chronic lymphocytic leukemia cells in a manner that is independent of their sensitivity to DAPT. The inhibition of Notch-2 signaling by gliotoxin was associated with down-regulated CD23 (FCER) expression and induction of apoptosis (1). **Target(s):** alcohol dehydrogenase (2); creatine kinase (3); NF-κB activation (4); protein farnesyltransferase (5-7); protein geranylgeranyltransferase I (7); proteasome chymotrypsin-like activity (8); RNA-directed RNA polymerase (9-11); RNA polymerase, viral (12-14). **1**. Hubmann, Hilgarth, Schnabl, *et al.* (2013) *Brit. J. Haematol.* **160**, 618. **2**. Waring, Sjaarda & Lin (1995) *Biochem. Pharmacol.* **49**, 1195. **3**. Hurne, Chai & Waring (2000) *J. Biol. Chem.* **275**, 25202. **4**. Umezawa, Ariga & Matsumoto (2000) *Anticancer Drug Des.* **15**, 239. **5**. Hara & Han (1995) *Proc. Natl. Acad. Sci. U.S.A.* **92**, 3333. **6**. Van der Pyl, Inokoshi, Shiomi, *et al.* (1992) *J. Antibiot.* **45**, 1802. **7**. Vigushin, Mirsaidi, Brooke, *et al.* (2004) *Med. Oncol.* **21**, 21. **8**. Paugam, Creuzet, Dupouy-Camet & Roisin (2002) *Parasitol. Res.* **88**, 785. **9**. Alaoui-Lsmaili, Hamel, L'Heureux, *et al.* (2000) *J. Hum. Virol.* **3**, 306. **10**. Hung, Gibbs & Tsiang (2002) *Antiviral Res.* **56**, 99. **11**. Ferrari, Wright-Minogue, Fang, *et al.* (1999) *J. Virol.* **73**, 1649. **12**. Ho & Walters (1968) *Antimicrobial Agents Chemother.* **8**, 68. **13**. Rodriguez & Carrasco (1992) *J. Virol.* **66**, 1971. **14**. Ferrari, Wright-Minogue, Fang, *et al.* (1999) *J. Virol.* **73**, 1649.

Gleevec, *See Imatinib*

Glipizide

This oral antidiabetic agent (FW = 445.54 g/mol; Photolabile; Phototoxic), also known as glydiazinamide, is a second-generation sulfonylurea drug that stimulates insulin secretion by binding to the sulfonylurea receptor and closing the K^+ channels in pancreatic β-cells, leading to the depolarization and causing voltage-dependent opening of Ca^{2+} channels and Ca^{2+} influx. **Key Pharmacokinetic Parameters:** *See* Appendix II in Goodman & Gilman's *THE PHARMACOLOGICAL BASIS OF THERAPEUTICS*, 12th Edition (Brunton, Chabner & Knollmann, eds.) McGraw-Hill Medical, New York (2011). **Target(s):** K^+ channel (1-4); transglutaminase (5). **1**. Zini, Ben-Ari & Ashford (1991) *J. Pharmacol. Exp. Ther.* **259**, 566. **2**. Raeburn & Brown (1991) *J. Pharmacol. Exp. Ther.* **256**, 480. **3**. Dorschner, Brekardin, Uhde, Schwanstecher & Schwanstecher (1999) *Mol. Pharmacol.* **55**, 1060. **4**. Wilson (1989) *J. Auton. Pharmacol.* **9**, 71. **5**. Gomis, Mathias, Lebrun, *et al.* (1984) *Res. Commun. Chem. Pathol. Pharmacol.* **46**, 331.

GKL003

This first-in-class symmetrical indole-based inhibitor (FW = 657.51 g/mol; Dissolve in DMSO; then mix with buffer; IUPAC Name: 2-(3-(4-bromophenyl)-4,6-dimethoxy-1*H*-indol-7-yl)-*N*'-(2-(3-(4-bromophenyl)-4,6-dimethoxy-1*H*-indol-7-yl)-2-oxoacetyl)-2-oxoacetohydrazide) is a pharmacophore model-based ligand that targets the bacterial RNA polymerase interaction site for *Thermus thermophilus* σ^A protein by binding to the so-called β'-CH region, thereby preventing initiation complex formation (K_i = 1.16 μM, based on a one-site binding model). **1**. Ma, Yang, Kandemir, *et al.* (2013) *Chem. Biol.* **8**, 1972.

Globomycin

This cyclic depsipeptide (FW = 655.82 g/mol; CAS = 67076-74-8), first isolated from a strain of *Streptomyces halstedii*, is a noncompetitive inhibitor of signal peptidase II and protein processing. Globomycin components are: glycine, L-serine, L-allothreonine, L-alloisoleucine, *N*-methyl-L-leucine, and 3-hydroxy-2-methylnonanoic acid. **1**. Sankaran (1998) in *Handb. Proteolytic Enzymes*, p. 982, Academic Press, San Diego. **2**. Sankaran & Wu (1995) *Meth. Enzymol.* **248**, 169. **3**. Tjalsma, Zanen, Bron & van Dijl (2002) *The Enzymes*, 3rd ed., **22**, 3. **4**. Kiho, Iwata, Kogen & Miyamoto (2003) *Drug Des. Discov.* **18**, 10. **5**. Dev, Harvey & Ray (1985) *J. Biol. Chem.* **260**, 5891. **6**. Hussain, Ozawa, Ichihara & Mizushima (1982) *Eur. J. Biochem.* **129**, 233. **7**. Dev & Ray (1990) *J. Bioenerg. Biomembr.* **22**, 271.

GLPG0634, See *Filgotinib*

GLPG 0974

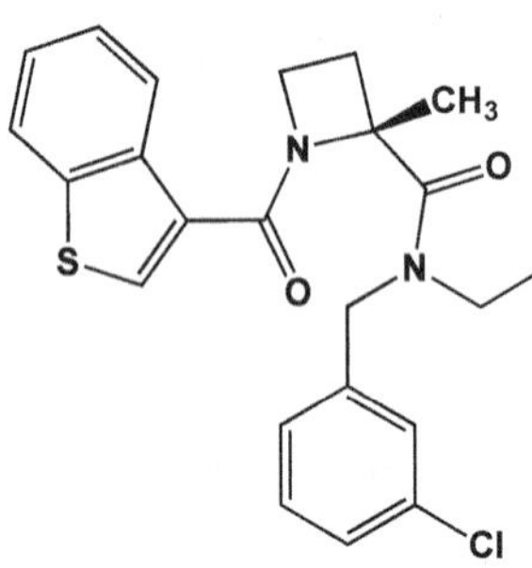

This FFA$_2$ antagonist (FW = 485.00 g/mol; CAS = 1391076-61-1; Soluble to 100 mM in DMSO and to 20 mM in 1 equivalent NaOH, with gentle warming), also known as 4-[[[(2R)-1-(benzo[*b*]thien-3-ylcarbonyl)-2-methyl-2-azetidinyl]carbonyl][(3-chlorophenyl)methyl]-amino]butanoic acid, targets free fatty acid receptor-2 (FFA$_2$), also called GPR43, is a G-protein coupled receptor for short chain fatty acids that is involved in mediating the inflammatory response. GLPG 0974 strongly inhibits acetate-induced neutrophil migration *in vitro*, demonstrating the ability to inhibit a neutrophil-based pharmacodynamic (PD) marker, CD11b activation-specific epitope [AE], in a human whole blood assay. **1.** Pizzonero, Dupont, Babel, *et al.* (2014) *J. Med. Chem.* 57, 10044.

D-Glucal

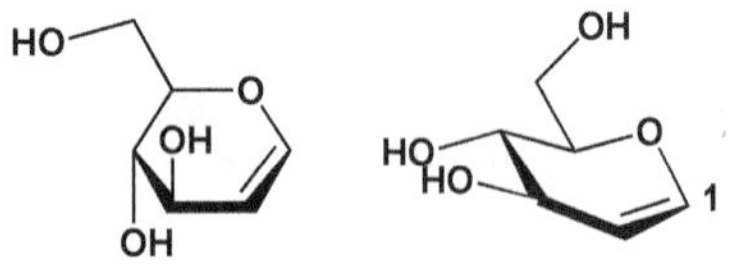

Haworth   Half-Chair
Projection   Configuration

This 1,2-anhydro derivative of glucose (FW = 146.14 g/mol; CAS 13265-84-4), also named 1,5-anhydro-2-deoxy-D-arabino-hex-1-enitol, contains a double bond in the ring, forcing a half-chair structure resembling an *oxa*-carbenium ion reaction intermediate in dissociative, *or* S_N1-type, nucleophilic substitution reactions.

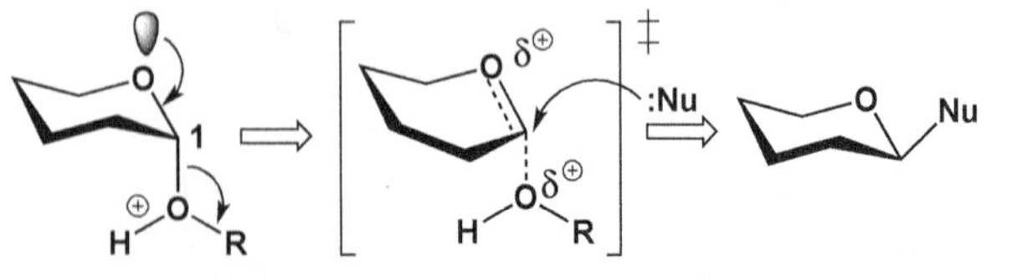

dissociation   oxa-carbenium ion   product/intermediate

Because some glycoside-hydrolyzing enzymes employ S_N1 reaction mechanisms, anhydro-sugars, such as D-glucal, often prove to be tight-binding transition-state mimics. In some glycosylases, the *oxa*-carbenium ion is attacked by an active-site carboxylate anion to form a covalently bound acylal species (also shown above). Incorrectly believing he had synthesized an aldehyde-containing product, the Nobel Laureate Emil Fischer assigned the name D-glucal to this hexa-1-enitol. (So great was his reputation that the term "glycal" was adopted as a general name for all sugars with a double-bond between carbon atoms 1 and 2.) *Note*: The term "glucal" is also unfortunately used to refer to the calcium salt of D-gluconic acid. **See** *D-Gluconate and D-gluconic Acid*. **Target(s):** glucose oxidase (1,2); α-glucosidase (3); β-glucosidase (3,4); glucosylceramidase (5); glycogen phosphorylase *a* (6); lactase/phlorizin hydrolase, *or* glycosylceramidase (7,8); α,α-trehalose phosphorylase configuration-retaining (9,10). **1.** Bright & Porter (1975) *The Enzymes*, 3rd ed., **12**, 421. **2.** Rogers & Brandt (1971) *Biochemistry* **10**, 4624 and 4636. **3.** Mooser (1992) *The Enzymes*, 3rd ed., **20**, 187. **4.** Dale, Ensley, Kern, Sastry & Byers (1985) *Biochemistry* **24**, 3530. **5.** Legler (1988) *NATO ASI Ser. A, Life Sci.* **150**, 63. **6.** Kasvinsky (1982) *J. Biol. Chem.* **257**, 10805. **7.** Arribas, Herrero, Martin-Lomas, Cañada, He & Withers (2000) *Eur. J. Biochem.* **267**, 6996. **8.** Hermida,

Corrales, Martinez-Costa, Fernandez-Mayoralas & Aragon (2006) *Clin. Chem.* **52**, 270. **9.** Eis, Watkins, Prohaska & Nidetzky (2001) *Biochem. J.* **356**, 757. **10.** Nidetzky & Eis (2001) *Biochem. J.* **360**, 727.

Glucantime

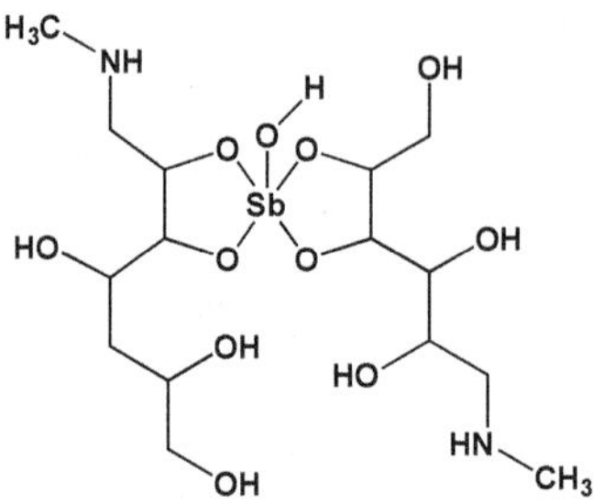

This antimony(V) compound (FW = indefinite composition; CAS = 133-51-7), also named meglumine antimoniate and (2R,3R,4R,5S)-6-methylaminohexane-1,2,3,4,5-pentol hydroxy-dioxostiborane, targets protozoal parasite *Leishmania* and is used in the treatment of leishmaniasis and a spectrum of related clinical diseases afflicting twelve million people worldwide (1-3). Its mechanism of action of is unknown, but may involve inhibition of the parasite's glycolytic and fatty acid oxidative pathways activity, resulting in decreased reducing equivalents for antioxidant defense and decreased synthesis of adenosine triphosphate (ATP), which is the energy required for the survival of the parasite. Osmolality measurements performed with meglumine antimonate solutions demonstrated an average of 1.43 antimony atoms per molecule of meglumine antimonite (2). Meglumine antimonite consists predominantly of aggregates of antimony and *N*-methyl-d-glucamine of the general compositions Sb$_n$NMG$_n$ and Sb$_n$NMG$_{n+1}$, with the forms containing one more *N*-methyl-d-glucamine moiety than antimony being in greater abundance for any value of *n* (1). **1.** Mishra, Saxena & Singh (2007) *Curr. Med. Chem.* **14**, 1153. **2.** Roberts, McMurray & Rainey (1998) *Antimicrob. Agents Chemother.* **42**, 1076. **3.** Berman. (1988) *Rev Infect Dis.* **10**, 560.

D-Glucaro-1,4-lactone

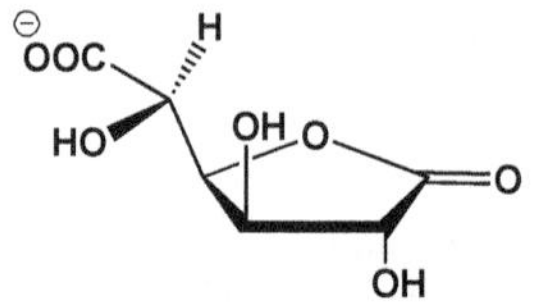

This D-glucarate lactone (FW$_{\text{free-acid}}$ = 192.13 g/mol; CAS = 61278-30-6; unstable above pH 6), also known as D-saccharo-1,4-lactone and D-glucosaccharo-1,4-lactone, possesses an ester carbonyl that imposes a half-chair conformation mimicking the geometry of *oxa*-carbenium ion or similar transition-state intermediates formed during glycosidase catalysis (*See D-Glucal for likely mechanistic implications*). **Target(s):** β-glucosidase (1); β-glucuronidase (2-22); hyaluronate lyase (23); hyaluronoglucosaminidase, *or* hyaluronidase (23); L-iduronidase (24). **1.** Park, Bae, Sung, Lee & Kim (2001) *Biosci. Biotechnol. Biochem.* **65**, 1163. **2.** Fishman & Bernfeld (1955) *Meth. Enzymol.* **1**, 262. **3.** Levvy & Conchie (1966) *Meth. Enzymol.* **8**, 571. **4.** Levvy & Marsh (1960) *The Enzymes*, 2nd ed., **4**, 397. **5.** Levvy (1952) *Biochem. J.* **52**, 464. **6.** Webb (1966) *Enzyme and Metabolic Inhibitors*, vol. 2, pp. 424, 427, Academic Press, New York. **7.** Levvy (1954) *Biochem. J.* **58**, 462. **8.** Levvy, Hay & Marsh (1957) *Biochem. J.* **65**, 203. **9.** Herd, Mayberry & Snell (1982) *Carbohydr. Res.* **99**, 33. **10.** Ho, Ho & Ho (1985) *Enzyme* **33**, 9. **11.** Ohtsuka & Wakabayashi (1970) *Enzymologia* **39**, 109. **12.** Salleh, Muellegger, Reid, *et al.* (2005) *Carbohydr. Res.* **341**, 49. **13.** Gupta & Singh (1983) *Biochim. Biophys. Acta* **748**, 398. **14.** Schulz & Weissenböck (1987) *Phytochemistry* **26**, 933. **15.** Diez & Cabezas (1979) *Eur. J. Biochem.* **93**, 301. **16.** Kim, Jin, Jung, Han & Kobashi (1995) *Biol. Pharm. Bull.* **18**, 1184. **17.** Fishman (1955) *Adv. Enzymol.* **16**, 361. **18.** Tsuchihashi, Yadome & Miyazaki (1984) *J. Biochem.* **96**, 1789. **19.** Kuroyama, Tsutsui, Hashimoto & Tsumuraya (2001) *Carbohydr. Res.* **333**, 27. **20.** Pereira, Cruz, Albuquerque, *et al.* (2005) *Protein Pept. Lett.* **12**, 695. **21.** Dean (1974) *Biochem. J.* **138**, 395. **22.** Nakamura, Takagaki, Majima, *et al.* (1990) *J. Biol. Chem.* **265**, 5390. **23.** Okorukwu & Vercruysse (2003) *J. Enzyme Inhib. Med. Chem.* **18**, 377. **24.** Weissmann & Santiago (1972) *Biochem. Biophys. Res. Commun.* **46**, 1430.

Glucokinase Regulatory Protein

This endogenous 625-residue protein (MW = 68701 g/mol; *Symbol*: GKRP) is a slow, tight-binding inhibitor of glucokinase (1-3). In mammals, GKRP's effect is modulated by fructose 6-phosphate, which reinforces the inhibition, and by fructose 1-phosphate which antagonizes it. **Mode of Inhibitory Action:** GKRP binds to the so-called super-open conformation of glucokinase, mainly through hydrophobic interactions that lock a small domain of GK and inhibit GK activity (4). GKRP releases GK in a sigmoidal manner in response to glucose concentration by restricting a structural rearrangement of the GK small domain via a single ion pair, showing that GKRP acts as an allosteric switch for GK. During fasting, GKRP binds glucokinase (GK), inactivating and sequestering the latter within the nucleus. GK remocal from the cytoplasm avoids futile cycling by phosphorylation/hydrolysis of glucose/glucose 6-P, as catalyzed by GK and glucose-6 phosphatase (1-3). Progress-curve analysis indicates that GKRP inhibition of GK is time-dependent, with apparent initial and final K_i values of 113 and 12.8 nM, respectively (1). This regulatory protein, a fructose 6-phosphate and fructose 1-phosphate sensitive GK inhibitor, appears to have arisen by the duplication of a gene similar to the bacterial *N*-acetylmuramate 6-phosphate etherase MurQ (2). GKRP inhibits glucokinase in the hepatocytes of mammals and lower vertebrates and its inhibitory action is potentiated by the presence of D-fructose 6-phosphate or D-sorbitol 6-phosphate and reversed by D-fructose 1-phosphate. These effects by fructose 1- and 6-phosphate are not observed for the regulatory protein from the livers of toad, turtle, and *Xenopus laevis*. **Inhibition of GK-GKRP Interactions:** AMG-1694 and AMG-3969 normalize blood glucose levels in several rodent models of diabetes by potently disrupting GK-GKRP interactions (5). These compounds reverse the inhibitory effect of GKRP on GK activity and promote GK translocation both *in vitro* (in isolated hepatocytes) and *in vivo* (in liver). A co-crystal structure of full-length human GKRP in complex with AMG-1694 revealed a previously unknown binding pocket in GKRP distinct from that of the phosphofructose-binding site. Furthermore, with AMG-1694 and AMG-3969 (but not GK activators), blood glucose lowering was restricted to diabetic and not normoglycemic animals (3). **Target(s):** glucokinase (6-9); hexokinase IV (6-9). **1.** Bourbonais, Chen, Huang, *et al.* (2012) *Biochem. J.* **441**, 881. **2.** Veiga-da-Cunha, Sokolova, Opperdoes & Van Schaftingen (2009) *Biochem. J.* **423**, 323. **3.** Brocklehurst, Davies & Agius (2004) *Biochem. J.* **378**, 693. **4.** Choi, Seo, Kyeong, Kim & Kim (2013) *Proc. Natl. Acad. Sci. U.S.A.* **110**, 10171. **5.** Lloyd, St Jean, Kurzeja, *et al.* (2013) *Nature* **504**, 437. **6.** Gloyn, Noordam, Willemsen, *et al.* (2003) *Diabetes* **52**, 2433. **7.** Grimsby, Sarabu, Corbett, *et al.* (2003) *Science* **301**, 370. **8.** Vandercammen & Van Schaftingen (1991) *Eur. J. Biochem.* **200**, 545. **9.** Gloyn, Odili, Zelent, *et al.* (2005) *J. Biol. Chem.* **280**, 14105.

D-Gluconate (D-Gluconic Acid)

This aldonic acid (FW$_{\text{free-acid}}$ = 196.16 g/mol; CAS = 526-95-4; M.P. = 132 °C; pK_a = 3.6 at 25°C), which is derived oxidatively from D-glucose, is widely used as a counter-ion in drug formulations. Gluconate is very soluble in water, forms complexes with metal ions, and readily forms lactones. Aqueous solutions contain a mixture of the free acid, the 1,4-lactone, and the 1,5-lactone. *Note:* The bioavailable calcium di-gluconate salt, $(Ca(C_6H_{11}O_7)_2)$, which is used to replenish calcium levels, is often by the commercial name GluCal. **Target(s):** L-fuconate dehydratase (1); β-glucosidase (2-4); lactate dehydrogenase (5); mannonate dehydratase (6,7); pantothenate synthetase, *or* pantoate:β-alanine ligase, moderately inhibited (8,9). **1.** Veiga & Guimaes (1991) *Arq. Biol. Tecnol.* **34**, 537. **2.** Webb (1966) *Enzyme and Metabolic Inhibitors*, vol. 2, p. 417, Academic Press, New York. **3.** Ezaki (1940) *Biochem. J.* **32**, 107. **4.** Dale, Ensley, Kern, Sastry & Byers (1985) *Biochemistry* **24**, 3530. **5.** Pirie (1934) *Biochem. J.* **28**, 411. **6.** Robert-Baudouy, Jimeno-Abendano & Stoeber (1982) *Meth. Enzymol.* **90**, 288. **7.** Robert-Baudouy, Jimeno-Abendano & Stoeber (1975) *Biochimie* **57**, 1. **8.** Miyatake, Nakano & Kitaoka (1979) *Meth. Enzymol.* **62**, 215. **9.** Miyatake, Nakano & Kitaoka (1978) *J. Nutr. Sci. Vitaminol. (Tokyo)* **24**, 243.

D-Glucono-1,4-lactone

This lactone (FW = 178.14 g/mol; M.P. = 133 135°C), also known as glucono-γ-lactone, possesses an ester carbonyl that imposes a half-chair conformation mimicking the sp^3 geometry of *oxa*-carbenium ion or similar transition-state intermediates often formed during glycosidase catalysis (*See D-Glucal for likely mechanistic implications*). Aqueous solutions also contain the free acid and the 1,5-lactone. **Target(s):** aryl-β-hexosidase (1); debranching enzyme (2,3); β-D-fucosidase, *or* β-D-galactosidase activity (4); α-glucosidase (5); β-glucosidase (2,5-13); lysozyme (14). **1.** Distler & Jourdian (1978) *Meth. Enzymol.* **50**, 524. **2.** Webb (1966) *Enzyme and Metabolic Inhibitors*, vol. 2, pp. 417-429, Academic Press, New York. **3.** Gunja, Manners & Maung (1961) *Biochem. J.* **81**, 392. **4.** Calvo, Santamaria, Melgar & Cabezas (1983) *Int. J. Biochem.* **15**, 685. **5.** Levvy & Conchie (1966) *Meth. Enzymol.* **8**, 571. **6.** Gatt (1969) *Meth. Enzymol.* **14**, 152. **7.** Heyworth & Walker (1962) *Biochem. J.* **83**, 331. **8.** Joshida, Kamada, Harada & Kato (1966) *Chem. Pharm. Bull. Tokyo* **14**, 583. **9.** Conchie, Gelman & Levvy (1967) *Biochem. J.* **103**, 609. **10.** Ezaki (1940) *J. Biochem.* **32**, 107. **11.** Conchie (1953) *Biochem. J.* **55**, xxi and (1954) *Biochem. J.* **58**, 552. **12.** Plant, Oliver, Patchett, Daniel & Morgan (1988) *Arch. Biochem. Biophys.* **262**, 181. **13.** Maret, Salvayre, Negre & Douste-Blazy (1983) *Eur. J. Biochem.* **133**, 283. **14.** Imoto, Johnson, North, Phillips & Rupley (1972) *The Enzymes*, 3rd ed., 7, 665.

D-Glucono-1,5-lactone

This lactone (FW = 178.14 g/mol; M.P. = 153°C), also known as D-glucono-δ-lactone and gluconolactone, possesses an ester carbonyl that imposes a half-chair conformation mimicking the geometry of *oxa*-carbenium ion or similar transition-state intermediates often formed during glycosidase catalysis (*See D-Glucal for likely mechanistic implications*). This lactone forms readily from gluconic acid, and aqueous solutions typically contain an equilibrium mixture of the 1,5- and 1,4-lactones, along with the acyclic gluconate. **Target(s):** β-*N*-acetylhexosaminidase (1,2); aldose 1-epimerase, *or* mutarotase (3); amygdalin β-glucosidase (4); amylo-α-1,6-glucosidase/4-α-glucanotransferase (glycogen debranching enzyme) (5); β-apiosyl-β-glucosidase (6); α-L-arabinofuranosidase (7); aryl-β-hexosidase (8); bacterial glycogen synthase (92); cellobiose (9); cellobiose phosphorylase (10,120); cellulase, *Phaseolus vulgaris* (11); coniferin β-glucosidase (12); dextransucrase (124; endo-1,3(4)-β-glucanase13; exo-(1→3)-β-glucanase (14,15); β-D-fucosidase16-18; α-L-fucosidase, weakly inhibited (19); β-galactosidase (20-22); glucan endo-1,6-β glucosidase (23); glucan 1,3-α-glucosidase, *or* glucosidase II, *or* processing mannosyl-oligosaccharide glucosidase II (24,25); glucan 1,3-β-glucosidase (26-34); glucan 1,4-α-glucosidase, *or* glucoamylase (35-37); glucan 1,4-β-glucosidase (38,39); 1,3-β-glucan synthase (119); α-glucosidase (40-47); β-glucosidase (40,48-80); glucosylceramidase (81-84); glycogen phosphorylase (85-89,125-127); glycogen synthase (90,91,123); isoamylase (93); isomaltulose synthase (94); lactase/phlorizin hydrolase, *or* glycosylceramidase (95,96); lactoylglutathione lyase, *or* glyoxalase I, weakly inhibited (97); licheninase (98); maltodextrin phosphorylase (89); maltose synthase (118); myrosinase, poor inhibitor (99); oligo-1,6 glucosidase (100,101); polygalacturonidase, *or* pectinase (102); protein-glucosylgalactosylhydroxylysine glucosidase (103); starch phosphorylase (104); starch synthase (92); (*S*)-strictosidine β-glucosidase (105,106); sucrose-phosphate synthase (121,122); sucrose phosphorylase (107); thioglucosidase, *or* myrosinase, *or* sinigrinase (108-111); trehalose (112); α,α-trehalose phosphorylase, configuration-retaining (113); vicianin β-glucosidase (114); xanthureate:UDP glucosyltransferase (115); xylan 1,4-β-xylosidase (116,117). **1.** Kapur & Gupta (1986) *Biochem. J.* **236**, 103. **2.** Khar & Anand (1977) *Biochim. Biophys. Acta* **483**, 141. **3.** Hucho & Wallenfels (1971) *Eur. J. Biochem.* **23**, 489. **4.** Petruccioli, Brimer,

Cicalini, Pulci & Federici (1999) *Biosci. Biotechnol. Biochem.* **63**, 805. **5.** Gillard & Nelson (1977) *Biochemistry* **16**, 3978. **6.** Hósel & Barz (1975) *Eur. J. Biochem.* **57**, 607. **7.** Hirano, Tsumuraya & Hashimoto (1994) *Physiol. Plant.* **92**, 286. **8.** Distler & Jourdian (1978) *Meth. Enzymol.* **50**, 524. **9.** Rapp, Knobloch & Wagner (1982) *J. Bacteriol.* **149**, 783. **10.** Sasaki (1988) *Meth. Enzymol.* **160**, 468. **11.** Durbin & Lewis (1988) *Meth. Enzymol.* **160**, 342. **12.** Dharmawardhana, Ellis & Carlson (1995) *Plant Physiol.* **107**, 331. **13.** Villa, Notario & Villanueva (1979) *Biochem. J.* **177**, 107. **14.** Ram, Romana, Shepherd & Sullivan (1984) *J. Gen. Microbiol.* **130**, 1227. **15.** Talbot & Vacquier (1982) *J. Biol. Chem.* **257**, 742. **16.** Calvo, Santamaria, Melgar & Cabezas (1983) *Int. J. Biochem.* **15**, 685. **17.** Chinchetru, Cabezas & Calvo (1983) *Comp. Biochem. Physiol.* **75**, 719. **18.** Srisomsap, Svasti, Surarit, *et al.* (1996) *J. Biochem.* **119**, 585. **19.** Reglero & Cabezas (1976) *Eur. J. Biochem.* **66**, 379. **20.** O'Mahony, Kelly & Fogarty (1988) *Biochem. Soc. Trans.* **16**, 183. **21.** Kelly, O'Mahony & Fogarty (1988) *Appl. Microbiol. Biotechnol.* **27**, 383. **22.** Konno, Yamasaki & Katoh (1986) *Plant Sci.* **44**, 97. **23.** Fleet & Phaff (1974) *J. Bacteriol.* **119**, 207. **24.** Presper & Heath (1983) *The Enzymes*, 3rd ed., **16**, 449. **25.** Burns & Touster (1982) *J. Biol. Chem.* **257**, 9991. **26.** Fleet & Phaff (1975) *Biochim. Biophys. Acta* **401**, 318. **27.** Molina, Cenamor, Sanchez & Nobela (1989) *J. Gen. Microbiol.* **135**, 309. **28.** Marshall & Grand (1975) *Arch. Biochem. Biophys.* **167**, 165. **29.** Nagasaki, Saito & Yamamoto (1977) *Agric. Biol. Chem.* **41**, 493. **30.** Talbot & Vacquier (1982) *J. Biol. Chem.* **257**, 742. **31.** Boucaud, Bigot & Devaux (1987) *J. Plant Physiol.* **128**, 337. **32.** Villa, Notario, Benitez & Villanueva (1976) *Can. J. Biochem.* **54**, 927. **33.** Bucheli, Durr, Buchala & Meier (1985) *Planta* **166**, 530. **34.** Hrmova, Harvey, Wang, *et al.* (1996) *J. Biol. Chem.* **271**, 5277. **35.** Ohnishi, Yamashita & Hiromi (1976) *J. Biochem.* **79**, 1007. **36.** Tanaka (1996) *Biosci. Biotechnol. Biochem.* **60**, 2055. **37.** Basaveswara Rao, Sastri & Subba Rao (1981) *Biochem. J.* **193**, 379. **38.** Rao & Mishra (1989) *Appl. Microbiol. Biotechnol.* **30**, 130. **39.** Wood & McCrae (1982) *Carbohydr. Res.* **110**, 291. **40.** Levvy & Conchie (1966) *Meth. Enzymol.* **8**, 571. **41.** Giblin, Kelly & Fogarty (1987) *Can. J. Microbiol.* **33**, 614. **42.** Suzuki, Yuki, Kishigami & Abe (1976) *Biochim. Biophys. Acta* **445**, 386. **43.** Shirai, Hung, Akita, *et al.* (2003) *Acta Crystallogr. Sect. D* **59**, 1278. **44.** Kelly, Giblin & Fogarty (1986) *Can. J. Microbiol.* **32**, 342. **45.** Thirunavukkarasu & Priest (1984) *J. Gen. Microbiol.* **130**, 3135. **46.** Silva & Terra (1995) *Insect Biochem. Mol. Biol.* **25**, 487. **47.** Chadalavada & Sivakami (1997) *Biochem. Mol. Biol. Int.* **42**, 1051. **48.** Santos & Terra (1985) *Biochim. Biophys. Acta* **831**, 179. **49.** Deshpande, Pettersson & Eriksson (1988) *Meth. Enzymol.* **160**, 126. **50.** Wood (1988) *Meth. Enzymol.* **160**, 221. **51.** Deshpande & Eriksson (1988) *Meth. Enzymol.* **160**, 415. **52.** Sadana, Patil & Shewale (1988) *Meth. Enzymol.* **160**, 424. **53.** Conchie (1953) *Biochem. J.* **55**, xxi, and Conchie (1954) *Biochem. J.* **58**, 552. **54.** Odoux, Chauwin & Brillouet (2003) *J. Agric. Food Chem.* **51**, 3168. **55.** Daniels, Coyle, Chiao, Glew & Labow (1981) *J. Biol. Chem.* **256**, 13004. **56.** Dale, Ensley, Kern, Sastry & Byers (1985) *Biochemistry* **24**, 3530. **57.** Feldwisch, Vente, Zettl, *et al.* (1994) *Biochem. J.* **302**, 15. **58.** Plant, Oliver, Patchett, Daniel & Morgan (1988) *Arch. Biochem. Biophys.* **262**, 181. **59.** Pitson, Seviour & McDougall (1997) *Enzyme Microb. Technol.* **21**, 182. **60.** Pocsi & Kiss (1988) *Biochem. J.* **256**, 139. **61.** Maret, Salvayre, Negre & Douste-Blazy (1983) *Eur. J. Biochem.* **133**, 283. **62.** Ferreira & Terra (1983) *Biochem. J.* **213**, 43. **63.** Glew, Peters & Christopher (1976) *Biochim. Biophys. Acta* **422**, 179. **64.** Riou, Salmon, Vallier, Gunata & Barre (1998) *Appl. Environ. Microbiol.* **64**, 3607. **65.** Sano, Amemura & Harada (1975) *Biochim. Biophys. Acta* **377**, 410. **66.** Hsieh & Graham (2001) *Phytochemistry* **58**, 995. **67.** Petruccioli, Brimer, Cicalini, Pulci & Federici (1999) *Biosci. Biotechnol. Biochem.* **63**, 805. **68.** Yeoh, Tan & Koh (1986) *Appl. Microbiol. Biotechnol.* **25**, 25. **69.** Lucas, Robles, Alvarez de Cienfuegos & Gálvez (2000) *J. Agric. Food Chem.* **48**, 3698. **70.** Parry, Beever, Owen, *et al.* (2001) *Biochem. J.* **353**, 117. **71.** Marana, Jacobs-Lorena, Terra & Ferreira (2001) *Biochim. Biophys. Acta* **1545**, 41. **72.** Kawai, Yoshida, Tani, *et al.* (2003) *Biosci. Biotechnol. Biochem.* **67**, 1. **73.** Belancic, Gunata, Vallier & Agosin (2003) *J. Agric. Food Chem.* **51**, 1453. **74.** Santos & Terra (1985) *Biochim. Biophys. Acta* **831**, 179. **75.** Yeoh (1989) *Phytochemistry* **28**, 721. **76.** Sanyal, Kundu, Dube & Dube (1988) *Enzyme Microb. Technol.* **10**, 91. **77.** Shewale & Sadana (1981) *Arch. Biochem. Biophys.* **207**, 185. **78.** Ait, Creuzet & Cattaneo (1982) *J. Gen. Microbiol.* **128**, 569. **79.** Mamma, Hatzinikolaou & Christakopoulos (2004) *J. Mol. Catal. B* **27**, 183. **80.** Yang, Ning, Shi, Chang & Huan (2004) *J. Agric. Food Chem.* **52**, 1940. **81.** Weinreb & Brady (1972) *Meth. Enzymol.* **28**, 830. **82.** van Weely, Brandsma, Strijland, Tager & Aerts (1993) *Biochim. Biophys. Acta* **1181**, 55. **83.** Reddy, Murray & Barranger (1985) *Biochem. Med.* **33**, 200. **84.** Legler (1988) *NATO ASI Ser. A, Life Sci.* **150**, 63. **85.** Tu, Jacobson & Graves (1971) *Biochemistry* **10**, 1229. **86.** Gold, Legrand & Sanchez (1971) *J. Biol. Chem.* **246**, 5700. **87.** Graves & Wang (1972) *The Enzymes*, 3rd ed., **7**, 435. **88.** Kasvinsky (1982) *J. Biol. Chem.* **257**, 10805. **89.** Tu, Jacobson & Graves (1971) *Biochemistry* **10**, 1229. **90.** Sundukov & Solov'eva (1990) *Biokhimiia* **55**, 1120 and 1287. **91.** Solov'eva & Sisse (1985) *Biokhimiia* **50**, 211. **92.** Fox, Kawaguchi, Greenberg & Preiss (1976) *Biochemistry* **15**, 849. **93.** Kitagawa, Amemura & Harada (1975) *Agric. Biol. Chem.* **39**, 989. **94.** Nagai, Sugitani & Tsuyuki (1994) *Biosci. Biotechnol. Biochem.* **58**, 1789. **95.** Malathi & Crane (1969) *Biochim. Biophys. Acta* **173**, 245. **96.** Kraml, Kolínská, Ellederová & Hirsová (1972) *Biochim. Biophys. Acta* **258**, 520. **97.** Douglas & Nadvi (1979) *FEBS Lett.* **106**, 393. **98.** Borriss (1981) *Z. Allg. Mikrobiol.* **21**, 7. **99.** Botti, Taylor & Botting (1995) *J. Biol. Chem.* **270**, 20530. **100.** Suzuki & Tomura (1986) *Eur. J. Biochem.* **158**, 77. **101.** Suzuki, Aoki & Hayashi (1982) *Biochim. Biophys. Acta* **704**, 476. **102.** Niture, Pant & Kumar (2001) *Eur. J. Biochem.* **268**, 832. **103.** Sternberg & Spiro (1979) *J. Biol. Chem.* **254**, 10329. **104.** Schwarz, Pierfederici & Nidetzky (2005) *Biochem. J.* **387**, 437. **105.** Hemscheid & Zenk (1992) *FEBS Lett.* **110**, 187. **106.** Luijendijk, Nowak & Verpoorte (1996) *Phytochemistry* **41**, 1451. **107.** Mieyal (1972) *Meth. Enzymol.* **28**, 935. **108.** Ohtsuru & Hata (1973) *Agric. Biol. Chem.* **37**, 2543. **109.** Ohtsuru, Tsuruo & Hata (1969) *Agric. Biol. Chem.* **33**, 1315. **110.** Tani, Ohtsuru & Hata (1974) *Agric. Biol. Chem.* **38**, 1623. **111.** Botti, Taylor & Botting (1995) *J. Biol. Chem.* **270**, 20530. **112.** Terra, Terra & Ferreira (1983) *Int. J. Biochem.* **15**, 143. **113.** Nidetzky & Eis (2001) *Biochem. J.* **360**, 727. **114.** Lizotte & Poulton (1988) *Plant Physiol.* **86**, 322. **115.** Real & Ferre (1990) *J. Biol. Chem.* **265**, 7407. **116.** Yasui & Matsuo (1988) *Meth. Enzymol.* **160**, 696. **117.** Gomez, Isorna, Rojo & Estrada (2001) *Biochimie* **83**, 961. **118.** Schilling (1982) *Planta* **154**, 87. **119.** López-Romero & Ruiz-Herrera (1978) *Antonie Van Leeuwenhoek* **44**, 329. **120.** Sasaki, Tanaka, Nakagawa & Kainuma (1983) *Biochem. J.* **209**, 803. **121.** Chen, Huang, Liu, *et al.* (2001) *Bot. Bull. Acad. Sin.* **42**, 123. **122.** Salerno & Pontis (1978) *Planta* **142**, 41. **123.** McVerry & Kim (1974) *Biochemistry* **13**, 3505. **124.** Kobayashi, Yokoyama & Matsuda (1986) *Agric. Biol. Chem.* **50**, 2585. **125.** Weinhausel, Griessler, Krebs, *et al.* (1997) *Biochem. J.* **326**, 773. **126.** Ariki & Fukui (1975) *J. Biochem.* **78**, 1191. **127.** Boeck & Schinzel (1996) *Eur. J. Biochem.* **239**, 150.

3-(β-D-Glucopyranosyl)-5-substituted-1,2,4-triazoles

These investigational hypoglycemic agents, 5-(4-aminophenyl)-3-(β-D-glucopyranosyl)-1,2,4-triazole (FW = 296.11 g/mol) and 3-(β-D-glucopyranosyl)-5-(2-naphthyl)-1,2,4-triazole (FW = 357.37 g/mol) target glycogen phosphorylase, with K_i values of 0.67 μM and 0.41 μM, respectively (1). In Type-2 diabetes mellitus (T2DM), hepatic glucose output is elevated, with glycogenolysis is an important contributor. In view of its key role in glycogen conversion to glucose 1-phosphate, glycogen phosphorylase (GP) may be a druggable target for treating Type 2 diabetes (T2DM), cardiac and cerebral ischemia, as well as certain cancers. The best glucose-based GP inhibitors are glucopyranosylidene-spiro-heterocycles (K_i = 0.16–0.63 μM) and *N*-acyl-*N*-β-D-glucopyranosyl ureas (K_i = 0.35–0.7 μM) (2). Glucopyranosylidene-spiro-thiohydantoin inhibits rat liver GP (K_i = 29.8 μM), with an attendant drop in blood sugar (3). An *N*-acyl-*N*'-β-D-glucopyranosyl urea derivative improves glucose tolerance in diabetic mice (4). **1.** Kun, Bokor, Varga, *et al.* (2014) *Europ. J. Med. Chem.* **76**, 567. **2.** Chrysina (2010) *Mini-Rev. Med. Chem.* **10**, 1093. **3.** Docsa, Czifrák, Hüse, Somsák, & Gergely (2011) *Molec. Med. Rep.* **4**, 477. **4.** Nagy, Docsa, Brunyánszki, *et al.* (2013) *PLoS One* **8**, e69420.

β-D-Glucosamine

This aminosugar ($FW_{free-base}$ = 179.17 g/mol; CAS = 3416-24-8; pK_a = 7.75), also known as 2-amino-2-deoxy-D-glucose and chitosamine, is a component of many polysaccharides, mucopolysacharides, mucoproteins, and glycoproteins: especially heparin, chitin, and chondroitin. The anomeric composition of the uncombined aminosugar is 36% α- and 64% β-D-glucosamine at 10°C in H_2O; however, the reported composition at 40°C in D_2O is 63% α- and 37% β-D-glucosamine. **Target(s):** β-N-acetylgalactosaminidase (1); β-N-acetylglucos-aminidase (2); β-N acetylhexosaminidase (1-7); α-amylase (8); arylamine N-acetyltransferase, or anthranilate N-acetyltransferase (39); cellulase (9); chitinase (10); chitosanase (11); fructokinase (12); β-D-fucosidase (13); glucan 1,4-α-glucosidase, or glucoamylase (14); glucokinase (15-17); glucose dehydrogenase (NAD^+) (18); β-glucosidase (19); glutamine synthetase, *Neurospora crassa*, mildly inhibits (20); glycogen synthase (16); hexokinase, also weak alternative substrate (21,22); laminaribiose phosphorylase (37); lipid-linked oligosaccharide assembly (23-26); lysozyme (27); α-mannosidase (28); procollagen glucosyltransferase (29); protein-glucosylgalactosylhydroxylysine glucosidase (30,31); protein glutamine γ-glutamyltransferase, or transglutaminase (38); thymidine kinase (32,33); α,α-trehalose phosphorylase, configuration-retaining (34); trypsin of *Glossina morsitans* midgut (35); UDP-N-acetylglucosamine 4-epimerase (36). **1.** Frohwein & Gatt (1967) *Biochemistry* **6**, 2775. **2.** Khar & Anand (1977) *Biochim. Biophys. Acta* **483**, 141. **3.** Frohwein & Gatt (1969) *Meth. Enzymol.* **14**, 161. **4.** Mommsen (1980) *Biochim. Biophys. Acta* **612**, 361. **5.** Potier, Teitelbaum, Melancon & Dallaire (1979) *Biochim. Biophys. Acta* **566**, 80. **6.** Jin, Jo, Kim, *et al.* (2002) *J. Biochem. Mol. Biol.* **35**, 313. **7.** Gers-Barlag, Bartz & Ruediger (1988) *Phytochemistry* **27**, 3739. **8.** Oosthuizen, Naude, Oelofsen, Muramoto & H. Kamiya (1994) *Int. J. Biochem.* **26**, 1313. **9.** Wood & McCrae (1978) *Biochem. J.* **171**, 61. **10.** Bhushan (2000) *J. Appl. Microbiol.* **88**, 800. **11.** Jo, Jo, Jin, *et al.* (2003) *Biosci. Biotechnol. Biochem.* **67**, 1875. **12.** Bueding & MacKinnon (1955) *J. Biol. Chem.* **215**, 495. **13.** Giordani & Lafon (1993) *Phytochemistry* **33**, 1327. **14.** Wong, Batt, Lee, Wagschal & Robertson (2005) *Protein J.* **24**, 455. **15.** Walker & Parry (1966) *Meth. Enzymol.* **9**, 381. **16.** Webb (1966) *Enzyme and Metabolic Inhibitors*, vol. 2, p. 382, Academic Press, New York. **17.** Balkan & Dunning (1994) *Diabetes* **43**, 1173. **18.** Pauly & Pfleiderer (1975) *Hoppe-Seyler's Z. Physiol. Chem.* **356**, 1613. **19.** Dale, Ensley, Kern, Sastry & Byers (1985) *Biochemistry* **24**, 3530. **20.** Kapoor & Bray (1968) *Biochemistry* **7**, 3583. **21.** Petit, Blázquez & Gancedo (1996) *FEBS Lett.* **378**, 185. **22.** Claeyssen, Wally, Matton, Morse & Rivoal (2006) *Protein Expr. Purif.* **47**, 329. **23.** Schwarz & Datema (1982) *Meth. Enzymol.* **83**, 432. **24.** Elbein (1983) *Meth. Enzymol.* **98**, 135. **25.** Elbein (1987) *Meth. Enzymol.* **138**, 661. **26.** Schwarz & Datema (1980) *Trends Biochem. Sci.* **5**, 65. **27.** Croux, Canard, Goma & Sucaille (1992) *Appl. Environ. Microbiol.* **58**, 1075. **28.** Mathur, Panneerselvam & Balasubramanian (1988) *Biochem. J.* **253**, 677. **29.** Barber & Jamieson (1971) *Biochim. Biophys. Acta* **252**, 533. **30.** Hamazaki & Hotta (1979) *J. Biol. Chem.* **254**, 9682. **31.** Hamazaki & Hotta (1980) *Eur. J. Biochem.* **111**, 587. **32.** Friedman, Kimball, Trotter & Skehan (1977) *Cancer Res.* **37**, 1068. **33.** Tesoriere, Tesoriere, Vento, Giuliano & Cantoro (1984) *Experientia* **40**, 705. **34.** Nidetzky & Eis (2001) *Biochem. J.* **360**, 727. **35.** Osir, Imbuga & Onyango (1993) *Parasitol. Res.* **79**, 93. **36.** Yamamoto, Kondo, Kumagai & Tochikura (1985) *Agric. Biol. Chem.* **49**, 603. **37.** Goldemberg, Maréchal & De Souza (1966) *J. Biol. Chem.* **241**, 45. **38.** Kim, Park, Jeong, *et al.* (2008) *Cancer Lett.* **273**, 243. **39.** Paul & Ratledge (1973) *Biochim. Biophys. Acta* **320**, 9.

D-Glucosamine 6-Phosphate

α-anomer   β-anomer

This metabolic intermediate (FW = 259.15 g/mol; CAS = 3616-42-0; pK_a values of ~1.0, 6.08, and 8.10 and is unstable above pH 7) is a product of the reactions catalyzed by glutamine:fructose-6-phosphate aminotransferase (isomerizing), hexokinase, glucosamine kinase, N-acetylglucosamine-6-phosphate deacetylase, and phosphoglucosamine mutase. **Target(s):** acetyl-CoA:α-glucosaminide N-acetyltransferase, or heparan-α-glucos-

aminide N-acetyltransferase (1); arylamine N-acetyltransferase, or anthranilate N-acetyltransferase (2); glucose-6 phosphate dehydrogenase (3-8); glucose-6-phosphate isomerase (4,9,10); glutamine:fructose-6 phosphate amidotransferase (11); glutamine synthetase (12-15); mannose-6-phosphate isomerase (9,16). **1.** Bame & Rome (1987) *Meth. Enzymol.* **138**, 607. **2.** Paul & Ratledge (1973) *Biochim. Biophys. Acta* **320**, 9. **3.** Noltmann & Kuby (1963) *The Enzymes*, 2nd ed., **7**, 223. **4.** Webb (1966) *Enzyme and Metabolic Inhibitors*, vol. **2**, p. 407 & 411, Academic Press, New York. **5.** Levy & Cook (1991) *Arch. Biochem. Biophys.* **291**, 161. **6.** Adediran (1991) *Biochimie* **73**, 1211. **7.** Bautista, Garrido-Pertierra & Soler (1988) *Biochim. Biophys. Acta* **967**, 354. **8.** Kanji, Toews & Carper (1976) *J. Biol. Chem.* **251**, 2258. **9.** Noltmann (1972) *The Enzymes*, 3rd ed., **6**, 271. **10.** Wolfe & Nakada (1956) *Arch. Biochem. Biophys.* **64**, 489. **11.** Broschat, Gorka, Page, *et al.* (2002) *J. Biol. Chem.* **277**, 14764. **12.** Rhee, Chock & Stadtman (1985) *Meth. Enzymol.* **113**, 213. **13.** Stadtman & Ginsburg (1974) *The Enzymes*, 3rd ed., **10**, 755. **14.** Wedler, Carfi & Ashour (1976) *Biochemistry* **15**, 1749. **15.** Dahlquist & Purich (1975) *Biochemistry* **14**, 1980. **16.** Gracey & Noltmann (1968) *J. Biol. Chem.* **243**, 5410.

D-Glucose

aldehydrol

H_2O

α-D-glucopyranose   open-chain   β-D-glucopyranose

D-glucofuranose

This common aldohexose (FW = 180.16 g/mol; CAS = 5996-10-1), is Nature's most abundant monosaccharide and is also a component of major disaccharides (sucrose, lactose, and maltose) and major polysaccharides (cellulose, amylose, amylopectin, and glycogen). The preponderance of glucose in living systems is best understood in terms of (a) its stereochemistry, where all hydroxyl groups as well as the hydroxymethyl occupy sterically preferred equatorial positions (especially the β-D-pyranosyl units of cellulose), (b) the favorable formation the pyranose (and the much scarcer furanose rings), thereby masking the otherwise reactive free aldehyde group, and (c) substantial hydration of the aldehyde group (to form the aldehydrol), further suppressing nonspecific reactions of glucose with amine-containing metabolites and proteins. NMR spectroscopy (conducted at 27°C in deuterium oxide) confirms that the β-D-pyranose form of glucose is most prevalent: 60.9% β-pyranose, 38.8% α-pyranose, 0.14% β-furanose, 0.15% α-furanose, 0.0045% aldehydrol, and only 0.0024% free aldehyde. At 5 mM glucose, the concentration of free aldehyde is ~10 μM. As a reducing sugar, the aldehyde group of open-chain glucose also reacts with protein amino groups. In the case of adult hemoglobin (HbA), glucose nonenzymatically reacts at the N-terminus of the β-chain, forming an imine that undergoes an Amadori rearrangement to 1-deoxyfructose. D-Glucose weakly inhibits scores of enzymes (K_i > 3-5 mM).

α-D-Glucose 1,6-Bisphosphate

This di-phosphorylated glucose ($FW_{free-acid}$ = 340.12 g/mol), formerly called α-D-glucose 1,6-diphosphate, is a key intermediate and reversibly bound cofactor in the reaction catalyzed by phosphoglucomutase. It was first isolated and its role with phosphoglucomutase characterized by Leloir and coworkers (1). It also serves as a cofactor in reactions catalyzed by phosphopentomutase, phosphomannomutase, and phospho-glucosamine mutase. Glucose 1,6-bisphosphate is an alternative activator of phosphofructokinase (2,3). **Target(s):** fructose-1,6-bisphosphatase (4,5); fructose-2,6-bisphosphate 2-phosphatase (6); glucose-1,6-bisphosphate synthase, by product inhibition (7-9); hexokinase (4,10-24); inositol polyphosphate 5-phosphatase, *or* inositol-1,4,5-trisphosphate 5-phosphatase (25); inositol-1,4,5 trisphosphate 5-phosphomonoesterase (26); 6-phosphofructokinase (27,28); 6-phosphogluconate dehydrogenase (4,29,30). **1.** Caputto, Leloir, Trucco, Cardini & Paladini (1948) *Arch. Biochem.* **18**, 201. **2.** Andres, Carreras & Cusso (1996) *Int. J. Biochem. Cell. Biol.* **28**, 1179. **3.** Espinet, Bartrons & Carreras (1988) *Comp. Biochem. Physiol. B* **90**, 453. **4.** Beitner (1979) *Trends Biochem. Sci.* **4**, 228. **5.** Marcus (1976) *J. Biol. Chem.* **251**, 2963. **6.** Villadsen & Nielsen (2001) *Biochem. J.* **359**, 591. **7.** Maliekal, Sokolova, Vertommen, Veiga-da-Cunha & van Schaftingen (2007) *J. Biol. Chem.* **282**, 31844. **8.** Rose, Warms & Kaklij (1975) *J. Biol. Chem.* **250**, 3466. **9.** Rose, Warms & Wong (1977) *J. Biol. Chem.* **252**, 4262. **10.** Crane & Sols (1955) *Meth. Enzymol.* **1**, 277. **11.** Easterby & Qadri (1982) *Meth. Enzymol.* **90**, 11. **12.** Beitner, Haberman & Livni (1975) *Biochim. Biophys. Acta* **397**, 355. **13.** Gerber, Preissler, Heinrich & Rapoport (1974) *Eur. J. Biochem.* **45**, 39. **14.** Rose & Warms (1975) *Arch. Biochem. Biophys.* **171**, 678. **15.** Bianchi, Serafini, Bartolucci, Giammarini & Magnani (1998) *Mol. Cell. Biochem.* **189**, 185. **16.** Beitner & Lilling (1984) *Int. J. Biochem.* **16**, 991. **17.** Webb (1966) *Enzyme and Metabolic Inhibitors*, vol. 2, p. 379, Academic Press, New York. **18.** de C. Cesar, Colepicolo, Rosa & Rosa (1997) *Comp. Biochem. Physiol. B* **118**, 395. **19.** Magnani, Stocchi, Serafini, *et al.* (1983) *Arch. Biochem. Biophys.* **226**, 377. **20.** Gao & Leary (2003) *J. Amer. Soc. Mass Spectrom.* **14**, 173. **21.** Bianchi, Serafini, Bartolucci, Giammarini & Magnani (1998) *Mol. Cell. Biochem.* **189**, 185. **22.** Lai, Behar, Liang & Hertz (1999) *Metab. Brain Dis.* **14**, 125. **23.** Stocchi, Magnani, Canestrari, Dachà & Fornaini (1982) *J. Biol. Chem.* **257**, 2357. **24.** Andreoni, Serafini, Laguardia & Magnani (2005) *Mol. Cell. Biochem.* **268**, 9. **25.** Milani, Volpe & Pozzan (1988) *Biochem. J.* **254**, 525. **26.** Rana, Sekar, Hokin & MacDonald (1986) *J. Biol. Chem.* **261**, 5237. **27.** Wegener, Schmidt, Leech & Newsholme (1986) *Biochem. J.* **236**, 925. **28.** Beinhauer, Klee, Schmist, G. Wegener & E. A. Newsholme (1987) *Biochem. Soc. Trans.* **15**, 378. **29.** Beitner & Nordenberg (1979) *Biochim. Biophys. Acta* **583**, 266. **30.** Nordenberg, Aviram, Beery, Stenzel & Novogrodsky (1984) *Cancer Lett.* **23**, 193.

α-D-Glucose 1-Phosphate

This non-reducing, glycolytic/glycogenolytic intermediate ($FW_{free-acid}$ = 260.14 g/mol; CAS = 59-56-3; Soluble in water; pK_a values = 1.11 and 6.13), also known as the Cori ester in honor of its discoverers (1), is produced by phosphorolysis of glycogen (by phosphorylase *a*) and isomerization of glucose 6-phosphate (by phosphoglucomutase). **Target(s):** aconitase (2); amylase (3); 1,5 anhydro-D-fructose reductase (4); fructose-1,6-bisphosphate aldolase (5,6); glucose dehydrogenase, weakly inhibited (7,8); glycerate kinase, weakly inhibited (9); glycogen synthase (10); isocitrate lyase (11); mannose-1-phosphate guanylyltransferase, *or* GDP-mannose pyrophosphorylase (12,13); mannose-6-phosphate isomerase, weakly inhibited (14); methylglyoxal synthase (15); NDP glucose:starch glucosyltransferase (16); β-phosphoglucomutase, mildly inhibited (17); [phosphorylase] phosphatase (18,19); pyruvate,orthophosphate dikinase, weakly inhibited (20); triose-phosphate isomerase, weakly inhibited (21); UDP-glucose 4-epimerase, partial inhibition of minor form (22). **1.** Cori, Colowick & Cori (1937) *J. Biol. Chem.* **121**, 465. **2.** Eprintsev, Semenova & Popov (2002) *Biochemistry (Moscow)* **67**, 795. **3.** Krishnamoorthy & Radha (1970) *Enzymologia* **39**, 26. **4.** Sakuma, Kametani & Akanuma (1998) *J. Biochem.* **123**, 189. **5.** Moorhead & Plaxton (1990) *Biochem. J.* **269**, 133. **6.** Bais, James, Rofe & Conyers (1985) *Biochem. J.* **230**, 53. **7.** Webb (1966) *Enzyme and Metabolic Inhibitors*, vol. 2, p. 410, Academic Press, New York. **8.** Metzger, Wilcox & Wick (1964) *J. Biol. Chem.* **239**, 1769. **9.** Saharan & Singh (1993) *Plant Physiol. Biochem.* **31**, 559. **10.** Zea & Pohl (2005) *Biopolymers* **79**, 106. **11.** Popov, Igamberdiev & Volvenkin (1996)

Biokhimiya **61**, 1898. **12.** Ning & Elbein (1999) *Arch. Biochem. Biophys.* **362**, 339. **13.** Szumilo, Drake, York & Elbein (1993) *J. Biol. Chem.* **268**, 17943. **14.** Noltmann (1972) *The Enzymes*, 3rd ed., **6**, 271. **15.** Murata, Fukuda, Watanabe, *et al.* (1985) *Biochem. Biophys. Res. Commun.* **131**, 190. **16.** Zea & Pohl (2005) *Biopolymers* **79**, 106. **17.** Nakamura, Shirokane & Suzuki (1998) *J. Ferment. Bioeng.* **85**, 350. **18.** Madsen (1986) *The Enzymes*, 3rd ed., **17**, 365. **19.** Wang (1999) *Biochem. J.* **341**, 545. **20.** Tjaden, Plagens, Dörr, Siebers & Hensel (2006) *Mol. Microbiol.* **60**, 287. **21.** Tomlinson & Turner (1979) *Phytochemistry* **18**, 1959. **22.** Ray & Bhaduri (1973) *Biochim. Biophys. Acta* **302**, 129.

α-D-Glucose 1-Phosphate 6-Vanadate

This phosphotransfer transition-state analogue ($FW_{free-acid}$ = 360.08 g/mol) combines with muscle phosphoglucomutase (K_i = 15 fM), mimicking a pentavalent intermediate thought to form during catalysis. The first step in the forward reaction is phosphoryl group transfer from the enzyme to glucose 1-phosphate to form glucose 1,6-bisphosphate, leaving the dephosphorylated enzyme, which then undergoes a rapid diffusional reorientation to position the 1-phosphate of the bisphosphate intermediate relative to the dephosphorylated enzyme. The dephosphorylated enzyme then facilitates phosphoryl group transfer from the glucose-1,6-bisP to phosphoglucomutase, again generating the phosphorylated and yielding glucose 6-phosphate. **1.** Percival, Doherty & Gresser (1990) *Biochemistry* **29**, 2764. **2.** Ray & Puvathingal (1990) *Biochemistry* **29**, 2790. **3.** Ray, Puvathingal & Liu (1991) *Biochemistry* **30**, 6875.

D-Glucose 6-Phosphate

This glycolytic and gluconeogenic pathway intermediate ($FW_{free-acid}$ = 260.14 g/mol; pK_a values of 0.94 and 6.11), which is unesterified at its anomeric carbon, is formed in the hexokinase, glucokinase, and phosphoglucomutase reactions. For mammalian brain hexokinase, potent inhibition by glucose 6-P ($K_i \approx 40$-50 μM (1)) occurs through its binding at the active-site (product site) and at a topologically remote (or allosteric) site. By contrast, glucose 6-P is a weak noncompetitive inhibitor of yeast hexokinase. **Target(s):** [acetyl-CoA carboxylase] kinase (2); *N*-acetylglucosamine-6-phosphate deacetylase (3,4); aconitase (5); ATP:citrate lyase, *or* ATP:citrate synthase (94,95); fructokinase (6,7); fructose-1,6-bisphosphatase (8); fructose-1,6-bisphosphate aldolase (9); fructose-2,6 bisphosphate 6-phosphatase, IC_{50} = 45 μM (10); glucokinase (11-13); glucosamine-6-phosphate deaminase (14); glucose dehydrogenase (15-17); glucose-1-phosphate adenylyltransferase (18); γ-glutamyl kinase, weakly inhibited (19); glycerate kinase, weakly inhibited (20); glycerol-1,2-cyclic-phosphate phosphodiesterase (21); glycerol kinase (22); glycogen phosphorylase (23,24,84-92); hexokinase (1,16,25-50); inositol-3-phosphate synthase, *or* inositol-1-phosphate synthase, for which the β-anomer is the natural substrate, and the α-anomer is an inhibitor (51); inositol-polyphosphate 5-phosphatase, *or* inositol-1,4,5-trisphosphate 5-phosphatase (52,53); isocitrate lyase (54,55); isopenicillin-N synthase (56); mannose-6-phosphate isomerase (57,58); methylglyoxal synthase (59); NADH peroxidase (60); peptidase K, *or* proteinase K (61); phosphoenolpyruvate carboxylase, note that the enzyme from some sources is activated (62); 6-phosphofructokinase, weakly inhibited (63); 6-phospho-β-galactosidase (64,65); phosphogluconate dehydrogenase (decarboxylating) (66); 6-phosphogluconolactonase (K_i = 0.3 mM (67); phosphopentomutase (16); phosphoprotein phosphatase (68); phosphorylase kinase (69-72); protein phosphatase 2A (68); pyrophosphatase (73); pyrophosphate:fructose-6-phosphate 1-phosphotransferase, weakly inhibited (74,75); pyruvate kinase (76,77); ribose-5-phosphate isomerase, weakly inhibited (16,78); starch phosphorylase (93); α,α-trehalose-phosphate synthase (UDP-forming) (79); triose-phosphate isomerase, weakly inhibited (80); UDP:hexose-1-phosphate uridylyltransferase (81); uridine nucleosidase (82,83).

1. Liu, Kim, Kurbanov, Honzatko & Fromm (1999) *J. Biol. Chem.* **274**, 31155. **2.** Heesom, Moule & Denton (1998) *FEBS Lett.* **422**, 43. **3.** Campbell, Laurent & Roden (1987) *Anal. Biochem.* **166**, 134. **4.** Weidanz, Campbell, Moore, L. J. deLucas, L. Roden, J. N. Thompson & P. Vezza (1996) *Brit. J. Haematol.* **95**, 645. **5.** Eprintsev, Semenova & Popov (2002) *Biochemistry (Moscow)* **67**, 795. **6.** Doelle (1982) *Eur. J. Appl. Microbiol. Biotechnol.* **14**, 241. **7.** Copeland, Harrison & Turner (1978) *Plant Physiol.* **62**, 291. **8.** Verhees, Akerboom, Schiltz, de Vos & van der Oost (2002) *J. Bacteriol.* **184**, 3401. **9.** Bais, James, Rofe & Conyers (1985) *Biochem. J.* **230**, 53. **10.** Plankert, Purwin & Holzer (1988) *FEBS Lett.* **239**, 69. **11.** Xu, Harrison, Weber & Pilkis (1995) *J. Biol. Chem.* **270**, 9939. **12.** Scopes, Testolin, Stoter, Griffiths-Smith & Algar (1985) *Biochem. J.* **228**, 627. **13.** Sener & Malaisse (1996) *Biochim. Biophys. Acta* **1312**, 73. **14.** Weidanz, Campbell, DeLucas, *et al.* (1995) *Brit. J. Haematol.* **91**, 72. **15.** Strecker (1955) *Meth. Enzymol.* **1**, 335. **16.** Webb (1966) *Enzyme and Metabolic Inhibitors*, vol. **2**, Academic Press, New York. **17.** Metzger, Wilcox & Wick (1964) *J. Biol. Chem.* **239**, 1769. **18.** Lapp & Elbein (1972) *J. Bacteriol.* **112**, 327. **19.** Baich (1970) *Biochem. Biophys. Res. Commun.* **39**, 544. **20.** Saharan & Singh (1993) *Plant Physiol. Biochem.* **31**, 559. **21.** Clarke & Dawson (1978) *Biochem. J.* **173**, 579. **22.** Kasinathan & Khuller (1984) *Lipids* **19**, 289. **23.** Fosset, Muir, Nielson & Fischer (1972) *Meth. Enzymol.* **28**, 960. **24.** Graves & Wang (1972) *The Enzymes*, 3rd ed., **7**, 435. **25.** McDonald (1955) *Meth. Enzymol.* **1**, 269. **26.** Crane & Sols (1955) *Meth. Enzymol.* **1**, 277. **27.** Chou & Wilson (1975) *Meth. Enzymol.* **42**, 20. **28.** Crane (1962) *The Enzymes*, 2nd ed., **6**, 47. **29.** P. Colowick (1973) *The Enzymes*, 3rd ed., **9**, 1. **30.** Copley & Fromm (1967) *Biochemistry* **6**, 3503. **31.** Purich & Fromm (1971) *J. Biol. Chem.* **246**, 3456. **32.** Aleshin, Malfois, Liu, *et al.* (1999) *Biochemistry* **38**, 8359. **33.** Sebastian, Wilson, Mulichak & Garavito (1999) *Arch. Biochem. Biophys.* **362**, 203. **34.** Hashimoto & Wilson (2002) *Arch. Biochem. Biophys.* **399**, 109. **35.** White & Wilson (1989) *Arch. Biochem. Biophys.* **274**, 375. **36.** Armstrong, Wilson & Shoemaker (1996) *Protein Expr. Purif.* **8**, 374. **37.** Wilson (2003) *J. Exp. Biol.* **206**, 2049. **38.** Cesar, Colepicolo, Rosa & Rosa (1997) *Comp. Biochem. Physiol. B* **118**, 395. **39.** Nemat-Gorgani & Wilson (1985) *Arch. Biochem. Biophys.* **236**, 220. **40.** Jangra, Malhotra, Saharan & Singh (2004) *Physiol. Mol. Biol. Plants* **10**, 217. **41.** Renz & Stitt (1993) *Planta* **190**, 166. **42.** Gao & Leary (2003) *J. Amer. Soc. Mass Spectrom.* **14**, 173. **43.** Dörr, Zaparty, Tjaden, Brinkmann & Siebers (2003) *J. Biol. Chem.* **278**, 18744. **44.** Bianchi, Serafini, Bartolucci, Giammarini & Magnani (1998) *Mol. Cell. Biochem.* **189**, 185. **45.** Aleshin, Zeng, Bourenkov, *et al.* (1998) *Structure* **6**, 39. **46.** Claeyssen & Rivoal (2007) *Phytochemistry* **68**, 709. **47.** Liu, Witkovsky & Yang (1983) *J. Neurochem.* **41**, 1694. **48.** Pabón, Cáceres, Gualdrón, *et al.* (2007) *Parasitol. Res.* **100**, 803. **49.** Stocchi, Magnani, Canestrari, Dachà & Fornaini (1982) *J. Biol. Chem.* **257**, 2357. **50.** Andreoni, G. Serafini, M. E. Laguardia & M. Magnani (2005) *Mol. Cell. Biochem.* **268**, 9. **51.** Wong & Sherman (1985) *J. Biol. Chem.* **260**, 11083. **52.** Fowler & Brännström (1990) *Biochem. J.* **271**, 735. **53.** Hansen, Johanson, Williamson & Williamson (1987) *J. Biol. Chem.* **262**, 17319. **54.** Tanaka, Nabeshima, Tokuda & Fukui (1977) *Agric. Biol. Chem.* **41**, 795. **55.** Popov, Igamberdiev & Volvenkin (1996) *Biokhimiya* **61**, 1898. **56.** Castro, Liras, Laiz, Cortes & Martin (1988) *J. Gen. Microbiol.* **134**, 133. **57.** Gracey & Noltmann (1968) *J. Biol. Chem.* **243**, 5410. **58.** Noltmann (1972) *The Enzymes*, 3rd ed., **6**, 271. **59.** Murata, Fukuda, Watanabe, *et al.* (1985) *Biochem. Biophys. Res. Commun.* **131**, 190. **60.** Badwey & Karnovsky (1979) *J. Biol. Chem.* **254**, 11530. **61.** Orstan & Gafni (1991) *Biochem. Int.* **25**, 657. **62.** Nakamura, Minoguchi & Izui (1996) *J. Biochem.* **120**, 518. **63.** Kombrink & Wöber (1982) *Arch. Biochem. Biophys.* **213**, 602. **64.** Calmes & Brown (1979) *Infect. Immun.* **23**, 68. **65.** Suzuki, Saito & Itoh (1996) *Biosci. Biotechnol. Biochem.* **60**, 708. **66.** Silverberg & Dalziel (1975) *Meth. Enzymol.* **41**, 214. **67.** Scopes (1985) *FEBS Lett.* **193**, 185. **68.** Dong, Ermolova & Chollet (2001) *Planta* **213**, 379. **69.** Madsen (1986) *The Enzymes*, 3rd ed. (), **17**, 365. **70.** Krebs, Love, Bratvold, *et al.* (1964) *Biochemistry* **3**, 1022. **71.** Carlson, Bechtel &

Graves (1979) *Adv. Enzymol. Relat. Areas Mol. Biol.* **50**, 41. **72.** Tu & Graves (1973) *Biochem. Biophys. Res. Commun.* **53**, 59. **73.** Gold & Veitch (1973) *Biochim. Biophys. Acta* **327**, 166. **74.** Bertagnolli, Younathan, Voll, Pittman & Cook (1986) *Biochemistry* **25**, 4674. **75.** Pfleiderer & Klemme (1980) *Z. Naturforsch. C* **35c**, 229. **76.** Lin, Turpin & Plaxton (1989) *Arch. Biochem. Biophys.* **269**, 228. **77.** Singh, Rogers & Singh (1991) *Biochem. Int.* **25**, 35. **78.** Axelrod (1955) *Meth. Enzymol.* **1**, 363. **79.** Murphy & Wyatt (1965) *J. Biol. Chem.* **240**, 1500. **80.** Tomlinson & Turner (1979) *Phytochemistry* **18**, 1959. **81.** Hill (1971) *J. Cell Physiol.* **78**, 419. **82.** Magni (1978) *Meth. Enzymol.* **51**, 290. **83.** Magni, Fioretti, Ipata & Natalini (1975) *J. Biol. Chem.* **250**, 9. **84.** Ercan-Fang, Taylor, Treadway, *et al.* (2005) *Amer. J. Physiol. Endocrinol. Metab.* **289**, E366. **85.** Schultz & Ankel (1970) *Biochim. Biophys. Acta* **215**, 39. **86.** Tanabe, Kobayashi & Matsuda (1987) *Agric. Biol. Chem.* **51**, 2465. **87.** Hata, Yokoyama, Suda, Hata & Matsuda (1987) *Comp. Biochem. Physiol. B Comp. Biochem.* **87**, 747. **88.** K. Mukundan & M. R. Nair (1977) *Fish. Technol.* **14**, 1. **89.** H. Schliselfeld (1973) *Ann. N.Y. Acad. Sci.* **210**, 181. **90.** L. Rath, P. K. Hwang & R. L. Fletterick (1992) *J. Mol. Biol.* **225**, 1027. **91.** Nader & J. U. Becker (1979) *Eur. J. Biochem.* **102**, 345. **92.** Vereb, A. Fodor & G. Bot (1987) *Biochim. Biophys. Acta* **915**, 19. **93.** Kokesh, Stephenson & Kakuda (1977) *Biochim. Biophys. Acta* **483**, 258. **94.** Evans & Ratledge (1985) *Can. J. Microbiol.* **31**, 1000. **95.** Shashi, Bachhawat & Joseph (1990) *Biochim. Biophys. Acta* **1033**, 23.

D-Glucuronate (D-Glucuronic Acid)

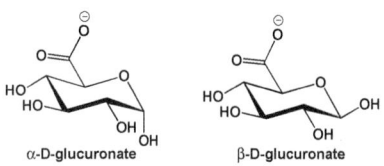

α-D-glucuronate   β-D-glucuronate

This uronic acid ($FW_{free-acid}$ = 194.14 g/mol) is a component of mucopolysaccharides (*e.g.*, hyaluronic acid) and many glucuronides. Glucuronides are produced in the detoxification of many hydroxyl-containing compounds. Note that aqueous solutions of glucuronic acid are mixtures of the acid and its stable lactones: however, equilibrium is reached slowly (*i.e.*, two months at room temperature). An equilibrium mixture consists of 20% lactone at room temperature and 60% lactone at 100°C. *See also* D-Glucuronolactone **Target(s):** acid phosphatase (1,2); aldose 1-epimerase (3); dextransucrase (4); β-glucosidase, weakly inhibited (5); α-glucuronidase (6,7); β-glucuronidase (6,8-14); limulin (*Limulus polyphemus* agglutinin (15); mannonate dehydratase (16); mannose isomerase (17); α-mannosidase, weakly inhibited (18); poly(ADP-ribose) glycohydrolase, weakly inhibited (19).

1. Nigam, Davidson & Fishman (1959) *J. Biol. Chem.* **234**, 1550. **2.** Kuo & Blumenthal (1961) *Biochim. Biophys. Acta* **52**, 13. **3.** Bailey, Fishman, Kusiak, Mulhern & Pentchev (1975) *Meth. Enzymol.* **41**, 471. **4.** Kobayashi, Yokoyama & Matsuda (1986) *Agric. Biol. Chem.* **50**, 2585. **5.** Harada, Tanaka, Fukuda, Hashimoto & Murata (2005) *Arch. Microbiol.* **184**, 215. **6.** Webb (1966) *Enzyme and Metabolic Inhibitors*, vol. **2**, pp. 424-427, Academic Press, New York. **7.** Nagy, Nurizzo, Davies, *et al.* (2003) *J. Biol. Chem.* **278**, 20286. **8.** Levvy & Marsh (1960) *The Enzymes*, 2nd ed., **4**, 397. **9.** Levvy (1952) *Biochem. J.* **52**, 464. **10.** Spencer & Williams (1951) *Biochem. J.* **48**, 538. **11.** Marsh & Levvy (1958) *Biochem. J.* **68**, 610. **12.** Ikegami, Matsunae, Hisamitsu, *et al.* (1995) *Biol. Pharm. Bull.* **18**, 1531. **13.** Diez & Cabezas (1979) *Eur. J. Biochem.* **93**, 301. **14.** Kim, Jin, Jung, Han & Kobashi (1995) *Biol. Pharm. Bull.* **18**, 1184. **15.** Barondes & Nowak (1978) *Meth. Enzymol.* **50**, 302. **16.** Robert-Baudouy & Stoeber (1973) *Biochim. Biophys. Acta* **309**, 473. **17.** Hey-Ferguson & Elbein (1970) *J. Bacteriol.* **101**, 777. **18.** Shepherd & Montgomery (1976) *Biochim. Biophys. Acta* **429**, 884. **19.** Tanuma, Sakagami & Endo (1989) *Biochem. Int.* **18**, 701.

D-Glutamate (D-Glutamic Acid)

This dicarboxylic acid (FW = 147.13 g/mol), also known as (*R*)-2-aminopentandioic acid, is the enantiomer of one of the twenty proteogenic amino acids. D-Glutamyl residues have been found in the cell walls of many bacteria. It is also found in some antibiotics such as bacitracin A. *See also* L-

Glutamic Acid **Target(s):** [citrate lyase] deacetylase, weakly inhibited (1,2); glutamate decarboxylase (3-5); glutamate dehydrogenase (6-9); glutamate formimidoyltransferase (10); glutaminase (9); glutamine synthetase, weakly inhibited and a poor substrate (9,11); glutamate-synthetase] adenylyltransferase (12); γ-glutamylcysteine synthetase (13-15); γ-D-glutamyl-*meso*-diaminopimelate peptidase (16); [pyridoxamine-phosphate aminotransferase, weakly inhibited (17). **1.** Giffhorn, Rode, Kuhn & Gottschalk (1980) *Eur. J. Biochem.* **111**, 461. **2.** Giffhorn & Gottschalk (1975) *J. Bacteriol.* **124**, 1052. **3.** Roberts & Frankel (1951) *J. Biol. Chem.* **190**, 505. **4.** Blindermann, Maitre, Ossola & Mandel (1978) *Eur. J. Biochem.* **86**, 143. **5.** Prabhakaran, Harris & Kirchheimer (1983) *Arch. Microbiol.* **134**, 320. **6.** Caughey, Smiley & Hellerman (1957) *J. Biol. Chem.* **224**, 591. **7.** Strecker (1955) *Meth. Enzymol.* **2**, 220. **8.** Smith, Austen, Blumenthal & Nyc (1975) *The Enzymes*, 3rd ed., **11**, 293. **9.** Webb (1966) *Enzyme and Metabolic Inhibitors*, vol. 2, Academic Press, New York. **10.** Miller & Waelsch (1957) *J. Biol. Chem.* **228**, 397. **11.** Ginsburg, Yeh, Hennig & Denton (1970) *Biochemistry* **9**, 633. **12.** Ebner, Wolf, Gancedo, Elsässer & Holzer (1970) *Eur. J. Biochem.* **14**, 535. **13.** Seelig & Meister (1985) *Meth. Enzymol.* **113**, 379. **14.** Griffith & Meister (1977) *Proc. Natl. Acad. Sci. U.S.A.* **74**, 3330. **15.** Sekura & Meister (1977) *J. Biol. Chem.* **252**, 2599. **16.** Arminjon, Guinand, Vacheron & Michel (1977) *Eur. J. Biochem.* **73**, 557. **17.** Tani, Ukita & Ogata (1972) *Agric. Biol. Chem.* **36**, 181.

L-Glutamic Acid

This dicarboxylic acid (FW$_{monosodium-salt}$ = 169.11 g/mol; pK_a values of 2.10, 4.07 (side chain), and 9.47 at 25°C; M.P.$_{decomposition}$ = 247-249°C; Specific rotation of + 31.8° at 25°C at the sodium D-line and 2.0 g/100 mL L-glutamic acid in 5 M HCl (+12.0° in distilled water)), also known as (*S*)-2-aminopentanedioic acid and 2-aminoglutaric acid and symbolized by Glu and E, is one of the twenty proteogenic (*i.e.*, protein-forming) amino acids. L-Glutamic acid has a "meaty" taste, whereas its enantiomer is tasteless. There are two codons: GAA and GAG. Glutamyl residues in proteins are almost always on the surface, and they serve as bases in enzyme active sites. Glutamyl residues favor α-helical structures more than aspartyl residues; in fact, glutamyl residues are frequently found in α-helices. Homologous proteins often have glutamyl residues interchanged with aspartyl, glutaminyl, alanyl, or, somewhat surprisingly, histidyl residues. Glutamate weakly inhibits many enzymes in the 3-5 mM range.

β-Glutamate (β-Glutamic Acid)

This β-amino acid (FW$_{monosodium-salt}$ = 169.11 g/mol; CAS: 1948-48-7 (free acid); Symbol: 3-AG), also known as 3-aminoglutaric acid and 3-aminoglutarate inhibits γ-glutamylcyclotransferase (1) and is a good alternative substrate for glutamine synthetase (2). 3-AG also inhibits glutamate-mediated neurotransmission both in primary neuronal cultures and in brain slices with more intact neural circuits (3). When assayed with the low affinity glutamate receptor antagonist γ-DGG, 3-AG significantly reduces the synaptic cleft glutamate concentration, suggesting that 3-AG may be a false transmitter that competes with glutamate during vesicle filling. In three different epileptic models (Mg^{2+}-free, 4-AP, and high K$^+$), 3-AG suppresses epileptiform activity both before and after its induction. Like L-glutamate, 3-AG is a high-affinity substrate for both the plasma membrane (EAATs) and vesicular (vGLUT) glutamate transporters. EAATs facilitate 3-AG entry into neuronal cytoplasm, and 3-AG competes with L-glutamate for transport into vesicles thus reducing glutamate content (4). In a synaptosomal preparation, 3-AG inhibits calcium-dependent endogenous L-glutamate release. Unlike L-glutamate, however, 3-AG had low affinity for both ionotropic (NMDA and AMPA) and G-protein coupled (mGlu1-8) receptors (4). Note also that the C-3 atom of 3-AG is prochiral, and sheep brain glutamine synthetase produces both β-glutamine stereoisomers, albeit with different yields. **1.** Griffith & Meister (1977) *Proc. Natl. Acad. Sci. U.S.A.* **74**, 3330. **2.** Allison, Todhunter & Purich (1977) *J. Biol. Chem.*

252, 6046. **3.** Wu, Foster, Staubli, *et al.* (2015) *Neuropharmacol.* **97**, 95. **4.** Foster, Chen, Runyan, *et al.* (2015) *Neuropharmacol.* **97**, 436.

L-Glutamate γ-Monohydroxamate

This aminoacyl-hydroxamate and glutamine analogue (FW = 162.15 g/mol; unstable in aqueous solutions) inhibits γ-glutamylcysteine synthetase and angiotensin-converting enzyme. This hydroxamate can be formed by the γ-glutamyl transferase activity of glutamine synthetase. **Target(s):** carbamoyl-phosphate synthetase (1); glutamate decarboxylase, weakly inhibited (2); glutamine:fructose-6-P aminotransferase (isomerizing) (3); glutaminyl-tRNA synthetase (4); γ-glutamylcysteine synthetase (5); 4-methyleneglutaminase (6); peptidyl-dipeptidase A, *or* angiotensin-converting enzyme (7); urease (8). **1.** Kaseman (1985) Ph.D. Dissertation, Department of Biochemistry, Cornell University Medical College, New York, New York. **2.** Tunnicliff (1990) *Int. J. Biochem.* **27**, 1235. **3.** Chmara, Andruszkiewicz & Borowski (1985) *Biochim. Biophys. Acta* **870**, 357. **4.** Lea & Fowden (1973) *Phytochemistry* **12**, 1903. **5.** Katoh, Hiratake & Oda (1998) *Biosci. Biotechnol. Biochem.* **62**, 1455. **6.** Ibrahim, Lea & Fowden (1984) *Phytochemistry* **23**, 1545. **7.** Liu, Lin & Hou (2004) *J. Agric. Food Chem.* **52**, 386. **8.** Davis & Shih (1984) *Phytochemistry* **23**, 2741.

L-Glutamate γ-Semialdehyde

This proline and ornithine precursor (FW = 131.13 g/mol), also known as L-glutamate 5-semialdehyde and (*S*)-2-amino-5-oxopentanoate, is formed by NADH-dependent reduction. Because its aldehyde group readily undergoes hydration to form an *sp*^{3}-hybridized carbon atom, glutamate semialdehyde is also a natural transition-state analogue inhibitor of *Escherichia coli* glutamine:fructose-6-phosphate transaminase (isomerizing). **Target(s):** CTP synthetase (1,2), most likely by binding to the catalytically essential thiol within the amidohydrolase active-site and inhibiting glutaminase activity; glutamine: fructose 6-P aminotransferase (isomerizing), *or* glucosamine-6-phosphate synthase (3,4). **1.** Bearne, Hekmat & MacDonnell (2001) *Biochem. J.* **356**, 223. **2.** Willemoes (2004) *Arch. Biochem. Biophys.* **424**, 105. **3.** Bearne & Wolfenden (1995) *Biochemistry* **34**, 11515. **4.** Badet-Denisot, Leriche, Massière & Badet (1995) *Bioorg. Med. Chem. Lett.* **5**, 815.

D-Glutamine

This amide-containing amino acid (FW = 146.15 g/mol), also known as (*R*)-2-amino-4-carbamoylbutanoate, is a component of the antibiotic tolaasin. **Target(s):** glutamate racemase (1); glutamate synthase (NADPH) (*Saccharomyces cerevisiae*) (2); [glutamine-synthetase] adenylyltransferase (3); γ-glutamyl transpeptidase (4,5); 4 methyleneglutaminase (6); UDP-*N*-acetylmuramoyl-L-alanine:D-glutamate ligase, weakly inhibited (7). **1.** Hwang, Cho, Kim, *et al.* (1999) *Nature Struct. Biol.* **6**, 422. **2.** Meister (1985) *Meth. Enzymol.* **113**, 327. **3.** Ebner, Wolf, Gancedo, Elsässer & Holzer (1970) *Eur. J. Biochem.* **14**, 535. **4.** Allison (1985) *Meth. Enzymol.* **113**, 419. **5.** Thompson & Meister (1977) *J. Biol. Chem.* **252**, 6792. **6.** Ibrahim, Lea & Fowden (1984) *Phytochemistry* **23**, 1545. **7.** Michaud, Blanot, Flouret & van Heijenoort (1987) *Eur. J. Biochem.* **166**, 631.

L-Glutamine

This proteogenic amino acid (FW = 146.15 g/mol; pK_a values of 2.17 and 9.13 at 25°C; Specific rotation of + 31.8° at 25°C at the sodium D-line and 2.0 g/100 mL L-glutamine in 5 M HCl; Symbols: Gln & Q; IUPAC: (S)-2-amino-4-carbamoylbutanoic acid) is one of the twenty proteogenic amino acids. There are two L-glutamine codons: CAA and CAG. Homologous proteins often have glutaminyl residues interchanged with glutamyl or histidyl residues. **Target(s):** acetylcholinesterase, weakly inhibited (1); adenylyl-[glutamine synthetase] hydrolase (adenylyl-[glutamate:ammonia ligase] hydrolase) (2-4); agaritine γ-glutamyltransferase, agaritine hydrolysis is inhibited (36); arginine decarboxylase (5); asparagine:oxo-acid aminotransferase (6,7); asparagine synthetase, Cys-1 mutants, ammonia-dependent activity inhibited (8); aspartate aminotransferase (9); aspartate 4-decarboxylase (10); carbamate kinase (11); carbamoyl-serine ammonia-lyase (12); [citrate lyase] deacetylase, weakly inhibited (13); glutamate dehydrogenase (14); γ-glutamyl kinase II, *or* glutamate 5-kinase II (15); γ-glutamyl transpeptidase, *or* γ-glutamyltransferase, also an alternative substrate (16,17,37-39); hemoglobin S polymerization (18); kynureninase, weakly inhibited (19); kynurenine:oxoglutarate aminotransferase (20-22); membrane alanyl aminopeptidase, *or* aminopeptidase N (23,24); 4-methyleneglutaminase (25,26); nitrogenase (27,28); ornithine carbamoyltransferase (40); [protein-P_{II}] uridylyltransferase (3,29,30); pyruvate kinase (31); saccharopine dehydrogenase (32); theanine hydrolase (33); UDP-N-acetylmuramoylalanine:D-glutamate ligase, inhibited at elevated concentrations (34); urease, weakly inhibited (35). **1.** Bergmann, Wilson & Nachmansohn (1950) *J. Biol. Chem.* **186**, 693. **2.** Shapiro (1970) *Meth. Enzymol.* **17A**, 936. **3.** Stadtman & Ginsburg (1974) *The Enzymes*, 3rd ed., **10**, 755. **4.** Shapiro (1969) *Biochemistry* **8**, 659. **5.** Li, Regunathan & Reis (1995) *Ann. N.Y. Acad. Sci.* **763**, 325. **6.** Maul & Schuster (1986) *Arch. Biochem. Biophys.* **251**, 585. **7.** Cooper (1977) *J. Biol. Chem.* **252**, 2032. **8.** Sheng, Moraga-Amador, van Heeke, *et al.* (1993) *J. Biol. Chem.* **268**, 16771. **9.** Rakhmanova & Popova (2006) *Biochemistry (Moscow)* **71**, 211. **10.** Wong (1985) *Neurochem. Int.* **7**, 351. **11.** Raijman & Jones (1973) *The Enzymes*, 3rd ed., **9**, 97. **12.** Cooper & Meister (1973) *Biochem. Biophys. Res. Commun.* **55**, 780. **13.** Giffhorn & Gottschalk (1975) *J. Bacteriol.* **124**, 1052. **14.** Strecker (1955) *Meth. Enzymol.* **2**, 220. **15.** Yoshinaga, Tsuchida & Okumura (1975) *Agri. Biol. Chem.* **39**, 1269. **16.** Allison (1985) *Meth. Enzymol.* **113**, 419. **17.** Thompson & Meister (1977) *J. Biol. Chem.* **252**, 6792. **18.** Rumen (1975) *Blood* **45**, 45. **19.** Jakoby & Bonner (1953) *J. Biol. Chem.* **205**, 709. **20.** Guidetti, Okuno & Schwarcz (1997) *J. Neurosci. Res.* **50**, 457. **21.** Okuno, Nishikawa & Nakamura (1996) *Adv. Exp. Med. Biol.* **398**, 455. **22.** Milart, Urbanska, Turski, Paszkowski & Sikorski (2001) *Placenta* **22**, 259. **23.** Lalu, Lampelo & Vanha-Perttula (1986) *Biochim. Biophys. Acta* **873**, 190. **24.** Bauvois & Dauzonne (2006) *Med. Res. Rev.* **26**, 88. **25.** Ibrahim, Lea & Fowden (1984) *Phytochemistry* **23**, 1545. **26.** Winter & Dekker (1991) *Plant Physiol.* **95**, 206. **27.** Jones & Monty (1979) *J. Bacteriol.* **139**, 1007. **28.** Vignais, Colbeau, Willison & Jouanneau (1985) *Adv. Microb. Physiol.* **26**, 155. **29.** Adler, Purich & Stadtman (1975) *J. Biol. Chem.* **250**, 6264. **30.** Bonatto, Couto, Souza, *et al.* (2007) *Protein Expr. Purif.* **55**, 293. **31.** Singh, Malhotra & Singh (2000) *Indian J. Biochem. Biophys.* **37**, 51. **32.** Fujioka & Nakatani (1972) *Eur. J. Biochem.* **25**, 301. **33.** Tsushida & Takeo (1985) *Agric. Biol. Chem.* **49**, 2913. **34.** Linnett & Tipper (1974) *J. Bacteriol.* **120**, 342. **35.** Davis & Shih (1984) *Phytochemistry* **23**, 2741. **36.** Gigliotti & Levenberg (1964) *J. Biol. Chem.* **239**, 2274. **37.** Suzuki, Kumagai & Tochikura (1986) *J. Bacteriol.* **168**, 1325. **38.** Castonguay, Lherbet & Keillor (2002) *Bioorg. Med. Chem.* **10**, 4185. **39.** Nakayama, Kumagai & Tochikura (1984) *J. Bacteriol.* **160**, 341. **40.** Lusty, Jilka & Nietsch (1979) *J. Biol. Chem.* **254**, 10030.

γ-(L-Glutamyl)-L-cysteine

This isopeptide (FW = 249.28 g/mol), the major thiol in aerobic phototrophic halobacteria (*e.g.*, *Halobacterium halobium*), is the direct precursor of glutathione in mammals. It is a product of the reaction catalyzed by γ-glutamylcysteine synthetase (1,2) and a substrate for glutathione synthetase (3). γ-Glutamyl-cysteine also an alternative product of γ-glutamyl transpeptidase (4) and glutathione S-transferase (5). **Target(s):** 5'-nucleotidase (6). **1.** Seelig & Meister (1985) *Meth. Enzymol.* **113**, 379. **2.** Seelig & Meister (1985) *Meth. Enzymol.* **113**, 390. **3.** Meister (1985) *Meth. Enzymol.* **113**, 393. **4.** Allison (1985) *Meth. Enzymol.* **113**, 419. **5.** Sugimoto, Kuhlenkamp, Ookhtens, *et al.* (1985) *Biochem. Pharmacol.* **34**, 3643. **6.** Liu & Sok (2000) *Neurochem. Res.* **25**, 1475.

L-Glutamyl-L-leucyl-L-aspartyl-[(2R,4S,5S)-5-amino-4-hydroxy-2,7-dimethyloctanoyl]-L-valyl-L-glutamyl-L-phenylalanine

This octapeptide analogue (FW = 935.06 g/mol; Sequence: ELDLΨAVEF, where Ψ denotes a nonhydrolyzable hydroxyethylene isostere (-(S)-CH(OH)CH₂-), replacing the peptide bond between the Leu-4 and Ala-5 residues), also known as OM00-3, inhibits memapsin-2 (β-secretase; K_i = 0.13 nM) and memapsin-1 (K_i = 0.18 nM). **1.** Turner, Loy, Nguyen, *et al.* (2002) *Biochemistry* **41**, 8742. **2.** Ghosh, Hong & Tang (2002) *Curr. Med. Chem.* **9**, 1135. **3.** Hong, Turner, Koelsch, *et al.* (2002) *Biochemistry* **41**, 10963.

L-Glutamyl-L-valyl-L-asparaginyl-[(2R,4S,5S)-5-amino-4-hydroxy-2,7-dimethyl-octanoyl]-L-alanyl-L-glutamyl-L-phenylalanine

This octapeptide analogue (FW = 892.00 g/mol; Sequence: ELDLΨAVEF, where Ψ denotes a nonhydrolyzable hydroxyethylene isostere (-(S)-CH(OH)CH₂-), replacing the peptide bond between the Leu-4 and Ala-5 residues), also known as OM99-2 and EVNLΨAAEF, is derived from the β-secretase site of Swedish β-amyloid precursor protein, in an aspartyl residue is replaced by an alanyl residue. OM99-2 inhibits the aspartyl proteases, memapsin-2 (*or* β-secretase), K_i = 1.9 nM (1-6), cathepsin D3, K_i = 48 nM (1), and memapsin 14 (2). **1.** Ghosh, Shin, Downs, *et al.* (2000) *J. Amer. Chem. Soc.* **122**, 3522. **2.** Ghosh, Bilcer, Harwood, *et al.* (2001) *J. Med. Chem.* **44**, 2865. **3.** Hong, Koelsch, Lin, *et al.* (2000) *Science* **290**, 150. **4.** Turner, Koelsch, Hong, *et al.* (2001) *Biochemistry* **40**, 10001. **5.** Ghosh, Hong & Tang (2002) *Curr. Med. Chem.* **9**, 1135. **6.** Hong, Turner, Koelsch, *et al.* (2002) *Biochemistry* **41**, 10963.

Glutarate (Glutaric Acid)

This dicarboxylic acid (FW$_{free-acid}$ = 132.12 g/mol; CAS = 110-94-1; M.P. = 97.5-98°C; pK_a values of 4.34 and 5.22 at 25°C; Solubility: 639 g/L H_2O 20°C), also known as pentanedioic acid and propane-1,3-dicarboxylic acid, is formed by the degradation of lysine and tryptophan. **Target(s):** 2-(acetamidomethylene)-succinate hydrolase (1); aerobactin synthase (2); alanine aminotransferase (3-6); aldehyde reductase (7); 2-aminoadipate aminotransferase, weakly inhibited (8); 4-aminobutyrate aminotransferase (9,10); aspartate aminotransferase (11-17); D-aspartate oxidase (18-21); γ-butyrobetaine dioxygenase, *or* γ-butyrobetaine hydroxylase, K_i = 0.32 mM (22); carbamoyl phosphate synthetase (23,24); carboxy-*cis,cis*-muconate cyclase (25); cyanate hydratase, *or* cyanase (26); cysteine sulfinate decarboxylase, slightly inhibited (27); fumarase (28,29); glucose-1,6-bisphosphate synthase (30); glucuronate reductase (7); glutamate decarboxylase (14,31-33); glutamate dihydro-genase (14,34-38); glutaryl-7-aminocephalosporanate acylase, by product inhibition (39-42); homocitrate synthase (55); 4-hydroxyglutamate aminotransferase (43); L-kynurenine aminotransferase, *or* kynurenine:2-oxoglutarate aminotransferase (44-46); leucine aminotransferase (47); lysine 2-monooxygenase (59); *N*-methylglutamate dehydrogenase (48); 3-oxoacid CoA-transferase, *or* 3-ketoacid CoA-transferase, weakly inhibited (49); procollagen-lysine 5-dioxygenase (56); procollagen-proline 3-dioxygenase (56); procollagen-proline 4-dioxygenase (56-58); pyrophosphatase (50); succinate dehydrogenase (14,51-54). **1.** Huynh & Snell (1985) *J. Biol. Chem.* **260**, 2379. **2.** Appanna & Viswanatha (1986) *FEBS Lett.* **202**, 107. **3.** Bulos & Handler (1965) *J. Biol. Chem.* **240**, 3283. **4.** Good & Muench (1992) *Plant Physiol.* **99**, 1520. **5.** Vedavathi, Girish & Kumar (2004) *Mol. Cell. Biochem.* **267**, 13. **6.** Vedavathi, Girish & Kumar (2006) *Biochemistry (Moscow)* **71** Suppl. 1, S105. **7.** Gabbay & Kinoshita (1975) *Meth. Enzymol.* **41**, 159. **8.** Deshmukh & Mungre (1989) *Biochem. J.* **261**, 761. **9.** Sytinsky & Vasilijev (1970) *Enzymologia* **39**, 1. **10.** Liu, Peterson, Langston, *et al.* (2005) *Biochemistry* **44**, 2982. **11.** Jenkins, Yphantis & Sizer (1959) *J. Biol. Chem.* **234**, 51. **12.** Velick & Vavra (1962) *The Enzymes*, 2nd ed., **6**, 219. **13.** Braunstein (1973) *The Enzymes*, 3rd ed., **9**, 379. **14.** Webb (1966) *Enzyme and Metabolic Inhibitors*, vol. **2**, Academic Press, New York. **15.** Bonsib, Harruff & Jenkins (1975) *J. Biol. Chem.* **250**, 8635. **16.** Tanaka, Tokuda, Tachibana, Taniguchi & Oi (1990) *Agric. Biol. Chem.* **54**, 625. **17.** Martins, Mourato & de Varennes (2001) *J. Enzyme Inhib.* **16**, 251. **18.** Dixon (1970) *Meth. Enzymol.* **17A**, 713. **19.** Dixon & Kenworthy (1967) *Biochim. Biophys. Acta* **146**, 54. **20.** de Marco & Crifò (1967) *Enzymologia* **33**, 325. **21.** Rinaldi (1971) *Enzymologia* **40**, 314. **22.** Ng, Hanauske-Abel & Englard (1991) *J. Biol. Chem.* **266**, 1526. **23.** Jones (1962) *Meth. Enzymol.* **5**, 903. **24.** Fahien, Schooler, Gehred & Cohen (1964) *J. Biol. Chem.* **239**, 1935. **25.** Thatcher & Cain (1975) *Eur. J. Biochem.* **56**, 193. **26.** Anderson, Johnson, Endrizzi, Little & Korte (1987) *Biochemistry* **26**, 3938. **27.** Heinamaki, Peramaa & Piha (1982) *Acta Chem. Scand. B* **36**, 287. **28.** Hill & Bradshaw (1969) *Meth. Enzymol.* **13**, 91. **29.** Massey (1953) *Biochem. J.* **55**, 172. **30.** Rose, Warms & Wong (1977) *J. Biol. Chem.* **252**, 4262. **31.** Fonda (1972) *Biochemistry* **11**, 1304. **32.** Stokke, Goodman & Moe (1976) *Clin. Chim. Acta* **66**, 411. **33.** Spink, Porter, Wu & Martin (1985) *Biochem. J.* **231**, 695. **34.** Caughey, Smiley & Hellerman (1957) *J. Biol. Chem.* **224**, 591. **35.** Frieden (1963) *The Enzymes*, 2nd ed., **7**, 3. **36.** Smith, Austen, Blumenthal & Nyc (1975) *The Enzymes*, 3rd ed., **11**, 293. **37.** Baker (1967) *Design of Active-Site-Directed Irreversible Enzyme Inhibitors*, Wiley, New York. **38.** Caughey, Smiley & Hellerman (1957) *J. Biol. Chem.* **224**, 591. **39.** Lee, Kim, Lee & Park (2000) *Biochim. Biophys. Acta* **1523**, 123. **40.** Battistel, Bianchi, Bortolo & Bonoldi (1998) *Appl. Biochem. Biotechnol.* **69**, 53. **41.** Aramori, Fukagawa, Tsumura, *et al.* (1992) *J. Ferment. Bioeng.* **73**, 185. **42.** Lee, Chang, Liu & Chu (1998) *Biotechnol. Appl. Biochem.* **28** (Part 2), 113. **43.** Goldstone & Adams (1962) *J. Biol. Chem.* **237**, 3476. **44.** Tanizawa, Asada & Soda (1985) *Meth. Enzymol.* **113**, 90. **45.** Tobes (1987) *Meth. Enzymol.* **142**, 217. **46.** Mason (1959) *J. Biol. Chem.* **234**, 2770. **47.** Pathre, Singh, Viswanathan & Sane (1987) *Phytochemistry* **26**, 2913. **48.** Hersh, Stark, Worthen & Fiero (1972) *Arch. Biochem. Biophys.* **150**, 219. **49.** Hersh & Jencks (1967) *J. Biol. Chem.* **242**, 3468. **50.** Naganna (1950) *J. Biol. Chem.* **183**, 693. **51.** Quastel & Woolbridge (1928) *Biochem. J.* **22**, 689. **52.** Massart (1950) *The Enzymes*, 1st ed., **1** (part 1), 307. **53.** Potter & Elvehjem (1937) *J. Biol. Chem.* **117**, 341. **54.** Thunberg (1933) *Biochem. Z.* **258**, 48. **55.** Qian, West & Cook (2006) *Biochemistry* **45**, 12136. **56.** Majamaa, Turpeenniemi-Hujanen, Latipää, *et al.* (1985) *Biochem. J.* **229**, 127. **57.** Majamaa, Hanauske-Abel, Günzler & Kivirikko (1984) *Eur. J. Biochem.* **138**, 239. **58.** Kaska, Günzler, Kivirikko & Myllylä (1987) *Biochem. J.* **241**, 483. **59.** Vandecasteele & Hermann (1972) *Eur. J. Biochem.* **31**, 80.

Glutaryl-CoA

This coenzyme A thioester (FW = 887.57 g/mol; CAS = 103192-48-9), an intermediate in lysine and tryptophan degradation, inhibits arylamine *N*-acetyltransferase (1), glutaconyl-CoA decarboxylase (2), 3-hydroxy-3 methylglutaryl-CoA synthase (3), methylmalonyl-CoA decarboxylase (4), methylmalonyl-CoA epimerase (5), (*S*)-methylmalonyl-CoA hydrolase (6).

1. Kawamura, Graham, Mushtaq, *et al.* (2005) *Biochem. Pharmacol.* **69**, 347. **2.** Haertel, Eckel, Koch, *et al.* (1993) *Arch. Microbiol.* **159**, 174. **3.** Middleton (1972) *Biochem. J.* **126**, 35. **4.** Galivan & Allen (1968) *Arch. Biochem. Biophys.* **126**, 838. **5.** Stabler, Marcell & Allen (1985) *Arch. Biochem. Biophys.* **241**, 252. **6.** Kovachy, Copley & Allen (1983) *J. Biol. Chem.* **258**, 11415.

Glutathione (Reduced Glutathione)

This redox-active tripeptide (FW = 307.32 g/mol; CAS = 70-18-8; pK_a values of 2.12, 3.53, 8.66, and 9.12), commonly abbreviated as GSH and chemically as L-γ-glutamyl-L-cysteinylglycine, is most abundant nonprotein thiol in living cells. (**See also** *Glutathione Disulfide*) Cellular glutathione concentrations are typically at 1–5 mM, with highest levels found within the eye. **Redox Properties:** GSH undergoes reversible oxidation to form glutathione disulfide (GSSG): $E° = -0.25$ V at pH 7. This value is based on measurements of the equilibrium poise of the glutathione reductase reaction (GS–SG + NADPH + H$^+$ ⇌ 2GSH + NADP$^+$), for which the equilibrium constant is $K_{eq} = γ$ typically obtained by absorption spectroscopy (changes in the NADPH spectrum, usually measured at 340 nm) or by NMR spectroscopy (1), the latter offering the advantage that several proton resonances may be measured simultaneously (2). **Roles in Cellular Metabolism:** Glutathione plays the following roles in cellular metabolism: detoxification of certain arenes; (a) reduction of Fe(III) in methemoglobin; (b) leukotriene and prostaglandin biosynthesis; (c) cofactor for a number of enzymes; (d) L-cystine translocation (as well as transport of certain other amino acids and dipeptides); (e) protein disulfide exchange reactions; and (f) trapping of hydrogen peroxide, organic peroxides, and free radicals. Glutathione biosynthesis occurs via two ATP-dependent reactions catalyzed by γ-glutamylcysteine synthetase followed by glutathione synthetase. **Inhibitory Targets:** Reduced glutathione binds to and/or reacts with disulfide-containing proteins, thereby reducing accessible disulfide linkages. Within cells, it suppresses the accumulation of disulfide-linked proteins arising from the oxidative action of molecular oxygen, peroxides, and various reactive oxygen species. Reduced glutathione also inhibits numerous extracellular enzymes that require intact disulfide linkages to remain catalytic active. γ-Glutamyl-cysteine synthetase is inhibited by glutathione under conditions similar to those prevailing *in vivo*, thus strongly suggesting a physiologically significant feedback mechanism (3). Inhibition is not allosteric, instead appearing to involve the binding of glutathione to the glutamate site of the enzyme as well as to another enzyme site; the latter requires a sulfhydryl group since ophthalmic acid (γ-glutamyl-α-aminobutyryl-glycine) is only a weak inhibitor. Such findings may explain observations on patients with 5-oxoprolinuria, who were shown to have a block in the γ-glutamyl cycle consisting of a marked deficiency of glutathione synthetase and consequently of glutathione (3). The glycine receptor (GlyR) is also inhibited by reduced glutathione (4); however, the affinity of strychnine or glycine for the GlyR is unaffected by such treatments. **1.** Veech, Eggleston & Krebs (1969) *Biochem. J.* **115**, 609. **2.** Millis, Weaver & Rabenstein (1993) *J. Org. Chem.* **58**, 4144. **3.** Richman & Meister (1975) *J. Biol. Chem.* **250**, 1422. **4.** Ruiz-Gómez, Fernández-Shaw, *et al.* (1991) *J. Neurochem.* **56**, 1690.

Glutathione Disulfide (Oxidized Glutathione)

This disulfide (FW = 612.60 g/mol; CAS = 70-18-8), frequently referred to as oxidized glutathione and abbreviated GSSG, is produced in many oxidation/reduction reactions. (**See** *Glutathione for details on redox chemistry*) Oxidized glutathione binds to and inhibits numerous intracellular enzymes that normally require thiol groups for full catalytic activity. **Target(s):** acid phosphatase (1); *S*-adenosylmethionine synthetase (2); adenylyl-sulfate kinase (3); arylamine glucosyltransferase, weakly inhibited (85); ATP:citrate lyase, or ATP:citrate synthase (4,5,87); cadmium-transporting ATPase (6); calpains (7); creatine kinase (8); deoxycytidine kinase (9); deoxyribopyrimidine photo-lyase (10); DNA-binding activity of c-Jun (11); DNA-directed DNA polymerase (12); DNA polymerase δ (12); eukaryotic initiation factor-2, *or* eIF-2 (4); *S* formylglutathione hydrolase (13); fructose-1,6-bisphosphatase (14,15); fructose-1,6-bisphosphate aldolase (16); 1,4-α-glucan branching enzyme (86); glucan endo-β-D-glucosidase (17); glutamine amino-transferase (18); γ-glutamyl cyclotransferase (19); γ-glutamylcysteine synthase (20); glutathione synthetase (21-23); glutathione *S*-transferase (24,83); glyceraldehyde-3-P dehydrogenase (25-28); glycogen synthase (4); guanidinoacetate *N*-methyltransferase (89,90); guanylate cyclase (4,29); hexokinase (4,30-33); cytomegalovirus protease (34); 3-hydroxy-3-methylglutaryl-CoA reductase (4,35,36); 15-hydroxyprosta-glandin dehydrogenase (37); 3α-hydroxy-steroid dehydrogenase (38); indoleamine *N*-acetyl-transferase (4); isocitrate lyase (5); maleylacetoacetate isomerase (39,40); methionine *S*-adenosyltransferase (84); 2-methyleneglutarate mutase (41); methylitaconate Δ-isomerase (41); 5-methyltetrahydropteoyltriglutamate: homocysteine *S*-methyl-transferase (88); methylumbelliferyl-acetate deacetylase (42); myosin ATPase (43); nucleotide diphosphatase (44); ornithine decarboxylase (45); pantetheine hydrolase (46, but ref. 47 reports no inhibition); peptidyl-dipeptidase A, *or* angiotensin-converting enzyme (48); 1-phosphatidylinositol 4-phosphate 5-kinase (49); phosphoadenylyl-sulfate reductase (93); phosphofructokinase (4); phosphogluco-mutase (50); phospho-inositide 5-phosphatase (51); phospholipase D (52); phospho-protein phosphatase types 1 and 2A (4,53-56); phospho-ribulokinase (57,58); [phosphorylase] phosphatase (4,54,59,60); phosphotyrosine protein phosphatase (61); prostaglandin Δ^{13} reductase (62); pyrophosphatase (63,64); pyruvate kinase (65); saccharolysin (66); sedoheptulose-bisphosphatase (15); sphingomyelin phosphodiesterase (67,68); stizolobate synthase (92); stizolobinate synthase (92); succinate dehydrogenase (27,69-73); succinyl-CoA synthetase (74); sucrose synthase (4); sulfur reductase (75); thiol oxidase (94,95); thiosulfate:thiol sulfurtransferase, product inhibition (76); transglutaminase (77); UDP-glucose dehydrogenase (78); xanthine dehydrogenase (91); xenobiotic transporting ATPase, multidrug-resistance protein (79-82). **1.** Terada (1997) *Int. J. Biochem. Cell Biol.* **29**, 985. **2.** Pajares, Duran, Corrales, Pliego & Mato (1992) *J. Biol. Chem.* **267**, 17598. **3.** Schriek & Schwenn (1986) *Arch. Microbiol.* **145**, 32. **4.** Gilbert (1984) *Meth. Enzymol.* **107**, 330. **5.** Spector (1972) *The Enzymes*, 3rd ed., **7**, 357. **6.** Li, Szczypka, Lu, Thiele & Rea (1996) *J. Biol. Chem.* **271**, 6509. **7.** Di Cola & Sacchetta (1987) *FEBS Lett.* **210**, 81. **8.** Reddy, Jones, Cross, Wong & Van Der Vliet (2000) *Biochem. J.* **347**, 821. **9.** Durham & Ives (1970) *J. Biol. Chem.* **245**, 2276. **10.** Saito & Werbin (1970) *Biochemistry* **9**, 2610. **11.** Klatt & Lamas (2002) *Meth. Enzymol.* **348**, 157. **12.** H. Malkas & Hickey (1996) *Meth. Enzymol.* **275**, 133. **13.** Uotila & Koivusalo (1981) *Meth. Enzymol.* **77**, 320. **14.** Nakashima & Ogino (1973) *J. Biochem.* **74**, 601. **15.** Nishizawa & Buchanan (1981) *J. Biol. Chem.* **256**, 6119. **16.** Offermann, McKay, Marsh & Bond (1984) *J. Biol. Chem.* **259**, 8886. **17.** Sanchez, Nombela, Villanueva & Santos (1982) *J. Gen. Microbiol.* **128**, 2047. **18.** T'ing-sêng (1961) *Biochemistry (U.S.S.R.)* [Engl. transl.] **25**, 870. **19.** Taniguchi & Meister (1978) *J. Biol. Chem.* **253**, 1799. **20.** Davis, Balinsky, Harington & Shepherd (1973) *Biochem. J.* **133**, 667. **21.** Gushima, Miya, Murata & Kimura (1983) *J. Appl. Biochem.* **5**, 210. **22.** Dennda & Kula (1986) *J. Biotechnol.* **4**, 143. **23.** Apontoweil & Berends (1975) *Biochim. Biophys. Acta* **399**, 1. **24.** Nishihara, Maeda, Okamoto, *et al.* (1991) *Biochem. Biophys. Res. Commun.* **174**, 580. **25.** Velick (1955) *Meth. Enzymol.* **1**, 401. **26.** Cotgreave, Gerdes, Schuppe-Koistinen & Lind (2002) *Meth. Enzymol.* **348**, 175. **27.** Boyer (1959) *The Enzymes*, 2nd ed., **1**, 511. **28.** Rapkine (1938) *Biochem. J.* **32**, 1729. **29.** Tsai, Adamik, Manganiello & Vaughan (1981) *Biochem. Biophys. Res. Commun.* **100**, 637. **30.** Fornaini, Dachà, Magnani & Stocchi (1982) *Meth. Enzymol.* **90**, 3. **31.** Magnani, Stocchi, Ninfali, Dacha & Fornaini (1980) *Biochim. Biophys. Acta* **615**, 113. **32.** Gao & Leary (2003) *J. Amer. Soc. Mass Spectrom.* **14**, 173. **33.** Liu, Witkovsky & Yang (1983) *J. Neurochem.* **41**, 1694. **34.** Baum, Ding, Siegel, *et al.* (1996) *Biochemistry* **35**, 5847. **35.** Gilbert & Stewart (1981) *J. Biol. Chem.* **256**, 1782. **36.** Dotan & Shechter (1983) *Arch. Biochem. Biophys.* **226**, 401. **37.** Chung, Fried, Williams-Ashman & Jarabak (1987) *Prostaglandins* **33**, 383. **38.** Terada, Nanjo, Shinagawa, *et al.* (1993) *J. Enzyme Inhib.* **7**, 33. **39.** Morrison, Wong & Seltzer (1976) *Biochemistry* **15**, 4228. **40.** Seltzer (1989) in *Coenzymes and Cofactors, Glutathione, Chem. Biochem. Med. Aspects Pt. A* (Dolphin, Poulson & Avromonic, eds.), vol. **3**, p. 733, Wiley, New York. **41.** Kung & Stadtman (1971) *J. Biol. Chem.* **246**, 3378. **42.** Lama, Lama, Mestriner & Mortari (1996) *Braz. J. Genet.* **19**, 243. **43.** Hartshorne & Morales (1965) *Biochemistry* **4**, 18. **44.** Krishnan & Rao (1972) *Arch. Biochem. Biophys.* **149**, 336. **45.** Guarnieri, Caldarera, Muscari, Flamigni & Caldarera (1982) *Ital. J. Biochem.* **31**, 404. **46.** Calvino & Barcia (2002) *J. Food Biochem.* **26**, 103. **47.** Ricci, Nardini, Chiaraluce, Dupre & Cavallini (1986) *Biochim. Biophys. Acta* **870**, 82. **48.** Hou, Chen & Lin (2003) *J. Agric. Food Chem.* **51**, 1706. **49.** Kai, Salway & Hawthorne (1968) *Biochem. J.* **106**, 791. **50.** Milstein (1961) *Biochem. J.* **79**, 591. **51.** Akhtar & Abdel-Latif (1978) *Biochim. Biophys. Acta* **527**, 159. **52.** Dai, Meij, Dhalla & Panagia (1995) *J. Lipid Mediat. Cell Signal.* **11**, 107. **53.** Rao & Clayton (2002) *Biochem. Biophys. Res. Commun.* **293**, 610. **54.** Ballou & Fischer (1986) *The Enzymes*, 3rd ed., **17**, 311. **55.** Nemani & Lee (1993) *Arch. Biochem. Biophys.* **300**, 24. **56.** Lee, Bai, Deguzman, *et al.* (1985) *Adv. Protein Phosphatases* **1**, 123. **57.** Michels, Wedel & Kroth (2005) *Plant Physiol.* **137**, 911. **58.** Lebreton, Graciet & Gontero (2003) *J. Biol. Chem.* **278**, 12078. **59.** Shimazu, Tokutake & Usami (1978) *J. Biol. Chem.* **253**, 7376. **60.** Gratecos, Detwiler, Hurd & Fischer (1977) *Biochemistry* **16**, 4812. **61.** Degl'Innocenti, Caselli, Rosati, *et al.* (1999) *IUBMB Life* **48**, 505. **62.** Sakuma, Fujimoto, Nishida, *et al.* (1992) *Prostaglandins* **43**, 435. **63.** Davis, Moses, Ndubuka & Ortiz (1987) *J. Gen. Microbiol.* **133**, 1453 **64.** Lathi & Raudaskoski (1983) *Folia Microbiol.* **28**, 371. **65.** Van Berkel, Koster & Hulsmann (1973) *Biochim. Biophys. Acta* **293**, 118. **66.** Achstetter, Ehmann & Wolf (1985) *J. Biol. Chem.* **260**, 4585. **67.** Fensome, Rodrigues-Lima, Josephs, Paterson & Katan (2000) *J. Biol. Chem.* **275**, 1128. **68.** Goñi & Alonso (2002) *FEBS Lett.* **531**, 38. **69.** Hopkins, Morgan & Lutwak-Mann (1938) *Biochem. J.* **32**, 1829. **70.** Hopkins & Morgan (1938) *Biochem. J.* **32**, 611. **71.** Schlenk (1951) *The Enzymes*, 1st ed., **2** (part 1), 316. **72.** Ames & Elvehjem (1944) *Proc. Soc. Exptl. Biol. Med.* **57**, 108. **73.** Slater (1949) *Biochem. J.* **45**, 130. **74.** Wider & Tigier (1971) *Enzymologia* **41**, 217. **75.** Zöphel, Kennedy, Beinert & Kroneck (1988) *Arch. Microbiol.* **150**, 72. **76.** Uhteg & Westley (1979) *Arch. Biochem. Biophys.* **195**, 211. **77.** Gomis, Arbos, Sener & Malaisse (1986) *Diabetes Res.* **3**, 115. **78.** Strominger & Mapson (1957) *Biochem. J.* **66**, 567. **79.** Loe, Stewart, Massey, Deeley & Cole (1997) *Mol. Pharm.* **51**, 1034. **80.** Mao, Deeley & Cole (2000) *J. Biol. Chem.* **275**, 34166. **81.** Paul, Breuninger & Kruh (1996) *Biochemistry* **35**, 14003. **82.** Paul, Breuninger, Tew, Shen & Kruh (1996) *Proc. Natl. Acad. Sci. U.S.A.* **93**, 6929. **83.** Willmore & Storey (2005) *FEBS J.* **272**, 3602. **84.** Corrales, Pérez-Mato, Sánchez del Pino, *et al.* (2002) *J. Nutr.* **132**, 2377S. **85.** Frear (1968) *Phytochemistry* **7**, 381. **86.** Praznik, Rammesmayer & Spies (1992) *Carbohydr. Res.* **227**, 171. **87.** Wells & Saxty (1992) *Eur. J. Biochem.* **204**, 249. **88.** Hondorp & Matthews (2004) *PLoS Biol.* **2**, e336. **89.** Konishi & Fujioka (1991) *Arch. Biochem. Biophys.* **289**, 90. **90.** Ogawa, Ishiguro & Fujioka (1983) *Arch. Biochem. Biophys.* **226**, 265. **91.** Hunt & Massey (1992) *J. Biol. Chem.* **267**, 21479. **92.** Saito & Komamine (1978) *Eur. J. Biochem.* **82**, 385. **93.** Lillig, Potamitou, Schwenn, Vlamis-Gardikis & Holmgren (2003) *J. Biol. Chem.* **278**, 22325. **94.** Neufeld, Green, Latterell & Weintraub (1958) *J. Biol. Chem.* **232**, 1093. **95.** Janolino & Swaisgood (1975) *J. Biol. Chem.* **250**, 2532.

Glyceraldehyde

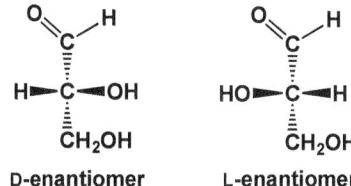

D-enantiomer   L-enantiomer

This aldotriose (FW = 90.08 g/mol; CAS = 56-82-6), of which the D-enantiomer is the form most commonly encountered in Nature, reacts with lysyl residues of many proteins to form imine adducts. **Target(s):** *N*-acetylneuraminate lyase, inhibited by DL-glyceraldehyde (1); aldose 1 epimerase, *or* mutarotase, inhibited by DL-glyceraldehyde (2); *N*-(5-amino-5-carboxypentanoyl)-L-cysteinyl-D-valine synthetase (3); formaldehyde transketolase, also an alternative substrate (4); galactonolactone dehydrogenase (5); glycerone kinase, inhibited by DL-glyceraldehyde, also weak alternative substrate (6); hemoglobin S polymerization, by DL-glyceraldehyde (7-9); phosphoribulokinase, inhibited by DL-glyceraldehyde (10); Cu,Zn-superoxide dismutase (11); transaldolase, inhibited by L-glyceraldehyde (12). **1.** Aisaka, Igarashi, Yamaguchi & Uwajima (1991) *Biochem. J.* **276**, 541. **2.** Hucho & Wallenfels (1971) *Eur. J. Biochem.* **23**, 489. **3.** Zhang & Demain (1992) *Arch. Microbiol.* **158**, 364. **4.** Kato, Higuchi, Sakazawa, *et al.* (1982) *Biochim. Biophys. Acta* **715**, 143. **5.** Mapson & Breslow (1958) *Biochem. J.* **68**, 395. **6.** Garcia-Alles, Siebold, Nyffeler, *et al.* (2004) *Biochemistry* **43**, 13037. **7.** Nigen & Manning (1977) *Proc. Natl. Acad. Sci. U.S.A.* **74**, 367. **8.** Nigen & Manning (1978) *J. Clin. Invest.* **61**, 11. **9.** Kark, Kale, Tarassoff, Woods & Lessin (1978) *J. Clin. Invest.* **62**, 888. **10.** Slabas & Walker (1976) *Biochem. J.* **153**, 613. **11.** Ukeda, Hasegawa, Ishi & Sawamura (1997) *Biosci. Biotechnol. Biochem.* **61**, 2039. **12.** Sprenger, Schörken, Sprenger & Sahm (1995) *J. Bacteriol.* **177**, 5930.

D-Glyceraldehyde 3-Phosphate

This key metabolic intermediate (FW_free-acid = 170.06 g/mol; CAS = 9001-50-7; Water-soluble; pK_a values of 1.42 and 6.45; Decomposes rapidly under basic pH conditions) is the structural isomer of dihydroxyacetone phosphate (*i.e.*, glycerone phosphate). Even neutral or acidic solutions should always be freshly prepared and used within 24 hours. May be stored frozen at −15°C. It is often advisable to generate D-glyceraldehyde 3-phosphate *in situ* from a protected precursor (*e.g.*, D-glyceraldehyde 3-phosphate diethyl acetal is stable at 4°C for at least a year). D-Glyceraldehyde 3-phosphate is a product of the reactions catalyzed by fructose-1,6 bisphosphate aldolase, triosephosphate isomerase, transketolase, transaldolase, formaldehyde transaldolase, triokinase, deoxyribose-phosphate aldolase, phosphoketolase, tryptophan synthase, glycerol-3-phosphate 1-dehydrogenase, and several other aldolases. It is a substrate for glyceraldehyde-3-phosphate dehydrogenases and glyceraldehyde-3-phosphate dehydrogenases (phosphorylating). It is reportedly a regulator of glucose-6-phosphate dehydrogenase (1). **Target(s):** acetyl-CoA synthetase, *or* acetate:CoA ligase, weakly inhibited (3); adenylate kinase (4); *N*-(5-amino-5-carboxypentanoyl)-L-cysteinyl-D-valine synthetase (5); aspartate aminotransferase (2,6); enolase (7); formate *C*-acetyltransferase, *or* pyruvate formate-lyase (18,30-32); D-glucosaminate dehydratase (8); glucose-6-phosphate isomerase (9); glutamine:fructose-6-phosphate aminotransferase (10); glycerol-3-phosphate acyltransferase, by the DL-mixture (11); glycerol-3-phosphate dehydrogenase (12,13); glycerone kinase, inhibited by the racemic mixture (14); methylglyoxal synthase (15); phosphoenolpyruvate phosphatase (16); 6-phosphogluconate dehydrogenase (17); pyruvate,water dikinase (19,20); ribose-phosphate diphosphokinase, *or* phosphoribosyl-pyrophosphate synthetase (21); ribose-5-phosphate isomerase (22,23); ribulose-bisphosphate carboxylase (24-26); ribulose-5-phosphate kinase (27); starch synthase (28); thymidylate synthase (29). **1.** Domagk & Chilla (1975) *Meth. Enzymol.* **41**, 205. **2.** Fitzgerald, Swearengin, Yeargans, *et al.* (2000) *J. Enzyme Inhib.* **15**, 79. **3.** O'Sullivan & Ettlinger (1976) *Biochim. Biophys. Acta* **450**, 410. **4.** McKellar, Charles & Butler (1980) *Arch. Microbiol.* **124**, 275. **5.** Zhang & Demain (1992) *Arch. Microbiol.* **158**, 364. **6.** Dubach & Hageman (1984) *Comp. Biochem. Physiol. B* **78**, 691. **7.** Niemczuk & Wolna (1987) *Biochem. Int.* **15**, 339. **8.** Iwamoto, Taniki, Koishi & Nakura (1995) *Biosci. Biotechnol. Biochem.* **59**, 408. **9.** Backhausen, Jöstingmeyer & Scheibe (1997) *Plant Sci.* **130**, 121. **10.** Kikuchi, Ikeda & Tsuiki (1972) *Biochim. Biophys. Acta* **289**, 303. **11.** Green & Bell (1984) *Biochim. Biophys. Acta* **795**, 348. **12.** Dawson & Thorne (1975) *Meth. Enzymol.* **41**, 254. **13.** Schryvers, Lohmeier & Weiner (1978) *J. Biol. Chem.* **253**, 783. **14.** Garcia-Alles, Siebold, Nyffeler, *et al.* (2004) *Biochemistry* **43**, 13037. **15.** Murata, Fukuda, Watanabe, *et al.* (1985) *Biochem. Biophys. Res. Commun.* **131**, 190. **16.** Duff, Lefebvre & Plaxton (1989) *Plant Physiol.* **90**, 734. **17.** Moritz, Striegel, De Graaf & Sahm (2000) *Eur. J. Biochem.* **267**, 3442. **18.** Takahashi, Abbe & Yamada (1982) *J. Bacteriol.* **149**, 1034. **19.** Cooper & Kornberg (1974) *The Enzymes*, 3rd ed., **10**, 631. **20.** Chulavatnatol & Atkinson (1973) *J. Biol. Chem.* **248**, 2712. **21.** Fox & Kelley (1972) *J. Biol. Chem.* **247**, 2126. **22.** Agosin & Aravena (1960) *Enzymologia* **22**, 281. **23.** Skrukrud, Gordon, Dorwin, *et al.* (1991) *Plant Physiol.* **97**, 730. **24.** Wishnick & Lane (1971) *Meth. Enzymol.* **23**, 570. **25.** Siegel & Lane (1975) *Meth. Enzymol.* **42**, 472.

26. Siegel, Wishnick & Lane (1972) *The Enzymes*, 3rd ed., **6**, 169. **27.** Hart & Gibson (1975) *Meth. Enzymol.* **42**, 115. **28.** Ghosh & Preiss (1965) *Biochemistry* **4**, 1354. **29.** Bures, Daron & Aull (1991) *Int. J. Biochem.* **23**, 733. **30.** Asanuma & Hino (2000) *Appl. Environ. Microbiol.* **66**, 3773. **31.** Asanuma & Hino (2002) *Appl. Environ. Microbiol.* **68**, 3352. **32.** Melchiorsen, Jokumsen, Villadsen, Israelsen & Arnau (2002) *Appl. Microbiol. Biotechnol.* **58**, 338.

.D-Glycerate (D-Glyceric Acid)

This (*R*)-hydroxy acid (FW_free-acid = 106.08 g/mol; CAS = 6000-40-4; pK_a = 3.52), also known as (*R*)-2,3-dihydroxypropanoic acid, is very soluble in water and ethanol. Note that glyceric acid will polymerize and produce an insoluble anhydride upon long standing. **Target(s):** acid phosphatase (1,2); phosphoenolpyruvate carboxykinase (GTP), weakly inhibited (3); phosphonate dehydrogenase, *or* phosphite dehydrogenase (4). **1.** Hollander (1971) *The Enzymes*, 3rd ed., **4**, 449. **2.** Kilsheimer & Axelrod (1957) *J. Biol. Chem.* **227**, 879. **3.** Guidinger & Nowak (1990) *Arch. Biochem. Biophys.* **178**, 131. **4.** Costas, White & Metcalf (2001) *J. Biol. Chem.* **276**, 17429.

DL-Glycerate (DL-Glyceric Acid)

This racemic hydroxy acid (FW_free-acid = 106.08 g/mol; pK_a = 3.55), also known as (*R,S*)-2,3-dihydroxypropanoic acid, is very soluble in water and ethanol. Surprisingly, the racemic mixture is less soluble in water than the pure enantiomer. Note that glyceric acid will polymerize to produce an insoluble anhydride upon long standing. **Target(s):** acid phosphatase (1,2); fructose-1,6-bisphosphatase (3); isocitrate lyase (4); L-lactate dehydrogenase (5); L-lactate dehydrogenase, *or* cytochrome (6); phosphoenolpyruvate carboxykinase (GTP), weakly inhibited (7); phosphoenolpyruvate carboxylase (8); phosphoglycerate phosphatase (9,10); sedoheptulose-bisphosphatase (3). **1.** Hollander (1971) *The Enzymes*, 3rd ed., **4**, 449. **2.** London, McHugh & Hudson (1958) *Arch. Biochem. Biophys.* **73**, 72. **3.** Schimkat, Heineke & Heldt (1990) *Planta* **181**, 97. **4.** McFadden (1969) *Meth. Enzymol.* **13**, 163. **5.** Yoshida & Freese (1975) *Meth. Enzymol.* **41**, 304. **6.** Nygaard (1963) *The Enzymes*, 2nd ed., **7**, 557. **7.** Guidinger & Nowak (1990) *Arch. Biochem. Biophys.* **178**, 131. **8.** Yoshida, Tanaka, Mitsunaga & Izumi (1995) *Biosci. Biotechnol. Biochem.* **59**, 140. **9.** Fallon & Byrne (1965) *Biochim. Biophys. Acta* **105**, 43. **10.** Pestka & Delwiche (1981) *Can. J. Microbiol.* **27**, 808.

Glycerol

This viscous water-soluble polyol (FW = 92.09 g/mol; CAS = 56-81-5; Density = 1.261 g/cm³; Melting Point = 17.8 °C) is a major component of saponifiable fats (mainly as mono-, di-, and tri-fatty acyl esters). Glycerol is often used as a cryoprotectant, stabilizing cells and and many proteins and enzymes. Indeed, it often promotes protein oligomerization and polymerization. For example, the presence of 2-4 M glycerol greatly promotes polymerization of chromatographically pure tubulin, enhancing microtubule self-assembly even in the absence of microtubule-stabilizing proteins like MAP2 and Tau. In glycerol-water mixtures (10-40 vol %), proteins give clear evidence that they are preferentially hydrated in ways that stabilize the native structure of globular proteins (1,2). Glycerol and other vicinal diols also react with propanediol dehydratase, typically resulting in conversion of enzyme-bound adenosylcobalamin to cob(II)alamin and formation of aldehyde or ketone derives from substrate (3,4). All are capable of effecting the irreversible inactivation of the enzyme. Glycerol is a very good diol dehydratase substrate as well as a potent inactivator, but does not induce cob(II)alamin formation to any detectable extent. In the presence of glycerol, diol dehydratase inactivation process accompanied by conversion of enzyme-bound adenosylcobalamin to an alkyl or thiol cobalamin, probably by substitution of an amino acid chain near the active site for the 5'-deoxy-5'-adenosyl ligand on the cobalamin (3). The inactivation reaction with glycerol exhibits a sizable deuterium isotope effect ($k_D/k_H = 14$), suggesting that hydrogen transfer is an important step in the inactivation mechanism. **1.** Gekko & Timasheff (1981) *Biochemistry* **20**, 4667. **2.** Gekko & Timasheff (1981) *Biochemistry* **20**, 4676. **3.** Bachovchin, Eagar, Moore & Richards (1977) *Biochemistry* **16**, 1082. **4.** Bachovchin, Moore & Richards (1978) *Biochemistry* **17**, 2218.

sn-Glycerol 1-Phosphate

This phosphorylated metabolite (FW$_{\text{free-acid}}$ = 172.07 g/mol; CAS = 57-03-4), also called (*S*)-glycerol 1-phosphate, L-glycerol 1-phosphate, D-α glycerophosphate, and D-glycerol 3-phosphate; is a structural isomer of the common component of many phospholipids, one that is only rarely encountered in the free form. Salts of *sn*-glycerol 1-phosphate are quite soluble in water. Solutions are relatively stable at 4°C. While chiral, glycerol-P has a barely measurable optical activity: $[\alpha]_D = -1.45°$ for the free acid (1). (*See also* *sn-Glycerol 3-Phosphate*) **Target(s):** 1-deoxy-D-xylulose-5-phosphate synthase, inhibited by DL-α glycerophosphate (2); fructose-2,6-bisphosphate 2-phosphatase (3); glyceraldehyde-3-phosphate dehydrogenase (4); glycerol-1,2-cyclic-phosphate phosphodiesterase (5); 6-phosphogluconate dehydratase (6); ribulose-phosphate 3-epimerase, inhibited by DL-α-glycerophosphate (7); triose phosphate isomerase (8). **1.** Baer (1952) *Biochem. Prep.* **2**, 31. **2.** Eubanks & Poulter (2003) *Biochemistry* **42**, 1140. **3.** Kurland & Pilkis (1995) *Protein Sci.* **4**, 1023. **4.** Dalziel (1975) *The Enzymes*, 3rd ed., **11**, 1. **5.** Clarke & Dawson (1978) *Biochem. J.* **173**, 579. **6.** Scopes & Griffiths-Smith (1984) *Anal. Biochem.* **136**, 530. 7. Chen, Hartman, Lu & Larimer (1998) *Plant Physiol.* **118**, 199. 8. Noltmann (1972) *The Enzymes*, 3rd ed., **6**, 271.

Glycerol 2-Phosphate

This phosphorylated metabolite (FW$_{\text{free-acid}}$ = 172.07 g/mol; CAS = 17181-54-3), also known as β-glycerophosphate, is a structural isomer of the common component of many phospholipids. Salts of glycerol 2-phosphate are quite soluble in water. **Target(s):** [acetyl-CoA carboxylase] kinase (1,2); 1-deoxy-D-xylulose-5-phosphate synthase (3); glycerol-1,2-phosphate-cyclic-phosphate phosphodiesterase (4); glycogen phosphorylase (5,6); glycogen synthase (7); 3-phosphoglycerate kinase (8); pyridoxamine:oxaloacetate aminotransferase (9); *Ustilago sphaerogena* ribonuclease (10); *Xenopus* ribosomal protein S6 kinase II (11). **1.** Ganzhorn & Chanal (1990) *Biochemistry* **29**, 6065. **2.** van Schaftingen, Davies & Hers (1982) *Eur. J. Biochem.* **124**, 143. **3.** Heesom, Moule & Denton (1998) *FEBS Lett.* **422**, 43. **4.** Heesom, Moule & Denton (1995) *Biochem. Soc. Trans.* **23**, 180S. **5.** Eubanks & Poulter (2003) *Biochemistry* **42**, 1140. **6.** Clarke & Dawson (1978) *Biochem. J.* **173**, 579. 7. Hassid, Doudoroff & Barker (1951) *The Enzymes*, 1st ed., **1** (Part 2), 1014. **8.** Cori, Cori & Green (1943) *J. Biol. Chem.* **151**, 39. **9.** Killilea & Whelan (1976) *Biochemistry* **15**, 1349. **10.** Szilágyi & Vas (1998) *Biochemistry* **37**, 8551. **11.** Wada & Snell (1962) *J. Biol. Chem.* **237**, 127. **12.** Glitz & Dekker (1964) *Biochemistry* **3**, 1391. **13.** Erikson, Maller & Erikson (1991) *Meth. Enzymol.* **200**, 252.

sn-Glycerol 3-Phosphate

This phosphorylated metabolite (FW$_{\text{free-acid}}$ = 172.07 g/mol; CAS = 57-03-4), also called (*R*)-glycerol 1-phosphate, L-glycerol 3-phosphate, and D-glycerol 1-phosphate, is a component of many phospholipids. Salts of *sn*-glycerol 3-phosphate are quite soluble in water. **Target(s):** [acetyl-CoA carboxylase] kinase (1); diglucosyl diacylglycerol synthase (2); fructose-1,6-bisphosphate aldolase, inhibited by DL-α-glycerophosphate (3); α-galactosidase (4); D-glucosaminate dehydratase (5); glucose-1,6-bisphosphate synthase (6); glutamine:fructose-6 phosphate transaminase (7); glyceraldehyde-3-phosphate dehydrogenase (8); glycerol-1,2-cyclic-phosphate phosphodiesterase (9); glycerone-phosphate *O*-acyltransferase, *or* dihydroxyacetone-phosphate *O*-acyltransferase (10,36-39); glycerol-phosphocholine phosphodiesterase (11); glycerophospho-inositol glycerophosphodiesterase (12); NADH peroxidase, inhibited by racemic mixture (13); 6-phosphofructo-2-kinase/ fructose 2,6-bisphosphatase (14-27); 6 phospho-β-glucosidase, inhibited by DL-α-glycerophosphate (28); 3-phosphoglycerate kinase (29); pyridoxamine:oxaloacetate aminotransferase (30); ribulose-phosphate 3-epimerase, inhibited by DL-α-glycerophosphate (31); triose-phosphate isomerase (32-35). **1.** Heesom, Moule & Denton (1998) *FEBS Lett.* **422**, 43. **2.** Vikström, Li & Wieslander (2000) *J. Biol. Chem.* **275**, 9296. **3.** Holden & Storey (1994) *Insect Biochem. Mol. Biol.* **24**, 265. **4.** Dey & Kaus (1981) *Phytochemistry* **20**, 45. **5.** Iwamoto, Taniki, Koishi & Nakura (1995) *Biosci. Biotechnol. Biochem.* **59**, 408. **6.** Maliekal, Sokolova, Vertommen, Veiga-da-Cunha & Van Schaftingen (2007) *J. Biol. Chem.* **282**, 31844. **7.** Kikuchi, Ikeda & Tsuiki (1972) *Biochim. Biophys. Acta* **289**, 303. **8.** Wold & Ballou (1967) *J. Biol. Chem.* **227**, 313. **9.** Clarke & Dawson (1978) *Biochem. J.* **173**, 579. **10.** Schlossman & Bell (1978) *J. Bacteriol.* **133**, 1368. **11.** Anfuso, Sipione, Lupo & Alberghina (1995) *Comp. Biochem. Physiol. B* **112**, 493. **12.** Dawson, Hemington, Richards & Irvine (1979) *Biochem. J.* **182**, 39. **13.** Badwey & Karnovsky (1979) *J. Biol. Chem.* **254**, 11530. **14.** García de Frutos & Baanante (1994) *Arch. Biochem. Biophys.* **308**, 461. **15.** Pilkis, Claus, Kountz & el-Maghrabi (1987) *The Enzymes*, 3rd ed., **18**, 3. **16.** Markham & Kruger (2002) *Eur. J. Biochem.* **269**, 1267. **17.** Villadsen & Nielsen (2001) *Biochem. J.* **359**, 591. **18.** Pilkis, Regen, Stewart, *et al.* (1984) *J. Biol. Chem.* **259**, 949. **19.** Rider (1987) *Biochem. Soc. Trans.* **15**, 988. **20.** Rider, Foret & Hue (1985) *Biochem. J.* **231**, 193. **21.** Larondelle, Mertens, van Schaftingen & Hers (1986) *Eur. J. Biochem.* **161**, 351. **22.** García de Frutos & Baanante (1995) *Arch. Biochem. Biophys.* **321**, 297. **23.** Martín-Sanz, Cascales & Boscá (1992) *Biochem. J.* **281**, 457. **24.** Loiseau, Rider & Hue (1987) *Biochem. Soc. Trans.* **15**, 384. **25.** Kountz, el-Maghrabi & Pilkis (1985) *Arch. Biochem. Biophys.* **238**, 531. **26.** Chevalier, Bertrand, Rider, *et al.* (2005) *FEBS J.* **272**, 3542. **27.** Francois, Van Schaftingen & Hers (1988) *Eur. J. Biochem.* **171**, 599. **28.** Wilson & Fox (1974) *J. Biol. Chem.* **249**, 5586. **29.** Szilágyi & Vas (1998) *Biochemistry* **37**, 8551. **30.** Wada & Snell (1962) *J. Biol. Chem.* **237**, 127. **31.** Chen, Hartman, Lu & Larimer (1998) *Plant Physiol.* **118**, 199. **32.** Krietsch (1975) *Meth. Enzymol.* **41**, 434. **33.** Gracy (1975) *Meth. Enzymol.* **41**, 442. **34.** Hartman & Norton (1975) *Meth. Enzymol.* **41**, 447. **35.** Wolfenden (1969) *Nature* **223**, 704. **36.** Hajra (1968) *J. Biol. Chem.* **243**, 3458. **37.** Jones & Hajra (1980) *J. Biol. Chem.* **255**, 8289. **38.** Schlossman & Bell (1977) *Arch. Biochem. Biophys.* **182**, 737. **39.** Declercq, Haagsman, Van Veldhoven, *et al.* (1984) *J. Biol. Chem.* **259**, 9064.

sn-Glycero-3-phosphocholine

This substituted glycerol (FW = 258.23 g/mol) is a component of many complex lipids, most notably phosphatidylcholine (*i.e.*, the diacylated derivative). Note that choline esters of glycerophosphate also isomerize in acid and alkaline solutions, resulting in partial racemization. **Target(s):** cholinesterase, acetylcholine esterase (1); Lysophospholipase (2); phosphatidylcholine desaturase (3); phosphatidylinositol transfer protein α (4); phosphoserine phosphatase (5); 4-phytase (6). **1.** Krvavica (1950) *Enzymologia* **14**, 39. **2.** Fallbrook, Turenne, Mamalias, Kish & Ross (1999) *Brain Res.* **834**, 207. **3.** Slack, Roughan & Browse (1979) *Biochem. J.* **179**, 649. **4.** Komatsu, Westerman, Snoek, Taraschi & Janes (2003) *Biochim. Biophys. Acta* **1635**, 67. **5.** Hawkinson, Acosta-Burruel, Ta & Wood (1997) *Eur. J. Pharmacol.* **337**, 315. **6.** Maiti, Majumder & Biswas (1974) *Phytochemistry* **13**, 1047.

Glycerophosphoinositol 4-Phosphate

This phosphoinositol metabolite (FW = 416.21 g/mol) inhibits the adenylyl cyclase activity (K_i = 1-4 µM) in FRTL5 membranes, when stimulated by the GTP-binding protein activator fluoroaluminate. Glycerophosphoinositol 4-phosphate exerts an inhibitory effect on the GTP-binding protein that stimulates the cyclase and serves as a Ras cascade messenger. **Target(s):** adenylate cyclase (1); glycerophosphoinositol glycerophospho-diesterase (2); inositol-polyphosphate 5-phosphatase, or inositol-1,4,5-trisphosphate 5-phosphatase (3); phosphoinositide phospholipase C (4). **1.** Iacovelli, Falasca, Valitutti, D'Arcangelo & Corda (1993) J. Biol. Chem. **268**, 20402. **2.** Zheng, Berrie, Corda & Farquhar (2003) Proc. Natl. Acad. Sci. U.S.A. **100**, 1745. **3.** Downes, Mussat & Michell (1982) Biochem. J. **203**, 169. **4.** Cruz-Rivera, Bennett & Crooke (1990) Biochim. Biophys. Acta **1042**, 113.

Glycidol Phosphate

This epoxide-containing glycerol-P/glyceraldehyde-P analogue (FW$_{free-acid}$ = 154.06 g/mol), also known as 1,2-epoxypropanol 3-phosphate, has R and S stereoisomers, and both inhibit enolase (1-3), 3-phosphoglycerate phosphatase (4), and triose-phosphate isomerase (3,5). **Primary Mode of Action:** Although similar concentrations of D- and L-glycidol-P give half-maximal rates of inactivate for triose phosphate isomerase, the maximum V_{inact} values is 10-times higher for D-glycidol-P. Although the same catalytic glutamate residue is esterified by addition to the C1 carbon of both enantiomers, the rate differences agree with inferences about active-site stereospecificity. **1.** Wold (1975) Meth. Enzymol. **41**, 120. **2.** Wold (1971) The Enzymes, 3rd. ed., **5**, 499. **3.** O'Connell & Rose (1977) Meth. Enzymol. **46**, 381. **4.** Randall & Tolbert (1971) J. Biol. Chem. **246**, 5510. **5.** Hartman (1972) Meth. Enzymol. **25**, 661. **6.** Hartman & Norton (1975) Meth. Enzymol. **41**, 447. **7.** Sigman & Mooser (1975) Ann. Rev. Biochem. **44**, 889. **8.** Hartman (1977) Meth. Enzymol. **46**, 130. **9.** Noltmann (1972) The Enzymes, 3rd ed., **6**, 271. **10.** Schray, O'Connell & Rose (1973) J. Biol. Chem. **248**, 2214. **11.** Waley, Miller, Rose & O'Connell (1970) Nature **227**, 181. **12.** Rose & O'Connell (1969) J. Biol. Chem. **244**, 6548.

Glycine

This α-amino acid (FW = 75.07 g/mol; CAS = 56-40-6; Symbols: Gly or G; pK_a values of 2.35 and 9.78 at 25°C), itself the simplest of the twenty proteogenic amino acids, has four codons: GGG, GGU, GGC, and GGA. Glycine residues are frequently found in reverse turns. They also tend to destabilize α-helical structures. Homologous proteins often have glycyl residues interchanged with alanyl, seryl, aspartyl, or asparaginyl residues. Because glycine is a cheap and widely used zwitterionic buffer, it should not be surprising that numerous papers report enzyme inhibition by glycine in vitro. Most studies are conducted at 50 mM glycine, and the inhibition is of dubious significance. **Target(s):** acetylcholinesterase, weakly inhibited (1); N^4-(β-N-acetylglucosaminyl)-L-asparaginase (2); agaritine γ-glutamyl-transferase (3); alanine aminotransferase, weakly inhibited) (4); L-alanine dehydrogenase (5); D-alanyl-D-alanine synthase, or D-alanine:D-alanine ligase (6); alkaline phosphatase (7-12); 2-aminohexano-6-lactam racemase (13); aminomalonate decarboxylase, now: aspartate β-decarboxylase (14); 8-amino-7-oxononanoate synthase (15-17); 5-aminovalerate aminotransferase (18); Arginase, weakly inhibited (19); asparaginase (20); asparagine:oxo-acid aminotransferase (21,22); asparagine synthetase,

weakly inhibited (23); bile acid acyl glucuronosyltransferase (24); carbamoyl-phosphate synthetase, glutamine-hydrolyzing (25); carbonic anhydrase (26,27); cystalysin (28); cystathionine β-lyase (29); cystathionine γ lyase, or cysteine desulfhydrase (28,30); 2-dehydro-3-deoxyglucarate aldolase, or 2-keto-3 deoxyglucarate aldolase (31); 2,2-dialkylglycine decarboxylase (pyruvate) (32); diphosphate:serine phosphotransferase, or pyrophosphate:serine phosphotransferase, weakly inhibited (33); formyltetrahydrofolate deformylase (34); fumarase (35); gelatinase A, matrix metalloproteinase 2 (36); glutamine synthetase (23,38-52); γ-glutamyl transpeptidase, or γ-glutamyltransferase (53-56); D-glycerate dehydrogenase (57); glycerol-2-phosphatase (58); polygalacturonase (59); haloacetate dehalogenase, weakly inhibited (60); histidine ammonia-lyase (61-65); formyltetrahydrofolate hydrolase (66,67); linoleate diol synthase (68); nucleotide diphosphatase (69); ornithine carbamoyltransferase (70); ornithine decarboxylase (71); pancreatic ribonuclease (72); pantothenase (73); phenylalanine ammonia-lyase, greater inhibition in the presence of phenol, o-cresol, or m cresol (74,75); phosphodiesterase I, 5'-exonuclease (76,77); 3-phosphoglycerate dehydrogenase (57); phosphoglycolate phosphatase (78); phosphoserine phosphatase (79-81); procollagen C endopeptidase, weakly inhibited (82); serine O-acetyltransferase (83-86); D-serine ammonia lyase (87,88); L-serine ammonia-lyase (89,90); serine hydroxymethyltransferase (91); tryptophan synthase (92); tyrosine decarboxylase, weakly inhibited (93); UDP-N-acetylmuramate:L-alanine ligase (6, 94). **1.** Bergmann, Wilson & Nachmansohn (1950) J. Biol. Chem. **186**, 693. **2.** Qian, Guan & Guo (2003) Structure **11**, 997. **3.** Gigliotti & Levenberg (1964) J. Biol. Chem. **239**, 2274. **4.** Saier & Jenkins (1967) J. Biol. Chem. **242**, 101. **5.** Yoshida & Freese (1965) Biochim. Biophys. Acta **96**, 248. **6.** Lugtenberg (1972) J. Bacteriol. **110**, 26. **7.** Massart & Vandendriessche (1945) Enzymologia **11**, 265. **8.** Bodansky (1946) J. Biol. Chem. **165**, 605. **9.** Massart (1950) The Enzymes, 1st ed., **1** (part 1), 307. **10.** Williams & Watson (1940) J. Biol. Chem. **135**, 337. **11.** Bodansky & Strachman (1948) J. Biol. Chem. **174**, 465. **12.** Bodansky & Strachman (1949) J. Biol. Chem. **179**, 81. **13.** Ahmed, Esaki, Tanaka & Soda (1983) Agric. Biol. Chem. **47**, 1887. **14.** Thanassi & Fruton (1962) Biochemistry **1**, 975. **15.** Izumi, Tani & Ogata (1979) Meth. Enzymol. **62**, 326. **16.** Der Garabedian (1986) Biochemistry **25**, 5507. **17.** Izumi, Morita, Tani & Ogata (1973) Agric. Biol. Chem. **37**, 1327. **18.** Izumi, Sato, Tani & Ogata (1973) Agric. Biol. Chem. **37**, 1335. **19.** Greenberg (1951) The Enzymes, 1st ed., **1** (part 2), 893. **20.** Chang & Farnden (1981) Arch. Biochem. Biophys. **208**, 49. **21.** Maul & Schuster (1986) Arch. Biochem. Biophys. **251**, 585. **22.** Cooper (1977) J. Biol. Chem. **252**, 2032. **23.** Schou, Grossowicz & Waelsch (1951) J. Biol. Chem. **192**, 187. **24.** Mano, Nishimura, Narui, Ikegawa & Goto (2002) Steroids **67**, 257. **25.** Mori & Tatibana (1978) Meth. Enzymol. **51**, 111. **26.** Bertini, Luchinat & Scozzafava (1977) Bioinorg. Chem. **7**, 225. **27.** Pocker & Stone (1968) Biochemistry **7**, 2936. **34.** Chu, Ebersole, Kurzban & Holt (1999) Clin. Infect. Dis. **28**, 442. **35.** Burnell & Whatley (1977) Biochim. Biophys. Acta **481**, 246. **36.** Fromageot & Grand (1942) Enzymologia **11**, 81. **32.** Wood (1972) The Enzymes, 3rd ed., **7**, 281. **33.** Sun, Zabinski & Toney (1998) Biochemistry **37**, 3865. **34.** Cagen & Friedmann (1972) J. Biol. Chem. **247**, 3382. **35.** Nagy, Marolewski, Benkovic & Zalkin (1995) J. Bacteriol. **177**, 1292. **36.** Massey (1953) Biochem. J. **55**, 172. **37.** Okada, Morodomi, Enghild, et al. (1990) Eur. J. Biochem. **194**, 721. **38.** Meister (1985) Meth. Enzymol. **113**, 185. **39.** Rhee, Chock & Stadtman (1985) Meth. Enzymol. **113**, 213. **40.** Shapiro & Stadtman (1970) Meth. Enzymol. **17A**, 910. **41.** Meister (1974) The Enzymes, 3rd ed., **10**, 699. **42.** Stadtman & Ginsburg (1974) The Enzymes, 3rd ed., **10**, 755. **43.** Orr & Haselkorn (1981) J. Biol. Chem. **256**, 13099. **44.** Dahlquist & Purich (1975) Biochemistry **14**, 1980. **45.** Kapoor & Bray (1968) Biochemistry **7**, 3583. **46.** Tate & Meister (1971) Proc. Natl. Acad. Sci. U.S.A. **68**, 781. **47.** Southern, Parker & Woods (1987) J. Gen. Microbiol. **133**, 2437. **48.** Hatanaka, Tachiki, Furukawa & Tochikura (1987) Agric. Biol. Chem. **51**, 425. **49.** Blanco, Alana, Llama & Serra (1989) J. Bacteriol. **171**, 1158. **50.** Krishnan, Singhal & Dua (1986) Biochemistry **25**, 1589. **51.** Ertan (1992) Arch. Microbiol. **158**, 35. **52.** Liaw, Pan & Eisenberg (1993) Proc. Natl. Acad. Sci. U.S.A. **90**, 4996. **54.** Thompson & Meister (1977) J. Biol. Chem. **252**, 6792. **55.** Nakayama, Kumagai & Tochikura (1984) J. Bacteriol. **160**, 341. **56.** Kimm, Kim & Park (1986) Korean J. Biochem. **18**, 135. **57.** Uhr & Sneddon (1971) FEBS Lett. **17**, 137. **58.** Morton (1955) Meth. Enzymol. **2**, 533. **59.** Kertesz (1955) Meth. Enzymol. **1**, 158. **60.** Goldman (1965) J. Biol. Chem. **240**, 3434. **61.** Rechler & Tabor (1971) Meth. Enzymol. **17B**, 63. **62.** Brand & Harper (1976) Biochemistry **15**, 1814. **63.** Hanson & Havir (1972) The Enzymes, 3rd ed., **7**, 75. **64.** Peterkofsky & Mehler (1963) Biochim. Biophys. Acta **73**, 159. **65.** Mehler & Tabor (1953) J. Biol. Chem.

201, 775. **66.** Zalkin (1997) *Meth. Enzymol.* **281**, 214. **67.** Nagy, Marolewski, Benkovic & Zalkin (1995) *J. Bacteriol.* **177**, 1292. **68.** Su, Brodowsky & Oliw (1995) *Lipids* **30**, 43. **69.** Yano, Funakoshi & Yamashina (1985) *J. Biochem.* **98**, 1097. **70.** Legrain & Stalon (1976) *Eur. J. Biochem.* **63**, 289. **71.** Morley & Ho (1976) *Biochim. Biophys. Acta* **438**, 551. **72.** Scheraga & Rupley (1962) *Adv. Enzymol.* **24**, 161. **73.** Airas (1983) *Anal. Biochem.* **134**, 122. **74.** Parkhurst & Hodgins (1972) *Arch. Biochem. Biophys.* **152**, 597. **75.** Alunni, Cipiciani, Fioroni & Ottavi (2003) *Arch. Biochem. Biophys.* **412**, 170. **76.** Yano, Funakoshi & Yamashina (1985) *J. Biochem.* **98**, 1097. **77.** López-Gómez, Costas, Meireles Ribeiro, *et al.* (1998) *FEBS Lett.* **421**, 77. **78.** Verin-Vergeau, Baldy & Cavalie (1980) *Phytochemistry* **19**, 763. **79.** Schmidt & Laskowski (1961) *The Enzymes*, 2nd ed., **5**, 3. **80.** Byrne (1961) *The Enzymes*, 2nd ed., **5**, 73. **81.** Bridgers (1967) *J. Biol. Chem.* **242**, 2080. **82.** Hojima, van der Rest & Prockop (1985) *J. Biol. Chem.* **260**, 15996. **83.** Hindson (2003) *Biochem. J.* **375**, 745. **84.** Hindson & Shaw (2003) *Biochemistry* **42**, 3113. **85.** Johnson, Huang, Roderick & Cook (2004) *Arch. Biochem. Biophys.* **429**, 115. **86.** Leu & Cook (1994) *Biochemistry* **33**, 2667. **87.** Federiuk & Shafer (1981) *J. Biol. Chem.* **256**, 7416. **88.** Schnackerz, Tai, Potsch & Cook (1999) *J. Biol. Chem.* **274**, 36935. **89.** Grabowski & Buckel (1991) *Eur. J. Biochem.* **199**, 89. **90.** Kubota, Yokozeki & Ozaki (1989) *J. Ferment. Bioeng.* **67**, 391. **91.** Schirch, Tatum & Benkovic (1977) *Biochemistry* **16**, 410. **92.** Ro & Miles (1999) *J. Biol. Chem.* **274**, 31189. **93.** Pamuk (1989) *Commun. Fac. Sci. Univ. Ank. Ser. C* **35**, 103. **94.** Liger, Blanot & van Heijenoort (1991) *FEMS Microbiol. Lett.* **80**, 111.

Glycochenodeoxycholate

This bile acid conjugate (FW$_{\text{free-acid}}$ = 449.63 g/mol; CAS = 640-79-9), also known as chenodeoxycholylglycine, is an important bile constituent. **Target(s):** alkaline phosphatase (1); carbonic anhydrase (2); β glucuronidase (3); γ-glutamyl transpeptidase (4); pepsin (5). **1.** Martins, Negrao, Hipolito-Reis & Azevedo (2000) *Clin. Biochem.* **33**, 611. **2.** Salomoni, Zuccato, Granelli, *et al.* (1989) *Scand. J. Gastroenterol.* **24**, 28. **3.** Ho (1985) *Biochim. Biophys. Acta* **827**, 197. **4.** Abbott & Meister (1983) *J. Biol. Chem.* **258**, 6193. **5.** Eto & Tompkins (1985) *Amer. J. Surg.* **150**, 564.

Glycocholate (Glycocholic Acid)

This bile acid conjugate (FW$_{\text{free-acid}}$ = 465.63 g/mol; FW$_{\text{sodium-salt}}$ = 523.64 g/mol (monohydrate); CAS = 207614-05-9), also known as cholylglycine, is the main bile constituent of herbivorous mammals. Its critical micelle

concentration (cmc) is 7.1 mM. **Target(s):** acetylcholinesterase, moderately inhibited (1); 1-acylglycerophosphocholine *O*-acyltransferase, *or* lysophosphatidylcholine *O*-acyltransferase (2); alkaline phosphatase (3); ATPase, Mg^{2+}-dependent (4); bile-acid-CoA:amino acid *N*-acyltransferase (5,6); carbonic anhydrase (7); β-carotene,15'-monooxygenase (8); esterase (9); γ-glutamyl transpeptidase (10,11); lipase, *or* triacylglycerol lipase (12,13); lysozyme (14); Na$^+$/K$^+$-exchanging ATPase (4); steryl-sulfatase, *or* arylsulfatase C, mildly inhibited (15). **1.** Sobotka & Antopol (1937) *Enzymologia* **4**, 189. **2.** Weltzien, Richter & Ferber (1979) *J. Biol. Chem.* **254**, 3652. **3.** Martins, Negrao, Hipolito-Reis & Azevedo (2000) *Clin. Biochem.* **33**, 611. **4.** Parkinson & Olson (1964) *Life Sci.* **3**, 107. **5.** Kimura, Okuno, Inada, Ohyama & Kido (1983) *Hoppe-Seyler's Z. Physiol. Chem.* **364**, 637. **6.** Czuba & Vessey (1980) *J. Biol. Chem.* **255**, 5296. **7.** Milov, Jou, Shireman & P. W. Chun (1992) *Hepatology* **15**, 288. **8.** Yang & Tume (1993) *Biochem. Mol. Biol. Int.* **30**, 209. **9.** Glick & King (1932) *J. Biol. Chem.* **97**, 675. **10.** Gardell & Tate (1983) *J. Biol. Chem.* **258**, 6198. **11.** Abbott & Meister (1983) *J. Biol. Chem.* **258**, 6193. **12.** Desnuelle (1972) *The Enzymes*, 3rd ed., **7**, 575. **13.** Schousboe (1976) *Biochim. Biophys. Acta* **450**, 165. **14.** Vantrappen, Ghoos & Peeters (1976) *Amer. J. Dig. Dis.* **21**, 547. **15.** Gniot-Szulzycka & Komoszynski (1972) *Enzymologia* **42**, 11.

Glycocyamine

This substituted guanidine (FW = 117.11 g/mol; CAS = 352-97-6), also known as *N*-amidinoglycine and guanidinoacetic acid, is very soluble in water. The *N*-phosphorylated derivative is a phosphagen in polychaete worms. **Target(s):** arginase (1); arginine decarboxylase (2,3); arginine deaminase (4); glycine amidinotransferase, product inhibition (5); 5-guanidino-2-oxopentanoate decarboxylase, *or* α-ketoarginine decarboxylase, weakly inhibited (6); Na$^+$/K$^+$-exchanging ATPase, possibly as a consequence of changes to membrane (7,8). **1.** Patchett, Daniel & Morgan (1991) *Biochim. Biophys. Acta* **1077**, 291. **2.** Boeker & Snell (1971) *Meth. Enzymol.* **17B**, 657. **3.** Blethen, Boeker & Snell (1968) *J. Biol. Chem.* **243**, 1671. **4.** Petrack, Sullivan & Ratner (1957) *Arch. Biochem. Biophys.* **69**, 186. **5.** Ratner & Rochovansky (1956) *Arch. Biochem. Biophys.* **63**, 296. **6.** Vanderbilt, Gaby, Rodwell & Bahler (1975) *J. Biol. Chem.* **250**, 5322. **7.** Zugno, Franzon, Chiarani, *et al.* (2004) *Int. J. Dev. Neurosci.* **22**, 191. **8.** Zugno, Stefanello, Streck, *et al.* (2003) *Int. J. Dev. Neurosci.* **21**, 183.

Glycodeoxycholate (Glycodeoxycholic Acid)

This conjugated bile salt (FW$_{\text{free-acid}}$ = 449.63 g/mol; CAS = 360-65-6; c.m.c. = 2.1 mM), also known as deoxy-cholylglycine, induces apoptosis in hepatocytes. **Target(s):** bile-acid-CoA:amino acid *N*-acyltransferase (1); β-glucuronidase (2); γ-glutamyl transpeptidase (3); lipase, *or* triacylglycerol lipase (4). **1.** Kimura, Okuno, Inada, Ohyama & Kido (1983) *Hoppe-Seyler's Z. Physiol. Chem.* **364**, 637. **2.** Ho (1985) *Biochim. Biophys. Acta* **827**, 197. **3.** Abbott & A. Meister (1983) *J. Biol. Chem.* **258**, 6193. **4.** Desnuelle (1972) *The Enzymes*, 3rd ed., **7**, 575.

Glycogen

This major storage polysaccharide (MW = 0.3-4 x 10^6 Da; CAS 9005-79-2) is similar in structure to amylopectin, *i.e.*, it is a polymer of D-glucopyranose with α-(1→4) linkages with α-(1→6) branches. (Branch points are more frequent in glycogen, as compared to amylopectin.) Because glycogen formation is localized within the cytosol of many cells, it is likely that glycogen particles trap (incorporate) cytosolic enzymes and may even alter their catalytic activity. In mammalian tissue, yeast, or bacteria, glycogen synthesis is initiated by protein autoglucosylation. In muscle, for example, this process is accomplished by Glycogenin-1. Anchorage of glycogenin most likely determines the subcellular location of glycogen. **Target(s):** casein kinase II; dual specificity protein phosphatase (2); [glycogen synthase-D] phosphatase (3,4); phosphoprotein phosphatase (5); [phosphorylase] phosphatase (6). **1.** Roach (1984) *Meth. Enzymol.* **107**, 81. **2.** Wang & Roach (2004) *Biochem. Biophys. Res. Commun.* **325**, 726. **3.** Stalmans & Hers (1973) *The Enzymes*, 3rd ed., **9**, 309. **4.** Thomas & Nakai (1973) *J. Biol. Chem.* **248**, 2208. **5.** Nakai & Thomas (1974) *J. Biol. Chem.* **249**, 6459. **6.** Wang (1999) *Biochem. J.* **341**, 545.

Glycolaldehyde

Monomer   **Dimer**

This hydroxyaldehyde $HOCH_2CHO$ (FW = 60.05 g/mol; CAS = 141-46-8), regarded by many as the simplest aldose, has a melting point of 97°C and is very soluble in water and hot ethanol. Glycolaldehyde dimer, also called 1,4-dioxane-2,5-diol and 2,5-dihydroxy-1,4-dioxane (FW = 120.11 g/mol), is commercially available. This dimer is actually a mixture of stereoisomers and melts between 80 and 90°C, depending on its actual diastereoisomeric composition. Freshly prepared aqueous solutions contain both the glycolaldehyde monomer and dimer; however, the solute becomes almost entirely monomeric after standing for 24 hours. **Target(s):** ethanolamine ammonia-lyase (1); glutamine:fructose-6-phosphate amino-transferase, isomerizing (2); glyceraldehyde-3-phosphate dehydrogenase (3); 4-hydroxy-2-oxoglutarate aldolase (4); propanediol dehydratase, inhibited by the hydrate (1,5,6); 3-propylmalate synthase (7); Cu,Zn superoxide dismutase (8,9); transketolase, as mechanism-based irreversible inhibitor (10). **1.** Abend, Bandarian, Reed & Frey (2000) *Biochemistry* **39**, 6250. **2.** Kikuchi, Ikeda & Tsuiki (1972) *Biochim. Biophys. Acta* **289**, 303. **3.** Morgan, Dean & Davies (2002) *Arch. Biochem. Biophys.* **403**, 259. **4.** Lane, Shapley & Dekker (1971) *Biochemistry* **10**, 1353. **5.** Abeles (1966) *Meth. Enzymol.* **9**, 686. **6.** Toraya & Fukui (1982) in B_{12} (Dolphin, ed.), vol. **2**, p. 233, Wiley, New York. **7.** Imai, Reeves & Ajl (1963) *J. Biol. Chem.* **238**, 3193. **8.** Ukeda, Hasegawa, Ishi & Sawamura (1997) *Biosci. Biotechnol. Biochem.* **61**, 2039. **9.** Ukeda, Hasegawa, Harada & Sawamura (2002) *Biosci. Biotechnol. Biochem.* **66**, 36. **10.** Chen, Baganz & Woodley (2007) *Chem. Eng. Sci.* **62**, 3178.

Glycolate (Glycolic Acid)

This hydroxy alkanoic acid (FW$_{\text{free-acid}}$ = 76.05 g/mol; CAS = 79-14-1; M.P. = 80° C; pK_a = 3.83; Soluble in water and ethanol), also known as hydroxyacetic acid, is a photosynthetic intermediate. **Target(s):** allantoate deaminase (1); allantoicase (2-4); creatine kinase, weakly inhibited (5); cystathionine γ-lyase, *or* cysteine desulfhydrase (6); glycerate kinase (7); isocitrate lyase (8-16); L-lactate dehydrogenase (17-19); malate synthase (20-29); malyl-CoA lyase (30); mandelonitrile lyase (31); pantothenate synthetase, *or* pantoate:β-alanine ligase, moderately inhibited (32,33); phosphoglycolate phosphatase (34-36); 3-propylmalate synthase (37); pyruvate kinase (38); succinate dehydrogenase, weakly inhibited (39); D-(–)-tartrate dehydratase, weakly inhibited (40); tartronic semialdehyde reductase, *or* 2-hydroxy-3-oxopropionate reductase (41,42); triose-phosphate isomerase (43); tyrosinase, *or* monophenol monooxygenase (44). **1.** Xu, Zhou & Huang (2004) *Zhi Wu Sheng Li Yu Fen Zi Sheng Wu Xue Xue Bao* **30**, 460. **2.** van der Drift & Vogels (1970) *Biochim. Biophys. Acta* **198**, 339. 3. Piedras, Munoz, Aguilar & Pineda (2000) *Arch. Biochem.*

Biophys. **378**, 340. **4.** S-Gravenmade, van der Drift & Vogels (1971) *Biochim. Biophys. Acta* **251**, 393. **5.** Watts (1973) *The Enzymes*, 3rd ed., **8**, 383. **6.** Fromageot & Grand (1942) *Enzymologia* **11**, 81. **7.** Saharan & Singh (1993) *Plant Physiol. Biochem.* **31**, 559. **8.** McFadden (1969) *Meth. Enzymol.* **13**, 163. **9.** Spector (1972) *The Enzymes*, 3rd ed., 7, 357. **10.** Nakamura, Amano, Nakadate & Kagami (1989) *J. Ferment. Bioeng.* **67**, 153. **11.** MacKintosh & Nimmo (1988) *Biochem. J.* **250**, 25. **12.** McFadden & Howes (1963) *J. Biol. Chem.* **238**, 1737. **13.** Hoyt, Robertson, Berlyn & Reeves (1988) *Biochim. Biophys. Acta* **966**, 30. **14.** Tanaka, Yoshida, Watanabe, Izumi & Mitsunaga (1997) *Eur. J. Biochem.* **249**, 820. **15.** Hoyt, Johnson & Reeves (1991) *J. Bacteriol.* **173**, 6844. **16.** Giachetti, Pinzauti, Bonaccorsi, Vincenzini & Vanni (1987) *Phytochemistry* **26**, 2439. **17.** Yoshida & Freese (1975) *Meth. Enzymol.* **41**, 304. **18.** Schwert & Winer (1963) *The Enzymes*, 2nd ed., 7, 142. **19.** Quastel & Woolbridge (1928) *Biochem. J.* **22**, 689. **20.** Dixon & Kornberg (1962) *Meth. Enzymol.* **5**, 633. **21.** Higgins, Kornblatt & Rudney (1972) *The Enzymes*, 3rd ed., 7, 407. **22.** Dixon, Kornberg & Lund (1960) *Biochim. Biophys. Acta* **41**, 217. **23.** Miernyk & Trelease (1981) *Phytochemistry* **20**, 2657. **24.** Munir, Hattori & Shimada (2002) *Biosci. Biotechnol. Biochem.* **66**, 576. **25.** Reinscheid, Eikmanns & Sahm (1994) *Microbiology* **140**, 3099. **26.** Smith, Huang, Miczak, *et al.* (2003) *J. Biol. Chem.* **278**, 1735. **27.** Beeckmans, Khan, Kanarek & Van Driessche (1994) *Biochem. J.* **303**, 413. **28.** Sundaram, Chell & Wilkinson (1980) *Arch. Biochem. Biophys.* **199**, 515. **29.** Durchschlag, Biedermann & Eggerer (1981) *Eur. J. Biochem.* **114**, 255. **30.** Hersh (1974) *J. Biol. Chem.* **249**, 5208. **31.** Jorns (1980) *Biochim. Biophys. Acta* **613**, 203. **32.** Miyatake, Nakano & Kitaoka (1979) *Meth. Enzymol.* **62**, 215. **33.** Miyatake, Nakano & Kitaoka (1978) *J. Nutr. Sci. Vitaminol.* **24**, 243. **34.** Verin-Vergeau, Baldy & Cavalie (1980) *Phytochemistry* **19**, 763. **35.** Rose, Grove & Seal (1986) *J. Biol. Chem.* **261**, 10996. **36.** Christeller & Tolbert (1978) *J. Biol. Chem.* **253**, 1791. **37.** Imai, Reeves & Ajl (1963) *J. Biol. Chem.* **238**, 3193. **38.** Fujii, Sato & Kaneko (1988) *Blood* **72**, 1097. **39.** Hatefi & Stiggall (1976) *The Enzymes*, 3rd ed., **13**, 175. **40.** Rode & Giffhorn (1982) *J. Bacteriol.* **150**, 1061. **41.** Kornberg & Gotto (1966) *Meth. Enzymol.* **9**, 240. **42.** Gotto & Kornberg (1961) *Biochem. J.* **81**, 273. **43.** Noltmann (1972) *The Enzymes*, 3rd ed., **6**, 271. **44.** Parvez, Kang, Chung & H. Bae (2007) *Phytother. Res.* **21**, 805.

Glycolithocholate (Glycolithocholic Acid)

This steroid (FW$_{\text{free-acid}}$ = 433.63 g/mol; CAS = 24404-83-9), also known as *N*-[(3α,5β)-3-hydroxycholan-24-oyl]glycine and lithocholylglycine, is one of the bile acid conjugates identified in the bile of man, bovine, rabbit, sheep, goat, and pig. **Target(s):** *N*-acetyllactosaminide α-2,3-sialyltransferase, IC$_{50}$ = 25 μM (1); glycochenodeoxycholate sulfotransferase (2). **1.** Chang, Lee, Chen & Li (2006) *Chem. Commun. (Cambridge)* **2006**, 629. **2.** Barnes, Burhol, Zander, *et al.* (1979) *J. Lipid Res.* **20**, 952.

Glycyl-(L-prolyl-)₅-glycyl-(L-prolyl-)₅-glycyl-(L-prolyl-)₄-L-proline

This octadecapeptide (FW = 1645.92 g/mol; *Sequence*: GPPPPP-GPPPPPGPPPPP) forms a one-to-one complex with profilin. an actin monomer-binding regulatory profilin (K_d = 84 μM), displacing the latter from its filament (+)-end-tracking proteins like vasodilator-stimulated phosphoprotein (VASP) and thereby inhibiting actin-based motility (*i.e.,* filopodial extension, lamellipodial protrusion, cell crawling, and intracellular motility of pathogen like *Listeria monocytogenes*, *Shigella flexneri*, *Vaccinia*, and *Rickettsia*). This peptide weakly inhibits exchange of actin-bound nucleotide in the absence or presence of profilin. **Target(s):** *Listeria* motility; profilin binding to vasodilator-stimulated phosphoprotein (VASP) and neural-Wiskott-Aldrich Syndrome protein (N-WASP). **1.** Kang, Laine, Bubb, Southwick & Purich (1997) *Biochemistry* **36**, 8384.

18β-Glycyrrhetinic Acid

This sweet, pentacyclic triterpene ($FW_{free-acid}$ = 470.69 g/mol), also known as enoxolone, is the active pharmacological ingredient of licorice (*Glycyrrhiza glabra*). It is the aglycon of glycyrrhizin and a number of saponins. **Target(s):** cAMP-dependent protein kinase (1); cAMP phosphodiesterase (2); complement component C2 (3); β-glucuronidase (4); glycogen phosphorylase (5); 3α-hydroxysteroid dehydrogenase (6); 3β-hydroxysteroid dehydrogenase (7); 3α/20β-hydroxysteroid dehydrogenase (8,9); 11β-hydroxysteroid dehydrogenase (7,10-14); 17β-hydroxysteroid dehydrogenase (15); Na⁺/K⁺-exchanging ATPase (16); protein kinase C (17); steroid 5β-reductase (10,18). **1.** Wang & Polya (1996) *Phytochemistry* **41**, 55. **2.** Birnbaum, Sapp & Tolman (1976) *J. Invest. Dermatol.* **67**, 235. **3.** Kroes, Beukelman, van den Berg, *et al.* (1997) *Immunology* **90**, 115. **4.** Shim, Kim & Kim (2000) *Planta Med.* **66**, 40. **5.** Wen, Sun, Liu, *et al.* (2008) *J. Med. Chem.* **51**, 3540. **6.** Akao, Akao, Hattori, Namba & Kobashi (1992) *Chem. Pharm. Bull. (Tokyo)* **40**, 1208. **7.** Akao, Terasawa, Hiai & Kobashi (1992) *Chem. Pharm. Bull. (Tokyo)* **40**, 3021. **8.** Ghosh, Erman, Pangborn, Duax & Baker (1992) *J. Steroid Biochem. Mol. Biol.* **42**, 849. **9.** Itoda, Takase & Nakajin (2002) *Biol. Pharm. Bull.* **25**, 1220. **10.** Latif, Conca & Morris (1990) *Steroids* **55**, 52. **11.** Monder, Stewart, Lakshmi, *et al.* (1989) *Endocrinology* **125**, 1046. **12.** Monder, Lakshmi & Miroff (1991) *Biochim. Biophys. Acta* **1115**, 23. **13.** Irie, Fukui, Negishi, Nagata & Ichikawa (1992) *Biochim. Biophys. Acta* **1160**, 229. **14.** Marandici & Monder (1993) *Steroids* **58**, 153. **15.** Krazeisen, Breitling, Moller & Adamski (2001) *Mol. Cell. Endocrinol.* **171**, 151. **16.** Itoh, Hara, Shiraishi, *et al.* (1989) *Biochem. Int.* **18**, 81. **17.** O'Brian, Ward & Vogel (1990) *Cancer Lett.* **49**, 9. **18.** Yoshida, Kuroki, Kobayashi & Tamaoki (1992) *J. Steroid Biochem. Mol. Biol.* **41**, 29.

Glycyrrhizin

This intensely sweet glycoside ($FW_{free-acid}$ = 822.94 g/mol; CAS = 1405-86-3), also known as glycyrrhizic acid, from licorice (*Glycyrrhiza glabra*) is amphiphilic and inhibits many enzymes. **Target(s):** arylamine *N*-acetyltransferase (1,2); β-glucuronidase (3-6); glycogen phosphorylase (7); HIV-1 reverse transcriptase (8,9); hyaluronoglucaminidase, *or* hyaluronidase (10); 3α/20β-hydroxysteroid dehydrogenase (11); 11β-hydroxysteroid dehydrogenase (12); Na⁺/K⁺-exchanging ATPase (13); phospholipase A₂ (14,15); polypeptide-dependent protein kinase, *or* protein kinase P, *or* casein kinase II (16-18); RNA-directed DNA polymerase, *or* reverse transcriptase (8,9); thrombin (19). GL also exerts a protective effect on ischemia-reperfusion injury in rat brains through the inhibition of inflammation, oxidative stress and apoptotic injury by antagonising the cytokine activity of High Mobility Group Box 1, *or* HMGB1 (20). **1.** Lo, Yen, Hsieh & Chung (1997) *J. Appl. Toxicol.* **17**, 385. **2.** Makarova (2008) *Curr. Drug Metab.* **9**, 538. **3.** Shim, Kim & Kim (2000) *Planta Med.* **66**, 40. **4.** Narita, Nagai, Hagiwara, *et al.* (1993) *Xenobiotica* **23**, 5. **5.** Kim,

Jin, Jung, Han & Kobashi (1995) *Biol. Pharm. Bull.* **18**, 1184. **6.** Sperker, Mürdter, Schick, *et al.* (1997) *J. Pharmacol. Exp. Ther.* **281**, 914. **7.** Wen, Sun, Liu, *et al.* (2008) *J. Med. Chem.* **51**, 3540. **8.** Nakashima, Matsui, Yoshida, *et al.* (1987) *Jpn. J. Cancer Res.* **78**, 767. **9.** Harada, Maekawa, Haneda, *et al.* (1998) *Biol. Pharm. Bull.* **21**, 1282. **10.** Furuya, Yamagata, Shimoyama, *et al.* (1997) *Biol. Pharm. Bull.* **20**, 973. **11.** Ghosh, Erman, Pangborn, Duax & Baker (1992) *J. Steroid Biochem. Mol. Biol.* **42**, 849. **12.** Idrus, Mohamad, Morat, Saim & Abdul Kadir (1996) *Steroids* **61**, 448. **13.** Itoh, Hara, Shiraishi, *et al.* (1989) *Biochem. Int.* **18**, 81. **14.** Okimasu, Moromizato, Watanabe, *et al.* (1983) *Acta Med. Okayama* **37**, 385. **15.** Shiki, Ishikawa, Shirai, Saito & Yoshida (1986) *Amer. J. Chin. Med.* **14**, 131. **16.** Ohtsuki & Iahida (1988) *Biochem. Biophys. Res. Commun.* **157**, 597. **17.** Ishikawa, Kanamaru, Wakui, Kanno & Ohtsuki (1990) *Biochem. Biophys. Res. Commun.* **167**, 876. **18.** Harada, Karino, Shimoyama, Shamsa & Ohtsuki (1996) *Biochem. Biophys. Res. Commun.* **227**, 102. **19.** Francischetti, Monteiro, Guimaraes & Francischetti (1997) *Biochem. Biophys. Res. Commun.* **235**, 259. **20.** Gong, Xiang, Yuan, *et al.* (2014) *PLoS One* **9**, e89450.

Glyoxal

This, the simplest of all dialdehydes (FW = 58.04 g/mol; CAS = 107-22-2; M.P.= 15°C; B.P. = 51°C) reacts under alkaline conditions with arginyl and lysyl residues in peptides and proteins. Extreme care should be exercised with this reagent, inasmuch as mixtures with air are known to explode. Glyoxal polymerizes quickly upon standing and reacts violently upon contact with water. Aqueous solutions of glyoxal (*i.e.*, three mol glyoxal per two mole of H_2O) are commercially available and relatively stable. **Target(s):** alcohol dehydrogenase (1); D-amino-acid oxidase (2); aminolevulinate aminotransferase (3); argininosuccinate synthetase (4); bacterial leucyl aminopeptidase (5); glutamate dehydrogenase (6); glutamine: fructose-6-phosphate aminotransferase (7); glutathione peroxidase (8); 4-hydroxy-2-oxoglutarate aldolase, *or* 2-keto-4-hydroxyglutarate aldolase (9); malate synthase (10); methylglyoxal reductase (NADPH-dependent) (11); propanediol dehydratase (12); 3-propylmalate synthase (13); pyruvate carboxylase (14); ribonuclease T₁ (15); subtilisin (16); Cu,Zn-superoxide dismutase (17); tartronate-semialdehyde reductase, *or* 2-hydroxy-3-oxopropionate reductase (18). **1.** Canella & Sodini (1975) *Eur. J. Biochem.* **59**, 119. **2.** Kotaki, Harada & Yagi (1966) *J. Biochem.* **60**, 592. **3.** Bajkowski & Friedmann (1982) *J. Biol. Chem.* **257**, 2207. **4.** Isashiki, Noda, Kobayashi, *et al.* (1989) *Protein Seq. Data Anal.* **2**, 283. **5.** Mäkinen, Mäkinen, Wilkes, Bayliss & Prescott (1982) *J. Biol. Chem.* **257**, 1765. **6.** Smith, Austen, Blumenthal & Nyc (1975) *The Enzymes*, 3rd ed., **11**, 293. **7.** Kikuchi, Ikeda & Tsuiki (1972) *Biochim. Biophys. Acta* **289**, 303. **8.** Park, Koh, Takahashi, *et al.* (2003) *Free Radic. Res.* **37**, 205. **9.** Grady, Wang & Dekker (1981) *Biochemistry* **20**, 2497. **10.** Munir, Hattori & Shimada (2002) *Biosci. Biotechnol. Biochem.* **66**, 576. **11.** Inoue, Rhee, Watanabe, Murata & Kimura (1988) *Eur. J. Biochem.* **171**, 213. **12.** Abeles (1966) *Meth. Enzymol.* **9**, 686. **13.** Imai, Reeves & Ajl (1963) *J. Biol. Chem.* **238**, 3193. **14.** Scrutton, Olmsted & Utter (1969) *Meth. Enzymol.* **13**, 235. **15.** Uchida & Egami (1971) *The Enzymes*, 3rd ed., **4**, 205. **16.** Genov & Idakieva (1983) *Int. J. Pept. Protein Res.* **21**, 536. **17.** Ukeda, Hasegawa, Ishi & Sawamura (1997) *Biosci. Biotechnol. Biochem.* **61**, 2039. **18.** Gotto & Kornberg (1961) *Biochem. J.* **81**, 273.

Glyoxylate

This naturally occurring carboxylic acid (FW = 74.04 g/mol; CAS = 298-12-4), also known as oxoacetic acid, glyoxalic acid, and oxoethanoic acid, is a key intermediate of the glyoxylate cycle occurring in some plants and microorganisms. The anhydrous acid is hygroscopic and has a melting point of 98°C. The hemihydrate melts at 70-75°C and will deliquesce rapidly, forming a syrup when exposed to the atmosphere. The common form of glyoxylic acid is the monohydrate which is also hygroscopic and melts at about 50°C. Glyoxylic acid is a relatively strong acid (pKa = 3.18 at 30°C) that is very soluble in water. The sodium salt is stable in neutral or slightly acidic conditions.

Glyphosate

This broad-spectrum herbicide (FW$_{\text{free-acid}}$ = 169.07; CAS = 1071-83-6; IUPAC Name: *N*-(phosphonomethyl)glycine), marketed in the U.S. as Round-Up™ and Tumbleweed™, is an extraordinarily potent homogentisate pathway inhibitor that is used on the farm and in the garden to eradicate weeds and other nuisance plants. Entire fields have been cleared and maintained in a weed-free state that facilitates highly efficient farming. Development of genetically modified (GM) cultivars that metabolize glyphosate allows farmers to suppress weeds without damaging the crop. Glyphosate's plant-killing power stems from its inhibitory action on 5-enolpyruvoyl-shikimate-3-phosphate synthetase (EPSPS), an essential enzyme for the formation of aromatic amino acids and plant vitality. The latter are precursors for certain secondary metabolites, including folates, ubiquinones and naphthoquines. While the shikimate pathway is absent in mammals and humans, glyphosate blocks phosphoenol pyruvate (PEP) binding to EPSPS, raising the concerning possibility that glyphosate may affect PEP-dependent reactions in other organisms. **Primary Mode of Action:** Various studies report glyphosate forms a tight ternary complex with EPSP synthase and shikimate 3-phosphate, prompting to the suggestion that in this complex, glyphosate functions as a transition-state analogue of the putative phosphoenolpyruvoyl oxonium ion (1,2). Despite this attractive suggestion, later studies (3) convincingly demonstrated that glyphosate is an uncompetitive inhibitor *versus* EPSP, K_{ii} = 54 µM. **Target(s):** adenosine kinase (4,5); 3-deoxy-7-phosphoheptulonate synthase, *or* 3-deoxy-D-*arabino*-heptulosonate-7-phosphate synthase (6,7); 3-phosphoshikimate 1-carboxyvinyltransferase, *or* 5-enolpyruvylshikimate-3-phosphate synthase (1-3,8-27); ribokinase (28). **1.** Anton, Hedstrom, Fish & Abeles (1983) *Biochemistry* **22**, 5903. **2.** Steinrücken & Amrhein (1984) *Eur. J. Biochem.* **143**, 351. **3.** Sammons, Gruys, Anderson, Johnson & Sikorski (1995) *Biochemistry* **34**, 6433. **4.** Park, Singh, Maj & Gupta (2004) *Protein J.* **23**, 167. **5.** Park, Singh & Gupta (2006) *Mol. Cell. Biochem.* **283**, 11. **6.** Ganson & Jensen (1988) *Arch. Biochem. Biophys.* **260**, 85. **7.** Jain & Bhalla-Sarin (2000) *Indian J. Biochem. Biophys.* **37**, 235. **8.** Du, Wallis & Payne (2000) *J. Enzyme Inhib.* **15**, 571. **9.** Lewendon & Coggins (1987) *Meth. Enzymol.* **142**, 342. **10.** Mousdale & Coggins (1987) *Meth. Enzymol.* **142**, 348. **11.** Steinrücken & Amrhein (1984) *Eur. J. Biochem.* **143**, 351. **12.** Steinrücken & Amrhein (1980) *Biochem. Biophys. Res. Commun.* **94**, 1207. **13.** Boocock & J. R. Coggins (1983) *FEBS Lett.* **154**, 127. **14.** Gruys, Walker & Sikorski (1992) *Biochemistry* **31**, 5534. **15.** Huynh (1991) *Arch. Biochem. Biophys.* **284**, 407. **16.** Sost, Chulz & Amrhein (1984) *FEBS Lett.* **173**, 238. **17.** Duncan, Lewendon & Coggins (1984) *FEBS Lett.* **165**, 121. **18.** Ream, Steinrücken, Porter & Sikorski (1988) *Plant Physiol.* **87**, 232. **19.** Huynh (1987) *Arch. Biochem. Biophys.* **258**, 233. **20.** He, Nie & Xu (2003) *Biosci. Biotechnol. Biochem.* **67**, 1405. **21.** Powell, Kerby, Rowell, Mousdale & Coggins (1992) *Planta* **188**, 484. **22.** Majumder, Selvapandiyan, Fattah, *et al.* (1995) *Eur. J. Biochem.* **229**, 99. **23.** Schönbrunn, Eschenburg, Shuttleworth, *et al.* (2001) *Proc. Natl. Acad. Sci. U.S.A.* **98**, 1376. **24.** Rubin, Gaines & Jensen (1984) *Plant Physiol.* **75**, 839. **25.** Mousdale & Coggins (1984) *Planta* **160**, 78. **26.** Du, Wallis, Mazzulla, Chalker, *et al.* (2000) *Eur. J. Biochem.* **267**, 222. **27.** Forlani, Parisi & Nielsen (1994) *Plant Physiol.* **105**, 1107. **28.** Park, van Koeverden, Singh & Gupta (2007) *FEBS Lett.* **581**, 3211.

GLYX-13

This NMDAR partial agonist (FW = 413.47 g/mol; CAS = 117928-94-6; *Sequence*: TPPT-NH$_2$) is a nootropic tetrapeptide that interacts at the receptor's glycine binding site and enhances learning and memory in young adult and learning-impaired aging rats (1,2). GLYX-13 readily crosses the blood brain barrier and increases long-term potentiation. It also induces gene expression of the NR1 subunit of *N*-methyl-D-aspartate receptor (NMDAR) within the hippocampus. GLYX-13 has a therapeutic index of 500 or more, onset within 20 minutes of administration, and produces antidepressant-like effects lasting approximately two weeks following administration (1). **Discovery:** The monoclonal antibody B6B21, which acts as a partial agonist at the glycine site of the NMDAR facilitated the discovery of a unique hippocampal neuron surface antigen (3). To create novel NMDAR-targetted drugs, the hyper-variable region of B6B21's light chain was cloned and sequenced, followed by screening of those peptides promoting binding of the NMDA-binding anticonvulsant MK-801 in the presence of 7-chlorokynurenic acid, a glycine site-specific competitive inhibitor of NMDAR (4). Peptides increasing MK-801 binding in a dose-dependent manner under such conditions were collectively termed "Glyxins". **Pharmacodynamics:** Inhibition of NMDA-mediated neurotransmission is known to provide anti-nociceptive actions in a number of animal models, and both partial agonists and antagonists of the NMDA-associated glycine site often exert antinociceptive actions at doses that are not ataxic. GLYX-13 is one such glycine site-specific, partial agonist NMDA receptor. Direct two-photon imaging of Schaffer collateral burst-evoked increases in [Ca^{2+}] within individual dendritic spines revealed that GLYX-13 selectively enhances burst-induced NMDAR-dependent spine Ca^{2+} influx (5). Examining the rate of MK-801 block of synaptic *versus* extra-synaptic NMDAR-gated channels revealed that GLYX-13 selectively enhances activation of burst-driven extra-synaptic NMDA receptors, an action that was blocked by the NR2B-selective NMDAR antagonist Ifenprodil. These findings suggest that GLYX-13 possesses unique a therapeutic potential, both as a learning-enhancing and memory-enhancing drug in view of its ability to simultaneously enhance long-term potentiation, while suppressing long-term depression, *or* LTD (5). **1.** Moskal, Kuo, Weiss, *et al.* (2005) *Neuropharmacol.* **49**, 1077. **2.** Moskal, Burch, Burgdorf, *et al.* (2013) *Expert Opin. Investig. Drugs* **23**, 243. **3.** Moskal & Schaffner (1986) *J. Neurosci.* **6**, 2045. **4.** Moskal, Yamamoto & Colley (2001) *Curr. Drug Targets* **2**. 331. **5.** Zhang, Sullivan, Moskal & Stanton (2008) *Neuropharmacol.* **55**, 1238.

GM-1489, *See N-[(2R)-2-(Carboxymethyl)-4-methylpentanoyl]-L-tryptophan-(S)-methyl-benzylamide*

GM6001

This broad-spectrum matrix metalloprotease inhibitor (FW = 388.46 g/mol; CAS = 142880-36-2), also known as Galardin, Ilomastat, and N^4-hydroxy-N^1-[(1*S*)-1-(1*H*-indol-3-ylmethyl)-2-(methylamino)-2-oxoethyl]-2-(2-methylpropyl)-(2*R*)-butanediamide, targets MMP-1 (K_i = 0.4 nM), MMP-2 (K_i = 0.5 nM), MMP-3 (K_i = 27 nM), MMP-7 (K_i = 3.7 nM), MMP-8 (K_i = 0.1 nM), MMP-9 (K_i = 0.2 nM), MMP-12 (K_i = 3.6 nM), MMP-14 (K_i = 13.4 nM), and MMP-26 (K_i = 0.36 nM). Hydroxamate-containing inhibitors rely on the hydroxamate moiety to interact with the active-site Zn^{2+} in matrix metalloprotease inhibitor, with specificity conferred by other side-chain groups. Penetration of epithelial cells into the submucosa, a process that requires digestion of the basal lamina and the surrounding extracellular matrix, is strongly inhibited by GM6001 (2). Compared to placebo, GM6001 significantly inhibited intimal hyperplasia and intimal collagen content, and it increased lumen area in stented arteries without effects on proliferation rates (3). **1.** Augé, Hornebeck, Decarme & Laronze (2003) *Bioorg. Med. Chem. Lett.* **13**, 1783. **2.** Tournier, Polette, Hinnrasky, *et al.* (1994) *J. Biol. Chem.* **269**, 25454. **3.** Li, Cantor, Nili (2002) *J. Am. Coll. Cardiol.* **39**, 1852.

GMX1778

This nicotinamide phosphoribosyltransferase-specific inhibitor (FW = 371.90 g/mol; CAS = 200484-11-3), also known as CHS 828 and N-[6-(4-chlorophenoxy)hexyl]-N'-cyano-N''-4-pyridinylguanidine) targets a rate-limiting enzyme required for NAD$^+$ synthesis from nicotinamide and 5'-phosphoribosyl-1'-pyrophosphate (1). GMX1778 increases intracellular reactive oxygen species (ROS) in cancer cells, while elevating the superoxide level and decreasing the cellular NAD$^+$ content. GMX1778-mediated ROS induction is p53-dependent, suggesting that the status of both p53 and NAPRT1 might affect tumor apoptosis (2). Significantly, GMX1778 is without effect on untransformed cells. GMX1778-induced ROS accumulation can be reduced by supplying cells with nicotinic acid, with the extent of reduction depending on active NAPRT1. Teglarinad chloride (FW = 672.60 g/mol; CAS = 432037-57-5; *Abbreviation*: GMX1777; IUPAC/Chemical name: (Z)-4-(3-(6-(4-chlorophenoxy)hexyl)-2-cyanoguanidino)-1-(3-oxo-2,4,7,10,13,16-hexaoxaheptadecyl)pyridin-1-ium chloride) is the water-soluble, intravenously administered pro-drug (IC$_{50}$ = 5-10 nM) that is taken up and rapidly converted into active drug (GMX1778) through hydrolytic cleavage of its carbonate ester bond (3). **1.** Hjarnaa, Jonsson, Latini, *et al.* (1999) *Cancer Res.* **59**, 5751. **2.** Cerna, Li, Flaherty, *et al.* (2012) *J. Biol. Chem.* **287**, 22408. **3.** Fuchs, Rodriguez, Eriksson, *et al.* (2010) *Int. J. Cancer* **126**, 2773.

GNF-7

This third-generation imatinib-based antineoplastic agent (FW = 547.54 g/mol; CAS = 839706-07-9; Solubility: 20 mg/mL DMSO; <1 mg/mL H$_2$O), also named N-[3-[1,4-dihydro-1-methyl-7-[(6-methyl-3-pyridinyl)amino]-2-oxopyrimido[4,5-d]pyrimidin-3(2H)-yl]-4-methylphenyl]-3-(trifluoromethyl)-benzamide, targets the protein kinase activity of Bcr-Abl$^{wild-type}$ (IC$_{50}$ < 5 nM), Bcr-Abl^{M351T} (IC$_{50}$ < 5 nM), Bcr-Abl^{T315I} (IC$_{50}$ = 61 nM), Bcr-Abl^{E255V} (IC$_{50}$ = 122 nM), Bcr-Abl^{G250E} (IC$_{50}$ = 136 nM), and c-Abl (IC$_{50}$ = 133 nM). Synthesized with the goal of bridging the ATP and allosteric binding sites with a linker that accommodates a larger gatekeeper residue, GNF-7 is among the first type-II PKI's to inhibit the T315I mutant. GNF-7 also displays excellent growth inhibitory activity against human colon cancer cell lines Colo205 and SW620, while having weaker action against the non-cancer HEK293T cell line. (***See also** Imatinib; Nilotinib; Dasatinib, and Bosutinib*) **1.** Choi, Ren, Adrian, *et al.* (2010) *J. Med. Chem.* **53**, 5439.

GNF-6702

This broad-spectrum antiprotozoal drug (FW = 429.42 g/mol), systematically named N-[4-fluoro-3-(6-pyridin-2-yl-[1,2,4]triazolo[1,5-a]pyrimidin-2-yl)phenyl]-2,4-dimethyl-1,3-oxazole-5-carboxamide, targets the kinetoplastid proteasomes of *Trypanosoma* cruzi, *Leishmania* spp. and *Trypanosoma brucei* spp., parasites now shownto share a common druggable target. GNF6702 inhibits the kinetoplastid proteasome noncompetitively, but does not inhibit the mammalian proteasome or the growth of mammalian cells. It is well-tolerated in mice. **1.** Khare, Nagle, Biggart, *et al.* (2016) *Nature* **537**, 229.

Gö 6976

This indolocarbazole (FW = 378.43 g/mol; CAS = 136194-77-9; IUPAC name: 12-(2-cyanoethyl)-6,7,12,13-tetrahydro-13-methyl-5-oxo-5H-indolo (2,3-a)pyrrolo(3,4-c)carbazole) inhibits protein kinase C (IC$_{50}$ = 7.9 nM). At higher concentrations, Gö 6976 will also inhibit protein kinase G (IC$_{50}$ = 6.2 μM) and [myosin light-chain] kinase (IC$_{50}$ = 5.8 μM). With respect to protein kinase C, Gö 6976 also exhibits some selectivity: the α-isozyme has an IC50 value of 2.3 nM and PKCβI IC$_{50}$ = 6.2 nM. The δ-, ε-, and ζ-isozymes are not inhibited, even at micromolar levels. Protein kinase Cμ has an IC$_{50}$ value of 20 nM. **Target(s):** polo kinase (1); protein kinase C (2-8). **1.** Johnson, Stewart, Woods, Giranda & Luo (2007) *Biochemistry* **46**, 9551. **2.** Martiny-Baron, Kazanietz, Mischak, *et al.* (1993) *J. Biol. Chem.* **268**, 9194. **3.** Qatsha, Rudolph, Marme, Schachtele & May (1993) *Proc. Natl. Acad. Sci. U.S.A.* **90**, 4674. **4.** Beaudry, Gendron, Guimond, Payet & Gallo-Payet (2006) *Endocrinology* **147**, 4263. **5.** Chen & Exton (2004) *J. Biol. Chem.* **279**, 22076. **6.** Beltowski, Marciniak, Jamroz-Wisniewska, Borkowska & Wojcicka (2004) *Acta Biochim. Pol.* **51**, 757. **7.** Harper & Poole (2007) *Biochem. Soc. Trans.* **35**, 1005. **8.** Johnson, Guptaroy, Lund, Shamban & Gnegy (2005) *J. Biol. Chem.* **280**, 10914.

Gonadotropin-Inhibitory Hormone

This hypothalamic neuropeptide (FW = 1390.66; Quail Sequence: SIKPSAYLPLRF-NH$_2$; Symbol: GnIH), which is secreted by the anterior hypophysis gland, inhibits gonadotropin secretion from the pituitary and plays a crucial role in vertebrate gonadal development and function. Its wide distribution within the brain also suggested roles in regulation of behavior. Melatonin acts directly on GnIH neurons through its receptor; to induce GnIH expression. In the quail, GnIH is also located in neurons of the paraventricular nucleus (*or* PVN) that terminate in the median eminence (*or* ME) (1-3). **1.** Tsutsui, Saigoh, Ukena, *et al.* (2000) *Biochem. Biophys. Res. Commun.* **275**, 661. **2.** Ukena, Ubuka & Tsutsui (2003) *Cell Tissue R*es. **312**, 73. 3. Ubuka, Ueno, Ukena & Tsutsui (2003) *J. Endocrinol.* **178**, 311.

Gonadotropin-Releasing Hormone

This peptide amide (FW = 1182.29 g/mol; CAS = 9034-40-6), also known as luteinizing hormone-releasing hormone (LHRH), luliberin, and gonadoliberin, stimulates the release of luteinizing hormone and follicle-stimulating hormone from the anterior pituitary. Human peptide sequence: L-pyroGlu-L-His-L-Try-L-Ser-L-Tyr-Gly-L-Leu-L-Arg-L-Pro-Gly-NH$_2$. **Target(s):** carboxypeptidase E (1); pyroglutamyl-peptidase II, K_i = 8.1 μM, with the salmon peptide (2-6). **1.** Hook & LaGamma (1987) *J. Biol. Chem.* **262**, 12583. **2.** Bauer (1998) in *Handb. Proteolytic Enzymes*, p. 1005, Academic Press, San Diego. **3.** Gallagher & O'Connor (1998) *Int. J. Biochem. Cell Biol.* **30**, 115. **4.** O'Connor & O'Cuinn (1985) *Eur. J. Biochem.* **150**, 47. **5.** Bauer, Nowak & Kleinkauf (1981) *Eur. J. Biochem.* **118**, 173. **6.** Taylor & Dixon (1978) *J. Biol. Chem.* **253**, 6934.

Gossypol

This toxic yellow secondary metabolite (FW = 518.56 g/mol; CAS = 303-45-7) from cotton (*Gossypium* sp.) inhibits sperm production in humans. Gossypol functions as a part of the cotton plant's defense system against pathogenic fungi and insects. **Target(s):** acid phosphatase (1); acrosomal proteinase, *or* acrosin (1,2); alcohol dehydrogenase (3); aldehyde dehydrogenase (3); aldose reductase (4,5); alkaline phosphatase (6); apyrase, *or* ATP diphosphohydrolase (7,8); arylsulfatase (1); azocoll proteinase (1); calcineurin (9); catechol *O*-methyltransferase (10); cathepsin L (11); cholesterol monooxygenase, side-chain cleavage (12); cytochrome *c* oxidase (13); DNA-directed DNA polymerase (14); DNA polymerase α (14); DNA topoisomerase II (15); glucose-6-phosphate dehydrogenase (16); β-glucuronidase (1); glutamate dehydrogenase (16,17); glutathione *S*-transferase (18); HIV-1 replication (19); hyaluronidase (1); α-hydroxyacid dehydrogenase (16,20); 3β-hydroxysteroid dehydrogenase (21); 11β-hydroxysteroid dehydrogenase (22,23); 17β-hydroxysteroid dehydrogenase, *potently* (41); inositol-polyphosphate multikinase (15); inositol-trisphosphate 3-kinase (15); lactate dehydrogenase (17,18), with K_i values of 1.9, 1.4, and 4.2 μM for LDH-A$_4$, LDH-B$_4$, and LDH-C$_4$, respectively (24); lactate dehydrogenase X (18,25,26); lipoxygenase (27); 5-lipoxygenase, *or* arachidonate 5-lipoxygenase (28); 12-lipoxygenase, *or* arachidonate 12-lipoxygenase (28); malate dehydrogenase (16,17,20,25); malic enzyme (16); maltase (6); [myosin light-chain] kinase (29); neuraminidase (1); phospholipase A$_2$ (30-32); procathepsin L self-processing, acrosomal (33); prostaglandin synthase (28); protein kinase C (29,34-36); ribonucleotide reductase (37); steroid Δ^5-Δ^4 isomerase (21); steroid 5α-reductase (38); steroid 11β-monooxygenase (21,12); succinate dehydrogenase (13); sucrase (6); telomerase (39); tubulin polymerization, microtubule assembly (40). **1.** Yuan, Shi & Srivastava (1995) *Mol. Reprod. Dev.* **40**, 228. **2.** Johnsen, Mas Diaz & Eliasson (1982) *Int. J. Androl.* **5**, 636. **3.** Messiha (1991) *Gen. Pharmacol.* **22**, 573. **4.** Deck, Vander Jagt & Royer (1991) *J. Med. Chem.* **34**, 3301. **5.** Deck, Chamblee, Royer, Hunsaker & Vander Jagt (1999) *Adv. Exp. Med. Biol.* **463**, 487. **6.** Baram, Cogan & Mokady (1987) *Agric. Biol. Chem.* **51**, 3437. **7.** Picher, Cote, Beliveau, Potier & Beaudoin (1993) *J. Biol. Chem.* **268**, 4699. **8.** Picher, Beliveau, Potier, *et al.* (1994) *Biochim. Biophys. Acta* **1200**, 167. **9.** Baumgrass, Weiwad, Erdmann, *et al.* (2001) *J. Biol. Chem.* **276**, 47914. **10.** Tang, Tsang, Lee & Wong (1982) *Contraception* **26**, 515. **11.** McDonald & Kadkhodayan (1988) *Biochem. Biophys. Res. Commun.* **151**, 827. **12.** Cuellar, Diaz-Sanchez & Ramirez (1990) *J. Steroid Biochem. Mol. Biol.* **37**, 581. **13.** Myers & Throneberry (1966) *Plant Physiol.* **41**, 787. **14.** Rosenberg, Adlakha, Desai & Rao (1986) *Biochim. Biophys. Acta* **866**, 258. **15.** Mayr, Windhorst & Hillemeier (2005) *J. Biol. Chem.* **280**, 13229. **16.** Gerez de Burgos, Burgos, Montamat, Rovai & Blanco (1984) *Biochem. Pharmacol.* **33**, 955. **17.** Burgos, Gerez de Burgos, Rovai & Blanco (1986) *Biochem. Pharmacol.* **35**, 801. **18.** Lee, Moon, Yuan & Chen (1982) *Mol. Cell. Biochem.* **47**, 65. **19.** Royer, Mills, Deck, Mertz & Vander Jagt (1991) *Pharmacol. Res.* **24**, 407. **20.** Montamat, Burgos, Gerez de Burgos, *et al.* (1982) *Science* **218**, 288. **21.** Wu, Chik, Albertson, Linehan & Knazek (1991) *Acta Endocrinol. (Copenhagen)* **124**, 672. **22.** Song, Lorenzo & Reidenberg (1992) *J. Lab. Clin. Med.* **120**, 792. **23.** Zhang, Lorenzo & Reidenberg (1994) *J. Steroid Biochem. Mol. Biol.* **49**, 81. **24.** Yu, Deck, Hunsaker, *et al.* (2001) *Biochem. Pharmacol.* **62**, 81. **25.** Olgiati & Toscano (1983) *Biochem. Biophys. Res. Commun.* **115**, 180. **26.** Giridharan, Bamji & Sankaram (1982) *Contraception* **26**, 607. **27.** Kulkarni, Cai & Richards (1992) *Int. J. Biochem.* **24**, 255. **28.** Hamasaki & Tai (1985) *Biochim. Biophys. Acta* **834**, 37. **29.** Jinsart, Ternai & Polya (1991) *Biol. Chem. Hoppe Seyler* **372**, 819. **30.** Glaser, Lock & Chang (1991) *Agents Actions* **34**, 89. **31.** Soubeyrand, Khadir, Brindle & Manjunath (1997) *J. Biol. Chem.* **272**, 222. **32.** Winkler, McCarte-Roshak, Huang, *et al.* (1994) *J. Lipid Mediat. Cell Signal.* **10**, 315. **33.** McDonald & Emerick (1995) *Arch. Biochem. Biophys.* **323**, 409. **34.** Ibrahim & Platt (1991) *J. Exp. Zool.* **260**, 202. **35.** Pelosin, Keramidas, Souvignet & Chambaz (1990) *Biochem. Biophys. Res. Commun.* **169**, 1040. **36.** Kimura, Sakurada & Katoh (1985) *Biochim. Biophys. Acta* **839**, 276. **37.** McClarty, Chan, Creasey & Wright (1985) *Biochem. Biophys. Res. Commun.* **133**, 300. **38.** Hiipakka, Zhang, Dai, Dai & Liao (2002) *Biochem. Pharmacol.* **63**, 1165. **39.** Mego (2002) *Bratisl. Lek. Listy* **103**, 378. **40.** Medrano & Andreu (1986) *Eur. J. Biochem.* **158**, 63 **41.** Hu, Zhou, Li, *et al.* (2009) *J. Steroid Biochem. Mol. Biol.* **115**, 14.

Gougerotin

This nucleoside antibiotic (FW = 443.42 g/mol; CAS = 2096-42-6), also known as aspiculamycin and asteromycin, is produced by *Streptomyces gougerotii* and inhibits both prokaryotic and eukaryotic protein biosynthesis (1-7) Gougerotin perturbs the relative positioning of the 3'-end of the P/P' site-bound tRNA and the peptidyltransferase loop region of 23S rRNA. **Target(s):** peptidyltransferase (1-10); peptidyl-tRNA hydrolase, *or* aminoacyl-tRNA hydrolase (11). **1.** Fernandez-Muñoz, Monro & Vazquez (1971) *Meth. Enzymol.* **20**, 481. **2.** Gottesman (1971) *Meth. Enzymol.* **20**, 490. **3.** Pestka (1974) *Meth. Enzymol.* **30**, 261. **4.** Carrasco, Battaner & Vazquez (1974) *Meth. Enzymol.* **30**, 282. **5.** Barbacid & Vazquez (1974) *Meth. Enzymol.* **30**, 426. **6.** Jiménez (1976) *Trends Biochem. Sci.* **1**, 28. **7.** Coutsogeorgopoulos (1967) *Biochemistry* **6**, 1704. **8.** Clark & Gunther (1963) *Biochim. Biophys. Acta* **76**, 636. **9.** Kirillov, Porse & Garrett (1999) *RNA* **5**, 1003. **10.** Spirin & Asatryan (1976) *FEBS Lett.* **70**, 101. **11.** Tate & Caskey (1974) *The Enzymes*, 3rd ed., **10**, 87.

GP 683

This diaryltubercidin analogue (FW = 402.44 g/mol), also named (2*R*,3*R*,4*S*,5*R*)-2-(4-anilino-5-phenylpyrrolo[2,3-*d*]pyrimidin-7-yl)-5-methyloxolane-3,4-diol, is a potent adenosine kinase inhibitor (ED$_{50}$ = 1.1 mg/kg) that maintains the adenosine concentration at seizure foci, thereby favoring activation of A$_1$ adenosine receptors and providing a measure of relief against further seizure. Because the use of adenosine receptor agonists is plagued by dose-limiting cardiovascular side effects, there is great interest in developing adenosine kinase inhibitors (AKIs) as a means for raising steady-state adenosine concentrations. GP683 also reduced epileptiform discharges induced by removal of Mg^{2+} in a rat neocortical preparation. This anti-seizure activity of GP683, both in *in vivo* and *in vitro* preparations, was markedly reversed by the adenosine receptor antagonists, such as theophylline and 8-(*p*-sulfophenyl)theophylline. GP683's ability to stabilize adenosine concentration surges promises to provide a measure of anti-seizure relief akin to 5'-amino-5'-deoxyadenosine, 5-iodotubercidin, and 5'-deoxy-5-iodotubercidin), but with fewer dose-limiting side effects. **1.** Wiesner, Ugarkar, Castellino (1999) *J. Pharmacol. Exp. Ther.* **289**, 1669.

GQ-177

This thiazolidinedione-based PPARγ partial agonist (FW = 452.34 g/mol) targets Peroxisome Proliferator-Activated Receptor-γ (*or* PPARγ), which regulates multiple pathways involved in the pathogenesis of obesity and atherosclerosis. GQ-177 mproves obesity-associated insulin resistance and dyslipidemia with atheroprotective effects in LDLr$^{-/-}$ mice. In a atherosclerosis mouse model, GQ-177 inhibits atherosclerotic lesion progression, increases plasma HDL and mRNA levels of PPARγ and ATP-binding cassette A1 in atherosclerotic lesions. ***See also*** *Rosiglitazone* **1.** Silva, César, de Oliveira, *et al.* (2016). *Pharmacol Res.* **104**. 49.

GR-20263, *See* Ceftazidime

GR 231118

This cyclic peptide amide (FW = 2352.77 g/mol; CAS = 158859-98-4; where Dpr = 2,3-Diaminopropionic acid), also known as 1229U91 and GW-1229, is a selective neuropeptide Y (NPY) Y_1-receptor antagonist ($K_d = 5 \times 10^{-9}$ M) and an NPY Y4 receptor agonist ($K_d = 1 \times 10^{-6}$ M). *In vitro*, GR-231118 also binds to neuropeptide FF (NPFF) receptors (K_i = 43-73 nM). **1**. Hegde, Bonhaus, Stanley, *et al.* (1995) *J. Pharmacol. Exp. Ther.* **275**, 1261. **2**. Parker, Babij, Balasubramaniam, *et al.* (1998) *Eur. J. Pharmacol.* **349**, 97. **3**. Ishihara, *et al.* (1998) *Am. J. Physiol.* 274, R1500. **4**. Mollereau, *et al.* (2001) *Br. J. Pharmacol.* **133**, 1.

GR38032F, *See* Ondansetron

Grazoprevir

This second generation, nonpeptide macrocycle and HCV protease inhibitor (FW = 766.90 g/mol; CAS = 1350514-68-9), also known by the code name MK-6170 and as 1*R*,18*R*,20*R*,24*S*,27*S*)-*N*-{(1*R*,2*S*)-1-[(cyclopropyl sulfonyl)carbamoyl]-2-vinylcyclopropyl}-7-methoxy-24-(2-methyl-2-propanyl)-22,25-dioxo-2,21-dioxa-4,11,23,26-tetraaza-pentacyclo[24.2.1.0^{3,12}. 0^{5,10}.0^{18,20}]nonacosa-3,5,7,9,11-pentaene-27-carboxamide, targets the Hepatitis C virus NS3/4a protease, exhibiting good activity against a range of HCV genotype variants, including some that are resistant to most currently used antiviral medications. Grazoprevir is used in combination with elbasvir, the resulting formulation, known as Zepatier®, is a highly effective, FDA-approved treatment of chronic HCV genotypes 1 or 4. A molecular modeling-derived strategy led to the design of this protease inhibitor which contains a conformationally constrained structure. This design arose from an analysis of the crystal structure of full-length NS3/4A with and without inhibitors docked in the active site. That strategy, when coupled with a modular synthetic approach expoiting key ring-closing metathesis reaction, allowed for the rapid exploration of these molecules. **1**. Harper S, McCauley JA, Rudd, *et al.* (2012) *ACS Med Chem Lett.* **3**, 332. **2**. Ali, Aydin, Gildemeister, *et al.* (2013) *ACS Chem. Biol.* **8**, 1469.

Gramicidins

H₂C(=O)–L–X–Gly–L–Ala–D–Leu–L–Ala–D–Val–L–Val–D–Val–L–Trp–D–Leu–L–Y–D–Leu–L–Trp–D–Leu–L–Trp–OCH₂CH₂NH₂

These linear membrane-penetrating antibiotics (FW$_{Val-GR-A}$ = 1882.30 g/mol; FW$_{Ile-GRA}$ = 1896.40 g/mol; FW$_{Val-GRB}$ = 1843.30 g/mol; FW$_{Ile-GRB}$ = 1857.30 g/mol; FW$_{Val-GRC}$ = 1859.30 g/mol; FW$_{Ile-GRC}$ = 1873.3 g/mol; CAS 1393-88-0 and 1405-97-6) are channel-forming ionophores that are active against Gram-positive microorganisms, hence their name (1-6). They efficiently uncouple oxidative phosphorylation and photophosphorylation by blocking formation of both the membrane potential and ΔpH. Truncated derivatives stimulate the ATPase without collapsing the membrane potential. Such findings suggest gamicidins interact directly with the H$^+$-ATPase (7). **Structural Features:** GRA, GRB, and GRC are linear 15-residue peptides. Key features: N-terminus = Formyl (H₂C(=O)–); Residue-X = Val and Ile residues in all three; Residue-Y = W in GRA, F in GRB, and Y in GRC; C-terminus = Ethanolamine (–OCH₂CH₂NH₂). They are formylated at the N-terminus and contain ethanolamine at the C-terminus. Each gramicidin consists of two chains, differing only in the residue at position one, where it is either an L-valyl residue or an L-isoleucyl residue. Note that the alternating stereochemical configurations of amino acid residues is an essential property for β-helix formation. The valine-containing ionophore is the principal form in Gramicidins A, B, and C, roughly amounting to 80-95%. Gramicidin D is a minor and more hydrophilic form. Commercial formulations from *Bacillus aneurinolyticus* (*Bacillus brevis*) consist of GRA (87.5%), GRB (7.1%), GRC (5.1%), and GRD (0.3% w/w). **1.** Dubos (1939) *J. Exp. Med.* **70**, 1. **2.** Dubos & Hotchkiss (1941) *J. Exp. Med.* **73**, 629. **3.** Gross & Witkop (1965) *Biochemistry* **4**, 2495. **4.** Reed (1979) *Meth. Enzymol.* **55**, 435. **5.** Slater (1967) *Meth. Enzymol.* **10**, 48. **6.** McCarty (1980) *Meth. Enzymol.* **69**, 719.n **7**. Izawa & Good (1972) *Meth. Enzymol.* **24**, 355. **7**. Rottenberg & Koeppe (1989) *Biochemistry* **28**, 4355.

Gramine

This alkaloid (FW$_{free-base}$ = 174.25 g/mol; CAS = 87-52-5), formed from L-tryptophan and found in relatively large amounts in barley, is a vasorelaxant and 5-HT$_{2A}$ receptor antagonist. **Target(s):** acetylcholinesterase (1); adenylate cyclase activity, dopamine-stimulated (2); 5-hydroxytryptamine 2A (5-HT$_{2A}$) receptor (3,4); NADH dehydrogenase, *or* Complex I (5); octopamine receptor (3,4). **1**. Lockhart, Closier, Howard, Steward & Lestage (2001) *Naunyn Schmiedebergs Arch. Pharmacol.* **363**, 429. **2**. Capasso, Creti, De Petrocellis, De Prisco & Parisi (1988) *Biochem. Biophys. Res. Commun.* **154**, 758. **3**. Froldi, Silvestrin, Dorigo & Caparrotta (2004) *Planta Med.* **70**, 373. **4**. Hiripi, Juhos & Downer (1994) *Brain Res.* **633**, 119. **5**. Niemeyer & Roveri (1984) *Biochem. Pharmacol.* **33**, 2973.

Granisetron

This serotonin 5-HT₃ receptor antagonist and anti-emetic (FW = 312.41 g/mol; CAS = 109889-09-0), also known as BRL 43694, Sancuso®, Kytril®, and 1-methyl-*N*-((1*R*,3*r*,5*S*)-9-methyl-9-azabicyclo[3.3.1]nonan-3-yl)-1*H*-indazole-3-carboxamide, is a FDA-approved drug (often administered as a transdermal patch) for the treatment nausea and vomiting associated with cancer chemotherapy. Often used in combination with other antiemetics, gransetron is specifically indicated for the prevention of both acute and delayed nausea and vomiting that attend the use of emetogenic chemotherapeutic agents, such as anthracycline and cyclophos-phamide. **Primary Mode of Inhibitor Action:** Gransetron affects both peripheral and central nervous systems, reducing vagus nerve activity and thereby deactivating the 5-HT₃ receptor-dense vomiting center located in the medulla oblongata. It is without effect on dopamine receptors or muscarinic receptors. BRL-43694 forms a high-affinity complex (K_d = 0.30 nM) with 5-HT₃ receptors, similar to that of quipazine. Scatchard analysis suggests binding occurs at a single population of receptors (1). There is also a direct correlation ($r = 0.98$, slope = 0.95) between BRL 43694 binding and observed inhibition of the 5-HT-evoked BJR, the so-called Bezold-Jarisch chemoreflex associated with combined bradycardia, hypotension, and peripheral vasodilation. (Other similarly acting 5-HT$_3$ antagonists include: tropisetron, ondansetron, dolasetron, and palonosetron. The effectiveness of each depends on particular variants of 5-HT$_3$ receptors expressed by the patient, including changes in promoters for the receptor genes.) **1**. Nelson & Thomas (1989) *Biochem. Pharmacol.* **38**, 1693.

Grepafloxacin

This orally bioavailable fluoroquinolone-class antibiotic (FW = 359.40 g/mol; CAS = 119914-60-2), also named Raxar® and (*RS*)-1-cyclopropyl-6-fluoro-5-methyl-7-(3-methylpiperazin-1-yl)-4-oxoquinoline-3-carboxylate, is a broad-spectrum systemic antibacterial agent that targets Type II DNA topoisomerases (gyrases), which are required for bacterial replication and transcription. Despite its effectiveness against Gram-positive and –negative bacteria (especially *Haemophilus influenzae*, *Streptococcus pneumoniae*, and *Moraxella catarrhalis*), grepafloxacin was withdrawn from the market, because it lengthens the QT interval, resulting in cardiac events and sudden death. For the prototypical member of this antibiotic class, *See Ciprofloxacin*

Griseofulvin

This orally active microtubule-directed antifungal agent from *Penicillium griseofulvum* and *P. janczewskii* (FW = 352.77 g/mol; CAS = 126-07-8; Solubility 50 mg/mL DMSO; <1 mg/mL H$_2$O; IUPAC: (2*S*,6'*R*)-7-chloro-2',4,6-trimethoxy-6'-methyl-3*H*,4'*H*-spiro[1-benzofuran-2,1'-cyclohex[2]-ene]-3,4'-dione) targets tubulin and microtubules (MTs), interfering with mitosis in *Trichophyton rubrum* and *Trichophyton mentagrophytes*, two pathogenic dermatophytes. Griseofulvin also targets centrosomal clustering that enables bipolar mitosis in cancer cell lines harboring supernumerary centrosomes. Griseofulvin induces multipolar spindles by inhibiting centrosome coalescence and mitotic arrest, with subsequent cell death in tumor cell lines, but not in diploid fibroblasts and keratinocytes having normal centrosome content. Inhibition of centrosome clustering by griseofulvin was not restricted to mitotic cells and occurs during interphase (1). While multipolar-spindle formation is dynein-independent, interphase MT depolymerization appears to be mechanistically involved in centrosomal declustering. SAR studies with the parent compound and 53 griseofulvin analogues strongly suggest that the mode-of-action of the compound class toward fungi and mammalian cancer cells is different (2). Treatment of *T. rubrum* fungal cells with griseofulvin decreased β-tubulin gene expression in a dose-dependent manner (3). Griseofulvin is used mainly to treat skin infections such as jock itch, athlete's foot, and ringworm; as well as fungal infections of the scalp, fingernails, and toenails. (*See also GF-15*) **Target(s):** ferrochelatase (rodent) (4); tubulin polymerization (microtubule assembly) (5-9); UDP-glucuronosyltransferase (10). **1**. Rebacz, Larsen, Clausen, *et al.* (2007) *Cancer Res.* **67**, 6342. **2**. Rønnest, Raab, Anderhub, *et al.* (2012) *J. Med. Chem.* **55**, 652. **3**. Zomorodian, Uthman, Tarazooie & Rezaie (2012) *J. Med. Chem.* **55**, 652. **4**. Cole & Marks (1984) *Mol. Cell. Biochem.* **64**, 127. **5**. Chaudhuri & Ludueña (1996) *Biochem. Pharmacol.* **51**, 903. **6**. Banks & Till (1975) *J. Physiol.* **252**, 283. **7**. Roobol, Gull & Pogson (1976) *FEBS Lett.* **67**, 248. **8**. Keates (1981) *Biochem. Biophys. Res. Commun.* **102**, 746. **9**. Sloboda, Van Blaricom, Creasey, Rosenbaum & Malawista (1982) *Biochem. Biophys. Res. Commun.* **105**, 882. **10**. Grancharov, H. Engelberg, Z. Naydenova, G. Muller, A. W. Rettenmeier & E. Golovinsky (2001) *Arch. Toxicol.* **75**, 609.

Griseoviridin

This antibiotic (FW = 477.53; CAS = 53216-90-3; λ_{max} = 221 nm (in methanol); $\varepsilon_{(1\%1cm)}$ = 870), obtained from a strain of *Streptomyces*, binds to the large subunit of *Escherichia coli* ribosomes, blocking the interaction of the 3' terminal end of peptidyl tRNA with the donor site of the peptidyltransferase. Note revised structure. **Target(s):** peptidyltransferase (1); prokaryotic protein biosynthesis (initiation and elongation) (2,3). **1**. Spirin & Asatryan (1976) *FEBS Lett.* **70**, 101. **2**. Jiménez (1976) *Trends Biochem. Sci.* **1**, 28. **3**. Barbacid, Contreras & Vazquez (1975) *Biochim. Biophys. Acta* **395**, 347.

GRN163L, *See Imetelstat*

GS 143

This β-TrCP1 ligase inhibitor (FW = 466.46 g/mol; CAS = 916232-21-8; Soluble to 100 mM in DMSO), systematically named 4-[4-[[5-(2-fluorophenyl)-2-furanyl]methylene]-4,5-dihydro-5-oxo-3-(phenylmethyl)-1*H*-pyrazol-1-yl]benzoic acid, blocks IκBα ubiquitination (IC$_{50}$ = 5.2 μM) without affecting proteasome activity. The inducible transcription factor NF-κB regulates divergent signaling pathways including inflammatory response and cancer development. NF-κB is canonically activated by preferential disposal of its inhibitory protein; IκB, which suppresses the nuclear translocation of NF-κB. Itself a major member of IκB family proteins, IκBα is phosphorylated by an IκB kinase (IKK) and subsequently undergoes polyubiquitylation by SCF(βTrCP1) ubiquitin-ligase in the presence of E$_1$ and E$_2$ and subsequent proteasomal degradation. GS143 suppresses IκBα ubiquitylation, markedly suppressing its destruction. The latter is stimulated by TNFα and a set of downstream responses coupled to NF-κB signaling, but not those of p53 and β-catenin *in vivo*. GS-143 also inhibits LPS-induced expression of inflammatory cytokines in human myelomonocytic cells. It suppresses antigen-induced NFκB expression, inflammation and mucus production in airways of OVA-sensitized mice. (Note: GS143 does not inhibit IκBα phosphorylation, MDM2-directed p53 ubiquitylation, and proteasome activity *in vitro*.) **1**. Hirose *et al.* (2008) *Biochem. Biophys. Res. Comm.* **374** 507. **2**. Nakajima, Fujiwara, Furuichi, Tanaka & Shimbara (2008) *Biochem. Biophys. Res. Comm.* **368**, 1007.

GS-1101

This orally bioavailable PI3Kδ inhibitor (FW = 415.43 g/mol), also known as CAL-101, Idelalisib®, and 5-fluoro-3-phenyl-2-([S])-1-[9H-purin-6-ylamino]propyl)-3H-quinazolin-4-one, targets the δ-isoform of the 110-kDa catalytic subunit of Class I phosphoinositide-3 kinase, showing potent immunomodulating and antineoplastic activities by through the inhibition of phosphatidylinositol-3,4,5-trisphosphate (PIP$_3$) formation. GS-1101 exhibits 40–300x selectivity over other PI3K isoforms (1-3). By preventing the activation of the PI3K signaling pathway, GS-1101 inhibits tumor cell proliferation, motility, and survival. Unlike other isoforms of PI3K, PI3K-δ is expressed primarily in hematopoietic lineages. GS-1101 also synergistically potentiates histone deacetylase inhibitor-induced proliferation inhibition and apoptosis through the inactivation of PI3K and extracellular signal-regulated kinase pathways (4). GS-1101 may also enjoy added therapeutic potential by targeting the infiltrative capacity of tumor-associated macrophages that is critical for their enhancement of tumor invasion and metastasis (5). **1**. Herman, Gordon, Wagner, *et al.* (2010) *Blood* **116**, 2078. **2**. Meadows, Vega, Kashishian *et al.* (2012) *Blood* **119**, 1897. **3**. Castillo, Furman & Winer (2012) *Expert Opin. Investig. Drugs* **21**, 15. **4**. Bodo, Zhao, Sharma, *et al.* (2013) *Brit. J. Haematol.* **163**, 72. **5**. Mouchemore, Sampaio, Murrey, *et al.* (2013) *FEBS J.* **280**, 5228.

GS-7977

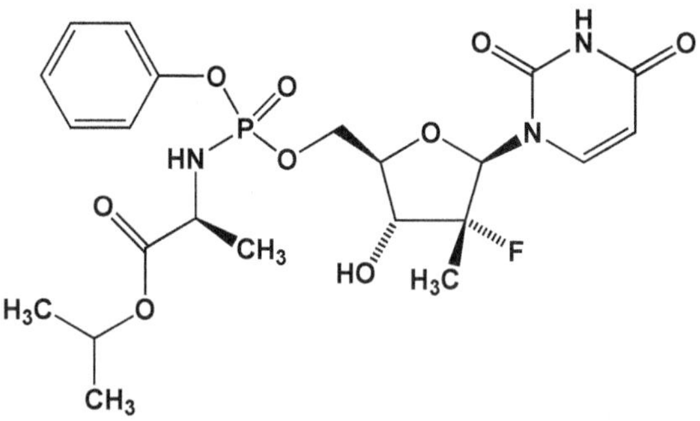

This antiviral pro-drug (FW = 387.44 g/mol; CAS = 1190307-88-0), also known as PPI-7977, Sofosbuvir®, and isopropyl (2S)-2-[[[(2R,3R,4R,5R)-5-(2,4-dioxopyrimidin-1-yl)-4-fluoro-3-hydroxy-4-methyltetrahydrofuran-2-yl]methoxyphenoxyphosphoryl]amino]propanoate, is metabolized into the active agent, 2'-deoxy-2'-α-fluoro-β-C-methyluridine-5'-monophosphate, which in turn inhibits Hepatis C virus RNA-dependent RNA polymerase. Clinical trials suggest Sofosbuvir plus Ribavirin for twelve weeks may be effective in previously untreated patients with HCV genotype 1, 2, or 3 infection (2). **1**. Sofia, Bao, Chang, *et al.* (2010) *J. Med. Chem.* **53**, 7202. **2**. Gane, Stedman, Hyland, et al. (2013) *New Engl. J. Med.* **368**, 34.

GS-8374

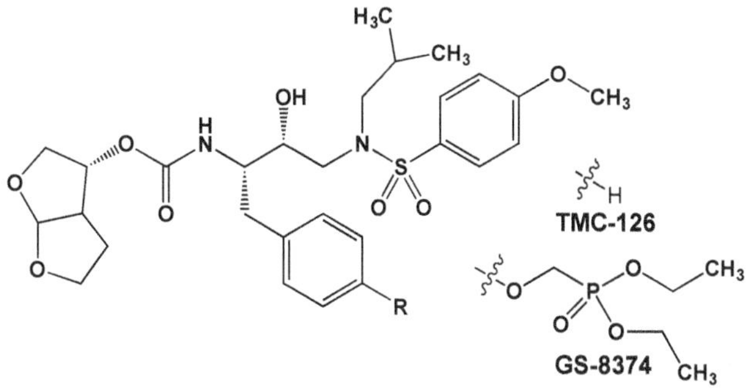

This prototypical phosphonate-containing protease inhibitor (FW = 728.79 g/mol), also named (3R,3aS,6aR)-hexahydrofuro[2,3-b]furan-3-yl-(2S,3R)-1-(4-((diethoxyphosphoryl)methoxy)-phenyl)-3-hydroxy-4-(N-isobutyl-4-methoxyphenylsulfonamido)butan-2-ylcarbamate, inhibits wild type HIV-1 protease (EC$_{50}$ = 0.01 nM) and antiviral-resistant forms (1). This diethylphosphonate derivative of TMC-126 (MW = 562.68 g/mol) (2,3) has a resistance profile that is superior to all clinically approved protease inhibitors. **1**. Grantz Šašková, Kožíšek, Stray, *et al.* (2014) *J. Virol.* **88**, 3586. **2**. Ghosh, Chapsal, Weber & Mitsuya (2008) *Acc. Chem. Res.* **41**, 78. **3**. Ghosh, Pretzer, Cho, Hussain & Duzgunes (2002) **54**, 29.

GS-9137, See *Elvitegravir*

GS-9350

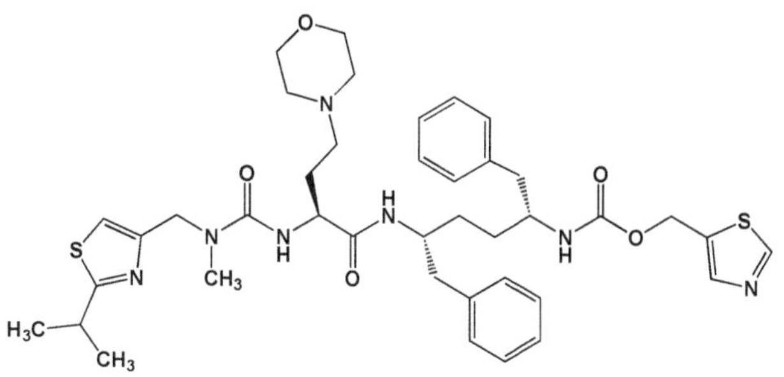

This potent mechanism-based inhibitor and pharmaco-enhancer (FW = 776.02 g/mol; CAS = 1004316-88-4; *Abbreviation*: COBI), also known as Cobicistat®, targets human CYP3A (*or* cytochrome P450 3A) isoforms (IC$_{50}$ = 30-285 nM), thereby boosting systemic exposure to co-administered agents that are otherwise deactivated by CYP3A (1). GS-9350 also displays weak action as an HIV proteinase inhibitor (IC$_{50}$ > 40-50 μM), allowing it to boost the action of higher-affinity anti-HIV drugs without risking selection of potential cobistat-resistant HIV variants. Cobistat also inhibits the intestinal efflux transporters P-glycoprotein and breast cancer resistance protein, thereby boosting the efficacy of HIV protease inhibitors atazanavir and darunavir and the lymphoid cell- and tissue-targeted pro-drug of the nucleotide analogue tenofovir (2). **1**. Mathias, German, Murray, *et al.* (2010) *Clin. Pharmacol. Ther.* **87**, 322. **2**. Lepist, Phan, Roy, *et al.* (2012) *Antimicrob. Agents Chemother.* **56**, 5409.

GSI-IX, See *DAPT*

GSK591

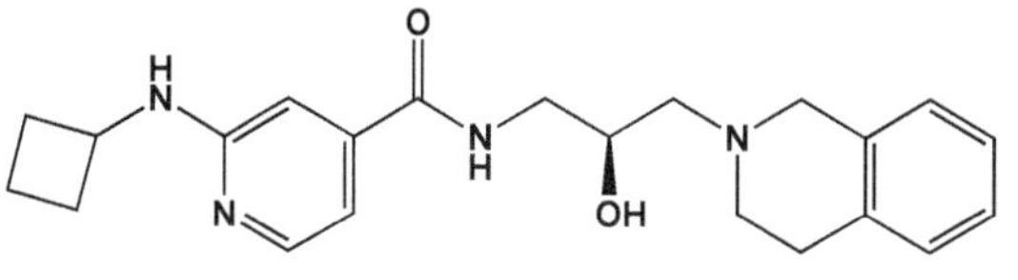

EPZ015866 (GSK3203591 or GSK591)

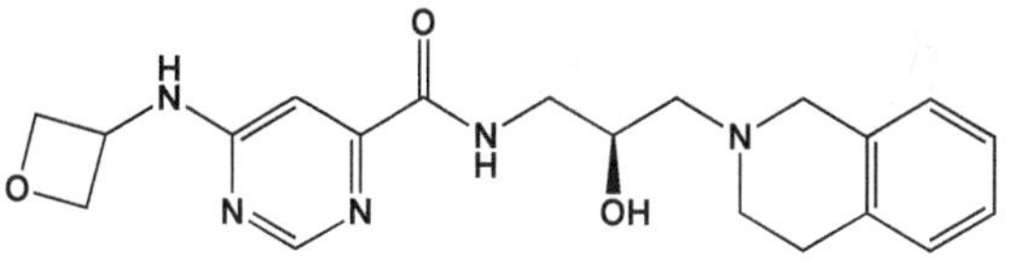

EPZ015666 (GSK3235025)

This PRMT5 inhibitor (FW = 380.49 g/mol; CAS = 1616391-87-7; Solubility: 79 mg/mL DMSO; < 1mg/mL H$_2$O), also known as (S)-2-(cyclobutylamino)-N-(3-(3,4-dihydroisoquinolin-2(1H)-yl)-2-hydroxypropyl)isonicotinamide, EPZ015866, GSK-3203591, GSK-591, and GSK 591, selectively targets Protein arginine methyltransferase-5 (IC$_{50}$ = 11 nM), an enzyme that catalyzes the symmetrical dimethylation of arginine residues in the small nuclear ribonucleoproteins SmD1 (SNRPD1) and SmD3 (SNRPD3), a step that is required in snRNP core particle assembly. PRMT5 also methylates SUPT5H and is likely to regulate its transcriptional elongation properties. PRMT5 also catalyzes the mono- and di-methylation of argininyl residues in myelin basic protein (MBP) *in vitro* and may play a role in cytokine-activated transduction pathways. PRMT5 also attenuates EGF signaling by monomethylating EGFR, RAF1 and probably BRAF. PRMT5 interacts with a number of binding partners that influence its substrate specificity. In an *in vitro* assay, GSK591 potently inhibits the PRMT5/MEP50 complex from methylating (histone) H4 (IC$_{50}$ = 11 nM). In Z-138 cells, GSK591 inhibits the symmetric arginine methylation of SmD3 (EC$_{50}$ = 56 nM). GSK591 is selective for PRMT5 (up to 50 μM) relative to a panel of methyltransferases. Reference-1 describes the design and optimization strategies using an initial hit compound with poor *in vitro* clearance to yield the *in vivo* tool compound EPZ015666 (GSK3235025) in addition to the potent *in vitro* tool molecule EPZ015866 (GSK-591). **1**. Duncan, Rioux, Boriak-Sjodin (2016) *ACS Med. Chem. Lett.* **7**, 162.

GSK621

This AMPK activator (FW = 489.91 g/mol; CAS 1346607-05-3; Solubility: 97 mg/mL DMSO; <1 mg/mL H_2O), also named 6-chloro-5-(2'-hydroxy-3'-methoxy[1,1'-biphenyl]-4-yl)-3-(3-methoxy-phenyl)-1H-pyrrolo[3,2-d]pyrimidine-2,4(3H,5H)-dione, potently and specifically targets Adenosine Monophosphate Kinase, doubling its basal activity at 1-2 µM and reaching nearly three-fold higher than basal at 8 µM. (Compare to corresponding values of 10 and 27 µM for A-769662) AMPK is a signal transduction kinase that acts either as a tumor suppressor or oncogene, depending on the context. GSK621 induces cytotoxicity by activating autophagy, independently of mTORC1 inhibition. Unexpectedly, activation of AMPK and constitutive mTORC1 signaling results in a synthetic lethal interaction across a range of acute myeloid leukemia (AML) primary samples and cell lines. 1. Sujobert, Poulain, Paubelle, et al. (2015) Cell Rep. 11, 1446.

GSK 2033

This potent cell-permeable LXR antagonist (FW = 591.66 g/mol; CAS = 1221277-90-2; Solubility: 20 mM in DMSO), also known as 2,4,6-Trimethyl-N-[[3'-(methylsulfonyl)[1,1'-biphenyl]-4-yl]methyl]-N-[[5-(trifluoromethyl)-2-furanyl]methyl]benzenesulfonamide, targets liver X receptor (K_i = 30 nM), a member of the nuclear receptor family of transcription factors that is closely related to nuclear receptors such as the PPARs, FXR and RXR (1,2). LXRs are important regulators of cholesterol, fatty acid, and glucose homeostasis. A ligand-bound cocrystal structure was determined which elucidated key interactions for high binding affinity (1). T-helper-17 (T_H17) cells, which are a subset of CD4$^+$ T cells characterized by the secretion of IL-17A, IL-17F, IL-21, and IL-22, play an important role in the regulation of immune responses against bacterial and fungal infections. These cells require the expression of both retinoic acid receptor-related orphan receptors α and γt (RORα and RORγt) for full differentiation. Although the LXR agonist T0901317 also displays high-affinity RORα and RORγt inverse activity, GSK-2003 was useful in demonstrating that it is T0901317's activity at RORα and γt, but not at LXR, that is facilitates the inhibition of T_H17 cell differentiation and function (2). 1. Zuercher, Buckholz, Campobasso, et al. (2010) J. Med. Chem. 53, 3412. 2. Solt, Kamenecka & Burris (2012) PLoS One 7, e46615.

GSK3787

This orally available PPARδ antagonist (FW = 392.78 g/mol; CAS = 188591-46-0; Solubility: 79 mg/mL DMSO; <1 mg/mL H_2O) selectively targets Peroxisome Proliferator-Activated Receptor δ (PPARδ), IC_{50} of 2.2 µM, without measurable affinity for hPPARα or hPPARγ. Mass spectral analysis confirms covalent binding to Cys249 within the PPARδ binding pocket (1). GW0742 induces up-regulation of Angptl4 as well as Adrp mRNA expression in wild-type mouse colon, but not in PPARβ/δ-null mouse colon, correlating well with reduced promoter occupancy of PPARβ/δ on genes for Angiopoietin-like 4, or Angptl4, and Adipose differentiation-related protein, Adrp. 1. Shearer, Wiethe, Ashe, et al. (2010) J. Med. Chem. 53, 1857. 2. Palkar, Borland, Naruhn, et al. (2010) Mol. Pharmacol. 78, 419.

GSK256066

This quinolone-based PDE inhibitor (FW = 518.55 g/mol), also named 6-({3-[(dimethylamino)carbonyl]phenyl}sulfonyl)-8-methyl-4-{[3-methyl-oxy)phenyl]amino}-3-quinolinecarboxamide, shows exceptionally high affinity (apparent IC_{50} = 3.2 pM; steady-state IC_{50} < 0.5 pM) for phosphodiesterase PDE4, abd exhibits >380,000-fold selectivity toward PDE4 over PDE1, PDE2, PDE3, PDE5, and PDE6, and >2500-fold against PDE7. 1. Woodrow, Ballantine, Barker, et al. (2009) Bioorg. Med. Chem. Lett. 19, 5261.

GSK-3203591, See GSK591

GSK 429286A

This potent PKI and vasodilator (FW = 432.37 g/mol for the monohydrochloride; CASs = 864082-47-3; Solubility: 90 mg/mL DMSO, <1 mg/mL H_2O), also known as N-(6-fluoro-1H-indazol-5-yl)-2-methyl-6-oxo-4-(4-(trifluoromethyl)phenyl)-1,4,5,6-tetra-hydropyridine-3-carbox-amide, targets ROCK1 and ROCK2, with IC_{50} values of 14 nM and 63 nM, respectively. RhoA/Rho kinases (abbreviated ROCK) play important roles in mediating vasoconstriction and vascular remodeling in the pathogenesis of pulmonary hypertension. GSK429286A slightly inhibits RSK and p70S6K with IC_{50} values of 0.78 µM and 1.94 µM, respectively. 1. Goodman, et al. (2007) J. Med. Chem. 50, 6. 2. Nichols, et al. (2009) Biochem. J. 424, 47.

GSK461364

This protein kinase inhibitor ($FW_{free-acid}$ = 543.60 g/mol; CASs = 929095-18-1 and 1000873-97-1; Solubility: 30 mg/mL DMSO; <1 mg/mL H_2O), also known systematically as 5-(6-((4-methylpiperazin-1-yl)-methyl)-1H-benzo[d]imidazol-1-yl)-3-((R)-1-(2-(trifluoromethyl)-phenyl)ethoxy)thiophene-2-carboxamide, targets the Polo-like kinase Plk1 (K_i = 2.2 nM), bringing about G_2/M cell-cycle arrest and cell death. RNA silencing of WT p53 increases the antiproliferative activity of GSK461364, suggesting that cancers with defective p53 would likewise be sensitive to co-administration of GSK461364 and DNA-damaging drugs. **1**. Gilmartin, et al. (2009) *Cancer Res.* **69**, 6969. **2**. Degenhardt, et al. (2010) *Mol. Cancer Ther.* **9**, 2079. **3**. Olmos, et al. (2011) *Clin. Cancer Res.* **17**, 3420.

GSK525762, *See I-BET-762*

GSK525762A, *See I-BET-762*

GSK690693

This pan-Akt inhibitor (F.W. = 425.48; CAS = 937174-76-0; Solubility (25°C): 39 mg/mL DMSO, <1 mg/mL H_2O), known systematically as 4-(2-(4-amino-1,2,5-oxadiazol-3-yl)-1-ethyl-7-((S)-piperidin-3-ylmethoxy)-1H-imidazo[4,5-c]-pyridin-4-yl)-2-methylbut-3-yn-2-ol, targets Akt1, Akt2 and Akt3 Protein Kinases (*or* PKB) with IC_{50} of 2 nM, 13 nM, and 9 nM, respectively. **1**. Rhodes, *et al.* (2008) *Cancer Res.* **68**, 2366. **2**. Levy, *et al.* (2009) *Blood* **113**, 1723. **3**. Altomare, *et al.* (2010) *Clin. Cancer Res.* 2010, **16**, 486.

GSK1059615

This potent and selective PI3Kα/PI4KIIIβ inhibitor (F.Wt. = 333.36; CAS = 958852-01-2, 1356195-42-0 (monohydrate, Na); Solubility (25°C): 2 mg/mL DMSO, <1 mg/mL Water), also known as (Z)-5-((4-(pyridin-4-yl)-quinolin-6-yl)methylene)thiazolidine-2,4-dione, has the following inhibitory profile: PI3Kα, IC_{50} = 0.4 nM; PI3Kβ, IC_{50} = 0.6 nM; PI3Kδ, IC_{50} = 2 nM; and PI3Kγ mTOR, IC_{50} = 0.4 nM. 1. Carnero (2009) *Expert Opin. Investig. Drugs* 18, 1265. 2. Knight, *et al.* (2010) *ACS Med. Chem. Lett.* 1, 39.

GSK1120212, *See Trametinib*

GSK1265744

This orally available strand-transfer inhibitor and potent antiviral (FW = 405.36 g/mol; CAS: 1051375-10-0), also known as GSK-1265744, S-265744, targets HIV integrase (with a low-nM IC_{50}), an essential enzyme that splices a viral DNA (vDNA) replica of its genome into host cell chromosomal DNA (hDNA) and is a promising druggable target for anti-AIDS agents. GSK1265744 monotherapy significantly reduces plasma HIV-1 RNA >100x below baseline to day-11 in HIV-1-infected subjects receiving 5 or 30 mg *versus* placebo. The mean plasma half-life is 31.5 hours, and a nanoparticle-based formulation has a half-life of 21–50 days. **1**. Spreen, Min, Ford, *et al.* (2013) *HIV Clin. Trials* **14**, 192.

GSK1322322

This first-in-class, orally bioavailable antibiotic (FW = 479.55 g/mol; CAS = 1152107-25-9) targets bacterial peptide deformylase (EC 3.5.1.88; *Reaction*: N-Formyl-L-methionyl peptide + H_2O ⇌ Formate + Methionyl peptide) which is required for protein maturation in prokaryotes. Intended for the treatment of complicated bacterial skin infections and hospitalized community-acquired pneumonia, GSK1322322 is active against such multidrug-resistant, Gram-negative pathogens as *Staphylococcus aureus* ATCC 29213 (MIC, 1-4 μg/mL), *Haemophilus influenzae* ATCC 49247 (MIC, 0.5-4 μg/mL), and *Streptococcus pneumoniae* ATCC 49619 (MIC, 0.12-0.5 μg/mL) (1). Upon single-dose administration, GSK1322322 is absorbed rapidly, with median time-to-maximum plasma concentration of 0.5 to 1.0 h (2). **1**. Ross, Scangarella-Oman, Miller, Sader & Jones (2011) *J. Clin. Microbiol.* **49**, 3928. **2**. Naderer, Dumont, Zhu, Kurtinecz & Jones (2005) *Antimicrob. Agents Chemother.* **57**, 2005.

GSK1363089, *See Foretinib*

GSK1562590

This orally active, high-affinity UT receptor antagonist ($FW_{HCl-Salt}$ = 617.95 g/mol; CAS = 1003878-07-6; Soluble to 100 mM in DMSO; N-[(1R)-1-[3'-(aminocarbonyl)-[1,1'-biphenyl]-4-yl]-2-(1-pyrrolidinyl)ethyl]-6,7-dichloro-2,3-dihydro-N-methyl-3-oxo-4H-1,4-benzoxazine-4-acetamide) selectively targets recombinant urotensin II receptors in monkey (pK_i = 9.14), human (pK_i = 9.28), mouse (pK_i = 9.34), cat (pK_i = 9.64) and rat (pK_i 9.66). GSK1562590 exhibits selectivity for UT receptors over a range of G protein-coupled receptors, ion channels, enzymes and neurotransmitter transporters. GSK1562590 also inhibits the hU-II-induced systemic pressor response in anaesthetized cats at a 10x lower dose than GSK1440115. The antagonistic effects GSK1562590 could not be reversed by washout in rat isolated aorta, suggesting a long receptor residence time. **1.** Behm, Aiyar, Olzinski, *et al.* (2010) *Brit. J. Pharmacol.* **161**, 207.

GSK1904529A

This orally available and reversible ATP-competitive inhibitor (FW = 851.96 g/mol; CAS = 1089283-49-7; Solubility: 170 mg/mL DMSO; <1 mg/mL H_2O), also known as *N*-(2,6-difluorophenyl)-5-(3-(2-(5-ethyl-2-methoxy-4-(4-(4-(methylsulfonyl)-piperazin-1-yl)-piperidin-1-yl)-phenyl-amino)pyrimidin-4-yl)*H*-imidazo[1,2-*a*]pyridin-2-yl)-2-methoxybenzamide, targets Insulin-like Growth Factor Receptor (IGF-IR) and Insulin Receptor (IR) with K_i values of 86 nM, blocking receptor autophosphorylation and downstream signaling, leading to cell cycle arrest. **1.** Sabbatini, Roland, Groy, *et al.* (2009) *Clin. Cancer Res.* **15**, 3058.

GSK2126458

This highly selective phospho-inositide-3-linase inhibitor (F.Wt. = 505.53; CAS = 1086062-66-9; Solubility (25°C): 100 mg/mL DMSO, <1 mg/mL Water), also known as 2,4-difluoro-*N*-(2-methoxy-5-(4-(pyridazin-4-yl)quinolin-6-yl)pyridin-3-yl)benzenesulfonamide, potently inhibits p110α, p110β, p110γ, p110δ, as well as mammalian target of rapamycin mTORC1 and mTORC2 with K_i of 0.019 nM, 0.13 nM, 0.024 nM, 0.06 nM, 0.18 nM and 0.3 nM, respectively. **1.** Knight, *et al.* (2010) *ACS Med. Chem. Lett.* **1**, 39. **2.** Greger, *et al.* (2012) *Mol. Cancer Ther.* **11**, 909.

GSK2190915

This orally bioavailable and potent 5-lipoxygenase-activating protein, *or* FLAP, inhibitor (FW = 637.84; CAS = 936350-00-4), also known by anther code name, AM803, and systematically as 3-[3-*tert*-butylsulfanyl-1-[4-(6-ethoxypyridin-3-yl)benzyl]-5-(5-methylpyridin-2-ylmethoxy)-1*H*-indol-2-yl]-2,2-dimethylpropionic acid, inhibits the synthesis of leukotrienes B₄ (LTB₄), IC₅₀ = 76 nM, as well as 5-oxo-6,8,11,14-eicosatetraenoic acid (5-oxo-ETE), and potency of 2.9 nM in FLAP binding assays. When rat lungs were challenged *in vivo* with calcium-ionophore, GSK2190915 inhibited LTB₄ formation and cysteinyl leukotriene (CysLT) production with ED₅₀ values of 0.12 mg/kg and 0.37 mg/kg, respectively (2). GSK2190915 also increased survival time in mice exposed to a lethal intravenous injection of platelet activating factor, *or* PAF (2). As such, GSK2190915 shows promise in the treatment of asthma, acute inflammation, and a model of lethal shock. GSK2190915 likewise inhibits the early asthmatic response (EAR) due to inhaled allergen. In nineteen huma subjects with mild asthma, GSK2190915 50 mg attenuated the EAR similarly to 100 mg of the same in our earlier study, suggesting 50 mg is already at the top of the dose–response curve (3). On the other hand, GSK2190915 10 mg is suboptimal. **1.** Stock, Bain, Zunic, *et al.* (2011) *J. Med. Chem.* **54**, 8013. **2.** Lorrain, Bain, Correa, *et al.* (2010) *Eur. J. Pharmacol.* **640**, 211. **3.** Singh, Boyce, Norris, Kent & Bentley (2013) *Int. J. Gen. Med.* **6**, 897.

GSK2248761A, See *Fosdevirine*

GSK2334470

This protein kinase inhibitor (FW = 462.59 g/mol; CAS = 1227911-45-6; Soluble in DMSO) targets 3-phosphoinositide-dependent protein kinase, *or* PDK1 (IC₅₀ ~10 nM), without affecting the activity of ninety-three other protein kinases (including Aurora, ROCK, p38 MAPK and PI3K), even at 5 μM, including thirteen PDK1-like kinases (1). PDK1 activates a subset of protein kinases A, C, and G, of which some play roles in cancer cell proliferation. Addition of GSK2334470 to HEK-293 (human embryonic kidney), U87 or MEF (mouse embryonic fibroblast) cells ablated T-loop phosphorylation and activation of Akt, but was more efficient at inhibiting Akt in response to stimuli such as serum that activated the PI3K (phosphoinositide 3-kinase) pathway weakly. GSK2334470 also inhibited activation of an Akt1 mutant lacking its pleckstrin homology (PH) domain more potently than full-length Akt1, suggesting that GSK2334470 may be more effective at inhibiting PDK1 substrates in the cytosol rather than those at the plasma membrane (1). Cancer cell resistance to this PDK1 inhibitor results from Akt being efficiently recruited to PDK1 via two alternative mechanisms: the first involves ability of Akt and PDK1 to mutually interact with the PI3K second messenger PtdIns(3,4,5)P₃, and the second involves PDK1 recruitment to Akt after its phosphorylation at Ser(473) by mTORC2, by means of a substrate-docking motif termed the PIF-pocket (2). **1.** Najafov, Sommer, Axten, *et al.* (2011) *Biochem.J.* **433**, 357. **2.** Najafov, Shpiro & Alessi (2012) *Biochem. J.* **448**, 285.

GSK2336805

This hepatitis C virus (HCV) inhibitor (FW = 465.95 g/mol) exhibits picomolar antiviral action against the standard genotypes (*i.e.*, 1a, 1b and 2a subgenomic replicons), but was inactive on twenty-two other RNA and DNA viruses tested. Changes in the N-terminal region of NS5A, a protein playing a key role in HCV replication, decreases the antiviral activity of GSK2336805. Such mutations in the genotype 1b replicon show modest shifts in compound activity (<13x), whereas mutations identified in the genotype 1a replicon have a more dramatic impact on potency. Combination and cross-resistance studies demonstrated that GSK2336805 could be used as a component of a multi-drug HCV regimen employing compounds with different mechanisms of action. **1.** Walker, Crosby, Wang, *et al.* (2013) *Antimicrob. Agents Chemother.* **58**, 38.

GSK2118436, *See Dabrafenib*

GSK2578215A

This potent, brain-penetrating Leucine-Rich Repeat Kinase-2, *or* LRRK2 inhibitor (FW = 399.42 g/mol; CAS = 1285515-21-0; Soluble to 100 mM in DMSO), also named 5-(2-fluoro-4-pyridinyl)-2-(phenylmethoxy)-*N*-3-pyridinylbenzamide, has IC_{50} values of 10 and 8 nM, respectively, for the wild type enzyme and its [Gly-2019-Ser]-mutant. GSK2578215A displays selectivity for LRRK2 *versus* 460 other protein kinases. GSK2578215A blocks Ser-910 and Ser-935 phosphorylation of both wild-type LRRK2 and G2019S mutant at a concentration of 0.3-1.0 µM in cells and in mouse spleen and kidney, but not in brain, following intraperitoneal injection of 100mg/kg. **1.** Reith, Bamborough & Jandu, *et al.* (2012) *Bioorg. Med. Chem. Lett.* **22**, 5625.

GSK 2606414

This orally available PERK inhibitor (FW = 451.44 g/mol; CAS = 1337531-36-8; Solubility: 100 mM in DMSO; 20 mM in H_2O for mono-HCl with gentle warming), also named 1-[5-(4-amino-7-methyl-7*H*-pyrrolo[2,3-*d*]pyrimidin-5-yl]-2,3-dihydro-1*H*-indol-1-yl]-2-[3-(trifluoromethyl)phenyl]ethanone, potently and selectively targets protein kinase R-like ER kinase, *or* PERK (IC_{50} = 0.4 nM), inhibiting thapsigargin-induced PERK phosphorylation in lung carcinoma A549 cells and attenuating subcutaneous pancreatic human tumor xenograft growth in mice.

GSK 2656157

This ATP-competitive protein kinase inhibitor (FW = 416.45 g/mol; CAS = 1337532-29-2; Solubility: 37 mg/mL DMSO, <1 mg/mL H_2O), also named 1-(5-(4-amino-7-methyl-7*H*-pyrrolo[2,3-*d*]pyrimidin-5-yl)-4-fluoroindolin-1-yl)-2-(6-methylpyridin-2-yl)ethanone, targets Protein kinase RNA-like Endoplasmic Reticulum Kinase (PERK), IC_{50} = 0.9 nM, showing 500x greater selectivity versus a panel of 300 kinases. PERK is an enzyme within the Unfolded Protein Response (UPR) signal transduction cascade that coordinates cellular adaptation to microenvironmental stresses that include hypoxia, nutrient deprivation, and change in redox status. GSK2656157 inhibits PERK activity in cells with an IC_{50} values of 10-30 nM, as shown by inhibition of stress-induced PERK autophosphorylation, eIF2α substrate phosphorylation, together with corresponding decreases in ATF4 and CAAT/enhancer binding protein homologous protein (CHOP) in multiple cell lines. Oral administration to mice shows both dose- and time-dependent pharmacodynamic responses in pancreas, as measured by PERK autophosphorylation. Twice daily dosing of GSK2656157 results in dose-dependent inhibition of multiple human tumor xenografts growth in mice. **1.** Atkins, Liu, Minthorn, *et al.* (2013) *Cancer Res.* **73**, 1993.

GSK J1

This JMJD3-selective inhibitor ($FW_{free-acid}$ = 388.45 g/mol; CAS = 1373422-53-7; λ_{max} = 253 nm), also named 3-((6-(4,5-dihydro-1*H*-benzo[*d*]azepin-3(2*H*)-yl)-2-(pyridin-2-yl)pyrimidin-4-yl)amino)propanoate, targets the histone H3 lysine-27 (*or* H3K27) demethylase known as JMJD3 (1). Jumonji family histone demethylases are Fe^{2+}/α-ketoglutarate-dependent oxygenases in regulatory transcriptional chromatin complexes. They play critical roles in transcriptional control of cell differentiation, development, inflammation, and cancer by silencing through the polycomb-repressive complex PRC1 or PRC2. GSK-J1's polar carboxylate restricts its cellular permeability, but its prodrug ethyl ester GSK-J4 is cell-penetrating. GSK-J1 shows significant activity (IC_{50} = 60 nM for human JMJD3) *in vitro* and in cells using an ester derivative (GSK-J4: 1 µM < IC_{50} < 10 µM). Another pyridine regio-isomer, known as GSK-J2, displays significantly less on-target activity (IC_{50} > 100 µM for human JMJD3) and can be used as control for target effects *in vitro*, and as ester derivative (GSK-J5) in cells. GSK-J1 also shows some activity (IC_{50} = 950 nM for Jarid1b, IC_{50} = 1.8 µM for Jarid1c) *versus* H3K4me3/2/1 demethylases. GSK-J1 has no effect on more than 100 different kinases or other unrelated proteins, including other chromatin-modifying enzymes such as histone deacetylases. *See also references 2-4 for comments on the selectivity of GSK-J1.* **1.** Kruidenier, Chung, Cheng, *et al.* (2012) *Nature* **488**, 404. **2.** Heinemann, Nielsen, Hudlebusch, *et al.* (2014) *Nature* **514**, E1. **3.** Kruidenier, Chung, Cheng, Liddle, *et al.* (2014) *Nature* **1514**, E2. **4.** http://www.thesgc.org/chemical-probes/GSKJ1

GSK J4

This GSK J1 pro-drug ($FW_{free-base}$ = 417.51 g/mol; CAS = 1373423-53-0; λ_{max} = 253 nm) that is cell-permeable and, once within cells, readily transformed into GSK J1 (*See GSK J1*), the first selective inhibitor of the H3K27 histone demethylase JMJD3 and UTX (IC_{50} = 60 nM), but is otherwise inactive against a panel of JMJ family demethylases. The jumonji (JMJ) family of histone demethylases are Fe^{2+}- and α-ketoglutarate-dependent oxygenases that are essential components of regulatory transcriptional chromatin complexes. These enzymes demethylate lysine residues in histones in a methylation-state and sequence-specific context. **1.** Kruidenier, Chung, Cheng, *et al.* (2012) *Nature* **488**, 404.

GSK-LSD1

This selective LSD1 inhibitor ($FW_{\text{free-base}}$ = 216.33 g/mol; FW_{2HCl} = 289.24 g/mol; CAS = 1431368-48-7(free-base)), also named *N*-((1*R*,2*S*)-2-phenylcyclopropyl)piperidin-4-amine, targets Lysine-Specific Demethylase 1, *or* LSD1 (IC_{50} = 16 nM), showing > 1000 fold selective over other closely related FAD utilizing enzymes (*i.e.*, LSD2, and monoamine oxidases MAO-A, MAO-B). LSD1 belongs to the flavin adenine dinucleotide (FAD)-dependent amine oxidase family that also includes monoamine oxidases (MAOs) and polyamine oxidase (PAO). LSD1 demethylates mono- and dimethylated histone H3 at Lysine-4, resulting in transcriptional repression ans controls p53 tumor suppressor activity of p53 by demethylation at Lysine-370. **1.** Shi, Lan, Matson, *et al.* (2004) *Cell* **119**, 941. **2.** Forneris, Binda, Vanoni, *et al.* (2005) *J. Biol. Chem.* **280**, 41360. **3.** Huang, Sengupta, Espejo, *et al.* (2007) *Nature* **449**, 105. **4. See also** http://www.thesgc.org/chemical-probes/LSD1

GsMTx4

GCLEFWWKCNPNDDKCCRPKLKCSKLFKLCNFSF-NH₂

This first-in-class, 35-residue peptide toxin (FW = 4095.86 g/mol; Soluble to 1 mg/mL in H_2O), first isolated from the venom of the tarantula *Grammostola spatulata*, is a cysteine knot family member that exploits its amphipathic structure to block cation-selective stretch-activated channels (SACs), K_d = 0.63 µM, based on the ratio of experimentally determined dissociation and association rate constants (1). The ability to bind to and burrow into phospholipid membranes is most likely a key to its mechanism of action, bending boundary lipids toward the channel and making the membrane appear thinner (2). In hypotonically swollen astrocytes, GsMTx-4 produces approximately 40% reduction in swelling-activated whole-cell current (1). Quite remarkably, a peptide (*en*GsMTx-4) of similar sequence, but containing D-amino acids, shows a similar K, suggesting that the binding of GsMTx4 is unlikely rely on stereochemical recognition (3). **1.** Suchyna, Johnson, Hamer, *et al.* (2000) *J. Gen. Physiol.* **115**, 583. (*Erratum in*: *J. Gen. Physiol.* (2001) **117**, 371.) **2.** Bowman, Gottlieb, Suchyna, Murphy & Sachs (2007) *Toxicon.* **49**, 249. **3.** Suchyna, Tape, Koeppe, 2nd, *et al.* (2004) *Nature* **430**, 235.

GTS-21

This orally bioavailable anabaseine analogue and partial nicotinic acetylcholine receptor agonist ($FW_{\text{free-base}}$ = 308.38 g/mol; $FW_{\text{diHCl-Salt}}$ = 381.31 g/mol; CAS = 156223-05-1), also named 3-[(2,4-dimethoxy-phenyl)methylene]-3,4,5,6-tetrahydro-2,3'-bipyridine and DMXBA, targets α_7-nicotinic acetylcholine receptors, showing cognition-enhancing effects in mammals including humans (1-4). Although GTS-21 selectively activates the α_7-nicotinic receptor, at higher concentrations, it is also an antagonist at $\alpha_4\beta_2$ and 5-HT$_3$ receptors. GTS-21 dose-dependently facilitates the induction of long-term potentiation (LTP) in the hippocampus (5). Sustained administration of GTS-21 improves deficient sensory inhibition

of repetitive auditory stimuli, a biomarker common in schizophrenics, without causing tachyphylaxis or receptor up-regulation in DBA/2 mice (6). The benzylidene substituent provides an extended π-electron bond system that suppresses dehydropiperidine ring opening (as occurs in anabaseine) and thereby maximizes the fraction of its pharmacologically active closed-ring form (7) (*See Anabasine & Anabaseine*). **1.** Woodruff-Pak, Li & Kem (1994) *Brain Res.* **645**, 309. **2.** De Fiebre, Meyer, Henry, Muraskin, Kem, Papke (1995) *Mol. Pharmacol.* **47**, 164. **3.** Arendash, Sengstock, Sanberg, Kem (1995) *Brain Res.* **674**, 252. **4.** Olincy, Harris, Johnson, Pender *et al.* (2006) *Arch. Gen. Psychiat.* **63**, 630. **5.** Hunter, de Fiebre, Papke, *et al.* (1994) *Neurosci. Lett.* **168**, 130. **6.** Stevens, Cornejo, Adams, *et al.* (2010) *Brain Res.* **1352**, **140**. **7.** Zoltewicz, Prokai-Tatrai, Bloom, Kem (1993) *Heterocycles* **35**, 171.

Guaiacol

This photosensitive phenol (FW = 124.14 g/mol; CAS = 90-05-1), also known as 2-methoxyphenol and *o*-hydroxyanisole, has been used as a hydrogen donor in assays of peroxidases. It is also an alternative substrate for laccase. Guaiacol is also produced as a pheromone within the gut of the desert locust, *Schistocerca gregaria*. (This compound should not be confused with the similarly named sesquiterpene, guaiol.) **Target(s):** acetaldehyde dehydrogenase (1); catalase (2); catechol 2,3-dioxygenase (3,4); linoleate diol synthase (5); 5-lipoxygenase, *or* arachidonate 5-lipoxygenase (6); 15-lipoxygenase, *or* arachidonate 15-lipoxygenase (6); prostaglandin H synthase (7); shikimate dehydrogenase (8,9). **1.** Gupta, Mat-Jan, Latifi & Clark (2000) *FEMS Microbiol. Lett.* **182**, 51. **2.** Garcia, Kaid, Vignaud & Nicolas (2000) *J. Agric. Food Chem.* **48**, 1050. **3.** Bertini, Briganti & Scozzafava (1994) *FEBS Lett.* **343**, 56. **4.** Arciero, Orville & Lipscomb (1985) *J. Biol. Chem.* **260**, 14035. **5.** Su, Sahlin & Oliw (1998) *J. Biol. Chem.* **273**, 20744. **6.** Dohi, Anamura, Shirakawa, Okamoto & Tsujimoto (1991) *Jpn. J. Pharmacol.* **55**, 547. **7.** Thompson & Eling (1989) *Mol. Pharmacol.* **36**, 809. **8.** Balinsky & Dennis (1970) *Meth. Enzymol.* **17A**, 354. **9.** Balinsky & Davies (1961) *Biochem. J.* **80**, 296.

Guaifenesin

This over-the-counter expectorant (FW = 198.22 g/mol; CAS = 93-14-1), also known as guaiphenesin, glyceryl guaiacolate, Mucinex®, and (*RS*)-3-(2-methoxyphenoxy)propane-1,2-diol, loosens mucus in the lungs, making it easier to cough up otherwise congestive mucus associated with chest colds. Guaifenesin increases the analgesic effects of acetaminophen and aspirin and increases the sedative effects of alcohol, tranquilizers, sedative, and anesthetics.

Guangxitoxins

EGECGGFWWKCGSGKPACCPLYVCSPKWGLCNFPMP

These peptide toxins (Symbol: GxTX) from the venom of the tarantula *Plesiophrictus guangxiensis* primarily inhibit outward voltage-gated $K_V2.1$ potassium channel currents. GxTX-1D (*Sequence*: DGECGGFWWKCG SGKPACCPLYVCSPKWGLCNFPMP; MW = 3922 g/mol). GxTX-1E (*Sequence*: EGECGGFWWKCGSGKPACCPLYVCSPKWGLCNFPMP; MW = 3936 g/mol). GxTX-2 (*Sequence*: ECRKMFGGCSVDSDCCAHL GCKPTLKYCAWDGT; MW = 3582 g/mol). The molecular structure of GxTX-1E is similar to those of tarantula toxins that target voltage sensors in Kv channels in that it contains an ICK motif, composed of beta-strands, and contains a prominent cluster of solvent-exposed hydrophobic residues surrounded by polar residues. Notably, inhibition of Kv2.x channels by

GxTX-1E or the small molecule inhibitor RY796 enhances glucose-stimulated insulin secretion in isolated wild-type mouse and human islets, but not in islets from Kv2.1(-/-) mice (1). In wild-type mice, neither inhibitor improved glucose tolerance *in vivo*. GxTX-1E and RY796 also enhance somatostatin release in isolated human and mouse islets and in situ perfused pancreata from WT and $K_V2.1$(-/-) mice (1). The NMR structure of the toxin reveals an amphipathic part and an inhibitor cystine knot motif (2). the structural architecture of GxTX-1E is also similar to that of JZTX-III, a tarantula toxin that interacts with $K_V2.1$ with low affinity (2). The most striking structural differences between GxTX-1E and JZTX-III are found in the orientation between the first and second cysteine loops and the C-terminal region of the toxins, suggesting that these regions of GxTX-1E are responsible for its high affinity (2). **1.** Li, Herrington, Petrov, *et al.* (2013) *J. Pharmacol. Exp. Ther.* **344**, 407. **2.** Lee, Milescu, Jung, *et al.* (2010) *Biochemistry* **49**, 5134.

γ-Guanidinobutyrate (γ-Guanidinobutyric Acid)

This guanidine-containing metabolite (FW = 145.16 g/mol; CAS = 463-00-3), which arises in marine invertebrates by amino acid oxidase action on arginine and in mammals by transamidination involving γ-aminobutyrate and an amidine donor, such as arginine, inhibits carboxypeptidase B, arginine 2-monooxygenase, and arginase. **Target(s):** Arginase (1,2); arginine decarboxylase (3); arginine deaminase (4); arginine 2-monooxygenase (5); carboxypeptidase B (6); guanidinoacetate kinase, weakly inhibited (7); 5-guanidino 2-oxopentanoate decarboxylase, *or* α-ketoarginine decarboxylase, K_i = 4.2 mM (8); putrescine carbamoyltransferase (9). **1.** Moreno-Vivian, Soler & Castillo (1992) *Eur. J. Biochem.* **204**, 531. **2.** Shimotohno, Iida, Takizawa & Endo (1994) *Biosci. Biotechnol. Biochem.* **58**, 1045. **3.** Smith (1979) *Phytochemistry* **18**, 1447. **4.** Park, Hirotani, Nakano & Kitaoka (1984) *Agric. Biol. Chem.* **48**, 483. **5.** Olomucki, Dangba & Nguyen Van Thoai (1964) *Biochim. Biophys. Acta* **85**, 480. **6.** Folk (1971) *The Enzymes*, 3rd ed., **3**, 57. **7.** Shirokane, Nakajima & Mizusawa (1991) *Agric. Biol. Chem.* **55**, 2235. **8.** Vanderbilt, Gaby, Rodwell & Bahler (1975) *J. Biol. Chem.* **250**, 5322. **9.** Stalon (1983) *Meth. Enzymol.* **94**, 339.

2-Guanidinoethylmercaptosuccinic Acid

This arginine isostere and potent enkephalin convertase inhibitor (FW = 235.26 g/mol; CAS = 77482-44-1), often abbreviated GEMSA), binds reversibly to membrane-bound and soluble forms. **Target(s):** carboxypeptidase B (1-6); carboxypeptidase D (7); carboxypeptidase E, *or* carboxypeptidase H, *or* enkephalin convertase, K_i = 8.8 nM (3,5,8-19); carboxypeptidase M (5,12,20-22); carboxypeptidase U (5,23-29); carboxypeptidase Z (5,30); lysine carboxypeptidase, *or* lysine(arginine) carboxypeptidase, *or* carboxypeptidase N (3,5,31-34); metallocarboxypeptidase D (8,35-40); thrombin activatable fibrinolysis inhibitor (41,42). **1.** Avilés & Vendrell (1998) in *Handb. Proteolytic Enzymes*, p. 1333, Academic Press, San Diego. **2.** McKay & Plummer, Jr. (1978) *Biochemistry* **17**, 401. **3.** Fricker, Plummer & Snyder (1983) *Biochem. Biophys. Res. Commun.* **111**, 994. **4.** Mackin & Noe (1987) *Endocrinology* **120**, 457. **5.** Mao, Colussi, Bailey, *et al.* (2003) *Anal. Biochem.* **319**, 159. **6.** Eaton, Malloy, Tsai, Henzel & Drayna (1991) *J. Biol. Chem.* **266**, 21833. **7.** Latchinian-Sadek & Thomas (1993) *J. Biol. Chem.* **268**, 534. **8.** Fricker (2002) *The Enzymes*, 3rd ed., **22**, 421. **9.** Bommer, Nikolarakis, Noble & Herz (1989) *Brain Res.* **492**, 305. **10.** Strittmatter, Lynch & Snyder (1984) *J. Biol. Chem.* **259**, 11812. **11.** Fricker & Snyder (1983) *J. Biol. Chem.* **258**, 10950. **12.** Deddish, Skidgel & Erdös (1989) *Biochem. J.* **261**, 289. **13.** Juvvadi, Fan, Nagle & Fricker (1997) *FEBS Lett.* **408**, 195. **14.** Stone, Li & Bernasconi (1994) *Arch. Insect Biochem. Physiol.* **27**, 193. **15.** Fricker (1988) *Ann. Rev. Physiol.* **50**, 309.

16. Davidson & Hutton (1987) *Biochem. J.* **245**, 575. **17.** Kumar (1996) *Adv. Exp. Med. Biol.* **410**, 319. **18.** Grimwood, Plummer & Tarentino (1989) *J. Biol. Chem.* **264**, 15662. **19.** Hook & Affolter (1988) *FEBS Lett.* **238**, 338. **20.** Skidgel (1998) in *Handb. Proteolytic Enzymes*, p. 1347, Academic Press, San Diego. **21.** Tan, Deddish & Skidgel (1995) *Meth. Enzymol.* **248**, 663. **22.** Tan, Balsitis, Black, *et al.* (2003) *Biochem. J.* **370**, 567. **23.** Hendriks (1998) in *Handb. Proteolytic Enzymes*, p. 1328, Academic Press, San Diego. **24.** Wang, Hendricks & Scharpe (1994) *J. Biol. Chem.* **269**, 15937. **25.** Bouma, Marx, Mosnier & Meijers (2001) *Thromb. Res.* **101**, 329. **26.** Boffa, Wang, Bajzar & Nesheim (1998) *J. Biol. Chem.* **273**, 2127. **27.** Von dem Borne, Bajzar, Meijers, Nesheim & Bouma (1997) *J. Clin. Invest.* **99**, 2323. **28.** Bajzar, Manuel & Nesheim (1995) *J. Biol. Chem.* **270**, 14477. **29.** Lazoura, Campbell, Yamaguchi, *et al.* (2002) *Chem. Biol.* **9**, 1129. **30.** Novikova & Fricker (1999) *Biochem. Biophys. Res. Commun.* **256**, 564. **31.** Skidgel & Erdös (1998) in *Handb. Proteolytic Enzymes*, p. 1344, Academic Press, San Diego. **32.** Plummer & Erdös (1981) *Meth. Enzymol.* **80**, 442. **33.** Skidgel (1995) *Meth. Enzymol.* **248**, 653. **34.** George, Ishikawa, Perryman & Roberts (1984) *J. Biol. Chem.* **259**, 2667. **35.** Song & Fricker (1995) *J. Biol. Chem.* **270**, 25007. **36.** Aloy, Companys, Vendrell, *et al.* (2001) *J. Biol. Chem.* **276**, 16177. **37.** Song & Fricker (1996) *J. Biol. Chem.* **271**, 28884. **38.** Eng, Novikova, Kuroki, Ganem & Fricker (1998) *J. Biol. Chem.* **273**, 8382. **39.** Sidyelyeva & Fricker (2002) *J. Biol. Chem.* **277**, 49613. **40.** Aloy, Companys, Vendrell, *et al.* (2001) *J. Biol. Chem.* **276**, 16177. **41.** Schneider & Nesheim (2003) *J. Thromb. Haemost.* **1**, 147. **42.** Bajzar, Manuel & Nesheim (1995) *J. Biol. Chem.* **270**, 14477.

Guanine

This purine base (FW = 151.13 g/mol; CAS = 73-40-5; Solubility: 4 mg/100 mL H_2O (more under alkaline or acidic solutions); λ_{max} = 246 nm at pH 7; ε = 10700 $M^{-1}cm^{-1}$; Symbol = G or Gua), systematically named 2-amino-6-hydroxypurine and 2-amino-1,7-dihydro-6*H*-purin-6-one, is one of the principal bases found in RNA and DNA. Guanine is an important structural element in nucleosides, nucleotides, and certain metabolites. Guanine derivatives are precursors in pterin, folate, and riboflavin biosynthesis. Spiders excrete guanine as their principal urotelic metabolite. **Target(s):** acid phosphatase (1); adenine phosphoribosyltransferase, weakly inhibited (2,3); dihydrofolate reductase, weakly inhibited (4); hydroxyacylglutathione hydrolase, glyoxalase II (5); 3-hydroxyanthranilate oxidase (6); lactoylglutathione lyase, glyoxalase I (5); *S*-methyl-5' thioadenosine phosphorylase (7,8); purine-nucleoside phosphorylase, *or* guanosine phosphorylase, by product inhibition (9,10); sepiapterin deaminase, weakly inhibited (11); starch phosphorylase (12,13); thiol oxidase (14); urate-ribonucleotide phosphorylase, weakly inhibited (15); xanthine dehydrogenase (16,17); xanthine oxidase (18); xanthine phosphoribosyltransferase, also alternative substrate (19-21). **1.** Belfield, Ellis & Goldberg (1972) *Enzymologia* **42**, 91. **2.** Tuttle & Krenitsky (1980) *J. Biol. Chem.* **255**, 909. **3.** Montero & Llorente (1991) *Biochem. J.* **275**, 327. **4.** Bertino, Perkins & Johns (1965) *Biochemistry* **4**, 839. **5.** Oray & Norton (1980) *Biochem. Biophys. Res. Commun.* **95**, 624. **6.** Nair (1972) *Arch. Biochem. Biophys.* **153**, 139. **7.** Garbers (1978) *Biochim. Biophys. Acta* **523**, 82. **8.** Fabianowska-Majewska, J. Duley, L. Fairbanks, A. Simmonds & Wasiak (1994) *Acta Biochim. Pol.* **41**, 391. **9.** Tarr (1967) *Meth. Enzymol.* **12A**, 113. **10.** Baker & Schaeffer (1971) *J. Med. Chem.* **14**, 809. **11.** Tsusue (1971) *J. Biochem.* **69**, 781. **12.** Kumar & Sanwal (1982) *Biochemistry* **21**, 4152. **13.** Kumar & Sanwal (1984) *Indian J. Biochem. Biophys.* **21**, 241. **14.** Sweisgood (1986) *Enzyme Microb. Technol.* **2**, 265. **15.** Laster & Blair (1963) *J. Biol. Chem.* **238**, 3348. **17.** Lyon & Garrett (1978) *J. Biol. Chem.* **253**, 2604. **18.** Sin (1975) *Biochim. Biophys. Acta* **410**, 12. **19.** Baker (1967) *J. Pharmaceut. Sci.* **56**, 959. **20.** Miller, Adamczyk, Fyfe & Elion (1974) *Arch. Biochem. Biophys.* **165**, 349. **21.** Naguib, Iltzsch, el Kouni, Panzica & el Kouni (1995) *Biochem. Pharmacol.* **50**, 1685. **22.** Krenitsky, Neil & Miller (1970) *J. Biol. Chem.* **245**, 2605.

Guanine Nucleotide Dissociation Inhibitor

These GoLoco (GL) motif-containing proteins regulate G-protein signaling by binding to G_α subunit and acting as guanine nucleotide dissociation inhibitors. GDI's bind to the GDP-bound form of Rho and Rab small GTPases, preventing exchange and keeping Rho or Rab in its off-state) and by blocking its localization to the membrane. Its inhibition can be reversed by the action of GDI displacement factors. GDI also inhibits cdc42 by binding to its tail, preventing cdc42 insertion into membrane and reducing its ability to trigger WASP-mediated nucleation of F-actin and associated motile events. The α-isoform of bovine Rab-GDI is a 50-kDa protein. (*Note*: The GoLoco motif is a 19-residue structural element within multidomain signaling regulators, such as Loco, RGS12, RGS14, and Rap1GAP, as well as in tandem arrays in proteins, such as AGS3, G18, LGN, Pcp-2/L7, and Partner of Inscuteable/Discs-large complex, the latter required to establish planar polarity during asymmetric cell division in *Drosophila*.) **1**. Geyer & Wittinghofer (1997) *Curr. Opin. Struct. Biol.* **7**, 786. **2**. Ueda, Kikuchi, Ohga, Yamamoto & Takai (1990) *J. Biol. Chem.* **265**, 9373. **3**. Hall (1998) *Science* **279**, 509. **4**. Hall (1994) *Ann. Rev. Cell Biol.* **10**, 31.

Guanosine

This ribonucleoside (FW = 283.24 g/mol; CAS = 118-00-3; λ_{max} = 253 nm at pH 6; ε = 13600 $M^{-1}cm^{-1}$), also known as 9-β-D-ribofuranosylguanine, is found in both the free form and as a component of nucleic acids and several enzyme cofactors. Guanosine is a substrate or product for a number of enzymes: for example, guanosine phosphorylase, guanosine deaminase, purine-nucleoside phosphorylase, purine nucleosidase, and diphosphate:purine nucleoside kinase. **Target(s):** adenosine deaminase (1-4); adenosine kinase (5,6); adenosylhomocysteinase (7); adenylosuccinate synthetase (8); alcohol dehydrogenase, yeast (9); cytidine deaminase, weakly inhibited (10); cytosine deaminase (11,12); deoxycytidine kinase, weakly inhibited (13); deoxyguanosine kinase (14); dihydroorotase (15); galactoside 2-α-L-fucosyltransferase, weakly inhibited (29); β-galactoside α-2,3-sialyltransferase (16); guanylate kinase (17); hydroxyacylglutathione hydrolase, ôr glyoxalase II (18); lactoylglutathione lyase, glyoxalase I (18); *S*-methyl-5'-deoxyadenosine phosphorylase (28); NMN nucleosidase (19); 5'-nucleotidase (20,21); 1-phosphatidylinositol 4-kinase (22); purine nucleosidase (23); ribonuclease T_2 (24); ribose-phosphate diphosphokinase, *or* phosphoribosyl-pyrophosphate synthetase (25); ribosylpyrimidine nucleosidase (26); starch phosphorylase (30); urate-ribonucleotide phosphorylase, weakly inhibited (27). **1**. Agarwal & Parks (1978) *Meth. Enzymol.* **51**, 502. **2**. Centelles, Franco & Bozal (1988) *J. Neurosci. Res.* **19**, 258. **3**. Harbison & Fisher (1973) *Arch. Biochem. Biophys.* **154**, 84. **4**. Akedo, Nishihara, Shinkai, Komatsu & Ishikawa (1972) *Biochim. Biophys. Acta* **276**, 257. **5**. Long & Parker (2006) *Biochem. Pharmacol.* **71**, 1671. **6**. Kidder (1982) *Biochem. Biophys. Res. Commun.* **107**, 381. **7**. Shimizu, Shiozaki, Ohshiro & Yamada (1984) *Eur. J. Biochem.* **141**, 385. **8**. Matsuda, Shimura, Shiraki & Nakagawa (1980) *Biochim. Biophys. Acta* **616**, 340. **9**. Sund & Theorell (1963) *The Enzymes*, 2nd ed., **7**, 25. **10**. Hosono & Kuno (1973) *J. Biochem.* **74**, 797. **11**. Ipata & Cercignani (1978) *Meth. Enzymol.* **51**, 394. **12**. Ipata, Marmocchi, Magni, Felicioli & Polidoro (1971) *Biochemistry* **10**, 4270. **13**. Krenitsky, Tuttle, Koszalka, *er al.* (1976) *J. Biol. Chem.* **251**, 4055. **14**. Yamada, Goto & Ogasawara (1982) *Biochim. Biophys. Acta* **709**, 265. **15**. Bresnick & Blatchford (1964) *Biochim. Biophys. Acta* **81**, 150. **16**. Sadler, Rearick, Paulson & Hill (1979) *J. Biol. Chem.* **254**, 4434. **17**. Anderson (1973) *The Enzymes*, 3rd ed., **9**, 49. **18**. Oray & Norton (1980) *Biochem. Biophys. Res. Commun.* **95**, 624. **19**. Foster (1981) *J. Bacteriol.* **145**, 1002. **20**. Webb (1966) *Enzyme and Metabolic Inhibitors*, vol. **2**, p. 472, Academic Press, New York. **21**. Klein (1957) *Z. Physiol. Chem.* **307**, 254. **22**. Hou, Z.-L. Zhang & H.-H. Tai (1988) *Biochim. Biophys. Acta* **959**, 67. **23**. Atkins, Shelp & Storer (1989) *J. Plant Physiol.* **134**, 447. **24**. Irie & Ohgi (1976) *J. Biochem.* **80**, 3925. Fox & Kelley (1972) *J. Biol. Chem.* **247**, 2126. **26**. Koszalka & Krenitsky (1979) *J. Biol. Chem.* **254**, 8185. **27**. Laster & Blair (1963) *J. Biol. Chem.* **238**, 3348. **28**. Koszalka & Krenitsky (1986) *Adv. Exp. Med. Biol.* **195B**, 559. **29**. Bella & Kim (1971) *Biochem. J.* **125**, 1157. **30**. Kumar & Sanwal (1984) *Indian J. Biochem. Biophys.* **21**, 241.

Guanosine 2',3'-Cyclic Monophosphate

This guanosine phosphodiester ($FW_{free-acid}$ = 345.21 g/mol), also called 2',3'-cyclic GMP and 2',3'-cGMP, is produced as an intermediate in the alkaline hydrolysis of RNA and in the enzyme-catalyzed hydrolysis of RNA by several ribonucleases (*e.g.*, ribonuclease T_1, ribonuclease T_2, ribonuclease U_2, *Bacillus subtilis* ribonuclease). 2',3'-cGMP is also a substrate for 2',3' cyclic-nucleotide 2'-phosphodiesterase and 2',3'-cyclic-nucleotide 3'-phosphodiesterase. A 2',3'-cyclic nucleoside phosphate termini is produced in the endonucleolytic cleavage of pre tRNA by tRNA-intron endonuclease. **Target(s):** 3',5'-cyclic-GMP phospho-diesterase, *or* cGMP phosphodiesterase (1); dCMP deaminase (2-4); galactoside 2-α-L-fucosyltransferase, weakly inhibited (5). **1**. Morishima (1975) *Biochim. Biophys. Acta* **410**, 310. **2**. Scarano, Bonaduce & de Petrocellis (1960) *J. Biol. Chem.* **235**, 3556. **3**. Webb (1966) *Enzyme and Metabolic Inhibitors*, vol. **2**; p. 469, Academic Press, New York. **4**. Rossi, Geraci & Scarano (1967) *Biochemistry* **6**, 3640. **5**. Bella & Kim (1971) *Biochem. J.* **125**, 1157.

Guanosine 3',5'-Cyclic Monophosphate

This key second messenger ($FW_{free-acid}$ = 345.21 g/mol; CAS = 7665-99-8; λ_{max} = 254 nm at pH 7; ε = 12950 $M^{-1}cm^{-1}$), also called 3',5'-cyclic GMP and 3',5'-cGMP, participates in many signal transduction pathways, including rhodopsin-mediate phototransduction, by means of cGMP-dependent protein kinases, cGMP-regulated channels, and cGMP-regulated cyclic nucleotide phosphodiesterases. 3',5'-cyclic GMP is produced from GTP by the action of guanylate cyclases and is hydrolyzed to GMP via cyclic-nucleotide phosphodiesterases. There is food evidence for accelerated cyclic nucleotide metabolic flux without changes in cell content of cGMP or cAMP in response to excitatory cellular stimuli. Sodium, potassium, and ammonium salts of cGMP are soluble in water, but the free acid is barely soluble. Its glycosidic bond is more stable in mineral acids than GMP. **Target(s):** adenylosuccinate synthetase (1-3); *Bacillus subtilis* ribonuclease (4); 3',5'-cyclic nucleotide phosphodiesterase, types 3A and 3B (5-12); cystinyl aminopeptidase (oxytocinase)13-15; glutaminase, phosphate-activated (16); poly(ADP-ribose) glycohydrolase (17,18). **1**. Spector, Jones & Elion (1979) *J. Biol. Chem.* **254**, 8422. **2**. Van der Weyden & Kelly (1974) *J. Biol. Chem.* **249**, 7282. **3**. Spector & Miller (1976) *Biochim. Biophys. Acta* **445**, 509. **4**. Kerjan & Szulmajster (1976) *Biochimie* **58**, 533. **5**. Manganiello, Degerman & Elks (1988) *Meth. Enzymol.* **159**, 504. **6**. Harrison, Beier, Martins & Beavo (1988) *Meth. Enzymol.* **159**, 685. **7**. Houslay, Pyne & Cooper (1988) *Meth. Enzymol.* **159**, 751. **8**. Harrison, Reifsnyder, Gallis, Cadd & Beavo (1986) *Mol. Pharmacol.* **29**, 506. **9**. Sudo, Tachibana, Toga, *et al.* (2000) *Biochem. Pharmacol.* **59**, 347. **10**. Hambleton, Krall, Tikishvili, *et al.* (2005) *J. Biol. Chem.* **280**, 39168. **11**. Grant & Colman (1984) *Biochemistry* **23**, 1801. **12**. Leroy, Degerman, Taira, *et al.* (1996) *Biochemistry* **35**, 10194. **13**. Mizutani, Tsujimoto & Nakazato (1998) in *Handb. Proteolytic Enzymes*, p. 1008, Academic Press, San Diego. **14**. Roy, Yeang & Karim (1982) *Prostaglandins Leukot. Med.* **8**, 173. **15**. Roy, Yeang & Karim (1981)

Prostaglandins Med. **6**, 577. **16**. Kvamme, Torgner & Svenneby (1985) *Meth. Enzymol.* **113**, 241. **17**. Sugimura, Yamada, Miwa, *et al.* (1973) *Biochem. Soc. Trans.* **1**, 642. **18**. Miwa, Tanaka, Matsushima & Sugimura (1974) *J. Biol. Chem.* **249**, 3475.

Guanosine 5'-Diphosphate

This purine nucleotide (FW$_{\text{free-acid}}$ = 443.20 g/mol; CAS = 146-91-8; pK_a for GDP are < 1, 2.9, 6.3, and 9.6) is a substrate for eukaryotic ribonucleotide reductases as well as nucleoside 5'-diphosphokinase (NDPK). It is found in either the free form or, more commonly, as a one-to-one complex with a divalent cation (*e.g.*, MgGDP). GDP is also a component of guanosinediphosphosugars. GDP is the product of many GTP-dependent enzymes and is a regulatory agent in many cellular processes. **Target(s):** α-*N*-acetylgalactosaminide α-2,6-sialyltransferase (1); *N*-acylneuraminate cytidylyl-transferase (2); α-*N*-acetylneuraminate α-2,8-sialyltransferase (97); adenine phosphoribosyltransferase (3,116-118); adenylosuccinate synthetase (4-11); adenylyl-sulfate kinase (12); alcohol dehydrogenase (13); amidophosphoribosyltransferase (14-17,102-114); D-amino-acid oxidase (18); AMP nucleosidase (19-22); asparagine synthetase, weakly inhibited (23); ATP diphosphatase (24); *Basidiobolus haptosporus* nuclease, *or* nuclease Bh1 (25,26); bis(5'-nucleosyl) tetraphosphatase, asymmetrical (27); CDP-diacylglycerol diphosphatase (28); cellulose synthase (141); chitobiosyldiphosphodolichol β-mannosyltransferase (29,125,126); [citrate lyase] deacetylase (30); cytosine deaminase (31,32); dCMP deaminase (4,33-35); deoxyguanosine kinase (36); diacylglycerol cholinephospho-transferase (37); dolichyl-phosphate β-D-mannosyltransferase (129-133); ethanolaminephosphotransferase (37); *N*-formylglutamate deformylase, *or* *N*-β-citrylglutamate deacylase (38); fructose-1,6-bisphosphatase (39); galactoside 2-α-L-fucosyltransferase (40,134,135); galactoside 3(4)-L-fucosyltransferase (1); β-galactoside α-2,3-sialyltransferase (1); 3-galactosyl-*N*-acetylglucosaminide 4-α-L-fucosyltransferase (138); GDP-mannose 4,6-dehydratase (41); GDP-mannose 6-dehydrogenase (42); glucokinase (43); glucosamine *N*-acetyltransferase (144); glucose-1-phosphate adenylyltransferase, *or* ADPglucose pyrophosphorylase (44,45); glucose-6-phosphate dehydrogenase (46); glycolipid 2-α-mannosyl-transferase (128); glucomannan 4-β-mannosyltransferase (140); glycolipid 3-α-mannosyltransferase (47,127); glycoprotein 3-α-L-fucosyl-transferase (120-124); glycoprotein 6-α-L-fucosyltransferase (136,137); GTP cyclohydrolase I (48); GTP cyclohydrolase IIa (49); hexokinase (50); hydroxyacylglutathione hydrolase, *or* glyoxalase II (51); IMP cyclohydrolase (52); IMP dehydrogenase (4,53-55); lactoylglutathione lyase, *or* glyoxalase I (51); mannose-1-phosphate guanylyltransferase (56); m^{7}G(5')pppN diphosphatase, slightly inhibited (57); micrococcal nuclease (58); *Neurospora* glutamate dehydrogenase (59); nicotinate phosphoribosyltransferase (115); NMN nucleosidase (60); nucleoside deoxyribosyltransferase (119); nucleoside-diphosphate kinase (61); 3'-nucleotidase (62); nucleotide diphosphatase (63,64); oligonucleotidase (65); orotate phosphoribosyltransferase (66); orotidylate decarboxylase (66); 1-phosphatidylinositol 4-kinase (67); phosphoenolpyruvate carboxykinase (GTP) (68,69); phosphoglycerate kinase (70); phosphomannan mannose-phosphotransferase (71); phosphoprotein phosphatase (72); 5-phosphoribosylamine synthetase, *or* ribose-5-phosphate:ammonia ligase (73,74); phosphoribosyl-pyrophosphate synthetase, *or* ribose-5-phosphate pyrophosphokinase (75); [phosphorylase] phosphatase (76); poly(A)-specific ribonuclease, weakly inhibited (77); poly(A) polymerase (78); polypeptide *N*-acetylgalactosaminyltransferase (139); polyphosphate kinase (79,80); protein-glutamine γ-glutamyltransferase, *or* transglutaminase (142,143); [protein-PII] uridylyltransferase (81); pyruvate,orthophosphate dikinase (82); ribonuclease, pancreatic, *or* ribonuclease A (83-85); ribonuclease T$_1$ (86); ribonuclease T$_2$ (87); ribose-phosphate diphosphokinase, *or* phosphoribosyl-pyrophosphate synthetase (88,89); ribose-5-phosphate isomerase (90); succinate dehydrogenase (14); threonine synthase, weakly inhibited (91); UDP-glucuronate 4-epimerase (92-94); undecaprenyl-phosphate galactose phosphotransferase, weakly inhibited

(95); uridine nucleosidase (96); xanthine phosphoribosyltransferase (98,99); 1,4-β-D-xylan synthase (100,101). **1**. Sadler, Beyer, Oppenheimer, *et al.* (1982) *Meth. Enzymol.* **83**, 458. **2**. Bravo, Barrallo, Ferrero, *et al.* (2001) *Biochem. J.* **358**, 585. **3**. Arnold & Kelley (1978) *Meth. Enzymol.* **51**, 568. **4**. Webb (1966) *Enzyme and Metabolic Inhibitors*, vol. **2**, Academic Press, New York. **5**. Wyngaarden & Greenland (1963) *J. Biol. Chem.* **238**, 1054. **6**. Spector, Jones & Elion (1979) *J. Biol. Chem.* **254**, 8422. **7**. Van der Weyden & Kelly (1974) *J. Biol. Chem.* **249**, 7282. **8**. Spector & Miller (1976) *Biochim. Biophys. Acta* **445**, 509. **9**. Matsuda, Shimura, Shiraki & Nakagawa (1980) *Biochim. Biophys. Acta* **616**, 340. **10**. Borza, Iancu, Pike, Honzatko & Fromm (2003) *J. Biol. Chem.* **278**, 6673. **11**. Matsuda, Ogawa, Fukutome, Shiraki & Nakagawa, (1977) *Biochem. Biophys. Res. Commun.* **78**, 766. **12**. Schwenn & Jender (1984) *Arch. Microbiol.* **138**, 9. **13**. Li & Vallee (1964) *J. Biol. Chem.* **239**, 792. **14**. Flaks & Lukens (1963) *Meth. Enzymol.* **6**, 52. **15**. Zalkin (1985) *Meth. Enzymol.* **113**, 264. **16**. Caskey, Ashton & Wyngaarden (1964) *J. Biol. Chem.* **239**, 2570. **17**. Hill & Bennett (1969) *Biochemistry* **8**, 122. **18**. McCormick, Chassy & Tsibris (1964) *Biochim. Biophys. Acta* **89**, 447. **19**. DeWolf, Fullin & Schramm (1979) *J. Biol. Chem.* **254**, 10868. **20**. Yoshino (1970) *J. Biochem.* **68**, 321. **21**. Ogasawara, Yoshino & Asai (1970) *J. Biochem.* **68**, 331. **22**. Yoshino & Takagi (1973) *J. Biochem.* **74**, 1151. **23**. Hongo, Matsumoto & Sato (1978) *Biochim. Biophys. Acta* **522**, 258. **24**. Torp-Pedersen, Flodgaard & Saermark (1979) *Biochim. Biophys. Acta* **571**, 94. **25**. Desai & Shankar (2001) *J. Biochem. Mol. Biol. Biophys.* **5**, 327. **26**. Desai & Shankar (2000) *Eur. J. Biochem.* **267**, 5123. **27**. Vallejo, Lobaton, Quintanilla, Sillero & Sillero (1976) *Biochim. Biophys. Acta* **438**, 304. **28**. Raetz, Hirschberg, Dowhan, Wickner & Kennedy (1972) *J. Biol. Chem.* **247**, 2245. **29**. Sharma, Lehle & Tanner (1982) *Eur. J. Biochem.* **126**, 319. **30**. Giffhorn & Gottschalk (1975) *J. Bacteriol.* **124**, 1052. **31**. Ipata & Cercignani (1978) *Meth. Enzymol.* **51**, 394. **32**. Ipata, Marmocchi, Magni, Felicioli & Polidoro (1971) *Biochemistry* **10**, 4270. **33**. Maley (1967) *Meth. Enzymol.* **12A**, 170. **34**. Scarano, Bonaduce & de Petrocellis (1960) *J. Biol. Chem.* **235**, 3556. **35**. Ellims, Kao & Chabner (1981) *J. Biol. Chem.* **256**, 6335. **36**. Green & Lewis (1979) *Biochem. J.* **183**, 547. **37**. Percy, Carson, Moore & Waechter (1984) *Arch. Biochem. Biophys.* **230**, 69. **38**. Miyake, Innami & Kakimoto (1983) *Biochim. Biophys. Acta* **760**, 206. **39**. Fujita & Freese (1979) *J. Biol. Chem.* **254**, 5340. **40**. Beyer & Hill (1980) *J. Biol. Chem.* **255**, 5373. **41**. Liao & Barber (1972) *Biochim. Biophys. Acta* **276**, 85. **42**. Roychoudhury, May, Gill, *et al.* (1989) *J. Biol. Chem.* **264**, 9380. **43**. Reeves, Montalvo & Sillero (1967) *Biochemistry* **6**, 1752. **44**. Preiss (1973) *The Enzymes*, 3rd ed., **8**, 73. **45**. Preiss, Shen, Greenberg & Gentner (1966) *Biochemistry* **5**, 1833. **46**. Horne, Anderson & Nordlie (1970) *Biochemistry* **9**, 610. **47**. Jensen & Schutzbach (1982) *J. Biol. Chem.* **257**, 9025. **48**. Yoo, Han, Ko & Bang (1998) *Arch. Pharm. Res.* **21**, 692. **49**. Graham, Xu & White (2002) *Biochemistry* **41**, 15074. **50**. Ning, Purich & Fromm (1969) *J. Biol. Chem.* **244**, 3840. **51**. Oray & Norton (1980) *Biochem. Biophys. Res. Commun.* **95**, 624. **52**. Szabados, Hindmarsh, Phillips, Duggleby & Christopherson (1994) *Biochemistry* **33**, 14237. **53**. Magasanik (1963) *Meth. Enzymol.* **6**, 106. **54**. Mager & Magasanik (1960) *J. Biol. Chem.* **235**, 1474. **55**. Hampton & Nomura (1967) *Biochemistry* **6**, 679. **56**. Szumilo, Drake, York & Elbein (1993) *J. Biol. Chem.* **268**, 17943. **57**. Nuss & Furuichi (1977) *J. Biol. Chem.* **252**, 2815. **58**. Cuatrecasas, Fuchs & Anfensen (1967) *J. Biol. Chem.* **242**, 1541, 3063, and 4759. **59**. Smith, Austen, Blumenthal & Nyc (1975) *The Enzymes*, 3rd ed., **11**, 293. **60**. Foster (1981) *J. Bacteriol.* **145**, 1002. **61**. Parks & Agarwal (1973) *The Enzymes*, 3rd ed., **8**, 307. **62**. Nguyen, Palcic & Hadziyev (1987) *J. Chromatogr.* **391**, 257. **63**. Kahn & Anderson (1996) *J. Biol. Chem.* **261**, 6016. **64**. Krishnan & Rao (1972) *Arch. Biochem. Biophys.* **149**, 336. **65**. Datta & Niyogi (1975) *J. Biol. Chem.* **250**, 7313. **66**. Jones, Kavipurapu & Traut (1978) *Meth. Enzymol.* **51**, 155. **67**. Hou, Zhang & Tai (1988) *Biochim. Biophys. Acta* **959**, 67. **68**. Chang, Maruyama, Miller & Lane (1966) *J. Biol. Chem.* **241**, 2421. **69**. Goto, Shimizu & Shukuya (1982) *J. Biochem.* **88**, 1239. **70**. Kuntz & Krietsch (1982) *Meth. Enzymol.* **90**, 103. **71**. Bretthauer, Kozak & Irwin (1969) *Biochem. Biophys. Res. Commun.* **37**, 820. **72**. Damuni & Reed (1987) *J. Biol. Chem.* **262**, 5133. **73**. Reem (1968) *J. Biol. Chem.* **243**, 5695. **74**. Westby & Tsai (1974) *J. Bacteriol.* **117**, 1099. **75**. Wong & Murray (1969) *Biochemistry* **8**, 1608. **76**. Detwiler, Gratecos & Fischer (1977) *Biochemistry* **16**, 4818. **77**. Martínez, Ren, Thuresson, Hellman, Åström & Virtanen (2000) *J. Biol. Chem.* **275**, 24222. **78**. Sillero, Socorro, Baptista, *et al.* (2001) *Eur. J. Biochem.* **268**, 3605. **79**. Lindner, Vidaurre, Willbold, Schoberth & Wendisch (2007) *Appl. Environ. Microbiol.* **73**, 5026. **80**. Tzeng & Kornberg (2000) *J. Biol. Chem.* **275**, 3977. **81**. Engleman & Francis (1978) *Arch. Biochem. Biophys.* **191**, 602. **82**. Ye, Wei, McGuire, *et al.* (2001) *J. Biol. Chem.* **276**, 37630. **83**. Zittle (1946) *J. Biol. Chem.* **162**, 287. **84**. Richards & H. W. Wyckoff (1971) *The Enzymes*,

3rd ed., **4**, 647. **85**. Zittle (1945) *J. Biol. Chem.* **160**, 527. **86**. Takahashi & Moore (1982) *The Enzymes*, 3rd ed., **15**, 435. **87**. Irie & Ohgi (1976) *J. Biochem.* **80**, 39. **88**. Fox & Kelley (1972) *J. Biol. Chem.* **247**, 2126. **89**. Arnvig, Hove-Jensen & Switzer (1990) *Eur. J. Biochem.* **192**, 195. **90**. Horitsu, Sasaki, Kikuchi, *et al.* (1976) *Agric. Biol. Chem.* **40**, 257. **91**. Giovanelli, Mudd, Datko & Thompson (1986) *Plant Physiol.* **81**, 577. **92**. Gaunt, Ankel & Schutzbach (1972) *Meth. Enzymol.* **28**, 426. **93**. Gaunt, Maitra & Ankel (1974) *J. Biol. Chem.* **249**, 2366. **94**. Munoz, Lopez, de Frutos & Garcia (1999) *Mol. Microbiol.* **31**, 703. **95**. Osborn & Tze-Yuen (1968) *J. Biol. Chem.* **243**, 5145. **96**. Raggi-Ranieri & Ipata (1971) *Ital. J. Biochem.* **20**, 27. **97**. Eppler, Morré & Keenan (1980) *Biochim. Biophys. Acta* **619**, 332. **98**. Miller, Adamczyk, Fyfe & Elion (1974) *Arch. Biochem. Biophys.* **165**, 349. **99**. Arent, Kadziola, Larsen, Neuhard & Jensen (2006) *Biochemistry* **45**, 6615. **100**. Dalessandro & Northcote (1981) *Planta* **151**, 61. **101**. Bailey & Hassid (1966) *Proc. Natl. Acad. Sci. U.S.A.* **56**, 1586. **102**. Messenger & Zalkin (1979) *J. Biol. Chem.* **254**, 3382. **103**. Liras, Argomaniz & Llorente (1990) *Biochim. Biophys. Acta* **1033**, 114. **104**. Zhou, Charbonneau, Colman & Zalkin (1993) *J. Biol. Chem.* **268**, 10471. **105**. King, Boounous & Holmes (1978) *J. Biol. Chem.* **253**, 3933. **106**. Holmes, McDonald, McCord, Wyngaarden & Kelley (1973) *J. Biol. Chem.* **248**, 144. **107**. Bera, Chen, Smith & Zalkin (2000) *J. Biol. Chem.* **182**, 3734. **108**. Chen, Tomchick, Wolle, *et al.* (1997) *Biochemistry* **36**, 10718. **109**. Itoh, Gorai, Usami & Tsushima (1979) *Biochim. Biophys. Acta* **581**, 142. **110**. Reynolds, Blevins & Randall (1984) *Arch. Biochem. Biophys.* **229**, 623. **111**. Wood & Seegmiller (1973) *J. Biol. Chem.* **248**, 138. **112**. Tsuda, Katunuma & Weber (1979) *J. Biochem.* **85**, 1347. **113**. Nagy, Reichert & Ribet (1974) *Biochim. Biophys. Acta* **370**, 85. **114**. Zhou, Smith & Zalkin (1994) *J. Biol. Chem.* **269**, 6784. **115**. Hayakawa, Shibata & Iwai (1984) *Agric. Biol. Chem.* **48**, 445. **116**. Sin & Finch (1972) *J. Bacteriol.* **112**, 439. **117**. Hirose & Ashihara (1983) *Z. Pflanzenphysiol.* **110**, 135. **118**. Nagy & Ribet (1977) *Eur. J. Biochem.* **77**, 77. **119**. Ghiorghi, Zeller, Dang & Kaminski (2007) *J. Biol. Chem.* **282**, 8150. **120**. de Vries, Storm, Rotteveel, *et al.* (2001) *Glycobiology* **11**, 711. **121**. Murray, Takayama, Schultz & Wong (1996) *Biochemistry* **35**, 11183. **122**. Campbell & Stanley (1984) *J. Biol. Chem.* **259**, 11208. **123**. Foster, Gillies & Glick (1991) *J. Biol. Chem.* **266**, 3526. **124**. Shinoda, Morishita, Sasaki, *et al.* (1997) *J. Biol. Chem.* **272**, 31992. **125**. Kaushal & Elbein (1987) *Biochemistry* **26**, 7953. **126**. Kaushal & Elbein (1986) *Arch. Biochem. Biophys.* **250**, 38. **127**. Jensen & Schutzbach (1981) *J. Biol. Chem.* **256**, 12899. **128**. Schutzbach, Springfield & Jensen (1980) *J. Biol. Chem.* **255**, 4170. **129**. Villagómez-Castro, Calvo-Méndez, Vargas-Rodríguez, Flores-Carreón & López Romero (1998) *Exp. Parasitol.* **88**, 111. **130**. Jensen & Schutzbach (1986) *Carbohydr. Res.* **149**, 199. **131**. Carlo & Villemez (1979) *Arch. Biochem. Biophys.* **198**, 117. **132**. Gasnier, Morelis, Louisot & Gateau (1987) *Biochim. Biophys. Acta* **925**, 297. **133**. Jensen & Schutzbach (1985) *Eur. J. Biochem.* **153**, 41. **134**. Bella & Kim (1971) *Biochem. J.* **125**, 1157. **135**. Bella & Kim (1971) *Arch. Biochem. Biophys.* **147**, 753. **136**. Ihara, Ikeda & Taniguchi (2006) *Glycobiology* **16**, 333. **137**. Kaminska, Wisniewska & Koscielak (2003) *Biochimie* **85**, 303. **138**. Prieels, Monnom, Dolmans, Beyer & Hill (1981) *J. Biol. Chem.* **256**, 10456. **139**. Takeuchi, Yoshikawa, Sasaki & Chiba (1985) *Agric. Biol. Chem.* **49**, 1059. **140**. Piro, Zuppa, Dalessandro & Northcote (1993) *Planta* **190**, 206. **141**. Haass, Hackspacher & Franz (1985) *Plant Sci.* **41**, 1. **142**. Ahvazi, Boeshans & Steinert (2004) *J. Biol. Chem.* **279**, 26716. **143**. Bergamini, Signorini & Poltronieri (1987) *Biochim. Biophys. Acta* **916**, 149. **144**. Piro, Buffo & Dalessandro (1994) *Physiol. Plant.* **90**, 181.

Guanosine 3'-Diphosphate 5'-Diphosphate (*or*, ppGpp)

This bacterial alarmone (FW = 588.17 g/mol), frequently abbreviated ppGpp and also called Magic Spot I, is elevated in microorganisms (*e.g.*, *Escherichia coli*) exhibiting the stringent response (*i.e.*, a metabolic stress response linked to poor nutrient and growth conditions such as a lack of

amino acids). ppGpp is an alternative product of the reaction catalyzed by GTP diphosphokinase (utilizing GDP as the acceptor substrate). Note that the 3'-substituent is more susceptible to hydrolysis that the 5'-pyrophosphate. ppGpp should be stored at –70°C. The half-life in 0.3 M KOH at 37°C is only 6.5 minutes. *For stability, concentration determinations, and metal ion binding properties, see Guanosine 5'-triphosphate.* **Target(s):** acetyl-CoA carboxylase (1); adenylosuccinate synthetase (2); ADPglucose pyrophosphorylase (3); ATP phosphoribosyltransferase, in the presence of L-histidine (4-6); 3 fructose-1,6-bisphosphatase (7); glycerol-3-phosphate dehydrogenase (8); hydroxydecanoyl-[acyl carrier-protein] dehydratase (9); hypoxanthine phosphoribosyltransferase (10); ornithine decarboxylase (11); phosphatidylglycerophosphate synthase (12); protein biosynthesis (elongation) (4,13,14); protein-synthesizing GTPase, elongation factor (4,13,14); RNA-directed RNA polymerase, *or* Qβ replicase (15,16); RNA polymerase (4). **1**. Polakis, Guchhait & Lane (1973) *J. Biol. Chem.* **248**, 7957. **2**. Stayton & Fromm (1979) *J. Biol. Chem.* **254**, 2579. **3**. Dietzler & Leckie (1977) *Biochem. Biophys. Res. Commun.* **77**, 1459. **4**. Bridger & Paranchych (1979) *Trends Biochem. Sci.* **4**, 176. **5**. Morton & Parsons (1977) *Biochem. Biophys. Res. Commun.* **74**, 172. **6**. Kleeman & Parsons (1977) *Proc. Natl. Acad. Sci. U.S.A.* **74**, 1535. **7**. Fujita & Freese (1979) *J. Biol. Chem.* **254**, 5340. **8**. Pieringer (1983) *The Enzymes*, 3rd ed., **16**, 255. **9**. Stein & Bloch (1976) *Biochem. Biophys. Res. Commun.* **73**, 881. **10**. Hochstadt (1978) *Meth. Enzymol.* **51**, 549. **11**. Paulin & Poso (1983) *Biochim. Biophys. Acta* **742**, 197. **12**. Merlie & Pizer (1973) *J. Bacteriol.* **116**, 355. **13**. Lucas-Lenard & Beres (1974) *The Enzymes*, 3rd ed., **10**, 53. **14**. Krab & Parmeggiani (1998) *Biochim. Biophys. Acta* **1443**, 1. **15**. Blumenthal (1982) *The Enzymes*, 3rd ed., **15**, 267. **16**. Blumenthal (1977) *Biochim. Biophys. Acta* **478**, 201.

Guanosine 5'-[β,γ-Imido]triphosphate

This GTP analogue (FW$_{\text{free-acid}}$ = 522.20 g/mol), also called GMPPNP, p(NH)ppG, and β,γ-imidoguanosine 5'-triphosphate, is an inhibitor of many GTP-dependent enzymes. It also serves as an alternative substrate for enzymes that hydrolyze GTP between the α and β phosphorus atoms: *e.g.*, snake venom phosphodiesterase hydrolyzes GMPPNP to produce GMP and imidodiphosphate (PNP). *Escherichia coli* alkaline phosphatase will hydrolyze the nucleotide to orthophosphate and the corresponding nucleoside diphosphate derivative. GMPPNP often bind to and reversibly activates G proteins. GMP-PNP is very soluble in water and binds metal ions with a greater affinity than GTP. It is unstable in acidic solutions, producing the corresponding phosphoramidate and orthophosphate. **Target(s):** adenylate cyclase (1); adenylosuccinate synthetase (2); dynamin GTPase (3); F$_1$ ATPase (4); GTPase (5); guanylate cyclase (6-8); [3-methyl-2-oxobutanoate dehydrogenase, *or* 2 methylpropanoyl-transferring)] phosphatase, *or* [branched-chain α-keto-acid dehydrogenase] phosphatase (9,10); phosphoprotein phosphatase (11); protein-synthesizing GTPase, elongation factor (12,13); ribosomal translocation (5); minor effect on tubulin polymerization, but substantial inhibition of depolymerization of p(NH)ppG-containing microtubule (13). **1**. Kirkham & Henley (1993) *Biochem. Pharmacol.* **46**, 1559. **2**. Jahngen & Rossomando (1984) *Arch. Biochem. Biophys.* **229**, 145. **3**. Shpetner & Vallee (1992) *Nature* **355**, 733. **4**. Jault, Divita, Allison & Di Pietro (1993) *J. Biol. Chem.* **268**, 20762. **5**. Eckstein, Kettler & Parmeggiani (1971) *Biochem. Biophys. Res. Commun.* **45**, 1151. **6**. Gorczyca (2000) *Meth. Enzymol.* **315**, 689. **7**. Gorczyca, Van Hooser & Palczewski (1994) *Biochemistry* **33**, 3217. **8**. Hayashi & Yamazaki (1991) *Proc. Natl. Acad. Sci. U.S.A.* **88**, 4746. **9**. Damuni & Reed (1988) *Meth. Enzymol.* **166**, 321. **10**. Damuni, Merryfield, Humphreys & Reed (1984) *Proc. Natl. Acad. Sci. U.S.A.* **81**, 4335. **11**. Damuni & Reed (1987) *J. Biol. Chem.* **262**, 5133. **12**. Uritani & Miyazaki (1988) *J. Biochem.* **103**, 522. **13**. Rodnina, Savelsbergh, Katunin & Wintermeyer (1997) *Nature* **385**, 37. **13**. Terry & Purich (1980) *J. Biol. Chem.* **255**, 10532.

Guanosine 2'-Monophosphate

This nucleotide (FW$_{\text{free-acid}}$ = 363.22 g/mol), also known as 2'-guanylate and abbreviated 2'-GMP, is produced from hydrolysis of guanosine 2',3'-cyclic monophosphate (as is 3'-GMP). The cyclic nucleotide is an intermediate in the alkaline hydrolysis of RNA. 2'-GMP is soluble in water and the *N*-glycosidic bond can be hydrolyzed at elevated temperatures in acidic conditions (guanine, ribose, and orthophosphate is produced in one hour in 1 M HCl at 100°C). **Target(s):** adenylosuccinate synthetase (1); 2',3'-cyclic-nucleotide 3'-phosphodiesterase (2); phosphogluconate dehydrogenase, decarboxylating (3); ribonuclease M (4); ribonuclease, pancreatic, ribonuclease A (5); ribonuclease T$_1$ (6-8); ribonuclease T$_2$ (4,9); ribonuclease Rh4, *or* ribonuclease U2 (10). **1**. Spector, Jones & Elion (1979) *J. Biol. Chem.* **254**, 8422. **2**. Díaz & Heredia (1996) *Biochim. Biophys. Acta* **1290**, 135. **3**. Silverberg & Dalziel (1975) *Meth. Enzymol.* **41**, 214. **4**. Irie & Ohgi (2001) *Meth. Enzymol.* **341**, 42. **5**. Richards & Wyckoff (1971) *The Enzymes*, 3rd ed., **4**, 647. **6**. Sato & Egami (1965) *Biochem. Z.* **342**, 437. **7**. Takahashi & Moore (1982) *The Enzymes*, 3rd ed., **15**, 435. **8**. Iida & Ooi (1969) *Biochemistry* **8**, 3897. **9**. Irie & Ohgi (1976) *J. Biochem.* **80**, 39. **10**. Sato & Uchida (1975) *Biochim. Biophys. Acta* **383**, 168.

Guanosine 3'-Monophosphate

This nucleotide (FW$_{\text{free-acid}}$ = 363.22 g/mol), also called 3'-guanylate, is readily produced by the alkaline hydrolysis of RNA (*e.g.*, by 0.3 M KOH at 37°C for 16 hours) and by the action of a number of ribonucleases (*e.g.*, spleen exonuclease, yeast ribonuclease, and ribonuclease T1). It is also a product of the reaction catalyzed by 2',3'-cyclic-nucleotide 2' phosphodiesterase. 3'-GMP is soluble in water. Note that the *N*-glycosidic bond can be hydrolyzed at elevated temperatures in acidic conditions. The ultraviolet spectrum has a λ_{max} of 252 nm (ε = 13,400 M^{-1}cm^{-1}) at pH 7. **Target(s):** adenylosuccinate synthetase (1); ribonuclease, pancreatic, *or* ribonuclease A (2); ribonuclease T$_1$ (3); ribonuclease T$_2$ (4); ribose-5-phosphate isomerase (5). **1**. Spector, Jones & Elion (1979) *J. Biol. Chem.* **254**, 8422. **2**. Richards & Wyckoff (1971) *The Enzymes*, 3rd ed., **4**, 647. **3**. Takahashi & Moore (1982) *The Enzymes*, 3rd ed., **15**, 435. **4**. Irie & Ohgi (1976) *J. Biochem.* **80**, 39. **5**. Horitsu, Sasaki, Kikuchi, *et al.* (1976) *Agric. Biol. Chem.* **40**, 257.

Guanosine 5'-Monophosphate

This common nucleotide (FW$_{\text{free-acid}}$ = 363.22 g/mol), abbreviated GMP and also known as guanylic acid, is produced from XMP via GMP synthases.

GMP can be hydrolyzed to guanine, orthophosphate and ribose in 1 M HCl at 100°C. GMP has a λ_{max} at pH 7 of 252 nm (ε = 13700 M^{-1}cm^{-1}).

Guanosine 5'-*O*-(2-Thiodiphosphate)

This nucleotide analogue (FW = 443.27 g/mol), abbreviated GDP-β-S, is a non-hydrolyzable GDP analogue that inhibits G-protein activation and the stimulation of adenylate cyclase by GTP and GTP analogues. **Target(s):** adenylate cyclase (1,2); diacylglycerol kinase (3); G protein (4,5). **1**. Svoboda, Furnelle, Eckstein & Christophe (1980) *FEBS Lett.* **109**, 275. **2**. Eckstein, Cassel, Levkovitz, Lowe & Selinger (1979) *J. Biol. Chem.* **254**, 9829. **3**. Wissing & Wagner (1992) *Plant Physiol.* **98**, 1148. **4**. Silk, Clejan & Witkom (1989) *J. Biol. Chem.* **264**, 21466. **5**. Burch & Axelrod (1987) *Proc. Natl. Acad. Sci. U.S.A.* **84**, 6374.

Guanosine 5'-*O*-(3-Thiotriphosphate)

This nucleotide (FW = 522.24 g/mol; Abbreviated: GTP-γ-S) is a non-hydrolyzable GTP analogue that inhibits many GTP-dependent reactions. It is also G-protein activator, stimulates adenyl cyclase, and will induce actin polymerization. **Target(s):** acetyl-CoA carboxylase (1); diacylglycerol kinase (2); DNA topoisomerase II (3); dynamin GTPase (4); GTP cyclohydrolase IIa (5); NAD$^+$ ADP-ribosyltransferase, *or* poly(ADP-ribose) polymerase (6); phospholipase A$_1$ (7); protein-glutamine γ-glutamyltransferase, *or* transglutaminase (8-11); protein-synthesizing GTPase, elongation factor (12,13); Rho ADP-ribosyltransferase (14); tubulin polymerization, microtubule assembly (15). **1**. Mick, Chun, VanderBloomer, Fu & McCormick (1998) *Biochim. Biophys. Acta* **1384**, 130. **2**. Kato & Takenawa (1990) *J. Biol. Chem.* **265**, 794. **3**. Arndt-Jovin, Udvardy, Garner, Ritter & Jovin (1993) *Biochemistry* **32**, 4862. **4**. Shpetner & Vallee (1992) *Nature* **355**, 733. **5**. Graham, Xu & White (2002) *Biochemistry* **41**, 15074. **6**. Kofler, Wallraff, Herzog, *et al.* (1993) *Biochem. J.* **293**, 275. **7**. Badiani, Lu & Arthur (1992) *Biochem. J.* **288**, 965. **8**. Siegel & Khosla (2007) *Pharmacol. Ther.* **115**, 232. **9**. Lee, Birckbichler & Patterson (1989) *Biochem. Biophys. Res. Commun.* **162**, 1370. **10**. Ahvazi, Boeshans & Steinert (2004) *J. Biol. Chem.* **279**, 26716. **11**. Zanetti, Ristoratore, Bertoni & Cariello (2004) *J. Biol. Chem.* **279**, 49289. **12**. Rodnina, Savelsbergh, Katunin & Wintermeyer (1997) *Nature* **385**, 37. **13**. Kisselev & Frolova (1995) *Biochem. Cell Biol.* **73**, 1079. **14**. Aktories & Just (1995) *Meth. Enzymol.* **256**, 184. **15**. Hamel & Lin (1984) *J. Biol. Chem.* **259**, 11060.

Guanosine 5'-Triphosphate

This purine nucleotide (FW$_{\text{free-acid}}$ = 523.18 g/mol; Abbreviation: GTP) is second only to ATP in its role as a phosphoryl-donor and source of Gibbs energy to drive mechanoenzyme (*or* energase) reactions. GTP is also a prominent metabolic regulator inasmuch as it binds to G-proteins. GTP is also a required substrate for RNA synthesis. GTP is produced by many substrate-level phosphorylation reactions, including nucleoside-diphosphate kinase (*Reaction*: MgATP^{2-} + GDP $\rightleftharpoons$ MgGTP^{2-} + ADP; K$_{eq}$ ≈ 1). **Stability:** Solid sodium and potassium salts of GTP are stable for months at –15°C and about a week at 0°C. However, it is always good procedure to freshly prepare solutions of GTP. In alkaline solutions, GTP degrades to pyrophosphate (or, diphosphate) and GMP even at 0°C. The phosphoanhydride bonds are very labile to acids (*e.g.*, 66% of total phosphorus produces orthophosphate in fifteen minutes at 100°C in 0.5 N H$_2$SO$_4$); the phosphate ester bond is not as labile. More drastic treatment (0.5 N H$_2$SO$_4$ at 100°C for one hour) will produce guanine, ribose 5-phosphate, ribose, and orthophosphate. Treatment with nitrous acid converts GTP xanthine 5'-triphosphate (XTP), via diazotization and hydrolysis of the diazonium salt. **Standardization of GTP Solutions:** Direct weighing is an unreliable way to calibrate GTP solutions. The actual concentration should be determined with the aid of a UV/visible spectrophotometer using a suitably diluted sample at a defined pH. At pH 7, λ$_{max}$ is 253 nm (ε = 13,700 M^{-1}cm^{-1}). At pH 1, λ$_{max}$ is 256 nm (ε = 12,400 M^{-1}cm^{-1}). At pH 11, λ$_{max}$ is 257 nm (ε = 11,900 M^{-1}cm^{-1}). **Metal Ion Binding:** GTP forms tight complexes with many metal ions. The stability constants are substantially the same as that for ATP. The stability constants for MgATP^{2-}, CaATP^{2-}, and MnATP^{2-} are 73000, 35000, and 100000 M^{-1}, respectively. Metal ion-free ATP is usually an inhibitor for MgATP-utilizing enzymes, and the same is undoubtedly true for GTP-utilizing enzymes. **Target(s):** α-*N*-acetylgalactosaminide α-2,6-sialyltransferase (1); *N*-acylneuraminate cytidylyltransferase (2,3); adenine phospho-ribosyltransferase (4,184); adenylate cyclase (5,6); adenylate dimethylallyltransferase (7); adenylylsulfatase (8-10); aldose-1-phosphate nucleotidyl-transferase (11); amidophosphoribosyl-transferase (12,180,181); 4-aminobutyrate amino-transferase (13); 5-aminolevulinate synthase (227); AMP deaminase (14-23); arylsulfate sulfotransferase (24); asparagine synthetase (25); aspartate ammonia-lyase (26); aspartate carbamoyltransferase (27-29,230-234); aspartate racemase (30); ATP:citrate lyase, *or* ATP:citrate synthase (211); *Bacillus subtilis* ribonuclease (31); *Basidobolus haptosporus* nuclease Bh1 (32); bis(5'-nucleosyl)-tetraphosphatase (asymmetrical) (33); casein kinase II (34); CDP-diacylglycerol:inositol 3-phosphatidyltransferase (35,36); cellulose synthase (UDP-forming) (210); chitobiosyldiphosphodolichol β-mannosyltransferase (37,191); chloroplast protein-transporting ATPase (38); citrate (*si*)-synthase, weakly inhibited (212); 3',5'-cyclic-GMP phosphodiesterase, *or* cGMP phosphodiesterase, slightly inhibited (39); 3',5'-cyclic-nucleotide phosphodiesterase (40); cysteinyl-tRNA synthetase (41); cytidine deaminase (42); cytosine deaminase (42,43); deoxycytidine kinase (44); deoxyguanosine kinase (45,46); dGTPase (47); diacylglycerol kinase (48-51); DNA topoisomerase II (52); dolichol kinase (53); dolichyl-phosphate β-D-mannosyltransferase (194-196); exopoly-phosphatase (54); 5'→3' exoribonuclease (55); exoribonuclease H (56); *N*-formylglutamate deformylase, *or* *N*-β-citryl-glutamate deacylase (57,58); β-fructofuranosidase, *or* invertase (59); fructose-1,6-bisphosphatase (60,61); fucokinase (62); fumarase (63); galactoside 2-α-L-fucosyltransferase (197-199); β-galactoside α-2,3-sialyltransferase (1,176); GDP-mannose 4,6-dehydratase (64,65); 1,3-β-glucan synthase (204,205); glucokinase, also weak alternative substrate (66,67); glucomannan 4-β-mannosyltransferase (206); glucosamine *N*-acetyltransferase (228); glucose-1 phosphatase (68); glucose-1-phosphate adenylyl-transferase (69); glucose-1-phosphate cytidylyltransferase (70); glucose-6-phosphate dehydrogenase (71,72); glucuronokinase (73); glutamate dehydrogenase (74-77); glutamine synthetase, *Neurospora crassa* (78); glycogen phosphorylase *b* (79); glycogen synthase (80-82); glycolipid 2-α-mannosyltransferase (193); glycolipid 3-α-mannosyltransferase (83, 192); glycoprotein 3-α-L-fucosyltransferase (187-189); glycoprotein 6-α-L fucosyl-transferase (200,201); guanylate kinase (84,85); hexokinase (86); hydroxyacylglutathione hydrolase, *or* glyoxalase II (87); 3-hydroxyanthranilate oxidasae, weakly inhibited (238); 5-hydroxypentanoate CoA-transferase, weakly inhibited (88); hypoxanthine(guanine) phosphoribosyltransferase (184-186); inosine kinase, *or* guanosine-inosine kinase (89); *myo*-inositol oxygenase (90,236,237); insulysin (91); D-lactate dehydrogenase (92,93); lactoylglutathione lyase, *or* glyoxalase I (87); membrane Dipeptidase, *or* renal Dipeptidase, *or* dehydropeptidase I (94,95); methionine *S*-adenosyltransferase (96-102); [3-methyl-2 oxobutanoate dehydrogenase, *or* 2-methylpropanoyl-transferring)] phosphatase, *or* [branched-chain α-keto-acid dehydrogenase] phosphatase (103-106); m$_7$G(5')pppN diphosphatase, slightly inhibited (107); mitochondrial processing peptidase (108); multiple inositol-polyphosphate phosphatase, *or* inositol-1,3,4,5-tetrakisphosphate 3-phosphatase (109);

NAD$^+$ diphosphatase (110); NADH peroxidase (111); NAD(P)H dehydrogenase (quinone) (112); nicotinate-nucleotide diphosphorylase (carboxylating) (178,179); nucleotidases (54); 3'-nucleotidase (54); 5'-nucleotidase (54,113-115); 2',5'-oligoadenylate synthetase (116); oligonucleotidase (117); ornithine carbamoyltransferase (229); palmitoyl-CoA hydrolase, *or* acyl-CoA hydrolase (118); phytanoyl-CoA dioxygenase (235); phosphatidate cytidylyl-transferase (119); phosphatidate phosphatase (120); 1-phosphatidylinositol 4 kinase (121-124); phosphodiesterase I, *or* 5'-exonuclease (125-127); phosphoenolpyruvate carboxylase (128-132); 1-phosphofructokinase (133,134); 6-phosphofructokinase (135-138); phosphoglucomutase (139); phosphogluconate dehydrogenase (140); 3-phosphoglycerate dehydrogenase (141); phosphoprotein phosphatase types 1 and 2A (106,142-147); phosphoribosyl-pyrophosphate synthetase, *or* ribose-5-phosphate pyrophosphokinase (148); phosphorylase kinase (149); phytoene synthase (150); polygalacturonate 4-α galacturonosyltransferase (202); polynucleotide adenylyltransferase, *or* poly(A) polymerase (151-157); polynucleotide 5'-hydroxyl-kinase (158,159); poly-peptide *N*-acetylgalactosaminyltransferase (203); porphobilinogen synthase (δ-aminolevulinate dehydratase) (160); protein-glutamine γ glutamyltransferase (transglutaminase) (161,213-226); [protein-PII] uridylyltransferase (162); pyruvate carboxylase (163); pyruvate dehydrogenase complex (164); pyruvate kinase (165-168); Rho ADP ribosyltransferase (169); ribonuclease H (56); ribose-phosphate diphosphokinase (phosphoribosyl pyrophosphate synthetase) (170,171); sterol 3β-glucosyltransferase (190); sucrose-phosphate synthase (207,208); sucrose synthase (209); thiamin-phosphate diphosphorylase (172,173); UDP-*N* acetylglucosamine:dolichyl-phosphate *N*-acetylglucosamine-phosphotransferase (174); uracil phosphoribosyltransferase, weakly inhibited (GTP is an activator in many organisms) (182,183); xanthine phosphoribosyltransferase (177). **1.** Sadler, Beyer, Oppenheimer, *et al.* (1982) *Meth. Enzymol.* **83**, 458. **2.** Rodríguez-Aparicio, Luengo, Gonzalez-Clemente & Reglero (1992) *J. Biol. Chem.* **267**, 9257. **3.** Bravo, Barrallo, Ferrero, *et al.* (2001) *Biochem. J.* **358**, 585. **4.** Arnold & Kelley (1978) *Meth. Enzymol.* **51**, 568. **5.** Ide (1971) *Arch. Biochem. Biophys.* **144**, 262. **6.** Yang & Epstein (1983) *J. Biol. Chem.* **258**, 3750. **7.** Takei, Sakakibara & Sugiyama (2001) *J. Biol. Chem.* **276**, 26405. **8.** Bailey-Wood, Dodgson & Rose (1970) *Biochim. Biophys. Acta* **220**, 284. **9.** Bayley-Wood, Dodgson & Rose (1969) *Biochem. J.* **112**, 257. **10.** Stokes, Denner & Dodgson (1973) *Biochim. Biophys. Acta* **315**, 402. **11.** Cabib & Carminatti (1966) *Meth. Enzymol.* **8**, 224. **12.** Hill & Bennett (1969) *Biochemistry* **8**, 122. **13.** Tunnicliff & Youngs (1989) *Proc. Soc. Exp. Biol. Med.* **192**, 11. **14.** Nathans, Chang & Deuel (1978) *Meth. Enzymol.* **51**, 497. **15.** Zielke & Suelter (1971) *The Enzymes*, 3rd ed., **4**, 47. **16.** Lushchak & Storey (1994) *Int. J. Biochem.* **26**, 1305. **17.** Lushchak, Smirnova & Storey (1998) *Comp. Biochem. Physiol. B* **119**, 611. **18.** Martini, Ranieri-Raggi, Sabbatini & Raggi (2001) *Biochim. Biophys. Acta* **1544**, 123. **19.** Ranieri-Raggi, Ronca, Sabbatini & Raggi (1995) *Biochem. J.* **309**, 845. **20.** Thakkar, Janero, Sharif, Hreniuk & Yarwood (1994) *Biochem. J.* **300**, 359. **21.** Yabuki & Ashihara (1992) *Phytochemistry* **31**, 1905. **22.** Yun & Suelter (1978) *J. Biol. Chem.* **253**, 404. **23.** Raffin, Izem & Thebault (1993) *Comp. Biochem. Physiol. B* **106**, 999. **24.** Konishi-Imamura, Kim, Koizumi & Kobashi (1995) *J. Enzyme Inhib.* **8**, 233. **25.** Hongo, Matsumoto & Sato (1978) *Biochim. Biophys. Acta* **522**, 258. **26.** Hanson & Havir (1972) *The Enzymes*, 3rd ed., **7**, 75. **27.** Jacobson & Stark (1973) *The Enzymes*, 3rd ed., **9**, 225. **28.** Webb (1966) *Enzyme and Metabolic Inhibitors*, vol. 2, p. 468, Academic Press, New York. **29.** Gerhart & Pardee (1962) *J. Biol. Chem.* **237**, 891. **30.** Shibata, Watanabe, Yoshikawa, *et al.* (2003) *Comp. Biochem. Physiol. B* **134**, 713. **31.** Yamasaki & Arima (1970) *Biochim. Biophys. Acta* **209**, 475. **32.** Desai & Shankar (2000) *Eur. J. Biochem.* **267**, 5123. **33.** Vallejo, Lobaton, Quintanilla, Sillero & Sillero (1976) *Biochim. Biophys. Acta* **438**, 304. **34.** Vorotnikov, Gusev, Hua, *et al.* (1993) *FEBS Lett.* **334**, 18. **35.** Antonsson & Klig (1996) *Yeast* **12**, 449. **36.** Antonsson (1994) *Biochem. J.* **297**, 517. **37.** Kaushal & Elbein (1986) *Arch. Biochem. Biophys.* **250**, 38. **38.** Subramanian, Ivey & Bruce (2001) *Plant J.* **25**, 349. **39.** Morishima (1975) *Biochim. Biophys. Acta* **410**, 310. **40.** Cheung (1967) *Biochemistry* **6**, 1079. **41.** Pan, Yu, Duh & Lee (1976) *J. Chin. Biochem. Soc.* **5**, 45. **42.** Ipata & Cercignani (1978) *Meth. Enzymol.* **51**, 394. **43.** Ipata, Marmocchi, Magni, Felicioli & Polidoro (1971) *Biochemistry* **10**, 4270. **44.** Kozai, Sonoda, Kobayashi & Sugino (1972) *J. Biochem.* **71**, 485. **45.** Yamada, Goto & Ogasawara (1982) *Biochim. Biophys. Acta* **709**, 265. **46.** Green & Lewis (1979) *Biochem. J.* **183**, 547. **47.** Kornberg, Lehman, Bessman, Simms & Kornberg (1958) *J. Biol. Chem.* **233**, 159. **48.** Kato & Takenawa (1990) *J. Biol. Chem.* **265**, 794. **49.** Wissing & Wagner (1992) *Plant Physiol.* **98**, 1148. **50.** Daleo, Piras & Piras (1976) *Eur. J. Biochem.*

68, 339. **51**. Kanoh & Akesson (1978) *Eur. J. Biochem.* **85**, 225. **52**. Arndt-Jovin, Udvardy, Garner, Ritter & Jovin (1993) *Biochemistry* **32**, 4862. **53**. Itami, Mueck & Keenan (1988) *Biochim. Biophys. Acta* **960**, 374. **54**. Proudfoot, Kuznetsova, Brown, *et al.* (2004) *J. Biol. Chem.* **279**, 54687. **55**. Slobin (2001) *Meth. Enzymol.* **342**, 282. **56**. Crouch & Dirksen (1982) *Cold Spring Harbor Monogr. Ser.* **14**, 211. **57**. Miyake, Innami & Kakimoto (1983) *Biochim. Biophys. Acta* **760**, 206. **58**. Asakura, Nagahashi, Hamada, *et al.* (1995) *Biochim. Biophys. Acta* **1250**, 35. **59**. Lee & Sturm (1996) *Plant Physiol.* **112**, 1513. **60**. Springgate & Stachow (1982) *Meth. Enzymol.* **90**, 378. **61**. Fujita & Freese (1979) *J. Biol. Chem.* **254**, 5340. **62**. Kilker, Shuey & Serif (1979) *Biochim. Biophys. Acta* **570**, 271. **63**. Hill & Teipel (1971) *The Enzymes*, 3rd. ed., **5**, 539. **64**. Liao & Barber (1972) *Biochim. Biophys. Acta* **276**, 85. **65**. Broschat, Chang & Serif (1985) *Eur. J. Biochem.* **153**, 397. **66**. Porter, Chassy & Holmlund (1982) *Biochim. Biophys. Acta* **709**, 178. **67**. Doelle (1982) *Eur. J. Appl. Microbiol. Biotechnol.* **14**, 241. **68**. Saugy, Farkas & MacLachlan (1988) *Eur. J. Biochem.* **177**, 135. **69**. Dietzler, Porter, Roth & Leckie (1984) *Biochem. Biophys. Res. Commun.* **122**, 289. **70**. Kimata & Suzuki (1966) *J. Biol. Chem.* **241**, 1099. **71**. Domagk, Chilla, Domschke, Engel & Sörensen (1969) *Hoppe Seyler's Z. physiol. Chem.* **350**, 626. **72**. Horne, Anderson & Nordlie (1970) *Biochemistry* **9**, 610. **73**. Gillard & Dickinson (1978) *Plant Physiol.* **62**, 706. **74**. Fisher (1985) *Meth. Enzymol.* **113**, 16. **75**. Fahien & Cohen (1970) *Meth. Enzymol.* **17A**, 839. **76**. Smith, Austen, Blumenthal & Nyc (1975) *The Enzymes*, 3rd ed., **11**, 293. **77**. Frieden (1962) *Biochim. Biophys. Acta* **59**, 484. **78**. Kapoor & Bray (1968) *Biochemistry* **7**, 3583. **79**. Madsen & Shechosky (1967) *J. Biol. Chem.* **242**, 3301. **80**. Stalmans & Hers (1973) *The Enzymes*, 3rd ed., **9**, 309. **81**. Rothman & Cabib (1967) *Biochemistry* **6**, 2107. **82**. Piras, Rothman & Cabib (1968) *Biochemistry* **7**, 56. **83**. Jensen & Schutzbach (1982) *J. Biol. Chem.* **257**, 9025. **84**. Anderson (1973) *The Enzymes*, 3rd ed., **9**, 49. **85**. le Floc'h & Lafleuriel (1990) *Plant Physiol. Biochem.* **28**, 191. **86**. Purich & Fromm (1971) *J. Biol. Chem.* **246**, 3456. **87**. Oray & Norton (1980) *Biochem. Biophys. Res. Commun.* **95**, 624. **88**. Eikmanns & Buckel (1990) *Biol. Chem. Hoppe-Seyler* **371**, 1077. **89**. Kawasaki, Shimaoka, Usuda & Utagawa (2000) *Biosci. Biotechnol. Biochem.* **64**, 972. **90**. Charalampous (1962) *Meth. Enzymol.* **5**, 329. **91**. Camberos, Pérez, Udrisar, Wanderley & Cresto (2001) *Exp. Biol. Med. (Maywood)* **226**, 334. **92**. LeJohn & Stevenson (1975) *Meth. Enzymol.* **41**, 293. **93**. LeJohn (1971) *J. Biol. Chem.* **246**, 2116. **94**. Harper, Rene & Campbell (1971) *Biochim. Biophys. Acta* **242**, 446. **95**. Ferren, Ward & Campbell (1975) *Biochemistry* **14**, 5280. **96**. Mudd (1973) *The Enzymes*, 3rd ed., **8**, 121. **97**. Chou & Talalay (1972) *Biochemistry* **11**, 1065. **98**. Schröder, Eichel, Breinig & Schröder (1997) *Plant Mol. Biol.* **33**, 211. **99**. Markham, Hafner, Tabor & Tabor (1980) *J. Biol. Chem.* **255**, 9082. **100**. Lu & Markham (2002) *J. Biol. Chem.* **277**, 16624. **101**. Chou & Talalay (1973) *Biochim. Biophys. Acta* **321**, 467. **102**. Porcelli, Cacciapuoti, Cartení-Farina & Gambacorta (1988) *Eur. J. Biochem.* **177**, 273. **103**. Damuni & Reed (1988) *Meth. Enzymol.* **166**, 321. **104**. Damuni & Reed (1987) *J. Biol. Chem.* **262**, 5129. **105**. Damuni, M. L. Merryfield, J. S. Humphreys & L. J. Reed (1984) *Proc. Natl. Acad. Sci. U.S.A.* **81**, 4335. **106**. Reed & Damuni (1987) *Adv. Protein Phosphatases* **4**, 59. **107**. Nuss & Furuichi (1977) *J. Biol. Chem.* **252**, 2815. **108**. Böhni, Daum & Schatz (1983) *J. Biol. Chem.* **258**, 4937. **109**. Höer, Höer & Oberdisse (1990) *Biochem. J.* **270**, 715. **110**. Nakajima, Fukunaga, Sasaki & Usami (1973) *Biochim. Biophys. Acta* **293**, 242. **111**. Badwey & Karnovsky (1979) *J. Biol. Chem.* **254**, 11530. **112**. Koli, Yearby, Scott & Donaldson (1969) *J. Biol. Chem.* **244**, 621. **113**. Madrid-Marina & Fox (1986) *J. Biol. Chem.* **261**, 444. **114**. Cercignani, Serra, Fini, *et al.* (1974) *Biochemistry* **13**, 3628. **115**. Newby, Luzio & Hales (1975) *Biochem. J.* **146**, 625. **116**. Ball (1982) *The Enzymes*, 3rd ed., **15**, 281. **117**. Datta & Niyogi (1975) *J. Biol. Chem.* **250**, 7313. **118**. Lee & Schulz (1979) *J. Biol. Chem.* **254**, 4516. **119**. Morii, Nishihara & Koga (2000) *J. Biol. Chem.* **275**, 36568. **120**. Berglund, I. Björkhem, B. Angelin & K. Einarsson (1989) *Biochim. Biophys. Acta* **1002**, 382. **121**. Vogel & Hoppe (1986) *Eur. J. Biochem.* **154**, 253. **122**. Steinert, Wissing & Wagner (1994) *Plant Sci.* **101**, 105. **123**. Hou, Zhang & Tai (1988) *Biochim. Biophys. Acta* **959**, 67. **124**. Yamakawa & T. Takenawa (1988) *J. Biol. Chem.* **263**, 17555. **125**. Luthje & A. Ogilvie (1985) *Eur. J. Biochem.* **149**, 119. **126**. Picher & Boucher (2000) *Amer. J. Respir. Cell Mol. Biol.* **23**, 255. **127**. Maruyama & Takashima (1993) *Cell Biochem. Funct.* **11**, 271. **128**. McDaniel & Siu (1972) *J. Bacteriol.* **109**, 385. **129**. Nakamura, Minoguchi & Izui (1996) *J. Biochem.* **120**, 518. **130**. Singal & Singh (1986) *Plant Physiol.* **80**, 369. **131**. Patel, Kraszewski & Mukhopadhyay (2004) *J. Bacteriol.* **186**, 5129. **132**. Sadaie, Nagano, Suzuki, Shinoyama & Fujii (1997) *Biosci. Biotechnol. Biochem.* **61**, 625. **133**. Van Hugo & Gottschalk (1974) *Eur. J. Biochem.* **48**, 455. **134**. Sapico & Anderson (1969) *J. Biol. Chem.* **244**, 6280. **135**. Cawood, F. C. Botha & J. G. C. Small (1988) *J. Plant Physiol.* **132**, 204. **136**. Isaac & Rhodes (1986) *Phytochemistry* **25**, 339. **137**. Sapico & Anderson (1969) *J. Biol. Chem.* **244**, 6280. **138**. Royt (1981) *Biochim. Biophys. Acta* **657**, 138. **139**. Maino & Young (1974) *J. Biol. Chem.* **249**, 5176. **140**. Silverberg & Dalziel (1975) *Meth. Enzymol.* **41**, 214. **141**. Slaughter & Davis (1975) *Meth. Enzymol.* **41**, 278. **142**. Ballou & Fischer (1986) *The Enzymes*, 3rd ed., **17**, 311. **143**. Damuni & L. J. Reed (1987) *J. Biol. Chem.* **262**, 5133. **144**. Brooks & Storey (1996) *Biochem. Mol. Biol. Int.* **38**, 1223. **145**. Mumby & J. A. Traugh (1980) *Biochim. Biophys. Acta* **611**, 342. **146**. Polya & Haritou (1988) *Biochem. J.* **251**, 357. **147**. Li (1982) *Curr. Top. Cell. Regul.* **21**, 129. **148**. Wong & Murray (1969) *Biochemistry* **8**, 1608. **149**. Chan & Graves (1982) *J. Biol. Chem.* **257**, 5948. **150**. Fraser, Schuch & Bramley (2000) *Planta* **211**, 361. **151**. Edmonds (1990) *Meth. Enzymol.* **181**, 161. **152**. Edmonds (1982) *The Enzymes*, 3rd ed., **15**, 217. **153**. Sarkar, Cao & Sarkar (1997) *Biochem. Mol. Biol. Int.* **41**, 1045. **154**. Pellicer, Salas & Salas (1978) *Biochim. Biophys. Acta* **519**, 149. **155**. Roggen & Slegers (1985) *Eur. J. Biochem.* 147, 225. **156**. Hunt, Meeks, Forbes, Das Gupta & Mogen (2000) *Biochem. Biophys. Res. Commun.* 272, 174. **157**. Tsiapalis, Dorson & Bollum (1975) *J. Biol. Chem.* **250**, 4486. **158**. Austin, Sirakoff, Roop & Moyer (1978) *Biochim. Biophys. Acta* **522**, 412. **159**. Levin & Zimmerman (1976) *J. Biol. Chem.* **251**, 1767. **160**. Tigier, Batlle & Locascio (1970) *Enzymologia* **38**, 43. **161**. Achyuthan & Greenberg (1987) *J. Biol. Chem.* **262**, 1901. **162**. Engleman & Francis (1978) *Arch. Biochem. Biophys.* **191**, 602. **163**. Scrutton & Utter (1965) *J. Biol. Chem.* **240**, 3714. **164**. Schwartz & Reed (1970) *Biochemistry* **9**, 1434. **165**. Iliffe-Lee & McClarty (2002) *Mol. Microbiol.* **44**, 819. **166**. Evans & Ratledge (1985) *Can. J. Microbiol.* **31**, 479. **167**. Singh, Malhotra & Singh (2000) *Indian J. Biochem. Biophys.* **37**, 51. **168**. Etges & Mukkada (1988) *Mol. Biochem. Parasitol.* **27**, 281. **169**. Aktories & Just (1995) *Meth. Enzymol.* **256**, 184. **170**. Tatibana, Kita, Taira, *et al.* (1995) *Adv. Enzyme Regul.* **35**, 229. **171**. Switzer & D. C. Sogin (1973) *J. Biol. Chem.* **248**, 1063. **172**. Penttinen (1979) *Meth. Enzymol.* **62**, 68. **173**. Kayama & Kawasaki (1973) *Arch. Biochem. Biophys.* **158**, 242. **174**. Zeng & Elbein (1995) *Eur. J. Biochem.* **233**, 458. **175**. Eppler, Morré & Keenan (1980) *Biochim. Biophys. Acta* **619**, 332. **176**. Kurosawa, Hamamoto, Inoue & Tsuji (1995) *Biochim. Biophys. Acta* **1244**, 216. **177**. Miller, Adamczyk, Fyfe & Elion (1974) *Arch. Biochem. Biophys.* **165**, 349. **178**. Shibata & Iwai (1980) *Agric. Biol. Chem.* **44**, 119. **179**. Taguchi & Iwai (1976) *Agric. Biol. Chem.* **40**, 385. **180**. Messenger & Zalkin (1979) *J. Biol. Chem.* **254**, 3382. **181**. Holmes, McDonald, McCord, Wyngaarden & Kelley (1973) *J. Biol. Chem.* **248**, 144. **182**. Carter, Donald, Roos & Ullman (1997) Mol. Biochem. Parasitol. 87, 137. **183**. Asai, Lee, Chandler & O'Sullivan (1990) *Comp. Biochem. Physiol. B Comp. Biochem.* 95, 159. **184**. Nagy & Ribet (1977) *Eur. J. Biochem.* 77, 77. **185**. Ohe & Watanabe (1980) *Agric. Biol. Chem.* **44**, 1999. **186**. Allen, Henschel, Coons, *et al.* (1989) *Mol. Biochem. Parasitol.* **33**, 273. **187**. de Vries, Storm, Rotteveel, *et al.* (2001) *Glycobiology* **11**, 711. **188**. Murray, Takayama, Schultz & Wong (1996) *Biochemistry* **35**, 11183. **189**. Shinoda, Morishita, Sasaki, *et al.* (1997) *J. Biol. Chem.* **272**, 31992. **190**. Madina, Sharma, Chaturvedi, Sangwan & Tuli (2007) *Biochim. Biophys. Acta* **1774**, 392. **191**. Kaushal & Elbein (1987) *Biochemistry* **26**, 7953. **192**. Jensen & Schutzbach (1981) *J. Biol. Chem.* **256**, 12899. **193**. Schutzbach, Springfield & Jensen (1980) *J. Biol. Chem.* **255**, 4170. **194**. Villagómez-Castro, Calvo-Méndez, Vargas-Rodríguez, Flores-Carreón & López Romero (1998) *Exp. Parasitol.* **88**, 111. **195**. Jensen & Schutzbach (1986) *Carbohydr. Res.* **149**, 199. **196**. Jensen & Schutzbach (1985) *Eur. J. Biochem.* **153**, 41. **197**. Bella & Kim (1971) *Biochem. J.* **125**, 1157. **198**. Scudder & Chantler (1981) *Biochim. Biophys. Acta* **660**, 128. **199**. Bella & Kim (1971) *Arch. Biochem. Biophys.* **147**, 753. **200**. Ihara, Ikeda & Taniguchi (2006) *Glycobiology* **16**, 333. **201**. Kaminska, Wisniewska & Koscielak (2003) *Biochimie* **85**, 303. **202**. Takeuchi & Tsumuraya (2001) *Biosci. Biotechnol. Biochem.* **65**, 1519. **203**. Takeuchi, Yoshikawa, Sasaki & Chiba (1985) *Agric. Biol. Chem.* **49**, 1059. **204**. Kamat, Garg & Sharma (1992) *Arch. Biochem. Biophys.* **298**, 731. **205**. Billon-Grand, Marais, Joseleau, *et al.* (1997) *Microbiology* **143**, 3175. **206**. Piro, Zuppa, Dalessandro & Northcote (1993) *Planta* **190**, 206. **207**. Sinha, Pathre & Sane (1997) *Phytochemistry* **46**, 441. **208**. Salerno & Pontis (1978) *Planta* **142**, 41. **209**. Hisajima & Ito (1981) *Biol. Plant.* **23**, 356. **210**. Haass, Hackspacher & Franz (1985) *Plant Sci.* **41**, 1. **211**. Antranikian, Herzberg & Gottschalk (1982) *J. Bacteriol.* 152, 1284. **212**. Okabayashi & Nakano (1979) *J. Biochem.* 85, 1061. **213**. Kawashima (1991) *Experientia* 47, 709. **214**. Siegel & C. Khosla (2007) *Pharmacol. Ther.* **115**, 232. **215**. Wada, Nakamura, Masutani, *et al.* (2002) *Eur. J. Biochem.* **269**, 3451. **216**. Wu, Lai & Tsai (2005) *Int. J. Biochem. Cell Biol.* **37**, 386. **217**. Carvajal-Vallejos, Campos, Fuentes-Prior, *et al.* (2007) *Biotechnol. Lett.* **29**, 1255. **218**. Mádi, Punyiczki, di Rao, Piacentini &

Fésüs (1998) *Eur. J. Biochem.* **253**, 583. **219.** Bergamini, Signorini & Poltronieri (1987) *Biochim. Biophys. Acta* **916**, 149. **220.** Murthy, Velasco & Lorand (1998) *Exp. Eye Res.* **67**, 273. **221.** Greenberg, Birckbichler & Rice (1991) *FASEB J.* **5**, 3071. **222.** Kang & Cho (1996) *Biochem. Biophys. Res. Commun.* **223**, 288. **223.** Hitomi, Yamagiwa, Ikura, Yamanishi & Maki (2000) *Biosci. Biotechnol. Biochem.* **64**, 2128. **224.** Dadabay & Pike (1989) *Biochem. J.* **264**, 679. **225.** Singh, Erickson & Cerione (1995) *Biochemistry* **34**, 15863. **226.** Zanetti, Ristoratore, Bertoni & Cariello (2004) *J. Biol. Chem.* **279**, 49289. **227.** Fanica-Gaignier & Clement-Metral (1973) *Eur. J. Biochem.* **40**, 19. **228.** Piro, Buffo & Dalessandro (1994) *Physiol. Plant.* **90**, 181. **229.** Ruepp, Müller, Lottspeich & Soppa (1995) *J. Bacteriol.* **177**, 1129. **230.** Achar, Savithri, Vaidyanathan & Rao (1974) *Eur. J. Biochem.* **47**, 15. **231.** Burns, Mendz & Hazell (1997) *Arch. Biochem. Biophys.* **347**, 119. **232.** Vickrey, Herve & Evans (2002) *J. Biol. Chem.* **277**, 24490. **233.** Durbecq, Thia-Toong, Charlier, *et al.* (1999) *Eur. J. Biochem.* **264**, 233. **234.** Rabinowitz, Hsiao, Gryncel, *et al.* (2008) *Biochemistry* **47**, 5881. **235.** Searls, Butler, Chien, *et al.* (2005) *J. Lipid Res.* **46**, 1660. **236.** Charalampous (1959) *J. Biol. Chem.* **234**, 220. **237.** Reddy, Pierzchala & Hamilton (1981) *J. Biol. Chem.* **256**, 8519. **238.** Nair (1972) *Arch. Biochem. Biophys.* **153**, 139.

(±)-Guatambuine

This alkaloid (FW = 264.37 g/mol; CAS = 2744-45-8) from *Aspidosperma subincanum* bark inhibits cyclopropane-fatty acyl-phospholipid synthase, IC$_{50}$ = 4 µM, with the inhibition patterns indicating that *Escherichia coli* CFAS operates via an ordered Bi Bi mechanism with S-adenosyl-L-methionine binding first. **1.** Gulanvarc'h, Guangqi, Drujon, Rey, Wang & Ploux (2008) *Biochim. Biophys. Acta* **1784**, 1652.

Guggulsterone

E-Guggulsterone

Z-Guggulsterone

This plant steroid and FXR antagonist (FW = 312.45 g/mol; CASs = 95975-55-6 (*E/Z*-isomer), 39025-24-6 (*E*-isomer), 39025-23-5 (*Z*-isomer)), also named pregna-4,17-diene-3,16-dione, from the resin of the guggul tree, *Commiphora mukul*, targets the Farnesoid X Receptor, acting as an antagonist in coactivator association assays, while enhancing FXR agonist-induced transcription of bile salt export pump (BSEP), a major hepatic bile acid transporter (1). Guggulipid, from which guggulsterone is derived, has been widely used in Asian folk medicine to treat hyperlipidemia. In HepG2 cells, in the presence of an FXR agonist, such as chenodeoxycholate or GW4064, GS enhances endogenous BSEP expression with a maximum induction of 400-500% that induced by an FXR agonist alone. This enhancement is also readily observed in FXR-dependent BSEP promoter activation using a luciferase reporter construct. GS is thus a selective bile acid receptor modulator that regulates expression of a subset of FXR targets. In rats, guggulipid treatment lowers serum triglyceride and raises serum high density lipoprotein levels. These data suggest that guggulsterone defines a novel class of FXR ligands characterized by antagonist activities in coactivator association assays but with the ability to enhance the action of agonists on BSEP expression *in vivo*. (*See also Dy 268*) **1.** Cui, Huang, Zhao, *et al.* (2003) *J. Biol. Chem.* **278**, 10214.

GW2016, *See Lapatinib*

GW2580

This oral cFMS kinase inhibitor (FW = 366.42 g/mol; CAS = 870483-87-7; Solubility: 48 mg/mL DMSO; <1 mg/mL H$_2$O), also named 5-[[3-methoxy-4-[(4-methoxyphenyl)methoxy]phenyl]methyl]-2,4-pyrimidinediamine, selectively targets monocyte growth and bone degradation by targeting cFMS kinase (1). Notably, GW-2580 inhibits the cellular homolog of the V-FMS oncogene product of the Susan McDonough strain of feline sarcoma virus, *or* c-FMS (IC$_{50}$ = 30 nM), about 150x to 500x lower concentration than its inhibition of b-Raf, CDK4, c-KIT, c-SRC, EGFR, ERBB2/4, ERK2, FLT-3, GSK3, ITK, and JAK2 kinases. GW-2580 selectively inhibits cFMS kinase compared with 186 other kinases *in vitro* and completely inhibited CSF-1-induced growth of rat monocytes (IC$_{50}$ = 0.2 µM) (2). At 1 µM, GW 2580 completely inhibits CSF-1-induced growth of mouse M-NFS-60 myeloid cells and human monocytes and completely inhibits bone degradation in cultures of human osteoclasts, rat calvaria, and rat fetal long bone (1). In contrast, GW2580 did not affect the growth of mouse NS0 lymphoblastoid cells, human endothelial cells, human fibroblasts, or five human tumor cell lines. GW2580 also did not affect lipopolysaccharide-(LPS-) induced TNF, IL-6, and prostaglandin E2 production in freshly isolated human monocytes and mouse macrophages. After oral administration, GW2580 blocked the ability of exogenous CSF-1 to increase LPS-induced IL-6 production in mice, inhibited the growth of CSF-1-dependent M-NFS-60 tumor cells in the peritoneal cavity, and diminished the accumulation of macrophages in the peritoneal cavity after thioglycolate injection (1). Use of GW2580 as a selective pharmacologic inhibitor of CSF1R signaling demonstrated that CSF-1 regulates the tumor recruitment of CD11b$^+$Gr-1(lo)Ly6C(hi) mononuclear Myeloid-Derived Suppressor Cells, *or* MDSCs (3). **1.** Conway, McDonald, Parham, *et al.* (2005) *Proc. Natl. Acad. Sci. U.S.A.* **102**, 16078. **2.** Conway, Pink, Bergquist, *et al.* (2008) *J. Pharmacol. Exp. Ther.* **326**, 41. **3.** Priceman, Sung, Shaposhnik, *et al.* (2010) *Blood* **115**, 1461.

GW5074

This potent signal transduction inhibtor (FW = 520.94 g/mol; CAS = 220904-83-6; Solubility: 100 mg/mL DMSO; <1 mg/mL H$_2$O), systematically named 3-[(3,5-dibromo-4-hydroxyphenyl)-methylene]-1,3-dihydro-5-iodo-2*H*-indol-2-one, selectively targets the proto-oncogene serine/threonine-protein kinase known as c-Raf (IC$_{50}$ = 9 nM), stimulating the Raf-MEK-ERK pathway, with little or no effect on the activities of JNK1, JNK2, JNK1, MEK1, MKK6, MKK7, CDK1, CDK2, c-Src, p38

MAP, VEGFR2 or c-Fms. GW5074 protects against the neurotoxic effects of MPP⁺ and methylmercury in cerebellar granule neurons, and glutathione depletion-induced oxidative stress in cortical neurons (1). GW-5074 also prevents neurodegeneration and improves behavioral outcome in an animal model of Huntington's disease (1). GW 5074 treatment (10 μM, 60 min) completely abolishes chronic morphine-mediated AC superactivation, suggesting that Raf-1 may have a crucial role in compensatory feedback regulation of cellular cAMP levels by clinically important opioid analgesics (2). Intraperitoneal administration of GW5074 significantly suppresses hyperresponsiveness of the airway contraction, whereas the airway epithelium-dependent relaxation is unaffected (3). **1**. Chin, Liu, Morrison, *et al.* (2004) *J. Neurochem.* **90**, 595. **2**. Yue, Varga, Stropova, *et al.* (2006) *Eur. J. Pharmacol.* **540**, 57. **3**. Lei, Cao, Xu, *et al.* (2008) *Respir. Res.* **9**, 71.

GW572016, *See Lapatinib*

GW788388

This orally active ALK5 inhibitor (FW = 425.28 g/mol; CAS = 452342-67-5; Solubility: 33 mg/mL DMSO; <1 mg/mL H_2O or Ethanol), also known as 4-(4-(3-(pyridin-2-yl)-1*H*-pyrazol-4-yl)pyridin-2-yl)-*N*-(tetrahydro-2*H*-pyran-4-yl)benzamide, potently and selectively targets the Activin receptor-Like Kinase activities of Transforming Growth Factor-β (TGF- β) type I receptor (*or* ALK5), with an IC_{50} of 18 nM, thereby blocking both TGF-β type II receptor and activin type II receptor kinase activities, but not that of the Bone Morphogenic Protein type II (BMP type II) receptor (1). When administered orally at 10 mg/kg once daily in a model of puromycin aminonucleoside-induced renal fibrosis, GW-788388 significantly reduces expression of Collagen IA1 (1). With db/db mice, which develop diabetic nephropathy, GW788388 (given orally for 5 weeks) significantly reduces renal fibrosis and decreases mRNA levels of key mediators of extracellular matrix deposition (2). Cardiomyocyte hypertrophy in MI hearts was likewise attenuated upon ALK5 inhibition by GW 788388 (3). In the bleomycin-induced fibrosis mouse model, treatment with GW788388 and bosentan prevents the fibrotic response, supporting the idea that the TGF-β /ET-1 axis has a role in wound repair and skin fibrosis (4). **1**. Gellibert, de Gouville, Woolven, *et al.* (2006) *J. Med. Chem.* **49**, 2210. **2**. Petersen, Thorikay, Deckers, *et al.* (2008) *Kidney Int.* **73**, 705. **3**. Tan, Zhang, Connelly, Gilbert & Kelly (2010) *Am. J. Physiol. Heart Circ. Physiol.* **298**, H1415. **4**. Lagares, García-Fernández, Jiménez, *et al.* (2010) *Arthritis Rheum.* **62**, 878.

GW823296, *See Orvepitant*

GW856553X, *See Losmapimod*

GX15-070, *See Obatoclax*

GYKI-14766, *See Efegatran*

GYKI 53773, *See Talampanel*

Gyromitrin

This toxin/carcinogen (FW = 100.12 g/mol; CAS = 16568-02-8), also known as acetaldehyde methylformylhydrazone, from the false morel *Gyromitra esculenta* hydrolyzes readily to form *N*-methyl-*N*-formylhydrazine, which then reacts with pyridoxal 5-phosphate, thereby depleting this cofactor within glutamic acid decarboxylase and reducing γ-aminobutyrate formation (1). Gyromitrin is a non-competitive inhibitor of human intestinal diamine oxidase, $ID_{50} = 1.6 \times 10^{-5}$ M (2). **1**. List & Luft (1968) *Arch. Pharm. Ber. Deutsch Pharm. Ges.* **301**, 294. **2**. Biegański, Braun & Kusche (1984) *Agents Actions* **14**, 351.

GZ-793A

This potent $VMAT_2$ inhibitor (FW = 464.04 g/mol; CAS = 1356447-90-9), also named (*R*)-3-[2,6-cis-di(4-methoxyphenethyl)piperidin-1-yl]propane-1,2-diol, selectively targets dopamine uptake by the vesicular monoamine transporter-2, inhibiting methamphetamine-evoked release of dopamine from striatal slices and methamphetamine self-administration in rats (1). (*See N-Methyl-α-methylphenethylamine* (*or methamphetamine*)). GZ-793A shows ~50x selectivity for $VMAT_2$ over the serotonin or dopamine transporters. GZ-793A's ability to evoke [³H]dopamine release and inhibit methamphetamine-evoked [³H]dopamine release from isolated striatal synaptic vesicles fit a two-site model, with high- and EC_{50} values of 15.5 nM and 29.3 μM (1). Tetrabenazine and reserpine completely inhibit GZ-793A-evoked [³H]dopamine release, but only at the High-affinity site. Low concentrations of GZ-793A that interact with the extra-vesicular dopamine uptake site and the High-affinity intra-vesicular DA release site also inhibited methamphetamine-evoked [³H]dopamine release from synaptic vesicles. These results support a hypothetical model of GZ-793A interaction at more than one site on the $VMAT_2$ protein, explaining its potent inhibition of dopamine uptake, dopamine release via a high-affinity tetrabenazine- and reserpine-sensitive site, dopamine release via a low-affinity tetrabenazine- and reserpine-insensitive site, and a low-affinity interaction with the dihydrotetrabenazine binding site on $VMAT_2$ (1). GZ-793A specifically decreases METH self-administration, without the development of tolerance (2). Increasing the unit dose of METH did not surmount the inhibition produced by GZ-793A on METH self-administration. GZ-793A did not serve as a substitute for self-administered METH. GZ-793A and related $VMAT_2$ inhibitors may be promising leads for reducing the risk of relapse to METH use following exposure to drug-associated cues (3). **1**. Horton, Nickell, Zheng, Crooks & Dwoskin (2013) *J. Neurochem.* **127**, 177. **2**. Beckmann, Denehy, Zheng, *et al.* (2012) *Psychopharmacol.* (Berlin). **220**, 395. **3**. Alvers, Beckmann, Zheng, *et al.* (2012) *Psychopharmacol.* (Berlin). **224**, 255.

GZD824

This novel, oral bioavailable Bcr-Abl inhibitor ($FW_{free-base}$ = 532.57 g/mol; $FW_{dimesylate-salt}$ = 724.77 g/mol; CAS = 1421783-64-3; Solubility: 100 mg/mL DMSO or H_2O), known systematically as 4-methyl-*N*-[4-[(4-methyl-1-piperazinyl)methyl]-3-(trifluoromethyl)phenyl]-3-[2-(1*H*-pyrazolo[3,4-*b*]-pyridin-5-yl)ethynyl]benzamide, methanesulfonate (1:2) targets both wild type Bcr-Abl (IC_{50} = 0.34 nM) and mutant Bcr-Abl^{T315I} (IC_{50} = 0.34 nM), the oncogenic kinases formed by fusion of the Breakpoint cluster (Bcr) and Abelson (Abl) genes and resulting in formation of the so-called Philadelphia chromosome found in more than 90% of chronic myelogenous leukemia (CML) cases. The Bcr-Abl^{T315I} mutant confers resistance to all FDA-approved Bcr-Abl inhibitors prior to ponatinib (*See Ponatinib*). GZD824 likewise potently suppresses proliferation of Bcr-Abl-positive K562 and Ku812 human CML cells with IC_{50} values of 0.2 and 0.13 nM, respectively. GZD824 displays an *in vivo* half-life of 10.6 hours. **1**. Ren, Pan, Zhang, *et al.* (2013) *J. Med. Chem.* **56**, 879.

– H –

H, *See* Histidine

H-77

This peptidomimetic (FW = 1011.24 g/mol; *Sequence*: D-His-L-Pro-L-Phe-L-His-L-Leu(CH$_2$NH)-L-Leu-L-Val-L-Tyr) inhibits canine (IC$_{50}$ = 24 nM), human (IC$_{50}$ = 1 µM), and rat renin (IC$_{50}$ = 0.6 µM) *in vitro*. This analogue corresponds to a sequence in the renin substrate, angiotensinogen, except that the peptide bond between Leu-5 and Leu-6 is replaced by a –CH$_2$NH– moiety. **Target(s):** cathepsin E, K_i = 50 nM (1-3); endothiapepsin, mildly inhibited (4); renin (5,6). **1.** Jupp, Richards, Kay, *et al.* (1988) *Biochem. J.* **254**, 895. **2.** Robinson, Lees, Kay & Cook (1992) *Biochem. J.* **284**, 407. **3.** Hill, Montgomery & Kay (1993) *FEBS Lett.* **326**, 101. **4.** Bailey & Cooper (1994) *Protein Sci.* **3**, 2129. **5.** Oldham, Arnstein, Major & Clough (1984) *J. Cardiovasc. Pharmacol.* **6**, 672. **6.** Szelke, Leckie, Tree, *et al.* (1982) *Hypertension* **4**, 59.

H 89

This Ser/Thr protein kinase inhibitor (FW = 519.28 g/mol; CAS 130964-39-5; Solubility: 25 mM in H$_2$O; 100 mM in DMSO), also named *N*-[2-[[3-(4-bromophenyl)-2-propenyl]amino]ethyl]-5-isoquinoline sulfonamide·HCl, targets ribosomal protein S6 kinase 1, *or* S6K1 (K_i = 80 nM), mitogen- and stress-activated protein kinase-1, *or* MSK1 (K_i = 120 nM), cAMP-stimulate protein kinase, *or* PKA (K_i = 135 nM), Rho-associated protein kinase, *or* ROCKII (K_i = 270 nM), Akt/PKBα (K_i = 2.6 µM), and Mitogen-Activated Protein Kinase-Activated protein kinase-2, *or* MAPKAP-K1b (K_i = 2.8 µM). On the unwarranted assumption that H-89 effect is mainly directed against PKA, H-89 has been widely used (in >1000 reports!) to evaluate the role of PKA in the heart, osteoblasts, hepatocytes, smooth muscle cells, neuronal tissue, epithelial cells, etc. Even so, H89 targets at least eight other kinases, a fact that compromises interpretation of earlier results (1). **Target(s):** forskolin-induced neurite outgrowth (2); phosphatidylcholine biosynthesis in HeLa cells (3); protein kinase A in LPS-induced activation of NF-κB proteins of J774 mouse macrophage-like cell line (4); incorporation of choline into phosphatidylcholine via inhibition of choline kinase (5); forskolin-stimulated cardiac L-type calcium current (6); protein kinase-mediated acetylcholine release from myenteric plexus of guinea pig ileum (7); collagenase induction by phorbol ester through a mechanism that does not involve protein kinase A (8). **1.** Lochner & Moolman (2006) *Cardiovasc. Drug Rev.* **24**, 261. **2.** Chijiwa, Mishima, Hagiwara, *et al.* (1990) *J. Biol. Chem.* **265**, 5267. **3.** Geilen, Wieprecht, Wieder & Reutter (1992) *FEBS Lett.* **309**, 381. **4.** Muroi & Suzuki (1993) *Cell Signal.* **5**, 289. **5.** Wieprecht, Wieder & Geilen (1994) *Biochem. J.* **297**, 241. **6.** Yuan & Bers (1995) *Am. J. Physiol.* **268**, C651. **7.** Takeuchi, Fukunaga, Hata & Yagasaki (1995) *J. Smooth Muscle Res.* **31**, 143. **8.** Shoshan, Ljungdahl & Linder (1996) *Cell Signal.* **8**, 191.

H-189

This synthetic peptidomimetic and human renin inhibitor (FW = 1268.53 g/mol; Sequence: Pro-His-Pro-Phe-His-Sta-(statinyl)-Val-Ile-His-Lys), itself an analogue of human angiotensinogen, inhibits endothiapepsin (K_i = 1 nM) and renin (K_i = 10 nM). Note that a tetrahedral statine moiety substitutes for the leucyl residue at the scissile bond. **Target(s):** cathepsin D, mildly inhibited (1); endothiapepsin (2-5); renin (1). **1.** Tree, Atrash,

Donovan, *et al.* (1983) *J. Hypertens.* **1**, 399. **2.** Coates, Erskine, Crump, Wood & Cooper (2002) *J. Mol. Biol.* **318**, 1405. **3.** Bailey, Cooper, Veerapandian, *et al.* (1993) *Biochem. J.* **289**, 363. **4.** Stultz & Karplus (2000) *Proteins* **40**, 258. **5.** Bailey & Cooper (1994) *Protein Sci.* **3**, 2129.

H-256

This reduced scissile-bond peptidomimetic (FW = 911.03 g/mol; *Sequence*: L-Pro-L-Thr-L-Glu-L-Phe-Ψ(CH$_2$NH)-L-Phe-L-Arg-L-Glu), is a potent inhibitor of a number of aspartic proteinases. K_i values for human pepsin, human gastricsin, human cathepsin D, penicillopepsin, and endothiapepsin are 30, 60, 150, 10, and 60 nM, respectively. **Target(s):** cathepsin D (1-3); cathepsin E (4); chymosin, weakly inhibited (1); endothiapepsin (1,2,5-8); gastricsin (2); penicillopepsin (1,2); pepsin (1,2); phytepsin (9). **1.** Hallett, Jones, Atrash, *et al.* (1985) in *Aspartic Proteinases and Their Inhibitors* (Kostka, ed.), Proc. of the F.E.B.S. Advanced Course 84/07, pp. 467-478, Walter de Gruyter & Co., Berlin. **2.** Foundling, Cooper, Watson, *et al.* (1987) *Nature* **327**, 349. **3.** Bonelli, Kay, Tessitore, *et al.* (1988) *Biol. Chem. Hoppe-Seyler* **369**, 323. **4.** Jupp, Richards, Kay, *et al.* (1988) *Biochem. J.* **254**, 895. **5.** Cooper, Foundling, Hemmings, *et al.* (1987) *Eur. J. Biochem.* **169**, 215. **6.** Stultz & Karplus (2000) *Proteins* **40**, 258. **7.** Coates, Erskine, Crump, Wood & Cooper (2002) *J. Mol. Biol.* **318**, 1405. **8.** Bailey & Cooper (1994) *Protein Sci.* **3**, 2129. **9.** Sarkkinen, Kalkkinen, Tilgmann, Siuro & Mikola (1992) *Planta* **186**, 317.

H-1152 & glycyl-H-1152

H 1152   Glycyl-H 1152

This cell-permeable isoquinoline (FW = 319.43 g/mol; CAS 871543-07-6; Photosensitive; Solubility: 100 mM in H$_2$O; 50 mM in DMSO), also named (*S*)-(+)-2-methyl-1-[(4-methyl-5-isoquinolinyl)sulfonyl]homopiperazine, inhibits G-protein Rho-associated kinase (K_i = 1.6 nM). The related Glycyl-H-1152 (FW = 449.40 g/mol; CAS 913844-45-8) displays improved ROCKII selectivity (*e.g.,* IC$_{50}$ values are 0.0118, 2.35, 2.57, 3.26, > 10 and >10 µM for ROCKII, Aurora A, CAMKII, PKG, PKA and PKC respectively). **Target(s):** ROCKII (IC$_{50}$ = 0.012 µM); CAMKII (IC$_{50}$ = 0.180 µM); PKG (IC$_{50}$ = 0.360 µM); Aurora A (IC$_{50}$ = 0.745 µM); PKA (IC$_{50}$ = 3.03 µM); PKC (IC$_{50}$ = 5.68 µM); and MLCK (IC$_{50}$ = 28.3 µM); polo kinase (1); protein kinase A (2,3); Rho-associated kinase (2,3). The structurally related Glycyl-H 1152 (FW = 319.43 g/mol; CAS 913844-45-8; Photosensitive), also named (*S*)-(+)-4-glycyl-2-methyl-1-[(4-methyl-5-isoquinolinyl)sulfonyl]hexahydro-1*H*-1,4-diazepine, displays improved selectivity toward ROCKII (4). **Target(s):** ROCKII (IC$_{50}$ = 0.0118 µM); Aurora A (IC$_{50}$ = 2.35 µM); CAMKII (IC$_{50}$ = 2.57 µM); PKG (IC$_{50}$ = 3.26 µM); PKA (IC$_{50}$ >10 µM); PKC (IC$_{50}$ >10 µM); and MLCK (IC$_{50}$ = 28.3 µM). **1.** Johnson, Stewart, Woods, Giranda & Luo (2007) *Biochemistry* **46**, 9551. **2.** Ikenoya, Hidaka, Hosoya, *et al.* (2002) *J. Neurochem.* **81**, 9. **3.** Sasaki, Suzuki & Hidaka (2002) *Pharmacol. Ther.* **93**, 225. **4.** Tamura, *et al.* (2005) *Biochim. Biophys. Acta* **1754**, 245.

HA14-1

This cell-permeable Bcl-2 inhibitor (FW = 409.24 g/mol; CAS 65673-63-4; Soluble to 100 mM in DMSO and to 100 mM in Ethanol), also named 2-amino-6-bromo-α-cyano-3-(ethoxycarbonyl)-4*H*-1-benzopyran-4-acetic acid ethyl ester, disrupts Bax/Bcl-2 interactions (IC$_{50}$ ~ 9 μM), thereby inducing apoptosis of tumor cells. *In vitro* binding studies demonstrated that HA14-1 interacts with a Bcl-2 surface pocket that is essential for its biological function (1). HA14-1 induces apoptosis of human acute myeloid leukemia (HL-60) cells overexpressing Bcl-2 protein associated with the decrease in mitochondrial membrane potential and activation of caspase-9, followed by caspase-3. Cytokine response modifier A, a potent inhibitor of Fas-mediated apoptosis, did not block apoptosis induced by HA14-1 (1). In acute myelogenous leukemia (AML) cell lines with constitutively activated MAPK, MAPK kinase (MEK) blockade by PD184352 strikingly potentiates apoptosis induced HA14-1 or by Bcl-2 antisense oligonucleotides (2). HA14-1 also binds to the antiapoptotic Blc-2 proteins, Bcl-XL and Bcl-w (3). **1**. Wang, Liu, Zhang, *et al.* (2000) *Proc. Natl. Acad. Sci. U.S.A.* **97**, 7124. **2**. Milella, Estrov, Kornblau, *et al.* (2002) *Blood* **99**, 3461. **3**. Doshi, Tian & Xing (2006) *J. Med. Chem.* **49**, 7731.

HA 130

This lipase inhibitor (FW = 463.29 g/mol; CAS 1229652-21-4; Solubility: 100 mM in DMSO), also named *B*-[3-[[4-[[3-[(4-fluorophenyl)methyl]-2,4-dioxo-5-thiazolidinylidene]methyl]phenoxy]methyl]phenyl]boronic acid, targets (IC$_{50}$ = 28 nM) ectonucleotide pyrophosphatase/phosphodiesterase-2, *or* E-NPP-2, *or* autotaxin (ATX), an extracellular enzyme catalyzes the conversion of lysophosphatidylcholine into the lipid mediator, lysophosphatidic acid (LPA) (1). Although autotaxin itself was mistakenly thought to be a tumor cell-motility-stimulating factor, its reaction product, LPA, is the actual signal molecule. ATX-LPA signaling has been implicated in angiogenesis, chronic inflammation, fibrotic diseases and tumor progression, making this system an attractive druggable target for therapy (2). The boronic acid moiety interacts with the active-site threonine in ATX, thereby increasing its affinity. **1**. Albers, van Meeteren, Egan, *et al.* (2010) *J. Med. Chem.* **53**, 4958. **2**. Albers, Dong, van Meeteren, *et al.* (2010) *Proc. Natl. Acad. Sci. USA* **107**, 7257.

(*R*)-(+)-HA-966

This brain-penetrant NMDA receptor antagonist/partial agonist (FW = 116.12 g/mol; CAS 123931-04-4; Soluble to 100 mM in H$_2$O), also named (*R*)-(+)-3-amino-1-hydroxypyrrolidin-2-one, is a centrally- and peripherally-acting antagonist (IC$_{50}$ ~ 150 μM), binding at the strychnine-insensitive glycine allosteric site of the *N*-methyl-D-aspartate receptor

ionophore complex (1) and reducing the affinity of glutamate for its recognition site on the NMDA receptor by 5x (2). By contrast, the *S*-enantiomer of HA-966 is much less potent, IC$_{50}$ > 1 mM. The IC$_{50}$ of (*R*)-HA-966 for displacing glycine binding is decreased in the presence of spermine, suggesting that spermine increases the affinity of (*R*)-HA-966 at the glycine binding site (3). **1**. Pullan, *et al.* (1990) *J. Neurochem.* **55**, 1346. **2**. Kemp & Priestley (1991) *Mol. Pharmacol.* **39**, 666. **3**. Pullan, *et al.* (1991) *Eur. J. Pharmacol.* **208**, 25.

HA-1077, See *Fasudil*

Hadacidin

This glycine derivative (FW$_{free-acid}$ = 119.08 g/mol; CAS 689-13-4), also called *N*-formyl-*N*-hydroxyglycine and *N*-formyl-*N* hydroxyaminoacetic acid, from *Penicillium frequentans* and *P. aurantio-violaceum* exhibits antitumor properties, inhibiting the growth of human adrenocarcinoma (1.2). Hadacidin also inhibits bacterial and mammalian purine nucleotide biosynthesis. **Target(s)**: adenylosuccinate synthetase (3-9); aspartate aminotransferase (10). **1**. Kaczka, Gitterman, Dulaney & Folkers (1962) *Biochemistry* **1**, 340. **2**. Stevens & Emery (1966) *Biochemistry* **5**, 74. **3**. Gale & Smith (1968) *Biochem. Pharmacol.* **17**, 2495. **4**. Jahngen & Rossomando (1984) *Arch. Biochem. Biophys.* **229**, 145. **5**. Poland, Lee, Subramanian, *et al.* (1996) *Biochemistry* **35**, 15753. **6**. Webb (1966) *Enzyme and Metabolic Inhibitors*, vol. **2**, p. 467, Academic Press, New York. **7**. Shigeura & Gordon (1962) *J. Biol. Chem.* **237**, 1937. **8**. Stayton, Rudolph & Fromm (1983) *Curr. Top. Cell. Regul.* **22**, 103. **9**. Matsuda, Shimura, Shiraki & Nakagawa (1980) *Biochim. Biophys. Acta* **616**, 340. **10**. Hatch (1973) *Arch. Biochem. Biophys.* **156**, 207.

Haemadin

This 57-residue exosite II-binding polypeptide (MW ≈ 5 kDa) from the land-living Indian leech *Haemadipsa sylivestris* has a mechanism of thrombin inhibition that is similar to that of hirudin (1,2). The N-terminal segment of haemadin binds within the thrombin active site, forming a parallel β-strand with residues Ser214-Gly216 of the proteinase. In contrast to hirudin, however, haemadin's markedly acidic C-terminal peptide does not bind the fibrinogen-recognition exosite, and interacts instead with the heparin-binding exosite of thrombin (3). (**See also** *Hirudin*) **1**. Strube, Kroger, Bialojan, Otte & Dodt (1993) *J. Biol. Chem.* **268**, 8590. **2**. Richardson, Fuentes-Prior, Sadler, Huber & Bode (2002) *Biochemistry* **41**, 2535. **3**. Richardson, Kröger, Hoeffken, *et al.* (2000) *EMBO J.* **19**, 5650.

Halichondrin B

This naturally occurring marine toxin (FW = 1125.36 g/mol; CAS 103614-76-2) from the marine sponge *Halichondria okadai*, is a tubulin-directed anti-mitotic agent with powerful anticancer properties (1). Its structurally less complicated analogue, eribulin, has already advanced to clinical trials, showing great promise as an antineoplastic agent that interrupts mitosis and induces apoptosis (**See** *Eribulin*). Halichondrin B is highly cytotoxic (IC$_{50}$ = 0.3 nM for L1210 murine leukemia cells), with arrest in mitosis (2). It also inhibits polymerization of purified brain tubulin in the absence or presence of microtubule-associated proteins (2). The total synthesis of Halichondrin B was accomplished in Kishi's laboratory at Harvard (3). **1**. Hirata & Uemura (1986) *Pure Appl. Chem.* **58**, 701. **2**. Bai, Paull, Herald, *et al.* (1991) *J. Biol. Chem.* **266**, 15882. **3**. Aicher, Buszek, Fang *et al.* (1992) *J. Am. Chem. Soc.* **114**, 3162.

Halofantrine

This antimalarial agent (FW$_{\text{free-base}}$ = 500.43 g/mol; CAS 69756-53-2), also known as 1,3-dichloro-α-[2-(dibutylamino)ethyl]-6-(trifluoromethyl)-9-phenanthrenemethanol, is effective against asexual forms of multidrug-resistant *Plasmodium falciparum* malaria. **Target(s):** cAMP-dependent protein kinase, *or* protein kinase A (1); CYP2D6, also as alternative substrate (2); hERG K$^+$ current (3-5). 1. Wang, Ternai & Polya (1994) *Biol. Chem. Hoppe Seyler* **375**, 527. 2. Halliday, Jones, Smith, Kitteringham & Park (1995) *Brit. J. Clin. Pharmacol.* **40**, 369. 3. Traebert, Dumotier, Meister, *et al.* (2004) *Eur. J. Pharmacol.* **484**, 41. 4. Mbai, Rajamani & January (2002) *Cardiovasc. Res.* **55**, 799. 5. Tie, Walker, Singleton, *et al.* (2000) *Brit. J. Pharmacol.* **130**, 1967.

Haloperidol

This powerful antipsychotic agent (FW$_{\text{free-base}}$ = 375.87 g/mol; CAS 52-86-8), also known as 4-[4-(4-chlorophenyl)-4-hydroxy-1-piperidinyl]-1-(4-fluoro-phenyl)-1-butanone, is a dopamine antagonist at D$_1$ (K_i = ~ 80 nM), D$_2$ (K_i = 8 nM), D$_3$ (K_i = ~ 7 nM), D$_4$ (K_i = 2 nM), and D$_5$ (K_i = ~ 100 nM) receptors and is frequently used to treat acute and chronic schizophrenia. While its precise mechanism of therapeutic action remains to be established, haloperidol appears to depress the CNS at the subcortex, midbrain, and brain stem. **Key Pharmacokinetic Parameters:** *See* Appendix II in Goodman & Gilman's THE PHARMACOLOGICAL BASIS OF THERAPEUTICS, 12th Edition (Brunton, Chabner & Knollmann, eds.) McGraw-Hill Medical, New York (2011). **Target(s):** Ca^{2+}-dependent cyclic-nucleotide phosphodiesterase (1); cholestenol Δ isomerase, *or* sterol Δ^{7,8}-isomerase (2); dopamine receptors (3,4); glycine transporter (5); monamine oxidase (6). 1. Hidaka, Inagaki, Nishikawa & Tanaka (1988) *Meth. Enzymol.* **159**, 652. 2. Moebius, Reiter, Bermoser, *et al.* (1998) *Mol. Pharmacol.* **54**, 591. 3. Leysen, Gommeren, Janssen, Van Gompel & Janssen (1988) *Psychopharmacol. Ser.* **5**, 12. 4. Nisoli, Tonello, Memo & Carruba (1992) *J. Pharmacol. Exp. Ther.* **263**, 823. 5. Williams, Mallorga, Conn, Pettibone & Sur (2004) *Schizophr. Res.* **71**, 103. 6. Giller, Hall, Reubens & Wojciechoswki (1984) *Biol. Psychiatry* 19. 517.

Haloxyfop

This aryloxyphenoxypropionate herbicide (FW$_{\text{free-acid}}$ = 361.70 g/mol; CAS 69806-34-4), also known as (*RS*)-2-{4-[3-chloro-5(trifluoromethyl)-2pyridyloxy]phenoxy}propionic acid, is a strong inhibitor of wheat acetyl-CoA carboxylase (IC$_{50}$ = 0.28 μM), an important target of aryloxyphenoxypropionate-class herbicides. Analysis of the interaction between the active site residues and (*R*)-haloxyfop shows the van der Waals interactions play a key role in binding. Note that the (*R*)-enantiomer is often referred to as haloxyfop-P. 1. Burton, Gronwald, Somers, *et al.* (1987) *Biochem. Biophys. Res. Commun.* **148**, 1039. 2. Rendina, Felts, Beaudoin, *et al.* (1988) *Arch. Biochem. Biophys.* **265**, 219. 3. Zhang, Tweel & Tong (2004) *Proc. Natl. Acad. Sci. U.S.A.* **101**, 5910. 4. Rendina, Craig-Kennard,

Beaudoin & Breen (1990) *J. Agric. Food Chem.* **38**, 1282. 5. Egli, Gengenbach, Gronwald, Somers & Wyse (1993) *Plant Physiol.* **101**, 499. 6. Zuther, Johnson, Haselkorn, McLeod & Gornicki (1999) *Proc. Natl. Acad. Sci. U.S.A.* **96**, 13387. 7. Zagnitko, Jelenska, Tevadze, Haselkorn & Gornicki (2001) *Proc. Natl. Acad. Sci. U.S.A.* **98**, 6617.

Halysin

This cysteine-rich, 71-residue disintegrin and hemorrhagic agent (MW = 7500 (SDS gel electrophoresis); CAS 137544-79-7) from the venom of the snake *Agkistrodon halys* contains an RGD loop that facilitates tight binding to integrins, resulting in the potent inhibition of platelet aggregation stimulated by ADP, thrombin, and collagen (IC$_{50}$ = 160–360 nM), without affecting platelet secretion. It was active in inhibiting platelet aggregation of platelet-rich plasma and whole blood. Halysin had no effect on thromboxane B$_2$ formation of platelets or intracellular Ca^{2+} mobilization of Quin 2-AM loaded platelets stimulated by thrombin. It inhibited the fibrinogen-induced aggregation of elastase-treated platelets. 1. Huang, Liu, Ouyang & Teng (1991) *Biochem. Pharmacol.* **42**, 1209.

HAMI3379

This potent receptor antagonist (FW = 595.70 g/mol; CAS 712313-35-4; IUPAC: 3-[[(3-carboxycyclohexyl)amino]carbonyl]-4-[3-[4-[4-(cyclohexyl-oxy) butoxy]phenyl]propoxy]benzoic acid) targets the G-protein-coupled leukotriene LTD$_4$- and LTC$_4$ receptor-mediated mobilization of intracellular calcium ion, with IC$_{50}$ values of 3.8 and 4.4 nM, respectively, with a far weaker effect on the Cys-LT$_1$ receptor (IC$_{50}$ = >10 μM) (1). Intracerebroventricular injection of HAMI 3379 protects against acute brain injury after focal cerebral ischemia in rats (2). 1. Wunder, Tinel, Kast, *et al.* (2010) *Br. J. Pharmacol.* **160**, 399. 2. Shi, Xiao, Zhao, *et al.* (2012) *Brain Res.* **1484**, 57.

Hammerhead Ribozyme-Mediated mRNA Knockdown

The smallest RNA endonucleases, hammerhead *trans*-acting ribozymes specifically bind and cleave RNA sequences at defined targets, often discriminating between those differing by a single nucleotide (1,2). Values for k_{cat} range from 3-15 min^{-1}, indicating an acceptable rate cleavage rate, and k_{cat}/K_m typically reaches values of 10^6 min^{-1} M^{-1}, a value comparable with those described for other biologically active ribozymes Mercatanti *et al.* (3) describe a method for predicting accessible target sites for hammerhead ribozymes within a given RNA sequence. Their approach maps all putative NUH cleavage sites (where N = A, C, G, U and H = A, C, U) and picks out short flanking regions as the binding domain for the corresponding ribozyme. The probabilistic level of unfolding, accessibility score (AS), is then calculated for each target region on the basis of a comparison of all folding structures obtained for the target RNA and arranged according to the Boltzmann's distribution. Application of a series of imposed limits then gives the best target sequences (*i.e.*, those endowed with likely accessibility and with a potentially active catalytic structure of the hammerhead sequence). 1. Phylactou, Tsipouras & Kilpatrick (1998) *Biochem Biophys Res Commun* **249**:804 2. Citti & Rainaldi (2005) *Curr. Gene Ther.* 5, 11. 3. Mercatanti, Rainaldi, Mariani, Marangoni & Citti (2009) *J. Comput. Biol.* **9**, 641.

Hanatoxin-1

This 34-residue channel-gating modifier (FW = 4120.71 g/mol; *Sequence:* ECRYLFGGCLTTSDCCKKHLGCKFRDKYCAWDFS; CAS 170780-00-4) from the venom of the Chilean rose tarantula (*Grammostola spatulata*) inhibits voltage-gated K$^+$ channels (*i.e.*, Kv2.1 voltage-activated K$^+$ channels) not by blocking the pore, but by altering the energetics of gating (1,2). HaTx1 adopts an inhibitor-cystine-knot motif, composed of two β-strands, presenting a hydrophobic patch and its surrounding charged residues that are principally responsible for the binding of this gating modifier toxins to the voltage-gated channel (3). **1**. Swartz & MacKinnon (1995) *Neuron* **15**, 941. **2**. Li-Smerin & Swartz (2000) *J. Gen. Physiol.* **115**, 673. **3**. Takahashi, Kim, Min, *et al.* (2000) *J. Mol. Biol.* **297**, 771.

Harmaline

This psychoactive alkaloid (FW$_{free-base}$ = 214.27 g/mol; CAS 304-21-2), also known as 3,4-dihydroharmine and harmidine, is found in a wide variety of plant groups, but was isolated from *Peganum harmala* in 1841 (1). Harmaline is a central nervous system stimulator that inhibits monoamine oxidase A, an enzyme that inactivates monoaminergic neurotransmitters, such as serotonin, melatonin, noradrenalin, and adrenalin. In view of their vital role in the inactivation of neurotransmitters, MAO dysfunctions are thought to underlie a number of psychiatric and neurological disorders. Harmaline also acts as a general inhibitor of Na$^+$-dependent transporters. **Target(s):** *N*-acetyltransferase (2); alanine transport, Na$^+$-dependent (3); aryl acylamidase (4); calcium ion channels (5); choline uptake (6); dicarboxylic acid transporter (7); DNA topoisomerase I (8); β-fructofuranosidase, *or* invertase (9); glutamine transport (10); histamine *N*-methyltransferase (11); *N* hydroxyarylamine *O*-acetyltransferase (12); γ-hydroxybutyrate transport (13); Mg^{2+}-dependent ATPase (14); monoamine oxidase, *or* amine oxidase (15-22); Na$^+$-dependent transport (6,23-26); Na$^+$/HCO$_3^-$ cotransporter (27); Na$^+$/K$^+$-exchanging ATPase (14,28,29); NhaB, a Na$^+$/H$^+$ antiporter (30); phenylalanine transport (23,25,31,32); sucrase (33); taurine transport (34). **1**. Manske, Perkin & Robinson (1927) *J. Chem. Soc.*, 1. **2**. Wright, Bird & Feldman (1979) *Res. Commun. Chem. Pathol. Pharmacol.* **24**, 259. **3**. Ganapathy, Mendicino & Leibach (1981) *J. Biol. Chem.* **256**, 118. **4**. Hsu (1984) *Res. Commun. Chem. Pathol. Pharmacol.* **43**, 223. **5**. Karaki, Kishimoto, Ozaki, *et al.* (1986) *Brit. J. Pharmacol.* **89**, 367. **6**. Smart (1981) *Eur. J. Pharmacol.* **75**, 265. **7**. Ogin & Grassl (1989) *Biochim. Biophys. Acta* **980**, 248. **8**. Sobhani, Ebrahimi & Mahmoudian (2002) *J. Pharm. Pharm. Sci.* **5**, 19. **9**. Gill, Kaur, Mahmood, Nagpaul & Mahmood (1998) *Indian J. Biochem. Biophys.* **35**, 86. **10**. Hundal, Mackenzie, Taylor & Rennie (1990) *Biochem. Soc. Trans.* **18**, 944. **11**. Cumming & Vincent (1992) *Biochem. Pharmacol.* **44**, 989. **12**. Yamamura, Sayama, Kakikawa, *et al.* (2000) *Biochim. Biophys. Acta* **1475**, 10. **13**. McCormick & Tunnicliff (1998) *Pharmacology* **57**, 124. **14**. Samarzija, Kinne-Saffran, Baumann & Fromter (1977) *Pflugers Arch.* **368**, 83. **15**. Blaschko (1963) *The Enzymes*, 2nd ed., **8**, 337. **16**. Udenfriend, Witkop, Redfield & Weissbach (1958) *Biochem. Pharmacol.* **1**, 160. **17**. Bouchaud, Couteaux & Gautron (1965) *C. R. Hebd. Seances Acad. Sci.* **260**, 348. **18**. Fuentes & Neff (1975) *Neuropharmacology* **14**, 819. **19**. Dugal (1977) *Biochim. Biophys. Acta* **480**, 56. **20**. Nelson, Herbet, Petillot, *et al.* (1979) *J. Neurochem.* **32**, 1817. **21**. Glover, Liebowitz, Armando & Sandler (1982) *J. Neural Transm.* **54**, 209. **22**. Kim, Sablin & Ramsay (1997) *Arch. Biochem. Biophys.* **337**, 137. **23**. Sepulveda & Robinson (1974) *Biochim. Biophys. Acta* **373**, 527. **24**. Sepulveda, Buclon & Robinson (1977) *Gastroenterol. Clin. Biol.* **1**, 87. **25**. Sepulveda & Robinson (1978) *J. Physiol.* **74**, 585. **26**. Aronson & Bounds (1980) *Amer. J. Physiol.* **238**, F210. **27**. Amlal, Wang, Burnham & Soleimani (1998) *J. Biol. Chem.* **273**, 16810. **28**. Robinson (1975) *Biochem. Pharmacol.* **24**, 2005. **29**. Canessa, Jaimovich & de la Fuente (1973) *J. Membr. Biol.* **13**, 263. **30**. Pinner, Padan & Schuldiner (1995) *FEBS Lett.* **365**, 18. **31**. Sepulveda & Robinson (1975) *Naunyn Schmiedebergs Arch. Pharmacol.* **291**, 201. **32**. Buclon, Sepulveda & Robinson (1977) *Naunyn Schmiedebergs Arch. Pharmacol.* **298**, 57. **33**. Mahmood & Alvarado (1977) *Biochim. Biophys. Acta* **483**, 367. **34**. Miyamoto, Tiruppathi, Ganapathy & Leibach (1989) *Amer. J. Physiol.* **257**, G65.

Harman

This antioxidant and lipid peroxidation inhibitor (FW$_{free-base}$ = 182.22 g/mol; CAS 486-84-0), also known as 1-methyl-9*H*-pyrido[3,4-*b*]indole and 2-methyl-β-carboline, is an a alkaloid first isolated from the seeds of *Peganum harmala*, but is also formed in cooked foods and cigarette smoke. Harman is also an I$_1$ imidazoline receptor agonist. Mammalian brain possesses a β-carboline 2-*N*-methyltransferase (perhaps phenylethanolamine *N*-methyl-transferase) that converts β-carbolines, such as norharman and harman, into 2-*N*-methylated β-carbolinium cations that are structural and functional analogs of the Parkinson Disease-inducing toxin 1-methyl-4-phenylpyridinium cation, *or* MPP$^+$ (1). **Target(s):** acetylcholinesterase (2); benzodiazepine receptor (3); cholesterol-5,6-oxide hydrolase (4); DNA-(apurinic or apyrimidinic site) lyase activity of the UV endonuclease induced by phage T$_4$ (5); DNA topoisomerase I (6,7); DNA topoisomerase II, weakly inhibited (7); monoamine oxidase A, amine oxidase (8-10); soluble epoxide hydrolase (11). **1**. Gearhart, Neafsey & Collins (2002) *Neurochem. Int.* **40**, 611. **2**. Skup, Oderfeld-Nowak & Rommelspacher (1983) *J. Neurochem.* **41**, 62. **3**. Rommelspacher, Nanz, Borbe, *et al.* (1980) *Naunyn Schmiedebergs Arch. Pharmacol.* **314**, 97. **4**. Palakodety, Vaz & Griffin (1987) *Biochem. Pharmacol.* **36**, 2424. **5**. Warner, Persson, Bensen, Mosbaugh & Linn (1981) *Nucl. Acids Res.* **9**, 6083. **6**. Sobhani, Ebrahimi & Mahmoudian (2002) *J. Pharm. Pharm. Sci.* **5**, 19. **7**. Funayama, Nishio, Wakabayashi, *et al.* (1996) *Mutat. Res.* **349**, 183. **8**. Udenfriend, Witkop, Redfield & Weissbach (1958) *Biochem. Pharmacol.* **1**, 160. **9**. Rommelspacher, May & Salewski (1994) *Eur. J. Pharmacol.* **252**, 51. **10**. May, Rommelspacher & Pawlik (1991) *J. Neurochem.* **56**, 490. **11**. Meijer & Depierre (1985) *Eur. J. Biochem.* **150**, 7.

Harringtonine

This alkaloid (FW$_{free-base}$ = 531.60 g/mol; CAS 26833-85-2), from the plum yew (*Cephalotaxus hainanensisanti*) indigenous to Hainan Island, is an antitumor agent has been used to treat acute or chronic myeloid leukemia. Harringtonine inhibits protein biosynthesis in HeLa cells, rabbit reticulocytes, and reticulocyte lysates. DNA synthesis is partially inhibited at 200 nM, but RNA synthesis is unaffected (1). These inhibitory effects on protein and DNA synthesis are partially reversible. Harringtonine induces breakdown of polyribosomes to monosomes with concomitant release of completed polypeptides. Its main effect is retarding the initiation of protein synthesis. **Target(s):** DNA biosynthesis (2); protein biosynthesis (1,3,4); telomerase (5). **1**. Huang (1975) *Mol. Pharmacol.* **11**, 511. **2**. Kato, Takamoto, Mizutani, Hato & Ota (1984) *Gan To Kagaku Ryoho* **11**, 2393. **3**. Fresno & Vázquez (1979) *Meth. Enzymol.* **60**, 566. **4**. Fresno, Jimenez & Vazquez (1977) *Eur. J. Biochem.* **72**, 323. **5**. Li, Zhang, Cao, *et al.* (2002) *J. Huazhong Univ. Sci. Technolog. Med. Sci.* **22**, 292 and 301.

HBI-8000, *See* Chidamide

HCT, *See* Hydrochlorothiazide

HC Toxin

This cyclopeptide mycotoxin (FW = 436.51 g/mol; CAS 83209-65-8) from the pathogenic plant fungus *Cochliobolus carbonium* (previously *Helminthosporium carbonium*, hence the "HC", named systematically as cyclo[D-prolyl-L-alanyl-D-alanyl-L-2-amino-8-oxo-9,10-epoxydecanoyl], is very toxic to plants, including maize, and is a potent histone deacetylase inhibitor (HDACi). The anti-neuroblastoma activity of HC toxin is superior to that of retinoic acids and all other HDACIs tested. Structurally, HC toxin adopts a rigid conformation that serves as a scaffold to mimic reverse-turns. **1**. Pope, Ciuffetti, Knoche, *et al.* (1983) *Biochemistry* **22**, 3502. **2**. Walton, Earle & Gibson (1982) *Biochem. Biophys. Res. Commun.* **107**, 785. **3**. Baidyaroy, Brosch, Graessle, Trojer & Walton (2002) *Eukaryot. Cell* **1**, 538. **4**. Brosch, Ransom, Lechner, Walton & Loidl (1995) *Plant Cell* **7**, 1941.

HCTZ, *See Hydrochlorothiazide*

Helenalin

This antitumor agent (FW = 262.31 g/mol; CAS 6754-13-8), first isolated from the sneezeweed *Helenium autumnale*, inhibits 5-lipoxygenase, undergoing Michael-type addition with the sulfhydryl groups of reduced glutathione and cysteine. **Target(s):** aminopyrine demethylase, cytochrome P450 (1); aniline hydroxylase, cytochrome P450 (1); ethoxyresorufin deethylase, cytochrome P450 (1); leukotriene C$_4$ synthase (2); 5-lipoxygenase, *or* arachidonate 5-lipoxygenase (2,3); NF-κB (4); oxidative phosphorylation (5); telomerase (6). **1**. Chapman, Holbrook, Chaney, Hall & Lee (1989) *Biochem. Pharmacol.* **38**, 3913. **2**. Tomhamre, Schmidt, Näsman-Glaser, Ericsson & Lindgren (2001) *Biochem. Pharmacol.* **62**, 903. **3**. Schneider & Bucar (2005) *Phytother. Res.* **19**, 81. **4**. Lyss, Schmidt, Merfort & Pahl (1997) *Biol. Chem.* **378**, 951. **5**. Narasimham, Kim & Safe (1989) *Gen. Pharmacol.* **20**, 681. **6**. Huang, Yeh & Wang (2005) *Cancer Lett.* **227**, 169.

Heliquinomycin

This antibiotic (FW = 714.59 g/mol; CAS 178182-49-5) from *Streptomyces* inhibits DNA helicase as well as the growth of HeLa S3, KB, LS180, K562, and HL60 human tumor cell lines (1). **Primary Mode of Action:** Heliquinomycin inhibits cellular DNA replication (IC$_{50}$ = 2.5 μM), without affecting level of chromatin-bound MCM4 (a DNA replication licensing factor) and without activating the DNA replication-stress checkpoint system, suggesting that the antibiotic perturbs DNA replication mainly by inhibiting the activity of replicative DNA helicase that unwinds DNA duplex at replication forks. Heliquinomycin inhibits the DNA helicase activity of MCM4/6/7 complex by stabilizing its interaction with single-stranded DNA. **1**. Chino, Nishikawa, Yamada, *et al.* (1998) *J. Antibiot.* **51**, 480. **2**. Sugiyama, Chino, Tsurimoto, Nozaki & Ishimi (2012) *J. Biochem.* **151**, 129.

Helvolic Acid

This steroid antibiotic (FW$_{free-acid}$ = 568.71 g/mol; CAS 29400-42-8), independently isolated from *Aspergillus fumigatus* in the early 1940s by Nobelists Waksman and Chain, is structurally related to fusidic acid and is effective against Gram-positive organisms. Helvolic acid forms a host-guest complex with γ-cyclodextrin. **Target(s):** elongation factor-2 (1); cholesteryl ester accumulation in macrophages (2). **1**. De Vendittis, De Paola, Gogliettino, *et al.* (2002) *Biochemistry* **41**, 14879. **2**. Shinohara, Hasumi & Endo (1993) *Biochim Biophys Acta* **1167**, 303.

Hematin

This Fe(III) porphyrin complex (FW$_{free-acid}$ = 633.51 g/mol; CAS 15489-90-4), also known as haematin, ferriprotoporphyrin IX hydroxide, and ferriheme hydroxide, is an oxidized form of heme that is the prosthetic group found in methemoglobin, metmyoglobin, catalase, cytochrome *b*, and plant peroxidases. Hematin also binds cooperatively to β$_1$-glycoglobulin and albumin, the latter a reservoir for excess heme that may enter the plasma. Hematin also activates the extra-hepatic P450 mono-oxygenases in tissue homogenates. Less stable than hemin (*i.e.*, ferriprotoporphyrin IX chloride), hematin can be reduced to heme by sodium dithionate (Na$_2$S$_2$O$_4$). Hematin is especially prone to self-assembly into dimers, oligomers and aggregates, depending on pH, solvent, temperature, concentration, ionic strength. This biocrystallization process converts hematin into insoluble and chemically inert paracrystals known as hemozoin, which consists of a centrosymmetric triclinic unit cell comprised of reciprocal head-to-tail dimeric units of heme bound through propionate O-Fe(III) (1). **Oligomerization in Plasmodium-infected Red Cells:** During its intraerythrocytic stages of infection, the malaria parasite consumes >80% of the red cell's hemoglobin, using the latter as a foodstuff for amino acids and releasing hematin, the latter in turn provoking membrane disruption and lysis. Left undeterred, the latter property would limit the parasite burden; however, to offset this effect, the parasite blocks hematin-induced lysis by oligomerization and paracrystal formation. Quinoline-type antimalarials (including chloroquine and amodiaquine) owe their antimalarial effects to the ability to inhibit hemozoin formation. (*See also Hemin*) **Target(s):** 5-aminolevulinate synthase (2,3); L-dopachrome isomerase (4); fatty-acid binding protein (5); glutathione *S*-transferase (6-19); 3-hydroxy-3-methylglutaryl-CoA reductase (20); malate dehydrogenase (acceptor) (21); porphobilinogen synthase, *or* 5-aminolevulinate dehydratase (22); prostaglandin-D synthase (23); succinate dehydrogenase (24-26). **1**. Pagola., Stephens, Bohle, Kosar & Madsen (2000) *Nature* **404**, 307. **2**. Jordan & Laghai-Newton (1986) *Meth. Enzymol.* **123**, 435. **3**. Burnham (1962) *Biochem. Biophys. Res. Commun.* **7**, 351. **4**. Pennock, Behnke, Bickle, *et al.* (1998) *Biochem. J.* **335**, 495. **5**. Stewart, Slysz, Pritting & Muller-Eberhard (1996) *Biochem. Cell Biol.* **74**, 249. **6**. Ahmad, Singh &

Awasthi (1991) *Lens Eye Toxic Res.* **8**, 431. **7**. Di Ilio, Aceto, Del Boccio, *et al.* (1988) *Eur. J. Biochem.* **171**, 491. **8**. Adewale & Afolayan (2004) *J. Biochem. Mol. Toxicol.* **18**, 332. **9**. Vander Jagt, Hunsaker, Garcia & Royer (1985) *J. Biol. Chem.* **260**, 11603. **10**. Papadopoulos, Polemitou, Laifi, Yiangou & Tananaki (2004) *Comp. Biochem. Physiol. C* **139C**, 93. **11**. Abdalla, El-Mogy, Farid & El-Sharabasy (2006) *Comp. Biochem. Physiol. B* **143B**, 76. **12**. Guthenberg, Warholm, Rane & Mannervik (1986) *Biochem. J.* **235**, 741. **13**. Stockman, McLellan & Hayes (1987) *Biochem. J.* **244**, 55. **14**. Awasthi & Singh (1984) *Biochem. Biophys. Res. Commun.* **125**, 1053. **15**. Warholm, Jensson, Tahir & Mannervik (1986) *Biochemistry* **25**, 4119. **16**. Alin, Jensson, Guthenberg, *et al.* (1985) *Anal. Biochem.* **146**, 313. **17**. Girardini, Amirante, Zemzoumi & Serra (2002) *Eur. J. Biochem.* **269**, 5512. **18**. Hamed, Abu-Shady, El-Beih, Abdalla & Afifi (2005) *Pol. J. Microbiol.* **54**, 153. **19**. Hiller, Fritz-Wolf, Deponte, *et al.* (2006) *Protein Sci.* **15**, 281. **20**. Usha Devi & Ramasarma (1987) *Mol. Cell. Biochem.* **77**, 103. **21**. Benziman (1969) *Meth. Enzymol.* **13**, 129. **22**. Coleman (1970) *Meth. Enzymol.* **17A**, 211. **23**. Thomson, Meyer & Hayes (1998) *Biochem. J.* **333**, 317. **24**. Bonner (1955) *Meth. Enzymol.* **1**, 722. **25**. Schlenk (1951) *The Enzymes*, 1st ed., 316. **26**. Singer & Kearney (1963) *The Enzymes*, 2nd ed., 7, 383.

Heme

This iron(II)-protoporphryn IX complex (FW$_{free-acid}$ = 616.50 g/mol; CAS 15489-90-4), also known as ferroprotoporphyrin IX and ferroheme, is the prosthetic group of myoglobin, hemoglobin, catalase, erythrocruorins, certain peroxidases, and cytochromes *b*. Absent the stabilizing effects of its protein binding partner, heme rapidly oxidizes; hence, heme solutions must be prepared under oxygen-free conditions. Heme iron can be removed by treatment with HCl. The presence of heme in blood samples is a confounding factor in the use of PCR in assays of bacterial and yeast sepsis, because heme inhibits Taq polymerase. Such variability can be minimized through the appropriate choice of another DNA polymerase (1). **Target(s):** aminolevulinate aminotransferase (2,3); δ-aminolevulinate synthase (4-7); fatty-acid binding protein (8); ferrochelatase (9-12); glutamyl-tRNA synthetase, plant (13); magnesium proptoporphyrin IX methyltransferase (14); nitric-oxide synthase (15); porphobilinogen synthase, *or* δ aminolevulinate dehydratase (16); protoporphyrinogen oxidase (17). **1**. Gosiewski, Brzychczy-Włoch, Pietrzyk, Sroka & Bulanda (2013) *Acta Biochim. Pol.* **60**, 603. **2**. McKinney & Ades (1991) *Int. J. Biochem.* **23**, 803. **3**. Sagar, Salotra, Bhatnagar & Datta (1995) *Microbiol. Res.* **150**, 419. **4**. Jordan & Shemin (1972) *The Enzymes*, 3rd ed., 7, 339. **5**. Burnham (1970) *Meth. Enzymol.* **17A**, 195. **6**. Burnham (1962) *Biochem. Biophys. Res. Commun.* **7**, 351. **7**. Volland & Felix (1984) *Eur. J. Biochem.* **142**, 551. **8**. Stewart, Slysz, Pritting & Muller-Eberhard (1996) *Biochem. Cell Biol.* **74**, 249. **9**. Dailey, Jones & Karr (1989) *Biochim. Biophys. Acta* **999**, 7. **10**. Dailey (1977) *J. Bacteriol.* **132**, 302. **11**. Dailey & Fleming (1983) *J. Biol. Chem.* **258**, 11453. **12**. Little & Jones (1976) *Biochem. J.* **156**, 309. **13**. Chang, Wegmann & Wang (1990) *Plant Physiol.* **93**, 1641. **14**. Gibson, Neuberger & Tait (1963) *Biochem. J.* **88**, 325. **15**. Cavicchi, Gibbs & Whittle (2000) *Gut* **47**, 771. **16**. Shemin (1972) *The Enzymes*, 3rd ed., 7, 323. **17**. Poulson & Polglase (1975) *J. Biol. Chem.* **250**, 1269.

Hemiasterlin

This cytotoxic oligopeptide (FW = 526.72 g/mol), isolated from the sponge genus *Auletta* and two collections of *Siphonochalina* sp., target tubulin, and, at low nanomolar toxic concentrations, produces abnormal mitotic spindles resembling those formed in the presence of taxol, nocodazole and vinblastine. At higher concentrations, hemiasterlin A does not cause microtubule bundling, as does taxol, but instead promotes microtubule depolymerization, much like nocodazole and vinblastine (1). Hemiasterlin induces the formation of ring-like tubulin aggregates with a diameter of about 40 nm, without producing any turbidity (2). A total synthesis of (–)-hemiasterlin has been accomplished in nine steps, with an overall yield of >35% (3). **1**. Anderson, Coleman, Andersen & Roberge (1997) *Cancer Chemother. Pharmacol.* **39**, 223. **2**. Bai, Durso, Sackett & Hamel (1999) *Biochemistry* **38**, 14302. **3**. Vedejs & Kongkittingam (2001) *J. Org, Chem.* **66**, 7355.

Hemicholinium-3 Dibromide

This symmetrical quaternary ammonium salt (FW = 574.35 g/mol; CAS 312-45-8; *Abbreviation*: HC-3) is a strong inhibitor of the Na$^+$-dependent choline uptake, resulting in an inhibition of acetylcholine synthesis. HC-3 is frequently used to deplete acetylcholine stores, and [^{3}H]-HC-3 hasd found use in locating HC-3 sites in rat brain. **Target(s):** acetylcholinesterase (1); carnitine *O*-octanoyltransferase (2); choline kinase (3-12); choline transport (13-16); ethanolamine kinase (6,8). **1**. Domino, Shellenberger & Frappier (1968) *Arch. Int. Pharmacodyn. Ther.* **176**, 42. **2**. Ramsay & Gandour (1999) *Adv. Exp. Med. Biol.* **466**, 103. **3**. Ishidate & Nakazawa (1992) *Meth. Enzymol.* **209**, 121. **4**. Ansell & Spanner (1974) *J. Neurochem.* **22**, 1153. **5**. Reinhardt, Wecker & Cook (1984) *J. Biol. Chem.* **259**, 7446. **6**. Spanner & Ansell (1977) *Biochem. Soc. Trans.* **5**, 164. **7**. Rodríguez-González, Ramírez de Molina, Benítez-Rajal & Lacal (2003) *Prog. Cell Cycle Res.* **5**, 191. **8**. Brophy, Choy, Toone & Vance (1977) *Eur. J. Biochem.* **78**, 491. **9**. Milanese, Espinosa, Campos, Gallo & Entrena (2006) *ChemMedChem* **1**, 1216. **10**. Ulane, Stephenson & Farrell (1978) *Biochim. Biophys. Acta* **531**, 295. **11**. Kinney & Moore (1988) *Arch. Biochem. Biophys.* **260**, 102. **12**. Choubey, Maity, Guha, *et al.* (2007) *Antimicrob. Agents Chemother.* **51**, 696. **13**. DiAugustine & Haarstad (1970) *Biochem. Pharmacol.* **19**, 559. **14**. Freeman, Macri, Choi & Jenden (1979) *J. Pharmacol. Exp. Ther.* **210**, 91. **15**. Guyenet, Lefresne, Rossier, Beaujouan & Glowinski (1973) *Mol. Pharmacol.* **9**, 630. **16**. Barker & Mittag (1975) *J. Pharmacol. Exp. Ther.* **192**, 86.

Hemin

This porphyrin complex (FW$_{free-acid}$ = 651.95 g/mol; CAS 16009-13-5), also known as ferriprotoporphyrin IX chloride and ferriheme chloride, is an oxidized form of heme, in which the iron is in the +3 oxidation state. Hemin is more stable than hematin (*i.e.*, ferriprotoporphyrin IX hydroxide), and the iron ion can be removed upon treatment with sulfuric acid. Hemin also regulates protein biosynthesis, by activation of a protein kinase (*i.e.*, the hemin-controlled translational repressor). The latter phosphorylates eIF-2, resulting in inhibition of polypeptide chain initiation and protein biosynthesis. (**See also** *Hematin; Heme*) **Target(s):** actin polymerization (1); aminolevulinate aminotransferase, *or* L-alanine:4,5-dioxovalerate aminotransferase (2,3,39-45); 5-aminolevulinate synthase (4-8,48-61); calpain (9); Ca^{2+}/Mg^{2+}-dependent ATPase (10); cAMP-dependent protein kinase, protein kinase A (11); casein kinase II (12); coproporphyrinogen III oxidase (13); cowpea mosaic virus 24-kDa proteinase (14,15); CYP1A2 (16); CYP3A4 (16); CYP2D6 (16); DNA polymerase (17); fatty-acid binding protein (18); ferrochelatase (19,20); glutamate racemase (21); glutathione *S*-transferase (22,46,47); guanylate cyclase (23); hemin-controlled repressor protein kinase (12); holocytochrome-*c* synthase (24); 3-hydroxy-3 methylglutaryl-CoA reductase (25); lactoylglutathione lyase, *or* glyoxalase I (26); lipoamidase (27); magnesium protoporphyrin IX methyltransferase (62,63); murine leukemia virus reverse transcriptase, *or* RNA-directed DNA polymerase (28); Na^+/K^+-exchanging ATPase (29); porphobilinogen synthase, *or* 5-aminolevulinate dehydratase (8,30); protein kinase (31); proteasome endopeptidase complex (32); protoporphyrinogen oxidase (33); pyruvate decarboxylase (34); ribonuclease (35); RNA-directed DNA polymerase (28); succinyl-CoA synthetase (36,37); uroporphyrinogen decarboxylase (38).

1. Avissar, Shaklai & Shaklai (1984) *Biochim. Biophys. Acta* **786**, 179. **2.** Shanker & Datta (1986) *Arch. Biochem. Biophys.* **248**, 652. **3.** Singh & Datta (1985) *Biochim. Biophys. Acta* **827**, 305. **4.** Irving & Elliott (1970) *Meth. Enzymol.* **17A**, 201. **5.** Jordan & Shemin (1972) *The Enzymes*, 3rd ed., **7**, 339. **6.** Scholnick, Hammaker & Marver (1972) *J. Biol. Chem.* **247**, 4132. **7.** Porra, Irving & Tennick (1972) *Arch. Biochem. Biophys.* **148**, 37. **8.** Ibrahim, Gruenspecht & Freedman (1978) *Biochem. Biophys. Res. Commun.* **80**, 722. **9.** Waxman (1981) *Meth. Enzymol.* **80**, 664. **10.** Leclerc, Vasseur, Bursaux, Marden & Poyart (1988) *Biochim. Biophys. Acta* **946**, 49. **11.** Scott, Kemp & Edwards (1985) *Biochim. Biophys. Acta* **847**, 301. **12.** Gonzatti-Haces & Traugh (1982) *J. Biol. Chem.* **257**, 6642. **13.** Rossi, Attwood & Garcia-Webb (1992) *Biochim. Biophys. Acta* **1135**, 262. **14.** Lomonossoff & Shanks (1998) in *Handb. Proteolytic Enzymes*, p. 716, Academic Press, San Diego. **15.** Bu & Shih (1989) *Virology* **173**, 348. **16.** Kim, Kim, Kim, *et al.* (2004) *Toxicol. Lett.* **153**, 239. **17.** Byrnes, Downey, Esserman & So (1975) *Biochemistry* **14**, 796. **18.** Stewart, Slysz, Pritting & Muller-Eberhard (1996) *Biochem. Cell Biol.* **74**, 249. **19.** Rossi, Attwood, Garcia-Webb & Costin (1990) *Biochim. Biophys. Acta* **1038**, 375. **20.** Dailey (1982) *J. Biol. Chem.* **257**, 14714. **21.** Choi, Esaki, Ashiuchi, Yoshimura & Soda (1994) *Proc. Natl. Acad. Sci. U.S.A.* **91**, 10144. **22.** Caccuri, Aceto, Piemonte, *et al.* (1990) *Eur. J. Biochem.* **189**, 493. **23.** elDeib, Parker & White (1987) *Biochim. Biophys. Acta* **928**, 83. **24.** Nicholson, Köhler & Neupert (1987) *Eur. J. Biochem.* **164**, 147. **25.** Usha Devi & Ramasarma (1987) *Mol. Cell. Biochem.* **77**, 103. **26.** Douglas & Sharif (1983) *Biochim. Biophys. Acta* **748**, 184. **27.** Oizumi & Hayakawa (1997) *Meth. Enzymol.* **279**, 202. **28.** Tsutsui & Mueller (1987) *Biochem. Biophys. Res. Commun.* **149**, 628. **29.** Yasuhara, Mori, Wakamatsu & Kubo (1991) *Biochem. Biophys. Res. Commun.* **178**, 95. **30.** Burnham, Pierce, Williams, Boyer & Kirby (1963) *Biochem. J.* **87**, 462. **31.** Hirsch & Martelo (1976) *Biochem. Biophys. Res. Commun.* **71**, 926. **32.** Akaishi, Sawada & Yokosawa (1996) *Biochem. Mol. Biol. Int.* **39**, 1017. **33.** Poulson & Polglase (1975) *J. Biol. Chem.* **250**, 1269. **34.** Cajori (1942) *J. Biol. Chem.* **143**, 357. **35.** Burka (1968) *Science* **162**, 1287. **36.** Bridger (1974) *The Enzymes*, 3rd ed., **10**, 581. **37.** Wider & Tigier (1971) *Enzymologia* **41**, 217. **38.** Jones & Jordan (1993) *Biochem. J.* **293**, 703. **39.** Prasad & Prasad (1989) *Biochem. Int.* **18**, 149. **40.** Sagar, Salotra, Bhatnagar & Datta (1995) *Microbiol. Res.* **150**, 419. **41.** Rhee, Murata & Kimura (1988) *J. Biochem.* **103**, 1045. **42.** Shanker & Datta (1985) *FEBS Lett.* **189**, 129. **43.** Tyagi & Datta (1993) *J. Biochem.* **113**, 557. **44.** Tyagi & Datta (1994) *Protein Expr. Purif.* **5**, 527. **45.** Singh, Tyagi & Datta (1991) *Eur. J. Biochem.* **198**, 581. **46.** Yeung & Gidari (1980) *Arch. Biochem. Biophys.* **205**, 404. **47.** Ahmad & Srivastava (2007) *Parasitol. Res.* **100**, 581. **48.** Yubisui & Yoneyama (1972) *Arch. Biochem. Biophys.* **150**, 77. **49.** Warnick & Burnham (1971) *J. Biol. Chem.* **246**, 6880. **50.** Scholnick, Hammaker & Marver (1972) *J. Biol. Chem.* **247**, 4132. **51.** Tait (1973) *Biochem. J.* **131**, 389. **52.** Ramaswamy & Nair (1973) *Biochim. Biophys. Acta* **293**, 269. **53.** Kreit (1981) *Biochimie* **63**, 439. **54.** Whiting & Granick (1976) *J. Biol. Chem.* **251**, 1340. **55.** Volland & Felix (1984) *Eur. J. Biochem.* **142**, 551. **56.** Whiting & Elliott (1972) *J. Biol. Chem.* **247**, 6818. **57.** Ades & Friedland (1988) *Int. J. Biochem.* **20**, 965. **58.** Ferreira & Gong (1995) *J. Bioenerg. Biomembr.* **27**, 151. **59.** Sato, Ishida, Mutsushika & Shimizu (1985) *Agric. Biol. Chem.* **49**, 3415. **60.** Paterniti & Beattie (1979) *J. Biol. Chem.* **254**, 6112. **61.** Murthy & Woods (1974) *Biochim. Biophys. Acta* **350**, 240. **62.** Gibson, Neuberger & Tait (1963) *Biochem. J.* **88**, 325. **63.** Henchigeri, Chan & Richards (1981) *Photosynthetica* **15**, 351.

Heparan Sulfate

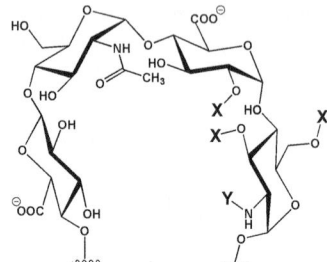

sample heparan sulfate sequence

This mucopolysaccharide (MW = 5-50 kDa for the indefinite biopolymer; CAS 9050-30-0; Note: X = H or SO_3^-, and Y = SO_3^- or CH_3CO), also known as heparitin sulfate, consists of repeating units of D-glucosamine and D-glucuronate (the latter predominating over L-iduronate), linked 1,4 and containing *O*- and *N*-sulfate groups. Heparan sulfate contributes to cell-cell adhesion. There are fewer sulfate groups in heparan sulfate compared to heparin: an average of one sulfate group per disaccharide unit is present in heparan sulfate. There is greater variability in structure and the polysaccharide tends to be longer in length: molecular weights vary between 5 and 50 kDa. In addition, heparan sulfate is typically found linked to a core protein. **Target(s):** *N*-acetylgalactosamine-6-sulfatase (1); α-*N*-acetylglucosaminidase (2); [β-adrenergic-receptor] kinase (3); casein kinase II (4); cathepsin G (5); cruzipain (6); β-glucuronidase (7); granzyme A (8); histone acetyltransferase, inhibited by heparan sulfate proteioglycan (9); hyalurono-glucosaminidase, *or* hyaluronidase (10-14); L-iduronidase (15); iduronate-2-sulfatase (16); nucleotide diphosphatase (17); phospholipase A_2 (18); procollagen N-endopeptidase (19-21). **1.** Glössl, Truppe & Kresse (1979) *Biochem. J.* **181**, 37. **2.** von Figura (1977) *Eur. J. Biochem.* **80**, 523. **3.** Benovic, Stone, Caron & Lefkowitz (1989) *J. Biol. Chem.* **264**, 6707. **4.** Hathaway, Lubben & Traugh (1980) *J. Biol. Chem.* **255**, 8038. **5.** Ledoux, Merciris, Barritault & Caruelle (2003) *FEBS Lett.* **537**, 23. **6.** Lima, Almeida, Tersariol, *et al.* (2002) *J. Biol. Chem.* **277**, 5875. **7.** Nakamura, Takagaki, Majima, *et al.* (1990) *J. Biol. Chem.* **265**, 5390. **8.** Vettel, Brunner, Bar-Shavit, Vlodavsky & Kramer (1993) *Eur. J. Immunol.* **23**, 279. **9.** Buczek-Thomas, Hsia, Rich, Foster & Nugent (2008) *J. Cell. Biochem.* **105**, 108. **10.** Mathews (1966) *Meth. Enzymol.* **8**, 654. **11.** Aronson & Davidson (1967) *J. Biol. Chem.* **242**, 441. **12.** Girish & Kemparaju (2005) *Biochemistry (Moscow)* **70**, 948. **13.** Toida, Ogita, Suzuki, Toyoda & Imanari (1999) *Arch. Biochem. Biophys.* **370**, 176. **14.** Afify, Stern, Guntenhöner & Stern (1993) *Arch. Biochem. Biophys.* **305**, 434. **15.** Schuchman, Guzman, Takada & Desnick (1984) *Enzyme* **31**, 166. **16.** Lissens, Zenati & Liebaers (1984) *Biochim. Biophys. Acta* **801**, 365. **17.** Vollmayer, Clair, Goding, Sano, Servos & Zimmermann (2003) *Eur. J. Biochem.* **270**, 2971. **18.** Yang, Farooqui & Horrocks (1996) *Adv. Exp. Med. Biol.* **416**, 309. **19.** Bruckner-Tuderman (1998) in *Handb. Proteolytic Enzymes*, p. 1244, Academic Press, San Diego. **20.** Hojima, McKenzie, van der Rest & Prockop (1980) *J. Biol. Chem.* **264**, 11336. **21.** Hojima, Morgelin, Engel, *et al.* (1994) *J. Biol. Chem.* **269**, 11381.

Heparin

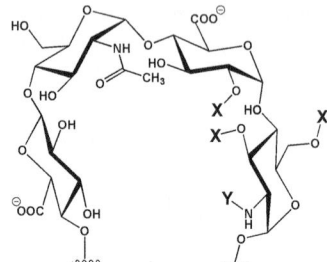

sample heparin sequence

This acid mucopolysaccharide and an anti-clotting agent (MW range = 5-40 kDa, $MW_{average}$ = 15 kDa, where R = H or SO_3^-, and R' = SO_3^- or CH_3CO) CAS 9005-49-6; Average Net Charge Per Polymer = –75) consists of repeating units of D-glucosamine and either D-glucuronate or L-iduronate (with L-iduronate comprising 90% of the uronic acids), joined by 1,4 linkages and containing *O*- and *N*-sulfate groups. On average, there are 2.7 sulfate groups per disaccharide unit, making heparin the most negatively charged polyelectrolyte in mammalian tissues. In addition, the amino group on D-glucosamine residues can be acetylated. Heparin inhibits the conversion of prothrombin to thrombin, activating antithrombin III. (**See also** *Antithrombin; Antithrombin III; Delteparin; Heparan Sulfate; Tinzaparin*) **Key Pharmacokinetic Parameters:** *See* Appendix II in Goodman & Gilman's THE *PHARMACOLOGICAL BASIS OF THERAPEUTICS*, 12th Edition (Brunton, Chabner & Knollmann, eds.) McGraw-Hill Medical, New York (2011). **Target(s):** *N*-acetylgalactosamine-4-sulfatase, *or* arylsulfatase B (1); *N*-acetylgalactosamine 6-sulfatase (2,3); α-*N*-acetyl-glucosaminidase (4); ADAMTS-4 endopeptidase, *or* aggrecanase (5); adenosine deaminase (RNA-adenosine deaminase) (6); adenylate cyclase (7); [β-adrenergic-receptor] kinase (8-17); alcohol dehydrogenase (18); alkaline phosphatase (19); α-amylase (20,21); β-amylase (20,21); arginyltransferase (199); Arylsulfatase (22); bovine spleen ribonuclease (23); [branched-chain α-keto-acid dehydrogenase] kinase, *or* [3-methyl-2-oxobutanoate dehydrogenase (acetyl-transferring)] kinase (24-27); [branched-chain-α-keto-acid dehydrogenase] phosphatase, *or* [3-methyl-2-oxobutanoate dehydrogenase] phosphatase (28-30); casein kinase-1 (31-33); casein kinase-2 (34-44); cathepsin G, also inhibited by low-molecular-weight heparin derivatives (45-47); chondroitin AC lyase (48); chondroitin 4-sulfotransferase (49); chondroitin 6-sulfotransferase (50); chymase (51); coagulation factor IXa (52); coagulation factor Xa (53); coagulation factor XIa (54); 2′,3′-cyclic-nucleotide 3′-phosphodiesterase (55-57); deoxyribonuclease I (58); DNA-directed DNA polymerase (59,60); DNA-directed RNA polymerase (59,61-64); DNA topoisomerase I (65-70); endopeptidase La (71); endopeptidase So (72); exopolyphosphatase (73-79); exo-α-sialidase, *or* neuraminidase, *or* sialidase (80,81); ferredoxin:NADP⁺ reductase (82); β-fructofuranosidase, *or* invertase, inhibited below pH 4.5 (20,83,84); fumarase (20,85); β-D-galactoside 2,3-sialyltransferase (86); β-glucuronidase (20,87-91); γ-glutamyl hydrolase (92,93); glycerophosphocholine phosphodiesterase (94); [glycogen synthase] kinase-I, *or* casein kinase I (31,32); [G-protein-coupled receptor] kinase (95-97); guanylate cyclase (98); [heparan sulfate]-glucosamine 3-sulfotransferase 1, inhibited by low affinity heparin (99); histone acetyltransferase (200); hyaluronoglucosaminidase, *or* hyaluronidase (20,100-116); L-iduronidase (117); inositol-trisphosphate 3 kinase (118); inositol-1,3,4-trisphosphate 5/6-kinase (119); leukocyte elastase (45,120,121); lipoprotein lipase (20,122); [low-density-lipoprotein receptor] kinase (123); lysozyme (20,124,125); myosin ATPase, competitively inhibited by heparins of molecular weights 1.75 to 11.6 kDa (126); [myosin heavy chain] kinase (127); nicotinamide phosphoribosyltransferase (128); non-specific serine/threonine protein kinase (31-44); nucleotide diphosphatase (129); pancreatic ribonuclease, *or* ribonuclease A (20,130-135); pappalysin-1 (136); 1-phosphatidylinositol-4-phosphate 5-kinase (137,138); phospholipase A₂ (139-141); phosphoprotein phosphatase (30,142-144); phosphorylase kinase (145); [phosphorylase] phosphatase (142-144); polynucleotide adenylyltransferase, *or* poly(A) polymerase (146); polynucleotide 5′-hydroxyl-kinase (147-150); polynucleotide phosphorylase, *or* polyribonucleotide nucleotidyl-transferase (151); procollagen N-endopeptidase (152-154); protein kinase C (155); protein tyrosine kinase (156); protein-tyrosine-phosphatase (157-167); protein xylosyltransferase (194-198); rhodopsin kinase (168-170); RNA-directed DNA polymerase, *or* reverse transcriptase (59,171); RNA directed RNA polymerase (172,173); [RNA-polymerase]-subunit kinase (174); RNA uridylyltransferase (175,176); *Schizophyllum commune* nuclease (177); simian immunodeficiency virus reverse transcriptase (171); squid ribonuclease (23); sterol esterase, *or* cholesterol esterase (178); thrombin (179-187); tRNA-pseudouridine synthase I (188); trypsin (20,189,190); tubulin-tyrosine carboxypeptidase (191); Vaccinia RNA polymerase (192); *Xenopus* ribosomal protein S₆ kinase II (193). **1.** Atsumi, Kawai, Seno & Anno (1972) *Biochem. J.* **128**, 983. **2.** Lim & Horwitz (1981) *Biochim. Biophys. Acta* **657**, 344. **3.** Glössl, Truppe & Kresse (1979) *Biochem. J.* **181**, 37. **4.** von Figura (1977) *Eur. J. Biochem.* **80**, 523. **5.** Sugimoto, Takahashi, Yamamoto, Shimada & Tanzawa (1999) *J. Biochem.* **126**, 449. **6.** Hough & Bass (1994) *J. Biol. Chem.* **269**, 9933. **7.** Deschodt-Lanckman, Robberecht & Christophe (1981) *Arch. Biochem. Biophys.* **208**, 1. **8.** Benovic (1991) *Meth. Enzymol.* **200**, 351. **9.** Benovic, Stone, Caron & Lefkowitz (1989) *J. Biol. Chem.* **264**, 6707. **10.** Kwatra, Benovic, Caron, Lefkowitz & Hosey (1989) *Biochemistry* **28**, 4543. **11.** Chuang, Sallese, Ambrosini, Parruti & De Blasi (1992) *J. Biol. Chem.* **267**, 6886. **12.** Parruti, Ambrosini, Sallese & De Blasi (1993) *Biochem. Biophys. Res. Commun.* **190**, 475. **13.** Onorato, Palczewski, Regan, *et al.* (1991)

Biochemistry **30**, 5118. **14.** Murga, Ruiz-Gómez, García-Higuera, *et al.* (1996) *J. Biol. Chem.* **271**, 985. **15.** Kim, Dion, Onorato & Benovic (1993) *Receptor* **3**, 39. **16.** Onorato, Gillis, Liu, Benovic & Ruoho (1995) *J. Biol. Chem.* **270**, 21346. **17.** Ungerer, Kessebohm, Kronsbein, Lohse & Richardt (1996) *Circ. Res.* **79**, 455. **18.** Paulikova, Molnarova & Podhradsky (1998) *Biochem. Mol. Biol. Int.* **46**, 887. **19.** Stinson & Chan (1987) *Adv. Protein Phosphatases* **4**, 127. **20.** Webb (1966) *Enzyme and Metabolic Inhibitors*, vol. **2**, Academic Press, New York. **21.** Myrbäck & Persson (1953) *Arkiv Kemi* **5**, 177 and 477. **22.** Farooqui & Hanson (1987) *Biochem. J.* **242**, 97. **23.** Irie (1998) *Pharmacol. Ther.* **81**, 77. **24.** Paxton (1988) *Meth. Enzymol.* **166**, 313. **25.** Randle, Patston & Espinal (1987) *The Enzymes*, 3rd ed., **18**, 97. **26.** Paxton & Harris (1984) *Arch. Biochem. Biophys.* **231**, 48. **27.** Reed, Damuni & Merryfield (1985) *Curr. Top. Cell. Regul.* **27**, 41. **28.** Damuni & Reed (1988) *Meth. Enzymol.* **166**, 321. **29.** Damuni, Merryfield, Humphreys & Reed (1984) *Proc. Natl. Acad. Sci. U.S.A.* **81**, 4335. **30.** Reed & Damuni (1987) *Adv. Protein Phosphatases* **4**, 59. **31.** Singh (1988) *Arch. Biochem. Biophys.* **260**, 661. **32.** Graves, Haas, Hagedorn, DePaoli-Roach & Roach (1993) *J. Biol. Chem.* **268**, 6394. **33.** Zhai, Graves, Robinson, *et al.* (1995) *J. Biol. Chem.* **270**, 12717. **34.** Hathaway & Traugh (1983) *Meth. Enzymol.* **99**, 317. **35.** Roach (1984) *Meth. Enzymol.* **107**, 81. **36.** Zhang, Jin & Roux (1993) *Plant Physiol.* **103**, 955. **37.** Hathaway, Lubben & Traugh (1980) *J. Biol. Chem.* **255**, 8038. **38.** Gounaris, Trangas & Tsiapalis (1987) *Arch. Biochem. Biophys.* **259**, 473. **39.** Vorotnikov, Gusev, Hua, *et al.* (1993) *FEBS Lett.* **334**, 18. **40.** Damuni & Reed (1988) *Arch. Biochem. Biophys.* **262**, 574. **41.** Mizoguchi, Yamaguchi-Shinozaki, Hayashida, Kamada & Shinozaki (1993) *Plant Mol. Biol.* **21**, 279. **42.** Hu & Rubin (1990) *J. Biol. Chem.* **265**, 5072. **43.** Kenyon, Homan, Storlie, Ikoma & Grose (2003) *J. Med. Virol.* **70** Suppl. 1, S95. **44.** Matsushita, Ohshima, Yoshioka, Nishiguchi & Nyunoya (2003) *J. Gen. Virol.* **84**, 497. **45.** Redini, Tixier, Petitou, Choay, Robert & Hornebeck (1988) *Biochem. J.* **252**, 515. **46.** Ledoux, Merciris, Barritault & Caruelle (2003) *FEBS Lett.* **537**, 23. **47.** Sissi, Lucatello, Naggi, Torri & Palumbo (2006) *Biochem. Pharmacol.* **71**, 287. **48.** Hong, Kim, Shin, *et al.* (2002) *Eur. J. Biochem.* **269**, 2934. **49.** Habuchi & Miyata (1980) *Biochim. Biophys. Acta* **616**, 208. **50.** Momburg, Stuhlsatz & Greiling (1972) *Hoppe-Seyler's Z. Physiol. Chem.* **353**, 1351. **51.** Kido, Fukusen & Katunuma (1985) *Arch. Biochem. Biophys.* **239**, 436. **52.** Lundblad & Davie (1964) *Biochemistry* **3**, 1720. **53.** Leadley (2001) *Curr. Top. Med. Chem.* **1**, 151. **54.** Ratnoff & Davie (1962) *Biochemistry* **1**, 677. **55.** Sprinkle (1989) *CRC Crit. Rev. Clin. Neurobiol.* **4**, 235. **56.** Nishizawa, Kurihara & Takahshi (1980) *Biochem. J.* **191**, 71. **57.** Sprinkle, Tippins & Kestler (1987) *Biochem. Biophys. Res. Commun.* **145**, 686. **58.** Napirei, Wulf, Eulitz, Mannherz & Kloeckl (2005) *Biochem. J.* **389**, 355. **59.** Schaffrath, Stuhlsatz & Greiling (1976) *Hoppe Seyler's Z. Physiol. Chem.* **357**, 499. **60.** Müller, Yamazaki & Zahn (1972) *Enzymologia* **43**, 1. **61.** Chamberlin (1974) *The Enzymes*, 3rd ed., **10**, 333. **62.** Boyer & Hallick (1998) *Plant Sci.* **137**, 13. **63.** Sethi (1971) *Prog. Biophys. Mol. Biol.* **23**, 67. **64.** Allan & Kropinski (1987) *Biochem. Cell Biol.* **65**, 776. **65.** Ishii, Futaki, Uchiyama, Nagasawa & Andoh (1987) *Biochem. J.* **241**, 111. **66.** Alkorta, Park, Kong, *et al.* (1999) *Arch. Biochem. Biophys.* **362**, 123. **67.** Vosberg (1985) *Curr. Top. Microbiol. Immunol.* **114**, 19. **68.** Yoshida, Nakata & Ichishima (1991) *Phytochemistry* **30**, 3885. **69.** Liu (1983) *Meth. Enzymol.* **100**, 133. **70.** Ishii, Hasegawa, Fujisawa & Andoh (1983) *J. Biol. Chem.* **258**, 12728. **71.** Waxman & Goldberg (1985) *J. Biol. Chem.* **260**, 12022. **72.** Chung & Goldberg (1983) *J. Bacteriol.* **154**, 231. **73.** Rodrigues, Ruiz, Vieira, Hill & Docampo (2002) *J. Biol. Chem.* **277**, 50899. **74.** Andreeva, Kulakovskaya & Kulaev (1998) *FEBS Lett.* **429**, 194. **75.** Andreeva, Kulakovskaya & Kulaev (2004) *Biochemistry (Moscow)* **69**, 387. **76.** Lichko, Kulakovskaya & Kulaev (2002) *Process Biochem.* **37**, 799. **77.** Kulakovskaya, Andreeva & Kualev (1997) *Biochemistry (Moscow)* **62**, 1051. **78.** Lichko, Kulakovskaya & Kulaev (2004) *Biochemistry (Moscow)* **69**, 270. **79.** Lichko, Kulakovskaya & Kulaev (2002) *Biochim. Biophys. Acta* **1599**, 102. **80.** Drzeniek (1966) *Nature* **211**, 1205. **81.** Iriyama, Takeuchi, Shiraishi, *et al.* (2000) *Comp. Biochem. Physiol. B* **126**, 561. **82.** Hosler & Yocum (1985) *Arch. Biochem. Biophys.* **236**, 473. **83.** Myrbäck (1960) *The Enzymes*, 2nd ed., **4**, 379. **84.** Astrup & Thorsell (1954) *Acta Chem. Scand.* **8**, 1859. **85.** Fischer & Herrmann (1937) *Enzymologia* **3**, 180. **86.** Schwarting, Gajewski, Carroll & DeWolf (1987) *Arch. Biochem. Biophys.* **256**, 69. **87.** Levvy & Marsh (1960) *The Enzymes*, 2nd ed., **4**, 397. **88.** Becker & Friedenwald (1949) *Arch. Biochem.* **22**, 101. **89.** Diez & Cabezas (1979) *Eur. J. Biochem.* **93**, 301. **90.** Tsuchihashi, Yadome & Miyazaki (1984) *J. Biochem.* **96**, 1789. **91.** Nakamura, Takagaki, Majima, *et al.* (1990) *J. Biol. Chem.* **265**, 5390. **92.** Silink, Reddel, Bethel & Rowe (1975) *J. Biol. Chem.* **250**, 5982. **93.** Elsenhans, Ahmad & Rosenberg (1984) *J. Biol. Chem.* **259**, 6364. **94.** Kusser & Fiedler (1984) *FEBS Lett.* **166**, 301. **95.** Fraga, Jose & Soares-da-

Silva (2004) *Amer. J. Physiol.* **287**, R772. **96.** Simon, Robin, Legrand & Cohen-Tannoudji (2003) *Endocrinology* **144**, 3058. **97.** Fraga, Luo, Jose, *et al.* (2006) *Cell Physiol. Biochem.* **18**, 347. **98.** Gorczyca, Van Hooser & Palczewski (1994) *Biochemistry* **33**, 3217. **99.** Razi & Lindahl (1995) *J. Biol. Chem.* **270**, 11267. **100.** Mathews (1966) *Meth. Enzymol.* **8**, 654. **101.** Fishman (1951) *The Enzymes*, 1st ed., **1** (part 2), 769. **102.** Meyer, Hoffman & Linker (1960) *The Enzymes*, 2nd ed., **4**, 447. **103.** Alburn & Whitley (1954) *Fed. Proc.* **13**, 330. **104.** Aronson & Davidson (1967) *J. Biol. Chem.* **242**, 441. **105.** Linker (1984) in *Methods in Enzymatic Analysis* (Bergmeyer, ed.), vol. 4, pp. 246-256, Verlag Chemie, Weinheim. **106.** Meyer (1947) *Physiol. Rev.* **27**, 335. **107.** Morey, Kiran & Gadag (2006) *Toxicon* **47**, 188. **108.** Meyer & Rapport (1952) *Adv. Enzymol.* **13**, 199. **109.** Krishnapillai, Taylor, Morris & Quantick (1999) *Food Chem.* **65**, 515. **110.** Girish & Kemparaju (2005) *Biochemistry (Moscow)* **70**, 948. **111.** Toida, Ogita, Suzuki, Toyoda & Imanari (1999) *Arch. Biochem. Biophys.* **370**, 176. **112.** Afify, Stern, Guntenhöner & Stern (1993) *Arch. Biochem. Biophys.* **305**, 434. **113.** Poh, Yuen, Chung & Khoo (1992) *Comp. Biochem. Physiol.* **101**, 159. **114.** Maksimenko, Schechilina & Tischenko (2003) *Biochemistry (Moscow)* **68**, 1055. **115.** Maksimenko, Petrova, Tischenko & Schechilina (2001) *Eur. J. Pharm. Biopharm.* **51**, 33. **116.** Ramanaiah, Parthasarathy & Venkaiah (1990) *Biochem. Int.* **20**, 301. **117.** Schuchman, Guzman, Takada & Desnick (1984) *Enzyme* **31**, 166. **118.** Takazawa & Erneux (1989) *Biochem. J.* **261**, 1059. **119.** Hughes, Kirk & Michell (1994) *Biochim. Biophys. Acta* **1223**, 57. **120.** Bieth (1998) in *Handb. Proteolytic Enzymes*, p. 54, Academic Press, San Diego. **121.** Jones, Elphick, Pettitt, Everard & Evans (2002) *Eur. Respir. J.* **19**, 1136. **122.** Korn (1962) *Meth. Enzymol.* **5**, 542. **123.** Kishimoto, Brown, Slaughter & Goldstein (1987) *J. Biol. Chem.* **262**, 1344. **124.** Imoto, Johnson, North, Phillips & Rupley (1972) *The Enzymes*, 3rd ed., **7**, 665. **125.** Kaiser (1953) *Nature* **171**, 607. **126.** Volpi, Bianchini & Bolognani (1991) *Biochem. Int.* **24**, 243. **127.** Murakami, Matsumura & Kumon (1984) *J. Biochem.* **95**, 651. **128.** Dietrich (1971) *Meth. Enzymol.* **18B**, 144. **129.** Vollmayer, Clair, Goding, *et al.* (2003) *Eur. J. Biochem.* **270**, 2971. **130.** McDonald (1955) *Meth. Enzymol.* **2**, 427. **131.** Scheraga & Rupley (1962) *Adv. Enzymol.* **24**, 161. **132.** Richards & Wyckoff (1971) *The Enzymes*, 3rd ed., **4**, 647. **133.** Zöllner & Fellig (1952) *Naturwissenschaften* **39**, 523. **134.** Zöllner & Fellig (1953) *Amer. J. Physiol.* **173**, 223. **135.** Stockx & Dierick (1959) *Enzymologia* **21**, 189. **136.** Parker, Gockerman, Busby & Clemmons (1995) *Endocrinology* **136**, 2470. **137.** Bazenet & Anderson (1992) *Meth. Enzymol.* **209**, 189. **138.** Cochet & Chambaz (1986) *Biochem. J.* **237**, 25. **139.** Diccianni, Mistry, Hug & Harmony (1990) *Biochim. Biophys. Acta* **1046**, 242. **140.** Yang, Farooqui & Horrocks (1996) *Adv. Exp. Med. Biol.* **416**, 309. **141.** Beghini, Toyama, Hyslop, *et al.* (2000) *J. Protein Chem.* **19**, 603. **142.** Cohen, Alemany, Hemmings, *et al.* (1988) *Meth. Enzymol.* **159**, 390. **143.** Ballou & Fischer (1986) *The Enzymes*, 3rd ed., **17**, 311. **144.** Chistova & Galabova (1998) *Z. Naturforsch. C* **53**, 951. **145.** Krebs, Love, Bratvold, *et al.* (1964) *Biochemistry* **3**, 1022. **146.** Ohyama, Fukami & Ohta (1980) *J. Biochem.* **88**, 337. **147.** Zimmerman & Pheiffer (1981) *The Enzymes*, 3rd ed., **14**, 315. **148.** Levin & Zimmerman (1976) *J. Biol. Chem.* **251**, 1767. **149.** Austin, Sirakoff, Roop & Moyer (1978) *Biochim. Biophys. Acta* **522**, 412. **150.** Pheiffer & Zimmerman (1979) *Biochemistry* **18**, 2960. **151.** Littauer & Soreq (1982) *The Enzymes*, 3rd ed., **15**, 517. **152.** Bruckner-Tuderman (1998) in *Handb. Proteolytic Enzymes*, p. 1244, Academic Press, San Diego. **153.** Hojima, McKenzie, van der Rest & Prockop (1980) *J. Biol. Chem.* **264**, 11336. **154.** Hojima, Morgelin, Engel, *et al.* (1994) *J. Biol. Chem.* **269**, 11381. **155.** Kikkawa & Nishizuka (1986) *The Enzymes*, 3rd ed., **17**, 167. **156.** Brunati & Pinna (1988) *Eur. J. Biochem.* **172**, 451. **157.** Tonks, Diltz & Fischer (1991) *Meth. Enzymol.* **201**, 427. **158.** Tonks, Diltz & Fischer (1991) *Meth. Enzymol.* **201**, 442. **159.** Ingebritsen (1991) *Meth. Enzymol.* **201**, 451. **160.** Harder, Owen, Wong, *et al.* (1994) *Biochem. J.* **298**, 395. **161.** Tonks, Diltz & Fischer (1988) *J. Biol. Chem.* **263**, 6731. **162.** Hernández-Hernández, Sánchez-Yagüe, Martín-Valmaseda & Llaniloo (1997) *Comp. Biochem. Physiol. B* **117**, 437. **163.** Dechert, Adam, Harder, Clark-Lewis & Jirik (1994) *J. Biol. Chem.* **269**, 5602. **164.** Hoppe, Berne, Stock, *et al.* (1994) *Eur. J. Biochem.* **223**, 1069. **165.** Tonks, Diltz & Fischer (1990) *J. Biol. Chem.* **265**, 10674. **166.** Zhao, Laroque, Ho, Fischer & Shen (1994) *J. Biol. Chem.* **269**, 8780. **167.** Daum, Zander, Morse, *et al.* (1991) *J. Biol. Chem.* **266**, 12211. **168.** Doza, Minke, Chorev & Selinger (1992) *Eur. J. Biochem.* **209**, 1035. **169.** Kikkawa, Yoshida, Nakagawa, Iwasa & Tsuda (1998) *J. Biol. Chem.* **273**, 7441. **170.** Palczewski, Ohguro, Premont & Inglese (1995) *J. Biol. Chem.* **270**, 15294. **171.** Lüke, Hoefer, Moosmayer, *et al.* (1990) *Biochemistry* **29**, 1764. **172.** Johnson, Sun, Hockman, *et al.* (2000) *Arch. Biochem. Biophys.* **377**, 129. **173.** Schiebel, Haas, Marinkovic, Klanner & Sänger (1993) *J. Biol. Chem.* **268**, 11851.

174. Guilfoyle (1989) *Plant Cell* **1**, 827. **175.** Zabel, Dorssers, Wernars & van Kammen (1981) *Nucl. Acids Res.* **9**, 2433. **176.** Bakalara, Simpson & Simpson (1989) *J. Biol. Chem.* **264**, 18679. **177.** Martin, Ullrich & Meyer (1986) *Biochim. Biophys. Acta* **867**, 67. **178.** Jamry, Sasser & Kumar (1995) *Int. J. Biochem. Cell Biol.* **27**, 415. **179.** Klein (1955) *Meth. Enzymol.* **2**, 139. **180.** Baughman (1970) *Meth. Enzymol.* **19**, 145. **181.** Magnusson (1970) *Meth. Enzymol.* **19**, 157. **182.** Lundblad, Kingdon & Mann (1976) *Meth. Enzymol.* **45**, 156. **183.** Olson, Björk & Shore (1993) *Meth. Enzymol.* **222**, 525. **184.** Hayakawa, Hayashi, Lee, *et al.* (2000) *Biochim. Biophys. Acta* **1543**, 86. **185.** Weitz (2003) *Thromb. Res.* **109** Suppl. 1, S17. **186.** De Cristofaro, Akhavan, Altomare, *et al.* (2004) *J. Biol. Chem.* **279**, 13035. **187.** Buchanan, Maclean & Brister (2001) *Thromb. Haemost.* **86**, 909. **188.** Green, Kammen & Penhoet (1982) *J. Biol. Chem.* **257**, 3045. **189.** Horwitt (1940) *Science* **92**, 89. **190.** Horwitt (1944) *J. Biol. Chem.* **156**, 427. **191.** Lopez, Arce & Barra (1990) *Biochim. Biophys. Acta* **1039**, 209. **192.** Gershon & Moss (1996) *Meth. Enzymol.* **275**, 208. **193.** Erikson, Maller & Erikson (1991) *Meth. Enzymol.* **200**, 252. **194.** Casanova, Kuhn, Kleesiek & Götting (2008) *Biochem. Biophys. Res. Commun.* **365**, 678. **195.** Götting, Kuhn, Tinneberg, Brinkmann & Kleesiek (2002) *Mol. Hum. Reprod.* **8**, 1079. **196.** Götting, Kuhn, Zahn, Brinkmann & Kleesiek (2000) *J. Mol. Biol.* **304**, 517. **197.** Götting, Müller, Schöttler, *et al.* (2004) *J. Biol. Chem.* **279**, 42566. **198.** Kuhn, Schnölzer, Schön, *et al.* (2005) *Biochem. Biophys. Res. Commun.* **333**, 156. **199.** Kato (1983) *J. Biochem.* **94**, 2015. **200.** Buczek-Thomas, Hsia, Rich, Foster & Nugent (2008) *J. Cell. Biochem.* **105**, 108.

Heparin Cofactor II

This human liver serpin and anticoagulant (MW = 65.6 kDa) is a natural serine protease with high specificity towards thrombin. HCII inactivates thrombin rapidly in the presence of dermatan sulfate (*i.e.*, heparin cofactor II is activated upon binding to cell surface glycosaminoglycans). The inhibition occurs by formation of a stable binary complex between the plasma glycoprotein and thrombin. Unlike antithrombin, which also inhibits other proteases (*e.g.*, coagulation factors IXa, Xa, XIa, XIIa, plasmin, and kallikerin), heparin cofactor II is highly selective. (*For a discussion of the likely mechanism of inhibitory action, **See** Serpins; also α_1-Antichymotrypsin*) **Target(s):** cathepsin G, weakly inhibited (1); chymotrypsin (2); thrombin (3-9).

1. Parker & Tollefsen (1985) *J. Biol. Chem.* **260**, 3501. **2.** Church, Noyes & Griffith (1985) *Proc. Natl. Acad. Sci. U.S.A.* **82**, 6431. **3.** Tollefsen (1997) *Adv. Exp. Med. Biol.* **425**, 35. **4.** Fortenberry, Whinna, Gentry, *et al.* (2004) *J. Biol. Chem.* **279**, 43237. **5.** Baglin, Carrell, Church, Esmon & Huntington (2002) *Proc. Natl. Acad. Sci. U.S.A.* **99**, 11079. **6.** Tollefsen, Majerus & Blank (1982) *J. Biol. Chem.* **257**, 2162. **7.** Hayakawa, Hayashi, Lee, *et al.* (2000) *Biochim. Biophys. Acta* **1543**, 86. **8.** Buchanan, Maclean & Brister (2001) *Thromb. Haemost.* **86**, 909. **9.** Maekawa, Sato & Tollefsen (2000) *Thromb. Res.* **100**, 443.

Hepatocyte Growth Factor Activator Inhibitor-1

This multidomain Type-I transmembrane glycoprotein (MW = 66 kDa; Symbol: HAI-1; Accession Number: AB000095) and its splice variant (HAI-1B) are novel Kunitz-type serine protease inhibitors, the latter containing an extra 16 residues adjacent to the C- terminus of its Kunitz-type domain (1). Its main target (IC_{50} = 30.5 nM) is Hepatocyte Growth Factor Activator (HGFA), a serum serine proteinase that specifically converts the inactive proform of Hepatocyte Growth Factor/Scatter Factor (*or pro-HGF/SF*) to its active form, HGF/SF, in response to tissue injury. Because HGF/SF is a pleiotropic factor with multifunctional roles in a variety of cells via its high-affinity receptor, c-Met receptor tyrosine kinase, HGFA is likely to play rolesin a number of pathophysiological phenomena *in vivo*. HAI-1 inhibits other serine proteases and has been shown to be essential for placental development by modulating the activities of extracellular proteases (1). **Structural Features:** The primary translation product consists of 513 residues, with the NH_2-terminal 35 residues serving as a signal peptide. The mature protein appears to be membrane-bound, because a hydrophobic region of about 20 amino acids is present at the COOH-terminal region. HAI has two well defined Kunitz domains, which contain around 60 residues and three disulfide bonds, thus resembling the functional domain of bovine pancreatic trypsin inhibitor. One or both of the Kunitz domains in HAI appear to be responsible for the inhibitory activity of the protein. **Secondary Targets:** matriptase (IC_{50} = 16.5 nM); plasma kallikrein (IC_{50} = 690 nM); plasmin (IC_{50} = 400 nM); and trypsin (IC_{50} = 2.4 nM). Attesting to its physologic significance is the finding that HAI-1-deficient mice die *in utero*, resulting from undifferentiated chorionic trophoblasts (1-3). HAI-1 plays a role in maintaining postnatal tissue homeostasis (*e.g.*, keratinization of the epidermis, hair development, and the integrity of the colonic epithelium integrity (4-6). Genetic studies in mice have established that improper inhibition of matriptase activity by HAI-1 can give rise to carcinogenesis (7). HAI-1 was first identified as an inhibitor co-purifying with hepatocyte growth factor activator, *or* HGFA-2 (8) and was later purified from human milk in complex with matriptase (9). Simultaneous genetic ablation of matriptase in mice rescues the placental defect caused by a lack of HAI-1 (3). HAI-1 also inhibits the glycophosphatidyl-inositol membrane-anchored serine protease known as prostasin (10). **1.** Tanaka, Nagaike, Takeda, *et al.* (2005) *Molec. Cellular Biol.* **25**, 5687. **2.** Fan, Brennan, Grant, *et al.* (2007) *Develop. Biol.* **303**, 222. **3.** Szabo, Molinolo, List & Bugge (2007) *Oncogene* **26**, 1546. **4.** Kawaguchi, Takeda, Hoshiko, *et al.* (2011) *Am. J. Pathol.* **179**, 1815. **5.** Nagaike, Kawaguchi, Takeda, *et al.* (2008) *Am. J. Pathol.* **173**, 1464. **6.** Szabo, Kosa, List & Bugge (2009) *Am. J. Pathol.* 174, 2015. 7. List, Szabo, Molinolo, *et al.* (2005) *Genes & Develop.* 19, 1934. **8.** Shimomura, Denda, Kitamura, *et al.* (1997) *J. Biol. Chem.* **272**, 6370. **9.** Lin, Anders, Johnson & Dickson (1999) *J. Biol. Chem.* **274**, 18237. **10.** Fan, Wu, Li & Kirchhofer (2005) *J. Biol. Chem.* **280**, 34513.

Hepcidin

This iron entry inhibitor and master regulator of iron homeostasis (FW = 2868.48 g/mol; Sequence: DTHFPICIFCCGCCHRSKCQMCCKT; Uniprot: P81172) binds to the iron export channel ferroportin on the outer basolateral surface of gut enterocytes and the plasma membrane of reticuloendothelial cells (macrophages). By inhibiting ferroportin, hepcidin prevents iron entry into the hepatic portal system, thereby reducing dietary iron absorption ans maintaining plasma iron concentrations in humans at 10-30 µM. Hepcidin is first expressed as its 84-residue pre-prohormone, which is processed into a 60-residue prohormone, and ultimately to the active 25-residue hormone. Hepatic hepcidin synthesis is transcriptionally regulated by extracellular and intracellular iron concentrations through a molecular complex of bone morphogenetic protein receptors and their iron-specific ligands, modulators and iron sensors. The hepcidin–ferroportin axis controls both extracellular iron concentrations and total body iron levels. It limits the size of the extracellular iron pool by binding ferroportin and mediating its degradation, thus preventing iron release from intracellular sources. **1.** Richardson (2005) *Curr. Med. Chem.* **12**, 2711. **2.** Ganz & Nemeth (2012) *Biochim. Biophys. Acta* **823**, 1434.

D-*manno*-2-Heptulose

This seven-carbon reducing monosaccharide (FW = 210.18 g/mol), most often isolated from the pulp of the avocado (1), is soluble in water and has a melting point of 151-152°C. More than 99% of the sugar is in the α-pyranose form at 22°C. **Target(s):** glucokinase (2,6); ketohexokinase, hepatic fructokinase, also as an alternative substrate (3); mannokinase, also as a weak alternative substrate (4); hexokinase (5). **1.** Shaw, Wilson & Knight (1980) *J. Agric. Food Chem.* **28**, 379. **2.** Reeves, Montalvo & Sillero (1967) *Biochemistry* **6**, 1752. **3.** Raushel & Cleland (1977) *Biochemistry* **16**, 2169. **4.** Sabáter, Sebastián & Asensio (1972) *Biochim. Biophys. Acta* **284**, 406. **5.** Claeyssen & Rivoal (2007) *Phytochemistry* **68**, 709. **6.** Sener & Malaisse (1996) *Biochim. Biophys. Acta* **1312**, 73.

2-*n*-Heptyl-4-hydroxyquinoline-*N*-oxide

This quinoline derivative (FW = 259.35 g/mol; Stock solutions prepared in 10 mM NaOH), often abbreviated HQNO and HOQNO, and also known as 2 heptyl-4-quinolinol 1-oxide, inhibits NADH and succinate oxidation and mitochondrial electron transport (at cytochrome *bc*₁ (1-5). It also inhibits

chloroplast electron transport (6). HQNO is also a streptomycin antagonist (7,8). **Target(s):** cytochrome *d* complex (9); cytochrome *o*-type oxidase (9-13); electron transport, chloroplast, and photosynthetic phosphorylation (6,14); electron transport, mitochondrial (2-5,7,15-19); hydrogen:quinone oxidoreductase (30); 5-lipoxygenase, *or* arachidonate 5-lipoxygenase (20); menaquinol oxidase21; methane monooxygenase (22); NADH dehydrogenase (1,15,23); plastoquinol:plastocyanin reductase (31); sulfur reductase (24); taurine dehydrogenase (25); ubiquinol:cytochrome-*c* reductase (Complex III) (2-4,18,19,26-28,32-34); ubiquinol oxidase, *Escherichia coli* (29). **1.** Nason & Vasington (1963) *Meth. Enzymol.* **6**, 409. **2.** Slater (1967) *Meth. Enzymol.* **10**, 48. **3.** Singer (1979) *Meth. Enzymol.* **55**, 454. **4.** Rapoport & Schewe (1977) *Trends Biochem. Sci.* **2**, 186. **5.** Izzo, Guerrieri & Papa (1978) *FEBS Lett.* **93**, 320. **6.** Izawa & Good (1972) *Meth. Enzymol.* **24**, 355. **7.** Lightbown & Jackson (1956) *Biochem. J.* **63**, 130. **8.** Machan, Taylor, Pitt, Cole & Wilson (1992) *J. Antimicrob. Chemother.* **30**, 615. **9.** Anraku & Gennis (1987) *Trends Biochem. Sci.* **12**, 262. **10.** Matsushita, Patel & Kaback (1986) *Meth. Enzymol.* **126**, 113. **11.** Kita, Kasahara & Anraku (1982) *J. Biol. Chem.* **257**, 7933. **12.** Matsushita, Patel & Kaback (1984) *Biochemistry* **23**, 4703. **13.** Withers & Bragg (1990) *Biochem. Cell Biol.* **68**, 83. **14.** Kamen (1963) *Meth. Enzymol.* **6**, 313. **15.** Jackson & Lightbown (1958) *Biochem. J.* **69**, 63. **16.** Brandon, Brocklehurst & Lee (1972) *Biochemistry* **11**, 1150. **17.** Izzo, Guerrieri & Papa (1978) *FEBS Lett.* **93**, 320. **18.** Papa, Izzo & Guerrieri (1982) *FEBS Lett.* **145**, 93. **19.** Halestrap (1982) *Biochem. J.* **204**, 49. **20.** Kitamura, Hashizume, Iida, *et al.* (1986) *J. Antibiot.* **39**, 1160. **21.** Lemma, Simon, Schagger & Kroger (1995) *Arch. Microbiol.* **163**, 432. **22.** Zahn & DiSpirito (1996) *J. Bacteriol.* **178**, 1018. **23.** Dancey, Levine & Shapiro (1976) *J. Biol. Chem.* **251**, 5911. **24.** Zöphel, Kennedy, Beinert & Kroneck (1988) *Arch. Microbiol.* **150**, 72. **25.** Kondo & Ishimoto (1987) *Meth. Enzymol.* **143**, 496. **26.** Hatefi (1978) *Meth. Enzymol.* **53**, 35. **27.** von Jagow & Link (1986) *Meth. Enzymol.* **126**, 253. **28.** Goldberger & Green (1963) *The Enzymes*, 2nd ed., **8**, 81. **29.** Kita, Konishi & Anraku (1986) *Meth. Enzymol.* **126**, 94. **30.** Ferber & Maier (1993) *FEMS Microbiol. Lett.* **110**, 257. **31.** Clark & Hind (1983) *J. Biol. Chem.* **258**, 10348. **32.** Hauska, Hurt, Gabellini & Lockau (1983) *Biochim. Biophys. Acta* **726**, 97. **33.** Yang & Trumpower (1986) *J. Biol. Chem.* **261**, 12282. **34.** Yu, Yang & Yu (1985) *J. Biol. Chem.* **260**, 963.

Herbimycin A

This ansamycin-class macrocyclic antibiotic (FW = 574.67 g/mol; CAS 70563-58-5; photosensitive; hygroscopic) was first isolated from species of *Streptomyces* (*e.g.*, *S. Hygroscopicus*). In view of its ability to bind to and inhibit Hsp90, treatment of cells with herbimycin A activates a heat shock response, even in the absence of heat shock. Pretreatment of cells with the Hsp90 inhibitors also delayed the rate of restoration of normal protein synthesis following a brief heat shock. Herbimycin also It selectively inactivates cytoplasmic tyrosine kinases by reacting irreversibly with the reactive SH group(s) and blocking ATP access to the active site. It profoundly decreases the total cellular activity of transmembrane tyrosine kinase receptors (including insulin-like growth factor, insulin, and epidermal growth factor receptors). This enhanced degradation is prevented by inhibitors of the 20S proteasome, whereas neither lysosomotropic agents nor general serine- and cysteine-protease inhibitors prevent receptor degradation. Herbimycin also inhibits p60v-*src* (IC₅₀ = 12 µM) and PDGF-induced ([platelet-derived growth factor]-induced) phospholipase D activation (IC₅₀ = 8 µg/mL). It is also reported to inhibit c-*src* related bone resorption (IC₅₀ = 70 nM). (**See also** *Geldanamycin, Radicicol, Deguelin, Derrubone, Macbecin*) **1.** Sepp-Lorenzino, Ma, Lebwohl, Venitsky & Rosen

(1995) *J. Biol. Chem.* **270**, 16580. **2**. Uehara & Fukazawa (1991) *Meth. Enzymol.* **201**, 370. **2**. Eriksson, Toivola, Sahlgren, Mikhailov & Härmälä-Braskén (1998) *Meth. Enzymol.* **298**, 542. **3**. Uehara, Hori, Takeuchi & Umezawa (1985) *Jpn. J. Cancer Res.* **76**, 672. **4**. Weinstein, Gold & DeFranco (1991) *Proc. Natl. Acad. Sci. U.S.A.* **88**, 4148. **5**. Park, Min & Rhee (1991) *J. Biol. Chem.* **266**, 24237. **6**. Uehara, Fukazawa, Murakami & Mizuno (1989) *Biochem. Biophys. Res. Commun.* **163**, 803.

Heroin

This often highly abused morphine derivative (FW = 369.41 g/mol; CAS 561-27-3), also known diacetylmorphine, morphine diacetate or diamorphine, is an opioid analgesic that is only occasionally administered in the U.S. and regularly in the UK (by subcutaneous, intramuscular, intrathecal or intravenous routes) for the treatment for acute pain associated with myocardial infarction, post-surgical pain as well as chronic pain, especially that associated with certain end-stage cancers. **Pharmacokinetics** Once within circulation, heroin is rapidly hydrolyzed to monoacetyl-morphine by serum cholinesterase (EC 3.1.1.8), *or* acylcholine acylhydrolase, *or* butyrylcholine esterase (*See Morphine*). The kinetics for 6-monoacetylmorphine formation coincide with the period over which users report experiencing a "rush", meaning an acute sense of a transcendent state of euphoria. Blood heroin and 6-acetylmorphine reach their maximal concentrations within minutes and are cleared rapidly. Upon intravenous injection or infusion, the mean half-life of heroin is only 3 min, and the mean clearance rate for heroin from the blood at apparent steady-state is 31 mL per kilogram of body weight per minute (1). Morphine levels rise more gradually, and morphine is cleared much more slowly. Oral administration of heroin results in measurable blood levels of morphine but not of heroin or 6-acetylmorphine. **Addictive Power:** Words alone cannot convey heroin's extreme addictiveness. The relative potency of heroin and its active metabolites (*e.g.*, 6-acetylmorphine and morphine) in lowering the threshold for reward stimulation of the medial forebrain bundle and raising the threshold for aversive stimulation of the mesencephalic reticular formation. Heroin is some 40-times more potent than morphine in lowering the reward threshold and approximately 6.5x more potent in raising the escape threshold (2). 6-Acetylmorphine and heroin were approximately equipotent in producing significant effects on the threshold in both rewarding and aversive brain stimulation. **1**. Inturrisi, Max, Foley, *et al.* (1984) *New Engl. J. Med.* **310**, 1213. **2**. Hubner & Kornetsky (1992) *J. Pharmacol. Exp. Ther.* 260, 562.

Hesperadin

This ATP-competitive Aurora kinase inhibitor (FW = 553.12 g/mol; CAS 422513-13-1: Solubility: 100 mM in DMSO; Desiccate at Room Temperature), also named *N*-[2,3-dihydro-2-oxo-3-[(3*Z*)-phenyl[[4-(1-piperidinylmethyl)phenyl]amino]methylene]-1*H*-indol-5-yl]ethanesulfon-amide hydrochloride, targets Aurora B kinase (IC$_{50}$ = 250 nM), blocking an inhibitor of chromosome alignment and segregation (1). Mammalian cells treated with Hesperadin enter anaphase in the presence of numerous monooriented chromosomes, many of which may have both sister kinetochores attached to one spindle pole (syntelic attachment). Hesperadin also causes cells arrested by taxol or monastrol to enter anaphase within <1 h, whereas cells in nocodazole stay arrested for 3-5 h. Such findings suggest that Aurora B is required to generate unattached kinetochores on monooriented chromosomes, which in turn could promote bipolar attachment as well as maintain checkpoint signaling (1). (*See similarly-sounding flavanones Hesperetin; Hesperidin*) **1**. Hauf, Cole, LaTerra, *et al.* (2003) *J. Cell Biol.* 161, 281.

Hesperetin

This flavanone (FW = 302.28 g/mol; CAS 520-33-2), also known as 3',5,7-trihydroxy-4-methoxyflavanone, is a potent antioxidant that is particularly effective against peroxynitrite. Hesperetin stimulates muscle phosphorylase kinase and is also demethylated by CYP1A1 and CYP1B1. (*See similarly-sounding flavanone Hesperidin*; *also similarly-sounding Aurora Kinase inhibitor Hesperadin*) **Target(s):** acid phosphatase, plant (1); aromatase, *or* CYP19 (2); 3',5'-cyclic-nucleotide phosphodiesterase (3); estrone sulfotransferase (4); fatty-acid synthase (5); flavanone 7-*O*-glucoside 2"-*O*-β-L-rhamnosyltransferase (6); 15-lipoxygenase, *or* arachidonate 15-lipoxygenase (7); phenol sulfotransferase (4); phosphodiesterase- 4 (3); xanthine oxidase (8). **1**. Axelrod (1947) *J. Biol. Chem.* **167**, 57. **2**. Jeong, Shin, Kim & Pezzuto (1999) *Arch. Pharm. Res.* **22**, 309. **3**. Ko, Shih, Lai, Chen & Huang (2004) *Biochem. Pharmacol.* **68**, 2087. **4**. Harris, Wood, Bottomley, *et al.* (2004) *J. Clin. Endocrinol. Metab.* **89**, 1779. **5**. Li & Tian (2004) *J. Biochem.* **135**, 85. **6**. Bar-Peled, Lewinsohn, Fluhr & Gressel (1991) *J. Biol. Chem.* **266**, 20953. **7**. Sadik, Sies & Schewe (2003) *Biochem. Pharmacol.* **65**, 773. **8**. Dew, Day & Morgan (2005) *J. Agric. Food Chem.* **53**, 6510.

Hesperidin

This flavone glycoside (FW = 610.57 g/mol), also known as cerantin, is abundant in lemons and sweet oranges (*e.g.*, approximately 8% of the dry weight of orange peel is hesperidin). Upon ingestion, hesperidin releases its aglycone, hesperetin (*See Hesperetin; See also similarly-sounding Aurora Kinase inhibitor Hesperadin*), which induces heme oxygenase. **Target(s):** Cyclooxygenase (1); β-glucuronidase, weakly inhibited (2); hyaluronidase, inhibited by acetylated, sulfonated, or phosphorylated hesperidin (3); lipase, *or* triacylglycerol lipase (4); 15-lipoxygenase, soybean (5); [myosin light-chain] kinase (6); nitric-oxide synthase (1); tyrosinase, *or* monophenol monooxygenase (7); xanthine oxidase, mildly inhibited, also inhibited by phosphorylated hesperidin (8). **1**. Sakata, Hirose, Qiao, Tanaka & Mori (2003) *Cancer Lett.* **199**, 139. **2**. Rodney, Swanson, Wheeler, Smith & Worrel (1950) *J. Biol. Chem.* **183**, 739. **3**. Beiler & Martin (1948) *J. Biol. Chem.* **174**, 31. **4**. Kawaguchi, Mizuno, Aida & Uchino (1997) *Biosci. Biotechnol. Biochem.* **61**, 102. **5**. Malterud & Rydland (2000) *J. Agric. Food Chem.* **48**, 5576. **6**. Jinsart, Ternai & Polya (1991) *Biol. Chem. Hoppe-Seyler* **372**, 819. **7**. Zhang, Lu, Tao, *et al.* (2007) *J. Enzyme Inhib. Med. Chem.* **22**, 91. **8**. Beiler & Martin (1951) *J. Biol. Chem.* **192**, 831.

Heteropodatoxins

DCGTIWHYCGTDQSECCEGWKCSRQLCKYVIDW
Heteropodatoxin-1

DDCGKLFSGCDTNADCCEGYVCRLWCKLDW
Heteropodatoxin-2

ECGTLFSGCSTHADCCEGFICKLWCRYERTW
Heteropodatoxin-3

These oligopeptides (Heteropodatoxin-1: FW_{H1} = 3917.41 g/mol; FW_{H2} = 3419.85 g/mol; Heteropodatoxin-3: FW_{H3} = 3606.11 g/mol) from venom of the spider *Heteropoda venatoria* inhibit Kv4.x potassium channels. **Target(s):** Ito1 (transient outer current; inhibited by heteropodatoxin-3 (1,2); potassium channel Kv4.x, inhibited by heteropodatoxins-1,-2, and -3 (2-4). **1**. Himmel, Wettwer, Q. Li & U. Ravens (1999) *Amer. J. Physiol.* **277**, H107. **2**. Sanguinetti, Johnson, Hammerland, *et al.* (1997) *Mol. Pharmacol.* **51**, 491. **3**. Ramakers & Storm (2002) *Proc. Natl. Acad. Sci. U.S.A.* **99**, 10144. **4**. Wang, Cheng, Tandan, *et al.* (2006) *J. Cardiovasc. Electrophysiol.* **17**, 298.

Hexachlorobiphenyls

These polychlorinated aromatics (FW = 360.88 g/mol), of which there are many isomer forms, are highly stable industrial chemicals that exert harmful effects on the environment. Their lipophilicity assures that they will readily penetrate tissues and will concentrate in fat-rich tissues and organs. **CAUTION:** PCB exposure gives rise to liver damage, nausea, jaundice, and decreased birth weight in the children of exposed pregnant women. Great care should be exercised in handling these and related compounds. Although some inhibitory targets are listed here, use of PCBs as metabolic inhibitors should be avoided. There are many better and safer alternatives. **Target(s):** L-aromatic amino acid decarboxylase (1); glycogen phosphorylase *a* and *b* (2); Ah receptor (3); uroporphyrinogen decarboxylase (4). **1**. Aarts, Denison, Cox, *et al.* (1995) *Eur. J. Pharmacol.* **293**, 463. **2**. Kawanishi, Seki & Sano (1983) *J. Biol. Chem.* **258**, 4285. **3**. Angus, Mousa, Vargas, *et al.* (1997) *Neurotoxicology* **18**, 857. **4**. Mead, Hart & Gamble (1982) *Biochim. Biophys. Acta* **701**, 173.

Hexachlorophene

This hexachlorinated bisphenol (FW = 406.91 g/mol; CAS 70-30-4), also known as 2,2'-methylenebis(3,4,6-trichlorophenol), is used as a disinfectant and flukicide. Because neurotoxic effects can occur upon hexachlorophene overexposure, wider use of this disinfectant was ababdoned in the 1970s. In rats, the oral LD_{50} is ~60 mg/kg, further evidence that it is relatively toxic. **Target(s):** acetylcholinesterase (1,2); adenylate cyclase (3); ATPase (3); chitin synthase (4); 3CL protease (5); cystathionine β-synthase, *or* serine sulfhydrase (6-8); enoyl-[acyl-carrier protein] reductase (9,10); glucose-6-phosphate dehydrogenase (11); glutathione *S*-transferase (12); homocysteine desulfhydrase (6); 3-oxoacyl-[acyl-carrier-protein] reductase (13); succinate dehydrogenase (14). **1**. Matsumura, Matsuoka, Igisu & Ikeda (1997) *Arch. Toxicol.* **71**, 151. **2**. Prasad, Indira & Rajendra (1987) *Bull. Environ. Contam. Toxicol.* **38**, 139. **3**. Mavier, Stengel & Hanoune (1976) *Biochem. Pharmacol.* **25**, 305. **4**. Pfefferle, Anke, Bross & Steglich (1990) *Agric. Biol. Chem.* **54**, 1381. **5**. Hsu, Kuo, Hsieh, *et al.* (2004) *FEBS Lett.* **574**, 116. **6**. Thong & Coombs (1987) *Exp. Parasitol.* **63**, 143. **7**.

Walker & Barrett (1992) *Exp. Parasitol.* **74**, 205. **8**. Papadopoulos, Walker & Barrett (1996) *Int. J. Biochem. Cell Biol.* **28**, 543. **9**. Marcinkeviciene, Jiang, Kopcho, *et al.* (2001) *Arch. Biochem. Biophys.* **390**, 101. **10**. Heath, Li, Roland & Rock (2000) *J. Biol. Chem.* **275**, 4654. **11**. Wang & Buhler (1981) *J. Toxicol. Environ. Health* **8**, 639. **12**. Liebau, Eckelt, Wildenburg, *et al.* (1997) *Biochem. J.* **324**, 659. **13**. Wickramasinghe, Inglis, Urch, *et al.* (2006) *Biochem. J.* **393**, 447. **14**. Lokanatha, Sailaja & Rajendra (1999) *J. Biochem. Mol. Toxicol.* **13**, 303.

3-[(*cis,cis*-7,10-Hexadecadienyl]-4-hydroxy-2-butenolide

This synthetic manoalide analogue (FW = 320.47 g/mol) inhibits lysophospholipase and phospholipase A_2. The closed and open forms of the α,β-unsaturated γ-lactone ring are in rapid equilibrium over the pH range from 4 to 9, with the cyclic form preferred at acidic pH values and the open *cis* form preferred at pH 9.5. When the pH is raised above 12, the α,β double bond isomerizes to form the *trans* butenolide. This *trans* isomer is stable at all pH values and does not recyclize to the γ-lactone ring. **Target(s):** lysophospholipase (1); phospholipase A_2 (2). **1**. Zhang & Dennis (1988) *J. Biol. Chem.* **263**, 9965. **2**. Deems, Lombardo, Morgan, Mihelich & Dennis (1987) *Biochim. Biophys. Acta* **917**, 258.

Hexadecylsulfonyl Fluoride

This long-chain alkyl sulfonyl fluoride (FW = 308.50 g/mol), also known as hexadecanesulfonyl fluoride and cetylsulfoyl fluoride, is an active-site directed irreversible inhibitor of a number of serine esterases. **Target(s):** acylglycerol lipase, *or* monoacylglycerol lipase (1-3); lipase, *or* triacylglycerol lipase (4,5); lipoprotein lipase (6-8); palmitoyl-protein thioesterase (9); phospholipase A, *Escherichia coli* (10,11). **1**. Dinh, Freund & Piomelli (2002) *Chem. Phys. Lipids* **121**, 149. **2**. Dinh, Carpenter, Leslie, *et al.* (2002) *Proc. Natl. Acad. Sci. U.S.A.* **99**, 10819. **3**. Saario, Savinainen, Laitinen, Järvinen & Niemi (2004) *Biochem. Pharmacol.* **67**, 1381. **4**. Tjeenk, Bulsink, Slotboom, *et al.* (1994) *Protein Eng.* **7**, 579. **5**. Konig, Jaeger, Sage, Vasil & Konig (1996) *Infect. Immun.* **64**, 3252. **6**. Olivecrona & Lookene (1997) *Meth. Enzymol.* **286**, 102. **7**. Skottova, Savonen, Lookene, Hultin & Olivecrona (1995) *J. Lipid Res.* **36**, 1334. **8**. Kokotos, Kotsovolou, Constantinou-Kokotou, Wu & Olivecrona (2000) *Bioorg. Med. Chem. Lett.* **10**, 2803. **9**. Das, Bellizzi, Tandel, Biehl, Clardy & Hofmann (2000) *J. Biol. Chem.* **275**, 23847. **10**. Horrevoets, Francke, Verheij & de Haas (1991) *Eur. J. Biochem.* **198**, 255. **11**. Horrevoets, Verheij & de Haas (1991) *Eur. J. Biochem.* **198**, 247.

1-Hexadecyl-3-trifluoroethylglycero-*sn*-2-phosphomethanol

This highly selective phospholipase inhibitor (FW = 493.50 g/mol (lithium salt)), also known as MJ33, targets Type-1B phospholipase A_2 (1-7) as well as 1-Cys peroxiredoxin (1). nt. MJ 33 has poor affinity for the Type II human synovial PLA_2 and has only moderate affinity toward macrophage lysosomal PLA_2. **1**. Chen, Dodia, Feinstein, Jain & Fisher (2000) *J. Biol. Chem.* **275**, 28421. **2**. Jain, Tao, Rogers, *et al.* (1991) *Biochemistry* **30**, 10256. **3**. Sekar, Eswaramoorthy, Jain & Sundaralingam (1997) *Biochemistry* **36**, 14186. **4**. Yu, Pan, Janssen, Bahnson & Jain (2005) *Biochemistry* **44**, 3369. **5**. Ghomashchi, Yu, Mihelich, Jain & Gelb (1991) *Biochemistry* **30**, 9559. **6**. Akiba, Dodia, Chen & Fisher (1998) *Comp.*

Biochem. Physiol. B **120**, 393. **7.** Wang, Dodia, Jain & Fisher (1994) *Biochem. J.* **304**, 131.

1-Hexanol

This primary alcohol (FW = 102.18 g/mol; M.P. = –51.6°C; B.P. = 157.5 °C), also known as *n*-hexyl alcohol, is a component of naturally occurring esters in certain fruits and seeds. **Target(s):** aryldialkylphosphatase, *or* paraoxonase, *or* arylesterase (1); arylesterase (1); bacterial luciferase, alkanal monooxygenase (FMN-linked), Ki = 0.369 mM (2); choline transport (3); cutinase (4); cytochrome *c* oxidase (5); diacylglycerol *O*-acyltransferase (6); esterase (7); firefly luciferase, *or Photinus*-luciferin 4-monooxygenase (8); glutaconyl-CoA decarboxylase (9); leucyl aminopeptidase (10,11); *N*-methyl-D-aspartate (NMDA) receptor (12); oxidative phosphorylation, uncoupling (13); peroxidase, horseradish (14); protein kinase C (15); pyroglutamyl-peptidase I (16). **1.** Debord, Dantoine, Bollinger, *et al.* (1998) *Chem. Biol. Interact.* **113**, 105. **2.** Curry, Lieb & Franks (1990) *Biochemistry* **29**, 4641. **3.** Deves & Krupka (1990) *Biochim. Biophys. Acta* **1030**, 32. **4.** Maeda, Yamagata, Abe, *et al.* (2005) *Appl. Microbiol. Biotechnol.* **67**, 778. **5.** Hasinoff & Davey (1989) *Biochem. J.* **258**, 101. **6.** Coleman & Bell (1976) *J. Biol. Chem.* **251**, 4537. **7.** Weber & King (1935) *J. Biol. Chem.* **108**, 131. **8.** Dickinson, Franks & Lieb (1993) *Biophys. J.* **64**, 1264. **9.** Buckel & Liedtke (1986) *Eur. J. Biochem.* **156**, 251. **10.** Smith & Hill (1960) *The Enzymes*, 2nd ed., **4**, 37. **11.** DeLange & Smith (1971) *The Enzymes*, 3rd ed., **3**, 81. **12.** Peoples & Ren (2002) *Mol. Pharmacol.* **61**, 169. **13.** Canton, Gennari, Luvisetto & Azzone (1996) *Biochim. Biophys. Acta* **1274**, 39. **14.** Biswas, Das, Pradhan, *et al.* (2008) *J. Phys. Chem. B* **112**, 6620.

Hexathionine Sulfoximine & Hexathionine Sulfoximine Phosphate

These derivatives (FW$_{HSX}$ = 250.36 g/mol; FW$_{HSX-P}$ = 330.34 g/mol) are tightly bound mechanism-based *reversible* inhibitors of γ-glutamylcysteine synthetase, which catalyzes the first step in glutathione biosynthesis. By mimicking the tetrahedral reaction intermediate, hexathionine sulfoximine is held firmly within the active site of glutamylcysteine synthetase. In contrast to buthionine sulfoximine, however, hexathionine sulfoximine causes substantial depletion of mitochondrial GSH. Hexathionine sulfoximine most likely perturbs mitochondrial membranes in such a way as to facilitate loss of GSH. Significantly, hexathionine sulfoximine produces substantial toxicity, whereas buthionine sulfoximine is relatively nontoxic and does not produce extensive loss of mitochondrial GSH (*See also* Buthionine Sulfoximine) **Target(s):** γ-glutamylcysteine synthetase (1,2). **1.** Seelig & Meister (1985) *Meth. Enzymol.* **113**, 379. **2.** Griffith (1987) *Meth. Enzymol.* **143**, 286.

Hexestrol

This diphenol (FW = 270.37 g/mol; CAS 5776-72-7), also known as dihydrodiethylstilbestrol and *meso*-3,4-bis(*p*-hydroxyphenyl) *n*-hexane, is a synthetic nonsteroidal estrogen. **Target(s):** ATPase (1,2); Ca^{2+}-transporting ATPase (3); 3α-hydroxysteroid dehydrogenase (4-6); 3-(or 17-)β-hydroxysteroid dehydrogenase (7-11); 3(20)α-hydroxysteroid dehydrogenase, indanol dehydrogenase (12,13); testosterone 17β-dehydrogenase (6); tyrosinase, monophenol monooxygenase (14); tubulin polymerization (microtubule assembly) (15,16). **1.** Lemker, Gruber, Schmid & Muller (2003) *FEBS Lett.* **544**, 206. **2.** Lemker, Ruppert, Stoger, Wimmers & Muller (2001) *Eur. J. Biochem.* **268**, 3744. **3.** Martinez-Azorin, Teruel, Fernandez-Belda & Gomez-Fernandez (1992) *J. Biol. Chem.* **267**, 11923. **4.** Sawada, Hara, Ohmura, Nakayama & Deyashiki (1991) *J. Biochem.* **109**, 770. **5.** Hara, Inoue, Nakagawa, Naganeo & Sawada (1988) *J. Biochem.* **103**, 1027. **6.** Hasebe, Hara, Nakayama, *et al.* (1987) *Enzyme* **37**, 109. **7.** Talalay (1963) *The Enzymes*, 2nd ed., **7**, 177. **8.** Webb (1966) *Enzyme and Metabolic Inhibitors*, vol. **2**, p. 449, Academic Press, New York. **9.** Talalay (1962) *Meth. Enzymol.* **5**, 512. **10.** Jarabak & Sack (1969) *Biochemistry* **8**, 2203. **11.** Talalay & Dobson (1953) *J. Biol. Chem.* **205**, 823. **12.** Hara, Nakagawa, Taniguchi & Sawada (1989) *J. Biochem.* **106**, 900. **13.** Hara, Mouri, Nakagawa, *et al.* (1989) *J. Biochem.* **106**, 126. **14.** Orenes-Pinero, Garcia-Carmona & Sanchez-Ferrer (2007) *J. Mol. Catal. B* **47**, 143. **15.** Sato, Murai, Oda, *et al.* (1987) *J. Biochem.* **101**, 1247. **16.** Chaudoreille, Peyrot, Braguer, Codaccioni & Crevat (1991) *Biochem. Pharmacol.* **41**, 685.

2-Hexylthioether-β,γ-methylene-ATP

This nonhydrolyzable ATP analogue (FW = 618.41 g/mol) targets NTPDase 2 (IC$_{50}$ = 20 μM), an ectonucleotidase catalyzing ATP hydrolysis and forming extracellular ADP, which binds to Type-2 purinergic receptors and controls multiple biological responses. The origin of inhibitor selectivity is its thiohexyl moiety, which is favorably located within a hydrophobic pocket of NTPDase 2, but, when bound to NTPDase 1, is exposed to solvent. **1.** Gillerman, Lecka, Simhaev, *et al.* (2014) *J. Med. Chem.* **57**, 5919.

5-Hexyne-1,4-diamine

This putrescine analogue (FW$_{free-base}$ = 112.17 g/mol; CAS 69355-11-9) is a potent irreversible inhibitor of ornithine decarboxylase. Prolonged incubation with dithiothreitol is ineffective in reactivating the enzyme. Kinetic isotope measurements using 4-deuterio-5-hexyne-1,4-diamine suggest covalent bond formation is rate-limiting. 5-Hexyne-1,4-diamine is also an alternative substrate of monoamine oxidase, producing 4-aminohex-5-ynoate, an irreversible inhibitor of 4-aminobutyrate aminotransferase. **1.** Metcalf, Bey, Danzin, *et al.* (1978) *J. Amer. Chem. Soc.* **100**, 2551. **2.** Danzin, Jung, Metcalf, Grove & Casara (1979) *Biochem. Pharmacol.* **28**, 627.

Higenamine

This benzyltetrahydroisoquinoline alkaloid (FW$_{free-base}$ = 271.32 g/mol; CAS 5843-65-2 (racemate), 106032-53-5 (R-enantiomer), 22672-77-1 (S-enantiomer)), also known as 1-[(4-hydroxyphenyl)methyl]-1,2,3,4-

tetrahydroisoquinoline-6,7-diol and isolated from the roots of *Aconitum japonicum*, inhibits dihydropteridine reductase. It also mediates cardiotonic, vascular relaxation and bronchodilator effects. It is also a pharmacologically active metabolite in *Nandina domestica Thunberg*, a plant long known to improve cough and breathing difficulties. Indeed, the isolated agent and chemically pure higenamine have equal potency in relaxing isolated guinea pig trachea (2). This action is the consequence of higenamine's actions as an agonist of β_2 adrenoceptors. Higenamine also inhibits iNOS expression by inhibiting nuclear factor κ_B activation by LPS + IFN-γ and may be beneficial in inflammatory diseases in which enhanced formation of NO is the main causative factor (3). Due to its positive inotropic action, higenamine may be more effective in conditions, where myocardial contractility is likely to be depressed, such as in septic shock and/or endotoxin-induced inflammatory disorders (3). It is also a new lead compound for treating erectile dysfunction (4). 1. Shen, Smith, Davis, Brubaker & Abell (1982) *J. Biol. Chem.* 257, 7294. 2. Tsukiyama, Ueki, Yasuda, *et al.* (2009) *Planta Med.* 75, 1393. 3. Kang, Lee, Lee, *et al.* (1999) *J. Pharmacol. Exp. Ther.* 291, 314. 4. Kam, Do, Choi, *et al.* (2012) *Int. J. Impot. Res.* 24, 77.

Himbacine

This alkaloid (FW = 345.53 g/mol; CAS 6879-74-9), also named (3S,3aR,4R,4aS,8aR,9aS)-4-[(1E)-2-[(2R,6S)-1,6-dimethyl-2-piperdinyl]-ethenyl]decahydro-3-methylnaphtho[2,3-c]furan-1(3H)-one, is a selective antagonist of M_2 and M_4 muscarinic acetylcholine receptors, with lower affinity for M_1 and M_3 forms. Its derivative, SCH 530348, is a thrombin receptor antagonist with potent anticoagulant properties. 1. Dorje, Wess, Lambrecht, *et al.* (1991) *J. Pharmacol. Exp. Ther.* 256, 727. 2. Gilani & Cobbin (1987) *Arch. Int. Pharmacodyn. Ther.* 290, 46.

Hippurate (Hippuric Acid)

This modified amino acid (FW$_{free-acid}$ = 179.18 g/mol; CAS 495-69-2; pK_a = 3.62), also known as N-benzoylglycine, can be found in the urine of herbivores and humans. As an oxidation product in toluene metabolism, high circulating levels of hippurate are a useful index of toluene toxicity. **Target(s):** acetylcholinesterase (1); adenylosuccinate synthetase, weakly inhibited (2); γ-glutamyl transpeptidase, transfer activity (3,4); glycine N-acyltransferase, product inhibition (5,6); glycine N-benzoyltransferase, product inhibition (5-7); lysine carboxypeptidase, *or* lysine(arginine) carboxypeptidase, *or* carboxypeptidase N (8); papain (9); peptidylglycine monooxygenase, also alternative substrate (10). 1. Bergmann, Wilson & Nachmansohn (1950) *J. Biol. Chem.* 186, 693. 2. Jahngen & Rossomando (1984) *Arch. Biochem. Biophys.* 229, 145. 3. Gardell & Tate (1983) *J. Biol. Chem.* 258, 6198. 4. Thompson & Meister (1979) *J. Biol. Chem.* 255, 2109. 5. van der Westhuizen, Pretorius & Erasmus (2000) *J. Biochem. Mol. Toxicol.* 14, 102. 6. Kelley & Vessey (1993) *J. Biochem. Toxicol.* 8, 63. 7. Kelley & Vessey (1990) *J. Biochem. Toxicol.* 5, 125. 8. Juillerat-Jeanneret, Roth & Bargetzi (1982) *Hoppe-Seyler's Z. Physiol. Chem.* 363, 51. 9. Sluyterman (1964) *Biochim. Biophys. Acta* 85, 316. 10. Francisco, Blackburn & Klinman (2003) *Biochemistry* 42, 1813.

Hirudin

This direct thrombin inhibitor and anticoagulant (MW ≈ 7 kDa; Sequence: VVYTDCTESGQNLCLEGSNVCGQGNKCILGSDGEKNQCVTGEGTP GPQSHNDGDFEEPEEYL; CAS 8001-27-2) from the saliva of the medicinal leech *Hirudo medicinalis* potently inhibits thrombin (K_i = 0.2

pM) by simultaneously occupying the fibrinogen site and the anionic exosite (1). Hirudin inhibits both free thrombin and fibrin clot-bound thrombin. X-ray structural analysis shows that, while thrombin's active site is blocked by hirudin, none of hirudin's residues are embedded at the S_1 subsite (2,3); its C-terminal residues bind to the fibrinogen's exosite. **Target(s):** coagulation factor IXa (4); complement component I (5); thrombin (1-4,6-15). 1. Markwardt (1994) *Thromb. Res.* 74, 1. 2. Grütter, Priestle, Rahuel, *et al.* (1990) *EMBO J.* 9, 2361. 3. Rydel, Ravichandran, Tulinsky, *et al.* (1990) *Science* 249, 277. 4. Bagby, Barabas, Gráf, Petersen & Magnusson (1976) *Meth. Enzymol.* 45, 669. 5. Tsiftsoglou & Sim (2004) *J. Immunol.* 173, 367. 6. Stone & Le Bonniec (1998) in *Handb. Proteolytic Enzymes*, p. 168, Academic Press, San Diego. 7. Baughman (1970) *Meth. Enzymol.* 19, 145. 8. Magnusson (1970) *Meth. Enzymol.* 19, 157. 9. Markwardt (1970) *Meth. Enzymol.* 19, 924. 10. Stone & Tapparelli (1995) *J. Enzyme Inhib.* 9, 3. 11. Lundblad, Kingdon & Mann (1976) *Meth. Enzymol.* 45, 156. 12. Stone & Maraganore (1993) *Meth. Enzymol.* 223, 312. 13. Magnusson (1971) *The Enzymes*, 3rd ed., 3, 277. 14. Prasad, Cantwell, Bush, *et al.* (2004) *J. Biol. Chem.* 279, 10103. 15. Weitz (2003) *Thromb. Res.* 109 Suppl. 1, S17.

Hispidin

This naturally occurring pyrone and antioxidant (FW = 246.22 g/mol; CAS 56070-89-4), also known as 6-(3,4-dihydroxystyryl)-4-hydroxy-2-pyrone, is a trypanocide that inhibits protein kinase Cβ, IC$_{50}$ = 2 μM (1). Hispidin also inhibits β-secretase, K_i = 8.4 μM (1) and prolyl endopeptidase, K_i = 24 μM (1). 1. Gonindard, Bergonzi, Denier, *et al.* (1997) *Cell Biol. Toxicol.* 13, 141. 2. Park, Jeon, Lee, Kim & Song (2004) *Planta Med.* 70, 143.

Histamine

This naturally occurring histidine metabolite (FW$_{free-base}$ = 111.15 g/mol; CAS 51-45-6; MP$_{free-base}$ = 83-84°C; Soluble in water and ethanol; pK_a values of 6.04 and 9.75), systematically named 1H-imidazole-4-ethanamine, is a potent vasodilator and H_1, H_2, H_3, and H_4 receptor agonist. Most histamine in the body is generated in the granules of mast cells and in basophils and eosinophils. Histamine release occurs when allergens bind to mast-cell-bound IgE antibodies. When released into circulation (or when injected intravenously), histamine causes most blood vessels to dilate, reducinh blood pressure. This mechanism is a key process in anaphylaxis. **Metabolism:** Histamine is formed by decarboxylation of the L-histidine, as catalyzed by the pyridoxal 5-phosphate-dependent histidine decarboxylase (HDC). Histamine is inactivated either by methylation of the imidazole ring, catalyzed by histamine N-methyltransferase (HMT) or by oxidative deamination of the primary amino group, catalyzed by diamine oxidase (DAO). Histamine also activates nitric-oxide synthase, causing smooth muscle and playing a role in controlling gastric secretions. Histamine is transported by the organic cation transporter (OCT), the plasma membrane monoamine transporter (PMAT), the vesicular monamine transporter (VMAT), and the endo-/exocytotic uptake/release in/from cellular vesicles or granules. **Histamine Receptors:** The H_1 receptor is a member of the family of Rhodopsin-like G-protein-coupled receptors and is expressed in smooth muscles, on the surface of vascular endothelial cells, as well as in the heart and central nervous system. (*See H$_1$- antagonists: Diphenhydramine; Loratadine; Cetirizine; Fexofenadine; Clemastine*) Histamine H$_2$ receptors are coupled to adenylate cyclase via G$_s$, forming CyclicAMP, which in turn activates Protein Kinase A. (*See H$_2$-antagonists: Ranitidine; Cimetidine; Famotidine; Nizatidine*) Histamine H$_3$ receptors are expressed in the CNS and to a lesser extent the peripheral nervous system, where they act as autoreceptors in presynaptic histaminergic neurons. They also control histamine turnover via feedback inhibition of histamine synthesis and release. (*See H$_3$-antagonists: Ciproxifan; Clobenpropit; Thioperamide*) Histamine H$_4$ receptors are highly expressed in bone marrow and white blood cells (also in colon, liver, lung, small intestine, spleen, testes, thymus, tonsils, and trachea) and regulates neutrophil release from bone marrow. It also acts on the histamine-gated chloride channel, producing fast inhibitory postsynaptic potentials in brain. (**See H₄-antagonists:** *Thioperamide; JNJ 7777120*) **Target(s):** amine oxidase, inhibited by elevated histamine concentrations (1); aminoimidazolase (2); chitinase (3,4); dipeptidyl-peptidase II (5); fibrin polymerization (6); glutaminyl peptide cyclotransferase (7); glycylpeptide *N*-tetradecanoyltransferase, *or* protein *N* myristoyltransferase (21,24); histidine ammonia-lyase (8-10); histidyl-tRNA synthetase (11); β-lactamase (penicillinase) (12); lysozyme (13-15); NAD⁺-diphthamide ADP-ribosyltransferase (16); ornithine decarboxylase (17); poly(ADP-ribose) polymerase (18); protein-glutamine γ-glutamyltransferase (*or* transglutaminase) (19,20); pyridoxal kinase (22); tryptase (23). **1.** Mondovì, Scioscia-Santoro, Rotilio & Costa (1965) *Enzymologia* **28**, 228. **2.** Rabinowitz & Pricer (1956) *J. Biol. Chem.* **222**, 537. **3.** Boller, Gehri, Mauch & Vögeli (1988) *Meth. Enzymol.* **161**, 479. **4.** Boller, Gehri, Mauch & Vögeli (1983) *Planta* **157**, 22. **5.** Struckhoff & Heymann (1986) *Biochem. J.* **236**, 215. **6.** Lorand & Jacobsen (1964) *Biochemistry* **3**, 1939. **7.** Schilling, Niestroj, Rahfeld, *et al.* (2003) *J. Biol. Chem.* **278**, 49773. **8.** Leuthardt (1951) *The Enzymes*, 1st ed. (Sumner & Myrbäck, eds.), **1** (part 2), 1156. **9.** Hanson & Havir (1972) *The Enzymes*, 3rd ed., **7**, 75. **10.** Brand & Harper (1976) *Biochemistry* **15**, 1814. **11.** Lepore, di Natale, Guarini & de Lorenzo (1975) *Eur. J. Biochem.* **56**, 369. **12.** Pollock (1960) *The Enzymes*, 2nd ed., **4**, 269. **13.** Audy, Trudel & Asselin (1988) *Plant Sci.* **58**, 43. **14.** Glazer, Barel, Howard & Brown (1969) *J. Biol. Chem.* **244**, 3583. **15.** Perin & Jolles (1973) *Mol. Cell. Biochem.* **2**, 189. **16.** Lee & Iglewski (1984) *Proc. Natl. Acad. Sci. U.S.A.* **81**, 2703. **17.** Ono, Inoue, Suzuki & Takeda (1972) *Biochim. Biophys. Acta* **284**, 285. **18.** Villamil, Podesta, Molina Portela & Stoppani (2001) *Mol. Biochem. Parasitol.* **115**, 29. **19.** Siegel & Khosla (2007) *Pharmacol. Ther.* **115**, 232. **20.** Yeh, Kao, Huang & Tsai (2006) *Biochim. Biophys. Acta* **1764**, 1167. **21.** Raju, Datla, Warrington & Sharma (1998) *Biochemistry* **37**, 14928. **22.** Gäng & von Collins (1973) *Int. J. Vitam. Nutr. Res.* **43**, 318. **23.** Alter & Schwartz (1989) *Biochim. Biophys. Acta* **991**, 426. **24.** Selvakumar, Lakshmikuttyamma, Shrivastav, *et al.* (2007) *Prog. Lipid Res.* **46**, 1.

Histargin

This fungal metabolite (FW = 355.40 g/mol; CAS 93361-66-1), also known as *N*-(2-((4-((amino(imino)methyl)amino)-1(*S*) carboxybutyl)amino) ethyl)-L-histidine, first isolated from a strain of *Streptomyces roseoviridis*, consists of L-histidine and L-arginine, with their α amino groups linked by an ethylene diamine bridge. **Target(s):** carboxypeptidase A and B (1,2); lysine carboxypeptidase (lysine(arginine) carboxypeptidase, *or* carboxy-peptidase N (1); neprilysin (3); peptidyl-dipeptidase A, *or* angiotensin-converting enzyme (1,3,4). **1.** Moriguchi, Umeda, Miyazaki, *et al.* (1988) *J. Antibiot.* **41**, 1823. **2.** Umezawa, Aoyagi, Ogawa, *et al.* (1984) *J. Antibiot.* **37**, 1088. **3.** Umezawa, Takeuchi, Aoyagi, Hamada & Ogawa (1984) Jpn. Kokai 216597. **4.** Wada, Aoyagi, Iinuma, *et al.* (1988) *Biotechnol. Appl. Biochem.* **10**, 435.

Histatin 5

This human salivary oligopeptide (FW = 3036.33 g/mol; CAS 104339-66-4; *Sequence*: DSHAKRHHGYKRKFHEKHHSHRGY) inhibits *Bacteroides gingivalis* protease, IC₅₀ = 55 nM (1) and clostripain, K_i = 10 nM (2). **1.**

Nishikata, Kanehira, Oh, *et al.* (1991) *Biochem. Biophys. Res. Commun.* **174**, 625. **2.** Gusman, Grogan, Kagan, Troxler & Oppenheim (2001) *FEBS Lett.* **489**, 97.

D-Histidine

This amino acid, (FW = 155.16 g/mol; CAS 351-50-8), systematically named (*R*)-2-amino-3-(1*H*-imidazol-4-yl)propanoic acid, is the enantiomer of the proteogenic amino acid. **Target(s):** alkaline phosphatase (1); alkylglycerophosphoethanolamine phospho-diesterase, *or* lysophospho-lipase D (2); 8-amino-7-oxononanoate synthase (3,20,21); D-arabinose isomerase, activates in presence of Mn²⁺; inhibits in absence of Mn²⁺ or presence of other ions (4); arylsulfatase A, *or* cerebroside-sulfatase) (5); carboxypeptidase A, K_i = 20 mM (6-9); L-fucose isomerase, activates in presence of Mn²⁺; inhibits in absence of Mn²⁺ or presence of other ions (4); histidine ammonia-lyase (10-17); histidyl-tRNA synthetase (18); L-serine ammonia-lyase, *or* L-serine dehydratase (19). **1.** Fishman & Sie (1971) *Enzymologia* **41**, 141. **2.** Clair, Koh, Ptaszynska, *et al.* (2005) *Lipids Health Dis.* **4**, 5. **3.** Izumi, Tani & Ogata (1979) *Meth. Enzymol.* **62**, 326. **4.** Izumori & Yamanaka (1979) *Agric. Biol. Chem.* **43**, 1993. **5.** Bleszynski & Leznicki (1967) *Enzymologia* **33**, 373. **6.** Pétra (1970) *Meth. Enzymol.* **19**, 460. **7.** Elkins-Kaufman & Neurath (1948) *J. Biol. Chem.* **175**, 893. **8.** Elkins-Kaufman, Neurath & De Maria (1949) *J. Biol. Chem.* **178**, 645. **9.** Hartsuck & Lipscomb (1971) *The Enzymes*, 3rd ed., **3**, 1. **10.** Wu, Kroening, White & Kendrick (1992) *Gene* **115**, 19. **11.** Brand & Harper (1976) *Biochemistry* **15**, 1814. **12.** Leuthardt (1951) *The Enzymes*, 1st ed., **1** (part 2), 1156. **13.** Hanson & Havir (1972) *The Enzymes*, 3rd ed., **7**, 75. **14.** Webb (1966) *Enzyme and Metabolic Inhibitors*, vol. **2**, p. 269, Academic Press, New York. **15.** Hernandez & Phillips (1993) *Protein Expression Purif.* **4**, 473. **16.** Wu, Kroening, White & Kendrick (1992) *Gene* **115**, 19. **17.** Allen, Clark & Phipps (1983) *Biochem. Soc. Trans.* **11**, 350. **18.** Kalousek & Konigsberg (1974) *Biochemistry* **13**, 999. **19.** Newman, Walker & Ziegler-Skylakis (1990) *Biochem. Cell Biol.* **68**, 723. **20.** Izumi, Morita, Tani & Ogata (1973) *Agric. Biol. Chem.* **37**, 1327. **21.** Izumi, Sato, Tani & Ogata (1973) *Agric. Biol. Chem.* **37**, 1335.

L-Histidine

This proteogenic amino acid (FW$_{neutral-species}$ = 155.16 g/mol; FW$_{HCl-Salt}$ = 191.62; CAS 71-00-1; pK_a values of 1.80 (–COOH), 6.04 (imidazole), and 9.33 (–NH₂) at 25°C), named systematically as (*S*)-2-amino-3-(1*H*-imidazol-4-yl)propanoic acid and symbolized by His and H, is the source of catalytic imidazole groups in many enzymes. L-Histidine is one of ten essential amino acids in mammals, and its two codons are CAU and CAC. Histidine is a dibasic amino acid that is frequently found on the surface of proteins and, because the imidazole ring has a pK_a value close to neutrality, often found as a proton acceptor/donor within the active sites of enzymes. They are often found in some α-helices and are more so in β-pleated sheets. Homologous proteins often have histidyl residues interchanged with asparaginyl or glutaminyl residues. The carbon atom situated between the two nitrogens in the ring can exchange its hydrogen atom slowly with the aqueous solvent. This exchange renders histidyl residues useful probes of protein structure in NMR experiments. Imidazole nitrogen atoms also have an affinity for metal ions and are often associated with iron, zinc, or copper ions. **Target(s):** acrosin (1); alkaline phosphatase (2-4); alkylglycerol-phosphoethanolamine phospho-diesterase, *or* lysophospholipase D (5); 2-aminohexano 6-lactam racemase (6,7); 8-amino-7-oxononanoate synthase (8,69,70); aminopeptidase B (9); D-arabinose isomerase, activates in presence of Mn²⁺; inhibits in absence of Mn²⁺ or presence of other ions (10); arginine kinase (11,12); argininosuccinate synthase (13); aspartate aminotransferase (50); ATP phosphoribosyltransferase (14-16,51-67); brachyuran, *or* *Novoden modestrus* collagenolytic protease (17); catalase (76); DNA nucleotidylexotransferase, *or* terminal deoxynucleotidyl transferase (18); dopamine β monooxygenase (72,73); L-fucose isomerase, activates in presence of Mn²⁺; inhibits in absence of Mn²⁺ or presence of other ions (10); gametolysin (19,20); gelatinase A, or matrix metalloproteinase-2 (21); gelatinase B, or matrix metalloproteinase-9 (21); α-glucosidase (22-24); glutamate decarboxylase, weakly inhibited (25); glutamine synthetase (26-31); glutaminyl-peptide cyclotransferase (68); γ-glutamyl transpeptidase (32,33); histidinol-phosphatase (34,35); homoserine kinase (36); iodide peroxidase, or thyroid peroxidase (74); kynureninase (37); microbial collagenase (38,39); ornithine carbamoyltransferase (71); ornithine decarboxylase (40); pancreatic ribonuclease (41); peroxidase (75); phenylalanine ammonia-lyase (42); phosphoprotein phosphatase (43); poly(ADP-ribose) polymerase (44); ribonuclease T_1 (45); L-serine ammonia-lyase, L-serine dehydratase (46); serine:pyruvate aminotransferase (47); Streptomyces griseus aminopeptidase (48); Vibrio collagenase (49).

1. Polakoski & McRorie (1973) J. Biol. Chem. 248, 8183. 2. Fernley (1971) The Enzymes, 3rd ed., 4, 417. 3. Bodansky & Strachman (1948) J. Biol. Chem. 174, 465. 4. Bodansky & Strachman (1949) J. Biol. Chem. 179, 81. 5. Clair, Koh, Ptaszynska, Bandle, et al. (2005) Lipids Health Dis. 4, 5. 6. Ahmed, Esaki, Tanaka & Soda (1983) Agric. Biol. Chem. 47, 1887. 7. Ahmed, Esaki, Tanaka & Soda (1985) Agric. Biol. Chem. 49, 2991. 8. Izumi, Tani & Ogata (1979) Meth. Enzymol. 62, 326. 9. Kawata, Takayama, Ninomiya & Makisumi (1980) J. Biochem. 88, 1601. 10. Izumori & Yamanaka (1979) Agric. Biol. Chem. 43, 1993. 11. Pereira, Alonso, Ivaldi, et al. (2003) J. Eukaryot. Microbiol. 50, 132. 12. Pereira, Alonso, Paveto, et al. (2000) J. Biol. Chem. 275, 1495. 13. Takada, Saheki, Igarashi & Katsunuma (1979) J. Biochem. 85, 1309. 14. Martin, Berberich, Ames, et al. (1971) Meth. Enzymol. 17B, 3. 15. Webb (1966) Enzyme and Metabolic Inhibitors, vol. 2, p. 351, Academic Press, New York. 16. Klungsøyr & Atkinson (1970) Biochemistry 9, 2021. 17. Kim, Park, Kim & Shahidi (2002) J. Biochem. Mol. Biol. 35, 165. 18. Bollum (1974) The Enzymes, 3rd ed., 10, 145. 19. Jaenicke, Kuhne, Spessert, Wahle & Waffenschmidt (1987) Eur. J. Biochem. 170, 485. 20. Matsuda, Saito, Yamaguchi & Kawase (1985) J. Biol. Chem. 260, 6373. 21. Upadhya & Strasberg (2000) Hepatology 31, 1115. 22. Halvorson (1966) Meth. Enzymol. 8, 559. 23. Halvorson & Ellias (1958) Biochem. Biophys. Acta 30, 28. 24. Suzuki, Yuki, Kishigami & Abe (1976) Biochim. Biophys. Acta 445, 386. 25. Roberts & Frankel (1951) J. Biol. Chem. 190, 505. 26. Rhee, Chock & Stadtman (1985) Meth. Enzymol. 113, 213. 27. Shapiro & Stadtman (1970) Meth. Enzymol. 17A, 910. 28. Stadtman & Ginsburg (1974) The Enzymes, 3rd ed., 10, 755. 29. Dahlquist & Purich (1975) Biochemistry 14, 1980. 30. Kapoor & Bray (1968) Biochemistry 7, 3583. 31. Southern, Parker & Woods (1987) J. Gen. Microbiol. 133, 2437. 32. Allison (1985) Meth. Enzymol. 113, 419. 33. Thompson & Meister (1977) J. Biol. Chem. 252, 6792. 34. Brady & Houston (1973) J. Biol. Chem. 248, 2588. 35. Millay & Houston (1973) Biochemistry 12, 2591. 36. Wormser & Pardee (1958) Arch. Biochem. Biophys. 78, 416. 37. Jakoby & Bonner (1953) J. Biol. Chem. 205, 709. 38. Seifter & Harper (1970) Meth. Enzymol. 19, 613. 39. Makinen & Makinen (1987) J. Biol. Chem. 262, 12488. 40. Guirard & Snell (1980) J. Biol. Chem. 255, 5960. 41. Scheraga & Rupley (1962) Adv. Enzymol. 24, 161. 42. Camm & Towers (1973) Phytochemistry 12, 961. 43. Bargoni, Fossa & Sisini (1963) Enzymologia 26, 65. 44. Villamil, Podesta, Molina Portela & Stoppani (2001) Mol. Biochem. Parasitol. 115, 249. 45. Takahashi (1966) J. Biochem. 60, 239. 46. Newman, Walker & Ziegler-Skylakis (1990) Biochem. Cell Biol. 68, 723. 47. Noguchi, Okuno & Kido (1976) Biochem. J. 159, 607. 48. Awad (1998) in Handb. Proteolytic Enzymes, p. 1431, Academic Press, San Diego. 49. Fukushima & Okuda (1998) in Handb. Proteolytic Enzymes, p. 1096, Academic Press, San Diego. 50. Xing & Whitman (1992) J. Bacteriol. 174, 541. 51. Ohta, Fujimori, Mizutani, et al. (2000) Plant Physiol. 122, 907. 52. Morton & Parsons (1976) Arch. Biochem. Biophys. 175, 677. 53. Martin (1963) J. Biol. Chem. 238, 257. 54. Kleeman & Parsons (1976) Arch. Biochem. Biophys. 175, 687. 55. Champagne, Sissler, Larrabee, Doublie & Francklyn (2005) J. Biol. Chem. 280, 34096. 56. Morton & Parsons (1977) Arch. Biochem. Biophys. 181, 643. 57. Tebar & Ballesteros (1976) Mol. Cell. Biochem. 11, 131. 58. Kryvi (1973) Biochim. Biophys. Acta 317, 123. 59. Champagne, Piscitelli & Francklyn (2006) Biochemistry 45, 14933. 60. Dall-Larsen, Kryvi & Klungsoyr (1976) Eur. J. Biochem. 66, 443. 61. Morton & Parsons (1977) Biochem. Biophys. Res. Commun. 74, 172. 62. Vega, Zou, Fernandez, et al. (2005) Mol. Microbiol. 55, 675. 63. Lohkamp, Coggins & Lapthorn (2000) Acta Crystallogr. Sect. D 56, 1488. 64. Cho, Sharma & Sacchettini (2003) J. Biol. Chem. 278, 8333. 65. Ames, Martin & Garry (1961) J. Biol. Chem. 236, 2019. 66. Voll, Appella & Martin (1967) J. Biol. Chem. 242, 1760. 67. Parsons & Koshland (1974) J. Biol. Chem. 249, 4104. 68. Schilling, Niestroj, Rahfeld, et al. (2003) J. Biol. Chem. 278, 49773. 69. Izumi, Morita, Tani & Ogata (1973) Agric. Biol. Chem. 37, 1327. 70. Izumi, Sato, Tani & Ogata (1973) Agric. Biol. Chem. 37, 1335. 71. Lusty, Jilka & Nietsch (1979) J. Biol. Chem. 254, 10030. 72. Colombo, Papadopoulos, Ash & Villafranca (1987) Arch. Biochem. Biophys. 252, 71. 73. Izumi, Hayakari, Kondo & Takemoto (1975) Hoppe-Seyler's Z. Physiol. Chem. 356, 1831. 74. Carvalho, Ferreira, Coelho, et al. (2000) Braz. J. Med. Biol. Res. 33, 355. 75. Suzuki, Honda, Mukasa & Kim (2006) Phytochemistry 67, 219. 76. Calandrelli, Gambacorta, Romano, Carratore & Lama (2008) World J. Microbiol. Biotechnol. 24, 2269.

L-Histidinol

This amino alcohol (FW = 141.17 g/mol; CAS 501-28-0), also known as 4-(2-amino-3-hydroxypropyl)imidazole, is an L-histidine precursor that strongly inhibits protein biosynthesis by targeting histidyl-tRNA synthetase, K_i = 1.3 μM for the Salmonella typhimurium enzyme (1-4). Quantitative structure-activity analysis, based on K_i values from the kinetics of ATP-PP$_i$ exchange and tRNA aminoacylation, led to the following conclusions: (a) the enzyme is stereospecific in its formation of aminoacyl-tRNA complexes, with D-histidine unable to influence either reaction; (b) the –COOH group is not required for binding, but bulky carboxy derivatives prevent binding to the enzyme; (c) the –NH$_2$ group is needed for good binding affinity; (d) the length of the side-chain plays a very important role in binding interactions; and (e) the heteroatoms within the five-membered ring also determine the inhibitory properties of L-histidine analogues (4). L-Histidinol has been gainfully employed in X-ray crystallographic studies as a nonreactive histidine analogue for exploring the active sites of Class II aminoacyl-tRNA synthetases. L-Histidinol dehydrogenase (HDH) appears to be essential for the survival of bacteria responsible for brucellosis and tuberculosis, suggesting that HDH-selective histidinol analogues may prove to be therapeutically useful. **Target(s):** glutaminyl-peptide cyclotransferase (5); glycylpeptide N-tetradecanoyltransferase, or protein N-myristoyltransferase (6,7); histidine ammonia-lyase (8,9); histidine decarboxylase (10); histidinol-phosphatase, weakly as a product inhibitor (11). 1. Chen & Somberg (1980) Biochim. Biophys. Acta 613, 514. 2. Freist, Verhey, Ruhlmann, Gauss & Arnez (1999) Biol. Chem. 380, 623. 3. Kalousek & Konigsberg (1974) Biochemistry 13, 999. 4. Lepore, di Natale, Guarini & de Lorenzo (1975) Eur. J. Biochem. 56, 369. 5. Schilling, Niestroj, Rahfeld, et al. (2003) J. Biol. Chem. 278, 49773. 6. Raju, Datla, Warrington & Sharma (1998) Biochemistry 37, 14928. 7. Selvakumar, Lakshmikuttyamma, Shrivastav, et al (2007) Prog. Lipid Res. 46, 1. 8. Brand & Harper (1976) Biochemistry 15, 1814. 9. Shibatani, Kakimoto & Chibata (1975) Eur. J. Biochem. 55, 263. 10. Gonchar, Petrii, Votrin & Debov (1981) Vopr. Med. Khim. 27, 815. 11. Brady & Houston (1973) J. Biol. Chem. 248, 2588.

Histogranin

This bioactive pentadecapeptide (FW = 1718.97 g/mol; CAS 150045-04-8; Sequence: MNYALKGQGRTLYGF), which co-purifies with bombesin-like immunoreactive peptides from bovine adrenal medulla, is concentrated within the pituitary (5.1 nmol/g) and adrenal glands (0.27 nmol/g), but it is also present in other tissues including the brain (1.6 pmol/g) and blood plasma (24 fmol/ml). A neuropeptide function for the adrenal peptide is suggested by its relative high concentration in chromaffin granules (42 fmol/mg protein as compared with 1 fmol/mg protein in cytosol) and its release from perfused bovine adrenal glands. In rat brain membrane preparations, synthetic histogranin displaces [^{3}H]CGP-39653, a specific ligand of N-methyl-D-aspartate (NMDA) receptor. The displacement curve was biphasic with IC$_{50}$ of 0.6 and 3955 nM, representing 33% and 67% of the binding sites, respectively. Intracerebroventricular (i.c.v.) injection of the peptide (5-100 nmol) in mice produced a dose-dependent protection against NMDA (0.5-1.0 nmol) -induced convulsions but not against (R,S)-α-amino-3-hydroxy-5-methyl-4-isoxazolepropionate (AMPA, 0.25-2.0 nmol), kainate (0.25-0.75 nmol) and bicuculline (1-10 nmol)-induced convulsions. Histogranin has 80% homology with residues 86–100 of histone H4 (2). However, because the structure of histone H4 is highly conserved, it is presumed that histogranin possesses its own precursor and that its gene is distinctly expressed (2). 1. Lemaire, Shukla, Rogers, et al. (1993) Eur. J. Pharmacol. 1993 245, 247. 2. Lemaire, Rogers, Dumont, et al. (1995) Life Sci. 56, 1233.

HKI-272, See Neratinib

HLM 006474

This E2F inhibitor (FW = 399.48 g/mol; CAS 353519-63-8; Soluble to 20 mM in DMSO; IUPAC: 7-[(4-ethoxy-3-methylphenyl)(2-pyridinylamino) methyl]-2-methyl-8-quinolinol), identified by computer-based virtual screen based on the crystal structure of the DNA-bound E2F4/DP2 heterodimer, reduces intracellular E2F4 DNA-binding activity in several cell lines, as measured by electrophoretic mobility shift assay. The E2F/Rb is critical in regulating the initiation of DNA replication, making it a reasonable target for developing chemotherapeutic agents for blocking the mammalian cell cycle. Overnight exposure to HLM006474 results in down-regulation of total E2F4 protein as well as known E2F targets. E2F4-null mouse embryonic fibroblasts are less sensitive than wild-type counterparts to the apoptosis-inducing activity of the compound, revealing its biological specificity. A375 cells, for example, are exquisitely sensitive to the apoptosis-inducing activity of HLM006474 in two-dimensional culture. It is also a potent inhibitor of melanocytes proliferation and subsequent invasion in a three-dimensional tissue culture model system. Importantly, HLM006474 is predicted to make hydrogen bonds with three residues that are absolutely conserved within the E2F family, suggesting that there is no reason to expect that HLM006474 is specific to E2F4 heterodimers. **1**. Ma, Kurtyka, Boyapalle, *et al.* (2008) *Cancer Res.* **68**, 6292.

HM30181

This P-gp inhibitor (FW = 688.74 g/mol; CAS 849675-66-7 (free base) and 849675-88-3 (hydrochloride)), targets Multi-Drug Resistance-1 (MDR1, ABCB1), also known as P-glycoprotein, thereby increasing the oral bioavailability and efficacy of paclitaxel. HM30181 is the most potent (IC$_{50}$ = 0.63 nM) of several MDR1 inhibitors (including cycloporin A, XR9576, and GF120918), and blocks (IC$_{50}$ = 35 nM) transepithelial transport of paclitaxel in MDCK monolayers (1). HM30181 does not inhibit MRP1 (ABCC1), MRP2 (ABCC2), and MRP3 (ABCC3), and only partially inhibits BCRP (ABCG2) at nonpharmacological levels (1). In humans, plasma concentrations peak at 14-42 hours and decline with $t_{1/2}$ of 76–169 hours after single administration; however, after multiple administrations peaks at 5.5-8 hours and declines with $t_{1/2}$ of 150-220 hours (2). After single administration, C_{max} and AUC$_\tau$ (*i.e.*, area under the concentration curve within dosing intervals) increase in a dose-dependent manner; however, after multiple administrations, C_{max} and AUC$_\tau$ dose-dependent increases were not observed (2). HM30181 does inhibit Pgp at the murine blood-brain barrier at clinically feasible doses, making [^{11}C]-HM30181 unsuitable as a PET tracer for visualise cerebral Pgp expression levels (3). **1**. Kwak, Lee, Lee, *et al.* (2010) *Eur. J. Pharmacol.* **627**, 92. **2**. Kim, Gu, Yoon, *et al.* (2012) *Clin. Ther.* **34**, 482. **3**. Bauer, Wanek, Mairinger, *et al.* (2012) *Eur. J. Pharmacol.* **696**, 18.

HMN-214

This protein kinase inhibitor (FW$_{free-acid}$ = 543.60 g/mol; CAS 173529-46-9; Solubility: 12 mg/mL DMSO; <1 mg/mL Water), also known as IVX-214 and systematically as 5-(6-((4-methylpiperazin-1-yl)methyl)-1*H*-benzo[d] imidazol-1-yl)-3-((*R*)-1-(2-(trifluoromethyl)-phenyl)-ethoxy)-thiophene-2-carboxamide, targets the Polo-like kinase Plk1 (IC$_{50}$ = 118 nM), bringing about G$_2$/M cell-cycle arrest and cell death. RNA silencing of WT p53 increases the antiproliferative activity of HMN-214, suggesting that cancers with defective p53 would likewise be sensitive to co-administration of HMN-214 and DNA-damaging drugs. **1**. Takagi, *et al.* (2003) *Invest. New Drugs*, **21**, 387. **2**. Tanaka, *et al.* (2003) *Cancer Res.* **63**, 6942. **2**. DiMaio, *et al.* (2009) *Mol. Cancer Ther.* **8**, 592.

HMR-1275, See Flavopiridol

HMR 1556

This potent channel blocker (FW = 411.44 g/mol; CAS 223749-46-0; Solubility: 100 mM in DMSO), also named *N*-[(3*R*,4*S*)-3,4-Dihydro-3-hydroxy-2,2-dimethyl-6-(4,4,4-trifluorobutoxy)-2*H*-1-benzopyran-4-yl]-*N*-methylmethanesulfonamide, targets the slowly activating, delayed I$_{Ks}$ rectifier potassium current (IC$_{50}$ = 10.5 and 34 nM, respectively, in canine and guinea pig ventricular myocytes). HMR-1556 selectively inhibits I$_{Ks}$ currents over I$_{Kr}$, I$_{Kl}$, I$_{to}$ and L-type Ca^{2+} channel currents, with little or no effect on K$_v$11.1 K$_v$1.5, K$_v$1.3, K$_{ir}$2.1 and HCN2 channel currents (1,2). While HMR1556 alone does not cause Torsade de pointes (TdP), a polymorphic ventricular tachycardia commonly associated with prolongation of the QT interval in ECGs, it does increase E-4031-induced TdP from 25 to 80% (3). **1**. Gerlach, Brendel, Lang, *et al.* (2001) *J. Med. Chem.* **44**, 3831. **2**. Thomas, Gerlach, & Antzelevitch (2003) *J. Cardiovasc. Pharmacol.* **41**, 140. **3**. Michael, Dempster, Kane & Coker (2007) *Brit. J. Pharmacol.* **152**, 1215.

HMR-1726, See Teriflunomide
HN-1, See bis(2-Chloroethyl)-ethylamine
HN-2, See Mechlorethamine

HO-3867

This curcumin-like synthetic antioxidant (FW = 468.59 g/mol), also named 1-[(1-oxyl-2,2,5,5-tetramethyl-2,5-dihydro-1*H*-pyrrol-3-yl)methyl]-(3*E*,5*E*) -3,5-bis(4-fluorobenzylidene)piperidin-4-one, exhibits potent anticancer efficacy towards human ovarian cancer cells and xenograft tumors, while exhibiting substantially lower toxicity towards noncancerous cells. In aerated solutions and cells, HO-3867 induced G_2/M cell-cycle arrest in A2780 cells as evidenced by a significant increase in the p53, p21, and p27 protein levels, but also a significant reduction in Cdk2 and cyclin A levels (1). HO-3867-mediated downregulation of *STAT3* gene expression, in both *in vitro* and *in vivo*, as a putative mechanism for increased Fas/CD95 expression. This is obvious from the substantial decrease in the level of Tyr705-phosphorylated STAT3, a major active form of activated STAT3. It is noteworthy that the expression level of Ser727-phosphorylated STAT3 was also clearly decreased *in vivo* (1). Ser727 phosphorylation is known to regulate the transcriptional activity of STAT3, and the attenuated phosphorylation is likely to account for the down-regulated transcriptional activity of STAT3 in the xenograft tumor treated with HO-3867. This agent also causes substantial inhibition of phospho-JAK1 (Tyr-1022/1023), suggesting it also inhibits constitutive activation of STAT3, which may be caused, at least in part, by pJAK1 inhibition. Depending on cell type, HO-3867 is also likely to inhibit STAT3 activation through JAK2, Src, Erb2, and EGFR, which are known to be implicated in STAT3 activation as well (1). Notably, the DAP compound HO-3867 selectively inhibited STAT3 phosphorylation, transcription and DNA binding without affecting the expression of other active STATs (2). Pharmacological analysis revealed higher bioabsorption/bioavailability of the active (cytotoxic) metabolites by cancer cells compared to normal cells. The selective cytotoxicity of HO-3867 appeared to be multifaceted, eliciting differential activation of the Akt pathway in normal versus cancer cells. RNAi attenuation experiments confirmed the requirement of STAT3 for HO-3867-mediated apoptosis in ovarian cancer cells. **1.** Selvendiran, Tong, Bratasz, *et al.* (2010) *Mol. Cancer Ther.* **9**, 1169. **2.** Rath, Naidu, Lata, *et al.* (2014) *Cancer Res.* **74**, 2316.

HOE-140

This peptidomimetic (FW = 1304.52 g/mol; Systematic Name: D-arginyl-L-arginyl-L-prolyl-(*trans*-4-hydroxy)-L-prolyl-glycyl-3-(2-thienyl)-L-alanyl-L-seryl-D-1,2,3,4-tetrahydro-3-isoquinolinecarbonyl-L-(2α,3β,7aβ)-octahydro-1*H*-indole-2-carbonyl-L-arginine, is a selective and longlasting B_2 bradykinin receptor antagonist, IC_{50} = 1 nM (1-3). Deletion of its C-terminal L-arg residue yields a potent B_1 bradykinin receptor antagonist (4,5). (*For its cyclized form, see* MEN 11270) **1.** Hock, Wirth, Albus, *et al.* (1991) *Brit. J. Pharmacol.* **102**, 769. **2.** Wirth, Hock, Albus, *et al.* (1991) *Brit. J. Pharmacol.* **102**, 774. **3.** Trifilieff, Da Silva, Landry & Gies (1992) *J. Pharmacol. Exp. Ther.* **263**, 1377. **4.** Rhaleb, Gobeil & Regoli (1992) *Life Sci.* **51**, PL125. **5.** Wirth, Wiemer & Schölkens (1992) *Agents Actions Suppl.* **38**, 406.

HOE-498, *See* Ramipril

HOE 642, *See* Cariporide

Homoarginine

This arginine homologue (FW = 188.23 g/mol; CAS 156-86-5), also known as 2-amino-6-guanidinohexanoic acid, and N^6-amidinolysine, is a strong inhibitor of alkaline phosphatases. **Target(s):** acrosin, by the L-isomer (1); alkaline phosphatase, by the L-isomer (2-15); aminopeptidase B (16); arginase (17,18); arginine decarboxylase, by the L-isomer (19-24); arginine deiminase, by the L-isomer (25-28); arginine kinase (29,30); arginine 2-

monooxygenase, by the L-isomer (31,32); arginyl tRNA synthetase (33); carboxypeptidase B, inhibited by racemic mixture (34); clostripain, by the L-isomer (35); Na^+/K^+-exchanging ATPase (36); peptidyl-Lys metalloendopeptidase (37,38); pyridoxal 5-phosphate hydrolase, alkaline phosphatase (2); transport system y^+ (39) **1.** Polakoski & McRorie (1973) *J. Biol. Chem.* **248**, 8183. **2.** Lumeng & Li (1979) *Meth. Enzymol.* **62**, 574. **3.** McDougall, Plumb, King & Hahnel (2002) *J. Histochem. Cytochem.* **50**, 415. **4.** Wada, Yagami, Niwa, Hayakawa & Tsuge (2001) *Exp. Anim.* **50**, 153. **5.** Weser & Kaup (1994) *Biochim. Biophys. Acta* **1208**, 186. **6.** Nair, Majeska & Rodan (1987) *Arch. Biochem. Biophys.* **254**, 18. **7.** Yora & Sakagishi (1986) *Comp. Biochem. Physiol. B.* **85**, 649. **8.** Lin & Fishman (1972) *J. Biol. Chem.* **247**, 3082. **9.** Lustig & Kellen (1971) *Enzymologia* **41**, 336. **10.** Herz (1985) *Experientia* **41**, 1357. **11.** Chakrabartty & Stinson (1983) *Biochim. Biophys. Acta* **839**, 174. **12.** Belland, Visser, Poppema & Stinson (1993) *Enzyme Protein* **47**, 73. **13.** Magnusson & Farley (2002) *Calcif. Tissue Int.* **71**, 508. **14.** Goldstein & Harris (1979) *Nature* **280**, 602. **15.** Wada, Yagami, Niwa, Hayakawa & Tsuge (2001) *Exp. Anim.* **50**, 153. **16.** Kawata, Takayama, Ninomiya & Makisumi (1980) *J. Biochem.* **88**, 1601. **17.** Colleluori & Ash (2001) *Biochemistry* **40**, 9356. **18.** Patchett, Daniel & Morgan (1991) *Biochim. Biophys. Acta* **1077**, 291. **19.** Ramakrishna & Adiga (1975) *Eur. J. Biochem.* **59**, 377. **20.** Boeker & Snell (1971) *Meth. Enzymol.* **17B**, 657. **21.** Blethen, Boeker & Snell (1968) *J. Biol. Chem.* **243**, 1671. **22.** Smith (1979) *Phytochemistry* **18**, 1447. **23.** Rosenfeld & Roberts (1976) *J. Bacteriol.* **125**, 601. **24.** Balbo, Patel, Sell, *et al.* (2003) *Biochemistry* **42**, 15189. **25.** Petrack, Sullivan & Ratner (1957) *Arch. Biochem. Biophys.* **69**, 186. **26.** Park, Hirotani, Nakano & Kitaoka (1984) *Agric. Biol. Chem.* **48**, 483. **27.** Knodler, Sekyere, Stewart, Schofield & Edwards (1998) *J. Biol. Chem.* **273**, 4470. **28.** Smith, Ganaway & Fahrney (1978) *J. Biol. Chem.* **253**, 6016. **29.** Pereira, Alonso, Ivaldi, *et al.* (2003) *J. Eukaryot. Microbiol.* **50**, 132. **30.** Pereira, Alonso, Paveto, *et al.* (2000) *J. Biol. Chem.* **275**, 1495. **31.** Nguyen van Thoai & Olomucki (1962) *Biochim. Biophys. Acta* **59**, 545. **32.** Flashner & Massey (1974) in *Mol. Mech. Oxygen Activ.* (Hayaishi, ed.) p. 245, Academic Press, New York. **33.** Charlier & Gerlo (1979) *Biochemistry* **18**, 3171. **34.** Folk (1971) *The Enzymes*, 3rd ed., **3**, 57. **35.** Mitchell & Harrington (1971) *The Enzymes*, 3rd ed., **3**, 699. **36.** da Silva, Parolo, Streck, *et al.* (1999) *Brain Res.* **838**, 78. **37.** Takio (1998) in *Handb. Proteolytic Enzymes*, p. 1538, Academic Press, San Diego. **38.** Nonaka, Ishikawa, Tsumuraya, *et al.* (1995) *J. Biochem.* **118**, 1014. **39.** McDonald, Rouhani, Handlogten, *et al.* (1997) *Biochim. Biophys. Acta* **1324**, 133.

L-Homocysteic Acid

This oxidized L-homocysteine metabolite (FW = 183.18 g/mol; CAS 1001-13-4; Abbreviation: L-HCA) is an L-glutamate analogue that inhibits a number of glutamate-dependent reactions, including the glutamate:cystine transport system. L-HCA occurs naturally in the mammalian CNS and is released from K^+-stimulated brain slices in a calcium-dependent manner. L-HCA is present in nerve terminals located in brain regions having a high density of NMDA receptors. L-HCA also appears to be an endogenous excitotoxic ligand at NMDA receptors. With the hypothesis that a high homocysteic acid level is a brain biomarker in Alzheimer's Disease brain and that increased levels may induce the accumulation of intraneuronal amyloid-β peptides, a recent study demonstrated a neuroprotective effect by anti-homocysteic acid antibodies in 3xTg-AD mouse model (1). Homocysteic acid is also an alternative substrate for folylpoly-γ-glutamate synthetase, and homocysteate is likely to prevent the incorporation of additional glutamate moieties, thereby reducing intracellular folate retention. *Note*: D-Homocysteic acid (CAS 56892-03-6) is likely to be a useful experimental control or reference molecule when probing the stereoselectivity of L-homocysteic acid. **Target(s):** acetylserotonin *O*-methyltransferase, *or* hydroxyindole *O*-methyltransferase (2); glutamate dehydrogenase (3); glutamate synthase (4); γ-glutamyl transpeptidase, *or* γ-glutamyltransferase; inhibited by the DL-mixture, K_i = 11 mM (5); kynurenine:oxoglutarate aminotransferase (6); methionine γ-lyase (7); methylaspartate mutase, *or* glutamate mutase (8); Na^+-dependent L-glutamate:L-cystine antiport exchanger (9); serine-sulfate ammonia-lyase, weakly inhibited by DL-mixture (10); sulfinoalanine decarboxylase, *or* cysteinesulfinate decarboxylase (11); tetrahydrofolate synthetase, chain termination inhibitor (12,13); UDP-*N*-acetylmuramoylalanine:D-glutamate ligase, inhibited by the DL-mixture; also an alternative substrate (14). **1.** Hasegawa, Mikoda, Kitazawa & LaFerla (2010) *PLoS ONE* **5**, e8593. **2.** Karahasanoglu & Ozand (1972) *J. Neurochem.* **19**, 411. **3.** Jolles-Bergeret (1967) *Biochim. Biophys. Acta* **146**, 45. **4.** Meister (1985) *Meth. Enzymol.* **113**, 327. **5.** Lherbet & Keillor (2004) *Org. Biomol. Chem.* **2**, 238. **6.** Kocki, Luchowski, Luchowska, *et al.* (2003) *Neurosci. Lett.* **346**, 97. **7.** Lockwood & Coombs (1991) *Biochem. J.* **279**, 675. **8.** Hartzoulakis & Gani (1994) *Proc. Indian Acad. Sci. Chem. Sci.* **106**, 1165. **9.** Patel, Warren, Rhoderick & Bridges (2004) *Neuropharmacology* **46**, 273. **10.** Tudball & Thomas (1973) *Eur. J. Biochem.* **40**, 25. **11.** Do & Tappaz (1996) *Neurochem. Int.* **28**, 363. **12.** Cichowicz & Shane (1987) *Biochemistry* **26**, 513. **13.** Shane & Cichowicz (1983) *Adv. Exp. Med. Biol.* **163**, 149. **14.** Pratviel-Sosa, Acher, Trigalo, *et al.* (1994) *FEMS Microbiol. Lett.* **115**, 223.

L-Homocysteine

This nonproteogenic thiol-containing L-amino acid (FW = 135.19 g/mol; CAS 6027-13-0; Symbol = Hcy; pK_a are 2.22 (–COOH), 8.87 (–SH), and 10.86 (–NH$_2$)) is a key intermediate in methionine metabolism as well as a proximal precursor in cysteine biosynthesis. Homocysteine is readily oxidized to homocystine. Homocysteine thiolactone forms within a few minutes in hot 20% HCl. **Target(s):** adenosylhomocysteinase (1-5); adenylylsulfatase (6); 1-aminocyclopropane-1 carboxylate synthase (7); 5-aminolevulinate aminotransferase (8,9,33); butyrylcholinesterase (10,11); carbamoyl-phosphate synthetase (12,13); catechol *O* methyltransferase (41); L-3-cyanoalanine synthase (14); cyclooxygenase (15); cystathionine β-lyase (16,17); cystathionine γ-lyase, *or* cysteine desulfhydrase (18); cystathionine γ-synthase (34); cysteine synthase, *or* *O*-acetylserine sulfhydrylase (16); cysteinyl-tRNA synthetase (19,20); cystine lyase (17); gelatinase A, *or* matrix metalloproteinase 2 (21); gelatinase B, *or* matrix metalloproteinase 9 (21); glutathione peroxidase (22,42); histidine ammonia-lyase (23); homoserine *O*-acetyltransferase (35); homoserine kinase (24); K$^+$ channels (25); methionine adenosyltransferase (26); methionyl-tRNA synthetase (27); Na$^+$/K$^+$-exchanging ATPase (10,28); serine *O*-acetyltransferase, weakly inhibited (36-38); L-serine ammonia-lyase, *or* L-serine dehydratase (29,30); *O*-succinylhomoserine (thiol)-lyase (31); thiopurine *S*-methyltransferase (39); tRNA (guanine-*N^2*-)-methyltransferase, weakly inhibited (40); tyrosine decarboxylase (32). **1.** Poulton & Butt (1976) *Arch. Biochem. Biophys.* **172**, 135. **2.** Walker & Duerre (1975) *Can. J. Biochem.* **53**, 312. **3.** Knudsen & Yall (1972) *J. Bacteriol.* **112**, 569. **4.** Fujioka & Takata (1981) *J. Biol. Chem.* **256**, 1631. **5.** Kajander, Eloranta & Raina (1976) *Biochim. Biophys. Acta* **438**, 522. **6.** Li & Schiff (1991) *Biochem. J.* **274**, 355. **7.** Nakajima & Imaseki (1986) *Plant Cell Physiol.* **27**, 969. **8.** Turner & Neuberger (1970) *Meth. Enzymol.* **17A**, 188. **9.** Braunstein (1973) *The Enzymes*, 3rd ed., **9**, 379. **10.** Stefanello, Franzon, Wannmacher, Wajner & Wyse (2003) *Metab. Brain Dis.* **18**, 273. **11.** Stefanello, Zugno, Wannmacher, Wajner & Wyse (2003) *Metab. Brain Dis.* **18**, 187. **12.** Kaseman & Meister (1985) *Meth. Enzymol.* **113**, 305. **13.** Kaseman (1985) Ph.D. Dissertation, Dept. Biochemistry, Cornell University Medical College, New York, New York. **14.** Warrilow & M. J. Hawkesford (2002) *J. Exp. Bot.* **53**, 439. **15.** Quere, Habib, Tobelem & Maclouf (1995) *Adv. Prostaglandin Thromboxane Leukot. Res.* **23**, 397. **16.** Burnell & Whatley (1977) *Biochim. Biophys. Acta* **481**, 246. **17.** Ramirez & Whitaker (1999) *J. Agric. Food Chem.* **47**, 2218. **18.** Fromageot (1951) *The Enzymes*, 1st ed., **1** (Part 2), 1237. **19.** Pan, Yu, Duh & Lee (1976) *J. Chin. Biochem. Soc.* **5**, 45. **20.** Burnell & Whatley (1977) *Biochim. Biophys. Acta* **481**, 266. **21.** Chavarria, Rodriguez-Nieto, Sanchez-Jimenez, Quesada & Medina (2003) *Biochem. Biophys. Res. Commun.* **303**, 572. **22.** Chen, Liu, Greiner & Holtzman (2000) *J. Lab. Clin. Med.* **136**, 58. **23.** Klee (1974) *Biochemistry* **13**, 4501. **24.** Burr, Walker, Truffa-Bachi & Cohen (1976) *Eur. J. Biochem.* **62**, 519. **25.** Shontz, Xu, Patel & Rozanski (2001) *J. Cardiovasc. Electrophysiol.* **12**, 175. **26.** Mudd (1973) *The Enzymes*, 3rd ed., **8**, 121. **27.** Hahn & Brown (1967) *Biochim. Biophys. Acta* **146**, 264. **28.** Streck, Zugno, Tagliari, *et al.* (2002) *Int. J. Dev. Neurosci.* **20**, 77. **29.** Suda & Nakagawa (1971) *Meth. Enzymol.* **17B**, 346. **30.** Pestaña, Sandoval & Sols (1971) *Arch. Biochem. Biophys.* **146**, 373. **31.** Kaplan & Guggenheim (1971) *Meth. Enzymol.* **17B**, 425. **32.** Webb (1966) *Enzyme and Metabolic Inhibitors*, vol. **2**, p. 307, Academic Press, New York. **33.** Neuberger & Turner (1963) *Biochim. Biophys. Acta* **67**, 342. **34.**

Guggenheim & Flavin (1969) *J. Biol. Chem.* **244**, 3722. **35.** Yamagata (1987) *J. Bacteriol.* **169**, 3458. **36.** Burnell & Whatley (1977) *Biochim. Biophys. Acta* **481**, 246. **37.** Smith & Thompson (1971) *Biochim. Biophys. Acta* **227**, 288. **38.** Nozaki, Shigeta, Saito-Nakano, Imada & Kruger (2001) *J. Biol. Chem.* **276**, 6516. **39.** Remy (1963) *J. Biol. Chem.* **238**, 1078. **40.** Taylor & Gantt (1979) *Biochemistry* **18**, 5253. **41.** Dhar & Rosazza (2000) *Appl. Environ. Microbiol.* **66**, 4877. **42.** Durmaz & Dikman (2007) *J. Enzyme Inhib. Med. Chem.* **22**, 733.

Homogentisic Acid

This metabolite (FW$_{free-acid}$ = 168.15 g.mol; CAS 451-13-8), also known as 2,5-dihydroxyphenylacetate, is an intermediate in the degradation of phenylalanine and tyrosine. Note that homogentisate will slowly oxidize to its quinone upon standing, particularly in alkaline pH, and this quinone will polymerize to produce a dark colored polymer. Alcaptonuria is an inborn error in metabolism in which there is a deficiency of homogentisate 1,2-dioxygenase and this dark polymer can be seen in the urine of individuals having this disorder. In addition, homogentisate can readily form the lactone (*i.e.*, 5-hydroxy-2-oxocoumaran). **Target(s):** imidazoleacetate 4-monooxygenase (1); tyrosine aminotransferase (2,3). **1.** Watanabe, Kambe, Imamura, *et al.* (1983) *Anal. Biochem.* **130**, 321. **2.** Ohisalo, Andersson & Pispa (1977) *Biochem. J.* **163**, 411. **3.** Schepartz (1970) *Enzymologia* **39**, 211.

Homoharringtonine

This toxic alkaloid (FW = 545.63 g/mol; CAS 26833-87-4), isolated from *Cephalotaxus hainanensisanti*, is an anticancer agent that has been used to treat acute or chronic myeloid leukemia. Homoharringtonine inhibits protein biosynthesis by blocking chain elongation. (*See also* Harringtonine) **1.** Tujebajeva, Graifer, Matasova, *et al.* (1992) *Biochim. Biophys. Acta* **1129**, 177. **2.** Tujebajeva, Graifer, Karpova & Ajtkhozhina (1989) *FEBS Lett.* **257**, 254. **3.** Fresno, Jimenez & Vazquez (1977) *Eur. J. Biochem.* **72**, 323.

Homoserine

L-homoserine

This γ-hydroxy amino acid (FW = 119.12 g/mol; CAS 672-15-1), a component in the antibiotics tolaasin and syringostatin, is an intermediate in the biosynthesis of L-threonine, L-methionine, and cystathionine. It is also the principal free amino acid in pea plants. Homoserine converts to the lactone upon standing at room temperature under acidic conditions Solutions should always be freshly prepared. **Target(s):** *O*-acetylhomoserine aminocarboxypropyltransferase, *or* *O*-acetylhomoserine sulfhydrylase (1-4); *O*-acetylserine/*O*-acetylhomoserine sulfhydrylase (5); 1-aminocyclopropane-1 carboxylate deaminase (6,7); asparagine:oxo-acid aminotransferase (8,9); aspartate kinase (10); cystathionine γ-lyase (11);

cysteine synthase (12,13); dihydrodipicolinate synthase, weakly inhibited by the L-enantiomer (14); diphosphate:serine phosphotransferase (15); hemoglobin S polymerization (16,17); hydroxylysine kinase (18); kynureninase, weakly inhibited by racemic mixture (19); methionine adenosyltransferase (20); methionine γ-lyase (21); methionyl-tRNA synthetase (22); muramoyltetrapeptide carboxy-peptidase, inhibited by D-homoserine, K_i = 43 mM (23); serine O-acetyltransferase, weakly inhibited (24,25); serine-phosphoethanolamine synthase (26); thetin:homocysteine methyltransferase, inhibited by the racemic mixture (27,28); threonine ammonia-lyase, or threonine dehydratase (29). **1**. Shiio & Ozaki (1987) *Meth. Enzymol.* **143**, 470. **2**. Murooka, Kakihara, Miwa, Seto & Harada (1977) *J. Bacteriol.* **130**, 62. **3**. Ozaki & Shiio (1982) *J. Biochem.* **91**, 1163. **4**. Yamagata (1971) *J. Biochem.* **70**, 1035. **5**. Yamagata (1987) *Meth. Enzymol.* **143**, 478. **6**. Minami, Uchiyama, Murakami, *et al.* (1998) *J. Biochem.* **123**, 1112. **7**. Honma, Shimomura, Shiraishi, Ichihara & Sakamura (1979) *Agric. Biol. Chem.* **43**, 1677. **8**. Maul & Schuster (1986) *Arch. Biochem. Biophys.* **251**, 585. **9**. Cooper (1977) *J. Biol. Chem.* **252**, 2032. **10**. Keng & Viola (1996) *Arch. Biochem. Biophys.* **335**, 73. **11**. Flavin & Segal (1964) *J. Biol. Chem.* **239**, 2220. **12**. Ascano & Nicholas (1977) *Phytochemistry* **16**, 889. **13**. Tamura, Iwasawa, Masada & Fukushima (1976) *Agric. Biol. Chem.* **40**, 637. **14**. Mazelis, Watley & Whatley (1977) *FEBS Lett.* **84**, 236. **15**. Cagen & Friedmann (1972) *J. Biol. Chem.* **247**, 3382. **16**. Kang & Benjamin (1975) *Exp. Mol. Pathol.* **22**, 220. **17**. Rumen (1975) *Blood* **45**, 45. **18**. Hiles & Henderson (1972) *J. Biol. Chem.* **247**, 646. **19**. Jakoby & Bonner (1953) *J. Biol. Chem.* **205**, 709. **20**. Mudd (1973) *The Enzymes*, 3rd ed., **8**, 121. **21**. Lockwood & Coombs (1991) *Biochem. J.* **279**, 675. **22**. Hahn & Brown (1967) *Biochim. Biophys. Acta* **146**, 264. **23**. Metz, Henning & Hammes (1986) *Arch. Microbiol.* **144**, 181. **24**. Burnell & Whatley (1977) *Biochim. Biophys. Acta* **481**, 246. **25**. Nozaki, Shigeta, Saito-Nakano, Imada & Kruger (2001) *J. Biol. Chem.* **276**, 6516. **26**. Allen & Rosenberg (1968) *Biochim. Biophys. Acta* **151**, 504. **27**. Webb (1966) *Enzyme and Metabolic Inhibitors*, vol. **2**, p. 353, Academic Press, New York. **28**. Fromm & Nordlie (1959) *Arch. Biochem. Biophys.* **81**, 363. **29**. Choi & Kim (1995) *J. Biochem. Mol. Biol.* **28**, 118.

trans-HR22C16

This cell-permeable, non-tubulin-interacting mitotic inhibitor (FW = 389.46 g/mol; CAS 462630-41-7) blocks mitosis (IC$_{50}$ = 800 nM) by binding to Eg5, a mitotic kinesin motor protein that plays an essential role in centrosome separation and formation of bipolar mitotic spindles. HR22C16 inhibits cell survival in both Taxol-sensitive and -resistant ovarian cancer cells with at least 15-fold greater efficacy than monastrol, the first generation Eg5 inhibitor (2). **1**. Hotha, Yarrow, Yang, *et al.* (2003) *Angew. Chem. Int. Ed, Engl.* **42**, 2379. **2**. Marcus, Peters, Thomas, *et al.* (2005) *J. Biol. Chem.* **280**, 11569.

HS-173

This potent PI3Kα inhibitor (FW = 422.26 g /mol; CAS 1276110-06-5; Solubility: 80 µg/mL DMSO, with warming; <1 µg/mL H$_2$O), also named ethyl-6-(5-(phenylsulfonamido)pyridin-3-yl)imidazo[1,2-*a*]pyridine-3-carboxylate, targets phosphatidyl-inositol 3-kinase α (IC$_{50}$ = 0.8 nM), a cell proliferation, growth, and survival signaling kinase is hyperactive in a wide variety of human cancers. HS-173 induces cell cycle arrest at the G$_2$/M phase as well as apoptosis (1). It also decreases the expression HIF-1α and VEGF, key factors in angiogenesis, evidenced by suppression of capillary tube formation and migration *in vitro* (1). Combined treatment of Panc-1 cells with HS-173 and Sorafenib synergistically inhibits viability, again with G$_2$/M arrest and apoptosis (2). **1**. Lee, Jung, Jeong, Hong & Hong (2013) *Cancer Lett.* **328**, 152. **2**. Yun, *et al.* (2013) *Cancer Lett.* **331**, 250. **3**. Son, *et al.* (2013) *Sci. Rep.* **3**, 3470.

HTH 01-015

This NUAK1 inhibitor (FW = 468.55 g/mol; CAS 1613724-42-7; Solubility: 100 mM in DMSO), systematically named 5,13-dihydro-4,5,13-trimethyl-2-[[1-(4-piperidinyl)-1*H*-pyrazol-4-yl]-amino]-6*H*-naphtho[2,3-*e*]pyrimido[5,4-*b*][1,4]diazepin-6-one, targets the NUAK family SNF1-like kinase-1, a serine/threonine-protein kinase regulating cell adhesion, ploidy, senescence, proliferation and Myc-driven tumor progression. HTH-01-015 inhibits only NUAK1 (IC$_{50}$ = 100 nM), displaying extreme selectivity and weak action against 139 other kinases that, including ten other AMP-activated protein kinase (AMPK) family members. In all cell lines tested, HTH-01-015 inhibits phosphorylation of the only well-characterized substrate, MYPT1 (myosin phosphate-targeting subunit 1), which is phosphorylated at Ser-445 by NUAK1. Attesting to its specificity, HTH-01-015 impairs the invasive potential of U2OS cells in a three-diensional cell invasion assay to the same extent as NUAK1 knockdown. **1**. Banerjee, Buhrlage, Huang, *et al.* (2004) *Biochem. J.* **457**, 215.

(±)-Huperzine A

This sesquiterpene alkaloid (FW = 242.32 g/mol; CAS 120786-18-7), also known as (±)-selagine, from the fir moss *Huperzia serrata*, inhibits acetylcholinesterase, with the (−)-stereoisomer more active (IC$_{50}$ = 47 nM). Huperzine A is also a NMDA receptor antagonist that readily crosses the blood-brain barrier. HupA also exerts effects on nicotinic acetylcholine receptors (mainly α$_7$ nAChRs and α$_4$β$_2$ nAChRs), evoking anti-inflammatory responses by decreasing IL-1β, TNF-α protein expression, and suppressing transcriptional activation of NF-κB signaling. HupA is classified as a dietary supplement in the U.S. **Target(s)**: acetylcholinesterase, both enantiomers inhibit (1-6); aryl-acylamidase (3); cholinesterase, butyrylcholinesterase, very weakly inhibited (1,3-5). **1**. Ashani, Peggins & Doctor (1992) *Biochem. Biophys. Res. Commun.* **184**, 719. **2**. Dvir, Jiang, Wong, *et al.* (2002) *Biochemistry* **41**, 10810. **3**. Costagli & Galli (1998) *Biochem. Pharmacol.* **55**, 1733. **4**. Luo, Yu, Zhan, *et al.* (2005) *J. Med. Chem.* **48**, 986. **5**. Giacobini (2003) *Neurochem. Res.* **28**, 515. **6**. Shi, Tai, McCammon, Taylor & Johnson (2003) *J. Biol. Chem.* **278**, 30905.

Huwentoxin XVI

CIGEGVPCDENDPRCCSGLVCLKPTLHGIWYKSYYCYKK

This EF-hand protein and potential analgesic agent (FW = 4437.13 g/mol; CAS 1600543-88-1; Disulfide Bridges: 1-16,8-21,15-36; Solubility: 1 mg/mL H$_2$O), one of many related polypeptide toxins from the poisonous venom of the Chinese bird spider, *or* tarantula, *Ornithoctonus huwena* (1), is a potent and reversible N-type Ca^{2+} channel blocker that, in rat dorsal root ganglion cells, gives an IC$_{50}$ ~60 nM (2). N-type calcium ion channels play important CNS roles in neurotransmission and pain signaling. Huwentoxin XVI does not block T-type Ca^{2+} channels, K$^+$ channels or Na$^+$ channels. **1.** Jiang, Zhang, Zhang, *et al.* (2010) *Comp. Biochem. Physiol. (Part D: Genomics Proteomics)* **5**, 81. **2.** Deng, Luo, Xiao, *et al.* (2014) *Neuropharmacol.* **79**, 657.

HWA-448, See *Torbafylline*

HY-14721, See *EMD-1214063*

HY-15185, See *PF-3084014*

Hyaluronic Acid

This unbranched, nonsulfated glycosaminoglycan (MW$_{average}$ ≈ 7 x 10^6 Da in synovial fluid of healthy subjects; MW$_{average}$ ≈ 4.8 x 10^6 in synovial fluid of rheumatoid arthritis patients; CAS 9004-61-9; *Symbol*: HA) is a viscous, extensively hydrated anionic polymer of disaccharide units. Each of the latter consists of an *N*-acetyl-D-glucosamine linked β1→4 to D-glucuronate, which is, in turn, linked β$_{1→3}$ to the next disaccharide unit. Found chiefly in the vitreous humor, umbilical cord, synovial fluid, and loose connective tissue in all vertebrates, most hyaluronic acid preparations also contain 1-2% of tightly associated protein. HA's distinct physicochemical properties commend its application as a viscoelastic and biocompatible substance for use as a joint lubricant and anti-inflammatory agent. Hyaluronic acid can also be cross-linked with 1,4-butanediol diglycidyl ether to form a nonabsorbable supramolecular substance suitable for use as a subdermal filler in cosmetic surgery. The broadly distributed cell surface protein known as CD44 is a likely cell adhesion receptor that binds to hyaluronate with high affinity. **Target(s):** *N*-acetylgalactosamine-6-sulfatase (1); acylphosphatase (2); cathepsin B (3); chondroitinase (4); β-fructofuranosidase (5,6); β-glucuronidase (7); γ-glutamyl hydrolase, weakly inhibited (8); histone acetyltransferase (9); lysozyme (5,10,11); phospholipase A$_2$ (12); ribonuclease (5,13); tissue kallikrein (14). **1.** Lim & Horwitz (1981) *Biochim. Biophys. Acta* **657**, 344. **2.** Koshland (1955) *Meth. Enzymol.* **2**, 555. **3.** Hayasaka, Shiono, Hara & Fukuyo (1983) *Invest. Ophthalmol. Vis. Sci.* **24**, 682. **4.** Thurston (1975) *Biochem. J.* **145**, 397. **5.** Webb (1966) *Enzyme and Metabolic Inhibitors*, vol. **2**, Academic Press, New York. **6.** Becker & Friedenwald (1949) *Arch. Biochem.* **22**, 101. **7.** Levvy & Marsh (1960) *The Enzymes*, 2nd ed., **4**, 397. **8.** Silink, Reddel, Bethel & Rowe (1975) *J. Biol. Chem.* **250**, 5982. **9.** Buczek-Thomas, Hsia, Rich, Foster & Nugent (2008) *J. Cell. Biochem.* **105**, 108. **10.** Skarnes & Watson (1955) *J. Bacteriol.* **70**, 110. **11.** Wang, Murao & Arai (1991) *Agric. Biol. Chem.* **55**, 1401. **12.** Yang, Farooqui & Horrocks (1996) *Adv. Exp. Med. Biol.* **416**, 309. **13.** Houck (1957) *Biochim. Biophys. Acta* **26**, 649. **14.** Forteza. Lauredo, Abraham & Conner (1999) *Amer. J. Respir. Cell Mol. Biol.* **21**, 666.

Hycanthone

This acid-sensitive thioxanthenone (FW = 356.49 g/mol; CAS 3105-97-3; Abbreviation: HC) is a widely used drug for schistosomiasis. Hycanthone also intercalate into DNA, preferring AT sequences (predominantly composed of alternating A and T residues), and act as a mutagen. **Primary Mode of Action:** HC exerts its antischistosomal activity by undergoing metabolic activation to a reactive ester which thenalkylates DNA. Resistant schistosomes are unaffected because they cannot convert HC to a reactive ester. **Target(s):** Apurinic endonuclease-1 (APE1), IC$_{50}$ = 80 nM, by binding to the enzyme, *not* DNA (1); DNase (2); RNA, DNA, and protein biosynthesis (3). **1.** Naidu, Agarwal, Pena, *et al.* (2011) *PLoS ONE* **6**. e23679. **2.** Bailly & Waring (1993) *Biochemistry* **32**, 5985. **3.** Pica-Mattoccia & Cioli (1985) *Amer. J. Trop. Med. Hyg.* **34**, 112.

Hydantocidin

This potent, nonselective phytotoxin and pro-herbicide (FW = 218.16 g/mol; CAS 130607-26-0), systematically named (2*S*)-4,5-dihydro-3α,4α-dihydroxy-5β-(hydroxymethyl)spiro[furan-2(3*H*),4'-imidazolidine]-2',5'-dione and (4*S*)-3'α,4'α-dihydroxy-5'β-(hydroxymethyl)-4',5'-dihydrospiro[imidazolidine-4,2'(3'*H*)-furan]-2,5-dione, first isolated from a submerged culture of *Streptomyces hygroscopicus* SANK 63584 (1), causes immediate cessation of plant growth, meristematic necrosis, and plant death in annuals as well as perennials, including monocotyledonous and dicotyledonous weeds. In *Arabidopsis thaliana*, adenine supplementation alleviates hydantocidin-induced symptoms, but guanine does not (2). In *A. thaliana* plants growing on agar plates, herbicidal effects were also reversed when the agar medium is supplemented with AMP, but not IMP or GMP, suggesting that hydantocidin blocks one or both of the two enzymes that convert IMP to AMP in the *de novo* purine biosynthesis pathway. Hydantocidin itself does not inhibit adenylosuccinate synthetase or adenylosuccinate lyase from *Zea mays*; however, *N*-acetyl-5'-phosphohydantocidin, a phosphorylated derivative of hydantocidin, is a potent inhibitor of the synthetase, but not the lyase. Such findings identify the site of action of hydantocidin and establish adenylosuccinate synthetase as a pro-herbicide target. With *Zea mays* adenylosuccinate synthetase, 5'-phosphohydantocidin is a time-dependent competitive inhibitor with respect to IMP (K_i = 22 nM), suggesting it bind to the IMP site as a tightly held dead-end complex (4). *See Hydantocidin 5'-Monophosphate* **1.** Nakajima, Itoi, Takamatsu, *et al.* (1991) *J. Antibiot. (Tokyo)* **44**, 293. **2.** Heim, Cseke, Gerwick, Murdoch & Green (1995) *Pesticide Biochem. Physiol.* **53**, 138. **3.** Siehl, Subramanian, Walters, *et al.* (1996) *Plant Physiol.* **110**, 753. **4.** Walters, Lee, Niderman, *et al.* (1997) *Plant Physiol.* **114**, 549. **(+)-**

Hydantocidin 5'-Monophosphate

This spironucleotide (FW$_{free-acid}$ = 298.15 g/mol), also known as 5'-phosphohydantocidin and HMP, is derived by metabolic phosphorylation of its pro-herbicide (*See Hydantocidin*). HMP inhibits adenylosuccinate synthetase, thereby blocking purine nucleotide biosynthesis (1-4). **1.** Fonné-Pfister, Chemla, Ward, *et al.* (1996) *Proc. Natl. Acad. Sci. U.S.A.* **93**, 9431. **2.** Poland, Lee, Subramanian, *et al.* (1996) *Biochemistry* **35**, 15753. **3.** Walters, Lee, Niderman, *et al.* (1997) *Plant Physiol.* **114**, 549. **4.** Siehl, Subramanian, Walters, *et al.* (1996) *Plant Physiol.* **110**, 753.

2-Hydrazino-3-(4-imidazolyl)propionic Acid

This histidine analogue competitively inhibits histidine ammonia-lyase and histidine decarboxylase. (*See also Histidine Hydrazide*) **Target(s):** histidine ammonia-lyase, inactivated by the D-stereomer; competitively inhibited by the L-isomer (1-3); histidine decarboxylase (4,5); histidyl-tRNA synthetase (6). **1.** Brand & Harper (1976) *Biochemistry* **15**, 1814. **2.** Allen, Clark & Phipps (1983) *Biochem. Soc. Trans.* **11**, 350. **3.** Brand & Harper (1976) *Arch. Biochem. Biophys.* **177**, 123. **4.** Snell & Guirard (1986) *Meth. Enzymol.* **122**, 139. **5.** Tanase, Guirard & Snell (1985) *J. Biol. Chem.* **260**, 6738. **6.** Lepore, di Natale, Guarini & de Lorenzo (1975) *Eur. J. Biochem.* **56**, 369.

Hydrochlorothiazide

This diuretic and most commonly prescribed antihypertensive (FW = 297.64 g/mol; CAS 58-93-5; *Abbreviations*: HCT, HCTZ, or HZT), also known as 6-chloro-1,1-dioxo-3,4-dihydro-2*H*-1,2,4-benzothiadiazine-7-sulfonamide is widely used in the treatment of congestive heart failure, diabetes insipidus, hypertension, renal tubular acidosis, and symptomatic edema (1,2). **Primary Mode of Action:** Hydrochlorothiazide reduces reabsorption of sodium and chloride ions at the beginning of the distal convoluted tubule, resulting in increased excretion of water and electrolytes (*e.g.*, sodium, potassium, chloride, and magnesium). It also exerts a natriuretic effect, mainly by decreasing sodium and chloride reabsorption in the cortical segment of the ascending limb of the loop of Henle by inhibition of a specific Na^+/Cl^- co-transporter. The antihypertensive effects of hydrochlorothiazide are still poorly understood. **Key Pharmacokinetic Parameters:** *See* Appendix II in Goodman & Gilman's *THE PHARMACOLOGICAL BASIS OF THERAPEUTICS*, 12th Edition (Brunton, Chabner & Knollmann, eds.) McGraw-Hill Medical, New York (2011). **Target(s):** Coenzyme Q_{10}-NADH-oxidase (3); cyclic 3',5'-AMP-dependent protein kinase (4); bone resorption in men despite experimentally elevated serum 1,25-dihydroxyvitamin D concentrations (5). **1.** Richterich (1959) *Klin Wochenschr.* **37**, 355. **2.** Kennedy & Crawford (1959) *Lancet* **1**, 66. **3.** Kishi, Kishi & Folkers (1975) *Res. Commun. Chem. Pathol. Pharmacol.* **12**, 533. **4.** Ferguson & Twite (1973) *Br. J. Pharmacol.* **49**, 288. **5.** Lemann, Gray, Maierhofer & Cheung (1985) *Kidney Int.* **28**, 951.

Hydrocinnamic Acid

This naturally occurring acid ($FW_{\text{free-acid}}$ = 150.18 g/mol; M.P. = 47-48°C), also known as 3-phenylpropionic acid and β-phenylpropionic acid, inhibits a surprising range of enzymes. **Target(s):** D-amino-acid oxidase (1,2); aromatic-amino-acid aminotransferase, K_i = 3.9 mM (3,57); carboxypeptidase A (1,4-21); carboxypeptidase B (14,22); carboxypeptidase C, *or* carboxypeptidase Y (23-26); catechol oxidase (27); chymotrypsin (1,6,28-37); elastase, pancreatic (38); enoate reductase (39); esterase (40); hemoglobin S polymerization (41); kynurenine aminotransferase, weakly inhibited (42); melilotate 3-monooxygenase (43); nicotinamidase (44); nonenzymatic hydroxylaminolysis of amino acid esters at –18°C (45); peptidyl-dipeptidase A, angiotensin I converting enzyme (46); phenylalanine dehydrogenase (47,48); phenylalanine 2-monooxygenase (49); phenylalanine racemase, *or* ATP-hydrolyzing (50); phenylalanyl-tRNA synthetase, weakly inhibited, Ki = 9.7 mM (51);

prephenate dehydratase, weakly inhibited (52); subtilisin (53); succinate dehydrogenase (54,55); tryptophan 2-monooxygenase (58); β-ureidopropionase, IC_{50} = 5.6 μM (56). **1.** Webb (1966) *Enzyme and Metabolic Inhibitors*, vol. 2, Academic Press, New York. **2.** Klein & Austin (1953) *J. Biol. Chem.* **205**, 725. **3.** Okamoto, Nakai, Hayashi, Hirotsu & Kagamiyama (1998) *J. Mol. Biol.* **280**, 443. **4.** Elkins-Kaufman, Neurath & De Maria (1949) *J. Biol. Chem.* **178**, 645. **5.** Pétra (1970) *Meth. Enzymol.* **19**, 460. **6.** Smith (1951) *The Enzymes*, 1st ed., **1** (part 2), 793. **7.** Neurath (1960) *The Enzymes*, 2nd ed., **4**, 11. **8.** Hartsuck & Lipscomb (1971) *The Enzymes*, 3rd ed., **3**, 1. **9.** Coleman & Vallee (1964) *Biochemistry* **3**, 1874. **10.** Quiocho & Richards (1966) *Biochemistry* **5**, 4062. **11.** Martinelli, Hanson, Thompson, *et al.* (1989) *Biochemistry* **28**, 2251. **12.** Riordan, Sokolovsky & Vallee (1967) *Biochemistry* **6**, 3609. **13.** Auld & Vallee (1970) *Biochemistry* **9**, 602. **14.** Gates & Travis (1973) *Biochemistry* **12**, 1867. **15.** Ludwig (1973) *Inorg. Biochem.* **1**, 438. **16.** Bodwell & Meyer (1981) *Biochemistry* **20**, 2767. **17.** Lacko & Neurath (1970) *Biochemistry* **9**, 4680. **18.** Spilburg, Bethune & Vallee (1977) *Biochemistry* **16**, 1142. **19.** Oshima & Nagasawa (1977) *J. Biochem.* **81**, 1285. **20.** Bradley, Naude, Muramoto, Yamauchi & Oeloefsen (1994) *Int. J. Biochem.* **26**, 555. **21.** Gettins (1986) *J. Biol. Chem.* **261**, 15513. **22.** Wintersberger, Cox & Neurath (1962) *Biochemistry* **1**, 1069. **23.** Hayashi (1976) *Meth. Enzymol.* **45**, 568. **24.** Bai & Hayashi (1979) *J. Biol. Chem.* **254**, 8473. **25.** Jonson & Aswad (1990) *Biochemistry* **29**, 4373. **26.** Bai, Hayashi & Hata (1975) *J. Biochem.* **77**, 81. **27.** Aerts & Vercauteren (1964) *Enzymologia* **28**, 1. **28.** Shaw (1970) *The Enzymes*, 3rd ed., **1**, 91. **29.** Wallace, Kurtz & Niemann (1963) *Biochemistry* **2**, 824. **30.** Kaufman & Neurath (1949) *J. Biol. Chem.* **180**, 181. **31.** Neurath & Gladner (1951) *J. Biol. Chem.* **188**, 407. **32.** Huang & Niemann (1952) *J. Amer. Chem. Soc.* **74**, 5963. **33.** Prahl & Neurath (1966) *Biochemistry* **5**, 2131. **34.** Wildnauer & Canady (1966) *Biochemistry* **5**, 2885. **35.** Kaufman & Neurath (1949) *J. Biol. Chem.* **181**, 623. **36.** Neurath & Gladner (1951) *J. Biol. Chem.* **188**, 407. **37.** Shiao (1970) *Biochemistry* **9**, 1083. **38.** Ardelt (1974) *Biochim. Biophys. Acta* **341**, 318. **39.** Buhler & Simon (1982) *Hoppe Seyler's Z. Physiol. Chem.* **363**, 609. **40.** Weber & King (1935) *J. Biol. Chem.* **108**, 131. **41.** Noguchi & Schechter (1978) *Biochemistry* **17**, 5455. **42.** Mason (1959) *J. Biol. Chem.* **234**, 2770. **43.** Husain, Schopfer & Massey (1978) *Meth. Enzymol.* **53**, 543. **44.** Albizati & Hedrick (1972) *Biochemistry* **11**, 1508. **45.** Grant & Alburn (1965) *Biochemistry* **4**, 1913. **46.** Oshima & Nagasawa (1977) *J. Biochem.* **81**, 57. **47.** Brunhuber, Thoden, Blanchard & Vanhooke (2000) *Biochemistry* **39**, 9174. **48.** Vanhooke, Thoden, Brunhuber, *et al.* (1999) *Biochemistry* **38**, 2326. **49.** Koyama (1984) *J. Biochem.* **96**, 421. **50.** Vater & Kleinkauf (1976) *Biochim. Biophys. Acta* **429**, 1062. **51.** Santi & Danenberg (1971) *Biochemistry* **10**, 4813. **52.** Dopheide, Crewther & Davidson (1972) *J. Biol. Chem.* **247**, 4447. **53.** Ottesen & Svendsen (1970) *Meth. Enzymol.* **19**, 199. **54.** Quastel & Woolbridge (1928) *Biochem. J.* **22**, 689. **55.** Massart (1950) *The Enzymes*, 1st ed., (Part 1), 307. **56.** Walsh, Green, Larrinua & Schmitzer (2001) *Plant Physiol.* **125**, 1001. **57.** Okamoto, Ishii, Hirotsu & Kagamiyama (1999) *Biochemistry* **38**, 1176. **58.** Emanuele, Heasley & Fitzpatrick (1995) *Arch. Biochem. Biophys.* **316**, 241.

Hydrogen Cyanide (Prussic Acid), *See Cyanide*

Hydrogen Peroxide

This common laboratory reagent and redox-active signaling metabolite (FW = 34.015 g/mol; CAS 7722-84-1) is a strong oxidant that reacts with cysteinyl and methionyl residues in proteins. In the presence of certain organic acids, ethers, or metal ions, hydrogen peroxide also attacks cystinyl, tryptophanyl, and tyrosyl residues. At 6×10^{-4} M, H_2O_2 causes a 90% inhibition of CO_2-fixation in isolated chloroplasts, but can be reversed by addition of catalase (2500 Units/mL) or DTT (10 mM). **Physical & Chemical Properties:** H_2O_2 is a nonplanar molecule with (twisted) C_2 symmetry, having a dihedral angle of 90.2° (compare to 104.5° for water). It is miscible in water as well as ethanol and diethyl ether. Commonly sold as a 3-90% (weight/volume) solution, preferably in a dark glass bottle, aqueous hydrogen peroxide decomposes with time, its the rate increased by metal ions, alkali, and even agitation. In dilute solutions, hydrogen peroxide is more acidic than water. H_2O_2 is also an extremely powerful oxidizer (*e.g.*, +1.8 V for the H_2O_2/H_2O couple). Through catalysis, H_2O_2 is converted into hydroxyl radical •OH, which is extremely reactive. **Enzymatic Reactions Involving Hydrogen Peroxide:** Catalase catalyzes the reaction: $2 H_2O_2 \rightarrow 2 H_2O + O_2$. Peroxidases (EC number 1.11.1.x) comprise a large family of enzymes that typically catalyze a reaction of the form: ROOR' + Electron donor (2 e⁻) + 2H⁺ → ROH + R'OH. D- and L-Amino oxidases also produce H₂O₂. **Target(s):** acetylacetone-cleaving enzyme (151,152); acetylcholinesterase (1-3); acetyl-CoA carboxylase (4); acetyl-CoA hydrolase (5); N-acetylneuraminate lyase (6); acid phosphatase (7-9); aconitase (aconitate hydratase) (10-14); adenosine nucleosidase (15); adenylate kinase (16,17); ADP ribosyl-[dinitrogen reductase] hydrolase (18); yeast alcohol dehydrogenase (19); aldehyde dehydrogenase (20); 1-alkylglycero-phosphocholine O-acetyltransferase (there are also reports of activation) (124); aminopeptidase I (21); β-amylase, weakly inhibited (22); arachidonyl-CoA synthetase, or arachidonate:CoA ligase (23); arylacetonitrilase (24); arylamine N-acetyltransferase (126-128); asclepain (25); L-ascorbate oxidase (26,27,176); aureusidin synthase, hydrogen peroxide required for the oxidation of 2',4,4',6'-tetrahydroxychalcone, but inhibits the oxidation of 2',3,4,4',6' pentahydroxychalcone (28,29); bacterial leucyl aminopeptidase (30); betaine:homocysteine S-methyltransferase (129); biphenyl-2,3-diol 1,2-dioxygenase (155); bromelain (25); γ-butyrobetaine dioxygenase (146,147); calcineurin (31); calotropain, *Calotropis gigantea* latex protease (32); carbamoyl phosphate synthetase (33,34); 3-carboxyethylcatechol 2,3-dioxygenase (160); catechol 1,2 dioxygenase (174); catechol 2,3-dioxygenase (35,172,173); cathepsin B (36); chlorite O₂-lyase (153); choline O-acetyltransferase (125); chymopapain (37,38); citrate (re)-synthase (123); clavaminate synthase (144); clostripain (39-41); complement system, guinea pig (42); cyclohexyl-isocyanide hydratase, ior isonitrile hydratase (43); CYP102 (138); cytochrome c₃ hydrogenase (175); cytochrome d complex (44); cytosine deaminase (Fe²⁺-enzyme inhibited (45); deoxyribo-dipyrimidine photo-lyase (46); 3,4 dihydroxyphenylacetate 2,3-dioxygenase (161-163); 2,5-dihydroxypyridine 5,6-dioxygenase (168,169); ficain, ior ficin (47); β-fructofuranosidase, or invertase (38,48); fructose-1,6-bisphosphate aldolase (49); galactose oxidase (50); gallate decarboxylase (51); gentisate 1,2-dioxygenase (52,171); glucose oxidase (38,53); glucose-6-phosphate dehydrogenase (49); glutathione S-transferase (117); glyceraldehyde-3-phosphate dehydrogenase (38,54); glycerol-3-phosphate dehydrogenase (55); glycerol-3-phosphate oxidase (56); glycerophosphocholine cholinephosphodiesterase (57); glyceryl-ether monooxygenase (136); histidine ammonia-lyase (38,58); histidine kinase (59); HIV-2 retropepsin (60,61); hydrogen lyase (62); 3-(hydroxyamino)phenol mutase (63); hydroxylamine oxidoreductase (64); 3-hydroxy-4-oxoquinoline 2,4-dioxygenase (154); 4-hydroxyphenylpyruvate dioxygenase (156-158); 4-hydroxyquinoline 3-monooxygenase (139); indoleamine 2,3-dioxygenase (150,166,167); inositol oxygenase (148,149); isocitrate lyase (65); β-lactamase (penicillinase) (66); lipoxygenase (164,165); lysozyme (67,68); maleate isomerase (69,70); 3-mercaptopyruvate sulfurtransferase (71); methionine S-adenosyltransferase (118,119); 2-methylcitrate synthase (122); methylisocitrate lyase (72); S-methyl-5'-thioadenosine phosphorylase (120); microbial collagenase (73); myosin ATPase (74-78); naphthalene 1,2 dioxygenase (143); nitric-oxide synthase (140); nitrile hydratase (79); pancreatic ribonuclease, ior ribonuclease A (80); pantetheine hydrolase (81); papain (82-84); pectin lyase (85); peptide-methionine (R)-S-oxide reductase (177); peptide-methionine (S)-S-oxide reductase (177,178); phenol 2-monooxygenase (141,142); phenylalanine 4-monooxygenase (86); 1-phosphatidylinositol-4-phosphate 5-kinase (87); phosphofructo-kinase (88,89); 6-phosphofructo-2-kinase/fructose-2,6-bisphosphatase (90); 6-phosphogluconate dehydratase (91); phospholipase D (92); phosphoribulokinase (116); phylloquinone monooxygenase (2,3-epoxidizing) (133); procollagen-proline 4-dioxygenase (145); protein phosphatase, or phosphoprotein phosphatase (93-95); protein phosphatase 2A (93); protein phosphatase 7 (95); protein-tyrosine phosphatase (96,97); protocatechuate 4,5-dioxygenase (170); pyruvate kinase (49); reticuline oxidase (98); ribonucleoside-diphosphate reductase (130); ribulose-bisphosphate carboxylase/oxygenase (99); selenide,water dikinase (100); selenophosphate synthetase (101); soluble epoxide hydrolase (102); spleen acid deoxyribonuclease (103); squalene monooxygenase, or squalene epoxidase (134); subtilisin (104); succinate dehydrogenase (105); sulfur dioxygenase (159); thiol oxidase (179,180); thiosulfate sulfurtransferase, or rhodanese (106); thymidylate synthase (107); tryptophan 2,3-dioxygenase (108); tryptophan 5-monooxygenase (137); tyrosinase, or monophenol monooxygenase (135); tyrosine 3-monooxygenase (109); urease (110); uricase, or urate oxidase (111,112); very-long-chain fatty acid α-hydroxylase (113); xanthine oxidase (114,115,131,132); xyloglucan:xyloglucosyl transferase (121). **1.** O'Malley, Mengel, Meriwether & Zirkle (1966) *Biochemistry* **5**, 40. **2.** Schallreuter, Elwary, Gibbons, Rokos & Wood (2004) *Biochem. Biophys. Res. Commun.* **315**, 502. **3.** Molochkina, Zorina, Fatkullina, Goloschapov & Burlakova (2005)

Chem. Biol. Interact. **157-158**, 401. **4.** Kozaki & Sasaki (1999) *Biochem. J.* **339**, 541. **5.** Nakanishi, Isohashi, Matsunaga & Sakamoto (1985) *Eur. J. Biochem.* **152**, 337. **6.** Uchida, Tsukada & Sugimori (1984) *J. Biochem.* **96**, 507. **7.** Schlosnagle, Sander, Bazer & Roberts (1976) *J. Biol. Chem.* **251**, 4680. **8.** Ketcham, Baumbach, Bazer & Roberts (1985) *J. Biol. Chem.* **260**, 5768. **9.** Vogel, Börchers, Marcus, *et al.* (2002) *Arch. Biochem. Biophys.* **401**, 164. **10.** Glusker (1971) *The Enzymes*, 3rd ed., **5**, 413. **11.** Bruchmann (1961) *Biochem. Z.* **335**, 199. **12.** Bruchmann (1961) *Naturwissenschaften* **48**, 79. **13.** Eprintsev, Semenova & Popov (2002) *Biochemistry (Moscow)* **67**, 795. **14.** Bulteau, Ikeda-Saito & Szweda (2003) *Biochemistry* **42**, 14846. **15.** Agarwala, Krishna Murti & Shrivastava (1954) *Enzymologia* **16**, 322. **16.** Colowick (1955) *Meth. Enzymol.* **2**, 598. **17.** Colowick & Kalckar (1943) *J. Biol. Chem.* **148**, 117. **18.** Nielsen, Bao, Roberts & Ludden (1994) *Biochem. J.* **302**, 801. **19.** Sund & Theorell (1963) *The Enzymes*, 2nd ed., **7**, 25. **20.** Duine (1990) *Meth. Enzymol.* **188**, 327. **21.** Röhm (1985) *Eur. J. Biochem.* **146**, 633. **22.** Hyun & Zeikus (1985) *Appl. Environ. Microbiol.* **49**, 1162. **23.** Hornberger & Patscheke (1990) *Eur. J. Biochem.* **187**, 175. **24.** Nagasawa, Mauger & Yamada (1990) *Eur. J. Biochem.* **194**, 765. **25.** Greenberg & Winnick (1940) *J. Biol. Chem.* **135**, 761. **26.** Stark & Dawson (1963) *The Enzymes*, 2nd ed., **8**, 297. **27.** Tokuyama & Dawson (1962) *Biochim. Biophys. Acta* **56**, 427. **28.** Nakayama, Yonekura-Sakakibara, Sato, *et al.* (2000) *Science* **290**, 1163. **29.** Sato, Nakayama, Kikuchi, *et al.* (2001) *Plant Sci.* **160**, 229. **30.** Chi, Chou, Wang, *et al.* (2004) *Antonie Leeuwenhoek* **86**, 355. **31.** Bogumil & Ullrich (2002) *Meth. Enzymol.* **348**, 271. **32.** Bose & Krishna (1958) *Enzymologia* **19**, 186. **33.** Kaseman & Meister (1985) *Meth. Enzymol.* **113**, 305. **34.** Kaseman (1985) Ph.D. Dissertation, Dept. Biochemistry, Cornell University Medical College, New York, New York. **35.** Milo, Duffner & Muller (1999) *Extremophiles* **3**, 185. **36.** Perona & Vallejo (1982) *Eur. J. Biochem.* **124**, 357. **37.** Kunimitsu & Yasunobu (1970) *Meth. Enzymol.* **19**, 244. **38.** Neumann (1972) *Meth. Enzymol.* **25**, 393. **39.** Mitchell & Harrington (1970) *Meth. Enzymol.* **19**, 635. **40.** Mitchell & Harrington (1971) *The Enzymes*, 3rd. ed., **3**, 699. **41.** Mitchell & Harrington (1968) *J. Biol. Chem.* **243**, 4683. **42.** Ecker, Pillemer, Martiensen & Wertheimer (1938) *J. Biol. Chem.* **123**, 351. **43.** Goda, Hashimoto, Shimizu & Kobayashi (2001) *J. Biol. Chem.* **276**, 23480. **44.** Anraku & Gennis (1987) *Trends Biochem. Sci.* **12**, 262. **45.** Porter & Austin (1993) *J. Biol. Chem.* **268**, 24005. **46.** Werbin (1977) *Photochem. Photobiol.* **26**, 675. **47.** Sugiura & Sasaki (1974) *Biochim. Biophys. Acta* **350**, 38. **48.** Sizer (1942) *J. Gen. Physiol.* **25**, 399. **49.** Palop, Rutherford & Marquis (1998) *Can. J. Microbiol.* **44**, 465. **50.** Malmström, Andréasson & Reinhammar (1975) *The Enzymes*, 3rd ed., **12**, 507. **51.** Zeida, Wieser, Yoshida, Sugio & Nagasawa (1998) *Appl. Environ. Microbiol.* **64**, 4743. **52.** Harpel & Lipscomb (1990) *Meth. Enzymol.* **188**, 101. **53.** Kleppe (1966) *Biochemistry* **5**, 139. **54.** Little & O'Brien (1969) *Eur. J. Biochem.* **10**, 533. **55.** Grant & Sargent (1961) *Biochem. J.* **81**, 206. **56.** Grant & Sargent (1960) *Biochem. J.* **76**, 229. **57.** Sok (1998) *J. Neurochem.* **70**, 1167. **58.** Shibatani, Kakimoto & Chibata (1975) *Eur. J. Biochem.* **55**, 263. **59.** Viaud, Fillinger, Liu, *et al.* (2006) *Mol. Plant Microbe Interact.* **19**, 1042. **60.** Davis, Newcomb, Moskovitz, *et al.* (2000) *Biochem. J.* **346**, 305. **61.** Davis, Newcomb, Moskovitz, *et al.* (2002) *Meth. Enzymol.* **348**, 249. **62.** Crewther (1953) *Australian J. Biol. Sci.* **6**, 205. **63.** Schenzle, Lenke, Spain & Knackmuss (1999) *J. Bacteriol.* **181**, 1444. **64.** Hooper & Terry (1977) *Biochemistry* **16**, 455. **65.** Rua, Soler, Busto & de Arriaga (2002) *Fungal Genet. Biol.* **35**, 223. **66.** Henry & Housewright (1947) *J. Biol. Chem.* **167**, 559. **67.** Imoto, Johnson, North, Phillips & Rupley (1972) *The Enzymes*, 3rd ed., **7**, 665. **68.** Meyer, Thompson, Palmer & Khorazo (1936) *J. Biol. Chem.* **113**, 303. **69.** Hatakeyama, Asai, Uchida, *et al.* (1997) *Biochem. Biophys. Res. Commun.* **239**, 74. **70.** Hatakeyama, Goto, Kobayashi, Terasawa & Yukawa (2000) *Biosci. Biotechnol. Biochem.* **64**, 1477. **71.** Nagahara & Katayama (2005) *J. Biol. Chem.* **280**, 34569. **72.** Tabuchi & Satoh (1977) *Agric. Biol. Chem.* **41**, 169. **73.** Seifter & Harper (1971) *The Enzymes*, 3rd ed., **3**, 649. **74.** Perry (1955) *Meth. Enzymol.* **2**, 582. **75.** Mehl (1944) *Science* **99**, 518. **76.** Ziff (1944) *J. Biol. Chem.* **153**, 25. **77.** Bailey & Perry (1947) *Biochim. Biophys. Acta* **1**, 506. **78.** Rainford, Hotta & Morales (1964) *Biochemistry* **3**, 1213. **79.** Nagasawa, Nanba, Ryuno, Takeuchi & Yamada (1987) *Eur. J. Biochem.* **162**, 691. **80.** Neumann, Moore & Stein (1962) *Biochemistry* **1**, 68. **81.** Ricci, Nardini, Chiaraluce, Dupre & Cavallini (1986) *Biochim. Biophys. Acta* **870**, 82. **82.** Smith (1951) *The Enzymes*, 1st ed., **1** (Part 2), 793. **83.** Bersin & Logemann (1933) *Z. Physiol. Chem.* **220**, 209. **84.** Purr (1935) *Biochem. J.* **29**, 5. **85.** Jong-Chon, Kim & Choi (1998) *J. Microbiol. Biotechnol.* **8**, 353. **86.** Shiman (1987) *Meth. Enzymol.* **142**, 17. **87.** Halstead, van Rheenen, Snel, *et al.* (2006) *Curr. Biol.* **16**, 1850. **88.** Lardy (1962) *The Enzymes*, 2nd ed., **6**, 67. **89.** Bloxham & Lardy (1973) *The Enzymes*, 3rd ed., **8**, 239. **90.**

Pilkis, Claus, Kountz & El-Maghrabi (1987) *The Enzymes*, 3rd ed., **18**, 3. **91**. Gardner & Fridovich (1991) *J. Biol. Chem.* **266**, 1478. **92**. Tang, Waksman, Ely & Liscovitch (2002) *Eur. J. Biochem.* **269**, 3821. **93**. Rao & Clayton (2002) *Biochem. Biophys. Res. Commun.* **293**, 610. **94**. Bargoni, Fossa & Sisini (1963) *Enzymologia* **26**, 65. **95**. Andreeva, Solov'eva, Kakuev & Kutuzov (2001) *Arch. Biochem. Biophys.* **396**, 65. **96**. Denu & Tanner (2002) *Meth. Enzymol.* **348**, 297. **97**. Raugei, Ramponi & Chiarugi (2002) *Cell. Mol. Life Sci.* **59**, 941. **98**. Steffens, Nagakura & Zenk (1985) *Phytochemistry* **24**, 2577. **99**. Marcus, Altman-Gueta, Finkler & Gurevitz (2003) *J. Bacteriol.* **185**, 1509. **100**. Lacourciere & Stadtman (1999) *Proc. Natl. Acad. Sci. U.S.A.* **96**, 44. **101**. Glass & Stadtman (1995) *Meth. Enzymol.* **252**, 309. **102**. Morisseau, Archelas, Guitton, D*et al.*(1999) *Eur. J. Biochem.* **263**, 386. **103**. Bernardi (1971) *The Enzymes*, 3rd ed., **4**, 271. **104**. Markland & Smith (1971) *The Enzymes*, 3rd ed., **3**, 561. **105**. Massart, Vandendreissche & Dufait (1940) *Enzymologia* **8**, 204. **106**. Nandi, Horowitz & Westley (2000) *Int. J. Biochem. Cell Biol.* **32**, 465. **107**. Dunlap (1978) *Meth. Enzymol.* **51**, 90. **108**. Knox & Mehler (1950) *J. Biol. Chem.* **187**, 419. **109**. Shiman & S. Kaufman (1970) *Meth. Enzymol.* **17A**, 609. **110**. Sumner (1951) *The Enzymes*, 1st ed., **1** (Part 2), 873. **111**. Holmberg (1939) *Biochem. J.* **33**, 1901. **112**. Truszkowski (1932) *Biochem. J.* **26**, 285. **113**. Kishimoto (1978) *Meth. Enzymol.* **52**, 310. **114**. Bray (1975) *The Enzymes*, 3rd ed., **12**, 299. **115**. Bernheim & Dixon (1928) *Biochem. J.* **22**, 113. **116**. Koyabashi, Tamoi, Iwaki, Shigeoka & Wadano (2003) *Plant Cell Physiol.* **44**, 269. **117**. Letelier, Martinez, Gonzalez-Lira, Faundez & Aracena-Parks (2006) *Chem. Biol. Interact.* **164**, 39. **118**. Avila, Corrales, Ruiz, *et al.* (1998) *Biofactors* **8**, 27. **119**. Corrales, Pérez-Mato, Sánchez del Pino, *et al.* (2002) *J. Nutr.* **132**, 2377S. **120**. Fernández-Irigoyen, Santamaría, Sánchez-Quiles, *et al.* (2008) *Biochem. J.* **411**, 457. **121**. Fry (1997) *Plant J.* **11**, 1141. **122**. Uchiyama & Tabuchi (1976) *Agric. Biol. Chem.* **40**, 1411. **123**. Goschalk & Dittbrenner (1970) *Hoppe-Seyler's Z. Physiol. Chem.* **351**, 1183. **124**. Hurst & Bazan (1997) *J. Ocul. Pharmacol. Ther.* **13**, 415. **125**. Zambrzycka, Alberghina & Strosznajder (2002) *Neurochem. Res.* **27**, 277. **126**. Makarova (2008) *Curr. Drug Metab.* **9**, 538. **127**. Dupret, Dairou, Atmane & Rodrigues-Lima (2005) *Meth. Enzymol.* **400**, 215. **128**. Sim, Westwood & Fullam (2007) *Expert. Opin. Drug Metab. Toxicol.* **3**, 169. **129**. Miller, Szegedi & Garrow (2005) *Biochem. J.* **392**, 443. **130**. Sato, Bacon & Cory (1984) *Adv. Enzyme Regul.* **22**, 231. **131**. De Renzo (1956) *Adv. Enzymol. Relat. Subj. Biochem.* **17**, 293. **132**. Machida & Nakanishi (1981) *Agric. Biol. Chem.* **45**, 425. **133**. Larson & Suttie (1978) *Proc. Natl. Acad. Sci. U.S.A.* **75**, 5413. **134**. M'Baya & Karst (1987) *Biochem. Biophys. Res. Commun.* **147**, 556. **135**. Kim & Uyama (2005) *Cell. Mol. Life Sci.* **62**, 1707. **136**. Rock, Baker, Fitzgerald & Snyder (1976) *Biochim. Biophys. Acta* **450**, 469. **137**. Friedman, Kappelman & Kaufman (1972) *J. Biol. Chem.* **247**, 4165. **138**. Eiben, Kaysser, Maurer, *et al.* (2006) *J. Biotech.* **124**, 662. **139**. Block & Lingens (1992) *Biol. Chem. Hoppe-Seyler* **373**, 249. **140**. Demady, Jianmongkol, Vuletich, Bender & Osawa (2001) *Mol. Pharmacol.* **59**, 24. **141**. Neujahr & Gaal (1973) *Eur. J. Biochem.* **35**, 386. **142**. Cadieux, Vrajmasu, Achim, Powlowski & Münck (2002) *Biochemistry* **41**, 10680. **143**. Lee (1999) *J. Bacteriol.* **181**, 2719. **144**. Salowe, Marsh & Townsend (1990) *Biochemistry* **29**, 6499. **145**. Koivu & Myllylä (1986) *Biochemistry* **25**, 5982. **146**. Blanchard, Englard & Kondo (1982) *Arch. Biochem. Biophys.* **219**, 327. **147**. Lindstedt (1967) *Biochemistry* **6**, 1271. **148**. Reddy, Pierzchala & Hamilton (1981) *J. Biol. Chem.* **256**, 8519. **149**. Naber, Swan & Hamilton (1986) *Biochemistry* **25**, 7201. **150**. Thomas & Stocker (1999) *Redox Rep.* **4**, 199. **151**. Lietgeb, Straganz & Nidetzky (2009) *Biochem. J.* **418**, 403. **152**. Straganz, Glieder, Brecker, Ribbons & Steiner (2003) *Biochem. J.* **369**, 573. **153**. Mehboob, Wolternink, Vermeulen, *et al.* (2009) *FEMS Microbiol. Lett.* **293**, 115. **154**. Bauer, Max, Fetzner & Lingens (1996) *Eur. J. Biochem.* **240**, 576. **155**. Yang, Xie, Zhang, Shi & Qian (2008) *Biochimie* **90**, 1530. **156**. Lindblad, Lindstedt, Lindstedt & Rundgren (1977) *J. Biol. Chem.* **252**, 5073. **157**. Lindstedt & Rundgren (1982) *J. Biol. Chem.* **257**, 11922. **158**. Lindstedt, Odelhög & Rundgren (1977) *Biochemistry* **16**, 3369. **159**. Suzuki & Silver (1966) *Biochim. Biophys. Acta* **122**, 22. **160**. Bugg (1993) *Biochim. Biophys. Acta* **1202**, 258. **161**. Gabello, Ferrer, Martín & Garrido-Pertierra (1994) *Biochem. J.* **301**, 145. **162**. Emerson, Kovaleva, Farquhar, Lipscomb & Que (2008) *Proc. Natl. Acad. Sci. U.S.A.* **105**, 7347. **163**. Ono-Kamimoto (1973) *J. Biochem.* **74**, 1049. **164**. Yamamoto, Fujii, Yasumoto & Mitsuda (1980) *Agric. Biol. Chem.* **44**, 443. **165**. Fornaroli, Petrussa, Braidot, Vianello & Macri (1999) *Plant Sci.* **145**, 1. **166**. Ferry, Ubeaud, Lambert, *et al.* (2005) *Biochem. J.* **388**, 205. **167**. Shimizu, Nomiyama, Hirata & Hayaishi (1978) *J. Biol. Chem.* **253**, 4700. **168**. Nozaki (1974) in *Mol. Mech. Oxygen Activ.* (Hayaishi, ed.), p. 135, Academic Press, New York. **169**. Gauthier & Rittenberg (1971) *J. Biol. Chem.* **246**, 3737. **170**. Yun,

Yun & Kim (2004) *J. Microbiol.* **42**, 152. **171**. Suárez, Ferrer & Martín (1996) *FEMS Microbiol. Lett.* **143**, 89. **172**. Viggiani, Siani, Notomista, *et al.* (2004) *J. Biol. Chem.* **279**, 48630. **173**. Chae, Kim, Bini & Zylstra (2007) *Biochem. Biophys Res. Commun.* **357**, 815. **174**. Gledraityte & Kalediene (2009) *Cent. Eur. J. Biochem.* **4**, 68. **175**. Yagi, Kimura, Daidoji, Sakai & Tamura (1976) *J. Biochem.* **79**, 661. **176**. White & Krupka (1965) *Arch. Biochem. Biophys.* **110**, 448. **177**. Schallreuter, Rübsam, Gibbons, *et al.* (2008) *J. Invest. Dermatol.* **128**, 808. **178**. Le, Chaffotte, Demey-Thomas, *et al.* (2009) *J. Mol. Biol.* **393**, 58. **179**. Lash & Jones (1983) *Arch. Biochem. Biophys.* **225**, 344. **180**. Lash & Jones (1986) *Arch. Biochem. Biophys.* **247**, 120.

Hydrogen Sulfide

This highly reactive metabolic signalling gas (FW = 34.08 g/mol; CAS 7783-06-4; M.P. = −85.5 °C; B.P. = −60.3 °C; pK_1 = 7.04, and pK_2 = 11.9) is formed endogenously and enzymatically (akin to the so-called gasotransmitters nitric oxide and carbon monoxide, reaching appreciable concentrations in brain, where it is thought to act as a neuromodulator and transmitter. H_2S also relaxes vascular smooth muscle cells, primes angiogenesis, promotes wound healing, inhibits platelet aggregation, and regulates *N*-methyl D-aspartate (NMDA) receptors in brain. The mechanisms of H_2S-mediated vasodilation include the activation of K_{ATP} channels, a variety of other channels, inhibition of phosphodiesterases, and synergy with NO (1) By stimulating K_{ATP} channels, H_2S causes vasorelaxation within a physiological concentration range. H_2S interacts with specific Cysteine-6 and Cysteine-26 on extracellular loop of SUR subunit of K_{ATP} channel complex in a novel posttranslational modification mechanism, i.e., protein *S*-sulfhydration (2). **H_2S Formation in Cells:** Hydrogen sulfide is synthesized in two pyridoxal phosphate (*or* PLP) requiring, enzyme-catalyzed reactions: (a) Cystathionine β-Synthase (*Major Reaction*: L-Serine + L-Homocysteine ⇌ L-Cystathionine + H_2O; *Minor H_2S-Producing Reaction*: L-Cysteine + L-Homocysteine ⇌ L-Cystathionine + H_2S), and (b) Cystathionine γ-Lyase (*Major Reaction*: L-Serine + L-Cystathionine ⇌ L-Cysteine + 2-Oxobutanoate + NH_3; (a) *Minor H_2S-Producing Elimination Reaction*: L-Cysteine ⇌ Pyruvate + NH_3 + H_2S; (b) *Minor H_2S-Producing Elimination Reaction*: L-Cystine ⇌ L-Cysteine + Pyruvate + H_2S). (The first enzyme requires heme, which may serve a regulatory role by interacting with NO and/or CO.) H_2S is involved in the regulation of vascular tone, myocardial contractility, neurotransmission, and insulin secretion. In vascular smooth muscle cells, neurons, cardiomyocytes and pancreatic β-cells, H_2S stimulates ATP-sensitive potassium channels (K_{ATP}) In addition, H_2S may react with reactive oxygen species (ROS) and/or reactive nitrogen species to limiting their toxic effects. Note that propargylglycine inhibits cystathionine γ-lyase-catalyzed formation H_2S from L-cysteine. **Generation of H_2S for Metabolic Studies:** In the laboratory, H_2S may be produced in a modified Kipp generator (3). Unless one uses a suitable chemical trap, it is advisable to avoid using such generators, which contain potentially toxic substances. Instead, one may directly administer 0.03–0.3 mM sodium hydrosulfide (Formula: NaHS; FW = 56.06 g/mol; CAS 16721-80-5), which is highly soluble in water. (**See also** *Sulfide Ions; Sulfur*) Nucleoside phosphorothioates, such as adenosine 5'-thiomonophosphate (*or* AMPS) and guanosine 5'-thiomonophosphate (*or* GMPS), are also converted to H_2S, allowing the latter to relax isolated kidney glomeruli *in vitro* and increasing glomerular filtration rate *in vivo* (4). AMPS and GMPS can be safely used as H_2S donors in experimental studies and possibly as H_2S-releasing drugs. **Slow-Releasing H_2S Donor:** Another method for generating H_2S relies on the use of GYY-4137 (FW = 377.48 g/mol; CAS 106740-09-4; IUPAC: 4-methoxyphenyl(morpholino) phosphinodithioate), a water-soluble H_2S donor with vasodilatory and anti-inflammatory properties (5,6). GYY4137 slowly releases appreciable levels of H_2S in the range of 10-250 μM, and the H_2S that is released is long-lived. **Toxicity:** Despite its recently discovered physiologic roles, H_2S gas is highly toxic, about 10-30x less so than hydrogen cyanide. Its distinctive odor of rotten eggs (perceptible over the 0.02-to-0.13 ppm range) provides some warning; however, at higher levels, olfactory fatigue occurs, making H_2S both inperceptible and hazardous. **Target(s):** D-amino acid oxidase (7,8); ascorbate oxidase (9); carbonic anhydrase (10-15); catalase (16-19); catechol oxidase (20); galactose oxidase (21); lactoperoxidase (22); maleylacetoacetate isomerase (23); thyroxine deiodinase, iodothyronine deiodinase (24); and tyrosinase (25). Sodium hydrosulfide (100 µM) attenuates β-amyloid (residues 1-42) (Aβ42) production in SH-SY5Y cells stimulated by 10-200 nM of the selective A_{2A} receptor agonist HENECA (25). NaHS also interferes with HENECA-stimulated production and post-translational modification of amyloid precursor protein (APP) by inhibiting its maturation. Measurement of the C-terminal APP fragments generated from its enzymatic cleavage by β-site amyloid precursor protein cleaving enzyme 1 (BACE1) showed that NaHS is without any significant effect on β-secretase activity (25). **1**. Wang (2012) *Physiol. Rev.* **92**, 791. **2**. Cohen (1915) *J. Am. Chem. Soc.* **37**, 14. **3**. **4**. Bełtowski, *et al.* (2014) *Pharmacol. Res.* **81**, 34. **5**. Li, Whiteman, Guan, *et al.* (2008) *Circulation* **117**, 2351. **6**. Li, Salto-Tellez, Tan, *et al.* (2009*) Free Radic. Biol. Med.* **47**, 103. **7**. Burton (1955) *Meth. Enzymol.* **2**, 199. **8**. Krebs (1951) *The Enzymes*, 1st ed., **2** (part 1), 499. **9**. Stark & Dawson (1963) *The Enzymes*, 2nd ed., **8**, 297. **10**. Keilin & Mann (1940) *Biochem. J.* **34**, 1163. **11**. Lindskog, Henderson, Kannan, *et al.* (1971) *The Enzymes*, 3rd. ed., **5**, 587. **12**. Schwimmer (1969) *Enzymologia* **37**, 163. **13**. Vullo, Franchi, Gallori, *et al.* (2003) *J. Enzyme Inhib. Med. Chem.* **18**, 403. **14**. Mangani & Hakansson (1992) *Eur. J. Biochem.* **210**, 867. **15**. Pocker & Stone (1968) *Biochemistry* **7**, 2936 and 3021. **16**. Nicholls & Schonbaum (1963) *The Enzymes*, 2nd ed., **8**, 147. **17**. Harris & Creighton (1915) *J. Biol. Chem.* **22**, 535. **18**. Senter (1903) *Zeit. für physiol. Chem.* **44**, 257. **19**. Malowan (1938) *Enzymologia* **5**, 89. **20**. Dawson & Magee (1955) *Meth. Enzymol.* **2**, 817. **21**. Malmström, Andréasson & Reinhammar (1975) *The Enzymes*, 3rd ed., (Boyer, ed.) **12**, 507. **22**. Elliott (1932) *Biochem. J.* **26**, 10. **10**. Seltzer (1972) *The Enzymes*, 3rd ed., **6**, 381. **23**. Nakagawa & Ruegamer (1967) *Biochemistry* **6**, 1249. **24**. Dawson & Tarpley (1951) *The Enzymes*, 1st ed. (Sumner & Myrbäck, eds.), **2** (part 1), 454. **25**. Nagpure & Bian (2014) *PLoS One* **9**, e88508.

(6*E*,8*Z*,11*Z*,14*Z*)-(5*S*)-5-Hydroperoxyeicosa-6,8,11,14-tetraenoic Acid

This unsaturated δ-hydroperoxy acid (FW$_{free-acid}$ = 336.47 g/mol), often abbreviated 5-HPETE, is an intermediate in arachidonate metabolism and a precursor in leukotriene and hydroxy-eicosatetraenoate biosynthesis. It is a product of the reaction catalyzed by 5-lipoxygenase (Note that 5-HPETE is also a strong product inhibitor of the 5-lipoxygenase.). Note that it is unstable and should be stored under an inert atmosphere. **Target(s)**: 5-lipoxygenase, *or* arachidonate 5-lipoxygenase (1-3); lipoxygenase, potato (4). **1**. Kim, Kim & Sok (1989) *Biochem. Biophys. Res. Commun.* **164**, 1384. **2**. Wiseman, Skoog, Nichols & Harrison (1987) *Biochemistry* **26**, 5684. **3**. Mirzoeva, Sud'ina, Pushkareva & Varfolomeev (1995) *FEBS Lett.* **377**, 306. **4**. Aharony, Redkar-Brown, Hubbs & Stein (1987) *Prostaglandins* **33**, 85.

(5*Z*,8*Z*,10*E*,14*Z*)-(12*S*)-12-Hydroperoxyeicosa-5,8,10,14-tetraenoic Acid

This unsaturated λ-hydroperoxy acid (FW$_{free-acid}$ = 336.47 g/mol), often abbreviated 12-HPETE, is an intermediate in arachidonate metabolism and a precursor in leukotriene and hydroxy-eicosatetraenoate biosynthesis.

Target(s): Ca^{2+}/calmodulin-dependent protein kinase II (1); prostacyclin synthase, *or* prostaglandin-I synthase (2); renin release (3); thromboxane-A synthase (4). **1**. Piomelli, Wang, Sihra, *et al.* (1989) *Proc. Natl. Acad. Sci. U.S.A.* **86**, 8550. **2**. Ham, Egan, Soderman, Gale & Kuehl (1979) *J. Biol. Chem.* **254**, 2191. **3**. Antonipillai (1990) *Proc. Soc. Exp. Biol. Med.* **194**, 224. **4**. Hammarstrom & Falardeau (1977) *Proc. Natl. Acad. Sci. U.S.A.* **74**, 3691.

(5*Z*,8*Z*,11*Z*,13*E*)-(15*S*)-15-Hydroperoxyeicosa-5,8,11,13-tetraenoic Acid

This unsaturated ξ–hydroperoxy acid (FW$_{free-acid}$ = 336.47 g/mol), often abbreviated 15-HPETE, is an intermediate in arachidonate metabolism and a precursor in leukotriene and hydroxy-eicosatetraenoate biosynthesis. **Target(s)**: 5-lipoxygenase, *or* arachidonate 5-lipoxygenase (1); 12-lipoxygenase, *or* arachidonate 12-lipoxygenase (2,3); lipoxygenase, potato (4); lipoxygenase, soybean (5,6); prostaglandin-D synthase, *or* prostaglandin H:prostaglandin D isomerase (7); prostaglandin-I synthase, *or* prostacyclin synthase (8-11); thromboxane-A synthase (12). **1**. Cashman, Lambert & Sigal (1988) *Biochem. Biophys. Res. Commun.* **155**, 38. **2**. Kühn, Heydeck, Brinckman & Trebus (1999) *Lipids* **34**, S273. **3**. Kishimoto, Nakamura, Suzuki, *et al.* (1996) *Biochim. Biophys. Acta* **1300**, 56. **4**. Kim, Kim & Sok (1989) *Biochem. Biophys. Res. Commun.* **164**, 1384. **5**. Kim & Sok (1991) *Arch. Biochem. Biophys.* **288**, 270. **6**. Huang, Kim & Sok (2008) *J. Agric. Food Chem.* **56**, 7808. **7**. Christ-Hazalhof & Nugteren (1982) *Meth. Enzymol.* **86**, 77. **8**. Salmon, Smith, Flower, Moncada & Vane (1978) *Biochim. Biophys. Acta* **523**, 250. **9**. Ham, Egan, Soderman, Gale & Kuehl (1979) *J. Biol. Chem.* **254**, 2191. **10**. Terashita, Nishikawa, Terao, Nakagawa & Hino (1979) *Biochem. Biophys. Res. Commun.* **91**, 72. **11**. Salmon & Flower (1982) *Meth. Enzymol.* **86**, 91. **12**. Shen & Tai (1986) *J. Biol. Chem.* **261**, 11592.

(9*Z*,11*E*)-(13*S*)-13-Hydroperoxyoctadeca-9,11-dienoic Acid

This unsaturated µ-hydroperoxy acid (FW$_{free-acid}$ = 312.45 g/mol), often abbreviated 13-HPODE, is produced by the action of lipoxygenase on linoleate. It induces moderate oxidative stress in cells and participates in the trapping of low-density lipoproteins in arterial subendothelial spaces. 13-HPODE also induces apoptosis and augments the epidermal growth factor receptor signaling pathway by attenuating receptor dephosphorylation. In addition, 13-HPODE inhibits prostacyclin biosynthesis. **Target(s)**: glucosamine synthetase (1); 15-hydroxyprostaglandin dehydrogenase (2); 15-lipoxygenase, *or* arachidonate 15-lipoxygenase (3); palmitoyl-CoA hydrolase, *or* acyl-CoA hydrolase (4); prostacyclin synthase, *or* prostaglandin I synthase (5); prostaglandin-I synthase, prostacyclin synthase (6-8). **1**. Fujita, Sakuma, Fudemoto, *et al.* (1997) *Res. Commun. Mol. Pathol. Pharmacol.* **96**, 209. **2**. Sakuma, Fujimoto, Miyata, *et al.* (1994) *Prostaglandins Leukot. Essent. Fatty Acids* **51**, 425. **3**. Kühn, Heydeck, Brinckman & Trebus (1999) *Lipids* **34**, S273. **4**. Sakuma, Fujimoto, Yoshimura, *et al.* (1999) *IUBMB Life* **48**, 539. **5**. Salmon, Smith, Flower, Moncada & Vane (1978) *Biochim. Biophys. Acta* **523**, 250. **6**. Pace-Asciak & Smith (1983) *The Enzymes*, 3rd ed., **16**, 543. **7**. DeWitt & Smith (1983) *J. Biol. Chem.* **258**, 3285. **8**. Smith, DeWitt & Day (1983) *Adv. Prostaglandin Thromboxane Leukotri. Res.* **11**, 87.

Hydroxocobalamin

This vitamin B_{12} derivative ($FW_{free-acid} = 1346.4$ g/mol; CAS 13422-51-0) is the conjugate base of aquacobalamin (pKa = 7.5) and a precursor in the biosynthesis of methylcobalamin. It binds tightly to the transcobalamins and is a nitric oxide scavenger. **Target(s):** ethanolamine ammonia-lyase (1-6); glycerol dehydratase (2,7); HIV-1 integrase (8); β-lysine 5,6-aminomutase (3); D-lysine-5,6-aminomutase (3); 2-methyleneglutarate mutase (9); propanediol dehydratase (2). **1.** Kaplan & Stadtman (1971) *Meth. Enzymol.* **17B**, 818. **2.** Abeles (1971) *The Enzymes*, 3rd ed., **5**, 481. **3.** Stadtman (1972) *The Enzymes*, 3rd ed., **6**, 539. **4.** Babior (1969) *J. Biol. Chem.* **244**, 2917. **5.** Bradbeer (1965) *J. Biol. Chem.* **240**, 4675. **6.** Blackwell & Turner (1978) *Biochem. J.* **175**, 555. **7.** Smiley & Sobolov (1962) *Arch. Biochem. Biophys.* **97**, 538. **8.** Weinberg, Shugars, Sherman, Sauls & Fyfe (1998) *Biochem. Biophys. Res. Commun.* **246**, 393. **9.** Barker (1972) *The Enzymes*, 3rd ed., **6**, 509.

N-Hydroxy-2-acetamidofluorene

This carcinogenic aminofluorene (FW = 387.25 g/mol) inactivates aromatic-hydroxylamine *O*-acetyltransferase and tyrosine-ester sulfotransferase. **Target(s):** aromatic-hydroxylamine *O*-acetyltransferase (1-3); arylamine *N*-acetyltransferase (4); tyrosine-ester sulfotransferase, *or* aryl sulfotransferase IV (5). **1.** Hanna (1994) *Adv. Pharmacol.* **27**, 401. **2.** Marhevka, Ebner, Sehon & Hanna (1985) *J. Med. Chem.* **28**, 18. **3.** Banks & Hanna (1979) *Biochem. Biophys. Res. Commun.* **91**, 1423. **4.** Sticha, Bergstrom, Wagner & Hanna (1998) *Biochem. Pharmacol.* **56**, 47. **5.** Ringer, Yerokun & Khan (1994) *Chem. Biol. Interact.* **92**, 343.

N-[2(R,S)-3-Hydroxyaminocarbonyl-2-benzyl-1-oxopropyl]glycine

This peptide analogue ($FW_{free-acid} = 280.28$ g/mol; Abbreviation: HACBO-Gly) inhibits neprilysin and thermolysin, with K_i values of 1.4 nM and 3.1 μM, respectively. **1.** Roques, Noble, Crine & Fournié-Zaluski (1995) *Meth. Enzymol.* **248**, 263. **2.** Le Moual, Devault, Roques, Crine & Boileau (1991) *J. Biol. Chem.* **266**, 15670. **3.** Waksman, Bouboutou, Devin, *et al.* (1985) *Biochem. Biophys. Res. Commun.* **131**, 262. **4.** Erdös & Skidgel (1989) *FASEB J.* **3**, 145. **5.** Waksman, Hamel, Delay-Goyet & Roques (1986) *EMBO J.* **5**, 3163. **6.** Marie-Claire, Ruffet, Tiraboschi & Fournié-Zaluski (1998) *FEBS Lett.* **438**, 215.

p-Hydroxyamphetamine

This adrenergic agent ($FW_{free-base} = 151.21$ g/mol; CAS 103-86-6), also known as 4-(2-aminopropyl)phenol, inhibits dopamine β-monooxygenase (1) and is a competitive inhibitor of brain monoamine oxidase (2-4). The DL-mixture inhibits *Escherichia coli* tyrosyl-tRNA synthetase, K_i = 10 μM (5). Both the free base and the hydrochloride salt are soluble in water and ethanol. **1.** Goldstein & Contrera (1962) *J. Biol. Chem.* **327**, 1898. **2.** Arai, Kim, Kinemuchi, *et al.* (1991) *Brain Res. Bull.* **27**, 81. **3.** Arai, Kim, Kinemuchi, Tadano, *et al.* (1990) *J. Neurochem.* **55**, 403. **4.** Huang & Eiduson (1977) *J. Biol. Chem.* **252**, 284. **5.** Santi & Peña (1973) *J. Med. Chem.* **16**, 273.

3-Hydroxyanthranilic Acid

This amino acid ($FW_{free-acid} = 153.14$ g/mol; CAS 548-93-6; λ_{max} at 235 and 298 nm (ε = 6400 and 3000 $M^{-1}cm^{-1}$, respectively) in 0.1 N HCl), also known as 2-amino-3-hydroxybenzoic acid, is an intermediate in L-tryptophan degradation. **Target(s):** *o*-aminophenol oxidase (1); anthranilate 3-monooxygenase (deaminating) (2); anthranilate phosphoribosyltransferase (3); kynureninase (4); pyridoxal kinase (5); *o*-pyrocatechuate decarboxylase (6,7); tryptophan 2,3-dioxygenase (8). **1.** Subba Rao & Vaidyanathan (1967) *Arch. Biochem. Biophys.* **118**, 388. **2.** Streeleela, SubbaRao, Premkumar & Vaidyanathan (1969) *J. Biol. Chem.* **244**, 2293. **3.** Bode & Birnbaum (1978) *Z. Allg. Mikrobiol.* **18**, 559. **4.** Tanizawa & Soda (1979) *J. Biochem.* **86**, 499. **5.** Karawya, Mostafa & Osman (1981) *Biochim. Biophys. Acta* **657**, 153. **6.** Subba Rao, Moore & Towers (1967) *Arch. Biochem. Biophys.* **122**, 466. **7.** Kamath, Dasgupta & Vaidyanathan (1987) *Biochem. Biophys. Res. Commun.* **145**, 586. **8.** Hitchcock & Katz (1988) *Arch. Biochem. Biophys.* **261**, 148.

3-Hydroxyaspartic Acid

This hydroxy amino acid (FW = 149.10 g/mol; CAS 71653-06-0; Symbol = Hya), also known as β-hydroxyaspartate, exists as four possible stereoisomeric formss. It is a glutamate uptake inhibitor and a substrate for excitatory amino acid transporters. It has also been identified in a number of proteins (such as in the vitamin K dependent protein S and in bovine factor X) as well as a number of special peptides (for example, in antimycotics and pyoverdins). **Target(s):** argininosuccinate synthetase (the *erythro* form is the inhibitor) (1); asparagine synthetase, mildly inhibited (2,3); aspartate aminotransferase, also a weak substrate (4-9); aspartate 1-decarboxylase (10,11); aspartate 4-decarboxylase, by both the *erythro* and *threo* forms (12-14); glutamate transporter (15-17); glycine:oxaloacetate aminotransferase (the *erythro*-L-form is the inhibitor) (5,18,19). **1**. Ratner (1970) *Meth. Enzymol.* **17A**, 298. **2**. Horowitz & Meister (1972) *J. Biol. Chem.* **247**, 6708. **3**. Lea & Fowden (1975) *Proc. R. Soc. Lond. B Biol. Sci.* **192**, 13. **4**. Sizer & Jenkins (1962) *Meth. Enzymol.* **5**, 677. **5**. Braunstein (1973) *The Enzymes*, 3rd ed., **9**, 379. **6**. Webb (1966) *Enzyme and Metabolic Inhibitors*, vol. 2, p. 355, Academic Press, New York. **7**. Czerlinski & Malkewitz (1965) *Biochemistry* **4**, 1127. **8**. Jenkins (1961) *J. Biol. Chem.* **236**, 1121. **9**. Martinez-Carrion, Tiemeier & Peterson (1970) *Biochemistry* **9**, 2574. **10**. Williamson (1985) *Meth. Enzymol.* **113**, 589. **11**. Williamson & Brown (1979) *J. Biol. Chem.* **254**, 8074. **12**. Miles & Sparrow (1970) *Meth. Enzymol.* **17A**, 689. **13**. Miles & Meister (1967) *Biochemistry* **6**, 1734. **14**. Wilson & Kornberg (1963) *Biochem. J.* **88**, 578. **15**. Sato, Inaba & Maede (1994) *Biochim. Biophys. Acta* **1195**, 211. **16**. Chatton, Shimamoto & Magistretti (2001) *Brain Res.* **893**, 46. **17**. Azzi, Chappell & Robinson (1967) *Biochem. Biophys. Res. Commun.* **29**, 148. **18**. Gibbs & Morris (1970) *Meth. Enzymol.* **17A**, 981. **19**. Gibbs & Morris (1966) *Biochem. J.* **99**, 27P.

4-Hydroxybenzaldehyde

This aromatic aldehyde (FW = 122.12 g/mol; CAS 123-08-0; M.P. = 116°C), also referred to a *p*-hydroxybenzaldehyde and 4-formylphenol, inhibits a number of oxidoreductases. 4-Hydroxybenzaldehyde is widely distributed in plants and is released upon mild oxidation of certain lignins (*e.g.*, monocotyledonous lignins) under alkaline conditions. **Target(s):** 4-aminobutyrate aminotransferase (1,2); *o*-aminophenol oxidase (3); chorismate lyase (4); cyanohydrin β-glucosyltransferase (5); dihydrodiol dehydrogenase (6); diphosphomevalonate decarboxylase (7); dopamine β-monooxygenase, K_i = 2.3 mM (8); β-glucosidase (9); 4-hydroxybenzoate 3-monooxygenase, weakly inhibited (10); 4-hydroxyphenylpyruvate dioxygenase (11); mandelonitrile lyase (12,13); phosphomevalonate kinase, weakly inhibited (7); succinate semialdehyde dehydrogenase (2,14,15); vomilenine glucosyltransferase (16). **1**. Tao, Xu & Yang (2006) *Bioorg. Med. Chem. Lett.* **16**, 3719. **2**. Tao, Yuan, Tang, Xu & Yang (2006) *Bioorg. Med. Chem. Lett.* **16**, 592. **3**. Suzuki, Furusho, Higashi, Ohnishi & Horinouchi (2006) *J. Biol. Chem.* **281**, 824. **4**. Holden, Mayhew, Gallagher & Vilker (2002) *Biochim. Biophys. Acta* **1594**, 160. **5**. Jones, Moller & Hoj (1999) *J. Biol. Chem.* **274**, 35483. **6**. Shinoda, Hara, Nakayama, Deyashiki & Yamaguchi (1992) *J. Biochem.* **112**, 840. **7**. Shama Bhat & Ramasarma (1979) *Biochem. J.* **181**, 143. **8**. Sirimanne, Herman & May (1987) *Biochem. J.* **242**, 227. **9**. Dale, Ensley, Kern, Sastry & Byers (1985) *Biochemistry* **24**, 3530. **10**. Sterjiades (1993) *Biotechnol. Appl. Biochem.* **17**, 77. **11**. Fellman (1987) *Meth. Enzymol.* **142**, 148. **12**. Yemm & Poulton (1986) *Arch. Biochem. Biophys.* **247**, 440. **13**. Xu, Singh & Conn (1986) *Arch. Biochem. Biophys.* **250**, 322. **14**. Albers & Koval (1961) *Biochim. Biophys. Acta* **52**, 29. **15**. Rivett & Tipton (1981) *Eur. J. Biochem.* **117**, 187. **16**. Ruyter & Stoeckigt (1991) *Helv. Chim. Acta* **74**, 1707.

m-Hydroxybenzoic Acid

This aromatic acid (FW$_{free-acid}$ = 138.12 g/mol; CAS 99-06-9), also known as 3-hydroxybenzoic acid, is a substrate for the reactions catalyzed by 3-hydroxybenzoate 4-monooxygenase, 3-hydroxybenzoate 6-monooxygenase, and 3-hydroxy-benzoate 2-monooxygenase. It is an intermediate in the degradation of *m*-cresol. **Target(s):** D-amino-acid oxidase (1,2); anthranilate 3-monooxygenase (deaminating) (3); L-ascorbate peroxidase, weakly inhibited (28); benzoate 4-monooxygenase, weakly (4); γ-butyrobetaine hydroxylase, *or* γ-butyrobetaine:2-oxoglutarate dioxygenase,

K_i = 0.02 mM (5); catechol 2,3-dioxygenase (27); catechol oxidase (29); cathepsin L, weakly inhibited (6); 4,5-dihydroxybenzoate decarboxylase (7); gentisate 1,2-dioxygenase (8); 3-hydroxy-anthranilate 3,4 dioxygenase (9); 4-hydroxybenzoate 4-*O*-β-glucosyltransferase (10); 4-hydroxybenzoate 3-monooxygenase (11,12); 4-hydroxybenzoate 3-monooxygenase (NAD(P)H) (13); 15-lipoxygenase, *or* arachidonate 15-lipoxygenase (14); melilotate 3-monooxygenase (15); polyphenol oxidase (29); protocatechuate 3,4-dioxygenase (16-19); protocatechuate 4,5-dioxygenase (20,21); *o*-pyrocatechuate decarboxylase (22,23); salicylate monooxygenase (24); serine-sulfate ammonia-lyase (25); tannase (26). **1**. Webb (1966) *Enzyme and Metabolic Inhibitors*, vol. 2, p. 341, Academic Press, New York. **2**. Bartlett (1948) *J. Amer. Chem. Soc.* **70**, 1010. **3**. Streeleela, SubbaRao, Premkumar & Vaidyanathan (1969) *J. Biol. Chem.* **244**, 2293. **4**. Reddy & Vaidyanathan (1976) *Arch. Biochem. Biophys.* **177**, 488. **5**. Ng, Hanauske-Abel & Englard (1991) *J. Biol. Chem.* **266**, 1526. **6**. Raghav & Singh (1993) *Indian J. Med. Res.* **98**, 188. **7**. Nakazawa & Hayashi (1978) *Appl. Environ. Microbiol.* **36**, 264. **8**. Harpel & Lipscomb (1990) *Meth. Enzymol.* **188**, 101. **9**. Nishizuka, Ichiyama & Hayaishi (1970) *Meth. Enzymol.* **17A**, 463. **10**. Katsumata, Shige & Ejiri (1989) *Phytochemistry* **28**, 359. **11**. Sterjiades (1993) *Biotechnol. Appl. Biochem.* **17**, 77. **12**. Hosokawa & Stanier (1966) *J. Biol. Chem.* **241**, 2453. **13**. Fujii & Kaneda (1985) *Eur. J. Biochem.* **147**, 97. **14**. Russell, Scobbie, Duthie & Chesson (2008) *Bioorg. Med. Chem.* **16**, 4589. **15**. Husain, Schopfer & Massey (1978) *Meth. Enzymol.* **53**, 543. **16**. Orville, Elango, Lipscomb & Ohlendorf (1997) *Biochemistry* **36**, 10039. **17**. Que & Epstein (1981) *Biochemistry* **20**, 2545. **18**. May, Phillips & Oldham (1978) *Biochemistry* **17**, 1853. **19**. Valley, Brown, Burk, *et al.* (2005) *Biochemistry* **44**, 11024. **20**. Arciero, Orville & Lipscomb (1985) *J. Biol. Chem.* **260**, 14035. **21**. Arciero & Lipscomb (1986) *J. Biol. Chem.* **261**, 2170. **22**. Santha, Rao & Vaidyanathan (1996) *Biochim. Biophys. Acta* **1293**, 191. **23**. Kamath, Dasgupta & Vaidyanathan (1987) *Biochem. Biophys. Res. Commun.* **145**, 586. **24**. Massey & Hemmerich (1975) *The Enzymes*, 3rd ed., **12**, 191. **25**. Tudball & Thomas (1973) *Eur. J. Biochem.* **40**, 25. **26**. Iibuchi, Minoda & Yamada (1972) *Agric. Biol. Chem.* **36**, 1553. **27**. Kobayashi, Ishida, Horiike, *et al.* (1995) *J. Biochem.* **117**, 614. **28**. Durner & Klessig (1995) *Proc. Natl. Acad. Sci. U.S.A.* **92**, 11312. **29**. Kanade, Suhas, Chandra & Gowda (2007) *FEBS J.* **274**, 4177.

p-Hydroxybenzoic Acid

This metabolite (FW$_{free-acid}$ = 138.12 g/mol; CAS 99-96-7; pK_a values = 4.67 and 9.37 at 25°C) is an intermediate in the biosynthesis of ubiquinone in mammals. (*See also p-Chloromercuribenzoate*) **Target(s):** γ-butyrobetaine hydroxylase, γ-butyrobetaine:2-oxoglutarate dioxygenase, K_i = 0.51 mM (1); catechol oxidase (34-36); cathepsin L, weakly inhibited (2); chorismate lyase (3); dihydrofolate synthase (4); dihydropteroate synthase (4); 4,5-dihydroxyphthalate decarboxylase, also alternative substrate (5); diphosphomevalonate decarboxylase (6); estrone sulfotranferase7; β-glucosidase (8); 3-hydroxyanthranilate 3,4-dioxygenase (9); imidazoleacetate 4-monooxygenase (10); 15-lipoxygenase, arachidonate 15-lipoxygenase (11); mandelonitrile lyase (12); phenylalanine ammonia lyase (13,14); phosphomevalonate kinase (6); polyphenol oxidase (34-36); prephenate dehydrogenase (15); procollagen-proline 4-dioxygenase (16,17); protocatechuate 3,4-dioxygenase (18-23); protocatechuate 4,5-dioxygenase (24,25); salicylate monooxygenase (26); shikimate dehydrogenase (27,28); tannase (29); thiopurine *S*-methyltransferase, IC$_{50}$ = 1.7 mM (30); tyrosinase, *or* monophenol monooxygenase, *or* polyphenol oxidase (31-33). **1**. Ng, Hanauske-Abel & Englard (1991) *J. Biol. Chem.* **266**, 1526. **2**. Raghav & Singh (1993) *Indian J. Med. Res.* **98**, 188. **3**. Holden, Mayhew, Gallagher & Vilker (2002) *Biochim. Biophys. Acta* **1594**, 160. **4**. Ortiz (1970) *Biochemistry* **9**, 355. **5**. Nakazawa & Hayashi (1978) *Appl. Environ. Microbiol.* **36**, 264. **8**. Dale, Ensley, Kern, Sastry & Byers (1985) *Biochemistry* **24**, 3530. **7**. Prusakiewicz, Harville, Zhang, Ackermann & Voorman (2007) *Toxicology* **232**, 248. **8**. Shama Bhat & Ramasarma (1979) *Biochem. J.* **181**, 143. **9**. Nishizuka, Ichiyama & Hayaishi (1970) *Meth. Enzymol.* **17A**, 463. **10**. Watanabe, Kambe, Imamura, *et al.* (1983) *Anal. Biochem.* **130**, 321. **11**. Russell, Scobbie, Duthie & Chesson (2008) *Bioorg. Med. Chem.* **16**, 4589. **12**. Yemm & Poulton (1986) *Arch.*

Biochem. Biophys. **247**, 440. **13**. Pridham & S. Woodhead (1974) *Biochem. Soc. Trans.* **2**, 1070. **14**. Jorrin & Dixon (1990) *Plant Physiol.* **92**, 447. **15**. Cotton & Gibson (1970) *Meth. Enzymol.* **17A**, 564. **16**. Majamaa, Günzler, Hanauske-Abel, Myllylä & Kivirikko (1986) *J. Biol. Chem.* **261**, 7819. **17**. Günzler, Hanauske-Abel, Myllylä, Mohr, Kivirikko (1987) *Biochem. J.* **242**, 163. **18**. Orville, Elango, Lipscomb & Ohlendorf (1997) *Biochemistry* **36**, 10039. **19**. Que & Epstein (1981) *Biochemistry* **20**, 2545. **20**. Hou, Lillard & Schwartz (1976) *Biochemistry* **15**, 582. **21**. May, Phillips & Oldham (1978) *Biochemistry* **17**, 1853. **22**. Durham, Sterling, Ornston & Perry (1980) *Biochemistry* **19**, 149. **23**. Valley, Brown, Burk, *et al.* (2005) *Biochemistry* **44**, 11024. **24**. Arciero, Orville & Lipscomb (1985) *J. Biol. Chem.* **260**, 14035. **25**. Arciero & J. D. Lipscomb (1986) *J. Biol. Chem.* **261**, 2170. **26**. Massey & Hemmerich (1975) *The Enzymes*, 3rd ed., **12**, 191. **27**. Balinsky & Dennis (1970) *Meth. Enzymol.* **17A**, 354. **28**. Balinsky & Davies (1961) *Biochem. J.* **80**, 296. **29**. Iibuchi, Minoda & Yamada (1972) *Agric. Biol. Chem.* **36**, 1553. **30**. Ames, Selassie, Woodson, *et al.* (1986) *J. Med. Chem.* **29**, 354. **31**. Webb (1966) *Enzyme and Metabolic Inhibitors*, vol. **2**, p. 298, Academic Press, New York. **32**. McIntyre & Vaughan (1975) *Biochem. J.* **149**, 447. **33**. Gawlik-Dziki, Zlotek & Swieca (2007) *Food Chem.* **107**, 129. **34**. Kanade, Suhas, Chandra & Gowda (2007) *FEBS J.* **274**, 4177. **35**. Wuyts, De Waele & Sweenen (2006) *Plant Physiol. Biochem.* **44**, 308. **36**. Xu, Zheng, Meguro & Kawachi (2004) *J. Wood Sci.* **50**, 260.

2-Hydroxybenzohydroxamoyl Adenylate

This reaction product analogue (FW = 482.35 g/mol) inhibits (2,3-dihydroxybenzoyl)adenylate synthase, $K_{i,app}$ = 37 nM. The free energy of binding is nearly equivalent to the sum of the corresponding values for adenosine 5'-phosphate and 2,3-dihydroxybenzoate. **1**. Callahan, Lomino & Wolfenden (2006) *Bioorg. Med. Chem. Lett.* **16**, 3802.

4-Hydroxybenzonitrile

This toxic aromatic nitrile (FW = 119.12 g/mol; CAS 767-00-0; M.P. = 110-113°C), also known as *p*-hydroxybenzonitrile and *p*-cyanophenol, is a substrate for a number of nitrilases. 4-Hydroxybenzonitrile readily forms the anion (pK_a = 7.9). **Target(s):** L-ascorbate oxidase (1); β-glucosidase (2); malate dehydrogenase, weakly inhibited (3); monoamine oxidase (4); toyocamycin nitrile hydratase (5). **1**. Gaspard, Monzani, Casella, *et al.* (1997) *Biochemistry* **36**, 4852. **2**. Dale, Ensley, Kern, Sastry & Byers (1985) *Biochemistry* **24**, 3530. **3**. Wedding, Hansch & Fukuto (1967) *Arch. Biochem. Biophys.* **121**, 9. **4**. Bright & Porter (1975) *The Enzymes*, 3rd ed. (P. D. Boyer, ed.), **12**, 421. **5**. Uematsu & Suhadolnik (1975) *Meth. Enzymol.* **43**, 759.

(3-Hydroxybenzyl)hydrazine

This hydrazine (FW$_{free-base}$ = 138.17 g/mol; CAS 637-33-2), also known as NSD 1015, inhibits aromatic-L-amino-acid decarboxylase and is a slow-binding inhibitor of γ-aminobutyrate aminotransferase. **Target(s):** 4-aminobutyrate aminotransferase (1,2); aromatic-L-amino-acid decarboxylase, *or* dopa decarboxylase (3-7); 4-hydroxybenzoate decarboxylase (8); monoamine oxidase (6); tyrosine decarboxylase (7). **1**. Lightcap &

Silverman (1996) *J. Med. Chem.* **39**, 686. **2**. Lightcap, Hopkins, Olson & Silverman (1995) *Bioorg. Med. Chem.* **3**, 579. **3**. Nissbrandt, Engberg, Wikstrom, Magnusson & Carlsson (1988) *Naunyn Schmiedebergs Arch. Pharmacol.* **338**, 148. **4**. Goodale & Moore (1976) *Life Sci.* **19**, 701. **5**. Kehr (1975) *Naunyn Schmiedebergs Arch. Pharmacol.* **290**, 347. **6**. Hunter, Rorie & Tyce (1993) *Biochem. Pharmacol.* **45**, 1363. **7**. Nagy & Hiripi (2002) *Neurochem. Int.* **41**, 9. **8**. Gallert & Winter (1992) *Appl. Microbiol. Biotechnol.* **37**, 119.

D-2-Hydroxy-3-butynoic Acid

This acetylenic analogue (FW$_{free-acid}$ = 100.07 g/mol; CAS 38628-65-8) of lactate and α-hydroxybutyrate acts as a suicide inhibitor for a number of enzymes. (*See also* 2-Keto-3-butynoate and 2-Keto-3-butynoic Acid) **Target(s):** D-amino-acid oxidase (1); cytochrome b_2 (2); D-α-hydroxyacid oxidase (1-3); D-lactate dehydrogenase (2-6). **1**. Olson, Massey, Ghisla & Whitfield (1979) *Biochemistry* **18**, 4724. **2**. Walsh, Cromartie, Marcotte & Spencer (1978) *Meth. Enzymol.* **53**, 437. **3**. Maycock (1980) *Meth. Enzymol.* **66**, 294. **4**. Kaczorowski, Kohn & Kaback (1978) *Meth. Enzymol.* **53**, 519. **5**. Sigman & Mooser (1975) *Ann. Rev. Biochem.* **44**, 889. **7**. Ghisla, Olson, Massey & Lhoste (1979) *Biochemistry* **18**, 4733.

3-*R*-Hydroxybutyr-(CH₂)-CoA

This hydroxybutyryl-*S*-coenzyme A analogue (FW = 837.53 g/mol; Symbol = HB-CH₂-CoA), in which the sulfur atom in 3-*R*-hydroxybutyrate coenzyme A (HB-*S*-CoA) is replaced with a CH₂ group, is a competitive inhibitor (K_{is} values of 40 and 14 μM, respectively) of a class I and a class III polyhydroxybutyrate (*or* PHB) synthases, enzymes that catalyze HB-*S*-CoA polymerization to produce polyoxoesters with molecular masses of 1–2 MDa. HB-CH₂-CoA was synthesized in 13 steps by a chemo-enzymatic approach (overall yield = 7.5%). To probe the elongation steps of the polymerization, HBCH₂CoA was incubated with a synthase already acylated with a [³H]-saturated trimer-CoA ([³H]-sTCoA). The reaction products were shown to be the methylene analogue of [³H]-sTCoA ([³H]-sT-CH₂-CoA), saturated dimer-([³H]-sD-CO₂H) and trimer-acid ([³H]-sT-CO₂H), distinct from the expected methylene analogue of [³H]-saturated tetramer-CoA ([³H]-sTet-CH₂-CoA). Detection of [³H]-sT-CH₂-CoA and its slow rate of formation suggest that HB-CH₂-CoA reports on the termination and re-priming process of the synthases, rather than elongation. **1**. Zhang, Shrestha, Buckley, *et al.* (2014) *ACS Chem. Biol.* **9**, 1773.

D-2-Hydroxybutyrate (D-2-Hydroxybutyric Acid)

This α-hydroxy acid (FW$_{free-acid}$ = 104.11 g/mol; CAS 20016-85-7), also known as (*R*)-2-hydroxybutanoic acid, inhibits D-3-hydroxybutyrate dehydrogenase (1-3), mandelate racemase, K_i = 17.3 mM (4), and D-threonine dehydrogenase (5). **1**. Krebs, Gawehn, Williamson & Bergmeyer (1969) *Meth. Enzymol.* **14**, 222. **2**. Delafield & Doudoroff (1969) *Meth. Enzymol.* **14**, 227. **3**. Nakada, Fukui, Saito, *et al.* (1981) *J. Biochem.* **89**, 625. **4**. Siddiqi, Bourque, Jiang, *et al.* (2005) *Biochemistry* **44**, 9013. **5**. Misono, Kato, Packdibamrung, Nagata & Nagasaki (1993) *Appl. Environ. Microbiol.* **59**, 2963.

D-3-Hydroxybutyrate (D-3-Hydroxybutyric Acid)

This β-hydroxyacid (FW$_{free-acid}$ = 104.11 g/mol; CAS 625-72-9; M.P. = 45.5-48°C; pK_a = 4.41 at 22°C; *Symbol*: BHB), also known as D(−)-β-hydroxybutyric acid and (*R*)-3-hydroxybutanoic acid, is one of three ketone bodies formed in ketogenesis (*See also Acetoacetate and Acetone*) (1-6). By providing an alternative substrate to fuel oxidative phosphorylation, ketone bodies support mammalian cell survival during periods of energy insufficiency. It is found in elevated concentrations in the urine of diabetics. The free acid is relatively unstable, forming esterification products; careless distillation will produce crotonic acid and water. Sodium and potassium salts are significantly more stable. The L-(+)-enantiomer, *or* (*S*)-3-hydroxybutyrate, is found as the CoA-thioester intermediate in fatty acid β-oxidation. The corresponding acyl-carrier protein intermediate in fatty acid biosynthesis is D-3-hydroxybutyryl-ACP. BHBA significantly increases levels of oxidation indicators (*e.g.*, MDA, NO and iNOS), but decreases levels of anti-oxidation indicators (*e.g.*, GSH-Px, CAT and SOD) in treated hepatocytes. IKKβ activity and phospho-IκBα (p-IκBα) are also increased in treated hepatocytes, suggesting that BHBA can induce hepatocyte inflammatory injury through the NF-κB signaling pathway and may be activated by oxidative stress (9). **Anti-inflammatory Mechanism(s):** BHB suppresses the activation of the NLRP3 inflammasome in response to urate crystals, ATP, and lipotoxic fatty acids, whereas acetoacetate, butyrate and acetate cannot (10). Significantly, BHB does not inhibit caspase-1 activation in response to pathogens that activate the NLR family, CARD domain-containing 4 (*or* NLRC4) or Absent in Melanoma-2 (*or* AIM2) inflammasome and does not affect non-canonical caspase-11-mediated inflammasome activation. BHB reduces the NLRP3 inflammasome response by preventing K$^+$ efflux and reducing the oligomerization of Apoptosis-associated Speck-like protein containing a Caspase-recruitment domain (*or* ASC) and speck formation (10). BHB's inhibitory effects on NLRP3 are independent of chirality and starvation-regulated mechanisms (*e.g.*, AMP-activated protein kinase (AMPK), reactive oxygen species (ROS), autophagy or glycolytic inhibition). BHB blocks the NLRP3 inflammasome without undergoing oxidation in the TCA cycle, and independently of Uncoupling Protein-2 (UCP2), Sirtuin-2 (SIRT2), the G-Protein-coupled Receptor GPR109A, or HydroCarboxylic Acid Receptor-2 (HCAR2). Instead, it reduces NLRP3 inflammasome-mediated interleukin (IL)-1β and IL-18 production in human monocytes (10). *In vivo*, BHB or a ketogenic diet attenuates caspase-1 activation and IL-1β secretion in mouse models of NLRP3-mediated diseases, suggesting the anti-inflammatory effects of caloric restriction or ketogenic diets may be linked to BHB-mediated inhibition of NLRP3 inflammasome (10). **Target(s):** aldehyde reductase, *or* aldose reductase (1); creatine kinase, weakly inhibited (2); 3-hydroxy-3-methylglutaryl-CoA reductase (3); palmitoyl-CoA hydrolase, *or* acyl-CoA hydrolase (4); pantothenate synthetase, *or* pantoate:β-alanine ligase, slightly inhibited (5); procollagen-proline 4-dioxygenase, weakly inhibited (6); L-threonine dehydrogenase (7,8). **1**. Hayman & Kinoshita (1965) *J. Biol. Chem.* **240**, 877. **2**. Watts (1973) *The Enzymes*, 3rd ed. (Boyer, ed.), **8**, 383. **3**. Fimognari & Rodwell (1965) *Biochemistry* **4**, 2086. **4**. Berge & Dossland (1979) *Biochem. J.* **181**, 119. **5**. Miyatake, Nakano & Kitaoka (1978) *J. Nutr. Sci. Vitaminol.* **24**, 243. **6**. Tuderman, Myllylä & Kivirikko (1977) *Eur. J. Biochem.* **80**, 341. **7**. Guerranti, Pagani, Neri, *et al.* (2001) *Biochim. Biophys. Acta* **1568**, 45. **8**. Pagani, Guerranti, Righi, *et al.* (1995) *Biochim. Biophys. Acta* **1244**, 49. **9**. Shi, Li, Li, *et al.* (2014) *Cell Physiol. Biochem.* **33**, 920. **10**. Youm, Nguyen, Grant, *et al.* (2015) *Nature Med.* **21**, 263.

10-Hydroxycamptothecin

This antitumor alkaloid (FW = 364.36 g/mol; CAS 67656-30-8) from the Chinese tree *Camptotheca acuminata* is a DNA intercalator that inhibits DNA topoisomerase (1-5). (S)-10-hydroxy-Camptothecin has strong anti-tumor activity against a wide range of experimental tumors including L1210 leukemia cells (IC$_{50}$ = 1.15 μM) (6). **1**. Xu & Ling (1985) *Am. J. Chin. Med.* **13**, 23. **2**. Ling, Tseng & Nelson (1991) *Differentiation* **46**, 135. **3**. Ling, Andersson & Nelson (1990) *Cancer Biochem. Biophys.* **11**, 23. **4**. Kingsbury, Boehm, Jakas, *et al.* (1991) *J. Med. Chem.* **34**, 98. **5**. Zhang, Li, Cai, *et al.* (1989) *Cancer Chemother. Pharmacol.* **41**, 257. **5**. Yu, Xia, Zhao, *et al.* (2012) *Biol. Pharmaceut. Bull.* **35**, 1295.

25-Hydroxycholesterol

This endogenous cholesterol metabolite, LXR agonist, and ER partial agonist (FW = 402.66 g/mol; CAS 2140-46-7; Symbol: 27HC), also known as cholest-5-ene-3β,25-diol, targets the Liver X Receptor (1) and Estrogen Receptor, respectively. The LXR is a member of the nuclear receptor family of transcription factors and is closely related to such nuclear receptors as the PPARs, FXR and RXR. LXRs are important regulators of cholesterol, fatty acid, and glucose homeostasis. While previously classified as orphan receptors, endogenous oxysterols were found to be their likely natural ligands. The most prevalent oxysterol in circulation, 27-hydroxycholesterol functions as a selective ER modulator (*or* SERM), the efficacy of which varies depends on the endpoints chosen (2). Importantly, 27HC positively regulates both gene transcription and cell proliferation in cellular models of breast cancer. Using combinatorial peptide phage display, we have determined that 27HC induces a unique conformational change in both ERα and ERβ, distinguishing it from 17β-estradiol and other SERMs (2). Hydroxycholesterol modulates hydroxymethylglutaryl-CoA reductase activity by rapidly increasing its degradation and decreasing its synthesis. Hydroxycholesterol suppresses β-estradiol-mediated breast cancer cell proliferation *in vitro*. It also increases α-synuclein levels and induces apoptosis in human neuroblastoma cells in vitro. 25-Hydroxycholesterol also acts within the Golgi compartment to induce the degradation of tyrosinase. **Target(s):** cholesterol efflux (3); cholesterol 7α-monooxygenase, *or* sterol 7α-monooxygenase (4); dolichol synthesis (5,7); hydroxymethylglutaryl-CoA reductase turnover (6-8). **1**. Fu, Menke, Chen, *et al.* (2001) *J. Biol. Chem.* **276**, 38378. **2**. DuSell, Umetani, Shaul, Mangelsdorf & McDonnell (2008) *Mol. Endocrinol.* **22**, 65. **3**. Kilsdonk, Morel, Johnson & Rothblat (1995) *J. Lipid Res.* **36**, 505. **4**. Schwartz & Margolis (1983) *J. Lipid Res.* **24**, 28. **5**. Elbein (1987) *Meth. Enzymol.* **138**, 661. **6**. Schwarz & Datema (1980) *Trends Biochem. Sci.* **5**, 65. **7**. Stange, Schneider, Preclik, Alavi & Ditschuneit (1981) *Lipids* **16**, 397. **8**. Tanaka, Edwards, Lan & Fogelman (1983) *J. Biol. Chem.* **258**, 13331.

2-Hydroxycitrate

This natural α-hydroxycitrate derivative (FW$_{free-acid}$ = 208.12 g/mol) is a potent inhibitor of ATP:citrate lyase. It is the active ingredient in the herb *Garcinia cambogia*. Note that there are four stereoisomers of 2 hydroxycitrate. The strongest inhibitor of ATP:citrate lyase is (2*S*,3*S*)-2-hydroxycitrate. This isomer is also an alternative activator of acetyl-CoA carboxylase and inhibits glucose-stimulated insulin secretion. **Target(s):** acetylcholine synthesis (1); acetyl-CoA hydrolase, inhibited by (–)-2-hydroxycitrate (2); ATP:citrate lyase, *or* ATP:citrate synthase, $K_i = 0.15$ μM (1,3-6); citrate (pro-*S*) lyase (7); citrate synthase, weakly inhibited (5); fatty acid biosynthesis (1); 6-phosphofructokinase (1,8); pyruvate dehydrogenase (1). **1**. Lowenstein & Brunengraber (1981) *Meth. Enzymol.* **72**, 486. **2**. Chabtree, Souter & Anderson (1989) *Biochem. J.* **257**, 673. **3**. Watson, Fang & Lowenstein (1969) *Arch. Biochem. Biophys.* **135**, 209. **4**. Spector (1972) *The Enzymes*, 3rd ed., **7**, 357. **5**. Sullivan, Singh, Srere & Glusker (1977) *J. Biol. Chem.* **252**, 7583. **6**. Li, Wang, Tino, *et al.* (2007) *Bioorg. Med. Chem. Lett.* **17**, 3208. **7**. Rokita & Walsh (1983) *Biochemistry* **22**, 2821. **8**. McCune, Foe, Kemp & Jurin (1989) *Biochem. J.* **259**, 925.

16-Hydroxycleroda-3,13-dien-16,15-olide

This clerodane diterpene (FW = 318.46 g/mol; *Abbreviation*: HCD) is non-cytotoxic antileishmanial lead, as evidenced by its effects on long-term survival (exceeding 6 months) of treated animals. Visceral leishmaniasis (or *Kala-azar*), the most devastating form of leishmaniasis, begins with invasion of the reticuloendothelial system (*e.g.*, spleen, liver and bone marrow) by the haemoflagellate protozoan parasite, *Leishmania donovani*. Since the 1940s, pentavalent antimony compounds have been a first-line treatment for all forms of leishmaniasis. Activity-guided fractionation of a crude ethanolic extract of the leaves of *Polyalthia longifolia* var. *pendula* led to the isolation of HCD (1), which arrests host cells in G_0-G_1 phase and promotes apoptosis. HCD decreases anti-apoptotic proteins, while simultaneously increasing pro-apoptotic proteins. Treated cells also formed filopodia, with reduced levels of FAK, pFAK, Rac1 and Cdc42. HCD inhibited the activity of MMP-2 and MMP-9. **1**. Koneni, Sashidhara, Singh, *et al.* (2010) *Brit. J. Pharmacol.* **159**, 1143. **2**. Thiyagarajan, Lin, Chia & Weng (2013) *Biochim. Biophys. Acta.* **1830**, 4091.

2-Hydroxydodecanoic Acid

This α-hydroxy acid (FW$_{free-acid}$ = 216.32 g/mol), also known as 2-hydroxylauric acid, inhibits butyryl-CoA synthetase, *or* butyrate:CoA ligase, *or* medium-chain-fatty-acyl-CoA synthetase, $K_i = 4.4$ μM. **1**. Kasuya, Yamaoka, Igarashi & Fukui (1998) *Biochem. Pharmacol.* **55**, 1769. **2**. Kasuya, Igarashi & Fukui (1999) *Chem. Biol. Interact.* **118**, 233. **3**. Kasuya, Hiasa, Kawai, Igarashi & Fukui (2001) *Biochem. Pharmacol.* **62**, 363. **4**. Morsczeck, Berger & Plum (2001) *Biochim. Biophys. Acta* **1521**, 59.

6-Hydroxy-DOPA

This allosteric RAD52 inhibitor (FW = 213.19 g/mol; CAS 21373-30-8; Soluble to 100 mM in 1 equuivalent HCl), also named 2,5-dihydroxy-DL-tyrosine, targets Rad52, a ring-like oligomeric protein encoded by a gene sharing similarity with *Saccharomyces cerevisiae* Rad52 and a key protein in DNA double-strand break repair and homologous recombination. Rad52 binds to the ends of single-stranded DNA (ssDNA), mediating the requisite DNA-DNA interactions for annealing complementary DNA strands. 6-OH-dopa) binds to and transforms RAD52 undecamer rings into dimers, abolishing ssDNA binding channel observed in crystal structures. 6-OH-Dopa also disrupts RAD52 heptamer and undecamer ring superstructures,

suppressing RAD52 recruitment and recombination activity in cells with negligible effects on other double-strand break repair pathways. 6-Hydroxy-DOPA selectively inhibits the proliferation of BRCA-deficient cancer cells, including those obtained from leukemia patients. **1**. Chandramouly, McDevitt, Sullivan, *et al.* (2015) *Chem. Biol.* **22**, 1491.

6-Hydroxydopamine

This neurotoxic agent (FW$_{free-base}$ = 169.18 g/mol; CAS 1199-18-4; photosensitive), also known as 2,4,5-trihydroxyphenethylamine, oxydopamine, and ODHA, destroy dopaminergic and noradrenergic neurons in the brains of experimental animals. Selective destruction of adrenergic nerve terminals by 6-hydroxydopamine appears to result from the covalent binding of its oxidation/photooxidation products to nucleophilic groups of biological macromolecules. The reaction seems to be nonspecific, and the high gelectivity of the destructive effect results from the efficient uptake of 6-hydroxydopamine into the adrenergic nerve terminals. **Target(s):** dihydropteridine reductase (1); glyceraldehyde-3-phosphate dehydrogenase (2); 5-lipoxygenase, *or* arachidonate 5-lipoxygenase, ID$_{50}$ = 26.7 μM (3); 12-lipoxygenase, *or* arachidonate 12-lipoxygenase, ID$_{50}$ = 45.3 μM (3); phenol sulfotransferase, *or* aryl sulfotransferase, also alternative substrate (4); prostaglandin synthase (ID$_{50}$ = 0.45 mM (3). **1**. S. Shen (1983) *Biochim. Biophys. Acta* **743**, 129. **2**. Hayes & Tipton (2002) *Toxicol. Lett.* **128**, 197. **3**. Hamasaki & Tai (1985) *Biochim. Biophys. Acta* **834**, 37. **4**. Yasuda, Liu, Suiko, Sakakibara & Liu (2007) *J. Neurochem.* **103**, 2679.

(5Z,8Z,11Z,13E)-(15S)-15-Hydroxyeicosa-5,8,11,13-tetraenoic Acid

This unsaturated ξ-hydroxyacid (FW = 320.47 g/mol), often abbreviated 15-HETE, is an arachidonate metabolite and lipoxygenase inhibitor. **Target(s):** 5-lipoxygenase, *or* arachidonate 5-lipoxygenase, I$_{50}$ values of 3 and 7.7 μM are reported (1-4); 12-lipoxygenase, *or* arachidonate 12-lipoxygenase (5,6). **1**. Coutts, Khandwala, Van Inwegen, *et al.* (1985) in *Prostaglandins, Leukotrienes, and Lipoxins* (Bailey, ed.), p. 627, Plenum Press, New York. **2**. Vanderhoek, Bryant & Bailey (1980) *J. Biol. Chem.* **255**, 10064. **3**. Kang & Vanderhoek (1995) *Biochim. Biophys. Acta* **1256**, 297. **4**. Cashman, Lambert & Sigal (1988) *Biochem. Biophys. Res. Commun.* **155**, 38. **5**. Nugteren (1982) *Meth. Enzymol.* **86**, 49. **6**. Pace-Asciak & Smith (1983) *The Enzymes*, 3rd ed., **16**, 543.

9-Hydroxyellipticine

This antitumor agent (FW = 262.31 g/mol) binds to DNA and stabilizes the DNA toposiomerase II:DNA cleavable complex. It is oxidized to reactive quinone-imine intermediates, inhibits telomerase, suppresses phosphorylation of the mutant p53 protein, and is likewise cytotoxic. **Target(s):** cyclin-dependent kinases (1); cytochrome P-448 activity, *or* ethoxyresorufin *O*-deethylase and biphenyl 2-hydroxylase (2); cytochrome P450 proteins (*e.g.*, benzo[*a*]pyrene hydroxylase activity) (3-5); DNA topoisomerase I (6); DNA topoisomerase II (7,8); p53 protein phosphorylation (9); telomerase (10). **1**. Meijer & Kim (1997) *Meth.*

Enzymol. **283**, 113. **2**. Delaforge, Ioannides & Parke (1982) *Chem. Biol. Interact.* **42**, 279. **3**. Lesca, Rafidinarivo, Lecointe & Mansuy (1979) *Chem. Biol. Interact.* **24**, 189. **4**. Sautereau & Lesca (1979) *Chem. Biol. Interact.* **27**, 269. **5**. Lesca, Beaune & Monsarrat (1981) *Chem. Biol. Interact.* **36**, 299. **6**. Riou, Douc-Rasy & Kayser (1986) *Biochem. Soc. Trans.* **14**, 496. **7**. Renault, Malvy, Venegas & Larsen (1987) *Toxicol. Appl. Pharmacol.* **89**, 281. **8**. Harding & Grummitt (2003) *Mini Rev. Med. Chem.* **3**, 67. **9**. Ohashi, Sugikawa & Nakanishi (1995) *Jpn. J. Cancer Res.* **86**, 819. **10**. Sato, Mizumoto, Kusumoto, *et al.* (1998) *FEBS Lett.* **441**, 318.

Bis(2-Hydroxyethyl) disulfide

This symmetrical disulfide (FW = g/mol; CAS 1892-29-1), also known as 2-hydroxyethyl disulfide and 2,2-dithiodiethanol, is formed by the oxidation of the protein thiol reductant β-mercaptoethanol (*See 2-mercaptoethanol*). *Bis*(2-hydroxyethyl) disulfide inactivates horse liver aldehyde dehydrogenase (EC 1.2.1.3) with formation of mixed-disulfides between protein sulfhydryl groups and β-mercaptoethanol. Upon extended incubation, the enzyme retained 30% of its initial catalytic activity, suggesting that 2-hydroxyethyl disulfide-treated aldehyde dehydrogenase retains catalytic activity. **1**. Brotherton & Rodwell (1980) *Physiol. Chem. Phys.* **12**, 483.

2-({[2-({3-[(2-Hydroxyethyl)amino]propyl}amino)phenyl] sulfonyl}amino)-5,6,7,8-tetrahydro-1-naphthalenecarboxylate

This substituted anthranilic acid sulfonamide (FW = 415.49 g/mol) inhibits methionyl aminopeptidase, IC$_{50}$ = 0.015 μM. **1**. Sheppard, Wang, Kawai, *et al.* (2006) *J. Med. Chem.* **49**, 3832.

(2'-Hydroxyethyl)cobalamin

This vitamin B$_{12}$ derivative (FW = 1374.42 g/mol) inhibits ethanolamine ammonia-lyase (1-3), glycerol dehydratase (4), and propanediol dehydratase (1). **1**. Abeles (1971) *The Enzymes*, 3rd ed., **5**, 481. **2**. Stadtman (1972) *The Enzymes*, 3rd ed., **6**, 539. **3**. Babior (1969) *J. Biol. Chem.* **244**, 2917. **4**. Poppe, Hull, Nitsche, *et al.* (1999) *Helv. Chim. Acta* **82**, 1250.

2-Hydroxyethyl Disulfide

This toxic reagent (FW = 154.25 g/mol; M.P. = 25-27°C), also known as 2,2'-dithiodiethanol, oxidized 2-mercaptoethanol, 2 mercaptoethanol disulfide, thioethylene glycol disulfide, dithiodiglycol, and diethylene glycol disulfide, is the disulfide of 2-mercaptoethanol. Although this reagent accumulates upon atmospheric exposure of 2-mercaptoethanol solutions used as a protein disulfide reductant, the far higher concentration of thiol likely reverses the formation of any mixed disulfides. (*See also 2-Mercaptoethanol*) **Target(s):** aldehyde dehydrogenase (1); ananain (2); bromelain (3); comosain (4); homocysteine desulfhydrase (5); 3-hydroxy-3-methylglutaryl-CoA reductase (6,7); iodothyronine 5'-deiodinase (8,9); L-methionone (*S*)-*S*-oxide reductase (10); ornithine carbamoyltransferase (11); pantetheine hydrolase, *or* pantetheinase (12); [phosphorylase] phosphatase (13); stem bromelain (3); thymidylate synthase (14). **1**. Brotherton & Rodwell (1980) *Physiol. Chem. Phys.* **12**, 483. **2**. Rowan (1998) in *Handb. Proteolytic Enzymes*, p. 569, Academic Press, San Diego. **3**. Rowan (1998) in *Handb. Proteolytic Enzymes*, p. 566, Academic Press, San Diego. **4**. *ibid*, p. 570. **5**. Thong & Coombs (1987) *Exp. Parasitol.* **63**, 143. **6**. Dotan & Shechter (1985) *J. Cell Physiol.* **122**, 14. **7**. Dotan & Shechter (1983) *Arch. Biochem. Biophys.* **226**, 401. **8**. Goswami & Rosenberg (1990) *Endocrinology* **126**, 2597. **9**. Goswami & Rosenberg (1989) *Biochem. Int.* **19**, 361. **10**. Black, Harte, Hudson & Wartofsky (1960) *J. Biol. Chem.* **235**, 2910. **11**. Marshall & Cohen (1972) *J. Biol. Chem.* **247**, 1669. **12**. Pitari, Maurizi, Ascenzi, Ricci & Dupre (1994) *Eur. J. Biochem.* **226**, 81. **13**. Gratecos, Detwiler, Hurd & Fischer (1977) *Biochemistry* **16**, 4812. **14**. Dunlap (1978) *Meth. Enzymol.* **51**, 90.

5'-(2-Hydroxyethylthio)immucillin-A

This nucleoside analogue (FW$_{HCl-Salt}$ = 373.87 g/mol), also known as (1*S*)-1-(9-deazaadenin-9-yl)-5-(2-hydroxyethylthio) 1,4-dideoxy-1,4-imino-D-ribitol, inhibits 5'-methylthioadenosine/*S*-adenosylhomocysteine nucleosidase (*K*$_i$' = 0.41 nM). It is also a tight, slow-binding inhibitor of human *S*-methyl-5' thioadenosine phosphorylase (*K*$_i$ = 14 nM). **Target(s):** adenosylhomocysteine nucleosidase (1); 5'-methylthioadenosine/*S*-adenosylhomocysteine nucleosidase (1); *S*-methyl-5'-thioadenosine phosphorylase (2). **1**. Singh, Evans, Lenz, *et al.* (2005) *J. Biol. Chem.* **280**, 18265. **2**. Evans, Furneaux, Schramm, Singh & Tyler (2004) *J. Med. Chem.* **47**, 3275.

α-Hydroxyfarnesylphosphonic Acid

This phosphonate (FW$_{free-acid}$ = 302.35 g/mol) is a strong competitive inhibitor of protein farnesyltransferase (IC$_{50}$ = 30 nM) as well as geranylgeranyltransferase I and II, albeit at significantly higher concentrations (IC$_{50}$ = 35.8 and 67 μM, respectively. Only 2*E*,6*E*-α-hydroxyfarnesyl-phosphonate induces alteration of Ras processing in intact human-derived leukemia cells and inhibits protein farnesyltransferase in enzyme assays. *Z*,*E*-α-farnesyl- and geranylphosphonates were inactive. **Target(s):** lipid-phosphate phosphatase, IC$_{50}$ = 73 μM (1); protein farnesyltransferase (2-7); rubber *cis*-polyprenyl*cis*transferase (8); soluble epoxide hydrolase's phosphatase activity, IC$_{50}$ = 73 μM (1). **1**. Tran, Aronov, Tanaka, *et al.* (2005) *Biochemistry* **44**, 12179. **2**. Terry, Long & Beese (2001) *The Enzymes*, 3rd ed., **21**, 19. **3**. Gibbs (2001) *The Enzymes*, 3rd ed., **21**, 81. **4**. Pompliano, Rands, Schaber, *et.* (1992) *Biochemistry* **31**, 3800. **5**. Gibbs, Pompliano, Mosser, *et al.* (1993) *J. Biol. Chem.* **268**, 7617. **6**. Hohl, Lewis, Cermak & Wiemer (1998) *Lipids* **33**, 39. **7**. Girgert, Hohnecker, Wittrock & Schweizer (1999) *Anticancer Res.* **19**, 2959. **8**. Mau, Garneau, Scholte, *et al.* (2003) *Eur. J. Biochem.* **270**, 3939.

Hydroxyfasudil

This potent protein kinase inhibitor and vasodilator (FW = 343.83 g/mol; CAS 155558-32-0), also known systematically as 1-(1-hydroxy-5-isoquinolinesulfonyl)homopiperazine (IC$_{50}$ = 10.7 µM), an enzyme (abbreviated ROCK) that plays a role in mediating vasoconstriction and vascular remodeling. Extending earlier findings suggesting a role of RhoA/Rho kinase in hippocampal-mediated human memory, Huentelman, Stephan, Talboom, *et. al.* (2009) demonstrated that peripheral administration of the ROCK inhibitor hydroxyfasudil improves spatial learning and working memory in a rodent learning model. *See Fasudil* **1**. Huentelman, Stephan, Talboom, *et al.* (2009) *Behav. Neurosci.* **123**, 218.

3-Hydroxyflavone

This flavone (FW = 238.24 g/mol; M.P. = 169-171°C), also known as flavonol, inhibits certain cytochrome P450 systems, sulfotransferase, and arachidonate metabolism. Note that the synonym flavanol also refers to this class of hydroxylated flavones and their derivatives. **Target(s):** carboxylase (1,2); cyclic AMP-dependent protein kinase (3); cyclooxygenase (4); CYP1A1 (5); CYP1A2 (5,6); CYP3A4, with respect to testosterone (7); 17β-hydroxysteroid dehydrogenase (8); lactoylglutathione lyase, *or* glyoxalase I (9); 5-lipoxygenase, *or* arachidonate 5-lipoxygenase (4); 15-lipoxygenase, *or* arachidonate 15-lipoxygenase (10); phenol sulfotransferase, *or* 7-hydroxyflavone sulfotransferase, *or* SULT1A1 (11,12); quercetin 2,3-dioxygenase (13). **1**. Karrer & Visconti (1947) *Helv. Chim. Acta* **30**, 268. **2**. Massart (1950) *The Enzymes*, 1st ed. (Sumner & Myrbäck, eds.), **1** (part 1), 307. **3**. Jinsart, Ternai & Polya (1992) *Biol. Chem. Hoppe Seyler* **373**, 205. **4**. Hoult, Moroney & Payá (1994) *Meth. Enzymol.* **234**, 443. **5**. Zhai, Dai, Friedman & Vestal (1998) *Drug Metab. Dispos.* **26**, 989. **6**. Zhai, Dai, Wei, Friedman & Vestal (1998) *Life Sci.* **63**, PL119. **7**. Schrag & Wienkers (2001) *Arch. Biochem. Biophys.* **391**, 49. **8**. Kristan, Krajnc, Konc, *et al.* (2005) *Steroids* **70**, 626 and 694. **9**. Allen, Lo & Thornalley (1993) *Biochem. Soc. Trans.* **21**, 535. **10**. Vasquez-Martinez, Ohri, Kenyon, Holman & Sepúlveda-Boza (2007) *Bioorg. Med. Chem.* **15**, 7408. **11**. Vietri, Pietrabissa, Spisni, Mosca & Pacifici (2002) *Xenobiotica* **32**, 563. **12**. Harris, Wood, Bottomley, *et al.* (2004) *J. Clin. Endocrinol. Metab.* **89**, 1779. **13**. Oka, Simpson & Krishnamurty (1972) *Can. J. Microbiol.* **18**, 493.

5-Hydroxyflavone

This flavone (FW = 238.24 g/mol; M.P. = 155-156°C), also known as primuletin, inhibits certain cytochrome P450 systems, certain matrix metalloproteinases, and a sulfotransferase. **Target(s):** CYP1A1 (1); CYP1A2 (1); gelatinase A, matrix metalloproteinase 2 (2); gelatinase B, *or* matrix metalloproteinase 9 (2); 17β-hydroxysteroid dehydrogenase (3); NADH oxidase (4); phenol sulfotransferase, *or* 7-hydroxyflavone sulfotransferase, *or* SULT1A1 (5,6); ribosyldihydronicotinamide dehydrogenase (quinone) (7). **1**. Zhai, Dai, Friedman & Vestal (1998)

Drug Metab. Dispos. **26**, 989. **2**. Ende & Gebhardt (2004) *Planta Med.* **70**, 1006. **3**. Kristan, Krajnc, Konc, *et al.* (2005) *Steroids* **70**, 626 and 694. **4**. Bohmont, Aaronson, Mann & Pardini (1987) *J. Nat. Prod.* **50**, 427. **5**. Vietri, Pietrabissa, Spisni, Mosca & Pacifici (2002) *Xenobiotica* **32**, 563. **6**. Harris, Wood, Bottomley, *et al.* (2004) *J. Clin. Endocrinol. Metab.* **89**, 1779. **7**. Boutin, Chatelain-Egger, Vella, Delagrange & Ferry (2005) *Chem.-Biol. Interact.* **151**, 213.

6-Hydroxyflavone

This flavone (FW = 238.24 g/mol; M.P. = 230-231°C) inhibits certain cytochrome P450 systems, sulfotransferases, and 3β-hydroxysteroid dehydrogenase. **Target(s):** CYP3A4 (*vs.* testosterone) (1); estrone sulfotransferase (2); 3β-hydroxysteroid dehydrogenase (3,4); phenol sulfotransferase, *or* aryl sulfotransferase, *or* SULT1A1 (2,5); steroid 11β-monooxygenase (3); steroid 17-monooxygenase (3); steroid 21-monooxygenase (3). **1**. Schrag & Wienkers (2001) *Arch. Biochem. Biophys.* **391**, 49. **2**. Harris, Wood, Bottomley, *et al.* (2004) *J. Clin. Endocrinol. Metab.* **89**, 1779. **3**. Ohno, Shinoda, Toyoshima, *et al.* (2002) *J. Steroid Biochem. Mol. Biol.* **80**, 355. **4**. Ohno, Matsumoto, Watanabe & Nakajin (2004) *J. Steroid Biochem. Mol. Biol.* **88**, 175. **5**. Vietri, Pietrabissa, Spisni, Mosca & Pacifici (2002) *Xenobiotica* **32**, 563.

7-Hydroxyflavone

This flavone (FW = 238.24 g/mol; M.P. = 240-241°C) inhibits certain cytochrome P450 systems, 17β-hydroxysteroid dehydrogenase, and alcohol dehydrogenase. **Target(s):** alcohol dehydrogenase (1); aromatase, *or* CYP19 (2-4); CYP1A1 (5); CYP1A2 (5); GLUT1 glucose transporter (6); 17β-hydroxysteroid dehydrogenase (2,7); NADH oxidase, weakly inhibited (8). **1**. Keung (1993) *Alcohol Clin. Exp. Res.* **17**, 1254. **2**. Le Bail, Laroche, Marre-Fournier & Habrioux (1998) *Cancer Lett.* **133**, 101. **3**. Ibrahim & Abul-Hajj (1990) *J. Steroid Biochem. Mol. Biol.* **37**, 257. **4**. Sanderson, Hordijk, Denison, *et al.* (2004) *Toxicol. Sci.* **82**, 70. **5**. Zhai, Dai, Friedman & Vestal (1998) *Drug Metab. Dispos.* **26**, 989. **6**. Martin, Kornmann & Fuhrmann (2003) *Chem. Biol. Interact.* **146**, 225. **7**. Le Lain, Nicholls, Smith & Maharlouie (2001) *J. Enzyme Inhib.* **16**, 35. **8**. Bohmont, Aaronson, Mann & Pardini (1987) *J. Nat. Prod.* **50**, 427.

3-Hydroxyglutamate (3-Hydroxyglutamic Acid)

This derivative of glutamic acid (FW$_{neutral-species}$ = 163.13 g/mol) pK_a values = 2.09, 4.18, and 9.20; Soluble in water and acetic acid; Insoluble in ethanol), also called β-hydroxyglutamate, has two chiral atoms and exists as four stereoisomers. The *threo* and *erythro* diastereoisomers of the L-compound are often used in neurochemical studies as well as in probes of a protein's specificity. **Target(s):** aspartate aminotransferase (Both the *erythro-* and *threo*-isomers are poor substrates, but are better inhibitors.) (1); glutamine synthetase (2); γ-glutamylcysteine synthetase, inhibited by the *threo*-β-hydroxy-DL-glutamate (It is also a weak alternative substrate.) (3). **1**. Braunstein (1973) *The Enzymes*, 3rd ed., **9**, 379. **2**. Webb (1966)

Enzyme and Metabolic Inhibitors, vol. **2**, p. 336, Academic Press, New York. **3**. Sekura & Meister (1977) *J. Biol. Chem.* **252**, 2599.

N-Hydroxy-DL-glutamate (N-Hydroxy-DL-glutamic Acid)

This modified amino acid (FW$_{free-acid}$ = 163.13 g/mol) irreversibly inhibits a number of pyridoxal-phosphate-dependent enzymes and reversibly inhibits glutamate dehydrogenase. **Target(s):** alanine aminotransferase (1); aspartate transaminase (1); glutamate decarboxylase (1); glutamate dehydrogenase (1); glutamate racemase (D-isomer also an alternative substrate) (2). **1**. Cooper & Griffith (1979) *J. Biol. Chem.* **254**, 2748. **2**. Glavas & Tanner (1997) *Bioorg. Med. Chem. Lett.* **7**, 2265.

2-Hydroxyglutarate (2-Hydroxyglutaric Acid)

This toxic and oncogenic α-hydroxy dicarboxylic acid (FW$_{free-acid}$ = 148.12 g/mol) arises by way of the slow NADH-dependent conversion of α-ketoglutarate, a side-reaction catalyzed by mitochondrial L-malate dehydrogenase (1) and by IDH1^{R132H} and IDH1^{R132C} mutants of isocitrate dehydrogenase isoform-1 (**See** *AGI-5198*). **L-2-Hydroxyglutaric Aciduria:** This disorder results from a defect in L-2-hydroxyglutarate dehydrogenase, a FAD-linked enzyme that catalyzes the irreversible conversion of L-2-hydroxyglutarate to α-ketoglutarate. L-2-hydroxyglutarate is formed as a NADH-dependent side-reaction by mitochondrial L-malate dehydrogenase, which normally interconverts oxaloacetate and L-malate. L-2-hydroxyglutarate dehydrogenase (L2HDGH) catalyzes the formation of α-ketoglutarate from L-2-hydroxyglutarate, and patients with this progressive neurodegenerative disease possess mutations in both alleles of the L2HDGH gene. **(R)-2HG's Oncometabolic Action:** A cardinal feature of several malignancies is the abundance of hydroxyglutarate. Isocitrate dehydrogenases IDH1 (cytosol) and IDH2 (mitochondrial matrix) catalyze the interconversion of isocitrate and α-ketoglutarate, a TCA cycle intermediate and an essential cofactor for JmjC domain-containing histone demethylases, TET 5-methylcytosine hydroxylases, and EglN prolyl-4-hydroxylases. Cancer-associated mutations IDH alter the enzymes such that they reduce α-ketoglutarate to the structurally similar metabolite (R)-2-hydroxyglutarate. These mutations are present in 5-20% of *de novo* NK-AML cases and in 10-20% of secondary AML cases resulting from "premalignant" myelodysplastic syndrome (MDS) and myeloproliferative neoplasm (MPN). IDH mutations are also present at a lower frequency (5-10%), in chronic-phase MDS and MP, but are rare in translocation-positive AML. (R)-2-Hydroxyglutarate levels in IDH mutant tumors are often greatly elevated, ranging from 1-30 mM. Because it is structurally and chemically very similar to α-ketoglutarate, one idea is that (R)-2-hydroxyglutarate transforms cells by competitively inhibiting α-ketoglutarate-dependent enzymes that function as tumor suppressors. Prolyl hydroxylation of Hypoxia-Inducible transcription Factor-α (HIFα) increases its affinity for the von Hippel-Lindau protein (pVHL), so designating it for degradation by the ubiquitin-proteasome system. Diminished HIF levels have the effect of enhancing cell proliferation and motility. **Target(s):** 2-(acetamidomethylene)-succinate hydrolase (2); carbamoyl-phosphate synthetase, ammonia-dependent (3); creatine kinase, inhibited by D- and L-isomers (4-7); cytochrome *c* oxidase, inhibited by D-isomer (8); glutamate decarboxylase, inhibited by both the D- and L-isomers (K_i = 3.4 and 4.4 mM, respectively) (9); glutamate dehydrogenase, inhibited by L-isomer (10,11). **1**. Van Schaftigen (2007) *Bull Mem Acad R Med Belg.* **162**, 451. **2**. Huynh & Snell (1985) *J. Biol. Chem.* **260**, 2379. **3**. Fahien, Schooler, Gehred & Cohen (1964) *J. Biol. Chem.* **239**, 1935. **4**. da Silva, Bueno, Schuck, *et al.* (2003) *Eur. J. Clin. Invest.* **33**, 840. **5**. da Silva, Bueno, Schuck, *et al.* (2004) *Neurochem. Int.* **44**, 45. **6**. da Silva, Bueno, Rosa, *et al.* (2003) *Neurochem. Res.* **28**, 1329. **7**. da Silva, Bueno, Schuck, *et al.* (2003) *Int. J. Dev. Neurosci.* **21**, 217. **8**. da Silva, Ribeiro, Leipnitz, *et al.* (2002) *Biochim. Biophys. Acta* **1586**, 81. **9**. Fonda (1972) *Biochemistry*

10, 1304. **9**. Bell, LiMuti, Renz & Bell (1985) *Biochem. J.* **225**, 209. **11**. Favilla & Bayley (1982) *Eur. J. Biochem.* **125**, 209.

12-Hydroxy-16-heptadecynoic Acid

This unsaturated λ-hydroxy acid (FW = 281.42 g/mol) irreversibly inhibits the cytochrome P450 system responsible for ω-oxidation of prostaglandins. The 12S-enantiomer is more active (K_i = 1.8 μM, $t_{1/2}$ = 0.7 min) than the 12R-isomer (K_i = 3.6 μM, $t_{1/2}$ = 0.8 min). **1**. Aitken, Roman, Loughran, de la Garza & Masters (2001) *Arch. Biochem. Biophys.* **393**, 329. **2**. Burger, Clark, Nishimoto, *et al.* (1993) *J. Med. Chem.* **36**, 1418. **3**. Muerhoff, Williams, Reich, *et al.* (1989) *J. Biol. Chem.* **264**, 749.

2-Hydroxyisonicotinate N-Oxide

This presumptive transition-state analogue (FW = 155.11 g/mol) inhibits protocatechuate 3,4-dioxygenase. Binding also results in a major change in the enzyme chromophore spectrum. **Target(s):** protocatechuate 2,3-dioxygenase (1); protocatechuate 3,4-dioxygenase (2-5); protocatechuate 4,5-dioxygenase (6). **1**. Wolgel & Lipscomb (1990) *Meth. Enzymol.* **188**, 95. **2**. May, Oldham, Mueller, Padgett & Sowell (1982) *J. Biol. Chem.* **257**, 12746. **3**. Whittaker & Lipscomb (1984) *J. Biol. Chem.* **259**, 4476 and 4487. **4**. Orville, Lipscomb & Ohlendorf (1997) *Biochemistry* **36**, 10052. **5**. May, Mueller, Oldham, Williamson & Sowell (1983) *Biochemistry* **22**, 5331. **6**. Arciero, Orville & Lipscomb (1985) *J. Biol. Chem.* **260**, 14035.

2-Hydroxy-4H-isoquinoline-1,3-dione

This isoquinoline derivative (FW = 177.16 g/mol) inhibits ribonucleases utilizing two divalent metal ions within their catalytic sites. **Target(s):** exoribonuclease H (1-3); HIV-1 reverse transcriptase (3,4); ribonuclease III (5); ribonuclease H (1-4); RNA-directed DNA polymerase (1,3,4). **1**. Hang, Rajendran, Yang, *et al.* (2004) *Biochem. Biophys. Res. Commun.* **317**, 321. **2**. Nakayama, Bingham, Tan & Maegley (2006) *Anal. Biochem.* **351**, 260. **3**. Klumpp, Hang, Rajendran, *et al.* (2003) *Nucl. Acids Res.* **31**, 6852. **4**. Hang, Li, Yang, *et al.* (2007) *Biochem. Biophys. Res. Commun.* **352**, 341. **5**. Sun, Pertzev & Nicholson (2005) *Nucl. Acids Res.* **33**, 807.

(S)-5-Hydroxy-4-ketonorvaline

This antifungal agent (FW = 147.13 g/mol), also known as (S)-2-amino-5-hydroxy-4-ketopentanoic acid and (S)-2 amino-5-hydroxy-4-oxopentanoic acid, inhibits homoserine dehydrogenase, forming a covalent adduct between C-5 of the agent and C-4 of the nicotinamide ring of NAD$^+$. Note that the (S)-stereoisomer is an L-aspartic acid analogue. **Target(s):** asparagine synthetase, weakly inhibited (1); homoserine dehydrogenase (2-4); porphobilinogen synthase, or δ-aminolevulinate dehydratase (5). **1.** Horowitz & Meister (1972) *J. Biol. Chem.* **247**, 6708. **2.** Jacques, Mirza, Ejim, *et al.* (2003) *Chem. Biol.* **10**, 989. **3.** Yamaki, Yamaguchi, Tsuruo & Yamaguchi (1992) *J. Antibiot. (Tokyo)* **45**, 750. **4.** Yamaki, Yamaguchi, Imamura, *et al.* (1990) *Biochem. Biophys. Res. Commun.* **168**, 837. **5.** Yamasaki & Moriyama (1971) *Biochim. Biophys. Acta* **227**, 698.

3-Hydroxykynurenine

L-isomer

This amino acid (FW = 224.22 g/mol), also known as 3-(3-hydroxyanthraniloyl)alanine, is an intermediate in tryptophan catabolism. Note that it is highly autooxidizable in air and will generate hydrogen peroxide. Aqueous solutions are yellow at alkaline pH values. **Target(s):** arylformamidase, inhibited by DL-mixture (1); cysteine-*S*-conjugate β-lyase (2); kynurenine: oxoglutarate aminotransferase (3,4); pyridoxal kinase (5); tryptophan 2,3-dioxygenase (6-8). **1.** Serrano & Nagayama (1991) *Comp. Biochem. Physiol. B Comp. Biochem.* **99**, 281. **2.** Stevens (1985) *J. Biol. Chem.* **260**, 7945. **3.** Guidetti, Okuno & Schwarcz (1997) *J. Neurosci. Res.* **50**, 457. **4.** Okuno, Nishikawa & Nakamura (1996) *Adv. Exp. Med. Biol.* **398**, 455. **5.** Karawya, Mostafa & Osman (1981) *Biochim. Biophys. Acta* **657**, 153. **6.** Hitchcock & Katz (1988) *Arch. Biochem. Biophys.* **261**, 148. **7.** Paglino, Lombardo, Arcà, Rizzi & Rossi (2008) *Insect Biochem. Mol. Biol.* **38**, 871. **8.** Schartau & Linzen (1976) *Hoppe-Seyler's Z. Physiol. Chem.* **357**, 41.

Hydroxylamine

This hygroscopic reagent (FW = 33.03 g/mol; pK_a = 7.97 at 20°C; Very soluble in water) is a powerful nucleophile at neutral to slightly alkaline pH, reacting nonenzymatically with acyl-phosphates, acid anhydrides, thiolesters, aldehydes, but not carboxyl groups. In its reaction with acetyl-P, hydroxylamine first forms the kinetically favored *O*-acetylhydroxamate (*i.e.*, $CH_3(C=O)$-O-NH_2); however, in the presence of excess NH_2OH, the *O*-acetylhydroxamate forms the thermodynamically favored $CH_3(C=O)$–NH-OH. Hydroxylamine an alternative substrate for most ammonia-dependent reactions, and suffices for ammonia in the glutamine synthetase reaction. NH_2OH also inhibits many enzymes, particularly those requiring pyridoxal 5'-phosphate. It is also a substrate of the reactions catalyzed by hydroxylamine reductase, hydroxylamine oxidase, hydroxylamine oxidoreductase, and aspartyltransferase. It is a product of hyponitrite reductase. It is also a chemical mutagen that deaminates cytosine residues within DNA, resulting in base pairing with an adenine residue instead of guanosine.

5-Hydroxylysine

L-5-hydroxylysine

D-5-hydroxylysine

allo-L-5-hydroxylysine

allo-D-5-hydroxylysine

This lysine derivative ($FW_{monohydrochloride}$ = 198.65 g/mol; Symbol = 5Hly or Hly), also called δ-hydroxylysine, 2,6-diamino-5-hydroxycaproate, and *erythro*-5-hydroxy-L-lysine, has two chiral centers, giving rise to four stereoisomers. Commercial sources typically contain all four isomers: D- and L-5-hydroxylysine (which are the *threo* isomers). The *allo*-isomers can be removed by chromatographic methods. L-5-Hydroxylysine has pKa values of 2.13, 8.85, and 9.83 (ε-amino). L-5-Hydroxylysyl residues are also found in the collagens and is an enzyme-catalyzed post-translational modification. **Target(s):** α-aminoadipate-semialdehyde dehydrogenase (1); glutamine synthetase (2,3); homocitrate synthase (4-6); homoserine kinase (7); D-lysine 5,6-aminomutase, inhibited by the DL mixture (8); L-lysine 6-aminotransferase, inhibited by the DL-5-hydroxylysine and DL-*allo*-5 hydroxylysine (9,10). **1.** Schmidt, Bode & Birnbaum (1990) *FEMS Microbiol. Lett.* **58**, 41. **2.** Webb (1966) *Enzyme and Metabolic Inhibitors*, vol. 2, p. 336, Academic Press, New York. **3.** Orr & Haselkorn (1981) *J. Biol. Chem.* **256**, 13099. **4.** Gaillardin, Poirier & Heslot (1976) *Biochim. Biophys. Acta* **422**, 390. **5.** Tucci & Ceci (1972) *Arch. Biochem. Biophys.* **153**, 742. **6.** Gray & Bhattacharjee (1976) *J. Gen. Microbiol.* **97**, 117. **7.** Thoen, Rognes & Aarnes (1978) *Plant Sci. Lett.* **13**, 103. **8.** Morley & Stadtman (1970) *Biochemistry* **9**, 4890. **9.** Soda & Misono (1971) *Meth. Enzymol.* **17B**, 222. **10.** Braunstein (1973) *The Enzymes*, 3rd ed., **9**, 379.

p-Hydroxymercuribenzoate

This sulfhydryl-reactive agent ($FW_{free-acid}$ = 338.71 g/mol; Abbreviation: PHMB) inhibits a large number of enzymes by modifying catalytically essential active-site thiols. The reactive species in aqueous solutions is the mercuric cation (^-O_3C–C_6H_4–Hg^+). Inhibition is often reversed by the addition of sulfhydryl-containing reagents such as cysteine, dithiothreitol, and glutathione. PHMB is water-soluble and the solubility increases with certain buffers (*e.g.*, Tris, glycylglycine, and pyrophosphate). **Note:** PHMB precipitates when combined with phosphate buffer. A similar result should be anticipated with nucleotides and other phosphorylated metabolites. PHMB forms readily upon hydrolysis of *p*-chloromercuribenzoate (PCMB). The sodium salt is prepared (PCMB is frequently dissolved in 200 mM NaOH followed by an adjustment to a lower pH [*e.g.*, pH 8]). **Targets:** PubMed and related literature searches indicate that >900 enzymes lose catalytic activity upon reaction with PHMB.

3-Hydroxy-4-methoxybenzaldehyde

This aromatic aldehyde (FW = 152.15 g/mol; M.P. = 113 115°C), also known as isovanillin, inhibits hemoglobin S polymerization. It is also a precursor in the biosynthesis of *p*'-*O* methylnorbellidine and lycorine. **Target(s):** aldehyde oxidase (1-3); hemoglobin S polymerization, mildly inhibited (4). **1.** Clarke, Harrell & Chenery (1995) *Drug Metab. Dispos.* **23**, 251. **2.** Austin, Baldwin, Cutler, *et al.* (2001) *Xenobiotica* **31**, 677. **3.** Panoutsopoulos & Beedham (2004) *Acta Biochim. Pol.* **51**, 649. **4.** Zaugg, Walder & Klotz (1977) *J. Biol. Chem.* **252**, 8542.

Nʸ-Hydroxy-Nʸ-methyl-L-arginine

This arginine derivative (FW = 205.24 g/mol) is both a reversible (K_i = 33.5 μM) and irreversible inhibitor (k_{inact} = 0.16 min⁻¹ and K_i = 26.5 μM) of nitric-oxide synthase. It also induces apoptosis in cell cultures and inhibited the growth of various transplantable mouse tumors. The demethylated N^γ-

hydroxy-L-arginine is an alternative substrate. **1**. Pufahl, Nanjappan, Woodard & Marletta (1992) *Biochemistry* **31**, 6822.

6-(1*S*-Hydroxy-3-methylbutyl)-7-methoxy-2*H*-chromen-2-one

This recently discovered coumarin (FW = 262.31 g/mol) is a novel mechanism-based inhibitor of carbonic anhydrase. Human carbonic anhydrase I, II, and IV have K_i values of 0.08, 0.06, and 3.8 μM, respectively. Unlike other anhydrase inhibitors, which coordinate directly to the active-site zinc ion, this agent undergoes slow conversion to its reaction product, *cis*-2-hydroxy-4-(1*S*-hydroxy 3-methylbutyl)-3-methoxycinnamic acid, which then remains firmly bound to the enzyme. The inhibition is time dependent, with maximum inhibition requiring 6 hours. The product adopts an extended conformation that plugs the entrance to the active site, while exhibiting no interaction with the zinc ion. **1**. Maresca, Temperini, Vu, *et al.* (2009) *J. Amer. Chem. Soc.* **131**, 3057.

3-Hydroxy-3-methylglutaric Acid

This β-hydroxy acid and isoprenoids/steroid precursor (FW$_{free-acid}$ = 126.11 g/mol; often abbreviated HMG), also known as 3-hydroxy-3 methylpentanedioic acid and dicrotalic acid, is a component of a key intermediate in ketogenesis and steroid biosynthesis. Note that the free di-acid is achiral; but becomes chiral upon formation of the CoA derivative. Only the (*S*)-isomer is biologically active. **Target(s):** aerobactin synthase (1); 3-hydroxy-3-methylglutaryl-CoA reductase (2-6); [3-hydroxy-3-methylglutaryl-CoA reductase (NADPH)]-phosphatase (7); kynurenine aminotransferase (8). **1**. Appanna & Viswanatha (1986) *FEBS Lett.* **202**, 107. **2**. Rodwell & Bensch (1981) *Meth. Enzymol.* **71**, 480. **3**. Russell (1985) *Meth. Enzymol.* **110**, 26. **4**. Sipat (1985) *Meth. Enzymol.* **110**, 40. **5**. Fimognari & Rodwell (1965) *Biochemistry* **4**, 2086. **6**. Kirtley & Rudney (1967) *Biochemistry* **6**, 230. **7**. Hegardt, Gil & Calvet (1983) *J. Lipid Res.* **24**, 821. **8**. Mason (1959) *J. Biol. Chem.* **234**, 2770.

(*S*)-3-Hydroxy-3-methylglutaryl-CoA

This thiolester (FW = 899.66 g/mol) is a key intermediate in the formation of steroids, ketone bodies, and terpenes. It is a product of the reactions catalyzed by hydroxymethylglutaryl-CoA synthase, methylglutaconyl-CoA hydratase, and succinate: hydroxymethylglutarate CoA-transferase. It is a substrate for hydroxymethylglutaryl-CoA lyase, hydroxymethylglutaryl-CoA reductase, and hydroxymethylglutaryl-CoA hydrolase. **Target(s):** [3-hydroxy-3-methylglutaryl-CoA reductase (NADPH)] kinase (1); [3-hydroxy 3-methylglutaryl-CoA reductase (NADPH)]-phosphatase (2); [3-methyl-2-oxobutanoate dehydrogenase, *or* 2-methylpropanoyl-transferring)]-phosphatase, *or* [branched-chain-α-keto-acid dehydrogenase] phosphatase (3,4); phosphoprotein phosphatase (2-4). **1**. Omkumar, Darnay & Rodwell (1994) *J. Biol. Chem.* **269**, 6810. **2**. Hegardt (1986) *Adv. Protein Phosphatases* **3**, 1. **3**. Damuni, Merryfield, Humphreys & Reed (1984) *Proc. Natl. Acad. Sci. U.S.A.* **81**, 4335. **4**. Reed & Damuni (1987) *Adv. Protein Phosphatases* **4**, 59.

9-(2-Hydroxymethylmorpholin-6-yl)adenine *O²*-Triphosphate

This nucleotide analogue (FW$_{free-acid}$ = 490.20 g/mol) acts as an alternative substrate for a number of DNA polymerases, resulting in chain termination. **Target(s):** DNA-directed DNA polymerase; DNA polymerase β; HIV-1 reverse transcriptase; Moloney murine leukemia virus reverse transcriptase (not a substrate); RNA directed DNA polymerase. **1**. Lebedeva, Seredina, Silnikov, *et al.* (2005) *Biochemistry (Moscow)* **70**, 1.

2-Hydroxy-1-naphthoate and 2-Hydroxy-1-naphthoic Acid

This naphthoate derivative (FW$_{free-acid}$ = 188.18 g/mol) inhibits 6-methylsalicylate decarboxylase, medium-chain acyl-CoA synthetase, and plant cinnamate 4-monooxygenase. **Target(s):** CYP73A family, *or* cinnamate 4-monooxygenase (1,2); medium-chain-fatty-acyl CoA synthetase, *or* butyryl-CoA synthetase, *or* butyrate:CoA ligase (3,4); 6-methylsalicylate decarboxylase (5). **1**. Schalk, Batard, Seyer, *et al.* (1997) *Biochemistry* **36**, 15253. **2**. Schoch, Nikov, Alworth & Werck-Reichhart (2002) *Plant Physiol.* **130**, 1022. **3**. Kasuya, Igarashi & Fukui (1996) *Biochem. Pharmacol.* **52**, 1643. **4**. Kasuya, Hiasa, Kawai, Igarashi & Fukui (2001) *Biochem. Pharmacol.* **62**, 363. **5**. Light & Vogel (1975) *Meth. Enzymol.* **43**, 530.

1-Hydroxy-2-nitro-1,3-propanedicarboxylic Acid

This α-hydroxy acid (FW$_{free-acid}$ = 193.11 g/mol), also known as nitroisocitrate, which is a mimic of an aconitase reaction-cycle intermediate mimic, displays slow-binding properties, with a K_i value of 0.7 nM (1). While the hydroxyl oxygen of both citrate and isocitrate are exchanged with solvent water via the action of aconitase, the hydroxyl oxygen of nitroisocitrate is not. **1**. Schloss, Porter, Bright & Cleland (1980) *Biochemistry* **19**, 2358. **2**. Ramsay, Dreyer, Schloss, *et al.* (1981) *Biochemistry* **20**, 7476. **3**. Werst, Kennedy, Beinert & Hoffman (1990) *Biochemistry* **29**, 10526. **4**. Telser, Emptage, Merkle, *et al.* (1986) *J. Biol. Chem.* **261**, 4840. **5**. Lauble, Kennedy, Beinert & Stout (1992) *Biochemistry* **31**, 2735.

(3-Hydroxy-2-nitropropyl)phosphonate

This transition-state inhibitor ($FW_{\text{free-acid}}$ = 185.07 g/mol) is an analogue of the intermediate in the reaction catalyzed by yeast enolase. The ionized (pK = 8.1) nitronate form, in the presence of 5 mM Mg^{2+}, has a K_i value of 6 nM. This finding supports a carbanion mechanism for enolase and suggests that the 3-hydroxyl of 2-phosphoglycerate is directly coordinated to Mg^{2+} prior to its elimination to yield phosphoenolpyruvate. **1.** Anderson, Weiss & Cleland (1984) *Biochemistry* **23**, 2779. **2.** Poyner, Cleland & Reed (2001) *Biochemistry* **40**, 8009.

4-Hydroxy-*trans*-2-nonenal

This neurotoxic aldehyde (FW = 156.22 g/mol; Abbreviation: HNE) inhibits a wide variety of enzymes and induces apoptosis. It is a major product of lipid peroxidation and is elevated during oxidative stress. Levels are also elevated in Alzheimer and Parkinson diseases. HNE can modify amino, sulfhydryl, and imidazole groups in proteins as well as the sulfhydryl groups of dihydrolipoyl groups in the pyruvate dehydrogenase and α-ketoglutarate dehydrogenase complexes. **Target(s):** adenylate cyclase (1); aldose reductase (2); bile-acid-CoA:amino acid *N*-acyltransferase (40); carbonic anhydrase (3); carnitine *O*-acetyltransferase (41,42); cathepsin B (4;) Ca^{2+}-transporting ATPase (5); creatine kinase (6); CYP1A1 (7,8); CYP1A2 (7,8); CYP3A6 (7); CYP2B1 (8); CYP2B4 (7,8); CYP2C3 (8); CYP2E1 (7,8); cytochrome-*c* oxidase (9,10,43); DNA-directed DNA polymerase (11); DNA polymerase α (11); ecto-ATPase (12); glucose-6-phosphatase (13); glucose-6-phosphate dehydrogenase (14,15); glutamate transport (16); glutathione-disulfide reductase (17); glutathione peroxidase (18,19); glutathione *S*-transferase (20); glyceraldehyde-3-phosphate dehydrogenase (21,22); glycine decarboxylase H protein (23); hsp72 mediated protein refolding (24); IκB kinase (25); interleukin-1β converting enzyme (26); isocitrate dehydrogenase (NADP$^+$) (27,28); α-ketoglutarate dehydrogenase complex (23,29); lipoxygenase-1, soybean (30); malic enzyme (23); Na^+/K^+-exchanging ATPase (31); 5'-nucleotidase (1); phospholipase D, sphingosine-stimulated (32); phospholipid-translocating ATPase ("flippase") (33); protein-disulfide isomerase (34); pyruvate dehydrogenase complex (23,28,35); succinate-semialdehyde dehydrogenase (36); thioredoxin reductase (37); ubiquitin thiolesterase (38); xanthine oxidase (39). **1.** Paradisi, Panagini, Parola, Barrera & Dianzani (1985) *Chem. Biol. Interact.* **53**, 209. **2.** Del Corso, Dal Monte, Vilardo, *et al.* (1998) *Arch. Biochem. Biophys.* **350**, 245. **3.** Uchida, Hasui & Osawa (1997) *J. Biochem.* **122**, 1246. **4.** Crabb, O'Neil, Miyagi, West & Hoff (2002) *Protein Sci.* **11**, 831. **5.** McConnell, Bittelmeyer & Raess (1999) *Arch. Biochem. Biophys.* **361**, 252. **6.** Eliuk, Renfrow, Shonsey, Barnes & Kim (2007) *Chem. Res. Toxicol.* **20**, 1260. **7.** Bestervelt, Vaz & Coon (1995) *Proc. Natl. Acad. Sci. U.S.A.* **92**, 3764. **8.** Kuo, Vaz & Coon (1997) *J. Biol. Chem.* **272**, 22611. **9.** Chen, Schenker, Frosto & Henderson (1998) *Biochim. Biophys. Acta* **1380**, 336. **10.** Musatov, Carroll, Liu, *et al.* (2002) *Biochemistry* **41**, 8212. **11.** Wawra, Zollner, Schaur, Tillian & Schauenstein (1986) *Cell Biochem. Funct.* **4**, 31. **12.** Foley (1999) *Neurochem. Res.* **24**, 1241. **13.** Koster, Slee, Montfoort, Lang & Esterbauer (1986) *Free Radic. Res. Commun.* **1**, 273. **14.** Szweda, Uchida, Tsai & Stadtman (1993) *J. Biol. Chem.* **268**, 3342. **15.** Ninfali, Ditroilo, Capellacci & Biagiotti (2001) *Neuroreport* **12**, 4149. **16.** Blanc, Keller, Fernandez & Mattson (1998) *Glia* **22**, 149. **17.** Vander Jagt, Hunsaker, Vander Jagt, *et al.* (1997) *Biochem. Pharmacol.* **53**, 1133. **18.** Kinter & Roberts (1996) *Free Radic. Biol. Med.* **21**, 457. **19.** Bosch-Morell, Flohe, Marin & Romero (1999) *Free Radic. Biol. Med.* **26**, 1383. **20.** van Iersel, Ploemen, Lo Bello, Federici & van Bladeren (1997) *Chem. Biol. Interact.* **108**, 67. **21.** Hiratsuka, Hirose, Saito & Watabe (2000) *Biochem. J.* **349**, 729. **22.** Ishii, Tatsuda, Kumazawa, Nakayama & Uchida (2003) *Biochemistry* **42**, 3474. **23.** Millar & Leaver (2000) *FEBS Lett.* **481**, 117. **24.** Carbone, Doorn, Kiebler, Sampey & Petersen (2004) *Chem. Res. Toxicol.* **17**, 1459. **25.** Ji, Kozak & Marnett (2001) *J. Biol. Chem.* **276**, 18223. **26.** Davis, Hamilton & Holian (1997) *J. Interferon Cytokine Res.*

17, 205. **27.** Yang, Yang & Park (2004) *Free Radic. Res.* **38**, 241. **28.** Benderdour, Charron, DeBlois, Comte & Des Rosiers (2003) *J. Biol. Chem.* **278**, 45154. **29.** Humphries & Szweda (1998) *Biochemistry* **37**, 15835. **30.** Gardner & Deighton (2001) *Lipids* **36**, 623. **31.** Siems, Hapner & van Kuijk (1996) *Free Radic. Biol. Med.* **20**, 215. **32.** Kiss, Crilly, Rossi & Anderson (1992) *Biochim. Biophys. Acta* **1124**, 300. **33.** Castegna, Lauderback, Mohmmad-Abdul & Butterfield (2004) *Brain Res.* **1004**, 193. **34.** Liu & Sok (2004) *Biol. Chem.* **385**, 633. **35.** Patel & Korotchkina (2002) *Methods Mol. Biol.* **186**, 255. **36.** Nguyen & Picklo (2003) *Biochim. Biophys. Acta* **1637**, 107. **37.** Yu, Moos, P. Cassidy, Wade & Fitzpatrick (2004) *J. Biol. Chem.* **279**, 28028. **38.** Nishikawa, Li, Kawamura, *et al.* (2003) *Biochem. Biophys. Res. Commun.* **304**, 176. **39.** Cighetti, Bortone, Sala & Allevi (2001) *Arch. Biochem. Biophys.* **389**, 195. **40.** Shonsey, Eliuk, Johnson, *et al.* (2008) *J. Lipid Res.* **49**, 282. **41.** Miyazawa, Ozasa, Furuta, Osumi & Hashimoto (1983) *J. Biochem.* **93**, 439. **42.** Liu, Killilea & Ames (2002) *Proc. Natl. Acad. Sci. U.S.A.* **99**, 1876. **43.** Kaplan, Tatarkova, Racay, *et al.* (2007) *Redox Rep.* **12**, 211.

erythro-9-(2-Hydroxy-3-nonyl)adenine

This hygroscopic purine nucleoside analogue ($FW_{\text{free-base}}$ = 277.37 g/mol), also known as EHNA and *erythro*-3-(adenin-9-yl)-2-nonanol, is a strong (slow-binding) inhibitor of adenosine deaminase, K_i = 4 nM (1-13), adenosine kinase (14), and cGMP-stimulated phosphodiesterase II, IC_{50} = 0.8 μM (15-19). EHNA also inhibits the enuculation of terminally differentiated erythroblasts, demonstrating a role for dynein in the formation of erythrocytes and suggesting that dynein helps to shift the nucleus to a position near the plasma membrane (20). EHNA dramatically inhibits 2D migration of the fibroblast, when cell contractility was blocked by Rho kinase or a myosin inhibitor, although EHNA itself did not affect cell migration (21). Cell migration in 3D soft collagen matrices, where the cell exerts a relatively low tractional force compared to that on a 2D stiff surface, is also profoundly inhibited by dynein intermediate chain (DIC) silencing regardless of the presence of myosin activity. **Target(s):** deoxycytidine kinase, K_i = 0.45 mM (22); dynein ATPase (23-26); GMP synthetase, weakly inhibited (27); 5'-methylthioadenosine nucleosidase (28,29); phosphodiesterase (14-18); phosphodiesterase 4 (18); protein-glutamate *O*-methyltransferase (24); purine-nucleoside phosphorylase (30). **1.** Agarwal, Spector & Parks (1977) *Biochem. Pharmacol.* **26**, 359. **2.** Agarwal & Parks (1978) *Meth. Enzymol.* **51**, 502. **3.** Morrison (1982) *Trends Biochem. Sci.* **7**, 102. **4.** Baker, Hanvey, Hawkins & Murphy (1981) *Biochem. Pharmacol.* **30**, 1159. **5.** Bessodes, Bastian, Abushanab, *et al.* (1982) *Biochem. Pharmacol.* **31**, 879. **6.** Schaeffer & Schwender (1974) *J. Med. Chem.* **17**, 6. **7.** O'Connell & Keller (1994) *Proc. Natl. Acad. Sci. U.S.A.* **91**, 10596. **8.** Philips, Robbins & Coleman (1987) *Biochemistry* **26**, 2893. **9.** Mardanyan, Sharoyan, Antonyan, Lupidi & Cristalli (2002) *Biochemistry (Moscow)* **67**, 770. **10.** Iwaki-Egawa & Watanabe (2002) *Comp. Biochem. Physiol. B* **133**, 173. **11.** Daddona, Wiesmann, Lambros, Kelley & Webster (1984) *J. Biol. Chem.* **259**, 1472. **12.** Lupidi, Marmocchi & Cristalli (1998) *Biochem. Mol. Biol. Int.* **46**, 1071. **13.** Wiginton, Coleman & Hutton (1981) *Biochem. J.* **195**, 389. **14.** Miller, Adamczyk, Miller, *et al.* (1979) *J. Biol. Chem.* **254**, 2346. **15.** Podzuweit, Nennstiel & Muller (1995) *Cell Signal* **7**, 733. **16.** Mery, Pavoine, Pecker & Fischmeister (1995) *Mol. Pharmacol.* **48**, 121. **17.** Sudo, Tachibana, Toga, *et al.* (2000) *Biochem. Pharmacol.* **59**, 347. **18.** O'Grady, Jiang, Maniak, *et al.* (2002) *J. Membr. Biol.* **185**, 137. **19.** Iffland, Kohls, Low, *et al.* (2005) *Biochemistry* **44**, 8312. **20.** Kobayashi, Ubukawa, Sugawara, *et al.* (2015) *Exp. Hematol.* **44**, 247. **21.** Kim, You & Rhee (2012) *Int. J. Mol. Med.* **29**, 440. **22.** Krenitsky, Tuttle, Koszalka, *et al.* (1976) *J. Biol. Chem.* **251**, 4055. **23.** Paschal, Shpetner & Vallee (1991) *Meth. Enzymol.* **196**, 181. **24.** Bouchard, Penningroth, Cheung, Gagnon & Bardin (1981) *Proc. Natl. Acad. Sci. U.S.A.* **78**, 1033. **25.** Saucier, Mariotti, Anderson & Purich (1985) *Biochemistry* **24**, 7581. **26.** Belles-Isles, Chapeau, White & Gagnon (1986) *Biochem. J.* **240**, 863. **27.** Spector & Beecham (1975) *J. Biol. Chem.*

250, 3101. **28**. Guranowski, Chiang & Cantoni (1983) *Meth. Enzymol.* **94**, 365. **29**. Guranowski, Chiang & Cantoni (1981) *Eur. J. Biochem.* **114**, 293. **30**. Lewis (1978) *Arch. Biochem. Biophys.* **190**, 662.

N^ω-Hydroxy-nor-L-arginine

This arginine derivative (FW = 176.18 g/mol) is a slow-binding inhibitor of arginase II, $K_i = 51$ nM at pH 7.5. Due to the reciprocal regulation between arginase and nitric oxide synthase, arginase inhibitors have therapeutic potential in treating nitric oxide-dependent smooth muscle disorders, such as erectile dysfunction. **1**. Colleluori & Ash (2001) *Biochemistry* **40**, 9356.

3-*endo*-8-*exo*-8-Hydroxy-2-oxabicyclo[3.3.1]non-6-ene-3,5-dicarboxylate

This bicyclic inhibitor (FW$_{free-acid}$ = 228.20 g/mol), also known as (1*R*,3*R*,5*S*)-3-carboxy-1-hydroxy-2-oxabicyclo[3.3.1]non-6-ene-5-carboxylate is a putative transition-state analogue for the reaction catalyzed by chorismate mutase, $K_i = 0.12$ μM for the *Escherichia coli* enzyme). **1**. Bartlett & Johnson (1985) *J. Amer. Chem. Soc.* **107**, 7792. **2**. Bartlett, Nakagawa, Johnson, Reich & Luis (1988) *J. Org. Chem.* **53**, 3195. **3**. Turnbull, Cleland & Morrison (1991) *Biochemistry* **30**, 7777. **4**. Mandal & Hilvert (2003) *J. Amer. Chem. Soc.* **125**, 5598. **5**. Gray, Eren & Knowles (1990) *Biochemistry* **29**, 8872. **6**. Sasso, Ramakrishnan, Gamper, Hilvert & Kast (2005) *FEBS J.* **272**, 375.

p-Hydroxyoxanilic Acid

This acid (FW$_{free-acid}$ = 181.14 g/mol) inhibits 4-hydroxyphenyl-pyruvate dioxygenase, $K_i = 6$ μM (1) and phenylpyruvate tautomerase, $K_i = 1.6$ mM (2), and exo-α-sialidase (*or* neuraminidase, *or* sialidase (3,4). **1**. Pascal, Oliver & Chen (1985) *Biochemistry* **24**, 3158. **2**. Pirrung, Chen, Rowley & McPhall (1993) *J. Amer. Chem. Soc.* **115**, 7103. **3**. Nok, Nzelibe & Yako (2003) *Z. Naturforsch. C* **58**, 594. **4**. Buratai, Nok, Ibrahim, Umar & Esievo (2005) *Cell Biochem. Funct.* **24**, 71.

4-Hydroxyphenylacetate

This plant phenolic (FW$_{free-acid}$ = 152.15 g/mol; M.P. = 149-151°C), found in dandelions (*Taraxacum officinalis*), is a product of reactions catalyzed by 4-hydroxyphenylacetaldehyde dehydrogenase and 4-hydroxyphenyl-pyruvate oxidase. (Do not confuse this anion with the ester, 4-hydroxyphenyl acetate, *i.e.*, *p*-HO–C₆H₄–O(C=O)CH₃.) **Target(s):** [branched-chain α-keto-acid dehydrogenase] kinase, *or* [3-methyl-2 oxobutanoate dehydrogenase (acetyl-transferring)] kinase (1); dihydropteridine reductase (2); 3,4 dihydroxyphenylacetate 2,3-dioxygenase (3); diiodotyrosine aminotransferase (4); diphosphomevalonate decarboxylase (5); β-glucosidase (6); glutamate decarboxylase (7);

homogentisate 1,2-dioxygenase (8); 4-hydroxybenzoate 3-monooxygenase (9); imidazoleacetate 4-monooxygenase (10); laccase (11); protocatechuate 3,4-dioxygenase (12,13); tyrosine aminotransferase (14,15). **1**. Paxton & Harris (1984) *Arch. Biochem. Biophys.* **231**, 58. **2**. Shen (1984) *Biochim. Biophys. Acta* **785**, 181. **3**. Que, Widom & Crawford (1981) *J. Biol. Chem.* **256**, 10941. **4**. Nakano (1967) *J. Biol. Chem.* **242**, 73. **5**. Shama Bhat & Ramasarma (1979) *Biochem. J.* **181**, 143. **6**. Dale, Ensley, Kern, Sastry & Byers (1985) *Biochemistry* **24**, 3530. **7**. Webb (1966) *Enzyme and Metabolic Inhibitors*, vol. **2**, p. 329, Academic Press, New York. **8**. Hudecová, Straková & Krizanova (1995) *Int. J. Biochem. Cell Biol.* **27**, 1357. **9**. Sterjiades (1993) *Biotechnol. Appl. Biochem.* **17**, 77. **10**. Watanabe, Kambe, Imamura, *et al.* (1983) *Anal. Biochem.* **130**, 321. **11**. Casella, Gullotti, Monzani, *et al.* (2006) *J. Inorg. Biochem.* **100**, 2127. **12**. Orville, Elango, Lipscomb & Ohlendorf (1997) *Biochemistry* **36**, 10039. **13**. Orville & Lipscomb (1989) *J. Biol. Chem.* **264**, 8791. **14**. Granner & Tomkins (1970) *Meth. Enzymol.* **17A**, 633. **15**. Jacoby & La Du (1964) *J. Biol. Chem.* **239**, 419.

4-Hydroxyphenylglyoxal

This α-ketoaldehyde (FW = 150.13 g/mol) is a more hydrophilic derivative of phenylglyoxal that also chemically modifies arginyl residues in proteins. By comparing the rates of modification with these two reagents, one may assess the hydrophobic *versus* hydrophilic environment surrounding essential arginyl residues. **Target(s):** amadoriase II (1); 5-enolpyruvylshikimate-3-phosphate synthase (2); β-1,3 galactosyl-*O*-glycosyl-glycoprotein β-1,6-*N*-acetylglucosaminyltransferase (3); glutamine synthetase (4); homoserine kinase (5); peptidyl dipeptidase IV (6); phosphate transporter (7); protein-*S* isoprenylcysteine *O*-methyltransferase (8); tryptophan synthase (9). **1**. Wu, Chen, Petrash & Monnier (2002) *Biochemistry* **41**, 4453. **2**. Padgette, Smith, Huynh & Kishore (1988) *Arch. Biochem. Biophys.* **266**, 254. **3**. Yang, Qin, Lehotay, *et al.* (2003) *Biochim. Biophys. Acta* **1648**, 62. **4**. Colanduoni & Villafranca (1985) *Biochem. Biophys. Res. Commun.* **126**, 412. **5**. Huo & Viola (1996) *Biochemistry* **35**, 16180. **6**. Lanzillo, Dasarathy & Fanburg (1989) *Biochem. Biophys. Res. Commun.* **160**, 243. **7**. McIntosh & Oliver (1994) *Plant Physiol.* **105**, 47. **8**. Bolvin, Lin & Béliveau (1997) *Biochem. Cell Biol.* **75**, 63. **9**. Eun & Miles (1984) *Biochemistry* **23**, 6484.

4-Hydroxyphenyllactic Acid

This tyrosine metabolite (FW$_{free-acid}$ = 182.18 g/mol), also known as 3-(4-hydroxyphenyl)-2-hydroxypropanoic acid, is soluble in water and ethanol. If heated to 100°C, hydroxyphenyllactic acid is converted to 4-hydroxycinnamic acid. **Target(s):** [branched-chain α-keto-acid dehydrogenase] kinase, *or* [3-methyl-2 oxobutanoate dehydrogenase acetyl-transferring)] kinase, weakly inhibited (1,2); dihydropteridine reductase (3); diiodotyrosine aminotransferase, weakly inhibited (4); diphospho-mevalonate decarboxylase (5); Na⁺/K⁺-exchanging ATPase (6); prephenate dehydrogenase (7,8); tyrosyl-tRNA synthetase, $K_i = 8.1$ mM for the *Escherichia coli* enzyme (9). **1**. Paxton (1988) *Meth. Enzymol.* **166**, 313. **2**. Paxton & Harris (1984) *Arch. Biochem. Biophys.* **231**, 58. **3**. Shen (1984) *Biochim. Biophys. Acta* **785**, 181. **4**. Nakano (1967) *J. Biol. Chem.* **242**, 73. **5**. Shama Bhat & Ramasarma (1979) *Biochem. J.* **181**, 143. **6**. Seda, Gove, Hughes & Williams (1984) *Clin. Sci. (London)* **66**, 415. **7**. Cotton & Gibson (1970) *Meth. Enzymol.* **17A**, 564. **8**. Christendat & Turnbull (1999) *Biochemistry* **38**, 4782. **9**. Santi & Peña (1973) *J. Med. Chem.* **16**, 273.

3-(3-Hydroxyphenyl)-N-propylpiperidine

These dopamine receptor modulators (FW$_{free-base}$ = 219.33 g/mol) exert opposing CNS effects. The R(+)-stereoisomer, also known as R(+)-N-(n-propyl)-3-(3-hydroxyphenyl)piperidine (or, R(+)-3-PPP), is a dopamine D$_2$ receptor agonist and σ$_1$ receptor antagonist. The S(–)-enantiomer, or preclamol, is a dopamine receptor antagonist and a dopamine autoreceptor agonist. **1**. Arnt, Bogeso, Christensen, et al. (1983) Psychopharmacology **81**, 199. **2**. Hjorth, Carlsson, Clark, et al. (1983) Psychopharmacology **81**, 89. **3**. Tokuyama, Hirata, Ide & Ueda (1997) Neurosci. Lett. **233**, 141. **4**. Bergeron & Debonnel (1997) Psychopharmacology **129**, 215.

4-Hydroxyphenylpyruvate

This tyrosine metabolite (FW$_{free-acid}$ = 180.16 g/mol), also known as 3-(4-hydroxyphenyl)-2-oxopropanoic acid, is a product of the reaction catalyzed by tyrosine aminotransferase and prephenate dehydrogenase. 4-Hydroxyphenylpyruvate is rapidly oxidized in alkaline solutions, yielding 4-hydroxybenzaldehyde. **Target(s):** aromatic-amino-acid decarboxylase, or dopa decarboxylase (1); [branched-chain α keto-acid dehydrogenase] kinase, or [3-methyl-2-oxobutanoate dehydrogenase (acetyl-transferring)] kinase, weakly inhibited (2,3); carnitine:acylcarnitine translocase, weakly inhibited (4); cathepsin D (5); choline acetyltransferase (6); cyclohexadienyl dehydrogenase (7); dihydropteridine reductase (8); D-dopachrome decarboxylase (9); histidine ammonia-lyase (10,11); homogentisate 1,2-dioxygenase (12); horseradish peroxidase (13); kynureninase14; Na$^+$/K$^+$-exchanging ATPase (15); peroxidase (13); phenylalanine 4-monooxygenase, activated at low concentrations (16); phenylpyruvate decarboxylase (17); prephenate dehydrogenase (18); pyruvate decarboxylase (1,19); rosmarinate synthase (20); transketolase (21,22); tyrosine 2,3-aminomutase (23); tyrosyl-tRNA synthetase, K_i = 9.9 mM for the Escherichia coli enzyme (24). **1**. Webb (1966) Enzyme and Metabolic Inhibitors, vol. **2**, Academic Press, New York. **2**. Paxton (1988) Meth. Enzymol. **166**, 313. **3**. Paxton & Harris (1984) Arch. Biochem. Biophys. **231**, 58. **4**. Parvin & Pande (1978) J. Biol. Chem. **253**, 1944. **5**. Woessner & Shamberger (1971) J. Biol. Chem. **246**, 1951. **6**. Nachmansohn & John (1945) J. Biol. Chem. **158**, 157. **7**. Xie, Bonner & Jensen (2000) Comp. Biochem. Physiol. C Toxicol. Pharmacol. **125**, 65. **8**. Shen (1984) Biochim. Biophys. Acta **785**, 181. **9**. Rosengren, Thelin, Aman, et al. (1997) Melanoma Res. **7**, 517. **10**. Rechler & Tabor (1971) Meth. Enzymol. **17B**, 63. **11**. Hug & Roth (1968) Biochem. Biophys. Res. Commun. **30**, 248. **12**. Fernández-Cañón & Peñalva (1997) Anal. Biochem. **245**, 218. **13**. Gamborg, Wetter & Neish (1961) Can. J. Biochem. Physiol. **39**, 1113. **14**. Shibata, Takeuchi, Tsubouchi, et al. (1991) Adv. Exp. Med. Biol. **294**, 523. **15**. Seda, Gove, Hughes & Williams (1984) Clin. Sci. (London) **66**, 415. **16**. Letendre, Dickens & Guroff (1975) J. Biol. Chem. **250**, 6672. **17**. Fujioka, Morino & Wada (1970) Meth. Enzymol. **17A**, 585. **18**. Cotton & Gibson (1970) Meth. Enzymol. **17A**, 564. **19**. Gale (1961) Arch. Biochem. Biophys. **94**, 236. **20**. Petersen (1991) Phytochemistry **30**, 2877. **21**. Solovjeva & Kochetov (1999) FEBS Lett. **462**, 246. **22**. Joshi, Singh, Kumar, et al. (2008) Mol. Biochem. Parasitol. **160**, 32. **23**. Kurylo-Borowska & Abramsky (1972) Biochim. Biophys. Acta **264**, 1. **24**. Santi & Peña (1973) J. Med. Chem. **16**, 273.

(S)-9-(3-Hydroxy-2-phosphonylmethoxypropyl)adenine

This acyclic adenylate analogue (FW$_{free-acid}$ = 303.21 g/mol), often abbreviated HPMPA, exhibits a broad-spectrum antiviral activity. The pseudonucleoside is phosphorylated in vivo to the diphosphate, which inhibits viral DNA phosphorylase. **Target(s):** herpes viruses (HSV-1 and HSV-2) replication (1,2); viral DNA polymerases (1,3,4). **1**. Prisbe & Chen (1996) Meth. Enzymol. **275**, 425. **2**. De Clercq, Holy, Rosenberg, et al. (1986) Nature **323**, 464. **3**. Merta, Votruba, Rosenberg, et al. (1990) Antiviral Res. **13**, 209. **4**. Arzuza, Garcia-Villalon, Tabares, Gil-Fernandez & De Clercq (1988) Biochem. Biophys. Res. Commun. **154**, 27.

5-Hydroxypipecolate (5-Hydroxypipecolic Acid)

This cyclic amino acid and 4-hydroxyproline "homologue" (FW = 145.16 g/mol; CAS 13096-31-6 (free base) and 824943-40-0 (HCl salt)), also called 5-hydroxy-2-piperidinecarboxylic acid and 5-hydroxypipecolinic acid, is readily prepared from 5-hydroxylysine. The L-enantiomer is found in a number of plants. The trans-isomer has been identified in Bocoa prouacensis, Candolleodendron brachystachyum, Xylia xylocarpa, and Swartzia macrosema. Both the cis- and trans-isomers have been identified in Gymnocladus dioicus. **Target(s):** hydroxylysine kinase (1); platelet aggregation (2). **1**. Hiles & Henderson (1972) J. Biol. Chem. **247**, 646. **2**. Mester, Szabados, Mester & Yadav (1979) Planta Med. **35**, 339.

17α-Hydroxyprogesterone

This steroid (FW = 330.47 g/mol), also known as 17α-hydroxypregn-4-ene-3,20-dione, is an intermediate in steroid hormone biosynthesis and a major metabolite of renal progesterone metabolism. It also exhibits antiglucocorticoid activity (1) and is an estrus regulator. **Target(s):** arylamine N-acetyltransferase (2); glucose-6-phosphate dehydrogenase (3); 20α-hydroxysteroid dehydrogenase (4); steroid 17α-hydroxylase-C17,20 lyase, hydroxylase activity inhibited (5); steroid 5α-reductase (6,7); steroid sulfatase (8); sterol O-acyltransferase, or cholesterol O-acyltransferase (ACAT) (9); sterol demethylase (10); UDP-glucuronosyltransferase (11). **1**. Scott & Tomkins (1975) Meth. Enzymol. **40**, 273. **2**. Kawamura,

Westwood, Wakefield, *et al.* (2008) *Biochem. Pharmacol.* **75**, 1550. **3.** Criss & McKerns (1969) *Biochim. Biophys. Acta* **184**, 486. **4.** Robertson, Frost, Hoyer & Weinkove (1982) *J. Steroid Biochem.* **17**, 237. **5.** Nakajin, Takahashi & Shinoda (1985) *Yakugaku Zasshi* **105**, 83. **6.** Schubert, Schumann, Rose, *et al.*(1978) *Endokrinologie* **72**, 141. **7.** Kinoshita (1981) *Endocrinol. Jpn.* **28**, 499. **8.** Notation & Ungar (1969) *Biochemistry* **8**, 501. **9.** Simpson & Burkhart (1980) *Arch. Biochem. Biophys.* **200**, 79. **10.** Gaylor, Chang, Nightingale, Recio & Ying (1965) *Biochemistry* **4**, 1144. **11.** Falany, Green, Swain & Tephly (1986) *Biochem. J.* **238**, 65.

4-Hydroxy-L-proline

cis-isomer   **trans-isomer**

This hydroxylated imino acid (FW = 131.13 g/mol; pK_a values of 1.82 and 9.66 at 25°C) has two stereoisomers, the most common being *trans*-4 hydroxy-L-proline (symbolized by 4Hyp). This nonproteogenic imino acid is abundant in collagen, arising by posttranslational modification catalyzed by prolyl hydroxylase. 4-Hydroxyproline can also be produced via the cyclization of γ-hydroxyglutamate. The *cis* form of L-proline (also called allo-4-hydroxy-L-proline) is found in free form in *Santalum album* and other plants. **Target(s):** alkaline phosphatase (1); arginase (1); hydroxylysine kinase (2); pyrroline-5 carboxylate dehydrogenase (3,4); pyrroline-5-carboxylate reductase (5). **1.** Bodansky & Strachman (1949) *J. Biol. Chem.* **179**, 81. **2.** Hunter & Downs (1945) *J. Biol. Chem.* **157**, 427. **3.** Hiles & Henderson (1972) *J. Biol. Chem.* **247**, 646. **4.** Strecker (1971) *Meth. Enzymol.* **17B**, 262. **5.** Webb (1966) *Enzyme and Metabolic Inhibitors*, vol. **2**, p. 355, Academic Press, New York.

3-Hydroxypyruvate

This α-keto acid ($FW_{free-acid}$ = 104.06 g/mol), also known as 3-hydroxy-2-oxopropanoic acid, is soluble in diethyl ether, acetone, and water. Aqueous solutions spontaneously dimerize to form 2,3-dihydroxy-2-hydroxymethyl-4-ketoglutarate. 3-Hydroxypyruvate is a product of the reactions catalyzed by glycerate dehydrogenase, serine 2-dehydrogenase, serine:glyoxylate aminotransferase, and serine:pyruvate aminotransferase. It is also a substrate for hydroxypyruvate reductase, hydroxypyruvate decarboxylase, pyruvate decarboxylase, and hydroxypyruvate isomerase. It is a suicide inhibitor of acetolactate synthase. **Target(s):** acetolactate synthase (1); *N*-acetylneuraminate lyase (2); *N*-acetylneuraminate lyase mutant with an increased dihydrodipicolinate synthase activity (3); acid phosphatase, weakly inhibited, K_i = 111 mM (4); glycerate kinase (5); 4-hydroxy-2-oxoglutarate aldolase (6,7); lactate dehydrogenase, also alternative substrate (8); phosphoglycerate phosphatase (9); pyruvate carboxylase (10,11); pyruvate decarboxylase, also alternative substrate (12); pyruvate dehydrogenase complex (13,14); pyruvate:ferredoxin oxidoreductase (15). **1.** Duggleby (2005) *J. Enzyme Inhib. Med. Chem.* **20**, 1. **2.** Lawrence, Barbosa, Smith, *et al.* (1997) *J. Mol. Biol.* **266**, 381. **3.** Joerger, Mayer & Fersht (2003) *Proc. Natl. Acad. Sci. U.S.A.* **100**, 5694. **4.** Kilsheimer & Axelrod (1957) *J. Biol. Chem.* **227**, 879. **5.** Yoshida, Fukuta, Mitsunaga, Yamada & Izumi (1992) *Eur. J. Biochem.* **210**, 849. **6.** Grady, Wang & Dekker (1981) *Biochemistry* **20**, 2497. **7.** Anderson, Scholtz & Schuster (1985) *Arch. Biochem. Biophys.* **236**, 82. **8.** Busch & Nair (1957) *J. Biol. Chem.* **229**, 377. **9.** Pestka & Delwiche (1981) *Can. J. Microbiol.* **27**, 808. **10.** Thomas, Diefenbach & Duggleby (1990) *Biochem. J.* **266**, 305. **11.** Charles & Willer (1984) *Can. J. Microbiol.* **30**, 532. **12.** Candy & Duggleby (1998) *Biochim. Biophys. Acta* **1385**, 323. **13.** Randall (1982) *Meth. Enzymol.* **89**, 408. **14.** Randall, Rubin & Fenko (1977) *Biochim. Biophys. Acta* **485**, 336. **15.** Williams, Leadlay & Lowe (1990) *Biochem. J.* **268**, 69.

8-Hydroxyquinoline

This laboratory reagent (FW = 145.16 g/mol; M.P. = 76°C), also known as 8-quinolinol, 8-oxyquinoline, and oxine, is a metal ion chelator that forms tight, water-insoluble complexes. Its solubility in diethyl ether, benzene, chloroform affords a way to deplete metal ion contaminants in aqueous solutions used in enzyme research. The copper complex has also been used as a fungicide. Chelators inhibit many metal ion-requiring enzymes as well as those enzymes, for which the active substrate must be complexed to a metal ion. **1.** Mellor & Maley (1947) *Nature* **159**, 370. **2.** Bjerrum (1941) *Metal Ammine Formation in Aqueous Solution*, P. Haase & Son, Copenhagen. **3.** Carlson, McReynolds & Verhoek (1945) J. Amer. Chem. Soc. **67**, 1334. **4.** Bjerrum & Anderson (1945) *Kgl. Danske Vidensk. Sel. Mat.-Fys. Medd.* **22**, No. 7 (1945). **5.** Pfeiffer, Thielert & Glaser (1939) *Prakt. Chem.* **152**, 145. **6.** Purich (2010) *Enzyme Kinetics: Catalysis and Control*, Academic Press-Elsevier, New York.

5-Hydroxytryptamine, *See* Serotonin

L-5-Hydroxytryptophan

This serotonin precursor (FW = 220.23 g/mol) is found in plants and animals. The L-enantiomer is a product of the reaction catalyzed by tryptophan 5-monooxygenase and substrate of L-aromatic amino-acid decarboxylase. 5-Hydroxytryptophan decomposes slowly when exposed to air and should be stored under nitrogen or argon. **Target(s):** alkaline phosphatase, by racemic mixture (1); anthranilate synthase (2); aromatic-L-amino-acid decarboxylase, *or* dopa decarboxylase, inhibited by the D-isomer (The L-enantiomer is an alternative substrate.) (3-6); catechol oxidase (7); chorismate mutase (8); 3-deoxy-7 phosphoheptulonate synthase, weakly inhibited (9); histidine decarboxylase (10); indoleamine 2,3 dioxygenase, also alternative substrate (11-13); nitronate monooxygenase (14); phenylalanine decarboxylase (15); thyroxine deiodinase, *or* iodothyronine deiodinase (16); tryptophan 2,3-dioxygenase (10,17-19); tryptophanyl-tRNA synthetase (20,21); tyrosinase, *or* monophenol monooxygenase (22); tyrosine aminotransferase (23); tyrosine 3-monooxygenase, moderately inhibited (24,25). **1.** Fishman & Sie (1971) *Enzymologia* **41**, 141. **2.** Widholm (1972) *Biochim. Biophys. Acta* **279**, 48. **3.** Voltattorni, Minelli & Dominici (1983) *Biochemistry* **22**, 2249. **4.** Jung (1986) *Bioorg. Chem.* **14**, 429. **5.** Mappouras, Stiakakis & Fragoulis (1990) *Mol. Cell. Biochem.* **94**, 147. **6.** Baxter & Slaytor (1972) *Phytochemistry* **11**, 2763. **7.** Pomerantz (1963) *J. Biol. Chem.* **238**, 2351. **8.** Hertel, Hieke & Gröger (1971) *Acta Biotechnol.* **11**, 39. **9.** Görisch & Lingens (1971) *Biochim. Biophys. Acta* **242**, 630. **10.** Webb (1966) *Enzyme and Metabolic Inhibitors*, vol. **2**, Academic Press, New York. **11.** Southan, Truscott, Jamie, *et al.* (1996) *Med. Chem. Res.* **6**, 343. **12.** Watanabe, Fujiwara, Yoshida & Hayaishi (1980) *Biochem. J.* **189**, 393. **13.** Ferry, Ubeaud, Lambert, *et al.* (2005) *Biochem. J.* **388**, 205. **14.** Kido, Soda & Asada (1978) *J. Biol. Chem.* **253**, 226. **15.** Lovenberg, Weissbach & Udenfriend (1962) *J. Biol. Chem.* **237**, 89. **16.** Nakagawa & Ruegamer (1967) *Biochemistry* **6**, 1249. **17.** Paglino, Lombardo, Arcà, Rizzi & Rossi (2008) *Insect Biochem. Mol. Biol.* **38**, 871. **18.** Matsumura, Osada & Aiba (1984) *Biochim. Biophys. Acta* **786**, 9. **19.** Feigelson & Brady (1974) in *Mol.*

Mech. Oxygen Activ. (O. Hayaishi, ed.) p. 87, Academic Press, New York. **20**. Davie (1962) *Meth. Enzymol.* **5**, 718. **21**. Sharon & Lipmann (1957) *Arch. Biochem. Biophys.* **69**, 219. **22**. Wittenberg & Triplett (1985) *J. Biol. Chem.* **260**, 12535. **23**. Jacoby & La Du (1964) *J. Biol. Chem.* **239**, 419. **24**. Moore & Dominic (1971) *Fed. Proc.* **30**, 859. **25**. McGeer, McGeer & Peters (1967) *Life Sci.* **6**, 2221.

Hydroxyurea

This free radical-scavenging antineoplastic agent (FW = 76.06 g/mol), also called *N*-hydroxyurea and hydroxycarbamide, inhibits eukaryotic ribonucleoside diphosphate reductase, quenching a catalytic tyrosyl radical (1). The resulting inhibition of DNA biosynthesis stalls cells in S-phase, often attended by cell death. This property has been exploited in the cell-cycle synchronization of animal, plant and yeast cells. Hydroxyurea is also used to treat sickle cell patients by inducing synthesis of fetal hemoglobin and reducing HbS polymerization and blocking adhesion of sickle cells to capillary membranes. **Key Pharmacokinetic Parameters:** *See* Appendix II in Goodman & Gilman's *THE PHARMACOLOGICAL BASIS OF THERAPEUTICS*, 12th Edition (Brunton, Chabner & Knollmann, eds.) McGraw-Hill Medical, New York (2011). **Target(s):** L-ascorbate peroxidase (2); carbonic anhydrase, human CAI, K_i = 0.1 mM (3); propanediol dehydratase (4); ribonucleotide reductase (ribonucleoside-diphosphate reductase), by reacting with and trapping an essential free radical catalytic intermediate (1,5-19); ribonucleotide reductase (ribonucleoside-triphosphate reductase) (20,21); urease, weak alternative substrate (22-28). **1**. Lassmann, Thelander & Graslund (1992) *Biochem. Biophys. Res. Commun.* **188**, 879. **2**. Mathews, Summers & Felton (1997) *Arch. Insect Biochem. Physiol.* **34**, 57. **3**. Scozzafava & Supuran (2003) *Bioorg. Med. Chem.* **11**, 2241. **4**. Hartmanis & Stadtman (1986) *Arch. Biochem. Biophys.* **245**, 144. **5**. Moore (1967) *Meth. Enzymol.* **12A**, 155. **6**. Scott & Tomkins (1975) *Meth. Enzymol.* **40**, 273. **7**. Lammers & Follmann (1983) *Struct. Bonding* **54**, 27. **8**. Hofmann, Feller, Pries & Follmann (1985) *Biochim. Biophys. Acta* **832**, 98. **9**. Cory, Sato & Lasater (1981) *Adv. Enzyme Regul.* **19**, 139. **10**. Larsen, Sjöberg & Thelander (1982) *Eur. J. Biochem.* **125**, 75. **11**. Stubbe (1990) *Adv. Enzymol. Relat. Areas Mol. Biol.* **63**, 349. **12**. Engström, Eriksson, Thelander & Akerman (1979) *Biochemistry* **18**, 2941. **13**. Elford, Van't Riet, Wampler, Lin & Elford (1981) *Adv. Enzyme Regul.* **19**, 151. **14**. Sato, Bacon & Cory (1984) *Adv. Enzyme Regul.* **22**, 231. **15**. Holmgren (1981) *Curr. Top. Cell. Regul.* **19**, 47. **16**. Averett, Lubbers, Elion & Spector (1983) *J. Biol. Chem.* **258**, 9831. **17**. Kucera & Paulus (1982) *Arch. Biochem. Biophys.* **214**, 114. **18**. Shao, Zhou, Zhu, *et al.* (2005) *Biochem. Pharmacol.* **69**, 627. **19**. Berglund & Sjöberg (1979) *J. Biol. Chem.* **254**, 253. **20**. Karp, Giles, Gojo, *et al.* (2008) *Leuk. Res.* **32**, 71. **21**. Yau & Wachsman (1973) *Mol. Cell. Biochem.* **1**, 101. **22**. Blakeley, Hinds, Kunze, Webb & Zerner (1969) *Biochemistry* **8**, 1991. **23**. Fishbein & Carbone (1965) *J. Biol. Chem.* **240**, 2407. **24**. Tanaka, Kawase & Tani (2004) *Bioorg. Med. Chem.* **12**, 501. **25**. Mahadevan, Sauer & Erfle (1977) *Biochem. J.* **163**, 495. **26**. Davis & Shih (1984) *Phytochemistry* **23**, 2741. **27**. Clemens, Lee & Horwitz (1995) *J. Bacteriol.* **177**, 5644. **28**. Nakano, Takenishi & Watanabe (1984) *Agric. Biol. Chem.* **48**, 1495.

Hydroxyzine

This diphenylmethane/piperazine-class antihistamine (FW = 374.90 g/mol; CAS 68-88-2), also known by the trade names Vistaril®, Atarax®, Equipose®, Masmoran®, and Paxistil® as well as the IUPAC name (±)-2-(2-

{4-[(4-chlorophenyl)-phenylmethyl]piperazin-1-yl}ethoxy)ethanol, exhibits anxiolytic, anti-obsessive, and antipsychotic properties. Hydroxyzine is as a potent inverse histamine H_1 receptor agonist, K_i = 2 nM (1-4). Hydroxyzine is also an antagonist for serotonin $5-HT_{2A}$ receptor (K_i = ~50 nM), dopamine D_2 receptor (K_i = 380 nM), and α_1-adrenergic (K_i = ~300 nM) receptor (2,3,5). Hydroxyzine's antiserotonergic effects likely underlie its usefulness as an anxiolytic (6), as other antihistamines without such properties are not effective in the treatment of anxiety (7). Upon oral and intramuscular administration, hydroxyzine is rapidly absorbed, distributed, and metabolized to cetirizine by the action of alcohol dehydrogenase. Drug effects are observed within one hour of administration, with a mean half-life of ~3 hours in adults. **Key Pharmacokinetic Parameters:** *See* Appendix II in Goodman & Gilman's *THE PHARMACOLOGICAL BASIS OF THERAPEUTICS*, 12th Edition (Brunton, Chabner & Knollmann, eds.) McGraw-Hill Medical, New York (2011). **1**. White & Boyajy (1960) *Arch. Internat. Pharmacodynam. Thérap.* **127**, 260. **2**. Kubo, Shirakawa, Kuno & Tanaka (March 1987) *Jpn. J. Pharmacol.* **43**, 277. **3**. Snowman & Snyder (1990) *J. Allergy Clin. Immunol.* **86**, 1025. **4**. Gillard, Van Der Perren, Moguilevsky, Massingham & Chatelain (2002) *Molec. Pharmacol.* **61**, 391. **5**. Haraguchi, Ito, Kotaki, Sawada & Iga (1997) in *Drug Metabolism and Disposition: the Biological Fate of Chemicals* **25**, 675. **6**. Rothbaum, Stein & Hollander, Eric (2009) *Textbook of Anxiety Disorders*, American Psychiatric Publishing, Inc. (ISBN 1-58562-254-0). **7**. Lamberty & Gower (2004) *Pharmacol. Biochem. Behav.* **79**, 119.

Hymenialdisine

10*E*-isomer   10*Z*-isomer

This marine sponge alkaloid (FW = 324.14 g/mol) inhibits nuclear factor-κB and protein tyrosine kinases. [Mitogen-activated protein kinase] kinase 1 is also inhibited by both the 10*E*- and 10*Z*-isomers, with IC_{50} values of 3 and 6 nM, respectively. **Target(s):** casein kinase 1 (1); cyclin-dependent kinase 2 (1); cyclin-dependent kinase 5/p35 (1); [glycogen synthase] kinase β (1); [mitogen-activated protein kinase] kinase 1 (2); nuclear factor-κB (3-5); protein-tyrosine kinase (5). **1**. Meijer, Thunnissen, White, *et al.* (2000) *Chem. Biol.* **7**, 51. **2**. Tasdemir, Mallon, Greenstein, *et al.* (2002) *J. Med. Chem.* **45**, 529. **3**. Breton & Chabot-Fletcher (1997) *J. Pharmacol. Exp. Ther.* **282**, 459. **4**. Roshak, Jackson, Chabot-Fletcher & Marshall (1997) *J. Pharmacol. Exp. Ther.* **283**, 955. **5**. Badger, Cook, Swift, *et al.* (1999) *J. Pharmacol. Exp. Ther.* **290**, 587.

Hyperforin

This photosensitive constituent (FW = 536.80 g/mol; M.P. = 79-80°C) from St John's wort (*Hypericum perforatum*) inhibits the activity of a number of cytochrome P450 systems (1-3). Hyperforin also inhibits the synaptosomal uptake of serotonin, norepinephrine, and dopamine, actions that may be responsible for the antidepressant activity of St. John's wort. Hyperforin acts as a protonophore, inducing TRPC6-independent H^+ currents in HEK-

293 cells, cortical microglia, chromaffin cells and lipid bilayers (4). **Target(s):** cyclooxygenase I (1); CYP1A2 (2); CYP3A4 (2,3); CYP2C9 (2,3); CYP2C19 (2); CYP2D6 (2,3); dopamine reuptake (5,6); 5-lipoxygenase (1,7); norepinephrine reuptake (5,6); serotonin reuptake (5,6). **1.** Albert, Zundorf, Dingermann, *et al.* (2002) *Biochem. Pharmacol.* **64**, 1767. **2.** Zou, Harkey & Henderson (2002) *Life Sci.* **71**, 1579. **3.** Obach (2000) *J. Pharmacol. Exp. Ther.* **294**, 88. **4.** Sell, Belkacemi, Flockerzi & Beck (2014) *Sci. Rep.* **4**, 7500. **5.** Chatterjee, Noldner, Koch & Erdelmeier (1998) *Pharmacopsychiatry* **31** Suppl. 1, 7. **6.** Muller, Singer, Wonnemann, *et al.* (1998) *Pharmacopsychiatry* **31** Suppl. 1, 16. **7.** Fischer, Szellas, Radmark, Steinhilber & Werz (2003) *FASEB J.* **17**, 949. **See also** *Hypericin*

Hypoestoxide

This bicyclo[9.3.1]pentadecane (FW = 375.49 g/mol), a diterpene from *Hypoestes rosea* (a shrub found in the African rain forest and constituent in Nigerian folk medicine), inhibits IκB kinase, thereby suppressing the synthesis of such pro-inflammatory cytokines as tumor necrosis factor α, interleukin 1β, and interleukin-6. Hypoestoxide also inhibits vascular endothelial growth factor induced cell proliferation in vitro, with an IC$_{50}$ of 28.6 μM. **1.** Ojo-Amaize, Kapahi, Kakkanaiah, *et al.* (2001) *Cell Immunol.* **209**, 149. **2.** Ojo-Amaize, Nchekwube, Cottam, *et al.* (2002) *Cancer Res.* **62**, 4007.

Hypoglycin A

This toxic amino acid (FW = 141.17 g/mol; CAS 156-56-9), also known as (αS,1R)-α-amino-2-methylenecyclopropanepropionic acid, from unripe ackee fruit (*Blighia sapida*) gives rise to methylenecyclopropyl)acetyl-CoA, which inhibits several acyl-CoA dehydrogenases, including butyryl-CoA dehydrogenase and isovaleryl-CoA dehydrogenase. (**See also** *(Methylenecyclopropyl)acetyl-CoA*) As indicated by its name, hypoglycin A also induces severe hypoglycemia. **1.** Billington & Sherratt (1981) *Meth. Enzymol.* **72**, 610.

Hypophosphite Ion

hypophosphite ion   formate ion

This conjugate base of hypophosphorus acid H$_3$PO$_2$ (FW$_{free-acid}$ = 64.99 g/mol; FW$_{NaSalt}$ = 87.98 g/mol; pK_a = 1.1) is a formate analogue that inhibits a number of formate-dependent reactions. It is a specific k_{cat} inhibitor of pyruvate formate-lyase, destroying the radical intermediate and forming enzyme-1-hydroxyethylphosphonate upon reaction with the acetyl-enzyme intermediate (1,2). **CAUTION:** Sodium hypophosphite (Formula: NaPO$_2$H$_2$·H$_2$O; FW = 105.99 g/mol; CAS 10039-56-2; DEA Listed (See Section 1310.02)), also known as sodium phosphinate, will explode upon exposure to strong oxidants. It should be kept in a cool, dry place that is isolated from oxidizing materials. When heated, it decomposes to form phosphine gas, a respiratory irritant. **Target(s):** formate C-acetyltransferase, pyruvate formate-lyase (1-5); formate dehydrogenase (6-10); formate hydrogen-lyase (9). **1.** Unkrig, Neugebauer & Knappe (1989)

Eur. J. Biochem. **184**, 723. **2.** Plaga, Frank & Knappe (1988) *Eur. J. Biochem.* **178**, 445. **3.** Brush, Lipsett & Kozarich (1988) *Biochemistry* **27**, 2217. **4.** Ulissi-DeMario, Brush & Kozarich (1991) *J. Amer. Chem. Soc.* **113**, 4341. **5.** Thauer, Kirchniawy & Jungermann (1972) *Eur. J. Biochem.* **27**, 282. **6.** Hemschemeier, Jacobs & Happe (2008) *Eukaryot. Cell* **7**, 518. **7.** Quayle (1966) *Meth. Enzymol.* **9**, 360. **8.** Ljungdahl & J. R. Andreesen (1978) *Meth. Enzymol.* **53**, 360. **9.** Crewther (1956) *Biochim. Biophys. Acta* **21**, 178. **10.** Deyhle & Barton (1977) *Can. J. Microbiol.* **23**, 125.

Hypotaurine

This sulfinic acid (FW$_{neutral}$ = 109.15; pK_a values of 2.16 and 9.56), also known as 2-aminoethanesulfinate, is an intermediate in taurine biosynthesis. Hypotaurine also exhibits antioxidant properties. It is a product of the reactions catalyzed by cysteamine dioxygenase and sulfinoalanine decarboxylase. **Target(s):** acetyl-L-carnitine uptake (1); β-alanine uptake (2); cystathionase, *or* cystathionine γ-lyase (3); mercaptopyruvate sulfurtransferase (4); Na$^+$-dependent γ-aminobutyrate uptake (5); Na$^+$/K$^+$-exchanging ATPase (6); taurine uptake (7,8). **1.** Burlina, Sershen, Debler & Lajtha (1989) *Neurochem. Res.* **14**, 489. **2.** Komura, Tamai, Senmaru, Terasaki, Sai & Tsuji (1996) *J. Neurochem.* **67**, 330. **3.** Cavallini, Mondovi, De Marco & Sciosciasantoro (1962) *Arch. Biochem. Biophys.* **96**, 456. **4.** Wing & Baskin (1992) *J. Biochem. Toxicol.* **7**, 65. **5.** Ramanathan, Brett & Giacomini (1997) *Biochim. Biophys. Acta* **1330**, 94. **6.** Mrsny & Meizel (1985) *Life Sci.* **36**, 271. **7.** Tamai, Senmaru, Terasaki & Tsuji (1995) *Biochem. Pharmacol.* **50**, 1783. **8.** Tayarani, Cloez, Lefauconnier & Bourre (1989) *Biochim. Biophys. Acta* **985**, 168.

Hypoxanthine

This purine (FW = 136.11 g/mol; CAS 68-94-0; Solubility: 0.07 g/100 mL H$_2$O; more soluble in alkaline or acidic solutions; λ$_{max}$ (pH 6) at 249.5 nm; ε = 10700 M^{-1}cm^{-1}; λ$_{max}$ (pH 6) at 259 nm (ε = 11100 M^{-1}cm^{-1}; Abbreviated Hyp), also called 6-hydroxypurine and purin-6(1H)-one, is an intermediate in the degradation of adenine and its nucleoside and nucleotide forms. Hyp can also be chemically prepared from adenine by the action of nitrous acid. In addition to its role as a metabolic intermediate, hypoxanthine is also found in certain tRNA molecules, presumably as a consequence of deamination. Hypoxanthine exists as tautomers; with the equilibrium favoring the lactam. Hypoxanthine is a substrate or product (and product inhibitor) for a number of enzymes, including hypoxanthine phosphoribosyltransferase, inosine nucleosidase, inosinate nucleosidase, DNA deoxyinosine glycosylase, and adenine deaminase. It is an important alternative substrate for xanthine oxidase. **Target(s):** adenine phosphoribosyltransferase (1); adenosine deaminase (2,3); alcohol dehydrogenase, yeast (4); D-amino-acid oxidase (5); 3',5'-cyclic-nucleotide phosphodiesterase (6); DNA-(apurinic or apyrimidinic site) lyase (7); guanine deaminase, weakly inhibited (8,9); inosine nucleosidase, product inhibition (10,11); NAD$^+$ ADP ribosyltransferase (poly(ADP-ribose) polymerase; IC$_{50}$ = 1.7 mM (12); nicotinamide phosphoribosyltransferase (13); pteridine oxidase (14); purine nucleosidase (15,16); purine-nucleoside phosphorylase (or guanosine phosphorylase; product inhibition (17,18); tRNA-guanine transglycosylase, weakly inhibited (19); urate oxidase, *or* uricase (20); xanthine phosphoribosyltransferase, also alternative substrate (21,22). **1.** Tuttle & Krenitsky (1980) *J. Biol. Chem.* **255**, 909. **2.** Centelles, Franco & Bozal (1988) *J. Neurosci. Res.* **19**, 258. **3.** Harbison & Fisher (1973) *Arch. Biochem. Biophys.* **154**, 84. **4.** Sund & Theorell (1963) *The Enzymes*, 2nd ed., **7**, 25. **5.** Burton (1951) *Biochem. J.* **48**, 458. **6.** Kemp & Huang (1974) *Meth. Enzymol.* **38**, 240. **7.** Kane & Linn (1981) *J. Biol. Chem.* **256**, 3405. **8.** Kimm, Park & Lee (1985) *Korean J. Biochem.* **17**, 139. **9.** Kimm, Park & Kim (1987) *Korean J. Biochem.* **19**, 39. **10.**

Yoshino & Tsukada (1988) *Int. J. Biochem.* **20**, 971. **11**. Yoshino, Tsukada & Tsushima (1978) *Arch. Microbiol.* **119**, 59. **12**. Rankin, Jacobson, Benjamin, Moss & Jacobson (1989) *J. Biol. Chem.* **264**, 4312. **13**. Preiss & Handler (1957) *J. Biol. Chem.* **225**, 759. **14**. Hong (1980) *Plant Sci. Lett.* **18**, 169. **15**. Parkin (1996) *J. Biol. Chem.* **271**, 21713. **16**. Parkin, Horenstein, Abdulah, Estupinan & Schramm (1991) *J. Biol. Chem.* **266**, 20658. **17**. Glantz & Lewis (1978) *Meth. Enzymol.* **51**, 525. **18**. Baker & Schaeffer (1971) *J. Med. Chem.* **14**, 809. **19**. Farkas, Jacobson & Katze (1984) *Biochim. Biophys. Acta* **781**, 64. **20**. Bergmann, Kwietny-Govrin, Ungar-Waron, Kalmus & Tamari (1963) *Biochem. J.* **86**, 567. **21**. Miller, Adamczyk, Fyfe & Elion (1974) *Arch. Biochem. Biophys.* **165**, 349. **22**. Naguib, Iltzsch, el Kouni, Panzica & el Kouni (1995) *Biochem. Pharmacol.* **50**, 1685.

– I –

I, *See* Inosine; Iodine and Iodide Ions; D-Isoleucine; L-Isoleucine

IA₃

This 68-residue vacuolar protease inhibitor (MW = 7707 Da), isolated from *Sacchaormyces cerevisiae*, inhibits yeast proteinase A (saccharopepsin or YPrA): K_i = 1.7 nM (1-7). The inhibitor appears to be completely selective in that only yeast aspartic proteinase A is inhibited to any significant extent. IA₃ thus represents the first example of a totally specific, naturally occurring, aspartic-proteinase inhibitor (1). The free inhibitor is intrinsically disordered in aqueous solution; however, upon binding to the protease, residues 2-34 form an α-helix. Multi-scale computer simulations were used to explore the inhibitor's detailed binding trajectories, suggesting important roles of non-native interactions in the initial encounter and binding steps preceding IA₃ folding (8). **1**. Dreyer, Valler, Kay, Charlton & Dunn (1985) *Biochem. J.* **231**, 777. **2**. Li, Phylip, Lees, *et al.* (2000) *Nat. Struct. Biol.* **7**, 113. **3**. Green, Ganesh, Perry, *et al.* (2004) *Biochemistry* **43**, 4071. **4**. Phylip, Lees, Brownsey, *et al.* (2001) *J. Biol. Chem.* **276**, 2023. **5**. Winther (1998) in *Handb. Proteolytic Enzymes*, p. 848, Academic Press, San Diego. **6**. Dunn (2002) *Chem. Rev.* **102**, 4431. **7**. Badasso, Read, Dhanaraj, *et al.* (2000) *Acta Crystallogr. Sect. D* **56**, 915. **8**. Wang, Wang, Chu, *et al.* (2011) *PLoS Comput. Biol.* **7**, e1001118.

IACFT, *See* Altropane

Ibandronate

This bone resorption inhibitor ($FW_{free-acid}$ = 308.23 g/mol; CAS 138926-19-9 (sodium salt)), also known by the trade names Boniva, Bondronat and Bonviva, and systematically as {1-hydroxy-3-[methyl(pentyl)-amino] propane-1,1-diyl}bis(phosphonic acid), inhibits farnesyl transferase and squalene synthase. The effects of nitrogen-containing bisphosphonates (N-BPs) on osteoclasts (Ocs) may differ with dose and regimen. **Key Pharmacokinetic Parameters:** *See* Appendix II in Goodman & Gilman's *THE PHARMACOLOGICAL BASIS OF THERAPEUTICS*, 12th Edition (Brunton, Chabner & Knollmann, eds.) McGraw-Hill Medical, New York (2011). **Target(s):** farnesyl*trans*transferase, or geranylgeranyl-diphosphate synthase (1); geranyl*trans*transferase, or farnesyl-diphosphate synthase (2-6); protein farnesyltransferase (7); protein geranylgeranyltransferase (7,8); squalene synthase (9). **1**. Szabo, Matsumura, Fukura, *et al.* (2002) *J. Med. Chem.* **45**, 2185. **2**. van Beek, Pieterman, Cohen, Löwik & Papapoulos (1999) *Biochem. Biophys. Res. Commun.* **255**, 491. **3**. Dunford, Thompson, Coxon, *et al.* (2001) *J. Pharmacol. Exp. Ther.* **296**, 235. **4**. Glickman & Schmid (2007) *Assay Drug Dev. Technol.* **5**, 205. **5**. Dunford, Kwaasi, Rogers, *et al.* (2008) *J. Med. Chem.* **51**, 2187. **6**. Lühe, Künkele, Haiker, *et al.* (2008) *Toxicol. In Vitro* **22**, 899. **7**. Luckman, Hughes, Coxon, *et al.* (1998) *J. Bone Miner. Res.* **13**, 581. **8**. Coxon, Helfrich, Van't Hof, *et al.* (2000) *J. Bone Miner. Res.* **15**, 1467. **9**. Amin, Cornell, Perrone & Bilder (1996) *Arzneimittel-forschung* **46**, 759.

Iberiotoxin

XFTDVDCSVSKECWSVCKDLFGVDRGKCMGKKCRCYQ

This 37-residue peptide (FW = 4221.90 g/mol; CAS 129203-60-7; *Sequence*: pEFTDVDCSVSKECWSVCLDLFGVDRGKCMGKKCRCYQ, where pE is a pyroglutamate; internal disulfides: Cys7-to-Cys28, Cys13-to-Cys33, and Cys17-to-Cys35; Solubility = ~0.7 mg/mL H₂O; Symbol: IbTx), from the venom of the Indian scorpion *Buthus tamulus*, is a selective and reversible high-affinity inhibitor of the high-conductance calcium ion-activated potassium channel, *or* PK,Ca (1). While charybdotoxin, noxiustoxin and iberiotoxin display a high degree of sequence homology, the first two are known to inhibit more than one class of channel (*i.e.*,

several Ca²⁺-activated and voltage-dependent K⁺ channels). Iberiotoxin (concentration range: 10-100 nM) is a selective inhibitor of PK, Ca, and tetraethylammonium ion that blocks PK,Ca at low (100-300 nM) concentrations, increasing the myogenic activity of bladder, but not portal vein (2). IbTX also causes a sustained contracture of aorta (2). **1**. Galvez, Gimenez-Gallego, Reuben, *et al.* (1990) *J. Biol. Chem.* **265**, 11083. **2**. Suarez-Kurtz, Garcia & Kaczorowski (1991) *J. Pharmacol. Exp. Ther.* **259**, 439.

I-BET-762

This BET protein-directed inhibitor (FW = 423.90 g/mol; CAS 1260907-17-2; Solubility: 80 mg/mL DMSO or Ethanol, 11 mg/mL H₂O), also known as GSK525762 and GSK525762A and systematically named 6-(4-chlorophenyl)-*N*-ethyl-8-methoxy-1-methyl-(4*S*)-4*H*-[1,2,4]triazolo[4,3-*a*][1,4]benzodiazepine-4-acetamide, targets Bromodomain and Extra-Terminal domain (BET) proteins (IC_{50} ~35 nM), interfering with their recognition of acetylated histones, disrupting formation of chromatin complexes essential for expression of pro-inflammatory genes, and suppressing production of such proteins by macrophages. I-BET-762 blocks acute inflammation and is highly selective over other bromodomain-containing proteins. I-BET-762 confers protection against lipopolysaccharide-induced endotoxic shock and bacteria-induced sepsis. Treatment of naïve CD4⁺ T cells with I-BET-762 during the first 2 days of differentiation has long-lasting effects on subsequent gene expression and cytokine production, with up-regulated expression of several anti-inflammatory gene products, including IL-10, Lag3, and Egr2, and down-regulated expression of several pro-inflammatory cytokines including GM-CSF and IL-17. The short 2-day treatment with I-BET-762 inhibits the ability of antigen-specific T cells (differentiated under Th1, but not Th17, conditions *in vitro*) to induce pathogenesis in an adoptive transfer model of experimental autoimmune encephalomyelitis. **1**. Nicodeme, Jeffrey, Schaefer, *et al.* (2010) *Nature* **468**, 1119. **2**. Bandukwala, Gagnon & Togher, *et al.* (2012) *Proc. Natl. Acad. Sci. USA* **109**, 14532.

IBMX, *See* 3-Isobutyl-1-methylxanthine

Ibodutant

This tachykinin/neurokinin, *or* NK₂, receptor antagonist (FW = 644.37 g/mol), also known by the code name MEN15596 and its IUPAC name [1-(2-phenyl-1R-[[1-(tetrahydropyran-4-ylmethyl)piperidin-4-ylmethyl]carbamoyl]ethylcarbamoyl)cyclopentyl]amide) is a candidate drug for the treatment of irritable bowel syndrome. In functional assays of phosphatidylinositol accumulation in human NK₂R-expressing CHO cells, ibodutant gives a $pK_{binding}$ of 10.6. Its antagonism is competitive and is reversed by the natural ligand neurokinin A. The same assay indicated that ibodutant attains equilibrium quickly and is released slowly from NK₂R (1). That ibodutant antagonizes NK₂R internalization in the human colon suggests it may be useful in treating hypermotility gut diseases, especially because NK₂ receptors are the predominant mediator of spasmogenic activity of tachykinins on enteric smooth muscle (2). When examined over the 0.3 to 100 nM range, ibodutant produces a concentration-dependent inhibition (calculated $pK_{binding}$ = 8.31) for the nonadrenergic-noncholinergic (NANC) contractile response, as induced by electrical field stimulation (EFS) of intrinsic airway nerves in main bronchi isolated from guinea pig

(3). **1**. Meini, Bellucci, Catalani, *et al.* (2009) *J. Pharmacol. Exp. Ther.* **329**, 486. **2**. Cipriani, Santicioli, Evangelista, *et al.* (2011) *Neurogastroenterol. Motil.* **23**, 96. **3**. Santicioli, Meini, Giuliani, Lecci & Maggi (2013) *Eur. J. Pharmacol.* **720**, 180.

Ibogaine

This alkaloid (FW$_{\text{free-base}}$ = 310.44 g/mol; CAS 83-74-9) from the African shrub *Tabernanthe iboga* possesses hallucinogenic properties and is banned in many countries. Ibogaine is a noncompetitive NMDA (*N*-methyl-D-aspartate) receptor antagonist, a σ$_2$ agonist, and also inhibits serotonin and dopamine transport. Ibogaine is an anti addictive drug that decreases withdrawal symptoms and drug craving over an extended periods of time after administration of a single dose. Extracts of the shrub are used by certain African natives when stalking game, enabling the hunter to remain motionless for long periods of time, while still remaining attentive and mentally alert. Ibogaine has a melting point of 152-153°C, is practically insoluble in water and soluble in ethanol and chloroform, and has λ$_{\text{max}}$ values of 226 and 298 nm (ε = 24,500 and 8,500 M^{-1}cm^{-1}, respectively). Its pK_a is 8.1 in 80% methyl cellosolve. **Primary Mechanism of Inhibitory Action:** Although ibogaine's structural similarity to serotonin had spurred early suggestions that it binds to the substrate site of serotonin transporters, *or* SERT, ibogaine binds to a distinct site, one that is accessible only from the cell's exterior. Ibogaine thereby inhibits both serotonin transport and serotonin-induced ionic currents (1). When added to the extracellular medium, ibogaine inhibits SERT substrate-induced currents, but does not when introduced into the cytoplasm by means of a patch electrode. Ibogaine inhibitiontt does not involve the formation of a long-lived complex with SERT; instead, it binds directly to the transporter, but only when the latter adopts an inward-open conformation. Ibogaine also noncompetitively inhibits the homologous dopamine transporter (DAT), blocking substrate-induced currents in DAT and increasing accessibility of the DAT cytoplasmic permeation pathway (1).   **Target(s):** cholinesterase (2); dopamine transport (3); NMDA receptor (4,5); serotonin transport (1,3). **1.** Bulling, Schicker, Zhang, *et al.* (2012) *J. Biol. Chem.* **287**, 18524. **2.** Augustinsson (1950) *The Enzymes*, 1st ed., **1** (Part 1), 443. **3.** Wells, Lopez & Tanaka (1999) *Brain Res. Bull.* **48**, 641. **4.** Leal, de Souza & Elisabetsky (2000) *Neurochem. Res.* **25**, 1083. **5.** Itzhak & Ali (1998) *Ann. N. Y. Acad. Sci.* **844**, 88.

Ibotenic acid

This naturally occurring neurotoxic 3-isoxazolol amino acid (FW = 158.11 g/mol; CAS 2552-55-8; Symbol: IA) from the exquisitely poisonous mushrooms *Amanita muscaria*, *A. pantherina*, and *A. strobiliformis*, is a powerful brain-lesioning, excitotoxic amino acid, which (like kainic acid) produces profound neuropathological and neurochemical changes in the basal ganglia. Lesioned animals show a behavioral syndrome reminiscent of Huntington's chorea, accompanied by a substantial increase in local cerebral metabolic activity in several striatal target structures within the extrapyramidal motor system (1). Injection results in demyelination, disrupting axonal transport in areas containing diffuse fiber systems, presumably as a consequence of a non-specific inflammatory response (2). As a conformationally restricted glutamate analogue that is closely related to muscimol (**See Muscimol**), ibotenic acid activates <u>I</u>nhibitory <u>G</u>lutamate <u>R</u>eceptors (IGluRs), a family of ion channel proteins closely related to ionotropic glycine and GABA receptors that are gated directly by glutamate; the open channel is permeable to chloride and sometimes potassium (3). *Note*: L-2-Amino-3-phosphonopropionate is a sterospecific inhibitor of ibotenic acid-stimulated phosphoinositide hydrolysis in rat brain slices, with little affinity for ionotropic receptors (4). L-2-Amino-3-phosphonopropionate's inhibitory effects are not readily reversed. **1.**

Isacson, Brundin, Kelly, Gage & Björklund (1984) *Nature* **311**, 458. **2.** Coffey, Perry, Allen, Sinden & Rawlins (1988) *Neurosci. Lett.* **84**,178. **3.** Cleland (1996) *Mol. Neurobiol.* **13**, 97. **4.** Schoepp, Johnson, Smith & McQuaid (1990) *Mol. Pharmacol.* **38**, 222.

Ibrutinib

This highly potent, orally available protein kinase inhibitor (FW = 450.50 g/mol; CAS 936563-96-1; Solubility: 88 mg/mL DMSO, <1 mg/mL H$_2$O), also known by the code name PCI-32765 and its systematic name 1-[(3*R*)-3-[4-amino-3-(4-phenoxyphenyl)pyrazole[3,4-*d*]pyrimidin-1-yl]piperidin-1-yl]prop-2-en-1-one, targets Bruton tyrosine kinase (BTK) autophosphorylation (IC$_{50}$ = 11 nM), phosphorylation of PLCγ, BTK's physiological substrate (IC$_{50}$ = 29 nM), and phosphorylation of ERK (IC$_{50}$ = 13 nM) (1,2). That the Bruton tyrosine kinase (Btk) is specifically required for BCR signaling was demonstrated by human and mouse mutations disrupting Btk function and likewise preventing B-cell maturation at steps requiring a functional B-cell antigen receptor (BCR) pathway (1-4). In contrast to conventional chemo-immunotherapy, ibrutinib is not myelo-suppressive (3,4). In chronic lymphocytic leukemia (*or* CLL), ibrutinib characteristically causes an early redistribution of tissue-resident CLL cells into the blood, with rapid resolution of enlarged lymph nodes, along with a surge in lymphocytosis (3,4). After weeks to months of continuous ibrutinib therapy, the growth- and survival-inhibitory activities of ibrutinib result in the normalization of lymphocyte counts and remissions in a majority of patients (3,4). Ibrutinib is also under development for the treatment of other B-cell malignancies, mantle cell lymphoma (MCL) and diffuse large B-cell lymphoma (DLBCL), as well as multiple myeloma (MM), follicular lymphoma (FL) and Waldenstrom's macroglobulinemia (WM). Actin-based motility of *Shigella flexneri* is regulated by Bruton's tyrosine kinase, and treatment with ibrutinib effectively impairs cell-to-cell spread of this pathogen (5). **1.** Honigberg, Smith, Sirisawad, *et al.* (2010) *Proc. Natl. Acad. Sci. U.S.A.* **107**, 13075. **2.** Herman, Gordon, Hertlein, *et al.* (2011) *Blood* **117**, 6287. **3.** Ponader, Chen, Buggy, *et al.* (2012) *Blood* **119**, 1182. **4.** Burger & Buggy (2013) *Leuk. Lymphoma.* **54**, 2385. **5.** Dragoi & Agaisse (2013) *Gut Microbes* **5**, 44.

Ibuprofen

(S)-(+)-ibuprofen

This orally active, nonsteroidal anti-inflammatory drug (NSAID) and anti-pyretic agent (FW$_{\text{free-acid}}$ = 206.28 g/mol; CAS 15687-27-1; M.P. = 75-77°C), also known as 2-(4-isobutylphenyl)propionic acid, was introduced in the UK in 1966 and in the U.S. in 1974. Ibuprofen was the first non-steroidal anti-inflammatory drug licensed for over-the-counter use in the U.K. in 1983 and in the U. S. in 1984. **Primary Mode of Action:** In 1971, Nobelist J. R. Vane first reported that aspirin and indomethacin inhibit prostaglandin biosynthesis, an observation that led to the concept that nonsteroidal anti-inflammatory drugs inhibit cyclooxygenases, thereby blocking the formation of cyclooxygenase pathway products. Ibuoprofen inhibits the cyclooxygenase activity of prostaglandin synthases 1 and 2, thereby reducing cytokine production, inhibiting associated signal transduction pathways, and modulating leucocyte activity. The (*S*)-(+)-stereoisomer is the more active, with IC$_{50}$ values of 8.9 and 7.2 μM for

COX-1 and COX-2, respectively. Ibuprofen also selectively activates 15-lipoxygenase activity in human PMNs at concentrations that completely inhibit 5-lipoxygenase as well as the human platelet cyclooxygenase pathways. In human platelets, they found that ibuprofen inhibited thromboxane B_2 production (IC$_{50}$ = 65 μM), whereas the lipoxygenase product 12-HETE was not appreciably affected, even at 5 mM ibuprofen. **Pharmacokinetics:** Absorption of ibuprofen is rapid and complete when given orally. Ibuprofen binds to and is carried by serum albumin, mainly at its high-affinity binding site (K_d for Site-II = 0.3 x 10^{-6} M). When displaced by naproxen, ibuprofen binds to serum albumin's low-affinity binding site (K_d for Site-I = 3 x 10^{-4} M), where it competes with aspirin and other agents. At doses >600 mg, a greater fraction of drug remains unbound, resulting to an increased ibuprofen clearance and a reduced AUC of total drug. Substantial concentrations of ibuprofen are attained in synovial fluid, the proposed site of action for most nonsteroidal anti-inflammatory drugs. Ibuprofen has a short elimination half-life ($t_{1/2}$ ~2 hours) and exceptional gastrointestinal tolerability. Ibuprofen is eliminated following biotransformation to glucuronide conjugates that are excreted in urine, with little of the drug eliminated in its original form. *S*-Ibuprofen is also metabolized by cytochrome P450 (CYP2C9), as are many other agents, including naproxen, celecoxib, sildenafil, warfarin, and tamoxifen. Voriconazole and fluconazole inhibit CYP2C9, thereby affecting ibuprofen turnover. **Key Pharmacokinetic Parameters:** *See* Appendix II in Goodman & Gilman's THE PHARMACOLOGICAL BASIS OF THERAPEUTICS, 12th Edition (Brunton, Chabner & Knollmann, eds.) McGraw-Hill Medical, New York (2011). **Target(s):** amidase (1,2); arylamine *N*-acetyltransferase (3); cyclooxygenase, *or* prostaglandin endoperoxide synthase (4-12); estrone sulfotransferase (13); fatty-acid amidohydrolase, *or* amidase (1,2); histidine decarboxylase (14); lysine decarboxylase (14); long-chain-fatty-acyl-CoA synthetase, *or* long chain-fatty-acid:CoA ligase, inhibited by the *R*-isomer (15); α-methylacyl-CoA epimerase (16,17); palmitoyl-CoA hydrolase, *or* acyl-CoA hydrolase (18); phenol sulfotransferase, *or* aryl sulfotransferase (13); phenyl-pyruvate tautomerase, weakly inhibited (19); phospho-ribosylamino-imidazolecarboxamide formyltransferase (20); thiopurine *S* methyltransferase (21). Ibuprofen also enhances the anticancer activity of cisplatin in lung cancer cells by inhibiting the heat shock protein 70 (22). **1.** Fowler, Holt & Tiger (2003) *J. Enzyme Inhib. Med. Chem.* **18**, 55. **2.** Holt, Nilsson, Omeir, Tiger & Fowler (2001) *Brit. J. Pharmacol.* **133**, 513. **3.** Makarova (2008) *Curr. Drug Metab.* **9**, 538. **4.** Pace-Asciak & Smith (1983) *The Enzymes*, 3rd ed., **16**, 543. **5.** Barnett, Chow, Ives, *et al.* (1994) *Biochim. Biophys. Acta* **1209**, 130. **6.** Cushman & Cheung (1976) *Biochim. Biophys. Acta* **424**, 449. **7.** Forghani, Ouellet, Keen, Percival & Tagari (1998) *Anal. Biochem.* **264**, 216. **8.** Kargman, Wong, Greig, *et al.* (1996) *Biochem. Pharmacol.* **52**, 1113. **9.** Meade, Smith & DeWitt (1993) *J. Biol. Chem.* **268**, 6610. **10.** Miller, Munster, Wasvary, *et al.*, (1994) *Biochem. Biophys. Res. Commun.* **201**, 356. **11.** Barnett, Chow, Ives, *et al.* (1994) *Biochim. Biophys. Acta* **1209**, 130. **12.** Patrignani, Panara, Sciulli, *et al.* (1997) *J. Physiol. Pharmacol.* **48**, 623. **13.** King, Ghosh & Wu (2006) *Curr. Drug Metab.* **7**, 745. **14.** Bruni, Dal Pra & Segre (1984) *Int. J. Tissue React.* **6**, 463. **15.** Knights & Jones (1992) *Biochem. Pharmacol.* **43**, 1465. **16.** Schmitz, Fingerhut & Conzelmann (1994) *Eur. J. Biochem.* **222**, 313. **17.** Schmitz, Albers, Fingerhut & Conzelmann (1995) *Eur. J. Biochem.* **231**, 815. **18.** Dixon, Osterloh & Becker (1990) *J. Pharm. Sci.* **79**, 103. **19.** Molnar & Garai (2005) *Int. Immunopharmacol.* **5**, 849. **20.** Ha, Morgan, Vaughn, Eto & Baggott (1990) *Biochem. J.* **272**, 339. **21.** Oselin & Anier (2007) *Drug Metab. Dispos.* **35**, 1452. **22.** Endo, Yano, Okumura & Kido (2014) *Cell Death Dis.* **5**, e1027.

IC261

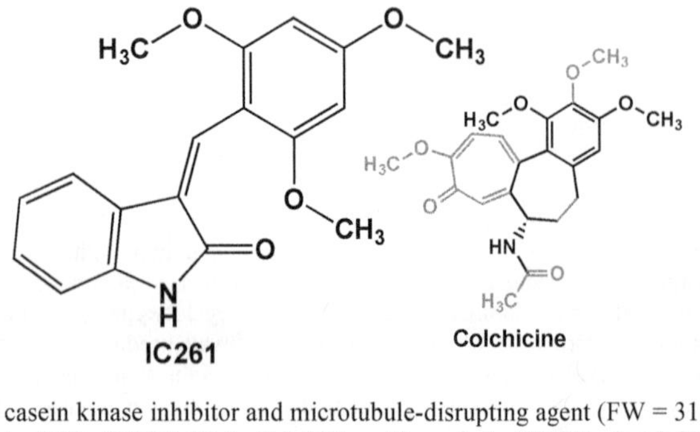

IC261   **Colchicine**

This casein kinase inhibitor and microtubule-disrupting agent (FW = 311.30 g/mol; CAS 186611-52-9; Solubility: 10 mg/mL DMSO; Insoluble in water) selectively targets casein kinase 1δ (IC$_{50}$ = 0.7-1.3 μM), and casein kinase 1ε (IC$_{50}$ = 0.6-1.4 μM), with weaker action against weakly inhibit casein kinase 1α$_1$ (IC$_{50}$ = 11-21 μM), PKA (IC$_{50}$ = >100 μM), p34/cdc2 (IC$_{50}$ = >100 μM), and Fyn, *or* p55 (IC$_{50}$ = >100 μM). X-ray structural analysis reveals that IC261 stabilizes casein kinase-1 in a conformation that is midway between nucleotide substrate-liganded and nonliganded conformations, with a delocalized network of side-chain interactions likely to account for the decreased dissociation rate for IC261 (1). Inhibition by is reversible and ATP-competitive. At low concentrations, IC261 causes centrosome amplification and multipolar mitosis, first suggesting a role for CK1δ and CK1ε isoforms in regulating centrosome or spindle function during mitosis (2). Later work, however, demonstrated that IC261 induces rapid and reversible microtubule depolymerization, an action that can be antagonized by pre-treatment of cells with taxol (3). (As shown above, IC261 has structural similarity to colchicine.) At lower IC261 concentrations, mitotic spindle microtubule dynamics are likewise affected, leading to cell-cycle arrest and, depending on the cellular background, to apoptosis in a dose-dependent manner. **1.** Mashhoon, DeMaggio, Tereshko, *et al.* (2000) *J. Biol. Chem.* **275**, 20052. **2.** Behrend, Milne, Stöter, *et al.* (2000) *Oncogene* **19**, 5303. **3.** Stöter, Krüger, Banting, Henne-Bruns & Knippschild (2014) *PLoS One* **9**, e100090.

IC-87114

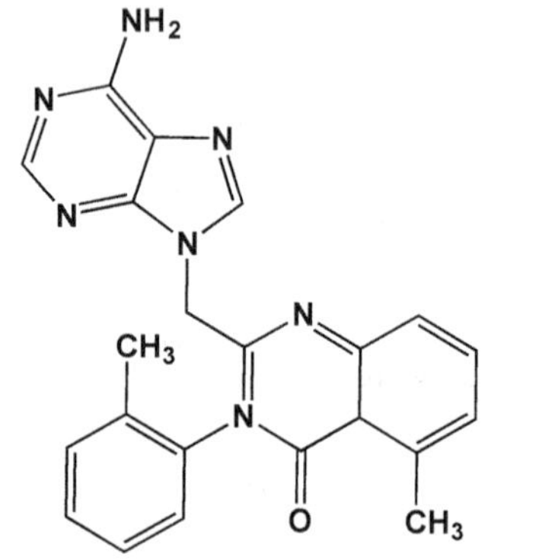

This ATP site-competitive (FW = 397.43; CAS 371242-69-2; Solubility (25°C): <1 mg/mL DMSO, <1 mg/mL Water), also known as 2-((6-amino-9*H*-purin-9-yl)methyl)-5-methyl-3-*o*-tolylquinazolin-4(3*H*)-one, targets the phosphoinositide-3-kinase- (PI3K-) directed signal-transduction pathway by selectively inhibiting PI3Kδ, IC$_{50}$ = 0.5 μM. Although Class-I phosphoinositide 3-kinases (PI3Ks) had been previously implicated in inflammatory responses, the use of non-selective inhibitors (*e.g.*, LY294002 and wortmannin) did not permit an assessment of the contribution of the individual PI3K isoforms in neutrophil activation. IC-87114 is a novel PI3K inhibitor that is for PI3Kδ, an isoform expressed predominantly in hematopoietic cells. (Notably, IC87114 does not inhibit any other protein kinases tested.) At 5 μM, IC-87114 blocked fMLP- and TNF1α-induced neutrophil generation of superoxide as well as the exocytosis of elastase. IC87114 also blocked TNF1α-stimulated elastase exocytosis from neutrophils in a mouse model for inflammation. IC-87114 also suppressed chemotaxis. When used in therapeutic mode, IC-87114 conferred prolonged protection from progression to overt diabetes in a number of animals, suggesting that PI3Kδ inhibitors could be useful for managing type 1 diabetes (3). **Targets:** PI3Kδ, IC$_{50}$ = 0.5 μM; PI3Kβ, IC$_{50}$ = 75 μM; and PI3Kγ, IC$_{50}$ = 29 μM. **1.** Sadhu, Masinovsky, Dick, Sowell & Staunton (2003) *J. Immunol.* **170**, 2647. **2.** Sadhu, Dick, Tino & Staunton (2003) *Biochem. Biophys. Res. Commun.* **308**, 764. **3.** Durant, Richer, Brenker, *et al.* (2013) *Autoimmunity* **46**, 62.

ICA 069673

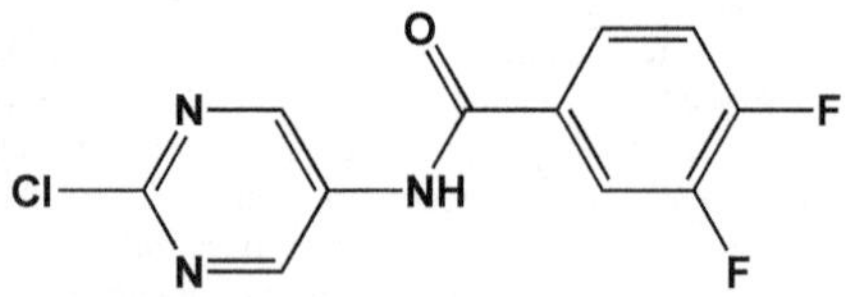

This orally available, potassium ion channel opener (FW = 269.63 g/mol; CAS 582323-16-8; Solubility: 100 mM in DMSO), also named *N*-(2-

chloro-5-pyrimidinyl)-3,4-difluorobenzamide, selectively targets the $K_V7.2/K_V7.3$ (or KCNQ2/3) potassium channel opener (EC_{50} = 0.69 μM), showing 20x greater selective for the $K_V7.2/K_V7.3$ channel over $K_V7.3/K_V7.5$ channels and showing little/no measurable activity over a panel of cardiac ion channels (i.e., IC_{50} >30 μM for $K_V11.1$ (hERG) channels, $Na_V1.5$ channels, L-type channels, and $K_V7.1$ channels). **1.** Amato, et al. (2011) ACS Med. Chem. Lett. **2**, 481.

Icatibant

This bradykinin peptidomimetic (FW = 1304.52 g/mol; CAS 130308-48-4), also known as Firazyr®, Hoe 140, and JE 049, is a bradykinin-competitive antagonist that binds to the kinin B_2 receptor, a G-protein-coupled receptor coupled to G_q and G_i, with G_q stimulating phospholipase C to increase intracellular free calcium and Gi inhibits adenylate cyclase. Icatibant's five unnatural amino acid residues allows it to resist the metalloproteases normally involved in kinin catabolism. Icatibant is used clinically to limit inflammatory and angioedema (i.e., vascular leakage) caused by an acute (uncontrolled) production of kinins and accumulation at the endothelium B_2 receptor. Angioedema is characterized by edematous attacks of subcutaneous and submucosal tissues, often causing painful intestinal consequences and can be life-threatening, if the larynx is affected. Icatibant is FDA-approved for the treatment of acute attacks of the hereditary bradykinin-mediated angioedema, due to C_1 inhibitor deficiency. **1.** Charignon, Späth, Martin & Drouet (2012) Expert Opin. Pharmacother. **13**, 2233.

ICE Inhibitor I

This cell-permeable peptidomimetic (FW = 1920.39 g/mol; Sequence: Ac-AAVALLPAVLLALLPYVAD-CHO, where CHO indicates a C-terminal aldehyde), also called interleukin-1β converting enzyme (ICE) inhibitor I, targets Caspase-1 and Caspase-4. The C-terminal YVAD-CHO sequence of this peptide is a highly specific and potent inhibitor of caspase-1 (K_i = 1 nM), most likely by reversibly forming a thiohemiacetal adduct with an active-site cysteine sulfhydryl group. The N-terminal sequence (i.e., residues 1–16) corresponds to the hydrophobic region (or H region) of the signal peptide of the Kaposi fibroblast growth factor (K-FGF), conferring cell permeability to the peptide. **1.** Garcia-Calvo, Peterson, Leiting, et al. (1998) J. Biol. Chem. **273**, 32608.

Ice-IX

This fictive material (Molecular Formula = $(H_2O)_n$; FW = indefinite), first mentioned in Kurt Vonnegut's novel Cat's Cradle, is a polymorph of water invented by one Dr. Felix Hoenikker. Rather than melting at 0 °C, Ice-IX melts at 45.8 °C, such that once it contacts liquid water at temperatures below 45.8 °C, Ice-IX seeds an immediate phase transition of liquid water, forming ever greater quantities of ice-nine. This unusual behavior anticipated the now well known behavior of prions in pathological amyloidigenesis.

ICG-001

This cell-permeable pyrazino-pyrimidine carboxamide (FW = 548.63 g/mol; CAS 780757-88-2; 847591-62-2) competes with β-catenin for CBP (or CREB-binding protein) binding, down-regulates β-catenin/T cell factor signaling (EC_{50} = 3 μM) by specifically binding to <u>C</u>yclic <u>A</u>MP <u>R</u>esponse <u>E</u>lement-<u>B</u>inding (or CREB) protein. ICG-001 selectively induces apoptosis in transformed cells but not in normal colon cells, reduces in vitro growth of colon carcinoma cells (2-5). Wnt signaling down-regulation by the inducible expression of Axin, ICAT, and dnTcf4E causes degeneration of hippocampal neurons, whereas Wnt signaling up-regulation by the inducible expression of Dvl and β-catenin has little effect. ICG-001 treatment results in hippocampal neuron degeneration, while the treatment with a JNK specific inhibitor is without effect. These data strongly support the idea that downregulation of Wnt/β-catenin signaling causes degeneration of hippocampal neurons in vivo and may be a cause of neurodegenerative diseases related to an anxiety related response. ICG-001 also selectively blocks CBP- (but not p300-) dependent TCF/β-catenin transcriptional regulation in vitro (IC_{50} = 5–25 μM) and in vivo (Dose = 5 mg/kg/day, by osmotic pump in mice), without affecting CBP-dependent transcriptional activities mediated by the AP-1 or CRE complex. ICG-001 also reactivates p300-/TCF-/β-catenin-mediated neuronal differentiation upon NGF stimulation of PC-12 cells expressing PS-1 Leu-286-Val mutant cultures. **1.** **Alternate Names:** β-Catenin/Tcf Inhibitor VI, Wnt Pathway Inhibitor XX, (6S,9aS)-hexahydro-6-((4-hydroxyphenyl)methyl)-8-(1-naphthalenyl-methyl)-4,7-dioxo-N-(phenylmethyl)-2H-pyrazino[1,2-a]-pyrimidine-1(6H)-carboxamide, (6S)-N-benzyl-6-(p-hydroxyphenylmethyl) -8-(1-naphthylmethyl)-4,7-dioxo-hexahydro-2H-pyrazino[1,2-a]-pyrimidine-1-carboxamide. **2.** Emami, Nguyen, Ma, et al. (2004) Proc. Natl. Acad. Sci. U.S.A. **101**, 12682. **3.** Eguchi, Nguyen, Lee & Kahn (2005) Med. Chem. **1**, 467. **4.** Yan, et al. (2012) J. Biol. Chem. **287**, 8598. **5.** Henderson, et al. (2010) Proc. Natl. Acad. Sci. USA **107**, 14309.

ICI-118551

This selective adrenoreceptor antagonist ($FW_{free-base}$ = 277.41 g/mol; CAS 72795-19-8 (free base); 72795-01-8(HCl salt)), also named (±)-erythro-(S*,S*)-1-[2,3-(dihydro-7-methyl-1H-inden-4-yl)oxy]-3-[(1-methylethyl)-amino]-2-butanol, targets the $β_2$-adrenergic receptor (K_i = 1.2 nM), with much greater affinity than the $β_1$ (K_i = 120 nM) or $β_3$ (K_i = 260 nM) subtypes. While specific antagonists of the beta 1-adrenoceptor, such as atenolol and betaxolol, had been widely available, ICI-118551 was the first-in-class $β_2$-selective antagonist (1). ICI 118,551 has no partial agonist activity but has a membrane-stabilising action similar to that of propranolol (1). ICI-118551, reverses the inhibitory effect of $β_2$-agonists (PC, salbutamol, and tulobuterol B) on N-formyl-methionyl-leucyl-phenylalanine-induced superoxide production in granulocytes (2). **1.** Bilski, Halliday, Fitzgerald & Wale (1983) J. Cardiovasc. Pharmacol. **5**, 430. **2.** Yasui, Kobayashi, Yamazaki, et al. (2006) Int. Arch. Allergy Immunol. **139**, 1.

ICI 125,211, See Tiotidine

ICI 182,780, See Fulvestrant

ICI 204,219, See Zafirlukast

ICI 204,636, See Quetiapine

ICI 230487

This *N*-methyl-quinolone (FW = 411.47 g/mol; IUPAC: 1-ethyl-6-{[3-fluoro-5-(4-methoxy-3,4,5,6-tetrahydro-2*H*-pyran-4-yl)phenoxy]methyl}-2-quinolone and ZM 230487, inhibits 5-lipoxygenase (IC$_{50}$ = 0.09 µM) and in turn leukotriene biosynthesis. **Target(s):** linoleate diol synthase (1); 5-lipoxygenase, *or* arachidonate 5-lipoxygenase (2-4). **1**. Su, Brodowsky & Oliw (1995) *Lipids* **30**, 43. **2**. Teixeira & Hellewell (1994) *Brit. J. Pharmacol.* **111**, 1205. **3**. Araico, Terencio, Alcaraz, *et al.* (2006) *Life Sci.* **78**, 2911. **4**. Young (1999) *Eur. J. Med. Chem.* **34**, 671.

ICRF-159, *See Razoxane*

Idarubicin

This anthracycline glycoside and antileukemic agent (FW$_{\text{free-base}}$ = 497.50 g/mol; CAS 58957-92-9), also known as Idamycin, is a structural analogue of daunorubicin. **Primary Mode of Inhibitory Action:** Like daunorubicin, darubicin binds to and intercalates into DNA, inducing DNA strand breaks, delaying cell cycle progression, and inhibiting DNA topoisomerase II. Its higher lipophilicity than daunorubicin results in faster accumulation within nuclei, superior DNA-binding capacity, and greater cytotoxicity. Severe myelosuppression occurs when idarubicin is used at an effective therapeutic dose. **Key Pharmacokinetic Parameters:** *See* Appendix II in Goodman & Gilman's *THE PHARMACOLOGICAL BASIS OF THERAPEUTICS*, 12th Edition (Brunton, Chabner & Knollmann, eds.) McGraw-Hill Medical, New York (2011). **Target(s):** *Ava*II restriction endonuclease (1); DNA ligase (ATP) (2); DNA topoisomerase II (3 5); *Eco*RI restriction endonuclease (1); *Hae*III restriction endonuclease (1); *Hha*I restriction endonuclease (1); *Hpa*II restriction endonuclease (1); *Sma*I restriction endonuclease (1). Idarubicin also inhibits phosphorylation of the main autophagy repressor mTOR and its downstream target p70S6 kinase (6). The treatment with the mTOR activator leucine prevented idarubicin-mediated induction of autophagy. Idarubicin-induced mTOR repression was associated with the activation of the mTOR inhibitor AMP-activated protein kinase and down-regulation of the mTOR activator Akt (6). **1**. Corneo, Pogliani, Biassoni & Tripputi (1988) *Ric. Clin. Lab.* **18**, 19. **2**. Ciarrocchi, Lestingi, Fontana, Spadari & Montecucco (1991) *Biochem. J.* **279**, 141. **3**. Larsson & Nygren (1994) *Cancer* **74**, 2857. **4**. Kellner, Rudolph & Parwaresch (2000) *Onkologie* **23**, 424. **5**. Insaf, Danks & Witiak (1996) *Curr. Med. Chem.* **3**, 437. **6**. Ristic, Bosnjak, Arsikin, *et al.* (2014) *Exp. Cell Res.* **326**, 90.

Idarucizumab, *See Dabigatran*

Idelalisib

This orally active P110δ inhibitor (FW = 415.43 g/mol; CAS 1146702-54-6), code named GS-1101 or CAL-101 and systematically named 5-fluoro-3-phenyl-2-[(1*S*)-1-(7*H*-purin-6-ylamino)-propyl]-4(3*H*)-quinazolin-one, is an antineoplastic agent that targets the δ-isoform of the enzyme phosphoinositide 3-kinase (IC$_{50}$ = 2.5nM; EC$_{50}$ = 8 nM), thereby blocking constitutive phosphatidylinositol-3-kinase signaling and resulting in decreased phosphorylation of Akt (and other downstream effectors), increased poly(ADP-ribose) polymerase, as well as induction of caspase cleavage and apoptosis. (1). By inhibiting the production of phosphatidylinositol-3,4,5-trisphosphate (PIP$_3$), idelalisib prevents activation of the PI3K signaling pathway, thus inhibiting tumor cell proliferation, motility, and survival. Idelalisib also induces apoptosis in B cell lines and primary B cells from patients with different B-cell malignancies, including chronic lymphoblastic leukemia, diffuse large B-cell lymphoma, multiple myeloma, and Hodgkin lymphoma. Idelalisib interferes with integrin-mediated CLL cell adhesion to EC and BMSC, providing a novel mechanism to explain idelalisib-induced redistribution of CLL cells from tissues into the blood (2). Idelalisib is metabolized by aldehyde oxidase to GS-563117, and to a lesser extent by cytochrome P450 (CYP) 3A and uridine 5'-diphospho (UDP)-glucuronosyltransferase (UGT) 1A4. *In vitro*, idelalisib inhibits P-glycoprotein (P-gp) and organic anion transporting polypeptides (OATP) 1B1 and 1B3, and GS-563117 is a time-dependent CYP3A inhibitor (3). **Target(s):** Idelalisib is 40-times to 300-times more selective for p110δ relative to other PI3K class I enzymes: p110α (IC$_{50}$ = 820 nM); p110β (IC$_{50}$ = 565 nM); and p110γ (IC$_{50}$ = 89 nM). Greater selectivity (400-times to 4000-times) is seen against related kinases C2β, hVPS34, DNA-PK, and mTOR, and no activity is observed against a panel of 402 diverse kinases, when assayed at 10µM (using Ambit KinomeScan). **1**. Lannutti, Meadows, Herman, *et al.* (2011) *Blood* **117**, 591. **2**. Fiorcari, Brown, McIntyre, *et al.* (2013) *PLoS One* **8**, e83830. **3**. Jin, Robeson, Zhou, *et al.* (2015) *J. Clin. Pharmacol.* **326**, 90.

IDN-6556, *See Emricasan*

IDX-899, *See Fosdevirine*

Z-IETD-fmk

This irreversible, cell-permeable peptide (FW = 654.03 g/mol; *Sequence*: Z-Ile-Glu(OMe)-Thr-Asp(OMe)-fluoromethylketone, where *Z* designates a N-terminal carbobenzoxy moiety; Solubility: 100 mM in DMSO; Use at 2-4 µM), also known as Caspase-8/FLICE inhibitor (fluoromethylketone), Granzyme B inhibitor (fluoromethyl-ketone), inhibits Caspase-8 and Granzyme B, with somewhat weaker action against caspase-10. The inhibitor possesses a Caspase-8 targeting IETD sequence and a methyl ester to facilitate cell permeability. Z-IETD-fmk inhibits Caspase-8 proteolysis, an essential step in TNF- and Fas-mediated apoptosis (1). To probe the underlying mechanism for phenylalanine-mediated neurotoxicity in phenylketonuria (PKU), Z-IETD-FMK (0.9 mM, 18 h) was found to strongly attenuate apoptosis in Phe-treated neurons, again suggesting involvement of the Fas receptor (FasR)-mediated cell death receptor pathway in Phe toxicity (2). **Targets:** Caspase-8 (*Z*-IETD-FMK), Caspase-9 (*Z*-LEHD-FMK), Pan-caspases (*Z*-VAD-FMK). **1**. Hatano, *et al.* (2000) *J. Biol. Chem.* **275**, 11814. **2**. Huang, Lu, Lv, *et al.* (2013) *PLoS One* **8**, e71553,

Ifenprodil

This subunit-selective NMDA antagonist (FW = 325.45 g/mol; CAS 23210-56-2), also named 4-[2-(4-benzylpiperidin-1-yl)-1-hydroxypropyl] phenol, targets those N-acetyl-D-glutamate receptors that contain NR2B. These glutamate-gated channels are the major excitatory neurotransmitters in the mammalian central nervous system (CNS) and are involved in numerous physiological and pathological processes including synaptic plasticity, chronic pain and psychosis. Ifenprodil shows efficacy in neuroprotection, anti-hyperalgesia, and anti-Parkinson models. Related NR2B-directed NMDA antagonists that are based on the ifenprodil pharmacophore include Eliprodil, CP-101,606, Ro25,6981 and PD0196860. 1. Mony, Kew, Gunthorpe & Paoletti (2009) *Brit. J. Pharmacol.* **157**, 1301.

IKK-16

This signal transduction inhibitor and antineoplastic agent (FW = 520.09 g/mol (HCl Salt); CAS 1186195-62-9), systematically named N-(4-Pyrrolidin-1-yl-piperidin-1-yl)-[4-(4-benzo[b]thiophen-2-yl-pyrimidin-2-ylamino)phenyl]carboxamide hydrochloride, targets IκB kinase (IKK), with IC_{50} values of 40, 70 and 200 nM for IKK-2, IKK complex, and IKK-1. IKK-16 inhibits TNF-α-stimulated IκB degradation and expression of adhesion molecules E-selectin, ICAM and VCAM, with IC_{50} values of 1.0, 0.5, 0.3 and 0.3 μM, respectively. IKK-16 inhibits LPS-induced TNF-α release in vivo and neutrophil extravasion in thioglycollate-induced peritonitis. 1. Waelchli, Bollbuck, Bruns, *et al.* (2006) *Bioorg. Med. Chem. Lett.* **16**, 108.

IL-1RA, *See Interleukin-1 Receptor Antagonist*

Ilicicolin H

This pyridinone (FW = 451.61 g/mol; CAS 12689-26-8), from *Cylindrocladium ilicicola* inhibits mitochondrial ubiquinol:cytochrome-c reductase (complex III), IC_{50} = 3-5 nM for yeast cytochrome c. Ilicicolin H displays complex kinetic behavior indicative of slow-binding inhibitors. Ilicicolin H also binds stoichiometrically to the yeast cytochrome bc_1 complex, but binds approximately two orders of magnitude less tightly to the bovine enzyme and is essentially non-inhibitory to the Paracoccus enzyme. 1. Rotsaert, Ding & Trumpower (2008) *Biochim. Biophys. Acta* **1777**, 211. 2. Ding, di Rago & Trumpower (2006) *J. Biol. Chem.* **281**, 36036. 3. Gutierrez-Cirlos, Merbitz-Zahradnik & Trumpower (2004) *J. Biol. Chem.* **279**, 8708.

Ilimaquinone

This sesquiterpenoid quinone (FW = 358.48 g/mol), obtained from the marine sponge *Hippiospongia metachromia*, induces reversible breakdown of Golgi membranes, inhibits protein processing and cellular secretions. Ilimaquinone also inhibiting gap-junction communication. **Targets:** S-adenosylhomo-cysteinase (1); gap junction communication (2,3); protein transport (4); pyruvate,orthophosphate dikinase (5,6). 1. Radeke & Snapper (1998) *Bioorg. Med. Chem.* **6**, 1227. 2. Feldman, Kim & Laird (1997) *J. Membr. Biol.* **155**, 275. 3. Cruciani, Leithe & Mikalsen (2003) *Exp. Cell Res.* **287**, 130. 4. Takizawa, Yucel, Veit, *et al.* (1993) *Cell* **73**, 1079. 5. Motti, Bourne, Burnell, *et al.* (2007) *Appl. Environ. Microbiol.* **73**, 1921. 6. Haines, Burnell, Doyle, *et al.* (2005) *J. Agric. Food Chem.* **53**, 3856.

Ilomastat

This substituted dipeptide (FW = 388.47 g/mol), also known as GM 6001, Galardin, and named systematically as N-[2R-(hydroxamidocarbonyl methyl)-4-methylpentanoyl]-L-tryptophanmethylamide, and N-[3-(N-hydroxycarbamoyl)-2-isobutylpropionyl]-L-tryptophanmethylamide, is a broad-spectrum matrix metalloproteinase (MMP) inhibitor, with action against interstitial collagenase (K_i = 0.4 nM; the value for the isobutyl diasteriosomer is 200 nM), gelatinase A (K_i = 0.5 nM), stromelysin 1 (K_i = 27 nM), neutrophil collagenase (K_i = 0.1 nM), and gelatinase B (K_i = 0.2 nM). It also inhibits *Pseudomonas aeruginosa* elastase (K_i = 20 nM), thermolysin (K_i = 20 nM), and peptide deformylase. **Targets:** gelatinase A, *or* matrix metalloproteinase 2 (1-6); gelatinase B, *or* matrix metalloproteinase 9 (1,2,4,5); interstitial collagenase, *or* matrix metalloproteinase 1 (2,4,7,8); membrane type matrix metalloproteinase 1 (matrix metalloproteinase 14 (9); neutrophil collagenase, *or* matrix metalloproteinase 8 (1,2,4,5); peptide deformylase (10); *Pseudomonas aeruginosa* elastase (7); stromelysin 1, *or* matrix metalloproteinase 3 (1,2,4,5,11); thermolysin (7). 1. Levy, Lapierre, Liang, *et al.* (1998) *J. Med. Chem.* **41**, 199. 2. Holleran, Galardy, Gao, *et al.* (1997) *Arch. Dermatol. Res.* **289**, 138. 3. Tournier, Polette, Hinnrasky, *et al.* (1994) *J. Biol. Chem.* **269**, 25454. 4. Galardy, Grobelny, Foellmer & Fernandez (1994) *Cancer Res.* **54**, 4715. 5. Galardy, Cassabonne, Giese, *et al.* (1994) *Ann. N. Y. Acad. Sci.* **732**, 315. 6. Cheng, Shen, Nan, Qian & Ye (2003) *Protein Expr. Purif.* **27**, 63. 7. Grobelny, Poncz & Galardy (1992) *Biochemistry* **31**, 7152. 8. Yamamoto, Tsujishita, Hori, *et al.* (1998) *J. Med. Chem.* **41**, 1209. 9. Rozanov, Ghebrehiwet, Postnova, *et al.* (2002) *J. Biol. Chem.* **277**, 9318. 10. Jayasekera, Kendall, Shammas, *et al.* (2000) *Arch. Biochem. Biophys.* **381**, 313. 11. Parker, Lunney, Ortwine, *et al.* (1999) *Biochemistry* **38**, 13592.

Imatinib

This first-in-class, rationally developed antineoplastic agent (FW$_{\text{free-base}}$ = 493.61 g/mol; CAS 152459-95-5; Symbol = IM; IUPAC Name: 4-[(4-methylpiperazin-1-yl)methyl]-*N*-(4-methyl-3-{[4-(pyridin-3-yl)pyrimidin-2-yl]amino}phenyl)benzamide), also known as CGP-57148 and STI-571 as well as by the proprietary names Gleevec® and Glivec®, potently inhibits protein-tyrosine kinases, showings high specificity for the Bcr-Abl (*or* Breakpoint Cluster Region/Abelson) oncoenzyme associated with chronic myelogenous leukemia (CML) and some forms of acute lymphoblastic leukemia (ALL) (1-7). CML is by marked myeloid proliferation, invariably terminating in acute leukemia. Prior to the advent of Gleevec, conventional therapeutic options included interferon-based regimens and stem-cell transplantation. **Rational Drug Discovery:** Chronic Myelogenous Leukemia is a clonal hematopoietic stem-cell disorder characterized by Chromosome 9:22 translocation. More than 90% of CML patients have hematopoietic cells showing evidence of a reciprocal translocation between Chromosomes-9 and 22. In this translocation, the c-abl gene, located on Chromosome-9q, is translocated to Chromosome-22, giving rise to a chimeric mRNA formed from exons at the 5'-end of the *bcr* gene and those from the 3'-end of the *c-abl* gene. The resulting Bcr-Abl oncoprotein exhibits elevated protein-tyrosine kinase activity, which is strongly implicated in the development of chronic myeloid leukemia. Imatinib inhibits Abl protein-tyrosine kinase *in vitro* and *in vivo* (1). Moreover, cellular proliferation and tumor formation by Bcr-Abl-expressing cells were specifically inhibited by this compound (2). Indeed, imatinib selectively inhibits the growth of BCR-ABL-positive cells (3,4). In the words of the Lasker Foundation, Druker, Lydon & Sawyers "seized upon the known molecular defect that underlies CML, formulated the idea of tackling this root cause of the disease, crafted a specific kinase inhibitor, and designed a second-generation inhibitor when drug resistance developed." In developing the first small molecule to target the activity of an oncoenzyme, they established a new paradigm for rational drug discovery. One can trace this historic development to work by Baltimore's lab on lymphocyte development, which used cells that has been immortalized by the Abelson mouse leukemia virus. While examining the virus's ability to cause cell transformation and cancer, they and Hunter's lab discovered that the oncoprotein was a growth-modulating protein tyrosine kinase. **Primary Mode of Inhibitory Action:** While many protein kinase inhibitors interact directly at the ATP-binding site, Gleevec is an example from a new class of allosteric inhibitors that alter protein kinase conformation to block productive ATP binding and/or transphosphorylation. The c-Abl protein kinase domain exists in active and inactive conformations, and, as shown by X-ray crystallography, imatinib binds only to the inactive enzyme. Imatinib is a competitive inhibitor relative to ATP, binding to a part of the ATP-binding site and extending into an adjacent hydrophobic region. Crystallographic analysis reveals that imatinib stabilizes the inactive, Aspartate-Phenylalanine-Glycine-out conformation of Abl, with the activation loop in a "closed" substrate-mimicking position and resultant distortion of the ATP binding loop. Imatinib prevents phosphoryl transfer to tyrosine and subsequent activation of phosphorylated targets, thereby blocking transmission of proliferative signals and inducing apoptosis in leukemic cells. **Off-Target Effects:** Beyond Bcr-Abl, imatinib is active against other tyrosine kinases, the inhibition of which has implications for treating other malignancies. For example, imatinib exhibits remarkable activity in gastrointestinal stromal tumors. Imatinib targets c-Kit receptor protein-tyrosine kinase (8-11), platelet-derived growth-factor receptor protein-tyrosine kinase (2,3), polo kinase (12), protein tyrosine kinase (1-7), receptor protein-tyrosine kinase (2,3,8-11,13,14), human quinone reductase 2, *or* NQO2 (15), ribosyldihydro-nicotinamide dehydrogenase (quinone) (16). It also stimulates low-density lipoprotein receptor-related protein 1-mediated ERK phosphorylation within insulin-producing β-cells (16).

Resistance: Imatinib resistance arises from mutations in residues of the Bcr-Abl kinase, mainly those making contact with the drug. Mutations at about 20 residues in the BCR-ABL kinase (particularly Bcr-Abl[E255K] and Bcr-Abl[T315I]) are associated with clinical resistance to imatinib in CML patients, most keeping the kinase from adopting a specific closed conformation stabilized by imatinib binding. A small number directly interfere with imatinib binding. The fusion gene *BCR/ABL* modulates cellular responses to DNA damage to promote genomic instability, a property that leads to resistance to imatinib, dasatinib and nilotinib, and likewise contributes to malignant disease progression. The former phenomenon is often caused by mutations in Bcr-Abl kinase, whereas the latter is associated with accumulation of additional genetic aberrations including chromosomal translocations, deletions, additional chromosomes, gene amplifications, and point mutations. **Treatment of Philadelphia-positive Acute Lymphoblastic Leukemia:** A thirteen-year follow-up study (17) on patients on hyper-CVAD (*i.e.,* cyclophosphamide, vincristine, adriamycin, plus dexamethasone) regimen, followed by imatinib-based consolidation/maintenance therapy, demonstrated: (a) that 42 of 45 patients treated for active disease achieved complete remission (one reaching full remission with incomplete recovery of platelets, 1 achieving partial remission, and one succumbing during induction); (b) the 5-year overall survival rate for all patients was 43%; (c) significant negative predictors of overall survival were: age >60 years, p190 molecular transcript, and active disease at enrollment; (d) while 16 patients underwent allogeneic stem cell transplantation, the median overall survival was no better; and (e) that patients with residual molecular disease at 3 months had improved complete remission duration with transplant. Hyper-CVAD and imatinib thus represents an effective regimen for treating Philadelphia-positive ALL. **Effect on Adriamycin/Doxorubicin-induced DNA Damage Checkpoint Arrest:** Src family kinases promote recovery from G$_2$ DNA damage checkpoint, and imatinib inhibits inactivation of ATM/ATR signaling pathway to suppress recovery from adriamycin/doxorubicin-induced DNA damage checkpoint arrest (18). Imatinib and pazopanib, two distinct inhibitors of PDGFR/c-Kit family kinases, delays recovery from checkpoint arrest and inhibited the subsequent S-G$_2$-M transition after adriamycin exposure. By contrast, imatinib and pazopanib does not delay the recovery from checkpoint arrest in the presence of caffeine, an ATM/ATR inhibitor. Imatinib consistently induces a persistent activation of ATR-Chk1 signaling (18). Maintenance of G$_2$ checkpoint arrest largely depends on ATR-Chk1 signaling. However, unlike Src inhibition, imatinib does not delay the recovery from checkpoint arrest in the presence of KU-55933, an ATM inhibitor. Imatinib induces a persistent activation of ATM-KAP1 signaling, suggesting a possible involvement of imatinib in an ATM-dependent DNA damage response. These results raise the likelihood that imatinib may inhibit resumption of tumor proliferation after chemo-and radiotherapy (18). **Key Pharmacokinetic Parameters:** *See* Appendix II in Goodman & Gilman's THE PHARMACOLOGICAL BASIS OF THERAPEUTICS, 12th Edition (Brunton, Chabner & Knollmann, eds.) McGraw-Hill Medical, New York (2011). **1.** Schindler, Bornmann, Pellicena, *et al.* (2000) *Science* **289**, 1938. **2.** Buchdunger, Zimmermann, Mett, *et al.* (1996) *Cancer Res.* **56**, 100. **3.** Carroll, Ohno-Jones, Tamura, *et al.* (1997) *Blood* **90**, 494. **4.** Sirvent, Boureux, Simon, Leroy & Roche (2007) *Oncogene* **26**, 7313. **5.** Tauchi & Ohyashiki (2006) *Int. J. Hematol.* **83**, 294. **6.** Cortes, Jabbour, Kantarjian, *et al.* (2007) *Blood* **110**, 4005. **7.** Anastasiadou & Schwaller (2003) *Curr. Opin. Hematol.* **10**, 40. **8.** Heinrich, Griffith, Druker, *et al.* (2000) *Blood* **96**, 925. **9.** Edling & Hallberg (2007) *Int. J. Biochem. Cell Biol.* **39**, 1995. **10.** Roskoski (2005) *Biochem. Biophys. Res. Commun.* **338**, 1307. **11.** Mol, Dougan, Schneider, *et al.* (2004) *J. Biol. Chem.* **279**, 31655. **12.** Johnson, Stewart, Woods, Giranda & Luo (2007) *Biochemistry* **46**, 9551. **13.** Ingram & Bonner (2006) *Curr. Mol. Med.* **6**, 409. **14.** Medinger & Drevs (2005) *Curr. Pharm. Des.* **11**, 1139. **15.** Winger, Hantschel, Superti-Furga & Kuriyan (2009) *BMC Struct. Biol.* **9**, 7. **16.** Fred, Boddeti, Lundberg & Welsh (2014) *Clin. Sci.* (London) **128**, 17. **17.** Daver, Thomas, Ravandi, *et al.* (2015) *Haematologica* **100**, 653. **18.** Morii, Fukumoto, Kubota, *et al.* (2015) *Cell Biol. Int.* **39**, 923.

Imazapic

This imidazolinone herbicide (FW = 275.30 g/mol), also known as Cadre, Plateau, and systematically as (±)-2-[4,5-dihydro-4-methyl-4-(1-methylethyl)-5-oxo-1H-imidazol-2-yl]-5-methyl-3-pyridinecarboxylic acid, inhibits acetolactate synthase (tobacco K_i = 1.2 μM). (*See Imazethapyr for mode of action*) Imazapac is a selective herbicide for both the pre- and post-emergent control of some annual and perennial grasses and some broadleaf weeds. Imazapic kills plants by inhibiting the production of branched chain amino acids essential for protein synthesis and cell growth. 1. Yoon, Chung, Chang, *et al.* (2002) *Biochem. Biophys. Res. Commun.* 293, 433. 2. Oh, Park, Yoon, Han & Choi (2001) *Biochem. Biophys. Res. Commun.* 282, 1237. 3. Le, Yoon, Kim & Choi (2005) *Biochim. Biophys. Acta* 1749, 103. 4. Shin, Chong, Chang & Choi (2000) *Biochem. Biophys. Res. Commun.* 271, 801. 5. Chong, Chang & Choi (1997) *J. Biochem. Mol. Biol.* 30, 274. 6. Le, Yoon, Kim & Choi (2005) *J. Biochem.* 138, 35.

Imazapyr

This imidazolinone herbicide (FW$_{free-acid}$ = 261.28 g/mol; CAS 81334-34-1), also known as AC 243997, inhibits acetolactate synthase and the biosynthesis of branched-chain amino acids (*Zea mays* K_i = 12 μM). Imazapyr is a non-selective herbicide used for the control of a broad range of weeds, including terrestrial annual and perennial grasses and broadleaved herbs, woody species, and riparian and emergent aquatic species. The isopropylamine salt is also known as Arsenal and Chopper. (*See Imazethapyr for mode of action*) 1. Zohar, Einav, Chipman & Barak (2003) *Biochim. Biophys. Acta* 1649, 97. 2. Chang & Duggleby (1997) *Biochem. J.* 327, 161. 3. Hill, Pang & Duggleby (1997) *Biochem. J.* 327, 891. 4. Southan & Copeland (1996) *Physiol. Plant.* 98, 824. 5. Singh, Stidham & Shaner (1988) *J. Chromatogr.* 444, 251. 6. Karim, Shim, Kim, *et al.* (2006) *Bull. Korean Chem. Soc.* 27, 549. 7. Muhitch, Shaner & Stidham (1987) *Plant Physiol.* 83, 451. 8. Shaner, Anderson & Stidham (1984) *Plant Physiol.* 76, 545.

Imazaquin

This imidazolinone herbicide (FW$_{free-acid}$ = 311.34 g/mol; CAS 81335-37-7), also known as AC 252214 and named systematically as 2-[(*RS*)-4-isopropyl-4-methyl-5-oxo-2-imidazolin-2-yl]quinoline-3-carboxylic acid, is a slow-binding inhibitor of acetolactate synthase, *or* acetohydroxyacid synthase (*Zea mays* K_i = 3.4 μM). (*See Imazethapyr for mode of action*) Imazaquin is often referred to by the proprietary name for the ammonium salt (Scepter). 1. Schloss & Van Dyk (1988) *Meth. Enzymol.* 166, 445. 2. Shaner & Singh (1993) *Plant Physiol.* 103, 1221. 3. Shaner, Anderson & Stidham (1984) *Plant Physiol.* 76, 545. 4. Zohar, Einav, Chipman & Barak (2003) *Biochim. Biophys. Acta* 1649, 97. 5. Southan & Copeland (1996) *Physiol. Plant.* 98, 824. 6. Yang & Kim (1997) *J. Biochem. Mol. Biol.* 30, 13.

Imazethapyr

This imidazolinone herbicide (FW$_{free-acid}$ = 289.33 g/mol; CAS 81335-77-5), also named 5-ethyl-2-[(*RS*)-4-isopropyl-4-methyl-5-oxo-2-imidazolin-2-yl]nicotinic acid, inhibits acetolactate synthase, *or* acetohydroxyacid synthase, which catalyzes the first reaction in branched-chain amino acid biosynthesis. A virtue of imazethapyr is that it is not toxic to animals. By starving grasses and dicotyledons of needed nutrients, acetohydroxyacid synthase inhibitors deprive cells of the ability to make key proteins and likewise inhibits DNA replication (1-4). Neither sulfonylurea- and imidazolinone-class herbicides are structural mimics of the substrates for the enzyme. They instead inhibit by blocking a channel through which substrates must pass to reach the active site. Sulfonylurea herbicides approach within 5 Å of the catalytic center, which is the C2 atom of the cofactor thiamin diphosphate, whereas imidazolinone herbicides are stationed at least 7 Å from this atom (5). The ammonium salt is known by the proprietary names Pivot and Pursuit. (*See also Imazapic, Imazapyr, and Imazaquin*) 1. Gaston, Zabalza, Gonzalez, *et al.* (2002) *Physiol. Plant* 114, 524. 2. Zohar, Einav, Chipman & Barak (2003) *Biochim. Biophys. Acta* 1649, 97. 3. Southan & Copeland (1996) *Physiol. Plant.* 98, 824. 4. McCourt, Pang, King-Scott, Guddat & Duggleby (2006) *Proc. Natl. Acad. Sci. U.S.A.* 103, 569.

IMC-1121B, *See Ramucirumab*

IMD 0354

This signal transduction inhibitor and antineoplastic agent (FW = 383.67 g/mol; CAS 978-62-1, 634914-41-3 (sodium salt); Solubility: 10 mg/mL DMSO; <1 mg/mL Water; Formulation: Dissolve in 5% carboxymethylcellulose in PBS), systematically named *N*-[3,5-bis(trifluoromethyl)phenyl]-5-chloro-2-hydroxybenzamide, targets IκB-2 (IC$_{50}$ < 5 μM), having the effect of selectively activating the NFκβ pathway and arresting the cell cycle at the G$_0$/G$_1$ transition (1-4). Constitutive activation of NF-κB also up-regulates cell cycle progression and contributes to the proliferation of human breast cancer cells, and IMD-0354 treatment inhibits growth by promoting apoptosis as well as cell-cycle transition from the G$_1$ phase to the S phase through regulation of D-type cyclins (5). IMD 0354 is also highly effective against cancer stem cells, which, owing to their higher drug efflux capability and stem cell-like properties, are concentrated in a side population of cells thought to be responsible for multi-drug resistance and tumor repopulation, events that cause breast cancer patients to succumb (6). Although typically used to control inflammation, NF-κB signaling pathway has been linked to proliferation, anti-apoptosis, angiogenesis, and immune tolerance in the tumor microenvironment, making this pathway an attractive target in cancer treatment. Significantly, targeted delivery of IMD 0354 through the use of a legumain inhibitor also enhances drug delivery under hypoxia, a hallmark of the tumor microenvironment, but not under normoxia. 1. Tanaka, *et al.* (2005) *Blood* 105, 2324. 2. Sugita, *et al.* (2009) *Int. Arch. Allergy Immunol.* 148, 186. 3. Kamon, *et al.* (2004) *Biochem Biophys Res Communs.* 323, 242. 4. Onai, *et al.* (2004) *Cardiovasc Res.* 63, 51. 5. Tanaka, Muto, Konno, Itai & Matsuda (2006) *Cancer Res.* 66, 419. 6. Gomez-Cabrero, Wrasidlo, Reisfeld, *et al.* (2013) *PLoS One* 8, e73607.

Imetelstat

(3'-5')d(3'-amino-3'-deoxy-*P*-thio)(O—T-A-G-G-G-T-T-A-G-A-C-A-A)

This first-in-class telomerase inhibitor (FW = 4610 g/mol; CAS 868169-64-6; *Sequence*: d(3'-amino-3'-deoxy-5'-[*O*-[2-hydroxy-3-[(1-oxohexadecyl) amino]propyl]-*P*-thio-TAGGGTTAGACAA)), also known as GRN163L, is a palmitoylated 13-mer oligonucleotide $N^{3'},P^{5'}$-thiophosphoramidate (*or* NPS oligonucleotide) that targets the template region of the human functional telomerase RNA (hTR) subunit (1,2). The essential role of telomerase in transformed cells makes telomere maintenance an attractive target for inhibition by GRN163L (3). Indeed, telomerase inhibition is observed with a wide range human tumor cells (including lung, breast, prostate, liver, brain, bladder, and hematological malignancies including multiple myeloma and lymphoma), in both cell culture systems and mouse xenograft models. Imetelstat is additive or synergistic when used in combination with several cancer drugs and/or radiation. GRN163L shows off-target effects, with loss of adhesion in A549 lung cancer cells as a result of decreased E-cadherin expression. There is also consequential disruption of the cytoskeleton, loss of cell adherence, cell-cycle arrest in G_1 phase, and decreased matrix metalloproteinase-2 (MMP-2) expression (4). Continuous exposure of CAPAN1 (IC_{50}=75 nM) and CD18 pancreatic cancer cells to GRN163L (IC_{50}=204 nM) initially results in rapid telomere shortening, followed by the maintenance of extremely short but stable telomeres (5). Further continuous exposure eventually led to crisis, with complete loss of viability after 47 (CAPAN1) and 69 (CD18) doublings. In these cells, crisis was accompanied by activation of a DNA damage response (γ-H2AX), senescence and apoptosis. **1.** Herbert, Gellert, Hochreiter, *et al* (2005) *Oncogene* **24**, 5262. **2.** Djojosubroto, Chin, Go, *et al.* (2005) *Hepatology* **42**, 1127. **3.** Harley (2008) *Nature Rev. Cancer.* 8, 167. **4.** Mender, Senturk, Ozgunes, *et al.* (2013) *Int. J. Oncol.* **42**, 1709. **5.** Burchett, Yan & Ouellette (2014) *PLoS One* **9**, e85155.

Imidacloprid

This insecticide (FW = 255.66 g/mol; CAS 138261-41-3), also known as *N*-{1-[(6-chloro-3-pyridyl)methyl]-4,5-dihydroimidazol-2-yl}nitramide, is a neonicotinoid that acts at pharmacologically distinct acetylcholine receptor (AChR) subtypes in the insect nervous system, but is ineffective on muscarinic receptors. By blocking nicotinergic neuronal pathways, treatment with imidacloprid results in paralysis and eventual death of insects. Its much weaker action on human and domestic animals and pets conmends its use by veterinarians to treat fleas and other insect pests (*See also Nitenpyram*). Monthly imidacloprid application of 7.5 to 10 mg/kg will rapidly kill existing and reinfesting flea infestations on dogs and break the flea life cycle by killing adult fleas before egg production begins (1). In single-channel patch clamp measurements on nicotinic acetylcholine receptors of rat pheochromocytoma (PC 12) cells, acetylcholine and imidacloprid induces single-channel currents of main conductance and subconductance states with conductances of 33.3 and 9.4 pS by acetylcholine and 30.4 and 9.8 pS by imidacloprid (2). However, the main conductance currents are generated predominantly by acetylcholine, whereas the subconductance currents are mainly generated by imidacloprid. **1.** Arther, Cunningham, Dorn, *et al.* (1997) *Am. J. Vet. Res.* **58**, 848. **2.** Nagata, Aistrup, Song & Narahashi (1996) *Neuroreport* **7**, 1025.

4-Imidazoleacetate (4-Imidazoleacetic Acid)

This metabolite (FW$_{free-base}$ = 125.11 g/mol; FW$_{hydrochloride}$ = 162.57 g/mol; CAS 3251-69-2 (hydrochloride)), also known as imidazole-4-acetic acid and IMA, is a product of histamine degradation and an antagonist at GABA$_C$ receptors. IMA also activates GABA$_A$ receptors. 4-Imidazole acetate is a substrate of the reactions catalyzed by imidazoleacetate 4-monooxygenase and imidazoleacetate:phosphoribosyldiphosphate ligase, and it is produced by the action of aldehyde dehydrogenase on imidazole 4-acetaldehyde. **Target(s):** dopamine β-monooxygenase, $K_i \approx 0.8$ mM (1); GABA$_C$ receptor (2,3); histidine ammonia-lyase (4); histidine decarboxylase (5-7). **1.** Townes, Titone & Rosenberg (1990) *Biochim. Biophys. Acta* **1037**, 240. **2.** Chebib, Mewett & Johnston (1998) *Eur. J. Pharmacol.* **357**, 227. **3.** Tunnicliff (1998) *Gen. Pharmacol.* **31**, 503. **4.** Webb (1966) *Enzyme and Metabolic Inhibitors*, vol. **2**, p. 353, Academic

Press, New York. **5.** Snell (1986) *Meth. Enzymol.* **122**, 128. **6.** Snell & Guirard (1986) *Meth. Enzymol.* **122**, 139. **7.** Rosenthaler, Guirard, Chang & Snell (1965) *Proc. Natl. Acad. Sci. U.S.A.* **54**, 152.

4-Imidazolepropionate (4-Imidazolepropionic Acid)

This histidine derivative (FW = 140.14 g/mol; CAS 1074-59-5), also known as 3-(4-imidazolyl)propionic acid, is a competitive inhibitor of urocanase (*Reaction*: L-Histidine $\rightleftharpoons$ Urocanate + NH4$^+$). **Target(s):** dimethyl-histidine *N*-methyltransferase, *or* histidine *N*-methyltransferase (1); histidine ammonia-lyase (2); histidine decarboxylase (3,4); histidyl-tRNA synthetase (5); urocanate hydratase, *or* urocanase (6-16). **1.** Ishikawa & Melville (1970) *J. Biol. Chem.* **245**, 5967. **2.** Brand & Harper (1976) *Biochemistry* **15**, 1814. **3.** Snell (1986) *Meth. Enzymol.* **122**, 128. **4.** Rosenthaler, Guirard, Chang & Snell (1965) *Proc. Natl. Acad. Sci. U.S.A.* **54**, 152. **5.** Lepore, di Natale, Guarini & de Lorenzo (1975) *Eur. J. Biochem.* **56**, 369. **6.** Magasanik, Kaminskas & Kimhi (1971) *Meth. Enzymol.* **17B**, 51. **7.** Hassall & Greenberg (1971) *Meth. Enzymol.* **17B**, 84. **8.** Matherly & Phillips (1980) *Biochemistry* **19**, 5814. **9.** Hug, O'Donnell & Hunter (1978) *Biochem. Biophys. Res. Commun.* **81**, 1435. **10.** Gerlinger & Retey (1987) *Z. Naturforsch. C* **42**, 349. **11.** Hug, O'Donnell & Hunter (1979) *J. Biol. Chem.* **253**, 7622. **12.** George & Phillips (1970) *J. Biol. Chem.* **245**, 528. **13.** O'Donnell & Hug (1989) *J. Photochem. Photobiol. B, Biol.* **3**, 429. **14.** Hunter & Hug (1989) *Pept. Res.* **2**, 240. **15.** Kaminskas, Kimhi & Magasanik (1970) *J. Biol. Chem.* **245**, 3536. **16.** Klepp, Fallert-Müller, Grimm, Hull & Retey (1990) *Eur. J. Biochem.* **192**, 669.

N-(2-(1*H*-Imidazol-4-yl)ethyl)-6-[4-(2-methoxy-phenoxy)-piperidin-1-yl]pyridazine-3-carboxamide

This potent, selective, orally bioavailable pyridazine (FW = 421.48 g/mol) inhibits stearoyl-CoA 9-desaturase (mouse, IC_{50} = 6 nM), catalyzes the committed step in the biosynthesis of monounsaturated fatty acids from saturated, long-chain fatty acids. Studies with stearoyl-CoA 9-desaturase knockout mice have established that these animals are lean and protected from leptin deficiency-induced and diet-induced obesity, with greater whole body insulin sensitivity than wild-type animals. **1.** Liu, Lynch, Freeman, *et al.* (2007) *J. Med. Chem.* **50**, 3086.

IMID-4F

This imidazoline derivative (FW$_{free-base}$ = 338.21 g/mol), also known as 2-[*N*-(2,6-dichlorophenyl)-*N*-(4 flurorobenzyl)amino]imidazoline, is a potent K$_{ATP}$ channel blocker. IMID-4F acts by interacting directly with the pore of the K$_{IR}$ channel, rather than through the sulphonylurea subunit of the K$_{ATP}$ channel complex. Note: IMID-4F is a *p*-fluorobenzyl derivative of clonidine. **1.** McPherson, Bell, Favaloro, Kubo & Standen (1999) *Brit. J. Pharmacol.* **128**, 1636. **2.** Bell, Favaloro, Khalil, Iskander & McPherson (2000) *Naunyn Schmiedebergs Arch. Pharmacol.* **362**, 145.

Imidobisphosphate (Imidobisphosphoric Acid)

This pyrophosphate analogue ($FW_{free-acid}$ = 175.98 g/mol), also known as imidodiphosphate, inhibits pyrophosphatase. It is also a valuable reagent for the synthesis of nucleotide analogues, such as p(NH)ppA and pp(NH)pA. **Target(s):** DNA-directed RNA polymerase (1); geranyl-diphosphate cyclase, *or* bornyl diphosphate synthase (2); guanylate cyclase (3,4); *trans*-octaprenyl*trans*transferase (5); phospho-enol-pyruvate carboxykinase (6,7); pyrophosphatase (8-12); pyrophosphate-dependent phosphofructokinase, *or* diphosphate:fructose-6 phosphate 1-phopho-transferase (13); pyruvate,orthophosphate dikinase (14). **1.** Komissarenko, Fomovskaia, Kolesnikova, Tarusova & Borisevich (1985) *Ukr. Biokhim. Zh.* **57**, 56. **2.** Wheeler & Croteau (1988) *Arch. Biochem. Biophys.* **260**, 250. **3.** Gorczyca (2000) *Meth. Enzymol.* **315**, 689. **4.** Gorczyca, Van Hooser & Palczewski (1994) *Biochemistry* **33**, 3217. **5.** Gotoh, Koyama & Ogura (1992) *J. Biochem.* **112**, 20. **6.** Wood, O'Brien & Michaels (1977) *Adv. Enzymol. Relat. Areas Mol. Biol.* **45**, 85. **7.** Willard & Rose (1973) *Biochemistry* **12**, 5241. **8.** Cooperman (1982) *Meth. Enzymol.* **87**, 526. **9.** Baltscheffsky & Nyrén (1986) *Meth. Enzymol.* **126**, 538. **10.** Schwarm, Vigenschow & Knobloch (1986) *Biol. Chem. Hoppe-Seyler* **367**, 119. **11.** Morita & Yasui (1978) *J. Biochem.* **83**, 719. **12.** Islam, Miyoshi, Kasuga-Aoki, *et al.* (2003) *Eur. J. Biochem.* **270**, 2814. **13.** Rowntree & Kruger (1992) *Plant Sci.* **86**, 183. **14.** Hiltpold, Thomas & Köhler (1999) *Mol. Biochem. Parasitol.* **104**, 157.

2-Iminobiotin

This guanidinium analogue of biotin (FW = 243.33 g/mol; CAS 13395-35-2), also known as guanidinobiotin, was first prepared in 1950 (1) and has been used in innumerable protein purification protocols and in binding investigations. 2-Iminobiotin is also a reversible inhibitor of murine inducible and rat neuronal nitric-oxide synthases, with K_i values of 21.8 and 37.5 μM, respectively (2) **1.** Hofmann & Axelrod (1950) *J. Biol. Chem.* **187**, 29. **2.** Sup, Green & Grant (1994) *Biochem. Biophys. Res. Commun.* **204**, 962.

Iminodiacetate (Iminodiacetic Acid)

This dicarboxylic acid ($FW_{free-acid}$ = 133.10 g/mol), also known as *N*-(carboxymethyl)glycine and diglycine, is a structural analogue of glutaric acid. It is soluble in water (2.4 g/100 mL at 5°C) and will readily form complexes with divalent cations (*e.g.*, Mn^{2+}). **Target(s):** acid phosphatase (1); glutamate dehydrogenase (2); phosphonoacetate hydrolase (3,4); pyrimidine-deoxynucleoside 2'-dioxygenase, *or* thymidine:2-oxoglutarate dioxygenase (5). **1.** Kawabe, Sugiura, Terauchi & Tanaka (1984) *Biochim. Biophys. Acta* **784**, 81. **2.** Smith, Austen, Blumenthal & Nyc (1975) *The Enzymes*, 3rd ed., **11**, 293. **3.** McGrath, Kulakova & Quinn (1999) *J. Appl. Microbiol.* **86**, 834. **4.** McMullan & Quinn (1994) *J. Bacteriol.* **176**, 320. **5.** Bankel, Lindstedt & Lindstedt (1972) *J. Biol. Chem.* **247**, 6128.

N^ϵ-(1-Iminoethyl)-L-lysine

This lysine derivative and arginine analogue (FW = 187.24 g/mol), often abbreviated L-NIL, is a highly selective inhibitor (IC_{50} = 3.3 μM) of inducible nitric-oxide synthase (iNOS). (The neuronal isoform has an IC_{50} value of 92 μM.) Studies using L-NIL suggest that total ablation of excess NO production does not improve cardiac graft survival, but partial inhibition does. This finding is consistent with the notion that NO may have both adverse and beneficial actions. **Target(s):** arginase (1); nitric-oxide synthase (2-6); transport system y^+ (7). **1.** Colleluori & Ash (2001) *Biochemistry* **40**, 9356. **2.** Griffith & Kilbourn (1996) *Meth. Enzymol.* **268**, 375. **3.** Pieper, Nilakantan, Hilton, *et al.* (2004) *Amer. J. Physiol. Heart Circ. Physiol.* **286**, H525. **4.** Moore, Webber, Jerome, *et al.* (1994) *J. Med. Chem.* **37**, 3886. **5.** Nilakantan, Halligan, Nguyen, *et al.* (2005) *J. Heart Lung Transplant.* **24**, 1591. **6.** Bodnárová & Moroz (2005) *J. Inorg. Biochem.* **99**, 922. **7.** McDonald, Rouhani, Handlogten, *et al.* (1997) *Biochim. Biophys. Acta* **1324**, 133.

N^δ-(1-Iminoethyl)-L-lysine

This ornithine derivative and arginine isostere (FW = 174.12 g/mol) inhibits of nitric-oxide synthase, with respective IC_{50} values of 20.2, 42.4, and 34.5 μM for the neuronal, inducible, and endothelial enzymes. It is an alternative substrate for arginine decarboxylase (1). **Target(s):** amino-acid *N*-acetyltransferase (*Escherichia coli* enzyme is mildly inhibited) (2); arginase (3); nitric-oxide synthase (4-11); transport system y^+ (12). **1.** Balbo, Patel, Sell, *et al.* (2003) *Biochemistry* **42**, 15189. **2.** Leisinger & Haas (1975) *J. Biol. Chem.* **250**, 1690. **3.** Colleluori & Ash (2001) *Biochemistry* **40**, 9356. **4.** Garvey, Furfine & Sherman (1996) *Meth. Enzymol.* **268**, 339. **5.** Griffith & Kilbourn (1996) *Meth. Enzymol.* **268**, 375. **6.** Moore, Webber, Jerome, *et al.* (1994) *J. Med. Chem.* **37**, 3886. **7.** McCall, Feelisch, Palmer & Moncada (1991) *Brit. J. Pharmacol.* **102**, 234. **8.** Knowles, Palacios, Palmer & Moncada (1990) *Biochem. J.* **269**, 207. **9.** Griffith & Stuehr (1995) *Ann. Rev. Physiol.* **57**, 707. **10.** Babu, Frey & Griffith (1999) *J. Biol. Chem.* **274**, 25218. **11.** Babu & Griffith (1998) *J. Biol. Chem.* **273**, 8882. **12.** McDonald, Rouhani, Handlogten, *et al.* (1997) *Biochim. Biophys. Acta* **1324**, 133.

Imipenem

This intravenously administered, semisynthetic β-lactam antibiotic (FW = 299.35; CAS 74431-23-5), also known as *N*-formimidoylthienamycin and Cilastatin, inhibits peptidoglycan biosynthesis and is a broad spectrum antibacterial agent. It is a stable derivative of thienamycin, produced by *Streptomyces cattleya*, and was the first of the carbapenems to be released for clinical use. Imipenem inhibits cell wall synthesis of various Gram-positive and Gram-negative bacteria and remains very stable, even in the presence of β-lactamase (*e.g.,* penicillinase and cephalosporinase). **Key Pharmacokinetic Parameters:** *See* Appendix II in Goodman & Gilman's *THE PHARMACOLOGICAL BASIS OF THERAPEUTICS*, 12th Edition (Brunton, Chabner & Knollmann, eds.) McGraw-Hill Medical, New York (2011). (***See*** *Carbapenem-Resistant Enterobacteriaceae; Ertapenem, Doripenem; Meropenem*) **Target(s):** D-alanine carboxypeptidase activities of penicillin-binding proteins 4 and 5 (1); β-lactamase (2-5); transglycosylase activity of penicillin-binding protein-1A (1); transpeptidase activities of penicillin-binding proteins 1A, 1B, and 2 (1). **1**. Hashizume, Ishino, Nakagawa, Tamaki & Matsuhashi (1984) *J. Antibiot.* **37**, 394. **2**. Beadle & Shoichet (2002) *Antimicrob. Agents Chemother.* **46**, 3978. **3**. Brown, Young & Amyes (2005) *Clin. Microbiol. Infect.* **11**, 15. **4**. Poirel, Brinas, Verlinde, Ide & Nordmann (2005) *Antimicrob. Agents Chemother.* **49**, 3743. **5**. Trehan, Beadle & Shoichet (2001) *Biochemistry* **40**, 7992.

Imipramine

This tricyclic antidepressant (FW$_{\text{free-base}}$ = 280.41 g/mol; CAS 50-49-7; Photosensitive), also known as Melipramine, Tofranil, and 3-(10,11-dihydro-5*H*-dibenzo[*b,f*]azepin-5-yl)-*N,N*-dimethylpropan-1-amine, binds tightly to the serotonin transporter, inhibiting 5-HT re-uptake. Imipramine has other notable effects: strong inhibition of norepinephrine re-uptake; inhibitory action on σ-receptors (weak, K_i > 500 nM); anticholinergic effects that limit its use in the elderly (especially suspected AD patients); action as a H$_1$ histamine receptor antagonist; and inhibitory effects on re-uptake and release at D$_1$ and D$_2$ receptors. (**See also** *Amitriptyline; Mianserin*) **Target(s):** Ca^{2+}-transporting ATPase (1); flavin-containing monooxygenase (2); glucuronosyltransferase (reported to be a substrate for UGT1A3 and UGT1A4 (3,4); K$^+$ channel, HERG (human ether-a-go-go related gene) (5); K$^+$ channel, transient outward (6); norepinephrine uptake (7); protein kinase C (8,9); serotonin transport (10,11); thiamin-triphosphatase (12). **1**. Hasselbach (1974) *The Enzymes*, 3rd ed., **10**, 431. **2**. Demirdoegen & Adall (2005) *Cell Biochem. Funct.* **23**, 245. **3**. Watanabe, Nakajima, Ohashi, Kume & Yokoi (2003) *Drug Metab. Dispos.* **31**, 589. **4**. Hara, Nakajima, Miyamoto & Yokoi (2007) *Drug Metab. Pharmacokinet.* **22**, 103. **5**. Teschemacher, Seward, Hancox & Witchel (1999) *Brit. J. Pharmacol.* **128**, 479. **6**. Casis & Sanchez-Chapula (1998) *J. Cardiovasc. Pharmacol.* **32**, 521. **7**. Linner, Arborelius, Nomikos, Bertilsson & Svensson (1999) *Biol. Psychiatry* **46**, 766. **8**. Kikkawa, Minakuchi, Takai & Nishizuka (1983) *Meth. Enzymol.* **99**, 288. **9**. Kikkawa & Nishizuka (1986) *The Enzymes*, 3rd ed., **17**, 167. **10**. Wood (1987) *Neuropharmacology* **26**, 1081. **11**. Hrdina (1989) *Int. J. Clin. Pharmacol. Res.* **9**, 119. **12**. Iwata, Baba, Matsuda & Terashita (1974) *Thiamine* (Proc. Pap. U.S.-Jpn. Semin, 2nd. Meeting), 213.

Immucillins

This novel class of rationally designed drugs maximizes binding affinity by replicating crucial stereoelectronic features of enzyme transition states, including, but not limited, to oxa-carbenium ion intermediates, attained during catalysis of various ribosyl transferase, nucleosidase, and nucleoside phosphorylase reactions. The systematic search for such high-affinity transition-state analogues by Schramm and associates is based on an approach termed Molecular Similarity (1-3). The overall goal is the rational design of extremely high-affinity inhibitors based on detailed reaction mechanism information that allows one to deduce an enzyme's transition-state configuration. One begins with the evaluation of multiple kinetic isotope effects arising from judiciously chosen heavy isotope substitiutions at or near the scissile bond. These data provide a glimpse of the most likely catalytic mechanisms. The process then exploits molecular mechanical modeling of the electrostatic surface potential of that transition-state configuration as well as numerous candidate analogues to obtain the analogue structures that best match the inferred transition-stae geometry. Subsequent testing of the inhibitory action of the most promising analogues provides an additional means for refining the investigator's structural model of the transition state. Vibrational modeling of a data set for experimentally determined kinetic isotope effects also allows one to obtain a geometric model of the ground state and transition state for the nucleoside hydrolase reaction. From these data, one creates an electrostatic potential surface for free inosine, the above-mentioned transition state, as well as the reaction products. In the Molecular Similarity Approach, a numerical index describes the similarity of geometric and electronic properties of an

analogue, a process intended to deduce the most likely transition-state configuration for an enzyme-catalyzed reaction. The strategy is in essence a tacit assertion of the Pauling postulate, namely that rate enhancement arises from the enzyme's ability to lower the energy input required to reach the transition state by selectively binding (and, hence, stabilizing) reactants in their transition-state configuration. That reaction transition states holds the secrets for the rational design of highly selective, if not specific, inhibitors for particular enzyme targets was first suggested by its principal proponent, Richard Wolfenden (5), who in turn credits Michael Polanyi's assertion in 1921 that to enhance the rate of any reaction, a catalyst must bind the altered substrate very tightly in the transition state. The approach for their design is best exemplified in studies of purine nucleoside phosphorylase, *or* PNP (*Reaction*: Inosine + P$_i$ ⇌ Hypoxanthine + Ribose-1-P), the inhibition of which has been exploited to treat leukemia, autoimmune disorders, and gout. Key features for tight binding are: (a) formation of ion-pair between bound phosphate (or its mimic) and inhibitor cation; (b) interaction of the leaving-group with N^1, O^6, and N^7 atoms of 9-deazahypoxanthine; (c) interaction between phosphate and inhibitor hydroxyl groups; and (d) interaction of His257 with the 5'-hydroxyl group (4). Four generations of ribocation transition-state mimics bind to PNP with roughly comparable affinity: (a) the first-generation drug known as Immucillin-H (K_i* = 58 pM) contains an iminoribitol cation with four asymmetric carbons; (b) the second-generation drug known as DADMe-Immucillin-H (K_i* = 9 pM) uses a methylene-bridged dihydroxypyrrolidine cation with two asymmetric centers; (c) the third-generation drug known as DATMe-Immucillin-H (K_i* = 9 pM) contains an open-chain amino alcohol cation with two asymmetric carbons; and (d) the fourth-generation drug known as SerMe-ImmH (K_i* = 5 pM) uses achiral dihydroxyaminoalcohol seramide as the ribocation mimic. Because enzymes evolve to best meet the demands of individual cell types and the needs of organisms, Molecular Similarity allows for the likelihood that isozymes (different polypeptides catalyzing the very same chemical reaction) may have different transition states, despite the fact that they catalyze the same reaction. The target enzyme most likely binds the inhibitor rapidly to form a loose-filling complex (Enz·I$_{\text{loose}}$) that ratchets through a succession of reversible, albeit slow, conformational isomerizations (*e.g.*, Enz + I ⇌ Enz·I$_{\text{loose}}$ ⇌ Enz·I$_1$ ⇌ Enz·I$_2$ ⇌ Enz·I$_3$ ⇌ · · · ⇌ Enz·I$_{\text{final}}$), ultimately resulting in extraordinarily tight binding that exploits multiple binding interactions and conformational adjustments. As pointed out by Ho *et al.* (4), despite the considerable chemical diversity in the structures of four generations of PNP transition-state analogues, the catalytic-site geometry is practically the same for all analogues, suggesting that, at least for human PNP, multiple solutions are available for converting the energy of transition-state stabilization into binding energy. **1**. Horenstein & Schramm (1993) *Biochemistry* **32**, 7089. **2**. Horenstein, Parkin, Estupinan & Schramm (1991) *Biochemistry* **30**, 10788. **3**. Schramm (1999) *Meth. Enzymol.* (Schramm & Purich, eds.) **308**, 301. **4**. Ho, Shi, Rinaldo-Matthis, Tyler, Evans, Clinch, Almo, & Schramm (2010) *Proc. Natl. Acad. Sci U.S.A.* **107**, 4805. **5**. Wolfenden (1969) *Nature* **223**, 704.

Immucillin-A

This adenosine nucleoside mimetic (FW = 264.27 g/mol) weakly inhibits 5'-methylthio-adenosine/*S*-adenosylhomo-cysteine nucleosidase. Immucillin-AP, which is the 5'-monophosphate derivative,s is a potent inhibitor of adenine phosphoribosyl-transferase (K_i = 200 nM). (**For Mode of Action, See** *Immucillin*) **Target(s):** adenine phosphoribosyl-transferase, inhibited by 5'-monophosphate (2); adenosyl-homocysteine nucleosidase, weakly inhibited (3); 5'-methylthioadenosine/*S*-adenosylhomocysteine nucleosidase, weakly inhibited (3); purine nucleosidase (4); purine-nucleoside phosphorylase (5). **1**. Ho, Shi, Rinaldo-Matthis (2010) *Proc Natl Acad Sci U S A*. **107**, 4805. **2**. Shi, Tanaka, Crother, *et al.* (2001) *Biochemistry* **40**, 10800. **3**. Singh, Evans, Lenz, *et al.* (2005) *J. Biol. Chem.* **280**, 18265. **4**. Versées, Barlow & Steyaert (2006) *J. Mol. Biol.* **359**, 331. **5**. Rinaldo-Matthis, Wing, Ghanem, *et al.* (2007) *Biochemistry* **46**, 659.

Immucillin-G

This guanosine nucleoside mimetic (FW = 281.27 g/mol), also known as (1*S*)-1-(9-deazaguanin-9-yl)-1,4-dideoxy-1,4-imino-D-ribitol, is an extremely high-affinity transition-state mimic for purine-nucleoside phosphorylase (*Toxoplasma gondii* K_i = 1.9 nM (1-4). (*For Mode of Action, See Immucillin*) 1. Miles, Tyler, Furneaux, Bagdassarian & Schramm (1998) *Biochemistry* **37**, 8615. 2. Bzowska, Kulikowska & Shugar (2000) *Pharmacol. Ther.* **88**, 349. 3. Rinaldo-Matthis, Wing, Ghanem, *et al.* (2007) *Biochemistry* **46**, 659. 4. Chaudhary, Ting, Kim & Roos (2006) *J. Biol. Chem.* **281**, 25652.

Immucillin-H

This purine nucleoside mimetic and novel T-cell selective immuno-suppressive agent (FW = 265.25 g/mol; CAS 209799-67-7), also known as , also known by its code name BCX-1777, its systematic name (1*S*)-1-(9-deazahypoxanthin-9-yl)-1,4-dideoxy-1,4-imino-D-ribitol, and its trade name Forodesine™, is an extremely high-affinity and orally effective transition-state mimic for purine nucleoside phosphorylase (PNP). The K_i values for human, bovine, and *Mycobacterium tuberculosis* PNP's are 73, 23, and 28 pM. (*For In Vitro Mode of Action, See Immucillin*) In the presence of 2'-deoxyguanosine (dGuo), Immucillin H inhibits human lymphocyte proliferation activated by various agents, including interleukin-2 (IL-2), mixed lymphocyte reaction and phytohemagglutinin. PNP catalyzes the phosphorolysis of dGuo to guanine (Gu) and 2'-deoxyribose-1-phosphate, whereas deoxycytidine kinase (dCK) converts dGuo to form dGMP, which undergoes metrabolic phosphorylation to dGTP. The affinity of dGuo is higher for PNP than for dCK. However, if PNP is blocked by forodesine, plasma dGuo is not cleaved to Gu, and instead is converted to dGTP by dCK, leading to inhibition of ribonucleotide reductase (RR), an enzyme required for DNA synthesis and cell replication. Blockade of the latter then triggers apoptosis. **Target(s):** hypoxanthine(guanine) phosphoribosyl-transferase (1); purine nucleosidase (2); purine-nucleoside phosphorylase (3-11). 1. Sauve, Cahill, Zech, *et al.* (2003) *Biochemistry* **42**, 5694. 2. Versées, Barlow & Steyaert (2006) *J. Mol. Biol.* **359**, 331. 3. Miles, Tyler, Furneaux, Bagdassarian & Schramm (1998) *Biochemistry* **37**, 8615. 4. Schramm (2002) *Biochim. Biophys. Acta* **1587**, 107. 5. Semeraro, Lossani, Botta, *et al.* (2006) *J. Med. Chem.* **49**, 6037. 6. Bzowska, Kulikowska & Shugar (2000) *Pharmacol. Ther.* **88**, 349. 7. Rinaldo-Matthis, Wing, Ghanem, *et al.* (2007) *Biochemistry* **46**, 659. 8. Murkin, Birck, Rinaldo-Matthis, *et al.* (2007) *Biochemistry* **46**, 5038. 9. Chaudhary, Ting, Kim & Roos (2006) *J. Biol. Chem.* **281**, 25652. 10. Canduri, Fadel, Basso, *et al.* (2005) *Biochem. Biophys. Res. Commun.* **327**, 646. 11. Nunez, Wing, Antoniou, Schramm & Schwartz (2006) *J. Phys. Chem. A* **110**, 463.

L-Immucillin-H

This immucillin-H enantiomer (FW = 265.25 g/mol) inhibits purine-nucleoside phosphorylase (K_i = 12 nM). (*See Immucillin for discussion of Primary Mode of Action*) 1. Rinaldo-Matthis, Murkin, Ramagopal, *et al.* (2008) *J. Amer. Chem. Soc.* **130**, 842.

Immunophilins

These clinically significant immunosuppressant-binding proteins include FK506-binding proteins (*or* FKBPs), which also bind rapamycin, as well as the 16 human cyclophilins, which possess peptidyl-prolyl *cis-trans* isomerase activity and also bind cyclosporin A. Immunosuppression occurs when the resulting Drug·Immunophilin complex binds to and inhibits calcineurin, the calcium-calmodulin-activated serine/threonine-specific phosphoprotein phosphatase abundant in the brain. Rapamycin also binds to FKBP, and the Rapamycin-FKBP complex binds to RAFT (<u>R</u>apamycin <u>A</u>nd <u>F</u>KBP-12 <u>T</u>arget, which in turn modulates translation by phosphorylating p70-S6 kinase. Once activated, this kinase phosphorylates ribosomal S6 protein and 4E-BP1, thereby inhibiting the initiation of translation. Immunophilin ligands as well as those that fail to suppress immunity because they fail to inhibit calcineurin, stimulate regrowth of damaged peripheral and central neurons, including those that are dopaminergic, serotoninergic, and cholinergic in intact animals. FKPB12 is physiologically associated with the ryanodine receptor and the inositol 1,4,5-trisphosphate (IP3) receptor, thereby regulating calcium ion flux. By influencing phosphorylation of neuronal nitric oxide synthase, FKBP12 also regulates formation of nitric oxide, which is reduced by FK506. In immune cells, formation of the Drug·Immunophilin·Calcineurin ternary complex blocks dephosphorylation-driven nuclear translocation of the nuclear factor of activated T cells (NFAT). This ultimately contributes to the failure of T cells to respond to antigenic stimuli. *See also Cyclosporin A; FK506; NIM811; Rapamycin*

IMPDHII (or IMPDH2)

This negative regulator of NF-κB activation (MW$_{monomer}$ = 55805 g/mol; Enzyme Commission Number = 1.1.1.205), also known as Inosine 5'-Monophosphate Dehydrogenase II, is a homodimer that decreases NF-κB activation by increasing SHP1 activity and subsequent dephosphorylation of the p85α subunit of PI3K. It serves as a negative regulator of the PI3K/Akt-dependent pathway of NF-κB transactivation (1). By diverting IMP away into the GMP synthesis from AMP synthesis pathway, IMPDHII also fuels lymphocyte proliferation. The *IMPDHII* gene is up-regulated in cancer, suggesting a role in malignant transformation. Its broader biologic function within innate immune signal pathways has yet to be established. In contrast to the ubiquitously expressed IMPDHI isozyme, IMPDHII expression is restricted to hematopoietic stem cells and is essential for B- and T-cell proliferation. IMPDHII may also contribute to the production of TNFα and nitric oxide following macrophage stimulation with LPS and interferon-γ through a guanosine-dependent mechanism (2). The rapid recruitment of IMPDH to lipid rafts was demonstrated by an unbiased proteomic analysis (1). Mounting evidence indicates that lipid rafts play a pivotal role in the fine regulation of the innate immune response through various mechanisms including the recruitment or exclusion of signaling molecules. 1. Toubiana, Rossi, Grimaldi, *et al.* (2011) *J. Biol.Chem.* **266**, 23319. 2. Weigel, Bertalanffy & Wolner (2002) *Mol. Pharmacol.* **62**, 453.

Imperatorin

This furanocoumarin (FW = 270.28 g/mol; M.P. = 102°C), isolated from *Imperatoria osthruthium*, *Angelica archangelica*, *A. dahurica*, and *Pastinaca sativa*, inhibits a number of cytochrome P450 systems. **Target(s):** acetylcholinesterase (1); 4-aminobutyrate aminotransferase (2); coumarin 7 monooxygenase (3); CYP1A1, moderately inhibited (4); CYP1B1 (5); CYP2B1, pentoxyresorufin *O*-dealkylase (6); electron transport (7,8). 1. Kim, Lim, Yang, *et al.* (2002) *Arch. Pharm. Res.* **25**, 856. 2. Choi, Ahn, Song, *et al.* (2005) *Phytother. Res.* **19**, 839. 3. Maenpaa, Sigusch, Raunio, *et al.* (1993) *Biochem. Pharmacol.* **45**, 1035. 4. Cai, Baer-Dubowska, Ashwood-Smith & DiGiovanni (1997) *Carcinogenesis* **18**, 215. 5. Mammen, Kleiner, DiGiovanni, Sutter & Strickland (2005) *Pharmacogenet. Genomics* **15**, 183. 6. Cai, Bennett, Nair, *et al.* (1993) *Chem. Res. Toxicol.* **6**, 872. 7. Kaiser, R. Kramar & Farkouth (1966) *Enzymologia* **30**, 65. 8. Olorunsogo, Uwaifo & Malomo (1990) *Chem. Biol. Interact.* **74**, 263.

Imperialine

This photosensitive alkaloid (FW = 429.64 g/mol; CAS 61825-98-7; photosensitive), also known as raddeamine and sipeimine, from the crown imperial (*Fritillaria imperialis*) as well as other species of *Fritillaria*, and *Petilium eduardi*. Imperialine resembles the alkaloid himbacine in terms of its pharmacological profile at muscarinic receptor subtypes in that it acts as an M_2-selective antagonist with respect to M_1 or M_3 sites. *Note*: This alkaloid also exists as a β-D-glucopyranoside, which is known as edpetiline. **1**. Eglen, Harris, Cox, *et al.* (1992) *Naunyn Schmiedebergs Arch. Pharmacol.* **346**, 144.

INCB-028050, *See* Baricitinib

INCB 024360-analogue

This potent, membrane-penetrating IDO inhibitor (FW = 271.64 g/mol; CAS 914471-09-3; Soluble to 100 mM in DMSO), also named 4-amino-*N'*-(3-chloro-4-fluorophenyl)-*N*-hydroxy-1,2,5-oxadiazole-3-carboximid-amide, targets indoleamine 2,3-dioxygenase, with a K_i of 19 nM in *in vitro* enzyme assays and an IC_{50} of 67 nM in HeLa cells, decreasing kynurenine levels in plasma and reducing tumor growth *in vivo* (1). The kynurenine pathway is responsible for the metabolism of >95% of dietary tryptophan and produces numerous bioactive metabolites. Indoleamine dioxygenase-1 (IDO1)inhibitors are currently in clinical trials for the treatment of cancer, and these agents may also have therapeutic utility in neurological disorders, including multiple sclerosis (2). **1**. Yue, Douty, Wayland, *et al.* (2009) *J. Med. Chem.* **52**, 7364. **2**. Dounay, Tuttle & Verhoest (2015) *J. Med. Chem.* **58**, 8762.

Indatraline

This substituted indanamine (FW_{free-base} = 292.21 g/mol), also known as Lu 19-005 and (±)-*trans*-3-(3,4-dichlorophenyl)-*N*-methyl-1-indanamine, is a strong inhibitor of dopamine, serotonin, and norepinephrine uptake, with K_i values of 1.7 nM, 0.42 nM, and 5.8 nM, respectively. Because DA, 5-HT, and NA all seem to be involved in depression, Lu 19-005's profile makes it an interesting experimental tool and a candidate antidepressant. **Target(s)**: dopamine uptake (1-3); norepinephrine uptake (1,2); serotonin uptake (1-3). **1**. Hyttel & Larsen (1985) *J. Neurochem.* **44**, 1615. **2**. Arnt, Christensen & Hyttel (1985) *Naunyn-Schmiedeberg's Arch. Pharmacol.* **329**, 101. **3**. Faraj, Olkowski & Jackson (1994) *Pharmacology* **48**, 320.

Indican

plant indican
indoxyl-β-D-glucoside

metabolic indican
3-indoxyl sulfate

This glucoside (FW = 295.29 g/mol; CAS 487-60-5), also known as plant indican and indoxyl-β-D-glucoside, is isolated from species of *Indigofera*, *Polygonum*, and *Isatis*. Indican is the precursor of indigo and reportedly stimulates ecdysone 20-monooxygenase activity. Metabolic indicant (FW_{free-acid} = 213.21 g/mol), also known as 3-indoxyl sulfate and indol-3-yl sulfate, is a uremic toxin found in the urine and blood plasma of mammals. **Target(s)**: *N*-acetylindoxyl oxidase, inhibited by 3-indoxyl sulfate (1); organic anion transporters, inhibited by 3-indoxyl sulfate (2); thyroxine transport, inhibited by 3-indoxyl sulfate (3). **1**. Beevers & French (1954) *Arch. Biochem. Biophys.* **50**, 427. **2**. Enomoto, Takeda, Taki, *et al.* (2003) *Eur. J. Pharmacol.* **466**, 13. **3**. Lim, Bernard, de Jong, *et al.* (1993) *J. Clin. Endocrinol. Metab.* **76**, 318.

Indinavir

This antiretroviral (FW = 613.80 g/mol), also known as L-735524, Crixivan®, and (2S)-1-[(2S,4R)-4-benzyl-2-hydroxy-4-{[(1S,2R)-2-hydroxy-2,3-dihydro-1*H*-inden-1-yl]carbamoyl}butyl]-*N-tert*-butyl-4-(pyridin-3-ylmethyl)piperazine-2-carboxamide, is an antiviral agent that inhibits Human Immunodeficiency Virus-1 protease (K_i = 0.52 nM) and HIV-2 protease (K_i = 3.3 nM) (1-11). Indinavir also inhibits the GLUT4 activity by binding reversibly to the glucose transporter's endofacial surface (12,13). Indinavir is effective against GLUT4 in its therapeutically relevant drug range (5-10 μM), but this range is without effect on the ubiquitously expressed erythrocyte transporter, GLUT1 (14). **Target(s)**: candidapepsin, a secreted aspartic proteinase of *Candida albicans*, weakly inhibited (15,16); glucuronosyltransferase (UDP-glucuronosyl-transferases: UGT1A1, UGT1A3, and UGT1A4 (17); HIV-1 retropepsin, *or* human immunodeficiency virus I protease (1-10); HIV-2 retropepsin, *or* human immunodeficiency virus 2 protease (11). **1**. Vacca, Dorsey, Schleif, *et al.* (1994) *Proc. Natl. Acad. Sci. U.S.A.* **91**, 4096. **2**. Clemente, Moose, Hemrajani, *et al.* (2004) *Biochemistry* **43**, 12141. **3**. Clemente, Hemrajani, Blum, Goodenow & Dunn (2003) *Biochemistry* **42**, 15029. **4**. Clemente, Coman, Thiaville, *et al.* (2006) *Biochemistry* **45**, 5468. **5**. Vacca (1994) *Meth. Enzymol.* **241**, 311. **6**. Wilson, Phylip, Mills, *et al.* (1997) *Biochim. Biophys. Acta* **1339**, 113. **7**. Liu, Boross, Wang, *et al.* (2005) *J. Mol. Biol.* **354**, 789. **8**. Surleraux, de Kock, Verschueren, *et al.* (2005) *J. Med. Chem.* **48**, 1965. **9**. Shuman, Haemaelaeinen & Danielson (2004) *J. Mol. Recognit.* **17**, 106. **10**. Liu, Kovalevsky, Louis, *et al.* (2006) *J. Mol. Biol.* **358**, 1191. **11**. Chen, Li, Chen, *et al.* (1994) *J. Biol. Chem.* **269**, 26344. **12**. Murata, Hruz & Mueckler (2000) *J. Biol. Chem.* **275**, 20251. **13**. Hresko & Hruz (2011) *PLoS One* **6**, e25237. **14**. Murata, Hruz & Mueckler (2002) *AIDS*

16, 859. **15.** Gruber, Speth, Lukasser-Vogl, *et al.* (1999) *Immunopharmacology* **41**, 227. **16.** Tossi, Benedetti, Norbedo, *et al.* (2003) *Bioorg. Med. Chem.* **11**, 4719. **17.** Zhang, Chando, Everett, *et al.* (2005) *Drug Metab. Dispos.* **33**, 1729.

Indirubin-3'-oxime

This CDK inhibitor (FW = 277.28 g/mol; CAS 160807-49-8; Symbol = IO), also known as indirubin-3'-monoxime, is an indicant dye that inhibits a number of kinases, particularly cyclin-dependent kinases by competing at their ATP sites. Indirubin-3'-oxime is also a component of Danggui Longhui Wan, a traditional Chinese medicine widely used for treating gynecological ailments, fatigue, mild anemia and high blood pressure. Indirubin-3'-oxime inhibits vascular smooth muscle cell (VSMC) proliferation and neointima formation *in vivo*, operating by a novel dual inhibitory mode, first by interfering with pro-migratory signaling in VSMC, and second by suppressing leukotriene biosynthesis in monocytes via direct inhibition of 5-lipoxygenase. 1-Methyl-indirubin-3'-oxime (MeIO) is a kinase-inactive analogue of IO that may be used as a negative control. **Target(s):** AMP-activated protein kinase, $IC_{50} = 0.22$ μM (1); cAMP-dependent protein kinase, *or* protein kinase A, weakly inhibited (1); cyclin-dependent kinase, *or* cyclin-dependent kinase 2/cyclin A, $IC_{50} = 0.59$ μM (1,2,6); cyclin-dependent kinase 1/cyclin B, $IC_{50} = 0.18$ μM (3,4); cyclin-dependent kinase 5/p25 (3,5); [glycogen synthase] kinase 3, *or* [tau protein] kinase; $IC_{50} = 0.19$ μM (1,3,5); Lck protein-tyrosine kinase (1); 5-lipoxygenase (7); lymphocyte kinase, $IC_{50} = 0.3$ μM (1); mitogen-activated protein kinase (1); non-specific serine/threonine protein kinase (1); pro-migratory signaling in vascular smooth muscle cells (7); protein kinase C, weakly inhibited (1); protein-tyrosine kinase, non-specific protein-tyrosine kinase (1); serum- and glucocorticoid-induced kinase, $IC_{50} = 0.38$ μM (1). **1.** Bain, McLauchlan, Elliott & Cohen (2003) *Biochem. J.* **371**, 199. **2.** Hoessel, Leclerc, Endicott, *et al.* (1999) *Nat. Cell Biol.* **1**, 60. **3.** Polychronopoulos, Magiatis, Skaltsounis, *et al.* (2004) *J. Med. Chem.* **47**, 935. **4.** Marko, Schatzle, Friedel, *et al.* (2001) *Brit. J. Cancer* **84**, 283. **5.** Leclerc, Garnier, Hoessel, *et al.* (2001) *J. Biol. Chem.* **276**, 251. **6.** Woodard, Li, Kathcart, *et al.* (2003) *J. Med. Chem.* **46**, 3877. **7.** Blazevic, Schaible, Weinhäupl, *et al.* (2013) *Cardiovasc. Res.* **101**, 522.

Indole

This coal tar aromatic and tryptophan metabolite (FW = 117.15 g/mol; M.P. = 52°C), also known as 2,3-benzopyrrole and 1*H*-benzo[*b*]pyrrole, is produced in the degradation of tryptophan. Indole is soluble in hot water and ho. Concentrated solutions have a unpleasant odor, but highly dilute solutions are relatively pleasant. **Target(s):** *N*-acetylindoxyl oxidase (1); chymotrypsin (2-12); 3'-demethylstaurosporine *O*-methyltransferase (13); histidine decarboxylase (14); indoleamine 2,3-dioxygenase, weakly inhibited (15); lysozyme (16); metridin (17,18); monoamine oxidase (19); myeloperoxidase (20); peroxidase (20); subtilisin (21-23); tryptophanase (24); tryptophan 2,3-dioxygenase (14,25,26). **1.** Beevers & French (1954) *Arch. Biochem. Biophys.* **50**, 427. **2.** Blow (1971) *The Enzymes*, 3rd ed., **3**, 185. **3.** Baker (1967) *Design of Active-Site-Directed Irreversible Enzyme Inhibitors*, Wiley, New York. **4.** Wallace, Kurtz & Niemann (1963) *Biochemistry* **2**, 824. **5.** Huang & Niemann (1953) *J. Amer. Chem. Soc.* **75**, 1395. **6.** Abrash & Niemann (1965) *Biochemistry* **4**, 99. **7.** Prahl & Neurath (1966) *Biochemistry* **5**, 2131. **8.** Shiao & Sturtevant (1969) *Biochemistry* **8**, 4910. **9.** Shiao (1970) *Biochemistry* **9**, 1083. **10.** Valenzuela & Bender (1970) *Biochemistry* **9**, 2440. **11.** Garel & Labouesse (1970) *J. Mol. Biol.* **47**, 41. **12.** Fasco & Fenton (1973) *Arch. Biochem. Biophys.* **159**, 802. **13.** Weidner, Kittelmann, Goeke, Ghisalba & Zähner (1998) *J. Antibiot.* **51**, 697. **14.** Webb (1966) *Enzyme and Metabolic Inhibitors*, vol. **2**, Academic Press, New York. **15.** Sono (1989)

Biochemistry **28**, 5400. **16.** Shinitzky, Katchalski, Grisaro & Sharon (1966) *Arch. Biochem. Biophys.* **116**, 332. **17.** Barrett (1998) in *Handb. Proteolytic Enzymes*, p. 24, Academic Press, San Diego. **18.** Gibson & Dixon (1969) *Nature* **222**, 753. **19.** McEwan, Sasaki & Jones (1969) *Biochemistry* **8**, 3952. **20.** Jantschko, Furtmüller, Zederbauer, *et al.* (2005) *Biochem. Pharmacol.* **69**, 1149. **21.** Ottesen & Svendsen (1970) *Meth. Enzymol.* **19**, 199. **22.** Glazer (1967) *J. Biol. Chem.* **242**, 433. **23.** Markland & Smith (1971) *The Enzymes*, 3rd ed., **3**, 561. **24.** Newton, Morino & Snell (1965) *J. Biol. Chem.* **240**, 1211. **25.** Matsumura, Osada & Aiba (1984) *Biochim. Biophys. Acta* **786**, 9. **26.** Feigelson & Brady (1974) in *Mol. Mech. Oxygen Activ.* (Hayaishi, ed.) p. 87, Academic Press, New York.

Indole-3-acetamide

This tryptophan metabolite (FW = 174.20 g/mol; M.P. = 148-150°C) inhibits aromatic-L-amino-acid decarboxylase (1), tryptophan 2,3-dioxygenase (2), tryptophan 2-monooxygenase, product inhibition (3-7). and tryptophan synthase (8). **1.** Nakazawa, Kumagai & Yamada (1987) *Agric. Biol. Chem.* **51**, 2531. **2.** Eguchi, Watanabe, Kawanishi, Hashimoto & Hayaishi (1984) *Arch. Biochem. Biophys.* **232**, 602. **3.** Kosuge (1970) *Meth. Enzymol.* **17A**, 446. **4.** Kosuge, Heskett & Wilson (1966) *J. Biol. Chem.* **241**, 3738. **5.** Emanuele, Heasley & Fitzpatrick (1995) *Arch. Biochem. Biophys.* **316**, 241. **6.** Sobrado & Fitzpatrick (2003) *Biochemistry* **42**, 13826 and 13833. **7.** Hutcheson & Kosuge (1985) *J. Biol. Chem.* **260**, 6281. **8.** Marabotti, Cozzini & Mozzarelli (2000) *Biochim. Biophys. Acta* **1476**, 287.

Indole-3-acetate (Indole-3-acetic Acid)

This plant hormone ($FW_{free-acid} = 175.19$ g/mol; CAS 6505-45-9; $pK_a = 4.75$; Sparingly soluble in water), also known as 3-indoleacetic acid, indol-3-ylacetic acid, (3-indolyl)acetic acid (IAA), and heteroauxin, is the main auxin of higher plants. IAA plays a key role in both root and shoot development. The hormone moves from one part of the plant to another by a designated importer *AUX1* and the efflux pumps *PIN1–7*. In peas, wounding or application of IAA induces the expression of ACS2, an IAA-responsive gene, suggesting that IAA may also accumulate at wound sites. So significant is IAA to plant physiology that there are at least four pathways for its synthesis. **Target(s):** adenylate kinase (1-3); alcohol dehydrogenase (4); D-amino-acid oxidase, $K_i = 2.3$ mM (5,6); aromatic-amino-acid aminotransferase, $K_d = 14$ mM (30); aromatic-L-amino-acid decarboxylase (7); carboxypeptidase A (8-16); catalase (39); cysteine-type carboxypeptidase (17); glucan 1,3-β-glucosidase, weakly inhibited (18); β-glucosidase (19); guanylate kinase (20); horseradish peroxidase (21,22); imidazoleacetate 4-monooxygenase, weakly inhibited (33,34); indoleamine 2,3-dioxygenase (38); mandelonitrile lyase, inactivation in light, not in dark (23); pectin lyase (24); peroxidase (21,22); tryptophanase (25,26); tryptophan 2,3-dioxygenase (25); tryptophan 2 monooxygenase (27,35-37); tryptophan synthase (28); tryptophanyl-tRNA synthetase (29); tyrosine aminotransferase (31); zeatin 9-aminocarboxy-ethyltransferase (32). **1.** Watanabe & Kubo (1982) *Eur. J. Biochem.* **123**, 587. **2.** Brownson & Spencer (1972) *Biochem. J.* **130**, 797. **3.** Ramotar & Pickard (1981) *Can. J. Microbiol.* **27**, 1053. **4.** Zikmanis & Kruce (1990) *Biofactors* **2**, 237. **5.** Webb (1966) *Enzyme and Metabolic Inhibitors*, vol. **2**, p. 342, Academic Press, New York. **6.** Klein & Austin (1953) *J. Biol. Chem.* **205**, 725. **7.** Baxter & Slaytor (1972) *Phytochemistry* **11**, 2763. **8.** Pétra (1970) *Meth. Enzymol.* **19**, 460. **9.** Neurath (1960) *The Enzymes*, 2nd ed., **4**, 11. **10.** Bradley, Naude, Muramoto, Yamauchi & Oelofsen (1994) *Int. J. Biochem.* **26**, 555. **11.** Giusti, Carrara, Cima & Borin (1985) *Eur. J. Pharmacol.* **116**,

287. **12.** Coleman & Vallee (1964) *Biochemistry* **3**, 1874. **13.** Rahmo & Fife (2000) *Bioorg. Chem.* **28**, 226. **14.** Auld & Vallee (1970) *Biochemistry* **9**, 602. **15.** Spilburg, Bethune & Vallee (1977) *Biochemistry* **16**, 1142. **16.** Bradley, Naude, Muramoto, Yamauchi & Oeloefsen (1994) *Int. J. Biochem.* **26**, 555. **17.** Greenbaum (1971) *The Enzymes*, 3rd ed., **3**, 475. **18.** Fleet & Phaff (1975) *Biochim. Biophys. Acta* **401**, 318. **19.** Feldwisch, Vente, Zettl, *et al.* (1994) *Biochem. J.* **302**, 15. **20.** Buccino & Roth (1969) *Arch. Biochem. Biophys.* **132**, 49. **21.** Fox, Purves & Nakada (1965) *Biochemistry* **4**, 2754. **22.** Reigh & Smith (1977) *Experientia* **33**, 1451. **23.** Petrounia, Goldberg & Brush (1994) *Biochemistry* **33**, 2891. **24.** Albersheim (1966) *Meth. Enzymol.* **8**, 628. **25.** Webb (1966) *Enzyme and Metabolic Inhibitors*, vol. **2**, Academic Press, New York. **27.** Gooder & Happold (1954) *Biochem. J.* **57**, 369. **28.** Wolf & Hoffmann (1974) *Eur. J. Biochem.* **45**, 269. **29.** Kisselev, Favorova & Kolaleva (1979) *Meth. Enzymol.* **59**, 174. **30.** Okamoto, Ishii, Hirotsu & Kagamiyama (1999) *Biochemistry* **38**, 1176. **31.** Jacoby & La Du (1964) *J. Biol. Chem.* **239**, 419. **32.** Parker, Entsch & Letham (1986) *Phytochemistry* **25**, 303. **33.** Watanabe, Kambe, Imamura, *et al.* (1983) *Anal. Biochem.* **130**, 321. **34.** Maki, Yamamoto, Nozaki & Hayaishi (1969) *J. Biol. Chem.* **244**, 2942. **35.** Kosuge, Heskett & Wilson (1966) *J. Biol. Chem.* **241**, 3738. **36.** Emanuele, Heasley & Fitzpatrick (1995) *Arch. Biochem. Biophys.* **316**, 241. **37.** Hutcheson & Kosuge (1985) *J. Biol. Chem.* **260**, 6281. **38.** Watanabe, Fujiwara, Yoshida & Hayaishi (1980) *Biochem. J.* **189**, 393. **39.** Chatterjee & Sanwal (1993) *Mol. Cell. Biochem.* **126**, 125.

Indole-3-acrylate (Indole-3-acrylic Acid)

This tryptophan metabolite (FW$_{\text{free-acid}}$ = 187.20 g/mol; CAS 29953-71-7) inhibits tryptophan synthase, tryptophan 2,3-dioxygenase, L-dopachrome isomerase, and xanthine oxidase. Substrate channeling in *Salmonella typhimurium* tryptophan synthase is regulated by allosteric interactions that are triggered by binding of ligand to the α-site and covalent reaction at the β-site. These interactions switch the enzyme between low-activity open conformations and high-activity closed conformations. This allosteriism has been demonstrated between the α-site and the external aldimine, α-aminoacrylate, and quinonoid forms of the β-site. **Target(s):** L-dopachrome isomerase (1); indoleamine 2,3-dioxygenase (2); tryptophanase, *or* tryptophan indole-lyase (3); tryptophan 2,3-dioxygenase (4,5); tryptophan synthase (6-8); xanthine oxidase (9). **1.** Suzuki, Sugimoto, Tanaka & Nishihira (1997) *J. Biochem.* **122**, 1040. **2.** Watanabe, Fujiwara, Yoshida & Hayaishi (1980) *Biochem. J.* **189**, 393. **3.** Snell (1975) *Adv. Enzymol. Relat. Areas Mol. Biol.* **42**, 287. **4.** Eguchi, Watanabe, Kawanishi, Hashimoto & Hayaishi (1984) *Arch. Biochem. Biophys.* **232**, 602. **5.** Webb (1966) *Enzyme and Metabolic Inhibitors*, vol. **2**, p. 325, Academic Press, New York. **6.** Matchett (1972) *J. Bacteriol.* **110**, 146. **7.** Wolf & Hoffmann (1974) *Eur. J. Biochem.* **45**, 269. **8.** Marabotti, Cozzini & Mozzarelli (2000) *Biochim. Biophys. Acta* **1476**, 287. **9.** Sheu & Chiang (1996) *Anticancer Res.* **16**, 311.

3-Indolebutyrate (3-Indolebutyric Acid)

This synthetic plant hormone (FW$_{\text{free-acid}}$ = 203.24 g/mol),, also known as 1*H*-indole-3-butanoic acid and 4-(3 indolyl)butyric acid (IBA; is an auxin that is used to initiate root growth (2). It is an endogenous auxin in

Arabidopsis. (**See also** *β-(3-Indolyl)butyrate and β-(3 Indolyl)butyric Acid*) **Target(s):** D-amino-acid oxidase, K_i = 1.1 mM (2); aromatic-amino-acid aminotransferase, K_d = 0.78 mM (3); carboxypeptidase A (4); chymotrypsin (5,6); tryptophan synthase (7,8); tyrosine aminotransferase (9). **1.** Hopkins (1999) *Introduction to Plant Physiology*, John Wiley & Sons. **2.** Klein & Austin (1953) *J. Biol. Chem.* **205**, 725. **3.** Okamoto, Ishii, Hirotsu & Kagamiyama (1999) *Biochemistry* **38**, 1176. **4.** Neurath (1960) *The Enzymes*, 2nd ed., **4**, 11. **5.** Webb (1966) *Enzyme and Metabolic Inhibitors*, vol. **2**, p. 370, Academic Press, New York. **6.** Neurath & Gladner (1951) *J. Biol. Chem.* **188**, 407. **7.** Meyer, Germershausen & Suskind (1970) *Meth. Enzymol.* **17A**, 406. **8.** Wolf & Hoffmann (1974) *Eur. J. Biochem.* **45**, 269. **9.** Jacoby & La Du (1964) *J. Biol. Chem.* **239**, 419.

Indole-3-carbinol

This tryptophan metabolite (FW = 147.17 g/mol; CAS 700-06-1; Symbol: I3C), a dietary compound found in cruciferous vegetables of the *Brassica* genus, such as broccoli and brussels sprouts, induces a G$_1$ cell-cycle arrest in LNCaP human prostate cancer cells, a process requiring activated p53 tumor suppressor protein (1). Western blots using phospho-specific p53 antibodies revealed that I3C treatment increases phosphorylation at Ser-15, Ser-37, and Ser-392, sites known to increase p53 protein stability as well as its transactivation potential. Ablation of p53 production using short interfering RNA (siRNA) prevents I3C-induced G$_1$ arrest and up-regulation of p21 expression. In female Sprague-Dawley rats, I3C significantly increases the extent of estradiol 2-hydroxylation (from 29% to 45%), indicating it may be chemopreventive in estrogen-dependent diseases (2). Indole-3-carbinol also activates AMP-activated protein kinase α (AMPKα), inhibiting fatty acid synthesis and blocking adipogenesis in 3T3-L1 cells (3) **1.** Hsu, Dev, Wing, *et al.* (2006) *Biochem. Pharmacol.* **72**, 1714. **2.** Michnovicz & Bradlow (1990) *J. Natl. Cancer Inst.* **82**, 947. **3.** Choi, Jeon, Lee & Lee (2014) *Biochimie* **104**, 127.

Indole-3-pyruvate (Indole-3-pyruvic Acid)

This α-keto acid (FW$_{\text{free-acid}}$ = 203.20 g/mol), also known as (indol-3-yl)pyruvic acid, is the product of the action of aromatic-amino-acid aminotransferase and L-amino-acid oxidase on L-tryptophan. It is also a product of the reactions catalyzed by indole-3-lactate dehydrogenase, tryptophan dehydrogenase, tryptophan aminotransferase, aromatic-amino-acid aminotransferase, and tryptophan:phenyl-pyruvate aminotransferase. It is a substrate of the reactions catalyzed by indole-3-pyruvate *C*-methyltransferase and indolepyruvate decarboxylase. It is also an intermediate in indolmycin, pigment, and auxin metabolism. **Target(s):** aromatic-amino-acid aminotransferase, product inhibition (1); cathepsin D (2); 3'-demethylstaurosporine *O*-methyltransferase (3); kynurenine: oxoglutarate aminotransferase (4); tryptophanase, *or* tryptophan indole-lyase (5); tryptophan 2-monooxygenase (6); tryptophan synthase (7); tryptophanyl-tRNA synthetase (8). **1.** Soto-Urzua, Xochinua-Corona, Flores-Encarnacion & Baca (1996) *Can. J. Microbiol.* **42**, 294. **2.** Woessner (1973) *J. Biol. Chem.* **248**, 1634. **3.** Weidner, Kittelmann, Goeke, Ghisalba & Zähner (1998) *J. Antibiot.* **51**, 697. **4.** Han & Li (2004) *Eur. J. Biochem.* **271**, 4804. **5.** Snell (1975) *Adv. Enzymol. Relat. Areas Mol. Biol.* **42**, 287. **6.** Sobrado & Fitzpatrick (2003) *Biochemistry* **42**, 13826 and 13833. **7.** Wolf & Hoffmann (1974) *Eur. J. Biochem.* **45**, 269. **8.** Kisselev, Favorova & Kolaleva (1979) *Meth. Enzymol.* **59**, 174.

Indolicidin

This tridecapeptide amide (FW = 1907.30 g/mol; *Sequence*: ILPWKWPWWPWRR-NH$_2$), from bovine neutrophils, exhibits bactericidal and fungicidal activity (1). Indolicidin binds to calmodulin and inhibits calmodulin-stimulated phosphodiesterase activity with IC$_{50}$ values in the nanomolar range (2). **Target(s):** aminoglycoside acetyltransferase (6')-II (3); aminoglycoside kinase (2') (3); calmodulin-stimulated phosphodiesterase (2); kanamycin kinase, *or* aminoglycoside phosphotransferase (3')-IIIa (3). **1**. Selsted, Novotny, Morris, *et al.* (1992) *J. Biol. Chem.* **267**, 4292. **2**. Sitaram, Subbalakshmi & Nagaraj (2003) *Biochem. Biophys. Res. Commun.* **309**, 879. **3**. Boehr, Draker, Koteva, *et al.* (2003) *Chem. Biol.* **10**, 189.

Indolmycin

This tryptophan metabolite and antibiotic (FW = 257.29 g/mol; CAS 21193-77-1), produced by *Streptomyces albus* (ATC12648) inhibits protein synthetase by targetting prokaryotic tryptophanyl-tRNA synthetase. Indolmycin is active against Gram-poisitive and negative microorganisms (1-3). Simultaneous binding of indolmycin and Mg^{2+}•ATP results in movement of the Tyr-125 and Gln-107 side-chains, opening of the mobile loop containing Lys-111; displacement of the Mg^{2+} ion by 0.4 Å, hexavalent metal coordination, stronger, nearly equivalent electrostatic interactions of Mg^{2+} with an oxygen from each phosphate group of ATP, and weaker coordination of phosphate group oxygen atoms by active site lysine residues (4). **Target(s):** tryptophan decarboxylase (1); tryptophan 2,3-dioxygenase, *or* tryptophan pyrrolase (1); tryptophanyl-tRNA synthetase (2,3). **1**. Werner & Reuter (1979) *Arzneimittelforschung* **29**, 59. **2**. Werner, Thorpe, Reuter & Nierhaus (1976) *Eur. J. Biochem.* **68**, 1. **3**. Retailleau, Huang, Yin, *et al.* (2003) *J. Mol. Biol.* **325**, 39. **4**. Williams, Yin & Carter (2015) *J. Biol. Chem.* **291**, 255.

3-(Indol-3-yl)propionate (3-(Indol-3-yl)propionic Acid)

This tryptophan analogue (FW$_{free-acid}$ = 189.21 g/mol), also known as 3-(3-indolyl)propionic acid, inhibits a number of enzymes, in some cases exploiting its hydroxyl radical-scavenging properties. **Target(s):** D-amino-acid oxidase, K_i = 2.3 mM (1); aromatic-amino-acid aminotransferase, K_d = 2.6 mM (2); carboxypeptidase A (3,4); chymotrypsin (4-7); cysteine-type carboxypeptidase (8); kynurenine aminotransferase, weakly inhibited (9); N-methyltryptophan oxidase (10); nicotinamidase (11); tryptophanase, *or* tryptophan indole-lyase (4,12-14); tryptophan 2,3-dioxygenase (4); tryptophan synthase (15,16); tryptophanyl-tRNA synthetase (17); tyrosine aminotransferase (18); zeatin 9-aminocarboxyethyltransferase (19). **1**. Klein & Austin (1953) *J. Biol. Chem.* **205**, 725. **2**. Okamoto, Ishii, Hirotsu & Kagamiyama (1999) *Biochemistry* **38**, 1176. **3**. Neurath (1960) *The Enzymes*, 2nd ed., **4**, 11. **4** . Webb (1966) *Enzyme and Metabolic Inhibitors*, vol. **2**, Academic Press, New York. **5**. Blow (1971) *The Enzymes*, 3rd ed., **3**, 185. **6**. Prahl & Neurath (1966) *Biochemistry* **5**, 2131. **7**. Neurath & Gladner (1951) *J. Biol. Chem.* **188**, 407. **8**. Greenbaum (1971) *The Enzymes*, 3rd ed., **3**, 475. **9**. Mason (1959) *J. Biol. Chem.* **234**, 2770. **10**. Khanna & Schuman Jorns (2001) *Biochemistry* **40**, 1441. **11**. Albizati & Hedrick (1972) *Biochemistry* **11**, 1508. **12**. Gooder & Happold (1954) *Biochem. J.* **57**, 369. **13**. Ikushiro & Kagamiyama (2000) *Biofactors* **11**, 97. **14**. Snell (1975) *Adv. Enzymol. Relat. Areas Mol. Biol.* **42**, 287. **15**. Meyer, Germershausen & Suskind (1970) *Meth. Enzymol.* **17A**, 406. **16**. Wolf & Hoffmann (1974) *Eur. J. Biochem.* **45**, 269. **17**. Kisselev, Favorova & Kolaleva (1979) *Meth. Enzymol.* **59**, 174. **18**. Jacoby & La Du (1964) *J.*

Biol. Chem. **239**, 419. **19**. Parker, Entsch & Letham (1986) *Phytochemistry* **25**, 303.

Indomethacin

This nonsteroidal anti-inflammatory drug (NSAID) and antipyretic (FW$_{free-acid}$ = 357.79 g/mol; CAS 53-86-1), also known as 1-(4-chlorobenzoyl)-5-methoxy-2-methyl-1*H*-indole-3-acetic acid, Indocin®, and Indo-Lemmon®, targets prostaglandin H synthase 1 and 2, with IC$_{50}$ values of 0.74 and 0.97 μM for COX-1 and COX-2 cyclooxygenases, respectively. Indomethacin is frequently used to relieve moderate/severe pain, tenderness, swelling, and stiffness caused by osteoarthritis, rheumatoid arthritis, and ankylosing spondylitis. It is also used to treat shoulder pain caused by bursitis and tendinitis. Indomethacin can also be used to treat acute gouty arthritis. **Pharmacokinetics:** Indomethacin enjoys high bioavailability (~98 %) and plasma protein (mainly albumin) binding (~90 %). Metabolized primarily in the liver, it has a half-life between 3–10 hours, with maximal plasma levels after 2 hours. Compared to naproxen and ibuprofen, indomethacin shows higher penetration of the blood-brain barrier. **Key Pharmacokinetic Parameters:** *See* Appendix II in Goodman & Gilman's *THE PHARMACOLOGICAL BASIS OF THERAPEUTICS*, 12th Edition (Brunton, Chabner & Knollmann, eds.) McGraw-Hill Medical, New York (2011). **Target(s):** alcohol dehydrogenase, NADP$^+$ (1); aldehyde reductase (1,2); 1-alkylglycerophosphocholine *O*-acetyl-transferase (38,39); 4-aminobutyrate aminotransferase, *or* GABA aminotransferase (3); ATPase, mitochondrial (4); carbonyl reductase (5,6); cathepsin B (7); cathepsin D (8); cathepsin L (9); cystinyl aminopeptidase, *or* oxytocinase (10,11); estrone sulfotransferase (36); glutamine:D-fructose-6-phosphate aminotransferase (12); glutathione S-transferase (13-16,37); hyaluronoglucosaminidase, *or* hyaluronidase (17,18); 15-hydroxyprostaglandin dehydrogenase (19,20); lactoperoxidase, weakly inhibited (54); leukotriene-C4 synthasem (16); lipoprotein lipase (21); 5-lipoxygenase, *or* arachidonate 5-lipoxygenase (52); multidrug resistance associated protein (15); peroxidase, weakly inhibited (54); phenol sulfotransferase, *or* aryl sulfotransferase (36); phospholipase A$_2$ (22-25); phospholipase C (26); prostaglandin D$_2$ 11-ketoreductase (27); prostaglandin-D synthase, IC$_{50}$ = 0.3 mM (28); prostaglandin-endoperoxide synthase, *or* cyclooxygenase activity, (15,19,29-33,40-51,53); prostaglandin E$_2$ 9-reductase (19,34,35); succinate-semialdehyde dehydrogenase (3). **Inhibition of Viral Protein Translation by Vesicular Stomatitis Virus Infection:** Beyond its potent anti-inflammatory and analgesic properties, indomethacin also possesses antiviral activity (55). During VSV infection, indomethacin activates double-stranded RNA-activated protein kinase (PKR) in an interferon- and dsRNA-independent manner, causing rapid (< 5 min) phosphorylation of eukaryotic initiation factor-2 α-subunit (eIF2α), thereby shutting-off viral protein translation and blocking viral replication (IC$_{50}$ = 2μM), while protecting host-cells from virus-induced damage. Indomethacin does not affect eIF2α kinases PERK and GCN2, and is unable to trigger eIF2α phosphorylation in the presence of 2-aminopurine, a PKR inhibitor. In addition, small-interfering RNA-mediated PKR-gene silencing dampened the antiviral effect in indomethacin-treated cells (55). Such results identify PKR as a critical target for the antiviral activity of indomethacin and indicate that eIF2α phosphorylation could be a key element in the broad-spectrum antiviral activity of the drug (55). **1**. Whittle & Turner (1981) *Biochem. Pharmacol.* **30**, 1191. **2**. Davidson & Murphy (1985) *Prog. Clin. Biol. Res.* **174**, 251. **3**. Whittle & Turner (1978) *J. Neurochem.* **31**, 1453. **4**. Chatterjee & Stefanovich (1976) *Arzneimittelforschung* **26**, 499. **5**. Wermuth (1981) *J. Biol. Chem.* **256**, 1206. **6**. Hara, Deyashiki, Nakagawa, Nakayama & Sawada (1982) *J. Biochem.* **92**, 1753. **7**. Yamamoto, Kamata & Kato (1984) *Jpn. J. Pharmacol.* **35**, 253. **8**. Yamamoto, Takeda & Kato (1988) *Nippon Yakurigaku Zasshi* **91**, 371. **9**. Raghav & Singh (1993) *Indian J. Med. Res.* **98**, 188. **10**. Roy, Yeang & Karim (1982) *Prostaglandins Leukot. Med.* **8**, 173. **11**. Roy, Yeang & Karim (1981) *Prostagladins Med.* **6**, 577. **12**. Chatterjee & Stefanovich (1976) *Arzneimittelforschung* **26**, 502. **13**. Danielson & Mannervik (1988) *Biochem. J.* **250**, 705. **14**. Di Ilio, Aceto,

Boccio, Casalone, Pennelli & Federici (1988) *Eur. J. Biochem.* **171**, 491.
15. Touhey, O'Connor, Plunkett, Maguire & Clynes (2002) *Eur. J. Cancer*
38, 1661. **16.** Söderström, Hammarström & Mannervik (1988) *Biochem. J.*
250, 713. **17.** Szary, Kowalczyk-Bronisz & Gieldanowski (1975) *Arch.
Immunol. Ther. Exp.* **23**, 131. **18.** Girish & Kemparaju (2005) *Biochemistry
(Moscow)* **70**, 708 and 948. **19.** Pace-Asciak & Smith (1983) *The Enzymes*,
3rd ed., **16**, 543. **20.** Hansen (1974) *Prostaglandins* **8**, 95. **21.** Majerus &
Prescott (1982) *Meth. Enzymol.* **86**, 11. **22.** Lobo & Hoult (1994) *Agents
Actions* **41**, 111. **23.** Bonney, Qizilbash & Franks (1988) *J. Endocrinol.*
119, 141. **24.** Jesse & Franson (1979) *Biochim. Biophys. Acta* **575**, 467.
25. Kaplan, Weiss & Elsbach (1978) *Proc. Natl. Acad. Sci. U.S.A.* **75**, 2955.
26. Shakir, Simpkins, Gartner, Sobel & Williams (1989) *Enzyme* **42**, 197.
27. Lovering, Ride, Bunce, *et al.* (2004) *Cancer Res.* **64**, 1802. **28.**
Thomson, Meyer & Hayes (1998) *Biochem. J.* **333**, 317. **29.** van der
Ouderaa & Buytenhek (1982) *Meth. Enzymol.* **86**, 60. **30.** Nugteren &
Christ-Hazelhof (1987) *Prostaglandins* **33**, 403. **31.** Laneuville, Breuer,
Dewitt, *et al.* (1994) *J. Pharmacol. Exp. Ther.* **271**, 927. **32.** Coutts,
Khandwala, Van Inwegen, *et al.* (1985) in *Prostaglandins, Leukotrienes,
and Lipoxins* (Bailey, ed.), p. 627, Plenum Press, New York. **33.** Stanford,
Roth, Shen & Majerus (1977) *Prostaglandins* **13**, 669. **34.** Tai & Yuan
(1982) *Meth. Enzymol.* **86**, 113. **35.** Yuan, Tai & Tai (1980) *J. Biol. Chem.*
255, 7439. **36.** King, Ghosh & Wu (2006) *Curr. Drug Metab.* **7**, 745. **37.**
Alin, Jensson, Guthenberg, *et al.* (1985) *Anal. Biochem.* **146**, 313. **38.**
Hurst & Bazan (1997) *J. Ocul. Pharmacol. Ther.* **13**, 415. **39.** Yamazaki,
Sugatani, Fujii, *et al.* (1994) *Biochem. Pharmacol.* **47**, 995. **40.** Tai, Tai &
Hollander (1976) *Biochem. J.* **154**, 257. **41.** Forghani, Ouellet, Keen,
Percival & Tagari (1998) *Anal. Biochem.* **264**, 216. **42.** Kargman, Wong,
Greig, *et al.* (1996) *Biochem. Pharmacol.* **52**, 1113. **43.** Meade, Smith &
DeWitt (1993) *J. Biol. Chem.* **268**, 6610. **44.** Doualla-Bell, Guay, Bourgoin
& Fortier (1998) *Biol. Reprod.* **59**, 1433. **45.** Miyamoto, Ogino, Yamamoto
& Hayaishi (1976) *J. Biol. Chem.* **251**, 2629. **46.** Reininger & Bauer
(2006) *Phytomedicine* **13**, 164. **47.** Barnett, Chow, Ives, *et al.* (1994)
Biochim. Biophys. Acta **1209**, 130. **48.** Patrignani, Panara, Sciulli, *et al.*
(1997) *J. Physiol. Pharmacol.* **48**, 623. **49.** Friedman, Lang & Burke (1975)
Biochim. Biophys. Acta **397**, 331. **50.** Kargman, Charleson, Cartwright, *et
al.* (1996) *Gastroenterology* **111**, 445. **51.** Parkes & Eling (1974)
Biochemistry **13**, 2598. **52.** Furakawa, Yoshimoto, Ochi & Yamamoto
(1984) *Biochim. Biophys. Acta* **795**, 458. **53.** Chi, Jong, Son, *et al.* (2001)
Biochem. Pharmacol. **62**, 1185. **54.** Nasralla, Ghoneim, Khalifa, Gad &
Abdel-Naim (2009) *Toxicol. Lett.* **191**, 347. **55.** Amici, La Frazia, Brunelli,
et al. (2015) *Cell Microbiol.* **17**, 1391.

Indomethacin 4-Methoxyphenyl Ester

This anti-inflammatory agent (FW = 463.92 g/mol), also named 1-(*p*-
chlorobenzoyl)-5-methoxy-2-methyl-1*H*-indole-3-acetic acid 4-
methoxyphenyl ester, is a selective inhibitor of cyclooxygenase-2. The IC_{50}
value for COX-2 is 40 nM, whereas the value for COX-1 is > 66 µM. **1.**
Kalgutkar, Marnett, Crews, Remmel & Marnett (2000) *J. Med. Chem.* **43**,
2860. **2.** Kalgutkar, Crews, Rowlinson, *et al.* (2000) *Proc. Natl Acad. Sci.
U.S.A.* **97**, 925.

Indospicine

This hepatotoxic arginine analogue (FW = 173.22 g/mol; CAS 16377-01-8),
also known as 6-amidino-2-aminohexanoic acid, is found in a tropical
legume (*Indigofera spicata*) known to be toxic to sheep and cows.
Target(s): amino-acid *N*-acetyltransferase (The *Escherichia coli* enzyme is
inhibited by L-indospicine.) (1); arginase (2,3); arginyl-tRNA synthetase
(4); nitric-oxide synthase (5). **1.** Leisinger & Haas (1975) *J. Biol. Chem.*

250, 1690. **2.** Madsen & Hegarty (1970) *Biochem. Pharmacol.* **19**, 2391. **3.**
Beruter, Colombo & Bachmann (1978) *Biochem. J.* **175**, 449. **4.** Madsen,
Christie & Hegarty (1970) *Biochem. Pharmacol.* **19**, 853. **5.** Pass, Arab,
Pollitt & Hegarty (1996) *Nat. Toxins* **4**, 135.

INDY & pro-INDY

This Dyrk1 inhibitor (FW = 235.30 g/mol; CAS 1169755-45-6; Solubility:
100 mM in DMSO; IUPAC: (1*Z*)-1-(3-ethyl-5-hydroxy-2(3*H*)-
benzothiazolylidene)-2-propanone, targets the Dual-specificity tyrosine-
phosphorylation regulated kinases Dyrk1A (IC_{50} = 0.24 µM) and Dyrk1B
(IC_{50} = 0.23 µM) by binding within the ATP-binding cleft. INDY reverses
aberrant tau-phosphorylation and rescues repressed calcineurin/Nuclear
Factor of Activated T cell (*or* NFAT) signaling. INDY impairs the self-
renewal capacity of subventricular zone neural stem cells. The cell-
permeable form is pro-INDY (FW = 277.34 g/mol; CAS 719277-30-2;
Solubility: 100 mM in DMSO), systematically named (1*Z*)-1-(5-acetyloxy-
3-ethyl-2(3*H*)-benzothiazolylidene)-2-propanone. Note that the β-carboline
alkaloid harmine is another highly selective and potent Dyrk1A inhibitor
that has proven to be very useful in cellular assays. **1.** Ogawa, Nonaka,
Goto, (2010) *Nature Commun.* **1**, 86. **2.** Pozo, Zahonero, Fernández, *et al.*
(2013) *J. Clin. Invest.* **123**, 2475.

Inhibin

This heterodimeric gonadal hormone ($MW_{alpha-subunit}$ = 20 kDa; $MW_{beta-subunit}$
= 12 kDa; CAS 57285-09-3) down-regulates Follicle-Stimulating Hormone
(FSH) production by exerting specific negative feedback actions on the
secretion of FSH from the gonadotropic cells of the pituitary gland. Inhibin
thus acts as a paracrine factor in the regulation of ovarian folliculogenesis as
well as steroidogenesis. Inhibin is a member of a much larger family of
glycoprotein hormones and growth factors that includes Müllerian-
inhibiting substance, transforming growth factor-β, erythroid differentiation
factor, and activin, the latter consisting of a dimer of two inhibin β-subunits.
Activin counteracts inhibin effects in pituitary cells. Inhibin α-subunit
knock-down (both transcriptionally and translationally) in mice primary
anterior pituitary cells promotes apoptosis by up-regulating Caspase-3, Bax
and Bcl-2 genes, but without effect on p53, markedly impairing progression
of G_1 cell-cycle phase and decreasing the fraction of cells in S phase (2).
Inhibin-α silencing results in up-regulation of mRNA and protein
expression of Gondotropin releasing hormone receptors [GnRHRs] and
down-regulates mRNA levels of β-glycans. Silencing of inhibin-α also
significantly increases activin-β concentration, without affecting FSH and
LH levels in anterior pituitary cells. Such findings suggest that up-
regulation of GnRH receptors by inhibin α-subunit gene silencing might
increase the concentration of activin-β in the culture medium (2). **1.**
Makanji, Zhu, Mishra, *et al.* (2014) *Endocr. Rev.* **35**, 747. **2.** Han, Wu,
Riaz, *et al.* (2013) *PLoS One* **8**, e74596.

Inhibitor & Antagonist Types

These organic or inorganic agents reduce the efficiency of an enzyme,
receptor, transporter/carrier, mechanoenzyme, and/or metabolic pathway.
When present at sufficient concentration, a **Total Inhibitor** reduces the
reaction velocity to zero, whereas a **Partial Inhibitor** cannot completely block
the action of a target enzyme, transporter, carrier, or metabolic process. By
resembling a substrate or reversible bound co-substrate (*e.g.*, NAD^+),
Classical reversible inhibitors are bound and released rapidly, attaining a
thermodynamic equilibrium that is defined by a true equilibrium constant.
Tight-Binding Inhibitors are high-affinity inhibitors that act without delay,
whereas **Slow Inhibitors** and **Slow, Tight-Binding Inhibitors** require additional
time to reach their fully inhibitory end-point, in most cases forming high-

affinity complexes. (Included under the latter are **Transition-State Inhibitors**, agents that mimic the configuration and/or charge of transitory intermediates formed during an enzyme reaction cycle. They initially form weak $E \cdot I$ complexes, but binding induces/triggers a subsequent series of conformational isomerizations (e.g., $E + I \rightleftharpoons (E \cdot I)_1 \rightleftharpoons (E \cdot I)_2 \rightleftharpoons (E \cdot I)_3 \rightleftharpoons \cdots (E \cdot I)_i \cdots \rightleftharpoons (E \cdot I)_{n-1} \rightleftharpoons (E \cdot I)_n$), ultimately resulting in tight binding and attended by total inhibition.) **Irreversible Inhibitors** form covalent bonds with enzymes, transporters/carriers, or mechanoenzymes, with attendant loss of catalytic activity. They first form weak $E \cdot I$ complexes, but binding induces/triggers formation of covalent bond with an active-site group (e.g., $E + I \rightleftharpoons (E \cdot I)_{reversible} \rightarrow (E \cdot I)_{irreversible}$), resulting total inhibition. **Mechanism-Based Inhibitors** (also called **Suicide Inhibitors**) are substrate-like molecules S' that undergo catalysis, most often forming reaction product P', but occasionally react with an active-site functional, resulting in total inhibition. The general scheme for their action is: Catalysis: $E + S' \rightleftharpoons E \cdot S'_{reversible} \rightleftharpoons E \cdot X'_{reactive} \rightleftharpoons E \cdot P'_{reversible} \rightarrow E + P'$; Enzyme Inactivation: $E \cdot X'_{reactive} \rightarrow E - P'_{irreversible}$) **Antagonists** are inhibitors of receptor-triggered signaling processes, as are **Partial Agonists**, the latter impairing receptor-triggered processes to a lesser degree at full site occupancy, and **Inverse Agonists**, agents that bind to the same receptor as an agonist but induce a pharmacological response opposite to that agonist. Many antagonsts are slw-binding, requiring time for full expression of their affinity for their target (e.g., $R + X \rightleftharpoons (R \cdot X)_1 \rightleftharpoons (R \cdot X)_2 \rightleftharpoons (R \cdot X)_3 \rightleftharpoons \cdots (R \cdot X)_i \cdots \rightleftharpoons (R \cdot X)_{n-1} \rightleftharpoons (R \cdot X)_n$) For a detailed discussion, consult Chapter 8 of Purich (2010) Enzyme Kinetics: Catalysis & Control, Academic Press (Elsevier).

INK-128, See Sapanisertib
INT-747, See Obeticholate (Obeticholic Acid)
JNJ-54781532, See Peficitinib
INK1197, See IPI-145

Inogatran

This unusual dipeptide (FW = 438.57 g/mol), systematically named 2-{[(2R)-1-[(2S)-2-[(4-carbamimidamidobutyl)carbamoyl]piperidin-1-yl]-3-cyclohexyl-1-oxopropan-2-yl]amino}acetic acid, is a potent inhibitor of thrombin (K_i = 15 nM). Stopped-flow data support the conclusion that kinetics are biphasic: the faster component accounted for the largest fraction of the change in the intrinsic fluorescence of thrombin induced by inhibitor binding and was dependent on the inhibitor concentration; the slower component was independent of inhibitor concentration. Inogatran was originally developed to treat arterial and venous thrombotic diseases. **1.** Nilsson, Sjoling-Ericksson & Deinum (1998) J. Enzyme Inhib. **13**, 11. **2.** Gustafsson, Elg, Lenfors, Borjesson & Teger-Nilsson (1996) Blood Coagulation Fibrinol. **7**, 69. **3.** Teger-Nilsson, Bylund, Gustafsson, Gyzander & Eriksson (1997) Thromb. Res. **85**, 133. **4.** Deinum, Gustavsson, Gyzander, et al. (2002) Anal. Biochem. **300**, 152. **5.** Dullweber, Stubbs, Musil, Stürzebecher & Klebe (2001) J. Mol. Biol. **313**, 593.

Inosine

This ribonucleoside (FW = 268.23 g/mol; CAS 58-63-9), also known as 9-β-D-ribofuranosyl-hypoxanthine and hypoxanthine 9-β-D-ribofuranoside, is found in both free form and as a component of the anticodon of some forms of tRNA. It is formed in the degradation of adenosine and is also an intermediate in the purine salvage pathway. Inosine, often abbreviated by I or Ino, is soluble in water (1.6 g/100 mL at 20°C) and is hydrolyzed by mineral acids to hypoxanthine and D-ribose. The reaction of adenosine with nitrous acid will produce inosine. The UV spectra has a λ_{max} at pH 6 of 248.5 nm (ε = 12300 $M^{-1}cm^{-1}$). **Target(s):** adenosine deaminase (1,2); adenosine nucleosidase (3); adenosine-phosphate deaminase (4); adenosylhomocysteinase (5-8); alcohol dehydrogenase, yeast (9); AMP nucleosidase (10); arginase (11); aspartate carbamoyltransferase, weakly inhibited (12); deoxycytidine kinase, weakly inhibited (13); dihydroorotase (14); GMP synthetase, weakly inhibited (15); guanine deaminase, weakly inhibited (16,17); hypoxanthine(guanine) phosphoribosyltransferase, weakly inhibited (18); IMP dehydrogenase, weakly inhibited (19); S-methyl-5'-deoxyadenosine phosphorylase, weakly inhibited (20); nucleoside-triphosphatase (21); 5'-nucleotidase (22-26); ribosylpyrimidine nucleosidase (27); urate-ribonucleotide phosphorylase, weakly inhibited (28); uridine nucleosidase (29). **1.** Centelles, Franco & Bozal (1988) J. Neurosci. Res. **19**, 258. **2.** Wolfenden, Sharpless & Allan (1967) J. Biol. Chem. **242**, 977. **3.** Burch & Stuchbury (1986) J. Plant Physiol. **125**, 267. **4.** Uchida, Narita, Masuda, Chen & Nomura (1995) Biosci. Biotechnol. Biochem. **59**, 2120. **5.** Guranowski, Montgomery, Cantoni & Chiang (1981) Biochemistry **20**, 110. **6.** Shimizu, Shiozaki, Ohshiro & Yamada (1984) Eur. J. Biochem. **141**, 385. **7.** Bozzi, Parisi & Martini (1993) J. Enzyme Inhib. **7**, 159. **8.** Kloor, Kurz, Fuchs, Faust & Osswald (1996) Kidney Blood Press. Res. **19**, 100. **9.** Sund & Theorell (1963) The Enzymes, 2nd ed., **7**, 25. **10.** DeWolf, Fullin & Schramm (1979) J. Biol. Chem. **254**, 10868. **11.** Rosenfeld, Dutta, Chheda & Tritsch (1975) Biochim. Biophys. Acta **410**, 164. **12.** Jacobson & Stark (1973) The Enzymes, 3rd ed., **9**, 225. **13.** Krenitsky, Tuttle, Koszalka, et al. (1976) J. Biol. Chem. **251**, 4055. **14.** Bresnick & Blatchford (1964) Biochim. Biophys. Acta **81**, 150. **15.** Spector & Beecham (1975) J. Biol. Chem. **250**, 3101. **16.** Kimm, Park & Lee (1985) Korean J. Biochem. **17**, 139. **17.** Kimm, Park & Kim (1987) Korean J. Biochem. **19**, 39. **18.** Olsen & Milman (1974) J. Biol. Chem. **249**, 4030. **19.** Nichol, Nomura & Hampton (1967) Biochemistry **6**, 1008. **20.** Koszalka & Krenitsky (1986) Adv. Exp. Med. Biol. **195B**, 559. **21.** Chern, McDonald & Morris (1969) J. Biol. Chem. **244**, 5489. **22.** Webb (1966) Enzyme and Metabolic Inhibitors, **2**, p. 472, Academic Press, New York. **23.** Klein (1957) Z. Physiol. Chem. **307**, 254. **24.** Segal & Brenner (1960) J. Biol. Chem. **235**, 471. **25.** Itoh, Mitsui & Tsushima (1967) Biochim. Biophys. Acta **146**, 151. **26.** Singh & Singh (1986) Phytochemistry **25**, 2267. **27.** Terada, Tatibana & Hayaishi (1967) J. Biol. Chem. **242**, 5578. **28.** Laster & Blair (1963) J. Biol. Chem. **238**, 3348. **29.** Kurtz, Exinger, Erbs & Jund (2002) Curr. Genet. **41**, 132.

Inosine 5'-Diphosphate

This nucleotide ($FW_{free-acid}$ = 428.19 g/mol; at pH 6.0, λ_{max} = 248.5 nm and ε = 12200 $M^{-1}cm^{-1}$), usually abbreviated IDP, is an analogue of both ADP and GDP, as evidenced by its ability to substitute for ADP and GDP in many nucleotide-requiring phosphotransfer reactions. IDP is generated by hydrolysis of the terminal phosphoryl group of inosine 5'-triphosphate (ITP), by phosphorylation of IMP, or by the nonenzymatic deamination of ADP, the latter achieved by the presence of nitrous acid. IDP is found in either the free form or, more commonly, associated with a divalent cation (1). **Target(s):** adenine phosphoribosyltransferase (2); adenylosuccinate synthetase (3,4); amidophosphoribosyl-transferase (5,6); cytochrome b_5 reductase (7); deoxyguanosine kinase (8); diacylglycerol kinase (9);; gluconokinase, weakly inhibited (10); glucose-6-phosphate dehydrogenase (11); glycoprotein 3-α-L-fucosyltransferase (12,13); hexokinase (14); hypoxanthine(guanine) phosphoribosyltransferase (15,16); isocitrate dehydrogenase (NAD^+) (17,18); nicotinate phosphoribosyltransferase (19); nicotinate nucleotide diphosphorylase (carboxylating), weakly inhibited (20); nucleoside-triphosphatase (21); 5'-nucleotidase (22); phosphoenolpyruvate carboxykinase (ATP) (23); [protein-PII] uridylyltransferase (24); ribose-5-phosphate adenylyltransferase, weakly inhibited (25); ribose phosphate diphosphokinase, *or* phosphoribosyl-pyrophosphate synthetase (26,27). **1.** Purich (2010) Enzyme Kinetics: Catalysis & Control, Academic Press, New York. **2.** Arnold & Kelley (1978) *Meth. Enzymol.* **51**, 568. **3.** Van der Weyden & Kelly (1974) *J. Biol. Chem.* **249**, 7282. **4.** Matsuda, Shimura, Shiraki & Nakagawa (1980) *Biochim. Biophys. Acta* **616**, 340. **5.** Hill & Bennett (1969) *Biochemistry* **8**, 122. **6.** Holmes, McDonald, McCord, Wyngaarden & Kelley (1973) *J. Biol. Chem.* **248**, 144. **7.** Strittmatter (1959) *J. Biol. Chem.* **234**, 2665. **8.** Yamada, Goto & Ogasawara (1982) *Biochim. Biophys. Acta* **709**, 265. **9.** Wissing & Wagner (1992) *Plant Physiol.* **98**, 1148. **10.** Coffee & Hu (1972) *Arch. Biochem. Biophys.* **149**, 549. **11.** Horne, Anderson & Nordlie (1970) *Biochemistry* **9**, 610. **12.** Murray, Takayama, Schultz & Wong (1996) *Biochemistry* **35**, 11183. **13.** Shinoda, Morishita, Sasaki, *et al.*(1997) *J. Biol. Chem.* **272**, 31992. **14.** Ning, Purich & Fromm (1969) *J. Biol. Chem.* **244**, 3840. **15.** Nagy &A.-M. Ribet (1977) *Eur. J. Biochem.* **77**, 77. **16.** Ohe & Watanabe (1980) *Agric. Biol. Chem.* **44**, 1999. **17.** Plaut (1969) *Meth. Enzymol.* **13**, 34. **18.** Chen & Plaut (1963) *Biochemistry* **2**, 1023. **19.** Hayakawa, Shibata & Iwai (1984) *Agric. Biol. Chem.* **48**, 445. **20.** Taguchi & Iwai (1976) *Agric. Biol. Chem.* **40**, 385. **21.** Chern, McDonald & Morris (1969) *J. Biol. Chem.* **244**, 5489. **22.** Madrid-Marina & Fox (1986) *J. Biol. Chem.* **261**, 444. **23.** Cymeryng, Cazzulo & Cannata (1995) *Mol. Biochem. Parasitol.* **73**, 91. **24.** Engleman & Francis (1978) *Arch. Biochem. Biophys.* **191**, 602. **25.** Stern & Avron (1966) *Biochim. Biophys. Acta* **118**, 577. **26.** Fox & Kelley (1972) *J. Biol. Chem.* **247**, 2126. **27.** Switzer & Sogin (1973) *J. Biol. Chem.* **248**, 1063.

Inosine 5'-Monophosphate

Amido Tautomer
(major species)

Imido Tautomer
(minor species)

This nucleotide (FW_{free-acid} = 316.21 g/mol; at pH 6.0, λ_{max} = 248.5 nm and ε = 12200 $M^{-1}cm^{-1}$), often abbreviated IMP and called inosinic acid and hypoxanthine riboside 5'-monophosphate, is the first purine nucleotide formed in the *de novo* biosynthesis, and it is thus the precursor of all other purine nucleotides. IMP may also be generated by the nonenzymatic deamination of ADP, the latter achieved by the presence of nitrous acid. An important property if IMP is that the base readily undergoes amido/imido tautomerization, allowing the hydroxyl form to undergo phosphorylation to generate an essential 6-*O*-phosphoryl-IMP intermediate in the adenylosuccinate synthetase reaction. Note that its keto group (in the amido form) would be a proton-acceptor in hydrogen bond formation, whereas its hydroxy group (in the imido form) would be a proton-donor in hydrogen bond formation. Such ambiguous base-bairing is likely to be mutagenic in DNA and RNA. **Target(s):** adenine phosphoribosyltransferase (1); adenylosuccinate synthetase (2); alcohol dehydrogenase (3); amidophospho-ribosyltransferase (4-7); D-amino-acid oxidase, weakly inhibited (8); AMP nucleosidase (9-13); asparagine synthetase, weakly inhibited (14); carbamoyl-phosphate synthetase (glutamine-hydrolyzing) (15-17); FAD diphosphatase (18); fructose-1,6-bisphosphate aldolase (19,20); glucokinase (21); glucose-6-phosphate dehydrogenase, weakly inhibited (22); GMP reductase (23); GTP cyclohydrolase IIa (24); guanylate kinase (25); hexokinase (26); 5'-nucleotidase (27,28); orotate phosphoribosyltransferase (29); orotidylate decarboxylase (29); pancreatic ribonuclease (30); phosphoglycerate kinase (31); 5-phosphoribosylamine synthetase, *or* ribose-5-phosphate:ammonia ligase (32); phosphoribosylamino-imidazole-succinocarboxamide synthetase (33); phosphoribosyl-pyrophosphate synthetase, *or* ribose-5-phosphate pyrophosphokinase (34); phosphorylase phosphatase (35); polyphosphate kinase (36); [protein-PII] uridylyltransferase (37,38); ribonuclease (30); ribose-phosphate diphosphokinase, *or* phosphoribosyl-pyrophosphate synthetase (39); starch phosphorylase, weakly inhibited (40); thiamin kinase (41); threonine synthase, weakly inhibited (41). **1.** Arnold & Kelley (1978) *Meth. Enzymol.* **51**, 568. **2.** Borza, Iancu, Pike, Honzatko & Fromm (2003) *J. Biol. Chem.* **278**, 6673. **3.** Li & Vallee (1964) *J. Biol. Chem.* **239**, 792. **4.** Flaks & Lukens (1963) *Meth. Enzymol.* **6**, 52. **5.** Baker (1967) *Design of*

Active-Site-Directed Irreversible Enzyme Inhibitors, Wiley, New York. **6.** Caskey, Ashton & Wyngaarden (1964) *J. Biol. Chem.* **239**, 2570. **7.** Hill & Bennett (1969) *Biochemistry* **8**, 122. **8.** McCormick, Chassy & Tsibris (1964) *Biochim. Biophys. Acta* **89**, 447. **9.** Heppel (1963) *Meth. Enzymol.* **6**, 117. **10.** DeWolf, Fullin & Schramm (1979) *J. Biol. Chem.* **254**, 10868. **11.** Yoshino (1970) *J. Biochem.* **68**, 321. **12.** Ogasawara, Yoshino & Asai (1970) *J. Biochem.* **68**, 331. **13.** Yoshino & Takagi (1973) *J. Biochem.* **74**, 1151. **14.** Hongo, Matsumoto & Sato (1978) *Biochim. Biophys. Acta* **522**, 258. **15.** Raushel, Thoden, Reinhart & Holden (1998) *Curr. Opin. Chem. Biol.* **2**, 624. **16.** Braxton, Mullins, Raushel & Reinhart (1999) *Biochemistry* **38**, 1394. **17.** Thoden, Raushel, Wesenberg & Holden (1999) *J. Biol. Chem.* **274**, 22502. **18.** Mistuda, Tsuge, Tomozawa & Kawai (1970) *J. Vitaminol.* **16**, 31. **19.** MacDonald & Storey (2002) *Arch. Biochem. Biophys.* **408**, 279. **20.** Bais, James, Rofe & Conyers (1985) *Biochem. J.* **230**, 53. **21.** Reeves, Montalvo & Sillero (1967) *Biochemistry* **6**, 1752. **22.** Horne, Anderson & Nordlie (1970) *Biochemistry* **9**, 610. **23.** Magasanik (1963) *Meth. Enzymol.* **6**, 106. **24.** Graham, Xu & White (2002) *Biochemistry* **41**, 15074. **25.** Anderson (1973) *The Enzymes*, 3rd ed., **9**, 49. **26.** Ning, Purich & Fromm (1969) *J. Biol. Chem.* **244**, 3840. **27.** Webb (1966) *Enzyme and Metabolic Inhibitors*, vol. 2, p. 471, Academic Press, New York. **28.** Hurwitz, Heppel & Horecker (1957) *J. Biol. Chem.* **226**, 525. **29.** Jones, Kavipurapu & Traut (1978) *Meth. Enzymol.* **51**, 155. **30.** Richards & Wyckoff (1971) *The Enzymes*, 3rd ed., **4**, 647. **31.** Kuntz & Krietsch (1982) *Meth. Enzymol.* **90**, 103. **32.** Westby & Tsai (1974) *J. Bacteriol.* **117**, 1099. **33.** Nelson, Binkowski, Honzatko & Fromm (2005) *Biochemistry* **44**, 766. **34.** Wong & Murray (1969) *Biochemistry* **8**, 1608. **35.** Rall & Sutherland (1962) *Meth. Enzymol.* **5**, 377. **36.** Lindner, Vidaurre, Willbold, Schoberth & Wendisch (2007) *Appl. Environ. Microbiol.* **73**, 5026. **37.** Engleman & Francis (1978) *Arch. Biochem. Biophys.* **191**, 602. **38.** Francis & Engleman (1978) *Arch. Biochem. Biophys.* **191**, 590. **39.** Fox & Kelley (1972) *J. Biol. Chem.* **247**, 2126. **40.** Kaziro (1959) *J. Biochem.* **46**, 1523. **41.** Giovanelli, Mudd, Datko & Thompson (1986) *Plant Physiol.* **81**, 577.

Inosine 5'-Triphosphate

This nucleotide (FW_{free-acid} = 508.08 g/mol; UV Spectrum: at pH 6.0, λ_{max} = 248.5 nm and ε = 12200 $M^{-1}cm^{-1}$), which is produced from IDP through the action of nucleoside-diphosphate kinase or by the action of nitrous acid on ATP, is an alternative substrate for many ATP- and GTP-utilizing enzymes. **Target(s):** adenine phosphoribosyltransferase (1); adenylate cyclase (2,3); adenylosuccinate synthetase (4); adenylylsulfatase (5-7); amidophospho-ribosyltransferase (8); asparagine synthetase (9); aspartate carbamoyltransferase (10); carbamate kinase, *Pyrococcus furiosus* (11); carbamoyl-phosphate synthetase (12); 3',5'-cyclic-nucleotide phosphodiesterase (13,14); exopolyphosphatase (15); GDP-mannose 4,6-dehydratase (16); glucose-6-phosphate dehydrogenase (17); GTP cyclohydrolase I (18); GTP cyclohydrolase II (19); guanylate cyclase (20-22); hexokinase (23); isocitrate dehydrogenase (NAD$^+$) (24-26); lactose synthase (27); D-lactate dehydrogenase (28); leucyl-tRNA synthetase29; nicotinamide mononucleotide adenylyltransferase26,30; nucleotidases (15); 3'-nucleotidase (15); 5'-nucleotidase (15,31,32); nucleotide diphosphatase (33); phosphoenolpyruvate carboxykinase (GTP) (34,35); phosphoprotein phosphatase (36); porphobilinogen synthase, *or* δ-aminolevulinate dehydratase (37); [protein-PII] uridylyltransferase (38); protein-synthesizing GTPase (elongation factor) (39); pyridoxal kinase, weakly inhibited (26,40); pyruvate carboxylase (41); pyruvate,orthophosphate dikinase (42-44); guanosine-triphosphate guanylyltransferase (45); *N*-acylneuraminate cytidylyltransferase (46); [glutamine-synthetase] adenylyltransferase (47); ribose-phosphate diphosphokinase, *or* phosphoribosyl-pyrophosphate synthetase (48,49); diacylglycerol kinase (50); 1-phosphofructokinase (51,52); glucuronokinase (53); pyruvate kinase (54); glucokinase (55,56); nicotinate-nucleotide diphosphorylase (carboxylating) (57,58); nicotinate phosphoribosyl-transferase (59); hypoxanthine(guanine) phosphoribosyltransferase (60); sucrose-phosphate synthase (61); *N*-acetylneuraminate 4-*O*-acetyltransferase (62). **1**. Arnold & Kelley (1978) *Meth. Enzymol.* **51**, 568. **2**. Ide (1971) *Arch. Biochem. Biophys.* **144**, 262. **3**. Yang & Epstein (1983) *J. Biol. Chem.* **258**, 3750. **4**. Van der Weyden & Kelly (1974) *J. Biol. Chem.* **249**, 7282. **5**. Bailey-Wood, Dodgson & Rose (1970) *Biochim. Biophys. Acta* **220**, 284. **6**. Bayley-Wood, Dodgson & Rose (1969) *Biochim. J.* **112**, 257. **7**. Stokes, Denner & Dodgson (1973) *Biochim. Biophys. Acta* **315**, 402. **8**. Hill & Bennett (1969) *Biochemistry* **8**, 122. **9**. Hongo, Matsumoto & Sato (1978) *Biochim. Biophys. Acta* **522**, 258. **10**. Jacobson & Stark (1973) *The Enzymes*, 3rd ed., **9**, 225. **11**. Uriarte, Marina, Ramón-Maiques, *et al.* (2001) *Meth. Enzymol.* **331**, 236. **12**. Durbecq, Legrain, Roovers, Piérard & Glansdorff (1997) *Proc. Natl. Acad. Sci. U.S.A.* **94**, 12803. **13**. Drummond & Yamamoto (1971) *The Enzymes*, 3rd ed., **4**, 355. **14**. Cheung (1967) *Biochemistry* **6**, 1079. **15**. Proudfoot, Kuznetsova, Brown, *et al.* (2004) *J. Biol. Chem.* **279**, 54687. **16**. Liao & Barber (1972) *Biochim. Biophys. Acta* **276**, 85. **17**. Horne, Anderson & Nordlie (1970) *Biochemistry* **9**, 610. **18**. Weisberg & O'Donnell (1986) *J. Biol. Chem.* **261**, 1453. **19**. Elstner & Suhadolnik (1975) *Meth. Enzymol.* **43**, 515. **20**. Hardman & Sutherland (1969) *J. Biol. Chem.* **244**, 6363. **21**. Krishnan, Fletcher, Chader & Krishna (1978) *Biochim. Biophys. Acta* **523**, 506. **22**. Durham (1976) *Eur. J. Biochem.* **61**, 535. **23**. Ning, Purich & Fromm (1969) *J. Biol. Chem.* **244**, 3840. **24**. Plaut (1969) *Meth. Enzymol.* **13**, 34. **25**. Chen & Plaut (1963) *Biochemistry* **2**, 1023. **26**. Webb (1966) *Enzyme and Metabolic Inhibitors*, vol. **2**, Academic Press, New York. **27**. Babad & Hassid (1966) *Meth. Enzymol.* **8**, 346. **28**. LeJohn & Stevenson (1975) *Meth. Enzymol.* **41**, 293. **29**. Marutzky, Flossdorf & Kula (1976) *Nucl. Acids Res.* **3**, 2067. **30**. Atkinson, Morton & Murray (1961) *Nature* **192**, 946. **31**. Madrid-Marina & Fox (1986) *J. Biol. Chem.* **261**, 444. **32**. Cercignani, Serra, Fini, *et al.* (1974) *Biochemistry* **13**, 3628. **33**. Krishnan & Rao (1972) *Arch. Biochem. Biophys.* **149**, 336. **34**. Vial, Oelckers, Rojas & Simpfendörfer (1995) *Comp. Biochem. Physiol. B* **112**, 451. **35**. Rohrer, Saz & Nowak (1986) *J. Biol. Chem.* **261**, 13049. **36**. Polya & Haritou (1988) *Biochem. J.* **251**, 357. **37**. Tigier, del C. Batlle & Locascio (1970) *Enzymologia* **38**, 43. **38**. Engleman & Francis (1978) *Arch. Biochem. Biophys.* **191**, 602. **39**. Masullp, De Vendittis & Bocchini (1994) *J. Biol. Chem.* **269**, 20376. **40**. Hurwitz (1953) *J. Biol. Chem.* **205**, 935. **41**. Scrutton & Utter (1965) *J. Biol. Chem.* **240**, 3714. **42**. South & Reeves (1975) *Meth. Enzymol.* **42**, 187. **43**. Milner, Michaels & Wood (1975) *Meth. Enzymol.* **42**, 199. **44**. Reeves (1971) *Biochem. J.* **125**, 531. **45**. Liu & McLennan (1994) *J. Biol. Chem.* **269**, 11787. **46**. Rodríguez-Aparicio, Luengo, Gonzalez-Clemente & Reglero (1992) *J. Biol. Chem.* **267**, 9257. **47**. Ebner, Wolf, Gancedo, Elsässer & Holzer (1970) *Eur. J. Biochem.* **14**, 535. **48**. Fox & Kelley (1972) *J. Biol. Chem.* **247**, 2126. **49**. Switzer & Sogin (1973) *J. Biol. Chem.* **248**, 1063. **50**. Wissing & Wagner (1992) *Plant Physiol.* **98**, 1148. **51**. Van Hugo & Gottschalk (1974) *Eur. J. Biochem.* **48**, 455. **52**. Sapico & Anderson (1969) *J. Biol. Chem.* **244**, 6280. **53**. Gillard & Dickinson (1978) *Plant Physiol.* **62**, 706. **54**. Etges & Mukkada (1988) *Mol. Biochem. Parasitol.* **27**, 281. **55**. Porter, Chassy & Holmlund (1982) *Biochim. Biophys. Acta* **709**, 178. **56**. Doelle (1982) *Eur. J. Appl. Microbiol. Biotechnol.* **14**, 241. **57**. Shibata & Iwai (1980) *Agric. Biol. Chem.* **44**, 119. **58**. Taguchi & Iwai (1976) *Agric. Biol. Chem.* **40**, 385. **59**. Hayakawa, Shibata & Iwai (1984) *Agric. Biol. Chem.* **48**, 445. **60**. Nagy & Ribet (1977) *Eur. J. Biochem.* **77**, 77. **61**. Sinha, Pathre & Sane (1997) *Phytochemistry* **46**, 441. **62**. Iwersen, Dora, Kohla, Gasa & Schauer (2003) *Biol. Chem.* **384**, 1035.

myo-Inositol

This polyol (FW = 180.16 g/mol; CAS 87-89-8; light-sensitive; Solubility: 14 g/100 mL H_2O at 25°C; *Abbreviation*: Ins) possesses an internal plane of symmetry and is accordingly optically inactive. When in the chair conformation, all but one hydroxyl group (designated the C2 position) are in the equatorial positions. *myo*-Inositol is a substrate in the following enzyme-catalyzed reactions: *myo*-inositol 2-dehydrogenase, inositol oxygenase, inositol 1-methyltransferase, inositol 3-methyltransferase, inositol 4-methyltransferase, indoleacetylglucose:inositol *O*-acyltransferase, inositol 3α-galactosyltransferase, inositol 3-kinase, and CDP-diacylglycerol:inositol 3-phosphatidyltransferase. It can be produced by the action of certain phosphatases on *myo* inositol 3-phosphate as well as the

reactions catalyzed by galactinol:raffinose galactosyl-transferase, galactinol: sucrose galactosyl-transferase, and inositol-1(or 4) monophosphatase. **Target(s):** alkaline phosphatase (1); CDP-diacylglycerol:glycerol-3-phosphate 3-phosphatidyl-transferase (2); CDP-diacylglycerol:serine *O*-phosphatidyl-transferase (3); β-fructofuranosidase, *or* invertase (4); α-galactosidase (5-10); galactosyl-galactosylglucosyl-ceramidase, weakly inhibited (11); glucose dehydrogenase (NAD^+) (12); α-glucosidase (13); β-glucosidase (14); glycerophosphoinositol inositol-phosphodiesterase, *or* 1,2-cyclic-inositol-phosphate phosphodiesterase (15); inositol-phosphate phosphatase, K_i = 400 mM (16,17); inositol-3 phosphate synthase, *or* inositol-1-phosphate synthase (18,19); phosphatidylinositol diacylglycerol lyase (20-22); phospholipase D (23). **1**. Ramakrishnan & Bhandari (1979) *Experientia* **35**, 994. **2**. Bleasdale & Johnston (1982) *Biochim. Biophys. Acta* **710**, 377. **3**. Carman & Bae-Lee (1992) *Meth. Enzymol.* **209**, 298. **4**. Isla, Vattuone, Gutierrez & Sampietro (1988) *Phytochemistry* **27**, 1993. **5**. Sharma (1971) *Biochem. Biophys. Res. Commun.* **43**, 572. **6**. Mujer, Ramirez & Mendoza (1984) *Phytochemistry* **23**, 1251. **7**. Bishop & Sweely (1978) *Biochim. Biophys. Acta* **525**, 399. **8**. Corchete & Guerra (1987) *Phytochemistry* **26**, 927. **9**. Lusis & Paigen (1976) *Biochim. Biophys. Acta* **437**, 487. **10**. Ueno, Ikami, Yamauchi & Kato (1980) *Agric. Biol. Chem.* **44**, 2623. **11**. Rietra, Tager & Borst (1972) *Biochim. Biophys. Acta* **279**, 436. **12**. Pauly & Pfleiderer (1975) *Hoppe Seylers Z. Physiol. Chem.* **356**, 1613. **13**. Thirunavukkarasu & Priest (1984) *J. Gen. Microbiol.* **130**, 3135. **14**. Langston, Sheehy & Xu (2006) *Biochim. Biophys. Acta* **1764**, 972. **15**. Ross & Majerus (1986) *J. Biol. Chem.* **261**, 11119. **16**. Miller, Bashir-Uddin Surfraz, Akhtar, Gani & Allemann (2004) *Org. Biomol. Chem.* **2**, 671. **17**. Nigou & Besra (2002) *Biochem. J.* **361**, 385. **18**. RayChaudhuri, Hait, das Gupta, *et al.* (1997) *Plant Physiol.* **115**, 727. **19**. Ju & Greenberg (2004) *Clin. Neurosci. Res.* **4**, 181. **20**. Shashidhar, Volwerk, Keana & Griffith (1990) *Biochim. Biophys. Acta* **1042**, 410. **21**. Feng, Wehbi & Roberts (2002) *J. Biol. Chem.* **277**, 19867. **22**. Wehbi, Feng & Roberts (2003) *Biochim. Biophys. Acta* **1613**, 15. **23**. Tang, Waksman, Ely & Liscovitch (2002) *Eur. J. Biochem.* **269**, 3821.

1D-*myo*-Inositol 2,4-Bisphosphate

This phosporylated inositol (FW$_{free-acid}$ = 340.12 g/mol; *Abbreviation*: Ins(2,4)P$_2$) inhibits fructose-1,6-bisphosphate aldolase. **1**. Koppitz, Vogel & Mayr (1986) *Eur. J. Biochem.* **161**, 421.

1D-*myo*-Inositol Hexasulfate

This sulfated inositol (FW$_{free-acid}$ = 660.53 g/mol), also known as 1D-*myo*-inositol hexakissulfate, inhibits phytase and certain protein kinases. **Target(s):** acid phosphatase (1); [β-adrenergic-receptor] kinase (2,3); casein kinase I (4); casein kinase II (5,6); L- and P-selectin (7); 3-phytase (10,14); 3- or 4-phytase (8,11-13); 4-phytase (9,10). **1**. Nagai, Funahashi & Egami (1963) *J. Biochem.* **54**, 191. **2**. Benovic (1991) *Meth. Enzymol.* **200**, 351. **3**. Benovic, Stone, Caron & Lefkowitz (1989) *J. Biol. Chem.* **264**, 6707. **4**. Hathaway, Tuazon & Traugh (1983) *Meth. Enzymol.* **99**, 308. **5**. Hathaway & Traugh (1983) *Meth. Enzymol.* **99**, 317. **6**. Dahmus (1981) *J. Biol. Chem.* **256**, 3319. **7**. Cecconi, Nelson, Roberts, *et al.* (1994) *J. Biol. Chem.* **269**, 15060. **8**. Chu, Guo, Lin, *et al.* (2004) *Structure* **12**, 2015. **9**. Ullah & Sethumadhavan (1998) *Biochem. Biophys. Res. Commun.* **251**, 260. **10**. Ullah, Sethumadhavan, Lei & Mullaney (2000) *Biochem. Biophys. Res. Commun.* **275**, 279. **11**. Ullah, Sethumadhavan, Mullaney, Ziegelhoffer &

Austin-Phillips (2002) *Biochem. Biophys. Res. Commun.* **290**, 1343. **12.** Ullah, Sethumadhavan, Mullaney, Ziegelhoffer & Austin-Phillips (2003) *Biochem. Biophys. Res. Commun.* **306**, 603. **13.** Ullah & Sethumadhavan (2003) *Biochem. Biophys. Res. Commun.* **303**, 463. **14.** Ullah, Sethumadhavan, Mullaney, Ziegelhoffer & Austin-Phillips (1999) *Biochem. Biophys. Res. Commun.* **264**, 201.

1D-*myo*-Inositol 1-Monophosphate

This optically active phosphoinositol (FW$_{free-acid}$ = 260.14 g/mol; *Abbreviation*: Ins(1)P) is a component of many phospholipids and inositides. Note that a mixture of 1- and 2-monophosphate derivatives are obtained upon treatment of 1D-*myo*-inositol 1-phosphate with hot dilute acid. 1D-*myo*-Inositol 1 monophosphate is produced from enzyme-catalyzed hydrolysis of phosphatidylinositol. (D-*myo*-Inositol 1-phosphate is identical to L-*myo*-inositol 3-phosphate, *i.e.*, D-*myo*-inositol 1-phosphate is the enantiomer of D-*myo*-inositol 3-phosphate). **Target(s):** fructose-1,6-bisphosphate aldolase (1); glycerophosphoinositol inositol-phosphodiesterase, *or* 1,2-cyclic-inositol-phosphate phosphodiesterase (2). **1.** Koppitz, Vogel & Mayr (1986) *Eur. J. Biochem.* **161**, 421. **2.** Ross & Majerus (1986) *J. Biol. Chem.* **261**, 11119.

1D-*myo*-Inositol 2-Monophosphate

This optically inactive phosphoinositol (FW$_{free-acid}$ = 260.14 g/mol; *Abbreviation*: Ins(2)P) is obtained upon treatment of 1D-*myo*-inositol 1-phosphate with hot dilute acid. It can also be generated by the action of phytase on phytic acid (*i.e.*, *myo*-inositol hexakisphosphate) and is stable in the presence of strong base. Migration of the phosphoryl moiety readily occurs, and 1-phosphate analogue is often formed. **Target(s):** glycerophosphoinositol inositolphospho-diesterase, *or* 1,2-cyclic-inositol phosphate phospho-diesterase (1); inositol-1-phosphatase (2). **1.** Ross & Majerus (1986) *J. Biol. Chem.* **261**, 11119. **2.** Hallcher & Sherman (1980) *J. Biol. Chem.* **255**, 10896.

1D-*myo*-Inositol (4-Palmitoyloxybutane-1-yl)-phosphonate

This nonhydrolyzable phosphodiester analogue (FW$_{free-acid}$ = CHOP g/mol) of 1-palmitoyl-lysophosphatidylinositol inhibits phosphatidylinositol-specific phospholipase C. **Target(s):** phosphatidylinositol-specific phospholipase C, *or* 1-phosphatidyl-inositol phosphodiesterase (1,2). **1.** Griffith, Volwerk & Kuppe (1991) *Meth. Enzymol.* **197**, 493. **2.** Shashidhar, Volwerk, Keana & Griffith (1990) *Biochim. Biophys. Acta* **1042**, 410.

1D-*myo*-Inositol 1,3,4,5,6-Pentakisphosphate

This phosphorylated inositol (FW$_{free-acid}$ = 580.06 g/mol; *Abbreviation*: Ins(1,3,4,5,6)P$_5$) modulates the binding of oxygen to avian and reptilian hemoglobins and also stimulates L-type Ca^{2+} channels. **Target(s):** fructose-1,6-bisphosphate aldolase (1); inositol-hexakisphosphate kinase (13); inositol-phosphate phosphatase (2); inositol-polyphosphate 5-phosphatase, *or* inositol 1,4,5 trisphosphate/1,3,4,5-tetrakisphosphate 5-phosphatase (3); inositol-tetrakisphosphate 1-kinase (4); inositol trisphosphate receptors (5); L- and P-selectin (6); multiple inositol-polyphosphate phosphatase, *or* 1,3,4,5-tetrakisphosphate 3-phosphatase, also as an alternative substrate (5,7-13); phosphatidylinositide 3-kinase/Akt signalling pathway (14). **1.** Koppitz, Vogel & Mayr (1986) *Eur. J. Biochem.* **161**, 421. **2.** Roth, Harkness & Isaacks (1981) *Arch. Biochem. Biophys.* **210**, 465. **3.** Höer & Oberdisse (1991) *Biochem. J.* **278**, 219. **4.** Tan, Bruzik & Shears (1997) *J. Biol. Chem.* **272**, 2285. **5.** Lu, Shieh & Chen (1996) *Biochem. Biophys. Res. Commun.* **220**, 637. **6.** Cecconi, Nelson, Roberts, *et al.* (1994) *J. Biol. Chem.* **269**, 15060. **7.** Nogimori, Hughes, Glennon, *et al.* (1991) *J. Biol. Chem.* **266**, 16499. **8.** Estrada-Garcia, Craxton, Kirk & Michell (1991) *Proc. R. Soc. Lond. B Biol. Sci.* **244**, 63. **9.** Hughes & Shears (1990) *J. Biol. Chem.* **265**, 9869. **10.** Höer & Oberdisse (1992) *Eicosanoids* **5** Suppl., S16. **11.** Höer, Höer & Oberdisse (1990) *Biochem. J.* **270**, 715. **12.** Craxton, Ali & Shears (1995) *Biochem. J.* **305**, 491. **13.** Piccolo, Vignati, Maffucci, *et al.* (2004) *Oncogene* **23**, 1754. **14.** Voglmaier, Bembenek, Kaplin, *et al.* (1996) *Proc. Natl. Acad. Sci. U.S.A.* **93**, 4305.

1D-*myo*-Inositol 1-Phosphate, *See* 1D-*myo*-Inositol 1-Monophosphate

1D-*myo*-Inositol 2-Phosphate, *See* 1D-*myo*-Inositol 2-Monophosphate

1D-*myo*-Inositol 4-Phosphate, *See* 1D-*myo*-Inositol 4-Monophosphate

1DL-*myo*-Inositol 1-Phosphate 4,5-Bisphosphorothioate

This sulfur-containing 1D-*myo*-inositol 1,4,5-trisphosphate isomer (FW = 516.22 g/mol) inhibits inositol-polyphosphate 5-phosphatase (K_i = 1.4 µM) and inositol-trisphosphate 3-kinase (K_i = 46 µM). **Target(s):** inositol-polyphosphate 5-phosphatase, *or* inositol-1,4,5-trisphosphate 5-phosphatase; inositol-trisphosphate 3-kinase. **1.** Safrany, Mills, Liu, *et al.* (1994) *Biochemistry* **33**, 10763.

myo-Inositol 1,2,3,5-Tetrakisphosphate

This phosphorylated inositol (FW$_{free-acid}$ = 500.08 g/mol; *Abbreviation*: Ins(1,2,3,5)P$_4$) inhibits inositol-trisphosphate 3 kinase. **1.** Choi, Chang, Chung & Choi (1997) *Bioorg. Med. Chem. Lett.* **7**, 2709.

1DL-*myo*-Inositol 1,2,4,5-Tetrakisphosphate

This 1D-*myo*-inositol tetrakisphosphate isomer (FW$_{free-acid}$ = 500.08 g/mol) inhibits inositol-polyphosphate 5-phosphatase, *or* inositol-1,4,5-

trisphosphate 5 phosphatase (K_i = 2.9 µM). **1**. Safrany, Mills, Liu, *et al.* (1994) *Biochemistry* **33**, 10763.

1D-*myo*-Inositol 1,2,5,6-Tetrakisphosphate

This phosphorylated inositol (FW$_{free\text{-}acid}$ = 500.08 g/mol; *Abbreviation*: Ins(1,2,5,6)P$_2$) is an intermediate in the inositol signaling pathway that inhibits inositol-tetrakisphosphate 1-kinase. **1**. Phillippy (1998) *Plant Physiol.* **116**, 291.

1D-*myo*-Inositol 1,3,4,5-Tetrakisphosphate

This polyphosphorylated inositol (FW$_{free\text{-}acid}$ = 500.08 g/mol; *Abbreviation*: Ins (1,3,4,5)P$_4$) regulates intracellular calcium ions by increasing Ca^{2+} passage across the plasma membrane. It is also an immediate precursor to inositol 1,3,4-trisphosphate. Note that it is an enantiomer of D-*myo*-inositol-3,4,5,6-tetrakisphosphate (*i.e.*, L-*myo*-inositol-1,3,4,5-tetrakisphosphate). **Target(s):** Ca^{2+}-transporting ATPase (1); inositol-hexakisphosphate kinase (2); inositol tetrakisphosphate 1-kinase (3,4); inositol-trisphosphate 3-kinase, by product inhibition (5-7); inositol 1,3,4-trisphosphate 5/6-kinase (8-10); inositol-polyphosphate 5-phosphatase, *or* inositol 1,4,5 trisphosphate 5-phosphatase, as alternative substrate for some isozymes and sources (11-13). **1**. Fraser & Sarnacki (1992) *Amer. J. Physiol.* **262**, F411. **2**. Voglmaier, Bembenek, Kaplin, *et al.* (1996) *Proc. Natl. Acad. Sci. U.S.A.* **93**, 4305. **3**. Phillippy (1998) *Plant Physiol.* **116**, 291. **4**. Tan, Bruzik & Shears (1997) *J. Biol. Chem.* **272**, 2285. **5**. Poinas, Backers, Riley, *et al.* (2005) *ChemBioChem* **6**, 1449. **6**. Nalaskowski, Bertsch, Fanick, *et al.* (2003) *J. Biol. Chem.* **278**, 19765. **7**. Choi, Chang, Chung & Choi (1997) *Bioorg. Med. Chem. Lett.* **7**, 2709. **8**. Abdullah, Hughes, Craxton, *et al.* (1992) *J. Biol. Chem.* **267**, 22340. **9**. Shears (1989) *J. Biol. Chem.* **264**. 19879. **10**. Hughes, Kirk & Michell (1994) *Biochim. Biophys. Acta* **1223**, 57. **11**. Kennedy, Sha'afi & Becker (1990) *J. Leukoc. Biol.* **47**, 535. **12**. Fowler & Brännström (1990) *Biochem. J.* **271**, 735. **13**. Connolly, Bansal, Bross, Irvine & Majerus (1987) *J. Biol. Chem.* **262**, 2146.

1D-*myo*-Inositol 1,3,4,6-Tetrakisphosphate

This phosphorylated inositol (FW$_{free\text{-}acid}$ = 500.08 g/mol; *Abbreviation*: Ins(1,3,4,6)P$_2$) is an intermediate in the inositol signaling pathway and stimulates intracellular Ca^{2+} release. It also inhibits inositol-1,3,4 trisphosphate 5/6-kinase and inositol-1,4,5,6-tetrakisphosphate 3-kinase. **Target(s):** inositol-tetrakisphosphate 1-kinase (1,2); inositol-1,4,5,6-tetrakisphosphate 3-kinase (3); inositol-polyphosphate 5-phosphatase, K_i = 7.7 µM (4); inositol-trisphosphate 3-kinase, K_i = 150 µM (4); inositol-1,3,4-trisphosphate 5/6-kinase (5,6). **1**. Q. Phillippy (1998) *Plant Physiol.* **116**, 291. **2**. Tan, K. S. Bruzik & S. B. Shears (1997) *J. Biol. Chem.* **272**, 2285. **3**. Craxton, C. Erneux & S. B. Shears (1994) *J. Biol. Chem.* **269**, 4337. **4**.

Safrany, Mills, Liu, *et al.* (1994) *Biochemistry* **33**, 10763. **5**. Abdullah, Hughes, Craxton, *et al.* (1992) *J. Biol. Chem.* **267**, 22340. **6**. Hughes, Kirk & Michell (1994) *Biochim. Biophys. Acta* **1223**, 57.

1D-*myo*-Inositol 3,4,5,6-Tetrakisphosphate

This phosphorylated inositol (FW$_{free\text{-}acid}$ = 500.08 g/mol; *Abbreviation*: Ins(3,4,5,6)P$_4$) inhibits the Ca^{2+}/calmodulin-dependent and kinase-regulated epithelial chloride channel. It has been observed in human astrocytoma cells, avian erythrocytes, and cultured murine macrophages. **Target(s):** calcium-activated Cl$^-$ currents (1-4); inositol-trisphosphate 3-kinase (5); inositol 1,3,4-trisphosphate 5/6-kinase (6,7); multiple inositol-polyphosphate phosphatase, *or* inositol 1,3,4,5-tetrakisphosphate 3-phosphatase (8-10). **1**. Nilius, Prenen, Voets, *et al.* (1998) *Pflugers Arch.* **435**, 637. **2**. Carew, Yang, Schultz & Shears (2000) *J. Biol. Chem.* **275**, 2690. **3**. Ho, Shears, Bruzik, Duszyk & French (1997) *Amer. J. Physiol.* **272**, C1160. **4**. Xie, Kaetzel, Bruzik, *et al.* (1996) *J. Biol. Chem.* **271**, 14092. **5**. Choi, Chang, Chung & Choi (1997) *Bioorg. Med. Chem. Lett.* **7**, 2709. **6**. Hughes, Kirk & Michell (1994) *Biochim. Biophys. Acta* **1223**, 57. **7**. Wilson, Sun, Cao & Majerus (2001) *J. Biol. Chem.* **276**, 40998. **8**. Höer, Höer & Oberdisse (1990) *Biochem. J.* **270**, 715. **9**. Hughes & Shears (1990) *J. Biol. Chem.* **265**, 9869. **10**. Nogimori, Hughes, Glennon, *et al.* (1991) *J. Biol. Chem.* **266**, 16499.

1D-*myo*-Inositol 1,3,4-Trisphosphate

This naturally-occurring phosphorylated cyclitol (FW$_{free\text{-}acid}$ = 420.10 g/mol; ; *Abbreviation*: Ins(1,3,4)P$_3$) inhibits inositol-3,4,5,6-tetrakisphosphate 1-kinase. **Target(s):** inositol-polyphosphate multikinase (1); inositol-tetrakisphosphate 1-kinase (2,3); multiple-inositol-polyphosphate phosphatase, *or* inositol-1,3,4,5-tetrakisphosphate 3 phosphatase (4-6). **1**. Riley, Deleu, Qian, *et al.* (2006) *FEBS Lett.* **580**, 324. **2**. Tan, Bruzik & Shears (1997) *J. Biol. Chem.* **272**, 2285. **3**. Craxton, Erneux & Shears (1994) *J. Biol. Chem.* **269**, 4337. **4**. Hughes & Shears (1990) *J. Biol. Chem.* **265**, 9869. **5**. Höer, Höer & Oberdisse (1990) *Biochem. J.* **270**, 715. **6**. Nogimori, Hughes, Glennon, *et al.* (1991) *J. Biol. Chem.* **266**, 16499.

1D-*myo*-Inositol 1,4,5-Trisphosphate

This phosphorylated inositol (FW$_{free\text{-}acid}$ = 420.10 g/mol; *Abbreviation*: Ins(1,4,5)P$_3$) mobilizes calcium ion concentrations. It is produced from phosphatidylinositol 4,5-bisphosphate via the action of phosphoinositase C. **Target(s):** Ca^{2+}-transporting ATPase (1,2); fructose-1,6-bisphosphate aldolase (3); guanylate cyclase (4); inositol-hexakisphosphate kinase (5); inositol-tetrakisphosphate 1-kinase (6,7); multiple inositol-polyphosphate phosphatase, *or* inositol-1,3,4,5-tetrakisphosphate 3-phosphatase (8-10); 3-phosphoglycerate kinase (11); sphingomyelin phosphodiesterase, mildly inhibited (12). **1**. Davis, Davis, Lawrence & Blas (1991) *FASEB J.* **5**, 2992. **2**. Kuo (1988) *Biochem.. Biophys.. Res. Commun.* **152**, 1111. **3**. Koppitz, Vogel & Mayr (1986) *Eur. J. Biochem.* **161**, 421. **4**. Janssens & de Jong (1988) *Biochem. Biophys. Res. Commun.* **150**, 405. **5**. Voglmaier,

Bembenek, Kaplin, *et al.* (1996) *Proc. Natl. Acad. Sci. U.S.A.* **93**, 4305. **6.** Phillippy (1998) *Plant Physiol.* **116**, 291. **7.** Tan, Bruzik & Shears (1997) *J. Biol. Chem.* **272**, 2285. **8.** Hughes & Shears (1990) *J. Biol. Chem.* **265**, 9869. **9.** Höer, Höer & Oberdisse (1990) *Biochem. J.* **270**, 715. **10.** Nogimori, Hughes, Glennon, *et al.* (1991) *J. Biol. Chem.* **266**, 16499. **11.** João & Williams (1993) *Eur. J. Biochem.* **216**, 1. **12.** Testai, Landek, Goswami, Ahmed & Dawson (2004) *J. Neurochem.* **89**, 636.

myo-Inosose-2

This optically inactive polyhydroxylated cyclohexanone (FW = 178.14 g/mol), also called *scyllo*-inosose and 2,4,6/3,5-pentahydroxy-cyclohexanone, is an intermediate in streptidine biosynthesis. *myo*-Inosose-2 is a substrate for glutamine:*scyllo*-inosose aminotransferase and *myo*-inosose-2 dehydratase. It decomposes slowly upon exposure to strong light and has a melting point of 199-202°C (with decomposition) (1). **Target(s):** CDP-diacylglycerol: inositol 3-phosphatidyltransferase (2); inositol 4-methyltransferase (3;) inositol oxygenase (4); D-xylose: NADP$^+$ 1-oxidoreductase, *or* D-xylose dehydrogenase, NADP$^+$-requiring (5). **1.** Posternak (1952) *Biochem. Prep.* **2**, 57. **2.** Takenawa & Egawa (1977) *J. Biol. Chem.* **252**, 5419. **3.** Wanek & Richter (1995) *Planta* **197**, 427. **4.** Naber, Swan & Hamilton (1986) *Biochemistry* **25**, 7201. **5.** Logemann & Wissler (1985) *Acta Med. Leg. Soc. (Liege)* **35**, 184.

Inostamycin

This polyether antibiotic (FW = 644.89 g/mol), first isolated from a species of *Streptomyces*, inhibits cytidine 5'-diphosphate 1,2-diacyl-sn-glycerol (CDP-DG): inositol transferase, caused a G$_1$-phase accumulation in the cell cycle of small cell lung carcinomas (1-3). Inostamycin, a novel polyether compound, reverses multidrug resistance, thereby allowing a dose-dependent accumulation of [^{3}H]-vinblastine in multidrug-resistant cell lines (2). **Target(s):** CDP-diacylglycerol:inositol 3-phosphatidyltransferase (1); P-glycoprotein, *or* multidrug resistance transporter (2); phosphatidylinositol turnover (3). **1.** Imoto, Taniguchi & Umezawa (1992) *J. Biochem.* **112**, 299. **2.** Kawada & Umezawa (1995) *Jpn. J. Cancer Res.* **86**, 873. **3.** Johtoh & Umezawa (1992) *Drugs Exp. Clin. Res.* **18**, 1.

S-Inosyl-L-homocysteine

This *S*-substituted homocysteine (FW = 385.40 g/mol), formed by the enzymatic deamination of *S*-adenosyl-L-homocysteine, weakly inhibits mRNA (nucleoside-2'-*O*-) methyltransferase and DNA (cytosine-5-)-methyltransferase. **Target(s):** DNA (cytosine-5-)-methyltransferase (1); *Hae*III DNA methyltransferase, K_i = 450 μM (1); *Hha*I DNA methyltransferase, K_i = 3.9 mM (1); mRNA (nucleoside-2'-*O*-) methyltransferase (2); tRNA (guanine-N^7-)-methyltransferase (3). **1.**

Cohen, Griffiths, Tawfik & Loakes (2005) *Org. Biomol. Chem.* **3**, 152. **2.** Pugh & Borchardt (1982) *Biochemistry* **21**, 1535. **3.** Paolella, Ciliberto, Traboni, Cimino & Salvatore (1982) *Arch. Biochem. Biophys.* **219**, 149.

S-Inosyl-L-(2-hydroxy-4-methylthio)butyrate

This *S*-adenosylmethionine analogue (FW = 385.42 g/mol) inhibits adenosylmethionine:8-amino-7-oxononanoate transaminase, *or* diamino-pelargonate aminotransferase (1), histone-arginine *N*-methyltransferase (2), and protein-arginine methyltransferase (3). **1.** Stoner & Eisenberg (1975) *J. Biol. Chem.* **250**, 4037. **2.** Gupta, Jensen, Kim & Paik (1982) *J. Biol. Chem.* **257**, 9677. **3.** Tuck & Paik (1984) *Meth. Enzymol.* **106**, 268.

Ins, *See myo*-Inositol

Ins(1,3)P$_2$, *See* 1D-*myo*-Inositol 1,3-Bisphosphate

Ins(1,4)P$_2$, *See* 1D-*myo*-Inositol 1,4-Bisphosphate

Ins(3,4)P$_2$, *See* 1D-*myo*-Inositol 3,4-Bisphosphate

Ins(1,3,4)P$_3$, *See* 1D-*myo*-Inositol 1,3,4-Trisphosphate

Ins(1,4,5)P$_3$, *See* 1D-*myo*-Inositol 1,4,5-Trisphosphate

Ins(1,3,4,5)P$_4$, *See* 1D-*myo*-Inositol 1,3,4,5-Tetrakisphosphate

Insensible

This novel nuclear inhibitor (MW = 19425.51 g; Accession Number = AE013599.5; Symbol = Insb) targets Notch signalling, which regulates a series of binary fate decisions in *Drosophila*, leading a single sensory organ precursor cell to form regularly spaced sensory organs through a series of asymmetric cell divisions. Insb over-expression leads to transcriptional repression of a Notch reporter and to phenotypes associated with Notch inhibition. While the complete loss of Insb activity showed no significant phenotype, it enhanced the bristle phenotype associated with reduced levels of Hairless, a nuclear protein acting as a co-repressor for Suppressor of Hairless. **1.** Coumailleau & Schweisguth (2014) *PLoS One* **9**, e98213.

Insulin degludec

This ultralong-acting insulin analogue (FW = 6103.97 g/mol; CAS 844439-96-9; Symbol: Ideg), also known as Tresiba®, provides a basal insulin level, reducing the need for frequent administration of fast-/short-acting bolus doses of insulin. The presence of the lysyl ε-amino-hexadecanediamide to Lysine-29 (B-chain) promotes self-assembly of insulin hexamers, which in turn form multi-hexamers, when deposited at appropriate levels within subcutaneous tissues. Insulin heximerization and polymerization serve to retard insulin dissolution and mobilization, thus providing for the nearly zero-order release of this hormone into the bloodstream. There is a continuous and highly predictable slow dissociation of insulin monomers from this depot; insulin levels rise immediately reaching t_{max} at 10-12 hours, followed by a slow decline, with a half-life of 17-21 hours, or roughly twice the duration of insulin glargine action. **Pharmacokinetics:** Onset of Action = 30–90 minutes (similar to insulin glargine and insulin detemir); Little or no peak in activity (due to the slow release into systemic circulation); Duration of Action = >24 hours; Time to Attain Steady-State = 2-3 days. Once-daily dosing produces a steady-state profile characterized by a near-constant effect, which varies little from injection to injection in a given patient. **1**. Rendell (2013) *Drugs Today* (*Barc*) **49**, 387. **2**. Goldman-Levine, Patel & Schnee (2013) *Ann. Pharmacother.* **47**, 269.

Inter-α-Trypsin Inhibitor (*or* Inter-α-Inhibitor)

This abundant plasma glycoprotein and trypsin inhibitor consists of three subunits (Heavy Chain 1, Heavy Chain 2, and Bikunin) joined covalently by a chondroitin sulfate chain originating at Ser-10 of bikunin. IαI interacts with the inflammation-associated Tumor Necrosis Factor-Inducible Gene 6 Protein (TNFAIP6 *or* TSG-6) within the extracellular matrix (ECM). This interaction leads to the transfer of the heavy chains from the chondroitin sulfate of inter-α-inhibitor to hyaluronan (HA), with consequential stabilization of the ECM. (*Note*: TNFAIP6 is itself secreted by various cells in response to stimulation by various inflammatory cytokines.) Hyaluronan and its binding proteins control inflammatory gene expression, inflammatory cell recruitment, and inflammatory cytokine release, thereby attenuating the course of inflammation and protecting against tissue damage. **1**. Salier, Rouet, Raguenez, & Daveau (1996) *Biochem. J.* **315** (Pt 1), 1. **2**. Zhuo, Hascall & Kimata (2004) *J. Biol. Chem.* **279**, 38079. **3**. Kobayashi (2006) *Biol Chem.* **387**, 1545.

Interleukin-1 Receptor Antagonist

This protein (MW = 17,258 g/mol; CAS 143090-92-0; Symbol: IL-1RA), which in humans is encoded by the *IL1RN* gene, binds non-productively to the Interleukin-1 Receptor (IL-1R), preventing IL-1 signal propagation to downstream targets. A member of the interleukin 1 cytokine family, IL1Rα is secreted by immune cells, epithelial cells, and adipocytes. It is a natural inhibitor of the pro-inflammatory effect of IL1β. This protein inhibits the activities of IL1Rα and IL1β, and thereby modulates a variety of IL1-related immune and inflammatory responses. It is commercially known as Anakinra (trade name Kineret®), is used to suppress inflammation and cartilage degradation associated with rheumatoid arthritis.

Intoplicine

This anticancer agent (FW_free-base = 348.45 g/mol; CAS 125974-72-3), also known as RP-60475, is a dual inhibitor of DNA topoisomerases I and II. Intoplicine induces both topoisomerase I- and II-mediated DNA strand breaks, and cleavage site patterns induced were different from those induced by camptothecin, *N*-[4-(9-acridinylamino)-3-methoxyphenyl] methanesulfonamide (m-AMSA), and other known topoisomerase inhibitors. Both topoisomerase I- and II-induced DNA breaks decreased at drug concentrations higher than 1 μM, a feature that is consistent with intoplicine's DNA-intercalating activity. 1. Nabiev, Chourpa, Riou, *et al.* (1994) *Biochemistry* **33**, 9013. 2. Poddevin, Riou, Lavelle & Pommier (1993) *Mol. Pharmacol.* **44**, 767.

Intrinsic Factor

This 417-residue glycoprotein (MW = 45,425 g; CAS 9008-12-2) is a vitamin B_{12}-binding glycoprotein produced by the parietal cells of the stomach. **Role in Vitamin B_{12} Absorption:** After dietary vitamin B_{12} is released from ingested proteins through the action of pepsin and stomach acid, it is rapidly bound by one of two vitamin B_{12}-binding proteins present in gastric juice. These binding proteins have a greater affinity for the vitamin than does intrinsic factor at acid pH. After transfer to the small intestine, pancreatic proteases digest the binding proteins, releasing vitamin B_{12}, which then becomes bound to intrinsic factor. Finally, there are receptors for intrinsic factor on the ileal mucosa, which bind the complex, allowing vitamin B_{12} to be enter portal circulation. **Physiology:** William Castle's

discovery of Intrinsic Factor in 1929 provided the first explanation for the pathogenesis and pathophysiology of pernicious anemia. He postulated that a factor was required to bind dietary vitamin B_{12} in a manner that permitted facilitated entry of B_{12} into circulation. The story is, of course, more complex. Cubilin (Mr = 215,000) is the intestinal receptor for the endocytosis of intrinsic factor-vitamin B_{12} complex. When intrinsic factor was complexed with various analogs of cyanocobalamin, the dissociation constant (K_i = 0.25 nM) of these complexes for ileal microvillous membranes were similar to that of intrinsic factor-bound cyanocobalamin. Mutations in *CUBN*, the gene for cubilin, is one cause of hereditary megaloblastic anemia and associated neurological deficits. Overexpression of an unstable intrinsic factor-cobalamin receptor occurs in Imerslund-Gräsbeck syndrome, an autosomal recessive, familial form of vitamin B_{12} deficiency caused by malfunction of the receptor located in the terminal ileum. **Mode of Inhibitory Action:** The dissociation constant for the intrinsic factor-cyanocobalamin complex was 0.066 nM. When added in excess, Intrinsic Factor tightly binds cobalamin derivatives, thereby depriving vitamin B_{12}-dependent enzymes of their essential cofactor. This inhibition can often be overcome by the addition of excess cobalamin or cobalamin analogue. **Target(s):** ethanolamine ammonia-lyase (1); leucine 2,3-aminomutase (2-4); β-lysine 5,6-aminomutase (1,5); D-lysine 5,6-aminomutase (1,6,7); β-methylaspartate mutase, *or* glutamate mutase (8,9); 2-methyleneglutarate mutase (10); methylmalonyl-CoA mutase (11); D-ornithine 4,5-aminomutase (1). **1**. Stadtman (1972) *The Enzymes*, 3rd ed., **6**, 539. **2**. Poston (1976) *J. Biol. Chem.* **251**, 1859. **3**. Poston (1977) *Science* **195**, 301. **4**. Poston (1978) *Phytochemistry* **17**, 401. **5**. Stadtman & Grant (1971) *Meth. Enzymol.* **17B**, 206. **6**. Stadtman & Grant (1971) *Meth. Enzymol.* **17B**, 211. **7**. Stadtman & Tsai (1967) *Biochem. Biophys. Res. Commun.* **28**, 920. **8**. Ellenbogen, Highley, Barker & Smyth (1960) *Biochem. Biophys. Res. Commun.* **3**, 178. **9**. Ohmori, Ishitani, Sato, Shimizu & Fukui (1971) *Biochem. Biophys. Res. Commun.* **43**, 156. **10**. Kung & Stadtman (1971) *J. Biol. Chem.* **246**, 3378. **11**. Poston (1978) *Phytochemistry* **17**, 401.

Iodine & Iodide

This group 17 (or VIIB) element (Symbol = I; Atomic Number = 53; Atomic Weight = 126.90) is a halogen located directly below bromine in the Periodic Table. In its elemental state, I_2 is a powerful oxidizing agent that, when applied as a dilute solution, finds wide utility as a disinfectant. Common oxidation states are I^{1-}, I^{5+}, and I^{7+}, for which the ionic radii are 2.20, 0.62, and 0.50 Å, respectively. Iodine has only one stable isotope (^{127}I), but several of its radioactive isotopes are used in biochemistry and biomedicine. Iodine-123 has a half-life of 13.2 hours and is used as a diagnostic imaging agent. Iodine-124 is a positron emitter with a half-life of 4.18 days and has been used in positron emission tomography. Iodine-125, which has a 59.4-day half-life for decay via electron capture (0.179 MeV), is frequently used to label macromolecules. Its emmisions require the experimenter to employ thick lead shielding. The frequently used iodine-131 has a half-life of 8.04 days (0.971 MeV). The thyroid gland is the critical organ in terms of radiation exposure, and investigators are strongly encouraged to adhere to institutional regulations regarding thyroid monitoring for localized radioiodine uptake. The effective biological half-life in humans is about 140 days, kept mainly as iodide monoanion and combined with thyroglobulin. Iodide anion (I^-) can react with one or more molecules of iodine (I_2) to produce polyiodide anions, of which I_3^- is found in aqueous systems. The triiodide ion is the most commonly used source for the iodination of proteins tyrosyl residues. Iodine is an important trace element required by a number of organisms. It is an essential element for vertebrates, Ascidians, corals, and possibly seed plants. It is accumulated in the mammalian thyroid gland (free iodide concentrations in the human thyroid are ~1 mM), in red and brown algae (and their marine predators), and a number of Porifera and Coelenterata. *See also Periodate*

Iodoacetamide

This alkylating agent (FW = 184.96 g/mol; M.P. = 93–96°C) is a potent irreversible inhibitor of enzymes requiring reactive thiols, ε-amino groups, and/or imidazole side chain groups. Nucleophilic displacement of the iodide by a enzyme sulfhydryl (*i.e.*, Enz–SH or Enz–S⁻) results in the formation of a stable thioether linkage (*i.e.*, Enz–S–CH₂C(=O)NH). In certain instances, methionyl residues also undergo alkylation. The reaction is typically a bimolecular process, rarely showing evidence of rate-saturation kinetics. Sulfhydryl groups are typically the most reactive protein functional moiety, with reactivity increasing with pH (optimal between pH 6 and 8.5). This toxic haloamide is soluble in water. Because iodoacetamide slowly decomposes in aqueous solutions, fresh stock solutions should be prepared. The reagent should also be stored away from light to minimize formation of iodine which often reacts with tyrosyl and histidyl residues. Iodoacetamide is usually slightly more reactive than iodoacetate.

Iodoacetate, *Similar in mode of action to* Iodoacetamide

3-Iodoacetol Phosphate

This analogue of dihydroxyacetone phosphate (FW$_{free-acid}$ = 279.95 g/mol), also known as 1-iodo-3-hydroxyacetone phosphate, inhibits a number of enzymes that utilize or produce dihydroxyacetone phosphate. Aqueous 3-iodoacetol phosphate is unstable (at pH 7, $t_{1/2}$ = 6.8 hours; at pH 8, $t_{1/2}$ = 3 hours). **Target(s):** fructose-1,6-bisphosphate aldolase (2,3); glycerol-phosphate dehydrogenase (2); methylglyoxal synthase (4); triose-phosphate isomerase (2,5-7). 1. Hartman (1970) *Biochemistry* 9, 1776. 2. Hartman (1972) *Meth. Enzymol.* 25, 661. 3. Hartman (1970) *Biochemistry* 9, 1783. 4. Yuan, Gracy & Hartman (1977) *Biochem. Biophys. Res. Commun.* 74, 1007. 5. Hartman & Norton (1975) *Meth. Enzymol.* 41, 447. 6. Hartman (1977) *Meth. Enzymol.* 46, 130. 7. Noltmann (1972) *The Enzymes*, 3rd ed., 6, 271.

5-Iodo-5'-amino-2',5'-dideoxyuridine-5'-*N'*-triphosphate

This phosphoramidate analogue (FW = 499.10 g/mol; *Abbreviation*: AIdUTP) of 5-iodo-2',5'-dideoxyuridine 5'-triphosphate reversibly inhibits thymidine kinase. Although AIdUTP is stable in alkaline solutions, below pH 8 it undergoes degradation by a novel phosphorylysis reaction which exhibits first order kinetics. AIdUTP is, however, 60-fold more potent as an allosteric inhibitor than is TTP 1. Chen & Prusoff (1978) *Meth. Enzymol.* 51, 354. 2. Chen, Ward & Prusoff (1976) *J. Biol. Chem.* 251, 4839. 3. Fischer, Fang, Lin, Hampton & Bruggink (1988) *Biochem. Pharmacol.* 37, 1293.

m-Iodobenzoate

This conjugate base of a substituted benzoic acid (FW$_{free-acid}$ = 248.02 g/mol; M.P. = 186-188°C) inhibits D-amino-acid oxidase (1,2), glutamate dehydrogenase (2-4), porphobilinogen synthase (5), and serine-sulfate ammonia-lyase (6). 1. Bartlett (1948) *J. Amer. Chem. Soc.* 70, 1010. 2. Webb (1966) *Enzyme and Metabolic Inhibitors*, vol. 2, Academic Press, New York. 3. Caughey, Smiley & Hellerman (1957) *J. Biol. Chem.* 224, 591. 4. Smith, Austen, Blumenthal & Nyc (1975) *The Enzymes*, 3rd ed., 11, 293. 5. del C. Batlle, Ferramola & Grinstein (1970) *Meth. Enzymol.* 17A, 216. 6. Tudball & Thomas (1973) *Eur. J. Biochem.* 40, 25.

o-Iodobenzoate (*o*-Iodobenzoic Acid)

This substituted benzoic acid (FW$_{free-acid}$ = 248.02 g/mol; M.P. = 162°C) inhibits aminopeptidase B (1); esterase (2); lysine 2-monooxygenase (3); macrocellulase complex (4); 6-phosphogluconate dehydrogenase (5); poly(A)-specific ribonuclease (6,7); propanediol dehydratase (8); and quinine oxidase (9). Note: Investigators should take care to not confuse *o*-iodobenzoate with *o*-iodosobenzoate. 1. Hopsu, Mäkinen & Glenner (1966) *Arch. Biochem. Biophys.* 114, 567. 2. Weber & King (1935) *J. Biol. Chem.* 108, 131. 3. Nakazawa (1971) *Meth. Enzymol.* 17B, 154. 4. Ljungdahl, Coughlan, Mayer, *et al.* (1988) *Meth. Enzymol.* 160, 483. 5. Noltmann & Kuby (1963) *The Enzymes*, 2nd ed., 7, 223. 6. Schröder, Zahn, Dose & Müller (1980) *J. Biol. Chem.* 255, 4535. 7. Schröder, Bachmann, Messer & Müller (1985) *Prog. Mol. Subcell. Biol.* 9, 53. 8. Ichikawa, Horike, Mori, Hosoi & Kitamoto (1985) *J. Ferment. Technol.* 63, 135. 9. Villela (1963) *Enzymologia* 25, 261.

m-Iodobenzylguanidine

This norepinephrine analogue (FW$_{free-base}$ = 275.09 g/mol; *Abbreviation*: MIBG), also known as iobenguane and often, interferes with cellular mono(ADP-ribosyl)transferases and acts as an alternative substrate for the ADP-ribosylating activity of cholera toxin and the mono(ADP-ribosyl)transferase of turkey erythrocyte membranes. MIBG is also used in the diagnosis and treatment of pheochromocytoma and inhibits mitochondrial respiration. **Target(s):** arginine-specific ADP-ribosyltransferase (1-4); monoamine oxidase (5); NADH dehydrogenase, *or* Complex I (6-8); ubiquinol:cytochrome *c* reductase, *or* Complex III (7,8). 1. Loesberg, van Rooij & Smets (1990) *Biochim. Biophys. Acta* 1037, 92. 2. Kharadia, Huiatt, Huang, Peterson & Graves (1992) *Exp. Cell Res.* 201, 33. 3. Smets, Loesberg, Janssen & Van Rooij (1990) *Biochim. Biophys. Acta* 1054, 49. 4. Deveze-Alvarez, Garcia-Soto & Martinez-Cadena (1997) *Adv. Exp. Med. Biol.* 419, 155. 5. Burba (1985) *Can. J. Physiol. Pharmacol.* 63, 166. 6. Cornelissen, Wanders, Van den Bogert, *et al.* (1995) *Eur. J. Cancer* 31A, 582. 7. Cornelissen, Wanders, Van Gennip, *et al.* (1995) *Biochem. Pharmacol.* 49, 471. 8. Cornelissen, Van Belzen, Van Gennip, Voute & Van Kuilenburg (1997) *Anticancer Res.* 17(1A), 265.

Iodocyanopindolol

This β$_1$-adrenoreceptor ligand (FW = 413.26 g/mol; CAS 85124-14-7; IUPAC: (*RS*)-4-[3-[(1,1-dimethylethyl)amino]-2-hydroxypropoxy]-3-iodo-1*H*-indole-2-carbonitrile) is a derivative of (*S*)-pindolol, a gold-standard β$_1$-adrenergic receptor antagonist (**See** *Pindolol*) and a partial-agonist/antagonist for 5-HT$_{1A}$ receptors. Its ^{125}I-labeled form has been widely exploited to map the distribution of β-adrenoreceptors, making it possible to directly determine the tissue concentration of β-adrenoceptors and the responsiveness of tissues to β-adrenergic stimulation (1). A major insight to emerge from radioligand-binding studies is that the tissue concentration of β-adrenoceptors is not a fixed number, but is instead dynamically regulated by a variety of drugs, hormones as well as physiological and pharmacological conditions. Moreover, because changes in β-adrenoceptors assessed in circulating lymphocytes mirror changes in density and functional responsiveness of the corresponding myocardial β-adrenoceptors, alterations in β-adrenoceptor function can be also examined in humans. Radioligand-binding studies also demonstrated the coexistence of $β_1$- and $β_2$-adrenoceptor subtypes in numerous tissues, including heart and lung. An example is the demonstration of the coexistence of $β_1$- and $β_2$-adrenoceptors in human right atrium by (+/-)-[^{125}I]iodocyanopindolol binding (2). 1. Brodde (1986) *J. Cardiovasc. Pharmacol.* 8 (Suppl. 4), S16. 2. Brodde, Karad, Zerkowski, Rohm & Reidemeister (1983) *Circulation. Res.* 53, 752.

2-Iodo-6-isopropyl-3-methyl-2',4,4'-trinitrodiphenyl Ether

This ether (FW = 487.21 g/mol; *Abbreviation*: DNP-INT), also known as 2,4-dinitrophenyl 2-iodo-4-isopropyl-6-nitrotoluen-3-yl ether and *O*-(2,4-dinitrophenyl)-2-iodo-4-nitrothymol, is a competitive inhibitor of the plastoquinol oxidation at the Qo site of the cytochrome b_6f complex. **Target(s):** photosynthetic electron transport (1); plastoquinol:plastocyanin reductase (*or* cytochrome b_6f complex) (2-10); ubiquinol:cytochrome-*c* reductase, *or* Complex III of *Spinacia oleracea* (11). 1. Trebst (1980) *Meth. Enzymol.* 69, 675. 2. Malkin (1988) *Meth. Enzymol.* 167, 341. 3. Delosme, Joliot & Trebst (1987) *Biochim. Biophys. Acta* 893, 1. 4. Hauska, Hurt, Gabellini & Lockau (1983) *Biochim. Biophys. Acta* 726, 97. 5. Krinner, Hauska, Hurt & Lockau (1982) *Biochim. Biophys. Acta* 681, 110. 6. Wynn, Bertsch, Bruce & Malkin (1988) *Biochim. Biophys. Acta* 935, 115. 7. Hurt & Hauska (1981) *Eur. J. Biochem.* 117, 591. 8. Oettmeier, Masson & E. Olschewski (1983) *FEBS Lett.* 155, 241. 9. Clark & Hind (1983) *J. Biol. Chem.* 258, 10348. 10. Malkin (1986) *FEBS Lett.* 208, 317. 11. Gabellini, Bowyer, Hurt, Melandri & Hauska (1982) *Eur. J. Biochem.* 126, 105.

5-Iodo-3-methoxy-tyrphostin AG538

This cell-permeable analogue of tyrphostin AG538 (FW = 437.19 g/mol) also known as α-cyano(3-methoxy-4-hydroxy-5-iodo)cinnamoyl(3',4' dihydroxyphenyl)ketone and tyrphostin I-OMe-AG538, inhibits insulin-like growth factor receptor protein-tyrosine kinase (IC_{50} = 3.4 μM). In computer modeling, the inhibitor docked in the kinase domain of the insulin receptor, sitting in place of tyrosines 1158 and 1162. The latter normally undergo autophosphorylation. Experimentally, it does not compete with ATP but instead competes with the IGF-1R substrate. 1. Blum, Gazit & Levitzki (2000) *Biochemistry* 39, 15705.

1-(5-Iodonaphthalene-1-sulfonyl)-1*H*-hexahydro-1,4-diazepine

This diazepine (FW$_{\text{free-base}}$ = 416.28 g/mol), commonly referred to as ML-7, is a selective [myosin light-chain] kinase inhibitor, IC_{50} = 0.3 μM. Other kinases are inhibited much less effectively (*e.g.*, protein kinase A, IC_{50} = 21

μM, and protein kinase C, IC_{50} = 42 μM). 1. Bain, McLauchlan, Elliott & Cohen (2003) *Biochem. J.* 371, 199. 2. Eriksson, Toivola, Sahlgren, Mikhailov & Härmälä-Braskén (1998) *Meth. Enzymol.* 298, 542. 3. Saitoh, Ishikawa, Matsushima, Naka & Hidaka (1987) *J. Biol. Chem.* 262, 7796. 4. Makishima, Honma, Hozumi, *et al.* (1991) *FEBS Lett.* 287, 175. 5. Tinsley, Yuan & Wilson (2004) *Trends Pharmacol. Sci.* 25, 64. 6. Lucero, Stack, Bresnick & Shuster (2006) *Mol. Biol. Cell* 17, 4093. 7. Behling-Kelly, McClenahan, Kim & Czuprynski (2007) *Infect. Immun.* 75, 4572. 8. Afonso, Ozden, Prevost, *et al.* (2007) *J. Immunol.* 179, 2576. 9. Wadgaonkar, Linz-McGillem, Zaiman & Garcia (2005) *J. Cell. Biochem.* 94, 351. 10. Niggli, Schmid & Nievergelt (2006) *Biochem. Biophys. Res. Commun.* 343, 602. 11. Grimm, Haas, Willipinski-Stapelfeldt, *et al.* (2005) *Cardiovasc. Res.* 65, 211. 12. Yanase, Ikeda, Ogata, *et al.* (2003) *Biochem. Biophys. Res. Commun.* 305, 223. 13. Deng, Williams & Schultz (2005) *Dev. Biol.* 278, 358. 14. Connell & Helfman (2006) *J. Cell Sci.* 119, 2269.

6β-Iodopenicillanic Acid

This substituted penicillanate (FW$_{\text{free-acid}}$ = 327.14 g/mol) reacts selectively with the active site of D-Ala-D-Ala carboxypeptidase to form an adduct that is much more stable than that formed with benzylpenicillin. The β-lactamases of *Bacillus cereus* and *Streptomyces cacaoi* are also inactivated, whereas the corresponding enzymes from *Streptomyces albus* and *Actinomadura* R39 catalyze the hydrolysis of the inactivator for multiple turnovers prior to inactivation (515 for the *S. albus* enzyme and 80 for *Actinomadura* R39). 6β-Iodopenicillanate is a substrate for class-B Zn^{2+}-containing β-lactamase II and inactivates lactamases from classes A, C, and D. **Target(s):** β-lactamase (1-7); zinc D-Ala-D-Ala carboxypeptidase (8-10). 1. Lenzini & Frere (1985) *J. Enzyme Inhib.* 1, 25. 2. Amicosante, Oratore, Joris, *et al.* (1988) *Biochem. J.* 254, 891. 3. Frere, Dormans, Duyckaerts & De Graeve (1982) *Biochem. J.* 207, 437. 4. Joris, De Meester, Galleni, *et al.* (1985) *Biochem. J.* 228, 241. 5. Ledent, Raquet, Joris, Van Beeumen & J. M. Frere (1993) *Biochem. J.* 292, 555. 6. Basu, Narayankumar, Van Beeumen & Basu (1997) *Biochem. Mol. Biol. Int.* 43, 557. 7. Caporale, Franceschini, Perilli, *et al.* (2004) *Antimicrob. Agents Chemother.* 48, 3579. 8. Ghuysen (1998) in *Handb. Proteolytic Enzymes*, p. 1354, Academic Press, San Diego. 9. Charlier, Dideberg, Jamoulle, *et al.* (1984) *Biochem. J.* 219, 763. 10. Joris, Jacques, Frere, Ghuysen & Van Beeumen (1987) *Eur. J. Biochem.* 162, 519.

p-Iodo-D-phenylalanine Hydroxamate

This hydroxamic acid (FW = 306.10 g/mol) inhibits *Vibrio* aminopeptidase. **Target(s):** bacterial leucyl aminopeptidase, *Vibrio* aminopeptidase, *or* *Aeromonas proteolytica* aminopeptidase. 1. Chevier & D'Orchymont (1998) in *Handb. Proteolytic Enzymes*, p. 1433, Academic Press, San Diego. 2. Chevrier, D'Orchymont, Schalk, Tarnus & Moras (1996) *Eur. J. Biochem.* 237, 393. 3. Schürer, Lanig & Clark (2004) *Biochemistry* 43, 5414.

(Z)-3-(4-Iodophenyl)-2-mercapto-2-propenoate

This selective calpain inhibitor, (FW = 306.12 g/mol; Photosensitive off-white solid) , also known as PD150606, is cell-permeable The K_i value for calpain-1 [μ-calpain] is 0.21 μM (noncompetitive) and 0.37 μM for calpain-2 [m-calpain]. It is directed to calpain's calcium binding sites and does not protect calpain against inactivation by the active-site-directed inhibiutor, *trans*-(epoxysuccinyl)-L-leucylamido-3-methylbutane. It also prevents dexamethasone-induced apoptosis in thymocytes. Note that there is some confusion in the literature concerning PD 150606. For example, in one of the references below, PD150606 is referred to as 3-(4-iodobenzyl)-2-mercapto-2 propenoate (*i.e.*, a substituted 2-butenoate) and not the inhibitor first reported in 1996. **1**. Edelstein, Shi & Schrier (1999) *J. Amer. Soc. Nephrol.* **10**, 1940. **2**. Lin, Chattopadhyay, Maki, *et al.* (1997) *Nat. Struct. Biol.* **4**, 539. **3**. Squier & Cohen (1997) *J. Immunol.* **158**, 3690. **4**. Edelstein, Yaqoob, Alkhunaizi, *et al.* (1996) *Kidney Int.* **50**, 1150. **5**. Wang, Nath, Posner, *et al.* (1996) *Proc. Natl. Acad. Sci. U.S.A.* **93**, 6687. **6**. Wang, Posner, Raser, *et al.* (1996) *Adv. Exp. Med. Biol.* **389**, 95.

5-Iodotubercidin

This adenosine analogue and genotoxic agent (FW = 392.15 g/mol; Symbol: Itu), also known as 7-iodo-7-deazaadenosine, NSC 113939, and 4-amino 5-iodo-7-(β-D-ribofuranosyl)pyrrolo[2,3-*d*]pyrimidine, inhibits adenosine kinase, reduces adenosine uptake, and stimulates glycogen synthesis (1-11). The K_i value for adenosine kinase is 30 nM. Because the use of adenosine receptor agonists is plagued by dose-limiting cardiovascular side effects, there is great interest in developing adenosine kinase inhibitors (AKIs) as a means for raising steady-state adenosine concentrations. However, 5-iodotubercidin also causes DNA damage, inducing DNA breaks and forming nuclear foci that are positive for γH2AX and TopBP1, activation of Atm and Chk2, and S15 phosphorylation and up-regulation of p53 (12). As such, 5-iodotubercidin induces G_2 cell-cycle arrest in a p53-dependent manner. It also induces cell death in p53-dependent and -independent modes. DNA breaks are likely generated by incorporation of a 5-iodotubercidin metabolite into DNA. Moreover, 5-iodotubercidin shows anti-tumor activity, reducing tumor size in carcinoma xenograft mouse models in p53-dependent and -independent ways (12). **See also Tubercidin** **Target(s):** adenosine kinase (1-11); adenosine transport (4,5); casein kinase 1 (13); [glycogen synthase] kinase (13,14); [3-hydroxy-3-methylglutaryl-CoA reductase (NADPH)] kinase (14); mitogen-activated protein kinase ERK2, K_i = 530 nM (15); phosphorylase kinase (13); protein kinase A (13); protein-tyrosine kinase, insulin receptor (13). **1**. Singh, Kumar & Sharma (2003) *J. Enzyme Inhib. Med. Chem.* **18**, 395. **2**. Ugarkar, DaRe, Kopcho, *et al.* (2000) *J. Med. Chem.* **43**, 2883. **3**. Cottam, Wasson, Shih, *et al.* (1993) *J. Med. Chem.* **36**, 3424. **4**. Parkinson & Geiger (1996) *J. Pharmacol. Exp. Ther.* **277**, 1397. **5**. Davies & Cook (1995) *Life Sci.* **56**, PL345. **6**. Long & Parker (2006) *Biochem. Pharmacol.* **71**, 1671. **7**. Palella, Andres & Fox (1980) *J. Biol. Chem.* **255**, 5264. **8**. Iqbal, Burbiel & Müller (2006) *Electrophoresis* **27**, 2505. **9**. Boyer, Ugarkar, Solbach, *et al.* (2005) *J. Med. Chem.* **48**, 6430. **10**. De Jong (1977) *Arch. Int. Physiol. Biochim.* **85**, 557. **11**. Long, Escuyer & Parker (2003) *J. Bacteriol.* **185**, 6548. **12**. Zhang, Jia, Liu,*et al.* (2013) *PLoS One* **8**, e62527./ **13**. Massillon, Stalmans, van de Werve & Bollen (1994) *Biochem. J.* **299**, 123. **14**. Musi, Hayashi, Fujii, *et al.* (2001) *Amer. J. Physiol.* **280**, E677. **15**. Fox, Coll, Xie, *et al.* (1998) *Protein Sci.* **7**, 2249.

3-Iodo-L-tyrosine

This tyrosine derivative (FW = 307.09 g/mol; Light-sensitive, with loss of iodide; pK_a values of 2.2 (–COOH), 8.7 (phenol), and 9.1 (–NH₂) at 25°C) is an alternative substrate for tubulin:tyrosine ligase. **Target(s):** aromatic-L-amino-acid decarboxylase, *or* dopa decarboxylase (1); cAMP phosphodiesterase (2); dihydropteridine reductase, weak inhibitor (3); tyrosine 3-monooxygenase (4-11). **1**. Nagy & Hiripi (2002) *Neurochem. Int.* **41**, 9. **2**. Law & Henkin (1984) *Res. Commun. Chem. Pathol.*

Pharmacol. **43**, 449. **3**. Shen (1983) *Biochim. Biophys. Acta* **743**, 129. **4**. Shiman & Kaufman (1970) *Meth. Enzymol.* **17A**, 609. **5**. Fitzpatrick (1991) *Biochemistry* **30**, 3658. **6**. Moore & Dominic (1971) *Fed. Proc.* **30**, 859. **7**. Bullard & Capson (1983) *Mol. Pharmacol.* **23**, 104. **8**. McGeer, McGeer & Peters (1967) *Life Sci.* **6**, 2221. **9**. Haavik (1997) *J. Neurochem.* **69**, 1720. **10**. Fitzpatrick (1988) *J. Bioil. Chem.* **263**, 16058. **11**. Neckameyer, Holt & Paradowski (2005) *Biochem. Genet.* **43**, 425.

5-Iodouracil

This substituted pyrimidine (FW = 237.98 g/mol; CAS 696-07-1) inactivates dihydropyrimidine dehydrogenase by forming an adduct with a cysteinyl residue. 5-Iodouracil also inhibits L-3-aminoisobutyrate aminotransferase and 4-aminobutyrate aminotransferase. **Target(s):** 4-aminobutyrate aminotransferase (1); (*R*)-3-amino-2 methylpropionate: pyruvate aminotransferase, *or* D-3-aminoisobutyrate aminotransferase (2); (*S*)-3-amino-2-methylpropionate: pyruvate aminotransferase, *or* L-3-aminoisobutyrate aminotransferase (3); dihydropyrimidine dehydrogenase (4,5). **1**. Tamaki, Kubo, Aoyama & Funatsuka (1983) *J. Biochem.* **93**, 955. **2**. Kaneko, Kontani, Kikugawa & Tamaki (1992) *Biochim. Biophys. Acta* **1122**, 45. **3**. Tamaki, Fujimoto Sakata & Matsuda (2000) *Meth. Enzymol.* **324**, 376. **4**. Dobritzsch, Ricagno, Schneider, Schnackerz & Lindqvist (2002) *J. Biol. Chem.* **277**, 13155. **5**. Porter, Chestnut, Taylor, Merrill & Spector (1991) *J. Biol. Chem.* **266**, 19988.

Ionophorous Antibiotics

These lipid-soluble molecules form stoichiometric complexes with alkali metal cations and transport bound ions across a variety of membranes (1-6) or into a bulk organic phase (3,6,7). Neutral sequestering agents, such as valinomycin (2,4,6), or the macrotetrolide homologues known as monactin, dinactin, tetranactin, and nonactin (2,4,7,8), form charged complexes with monovalent cations while the negatively charged carboxylic acid ionophores such as nigericin (3,4,9,10) or the monensins (11,12) form a neutral metal-antibiotic species. The carboxylic acid X537A binds barium (13), other divalent cations, and lanthanum (14) as well as alkali metal cations (4,5). While the neutral sequestering ionophores induce active, electrogenic accumulation of monovalent cations by respiring mitochondria of rat liver (2,4-6,8), the carboxylic acid antibiotics catalyze an electroneutral exchange of accumulated or endogenous alkali metal cations for extramitochondrial protons (3,6,10-12). A23187 is a carboxylic acid antibiotic that affects mitochondrial function by perturbing endogenous calcium and magnesium content. (**See also Valinomycin; Nactin; Monensin; Nigericin; X537A; A23187**) **1**. Reed & Lardy (1972) *J. Biol. Chem.* **247**, 6970. **2**. Pressman (1965) *Proc. Nat. Acad. Sci. USA.* **63**, 1076. **3**. Pressman, Harris, Jagger & Johnson (1967) *Proc. Nat. Acad. Sci. USA* **68**, 1949. **4**. Lardy, Graven & Estrada (1967) *Fed. Proc.* **26**, 1355. **5**. Henderson, McGivan & Chappell (1969) *Biochem. J.* **111**, 521. **6**. Pressman (1969) *Ann. N. Y. Acad. Sci.* **147**, 829. **7**. Eisenman, Ciana, Szabo (1969) *J. Membrane Biol.* **1**, 294. **8**. Graven, Lardy & Estrada (1967) *Biochemistry* **6**, 365. **9**. Lardy, Johnson & McMurray (1958) *Arch. Biochem. Biophys.* **76**, 587. **10**. Graven, Estrada & Lardy (1966) *Proc. Nat. Acad. Sci. USA* **66**, 654. **11**. Estrada, Rightmire & Lardy (1967) *Antimicrob. Agents Chemother.* p. 279. **12**. Estrada & Calder (1970) *Biochemistry* **9**, 2092. **13**. Johnson, Herrin, Liu & Paul (1970) *J. Amer. Chem. Soc.* **92**, 4428.

IP7e

This orally bioavailable, brain-penetrant Nurr1 activator (FW = 390.43 g/mol; CAS 500164-74-9; Solubility: 100 mM in DMSO; 20 mM in H_2O, the latter with warming), also named 6-[4-[(2-methoxyethoxy)methyl]phenyl]-5-methyl-3-phenylisoxazolo[4,5-c]pyridin-4(5H)-one, targets the <u>N</u>uclear <u>r</u>eceptor-<u>r</u>elated 1 (EC_{50} = 3.9 nM), a protein that resembles steroid nuclear hormone receptor, but is not regulated by ligands and is instead locked in a constitutively active form. In Nurr1-knockout mice, complete midbrain dopamine cell agenesis is apparent at birth, and point mutations in Nurr1 gene are associated with a familiar form of Parkinson Disease. Reduced Nurr1 expression levels is observed in brains of aged individuals. Nurr1 also plays an anti-inflammatory role by inhibiting the expression of inflammatory genes in microglia and astrocytes, and Nurr1 helps to repress the activity of the pro-inflammatory transcription factor NF-κB. IP7e reduces the incidence and severity of experimental autoimmune encephalomyelitis (EAE) in mice, attenuating inflammation and neurodegeneration in spinal cords of EAE mice by an NF-κB pathway-dependent process. **1**. Montarolo, Raffaele, Perga, *et al*. (2014) *PLoS One* **9**, e108791. **2**. Hintermann, Chiesi, von Krosigk, *et al*. (2007) *Bioorg. Med. Chem. Lett*. **17**, 193.

IPI-145

This novel sinalling pathway protein kinase inhibitor (FW = 416.86 g/mol; CAS 1201438-56-3; Solubility = 84 mg/mL DMSO; <1 mg/mL H_2O), also known as Duvelisib, INK1197, and 8-chloro-2-phenyl-3-[(1S)-1-(9H-purin-6-ylamino)ethyl]-1(2H)-isoquinolinone, selectively targets PI3K-δ (K_i = 23 pM (calculated from k_{on} = 15.2 x 10^6 $M^{-1}s^{-1}$ and k_{off} = 0.9 hour^{-1}); IC_{50} = 1 nM), showing lesser affinity to PI3K-γ (K_i = 243 pM (calculated from k_{on} = 3.4 x 10^6 $M^{-1}s^{-1}$ and k_{off} = 3 hour^{-1}); IC_{50} = 50 nM) and little inhibition of PI3K-β (K_i = 15.5 pM (calculated from k_{on} = 2.7 x 10^6 $M^{-1}s^{-1}$ and k_{off} = 2.7 hour^{-1}) and PI3K-α (K_i = 25.9 μM (calculated from k_{on} = 4.2 x 10^6 $M^{-1}s^{-1}$ and k_{off} = >1800 hour^{-1}). IPI-145 was a weaker inhibitor (K_i > 1 μM) with 440 other protein kinases. This inhibitor inhibits human B and T cell proliferation, with EC_{50} values of 0.5 and 9.5 nM. **1**. "Targeting PI3K-δ and PI3K-γ in Inflammation", by Palombella, Infinity Pharmaceuticals, Inc., *New York Acad. Sci. Conference*, June 27, 2012 (http://www.infi.com/product-candidates-publications.asp)

Ipilimumab

This CTLA-4 function-blocking drug (MW = 148634.91 g/mol; CAS 477202-00-9), also known by the code names MDX-010 and MDX-101 and by its trade name Yervoy®, is a humanized IgG1 monoclonal antibody against the extracellular domain of the Cytotoxic T-Lymphocyte Antigen-4, a protein that activates the immune system by targeting the CTLA-4 receptor, inhibiting inappropriate or prolonged T-cell activation. Ipilimumab and like-acting anti-CTLA-4 antibodies bind to CTLA-4 on the cell surface, blocking its interaction with the costimulatory molecules B7.1 and B7.2 and interrupting the negative signalling mediated by CTLA-4 and blunting T-cell responses involved in immune tolerance against self antigens. Ipilimumab activates tumor-reactive T cells, while removing inhibitory activity with CTLA-4 blockade, a drug action that commends its use combination with other cancer immunotherapies to improve clinical outcomes, particularly in melanoma treatment. **Formulation:** Ipilimumab is diluted to 1–2 mg/mL in normal saline or 5% dextrose in water and administered intravenously over 90 minutes using an in-line filter. Among previously untreated patients with metastatic melanoma, nivolumab alone or combined with ipilimumab resulted in significantly longer progression-free survival than ipilimumab alone (3). *See Nivolumab* **1**. Maker, Phan, Attia, *et al*. (2005) *Ann. Surg. Oncol*. **12**, 1005. **2**. Acharya & Jeter (2013) *Clin. Pharmacol*. **5**, 21. **3**. Larkin, Chiarion-Sileni, Gonzalez, *et al*. (2015) *New Engl. J. Med*. **373**, 23.

Ipragliflozin

This orally active benzothiophene C-glucoside (FW = 404.45 g/mol; CAS 761423-87-4), also known by its code name ASP1941 and its systematic name (1S)-1,5-Anhydro-1-C-[3-[(1-benzothiophen-2-yl)methyl]-4-fluoro-phenyl]-D-glucitol, a potent and selective sodium glucose co-transporter 2 (SGLT2) inhibitor for the treatment of Type 2 Diabetes Mellitus (T2DM). **Rationale:** SGLT2 is a high-capacity, low-affinity transport system primarily expressed in the renal proximal tubules, where it plays an important role in the regulation of glucose levels. Ipragliflozin induces sustained increases in urinary glucose excretion by inhibiting renal glucose reabsorption, with subsequent antihyperglycemic effect and a low risk of hypoglycemia. Therefore, ipragliflozin has the therapeutic potential for treating hyperglycemia in diabetes by increasing glucose excretion into urine. **1**. Imamura, Nakanishi, Suzuki, *et al*. (2012) *Bioorg. Med. Chem*. **20**, 3263. **2**. Tahara, Kurosaki, Yokono, *et al*. (2012) *Naunyn. Schmiedebergs. Arch. Pharmacol*. **385**, 423 **2**. Schwartz, Akinlade, Klasen, *et al*. (2011) *Diabetes Technol. Ther*. **13**, 1219.

Iproniazid

This non-selective irreversible MAO inhibitor ($FW_{free-base}$ = 179.22 g/mol; CAS 54-92-2), also called isonicotinic acid 2-isopropylhydrazide, 1-isonicotinyl-2-isopropylhydrazine, and Marsilid® (iproniazid's phosphate salt; FW = 277.21 g/mol; CAS 305-33-9), is a first-generation isoniazid-based antidepressant. Initially developed to treat tuberculosis, iproniazid became the first antidepressant ever marketed, after its mood elevating properties were noted in the early 1950s. Hepatotoxicity eventually led to its withdrawal from the market. Once heralded as a "psychic energizer", iproniazid's significance today is as the drug that awakened wider interest in monoamine oxidase as a gateway target in the development of numerous antidepressants for treating mood and anxiety spectrum disorders. **Target(s):** ethanolamine oxidase (1); 5-hydroxytryptophan decarboxylase, aromatic amino-acid decarboxylase (2); isopenicillin-N epimerase (3,4); monoamine oxidase (5-21); putrescine oxidase (22); taurine dehydrogenase (23). **1**. Narrod & Jakoby (1966) *Meth. Enzymol*. **9**, 354. **2**. Uteshev (1964) *Farmakol. Toksikol*. **27**, 293. **3**. Usui & Yu (1989) *Biochim. Biophys. Acta* **999**, 78. **4**. Jensen, Westlake & Wolfe (1983) *Can. J. Microbiol*. **29**, 1526. **5**. Tabor, Tabor & Rosenthal (1955) *Meth. Enzymol*. **2**, 390. **6**. McEwen (1971) *Meth. Enzymol*. **17B**, 692. **7**. Yamada & Adachi (1971) *Meth. Enzymol*. **17B**, 705. **8**. Yasunobu & Gomes (1971) *Meth. Enzymol*. **17B**, 709. **9**. Yamada, Kumagai & Uwajima (1971) *Meth. Enzymol*. **17B**, 722. **10**. A. Zeller (1963) *The Enzymes*, 2nd ed., **8**, 313. **11**. Blaschko (1963) *The Enzymes*, 2nd ed., **8**, 337. **12**. Smith, Weissbach & Udenfriend (1963) *Biochemistry* **2**, 746. **13**. Blaschko (1963) *The Enzymes*, 2nd ed., **8**, 337. **14**. Udenfriend, Witkop, Redfield & Weissbach (1958) *Biochem. Pharmacol*. **1**, 160. **15**. Davison (1957) *Biochem. J*. **67**, 316. **16**. Verdiani (1958) *Boll. Chim. Farm*. **97**, 307. **17**. Gomes (1983) *Indian J. Biochem. Biophys*. **20**, 96. **18**. Zeller & Sarkar (1962) *J. Biol. Chem*. **237**, 2333. **19**. Rucker & Goettlich-Riemann (1972) *Enzymologia* **43**, 33. **20**. Tabor, Tabor & Rosenthal (1954) *J. Biol. Chem*. **208**, 645. **21**. McEwan, Sasaki & Jones (1969) *Biochemistry* **8**, 3963. **22**. Yamada (1971) *Meth. Enzymol*. **17B**, 726. **23**. Kondo & Ishimoto (1987) *Meth. Enzymol*. **143**, 496.

IPSU

This selective OX$_2$R antagonist (FW = 405.49 g/mol; CAS 1373765-19-5; Solubility: 100 mM in DMSO; 100 mM in H$_2$O, as mono-HCl salt), also named 2-((1*H*-indol-3-yl)methyl)-9-(4-methoxypyrimidin-2-yl)-2,9-diazaspiro[5.5]undecan-1-one, targets orexin OX$_1$ receptors (IC$_{50}$ = 11 nM), exhibiting ~6-fold selectivity versus OX$_1$ receptors. Orexin receptor antagonists represent attractive targets for the development of drugs for the treatment of insomnia, inasmuch as the receptors are crucial for the stability of wake and sleep states. OX$_1$R and OX$_2$R are G protein-coupled receptors with 64% sequence identity in human. Both couple to G$_q$ and mobilize intracellular Ca^{2+} via activation of phospholipase C, while OX$_2$R also couples G$_i$/G$_o$ and inhibit cAMP production via inhibition of adenylate cyclase. In non-neuronal cells OX$_2$R is capable of extracellular signal-regulated kinase activation via G$_s$, G$_q$, and G$_i$. *See Orexins A & B; Almorexin 1.* Callander, Olorunda, Monna, *et al.* (2013) *Front. Neurosci.* **7**, 230.

IRAK1/4 Inhibitor I

This protein kinase inhibitor (FW = 395.41 g/mol; CAS 509093-47-4; Solubility: 20 mM in DMSO), also named *N*-[1-[2-(4-Morpholinyl)ethyl]-1*H*-benzimidazol-2-yl]-3-nitrobenzamide, targets interleukin-1 receptor-associated kinases IRAK-1 (IC$_{50}$ = 0.3 μM) and IRAK-4 (IC$_{50}$ = 0.2 μM), serine/threonine kinases that promote protein phosphorylation, ubiquitination, and degradation. Targeted inhibition of IRAK1 and IRAK4, either with shRNA or with a pharmacological IRAK1/4 inhibitor, dramatically impeded proliferation of T acute lymphoblastic leukemia (T-ALL) cells isolated from patients and T-ALL cells in a murine leukemia model; however, IRAK1/4 inhibition had little effect on cell death (2). **1.** Powers, Li, Jaen, *et al.* (2006) *Bioorg. Med. Chem. Lett.* **16**, 2842. **2.** Li, Younger, *et al.* (2015) *J. Clin. Invest.* **125**, 1081.

Irbesartan

This oral antihypertensive (FW = 428.53 g/mol; CAS 138402-11-6; Soluble to 100 mM in DMSO), also named 2-butyl-3-({4-[2-(2*H*-1,2,3,4-tetrazol-5-yl)phenyl]phenyl}methyl)-1,3-diazaspiro[4.4]non-1-en-4-one and marketed under the trade names Aprovel®, Karvea®, and Avapro®, is a nonpeptide angiotensin II receptor antagonist (IC$_{50}$ = 1.2 nM), reportedly providing a blockade of the actions of angiotensin II other than inhibition of renin or angiotensin-converting enzyme (ACE). (*See also Candesartan; Eprosartan; Losartan; Olmesartan; Telmisartan; Valsartan*) **Key Pharmacological Parameters:** Bioavailability = 7 %; Food Effect? NO; Drug $t_{1/2}$ = 11–15 hours; Metabolite $t_{1/2}$ = not determined; Drug's Protein Binding = 90-95 %; Metabolite's Protein Binding = – %; Route of Elimination = 33 % Renal and 66 % Hepatic. **1.** Massie, Carson, McMurray, *et al.* (2008) *New Engl. J. Med.* **359**, 2456.

Irinotecan

This antineoplastic pro-drug (FW$_{free-base}$ = 586.69 g/mol), also known as CPT-11 and 7-ethyl-10-[4-(1-piperidino)-1-piperidinocarbonyloxycampto-hecin, is a water-soluble analogue of camptothecin and inhibits DNA topoisomerase I. Cancer cell deaths are reportedly due to DNA strand breaks caused by the formation of cleavable complexes. Irinotecan is deesterified *in vivo* to 7-ethyl-10-hydroxycamptothecin (SN-38). **Key Pharmacokinetic Parameters:** *See* Appendix II in Goodman & Gilman's *THE PHARMACOLOGICAL BASIS OF THERAPEUTICS*, 12th Edition (Brunton, Chabner & Knollmann, eds.) McGraw-Hill Medical, New York (2011). **1.** Chabot (1997) *Clin. Pharmacokinet.* **33**, 245. **2.** Kawato, Aonuma, Hirota, Kuga & Sato (1991) *Cancer Res.* **51**, 4187. **3.** Iyer & Ratain (1998) *Cancer Chemother. Pharmacol.* **42** Suppl., S31.

Isatin

This orange-colored indole derivative (FW = 147.13 g/mol; M.P. = 203.5°C), also known as indole-2,3-dione, is formed in the metabolism of many indole-containing compounds and has been used as a reagent in the detection of amino acids, mercaptans, cuprous ions, and thiophenes. It is also an intermediate in the chemical synthesis of indigo and other dyes. **Target(s):** acetylcholinesterase (1); *N*-acetylindoxyl oxidase, weakly inhibited (2); acid phosphatase (3); alkaline phosphatase (4-10); amine oxidase, *or* monoamine oxidase (1,11-13); aryldialkylphosphatase, *or* paraoxonase (14); arylesterase (14); glucokinase (15); D-glucose transport, Na$^+$-dependent (16-18); guanylate cyclase, *or* guanylyl cyclase (19); guanylate cyclase-coupled atrial natriuretic peptide receptors (20); hyaluronidase (21); lysine transport, Na$^+$-dependent (22); Na$^+$/K$^+$-exchanging ATPase (1); sucrase (23,24); xanthine oxidase (25,26). **1.** Kumar, Bansal & Mahmood (1994) *Indian J. Med. Res.* **100**, 246. **2.** Beevers & French (1954) *Arch. Biochem. Biophys.* **50**, 427. **3.** Singh, Sharma, Sareen & Sohal (1977) *Enzyme* **22**, 256. **4.** Yasuda, Ihjima, Koshikawa & Watanabe (1957) *Okajimas Folia Anat. Jpn.* **29**, 317. **5.** Kumar, Nagpaul, Singh, Bansal & Sharma (1978) *Experientia* **34**, 434. **6.** Singh, Kumar, Bansal, Nagpaul & Sharma (1978) *Enzyme* **23**, 22. **7.** Singh, Kumar, Bansal, Nagpaul & Sharma (1977) *Indian J. Biochem. Biophys.* **14**, 88. **8.** Kumar (1979) *Enzyme* **24**, 152. **9.** Singh, Sharma, Nagpaul, Bansal & Kumar (1979) *Enzyme* **24**, 67. **10.** E. Sarciron, Hamoud, Azzar & Petavy (1991) *Comp. Biochem. Physiol. B* **100**, 253. **11.** Glover, Halket, Watkins, Clow, Goodwin & Sandler (1988) *J. Neurochem.* **51**, 656. **12.** Sandler, Glover, Clow & Elsworth (1985) *Prog. Clin. Biol. Res.* **192**, 359. **13.** Medvedev, Ivanov, Kamyshanskaya, *et al.* (1995) *Biochem. Mol. Biol. Int.* **36**, 113. **14.** Billecke, Draganov, Counsell, *et al.* (2000) *Drug Metab. Dispos.* **28**, 1335. **15.** Lenzen, Brand & Panten (1988) *Brit. J. Pharmacol.* **95**, 851. **16.** Nagpal, Wali & Mahmood (1983) *Biochem. Med.* **29**, 46. **17.** Nagpal, Wali, Singh, *et al.* (1985) *Biochem. Med.* **34**, 207. **18.** Patil, Prakash & Hegde (1985) *Indian J. Biochem. Biophys.* **22**, 249. **19.** Medvedev, Sandler & Glover (1999) *Eur. J. Pharmacol.* **384**, 239. **20.** Glover, Medvedev & Sandler (1995) *Life Sci.* **57**, 2073. **21.** Kumar, Trehan, Sareen & Dani (1978) *Indian J. Med. Res.* **67**, 73. **22.** Gargari, Bansal, Mahmood & Mahmood (1996) *Indian J. Biochem. Biophys.* **33**, 519. **23.** Mousavi, Bansal, Singh & Mahmood (1994) *Indian J. Exp. Biol.* **32**, 612. **24.** Gargari, Siddiqui, Bansal, Singh & Mahmood (1994) *Indian J. Biochem. Biophys.* **31**, 191. **25.** Susheela, Singh, Dani, Amma & Sareen (1969) *Enzymologia* **37**, 325. **26.** De Renzo (1956) *Adv. Enzymol. Relat. Subj. Biochem.* **17**, 293.

Isavuconazole & Isavuconazonium sulfate

Isavuconazonium sulfate

Isavuconazole

This potent oral/intravenous triazole-class antifungal (FW = 814.83 g/mol; CAS 946075-13-4), also named BAL8557, Cresemba®, and [2-[1-[1-[(2R,3R)-3-[4-(4-cyanophenyl)-1,3-thiazol-2-yl]-2-(2,5-difluorophenyl)-2-hydroxybutyl]-1,2,4-triazol-4-ium-4-yl]ethoxycarbonylmethylamino] pyridin-3-yl]methyl 2-(methylamino)acetate, is a prodrug that is metabolized to the active drug isavuconazole (FW = 437.47 g/mol; CAS 241479-67-4; IUPAC: 4-{2-[(1R,2R)-(2,5-difluorophenyl)-2-hydroxy-1-methyl-3-(1H-1,2,4-triazol-1-yl)propyl]-1,3-thiazol-4-yl}benzonitrile), the latter inhibiting lanosterol 14 α-demethylase, which converts lanosterol into ergosterol. Depletion of ergosterol in the fungal cell membrane leads to a build-up of lanosterol, compromising fungal membrane integrity. Selectivity toward fungi is achieved, because mammalian cells are resistant to demethylation inhibition by azoles. BAL4815 shows broad-spectrum of activity against *Aspergillus spp.*, *Candida spp.*, *Cryptococcus spp.*, Mucorales (Order: *Zygomycete*), black yeasts and filamentous relatives. It is likewise cytotoxic against true pathogenic fungi, including *Histoplasma capsulatum* and *Blastomyces dermatitidis*. See (1) for a discussion of the pharmacodynamics of isavuconazole and its prodrug in an *Aspergillus fumigatus* mouse infection model. **1.** Seyedmousavi, Brüggemann, Meis, *et al.* (2015) *Antimicrob. Agents Chemother.* **59**, 2855.

3-Isoadenosine 5'-Monophosphate

This AMP analogue (FW = 347.22 g/mol; CAS 2862-19-3; Symbol: piA) has the ribofuranosyl ring linked to the N^3-atom of the pyridine ring (instead of the imidazole ring) of the purine. Iso-AMP is a strong inhibitor of AMP deaminase (1-3). It is also an activator of threonine dehydratase, acting in place of AMP. **1.** Zielke & Suelter (1971) *The Enzymes*, 3rd ed., **4**, 47. **2.** Setlow & Lowenstein (1968) *J. Biol. Chem.* **243**, 3409 and 6216. **3.** Zielke & Suelter (1969) *Fed. Proc.* **28**, 728.

Isoascorbate

This ascorbic acid epimer (FW = 176.13 g/mol), also known as D-isoascorbate, can substitute for ascorbate in a number of enzyme-catalyzed reactions (*e.g.*, procollagen-proline 4 monooxygenase) and is frequently used as an antioxidant. Isoascorbate exhibits about 5% of vitamin equvalence of L-ascorbate. **Target(s):** alkaline phosphatase (1); catechol oxidase (2,3); dihydrodiol dehydrogenase (4,5); hyaluronate lyase (6); hyaluronoglucosaminidase, hyaluronidase (6); laccase (7); lactoylglutathione lyase, *or* glyoxalase I (8); polyphenol oxidase (2,3); tyrosine monooxygenase (9). **1.** Miggiano, Mordente, Martorana, Meucci & Castelli (1984) *Biochim. Biophys. Acta* **789**, 343. **2.** Motoda (1979) *J. Ferment. Technol.* **57**, 79. **3.** Yamamoto, Yoshitama & Teramoto (2002) *Plant Biotechnol.* **19**, 95. **4.** Shinoda, Hara, Nakayama, Deyashiki & Yamaguchi (1992) *J. Biochem.* **112**, 840. **5.** Hara, Shinoda, Kanazu, *et al.* (1991) *Biochem. J.* **275**, 121. **6.** Okorukwu & Vercruysse (2003) *J. Enzyme Inhib. Med. Chem.* **18**, 377. **7.** Ishigami, Hirose & Yamada (1988) *J. Gen. Appl. Microbiol.* **34**, 401. **8.** Douglas & Nadvi (1979) *FEBS Lett.* **106**, 393. **9.** Wilgus & Roskoski (1988) *J. Neurochem.* **51**, 1232.

Isobongkrekic Acid

This geometric isomer of bongkrekic acid (FW_{free-acid} = 486.61 g/mol; CAS 60132-21-0), from *Pseudomonas cocovenenans*, can be generated by exposing bongkrekic acid to alkaline pH. It is an uncompetitive inhibitor of mitochondrial adenine nucleotide transporter, *or* ADP/ATP carrier. (***See also*** *Bongkrekate and Bongkrekic Acid*) **Target(s): 1.** Vignais, Block, Boulay, Brandolin & Lauquin (1982) in *Membranes and Transport*, (Martonosi, ed.) vol. 1, 405, Plenum Press, New York. **2.** Fiore, Trezeguet, Le Saux, *et al.* (1998) *Biochimie* **80**, 137. **3.** Lauquin, Duplaa, Klein, Rousseau & Vignais (1976) *Biochemistry* **15**, 2323.

Isobutyl Alcohol

This polar organic solvent (FW = 74.12 g/mol; CAS 78-83-1), also known as 2-methyl-1-propanol and isobutanol, is a flammable liquid (M.P = – 108°C; B.P. = 108°C) with a dipole moment of 1.79 D at 25°C and a dielectric constant of 16.68 at 20°C. This solvent is often used in organic synthesis, extractions, chromatography, spectrophotometry, and in toxicology investigations. **Target(s):** aryldialkyl-phosphatase, *or* paraoxonase, *or* arylesterase (2); arylesterase (2); carbonic anhydrase (3); glucan 1,3-α-glucosidase, *or* glucosidase II, *or* mannosyl oligosaccharide glucosidase II (4); hydroxynitrilase, *or* hydroxynitrile lyase (5); pepsin (6); pyroglutamyl peptidase 1 (7); pyrophosphatase (8); retinol isomerase (9); thermolysin (10). **1.** Gopalan, Glew, Libell & DePetro (1989) *J. Biol. Chem.* **264**, 15418. **2.** Debord, Dantoine, Bollinger, *et al.* (1998) *Chem. Biol. Interact.* **113**, 105. **3.** Pocker & Dickerson (1968) *Biochemistry* **7**, 1995. **4.** Santa-Cecilia, Alonso & Calvo (1991) *Biol. Chem.* **372**, 373. **5.** Chueskul & Chulavatnatol (1996) *Arch. Biochem. Biophys.* **334**, 401. **6.** Tang (1965) *J. Biol. Chem.* **240**, 3810. **7.** Exterkate (1979) *FEMS Microbiol. Lett.* **5**, 111. **8.** Balocco (1962) *Enzymologia* **24**, 275. **9.** Stecher & Palczewski (2000) *Meth. Enzymol.* **316**, 330. **10.** Muta & K. Inouye (2002) *J. Biochem.* **132**, 945.

Isobutylamine

This toxic primary amine (FW$_{free-base}$ = 73.14 g/mol; CAS 78-81-9), also known as 2-methyl-1-propylamine and 1-amino-2 methylpropane, will cause skin blistering and headaches when inhaled. The free base is a liquid (M.P. = –85°C; B.P.= 68-69°C) at room temperature and is miscible with water. **Target(s):** δ-aminolevulinate synthase (1); glutamate:ethylamine ligase, *or* theanine synthetase (2); lysyl endopeptidase (3); spermidine synthase (4); valine decarboxylase, *or* leucine decarboxylase (5,6). **1.** Jordan & Shemin (1972) *The Enzymes*, 3rd ed., **7**, 339. **2.** Sasaoka, Kito & Onishi (1965) *Agric. Biol. Chem.* **29**, 984. **3.** Sakiyama & Masaki (1994) *Meth. Enzymol.* **244**, 126. **4.** Shirahata, Morohohi, Fukai, Akatsu & Samejima (1991) *Biochem. Pharmacol.* **41**, 205. **5.** Webb (1966) *Enzyme and Metabolic Inhibitors*, vol. **2**, p. 352, Academic Press, New York. **6.** Sutton & King (1962) *Arch. Biochem. Biophys.* **96**, 360.

N-{4-[N-(1-Isobutyl-1H-imidazol-5-yl)methylamino]-2-phenylbenzoyl}-L-methionine

This imidazole-containing peptidomimetic agent (FW = 479.63 g/mol), also called FTI-2239, inhibits mouse protein farnesyltransferase (IC$_{50}$ = 36 nM). It is also highly selective for PFTase over PGGTase-I both *in vitro* and in intact cells. **1.** Ohkanda, Strickland, Blaskovich, *et al.* (2006) *Org. Biomol. Chem.* **4**, 482.

3-Isobutyl-8-methoxymethyl-1-methylxanthine

This purine (FW = 266.30 g/mol; *Abbreviation*: MIMX), also named 8-methoxymethyl-1-methyl-3-(2-methylpropyl)xanthine, inhibits phosphodiesterase 1 (IC$_{50}$ = 4 µM). **Target(s):** 3′,5′-cyclic-GMP phosphodiesterase, *or* cGMP phosphodiesterase (1,2); 3′,5′ cyclic-nucleotide phosphodiesterase (2-8); phosphodiesterase 1 (2-5,7,8); phosphodiesterase 2 (2); phosphodiesterase 5 (1,2). **1.** Thomas, Francis & Corbin (1990) *J. Biol. Chem.* **265**, 14964. **2.** O'Grady, Jiang, Maniak, *et al.* (2002) *J. Membr. Biol.* **185**, 137. **3.** Ahn, Crim, Romano, Sybertz & Pitts (1989) *Biochem. Pharmacol.* **38**, 3331. **4.** Schneller, Ibay, Martinson & Wells (1986) *J. Med. Chem.* **29**, 972. **5.** Haddad, Land, Tarnow-Mordi, *et al.* (2002) *J. Pharmacol. Exp. Ther.* **300**, 567. **6.** Yu, Wolda, Frazier, *et al.* (1997) *Cell. Signal.* **9**, 519. **7.** Clapham & Wilderspin (1996) *Biochem. Soc. Trans.* **24**, 320. **8.** Sairenji, Satoh & Sugiya (2006) *Biomed. Res.* **27**, 37.

3-Isobutyl-1-methylxanthine

This purine derivative (FW = 222.25 g/mol; CAS 28822-58-4; M.P. = 200-201°C; Solubility = 75 mM in DMSO; *Abbreviations*: IBMX and MIX), also known as 1-methyl-3-isobutylxanthine and 3-isobutyl-1-methyl-2,6(1H,3H)-purinedione, is a nonselective cyclic-nucleotide phosphodiesterase inhibitor, with IC$_{50}$ values of 2–50 µM. IBMX does not inhibit PDE-8 or PDE-9. IBMX also acts as a nonselective adenosine receptor antagonist. **Target(s):** α$_2$-adrenoceptor-mediated 5-hydroxytryptamine release (1); Ca^{2+}-channels, L type (2); 3′,5′-cyclic-GMP phosphodiesterase, *or* cGMP phosphodiesterase (3-15); 3′,5′-cyclic nucleotide phosphodiesterase (4-7,9,14,16-24); cytokinin 7β-glucosyltransferase (29); guanylate cyclase (25); NAD$^+$ ADP-ribosyltransferase (poly(ADP-ribose) polymerase; IC$_{50}$ = 3.1 mM (28); phosphodiesterase 1 (7); phosphodiesterase 2 (7); phosphodiesterase 3 (7); phosphodiesterase 4 (7,14); phosphodiesterase 5 (3,4,6-8,12,14,15); phosphodiesterase 6 (8); phosphodiesterase 7 (26); phosphodiesterase 9 (27); phosphodiesterase 10 (24); phosphodiesterase 11, slightly inhibited (17,18). **1.** Freitag, Wessler & Racke (1998) *Eur. J. Pharmacol.* **354**, 67. **2.** Fearon, Palmer, Balmforth, *et al.* (1998) *Eur. J. Pharmacol.* **342**, 353. **3.** Liebman & Evanczuk (1982) *Meth. Enzymol.* **81**, 532. **4.** Wells & Miller (1988) *Meth. Enzymol.* **159**, 489. **5.** Manganiello, Degerman & Elks (1988) *Meth. Enzymol.* **159**, 504. **6.** Hamet & Tremblay (1988) *Meth. Enzymol.* **159**, 710. **7.** Coste & Grondin (1995) *Biochem. Pharmacol.* **50**, 1577. **8.** Zhang, Kuvelkar, Wu, *et al.* (2004) *Biochem. Pharmacol.* **68**, 867. **9.** Methven, Lemon & Bhoola (1980) *Biochem. J.* **186**, 491. **10.** Thomas, Francis & Corbin (1990) *J. Biol. Chem.* **265**, 14964. **11.** Srivastava, Fox & Hurwitz (1995) *Biochem. J.* **308**, 653. **12.** Wang, Wu, Myers, *et al.* (2001) *Life Sci.* **68**, 1977. **13.** Kuwayama, Snippe, Derks, Roelofs & Van Haastert (2001) *Biochem. J.* **353**, 635. **14.** Huai, Liu, Francis, Corbin & Ke (2004) *J. Biol. Chem.* **279**, 13095. **15.** Su & Vacquier (2006) *Mol. Biol. Cell* **17**, 114. **16.** Harrison, Beier, Martins & Beavo (1988) *Meth. Enzymol.* **159**, 685. **17.** Sharma (2003) *Ind. J. Biochem. Biophys.* **40**, 77. **18.** Grant & Colman (1984) *Biochemistry* **23**, 1801. **19.** Yu, Wolda, Frazier, *et al.* (1997) *Cell. Signal.* **9**, 519. **20.** Yuasa, Ohgaru, Asahina & Omori (2001) *Eur. J. Biochem.* **268**, 4440. **21.** D'Andrea, Qiu, Haynes-Johnson, *et al.* (2005) *J. Histochem. Cytochem.* **53**, 895. **22.** D'Angelo, Garzia, Andre, *et al.* (2004) *Cancer Cell* **5**, 137. **23.** Kunz, Oberholzer & Seebeck (2005) *FEBS J.* **272**, 6412. **24.** Fujishige, Kotera & Omori (1999) *Eur. J. Biochem.* **266**, 1118. **25.** Gorczyca (2000) *Meth. Enzymol.* **315**, 689. **26.** Sudo, Tachibana, Toga, *et al.* (2000) *Biochem. Pharmacol.* **59**, 347. **27.** Wang, Wu, Egan & Billah (2003) *Gene* **314**, 15. **28.** Rankin, Jacobson, Benjamin, Moss & Jacobson (1989) *J. Biol. Chem.* **264**, 4312. **29.** Hou, Lim, Higgins & Bowles (2004) *J. Biol. Chem.* **279**, 47822.

Isobutyryl-CoA

This thiolester (FW = 855.65 g/mol; CAS 15621-60-0), also known as 2-methylpropanoyl-CoA, is an intermediate in the degradation of valine and a substrate of the reactions catalyzed by isobutyryl CoA mutase and dihydrolipoyllysine-residue (2-methylpropanoyl)transferase and an alternative substrate for phloroisovalero-phenone synthase. It is also a product of the reactions catalyzed by 2 oxoisovalerate dehydrogenase (acylating) and 3-methyl-2-oxobutanoate dehydrogenase (ferredoxin). **Target(s):** [branched-chain α-keto-acid dehydrogenase] kinase, *or* [3-methyl-2 oxobutanoate dehydrogenase (acetyl-transferring)] kinase (1,2); glycine cleavage complex, *or* glucine synthase (3); 3-hydroxyisobutyryl-CoA hydrolase (4,5); β-ketoacyl-[acyl-carrier-protein] synthase III (6); 3-

methylcrotonoyl-CoA carboxylase (7); [3-methyl-2-oxobutanoate dehydrogenase, *or* 2-methylpropanoyl-transferring)] phosphatase, *or* [branched-chain-α-keto-acid dehydrogenase] phosphatase (8-10); phosphoprotein phosphatase (10,11); pyruvate dehydrogenase (12). **1.** Paxton (1988) *Meth. Enzymol.* **166**, 313. **2.** Paxton & Harris (1984) *Arch. Biochem. Biophys.* **231**, 48. **3.** Kolvraa (1979) *Pediatr. Res.* **13**, 889. **4.** Shimomura, Murakami, Nakai, *et al.* (2000) *Meth. Enzymol.* **324**, 229. **5.** Shimomura, Murakami, Fujitsuka, *et al.* (1994) *J. Biol. Chem.* **269**, 14248. **6.** Smirnova & Reynolds (2001) *J. Bacteriol.* **183**, 2335. **7.** Diez, Wurtele & Nikolau (1994) *Arch. Biochem. Biophys.* **310**, 64. **8.** Damuni & Reed (1987) *J. Biol. Chem.* **262**, 5129. **9.** Damuni, Merryfield, Humphreys & Reed (1984) *Proc. Natl. Acad. Sci. U.S.A.* **81**, 4335. **10.** Reed & Damuni (1987) *Adv. Protein Phosphatases* **4**, 59. **11.** Damuni & L. J. Reed (1987) *J. Biol. Chem.* **262**, 5133. **12.** Martin-Requero, Corkey, Cerdan, *et al.* (1983) *J. Biol. Chem.* **258**, 3673.

Isocaproate (Isocaproic Acid)

This branched chain organic acid (FW$_{\text{free-acid}}$ = 116.16 g/mol; CAS 646-07-1; B.P. = 199-201°C), also known as 4-methylvaleric acid and 4-methypentanoic acid, inhibits aminoacylase (1); branched-chain-amino-acid transaminase (2); leucyl aminopeptidase (3-5); ornithine aminotransferase (6). **1.** Park & Fox (1960) *J. Biol. Chem.* **235**, 3193. **2.** Koide, Honma & Shimomura (1977) *Agric. Biol. Chem.* **41**, 1171. **3.** Smith & Hill (1960) *The Enzymes*, 2nd ed., **4**, 37. **4.** DeLange & Smith (1971) *The Enzymes*, 3rd ed., **3**, 81. **5.** Ledeme, Hennon, Vincent-Fiquet & Plaquet (1981) *Biochim. Biophys. Acta* **660**, 262. **6.** Kalita, Kerman & Strecker (1976) *Biochim. Biophys. Acta* **429**, 780.

Isocitrate (Isocitric Acid)

This monohydroxy tricarboxylic acid (FW$_{\text{free-acid}}$ = 192.13 g/mol; CAS 320-77-4; pK_a values of 3.29, 4.71, and 6.40 at 25°C), also known as (1R,2S)-1-hydroxypropane-1,2,3-tricarboxylic acid and formerly called *threo*-DS-isocitric acid, are key intermediates in the tricarboxylic acid and glyoxylate cycles. **Target(s):** aconitate decarboxylase (1); aspartate aminotransferase (30); ATP:citrate lyase, *or* ATP:citrate synthase (31); glucose-1,6-bisphosphate synthase (2); glucose-1-phosphate adenylyltransferase (3); glutamate decarboxylase, inhibited by *dl*-isocitrate (4); [isocitrate dehydrogenase (NADP+)] kinase (5-7); L-lactate dehydrogenase (8); 2-methylcitrate dehydratase (9); 2 methylisocitrate dehydratase (10); oxaloacetate tautomerase, inhibited by *dl*-isocitrate (11); phosphoenolpyruvate carboxykinase (GTP), weakly inhibited by *dl*-isocitrate (12); phosphoenolpyruvate carboxylase (13,14); phosphoenolpyruvate mutase (15); 6-phosphofructokinase (16-20); phosphoglucomutase, weakly inhibited (21); procollagen-proline 4-dioxygenase (33); pyruvate kinase (22-28); thiosulfate sulfurtransferase, rhodanese; inhibited by *dl*-isocitrate (29); trimethyllysine dioxygenase (32). **1.** Bentley & Thiessen (1957) *J. Biol. Chem.* **226**, 703. **2.** Rose, Warms & Wong (1977) *J. Biol. Chem.* **252**, 4262. **3.** Preiss, Crawford, Downey, Lammel & Greenberg (1976) *J. Bacteriol.* **127**, 193. **4.** Gerig & Kwock (1979) *FEBS Lett.* **105**, 155. **5.** Nimmo (1984) *Trends Biochem. Sci.* **9**, 475. **6.** Nimmo & Nimmo (1984) *Eur. J. Biochem.* **141**, 409. **7.** Miller, Chen, Karschnia, *et al.* (2000) *J. Biol. Chem.* **275**, 833. **8.** Yoshida & Freese (1975) *Meth. Enzymol.* **41**, 304. **9.** Aoki & Tabuchi (1981) *Agric. Biol. Chem.* **45**, 2831. **10.** Tabuchi, Umetsu, Aoki & Uchiyama (1995) *Biosci. Biotechnol. Biochem.* **59**, 2013. **11.** Belikova, Kotlyar & Vinogradov (1989) *FEBS Lett.* **246**, 17. **12.** Hebda & Nowak (1982) *J. Biol. Chem.* **257**, 5503. **13.** Moraes & Plaxton (2000) *Eur. J. Biochem.* **267**, 4465. **14.** Owttrim & Colman (1986) *J. Bacteriol.* **168**, 207. **15.** Seidel & Knowles (1994) *Biochemistry* **33**, 5641. **16.** Sols & Salas (1966) *Meth. Enzymol.* **9**, 436. **17.** Bloxham & Lardy (1973) *The Enzymes*, 3rd ed., **8**, 239. **18.** Webb (1966) *Enzyme and Metabolic Inhibitors*, vol. **2**, p. 385, Academic Press, New York. **19.** Passonneau & Lowry (1963) *Biochem. Biophys. Res. Commun.* **13**, 372. **20.** Van Praag, Zehavi & Goren (1999)

Biochem. Mol. Biol. Int. **47**, 749. **21.** Popova, Matasova & Lapot'ko (1998) *Biochemistry (Moscow)* **63**, 697. **22.** Singh, Malhotra & Singh (2000) *Indian J. Biochem. Biophys.* **37**, 51. **23.** Singh, Malhotra & Singh (1998) *Indian J. Biochem. Biophys.* **35**, 346. **24.** Wu & Turpin (1992) *J. Phycol.* **28**, 472. **25.** Lin, Turpin & Plaxton (1989) *Arch. Biochem. Biophys.* **269**, 228. **26.** Turner, Knowles & Plaxton (2005) *Planta* **222**, 1051. **27.** Andre, Froehlich, Moll & Benning (2007) *Plant Cell* **19**, 2006. **28.** Evans & Ratledge (1985) *Can. J. Microbiol.* **31**, 479. **29.** Oi (1975) *J. Biochem.* **78**, 825. **30.** Rakhmanova & Popova (2006) *Biochemistry (Moscow)* **71**, 211. **31.** Antranikian, Herzberg & Gottschalk (1982) *J. Bacteriol.* **152**, 1284. **32.** Henderson, Nelson & Henderson (1982) *Fed. Proc.* **41**, 2843. **33.** Tuderman, Myllylä & Kivirikko (1977) *Eur. J. Biochem.* **80**, 341.

threo-Ls-Isocitrate (*threo*-Ls-Isocitric Acid)

This citrate stereoisomer (FW$_{\text{free-acid}}$ = 192.13 g/mol) inhibits aconitase (1); fumarase (2); isocitrate dehydrogenase, NADP$^+$-requiring (3); 2-methylisocitrate dehydratase (4); phosphoenolpyruvate mutase (5). **1.** Glusker (1971) *The Enzymes*, 3rd ed., **5**, 413. **2.** Hill & Teipel (1971) *The Enzymes*, 3rd. ed., **5**, 539. **3.** Seelig & Colman (1978) *Arch. Biochem. Biophys.* **188**, 394. **4.** Tabuchi, Umetsu, Aoki & Uchiyama (1995) *Biosci. Biotechnol. Biochem.* **59**, 2013. **5.** Seidel & Knowles (1994) *Biochemistry* **33**, 5641.

Isocytosine

This structural isomer of cytosine (FW = 111.10 g/mol; CAS 108-53-2), also known as 2-amino-4-hydroxypyrimidine, inhibits cytosine deaminase (1), hypoxanthine(guanine) phosphoribosyltransferase (2,3), tRNA-guanine transglycosylase (4), and xanthine phosphoribosyltransferase (2). **1.** Kream & Chargaff (1952) *J. Amer. Chem. Soc.* **74**, 5157. **2.** Naguib, Iltzsch, el Kouni, Panzica & el Kouni (1995) *Biochem. Pharmacol.* **50**, 1685. **3.** Miller, Ramsey, Krenitsky & Elion (1972) *Biochemistry* **11**, 4723. **4.** Farkas, Jacobson & Katze (1984) *Biochim. Biophys. Acta* **781**, 64.

Isofagomine

This piperidine (FW$_{\text{free-base}}$ = 147.17 g/mol; CAS 161302-93-8), also known as (3R,4R,5R)-5-(hydroxymethyl)piperidine-3,4-diol, inhibits liver glycogen phosphorylase (IC$_{50}$ = 0.7 μM) and is a slow-binding inhibitor of almond β-glucosidase. **Target(s):** glucoamylase (1); α-glucosidase (1); β-glucosidase (1-4); glycogen phosphorylase (5,6); oligo-1,6-glucosidase, *or* isomaltase (1,2); α,α-trehalose phosphorylase, *or* configuration-retaining (7). **1.** Dong, Jespersen, Bols, Skrydstrup & Sierks (1996) *Biochemistry* **35**, 2788. **2.** Lohse, Hardlei, Jensen, Plesner & Bols (2000) *Biochem. J.* **349**, 211. **3.** Bulow, Plesner & Bols (2001) *Biochim. Biophys. Acta* **1545**, 207. **4.** Zechel, Boraston, Gloster, *et al.* (2003) *J. Amer. Chem. Soc.* **125**, 14313. **5.** Jakobsen, Lundbeck, Kristiansen, *et al.* (2001) *Bioorg. Med. Chem.* **9**, 733. **6.** S. Waagepetersen, N. Westergaard & A. Schousboe (2000) *Neurochem. Int.* **36**, 435. **7.** Nidetzky & C. Eis (2001) *Biochem. J.* **360**, 727.

Isoferulate (Isoferulic Acid)

This α,β-unsaturated acid (FW$_{free-acid}$ = 194.19 g/mol) inhibits diphospho-mevalonate decarboxylase (1); phosphomevalonate kinase (1), tyrosinase, *or* monophenol monooxygenase, weakly inhibited (2), xanthine oxidase (3,4). **1**. Shama Bhat & Ramasarma (1979) *Biochem. J.* **181**, 143. **2**. Karioti, Protopappa, Magoulas & Skaltsa (2007) *Bioorg. Med. Chem.* **15**, 2708. **3**. Chan, Wen & Chiang (1995) *Anticancer Res.* **15**, 703. **4**. Chang, Lee, Chen, *et al.* (2007) *Free Rad. Biol. Med.* **43**, 1541.

Isoflurane

This anesthetic (FW = 184.49 g/mol; CAS 26675-46-7; Nonflammable Liquid; B.P. = 48.5°C; Miscible with many organic solvents and usually co-administered by mask with O_2 or air), also known as 2-chloro-2-(difluoromethoxy)-1,1,1-trifluoroethane, appears to depress CA1 synapses at presynaptic sites downstream from sodium ion channels, as indicated by the increased facilitation that accompanies isoflurane depression of excitatory postsynaptic potential (1-3). The latter refers to the temporary depolarization of postsynaptic membrane potential caused by the flow of sodium ions into postsynaptic cells, after the opening of ligand-gated ion channels. **Targets:** Volatile anesthetics are known to depress glutamatergic synaptic transmission, although the molecular mechanisms are unclear. At clinically relevant concentrations, isoflurane inhibited the L-, N- and P/Q-types, but not toxin-resistant, Ca^{2+} channel currents. The respective IC$_{50}$ values for the L-, N- and P/Q-type neuronal currents were 0.7%, 1.3% and 3.0%, corresponding to 0.35, 0.68 and 1.46 mM in the aqueous phase) (3). In the nervous system, the precise physiological roles of these calcium currents remain to be elucidated, but N- and P/Q-types have important roles in synaptic transmission at the neuromuscular junction. Inhibition of such channels in the CNS may lead to loss of consciousness, loss of sensation, as well as circulatory and respiratory depression. Isoflurane also inhibits endothelium-dependent relaxation, apparently by inhibiting the nitric oxide-guanylyl cyclase signaling pathway. While isoflurane inhibits brain NO synthase *in vitro*, it has no effect at clinically relevant concentrations (4). Plasma membranes Ca^{2+}-ATPase is significantly inhibited, in a dose-related manner, by clinically relevant partial pressures of halothane, isoflurane, xenon, and nitrous oxide (5). **1**. Study (1994) *Anesthesiology* **81**, 104. **2**. Puil, Hutcheon & Reiner (1994) *Neurosci. Lett.* **176**, 63. **3**. Kameyama, Aono & Kitamura (1999) *Brit. J. Anaesth.* **82**, 402. **4**. Tobin, Martin, Breslow & Traystman (1994) *Anesthesiology* **81**, 1264. **5**. Franks, Horn, Janicki & Singh (1995) *Anesthesiology* **82**, 108.

Isoglutamine

L-isoglutamine   D-isoglutamine

This α-amide of glutamic acid (FW = 146.15 g/mol; pK$_a$ values of 3.81 and 7.88), also known as γ-aminoglutaramic acid and 4-amino-4-carbamoylbutanoic acid The L-stereoisomer has been observed in some proteins, whereas the D-isomer is a component of the peptidoglycan of a number of organisms (*e.g.*, *Staphylococcus aureus*). **Target(s):** arginyltransferase (1); glutamate decarboxylase, inhibited by L-stereoisomer, K_i = 95 mM (2). **1**. Soffer (1973) *J. Biol. Chem.* **248**, 2918. **2**. Fonda (1972) *Biochemistry* **11**, 1304.

Isoguanine

This rare purine base (FW = 151.13 g/mol; CAS 3373-53-3) also known as *iso*guanine, 2-hydroxy-6-aminopurine, 2-oxoadenine, and 2-hydroxyadenine, was originally isolated from croton bean. It has also been reported in the nudibranch mollusk *Diaulula sandiegensis*. Isoguanine does not deaminate when treated with nitrous acid. Insertion of isoguanine into DNA increases the rate of mutagenesis. Both polar solvents and the DNA microenvironment dramatically change the intrinsic tautomeric properties of this purine (1). **Target(s):** hypoxanthine(guanine) phosphoribosyl-transferase (2); xanthine phosphoribosyltransferase (2,3). **1**. Blas, Luque & Orozco (2004) *J. Amer. Chem. Soc.* **126**, 154. **2**. Naguib, Iltzsch, el Kouni, Panzica & el Kouni (1995) *Biochem. Pharmacol.* **50**, 1685. **3**. Miller, Adamczyk, Fyfe & Elion (1974) *Arch. Biochem. Biophys.* **165**, 349.

Isoguanosine

This rare nucleoside (FW = 283.24 g/mol), also known as crotonoside and 2-hydroxyadenosine, has been reported in croton bean and the nudibranch mollusk *Diaulula sandiegensis*. Isoguanosine incorporation into DNA increases the rate of mutagenesis. Both polar solvents and the DNA microenvironment dramatically change the intrinsic tautomeric properties of this nucleoside. **Target(s):** adenosine kinase (1); deoxycytidine kinase, weakly inhibited (2). **1**. Long & Parker (2006) *Biochem. Pharmacol.* **71**, 1671. **2**. Krenitsky, Tuttle, Koszalka, *et al.* (1976) *J. Biol. Chem.* **251**, 4055.

L-Isoleucine

This nutritionally essential branched-chain amino acid (FW = 131.18 g/mol; pK$_a$ values of 2.32 (R–COOH) and 9.76 (R–NH$_2$) at 25°C; Specific Rotation = + 39.5° at 25°C at the sodiun D-line; Solubility: 4.12 g/100 mL H$_2$O), also named (2S,3S)-2-amino-3-methylpentanoic acid, is one of two of proteogenic amino acids containing two chiral centers. There are three codons for isoleucine: AUU, AUC, and AUA. With its hydrophobic side-chain, isoleucine residues are usually located in the interior of proteins. They do not favor α-helical structures, but instead appear in β-pleated sheets. Homologous proteins often have isoleucyl residues interchanged with valyl, leucyl, and threonyl residues. Because of the second chiral center, there are two diastereoisomers of D- and L-isoleucine: D- and L-alloisoleucine. L-isoleucine is a physiologic feedback inhibitor of *Escherichia coli* threonine deaminase and aspartokinase. Low isoleucine levels during amino acid starvation result in ppGpp ("Magic Spot") formation and interruption of ribosomal protein synthesis. **Target(s):** acetolactate synthase (1,50-61); aminopeptidase B (2); arginase, weakly inhibited (2-8); argininosuccinate synthetase (9); arylformamidase, weakly inhibited (10); aspartate kinase (11,12); bacterial leucyl aminopeptidase (13); (*R*)-citramalate synthase (48); creatine kinase (14); cytosol nonspecific Dipeptidase (15); dihydrodipicolinate synthase (16); 2-ethylmalate synthase, weakly inhibited (17); glutamine synthetase, weakly inhibited (17,18); γ-glutamyl transpeptidase, weakly inhibited (19,20); homoserine kinase (21-24); 2-isopropylmalate synthase (25); ornithine aminotransferase (44-46); ornithine *N*-benzoyltransferase, when arginine is the acceptor substrate (49); ornithine carbamoyltransferase (62,63); saccharopine dehydrogenase (26); threonine ammonia-lyase, *or* threonine dehydratase, *or* threonine deaminase (33-48); Xaa-Pro dipeptidase, *or* prolidase isozyme I (43). **1**. Barak, Calvo & Schloss (1988) *Meth. Enzymol.* **166**, 455. **2**. Kawata, Takayama, Ninomiya & Makisumi (1980) *J. Biochem.* **88**, 1601.

3. Hunter & Downs (1945) *J. Biol. Chem.* **157**, 427. **4**. Webb (1966) *Enzyme and Metabolic Inhibitors*, vol. **2**, p. 337, Academic Press, New York. **5**. Mohamed, Fahmy, Mohamed & Hamdy (2005) *Comp. Biochem. Physiol. B Biochem. Mol. Biol.* **142**, 308. **6**. Patchett, Daniel & Morgan (1991) *Biochim. Biophys. Acta* **1077**, 291. **7**. Kaysen & Strecker (1973) *Biochem. J.* **133**, 779. **8**. Singh & Singh (1990) *Arch. Int. Physiol. Biochim.* **98**, 411. **9**. Takada, Saheki, Igarashi & Katsunuma (1979) *J. Biochem.* **85**, 1309. **10**. Serrano & Nagayama (1991) *Comp. Biochem. Physiol. B Comp. Biochem.* **99**, 281. **11**. Truffa-Bachi (1973) *The Enzymes*, 3rd ed., **8**, 509. **12**. Cahyanto, Kawasaki, Nagashio, Fujiyama & Seki (2006) *Microbiology* **152**, 105. **13**. Baker, Wilkes, Bayliss & Prescott (1983) *Biochemistry* **22**, 2098. **14**. Pilla, Cardozo, Dornelles, *et al.* (2003) *Int. J. Dev. Neurosci.* **21**, 145. **15**. Wilcox & Fried (1963) *Biochem. J.* **87**, 192. **16**. Wallsgrove & Mazelis (1981) *Phytochemistry* **20**, 2651. **17**. Southern, Parker & Woods (1987) *J. Gen. Microbiol.* **133**, 2437. **18**. Singh & Singh (1992) *Arch. Int. Physiol. Biochim. Biophys.* **100**, 203. **19**. Allison (1985) *Meth. Enzymol.* **113**, 419. **20**. Thompson & Meister (1977) *J. Biol. Chem.* **252**, 6792. **21**. Wormser & Pardee (1958) *Arch. Biochem. Biophys.* **78**, 416. **22**. Burr, Walker, Truffa-Bachi & Cohen (1976) *Eur. J. Biochem.* **62**, 519. **23**. Thoen, Rognes & Aarnes (1978) *Plant Sci. Lett.* **13**, 103. **24**. Baum, Madison & Thompson (1983) *Phytochemistry* **22**, 2409. **25**. Gross (1970) *Meth. Enzymol.* **17A**, 777. **26**. Fujioka & Nakatani (1972) *Eur. J. Biochem.* **25**, 301. **27**. Burns (1971) *Meth. Enzymol.* **17B**, 555. **28**. Hatfield & Umbarger (1971) *Meth. Enzymol.* **17B**, 561. **29**. Datta (1971) *Meth. Enzymol.* **17B**, 566. **30**. Davis & Metzler (1972) *The Enzymes*, 3rd ed., **7**, 33. **31**. Maeba & Sanwal (1966) *Biochemistry* **5**, 525. **32**. Eisenstein (1991) *J. Biol. Chem.* **266**, 5801. **33**. Kagan, Sinelnikova & Kretovich (1969) *Enzymologia* **36**, 335. **34**. Desai, Laub & Anita (1972) *Phytochemistry* **11**, 277. **35**. Wessel, Graciet, Douce & Dumas (2000) *Biochemistry* **39**, 15136. **36**. Laakmann-Ditges & Klemme (1988) *Arch. Microbiol.* **149**, 249. **37**. Bode, Schult & Birnbaum (1986) *J. Basic Microbiol.* **26**, 443. **38**. Cohn & Phillips (1974) *Biochemistry* **13**, 1208. **39**. Dougall (1979) *Phytochemistry* **9**, 959. **40**. Eisenstein (1995) *Arch. Biochem. Biophys.* **316**, 311. **41**. Calhoun, Rimerman & Hatfield (1973) *J. Biol. Chem.* **248**, 3511. **42**. Sharma & Mazumder (1970) *J. Biol. Chem.* **245**, 3008. **43**. Liu, Nakayama, Sagara, *et al.* (2005) *Clin. Biochem.* **38**, 625. **44**. Goldberg, Flescher & Lengy (1979) *Exp. Parasitol.* **47**, 333. **45**. Kalita, Kerman & Strecker (1976) *Biochim. Biophys. Acta* **429**, 780. **46**. Matsuzawa (1974) *J. Biochem.* **75**, 601. **47**. Rabin, Salomon, Bleiweis, Carlin & Ajl (1968) *Biochemistry* **7**, 377. **48**. Xu, Zhang, Guo, *et al.* (2004) *J. Bacteriol.* **186**, 5400. **49**. Seymour, Millburn & Tait (1987) *Biochem. Soc. Trans.* **15**, 1108. **50**. Oda, Nakano & Kitaoka (1982) *J. Gen. Microbiol.* **128**, 1211. **51**. Chong, Chang & Choi (1997) *J. Biochem. Mol. Biol.* **30**, 274. **52**. Bekkaoui, Schorr & Crosby (1993) *Physiol. Plant.* **88**, 475. **53**. Southan & Copeland (1996) *Physiol. Plant.* **98**, 824. **54**. Chin & Trela (1973) *J. Bacteriol.* **114**, 674. **55**. Blatt, Pledger & Umbarger (1972) *Biochem. Biophys. Res. Commun.* **48**, 444. **56**. DeFelice, Lago, Squires & Calvo (1982) *Ann. Microbiol.* **133A**, 251. **57**. Xing & Whitman (1987) *J. Bacteriol.* **169**, 4486. **58**. Arfin & Koziell (1973) *Biochim. Biophys. Acta* **321**, 348. **59**. McDonald, Satyanarayana & Kaplan (1973) *J. Bacteriol.* **114**, 332. **60**. Yang & Kim (1993) *Biochim. Biophys. Acta* **1157**, 178. **61**. Singh, Stidham & Shaner (1988) *J. Chromatogr.* **444**, 251. **62**. Legrain & Stalon (1976) *Eur. J. Biochem.* **63**, 289. **63**. Lusty, Jilka & Nietsch (1979) *J. Biol. Chem.* **254**, 10030.

L-Isoleucinol

This reduced isoleucine derivative (FW = 117.19 g/mol; *Symbol* = Ile-ol) competitively inhibits isoleucine interactions with isoleucyl-tRNA synthetase. Isoleucyl-tRNA synthetase from *Escherichia coli* catalyzes the activation of $[^{18}O_2]$isoleucine by adenosine 5'-$[(R)$-α-$^{17}O]$triphosphate with inversion of configuration at phosphorus. Moreover, isoleucyl-tRNA synthetase does not catalyze positional isotope exchange in adenosine 5'-$[\beta$-$^{18}O]$triphosphate in the absence of isoleucine or in the presence of the competitive inhibitor isoleucinol, which effectively eliminates the possibility of either adenylyl-enzyme or adenosine metaphosphate intermediates being involved. These observations require that isoleucyl-tRNA synthetase catalyzes the activation of isoleucine by associative "in-

line" nucleotidyl transfer. **1**. Lowe, Sproat, Tansley & Cullis (1983) *Biochemistry* **22**, 1229. **2**. Freist & Cramer (1983) *Eur. J. Biochem.* **131**, 65. **3**. Rainey, Hammer-Raber, Kula & Holler (1977) *Eur. J. Biochem.* **78**, 239.

L-Isoleucinol Adenylate

This aminoalkyl adenylate (FW = 326.29 g/mol; *Abbreviation*: Ile-ol-AMP) is formed by the condensation of the amino alcohol analogue of isoleucine and AMP. Ile-ol-AMP is a structural analogue of the aminoacyl adenylate intermediate of the reaction catalyzed by isoleucyl-tRNA synthetase, and, as such, is a strong inhibitor of that enzyme (1-4). Ile-ol-AMP weakly inhibits valyl-tRNA synthetase and has been used to investigate kinetic proofreading mechanisms in the catalysis of aminoacyl-tRNA synthetases (1). **1**. Cassio, Lemoine, Waller, Sandrin & Boissonnas (1967) *Biochemistry* **6**, 827. **2**. Pope, Lapointe, Mensah, *et al.* (1998) *J. Biol. Chem.* **273**, 31680. **3**. Pope, Moore, McVey, *et al.* (1998) *J. Biol. Chem.* **273**, 31691. **4**. Pope, McVey, Fantom & Moore (1998) *J. Biol. Chem.* **273**, 31702.

L-Isoleucyl-L-prolyl-L-isoleucine

This tripeptide (FW = 341.45 g/mol; CAS 90614-48-5), also known as diprotin A and IPI, inhibits dipeptidyl-peptidase IV (*or* dipeptidyl-aminopeptidase IV), K_i = 4.6 μM. This tripeptide is actually an slowly hydrolyzed alternative substrate. **Target(s)**: dipeptidyl-peptidase II (2); dipeptidyl-peptidase IV (1-15); prolyl oligopeptidase, *Pyrococcus furiosus* (6,17); tripeptidyl-peptidase I (18); tripeptidyl-peptidase II, K_i = 1 μM (19); Xaa-Pro dipeptidyl-peptidase, K_i = 2 μM (5). **1**. Hiramatsu, Yamamoto, Kyono, *et al.* (2004) *Biol. Chem.* **385**, 561. **2**. Leiting, Pryor, Wu, *et al.* (2003) *Biochem. J.* **371**, 525. **3**. Umezawa, Aoyagi, Ogawa, *et al.* (1984) *J. Antibiot.* **37**, 422. **4**. Rahfeld, Schierhorn, Hartrodt, Neubert & Heins (1991) *Biochim. Biophys. Acta* **1076**, 314. **5**. Rigolet, Xi, Rety & Chich (2005) *FEBS J.* **272**, 2050. **6**. Jobin, Martinez, Motard, Gottschalk & Grenier (2005) *J. Bacteriol.* **187**, 795. **7**. Malik, Busek, Mares, *et al.* (2003) *Adv. Exp. Med. Biol.* **524**, 95. **8**. De Meester, Vanhoof, Hendriks, *et al.* (1992) *Clin. Chim. Acta* **210**, 23. **9**. Ohkubo, Huang, Ochiai, Takagaki & Kani (1994) *J. Biochem.* **116**, 1182. **10**. Davy, Thomsen, Juliano, *et al.* (2000) *Plant Physiol.* **122**, 425. **11**. Shibuya-Saruta, Kasahara & Hashimoto (1996) *J. Clin. Lab. Anal.* **10**, 435. **12**. Kikuchi, Fukuyama & Epstein (1988) *Arch. Biochem. Biophys.* **266**, 369. **13**. Kudo, Nakamura & Koyama (1985) *J. Biochem.* **97**, 1211. **14**. Cohen, Fruitier-Arnaudin & Piot (2004) *Biochimie* **86**, 31. **15**. Thoma, Löffler, Stihle, *et al.* (2003) *Structure* **11**, 947. **16**. Harwood & Schreier (2001) *Meth. Enzymol.* **330**, 445. **17**. Harwood, Denson, Robinson-Bidle & Schreier (1997) *J. Bacteriol.* **179**, 3613. **18**. Du, Kato, Li, *et al.* (2001) *Biol. Chem.* **382**, 1715. **19**. Ganellin, Bishop, Bambal, *et al.* (2000) *J. Med. Chem.* **43**, 664.

1-(L-Isoleucyl)pyrrolidone

This dipeptide analogue (FW$_{free-base}$ = 184.28 g/mol), also known as L-isoleucine-pyrrolidide, inhibits dipeptidyl peptidases II and IV. **1**. Stöckel-Maschek, Mrestani-Klaus, Stiebitz, Demuth & Neubert (2000) *Biochim. Biophys. Acta* **1479**, 15. **2**. Lambeir, Rea, Fulop, *et al.* (2003) *Adv. Exp. Med. Biol.* **524**, 29. **3**. Stöckel-Maschek, Stiebitz, Born, *et al.* (2000) *Adv. Exp. Med. Biol.* **477**, 117. **4**. Stöckel-Maschek, Stiebitz, Faust, *et al.* (2003)

Adv. Exp. Med. Biol. **524**, 69. **5**. Balzarini, Andersson, Schols, *et al.* (2004) *Int. J. Biochem. Cell Biol.* **36**, 1848.

5'-*O*-[*N*-(L-Isoleucyl)sulfamoyl]adenosine

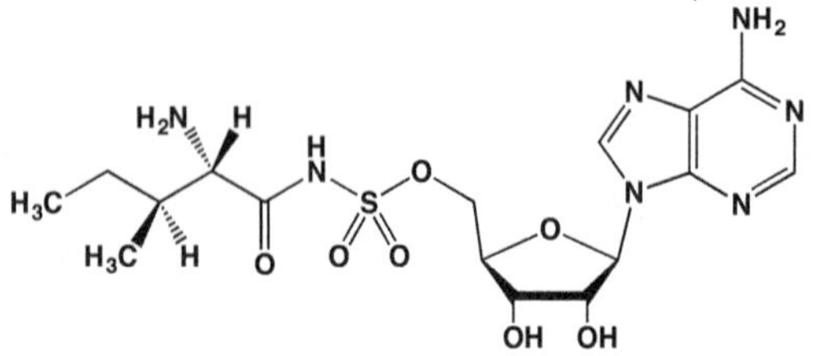

This isoleucyl adenylate analogue (FW$_{\text{free-base}}$ = 459.48 g/mol) inhibits isoleucyl-tRNA synthetase. **1**. Lee, Kim, Lee, *et al.* (2003) *Bioorg. Med. Chem. Lett.* **13**, 1087. **2**. Pope, Moore, McVey, *et al.* (1998) *J. Biol. Chem.* **273**, 31691. **3**. Nakama, Nureki & Yokoyama (2001) *J. Biol. Chem.* **276**, 47387.

3-(L-Isoleucyl)thiazolidine

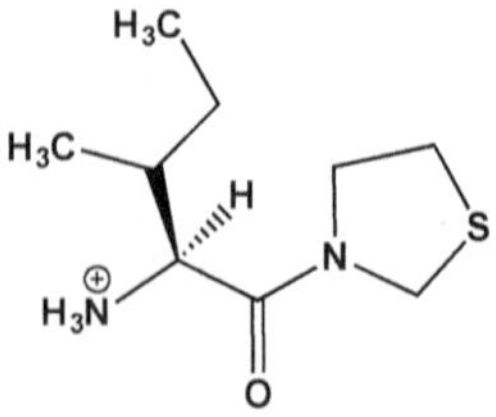

This dipeptide analogue (FW$_{\text{free-base}}$ = 202.32 g/mol), also known as isoleucyl-thiazolidide, inhibits cytosol nonspecific Dipeptidase, *or* prolyl dipeptidase (1), dipeptidyl-peptidase II (1-3), and dipeptidyl-peptidase IV (1-8). **1**. Jiaang, Chen, Hsu, *et al.* (2005) *Bioorg. Med. Chem. Lett.* **15**, 687. **2**. Leiting, Pryor, Wu, *et al.* (2003) *Biochem. J.* **371**, 525. **3**. Stöckel-Maschek, Mrestani-Klaus, Stiebitz, Demuth & Neubert (2000) *Biochim. Biophys. Acta* **1479**, 15. **4**. Malík, Busek, Mares, *et al.* (2003) *Adv. Exp. Med. Biol.* **524**, 95. **5**. Stöckel-Maschek, Stiebitz, Born, *et al.* (2000) *Adv. Exp. Med. Biol.* **477**, 117. **6**. Schön, Born, Demuth, *et al.* (1991) *Biol. Chem. Hoppe-Seyler* **372**, 305. **7**. Stöckel-Maschek, Stiebitz, Faust, *et al.* (2003) *Adv. Exp. Med. Biol.* **524**, 69. **8**. Baer, Gerhartz, Hoffmann, Rosche & Demuth (2003) *Adv. Exp. Med. Biol.* **524**, 103.

Isolinderalactone

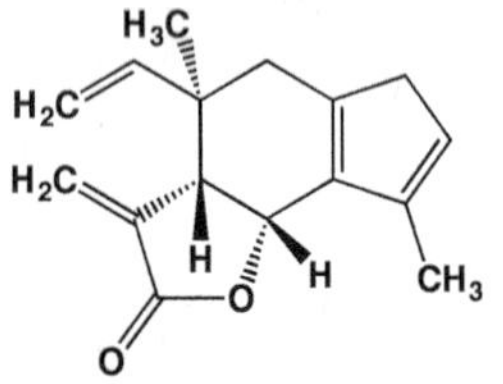

This naturally occurring sesquiterpene (FW = 242.32 g/mol), obtained from the roots of *Lindera strychnifolia*, competitively inhibits prolyl oligopeptidase, (PEP, EC 3.4.21.26), an enzyme that plays a role in degradation of proline-containing neuropeptides (*e.g.*, vasopressin, substance P, and thyrotropin-releasing hormone, *or* TRH) involved in the processes of learning and memory. This oligopeptidase is also inhibited noncompetitively by two tannins (epicatechin and aesculitannin) and competitively by three other sesquiterpenes (linderene and linderene acetate, and linderalactone). Isolinderalactone readily isomerizes to lideralactone via a Cope rearrangement. **1**. Kobayashi, Miyase, Sano, *et al.* (2002) *Biol. Pharm. Bull.* **25**, 1049.

Isoliquiritigenin

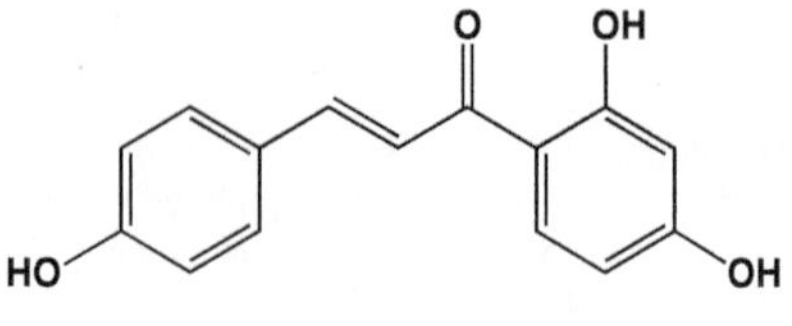

This substituted chalcone (FW = 256.26 g/mol), also known as 4,2',4'-trihydroxychalcone and (*E*)-1-(2,4-dihydroxyphenyl)-3-(4-hydroxyphenyl)-2-propen-1-one, which is found in licorice roots and other plants, is a guanylate cyclase activator that exhibits vasorelaxant effects. **Target(s):** aldose reductase (1,2); aryl hydrocarbon hydroxylase, cytochrome P450 (3); cAMP phosphodiesterase (4,5); catechol oxidase (6); fatty-acid synthase (7); glutathione *S*-transferase, *Triticum aestivum* (8); GLUT1 glucose transporter (9); monoamine oxidase (10); [myosin light-chain] kinase (11); naringenin-chalcone synthase (12); tyrosinase, *or* monophenol monooxygenase (6,13,14); xanthine oxidase (15). **1**. Tawata, Aida, Noguchi, *et al.* (1992) *Eur. J. Pharmacol.* **212**, 87. **2**. Aida, Tawata, Shindo, *et al.* (1990) *Planta Med.* **56**, 254. **3**. Friedman, West, Sugimura & Gelboin (1985) *Pharmacology* **31**, 203. **4**. Abdollahi, Chan, Subrahmanyam & O'Brien (2003) *Mol. Cell. Biochem.* **252**, 205. **5**. Kusano, Nikaido, Kuge, *et al.* (1991) *Chem. Pharm. Bull.* **39**, 930. **6**. Kim & Uyama (2005) *Cell. Mol. Life Sci.* **62**, 1707. **7**. Li, Ma, Wang & Tian (2005) *J. Biochem.* **138**, 679. **8**. Cummins, O'Hagan, Jablonkai, Cole, Hehn, Werck-Reichhart & Edwards (2003) *Plant Mol. Biol.* **52**, 591. **9**. Martin, Kornmann & Fuhrmann (2003) *Chem. Biol. Interact.* **146**, 225. **10**. Pan, Kong, Zhang, Cheng & Tan (2000) *Acta Pharmacol. Sin.* **21**, 949. **11**. Jinsart, Ternai & Polya (1991) *Biol. Chem. Hoppe-Seyler* **372**, 819. **12**. Whitehead & Dixon (1983) *Biochim. Biophys. Acta* **747**, 298. **13**. Nerya, Musa, Khatib, Tamir & Vaya (2004) *Phytochemistry* **65**, 1389. **14**. Nerya, Vaya, Musa, *et al.* (2003) *J. Agric. Food Chem.* **51**, 1201. **15**. Kong, Zhang, Pan, Tan & Cheng (2000) *Cell. Mol. Life Sci.* **57**, 500.

Isomalathion

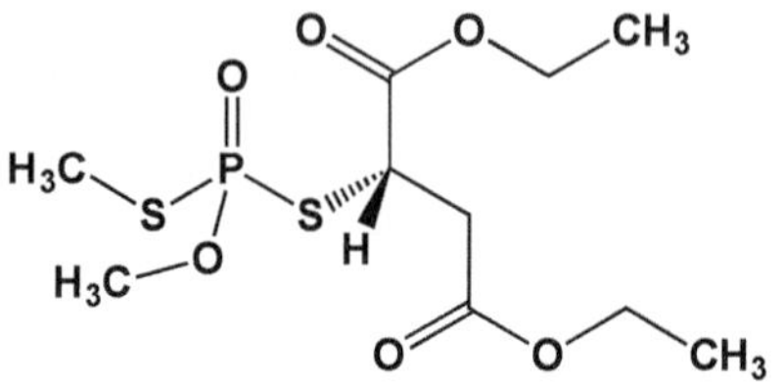

This malathion isomer (FW = 330.36 g/mol) inhibits acetylcholinesterases and butyrylcholinesterases. The (1*R*,3*R*)-isomer (shown above) displays greater potency than (1*S*,3*S*)-isomalathion. *Torpedo californica* acetylcholine esterase inhibited by either of the (1*R*)-isomers can be readily reactivated (1). The enzyme is significantly more difficult to reactivate following inhibition by either of the (1*S*)-isomers. The primary leaving group for the (1*R*)-isomers is diethyl thiosuccinyl, whereas it is thiomethyl for (1*S*)-isomalathions: the (1*R*)-isomalathion produces an *O,S*-dimethyl phosphate enzyme adduct whereas (1*S*)-isomalathions generated an *O*-methyl phosphate adduct (2). **Target(s):** acetylcholinesterase (1,2); butyrylcholine sterase (3); cholesterol esterase (3). **1**. Berkman, Quinn & Thompson (1993) *Chem. Res. Toxicol.* **6**, 724. **2**. Doorn, Thompson, Christner & Richardson (2003) *Chem. Res. Toxicol.* **16**, 958. **3**. Doorn, Talley, Thompson & Richardson (2001) *Chem. Res. Toxicol.* **14**, 807.

Isomaltose

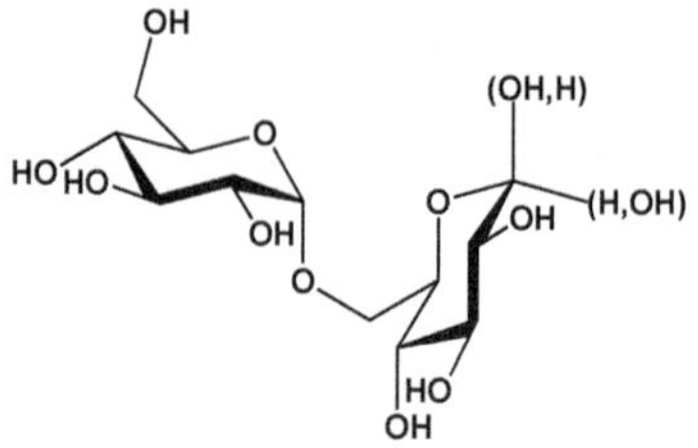

This reducing disaccharide (FW = 342.30 g/mol), also known as brachyose, consists of two D-glucopyranosyl residues joined by α1,6 linkages. Isomaltose is a structural component of glycogen and amylopectin as well as certain dextrans. It is a substrate for sucrase-isomaltase. **Target(s):** concanavalin A (1); α-dextrin endo-1,6-α-glucosidase (2,3); glucan 1,6-α-0isomaltosidase (4); α-glucosidase (2,5-7); isoamylase (8). **1**. Goldstein, Hollerman & Smith (1965) *Biochemistry* **4**, 876. **2**. Webb (1966) *Enzyme and Metabolic Inhibitors*, vol. 2, pp. 416-423, Academic Press, New York. **3**. Lukomskaia & Rozenfel'd (1958) *Biochemistry (U.S.S.R.)* **23**, 244. **4**. Tonozuka, Suzuki, Ikehara, *et al.* (2004) *J. Appl. Glycosci.* **51**, 27. **5**. Brown, Brown & Jeffrey (1972) *Meth. Enzymol.* **28**, 805. **6**. Dahlqvist (1959) *Acta Chem. Scand.* **13**, 2156. **7**. Halvorson & Ellias (1958) *Biochem. Biophys. Acta* **30**, 28. **8**. Kitagawa, Amemura & Harada (1975) *Agric. Biol. Chem.* **39**, 989.

Isomazole

This vasodilator (FW$_{free-base}$ = 287.34 g/mol; Hygroscopic; Photosensitive), also known as EMD 52750, LY 175326, and 2-[2-methoxy-4-(methylsulfinyl)phenyl]-1H-imidazo[4,5-c]pyridine, inhibits cGMP phosphodiesterase III, IC$_{50}$ = 42 μM for guinea pig enzyme (1-3) and cGMP phosphodiesterase V, IC$_{50}$ = 58 μM for pig enzyme (1). Isomazole is commercially available as the dihydrochloride. **1.** Coates, Connolly, Dhanak, Flynn & Worby (1993) *J. Med. Chem.* **36**, 1387. **2.** Bethke, Klimkiewicz, Kohl, *et al.* (1991) *J. Cardiovasc. Pharmacol.* **18**, 386. **3.** Hayes, Bowling, Pollock & Robertson (1986) *J. Pharmacol. Exp. Ther.* **237**, 18.

Isoniazid

This antibiotic (FW = 137.14 g/mol; CAS 54-85-3), also known as and marketed under other names including Isonicotinylhydrazide (INH) and Hydra®, Sovit®, Laniazid®, and Nydrazid®, is a first-line drug for preventing and treating both latent and active tuberculosis (*See also Ethionamide*). Isoniazid is effective against *Mycobacterium tuberculosis* as well as atypical strains like *M. kansasii* and *M. xenopi* (1). Itself a prodrug, isoniazid first must be activated by KatG, a mycobacterium catalase-peroxidase that oxidatively forms isonicotinic acyl-NADH (2). The latter binds to and inhibits InhA, an enoyl-acyl carrier protein reductase, thereby blocking the synthesis of mycolic acid, which is required for the mycobacterial cell membrane. The presence of a tyrosyl-like radical in KatG during catalase turnover has been demonstrated by RFQ-EPR spectroscopy. The persistence of this radical correlates with the amount of H$_2$O$_2$ present during catalase turnover, and therefore it can be considered a catalytically competent species in the catalase cycle (2). Despite earlier evidence for a common site of action, Isoniazid and Ethionamide (ETA) are activated by different mechanisms as resistance to INH does not confer resistance to ETA. **1.** Berning SE, Peloquin (1998) "Antimycobacterial agents: Isoniazid," In: *Antimicrobial Therapy and Vaccines* (Yu, Merigan & Barriere, eds.), Williams and Wilkins, Baltimore 1998. **2.** Suarez, Ranguelova, Jarzecki, *et al.* (2009) *J. Biol. Chem.* **284**, 7017.

Isonicotinate (Isonicotinic Acid)

This nicotinic acid isomer (FW$_{free-acid}$ = 123.11 g/mol; CAS 16887-79-9; Solubility: 0.50 g/100 mL H$_2$O), also known as 4-pyridinecarboxylic acid and 4-picolinic acid, is a secondary metabolite in plants and microorganisms. **Target(s):** acetoin dehydrogenase (1); alcohol dehydrogenase (NADP$^+$) (1); 5-aminolevulinate aminotransferase (2); γ-butyrobetaine dioxygenase, *or* γ-butyrobetaine hydroxylase, K_i = 0.08 mM (3); carbonyl reductase (NADPH) (1); *erythro*-3-hydroxyaspartate ammonia-lyase (4); NADase, weakly inhibited (5); nicotinamide-nucleotide amidase (6); nicotinate glucosyltransferase (7); nicotinate N-methyltransferase (8); nicotinate phosphoribosyltransferase (9); procollagen-lysine 5-dioxygenase (10); procollagen-proline 3-dioxygenase (10); procollagen-proline 4-dioxygenase (10,11); tryptophan 2-C-methyltransferase (12). **1.** Hara, Seiriki, Nakayama & Sawada (1985) *Prog. Clin. Biol. Res.* **174**, 291. **2.** Turner & Neuberger (1970) *Meth. Enzymol.* **17A**, 188. **3.** Ng, Hanauske-Abel & Englard (1991) *J. Biol. Chem.* **266**, 1526. **4.** Gibbs & Morris (1965) *Biochem. J.* **97**, 547. **5.** Zatman, Kaplan, Colowick & Ciotti (1954) *J. Biol. Chem.* **209**, 453. **6.** Imai (1973) *J.*

Biochem. **73**, 139. **7.** Taguchi, Sasatani, Nishitani & Okumura (1997) *Biosci. Biotechnol. Biochem.* **61**, 720. **8.** Taguchi, Nishitani, Okumura, Shimabayashi & Iwai (1989) *Agri. Biol. Chem.* **53**, 2867. **9.** Gaut & Solomon (1971) *Biochem. Pharmacol.* **20**, 2903. **10.** Majamaa, Turpeenniemi-Hujanen, Latipää, *et al.* (1985) *Biochem. J.* **229**, 127. **11.** Majamaa, Hanauske-Abel, Günzler & Kivirikko (1984) *Eur. J. Biochem.* **138**, 239. **12.** Speedie (1987) *Meth. Enzymol.* **142**, 235.

Isonicotinic Acid Hydrazide

This hepatotoxic hydrazide (FW = 137.14 g/mol), also called isoniazid and 1-isonicotinylhydrazine, is an tuberculostatic agent. This drug was first prepared in 1912, but was patented for its therapeutic effects by Hoffmann-La Roche in 1952. **Primary Mode of Action:** Oxidation of this hydrazide leads to the generation of reactive intermediates that inactivate an NADH-specific enoyl reductase of *Mycobacterium tuberculosis*, reacting with an active-site cysteinyl residue1. The reductase undergoes slow-binding inhibition by an isoniazid-NAD$^+$ adduct (K_i = 0.75 nM). **Key Pharmacokinetic Parameters:** *See* Appendix II in Goodman & Gilman's THE PHARMACOLOGICAL BASIS OF THERAPEUTICS, 12th Edition (Brunton, Chabner & Knollmann, eds.) McGraw-Hill Medical, New York (2011). **Target(s):** Nicotinic acid hydrazide is a pyridoxine antagonist in some organisms, reacting with the pyridoxal 5-phosphate (PLP) cofactor of nearly every transaminase tested. One may anticipate similar inhibitory effects on PLP-dependent racemases, decarboxylase, dehydratases, *etc.* **1.** Johnsson, King & Schultz (1995) *J. Amer. Chem. Soc.* **117**, 5009. **2.** Gupta, Hollis, Patel, Patrick & Bell (2004) *J. Bone Miner. Res.* **19**, 680.

N^6-(Δ^2-Isopentenyl)adenine

This purine derivative (FW = 203.25 g/mol), also called N^6-(γ.γ-dimethylallylamino)purine, 6-Δ^2-isopentenylaminopurine (IPA), and bryokinin, inhibits cyclin-dependent kinases (IC$_{50}$ = 55 μM). It is also a rare base in certain tRNA and has been reported to act as a cytokinin (*e.g.*, in callus cells of moss sporophytes). IPA has a λ$_{max}$ of 269 nm at pH 7 (ε = 19400 M^{-1}cm^{-1}). **Target(s):** alcohol dehydrogenase (1); cyclin-dependent kinases (2); NADH dehydrogenase (3); xanthine oxidase (4); *cis*-zeatin O-β-D-glucosyltransferase (5). **1.** Zikmanis & Kruce (1990) *Biofactors* **2**, 237. **2.** Meijer & Kim (1997) *Meth. Enzymol.* **283**, 113. **3.** Sue, Miyoshi & Iwamura (1997) *Biosci. Biotechnol. Biochem.* **61**, 1806. **4.** Sheu, Lin & Chiang (1996) *Anticancer Res.* **16**, 3571. **5.** Veach, Martin, Mok, *et al.* (2003) *Plant Physiol.* **131**, 1374.

N^6-(Δ^2-Isopentenyl)adenosine

This rare, naturally occurring nucleoside (FW = 335.36 g/mol), also known as 6-(γ,γ dimethylallylamino)purine-9-riboside, 6-(3-methyl-2-butenyl amino)-9-β-D-ribofuranosylpurine, and 6-(3-methyl-2-butenylamino)purine -9-riboside, is a component of certain tRNA's and is produced by tRNA isopentenyltransferase. It also exhibits cytokinin activity and can induce apoptosis. This nucleoside has a λ_{max} at 269 nm at pH 7 (ε = 20,000 $M^{-1}cm^{-1}$). **Target(s):** adenosine kinase (1,2); adenosine nucleosidase (3); 3',5'-cyclic-nucleotide phosphodiesterase (4); cytidine transport (5); GMP synthetase (6); methionine S-adenosyltransferase (7); NADH dehydrogenase (8); tRNA methyltransferases (9); uridine transport (5). **1.** Palella, Andres & Fox (1980) J. Biol. Chem. **255**, 5264. **2.** Faye & Le Floc'h (1997) Plant Physiol. Biochem. **35**, 15. **3.** Burch & Stuchbury (1986) J. Plant Physiol. **125**, 267. **4.** Hecht, Faulkner & Hawrelak (1974) Proc. Natl. Acad. Sci. U.S.A. **71**, 4670. **5.** Hakala, Slocum & Gryko (1975) J. Cell Physiol. **86**, 281. **6.** Spector & Beecham III (1975) J. Biol. Chem. **250**, 3101. **7.** Berger & Knodel (2003) BMC Microbiol. **3**, 12. **8.** Sue, Miyoshi & Iwamura (1997) Biosci. Biotechnol. Biochem. **61**, 1806. **9.** Kerr & Borek (1973) The Enzymes, 3rd ed., **9**, 167.

Isopentenyl Diphosphate

This activated five-carbon building block ($FW_{free-acid}$ = 246.09 g/mol), also known as isopentenyl pyrophosphate (IPP), is a key intermediate in the biosynthesis of the sterols and other terpenes. Note that it is a geometric isomer of dimethylallyl diphosphate with which it is interconvertible via isopentenyl diphosphate Δ-isomerase. **Target(s):** aspulvinone dimethylallyltransferase (1); dolichyl-diphosphatase (2); geranylgeranyl-diphosphatase (3); 4-hydroxy-benzoate nonaprenyltransferase (4); mevalonate kinase (5). **1.** Takahashi, Ojima, Ogura & Seto (1978) Biochemistry **17**, 2696. **2.** Scher & Waechter (1984) J. Biol. Chem. **259**, 14580. **3.** Nah, Song & Back (2001) Plant Cell Physiol. **42**, 864. **4.** Kawahara, Koizumi, Kawaji, et al. (1991) Agric. Biol. Chem. **55**, 2307. **5.** Hinson, Chambliss, Toth, Tanaka & Gibson (1997) J. Lipid Res. **38**, 2216.

N-Isopentyl-6-[4-(2-(trifluoromethyl)phenoxy)piperidin-1-yl]pyridazine-3-carboxamide

This pyridazine derivative (FW = 436.46 g/mol) inhibits stearoyl-CoA 9-desaturase (mouse, IC_{50} = 34 nM). Studies with stearoyl-CoA 9-desaturase knockout mice have established that these animals are lean and protected from leptin deficiency-induced and diet-induced obesity, with greater whole body insulin sensitivity than wild-type animals. **1.** Liu, Lynch, Freeman, et al. (2007) J. Med. Chem. **50**, 3086.

Isophthalate (Isophthalic Acid)

This aromatic dicarboxylic acid ($FW_{free-acid}$ = 166.13 g/mol) is an unlikely glutamate analogue that inhibits many glutamate-dependent reactions, particularly the reactions catalyzed by glutamate dehydrogenase and glutamate decarboxylase. **Target(s):** γ-butyrobetaine dioxygenase, or γ-butyrobetaine hydroxylase; K_i = 3.8 mM (1); dihydrodipicolinate reductase (2); glutamate decarboxylase (3,4); glutamate dehydrogenase (5-14); α-ketoglutarate dehydrogenase (15); kynurenine aminotransferase (16); procollagen-lysine 5-dioxygenase (15); procollagen-proline 3-dioxygenase (15); procollagen-proline 4-dioxygenase (15,17-19); serine-sulfate ammonia-lyase (20). **1.** Ng, Hanauske-Abel & Englard (1991) J. Biol. Chem. **266**, 1526. **2.** Tamir (1971) Meth. Enzymol. **17B**, 134. **3.** Fonda (1972) Biochemistry **11**, 1304. **4.** Youngs & Tunnicliff (1991) Biochem. Int. **23**, 915. **5.** Caughey, Smiley & Hellerman (1957) J. Biol. Chem. **224**, 591. **6.** Frieden (1963) The Enzymes, 2nd ed., **7**, 3. **7.** Shaw (1970) The Enzymes, 3rd ed., **1**, 91. **8.** Smith, Austen, Blumenthal & Nyc (1975) The Enzymes, 3rd ed., **11**, 293. **9.** Baker (1967) Design of Active-Site-Directed

Irreversible Enzyme Inhibitors, Wiley, New York. **10.** Webb (1966) Enzyme and Metabolic Inhibitors, vol. **2**, p. 330, Academic Press, New York. **11.** O'Donnell, Maher & Griffin (1995) Biochem. Soc. Trans. **23**, 364S. **12.** Youngs & Tunnicliff (1991) Biochem. Int. **23**, 915. **13.** Stevens, Duncan & Robertson (1989) FEMS Microbiol. Lett. **48**, 173. **14.** Boots, Franklin, Dunlavey, et al. (1976) Proc. Soc. Exp. Biol. Med. **151**, 316. **15.** Majamaa, Turpeenniemi-Hujanen, Latipää, et al. (1985) Biochem. J. **229**, 127. **16.** Mason (1959) J. Biol. Chem. **234**, 2770. **17.** Tschank, Hanauske-Abel & Peterkofsky (1988) Arch. Biochem. Biophys. **261**, 312. **18.** Majamaa, Hanauske-Abel, Günzler & Kivirikko (1984) Eur. J. Biochem. **138**, 239. **19.** Kaska, Günzler, Kivirikko & Myllylä (1987) Biochem. J. **241**, 483. **20.** Tudball & Thomas (1973) Eur. J. Biochem. **40**, 25.

Isophthalic Acid 2(S)-Acetylpyrrolidine N-(L-Prolyl)-2(S)-(hydroxyacetyl)pyrrolidine Diamide

This substrate analogue (FW = 469.54 g/mol) inhibits pig brain prolyl oligopeptidase (IC_{50} = 1.2 nM). **1.** Wallén, Christiaans, Jarho, et al. (2003) J. Med. Chem. **46**, 4543.

4-Isopropylcatechol

This substituted pyrocatechol (FW = 152.19 g/mol), also known as 1,2-dihydroxy-4-isopropylbenzene, inhibits steroid oxygenase and is a potent irreversible, cutaneous depigmenting agent. **Target(s):** 3,4-dihydroxy-9,10-secoandrosta-1,3,5(10)-triene-9,17-dione 4,5-dioxygenase, or steroid oxygenase (1,2). **1.** Hayaishi, Nozaki & Abbott (1975) The Enzymes, 3rd ed., **12**, 119. **2.** Nozaki (1974) in Mol. Mech. Oxygen Activ. (Hayaishi, ed.), p. 135, Academic Press, New York.

Isopropylmalate (Isopropylmalic Acid)

2-isopropylmalate   **3-isopropylmalate**

This metabolite ($FW_{free-acid}$ = 176.17 g/mol) refers to either (2S)-2- or (2R,3S)-3-isopropylmalic acid (i.e., 3-carboxy-3-hydroxy-4-methylpentanoic acid and 3-carboxy-2-hydroxy-4-methylpentanoic acid. Both are intermediates in L-leucine biosynthesis, and they are interconverted in the reaction catalyzed by 3-isopropylmalate dehydratase. 2-Isopropylmalate is a product of 2-isopropylmalate synthase, and 3-isopropylmalate is a substrate of the reaction catalyzed by 3-isopropylmalate dehydrogenase. **Target(s):** isocitrate dehydrogenase, inhibited by 3-isopropylmalate, particularly the N99L enzyme (1); 3-isopropylmalate dehydratase, inhibited by racemic mixture (2). **1.** Miyazaki, Yaoi & Oshima (1994) Eur. J. Biochem. **221**, 899. **2.** Gross, Burns & Umbarger (1963) Biochemistry **2**, 1046.

2-Isopropyl-4-pentenamide

This mechanism-based inhibitor (FW = 141.21 g/mol; Abbreviation: AIA), also known as allylisopropylacetamide, inactivates cytochrome P450 systems. It also induces the synthesis of δ-aminolevulinate synthase. **1.**

Ator & Ortiz de Montellano (1990) *The Enzymes*, 3rd ed., **19**, 213. **2.** Davies, Britt & Pohl (1986) *Chem. Biol. Interact.* **58**, 345. **3.** Davies, Britt & Pohl (1986) *Arch. Biochem. Biophys.* **244**, 387. **4.** Ortiz de Montellano, Yost, Mico, *et al.* (1979) *Arch. Biochem. Biophys.* **197**, 524.

Isopropyl 1-thio-β-D-galactopyranoside

This synthetic thioglycoside (FW = 238.30 g/mol; M.P. = 174-175°C; *Abbreviation*: IPTG) is a nonmetabolizable galactose analogue that is a convenient and widely used inducer of β-galactosidase and β-galactoside permease. The labeled galactopyranoside has also been used to isolate the Lac repressor. **Target(s):** β-galactosidase (1-3); protein-$N\pi$-phosphohistidine:sugar phosphotransferase (4). **1.** Shaikh, Khire & Khan (1999) *Biochim. Biophys. Acta* **1472**, 314. **2.** Huber, Roth & Bahl (1990) *J. Protein Chem.* **15**, 621. **3.** Nakao, Harada, Kodama, *et al.* (1994) *Appl. Microbiol. Biotechnol.* **40**, 657. **4.** Peters, Frank & Hengstenberg (1995) *Eur. J. Biochem.* **228**, 798.

2'-Isopropyl-4'-(trimethylammonium chloride)-5'-methylphenyl piperidine-1-carboxylate

This plant growth retardant (FW = 354.92 g/mol), also known as Amo 1618 and (5 hydroxycarvacryl)trimethylammonium chloride 1-piperidinecarboxylate, is a choline analogue that blocks gibberellin biosynthesis via inhibition of the copalyl diphosphate synthase activity. **Target(s):** beyerene biosynthesis (1); cholinesterase, *or* butyrylcholinesterase (2,3); *ent*-copalyl diphosphate synthase (1,4-9); *ent*-kaurene synthase (1,4-9); 2,3-oxidosqualene-lanosterol cyclase, *or* lanosterol synthase (10); phospholipase D (11); sandaracopimaradiene biosynthesis (1); squalene:hopene cyclase (12,13); trachylobane biosynthesis (1). **1.** Robinson & West (1970) *Biochemistry* **9**, 80. **2.** Fluck & Jaffe (1975) *Biochim. Biophys. Acta* **410**, 130. **3.** Riov & Jaffe (1973) *Plant Physiol.* **51**, 520. **4.** West & Upper (1969) *Meth. Enzymol.* **15**, 481. **5.** Rademacher (2000) *Annu. Rev. Plant Physiol. Plant Mol. Biol.* **51**, 501. **6.** Kawaide, Imai, Sassa & Kamiya (1997) *J. Biol. Chem.* **272**, 21706. **7.** Fall & West (1971) *J. Biol. Chem.* **246**, 6913. **8.** Hedden, Phinney, MacMillan & Sponsel (1977) *Phytochemistry* **16**, 1913. **9.** Frost & West (1977) *Plant Physiol.* **59**, 22. **10.** Ono & Bloch (1975) *J. Biol. Chem.* **250**, 1571. **11.** Herman & Chrispeels (1980) *Plant Physiol.* **66**, 1001. **12.** Seckler & Poralla (1986) *Biochim. Biophys. Acta* **881**, 356. **13.** Ochs, Tappe, Gartner, Kellner & Poralla (1990) *Eur. J. Biochem.* **194**, 75.

Isoquercitrin

This flavonoid glucoside (FW = 464.38 g/mol), also known as quercetin 3-*O*-β-D-glucopyranoside, has been isolated from the flowers of *Gossypium herbaceum* (Levant cotton), *Eucommia ulmoides* (Hardy rubber tree), *Aesculus hippocastanum* (horse chestnut), *Tropaeolum majus* (garden nasturnium), *Arnica montana* (Leopard's bane and Wolf's bane), and many other plants. It is a yellow solid with a low solubility in water. **Target(s):** aldose reductase (1); fatty-acid synthase (2); 5-lipoxygenase, *or* arachidonate 5 lipoxygenase (3); monoamine oxidase B (4); Na+/glucose cotransporter, *or* SGLT-1 (5); neprilysin (6); peptidyl-dipeptidase A, *or* angiotensin I-converting enzyme (6,7); phosphatidylinositol 3-kinase (8); ribosyl-dihydronicotinamide dehydrogenase (quinone), IC$_{50}$ = 1.5 μM (9); xanthine oxidase (10). **1.** Sakai, Izumi, Murano, *et al.* (2001) *Jpn. J. Pharmacol.* **85**, 322. **2.** Wang, Zhang, Ma & Tian (2006) *J. Enzyme Inhib. Med. Chem.* **21**, 87. **3.** Schneider & Bucar (2005) *Phytother. Res.* **19**, 81. **4.** Lee, Lin, Shen, *et al.* (2001) *J. Agric. Food Chem.* **49**, 5551. **5.** Ader, Block, Pietzsch & Wolffram (2001) *Cancer Lett.* **162**, 175. **6.** Kiss, Kowalski & Melzig (2004) *Planta Med.* **70**, 919. **7.** Oh, Kang, Kwon, *et al.* (2004) *Biol. Pharm. Bull.* **27**, 2035. **8.** Shibasaki, Fukui & Takenawa (1993) *Biochem. J.* **289**, 227. **9.** Boutin, Chatelain-Egger, Vella, Delagrange & Ferry (2005) *Chem.-Biol. Interact.* **151**, 213. **10.** Dew, Day & Morgan (2005) *J. Agric. Food Chem.* **53**, 6510.

1-(5-Isoquinolinesulfonyl)-2-methylpiperazine

This substituted isoquinoline, (FW = 291.37 g/mol), commonly referred to as H-7, is a strong inhibitor of a number of protein kinases: *e.g.*, cAMP-dependent protein kinase (protein kinase A; K_i = 3 μM), cGMP-dependent protein kinase (protein kinase G; K_i = 5.3 μM), [myosin light chain] kinase (K_i = 97 μM), protein kinase C (K_i = 6 μM), and casein kinase I (IC$_{50}$ = 0.1 mM). It also inhibits telomerase activity in quercetin-, H-89-, or herbimycin A-treated NPC-076 cells. **Target(s):** [β-adrenergic-receptor] kinase, weakly inhibited (1,2); cAMP-dependent protein kinase, *or* protein kinase A (3-6); casein kinase I, mildly inhibited (4); cGMP-dependent protein kinase, *or* protein kinase G (3,4); microtubule-associated protein-2 kinase, weakly inhibited (7); [myosin light-chain] kinase (3,4); non-specific serine/threonine protein kinase (4,8); PknH protein kinase (8); protein kinase C (3,4,9-13); rhodopsin kinase (K_i = 46 μM (14); [RNA-polymerase]-subunit kinase, *or* CTD kinase (15,16); telomerase (17). **1.** Benovic (1991) *Meth. Enzymol.* **200**, 351. **2.** Chuang, Sallese, Ambrosini, Parruti & De Blasi (1992) *J. Biol. Chem.* **267**, 6886. **3.** Quick, Ware & Driedger (1992) *Biochem. Biophys. Res. Commun.* **187**, 657. **4.** Hidaka, Inagaki, Kawamoto & Sasaki (1984) *Biochemistry* **23**, 5036. **5.** Engh, Girod, Kinzel, Huber & Bossemeyer (1996) *J. Biol. Chem.* **271**, 26157. **6.** Langer, Vogtherr, Elshorst, *et al.* (2004) *ChemBioChem* **5**, 1508. **7.** Schanen & Landreth (1992) *Mol. Brain Res.* **14**, 43. **8.** Sharma, Chandra, Gupta, *et al.* (2004) *FEMS Microbiol. Lett.* **233**, 107. **9.** Hidaka, Watanabe & Kobayashi (1991) *Meth. Enzymol.* **201**, 328. **10.** Kikkawa & Nishizuka (1986) *The Enzymes*, 3rd ed., **17**, 167. **11.** Kawamoto & Hidaka (1984) *Biochem. Biophys. Res. Commun.* **125**, 258. **12.** Nishikawa, Uemura, Hidaka & Shirakawa (1986) *Life Sci.* **39**, 1101. **13.** Sakurai, Onishi, Tanimoto & Kizaki (2001) *Biol. Pharm. Bull.* **24**, 973. **14.** Palczewski, Kahn & Hargrave (1990) *Biochemistry* **29**, 6276. **15.** Medlin, Uguen, Taylor, Bentley & Murphy (2003) *EMBO J.* **22**, 925. **16.** Jacobs, Ogiwara & Weiner (2004) *Mol. Cell. Biol.* **24**, 846. **17.** Ku, Cheng & Wang (1997) *Biochem. Biophys. Res. Commun.* **241**, 730.

Isosteviol

This diterpene (FW$_{\text{free-acid}}$ = 318.46 g/mol; M.P. = 234°C), also known as *ent*-16-oxobeyeran-19-oic acid, is produced by acid hydrolysis of steviol via a Wagner-Meerwein rearrangement. **Target(s):** ATPase (1); DNA-directed DNA polymerase (2); DNA polymerase α (2); DNA topoisomerase II (2);

glucose-6-phosphatase (3); glutamate dehydrogenase (1); oxidative phosphorylation (1); succinate dehydrogenase (1). **1**. Kelmer Bracht, Alvarez & Bracht (1985) *Biochem. Pharmacol.* **34**, 873. **2**. Mizushina, Akihisa, Ukiya, *et al.* (2005) *Life Sci.* **77**, 2127. **3**. Ishii & Bracht (1987) *Braz. J. Med. Biol. Res.* **20**, 837.

Isovalerate (Isovaleric Acid)

This branched-chain carboxylic acid (FW$_{free-acid}$ = 102.13 g/mol; M.P. = – 37°C; B.P. = 175-177°C; pK_a = 4.78 at 25°C), also known a 3-methylbutanoic acid and delphinic acid. Is found in urine at elevated levels in individuals with isovaleryl-CoA dehydrogenase deficiency. **Targets:** betaine:homocysteine *S*-methyltransferase (1-3); dimethylglycine *N*-methyltransferase, weakly inhibited (4); homoserine kinase (5); 2-isopropyl-malate synthase (6); 2 ketoisovalerate dehydrogenase (7); 3-methyl-2-ketobutanoate Hydroxymethyltransferase, *or* 3-methyl 2-oxobutanoate Hydroxymethyltransferase (8); ornithine aminotransferase (9); sarcosine/ dimethylglycine *N*-methyltransferase, weakly inhibited (4); valine decarboxylase, leucine decarboxylase (10,11). **1**. Skiba, Wells, Mangum & Awad (1987) *Meth. Enzymol.* **143**, 384. **2**. Awad, Whitney, Skiba, Mangum & Wells (1983) *J. Biol. Chem.* **258**, 12790. **3**. Skiba, Taylor, Wells, Mangum & Awad, Jr. (1982) *J. Biol. Chem.* **257**, 14944. **4**. Waditee, Tanaka, Aoki, *et al.* (2003) *J. Biol. Chem.* **278**, 4932. **5**. Wormser & Pardee (1958) *Arch. Biochem. Biophys.* **78**, 416. **6**. Allison, Baetz & Wiegel (1984) *Appl. Environ. Microbiol.* **48**, 1111. **7**. Connelly, Danner & Bowden (1970) *Meth. Enzymol.* **17A**, 818. **8**. Powers & Snell (1976) *J. Biol. Chem.* **251**, 3786. **9**. Kalita, Kerman & Strecker (1976) *Biochim. Biophys. Acta* **429**, 780. **10**. Webb (1966) *Enzyme and Metabolic Inhibitors*, vol. **2**, p. 352, Academic Press, New York. **11**. Sutton & King (1962) *Arch. Biochem. Biophys.* **96**, 360.

Isovaleryl-CoA

This Coenzyme A thiolester (FW = 851.67 g/mol), itself an intermediate in leucine degradation, inhibits [branched-chain α-keto-acid dehydrogenase] kinase and pyruvate dehydrogenase. It is also substrate of the reactions catalyzed by isovaleryl-CoA dehydrogenase and phloroisovalero-phenone synthase. **Target(s):** amino acid acetyltransferase (1); [branched-chain α-keto-acid dehydrogenase] kinase, *or* [3-methyl-2-oxobutanoate dehydrogenase (acetyl-transferring)] kinase) (2,3); [3-methyl-2-oxobutanoate dehydrogenase, *or* 2-methylpropanoyl-transferring)]-phosphatase, *or* [branched-chain-α keto-acid dehydrogenase] phosphatase (4-6); phosphoprotein phosphatase (6,7); pyruvate dehydrogenase (8,9). **1**. Coude, Sweetman & Nyhan (1979) *J. Clin. Invest.* **64**, 1544. **2**. Paxton (1988) *Meth. Enzymol.* **166**, 313. **3**. Paxton & Harris (1984) *Arch. Biochem. Biophys.* **231**, 48. **4**. Damuni & Reed (1987) *J. Biol. Chem.* **262**, 5129. **5**. Damuni, Merryfield, Humphreys & Reed (1984) *Proc. Natl. Acad. Sci. U.S.A.* **81**, 4335. **6**. Reed & Damuni (1987) *Adv. Protein Phosphatases* **4**, 59. **7**. Damuni & Reed (1987) *J. Biol. Chem.* **262**, 5133. **8**. Martin-Requero, Corkey, Cerdan, *et al.* (1983) *J. Biol. Chem.* **258**, 3673. **9**. Gregersen (1981) *Biochem. Med.* **26**, 20.

N$^\alpha$-Isovaleryl-L-histidyl-L-prolyl-L-phenylalanyl-L-histidyl-statinyl-L-isoleucyl-L-phenylalaninamide

This peptidomimetic inhibitor (FW = 1037.28 g/mol), also known as ISCRIP and U-77455E, derives its high-affinity action on target aspartic proteases by virtue of its statine (*or* (3*S*,4*S*)-4-amino-3-hydroxy-6 methylheptanoyl) moiety., which resembles a tetrahedral reaction intermediate **Target(s):** cathepsin D (1,2); cathepsin E (1); endothiapepsin (1); gastricsin (1); pepsin, IC$_{50}$ = 0.87 μM (1,2); renin, IC$_{50}$ = 2 nM (1-3). **1**. Jupp, Dunn, Jacobs, *et al.* (1990) *Biochem J.* **265**, 871. **2**. Boger, Lohr, Ulm, *et al.* (1983) *Nature* **303**, 81. **3**. Epps, Cheney, Schostarez, *et al.* (1990) *J. Med. Chem.* **33**, 2080.

Isovaline

This valine isomer (FW = 117.15 g/mol; CAS 465-58-7), also known as 2-amino-2-methylbutyric acid, which is present in a number of antibiotics (*e.g.*, suzukacillin-A and trichotoxins), inhibits D-amino-acid oxidase (1), 1-aminocyclopropanecarboxylate malonyltransferase, with the *R*-stereoisomer inhibiting more than the *S*-isomer (2). **1**. Horowitz (1944) *J. Biol. Chem.* **154**, 141. **2**. Liu, Su & Yang (1984) *Arch. Biochem. Biophys.* **235**, 319.

Isoxaflutole

Isoxaflutole

Active Form

This commercial isoxazole-based herbicide (FW = 359.32 g/mol; CAS 141112-29-0), also named 5-cyclopropyl-4-[2-(methylsulfonyl)-4-(trifluoromethyl)benzoyl]isoxazole, rearranges to diketonitrile, a slow, tight-binding inhibitor of plant 4-hydroxyphenylpyruvate dioxygenase (K_i = 6 nM). Isoxaflutole prevents the biosynthesis of carotenoid pigments, which protect a chlorophyll from photodecomposition, such that carotenoid-depleted plants are damaged by the sun and eventually die. Isoxaflutole is the active ingredient in BALANCE®, a soil-applied pre-emergent herbicide that is spred before field corn begins growing in a field. Isoxaflutole's carcinogenic risk is estimated to be 9.3 x 10^{-8}. **1**. Yang, Pflugrath, Camper, *et al.* (2004) *Biochemistry* **43**, 10414. **2**. Garcia, Job & Matringe (2000) *Biochemistry* **39**, 7501.

ISP-1

This immunosuppressant (FW = 401.54 g/mol), also known as myriocin, thermozymocidin, and 2-amino-3,4-dihydroxy-2-hydroxymethyl-14-oxoeicos-6-enoate, is a potent inhibitor of fungal serine *C*-palmitoyltransferase. It is produced by *Myriococcum albomyces*, *Mycelia sterilia*, and *Isaria* (*or* *Cordyceps*) *sinclairii* (a parasite on moths). **1**. Dickson, Lester & Nagiec (2000) *Meth. Enzymol.* **311**, 3. **2**. Miyake, Kozutsumi, Nakamura, Fujita & Kawasaki (1995) *Biochem. Biophys. Res. Commun.* **211**, 396. **3**. Riley & Plattner (2000) *Meth. Enzymol.* **311**, 348. **4**. Yamaji-Hasegawa, Takahashi, Tetsuka, Senoh & Kobayashi (2005)

Biochemistry **44**, 268. **5**. Hanada, Nishijima, Fujita & Kobayashi (2000) *Biochem. Pharmacol.* **59**, 1211. **6**. Weiss & Stoffel (1997) *Eur. J. Biochem.* **249**, 239. **7**. Perry (2002) *Biochim. Biophys. Acta* **1585**, 146. **8**. Ikushiro, Hayashi & Kagamiyama (2004) *Biochemistry* **43**, 1082.

Ispinesib

This potent, reversible allosteric inhibitor (FW = 517.06; CAS 336113-53-2, 514820-03-2 (methanesulfonate), 514820-04-3 (HCl); Solubility: 100 mg/mL DMSO; <1 mg/mL H_2O), also known as SB-715992, CK0238273, and (*R*)-*N*-(3-aminopropyl)-*N*-(1-(3-benzyl-7-chloro-4-oxo-3,4-dihydro-quinazolin-2-yl)-2-methylpropyl)-4-methylbenzamide, specifically inhibits kinesin KSP (K_i = 1.7 nM), also known as HsEg5, which is essential for the formation of a bipolar mitotic spindle and progression through mitosis (1). Ispinesib alters the ability of KSP to bind to microtubules (MTs) and inhibits its movement by preventing the release of ADP without preventing KSP·ADP dissociation from the MT·KSP·ADP complex. Ispinesib and monastrol share a common mode of inhibition. SB-715992 (6–10 mg/kg) also inhibits murine solid tumors, including Madison 109 lung carcinoma, M5076 sarcoma, as well as L1210 and P388 leukemias (2). **1**. Lad, Luo, Carson, *et al.* (2008) *Biochemistry* **47**, 3576. **2**. Davis, Sarkar, Hussain & Sarkar (2006) *BMC Cancer* **6**, 22.

Isradipine

This L-class calcium channel blocker (FW = 371.39 g/mol; CAS 75695-93-1), also known by its trade names DynaCirc® and Prescal® as well as its systematic name 3-methyl 5-propan-2-yl 4-(2,1,3-benzoxadiazol-4-yl)-2,6-dimethyl-1,4-dihydropyridine-3,5-dicarboxylate, selectively inhibits the sinus node, but not atrioventricular, conduction, making it a potent vasodilator without deleterious effects on blood pressure either at rest or during exercise. Isradipine is usually prescribed for the treatment of high blood pressure in order to reduce the risk of stroke and heart attack. When administered at doses that are relevant for antihypertensive treatment, animal studies indicate that isradipine has powerful anti-atherosclerotic effects as well as a brain tissue-preserving effect after experimentally induced stroke. Recent rodent studies suggest that isradipine may be useful in treating Parkinson's disease, because isradipine appears to force dopaminergic neurons in to revert to a juvenile, Ca^{2+}-independent mechanism to generate autonomous activity. **1**. Surmeier (2007) *Lancet Neurol.* **6**, 933.

Itaconate (Itaconic Acid)

This naturally occurring dicarboxylic acid ($FW_{free-acid}$ = 130.10 g/mol; pK_a values of 3.63 and 5.00; Water-soluble in water (8.3 g/100 g at 20°C), also known as methylenebutanedioic acid and methylenesuccinic acid, is a product of the reaction catalyzed by aconitate decarboxylase and is an alternative substrate of succinate:citramalate CoA-transferase. **Target(s):** 2-(acetamidomethylene)-succinate hydrolase (1); carboxy-*cis,cis*-muconate cyclase (2); fumarase (3,4); glutamate decarboxylase (5); 2-(hydroxymethyl)-3 (acetamidomethylene)-succinate hydrolase (1); isocitrate lyase, slow-binding inhibition (6-22); kynurenine aminotransferase (23);

methylaspartate mutase, *or* glutamate mutase (24,25); 2-methyleneglutarate mutase (26-28); phosphofructokinase (29); succinate dehydrogenase, *or* succinate oxidase (30-34). **1**. Huynh & Snell (1985) *J. Biol. Chem.* **260**, 2379. **2**. Thatcher & Cain (1975) *Eur. J. Biochem.* **56**, 193. **3**. Webb (1966) *Enzyme and Metabolic Inhibitors*, vol. 2, p. 279, Academic Press, New York. **4**. Jacobsohn (1953) *Enzymologia* **16**, 113. **5**. Fonda (1972) *Biochemistry* **11**, 1304. **6**. Daron & Gunsalus (1963) *Meth. Enzymol.* **6**, 622. **7**. McFadden (1969) *Meth. Enzymol.* **13**, 163. **8**. Morrison (1982) *Trends Biochem. Sci.* **7**, 102. **9**. Munir, Hattori & Shimada (2002) *Arch. Biochem. Biophys.* **399**, 225. **10**. Spector (1972) *The Enzymes*, 3rd ed., **7**, 357. **11**. Runquist & Kruger (1999) *Plant J.* **19**, 423. **12**. Patel & McFadden (1978) *Exp. Parasitol.* **44**, 262. **13**. McFadden & Purohit (1977) *J. Bacteriol.* **131**, 136. **14**. Rittenhouse & McFadden (1974) *Arch. Biochem. Biophys.* **163**, 79. **15**. Tsukamoto, Ejiri & Katsumata (1986) *Agric. Biol. Chem.* **50**, 409. **16**. Höner zu Bentrup, Miczak, Swenson & Russell (1999) *J. Bacteriol.* **181**, 7161. **17**. Schloss & Cleland (1982) *Biochemistry* **21**, 4420. **18**. Hoyt, Johnson & Reeves (1991) *J. Bacteriol.* **173**, 6844. **19**. Giachetti, Pinzauti, Bonaccorsi, Vincenzini & Vanni (1987) *Phytochemistry* **26**, 2439. **20**. DeLucas, Amor, Diaz, Turner & Laborda (1997) *Mycol. Res.* **101**, 410. **21**. Chell, Sundaram & Wilkinson (1978) *Biochem. J.* **173**, 165. **22**. Hönes, Simon & Weber (1991) *J. Basic Microbiol.* **31**, 251. **22**. Mason (1959) *J. Biol. Chem.* **234**, 2770. **23**. Hartzoulakis & Gani (1994) *Proc. Indian Acad. Sci. Chem. Sci.* **106**, 1165. **24**. Hartzoulakis & Gani (1994) *Proc. Indian Acad. Sci. Chem. Sci.* **106**, 1165. **25**. Leutbecher, Bocher, Linder & Buckel (1992) *Eur. J. Biochem.* **205**, 759. **26**. Barker (1972) *The Enzymes*, 3rd ed., **6**, 509. **27**. Zelder & Buckel (1993) *Biol. Chem. Hoppe-Seyler* **374**, 85. **28**. Kung & Stadtman (1971) *J. Biol. Chem.* **246**, 3378. **29**. Passonneau & Lowry (1963) *Biochem. Biophys. Res. Commun.* **13**, 372. **30**. Booth, Taylor, Wilson & DeEds (1952) *J. Biol. Chem.* **195**, 697. **31**. Lang & Bässler (1953) *Biochem. Z.* **323**, 456. **32**. Veeger, DerVartanian & Zeylemaker (1969) *Meth. Enzymol.* **13**, 81. **33**. Hatefi (1978) *Meth. Enzymol.* **53**, 27. **34**. Hatefi & Stiggall (1976) *The Enzymes*, 3rd ed., **13**, 175.

ITF2357, *See* Givinostat

ITMN-191, *See* Danoprevir

Ivabradine

This first-in-class HCN channel blocker (FW = 468.59 g/mol; CAS 155974-00-8), also named 3-[3-({[(7*S*)-3,4-dimethoxy-bicyclo[4.2.0]octa-1,3,5-trien-7-yl]methyl}(methyl)amino)propyl]-7,8-dimethoxy-2,3,4,5-tetrahydro-1*H*-3-benzazepin-2-one, S-16257, and Corlanor®, is a bradycardic agent that acts on the I_f (*or* I_q) current, where subscripts f and q stand for "funny" or "queer" to connote the channel's unusual properties, as compared to already characterized ion currents. The Hyperpolarization-activated Cyclic Nucleotide-gated (*or* HCN) I_h channel is highly expressed in the sinoatrial node (SAN), where its regulating pacemaker activity. HCN channels are of voltage-gated K^+ (K_V) and cyclic nucleotide-gated (CNG) channels that consist of four identical or four non-identical subunits that become integrally embedded in the cell membrane, creating a potassium ion-conducting pore. HCN subunits have six membrane-spanning domains (numbered S1 to S6) that include: a putative voltage sensor (at S4) and a pore region (between S5 and S6) and a cyclic nucleotide-binding domain (CNBD) in the C-terminus. Within the SAN, the most prominently expressed I_h channel is HCN4, with HCN2 and HCN1 present at moderate and low levels. When tested in rabbit isolated SAN tissue, S-16257 and UL-FS 49 (1, 3, and 10 µM) were equipotent in slowing spontaneous APs firing predominantly by decreasing the rate of diastolic depolarization (at 3 µM, –24% and –28 %, respectively) (1). Ivabradine selectively inhibits the pacemaker I_h current in a dose-dependent manner, slowing heart rate and allowing more time for blood to flow to the myocardium. This anti-anginal and anti-ischemic agent selectively and specifically inhibits the I_f current in the sino-atrial node, provides pure heart rate reduction without altering other cardiac parameters, such as conduction, and without affecting other hemodynamic parameters. It is approved for the treatment of coronary artery disease and heart failure. Recent work (3) demonstrated that ivabradine can also interact with KCNH2-encoded human Ether-à-go-go-Related Gene (or hERG) potassium channels that strongly influence ventricular repolarization and susceptibility to torsades de pointes arrhythmia. Indeed, ivabradine does not discriminate between hERG and HCN channels: it inhibits Ih ERG with similar potency to that reported for native I_f and HCN channels, with S6 binding determinants resembling those observed for HCN4 (3). In 2015, the U.S. Food and Drug Administration (FDA) approved Corlanor to reduce the risk of hospitalization for worsening heart failure in patients with stable, symptomatic chronic heart failure with left ventricular ejection fraction (LVEF) ≤35 percent, who are in sinus rhythm with resting heart rate ≥70 beats per minute, and either are on maximally tolerated doses of β-blockers or have a contraindication to β-blocker use. **1**. Thollon, Cambarrat, Vian, *et al.* (1994) *Brit. J. Pharmacol.* **112**, 37. **2**. Deedwania (2013) *Drugs* **73**, 1569. **3**. Melgari, Brack, Zhang, *et al.* (2015) *J. Am. Heart Assoc.* **4**, e001813.

Ivermectin

Ivermectin B$_{1a}$ (R = CH$_3$)
Ivermectin B$_{1b}$ (R = H)

This semi-synthetic anthelmintic and insecticide consists primarily of two components, 80% 22,23-dihydroabamectin B$_{1a}$ (FW = 875.11 g/mol; CAS 70288-86-7) and 20% 22,23-dihydroabamectin B$_{1b}$ (FW = 861.08 g/mol; CAS 71827-03-7). Both are slightly soluble in water (~4 µg/mL) and are very soluble in methyl ethyl ketone and propylene glycol. The drug binds to glutamate-gated chloride channels (GluCls) in the membranes of invertebrate nerve and muscle cells, causing increased permeability to chloride ions, resulting in cellular hyper-polarization, followed by paralysis and death. Ivermectin has been used to treat heartworm in dogs and human filarial infections, particularly as a donated product for onchocerciasis and lymphatic filariasis. Ivermectin augments immune responses and impairs the neuromuscular function of the parasites, leading to paralysis. It is also an agonist of the glycine receptor chloride channel and is a substrate of CYP3A4. **Key Pharmacokinetic Parameters:** *See* Appendix II in Goodman & Gilman's THE PHARMACOLOGICAL BASIS OF THERAPEUTICS, 12th Edition (Brunton, Chabner & Knollmann, eds.) McGraw-Hill Medical, New York (2011). **Target(s):** Ca2+-transporting ATPase (1); P-glycoprotein, multidrug resistance factor (2). **1**. Bilmen, Wootton & Michelangeli (2002) *Biochem. J.* **366**, 255. **2**. Didier & Loor (1996) *Anticancer Drugs* **7**, 745.

Ivosidenib

This first-in-class orally active IDH1 inhibitor (FW = 582.96 g/mol; CAS 1448347-49-6; Solubility: 100 mg/mL DMSO; < 1 mg/mL H$_2$O), also known as AG-120 and (*S*)-*N*-((*S*)-1-(2-chlorophenyl)-2-((3,3-difluorocyclobutyl)-amino)-2-oxoethyl)-1-(4-cyano-pyridin-2-yl)-*N*-(5-fluoro-pyridin-3-yl)-5-oxopyrrolidine-2-carboxamide, targets a mutant form

of Isocitrate Dehydrogenase Type-1 located within the cytoplasm, which inhibits formation of the oncometabolite, 2-hydroxyglutarate (2HG), inducing cellular differentiation and inhibiting cellular proliferation in IDH1-expressing tumor cells. **1**. Agresta, *et al.* (2015) PCT Int. Appl. (2015), WO 2015127172, A1 20150827.

IVX-214, *See* HMN-214

IWP-2

This novel Wnt pathway inhibitor (FW = 466.60 g/mol; CAS 686770-61-6; Solubility: <1 mg/mL DMSO, Ethanol, or H$_2$O), also named *N*-(6-methyl-2-benzothiazolyl)-2-[(3,4,6,7-tetrahydro-4-oxo-3-phenylthieno[3,2-*d*]pyrimidin-2-yl)thio]acetamide, targets Wnt processing and secretion (IC$_{50}$ = 27 nM), selectively blocking Porcupine-mediated Wnt palmitoylation, without affecting Wnt/β-catenin generally and without inhibiting Wnt-stimulated cellular responses (1). IWP-2 also reduces secretion of functionally active Wnt5a, suppressing both bacterial phagocytosis and the secretion of proinflammatory cytokines, and also accelerating the rate of bacterial killing (2). Suppression of Wnt secretion from Mesenchymal stem cells (MSCs) by IWP-2 (5 µmol/L) reduces the efficacy of tyrode (*or* ConT) in inducing phospho-LRP6 and to increase cardiac conduction (3). Such findings support the hypothesis that MSC-mediated Cx43 upregulation is partly mediated through the canonical Wnt signaling pathway (3). IWP-2 decreases MKN28 gastric cancer cell proliferation, migration and invasion, as well as elevates caspase 3/7 activity (4). IWP-2 downregulates the transcriptional activity of the Wnt/β-catenin signaling pathway and downregulated the expression levels of downstream Wnt/β-catenin target genes in MKN28 cells (4). (***See*** *IWP-4; IWP-L6; IWR-1-endo*) **1**. Chen, Dodge, Tang, *et al.* (2009) *Nature Chem. Biol.* **5**, 100. **2**. Maiti, Naskar & Sen (2012) *Proc. Natl. Acad. Sci. USA* **109**, 16600. **3**. Mureli, Gans, Bare, *et al.* (2013) *Am. J. Physiol. Heart Circ. Physiol.* **304**, H600. **4**. Mo, Li, Chen, *et al.* (2013) *Oncol. Lett.* **5**, 1719.

IWP-4

This novel Wnt pathway inhibitor (FW = 496.62 g/mol; CAS 686772-17-8; IUPAC Name: *N*-(6-methyl-2-benzothiazolyl)-2-[(3,4,6,7-tetrahydro-3-(2-methoxyphenyl)-4-oxothieno[3,2-*d*]pyrimidin-2-yl)thio]acetamide) has a similar mode of action to IWP-2 and IWP-L6, inhibiting Wnt/β-catenin signaling (IC$_{50}$ = 25 nM), with minimal effect on Notch and Hedgehog signaling pathways. (***See*** *IWP-2; IWP-L6; IWR-1-endo*) **1**. Chen, *et al.* (2009) *Nature Chem. Biol.* **5**, 100. **2**. Lian, *et al.* (2012) *Proc. Natl. Acad. Sci. U.S.A.* **109**, 1848. **3**. Frith, *et al.* (2013) *PLoS One* **8**, e82931.

IWP-L6

This potent Wnt signal transduction pathway inhibitor (FW = 472.58 g/mol; CAS 1427782-89-5; Solubility: 25 mg/mL DMSO; < 1 mg/mL H_2O; *N*-(5-phenyl-2-pyridinyl)-2-[(3,4,6,7-tetrahydro-4-oxo-3-phenylthieno[3,2-*d*]-pyrimidin-2-yl)thio]acetamido) targets Porcupine, *or* Porcn (EC_{50} = 0.5 nM), a membrane-bound *O*-acyltransferases that catalyzes palmitoylation of Wnt proteins, which is required for their secretion and activity (1). Dysregulated Porcn commonly results in developmental disorders, such as focal dermal hypoplasia (Goltz syndrome), and Porcn hyperactivity is often associated with cancerous cell growth. (*See IWP-2; IWP-4; IWR-1-endo*) **1**. Wang, Moon, Dodge, *et al.* (2013) *J. Med. Chem.* **56**, 2700.

Ixabepilone

This novel epothilone B analogue (FW = 506.70 g/mol; CAS 219989-84-1), also known as azaepothilone B, its code name BMS-247550, its trade name Ixempra®, and the IUPAC name (1*R*,5*S*,6*S*,7*R*,10*S*,14*S*,16*S*)-6,10-dihydroxy-1,5,7,9,9-pentamethyl-14-[(*E*)-1-(2-methyl-1,3-thiazol-4-yl)-prop-1-en-2-yl]-17-oxa-13-azabicyclo[14.1.0]heptadecane-8,12-dione), inhibits microtubule disassembly and demonstrates antineoplastic properties that showed greater promise than paclitaxel (*e.g.*, requiring a less frequent treatment schedule than taxol and allowing for oral administration). *In vitro*, ixabepilone is twice as potent as paclitaxel in inducing tubulin polymerization (1). Like paclitaxel, BMS-247550 is a highly potent cytotoxic agent capable of killing cancer cells at low nanomolar concentrations; however, it retains its antineoplastic activity against human cancers that are naturally insensitive to paclitaxel or that have developed resistance to paclitaxel, both *in vitro* and *in vivo* (1). This enhanced cytotoxic effect has been attributed to its poor substrate properties with p-glycoprotein drug multidrug resistance protein as well as its high affinity for various β-tubulin isoforms. **Background:** The epothilones constitute a novel class of non-taxane microtubule-stabilizing agents isolated from the cellulose degrading myxobacteria, *Sorangium cellulosum*. Their chemical structures are distinct from taxanes, stereochemically less complex, and therefore more amenable to synthetic modification (2). Other epothilones include patupilone (epothilone B), BMS 310705, sagopilone (ZK-EPO), KOS-862 (epothilone D), and KOS-1584. **Pharmacodynamics & Pharmacokinetics:** Ixabepilone induces cell cycle arrest at the G_2-M phase transition, followed by apoptotic cell death of paclitaxel-resistant MDA-MB-468 cells (3). Treatment with EpoB triggers a conformational change in the Bax protein and its translocation from the cytosol to the mitochondria, which is accompanied by cytochrome *c* release from the inter-membrane space of mitochondria into the cytosol (3). In patients treated at 40 mg/m² (administered as a 1-hour intravenous infusion every 3 weeks), Plasma ixabepilone concentration and the severity of neutropenia correlate with the level of peak microtubule bundle formation (MBF) in peripheral blood mononuclear cells (PBMCs). When used in monotherapy or in combination with capecitabine, Ixempra is indicated for the treatment of metastatic or locally advanced breast cancer in patients after failure of anthracycline and a paclitaxel. **1**. Lee, Borzilleri, Fairchild, *et al.* (2001) *Clin Cancer Res.* **7**, 1429. **2**. Kit, Cheng, Bradley, & Budman (2008) *Biologics* **2**, 789. **3**. Yamaguchi, Paranawithana, Lee, *et al.* (2002) *Cancer Res.* **62**, 466.

Ixazomib

This orally bioavailable, second-generation proteasome inhibitor (FW = 361.03 g/mol; CAS 1072833-77-2; Solubility: 72 mg/mL DMSO), codenamed MLN2238 and known as (*R*)-1-(2-(2,5-dichlorobenz-amido)acetamido)-3-methylbutylboronic acid, is a boronate peptidomimetic that targets the chymotrypsin-like activity at the β5 proteolytic site of the 20S proteasome (IC_{50} = 3.4 nM; K_i = 0.93 nM) as well as the caspase-like activity at the β1 site (IC_{50} = 31 nM) and trypsin-like activity at the β2 site (IC_{50} = 3.5 μM). Ixazomib has greater pharmacodynamic effects in tissues than bortezomib, showing activity in both solid tumor and hematologic preclinical xenograft models. (*See also Bortezomib; Carfilszomib; Oprozomib*) **1**. Kupperman, Lee, Cao, *et al.* (2010) *Cancer Res.* **70**, 1970. *Erratum: Cancer Res.* (2010) **70**, 3853. **2**. Lee, Fitzgerald, Bannerman, *et al.* (2011) *Clin Cancer Res.* **17**, 7313.

J 104129

This potent, orally bioavailable muscarinic acetylcholine receptor antagonist (FW = 500.63 g/mol; CAS 257603-40-0; Solubility: 100 mM in DMSO), also named (αR)-α-cyclopentyl-α-hydroxy-N-[1-(4-methyl-3-pentenyl)-4-piperidinyl]benzeneacetamide, targets M_3 muscarinic acetylcholine receptors (K_i = 4.2 nM), displaying 5x and ~110x selectivity over M_1 (K_i = 19 nM) and M_2 (K_i = 490 nM) receptors. J 104129 also exhibits >250x selectivity in its inhibition of acetylcholine-induced bronchoconstriction (K_B = 3.3 nM) over acetylcholine-induced bradycardia. **1.** Mitsuya, Mase, Tsuchiya, *et al.* (1999) *Bioorg. Med. Chem.* **7**, 2555. **2.** Mitsuya, Ogino, Kawakami, *et al.* (2000) *Bioorg. Med. Chem.* **8**, 825. **3.** Mitsuya, Kawakami, Ogino, Miura & Mase (1999) *Bioorg. Med. Chem.* **9**, 2037.

J-104871

This potent inhibitor (FW$_{free-acid}$ = 708.68 g/mol) of rat brain protein farnesyltransferase (IC$_{50}$ = 3.9 nM) is without effect on squalene synthase. J-104871 weakly inhibits protein geranylgeranyltransferase I and Ras processing in activated H-*ras*-transformed NIH 3T3 cells. **Target(s):** protein farnesyltransferase (1,2). **1.** Gibbs (2001) *The Enzymes*, 3rd ed. (Tamanoi & Sigman, eds), **21**, 81. **2.** Yonemoto, Satoh, Arakawa, *et al.* (1998) *Mol. Pharmacol.* **54**, 1.

J-107088, See Edotecarin

(±)-J 113397

This potent and selective non-peptidyl NOP receptor antagonist (FW = 399.58 g/mol; CAS 217461-40-0; Solubility: 50 mM in DMSO; 50 mM in Ethanol), a racemic mixture also named (±)-1-[(3R*,4R*)-1-(cyclooctylmethyl)-3-(hydroxymethyl)-4-piperidinyl]-3-ethyl-1,3-dihydro-2H-benzimidazol-2-one, targets the nociceptin receptor (*or* NOP), also known as the orphanin FQ receptor *or* κ-type 3 opioid receptor, (IC$_{50}$ = 2.3 nM), κ-opioid receptor (IC$_{50}$ = 1400 nM), μ-opioid receptor (IC$_{50}$ 2200 nM) and δ-opioid receptor (IC$_{50}$ > 10000 nM). **1.** Kawamoto, Ozaki, Itoh, *et al.* (1999) *J. Med. Chem.* **42**, 5061.

J 113863

This potent chemokine receptor antagonist (FW$_{iodide-salt}$ = 655.44 g/mol; CAS 353791-85-2; Solubility: 100 mM in DMSO; 50 mM in Ethanol), also named 1,4-*cis*-1-(1-cyclooocten-1-ylmethyl)-4-[[(2,7-dichloro-9H-xanthen-9-yl)carbonyl]amino]-1-ethylpiperidinium iodide, targets CCR$_1$ receptors (IC$_{50}$ = 0.9 nM for human; IC$_{50}$ = 5.8 nM for mouse), displaying high selectivity for human CCR$_3$ receptors (IC$_{50}$ = 0.58 nM), but not mouse CCR3 receptors (IC$_{50}$ = 460 nM) (1). J-113863 improves paw inflammation, joint damage and dramatically reduces cell infiltration into joints in a mouse collagen-induced arthritis model (2). (*Note*: J-113863 is an isomer of the chemokine CCR$_1$ and CCR$_3$ receptor antagonist, UCB 35625.) **1.** Naya, Sagara, Ohwaki, *et al.* (2001) *J. Med. Chem.* **44**, 1429. **2.** Amat, Benjamin, Williams, *et al.* (2006) *Brit. J. Pharmacol.* **149**, 666.

JA-2, See N-(1-(RS)-Carboxy-3-phenylpropyl)-L-alanyl-α-aminoisobutyryl-L-tyrosine-p-aminobenzoate

Jaceosidin

This flavonoid (FW = 330.29 g/mol; CAS 18085-97-7), systematically named 5,7-dihydroxy-2-(4-hydroxy-3-methoxyphenyl)-6-methoxy-4H-1-benzopyran-4-one, from *Artemisia princeps* (Japanese mugwort), is more potent than cisplatin as an inhibitor of human endometrial cancer cells, increasing the phosphorylation of Cdc25C, ATM-Chk1/2, and ERK, leading to the inactivation of the Cdc2-cyclin B$_1$ complex, and resulting in G$_2$/M cell cycle arrest. **1.** Lee, Kim, Ahn, *et al.* (2013) *Food Chem. Toxicol.* **55**, 214.

JAK3 Inhibitor I, See WHI-P131
JAK3 Inhibitor II, See WHI-P154
JAK3 Inhibitor III, See WHI-P97
JAK3 Inhibitor IV, See ZM 39923
JAK3 Inhibitor V, See ZM 429829

JAK3 Inhibitor VI

This cell-permeable, selective JAK3 inhibitor (FW$_{Mesylate-Salt}$ = 383.42 g/mol; CAS 856436-16-3; Solubility: 5 mg/mL in DMSO and 1 mg/mL in H$_2$O), also named (3Z)-5-(3-pyridinyl)-3-(1H-pyrrol-2-ylmethylene)-1,3-dihydro-2H-indol-2-one(methanesulfonate salt)), targets JAK3 tyrosine kinase, IC$_{50}$

= 27 nM) in vitro, with action against IL-2-induced cell proliferation (IC$_{50}$ = 760 nM in mouse CTLL) and IL-2-induced cell proliferation (IC$_{50}$ = 250 nM in human T-cells) (1,2) It is also a mutant-specific inhibitor for gatekeeper mutant of Epidermal Growth Factor Receptor, selectively inhibiting EGFR T790M/L858R, with weaker action against wild-type EGFR *in vitro* (3). JAK3 inhibitor VI also specifically reduced autophosphorylation of EGFR T790M/L858R in NCI-H1975 cells upon EGF stimulation, but did not show the inhibitory effect on wild-type EGFR in A431 cells. Furthermore, JAK3 inhibitor VI suppressed the proliferation of NCI-H1975 cells, but showed limited inhibitory effects on the wild-type EGFR-expressing cell lines A431 and A549 (2). A docking simulation between JAK3 inhibitor VI and the ATP-binding pocket of EGFR T790M/L858R predicted a potential binding status with hydrogen bonds. Estimated binding energy of JAK3 inhibitor VI to EGFR T790M/L858R was more stable than its binding energy to the wild-type EGFR (2). **1**. Adams, Aldous, Amendola, *et al.* (2003) *Bioorg. Med. Chem. Lett.* **13**, 3105. **2**. Chen, Thakur, Clark, *et al.* (2006) *Bioorg. Med. Chem. Lett.* **16**, 5633. **3**. Nishiya, Sakamoto, Oku, Nonaka & Uehara (2015) *World J. Biol. Chem.* **6**, 409.

JANEX-1 *See WHI-P131*

Janus Green B

This supravital mitochondria-staining dye (FW$_{HCl-Salt}$ = 511.07 g/mol; CAS 2869-83-2), also named 3-(diethylamino)-7-{4-(dimethylamino)phenyl]-azo}-5-phenylphenazinium chloride, interacts with double-stranded DNA and is also incorporated in mitochondria (1). The latter property is routinely exploited in histology. **Target(s):** β-fructo-furanosidase, *or* invertase (2,3), fumarase (4), and urease (5). **1**. Huang, Li, Huang & Li (2000) *Analyst* **125**, 1267. **2**. Quastel & Yates (1936) *Enzymologia* **1**, 60. **3**. Massart (1950) *The Enzymes*, 1st ed. (Sumner & Myrbäck, eds), **1** (part 1), 307. **4**. Quastel (1931) *Biochem. J.* **25**, 898. 5. Quastel (1932) *Biochem. J.* **26**, 1685.

Jasplakinolide

This cytotoxic, cell-permeant cyclodepsipeptide (FW = 695.66 g/mol; CAS 102396-24-7; Soluble to 2 mg/mL in DMSO), also named cyclo[(3*R*)-3-(4-hydroxyphenyl)-β-alanyl-(2*S*,4*E*,6*R*,8*S*)-8-hydroxy-2,4,6-trimethyl-4-non-enoyl-L-alanyl-2-bromo-*N*-methyl-D-tryptophyl], from the marine sponge *Jaspis johnstoni* binds to monomeric actin (K$_d$ = 15 nM; competitive with phalloidin) and promotes actin filament self-assembly by inhibiting actin depolymerization (1). Jasplakinolide exhibits both antifungal and antiproliferative activity. (Microtubules, intermediate filaments, and their associated proteins and processes are unaffected.) Cells allowed to recover from jasplakinolide treatment continue to grow, but show severe changes in the cell pattern and displacement of organelles, suggesting that even after removal of the drug, some basic morphogenetic features remain altered.

Jasplakinolide also blocks store-mediated Ca^{2+} entry into cells as well as enhances the induction of apoptosis by interleukin-2 deprivation. Jasplakinolide has a much greater effect on Mg^{2+}-actin than on Ca^{2+}-actin (2). Competitive binding studies using rhodamine-phalloidin suggest that jasplakinolide binds to F-actin competitively with phalloidin with a dissociation constant of ~15 nM. This compares favorably to the previously reported IC$_{50}$ of 35 nM for the antiproliferative effect of jasplakinolide on PC3 prostate carcinoma cells. Somewhat paradoxically, jasplakinolide stabilizes actin filaments *in vitro*, but disrupts actin filaments *in vivo*, where it also induces amorphous aggregation of monomeric actin. The answer to this paradox lies in the ability of this agent to enhances the rate of nucleation markedly by switching the critical nucleus size from four to approximately three subunits, an action that is mechanistically consistent with the localization of the jasplakinolide-binding site at an interface of three actin subunits (3). Any or all of the above actions may account for jasplakinolide's potent antifungal and antitumor properties. **1**. Holzinger (2001) *Methods Mol. Biol.* **161**, 109. **2**. Bubb, Senderowicz, Sausville, Duncan & Korn (1994) *J. Biol. Chem.* **269**, 14869. **3**. Bubb, Spector, Beyer & Fosen (2000) *J. Biol. Chem.* **275**, 5163.

Jatrophone

This antibiotic (FW = 300.40 g/mol; CAS 29444-03-9; λ$_{max}$ = 285 nm), diterpene of *Jatropha isabelli* (*Euphorbiaceae*), is active against *Leischmania amazonensis* and *Trypanosoma cruzi* and cancer of the nasalpharyx (1,2). Jatrophone also binds tightly to RNA, greatly increasing its thermal stability. In the isolated longitudinal muscle from guinea-pig ileum, 1.6–12.8 μM jatrophone produces a concentration-dependent inhibition of acetylcholine, bradykinin and histamine-induced contractions. In this tissue, jatrophone was about 5 to 8-fold more potent in inhibiting acetylcholine and bradykinin than in uterine muscle. **1**. Akendengue-Milama, Laurens & Hocquemiller (1999) *Parasite* **6**, 3. **2**. Gutiérrez (2007) *Handbook of Compounds with Antiprotozoal Activity from Plants*, p. 86, Nova Science, Hauppauge, NY. **3**. Calixo & Sant'Ana (1987) *Phytotherap. Res.* **1**, 122.

JE 049, *See Icatibant*

Jervine

This signal transduction inhibitor (FW = 425.61 g/mol; CAS 469-59-0; Solubility: 50 mM in Ethanol; 20 mM in DMSO), also named 2'*R*,3*S*,3'*R*,3'*aS*,6'*S*,6*aS*,6*bS*,7'*aR*,11*aS*,11*bR*)-2,3,3'*a*,4,4',5',6,6',6*a*,6*b*,7,7', 7'*a*,8,11*a*,11*b*-hexadecahydro-3-hydroxy-3',6',10,11*b*-tetramethyl-spiro[9*H*-benzo[*a*]fluorene-9,2'(3'*H*)-furo[3,2-*b*]pyridin]-11(1*H*)-one, from the white hellebore (*Veratrum album*) and green hellebore (*V. viride*), binds directly to Smoothened, *or* Smo (IC$_{50}$ = 500-700 nM), targeting signal transduction via Sonic hedgehog (1) as well as the modification of proteins by cholesterol (2). **1**. Cooper, Porter, Young & Beachy (1998) *Science* **280**, 1603. **2**. Mann & Beachy (2000) *Biochim. Biophys. Acta* **1529**, 188.

JFD01307SC

This antimetabolite (FW = 193.22 g/mol; CAS 51070-56-5), also known as *N*-(tetrahydro-1,1-dioxido-3-thienyl)glycine, is a methionine sulfoximine analogue that targets *Mycobacterium tuberculosis* glutamine synthetase, presumably by mimicking the tetrahedral ammonia adduct formed upon attack of ammonia on the enzyme's catalytically essential γ-glutamyl-phosphate intermediate. JFD01307SC shows activity against *M. tuberculosis*, with a minimal inhibitory concentration (MIC) in the 8–16 μg/mL range. **1**. Lamichhane, Freundlich, Ekins, *et al.* (2011) *mBio.* **2**, e00301.

JH-II-127

This highly potent and selective, brain-penetrating LRRK2 inhibitor (FW = 416.87 g/mol), also named 2-anilino-4-methylamino-5-chloropyrrolo-pyrimidine, targets Leucine-Rich Repeat Kinase-2, showing an IC_{50} value of 6.2 nM for the wild-type kinase and IC_{50} values of 2.2 nM and 47.7 nM, respectively, for the Gly-2019-Ser and Ala-2016-Thr mutants. JH-II-127 substantially inhibits Ser-910 and Ser-935 phosphorylation of both wild-type LRRK2 and G2019S mutant at a concentration of 0.1–0.3 μM in a variety of cell types. It also inhibits phosphorylation at Ser-935 in mouse brain after oral delivery of doses as low as 30 mg/kg. The Gly-2019-Ser mutation, which enhances catalytic activity, is one of a small number of LRRK2 mutations likely to cause Parkinson's disease, especially the sporadic type. Notably, the Ala-2016-Thr catalytically is active, but resists inhibition by H-1152, Y-27632, and sunitinib (2). **1**. Hatcher, Zhang, Choi, *et al.* (2015) *ACS Med. Chem. Lett.* **6**, 584. **2**. Nichols, Dzamko, Hutti, *et al.* (2009) *Biochem J.* **424**, 47.

JHW 007

This potent dopamine uptake inhibitor ($FW_{HCl-Salt}$ = 421.95 g/mol; CAS = 202645-74-7; Solubility: 100 mM in DMSO), also named (3-*endo*)-3-[bis(4-fluorophenyl)methoxy]-8-butyl-8-azabicyclo[3.2.1]octane hydrochloride, displays high affinity for the dopamine transporter, *or* DAT (K_i = 25 nM). JHW-007 shows substantially weaker action against the serotonin transporter, *or* SERT (K_i = 1.73 μM) as well as the norepinephrine transporter, *or* NET (K_i = 1.33 μM), the latter also known as Solute Carrier family 6 member 2, *or* SLC6A2. JHW-007 suppresses cocaine's dopamine

effects in a dose-dependent manner. **1**. Kopajtic, *et al.* (2010) *J. Pharmacol. Exp. Ther.* **335**, 703. **2**. Velaquez-Sanchez, *et al.* (2010) *Eur. Neuropsychopharmacol.* **20**, 501. **3**. Kristensen, *et al.* (2011) *Pharm. Rev.* **63**, 585.

JIB 04

This pan Jumonji histone demethylase inhibitor ($FW_{free-base}$ = 308.77 g/mol; CAS 199596-05-9; Solubility: 100 mM DMSO; 50 mM in 1 equivalent of HCl), also named 5-chloro-2-[(*E*)-2-[phenyl(pyridin-2-yl)methyl-idene]hydrazin-1-yl]pyridine, targets $JARID_{1A}$ (K_i = 230 nM), $JMJD_{2E}$ (K_i = 340 nM), $JMJD_{2B}$ (K_i = 435 nM), $JMJD_{2A}$ (K_i = 445 nM), $JMJD_3$ (K_i = 855 nM) and $JMJD_{2C}$ (K_i = 1100 nM) demethylases (1,2). Grunicke & Hofmann (1997) *J. Med. Chem.* **40**, 4420. **2**. Wang, *et al.* (2013) *Nat. Commun.* **4**, 2035.

Jingzhaotoxin-I

DGECGGFWWKCGRGKPPCCKGYACSKTWGWCAVEAP

This venom-ed derived 33-residue polypeptide (MW = 3497 g/mol; Symbol: JZTX-I; Solubility: 1 mg/mL in H_2O), from the Chinese tarantula *Chilobrachys jing-zhao*, interacts with sodium channels, showing high affinity (IC_{50} = 32 nM) for tetrodotoxin-resistant subtype found in mammalian cardiac myocytes. Because it neither modifies the current-voltage relationships nor shifts the steady-state inactivation of sodium channels, JZTX-I defines a new subclass of spider toxins. It binds to the $Na_V1.5$ S3-S4 linker of domain II, thereby discriminating its action at cardiac sodium channels from inhibition of the presumably functionally similar, tetrodotoxin-resistant, neuronal isoforms. **1**. Xiao, Tang, Hu, *et al.* (2005) *J. Biol. Chem.* **280**, 12069.

JIP-1 (153-163)

This cJNK-selective peptide inhibitor (FW = 1341.6 g/mol; Sequence: RPKRPTTLNLF-NH2; CAS 438567-88-5; Solubility: 1 mg/mL in H_2O), corresponding to the C-terminal eleven residues of I-JIP (Inhibitor of JNK-based on JIP-1), binds directly to and inhibits recombinant c-Jun N-terminal kinase, but not its substrates, based on surface plasmon resonance data. (JNK-interacting protein-1 (JIP1) is a scaffolding protein that enhances JNK signaling by creating a proximity effect between JNK and upstream kinases.) The JNK subfamily of the mitogen-activated protein kinases (MAPKs) transduce extracellular signals to inform cytoplasmic and nuclear targets. Four residues (*e.g.*, Arg-156, Pro-157, Leu-160, or Leu-162) are critical for inhibition (1), a finding that extends an earlier report that peptides corresponding to residues 143–163 inhibit cJNK (2). **1**. Barr, Kendrick & Bogoyevitch (2002) *J. Biol. Chem.* **277**, 10987. **2**. Bonny, Oberson, Negri, Sauser & Schorderet (2001) *Diabetes* **50**, 77.

JJKK 048

This potent and selective MAGL inhibitor (FW = 434.44 g/mol; CAS 1515855-97-6; Soluble to 100 mM in DMSO), also named 4-[bis(1,3-benzodioxol-5-yl)methyl]-1-piperidinyl]-1H-1,2,4-triazol-1-yl-methanone, targets monoacylglycerol lipase (IC$_{50}$ = 0.4 nM) and exhibits >13,000x and ~630x selectivity over FAAH and ABHD6, respectively (1,2). MAGL catalyzes the hydrolysis of 2-arachidonoylglycerol (or 2-AG), thereby terminating the signaling by this endocannabinoid. 2-AG hydrolysis liberates arachidonic acid, the principal substrate for the neuroinflammatory prostaglandins. MAGL also redirects lipid stores toward protumorigenic signaling lipids (1). Thus MAGL inhibitors may have great therapeutic potential. **1**. Aaltonen, et al. (2013) Chem. Biol. **20**, 379. **2**. Laitinen, et al. (2014) Mol. Pharmacol. **85**, 510.

JK 184

This signal transduction modulator (FW = 350.44 g/mol; CAS 315703-52-7; Solubility: 100 mM in DMSO; 25 mM in Ethanol), also named N-(4-ethoxyphenyl)-4-(2-methylimidazo[1,2-a]pyridin-3-yl)-2-thiazolamine, is a alcohol dehydrogenase-7 (Adh7) inhibitor (IC$_{50}$ = 210 nM), a potent downstream hedgehog (Hh) signaling inhibitor that prevents Gli-dependent transcriptional activity (IC$_{50}$ = 30 nM), and a microtubule depolymerizing agent in vitro. JK-184 also exhibits antiproliferative activity in a range of cancer cell lines (GI$_{50}$ = 3-21 nM) and human cancer cell xenografts in vivo. **1**. Lee, et al. (2007) Chembiochem. **8**, 1916. **2**. Cupido, et al. (2009) Angew. Chem. Int. Ed. Engl. **48**, 2321.

JKC 301

This cyclic pentapeptide (FW$_{free-acid}$ = 624.74 g/mol; Sequence: cyclo[D-Asp-L-Pro-D-Ile-L-Leu-D-Trp]; CAS 136553-96-3) is a selective endothelin (ET-A) receptor antagonist (1,2). Selective ET-A antagonists (e.g., BQ123, BMS-182,874 and JKC-301) all display low affinities at the endothelin receptors, whereas the selective ET-B antagonist BQ-788 showed moderate affinity, effective in the low nanomolar range (3) **1**. Fekete, Yang, Kimura & Aviv (1995) J. Pharmacol. Exp. Ther. **275**, 215. **2**. Nambi, Pullen, Kincaid, et al. 1997) Mol. Pharmacol. **52**, 582. **3**. Widdowson & Kirk (1996) Br. J. Pharmacol. **118**, 2126.

JKC 363

This potent and selective MC$_4$ receptor antagonist (FW = 1506.72 g/mol; CAS 436083-30-6; Sequence: XEHXRWGCPPKD, with a disulfide bridge between residues 1 - 8, where X-1 = Mpr, X-4 = D-2-Nal, Asp-12 = C-terminal amide); Solubility: 1 mg/mL in H$_2$O) targets the melanocortin MC$_4$ receptor (IC$_{50}$ = 0.5 nM), antagonizing the effects of α-MSH, suppressing thyrotropin-releasing hormone (TRH) release, attenuating food intake, and reducing formalin-induced pain in vivo (1-3). JKC 363 shows weaker action against MC$_3$ (IC$_{50}$ = 45 nM). **1**. Kim, Small, Russell, et al. (2002) J. Neuroendocrinol. **14**, 276. **2**. Verty, McFarlane, McGregor & Mallet (2004) Endocrinol. **145**, 3224.

JLK 6

This γ-secretase-mediated β-APP processing inhibitor (FW = 225.63 g/mol; CAS 62252-26-0; Solubility: 100 mM in DMSO; 25 mM in Ethanol), also named 7-amino-4-chloro-3-methoxy-1H-2-benzopyran, selectively targets β-Amyloid Precursor Protein cleavage without affecting other γ-secretase-mediated pathways (1). JLK-6 prevents recovery of Aβ40 and Aβ42 from HEK293 cell that overexpress wild-type or Swedish-mutated βAPP (IC$_{50}$ ~ 30 μM), while displaying no effect on Notch cleavage and Notch-mediated intracellular signaling. JLK-6 displays no activity on BACE1, BACE2, α-secretase, proteasome activity, or GSK3β. **1**. Petit, et al. (2001) Nature Cell. Biol. **3**, 507. **2**. Petit, et al. (2003) J. Neurosci. Res. **74**, 370. **3**. Hellstrom, et al. (2007) Nature **445**, 776.

JLKG See 7-Amino-4-chloro-3-methoxyisocoumarin

JMV 390-1

This neurotensin inhibitor and neuromedin N-degrading enzyme inhibitor (FW = 449.55 g/mol; CAS 148473-36-3; Solubility: 0.80 mg/mL in 0.01M sodium bicarbonate), also named N-[(2R)-4-hydroxyamino)-1,4-dioxo-2-(phenylmethyl)butyl]-L-isoleucyl-L-leucine, targets the major neurotensin and neuromedin N degrading enzymes, with respective IC$_{50}$ values are 31, 40, 52, and 58 nM for endopeptidase 24.15, endopeptidase 24.11, leucine aminopeptidase and endopeptidase 24.16. Hydroxamate-based inhibitors are often highly effective against metalloproteinases. Cytotoxic activity of Bacillus anthracis lethal factor is inhibited by leukotriene A$_4$ hydrolase and this metallopeptidase inhibitor. JMV 390-1 exhibits analgesic effects in various nociception assays. **1**. Kitabgi, et al. (1992) Neurosci. Lett. **142**, 200. **2**. Doulut, et al. (1993) J. Med. Chem. **36**, 1369. **3**. Menard, et al. (1996) Biochem. J. **320**, 687.

JMV 449

This potent NTR agonist (FW = 746.96 g/mol; CAS 139026-66-7; Solubility: 2 mg/mL acetonitrile (20% vol/vol) in acetic acid (0.1% vol/vol)), is a pseudopeptide corresponding neurotensin (NT) residues 8-13 (KKPYIL), in which the Lys-Lys peptide bond is replaced by the nonhydrolyzable pseudopeptide moiety (or Ψ(CH$_2$-NH)). JMV449 blocks

[^{125}I]-neurotensin binding (IC$_{50}$ = 0.15 nM) to the neurotensin receptor (NTR) in neonatal mouse brain and inhibits the contraction of guinea pig ileum (EC$_{50}$ = 1.9 nM). JMV-449 also produces long-lasting hypothermic, neuroprotective, and analgesic effects in mice, when administered centrally. **1**. Lugrin, Vecchini, Doulut, *et al.* (1991) *Eur. J. Pharmacol.* **205**, 191. **2**. Dubuc, Costentin, Doulut, *et al.* (1992) *Eur. J. Pharmacol.* **219**, 327. **3**. Torup, Borsdall & Sager (2003) *Neurosci. Lett.* **351**, 173.

JMV 2959

This small-molecule ghrelin receptor antagonist (FW$_{free-base}$ = 508.63 g/mol; FW$_{HCl-Salt}$ = 545.08 g/mol; Solubility (hydrochloride): 100 mg/mL DMSO = 5 mg/mL H$_2$O), known systematically as (*R*)-*N*-(2-(1*H*-indol-3-yl)-1-(4-(4-methoxybenzyl)-5-phenethyl-4*H*-1,2,4-triazol-3-yl)ethyl)-2-aminoacetamide, inhibits (IC$_{50}$ = 32 nM) [^{125}I-His^9]-ghrelin binding to the human ghrelin receptor-1a, *or* hGHS-1a, receptor (transiently expressed on LLC-PK1 cells), but does not induce any intracellular calcium mobilization (1,2). The dissociation constant of the receptor-JMV-2959 complex was determined by the Schild method. This experiment revealed a dissociation constant K_b of 19 ± 6 nM. **1**. Moulin, Demange, Bergé, *et al.* (2007) *J. Med. Chem.* **50**, 5790. **2**. Moulin, Brunel, Boeglin, *et al.* (2012) *Forum Amino Acid, Peptide & Protein Research* **12**, 1355.

JNJ 303

This I$_{Ks}$ blocker (FW = 440.98 g/mol; CAS 878489-28-2; Solubility: 25 mM in DMSO), also named 2-(4-chlorophenoxy)-2-methyl-*N*-[5-[(methylsulfonyl)amino]tricyclo[3.3.1.13,7]dec-2-yl]propanamide, potently and selectively targets the cardiac I$_{Ks}$ channel (IC$_{50}$ = 64 nM), displaying no effects on other cardiac channels (*i.e.*, IC$_{50}$ values are 3.3, >10, 11.1 and 12.6 μM for I$_{Na}$, I$_{Cas}$, I$_{to}$ and I$_{Kr}$ currents, respectively). The I$_{Ks}$ channel is a major repolarization current in the heart and responds rapidly to sympathetic nervous system stimulation, ensuring an adequate diastolic filling time as the heart rate increases (2). In cardiac myocytes, the I$_{Ks}$ channel is a macromolecular complex composed of a pore-forming α (KCNQ1) subunit and modulatory β (KCNE1) subunit, as well as intercellular proteins critical for controlling the phosphorylation state of the complex (2). **1**. Towart et al (2009) *J. Pharm. Toxicol. Meth.* **60**, 1. **2**. Osteen, Sampson & Kass (2010) *Proc. Natl. Acad. Sci. USA* **107**, 18751.

JNJ 5207787

This brain-penetrant and selective NPY antagonist (FW = 510.67 g/mol; CAS 683746-68-1; Solubility: 10 mM in DMSO (gentle warming); 10 mM in Ethanol (gentle warming)), also named (2*E*)-*N*-(1-acetyl-2,3-dihydro-1*H*-indol-6-yl)-3-(3-cyanophenyl)-*N*-[1-(2-cyclopentylethyl)-4-piperidinyl]-2-propenamide, displays 100x selectivity for the neuropeptide Y Y$_2$ receptor (IC$_{50}$ = 0.1 μM) over NPY Y$_1$, NPY Y$_4$ and Y$_5$ receptors and much greater selectivity versus a panel of fifty other receptors, ion channels and

transporters. **1**. Bonaventure, *et al.* (2004) *J. Pharmacol. Exp. Ther.* **308**, 1130. **2**. Jablonowski, *et al.* (2004) *Bioorg. Med. Chem. Lett.* **14**, 1239.

JNJ-7706621

This cyclin kinase-directed inhibitor (FW = 394.36 g/mol; CAS 443797-96-4; Solubility: 80 mg/mL DMSO; <1 mg/mL Water;; Formulation: Dissolve in 0.5% methylcellulose containing 0.1% polysorbate 80 in sterile water), systematically named 4-[5-amino-1-(2,6-difluorobenzoyl)-1*H*-[1,2,4]triazol-3-ylamino]benzenesulfonamide, targets CDK1/cyclin B, CDK2/cyclin A, CDK2/cyclin E, Aurora A kinase and Aurora B kinase with an IC$_{50}$ value of 9 nM, 4 nM, 3 nM, 11 nM and 15 nM, respectively. Flow cytometric analysis of DNA content showed that JNJ-7706621 delayed progression through G$_1$ and arrested the cell cycle at the G2-M phase. **Cyclin Target Selectivity:** Cdk1 (+++), Cdk2 (+++), Cdk3 (weak, if any), Cdk4 (weak, if any), Cdk5 (weak, if any), Cdk6 (weak, if any), Cdk7 (weak, if any), Cdk8 (weak, if any), Cdk9 (weak, if any), Cdk10 (weak, if any), Aurora A Kinase (++), Aurora B Kinase (++). **1**. Emanuel, *et al.* (2005) *Cancer Res.* **65**, 9038. **2**. Seamon, *et al.* (2006) *Mol. Cancer Ther.* **5**, 2459. **3**. Lin, *et al.* (2005) *J. Med. Chem.* **48**, 4208.

JNJ 5207852

This high-affinity, brain-penetrant and orally active histamine receptor neutral antagonist (FW = 389.40 g/mol; CAS 398473-34-2; Solubility: 50 mM in water; 10 mM in DMSO), also named 1-[3-[4-(1-piperidinylmethyl)phenoxy]propyl]piperidine, targets the H$_3$ receptor (human, pK$_i$ = 8.9; rat, pK$_i$ = 9.2), showing 100x and 3x greater affinity than thioperamide, respectively. JNJ 5207852 suppresses slow-wave sleep and exhibits wake-promoting effects in rodent arousal models. **1**. Barbier et al (2004) *Brit. J. Pharmacol.* **143**, 649. **2**. Jia, *et al.* (2005) *Neuropharmacol.* **50**, 404. **3**. Le, *et al.* (2008) *J. Pharmacol. Exp. Ther.* **325**, 902.

JNJ 7777120a

This anti-imflammatory (FW$_{free-base}$ = 277.75 g/mol; CAS 459168-41-3) was among the first potent, non-imidazole, human H$_4$ histamine receptor antagonists (1). JNJ7777120 also recruits β-arrestin to the H$_4$ receptor, independent of G protein activation (2). **1**. Jablonowski, Grice, Chai, *et al.* (2003) *J. Med. Chem.* **46**, 3957. **2**. Osethorne & Charlton (2011) *Mol. Pharmacol.* **79**, 749.

JNJ 10329670

This immunosuppressive agent (FW = 651.15 g/mol; CAS 398473-34-2), known systematically as 1-[3-[4-(6-chloro-2,3-dihydro-3-methyl-2-oxo-1H-benzimidazol-1-yl)-1-piperidinyl]propyl]-4,5,6,7-tetrahydro-5-(methylsulfonyl)-3-[4-(trifluoromethyl)phenyl]-1H-pyrazolo[4,3-c]pyridine, inhibits human the cysteine-protease cathepsin S (K_i = 34 nM). Dialysis against 10 mM dithiothreitol fully reverses all inhibition. Treatment of human B cell lines and primary human dendritic cells with JNJ 10329670 resulted in the accumulation of the p10 fragment of the invariant chain (IC$_{50}$ of 1μM). In contrast, inhibition of invariant chain proteolysis was much less effective in a human monocytic cell line, suggesting that other enzymes may degrade the invariant chain. **Target(s):** JNJ-10329670 is much less active against the mouse (K_i = 2364 nM), dog (K_i = 124 nM), monkey (K_i = 266 nM), and bovine (K_i = 411 nM) enzymes. The compound is inactive against other proteases, including the closely related cathepsins L, F, and K. Such selectivity suggests that JNJ 10329670 would be excellent tool for exploring the role of cathepsin S in human systems. **1.** Thurmond, Sun, Sehon, *et al.* (2004) *J. Pharmacol. Exp. Ther.* **308**, 268.

JNJ 17203212

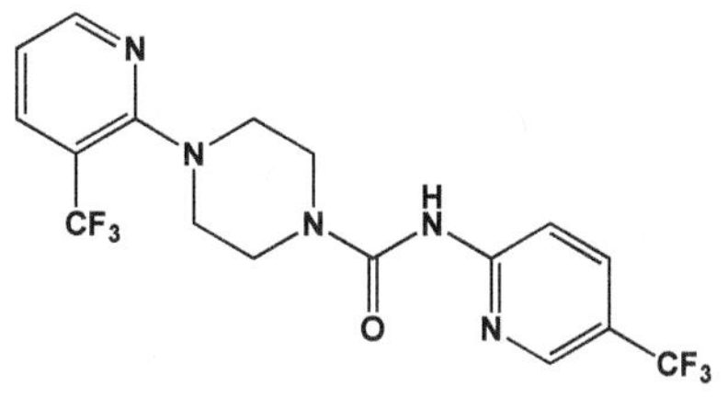

This reversible, competitive and potent TRPV$_1$ antagonist (FW = 419.32 g/mol; CAS 821768-06-3; Solubility: 100 mM in DMSO; 100 mM in ethanol), also named 4-[3-(trifluoromethyl)-2-pyridinyl]-N-[5-(trifluoromethyl)-2-pyridinyl]-1-piperazinecarboxamide, targets the Transient Receptor Potential cation channel subfamily V member 1 (TrpV1), also known as the capsaicin receptor and the vanilloid receptor 1, (pK_i values are 6.5, 7.1 and 7.3 at rat, guinea pig and human TRPV1 respectively). JNJ 17203212 inhibits capsaicin- and H$^+$-induced channel activation (pIC$_{50}$ values are 6.32 and 7.$_{23}$ respectively), exhibiting antitussive and analgesic activity *in vivo*. **1.** Ghilardi, *et al.* (2005) *J. Neurosci.* **25**, 3126. **2.** Swanson, *et al.* (2005) *J. Med. Chem.* **48**, 1857. **3.** Bhattacharya, *et al.* (2007) *J. Pharmacol. Exp. Ther.* **323**, 665.

JNJ 28871063

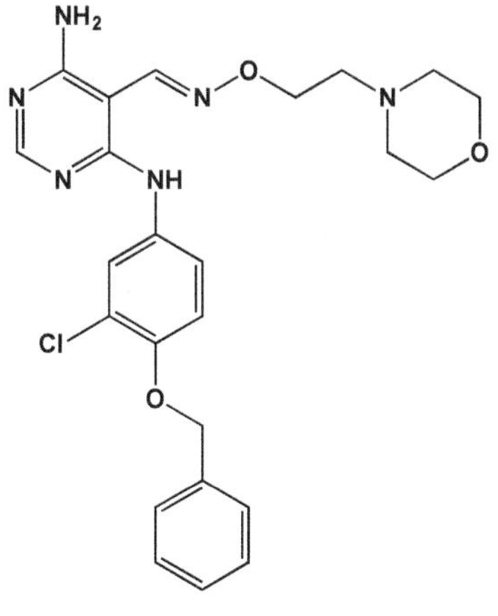

This ErbB receptor inhibitor (FW$_{free-base}$ = 482.97 g/mol; CAS 944342-90-9; Solubility: 50 mM in DMSO), also named 5E-4-amino-6-(4-benzyloxy-3-chlorophenylamino)pyrimidine-5-carboxaldehyde N-(2-morpholin-4-ylethyl) oxime, is a potent and highly selective pan-ErbB kinase inhibitor that blocks the proliferation of epidermal growth factor receptor (EGFR; ErbB1)- and ErbB2-overexpressing cells, but does not affect the growth of non-ErbB-overexpressing cells. Treatment of human cancer cells with JNJ-28871063 inhibited phosphorylation of key tyrosine residues in both EGFR and ErbB2, blocking downstream signal transduction pathways responsible for proliferation and survival. A single dose of compound reduced phosphorylation of ErbB2 receptors in tumor-bearing mice, demonstrating target suppression *in vivo*. **1.** Emanuel, Hughes, Adams, *et al.* (2008) *Mol. Pharmacol.* **73**, 338.

JNJ-31020028

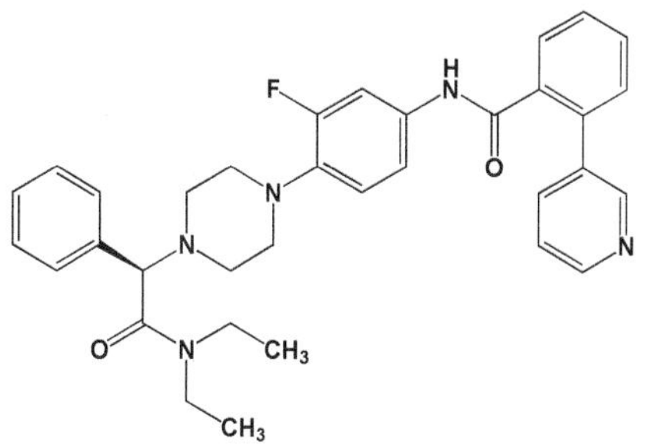

(R)-(–)-JNJ-31020028

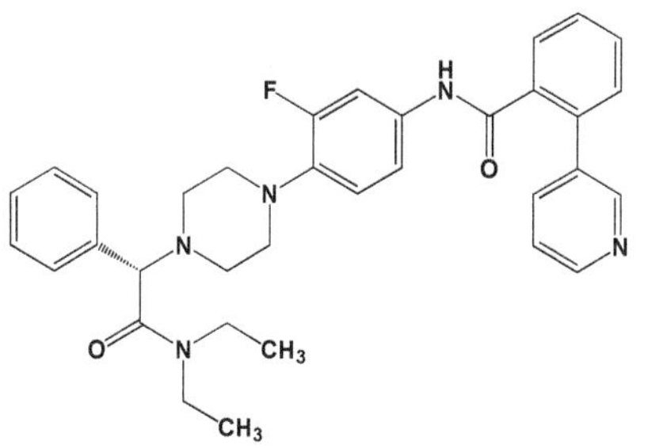

(S)-(+)-JNJ-31020028

This brain-penetrating, neuropeptide Y$_2$-receptor antagonist (FW = 565.69 g/mol; CAS 1094873-14-9), systematically named, N,N-diethyl-4-[2-fluoro-4-[[2-(3-pyridinyl)benzoyl]amino]phenyl]-α-phenyl-1-piperazine acetamide, inhibits binding of neuropeptide Y to human Y$_2$ receptors. Its pharmacologic action was demonstrated in KAN-Ts cells and rat Y$_2$ receptors in rat hippocampus, where it blocks PYY-stimulated calcium responses in cells expressing a chimeric G-protein Gqi5 and in the rat vas deferens, the latter a prototypical Y$_2$ bioassay. JNJ-31020028 bound with high affinity (pIC$_{50}$ = 8.07, human, and pIC$_{50}$ = 8.22, rat) and was >100x more selective than with human Y$_1$, Y$_4$, and Y$_5$ receptors (1). JNJ-31020028 also crosses the blood brain barrier and occupies the receptor after subcutaneous dosing. Chiral resolution demonstrated that the IC$_{50}$ values for the R- and S-enantiomers are 9.2 and 14.3 nM, respectively (2). **1.** Shoblock, Welty, Nepomuceno, *et al.* (2010) *Psychopharmacology* (Berlin) **208**, 265. **3.** Swanson, Wong, Jablonowski, *et al.* (2011) *Bioorg. & Med. Chem. Lett.* **21**, 5552.

JNJ-38877605

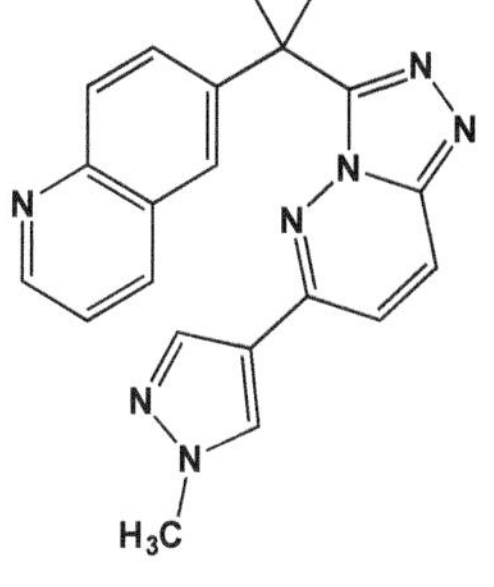

This ATP-competitive protein kinase inhibitor (FW = 377.36 g/mol; CAS 943540-75-8; Solubility: 37 mg/mL DMSO; < 1 mg/mL H$_2$O), also named 6-(difluoro(6-(1-methyl-1H-pyrazol-4-yl)[1,2,4]triazolo[4,3-b]pyridazin-3-yl)methyl)quinoline, targets c-Met (IC$_{50}$ = 4 nM), showing 600-times selectivity for *versus* 200 other tyrosine and serine-threonine kinases. MET silencing by siRNA or inhibition of its kinase activity by JNJ-38877605 counteracts radiation-induced invasiveness, promotes apoptosis, and prevents cells from resuming proliferation after irradiation *in vitro*. **1.** De Bacco F, Luraghi P, Medico, *et al.* (2011) *J. Natl. Cancer Inst.* **103**, 645.

JNJ DGAT2-A

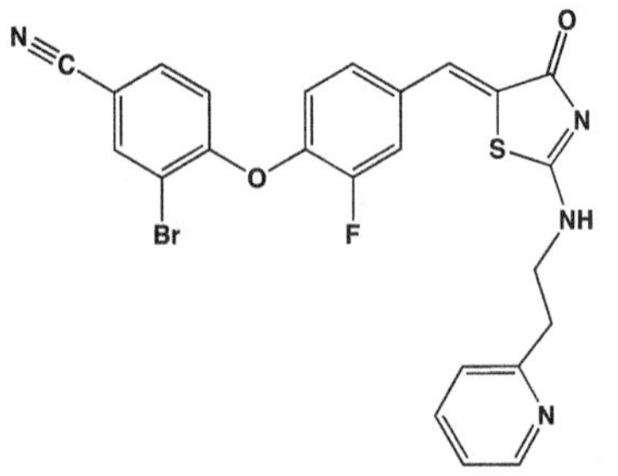

This potent and selective DGAT2 inhibitor (FW = 523.38 g/mol; Soluble to 100 mM in DMSO), also named 3-bromo-4-[2-fluoro-4-[[4-oxo-2-[[2-(pyridin-2-yl)ethyl]amino]-1,3-thiazol-5-(4H)-ylid ene]methyl]phenoxy]benzonitrile, targets isoform-2 of diacylglycerol acyltransferase (DGAT), the enzyme catalyzing the final step in triglyceride (TG) synthesis. There are two enzyme isoforms, DGAT1 and DGAT2, each with different physiological functions, indicating the utility of differential inhibition of DGAT1 and DGAT2 to probe hepatic TG synthesis. Moreover, hyperglycemia and hypertriglyceridaemia are both characteristic of pre-diabetes conditions, and DGAT2 catalyzes the de novo synthesis of triacylglycerols from newly synthesized fatty acids and nascent diacylglycerols. This property identifies DGAT2 as a likely link between hyperglycemia and hypertriglyceridaemia. JNJ DGAT2-A inhibits DGAT2 (IC_{50} = 140 nM), showing >70-fold selectivity for DGAT2 over DGAT1 and MGAT2. An intriguing finding is that DGAT1 inhibitors (JNJ-DGAT1-A and A-92250) block incorporation of exogenously added oleate, but do not tdisrupt incorporation of exogenously added glycerol into TG; the opposite is true for the selective DGAT2 inhibitor, JNJ DGAT2-A. In other words, DGAT2 preferentially esterifies exogenously added glycerol to endogenously synthesized FAs, and DGAT1 preferentially esterifies exogenously added oleic acid to already-existing diacylglycerol. 1. Qi, Lang, Geisler, et al. (2012) J. Lipid Res. 53, 1106.

JNK Inhibitor I

This cell-penetrating peptide (MW = 3920 g/mol; pI = 12.5; Sequence: GRKKRRQRRRPPRPKRPTTLNLFPQVPRSQ-DT-NH₂), also known as SAPK Inhibitor I, diminishes c-Jun amino-terminal kinase (JNK) signaling by blocking the activation of the transcription factor c-jun. JNK Inhibitor 1 is a fusion peptide consisting of the C-terminal sequence of the JNK binding protein attached to the HIV-TAT pentratin domain (residues 48-57). The HIV-TAT sequence permits entry into the cytoplasm by means of micropinocytosis (2). 1. Barr, Kendrick & Bogoyevitch (2002) J. Biol. Chem. 277, 10987. 2. Chauhan, Tikoo, Kapur & Singh (2007) J. Control Release 117, 148.

JNK Inhibitor II, See SP 600125

JO146

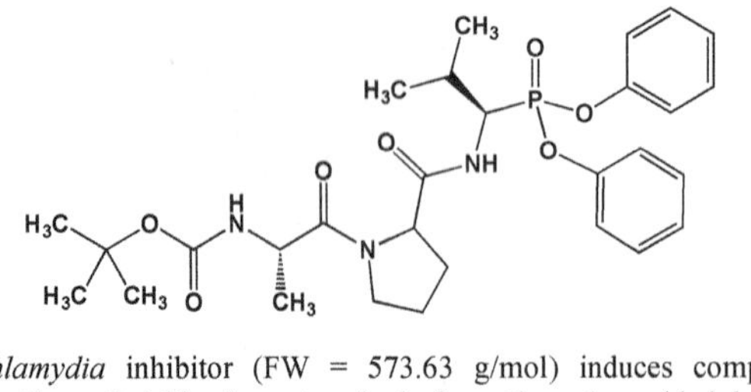

This Chlamydia inhibitor (FW = 573.63 g/mol) induces complete or significant loss of viable elementary body formation when added during the replicative phase of Chlamydia trachomatis, an obligate intracellular bacterial pathogens with significant clinical importance inasmuch as it is now the most common sexually transmitted bacterial infection world wide and ocular-infecting serovars are the most common cause of preventable blindness worldwide. JO146 possesses a C-terminal phosphonate 'warhead' electrophile that reacts irreversibly with the protease active-site serine residue. When treated during the replicative phase of development, Chlamydia lose their distinctive morphology, with ultimate loss of inclusions from infected host cells. JO146 completely prevented the formation of viable Chlamydia elementary bodies. 1. Gloeckl, Ong, Patel, et al. (2013) Mol. Microbiol. 89, 676.

Joro Spider Toxin

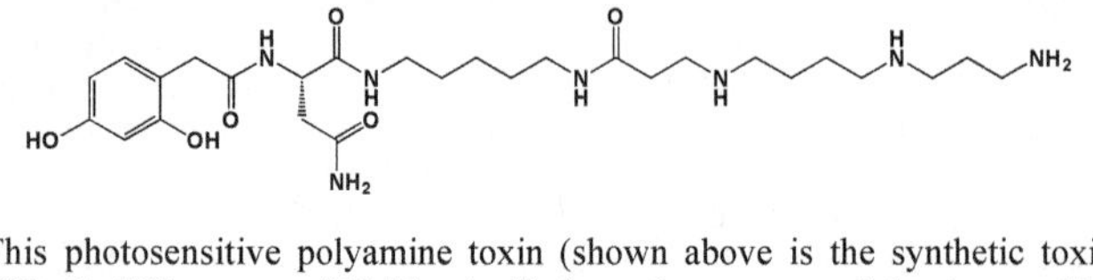

This photosensitive polyamine toxin (shown above is the synthetic toxin JSTx-3, $FW_{free-base}$ = 565.71 g/mol) from the venom of the joro spider (Nephila clavata) irreversibly blocks excitatory postsynaptic potentials without affecting inhibitory postsynaptic potentials. JST not only inhibits L-glutamate binding to synaptic membrane receptors, but also L-glutamate uptake by synaptosomes. In fact, it joro spider toxins are one of the most potent antagonists of glutamatergic AMP_A/K_A (or α-amino-3-hydroxy-5-methylisoxazole-4-propionate/kainate) receptor channels in invertebrates and vertebrates. (See also 1-Naphthylacetylspermine) Target(s): AMP_A receptors (1); glutamate receptor (2) 1. Van Den Bosch, Vandenberghe, Klaassen, Van Houtte & Robberecht (2000) J. Neurol. Sci. 180, 29. 2. Pan-Hou, Suda, Sumi, Yoshioka & Kawai (1989) Brain Res. 476, 354.

JP 1302

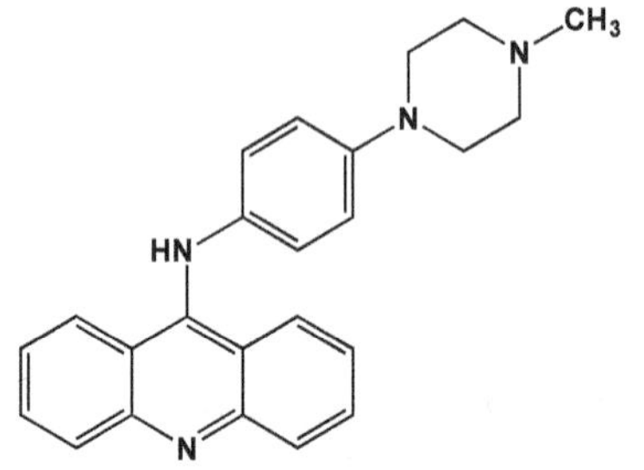

This $α_{2C}$ adrenoceptor antagonist (FW = 441.41 g/mol; CAS 1259314-65-2; Solubility: 100 mM in water (as dihydrochloride) and to 5 mM in DMSO), also named N-[4-(4-Methyl-1-piperazinyl)phenyl]-9-acridinamine, selectively targets the human $α_{2C}$ subtype (K_i = 28 nM), with much weaker action against $α_{2B}$ (K_i = 1470 nM), $α_{2D}$ (K_i = 1700 nM) and $α_{2A}$ (K_i = 3150 nM) receptor subtypes. JP-1302 also antagonizes epinephrine-stimulated $GTP_γS$ binding in vitro (K_B = 16 nM) and produces antidepressant and antipsychotic-like effects in vivo. 1. Sallinen, et al. (2007) Brit. J. Pharmacol. 150, 391.

(+)-JQ1

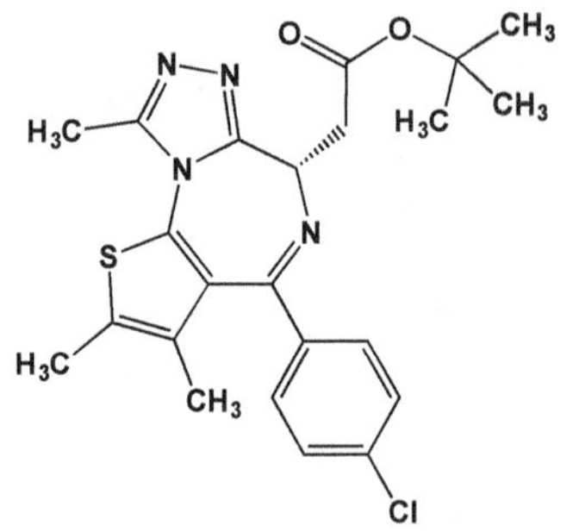

This potent, selective and cell-permeable BET bromodomain inhibitor (FW = 456.99 g/mol; CAS 1268524-70-4; Solubility: 100 mM in DMSO; 100 mM in Ethanol), also named (6S)-4-(4-chlorophenyl)-2,3,9-trimethyl-6H-thieno[3,2-f][1,2,4]triazolo[4,3-a][1,4]diazepine-6-acetic acid 1,1-dimethylethyl ester, targets BRD2 (N-terminal), IC_{50} = 18 nM, K_d = 128 nM; BRD3 (N-terminal), K_d = 60 nM; BRD3 (C-terminal), K_d = 82 nM; BRD4 (C-terminal), IC_{50} = 33 nM, K_d = 90 nM; BRD4 (N-terminal), IC_{50} = 77 nM, K_d = 49 nM; and CREBBP, IC_{50} = 12942 nM; and BRDT (N-terminal), K_d = 190 nM. Recurrent translocation of BRD4 is observed in a genetically-defined, incurable subtype of human squamous carcinoma. Competitive binding by JQ1 displaces the BRD4 fusion oncoprotein from chromatin, prompting squamous differentiation and specific antiproliferative effects in BRD4-dependent cell lines and patient-derived xenograft models (1). JQ1 also exerts major anti-leukemic effects in a broad range of human AML subtypes, including relapsed and refractory patients and all relevant stem- and progenitor cell compartments, including CD34⁺/CD38⁻ and CD34⁺/ CD38⁺ AML cells (2). Treatment of mice with JQ1 reduces seminiferous tubule area, testis size, and spermatozoa number and motility, without affecting hormone levels (3). JQ1 exerts a complete and reversible contraception, making it a lead compound for germ cell-based contraception. JQ1 impairs LPS-induced human monocyt-derived dentritic cell maturation by inhibiting STAT5 activity, thereby generating cells that can only weakly stimulate an adaptive-immune response and suggesting JQ1 may have beneficial effects in treating T cell-mediated inflammatory diseases (4). **1**. Filippakopoulos, Qi, Picaud, *et al.* (2010) *Nature* **468**, 1067. **2**. Herrmann, Blatt, Shi, *et al.* (2012) *Oncotarget* **3**, 1588. **3**. Matzuk, McKeown, Filippakopoulos, *et al.* (2012) *Cell* **150**, 673. **4**. Toniolo, Liu, Yeh, *et al.* (2015) *J. Immunol.* **194**, 3180.

JTC 801

This potent, selective and orally bioavailable NOP receptor antagonist (FW = 447.96 g/mol; CAS 244218-51-7; Solubility: 5 mM in H_2O; 100 mM in DMSO), also named *N*-(4-amino-2-methyl-6-quinolinyl)-2-[(4-ethylphenoxy)methyl]benzamide, targets the nociceptin *or* NOP receptor (K_i = 8.2 nM), also known as the orphanin FQ receptor or κ-type-3 opioid receptor, displaying 12x, 130x and 1000x selectivity over human μ-, κ- and δ-opioid receptors, respectively (1-3). TC-801 inhibits (K_i = 45 nM) [^{3}H]-nociceptin binding to human ORL$_1$ receptors expressed in HeLa cells. It also completely antagonizes the suppression of nociceptin on forskolin-induced accumulation of cyclic AMP (IC_{50} = 2.6 μM) in ORL$_1$ receptor-expressing HeLa cells *in vitro* (2). TC-801's anti-nociceptive action of JTC-801 was not inhibited by naloxone (2). JTC-801 exerted anti-allodynic and anti-hyperalgesic effects in rats, suggesting that N/OFQ system might be involved in the modulation of neuropathic pain and inflammatory hyperalgesia (3). **1**. Shinkai, Ito, Iida *et al.* (2000) *J. Med. Chem.* **43**, 4667. **2**. Yamada, Nakamoto, Suzuki, Ito & Aisaka (2002) *Brit. J. Pharmacol.* **135**, 323. **3**. Tamai, Sawamura, Takeda, Orii & Hanaoka (2005) *Eur. J. Pharmacol.* **510**, 223.

JTE 013

This S1P$_2$ receptor antagonist (FW = 408.29 g/mol; CAS 547756-93-4; Solubility: 100 mM in DMSO *or* Ethanol), also named 1-[1,3-dimethyl-4-(2-methylethyl)-1*H*-pyrazolo[3,4-*b*]pyridin-6-yl]-4-(3,5-dichloro-4-pyridinyl)semicarbazide, selectively targets the sphingosine-1-phosphate S1P$_2$ receptors (IC_{50} = 17.6 nM), with little effect (~4% inhibition at 10 μM JTE-013) on S1P$_3$ receptors and none on S1P$_1$ receptors (1-3). JTE-013 enhances S1P-induced angiogenesis *in vivo*. **1**. Ohmori, *et al.* (2003) *Cardiovasc. Res.* **58**, 170. **2**. Parrill, *et al.* (2004) *Semin. Cell. Dev. Biol.* **15**, 467. **3**. Inoki, *et al.* (2006) *Biochem. Biophys. Res. Comm.* 346 293.

JTE 907

This cannabinoid receptor inverse agonist (FW = 436.51 g/mol; CAS 282089-49-0; Solubility: 100 mM in DMSO; 10 mM in Ethanol), also named *N*-(1,3-benzodioxol-5-ylmethyl)-1,2-dihydro-7-methoxy-2-oxo-8-(pentyloxy)-3-quinolinecarboxamide, targets rat, mouse and human CB$_2$ receptors with K_i values of 0.38, 1.55 and 35.9 nM respectively, producing anti-inflammatory effects *in vivo* (1,2). JTE 907 also suppresses spontaneous itch-associated responses of NC mice, a model of atopic dermatitis (3) **1**. Iwamura, et al. (2001) *J. Pharmacol. Exp. Ther.* **296**, 420. **2**. Ueda, *et al.* (2005) Eur. J. Pharmacol. **520**, 164. **3**. Maekawa et al (2006) *Eur. J. Pharmacol.* **542**, 179.

JTK-303, *See Elvitegravir*

JTK-853

This novel, non-nucleoside inhibitor (FW = 650.47 g/mol) targets the palm binding site of hepatitis C virus (HCV) polymerase and demonstrates antiviral activity in HCV-infected patients (1-3). JTK-853 has a high genetic barrier to resistance, suggesting that combination therapies including this agent will be potent in suppressing the emergence of drug resistance in HCV-infected patients. Amino acid substitutions in the polymerase (*e.g.,* Met-414-Thr, Cys-445-Arg, Tyr-448-Cys/His, and Leu-466-Phe) confer viral resistance in HCV-infected patients on 3-day JTK-853 monotherapy (3). **1**. Ando, Ogura, Toyonaga, *et al.* (2013) *Intervirology* **56**, 302. **2**. Ogura, Toyonaga, Ando, *et al.* (2012) *Antimicrob. Agents Chemother.* **57**, 436. **3**. Ando, Adachi, Ogura, *et al.* (2012) *Antimicrob. Agents Chemother.* **56**, 4250.

JTP-4819

This memory-enhancing drug (FW = 359.43 g/mol; CAS 68497-62-1), also named (−)-(2*S*)-1-benzylaminocarbonyl-[(2*S*)-2-glycoloylpyrrolidinyl]-2-pyrrolidinecarboxamide, increases substance P-like immunoreactivity in the cerebral cortex and hippocampus through its novel action as a prolyl endopeptidase inhibitor, IC_{50} = 0.7–0.8 nM (1). In studies on the isolated prolyl endopeptidase, JTP-4819 behaves as a tight-binding competitive inhibitor, displaying a K_{ic} value of 0.045 nM (2). JTP-4819 inhibits pig brain prolyl oligopeptidase, IC_{50} = 0.2 nM (3). (*See Baicalin and Pramiracetam*) **1**. Toide, Okamiya, Iwamoto & Kato (1995) *J. Neurochem.* **65**, 234. **2**. Venäläinen, Juvonen, Forsberg (2002) *Biochem. Pharmacol.* **64**, 463. **3**. Wallén, Christiaans, Saario, *et al.* (2002) *Bioorg. Med. Chem.* **10**, 2199.

JTP-74057, *See Trametinib*

JTT-130

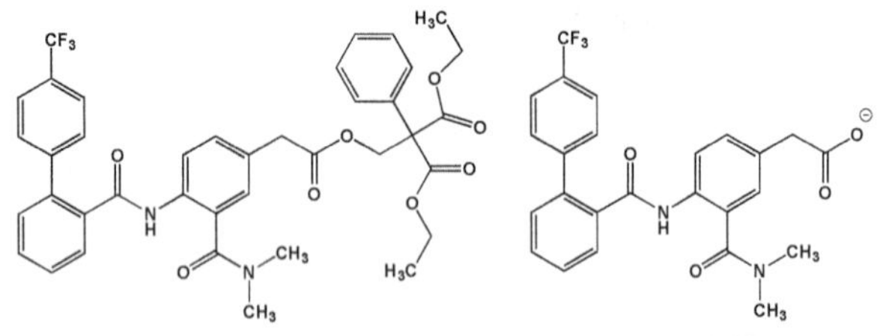

JTT-130   **Metabolite M1**

This microsomal triglyceride transfer inhibitor (FW = 718.73 g/mol) inhibits enterocyte Microsomal Triglyceride-transfer Protein (MTP), a carrier playing a central role in intestinal lipid absorption and a druggable target for treating metabolic syndrome. JTT-130 potently inhibited lipid transfer by MTP derived from human small intestine in a concentration-dependent manner for transfer of triglycerides (IC_{50} = 0.83 nM) and cholesteryl ester (IC_{50} = 0.74 nM). **Rationale fo Enterocyte-Directed Action:** Localized within the endoplasmic reticulum, MTP participates in the assembly of TG-rich lipoproteins, such as chylomicron particles in the small intestine, and very low-density lipoprotein (VLDL) particles in the liver. In view of reported undesireable *in vivo* effects of MTP inhibitors, it was evident that inhibition of hepatic MTP could result in the potent blockade of VLDL release, resulting in reduced plasma lipids and inducing fatty liver and hepatic dysfunction. In a strategy designed to minimize any tendency of enteric MTP inhibitors to interfere with hepatic MTP, JTT-130 was designed to undergo rapid metabolic hydrolysis after oral administration to its inactive metabolite M1 (shown above). **1.** Mera, Odani, Kawai, *et al.* (2011) *J. Pharmacol. Exp. Ther.* **336**, 321.

JTT-705, *See Dalcetrapib*

JTV-803

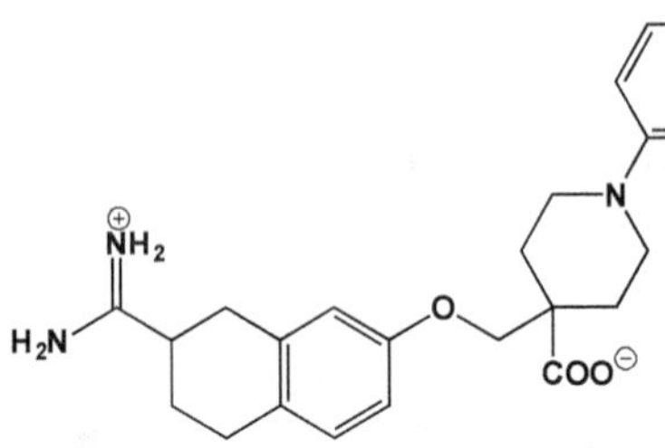

This anticoagulant (FW = 409.49 g/mol; CAS 247131-79-9) potently inhibits coagulation factor Xa (K_i = 19 nM) but weakly inhibits trypsin (K_i = 13.6 µM) and plasmin (K_i = 78 µM). JTV-803 was dose-dependently effective against disseminated intravascular coagulation in both TF-induced and LPS-induced rat models **1.** Hayashi, Matsuo, Nakamoto & Aisaka (2001) *Eur. J. Pharmacol.* **412**, 61. **2.** Ieko, Tarumi, Takeda, *et al.* (2004) *J. Thromb. Haemost.* **2**, 612.

Juglone

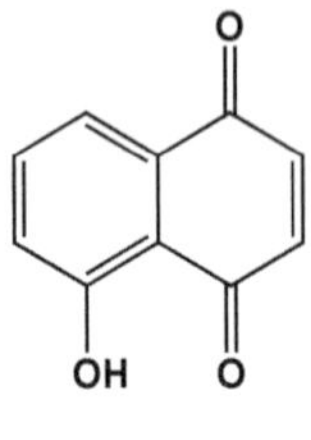

This pH-indicating, yellow-brown naphthoquinone (FW = 174.16 g/mol; CAS 481-39-0), also known as 5-hydroxy-1,4-naphthoquinone, from the walnut shells, leaves, and roots, potently irreversibly inhibits sarcoplasmic reticulum Ca^{2+}-transporting ATPase activity (1). Another naphthoquinone, 5-OH-2-CH_3-1,4-naphthoquinone (plumbagine) also inhibits irreversibly, whereas 2-CH_3-1,4-naphthoquinone (menadione) and 2,3-$(OCH_3)_2$-1,4-naphthoquinone (2,3diOmeNQ) do not. Such findings suggest that the C5 hydroxyl is essential for juglone's direct interaction with a catalytic SH group. (**See also** *Lawsone*) **Target(s):** Ca^{2+}-transporting ATPase (1); ζ-crystallin's lactate dehydrogenase activity (2,3); dihydroxy-acid dehydratase (4); ecdysone 20-monooxygenase (5); 3α-hydroxysteroid transhydrogenase (6); peptidyl-prolyl isomerase, irreversibly modifying

sulfhydryl groups by a Michael-type addition (7,8); pyruvate decarboxylase (9); ribonuclease H activity associated with HIV-I reverse transcriptase (10); urease (11); acetyl-choline synthesis (12). **1.** Floreani, Forlin, Bellin & Carpenedo (1995) *Biochem. Mol. Biol. Int.* **37**, 757. **2.** Duhaiman (1996) *Biochem. Biophys. Res. Commun.* **218**, 648. **3.** Bazzi (2001) *Arch. Biochem. Biophys.* **395**, 185. **4.** Babu & Brown (1995) *Microbios* **82**, 157. **5.** Mitchell & Smith (1988) *Experientia* **44**, 990. **6.** Koide (1962) *Biochim. Biophys. Acta* **59**, 708. **7.** Chao, Greenleaf & Price (2001) *Nucl. Acids Res.* **29**, 767. **8.** Hennig, Christner, Kipping, *et al.* (1998) *Biochemistry* **37**, 5953. **9.** Kuhn & Beinert (1947) *Ber.* **80**, 101. **10.** Min, Miyashiro & Hattori (2002) *Phytother. Res.* **16** Suppl. 1, S57. **11.** Schopfer & Grob (1949) *Helv. Chim. Acta* **32**, 829. **12.** Haubrich & Wang (1976) Biochem Pharmacol. **25**, 669.

Justicidins

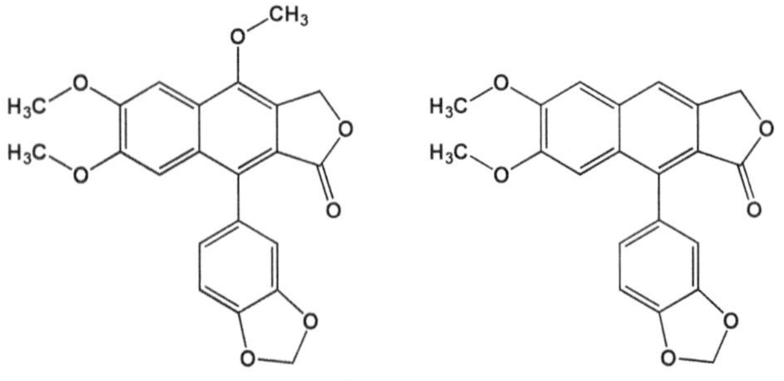

Justicidin A   **Justicidin B**

This class of lignans ($FW_{Justicidin-A}$ = 394.38 g/mol; CAS 25001-57-4; $FW_{Justicidin-B}$ = 364.35 g/mol; CAS 17951-19-8) from species of *Justicia*, *Sesbania*, and *Acanthaceae* is both cytotoxic and antiviral. They also inhibit platelet aggregation (1). Justicidin A inhibits the transport of TNF-α to cell surface in lipopolysaccharide-stimulated RAW 264.7 macrophages (2). **1.** Chen, Hsin, Ko, *et al.* (1996) *J. Nat. Prod.* **59**, 1149. **2.** Tsao, Lin & Wang (2004) *Mol. Pharmacol.* **65**, 1063.

Juvenile Hormones

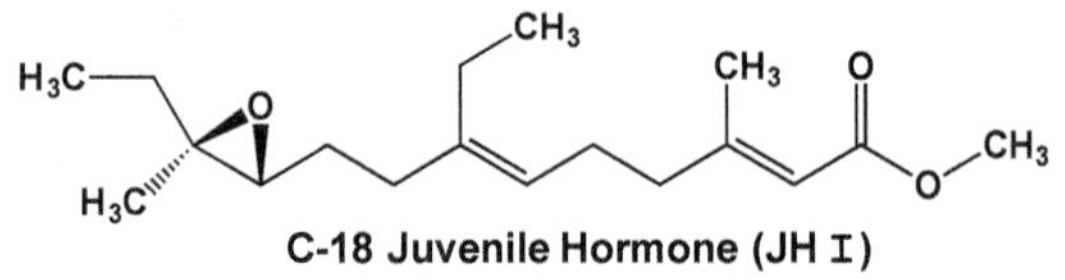

C-18 Juvenile Hormone (JH I)

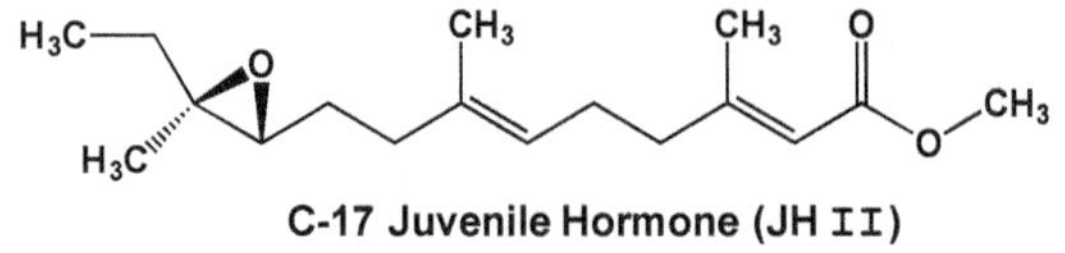

C-17 Juvenile Hormone (JH II)

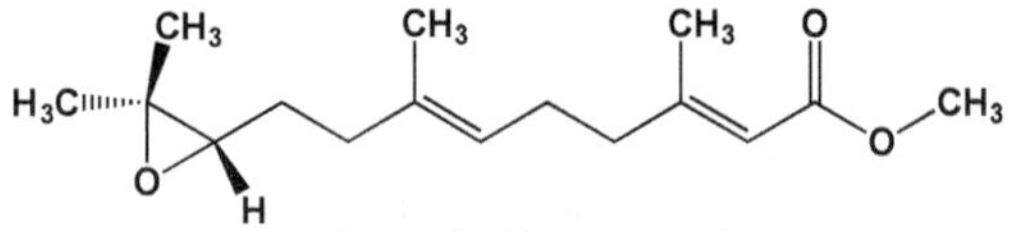

Juvenile Hormone III

These insect hormones (JH) function to maintain the larval stage, thereby preventing maturation. As such, they have been exploited as insect-controlling agents (1,2). The best characterized examples are those of the silk moth: C-18 JH (JH I; *cis*-10,11-epoxy-7-ethyl-3,11 dimethyl-*trans,trans*-2,6-tridecadienoic acid methyl ester; FW = 294.43 g/mol) and C-17 JH (JH II; *cis*-10,11-epoxy-3,7,11-trimethyl-*trans,trans*-2,6-tridecadienoic acid methyl ester; FW = 280.41 g/mol). JH III is the smaller *cis*-10,11-epoxy-3,7,11-trimethyl-*trans,trans*-2,6 dodecadienoic acid methyl ester (FW = 266.38 g/mol). C-18 JH was stereochemically synthesized by Corey and coworkers in 1968 (3). **Target(s):** microsomal epoxide hydrolase (JH-I inhibition is competitive, K_i = 0.27 mM; JH-I is a very poor substrate for *Trypanosoma cruzi* epoxide hydrolase) (4). **1.** Pratt & Brooks, eds (1981) *Juvenile Hormone Biochemistry*, Elsevier-North Holland, Amsterdam. **2.** Davey (2000) *Insect Biochem. Mol. Biol.* **30**, 663. **3.** Corey, Katzenellenbogen, Gilman, *et al.* (1968) *J. Amer. Chem. Soc.* **90**, 5618. **4.** Yawetz & Agosin (1979) *Biochim. Biophys. Acta* **585**, 210.

JW55

This tankyrase inhibitor (FW = 434.48 g/mol; CAS 664993-53-7; Solubility: 100 mM in DMSO), also named *N*-[4-[[[[tetrahydro-4-(4-methoxyphenyl)-2*H*-pyran-4-yl]methyl]amino]carbonyl]phenyl]-2-furan carboxamide, bids selectively within the poly(ADP-ribose) polymerase (*or* PARP) domain of Tankyrases 1 and 2 (TNKS1/2), resulting in stabilization of Axin2 and increased degradation of β-catenin. JW55 inhibits the β-catenin signaling pathway, decreasing canonical Wnt signaling in SW480 and HCT-15 colon carcinoma cell lines and reducing cell cycle progression and proliferation in SW480 cells *in vitro*. **1.** Waaler, Machon, Tumova, *et al.* (2012) *Cancer Res.* **72**, 2822.

JW67

This canonical Wnt signaling pathway inhibitor (FW = 394.38 g/mol; CAS 442644-28-2; Solubility: 100 mM in DMSO), also named tri-spiro[3*H*-indole-3,2'-[1,3]dioxane-2'',3'''-[3*H*]indole]-2,2'''(1*H*,1'''*H*)-dione, targets (IC$_{50}$ = 1.17 μM) the β-catenin destruction complex (*i.e.*, GSK-3β/AXIN/APC) to induce β-catenin degradation, blocking cell cycle progression at G$_1$/S in colorectal cancer cell lines (GI$_{50}$ = 7.8 μM). JW-67 is selective for the canonical Wnt pathway over the Sonic hedgehog (Shh) and NF-κB pathways. **1.** Waaler, *et al.* (2011) *Cancer Res.* **71**, 197. **2.** Shultz, *et al.* (2012) *J. Med. Chem.* **55**, 1127.

JW74

This tankyrase inhibitor (FW = 456.52 g/mol; CAS 863405-60-1; Solubility: 20 mg/mL DMSO; IUPAC Name: 4-[4-(4-methoxyphenyl)-5-[[[3-(4-methylphenyl)-1,2,4-oxadiazol-5-yl]methyl]thio]-4*H*-1,2,4-triazol-3-yl]pyridine) is an efficient and specific inhibitor of the canonical Wnt signaling, both *in vitro* and *in vivo*. A major regulator of stem cell self-renewal and differentiation, the Wnt/β-catenin signaling pathway is aberrantly activated in a several cancers, including osteosarcoma. JW74 rapidly reduces active β-catenin, attended subsequently by downregulation of the Wnt target genes AXIN2, SP5, and NKD1. AXIN2 protein levels were strongly increased after compound exposure. Long-term treatment with JW74 inhibited the growth of osteosarcoma tumor cells in both a mouse xenograft model of CRC and in Apc(Min) mice. (*Note:* Min = Multiple intestinal neoplasias) *See also G007-LK* **1.** Waaler, Machon, von Kries, *et al.* (2011) *Cancer Res.* **71**, 197. **2.** Wessel, Daffinrud, Munthe, *et al.* (2013) *Cancer Med.* **3**, 36.

JW480

This KIAA1363 inhibitor (FW = 333.43 g/mol; CAS 1354359-53-7; Solubility: 100 mM in DMSO *or* Ethanol), also named 2-Isopropylphenyl(2-(naphthalen-2-yl)ethyl)carbamate, potently and selectively targets (IC$_{50}$ = 20 nM in mouse brain) Neutral Cholesterol Ester Hydrolase 1 (NCEH), also known as Arylacetamide deacetylase-like-1 (AADACL1) or KIAA1363, a so-called serine hydrolase Impairs migration, invasion, survival and tumor growth of prostate cancer cell lines *in vivo*. Reduces monoalkylglycerol ether (MAGE) levels. NCEH hydrolyzes 2-acetyl monoalkylglycerol ether in the pathway regulating platelet activating factor (*or* =PAF) levels. **1.** Chang, Nomura & Cravatt (2011) *Chem Biol.* **18**, 476.

JW642

This MAGL inhibitor (FW = 462.39 g/mol; CAS 1416133-89-5; Solubility: 100 mM in DMSO) potently and selectively targets monoacylglycerol lipase, *or* MAGL, (IC$_{50}$ = 3.7 nM), displaying a remarkable 6000x selectivity for MAGL over fatty acid amide hydrolase, *or* FAAH, (IC$_{50}$ = 20.6 μM). Because the endocannabinoids 2-arachidonoyl glycerol (2-AG) and *N*-arachidonoyl ethanolamine (*or* anandamide) are mainly degraded by monoacylglycerol lipase (MAGL) and fatty acid amide hydrolase (FAAH), respectively, JW642 selectively blocks a decrease in 2-AG without altering anandamide levels. **1.** Chang, Niphakis, Lum, *et al.* (2012) *Chem. Biol.* **19**, 579.

JWH-007

This naphthoylindole family analgesic (FW = 355.47 g/mol; CAS 155471-10-6; IUPAC Name: 1-pentyl-2-methyl-3-(1-naphthoyl)indole) acts as a cannabinoid CB$_1$ and CB$_2$ receptor agonist, with K_d values of 9.5 and 2.9 nM, respectively, or around 4-5x tighter than for Δ^9-THC (1). JHW-007 exhibits high affinity binding to dopamine transporters (DAT), but lacks cocaine-like behavioral effects and antagonizes the effects of cocaine (2). JHW-007 shifts the cocaine dose-effect curve 3.07x to the right, a behavior that is suggestive of competitive antagonism (2). JHW-007 blocks cocaine-induced reward, locomotor stimulation, and sensitization (3). This atypical dopamine transport inhibitor also prevents amphetamine-induced sensitization and synaptic reorganization within the nucleus accumbens (4). Because DAT is considered the biological target responsible for the abuse liability of cocaine, DAT ligands like JWH-007 may be viewed as potential lead molecules in searches for suitable cocaine antagonists. **1.** Huffman,

Dai, Martin, *et al.* (1994) *Bioorg. Med. Chem. Lett.* **4**, 563. **2**. Desai, Kopajtic, Koffarnus, Newman & Katz (2005) *J. Neurosci.* 25, 1889. **3**. Velázquez-Sánchez, Ferragud, Murga, Cardá & Canales (2010) *Eur. Neuropsychopharmacol.* **20**, 501. **4**. Velázquez-Sánchez, García-Verdugo, Murga & Canales (2013) *Prog. Neuropsychopharmacol. Biol. Psychiatry* **44**, 73.

JX 401

This potent p38 MAP kinase inhibitor (FW = 355.50 g/mol; CAS 349087-34-9; Solubility: 75 mM in DMSO), also named 1-[2-methoxy-4-(methylthio)benzoyl]-4-(phenylmethyl)piperidine, targets p38α (IC$_{50}$ = 32 nM), a stress-activated protein kinase that negatively regulates malignant transformation induced by oncogenic H-Ras. p38α is also hyperactive in inflammatory diseases, and various indications suggest that its inhibition would reverse inflammation. JX-401 inhibits the differentiation of myoblasts to myotubes in mammalian cells in culture. JX-401 also inhibits the p38γ isoform, but only weakly (IC$_{50}$ > 10 μM). **1**. Friedmann, Shriki, Bennett, *et al.* (2006) *Mol. Pharmacol.* **70**, 1395.

JZL 184

This potent and selective endocannabinoid hydrolysis inhibitor (FW = 520.50 g/mol; CAS 1101854-58-3; λ$_{max}$ = 284 nm), also named 4-nitrophenyl-4-(dibenzo[*d*][1,3]dioxol-5-yl(hydroxy)methyl)piperidine-1-carboxylate, targets monoacylglycerol lipase (MAGL) and fatty acid amide hydrolase (FAAH) in mouse brain membranes, with respective IC$_{50}$ values of 8 nM and 4 μM. MAGL catalyzes the hydrolysis of the endocannabinoid 2-arachidonoyl glycerol (2-AG) to arachidonic acid and glycerol, thus terminating its biological function (1). FAAH also catalyzes the hydrolysis of the endocannabinoid arachidonoyl ethanolamide (AEA). When administered to mice at 16 mg/kg, intraperitoneally, JZL184 reduces MAGL activity by 85%, elevates brain 2-AG levels by 8x, and elicits analgesic activity in a variety of pain assays that qualitatively mimics direct central cannabinoid (CB$_1$) agonists. **1**. Long, Li, Booker, *et al.* (2009) *Nature Chem. Biol.* **5**, 37.

K, **See** *Lysine; Potassium and Potassium Ions*

K-115, See *Ripasudil*

K-252a

This cell-permeable staurosporine analogue (FW = 467.48 g/mol; Pale yellow powder; Soluble in methylene chloride, *N,N*,-dimethylformamide, and dimethyl sulfoxide) is a broad-spectrum protein kinase inhibitor, acting on CaM kinase II (K_i = 1.8 nM), protein kinase A (K_i = 18-25 nM), protein kinase C (K_i = 25 nM), protein kinase G (K_i = 20 nM), phosphorylase kinase (IC$_{50}$ = 1.7 nM), and myosin-light-chain kinase (K_i = 17 nM). K-252a also inhibits the protein-tyrosine kinase activity of the NGF receptor gp140trk (IC$_{50}$ = 3 nM) and induces apoptosis and cell cycle arrest by inhibiting Cdc2 and Cdc25c. (**See also** *KT5720; KT5823; KT5926; K-252b*) **Target(s):** AfsK serine/threonine protein kinase (1); Ca^{2+}/calmodulin-dependent kinase II, *or* calmodulin kinase II (2,3); cAMP-dependent protein kinase, *or* protein kinase A (4); cGMP dependent protein kinase, *or* protein kinase G (4); cyclin-dependent kinases (5); heat-shock protein (HSP25) kinase (6); mitogen-activated protein kinase activated protein kinase-27, *or* [myosin light chain] kinase (8); NGF receptor (9-13); non-specific serine/threonine protein kinase (1); phosphorylase kinase (14); protein kinase C (4,15-17); trk receptor-linked protein-tyrosine kinase (12). **1**. Horinouchi (2003) *J. Ind. Microbiol. Biotechnol.* **30**, 462. **2**. Hashimoto, Nakayama, Teramoto, *et al.* (1991) *Biochem. Biophys. Res. Commun.* **181**, 423. **3**. Rodriguez-Mora, LaHair, *et al.* (2005) *Exp. Opin. Ther. Targets* **9**, 791. **4**. Kase, Iwahashi, Nakanishi, *et al.* (1987) *Biochem. Biophys. Res. Commun.* **142**, 436. **5**. Chin, Murray, Doherty & Singh (1999) *Cancer Invest.* **17**, 391. **6**. Hayess & Benndorf (1997) *Biochem. Pharmacol.* **53**, 1239. **7**. Schindler, Godbey, Hood, *et al.* (2002) *Biochim. Biophys. Acta* **1598**, 88. **8**. Nakanishi, Yamada, Kase, Nakamura & Nonomura (1988) *J. Biol. Chem.* **263**, 6215. **9**. Koizumi, Contreras, Matsuda, *et al.* (1988) *J. Neurosci.* **8**, 715. **10**. Hashimoto (1988) *J. Cell Biol.* **107**, 1531. **11**. Tapley, Lamballe & Barbacid (1992) *Oncogene* **7**, 371. **12**. Angeles, Yang, Steffler & Dionne (1998) *Arch. Biochem. Biophys.* **349**, 267. **13**. Muroya, Hashimoto, Hattori & Nakamura (1992) *Biochim. Biophys. Acta* **1135**, 353. **14**. Elliott, Wilkinson, Sedgwick, *et al.* (1990) *Biochem. Biophys. Res. Commun.* **171**, 148. **15**. Mizuno, Saido, Ohno, Tamaoki & Suzuki (1993) *FEBS Lett.* **330**, 114. **16**. Gschwendt, Dieterich, Rennecke, *et al.* (1996) *FEBS Lett.* **392**, 77. **17**. Ranganathan, Liu, Migliorini, *et al.* (2004) *J. Biol. Chem.* **279**, 40536.

K-252b

This cell-permeable general protein kinase inhibitor (FW = 453.45 g/mol; Photosensitive; Pale yellow powder) is a staurosporine analogue that inhibits protein kinase A (K_i = 90 nM), protein kinase C (K_i = 20 nM), protein kinase G (K_i = 100 nM), and myosin-light-chain kinase (K_i = 147 nM). K 252b (FWfree acid = 453.45 g/mol) is soluble in soluble in *N,N*-dimethylformamide and dimethyl sulfoxide. The corresponding ester K-252a is a more potent and cytotoxic agent on intact cells than K-252b, and the former dissolves more effectively into the hydrophobic interior of cell membranes (1). **Target(s):** cAMP-dependent protein kinase (2); casein kinase II-like ectokinase (3); cGMP dependent protein kinase, *or* protein kinase G (2); myosin-light-chain kinase (2); protein kinase C (2,4). **1**. Ross, McKinnon, Daou, Ratliff & Wolf (1995) *J. Neurochem.* **65**, 2748. **2**. Kase, Iwahashi, Nakanishi, *et al.* (1987) *Biochem. Biophys. Res. Commun.* **142**, 436. **3**. Teshima, Onose, Saito, *et al.* (1999) *Immunol. Lett.* **68**, 369. **4**. Nakanishi, Matsuda, Iwahashi & Kase (1986) *J. Antibiot. (Tokyo)* **39**, 1066.

K-579

This long-lasting hypoglycemic agent (FW$_{free-base}$ = 328.41 g/mol; CAS 440100-64-1; Soluble to 50 mM as mono-HCl salt; 100 mM in DMSO as the free base), also named (2*S*)-1-[[[4-methyl-1-(2-pyrimidinyl)-4-piperidinyl]amino]acetyl]-2-pyrrolidinecarbonitrile, is a slow-binding inhibitor rat, canine, human and monkey dipeptidyl peptidase IV, with IC$_{50}$ values of 3, 5, 8, and 8 nM, respectively. K-579 preserves the endogenously secreted active forms of glucagon-like peptide-1 (Glp-1), augments insulin responsiveness, and ameliorates the hypergycemic excursions during oral glucose tolerance test in rats. It likewise attenuates hyperglycemic excursion following glucose loading in Zucker fatty rats (1). In glibenclamide-pretreated rats, K579 significantly suppresses blood glucose elevation without excessive hypoglycemia, and continues to inhibit plasma DPP-IV even 8 h after administration (2). K579's sustained inhibitory action is due to the presence of K579 metabolites that also inhibits dipeptidyl peptidase IV (3). **1**. Takasaki, Iwase, Nakajima, *et al.* (2004) *Eur. J. Pharmacol.* **486**, 335. **2**. Takasaki, Nakajima, Ueno, Nomoto & Higo (2004) *J. Pharmacol. Sci.* **95**, 291. **3**. Takasaki, Takada, Nakajima, *et al.* (2004) *Eur. J. Pharmacol.* **505**, 237.

K858

This mitotic drug (FW = 277.34 g/mol; CAS 910634-41-2), also known by the code names LY2523355 and KF 89617 as well as its systematic name, targets kinesin Eg5, a mitotic motor protein that plays a key role in the M-phase of the cell cycle. Long-term continuous treatment of cancer cells with K858 results in antiproliferative effects through the induction of mitotic cell death, polyploidy, and senescence. In contrast, treatment of nontransformed cells results in mitotic slippage without cell death, but leads to cell-cycle arrest in G$_1$ phase in a tetraploid state. (**See also** *Litronesib (structurally related)*, *Monastrol, Terpendole E, S-Trityl-L-cysteine*) **1**. Nakai, Iida, Takahashi, *et al.* (2009) *Cancer Res.* **69**, 3901.

K03861

This type II CDK2 inhibitor (FW = 501.50 g/mol; CAS 853299-07-7; Solubility: 100 mg/mL DMSO), also named *N*-[4-[(2-amino-4-pyrimidinyl) oxy]phenyl]-N'-[4-[(4-methyl-1-piperazinyl)methyl]-3-(trifluoro-methyl) phenyl]urea, targets Cyclin-Dependent Kinase-2 CDK2^{WT} (K_d = 50 nM), CDK2^{C118L} (K_d = 18.6 nM), CDK2^{A144C} (K_d = 15.4 nM), and CDK2$^{C118L/A144C}$ (K_d = 9.7 nM). **1**. Alexander, Möbitz, Drueckes, *et al.* (2015) *ACS Chem Biol.* **10**, 2116.

K-7174

This first-in-class, orally active proteasome inhibitor and novel cell adhesion inhibitor (FW = 568.76 g/mol), also known systematically as N,N'-bis-(E)-[5-(3,4,5-trimethoxyphenyl)-4-pentenyl]homopiperazine, targets all three proteasome catalytic subunits (β1, β2 and β5), whereas bortezomib and other proteasome inhibitors mainly act on the β5 subunit (1). Perhaps, more significantly, K-7174 inhibits the growth of bortezomib-resistant myeloma cells carrying a β5-subunit mutation, and also acts additively, when used in combination with bortezomib, to inhibit proteasome action and induce apoptosis in myeloma cells (1). K-7174 induces transcriptional repression of Class I histone deacetylases HDAC1, 2 and 3 via caspase-8-dependent degradation of Sp1, the most potent transactivator of Class-I HDAC genes (2). HDAC1 overexpression ameliorates K-7174's cytotoxic effect and abrogates histone hyperacetylation without affecting the accumulation of ubiquitinated proteins in treated myeloma cells. K-7174 is also more effective when administered by oral, rather than an intravenous route. K-7174 is also novel cell adhesion inhibitor that blocks adhesion of monocytes to cytokine-stimulated endothelial cells by reducing endothelial VCAM-1 induction by binding to the GATA motifs in the VCAM-1 gene promoter region (3). Importantly, K-7174 did not influence the binding to any of the following binding motifs: octamer binding protein, AP-1, SP-1, ets, NFκB, or interferon regulatory factor, suggesting that the regulation of GATA binding may be a new druggable target for anti-inflammatory agents that act through a NFκB-independent mechanism (3). **1**. Kikuchi, Shibayama, Yamada, *et al.* (2013) *PLoS One* **8**, e6064. **2**. Kikuchi, Yamada, Koyama *et al.* (2013) *J. Biol. Chem.* **288**, 25593. **3**. Umetani, Nakao, Doi, *et al.* (2000) *Biochem. Biophys. Res Commun.* **272**, 370.

K41448

This highly selective and peripherally acting CRF$_2$ antagonist (MW = 3632.26 g/mol; CAS 434938-41-7; Sequence: D-Phe-His-Leu-Leu-Arg-Lys-Nle-Ile-Glu-Ile-Glu-Lys-Gln-Gln-Ala-Ala-Asn-Asn-Arg-Leu-Leu-Leu-Asp-Thr-Ile-NH$_2$; Solubility: 5 mg/mL H$_2$O) targets human Corticotropin-Releasing hormone (CRH) Receptor-2 subtypes CRF$_{2\alpha}$ and CRF$_{2\beta}$, with K_i values of 0.66 and 0.62 nM, respectively, but with a much higher value of 425 nM for CRF$_1$. (**See also** *Antisauvagine-30; Astressin*) Found in plasma membranes of hormone-sensitive cells, these Type-2 GPCRs bind Corticotropin-Releasing Hormone (CRH), a neurotransmitter peptide involved in stress responses. Produced by parvocellular neuroendocrine cells in the paraventricular nucleus of the hypothalamus, CRH is released into the primary capillary plexus of the hypothalamo-hypophyseal portal system. CRH is also synthesized by the placenta and appears to play a role in parturition. An analogue of antisauvagine-30, K41448 inhibits sauvagine-stimulated cAMP accumulation in hCRF$_{2\alpha}$- and hCRF$_{2\beta}$-expressing cells. In rats, K41448 blocks urocortin-induced hypotension after systemic administration. **1**. Lawrence, *et al.* (2002) *Brit. J. Pharmacol.* **136**, 896. **2**. Ruhmann, *et al.*, (2002) *Peptides* **23**, 453.

Kahweol

This anticarcinogenic agent (FW = 314.42 g/mol; CAS 6894-43-5; (3bS,5aS,7R,8R,10aR,10bS)-3b,4,5,6,7,8,9,10,10a,10b-decahydro-7-hydroxy-10b-methyl-5a,8-methano-5aH-cyclohepta(5,6)naphtho(2,1-b) furan-7-methanol, is an antioxidant coffee diterpenoid that, among its many actions, inhibits cytochrome P450 enzymes (1-3). Kahweol prevents osteoclastogenesis by impairing NFATc1 expression and by blocking Erk phosphorylation (4). It also suppresses cell proliferation by inducing proteasomal degradation of cyclin D$_1$ via ERK1/2-, JNK- and GKS3β-dependent threonine-286 phosphorylation in human colorectal cancer cells (5). Indeed, epidemiologic evidence suggests a positive correlation between coffee consumption and a reduced risk of cancer and cirrhosis. Among the mechanisms likely to account for these chemoprotective effects are induction of conjugating enzymes (*e. g.*, glutathione *S*-transferases, glucuronosyl *S*-transferases), increased expression of cellular defense enzymes (*e. g.*, γ-glutamyl cysteine synthetase and heme oxygenase-1), and reduced CYP expression, perhaps decreasing carcinogen activation (6). **Targets:** 4-aminobiphenyl hydroxylation (2); CYP1A1 (1); CYP2B6 (1,3). **1**. Cavin, Holzhaeuser, Scharf, *et al.* (2002) *Food Chem. Toxicol.* **40**, 1155. **2**. Hammons, Fletcher, Stepps, *et al.* (1999) *Nutr. Cancer* **33**, 46. **3**. Cavin, Mace, Offord & Schilter (2001) *Food Chem. Toxicol.* **39**, 549. **4**. Fumimoto, Sakai, Yamaguchi, *et al.* (2012) *J. Pharmacol. Sci.* **118**, 479. **5**. Park, Song & Jeong (2016) *Food Chem. Toxicol.* **95**, 142. **6**. Cavin, Holzhaeuser, Scharf, *et al.* (2002) *Food Chem. Toxicol.* **40**, 1155.

Kainate (Kainic Acid)

This cyclic analogue of L-glutamate and marine natural product (FW$_{free-acid}$ = 213.23 g/mol; CAS 487-79-6), also known as (2S,3S,4S)-2-carboxy-4-(1-methylethenyl)-3-pyrrolidineacetic acid and digenic acid, is an agonist for the kainate class of glutamate receptors, inducing seizures and neurodegeneration. Kainate receptor agonists depress transmitter release at synapses in the hippocampus. Distinct mechanisms appear to underlie this phenomenon at different synapses. Pre-synaptic kainate receptors can also potentiate the release of both GABA and glutamate. **Target(s):** choline *O*-acetyltransferase. **1**. Loureiro-Dos-Santos, Reis, Kubrusly, *et al.* (2001) *J. Neurochem.* **77**, 1136.

Kallistatin

This serpin (MW = 48.5 kDa), also known as protease inhibitor 4, is an acidic protein (Isoelectric Point = 4.6–5.2) that forms a 1:1 covalent complex with tissue kallikrein. Heparin blocks kallistatin: kallikrein complex formation. (*For a discussion of the likely mechanism of inhibitory action,* **See** *Serpins; also* α_1-*Antichymotrypsin*) **1**. Chao (1998) in *Handb. Proteolytic Enzymes*, p. 97, Academic Press, San Diego. **2**. Zhou, Chao & Chao (1992) *J. Biol. Chem.* **267**, 25873. **3**. Chao, Chai, Chen, *et al.* (1990) *J. Biol. Chem.* **265**, 16394. **4**. Chao, Chai & Chao (1996) *Immunopharmacology* **32**, 67. **5**. Chao & Chao (1995) *Biol. Chem. Hoppe Seyler* **376**, 705. **6**. Chen, Chao & Chao (2000) *J. Biol. Chem.* **275**, 38457 and 40371.

Kanamycins

Kanamycin A

Kanamycin B

Kanamycin C

These aminoglycoside antibiotics produced by *Streptomyces kanamyceticus* include the major component Kanamycin A (FW$_{free-base}$ = 484.50 g/mol),

Kanamycins B (FW$_{\text{free-base}}$ = 483.52 g/mol), and Kanamycins C (FW$_{\text{free-base}}$ = 484.50 g/mol) varying only in the 6-glucosamine component . All are soluble in water (particularly as the sulfate salts) and each has antibacterial activity. (Numbers adjacent to the amino groups in kanamycin B correspond to the pK_a values.) **Primary Mode of Action:** Kanamycins bind to both the 50S and 70S ribosomal subunits, inhibiting initiation and elongation and thus protein biosynthesis (1,2) The peptidyl-tRNA is fixed to the A-site of the ribosome. In addition, the kanamycins also inhibit the release of tRNA. Binding to the 16S component (near nucleotide 1400 of *Escherichia coli* 16S, with two adenine residues, 1408 and 1493 playing essential roles in binding) decreases the rate of dissociation of aminoacyl-tRNA, thus inhibiting processivity and resulting in miscoding. **Target(s):** cellulose 1,4-β-cellobiosidase (3); DNA-directed DNA polymerase (Klenow DNA polymerase) (4); gentamicin 3'-*N*-acetyltransferase, inhibited by kanamycins A and C (5); kanamycin 6'-acetyltransferase, by kanamycin C, but not kanamycin A (6); ornithine decarboxylase (7); phosphatidylinositol-specific phospholipase C (8); phospholipase D (9); protein biosynthesis (initiation and elongation) (1,2); poly(A)-specific ribonuclease (4); protein kinase C (10); ribonuclease P (11,12); ribozymes, group I (13); streptomycin 3''-adenylyltransferase (14). **1.** Pestka (1974) *Meth. Enzymol.* **30**, 261. **2.** Jiménez (1976) *Trends Biochem. Sci.* **1**, 28. **3.** Singh, Agrawal, Abidi & Darmwal (1990) *J. Gen. Appl. Microbiol.* **36**, 245. **4.** Ren, Martínez, Kirsebom & Virtanen (2002) *RNA* **8**, 1393. **5.** Williams & Northrop (1978) *J. Biol. Chem.* **253**, 5908. **6.** Haas & Dowding (1975) *Meth. Enzymol.* **43**, 611. **7.** Henley, Mahran & Schacht (1988) *Biochem. Pharmacol.* **37**, 1679. **8.** Lipsky & Lietman (1982) *J. Pharmacol. Exp. Ther.* **220**, 287. **9.** Liscovitch, Chalifa, Danin & Eli (1991) *Biochem. J.* **279**, 319. **10.** Hagiwara, Inagaki, Kanamura, Ohta & Hidaka (1988) *J. Pharmacol. Exp. Ther.* **244**, 355. **11.** Tekos, Tsagla, Stathopoulos & Drainas (2000) *FEBS Lett.* **485**, 71. **12.** Mikkelsen, Brannvall, Virtanen & Kirsebom (1999) *Proc. Natl. Acad. Sci. U.S.A.* **96**, 6155. **13.** von Ahsen, Davies & Schroeder (1991) *Nature* **353**, 368. **14.** Kim, Hesek, Zajicek, Vakulenko & Mobashery (2006) *Biochemistry* **45**, 8368.

Katanosins

Katanosin A

These cyclic depsipeptide antibiotics (FW$_{\text{Katanosin-A}}$ = 1262.47 g/mol; CAS 116103-86-7; (FW$_{\text{Katanosin-B}}$ = 1276.49 g/mol; CAS 116340-02-4) from the culture broth of a strain related to the genus *Cytophaga* exhibit strong antimicrobial activity against methicillin-resistant *Staphylococcus aureus* and VanA-type vancomycin-resistant *enterococci*, with MICs ranging from 0.39 to 3.13 µg/mL, as well as against other Gram-positive bacteria (1). They inhibit incorporation of *N*-acetylglucosamine, a precursor of cell wall synthesis, into peptidoglycan of *S. aureus* whole cells at concentrations close to their MICs. *Note:* In katanosin B, the Val residue is replaced by Ile. **1.** Maki, Miura & Yamano (2001) *Antimicrob. Agents Chemother.* 45, 1823.

Kasugamycin

This aminoglycoside antibiotic (FW = 379.37 g/mol) from a strain of *Streptomyces kasugaensis* inhibits prokaryotic protein biosynthesis (1,2) by binding to the 30S ribosomal subunit. It inhibits initiation-complex formation. In addition, kasugamycin will block translation in fungal systems at pH 4–5. Interestingly, translation of leaderless mRNA continues in the presence of kasugamycin (3). The amino sugar kasugamine (*i.e.*, 2,4-diamino-2,3,4,6-tetradeoxy-D-*arabino*-hexose) is a component. Kasugamycin should be stored dry and at 4°C. **1.** Pestka (1974) *Meth. Enzymol.* **30**, 261. **2.** Jiménez (1976) *Trends Biochem. Sci.* **1**, 28. **3.** Moll & Blasi (2002) *Biochem. Biophys. Res. Commun.* **297**, 1021. **4.** Okuyama, Machiyama, Kinoshita & Tanaka (1971) *Biochem. Biophys. Res. Commun.* **43**, 196. **5.** Masukawa, Tanaka & Umezawa (1968) *J. Antibiot. (Tokyo)* **21**, 73. **6.** Tanaka, Yoshida, Sashikata, Yamaguchi & Umezawa (1966) *J. Antibiot. (Tokyo)* **19**, 65.

KB-R7943

This potent and selective inhibitor (FW = 427.29 g/mol; CAS 182004-65-5; Solubility: 100 mM in DMSO; Stable at room temperature), also named 2-[2-[4-(4-nitrobenzyloxy)phenyl]ethyl]isothiourea mesylate, targets reverse mode of the Na$^+$/Ca^{2+} exchanger, *or* NCX, IC$_{50}$ = 0.7 µM (1). KB-R7943 significantly improves recovery of population spike amplitudes in rat hippocampal slices by inhibiting the Na$^+$/Ca^{2+} exchanger, operating in reverse mode, contributes to hypoxia/hypoglycemia-induced injury in CA1 neurons (2). KB-R7943 also suppresses ouabain-induced arrhythmias by inhibiting the reverse-mode NCX (3). KB-R7943 also potently inhibits the mitochondrial Ca^{2+} uniporter (MCU). In permeabilized HeLa cells, KB-R7943 inhibits mitochondrial Ca^{2+} uptake (K_i = 5.5 µM). In intact cells, KB-R7943 (10 µM) reduces the histamine-induced mitochondrial [Ca^{2+}] peak by 80%. KB-R7943 does not modify the mitochondrial membrane potential and is without effect on the mitochondrial Na$^+$/Ca^{2+} exchanger (4). KB-R7943 inhibits histamine-induced ER-Ca^{2+} release in intact cells, but not in cells loaded with a Ca^{2+}-chelator to dampen cytosolic [Ca^{2+}] changes (4). Inhibition of ER-Ca^{2+}-release by KB-R7943 was probably due to the increased feedback Ca^{2+}-inhibition of inositol 1,4,5-trisphosphate receptors after MCU block. **1.** Hoyt, Arden, Aizenman & Reynolds (1998) *Mol. Pharmacol.* **53**, 742. **2.** Schröder et al (1999) *Neuropharmacology* **38**, 319. **3.** Watano, Harada, Harada & Nishimura (1999) *Brit. J. Pharmacol.* **127**, 1846. **4.** Santo-Domingo et al (2007) *Brit. J. Pharmacol.* **151**, 647.

Kelatorphan

This potent neprilysin inhibitor (FW$_{\text{free-acid}}$ = 294.31 g/mol), also known as [(*R*)-3-(*N*-hydroxy)carboxamide-2 benzylpropanoyl]-L-alanine, is the first complete inhibitor of enkephalin metabolism and the only compound with analgesic activity exceeding the combined action of thiorphan and bestatin. **Target(s):** dactylysin (1); leukotriene-A$_4$ hydrolase (2,3); membrane alanyl aminopeptidase, also known as aminopeptidase N, K_i = 380 nM (4,5); neprilysin, K_i = 1.8 nM (4,5). **1.** Carvalho, Joudiou, Boussetta, Leseney & Cohen (1992) *Proc. Natl. Acad. Sci. U.S.A.* **89**, 84. **2.** Penning (2001) *Curr. Pharm. Des.* **7**, 163. **3.** Haeggström, Kull, Rudberg, Tholander & Thunnissen (2002) *Prostaglandins* **68 69**, 495. **4.** Roques, Noble, Crine & Fournié-Zaluski (1995) *Meth. Enzymol.* **248**, 263. **5.** Bauvois & Dauzonne (2006) *Med. Res. Rev.* **26**, 88.

Kenpaullone

This relatively nonselective paullone-derived protein kinase inhibitor (FW = 327.18 g/mol; CAS 142273-20-9; Solubility: 65 mg/mL DMSO; <1 mg/mL H₂O), also known as 9-bromo-7,12 dihydroindolo[3,2-d][1]benzazepin-6(5H)-one and NSC-664704, potently inhibits cyclin-dependent protein kinases, glycogen synthase kinase-3β, c-Src, casein kinase II, ERK1 (extracellular-regulated kinase-1), and ERK2. **Target(s):** casein kinase 2 (1); c-Src (1); cyclin-dependent kinase, *or* cyclin-dependent protein kinase 1/cyclin B, IC_{50} = 0.4 μM (1-3); cyclin-dependent protein kinase 2/cyclin A, IC_{50} = 0.68 μM (1,4); cyclin-dependent protein kinase 2/cyclin E, IC_{50} = 7.5 μM (1); cyclin-dependent protein kinase 5, IC_{50} = 0.85 μM (1); ERK1, *or* extracellular-regulated kinase 1 (1); ERK2, *or* extracellular-regulated kinase 2 (1); glycogen synthase kinase 3β, *or* [tau protein] kinase, IC_{50} = 0.23 μM (4,5); Lck protein-tyrosine kinase (4); lymphocyte kinase, IC_{50} = 0.47 μM (4); mitogen-activated protein kinase (4); non-specific serine/threonine protein kinase (1,4); protein-tyrosine kinase, *or* nonspecific protein-tyrosine kinase (4). **1**. Zaharevitz, Gussio, Leost, *et al.* (1999) *Cancer Res.* **59**, 2566. **2**. Schultz, Link, Leost, *et al.* (1999) *J. Med. Chem.* **42**, 2909. **3**. Woodard, Li, Kathcart, *et al.* (2003) *J. Med. Chem.* **46**, 3877. **4**. Bain, McLauchlan, Elliott & Cohen (2003) *Biochem. J.* **371**, 199. **5**. Cole, Frame & Cohen (2004) *Biochem. J.* **377**, 249.

Ketamine

(2R,6S)-Ketamine

This once commonly used dissociative anesthetic ($FW_{ketamine}$ = 237.73 g/mol; CAS 6740-88-1; IUPAC Name: (*RS*)-2-(2-chlorophenyl)-2-(methylamino)cyclohexanone; HCl salt is very soluble, 20 g/100 mL H₂O) is a noncompetitive NMDA (*N*-methyl-D-aspartate) receptor antagonist. When administered at the higher doses required for anesthesia, ketamine is hallucinogenic, inducing schizophrenia-like symptoms, amnesia, loss of consciousness, and immobility. These properties explain why ketamine is not typically used as a primary anesthetic (although it becomes the anesthetic of choice when a reliable ventilator is unavailable) (1). **Mechanism of Ketamine Action:** Ketamine acts on the Central Nervous System, exerting both general anf local anaesthetic properties that are mediated by noncompetitive antagonism of the NMDA receptor's Ca²⁺ channel pore. Channel blockade appears to be the primary source of its anaesthetic and analgesic properties. Ketamine also reduces presynaptic glutamate release. The S-(+) enantiomer has a three- to four-fold greater affinity for the NMDA receptor than the R(−) form The S(+)-enantiomer inhibits more effectively than R(−)-isomer and shows fewer side effects. Ketamine also increases the phosphorylation of the transcription factor cAMP response element-binding protein (pCREB) and neurotrophic factor/tropomyosin related kinase B receptor (pTrKB) in the prefrontal cortex (PFC), and decreased pCREB in the hippocampus (2). The MEK inhibitor PD184161 abolishes ketamine's effects in the hippocampus (2). (*See also Phenclycidine*) **Pharmacokinetics:** In humans, plasma ketamine concentration-time curves fit a two-compartment open model with a terminal $t_{1/2}$ of 186 min (3). Absorption after intramuscular injection was rapid and the bioavailability was 93%; however, only 17% of an oral dose was absorbed because of extensive first-pass metabolism (3). **Metabolism:** Ketamine's immediate metabolic products are norketamine (FW = 223.70 g/mol; CASs = 35211-10-0 (free base); 79499-59-5 (HCl salt); IUPAC = 2-amino-2-(2-chlorophenyl)cyclohexan-1-one); and *cis*-6-hydroxy-

norketamine (FW = 239.70 g/mol; CAS 81395-70-2; IUPAC: 2-amino-2-(2-chlorophenyl)-6-hydroxycyclohexan-1-one).

(2R,6S)-Ketamine

(2R,6S)-Norketamine

(2S,6S) & (2R,6R)-Hydroxynorketamine

The former is a potent non-competitive NMDA receptor antagonist (K_i = 3.6 μM), whereas the latter increases AMPA receptor-mediated excitatory post-synaptic potentials in hippocampal slices. **Ketamine Abuse:** Ketamine is listed as a controlled substance under the United States Controlled Substance Act. Given the relative ease of its synthesis, ketamine is a widely abused recreational. Known by the street names "K", Special K, Vitamin K, or Lady K, ketamine displays potent PCP-and DXM-like qualities. Broadly classified as a dissociative anesthetic, ketamine distorts sight and sound perception, producing feelings of detachment and often inducing hallucinations. The onset of its effects depends on the mode of ingestion: 1-5 minutes, upon injection; 5-15 minutes, upon snorting the dry powder; 5-30 minutes, orally. A dose of 1-2 mg/kg body-weight produces an intense experience lasting ~1 hour. Side effects include flashbacks, amnesia, impaired motor functioning, hallucinations, disorientation, tachycardia, muscle rigidity, loss of coordination, sense of invulnerability, aggressive and/or violent behavior, and, in cases of overdose, death. Chronic use of ketamine alters apoptotic markers in the prefrontal cortex and causes abnormal walking, jumping, and climbing. **Target(s):** acetylcholinesterase (4); cholinesterase (5); K⁺ channels, neuronal (6); Na⁺ channels, voltage-gated (7); Na⁺/K⁺-exchanging ATPase (8); nicotinic acetylcholine receptors, neuronal (9); NMDA receptor (10). Ketamine also suppresses proinflammatory cytokine production in human whole blood *in vitro* (11). Curiously, a single dose of ketamine produces rapid (commencing within 2 hours) and sustained (lasting >7 days) antidepressant efficacy in treatment-resistant patients (12). **1**. Pai & Heining (2007) *Contin. Educ. Anaest. Crit. Care & Pain* **7**, 59. **2**. Réus, Abaleira, Titus, *et al.* (2016) *Pharmacol. Rep.* **68**, 177. **3**. Clements, Nimmo, & Grant (1982) *J Pharm Sci.* **71**, 539. **4**. Cohen, Chan, Bhargava & Trevor (1974) *Biochem. Pharmacol.* **23**, 1647. **5**. Schuh (1975) *Brit. J. Anaesth.* **47**, 1315. **6**. Friederich, Benzenberg & Urban (2001) *Eur. J. Anaesthesiol.* **18**, 177. **7**. Wagner, Gingrich, Kulli & Yang (2001) *Anesthesiology* **95**, 1406. **8**. Rabin & Acara (1993) *Biochem. Pharmacol.* **45**, 1653. **9**. Yamakura, Chavez-Noriega & Harris (2000) *Anesthesiology* **92**, 1144. **10**. Elliott, Kest, Man, Kao & Inturrisi (1995) *Neuropsychopharmacology* **13**, 347. **11**. Kawasaki, Ogata, Kawasaki, *et al.* (1999) *Anesth. Analg.* **89**, 665. **12**. Wang, Jing, Toledo-Salas & Xu (2014) *Neurosci. Bull.* **31**, 75.

Ketene

$$H_2C{=}C{=}O$$

This powerful acetylating agent (FW = 42.04 g/mol; M.P. = −150°C; B.P. = −56°C), which is prepared by the thermal decoposition of acetone, diketene, or acetic anhydride, is an acylating agent that reacts with nucleophilic groups in proteins. Ketene is toxic, and all work with it should be conducted in a fume hood. (*See also Diketene*) **Target(s):** alkaline phosphatase (1,2); amylase, pancreatic (3); chymotrypsin (4); diphtheria toxin (5); papain (6); pepsin, with *O*-acetylation (not *N*-acetylation) of the latter leading to inactivation (7-10). **1**. Fernley (1971) *The Enzymes*, 3rd ed., **4**, 417. **2**.

Gould (1944) *J. Biol. Chem.* **156**, 365. **3.** Little & Caldwell (1942) *J. Biol. Chem.* **142**, 585. **4.** Sizer (1945) *J. Biol. Chem.* **160**, 547. **5.** Pappenheimer (1938) *J. Biol. Chem.* **125**, 201. **6.** Greenberg & T. Winnick (1940) *J. Biol. Chem.* **135**, 761. **7.** Smith (1951) *The Enzymes*, 1st ed., **1** (part 2), 793. **8.** Herriott & Northrop (1934) *J. Gen. Physiol.* **18**, 35. **9.** Bovey & Yanari (1960) *The Enzymes*, 2nd ed., **4**, 63. **10.** Fruton (1971) *The Enzymes*, 3rd ed., **3**, 119.

2-Ketoadipate (2-Oxoadipate)

This α-keto acid (FW$_{\text{free-acid}}$ = 160.13 g/mol), also called α-ketoadipic acid, 2-oxoadipic acid, and 2-oxohexanedioic acid, is an intermediate in the degradation of L-lysine and L-tryptophan. It is also an intermediate in the pathway for L-lysine biosynthesis in fungi. **Target(s):** 4-aminobutyrate aminotransferase (1); [branched-chain α-keto-acid dehydrogenase] kinase, *or* 3-methyl-2-oxobutanoate dehydrogenase (acetyl-transferring)] kinase (2-4); γ-butyrobetaine dioxygenase, *or* γ-butyrobetaine hydroxylase, K_i = 0.17 mM (5); 4-hydroxy-4 methyl-2-oxoglutarate aldolase (6); kynurenine:oxoglutarate aminotransferase (7,8); lactate dehydrogenase (9); 3-oxoadipate CoA-transferase (10); phospho-enolpyruvate carboxylase, weakly inhibited (11); procollogen-lysine 5-dioxygenase (12); procollagen-proline 3-dioxygenase (12); procollagen-proline 4-dioxygenase (12-14); pyrimidine-deoxynucleoside 2'-dioxygenase, *or* thymidine,α-ketoglutarate dioxygenase (15,16); pyruvate carboxylase (17); saccharopine dehydrogenase (18). **1.** Schousboe, Wu & Roberts (1974) *J. Neurochem.* **23**, 1189. **2.** Paxton (1988) *Meth. Enzymol.* **166**, 313. **3.** Espinal, Beggs & Randle (1988) *Meth. Enzymol.* **166**, 166. **4.** Randle, Patston & Espinal (1987) *The Enzymes*, 3rd ed., **18**, 97. **5.** Ng, Hanauske-Abel & Englard (1991) *J. Biol. Chem.* **266**, 1526. **6.** Maruyama (1990) *J. Biochem.* **108**, 327. **7.** Tobes & Mason (1978) *Life Sci.* **22**, 793. **8.** Takeuchi, Otsuka & Shibata (1983) *Biochim. Biophys. Acta* **743**, 323. **9.** Okabe, Hayakawa, Hamada & Koike (1968) *Biochemistry* **7**, 79. **10.** Kaschabek, Kuhn, Müller, Schmidt & Reineke (2002) *J. Bacteriol.* **184**, 207. **11.** Mori & I. Shiio (1985) *J. Biochem.* **98**, 1621. **12.** Majamaa, Turpeenniemi-Hujanen, Latipää, *et al.* (1985) *Biochem. J.* **229**, 127. **13.** Majamaa, Hanauske-Abel, Günzler & Kivirikko (1984) *Eur. J. Biochem.* **138**, 239. **14.** Kaska, Günzler, Kivirikko & R. Myllylä (1987) *Biochem. J.* **241**, 483. **15.** Bankel, Lindstedt & Lindstedt (1972) *J. Biol. Chem.* **247**, 6128. **16.** Hayaishi, Nozaki & Abbott (1975) *The Enzymes*, 3rd ed., **12**, 119. **17.** Osmani & Scrutton (1985) *Eur. J. Biochem.* **147**, 119. **18.** Fujioka & Nakatani (1972) *Eur. J. Biochem.* **25**, 301.

3-Ketoadipate (3-Oxoadipate)

This β-keto acid (FW$_{\text{free-acid}}$ = 160.13 g/mol), also called β-ketoadipic acid, 3-oxoadipic acid, and 3-oxohexanedioic acid, is an intermediate in the degradation of a number of aromatic compounds by bacteria. It has a melting point of 124-125°C and is slightly soluble in water. It generates a violet color with FeCl$_3$. 3-Ketoadipate is a product of the reactions catalyzed by 3 oxoadipate enol-lactonase and maleylacetate reductase. It is also substrate for 3-oxoadipate CoA transferase. **Target(s):** deacetoxycephalosporin-C synthase (1); kynurenine aminotransferase (2); 3 oxoadipate enol-lactonase, *or* β-ketoadipate enol-lactonase (3); porphobilinogen synthase, *or* δ-aminolevulinate dehydratase (4); pyridoxamine:oxaloacetate aminotransferase (5); pyrimidine deoxynucleoside 2'-dioxygenase, *or* thymidine,α-ketoglutarate dioxygenase (6,7). **1.** Cortés, Martín, Castro, Láiz & Liras (1987) *J. Gen. Microbiol.* **133**, 3165. **2.** Mason (1959) *J. Biol. Chem.* **234**, 2770. **3.** Yeh & Ornston (1984) *Arch. Microbiol.* **138**, 102. **4.** Tschudy, Collins, Caughey & Kleinspehn (1960) *Science* **131**, 1380. **5.** Wu & Mason (1964) *J. Biol. Chem.* **239**, 1492. **6.** Hayaishi, Nozaki & Abbott (1975) *The Enzymes*, 3rd ed., **12**, 119. **7.** Bankel, Lindstedt & Lindstedt (1972) *J. Biol. Chem.* **247**, 6128.

2-Keto-3-butynoate (2-Oxo-3-butynoate)

This acetylenic α-keto acid (FW$_{\text{free-acid}}$ = 98.06 g/mol) is an mechanism-based inhibitor that irreversibly inactivates a number of flavoproteins. 2-Hydroxy-3-butynoate is both a substrate and an irreversible inactivator of the flavoenzyme L-lactate oxidase. The partitioning between catalysis and inactivation is determined by the concentration of the secoGnd substrate, dioxygen. The common intermediate that forms in these two pathways is a charge-transfer complex between enzyme-reduced flavin and 2-keto-3-butynoate. Inactivation occurs by covalent modification of the enzyme-bound cofactor and does not involve labeling of the apoprotein. The spectrum of the enzyme bound adduct suggests a 4a,5-adduct. **Target(s):** flavocytochrome b_2, *or* L-lactate dehydrogenase (cytochrome) (1,2); L-hydroxyacid oxidase (3,4); lactate oxidase (5,6); 4-oxalocrotonate tautomerase (7); pyruvate decarboxylase (8); pyruvate dehydrogenase (8); pyruvate oxidase (8). **1.** Pompon & Lederer (1982) *Eur. J. Biochem.* **129**, 143. **2.** Pompon & Lederer (1985) *Eur. J. Biochem.* **148**, 145. **3.** Cromartie & Walsh (1975) *Biochemistry* **14**, 3482. **4.** Cromartie & Walsh (1975) *Biochemistry* **14**, 2588. **5.** Ator & Ortiz de Montellano (1990) *The Enzymes*, 3rd ed., **19**, 213. **6.** Schonbrunn, Abeles, Walsh, *et al.* (1976) *Biochemistry* **15**, 1798. **7.** Taylor, Czerwinski, Johnson, Whitman & Hackert (1998) *Biochemistry* **37**, 14692. **8.** Brown, Nemeria, Yi, *et al.* (1997) *Biochemistry* **36**, 8071.

2-Ketobutyrate (2-Oxobutyrate)

This α-keto acid (FW$_{\text{free-acid}}$ = 102.09 g/mol; Water-soluble; pK_a = 2.5), also known as α-ketobutyric acid, 2-oxobutyric acid, and 2-oxobutanoic acid, is formed from L-threonine by the action of threonine dehydratase. It is also produced in the degradation of L-methionine. 2-Ketobutyrate is the substrate of the reactions catalyzed by 2-ethylmalate synthase. It is an alternative substrate for α-keto-acid dehydrogenase. **Target(s):** acetolactate synthase, *or* acetohydroxy acid synthase (1-3,25,26); aldehyde reductase, *or* aldose reductase (4); D-amino-acid oxidase (5); [branched-chain α-keto acid dehydrogenase] kinase (6); γ-butyrobetaine dioxygenase, *or* γ-butyrobetaine hydroxylase, K_i = 1.3 mM (7); carbamoyl-phosphate synthetase I (8); deacetoxycephalosporin-C synthase (27); 2-dehydro-3-deoxy-L-arabinonate dehydratase (9); dihydrodipicolinate synthase (10); fumarylacetoacetate (11); 4-hydroxy-4-methyl-2-oxoglutarate aldolase (12); 4-hydroxy-2-oxoglutarate aldolase (13); malic enzyme (14); 3-mercaptopuruvate sulfurtransferase (15); methylmalonyl-CoA carboxyltransferase, *or* transcarboxylase (16,17); ornithine δ-aminotransferase (18); phosphoenolpyruvate carboxykinase (19); phosphoenolpyruvate carboxykinase (GTP) (19); phosphoenol-pyruvate:glucose phosphotransferase transport system (20); procollagen-proline 3-dioxygenase (28); procollagen-proline 4-dioxygenase (28,29); pyruvate dehydrogenase, as alternative substrate (21,22); [pyruvate dehydrogenase (acetyl transferring)] kinase (23); saccharopine dehydrogenase (24). **1.** Shaw & Berg (1980) *J. Bacteriol.* **143**, 1509. **2.** Tse & Schloss (1993) *Biochemistry* **32**, 10398. **3.** DeFelice, Squires & Levinthal (1978) *Biochim. Biophys. Acta* **541**, 9. **4.** Hayman & Kinoshita (1965) *J. Biol. Chem.* **240**, 877. **5.** Dixon & Kleppe (1965) *Biochim. Biophys. Acta* **96**, 368. **6.** Paxton (1988) *Meth. Enzymol.* **166**, 313. **7.** Ng, Hanauske-Abel & Englard (1991) *J. Biol. Chem.* **266**, 1526. **8.** Rabier, Briand, Petit, Kamoun & Cathelineau (1986) *Biochimie* **68**, 639. **9.** Portsmouth, Stoolmiller & Abeles (1967) *J. Biol. Chem.* **242**, 2751. **10.** Karsten (1997) *Biochemistry* **36**, 1730. **11.** Braun & Schmidt (1973) *Biochemistry* **12**, 4878. **12.** Maruyama (1990) *J. Biochem.* **108**, 327. **13.** Lane, Shapley & Dekker (1971) *Biochemistry* **10**, 1353. **14.** Mallick, Harris & Cook (1991) *J. Biol. Chem.* **266**, 2732. **15.** Porter & Baskin (1996) *J. Biochem. Toxicol.* **11**, 45. **16.** Wood, Jacobson, Gerwin & Northrop (1969) *Meth. Enzymol.* **13**, 215. **17.** Wood (1972) *The Enzymes*, 3rd ed., **6**, 83. **18.**

Strecker (1965) *J. Biol. Chem.* **240**, 1225. **19.** Guidinger & Nowak (1990) *Arch. Biochem. Biophys.* **278**, 131. **20.** Daniel, Joseph & Danchin (1984) *Mol. Gen. Genet.* **193**, 467. **21.** Lapointe & Olson (1985) *Arch. Biochem. Biophys.* **242**, 417. **22.** Bisswanger (1981) *J. Biol. Chem.* **256**, 815. **23.** Hucho, Randall, Roche, *et al.* (1972) *Arch. Biochem. Biophys.* **151**, 328. **24.** Fujioka & Nakatani (1972) *Eur. J. Biochem.* **25**, 301. **25.** Oda, Nakano & Kitaoka (1982) *J. Gen. Microbiol.* **128**, 1211. **26.** Glatzer, Eakin & Wagner (1972) *J. Bacteriol.* **112**, 453. **27.** Cortés, Martín, Castro, L. Láiz & P. Liras (1987) *J. Gen. Microbiol.* **133**, 3165. **28.** Majamaa, Turpeenniemi-Hujanen, Latipää, *et al.* (1985) *Biochem. J.* **229**, 127. **29.** Majamaa, Hanauske-Abel, Günzler & Kivirikko (1984) *Eur. J. Biochem.* **138**, 239.

2-Ketocaproate (2-Oxocaproate)

This α-keto acid (FW$_{\text{free-acid}}$ = 130.14 g/mol), also known as α-ketocaproic acid, 2-oxocaproic acid, and 2-oxohexanoic acid, is an alternative substrate for a number of aminotransferases and inhibits [branched-chain α-keto-acid dehydrogenase] kinase. **Target(s):** aldehyde reductase, *or* aldose reductase (1); branched-chain α-keto-acid dehydrogenase (2); [branched-chain α-keto-acid dehydrogenase] kinase, *or* [3-methyl-2-oxobutanoate dehydrogenase (acetyl-transferring)] kinase (3); ornithine δ-aminotransferase (4); saccharopine dehydrogenase (5). **1.** Hayman & Kinoshita (1965) *J. Biol. Chem.* **240**, 877. **2.** Paxton & Harris (1984) *Arch. Biochem. Biophys.* **231**, 48. **3.** Paxton (1988) *Meth. Enzymol.* **166**, 313. **4.** Strecker (1965) *J. Biol. Chem.* **240**, 1225. **5.** Fujioka & Nakatani (1972) *Eur. J. Biochem.* **25**, 301.

Ketoconazole

This synthetic oral antifungal and anti-androgen (FW = 531.43 g/mol; CAS 65277-42-1; *Symbol*: KCZ), also known as Nizoral™, Sebizole™, and Ketomed™, and systematically as 1-[4-(4-{[(2R,4S)-2-(2,4-dichlorophenyl)-2-(1H-imidazol-1-ylmethyl)-1,3-dioxolan-4-yl]methoxy}phenyl)piperazin-1-yl]ethan-1-one, is widely used clinically to treat fungal skin infections in immunocompromised patients, especially those with HIV-AIDS. **Primary Mode(s) of Action:** Ketoconazole interferes with the fungal ergosterol synthesis, thereby depriving fungal membranes of an essential constituent. Like fluconazole and itraconazole, ketoconazole inhibits the enzyme cytochrome P450 14-α-demethylase (*or* P45014DM) enzyme, which catalyzes the conversion of lanosterol into ergosterol (1). As an anti-androgen, ketoconazole blocks both testicular and adrenal androgen biosynthesis and is also a weaker androgen receptor antagonist (2). A third antifungal action is ketoconazole inhibition of mitochondrial NADH oxidase (3). Ketoconazole inhibition of triazolam α- and 4-hydroxylation, midazolam α-hydroxylation, testosterone 6β-hydroxylation, and nifedipine oxidation appeared to be a mixed competitive-noncompetitive process, with the noncompetitive component being dominant but not exclusive (4). Quantitative estimates of K_i were in the low-nanomolar range for all four substrates (4). **Target(s):** cortisol secretion of an adrenal adenoma *in vivo* and *in vitro* (5); adrenal steroidogenesis by inhibiting cytochrome P450-dependent enzymes (6); adrenocorticotropic hormone stimulated cyclicAMP production (6); leukotriene biosynthesis *in vitro* and *in vivo* (7); Cytochrome P450 CYP27-catalyzed oxidation of C27-steroid into C27-acid (8). A recent study found that KCZ inhibits the following P450 steroidogenic enzymes in a selective and dose-dependent manner: cholesterol side-chain cleavage cytochrome P450, *or* CYP11A1/P450scc (IC$_{50}$ = 0.56 μM), 17α-hydroxylase activity of CYP17A1/P450c17 (IC$_{50}$ = 3.36 μM), and CYP19A1/P450arom (IC$_{50}$ = 0.56 μM). At 10 μg KCZ per mL, the folowing enzymes were unaffected: CYP17A1's 17,20 lyase

activity and five non-cytochrome steroidogenic enzymes (3β-hydroxysteroid dehydrogenase-Δ$^{5\text{-}4}$ isomerase type 1, *or* 3βHSD1; 5α-reductase; 20α-hydroxysteroid dehydrogenase, *or* 20α-HSD; 3α-hydroxysteroid dehydrogenase, *or* 3α-HSD; and 17β-hydroxysteroid dehydrogenase, type 1, *or* 17HSD1 (9). **1.** Reinhardt & Horst (1989) *Arch. Biochem. Biophys.* **272**, 459. **2.** Engelhardt, Mann, Hörmann, Braun & Karl (1983) *Klin Wochenschr.* **61**, 373. **3.** Rodriguez & Acosta (1996) *J. Biochem. Toxicol.* **11**, 127. **4.** Greenblatt, Zhao, Venkatakrishnan, *et al.* (2011) *J. Pharm. Pharmacol.* **63**, 214. **5.** Grosso, Boyden, Pamenter; *et al.* (1983) *Antimicrob. Agents Chemother.* **23**, 207. **6.** Loose, Kan, Hirst, Marcus & Feldman (1983) *J. Clin. Invest.* **71**, 1495. **7.** Beeten, Loots, Somers, Coene & De Clerck (1986) *Biochem. Pharmacol.* **35**, 883. **8.** Betsholtz & Wikvall (1995) *J. Steroid Biochem. Mol. Biol.* **55**, 115. **9.** Gal & Orly (2014) *Clin. Med. Insights Reprod. Health* **8**, 15.

2-Keto-D-gluconate (2-Oxo-D-gluconate)

This ketoadonic acid (FW$_{\text{free-acid}}$ = 194.14 g/mol), also called 2-dehydro-D-gluconic acid, 2-oxo-D-gluconic acid, D-glucosonic acid, fructuronic acid, and D-*arabino*-hexalusonic acid, is a hygroscopic solid that is soluble in water. **Target(s):** cellobiose phosphorylase (1,2); 2-dehydro-3-deoxyglucarate aldolase (3); tagaturonate reductase (4). **1.** Sasaki (1988) *Meth. Enzymol.* **160**, 468. **2.** Sasaki, Tanaka, Nakagawa & Kainuma (1983) *Biochem. J.* **209**, 803. **3.** Fish & Blumenthal (1966) *Meth. Enzymol.* **9**, 529. **4.** Portalier & Stoeber (1982) *Meth. Enzymol.* **89**, 210.

α-Ketoglutarate (α-Oxoglutarate)

This common α-keto-dicarboxylic acid (FW$_{\text{free-acid}}$ = 146.10 g/mol; Symbol: α-KG, 2-OG), also called 2-oxoglutaric acid and 2-oxopentanedicarboxylic acid, is a key intermediate in the tricarboxylic acid cycle and is a co substrate/product for many aminotransferases and other enzymes. It is also produced by oxidative deamination of glutamate. α-Ketoglutaric acid is soluble in water and ethanol and slightly soluble in diethyl ether. It melts at 113.5°C and has pKa values of 2.47 and 4.68 (at 25°C). Note that this α-keto acid is in equilibrium with the *gem*-diol and a cyclic lactol, *i.e.*, 5-hydroxy-2-oxo 5-tetrahydrofurancarboxylic acid (1). ^{13}C-NMR indicates that the α-keto acid is the predominant form at neutral pH with 7% *gem*-diol and a small amount of the lactol present (2). The relative amount of lactol increases with lower pH. With both carboxyl groups protonated there are relative amounts are 35% α-keto acid, 30% lactol, and 35% *gem*-diol.2 Under slightly different conditions at pH 0.5 in D2O, the reported values are 31%, 16%, and 53% (1). **Target(s):** 2-(acetamidomethylene)-succinate hydrolase (3); acetoacetate decarboxylase (4); alanopine dehydrogenase (5); aldehyde reductase, *or* aldose reductase (6); aminocyclo-propanecarboxylate oxidase (129); aminolevulinate aminotransferase (117-120); 5-aminolevulinate synthase (7); asparagine synthetase (8,9); D-aspartate oxidase (10); carbamoyl phosphate synthetase (11-13); choline acetyltransferase (14-16); citrate lyase, weakly inhibited (17); [citrate lyase] deacetylase, weakly inhibited (105); citrate (*re*)-synthase (123); citrate (*si*) synthase (18,19,124-128); cyanate hydratase (cyanase)20; cystathionine γ-lyase21; cysteine dioxygenase133; 3,4-dihydroxyphenylalanine oxidative deaminase (131); exo-α-sialidase, *or* neuraminidase, *or* sialidase (103); fumarase, *or* fumarate hydratase (22); fructose-1,6-bisphosphate aldolase (23); fumarylacetoacetase (24); glucose-6-phosphatase (104); glucose-6-phosphate dehydrogenase (25); glucose-6-phosphate isomerase, weakly inhibited (26); glutamate carboxypeptidase II (27); glutamate decarboxylase, K_i = 0.63 mM (28-32); glutamate racemase (33); glutaminase (34-36); glutamine synthetase (37-39); [glutamine-synthetase] adenylyltransferase (37,40,41,110-112); D-2-hydroxyacid dehydrogenase (42,43); 4-hydroxy-4-methyl-2 ketoglutarate aldolase, *or* 4-hydroxy-4-

methyl-2-oxoglutarate aldolase (44-46); 4-hydroxy-2 oxoglutarate aldolase, *or* 2-keto-4-hydroxyglutarate aldolase (47,48); indolepyruvate decarboxylase (49); inositol oxygenase (130); isocitrate dehydrogenase (NAD⁺) (51); [isocitrate dehydrogenase (NADP⁺)] kinase (50,106); isocitrate lyase (52-56); kynurenine 3-monooxygenase (57); lactase-phlorizin hydrolase, *or* glycosylceramidase (102); D-lactate dehydrogenase (cytochrome) (58); L-lactate dehydrogenase (42,59); lysine 2-monooxygenase (132); malate dehydrogenase (60,61); 3-mercaptopyruvate sulfurtransferase (62); methylaspartate mutase, *or* glutamate mutase (63); *N*-methylglutamate dehydrogenase (64); nicotinate nucleotide diphosphorylase (carboxylating) (121,122); oxaloacetate decarboxylase, weakly inhibited (65); oxaloacetate tautomerase (66); phosphoenolpyruvate carboxykinase, *or weakly inhibited* (67); phosphoenolpyruvate carboxykinase (GTP) (67-71); phosphoenolpyruvate carboxylase (72-79); phosphoenolpyruvate mutase (80); 6-phosphofructokinase (81-83,115,116); pyridoxamine: oxaloacetate aminotransferase (84); pyridoxamine-5-phosphate oxidase (85); pyruvate carboxylase (86-92); pyruvate decarboxylase, weakly inhibited (42,93); pyruvate kinase (94,113,114); pyruvate,water dikinase (107-109); ribulose-bisphosphate carboxylase/oxygenase (97); succinyl-CoA synthetase (ADP-forming), *or* succinate:CoA ligase (ADP-forming) (98); thiosulfate sulfurtransferase, *or* rhodanese (95,96); threonine ammonia-lyase, *or* threonine dehydratase (99,100); tyrosine 2,3 aminomutase (101). **1.** Cooper & Redfield (1975) *J. Biol. Chem.* **250**, 527. **2.** Viswanathan, R. E. Johnson & H. F. Fischer (1982) *Biochemistry* **21**, 339. **3.** Huynh & Snell (1985) *J. Biol. Chem.* **260**, 2379. **4.** Davies (1943) *Biochem. J.* **37**, 230. **5.** Fields & Hochachka (1981) *Eur. J. Biochem.* **114**, 615. **6.** Hayman & Kinoshita (1965) *J. Biol. Chem.* **240**, 877. **7.** Jordan & Shemin (1972) *The Enzymes*, 3rd ed., **7**, 339. **8.** Lea & Fowden (1975) *Proc. R. Soc. Lond. B Biol. Sci.* **192**, 13. **9.** Rognes (1980) *Phytochemistry* **19**, 2287. **10.** Dixon & Kenworthy (1967) *Biochim. Biophys. Acta* **146**, 54. **11.** Jones (1962) *Meth. Enzymol.* **5**, 903. **12.** Fahien, Schooler, Gehred & Cohen (1964) *J. Biol. Chem.* **239**, 1935. **13.** Garcia-Espana, Alonso & Rubio (1991) *Arch. Biochem. Biophys.* **288**, 414. **14.** Augustinsson (1952) *The Enzymes*, 1st ed., **2** (part 2), 906. **15.** Nachmansohn & John (1945) *J. Biol. Chem.* **158**, 157. **16.** Nachmansohn & Weiss (1948) *J. Biol. Chem.* **172**, 677. **17.** Giffhorn & Gottschalk (1975) *J. Bacteriol.* **124**, 1052. **18.** Weitzman (1969) *Meth. Enzymol.* **13**, 22. **19.** Spector (1972) *The Enzymes*, 3rd ed., **7**, 357. **20.** Anderson, Johnson, Endrizzi, Little & Korte (1987) *Biochemistry* **26**, 3938. **21.** Metaxas & Delwiche (1955) *J. Bacteriol.* **69**, 735. **22.** Behal & Oliver (1997) *Arch. Biochem. Biophys.* **348**, 65. **23.** Ujita & Kimura (1982) *Meth. Enzymol.* **90**, 235. **24.** Braun & Schmidt (1973) *Biochemistry* **12**, 4878. **25** Sokolov & Y. A. Trotsenko (1990) *Meth. Enzymol.* **188**, 339. **26.** Sangwan & Singh (1989) *J. Biosci.* **14**, 47. **27.** Robinson, Blakely, Couto & Coyle (1987) *J. Biol. Chem.* **262**, 14498. **28.** Fonda (1972) *Biochemistry* **11**, 1304. **29.** Gerig & Kwock (1979) *FEBS Lett.* **105**, 155. **30.** Roberts & Frankel (1951) *J. Biol. Chem.* **190**, 505. **31.** Blindermann, Maitre, Ossola & Mandel (1978) *Eur. J. Biochem.* **86**, 143. **32.** Denner, Wei, Lin, Lin & Wu (1987) *Proc. Natl. Acad. Sci. U.S.A.* **84**, 668. **33.** Ayengar & Roberts (1952) *J. Biol. Chem.* **197**, 453. **34.** Curthoys & Shapiro (1982) *Contrib. Nephrol.* **31**, 71. **35.** Ardawi & Newsholme (1984) *Biochem. J.* **217**, 289. **36.** Huerta-Saquero, Calderon-Flores, Diaz-Villasenor, Du Pont & Duran (2004) *Biochim. Biophys. Acta* **1673**, 201. **37.** Shapiro & Stadtman (1970) *Meth. Enzymol.* **17A**, 910. **38.** Donohue & Bernlohr (1981) *J. Bacteriol.* **147**, 589. **39.** Bhandari & Nicholas (1981) *Aust. J. Biol. Sci.* **34**, 527. **40.** Ebner, Gancedo & Holzer (1970) *Meth. Enzymol.* **17A**, 922. **41.** Stadtman & Ginsburg (1974) *The Enzymes*, 3rd ed., **10**, 755. **42.** Webb (1966) *Enzyme and Metabolic Inhibitors*, vol. **2**, Academic Press, New York. **43.** Cremona (1964) *J. Biol. Chem.* **239**, 1457. **44.** Wood (1972) *The Enzymes*, 3rd ed., **7**, 281. **45.** Maruyama (1990) *J. Biochem.* **108**, 327. **46.** Shannon & Marcus (1962) *J. Biol. Chem.* **237**, 3342. **47.** Grady, Wang & Dekker (1981) *Biochemistry* **20**, 2497. **48.** Lane, Shapley & Dekker (1971) *Biochemistry* **10**, 1353. **49.** Koga, Adachi & Hidaka (1992) *J. Biol. Chem.* **267**, 15823. **50.** Nimmo (1984) *Trends Biochem. Sci.* **9**, 475. **51.** Cook & Sanwal (1969) *Meth. Enzymol.* **13**, 42. **52.** McFadden (1969) *Meth. Enzymol.* **13**, 163. **53.** Tanaka, Nabeshima, Tokuda & Fukui (1977) *Agric. Biol. Chem.* **41**, 795. **54.** MacKintosh & Nimmo (1988) *Biochem. J.* **250**, 25. **55.** Takao, Takahashi, Tanida & Takahashi (1984) *J. Ferment. Technol.* **62**, 577. **56.** McFadden & Howes (1963) *J. Biol. Chem.* **238**, 1737. **57.** Shin, Sano & Umezawa (1982) *J. Nutr. Sci. Vitaminol. (Tokyo)* **28**, 191. **58.** Nygaard (1963) *The Enzymes*, 2nd ed., **7**, 557. **59.** Boeri, Cremona & Singer (1960) *Biochem. Biophys. Res. Commun.* **2**, 298. **60.** Fahien, Kmiotek, MacDonald, Fibich & Mandic (1988) *J. Biol. Chem.* **263**, 10687. **61.** Sprott, McKellar, Shaw, Giroux & Martin (1979) *Can. J. Microbiol.* **25**, 192. **62.** Porter & Baskin (1996) *J. Biochem. Toxicol.* **11**, 45. **63.**

Roymoulik, Chen & Marsh (1999) *J. Biol. Chem.* **274**, 11619. **64.** Hersh, Stark, Worthen & Fiero (1972) *Arch. Biochem. Biophys.* **150**, 219. **65.** Sender, Martin, Peiru & Magni (2004) *FEBS Lett.* **570**, 217. **66.** Annett & Kosicki (1969) *J. Biol. Chem.* **244**, 2059. **67.** Hebda & Nowak (1982) *J. Biol. Chem.* **257**, 5503. **68.** Pönsgen-Schmidt, Schneider, Hammer & Betz (1988) *Plant Physiol.* **86**, 457. **69.** Bentle & Lardy (1977) *J. Biol. Chem.* **252**, 1431. **70.** Titheradge, Picking & Haynes (1992) *Biochem. J.* **285**, 767. **71.** Mukhopadhyay, Concar & Wolfe (2001) *J. Biol. Chem.* **276**, 16137. **72.** Mori & Shiio (1985) *J. Biochem.* **98**, 1621. **73.** Schwitzguebel & Ettlinger (1979) *Arch. Microbiol.* **122**, 109. **74.** Iglesias, Gonzalez & Andreo (1986) *Planta* **168**, 239. **75.** Rivoal, Plaxton & Turpin (1998) *Biochem. J.* **331**, 201. **76.** Owttrim & Colman (1986) *J. Bacteriol.* **168**, 207. **77.** Schuller, Turpin & Plaxton (1990) *Plant Physiol.* **94**, 1429. **78.** Marczewski (1989) *Physiol. Plant* **76**, 539. **79.** Schuller, Plaxton & Turpin (1990) *Plant Physiol.* **93**, 1303. **80.** Seidel & Knowles (1994) *Biochemistry* **33**, 5641. **81.** Bloxham & Lardy (1973) *The Enzymes*, 3rd ed., **8**, 239. **82.** Webb (1966) *Enzyme and Metabolic Inhibitors*, vol. **2**, p. 385, Academic Press, New York. **83.** Passonneau & Lowry (1963) *Biochem. Biophys. Res. Commun.* **13**, 372. **84.** Wu & Mason (1964) *J. Biol. Chem.* **239**, 1492. **85.** Pogell (1963) *Meth. Enzymol.* **6**, 331. **86.** Osmani, Mayer, Marston, Selmes & Scrutton (1984) *Eur. J. Biochem.* **139**, 509. **87.** Scrutton & White (1974) *J. Biol. Chem.* **249**, 5405. **88.** Osmani, Marston, Selmes, Chapman & Scrutton (1981) *Eur. J. Biochem.* **118**, 271. **89.** Libor, Sundaram & Scrutton (1978) *Biochem. J.* **169**, 543. **90.** Mukhopadhyay, Stoddard & Wolfe (1998) *J. Biol. Chem.* **273**, 5155. **91.** Modak & Kelly (1995) *Microbiology* **141**, 2619. **92.** Jitrapakdee & Wallace (1999) *Biochem. J.* **340**, 1. **93.** Gale (1961) *Arch.Biochem. Biophys.* **94**, 236. **94.** Guderley & P. W. Hochachka (1980) *J. Exp. Zool.* **212**, 461. **95.** Oi (1974) *J. Biochem.* **76**, 455. **96.** Oi (1975) *J. Biochem.* **78**, 825. **97.** Wang & Tabita (1992) *J. Bacteriol.* **174**, 3593. **98.** Danson, Black, Woodland & Wood (1985) *FEBS Lett.* **179**, 120. **99.** Maeba & Sanwal (1966) *Biochemistry* **5**, 525. **100.** Choi & Kim (1995) *J. Biochem. Mol. Biol.* **28**, 118. **101.** Kurylo-Borowska & Abramsky (1972) *Biochim. Biophys. Acta* **264**, 1. **102.** Hamaswamy & Radhakrishnan (1973) *Biochem. Biophys. Res. Commun.* **54**, 197. **103.** Engstler, Reuter & Schauer (1992) *Mol. Biochem. Parasitol.* **54**, 21. **104.** Mithieux, Vega & Riou (1990) *J. Biol. Chem.* **265**, 20364. **105.** Giffhorn & Gottschalk (1975) *J. Bacteriol.* **124**, 1052. **106.** Nimmo & Nimmo (1984) *Eur. J. Biochem.* **141**, 409. **107.** Cooper & Kornberg (1974) *The Enzymes*, 3rd ed., **10**, 631. **108.** Chulavatnatol & Atkinson (1973) *J. Biol. Chem.* **248**, 2712. **109.** Tjaden, Plagens, Dörr, Siebers & Hensel (2006) *Mol. Microbiol.* **60**, 287. **110.** Caban & Ginsburg (1976) *Biochemistry* **15**, 1569. **111.** Wolf & Ebner (1972) *J. Biol. Chem.* **247**, 4208. **112.** Jonsson, Teixeira & Nordlund (2007) *FEBS J.* **274**, 2449. **113.** Plaxton, Smith & Knowles (2002) *Arch. Biochem. Biophys.* **400**, 54. **114.** Knowles, Smith, Smith & Plaxton (2001) *J. Biol. Chem.* **276**, 20966. **115.** Van Praag, Zehavi & Goren (1999) *Biochem. Mol. Biol. Int.* **47**, 749. **116.** Botha & Turpin (1990) *Plant Physiol.* **93**, 871. **117.** Varticovski, Kushner & Burnham (1980) *J. Biol. Chem.* **255**, 3742. **118.** Shioi, Nagamine & Sasa (1984) *Arch. Biochem. Biophys.* **234**, 117. **119.** Bajkowski & Friedmann (1982) *J. Biol. Chem.* **257**, 2207. **120.** Shanker & Datta (1986) *Arch. Biochem. Biophys.* **248**, 652. **121.** Shibata & Iwai (1980) *Biochim. Biophys. Acta* **611**, 280. **122.** Iwai, Shibata & Taguchi (1977) *Agric. Biol. Chem.* **43**, 351. **123.** Goschalk & Dittbrenner (1970) *Hoppe-Seyler's Z. Physiol. Chem.* **351**, 1183. **124.** Tabrett & Copeland (2000) *Arch. Microbiol.* **173**, 42. **125.** Lee, Park & Yim (1997) *Mol. Cells* **7**, 599. **126.** Robinson, Easom, Danson & Weitzman (1983) *FEBS Lett.* **154**, 51. **127.** Pereira, Donald, Hesfield & Duckworth (1994) *J. Biol. Chem.* **269**, 412. **128.** Danson, Black, Woodland & Wood (1985) *FEBS Lett.* **179**, 120. **129.** Vioque & Castellano (1998) *J. Agric. Food Chem.* **46**, 1706. **130.** Reddy, Pierzchala & Hamilton (1981) *J. Biol. Chem.* **256**, 8519. **131.** Ranjith, Ramana & Sasikala (2008) *Can. J. Microbiol.* **54**, 829. **132.** Vandecasteele & Hermann (1972) *Eur. J. Biochem.* **31**, 80. **133.** Chai, Bruyere & Maroney (2006) *J. Biol. Chem.* **281**, 15774.

3-Ketoglutarate (3-Oxtoglutarate)

This β-keto dicarboxylic acid (FWfree acid = 146.10 g/mol), also known as 3-oxoglutaric acid and acetonedicarboxylic acid, is an inhibitor of γ-butyrobetaine hydroxylase. Note that this water soluble acid decomposes in hot water, acids, and alkalies to acetone and carbon dioxide. **Target(s):** acetoacetate decarboxylase (1); aldehyde reductase, *or* aldose reductase (2);

γ-butyrobetaine dioxygenase, *or* γ-butyrobetaine hydroxylase; K_i = 0.29 mM (3); hyoscyamine (6S)-dioxygenase (4); procollagen-lysine 5-dioxygenase (5); procollagen-proline 3-dioxygenase (5); procollagen-proline 4-dioxygenase (5,6); pyrimidine-deoxynucleoside 2'-dioxygenase, *or* thymidine:2 oxoglutarate dioxygenase (7). **1.** Davies (1943) *Biochem. J.* **37**, 230. **2.** Hayman & Kinoshita (1965) *J. Biol. Chem.* **240**, 877. **3.** Ng, Hanauske-Abel & Englard (1991) *J. Biol. Chem.* **266**, 1526. **4.** Hashimoto & Yamada (1987) *Eur. J. Biochem.* **164**, 277. **5.** Majamaa, Turpeenniemi-Hujanen, Latipää, *et al.* (1985) *Biochem. J.* **229**, 127. **6.** Majamaa, Hanauske-Abel, Günzler & Kivirikko (1984) *Eur. J. Biochem.* **138**, 239. **7.** Bankel, Lindstedt & Lindstedt (1972) *J. Biol. Chem.* **247**, 6128.

2-Ketoisocaproate (2-Oxoisocaproate)

This α-keto acid (FW$_{\text{free-acid}}$ = 130.14 g/mol; M.P. = –2 °C; B.P. = 84-85 °C at 15 mm Hg), also known as α-ketoisocaproic acid, 2-oxoisocaproic acid, 4-methyl-2 oxovaleric acid, and 4-methyl-2-oxopentanoic acid, is an intermediate in both the biosynthetic and degradation pathways of L-leucine. This α-keto acid analogue of leucine is prepared from leucine by the action of D- or L-amino acid oxidase (1). While relatively stable at room temperature, it decompose slowly, even at 0 °C (1). **Target(s):** aldehyde reductase, *or* aldose reductase (2); branched-chain α-keto-acid dehydrogenase (3); [branched-chain α-keto-acid dehydrogenase] kinase, *or* 3-methyl-2-oxobutanoate dehydrogenase (acetyl-transferring)] kinase (4-10); glutamine:pyruvate aminotransferase (11); 2-isopropylmalate synthase (12-16); K$^+$ channel, ATP-regulated (17); α-ketoglutarate decarboxylation, presumably by pyruvate dehydrogenase in rat liver (18); kynurenine 3-monooxygenase (19); pyruvate carrier (20); pyruvate decarboxylation, presumably by pyruvate dehydrogenase in rat brain and liver (18,21). **1.** Meister (1953) *Biochem. Prep.* **3**, 66. **2.** Hayman & Kinoshita (1965) *J. Biol. Chem.* **240**, 877. **3.** Paxton & Harris (1984) *Arch. Biochem. Biophys.* **231**, 48. **4.** Roach (1984) *Meth. Enzymol.* **107**, 81. **5.** Harris, Kuntz & Simpson (1988) *Meth. Enzymol.* **166**, 114. **6.** Paxton (1988) *Meth. Enzymol.* **166**, 313. **7.** Randle, Patston & Espinal (1987) *The Enzymes*, 3rd ed., **18**, 97. **8.** Paxton & Harris (1984) *Arch. Biochem. Biophys.* **231**, 48. **9.** Lau, Fatania & Randle (1982) *FEBS Lett.* **144**, 57. **10.** Murakami, Matsuo, Shimizu & Shimomura (2005) *J. Nutr. Sci. Vitaminol.* **51**, 48. **11.** Cooper & Meister (1973) *J. Biol. Chem.* **248**, 8489. **12.** Kohlhaw (1988) *Meth. Enzymol.* **166**, 414. **13.** Wiegel & Schlegel (1977) *Arch. Microbiol.* **114**, 203. **14.** Wiegel & Schlegel (1977) *Arch. Microbiol.* **112**, 247. **15.** Wiegel (1981) *Arch. Microbiol.* **130**, 385. **16.** Kohlhaw, Leary & Umbarger (1969) *J. Biol. Chem.* **244**, 2218. **17.** Branstrom, Efendic, Berggren & Larsson (1998) *J. Biol. Chem.* **273**, 14113. **18.** Bowden, Brestel, Cope, *et al.* (1970) *Biochem. Med.* **4**, 69. **19.** Shin, Sano & Umezawa (1982) *J. Nutr. Sci. Vitaminol. (Tokyo)* **28**, 191. **20.** Paradies & Papa (1975) *Boll. Soc. Ital. Biol. Sper.* **51**, 1252. **21.** Bowden, McArthur & Fried (1971) *Biochem. Med.* **5**, 101.

2-Ketoisovalerate (2-Oxoisovalerate)

This α-keto acid (FW$_{\text{free-acid}}$ = 116.12 g/mol), also known as α-ketoisovaleric acid, 2-oxoisovaleric acid, 3-methyl-2 oxobutanoic acid, and ketovaline, is an intermediate in both the biosynthetic and degradation pathways of L-valine. It is also an intermediate in the biosynthesis of L-leucine and pantothenate. 2-ketoisovalerate is α-keto acid product of valine transamination. **Target(s):** aldehyde reductase, *or* aldose reductase (1); branched-chain α-keto-acid dehydrogenase (2); [branched-chain α-keto-acid dehydrogenase] kinase, *or* [3-methyl-2-oxobutanoate dehydrogenase (acetyl-transferring)] kinase (3-9); 2-dehydropantolactone reductase (10); glutamine:pyruvate aminotransferase (11); ketol-acid reductoisomerase (12); kynurenine 3-monooxygenase (13); ornithine δ-aminopeptidase (14); pyruvate dehydrogenase (15). **1.** Hayman & Kinoshita (1965) *J. Biol. Chem.* **240**, 877. **2.** Paxton & Harris (1984) *Arch. Biochem. Biophys.* **231**, 48. **3.** Roach (1984) *Meth. Enzymol.* **107**, 81. **4.** Harris, Kuntz & Simpson (1988) *Meth. Enzymol.* **166**, 114. **5.** Paxton (1988) *Meth. Enzymol.* **166**,

313. **6.** Randle, Patston & Espinal (1987) *The Enzymes*, 3rd ed., **18**, 97. **7.** Paxton & Harris (1984) *Arch. Biochem. Biophys.* **231**, 48. **8.** Lau, Fatania & Randle (1982) *FEBS Lett.* **144**, 57. **9.** Murakami, Matsuo, Shimizu & Shimomura (2005) *J. Nutr. Sci. Vitaminol.* **51**, 48. **10.** Wilken, King & Dyar (1979) *Meth. Enzymol.* **62**, 209. **11.** Cooper & A. Meister (1973) *J. Biol. Chem.* **248**, 8489. **12.** Arfin (1970) *Meth. Enzymol.* **17A**, 751. **13.** Shin, Sano & Umezawa (1982) *J. Nutr. Sci. Vitaminol. (Tokyo)* **28**, 191. **14.** Strecker (1965) *J. Biol. Chem.* **240**, 1225. **15.** Jackson & Singer (1983) *J. Biol. Chem.* **258**, 1857.

Ketomalonate (Oxomalonate)

This α-keto dicarboxylic acid (FW$_{\text{free-acid}}$ = 118.05 g/mol), also known as oxomalonic acid, and oxopropanedioic acid, is a homologue of oxaloacetate and α-ketoglutarate that is found in beet molasses. It is an alternative substrate for a number of enzymes: for example, serine:glyoxylate aminotransferase. Ketomalonate is typically supplied as the monohydrate, which is found mainly in the *gem*-diol form (i.e., dihydroxymalonic acid; FW$_{\text{free-acid}}$ = 136.06 g/mol). Malic enzyme slowly reduces ketomalonate in the presence of NADPH, indicating formation of the ternary complex Enzyme·NADPH·Ketomalonate (1). The *gem*-diol binds to form the enzyme·NADP$^+$ binary complex. **Target(s):** acid phosphatase, weakly inhibited (2); D-aspartate oxidase (K_i = 10 mM) (3,4;) homocitrate synthase (5); α-ketoglutarate dehydrogenase (6); L-lactate dehydrogenase (7); malic enzyme, *or* malate dehydrogenase (decarboxylating) (1,8,9); oxaloacetate decarboxylase (Na$^+$-pumping) (10-13); procollagen-proline 4-dioxygenase (14); pyridoxamine: oxaloacetate aminotransferase (15); pyruvate decarboxylase (16-18). **1.** Schimerlik & Cleland (1977) *Biochemistry* **16**, 565. **2.** Kilsheimer & Axelrod (1957) *J. Biol. Chem.* **227**, 879. **3.** de Marco & Crifò (1967) *Enzymologia* **33**, 325. **4.** Rinaldi (1971) *Enzymologia* **40**, 314. **5.** Qian, West & Cook (2006) *Biochemistry* **45**, 12136. **6.** Bunik Pavlova (1997) *Biochemistry (Moscow)* **62**, 1012. **7.** Yoshida & Freese (1975) *Meth. Enzymol.* **41**, 304. **8.** Kun (1963) *The Enzymes*, 2nd ed., **7**, 149. **9.** Stickland (1959) *Biochem. J.* **73**, 646 and 654. **10.** Dimroth (1986) *Meth. Enzymol.* **125**, 530. **11.** Dimroth (1981) *Eur. J. Biochem.* **115**, 353. **12.** Dimroth & Thomer (1986) *Eur. J. Biochem.* **156**, 157. **13.** Dahinden, Auchli, Granjon, *et al.* (2005) *Arch. Microbiol.* **183**, 121. **14.** Hutton, Tappel & Udenfriend (1967) *Arch. Biochem. Biophys.* **118**, 231. **15.** Wu & Mason (1964) *J. Biol. Chem.* **239**, 1492. **16.** Gale (1961) *Arch.Biochem. Biophys.* **94**, 236. **17.** Webb (1966) *Enzyme and Metabolic Inhibitors*, vol. **2**, p. 431, Academic Press, New York. **18.** Scrutton, Olmsted & Utter (1969) *Meth. Enzymol.* **13**, 235.

2-Keto-3-methylvalerate (2-Oxo-3-methylvalerate)

This α-keto acid (FW$_{\text{free-acid}}$ = 130.14 g/mol; M.P. = 38-40°C; B.P. = 73°C at 10 mm Hg; pK_a = 2.3 at 25°C), also known as α-keto-β-methylvaleric acid, 2-oxo-3-methylvaleric acid, 3-methyl-2-oxopentanoic acid, and ketoisoleucine, is the α-keto acid analogue of isoleucine and is an intermediate in both the biosynthetic and degradation pathways of L-isoleucine. *Note*: There are two possible stereoisomers: the natural isomer is the D-enantiomer (i.e., the 3S-isomer). **Target(s):** branched-chain α-keto-acid dehydrogenase (1); [branched-chain α-keto-acid dehydrogenase] kinase, *or* [3-methyl-2-oxobutanoate dehydrogenase (acetyl-transferring)] kinase (2-6); kynurenine 3-monooxygenase (7); pyruvate carrier (8). **1.** Paxton & Harris (1984) *Arch. Biochem. Biophys.* **231**, 48. **2.** Roach (1984) *Meth. Enzymol.* **107**, 81. **3.** Harris, Kuntz & Simpson (1988) *Meth. Enzymol.* **166**, 114. **4.** Paxton (1988) *Meth. Enzymol.* **166**, 313. **5.** Randle, Patston & Espinal (1987) *The Enzymes*, 3rd ed., **18**, 97. **6.** Paxton & Harris (1984) *Arch. Biochem. Biophys.* **231**, 48. **7.** Shin, Sano & Umezawa (1982) *J. Nutr. Sci. Vitaminol. (Tokyo)* **28**, 191. **8.** Paradies & Papa (1975) *Boll. Soc. Ital. Biol. Sper.* **51**, 1252.

7-Keto-20-oxacholesterol (7-Oxo-20-oxacholesterol)

This 20-oxa analogue of 7-ketocholesterol (FW = 388.59 g/mol), also known as 7-oxo-20-oxacholesterol and SC 31769, inhibits ACAT, *or* sterol acyltransferase (cholesterol acyltransferase). **1**. Billheimer (1985) *Meth. Enzymol.* **111**, 286.

(2S,4E)-3-Keto-N-palmitoylceramide ((2S,4E)-3-Oxo-N-palmitoylceramide)

This ceramide analogue (FW = 535.90 g/mol) inhibits ceramidase. For inhibition, the enzyme requires the primary and secondary hydroxyl groups, the C^4-C^5 double bond, the trans configuration of this double bond, and the NH-protons from either the amide of ceramide or the amine of sphingosine. 1. Usta, El Bawab, Roddy, *et al.* (2001) *Biochemistry* **40**, 9657.

2-Ketopelargonate (2-Oxopelargonate)

This α-keto acid (FW_free-acid = 172.22 g/mol), also known as 2-ketononanoic acid and 2-oxononanoic acid, inhibits lens aldose reductase, an enzyme that catalyzes the reduction of glucose to sorbitol, which cannot be exported from cells. These properties lead to the accumulation of this osmolyte (especially during hyperglycemic episodes), leading to osmotic damage to retinal neurons. 1. Hayman & Kinoshita (1965) *J. Biol. Chem.* **240**, 877.

S-(2-Ketopentadecyl)-CoA (S-(2-Oxopentadecyl)-CoA)

This acyl-CoA analogue (FW = 991.93 g/mol), also known as S-(2-oxopenta-decyl)-CoA, is a strong inhibitor of glycylpeptide N-tetradecanoyltransferase, *or* protein N-myristoyltransferase (1-5). **1**. Glover, Tellez, Guziec & Felsted (1991) *Biochem. Pharmacol.* **41**, 1067. **2**. Selvakumar, Lakshmikuttyamma, Shrivastav, *et al.* (2007) *Prog. Lipid Res.* **46**, 1. **3**. Paige, Zheng, DeFrees, Cassady & Geahlen (1990) *Biochemistry* **29**, 10566. **4**. Rudnick, McWherter, Rocaque, *et al.* (1991) *J. Biol. Chem.* **266**, 9732. **5**. Zhang, Jackson-Machelski & Gordon (1996) *J. Biol. Chem.* **271**, 33131.

2-Keto-3-pentynoate (2-Oxo-3-pentynoate)

This acetylenic α-keto acid (FW_free-acid = 112.08 g/mol) is an irreversible inhibitor of 4 oxalocrotonate tautomerase. The enzyme is covalently modified at Pro1 by the resulting 2-oxo-3 pentenoate adduct. The ε and η nitrogens of the guanidinium side chain of Arg39" from a neighboring dimer interact with the C2 carbonyl oxygen and one C1 carboxylate oxygen of the adduct, respectively. The side-chain of Arg-61' from the same dimer as the modified Pro-1 interacts with the C1 carboxylate group in a bidentate fashion. Two ordered water molecules within the active site interact as well with the 2-carbonyl group of the adduct (1). **Target(s):** 4-oxalocrotonate tautomerase (1,2); pyruvate decarboxylase (3); pyruvate dehydrogenase (3); pyruvate oxidase (3); tryptophan 2-monooxygenase (4,5). **1**. Taylor, Czerwinski, Johnson, Whitman & Hackert (1998) *Biochemistry* **37**, 14692. **2**. Johnson, Czerwinski, Fitzgerald & Whitman (1997) *Biochemistry* **36**, 15724. **3**. Brown, Nemeria, Yi, *et al.* (1997) *Biochemistry* **36**, 8071. **4**. Sobrado & Fitzpatrick (2002) *Arch. Biochem. Biophys.* **402**, 24. **5**. Gadda, Dangott, Johnson, Whitman & Fitzpatrick (1999) *Biochemistry* **38**, 5822.

2-Ketopimelate (2-Oxopimelate)

This α-keto dicarboxylic acid (FW_free-acid = 174.15 g/mol), also known as 2-oxopimelic acid and 2-oxoheptanedioic acid, inhibits dihydrodipicolinate synthase and is a weak inhibitor of pyrimidine-deoxynucleoside 2'-dioxygenase. **Target(s):** dihydrodipicolinate synthase (1); 2,3,4,5-tetrahydropyridine-2,6-dicarboxylate N-succinyltransferase (2); pyrimidine-deoxynucleoside 2'-dioxygenase, *or* thymidine:2-oxoglutarate dioxygenase, weakly inhibited (3). **1**. Blickling, Renner, Laber, *et al.* (1997) *Biochemistry* **36**, 24. **2**. Berges, DeWolf, Dunn, *et al.* (1986) *J. Biol. Chem.* **261**, 6160. **3**. Bankel, Lindstedt & Lindstedt (1972) *J. Biol. Chem.* **247**, 6128.

(S)-Ketoprofen

This nonsteroidal anti-inflammatory agent (FW_free-acid = 254.29 g/mol; M.P. = 94°C) is a strong, nonselective inhibitor of prostaglandin-endoperoxide synthase (cyclooxygenase). The (S)-enantiomer is about 100-500-fold more potent than the (R)-stereoisomer. Ketoprofen is slightly soluble in water and soluble in dimethylformamide, ethanol, acetone, and ethyl acetate. **Target(s):** anandamide amidohydrolase (1); arylamine N-acetyltransferase (2); cyclooxygenase, *or* prostaglandin-endoperoxide synthase; COX-1 and COX-2 (3-18); CYP2C10 (19); glutathione S transferase-π (20); neutrophil elastase (21); palmitoyl-CoA ligase, *or* long-chain-fatty-acyl-CoA synthetase, by R-isomer (22,23); phenol sulfotransferase (24); phospholipase A₂, weakly inhibited (25); thiopurine S-methyltransferase (26). **1**. Fowler, Tiger & Stenstrom (1997) *J. Pharmacol. Exp. Ther.* **283**, 729. **2**. Makarova (2008) *Curr. Drug Metab.* **9**, 538. **3**. Moore, Tramer, Carroll, Wiffen & McQuay (1998) *Brit. Med. J.* **316**, 333. **4**. Evans (1996) *J. Clin. Pharmacol.* **36**, 7S. **5**. Parsadaniantz, Lebeau, Duval, *et al.* (2000) *J. Neuroendocrinol.* **12**, 766. **6**. Kay-Mugford, Benn, LaMarre & Conlon (2000) *Amer. J. Vet. Res.* **61**, 802. **7**. Cryer & Feldman (1998) *Amer. J. Med.* **104**, 413. **8**. Carabaza, Cabre, Garcia, *et al.* (1997) *Chirality* **9**, 281. **9**. Carabaza, Cabre, Rotllan, *et al.* (1996) *J. Clin. Pharmacol.* **36**, 505. **10**. Grossman, Wiseman, Lucas, Trevethick & Birch (1995) *Inflamm. Res.* **44**, 253. **11**. Suesa, Fernandez, Gutierrez, *et al.* (1993) *Chirality* **5**, 589. **12**. Hayball, Nation & Bochner (1992) *Chirality* **4**, 484. **13**. Miyazawa, Iimori, Makino, Mikami & Miyasaka (1985) *Jpn. J. Pharmacol.* **38**, 199. **14**. Humes, Winter, Sadowski & Kuehl, (1981) *Proc. Natl. Acad. Sci. U.S.A.* **78**, 2053. **15**. Tachizawa, Tsukada, Saito & Akimoto (1977) *Jpn. J. Pharmacol.* **27**, 351. **16**. Streppa, Jones & Budsberg (2002) *Amer. J. Vet. Res.* **63**, 91. **17**. Barnett, Chow, Ives, *et al.* (1994) *Biochim. Biophys. Acta* **1209**, 130. **18**. Patrignani, Panara, Sciulli, *et al.* (1997) *J. Physiol.*

Pharmacol. **48**, 623. **19**. Kappers, Groene, Kleij, *et al.* (1996) *Xenobiotica* **26**, 1231. **20**. Goto, Iida, Cho, *et al.* (1999) *Free Radic. Res.* **31**, 549. **21**. Kang, Bae, Kim, *et al.* (2000) *Exp. Mol. Med.* **32**, 146. **22**. Roberts & Knights (1992) *Biochem. Pharmacol.* **44**, 261. **23**. Knights & Jones (1992) *Biochem. Pharmacol.* **43**, 1465. **24**. Vietri, De Santi, Pietrabissa, Mosca & Pacifici (2000) *Eur. J. Clin. Pharmacol.* **56**, 81. **25**. Makela, Kuusi & Schroder (1997) *Scand. J. Clin. Lab. Invest.* **57**, 401. **26**. Oselin & Anier (2007) *Drug Metab. Dispos.* **35**, 1452.

Ketorolac

This anti-inflammatory agent (FW$_{free-acid}$ = 255.27 g/mol; pK_a = 3.49; M.P. = 160-161 °C), also known as 5-benzoyl-2,3-dihydro-1*H*-pyrrolizine-1-carboxylic acid, inhibits the induction of nitric-oxide synthase by lipopolysaccharides. The (*S*)-stereoisomer is the more active enantiomer. **Target(s):** anandamide amidohydrolase (1); prostaglandin H synthase, *or* cyclooxygenase 1 and 2 (2-5). **1**. Fowler, Janson, Johnson, *et al.* (1999) *Arch. Biochem. Biophys.* **362**, 191. **2**. Dionne, Khan & Gordon (20 01) *Clin. Exp. Rheumatol.* **19** (6 Suppl. 25), S63. **3**. Jett, Ramesha, Brown, *et al.* (1999) *J. Pharmacol. Exp. Ther.* **288**, 1288. **4**. Carabaza, Cabre, Rotllan, *et al.* (1996) *J. Clin. Pharmacol.* **36**, 505. **5**. Litvak & McEvoy (1990) *Clin. Pharm.* **9**, 921.

KF 89617, *See* Litronesib

KG 5

This allosteric inhibitor (FW = 459.45 g/mol; CAS 877874-85-6), also named 2-(methylthio)-6-[4-[5-[[3-(trifluoromethyl)phenyl]amino]-1*H*-1,2,4-triazol-3-yl]phenoxy]-4-pyrimidinamine, targets PDGFRβ and B-Raf, two important targets for pericyte recruitment and endothelial cell survival, respectively. KG-5 targets the allosteric site, binding to the inactive kinase conformation (type II inhibitors) and providing an alternative for selective inhibitors that are physiologically active. **1**. Murphy, Shields, Stoletov, *et al.* (2010) *Proc. Natl. Acad. Sci. USA* **107**, 4299.

KGA-2727

This phlorizin-like pyrazole-*O*-glucoside (FW = 489.59 g/mol), also named 3-(3-{4-[3-(β-D-glucopyranosyloxy)-5-isopropyl-1*H*-pyrazol-4-ylmethyl]-3-methylphenoxy}propylamino)propionamide, is a novel and selective inhibitor of high-affinity sodium glucose cotransporter-1, *or* SGLT1, exhibiting antidiabetic efficacy in rodent models. In the oral glucose tolerance test with streptozotocin-induced diabetic rats, KGA-2727 reduced plasma glucose after glucose loading, indicating that KGA-2727 improved postprandial hyperglycemia. In the Zucker diabetic fatty (ZDF) rat, chronic treatments with KGA-2727 reduced plasma glucose as well as glycated hemoglobin. **1**. Shibazaki, Tomae, Ishikawa-Takemura (2012) *J. Pharmacol. Exp. Ther.* **342**, 288.

KGP94

This protease inhibitor (FW = 350.23 g/mol; dissolve in sterile DMSO to obtain 10-25 mM stock solutions and dilute 1000x in cell culture media), also named 3-bromophenyl-3-hydroxyphenyl-ketone thiosemicarbazone, targets Cathepsin L (1), a protease that plays a crucial role in remodeling the tumor microenvironment and freeing cells to disseminate, resulting in therapeutic failure, poor prognosis, and high mortality (2). Acute exposures to hypoxic or acidic conditions are known to significantly elevate secreted Cathepsin L concentration, either through an increase in intracellular Cathepsin L levels or through activation of lysosomal exocytosis, or both, depending on the tumor type. Increases in Cathepsin L secretion also closely parallel enhanced tumor cell migration and invasion suggesting that this protease could be an essential factor in tumor microenvironment triggered metastasis. KGP94 treatment markedly attenuates tumor cell invasion and migration under both normal and aberrant microenvironmental conditions, suggesting that Cathepsin L inhibitors may have significant utility as an anti-metastatic agent. **1**. Kishore-Kumar, Chavarria, Charlton-Sevcik, *et al.* (2010) *Bioorg. Med. Chem. Lett.* **20**, 1415. **2**. Sudhan & Siemann (2013) *Clin. Exp. Metastasis* **30**, 891. **3**. Sudhan & Siemann (2015) *Pharmacol. Ther.* **155**, 105.

Khellin

This coronary vasodilator (FW = 260.25 g/mol; CAS 82-02-0; Solubility = 0.025 g/100 mL H_2O at 25°C; M.P. = 154-155°C), systematically named 4,9-dimethoxy-7-methyl-5*H*-furo[3,2-g][1]benzopyran-5-one, is a furocoumarin that is abundant in the plant *Ammi visnaga*. It elicits an intracellular response by $G_{i\alpha}$ protein, the inhibitory G-protein component that blocks adenylate cyclase (1). Khellin is structurally related to other furocoumarins and, with its photoreactive quinoidal structure, is a phototherapeutic agent that has been used to treat psoriasis and vitiligo. It forms furan and pyrone mono-photoadducts with thymine bases in DNA. Khellin is also a active principle in the traditional Egyptian medicine known as khella, which is prepared as an aqueous extract of the dried ripe fruit, and, owing to khelin's vasodilating properties, is used to treat respiratory conditions including asthma, bronchitis, persistent cough, as well as whooping cough. Its relatively low toxicity commends its use as an antilitogenic agent for treating calcium oxalate kidney stone formation. **Target(s):** adenylyl cyclase system (1); cAMP phosphodiesterase, weakly inhibited (2). **1**. Di Stefano, La Gaetana, Bovalini, Lusini & Martelli (1996) *Biochim. Biophys. Acta* **1314**, 105. **2**. Bovalini, Lusini, Simoni, *et al.* (1987) *Z. Naturforsch. [C]* **42**, 1009.

Kifunensine

This hygroscopic cyclic oxamide analogue (FW = 232.19 g/mol) of the alkaloid 1-aminomannojirimycin, produced by the actinomyces

Kitasatosporia kifunense strain 9482, is a potent inhibitor of the plant glycoprotein processing enzyme, mannosidase I (IC$_{50}$ = 20-50 nM), but is virtually inactive toward mannosidase II (K_i = 5.2 mM). Kifunensine inhibits rat liver mannosidase I, but not the endoplasmic reticulum mannosidase. When tested *in vitro* for its ability to inhibit influenza virus glycoprotein processing in canine kidney cells, 4 μM kifunensine completely shifted the structure *N*-linked oligosaccharide from complex chain to Man$_9$(GlcNAc)$_2$ structure, again in keeping its selected action on mannosidase I. **Target(s):** glycoprotein processing (1-3); α-mannosidase (4,5); mannosyl-oligosaccharide (1,2); α-mannosidase, *or* mannosidase I, *or* α1,2-mannosidase (1,2,6-14); mannosyl-oligosaccharide 1,3-1,6-α mannosidase, *or* mannosidase II, weakly (6). **1**. Kaushal & Elbein (1994) *Meth. Enzymol.* **230**, 316. **2**. Elbein, Tropea, Mitchell & Kaushal (1990) *J. Biol. Chem.* **265**, 15599. **3**. Elbein, Kerbacher, Schwartz & Sprague (1991) *Arch. Biochem. Biophys.* **288**, 177. **4**. Kayakiri, Takese, Shibata, *et al.* (1989) *J. Org. Chem.* **54**, 4015. **5**. Woo, Miyazaki, Hara, Kimura & Kimura (2004) *Biosci. Biotechnol. Biochem.* **68**, 2547. **6**. Shah, Kuntz & Rose (2003) *Biochemistry* **42**, 13812. **7**. Elbein (1991) *FASEB J.* **5**, 3055. **8**. Akao, Yamaguchi, Yahara, *et al.* (2006) *Biosci. Biotechnol. Biochem.* **70**, 471. **9**. Gonzalez, Karaveg, Vandersall-Nairn, Lal & Moremen (1999) *J. Biol. Chem.* **274**, 21375. **10**. Lal, Pang, Kalelkar, *et al.* (1998) *Glycobiology* **8**, 981. **11**. Pan, Kaushal, Papandreou, Ganem & Elbein (1992) *J. Biol. Chem.* **267**, 8313. **12**. Bonay & Fresno (1999) *Glycobiology* **9**, 423. **13**. Yoshida, Kato, Asada & Nakajima (2001) *Glycoconjugate J.* **17**, 745. **14**. Tempel, Karaveg, Liu, *et al.* (2004) *J. Biol. Chem.* **279**, 29774.

Kinetensin

This nonapeptide (MW = 1172.40 g/mol; *Sequence*: IARRHPYFL; Isoelectric Point = 19.84), isolated from pepsin-treated human plasma, exhibits neurotensin-like immunoreactivity. Kinetensin also stimulates histamine release. It is an alternative substrate for tryptase, chymase, and mast cell carboxypeptidase. It is not a substrate of neurolysin. **1**. Vincent, Dauch, Vincent & Checler (1997) *J. Neurochem.* **68**, 837. . Barelli, Vincent & Checler (1993) *Eur. J. Biochem.* **211**, 79.

Kinetin

This adenine derivative (FW = 215.21 g/mol; λ$_{max}$ in ethanol at 268 nm; ε = 18650 M^{-1}cm^{-1})), also known as 6-furfurylaminopurine and *N*^{6}-furfuryladenine, is a synthetic cytokinin, exhibiting activity similar to that of *N*^{6}-benzyladenine. Kinetin stimulates cell division in the presence of auxin. Kinetin was originally obtained in 1954 from the hydrolysis of herring sperm DNA and was soon found to stimulate cell division in tobacco pith cells (1) Note: A commercial form of hyaluronidase also goes by the name Kinetin. **Target(s):** Hill activity decay (2); 5-hydroxyfuranocoumarin 5-*O*-methyltransferase (3); 8-hydroxyfuranocoumarin 8-*O*-methyltransferase (3); indole-3-acetate β-glucosyltransferase, weakly inhibited (4); protein oxidation and glycoxidation (5); xanthine dehydrogenase (6); *cis*-zeatin *O*-β-D-glucosyltransferase (7). **1**. Miller, Skoog, Von Saltza & Strong (1955) *J. Amer. Chem. Soc.* **77**, 1392. **2**. Banerji & Kumar (1975) *Biochem. Biophys. Res. Commun.* **65**, 940. **3**. Thompson, Sharma & Brown (1978) *Arch. Biochem. Biophys.* **188**, 272. **4**. Leznicki & Bandurski (1988) *Plant Physiol.* **88**, 1481. **5**. Verbeke, Siboska, Clark & Rattan (2000) *Biochem. Biophys. Res. Commun.* **276**, 1265. **6**. Sauer, Frebortova, Sebela, *et al.* (2002) *Plant Physiol. Biochem.* **40**, 393. **7**. Veach, Martin, Mok, *et al.* (2003) *Plant Physiol.* **131**, 1374.

Kingianin A

This cytotoxic secondary metabolite and apoptosis-promoting agent (FW = 682.90 g/mol) from the bark of the Malaysian plant, *Endiandra kingiana* (Lauraceae), displays significant affinity for the protein Bcl-X$_L$, K_i = 1-3 μM, thereby blocking its anti-apoptotic actions (1). Kingianin A was recently synthesized by a 12-step synthesis based on a radical cation Diels-Alder (RCDA) reaction that is believed to be biomimetic (2). **1**. Leverrier, Awang, Guéritte & Litaudon (2011) *Phytochemistry* **72**, 1443. **2**. Lim & Parker (2013) *Org. Lett.* **15**, 398.

Kirromycin

This antibiotic (FW = 796.96 g/mol), also called mocimycin and delvomycin, prevents the release of elongation factor-Tu (EF-Tu) from the bacterial ribosome and thus inhibits protein biosynthesis. Translocation is inhibited by kirromycin stabilizatio of aminoacyl-tRNA bound to the ribosomal A-site. Kirromycin also inhibits the phosphorylation of EF-Tu. **Target(s):** elongation factor EF-Tu (1-4); RNA-directed RNA polymerase (Qb replicase), weakly inhibited (5). **1**. Young & Neidhardt (1978) *J. Bacteriol.* **135**, 675. **2**. Wolf, Chinali & Parmeggiani (1977) *Eur. J. Biochem.* **75**, 67. **3**. Hogg, Mesters & Hilgenfeld (2002) *Curr. Protein Pept. Sci.* **3**, 121. **4**. Parmeggiani & Swart (1985) *Annu. Rev. Microbiol.* **39**, 557. **5**. Brown & Blumenthal (1976) *J. Biol. Chem.* **251**, 2749.

KN-93

KN-93

KN-92

This cell-permeable, dual-action CaM kinase inhibitor and K$_V$ channel blocker (FW = 501.04 g/mol; CAS 139298-40-1; Solubility: 100 mM in DMSO with gentle warming), also named *N*-[2-[[[3-(4-chlorophenyl)-2-propenyl]methylamino]methyl]phenyl]-*N*-(2-hydroxyethyl)-4-methoxybenzenesulfonamide, has two modes of action. **Mechanism of Action as a CaM Kinase Inhibitor:** KN-73 selectively and potently targets the Ca^{2+}/calmodulin-stimulated protein kinase CaM kinase II (IC$_{50}$ = 0.37 μM). KN-73 has no significant effect on cAMP-dependent protein kinase (PKA), Ca^{2+}/phospholipid dependent protein kinase, myosin light chain kinase, and Ca^{2+}-phosphodiesterase (1). A significant clinical finding is that persistently elevated CaMKII activity (by β-adrenergic stimulation or rapid heart rate, by reactive oxidative species (ROS), by angiotensin II signaling, or by GlcNAcylation during hyperglycemia) appears to be a key factor in heart failure, arrhythmia, and other forms of heart disease. KN-93 also inhibits autophosphorylation of both CaMKII α- and β-subunits. KN-93 inhibits CaMKII competitively relative to calmodulin (1). In rabbit hearts pretreated for 10 min with 0.5 μM KN-93 or a similar level of its inactive analogue KN-92 (FW = 554.98 g/mol; CAS 1135280-28-2; IUPAC Name: 2-[*N*-(4-methoxybenzenesulfonyl)]-amino-*N*-(4-chlorocinnamyl)-*N*-methylbenzyl amine) before clofilium exposure, early after-depolarizations (EADs) are significantly suppressed by KN-93 (4 out of 10 hearts) compared to KN-92 (10 out of 11 hearts) (2). KN-93 was also instrumental in demonstrating calcium/calmodulin-dependent phosphorylation and activation of human Cdc25-C at the G$_2$/M phase transition in HeLa cells (3). **Mechanism of Action as a K$_V$ Channel Blocker:** KN-93 is also a direct extracellular blocker of a wide range of cloned voltage-gated potassium (K$_V$) channels from a number of different subfamilies (4). In all channels tested, this effect is CaMK II-independent, as indicated by the fact that KN-92 (1 µM) has a similar inhibitory action on these ion currents (4). Importantly, KN-93 is ineffective as a blocker, when applied intracellularly, suggesting that CaMK II-independent effects of KN-93 on K$_V$ channels can be circumvented by intracellular application of KN-93 (4). **Recommendation:** The dual action of KN-93 on CaMKII and ion channels requires that the effects of KN-93 and KN-92 must be separately evaluated before assigning the most likely mechanism of action. For example, one study showed that, with injection of 1, 3 and 6 µg KN93 into the nucleus accumbens, there is a dose-dependent increase in hindpaw withdrawal latency (HWL) upon exposure to noxious thermal and mechanical stimulation in mononeuropathic rats (5). Given the likely significance of both CaMKII and K$_V$ channels in the nucleus accumbens, however, one cannot unambiguously conclude, as the authors did, that the observed KN-93 effects are directed to CaMKII. **1.** Sumi, Kiuchi, Ishikawa, *et al.* (1991) *Biochem. Biophys. Res. Commun.* **181**, 968. **2.** Anderson, Braun, Wu, *et al.* (1998) *J. Pharmacol. Exp. Ther.* **287**, 996. **3.** Patel, Holt, Philipova, *et al.* (1999) *J. Biol. Chem.* **274**, 7958. **4.** Rezazadeh, Claydon & Fedida (2006) *J. Pharmacol. Exp. Ther.* **317**, 292. **5.** Bian & Yu, *et al.* (2014) *Neurosci. Lett.* **583**, 6.

K-(D-1-Nal)-FWLL-NH$_2$

This high affinity inverse agonist (FW = 902.13 g/mol; CAS 1394288-22-2; Soluble to 2 mg/mL in water), also named lysyl-D-naphthylamido-phenylalanyl-tryptophanyl-leucyl-leucylamide, targets the ghrelin receptor, a constitutive receptor activity that regulates appetite and food intake. Els, Schild, Petersen, *et al.* (1) showed that subtle changes in the core WFW binding motif of the ghrelin receptor ligand K^{1}W^{2}F^{3}W^{4}L^{5}L^6-NH$_2$ determine whether a resulting peptide analogue shows inverse agonism or agonism. Introduction of β-(3-benzothienyl)-D-alanine (D-Bth), 3,3-diphenyl-D-alanine (D-Dip) and 1-naphthyl-D-alanine (D-1-Nal) at Position-2 resulted in highly potent and efficient inverse agonists, whereas the substitution of D-tryptophan at Position-4 with 1-naphthyl-D-alanine (D-1-Nal) and 2-naphthyl-d-alanine (D-2-Nal) induces agonism in functional assays. **1.** Els, Schild, Petersen, *et al.* (2012) *J. Med. Chem.* **55**, 7437.

KNI 272

This peptide analogue (FW = 667.85 g/mol), also known as kynostatin-272, contains allophenylnorstatine and inhibits HIV-1 protease: K_i = 9.8 pM. **1.** Clemente, Hemrajani, Blum, Goodenow & Dunn (2003) *Biochemistry* **42**,

15029. **2.** Huang, Xu, Luo, *et al.* (2002) *J. Med. Chem.* **45**, 333. **3.** Baldwin, Bhat, Gulnik, *et al.* (1995) *Structure* **3**, 581. **4.** Mimoto, Imai, Kisanuki, *et al.* (1992) *Chem. Pharm. Bull. (Tokyo)* **40**, 2251. **5.** Kageyama, Mimoto, Murakawa, *et al.* (1993) *Antimicrob. Agents Chemother.* **37**, 810. **6.** Gulnik, Suvorov, Liu, *et al.* (1995) *Biochemistry* **34**, 9282. **7.** Hamada, Matsumoto, Yamaguchi, *et al.* (2004) *Bioorg. Med. Chem.* **12**, 159.

Ko 1173, *See Mexiletine*

Kobe0065

This orally active H-Ras-cRaf1 interaction inhibitor (FW = 449.79 g/mol; CAS 436133-68-5; Solubility: 5 mM in H$_2$O, with gentle warming), also named *N*-(3-chloro-4-methylphenyl)-2-[2,6-dinitro-4-(trifluoromethyl) phenyl]hydrazinecarbothioamide, binds to the GTP·Ras complex, disrupting its interactions with c-Raf-1 (K_i = 10^{-6}–10^{-5} M). Because mutational activation of H-Ras, K-Ras, and N-Ras is often observed in human cancers, protein interaction disruptors are promising anticancer drug targets. When added to the culture medium at 2 and 20 µM, respectively, Kobe0065 and its analogue Kobe2602 inhibit both anchorage-dependent and anchorage-independent growth and induce apoptosis of H-ras^{G12V}-transformed NIH 3T3 cells, accompanied by down-regulation of downstream molecules such as MEK/ERK, Akt, and RalA as well as the upstream signaling molecule, Son of sevenless (SOS). Moreover, they exhibit antitumor activity on a xenograft of human colon carcinoma SW480 cells carrying the K-ras^{G12V} gene by oral administration. The NMR structure of a complex of the compound with GTP·H-Ras^{T35S} confirms its insertion into one of the surface pockets and provides a molecular basis for binding inhibition toward multiple Ras·GTP-interacting molecules. **1.** Shima, Yoshikawa, Ye, *et al.* (2013) *Proc. Natl. Acad. Sci. U.S.A.* **110**, 8182.

Kodaistatin A

This secondary fungal metabolite (FW = 628.63 g/mol), isolated from *Aspergillus terreus thom* DSM 11247, is a potent inhibitor of glucose-6-phosphate translocase (IC$_{50}$ = 80 nM). A hydroxylated kodaistatin A derivative, kodaistatin C is a slightly weaker inhibitor of the translocase (IC$_{50}$ = 130 nM). **1.** Van Schaftingen & Gerin (2002) *Biochem. J.* **362**, 513. **2.** Vértesy, Burger, Kenja, *et al.* (2000) *J. Antibiot.* **53**, 677.

Kojibiose

This α1,2-linked glucose disaccharide FW = 342.30 g/mol, named (2-*O*-α-D-glucopyranosyl-D-glucopyranose, is a substrate for a number of α-glucosidases. Kojibiose is often found linked to teichoic acids in Gram-positive bacteria. **Target(s):** concanavalin A (1); mannosyl-oligosaccharide glucosidase, *or* glucosidase I (2-11). **1.** Goldstein, Hollerman & Smith (1965) *Biochemistry* **4**, 876. **2.** Kaushal & Elbein (1989) *Meth. Enzymol.* **179**, 452. **3.** Zeng & Elbein (1998) *Arch. Biochem. Biophys.* **355**, 26. **4.** Presper & Heath (1983) *The Enzymes*, 3rd ed., **16**, 449. **5.** Shailubhai, Pratta & Vijay (1987) *Biochem. J.* **247**, 555. **6.** Szumilo, Kaushal & Elbein (1986) *Arch. Biochem. Biophys.* **247**, 261. **7.** Ugalde, Staneloni & Leloir (1980) *Eur. J. Biochem.* **113**, 97. **8.** Bause, Erkens, Schweden & Jaenicke (1986) *FEBS Lett.* **206**, 208. **9.** Bause, Schweden, Gross & Orthen (1989) *Eur. J. Biochem.* **183**, 661. **10.** Herscovics (1999) *Biochim. Biophys. Acta* **1426**, 275. **11.** Dhanawansa, Faridmoayer, van der Merwe, Li & Scaman (2002) *Glycobiology* **12**, 229.

Kojic Acid

This pyranone derivative (FW = 142.11 g/mol; M.P. = 153-154°C; Soluble in water, ethanol, methanol, and acetone), also known as 5-hydroxy-2-(hydroxymethyl)-4*H*-pyran-4-one, is an antibiotic that is produced by a number of microorganisms (*e.g.*, *Aspergillus oryzae*, *A. terreus*, and *Penicillium* sp.). It is a slow-binding inhibitor of tyrosinase and inhibits melanin production. It has been used as a whitening agent and food additive. **Target(s):** alcohol dehydrogenase, yeast (1); D-amino-acid oxidase (2-5); L-amino-acid oxidase (3); catechol oxidase (6,23,35,36,38,39); laccase (7-11); phenol oxidases, insect (12); polyphenol oxidase (13,35,36); quercetin 2,3-dioxygenase (14,15); tyrosinase, *or* monophenol monooxygenase (6,12,13,16-35,37); xanthine oxidase (4). **1.** Sund & Theorell (1963) *The Enzymes*, 2nd ed., **7**, 25. **2.** Burton (1955) *Meth. Enzymol.* **2**, 199. **3.** Webb (1966) *Enzyme and Metabolic Inhibitors*, vol. **2**, Academic Press, New York. **4.** Klein & Olsen (1947) *J. Biol. Chem.* **170**, 151. **5.** Klein & Austin (1953) *J. Biol. Chem.* **205**, 725. **6.** Battaini, Monzani, Casella, Santagostini & Pagliarin (2000) *J. Biol. Inorg. Chem.* **5**, 262. **7.** Salony, Garg, Baranwal, *et al.* (2008) *Biochim. Biophys. Acta* **1784**, 259. **8.** Baldrian (2004) *Appl. Microbiol. Biotechnol.* **63**, 560. **9.** Saito, Hong, Kato, *et al.* (2003) *Enzyme Microb. Technol.* **33**, 520. **10.** Iyer & Chattoo (2003) *FEMS Microbiol. Lett.* **227**, 121. **11.** Niladevi, Jacob & Prema (2008) *Process Biochem.* **43**, 654. **12.** Dowd (1999) *Nat. Toxins* **7**, 337. **13.** Perez-Gilabert, Morte, Honrubia & Garcia-Carmona (2001) *J. Agric. Food Chem.* **49**, 1922. **14.** Steiner, Kooter & Dijkstra (2002) *Biochemistry* **41**, 7955. **15.** Barney, Schaab, LoBrutto & Francisco (2004) *Protein Expr. Purif.* **35**, 131. **16.** Koketsu, Choi, Ishihara, B*et al.* (2002) *Chem. Pharm. Bull.* **50**, 1594. **17.** Tepper, Bubacco & Canters (2002) *J. Biol. Chem.* **277**, 30436. **18.** Sakuma, Ogawa, Sugibayashi, Yamada & Yamamoto (1999) *Arch. Pharm. Res.* **22**, 335. **19.** Curto, Kwong, Hermersdorfer, *et al.* (1999) *Biochem. Pharmacol.* **57**, 663. **20.** Cabanes, Chazarra & Garcia-Carmona (1994) *J. Pharm. Pharmacol.* **46**, 982. **21.** Parvez, Kang, Chung & Bae (2007) *Phytother. Res.* **21**, 805. **22.** Abdel-Halim, Marzouk, Mothana & Awadh (2008) *Pharmazie* **63**, 405. **23.** Kim & Uyama (2005) *Cell. Mol. Life Sci.* **62**, 1707. **24.** Criton & Le Mellay-Hamon (2008) *Bioorg. Med. Chem. Lett.* **18**, 3607. **25.** Shiino, Watanabe & Umezawa (2008) *J. Enzyme Inhib. Med. Chem.* **23**, 16. **26.** Selinheimo, Ni Eidhin, Steffensen, *et al.* (2007) *J. Biotech.* **130**, 471. **27.** Kong, Hong, Choi, Kim & Cho (2000) *Biotechnol. Appl. Biochem.* **31**, 113. **28.** Flurkey, Cooksey, Reddy, *et al.* (2008) *J. Agric. Food Chem.* **56**, 4760. **29.** Karioti, Protopappa, Magoulas & Skaltsa (2007) *Bioorg. Med. Chem.* **15**, 2708. **30.** Kong, Park, Hong & Cho (2000) *Comp. Biochem. Physiol. B Biochem. Mol. Biol.* **125**, 563. **31.** Ito & Oda (2000) *Biosci. Biotechnol. Biochem.* **64**, 261. **32.** Yoshimoto, Yamamoto & Tsuru (1985) *J. Biochem.* **97**, 1745. **33.** Momtaz, Mapunya, Houghton, *et al.* (2008) *J. Ethnopharmacol.* **119**, 507. **34.** Zhang, Lu, Tao, *et al.* (2007) *J. Enzyme Inhib. Med. Chem.* **22**, 91. **35.** Khan, Mughal, Khan, *et al.* (2006) *Bioorg. Med. Chem.* **14**, 6027. **36.** Yamamoto, Yoshitama & Teramoto (2002) *Plant Biotechnol.* **19**, 95. **37.** Khan, Maharvi, Khan, *et al.* (2006) *Bioorg. Med. Chem.* **14**, 344. **38.** Jaenicke & Decker (2008) *FEBS J.* **275**, 1518. **39.** Olianas, Sanjust, Pellegrini & Rescigno (2005) *J. Comp. Physiol. B* **175**, 405.

KP1212, See *5-Aza-5,6-dihydro-2'-deoxycytidine*
KPT-330, See *Selinexor*

KPT-335, See *Verdinexor*

K-Ras(G12C) inhibitor 6

This synthetic allosteric-type irreversible G-protein inhibitor (FW = 405.34 g/mol; Solubility: 80 mg/mL DMSO; <1 mg/mL H₂O), also named *N*-(1-(2-(2,4-dichlorophenoxy)acetyl)piperidin-4-yl)-4-mercapto-butanamide, targets the frequently oncogenic Gly-to-Cys mutant form of K-Ras, blocking K-Ras(G12C) interactions with its proliferation-promoting binding partners. Oncogenic mutations result in functional activation of Ras family proteins by impairing GTP hydrolysis. With diminished regulation by GTPase activity, the nucleotide state of Ras becomes more dependent on relative nucleotide affinity and concentration, allowing GTP to outwegn GDP and increasing the proportion of active GTP-bound Ras. Binding of the inhibitor to K-Ras(G12C) disrupts both Switch-I and Switch-II, subverting the native nucleotide preference to favor GDP over GTP and impairing Ras binding to Raf. (**See also** *K-Ras(G12C) Inhibitor 9*) **Role in Modulating the Subcellular Localization of p21:** The nuclear exporter, known as exportin (XPO1; CRM1), controls the nucleo-cytoplasmic localization of more than two hundred Nuclear Export Signal (NES)-containing proteins, many of which are tumor suppressor proteins (TSPs), including p21. The latter was originally described as a cyclin-dependent-kinase inhibitor (CKI) of cyclin-CDK2, cyclin-CDK1, and cyclin-CDK4/6 complexes whose expression is classically regulated by p53. PO1 inhibition by KPT-330 attenuates renal cell carcinoma (RCC) viability through cell cycle arrest as well as induction of apoptosis, and that increased nuclear p21 by XPO1 inhibition plays a major role in the efficacy of KPT-330 in RCC (2). Although normally localized within the nucleus, p21 binds to cyclin-CDK complexes thereby inhibiting their function in cell cycle progression, resulting in cell cycle arrest. However, when p21 localizes to the cytosolic compartment, it inhibits apoptosis by complexing with pro-apoptotic proteins such as pro-caspase-3 or ASK We show that KPT-330 potentiates the antitumor activity of sunitinib, resulting in complete inhibition of RCC tumor growth in vivo with this combination. KPT-330 resistant cells remains sensitive to sunitinib, sorafenib, everolimus, and temsirolimus, suggesting that patients who fail KPT-330 or for whom resistance develops, could still respond to these older drugs (2). Finally, given the aforementioned binding of KPT-330 to small G-proteins like K-Ras(G12C), it seems likely that KPT-330 binds directly to p21, thereby altering its capacity to be processed appropriately by XPO1. **1.** Ostrem, Peters, Sos, Wells & Shokat (2013) *Nature* **503**, 548. **2.** Wettersten, Landesman, Friedlander, *et al.* (2014) *PLoS One* **9**, e113867.

K-Ras(G12C) Inhibitor 9

This synthetic allosteric-type irreversible G-protein inhibitor (FW = 513.78 g/mol; CAS 1469337-91-4; Solubility: 53 mg/mL DMSO; <1 mg/mL H₂O or Ethanol), also named *N*-[1-[2-[(4-chloro-5-iodo-2-methoxyphenyl)amino]acetyl]-4-piperidinyl]ethenesulfonamide, targets the frequently oncogenic Gly-to-Cys mutant form of K-Ras, blocking K-Ras(G12C) interactions with its proliferation-promoting binding partners. This compound relies on the mutant cysteine SH group for reactivity and does not affect the wild-type protein. Crystallographic studies reveal the formation of a new pocket that is not apparent in previous structures of Ras, beneath the effector binding switch-II region. Binding of this inhibitor to K-Ras(G12C) disrupts both switch-I and switch-II, favoring the binding of GDP over GTP and impairing K-Ras(G12C) binding to Raf. (**See also** *K-Ras(G12C) Inhibitor 6*) **1.** Ostrem, Peters, Sos, Wells & Shokat (2013) *Nature* **503**, 548.

KRN-951, See *Tivozanib*
KRX-0401, See *Perifosine*

KT-611, See *Naftopidil*

KT-5555, See *Lestaurtinib*

KT-5720

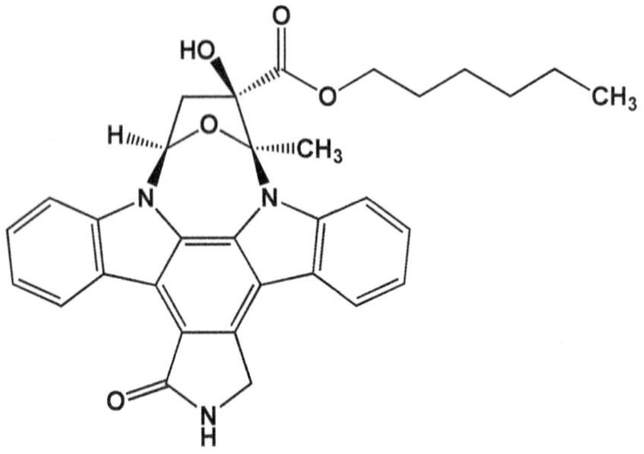

This potent and selective protein kinase inhibitor (FW = 537.61 g/mol; CAS 108068-98-0; Soluble to 50 mM in DMSO), also named (9R,10S,12S)-2,3,9,10,11,12-hexahydro-10-hydroxy-9-methyl-1-oxo-9,12-epoxy-1H-diindolo[1,2,3-fg:3',2',1'-kl]pyrrolo[3,4-i][1,6]benzodiazocine-10-carboxylate, hexyl ester, targets protein kinase A (K_i = 60 nM), with no effect on PKG (K_i > 2 μM) or PKC (K_i > 2 μM). KT5720 reversibly arrests human skin fibroblasts, holding them in G$_1$ phase. PKA inhibition by KT-5720 results in altered astrocyte morphology, attended by a reduction in Profilin-2 levels (4). **1.** Kase, *et al.* (1987) *Biochem. Biophys. Res. Commun.* **142**, 436. **2.** Gadbois, *et al.* (1992) *Proc. Natl. Acad. Sci. U.S.A.* **89**, 8626. **3.** Cabell, *et al.* (1993) *Int. J. Dev. Neurosci.* **11**, 357. **4.** Schweinhuber, Meßerschmidt, Hänsch, Korte & Rothkegel (2015) *PLoS One* **10**, e0117244.

KT5823

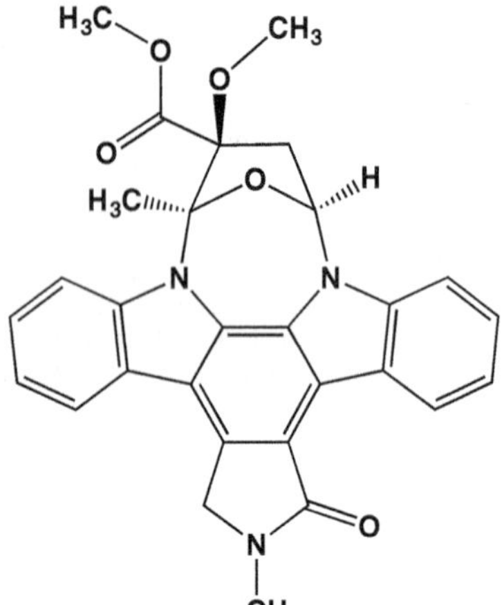

This cell-permeable reagent (FW = 495.53 g/mol; Photosensitive), also known as (9R,10S,12S)-2,3,9,10,11,12 hexahydro-10-methoxy-9-methyl-1-oxo-9,12-epoxy-1H-diindolo[1,2,3-fg:3,2,1-k]-N-methylpyrrolo[3,4-l][1,6]benzodiazocine-10-carboxylic acid methyl ester, is a strong inhibitor of protein kinase G (cGMP-dependent protein kinase; K_i = 230 nM). It will also inhibit protein kinase C (Ki = 4.0 μM) and weakly inhibit protein kinase A (K_i > 10 μM). It is a modified derivative of K-252a. Inhibition of p38-regulated/activated kinase by KT5823 varies from batch to batch, suggesting the actual inhibitory agent may vary from batch to batch (1). **Target(s):** cGMP-dependent protein kinase, *or* protein kinase G (1-7); [glycogen-synthase] kinase 3β, *or* [tau protein] kinase (1); p38-regulated/activated kinase (1); protein kinase A, weakly inhibited (1); protein kinase C, weakly inhibited (1). **1.** Bain, McLauchlan, Elliott & Cohen (2003) *Biochem. J.* **371**, 199. **2.** Kase, Iwahashi, Nakanishi, *et al.* (1987) *Biochem. Biophys. Res. Commun.* **142**, 436. **3.** Eriksson, Toivola, Sahlgren, Mikhailov & Härmälä-Braskén (1998) *Meth. Enzymol.* **298**, 542. **4.** Dawson-Scully, Armstrong, Kent, Robertson & Sokolowski (2007) *PLoS ONE* **2**, e773. **5.** Li, Zhang, Marjanovic, Ruan & Du (2004) *J. Biol. Chem.* **279**, 42469. **6.** Schlossmann & Hofmann (2005) *Drug Discov. Today* **10**, 627.

KT5926

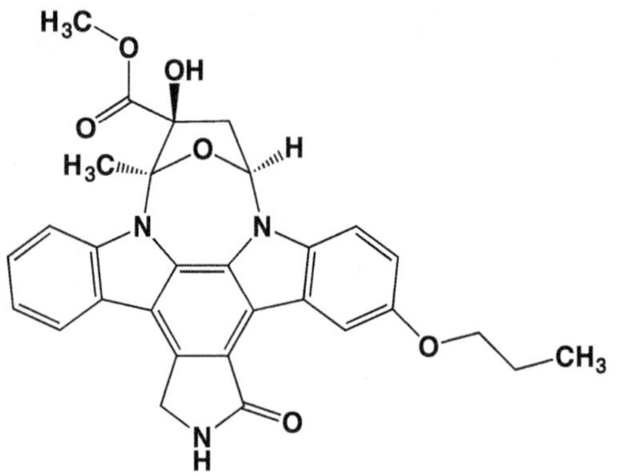

This potent and selective protein kinase inhibitor (FW = 525.56 g/mol), also known as ((8R*,9S*,11S*)-(–)-9-hydroxy-9-methoxycarbonyl-8-methyl-14-n-propoxy-2,3,9,10-tetrahydro-8,11-epoxy,1H,8H,11H-2,7β,11α-triaza-dibenzo[a,g]cycloocta[cde]trinden-1-one, inhibits [myosin light-chain] kinase competitively relative to ATP (K_i = 18 nM) and noncompetitively relative to the myosin light chain (K_i = 12 nM). KT5926 exhibits lower affinity for protein kinase C (K_i = 723 nM), cAMP dependent protein kinase (K_i = 1.2 μM), and cGMP-dependent protein kinase (K_i = 158 nM). Notably, Ca^{2+}-ATPase, Na$^+$/K$^+$-exchanging ATPase, hexokinase, and 5'-nucleotidase were not inhibited by KT5926 below 10 μM. This agent inhibited serotonin secretion induced by platelet-activating factor, but its potency was significantly less than that of K-252a. The K_i value for Ca^{2+}/calmodulin-dependent protein kinase II was 4.4 nM. (**See also** *K-252a*) **Target(s):** Ca^{2+}/calmodulin-dependent protein kinase II (1); cAMP-dependent protein kinase (2); cGMP-dependent protein kinase, *or* protein kinase G (2,3,7,8); heat-shock protein (HSP25) kinase (5); [myosin light-chain] kinase (2,4,6); protein kinase C (2). **1.** Hashimoto, Nakayama, Teramoto, *et al.* (1991) *Biochem. Biophys. Res. Commun.* **181**, 423. **2.** Nakanishi, Yamada, Iwahashi, Kuroda & Kase (1990) *Mol. Pharmacol.* **37**, 482. **3.** Baker & Deng (2005) *Front. Biosci.* **10**, 1229. **4.** Tinsley, Yuan & Wilson (2004) *Trends Pharmacol. Sci.* **25**, 64. **5.** Hayess & Benndorf (1997) *Biochem. Pharmacol.* **53**, 1239. **6.** Wadgaonkar, Linz-McGillem, Zaiman & Garcia (2005) *J. Cell. Biochem.* **94**, 351. **7.** Murthy & Zhou (2003) *Amer. J. Physiol. Gastrointest. Liver Physiol.* **284**, G221. **8.** Yamamoto & Suzuki (2005) *J. Biol. Chem.* **280**, 16979.

KU-55933

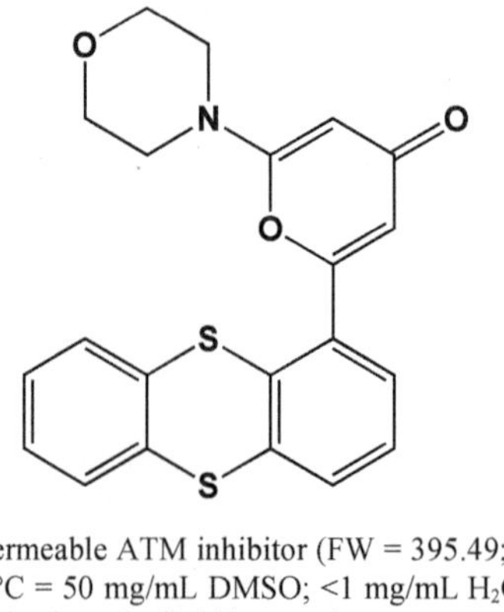

This potent, cell-permeable ATM inhibitor (FW = 395.49; CAS 587871-26-9; Solubility at 25°C = 50 mg/mL DMSO; <1 mg/mL H$_2$O), also known as 2-morpholino-6-(thianthren-1-yl)-4H-pyran-4-one, targets Ataxia Telangiectasia-Mutated (IC$_{50}$ = 13 nM; K_i = 2.2 nM), a nonselective serine/threonine protein kinase that is recruited to and activated by DNA double-strand breaks, where it phosphorylates such downstream proteins as p53, CHK2, NBS1, and BRCA1 (1). **Background:** Ataxia-telangiectasia syndrome (also known as Louis–Bar syndrome) is a rare, neurodegenerative, autosomal recessive disability. ATM was first identified as an syndrome-associated protein that likely gave rise to the AT phenotype, the latter characterized by higher degrees of chromosome breakage, increased sensitivity to ionizing radiation and radiomimetic drugs, and defective cell-cycle checkpoints in response to genome damage. Cellular inhibition of ATM by KU-55933 is demonstrated by the ablation of ionizing radiation-dependent phosphorylation of a range of ATM targets, including p53, γH2AX, NBS1, and SMC1. KU-55933 does not show inhibition of UV light DNA damage induced cellular phosphorylation events (1). Indeed, exposure to KU-55933 significantly sensitizes cells to the cytotoxic effects of ionizing radiation as well as to DNA double-strand break-inducing chemotherapeutic agents (*e.g.*, etoposide, doxorubicin, and camptothecin). **Effects on Deoxynucleoside Homeostasis:** Inhibition of ATM by KU-55933 also causes loss of ionizing radiation-induced cell cycle arrest linked to changes in deoxynucleotide metabolism (1-4). Human cells use two different pathways to generate the 2'-deoxy-ribonucleoside triphosphates (dNTPs) needed for DNA replication and repair. While the main pathway involves *de novo* reduction of NDPs by ribonucleotide reductase (RNR) to their corresponding dNDPs, a second pathway salvages deoxynucleosides (dNs) through deoxynucleoside kinases. Cells coordinate *de novo* and salvage pathways to assure that the dNTP supply is sufficient for replication and repair, thereby avoiding deleterious genotoxic stress. In the salvage pathway, deoxynucleoside kinases are rate-limiting for dNTP supply, with thymidine kinase (TK1) and/or deoxycytidine kinase (dCK) in the cytosol and TK2 and deoxyguanosine kinase (dGK) in mitochondria. TK1 is S-phase specific through a mechanism similar to that of RNR. (The other three deoxynucleoside kinases are constitutively active across the cell cycle.) Deoxynucleoside kinases are variably expressed in human normal and cancer tissues, with TK1 being elevated in cervical cancers. The substrates of these salvage enzymes, deoxynucleosides, enter cells and mitochondria passively by plasma membrane-equilibrative nucleoside transporters. ATM integrates the DNA damage-response by catalyzing the phosphorylation of deoxy-cytidine kinase (dCK) at Ser^{74} in response to cell exposure to ionizing radiation (IR). Deoxy-cytidine kinase catalyzes the phosphorylation pyrimidine (dC) as well as purine (dA, dG) deoxynucleosides, and a phosphomimetic Ser-74-Glu mutation increases catalysis of phosphorylation of dC, but not dA and dG (4). ATM-dependent phosphorylation of dCK at Ser^{74} after ionizing radiation preferentially increases dCMP formation over dGMP and dAMP. dG and dA are not exclusively phosphorylated by dCK, and the rates of dA and dG phosphorylation remained unchanged. Moreover, Ser-74-Ala substitution decreased dCK activity for dC, but had little effect on dA and dG phosphorylation. dCK^{S74E} tripled the rate of dC phosphorylation, but decreased dG and dA phosphorylation rates by 50%. Such findings suggest that IR-induced ATM-dependent phosphorylation of dCK at Ser-74 increases the flux of deoxy-cytidine through the salvage pathway (4). KU-55933 treatment increases intracellular dCTP pools post-IR and enhances the DNA repair rate. Mutation of Ser^{74} profoundly affects murine T and B-lymphocyte development, suggesting posttranslational regulation of dCK is vitally for genomic stability during hematopoiesis (4). **Other Targets:** DNA-PK, IC$_{50}$ = 2.5 µM (1); mTOR PK, IC$_{50}$ = 9.3 µM (1); PI3K, IC$_{50}$ = 16.6 µM (1); ATM-mediated phosphorylation of AMP-activated protein kinase (AMPK) in HeLa and A29 cells (5); ATM signaling in primary A-T fibroblasts, resulting in global dysregulation of R$_1$, R$_2$, and p53R$_2$ subunits of RNR, abrogation of RNR-dependent up-regulation of *mt*-DNA in response to ionizing radiation, increase in mitochondrial transcription factor A to *mt*-DNA ratios, and increased resistance to inhibitors of mitochondrial respiration and translation (6); inhibition of phosphatidylinositol 3-kinase and autophagy at much higher KU-55933 concentrations (2 µM) (7); many other cellular processes, owing to ATM's notoriously promiscuous phosphorylation of numerous targets, including Abl gene, BRCA1, Bloom syndrome protein, DNA-PKcs, FANCD2, MRE11A, Nibrin, p53, RAD17, RAD51, RBBP8, RHEB, RRM2B, SMC1A TERF1, and TP53BP1. **Other ATM Inhibitors:** KU-60019 (IC$_{50}$ = 6.3 nM), Wortmannin (IC$_{50}$ = 150 nM), CP-466722 (IC$_{50}$ = 450 nM), Torin 2 (IC$_{50}$ = 28 nM), and CGK733 (IC$_{50}$ = 200 nM). **1**. Hickson, *et al.* (2004) *Cancer Res.* 64, 9152. **2**. Soleimani, *et al.* (2011) *Aging* 3, 782. **3**. Ivanov, *et al.* (2009) *Cancer Res.* 69, 3510. **4**. Bunimovich, Nair-Gill, Riedinger, *et al.* (2014) *PLoS One* 9, e104125. **5**. Sun, Connors & Yang (2007) *Mol. Cell Biochem.* 306, 239. **6**. Eaton, Lin, Sartorelli, Bonawitz & Shadel (2007) *J. Clin. Invest.* 117, 2723. **7**. Farkas, Daugaard & Jäättelä (2011) *J. Biol. Chem.* 286, 38904.

KU-60019

This potent and specific ATM inhibitor (F.Wt. = 547.67; CAS 587871-26-9; Solubility (25°C): 40 mg/mL DMSO, <1 mg/mL Water), also known as 2-((2*S*,6*R*)-2,6-dimethylmorpholino)-*N*-(5-(6-morpholino-4-oxo-4*H*-pyran-2-yl)-9*H*-thioxanthen-2-yl)acetamide, targets Ataxia telangiectasia mutated, *or* ATM (*IC$_{50}$* = 6.3 nM), the curiously named serine/threonine protein kinase recruited to and activated by DNA double-strand breaks (1). While similar in its action to KU-55933, KU-60019 has improved water-solubility. KU-60019 has little activity against DNA-PK's and ATR with *IC$_{50}$* values of 1.7 µM and >10 µM, respectively, as well as 229 other protein kinases, including PI3K, mTOR and mTOR/FKBP12. **1**. Golding, *et al.* (2009) *Mol. Cancer Ther.* 8, 2894.

KU-60648

This potent DNA-PK inhibitor (FW = 582.71 g/mol; CAS 881375-00-4), also known as 4-ethyl-*N*-[4-[2-(4-morpholinyl)-4-oxo-4*H*-1-benzopyran-8-yl]-1-dibenzothienyl]-1-piperazineacetamide, targets DNA-dependent protein kinase (IC$_{50}$ = 8.6 nM) as well as its autophosphorylation (IC$_{50}$ = 20-40 nM), depending on cell type. In SW620 cells, 1 µM KU-0060648 enhances etoposide cytotoxicity more than 100 times and doxorubicin cytotoxicity by more than 10 times (1). Although radiation and genotoxic drugs are used widely used in the clinic to treat cancer, DNA repair often limits their cytotoxic action. To increase the efficacy of such treatments, DNA-PK inhibitors promise to enhance cancer cell chemo- and/or radio-sensitivity. **1**. Furgason & Bahassi (2013) *Pharmacol. Ther.* 137, 298. **2**. Munck, Batey, Zhao, *et al.* (2012) *Mol. Cancer Ther.* 11, 1789.

Kunitzins

AAKIILNPKFRCKAAFC

Kunitzin-RE

AVNIPFKVHLRCKAAFC

Kunitzin-OS

These novel structurally related protease inhibitors, Kunitzin-RE (MW = 1894.37 g/mol; Sequence: AAKIIL-NPKFRCKAAFC; Isoelectric Point = 9.85) and Kunitzin-OS (MW = 1917.36 g/mol; Sequence: AVNIPFKVHLRCK-AAFC; Isoelectric Point = 9.51), from the lyophilized skin secretions of the European frog *Rana esculenta* and the Chinese frog *Odorrana schmackeri*, target serine proteases. Due to the presence of the canonical Kunitz-type protease inhibitor motif (CKAAFC), these peptides were named kunitzin-RE and kunitzin-OS. Both potently inhibit trypsin (K$_i$ = 5.56 µM and 7.56 µM), representing prototypes of a novel class of highly attenuated amphibian skin protease inhibitors. Substitution of Lys-13, the predicted residue occupying the P$_1$ position within the inhibitory loop, with Phe resulted in decrease in trypsin inhibitor effectiveness and antimicrobial activity against *Escherichia coli*, but exhibits a potential inhibition activity against chymotrypsin. **1**. Chen, Wang, Shen, *et al.* (2016) *Biochem. Biophys. Res. Commun.* 477, 534.

Kurtoxin

KIDGYPVDYWNCKRICWYNNKYCNDLCKGLKADSGYCWGWTLSCYCQGLPDNARIKRSGRCRA

This 63-residue oligopeptide (MW = 7394.48 g/mol) from the venom of the scorpion *Parabuthus transvaalicus*, blocks Ca^{2+} channels, particularly low-threshold (T-type) channels. **1**. Sidach & Mintz (2002) *J. Neurosci.* 22, 2023. **2**. Nishio, Nishiuchi, Ishimaru & T. Kimura (2003) *Lett. Pept. Sci.* 10, 589. **3**. Chuang, Jaffe, Cribbs, Perez-Reyes & Swartz (1998) *Nat. Neurosci.* 1, 668.

KVX-478, See *Amprenavir*

KW-2478

This nonansamycin Hsp90 inhibitor (FW$_{free-base}$ = 574.66 g/mol; CAS 819812-04-9; FW$_{HCl-Salt}$ = 610.13 g/mol; CAS 819812-18-5), also known as 2-(2-ethyl-3,5-dihydroxy-6-(3-methoxy-4-(2-morpholinoethoxy)benzoyl)phenyl)-*N,N*-bis(2-methoxyethyl)acetamide, targets Heat Shock Protein-90 (IC$_{50}$ = 3.8 nM), an ATP-dependent molecular chaperone that stabilizes various oncogenic kinases, including HER2, EGFR, BCR-ABL, B-Raf and EML4-ALK, all of which are essential for tumor growth. In a novel orthotopic multiple myeloma (MM) model of i.v. inoculated OPM-2-GFP protein, KW-2478 showed a significant reduction of both serum M protein and MM tumor burden in the bone marrow (1). (**See also** *Alvespimycin; Ansamycin; Deguelin; Derrubone; Ganetespib; Geldanamycin; Herbimycin; Macbecin; Radicicol; Tanespimycin*) **1.** Nakashima, Ishii, Tagaya, *et al.* (2010) *Clin. Cancer Res.* **16**, 2792.

KW-3049, See *Benidipine*

KX2-391

This orally bioavailable, non-ATP competitive inhibitor (FW$_{free-base}$ = 431.54 g/mol; CAS 897016-82-9 (dihydrochloride); Solubility of free-base is 85 mg/mL DMSO, <1 mg/mL H$_2$O or Ethanol), known also by its alternate code name KX-01 and systematically as *N*-benzyl-2-(5-(4-(2-morpholinoethoxy)phenyl)pyridin-2-yl)acetamide, binds in the target protein substrate pocket of Src kinase, a proto-oncogene protein kinase that is up-regulated in many cancer cell lines and exerts control over multiple genetic and signaling pathways involved in the proliferation, survival, angiogenesis, invasion, and migration of various types of cancer cells. By targeting the peptide substrate rather than the ATP binding site, this inhibitor has a mode of action that is distinct from other Src kinase inhibitors. The KX2-391 binding site on tubulin (IC$_{50}$ = 0.5 µM) is distinct from sites for well-known tubulin inhibitors. KX2-391 nominally exhibits a GI$_{50}$ of 9-60 nM for the growth inhibition of various liver cancer cell lines (*e.g.*, Huh7, PLC/PRF/5, Hep3B, and HepG2). **1.** Fallah-Tafti, Foroumadi, Tiwari, *et al.* (2011) *Eur. J. Med. Chem.* **46**, 4853.

KY02111

This (FW = 376.86 g/mol; CAS 1118807-13-8; Solubility: 15 mg/mL DMSO; <1 mg/mL H$_2$O), systematically named *N*-(6-chloro-2-benzothiazolyl)-3,4-dimethoxybenzenepropanamide, promotes human pluripotent stem cell (hPSC) differentiation into cardiomyocytes by inhibiting Wnt signaling (renamed after *Drosophila* Wingless/int family of secreted lipid-modified signaling glycoproteins, with int1 becoming Wnt1). Although the direct target of KY02111 remains unknown, it is believed to

act downstream of APC and GSK3β. **1.** Minami, Yamada, Otsuji, *et al.* (2012) *Cell Rep.* **2**, 1448.

Kynurenate (Kynurenic Acid)

This quinaldate biosynthesis intermediate (FW$_{free-acid}$ = 189.17 g/mol; M.P. = 282-283°C; Yellow solid; Slightly soluble in water (0.9% at 100°C); Insoluble in diethyl ether; λ$_{max}$ at 332 and 344 nm at pH 7; ε = 9800 and 7920 M^{-1}cm^{-1}, respectively), also known as 4-hydroxyquinaldate, 4-hydroxy-2-quinoline-carboxylate, and 4 oxo-4*H*-quinoline-2-carboxylate, is a nonselective NMDA and AMPA/kainate receptor antagonist. **Target(s):** acetyl-CoA carboxylase (1,2); aldehyde reductase, *or* aldose reductase (3); 2 aminoadipate aminotransferase (4); AMPA/kainate receptors (5,6); dihydrolipoamide dehydrogenase (7); glutamate dehydrogenase (8-10); glyceraldehyde-3-phosphate dehydrogenase (10); kynureninase (11); kynurenine 7,8-hydroxylase (12); kynurenine 3-monooxygenase, weakly inhibited (13); kynurenine: oxoglutarate aminotransferase (14); lactate dehydrogenase (8-10); malate dehydrogenase (10); NAD$^+$ ADP-ribosyltransferase, *or* poly(ADP-ribose) polymerase (15); NMDA receptor (5,6,16); tryptophan 2,3-dioxygenase (17). **1.** Tanabe, Nakanishi, Hashimoto, *et al.* (1981) *Meth. Enzymol.* **71**, 5. **2.** Hashimoto, Isano, Iritani & Numa (1971) *Eur. J. Biochem.* **24**, 128. **3.** Kador, Sharpless & Goosey (1982) *Prog. Clin. Biol. Res.* **114**, 243. **4.** Deshmukh & Mungre (1989) *Biochem. J.* **261**, 761. **5.** Birch, Grossman & Hayes (1988) *Eur. J. Pharmacol.* **151**, 313. **6.** Weber, Dietrich, Grasel, *et al.* (2001) *J. Neurochem.* **77**, 1108. **7.** Furuta & Hashimoto (1982) *Meth. Enzymol.* **89**, 414. **8.** Baker (1967) *Design of Active-Site-Directed Irreversible Enzyme Inhibitors*, Wiley, New York. **9.** Baker, Lee, Tong, Ross & AMartinez (1962) *J. Theoret. Biol.* **3**, 446. **10.** Baker & Bramhall (1972) *J. Med. Chem.* **15**, 230. **11.** Tanizawa & Soda (1979) *J. Biochem.* **86**, 499. **12.** Taniuchi & Hayaishi (1963) *J. Biol. Chem.* **238**, 283. **13.** Saito, Quearry, Saito, *et al.* (1993) *Arch. Biochem. Biophys.* **307**, 104. **14.** Takeuchi, H. Otsuka & Y. Shibata (1983) *Biochim. Biophys. Acta* **743**, 323. **15.** Banasik, Komura, Shimoyama & Ueda (1992) *J. Biol. Chem.* **267**, 1569. **16.** Ganong & Cotman (1986) *J. Pharmacol. Exp. Ther.* **236**, 293. **17.** Paglino, Lombardo, Arcà, Rizzi & Rossi (2008) *Insect Biochem. Mol. Biol.* **38**, 871.

L-Kynurenine

This amino acid (FW = 208.22 g/mol; (λ$_{max}$ at 230, 257, and 360 nm at pH 7; ε = 18900, 7500, and 4500 M^{-1}cm^{-1}, respectively), also called 3-anthraniloyl-L-alanine, is an intermediate in L-tryptophan degradation and in the biosynthesis of nicotinate from L-tryptophan (1,2). **Target(s):** arylformamidase (3-5); cysteine-*S*-conjugate β-lyase (6); tryptophanase (7,8); tryptophan 2,3-dioxygenase (9); xanthine oxidase (9). **1.** Hayaishi & Stanier (1951) *J. Bacteriol.* **62**, 691. **2.** Hayaishi (1953) *Biochem. Prep.* **3**, 108. **3.** Serrano & Nagayama (1991) *Comp. Biochem. Physiol. B Comp. Biochem.* **99**, 281. **4.** Shinohara & Ishiguro (1977) *Biochim. Biophys. Acta* **483**, 409. **5.** Menge (1979) *Hoppe-Seyler's Z. Physiol. Chem.* **360**, 185. **6.** Stevens (1985) *J. Biol. Chem.* **260**, 7945. **7.** Morino & Snell (1970) *Meth. Enzymol.* **17A**, 439. **8.** Hoch, Simpson & DeMoss (1966) *Biochemistry* **5**, 2229. **9.** Paglino, Lombardo, Arcà, Rizzi & Rossi (2008) *Insect Biochem. Mol. Biol.* **38**, 871. **10.** Kaszaki, Palásthy, Érczes, *et al.* (2008) *Neurogastroenterol. Motil.* **20**, 53.

KYP-2047

This brain-penetrating proline/cyanoproline-containing peptide derivative (FW = 339.43 g/mol; CAS 796874-99-2), also named (2S)-1-[[(2S)-1-(1-oxo-4-phenylbutyl)-2-pyrrolidinyl]carbonyl]-2-pyrrolidinecarbonitrile, modulates inflammation, angiogenesis and neurodegeneration by virtue of its extreme inhibition of prolyl oligopeptidase, *or* POP (K_i = 23 pM), also known as prolyl endopeptidase (1). KYP-2047 rapidly penetrates mouse brain (t_{max} ≤10 min), achieving pharmacologically active concentrations after a single dose of 15 or 50 µmol/kg (i.p.) (4). The brain/blood AUC ratio was 0.050 and 0.039 after 15 and 50 µmol/kg (i.p.), respectively (2). KYP-2047 was equally distributed between the cortex, hippocampus and striatum. KYP-2047 clears the α-synuclein aggregates induced by oxidative stress in in cell lines overexpressing wild-type or A30P/A53T mutant human α-syn and in the brains of two A30P α-synuclein transgenic mouse strains (3). α-Synuclein is the main component of Lewy bodies in Parkinson Disease that cause neuronal cell death and dementia (3). KYP-2047 also inhibits AcSDKP formation from the actin monomer-sequestering protein thymosin β4 (*Sequence*: MSDKPDMAEIEKFDKSKLKKTETQEKKNPL-PSKETIEQEKQAGES). Ac-SDKP is a pro-angiogenic and antifibrogenic agent that is involved in hemopoietic stem cell differentiation (4). **1.** Jalkanen, Puttonen, Venäläinen, *et al.* (2007) *Basic Clin. Pharmacol. Toxicol.* **100**, 132. **2.** Jalkanen, Leikas & Forsberg (2013) *Basic Clin. Pharmacol. Toxicol.* **114**, 460. **3.** Myöhänen, Hannula & Van Elzen, *et al.* (2012) *Brit. J. Pharmacol.* **166**, 1097. **4.** Myöhänen, Tenorio-Laranga, Jokinen, *et al.* (2011) *Brit. J. Pharmacol.* **163**, 1666.

– L –

L, *See* L-Leucine

L803-mts

This selective, cell-permeable peptide inhibitor (FW = 1464.64 g/mol; *Sequence*: Myr-GKEAPPAPPQpSP-NH$_2$, with modifications at: Gly-1 = *N*-Myristoyl-Gly, Ser-11 = pSer, and Pro-12 = C-terminal amide; CAS 1043881-55-5) targets glycogen synthase kinase-3, *or* GSK-3 (IC$_{50}$ = 40 μM), but is without effect on Cdc2, PKB, and PKC. This phosphorylated peptide inhibitor is protein substrate site-competitive, in sharp contrast to protein kinase inhibitors directed to ATP binding sites. L803-mts mimics insulin action, activating glycogen synthase activity (~2.5-fold) in HEK293 cells and improving glucose tolerance in diabetic mice. **1**. Plotkin, Kaidanovich, Talior & Eldar-Finkelman (2003) *J. Pharm. Exp. Ther.* **305**, 974.

L2-401

This thiadiazolidinone (FW = 304.34 g/mol), systematically named 4-[(4-fluorophenyl)methyl]-2-(4-methylphenyl)-1,2,4-thiadiazolidine-3,5-dione, is a slow binding inhibitor of the pyridoxal 5′-phosphate-dependent enzyme alanine racemase (K_i = 0.2 μM), which catalyzes the racemization of L-alanine to D-alanine. The latter is an essential precursor for the synthesis of the pentapeptide cross-linkers in cell wall peptidoglycan of methicillin-resistant *Staphylococcus aureus*. When presented on a molar basis, L2-401 and D-cycloserine showed comparable inhibitory effects *in vitro* against methicillin-susceptible and methicillin-resistant *S. aureus* strains, as revealed by the MIC required to inhibit bacterial growth. L2-401 also inhibits HeLa cell proliferation. Thiadiazolinediones are known to inhibit Glycogen Synthase Kinase-3 β, but it is not known whether GSK-3β inhibition is responsible for L2-401's growth-inhibiting effect on HeLa cells. **1**. Ciusteaa, Mootiena, Rosato (2012) *Biochem. Pharmacol.* **83**, 368.

L-163,191, *See* MK 0677

L 216140

This antibiotic (FW = 308.33 g/mol) competitively inhibits UDP-3-*O*-(*R*-3-hydroxymyristoyl)-GlcNAc deacetylase, *or* LpxC (K_i = 50 nM), an enzyme that catalyzes the committed reaction in the biosynthesis of Lipid A, an essential molecule for bacterial proliferation. LpxC is thus a prime target for the development of novel antibiotics (1). By using L-161,240-insensitive LpxC activity from *P. aeruginosa* expressed from a low copy plasmid, LpxC was shown to be the primary target of L-161,240 in *E. coli* (2). When challenged with a lethal dose of *E. coli*, L-161–240 protects mice against septicemia; however, it is ineffective against *Pseudomonas aeruginosa* or *Serratia marcescens*. **1**. Onishi, Pelak, Gerckens, *et al.* (1996) *Science* **274**, 980. **2**. Mdluli, Witte, Kline, *et al.* (2006) *Antimicrob. Agents Chemother.* **50**, 2178.

L-368899

This orally bioavailable, nonpeptide oxytocin receptor antagonist (FW$_{free-base}$ = 554.78 g/mol; CAS 148927-60-0; Soluble to 100 mM in water and to 100 mM in DMSO), also known as 1-((7,7-dimethyl-2(*S*)-(2(*S*)-amino-4-(methylsulfonyl)butyramido)bicyclo[2.2.1]-heptan-1(*S*)-yl)methyl)sulfonyl)-4-(2-methylphenyl)piperazine, exhibits good selectivity (IC$_{50}$ = 8.9 nM), with > 40-fold selectivity over vasopressin V$_{1a}$ (IC$_{50}$ = 370 nM) and V$_2$ (IC$_{50}$ = 570 nM) receptors. L-368899 antagonizes oxytocin-induced uterine contractions *in vitro* and *in vivo* (1). Unlike peripherally selective L-371257, brain penetration by L-368899 rapid and selectively accumulates in areas of the limbic system, making it a useful tool for investigating the centrally mediated roles of oxytocin, including social behaviour and pair bonding. L-368899 reduces sexual activity, food-sharing, and related mothering instincts. Treatment with L-368899 significantly reduced episodes luteolytic PGF$_{2\alpha}$ release in ewes, suggesting it may be have value in reducing early pregnancy failure (2). **1**. Williams, Anderson, Ball, *et al.* (1994) *J. Med. Chem.* **37**, 565. **2**. Mann, Lamming, Scholey, Hunter & Pettibone (2003) **25**, 255.

L-651,582, *See* Carboxyamidotriazole

L-655238

This substituted quinolone (FW = 335.45 g/mol; CAS 101910-24-1), also known as α-pentyl-4-(2-quinolinylmethoxy)benzenemethanol and REV 5901, is a potent inhibitor of the 5-lipoxygenase activating protein, *or* FLAP (IC$_{50}$ = 100 nM) and a competitive leukotriene D$_4$ antagonist. **Target(s):** leukotriene D$_4$ (1); 5-lipoxygenase-activating protein (2). **1**. Datta, Rathod, Manning, Turnbull & McNeil (1996) *J. Endocrinol.* **149**, 269. **2**. Evans, Leville, Mancini, *et al.* (1991) *Mol. Pharmacol.* **40**, 22.

L-658758

This cephalosporin derivative (FW$_{free-acid}$ = 414.41 g/mol), also known as 1-[[3-(acetoxymethyl)-7α-methoxy-8-oxo-5-thia-1-azabicyclo[4.2.0]oct-2-en-2-yl]carbonyl]proline *S,S*-dioxide, irreversibly inhibits leukocyte elastase, with acylation of an active-site serine hydroxyl. **Target(s):** leukocyte elastase, *or* neutrophil elastase (1-4); myeloblastin, *or* proteinase 3 (4). **1**. Davies, Ashe, Bonney, *et al.* (1991) *Ann. N. Y. Acad. Sci.* **624**, 219. **2**. Knight, Maycock, Green, *et al.* (1992) *Biochemistry* **31**, 4980. **3**. Pacholok,

Davies, Dorn, *et al.* (1995) *Biochem. Pharmacol.* **49**, 1513. **4**. Rees, Brain, Wohl, Humes & Mumford (1997) *J. Pharmacol. Exp. Ther.* **283**, 1201.

L-659,699, See *F-244*

L-660631

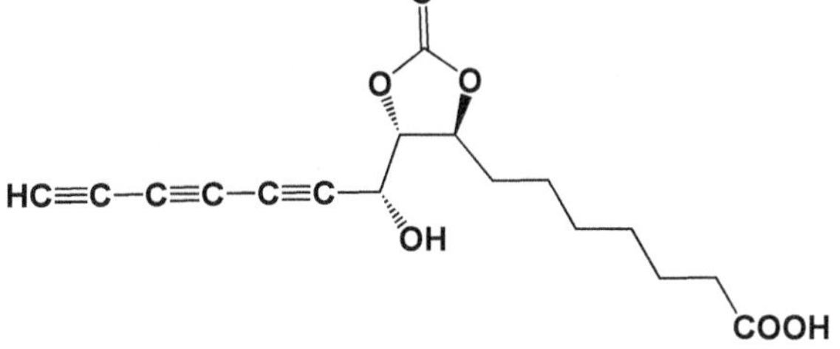

This triyne carboxylate natural product (FW = 318.32 g/mol; CAS 115216-83-6), also known as 2-oxo-5-(1-hydroxy-2,4,6-heptatriynyl)-1,3-dioxolone-4-heptanoic acid), first isolated from a culture of *Actinomyces*, targets rat liver cytosolic acetyl-CoA *C*-acetyltransferase, thereby inhibiting sterol biosynthesis. L=660631 also regulates the growth of *Candida albicans*. The inhibitor had no effect on other sulfhydryl containing enzymes of lipid synthesis such as HMG-CoA synthase, HMG-CoA reductase, and fatty acid synthase. **1**. Greenspan, Yudkovitz, Chen, *et al.* (1989) *Biochem. Biophys. Res. Commun.* **163**, 548.

L-660711, See *MK-0571*

L-663,536, See *MK-886*

L-671,152, See *Dorzolamide*

L-687908

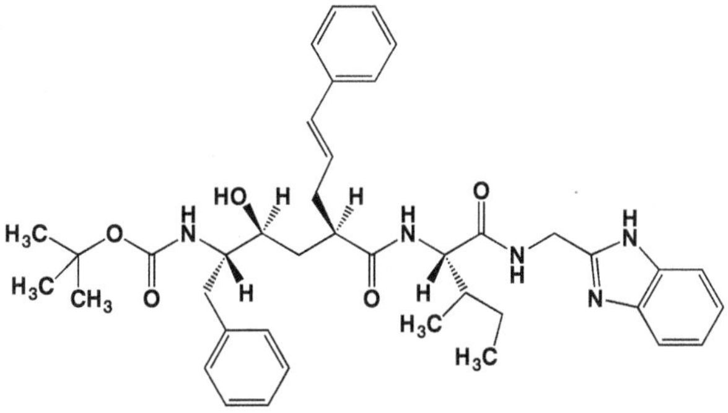

This potent hydroxyethylene-containing peptidomimetic (FW = 681.88 g/mol) inhibits human immunodeficiency virus-I protease (IC$_{50}$ = 30 pM). **1**. Vacca (1994) *Meth. Enzymol.* **241**, 311. **2**. Vacca, Guare, deSolms, *et al.* (1991) *J. Med. Chem.* **34**, 1225.

L-700,462, See *Tirofiban*

L-739749

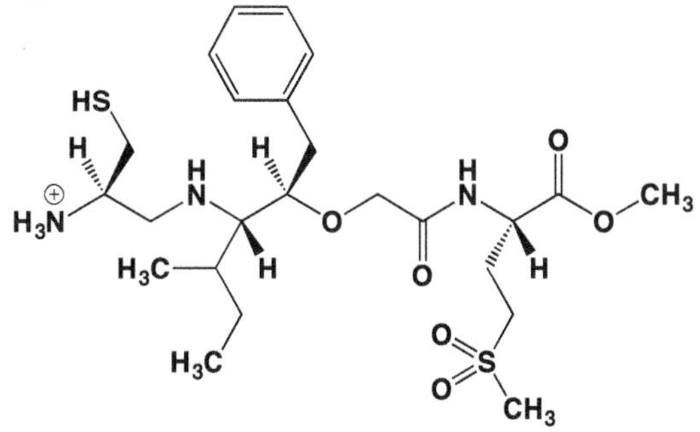

This peptidomimetic (FW = 531.74 g/mol), also known as 2(*S*)-[2(*S*)-[2(*R*)-amino-3-mercapto]propylamino-3(*S*)-methyl]pentyloxy-3-phenylpropionyl-methioninesulfone methyl ester, is L-cysteinyl-L-isoleucyl-L-phenylalanyl-L-methionine methyl ester analogue and a prodrug for the early-stage inhibitor of protein farnesyltransferase L-739750 (1-3). L-739749 principally interferes with RhoB and suppresses Ras transformation. The half-life of RhoB was found to be ~2 h, and can be functionally depleted

during the 18-h period required by L-739749 to induce reversion. Cell treatment with L-739749 disrupted the vesicular localization of RhoB but did not effect the localization of the closely related RhoA protein. Ras-transformed Rat1 cells ectopically that express *N*-myristylated forms of RhoB (*e.g.,* Myr-RhoB, whose vesicular localization was unaffected by L-739749) were resistant to drug treatment (2). **1**. Gibbs (2001) *The Enzymes*, 3rd ed., **21**, 81. **2**. Prendergast, Davide, deSolms, *et al.* (1994) *Mol. Cell. Biol.* **14**, 4193. **3**. Kohl, Wilson, Mosser, *et al.* (1994) *Proc. Natl. Acad. Sci. U.S.A.* **91**, 9141.

L-739750

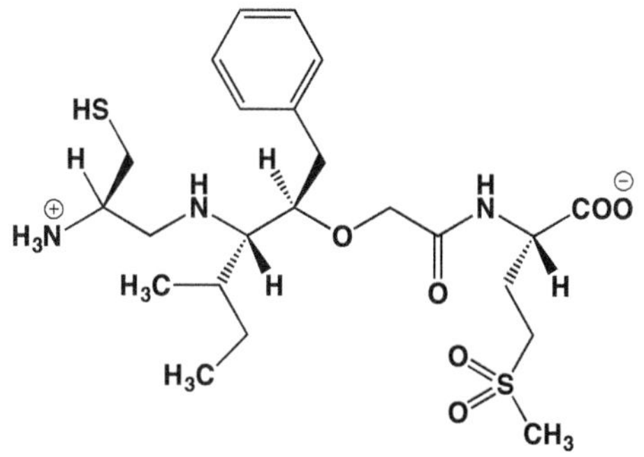

This Cys-Ile-Phe-Met peptidomimetic (FW = 501.71 g/mol) is the metabolic product of the prodrug L-739749 and the similarly acting prodrug L-744832. It inhibits protein farnesyltransferase, which catalyzes posttranslational modification of signal-transduction G-proteins, including Ras (1-3). L-739750 suppresses the anchorage-independent growth of Rat1 cells transformed with viral H-ras. (**See** *L-739749 above for additional details.*) **1**. Terry, Long & Beese (2001) *The Enzymes*, 3rd ed., **21**, 19. **2**. Kohl, Wilson, Mosser, *et al.* (1994) *Proc. Natl. Acad. Sci. U.S.A.* **91**, 9141. **3**. Guida, Hamilton, Crotty & Sebti (2006) *J. Comp.-Aided Mol. Des.* **19**, 871.

L-743,726, See *Efavirenz*

L-745870

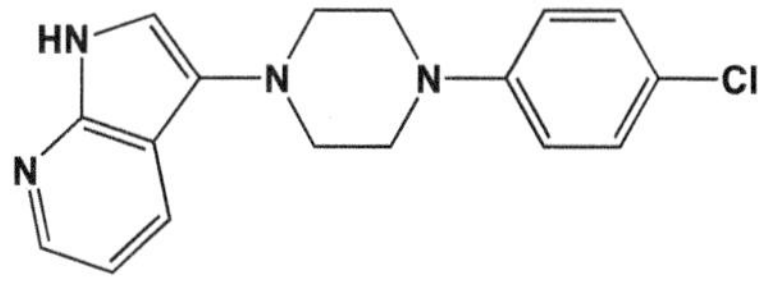

This pyrrolopyridine (FW$_{free-base}$ = 326.83 g/mol), also known as 3-([4-(4-chlorophenyl)piperazin-1-yl]methyl)-1*H*-pyrrolo[2,3-*b*]pyridine, is a selective D$_4$ dopamine receptor antagonist. It displaces the specific binding of 0.2 nM spiperone to cloned human dopamine D4 receptors with a *K*i value of 0.43 nM. **1**. Kulagowski, Broughton, Curtis, *et al.* (1996) *J. Med. Chem.* **39**, 1941. **2**. Patel, Freedman, Chapman, *et al.* (1997) *J. Pharmacol. Exp. Ther.* **283**, 636.

L-754030, See *Aprepitant*

L868275, See *Flavopiridol*

Laccaic Acid A

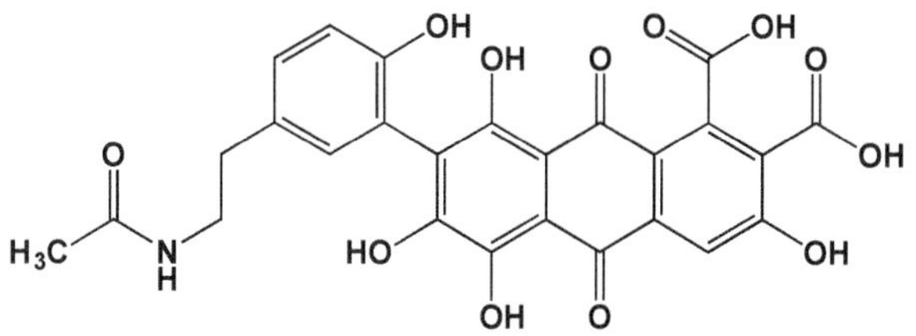

This complex anthraquinone and widely used food colorant (FW = 537.43 g/mol; CAS 15979-35-8; *Abbreviation*: LCA) is a novel, first-in-class direct, DNA-competitive inhibitor (*K*$_i$ = 310 nM) of DNA methyltransferase 1 (*or* DNMT1), a key enzyme in the methylation of cytosines within CpG dinucleotide during epigenetic marking of vertebrate DNA, thereby repressing transcription. This enzyme binds to DNA replication sites in S-phase and acts to maintain the methylation pattern in the newly synthesized strand, a condition that is essential for epigenetic inheritance. It also associates with chromatin during G_2 and M phases to maintain DNA methylation independently of replication. Significantly, LCA exhibited mixed inhibition with respect to *S*-adenosyl-methionine (1). Given its polycyclic aromatic structure, LCA is a likely DNA intercalator. Significantly, LCA and 5-aza-dC act synergistically in activating expression of silenced tumor suppressor genes, thus demonstrating the value of using agents having complementary mechanisms of action. Laccaic acid is a potent inhibitor of the autoactivation of the plasma hyaluronan-binding protein (PHBP) proenzyme, $IC_{50} \approx 1$ µM, as well as the PHB's catalytic activity of the active enzyme, $IC_{50} \approx 1$ µM (2). **1**. Fagan, Cryderman, Kopelovich, Wallrath & Brenner (2013) *J. Biol. Chem.* **288**, 23858. **2**. Sekido, Nishimura, Takai & Hasumi (2010) *Biosci. Biotechnol. Biochem.* **74**, 2320.

Lacosamide

This functionalized amino acid (FW = 250.29 g/mol; CAS 175481-36-4; *Abbreviation*: LCM), also known as Vimpat® and systematically named N^2-acetyl-*N*-benzyl-D-homoserinamide, is an orally available anticonvulsant that selectively enhances the slow inactivation of sodium channels and interacts with the neuroplasticity-relevant target, collapsin-response mediator protein-2 (CRMP-2), a cytosolic phosphoprotein involved in neuronal differentiation and axonal guidance. Pharmacokinetic studies show that it is renally excreted, minimally bound to plasma proteins, with no known drug-drug interactions. LCM inhibits CRMP2-mediated neurite outgrowth, an effect that is phenocopied by CRMP2 knockdown (1). At 10 µM, LCM does not bind to AMPA receptors; Kainate receptors; NMDA receptors; $GABA_A$ (muscimol/benzodiazepine) receptor; $GABA_B$ receptor; adenosine A_1 receptor, adenosine A_2 receptor, adenosine A_3 receptor; α_1-receptor, α_2-receptor; β_1-receptor, β_2-receptor; muscarinic M_1, M_2, M_3, M_4, and M_5 receptors, histamine H_1, H_2, and H_3 receptors, dopamine D_1, D_2, D_3, D_4, and D_5 receptors; serotonin HT_{1A}, HT_{1B}, HT_{2A}, HT_{2C}, HT_3, HT_{5A}, HT_6, and HT_7 receptors, and K_{ATP} receptor, indicating that LCM does not act by means of a high-affinity interaction with an acknowledged recognition site on a target for existing antiepileptic drugs (2). Instead, lacosamide affects voltage-gated sodium channels in a novel way by enhancing the slow-inactivating 'braking' state of these channels (3). **1**. Wilson SM, Xiong W, Wang (2012) *Neuroscience* **210**, 451. **2**. Errington, Coyne, Stöhr, Selve & Lees (2006) *Neuropharmacology* **50**, 1016. **3**. Bee & Dickenson (2009) *Neuropharmacology* **57**, 472.

Lactacystin

Lactacystin   **Lactalysin Lactone**

This cell-permeable, pro-inhibitor ($FW_{\text{free-acid}}$ = 376.43 g/mol), a *Streptomyces lactacystinaeus* metabolite that spontaneously converts to its *clasto*-lactacystin β-lactone (FW = 213.23 g/mol), also known as omuralide, which irreversibly inhibits mammalian proteasomes, blocking both chymotryptic- and tryptic-like activities (K_i = 4 nM). It also inhibits NF-κB activation. Lactacystin is also an inducer of neurite outgrowth in mouse neuroblastoma cells. **Target(s)**: carboxypeptidase C, *or* cathepsin A (1-3); HslVU protease (4); proteasome ATPase (5); proteasome endopeptidase complex (6-12). **1**. Kozlowski, Stoklosa, Omura, *et al.* (2001) *Tumour Biol.* **22**, 211. **2**. Ostrowska, Wojcik, Wilk, *et al.* (2000) *Int. J. Biochem. Cell Biol.* **32**, 747. **3**. Satoh, Kadota, Oheda, *et al.* (2004) *J. Antibiot.* **57**, 316. **4**. Rohrwild, Huang, Goldberg, Yoo & Chung (1998) in *Handb. Proteolytic Enzymes*, p. 492, Academic Press, San Diego. **5**. Certad, Abrahem &

Georges (1999) *Exp. Parasitol.* **93**, 123. **6**. Seemüller, Dolenc & Lupas (1998) in *Handb. Proteolytic Enzymes*, p. 495, Academic Press, San Diego. **7**. Grisham, Palombella, Elliott, *et al.* (1999) *Meth. Enzymol.* **300**, 345. **9**. Dick, Cruikshank, Grenier, *et al.* (1996) *J. Biol. Chem.* **271**, 7273. **10**. Craiu, Gaczynska, Akopian, *et al.* (1997) *J. Biol. Chem.* **272**, 13437. **11**. Corey & Li (1999) *Chem. Pharm. Bull. (Tokyo)* **47**, 1. **12**. Wojcikiewicz, Xu, Webster, Alzayady & Gao (2003) *J. Biol. Chem.* **278**, 940.

β-Lactam Antibiotics

These naturally occurring, fungus-derived antibiotics target the transpeptidases that catalyze a key step in the cross-linking of the peptidoglycan required in bacterial cell wall formation. These natural products are divided into five classes, the core structures of which are as follows.

Penicillins

Cephalosporins

Carbapenams

Monobactams

Clavams

The common structural element within these antibiotics is the central four-member β-lactam nucleus, a highly strained structure that intercepts the transpeptidase (TP) and activates covalent formation of *O*-acyl-TP. These antibiotics likewise exclude/diminish water attack on the *O*-acyl-TP, thereby preventing hydrolysis. The net effect is that β-lactam antibiotics successfully interrupt cell-wall formation. **See specific agent; For details on mode of action, See** *Penicillins*

D-Lactate (D-Lactic Acid)

This α-hydroxyacid (FW = 90.08 g/mol; pK_a = 3.86 at 25°C; Soluble in water, ethanol, and acetone; M.P. = 52.8°C), systematically named (*R*)-2-hydroxypropanoic acid, is produced by some microorganisms, fungi, plants, as well such microbes as *Escherichia coli* and *Lactobacillus leishmanii..* The solid is hygroscopic and concentrated solutions tend to oligomerize. (*See also* DL-*Lactate;* L-*Lactate*) **Target(s)**: ε-crystallin L-lactate dehydrogenase (1); 3-hydroxybutyrate dehydrogenase (2-4); L-lactate dehydrogenase (1,5-7); L-lactate 2-monooxygenase (8-13); D-lactate-2-sulfatase (14); proline dehydrogenase (15); propionate CoA-transferase, also as an alternative substrate (16). **1**. Chang, Huang & Chiou (1991) *Arch. Biochem. Biophys.* **284**, 285. **2**. Krebs, Gawehn, Williamson & Bergmeyer (1969) *Meth. Enzymol.* **14**, 222. **3**. Bergmeyer, Gawehn, Klotzsch, Krebs & Williamson (1967) *Biochem. J.* **102**, 423. **4**. Delafield, Cooksey & Doudoroff (1965) *J. Biol. Chem.* **240**, 4023. **5**. Snoswell (1966)

Meth. Enzymol. **9**, 321. **6**. Yoshida & Freese (1975) *Meth. Enzymol.* **41**, 304. **7**. Dikstein (1959) *Biochim. Biophys. Acta* **36**. 397. **8**. Xu, Yano, Yamamoto, Suzuki & Kumagai (1996) *Appl. Biochem. Biotechnol.* **56**, 277. **9**. Durfor & Cromartie (1981) *Arch. Biochem. Biophys.* **210**, 710. **10**. Takemori, Nakai, Nakazawa, Katagiri & Nakamura (1973) *Arch. Biochem. Biophys.* **154**, 137. **11**. Takemori & Katagiri (1975) *Meth. Enzymol.* **41B**, 329. **12**. Sun, Williams & Massey (1996) *J. Biol. Chem.* **271**, 17226. **13**. Müh, Williams & Massey (1994) *J. Biol. Chem.* **269**, 7989. **14**. Crescenzi, Dodgson & White (1984) *Biochem. J.* **223**, 487. **15**. Scarpulla & Soffer (1978) *J. Biol. Chem.* **253**, 5997. **16**. Schweiger & Buckel (1984) *FEBS Lett.* **171**, 79.

L-Lactate (L-Lactic Acid)

This α-hydroxy acid (FW = 90.08 g/mol; pK_a = 3.86 at 25°C; Soluble in water, ethanol, and acetone; M.P. = 52.8°C), systematically named (*S*)-2-hydroxypropanoic acid, is the enantiomer that is most often encountered in Nature and is formed via anaerobic glycolysis. L-Lactate accumulates in muscle cells and blood following strenuous exercise. **Target(s):** aminoacylase (1); catalase (2; creatine kinase (3,4); cutinase (5); cystathionine γ-lyase, *or* cysteine desulfhydrase (6); D-2-hydroxy-acid dehydrogenase (7); D-lactate dehydrogenase (8); nicotinate-nucleotide diphosphorylase (carboxylating) (9); oxaloacetate tautomerase (10); phosphoenolpyruvate carboxylase (11); proline dehydrogenase (12); propionate CoA-transferase, also as an alternative substrate (13); pyrophosphatase (14); pyruvate kinase (15,16); tauropine dehydrogenase (17); tyrosinase, *or* monophenol monooxygenase (18); tyrosine decarboxylase (19); UDP-*N*-acetylmuramate:L-alanine ligase (20). **1**. Tamura, Oki, Yoshida, *et al.* (2000) *Arch. Biochem. Biophys.* **379**, 261. **2**. Chatterjee & Sanwal (1993) *Mol. Cell. Biochem.* **126**, 125. **3**. Watts (1973) *The Enzymes*, 3rd ed., **8**, 383. **4**. Tang, Ou & Zhou (2003) *Biochem. Cell Biol.* **81**, 1. **5**. Maeda, Yamagata, Abe, *et al.* (2005) *Appl. Microbiol. Biotechnol.* **67**, 778. **6**. Fromageot & Grand (1942) *Enzymologia* **11**, 81. **7**. Cammack (1975) *Meth. Enzymol.* **41**, 323. **8**. Snoswell (1966) *Meth. Enzymol.* **9**, 321. **9**. Shibata & Iwai (1980) *Biochim. Biophys. Acta* **611**, 280. **10**. Annett & Kosicki (1969) *J. Biol. Chem.* **244**, 2059. **11**. Miziorko, Nowak & Mildvan (1974) *Arch. Biochem. Biophys.* **163**, 378. **12**. Scarpulla & Soffer (1978) *J. Biol. Chem.* **253**, 5997. **13**. Schweiger & Buckel (1984) *FEBS Lett.* **171**, 79. **14**. Naganna (1950) *J. Biol. Chem.* **183**, 693. **15**. Plaxton & Storey (1984) *Eur. J. Biochem.* **143**, 257. **16**. Chrispeels & Gaede (1985) *Comp. Biochem. Physiol. B Comp. Biochem.* **82**, 162. **17**. Gade (1986) *Eur. J. Biochem.* **160**, 311. **18**. Kong, Park, Hong & Cho (2000) *Comp. Biochem. Physiol. B Biochem. Mol. Biol.* **125**, 563. **19**. Moreno-Arribas & Lonvaud-Funel (1999) *FEMS Microbiol. Lett.* **180**, 55. **20**. Liger, Blanot & van Heijenoort (1991) *FEMS Microbiol. Lett.* **80**, 111.

Lacticin 481

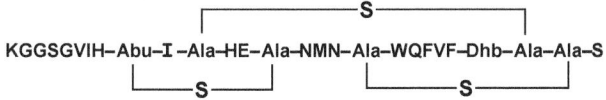

This 1.7-kDa lanthionine-containing peptide antibiotic, produced by *Lactococcus lactis* CNRZ 481, is a bacteriocin that acts adversely on certain other microorganisms via pore formation (1-3). The sequence is shown above where single uppercased letters refer to standard aminoacyl residues and dhB, Abu–S–Ala, and Ala–S–Ala refer to didehydrobutyrine, β-methyllanthionine, and lanthionine, respectively. Lacticin 481 is synthesized on ribosomes as a prepeptide (LctA) and posttranslationally modified to its mature form. These modifications include dehydration of serines and threonines, followed by intramolecular addition of cysteines to the unsaturated amino acids, which generates cyclic thioethers. This process breaks eight chemical bonds and forms six newbonds and is catalyzed by one enzyme, LctM (4). **1**. Piard, Delorme, Novel, Desmazeaud & Novel (1993) *FEMS Microbiol. Lett.* **112**, 313. **2**. Piard, Kuipers, Rollema, Desmazeaud & de Vos (1993) *J. Biol. Chem.* **268**, 16361. **3**. Piard, Muriana, Desmazeaud & Klaenhammer (1992) *Appl. Environ. Microbiol.* **58**, 279. 4. Xie, Miller, Chatterjee, *et al.* (2004) *Science* **303**, 679.

Lactitol

This hygroscopic reduced sugar (FW = 344.32 g/mol; CAS 585-86-4), also known as 4-*O*-β-D-galactopyranosyl-D-glucitol and lacitol, is as artificial sweetener and as a laxative. **Target(s):** 3-galactosyl-*N*-acetylglucosaminide 4-α-L-fucosyltransferase, *or* galactoside 3(4)-L-fucosyltransferase, K_i = 17 mM (1,2); *trans*-sialidase (3). **1**. Sadler, Beyer, Oppenheimer, *et al.* (1982) *Meth. Enzymol.* **83**, 458. **2**. Prieels, Monnom, Dolmans, Beyer & Hill (1981) *J. Biol. Chem.* **256**, 10456. **3**. Agusti, Paris, Ratier, Frasch & de Lederkremer (2004) *Glycobiology* **14**, 659.

Lactocin S

dhB–NLA(DA)VAV(DA)MELLPTA(DA)VLY–Ala–DVAG–Ala–PKY–Ala–AKHH–Ala

This lanthionine-containing peptide antibiotic (CAS 125387-34-0), from *Lactobacillus sake*, is a bacteriocin that exhibits potent antimicrobial activity against a broad range of Gram-positive organisms. The sequence is shown above where single uppercased letters refer to standard aminoacyl residues, and dhB and Ala–S–Ala refer to didehydrobutyrine and lanthionine, respectively. Note that lactocin S also contains D-alanyl residues (DA). This bacteriocin is the first prokaryotic exception to the rule that only L-amino acids are included in ribosomally synthesized peptides. **1**. Skaugen, Nissen-Meyer, Jung, *et al.* (1994) *J. Biol. Chem.* **269**, 27183.

Lactose

α-anomer   **β-anomer**

This disaccharide (FW = 342.30 g/mol; CAS 63-42-3), also known as 4-*O*-β-D-galactopyranosyl-D-glucose, milk sugar, and lactobiose, is a yeast-fermentable reducing sugar. Both the α- and β-anomers are commercially available, and α-lactose is less soluble in water (1 g/2.6 mL boiling water vs. 1g/1.1 mL for the β-anomer) and not as sweet as β-lactose. **Target(s):** β-*N*-acetylhexosaminidase (19,20); aldose 1-epimerase (1,2); arabino-galactan endo 1,4-β-galactosidase (3); aralin (4); *Bauhinia purpurea* agglutinin (5); cellulase (25); cellulose 1,4-β cellobiosidase (6-9); discoidins II (*Dictyostelium discoideum* agglutinins) (10); electrolectin (11); β-fructofuranosidase, *or* invertase, weakly inhibited (24); β-D-fucosidase (22,23); 1,2-α-L-fucosidase (12); α-galactosidase (13); galactoside 2-α-L-fucosyltransferase (26); galactosylceramidase (21); glucuronosyltransferase (27); isomaltulose synthase (14); levansucrase (15); pallidin (*Polysphondylium pallidum* agglutinin) (16); phloretin hydrolase (17); rRNA *N*-glycosylase (4); *Sophora japonica* hemagglutinin (18). **1**. Bentley (1962) *Meth. Enzymol.* **5**, 219. **2**. Fishman, Pentchev & Bailey (1975) *Meth. Enzymol.* **41**, 484. **3**. Nakano, Takenishi & Watanabe (1985) *Agric. Biol. Chem.* **49**, 3445. **4**. Tomatsu, Kondo, Yoshikawa, *et al.* (2004) *Biol. Chem.* **385**, 819. **5**. Osawa, Irimura & Kawaguchi (1978) *Meth. Enzymol.* **50**, 367. **6**. Fracheboud & Canevascini (1989) *Enzyme Microb. Technol.* **11**, 220. **7**. Ubhayasekera, Muñoz, Vasella, Ståhlberg & Mowbray (2005) *FEBS J.* **272**, 1952. **8**. Kruus, Andreacchi, Wang & Wu (1995) *Appl. Microbiol. Biotechnol.* **44**, 399. **9**. Tuohy, Walsh, Murray, *et al.* (2002) *Biochim. Biophys. Acta* **1596**, 366. **10**. Barondes, Rosen, Frazier, Simpson & Haywood (1978) *Meth. Enzymol.* **50**, 306. **11**. Teichberg (1978) *Meth. Enzymol.* **50**, 291. **12**. Miura, Okamoto & Yanase (2005) *J. Biosci. Bioeng.* **99**, 629. **13**. Pederson & Goodman (1980) *Can. J. Microbiol.* **26**, 978. **14**. Nagai, Sugitani & Tsuyuki (1994) *Biosci. Biotechnol. Biochem.* **58**, 1789. **15**. Pabst (1977) *Infect. Immun.* **15**, 518. **16**. Barondes, Simpson, Rosen & Haywood (1978) *Meth. Enzymol.* **50**, 312. **17**. Mackey, Henderson &

Gregory (2002) *J. Biol. Chem.* **277**, 26858.   **18**. Poretz (1972) *Meth. Enzymol.* **28**, 349.   **19**. Keyhani & Roseman (1996) *J. Biol. Chem.* **271**, 33425.   **20**. Sakai, Narihara, Kasama, Wakayama & Moriguchi (1994) *Appl. Environ. Microbiol.* **60**, 2911.   **21**. Rushton & Dawson (1975) *Biochim. Biophys. Acta* **388**, 92.   **22**. Nunoura, Ohdan, Yano, Yamamoto & Kumagai (1996) *Biosci. Biotechnol. Biochem.* **60**, 188.   **23**. Colas (1980) *Biochim. Biophys. Acta* **613**, 448.   **24**. Isla, Vattuone, Gutierrez & Sampietro (1988) *Phytochemistry* **27**, 1993.   **25**. Mori (1992) *Biosci. Biotechnol. Biochem.* **56**, 1198.   **26**. Beyer & Hill (1980) *J. Biol. Chem.* **255**, 5373.   **27**. Kakuda, Oka & Kawasaki (2004) *Protein Expr. Purif.* **35**, 111.

Lactulose

This nonabsorbable disaccharide (FW = 342.30 g/mol; CAS 4618-18-2), also known by the brand names Constulose®, Enulose®, Generlac®, Cholac®, and Constilac®, as well as its IUPAC name 4-*O*-β-D-galactopyranosyl-β-D-fructofuranose, is an osmolyte that, when taken orally (typical dose = 10 g/15 mL), remains within the digestive tract, thereby increasing colon water content and relieving constipation within one day. Lactulose may also be taken orally or rectally to treat hepatic encephalopathy arising from hyperammonemia. In this case, lactulose is metabolized by gut flora, forming lactic acid and acetic acid and acidifying the colon. The latter converts otherwise freely diffusible NH_3 into NH_4^+ ion, which cannot diffuse back into the bloodstream and is instead eliminated. Lactulose also stimulates the growth of health-promoting bacteria (*e.g.*, bifidobacteria and lactobacilli) in the GI tract, while inhibiting growth of pathogenic bacteria, including *Salmonella*. Pharmacological inhibition of galectin-1 by lactulose alleviates weight gain in diet-induced obese rats (1). Notably, lactulose reduces adipogenesis and fat accumulation *in vitro* by down-regulating such adipogenic transcription factors such as C/EBPα and PPARγ. *In vivo* treatment of lactulose in 5-week-old Sprague-Dawley male rats significantly alleviates High-Fat-Diet-induced body weight gain and food efficiency as well as improved plasma and other metabolic parameters. Lactulose treatment also down-regulates major adipogenic marker proteins (*e.g.*, C/EBPα and PPARγ) in adipose tissue as well as stimulated expression of proteins involved in energy expenditure and lipolysis (*e.g.*, ATP5B, COXIV, HSL, and CPT1). Lactulose was first identified in milk that was heat-treated, and it can now be prepared in highly efficient chemical and enzymatic syntheses.   **1**. Mukherjee & Yun (2016) *Life Sci.* **148**, 112.

Ladostigil

This orally bioavailable AChE inhibitor and novel multimodal neuroprotective drug (FW = 272.34 g/mol; CAS 209349-27-4), also codenamed TV-3,326 and known systematically as [(3*R*)-3-(prop-2-ynylamino)indan-5-yl]-*N*-propylcarbamate, reversibly targets both acetylcholinesterase (AChE) as well as butyrylcholinesterase (BuChE) in rats after oral doses of 10-100 mg/kg (1) and irreversibly inhibits monoamine oxidase B (2). Through the combination of the pharmacophore of rasagiline with the carbamate moiety of the cholinesterase inhibitor rivastigmine, ladostigil is a versatile inhibitor that shows promise for treating neurodegenerative disorders like Alzheimer's disease, Lewy body disease, and Parkinson's disease. (***See also*** *Rasagiline; Rivastigmine*) After chronic but not acute treatment, it inhibits MAO-A and -B in the brain by more than 70% but has almost no effect on these enzymes in the small intestine in rats and rabbits (1). The brain selectivity results in minimal potentiation of the pressor response to oral tyramine. Ladostigil acts like other antidepressants in the forced swim test in rats, indicating a potential for antidepressant activity. Chronic treatment of mice with ladostigil (26 mg/kg) prevents the destruction of nigrostriatal neurons by the neurotoxin *N*-methyl-4-phenyl-1,2,3,6-tetrahydropyridine (*or* MPTP) (2). An additional

and new neuroprotective effect, shared by propargylamine compounds, relates to their ability to regulate the processing of amyloid-β protein precursor (AβPP) by the non-amyloidogenic α-secretase pathway. Such effects are linked to the p42/44 mitogen-activated protein kinase (MAPK) and protein kinase C (PKC) signaling pathway (2). Moreover, by enhancing the expression of Glia-Derived Neurotrophic Factor (GDNF) and Brain-Derived Neurotrophic Factor (BDNF), ladostigil may reverse damaging effects of neurodegenerative diseases by inducing neurogenesis (3). Ladostigil has antidepressant effects, and may prove useful as a way to reduce comorbid depression and anxiety in such disorders (4,5). Recent work indicates that ladostigil prevents gliosis and oxidative-nitrative stress and reduced the deficits in episodic and spatial memory induced by intracerebroventricular injection of streptozotocin in rats (6). It also possesses potent anti-apoptotic and neuroprotective activities *in vitro* and in various neurodegenerative rat models, (*e.g.*, hippocampal damage induced by global ischemia in gerbils and cerebral oedema induced in mice by closed head injury). These neuroprotective activities involve regulation of amyloid precursor protein processing; activation of protein kinase C and mitogen-activated protein kinase signaling pathways; inhibition of neuronal death markers; prevention of the fall in mitochondrial membrane potential and upregulation of neurotrophic factors and antioxidative activity. Recent findings demonstrated that the major metabolite of ladostigil, hydroxy-1-(*R*)-aminoindan has also a neuroprotective activity and thus, may contribute to the overt activity of its parent compound (6).   **1**. Weinstock, Gorodetsky, Poltyrev, *et al.* (2003) *Prog. Neuropsychopharmacol. Biol. Psychiatry* **27**, 555.   **2**. Bar-Am, Amit, Weinreb, Youdim & Mandel (2010) *J. Alzheimer's Dis.* **21**, 361.   **3**. Weinreb, Amit, Bar-Am & Youdim (2007) *Ann. N. Y. Acad. Sci.* **1122**, 155.   **4**. Weinstock, Poltyrev, Bejar & Youdim (2002) *Psychopharmacol.* **160**, 318.   **5**. Weinstock, Gorodetsky, Poltyrev, *et al.* (2003) *Prog. Neuropsychopharmacology & Biological Psychiatry* **27**, 555.   **6**. Weinreb, Amit, Bar-Am & Youdim (2012) *Curr. Drug Targets* **13**, 483.

Lambrolizumab

This humanized monoclonal IgG$_4$κ isotype antibody (MW = 146.3 kDa; CAS 1374853-91-4), also known as MK-3475, targets the programmed death-1 receptor (PD-1), a negative regulator of T-cell effector mechanisms that limits immune responses against cancer. The variable region sequences of a very-high-affinity mouse antihuman PD-1 antibody (K_d = 28 pM) were grafted into a human IgG$_4$ immunoglobulin with a stabilizing S228P Fc alteration (1). The IgG4 immunoglobulin subtype does not engage Fc receptors or activate complement, thus avoiding cytotoxic effects of the antibody when it binds to the T cells that it is intended to activate. In advanced stage melanoma patients, including those with disease progression while on ipilimumab, lambrolizumab treatment resulted in a high rate of durable tumor regression (1). Abrogating an immune-system checkpoint to limit the antitumor activity of preexisting tumor-specific cytotoxic T cells underscores the importance of focusing on immune regulatory events for cancer therapy (1). In April, 2013, Lambrolizumab was granted breakthrough therapy status to the experimental drug.   **1**. Hamid, Robert, Daud, *et al.* (2013) *New Engl. J. Med.* **369**, 134.

LamB Signal Peptide

This basic oligopeptide (MW = 2545.22 g/mol; *Sequence*: MMITLRKLPLAVAVAAGVMSAQAMA; Isoelectric Point = 11.0), corresponds to the signal sequence of the *Escherichia coli* outer membrane protein, LamB, inhibits the signal-recognition-particle GTPase activity (K_i = 5-8 µM), as does MMITLRKRRRKLPLAVAVAAGVMSAQAMA (MW = 2985.77; Isoelectric Point = 12.3; K_i =3 µM). With fully a third of the *E. coli* proteome comprised of proteins making up the cytoplasmic membrane, periplasm, or outer membrane, these peptides are useful probes of whether an *E. coli* protein is processed by the general secretion (Sec) pathway.   **1**. Swain & Gierasch (2001) *J. Biol. Chem.* **276**, 12222.

Lamivudine

This nucleoside RTI and antiviral (FW = 229.26 g/mol; CAS 134678-17-4; Symbol: 3TC; IUPAC Name: 4-amino-1-[(2*R*,5*S*)-2-(hydroxy-methyl)-1,3-oxathiolan-5-yl]-1,2-dihydropyrimidin-2-one), also known as Epivir® and Epivir-HBV®, the (–)-enantiomer of 2'-deoxy-3'-thiacytidine, inhibits HIV reverse transcriptase and prevents hepatitis B virus (HBV) infection by likewise targeting the HBV reverse transcriptase (1-7). Within the cell, lamivudine is converted to its triphosphate, *or* 3TC-TP (*See Lamivudine 5'-Triphosphate*), which can act (a) as a reverse transcriptase chain terminator, after incorporation into the growing HBV DNA chain and (b) as a competitive inhibitor of deoxycytidine triphosphate (dCTP) incorporation at the level of the DNA polymerase. 3TC-TP inhibits viral DNA synthesis, but without effect on mitochondrial DNA synthesis. Lamivudine also interrupts the recycling of virions to the nucleus and suppress formation of covalently closed circular DNA (*or* cccDNA). When receiving lamivudine for 3 years to treat chronic hepatitis B, 67-75% of patients develop drug-resistant B-domain (L528M), C-domain (M552I), or M552V mutations in HBV polymerase (8). *In vitro* studies show resistance in single-drug exposure of L528M-, M552I-, M552V-, L528M/M552I-, or L528M/M552V-containing hepatoma cells to lamivudine, adefovir, entecavir, racivir, emtricitabine, L-D4FC, clevudine, D-DAPD, or (-)-carbovir (8). Significantly, only adefovir and entecavir are effective against all five HBV mutants, and higher doses of these compounds were necessary to inhibit the double mutants compared with the single mutants. The B-domain mutation (L528M) of HBV polymerase not only restores the replication competence of C-domain mutants, but also increases resistance to nucleoside analogues. (**See also** *Adefovir & Adefovir Dipivoxil; Entecavir; Telbivudine; /Tenofovir; Emtricitabine*) **Target(s):** deoxycytidine kinase (1); deoxynucleoside kinase (1); feline immunodeficiency virus reverse transcriptase, via the 5'-triphosphate (2); feline leukemia virus reverse transcriptase, via the 5'-triphosphate (2); HBV reverse transcriptase, via the 5'-triphosphate (3,4); HIV reverse transcriptase, via the 5'-triphosphate (5-7); RNA-directed DNA polymerase, via the 5' triphosphate (2-7). **1**. Johansson, Van Rompay, Degrève, Balzarini & Karlsson (1999) *J. Biol. Chem.* **274**, 23814. **2**. Operario, Reynolds & Kim (2005) *Virology* **335**, 106. **3**. Keam & Scott (2002) *Paediatr. Drugs* **4**, 687. **4**. Chang, Skalski, Zhou & Cheng (1992) *J. Biol. Chem.* **267**, 22414. **5**. Balzarini & De Clercq (1996) *Meth. Enzymol.* **275**, 472. **6**. Coates, Cammack, Jenkinson, *et al.* (1992) *Antimicrob. Agents Chemother.* **36**, 733. **7**. El Safadi, Vivet-Boudou & Marquet (2007) *Appl. Microbiol. Biotechnol.* **75**, 723. **8**. Ono, Kato, Shiratori, *et al.* (2001) *J. Clin. Invest.* **107**, 449.

Lamivudine 5'-Triphosphate

This nucleotide analogue (FW = 470.21 g/mol), the (–)-enantiomer of 2'-deoxy-3'-thiacytidine 5'-triphosphate, inhibits DNA polymerase (1), feline immunodeficiency virus reverse transcriptase (2), feline leukemia virus reverse transcriptase (2), HIV reverse transcriptase (1,3), and RNA-directed DNA polymerase (1-3). 3TC 5'-triphosphate is a competitive inhibitor with respect to dCTP in the RNA-dependent DNA polymerase activity (K_i = 10.6 to 1.24 μM, depending on the template and primer used). The DNA-dependent DNA polymerase activity is 50% inhibited at 23 μM 3TC 5'-triphosphate concentration (1). **1**. Hart, Orr, Penn, *et al.* (1992) *Antimicrob. Agents Chemother.* **36**, 1688. **2**. Operario, Reynolds & Kim (2005) *Virology* **335**, 106. **3**. El Safadi, Vivet-Boudou & Marquet (2007) *Appl. Microbiol. Biotechnol.* **75**, 723.

Lamotrigine

This triazine (FW = 256.09 g/mol; CAS 84057-84-1), named systematically as 6-(2,3-dichlorophenyl)-1,2,4-triazine-3,5-diamine, is an anticonvulsant drug that has been used in the treatment of bipolar disorders. **Primary Mode of Action:** Lamotrigine acts presynaptically on voltage-gated sodium channels to decrease glutamate release. Its antiepileptic effects are, at least in part, due to use- and voltage-dependent modulation of fast voltage-

dependent sodium currents. **Key Pharmacokinetic Parameters:** *See* Appendix II in Goodman & Gilman's THE PHARMACOLOGICAL BASIS OF THERAPEUTICS, 12th Edition (Brunton, Chabner & Knollmann, eds.) McGraw-Hill Medical, New York (2011). **Target(s):** Ca²⁺ channels (1-3); glutamate release (4-6); Na⁺ channels (3,4,7,8). **1**. Stefani, Spadoni, Siniscalchi & Bernardi (1996) *Eur. J. Pharmacol.* **307**, 113. **2**. Wang, Huang, Hsu, Tsai & Gean (1996) *Neuroreport* **7**, 3037. **3**. Stefani, Spadoni & Bernardi (1997) *Exp. Neurol.* **147**, 115. **4**. Cheung, Kamp & Harris (1994) *Epilepsy Res.* **13**, 107. **5**. Leach, Marden & Miller (1986) *Epilepsy* **27**, 490. **6**. Wang, Sihra & Gean (2001) *Neuroreport* **12**, 2255. **7**. Kuo & Lu (1997) *Brit. J. Pharmacol.* **121**, 1231. **8**. Kuo (1998) *Mol. Pharmacol.* **54**, 712.

Laninamivir Octanoate

This viral neuraminidase (NA) -directed prodrug ($FW_{octanoyl-ester}$ = 514.62 g/mol; $FW_{alcohol}$ = 472.53 g/mol CAS 203120-17-6), also known as CS-8958 and systematically named (4S,5R,6R)-5-acetamido-4-carbamimidamido-6-[(1R,2R)-3-(*O*)-octanoyl-2-methoxypropyl]-5,6-dihydro-4*H*-pyran-2-carboxylic acid, is the bioavailable ester that is metabolically hydrolyzed (▷ indicates the scissile bond) by *S*-formylglutathione hydrolase (esterase D) and acyl-protein thioesterase 1 (APT1) to generate the active drug used in treatment and prophylaxis of Influenza virus A and Influenza virus B infections. These inhibitors bind to the NA surface glycoprotein on newly formed virus particles, preventing their release from the host cell. The two widely used NA inhibitors are the inhaled drug zanamivir, *or* Relenza® (10 mg/dose, twice daily) and the oral drug oseltamivir phosphate, *or* Tamiflu® (75 mg/dose, twice daily). By contrast, single daily administration of CS-8958 is highly effective against Influenza viruses A and B. **1**. Yamashita, Tomozawa, Kakuta, *et al.* (2009) *Antimicrob. Agents Chemother.* **53**, 186. **2**. Kubo, Tomozawa, Kakuta, Tokumitsu & Yamashita (2010) *Antimicrob. Agents Chemother.* **54**, 1256.

Lankamycin

This macrolide antibiotic (FW = 920.53 g/mol; CAS 30042-37-6) from *Streptomyces violaceoniger* from the tropical forests of Sri Lanka, also known as kujimycin B, inhibits protein biosynthesis (1). Lankamycin binds to ribosomes in a similar manner to erythromycin; however, when in complex with lankacidin, lankamycin is posed so that it interacts with lankacidin in the adjacent ribosomal binding site. When paired, lankacidin and lankamycin inhibit the peptidyl transferase center and nascent peptide exit tunnel, respectively (2). **1**. Pestka (1974) *Meth. Enzymol.* **30**, 261. **2**. Belousoff, Shapira, Bashan, *et al.* (2011) *Proc. Natl. Acad. Sci. U.S.A.* **108**, 2717.

Lanosterol

This cholesterol precursor (FW = 426.73 g/mol; M.P. = 138-140°C), also known as kryptosterol and systematically named (3β)-lanosta-8,24-dien-3-ol, and 4,4,14α-trimethylzymosterol, is also required for the synthesis of other steroids in animals. (In plants, lanosterol's role is fulfilled by cycloartenol.). Lanosterol is present at high concentrations in lanolin. **Target(s):** 3-hydroxy-3-methylglutaryl-CoA reductase (1); sterol acyltransferase, cholesterol acyltransferase (2). **1.** Berg, Draber, von Hugo, Hummel & Mayer (1981) *Z. Naturforsch. [C]* **36**, 798. **2.** Billheimer (1985) *Meth. Enzymol.* **111**, 286.

Lanthanum & Lanthanum Ions

Ions of this group 3 (or group IIIA) element (Symbol = La; Atomic Number = 57; Atomic Weight = 138.905), have valences of 1 of 3 (ionic radii for La$^+$ and La^{3+} are 1.39 and 1.061 Å, respectively). There are two naturally occurring isotopes of lanthanum: ^{138}La (natural abundance = 0.0902%) and ^{139}La (99.9098%). Isotope-138 is radioactive, with a half-life of 1.06 x 10^{11} years. Another radioisotope, ^{137}La, has a half-life of 8 x 10^4 years. With an ionic radius near that of Ca^{2+}, tervalent lanthanum is often used as calcium channel blockers in the study of a number of calcium-dependent processes. Lanthanum ions also alter the stability of phospho-enzyme intermediates and compete with magnesium ions, inhibiting certain ATPases. When exposed to elevated levels of lanthanum ions, mammals tend to accumulate the metal in their bones. In addition, the tree *Carya* spp. has been reported to be a lanthanum accumulator. **Target(s):** acetylcholinesterase (1); acid phosphatase (2); aryldialkyl-phosphatase, *or* paraoxonase (3,4); arylesterase (5); calcium channels (6-9); catechol *O*-methyltransferase (10); Ca^{2+}-transporting ATPase (9,11-15); diphosphate:serine phosphotransferase (16); ferroxidase (35,36); β-fructofuranosidase (17); β-D-fucosidase (18); fumarylacetoacetase (22); β-glucosidase (19); glutamate formimidoyl-transferase (34); histamine release (20); 3-ketovalidoxylamine C-N-lyase (21); micrococcal nuclease (23); Na$^+$/K$^+$-exchanging ATPase (24); norepinephrine transport (25); phosphatidylinositol diacylglycerol-lyase, *or* phosphatidylinositol-specific phospholipase C (26); 1-phosphatidylinositol 4-kinase (27); phospholipase C (28); phospholipid-translocating ATPase, *or* flippase (29); 3-phosphoshikimate 1-carboxyvinyltransferase, *or* 3-enolpyruvoylshikimate-5 phosphate synthase (30); protein-glutamine γ-glutamyltransferase, *or* transglutaminase (31); xyloglucan:xyloglucosyl transferase (32,33). **1.** K. Marquis & E. E. Black (1985) *Biochem. Pharmacol.* **34**, 533. **2.** Lin, Lee, Li & Chu (1983) *Biochemistry* **22**, 1055. **3.** Gil, Pla, Gonzalvo, Hernández & Villanueva (1993) *Chem. Biol. Interact.* **87**, 149. **4.** Gil, Gonzalvo, Hernández, Villanueva & Pla (1994) *Biochem. Pharmacol.* **48**, 1559. **5.** Lorentz, Wirtz & Weiss (2001) *Clin. Chim. Acta* **308**, 69. **6.** Mliner & Enyeart (1993) *J. Physiol.* **469**, 639. **7.** Beedle, Hamid & Zamponi (2002) *J. Membr. Biol.* **187**, 225. **8.** Mela (1969) *Biochemistry* **8**, 2481. **9.** Carafoli (1992) *J. Biol. Chem.* **267**, 2115. **10.** R. Quiram & R. M. Weinshilboum (1976) *Biochem. Pharmacol.* **25**, 1727. **11.** J. Herscher & A. F. Rega (1997) *Ann. N. Y. Acad. Sci.* **834**, 407. **12.** Nakazawa, Liu, Inoue & Ohno (1997) *Eur. J. Pharmacol.* **325**, 237. **13.** Herscher & Rega (1996) *Biochemistry* **35**, 14917. **14.** Gruner, Diezel, Strunk, *et al.* (1991) *J. Invest. Dermatol.* **97**, 478. **15.** Caroni & Carafoli (1981) *J. Biol. Chem.* **256**, 3263. **16.** Cagen & Friedmann (1972) *J. Biol. Chem.* **247**, 3382. **17.** Massart (1950) *The Enzymes*, 1st ed. (J. B. Sumner & K. Myrbäck, eds.), **1** (part 1), 307. **18.** Giordani & Lafon (1993) *Phytochemistry* **33**, 1327. **19.** Plant, Oliver, Patchett, Daniel & Morgan (1988) *Arch. Biochem. Biophys.* **262**, 181. **20.** Gruner, Sehrt, Muller, *et al.* (1992) *Agents Actions* **36**, 207. **21.** Takeuchi, Asano, Kameda & Matsui (1985) *J. Biochem.* **98**, 1631. **22.** Connors & Stotz (1949) *J. Biol. Chem.* **178**, 881. **23.** Cuatrecasas, Fuchs & Anfinsen (1967) *J. Biol. Chem.* **242**, 1541. **24.** Knudsen, Berthelsen & Johansen (1990) *Brit. J. Pharmacol.* **100**, 453. **25.** Bryan-Lluka & Bonisch (1997) *Naunyn Schmiedebergs Arch. Pharmacol.* **355**, 699. **26.** Abdel-Latif, Luke & Smith (1980) *Biochim. Biophys. Acta* **614**, 425. **27.** Collins & Wells (1983) *J. Biol. Chem.* **258**, 2130. **28.** McDonald & Mamrack (1989) *Biochemistry* **28**, 9926. **29.**

Auland, Morris & Roufogalis (1994) *Arch. Biochem. Biophys.* **312**, 272. **30.** Steinrücken & Amrhein (1984) *Eur. J. Biochem.* **143**, 341. **31.** Achyuthan, Mary & Greenberg (1989) *Biochem. J.* **257**, 331. **32.** Takeda & Fry (2004) *Planta* **219**, 722. **33.** Fry, Smith, Renwick, *et al.* (1992) *Biochem. J.* **282**, 821. **34.** Miller & Waelsch (1957) *J. Biol. Chem.* **228**, 397. **35.** Frieden & Hsieh (1976) *Adv. Enzymol. Rel. Areas Mol. Biol.* **44**, 187. **36.** Huber & Frieden (1970) *J. Biol. Chem.* **245**, 3979.

Lansoprazole

This proton-pump inhibitor prodrug, *or* PPI (FW = 369.36 g/mol; CAS 103577-45-3), also named AG-1749, Prevacid® and (*RS*)-2-([3-methyl-4-(2,2,2-trifluoroethoxy)pyridin-2-yl]methylsulfinyl)-1*H*-benzo[*d*]imidazole, reduces gastric acid production by targeting (H$^+$,K$^+$)-adenosine triphosphatase, *or* HK-ATPase (1). Lansoprazole inhibits acid formation after being transformed into its active forms within the acidic compartment of the parietal cells.

Lansoprazole

Inhibition of acid formation is reversible in parietal cells, and *de novo* synthesis of HK-ATPase does not participate in the recovery from the inhibition. Endogenous glutathione is involved in the recovery process and also affects the inhibitory action. Both the (+)-and the (-)-enantiomer of lansoprazole inhibit the acid formation stimulated by dibutyryl cyclic AMP (db-cAMP) in isolated canine parietal cells in a concentration-dependent manner with IC$_{50}$ values of 59 and 82 nM, respectively (3). The enantiomers showed concentration-dependent inhibition of HK-ATPase with IC$_{50}$ values of 4.2 and 5.2 µM, respectively (3). On the other hand, the IC$_{50}$ values of lansoprazole for db-cAMP-stimulated acid formation and HK-ATPase were 59 nM and 2.1 µM, respectively (3). Such results suggest that the two enantiomers of lansoprazole have antisecretory action due to inhibition of HK-ATPase. Lansoprazol's plasma elimination half-life (1.5 hours) is not proportional to the duration of its effects on gastric acid suppression. Indeed, when used for a day or more, its effects last for over 24 hours. **1.** Nagaya, Satoh, Kubo & Maki (1989) *J. Pharmacol. Exp. Ther.* **248**, 799. **2.** Nagaya, Satoh & Maki (1990) *J. Pharmacol. Exp. Ther.* **252**, 1289. **3.** Nagaya, Inatomi, Nohara & Satoh (1991) *Biochem. Pharmacol.* **42**, 1875.

β-Lapachone

This cytotoxic agent (FW = 242.27 g/mol; CAS 4707-32-8; Solubility: >25 mg/mL DMSO at 5 °C), systematically named 3,4-dihydro-2,2-dimethyl-2*H*-naphtho[1,2-b]pyran-5,6-dione, is a DNA topoisomerase I inhibitor.

Generation of Reactive Oxygen Species: β-lapachone is redox-active, and its structural similarities to menadione suggest a common mechanism of action (1) that targets NQO1 NAD(P)H quinone dehydrogenase 1 and alters the NAD(P)H metabolism through the formation hydroquinone and semiquinone species as well as reactive oxygen species (ROS):

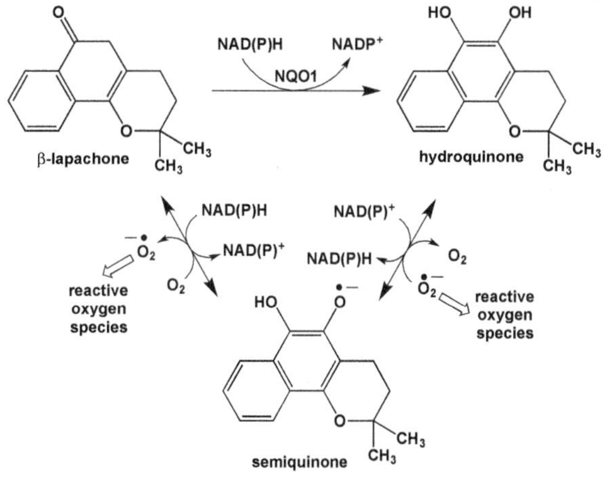

Biochemical studies suggest that reduction of β-lapachone by NQO1 leads to a futile cycling between the quinone and hydroquinone forms, with a concomitant loss of reduced NAD(P)H. **DNA Damage:** Both quinones form adducts with mercaptoethanol, and β-lapachone is 10-times more reactive, and there is an apparent correlation between the rates of the adduct formation with thiols and of the topoisomerase II-poisoning activity of the aforementioned quinones (1). β-Lapachone induces protein-linked DNA breaks in the presence of purified human DNA topoisomerase IIα, without intercalating into DNA or inhibiting topoisomerase II or ligation by mammalian or T4 DNA ligases. Its action is ATP-independent and involves the formation of reversible cleavable complexes. A 4-hour post-treatment with 4 μM β-lapachone was previously shown to enhance the lethality of X-rays against human laryngeal epidermoid carcinoma (HEp-2) cells (2). It was subsequently demonstrated that β-lapachone (a) activates the DNA-unwinding activity of topoisomerase I, (b) inhibits the fast component of Potentially Lethal Damage Repair (PLDR) carried out by HEp-2 cells when present during or immediately following X-irradiation, (c) specifically and synergistically enhances the cytotoxic effects of DNA-damaging agents which induce DNA strand incisions, such as neocarzinostatin or X-rays, against a radioresistant human malignant melanoma (U1-Mel) cell line, (d) does not synergistically potentiate melphalan-induced lethality against U1-Mel cells but inhibits survival recovery and increases sister chromatid exchanges, and (e) does not further enhance the lethal effects of X-rays following prolonged drug exposures, indicating that β-lapachone modifies initially created DNA lesions or inhibits lesion repair, but does not create lethal lesions by itself (3). NQO1 enhances the cytotoxic effects of β-lapachone, and NQO1 gene expression levels directly correlate with sensitivity to a 4-hour pulse of β-lapachone in a panel of breast cancer cell lines (4). β-Lapachone accelerates the DNA-unwinding activity of topoisomerase I derived from avian erythrocytes, calf thymus, or HEp-2 cells. β-Lapachone activates apoptosis in a number of cell lines (4). **Dicoumarol Cytoprotection:** The NQO1 inhibitor, dicoumarol, significantly protects NQO1-expressing cells from all aspects of β-lapachone toxicity. Stable transfection of the NQO1-deficient cell line, MDA-MB-468, with an NQO1 expression plasmid increases apoptotic responses and lethality after β-lapachone exposure. Dicoumarol also blocks apoptotic responses and lethality. In addition, the activation of a cysteine protease, which has characteristics consistent with the neutral calcium-dependent protease, calpain, is observed after β-lapachone treatment (5). NQO1 expression directly correlated with sensitivity to a 4-hour pulse of β-lapachone in a panel of breast cancer cell lines, and the NQO1 inhibitor, dicoumarol, significantly protected NQO1-expressing cells from all aspects of β-lapachone toxicity. Stable transfection of the NQO1-deficient cell line, MDA-MB-468, with an NQO1 expression plasmid increases apoptotic responses and lethality after β-lapachone exposure. Dicoumarol blocks both the apoptotic responses and lethality. This finding is the first definitive elucidation of an intracellular target for β-lapachone in tumor cells. Such studies suggest that NQO1 may be exploited for gene therapy, radiotherapy, and/or chemopreventive interventions, since the enzyme is elevated in a number of tumor types (*i.e.*, breast and lung) and during neoplastic transformation. **NAD(P) Depletion by β-Lapachone:** The aforementioned β-lapachone-induced DNA damage also hyperactivates poly(ADP-ribosyl)transferase-1 (PARP1), an enzyme that regulates molecular events involved in cell recovery. As a consequence, cellular stores of NAD$^+$ are greatly depleted by β-lapachone treatment. Finally, β-Lapachone is also a specific probe for ligand binding to the glucocorticoid receptor (6). Although structurally unlike any glucocorticoid, β-lapachone interacts rapidly and competitively with the ligand-binding site of unpurified and highly purified glucocorticoid receptor (6). 1. Frydman, Marton, Sun, *et al.* (1997) *Cancer Res.* **57**, 620. 2. Boothman, Greer & Pardee (1988) *Cancer Res.* **47**, 5361. 3. Boothman, Trask & Pardee (1989) *Cancer Res.* **49**, 605. 4. Pink, Planchon, Tagliarino, *et al.* (2000) *J. Biol. Chem.* **275**, 5416. 5. Pink, Planchon, Tagliarino, *et al.* (2000) *J. Biol. Chem.* **275**, 5416. 6. Schmidt, Miller-Diener & Litwack (1984) *J. Biol. Chem.* **259**, 9536.

Lapatinib

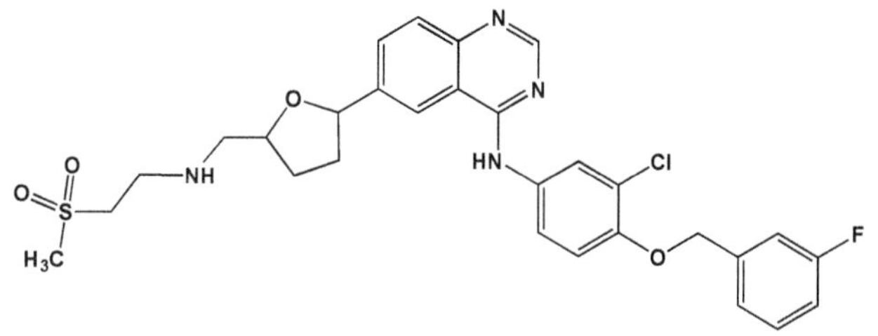

This orally EGFR/HER2 tyrosine kinase inhibitor (F.Wt. = 925.46; CAS 388082-77-7, 231277-92-2 (free-base), 1187538-35-7 (4-methylbenzene-sulfonate salt); Solubility: 185 mg/mL DMSO, <1 mg/mL H$_2$O), also known as GW572016, GW2016, Tykerb®, Tyverb®, and *N*-(4-(3-fluorobenzyloxy)-3-chlorophenyl)-6-(5-((2-(methyl-sulfonyl)ethylamino)methyl)furan-2-yl)quinazolin-4-amine, di-4-methylbenzene sulfonate, blocks phosphorylation of purified EGFR and ErbB-2 tyrosine kinase domains, with IC$_{50}$ of 10.2 and 9.8 nM, respectively (1). **Mode of Action:** GW2016 treatment inhibited tumor xenograft growth of the HN5 and BT474 cells in a dose-responsive manner at 30 and 100 mg/kg orally, twice daily, with complete inhibition of tumor growth at the higher dose. Lapatinib markedly reduces tyrosine phosphorylation of EGFR and erbB2, inhibiting the activation of Erk1/2 and AKT which are important downstream effectors of proliferation and cell survival, respectively. Complete inhibition of activated AKT in erbB2 overexpressing cells correlated with a 23-fold increase in apoptosis compared with vehicle controls. EGF, often elevated in cancer patients, did not reverse the inhibitory effects of GW572016 (2). Client protein kinases are recruited to the Hsp90 molecular chaperone by Cdc37, which simultaneously binds Hsp90 and kinases as well as regulates the Hsp90 chaperone cycle (3). Significantly, Cdc37 binding to protein kinases is itself antagonized by ATP-competitive kinase inhibitors, such as lapatinib, and in cancer cells, such inhibitors deprive access of oncogenic kinases, such as B-Raf and ErbB2, to Hsp90-Cdc37, resulting in degradation of these protein kinases (3). Lapatinib-induced cell growth inhibition and G$_1$ cell-cycle arrest in HER2-overexpressing human breast cancer cells depends on upregulated expression of p27 (Kip1), which is mediated through both transcriptional and post-translational mechanisms (4). **Metabolism & Distribution:** Lapatinib is metabolized by CYP3A4/5 to yield an *O*-debenzylated metabolite that undergoes further oxidation to a reactive quinone imine (5). The *O*-debenzylated metabolite is significantly more cytotoxic than lapatinib, and pre-treatment with L-buthionine sulfoximine (25 μM) to deplete intracellular glutathione markedly enhances lapatinib cytotoxicity. In mice, lapatinib levels were 4x higher in tumor than blood, with a 4x longer half-life (6). Tumor concentrations exceeded the in vitro IC$_{90}$ (~ 900 nM or 500 ng/mL) for inhibition of HER2 phosphorylation throughout the 12-hour dosing interval. In patients, tumor levels were 6- and 10-fold higher with QD and BID dosing, respectively, compared to plasma trough levels. The relationship between tumor and plasma concentration was complex, indicating multiple determinants (6). 1. Rusnak, Lackey, Affleck, *et al.* (2001) *Mol. Cancer Ther.* **1**, 85. 2. Xia, Mullin, Keith, *et al.* (2002) *Oncogene* **21**, 6255. 3. Polier, Samant, Clarke, *et al.* (2013) *Nature Chem. Biol.* **9**, 307. 4. Tang, Wang, Strom, Gustafsson & Guan (2013) *Cell Cycle* **12**, 2665. 5. Hardy, Wahlin, Papageorgiou, *et al.* (2013) *Drug Metab. Dispos.* **42**, 162. 6. Spector, Robertson, Bacus, *et al.* (2015) *PLoS One* **10**, e0142845.

Largazole

This novel cytotoxic cyclodepsipeptide (FW = 608.83 g/mol; CAS 1009815-87-5) from a marine cyanobacterium (*Symploca* sp., collected in the Florida Keys) exhibits potent antiproliferative activity. In studies with transformed and nontransformed epithelial cells, largazole gave respective growth-inhibition (GI$_{50}$) values of 7.7 nM and 122 nM; similarly, with transformed and nontransformed fibroblast cells, it gave GI$_{50}$ values of 55 nM and 480 nM, respectively (1). Largazole possesses a densely assembled structure, with a rare 4-methylthiazoline linearly fused to a thiazole in its cyclic core and a hitherto undescribed 3-hydroxy-7-mercaptohept-4-enoic acid unit incorporated in an ester, thioester, and amide framework (1). Enantioselective total synthesis of (+)-largazole has been reported (2). Largazole is a potent inhibitor of class I histone deacetylases (HDACs) (1,3), a potent inducer of gene expression in latent Epstein-Bar Virus (EBV) (3), and a sensitizing agent that increases the susceptibility of lymphoma cells to nucleoside antiviral agents (3). **1**. Taori, Paul & Luesch (2008) *J. Am. Chem. Soc.* **130**, 1806. **2**. Ghosh & Kulkarni (2008) *Org. Lett.* **10**, 3907. **3**. Ghosh, Perrine, Williams & Faller (2012) *Blood* **119**, 1008.

Lasalosid A

This ionophorus antibiotic (FW$_{free-acid}$ = 590.80 g/mol), also known as X-537A, from certain strains of *Streptomyces lasaliensis* facilitates the exchange of monovalent and divalent cations for protons. Nonovalent cations are transported about ten-times more effectively than divalent cations (1). Lasalosid A also will transport catecholamines across membranes. The Ca^{2+}-dependent ATPase of the sarcoplasmic reticulum is strongly inhibited by lasolosid A. **Target(s):** ATPase (1,2); Ca^{2+}-dependent ATPase (3-5); oxidative phosphorylation (2); protein tyrosine kinase (6). **1**. Pressman (1976) *Ann. Rev. Biochem.* **45**, 501. **2**. Reed (1979) *Meth. Enzymol.* **55**, 435. **3**. Kawashima, Hara & Kanazawa (1990) *J. Biol. Chem.* **265**, 10993. **4**. Hasselbach, Ludi & Migala (1983) *Eur. J. Biochem.* **132**, 9. **5**. Hasselbach, Ludi & Migala (1982) *Z. Naturforsch. [C]* **37**, 1290. **6**. Petukhov, Chibalin, Kovalenko, Bulargina & Severin (1991) *Biokhimiia* **56**, 2077.

Lasofoxifene

This orally active selective estrogen receptor modulator, *or* SERM (FW = 413.55 g/mol; CAS 180916-16-9), also named (5R,6S)-6-phenyl-5-[4-(2-pyrrolidin-1-ylethoxy)phenyl]-5,6,7,8-tetrahydronaphthalen-2-ol, is a tamoxifen-related drug that competitively inhibits estradiol binding to both ERα and ERβ with high affinity. Its IC$_{50}$ for ERα is 1.5 nM, similar to that of estradiol (4.8 nM) and at least 10-fold tighter than tamoxifen and raloxifene. Lasofoxifene also shows remarkably improved oral bioavailability relative to tamoxifen and raloxifene, a property that may account for its greater potency. *See also Tamoxifen; Toremifene*

LASSBio-1135

This weak COX-2 inhibitor (FW = 286.34 g/mol) blocks capsaicin-elicited currents noncompetitively (IC$_{50}$ = 580 nM) as well as low pH-induced current at 50 μM. LASSBio-1135 also inhibits TNFα release in these cells stimulated by lipopolysaccharide with an IC$_{50}$ of 546 nM by reducing p38 MAPK phosphorylation. **1**. Lima, Silva, Lacerda, *et al.* (2014) *PLoS One* **9**, e99510.

Lassomycin

This ribosomally encoded cyclic peptide antibiotic (FW = 1880.22 g/mol; Isoelectric Point = 12), which possesses an unusual structural fold partially resembling other lasso peptides, binds to a highly acidic region of the ClpC1 ATPase complex, thereby markedly stimulating its ATPase activity without stimulating ClpP1P2-catalyzed protein breakdown. The latter is essential for viability of *Mycobacterium tuberculosis*. Uncoupling of ATPase activity from proteolytic activity accounts for its bactericidal activity. **1**. Gavrish, Sit, Cao, *et al.* (2014) *Chem. Biol.* **21**, 509.

Latrunculin A

This cell-permeant, marine natural product and actin polymerization inhibitor (FW = 421.56 g/mol; CAS 76343-93-6; Abbreviation: LAT-A) from the Red Sea sponge *Latrunculia magnifica* alters eukaryotic cell shape by disrupting microfilament organization and inhibiting the some microfilament-associated processes during fertilization and early development (1-4). **Primary Mode of Action:** Latrunculin-A inhibits the self-assembly of actin filaments (*or* microfilaments) by forming a one-to-one complex with monomeric actin, thereby hindering polymerization-requiring conformational rearrangements in Domains 1 and 2 (5). Latrunculin A has a K_d of 180 nM–220 nM, as determined by binding to pyrene-labeled rabbit skeletal muscle actin. **Preparation:** Stock solutions of Latrunculin-A are prepared at 2-10 mM DMSO and then diluted to 100 μM in Actin Monomer-Stabilizing (*or* "G") Buffer (5 mM Tris (pH 7.8), 2 mM MgCl$_2$, 0.1 mM CaCl$_2$, 0.2 mM ATP, 0.2 mM DTT, and 0.01-0.1% sodium azide). **Cellular Effects:** Exposure of most cells to Latrunculin A or Latrunculin B (0.03 μg/mL for 1-3 hours) induces concentration-dependent changes in cell shape and actin organization, causing complete rounding-up of all cells. Latrunculins also inhibit cytokinesis in synchronized cells, with Latrunculin-A having longer-lasting effects (2,3). Latrunculin-A causes complete disruption of the yeast actin cytoskeleton within 2-5 min, suggesting that although yeast are nonmotile, their actin filaments undergo rapid cycles of assembly and disassembly *in vivo* (6). Latrunculin-A potently inducees PMN leukocyte aggregation in a dose- and time-dependent manner over the 12 nM to 60 nM concentration range. At 120 nM, Latrunculin-A causes a 50% loss of actin filaments in PMN leukocyte after 5-min incubation (7). (The effects of Latrunculin-B were similar to those of Latrunculin-A, albeit slightly less potent and effective for shorter periods. Some cells may possess the ability to deactivate Latrunculin-B, most likely by ATP-dependent export.) **1.** Ayscough (1998) *Meth. Enzymol.* **298**, 18. **2.** Coue, Brenner, Spector & Korn (1987) *FEBS Lett.* **213**, 316. **3.** Morton, Ayscough & McLaughlin (2000) *Nature Cell Biol.* **2**, 376. **4.** de Oliveira & Mantovani (1988) *Life Sci.* **43**, 1825. **5.** Rennebaum & Caflisch (2012) *Proteins* **80**, 1998 **6**. Ayscough, Stryker, Pokala *et al.* (1997) *J. Cell. Biol.* **137**, 399. [*Erratum*: *J. Cell. Biol.* **146**, following p. 1201 (1999)] **7.** Oliveira, Chedraoui & Mantovani (1997) *Life Sci.* **61**, 603.

Latrunculin B, *Similar in action to* Latrunculin A

Laulimalide

This marine macrolide and cytoskeleton-directed drug (FW = 514.66 g/mol), originally isolated from the marine sponges *Cacospongia mycofijiensis* from Fiji (1) and a *Hyattella* species in Indonesia (2), is a microtubule-stabilizing agent that inhibits cell-cycle progression (IC$_{50}$ ≈ 0.1 μM). Laulimalide promotes tubulin polymerization by preferentially stabilizing assembled microtubules and is quantitatively comparable to paclitaxel (taxol). Even so, laulimalide does not bind in the taxoid site (3). Significantly, laulimalide enhances assembly synergistically with paclitaxel, as would be expected if these drugs bound at different microtubule sites (4). Stoichiometric amounts of laulimalide and paclitaxel cause extensive tubulin polymerization, with maximum synergy observed at lower temperatures under conditions where paclitaxel is relatively inactive (5). Laulimalide-induced assembly, like paclitaxel-induced assembly, is inhibited by drugs that inhibit tubulin assembly by binding at either the colchicine- or vinblastine-binding site (4). When radiolabeled GTP is present in a reaction mixture with either laulimalide or paclitaxel, nucleotide hydrolysis occurs with incorporation of radiolabeled GDP into polymer (4). Laulimalide is also active in cells over-expressing multi-drug resistance (MDR) P-glycoprotein. Remarkably, eight different total syntheses of (–)-laulimalide reported in the span of less than two years (6). **1.** Quinoa, Kakou & Crews (1988) *J. Org. Chem.* **53**, 3642. **2.** Corley, Herb, Moore, Scheuer & Paul (1988) *J. Org. Chem.* **53**, 3644. **3.** Pryor, O'Brate, Bilcer, *et al.* (2002) *Biochemistry* **41**, 9109. **4.** Gapud, Bai, Ghosh & Hamel (2004) *Mol. Pharmacol.* **66**, 113. **5.** Thompson, Wilson & Purich (1981) *Cell Motil.* **1**, 445. **6.** Crimmins (2002) *Curr. Opin. Drug Discov. & Devel.* **5**, 944.

Laurate (Lauric Acid)

This medium-chain fatty acid (FW$_{free-acid}$ = 200.32 g/mol; pK_a = 4.8; M.P. = 44°C; B.P. = 298.9°C; Solubility: 0.0055 g/100 mL H$_2$O), also called *n*-dodecanoic acid, is widely distributed in animals and plants. (Note: When heated to 60 °C in the presence of sufficient concentaration (0.1% w/v) of Sodium Dodecyl Sulfate (*or* SDS), many proteins are denatured, with consequential loss of enzyme catalysis or biological function.) **Target(s):** acetyl-CoA carboxylase (1,2); alcohol dehydrogenase (3,4); D-amino-acid oxidase (5,6); ceramidase (at pH 5) (7); chondroitin AC lyase, weakly inhibited (8); chondroitin B lyase, weakly inhibited (8); cyclooxygenase (9); cytosol alanyl aminopeptidase (30); esterase (10); 7-ethoxycoumarin *O*-deethyase (cytochrome P450) (11); firefly luciferase, IC$_{50}$ = 1.2 μM (19,20,35,36); formaldehyde dehydrogenase (12); β-galactoside α-2,6 sialyltransferase (13,33); glucokinase (14); glutathione *S*-transferase (15); glycylpeptide *N* tetradecanoyltransferase (34); hexokinase (14,16,17,32); hyaluronate lyase (8); *S* (hydroxymethyl)glutathione dehydrogenase (12); lecithin:cholesterol acyltransferase, *or* phosphatidylcholine:cholesterol acyltransferase (18); lysozyme (21); monoamine oxidase (22);

pectinesterase, weakly inhibited (23); peptidyl-dipeptidase A, *or* angiotensin I-converting enzyme (24); phosphofructokinase (14,16,17); pyruvate kinase (17,25); sphingomyelin phosphodiesterase, *or* sphingomyelinase (31); steroid 5α-reductase (26-28); L-threonine dehydrogenase (29). **1.** Miller & Levy (1975) *Meth. Enzymol.* **35**, 11. **2.** Levy (1963) *Biochem. Biophys. Res. Commun.* **13**, 267. **3.** Sund & Theorell (1963) *The Enzymes*, 2nd ed., 7, 25. **4.** Winer & Theorell (1960) *Acta Chem. Scand.* **14**, 1729. **5.** Dixon & Kleppe (1965) *Biochim. Biophys. Acta* **96**, 368. **6.** Brachet, Carreira & Puigserver (1990) *Biochem. Int.* **22**, 837. **7.** Gatt (1966) *J. Biol. Chem.* **241**, 3724. **8.** Suzuki, Terasaki & Uyeda (2002) *J. Enzyme Inhib. Med. Chem.* **17**, 183. **9.** Henry, Momin, Nair & Dewitt (2002) *J. Agric. Food Chem.* **50**, 2231. **10.** Weber & King (1935) *J. Biol. Chem.* **108**, 131. **11.** Muller-Enoch, Fintelmann, Nicolaev & Gruler (2001) *Z. Naturforsch.* **56**, 1082. **12.** Sanghani, Robinson, Bosron & Hurley (2002) *Biochemistry* **41**, 10778. **13.** Bador, Morelis & Louisot (1984) *Biochimie* **66**, 223. **14.** Lea & Weber (1968) *J. Biol. Chem.* **243**, 1096. **15.** Young & Briedis (1989) *Biochem. J.* **257**, 541. **16.** Lea, Weber & Morris (1976) *Oncology* **33**, 205. **17.** Brown & Smith (1977) *Enzyme* **22**, 357. **18.** Bonelli & Jonas (1993) *Biochim. Biophys. Acta* **1166**, 92. **19.** Ueda & Suzuki (1998) *Biophys. J.* **75**, 1052. **20.** Takehara, Kamaya & Ueda (2005) *Biochim. Biophys. Acta* **1721**, 124. **21.** Jollès (1960) *The Enzymes*, 2nd ed., **4**, 431. **22.** McEwan, Sasaki & Jones (1969) *Biochemistry* **8**, 3952. **23.** Miller & McColloch (1959) *Biochem. Biophys. Res. Commun.* **1**, 91. **24.** Oshima & Nagasawa (1977) *J. Biochem.* **81**, 57. **25.** Kayne (1973) *The Enzymes*, 3rd ed., **8**, 353. **26.** Weisser, Tunn, Behnke & Krieg (1996) *Prostate* **28**, 300. **27.** Niederprum, Schweikert, Thuroff & Zanker (1995) *Ann. N. Y. Acad. Sci.* **768**, 227. **28.** Raynaud, Cousse & Martin (2002) *J. Steroid Biochem. Mol. Biol.* **82**, 233. **29.** Guerranti, Pagani, Neri, *et al.* (2001) *Biochim. Biophys. Acta* **1568**, 45. **30.** Garner & Behal (1977) *Arch. Biochem. Biophys.* **182**, 667. **31.** Barnholz, Roitman & Gatt (1966) *J. Biol. Chem.* **241**, 3731. **32.** Morris, DeBruin, Yang, *et al.* (2006) *Eukaryot. Cell* **5**, 2014. **33.** Hickman, Ashwell, Morell, van den Hamer & Scheinberg (1970) *J. Biol. Chem.* **245**, 759. **34.** Selvakumar, Lakshmikuttyamma, Shrivastav, *et al.* (2007) *Prog. Lipid Res.* **46**, 1. **35.** Matsuki, Suzuki, Kamaya & Ueda (1999) *Biochim. Biophys Acta* **1426**, 143. **36.** Takehara, Kamaya & Ueda (2005) *Biochim. Biophys. Acta* **1721**, 124.

Lauroyl-CoA

This thiolester (FW = + 767.54), also known as dodecanoyl-CoA, is a substrate for many acyltransferases. **Target(s):** acetyl-CoA carboxylase, K_i = 74 μM for rat liver enzyme (1); 1-alkyl-*sn* glycero-3-phosphate acetyltransferase (2); ATP:citrate lyase, *or* ATP-citrate synthase (3,4); carnitine *O*-octanoyl-transferease, also alternative substrate (5); citrate (*si*)-synthase (6); dodecenoyl-CoA Δ-isomerase (7); glucose-6-phosphatase (8,9); glucose-6-phosphate dehydrogenase (10); glycylpeptide *N*-tetradecanoyl-transferase, *or* protein *N*-myristoyltransferase (11); phosphoketolase (12); stearoyl-CoA 9-desaturase (13,14). **1.** Bortz & Lynen (1963) *Biochem. Z.* **337**, 505. **2.** Baker & Chang (1995) *J. Neurochem.* **64**, 364. **3.** Shashi, Bachhawat & Joseph (1990) *Biochim. Biophys. Acta* **1033**, 23. **4.** Boulton & Ratledge (1983) *J. Gen. Microbiol.* **129**, 2863. **5.** Solberg (1974) *Biochim. Biophys. Acta* **360**, 101. **6.** Måhlén (1972) *Eur. J. Biochem.* **29**, 60. **7.** Stoffel & Ecker (1969) *Meth. Enzymol.* **14**, 99. **8.** Fulceri, Gamberucci, Scott, *et al.* (1995) *Biochem. J.* **307**, 391. **9.** Methieux & Zitoun (1996) *Eur. J. Biochem.* **235**, 799. **10.** Eger-Neufeldt, Teinzer, Weiss & Wieland (1965) *Biochem. Biophys. Res. Commun.* **19**, 43. **11.** Glover, Goddard & Felsted (1988) *Biochem. J.* **250**, 485. **12.** Whitworth & Ratledge (1977) *J. Gen. Microbiol.* **102**, 397. **13.** Joshi, Prasad & Sreekrishna (1981) *Meth. Enzymol.* **71**, 252. **14.** Prasad & Joshi (1979) *J. Biol. Chem.* **254**, 6362.

N-Lauroylsarcosine and *N*-Lauroylsarcosinate

This anionic surfactant (FW$_{sodium-salt}$ = 293.38 g/mol), also known as Sarcosyl, Sarkosyl, Gardol, and *N*-dodecanoylsarcosine, is a milder denaturing agent than sodium dodecyl sulfate and is frequently used to isolate DNA and RNA as well as to dissociate nucleosomes and ribosomes.

Target(s): ceramide kinase (1); 1,3-β-glucan synthase (2); α-glucosaminide *N*-acetyltransferase (3); guanylate cyclase (4); herpes simplex virus-1 (5); IgA-specific metalloendopeptidase (6); lipase, *or* triacylglycerol lipase (7); phenylacetone monooxygenase (8); phorbol diester hydrolase (9,10); protein xylosyltransferase (11); RNA polymerase II (12-15); trehalose phosphatase (16). **1.** Van Overloop, Gijsbers & Van Veldhoven (2006) *J. Lipid Res.* **47**, 268. **2.** Beaulieu, Tang, Yan, *et al.* (1994) *Antimicrob. Agents Chemother.* **38**, 937. **3.** Baum & Rome (1987) *Meth. Enzymol.* **138**, 607. **4.** Fleischman & Denisevich (1979) *Biochemistry* **18**, 5060. **5.** Piret, Roy, Gagnon, *et al.* (2002) *Antimicrob. Agents Chemother.* **46**, 2933. **6.** Simpson, Hausinger & Mulks (1988) *J. Bacteriol.* **170**, 1866. **7.** Saxena, Davidson, Sheoran & B. Giri (2003) *Process Biochem.* **39**, 239. **8.** Fraaije, Kamerbeek, Heidekamp, Fortin & Janssen (2004) *J. Biol. Chem.* **279**, 3354. **9.** Shoyab, Warren & Todaro (1981) *J. Biol. Chem.* **256**, 12529. **10.** Kadner, Katz, Levitz & Finlay (1985) *J. Biol. Chem.* **260**, 15604. **11.** Casanova, Kuhn, Kleesiek & Götting (2008) *Biochem. Biophys. Res. Commun.* **365**, 678. **12.** Szentirmay & Sawadogo (1994) *Nucl. Acids Res.* **22**, 5341. **13.** Wu, Jupp, Stenberg, J. A. Nelson & P. Ghazal (1993) *J. Virol.* **67**, 7547. **14.** Kephart, Marshall & Price (1992) *Mol. Cell. Biol.* **12**, 2067. **15.** Izban & Luse (1991) *Genes Dev.* **5**, 683. **16.** Klutts, Pastuszak, Edavana, *et al.* (2003) *J. Biol. Chem.* **278**, 2093.

Lavendustin A

This salicylic acid derivative ($FW_{\text{free-acid}}$ = 381.39 g/mol), also called RG14355 and 5-amino[*N*-(2,5-dihydroxybenzyl)-*N*-(2-hydroxybenzyl)] salicylic acid, from *Streptomyces griseolavendus* and is a cell-permeable inhibitor of a number of protein-tyrosine kinases (1-5). Epidermal growth factor receptor protein-tyrosine kinase, IC_{50} = 11 nM (2) and the p60c-*src* kinase, IC_{50} = 500 nM, are inhibited with little effect on protein kinase A or protein kinase C (IC_{50} > 200 mM). It also inhibits *N*-methyl-D-aspartate-stimulated cGMP production, IC_{50} = 30 nM (6). Lavendustin A exhibits anti-proliferative properties and suppresses the angiogenic action of vascular endothelial growth factor in rats (7). **Target(s):** inositol-trisphosphate 3-kinase (8); protein-tyrosine kinase, non-specific protein-tyrosine kinase (1-5); receptor protein-tyrosine kinase (9). **1.** Eriksson, Toivola, Sahlgren, Mikhailov & Härmälä-Braskén (1998) *Meth. Enzymol.* **298**, 542. **2.** Hsu, Persons, Spada, *et al.* (1991) *J. Biol. Chem.* **266**, 21105. **3.** O'Dell, Kandel & Grant (1991) *Nature* **353**, 558. **4.** Onoda, Iinuma, Sasaki, *et al.* (1989) *J. Nat. Prod.* **52**, 1252. **5.** Ortega, Velasquez & Perez (2005) *Biol. Res.* **38**, 89. **6.** Rodriguez, Quignard, Fagni, Lafon-Cazal & Bockaert (1994) *Neuropharmacology* **33**, 1267. **7.** Hu & Fan (1995) *Brit. J. Pharmacol.* **114**, 262. **8.** Mayr, Windhorst & Hillemeier (2005) *J. Biol. Chem.* **280**, 13229. **9.** Mergler, Dannowski, Bednarz, *et al.* (2003) *Exp. Eye Res.* **77**, 485.

Lazabemide

This pyridinecarboxamide ($FW_{\text{hydrochloride}}$ = 236.10 g/mol), also known as *N*-(2-aminoethyl)-5-chloro-2-pyrinidinecarboxamide and Ro 19-6327, is a potent and selective inhibitor of monoamine oxidase B. **1.** Da Prada, Kettler, Keller, *et al.* (1990) *J. Neural Transm. Suppl.* **29**, 279. **2.** Fernandes & Soares-da-Silva (1990) *J. Pharmacol. Exp. Ther.* **255**, 1309. **3.** Pestana & P. Soares-da-Silva (1994) *Brit. J. Pharmacol.* **113**, 1269. **4.** Henriot, Kuhn, Kettler & Da Prada (1994) *J. Neural Transm. Suppl.* **41**, 321. **5.** Cesura, Galva, Imhof, *et al.* (1989) *Eur. J. Pharmacol.* **162**, 457.

LB30870

This extremely potent anticoagulant (FW = 564.35 g/mol) is a direct benzamidine-based thrombin inhibitor (K_i = 15 pM), with oral bioavailabilities of 43%, 42%, and 15% in rats, dogs, and monkeys, respectively (1). On a gravimetric basis, LB30870 is more effective than enoxaparin, a low-molecular-weight heparin, in the venous thrombosis models in rats and rabbits. ED_{50} values for LB30870, melagatran and enoxaparin are 50 μg/kg/min, 35 μg/kg/min and 200 μg/kg/min, respectively, showing that LB30870 has promising properties for providing effective and safe prophylaxis for venous and arterial thrombosis (2). No significant bleeding was observed with LB30870 at the dose up to two times ED_{80} in rats. LB30889, a double prodrug of LB30870, showed species difference in pharmacokinetics, and oral bioavailability in rats or dogs was no better than LB30870 (2). When studied in healthy men (at oral doses of 5–240 mg under fasting conditions), C_{max} occurred at 1.3–3.0 hours post-dose, with a mean apparent terminal $t_{1/2}$ of 2.8–4.1 hours (3). AUC after doses above 15 mg appeared greater than dose-proportional. In fed state, however, AUC data showed 80% reduction relative to fasting condition, suggesting that this food effect must be overcome for LB30870 to become a successful oral anti-coagulant (3). **1.** Lee, Park, Jung, *et al.* (2003) *J. Med. Chem.* **46**, 3612. **2.** Park, Lee, Kim, *et al.* (2013) *Biorg. & Med. Chem. Lett.* **23**, 4779. **3.** Kim, Lee, Boyce, *et al.* (2015) *Xenobiotica* **45**, 663.

LBH589, *See Panobinostat*

LBQ657

This NEP inhibitor (FW = 383.44 g/mol; CAS 149709-44-4), also named (2*R*,4*S*)-5-(biphenyl-4-yl)-4-((3-carboxypropionyl)amino)-2-methyl-pentan-oic acid, targets neprilysin (IC_{50} = 5nM), the enzyme responsible for degrading atrial and brain natriuretic peptides, two blood pressure-lowering peptides that mainly work by reducing blood volume. LBQ-657 is formed from the prodrug AHU-377 by intracellular esterases (**See** *AHU-377; LCZ696*). **1.** Ksander, Ghai, de Jesus, *et al.* (1995) *J. Med. Chem.* **38**, 1689.

LC-44, *See Flupenthixol*

LCB 03-0110

This protein kinase inhibitor (FW = 490.45 g/mol; Soluble to 100 mM in H_2O; 100 mM in DMSO), also named 3-[[2-[3-(4-morpholinylmethyl) phenyl]thieno[3,2-*b*]pyridin-7-yl]amino]phenol, targets c-Src kinase inhibitor (IC_{50} = 1.3 nM) as well as discoidin domain receptor 2 (DDR2) family tyrosine kinases, BTK and Syk. LCB 03-0110 suppresses the proliferation and migration of primary dermal fibroblasts induced by transforming growth factor β1 and type I collagen, and this result correlates with the inhibition ability of the compound against enhanced expression of α-smooth muscle actin and activation of Akt1 and focal adhesion kinase. In J774A.1 macrophage cells activated by lipopolysaccharide LCB 03-0110 inhibits cell migration and nitric oxide, inducible nitric-oxide synthase, cyclooxygenase 2, and tumor necrosis factor-α synthesis. When applied topically to full excisional wounds on rabbit ears, LCB 03-0110 suppresses the accumulation of myofibroblast and macrophage cells in the healing wound and reduces hypertrophic scar formation after wound closing, without delaying the wound closing process. **1.** Sun, Phan, Jung, *et al.* (2012) *J. Pharmacol. Exp. Ther.* **340**, 510.

LCZ696

This novel binary drug consists of Valsartan, the angiotensin II receptor Type-1 (AT_1) antagonist that causes vasodilation and increases renal excretion of sodium ion and water by reducing aldosterone production, and AHU-377, a prodrug that is converted to LBQ657, the latter inhibiting neprilysin. **See** *Valsartin; AHU377 (and LBQ657)* **1.** McMurray, Packer, Desai, *et al.* (2014) *New Engl. J. Med.* **371**, 993.

LDE225, *See Sonidegib (Erismodegib)*

LDN 27219

This reversible transglutaminase inhibitor (FW = 408.50 g/mol; CAS 312946-37-5; Soluble in DMSO), named systematically as 2-[(3,4-dihydro-4-oxo-3,5-diphenylthieno[2,3-*d*]pyrimidin-2-yl)thio]acetylhydrazide, targets tissue transglutaminase (TGase), a Ca^{2+}-dependent enzyme that catalyzes the cross-linking of intracellular proteins through a mechanism involving isopeptide bond formation between Gln and Lys residues and that is allosterically regulated by GTP. The IC_{50} value is 0.25 μM. LDN-27219 is a slow-binding inhibitor that appears to bind, not at the enzyme's active site, but instead its GTP regulatory site (or at a site that regulates GTP binding). The potency and kinetics of inhibiton are dependent on substrate, suggest a novel inhibitory mechanism involving differential binding of LDN-27219 to multiple conformational states of this enzyme. **1.** Case & Stein (2007) *Biochemistry* **46**, 1106.

LDN-57444

This proteasome inhibitor (FW = 397.64 g/mol; CAS 668467-91-2; Solubility: 11 mg/mL DMSO; <1 mg/mL H_2O or Ethanol), also known as 5-chloro-1-[(2,5-dichlorophenyl)methyl]-1*H*-indole-2,3-dione 3-(*O*-acetyloxime), competitively and reversibly targets ubiquitin C-terminal hydrolase, or Uch-L1 (IC_{50} = 0.88 μM), showing 28x selectivity over isoform Uch-L3

(1-4). Abnormal accumulation and aggregation of α-synuclein (α-syn) within neurons, and mutations in the α-syn and UCH-L1 genes have been shown to play a role in the pathogenesis of Parkinson Disease. In typical experiments, cultured neurons are treated with 10 μM LDN-57444 for 24 hr, a dose assuring UCH-L1-specific action. For short-term *in vivo* experiments, mice are injected intraperitoneally (0.5 mg LDN-57444 per kg), at time zero and again with the same dosage four hours after the first injection. **1.** Liu, Lashuel, Choi, *et al.* (2003) *Chem. Biol.* **10**, 837. **2.** Cartier, Ubhi, Spencer, *et al.* (2012) *PLoS One* **7**, e34713. **3.** Gong, Cao, Zheng, *et al.* (2006) *Cell* **126**, 775.

LDN 193188

This PC-TP inhibitor (FW = 494.35 g/mol; CAS 1267610-30-9; Soluble to 100 mM in DMSO), also named 2,4-dichloro-*N*-[[[4-[[(4,6-dimethyl-2-pyrimidinyl)amino]sulfonyl]phenyl]amino]carbonyl]benzamide, targets Phosphatidylcholine Transfer Protein, (PC-TP, also referred to as StarD2) is a highly specific intracellular lipid-binding protein that catalyzes the transfer of phosphatidylcholines between membranes *in vitro*. Recent studies have suggested that PC-TP *in vivo* functions to regulate fatty acid and glucose metabolism, possibly via interactions with selected other proteins. LDN 193188 binds to PC-TP, displacing phosphatidylcholines from its lipid binding site and increasing the thermal stability of the protein. **1.** Wagle, Xian, Shishova, *et al.* (2008) *Anal. Biochem.* **383**, 85. **2.**

LDN-212854

This potent BMP receptor inhibitor (FW = 406.48 g/mol; CAS 1432597-26-6; Solubility: 80 mg/mL DMSO (with heating); <1 mg/mL H_2O or Ethanol), also named 5-[6-4-(1-piperazinyl)phenyl]pyrazolo[1,5-*a*]pyrimidin-3-yl]quinoline, selectively targets Bone Morphogenetic Protein receptors, with IC_{50} of 1.3 nM for ALK2, about 2x, 66x, 1640x, and 7135x selectivity over ALK1, ALK3, ALK4, and ALK5, respectively. In contrast to previously described BMP receptor kinase inhibitors, LDN-212854 exhibits nearly four orders of selectivity versus closely related TGF-β and Activin Type I receptors. LDN-212854 potently inhibits heterotopic ossification in an inducible transgenic mutant ALK2 mouse model of fibrodysplasia ossificans progressive, allowing selective targeting individual of the highly homologous members of the BMP type I receptor family. **1.** Mohedas, Xing, Armstrong, *et al.* (2013) *ACS Chem. Biol.* **8**, 1291.

LDN-214117

This potent and selective BMP type I receptor kinase inhibitor (FW = 419.52 g/mol; CAS 1627503-67-6; Solubility: 83 mg/mL DMSO; < 1 mg/mol H_2O), also named 1-(4-(6-methyl-5-(3,4,5-trimethoxyphenyl) pyridin-3-yl)phenyl)piperazine, targets Bone Morphogenetic Protein receptor's Activin-Like Kinase 2, or ALK2 (IC_{50} = 24 nM). Contrary to the notion that activating mutations of ALK2 might alter inhibitor efficacy due to potential conformational changes in the ATP-binding site, the action of LDN-214117 demonstrates consistent binding to a panel of mutant and wild-type ALK2 proteins. **1**. Mohedas, Wang, Sanvitale, *et al.* (2014) *J. Med. Chem.* **57**, 7900.

Lead

This Group 14 (*or* IVB) heavy metal element (Atomic Weight = 207.2; Symbol = Pb; Atomic Number = 82; Located beneath Tin in the Periodic Table) forms bioreactive ions and organic compounds that inhibit numerous proteins, enzymes and vital cellular processes. **Lead Ions:** Typical valences of lead are +2 and +4, having ionic radii of 1.20 and 0.84 Å, respectively. Although the exact solvation number *n* of the $[Pb(H_2O)_n]^{2+}$ ion is unknown for dilute Pb^{2+} solutions, the hexaaquo species is likely. Aquated lead ions bind tightly to proteins, most often reacting with thiol groups (*Reaction:* Protein-SH + $[Pb(H_2O)_n]^{2+} \rightleftharpoons$ Protein-S-$Pb(H_2O)_{(n-1)}^{1+}$ + H^+), and less so with carboxyl groups (*Reaction:* Protein-C(=O)-O^- + $[Pb(H_2O)_n]^{2+} \rightleftharpoons$ Protein-C(=O)-O-$Pb(H_2O)_{(n-1)}^{1+}$ + H^+), and neutral amino groups (*Reaction:* Protein-NH_2 + $[Pb(H_2O)_n]^{2+} \rightleftharpoons$ Protein-NH-$Pb(H_2O)_{(n-1)}^{1+}$ + H^+). Pb^{2+} displaces Ca^{2+} and Zn^{2+} present in many proteins, binding oxygen atoms from amino acids or water as a major ligand, followed by sulfur and nitrogen. Around one-third of the lead binding sites are due to zinc or calcium ionic displacement, whereas two-thirds were opportunistic (1). As a heavy metal ion, Pb^{2+} is an environmentally persistent toxin that impairs circulatory, hematological, gastrointestinal, neurological, reproductive, and immunological processes. It also catalyzes oxidation reactions, leading to the formation and accumulation of reactive oxygen species (ROS) that damage cells and tissues. With many divalent cation transporters in humans, there is still no clear view of which pump Pb^{2+}. Given the similar radii of Ca^{2+} and Pb^{2+} ions, the ability of the latter to cross the blood-brain barrier is likely facilitated by calcium ion transporters, and, once within the brain, lead-induced damage can be observed in the prefrontal cerebral cortex, hippocampus, and cerebellum. CNS lead accumulation also affects the hypothalamic-pituitary axis, blunting responses to thyrotropin-releasing factor (TRF or TRH), growth-hormone-releasing factor (GRF or GHRH), and gonadotropin-releasing hormone (GnRH). Pb^{2+} likewise binds to calmodulin, decreasing the efficiency of calmodulin in activating CAM-dependent protein kinases and reducing the phosphorylation of brain membrane proteins. Less appreciated, however, is that Pb^{2+} also readily binds to the negatively charged phosphodiesters of RNA and DNA, thereby interfering with nucleic acid regulatory interactions and worse still catalyzing nonenzymatic strand cleavage (2). The latter reactions are illustrated below, with bound lead ion acting as a Lewis acid that mediates general and specific acid catalysis, respectively, during DNA and RNA phosphodiester hydrolysis.

Lewis Acid Catalysis of DNA and RNA Cleavage by Bound Pb(II)

These inherently mutagenic reactions cause damage requiring prompt repair, but the exact enzymes responsible for removing bound led ions and executing such strand breakage repair are still unclear. **Organolead:** The term organolead implies that a covalent bond joins a carbon atom to a lead atom within an organic molecule, a criterion that is different from O–Pb and N–Pb bonds found in chelate complexes with Pb^{2+} or those compounds having S–Pb and Se-Pb bonds in sulfur- and selenium-containing molecules fixed to lead atoms. That said, the lead atoms in Pb-containing organic molecules often form strong coordinate covalent bonds with thiol-containing side-chain groups within enzymes, often resulting in loss of

catalytic activity. Moreover, whereas inorganic compounds tend to have Pb(II) centers, the dominant species is Pb(IV) in organolead compounds. Inorganic lead compounds form with elements, such as nitrogen, oxygen and the halides, which have have higher electronegativity than lead, and the partial positive charge on lead leads to a stronger contraction of the 6*s*-orbital than the 6*p*-orbital, making the 6*s*-orbital inert (*i.e.,* the so-called "inert pair" effect). The biotransformation and metabolic impact of organolead is the subject of a monograph by Grandjean (3). Starting in 1920, tetraethyllead was added to gasoline (0.5 g per gallon) to suppress engine knocking (***See** Tetraethyllead*). Much was oxidized, yielding CO_2, H_2O, and lead oxide, and, given the many billions of gallons used in automobiles over the ensuing fifty years, the amount of lead from auto exhaust is beyond reckoning. **Heme Biosynthesis:** Lead ions strongly inhibit heme biosynthesis in mammals, targeting δ-aminolevulinate synthetase (*or* ALA synthase), δ-aminolevulinate dehydrase (*or* porphobilinogen synthase), and the Fe(II)-inserting enzyme, ferrochelatase (4,5). As the only cytosolic enzyme, δ-aminolevulinate dehydrase is most affected, and the overall rate of heme synthesis remains unaffected until 80-90% of this enzyme is inhibited (occurring at 50 μg/dL plasma). Notably, lead inhibition of ferrochelatase results in the incorporation of divalent zinc into protoporphryn IX, with consequential loss of the heme needed for oxygen transport and heme redox chemistry. **Transcription Factor Failure:** As assayed by DNase I protection, the interaction of Cys_2His_2 zinc finger protein transcription factor IIIA (TFIIIA) with the 50-bp internal control region of the 5S ribosomal gene was completely inhibited at 10 μM Pb^{2+} (or 300 μg/dL) an effect that cannot be reversed by addition of 2-mercaptoethanol (6). Moreover, prior incubation of free TFIIIA with lead resulted in DNA-binding inhibition, whereas prior incubation of a TFIIIA/5S RNA complex with lead did not. Inhibition of Cys_2His_2 zinc finger transcription factors by lead ions at concentrations near those known to have deleterious physiological effects points to new molecular mechanisms for lead toxicity in promoting disease. **Lead Toxicity:** Lead exposure also induces certain stress-related proteins (7) as well as epoxide hydrolase (8) and the glutathione transferase P gene (9). Lead ions can replace Ca^{2+} at their sites of calcium ion-binding proteins. For nearly two hundred years, manufacturers added high amounts of lead tetraoxide as a base pigment for white and lightly colored paints. Lead was also used to join cast iron pipes and to solder copper pipes and wires. Chronic lead exposure results in accumulation in mammalian bone as well as the roots of many plants. The Biological Exposure Index (BEI) for adult blood lead levels is 30 μg/dL. Lead poisoning is insidious, gradual at first, and then followed by its full-blown manifestations. Unfortunately, early evidence of lead poisoning can be elusive. Even seemingly healthy individuals can have high circulating lead levels. Signs and symptoms aren't apparent until dangerous amounts have accumulated. In children, they include irritability, loss of appetite, unaccountable weight loss, sluggishness and fatigue, unaccountable abdominal pain, vomiting, constipation, and learning difficulties. For children, a blood lead level (BLL) of 10 μg/dL is considered grounds for intervention (10,11). In fact, studies have repeatedly linked chronic exposure to BLLs < 10 μg/dL with impairments in cognitive function and behavior in young children despite the absence of overt signs of toxicity (10). The accumulation and efflux of lead depends in part on the ATP-dependent efflux pump, *or* Multidrug Resistance Protein-1, and requires glutathione, with the latter most likely forming a conjugate with lead ion, followed by export of that conjugate (12). **Abatement & Prophylaxis:** An imperative for lead abatement is the removal of flaking lead house paint from older homes (built prior to 1977 in the U.S.). Chelation therapy using EDTA has been an FDA-approved treatment for lead poisoning for more than forty years. *N*-Acetylcysteine, α-lipoic acid, vitamin E, quercetin and a few herbal extracts also show helpful prophylactic benefits against the majority of lead-mediated injuries, both *in vitro* and *in vivo*. **Enzyme Targets:** Over 1100 proteins and enzymes are inactivated upon even brief exposure to ionic and organic lead compounds – far too many to be conveniently listed in this monograph. Because metal ion exchange most often comports to the Eigen-Tamm mechanism, in which the ligation kinetics is dominated by ligand dissociation rates, the binding of lead ions can be complex. Typical targets have essential thiol groups in their active-sites or as functional groups in their substrates (*e.g.,* cysteine, glutathione, thiosulfate, *etc.*) and coenzymes (*e.g.,* CoA-SH, lipoic acid, and acyl carrier protein). Pb^{2+} also replaces metal ions in enzyme active sites. Significantly, Pb^{2+} also activates calmodulin and, in fact, has a higher affinity than Ca^{2+} (13). One must not ignore that the heavy metal ion binds firmly to the cysteine-rich zinc finger proteins that bind to and modify DNA structure and function. **1**. Kirberger & Yang (2008) *J. Inorg. Biochem.* **102**,

1901. **2**. Brown, Hingerty, Dewan & Klug (1983) *Nature* **303**, 543. **3**. Grandjean (1984) *Biological Effects of Organolead Compounds*, 288 pp., CRC Press, Boca Raton. **4**. Nieburg, Weiner, Oski & Oski (1974) *Am. J. Dis. Child.* **127**, 348 **5**. Campbell, Brodie, Thompson, Meredith, Moore & Goldberg (1977) *Clin Sci Mol Med.* **53**, 335. **6**. Hanas, Rodgers, Bantle & Cheng (1999) *Molec. Pharmacol.* **56**, 982. **7a**. Ahamed, Verma, Kumar (2005) *Sci. Total Environ.* **346**, 48. **7b**. Giusi, Alò, Crudo, Facciolo & Canonaco (2008) *Toxicol. Appl. Pharmacol.* **227**, 248. **8**. Ahotupa, Hartiala & Aitio (1979) *Acta Pharmacol. Toxicol. (Copenhagen)* **44**, 359. **9**. Cummings & Cavlock (2004) *Crit. Rev. Toxicol.* **34**, 461. **10**. *CDC: Preventing Lead Poisoning in Young Children.* (1991) Centers for Disease Control and Prevention, Atlanta, GA. **11**. Bellinger & Needleman (2003) *New Engl. J. Med.* **349**, 500. **12**. Huang, Ye, Yu, *et al.* (2014) *Toxicol. Lett.* **226**, 277. **13**. Ouyang & Vogel (1998) *Biometals* **11**, 213.

LEDGIN-6

This HIV inhibitor (FW = 353.85 g/mol; off-white crystalline solid: M.P. = 215-217 °C), known systematically as 2-(6-chloro-2-methyl-4-phenylquinolin-3-yl)pentanoic acid, targets the multifunctional HIV-1 enzyme integrase, which interacts with viral DNA and its key cellular cofactor LEDGF to efficiently integrate the reverse transcript into a host cell chromosome (1,2). LEDGIN-6 and the structurally related 2-(quinolin-3-yl) acetate, BI-1001, selectively inhibit integrase-LEDGF interactions *in vitro*, with similar IC$_{50}$ values, defining each as an allosteric inhibitor of integrase function. In blocking the formation of the stable synaptic complex between integrase and viral DNA, they stabilize an inactive multimeric form of integrase (2). These compounds also inhibit LEDGF binding to the stable synaptic complex, resulting in cooperative inhibition of the concerted HIV integration and replication in cell culture. **1**. Christ, Voet, Marchand, *et al.* (2010) *Nat. Chem. Biol.* **6**, 442. **2**. Kessl, Jena, Koh, *et al.* (2012) *J. Biol. Chem.* **287**, 16801.

Ledipasvir

This hepatitis C drug (FW = 889.00 g/mol; CAS 1256388-51-8), also named GS-5885 and methyl *N*-[(2*S*)-1-[(6*S*)-6-[5-[9,9-difluoro-7-[2-[(1*S*,2*S*,4*R*)-3-[(2*S*)-2-(methoxycarbonylamino)-3-methylbutanoyl]-3-aza-bicyclo[2.2.1]heptan-2-yl]-3*H*-benzimidazol-5-yl]fluoren-2-yl]-1*H*-imidazol-2-yl]-5-azaspiro[2.4]heptan-5-yl]-3-methyl-1-oxobutan-2-yl]carbamate, inhibits the viral non-structural protein 5A, *or* NS5A, a Zn(II)-binding and proline-rich hydrophilic viral phosphoprotein that is involved in viral replication, assembly, and secretion (1,2). NS5A also modulates the polymerase activity of NS5B, an RNA-dependent RNA polymerase (RdRp). Subject-derived Met-28-Thr, Gln-30-Arg, Leu-31-Met, and Tyr-93-Cys mutations all confer >30x reductions in GS-5885 and daclatasvir susceptibilities *in vitro* (3) Site-directed NS5A mutants also show reduced susceptibility. However, all NS5A mutants tested remained fully susceptible to other classes of direct-acting antivirals, interferon α, and ribavirin.Moreover, the nonoverlapping resistance profiles and high potency of ledipasvir favor its use with other antivirals for treating chronic HCV (3). In 2014, the FDA approved Harvoni®, a combination drug consisting of ledipasvir (90 mg) and sofosbuvir (400 mg). *See also Sofosbuvir* **1**. Afdhal, Zeuzem, Kwo, *et al.* (2014). *New Engl. J. Med.* **370**, 1889. **2**. Mizokami,

Yokosuka, Takehara, *et al.* (2015) *Lancet Infect. Dis.* **15**, 645. **3**. Wong, Worth, Martin, *et al.* (2013) *Antimicrob. Agents Chemother.* **57**, 6333.

LEE011

This orally available and highly specific CDK4/6 inhibitor. (FW = 434.54 g/mol; CAS 1211441-98-3; Solubility: 7 mg/mL (16 mM) DMSO; <1 mg/mL H$_2$O or Ethanol), also named 7-cyclopentyl-*N*,*N*-dimethyl-2-((5-(piperazin-1-yl)pyridin-2-yl)amino)-7*H*-pyrrolo[2,3-*d*]pyrimidine-6-carbox amide, greatly reduces proliferation in 12 of 17 human neuroblastoma-derived cell lines by inducing cytostasis at (mean IC$_{50}$ = 307 ± 68 nM in sensitive lines). LEE011 causes cell-cycle arrest and cellular senescence attributable to dose-dependent decreases in phosphorylated RB and FOXM1, respectively. **1**. Rader, Russell, Hart, *et al.* (2013) *Clin. Cancer Res.* **19**, 6173.

Leech-derived Tryptase Inhibitor

This 46-residue Kazal-type protein (MW = 4738 Daltons), often abbreviated LDTI and first isolated from the leech *Hirudo medicinalis*, is a potent inhibitor of human tryptase (K_d = 1.4 nM). **Target(s):** chymotrypsin (1); trypsin (1-3); tryptase (1-4). **1**. Sommerhoff, Söllner, Mentele, *et al.* (1994) *Biol Chem Hoppe-Seyler.* **375**, 685. **2**. Erba, Fiorucci, Sommerhoff, Coletta & Ascoli (2000) *Biol. Chem.* **381**, 1117. **3**. Stubbs, Morenweiser, Stürzebecher, *et al.* (1997) *J. Biol. Chem.* **272**, 19931. **4**. Pereira, Bergner, Macedo-Ribeiro, *et al.* (1998) *Nature* **392**, 306.

Leelamine

This chiral diterpene amine and investigational chemotherapeutic agent (FW = 285.48 g/mol; CAS 1446-61-3; from *leela* (Sanskrit for "play"), also named as 1*R*,2,3,4,4*aS*,9,10,10*aR*-octahydro-1,4*a*-dimethyl-7-(1-methylethyl)-1-phenanthrenemethanamine, is a lysosomotropic, intracellular cholesterol transport inhibitor, resulting in the inhibition of autophagic flux and induction of cholesterol accumulation in lysosomal-endosomal cell compartments (1). Treatment of cells with leelamine results in the accumulation of cholesterol within acidic organelles such as lysosomes, disrupting autophagic flux and intracellular cholesterol trafficking as well as receptor-mediated endocytosis (1). Leelamine inhibited the growth of preexisting xenografted melanoma tumors by an average of 60% via targeting of the PI3K, MAPK and STAT3 pathways, without affecting animal body weight or blood markers of major organ function (2). Leelamine inhibits pyruvate dehydrogenase kinase, IC$_{50}$ = 9.5 μM (3). Leelamine also shows weak cannabinoid receptor affinity, displacing 20% of [^{3}H]-CP55940 from central CB$_1$ and the peripheral CB$_2$ receptors, when present at 10 μM. **1**. Kuzu, Gowda, Sharma & Robertson (2014) *Mol. Cancer Ther.* **13**, 1690. **2**. Gowda, Madhanapantula, Kuzu, Sharma & Robertson (2014) *Mol. Cancer Ther.* **13**, 1679. **3**. Cadoudal, Distel, Durant, *et al.* (2008) *Diabetes* **57**, 2272.

Leflunomide

Leflunomide   A771726

This antirheumatic (FW = 270.21 g/mol; CAS 75706-12-6), also known as *N*-[4-(trifluoromethyl)phenyl]-5 methylisoxazole-4-carboxamide, Arava, and SU 101, is metabolized to an open-ring derivative A771726 (**See Teriflunomide**) that is a potent, reversible inhibitor of mammalian dihydroorotate dehydrogenase, the rate-limiting step in *de novo* pyrimidine biosynthesis. (The bacterial enzyme is unaffected.) Because activated white blood cells have an elevated demand for purines and pyrimidines, they rely on both salvage and *de novo* synthesis pathways. In the presence of A771726, lymphocytes cannot accumulate sufficient stores of pyrimidine nucleotides to support DNA biosynthesis. Leflunomide prevents proliferative expansion of activated and autoimmune lymphocytes by interfering with their cell-cycle progression. By contrast, nonlymphoid cells rely on pyrimidine salvage pathway, making them less dependent on *de novo* synthesis. While the immunoregulatory effects of A771726 are observed at the lower concentrations, higher doses inhibit protein-tyrosine kinases responsible for early T-cell and B-cell signalling in the G_0/G_1 phases are also inhibited (1). Leflunomide inhibits platelet-derived-growth factor-mediated signaling events, including receptor tyrosine phosphorylation, DNA synthesis, cell cycle progression, and cell proliferation (2). It also inhibits dihydroorotate dehydrogenase (DHODH), an enzyme that is involved in the biosynthesis of uridine (**See also 4SC-101**) (3). A77-1726, a pharmacologically active metabolite of leflunomide, attenuates lupus nephritis by promoting the development of regulatory T cells and inhibiting IL-17-producing double negative T cells (4). **Key Pharmacokinetic Parameters:** See Appendix II in Goodman & Gilman's *THE PHARMACOLOGICAL BASIS OF THERAPEUTICS*, 12th Edition (Brunton, Chabner & Knollmann, eds.) McGraw-Hill Medical, New York (2011). **1**. Xu, Williams, Gong, Finnegan & Chong (1996) *Biochem. Pharmacol.* **52**, 527. **2**. Shawver, Schwartz, Mann, *et al.* (1997) *Clin. Cancer Res.* **3**, 1167. **3**. Leban & Vitt (2011) *Arzneimittelforschung* **61**, 66. **4**. Qiao, Yang, Li, Williams & Zhang (2015) *Clin. Immunol.* **157**, 166..

Leiurotoxin I

This channel blocker (FW = 3429.18 g/mol; CAS 142948-19-4; *Sequence*: AFCNLRMCQLSCRSLGLLGKCIGDKCECVKH; *Symbol*: LeTx1), also known as Scyllatoxin, is a peptide inhibitor of the small conductance, Ca^{2+}-activated K^+ channels. Leiurotoxin I comprises <0.02% of the protein in the venom of the scorpion (*Leiurus quinquestriatus hebraeus*). Leiurotoxin shares some homology to other scorpion toxins such as charybdotoxin, noxiustoxin, and neurotoxin P_2. Leiurotoxin I completely inhibits ^{125}I-apamin binding to rat brain synaptosomal membranes (K_i = 75 pM), but is some 10-20x less potent than apamin (**See Apamin**). Like apamin, leiurotoxin I blocks the epinephrine-induced relaxation of guinea pig smooth muscle (ED_{50} = 6.5 nM), but is without effect on the rate or force of contraction in guinea pig atria or rabbit portal vein preparations. The cysteine spacing and disulfides found in all scorpion toxins (and preserved in LeTx1) may play an active role in folding, with only two native disulfide bonds in LeTx1 sufficient to preserve its active conformation (3). **1**. Chicchi, Gimenez-Gallego, Ber, *et al.* (1988) *J. Biol Chem.* **263**, 10192. **2**. Auguste, Hugues, Gravé, *et al.* (1990) *J. Biol. Chem.* **265**, 4753. **3**. Zhu, Liang, Martin, *et al.* (2002) *Biochemistry* **41**, 11488.

Lenalidomide

This thalidomide derivative (FW = 259.26 g/mol; CAS 191732-72-6; Solubility: 50 mg/mL DMSO, <1 mg/mL H_2O), also known by the trade name Revlimid® and its systematic name (*RS*)-3-(4-amino-1-oxo-1,3-dihydro-2*H*-isoindol-2-yl)piperidine-2,6-dione, is a TNFα secretion inhibitor (IC_{50} = 13 nM) and is indicated for the treatment of patients with mantle cell lymphoma (MCL) whose disease has relapsed or progressed after two prior therapies, one of which included bortezomib. Although originally intended for treating multiple sclerosis, lenalidomide has proven to be an effective immunomodulatory agent that stimulates production of the cytokyne IL-2 and induces the biosynthesis of interleukin-2 receptor (IL-2R), a heterotrimeric protein expressed on the outer surface of certain immune cells, including lymphocytes. **1**. Muller, *et al. Bioorg. Med. Chem.*

Lett. (1999) **9**, 1625. **2**. Zangari M, *et al.* (2005) *Expert Opin. Investig. Drugs* 14, 1411. **3**. Ramsay, *et al.* (2012) *Blood* 120, 1412.

Lenvatinib

This TKI and angiogenesis inhibitor (FW = 426.86 g/mol; CAS 417716-92-8; Solubility: 40 mg/mL DMSO; <1mg/mL H_2O), also named 1-(4-(6-carbamoyl-7-methoxyquinolin-4-yloxy)-2-chlorophenyl)-3-cyclopropyl-urea, E7080 and Lenvima®, shows potent action against Stem Cell Factor- (SCF-) induced angiogenesis, both *in vitro* and in SCF-producing human small cell lung carcinoma H146 cells *in vivo* (1). It is specifically indicated for treating locally recurrent or progressive metastatic radioiodine-refractory thyroid cancer. Lenvatinib inhibits SCF-driven tube formation of HUVEC, which express SCF receptor, KIT (IC_{50} = 5.2 nM) and is almost identical for VEGF (IC_{50} = 5.1 nM). Imatinib, by contrast, do not show the potent antitumor activity against H146 cells *in vitro* (IC_{50} = 2.2 μM), because these cells do not express KIT. (**See Imatinib**) However, when administered orally at 160 mg/kg, imatinib slows tumor growth of H146 cells in nude mice, attended by decreased microvessel density. Oral administration of lenvatinib inhibits H146 cells at doses of 30 and 100 mg/kg in a dose-dependent manner, causing tumor regression at 100 mg/kg. While anti-VEGF antibody also slows tumor growth, it fails to cause tumor regression. Such results indicate that KIT signaling plays a role in tumor angiogenesis of SCF-producing H146 cells, and lenvatinib causes regression of H146 tumors as a result of antiangiogenic activity mediated by inhibition of both KIT and VEGF receptor signaling (1). VEGF-R3 kinase inhibition by E7080 also decreases lymphatic vessel density (LVD) within MDA-MB-231 tumors that express VEGF-C. Simultaneous inhibition of both VEGF-R2 and VEGF-R3 kinases by lenvatinib may be a promising new strategy to control regional lymph node and distant lung metastases (2). Lenvatinib inhibits VEGF- and FGF-driven angiogenesis and showed a broad spectrum of antitumor activity with a wide therapeutic window. MVD and pericyte-coverage of tumor vasculature might be biomarkers for lenvatinib-responsive therapies (3). **1**. Matsui, Yamamoto, Funahashi, *et al.* (2008) *Int. J. Cancer* 122, 664. **2**. Matsui, Funahashi, Uenaka, *et al.* (2008) *Clin. Cancer Res.* 14, 5459. **3**. Yamamoto, Matsui, Matsushima, *et al.* (2014) *Vasc. Cell.* 6, 18.

Leptomycin B

This unusual unsaturated fatty acid ($FW_{free-acid}$ = 540.74 g/mol) inhibits the nuclear export of proteins in eukaryotic cells. Leptomycin B binds covalently via its αβ-unsaturated δ-lactone to the sulfhydryl group of a cysteinyl residues of CRM1 (*or* Exportin 1), a receptor for the nuclear export signal of proteins. **1**. Nishi, Yoshida, Fujiwara, *et al.* (1994) *J. Biol. Chem.* **269**, 6320. **2**. Fukuda, Asano, Nakamura *et al.* (1997) *Nature* **390**, 308. **3**. Kudo, Matsumori, Taoka, *et al.* (1999) *Proc. Natl. Acad. Sci. U.S.A.* **96**, 9112. **4**. Kudo, Wolff, Sekimoto, *et al.* (1998) *Exp. Cell Res.* **242**, 540.

Lestaurtinib

This staurosporine-like multi-tyrosine kinase inhibitor, *or* TKI (FW = 439.47 g/mol; CAS 111358-88-4), also known as CEP-701 and KT-5555, potently targets the tyrosine kinase activity of the neurotrophin-specific Trk receptors (IC$_{50}$ = 4 nM), FMS-like tyrosine kinase-3 (FLT3), and JAK2, STAT5 and STAT3. Trk receptor ligands are neurotrophins, a family of neuronal growth factors, including NGR, that activate Trks, which in turn promote cell survival. With aberrant expression of the Trk receptors (Trk A, B, C) in pancreatic ductal adenocarcinoma (PDAC), lestaurtinib is active against xenograft models using the poorly differentiated PDAC cell line, Panc1, inhibiting both growth and invasion (1) and exerts similar effects on both primary and metastatic prostate cancer cells (2). Internal tandem duplication of FMS-like tyrosine kinase-3 (FLT3) is associated with poor outcomes in acute myelogenous leukemia and Hodgkin Lymphoma (HL), and lestaurtinib potently inhibits autophosphorylation of FLT3 and JAK2 *in vitro* (3-5). (***See also*** *Staurosporine*) **IUPAC:** (9*S*,10*S*,12*R*)-2,3,9,10,11,12-hexahydro-10-hydroxy-10-(hydroxymethyl)-9-methyl-9,12-epoxy-1*H*-diindolo[1,2,3-*fg*:3',2',1'-*kl*]pyrrolo[3,4-*i*][1,6]benzodiazocin-1-one. **1.** Miknyoczki, Chang, Klein-Szanto, Dionne & Ruggeri (1999) *Clin. Cancer Res.* **5**, 2205. **2.** Weeraratna, Dalrymple, Lamb, *et al.* (2001) *Clin. Cancer Res.* **7**, 2237. **3.** Fath & Lewis (2009) *Expert Rev. Hematol.* **2**, 17. **4.** Sanz, Burnett, Lo-Coco & Löwenberg (2009) *Curr. Opin. Oncol.* **21**, 594. **5.** Diaz, Navarro, Ferrer, *et al.* (2011) *PLoS One* **6**, e18856.

Letrozole

This orally bioavailable, non-steroidal anti-estrogen agent (FW = 285.30 g/mol; CAS 112809-51-5; Symbol: LZ), known by the trade name Femara™ and systematically as 4,4'-((1*H*-1,2,4-triazol-1-yl)methylene)-dibenzonitrile, targets aromatase, *or* CYP19A1 (IC$_{50}$ = 50-100 nM), a member of the cytochrome P450 superfamily (EC 1.14.14.1). This monooxygenase catalyzes numerous reactions in steroidogenesis and is responsible for the conversion of androgens into estrogens. Letrozole is widely used to treat post-surgical, hormonally-responsive breast cancer (1). Early studies showed that concurrent treatment of MCF-7 cells with 17-β-estradiol in the presence of LZ significantly suppresses the estradiol-induced stimulation of matrix metalloproteinases (MMP) levels. These results indicate that letrozole is a potent inhibitor of *in vitro* cell proliferation and of type IV collagenase expression in ER-positive MCF-7 cells (2). Letrozole likewise suppresses breast tumor growth and invasiveness (2). LZ treatment of female germ cells brings about sex inversion in yellow catfish (*Pelteobagrus fulvidraco*), a species displaying sexual dimorphic growth. These results suggest that aromatase activity plays a vital role in sex differentiation. **1.** Bhatnagar, Häusler, Schieweck, Lang & Bowman (1990) *J. Steroid Biochem. Mol. Biol.* **37**, 1021. **2.** Mitropoulou, Tzanakakis, Kletsas, Kalofonos & Karamanos (2003) *Int. J. Cancer* **104**, 155. **3.** Shen, Fan, Yang, *et al.* (2013) *Biol. Bull.* **225**, 18.

L-Leucinal

This leucine derivative (FW$_{\text{free-base}}$ = 103.17 g/mol), also known as 2(*S*)-amino-4-methylpentanal, inhibits tripeptide aminopeptidase, leucyl aminopeptidase, and aminopeptidase N. **Target(s):** leucyl aminopeptidase (1-3); membrane alanyl aminopeptidase, *or* aminopeptidase N (1); tripeptide aminopeptidase (4). **1.** Bauvois & Dauzonne (2006) *Med. Res. Rev.* **26**, 88. **2.** Sträter & Lipscomb (2004) in *Handbook of Metalloproteins*, **3**, 199, Wiley, New York. **3.** Andersson, MacNeela & Wolfenden (1985) *Biochemistry* **24**, 330. **4.** Frick & Wolfenden (1985) *Biochim. Biophys. Acta* **829**, 311.

D-Leucine

This branched chain amino acid (FW = 131.175 g/mol), (*R*)-2-amino-4-methylpentanoic acid, is the enantiomer of the proteogenic amino acid, L-leucine. D-Leucyl residues are present in a number of antibiotics: for example, gramicidin A, polymixin D$_1$, and tolaasin. (***See also*** *L-Leucine*) **Target(s):** alkaline phosphatase, weakly inhibited (1); carboxypeptidase A, weakly inhibited (2); 2-ethylmalate synthase, weakly inhibited (3); gramicidin S synthase (4); 2-isopropylmalate synthase, weakly inhibited (5); leucyl aminopeptidase (6); system L amino acid transport (7). **Key Pharmacokinetic Parameters:** *See* Appendix II in Goodman & Gilman's *THE PHARMACOLOGICAL BASIS OF THERAPEUTICS*, 12th Edition (Brunton, Chabner & Knollmann, eds.) McGraw-Hill Medical, New York (2011). **1.** Hoylaerts, Manes & Millán (1992) *Biochem. J.* **286**, 23. **2.** Coleman & Vallee (1964) *Biochemistry* **3**, 1874. **3.** Rabin, Salomon, Bleiweis, Carlin & Ajl (1968) *Biochemistry* **7**, 377. **4.** Zimmer & Laland (1975) *Meth. Enzymol.* **43**, 567. **5.** Ulm, Böhme & Kohlhaw (1972) *J. Bacteriol.* **110**, 1118. **6.** Machuga & Ives (1984) *Biochim. Biophys. Acta* **789**, 26. **7.** Barker, Wilkins, Golding & Ellory (1999) *J. Physiol.* **514**, 795.

L-Leucine

This branched-chain amino acid (FW = 131.18 g/mol; pK_1 = 2.33 (α-COOH), pK_2 = 9.74 (α-NH$_2$) at 25°C; dpK_{a2}/dT = –0.026; Specific Rotation = + 16.0° at 25°C; Symbol = Leu or L), named systematically as (*S*)-2-amino-4-methylpentanoic acid, is a proteogenic amino acid that is nutritionally essential in mammals. Leucine was identified early on in biochemical investigations of protein components (1-5). The six codons for Leucine are UUA, CUU, UUG, CUA, CUC, and CUG. One of the most hydrophobic amino acids, leucine residues are usually buried in the apolar interior, where they are found in both α-helices and β-pleated sheets. Leucine residues are also found in coiled-coil regions and most famously in so-called "leucine zipper" proteins. Homologous proteins often have leucine residues interchanged with valine, isoleucine, methionine, or phenylalanine residues. **Target(s):** acetolactate synthase (6,7,83-93); acetylcholinesterase8; alkaline phosphatase (9,10,52-55); D-amino-acid oxidase (11-14); 2-aminohexanoate transaminase, also weaker alternative substrate (59); 2-aminohexano-6-lactam racemase, weakly inhibited (15); arginase (16-19); asparagine:oxo-acid aminotransferase (61,62); aspartate kinase (20,57); ATP:citrate lyase, *or* ATP:citrate synthase, weakly inhibited (81); bacterial leucyl aminopeptidase, product inhibition (45); creatine kinase (56);

cystinyl aminopeptidase, *or* oxytocinase (46,48); cytosol alanyl aminopeptidase (43,44); cytosol nonspecific dipeptidase (21,40,41); dipeptidyl-peptidase IV (22); 2-ethylmalate synthase (82); 2-glucose-6-phosphatase (23); glutamine synthetase (24); homoserine kinase (58); (*S*)-2-hydroxy-acid oxidase (25); isoleucine incorporation into proteins (26); 2-isopropylmalate synthase (27-32,68-80); kynurenine: oxoglutarate aminotransferase (67); lactate 2-monooxygenase (97); leucyl aminopeptidase (51); membrane alanyl aminopeptidase, *or* aminopeptidase N (33,46,49,50); ornithine aminotransferase (63-66); ornithine carbamoyl-transferase (34,94-96); saccharopine dehydrogenase (35); theanine hydrolase, *or* weakly inhibited (36); threonine ammonia-lyase, *or* threonine dehydratase (37,38); tripeptide aminopeptidase (39,47); valine:pyruvate transaminase (60); Xaa-Pro dipeptidase, *or* prolidase isozyme I (42). **1.** Proust (1819) *Ann. chim. phys.* [2] **10**, 29. **2.** Braconnot (1820) *Ann. chim. phys.* [2] **13**, 113. **3.** Mulder (1839) *J. prakt. Chem.* **16**, 290 and **17**, 57. **4.** Laurent & C. Gerhardt (1848) *Compt. rend.* **27**, 256-258 and *Ann. chim. phys.* [3] **24**, 321. **5.** Schulze & Likiernik (1891) *Ber.* **24**, 669-673 and (1893) *Z. physiol. Chem.* **17**, 513. **6.** Gollop, Chipman & Barak (1983) *Biochim. Biophys. Acta* **748**, 34. **7.** Barak, Calvo & Schloss (1988) *Meth. Enzymol.* **166**, 455. **8.** Bergmann, Wilson & Nachmansohn (1950) *J. Biol. Chem.* **186**, 693. **9.** Hoylaerts, Manes & Millan (1992) *Biochem. J.* **286**, 23. **10.** Bounias & Pacheco (1974) *C. R. Acad. Sci. Hebd. Seances Acad. Sci. D* **279**, 691. **11.** Burton (1955) *Meth. Enzymol.* **2**, 199. **12.** Dixon & Kleppe (1965) *Biochim. Biophys. Acta* **96**, 368. **13.** Bright & Porter (1975) *The Enzymes*, 3rd ed., **12**, 421. **14.** Webb (1966) *Enzyme and Metabolic Inhibitors*, vol. **2**, p. 340, Academic Press, New York. **15.** Ahmed, Esaki, Tanaka & Soda (1983) *Agric. Biol. Chem.* **47**, 1887. **16.** Hunter & Downs (1945) *J. Biol. Chem.* **157**, 427. **17.** Patchett, Daniel & Morgan (1991) *Biochim. Biophys. Acta* **1077**, 291. **18.** Kaysen & Strecker (1973) *Biochem. J.* **133**, 779. **19.** Singh & Singh (1990) *Arch. Int. Physiol. Biochim.* **98**, 411. **20.** Truffa-Bachi (1973) *The Enzymes*, 3rd ed., **8**, 509. **21.** Wilcox & Fried (1963) *Biochem. J.* **87**, 192. **22.** Shibuya-Saruta, Kasahara & Hashimoto (1996) *J. Clin. Lab. Anal.* **10**, 435. **23.** Nordlie (1971) *The Enzymes*, 3rd ed., **4**, 543. **24.** Singh & Singh (1990) *Arch. Int. Physiol. Biochim.* **98**, 95. **25.** Robinson, Keay, Molinari & Sizer (1962) *J. Biol. Chem.* **237**, 2001. **26.** So & Davie (1965) *Biochemistry* **4**, 1973. **27.** Kohlhaw & Leary (1970) *Meth. Enzymol.* **17A**, 771. **28.** Gross (1970) *Meth. Enzymol.* **17A**, 777. **29.** Kohlhaw (1988) *Meth. Enzymol.* **166**, 414. **30.** Higgins, Kornblatt & Rudney (1972) *The Enzymes*, 3rd ed., **7**, 407. **31.** Webster & Gross (1965) *Biochemistry* **4**, 2309. **32.** Allison, Baetz & Wiegel (1984) *Appl. Environ. Microbiol.* **48**, 1111. **33.** Wachsmuth, Fritze & Pfleiderer (1966) *Biochemistry* **5**, 175. **34.** Lusty, Jilka & Nietsch (1979) *J. Biol. Chem.* **254**, 10030. **35.** Fujioka & Nakatani (1972) *Eur. J. Biochem.* **25**, 301. **36.** Tsushida & Takeo (1985) *Agric. Biol. Chem.* **49**, 2913. **37.** Nath & Sanwal (1972) *Arch. Biochem. Biophys.* **151**, 420. **38.** Oda, Nakano & Kitaoka (1983) *J. Gen. Microbiol.* **129**, 57. **39.** Doumeng & Maroux (1979) *Biochem. J.* **177**, 801. **40.** Lenney (1990) *Biol. Chem. Hoppe-Seyler* **371**, 433. **41.** Wang, Liu, Yamashita, Manabe & Kodama (2004) *Clin. Chem. Lab. Med.* **42**, 1102. **42.** Liu, Nakayama, Sagara, *et al.* (2005) *Clin. Biochem.* **38**, 625. **43.** Garner & Behal (1977) *Arch. Biochem. Biophys.* **182**, 667. **44.** Garner & Behal (1975) *Biochemistry* **14**, 3208. **45.** Baker, Wilkes, Bayliss & Prescott (1983) *Biochemistry* **22**, 2098. **46.** Lalu, Lampelo & Vanha-Perttula (1986) *Biochim. Biophys. Acta* **873**, 190. **47.** Sachs & Marks (1982) *Biochim. Biophys. Acta* **706**, 229. **48.** Krishna & Kanagasabapathy (1989) *J. Endocrinol.* **121**, 537. **49.** Bauvois & Dauzonne (2006) *Med. Res. Rev.* **26**, 88. **50.** McCaman & Villarejo (1982) *Arch. Biochem. Biophys.* **213**, 384. **51.** Machuga & Ives (1984) *Biochim. Biophys. Acta* **789**, 26. **52.** Al-Saleh (2002) *J. Nat. Toxins* **11**, 357. **53.** Le Du, Stigbrand, Taussig, Menez & Stura (2001) *J. Biol. Chem.* **276**, 9158. **54.** Wennberg, Kozlenkov, Di Mauro, *et al.* (2002) *Hum. Mutat.* **19**, 258. **55.** Hoylaerts, Manes & Millán (1992) *Biochem. J.* **286**, 23. **56.** Pilla, Cardozo, Dornelles, *et al.* (2003) *Int. J. Dev. Neurosci.* **21**, 145. **57.** Curien, Ravanel, Robert & Dumas (2005) *J. Biol. Chem.* **280**, 41178. **58.** Thoen, Rognes & Aarnes (1978) *Plant Sci. Lett.* **13**, 103. **59.** Der Garabedian & Vermeersch (1987) *Eur. J. Biochem.* **167**, 141. **60.** Berg, Whalen & Archambault (1983) *J. Bacteriol.* **155**, 1009. **61.** Maul & Schuster (1986) *Arch. Biochem. Biophys.* **251**, 585. **62.** Cooper (1977) *J. Biol. Chem.* **252**, 2032. **63.** Strecker (1965) *J. Biol. Chem.* **240**, 1225. **64.** Yasuda, Tanizawa, Misono, Toyama & Soda (1981) *J. Bacteriol.* **148**, 43. **65.** Kalita, Kerman & Strecker (1976) *Biochim. Biophys. Acta* **429**, 780. **66.** Matsuzawa (1974) *J. Biochem.* **75**, 601. **67.** Wejksza, Rzeski, Okuno, *et al.* (2005) *Neurochem. Res.* **30**, 963. **68.** Wiegel & Schlegel (1977) *Arch. Microbiol.* **112**, 247. **69.** Wiegel & Schlegel (1977) *Arch. Microbiol.* **114**, 203. **70.** Wiegel (1981) *Arch. Microbiol.* **130**, 385. **71.** Kohlhaw, Leary & Umbarger (1969) *J. Biol. Chem.* **244**, 2218. **72.** Hill & Schlegel (1969) *Arch.*

Mikrobiol. **68**, 1. **73.** Ulm, Böhme & Kohlhaw (1972) *J. Bacteriol.* **110**, 1118. **74.** Bode & Birnbaum (1991) *J. Basic Microbiol.* **31**, 21. **75.** de Kraker, Luck, Textor, Tokuhisa & Gershenzon (2007) *Plant Physiol.* **143**, 970. **76.** de Carvalho, Argyrou & Blanchard (2005) *J. Amer. Chem. Soc.* **127**, 10004. **77.** Stieglitz & Calvo (1974) *J. Bacteriol.* **118**, 935. **78.** Beltzer, Morris & Kohlhaw (1988) *J. Biol. Chem.* **263**, 368. **79.** Teng-Leary & Kohlhaw (1973) *Biochemistry* **12**, 2980. **80.** Hagelstein & Schultz (1993) *Biol. Chem. Hoppe-Seyler* **374**, 1105. **81.** Adams, Dack, Dickinson & Ratledge (2002) *Biochim. Biophys. Acta* **1597**, 36. **82.** Rabin, Salomon, Bleiweis, Carlin & Ajl (1968) *Biochemistry* **7**, 377. **83.** Oda, Nakano & Kitaoka (1982) *J. Gen. Microbiol.* **128**, 1211. **84.** Chong, Chang & Choi (1997) *J. Biochem. Mol. Biol.* **30**, 274. **85.** Singh, Szamosi, Hand & Misra (1992) *Plant Physiol.* **99**, 812. **86.** Bekkaoui, Schorr & Crosby (1993) *Physiol. Plant.* **88**, 475. **87.** Southan & Copeland (1996) *Physiol. Plant.* **98**, 824. **88.** Blatt, Pledger & Umbarger (1972) *Biochem. Biophys. Res. Commun.* **48**, 444. **89.** DeFelice, Lago, Squires & Calvo (1982) *Ann. Microbiol.* **133A**, 251. **90.** Xing & Whitman (1987) *J. Bacteriol.* **169**, 4486. **91.** Arfin & Koziell (1973) *Biochim. Biophys. Acta* **321**, 348. **92.** Yang & Kim (1993) *Biochim. Biophys. Acta* **1157**, 178. **93.** Singh, Stidham & Shaner (1988) *J. Chromatogr.* **444**, 251. **94.** Legrain & Stalon (1976) *Eur. J. Biochem.* **63**, 289. **95.** Marshall & Cohen (1972) *J. Biol. Chem.* **247**, 1654. **96.** Pierson, Cox & Gilbert (1977) *J. Biol. Chem.* **252**, 6464. **97.** Xu, Yano, Yamamoto, Suzuki & Kumagai (1996) *Appl. Biochem. Biotechnol.* **56**, 277.

5-Leucine Enkephalin

This pentapeptide (FW = 553.62 g/mol; *Sequence*: YGGFL; Abbreviation: [Leu]Enkephalin) is an endogenous opioid neurotransmitter and neuromodulator that functions as an agonist at μ and δ opioid receptors. [Leu]Enkephalin is itself a substrate for dipeptidyl-peptidase III. **Target(s):** Ca^{2+}-dependent ATPase (1); carboxypeptidase E, K_i = 6.5 mM (2,3); dactylysin, K_i = 250 μM (The amide also inhibits.) (4,5); protein-disulfide isomerase (6). **1.** Yamasaki & Way (1985) *Neuropeptides* **5**, 359. **2.** Hook (1990) *Life Sci.* **47**, 1135. **3.** Hook & LaGamma (1987) *J. Biol. Chem.* **262**, 12583. **4.** Joudiou, Carvalho, Camarao, Boussetta & Cohen (1993) *Biochemistry* **32**, 5959. **5.** Carvalho, Joudiou, Boussetta, Leseney & Cohen (1992) *Proc. Natl. Acad. Sci. U.S.A.* **89**, 84. **6.** Morjana & Gilbert (1991) *Biochemistry* **30**, 4985.

L-Leucine Hydroxamate

This hydroxamate (FW = 146.19 g/mol; CAS 31982-78-2) inhibits *Vibrio* aminopeptidase and thermolysin (K_i = 14 μM). A substituted β-mercaptoketone is likely to inhibit in a similar manner (K_i = 1 μM), suggesting a mechanism in which a zinc-bound hydroxide ion participates in concerted proton-transfer processes and where the coordination and charge field at the zinc atom remain unchanged. **Target(s):** atrolysin A (1); bacterial leucyl aminopeptidase, as alternative substrate (2,3); coccolysin (4); leucyl aminopeptidase, as alternative substrate (3,4); membrane alanyl aminopeptidase (3); thermolysin, K_i = 190 μM (6-9); *Vibrio* aminopeptidase (10); vibriolysin (11). **1.** Fox, Campbell, Beggerly & Bjarnason (1986) *Eur. J. Biochem.* **156**, 65. **2.** Baker, Wilkes, Bayliss & Prescott (1983) *Biochemistry* **22**, 2098. **3.** Wilkes & Prescott (1983) *J. Biol. Chem.* **258**, 13517. **4.** Xu, Shawar & Dresden (1990) *Exp. Parasitol.* **70**, 124. **5.**

Mäkinen, Clewell, An & Mäkinen (1989) *J. Biol. Chem.* **264**, 3325. **6.** Beynon & Beaumont (1998) in *Handb. Proteolytic Enzymes*, p. 1037, Academic Press, San Diego. **7.** Wasserman & Hodge (1996) *Proteins* **24**, 227. **8.** Nishino & Powers (1978) *Biochemistry* **17**, 2846. **9.** Holmes & Matthews (1981) *Biochemistry* **20**, 6912. **10.** Chevier & D'Orchymont (1998) in *Handb. Proteolytic Enzymes*, p. 1433, Academic Press, San Diego. **11.** Wilkes, Bayliss & Prescott (1988) *J. Biol. Chem.* **263**, 1821.

Leucinostatins

This family of Class-II cystatins and antibiotic/phytotoxic nonapeptides (FW$_{\text{Leukostatin-A}}$ = 1218.66; CAS 76600-38-9) are produced by a number of different microorganisms, among them: *Paecilomyces marquandii, P. lilacinus*, and species of *Acremonium*). These agents also go by the synonyms Paecilotoxin, LS-87870, AC1O51DJ, 39405-64-6, and Antibiotic A20668. By far the most abundant family member, Leucinostatin A inhibits mitochondrial (1-8) and chloroplast (9) ATP synthases as well as prostate cancer growth by reducing insulin-like growth factor-I expression in prostate stromal cells (10). **1.** Linnett & Beechey (1979) *Meth. Enzymol.* **55**, 472. **2.** Shima, Fukushima, Arai & Terada (1990) *Cell Struct. Funct.* **15**, 53. **3.** Unitt & Lloyd (1981) *J. Gen. Microbiol.* **126**, 261. **4.** Higa & Cazzulo (1981) *Mol. Biochem. Parasitol.* **3**, 357. **5.** Yarlett & Lloyd (1981) *Mol. Biochem. Parasitol.* **3**, 13. **6.** Clarke, Fuller & Morris (1979) *Eur. J. Biochem.* **98**, 597. **7.** Bowman, Mainzer, Allen & Slayman (1978) *Biochim. Biophys. Acta* **512**, 13. **8.** Lardy, Reed & Lin (1975) *Fed. Proc.* **34**, 1707. **9.** Lucero, Ravizzine & Vallejos (1976) *FEBS Lett.* **68**, 141. **10.** Kawada, Inoue, Ohba, *et al.* (2010) *Int. J. Cancer* **126**, 810.

Leukemia Inhibitory Factor

This 202-residue interleukin-6 class cytokine (MW = 22007.74 g; NCBI Reference Sequence: NP 002300.1; Symbol: LIF) controls the induction of hematopoietic differentiation in normal myeloid and leukemia cells, the promotion of neuron and muscle cell differentiation, the regulation of mesenchymal-to-epithelial conversion, as well as trophoblast implantation (1-3). To activate signaling cascades in stromal cells, chondrocytes, osteoblasts, osteocytes, adipocytes, and synovial fibroblasts, LIF signals through cytokine-specific LIF receptor, *or* LIFR (also known as also known as Cluster of Differentiation-118, *or* CD118), which is complexed to the cytokine co-receptor Glycoprotein-130 (3). This same LIFR/Gp-130 receptor complex is also used by Ciliary Neurotrophic Growth Factor (CNTF), Oncostatin M, Cardiotrophin-1 (CT1) and Cardiotrophin-like Cytokine (CLC). **Role in Maintaining Embryonic Stem Cell Self-Renewal & Pluripotency:** Pluripotency is a transient *in vivo*, but cultured pluripotent cells can be indefinitely maintained in an artificially induced state of self-renewal by supplementing them with appropriate exogenous cues (2). An indispensable agent for the pluripotency of mouse embryonic stem cells (mESCs), LIF binds to the LIF receptor, thereby activating JAK/STAT3, PI3K/AKT and SHP2/MAPK signaling pathways to choreograph the gene expression pattern that is specific to mESCs (3). When LIF is withdrawn from culture medium, the activity of mTOR (mammalian Target of Rapamycin) increases rapidly, as indicated by the phosphorylation of its targets, ribosomal protein S6 and translation factor 4EBP1. Suppression of phosphorylation at Tyr-705 in STAT3 by WP1066 (IC$_{50}$ = 2.4 µM) also activates phosphorylation of the mTOR target S6 ribosomal protein (*See also WP1066*). LIF removal strongly activates ERK activity, either by direct phosphorylation of mTOR or by phosphorylation of upstream negative regulators of mTOR, such as TSC1 and TSC2 proteins (4). LIF withdrawal leads to phosphorylation of TSC2 protein, relieving its inhibition of mTOR activity and thereby decreasing pluripotent gene expression Oct-4, Nanog, Sox2, while augmenting expression of the fgf5 gene, a marker of post-implantation epiblasts. Such findings indicate that LIF-depleted mouse ESCs undergo a transition from the LIF/STAT3-supported pluripotency to the FGFR/ERK-committed, primed-like state that attends expression of early differentiation markers, as mediated by mTOR signaling (5). Although LIF knockdown suppresses proliferation and invasion of osteosarcoma by

blocking the STAT3 signal pathway, treatment with recombinant LIF significantly promotes osteosarcoma growth and invasion by enhancing STAT3 phosphorylation, an effect that can be partially neutralized by the STAT3 inhibitor, HO-3867 (5). See Reference-6 for details on the molecular cloning and expression of murine myeloid LIF cDNA. **1.** Granger, Moore, White, *et al.* (1970) *J. Immunol.* **104**, 1476. **2.** Weinberger, Ayyash, Novershstern & Hanna (2016) *Nature Reviews: Molecular Cell Biol.* **17**, 155. **3** Sims & Johnson (2012) *Growth Factors* **30**, 76. **4.** Cherepkova, Sineva & Pospelov (2016) *Cell Death Dis.* **7**, e2050. **5.** Liu, Lu, Li, *et al.* (2015) *Acta Pathol. Microbiol. Immunol. Scand.* **123**, 837. **6.** Gearing, Gough, King, *et al.*, (1987) *EMBO J.* **6**, 3995.

Leukotriene A₃ & Leukotriene A₃ Methyl Ester

This 5-hydroperoxyeicosatrienoate metabolite (FW$_{\text{free-acid}}$ = 320.48 g/mol; IUPAC: 5S-*trans*-5,6-oxido-7E,9E,11Z-eicosatrienoic acid; *Abbreviation* = LTA₃) and its cell-permeable methyl ester (FW$_{\text{ester}}$ = 334.49 g/mol; CAS 83851-38-1) are natural and synthetic precursors, respectively, of leukotriene C₃. Leukotriene A₃ is an analogue of leukotriene A₄, but lacks the C14-C15 double bond. As the free acid, LTA₃ is highly unstable, and the methyl ester is preferred in biochemical experiments that rely on endogenous esterases to generate the free acid. LTA₃ is a potent suicide inhibitor of leukotriene A₄ hydrolase, which catalyzes the conversion of the unstable epoxide LTA₄ into pro-inflammatory LTB₄, by reacting covalently with an active-site residue (1-3). Mass spectroscopic analysis of proteolytic fragments are consistent with nucleophilic attack by the hydroxyl group of Tyrosine-383 at the conjugated triene epoxide of LTA₃, thereby forming a triene ether carbinol covalent adduct (4). **1.** Ohishi, Izumi, Seyama & Shimizu (1990) *Meth. Enzymol.* **187**, 286. **2.** Evans, Nathaniel, Zamboni & Ford-Hutchinson (1985) *J. Biol. Chem.* **260**, 10966. **3.** Evans, Nathaniel, Charleson, *et al.* (1986) *Prostaglandins Leukot. Med.* **23**, 167. **4.** Mancini, Waugh, Thompson, *et al.* (1998) *Arch. Biochem. Biophys.* **354**, 117.

Leukotriene A₃

This 5-hydroperoxyeicosatrienoate metabolite (FW = 304.46 g/mol; CAS 83851-38-1 (cell-permeable methyl ester); *Abbreviation* = LTA₃) is a precursor of leukotriene C₃. Leukotriene A₃ is an analogue of leukotriene A₄, lacking the C14 double bond. LTA₃ is a potent suicide inhibitor of leukotriene A₄ hydrolase, with an active-site tyrosine residue in the hydrolase acting as the nucleophile and forming a carbinol adduct. **1.** Ohishi, Izumi, Seyama & Shimizu (1990) *Meth. Enzymol.* **187**, 286. **2.** Mancini, Waugh, Thompson, *et al.* (1998) *Arch. Biochem. Biophys.* **354**, 117. **3.** Evans, Nathaniel, Zamboni & Ford-Hutchinson (1985) *J. Biol. Chem.* **260**, 10966. **4.** Evans, Nathaniel, Charleson, *et al.* (1986) *Prostaglandins Leukot. Med.* **23**, 167.

Leukotriene A₄ & Leukotriene A₄ Methyl Ester

This unstable eicosatetraenoate (FW = 317.45 g/mol; IUPAC: 5(S)-trans-5,6-oxido-11,14-*cis*-eicosatetraenoate; *Abbreviation* = LTA₄; FW$_{\text{ester}}$ = 332.48 g/mol; CAS 73466-12-3) is synthesized in mast cells, eosinophils, and neutrophils from arachidonic acid by 5-lipoxygenase (5-LO), which exhibits both lipoxygenase and LTA₄ synthase activities. It is a substrate of leukotriene A₄ hydrolase, which catalyzes the conversion of LTA₄ into pro-inflammatory leukotriene B_4 (or LTB_4). During catalysis, this hydrolase, which also exhibits peptidase activity, is inactivated through covalent binding to an active-site tyrosyl residue (1-17). This process blocks both enzyme activities and may be of importance for the overall regulation of LTB_4 biosynthesis. In addition, leukotriene A_4, which is produced by 5-lipoxygenase, irreversibly inactivates the lipoxygenase, representing an unusual case of product inactivation (18,19). Note also that leukotriene A_4 is also a substrate of a glutathione S-transferase, producing leukotriene C_4. **1.** Haeggström (1998) in *Handb. Proteolytic Enzymes*, p. 1022, Academic Press, San Diego. **2.** Ohishi, Izumi, Seyama & Shimizu (1990) *Meth. Enzymol.* **187**, 286. **3.** Haeggström (1990) *Meth. Enzymol.* **187**, 324. **4.** Mancini, Waugh, Thompson, *et al.* (1998) *Arch. Biochem. Biophys.* **354**, 117. **5.** Kull, Ohlson, Lind & Haeggström (2001) *Biochemistry* **40**, 12695. **6.** Evans, Nathaniel, Zamboni & Ford-Hutchinson (1985) *J. Biol. Chem.* **260**, 10966. **7.** McGee & Fitzpatrick (1985) *J. Biol. Chem.* **260**, 12832. **8.** Mueller, Wetterholm, Blomster, *et al.* (1995) *Proc. Natl. Acad. Sci. U.S.A.* **92**, 8383. **9.** Mueller, Blomster, Oppermann, *et al.* (1996) *Proc. Natl. Acad. Sci. U.S.A.* **93**, 5931. **10.** Orning, Jones & Fitzpatrick (1990) *J. Biol. Chem.* **265**, 14911. **11.** Ohishi, Izumi, Minami, *et al.* (1987) *J. Biol. Chem.* **262**, 10200. **12.** Haeggström, Kull, Rudberg, Tholander & Thunnissen (2002) *Prostaglandins* **68 69**, 495. **13.** Kull, Ohlson & Haeggström (1999) *J. Biol. Chem.* **274**, 34683. **14.** Orning, Krivi & Fitzpatrick (1991) *J. Biol. Chem.* **266**, 1375. **15.** Samuelsson & Funk (1989) *J. Biol. Chem.* **264**, 19469. **16.** Orning & Fitzpatrick (1999) *Arch. Biochem. Biophys.* **368**, 131. **17.** Mueller, Andberg & Haeggström (1998) *J. Biol. Chem.* **273**, 11570. **18.** Arai, Shimoji, Konno, *et al.* (1983) *J. Med. Chem.* **26**, 72. **19.** Lepley & Fitzpatrick (1994) *J. Biol. Chem.* **269**, 2627.

14,15-Leukotriene A_4

This unstable fatty acid derivative ($FW_{free-acid}$ = 318.46 g/mol; *Abbreviation* = 14,15-LTA_4), also known as 14,15-oxido-5,8-*cis*-10,12-*trans* eicosatetraenoic acid, is an analogue of leukotriene A_4 (1,2). 14,15-LTA_4 is not a substrate of leukotriene A_4 hydrolase. It is produced from 15 hydroperoxyeicosatetraenoate in anaerobic incubations. **1.** Ohishi, Izumi, Seyama & Shimizu (1990) *Meth. Enzymol.* **187**, 286. **2.** Ohishi, Izumi, Minami, *et al.* (1987) *J. Biol. Chem.* **262**, 10200.

Leukotriene A_5

This leukotriene A_4 analogue (FW = 316.44 g/mol; CAS 90121-06-5 (membrane-penetrating methyl ester); *Abbreviation* = LTA_5), also known as 5(*S*)-*trans*-5,6-oxido-7,9-*trans*-11,14,17-*cis* eicosapentaenoic acid, is also a substrate of leukotriene A_4 hydrolase, albeit weaker, and also acts as a suicide substrate inhibitor. **1.** Ohishi, Izumi, Seyama & Shimizu (1990) *Meth. Enzymol.* **187**, 286. **2.** Nathaniel, Evans, Leblanc, *et al.* (1985) *Biochem. Biophys. Res. Commun.* **131**, 827. **3.** Evans, Nathaniel, Charleson, *et al.* (1986) *Prostaglandins Leukot. Med.* **23**, 167.

Leukotriene B_4

This unstable leukotriene ($FW_{free-acid}$ = 336.47 g/mol; CAS71160-24-2; *Abbreviation* = LTB_4), also known as 5*S*,12*R*-dihydroxy-6*Z*,8*E*,10*E*,14*Z*-eicosatetraenoic acid, is a potent chemotactic agent for polymorphonuclear leukocytes. It increases vascular permeability and functions an activator of inflammatory cells. LTB_4 also stimulated c-Fos and c-Jun proto-oncogene transcription in human monocytes. **Target(s):** leukotriene E4 20-

monooxygenase (1). **1.** Huwyler, Jedlitschky, Keppler & Gut (1992) *Eur. J. Biochem.* **206**, 869.

Leukotriene C_4

This leukotriene (FW = 625.77g/mol; CAS 72025-60-6; *Abbreviation* = LTC_4) is formed by the conjugation of glutathione to leukotriene A_4, via leukotriene C_4 synthase and glutathione S-transferases. It induces renal vasoconstriction and broncho-constriction. Leukotriene C_4 is an important mediator of vascular tone, permeability, and smooth muscle contractility. It is also a very good substrate (K_m = 6 µM) in the γ–glutamyl transpeptidase reaction. **Target(s):** N-acetylgalactosamine-4-sulfatase, *or* arylsulfatase B (2); glutathione S transferase-1 (3); 15-hydroxyprostaglandin dehydrogenase (NADP+) (4); xenobiotic-transporting ATPase, multidrug-resistance protein (5,6). **1.** Anderson, Allison & Meister (1982) *Proc. Natl. Acad. Sci. U.S.A.* **79**, 1088. **2.** Weller, Corey, Austen & Lewis (1986) *J. Biol. Chem.* **261**, 1737. **3.** Bannenberg, Dahlen, Luijerink, Lundqvist & Morgenstern (1999) *J. Biol. Chem.* **274**, 1994. **4.** Chung, Harvey, Armstrong & Jarabak (1987) *J. Biol. Chem.* **262**, 12448. **5.** Loe, Stewart, Massey, Deeley & Cole (1997) *Mol. Pharm.* **51**, 1034. **6.** Paul, Breuninger, Tew, Shen & Kruh (1996) *Proc. Natl. Acad. Sci. U.S.A.* **93**, 6929.

Leukotriene D_4

This leukotriene (FW = 496.66g/mol; CAS 73836-78-9; *Abbreviation* = LTD4), produced by γ-glutamyl transpeptidase on leukotriene C_4 (1), is a potent broncho-constrictor and participates in inflammatory processes. **Target(s):** N-acetylgalactosamine-4-sulfatase, *or* arylsulfatase B (2); leukotriene-C_4 synthase (3). **1.** Anderson, Allison & Meister (1982) *Proc. Natl. Acad. Sci. U.S.A.* **79**, 1088. **2.** Weller, Corey, Austen & Lewis (1986) *J. Biol. Chem.* **261**, 1737. **3.** Nicholson, Klemba, Rasper, Metters & Zamboni (1992) *Eur. J. Biochem.* **209**, 725.

Leukotriene E_4

This leukotriene (FW = 439.61g/mol; CAS 75715-89-8; *Abbreviation* = LTE_4), produced from leukotriene D_4 via the action of a dipeptidase (1), mediates anaphalaxis and inflammatory processes. **Target(s):** N-acetylgalactosamine-4-sulfatase, *or* arylsulfatase B (2); leukotriene-C_4 synthase (3). **1.** Anderson, Allison & Meister (1982) *Proc. Natl. Acad. Sci. U.S.A.* **79**, 1088. **2.** Weller, Corey, Austen & Lewis (1986) *J. Biol. Chem.*

261, 1737. **3**. Nicholson, Klemba, Rasper, Metters & Zamboni (1992) *Eur. J. Biochem.* **209**, 725.

Leupeptin

This widely used, water-soluble tripeptide aldehyde (FW = 426.55 g/mol; CAS 55123-66-5; *Sequence*: *N*-Ac-LLR-CHO), originally isolated from an *Actinomycetes* broth (1,2), is a transition-state mimic (as thye hydrated aldehyde) and potent competitive inhibitor of many proteinases, including trypsin, plasmin, cathepsin B, and papain. Leupeptin is an ineffective against chymotrypsin, pepsin, thrombin, elastase, cathepsin A, and cathepsin D. Leupeptin is itself an alternative substrate for some proteinases, such as thermolysin. Note that the terminal aldehyde group, which in aqueous solutions is either hydrated or rearranged into cyclic carbinolamine, is essential for inhibition. **Practical Considerations:** Due to its limited specificity, leupeptin is frequently used in conjunction with other protease inhibitors during the purification of protease-sensitive proteins. The effective inhibitory concentration range is 10-100 μM. Stock solutions are prepared as a 5-10 mM aqueous solution. If needed, higher stock concentrations are achieved in *N*,*N*-dimethylformamide or ethanol. Aqeous stock solutions are stable for hours at room temperature, for days in an ice-bath, and for months at –20 °C. **1**. Aoyagi, Takeuchi, Matsuzaki, *et al.* (1969) *J. Antibiotics (Tokyo)*, Series A **22**, 283. **2**. Aoyagi, Miyata, Nanbo, *et al.* (1969) *J. Antibiotics (Tokyo)*, Series A **22**, 558.

Leuprolide

This pituitary GnRH receptor densensitizer and luteinizing hormone-releasing hormone (LHRH) agonist (FW = 1209.40 g/mol; CAS 53714-56-0; Sequence: Pyro-Glu-His-Trp-Ser-Tyr-D-Leu-Leu-Arg-Pro-NHC$_2$H$_5$), also known as Leuprorelin and Lupron®, acts primarily on the anterior pituitary, inducing a transient early rise in gonadotrophin release. With continued use, leuprorelin treatment results in pituitary desensitization, indirectly down-regulating secretion of gonadotropins luteinizing hormone (LH) and follicle-stimulating hormone (FSH) and dramatically reducing estradiol and testosterone levels in both sexes. **1**. Plosker & Brogden (1994) *Drugs* **48**, 930.

Levamisole

This antihelminthic (FW$_{free-base}$ = 204.30 g/mol; M.P. = 60-61.5°C), also called (–)-2,3,5,6-tetrahydro-6-phenylimidazo[2,1-*b*]thiazole, is the 6*S* stereoisomer of tetramisole (*i.e.*, an equal mixture of levamisole, the (–)-stereoisomer, and dexamisole, the (+)-enantiomer). Some literature sources use the terms tetramisole and levamisole synonymously. Levamisole inhibits liver, bone, and kidney alkaline phosphatases; however, intestinal alkaline phosphatase is only weakly inhibited. (***See also*** *Tetramisole*) **Target(s):** alkaline phosphatase (1-12); *Ascaridia galli* ATPase (13); *Ascaridia galli* acid phosphatase (13); Ca^{2+}-dependent ATPase, *Setaria cervi* (14,15); ecto-alkaline phosphatase (3,4); membrane-specific ATPase (16); Na$^+$/K$^+$-exchanging ATPase (15,17); organic cation transporter rOCT1 (18); phosphotyrosine phosphatase (19); pyridoxal phosphatase (20). **1**. Van Belle (1972) *Biochim. Biophys. Acta* **289**, 158. **2**. Cyboron & R. E. Wuthier (1981) *J. Biol. Chem.* **256**, 7262. **3**. Anagnostou, Plas, Nefussi & Forest (1996) *J. Cell Biochem.* **62**, 262. **4**. Anagnostou, C. Plas & N. Forest (1996) *J. Cell Biochem.* **60**, 484. **5**. Stagni, Vittur & B. de Bernard (1983) *Biochim. Biophys. Acta* **761**, 246. **6**. Herz (1985) *Experientia* **41**, 1357. **7**. Stinson & Chan (1987) *Adv. Protein Phosphatases* **4**, 127. **8**. Magnusson & Farley (2002) *Calcif. Tissue Int.* **71**, 508. **9**. Kutzler, Solter, Hoffman & Volkmann (2003) *Theriogenology* **60**, 299. **10**. Champion, Glazier, Greenwood, *et al.* (2003) *Placenta* **24**, 453. **11**. Jahan & Butterworth (1984) *Biochem. Soc. Trans.* **12**, 792. **12**. Mussarat & Butterworth (1985) *Biochem. Soc. Trans.* **13**, 773. **13**. Aggarwal, Sanyal & Khera (1992) *Acta Vet. Hung.* **40**, 243. **14**. Agarwal, Shukla & Tekwani (1992) *Int. J. Biochem.* **24**, 1447. **15**. Agarwal, Tekwani, Shukla & Ghatak (1990) *Indian J. Exp. Biol.* **28**, 245. **16**. Pizauro, Demenis, Ciancaglini & Leone (1998) *Biochim. Biophys. Acta* **1368**, 108. **17**. Skobis & Bereiter-Hahn (1991) *Naturwissenschaften* **78**, 226. **18**. Martel, Ribeiro, Calhau & Azevedo (1999) *Pharmacol. Res.* **40**, 275. **19**. Burch, Hamner & Wuthier (1985) *Metabolism* **34**, 169. **20**. Tazoe, Ichikawa & Hoshino (2005) *Biosci. Biotechnol. Biochem.* **69**, 2277.

Levetiracetam

This orally available nonanaleptic drug (FW = 170.21 g/mol; CAS 102767-28-2; the *S*-enantiomer of etiracetam), also known as UCB L059, Keppra® and (*S*)-2-(2-oxopyrrolidin-1-yl)butanamide, is a high-affinity synaptic vesicle protein 2A ligand that is indicated for the treatment of partial (focal), myoclonic, and tonic-clonic seizures. Absorption is rapid and essentially complete, putting levetiracetam's bioavailability at ~100 %, with no effect of food on absorption. The volume of distribution of levetiracetam is similar to total body water. It is not extensively metabolized, and any metabolites that do form are pharmacologically inactive. The plasma half-life in adults is 6-8 hours. Levetiracetam is eliminated primarily by the kidneys, with ~66 percent of the drug unchanged. The drug is active, even when memory deficits are induced by electroconvulsive shock, undertraining, or by a long training-to-test interval (1). Levetiracetam induces alterations in GABA metabolism and turnover in discrete areas of rat brain and reduces neuronal activity in substantia nigra pars reticulata (2). It also prevents perforin-mediated neuronal injury induced by acute cerebral ischemia reperfusion (3). LEV inhibits voltage-gated presynaptic Ca$_V$ channels, reducing neuronal excitability (4). Levetiracetam reverses the inhibition by negative allosteric modulators of neuronal GABA- and glycine-gated currents (5). **1**. Sara (1980) *Psychopharmacol.* (Berl.) **68**, 235. **2**. Löscher, Hönack & Bloms-Funke (1996) *Brain Res.* **735**, 208. **3**. Zhang, Li, Zuo, *et al.* (2015) *Mol. Neurobiol.* Doi:10.1007/s12035-015-9467-9. **4**. Vogl, Mochida, Wolff, Whalley & Stephens (2012) *Mol. Pharmacol.* **82**, 199. **5**. Rigo, Hans, Nguyen, *et al.* (2002) *Brit. J. Pharmacol.* **136**, 659.

Levofloxacin

This fluoroquinolone-class antibiotic (FW = 361.37 g/mol; CAS 100986-85-4; IUPAC Name: (*S*)-9-fluoro-2,3-dihydro-3-methyl-10-(4-methylpiperazin-1-yl)-7-oxo-7*H*-pyrido[*1,2,3-de*]-1,4-benzoxazine-6-carboxylate), also trade-named Levaquin®, Tavanic®, and Iquix®, is a broad-spectrum systemic antibacterial that targets Type II DNA topoisomerases (gyrases), enzymes that are required for bacterial replication and transcription. Levofloxacin shows greater action against the respiratory pathogen *Streptococcus pneumoniae* than ciprofloxacin. It is also used to treat urinary tract infections, prostatitis, anthrax, endocarditis, meningitis, pelvic inflammatory disease, traveler's diarrhea, tuberculosis, and bubonic plague. For the prototypical member of this antibiotic class, ***See*** *Ciprofloxacin*

Levomilnacipran

This FDA-approved antidepressant (FW = 246.35 g/mol; CAS 96847-55-1; *Symbol*: LVM), also known by its code name F2695, its proprietary name Fetzima®, and (*1S,2R*)-2-(aminomethyl)-*N*,*N*-diethyl-1-phenylcyclopropanecarboxamide, is the more active enantiomer of milnacipran, displaying therefore similar pharmacological action as a serotonin-norepinephrine reuptake inhibitor (***See*** *Milnacipran*). LVM exhibits high affinity for human norepinephrine transporter (*K*$_i$ = 92.2 nM), 5-HT transporter (K_i = 11.2 nM), and potently inhibited norepinephrine (IC$_{50}$ = 10.5 nM) and 5-HT (IC$_{50}$ = 19.0 nM) reuptake (human transporter) *in vitro*. LVM had 2-fold greater potency for norepinephrine relative to serotonin reuptake inhibition (i.e. NE/5-HT potency ratio: 0.6) and 17 and 27 times higher selectivity for NE reuptake inhibition compared with venlafaxine and duloxetine, respectively. LVM exhibit inconsequential affinity for 23 off-target receptors. **1**. Auclair, Martel, Assié, *et al.* (2013) *Neuropharmacol.* **70**, 338. **2**. Citrome (2013) *Int. J. Clin. Pract.* **67**, 1089. **3**. Saraceni, Venci & Gandhi (2013) *J. Pharm. Pract.* **27**, 389.

Levonorgestrel

This orally available second-generation synthetic progestogen and emergency contraceptive, *or ECP* (FW = 312.45 g/mol; CAS 797-63-7; IUPAC: (–)-13-ethyl-17-ethynyl-17-hydroxy-1,2,6,7,8,9,10,11,12,13,14,15, 16,17-tetradecahydrocyclopenta[*a*]phenanthren-3-one) is an active ingredient in several contraceptives (including combined oral contraceptive pills, progestogen-only pills, intrauterine devices and as contraceptive implants) as well as in hormone-replacement therapy. Levonorgestrel's action as an emergency contraceptive is mainly by preventing ovulation or fertilization by altering tubal transport of sperm and/or egg. It may also inhibit implantation by altering the endometrium. Efficacy is higher if taken as soon as possible after unprotected intercourse. Moreover, levonorgestrel can be used at any time during the menstrual cycle. Levonorgestrel is ineffective after implantation has commenced. **Pharmacokinetics:** *Absorption*: After administration of a single dose of Plan B (0.75 mg) under fasting conditions, maximum serum concentrations of levonorgestrel are 14.1 ± 7.7 ng/mL for an average of 1.6 ± 0.7 hours. *Distribution*: Serum levonorgestrel is primarily protein bound, with ~50% bound to albumin and 47.5% bound to Sex Hormone Binding Globulin (*or SHBG*). *Metabolism*: Following a single oral dosage, levonorgestrel does not appear to be extensively metabolized by the liver. The primary metabolites are 3α,5β- and 3α,5α-tetrahydrolevonorgestrel, with 16β-hydroxynorgestrel also identified. Together, they account for less than 10% of parent plasma levels. Urinary metabolites are hydroxylated at the 2α-and 16β-positions. Small amounts of these metabolites undergo sulfation and glucuronidation. *Excretion*: The elimination half-life of levonorgestrel following single dose administration as Plan B (0.75 mg) is 24.4 ± 5.3 hours. ***See also Mifepristone and Ulipristal acetate*** **1**. Brand Names of Levonorgestrel-Only ECPs: Escapelle®, Plan B®, Levonelle®, Glanique®, NorLevo®, Postinor-2®, i-Pill®, Next Choice®, and 72-HOURS®.

Levulinate (Levulinic Acid)

This photosensitive γ-keto acid (FW = 116.12 g/mol; Soluble in water and ethanol; M.P. = 33-35°C), also known as 4-ketopentanoic acid and 4-oxopentanoic acid, which is produced by boiling starch or cellulose in aqueous HCl, inhibits porpholinogen synthase by forming a Schiff base with the ε-amino groups of an active-site lysyl residue. **Target(s):** γ-butyrobetaine dioxygenase, *or* γ-butyrobetaine hydroxylase, K_i = 13 mM (1); porphobilinogen synthase, *or* δ-aminolevulinate synthase (2-10); procollagen-proline 3-dioxygenase (11). **1**. Ng, Hanauske-Abel & England (1991) *J. Biol. Chem.* **266**, 1526. **2**. Shemin (1970) *Meth. Enzymol.* **17A**, 205. **3**. Erskine, Newbold, Brindley, *et al.* (2001) *J. Mol. Biol.* **312**, 133. **4**. Erskine, Newbold, Roper, *et al.* (1999) *Protein Sci.* **8**, 1250. **5**. Frankenberg, Erskine, Cooper, *et al.* (1999) *J. Mol. Biol.* **289**, 591. **6**. Erskine, Norton, Cooper, *et al.* (1999) *Biochemistry* **38**, 4266. **7**. Barnard, Itoh, Hohberger & Shemin (1977) *J. Biol. Chem.* **252**, 8965. **8**. Nandi & Shemin (1968) *J. Biol. Chem.* **243**, 1236. **9**. Yamasaki & Moriyama (1971) *Biochim. Biophys. Acta* **227**, 698. **10**. Van Heyningen & Shemin (1971)

Biochem. J. **124**, 68P. **11**. Majamaa, Turpeenniemi-Hujanen, Latipää, *et al.* (1985) *Biochem. J.* **229**, 127.

LGD1069, *See Bexarotene*

LGK-974

This potent and specific PORCN inhibitor (FW = 396.44 g/mol; CAS 1243244-14-5; Solubility: <1 mg/mL DMSO, Ethanol, H$_2$O), also known 2',3-dimethyl-N-[5-(2-pyrazinyl)-2-pyridinyl]-[2,4'-bipyridine]-5-acetamide, targets Wnt signaling with IC$_{50}$ of 0.4 nM. LGK974 is potent and efficacious in multiple tumor models at well-tolerated doses in vivo, including murine and rat mechanistic breast cancer models driven by MMTV-Wnt1 and a human head and neck squamous cell carcinoma model (1,2). Mutational inactivation of RNF43 in pancreatic adenocarcinoma confers Wnt dependency. **1**. Liu J, et al. (2013) *Proc. Natl. Acad. Sci. U.S.A.* **110**, 20224. **2**. Jiang X, et al. (2013) *Proc. Natl. Acad. Sci. U.S.A.* **110**, 12649.

Lidocaine

This rapid-acting local anesthetic (FW$_{free-base}$ = 234.34 g/mol; CAS 137-58-6; M.P. = 68-69°C; water-insoluble; Systematic Name: 2-(diethylamino)-N-(2,6-dimethylphenyl)acetamide; HCl monohydrate (FW$_{HCl-H2O}$ = 288.81 g/mol; CAS 6108-05-0), highly water-soluble; pK_a = 7.8), also known by the trade names Xylocaine®, Recticare®, Lidoderm®, Solarcaine®, and Lmx4®, is a membrane-stabilizing agent and Na$^+$ channel blocker. Lidocaine finds broad utility in spinal anesthesia, intravenous regional anesthesia, surface anesthesia, emergency treatment of ventricular arrhythmias, and as a peripheral nerve block. In the heart, lidocaine reduces Phase-4 depolarization and automaticity. Lidocaine patches relieve the pain of post-herpetic neuralgia associated with shingles. **Primary Mode of Action:** Lidocaine stabilizes neuronal membranes by inhibiting the sodium ion fluxes needed to initiate nerve impulses, resulting in local anesthetic action. Most local anesthetics are hydrophobic and access the cytoplasmic-face of the sodium channel by diffusing through the peripheral membrane, thereby diminishing the excitability of all neurons, not just sensory neurons. The membrane-impermeant lidocaine derivative QX-314, however, relies on the TRPV1 (transient receptor potential vanilloid type-1) channel, to target primary sensory nociceptor neurons selectively, thereby eliciting its pain-specific local anesthetic effect on sodium channels (1) Moreover, capsaicin activates TRPV1, allowing more QX-314 (lidocaine) to enter primary sensory nociceptor neurons, thereby synergizing lidocaine anesthesia. Lidocaine also induces ROCK-dependent membrane blebbing and subsequent cell death in rabbit articular chondrocytes (2). **Target(s):** acetylcholinesterase (3); actomyosin motility (4); Ca^{2+}-dependent cyclic-nucleotide phosphodiesterase (5); histamine *N*-methyltransferase (6); 15-hydroxyprostaglandin dehydrogenase (7); kinesin (8); M$_3$ muscarinic acetylcholine receptors (9); Na$^+$ channels (10-12); Na$^+$/K$^+$-exchanging ATPase, weakly inhibited (13); protein-tyrosine kinase (14); sterol acyltransferase, *or* cholesterol acyltransferase, *or* ACAT (15-17). **1**. Binshtok, Bean & Woolf (2007) *Nature* **449**, 607. **2**. Maeda, Toyoda, Imai, *et al.* (2016) *J. Orthop. Res.* **34**, 754. **3**. Hannesson & DeVries (1990) *J. Neurosci. Res.* **27**, 84. **4**. Tsuda, Mashimo, Yoshiya, *et al.* (1996) *Biophys. J.* **71**, 2733. **5**. Hidaka, Inagaki, Nishikawa & Tanaka (1988) *Meth. Enzymol.* **159**, 652. **6**. Thithapandha & Cohn (1978) *Biochem. Pharmacol.* **27**, 263. **7**. Pace-Asciak & Smith (1983) *The Enzymes*, 3rd ed., **16**, 543. **8**. Miyamoto, Muto, Mashimo, *et al.* (2000) *Biophys. J.* **78**, 940. **9**. Hollmann, Ritter, Henle, *et al.* (2001) *Brit. J. Pharmacol.* **133**, 207. **10**. Charpentier (2002) *Gen. Physiol. Biophys.* **21**, 355. **11**. Sheets & Hanck (2003) *J. Gen. Physiol.* **121**, 163. **12**. Courtney (1975) *J. Pharmacol. Exp. Ther.* **195**, 225. **13**. Kutchai & Geddis (2001) *Pharmacol. Res.* **43**, 399. **14**. Hirata,

Sakaguchi, Mochida, *et al.* (2004) *Anesthesiology* **100**, 1206. **15.** Billheimer (1985) *Meth. Enzymol.* **111**, 286. **16.** Chang & Doolittle (1983) *The Enzymes*, 3rd ed., **16**, 523. **17.** Bell & Hubert (1980) *Biochim. Biophys. Acta* **619**, 302.

LIMKi 3

This potent LIMK inhibitor (FW = 431.29 g/mol; CAS Regtistry Number = 1338247-35-0; Solubility: 100 mM in DMSO), also named BMS 5 and *N*-[5-[1-(2,6-dichlorophenyl)-3-(difluoromethyl)-1*H*-pyrazol-5-yl]-2-thiazolyl]-2-methylpropanamide, targets LIM kinases LIMK1 (IC_{50} = 7 nM) and LIMK2 (IC_{50} = 8 nM), inhibiting phosphorylation of ADF/cofilin in MDA-MB-231 breast cancer cells and reducing MDA-MB-231 tumor cell invasion in a 3D matrigel invasion assay (1-3). LIMK1 and LIMK2 regulate the actin cytoskeleton by phosphorylating (and thereby inactivating) the cofilin family of actin-depolymerizing factors. LIMK1 also acts to destabilize microtubules and regulates cell motility, including tumor metastasis. **1.** Ross-Macdonald, de Silva, Guo, *et al.* (2008) *Mol. Cancer Ther.* **7**, 3490. **2.** Scott, Hooper, Crighton, *et al.* (2010) *J. Cell Biol.* **191**, 169. **3.** Sparrow, Manetti, Bott, *et al.* (2012) *J. Neurosci.* **32**, 5284.

Linaclotide

This oral tetradecapeptide drug (FW =1526.74 g/mol; CAS 851199-59-2), also known by the trade names Linzess® and Constella®, activates guanylate cyclase-C (GC-C) receptors on the luminal membrane surface to increase chloride and bicarbonate secretions into the intestine and to inhibit sodium ion absorption (1). By increasing the secretion of water into the lumen and improving defecation, linaclotide is an FDA-approved drug for the treatment of constipation-predominant irritable bowel syndrome (IBS-C) as well as chronic constipation. Linaclotide is minimally absorbed into systemic circulation. In mice, linaclotide inhibits colonic nociceptors with greater efficacy during chronic visceral hypersensitivity (2). Intra-colonic administration of linaclotide reduces signaling of noxious colorectal distention to the spinal cord. The colonic mucosa, but not neurons, expresses linaclotide's target, GC-C, and the latter's downstream effector cGMP is released after administration of linaclotide and also inhibited nociceptors. The linaclotide effects are lost in Gucy2c⁻/⁻ knock-out mice and is likewise prevented by inhibiting cGMP transporters or removing the mucosa (2). **1.** Lee & Wald (2011) *Expert Opin. Drug Metab. Toxicol.* **7**, 651. **2.** Castro, Harrington, Hughes, *et al.* (2013) *Gastroenterology.* **145**, 1334.

Linezolid

This orally available 1,3-oxazolidinone-containing antibiotic (FW = 337.35 g/mol; CAS 165800-03-3), also known by the trade names Zyvox® (U.S., UK, Australia), Zyvoxid® (Europe), and Zyvoxam® (Canada & Mexico), Linospan® (India) and Arlin® (Bangladesh) as well as by its systematic name (*S*)-*N*-({3-[3-fluoro-4-(morpholin-4-yl)phenyl]-2-oxo-1,3-oxazolidin-5-yl}methyl)acetamide, targets translation at its initiation stage, most likely by preventing the formation of the initiation complex (1-3). Linezolid is active against most disease-causing Gram-positive bacteria, including

vancomycin-resistant enterococci (VRE) and methicillin-resistant *Staphylococcus aureus* (MRSA). Due to its unique mechanism of action, cross-resistance between linezolid and other protein synthesis inhibitors is highly infrequent or nonexistent. **Mode of Antibiotic Action:** Initiation of protein synthesis requires the simultaneous presence of *N*-formyl-methionyl-tRNA, 30S ribosome subunit, mRNA, GTP, and initiation factors IF1, IF2, and IF3. By binding at the 23S portion of the 50S subunit (near sites for chloramphenicol and lincomycin), linezolid distorts the local structure, preventing formation of the initiation complex from 30S and 50S units, tRNA, and mRNA. Linezolid does not inhibit the independent binding of either mRNA or *N*-formylmethionyl-tRNA to *E. coli* 30S ribosomal subunits, nor does it prevent the formation of the IF2-*N*-formylmethionyl-tRNA binary complex (3). Oxazolidinone antibiotics therefore inhibit the formation of the initiation complex in bacterial translation systems by preventing formation of *N*-formylmethionyl-tRNA·Ribosome·mRNA ternary complex (3). Linezolid is also a potent inhibitor of cell-free transcription-translation in *E. coli* (IC_{50} = 1.8 μM). (**See also** *Tedizolid*) In view of structural similarities between bacterial and mitochondrial ribosomes, linezolid is apt to inhibit vital cellular processes that may account for neuropathies noted with long-term linezolid therapy. **Key Pharmacokinetic Parameters:** *See* Appendix II in Goodman & Gilman's THE PHARMACOLOGICAL BASIS OF THERAPEUTICS, 12ᵗʰ Edition (Brunton, Chabner & Knollmann, eds.) McGraw-Hill Medical, New York (2011). **1.** Spangler, Jacobs & Appelbaum (1996) *Antimicrob. Agents Chemother.* **40**, 481. **2.** Brickner, Hutchinson, Barbachyn, *et al.* (1996) *J. Med. Chem.* **39**, 673. **3.** Swaney, Aoki, Ganoza & Shinabarger (1998) *Antimicrob. Agents Chemother.* **42**, 3251.

Lincomycin

This antibiotic (FW = 406.54 g/mol), produced by *Streptomyces lincolnensis*, inhibits protein biosynthesis in Gram-positive microorganisms, but is less effective against Gram-negative bacteria. Lincomycin belongs to a class of antibiotics that inhibits the binding of sRNA to the ribosome-messenger complex. In view of its adverse effects and toxicity, lincomycin is now rarely used in the clinic. **Target(s):** peptidyltransferase (1-6); peptidyl-tRNA hydrolase, *or* aminoacyl-tRNA hydrolase (7); protein biosynthesis, initiation and elongation (1-4,8-12). **1.** Jiménez (1976) *Trends Biochem. Sci.* **1**, 28. **2.** Fernandez-Muñoz, Monro & Vazquez (1971) *Meth. Enzymol.* **20**, 481. **3.** Pestka (1974) *Meth. Enzymol.* **30**, 261. **4.** Lucas-Lenard & Beres (1974) *The Enzymes*, 3rd ed., **10**, 53. **5.** Kallia-Raftopoulos, Kalpaxis & Coutsogeorgopoulos (1992) *Arch. Biochem. Biophys.* **298**, 332. **6.** Spirin & Asatryan (1976) *FEBS Lett.* **70**, 101. **7.** Tate & Caskey (1974) *The Enzymes*, 3rd ed., **10**, 87. **8.** Chang & Weisblum (1967) *Biochemistry* **6**, 836. **9.** Chang, Sih & Weisblum (1966) *Proc. Natl. Acad. Sci. U.S.A.* **55**, 431. **10.** Kallia-Raftopoulos, Kalpaxis & Coutsogeorgopoulos (1994) *Mol. Pharmacol.* **46**, 1009. **11.** Josten & Allen (1964) *Biochem. Biophys. Res. Commun.* **14**, 241. **12.** Cundliffe (1969) *Biochemistry* **8**, 2063.

Lindane

This toxic pesticide and carcinogen (FW = 290.83 g/mol; M.P. = 112.5°C), also known as (1α,2α,3β,4α,5α,6β)-1,2,3,4,5,6-hexachlorocyclohexane, is an inhibitor of the γ-aminobutyrate receptor/chloride ionophore complex. Its resistance to hydrolysis accounts for its long biological half-life and the considerable concerns about ground water contamination. Another hazardous feature is the propensity of lindane to concentrate within membranes and fat depots. **Target(s):** γ-aminobutyrate receptor/chloride ionophore complex (1,2); α-amylase (3,4); Ca^{2+}-transporting ATPase (5); CDP-diacylglycerol:inositol 3-phosphatidyl-transferase, inhibited by several hexachlorocyclohexane stereoisomers (6-8); gap junction formation (9). **1.** Bloomquist (2003) *Arch. Insect Biochem. Physiol.* **54**, 145. **2.** Narahashi (1996) *Pharmacol. Toxicol.* **79**, 1. **3.** Lane & Williams (1948) *Arch. Biochem.* **19**, 329. **4.** Schwimmer & Balls (1949) *J. Biol. Chem.* **179**, 1063. **5.** Jones, Froud & Lee (1985) *Biochim. Biophys. Acta* **812**, 740. **6.** Antonsson (1997) *Biochim. Biophys. Acta* **1348**, 179. **7.** Parries & Hokin-Neaverson (1985) *J. Biol. Chem.* **260**, 2687. **8.** Imoto, Taniguchi & Umezawa (1992) *J. Biochem.* **112**, 299. **9.** Li & Mather (1997) *Endocrinology* **138**, 4477.

Linoleate (Linoleic Acid)

This all-*cis* unsaturated fatty acid (FW$_{\text{free-acid}}$ = 280.45 g/mol; M.P. = – 12°C), also known as (Z,Z)-9,12-octadecadienoic acid, is a nutritionally essential fatty acid in mammals. It is commonly found esterified to glycerol and in many vegetable oils. **Target(s):** 2-acylglycerol *O*-acyltransferase, *or* monoacylglycerol *O*-acyltransferase (32); alcohol *O*-acetyltransferase (30,31); arachidonyl-CoA synthetase, *or* arachidonate:CoA ligase (1); chondroitin AC lyase (2); chondroitin B lyase (2); chymase (3,4); 3',5'-cyclic-nucleotide phosphodiesterase (5); cyclooxygenase, *or* prostaglandin-endoperoxide synthase (6,33); DNA polymerase (7-11); DNA topoisomerase I (11); DNA topoisomerase II (10,11); ferrochelatase (12); 1,3-β glucan synthase (28,29); hexokinase (13); hyaluronate lyase (2); 3-hydroxy-3-methylglutaryl-CoA synthase (14); leukocyte elastase (15); 5-lipoxygenase, *or* arachidonate 5-lipoxygenase (34); microbial collagenase (15); [myosin light-chain] kinase (16); NAD$^+$ ADP-ribosyltransferase, *or* poly(ADP-ribose) polymerase (17); Na$^+$/K$^+$-exchanging ATPase (18); palmitoyl-CoA hydrolase, *or* acyl-CoA hydrolase (19); peroxidase (36); phosphatidate phosphatase (20); phospholipid-hydroperoxide glutathione peroxidase (35); prostaglandin D$_2$ dehydrogenase (21); prostaglandin D synthetase (21); prostaglandin E$_2$ 9-reductase, *or* 9-hydroxyprostaglandin dehydrogenase (22); sphingomyelin phosphodiesterase (23); steroid 5α-reductase (24,25); terminal deoxynucleotidyltransferase (7); tyrosine kinase (26); uroporphyrinogen decarboxylase (27). **1.** Wilson, Prescott & Majerus (1982) *J. Biol. Chem.* **257**, 3510. **2.** Suzuki, Terasaki & Uyeda (2002) *J. Enzyme Inhib. Med. Chem.* **17**, 183. **3.** Kido, Fukusen & Katunuma (1984) *Arch. Biochem. Biophys.* **230**, 610. **4.** Kido, Fukusen & Katunuma (1985) *Arch. Biochem. Biophys.* **239**, 436. **5.** Orellana, Jedlicki, Allende & Allende (1984) *Arch. Biochem. Biophys.* **231**, 345. **6.** Ringbom, Huss, Stenholm, *et al.* (2001) *J. Nat. Prod.* **64**, 745. **7.** Mizushina, Tanaka, Yagi, *et al.* (1996) *Biochim. Biophys. Acta* **1308**, 256. **8.** Mizushina, Yoshida, Matsukage & Sakaguchi (1997) *Biochim. Biophys. Acta* **1336**, 509. **9.** Mizushina, Ohkubo, Date, *et al.* (1999) *J. Biol. Chem.* **274**, 25599. **10.** Mizushina, Sugawara, Iida & Sakaguchi (2000) *J. Mol. Biol.* **304**, 385. **11.** Mizushina, Tsuzuki, Eitsuka, *et al.* (2004) *Lipids* **39**, 977. **12.** Hanson & Dailey (1984) *Biochem. J.* **222**, 695. **13.** Stewart & Blakely (2000) *Biochim. Biophys. Acta* **1484**, 278. **14.** Kurodam & Endo (1976) *Biochim. Biophys. Acta* **486**, 70. **15.** Rennert & Melzig (2002) *Planta Med.* **68**, 767. **16.** Kigoshi, Uchida, Kaneko, *et al.* (1990) *Biochem. Biophys. Res. Commun.* **171**, 369. **17.** Banasik, Komura, Shimoyama & Ueda (1992) *J. Biol. Chem.* **267**, 1569. **18.** Miller & Woodhouse (1977) *Aust. J. Exp. Biol. Med. Sci.* **55**, 741. **19.** Joyard & Stumpf (1980) *Plant Physiol.* **65**, 1039. **20.** Elabbadi, Day, Virden & Yeaman (2002) *Lipids* **37**, 69. **21.** Osama, Narumiya, Hayaishi, *et al.* (1983) *Biochim. Biophys. Acta* **752**, 251. **22.** Tai & Yuan (1982) *Meth. Enzymol.* **86**, 113. **23.** Liu, Nilsson & Duan (2002) *Lipids* **37**, 469. **24.** Liang & Liao (1992) *Biochem. J.* **285**, 557. **25.** Raynaud, Cousse & Martin (2002) *J. Steroid Biochem. Mol. Biol.* **82**, 233. **26.** Tomaska & Resnick (1993) *J. Biol. Chem.* **268**, 5317. **27.** Seki, Kawanishi & Sano (1986) *Meth. Enzymol.* **123**, 415. **28.** Kauss & Jeblick (1986) *Plant Physiol.* **80**, 7. **29.** Ko, Frost, Ho, Ludescher & Wasserman (1994) *Biochim. Biophys. Acta* **1193**, 31. **30.** Yoshioka & Hashimoto (1981) *Agric. Biol. Chem.* **45**, 2183. **31.** Minetoki, Bogaki, Iwamatsu, Fujii & Hamachi (1993) *Biosci. Biotechnol. Biochem.* **57**, 2094. **32.** Bhat, Wang & Coleman (1994) *J. Biol. Chem.* **269**, 13172. **33.** Reininger & Bauer (2006) *Phytomedicine* **13**, 164.

34. Schneider & Bucar (2005) *Phytother. Res.* **19**, 81. **35.** Maiorino, Roveri, Gregolin & Ursini (1986) *Arch. Biochem. Biophys.* **251**, 600. **36.** Suzuki, Honda, Mukasa & Kim (2006) *Phytochemistry* **67**, 219.

Linolenate (Linolenic Acid)

This all-*cis* unsaturated fatty acid (FW$_{\text{free-acid}}$ = 278.44 g/mol; M.P. = – 14.5°C), also known as (Z,Z,Z)-9,12,15-octadecadienoic acid and α linolenic acid, is a nutritionally essential fatty acid in mammals. Linolenate is commonly found esterified to glycerol and in many vegetable oils. **Target(s):** 2-acylglycerol *O*-acyltransferase, *or* monoacylglycerol *O*-acyltransferase (20); alcohol *O*-acetyltransferase (18,19); chondroitin AC lyase (1); chondroitin B lyase (1); chymase (2,3); cyclooxygenase (4); DNA-directed DNA polymerase (5,6); DNA polymerases α, γ, δ, and ε (5,6); DNA topoisomerase II (6); 1,3-β-glucan synthase (7); hexokinase (8); hyaluronate lyase (1); leukocyte elastase (9); microbial collagenase (9); NAD$^+$ ADP-ribosyltransferase, *or* poly(ADP-ribose) polymerase (10); palmitoyl-CoA hydrolase, *or* acyl-CoA hydrolase (11); peroxidase (23); phospholipid-hydroperoxide glutathiome peroxidase (22); photophosphorylation, chloroplast (uncoupler) (12); photosystem II (13); platelet-derived growth factor receptor protein-tyrosine kinase (14); prostaglandin D$_2$ dehydrogenase (15); prostaglandin D synthetase (15); prostaglandin-endoperoxide synthase, cyclooxygenase (21); protein-tyrosine kinase (14); sphingomyelin phosphodiesterase (16); steroid 5α-reductase (17). **1.** Suzuki, Terasaki & Uyeda (2002) *J. Enzyme Inhib. Med. Chem.* **17**, 183. **2.** Kido, Fukusen & Katunuma (1984) *Arch. Biochem. Biophys.* **230**, 610. **3.** Kido, Fukusen & Katunuma (1985) *Arch. Biochem. Biophys.* **239**, 436. **4.** Ringbom, Huss, Stenholm, *et al.* (2001) *J. Nat. Prod.* **64**, 745. **5.** Mizushina, Tanaka, Yagi, *et al.* (1996) *Biochim. Biophys. Acta* **1308**, 256. **6.** Mizushina, Tsuzuki, Eitsuka, *et al.* (2004) *Lipids* **39**, 977. **7.** Kauss & Jeblick (1986) *Plant Physiol.* **80**, 7. **8.** Stewart & Blakely (2000) *Biochim. Biophys. Acta* **1484**, 278. **9.** Rennert & Melzig (2002) *Planta Med.* **68**, 767. **10.** Banasik, Komura, Shimoyama & Ueda (1992) *J. Biol. Chem.* **267**, 1569. **11.** Joyard & Stumpf (1980) *Plant Physiol.* **65**, 1039. **12.** Izawa & Good (1972) *Meth. Enzymol.* **24**, 355. **13.** Trebst (1980) *Meth. Enzymol.* **69**, 675. **14.** Tomaska & Resnick (1993) *J. Biol. Chem.* **268**, 5317. **15.** Osama, Narumiya, Hayaishi, *et al.* (1983) *Biochim. Biophys. Acta* **752**, 251. **16.** Liu, Nilsson & Duan (2002) *Lipids* **37**, 469. **17.** Liang & Liao (1992) *Biochem. J.* **285**, 557. **18.** Yoshioka & Hashimoto (1981) *Agric. Biol. Chem.* **45**, 2183. **19.** Minetoki, Bogaki, Iwamatsu, Fujii & Hamachi (1993) *Biosci. Biotechnol. Biochem.* **57**, 2094. **20.** Bhat, Wang & Coleman (1994) *J. Biol. Chem.* **269**, 13172. **21.** Reininger & Bauer (2006) *Phytomedicine* **13**, 164. **22.** Maiorino, Roveri, Gregolin & Ursini (1986) *Arch. Biochem. Biophys.* **251**, 600. **23.** Suzuki, Honda, Mukasa & Kim (2006) *Phytochemistry* **67**, 219.

(6,9,12)-Linolenate ((6,9,12)-Linolenic Acid)

This all-*cis* unsaturated fatty acid (FW$_{\text{free-acid}}$ = 278.44 g/mol), also known as (Z,Z,Z)-6,9,12-octadecadienoic acid and γ-linolenic acid, is a component of plants fatty esters, such as the evening primrose (*Oenothera* sp.), black current oil, and borage oil. It is a precursor in the biosynthesis of arachidonate. **Target(s):** carnitine *O*-palmitoyltransferase (1); DNA-directed DNA polymerase (2); DNA polymerase α (2); leukocyte elastase (3); NAD$^+$ ADP-ribosyltransferase, *or* poly(ADP-ribose) polymerase (4); steroid 5α-reductase (5). **1.** Colquhoun (2002) *Biochim. Biophys. Acta* **1583**, 74. **2.** Mizushina, Tanaka, Yagi, *et al.* (1996) *Biochim. Biophys. Acta* **1308**, 256. **3.** Rennert & Melzig (2002) *Planta Med.* **68**, 767. **4.** Banasik,

Komura, Shimoyama & Ueda (1992) *J. Biol. Chem.* **267**, 1569. **5.** Liang & Liao (1992) *Biochem. J.* **285**, 557.

Linoleoyl-CoA

This acylated coenzyme A derivative (FW = 1029.98 g/mol) inhibits acetyl-CoA carboxylase, cyclooxygenase, and lipoxygenase. **Target(s):** acetyl-CoA carboxylase (1); cyclooxygenase (2); hexokinase (3); 3-hydroxy-3 methylglutaryl-CoA reductase (4); lipoxygenase (2); [pyruvate dehydrogenase (acetyl-transferring)] kinase (5). **1.** Tanabe, Nakanishi, Hashimoto, *et al.* (1981) *Meth. Enzymol.* **71**, 5. **2.** Fujimoto, Tsunomori, Sumiya, *et al.* (1995) *Prostaglandins Leukot. Essent. Fatty Acids* **52**, 255. **3.** Thompson & Cooney (2000) *Diabetes* **49**, 1761. **4.** Faas, Carter & Wynn (1978) *Biochim. Biophys. Acta* **531**, 158. **5.** Rahmatullah & Roche (1985) *J. Biol. Chem.* **260**, 10146.

Linuron

This urea derivative (FW = 249.10 g/mol; CAS 330-55-2), also known as Lorox® and *N*'-(3,4-dichlorophenyl)-*N*-methoxy-*N*-methylurea, is a selective pre- and post-emergence herbicide, targeting photosynthesis and photosynthetic electron transport. Linuron is used to control a wide variety of annual and perennial broadleaf and grassy weeds on both crop and noncrop sites. Linuron is also an anti-androgenic agent that produces reproductive malformations in male rats. **1.** Hulsen, Minne, Lootens, Vandecasteele & Hofte (2002) *Environ. Microbiol.* **4**, 327. **2.** Maier-Bode & Hartel (1981) *Residue Rev.* **77**, 1. **3.** Dodge (1977) *Spec. Publ., Chem. Soc.* **29**, 7.

Lipiarmycins

Lipiarmycin A₃

Lipiarmycin A₄

rest is identical to A₃

rest is identical to A₄

Lipiarmycin B₃

Lipiarmycin B₄

Consisting of a complex mixture of lipiarmycins A₃ and A₄ and two minor components, these antibiotics ($FW_{Lipiarmycin-A3}$ = 1060.06 g/mol (CAS 56645-60-4); $FW_{Lipiarmycin-A4}$ = 1046.04 g/mol; $FW_{Lipiarmycin-B3}$ = 1060.06 g/mol;

$FW_{Lipiarmycin-B4}$ = 1034.02 g/mol) amusingly draw their name from an *Actinoplanes* strain isolated on the leap year date, February 29, 1972. The B series of lipiarmycins were isolated from *Actinoplanes deccanensis*. Note that clostomicins B₁ and B₂ of *Micromonospora echinospora* correspond to lipiarmycins A₃ and B₃, respectively. **Target(s):** DNA-directed RNA polymerase (1-8). **1.** Chamberlin & Ryan (1982) *The Enzymes*, 3rd ed., **15**, 87. **2.** Osburne & Sonenshein (1980) *J. Virol.* **33**, 945. **3.** Sergio, Pirali, White & Parenti (1975) *J. Antibiot.* **28**, 543. **4.** Parenti, Pagani & Beretta (1975) *J. Antibiot.* **28**, 247. **5.** Coronelli, White, Lancini & Parenti (1975) *J. Antibiot.* **28**, 253. **6.** Talpaert, Campagnari & Clerici (1975) *Biochem. Biophys. Res. Commun.* **63**, 328. **7.** Cavalleri, Arnone, Di Modugno, Nasini & Goldstein (1988) *J. Antibiot.* **41**, 308. **8.** Chopra (2007) *Curr. Opin. Invest. Drugs* **8**, 600.

α-Lipoate (α-Lipoic Acid)

This acyl-transfer and redox coenzyme ($FW_{free-acid}$ = 206.33 g/mol; CAS 62-46-4), also known as 6-thioctic acid and (*R*)-1,2-dithiolane-3 pentanoic acid, has a melting point of 46-48°C and a p*K*a of 4.76 at 25°C. contains two sulfhydryl groups that form a dithiolane ring in the oxidized (disulfide) form. The redox potential at pH 7 is –0.29 volts. Lipoic acid is attached to the ε-amino group of lysyl residues of transacetylases (subunits of α-ketoacid dehydrogenase complexes; *i.e.*, the dihydrolipoyl acyltransferase subunit), thereby permitting acyl transfer as a swinging arm akin to that of carboxybiocytin. Note that free lipoate will not act as a substrate for the E₁ component (*i.e.*, α-keto acid decarboxylase) of the α-ketoacid dehydrogenase complexes. (**See also** ✦*Dihydrolipoate and* ✦*Dihydrolipoic Acid*) **Target(s):** acetyl-CoA hydrolase (1); choline *O*-acetyltransferase (2); dihydrolipoamide acetyltransferase (3); dihydrolipoamide dehydrogenase (3); glucose-6-phosphate dehydrogenase, moderately inhibited (4); glutamate dehydrogenase (5); glyceraldehyde-3-phosphate dehydrogenase, moderately inhibited (4); glycogen phosphorylase *b* (6); HIV-1 replication (7); (*S*)-2-hydroxy-acid oxidase (8); isocitrate dehydrogenase, moderately inhibited (4); lactate dehydrogenase, moderately inhibited (4); luciferase, firefly (*Photinus*-luciferin 4-monooxygenase (ATP-hydrolyzing), K_i = 0.026 μM (17); malate dehydrogenase (4,9); NADPH:cytochrome P450 reductase (10); 6-phosphogluconate dehydrogenase, moderately inhibited (4); poly(ADP-ribose) synthetase (11); proline racemase (12,13); pyruvate dehydrogenase (acetyl-transferring) complex (3); [pyruvate dehydrogenase (acetyl-transferring)] kinase (14); thiosulfate sulfur transferase, *or* rhodanese (15-16).). **1.** Robinson, Mahan & Koeppe (1976) *Biochem. Biophys. Res. Commun.* **71**, 959. **2.** Haugaard & Levin (2000) *Mol. Cell. Biochem.* **213**, 61. **3.** Hong, Jacobia, Packer & Patel (1999) *Free Radic. Biol. Med.* **26**, 685. **4.** Henderson & Eakin (1960) *Biochem. Biophys. Res. Commun.* **3**, 169. **5.** Yanagawa & Egami (1975) *J. Biochem.* **78**, 1153. **6.** Klinova, Klinov, Kurganov, Mikhno & Baliakina (1988) *Bioorg. Khim.* **14**, 1520. **7.** Baur, Harrer, Peukert, *et al.* (1991) *Klin. Wochenschr.* **69**, 722. **8.** Robinson, Keay, Molinari & Sizer (1962) *J. Biol. Chem.* **237**, 2001. **9.** Foster & Harrison (1975) *Biochem. Biophys. Res. Commun.* **64**, 528. **10.** Slepneva, Sergeeva & Khramtsov (1995) *Biochem. Biophys. Res. Commun.* **214**, 1246. **11.** Banasik, Komura & Ueda (1990) *FEBS Lett.* **263**, 222. **12.** Stadtman (1962) *Meth. Enzymol.* **5**, 875. **13.** Stadtman & Elliott (1957) *J. Biol. Chem.* **228**, 983. **14.** Korotchkina, Sidhu & Patel (2004) *Free Radic. Res.* **38**, 1083. **15.** Turkowsky, Blotevogel & Fischer (1991) *FEMS Microbiol. Lett.* **81**, 251. **16.** Pagani, Bonomi & Cerletti (1983) *Biochim. Biophys. Acta* **742**, 116. **17.** Inouye (2010) *Cell. Mol. Life Sci.* **67**, 387.

Lipstatin

This potent lipase irreversible pancreatic lipase inhibitor (FW = 505.74 g/mol; CAS 96829-59-3), also known as [(2*S*,4*E*,7*E*)-1-(3-hexyl-4-

oxooxetan-2-yl)trideca-4,7-dien-2-yl](2S)-2-formamido-3-methylpentano-ate, a natural product from the actinobacterium, *Streptomyces toxytricin* (1). Lipstatin contains a reactive small-ring β-lactone structure that probably accounts for the irreversible inhibition of pancreatic lipase (IC$_{50}$ = 0.14 μM). In mice, triolein absorption is dose-dependently inhibited by lipstatin, whereas oleic acid is absorbed normally. Other pancreatic enzymes, such as phospholipase A$_2$ and trypsin, are not inhibited, even at an inhibitor concentration of 200 μM (1). Lipstatin is closely related structurally to the known esterase inhibitor, esterastin; the former contains a *N*-formyl-L-leucine side-chain and the latter an *N*-acetyl-L-asparagine side-chain (2). (**See also** *Orlistatin*) **1**. Weibel, Hadvary, Hochuli, Kupfer & Lengsfeld (1987) *J. Antibiot.* (Tokyo). **40**, 1081, **2**. Hochuli, Kupfer, Maurer, *et al.* (1987) *J. Antibiot.* (Tokyo). **40**, 1086.

Liraglutide

This long-acting, intravenously administered GLP-1 Receptor agonist and Type-2 Diabetes drug (FW = 3751.20 g/mol; CAS 204656-20-2; Sequence: HAEGTFTSDVSSYLEGQAAK(γ-E-palmitoyl)EFIAWLV-RGRG), which is marketed under the trade names Saxenda® and Victoza®, binds in place of the Glucagon-Like Peptide-1 (Sequence: HAEGTFTSDVSSYLEG QAAKEFIAWLVKG-RG), the endogenous metabolic hormone that stimulates insulin secretion. By 12 hours, liraglutide begins to reduce fasting as well as meal-related glycemia by modifying insulin secretion, delaying gastric emptying, and suppressing prandial glucagon secretion. Liraglutide does so by activating the pancreatic β-cell GLP-1 receptor, the latter a G-protein-coupled receptor that increases intracellular 3',5'-cyclicAMP, thereby stimulating insulin release in the presence of elevated glucose (1,2). Experiments carried out both *in vitro* and in animals demonstrate that GLP-1 increases β-cell mass by stimulating islet cell neogenesis, while inhibiting islet apoptosis. Liraglutide has an 11-15 hour half-life upon subcutaneous injection, compared to 30 min for naturally occurring GLP-1. Liraglutide reduces body weight and food intake in obese candy-fed rats, whereas a dipeptidyl peptidase-IV inhibitor, vildagliptin, does not (2). In studies on primary neonatal rat islets, liraglutide inhibits both cytokine- and free fatty acid-induced apoptosis in a dose-dependent manner (3). This anti-apoptotic effect was mediated by the GLP-1 receptor, inasmuch as the specific GLP-1 receptor antagonist, exendin, blocks liraglutide's effect. The adenylate cyclase activator, forskolin, exerts a similar anti-apoptotic effect, indicating that liraglutide's effect is cAMP-mediated. Liraglutide's actions are inhibited by locking the PI3 kinase pathway with wortmannin, whereas inhibition of MAP kinase pathways by PD98059 is without effect. **1**. Larsen, Fledelius, Knudsen & Tang-Christensen (2001) *Diabetes* **50**, 2530. **2**. Raun, von Voss, Gotfredsen, *et al.* (2007) *Diabetes* **56**, 8. **3**. Bregenholt, Møldrup, Blume, *et al.* (2006) *Biochem. Biophys. Res. Commun.* **330**, 577.

Lisinopril

This antihypertensive agent (FW = 405.50 g/mol; CAS 83915-83-7), also known as Prinivil® and *N*$^\alpha$-[(S)-1-carboxy-3-phenylpropyl]-L-lysyl-L-proline, is an orally active inhibitor of human angiotensin I-converting enzyme (IC$_{50}$ = 4.7 nM) (1-17). Lisinopril treatment increases expression of eNOS, SK$_{Ca}$ and BK(Ca) proteins (18). In small mesenteric arteries, SK$_{Ca}$ and IK$_{Ca}$ channels play a predominant role in ACh mediated-relaxation compared to NO. Lisinopril treatment increased expression of eNOS, SK$_{Ca}$, BK$_{Ca}$ channel protein and increased the contribution of NO to ACh-mediated relaxation (18). **Key Pharmacokinetic Parameters:** *See* Appendix II in Goodman & Gilman's *THE PHARMACOLOGICAL BASIS OF THERAPEUTICS*, 12th Edition (Brunton, Chabner & Knollmann, eds.) McGraw-Hill Medical, New York (2011). **Target(s):** lysine carboxypeptidase, *or* lysine(arginine) carboxy-peptidase, *or* carboxypeptidase N (19); Xaa-Trp aminopeptidase, *or* aminopeptidase W (20). **1**. Corvol & Williams (1998) in *Handb. Proteolytic Enzymes*, p. 1066, Academic Press, San Diego. **2**. Isaac & Coates (1998) in *Handb. Proteolytic Enzymes*, p. 1076, Academic Press, San Diego. **3**. Corvol, Williams & Soubrier (1995) *Meth. Enzymol.* **248**,

283. **4**. Millar, Derkx, McLean & Reid (1982) *Brit. J. Clin. Pharmacol.* **14**, 347. **5**. Arzubiaga & Beck (1992) *Amer. J. Med. Sci.* **303**, 340. **6**. Nagamori, Fujishima & Okada (1990) *Agric. Biol. Chem.* **54**, 999. **7**. Kawamura, Oda & Muramatsu (2000) *Comp. Biochem. Physiol. B* **126**, 29. **8**. Kawamura, Kikuno, Oda & Muramatsu (2000) *Biosci. Biotechnol. Biochem.* **64**, 2193. **9**. Van Dyck, Novakova, Van Schepdael & Hoogmartens (2003) *J. Chromatogr. A* **1013**, 149. **10**. Garats, Nikolskaya, Binevski, Pozdnev & Kost (2001) *Biochemistry (Moscow)* **66**, 429. **11**. Binevski, Sizova, Pozdnev & Kost (2003) *FEBS Lett.* **550**, 84. **12**. Lanzillo, Stevens, Dasarathy, Yotsumoto & Fanburg (1985) *J. Biol. Chem.* **260**, 14938. **13**. Wei, Clauser, Alhenc-Gelas & Corvol (1992) *J. Biol. Chem.* **267**, 13398. **14**. Hooper & Turner (1987) *Biochem. J.* **241**, 625. **15**. Bull, Thornberry & Cordes (1985) *J. Biol. Chem.* **260**, 2963. **16**. Sturrock, Natesh, van Rooyen & Acharya (2004) *Cell. Mol. Life Sci.* **61**, 2677. **17**. Brunner, Desponds, Biollaz, *et al.* (1981) *Brit. J. Clin. Pharmacol.* **11**, 461 and (2004) **58**, S778. **18**. Albarwani, Al-Siyabi, Al-Husseini, *et al.* (2014) *Physiol Res.* **64**, 39. **19**. Skidgel, Weerasinghe & Erdös (1989) *Adv. Exp. Med. Biol.* **247A**, 325. **20**. Tieku & Hooper (1992) *Biochem. Pharmacol.* **44**, 1725.

Listeriolysin O

This thiol-activated, four-domain, pore-forming hemolysin (MW = 56-kDa; CAS 72270-41-8) is a toxic virulence factor produced by *Listeria monocytogenes*, the intracellular pathogen primarily responsible for listeriosis, a serious infection often caused by eating deli food contaminated with the *Listeria* bacterium. This disease primarily affects those with attenuated immunity, *e.g.*, older adults, pregnant women, newborns, and adults with HIV-AIDS (1). Bacteremia, meningitis, and rhabdomyolysis are more serious manifestations of listeriosis. Listeriolysin O binds to cholesterol-rich membranes, assembling into pore-forming aggregates that create 10-20-nm interior holes (2-5). Similarly acting thiol-activated toxins include Streptolysin O, Alveolysin, and Theta toxin. **Role in** *Listeria* **Pathogenesis:** Upon phagocytosis of Listeria bacteria by a nonprofessional phagocyte, the latter's phagosome fuses with a lysosome to form a phagolysosome. Acidification of that compartment to pH ~5.5 initiates what will prove to be a futile attempt to initiate bacterial lysis. Paradoxically, that same pH change activates listerolysin O (optimally active at pH 5.5-5.6), which undergoes a conformational change that potentiates its pore-forming and membrane-fragmenting power. The latter action destroys the phagolysosome, freeing the pathogen, which takes up residence within the cytoplasm, where it thrives and multiplies. *L. monocytogenes* eventually deploys its ActA surface protein to initiate actin-based Listeria motility, thereby promoting the pathogen's cell-to-cell spread before the host protein inevitably undergoes listeriolysin-induced autophagy, apoptosis and destruction (1). **1**. Southwick & Purich (1995) *New. Engl. J. Med.* **334**, 770. **2**. Cossart (1988) *Infection.* **16** (Suppl. 2), S157. **3**. Dramsi & Cossart (2002) *J. Cell Biol.* **156**, 943. **4**. Palmer (2001) *Toxicon* **39**, 1681. **5**. Gedde, Higgins, Tilney & Portnoy (2000) *Infect. Immun.* **68**, 999.

Lithium Ion

This monovalent ion of the smallest Group 1 (or Group IA) element (*Symbol* = Li; *Atomic Weight* = 6.941) is itself so small (*Ionic Radius* = 0.68 Å, with refined estimates of 0.60 and 0.79 Å for four- and six-coordinate complexes) that its inner coordination sphere is very crowded, often containing as few as 3.6 water molecules in the first coordination sphere. Located directly above sodium in the Periodic Table, lithium is an alkali metal that readily oxidizes to Li$^+$, which inhibits a range of biochemical processes, often by competing favorably with K$^+$ and Na$^+$ for monovalent cation sites within proteins, nucleic acids, and charged polysaccharides (1). **Clinical Aspects:** Marketed as Eskalith® and Lithobid®, lithium carbonate (FW = 73.89 g/mol; CAS 554-13-2) is widely used to treat bipolar disorder. This mood stabilizer influences behavior and Akt/GSK-3 signaling in mice and many antipsychotics have been shown to more potently antagonize the activity of the β-arrestin-2 pathway relative to the G protein-dependent pathway. It reduces the severity and frequency of mania and is often prescribed in combination with mildly anti-psychotic agents. In the context of mood disorders and neurodegenerative diseases, lithium ion is thought to targets inositol monophosphatase in the phosphatidylinositol signalling pathway as well as the protein kinase glycogen synthase kinase-3 (2). Neuroprotection is the most likely outcome of lithium treatment in both preclinical and clinical models of bipolar disorder. Based on comprehensive reviews and meta-analysis of patient records and randomized clinical trials, McKnight *et al.* (3) report that patients taking lithium have a slightly reduced glomerular filtration rate, a slightly higher risk of renal failure, a tendency toward clinical hypothyroidism, slightly increased blood calcium and parathyroid hormone, as well as greater weight gain. Other studies show a slight increase in suicide risk. Lithium ion is also a suspected teratogen (4). **Nonspecific Inhibition of Glycogen Sythase Kinase:** LiCl is an ATP noncompetitive inhibitor of GSK-3β activity ($K_i \approx 2$ mM), and, although its has been commonly used to investigate the effects of GSK-3β inhibition, it is notoriously nonspecific and is unlikely to produce results strictly attributable to GSK inhibition. One should instead consider the use of SB415286, a far more potent inhibitor that is highly specific for GSK-3β (5). Other ways to inhibit cellular GSK-3 activity include antisense targeting of GSK-3 and overexpression of a dominant negative form of the kinase. **Removal from Commercial Nucleotides & Coenzymes:** By binding tightly to the phospates of ATP, GTP and many other metabolites, lithium ions help to stabilize these molecules against hydrolysis, thus facilitating long-term storage. Nonetheless, 1 mM lithium salt of ATP contains 4 mM lithium ion, leading to the possibility that this monovalent ion will be a confounding variable in enzyme assays or metabolic experiments. To avoid such ambiguity, a good practice is to acidify commercial lithium salts by the addition of dilute acetic acid to pH 5 (thereby weakening lithium ion binding by protonating the phosphates), followed by passage over a small Dowex-50 column that has been previously equilibrated with sodium or potassium ion and thoroughly washed with deionized water. (*Best Source Recommendation*: AG® 50W and AG® MP-50 Cation Exchange Resins). Because ion exchange kinetics is rapid, this procedure can be conveniently accomplished by use of a Centricon Plus-20 PLGC centrifuge-tube filter. **Lithium Effect on Circadian Rhythm:** Lithium also appears to exert a novel amplitude-enhancing effect on the PER2 protein rhythms in the central and peripheral circadian clockwork, an action that may involve the glycogen synthase kinase-3-mediated signalling pathway (6). **Key Pharmacokinetic Parameters:** *See* Appendix II in Goodman & Gilman's THE PHARMACOLOGICAL BASIS OF THERAPEUTICS, 12th Edition (Brunton, Chabner & Knollmann, eds.) McGraw-Hill Medical, New York (2011). **Target(s):** Lithium has many metabolic effects, inhibiting glucose synthesis in the liver, in part through effects on gene expression (7). Also inhibited are glycogen synthase kinase-3, *or* Tau kinase (8) and *myo*-inositol-1-phosphatase, the latter showing uncompetitive inhibition (9). When viewed in the context of the phosphatidyl-inositide cycle, lithium blocks conversion of inositol-1-phosphate (IP_1) to myoinositol, resulting in the accumulation of inositol monophosphates and depletion of myoinositol, reducing PIP_2 as well as its cleavage products, IP_3 and DAG. Lithium is also a mixed-type noncompetitive inhibitor of sodium ion/dicarboxylate-1 (NaDC-1), an electrogenic anion transporter accepting either Na^+ or Li^+ as coupling cations (10). NaDC-1 possesses a single high-affinity binding site for Li^+ that, when occupied, inhibits dicarboxylate transport. In studies of fructose 1,6-bisphosphatase, Li^+ was found to be a linear noncompetitive inhibitor with respect to Mg^{2+} (11). Li^+ is believed to inhibit FBPase, either by distorting the geometry of the active site or by retarding catalytic turnover or product release. **1.** Garrett (2004) *Handbook of Lithium and Natural Calcium Chloride: Their Deposits, Processing, Uses and Properties*, Elsevier. (ISBN: 978-0-12-276152-2). **2.** Brown & Tracy (2013) *Ther. Adv. Psychopharmacol.* **3**, 163. **3.** McKnight, Adida, Budge (2012) *Lancet* **379**, 721. **4.** Yacobi & Ornoy (2008) *Isr. J. Psychiatry Relat. Sci.* **45**, 95. **5.** Coghlan, Culbert, Cross, *et al.* (2000) *Chem. Biol.* **7**, 793. **6.** Li, Lu, Beesley, Loudon and Meng (2012) PLoS ONE **7**, e33292. **7.** Lewitt, Brismar, Ohlson, & Hartman (2001) *J. Endocrinol.* **171**, R11. **8.** Freland & Beaulieu (2012) *Front. Mol. Neurosci.* **5**, 14. **9.** Inhorn & Majerus (1987) *J. Biol. Chem.* **262**, 15946. **10.** Pajor, Hirayama, Loo (1998)) *J. Biol. Chem.* **273**, 18923. **11.** Zhang, Villeret, Lipscomb & Fromm (1996) *Biochemistry* **45**, 3038.

Lithocholate (Lithocholic Acid)

This steroid ($FW_{\text{free-acid}} = 376.58$ g/mol; CAS 434-13-9; very low water-solubility and more soluble in diethyl ether), also known as (3α,5β)-3-hydroxycholan-24-oic acid, is one of the bile acids identified in human, bovine, rabbit, sheep, goat, and porcine bile. Lithocholate is typically found is the taurine or glycine conjugate. **Target(s):** α-*N*-acetyllactosaminide α-2,3-sialyltransferase (1); α-amylase (2); cAMP dependent protein kinase (3); carbonic anhydrase (4,5); choloyl-CoA synthetase, *or* cholate:CoA ligase (6,7); DNA polymerase β (8-10); DNA topoisomerase II (8); glutathione *S*-transferase (11-15); pepsin (16); steryl-sulfatase, *or* arylsulfatase C, mildly inhibited (17); threonine ammonia-lyase, *or* threonine dehydratase (18); UDP-glucuronosyltransferases (19). **1.** Chang, Lee, Chen & Li (2006) *Chem. Commun. (Cambridge)* **2006**, 629. **2.** O'Donnell, McGeeney & FitzGerald (1975) *Enzyme* **19**, 129. **3.** Wang & Polya (1996) *Phytochemistry* **41**, 55. **4.** Milov, Jou, Shireman & Chun (1992) *Hepatology* **15**, 288. **5.** Milov, Shireman & Chun (1990) *FASEB J.* **4**, A731. **6.** Wheeler, Shaw & Barnes (1997) *Arch. Biochem. Biophys.* **348**, 15. **7.** Falany, Xie, Wheeler, *et al.* (2002) *J. Lipid Res.* **43**, 2062. **8.** Mizushina, Kasai, Sugawara, *et al.* (2001) *J. Biochem.* **130**, 657. **9.** Mizushina, Ohkubo, Sugawara & Sakaguchi (2000) *Biochemistry* **39**, 12606. **10.** Ogawa, Murate, Suzuki, Nimura & Yoshida (1998) *Jpn. J. Cancer Res.* **89**, 1154. **11.** Singh, Leal & Awasthi (1988) *Toxicol. Appl. Pharmacol.* **95**, 248. **12.** Takikawa, Sugiyama & Kaplowitz (1986) *J. Lipid Res.* **27**, 955. **13.** Hayes & Mantle (1986) *Biochem. J.* **233**, 407. **14.** Lyon & Atkins (2002) *Biochemistry* **41**, 10920. **15.** Liebau, Eckelt, Wildenburg, *et al.* (1997) *Biochem. J.* **324**, 659. **16.** Tompkins & Tompkins (1982) *Amer. J. Surg.* **144**, 300. **17.** Gniot-Szulzycka & Komoszynski (1972) *Enzymologia* **42**, 11. **18.** Pagani, Leoncini, Terzuoli, Guerranti & Marinello (1990) *Enzyme* **43**, 122. **19.** Falany, Green, Swain & Tephly (1986) *Biochem. J.* **238**, 65.

Litronesib

This mitotic drug and cytostatic agent (FW = 511.70 g/mol; CAS 910634-41-2), also known as LY2523355, KF 89617 and its systematic name (*R*)-*N*-(5-((2-(ethylamino)ethylsulfonamido)methyl)-5-phenyl-4,5-dihydro-1,3,4-thiadiazol-2-yl)pivalamide, targets kinesin Eg5, a mitotic motor protein that plays a key role in mitosis including chromosome positioning, separating the centrosome, and establishing a bipolar spindle in the M-phase of the cell cycle. Litronesib is an ATP-noncompetitive, allosteric, reversible inhibitor with no effect on microtubule formation. Litronesib has been investigated in clinical trials for the treatment of multiple cancers as well as collagen- and adjuvant-induced arthritis. Litronesib racemizes upon storage in aqueous solution, because its substituted 1,3,4-thiadiazoline ring forms a positively charged trigonal intermediate through a relay mechanism involving intramolecular deprotonation, attack of the resulting sulfonamide anion on C^5, with cleavage of the C^5–S bond to form an aziridine that suffers heterolytic dissociation to yield a reactive ylide (2). (*See also K858, Monastrol, Terpendole E, S-Trityl-L-cysteine*) **1.** Wakui, Yamamoto, Kitazono, *et al.* (2014) *Cancer Chemother. Pharmacol.* **74**, 15. **2.** Baertschi, Jansen, Montgomery, *et al.* (2014) *J. Pharm. Sci.* **103**, 2797.

Lividomycins

These 3'-deoxyaminoglycoside antibiotics (FW = 761.78 g/mol for lividomycin A (shown above)) are alternative substrate for a number of aminoglycoside-dependent enzyme (*e.g.*, aminoglycoside 3' phosphotransferase III). **Target(s):** DNA-directed DNA polymerase, Klenow fragment (1); gentamicin 3'-*N* acetyltransferase, inhibited by lividomycin B, K_i = 2 µM (2); kanamycin 6'-acetyltransferase, inhibited by lividomycin A (3); poly(A)-specific ribonuclease (1). **1.** Ren, Martínez, Kirsebom & Virtanen (2002) *RNA* **8**, 1393. **2.** Williams & Northrop (1978) *J. Biol. Chem.* **253**, 5908. **3.** Haas & Dowding (1975) *Meth. Enzymol.* **43**, 611.

LJC10,627, *See* Biapenem

LJH685

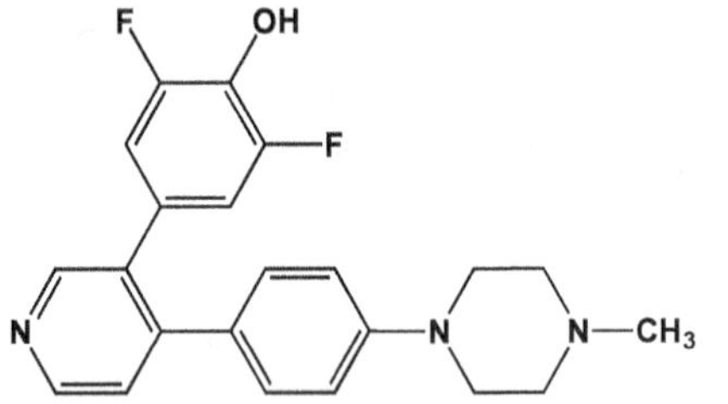

This potent pan-RSK inhibitor (FW = 381.43 g/mol; CAS 1627710-50-2; Solubility: 70 mg/mL DMSO; < 1 mg/mL H$_2$O), also named 2,6-difluoro-4-[4-[4-(4-methyl-1-piperazinyl)phenyl]-3-pyridinyl]phenol, targets p90 ribosomal S6 kinases RSK1 (IC$_{50}$ = 6 nM), RSK2 (IC$_{50}$ = 5 nM), and RSK3 (IC$_{50}$ = 4 nM). Based on X-ray crystallography, the nonplanar shape of LJH685 and its conformational properties distinct from the currently available and commonly used inhibitor dihydropteridinone (BI-D1870) contribute to its significantly improved selectivity profile. LJH685 RSK inhibition correlates with antiproliferative effects in MAPK pathway–dependent cancer cell lines only in anchorage-independent growth setting. For further details on LJH685's inhibitory action and its morpholino analogue, *See LJI308*. **1.** Aronchik, Appleton, Basham, *et al.* (2014) *Mol. Cancer Res.* **12**, 803.

LJI308

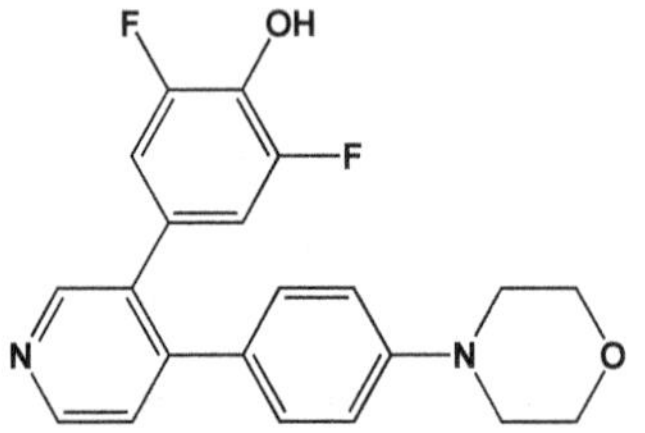

This potent pan-RSK inhibitor (FW = 368.38 g/mol; CAS 1627709-94-7; Solubility: 73 mg/mL DMSO; < 1 mg/mL H$_2$O), also named 2,6-difluoro-4-[4-[4-(4-morpholinyl)phenyl]-3-pyridinyl]phenol, targets p90 ribosomal S6 kinases RSK1 (IC$_{50}$ = 6 nM), RSK2 (IC$_{50}$ = 4 nM), and RSK3 (IC$_{50}$ = 13 nM). The p90 ribosomal S6 kinase (RSK) comprises a family of four closely related proteins that are widely expressed in cancer cell lines and tissues and are activated in response to a number of growth factors and hormones. RSK kinases have a unique structure containing two nonidentical kinase domains, N-terminal and C-terminal, separated by a linker region. One physiologic RSK substrate is Y-box–binding protein 1 (YB1), which regulates transcription and translation by binding to its recognition motifs in both DNA and RNA to regulate key cell processes such as proliferation, motility, and stemness. although phosphorylation of YB1 at Ser102 was well inhibited by LJH685 and LJI308, this did not closely correlate with widespread functional consequences. These results were surprising given that earlier work linked YB1 to important physiologic roles in cancer progression, including the expression of fibroblast growth factor receptor (FGFR), proliferating cell nuclear antigen (PCNA), MAP Kinase interacting serine/threonine kinase (MKNK1), and matrix metalloproteinase 2 (MMP2). Although earlier reports indicated that YB1 is critical for these widespread roles, results with LJH685 and LJI308 suggest that the importance of phosphorylation at Ser102 may be more nuanced. (*See also* LJH685) **1.** Aronchik, Appleton, Basham, *et al.* (2014) *Mol. Cancer Res.* **12**, 803.

LLY-507

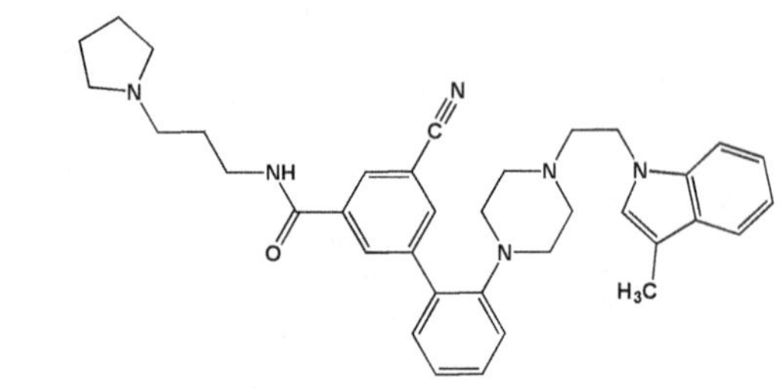

This potent methylation inhibitor (FW = 574.77 g/mol) targets the SMYD2, an enzyme that methylates histone H3 at Lysine-36, IC$_{50}$ ~ 30 nM (1). LLY-507 exhibits >100-fold selective for SMYD2 over a broad range of other methyltransferase and non-methyltransferase targets. The SMYD methylase family, consisting of SMYD1 through SMYD5, possess a distinctive SET domain that is split by a MYND domain. The SET domain is catalytic, and MYND domain is a zinc finger that facilitates binding to certain proline-rich sequences. SMYDs 1-5 also contain a C-terminal domain (CTD) that regulates histone methylation; and binding of Heat Shock Protein 90 (Hsp90) causes the CTD to swing out and fully expose the methylase active site. A 1.63Å resolution crystal structure of a SMYD2·LLY-507 complex shows the inhibitor occupying the binding pocket for the substrate peptide (1). LLY-507 is active in cells, as measured by reduction of SMYD2-induced mono-methylation of p53 K370 at sub-micromolar doses (1). Mass spectrometry-based proteomics revealed that cellular global histone methylation levels are not significantly affected by SMYD2 inhibition with LLY-507, and subcellular fractionation studies indicate that SMYD2 is primarily cytoplasmic, suggesting SMYD2 targets a very small subset of histones at specific chromatin loci, and/or non-histone substrates (1). Unlike LLY-507, a previously disclosed SMYD2 inhibitor shows weak cellular activity (2). **1.** Nguyen, Allali-Hassani, Antonysamy, *et al.* (2015) *J. Biol. Chem.* **290**, 13641. **2.** Ferguson, Larsen, Howard, *et al.* (2011) *Structure* **19**, 1262.

LLY-2707

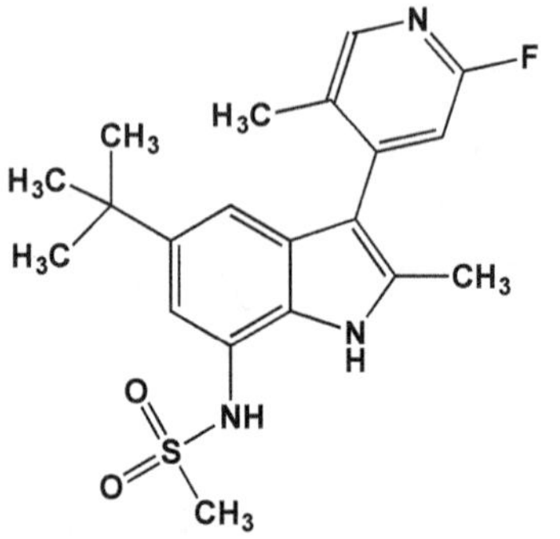

This novel non-steroidal glucocorticoid receptor antagonist (FW = 389.16 g/mol), named systematically as *N*-(5-(*tert*-butyl)-3-(2-fluoro-5-methylpyridin-4-yl)-2-methyl-1*H*-indol-7-yl)-methanesulfonamide, reduces weight gain in female rats treated with atypical antipsychotic drugs (*or* AAPDs), such as olanzapine (*or* Zyprexa®). In humans, clinically significant weight gain (~0.9 kg/month) results in a 6-10 kg body weight gain after one year on olanzapine (2), and LLY-2707 may represent a novel way for preventing weight gain and diabetes during AAPD treatment. (*See Olanzapine*) **1.** Sindelar, Carson, Morin, *et al.* (2013) *J. Pharmacol. Exp. Ther.* **348**, 192. **2.** Nemeroff (1997) *J. Clin. Psychiatry* **58** (Suppl 10) 45.

LM-10

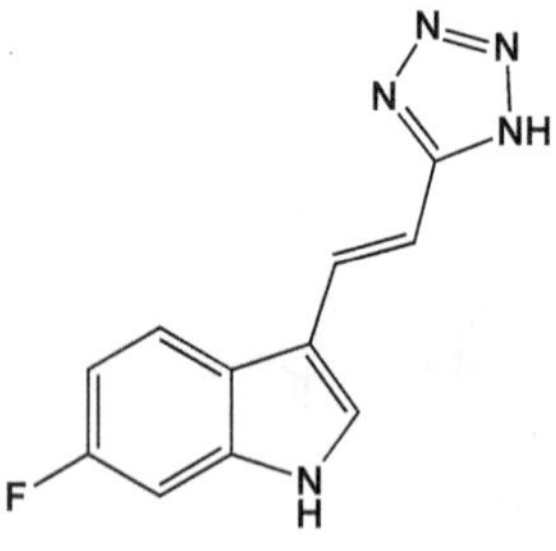

This indole (FW = 229.22 g/mol; CAS 1316695-35-8), also named 6-fluoro-3-[(1*E*)-2-(2*H*-tetrazol-5-yl)ethenyl]-1*H*-indole, selectively inhibits tryptophan 2,3-dioxygenase (*Reaction*: L-tryptophan + O_2 $\rightleftharpoons$ *N*-formyl-L-kynurenine), with respective IC_{50} values of 0.62 and 2 μM for the human and mouse enzymes, and exhibiting lower action against IDO, MAO-A, MAO-B, and a panel of receptors and transporters. LM 10 educes growth of TDO-expressing P815 mastocytoma tumors in mice. **1**. Pilotte, *et al.* (2012) *Proc. Natl. Acad. Sci. U.S.A.* **109**, 2497. **2**. Dolusić, *et al.* (2011) *J. Med. Chem.* **54**, 5320.

LN-1-255

This 6-alkylidene-2′-substituted penicillin sulfone (FW = 487.46 g/mol) targets Class C and Class A serine β-lactamases, thereby improving the antibiotic action of carbapenem antibiotics against otherwise resistant bacterial strains (1). By possessing a catecholic functionality resembling a natural bacterial siderophore, LN-1-255 is unique among β-lactamase inhibitors (2). In combination with LN-1-255, piperacillin is more potent against *Escherichia coli* DH10B strains bearing *bla*$_{SHV}$ extended-spectrum and inhibitor-resistant β-lactamases than an equivalent amount of tazobactam and piperacillin (2). LN-1-255 also significantly enhances the activity of ceftazidime and cefpirome against extended-spectrum cephalosporin- and Sme-1-containing carbapenem-resistant clinical strains (2). LN-1-255 inhibited SHV-1 and SHV-2 β-lactamases (K_i = 110 nM and 100 nM, respectively). When LN-1-255 inactivates SHV β-lactamases, mass spectrometry reveals a single bicyclic aromatic intermediate that X-ray crystallography shows to have its carbonyl oxygen pointing out of the oxyanion hole and forming hydrogen bonds with Lys-234 and Ser-130 within the active site (2). (Covalent docking of LN-1-255 was applied between the catalytic Ser-81 (oxygen atom side chain) and the carboxylate group (oxygen atom) of the indolizine obtained after covalent modification of OXA-24/40 from *A. baumannii*.) LN-1-255 also inhibits OXA-48, the class D carbapenemase that presents unique challenges, inasmuch as it is resistant to most β-lactam inhibitors (3). Both OXA-48-producing clinical and transformant strains of *Klebsiella pneumoniae* and *Escherichia coli* display enhanced susceptibility to carbapenem antibiotics in the presence of 4 mg/L LN-1-255 (*i.e.*, 2-32x greater) and 16 mg/L LN-1-255 (*i.e.*, 4-64x greater). Kinetic measurements demonstrate that LN-1-255 inhibits OXA-48 with an acylation efficiency *(k_2/K_m)* of 10 x10^4 $M^{-1}s^{-1}$ and a slow deacylation rate k_{off} of 7 x 10^{-4} s^{-1} (3). The IC_{50} is 3 nM for LN-1-255, whereas an IC_{50} of 1.5 μM is obtained for tazobactam, largely because k_{cat}/k_{inact} is 500x lower for LN-1-255 compared to tazobactam. Combined use of carbapenem antibiotics and LN-1-255 is highly effective against OXA-48, affording a promising route for circumventing antibiotic resistance (3). **See also** Piperacillin; Tazobactam **1**. Buynak, Rao, Doppalapudi, *et al.* (1999) *Bioorg. Med. Chem. Lett.* **9**, 1997. **2**. Pattanaik, Bethel, Hujer, *et al.* (2009) *J. Biol. Chem.* **284**, 945. **3**. Vallejo, Martínez-Guitián, Vázquez-Ucha, *et al.* (2016) **71**, 2171.

LOC14

This neuroprotective PDI modulator (FW = 316.41 g/mol; CAS 877963-94-5), also named 2-[[4-(cyclopropyl-carbonyl)-1-piperazinyl]methyl]-1,2-benzisothiazol-3(2*H*)-one, targets Protein Disulfide Isomerase (K_d = 62 nM) by adopting an oxidized conformation and suppressing its activty. An endoplasmic reticulum enzyme in eukaryotes, PDI catalyzes the formation/breakage of disulfide bonds between cysteine residues within proteins as they fold. **1**. Kaplan, Gaschler, Dunn, *et al.* (2015) *Proc. Natl. Acad. Sci. USA* **112**, :E2245.

Lometrexol

This antifolate (FW = 443.65 g/mol; CAS 106400-81-1; Solubility: 5 mg/mL DMSO), also named *N*-[4-[2-[(6*R*)-2-amino-3,4,5,6,7,8-hexahydro-4-oxopyrido-[2,3-d]pyrimidin-6-yl]ethyl]-benzoyl]-L-glutamate, 5,10-dideazatetrahydrofolate, and LY264618, targets glycinamide ribonucleotide formyltransferase (GARFTase Reaction: 10-Formyl-tetrahydrofolate + N^1-(5-Phospho-D-ribosyl)glycinamide $\rightleftharpoons$ Tetrahydrofolate + N^2-Formyl-N^1-(5-phospho-D-ribosyl)-glycinamide), a folate-requiring enzyme in *de novo* purine synthesis. Although a potent GARFTase inhibitor, lometrexol does not interfere with the enzymes involved in folate synthesis. Lometrexerol has been evaluated clinically for the treatment of various cancers as an anti-folate like agent, similar to methotrexate. Treatment of cells with lometrexol rapidly decreases ATP and GTP levels, resulting in cell cycle arrest and apoptosis. Although depletion of nucleotide pools induces p53 expression, lometrexol is cytotoxic in both wild type and mutant p53-expressing tumor cells. Lometrexol is cytotoxic in CCRF-CEm leukemia cells (IC_{50} = 2.9 nM). **1**. Sessa, de Jong, D'Incalci, *et al.* (1996) *Clin. Cancer Res.* **2**, 1123.

Lomustine

This nitrosourea derivative (FW = 233.70 g/mol; CAS 13010-47-4), also known as 1-(2-chloroethyl)-3-cyclohexyl-1-nitrosourea (CCNU), is an alkylating reagent and exhibits antitumor activity in a manner resembling carmustine and nimustine. This nitrosourea undergoes hydrolysis *in vivo* to form reactive metabolites. With ability to cross the blood-brain barrier, lomustine is often used to treat brain tumors. (*For other mechanistic details, see* Nitrosoureas) CCNU is not detected in the plasma of patients, due to its complete conversion to monohydroxylated metabolites during the 'first pass' through liver and gut (1). Two monohydroxylated metabolites, *trans*-4-hydroxy CCNU and *cis*-4-hydroxy CCNU reach peak concentrations (0.8-0.9 μg/mL) within 2-4 hours. These metabolites are also detected in a tumor biopsy. Plasma clearance half-lives of the two metabolites are similar in each patient but showed a two-fold variation between patients, from 1.3-2.9 hours (1). **Target(s):** arylamine *N*-acetyltransferase (2); chymotrypsin, inhibited by metabolite of CCNU (3); CYP2D6, *or* dextromethorphan *O*-demethylase (4); DNA polymerase (5); esterase (6); glutathione-disulfide reductase (7); ribonucleotide reductase (7); thioredoxin reductase (7); transglutaminase, *or* protein-glutamine γ-glutamyltransferase (8); tyrosinase (9). **1**. Lee, Workman, Roberts & Bleehen (1985) *Cancer Chemother. Pharmacol.* **14**, 125. **2**. Hung (2000) *Neurochem. Res.* **25**, 845. **3**. Babson, Reed & Sinkey (1977) *Biochemistry* **16**, 1584. **4**. Le Guellec, Lacarelle, Catalin & Durand (1993) *Cancer Chemother. Pharmacol.* **32**, 491. **5**. Chuang, Laszlo & Keller (1976) *Biochim. Biophys. Acta* **425**, 463. **6**. Dive, Workman & Watson (1988) *Biochem. Pharmacol.* **37**, 3987. **7**. Schallreuter, Gleason & Wood (1990) *Biochim. Biophys. Acta* **1054**, 14. **8**. Tarantino & Thompson (1979) *Cancer Biochem. Biophys.* **4**, 33. **9**. Rachkova, Raikova & Raikov (1991) *Cancer Biochem. Biophys.* **12**, 59.

Lonafarnib

This orally bioavailable farnesyltransferase inhibitor, *or* FTI (FW = 638.83 g/mol; CAS 193275-84-2; Solubility: 125 mg/mL DMSO; <1 mg/mL H_2O), also named SCH66336 and 4-[2-[4-[(11*R*)-3,10-dibromo-8-chloro-6,11-dihydro-5*H*-benzo[5,6]cyclohepta[1,2-*b*]pyridin-11-yl]-1-piperidinyl]-2-oxoethyl]-1-piperidinecarboxamide, blocks farnesylation of H-ras (IC_{50} = 1.9 nM), K-ras-4B (IC_{50} = 5.2 nM) and N-ras (IC_{50} = 2.8 nM), inhibiting anchorage-independent growth of many human tumor lines lacking an activated ras oncogene. In nude mouse xenograft models for human tumors of colon, lung, pancreas, prostate, and urinary bladder origin, SCH 66336 demonstrated potent oral activity (1). Enhanced *in vivo* efficacy is observed when combined with other cytotoxic agents, *e.g.*, cyclophosphamide, 5-fluorouracil, and vincristine (1). Lonafarnib suppresses protein kinase B/Akt activity as well as the phosphorylation of the Akt substrates glycogen synthase kinase (GSK)-3β, forkhead transcription factor, and BAD (2). Infection of SqCC/Y1 cells with an adenovirus that contained a constitutively active form of Akt rescued cells from lonafarnib-induced apoptosis, suggesting it is a potent apoptosis inducer in HNSCC cells that acts by suppressing the Akt pathway. Lonafarnib treatment of tumor cells *in vitro* is proapoptotic under certain cell culture conditions, and induction of apoptosis generally requires a second death-promoting signal. At achievable clinical concentration ranges, lonafarnib effectively inhibits the growth of ten different nonsmall-cell lung cancer (NSCLC) cell lines, particularly after a prolonged treatment, *regardless* of Ras mutational status (3). Lonafarnib arrested cells growth at G_1 or G_2M phase in the majority tested cell lines, but it also induced apoptosis when cells were cultured in a low serum (0.1%) medium. The majority of NSCLC cell lines expressed undetectable level of phosphorylated Akt (p-Akt). Lonafarnib at up to 10 µM did not decrease either total Akt or p-Akt levels in tested cell lines, even after a 48-hour treatment (3). It even increased p-Akt level in one cell line, although it was as sensitive as others to lonafarnib treatment and underwent lonafarnib arrest. Lonafarnib thus appears to be effective in inhibiting NSCLC cells via growth arrest at G_2M or by induction of apoptosis, and, in some cases, does so without down-regulation of Akt. **1.** Liu, Bryant, Chen, *et al.* (1998) *Cancer Res.* **58**, 4947. **2.** Chun, Lee, Hassan, *et al.* (2003) *Cancer Res.* **63**, 4796. **3.** Sun, Zhou, Wang, Fu & Khuri (2004) *Cancer Biol. Ther.* **3**, 1092.

Lonidamine

This antineoplastic agent ($FW_{free-acid}$ = 321.16 g/mol; M.P. = 207°C), also known as 1-(2,4-dichlorobenzyl)-1*H*-indazole-3-carboxylic acid, also functions as an antispermatogenic drug. Lonidamine is particularly effective in selectively sensitizing tumors to chemotherapy, hyperthermia and radiotherapy. LND potently inhibits mitochondrial pyruvate carrier (MPC) in isolated rat liver mitochondria (K_i = 2.5 µM) and co-operatively inhibits L-lactate transport by MCT1, MCT2 and MCT4 expressed in *Xenopus laevis* oocytes ($K_{0.5}$ = 36-40 µM; Hill coefficient = 1.65-1.85) 1). In rat heart mitochondria LND inhibited the MPC with similar potency and uncoupled oxidation of pyruvate was inhibited more effectively (IC_{50} = ~7 µM) than other substrates including glutamate (IC_{50} = ~20 µM). In isolated DB-1 melanoma cells 1-10 µM, LND increases L-lactate output, consistent with MPC inhibition, but higher concentrations (150 µM) decreased L-lactate

output whereas increasing intracellular [L-lactate] > 5-fold, consistent with MCT inhibition (1). **Target(s):** cystic fibrosis transmembrane conductance regulator (CFTR) chloride channel (2); electron transport (3-6); hexokinase (3,7,8); lactate transport (9). **1.** Nancolas, Guo, Zhou, *et al.* (2016) *Biochem. J.* **473**, 929. **2.** Gong, Burbridge, Lewis, Wong & Linsdell (2002) *Brit. J. Pharmacol.* **137**, 928. **3.** Gatto, Tita, Artico & Saso (2002) *Contraception* **65**, 277. **4.** Floridi, D'Atri, Bellocci, *et al.* (1984) *Exp. Mol. Pathol.* **40**, 246. **5.** Floridi & Lehninger (1983) *Arch. Biochem. Biophys.* **226**, 73. **6.** Turrens (1986) *Mol. Biochem. Parasitol.* **20**, 237. **7.** Floridi, Paggi, Marcante, *et al.* (1981) *J. Natl. Cancer Inst.* **66**, 497. **8.** Nista, De Martino, Malorni, *et al.* (1985) *Exp. Mol. Pathol.* **42**, 194. **9.** Ben-Horin, Tassini, Vivi, Navon & Kaplan (1995) *Cancer Res.* **55**, 2814.

Lonomycins

These polyether antibiotics from *Streptomyces ribosidificus*, are ionophores that uncouple oxidative phosphorylation. The major component is lonomycin A (FW = 829.08 g/mol; CAS 58845-80-0), which was originally called simply lonomycin: lonomycins B and C are minor components. Lonomycin A more readily forms complexes with K^+ than with NH_4^+, Rb^+, or Na^+. Complexes were not reported with Li^+, Cs^+, or Ca^{2+}. **1.** Reed (1979) *Meth. Enzymol.* **55**, 435.

Loperamide

This widely used antidiarrheal agent ($FW_{free-base}$ = 477.05 g/mol; CASs: 53179-11-6 (free base) and 34552-83-5 (HCl salt)), also known as Imodium® and 4-(4-chlorophenyl)-4-hydroxy-*N*,*N*-dimethyl-α,α-diphenyl-1-piperidinebutanamide, is a nonselective Ca^{2+} channel blocker and opioid agonist. While nanomolar concentrations of loperamide activate opioid receptors in the gastrointestinal tract, somewhat higher concentrations are required to block calmodulin activity, calcium channels, *N*-methyl-D-aspartate-receptor channels, and maitotoxin elicited calcium influx. P-glycoprotein-mediated efflux prevents loperamide from crossing the blood-brain barrier, thereby minimizing CNS effects. **1.** Daly & Harper (2000) *Cell Mol. Life Sci.* **57**, 149. **2.** Ooms, Degryse & Janssen (1984) *Scand. J. Gastroenterol. Suppl.* **96**, 145. **3.** Church, Fletcher, Abdel-Hamid & MacDonald (1994) *Mol. Pharmacol.* **45**, 747.

Lophotoxin

This marine neurotoxin (FW = 416.43 g/mol; CAS 78697-56-0) from sea fans and whips (*e.g.*, species of *Lophogorgia*) irreversibly inactivates nicotinic acetylcholine receptors, targeting a tyrosyl residue within the α subunit (1-4) (**See also** *Bipinnatins*). A total of 12 active lophotoxin analogues have been isolated from various species of the soft corals *Lophogorgia* and *Pseudopterogorgia* (Fenical *et al.*, 1981; Culver *et al.*, 1985; Abramson *et al.*, 1989; Wright *et al.*, 1989). A radiolabeled analog was used to show that the lophotoxins irreversibly inhibit nicotinic receptors by forming a covalent bond with Tyr^{190} in the α-subunits (αTyr^{190}) of the receptor. Lophotoxins are unique among naturally occurring neurotoxins in their selective inhibition of a neurotransmitter receptor by irreversible, covalent modification. Structural comparison of lophotoxin and bipinnatins suggests that the greater stability of lophotoxin results from substituents at positions R_1 and R_2 (3). At position R_2, lophotoxin lacks an acetate ester present in bipinnatins. Spontaneous hydrolysis of the acetate ester at position R_2 may be an initiating event in activation of the bipinnatins (3). Alternatively, the presence of an aldehyde at position R_1 in lophotoxin may account for the increased activity of lophotoxin as well as its greater stability. **1.** Smythies (1983) *Med. Hypotheses* **10**, 465. **2.** Hann, Pagan, Gregory, *et al.* (1998) *J. Pharmacol. Exp. Ther.* **287**, 253. **3.** Groebe & Abramson (1995) *J. Biol. Chem.* **270**, 281. **4.** Abramson, Culver, Kline, *et al.* (1988) *J. Biol. Chem.* **263**, 18568.

Lopinavir

This antiviral agent (FW = 628.81 g/mol), also known as ABT-378, is a strong inhibitor of HIV-1 protease. Note that lopinavir is an effective inhibitor of ritonavir-resistant HIV strains, with a K_i of 1–4 pM for the Val-82-Phe viral proteinase (1-5). **Key Pharmacokinetic Parameters:** *See* Appendix II in Goodman & Gilman's *THE PHARMACOLOGICAL BASIS OF THERAPEUTICS*, 12th Edition (Brunton, Chabner & Knollmann, eds.) McGraw-Hill Medical, New York (2011). **Target(s):** UGT1A1, UGT1A3, and UGT1A4 glucuronosyltransferase (6). **1.** Carrillo, Stewart, Sham, *et al.* (1998) *J. Virol.* **72**, 7532. **2.** Sham, Kempf, Molla, *et al.* (1998) *Antimicrob. Agents Chemother.* **42**, 3218. **3.** Clemente, Coman, Thiaville, *et al.* (2006) *Biochemistry* **45**, 5468. **4.** Surleraux, de Kock, Verschueren, *et al.* (2005) *J. Med. Chem.* **48**, 1965. **5.** Shuman, Haemaelaeinen & Danielson (2004) *J. Mol. Recognit.* **17**, 106. **6.** Zhang, Chando, Everett, *et al.* (2005) *Drug Metab. Dispos.* **33**, 1729.

Loratadine

Loratadine

Desloratadine

3-Hydroxy-Desloratadine

This long-acting antihistamine pro-drug and antiallergic agent (FW = 382.89 g/mol; CAS 79794-75-5; IUPAC: ethyl 4-(8-chloro-5,6-dihydro-11*H*-benzo[5,6]cyclohepta[1,2-*b*]pyridin-11-ylidene)-1-piperidinecarboxylate; Soluble to 20 mM in DMSO), also named SCH 29851, Claritin®, and Clarityn®, is metabolized to descarboxyethoxyloratadine, *or* desloratadine (FW = 310.82 g/mol; CAS 100643-71-8; IUPAC: 8-chloro-6,11-dihydro-11-(4-piperdinylidene)-5*H*-benzo[5,6]cyclohepta[1,2-*b*]pyridine; Soluble to 20 mM in DMSO), eventually forming the more active agent, 3-hydroxydesloratadine (FW = 326.82 g/mol; CAS 119410-08-1; IUPAC: 8-chloro-6,11-dihydro-11-(4-piperidinylidene)-5*H*-benzo[5,6]cyclohepta[1,2-*b*]pyridin-3-ol). UGT2B10 catalyzes glucuronidation of desloratadine, followed by CYP2C8-catalyzed oxidation and a de-conjugation to form 3-hydroxydesloratadine (6). Desloratadine is an inverse agonist (K_i = 0.87 nM) of histamine H_1 receptor, devoid of CNS effects (1-5). (Compare to a K_i value of 35 nM for loratidine.) Unlike most classical antihistamines, loratadine (ED_{50} = 0.40 mg/kg, p.o.) is a peripheral H_1 antagonist that lacks central nervous system depressing effects, including drowsiness. In anti-immunoglobulin-triggered human basophils and 2,4-dinitrophenyl-triggered rat basophilic leukemia cells, dose-dependent inhibition of histamine release occurs at 2 μM descarboxyethoxy-loratadine or at 7 μM loratadine (7). By using the whole-cell patch clamp technique, the effects of loratadine on the transient outward K current (I_{to}), sustained current (I_{sus}), and current measured at -100 mV (I_{K1} and I_{ns}), the major inward and outward potassium currents present in human atrial myocytes, Crumb (8) assessed the possible molecular mechanisms for the observed atrial arrhythmias reported with loratadine use. Loratadine rate-dependently inhibited I_{to} at therapeutic concentrations with 10 nM loratadine reducing I_{to} amplitude at a pacing rate of 2 Hz by 35 %. In contrast, loratadine had no effect on either I_{sus} or the current measured at -100 mV. Such results provide a possible explanation for the incidences of supraventricular arrhythmias reported with the use of loratadine. inhibition of brain PgP multi-drug resistance factor by verapamil can convert desloratadine to a sedating antihistamine in mice (9). BCR/ABL promotes histamine production in CML cells and desloratidine exert antileukemic effects on CML cells (10). **See also** *Terfenadine, the first FDA-approved nonsedating H1-directed antihistamine*. *Note*: Loratadine is distantly related to quetiapine, an atypical antipsychotic. **Key Pharmacokinetic Parameters:** *See* Appendix II in Goodman & Gilman's *THE PHARMACOLOGICAL BASIS OF THERAPEUTICS*, 12th Edition (Brunton, Chabner & Knollmann, eds.) McGraw-Hill Medical, New York (2011). **Target(s):** aldehyde oxidase (1); CYP3A4, as alternative substrate (2); CYP2C19 (2); CYP2D6 (2); glucuronidation reactions (2); histamine H_1 receptor (3,4); P-glycoprotein, *or* xenobiotic transporting ATPase, multidrug-resistance protein (5). When present at 10 μM, H_1 antihistamines inhibit oxidative burst of human neutrophils (Rank Order of Potency: Dithiaden > Loratadine > Brompheniramine > Chlorpheniramine > Pheniramine) (11). **1.** Obach, Huynh, Allen & Beedham (2004) *J. Clin. Pharmacol.* **44**, 7. **2.** Nicolas, Whomsley, Collart & Roba (1999) *Chem. Biol. Interact.* **123**, 63. **3.** Haria, Fitton & Peters (1994) *Drugs* **48**, 617. **4.** Barnett, Iorio, Kreutner, *et al.* (1984) *Agents Actions* **14**, 590. **5.** Wang, Casciano, Clement & Johnson (2001) *Drug Metab. Dispos.* **29**, 1080. **6.** Kazmi, Barbara, Yerino & Parkinson (2015) *Drug Metab. Dispos.* **43**, 523. **7.** Berthon, Taudou, Combettes, *et al.* (1994) *Biochem. Pharmacol.* **47**, 789. **8.** Crumb (1999) *Brit. J. Pharmacol.* **126**, 575. **9.** Katta, Dhananjeyan, Bykowski, *et al.* (2007) *Drug Metab. Lett.* **1**, 7. **10.** Aichberger, Mayerhofer, Vales, *et al.* (2006) *Blood* **108**, 3538. **11.** Nosá, Jančinová, Drábiková & Perečko (2015) *Gen. Physiol. Biophys.* **34**, 209.

Lorazepam

This a high-potency, intermediate-duration, 3-hydroxybenzodiazepine anticonvulsant/antianxiety agent (FW = 321.16 g/mol; CAS 846-49-1) binds to $GABA_A$ receptors in the CNS, exhibiting anxiolytic effects, amnestic effects, sedative/hypnotic effects, anticonvulsant effects, antiemetic effects, and muscle relaxant effects. **Key Pharmacokinetic Parameters:** *See* Appendix II in Goodman & Gilman's *THE PHARMACOLOGICAL BASIS OF THERAPEUTICS*, 12th Edition (Brunton, Chabner & Knollmann, eds.)

McGraw-Hill Medical, New York (2011). **Target(s):** glucuronosyl-transferase (1); Na$^+$/K$^+$-exchanging ATPase (2); adenosine deaminase (3). **1**. Hara, Nakajima, Miyamoto & Yokoi (2007) *Drug Metab. Pharmacokinet.* **22**, 103. **2**. Saha, Sengupta, Dutta, Sirkar & Sengupta (1989) *Indian J. Exp. Biol.* **27**, 44. **3**. Moosavi-Movahedi, Sepassi Tehrani, Amanlou, *et al.* (2010) *Protein Pept Lett.* **17**, 197.

Lorcaserin

This centrally acting, selective serotonin receptor agonist and anti-obesity drug (FW = 195.68 g/mol; CAS 616202-92-7), also named Belviq® and (1R)-8-chloro-1-methyl-2,3,4,5-tetrahydro-1H-3-benzazepine, targets 5-HT$_{2c}$ receptors (EC$_{50}$ = 39 nM; K_i = 13 nM), with weaker action against 5-HT$_{2a}$ receptors (EC$_{50}$ = 550 nM; K_i = 90 nM) and 5-HT$_{2b}$ receptors (EC$_{50}$ = 2400 nM; K_i = 150 nM). Lorcaserin does not compete for binding of ligands to serotonin, dopamine, and norepinephrine transporters. 5-HT$_{2c}$ receptors are located in the choroid plexus, cortex, hippocampus, cerebellum, amygdala, thalamus, and hypothalamus. Although lorcaserin's mechanism of appetite suppression is unclear, 5-HT$_{2c}$ receptor activation in the hypothalamus is thought to activate pro-opiomelanocortin (POMC), thereby promoting satiety and modest weight loss (3). **1**. Smith, Smith, Tsai, *et al.* (2008) *J. Med. Chem.* **51**, 305. **2**. Thomsen, Grottick, Menzaghi, *et al.* (2008) *J. Pharmacol. Exp. Ther.* **325**, 577. **3**. Chan, He, Chui, *et al.* (2013) *Obesity Rev.* **14**, 383.

Losmapimod

This orally bioavailable protein kinase inhibitor and anti-depresion drug (FW = 383.46 g/mol; CAS 585543-15-3), also named GW856553X and 6-[5-(cyclopropylcarbamoyl)-3-fluoro-2-methylphenyl]-N-(2,2-dimethylpropyl)pyridine-3-carboxamide, selectively targets p38 mitogen-activated protein kinases that serve as mediators of inflammation (1). Inhibition of these enzymes produces antidepressant and antipsychotic effects in mechanism likely to involve release of bone-derived nerve factor(BDNF) release. It also reduces plasma fibrinogen in patients with chronic obstructive pulmonary disease (2). **1**. Aston, Bamborough, Buckton, *et al.* (2009) *Journal of Medicinal Chemistry* **52**, 6257. **2**. Lomas, Lipson, Miller, *et al.* (2012) *J. Clin. Pharmacol.* **52**, 416.

Losartan

This oral antihypertensive drug (FW = 422.92 g/mol; CAS 114798-26-4), also named (2-butyl-4-chloro-1-{[2'-(1H-tetrazol-5-yl)biphenyl-4-yl]methyl}-1H-imidazol-5-yl)methanol and marketed under the trade name Cozaar, is a nonpeptide angiotensin II receptor antagonist (IC$_{50}$ = 19 nM) that reportedly provides a more specific and complete blockade of angiotensin II actions than inhibition of renin or angiotensin-converting enzyme (ACE). Indeed, All of the physiological effects of angiotensin II, including stimulation of release of aldosterone, are antagonized in the presence of losartan. Reduction in blood pressure occurs independently of

the status of the renin-angiotensin system. Losartan is the recommended first-line treatment in patients under age 55, who cannot tolerate an ACE inhibitor. (***See also*** *Candesartan; Eprosartan; Irbesartan; Olmesartan; Telmisartan; Valsartan*) **Key Pharmacokinetic Parameters:** Bioavailability = 33 %; Food Effect? NO; Drug $t_{1/2}$ = 2 hours; Metabolite $t_{1/2}$ = 6–9 hours; Drug's Protein Binding = 98.7 %; Metabolite's Protein Binding = 99.8 %; Route of Elimination = 35 % Renal, 60 % Liver. *See also* Appendix II in Goodman & Gilman's THE PHARMACOLOGICAL BASIS OF THERAPEUTICS, 12th Edition (Brunton, Chabner & Knollmann, eds.) McGraw-Hill Medical, New York (2011). **1**. Chiu, McCall, Price, *et al.* (1990) *J. Pharmacol. Exp. Ther.* **252**, 711. **2**. Wong, Price, Chiu, *et al.* (1990) *J. Pharmacol. Exp. Ther.* **252**, 719 and 726.

Lovastatin

Lovastatin Lactone   **Sodium Salt**

This serum cholesterol-reducing drug (FW$_{lactone}$ = 404.55 g/mol), also known as MK-803, mevinolin, and monacolin K, is a fungal metabolite from *Aspergillus terreus* and *Monascus ruber* that inhibits 3-hydroxy-3-methyl-glutaryl-coenzyme A (*or* HMG-CoA) reductase. When taken at pharmacologic doses, this tight-binding competitive inhibitor blocks mevalonate biosynthesis and induces the expression of LDL receptors in the liver, which in turn increases the catabolism of plasma LDL and lowers the plasma concentration of cholesterol, an important determinant of atherosclerosis. When taken along with a low-fat diet, lovastatin reduces the risk of heart attack, stroke, certain types of heart surgeries, and chest pain in patients with heart disease or subjects having several common heart disease risk factors (*i.e.*, family history of heart disease, high blood pressure, age, low HDL cholesterol, and smoking). **Primary Mode of Action:** Lovastatin is a potent inhibitor of 3-hydroxy-3-methylglutaryl-CoA reductase (K_i = 0.5 nM) and is highly effective in the treating hypercholesterolemia. Lovastatin received approved by the Food and Drug Administration in 1987. The lactone is very stable and is soluble in organic solvents, with lower solubility in water (0.4 x 10^{-3} mg/mL at room temperature). The sodium salt is unstable at room temperature and should be prepared only immediately prior to use (1). Note that mevastatin is the 6-demethylated derivative of lovastatin. *See Mevastatin; Statin Side-Effects* **Key Pharmacokinetic Parameters:** *See* Appendix II in Goodman & Gilman's THE PHARMACOLOGICAL BASIS OF THERAPEUTICS, 12th Edition (Brunton, Chabner & Knollmann, eds.) McGraw-Hill Medical, New York (2011). Note: Lovastatin is a CYP3A4-metabolized statin, and co-administration of a CYP3A4 inhibitors can result in adverse drug interactions. **Target(s):** acetylcholinesterase (2); cholinesterase, *or* butyrylcholinesterase) (2); 3-hydroxy-3 methylglutaryl-CoA reductase (NADPH) (1,3-6). **1**. Endo (1981) *Meth. Enzymol.* **72**, 684. **2**. Chiou, Lai, Tsai, *et al.* (2005) *J. Chin. Chem. Soc.* **52**, 843. **3**. Endo (1981) *Trends Biochem. Sci.* **6**, 10. **4**. Chang (1983) *The Enzymes*, 3rd ed., **16**, 491. **5**. Gibson & Parker (1987) *The Enzymes*, 3rd ed., **18**, 179. **6**. Endo (1980) *J. Antibiot.* **33**, 334.

Loxapine

This dibenzoxazepine-based antipsychotic agent (FW = 327.81 g/mol; CAS 27833-64-3; Solubility: 90 mg/mL DMSO), also named 2-chloro-11-(4-methyl-1-piperazinyl)dibenz[b,f][1,4]oxazepine, is a potent D$_2$/D$_4$-dopamine receptor and 5-HT$_{2A}$/5-HT$_{2B}$/5-HT$_7$ serotonergic receptor antagonist that is frequently used in the treatment of schizophrenia (1-4). In studies with human receptors, its rank-order potency is: 5-HT$_2$ receptor ≥ D$_4$ receptor >>> D$_1$ receptor > D$_2$ receptor (5). **1**. Roth, Tandra, Burgess, Sibley & Meltzer (1995) *Psychopharmacology* **120**, 365. **2**. Shen, Monsma, Metcalf, *et al.* (1993) *J. Biol. Chem.* **268**, 18200. **3**. Fiorella, Rabin & Winter (1995) *Psychopharmacology* **121**, 347. **4**. Glusa & Pertz (2000) *Brit. J. Pharmacol.* **130**, 692. **5**. Singh, Barlas, Singh, Franks & Mishra (1996) *J. Psychiatry Neurosci.* **21**, 29.

Loxoprofen

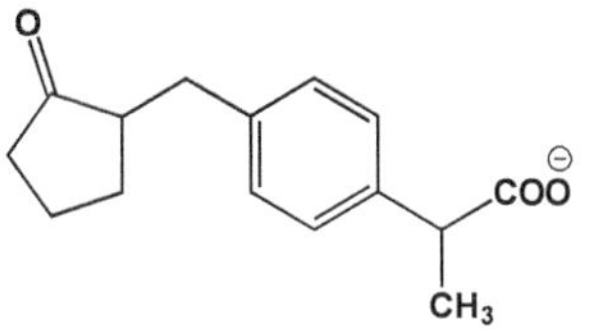

Loxoprofen 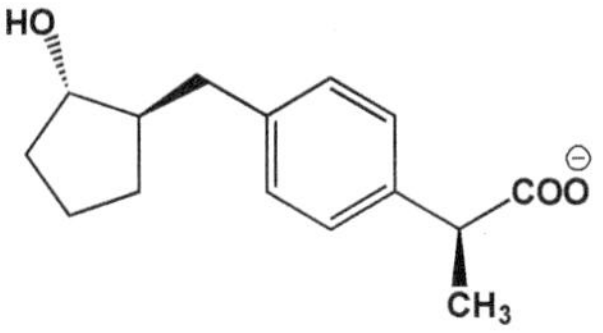   **Most Potent Metabolite**

This anti-inflammatory agent (FW$_{free-acid}$ = 246.31 g/mol), also named CS-600 and 2-[4-(2 oxocyclo-pentylmethyl)phenyl]propionic acid, is the pro-drug of an inhibitor of prostaglandin synthesis (*i.e.*, 2-[4-(*trans*-2-hydroxycyclopentylmethyl)phenyl]propionic acid). The (2*S*,1'*R*,2'*S*)-*trans*-alcohol isomer was found to be the most active metabolite. **Target(s):** carbonyl reductase (1); cyclooxygenase-1 and -2 (2,3). **1**. Imamura, Koga, Higuchi, *et al.* (1997) *J. Enzyme Inhib.* **11**, 285. **2**. Matsuda, Tanaka, Ushiyama, Ohnishi & Yamazaki (1984) *Biochem. Pharmacol.* **33**, 2473. **3**. Riendeau, Salem, Styhler, *et al.* (2004) *Bioorg. Med. Chem. Lett.* **14**, 1201.

Lq2

This oligopeptides (MW = 4250; *Sequence:* ZFTNESCTASNQCWSIC KRLHNTNRGKCMNKKCRCYS, where "Z" refers to a pyroglutamyl residue), also known as ChTX-Lq2, is a charybdotoxin homologue (**See** *Charybdotoxin*) from the venom of the Israeli scorpion *Leiurus quinquestriatus hebraeus*. Lq2 binds to the voltage-activated, the Ca^{2+}-activated and the inward-rectifier potassium channels, with high affinity for the latter class, especially the Renal Outer Medullary Potassium channel ROMK1 that regulates the resting membrane potential. Note that Lq2 has disulfide bonds between Cys7–Cys28, Cys13–Cys33, and Cys17–Cys35. **1**. Lucchesi, Ravindran, Young & Moczydlowski (1989) *J. Membr. Biol.* **109**, 269. **2**. Lu & MacKinnon (1997) *Biochemistry* **36**, 6936.

LS-187118, See *Sarpogrelate*

LS-193,855, See *RPL-554*

LSD, See *Lysergic Acid Diethylamide*

LspA Virulence Factor YL2 Domain

This protein domain (MW = 37 kDa) allows *Haemophilus ducreyi*, a microorganism that causes the sexually transmitted infection known as chancroid, to elude phagocytosis and eventual destruction. The YL2 domain of the 456-kDa LspA1 also inhibits phagocytosis of IgG-opsonized targets. When expressed in mammalian cells, YL2 decreases levels of active Src tyrosine kinases (SFKs). Tyrosine phosphorylation sites in two EPIYG motifs within YL2 are required, and purified Csk phosphorylates these motifs, in turn increasing Csk catalytic activity in a positive feedback loop. A dominant-negative Csk rescues phagocytosis in the presence of YL2, suggesting that Csk is the main host protein interacting with YL2. *Helicobacter pylori* CagA protein also inhibits phagocytosis in a Csk-dependent manner, suggesting a possibly general mechanism to subvert Fcγ receptor (FcγR)-mediated phagocytosis, the latter a crucial component of the innate immune response. **1**. Dodd, Worth, Rosen, *et al.* (2014) *MBio.* **5**, e01178.

LU 1631, See *Amezinium*

Lu 3-010, See *Talopram*

Lu-10-171, See *Escitalopram*

Lu 19-005, See *Indatraline*

LU-208075, See *Ambrisentan*

Lu AA21004, See *Vortioxetine*

Lu AA 47070

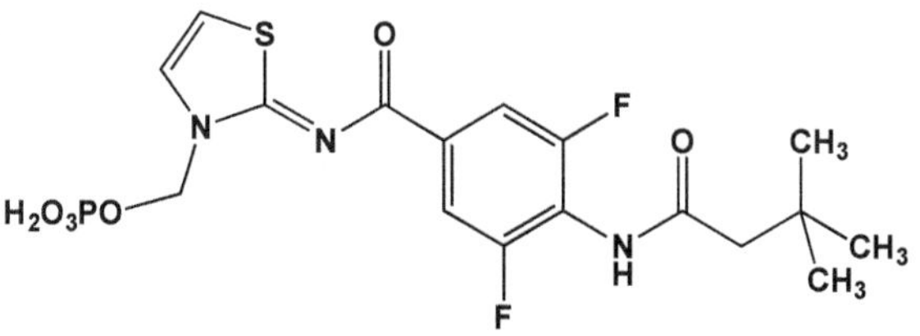

Lu AA 47070 (prodrug)

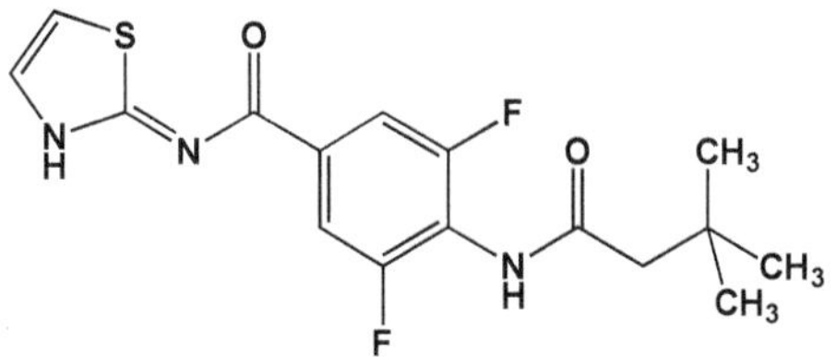

Active Drug

This orally bioavailable prodrug (FW = 463.39 g/mol; CAS 913842-25-8; Soluble to 100 mM in DMSO), also named 4-[(3,3-dimethyl-1-oxobutyl)amino]-3,5-difluoro-*N*-[3-[(phosphonooxy)methyl]-2(3*H*)-thiazo-lylidene]benzamide, is metabolically converted into potent and selective adenosine A$_{2A}$ receptor antagonist (K_i = 5.9 nM), with much weaker action against the A$_{2B}$ (K_i = 260 nM), A$_1$ (K_i = 410 nM), and A$_3$ receptors (K_i < 10,000 nM) (1). The active drug had acceptable ADME properties; however, its low intrinsic solubility limited its development as a therapeutic agent, because oral bioavailability from dosing in suspension was significantly lower than the oral bioavailability from solution dosage. When administered intravenously, Lu AA47070 (dose: 3.75-30 mg/kg) reverses the effects of the D$_2$ receptor antagonist pimozide (1.0 mg/kg, administered intravenously) using several measures of motor impairment (*e.g.*, catalepsy, locomotion, and tremulous jaw movements) in a rodent model of Parkinsonian tremor (2) **1**. Sams, Mikkelsen, Larsen, *et al.* (2011) *J. Med. Chem.* **54**, 751. **2**. Collins, Sager, Sams, et al. (2012) *Pharmacol. Biochem. Behav.* **100**, 498.

Lu AF64280

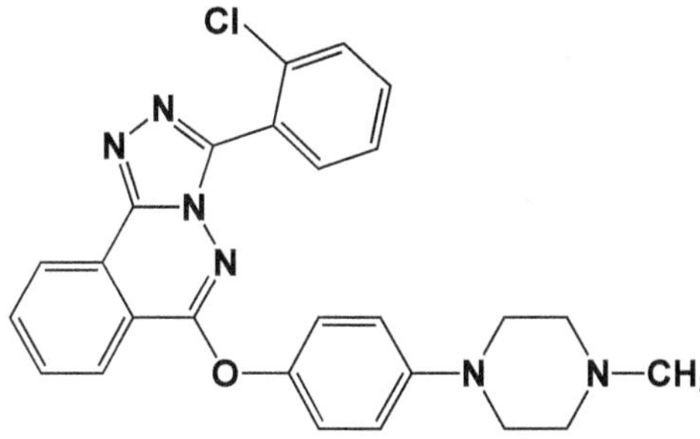

This brain-penetrating PDE inhibitor (FW = 470.96 g/mol) targets phosphodiesterase PDE2a, both *in vitro* and *in vivo*, and is suitable for assessing roles played by PDE2A in synaptic plasticity under normal and perturbed conditions. Lu AF64280 attenuates spontaneous P20-N40 gating deficits in DBA/2 mice, an inbred mouse strain proposed to model P50 gating deficits observed in some schizophrenic patients. Lu AF64280 significantly increases cGMP levels in the hippocampus, attenuates sub-chronic phencyclidine-induced deficits in novel object exploration in rats, blocks early postnatal phencyclidine-induced deficits in the intradimensional/extradimensional shift task in rats and attenuates spontaneous P20-N40 auditory gating deficits in DBA/2 mice. At any dose tested, Lu AF64280 fails to attenuate phencyclidine-induced hyperactivity in mice, and is devoid of antipsychotic-like activity in the conditioned avoidance response paradigm in rats. **Target(s):** Lu AF64280 inhibits full-length human PDE2A (K_i = 20 nM) and moderately inhibits both human PDEs 9A (K_i = 1,000 nM) and 10A (K_i = 1,800 nM). Lu AF64280 displays a K_i above 5,000 nM against all other tested members of the PDE family, revealing >250x selectivity window towards other PDEs. **1**. Redrobe, Jørgensen, Christoffersen, *et al.* (2014) *Psychopharmacol.* **231**, 3151.

Lubiprostone

This bowel movement-promoting medication (FW = 390.46 g/mol; CAS 136790-76-6; IUPAC: 7-[(1*R*,3*R*,6*R*,7*R*)-3-(1,1-difluoropentyl)-3-hydroxy-8-oxo-2-oxabicyclo[4.3.0]non-7-yl]heptanoic acid) is used to treat chronic idiopathic constipation, irritable bowel syndrome-associated constipation, as well as opioid-induced constipation. This bicyclic fatty acid, itself derived from prostaglandin E_1, activates the ClC-2 voltage-gated Cl⁻ channel on gastrointestinal epithelial cells, producing a chloride-rich fluid secretion that softens the stool, increases motility, and promotes spontaneous bowel movements. **1**. Müller-Lissner (2013) *Expert Opin. Drug Metab. Toxicol.* **9**, 391. **2**. Raschi & De Ponti (2014) *Expert Opin. Drug Metab. Toxicol.* **10**, 293.

Lucanthone

This anthelmintic and bacterial antibiotic (FW$_{free-base}$ = 340.49 g/mol; FW$_{hydrochloride}$ = 376.49; CAS; IUPAC Name: 1-[[2-(diethylamino)ethyl]amino]-4-methyl-9*H*-thioxanthen-9-one hydrochloride), also known by the code names MS-752 and RP-3735 as well as the trade names Miracil D™, Nilodin™, Miracol™ and Tixantone™, is used to treat parasitic infections, particularly by *Schistosoma mansoni* (1). When administered at 30-60 µM, lucanthone is toxic to *Escherichia coli* and *Bacillus subtilis*. **Schistosomiasis:** Although these flatworms are not found in the United States, they infect more than 200 million people worldwide, making schistosomiasis a major health challenge. The causative organism is found throughout Africa, parts of South America (particularly Brazil, Venezuela, Surinam, and Guyana), and the Carribean islands (mainly Peurto Rico, St. Lucia, and Martinique), where it flourishes in certain freshwater snails. **Primary Mode(s) of Action:** Lucanthone is a DNA intercalator that inhibits DNA and/or RNA biosynthesis inhibitor, DNA topoisomerase, and base-excision repair. Lucanthone inhibits the base-excision endonuclease activity of APE1, without affecting its redox activity, IC$_{50}$ = 5 µM (2). X-ray structures of APE1 reveal a hydrophobic pocket where hydrophobic small molecules like lucanthone can bind. Exposure of cells to lucanthone results in processing and recruitment of Microtubule-Associated Protein-1 light chain 3 (MAP1-LC3) to autophagosomes, but impairs autophagic degradation (3,4). **Target(s):** ribosomal RNA synthesis (5); butyrylcholinesterase (6). **1**. Hirschberg (1975) in *Antibiotics* (Corcoran & Hahn, eds), Springer-Verlag, New York, p. 274. **2**. Naidu, Agarwal, Pena, *et al.* (2011) *PLoS One* **6**, e23679. **3**. Carew, Espitia, Esquivel, *et al.* (2011) *J. Biol. Chem.* **286**, 6602. **4**. Fishel & Kelley (2007) *Mol. Aspects Med.* **28**, 375. **5**. Bases & Mendez (1969) *J. Cell Physiol.* **74**, 283. **6**. Verdier & Wolfe (1986) Biochem Pharmacol. **35**, 1605.

Lufaxin

This novel FXa inhibitor (MW = 32.4 kDa), from the salivary gland of the sand fly *Lutzomyia longipalpis*, blocks Protease-Activated Receptor 2 (PAR2) activation and inhibits inflammation and thrombosis *in vivo*. Also known as Coagulation Factor II (thrombin) Receptor-like 1 (F2RL1), PAR2

modulates inflammatory responses and acts as a sensor for proteolytic enzymes generated during infection. Lufaxin inhibits thrombin formation by prothrombinase, which consists of the serine proteins, Factor Xa, and the protein cofactor, Factor Va. It was identified as a slow, tight-binding noncompetitive (reversible) inhibitor of coagulation Factor Xa. Surface plasmon resonance revealed that FXa forms a high-affinity complex with Lufaxin (K_d ~ 3 nM), and isothermal titration calorimetry demonstrated the binding stoichiometry was 1:1. Lufaxin also prevents PAR2 activation by FXa in the MDA-MB-231 cell line and abrogates edema formation triggered by injection of FXa in the paw of mice. Moreover, Lufaxin prevents FeCl$_3$-induced carotid artery thrombus formation and prolongs aPTT *ex vivo*, implying that it works as an anticoagulant *in vivo*. **1**. Collin, Assumpção, Mizurini, *et al.* (2012) *Arterioscler. Thromb. Vasc. Biol.* **32**, 2185.

Luminespib

This highly potent HSP90 inhibitor (FW = 465.54 g/mol; CAS 747412-49-3; Solubility: 90 mg/mL DMSO; <1 mg/mL H$_2$O), also known as AUY922 and NVP-AUY922, targets Heat Shock Protein 90 (IC$_{50}$ = 13 nM and 21 nM for HSP90α and HSP90β) and human tumor cell proliferation with GI$_{50}$ values of 2–40 nM, inducing G$_1$-G$_2$ arrest and apoptosis. Its action on is characterized by, respectively (1-2). NVP-AUY922 induces HSP70, the binding of which to client proteins (*e.g.*, HER-2, Akt and thymidylate synthase) results in their proteasomal degradation (3), with attendant cell growth inhibition. When used in combination with a PKC-9 inhibitor, AUY922 (15 nM) almost completely suppresses HIV-1 reactivation, with no apparent cytotoxicity (4). Indeed, Hsp90 is required downstream of PKCs but not for activation of mitogen-activated protein kinase. Inhibition of Hsp90 reduces degradation of IkBα and blocks nuclear translocation of transcription factor p65/p50, suppressing the NF-κB pathway. Coimmunoprecipitation experiments show that Hsp90 interacts with inhibitor of nuclear factor kappa-B kinase (IKK) together with cochaperone Cdc37, which is critical for the activity of several kinases. Targeting of Hsp90 by AUY922 dissociates Cdc37 from the complex. Hsp90 thus controls HIV-1 reactivation from latency by keeping the IKK complex functional and connects T-cell activation with HIV-1 replication (4). AUY922 treatment also interrupts HSP90 interactions with its protein clients Bcr-Abl or Bcr-Abl kinase domain mutants (T315I and E255K), leading to Bcr-Abl degradation, nhibiting cell proliferation and inducing apoptosis of both imatinib-sensitive 32Dp210 (IC$_{50}$ = 6 nM) and imatinib-resistant 32Dp210T315I cells (IC$_{50}$ ≈ 6 nM) and human KBM-5R/KBM-7R cell lines (IC$_{50}$ = 50 nM) (5). **1**. Brough, Aherne, Barril, *et al.* (2008) *J. Med. Chem.* **51**, 196. **2**. Eccles, Massey, Raynaud, *et al.* (2008) *Cancer Res.* **68**, 2850. **3**. Lee, Lee, Han, *et al.* (2011) *Cancer Sci.* **102**, 1388. **4**. Anderson, Low, Weston, *et al.* (2014) *Proc. Natl. Acad. Sci. U.S.A.* **111**, E1528. **5**. Tao, Chakraborty, Leng, Ma & Arlinghaus (2015) *Genes Cancer* **6**, 19.

Lurasidone

This orally active, atypical antipsychotic (FW = 491.69 g/mol; CAS 367514-87-2), also known by the trade Latuda®, the developmental code name SM-13496, and the systematic name (3aR,4S,7R,7aS)-2-{(1R,2R)-2-[4-(1,2-benzisothiazol-3-yl)piperazin-1-ylmethyl]cyclohexylmethyl}hexahydro-4,7-methano-2H-isoindole-1,3-dione, is an FDA-approved drug for the treatment of depressive episodes associated with bipolar type I disorder as well as bipolar type II disorder in adults, either alone or in combination with lithium or valproate (1,2). Lurasidone has negligible affinity for histamine H_1 and muscarinic M_1 receptors, which are thought to contribute to side effects such as weight gain, sedation, and worsening of cognitive deficits (3). Structural models of lurasidone-GPCR complexes were determined by homology modeling of receptors, exhaustive ligand docking, and molecular dynamics simulation-based refinement of these complexes. This modeling gave reliable structural models for D_2, 5-HT_{2A}, and 5-HT_7 recptors with overall structural complementarities with a salt bridge anchor at the center of the lurasidone molecule, but not for H_1 and M_1 receptors, owing to steric hindrance between the norbornane-2,3-dicarboximide and/or cyclohexane part of lurasidone and both receptors. By comparison with the structural models of olanzapine-GPCRs and ziprasidone-GPCRs using the same computational protocols, it was suggested that the bulkiness of the norbornane-2,3-dicarboximide part and the rigidity and the bulkiness of the cyclohexyl linker gave lurasidone high selectivity for the desired aminergic GPCRs (3). **Targets:** α_1-adrenergic receptor, as antagonist (K_i = 48 nM); α_{2A}-adrenergic receptor, as antagonist (K_i = 1.6 nM); α_{2C}-adrenergic receptor, as antagonist (K_i = 10.8 nM); dopamine D_1 receptor, as antagonist (K_i = 262 nM); dopamine D_2 receptor, as antagonist (K_i = 1.7 nM); serotonin 5-HT_{1A} receptor, as partial agonist (K_i = 6.8 nM); serotonin 5-HT_{2A} receptor, as antagonist (K_i = 2.0 nM); serotonin 5-HT_{2C} receptor, as antagonist (K_i = 415 nM); and serotonin 5-HT_7 receptor, as antagonist (K_i = 0.5 nM). **1.** Meyer, Loebel & Schweizer (2009) *Expert Opin. Investigatnl. Drugs* **18**, 1715. **2.** Ishibashi, Horisawa, Tokuda, (2010) J. Pharmacol. Exper. Therapeut. **334**, 171. **3.** Ichikawa, Okazaki, Nakahira, *et al.* (2012) *Neurochem. Int.* **61**, 1133.

Luseogliflozin

This orally active and highly selective, second generation SGLT2 inhibitor (FW = 452.56 g/mol; CAS 898537-18-3), also known as (1S)-1,5-anhydro-1-[5-(4-ethoxybenzyl)-2-methoxy-4-methylphenyl]-1-thio-D-glucitol and TS-071, is a Type-2 diabetes drug that targets the sodium-glucose transporter, Subtype 2 (IC_{50} = 2.26 nM), increasing urinary glucose excretion independently of insulin secretion/action. Chiefly expressed in the proximal tubules, the SGLT2 is a low-affinity, high-capacity glucose transporter that is responsible for >90% of kidney glucose reabsorbed in hyperglycemic patients. Because its inhibitory action is insulin-independent, luseogliflozin provides steady glucose control, without risk for hypoglycemia and with favorable reductions in both body weight and blood pressure. Other SGLT2 inhibitors include canagliflozin, dapagliflozin, empagliflozin, and tofogliflozin. **1.** Kojima, Williams, Takahashi, Miyata & Roman (2013) *J. Pharmacol. Exp. Ther.* **345**, 464. **2.** Kurosaki & Ogasawara (2013) *Pharmacol. Ther.* **139**, 51.

Luteolin 7-*O*-Glucoside

This yellow glycoside (FW = 448.38 g/mol), also known as cynaroside, from many plants (*e.g.*, *Achillea millefolium* and *Ixeris chinensis*) inhibits α-glucosidase and α-amylase. **Target(s):** α-amylase (1,2); catechol oxidase (3); complement pathway (4); 3',5'-cyclic nucleotide phosphodiesterase (5); α-glucosidase (1); phosphodiesterase 2 (5); phosphodiesterase 4 (5); thyroxine 5-deiodinase, *or* iodothyronine deiodinase (6,7); tyrosinase, *or* monophenol monooxygenase (3,8); xanthine oxidase (9). **1.** Kim, Kwon & Son (2000) *Biosci. Biotechnol. Biochem.* **64**, 2458. **2.** Funke & Melzig (2005) *Pharmazie* **60**, 796. **3.** Kim & Uyama (2005) *Cell. Mol. Life Sci.* **62**, 1707. **4.** Pieroni, Pachaly, Huang, Van Poel & Vlietinck (2000) *J. Ethnopharmacol.* **70**, 213. **5.** Ko, Shih, Lai, Chen & Huang (2004) *Biochem. Pharmacol.* **68**, 2087. **6.** Auf'mkolk, Kohrle, Gumbinger, Winterhoff & Hesch (1984) *Horm. Metab. Res.* **16**, 188. **7.** Auf'mkolk, Koehrle, Hesch & Cody (1986) *J. Biol. Chem.* **261**, 11623. **8.** Karioti, Protopappa, Magoulas & Skaltsa (2007) *Bioorg. Med. Chem.* **15**, 2708. **9.** Nguyen, Awale, Tezuka, *et al.* (2006) *Planta Med.* **72**, 46.

LW8

This HIF-1α modulator (FW = 435.51 g/mol; CAS 934593-90-5; Soluble in DMSO, but not H_2O) inhibits the accumulation of Hypoxia-Induced Factor-1α, promoting the degradation of wild-type HIF-1α, but not of a DM-HIF-1α, having P402A and P564A modifications, at hydroxylation sites within the O_2-Dependent Degradation Domain (ODDD). In the presence of LW8, VHL knockdown does not abolish HIF-1α protein accumulation, indicating that LW8 promotes HIF-1α degradation by regulating expression of von Hippel-Lindau tumor suppressor (also known as pVHL). In mice carrying xenografts of human colon cancer HCT116 cells, LW8 shows strong anti-tumor efficacy in vivo, decreasing HIF-1α expression in frozen-tissue immuno-histochemical staining. **1.** Lee, Kang, Park, *et al.* (2010) *Biochem. Pharmacol.* **80**, 982.

LY5

This non-peptide cell-permeable inhibitor (FW = 329.33 g/mol) targets STAT3 dimerization, blocking its activation with low IC_{50} values in the 0.5–1.4 μM range. Strong binding affinity to the STAT3 SH2 domain was confirmed by fluorescence polarization. The IC_{50} values for the inhibition of viability are 0.364, 0.318, and 0.488 μm, respectively, in UW288-1, UW426, and DAOY cells. Its ability to significantly inhibit STAT3 phosphorylation at the lower drug concentration in medulloblastoma expressing constitutive STAT3 indicates its high potency. Several known STAT3 downstream targets, such as cyclin-D1, survivin, and bcl-XL, were decreased by LY5 treatment, as demonstrated by RT-PCR analysis and angiogenesis array, again supporting the idea of LY5 is a potent STAT3 inhibitor. miR-21 gene expression was down-regulated by LY5, another indication of the inhibitory effect of LY5 on STAT3. The fact that LY5 inhibited STAT3 phosphorylation stimulated by cytokines and growth factors including IL-6, IGF-1, IGF-2, and LIF, but had no effect on STAT1 or STAT5 phosphorylation induced by IFN-γ or EGF, suggests that it selectively suppresses STAT3 activation. LY5 shows anti-tumor action by inducing apoptosis in medulloblastoma cells, via STAT3 inhibition. That LY5 blocks tumor cell migration and angiogenesis *in vitro* through inhibition of STAT3 supports this suggestion. Treatment with LY5 and cisplatin or X-radiation produced a statistically greater inhibitory effect than single treatment. **1.** Xiao, Bid, Jou, *et al.* (2015) *J. Biol. Chem.* **290**, 3418.

LY-146032, *See Daptomycin*

LY139603, *See Atomoxetine*

LY 210448, *See Dapoxetine*

LY 215840

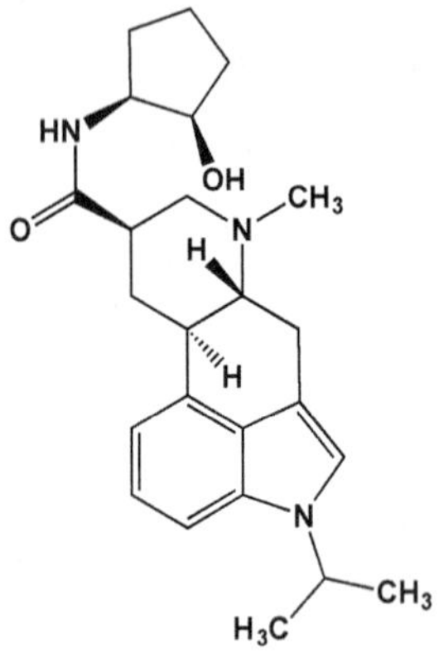

This serotonin receptor antagonist (FW = 395.26 g/mol; CAS 137328-52-0; Solubility: 100 mM in DMSO), also named [8β(1S,2R)]-N-(2-hydroxycyclopentyl)-6-methyl-1-(1-methylethyl)ergoline-8-carboxamide, targets 5-HT$_2$ and 5-HT$_7$ receptors, reducing the stimulatory effects of serotonin on aldosterone secretion and reducing 5-HT-induced cAMP formation (pK_B = 8.26). **1**. Cushing, *et al.* (1996) *J. Pharmacol. Exp. Ther.* **277**, 1560. PMID: 8667223. **2**. Lenglet, *et al.* (2002) *Endocrinology* **143**, 1748. **3**. Louiset, *et al.* (2006) *J. Clin. Endocrinol. Metab.* **91**, 4578.

LY 225910

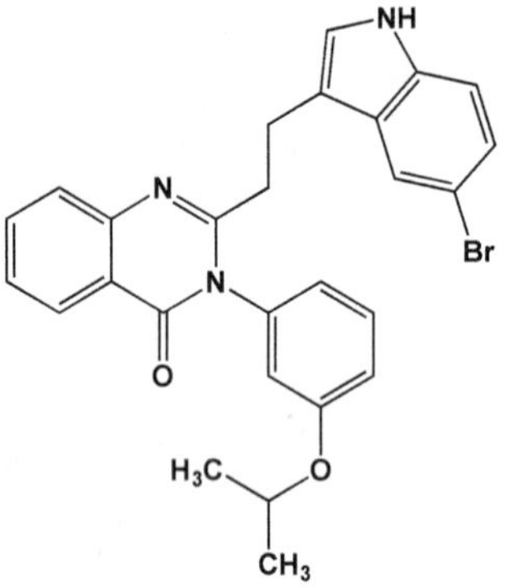

This potent CCK receptor antagonist (FW = 502.41 g/mol; CAS 133040-77-4; Solubility: 100 mM in DMSO), also named 2-[2-(5-bromo-1H-indol-3-yl)ethyl]-3-[3-(1-methylethoxy)phenyl]-4(3H)-quinazoline, targets the Cholecystokinin CCK$_2$ receptor, a G-protein coupled receptor that binds the peptide hormones cholecystokinin or gastrin. LY-225910 has an IC$_{50}$ value of 9.3 nM for the inhibition of binding of [^{125}I]-CCK-8 sulfate at mouse brain receptors. **1**. Yu, *et al.* (1991) *J. Med. Chem.* **34**, 1505. **2**. Suman-Chauhan, *et al.* (1996) *Regul. Pept.* **65**, 37.

LY 223053

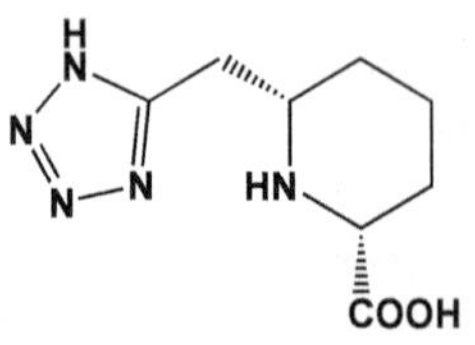

This NMDA receptor antagonist (FW = 211.22 g/mol; CAS 125546-04-5; Solubility: 100 mM in H$_2$O or DMSO), also named (2R*,4S*)-4-(1H-tetrazol-5-ylmethyl)-2-piperidinecarboxylic acid, targets N-methyl-D-aspartate receptor (IC$_{50}$ = 7 nM), but displays no affinity for AMPA or kainate receptors, even at 10 μM. LY-223052 inhibits NMDA-induced neuronal degeneration and also protects from NMDA-induced convulsions in neonatal rats. **1**. Schoepp, *et al.* (1990) *J. Pharmacol. Exp. Ther.* **255**, 1301. **2**. Madden, *et al.* (1992) *J. Neurosurg.* **76**, 106. **3**. Madden, *et al.* (1993) *Stroke* **24**, 1068.

LY 264618, *See Lometrexol*

LY248686, *See Duloxetine*

LY 255283

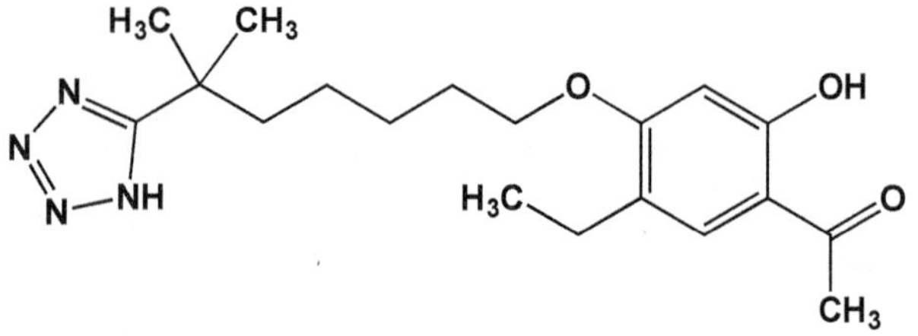

This competitive BLT antagonist (FW = g/mol; CAS 117690-79-6; Solubility: 100 mM in DMSO), also named 1-[5-ethyl-2-hydroxy-4-[[6-methyl-6-(1H-tetrazol-5-yl)heptyl]oxy]-phenyl]-ethanone, selectively targets human recombinant Leukotriene B-2, or BLT$_2$ receptors (IC$_{50}$ ~ 1 nM), with an IC$_{50}$ value > 10 μM ahainst BLT$_1$ receptors. LY 255283 inhibits LTB$_4$-induced contraction of lung parenchyma (pA$_2$ = 7.2), and reduces LTB$_4$-mediated airway obstruction in guinea pigs following intravenous and oral administration. **1**. Herron, *et al.* (1992) *J. Med. Chem.* **35**, 1818. **2**. Silbaugh, *et al.* (1992) *Eur. J. Pharmacol.* **223**, 57. **3**. Yokomizo, *et al.* (2001) *J. Biol. Chem.* **276**, 12454.

LY 266097

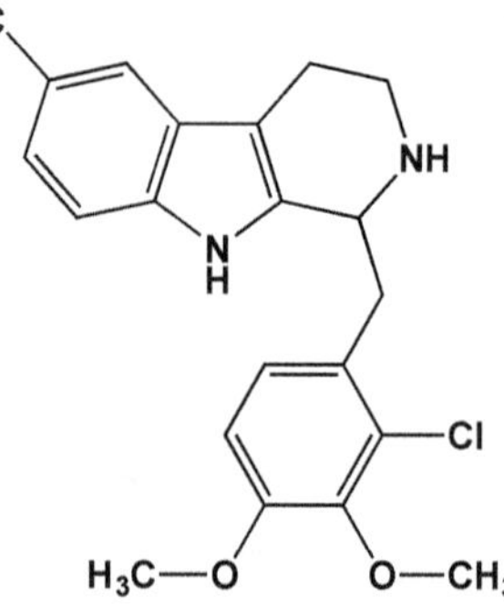

This serotonin receptor antagonist (FW = 370.88 g/mol; CAS 172895-39-5; Solubility: 100 mM in DMSO or Ethanol), also named 1-[(2-chloro-3,4-dimethoxyphenyl)methyl]-2,3,4,9-tetrahydro-6-methyl-1H-pyrido[3,4-b]indole, targets 5-HT$_{2B}$ receptor (pK_i = 9.3), with >100x selectivity over 5-HT$_{2A}$ and 5-HT$_{2C}$ receptors. LY 266097 attenuates amphetamine-induced locomotion in the rat. (*See LY 272015*) **1**. Audia, Evrard, Murdoch, *et al.* (1996) *J. Med. Chem.* **39**, 2773. **2**. Auclair, Cathala, Sarrazin, *et al.* (2010) *J. Neurochem.* **114**, 1323.

LY 272015

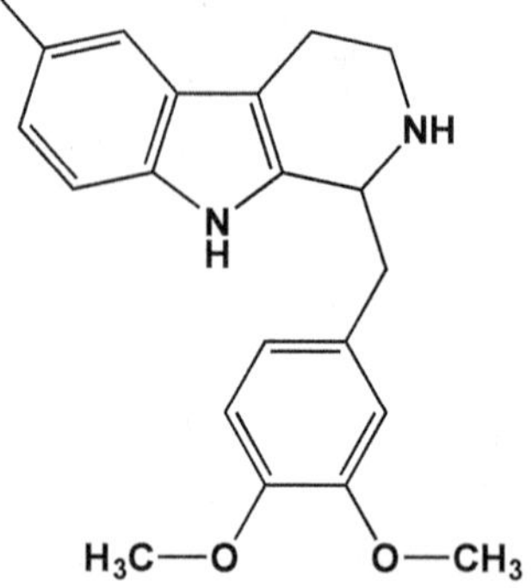

This orally available serotonin receptor antagonist (FW = 336.74 g/mol; CAS 172895-15-7; Solubility: 100 mM in DMSO; 10 mM in Ethanol), also named 1-[(3,4-dimethoxyphenyl)methy]-2,3,4,9-tetrahydro-6-methyl-1H-pyrido[3,4-b]indole, targets 5-HT$_{2B}$ receptors (K_i = 0.75 nM), displays selectivity over 5-HT$_{2C}$ (K_i = 22 nM) and 5-HT$_{2A}$ receptors (K_i = 29 nM). (*See LY 266097*) **1**. Cohen, *et al.* (1996) *Serotonin Res.* **3**, 131. **2**. Sevoz-Couche, *et al.* (2000) *Br. J. Pharmacol.* **131**, 1445. **3**. Niebert, *et al.* (2011) *PLoS One* **6**, e21395.

LY 288513

This (FW = 436.31 g/mol; CAS 147523-65-7; Solubility = 100 mM in DMSO or Ethanol), also named (4S,5R)-N-(4-bromophenyl)-3-oxo-4,5-diphenyl-1-pyrazolidinecarboxamide, targets the Cholecystokinin CCK₂ receptor (IC$_{50}$ = 16 nM), a G-protein coupled receptor that binds the peptide hormones cholecystokinin or gastrin. It is a much weaker antagonist (IC$_{50}$ > 30,000 nM) for CCK₁ receptors). LY 288513 also displays anxiolytic and antipsychotic properties *in vivo*. **1** Rasmussen, *et al.* (1992) *J. Pharmacol. Exp. Ther.* **264**, 480. **2.** Rasmussen, *et al.* (1994) *Ann. N.Y. Acad. Sci.* **713**, 300. **3.** Helton, *et al.* (1996) *Pharmacol. Biochem. Behav.* **53**. 493.

LY 294002

LY 294002   **LY 303511**

This chromone (FW = 307.34; CASs = 957054-30-7 and 957054-33-0 (hydrochloride); Solubility (25°C): 36 mg/mL DMSO, < 1 mg/mL Water), also known as 2-(4-morpholinyl)-8-phenyl-4H-1-benzopyran-4-one, completely inhibits phosphatidylinositol 3-kinase activity (IC$_{50}$ = 1.4 μM) and voltage-dependent K$^+$ channels, but does not inhibit phosphatidylinositol 4-kinase, protein kinase A, protein kinase C, diacylglycerol kinase, and EGF receptor protein-tyrosine kinase. Note that the structural analogue LY 303511 (*i.e.*, 2-(4-piperazinyl)-8-phenyl-4H-1-benzopyran-4-one; FW$_{free-base}$ = 306.36 g/mol) is inactive as an inhibitor of phosphatidylinositol 3-kinase but will inhibit voltage-dependent K$^+$ channels. **Target(s):** casein kinase 2 (1); DNA-dependent protein kinase (2); phosphatidylinositol-4,5 bisphosphate 3-kinase (3-5); phosphatidylinositol 3-kinase (1,6-15); phosphatidylinositol-4-phosphate 3-kinase (16,17); polo kinase (18); voltage-dependent K$^+$ channel (19). **1.** Davies, Reddy, Caivano & Cohen (2000) *Biochem. J.* **351**, 95. **2.** Smith, Divecha, Lakin & Jackson (1999) *Biochem. Soc. Symp.* **64**, 91. **3.** Vanhaesebroeck, Leevers, Ahmadi, *et al.* (2001) *Ann. Rev. Biochem.* **70**, 535. **4.** Metzner, Heger, Hofmann, Czech & Norgauer (1997) *Biochem. Biophys. Res. Commun.* **232**, 719. **5.** Vanhaesebroeck, Welham, Kotani, *et al.* (1997) *Proc. Natl. Acad. Sci. U.S.A.* **94**, 4330. **6.** Vlahos, Matter, Hui & Brown (1994) *J. Biol. Chem.* **269**, 5241. **7.** Alaimo, Knight & Shokat (2005) *Bioorg. Med. Chem.* **13**, 2825. **8.** Gray, Olsson, Batty, Priganica & Downes (2003) *Anal. Biochem.* **313**, 234. **9.** Hardy, Langelier & Prentki (2000) *Cancer Res.* **60**, 6353. **10.** Anderson & Jackson (2003) *Int. J. Biochem. Cell Biol.* **35**, 1028. **11.** Cataldi, Di Pietro, Centurione, *et al.* (2000) *Cell. Signal.* **12**, 667. **12.** Wang & Sul (1998) *J. Biol. Chem.* **273**, 25420. **13.** Huang, Dedousis, Bhatt & O'Doherty (2004) *J. Biol. Chem.* **279**, 21695. **14.** Leblais, Jo, Chakir, *et al.* (2004) *Circ. Res.* **95**, 1183. **15.** Tang & Downes (1997) *J. Biol. Chem.* **272**, 14193. **16.** Arcaro, Khanzada, Vanhaesebroeck, *et al.* (2002) *EMBO J.* **21**, 5097. **17.** Domin, Pages, Volinia, *et al.* (1997) *Biochem. J.* **326**, 139. **18.** Johnson, Stewart, Woods, Giranda & Luo (2007) *Biochemistry* **46**, 9551. **19.** El-Kholy, Macdonald, Lin, *et al.* (2003) *FASEB J.* **17**, 720.

LY 294468, **See** *Efegatran*
LY 300164, **See** *Talampanel*

LY 311727

This selective phospholipase inhibitor (FW = 430.43 g/mol; CAS 164083-84-5; Solubility: 100 mM in DMSO; 100 mM in 1eq. NaOH), also named [3-[[3-(2-amino-2-oxoethyl)-2-ethyl-1-(phenylmethyl)-1H-indol-5-yl]oxy] propyl]phosphonic acid, targets secreted phospholipase A₂ (sPLA₂) inhibitor (IC$_{50}$ < 1 μM for group IIA PLA₂; IC$_{50}$ <50 μM for group V PLA₂), attenuating synthesis of VEGF-mediated platelet-activating factor (*or* PAF) in both Bovine Aortic Endothelial Cells (BAEC) and Human Umbilical Vein Endothelial Cells (HUVEC) BAEC at 100 μM. **1.** Schevitz, *et al.* (1995) *Nat. Struct. Biol.* **2**, 429. **2.** Bernatchez et al (2001) *Br. J. Pharmacol.* **134**, 197.

LY 315920

This orally available selective sPLA₂ inhibitor (FW = 380.39 g/mol; CAS 172732-68-2; Solubility: 100 mM DMSO), also known as Varespladib and 2-[[3-(2-amino-2-oxoacetyl)-2-ethyl-1-(phenylmethyl)-1H-indol-4-yl]oxy]-acetate, targets secreted phospholipase A₂ (IC$_{50}$ = 50-750 nM for a range of sPLA₂ isoforms), but has no effect on cytosolic PLA₂, COX-1, or COX-2. LY 315920 attenuates sPLA₂-induced thromboxane formation in bronchoalveolar lavage (BAL) cells. It also suppresses sPLA₂-induced contractions of lung pleural strips studied *ex vivo*. **1.** Snyder, *et al.* (1999) *J. Pharmacol. Exp. Ther.* **288**, 1117. **2.** Oslund, *et al.* (2008) *J. Med. Chem.* **51**, 4708.

LY 320135

This potent cannabinoid receptor antagonist/inverse-agonist (FW = 383.40 g/mol; CAS 176977-56-3; Solubility: 30 mM in DMSO), also named 4-[[6-methoxy-2-(4-methoxyphenyl)-3-benzofuranyl]carbonyl]benzonitrile, targets CB₁ receptors (K$_i$ = 141 nM), showing >70x selectivity over CB₂ receptors (K$_i$ > 10 μM). LY-320135 shows weak binding to both 5-HT₂ (K$_i$ = 6.4 μM) and muscarinic receptors (K$_i$ = 2.1 μM). **1.** Felder, *et al.* (1998) *J. Pharmacol. Exp. Ther.* **284**, 291. **2.** Holland, *et al.* (1999) *Brit. J. Pharmacol.* **128**, 597. **3.** Pertwee (2005) *Life Sci.* **76**, 1307.

LY333328, **See** *Oritavancin*

LY 333531

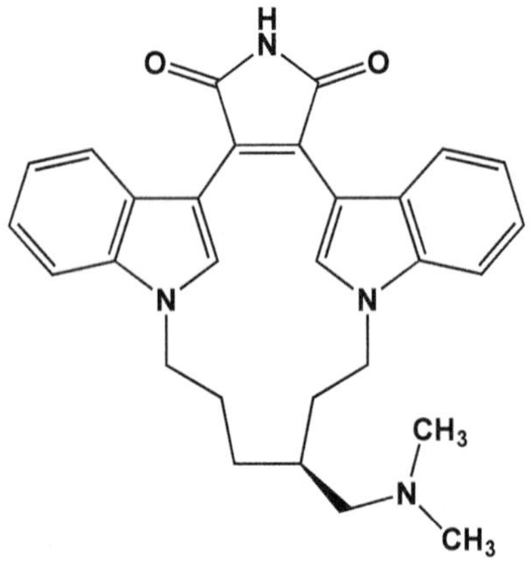

This isozyme-selective PKC inhibitor (FW = 466.58 g/mol; CAS 169939-93-9; Solubility: 20 mM in DMSO) competitively and reversibly inhibits protein kinase-βI, *or* PKCβI (IC$_{50}$ = 4.7 nM) and PKCβII (IC$_{50}$ = 5.9 nM), with strong selectivity for PKCβ with respect to isozymes PKC-η (IC$_{50}$ = 0.052 μM), PKC-δ (IC$_{50}$ = 0.25 μM), PKC-γ (IC$_{50}$ = 0.30 μM), PKC-α (IC$_{50}$ = 0.36 μM), PKC-ε (IC$_{50}$ = 0.60 μM), and PKC-ζ (IC$_{50}$ = >100 μM). **1.** Jirousek, *et al.* (1996) *J. Med. Chem.* **39**, 2664. **2.** Samokhin, *et al.* (1999) *Eur. J. Pharmacol.* **386**, 297. **3.** Faul, *et al.* (2003) *Bioorg. Med. Chem. Lett.* **13**, 1857.

LY335979, See *Zosuquidar*

LY 341495

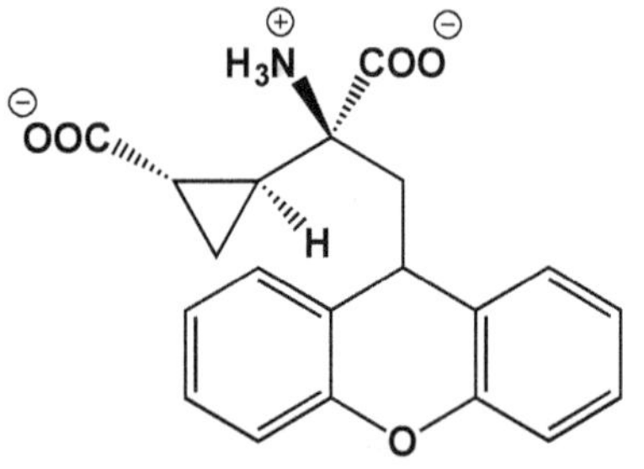

This highly potent and selective group II metabotropic glutamate receptor antagonist (FW = 353.37 g/mol; CAS 201943-63-7; Solubility: 10 mM in DMSO with gentle warming; 100 mM in 1.2 equivalents NaOH), also named (2*S*)-2-amino-2-[(1*S*,2*S*)-2-carboxycycloprop-1-yl]-3-(xanth-9-yl)propanoic acid, targets human mGlu$_2$ receptors (IC$_{50}$ = 2.3 nM), mGlu$_3$ receptors (IC$_{50}$ = 1.3 nM), mGlu$_8$ receptors (IC$_{50}$ = 173 nM), mGlu$_{7a}$ receptors (IC$_{50}$ = 990 nM), mGlu$_{1a}$ receptors (IC$_{50}$ = 6800 nM), mGlu$_{5a}$ receptors (IC$_{50}$ = 8200 nM) and mGlu$_{4a}$ receptors (IC$_{50}$ = 22000 nM) (1-3). Although selective at low-nM concentrations, LY 341495 can be used in higher concentrations to block all hippocampal mGlu receptors (2). LY341495 competitively antagonizes DHPG-stimulated phosphatidyl-inositol (PI) hydrolysis in AV12-664 cells expressing either human mGlu$_1$ or mGlu$_5$ receptors with K_i-values of 7.0 and 7.6 μM, respectively (2). When tested against 10 μM L-glutamate-stimulated Ca^{2+} mobilisation in rat mGlu$_5$-expressing CHO cells, it produces substantial or complete block at a concentration of 100 μM. In rat hippocampal slices, LY341495 eliminates 30 μM DHPG-stimulated PI hydrolysis and 100 μM (1*S*,3*R*)-ACPD-inhibition of forskolin-stimulated cAMP formation at concentrations of 100 and 0.03 μM, respectively (2). In area CA1 (*Cornu Ammon, or* Ammon's Horn) of the mammalian hippocampus in the medial temporal lobe, it antagonizes DHPG-mediated potentiation of NMDA-induced depolarisations and DHPG-induced long-lasting depression of AMPA receptor-mediated synaptic transmission (2). LY341495 also blocks NMDA receptor-independent depotentiation and setting of a molecular switch involved in the induction of Long-Term Potentiation (LTP), effects previously shown to be blocked by (*S*)-MCPG. **1.** Kingston, Ornstein, Wright, *et al.* (1998) *Neuropharmacol.* **37**, 1. **2.** Fitzjohn, Bortolotto, Palmer, *et al.* (1998) *Neuropharmacol.* **37**, 1445. **3.** Johnson, Wright, Arnold, *et al.* (1999) *Neuropharmacol.* **38**, 1519.

LY 367265

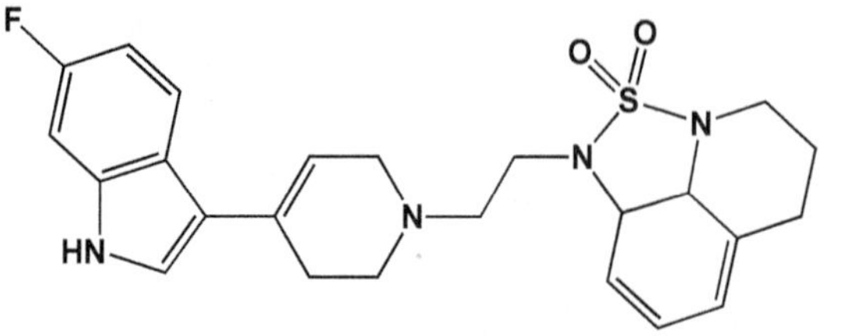

This indole-based antidepressant (FW = 454.56 g/mol), also named (1-[2-[4-(6-fluoro-1*H*-indol-3-yl)-3,6-dihydro-1(2*H*)-pyridinyl]ethyl]-5,6-dihydro-1*H*,4*H*-[1,2,5]thiadiazolo[4,3,2-*ij*]quinoline-2,2-dioxide, has a high affinity for the 5-hydroxytryptamine transporter (K_i = 2.3 nM) and is a 5-hydroxytryptamine (HT$_{2A}$) receptor antagonist (K_i = 0.81 nM). LY 367265 also presynaptically decreases nerve terminal excitability, which subsequently attenuates the Ca^{2+} entry through voltage-dependent Ca^{2+} channels to cause a decrease in evoked glutamate release (2). **1.** Pullar, Carney, Colvin, *et al.* (2000) *Eur. J. Pharmacol.* **407**, 39. **2.** Wang (2005) *Synapse* **55**, 156.

LY 367385

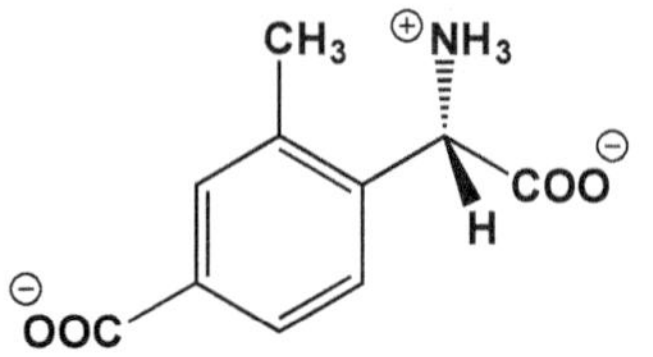

This neuroprotective metabolotropic glutamate receptor antagonist (FW = 209.2 g/mol; CAS 198419-91-9; Solubility: 100 mM in 1.1 equivalents of NaOH), also named (*S*)-(+)-α-amino-4-carboxy-2-methylbenzeneacetic acid, selectively targets mGlu$_{1a}$ receptors (IC$_{50}$ = 8.8 μM for quisqualate-induced phosphoinositide hydrolysis), as contrasted with an IC$_{50}$ value >100 μM for mGlu$_{5a}$. LY 367385 shows negligible effects on group II and III metabolotropic glutamate receptors (1). Activation of mGlu$_1$ receptors appears to contribute to a variety of epileptic syndromes, and antagonists of mGlu$_1$ receptors show promise as anticonvulsants (2). **1.** Bruno, *et al.* (1999) *Neuropharmacol.* **38**, 199. **2.** Chapman, *et al.* (1999) *Eur. J. Pharmacol.* **368**, 17.

(±)-LY 395756

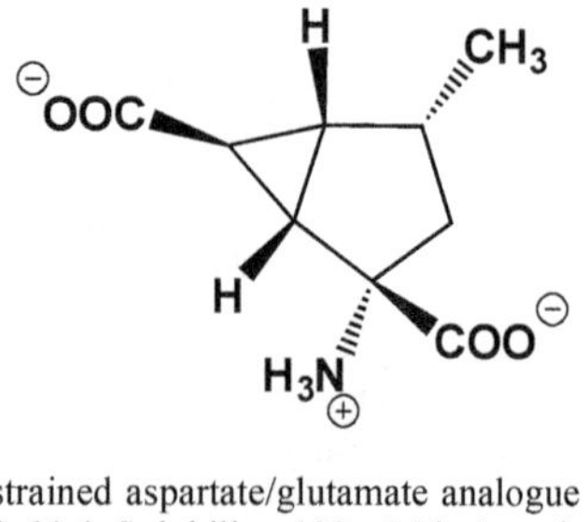

This sterically constrained aspartate/glutamate analogue (FW$_{neutral}$ = 199.21 g/mol; CAS 852679-66-4; Solubility: 100 mM in 1 equivalent of HCl and to 100 mM in 1 equivalent of NaOH), also named (1*S*,2*S*,4*R*,5*R*,6*S*)-rel-2-amino-4-methylbicyclo[3.1.0]hexane-2,6-dicarboxylic acid, selectively targets mGlu$_2$ receptors (as agonist, K_a = 0.165 μM) and mGlu$_3$ receptors (as antagonist, K_i = 0.302 μM). **1.** Dominguez, Prieto, Valli, *et al.* (2005) *J. Med. Chem.* **48**, 3605.

LY-450139, See *Semagacestat*

LY 456236

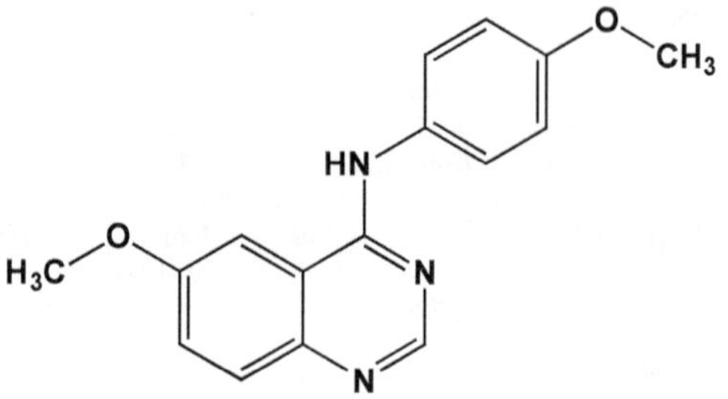

This metabolotropic glutamate receptor antagonist (FW = 218.32 g/mol; CAS 338736-46-2; Solubility: 20 mM in DMSO), also named 6-methoxy-N-(4-methoxyphenyl)-4-quinazolinamine, selectively targets mGlu$_1$ receptors (IC$_{50}$ = 143 nM), with an IC$_{50}$ > 10 μM for mGlu$_5$ receptors. LY 456246 reduces hyperalgesic behavior induced by formalin in both mouse (ED$_{50}$ = 28 mg/kg) and rat (ED$_{50}$ = 16 mg/kg). In male and female DBA/2 mice, LY456236 produced a dose-related inhibition of sound-induced clonic-tonic seizures. In male CF1 mice, LY456236 produced a dose-related inhibition of tonic extensor seizures in the threshold electroshock model, and limbic seizures in the 6-Hz focal seizure model. **1**. Shannon, Peters & Kingston (2005) *Neuropharmacol.* **49** (Suppl 1), 188.

LY2090314

This potent intravenous GSK-3 inhibitor (FW = 512.53 g/mol; CAS 603288-22-8), also named 3-(9-fluoro-2-(piperidine-1-carbonyl)-1,2,3,4-tetrahydro-[1,4]diazepino-[6,7,1-hi]indol-7-yl)-4-(imidazo[1,2-a]pyridin-3-yl)-1H-pyrrole-2,5-dione, targets both GSK-3α (IC$_{50}$ = 1.5 nM) and GSK-3β (IC$_{50}$ = 0.9 nM), with properties that promise to improve the efficacy of platinum-based chemotherapy regimens. In rats and humans, LY2090314 is rapidly cleared by extensive metabolism, with negligible circulating metabolite exposures due to biliary excretion of metabolites into feces and with no apparent intestinal reabsorption (1). While not a time-dependent CYP2B6 inhibitor, LY2090314 significantly reduces CYP2B6 mRNA levels, which significantly correlate with observed changes in catalytic activity (2). Cytotoxicity assays reveal that melanoma cell lines are very sensitive to LY2090314 *in vitro* (IC$_{50}$ ~10 nM after 72hr of treatment) in contrast to other solid tumor cell lines (IC$_{50}$ >10 uM), as evidenced by caspase activation and PARP cleavage (3). Cell lines harboring mutant B-RAF or N-RAS were equally sensitive to LY2090314 as were those with acquired resistance to the BRAF inhibitor vemurafenib (3). **1**. Zamek-Gliszczynski, Abraham, Alberts, *et al.* (2013) *Drug Metab. Dispos.* **41**, 714. **2**. Zamek-Gliszczynski, Mohutsky, Rehmel & Ke (2014) *Metab. Dispos.* **42**, 1008. **3**. Atkinson, Rank, Zeng, *et al.* (2015) *PLoS One* **10**, :e0125028.

LY 2109761

This TGF-β receptor inhibitor (FW = 441.52 g/mol; CAS 700874-71-1; Solubility: <1 mg/mL DMSO, H$_2$O, or Ethanol), also named 7-(2-morpholinoethoxy)-4-(2-(pyridin-2-yl)-5,6-dihydro-4H-pyrrolo[1,2-b]pyrazol-3-yl)quinoline, is a membrane-permeable synthetic compound that selectively targets TGF-β-I (K_i = 38 nM) and TGF-β-II (K_i = 300 nM) receptors, thereby reducing Smad2 phosphorylation (1-3). LY2109761 significantly inhibited the L3.6pl/GLT soft agar growth, suppressed both basal and TGF-β1-induced cell migration and invasion, and induced anoikis. When administered in combination with gemcitabine, LY2109761significantly reduced the tumor burden, prolonged survival, and reduced spontaneous abdominal metastases (1). LY2109761 likewise blocks migration and invasion of hepatocellular carcinoma (HCC) cells by up-regulating E-cadherin (2). In an orthotopic intracranial model for glioblastoma multiforme, *or* GBM, LY2109761 significantly reduces tumor

growth, prolongs survival, and extends the prolongation of survival induced by radiation (3. Histologic analyses revealed that LY2109761 inhibited tumor invasion promoted by radiation, reduced tumor microvessel density, and attenuated mesenchymal transition (3). **1**. Melisi, Ishiyama, Sclabas, *et al.* (2008) *Mol. Cancer Ther.* **7**, 829. **2**. Fransvea, Angelotti, Antonaci & Giannelli (2008) *Hepatology* **47**, 1557. **3**. Zhang, Kleber, Röhrich, et al. (2011) *Cancer Res.* **71**, 7155.

LY 2157299

This novel and orally available TGF-β signal transduction inhibitor (FW = 369.44 g/mol; CAS 700874-72-2; λ$_{max}$ at 238 and 282 nm), 4-[5,6-dihydro-2-(6-methyl-2-pyridinyl)-4H-pyrrolo[1,2-b]pyrazol-3-yl]-6-quinolinecarboxamide, targets TGF-β receptor type 1 kinase (IC$_{50}$ = 56 nM) (1) and chemotherapy-induced expansion of cancer stem-like cells in triple negative breast cancer cell lines and xenografts (2). By disrupting Smad-2 phosphorylation, LY2157299 inhibits migration and growth of hepatocellular carcinoma cells (3). TGF-β signaling helps to orchestrate a favorable microenvironment for tumor cell growth and to promote epithelial-mesenchymal transition (EMT) (4). TGF-β promotes hepatocellular carcinoma progression by an intrinsic activity as an autocrine/paracrine growth factor and also by an extrinsic activity by inducing microenvironment changes, including cancer-associated fibroblasts, T regulatory cells, and inflammatory mediators. Intriguingly, TGF-β signaling inhibitors can actually promote tumor progression under some conditions (4). Briefly, when Panc-1 pancreatic carcinoma cells and normal fibroblasts are placed atop a three-dimensional (3D) collagen gel, tumor cells scatter into the fibroblast layer, apparently undergoing epithelial-mesenchymal transition. However, when fibroblasts are first placed within collagen gel, Panc-1 cells actively invade the gel, extending long microtubule-based protrusions. Although TGF-β and HGF individually stimulate tumor cell invasion into collagen gel without fibroblasts, LY-2157299 significantly enhances Panc-1 cell invasion into fibroblast co-cultures. **1**. Bueno, de Alwis, Pitou, *et al.* (2008) *Eur. J. Cancer.* **44**, 142. **2**. Bhola, Balko, Dugger, *et al.* (2013) *J. Clin. Invest.* **123**, 1348. **3**. Dituri, Mazzocca, Peidrò,*et al.* (2013) *PLoS One* **8**, e67109. **3**. Giannelli, Villa, Lahn, *et al.* (2014) *Cancer Res.* **74**, 1890. **4**. Oyanagi, Kojima, Sato, *et al.* (2014) *Exp. Cell Res.* **27**, 167.

LY2183240

This nonspecific endocannabinoid transport inhibitor (FW = 307.35 g/mol; CAS 874902-19-9; IUPAC Name: N,N-dimethyl-5-[(4-biphenyl)methyl] tetrazole-1-carboxamide) disrupts cellular uptake (IC$_{50}$ = 270 pM) of anandamide, an endogenous cannabinoid (EC) neurotransmitter, thereby giving rise to its analgesic and anxiolytic effects. LY2183240 covalently inactivates the EC-degrading enzyme fatty acid amide hydrolase (FAAH) by carbamylation of Serine-241, an active-site nucleophile (1). Activity-based proteomic screening of LY2183240 identified forty other serine hydrolase targets. Thus, while the blockade of anandamide uptake by LY2183240 may be due to FAAH inactivation, this inhibitor's promiscuity greatly limits its potential as a drug. **1**. Alexander & Cravatt (2006) *J. Am. Chem. Soc.* **128**, 9699.

LY 2365109

This potent GlyT₁ inhibitor (FW = g/mol; CAS 868265-28-5; Solubility: 100 mM in DMSO; 50 mM in Ethanol), also named *N*-[2-[4-(1,3-benzodioxol-5-yl)-2-(1,1-dimethylethyl)phenoxy]ethyl]-*N*-methylglycine, selectively targets the glycine transporter 1 (GlyT₁) inhibitor (IC₅₀ = 15.8 nM), with far weaker action on GlyT₂ (IC₅₀ > 30,000 nM). LY 2365109 induces a dose-dependent elevation in CSF levels of glycine, and enhances acetylcholine and dopamine release in the striatum and prefrontal cortex, respectively, profoundly impairintg locomotor and respiratory function at higher doses. **1**. Perry, Falcone, Fell, *et al.* (2008) *Neuropharmacol.* **55**, 743.

LY 2389575

This selective negative allosteric mGlu modulator (FW = 402.12 g/mol; CAS 885104-09-6; Solubility: 20 mM in DMSO), also named (3*S*)-*N*-(2,4-dichlorobenzyl)-1-(5-bromopyrimidin-2-yl)pyrrolidinyl-3-amine, targets the metabolotrophic glutamate receptor, *or* mGlu₃ (IC₅₀ = 190 nM), exhibiting > 65x weaker action against other mGlu receptors. LY 2389575 abolishes the neuroprotective action of LY 379268 against amyloid-β toxicity in mixed cortical neuronal and astrocyte cell cultures. **1**. Caraci, Molinaro, Battaglia, *et al.* (2011) *Mol. Pharmacol.* **79**, 618.

LY2523355, *See* Litronesib

LY2584702

This potent and highly selective ATP-competitive protein kinase inhibitor (FW₍free-base₎ = 445.43 g/mol; FW₍tosylate-salt₎ = 617.62 g/mol; CAS 1082949-68-5), also named 4-{4-[4-(4-fluoro-3-trifluoromethylphenyl)-1-methyl-1*H*-imidazol-2-yl]piperidin-1-yl}-1*H*-pyrazolo[3,4-*d*]pyrimidine *p*-toluene sulfonate, targets p70S6K (IC₅₀ = 4 nM), a serine/threonine kinase that acts downstream of PIP₃ and phosphoinositide-dependent kinase-1 in the PI₃ kinase pathway. Its target substrate is the S6 ribosomal protein, the phosphorylation of which induces protein synthesis.

LY2784544

This potent, selective and ATP-competitive tyrosine kinase inhibitor (FW = 469.95 g/mol; CAS 1229236-86-5; Solubility: 94 mg/mL in DMSO), also named 3-(4-chloro-2-fluorobenzyl)-2-methyl-*N*-(3-methyl-1*H*-pyrazol-5-yl)-8-(morpholinomethyl)imidazo-[1,2-*b*]pyridazin-6-amine, targets Janus Kinase-2, *or* JAK2, IC₅₀ = 3 nM, including its Val-617-Phe mutant, with 8x and 20x selectivity in its action against JAK1 and JAK3. **1**. Ma, Clayton, Walgren, *et al.* (2013) *Blood Cancer J.* **3**, e109.

LY2835219, *See* Abemaciclib

LY2940680

This synthetic inhibitor (FW = 512.50 g/mol; CAS 1258861-20-9), systematically named 4-fluoro-*N*-methyl-*N*-(1-(4-(1-methyl-1*H*-pyrazol-5-yl)phthalazin-1-yl)piperidin-4-yl)-2-(trifluoromethyl)benzamide, targets the smoothened (SMO) receptor, a key component in the hedgehog (Hh) signalling transduction pathway that helps maintain normal embryonic development and is also likely to play a role in carcinogenesis. LY2940680 potently inhibits Hh signaling in Daoy, a human medulloblastoma tumor cell line, and C3H10T1/2, a mouse mesenchymal cell line (1). It also inhibits the functional activity of resistant Smo mutant (Asp-473-His) produced by treatment with GDC-0449, another Smo receptor antagonist. LY2940680 induces Caspase-3 activity and reduces proliferation of medulloblastoma tumors. **Mode of Inhibitor Action:** LY2940680 binds to the transmembrane domain of the human SMO receptor, which shares the seven-transmembrane helical fold but lacks most of the conserved motifs for Class A GPCRs (2). Receptor-bound LY2940680 has a canonical GPCR 7TM bundle fold with a short helix VIII packed parallel to the membrane bilayer. The ligand binds at the extracellular end of the seven-transmembrane-helix bundle and forms extensive contacts with the loops. **1**. Bender, Hipskind, Capen, *et al.* (2011) *Cancer Res.* **71**, (Supplement 1) Abstract 2819. **2**. Wang, Wu, Katritch, *et al.* (2013) *Nature* **497**, 338.

LY3009104, *See* Baricitinib

Lycorine

This photosensitive alkaloid (FW = 287.32 g/mol), isolated from *Lycoris radiata*, *Narcissus tazwtta*, *Amaryllis belladonna*, and other members of the *Amaryllidaceae*, is biosynthesized from L-tyrosine and L-phenylalanine. Lycorine is an inhibitor of the peptidyltransferase center and also inhibits ascorbate biosynthesis *in vivo*. **Target(s):** acetylcholinesterase, weakly inhibited (1); L-galactono-γ-lactone dehydrogenase (2,3); peptidyltransferase (4); protein biosynthesis, elongation), eukaryotic (4-7). Lycorine also inhibits lipopolysaccharide-induced up-regulation of iNOS and COX-2 by suppressing P38 and STATs activation, thereby increasing the survival rate of mice after LPS challenge (8). **1.** Elgorashi, Stafford & Van Staden (2004) *Planta Med.* **70**, 260. **2.** Imai, Karita, Shiratori, *et al.* (1998) *Plant Cell Physiol.* **39**, 1350. **3.** Arrigoni, Paciolla & De Gara (1996) *Boll. Soc. Ital. Biol. Sper.* **72**, 37. **4.** Kukhanova, Victorova & Krayevsky (1983) *FEBS Lett.* **160**, 129. **5.** Jiménez (1976) *Trends Biochem. Sci.* **1**, 28. **6.** Vrijsen, Vanden Berghe, Vlietinck & Boeye (1986) *J. Biol. Chem.* **261**, 505. **7.** Jiménez, Santos, Alonso & Vazquez (1976) *Biochim. Biophys. Acta* **425**, 342. **8.** Kang J, Zhang Y, Cao X (2012) *Int. Immunopharmacol.* **12**, 249.

Lyn Peptide Inhibitor
This stearoylated signal transduction inhibitor (FW = 2370.91 g/mol; Sequence: *N*-Octadecanoyl-YGYRLRRK-WEEKIPNP; CAS 222018-18-0; Solubility: 10 mg/mL in H_2O) is cell-permeable and blocks Lyn kinase-dependent effects of the IL-5 receptor (1,2). The IL-5, IL-3, and GM-CSF cytokines regulate hematopoietic cell growth, differentiation, and survival. While GM-CSF and IL-3 have broader overlapping hematopoietic activities, IL-5 is eosinophil-specific and plays a central role in eosinophilic inflammation in atopy and asthma (2). Receptors for all three cytokines are composed of a ligand-specific α chain and a dimer of the common β subunit (or βc) that is shared by all three cytokines. Each specific α chain forms a low-affinity receptor, when bound by its respective ligand, but association with βcleads to high-affinity binding. Association of α and βc subunits, as mediated by ligand binding, initiates signaling through the JAK/STAT, Ras/MAPK, and PI3-K signaling pathways (2). Lyn peptide inhibitor blocks the association of Lyn, but not with the βc receptor *in situ*, suggest that the N-terminal unique domain of Lyn kinase is important for binding to the βc receptor (1). This novel Lyn-binding peptide inhibitor blocks eosinophil differentiation, survival, and airway eosinophilic inflammation (1). **1.** Adachi, et al (1999) *J. Immunol.* **163**, 939. **2.** Martinez-Moczygemba & Huston (2001) *J. Clin. Invest.* **108**, 1797.

Lysergic Acid Diethylamide (*or* LSD)

This potent hallucinogen and widely abused psychedelic agent (FW = 323.44 g/mol; CAS 50-37-3; M.P. = 80-85°C; Symbol: LSD or LSD-25), systematically named (6aR,9R)-*N*,*N*-diethyl-7-methyl-4,6,6a,7,8,9-hexahydroindolo[4,3-*fg*]quinoline-9-carboxamide, is a semisynthetic product of lysergic acid, the latter a secondary metabolite derived from tryptophan in the parasitic rye fungus *Claviceps purpurea*. First synthesized in 1938 by Albert Hofmann in his search for pharmacologically useful lysergic acid derivatives, its profound psychoactive properties were reported five years later (1). Marketed as Delysid by Sandoz in the 1950s, LSD's psychoactive properties were touted in the popular press as a "mind-expanding" synesthetic drug, wherein stimulation of one sensory or cognitive pathway elicits automatic, involuntary experiences in other sensory and/or cognitive pathways. Already widely abused by the early 1960s, its actions soon became the topic of New Age poetry and song. **Pharmacology:** LSD's action is complex, with its mechanisms still incompletely understood (2,3). Although distributed throughout the body, where it alters phosphate

excretion, its main effects are central, altering dopamine excretion, but without effect on the metabolism or retention of other neurotransmitters (4). In humans, metabolism is rapid, and LSD undergoes hydroxylation, oxidation, and glucuronidation. LSD and its metabolites have been quantified in human biological fluids by HPLC and mass spectrometry (4). Numerous studies report nM-affinity for serotonin receptors, affecting nearly all, but showing strongest effects on 5-HT_{1A} receptors (K_i = 1.1 nM) and 5-HT_{1B} receptors (K_i = 4 nM) (3). **LSD-induced Euphoria:** At moderate doses (75–150 μg p.o.), LSD alters one's perceptions, as indicated by euphoria, enhanced introspection, hypnagogic experiences, lost sense of time, and often wild and colorful dreams/hallucinations. First effects (dilated pupils, altered heart rate and blood pressure, and fluctuations in body temperature, including sweats or chills) typically occur within 30–90 minutes, soon impairing objectivity and judgement. Hallucinations often ensue, with visual perception of objects becoming distorted in size and shape, movements accentuated, colors more vivid, and sounds amplified. Some sensations are said to be strangely pleasurable (encouraging further use), but many individuals "have a bad trip", attended by mood swings, profound confusion, fright, extreme paranoia, and delirious flashbacks. Users quickly develop a high degree of tolerance and need higher doses to induce LSD's much-vaunted euphoria. **Target(s):** acetylcholinesterase (3,5); aryl-acylamidase (5,6); cholinesterase (7,8); serotonin receptors (9-11). **1.** Stoll & Hofmann (1943) *Helv. Chim. Acta* **26**, 944. **2.** Bajgar, Patocka & Zizkovsky (1971) *Sb. Ved. Pr. Lek. Fak. Karlovy Univerzity Hradci Kralove Suppl.* **14**, 417. **3.** Passie Halpern, Stichtenoth, Emrich & Hintzen (2008) CNS Neurosci Ther. **14**, 295. **4.** Canezin, Cailleux, Turcant, *et al.* (2001) *J. Chromatogr. B Biomed. Sci. Appl.* **765**, 15. **5.** Paul & Halaris (1976) *Biochem. Biophys. Res. Commun.* **70**, 207. **6.** Hsu, Paul, Halaris & Freedman (1977) *Life Sci.* **20**, 857. **7.** Zehnder & Cerletti (1956) *Helv. Physiol. Pharmacol. Acta* **14**, 264. **8.** Thompson, Tickner & Webster (1954) *Process Biochem.* **58**, xix. **9.** Peroutka, Sleight, McCarthy, *et al.* (1989) *J. Neuropsychiatry Clin. Neurosci.* **1**, 253. **10.** Boess & Martin (1994) *Neuropharmacol.* **33**, 275. **11.** Egan, Herrick-Davis, Miller, Glennon & Teitler (1998) *Psychopharmacol.* **136**, 409.

δ-Lysin
This 26-residue, *N*-formylated oligopeptide (MW = 3048.63 g; *Sequence*: *N*-formyl-MAQEIISTIGELVKWIIETVNKF-TKK; Soluble in both aqueous and organic solvents such as chloroform/methanol) is a hemolytic peptide toxin secreted by *Staphylococcus aureus* that forms an amphipathic helix upon binding to lipid bilayers. alters the membrane permeability of those bilayers. δ- δ-Lysin aggregates in aqueous solution at concentrations ≥1 μM, forming dimers, trimers and tetramers. Lysin insertion is strongly dependent on the peptide-to-lipid ratio, suggesting that association of a critical number of monomers on/within the membrane is required for activity. There is no evidence of a stable membrane-inserted pore; instead, the peptide appears to cross the membrane rapidly and reversibly, mediating the release of the lipid vesicle contents in the process. **1.** Pokorny, Birkbeck &Almeida (2002) *Biochemistry* **41**, 11044.

L-Lysine

This dibasic amino acid ($FW_{free-base}$ = 146.19 g/mol; Symbols: Lys and K; pK_a values = 2.16 (α-COOH), 9.06 (α-NH_2), and 10.54 (ε-amino group) at 25°C; Specific Rotation = + 25.9° at 25°C for the D-line), systematically named (*S*)-2,6-diaminohexanoic acid is one of the twenty proteogenic amino acids and is nutritionally essential for mammals. There are two codons: AAA and AAG. The side-chain of lysine consists of a hydrophobic chain of four methylene groups, ending with a charged amino group. Hence, most lysyl residues in proteins are on the protein surface, but the alkyl chain can make hydrophobic interactions. The ε-amino group is protonated under most physiological conditions. A small fraction is always unprotonated and can act as a nucleophile. This amino group can thus become acylated, alkylated, arylated, and amidated. Lysines are found frequently in α-helices. Those in collagens are often hydroxylated at the C5 position. In homologous proteins, lysyl residues are often interchanged with arginyl, asparaginyl, threonyl, seryl, and glutaminyl residues. **Target(s):** acetylcholinesterase (weakly inhibited) (1); alkaline phosphatase (2); α-aminoadipate-semialdehyde dehydrogenase (3,4); 2-aminohexano-6-lactam racemase (5); aminolevulinate aminotransferase, weakly inhibited (93); aminopeptidase B (6,70-72); 5 aminovalerate aminotransferase (93); D-arginase (7); arginase (8-20); arginine decarboxylase (21-24); arginine deaminase (25); arginine kinase, weakly inhibited (74); aspartate kinase (13,26,27,75-88); carboxypeptidase B (28); carboxypeptidase E (28); coagulation factor VIIa (30); L-cystine transport (31); 2-dehydro-3-deoxyglucarate aldolase, or 2-keto-3-deoxyglucarate aldolase (32); diaminopimelate decarboxylase (33); dihydrodipicolinate synthase (34-42); dipeptidyl-peptidase IV (43); GfMEP, or thermostable lysine-specific metalloendopeptidase (44,45); glutamate decarboxylase, mildly inhibited (56); glutamine synthetase (47); γ-glutamyl transpeptidase (48,49); histidine ammonia-lyase, weakly inhibited (13); homocitrate synthase (50,51,95-103); homoserine kinase (52,91); hydroxylysine kinase (89,90); kynureninase (53-55); leucyl aminopeptidase (73); leucyltransferase (105); 15-lipoxygenase, or arachidonate 15-lipoxygenase (109); lipoxygenase, soybean, weakly inhibited (109); lysine carboxypeptidase, or lysine(arginine) carboxypeptidase, or carboxypeptidase N (56,57); D-ornithine 4,5 aminomutase, weakly inhibited (58); ornithine aminotransferase (94); ornithine carbamoyltransferase (59,106,107); ornithine decarboxylase (60-63); penicillopepsin (64); peptidyl-Lys metalloendopeptidase (Armillaria mellea protease) (44,45,65); procollagen C-endopeptidase (66-68); protein-glutamine γ-glutamyltransferase, or transglutaminase (104); [ribulose-bisphosphate carboxylase]-lysine N-methyltransferase (108); Xaa-Arg Dipeptidase, or aminoacyl-lysine dipeptidase (69).

1. Bergmann, Wilson & Nachmansohn (1950) *J. Biol. Chem.* **186**, 693. **2**. Bodansky & Strachman (1948) *J. Biol. Chem.* **174**, 465. **3**. Schmidt, R. Bode & D. Birnbaum (1990) *FEMS Microbiol. Lett.* **58**, 41. **4**. Affenzeller, W. M. Jaklitsch, C. Honlinger & C. P. Kubicek (1989) *FEMS Microbiol. Lett.* **49**, 293. **5**. Ahmed, Esaki, Tanaka & Soda (1985) *Agric. Biol. Chem.* **49**, 2991. **6**. Kawata, Takayama, Ninomiya & Makisumi (1980) *J. Biochem.* **88**, 1601. **7**. Nadai (1958) *J. Biochem.* **45**, 1011. **8**. Prasad, Lokanatha, Sreekanth & Rajendra (1997) *J. Enzyme Inhib.* **12**, 255. **9**. Hunter & Downs (1945) *J. Biol. Chem.* **157**, 427. **10**. Greenberg (1951) *The Enzymes*, 1st ed., **1** (part 2), 893. **11**. Greenberg (1960) *The Enzymes*, 2nd ed., **4**, 257. **12**. Ameen & Palmer (1987) *Biochem. Int.* **14**, 395. **13**. Webb (1966) *Enzyme and Metabolic Inhibitors*, vol. **2**, Academic Press, New York. **14**. Hunter & Downs (1945) *J. Biol. Chem.* **157**, 427. **15**. Mohamed, Fahmy, Mohamed & Hamdy (2005) *Comp. Biochem. Physiol. B Biochem. Mol. Biol.* **142**, 308. **16**. Patchett, Daniel & Morgan (1991) *Biochim. Biophys. Acta* **1077**, 291. **17**. Shimotohno, Iida, Takizawa & Endo (1994) *Biosci. Biotechnol. Biochem.* **58**, 1045. **18**. Kaysen & Strecker (1973) *Biochem. J.* **133**, 779. **19**. Orellana, Lopez, Uribe, et al. (2002) *Arch. Biochem. Biophys.* **403**, 155. **20**. Nakamura, Fujita & Kimura (1973) *Agric. Biol. Chem.* **37**, 2827. **21**. Boeker & Snell (1971) *Meth. Enzymol.* **17B**, 657. **22**. Blethen, Boeker & Snell (1968) *J. Biol. Chem.* **243**, 1671. **23**. Rosenfeld & Roberts (1976) *J. Bacteriol.* **125**, 601. **24**. Li, Regunathan & Reis (1995) *Ann. N.Y. Acad. Sci.* **763**, 325. **25**. Smith, Ganaway & Fahrney (1978) *J. Biol. Chem.* **253**, 6016. **26**. Truffa-Bachi (1973) *The Enzymes*, 3rd ed., **8**, 509. **27**. Wampler & Westhead (1968) *Biochemistry* **7**, 1661. **28**. Wolff, Schirmer & Folk (1962) *J. Biol. Chem.* **237**, 3094. **36**. Hook (1990) *Life Sci.* **47**, 1135. **30**. Radcliffe & Heinze (1978) *Arch. Biochem. Biophys.* **189**, 185. **31**. Biber, Stange, Stieger & Murer (1983) *Pflugers Arch.* **396**, 335. **32**. Wood (1972) *The Enzymes*, 3rd ed., **7**, 281. **33**. White (1971) *Meth. Enzymol.* **17B**, 140. **34**. Shedlarski (1971) *Meth. Enzymol.* **17B**, 129. **35**. Karsten (1997) *Biochemistry* **36**, 1730. **36**. Laber, Gomis-Rueth, Romao & Huber (1992) *Biochem. J.* **288**, 691. **37**. Kumpaisal, Hashimoto & Yamada (1987) *Plant Physiol.* **85**, 145. **38**. Wallsgrove & Mazelis (1981) *Phytochemistry* **20**, 2651. **39**. Mazelis, Watley & Whatley (1977) *FEBS Lett.* **84**, 236. **40**. Bakhiet, Forney, Stahly & Daniels (1984) *Curr. Microbiol.* **10**, 195. **41**. Blickling, Beisel, Bozic, et al. (1997) *J. Mol. Biol.* **274**, 608. **42**. Frisch, Gengenbach, Tommy, et al. (1991) *Plant Physiol.* **96**, 444. **43**. Shibuya-Saruta, Kasahara & Hashimoto (1996) *J. Clin. Lab. Anal.* **10**, 435. **44**. Takio (1998) in *Handb. Proteolytic Enzymes*, p. 1538, Academic Press, San Diego. **45**. Nonaka, Ishikawa, Tsumuraya, et al. (1995) *J. Biochem.* **118**, 1014. **46**. Roberts & Frankel (1951) *J. Biol. Chem.* **190**, 505. **47**. Blanco, Alana, Llama & Serra (1989) *J. Bacteriol.* **171**, 1158. **48**. Allison (1985) *Meth. Enzymol.* **113**, 419. **49**. Thompson & Meister (1977) *J. Biol. Chem.* **252**, 6792. **50**. Jaklitsch & Kubicek (1990) *Biochem. J.* **269**, 247. **51**. Malik (1980) *Trends Biochem. Sci.* **5**, 68. **52**. Wormser & Pardee (1958) *Arch. Biochem. Biophys.* **78**, 416. **53**. Jakoby & Bonner (1953) *J. Biol. Chem.* **205**, 709. **54**. Tanizawa & Soda (1979) *J. Biochem.* **86**, 499. **55**. Soda & Tanizawa (1979) *Adv. Enzymol. Relat. Areas Mol. Biol.* **49**, 1. **56**. Ryan (1988) *Meth. Enzymol.* **163**, 186. **57**. Juillerat-Jeanneret, Roth & Bargetzi (1982) *Hoppe-Seyler's Z. Physiol. Chem.* **363**, 51. **58**. Somack & Costilow (1973) *Biochemistry* **12**, 2597. **59**.

Paik, Pearson, Nochumson & Kim (1977) *Inter. J. Biochem.* **8**, 317. **60**. Guirard & Snell (1980) *J. Biol. Chem.* **255**, 5960. **61**. Schaeffer & Donatelli (1990) *Biochem. J.* **270**, 599. **62**. Pantazaki, Anagnostopoulos, Lioliou & Kyriakidis (1999) *Mol. Cell. Biochem.* **195**, 55. **63**. Flamigni, Guarnieri & Caldarera (1984) *Biochim. Biophys. Acta* **802**, 245. **64**. Hofmann & Shaw (1964) *Biochim. Biophys. Acta* **92**, 543. **65**. Lewis, Basford & Walton (1978) *Biochim. Biophys. Acta* **522**, 551. **66**. Kessler (1998) in *Handb. Proteolytic Enzymes*, p. 1236, Academic Press, San Diego. **67**. Hojima, van der Rest & Prockop (1985) *J. Biol. Chem.* **260**, 15996. **68**. Kessler, Takahara, Biniaminov, Brusel & Greenspan (1971) *Science* **271**, 360. **69**. Kumon, Matsuoka, Kakimoto, Nakajima & Sano (1970) *Biochim. Biophys. Acta* **200**, 466. **70**. Goldstein, Nelson, Kordula, Mayo & Travis (2002) *Infect. Immun.* **70**, 836. **71**. Lauffart, McDermott, Jones & Mantle (1988) *Biochem. Soc. Trans.* **16**, 849. **72**. Kawata, Takayama, Ninomiya & Makisumi (1980) *J. Biochem.* **88**, 1601. **73**. Machuga & Ives (1984) Biochim. Biophys. Acta **789**, 26. **74**. Pereira, Alonso, Paveto, et al. (2000) *J. Biol. Chem.* **275**, 1495. **75**. Hamano, Nicchu, Shimizu, et al. (2007) *Appl. Microbiol. Biotechnol.* **76**, 873. **76**. Paulus & Gray (1967) *J. Biol. Chem.* **242**, 4980. **77**. Wong & Dennis (1973) *Plant Physiol.* **51**, 322. **78**. Moir & Paulus (1977) *J. Biol. Chem.* **252**, 4648. **79**. Kochhar, Kochhar & Sane (1998) *Biochem. Mol. Biol. Int.* **44**, 795. **80**. Kochhar, Kochhar & Sane (1986) *Biochim. Biophys. Acta* **880**, 220. **81**. Relton, Bonner, Wallsgrove & Lea (1988) *Biochim. Biophys. Acta* **953**, 48. **82**. Zhang, Jiang, Zhao, Yang & Chiao (2000) *Appl. Microbiol. Biotechnol.* **54**, 52. **83**. Cahyanto, Kawasaki, Nagashio, Fujiyama & Seki (2006) *Microbiology* **152**, 105. **84**. Dotson, Somers & Gengenbach (1989) *Plant Physiol.* **91**, 1602. **85**. Dungan & Datta (1973) *J. Biol. Chem.* **248**, 8534. **86**. Ferreira, Meinhardt & Azevedo (2006) *Ann. Appl. Biol.* **149**, 77. **87**. Curien, Laurencin, Robert-Genthon & Dumas (2007) *FEBS J.* **274**, 164. **88**. Lugli, Gaziola & Azevedo (2000) *Plant Sci.* **150**, 51. **89**. Hiles & Henderson (1972) *J. Biol. Chem.* **247**, 646. **90**. Chang (1977) *Enzyme* **22**, 230. **91**. Thoen, Rognes & Aarnes (1978) *Plant Sci. Lett.* **13**, 103. **92**. Der Garabedian (1986) *Biochemistry* **25**, 5507. **93**. Neuberger & Turner (1963) *Biochim. Biophys. Acta* **67**, 342. **94**. Yasuda, Tanizawa, Misono, Toyama & Soda (1981) *J. Bacteriol.* **148**, 43. **95**. Gray & Bhattacharjee (1976) *Can. J. Microbiol.* **22**, 1664. **96**. Wulandari, Miyazaki, Kobashi, et al. (2002) *FEBS Lett.* **522**, 35. **97**. Luengo, Revilla, Lopez, Villanueva & Martin (1980) *J. Bacteriol.* **144**, 869. **98**. Schmidt, Bode, Lindner & Birnbaum (1985) *J. Basic Microbiol.* **25**, 675. **99**. Gaillardin, Poirier & Heslot (1976) *Biochim. Biophys. Acta* **422**, 390. **100**. Takenouchi, Tanaka & Soda (1981) *J. Ferment. Technol.* **59**, 429. **101**. Tucci & Ceci (1972) *Arch. Biochem. Biophys.* **153**, 742. **102**. Demain & Masurekar (1974) *J. Gen. Microbiol.* **82**, 143. **103**. Feller, Ramos, Pierard & Dubois (1999) *Eur. J. Biochem.* **261**, 163. **104**. Kumazawa, Ohtsuka, Ninomiya & Seguro (1997) *Biosci. Biotechnol. Biochem.* **61**, 1086. **105**. Soffer (1973) *J. Biol. Chem.* **248**, 8424. **106**. Legrain & Stalon (1976) *Eur. J. Biochem.* **63**, 289. **107**. Lusty, Jilka & Nietsch (1979) *J. Biol. Chem.* **254**, 10030. **108**. Trievel, Flynn, Houtz & Hurley (2003) *Nat. Struct. Biol.* **10**, 545. **109**. Schurink, van Berkel, Wichers & Boeriu (2007) *Peptides* **28**, 2268.

L-Lysyl-D-amphetamine

This fatigue-reducing amphetamine prodrug (FW = 263.38 g/mol; CAS 608137-32-2), also named L-lysine-dextroamphetamine, (2S)-2,6-diamino-N-[(2S)-1-phenylpropan-2-yl]hexanamide, lisdexamfetamine, and Vyvanse, is metabolized to D-amphetamine (or dextroamphetamine), a CNS stimulant. L-Lysine-dextroamphetamine is prescribed for the treatment of Attention Deficit Hyperactivity Disorder (ADHD) and narcolepsy. Amphetamine's augmentation of synaptic monoamine concentrations is mediated mainly by monoamine release, an effect that is complemented by reuptake inhibition as well as inhibition of monoamine oxidase (MAO). Given the need to set a narrow concentration range for circulating lisdexamfetamine, Vyvanse® is available 10-mg, 20-mg, 30-mg, 40-mg, 50-mg, 60-mg, and 70 mg doses. As an amphetamine precursor, lisdexamfetamine is a Class B/Schedule II substance in the United Kingdom and a Schedule II controlled substance in the United States (DEA Number: 1205). **See also** *Methamphetamine*

Lysophosphatidate (Lysophosphatidic Acid)

1-Palmitoyl-*sn*-glycerol 3-phosphate

These fatty acetyl-glycerophosphate derivatives (FW = 409.48 g/mol for 1-palmitoyl-*sn*-glycerol 3-P), also called acylglycerophosphates, in which one of the hydroxyl groups is esterified to a fatty acid. Note that the term does not stipulate which hydroxyl group is acylated nor the location of the phosphoryl group. Nevertheless, unless otherwise stated below, lysophosphatidate is assumed to refer to 1-monoacyl-*sn*-glycerol 3-P. (**See also** *specific compound*) **Target(s):** adenylyl cyclase, inhibited by 1-oleoyl-*sn*-glycerol 3-phosphate (1); alkylglycerophospho-ethanolamine phosphodiesterase, lysophospholipase D (2); carboxylesterase (3); ceramidase (4); DNA-directed DNA polymerase (5); DNA polymerase α (5); dolichyl-phosphatase (6,7); glycerol-3-phosphate *O*-acyltransferase (8); glycosylphosphatidylinositol-specific phospholipase D (9); Lysophospholipase (10,11); Na^+/K^+-exchanging ATPase (12); palmitoyl-CoA hydrolase, *or* acyl-CoA hydrolase (13); phosphatidylinositol 3-kinase (14,15); polar-amino-acid-transporting ATPase, *or* histidine permease (16); sphingomyelin phosphodiesterase (17). **1.** Proll, Clark & Butcher (1985) *Mol. Pharmacol.* **28**, 331. **2.** van Meeteren, Ruurs, Christodoulou, *et al.* (2005) *J. Biol. Chem.* **280**, 21155. **3.** Mukherjee, Jay & Choy (1993) *Biochem. J.* **295**, 81. **4.** Yada, Higuchi & Imokawa (1995) *J. Biol. Chem.* **270**, 12677. **5.** Murakami-Murofushi, Shioda, Kaji, Yoshida & Murofushi (1992) *J. Biol. Chem.* **267**, 21512. **6.** Rip, Rupar, Chaudhary & Carroll (1981) *J. Biol. Chem.* **256**, 1929. **7.** Ravi, Rip & Carroll (1983) *Biochem. J.* **213**, 513. **8.** Haldar & Vancura (1992) *Meth. Enzymol.* **209**, 64. **9.** Low & Huang (1993) *J. Biol. Chem.* **268**, 8480. **10.** Baker & Chang (2000) *Biochim. Biophys. Acta* **1483**, 58. **11.** Sugimoto & Yamashita (1994) *J. Biol. Chem.* **269**, 6252. **12.** Nishikawa, Tomori, Yamashita & Shimizu (1988) *Jpn. J. Pharmacol.* **47**, 143. **13.** Mukherjee, Jay & Choy (1993) *Biochem. J.* **295**, 81. **14.** Lavie & Agranoff (1996) *J. Neurochem.* **66**, 811. **15.** Carpenter, Duckworth, Auger, *et al.* (1990) *J. Biol. Chem.* **265**, 19704. **16.** Nikaido, Liu & Ames (1997) *J. Biol. Chem.* **272**, 27745. **17.** Testai, Landek, Goswami, Ahmed & Dawson (2004) *J. Neurochem.* **89**, 636.

Lysophosphatidylethanolamine

1-Palmitoyl-*sn*-glycerol 3-phosphoethanolamine

This naturally occurring lysophospholipid, *or* LPL ($FW_{palmitoyl-ester}$ = 453.56 g/mol), itself a product of the phospholipase A action on other membrane-associated phospholipids, functions as a mediator through G-protein-coupled receptors. It also retards the senescence of leaves, flowers, and postharvest fruits. **Target(s):** α-*N*-acetylneuraminate α-2,8-sialyltransferase (1); diacylglycerol *O* acyltransferase (2); 1,3-β-glucan synthase (3); palmitoyl-CoA hydrolase, or acyl-CoA hydrolase, weakly inhibited (4); phosphatidate phosphatase (5); phospholipase D (noncompetitive) (6); steryl-β-glucosidase (7). **1.** Eppler, Morré & Keenan (1980) *Biochim. Biophys. Acta* **619**, 318. **2.** Parthasarathy, Murari, Crilly & Baumann (1981) *Biochim. Biophys. Acta* **664**, 249. **3.** Ko, Frost, Ho, Ludescher & Wasserman (1994) *Biochim. Biophys. Acta* **1193**, 31. **4.** Sanjanwala, Sun & MacQuarrie (1987) *Arch. Biochem. Biophys.* **258**, 299. **5.** Humble & Berglund (1991) *J. Lipid Res.* **32**, 1869. **6.** Ryu, Karlsson, Ozgen & Palta (1997) *Proc. Natl. Acad. Sci. U.S.A.* **94**, 12717. **7.** Kalinowska & Wojciechowski (1986) *Phytochemistry* **25**, 45.

N-[*N*-(L-Lysyl-L-prolyl-L-leucyl-L-glutamyl)-(2-amino-3-phenylpropyl)]-L-tyrosyl-L-arginyl-L-valine

This peptidomimetic inhibitor (FW = 1066.34 g/mol; *Sequence*: KPLEF ΨYRV, containing a nonhydrolyzable methyleneamino component, designated Ψ) inhibits cathepsin D, K_i = 8.5 nM, plasmepsin-2, K_i = 19.5 μM, and plasmepsin-4 (K_i = 0.48 nM). **1.** Beyer, Johnson, Chung, *et al.* (2005) *Biochemistry* **44**, 1768.

D-Lyxose

This reducing aldopentose (FW = 150.13 g/mol; *Abbreviation* = Lyx), also known as D-*lyxo*-pentose, is a rare sugar and epimer of D-xylose, D-arabinose, and L-ribose. Note that it is a structural homologue of D-mannose. It is a substrate of the reaction catalyzed by D-lyxose ketol-isomerase and is an alternative substrate of mannose isomerase. Approximately 70% of D-lyxose is present as the α-pyranose structure at 31°C in D_2O, with 28% a the β-pyranose, 1.5% as the α-furanose, and 0.5% as the β-furanose. **Target(s):** β-galactosidase (1-3); glucokinase (4,5); α-glucosidase, weakly inhibited (6); β-glucosidase (7); hexokinase (8-10), where lyxose not only competes with hexose substrates, but also induces a conformational change in yeast hexokinase that greatly activates the enzyme's intrinsic ATPase activity (9); D-xylose dehydrogenase (11); xylose isomerase (12-15). **1.** Webb (1966) *Enzyme and Metabolic Inhibitors*, vol. 2, p. 418, Academic Press, New York. **2.** Huber & Brockbank (1987) *Biochemistry* **26**, 1526. **3.** Huber, Roth & Bahl (1990) *J. Protein Chem.* **15**, 621. **4.** Walker & Parry (1966) *Meth. Enzymol.* **9**, 381. **5.** Parry & Walker (1966) *Biochem. J.* **99**, 266. **6.** Yao, Mauldin & Byers (2003) *Biochim. Biophys. Acta* **1645**, 22. **7.** Dale, Ensley, Kern, Sastry & Byers (1985) *Biochemistry* **24**, 3530. **8.** Sols, de la Fuenta, Villar-Palasí & Asensio (1958) *Biochim. Biophys. Acta* **30**, 92. **9.** Rudolph & Fromm (1971) *J. Biol. Chem.* **246**, 2104. **10.** Ohning & Neet (1983) *Biochemistry* **22**, 2986. **11.** Zepeda, Monasterio & Ureta (1990) *Biochem. J.* **266**, 637. **12.** Yamanaka (1975) *Meth. Enzymol.* **41**, 466. **13.** Noltmann (1972) *The Enzymes*, 3rd ed., **6**, 271. **14.** Yamanaka (1969) *Arch. Biochem. Biophys.* **131**, 502. **15.** Lehmacher & Bisswanger (1990) *Biol. Chem. Hoppe-Seyler* **371**, 527.

– M –

M, *See Methionine; Also used to represent a Methyl group, when indicating an epigenetic mark.*

M 25

This Smoothened receptor antagonist (FW = 379.50 g/mol; Solubility: 100 mM in DMSO; 100 mM in Ethanol), also named *N*-[[1-(2-methoxyphenyl)-1*H*-indazol-5-yl]methyl]-2-propylpentanamide, targets the Hedgehog signaling pathway, inhibiting Shh-induced Gli luciferase reporter activity in Shh-light 2 cells (IC$_{50}$ = 5 nM) (1) and antagonizing GSA-10 (IC$_{50}$ = 69 nM) and GSA-10 (*or* propyl 4-(1-hexyl-4-hydroxy-2-oxo-1,2-dihydro-quinoline-3-carboxamido) benzoate) (IC$_{50}$ = 115 nM) -mediated differentiation of the C3H10T1/2 cell line (2). **1**. Dessole, Branca, Ferrigno, *et al.* (2009) *Bioorg. Med. Chem. Lett.* **19**, 4191. **2**. Gorojankina, Hoch, Faure, *et al.* (2013) *Mol. Pharmacol.* **83**, 1020.

M 084

This TRPC4/5 channel blocker (FW$_{HCl-Salt}$ = 225.72 g/mol; Solubility: 100 mM in water; 100 mM in DMSO), also named *N*-butyl-1*H*-benzimidazol-2-amine, targets Transient Receptor Potential Canonical (TRPC) channels TRPC4 (IC$_{50}$ = 3.7-10.3 μM) TRPC5 (IC$_{50}$ = 8.2 μM), while weakly blockong TRPC3 channels (1). These channels are non-selective cation channels that mediate Na$^+$ and Ca^{2+} entry into cells, leading to membrane depolarization and an increase in intracellular Ca^{2+} concentration ([Ca^{2+}]$_i$). A rise in Na$^+$ level at the cytoplasmic side may also be an important factor for Na$^+$-dependent transport. Because PLC activation is commonly achieved by the stimulation of GPCRs or receptor tyrosine kinases, the TRPC channels are often referred to as receptor-operated channels. In mice subjected to chronic unpredictable stress, M084 treatment reverses the enhanced immobility time in forced swim test and decreases the latency to feed in novelty suppressed feeding test (2). M084 treatment increases BDNF expression in both mRNA and protein levels, as well as phosphorylation levels of AKT and ERK, in prefrontal cortex. Such findings suggest that M084 exerts rapid antidepressant and anxiolytic-like effects at least in part by acting on BDNF and its downstream signaling. **1**. Shu, Lu, Qu, *et al.* (2015) *Br J Pharmacol.* **172**, 3495. **2**. Yang, Jiang, Wu, *et al.* (2015) *PLoS One* **10**, e0136255.

M344

This HDAC inhibitor and antiproliferative agent (FW = 307.19 g/mol; CAS 251456-60-7; Solubility: 100 mM in DMSO), also named 4-(diethylamino)-*N*-[7-(hydroxyamino)-7-oxoheptyl]benzamide, targets histone deacetylase (IC$_{50}$ = 100 nM), inducing terminal cell differentiation. M344 up-regulates SMN2 protein expression in fibroblast cells derived from SMA patients up to 7x after 64 h of treatment (1). Moreover, M344 significantly raises the total number of gems/nucleus as well as the number of nuclei that contain gems. Treatment with M-344 give rise to hyperacetylated histone H4. M-344 suppresses the growth of human endometrial and ovarian cancer cells by inducing cell cycle arrest and apoptosis (2). **1**. Riessland, Brichta, Hahnen & Wirth (2006) *Hum. Genet.* **120**, 101. **2**. Takai, Ueda, Nishida, Nasu & Narahara (2006) *Gynecol. Oncol.* **101**, 108.

M617

This selective GAL$_1$ agonist (FW = 2361.68 g/mol; CAS 1172089-00-7; Sequence: GWTLNSAGYLLGPQPPGFSPFR-NH$_2$; Soluble to 1 mg/mL in H$_2$O), also named Galanin(1-13)-Gln^{14}-bradykinin(2-9)amide, targets galanin GAL$_1$ receptor (*K*$_i$ = 0.23 nM), with lowere activity against GAL$_2$ receptors (*K*$_i$ = 5.71 nM). M617 enhances food consumption in rats, following intracerebroventricular administration, and reduces CAP-induced inflammatory pain. **1**. Lundstrom, *et al.* (2005) *Int. J. Pept. Res. Ther.* **11**, 17. **2**. Jimenez-Andrade, *et al.* (2006) *Pharmacol. Biochem. Behav.* **85**, 273. **3**. Mazarati, *et al.* (2006) *J. Pharmacol. Exp. Ther.* **318**, 700.

M871

This GAL$_2$ antagonist (FW = 2287.64 g/mol; CAS 908844-75-7; Sequence: WTLNSAGYLLGPEHPPPALALA-NH$_2$; Solubility: 1 mg/mL in acetonitrile:H$_2$O (30% vol/vol)), also named Galanin-(2-13)-Glu-His-(Pro)$_3$-(Ala-Leu)$_2$-Ala-amide, selectively targets the galanin GAL$_2$ receptor antagonist (*K*$_i$ = 13 nM) with much weaker action against GAL$_1$ receptors (*K*$_i$ = 420 nM). M871 blocks the pro-nociceptive effect of GAL$_2$ receptor agonists. **1**. Jimenez-Andrade, *et al.* (2006) *Pharmacol. Biochem. Behav.* **85**, 273. **2**. Sollenburg, *et al.* (2006) *Int. J. Pept. Res. Ther.* **12**, 115.

M1145

This GAL$_2$ agonist (FW = 2774.26 g/mol; CAS 860790-38-1; Sequence: RGRGNWTLNSAGYLLGPVLPPPALALA-NH$_2$; Soluble to 5 mg/mL in H$_2$O) potently and selectively targets the galanin receptor-2 (EC$_{50}$ = 38 nM), with *K*$_i$ values of 6.55, 497 and 587 nM for GAL$_2$, GAL$_3$ and GAL$_1$ respectively. M 1145 yields an agonistic effect *in vitro*, observable as an increase in inositol phosphate (IP) accumulation in the absence or presence of galanin. **1**. Runesson, Saar, Lundström, Järv & Langel (2009) *Neuropeptides* **43**, 187.

MA 2029

This orally active, potent, and competitive motilin receptor antagonist (FW = 556.34 g/mol; CAS 287206-61-5; Solubility: 20 mM in 1eq. HCl; 100 mM in DMSO), also named 4-fluoro-*N*-methyl-L-phenylalanyl-*N*-methyl-L-valyl-3-(1,1-dimethylethyl)-*N*-ethyl-L-tyrosinamide, selectively targets the motilin receptor (IC$_{50}$ = 4.9 nM) over a range of other receptors and ion channels. Inhibits motilin-induced duodenal muscle contractions *in vitro*. MA-2029 also inhibits motilin-induced colonic and abdominal contractions *in vivo*, but contractile responses to acetylcholine and substance P were unaffected even at 1 μM MA-2029. **1**. Sudo, Yoshida, Ozaki, *et al.* (2008) *Eur. J. Pharmacol.* **581**, 296.

mAb100 & mAb114

These enzyme-inhibiting human monoclonal antibodies neutralize Ebola virus, preventing all signs of infection in nonhuman primates exposed to this notoriously contagious virus. Ebola virus glycoprotein (GP) is a Class-I fusion protein comprising disulfide-linked GP1 and GP2 subunits. GP1 binds to the Niemann-Pick C1 receptor, an event that allows GP2-mediated fusion of the viral and host-cell membranes. mAb100 binds to an ectodomain at the base of the chalice-shaped Ebola virus GP trimer, blocking access of cathepsin to its cleavage loop, thereby preventing GP proteolysis that is required for virus entry into host cells. By contrast, monoclonal mAb114 interacts with a different ectodomain on the glycan cap and inner chalice of the glycoprotein, remaining associated after proteolytic removal of the cap and inhibiting GP binding to its receptor. **1**. Misasi, Gilman, Kanekiyo, *et al.* (2016) *Science* **351**, 1343.

Macbecin I

This geldanamycin derivative (FW = 558.67 g/mol; CAS 73341-72-7; Soluble in DMSO), also named (15R)-6,17-didemethoxy-15-methoxy-6-methyl-11-O-methylgeldanamycin, is an ansamycin-related antibiotic compound that inhibits Hsp90 activity (IC$_{50}$ = 2 µM) by binding to the chaperonin ATP-binding site (1-3). Cytocidal changes induced by Macbecin I were also observed in cells which were temporarily prevented from entering mitosis by treatment with known antitumor agents such as 5-fluorouracil, cyclophosphamide, and neocarzinostatin, whereas such cytolysis was not observed in cells which were arrested in metaphase by treatment with ansamitocin P-3 (1). Cytotoxicity is observed at ≥ 0.1 µg Macbecin I per mL. Macbecin I exhibits antitumor and cytocidal activities (IC$_{50}$ ~ 0.4 µM) by blocking the folding of key oncogenic client proteins, including ErbB2 and cRaf1. Macbecin I displays higher affinity and potency than geldanamycin (2). (**See also** Alvespimycin; Ansamycin; Deguelin; Derrubone; Ganetespib; Geldanamycin; Herbimycin; Radicicol; Tanespimycin; and the nonansamycin Hsp90 inhibitor KW-2478) **1**. Ono, Kozai & Ootsu (1982) Gann. **73**, 938. **2**. Martin, Gaisser, Challis, et al. (2008) J. Med. Chem. **51**, 2853.

Macitentan

Macitentan (R = CH$_3$CH$_2$CH$_2$–) and ACT-132577 (R = H–)

This orally active, tissue-targeting, dual endothelin receptor antagonist (FW = 585.96 g/mol; CAS 441798-33-0; n-octanol/H$_2$O distribution coefficient = 800), also called Actelion-1, ACT-064992, Opsumit®, N-[5-(4-bromophenyl)-6-(2-(5-bromopyrimidin-2-yloxy)ethoxy)pyrimidin-4-yl]-N'-propylaminosulfonamide, is metabolized into the pharmacologically active major metabolite, ACT-132577 (FW = 546.19 g/mol) (1,2). Macitentan and ACT-132577 antagonize the specific binding of ET-1 on membranes of cells overexpressing ET$_A$ and ET$_B$ receptors and blunt ET-1-induced calcium mobilization in various natural cell lines, with IC$_{50}$ values of 0.2 nM for ET$_A$ and 381 nM for ET$_B$ (1,2). For ACT-132577, IC$_{50}$ values were 3.4 nM for ET$_A$ and 987 nM for ET$_B$. In functional assays, both macitentan and ACT-132577 inhibited ET-1-induced contractions in isolated endothelium-denuded rat aorta (ET$_A$ receptors) and sarafotoxin S6c-induced contractions in isolated rat trachea (ET$_B$ receptors) (3). In October 2013, Opsumit was approved by the U.S. FDA for the treatment of pulmonary arterial hypertension. **1**. Iglarz, Binkert, Morrison, et al. (2008) J. Pharmacol. Expmtl. Ther. **327**, 736. **2**. Sidharta, van Giersbergen & Dingemanse (2013) J. Clin. Pharmacol. **53**, 1131. **3**. Bolli, Boss, Binkert, et al. (2012) J. Med. Chem. **55**, 7849.

Macrophage Migration Inhibitory Factor

This pluripotent catalytic cytokine and upstream mediator of the inflammatory pathway (MW = 12,476 g; Sequence: MPMFIVNTNV PRASVPDGFLSELTQQLAQATGKPPQYIAVHVVPDQLMAFGGSSEP CALCSLHSIGKIGGAQNRSYSKLLCGLLAERLRISPDRVYINYYDMN AANVGWNNSTFA), also known as MIF and Glycosylation-Inhibiting Factor (or GIF), plays a pivotal role in systemic as well as local inflammatory and immune responses. First identified as a T cell-derived lymphokine that inhibits the random migration of macrophages out of capillary tubes in vitro (1,2) MIF enhances adherence, phagocytosis, and tumoricidal activity (3–6). MIF is also expressed in a variety other cells, indicating its broader involvement within and beyond the immune system (7,8). MIF functions as an anterior pituitary-derived hormone and a glucocorticoid-induced immunomodulator (9). MIF inhibitory action is itself inhibited by p425 (See p425). While lacking any significant primary

sequence homology, MIF shares homologous three-dimensional features with several catabolic enzymes involved in the degradation of aromatic compounds. MIF also catalyzes tautomerization of 2-carboxy-2,3-dihydroindole-5,6-quinone to form dihydroxyindole carboxylic acid (10). A tautomerase-null MIF gene knock-in mouse model demonstrated that protein-protein interactions, and not enzymatic activity, are most likely required for MIF's role in cell growth regulation (11). **Role in Atherosclerosis:** MIF expression within atheromatous plaques has been closely associated with progression and instability in human disease, and MIF deficiency is known to significantly impair atheroma development in low-density lipoprotein receptor–deficient mice, and inhibition of MIF activity (using an anti-MIF antibody) prevents atherosclerosis in ApoE$^{-/-}$ mice (12). Proatherogenic stimuli that promote MIF expression by smooth musce cells (SMCs) in vivo via a Tissue Factor-dependent mechanism. Inhibition of this process completely inhibits MIF secretion and subsequent development of atherosclerosis in mouse models (12). **See also** Tissue Factor Pathway Inhibitor **1**. David (1966) Proc. Natl. Acad. Sci. USA **56**, 72. **2**. Bloom & Bennett (1966) Science **153**, 80. **3**. Nathan, Karnovsy & David (1971) J. Exp. Med. **133**, 1356. **4**. Nathan, Remold & David (1973) J. Exp. Med. **137**, 275. **5**. Churchhill, Piessens, Sulis, et al. (1975) J. Immunol. **115**, 781. **6**. Herriott, Jiang, Stewart et al. (1993) J. Immunol. **150**, 4524. **7**. Bucala (1996) FASEB J. **7**, 19. **8**. Nishihira (2000) J. Interferon Cytokine Res. **20**, 751. **9**. Calandra, Bernhagen, Metz, et al. (1995) Nature **377**, 68. **10**. Rosengren, Bucala, Aman, et al. (1996) Mol. Med. **2**, 143. **11**. Fingerle-Rowson, Kaleswarapu, Schlander, et al. (2009) Mol. Cell. Biol. **29**, 1922. **12**. Chen, Xia, Hayford, et al. (2015) Circulation **131**, 1350.

Macroautophagy Inhibitors

Macroautophagy is the main pathway that cells use to remove damaged cell organelles and/or unused proteins. It is characterized by the formation of a double membrane around the organelle, and this structure is termed an autophagosome. This universal cellular process is cytoprotective, allowing renewal by recycling degraded elements of organelles and proteins. In the canonical starvation-induced macroautophagic pathway, autophagosome formation is induced by class 3 phosphoinositide-3-kinases as well as GTP-regulatory proteins. Indeed, transfection of cells with specific class-specific PI3K antisense oligonucleotides greatly inhibits macroautophagy. Macroautophagy is also inhibited by small interfering RNA targeting key proteins (e.g., Atg5, Atg6/Beclin 1-1, Atg10, or Atg12) and pharmacologically with 3-methyladenine, hydroxychloroquine, bafilomycin A$_1$, Gö 6976, monensin, or the ATM kinase nhibitor KU-55933. Cell death occurs by apoptosis, but the extent of cell death can be reduced by stabilizing mitochondrial membranes with Bcl-2, vMIA (a cytomegalovirus-derived gene) or by caspase inhibition (1). Suppression of the ubiquitin–proteasome system by proteasome inhibitors induces macroautophagy through multiple pathways, including: (a) accumulation of ubiquitinated proteins; (b) activation of the IRE1-JNK pathway; (c) proteasomal stabilization of ATF4; (d) inhibition of mTOR signaling; and (e) reduced proteasomal degradation of LC3 (2), **1**. Boya, González-Polo, Casares, et al. (2005) Mol. Cell Biol. **25**, 1025. **2**. Wua, Sakamotob, Milanic et al. (2010) Drug Resistance Updates **13**, 87.

MAFP

This FAAH inhibitor (FW = 370.24 g/mol; CAS 188404-10-6; Soluble in methyl acetate), also named methyl arachidonyl fluorophosphonate and (5Z,8Z,11Z,14Z)-5,8,11,14-eicosatetraenyl-methyl ester phosphonofluoridic acid, targets fatty acid amide hydrolase, or FAAH, an anandamide amidase (IC$_{50}$ = 1-3 nM), acting at about 1000x lower concentrations than required for phospholipase A$_2$ inhibition (1). As a phosphofluoridate-based inhibitor, MAPF irreversibly modifies an active-site nucleophile in FAAH, with permanent loss of catalytic activity. As such, MAPF is among most potent anandamide amidohydrolase inhibitor described thus far. FAAH is responsible for anandamide hydrolysis (IC$_{50}$ = 2.5 nM), making MAFP an important inhibitor in endocannabinoid metabolism. MAPF also binds irreversibly to CB$_1$ cannabinoid receptors (IC$_{50}$ = 20 nM). In intact neuroblastoma cells, MAFP was also around 1000-fold more potent than arachidonyl trifluoromethyl ketone (or Arach-CF3), previously the best inhibitor (2). MAFP demonstrates selectivity towards anandamide amidase for which it was approximately 3000 and 30,000-fold more potent than towards chymotrypsin and trypsin, respectively (1). MAFP displaced [^{3}H]CP-55940 binding to the CB1 cannabinoid receptor with an IC$_{50}$ of 20 nM versus 40 nM for anandamide (1). It binds irreversibly and prevents subsequent binding of the cannabinoid radioligand [H]CP-55940. Subch findings suggest MAFP is a potent and specific inhibitor of anandamide amidase and can interact with the cannabinoid receptors at the cannabinoid binding site. **1.** Deutsch, Omeir, Arreaza, *et al.* (1997) *Biochem. Pharmacol.* **53**, 255. **2.** Koutek, Prestwich, Howlett, *et al.* (1997) *J. Biol. Chem.* **269**, 22937.

Magnolol

This biphenyl derivative and free-radical scavenger (FW = 266.34 g/mol; CAS 528-43-8) is an anti-inflammatory agent found in the bark of *Magnolia officinalis*. **Target(s):** 1-alkylglycerophosphocholine *O*-acetyltransferase (1); Ca^{2+} channels (2); cyclooxygenase (3,4); histamine release (5); 11β-hydroxysteroid dehydrogenase (6,7); leukotriene A$_4$ hydrolase (8); leukotriene C$_4$ synthase (8); 5-lipoxygenase (3,8); phospholipase A$_2$ (8); sterol *O*-acyltransferase, *or* cholesterol *O*-acyltransferase, *or* ACAT (9); streptococcal glucosyltransferases (10,11). **1.** Yamazaki, Sugatani, Fujii, *et al.* (1994) *Biochem. Pharmacol.* **47**, 995. **2.** Teng, Yu, Chen, Huang & Huang (1990) *Life Sci.* **47**, 1153. **3.** Hsu, Lu, Tsao, *et al.* (2004) *Biochem. Pharmacol.* **67**, 831. **4.** Schühly, Khan & Fischer (2009) *Inflammopharmacology* **17**, 106. **5.** Ikarashi, Yuzurihara, Sakakibara, *et al.* (2001) *Planta Med.* **67**, 709. **6.** Homma, Oka, Niitsuma & Itoh (1994) *J. Pharm. Pharmacol.* **46**, 305. **7.** Horigome, Homma, Hiran, *et al.* (2001) *Planta Med.* **67**, 33. **8.** Hamasaki, Kobayashi, Zaitu, *et al.* (1999) *Planta Med.* **65**, 222. **9.** Kwon, Kim, Lee, *et al.* (1997) *Planta Med.* **63**, 550. **10.** Jun, Jifang, Miaoquan, Yingyan & Xihong (2004) *Lett. Appl. Microbiol.* **39**, 459. **11.** Huang, Fan, Wang, *et al.* (2006) *Arch. Oral Biol.* **51**, 899.

Maitotoxin

This huge amphipathic polyether toxin (MW = 3425.91; CAS 59392-53-9; *Abbreviation*: MTX) from the marine, red-tide dinoflagellate (*Gambierdiscus toxicus*) is a pore-former that increases calcium ion uptake, thereby mobilizing internal calcium stores and enhancing Ca^{2+}-dependent

phosphoinositide turnover and activating calpain (1,2). With 98 chiral centers, maitoxin presents a daunting challenge for synthetic chemists. **Primary Mode of Action:** One of Nature's exquisitely potent toxins (LD$_{50}$ = 25 pg/g in mice), maitotoxin's toxicity remained unclear for decades. One clue to MTX action was that its toxic effects are blocked by SKF 96365, typically an inhibitor of receptor-mediated calcium entry (3). It now appears that MTX binds to the plasmalemmal Ca^{2+}-ATPase (PMCA) pump, converting the latter into an open calcium ion channel (4). MTX-induced calcium ion currents were significantly reduced in PMCA1-knockdown cells and in fibroblasts from PMCA1(+/−) and PMCA4(−/−) transgenic mice. (*See Palytoxin*) **Targets:** actin cytoskeleton, significant reduction *in vivo* (5); cell cycle progression through the G$_1$/S and G$_2$/M transitions (6); G-protein α-subunit guanine nucleotide exchange (7). **1.** Choi, Padgett, Nishizawa, *et al.* (1990) *Mol. Pharmacol.* **37**, 222. **2.** Gusovsky & Daly (1990) *Biochem. Pharmacol.* **39**, 1633. **3.** Soergel, Yasumoto, Daly & Gusovsky (1992) *Molec. Pharmacol.* **41**, 487. **4.** Sinkins, Estacion, Prasad, Goel, Shull, Kunze, Schilling WP. (2009) *Am. J. Physiol. Cell Physiol.* **297**, C1533 **5.** Ares, Louzao, Vieytes, *et al.* (2005) *J. Exp. Biol.* **208**, 4345. **6.** Van Dolah & Ramsdell (1996) *J. Cell Physiol.* **166**, 49. **7.** Khan, Tall, Rebecchi, *et al.* (2001) *Int. J. Toxicol.* **20**, 39.

Malachite Green

This cytotoxic, triphenylmethane dye (FW$_{chloride}$ = 364.92 g/mol; CAS 569-64-2; Highly soluble in water, forming a blue-green solution), also known as aniline green, basic green 4 and China green, has green color reminiscent of the mineral malachite. Solutions are yellow below pH 2 and colorless at pH 14. Malachite green dissipates mitochondrial membrane potential and oxidative phosphorylation. The reduced metabolite is called leucomalachite green. Malachite green is used to treat and prevent fungal and parasitic infections in the aquaculture industry. **Target(s):** alcohol dehydrogenase, yeast (1); cholinesterase, *or* butyrylcholinesterase (2); cytochrome P450 monooxygenases (3); DNA-directed DNA polymerase (4); DNA polymerase I (4); F$_1$ ATPase (5,6); fumarase (7); glutathione *S*-transferase (8,9); oxidative phosphorylation (10,11); urease (12). **1.** Sund & Theorell (1963) *The Enzymes*, 2nd ed., **7**, 25. **2.** Kücükkilinc & Ozer (2005) *Arch. Biochem. Biophys.* **440**, 118. **3.** Beyhl (1981) *Experientia* **37**, 943. **4.** Wolfe (1977) *Biochemistry* **16**, 30. **5.** Bullough, Ceccarelli, Roise & Allison (1989) *Biochim. Biophys. Acta* **975**, 377. **6.** Mai & Allison (1983) *Arch. Biochem. Biophys.* **221**, 467. **7.** Quastel (1931) *Biochem. J.* **25**, 898. **8.** Glanville & Clark (1997) *Life Sci.* **60**, 1535. **9.** Debnam, Glanville & Clark (1993) *Biochem. Pharmacol.* **45**, 1227. **10.** Werth & Boiteux (1968) *Arch. Toxikol.* **23**, 82. **11.** Werth & Boiteux (1967) *Arzneimittelforschung* **17**, 1231. **12.** Quastel (1932) *Biochem. J.* **26**, 1685.

Malaoxon

This strong acetyl- and butyryl-cholinesterase inhibitor (FW = 314.30 g/mol; CAS 1634-78-2; IUPAC Name: diethyl 2-(dimethoxyphosphoryl sulfanyl)butanedioate) is the phosphorothioate that is formed upon hydrolysis of malathion, a once well known insecticide. **Target(s):** acetylcholinesterase (1-5,12-16); acylaminoacylpeptidase, moderately inhibited (6); cholinesterase, *or* butyrylcholinesterase (3,5,7-9,12); procollagen-lysine 5-dioxygenase, *or* lysyl hydrolase (10,11). **1.** Bozsik, Francis, Gaspar & Haubruge (2002) *Meded. Rijksuniv. Gent. Fak. Landbouwkd. Toegep. Biol. Wet.* **67**, 671. **2.** Gao, Rao, Wilde & Zhu (1998) *Arch. Insect Biochem. Physiol.* **39**, 118. **3.** Rodriguez, Muth, Berkman, Kim & Thompson (1997) *Bull. Environ. Contam. Toxicol.* **58**,

171. **4**. Main & Dauterman (1967) *Can. J. Biochem.* **45**, 757. **5**. Main (1969) *J. Biol. Chem.* **244**, 829. **6**. Richards, Johnson & Ray (2000) *Mol. Pharmacol.* **58**, 577. **7**. Baker (1967) *Design of Active-Site-Directed Irreversible Enzyme Inhibitors*, Wiley, New York. **8**. Main (1964) *Science* **144**, 992. **9**. Shaw (1970) *The Enzymes*, 3rd ed., **1**, 91. **10**. Samimi & Last (2001) *Toxicol. Appl. Pharmacol.* **176**, 181. **11**. Samimi & Last (2001) *Toxicol. Appl. Pharmacol.* **172**. 203. **12**. Schopfer, Voelker, Bartels, Thompson & Lockridge (2005) *Chem. Res. Toxicol.* **18**, 747. **13**. Frasco, Fournier, Carvalho & Guilhermino (2006) *Aquat. Toxicol.* **77**, 412. **14**. Shaonan, Xianchuan, Guonian & Yajun (2004) *Aquat. Toxicol.* **68**, 293. **15**. Walsh, Dolden, Moores, *et al.* (2001) *Biochem. J.* **359**, 175. **16**. Boublik, Saint-Aguet, Lougarre, *et al.* (2002) *Protein Eng.* **15**, 43.

D-Malate (D-Malic Acid)

This dicarboxylic acid (FW$_{\text{free-acid}}$ = 134.09 g/mol; CAS 636-61-3), also known as D-(+)-monohydroxysuccinic acid and (*R*)-hydroxybutanedioic acid, is not the stereoisomer that participates in the tricarboxylic acid and glyoxylate cycles. The free acid is soluble in water and ethanol, has a melting point of 101°C, and pK_a values of 3.5 and 5.1. (*See also* DL-*Malate and* DL-*Malic Acid;* L-*Malate and* L-*Malic Acid*) **Target(s):** acid phosphatase, weakly inhibited (1-4); aspartate ammonia-lyase, K_i = 0.66 mM (5,6); D-aspartate oxidase (7-9); cyanate hydratase, *or* cyanase (10); 2,3-diketogulonate reductase (11); fumarase (2,12,13); glutamate decarboxylase, K_i = 15 mM (14); L-lactate dehydrogenase (cytochrome) (15); L-malate dehydrogenase (16); L-malate transport, as alternative substrate (17); maleate hydratase, as product inhibitor (18,19); malic enzyme (oxaloacetate decarboxylating activity) (20-22); (*S*)-2-methylmalate dehydratase (23,24); sinapoylglucose:malate *O*-sinapoyltransferase (25); tartrate decarboxylase (26); D-(−)-tartrate dehydratase (27-29); L-(+)-tartrate dehydratase (30); *meso*-tartrate dehydratase (27). **1**. Hollander (1971) *The Enzymes*, 3rd ed., **4**, 449. **2**. Webb (1966) *Enzyme and Metabolic Inhibitors*, vol. **2**, Academic Press, New York. **3**. London, McHugh & Hudson (1958) *Arch. Biochem. Biophys.* **73**, 72. **4**. Kilsheimer & Axelrod (1957) *J. Biol. Chem.* **227**, 879. **5**. Falzone, Karsten, Conley & Viola (1988) *Biochemistry* **27**, 9089. **6**. Viola (2000) *Adv. Enzymol. Relat. Areas Mol. Biol.* **74**, 295. **7**. Dixon & Kenworthy (1967) *Biochim. Biophys. Acta* **146**, 54. **8**. de Marco & Crifò (1967) *Enzymologia* **33**, 325. **9**. Rinaldi (1971) *Enzymologia* **40**, 314. **10**. Anderson, Johnson, Endrizzi, Little & Korte (1987) *Biochemistry* **26**, 3938. **11**. Forouhar, Lee, Benach, *et al.* (2004) *J. Biol. Chem.* **279**, 13148. **12**. Hill & Bradshaw (1969) *Meth. Enzymol.* **13**, 91. **13**. Massey (1953) *Biochem. J.* **55**, 172. **14**. Fonda (1972) *Biochemistry* **11**, 1304. **15**. Nygaard (1963) *The Enzymes*, 2nd ed., **7**, 557. **16**. Yoshida (1969) *Meth. Enzymol.* **13**, 141. **17**. Sousa, Mota & Leao (1992) *Yeast* **8**, 1025. **18**. Van der Werf, Van der Tweel & Hartmans (1993) *Appl. Environ. Microbiol.* **59**, 2823. **19**. Van der Werf, Hartmans & Van der Tweel (1995) *Enzyme Microb. Technol.* **17**, 430. **20**. Ochoa (1955) *Meth. Enzymol.* **1**, 739. **21**. Hsu, Mildvan, Chang & Fung (1976) *J. Biol. Chem.* **251**, 6574. **22**. Veiga Salles & Ochoa (1950) *J. Biol. Chem.* **187**, 849. **23**. Wang & Barker (1969) *Meth. Enzymol.* **13**, 331. **24**. Wang & Barker (1969) *J. Biol. Chem.* **244**, 2516. **25**. Gräwe, Bachhuber, Mock & Strack (1992) *Planta* **187**, 236. **26**. Furuyoshi, Nawa, Kawabata, Tanaka & Soda (1991) *J. Biochem.* **110**, 520. **27**. Shilo (1957) *J. Gen. Microbiol.* **16**, 472. **28**. Rode & Giffhorn (1982) *J. Bacteriol.* **150**, 1061. **29**. Rode & Giffhorn (1982) *J. Bacteriol.* **151**, 1602. **30**. Hurlbert & Jakoby (1965) *J. Biol. Chem.* **240**, 2772.

L-Malate (L-Malic Acid)

This dicarboxylic acid (FW$_{\text{free-acid}}$ = 134.09 g/mol; CAS 97-67-6; free acid is soluble in water and ethanol; M.P. = 101°C; pK_1 = 3.40; pK_2 = 5.13), also known as L-(−)-monohydroxysuccinic acid and (*S*)-hydroxybutanedioic acid, is the stereoisomer that participates in the tricarboxylic acid and glyoxylate cycles. Surprisingly, L-malic acid is actually dextrorotatory at very high concentrations and 20°C. (*See also* D-*Malate and* D-*Malic Acid;* DL-

Malate and DL-*Malic Acid*) Malate is a substrate for reactions catalyzed by malate dehydrogenase, malate dehydrogenase (oxaloacetate decarboxylating) (known widely as the malic enzyme), malate dehydrogenase (decarboxylating), malate dehydrogenase (oxaloacetate decarboxylating) (NADP$^+$), malate dehydrogenase (NADP$^+$), malate oxidase, malate dehydrogenase (acceptor), sinapoylglucose:malate *O*-sinapoyltransferase, and malate:CoA ligase (malyl-CoA synthetase). It is a product of the reactions catalyzed by fumarate hydratase (fumarase), lactate:malate transhydrogenase, and malate synthase. **Target(s):** acid phosphatase, weakly inhibited (1,67); aspartate aminotransferase (78-80); aspartate ammonia-lyase, very weakly inhibited (2); aspartate carbamoyltransferase (3,84); aspartate kinase (68); D-aspartate oxidase (4); carbamoyl-phosphate synthetase I (5); carboxy-*cis,cis* muconate cyclase (6); catalase (86); citrate (*si*)-synthase, weakly inhibited (83); cyanate hydratase, *or* cyanase (7); glucose-1,6-bisphosphate synthase, weakly inhibited (69); glutamate decarboxylase, K_i = 13 mM (8-10); glutamate synthase (11); D-2-hydroxyacid dehydrogenase (12,13); isocitrate lyase (14,15); lactate dehydrogenase (12,16); D-lactate dehydrogenase (17,18); L-lactate dehydrogenase (cytochrome) (17); maleate hydratase (19); malic enzyme (oxaloacetate decarboxylating activity) (20,21); malic enzyme, NAD+-dependent (oxaloacetate decarboxylating activity) (22); *N*-methylglutamate dehydrogenase (23); 2-methyleneglutarate mutase (24,25); nicotinate-nucleotide diphosphorylase (carboxylating) (81,82); nucleotidases (26); oxaloacetate decarboxylase (27-31); pantothenase (32); phosphoenolpyruvate carboxykinase (ATP) (33,34); phosphoenolpyruvate carboxykinase (GTP) (35); phosphoenolpyruvate carboxykinase (36); phosphoenolpyruvate carboxylase (37-51); 6-phosphofructokinase (12,52,53,76,77); phosphoprotein phosphatase (66); phosphoribulokinase (75); procollagen-proline 4-dioxygenase (85); protein phosphatase 2A (66); pyrophosphatase, *or* inorganic diphosphatase (54,55); pyrophosphate-dependent phosphofructokinase, *or* diphosphate:fructose-6-phosphate 1-phosphotransferase (70); pyruvate carboxylase (56,57); pyruvate kinase (71-74); succinate dehydrogenase (12,59,60); tartrate decarboxylase, less inhibited than with D-malate (61); D-(−)-tartrate dehydratase, weakly inhibited (62,63); L-(+) tartrate dehydratase (64); tartronate-semialdehyde reductase, *or* 2-hydroxy-3-oxopropionate reductase, weakly inhibited (65); thiosulfate sulfurtransferase, *or* rhodanese (58). **1**. London, McHugh & Hudson (1958) *Arch. Biochem. Biophys.* **73**, 72. **2**. Falzone, Karsten, Conley & Viola (1988) *Biochemistry* **27**, 9089. **3**. Jacobson & Stark (1973) *The Enzymes*, 3rd ed., **9**, 225. **4**. Dixon & Kenworthy (1967) *Biochim. Biophys. Acta* **146**, 54. **5**. Garcia-Espana, Alonso & Rubio (1991) *Arch. Biochem. Biophys.* **288**, 414. **6**. Thatcher & Cain (1975) *Eur. J. Biochem.* **56**, 193. **7**. Anderson, Johnson, Endrizzi, Little & Korte (1987) *Biochemistry* **26**, 3938. **8**. Zhou, Zheng & Zhang (1996) *Yao Xue Xue Bao* **31**, 897. **9**. Fonda (1972) *Biochemistry* **11**, 1304. **10**. Gerig & Kwock (1979) *FEBS Lett.* **105**, 155. **11**. Matsuoka & Kimura (1986) *J. Biochem.* **99**, 1087. **12**. Webb (1966) *Enzyme and Metabolic Inhibitors*, vol. **2**, Academic Press, New York. **13**. Cremona (1964) *J. Biol. Chem.* **239**, 1457. **14**. Kleber (1975) *Acta Biol. Med. Ger.* **34**, 723. **15**. McFadden & Howes (1963) *J. Biol. Chem.* **238**, 1737. **16**. Boeri, Cremona & Singer (1960) *Biochem. Biophys. Res. Commun.* **2**, 298. **17**. Nygaard (1963) *The Enzymes*, 2nd ed., **7**, 557. **18**. Strasser de Saad, Pesce de Ruiz Holgado & Oliver (1986) *Biotechnol. Appl. Biochem.* **8**, 370. **19**. Van der Werf, Van der Tweel & Hartmans (1993) *Appl. Environ. Microbiol.* **59**, 2823. **20**. Ochoa (1955) *Meth. Enzymol.* **1**, 739. **21**. Veiga Salles & Ochoa (1950) *J. Biol. Chem.* **187**, 849. **22**. Korkes, del Campillo & Ochoa (1950) *J. Biol. Chem.* **187**, 891. **23**. Hersh, Stark, Worthen & Fiero (1972) *Arch. Biochem. Biophys.* **150**, 219. **24**. Barker (1972) *The Enzymes*, 3rd ed., **6**, 509. **25**. Kung & Stadtman (1971) *J. Biol. Chem.* **246**, 3378. **26**. Arsenis & Touster (1978) *Meth. Enzymol.* **51**, 271. **27**. Dimroth (1986) *Meth. Enzymol.* **125**, 530. **28**. Ochoa & Weisz-Tabori (1948) *J. Biol. Chem.* **174**, 123. **29**. Plaut & Lardy (1949) *J. Biol. Chem.* **180**, 13. **30**. Dimroth (1981) *Eur. J. Biochem.* **115**, 353. **31**. Benziman, Russo, Hochmann & Weinhouse (1978) *J. Bacteriol.* **134**, 1. **32**. Airas (1976) *Biochim. Biophys. Acta* **452**, 201. **33**. Cannata & Stoppani (1963) *J. Biol. Chem.* **238**, 1919. **34**. Chinthapalli, Raghavan, Blasing, Westhoff & Raghavendra (2000) *Photosynthetica* **38**, 415. **35**. Hebda & Nowak (1982) *J. Biol. Chem.* **257**, 5503. **36**. Wood, Davies & Willard (1969) *Meth. Enzymol.* **13**, 297. **37**. Utter & Kolenbrander (1972) *The Enzymes*, 3rd ed., **6**, 117. **38**. Zhang, Li & Chollet (1995) *Plant Physiol.* **108**, 1561. **39**. Frank, Clarke, Vater & Holzwarth (2001) *Biophys. Chem.* **92**, 53. **40**. Moraes & Plaxton (2000) *Eur. J. Biochem.* **267**, 4465. **41**. Echevarria, Pacquit, Bakrim, *et al.* (1994) *Arch. Biochem. Biophys.* **315**, 425. **42**. Yoshinaga (1977) *J. Biochem.* **81**, 665. **43**. Daniel, Bryant & Woodward (1984) *Biochem. J.* **218**, 387. **44**. Munoz, Escribano & Merodio (2001) *Phytochemistry* **58**, 1007. **45**. Hoban

& Lyric (1975) *Can. J. Biochem.* **53**, 875. **46.** Andreo, Gonzalez & Iglesias (1987) *FEBS Lett.* **213**, 1. **47.** Pays, Jones, Wilkins, Fewson & Malcolm (1980) *Biochim. Biophys. Acta* **614**, 151. **48.** Takahashi-Terada, Kotera, Ohshima, *et al.* (2005) *J. Biol. Chem.* **280**, 11798. **49.** Nhiri, Bakrim, El Hachimi-Messouak, Echevarria & Vidal (2000) *Plant Sci.* **151**, 29. **50.** Gayathri, Parvathi & Raghavendra (2000) *Photosynthetica* **38**, 45. **51.** Tripodi, Turner, Gennidakis & Plaxton (2005) *Plant Physiol.* **139**, 969. **52.** Bloxham & Lardy (1973) *The Enzymes*, 3rd ed., **8**, 239. **53.** Passonneau & Lowry (1963) *Biochem. Biophys. Res. Commun.* **13**, 372. **54.** Krishnan & Gnanam (1988) *Arch. Biochem. Biophys.* **260**, 277. **55.** Naganna (1950) *J. Biol. Chem.* **183**, 693. **56.** Scrutton, Olmsted & Utter (1969) *Meth. Enzymol.* **13**, 235. **57.** Scrutton & Mildvan (1968) *Biochemistry* **7**, 1490. **58.** Oi (1975) *J. Biochem.* **78**, 825. **59.** Potter & Elvehjem (1937) *J. Biol. Chem.* **117**, 341. **60.** Swingle, Axelrod & Elvehjem (1942) *J. Biol. Chem.* **145**, 581. **61.** Furuyoshi, Nawa, Kawabata, Tanaka & Soda (1991) *J. Biochem.* **110**, 520. **62.** Rode & Giffhorn (1982) *J. Bacteriol.* **150**, 1061. **63.** Rode & Giffhorn (1982) *J. Bacteriol.* **151**, 1602. **64.** Hurlbert & Jakoby (1965) *J. Biol. Chem.* **240**, 2772. **65.** Gotto & Kornberg (1961) *Biochem. J.* **81**, 273. **66.** Dong, Ermolova & Chollet (2001) *Planta* **213**, 379. **67.** Kuo & Blumenthal (1961) *Biochim. Biophys. Acta* **52**, 13. **68.** Keng & Viola (1996) *Arch. Biochem. Biophys.* **335**, 73. **69.** Rose, Warms & Wong (1977) *J. Biol. Chem.* **252**, 4262. **70.** Stitt (1989) *Plant Physiol.* **89**, 628. **71.** Wu & Turpin (1992) *J. Phycol.* **28**, 472. **72.** Lin, Turpin & Plaxton (1989) *Arch. Biochem. Biophys.* **269**, 228. **73.** Guderley & Hochachka (1980) *J. Exp. Zool.* **212**, 461. **74.** Turner, Knowles & Plaxton (2005) *Planta* **222**, 1051. **75.** Marsden & Codd (1984) *J. Gen. Microbiol.* **130**, 999. **76.** Van Praag, Zehavi & Goren (1999) *Biochem. Mol. Biol. Int.* **47**, 749. **77.** Botha & Turpin (1990) *Plant Physiol.* **93**, 871. **78.** Owen & Hochachka (1974) *Biochem. J.* **143**, 541. **79.** Martins, Mourato & de Varennes (2001) *J. Enzyme Inhib.* **16**, 251. **80.** Hatch (1973) *Arch. Biochem. Biophys.* **156**, 207. **81.** Shibata & Iwai (1980) *Biochim. Biophys. Acta* **611**, 280. **82.** Iwai, Shibata & Taguchi (1979) *Agric. Biol. Chem.* **43**, 351. **83.** Stern (1961) *The Enzymes*, 2nd ed. **5**, 367. **84.** Jacobson & Stark (1975) *J. Biol. Chem.* **250**, 6852. **85.** Tuderman, Myllylä & Kivirikko (1977) *Eur. J. Biochem.* **80**, 341. **86.** Chatterjee & Sanwal (1993) *Mol. Cell. Biochem.* **126**, 125.

Maleate (Maleic Acid)

This unsaturated dicarboxylic acid (FW$_{free-acid}$ = 116.07 g/mol; CAS 110-16-7; pK_a values of 1.97 and 6.24), also known as (*Z*)-butenedioic acid and rarely as toxilic acid, is the *cis*-isomer of fumaric acid. The free acid has a melting point of 130-131°C, significantly lower than that of fumaric acid (287°C). Maleic acid isomerizes to fumaric acid, when heated above its melting point. It is soluble in water (78.8 g/100 mL at 25°C), diethyl ether (7.6 g/100 g), and benzene (0.024 g/100 g). Sodium hydrogen maleate and potassium hydrogen maleate are common pH buffers. **Target(s):** 2-(acetamido-methylene)succinate hydrolase (1); acid phosphatase (2,3); aconitase, *or* aconitate hydratase (4); adenosine-phosphate deaminase (5,6); adenylosuccinate synthetase (7,8); alanine aminotransferase (10,11,113-115); alcohol dehydrogenase (12); aldose reductase (13); D-amino-acid aminotransferase, *or* D-alanine aminotransferase (9,109); 4-aminobutyrate aminotransferase (110); AMP deaminase (5); α-amylase, weakly inhibited (14); arylsulfatase (15); asclepain (16,17); aspartate aminotransferase (2,18-25,116-119); aspartate carbamoyltransferase (26-28,127-130); aspartate 4-decarboxylase (29-31); -aspartate oxidase (32); bromelain, weakly inhibited (16); calotropain, *or* Calotropis gigantea latex protease (17); carbamoyl-phosphate synthetase (26); carboxy-*cis,cis* muconate cyclase (33); catalase (134); cellulose polysulfatase (34); ceruloplasmin (35); cholinesterase (36,37); citrate (*si*)-synthase, weakly inhibited (124); *p*-coumarate decarboxylase (38); cyanate hydratase, *or* cyanase (39); cysteine aminotransferase (112); cysteine desulfurase, *or* cystathionine γ-lyase (40); dehydro-L-gulonate decarboxylase (41); fatty-acid synthase (42,43); formate hydrogen-lyase (44); 5-formyltetrahydrofolate cyclo-ligase, *or* 5,10-methenyltetrahydrofolate synthetase (45); β-fructofuranosidase, *or* invertase (105); fumarase (2,46,47); glucose dehydrogenase, weakly inhibited (48); glucose-6-phosphate isomerase, moderately inhibited (49); glutamate decarboxylase, weakly inhibited, K_i = 18 mM (50,51); glutamate formimidoyltransferase (131); γ-glutamyl transpeptidase, *or* γ-glutamyltransferase, hydrolase activity is stimulated and transpeptidase activity is inhibited (52 56,125,126); glutathione peroxidase (133); β-glycerophosphatase, *or* glycerol-2-phosphatase (3); glycine amino-

transferase (57); glycine:oxaloacetate aminotransferase (21,58); glycogen synthase (59); glyoxalase (60); *erythro*-3-hydroxyaspartate ammonia-lyase (58,61); 4-hydroxybenzoate 3-monooxygenase, weakly inhibited (132); 4-hydroxyglutamate aminotransferase (108); [3-hydroxy-3 methylglutaryl-CoA reductase (NADPH)]-phosphatase (106); isocitrate lyase (62-69); α-ketoglutarate oxidase (70); lactase-phlorizin hydrolase, *or* glycosylceramidase (104); lactate dehydrogenase, weakly inhibited (71); leucine aminotransferase (111); malate dehydrogenase (70,72); malic enzyme (malate dehydrogenase (decarboxylating), weakly inhibited (73,74); 2-methyleneglutarate mutase (75); myrosulfatase (15); nicotinate-nucleotide diphosphorylase (carboxylating) (120,121); oxaloacetate keto enol tautomerase (76); oxaloacetate tautomerase (77,78); papain (16,60,79); pepsin (80); phenylpyruvate enol-keto tautomerase (81); phosphoenolpyruvate carboxykinase, pyrophosphate-requiring (82); phosphoenolpyruvate carboxylase (83,84); phosphofructokinase (85); phosphoribosylaminoimidazole-succinocarboxamide synthetase (86); polynucleotide 5'-hydroxyl kinase (107); progesterone 11α-hydroxylase (87); progesterone 11β-hydroxylase (87); proline racemase (88,89); pyrophosphatase, *or* inorganic diphosphatase (90); pyruvate decarboxylase (91); pyruvate dehydrogenase (92); pyruvate oxidase (92); sterol 3β-glucosyltransferase (122); succinate dehydrogenase (2,60,93-100); succinate oxidase (92,101); sucrose-phosphate synthase (123); triose phosphate isomerase (102); urease (103). **1.** Huynh & Snell (1985) *J. Biol. Chem.* **260**, 2379. **2.** Webb (1966) *Enzyme and Metabolic Inhibitors*, vol. 2, Academic Press, New York. **3.** Nigam, Davidson & Fishman (1959) *J. Biol. Chem.* **234**, 1550. **4.** Treton & Heslot (1978) *Agric. Biol. Chem.* **42**, 1201. **5.** Zielke & Suelter (1971) *The Enzymes*, 3rd ed., **4**, 47. **6.** Yates (1969) *Biochim. Biophys. Acta* **171**, 299. **7.** Markham & Reed (1977) *Arch. Biochem. Biophys.* **184**, 24. **8.** Gorrell, Wang, Underbakke, *et al.* (2002) *J. Biol. Chem.* **277**, 8817. **9.** Martinez-Carrion & Jenkins (1970) *Meth. Enzymol.* **17A**, 167. **10.** Segal & Matsuzawa (1970) *Meth. Enzymol.* **17A**, 153. **11.** Bulos & Handler (1965) *J. Biol. Chem.* **240**, 3283. **12.** Lutwak-Mann (1938) *Biochem. J.* **32**, 1364. **13.** Hargitai & Erdo (1997) *Neurobiology (Bp)* **5**, 453. **14.** Di Carlo & Redfern (1947) *Arch. Biochem.* **15**, 343. **15.** Takahashi (1960) *J. Biochem.* **47**, 230. **16.** Greenberg & Winnick (1940) *J. Biol. Chem.* **135**, 761. **17.** Bose & Krishna (1958) *Enzymologia* **19**, 186. **18.** Jenkins, Yphantis & Sizer (1959) *J. Biol. Chem.* **234**, 51. **19.** Velick & Vavra (1962) *The Enzymes*, 2nd ed., **6**, 219. **20.** Vessal & Taher (1995) *Comp. Biochem. Physiol. B Biochem. Mol. Biol.* **110**, 431. **21.** Braunstein (1973) *The Enzymes*, 3rd ed., **9**, 379. **22.** Goldstone & Adams (1962) *J. Biol. Chem.* **237**, 3476. **23.** Velick & Vavra (1962) *J. Biol. Chem.* **237**, 2109. **24.** Jager, Moser, Sauder & Jansonius (1994) *J. Mol. Biol.* **239**, 285. **25.** Bonsib, Harruff & Jenkins (1975) *J. Biol. Chem.* **250**, 8635. **26.** Jones (1962) *Meth. Enzymol.* **5**, 903. **27.** Gerhart & Pardee (1964) *Fed. Proc.* **23**, 727. **28.** Burns, Mendz & Hazell (1997) *Arch. Biochem. Biophys.* **347**, 119. **29.** Miles & Sparrow (1970) *Meth. Enzymol.* **17A**, 689. **30.** Wilson & Kornberg (1963) *Biochem. J.* **88**, 578. **31.** Shibatani, Kakimoto, Kato, Nishimura & Chibata (1974) *J. Ferment. Technol.* **52**, 886. **32.** Dixon & Kenworthy (1967) *Biochim. Biophys. Acta* **146**, 54. **33.** Thatcher & Cain (1975) *Eur. J. Biochem.* **56**, 193. **34.** Takahashi (1960) *J. Biochem.* **48**, 691. **35.** Curzon (1960) *Biochem. J.* **77**, 66. **36.** Nachmansohn & Lederer (1939) *Bull. Soc. Chim. Biol.* **21**, 797. **37.** Mounter & Whittaker (1953) *Biochem. J.* **53**, 167. **38.** Harada & Mino (1976) *Can. J. Microbiol.* **22**, 1258. **39.** Anderson, Johnson, Endrizzi, Little & Korte (1987) *Biochemistry* **26**, 3938. **40.** Fromageot & Grand (1944) *Enzymologia* **11**, 235. **41.** Kagawa & Shimazono (1970) *Meth. Enzymol.* **18A**, 46. **42.** Muesing & Porter (1975) *Meth. Enzymol.* **35**, 45. **43.** Hsu & Wagner (1970) *Biochemistry* **9**, 245. **44.** Crewther (1953) *Australian J. Biol. Sci.* **6**, 205. **45.** Grimshaw, Henderson, Soppe, *et al.* (1984) *J. Biol. Chem.* **259**, 2728. **46.** Hill & Bradshaw (1969) *Meth. Enzymol.* **13**, 91. **47.** Massey (1953) *Biochem. J.* **55**, 172. **48.** Nakamura (1954) *J. Biochem.* **41**, 67. **49.** Sangwan & Singh (1989) *J. Biosci.* **14**, 47. **50.** Ohno & Okunuki (1962) *J. Biochem.* **51**, 313. **51.** Fonda (1972) *Biochemistry* **11**, 1304. **52.** Tate & Meister (1974) *Proc. Natl. Acad. Sci. U.S.A.* **71**, 3329. **53.** Tate & Meister (1975) *J. Biol. Chem.* **250**, 4619. **54.** Thompson & Meister (1979) *J. Biol. Chem.* **254**, 2956. **55.** Thompson & Meister (1979) *J. Biol. Chem.* **255**, 2109. **56.** Prusiner & Prusiner (1978) *J. Neurochem.* **30**, 1261. **57.** Nakada (1964) *J. Biol. Chem.* **239**, 468. **58.** Gibbs & Morris (1970) *Meth. Enzymol.* **17A**, 981. **59.** Rothman & Cabib (1967) *Biochemistry* **6**, 2098 and 2107. **60.** Morgan & Friedman (1938) *Biochem. J.* **32**, 862. **61.** Gibbs & Morris (1965) *Biochem. J.* **97**, 547. **62.** Spector (1972) *The Enzymes*, 3rd ed., **7**, 357. **63.** McFadden & Howes (1963) *J. Biol. Chem.* **238**, 1737. **64.** Daron, Rutter & Gunsalus (1966) *Biochemistry* **5**, 895. **65.** Tsukamoto, Ejiri & Katsumata (1986) *Agric. Biol.*

Chem. **50**, 409. **66.** McFadden & W. V. Howes (1963) *J. Biol. Chem.* **238**, 1737. **67.** Schloss & Cleland (1982) *Biochemistry* **21**, 4420. **68.** Giachetti, Pinzauti, Bonaccorsi, Vincenzini & Vanni (1987) *Phytochemistry* **26**, 2439. **69.** Chell, Sundaram & Wilkinson (1978) *Biochem. J.* **173**, 165. **70.** Angielski & Rogulski (1962) *Acta Biochim. Polon.* **9**, 357. **71.** Hopkins, Morgan & Lutwak-Mann (1938) *Biochem. J.* **32**, 1829. **72.** Scholefield (1955) *Biochem. J.* **59**, 177. **73.** Stickland (1959) *Biochem. J.* **73**, 646. **74.** Schimerlik & Cleland (1977) *Biochemistry* **16**, 565. **75.** Kung & Stadtman (1971) *J. Biol. Chem.* **246**, 3378. **76.** Belikova, Burov & Vinogradov (1988) *Biochim. Biophys. Acta* **936**, 10. **77.** Vinogradov, Kotlyar, Burov & Belikova (1989) *Adv. Enzyme Regul.* **28**, 271. **78.** Belikova, Burov & Vinogradov (1988) *Biochim. Biophys. Acta* **936**, 10. **79.** Ganapathy & Sastri (1939) *Biochem. J.* **33**, 1175. **80.** Crippa & Maffei (1940) *Gazz. Chim. Ital.* **70**, 212. **81.** Knox & Pitt (1957) *J. Biol. Chem.* **225**, 675. **82.** Wood, Davies & Willard (1969) *Meth. Enzymol.* **13**, 297. **83.** Owttrim & Colman (1986) *J. Bacteriol.* **168**, 207. **84.** Gold & Smith (1974) *Arch. Biochem. Biophys.* **164**, 447. **85.** Passonneau & Lowry (1963) *Biochem. Biophys. Res. Commun.* **13**, 372. **86.** Nelson, Binkowski, Honzatko & Fromm (2005) *Biochemistry* **44**, 766. **87.** Abdel-Fattah & Badawi (1978) *Zentralbl. Bakteriol. Naturwiss.* **133**, 733. **88.** Cardinale & Abeles (1968) *Biochemistry* **7**, 3970. **89.** Reina-San-Martin, Degrave, Rougeot, *et al.* (2000) *Nature Med.* **6**, 890. **90.** Naganna (1950) *J. Biol. Chem.* **183**, 693. **91.** Talarico, Ingram & Maupin-Furlow (2001) *Microbiology* **147**, 2425. **92.** Peters & Wakelin (1946) *Biochem. J.* **40**, 513. **93.** Veeger, DerVartanian & Zeylemaker (1969) *Meth. Enzymol.* **13**, 81. **94.** Hatefi (1978) *Meth. Enzymol.* **53**, 27. **95.** Hopkins, Morgan & Lutwak-Mann (1938) *Biochem. J.* **32**, 1829. **96.** Hopkins & Morgan (1938) *Biochem. J.* **32**, 611. **97.** Hatefi & Stiggall (1976) *The Enzymes*, 3rd ed., **13**, 175. **98.** Dervartanian & Veeger (1962) *Biochem. J.* **84**, 65p. **99.** Takeya, Sawada & Matsuyama (1953) *Igaku To Seibutsugaku* **27**, 114. **100.** Morgan & Friedmann (1938) *Biochem. J.* **32**, 862. **101.** Potter & DuBois (1943) *J. Gen. Physiol.* **26**, 391. **102.** Noltmann (1972) *The Enzymes*, 3rd ed., **6**, 271. **103.** Kistiakowsky, Mangelsdorf, Rosenberg & Shaw (1952) *J. Amer. Chem. Soc.* **74**, 5015. **104.** Hamaswamy & Radhakrishnan (1973) *Biochem. Biophys. Res. Commun.* **54**, 197. **105.** Prado, Vattuone, Fleischmacher & Sampietro (1985) *J. Biol. Chem.* **260**, 4952. **106.** Hegardt, Gil & Calvet (1983) *J. Lipid Res.* **24**, 821. **107.** Novogrodsky, Tal, Traub & Hurwitz (1966) *J. Biol. Chem.* **241**, 2933. **108.** Goldstone & Adams (1962) *J. Biol. Chem.* **237**, 3476. **109.** Martinez-Carrion & Jenkins (1965) *J. Biol. Chem.* **240**, 3547. **110.** Liu, Peterson, Langston, *et al.* (2005) *Biochemistry* **44**, 2982. **111.** Pathre, Singh, Viswanathan & Sane (1987) *Phytochemistry* **26**, 2913. **112.** Thibert & Schmidt (1977) *Can. J. Biochem.* **55**, 958. **113.** Agarwal (1985) *Indian J. Biochem. Biophys.* **22**, 102. **114.** Vedavathi, Girish & Kumar (2004) *Mol. Cell. Biochem.* **267**, 13. **115.** Vedavathi, Girish & Kumar (2006) *Biochemistry (Moscow)* **71** Suppl. 1, S105. **116.** Nobe, Kawaguchi, Ura, *et al.* (1998) *J. Biol. Chem.* **273**, 29554. **117.** Tanaka, Tokuda, Tachibana, Taniguchi & Oi (1990) *Agric. Biol. Chem.* **54**, 625. **118.** Martins, Mourato & de Varennes (2001) *J. Enzyme Inhib.* **16**, 251. **119.** Vessal & Taher (1995) *Comp. Biochem. Physiol. B* **110**, 431. **120.** Shibata & Iwai (1980) *Biochim. Biophys. Acta* **611**, 280. **121.** Iwai, Shibata & Taguchi (1979) *Agric. Biol. Chem.* **43**, 351. **122.** Forsee, Laine & Elbein (1974) *Arch. Biochem. Biophys.* **161**, 248. **123.** Salerno & H. G. Pontis (1978) *Planta* **142**, 41. **124.** Stern (1961) *The Enzymes*, 2nd ed. **5**, 367. **125.** Das & Shichi (1979) *Exp. Eye Res.* **29**, 109. **126.** Miller, Awasthi & Srivastava (1976) *J. Biol. Chem.* **251**, 2271. **127.** Achar, Savithri, Vaidyanathan & N. A. Rao (1974) *Eur. J. Biochem.* **47**, 15. **128.** Burns, Mendz & Hazell (1997) *Arch. Biochem. Biophys.* **347**, 119. **129.** Masood & Venkitasubramanian (1988) *Biochim. Biophys. Acta* **953**, 106. **130.** Chang & Jones (1974) *Biochemistry* **13**, 638. **131.** Beaudet & Mackenzie (1975) *Biochim. Biophys. Acta* **410**, 252. **132.** Shoun, Arima & Beppu (1983) *J. Biochem.* **93**, 169. **133.** Flohe (1989) in *Coenzymes and Cofactors, Glutathione, Chem. Biochem. Med. Aspects, Pt. A*, p. 643, Wiley, New York. **134.** Chatterjee & Sanwal (1993) *Mol. Cell. Biochem.* **126**, 125.

Malformin A₁

D-Cys-D-Cys-L-Val-D-Leu-L-Ile

This cyclic peptide antibiotic (FW = 529.73 g/mol; CAS 3022-92-2) from the mycelium of *Aspergillus niger* promotes cell elongation in higher plants as well as malformations of stems and petioles. **Target(s):** interleukin-1β binding to human monocytes, IC_{50} = 250 nM (1); protein biosynthesis (2).

1. Herbert, Savi, Lale, *et al.* (1994) *Biochem. Pharmacol.* **48**, 1211. **2.** Bannon, Dawes & Dean (1994) *Thromb. Haemost.* **72**, 482.

Malonate (Malonic Acid)

This dicarboxylic acid (FW$_{free-acid}$ = 104.06 g/mol; CAS 141-82-2; decomposes ~135°C), is the first agent ever identified as a reversible competitive inhibitor (for succinate dehydrogenase). The free acid is very soluble in water (73.5 g/100 g at 20°C) and ethanol (57 g/100 g at 20°C). The pK_a values are 2.58 and 5.17 at 37°C. Malonate is produced in the reactions catalyzed by malonate-semialdehyde dehydrogenase and barbiturase. It is also a substrate of the reactions catalyzed by malonate CoA-transferase and malonate decarboxylase. **Target(s):** 2-(acetamidomethylene)succinate hydrolase (1); acetoacetate decarboxylase (2); acid phosphatase (3,4); allophanate hydrolase (5); D-amino-acid oxidase (6); aspartate carbamoyltransferase (7,8); aspartate 4-decarboxylase (9); aspartate kinase (121); D-aspartate oxidase (10,11); γ-butyrobetaine dioxygenase, *or* γ-butyrobetaine hydroxylase, K_i = 7.0 mM (12); carbamoyl-phosphate synthetase (13); carboxy-*cis*,*cis*-muconate cyclase (14); choline acetyltransferase (15); creatine kinase (16-18); cyanate hydratase, *or* cyanase (19,20); diamine oxidase (21); diisopropylfluorophosphatase (22); dopamine β-monooxygenase (123); fructose-1,6 bisphosphatase (23); 1,3-α-L-fucosidase (118); fumarase, *or* fumarate hydratase (24,25); fumarylacetoacetase (26); glucose-6-phosphate isomerase, weakly inhibited (27); β-glucuronidase (28); glutamate decarboxylase, weakly inhibited, K_i = 36 mM (29); γ-glutamyl transpeptidase (30); glycogen synthase (31); D-2-hydroxyacid dehydrogenase (3,32); (*S*)-2-hydroxy-acid oxidase (33,34); 3-hydroxybutyrate dehydrogenase (35); 2-(hydroxymethyl)-3-(acetamido-methylene)succinate hydrolase (1); [3-hydroxy-3-methylglutaryl-CoA reductase (NADPH)]-phosphatase (119); IMP dehydrogenase (36); isocitrate lyase (37-52); kynurenine:2-oxoglutarate aminotransferase (53); laccase (127); D-lactate dehydrogenase (cytochrome) (54); L-lactate dehydrogenase (55-59); lactate 2 monooxygenase (60); lipase (61); malate dehydrogenase (59,62-65); malic enzyme, NAD⁺-dependent (decarboxylating), *or* malate dehydrogenase, NAD⁺ (66); malic enzyme (oxaloacetate-decarboxylating activity) (67-70); malonyl-*S*-ACP:biotin-protein carboxyltransferase (122); *N*-methylglutamate dehydrogenase (71); NADH oxidase (72); nucleotidases, weakly inhibited (120); oxaloacetate decarboxylase (62,70,73-76); oxaloacetate tautomerase (77,78); 3-oxoacid CoA-transferase, *or* 3-ketoacid CoA-transferase, weakly inhibited (79,80); peptidyl-dipeptidase A, *or* angiotensin I-converting enzyme (117); phosphoenolpyruvate carboxykinase (GTP), weakly inhibited (81); phosphoenolpyruvate carboxylase (82-84); phosphonoacetate hydrolase, weakly inhibited (85); procollagen-lysine 5-dioxygenase (125); procollagen-proline 3-dioxygenase (125); procollagen-proline 4-dioxygenase (86,125,126); pyrophosphatase (87,88); pyruvate carboxylase (89,90); pyruvate kinase (91); ribose-5-phosphate isomerase (92); 3-serine dehydrogenase (93); succinate dehydrogenase (3,56,57,94-110); succinate:3-hydroxy-3-methylglutarate CoA-transferase (Note: Malonate is an alternative product/substrate (111); succinyl-CoA synthetase (ADP-forming), *or* succinate:CoA ligase (ADP) (112); D-threonine dehydrogenase (113); L-threonine dehydrogenase (114); trimethyllysine dioxygenase (124); triose-phosphate isomerase (115); urease (116). **Malonate Neurotoxicity:** Malonate causes dose-dependent neurotoxicity, both *in vivo* and *in vitro*, via inhibition of succinate dehydrogenase and depletion of striatal ATP stores (128-131) resulting in neuronal depolarization and secondary excitotoxicity (132,133). Malonate also induces mitochondrial potential collapse, swelling, release of cytochrome *c* (Cyt *c*), depletion of glutathione (GSH), and reduction in nicotinamide adenine dinucleotide coenzyme (NAD(P)H) within brain-isolated mitochondria (134). Mitochondrial potential collapse occurs almost immediate after malonate addition, and mitochondrial swelling is evident after 15 min of drug addition (134). The latter effect is blocked by cyclosporin A (CsA), Ruthenium Red (RR), magnesium, catalase, GSH and vitamin E. When added to SH-SY5Y cell cultures, malonate produces a marked loss of cell viability together with the release of Cytochrome *c* and depletion of GSH and NAD(P)H concentrations. All these effects were not apparent in SH-SY5Y cells overexpressing Bcl-xL (134). Taken together, these data suggest that malonate causes a rapid collapse of mitochondrial potential and production of reactive oxygen species, overwhelming mitochondrial antioxidant capacity and leads to mitochondrial swelling (134). **1**. Huynh & Snell (1985) *J. Biol. Chem.* **260**, 2379. **2**. Davies (1943) *Biochem. J.* **37**, 230. **3**. Webb (1966) *Enzyme and Metabolic Inhibitors*, vol. 2, Academic Press, New York. **4**. Nigam, Davidson & Fishman (1959) *J. Biol. Chem.* **234**, 1550. **5**. Maitz, Haas & Castric (1982) *Biochim. Biophys. Acta* **714**, 486. **6**. Frisell, Lowe & Hellerman (1956) *J. Biol. Chem.* **223**, 75. **7**. Jacobson & Stark (1973) *The Enzymes*, 3rd ed., **9**, 225. **8**. Gouaux & Lipscomb (1990) *Biochemistry* **29**, 389.26. **9**. Shibatani, Kakimoto, Kato, Nishimura & Chibata (1974) *J. Ferment. Technol.* **52**, 886. **10**. Dixon & Kenworthy (1967) *Biochim. Biophys. Acta* **146**, 54. **11**. Rinaldi (1971) *Enzymologia* **40**, 314. **12**. Ng, Hanauske-Abel & Englard (1991) *J. Biol. Chem.* **266**, 15. **13**. Jones (1962) *Meth. Enzymol.* **5**, 903. **14**. Thatcher & Cain (1975) *Eur. J. Biochem.* **56**, 193. **15**. Korey, de Braganza & Nachmansohn (1951) *J. Biol. Chem.* **189**, 705. **16**. Kuby, Noda & Lardy (1954) *J. Biol. Chem.* **210**, 65. **17**. Kuby & Noltmann (1962) *The Enzymes*, 2nd ed., **6**, 515. **18**. Watts (1973) *The Enzymes*, 3rd ed., **8**, 383. **19**. Anderson, Johnson, Endrizzi, Little & Korte (1987) *Biochemistry* **26**, 3938. **20**. Anderson & Little (1986) *Biochemistry* **25**, 1621. **21**. Zeller (1938) *Helv. Chim. Acta* **21**, 1645. **22**. Mounter & Chanutin (1953) *J. Biol. Chem.* **204**, 837. **23**. Aubel, Rosenberg & Szulmajster (1950) *Biochim. Biophys. Acta* **5**, 228. **24**. Hill & Bradshaw (1969) *Meth. Enzymol.* **13**, 91. **25**. Massey (1953) *Biochem. J.* **55**, 172. **26**. Connors & Stotz (1949) *J. Biol. Chem.* **178**, 881. **27**. Sangwan & Singh (1989) *J. Biosci.* **14**, 47. **28**. Mills, Paul & Smith (1953) *Biochem. J.* **53**, 232. **29**. Fonda (1972) *Biochemistry* **11**, 1304. **30**. Prusiner & Prusiner (1978) *J. Neurochem.* **30**, 1261. **31**. Rothman & Cabib (1967) *Biochemistry* **6**, 2098. **32**. Cremona (1964) *J. Biol. Chem.* **239**, 1457. **33**. Zelitch (1955) *Meth. Enzymol.* **1**, 528. **34**. Jorns (1975) *Meth. Enzymol.* **41**, 337. **35**. Bergmeyer, Gawehn, Klotzsch, Krebs & Williamson (1967) *Biochem. J.* **102**, 423. **36**. Hall, Barnes, Ward, Wheaton & Izydore (2001) *J. Pharm. Pharmacol.* **53**, 749. **37**. Daron & Gunsalus (1963) *Meth. Enzymol.* **6**, 622. **38**. McFadden (1969) *Meth. Enzymol.* **13**, 163. **39**. Spector (1972) *The Enzymes*, 3rd ed., **7**, 357. **40**. Daron, Rutter & Gunsalus (1966) *Biochemistry* **5**, 895. **41**. Rao & McFadden (1965) *Arch. Biochem. Biophys.* **112**, 294. **42**. Nakamura, Amano, Nakadate & Kagami (1989) *J. Ferment. Bioeng.* **67**, 153. **43**. McFadden & Howes (1963) *J. Biol. Chem.* **238**, 1737. **44**. Hoyt, Robertson, Berlyn & Reeves (1988) *Biochim. Biophys. Acta* **966**, 30. **45**. Hoyt, Johnson & Reeves (1991) *J. Bacteriol.* **173**, 6844. **46**. Giachetti, Pinzauti, Bonaccorsi, Vincenzini & Vanni (1987) *Phytochemistry* **26**, 2439. **47**. DeLucas, Amor, Diaz, Turner & Laborda (1997) *Mycol. Res.* **101**, 410. **48**. Khan, Sallemuddin, Siddiqi & McFadden (1977) *Arch. Biochem. Biophys.* **183**, 13. **49**. Vicenzini, Nerozzi, Vincieri & Vanni (1980) *Phytochemistry* **19**, 769. **50**. Chell, Sundaram & Wilkinson (1978) *Biochem. J.* **173**, 165. **51**. Vanni, Vicenzini, Nerozzi & Sinha (1979) *Can. J. Biochem.* **57**, 1131. **52**. Reiss & Rothstein (1974) *Biochemistry* **13**, 1796. **53**. Tobes (1987) *Meth. Enzymol.* **142**, 217. **54**. Nygaard (1963) *The Enzymes*, 2nd ed., **7**, 557. **55**. Torres-da Matta, Nery da Matta & Hasson-Voloch (1976) *An. Acad. Bras. Cienc.* **48**, 145. **56**. Quastel & Woolbridge (1928) *Biochem. J.* **22**, 689. **57**. Massart (1950) *The Enzymes*, 1st ed., 1 (part 1), 307. **58**. Schwert & Winer (1963) *The Enzymes*, 2nd ed., **7**, 142. **59**. Das (1937) *Biochem. J.* **31**, 1124. **60**. Ghisla & Massey (1977) *J. Biol. Chem.* **252**, 6729. **61**. Gajdos (1939) *Compt. Rend. Soc. Biol.* **130**, 1566. **62**. Lwoff, Audereau & Cailleau (1947) *Compt. Rend.* **224**, 303. **63**. Yamamura, Kusunose, Nagai & Kusunose (1954) *J. Biol. Chem.* **41**, 513. **64**. Scholefield (1955) *Biochem. J.* **59**, 177. **65**. Green (1936) *Biochem. J.* **30**, 2095. **66**. Korkes, del Campillo & Ochoa (1950) *J. Biol. Chem.* **187**, 891. **67**. Ochoa (1955) *Meth. Enzymol.* **1**, 739. **68**. Kun (1963) *The Enzymes*, 2nd ed., **7**, 149. **69**. van Heyningen & Pirie (1953) *Biochem. J.* **53**, 436. **70**. Veiga Salles & Ochoa (1950) *J. Biol. Chem.* **187**, 849. **71**. Hersh, Stark, Worthen & Fiero (1972) *Arch. Biochem. Biophys.* **150**, 219. **72**. Eichel (1959) *Biochem. J.* **71**, 106. **73**. Evans, Vennesland & Slotin (1943) *J. Biol. Chem.* **147**, 771. **74**. Liebecq & Peters (1949) *Biochim. Biophys. Acta* **3**, 215. **75**. Sender, Martin, Peiru & Magni (2004) *FEBS Lett.* **570**, 217. **76**. Morinaga & Shirakawa (1971) *Agric. Biol. Chem.* **35**, 1166. **77**. Annett & Kosicki (1969) *J. Biol. Chem.* **244**, 2059. **78**. Belikova, Burov & Vinogradov (1988) *Biochim. Biophys. Acta* **936**, 10. **79**. Fenselau & Wallis (1974) *Biochemistry* **13**, 3884. **80**. Hersh & Jencks (1967) *J. Biol. Chem.* **242**, 3468. **81**. Guidinger & Nowak (1990) *Arch. Biochem. Biophys.* **178**, 131. **82**. Raghavendra & Das (1975) *Biochem. Biophys. Res. Commun.* **66**, 160. **83**. Owttrim & Colman (1986) *J. Bacteriol.* **168**, 207. **84**. Pays, Jones, Wilkins, Fewson & Malcolm (1980) *Biochim. Biophys. Acta* **614**, 151. **85**. McMullan & Quinn (1994) *J. Bacteriol.* **176**, 320. **86**. Hutton, Tappel & Udenfriend (1967) *Arch. Biochem. Biophys.* **118**, 231. **87**. Naganna & Menon (1948) *J. Biol. Chem.* **174**, 501. **88**. Naganna (1950) *J. Biol. Chem.* **183**, 693. **89**. Scrutton, Olmsted & Utter (1969) *Meth. Enzymol.* **13**, 235.

90. Scrutton & Mildvan (1968) *Biochemistry* **7**, 1490. **91**. Fawaz & Fawaz (1962) *Biochem. J.* **83**, 438. **92**. Woodruff III & Wolfenden (1979) *J. Biol. Chem.* **254**, 5866. **93**. Chowdhury, Higuchi, Nagata & Misono (1997) *Biosci. Biotechnol. Biochem.* **61**, 152. **94**. Bonner (1955) *Meth. Enzymol.* **1**, 722. **95**. Bernath & Singer (1962) *Meth. Enzymol.* **5**, 597. **96**. Slater (1967) *Meth. Enzymol.* **10**, 48. **97**. Veeger, DerVartanian & Zeylemaker (1969) *Meth. Enzymol.* **13**, 81. **98**. Hatefi (1978) *Meth. Enzymol.* **53**, 27. **99**. Schäfer, Moll & Schmidt (2001) *Meth. Enzymol.* **331**, 369. **100**. Potter & DuBois (1943) *J. Gen. Physiol.* **26**, 391. **101**. Schlenk (1951) *The Enzymes*, 1st ed., **2** (part 1), 316. **102**. Singer & Kearney (1963) *The Enzymes*, 2nd ed., **7**, 383. **103**. Hatefi & Stiggall (1976) *The Enzymes*, 3rd ed., **13**, 175. **104**. Thorn (1953) *Biochem. J.* **54**, 540. **105**. Quastel & Wheatley (1931) *Biochem. J.* **25**, 117. **106**. Potter & Elvehjem (1937) *J. Biol. Chem.* **117**, 341. **107**. Kun & Abood (1949) *J. Biol. Chem.* **180**, 813. **108**. Chakravorty & Das (1966) *Enzymologia* **31**, 51. **109**. Das (1937) *Biochem. J.* **31**, 1124. **110**. Hopkins, Morgan & Litwak-Mann (1938) *Biochem. J.* **32**, 1829. **111**. Deana, Rigoni, Deana & Galzigna (1981) *Biochim. Biophys. Acta* **662**, 119. **112**. Palmer & Wedding (1966) *Biochim. Biophys. Acta* **113**, 167. **113**. Misono, Kato, Packdibamrung, Nagata & Nagasaki (1993) *Appl. Environ. Microbiol.* **59**, 2963. **114**. Guerranti, Pagani, Neri, *et al.* (2001) *Biochim. Biophys. Acta* **1568**, 45. **115**. Noltmann (1972) *The Enzymes*, 3rd ed., **6**, 271. **116**. Kistiakowsky, Mangelsdorf, Rosenberg & Shaw (1952) *J. Amer. Chem. Soc.* **74**, 5015. **117**. Oshima & Nagasawa (1977) *J. Biochem.* **81**, 57. **118**. Imbert, Glasgow & Pizzo (1982) *J. Biol. Chem.* **257**, 8205. **119**. Hegardt, Gil & Calvet (1983) *J. Lipid Res.* **24**, 821. **120**. Tjernshaugen (1978) *Biochem. J.* **169**, 597. **121**. Keng & Viola (1996) *Arch. Biochem. Biophys.* **335**, 73. **122**. Hilbi & Dimroth (1994) *Arch. Microbiol.* **162**, 48. **123**. Colombo, Papadopoulos, Ash & Villafranca (1987) *Arch. Biochem. Biophys.* **252**, 71. **124**. Henderson, Nelson & Henderson (1982) *Fed. Proc.* **41**, 2843. **125**. Majamaa, Turpeenniemi-Hujanen, Latipää, *et al.* (1985) *Biochem. J.* **229**, 127. **126**. Majamaa, Hanauske-Abel, Günzler & Kivirikko (1984) *Eur. J. Biochem.* **138**, 239. **127**. Lorenzo, Moldes, Rodríguez Couto & Sanromán (2005) *Chemosphere* **60**, 1124. **128**. Beal, Brouillet, Jenkins, *et al.* (1993) *J. Neurochem.* **61**, 1147. **129**. Greene & Greenamyre (1995) *J. Neurochem.* **64**, 430. **130**. Stokes, Bernard, Nicklas & Zeevalk (2001) *J. Neurosci. Res.* **64**, 43. **131**. Van Westerlaak, Joosten, Gribnau, *et al.* (2001) *Brain Res.* **922**, 243. **132**. Henshaw, Jenkins, Schulz, *et al.* (1994) *Brain Res.* **647**, 161. **133**. Greene & Greenamyre (1996). *J. Neurochem.* **66**, 637. **134**. Fernandez-Gomez, Galindo, Gómez-Lázaro, *et al.* (2005) *Br. J. Pharmacol.* **144**, 528.

Malonate Semialdehyde (Malonic Acid Semialdehyde)

This acid aldehyde (FW$_{\text{free-acid}}$ = 88.06 g/mol), an intermediate in the pyrimidine degradation, is also a product of the reactions catalyzed by β-alanine aminotransferase (*or* β-alanine:pyruvate transaminase), 3-hydroxypropionate dehydrogenase, 3-*aci*-nitropropanoate oxidase, acetylenecarboxylate hydratase, and *cis*- and *trans*-3-chloroacrylate dehalogenase. Malonate semialdehyde is also a substrate for malonate-semialdehyde dehydrogenase, malonate-semialdehyde dehydrogenase (acetylating), and malonate-semialdehyde dehydratase. **Target(s):** phosphonoacetaldehyde hydrolase (1); succinate-semialdehyde dehydrogenase (NAD(P)$^+$) (2). **1**. Olsen, Hepburn, Lee, *et al.* (1992) *Arch. Biochem. Biophys.* **296**, 144. **2**. Jakoby (1962) *Meth. Enzymol.* **5**, 765.

Malondialdehyde

main species in H$_2$O   main species in organic solvents

This mutagenic dialdehyde (FW = 72.06 g/mol; CAS 5578-67-6) is a quantifiable lipid peroxidation end-product of cellular oxidative stress. Lipid peroxidation is a process in which oxidants, such as free radicals, attack unsaturated lipids, especially polyunsaturated fatty acids. The main primary products of lipid peroxidation are lipid hydroperoxides (LOOH). Among the different aldehydes that form as secondary products of lipid peroxidation are malondialdehyde (MDA), propanal, hexanal, and 4-hydroxynonenal (4-HNE). MDA has been widely used for many years as a convenient biomarker for lipid peroxidation of omega-3 and omega-6 fatty acids because of its facile reaction with 2-thiobarbituric acid (TBA). At low pH and elevated temperature, MDA reacts with TBA, generating the red-colored 1MDA:2TBA ternary adduct (shown below), the fluorescence of which may be used to measure malonaldehyde production.

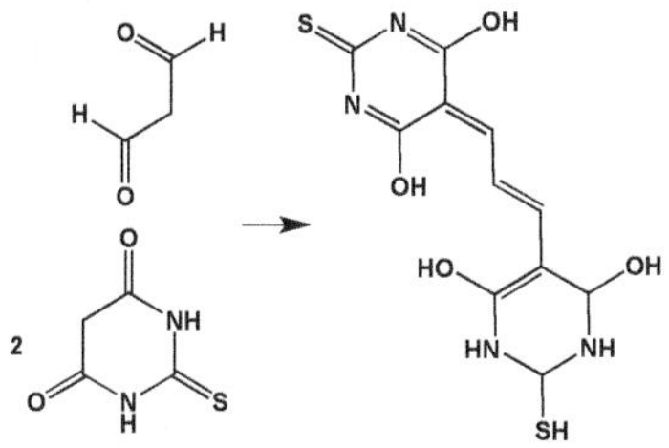

That said, thiobarbituric acid is notoriously nonspecific, leading to controversy concerning its use in MDA quantification for *in vivo* samples. Better methods for determining free and total MDA include gas chromatography-mass spectrometry (GC-MS), liquid chromatography-mass spectrometry (LC-MS), and other derivitization techniques. Malondialdehyde also reacts with (a) guanidinium groups of arginyl residues in peptides and proteins under strongly acidic conditions and (b) amino groups of deoxyadenosine and guanosine within DNA. **Target(s):** aldehyde dehydrogenase (1); carnitine *O*-acetyltransferase (2); cytochrome P450 (3); glucose-6-phosphate dehydrogenase (4); pancreatic ribonuclease, *or* ribonuclease A (5,6); xanthine oxidase (7); xanthine oxidoreductase (7). Given the reactivity of this metabolite, this list is likely to under-represent the range of proteins and enzymes that react with malondialdehyde. **1.** Hjelle, Grubbs & Petersen (1982) *Toxicol. Lett.* **14**, 35. **2.** Liu, Killilea & Ames (2002) *Proc. Natl. Acad. Sci. U.S.A.* **99**, 1876. **3.** Buko, Artsukevich & Ignatenko (1999) *Exp. Toxicol. Pathol.* **51**, 294. **4.** Ganea & Harding (2000) *Biochim. Biophys. Acta* **1500**, 49. **5.** Chio & Tappel (1969) *Biochemistry* **8**, 2827. **6.** King (1966) *Biochemistry* **5**, 3454. **7.** Cighetti, Bortone, Sala & Allevi (2001) *Arch. Biochem. Biophys.* **389**, 195.

Malonoben

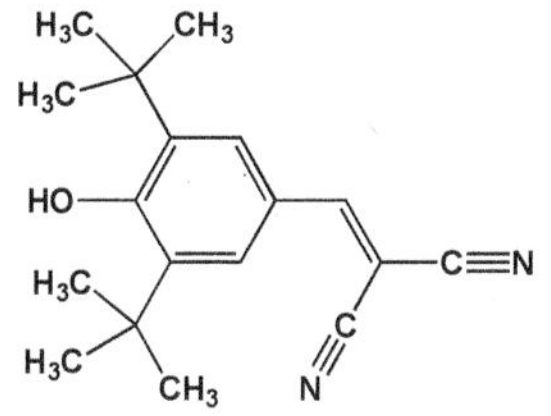

This protonophore (FW = 282.38 g/mol; CAS 10537-47-0; Solubility: 75 mM DMSO), also named SF 6847 and 3,5-di-*tert*-butyl-4-hydroxy-benzylidenemalononitrile, is the most powerful known uncoupler of oxidative phosphorylation. See reference (2) for pH-dependence of uncoupling activity, binding to mitochondria, as well as spectral properties in the presence of different types of liposomes, biopolymers and mitochondria, and model membranes. **1.** Muraoka & Terada (1972) *Biochim. Biophys. Acta* **275**, 271. **2.** Terada (1975) *Biochim. Biophys. Acta* **387**, 519. **3.** Terada (1981) *Biochim. Biophys. Acta* **639**, 225.

Malonyl-CoA

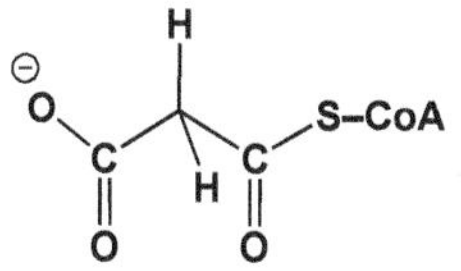

This hygroscopic thioester (FW = 753.59 g/mol; CAS 524-14-1), itself a key intermediate in fatty-acid biosynthesis, is a product of the reactions catalyzed by acetyl-CoA carboxylase and malonate CoA-transferase. nhibits the rate-limiting step in β-oxidation of fatty acids. It also regulating the enzyme carnitine acyltransferase, inhibiting fatty acids from combining with carnitine and thereby preventing their entry into mitochondria, Malonyl-CoA is also a substrate of the fatty acid chain-extending reactions catalyzed by fatty-acid synthase and fatty acyl-CoA synthase (*i.e.*, [acyl-carrier protein] *S*-malonyltransferase). Malonyl CoA:Acyl Carrier Protein

transacylase (MCAT) transfers malonate from malonyl-CoA to the terminal thiol of holo-acyl carrier protein (ACP) Malonyl-CoA is also a substrate for naringenin chalcone synthase, erythronolide synthase, trihydroxystilbene synthase, D-tryptophan *N*-malonyltransferase, anthranilate *N*-malonyltransferase, 3,4-dichloroaniline *N*-malonyl-transferase, isoflavone-7-*O*-β-glucoside 6''-*O*-malonyl-transferase, flavonol-3-*O*-β-glucoside *O*-malonyltransferase, icosanoyl-CoA synthase, pinosylvin synthase, benzophenone synthase, phloroisovalerophenone synthase, acridone synthase, lovastatin nonaketide synthase, 6-methylsalicylic acid synthase, and malonyl-CoA decarboxylase. **Target(s):** acetyl-CoA carboxylase (1,2); arylamine *N*-acetyltransferase (53); ATP:citrate lyase, *or* ATP:citrate synthase (3); [branched-chain α-keto-acid dehydrogenase] kinase, *or* [3-methyl-2 oxobutanoate dehydrogenase (acetyl-transferring)] kinase (4,5); carnitine *O*-acyltransferase (6,7,52); carnitine *O*-octanoyltransferase (8-15); carnitine *O*-palmitoyltransferase (7,13,16,31-51); enoyl-[acyl carrier-protein] reductase (17); 3-hydroxyisobutyryl-CoA hydrolase (18,19); methylmalonyl-CoA decarboxylase (20); methylmalonyl-CoA epimerase (21); (*S*)-methylmalonyl-CoA hydrolase (22); pantothenate kinase (23-29); propionyl-CoA carboxylase (1); pyruvate carboxylase (30). **1.** Hügler, Krieger, Jahn & Fuchs (2003) *Eur. J. Biochem.* **270**, 736. **2.** Thampy (1989) *J. Biol. Chem.* **264**, 17631. **3.** Adams, Dack, Dickinson & Ratledge (2002) *Biochim. Biophys. Acta* **1597**, 36. **4.** Paxton (1988) *Meth. Enzymol.* **166**, 313. **5.** Paxton & Harris (1984) *Arch. Biochem. Biophys.* **231**, 48. **6.** McGarry, Mannaerts & Foster (1977) *J. Clin. Invest.* **60**, 265. **7.** Bieber & Farrell (1983) *The Enzymes*, 3rd ed., **16**, 627. **8.** Chung & Bieber (1993) *J. Biol. Chem.* **268**, 4519. **9.** Hegardt, Bach, Asins, *et al.* (2001) *Biochem. Soc. Trans.* **29**, 316. **10.** Lilly, Bugaisky, Umeda & Bieber (1990) *Arch. Biochem. Biophys.* **280**, 167. **11.** Saggerson & Carpenter (1982) *FEBS Lett.* **137**, 124. **12.** Bird, Munday, Saggerson & Clark (1985) *Biochem. J.* **226**, 323. **13.** Morillas, Gómez-Puertas, Roca, *et al.* (2001) *J. Biol. Chem.* **276**, 45001. **14.** Nic A'Bháird & Ramsay (1992) *Biochem. J.* **286**, 637. **15.** Morillas, Gómez-Puertas, Bentebibel, *et al.* (2003) *J. Biol. Chem.* **278**, 9058. **16.** McGarry, Leatherman & Foster (1978) *J. Biol. Chem.* **253**, 4128. **17.** Wakil & Stoops (1983) *The Enzymes*, 3rd ed., **16**, 3. **18.** Shimomura, Murakami, Nakai, *et al.* (2000) *Meth. Enzymol.* **324**, 229. **19.** Shimomura, Murakami, Fujitsuka, *et al.* (1994) *J. Biol. Chem.* **269**, 14248. **20.** Galivan & Allen (1968) *Arch. Biochem. Biophys.* **126**, 838. **21.** Stabler, Marcell & Allen (1985) *Arch. Biochem. Biophys.* **241**, 252. **22.** Kovachy, Copley & Allen (1983) *J. Biol. Chem.* **258**, 11415. **23.** Falk & Guerra (1993) *Arch. Biochem. Biophys.* **301**, 424. **24.** Vallari, Jackowski & Rock (1987) *J. Biol. Chem.* **262**, 2468. **25.** Zhang, Rock & Jackowski (2005) *J. Biol. Chem.* **280**, 32594. **26.** Lehane, Marchetti, Spry, *et al.* (2007) *J. Biol. Chem.* **282**, 25395. **27.** Halvorsen & Skrede (1982) *Eur. J. Biochem.* **124**, 211. **28.** Zhang, Rock & Jackowski (2006) *J. Biol. Chem.* **281**, 107. **29.** Kotzbauer, Truax, Trojanowski & Lee (2005) *J. Neurosci.* **25**, 689. **30.** Scrutton, Olmsted & Utter (1969) *Meth. Enzymol.* **13**, 235. **31.** Ramsay, Derrick, Friend & Tubbs (1987) *Biochem. J.* **244**, 271. **32.** Ramsay, Mancinelli & Arduini (1991) *Biochem. J.* **275**, 685. **33.** Declercq, Falck, Kuwajima, *et al.* (1987) *J. Biol. Chem.* **262**, 9812. **34.** Miyazawa, Ozaka, Osumi & Hashiomoto (1983) *J. Biochem.* **94**, 529. **35.** Kashfi, Mynatt & Cook (1994) *Biochim. Biophys. Acta* **1212**, 245. **36.** Zierz & Engel (1987) *Biochem. J.* **245**, 205. **37.** Bentebibel, Sebastian, Herrero, *et al.* (2006) *Biochemistry* **45**, 4339. **38.** Murthy & Pande (1987) *Biochem. J.* **248**, 727. **39.** Woeltje, Esser, Weis, *et al.* (1990) *J. Biol. Chem.* **265**, 10714. **40.** Weis, Cowan, Bro wn, Foster & McGarry (1994) *J. Biol. Chem.* **269**, 26443. **41.** Kuhajda & Ronnett (2007) *Curr. Opin. Invest. Drugs* **8**, 312. **42.** Colquhoun (2002) *Biochim. Biophys. Acta* **1583**, 74. **43.** Murthy & Pande (1990) *Biochem. J.* **268**, 599. **44.** Hertel, Gellerich, Hein & Zierz (1999) *Adv. Exp. Med. Biol.* **466**, 87. **45.** Brindle, Zammit & Pogson (1985) *Biochem. J.* **232**, 177. **46.** Ramsay (1988) *Biochem. J.* **249**, 239. **47.** Kerner & Bieber (1990) *Biochemistry* **29**, 4326. **48.** Brown, Hill, Esser, *et al.* (1997) *Biochem. J.* **327**, 225. **49.** Brindle, Zammit & Pogson (1985) *Biochem. Soc. Trans.* **13**, 880. **50.** Zhu, Shi, Cregg & Woldegiorgis (1997) *Biochem. Biophys. Res. Commun.* **239**, 498. **51.** Lund (1987) *Biochim. Biophys. Acta* **918**, 67. **52.** Farrell, Fiol, Reddy & Bieber (1984) *J. Biol. Chem.* **259**, 13089. **53.** Kawamura, Graham, Mushtaq, *et al.* (2005) *Biochem. Pharmacol.* **69**, 347.

Maltitol

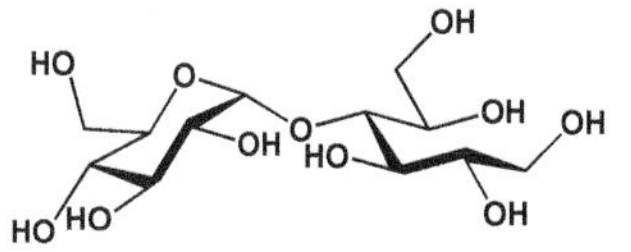

This reduced disaccharide and sweetening agent (FW = 344.31 g/mol; CAS 585-88-6), also known as 4-*O*-α-D-glucopyranosyl-D-sorbitol, inhibits glucoamylase and reportedly activates cyclomaltodextrin glucanotransferase. Maltitol is a high-sweetness, FDA-approved food additive that resists bacterial degradation, ideal for low-calorie hard candy and chewing gum. Many oral pharmaceuticals are likewise sweetened by maltitol. Excessive consumption may have a laxative effect. **Target(s):** α-amylase (1); β-amylase (2); cyclomaltodextrin glucanotransferase (3); glucan 1,4-α glucosidase, *or* glucoamylase (1,4,5); α-1,4-glucan lyase (6); glucoamylase/maltase complex, as a slow alternative substrate (7). 1. De Mot & Verachtert (1987) *Eur. J. Biochem.* 164, 643. 2. Ye, Miyake, Tatsumi, Nishimura & Nitta (2004) *J. Biochem.* 135, 355. 3. Bovetto, Villette, Fontaine, Sicard & Bouquelet (1992) *Biotechnol. Appl. Biochem.* 15, 59. 4. Wursch & Del Vedovo (1981) *Int. J. Vitam. Nutr. Res.* 51, 161. 5. Fogarty & Benson (1983) *Eur. J. Appl. Microbiol. Biotechnol.* 18, 271. 6. Yu, Christensen, Kragh, Bojsen & Marcussen (1997) *Biochim. Biophys. Acta* 339, 311. 7. Gunther, Wehrspaun & Heymann (1996) *Arch. Biochem. Biophys.* 327, 295.

Maltose

α-Maltose

β-Maltose

This reducing disaccharide (FW = 342.30 g/mol; CAS 69-79-4), also known as 4-*O*-α-D-glucopyranosyl-D-glucose, maltobiose, and malt sugar, is very soluble in water (108 g/100 g at 20°C) and is typically supplied as the monohydrate. Maltose is a substrate of the reactions catalyzed by maltose *O*-acetyltransferase, maltose phosphorylase, sucrose α-glucosidase, maltose-transporting ATPase, maltase (α-glucosidase), maltose epimerase, maltose α-D-glucosyltransferase, and exo-(1,4)-α-D-glucan lyase. It is product of the reactions catalyzed by maltose synthase, β-amylase, glucan 1,4-α maltohydrolase, pullulanase, and limit dextrinase. It is an alternative substrate of hexose oxidase and aldose 1-epimerase. **Target(s):** acetyl-CoA:α-glucosaminide *N*-acetyltransferase, *or* heparan-α-glucosaminide *N* acetyltransferase (1); β-*N*-acetylhexosaminidase (28); aldose 1-epimerase, *or* mutarotase (2-4); α-amylase (5,6,43,44); β-amylase (7,8,39-42); amylo-α-1,6-glucosidase/4-α-glucanotransferase (glycogen debranching enzyme (30); concanavalin A (9); cyclomaltodextrin glucanotransferase (55,56); dextransucrase, also as a weak alternative substrate (57,58); dextrin Dextranase (59); exo-(1→4)-α-D-glucan lyase (10); β-fructofuranosidase, *or* invertase (11,31); β-D-fucosidase (29); β-galactosidase (32-36); GDP mannose 6-dehydrogenase (12); glucan 1,3-α-glucosidase, *or* glucosidase II, *or* mannosyl-oligosaccharide glucosidase II (13-16); glucan 1,6-α-isomaltosidase (17); glucokinase (53); β-glucosidase (18,19,37,38); D-glucosyltransferases, *Streptococcus* (20); glycogenin glucosyltransferase, weakly inhibited (54); isoamylase (21); limit dextrinase (22); α-mannosidase (18,23); membrane-oligosaccharide glycerophosphotransferase (51); mycodextranase (24); phloretin hydrolase (25); riboflavin phosphotransferase (52); sucrose α-glucosidase, *or* sucrase-isomaltase (29); sucrose-phosphatase (45-50); thioglucosidase, *or* myrosinase (26,27). 1. Bame & Rome (1987) *Meth. Enzymol.* 138, 607. 2. Bentley (1962) *Meth. Enzymol.* 5, 219. 3. Bailey, Fishman, Kusiak, Mulhern & Pentchev (1975) *Meth. Enzymol.* 41, 471. 4. Fishman, Pentchev & Bailey (1975) *Meth. Enzymol.* 41, 484. 5. De Mot & Verachtert (1987) *Eur. J. Biochem.* 164, 643. 6. Schwimmer (1950) *J. Biol. Chem.* 186, 181. 7. Thoma, Spradlin & Dygert (1971) *The Enzymes*, 3rd ed., 5, 115. 8. Mikami, Nomura & Morita (1983) *J. Biochem.* 94, 107. 9. Goldstein, Hollerman & Smith (1965) *Biochemistry* 4, 876. 10. Yu, Christensen, Kragh, Bojsen & Marcussen (1997) *Biochim. Biophys. Acta* 1339, 311. 11. Ruchti & McLaren (1964) *Enzymologia* 27, 185. 12. Roychoudhury, May, Gill, *et al.* (1989) *J. Biol. Chem.* 264, 9380. 13. Presper & Heath (1983) *The Enzymes*, 3rd ed., 16, 449. 14. Saxena, Shailubhai, Dong-Yu & Vijay (1987) *Biochem. J.* 247, 563. 15. Burns & Touster (1982) *J. Biol. Chem.* 257, 9991. 16. Brada &

Dubach (1984) *Eur. J. Biochem.* 141, 149. 17. Tonozuka, Suzuki, Ikehara, *et al.* (2004) *J. Appl. Glycosci.* 51, 27. 18. Webb (1966) *Enzyme and Metabolic Inhibitors*, vol. 2, Academic Press, New York. 19. Ezaki (1940) *J. Biochem.* 32, 107. 20. McAlister, Doyle & Taylor (1989) *Carbohydr. Res.* 187, 131. 21. Kitagawa, Amemura & Harada (1975) *Agric. Biol. Chem.* 39, 989. 22. Dunn & Manners (1975) *Carbohydr. Res.* 39, 283. 23. Hockenhull, Ashton, Fantes & Whitehead (1954) *Biochem. J.* 57, 93. 24. Tung, Rosenthal & Nordin (1971) *J. Biol. Chem.* 246, 6722. 25. Mackey, Henderson & Gregory (2002) *J. Biol. Chem.* 277, 26858. 26. Tani, Ohtsuru & Hata (1974) *Agric. Biol. Chem.* 38, 1623. 27. Tsuruo & Hata (1968) *Agric. Biol. Chem.* 32, 1420. 28. Keyhani & Roseman (1996) *J. Biol. Chem.* 271, 33425. 29. Chinchetru, Cabezas & Calvo (1983) *Comp. Biochem. Physiol.* 75, 719. 30. Liu, Madsen, Fan, *et al.* (1995) *Biochemistry* 34, 7056. 31. Isla, Vattuone, Gutierrez & Sampietro (1988) *Phytochemistry* 27, 1993. 32. Ohtsuka, Tanoh, Ozawa, *etr al.* (1990) *J. Ferment. Bioeng.* 70, 301. 33. Tanaka, Kagamiishi, Kiuchi & Horiuchi (1975) *J. Biochem.* 77, 241. 34. Choi, Kim, Lee & Lee (1995) *Biotechnol. Appl. Biochem.* 22, 191. 35. Itoh, Suzuki & Adachi (1982) *Agric. Biol. Chem.* 46, 899. 36. Levin & Mahoney (1981) *Antonie Leeuwenhoek* 47, 53. 37. Ferreira & Terra (1983) *Biochem. J.* 213, 43. 38. Seidle, Marten, Shoseyov & Huber (2004) *Protein J.* 23, 11. 39. Oyama, Miyake, Kusunoki & Nitta (2003) *J. Biochem.* 133, 467. 40. Lizotte, Henson & Duke (1990) *Plant Physiol.* 92, 615. 41. Van Damme, Hu, Barre, *et al.* (2001) *Eur. J. Biochem.* 268, 6263. 42. Doehlert, Duke & Anderson (1982) *Plant Physiol.* 69, 1096. 43. Bhella & Altosaar (1985) *Can. J. Microbiol.* 31, 149. 44. D'Amico, Sohier & Feller (2006) *J. Mol. Biol.* 358, 1296. 45. Hawker (1967) *Biochem. J.* 102, 401. 46. Echeverria & Salerno (1994) *Plant Sci.* 96, 15. 47. Hawker & Smith (1984) *Phytochemistry* 23, 245. 48. Echeverria & Salerno (1993) *Physiol. Plant.* 88, 434. 49. Whitaker (1984) *Phytochemistry* 23, 2429. 50. Hawker (1971) *Phytochemistry* 10, 2313. 51. Goldberg, Rumley & Kennedy (1981) *Proc. Natl. Acad. Sci. U.S.A.* 78, 5513. 52. Katagiri, Yamada & Imai (1959) *J. Biochem.* 46, 1119. 53. Goward, Hartwell, Atkinson & Scawen (1986) *Biochem. J.* 237, 415. 54. Meezan, Manzella & Roden (1995) *Trends Glycosci. Glycotechnol.* 7, 303. 55. Bovetto, Villette, Fontaine, Sicard & Bouquelet (1992) *Biotechnol. Appl. Biochem.* 15, 59. 56. Martins & Hatti-Kaul (2003) *Enzyme Microb. Technol.* 33, 819. 57. Kobayashi, Yokoyama & Matsuda (1986) *Agric. Biol. Chem.* 50, 2585. 58. Kitaoka & Robyt (1999) *Carbohydr. Res.* 320, 183. 59. Suzuki, Unno & Okada (2000) *J. Appl. Glycosci.* 47, 27.

Mambalgin 1

LKCYQHGKVVTCHRDMKFCYHNTGMPFRNLKLILQGCSSSCSETENNKCCSTDRCNK

This analgesic (FW = 6554.51 g/mol; CAS 1609937-15-6; Disulfide bridges: Cys3-Cys19, Cys12-Cys37, Cys41-Cys49, Cys50-Cys55; Soluble to 1 mg/mL H$_2$O) is a snake venom toxin from the Black Mamba (*Dendroaspis polylepis*), a deadly African snake from the *Elapidae* family. Mamalgin-1 selectively targets Acid-Sensing Ion Channel-1a, a neuronal voltage-insensitive cationic channel that activated by extracellular protons. Mambalgin-1 binds to closed (inactive) channel, with IC$_{50}$ values of 192 and 72 nM for human ASIC1a and ASIC1a/1b dimer, respectively. Mambalgin 1 is selective for ASIC1a *versus* ASIC2a, ASIC3, TRPV1, P2X$_2$, 5-HT$_3$, Na$_v$1.8, Ca$_v$3.2 and K$_v$1.2 channels. It increases latency of tail and paw withdrawal in mouse tail-flick and paw-flick tests. 1. Diochot, Baron, Salinas, *et al.* (2012) *Nature* 490, 552.

Mandelate (Mandelic Acid)

L-(S)-(+)-Mandelate   D-(S)-(−)-Mandelate

These naturally occurring stereoisomers (FW$_{free-acid}$ = 152.15 g/mol; CAS 611-71-2), also known as α-hydroxybenzeneacetic acid and 2-hydroxy-2-phenylacetic acid, exists naturally. Mandelic acid is slightly photosensitive and has a pKa value of 3.37 at 25°C. The free acid is soluble in water (1 g/6.3 mL). Mandelate is a substrate of mandelate 4-monooxygenase and mandelate racemase. The (*R*)-enantiomer is also produced by mandelamide amidase. **Target(s):** L-amino-acid oxidase (1-3); benzoylformate decarboxylase, inhibited by (*R*)- and (*S*)-isomers (4,5); butyryl-CoA synthetase, *or* butyrate:CoA ligase, *or* medium-chain-fatty-acyl-CoA synthetase (6); carboxypeptidase A (7,8); L-lactate dehydrogenase (9,10); L-lactate dehydrogenase (cytochrome) (11-13); D-mandelate dehydrogenase, inhibited by L-mandelate (14); mandelonitrile lyase (15); prunasin β-glucosidase, weakly inhibited (16); pyrophosphatase (17); succinate dehydrogenase, enantiomer not specified (18,19). **1**. Ratner (1955) *Meth. Enzymol.* **2**, 204. **2**. Zeller & Maritz (1945) *Helv. Chim. Acta* **28**, 365. **3**. Krebs (1951) *The Enzymes*, 1st ed., **2** (part 1), 499. **4**. Polovnikova, McLeish, Sergienko, *et al.* (2003) *Biochemistry* **42**, 1820. **5**. Weiss, Garcia, Kenyon, Cleland & Cook (1988) *Biochemistry* **27**, 2197. **6**. Kasuya, Yamaoka, Igarashi & Fukui (1998) *Biochem. Pharmacol.* **55**, 1769. **7**. Pétra (1970) *Meth. Enzymol.* **19**, 460. **8**. Hartsuck & Lipscomb (1971) *The Enzymes*, 3rd ed., **3**, 1. **9**. Schwert & Winer (1963) *The Enzymes*, 2nd ed., **7**, 142. **10**. Lehmann (1938) *Skand. Arch. Physiol.* **80**, 237. **11**. Nygaard (1963) *The Enzymes*, 2nd ed., **7**, 557. **12**. Gondry, Dubois, Terrier & Lederer (2001) *Eur. J. Biochem.* **268**, 4918. **13**. Dikstein (1959) *Biochim. Biophys. Acta* **36**. 397. **14**. Allison, O'Donnell & Fewson (1985) *Biochem. J.* **231**, 407. **15**. Yemm & Poulton (1986) *Arch. Biochem. Biophys.* **247**, 440. **16**. Kuroki & Poulton (1987) *Arch. Biochem. Biophys.* **255**, 19. **17**. Naganna (1950) *J. Biol. Chem.* **183**, 693. **18**. Quastel & Woolbridge (1928) *Biochem. J.* **22**, 689. **19**. Massart (1950) *The Enzymes*, 1st ed., **1** (Part 1), 307.

Mandelonitrile

This aromatic cyanohydrin (FW = 133.15 g/mol; CAS 532-28-5; almost insoluble in water), also known as α-hydroxybenzeneacetonitrile, is a product of the reactions catalyzed by prunasin β-glucosidase and vicianin β-glucosidase as well as a substrate of mandelonitrile lyase, hydroxynitrilase, and cyanohydrin β-glucosyltransferase. The racemic mixture is an oily liquid (M.P. = −10°C), while the optically pure isomers melt at 28-30°C. Both enantiomers, their racemate, and their β-glycosides are naturally occurring (*e.g.*, mandelonitrile is a component of amygdalin). **See also Prunasin** **Target(s):** cyanohydrin β-glucosyltransferase (1); α,α-trehalase (2,3). **1**. Jones, Moller & Hoj (1999) *J. Biol. Chem.* **274**, 35483. **2**. Silva, Terra & Ferreira (2006) *Comp. Biochem. Physiol. B Biochem. Mol. Biol.* **143**, 367. **3**. Silva, Terra & Ferreira (2004) *Insect Biochem. Mol. Biol.* **34**, 1089.

Manganese Ion

Hexaaquo-Mn(II) Cation

This divalent cation (Symbol = Mn^{2+}) is the principal ionic form of the Group 7 (or Group VIIA) metal (Atomic Number = 25; Atomic Weight = 54.94), an element, located directly above technetium in the Periodic Table. The ground state electronic configuration of the neutral atom is $1s^2 2s^2 2p^6 3s^2 3p^6 4s^2 3d^6$). The ionic radii for Mn^{2+}, Mn^{3+}, Mn^{4+}, and Mn^{7+} are 0.80, 0.66, 0.60, and 0.46 Å, respectively. Mn^{2+} is an essential trace element for living organisms and is an important agent in photosynthesis. In mammals, Mn^{2+} is transported by DCT1, a divalent cation transporter with unusually broad substrate range that also includes Fe^{2+}, Zn^{2+}, Co^{2+}, Cd^{2+}, Cu^{2+}, Ni^{2+} and Pb^{2+}. DCT1-mediated active transport is proton-coupled and depends on the cell membrane potential. Mn^{2+} substitutes for Mg^{2+} in numerous nucleotide-requiring enzyme-catalyzed reactions. It can also replace Zn^{2+} in certain zinc-dependent enzymes. The paramagnetic properties of Mn^{2+} have exploited it as an atomic-scale, distance-surveying probe of enzyme active sites. **1**. Mildvan (1970) Metals in enzyme catalysis, in *The Enzymes*, 3rd ed., vol. 2, p. 445. **2**. Scrutton, Reed & Mildvan (1973) *Adv. Exp. Med. Biol.* **40**, 79. **3**. Dahlquist & Purich (1975) *Biochemistry* **14**, 1980. **4**. Purich (2010) *Enzyme Kinetics: Catalysis & Control*, Elsevier, New York.

α-Mangostin

This xanthone (FW = 410.47 g/mol; CAS 6147-11-1), also known as 1,3,6-trihydroxy-7-methoxy-2,8-bis(3-methyl-2-butenyl)-9*H* xanthen-9-one and NSC30552, is isolated from various parts, including the fruit, of the mangosteen tree (*Garcinia mangostana*). It is a yellow solid that is practically insoluble in water. **Target(s):** Ca^{2+}-transporting ATPase (1); cyclic AMP-dependent protein kinase (2); histamine H_1 receptor (3); HIV-1 protease (4); 12-lipoxygenase, *or* arachidonate 12-lipoxygenase (5); 15-lipoxygenase, *or* arachidonate 15-lipoxygenase (5); myosin light-chain kinase (2); sphingomyelin phosphodiesterase, *or* acid sphingomyelinase (6,7). **1**. Furukawa, Shibusawa, Chairungsrilerd, *et al.* (1996) *Jpn. J. Pharmacol.* **71**, 337. **2**. Jinsart, Ternai, Buddhasukh & Polya (1992) *Phytochemistry* **31**, 3711. **3**. Chairungsrilerd, Furukawa, Ohta, Nozoe & Ohizumi (1996) *Eur. J. Pharmacol.* **314**, 351. **4**. Chen, Wan & Loh (1996) *Planta Med.* **62**, 381. **5**. Deschamps, Gautschi, Whitman, *et al.* (2007) *Bioorg. Med. Chem.* **15**, 6900. **6**. Okudaira, Ikeda, Kondo, *et al.* (2000) *J. Enzyme Inhib.* **15**, 129. **7**. Goñi & Alonso (2002) *FEBS Lett.* **531**, 38.

D-Mannitol

This sugar alcohol (FW = 182.17 g/mol; CAS 69-65-8; M.P. = 166-168°C), also known as mannite and manna sugar, is found in many microorganisms, algae, and plants, including the manna ash tree (*Fraxinus ornus*) and the edible Tamogi-take mushroom (*Pleurotus cornucopiae*). D-Mannitol has a sweet taste and is very soluble in water (15.6 g/100 mL at 18°C) and glycerol (1 g/18 mL). D-Mannitol is a substrate of the reactions catalyzed by mannitol 2-dehydrogenase, mannitol 2-dehydrogenase ($NADP^+$), mannitol 1-dehydrogenase, mannitol 2-dehydrogenase (cytochrome), and D-mannitol oxidase. It is a product of the reaction catalyzed by mannitol-1-phosphatase. **Role in Enhancing Antibiotic Susceptibility in Persistor-Containing Biofilms:** Antibiotic failure in clearing *Pseudomonas aeruginosa* lung infection, the key mortality factor for cystic fibrosis patients, is attributable, at least in part, to the high tolerance of *P. aeruginosa* biofilms. At 10-40 mM, mannitol restores aminoglycoside (tobramycin) sensitivity of *P. aeruginosa* persister cells by generating a proton-motive force (PMF), thereby enhancing tobramycin sensitivity up to 1,000x (1). **Target(s):** aldose 1-epimerase, *or* mutarotase (2); α-amylase (3); inositol-3-phosphate synthase (4); lipoxygenase, soybean (5); mannitol-1-phosphatase (6); mannonate dehydratase (7,8); mannose isomerase (9); Na^+/K^+-exchanging ATPase (10); peptidyl-dipeptidase A, *or* angiotensin I converting enzyme (11); α,α-phosphotrehalase, *or* trehalose-6-phosphate hydrolase (12); sulfite oxidase (13); α,α-trehalase, weakly inhibited (14); xylose isomerase (15,16). **1**. Barraud, Buson, Jarolimek & Rice (2013) *PLoS One* **8**, e84220. **2**. Bailey, Fishman, Kusiak, Mulhern & Pentchev (1975) *Meth. Enzymol.* **41**, 471. **3**. Ali & Abdel-Moneim (1989) *Zentralbl. Mikrobiol.* **144**, 615. **4**. RayChaudhuri, Hait, das Gupta, *et al.* (1997) *Plant Physiol.* **115**, 727. **5**. Morrison, Winokur & Brown (1982) *Biochem. Biophys. Res. Commun.* **108**, 1757. **6**. Iwamoto, Kawanobe, Shiraiwa & Ikawa (2001) *Mar. Biotechnol.* **3**, 493. **7**. Robert-Baudouy, Jimeno-Abendano & Stoeber (1982) *Meth. Enzymol.* **90**, 288. **8**. Robert-Baudouy & Stoeber (1973) *Biochim. Biophys. Acta* **309**, 473. **9**. Hirose, Maeda, Yokoi & Takasaki (2001) *Biosci. Biotechnol. Biochem.* **65**, 658. **10**. Whikehart, Angelos & Montgomery (1995) *Cornea* **14**, 295. **11**. Hagiwara, Takahashi, Shen, *et al.* (2005) *Biosci. Biotechnol. Biochem.* **69**, 1603. **12**. Helfert, Gotsche & Dahl (1995) *Mol. Microbiol.* **16**, 111. **13**. Cohen & Fridovich (1971) *J. Biol. Chem.* **246**, 359. **14**. Lopez & Torrey (1985) *Arch. Microbiol.* **143**, 209. **15**. Noltmann (1972) *The Enzymes*, 3rd ed., **6**, 271. **16**. Danno (1970) *Agric. Biol. Chem.* **34**, 1805.

D-Mannitol 1-Phosphate

This phosphorylated sugar alcohol (FW$_{free-acid}$ = 262.15 g/mol; CAS 15806-48-1), also called D-mannitol 6-phosphate, is a product of the reaction catalyzed by mannose-6-phosphate 6-reductase and is a substrate in the reactions catalyzed by mannitol 1-phosphate 5-dehydrogenase and mannitol 1-phosphatase. **Target(s):** glucose-6-phosphate isomerase (1,2); inositol-3-phosphate synthase (3); D-mannitol dehydrogenase (4,5), mannose-6-phosphate isomerase (1,2,6,7). **1.** Noltmann (1972) *The Enzymes*, 3rd ed., **6**, 271. **2.** Lee & Matheson (1984) *Phytochemistry* **23**, 983. **3.** Barnett, Rasheed & Corina (1973) *Biochem. Soc. Trans.* **1**, 1267. **4.** Ueng & McGuinness (1977) *Biochemistry* **16**, 107. **5.** Ueng, Hartanowicz, Lewandoski, *et al.* (1976) *Biochemistry* **15**, 1743. **6.** Gracey & Noltmann (1968) *J. Biol. Chem.* **243**, 5410. **7.** Gao, Yu & Leary (2005) *Anal. Chem.* **77**, 5596.

D-Mannonolactam

This mannopyranosyl cation (FW = 176.17 g/mol; CAS 62362-63-4) mimic inhibits several mannosidases. **Target(s):** α-mannosidase, IC$_{50}$ = 400 nM for the jack bean enzyme; β-mannosidase, moderately inhibited, IC$_{50}$ = 150 μM); mannosyl-oligosaccharide 1,2-α mannosidase (mannosidase I; IC$_{50}$ = 4 μM); mannosyl-oligosaccharide 1,3-1,6-α-mannosidase (mannosidase II; IC$_{50}$ = 90-100 nM). **1.** Pan, Kaushal, Papandreou, Ganem & Elbein (1992) *J. Biol. Chem.* **267**, 8313.

D-Mannono-1,5-Lactone

This putative transition-state analogue (FW = 178.14 g/mol; CAS 10366-75-3) adopts an sp^2 hybrid orbital that mimics the configuration of oxacarbenium ion intermediates. D-Mannono-1,5-lactone targets 2-deoxyglucosidase (1), α-mannosidase, weakly inhibited (2-11), β-mannosidase (2,8), and mannosyl-oligosaccharide 1,2-α-mannosidase (11). **1.** Canellakis, Bondy, May, Myers-Robfogel & Sartorelli (1984) *Eur. J. Biochem.* **143**, 159. **2.** Sukeno, Tarentino, Plummer & Maley (1972) *Meth. Enzymol.* **28**, 777. **3.** Matta & Bahl (1972) *Meth. Enzymol.* **28**, 749. **4.** Opheim & Touster (1978) *Meth. Enzymol.* **50**, 494. **5.** Li (1967) *J. Biol. Chem.* **242**, 5474. **6.** Levvy, Hay & Conchie (1964) *Biochem. J.* **91**, 378. **7.** Levvy & Conchie (1966) *Meth. Enzymol.* **8**, 571. **8.** Sukeno, Tarentino, Plummer & Maley (1972) *Biochemistry* **11**, 1493. **9.** Matta & Bahl (1972) *J. Biol. Chem.* **247**, 1780. **10.** Waln & Poulton (1987) *Plant Sci.* **53**, 1. **11.** Swaminathan, Matta, Donoso & Bahl (1972) *J. Biol. Chem.* **247**, 1775.

D-Mannosamine

This reducing amino sugar (FW$_{hydrochloride}$ = 215.63 g/mol; CAS 179-1711-2; Soluble in water and methanol; pK_a = 7.28), also known as 2-amino-2-deoxy-D-mannose, is a constituent of many glycoproteins and mucolipids. The anomeric composition of D-mannosamine is 43% α-anomer and 57%

β-anomer in D$_2$O at 40°C. **Target(s):** β-*N*-acetylglucosaminidase (1,2); β-*N*-acetylhexosaminidase (17,18); amine oxidase, semicarbazide-sensitive (*or* diamine oxidase; amine oxidase (copper-containing) (3); glucokinase (4); glycosylphosphatidylinositol anchor biosynthesis (5-9); mannokinase, as weak alternative substrate (22,23); α-mannosidase (10,11); β-mannosidase (19-21); mannosyl-oligosaccharide 1,2-α mannosidase, *or* mannosidase I (12,16); 1,2-α-mannosyltransferase (9); protein and peptide glycosylation (12-14); protein-glucosylgalactosyl-hydroxylysine glucosidase (15). **1.** Kapur & Gupta (1986) *Biochem. J.* **236**, 103. **2.** Reglero (1979) *Int. J. Biochem.* **10**, 285. **3.** O'Sullivan, O'Sullivan, Tipton, *et al.* (2003) *Biochim. Biophys. Acta* **1647**, 367. **4.** Reeves, Montalvo & Sillero (1967) *Biochemistry* **6**, 1752. **5.** Ralton, Milne, Guther, Field & Ferguson (1993) *J. Biol. Chem.* **268**, 24183. **6.** Field, Medina-Acosta & Cross (1993) *J. Biol. Chem.* **268**, 9570. **7.** Pan, Kamitani, Bhuvaneswaran, *et al.* (1992) *J. Biol. Chem.* **267**, 21250. **8.** Lisanti, Field, Caras, Menon & Rodriguez-Boulan (1991) *EMBO J.* **10**, 1969. **9.** Sevlever & Rosenberry (1993) *J. Biol. Chem.* **268**, 10938. **10.** Tulsiani, Skudlarek & Orgebin-Crist (1989) *J. Cell Biol.* **109**, 1257. **11.** Lucas, Martin-Barrientos & Cabezas (1984) *Int. J. Biochem.* **16**, 207. **12.** Kaushal & Elbein (1989) *Meth. Enzymol.* **179**, 452. **13.** Elbein (1983) *Meth. Enzymol.* **98**, 135. **14.** Elbein (1987) *Meth. Enzymol.* **138**, 661. **15.** Hamazaki & Hotta (1979) *J. Biol. Chem.* **254**, 9682. **16.** Szumilo, Kaushal, Hori & Elbein (1986) *Plant Physiol.* **81**, 383. **17.** Garcia-Alonso, Reglero & Cabezas (1990) *Int. J. Biochem.* **22**, 645. **18.** Mian, Herries, Cowen & Batte (1979) *Biochem. J.* **177**, 319. **19.** Houston, Latimer & Mitchell (1974) *Biochim. Biophys. Acta* **370**, 276. **20.** Ouellette & Bewley (1986) *Planta* **169**, 333. **21.** Noeske & Mersmann (1983) *Hoppe-Seyler's Z. Physiol. Chem.* **364**, 1645. **22.** Sabater, Sebastián & Asensio (1972) *Biochim. Biophys. Acta* **284**, 406. **23.** Sebastián & Asensio (1972) *Arch. Biochem. Biophys.* **151**, 227.

D-Mannose

α-D-mannopyranose   β-D-mannopyranose

This reducing aldohexose (FW = 180.16 g/mol; CAS 3458-28-4) is a key component of many glycoproteins, hemicelluloses, and mannans. Mannose is very soluble in water (248 g/100 mL at 17°C) and sparingly soluble in ethanol and methanol. The α-pyranose is the more abundant anomer (68%) at 21°C in D$_2$O. The structure of mannose was determined by Nobelist Emil Fischer (1). **Target(s):** β-*N*-acetylhexosaminidase (22-24); β-amylase (42); concanavalin A$_4$; cyclomaltodextrin glucanotransferase (50); fructokinase (44); 1,2-α-L-fucosidase (21); α-galactosidase (37); glucan 1,3-α-glucosidase, *or* glucosidase II, *or* mannosyl-oligosaccharide glucosidase II (9); glucokinase (5,45); glucose oxidase (6); α-glucosidase (7,8); β-glucosidase (38-41); glycogen phosphorylase, weakly inhibited (52); hexokinase, as an alternative substrate (10,45-48); laminaribiose phosphorylase (49); levansucrase (51); α-mannosidase (11,12,18,29-36); β-mannosidase (13,26-28); mannosyl glycoprotein endo-β-*N*-acetyl-glucosamidase (2,3,14); mannosyl-oligosaccharide glucosidase, *or* glucosidase I (19,20); protein:*N*$^{\pi}$-phosphohistidine:sugar phospho-transferase (43); thioglucosidase, *or* myrosinase (15,16); α,α-trehalase (25); xylose isomerase, weakly inhibited (17). **1.** Fischer & Hirschberger (1888) *Ber.* **21**, 1805; (1889) *Ber.* **22**, 1155; and (1889) *Ber.* **22**, 3218. **2.** Koide & Muramatsu (1975) *Biochem. Biophys. Res. Commun.* **66**, 411. **3.** Reglero (1979) *Int. J. Biochem.* **10**, 285. **4.** Goldstein, Hollerman & Smith (1965) *Biochemistry* **4**, 876. **5.** Parry & Walker (1966) *Biochem. J.* **99**, 266. **6.** Schepartz & Subers (1964) *Biochim. Biophys. Acta* **85**, 228. **7.** Webb (1966) *Enzyme and Metabolic Inhibitors*, vol. **2**, pp. 416-417, Academic Press, New York. **8.** Michaelis & Rona (1914) *Biochem. Z.* **60**, 62. **9.** Presper & Heath (1983) *The Enzymes*, 3rd ed., **16**, 449. **10.** Copley & Fromm (1967) *Biochemistry* **6**, 3503. **11.** Opheim & Touster (1978) *Meth. Enzymol.* **50**, 494. **12.** Tulsiani & Touster (1978) *Meth. Enzymol.* **50**, 500. **13.** McCleary (1988) *Meth. Enzymol.* **160**, 589. **14.** Muramatsu (1978) *Meth. Enzymol.* **50**, 555. **15.** Tani, M. Ohtsuru & Hata (1974) *Agric. Biol. Chem.* **38**, 1623. **16.** Tsuruo & Hata (1968) *Agric. Biol. Chem.* **32**, 1420. **17.** Noltmann (1972) *The Enzymes*, 3rd ed., **6**, 271. **18.** Tulsiani, Opheim & Touster (1977) *J. Biol. Chem.* **252**, 3227. **19.** Grinna & Robbins (1979) *J. Biol. Chem.* **254**, 8814. **20.** Grinna & Robbins (1980) *J. Biol. Chem.* **255**, 2255. **21.** Miura, Okamoto & Yanase (2005) *J. Biosci. Bioeng.* **99**, 629. **22.** Calvo, Reglero & Cabezas (1978) *Biochem. J.* **175**, 743. **23.** Garcia-

Alonso, Reglero & Cabezas (1990) *Int. J. Biochem.* **22**, 645. **24**. Sakai, Narihara, Kasama, Wakayama & Moriguchi (1994) *Appl. Environ. Microbiol.* **60**, 2911. **25**. Lopez & Torrey (1985) *Arch. Microbiol.* **143**, 209. **26**. Wan, Muldrey, Li & Li (1976) *J. Biol. Chem.* **251**, 4384. **27**. Kulminskaya, Eneiskaya, Isaeva-Ivanova, *et al.* (1999) *Enzyme Microb. Technol.* **25**, 372. **28**. Bouquelet, Spik & Montreuil (1978) *Biochim. Biophys. Acta* **522**, 521. **29**. Plant & Moore (1982) *Phytochemistry* **21**, 985. **30**. Gaikwald, Keskar & Khan (1995) *Biochim. Biophys. Acta* **1250**, 144. **31**. Einhoff & Rudiger (1988) *Biol. Chem. Hoppe-Seyler* **369**, 165. **32**. Tulsiani, Skudlarek, Nagdas & Orgebin-Crist (1993) *Biochim. J.* **290**, 427. **33**. Tulsiani, Opheim & Touster (1977) *J. Biol. Chem.* **252**, 3227. **34**. Opheim & Touster (1978) *J. Biol. Chem.* **253**, 1017. **35**. Opheim (1978) *Biochim. Biophys. Acta* **524**, 121. **36**. Vega & Dominguez (1988) *J. Basic Microbiol.* **28**, 371. **37**. Dey, Naik & Pridham (1986) *Planta* **167**, 114. **38**. Park, Bae, Sung, Lee & Kim (2001) *Biosci. Biotechnol. Biochem.* **65**, 1163. **39**. Seidle, Marten, Shoseyov & Huber (2004) *Protein J.* **23**, 11. **40**. Langston, Sheehy & Xu (2006) *Biochim. Biophys. Acta* **1764**, 972. **41**. Harada, Tanaka, Fukuda, Hashimoto & Murata (2005) *Arch. Microbiol.* **184**, 215. **42**. Van Damme, Hu, Barre, *et al.* (2001) *Eur. J. Biochem.* **268**, 6263. **43**. Hüdig & Hengstenberg (1980) *FEBS Lett.* **114**, 103. **44**. Bueding & MacKinnon (1955) *J. Biol. Chem.* **215**, 495. **45**. Fernández, Herrero & Moreno (1985) *J. Gen. Microbiol.* **131**, 2705. **46**. Claeyssen, Wally, Matton, Morse & Rivoal (2006) *Protein Expr. Purif.* **47**, 329. **47**. Mulcahy, O'Flaherty, Jennings & Griffin (2002) *Anal. Biochem.* **309**, 279. **48**. Röber, Stolle & Reuter (1984) *Z. Allg. Mikrobiol.* **24**, 619. **49**. Goldemberg, Maréchal & De Souza (1966) *J. Biol. Chem.* **241**, 45. **50**. Bovetto, Villette, Fontaine, Sicard & Bouquelet (1992) *Biotechnol. Appl. Biochem.* **15**, 59. **51**. Lyness & Doelle (1983) *Biotechnol. Lett.* **5**, 345. **52**. Tanabe, Kobayashi & Matsuda (1988) *Agric. Biol. Chem.* **52**, 757.

L-Mannose

α-L-mannopyranose   β-L-mannopyranose

This reducing aldohexose (FW = 180.16 g/mol; CAS 10030-80-5; very soluble in water and sparingly soluble in ethanol and methanol) is the enantiomer of a key component of many glycoproteins, hemicelluloses, and mannans. L-Mannose was first synthesized in the laboratory by Nobelist Emil Fischer (1). **Target(s):** β-glucosidase, K_i = 550 mM for the sweet almond enzyme (2); α-L-rhamnosidase (3). **1**. Fischer (1890) *Ber.* **23**, 370 and (1891) *Ber.* **24**, 2683. **2**. Dale, Ensley, Kern, Sastry & Byers (1985) *Biochemistry* **24**, 3530. **3**. Jang & Kim (1996) *Biol. Pharm. Bull.* **19**, 1546.

α-D-Mannose 1-Phosphate

This phosphorylated hexose (FW$_{free-acid}$ = 260.14; CAS 27251-84-9), a key intermediate in the formation of GDP-mannose, glycoproteins, and glucomannans, is more acid labile than D-mannose 6-phosphate. In 0.95 M acid at 30°C, the $t_{1/2}$ for α-D-mannose 1-phosphate hydrolysis is six hours. **Target(s):** galactose-1-phosphate uridylyltransferase (1); glucose-1-phosphate guanylyltransferase (2); glycolipid 3-α-mannosyl-transferase, *or* α-1,3-mannosyltransferase (3); starch phosphorylase (4); α,α-trehalose phosphorylase, configuration-retaining (5). **1**. Lang, Groebe, Hellkuhl & von Figura (1980) *Pediatr. Res.* **14**, 729. **2**. Danishefsky & Heritier-Watkins (1967) *Biochim. Biophys. Act*a **139**, 349. **3**. Doering (1999) *J. Bacteriol.* **181**, 5482. **4**. Hsu, Yang, Su & Lee (2004) *Bot. Bull. Acad. Sin.* **45**, 187. **5**. Eis & Nidetzky (2002) *Biochem. J.* **363**, 335.

D-Mannose 6-Phosphate

This phosphorylated hexose (FW$_{free-acid}$ = 260.14; CAS 672-15-9), an important intermediate in mannose metabolism, is also a precursor in the formation of GDP-mannose, glycoproteins, and glucomannans. D-Mannose 6-phosphate is more stable in acid than D-mannose 1-phosphate: 50% is hydrolyzed in 16.6 hour in 1.0 M HCl at 100°C *versus* 6 hours in 0.95 M HCl at 30°C. D-Mannose 6-phosphate is a product of the reaction catalyzed by mannokinase and is a substrate of the reactions catalyzed by mannose-6-phosphate 6-reductase, mannose-6-phosphate isomerase, and phosphomannomutase. It can also be synthesized by the action of hexokinase or acyl-phosphate:hexose phosphotransferase. Mannose 6-phosphate receptors are located in the Golgi apparatus, where they bind proteins containing phosphomannosyl residues and function in targeting proteins to lysosomes. **Target(s):** *N*-acetylglucosamine-1-phosphodiester α-*N*-acetylglucosaminiidase (1,2); glucosamine-6-phosphate isomerase (3); glucose-6-phosphate isomerase (4-6); inositol-3-phosphate synthase, *or* *myo*-inositol-1-phosphate synthase (7-9); mannose-6-phosphate isomerase, weakly inhibited by the α-anomer, whereas the β-anomer is the substrate (10). **1**. Varki & Kornfeld (1981) *J. Biol. Chem.* **256**, 9937. **2**. Lee & Pierce (1995) *Arch. Biochem. Biophys.* **319**, 413. **3**. Friedman & Benson (1975) *Meth. Enzymol.* **41**, 400. **4**. Noltmann (1972) *The Enzymes*, 3rd ed., **6**, 271. **5**. Concepcion, Chataing & Dubourdieu (1999) *Comp. Biochem. Physiol. B Biochem. Mol. Biol.* **122**, 211. **6**. Jeong, Fushinobu, Ito, *et al.* (2003) *FEBS Lett.* **535**, 200. **7**. Wong, Mauck & Sherman (1982) *Meth. Enzymol.* **90**, 309. **8**. Wong & Sherman (1985) *J. Biol. Chem.* **260**, 11083. **9**. Naccarato, Ray & Wells (1974) *Arch. Biochem. Biophys.* **164**, 194. **10**. Rose, O'Connell & Schray (1973) *J. Biol. Chem.* **248**, 2232.

Mannostatin A

This metabolite (FW$_{free-base}$ = 179.24 g/mol; CAS 102822-56-0) is produced by *Streptoverticillium verticillus* and is a potent inhibitor of α-mannosidases and was the first nonalkaloidal glycoprotein processing inhibitor. **Target(s):** arylmannosidase (1-3); α-mannosidase (1,3-6); mannosyl-oligosaccharide 1,3-1,6-α mannosidase, *or* mannosidase II, K_i = 36 nM (1,3,6-8). **1**. Kaushal & Elbein (1994) *Meth. Enzymol.* **230**, 316. **2**. Pastuszak, Kaushal, Wall, *et al.* (1990) *Glycobiology* **1**, 71. **3**. Tropea, Kaushal, Pastuszak, *et al.* (1990) *Biochemistry* **29**, 10062. **4**. Aoyagi, Yamamoto, Kojiri, *et al.* (1989) *J. Antibiot.* **42**, 883. **5**. Ogawa & Yu (1995) *Bioorg. Med. Chem.* **3**, 939. **6**. Li, Kawatkar, George, *et al.* (2004) *ChemBioChem* **5**, 1220. **7**. Elbein (1991) *FASEB J.* **5**, 3055. **8**. Kawatkar, Kuntz, Woods, Rose & Boons (2006) *J. Amer. Chem. Soc.* **128**, 8310.

Manoalide

This amphipathic marine sesterterpenoid (FW = 416.55 g/mol; CAS 75088-80-1) from the sponge *Luffariella variabilis* is an irreversible inhibitor of phospholipase A$_2$ by modifying several lysyl residues. At elevated pH, the ring opens to generate a molecular entity with α,β unsaturated aldehydes. Related inhibitors of *L. variabilis* PLA$_2$ include secomanoalide and luffariellins A, B, C, and D. Given the diverse isoforms and roles of PLA$_2$, manoalide-like molecules hold considerable promise in pharmacology, especially as immune modulators. **Target(s):** Ca^{2+} channels (1); 5-lipoxygenase (2); phosphatidylinositol-specific phospholipase C, *or* 1-phosphatidylinositol phosphodiesterase (3,4); phospholipase A$_2$ (5-13). **1.** Wheeler, Sachs, De Vries, *et al.* (1987) *J. Biol. Chem.* **262**, 6531. **2.** De Vries, Amdahl, Mobasser, Wenzel & Wheeler (1988) *Biochem. Pharmacol.* **37**, 2899. **3.** Bennett, Angioli & Crooke (1991) *Meth. Enzymol.* **197**, 526. **4.** Bennett, Mong, Wu, *et al.* (1987) *Mol. Pharmacol.* **32**, 587. **5.** Folmer, Jaspars, Schumacher, Dicato & Diederich (2010) *Biochem Pharmacol.* **80**, 1793. **6.** Reynolds, Mihelich & Dennis (1991) *J. Biol. Chem.* **266**, 16512. **6.** Reynolds & Dennis (1991) *Meth. Enzymol.* **197**, 359. **8.** Lombardo & Dennis (1985) *J. Biol. Chem.* **260**, 7234. **9.** Jacobson, Marshall, Sung & Jacobs (1990) *Biochem. Pharmacol.* **39**, 1557. **10.** Mayer, Glaser & Jacobs (1988) *J. Pharmacol. Exp. Ther.* **244**, 871. **11.** Glaser, Vedvick & Jacobs (1988) *Biochem. Pharmacol.* **37**, 3639. **12.** Glaser & Jacobs (1987) *Biochem. Pharmacol.* **36**, 2079. **13.** Deems, Lombardo, Morgan, Mihelich & Dennis (1987) *Biochim. Biophys. Acta* **917**, 258.

Manumycin A

This unusual antibiotic (FW = 550.65 g/mol; CAS 52665-74-4), also called UCF1-C, inhibits protein farnesyltransferases (IC$_{50}$ = 5 μM) and protein geranylgeranyltransferase type I (IC$_{50}$ = 180 μM). **Target(s):** protein farnesyltransferase (1-5); protein geranylgeranyltransferase type I1, *or* sphingomyelin phosphodiesterase, *or* neutral sphingomyelinase (6,7). **1.** Mitsuzawa & Tamanoi (1995) *Meth. Enzymol.* **250**, 43. **2.** Gibbs (2001) *The Enzymes*, 3rd ed., **21**, 81. **3.** Hara, Akasaka, Akinaga, *et al.* (1993) *Proc. Natl. Acad. Sci. U.S.A.* **90**, 2281. **4.** Courdavault, Thiersault, Courtois, *et al.* (2005) *Plant Mol. Biol.* **57**, 855. **5.** Qian, Zhou, Ju, Cramer & Yang (1996) *Plant Cell* **8**, 2381. **6.** Arenz, Thutewohl, Block, *et al.* (2001) *Chembiochem* **2**, 141. **7.** Goñi & Alonso (2002) *FEBS Lett.* **531**, 38.

MAP-2 Fragment 1705-1722

This octadecapeptide fragment (FW = 1867.16; Sequence: VTSKCGSLKNIRHRPGGG; Isoelectric Point: 11.01) corresponding to residues 1705-1722 within the microtubule-binding domain of Microtubule-Associated Protein-2 (MAP2), stimulates polymerization of MAP-free tubulin and also inhibits the binding of microtubule-associated proteins (*e.g.*, Tau proteins and MAP2) to microtubules. This peptide also greatly reduces microtubule dynamics. **1.** Joly & Purich (1990) *Biochemistry* **29**, 8916. **2.** Yamauchi, Flynn, Marsh & Purich (1993) *J. Neurochem.* **60**, 817.

MAP4, *See* (S)-2-Amino-2-methyl-4-phosphonobutyrate

Maprotiline

This antidepressant (FW$_{hydrochloride}$ = 313.87 g/mol; CAS 10262-69-8; Solubility: 10 mM in H$_2$O), also named 1-(3-methylaminopropyl)dibenzo[*b,e*]bicyclo[2.2.2]octadiene hydrochloride, is a selective norepinephrine uptake inhibitor. Anxiolytic-like effects are also observed with a low-dose maprotiline (0.5 mg/kg), but are lost at higher doses. **Target(s):** aldehyde oxidase (1); diamine oxidase (2); monoamine oxidase (3,4); norepinephrine uptake (5,6). **1.** Obach, Huynh, Allen & Beedham (2004) *J. Clin. Pharmacol.* **44**, 7. **2.** Obata & Yamanaka (2000) *Neurosci. Lett.* **286**, 131. **3.** Egashira, Takayama & Yamanaka (1999) *Jpn.*

J. Pharmacol. **81**, 115. **4.** Ask, Fagervall & Ross (1983) *Naunyn Schmiedebergs Arch. Pharmacol.* **324**, 79. **5.** Hosli & Hosli (1995) *Int. J. Dev. Neurosci.* **13**, 897. **6.** Wirz-Justice & Lichtsteiner (1976) *J. Pharm. Pharmacol.* **28**, 172.

Maraviroc

This synthetic antiretroviral drug (FW = 513.67 g/mol; CAS 376348-65-1), known also by its trade names, Selzentry® and Celsentri®, the code name UK-427,857, as well as 4,4-difluoro-*N*-{(1*S*)-3-[3-(3-isopropyl-5-methyl-4*H*-1,2,4-triazol-4-yl)-8-azabicyclo[3.2.1]oct-8-yl]-1-phenylpropyl}cyclo-hexanecarboxamide, is CCR5 receptor antagonist that binds to a hydrophobic pocket formed by the receptor's transmembrane helices, thus preventing HIV-1 from entering and infecting CD4$^+$ immune cells. CCR5 is a member of G-protein-coupled, seven transmembrane segment receptors, also known as C-C motif chemokine receptor-5. Because maraviroc binds at a site that is topologically distinct from the HIV-1 binding site, it is said to be an allosteric inhibitor. With its borderline physicochemical properties (*i.e.*, moderate lipophilicity, slightly basic pK_a of 7.3, and eight H-bonding groups, and MW > 500), maraviroc exhibits relatively poor membrane permeability, but its transcellular flux is enhanced in the presence of P-glycoprotein inhibitors (2). High-level resistance to maraviroc occurs when primary isolates are passaged in peripheral blood lymphocytes (3). Other similarly acting viral entry inhibitors include: aplaviroc, vivriviroc, TAK-779, SCH C (*or* SCH 351125). By contrast, the entry inhibitor PRO 140 is a human monoclonal antibody targeting the CCR5 receptor. **Key Pharmacokinetic Parameters:** *See* Appendix II in Goodman & Gilman's *THE PHARMACOLOGICAL BASIS OF THERAPEUTICS*, 12th Edition (Brunton, Chabner & Knollmann, eds.) McGraw-Hill Medical, New York (2011). **1.** Wood & Armour (2005) *Prog. Med. Chem.* **43**, 239. **2.** Walker, Abel, Comby, *et al.* (2005) *Drug Metab. Dispos.* **33**, 587.

Marbofloxacin

This orally bioavailable third-generation fluoroquinolone-class antibiotic (FW = 362.36 g/mol; CAS 115550-35-1; IUPAC: 9-fluoro-2,3-dihydro-3-methyl-10-(4-methyl-1-piperazinyl)-7-oxo-7*H*-pyrido[3,2,1-*ij*][4,1,2]benz-oxadiazine-6-carboxylate, is a broad-spectrum systemic antibacterial targeting Type II DNA topoisomerases (gyrases), thereby blocking bacterial replication and transcription in a wide range of Gram-positive and Gram-negative bacterial pathogens (*e.g.*, Aeromonas, Brucella, Campylobacter, Chlamydia trachomatis, Enterobacter, Escherichia coli, Haemophilus, Klebsiella spp, Mycobacterium, Mycoplasma, Proteus, Pseudomonas aeruginosa, Salmonella, Serratia, Shigella, Staphylococci (including penicillinase-producing and methicillin-resistant strains), Vibrio, and Yersinia). Most often used as a monotherapy in veterinary medicine (where pathogen prevention and control are constant challenges, especially in densely populated stock) under the trade names Forcyl®, Marbocyl®, and Zeniquin®, marbofloxacin has also been combined with clotrimazole and dexamethasone under the trade name Aurizon®. For the prototypical member of this antibiotic class, **See** *Ciprofloxacin*

Marimastat

This peptidomimetic agent (FW = 331.41 g/mol; CAS 154039-60-8) is a strong inhibitor of several matrix metalloproteinases, among the interstitial collagenase (K_i = 0.7 nM), gelatinase A (K_i = 0.6 nM), stromelysin 1 (K_i = 8.9 nM), matrilysin (IC$_{50}$ = 20 nM), neutrophil collagenase (IC$_{50}$ = 2 nM), gelatinase B (IC$_{50}$ = 3 nM), and ADAM 17 endopeptidase (K_i = 6.3 nM). **Target(s):** ADAM 17 endopeptidase, tumor necrosis factor-α converting enzyme, *or* TACE (1); gelatinase A, *or* MMP 2 (1,2); gelatinase B , or MMP 9 (2,3); interstitial collagenase MMP 1 (1,2); matrilysin, *or* MMP 7 (2); neutrophil collagenase , *or* MMP 8 (2); membrane-type MMP 1 autocatalytic processing (*or* MMP 14 autocatalytic processing) (4); stromelysin 1, *or* MMP 3 (1,2). **1.** Kottirsch, Koch, Feifel & Neumann (2002) *J. Med. Chem.* **45**, 2289. **2.** Whittaker & Brown (1998) *Curr. Opin. Drug Discov. Dev.* **1**, 157. **3.** Underwood, Min, Lyons & Hambley (2003) *J. Inorg. Biochem.* **95**, 165. **4.** Toth, Hernandez-Barrantes, Osenkowski, *et al.* (2002) *J. Biol. Chem.* **277**, 26340.

Maritoclax

This axially dissymmetric marine-derived *Streptomycetes* natural product (MW = 510.15 g/mol), also known as Marinopyrrole A, is active against methicillin-resistant *Staphylococcus aureus*, with an MIC$_{90}$ of 2 μM (1). Its structure was assigned by X-ray crystallography, and, although configurationally stable at room temperature, the naturally occurring *M*-(–)-enantiomer racemizes to the non-natural *P*-(+)-*atropo*-enantiomer at elevated temperatures (1). Maritoclax also binds to Mcl-1, but not Bcl-X$_L$, a related anti-apoptotic factor in the Bcl-2 protein family (2). In doing so, Maritoclax disrupts the interaction between Bim and Mcl-1, inducing the latter's degradation by the proteasome system, an action associated with Maritoclax's pro-apoptotic activity. Importantly, Maritoclax selectively kills Mcl-1-dependent, but not Bcl-2- or Bcl-X$_L$-dependent, leukemia cells and markedly enhances the efficacy of ABT-737 against hematologic malignancies, including K562, Raji, and multidrug-resistant HL60/VCR, by ~60- to 2000-fold at 1-2 μM (2). **1.** Hughes, Prieto-Davo, Jensen & Fenical (2008) *Org. Lett.* **10**, 629. **2.** Doi, Li, Sung, *et al.* (2012) *J. Biol. Chem.* **287**, 10224.

MARTA2 Dominant Negative Inhibitor

Dendritic targeting of mRNAs encoding the microtubule-associated protein 2 (MAP2) in neurons involves high-affinity and high-specificity binding of its *cis*-acting dendritic targeting element to MARTA2 (*or* MAP2-RNA trans-acting) protein. MARTA2, bind to the cis-element with both high affinity and specificity. MARTA2 is an orthologue of human far-upstream element binding protein 3, which contains four centrally located KH domains that bind to single-stranded DNA and RNA. (KH domains were first identified in the human heterogeneous nuclear ribonucleoprotein (hnRNP) K.) In neurons, MARTA2 resides in somatodendritic granules and dendritic spines, where it associates with MAP2 mRNAs. Expression of a dominant-negative MARTA2-KH-eGFP variant (containing only the KH domains and an enhanced-GFP reporter) disrupts dendritic targeting of endogenous MAP2 mRNAs, without noticeably disturbing the level and subcellular distribution of polyadenylated mRNAs. MAP2 transcripts also associate with the microtubule-based motor KIF5 and inhibition of KIF5, but not cytoplasmic dynein, disrupts extrasomatic trafficking of MAP2

mRNA granules. Such findings indicate that MARTA2 is a key *trans*-acting factor that factors in KIF5-mediated dendritic targeting of MAP2 mRNAs. **1.** Zivraj, Rehbein, Ölschläger-Schütt, *et al.* (2013) *J. Neurochem.* **124**, 670. **2.** Zivraj (2005) *Regulation of Dendritic MAP2 Targeting by MARTA2 in Rattus norvegicus*, a Dissertation submitted to the University of Hamburg.

Martentoxin

This secreted high-affinity potassium channel antagonist (MW = 4056.3 g/mol; (*Sequence*: CSISICTEAFGLIDVKCFASSECWTA), also known as potassium channel toxin α-KTx16.2, BmK 622, and BmTx3B, from the venom of the East-Asian scorpion *Buthus martensi*, inhibits the voltage-dependent, large-conductance Ca^{2+}-activated K$^+$ channel, often referred to the BK channel (IC$_{50}$ = 80 nM). This toxin appears to block channel activity by a simple one-step binding reaction exhibiting a fast association to and slow dissociation from channels on rat brain synaptosomes. **Target(s):** BK$^+$ channel (1,2); delayed rectifying K$^+$ channel (IC$_{50}$ = 50 μM) (1,2). **1.** Wang, Chen, Xhang, Wu & Wu (2005) *Proteins* **58**, 489. **2.** Ji, Wang, Ye, *et al.* (2003) *Neurochem.* **84**, 325.

Maslinate (Maslinic Acid)

This triterpene acid (FW$_{free-acid}$ = 472.71 g/mol; CAS 4373-41-5) from a number of plants, including olive pomace (*Olea europaea*) is an antioxidant that also displays a broad spectrum of inhibitory actions. **Target(s):** DNA-directed DNA polymerase (1); DNA polymerase α (1); DNA polymerase β (1); DNA topoisomerase II (1); glycogen phosphorylase (2-4); HIV-1 retropepsin (5); sterol *O*-acyltransferase, *or* acyl-CoA:cholesterol *O*-acyltransferase, *or* ACAT (6); tyrosinase, *or* monophenol monooxygenase, IC$_{50}$ = 1.7 μM (7). **1.** Mizushina, Ikuta, Endoh, *et al.* (2003) *Biochem. Biophys. Res. Commun.* **305**, 365. **2.** Wen, Sun, Liu, *et al.* (2005) *Bioorg. Med. Chem. Lett.* **15**, 4944. **3.** Wen, Zhang, Liu, *et al.* (2006) *Bioorg. Med. Chem. Lett.* **16**, 722. **4.** Wen, Sun, Liu, *et al.* (2008) *J. Med. Chem.* **51**, 3540. **5.** Xu, Zeng, Wan & Sim (1996) *J. Nat. Prod.* **59**, 643. **6.** Kim, Han, Chung, *et al.* (2005)*Arch. Pharm. Res.* **28**, 550. **7.** Ullah, Hussain, Hussain, *et al.* (2007) *Phytother. Res.* **21**, 1976.

Masitinib

This novel, orally bioavailable, ATP-competitive tyrosine protein kinase inhibitor (FW = 498.65 g/mol; CAS 790299-79-5; Solubility: 100 mg/mL DMSO, <1 mg/mL H$_2$O), also named AB1010 and *N*-(4-methyl-3-(4-(pyridin-3-yl)thiazol-2-ylamino)phenyl)-4-((4-methylpiperazin-1-yl)-methyl)benzamide, targets the stem cell factor receptor Kit (IC$_{50}$ = 200 nM) and platelet-derived growth factor receptors α and β, *or* PDGFRα (IC$_{50}$ = 540 nM) and PDGFRβ (IC$_{50}$ = 800 nM), with weaker inhibition to ABL and c-Fms (1). When docked into the KIT binding site, masitinib's aminothiazole hydrogen bonds with the side-chain of the gatekeeper residue Thr-670. Its amide NH forms an H-bond to the Glu-640 side-chain, and the meta-nitrogen of the pyridine ring interacts with the backbone NH of Cys-673. For the methylpiperazine group, an additional H-bond occurs between the protonated CH$_3$-NH and the backbone C=O of His-790. Its thiazole ring packs loosely between the aliphatic portions of the side-chains of Ala-621, Leu-799, Cys-809, and Phe-811 (1). Although c-Kit is required for early amplification of erythroid progenitors, it must disappear from cell surface for the cell to enter the final steps of maturation in an erythropoietin-dependent manner. Both imatinib and masitinib block this internalization process (2) Importantly, masitinib was inactive against Flt3 (>10 μM), but moderately inhibited c-Fms in both cell proliferation and recombinant protein kinase assays, IC_{50} of 1.0 μM and 1.5 μM, respectively (1). Such selectivity is also consistent with the absence of masitinib-induced cardiotoxicity or genotoxicity in animal studies. Masitinib is an investigational drug for rheumatoid arthritis (3), progressive multiple sclerosis (4,5), Alzheimer disease (6), corticosteroid-dependent asthma (7), and gastrointestinal stromal tumors, *or* GIST (8). Masitinib mesylate (dose = 12.5mg/kg, once daily) is also the first anticancer therapy approved in veterinary medicine for the treatment of unresectable canine mast cell tumors (CMCTs), harboring activating c-KitR mutations (9). **1.** Dubreuil, Letard, Ciufolini, *et al.* (2009) *PLoS One* **4**, e7258. **2.** D'allard, Gay, Descarpentries, *et al.* (2013) *PLoS One* **8**, e60961. **3.** Tebib, Mariette, Bourgeois, *et al.* (2009) *Arthritis Res. Ther.* 11, R95. **4.** Vermersch, Benrabah, Schmidt , *et al.* (2012) *BMC Neurol.* **12**, 36. **5.** Rommer & Stüve (2013) *Curr. Treat. Options Neurol.* **15**, 241. **6.** Piette, Belmin, Vincent, *et al.* (2011) *Alzheimers Res. Ther.* **3**, 16. **7.** Humbert, de Blay, Garcia, *et al.* (2009) *Allergy* 64, 1194. **8.** Kim & Zalupski (2011) *Surg. Oncol.* **104**, 901. **9**. **1.** Marech, Patruno, Zizzo, *et al.* (2013) *Crit. Rev. Oncol. Hematol.* **91**, 98.

3-MATIDA

This sterically constrained glutamate analogue (FW = 215.22 g/mol; CAS 518357-51-2; Soluble to 100 mM in 1eq. NaOH and to 50 mM in DMSO) targets the metabotropic glutamate $mGlu_1$ receptor antagonist (IC_{50} = 6.3 μM, rat $mGlu_{1a}$) and displays ≥ 40x selectivity over $mGlu_5$, $mGlu_2$, $mGlu_{4a}$ (each with IC_{50} > 300 μM) as well as NMDA and AMPA receptors (IC_{50} = 250 μM). 3-Mathida is neuroprotective in cultured murine cortical cells and rat hippocampal slice cultures *in vitro*. It also reduces the size of ischemia-induced brain infarcts in rats. **1.** Cozzi, Meli, Carlà, *et al.* (2002) *Neuropharmacol.* **43**, 119. **2.** Moroni, Attucci, Cozzi, *et al.* (2002) *Neuropharmacol.* **42**, 741.

Maytansine

This microtubule-disrupting alkaloid (FW = 692.21 g/mol; CAS 35846-53-8) from the plant *Maytenus ovatus* is a potent antitumor agent that inhibits tubulin polymerization and is a potent inhibitor of mitosis. Although earlier work suggested that aytansine appears to compete with the *Catharanthus* alkaloids (vinblastine, vincristine, and vinleursine) for a common or overlapping site on the tubulin dimer (1), new structural data shows that this macrocyclic drug binds to a site on β-tubulin that is distinct from the vinca domain and that blocks the formation of longitudinal tubulin interactions in microtubules (2). The total synthesis of (–)-maytansine was realized in 1980 by Corey and coworkers (3). **Target(s):** calmodulin (4); tubulin polymerization, *or* microtubule self-assembly (1,2,4,5); vinblastine and colchicine binding to tubulin (6). **1.** Huang, Lin & Hamel (1985) *Biochem. Biophys. Res. Commun.* **128**, 1239. **2.** Prota, Bargsten, Diaz, *et al.* (2014)

Proc. Natl. Acad. Sci. USA **111**, 13817. **3.** Corey, Weigel, Chamberlin, Cho & Hua (1980) *J. Amer. Chem. Soc.* **102**, 6613. **4.** Watanabe & West (1982) *Fed. Proc.* **41**, 2292. **5.** Remillard, Rebhun, Howie & Kupchan (1975) *Science* **189**, 1002. **6.** Purich & Kristofferson (1984) *Adv. Protein Chem.* **36**, 133. **7.** Lin, Hamel & Wolpert-DeFilippes (1981) *Res. Commun. Chem. Pathol. Pharmacol.* **31**, 443.

MC 1742

This cell-permeable and potent Class I and IIb HDAC inhibitor (FW = 395.47 g/mol; CAS 1776116-74-5; Solubiity: 100 mM in DMSO), also known as 5-[(4-[1,1'-biphenyl]-4-yl-1,6-dihydro-6-oxo-2-pyrimidinyl)thio]-*N*-hydroxypentanamide, targets histone deacetylases, with respective IC_{50} values of 7, 20, 40, 100, 110 and 610 nM for HDAC6, HDAC3, HDAC10, HDAC1, HDAC2 and HDAC8. MCC-1742 suppresses cell proliferation and induces apoptosis of sarcoma cancer stem cells (CSCs) when used at >500 nM. It also induces osteogenesis in sarcoma CSC cells in the 25–500 nM range. **1.** Di Pompo, *et al.* (2015) *J. Med. Chem.* **58**, 4073. **2.** Mai, *et al.* (2008) *Bioorg. Med. Chem. Lett.* **18**, 2530. **3.** Mai, *et al.* (2006) *J. Med. Chem.* **49**, 6046.

MCI-9042, *See* Sarpogrelate

MC-LR, *See* Micrococcin P_1

McN-4853, *See* Topiramate

MD-805, *See* Argatroban

MDC-1016

This farnesyl analogue ($FW_{free-acid}$ = 510.63 g/mol; *Symbol*: PFTS), also known as phospho-farnesylthiosalicylate, inhibits the growth of human pancreatic cancer cells in culture and in a xenograft model in a concentration- and time-dependent manner. MDC-1016 prevents pancreatitis-accelerated acinar-to-ductal metaplasia in mice with activated K-ras. After oral administration, PFTS is rapidly absorbed, metabolized to FTS and FTS glucuronide, which are distributed through the blood to body organs. PFTS inhibits Ras-GTP, the active form of Ras, both *in vitro* and *in vivo*, leading to the inhibition of downstream effector pathways involving c-RAF/mitogen-activated protein-extracellular signal-regulated kinase (ERK) kinase, (MEK)/ERK1/2 kinase, and phosphatidylinositol 3-kinase/AKT. **1.** Mackenzie, Bartels, Xie, *et al.* (2013) *Neoplasia* **15**, 1184.

mdivi-1

This mitochondrial division inhibitor (FW = 353.22 g/mol; CAS 338967-87-6; Soluble to 100 mM in DMSO), systematically named 3-(2,4-dichloro-5-methoxyphenyl)-2,3-dihydro-2-thioxo-4(1H)-quinazolinone, attenuates mitochondrial division in yeast and mammalian cells by selectively inhibiting the mechano-protein Dynamin-1, or Dnm1/Drp1 GTPase (IC$_{50}$ = 1-10 μM). In cells, mdivi-1 retards apoptosis by inhibiting mitochondrial outer membrane permeabilization. *In vitro*, mdivi-1 potently blocks Bid-activated Bax/Bak-dependent cytochrome *c* release from mitochondria. These findings indicate that dynamin directly regulates mitochondrial outer membrane permeabilization independent of Drp1-mediated division. **1.** Cassidy-Stone, Chipuk, Ingerman, *et al.* (2008) *Dev. Cell* **14**, 193. **2.** Tanaka & Youle (2008) *Mol. Cell* **29**, 409. **3.** Lackner & Nunnari (2010) *Chem. Biol.* **17**, 578.

MDL 11,939

This orally active and centrally acting serotonin receptor antagonist (FW = 295.42 g/mol; CAS 107703-78-6; Solubility: 50 mM in DMSO), also named α-phenyl-1-(2-phenylethyl)-4-piperidinemethanol, selectively targets rabbit 5-HT$_{2A}$ (K_i = 0.54 nM) and human 5-HT$_{2A}$ receptors (K_i = 2.5 nM), with much weaker action against rabbit 5-HT$_{2C}$ receptors (K_i = 82 nM) and human 5-HT$_{2C}$ receptors (K_i = ~10,000 nM). **1.** Dudley, *et al.* (1988) *Drug Dev. Res.* **13**, 29.

MDL 12330A

This reagent (FW$_{free-base}$ = 340.55 g/mol; FW$_{HCl-Salt}$ = 377.01 g/mol; CAS 40297-09-4; Solubility = 20 mg/mL DMSO), also known as *N*-(*cis*-2-phenylcyclopentyl)azacyclotridec-1-en-2-amine and RMI 12330A, inhibits adenylyl cyclase, *or* AC (1-3) and cAMP phosphodiesterase (4). MDL-12330A has a non-specific effect on glycine transport in Müller cells from the retina (5). Importantly, MDL-12,330A also enhances [Ca^{2+}]$_i$ and insulin secretion via inhibition of KV-channels rather than AC antagonism in β-cells, suggesting that one must recognize the possibility of non-specific side effects when evaluating the action of this agent (5). MDL-12330A blocks slow extracellular and store-operated Ca^{2+} entry into cells. **1.** Lippe & Ardizzone (1991) *Comp. Biochem. Physiol. C* **99**, 209. **2.** Chen, Lane, Bock, Leinders-Zufall & Zufall (2000) *J. Neurophysiol.* **84**, 575. **3.** Ferretti, Sonetti, Pareschi, *et al.* (1996) *Neurosci. Lett.* **207**, 191. **4.** Biondi, Campi, Pareschi, Portolan & Ferretti (1990) *Neurosci. Lett.* **113**, 322. **5.** Gadeaa, Lópeza & López-Colomé (1999) *Brain Res.* **838**, 200. **5.** Li, Guo, Gao, *et al.* (2013) *PLoS One* **8**, e77934.

MDL 25637

This putative transition-state analogue (FW$_{free-base}$ = 355.34 g/mol; CAS 104343-33-1), also known as 2,6-dideoxy-2,6-imino-7-*O*-(β-D glucopyranosyl)-D-*glycero*-L-*gulo*-heptitol and α-homonojirimycin-7-*O*-β-D-glucopyranoside, is a slow, tight-binding inhibitor of a variety of glycohydrolases. **Target(s):** glucoamylase (1); α-glucosidase (2); glucan 1,3-α-glucosidase, *or* glucosidase II, *or* mannolsyl-oligosaccharide

glucosidase II (2); isomaltase (1); maltase (1); mannosyl-oligosaccharide glucosidase, *or* glucosidase I (2); sucrase (1); α,α-trehalase (1,3-5). **1.** Rhinehart, Robinson, Liu, *et al.* (1987) *J. Pharmacol. Exp. Ther.* **241**, 915. **2.** Kaushal & Elbein (1994) *Meth. Enzymol.* **230**, 316. **3.** Salleh & Honek (1990) *FEBS Lett.* **262**, 359. **4.** Anzeveno, Creemer, Daniel, King & Liu (1989) *J. Org. Chem.* **54**, 2539. **5.** Temesvari & Cotter (1997) *Biochimie* **79**, 229.

MDL 28170, *See* N-Benzyloxycarbonyl-L-valyl-L-phenylalaninal

MDL 28815, *See* N-(1,5,9-Trimethyldecyl)-4α,10-dimethyl-8-aza-trans-decal-3β-ol

MDL 72222, *See* 3-Tropanyl 3,5-Dichlorobenzoate

MDL 72527, *See* N,N'-Bis(2,3-butadienyl)-1,4-butanediamine

MDL-72,974, *See* Mofegiline

MDL 73811, *See* 5'-([(Z)-4-Amino-2-butenyl]methylamino)-5'-deoxyadenosine

MDL100907

This potent serotonin (5-hydroxytryptamine) receptor antagonist and antipsychotic agent *in vivo* (FW = 373.46 g/mol; CAS 139290-65-6), systematically named (*R*)-(+)-α-(2,3-dimethoxyphenyl)-1-[2-(4-fluorophenyl)ethyl]-4-piperinemethanol, selectively targets 5-HT$_{2A}$ (K_i = 0.36 nM) , in behavioral, electrophysiological and neurochemical models of antipsychotic activity and extrapyramidal side-effect liability (1). Scatchard analysis shows high-affinity binding, single-site binding of [^{3}H]-MDL100907 to 5-HT$_{2A}$ receptors in rabbit cortical membranes, with a mean B$_{max}$ was 8.5 fmol/mg tissue and a mean K_d was 33 pM (2). MDL 100907 exhibits >80x selectivity over other 5-HT serotonergic receptor subtypes (1). In mice, MDL 100,907 blocked amphetamine-stimulated locomotion at doses that did not significantly affect apomorphine-stimulated climbing behavior (1). MDL 100,907 has a clozapine-like profile of potential antipsychotic activity with low extrapyramidal side-effect liability (1). Using inhibition of D-amphetamine-stimulated locomotion in mice as a measure of potential antipsychotic efficacy, MDL 100,907 shows a superior CNS safety index relative to reference compounds (*e.g.*, haloperidol, clozapine, risperidone, ritanserin, and amperozide) in each of five tests for side effect potential, including measures of ataxia, general depressant effects, α$_1$-adrenergic antagonism, striatal D$_2$ receptor antagonism, and muscle relaxation (3). MDL 100,907 does not antagonize apomorphine-induced stereotypes in rats, suggesting that it may lack extrapyramidal side effect liability. MDL 100,907 shows selectivity as a potential antipsychotic in that it lacked consistent activity in selected rodent models of anticonvulsant, antidepressant, analgesic, or anxiolytic activity (3). **1.** Sorensen, Kehne, Fadayel, *et al.* (1993) *J. Pharmacol. Exp. Ther.* **266**, 684. **2.** Aloyo & Harvey (2000) *Eur. J. Pharmacol.* **406**, 163. **3.** Kehne, Baron, Carr, *et al.* (1996) *J. Pharmacol. Exp. Ther.* **277**, 968.

MDL 101146

This orally active pentafluoroethyl ketone (FW = 632.62 g/mol; CAS 149859-17-6), also named *N*-[4-(4-morpholinylcarbonyl)benzoyl]-L-valyl-*N*-[3,3,4,4,4-pentafluoro-1-(1-methylethyl)-2-oxobutyl]   L-prolinamide, inhibits both porcine pancreatic elastase (PPE) and human leukocyte (neutrophil) elastase (HNE). Its modeling-based design was corroborated with X-ray crystallographic analysis of its complex with PPE and subsequently HNE. **1**. Cregge, Durham, Farr, *et al.* (1998) *J. Med. Chem.* **41**, 2461. **2**. Durham, Hare, Angelastro, *et al.* (1994) *J. Pharmacol. Exp. Ther.* **270**, 185.

MDL 27,088, See *[4S-[4α,7α(R*),12bβ]]-7-[S-(1-Carboxy-3-phenylpropyl)amino]-1,2,3,4,6,7,8,12b-octahydro-6-oxopyrido[2,1-a]-[2]benzazepine-4-carboxylate*

MDL 201449A, See *9N-[(1'R,3'R)-trans-3'-Hydroxycyclopentanyl]adenine*

MDMA, See *3,4-Methylenedioxymethamphetamine*

MDMP, See *2-(4-Methyl-2,6-dinitroanilino)-N-methylpropionamide*

MDP, See *N-Acetylmuramyl-L-alanyl-D-isoglutamine; Methylenediphosphonate*

MDX-010, MDX-101, See *Ipilimumab*

MDX1106, See *Nivoluma*

R(–)-Me5, See *R(–)-1-(2,6-Dimethylphenoxy)-3-methyl-2-butanamine*

ME0328

This potent PARP inhibitor (FW = 321.37 g/mol; CAS 1445251-22-8; Solubility: <1 mg/mL DMSO, H_2O, or Ethanol), also named 3,4-dihydro-4-oxo-*N*-[(1*S*)-1-phenylethyl]-2-quinazolinepropanamide, selectively targets ADP-ribosyltransferase-3/poly(ADP-ribose) polymerase-3 (ARTD3), a druggable regulator of DNA repair and mitotic progression. *In vitro* profiling against 12 PARP family members suggests that ME0328 is selective for ARTD3, and X-ray crystal structures provide a basis for such selectivity. ME0328 is active in cells, where it elicits ARTD3-specific effects at submicromolar concentration. **1**. Lindgren, Karlberg, Thorsell, *et al.* (2013) *ACS Chem Biol.* **8**, 1698.

Mead Acid, See *5Z,8Z,11Z-Eicosatrienoate and 5Z,8Z,11Z-Eicosatrienoic Acid*

Meayamycin

This FR901464 analogue (FW = 489.65) shows picomolar anti-proliferative activity of against various cancer cell lines and multidrug resistant cells. The observed time-dependence suggests that meayamycin may form a covalent bond with its protein target(s). Like FR901464 and spliceostatin A, meayamycin inhibits pre-mRNA splicing of the AdML pre-mRNA substrate in a concentration-dependent manner, with 50 nM sufficient for complete inhibition. Incubation of the same nuclear extract with up to 10 µM of the non-epoxide analogue was without effect upon splicing. Unlike FR901464, meayamycin is very stable in phosphate buffer at pH 7.4 and is likewise relatively stable in cell culture medium. (*See also E7107; FR901464; Pladienolide B; Spliceomycin A*) **1**. Albert, McPherson, O'Brien, *et al.* (2009;) *Mol. Cancer Ther.* **8**, 2308.

Mebendazole

This benzimidazole-containing anthelminthic (FW = 295.30 g/mol; CAS 31431-39-7), also known as methyl-(5-benzoyl-1*H*-benzimidazol-2-yl)carbamate, is a widely used veterinary medicine for treating roundworm, whipworm, threadworm, and hookworm infestation. By binding to the colchicine-sensitive site of tubulin and inhibiting the assembly of microtubules, mebendazole brings about degenerative changes in the tegument and intestinal cells of the worm. **Target(s):** malate dehydrogenase (1,2); tubulin (3-5). **1**. Vessal & Tabei (1996) *Comp. Biochem. Physiol. B Biochem. Mol. Biol.* **113**, 757. **2**. Tejada, Sanchez-Moreno, Monteoliva & Gomez-Banqueri (1987) *Vet. Parasitol.* **24**, 269. **3**. Ochola, Prichard & Lubega (2002) *J. Parasitol.* **88**, 600. **4**. Sasaki, Ramesh, Chada, *et al.* (2002) *Mol. Cancer Ther.* **1**, 1201. **5**. Laclette, Guerra & Zetina (1980) *Biochem. Biophys. Res. Commun.* **92**, 417.

Mecamylamine

This noncompetitive nicotinic acetylcholine receptor antagonist (FW$_{free-base}$ = 167.29 g/mol; CAS 826-39-1), also known as Inversine™, was the first orally available antihypertensive agent. While popular in the 1950s, it is now rarely used for hypertension owing to its widespread ganglionic side-effects. At lower doses, it blocks the physiological effects of nicotine, improving abstinence rates in those seeking to stop smoking. **1**. Damaj, Glassco, Dukat & Martin (1999) *J. Pharmacol. Exp. Ther.* **291**, 1284. **2**. Bissada, Welch & Finkbeiner (1978) *Urology* **11**, 425. **3**. Varanda, Aracava, Sherby, *et al.* (1985) *Mol. Pharmacol.* **28**, 128.

Mechlorethamine

This nitrogen mustard (FW$_{free-base}$ = 156.05 g/mol; M.P. = –60°C; Slightly soluble in water; Soluble in many organic solvents; HCl Salt (M.P. = 109 111°C; CAS 55-86-7) is water-soluble), also known as HN-2 and *N*-bis(2-chloroethyl)methylamine, is a chemical warfare agent, acting as cholinesterase poison. It is also DNA and RNA alkylating agent, the latter actions inhibiting both transcription and translation (1) The hydrochloride is used to treat Hodgkin's disease. **CAUTION:** As the free base, HN-2 is a powerful vesicant, requiring use of a chemical hood or gas mask. Avoid contact with skin and wear protective gown and thick chemical rubber gloves. **History Note:** Although Germany and the United States produced nitrogen mustard agents in 1941 and 1943, respectively, there are no indications that they were ever used on the battlefield. Several nations currently maintain stockpiles of sulfur and nitrogen mustards. (*See also 2,2'-Dichloro-N-triethylamine; Tris(2 chloroethyl)amine*) **Target(s):** acetyl-cholinesterase (2-6); betaine-aldehyde dehydrogenase (7); choline acetyltransferase (8); choline dehydrogenase (9-12); choline oxidase (3); hemoglobin S polymerization (13-15); hexokinase (16,17); Na$^+$/K$^+$/Cl$^-$-cotransporter (18). **1**. Masta, Gray & Phillips (1995) *Nucl. Acids Res.* **23**, 3508. **2**. Fujii, Ohnoshi, Namba & Kimura (1982) *Gan To Kagaku Ryoho* **9**, 831. **3**. Gilman & Phillips (1946) *Science* **103**, 409. **4**. Peters (1947) *Nature* **159**, 149. **5**. Augustinsson (1950) *The Enzymes*, 1st. ed., **1** (Part 1), 443. **6**. Adams & Thompson (1948) *Biochem. J.* **42**, 170. **7**. Rothschild & Barron (1954) *J. Biol. Chem.* **209**, 511. **8**. Augustinsson (1952) *The Enzymes*, 1st ed., **2** (Part 2), 906. **9**. Kimura & Singer (1962) *Meth. Enzymol.* **5**, 562. **10**. Singer (1963) *The Enzymes*, 2nd ed., **7**, 345. **11**. Hatefi & Stiggall (1976) *The Enzymes*, 3rd ed., **13**, 175. **12**. Sivak, Mahoney & Rogers (1967) *Biochem. Pharm.* **16**, 1919. **13**. Roth, Nagel, Bookchin & Grayzel (1972) *Biochem. Biophys. Res. Commun.* **48**, 612. **14**. Fung, Ho, Roth & Nagel (1975) *J. Biol. Chem.* **250**, 4786. **15**. Harrington & Nagel (1977) *J. Lab. Clin. Med.* **90**, 863. **16**. McDonald (1955) *Meth. Enzymol.* **1**, 269. **17**. Colowick (1951) *The Enzymes*, 1st ed., **2** (Part 1), 114. **18**. Wilcock, Chahwala & Hickman (1988) *Biochim. Biophys. Acta* **946**, 368.

Meclofenamate (Meclofenamic Acid)

This nonsteroidal anti-inflammatory, *or* NDAID, and "double" inhibitor (FW$_{\text{free-acid}}$ = 296.15 g/mol; CAS 644-62-2), also known as Meclomen[®] and 2-[(2,6-dichloro-3-methylphenyl)amino]benzoic acid, targets the cyclooxygenase and lipoxygenase pathways of arachidonate metabolism. Meclomen also interferes with leukotriene receptors and post-receptor mechanisms. Meclomen inhibits leukotriene C_4 and leukotriene D_4 induced contractions, but had no effect on histamine induced contractions. Ibuprofen, naproxen and indomethacin had no effect on lipoxygenases or on leukotriene receptors but were powerful inhibitors of cyclooxygenases. It is FDA-approved for the treatment of joint, muscular pain, arthritis and dysmenorrhea. **Target(s):** cyclooxygenase, *or* prostaglandin-endoperoxide synthase (1-4); estrone sulfotransferase (5); hKv2.1, a major human neuronal voltage-gated K^+ channel (6); 5-lipoxygenase (1,2); 15-lipoxygenase (2); phenol sulfotransferase, *or* aryl sulfotransferase (5). **1.** Boctor, Eickholt & Pugsley (1986) *Prostaglandins Leukot. Med.* **23**, 229. **2.** Stadler, Kapui & Ambrus (1994) *J. Med.* **25**, 371. **3.** Tai, Tai & Hollander (1976) *Biochem. J.* **154**, 257. **4.** Meade, Smith & DeWitt (1993) *J. Biol. Chem.* **268**, 6610. **5.** King, Ghosh & Wu (2006) *Curr. Drug Metab.* **7**, 745. **6.** Lee & Wang (1999) *Eur. J. Pharmacol.* **378**, 349.

MEDI-4736, *See* Durvalumab

MEDICA 16

This nonmetabolizable long-chain fatty acid and FFA receptor agonist (FW = 342.51 g/mol; CAS 87272-20-6; Origin of Name: alpha, oMEga-DICarboxylic Acid; Solubility: 100 mM in DMSO or Ethanol), also named 3,3,14,14-tetramethylhexadecanedioate and β,β′-tetramethyl-hexadecanedioate, is a rotationally symmetric fatty acid analogue that targets Free Fatty Acid (*or* FFA) receptors, exhibiting selectivity for FFA$_1$ (GPR40) over GPR120. The latter are G-protein-coupled receptors (whose endogenous ligands are medium- and long-chain free fatty acids) thought to play an important physiological role in insulin release. MEDICA16 selectively activates [Ca^{2+}]$_i$ response in GPR40-expressing cells, but not in GPR120-expressing cells (1). Medica-16 also inhibits ATP citrate lyase, limiting the supply of acetyl-CoA required by acetyl-CoA carboxylase (ACC), the rate-limiting liver enzyme in fatty acid biosynthesis and a key regulator of muscle fatty acid oxidation (2,3). Rat liver acetyl-CoA carboxylase activity was inhibited by the free as well as the CoA monothioester of β,β′-tetramethylhexadecanedioate (1). The CoA monothioester of MEDICA 16 is a dead-end inhibitor, with K_i values of 2 μM and 58 μM for carboxylated and noncarboxylated ACC (3). MEDICA 16-CoA binding is not mutually exclusive with that of citrate and does not affect the avidin-resistance of rat liver acetyl-CoA carboxylase. The free dioic acid of MEDICA 16 is competitive to citrate (K_i = 70 μM), compared to a K_a of 2-8 mM for citrate as an activator. Inhibition of the carboxylase by the free dioic acid of MEDICA 16 is accompanied by an increase in its avidin resistance (3). The resultant inhibition of acetyl-CoA carboxylase by MEDICA 16 and its CoA thioester, together with the previously reported citrate-competitive inhibition of ATP-citrate lyase by MEDICA 16, may account for the drug's observed hypolipidemic effect under dietary conditions where liver lipogenesis constitutes a major flux of lipid synthesis (3). **1.** Hara, Hirasawa, Sun, *et al.* (2009) *Naunyn. Schmiedebergs. Arch. Pharmacol.* **380**, 247. **2.** Bar-Tana, Rose-Kahn & Srebnik (1985) *J. Biol. Chem.* **260**, 8404. **3.** Rose-Kahn & Bar-Tana (1990) *Biochim. Biophys. Acta* **1042**, 259.

Mefloquine

This antimalarial agent (FW$_{\text{free base}}$ = 378.32 g/mol; CAS 53230-10-7) interacts with DNA, raises the intravascular pH in the acid vesicles of the parasite (*Plasmodium falciparum*), and blocks gap junction channels Cx36 and Cx50. Note that mefloquine is structurally related to quinine. It reportedly stimulates Na^+/K^+-exchanging ATPase, Ca^{2+}/Mg^{2+}-ATPase activity, and the choline/Mg^{2+} antiport system. **Key Pharmacokinetic Parameters:** *See* Appendix II in Goodman & Gilman's *THE PHARMACOLOGICAL BASIS OF THERAPEUTICS*, 12th Edition (Brunton, Chabner & Knollmann, eds.) McGraw-Hill Medical, New York (2011). **Target(s):** acetylcholinesterase (1,2); aromatase, *or* CYP19 (3); Ca^{2+} currents4; cholinesterase, *or* butyrylcholinesterase (1,2,5); Cl^- channels, volume-regulated and Ca^{2+}-activated (6); dolichol kinase (7); gap junction channels Cx36 and Cx508; halofantrine dealkylase (9); K^+ channels (10-12); ornithine decarboxylase, weakly inhibited (13); *Plasmodium falciparum* aminopeptidase (14); protein kinase C (15,16); ribosyldihydronicotinamide dehydrogenase (quinone) (17); tolbutamide hydroxylase (cytochrome P450) (18). **1.** Lim & Go (1985) *Clin. Exp. Pharmacol. Physiol.* **12**, 527. **2.** Ngiam & Go (1987) *Chem. Pharm. Bull.* **35**, 409. **3.** Ayub & Scott (1988) *J. Steroid Biochem.* **29**, 149. **4.** Nori & Barry (1997) *J. Comp. Physiol. [A]* **180**, 473. **5.** McArdle, Sellin, Coakley, *et al.* (2005) *Neuropharmacology* **49**, 1132. **6.** Maertens, Wei, Droogmans & Nilius (2000) *J. Pharmacol. Exp. Ther.* **295**, 29. **7.** Walter (1986) *Exp. Parasitol.* **62**, 356. **8.** Cruikshank, Hopperstad, Younger, *et al.* (2004) *Proc. Natl. Acad. Sci. U.S.A.* **101**, 12364. **9.** Baune, Furlan, Taburet & Farinotti (1999) *Drug Metab. Dispos.* **27**, 565. **10.** Gribble, Davis, Higham, Clark & Ashcroft (2000) *Brit. J. Pharmacol.* **131**, 756. **11.** Kang, Chen, Wang & Rampe (2001) *J. Pharmacol. Exp. Ther.* **299**, 290. **12.** Traebert, Dumotier, Meister, *et al.* (2004) *Eur. J. Pharmacol.* **484**, 41. **13.** Konigk & Putfarken (1983) *Tropenmed. Parasitol.* **34**, 1. **14.** Vander Jagt, Baack & Hunsaker (1984) *Mol. Biochem. Parasitol.* **10**, 45. **15.** Agarwal & Walter (1988) *Gen. Pharmacol.* **19**, 219. **16.** el Benna, Hakim & Labro (1992) *Biochem. Pharmacol.* **43**, 527. **17.** Graves, Kwiek, Fadden, *et al.* (2002) *Mol. Pharmacol.* **62**, 1364. **18.** Karbwang, Back, Bunnag & Breckenridge (1988) *Southeast Asian J. Trop. Med. Public Health* **19**, 235.

MEK162, *See* Binimetinib

Melagatran

This antithrombotic agent (FW = 429.52 g/mol; CAS 159776-70-2) inhibits thrombin, K_i = 2 nM (1-5). The prodrug is known as ximelagatran. In early 2006, AstraZeneca withdrew the anticoagulant Exanta™ (melagatran/ximelagatran) from the market; although effective for short-term prevention of venous thromboembolism following orthopedic surgery, Exanta showed a higher risk of liver injury. **1.** Nilsson, Sjoling-Ericksson & Deinum (1998) *J. Enzyme Inhib.* **13**, 11. **2.** Gustafsson, Antonsson, Bylund, *et al.* (1998) *Thromb. Haemost.* **79**, 110. **3.** Weitz (2003) *Thromb. Res.* **109** Suppl. 1, S17. **4.** Deinum, Gustavsson, Gyzander, *et al.* (2002) *Anal. Biochem.* **300**, 152. **5.** Dullweber, Stubbs, Musil, Stürzebecher & Klebe (2001) *J. Mol. Biol.* **313**, 593.

Melanostatin (or Melanotropin Release-inhibiting Hormone)

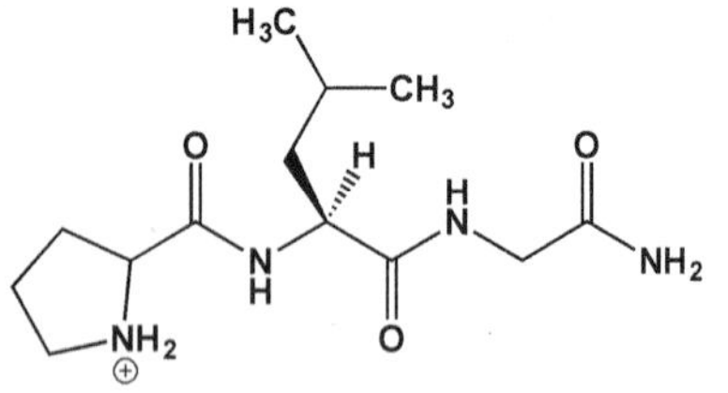

This peptide amide (FW = 284.37 g/mol; CAS 2002-44-0; *Sequence*: PLG-NH$_2$), which is derived by proteolysis of oxytocin (*Sequence*: CYIQNCPLG-NH$_2$), inhibits anterior pituitary synthesis/secretion of melanotropin. This peptide blocks the effects of opioid receptor activation (1). Melanostatin is also an allosteric activator of the D$_2$ and D$_4$ dopamine receptor subtypes. (*Note*: Neuropeptide Y has also been called melanostatin.) **1.** Dickinson & Slater (1980) *Peptides* **1**, 293.

Melarsen Oxide

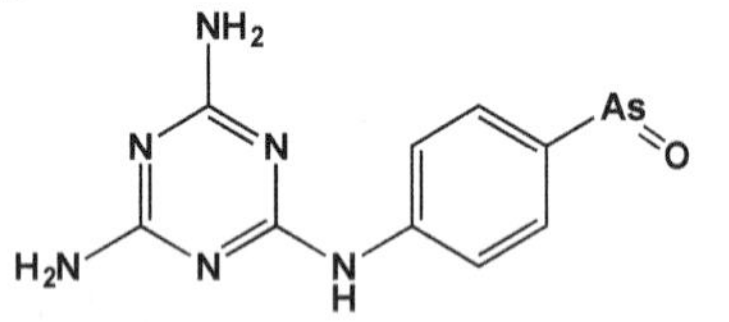

This organoarsenical (FW = 292.13 g/mol; CAS 21840-08-4), also named *p*-(4,6-diamino-1,3,5-triazin-2-yl)aminophenylarsene oxide, is the active form of the trypanocidal drugs melarsoprol and cymelarsan. Melarsen oxide forms tight complexes with DL-dihydrolipoamide and DL-dihydrolipoate with association constants of 5.47 x 10^9 and 4.51 x 10^9 M^{-1}, respectively. It also forms a stable adduct with trypanothione (association constant = 1.05 x 10^7 M^{-1}). (*Note*: In the older literature (*e.g.*, in ref. 2 below), the oxide has also been referred to simply as melarsen.) (*See also Pentamidine, Suramin, and Eflomithine*) **Target(s):** aldehyde dehydrogenase (1); calcineurin (2,3); dihydrolipoamide dehydrogenase (4); fructose-2,6-bisphosphatase (5); fumarase (6); glutamate dehydrogenase (7); glutathione-disulfide reductase (8,9); glycerol-3-phosphate dehydrogenase (10-12); malate dehydrogenase (7); malic enzyme (7); phosphoenolpyruvate carboxylase (13); 6-phosphofructo-2-kinase (5); pyruvate carboxylase (7); pyruvate kinase. weakly inhibited (5,7); succinate dehydrogenase (14); trypanothione reductase (9). **1.** Schwarcz & Stoppani (1970) *Enzymologia* **38**, 269. **2.** Bogumil & Ullrich (2002) *Meth. Enzymol.* **348**, 271. **3.** Bogumil, Namgaladze, Schaarschmidt, *et al.* (2000) *Eur. J. Biochem.* **267**, 1407. **4.** Fairlamb, Smith & Hunter (1992) *Mol. Biochem. Parasitol.* **53**, 223. **5.** Van Schaftingen, Opperdoes & Hers (1987) *Eur. J. Biochem.* **166**, 653. **6.** Favelukes & Stoppani (1958) *Biochim. Biophys. Acta* **28**, 654. **7.** Mottram & Coombs (1985) *Exp. Parasitol.* **59**, 151. **8.** Muller, Walter & Fairlamb (1995) *Mol. Biochem. Parasitol.* **71**, 211. **9.** Cunningham, Zvelebil & Fairlamb (1994) *Eur. J. Biochem.* **221**, 285. **10.** Suresh, Turley, Opperdoes, Michels & Hol (2000) *Structure Fold Des.* **8**, 541. **11.** Marche, Michels & Opperdoes (2000) *Mol. Biochem. Parasitol.* **106**, 83. **12.** Grant & Sargent (1961) *Biochem. J.* **81**, 206. **13.** Cannata & Stoppani (1963) *J. Biol. Chem.* **238**, 1208 and 1919. **14.** Stoppani & Brignone (1956) *Biochem. J.* **64**, 196.

Melatonin

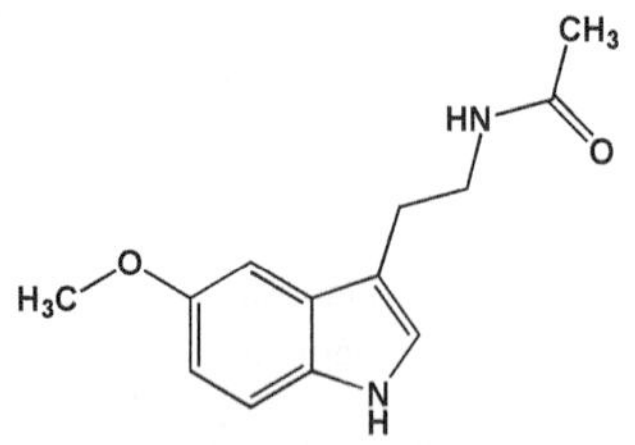

This photosensitive pineal hormone (FW = 232.28 g/mol; CAS 73-31-4; M.P. = 116-118°C; λ$_{max}$ at 223 nm (in 95% ethanol), with ε = 27550 M^{-1}cm^{-1}, and 278 nm, with ε = 6300 M^{-1}cm^{-1}), also known as *N*-acetyl-5-methoxytryptamine, mediates photoperiodicity and serves as a peroxynitrite scavenger. It also stimulates the aggregation of melanosomes in amphibians, affects locomotor activity in sparrows, and participates in thermoregulation of some ectotherms. Melatonin will lighten skin color by reversing the effect of melanocyte-stimulating hormone. **Target(s):**

aralkylamine *N*-acetyltransferase (1,2); nitric-oxide synthase (3,4); ribosydihydronicotinamide dehydrogenase (quinone) (5-8). **1.** Ferry, Loynel, Kucharczyk, *et al.* (2000) *J. Biol. Chem.* **275**, 8794. **2.** Zilberman-Peled, Benhar, Coon, Ron & Gothilf (2004) *Gen. Comp. Endocrinol.* **138**, 139. **3.** Pozo, Reiter, Calvo & Guerrero (1994) *Life Sci.* **55**, PL455. **4.** Pozo, Reiter, Calvo & Guerrero (1997) *J. Cell Biochem.* **65**, 430. **5.** Maiti, Reddy, Sturdy, *et al.* (2009) *J. Med. Chem.* **52**, 1873. **6.** Mailliet, Ferry, Vella, *et al.* (2005) *Biochem. Pharmacol.* **71**, 74. **7.** Nosjean, Ferro, Cogé, *et al.* (2000) *J. Biol. Chem.* **275**, 31311. **8.** Calamini, Santarsiero, Boutin & Mesecar (2008) *Biochem. J.* **413**, 81.

Meldonium

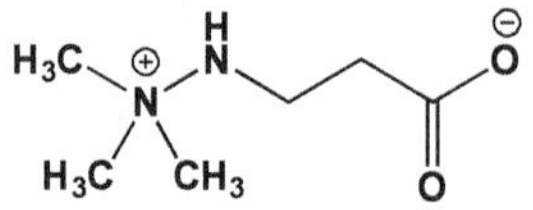

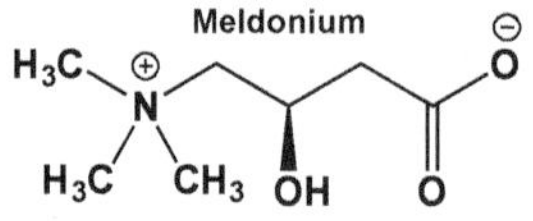

This carnitine analogue and anti-ischemic drug (FW = 147.19 g/mol; CAS 86426-17-7; Solubility: >40 mg/mL), also known as Mildronate, Mildronāts, MET-88, and Quaterine, as well as 2-(2-carboxyethyl)-1,1,1-trimethylhydrazinium, is listed (as of January, 2016) by the World Anti-Doping Agency (WADA) as a banned substance in athletic competition. Meldonium was originally developed as a performance-enhancing drug by selecting agents that increased the work capacity of previously trained animals (1). Treatment of guinea pigs with meldonium results in a 50% decline in myocardial carnitine content, from 11 to 5.6 μmol/g dry weight (2). Meldonium is a noncompetitive inhibitor of butyrobetaine hydroxylase and inhibits fatty acid oxidation (3). Meldonium inhibits carnitine acetyl transferase, thereby contributing to an increase in acetyl-CoA availability for intramitochondrial metabolic pathways. It also prevents L-carnitine-induced stimulation of [U-^{14}C]-palmitic acid oxidation *in vitro*. During administration as well as after addition into incubation medium, meldonium shows no significant effects on the oxidation of [1-^{14}C]-palmitoyl-L-carnitine, but decreased the exogenous L-carnitine induced oxidation of the substrate *in vitro*. It does not affect the activity of carnitine palmitoyl transferase I (3). Meldonium inhibits carnitine acetyltransferase competitively with respect to carnitine (K_i = 1.6 mM), and NMR spectroscopy and molecular docking show that its bound conformation closely resembles that of carnitine, except for the orientation of the (CH$_3$)$_3$N$^+$ group of meldonium is exposed to the solvent (4). It now appears that, by reducing fatty acid oxidation, meldonium increases the amount of O$_2$ available for respiration and mitochondrial ATP synthesis, thereby giving an aerobic advantage to athletes using this agent. On March 7, 2016, tennis star Maria Sharapova admitted to having tested positive for meldonium and to having taken this drug for health reasons over the previous decade. During that interval, she had won four of her five Grand Slam tournament championships. **1.** Seĭfulla, Trevisani, Morozov, *et al.* (1993) *Eksp. Klin. Farmakol.* **56**, 34. **2.** Dhar, Grupp, Schwartz, Grupp & Matlib (1996) *J. Cardiovasc. Pharmacol. Ther.* **1**, 235. **3.** Shutenko, Simkhovich, Meĭrena, *et al.* (1989) *Vopr. Med. Khim.* **35**, 59. **4.** Jaudzems, Kuka, Gutsaits, *et al.* (2009) *J. Enzyme Inhib. Med. Chem.* **24**, 1269)

Melezitose

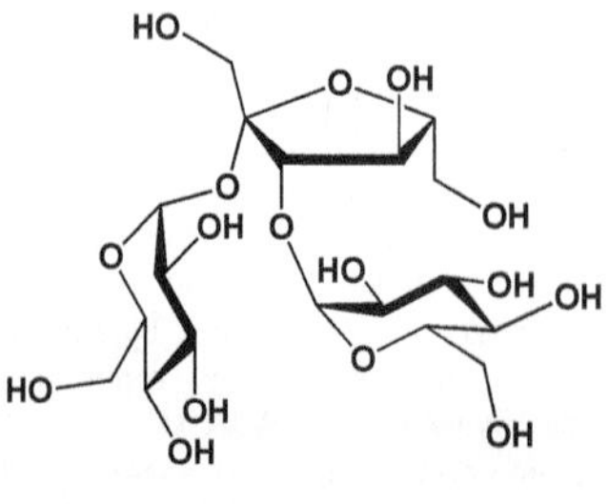

This nonreducing tri-saccharide (FW = 504.44 g/mol; CAS 597-12-6), also known as *O*-α-D-glucopyranosyl-(1→3)-*O*-β-D-fructofuranosyl-(2→1)-α-D-glucopyranoside, is produced by many plants and is found in a manna produced by Douglas fir and certain pine trees. Bees that have collected manna from these trees during times of drought produce honey containing elevated concentrations of melezitose, which is nutritionally unsuitable for maintenance of the hive. **Target(s):** concanavalin A (1); β-fructofuranosidase, *or* invertase, weakly inhibited (2); α-glucosidase, weakly inhibited (3); inulinase (4); sucrose-phosphatase (5-8). **1.** Goldstein, Hollerman & Smith (1965) *Biochemistry* **4**, 876. **2.** Isla, Vattuone, Gutierrez & Sampietro (1988) *Phytochemistry* **27**, 1993. **3.** Webb (1966) *Enzyme and Metabolic Inhibitors*, vol. **2**, p. 416, Academic Press, New York. **4.** Avigad & Bauer (1966) *Meth. Enzymol.* **8**, 621. **5.** Hawker & Hatch (1975) *Meth. Enzymol.* **42**, 341. **6.** Hawker (1967) *Biochem. J.* **102**, 401. **7.** Echeverria & Salerno (1993) *Physiol. Plant.* **88**, 434. **8.** Hawker (1971) *Phytochemistry* **10**, 2313.

β-D-Melibiose

This reducing disaccharide (FW = 342.30 g/mol; CAS 585-99-9), also known as 6-*O*-α-D-galactopyranosyl-D-glucose, is produced by a number of plants and is a component of the structure of raffinose. It is an epimer of isomaltose, is very soluble in water, and typically supplied as a mixture of the α- and β-anomers. Dilute acids will hydrolyze melibiose to the constituent monosaccharides. It is also a substrate for many α-galactosidases. **Target(s):** aldose 1-epimerase (1); aralin (2); *Bauhinia purpurea* agglutinin (3); β-Fructofuranosidase, *or* invertase, weakly inhibited (4); galactinol:raffinose galactosyltransferase, *or* stachyose synthase, weakly inhibited (5); α-galactosidase (6-8); β-galactosidase (6,9,10); β-glucosidase (11); rRNA *N*-glycosylase (2). **1.** Bailey, Fishman, Kusiak, Mulhern & Pentchev (1975) *Meth. Enzymol.* **41**, 471. **2.** Tomatsu, Kondo, Yoshikawa, *et al.* (2004) *Biol. Chem.* **385**, 819. **3.** Osawa, Irimura & Kawaguchi (1978) *Meth. Enzymol.* **50**, 367. **4.** Isla, Vattuone, Gutierrez & Sampietro (1988) *Phytochemistry* **27**, 1993. **5.** Gaudreault & Webb (1981) *Phytochemistry* **20**, 2629. **6.** Webb (1966) *Enzyme and Metabolic Inhibitors*, vol. **2**, p. 418, Academic Press, New York. **7.** Sheinin & Crocker (1961) *Can. J. Biochem. Physiol.* **39**, 55. **8.** Pederson & Goodman (1980) *Can. J. Microbiol.* **26**, 978. **9.** Lester & Bonner (1952) *J. Bacteriol.* **63**, 759. **10.** Ikura & Horikoshi (1979) *Agric. Biol. Chem.* **43**, 1359. **11.** Ferreira & Terra (1983) *Biochem. J.* **213**, 43.

Melittin

This oligopeptideamide apitoxin and membranolytic agent (MW = 2850 g; CAS 37231-28-0; *Sequence*: GIGAVLKVLTTGLPALISWIKRKRQQ-NH$_2$, with N-terminal glycyl residue often formylated; Solubility: 1 mg/mL acetonitrile-H$_2$O, 20% vol/vol) is the main hemolytic component in honey bee (*Apis mellifica*) venom, comprising ~40-50% of the dry venom by weight. Melittin forms transmembrane pores that alter permeability barriers. When adsorbed onto a membrane surface, melittin self-associates and perturbs bilayer structure. Melittin also activates many phospholipases and binds to calmodulin and calmodulin-dependent proteins. Its mouse LD$_{50}$ is about 4 μg/g body weight. **Structural Features:** Like other pore-forming peptides, melittin is amphipathic. Its hydrophobic and positively charged groups arrange on opposite surfaces of an α-helix, as indicated by helical wheel diagrams and its solubility (*i.e.*, >250 mg/mL H$_2$O and ~20 mg/mL CH$_3$OH). In the presence of a phospholipid-rich bilayer, melittin molecules bind within milliseconds and orient either parallel or perpendicular with respect to the plane of the membrane. Those molecules adopting the perpendicular orientation are believed to self-assemble into transmembrane pores, whereas those in the parallel arrangement are inactive (1). The kinetics of the overall pore-forming can be studied monitoring hemoglobin leakage from melittin-treated red blood cells. **Target(s):** acetylcholinesterase (2); calmodulin-lysine *N*-methyltransferase (3,4); Ca^{2+}-transporting ATPase (5-10); epidermal growth factor-induced protein-tyrosine kinase (11); F$_1$F$_o$ ATPase, *or* F$_1$F$_o$ ATP synthase (12); G$_s$ activity of guanine nucleotide-dependent adenylyl cyclase activity in synaptic membranes of the rat cerebral cortex by inhibition of GDP release in exchange for added guanine nucleotides or the association of guanine nucleotides (13); H$^+$/K$^+$-dependent ATPase (14-16); Na$^+$/K$^+$-exchanging ATPase (6,14,17-20); phosphorylase kinase (21); photophosphorylation (uncoupled) (22); protein kinase C (17,23-28); protein-tyrosine phosphatase

(29); Rho ADP-ribosyltransferase (30,31). **1.** Guzmán, Mika & Poolman (2008) *J. Biol. Chem.* **283**, 33854. **2.** Mitchell, Lowy, Sarmiento & Dickson (1971) *Arch. Biochem. Biophys.* **145**, 344. **3.** Pech & Nelson (1994) *Biochim. Biophys. Acta* **1199**, 183. **4.** Wright, Bertics & Siegel (1996) *J. Biol. Chem.* **271**, 12737. **5.** Shorina, Mast, Lopina & Rubtsov (1997) *Biochemistry* **36**, 13455. **6.** Murtazina, Mast, Rubtsov & Lopina (1997) *Biochemistry (Moscow)* **62**, 54. **7.** Baker, East & Lee (1995) *Biochemistry* **34**, 3596. **8.** Voss, Mahaney & Thomas (1995) *Biochemistry* **34**, 930. **9.** Voss, Birmachu, Hussey & Thomas (1991) *Biochemistry* **30**, 7498. **10.** Mahaney & Thomas (1991) *Biochemistry* **30**, 7171. **11.** Errasfa & Stern (1994) *Biochim. Biophys. Acta* **1222**, 471. **12.** Bullough, Ceccarelli, Roise & Allison (1989) *Biochim. Biophys. Acta* **975**, 377. **13.** Fukushima, Kohno, Kato, *et al.* (1998) *Peptides* **19**, 811. **14.** Yang & Carrasquer (1997) *Zhongguo Yao Li Xue Bao* **18**, 3. **15.** Cuppoletti & Malinowska (1992) *Mol. Cell Biochem.* **114**, 57. **16.** Cuppoletti, Blumenthal & Malinowska (1989) *Arch. Biochem. Biophys.* **275**, 263. **17.** Raynor, Zheng & Kuo (1991) *J. Biol. Chem.* **266**, 2753. **18.** Cuppoletti & Abbott (1990) *Arch. Biochem. Biophys.* **283**, 249. **19.** Yang, Zhang & Jiang (2001) *Acta Pharmacol. Sin.* **22**, 279. **20.** Chen & Lin-Shiau (1985) *Biochem. Pharmacol.* **34**, 2335. **21.** Paudel, Xu, Jarrett & Carlson (1993) *Biochemistry* **32**, 11865. **22.** Davies & Berg (1981) *Arch. Biochem. Biophys.* **211**, 297. **23.** Kikkawa & Nishizuka (1986) *The Enzymes*, 3rd ed., **17**, 167. **24.** Baudier, Mochly-Rosen, Newton, *et al.* (1987) *Biochemistry* **26**, 2886. **25.** Katoh (2002) *J. Vet. Med. Sci.* **64**, 779. **26.** Gravitt, Ward & O'Brian (1994) *Biochem. Pharmacol.* **47**, 425. **27.** O'Brian & Ward (1989) *Mol. Pharmacol.* **36**, 355. **28.** Katoh, Raynor, Wise, *et al.* (1982) *Biochem. J.* **202**, 217. **29.** Errasfa & Stern (1993) *Eur. J. Pharmacol.* **247**, 73. **30.** Aktories & Just (1995) *Meth. Enzymol.* **256**, 184. **31.** Koch, Habermann, Mohr, Just & Aktories (1992) *Eur. J. Pharmacol.* **226**, 87.

Meloxicam

This non-steroidal anti-inflammatory drug (FW = 351.41 g/mol; CAS 71125-38-7) inhibits cyclooxygenase-2, the inducible isozyme associated with inflammatory responses (IC$_{50}$ = 4.7 μM) with a selectivity for cyclooxygenase-1 (IC$_{50}$ = 37 μM). In clinical studies, meloxicam has exhibited reliable efficacy against rheumatoid arthritis, osteoarthritis, lumbago, scapulohumeral periarthritis, and neck-shoulder-arm syndrome. **Target(s):** collagenase (1); cyclooxygenase, *or* prostaglandin-endoperoxide synthase (2-5); oxidative phosphorylation, as an uncoupler (6). Meloxicam also exerts antitumor effects by targeting the COX-2/MMP-2/E-cadherin, AKT, apoptotic and autophagic pathways in COX-2-dependent and COX-2-independent pathways, and by inhibiting cell autophagy (7). **1.** Barracchini, Franceschini, Amicosante, *et al.* (1998) *J. Pharm. Pharmacol.* **50**, 1417. **2.** Noble & Balfour (1996) *Drugs* **51**, 424. **3.** Engelhardt (1996) *Brit. J. Rheumatol.* **35** (Suppl. 1), 4. **4.** Patrignani, Panara, Sciulli, *et al.* (1997) *J. Physiol. Pharmacol.* **48**, 623. **5.** Sud'ina, Pushkareva, Shephard & Klein (2008) *Prostaglandins Leukot. Essent. Fatty Acids* **78**, 99. **6.** Moreno-Sanchez, Bravo, Vasquez, *et al.* (1999) *Biochem. Pharmacol.* **57**, 743. **7.** Dong, Li, Xiu, *et al.* (2014) *PLoS One* **9**, e92864.

Melphalan

This antineoplastic agent (FW = 305.20 g/mol; CAS 148-82-3), also known as L-sarcolysine, 4-[*bis*(2-chloroethyl)amino]-L-phenylalanine, and L-phenylalanine mustard, forms inter-strand cross-links in duplex DNA. The mutation-induced alkylations result primarily in transversions and occur preferentially at GGC sequences. The agent selectively forms a two base-staggered inter-strand cross-link between the 5'-G and the G-opposite-C in the 5'-GGC sequence (1). (*Note*: The term melphalan refers specifically to the L-enantiomer. The D-isomer is known as medphalan, and the racemic mixture is merphalan.) **Target(s):** p34-cdc2 kinase (1,2); RNA polymerase

(2,3); thioredoxin reductase (4). **1**. Bauer & Povirk (1997) *Nucl. Acids Res.* **25**, 1211. **2**. Orlandi, Zaffaroni, Bearzatto & Silvestrini (1996) *Brit. J. Cancer* **74**, 1924. **3**. Hill (1977) *Cancer Biochem. Biophys.* **2**, 43. **4**. Witte, Anestal, Jerremalm, Ehrsson & Arner (2005) *Free Radic. Biol. Med.* **39**, 696.

Memantine

This voltage-dependent, low-affinity noncompetitive NMDA receptor antagonist (FW = 215.76 g/mol; CAS 41100-52-0, for the HCl salt; Soubility: 100 mM in H_2O for HCL salt), also known by the code name D145, trade names Axura™, Akatinol™, Namenda™, Ebixa™, Abixa™ and Memox™, and its systematic name 1,3-dimethyl-5-aminoadamantan or 3,5-dimethyl-tricyclo-[3.3.1.13,7]decan-1-amine, is a neuroprotective drug used in the symptomatic treatment of Alzheimer's Disease and mild to moderate vascular dementia. By binding with more tightly to the NMDA receptor than Mg^{2+}, memantine inhibits prolonged Ca^{2+} influx, thereby forming the basis for its action in ameliorating neuronal excitotoxicity (1-3). This FDA-approved AD drug, which has been shown to modestly benefit cognition, function, and global outcome in patients with moderate to severe AD, is also a a noncompetitive seritonin 5HT₃ receptor antagonist (4). A sensitive and rapid HPLC method for the determination of memantine in human plasma has been reported (5). It should also be noted that memantine is a non-competitive antagonist of the type-3 5-hydroxytryptamine (serotonin) receptor, *or* 5-HT3 receptor and a non-competitive antagonist of different neuronal nicotinic acetylcholine receptors (nAChRs), with potencies comparable to NMDA and 5-HT3 receptors. **Key Pharmacokinetic Parameters:** *See* Appendix II in Goodman & Gilman's *THE PHARMACOLOGICAL BASIS OF THERAPEUTICS*, 12th Edition (Brunton, Chabner & Knollmann, eds.) McGraw-Hill Medical, New York (2011). **1**. Wesemann, Sontag & Maj (1983) *Arzneimittelforschung.* **33**, 1122. **2**. Rogawski & Wenk (2003) *CNS Drug Rev.* **9**, 275. **3**. Cacabelos, Takeda, Winblad (1999) *Int. J. Geriatr. Psychiatry* **14**, 3. **4**. Rammes, Rupprecht, Ferrari *et al.* (2001) *Neurosci. Lett.* **306**, 81. **5**. Zarghi, Shafaati, Foroutin *et al.* (2010) *Sci. Pharm.* **78**, 847.

MEN 10376

This potent and selective NK receptor antagonist (FW = 1081.24 g/mol; CAS 135306-85-3; Peptide Sequence: L-Asp-L-Tyr-D-Trp-L-Val-D-Trp-D-Trp-L-Lys-NH₂) targets Tachykinin NK₂ receptors (K_d = 9 nM, rabbit pulmonary artery). MEN 10376 shows >250x selectivity over NK₁ (K_d = 0.4 µM, guinea pig ileum) and NK₃ (K_d = 1 mM, guinea pig brain) (1-3). Tachykinin receptors regulate both excitatory and inhibitory intestinal motor functions as well as secretions, inflammation and visceral sensitivity. In particular, the NK₂ receptor stimulation inhibits intestinal motility by activating sympathetic extrinsic pathways or NANC intramural inhibitory components (4). MEN 10,376 is devoid of significant agonist activity (intravenous dose: 1-3 µmol/kg), selectively antagonizing the effects of an NK-2 agonist [β-Ala8]-Neurokinin A(4-10) in bladder contraction in rats and bronchoconstriction in guinea pigs, without affecting the response to the NK-1 agonist, [Sar9]-substance P sulfone on hypotension, salivation and bladder contraction in rats, bronchoconstriction in guinea pigs (1). **1**. Maggi, Giuliani, Ballati, *et al.* (1991) *J. Pharmacol. Exp. Ther.* **257**, 1172. **2**. Bartho, Santicioli, Patacchini & Maggi (1992) *Brit. J. Pharmacol.* **105**, 805. **3**. Subramanian N1, Ruesch C, Bertrand (1994) *Biochem. Biophys. Res. Commun.* **200**, 1512. **4**. Lecci, Capriati &Maggi (2004) *Brit. J. Pharmacol.* **141**, 1249.

MEN 11270

This conformationally constrained bradykinin receptor antagonist (FW = 1299.56 g/mol; CAS 235082-52-7; Structure: RRPX₁GX₂X₃X₄X₅R, with modifications: first Arg = D-Arg, X₁ = Hyp, X₂ = Thi, X₃ = Dab, X₄ = D-Tic, X₅ = Oic, cyclized 7γ-10α; Solubility: 1 mg/mL in acetonitrile/water (30% vol/vol)) targets the B₂ bradykinin receptor (K_i = 0.2 nM). *Note*: MEN 11270 is the cyclized analogue of HOE 140 (**See** *HOE 140*), and blocks hypotension and bronchoconstriction *in vivo*. MEN 11270 displays selectivity for B₂ over tenty-nine other receptors and ion channels (pIC₅₀ < 5.5). **1**. Meini, *et al.* (1999) *J. Pharmacol. Exp. Ther.* **289**, 1250. **2**. Tramontana, *et al.* (2001) *J. Pharmacol. Exp. Ther.* **296**, 1051. **3**. Cucchi, *et al.* (2002) *Peptides* **23**, 1457.

MEN15596, *See* Ibodutant

Menadione

This photosensitive dione (FW = 172.18 g/mol; CAS 58-27-5; M.P. = 105-107°C), also known as 2-methyl-1,4-naphthalenedione and vitamin K₃, is a synthetic quinone that exhibits the physiological properties and characteristics of vitamin K. Menadione functions as an electron carrier and participates in a number of oxidation-reduction reactions. Menadione is also a thiol reactive reagent. (**See also** *Menadiol; Menadione Bisulfite*) **Target(s):** aldehyde oxidase (1,2); aldose reductase, *or* aldehyde reductase (3,4); alkane 1-monooxygenase (37); aminopyrine *N*-demethylase (cytochrome P450) (5); aniline-*p*-hydroxylase (cytochrome P450) (5); (*S*)-canadine synthase (34); Ca²⁺-transporting ATPase (6); choline acetyltransferase (7,8); cholinesterase (7,9); *trans*-cinnamate 4-monooxygenase (39,40); glutathione disulfide reductase (10); α-glycerophosphatase (11); histidine kinase (12); horseradish peroxidase (13); β-hydroxybutyrate dehydrogenase. weakly inhibited (14); 3α-hydroxysteroid transhydrogenase (15); indoleamine 2,3-dioxygenase (41,42); *myo*-inositol oxygenase (16,17); iodide peroxidase, *or* thyroid peroxidase (44); lactate dehydrogenase (18); 5-lipoxygenase, *or* arachidonate 5-lipoxygenase (43); 3-mercaptopyruvate sulfurtransfrase (19); [mitogen-activated protein kinase kinase] kinase (20); NAD⁺ ADP-ribosyltransferase, *or* poly(ADP-ribose) polymerase (32); NADH dehydrogenase (21); nitrate reductase (22); peroxidase (13); protein-glutamine γ-glutamyltransferase, *or* transglutaminase (33); pyruvate decarboxylase (23); quinine oxidase (24); ribonuclease H activity associated with HIV-I reverse transcriptase (25); stearoyl-CoA Δ⁹-desaturase (35,36); steroid 9α-monooxygenase (26); sterol 14-demethylase, *or* CYP51 (38); succinate dehydrogenase (27); thiosulfate sulfurtransferase, rhodanese (19); ubiquinone thiolesterase (28); urease (29,30); vitamin K-dependent carboxylase, *or* protein-glutamate carboxylase (31). **1**. Rajagopalan, Fridovich & Handler (1962) *J. Biol. Chem.* **237**, 922. **2**. Lakshmanan, Vaidyanathan & Cama (1964) *Biochem. J.* **90**, 569. **3**. Bhatnagar, Liu, Petrash & Srivastava (1992) *Mol. Pharmacol.* **42**, 917. **4**. Davidson & Murphy (1985) *Prog. Clin. Biol. Res.* **174**, 251. **5**. Floreani & Carpenedo (1990) *Toxicol. Appl. Pharmacol.* **105**, 333. **6**. Floreani & Carpenedo (1989) *Arch. Biochem. Biophys.* **270**, 33. **7**. Torda & Wolff (1944) *Proc. Soc. Exptl. Biol. Med.* **57**, 236. **8**. Nachmansohn & Berman (1946) *J. Biol. Chem.* **165**, 551. **9**. Taylor, Weller & Hastings (1952) *Amer. J. Physiol.* **168**, 658. **10**. Luond, McKie, Douglas, Dascombe & Vale (1998) *J. Enzyme Inhib.* **13**, 327. **11**. Hoffmann-Ostenhof & Putz (1948) *Monatsh. Chem.* **79**, 421. **12**. Viaud, Fillinger, Liu, *et al.* (2006) *Mol. Plant Microbe Interact.* **19**, 1042. **13**. Klapper & Hackett (1963) *J. Biol. Chem.* **238**, 3736. **14**. Martius & Nitz-Litzow (1953) *Biochim. Biophys. Acta* **12**, 134. **15**. Koide (1962) *Biochim. Biophys. Acta* **59**, 708. **16**. Charalampous (1962) *Meth. Enzymol.* **5**, 329. **17**. Charalampous (1959) *J. Biol. Chem.* **234**, 220. **18**. Snoswell (1959) *Australian J. Exptl. Biol. Med. Sci.* **37**, 49. **19**. Wrobel & Jurkowska (2007) *Acta Biochim. Pol.* **54**, 407. **20**. Cross & Templeton (2004) *Biochem. J.* **381**, 675. **21**. Frimmer (1960) *Biochem. Z.* **332**, 522. **22**. Sadana & McElroy (1957) *Arch. Biochem. Biophys.* **67**, 16. **23**. Kuhn & Beinert (1947) *Ber.* **80**, 101. **24**. Villela (1963) *Enzymologia* **25**, 261. **25**. Min, Miyashiro & Hattori (2002) *Phytother. Res.* **16** Suppl. 1, S57. **26**. Chang & Sih (1964) *Biochemistry* **3**, 1551. **27**. Potter & DuBois (1943) *J. Gen. Physiol.* **26**, 391. **28**. Liu, Lashuel, Choi, *et al.* (2003) *Chem. Biol.* **10**, 837. **29**. Grant & Kinsey (1946) *J. Biol. Chem.* **165**, 485. **30**. Schopfer & Grob (1949) *Helv. Chim. Acta* **32**, 829. **31**. Johnson (1980) *Meth. Enzymol.*

67, 165. **32**. Banasik, Komura, Shimoyama & Ueda (1992) *J. Biol. Chem.* **267**, 1569. **33**. Lai, Liu, Tucker, *et al.* (2008) *Chem. Biol.* **15**, 969. **34**. Rueffer & Zenk (1994) *Phytochemistry* **36**, 1219. **35**. Fulco & Bloch (1964) *J. Biol. Chem.* **239**, 993. **36**. Wilson & Miller (1978) *Can. J. Biochem.* **56**, 1109. **37**. Nakayama & Shoun (1994) *Biochem. Biophys. Res. Commun.* **202**, 586. **38**. Aoyama, Yoshida & Sato (1984) *J. Biol. Chem.* **259**, 1661. **39**. Potts, Weklych & Conn (1974) *J. Biol. Chem.* **249**, 5019. **40**. Billett & Smith (1978) *Phytochemistry* **17**, 1511. **41**. Kumar, Malachowski, DuHadaway, *et al.* (2008) *J. Med. Chem.* **51**, 1706. **42**. Macchiarulo, Camaioni, Nuti & Pellicciari (2009) *Amino Acids* **37**, 219. **43**. Young (1999) *Eur. J. Med. Chem.* **34**, 671. **44**. Sugawara, Sugawara, Wen & Giulivi (2002) *Exp. Biol. Med. (Maywood)* **227**, 141.

O-Menthyl Hexylphosphonochloridate

This phosphonochloridate (FW = 322.86 g/mol) is an irreversible inhibitor of *Candida rugosa* lipase (*or* triacylglycerol lipase, reacting covalently with an active-site serine residue. While both *S*- and *R*-stereoisomers inhibit, the *R*-isomer is more efficient. **1**. Ransac, Gargouri, Marguet, *et al.* (1997) *Meth. Enzymol.* **286**, 190. **2**. Cygler, Grochulski, Kazlauskas, *et al.* (1994) *J. Amer. Chem. Soc.* **116**, 3180.

Mepacrine, *See* Quinacrine

Mepanipyrim

This substituted pyrimidine (FW = 223.28 g/mol; CAS 110235-47-7; Solubility: 5.58 mg/L H$_2$O at 20°C and soluble in most organic solvents), also known as *N*-(4-methyl-6-prop-1-ynylpyrimidin-2-yl)aniline, is a fungicide that inhibits the secretion of host cell wall-degrading enzymes. Mepanipyrim does not inhibit the cell wall-degrading enzymes (*e.g.,* cutinase, pectinase, and cellulases). Mepanipyrim also inhibits secretion of lipases, proteases, and invertase. **Target(s):** intracellular transport of very-low-density lipoprotein, *or* VLDL (1); retrograde Golgi-to-ER trafficking (2); secretion of host-cell wall degrading enzymes (3). **1**. Terada, Mizuhashi, Murata & Tomita (1999) *Toxicol. Appl. Pharmacol.* **154**, 1. **2**. Nakamura, Kono & Takatsuki (2003) *Biosci. Biotechnol. Biochem.* **67**, 139. **3**. Miura, Kamakura, Maeno, *et al.* (1994) *Pest. Biochem. Physiol.* **48**, 222.

Meperidine

This analgesic and μ-opioid receptor agonist (FW$_{free-base}$ = 247.34 g/mol; CAS 57-42-1), is known in the U.S. by the trade name Demerol™ (as the hydrochloride) and elsewhere as pethidine, Isonipecaine™, Lidol™, Pethanol™, Piridosal™, Algil™, Alodan™, Centralgin™, Dispadol™, Dolantin™, and Mialgin™. When compared with morphine, demerol is safer, less addictive, and, due to its anticholinergic effects, is also superior in the treatment of pain associated with biliary spasm or renal colic. **Pharmacokinetic Parameters:** Meperidine is rapidly hydrolysed in the liver

to pethidinic acid and is also demethylated (by CYP2B6, CYP3A4, and CYP2C19) to norpethidine, the latter having half its analgesic activity but a longer elimination half-life (8–12 hours). Meperidine is also de-esterified by hepatic carboxyesterase to form Pethidinic acid (inactive), which is then demethylated (by CYP2B6, CYP3A4, and CYP2C19) to Norpethidinic acid (inactive). Meperidine, Norpethidine, Pethidinate and Norpethidinate all undergo glucuronidation and subsequent renal excretion. *See also* Appendix II in Goodman & Gilman's THE PHARMACOLOGICAL BASIS OF THERAPEUTICS, 12th Edition (Brunton, Chabner & Knollmann, eds.) McGraw-Hill Medical, New York (2011). **Target(s):** K$^+$ channel (1); Na$^+$ channel (1,2); NADH dehydrogenase (*or* complex I *or* NADH:ubiquinone reductase) (3-5); NMDA (*N*-methyl-D-aspartate) receptor (6). **1**. Brau, Koch, Vogel & Hempelmann (2000) *Anesthesiology* **92**, 147. **2**. Wagner, Eaton, Sabnis & Gingrich (1999) *Anesthesiology* **91**, 1481. **3**. Hatefi & Stiggall (1978) *Meth. Enzymol.* **53**, 5. **4**. Hatefi (1978) *Meth. Enzymol.* **53**, 11. **5**. Hatefi & Stiggall (1976) *The Enzymes*, 3rd ed., **13**, 175. **6**. Yamakura, Sakimura & Shimoji (2000) *Anesth. Analg.* **90**, 928.

Mephedrone

This CNS-penetrant, serotonin reuptake inhibitor and "designer" stimulant (FW$_{HCl-Salt}$ = 213.69 g/mol; CAS 1189726-22-4; Soluble to 100 mM in H$_2$O; 100 mM in DMSO), also named (±)-2-methylamino-1-(4-methylphenyl)propan-1-one hydrochloride and 4-methylmethcathinone, targets striatal dopamine uptake (IC$_{50}$ = 467 nM) and hippocampal 5-HT uptake into synaptosomes (IC$_{50}$ = 558 nM). Administration of Mephedrone delpetes dopamine and reduces 5-HT transporter function. Mephedrone has a unique pharmacological profile with both abuse liability and neurotoxic potential. **1**. Hadlock, Webb, McFadden, *et al.* (2011) *J. Pharmacol. Exp. Ther.* **339**, 530.

Mepivacaine

This local anesthetic (FW = 246.35 g/mol; CAS 22801-44-1) reversibly blocks the transient inward sodium ion current (IC$_{50}$ = 0.3 mM for rat dorsal horn neurons of the spinal cord) and the steady-state potassium ion outward current (IC$_{50}$ = 0.2 mM for rat dorsal horn neurons of the spinal cord). **Target(s):** cholinesterase (1); hKv1.5 potassium channel (2); K$^+$ channel (3,4); Na$^+$ channel (3,4); TASK tandem pore baseline K$^+$ channel (5). **1**. Perez-Guillermo, Delgado & Vidal (1987) *Biochem. Pharmacol.* **36**, 3593. **2**. Longobardo, Delpon, Caballero, Tamargo & Valenzuela (1998) *Mol. Pharmacol.* **54**, 162. **3**. Brau, Vogel & Hempelmann (1998) *Anesth. Analg.* **87**, 885. **4**. Olschewski, Hempelmann, Vogel & Safronov (1998) *Anesthesiology* **88**, 172. **5**. Kindler, Yost & Gray (1999) *Anesthesiology* **90**, 1092.

Meprobamate

This anxiolytic agent and muscle relaxant (FW = 218.25 g/mol; CAS 57-53-4; M.P = 104-106°C), also known as 2-methyl-2-propyl-1,3-propanediol dicarbamate, Miltown™, and Equanil™, is not as effective as benzodiazepines and is often used for short-term relief of anxiety or nervousness (1). The drug can also be delivered as a pro-drug (Tybamate). **Target(s):** carbonic anhydrase (2); ecto-ATPase (3); protein-binding of histamine and serotonin (4). **1**. Ludwig & Piech (1951) *J. Amer. Chem. Soc.* **73**, 5779. **2**. Parr & Khalifah (1992) *J. Biol. Chem.* **267**, 25044. **3**. Melzig, Michalski & Teuscher (1989) *Biomed. Biochim. Acta* **48**, 431. **4**. Di Maggio, Cannavà & Ciaceri (1968) *Boll. Soc. Ital. Biol. Sper.* **44**, 2096.

Mepyramine

This selective histamine receptor inverse agonist (FW = 401.46 g/mol; CAS 59-33-6; Solubility: 10 mM in H_2O), also named Pyrilamine and 2-((2-(dimethylamino)ethyl)(*p*-methoxybenzyl)amino)pyridine maleate, targets H_1 receptors, inhibiting (log EC_{50} = –7.94) histamine-induced inositol phosphate production and intracellular calcium mobilization. [^{3}H]-Mepyramine labels the H_1 receptor as well as a liver [^{3}H]-Mepyramine-binding protein (MBP) related to debrisoquine 4-hydroxylase, *or* cytochrome P450IID (1). Binding of [^{3}H]-Mepyramine to cloned H_1 receptor is unaffected by 1 μM quinine, an inhibitor of debrisoquine 4-hydroxylase (1). H_1 receptors represent the majority of [^{3}H]-Mepyramine binding sites in the cerebral cortex, thalamus and hypothalamus, hippocampus, heart, aorta, lung and spleen (2). Mepyramine sequesters $G_{q/11}$, reducing its availability to other unrelated receptors within the same signaling pathway (2). **1**. Liu, Horio, Fujimoto & Fukui (1994) *J. Pharmacol. Exp. Ther.* **268**, 959. **2**. Fitzsimons, Monczor, Fernández, Shayo & Davio (2004) *J. Biol. Chem.* **279**, 34431.

Merbarone

This antitumor agent (FW = 263.28 g/mol; CAS 97534-21-9), also known as 5-(*N*-phenylcarboxamido)-2-thiobarbituric acid and NSC 336628, inhibits DNA topoisomerase II and blocks DNA cleavage (1-4). **1**. Andersen, Bendixen & Westergaard (1996) *DNA Replication in Eucaryotic Cells, Cold Spring Harbor Laboratory Press*, p. 587. **2**. Drake, Hofmann, Mong, *et al.* (1989) *Cancer Res.* **49**, 2578. **3**. Hammonds, Maxwell & Jenkins (1998) *Antimicrob. Agents Chemother.* **42**, 889. **4**. Insaf, Danks & Witiak (1996) *Curr. Med. Chem.* **3**, 437,

2-Mercaptobenzothiazole

This thiol (FW = 167.26 g/mol; CAS 149-30-4; *Abbreviation*: MBT), also called Mercap, Captax, and 2(3*H*)-benzothiazolethione, is a pale yellow solid that binds copper ions and inhibits some copper-dependent enzymes. Note that 2-mercaptobenzothiazole has a very low solubility in water, but is soluble in ethanol (2 g/100 mL at 25°C). 2-Mercaptobenzothiazole can also be used as a reducing reagent. Its primary commercial use is as a vulcanization accelerator in rubber production and as an inhibitor the corrosion of copper immersed in water. **Target(s):** catechol oxidase (1); dopamine β-monooxygenase (2); laccase (3); polyphenol oxidase (1,4); tyrosinase (5). **1**. Motoda (1979) *J. Ferment. Technol.* **57**, 79. **2**. Johnson, Boukma & Platz (1970) *J. Pharm. Pharmacol.* **22**, 710. **3**. Ishigami, Hirose & Yamada (1988) *J. Gen. Appl. Microbiol.* **34**, 401. **4**. Palmer & Roberts (1967) *Science* **157**, 200. **5**. Zhang, van Leeuwen, Wichers & Flurkey (1999) *J. Agric. Food Chem.* **47**, 374.

2-Mercaptoethanol

This reducing reagent (FW = 78.14 g/mol; CAS 60-24-2; Density = 1.114 g/mL; Concentration = 14.3 M), commonly known as β-mercaptoethanol (*or*, βME), reduces disulfide groups, thereby restoring the activity of many enzymes with catalytically essential thiol groups. When present in large molar excess relative to a disulfide-containing peptide or protein, 2-mercaptoethanol reduces accessible disulfide bonds with stoichiometric formation of the disulfide $HOCH_2CH_2-S-S-CH_2CH_2OH$ (*See Bis(2-hydroxyethyl) disulfide*). Mercaptoethanol is also used to maintain reduced thiols in coenzyme A and lipoic acid. The equilibrium constant for the thiol-disulfide exchange with glutathione disulfide (thus, 2 βME + GS–SG ⇌ βME–S–S–SβME + 2 GSH) is approximately one, whereas the corresponding value for disulfide bonds in intact proteins is often considerably less than one. For this reason, a large excess of 2-mercaptoethanol is required in order to reduce recalcitrant disulfides. Note that 2-mercaptoethanol is easily air oxidized, requiring βME solutions to be freshly prepared. Those enzymes and cofactors requiring disulfides to maintain their active conformation may be inhibited by 2-mercaptoethanol. βME complexation of metal ions can also result in inhibition.

2-Mercaptoethanol Disulfide, *See 2-Hydroxyethyl Disulfide*

Mercaptoethylguanidine

This peroxynitrite scavenger (FW = 119.19 g/mol) inhibits the inducible isoform of nitric-oxide synthase. **Target(s):** cysteamine dioxygenase (1); dopamine β-hydroxylase (2); nitric-oxide synthase (3-5). **1**. Richerson & Ziegler (1987) *Meth. Enzymol.* **143**, 410. **2**. Diliberto, DiStefano & Smith (1973) *Biochem. Pharmacol.* **22**, 2961. **3**. Szabo, Ferrer-Sueta, Zingarelli, *et al.* (1997) *J. Biol. Chem.* **272**, 9030. **4**. Zingarelli, Cuzzocrea, Szabo & Salzman (1998) *J. Pharmacol. Exp. Ther.* **287**, 1048. **5**. Southan, Zingarelli, O'Connor, Salzman & Szabo (1996) *Brit. J. Pharmacol.* **117**, 619.

8-Mercaptoguanine

This antineoplastic guanine derivative (FW = 362.35 g/mol; CAS 6324-72-7) inhibits hypoxanthine(guanine) phosphoribosyltransferase (1). The essential folate biosynthetic enzyme, 6-hydroxymethyl-7,8-dihydropterin pyrophosphokinase, which catalyzes the Mg^{2+}-dependant transfer of pyrophosphate from ATP to 6-hydroxymethyl-7,8-dihydropterin binds 8-mercaptoguanine at the substrate site (K_d ~13 μM) and inhibits *Staphylococcus aureus* enzyme HPPK (IC_{50} ~ 41 μM) (2). 8-Mercaptoguanine is also a useful starting reagent for the synthesis of various C-8 derivatives of purines. **1**. Miller, Ramsey, Krenitsky & Elion (1972) *Biochemistry* **11**, 4723. **2**. Chhabra, Barlow, Dolezal, *et al.* (2013) *PLoS One* **8**, e59535.

N-(2(*R*,*S*)-Mercaptoheptanoyl)-L-phenylalanyl-L-alanine

This α-mercaptoacyldipeptide (FW$_{\text{free-acid}}$ = 380.51 g/mol), also known as N-(2(R,S)-sulfanylheptanoyl)-L-phenylalanyl-L-alanine, inhibits neprilysin and thermolysin: K_i = 2.9 and 48 nM, respectively. Crystal structure analysis suggests that the mercaptoacyl moieties act as bidentates, with Zn–S and Zn–O distances of 2.3 and 2.4 Å, respectively. The side chains fit within the S$_1$, S$_1$', and S$_2$' pockets. Moreover, a distance of 3.1 Å between the sulfur atom and the OE1 of Glu^{143} suggests that they are H-bonded and that one of these atoms is protonated. This H-bond network involving Glu^{143}, the mercaptoacyl group of the inhibitor, and the Zn ion could be considered a "modified" transition state mimic of the peptide bond hydrolysis. **1**. Gaucher, Selkti, Tiraboschi, *et al.* (1999) *Biochemistry* **38**, 12569.

3-Mercapto-2-ketoglutarate

This α-ketoglutarate derivative (FW$_{\text{free-acid}}$ = 178.17 g/mol), also known as 3-mercapto-2-oxoglutarate and β-mercapto-α-ketoglutarate, potently inhibits isocitrate dehydrogenase and is an alternative substrate of α-ketoglutarate dehydrogenase. For porcine mitochondrial NADP$^+$-specific isocitrate dehydrogenase, inhibition is competitive with respect to isocitrate, K_i = 5 nM (1). **Target(s):** isocitrate dehydrogenase, NAD$^+$-specific (1); isocitrate dehydrogenase NADP$^+$-specific (1); γ-butyrobetaine dioxygenase, thereby blocking L-carnitine synthesis (2). **1**. Plaut, Aogaichi & Gabriel (1986) *Arch. Biochem. Biophys.* **245**, 114. **2**. Wehbie, Punekar & Lardy (1988) *Biochemistry* **27**, 2222.

Mercaptomerin

This diuretic (FW$_{\text{disodium-salt}}$ = 606.01 g/mol), also known as thiomerin, reportedly inhibits ATPase, oubain-sensitive (1), glutamine synthesis (2), histamine N-methyltransferase, weakly inhibited (3), and Na$^+$/K$^+$-exchanging ATPase (4). **1**. Nechay, Nelson, Contreras, *et al.* (1975) *J. Pharmacol. Exp. Ther.* **192**, 303. **2**. Richterich-Van Baerle, Goldstein & Dearborn (1957) *Enzymologia* **18**, 327. **3**. Thithapandha & Cohn (1978) *Biochem. Pharmacol.* **27**, 263. **4**. Czackes, Wald & Gutman (1977) *Res. Commun. Chem. Pathol. Pharmacol.* **17**, 133.

DL-2-Mercaptomethyl-3-guanidinoethylthiopropanoic Acid

This thiol-containing arginine analogue (FW = 237.35 g/mol), also known as MGTA, MERGETPA, and Plummer's inhibitor, is a potent inhibitor of carboxypeptidase N, which catalyzes the hydrolysis of bradykinin. **Target(s):** carboxypeptidase B (1,12); carboxypeptidase D (2); carboxypeptidase E, *or* carboxypeptidase H (2,3,12); carboxypeptidase M (4-9,12); carboxypeptidase U (10-12); lysine carboxypeptidase, *or* lysine(arginine) carboxypeptidase, *or* carboxypeptidase N, *or* arginine carboxypeptidase, K_i = 2 nM (12-19); metallocarboxypeptidase D (20); Xaa-Pro aminopeptidase, *or* aminopeptidase P (21-23). **1**. Muto, Suzuki, Sato & Ishii (2003) *Eur. J. Pharmacol.* **461**, 181. **2**. Fricker (2002) *The Enzymes*, 3rd ed., **22**, 421. **3**. Juvvadi, Fan, Nagle & Fricker (1997) *FEBS Lett.* **408**, 195. **4**. Skidgel (1998) in *Handb. Proteolytic Enzymes*, p. 1347, Academic Press, San Diego. **5**. Nagae, Abe, Becker, *et al.* (1993) *Amer. J. Respir. Cell Mol. Biol.* **9**, 221. **6**. Nagae, Deddish, Becker, *et al.* (1992) *J. Neurochem.* **59**, 2201. **7**. Desmazes, Lockhart, Lacroix & Dusser (1992) *Amer. J. Respir. Cell Mol. Biol.* **7**, 477. **8**. Tan, Deddish & Skidgel (1995) *Meth. Enzymol.* **248**, 663. **9**. Skidgel, Davis & Tan (1989) *J. Biol. Chem.* **264**, 2236. **10**. Hendriks (1998) in *Handb. Proteolytic Enzymes*, p. 1328, Academic Press, San Diego. **11**. Bouma, Marx, Mosnier & Meijers (2001) *Thromb. Res.* **101**, 329. **12**. Mao, Colussi, Bailey, *et al.* (2003) *Anal. Biochem.* **319**, 159. **13**. Plummer & Ryan (1981) *Biochem. Biophys. Res. Commun.* **98**, 448. **14**. Skidgel & Erdös (1998) in *Handb. Proteolytic*

Enzymes, p. 1344, Academic Press, San Diego. **15**. Masuda, Yoshioka, Hinode & Nakamura (2002) *Infect. Immun.* **70**, 1807. **16**. Fisher, Ryan, Chung & Plummer (1986) *Adv. Exp. Med. Biol.* **198** Pt. A, 405. **17**. Ryan (1988) *Meth. Enzymol.* **163**, 186. **18**. Skidgel (1995) *Meth. Enzymol.* **248**, 653. **19**. Barabé & Huberdeau (1991) *Biochem. Pharmacol.* **41**, 821. **20**. Novikova, Eng, Yan, Qian & Fricker (1999) *J. Biol. Chem.* **274**, 28887. **21**. Ryan, Valido, Berryer, Chung & Ripka (1992) *Biochim. Biophys. Acta* **1119**, 140. **22**. Orawski & Simmons (1995) *Biochemistry* **34**, 11227. **23**. Simmons & Orawski (1992) *J. Biol. Chem.* **267**, 4897.

2-Mercaptomethyl-5-guanidinopentanoic Acid

This thiol-containing arginine analogue (FW = 205.28 g/mol), also known as SQ 24798, is a carboxypeptidase B inhibitor (K_i = 0.42 nM). The inhibitor potency is attributed to the combined action of substrate specificity-conferring features and thiol-mediated chelation of an active-site zinc ion. It is a weaker inhibitor of carboxypeptidase A (K_i = 11.6 μM) and angiotensin-converting enzyme (K_i = 288 μM). **Target(s):** carboxypeptidase A (1); carboxypeptidase B (1,2); lysine carboxypeptidase *or* carboxypeptidase N, *or* lysine(arginine) carboxypeptidase (3); peptidyl-dipeptidase A, *or* angiotensin converting enzyme (1). **1**. Ondetti, Condon, Reid, *et al.* (1979) *Biochemistry* **18**, 1427. **2**. Avilés & Vendrell (1998) in *Handb. Proteolytic Enzymes*, p. 1333, Academic Press, San Diego. **3**. Plummer & Ryan (1981) *Biochem. Biophys. Res. Commun.* **98**, 448.

3-[(R)-1-(Mercaptomethyl)-2-phenylethyl]amino-3-oxopropanoate

This thiorphan analogue (FW = 253.32 g/mol), also known as *retro*-thiorphan, inhibits neprilysin; the R-stereoisomer has a K_i value of 2.3 nM. whereas the value for the S-isomer is 210 nM (1-3). The K_i value for thremolysin is 0.8 μM (3-5). (*See also Thiorphan*) **1**. Roques, Noble, Crine & Fournié-Zaluski (1995) *Meth. Enzymol.* **248**, 263. **2**. Fournié-Zaluski, Lucas-Soroca, Devin & Roques (1986) *J. Med. Chem.* **29**, 751. **3**. Marie-Claire, Ruffet, Antonczak, *et al.* (1997) *Biochemistry* **36**, 13938. **4**. Beynon & Beaumont (1998) in *Handb. Proteolytic Enzymes*, p. 1037, Academic Press, San Diego. **5**. Roderick, Fournie-Zaluski, Roques & Matthews (1989) *Biochemistry* **28**, 1493.

3-Mercaptopicolinate (3-Mercaptopicolinic Acid)

This hypoglycemic agent (FW$_{\text{free-acid}}$ = 155.18 g/mol; CAS 14623-54-2 (HCl salt)) inhibits gluconeogenesis, binds metal ions (particularly iron), and inhibits phosphoenolpyruvate carboxykinase. **Target(s):** glucose-6-phosphatase (1,2); glucose-6-phosphate translocase (3); phosphoenol-pyruvate carboxykinase (ATP) (4,5); phosphoenolpyruvate carboxykinase (GTP) (6-14). **1**. Bode, Foster & Nordlie (1993) *Biochem. Cell Biol.* **71**, 113. **2**. Foster, Bode & Nordlie (1994) *Biochim. Biophys. Acta* **1208**, 222. **3**. Van Schaftingen & Gerin (2002) *Biochem. J.* **362**, 513. **4**. Wingler, Walker, Chen & Leegood (1999) *Plant Physiol.* **120**, 539. **5**. Urbina, Orsono & Rojas (1990) *Arch. Biochem. Biophys.* **282**, 91. **6**. Rognstad (1979) *J. Biol. Chem.* **254**, 1875. **7**. Jomain-Baum, Schramm & Hanson (1976) *J. Biol. Chem.* **251**, 37. **8**. MacDonald & Grewe (1981) *Biochim. Biophys. Acta* **663**, 302. **9**. Watford, Vinay, Lemieux & Gougoux (1980) *Can. J. Biochem.* **58**, 440. **10**. Makinen & Nowak (1983) *J. Biol. Chem.* **258**, 11654. **11**. MacDonald (1978) *Biochim. Biophys. Acta* **526**, 293. **12**. MacDonald (1979) *Biochem. Biophys. Res. Commun.* **90**, 741. **13**. Foley,

Wang, Dunten, *et al.* (2003) *Bioorg. Med. Chem. Lett.* **13**, 3607. **14**. Harlocker, Kapper, Greenwalt & Bishop (1991) *J. Exp. Zool.* **257**, 285.

3-Mercaptopropionate (3-Mercaptopropionic Acid)

This toxic β-thiol acid and coenzyme M analogue (FW$_{\text{free-acid}}$ = 106.15 g/mol; CAS 107-96-0; M.P. = 17-19°C), also known as 3-mercaptopropanoic acid, is a convulsion-inducing agent. **Target(s):** 4-aminobutyrate aminotransferase (1,2); aspartate aminotransferase (3); aspartate carbamoyltransferase, weakly inhibited (4); butyrate uptake (5); cathepsin B (6); cystathionine γ synthase, also alternative substrate (7,8); cysteine dioxygenase, weakly inhibited (9-11); cysteinesulfinic acid decarboxylase (12); dihydroorotase (13); glutamate decarboxylase (14-23); guanidinobutyrase (24); long-chain acyl-CoA dehydrogenase (25); 3-mercaptopyruvate sulfurtransferase (26); phosphoenol-pyruvate carboxylase (27); *O*-succinylhomoserine (thiol)-lyase (28); urocanate hydratase, *or* urocanase (29); Zn^{2+}-type D-alanyl-D-alanine-cleaving carboxypeptidase (30,31). **1**. Gale (1984) *Life Sci.* **34**, 701. **2**. Schousboe, Wu & Roberts (1974) *J. Neurochem.* **23**, 1189. **3**. Braunstein (1973) *The Enzymes*, 3rd ed., **9**, 379. **4**. Foote, Lauritzen & Lipscomb (1985) *J. Biol. Chem.* **260**, 9624. **5**. Stein, Schroder, Milovic & Caspary (1995) *Gastroenterology* **108**, 673. **6**. Krepela, Prochazka & Karova (1999) *Biol. Chem.* **380**, 541. **7**. Posner (1972) *Biochim. Biophys. Acta* **276**, 277. **8**. Kaplan & Guggenheim (1971) *Meth. Enzymol.* **17B**, 425. **9**. Dominy, Simmons, Karplus, Gehring & Stipanuck (2006) *J. Bacteriol.* **188**, 5561. **10**. Bruland, Wübbeler & Steinbüchel (2009) *J. Biol. Chem.* **284**, 660. **11**. Chai, Bruyere & Maroney (2006) *J. Biol. Chem.* **281**, 15774. **12**. Heinama, Peramaa & Piha (1982) *Acta Chem. Scand. B* **36**, 287. **13**. Christopherson & Jones (1980) *J. Biol. Chem.* **255**, 3358. **14**. Netopilova, Drsata, Haugvicova, Kubova & Mares (1997) *Neurosci. Lett.* **226**, 68. **15**. Roberts, Taberner & Hill (1978) *Neuropharmacology* **17**, 715. **16**. Netopilova, Drsata, Kubova & Mares (19995) *Epilepsy Res.* **20**, 179. **17**. Youngs & Tunnicliff (1991) *Biochem. Int.* **23**, 915. **18**. Blindermann, Maitre, Ossola & Mandel (1978) *Eur. J. Biochem.* **86**, 143. **19**. Lamar (1970) *J. Neurochem.* **17**, 165. **20**. Tunnicliff (1990) *Int. J. Biochem.* **27**, 1235. **21**. Denner, Wei, Lin, Lin & Wu (1987) *Proc. Natl. Acad. Sci. U.S.A.* **84**, 668. **22**. Youngs & Tunnicliff (1991) *Biochem. Int.* **23**, 915. **23**. Satyanarayan & Nair (1985) *Eur. J. Biochem.* **150**, 53. **24**. Yorifuji, Shimizu, Hirata, *et al.* (1992) *Biosci. Biotechnol. Biochem.* **56**, 773. **25**. Yamamoto & Nakamura (1994) *Gastroenterology* **107**, 517. **26**. Porter & Baskin (1995) *J. Biochem. Toxicol.* **10**, 287. **27**. Gold & Smith (1974) *Arch. Biochem. Biophys.* **164**, 447. **28**. Kaplan & Guggenheim (1971) *Meth. Enzymol.* **17B**, 425. **29**. Hug, O'Donnell & Hunter (1978) *Biochem. Biophys. Res. Commun.* **81**, 1435. **30**. Charlier, Dideberg, Jamoulle, *et al.* (1984) *Biochem. J.* **219**, 763. **31**. Ghuysen (1998) in *Handb. Proteolytic Enzymes*, p. 1354, Academic Press, San Diego.

N-(2-Mercaptopropionyl)glycine, *See* *Tiopronin*

2-(3-Mercaptopropyl)pentanedioic acid

This orally bioavailable substituted dicarboxylic acid (FW = 206.26 g/mol; CAS 254737-29-6; Solubility: 100 mM in H$_2$O or DMSO), also known as 2-MPPA, is a selective inhibitor (IC$_{50}$ = 90 nM) of glutamate carboxypeptidase II (GCP II), also variously known as *N*-acetyl-L-aspartyl-L-glutamate peptidase I (*or* NAALADase I), NAAG peptidase, or Prostate specific membrane antigen (*or* PSMA) (1-3). This zinc metalloenzyme catalyzes the formation of glutamate and *N*-acetyl-aspartate from *N*-acetyl-aspartyl-glutamate (*or* NAAG), the latter one of the three most abundant neurotransmitters in the central nervous system. By reducing the amount of glutamate formed, selective GCPII inhibitors protect motor neurons (MN) against glutamate excitotoxicity, a process believed to be a factor in both sporadic (ALS) and familial (FALS) forms of amyotrophic lateral sclerosis. 2-MPPA prevents MN cell death in both of these systems because of the resultant decrease in glutamate levels (1,2). When administered at 60 mg/kg (but not 10 or 30 mg/kg), 2-MPPA prevented morphine tolerance without

affecting acute morphine antinociception (3). **1**. Ghadge, Slusher, Bodner, *et al.* (2003) *Proc. Natl. Acad. Sci. U.S.A.* **100**, 9554. **2**. Majer, Jackson, Delahanty, *et al.* (2003) *J. Med. Chem.* **46**, 1989. **3**. Kozela, Wrobel, Kos, *et al.* (2005) *Psychopharmacology* (Berlin) **183**, 275.

2-Mercaptopurine

This substituted purine (FW = 152.18 g/mol; CAS 28128-19-0), also known as 2-purinethiol and 2-thiopurine, inhibits 6-phosphofructo-2-kinase (1), transaminidase (2), and xanthine phoshoribosyltransferase (3). **Key Pharmacokinetic Parameters:** *See* Appendix II in Goodman & Gilman's *THE PHARMACOLOGICAL BASIS OF THERAPEUTICS*, 12th Edition (Brunton, Chabner & Knollmann, eds.) McGraw-Hill Medical, New York (2011). **1**. Mojena, Bosca, Rider, Rousseau & Hue (1992) *Biochem. Pharmacol.* **43**, 671. **2**. Karelin (1974) *Vopr. Med. Khim.* **20**, 406. **3**. Miller, Adamczyk, Fyfe & Elion (1974) *Arch. Biochem. Biophys.* **165**, 349.

6-Mercaptopurine

This toxic hypoxanthine analogue (FW = 152.18 g/mol; CAS 50-44-2; yellow, with λ$_{\text{max}}$ at 222 (9240 M^{-1}cm^{-1}) and 327 nm (21300 M^{-1}cm^{-1}) in 0.1 N HCl, and λ$_{\text{max}}$ values of 230 (14000 M^{-1}cm^{-1}) and 312 nm (19600 M^{-1}cm^{-1}) in 0.1 N NaOH), also called 6-thiopurine and thiohypoxanthine, was first synthesized in the 1950s by Gertrude Elion and George Hitchings, the 1988 Nobel Laureates in Medicine/Physiology. (*See also* prodrug *Azathioprine*) In humans, 6-MP transport involves sodium-dependent, purine-selective transporters SLC28A2, SLC28A3, SLC29A1 and SLC29A2. As an alternative substrate for the salvage enzyme, hypoxanthine:guanine phosphoribosyltransferase (*Physiologic Reaction*: Guanine (or Hypoxanthine) + PRPP ⇌ GMP (or IMP) + PP$_i$), 6-mercaptopurine is converted to 6-mercaptopurine-β-D-ribose 5'-monophosphate, *or* 6-thio-IMP (*Drug Reaction*: 6-Thiopurine + PRPP ⇌ 6-thio-IMP + PP$_i$). (*See also 6-Thioinosine 5'-Triphosphate; 6-Thioinosine 5'-Triphosphate*). 6-Thio-IMP is a potent inhibitor (slow alternative substrate) for IMP dehydrogenase (*Physiologic Reaction*: IMP + NAD$^+$ + H$_2$O ⇌ XMP + NADH + H$^+$; *Thio-IMP Reaction*: 6-Thio-IMP + NAD$^+$ + H$_2$O ⇌ 6-Thio-XMP + NADH + H$^+$), and 6-Thio-XMP is then converted to 6-Thio-GMP by GMP synthase. 6-Thio-IMP is also a competitive inhibitor for adenylosuccinate synthetase (*Reaction*: IMP + Aspartate + GTP ⇌ Adenylosuccinate + GDP + P$_i$). 6-Thio-IMP is likewise an feedback inhibitor for glutamine:phosphoribosylamine synthetase (*Reaction*: Glutamine + PRPP ⇌ Glutamate + Phosphoribosylamine + PP$_i$), the first committed step and pacemaker in *de novo* purine nucleotide biosynthesis. These properties commend 6-mercaptopurine as a powerful immunosuppressant and antileukemic agent that throttles purine biosynthesis as well as cell proliferation. In humans, 6-thio-IMP is converted to thio-guanosine nucleotides, the cytotoxicity of which occurs through 6-mercapto-dGTP incorporation into DNA, thereby impairing enzymes involved in DNA replication and repair and inducing DNA damage (*e.g.*, single strand-breaks, DNA–protein cross-links, and chromatid exchanges). Incorporation of 6-mercapto-guanosine triphosphate into RNA also impairs mRNA structure and subsequent translation. (*See also 6-Thioguanine; 6-Thioguanosine 5'-Monophosphate*) Note also that 6-mercaptopurine is a substrate for thiopurine methyltransferase (*Reaction*: 6-Mercaptopurine + *S*-Adenosyl-Methionine ⇌ *S*-CH$_3$-Mercaptopurine + *S*-Adenosyl-homocysteine), with *S*-methyl-thiopurine preventing the formation of thio-guanosine nucleotides. Moreover, because xanthine oxidase metabolizes 6-mercaptopurine, patients taking xanthine oxidase inhibitors (*e.g.*, allopurinol, oxypurinol, tisopurine, febuxostat, topiroxostat, and phytic acid) to prevent gout are therefore at elevated risk for mercaptopurine toxicity. *See other IMPDH inhibitors: Merimepodib; Mycophenolate; Mycophenolate mofetil* **Target(s):** adenine phosphoribosyl-transferase (1); cyclooxygenase (2); DNA methylation

(3,4); hypoxanthine(guanine) phosphoribosyltransferase, also as an alternative substrate (5-8); methionine S-adenosyltransferase (9); 6-phosphofructo-2-kinase (10); purine-nucleoside phosphorylase, or guanosine phosphorylase, as product inhibitor (11,12); succinate dehydrogenase (13); urate oxidase, or uricase (14); xanthine dehydrogenase (15); xanthine oxidase (16); xanthine phosphoribosyltransferase (6,17). 1. Queen, Vander Jagt & Reyes (1989) Biochim. Biophys. Acta 996, 160. 2. Homo-Delarche, Bach & Dardenne (1988) Prostaglandins 35, 479. 3. Lambooy, Leegwater, van den Heuvel, Bokkerink & De Abreu (1998) Clin. Chem. 44, 556. 4. De Abreu, Lambooy, Stet, Vogels-Mentink & Van den Heuvel (1995) Adv. Enzyme Regul. 35, 251. 5. Baker (1967) Design of Active-Site-Directed Irreversible Enzyme Inhibitors, Wiley, New York. 6. Naguib, Iltzsch, el Kouni, Panzica & el Kouni (1995) Biochem. Pharmacol. 50, 1685. 7. Walter & Königk (1974) Tropenmed. Parasitol. 25, 227. 8. Miller, Ramsey, Krenitsky & Elion (1972) Biochemistry 11, 4723. 9. Berger & Knodel (2003) BMC Microbiol. 3, 12. 10. Mojena, Bosca, Rider, Rousseau & Hue (1992) Biochem. Pharmacol. 43, 671. 11. Koch (1956) J. Biol. Chem. 223, 535. 12. Baker & Schaeffer (1971) J. Med. Chem. 14, 809. 13. Dombrad & Csizmadia (1964) Enzymologia 26, 321. 14. Bergmann, Kwietny-Govrin, Ungar-Waron, Kalmus & Tamari (1963) Biochem. J. 86, 567. 15. Schräder, Rienhöfer & Andreesen (1999) Eur. J. Biochem. 264, 862. 16. Webb (1966) Enzyme and Metabolic Inhibitors, vol. 2, p. 282, Academic Press, New York. 17. Miller, Adamczyk, Fyfe & Elion (1974) Arch. Biochem. Biophys. 165, 349.

3-Mercaptopyruvate (3-Mercaptopyruvic Acid)

This naturally occurring pyruvate derivative (FW$_{free-acid}$ = 120.13 g/mol; CAS 2464-23-5) is an intermediate in the catabolism of cysteine in a number of organisms. It is formed in the cysteine aminotransferase reaction and a substrate for 3-mercaptopyruvate sulfurtransferase. In some organisms, 3-mercaptopyruvate is an alternative substrate of tRNA sulfurtransferase. It is also an alternative substrate of lactate dehydrogenase and a mechanism-based inactivator of pyruvate formate-lyase. Note that aqueous solutions of 3-mercaptopyruvic acid readily dimerize to form 2,5-dihydroxy-1,4-dithiane-2,5-dicarboxylic acid (1). The equilibrium constant favors the cyclic form at pH 5-7. In neutral or alkaline solutions, 3-mercaptopyruvate irreversibly forms the acyclic aldol dimer ($^-$OOCCOCH(SH)C(OH)(CH$_2$SH)COO$^-$), which then undergoes further oligomerization. **Target(s):** glutamine:pyruvate aminotransferase (2,3); 4-hydroxy-2-oxoglutarate aldolase (4); pyruvate formate-lyase (5); tRNA sulfurtransferase (6). 1. Cooper, Haber & Meister (1982) J. Biol. Chem. 257, 816. 2. Cooper & Meister (1977) CRC Crit. Rev. Biochem. 4, 281. 3. Cooper & Meister (1972) Biochemistry 11, 661. 4. Anderson, Scholtz & Schuster (1985) Arch. Biochem. Biophys. 236, 82. 5. Parast, Wong, Kozarich, Peisach & Magliozzo (1995) Biochemistry 34, 5712. 6. Wong, Weiss, Eliceiri & Bryant (1970) Biochemistry 9, 2376.

Mercuric Ion

This heavy metal cation (Symbol: Hg^{2+}; Atomic Number: 80; Atomic Weight: 200.59), a Group 12 element (or group IIB), reacts rapidly with sulfhydryl groups, accounting for its frequent use as an enzyme inhibitor as well as an analytical reagent for detecting and quantifying cysteinyl residues and other thiol-containing substances. Because mercuric ion is divalent, it reacts with up to two sulfhydryl groups. At low concentrations, a simple mercaptide is generated (Reaction: R-SH + Hg^{2+} → R-S-Hg$^+$ + H$^+$). This mercaptide reacts with a second thiol to produce a bridged structure (Reaction: R-SHg$^+$ + R'-SH → R-S-Hg-S-R' + H$^+$). Excess mercuric ion favors reformation of the mercaptide from this bridged structure (Reaction: R-S-Hg-S-R' + Hg^{2+} → R-S-Hg$^+$ + R'-S-Hg$^+$). Compounds containing Hg(II) are also common. The majority of organomercurials have either the general structure R-HgX or R$_2$-Hg. The CH$_3$-Hg$^+$ ion, for example, which is formed by methylation of metallic mercury by methylcobalamin, binds firmly to sulfur in biomolecules. In addition, mercuric ion can denature DNA and damage chromosomes. **Toxicity:** In humans and mammals, mercury ions have toxic effects on number organs, and the proximal kidney tubule is especially vulnerable. Because cysteines in most cell-surface proteins/receptors are mainly disulfides, mercuric ions rarely inhibit them. Even so, circulating mercury ions combine with albumin, metallothionein,

glutathione, and cysteine, and these mercurated substances have been implicated in mechanisms involved in the proximal tubular uptake, accumulation, transport, and toxicity of mercuric ions. Ionic mercury and certain organomercurials also react with aquaporins, inhibiting water transport (3). 1. Zalups (2000) Pharmacol. Rev. 52, 113. 2. Broussard, Hammett-Stabler, Winecker & Ropero-Miller (2002) Laboratory Medicine 33, 614. 3. Macey (1984) Am. J Physiol. 246, C195.

Merimepodib

This orally bioavailable IMPDH inhibitor (FW = 452.17 g/mol; CAS 198821-22-6), also known as VX-497, VX497, VI21497, and (S)-tetrahydrofuran-3-yl 3-(3-(3-methoxy-4-(oxazol-5-yl)phenyl)ureido)benzyl-carbamate, targets IMP dehydrogenase and inhibits the proliferation of primary human, mouse, rat, and dog lymphocytes at concentrations of ~100 nM (1). Its inhibitory effect on lymphocytes is reversed in the presence of exogenous guanosine, but not in the presence of adenosine or uridine, confirming that the antilymphocytic activity of VX-497 is specifically due to inhibition of IMPDH. In mice, oral administration of merimepodib inhibits the primary IgM antibody response in a dose-dependent manner (ED$_{50}$ = 30-35 mg/kg) (1). Merimepodib is an uncompetitive inhibitor, K_i = 6 nM for IMPDH-II (2). It inhibits proliferation of mitogen-stimulated primary human lymphocytes (IC$_{50}$ = ~80 nM). Merimepodib does not inhibit proliferation of nonlymphoid cell types, such as fibroblasts, indicating selectivity for IMPDH inhibition (2). It also prolongs skin graft survival and improves graft versus host disease in mice (3). 1. Jain, Almquist, Shlyakhter & Harding (2001) J. Pharm. Sci. 90, 625. 2. Jain, Almquist, Heiser, et al. (2002) J. Pharmacol. Exp. Ther. 302, 1272. 3. Decker, Heiser, Chaturvedi, et al. (2001) Drugs Exp. Clin. Res. 27, 89.

Meropenem

This ultra-broad-spectrum, injectable, carbapenem-class antibiotic (FW = 383.46 g/mol; CAS 119478-56-7), also named SM-7338, Merofit® (India), Meronem® (Germany), 3-[5-(dimethyl-carbamoyl)pyrrolidin-2-yl]sulfanyl-6-(1-hydroxyethyl)-4-methyl-7-oxo-1-azabicyclo[3.2.0]hept-2-ene-2-carboxylate, inhibits bacterial wall synthesis like other β-lactam antibiotics, but enjoys high resistance to enzymatic degradation by β-lactamases or cephalosporinases. Meropenem shows potent antibacterial activity against a broad spectrum of aerobes, including Staphylococcus aureus, β-hemolytic streptococci, Streptococcus pneumoniae, Haemophilus influenzae, Neisseria spp., as well as members of the family Enterobacteriaceae, Pseudomonas spp., as well as Gram-positive and Gram-negative anaerobes in a collection of 1,102 unselected clinical isolates (1). Clinical meropenem response in vitro is defined as: susceptible (MICs ≤ 4 µg/mL); intermediate susceptibility (MICs = 8 µg/mL); and resistance (MICs ≥ 16 µg/mL) (2). In six volunteers given a single 1-g intravenous dose, the mean elimination $t_{1/2}$ in plasma was 1.1 hours, with a mean plasma concentration of 24 µg/mL at 1 hour to 0.7 µg/mL at 6 hours (3). Inflammatory fluid penetration is rapid (reaching maximum concentration of drug in serum in 45 minutes) with a penetration of 111% (3). (See Carbapenem-Resistant Enterobacteriaceae; Ertapenem; Doripenem; Imipenem) 1. Edwards, Turner, Wannop, et al. (1989) Antimicrob. Agents Chemother. 33, 215. 2. Edwards (1995) J. Antimicrob. Chemother. 36 (Supplement A), 1. 3. Wise, Logan, Cooper, Ashby & Andrews (1990) Antimicrob. Agents Chemother. 34, 1515.

Mersalyl

This sulfhydryl-reactive agent (FW$_{\text{sodium-salt}}$ = 505.85 g/mol; CASs = 492-18-2, 486-67-9 (mersalyl acid); photosensitive), also known as Salyrgan and [3-[[2-(carboxylatomethoxy)benzoyl]amino]-2-methoxypropyl]hydroxymercurate, was originally adapted for use as a diuretic from calomel (HgCl), itself a diuretic first discovered by the German-Swiss Renaissance physician Paracelsus. In mice, mersalyl exhibits LD$_{50}$ values of 400, 500, and 1400 mg/Kg for intravenous, subcutaneous, and oral routes. Because the cysteine sulfur atoms of cell-surface proteins and receptors are present as disulfides, mersalyl rarely inhibits these processes. Moreover, although many thiol-requiring enzymes are inhibited *in vitro*, mersalyl does not cross biomembranes and rarely inhibits enzymes *in vivo*. In addition to the enzymes listed below, mersalyl inhibits transporters for orthophosphate, sulfate, glutamine, and carnitine/acylcarnitine. **Target(s):** 4-acetamidobutyrate deacetylase (1); α-*N*-acetylneuraminyl-2,3-β-galactosyl 1,3-*N*-acetyl-galactosamide 6-α-sialyltransferase (79); aconitase (aconitate hydratase) (2); actin polymerization (3,4); adenine nucleotide carrier (5); adenylate cyclase (6); alcohol dehydrogenase, yeast (7); amine oxidase (8); aquacobalamin reductase (10); aquacobalamin reductase (NADPH) (9,10,83,84); arginine deaminase (11); L-ascorbate peroxidase (89); aspartate aminotransferase (76); aspartate carbamoyltransferase (12); benzylamine oxidase (8); benzylamine/putrescine oxidase (8); branched-chain amino-acid aminotransferase (74,75); Ca^{2+}-dependent ATPase (13,14); carbamoyl-phosphate synthetase (15); carbon monoxide dehydrogenase (16); catechol 1,2-dioxygenase (17); cathepsin L (18); chloride/bicarbonate-dependent ATPase (19); CoB-CoM heterodisulfide reductase (90); 3-cyanoalanine hydratase (20); cyanoalanine nitrilase (20); cytidine deaminase (21-23); cytochrome *b$_5$* reductase (24); cytosine deaminase (25); dynein ATPase (26); ecto-ATPase (27); endopeptidase La (68); 2-enoate reductase (28); ferredoxin hydrogenase (88); ferredoxin:nitrite reductase (29); formyltetrahydrofolate synthetase (30); glutaminase (31,32); glutathione dehydrogenase (ascorbate) (91); glutathione *S*-transferase (77); glycerol kinase (73); glycoprotein 6-α-L-fucosyltransferase (81); 9-hydroxyprostaglandin dehydrogenase (33); imidazoleacetate 4-monooxygenase (86); levansucrase (82); lysine 2-monooxygenase (34,87); lysyl aminopeptidase, bacterial (35); macropain (proteasome endopeptidase complex) (36-38); methylamine oxidase (8); methylglutaconyl-CoA hydratase (39); microsomal epoxide hydrolase, weakly inhibited (70); NADH dehydrogenase (40-45); NADH: ferricyanide reductase (46); NADH:hexacyanoferrate(III) reductase (47); NADPH:cytochrome *c* reductase (46,48); NADPH:cytochrome P450 reductase, *or* NADPH: ferrihemoprotein reductase (24,46); NAD(P)H dehydrogenase (49,50); NADPH oxidase (51); Na$^+$/K$^+$-exchanging ATPase (52,53); peroxidase, gastric (54,55); 1-phosphatidylinositol 4-kinase (72); polygalacturonidase, *or* pectinase (71); proline dehydrogenase, *or* proline oxidase (56); prolyl aminopeptidase (69); prostaglandin E$_2$ 9-reductase (57,58); proteasome endopeptidase complex (36-38); pyrophosphatase, *or* inorganic diphosphatase (59); ribose-5-phosphate isomerase (60,61); rubredoxin:NAD$^+$ reductase (62); steroid 21-monooxygenase, *or* CYP21A1 (63,64,85); steroid 5α-reductase, rat (65); sterol 3β-glucosyltransferase (80); succinate dehydrogenase (66); thiamin pyridinylase, *or* thiaminase I (78); threonine dehydratase (67). **1.** Haywood & Large (1986) *J. Gen. Microbiol.* **132**, 7. **2.** Suzuki, Akiyama, Fujimoto, *et al.* (1976) *J. Biochem.* **80**, 799. **3.** Drabikowski, Kuehl & Gergely (1961) *Biochem. Biophys. Res. Commun.* **5**, 389. **4.** Lusty & Fasold (1969) *Biochemistry* **8**, 2933. **5.** Spagnoletta, De Santis, Palmieri & Genchi (2002) *J. Bioenerg. Biomembr.* **34**, 465. **6.** Mavier & Hanoune (1975) *Eur. J. Biochem.* **59**, 593. **7.** Sund & Theorell (1963) *The Enzymes*, 2nd ed., **7**, 25. **8.** Large & Haywood (1990) *Meth. Enzymol.* **188**, 427. **9.** Watanabe & Nakano (1997) *Meth. Enzymol.* **281**, 289. **10.** Watanabe & Nakano (1997) *Meth. Enzymol.* **281**, 295. **11.** Park, Hirotani, Nakano & Kitaoka (1984) *Agric. Biol. Chem.* **48**, 483. **12.** Chang, Prescott & Jones (1978) *Meth. Enzymol.* **51**, 41. **13.** Thompson &

Nechay (1981) *J. Toxicol. Environ. Health* **7**, 901. **14.** Michaelson, Ophir & Angel (1980) *J. Neurochem.* **35**, 116. **15.** Jones (1962) *Meth. Enzymol.* **5**, 903. **16.** Ragsdale & Wood (1985) *J. Biol. Chem.* **260**, 3970. **17.** Nakazawa & Nakazawa (1970) *Meth. Enzymol.* **17A**, 518. **18.** McDonald & Kadkhodayan (1988) *Biochem. Biophys. Res. Commun.* **151**, 827. **19.** Gassner & Komnick (1981) *Eur. J. Cell Biol.* **25**, 108. **20.** Yanase, Sakai & Tonomura (1983) *Agric. Biol. Chem.* **47**, 473. **21.** Frick, Yang, Marquez & Wolfenden (1989) *Biochemistry* **28**, 9423. **22.** Hosono & Kuno (1973) *J. Biochem.* **74**, 797. **23.** Ashley & Bartlett (1984) *J. Biol. Chem.* **259**, 13615. **24.** Williams (1976) *The Enzymes*, 3rd ed., **13**, 89. **25.** Katsuragi, Sakai, Matsumoto & Tonomura (1986) *Agric. Biol. Chem.* **50**, 1721. **26.** Yamin & Tamm (1982) *J. Cell Biol.* **95**, 589. **27.** Manery, Dryden, Still & Madapallimattam (1984) *Can. J. Biochem. Cell Biol.* **62**, 1015. **28.** Tischer, Bader & Simon (1979) *Eur. J. Biochem.* **97**, 103. **29.** Vega, Cárdenas & Losada (1980) *Meth. Enzymol.* **69**, 255. **30.** Buttlaire (1980) *Meth. Enzymol.* **66**, 585. **31.** Ardawi & Newsholme (1984) *Biochem. J.* **217**, 289. **32.** Kvamme & Olsen (1979) *FEBS Lett.* **107**, 33. **33.** Yuan, Tai & Tai (1980) *J. Biol. Chem.* **255**, 743. **34.** Nakazawa (1971) *Meth. Enzymol.* **17B**, 154. **35.** Klein & Henrich (1998) in *Handb. Proteolytic Enzymes*, p. 1018, Academic Press, San Diego. **36.** Clark, Ilgen, Haire & Mykles (1991) *Comp. Biochem. Physiol. B* **99**, 413. **37.** McGuire & DeMartino (1986) *Biochim. Biophys. Acta* **873**, 279. **38.** Dahlmann, Kuehn, Rutschmann & Reinauer (1985) *Biochem. J.* **228**, 161. **39.** Hilz, Knappe, Ringelmann & Lynen (1958) *Biochem. Z.* **329**, 476. **40.** Galante & Hatefi (1978) *Meth. Enzymol.* **53**, 15. **41.** Palmer & Møller (1982) *Trends Biochem. Sci.* **7**, 258. **42.** Frimmer (1960) *Biochem. Z.* **332**, 522. **43.** Knudten, Thelen, Luethy & Elthon (1994) *Plant Physiol.* **106**, 1115. **44.** del Castillo-Olivares, Medina, Nunez de Castro & Marquez (1996) *Biochem. J.* **314**, 587. **45.** Gutman, Mersmann, Luthy & Singer (1970) *Biochemistry* **9**, 2678. **46.** Crowder & Brady (1979) *J. Biol. Chem.* **254**, 408. **47.** Berczi, Fredlund & Moller (1995) *Arch. Biochem. Biophys.* **320**, 65. **48.** Laporte, Doussiere, Mechin & Vignais (1991) *Eur. J. Biochem.* **196**, 59. **49.** Luethy, Thelen, Knudten & Elthon (1995) *Plant Physiol.* **107**, 443. **50.** Rasmusson & Moller (1991) *Eur. J. Biochem.* **202**, 617. **51.** Doussiere & Vignais (1985) *Biochemistry* **24**, 7231. **52.** Anner, Moosmayer & Imesch (1992) *Amer. J. Physiol.* **262**, F830. **53.** Nechay, Nelson, Contreras, *et al.* (1975) *J. Pharmacol. Exp. Ther.* **192**, 303. **54.** De & Banerjee (1986) *Eur. J. Biochem.* **160**, 319. **55.** Banerjee & Datta (1981) *Acta Endocrinol. (Copenhagen)* **96**, 208. **56.** Lang & Lang (1958) *Biochem. Z.* **329**, 577. **57.** Tai & Yuan (1982) *Meth. Enzymol.* **86**, 113. **58.** Pace-Asciak & Smith (1983) *The Enzymes*, 3rd ed., **16**, 543. **59.** Gordon-Weeks, Steele & Leigh (1996) *Plant Physiol.* **111**, 195. **60.** Bruns, Noltman & Vahlhaus (1958) *Biochem. Z.* **330**, 483. **61.** Middaugh & MacElroy (1976) *J. Biochem.* **79**, 1331. **62.** Ueda & Coon (1972) *J. Biol. Chem.* **247**, 5010. **63.** Hayano & Dorfman (1962) *Meth. Enzymol.* **5**, 503. **64.** Rosenthal & Narasimhulu (1969) *Meth. Enzymol.* **15**, 596. **65.** Liang, Cascieri, Cheung, Reynolds & Rasmusson (1985) *Endocrinology* **117**, 571. **66.** Hatefi & Stiggall (1976) *The Enzymes*, 3rd ed., **13**, 175. **67.** Nakazawa (1971) *Meth. Enzymol.* **17B**, 571. **68.** Desautels & Goldberg (1982) *J. Biol. Chem.* **257**, 11673. **69.** Ehrenfreud, Mollay & Kreil (1992) *Biochem. Biophys. Res. Commun.* **184**, 1250. **70.** Oesch & Daly (1971) *Biochim. Biophys. Acta* **227**, 692. **71.** Sakai, Okushima & Sawada (1982) *Agric. Biol. Chem.* **46**, 2223. **72.** Hou, Zhang & Tai (1988) *Biochim. Biophys. Acta* **959**, 67. **73.** Bergmeyer, Holz, Kauder, Möllering & Wieland (1961) *Biochem. Z.* **333**, 471. **74.** Hall, Wallin, Reinhart & Hutson (1993) *J. Biol. Chem.* **268**, 3092. **75.** Taylor & Jenkins (1966) *J. Biol. Chem.* **241**, 4406. **76.** Ryan, Bodley & Fottrell (1972) *Phytochemistry* **11**, 957. **77.** Letelier, Martinez, Gonzalez-Lira, Faundez & Aracena-Parks (2006) *Chem. Biol. Interact.* **164**, 39. **78.** McCleary & Chick (1977) *Phytochemistry* **16**, 207. **79.** Baubichon-Cortay, Broquet, George & Louisot (1989) *Glycoconjugate J.* **6**, 115. **80.** Ullmann, Ury, Rimmele, Benveniste & Bouvier-Navé (1993) *Biochimie* **75**, 713. **81.** Kaminska, Wisniewska & Koscielak (2003) *Biochimie* **85**, 303. **82.** Yanase, Iwata, Nakahigashi, *et al.* (1992) *Biosci. Biotechnol. Biochem.* **56**, 1335. **83.** Watanabe, Oki, Nakano & Kitaoka (1987) *J. Biol. Chem.* **262**, 11514. **84.** Saido, Watanabe, Tamura, *et al.* (1993) *J. Nutr.* **123**, 1868. **85.** Ryan & Engel (1957) *J. Biol. Chem.* **225**, 103. **86.** Okamoto, Nozaki & Hayaishi (1968) *Biochem. Biophys. Res. Commun.* **32**, 30. **87.** Takeda, Yamamoto, Kojima & Hayaishi (1969) *J. Biol. Chem.* **244**, 2935. **88.** Nakos & Mortenson (1971) *Biochemistry* **10**, 2442. **89.** De Leonardis, Dipierro & Dipierro (2000) *Plant Physiol. Biochem.* **38**, 773. **90.** Murakami, Deppenmeier & Ragsdale (2001) *J. Biol. Chem.* **276**, 2432. **91.** Dipierro & Borraccino (1991) *Phytochemistry* **30**, 427.

Merthiolate, See *Thimerosal*

Mesaconate (Mesaconic Acid)

This dicarboxylic acid (FW$_{free-acid}$ = 130.10 g/mol; CAS 498-24-8), also referred to as methylfumaric acid and (E)-2-methyl-2-butenedioic acid, is an intermediate in the catabolism of glutamate in a number of organisms (e.g., Clostridium tetanomorphum). The free acid is soluble in water (2.7 g/100 g at 18°C) and has pK_a values of 3.07 and 4.82 at 25°C. Mesaconate is an activator of dopamine β-monooxygenase. **Target(s):** carboxy-cis,cis-muconate cyclase (1); fumarase (2-4); fumarate reductase activity of Shewanella putrefaciens flavocytochrome c (5); glutamate decarboxylase, K_i = 4.1 mM (6); 2-methyleneglutarate mutase (7,8). **1.** Thatcher & Cain (1975) Eur. J. Biochem. **56**, 193. **2.** Hill & Bradshaw (1969) Meth. Enzymol. **13**, 91. **3.** Massey (1953) Biochem. J. **55**, 172. **4.** Webb (1966) Enzyme and Metabolic Inhibitors, vol. 2, p. 275, Academic Press, New York. **5.** Morris, Black, Pealing, et al. (1994) Biochem. J. **302**, 587. **6.** Fonda (1972) Biochemistry **11**, 1304. **7.** Barker (1972) The Enzymes, 3rd ed., **6**, 509. **8.** Kung & Stadtman (1971) J. Biol. Chem. **246**, 3378.

Mesalamine

This anti-inflammatory aminosalicylate (FW = 153.14 g/mol; CAS 89-57-6), also known 5-aminosalicylic acid (or 5-ASA) and 5-amino-2-hydroxybenzoate, is a mainstay drug in the treatment of inflammatory bowel disease, such as ulcerative colitis. Mesalazine is bowel-specific, acting locally in the gut and showing few systemic side effects. Mesalamine is the active moiety in sulfasalazine (FW = 398.41 g/mol; CAS 599-79-1; formed by linking 5-ASA with sulfapyridine; IUPAC: 2-hydroxy-5-[(E)-2-{4-[(pyridin-2-yl)sulfamoyl]phenyl}diazen-1-yl]benzoic acid), which was initially developed for the treatment of rheumatoid arthritis more than 60 years ago. Mesalamine dose-dependently inhibits IL-1-stimulated NF-κB-dependent transcription without preventing IκB degradation or nuclear translocation and DNA binding of the transcriptionally active NF-κB proteins, RelA, c-Rel, or RelB. It also inhibits IL-1-stimulated RelA phosphorylation, suggesting that pharmacologic modulation of the phosphorylation status of RelA regulates the transcriptional activity of NF-κB in a manner that is independent of nuclear translocation and DNA binding (1). NCX-456, the nitric oxide-releasing derivative of mesalamine is significantly more effective than mesalamine in reducing the severity of colitis (i.e., damage and granulocyte infiltration) (2). Unlike mesalamine, NO-mesalamine significantly suppresses leukocyte adherence to the vascular endothelium in vivo. NO-mesalamine inhibited IL-1β and IFN-γ release and caspase 1 activity in splenocytes; such effects were not found in the inflamed colon (2). **1.** Egan, Mays, Huntoon, et al. (1999) J. Biol. Chem. **274**, 26448. **2.** Wallace, Vergnolle, Muscará, et al. (1999) Gastroenterol. **117**, 557.

Mescaline

This alkaloid and hallucinogenic tyrosine derivative (FW = 211.26 g/mol; CAS 832-92-8 (hydrochloride)), from the flowering buds of the mescal peyote cactus Lophophora williamsii, is a partial agonist at 5-HT$_{2A}$ and 5-HT$_2$ receptors, subtypes of the serotonin G protein-coupled receptor family. Its free base (M.P. = 35-36°C) is moderately water-soluble, but the hydrochloride is far more so. Psycodelic effects often include altered thinking, distorted sense of time and self-awareness, and vivid visual hallucinations. **Target(s):** diamine N-acetyltransferase (1); weakly inhibits microtubule self-assembly, or tubulin polymerization (2). Note also mescaline's structural similarity with the trimethoxyphenyl ring and amino group location in colchicine. Native North Americans used peyote for spiritual purposes for nearly six millennia. **1.** Seiler & al-Therib (1974) Biochim. Biophys. Acta **354**, 206. **2.** Andreu & Timasheff (1982) Biochemistry **21**, 534.

Mesopram

This orally active PDE inhibitor (FW = 265.31 g/mol; CAS 189940-24-7; Solubility: 100 mM in DMSO) targets type IV phosphodiesterase (PDE IV), inhibiting type 1 helper T (Th1) cell proliferation and markedly decreasing production of IFN-γ, TNF-α, IL-10 and iNOS in vitro (1,2). Pharmacological inhibition of PDE IV by mesopram triggers ovulation in follicle-stimulating hormone-primed rats (3). It also exhibits efficacy against experimental autoimmune encephalomyelitis and murine colitis in vivo. **1.** Dinter, Tse, Halks-Miller, et al. (2000) J. Neuroimmunol. **108**, 136. **2.** Loher, Schmall, Freytag (2003) J. Pharmacol. Exp. Ther. **305**, 549. **3.** McKenna et al (2005) Endocrinology **146**, 208.

Mesoxalonitrile 3-Chlorophenylhydrazone

This protonophore (FW = 204.62 g/mol; CAS 555-60-2; Abbreviation: CCCP), also named carbonyl cyanide m-chlorophenylhydrazone, uncouples oxidative phosphorylation and photophosphorylation. (**See also** p-Nitrophenol, FCCP, Protonophore) **Target(s):** ATP synthase (1); chloride/proton symporter (2); H$^+$-transporting ATPase, lysosomal (3); L-lysine 6-monooxygenase (NADPH) (4); nickel-transporting ATPase (5); oxidative and photophosphorylation (uncoupler) (1,6-13). **1.** Stiggall, Galante & Hatefi (1979) Meth. Enzymol. **55**, 308. **2.** Alvarado & Vasseur (1998) Amer. J. Physiol. **274**, C481. **3.** Jonas, Smith, Allison, et al. (1983) J. Biol. Chem. **258**, 11727. **4.** Thariath, Fatum, Valvano & Viswanatha (1993) Biochim. Biophys. Acta **1203**, 27. **5.** Yang, Daniel, Hsu & Drake (1989) Appl. Environ. Microbiol. **55**, 1078. **6.** Slater (1967) Meth. Enzymol. **10**, 48. **7.** Izawa & Good (1972) Meth. Enzymol. **24**, 355. **8.** Heytler (1979) Meth. Enzymol. **55**, 462. **9.** Hanstein (1976) Trends Biochem. Sci. **1**, 65. **10.** Heytler (1963) Biochemistry **2**, 357. **11.** Goldsby & Heytler (1963) Biochemistry **2**, 1142. **12.** Kimimura, Katoh, Ikegami & Takamiya (1971) Biochim. Biophys. Acta **234**, 92. **13.** Heytler & Prichard (1962) Biochem. Biophys. Res. Commun. **7**, 272.

Mesulergine hydrochloride

This receptor antagonist (FW = 398.95 g/mol; CAS 72786-12-0; Solubility: 100 mM in DMSO and to 5 mM in H_2O with gentle warming), also named N'-[(8-α1,6-dimethylergolin-8-yl]-N,N-dimethylsulfamide hydrochloride, targets 5-HT$_{2A}$ (pA$_2$ = 9.1) and HT$_{2C}$ receptor antagonist (pA$_2$ = 9.1) (1-3). Mersulergine is also D$_2$-like dopamine receptor partial agonist (K_i = 8 nM). Mersulergine displays antiprolactin and antiparkinsonian effects *in vivo*. Like mesulergine, its 1,20-N,N-bidemethylated metabolite interact directly with D1- and D$_2$-receptors. **1**. Markstei (1983) Mesulergine and its 1,20-N,N-bidemethylated metabolite interacts directly with D$_1$- and D$_2$-receptors. *Eur.J.Pharmacol.* **95**, 101. **2**. Marko (1984) *Eur. J. Pharmacol.* **101**, 263. 3. Hoyer, *et al.* (1994) *Pharmacol. Rev.* **46**, 157.

Metformin

This oral, biguanide-class antidiabetic drug (FW = 129.17 g/mol; CAS 657-24-9), also named N,N'-dimethyl-biguanide, is the longstanding first-line treatment for Type 2 diabetes, decreasing hepatic glucose production, increasing insulin sensitivity (without stimulating endogenous insulin secretion), and enhancing peripheral glucose uptake (1,2). Metformin also transiently inhibits the mitochondrial respiratory chain Complex-1, decreasing hepatic energy state and thereby activating the AMP protein kinase (AMPK); however, this idea has been challenged on the basis of loss-of-function experiments. Other evidence suggests metformin inhibits hepatic glucose output, reducing cellular energy status of adenine nucleotides instead of directly affecting gluconeogenic gene expression. **Role in Cancer Therapy:** Metformin significantly inhibits cell proliferation and apoptosis in all pancreatic cell lines. In these and other human cancer cells, metformin inhibits mitochondrial Complex I (NADH dehydrogenase) as well as respiration (3). It likewise inhibits cellular proliferation in the presence of glucose, but induces cell death upon glucose deprivation. These findings suggest that cancer cells rely exclusively on glycolysis for survival in the presence of metformin. Metformin also reduces hypoxic activation of hypoxia-inducible factor 1 (HIF-1). All effects of metformin were reversed when the metformin-resistant *Saccharomyces cerevisiae* NADH dehydrogenase (NDI1) is overexpressed. When tested in mice, metformin inhibits the growth of control human cancer cells, but not those expressing NDI1. [*Note:* The related biguanides phenformin (*or,* 2-(*N*-phenethyl-carbamimidoyl)guanidine) and buformin (*or, N*-butylimidocarbonimidic diamide) were withdrawn from the market in most countries as a consequence of toxicity, mainly severe lactic acidosis.] Metformin-treated cells exhibited significantly lower viability and proliferation and significantly more cell cycle arrest in G$_1$ and G$_2$/M than control cells (4). These cells also exhibited significantly more apoptosis via both intrinsic and extrinsic pathways. In addition, metformin treatment induced autophagy. Inhibition of autophagy, either by Beclin1 knockdown or by 3-methyladenine-mediated inhibition of Caspase-3/7, suppressed the anti-proliferative effects of metformin on endometrial cancer cells. **Pharmacokinetics:** Intestinal metformin absorption is most likely mediated by the plasma membrane monoamine transporter (PMAT), whereas uptake into epithelial cells from circulation is primarily facilitated by OCT2. Metformin is not metabolized and is excreted unmodified in urine ($t_{1/2}$ ~5 hours), with a renal clearance (CL$_r$) of 510 mL/min. The Organic Cation Transporter 1 (OCT1) plays a key role in cellular uptake of metformin, and OCT1 is strongly inhibited by verapamil (5). **Key Pharmacokinetic Parameters:** *See* Appendix II in Goodman & Gilman's THE *PHARMACOLOGICAL BASIS OF THERAPEUTICS*, 12th Edition (Brunton, Chabner & Knollmann, eds.) McGraw-Hill Medical, New York (2011). **Target(s):** spermidine uptake, competitively (6); basal glycogenesis, but restores insulin-stimulated glycogenesis in cholesterol hemisuccinate-treated cells (6); leptin secretion in cultured rat adipose tissue (7); non-enzymatic glycation in diabetes (8); also binds α-dicarbonyls, such as methylglyoxal and 3-deoxyglucosone, thereby lowering nonenzymatic protein modification (8); tyrosine phosphatase activity, thereby stimulating insulin receptor tyrosine kinase (9); hepatic gluconeogenesis via AMP-activated protein kinase-dependent regulation of the orphan nuclear receptor SHP (10); Stat3 activation by phosphorylation at Tyr-705 and Ser-727 residues and Stat3-downstream signaling, thereby inhibiting cell growth and inducing apoptosis in triple-negative breast cancers (11); AMP deaminase, thereby activating AMP-stimulated protein kinase, *or* AMPK (12); mTOR, with cell-cycle arrest mediated through REDD1 (13). **1**. Edelman (1998)

Adv. Intern. Med. **43**, 449. **2**. Dagogo-Jack & Santiago (1997) *Arch. Intern. Med.* **157**, 1802. **3**. Wheaton, Weinberg, Hamanaka, *et al.* (2014) *Elife* **3**, e02242. **3**. Khan, Wiernsperger, Quemener, Havouis & Moulinoux (1992) *J. Cell Physiol.* **152**, 310. **4**. Takahashi, Kimura, Yamanaka, *et al.* (2014) *Cancer Cell Int.* **14**, 53. **5**. Cho, Kim, Park & Chung (2014) *Brit. J. Clin. Pharmacol.* **78**, 1426. **6**. Mick, Wang, Ling Fu & McCormick (2000) *Biochim. Biophys. Acta.* **1502**, 426. **7**. Beisswenger & Ruggiero-Lopez (2003) *Diabetes Metab.* **29**, 6S95. **8**. Holland, Morrison, Chang, Wiernsperger & Stith (2004) *Biochem. Pharmacol.* **67**, 2081. **9**. Kim, Park, Lee, *et al.* (2008) *Diabetes* **57**, 306. **10**. Deng, Wang, Deng, *et al.* (2012) *Cell Cycle* **11**, 367. **11**. Ouyang, Parakhia & Ochs (2011) *J. Biol. Chem.* **286**, 1. **12**. Ben Sahra, Regazzetti, Robert, *et al.* (2011) *Cancer Res.* **71**, 4366. **13**. Dowling, Zakikhani, Fantus, Pollak & Sonenberg (2007) *Cancer Res.* **67**, 10804.

Metavanadate

This vandate ion (FW$_{sodium-salt}$ = 121.93 g/mol; CAS 16389-35-8; Oxidation State = 5) is a metaphosphate ion analogue that forms indefinite chains as well as cyclic structures. Investigators should be certain that the correct form of vanadate is used in investigations. (***See*** *Orthovanadate; Vanadium and Vanadium Ions*) **Target(s):** acid phosphatase (1); alkaline phosphatase (2); bisphosphoglycerate phosphatase (3); carbonic anhydrase (4); dynein ATPase (5); frictose-1,6-bisphosphatase (3); hexokinase (3); phosphoglucomutase (3); phosphoglycerate kinase (3); phosphoglycerate mutase (3,6,7); sabinene-hydrate synthase (8); succinyl-CoA synthetase (ADP-forming) (9,10); succinyl-CoA synthetase (GDP-forming) (9); UDP-*N*-acetylglucosamine 2-epimerase (11). **1**. Andrews & C. Pallavicini (1973) *Biochim. Biophys. Acta* **321**, 197. **2**. Srinivas, Jayalakshmi, Sreeramulu, Sherman & Rao (2006) *Biochim. Biophys. Acta* **1760**, 310. **3**. Climent, Bartrons, Pons & Carreras (1981) *Biochem. Biophys. Res. Commun.* **101**, 570. **4**. Innocenti, Vullo, Scozzafava & Supuran (2005) *Bioorg. Med. Chem. Lett.* **15**, 567. **5**. Shimizu (1981) *Biochemistry* **20**, 4347. **6**. White, Nairn, Price, *et al.* (1992) *J. Bacteriol.* **174**, 434. **7**. Chander, Setlow & Setlow (1998) *Can. J. Microbiol.* **44**, 759. **8**. Hallahan & Croteau (1988) *Arch. Biochem. Biophys.* **264**, 618. **9**. Krivanek & Novakova (1992) *Physiol. Res.* **41**, 345. **10**. Krivanek & Novakova (1991) *Gen. Physiol. Biophys.* **10**, 71. **11**. Zeitler, J. P. Banzer, C. Bauer & W. Reutter (1992) *Biometals* **5**, 103.

Metergoline

This serotonin (HT) antagonist (FW$_{free-base}$ = 403.52 g/mol; CAS 17692-51-2; Soluble to 100 mM in DMSO), also named [(8β)-1,6-dimethylergolin-8-yl)methyl]carbamic acid phenylmethyl ester, targets 5-HT$_1$/5-HT$_2$ with activity at 5-HT$_{1D}$. Specific binding site for [^{3}H]-metergoline characterized by a K_d of 0.5-1.0 nM was detected in microsomal and synaptic plasma membranes from various areas of the adult rat brain. Metergoline has moderate affinity for 5-HT$_6$ and high affinity for 5-HT$_7$. It is also a full agonist at 5-HT$_{1Dβ}$ receptors (2). **1**. Hamon, Mallat, Herbet, *et al.* (1981) *J. Neurochem.* **36**, 613. **2**. Miller, King, Demchyshyn, Niznik & Teitler M (1992) *Eur. J. Pharmacol.* **227**, 99.

Methadone

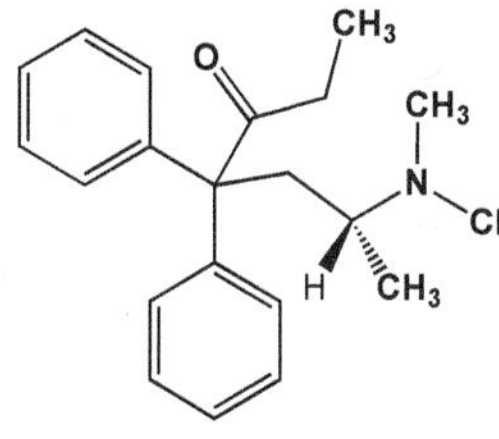

This orally available narcotic analgesic and full opioid agonist (FW = 309.45 g/mol; CAS 76-99-3), also known as 6-dimethylamino-4,4-diphenyl-3-heptanone and amidon, is widely used to treat individuals with moderate to severe opioid dependency. It is a μ-opioid receptor agonist that, upon chronic exposure, desensitizes both the μ- and δ-receptors. Both (R)- and (S)-methadone are N-methyl-D-aspartate (NMDA) receptor antagonists, but the latter is strongest. **Methadone Maintenance Therapy:** MMT is prescribed to individuals who wish to abstain from illicit drug use but have failed to maintain abstinence from opiates for significant periods. Therapeutic goals mainly focus on reducing withdrawal symptoms and managing psychosocial problems and psychiatric co-morbidity. The duration of methadone maintenance can be for months or even years. Side-effects include cardiac dysrhythmia, hypotension, diaphoresis, constipation, nausea, vomiting, dizziness, and sedation. Methadone is hardly innocuous: ~5,000 U.S. drug users die annually as a consequence of methadone overdose. **Pharmacokinetics:** Pharmacologically active (R)-methadone (shown above) is metabolized slowly and has very high fat solubility, lasting longer than morphine-based drugs. Its mean bioavailability is around 75%, and its elimination half-life of 15-60 hours, with mean of ~22 hours. Blood concentrations fluctuate greatly between dosing. Cytochrome P450 3A4 and 2D6 are involved in its hepatic metabolism. **Key Pharmacokinetic Parameters:** *See* Appendix II in Goodman & Gilman's THE PHARMACOLOGICAL BASIS OF THERAPEUTICS, 12th Edition (Brunton, Chabner & Knollmann, eds.) McGraw-Hill Medical, New York (2011). **Target(s):** aldehyde oxidase, mixed non-competitive type inhibition, K_{is} = 0.03 μM, K_{ii} = 0.57 μM (1); calcium channels, L-type (2); cholinesterase (3,4); CYP2D6, *or* cytochrome P450 2D6 (5); glycogen phosphorylase (6); glycogen synthase (6); hexokinase (7); N-methyl D-aspartate (NMDA) receptor (8); rhodamine123 transport (9). **1.** Robertson & Gamage (1994) *Biochem. Pharmacol.* **47**, 584. **2.** Yang, Shan, Ng & Pang (2000) *Brain Res.* **870**, 199. **3.** Augustinsson (1950) *The Enzymes*, 1st. ed., **1** (part 1), 443. **4.** Greig & Howell (1948) *Proc. Soc. Exp. Biol. Med.* **68**, 352. **5.** Wu, Otton, Sproule, *et al.* (1993) *Brit. J. Clin. Pharmacol.* **35**, 30. **6.** Gourley & Schwarzmeier (1973) *Life Sci.* **13**, 1353. **7.** Greic (1948) *Arch. Biochem.* **17**, 129. **8.** Gorman, Elliott & Inturrisi (1997) *Neurosci. Lett.* **223**, 5. **9.** Stormer, Perloff, von Moltke & Greenblatt (2001) *Drug Metab. Dispos.* **29**, 954.

Methamphetaminse, *See* *N-Methyl-α-methylphenethylamine*

Methaqualone

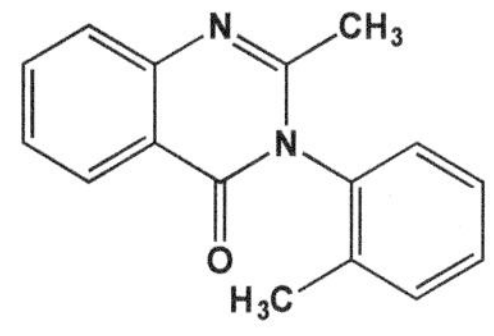

This synthetic, barbiturate-like, central nervous system depressant and anticonvulsant (FW = 250.30 g/mol; CAS 72-44-6), also known as 2-methyl-3-o-tolyl-4(3H)-quinazolinone and Quaalude™, is a controlled (often abused) substance with hypnotic and sedative properties. Its principal mode of action centers on GABA receptor blockade, rather than effects on glycine receptors. In the 1980's, it gained popularity as a euphoriant among casual recreational drug users. Methaqualone also induces the activities of cytochrome P450-dependent systems and UDPglucuronosyltransferases. Methaqualone significantly stimulates the activity of hepatic δ-aminolevulinic acid (ALA) synthetase, the first committed step and pacemaking enzyme in heme biosynthesis in mammals. Methqualone is a noncompetitive antagonist for the α-amino-3-hydroxy-5-methyl-4-isoxazolepropionate, *or* AMPA, receptor (1). **1.** Chenard, Menniti, Pagnozzi, *et al.* (2000) *Bioorg. Med. Chem. Lett.* **10**, 1203.

Methicillin

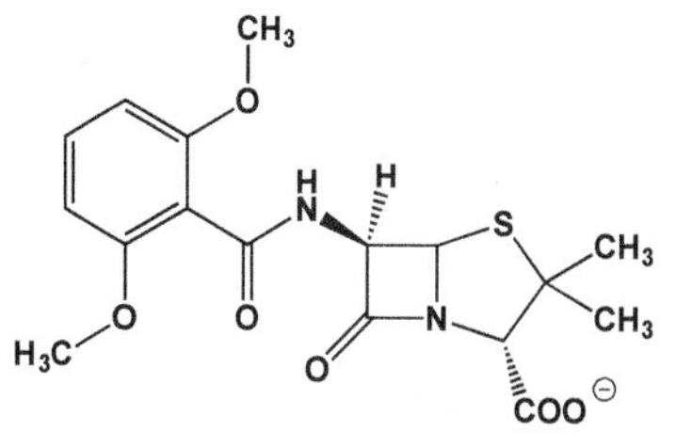

This β-lactam antibiotic (FW$_{free-acid}$ = 380.42 g/mol; CAS 61-32-5), also known as 6-(2,6-dimethoxybenzamido)penicillanic acid and meticillin, is a semisynthetic, narrow-spectrum penicillin that is bacteriostatic by virtue of its inhibition of peptidoglycan biosynthesis by binding to and competitively inhibiting the transpeptidase enzyme used by bacteria to cross-link the peptide (D-alanyl-D-alanine) required in peptidoglycan synthesis. A defining quality of methicillin is its resistance to the action of many β-lactamases. In the clinic, methicillin has been largely replaced by flucloxacillin and dicloxacillin. The term MRSA refers any *Staphylococcus aureus* strain that resists β-lactam antibiotics, including the penicillins (*e.g.*, methicillin, dicloxacillin, nafcillin, oxacillin, *etc.*) as well as the cephalosporins. **Target(s):** alanyl aminopeptidase, *or* aminopeptidase N (1,2); aminopeptidase D (3); cytosol alanyl aminopeptidase (1,2); glycopeptide transpeptidase (4); β-lactamase, *or* penicillinase (5-11); lecithin:cholesterol acyltransferase, *or* LCAT (12); lipase (13); membrane alanyl aminopeptidase, aminopeptidase N (2); muramoylpentapeptide carboxypeptidase (4,14,15); serine-type D-Ala-D-Ala carboxypeptidase (16). **1.** Starnes, Szechinski & Behal (1982) *Eur. J. Biochem.* **124**, 363. **2.** McClellan & Garner (1980) *Biochim. Biophys. Acta* **613**, 160. **3.** Starnes, Desmond & Behal (1980) *Biochim. Biophys. Acta* **616**, 290. **4.** Izaki, Matsuhashi & Strominger (1966) *Meth. Enzymol.* **8**, 487. **5.** Ross (1975) *Meth. Enzymol.* **43**, 678. **6.** Hamilton-Miller & Smith (1964) *Nature* **201**, 999. **7.** Fink, Behner & Tan (1987) *Biochemistry* **26**, 4248. **8.** Citri & Garber (1961) *Biochem. Biophys. Res. Commun.* **4**, 143. **9.** Garber & Citri (1962) *Biochim. Biophys. Acta* **62**, 385. **10.** Fujii-Kuriyama, Yamamoto & Sugawara (1977) *J. Bacteriol.* **131**, 726. **11.** Matthew & Sykes (1977) *J. Bacteriol.* **132**, 341. **12.** Bojesen (1978) *Scand. J. Clin. Lab. Invest.* Suppl. **150**, 26. **13.** Geitman, Kardash & Kivman (1977) *Vopr. Med. Khim.* **23**, 632. **14.** Davis & Salton (1975) *Infect. Immun.* **12**, 1065. **15.** Diaz-Mauriño, Nieto & Perkins (1974) *Biochem. J.* **143**, 391. **16.** Coyette, Ghuysen & Fontana (1978) *Eur. J. Biochem.* **88**, 297.

Methimazole

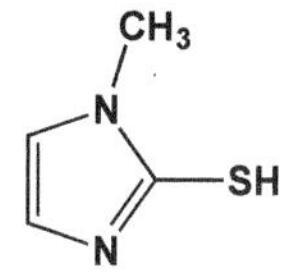

This widely used antihyperthyroid agent (FW = 114.17 g/mol; CAS 60-56-0), also known as 1-methyl-2-mercaptoimidazole and 1-methylimidazole-2-thiol, binds to thyroid peroxidase and blocks the conversion of iodide to iodine, thereby limiting thryoglobulin iodination. **Target(s):** albendazole monooxygenase (31); catechol oxidase (31); chlorophenol 4-monooxygenase (2); CYP3A4 (3); CYP2B6 (3); CYP2C9 (3); CYP2C19 (3); cysteine conjugate S-oxidase (4); cytochrome P450 (3,5); dopamine β-monooxygenase (6-9); flavin-containing monooxygenase, also as an alternative substrate (10-14); iodide peroxidase, *or* thyroid peroxidase (18,26-29,33,34); aactoperoxidase (15-18,35); myeloperoxidase (19-21); peroxidase (15-22,35); prostaglandin H synthase (23-25); tyrosinase, *or* monophenol monooxygenase (30-32). **1.** Moroni, Buronfosse, Longin-Sauvageon, Delatour & Benoit (1995) *Drug Metab. Dispos.* **23**, 160. **2.** Martin-Le Garrec, Artaud & Capeillere-Blandin (2001) *Biochim. Biophys. Acta* **1547**, 288. **3.** Guo, Raeissi, White & Stevens (1997) *Drug Metab. Dispos.* **25**, 390. **4.** Sausen & Elfarra (1990) *J. Biol. Chem.* **265**, 6139. **5.** Nnane & Damani (2003) *Life Sci.* **73**, 359. **6.** Rosenberg (1983) *Biochim. Biophys. Acta* **749**, 276. **7.** Hidaka & Nagasaka (1977) *Biochem. Pharmacol.* **26**, 1092. **8.** Fuller, Ho, Matsumoto & Clemens (1976) *Adv. Enzyme Regul.* **15**, 267. **9.** Stolk & Hanlon (1973) *Life Sci. II* **12**, 417. **10.** Lee, Shin, Lee, *et al.* (2003) *Toxicol. Lett.* **136**, 163. **11.** Kousba, Soll, Yee & Martin (2007) *Drug Metab. Dispos.* **35**, 2242. **12.** Reid, Walker, Miller, *et al.* (2004) *Clin. Cancer Res.* **10**, 1471. **13.** Chiba, Kobayashi, Itoh, *et al.*

(1995) *Eur. J. Pharmacol.* **293**, 97. **14.** Peters, Livingstone, Shenin-Johnson, Hines & Schlenk (1995) *Xenobiotica* **25**, 121. **15.** Bandyopadhyay, Biswas & Banerjee (2002) *Toxicol. Lett.* **128**, 117. **16.** Bandyopadhyay, Bhattacharyya, Chatterjee & Banerjee (1995) *Biochem. J.* **306**, 751. **17.** Doerge (1986) *Biochemistry* **25**, 4724. **18.** Magnusson, Taurog & Dorris (1984) *J. Biol. Chem.* **259**, 197. **19.** Wagner, Buettner, Oberley, Darby & Burns (2000) *J. Biol. Chem.* **275**, 22461. **20.** Jacquet, Deby, Mathy, *et al.* (1991) *Arch. Biochem. Biophys.* **291**, 132. **21.** Pincemail, Deby, Thirion, de Bruyn-Dister & Goutier (1988) *Experientia* **44**, 450. **22.** Bandyopadhyay, Bhattacharyya & Banerjee (1993) *Biochem. J.* **296**, 79. **23.** Lagorce, Moulard, Rousseau, *et al.* (1997) *Pharmacology* **55**, 173. **24.** Petry & Eling (1987) *J. Biol. Chem.* **262**, 14112. **25.** Zelman, Rapp, Zenser, Mattammal & Davis (1984) *J. Lab. Clin. Med.* **104**, 185. **26.** Coval & Taurog (1967) *J. Biol. Chem.* **242**, 5510. **27.** Engler, Taurog & Nakashima (1982) *Biochem. Pharmacol.* **31**, 3801. **28.** Davidson, Soodak, Neary, *et al.* (1978) *Endocrinology* **103**, 871. **29.** Nagasaka & Hidaka (1976) *J. Clin. Endocrinol. Metab.* **43**, 152. **30.** Andrawis & Kahn (1986) *Biochem. J.* **235**, 91. **31.** Kim & Uyama (2005) *Cell. Mol. Life Sci.* **62**, 1707. **32.** Flurkey, Cooksey, Reddy, *et al.* (2008) *J. Agric. Food Chem.* **56**, 4760. **33.** Schmutzler, Bacinski, Gotthardt, *et al.* (2007) *Endocrinology* **148**, 2835. **34.** Carvalho, Ferreira, Coelho, *et al.* (2000) *Braz. J. Med. Biol. Res.* **33**, 355. **35.** Bhuyan & Mugesh (2008) *Inorg. Chem.* **47**, 6569.

D-Methionine

This naturally occurring, sulfide-containing amino acid (FW = 149.21 g/mol; CAS 348-67-4; pK_a values of 2.13 and 9.28 at 25°C), the enantiomer of the proteogenic amino acid L-methionine, is a component in the cyanobacterial cyclic peptide, nostocyclamide M (1). **Target(s):** D-alanine aminotransferase (2); homoserine *O*-succinyltransferase (3); methionine *S*-adenosyltransferase (4); methionine decarboxylase (5,6); methionine *S*-methyltransferase, weakly inhibited (7); methionyl-tRNA synthetase (8); murein tetrapeptide LD-carboxypeptidases (9); tryptophan 2-monooxygenase, weakly inhibited (10). **1.** Juttner, Todorova, Walch & von Philipsborn (2001) *Phytochemistry* **57**, 613. **2.** Martinez-Carrion & Jenkins (1965) *J. Biol. Chem.* **240**, 3547. **3.** Mares, Urbanowski & Stauffer (1992) *J. Bacteriol.* **174**, 390. **4.** Yarlett, Garofalo, Goldberg, *et al.* (1993) *Biochim. Biophys. Acta* **1181**, 68. **5.** Misono, Kawabata, Toyosato, Yamamoto & Soda (1980) *Bull. Inst. Chem. Res. Kyoto Univ.* **58**, 323. **6.** Stevenson, Akhtar & Gani (1990) *Biochemistry* **29**, 7631. **7.** James, Nolte & Hanson (1995) *J. Biol. Chem.* **270**, 22344. **8.** Burnell & Whatley (1977) *Biochim. Biophys. Acta* **481**, 266. **9.** Templin & Höltje (1998) in *Handb. Proteolytic Enzymes*, p. 1574, Academic Press, San Diego.

L-Methionine

This nutritionally essential, sulfide-containing amino acid (FW = 149.21 g/mol; CAS 63-68-3; pK_a values of 2.13 and 9.28 at 25°C; Codon = AUG (AUA in certain mitochondria; Symbol = Met and M), also known as (*S*)-2-amino-4-(methylthio)butanoic acid, is one of the twenty proteogenic amino acids and is the initiating, N-terminal aminoacyl residue in translation. In prokaryotes, the methionyl residue is *N*-formylated, and the formyl group is usually removed post-translationally via an enzyme-catalyzed reaction. For other proteins, the methionyl residue is removed by a specific aminopeptidase. Methionine is regarded as a nonpolar amino acid in protein and peptide structures. Methionyl residues are frequently found in α-helices as well as β-pleated sheets in proteins. In homologous proteins, methionyl residues are often interchanged with leucyl and isoleucyl residues. The sulfur atom of the side-chain can act as a nucleophile, even under acidic conditions, and reacts famously with cyanogen bromide. The resulting acid-catalyzed cleavage of the polypeptide chain converts the methionyl residue to L-homoserine lactone. Mueller (1,2) first its presence in meat infusion was essential for the growth of hemolytic streptococcus. After he announced the discovery of a new amino acid in the hydrolysate of casein (3,4). Barger and Coyne (5) synthesized γ-methylthiol-α-aminobutyric acid, naming it methionine. **Target(s):** *O*-acetylhomoserine aminocarboxy-propyltransferase *or* *O*-acetylhomoserine sulfhydrylase (6,7,54-63); *O*-acetylserine/*O*-acetylhomoserine sulfhydrylase, *oe* cysteine synthase (8,9,63); *S*-adenosylmethionine cyclotransferase, weakly inhibited (68); alkaline phosphatase, mildly inhibited by the racemic mixture (10); D-amino-acid oxidase (11); 2-aminohexano-6-lactam racemase (12); arginase (13,14); aromatic-amino-acid aminotransferase (51); arylformamidase, weakly inhibited (15); asparagine:oxo-acid aminotransferase (52,53); aspartate kinase (43-46); choline kinase (16,49); L-3-cyanoalanine synthase (17); cystathionine β-lyase (18); cysteine synthase (8,9,63-67); cystinyl aminopeptidase, *or* oxytocinase (36-38); cytosol alanyl aminopeptidase (19,20); diaminopimelate decarboxylase (21); dimethylglycine *N*-methyltransferase, weakly inhibited (81); dipeptidyl-peptidase IV (22); glutamine synthetase (23); [glutamine-synthetase] adenylyltransferase (41); γ-glutamyl transpeptidase, *or* γ-glutamyltransferase, weakly inhibited (24,25,70,71); homoserine *O*-acetyltransferase (26,27,75); homoserine kinase (47,48); homoserine *O*-succinyltransferase (72-74); iodide peroxidase, thyroid peroxidase (84); kynureninase, weakly inhibited by racemic mixture (28); membrane alanyl aminopeptidase, *ore* aminopeptidase N (36,39,40); methionyl aminopeptidase, as product inhibitor (29,30); *S*-methyl-5'-thioadenosine phosphorylase, weakly inhibited (69); ornithine carbamoyltransferase (78,79); phenylalanine(histidine) aminotransferase (50); phenylalanine 4-monooxygenase (83); phosphoprotein phosphatase, mildly inhibited (31); sarcosine/dimethylglycine *N*-methyltransferase, weakly inhibited (81); serine *O* acetyltransferase (76,77); serine hydroxymethyltransferase, *or* glycine hydroxymethyltransferase (80); sulfate adenylyltransferase, weakly inhibited (42); sulfinoalanine decarboxylase, *or* cysteinesulfinate decarboxylase (32); tRNA (uracil-5-)-methyltransferase, weakly inhibited (82); tryptophanase, *or* tryptophan indole-lyase (33-35). **1.** Mueller (1921) *Proc. Soc. Exper. Biol. Med.* **18**, 14. **2.** Mueller (1921) *Proc. Soc. Exper. Biol. Med.* **18**, 225. **3.** Mueller (1922) *Proc. Soc. Exper. Biol. Med.* **19**, 161. **4.** Mueller (1923) *J. Biol. Chem.* **56**, 157. **5.** Barger & Coyne (1928) *Biochem. J.* **22**, 1417. **6.** Yamagata (1987) *Meth. Enzymol.* **143**, 465. **7.** Shiio & Ozaki (1987) *Meth. Enzymol.* **143**, 470. **8.** Yamagata (1987) *Meth. Enzymol.* **143**, 478. **9.** Burnell & Whatley (1977) *Biochim. Biophys. Acta* **481**, 246. **10.** Fishman & Sie (1971) *Enzymologia* **41**, 141. **11.** Dixon & Kleppe (1965) *Biochim. Biophys. Acta* **96**, 368. **12.** Ahmed, Esaki, Tanaka & Soda (1983) *Agric. Biol. Chem.* **47**, 1887. **13.** Hunter & Downs (1945) *J. Biol. Chem.* **157**, 427. **14.** Kaysen & Strecker (1973) *Biochem. J.* **133**, 779. **15.** Serrano & Nagayama (1991) *Comp. Biochem. Physiol. B Comp. Biochem.* **99**, 281. **16.** Ishidate & Nakazawa (1992) *Meth. Enzymol.* **209**, 121. **17.** Macadam & Knowles (1984) *Biochim. Biophys. Acta* **786**, 123. **18.** Morinaga, Tani & Yamada (1983) *Agric. Biol. Chem.* **47**, 2855. **19.** Garner & Behal (1977) *Arch. Biochem. Biophys.* **182**, 667. **20.** Garner & Behal (1975) *Biochemistry* **14**, 3208. **21.** Selli, Crociani, Di Gioia, *et al.* (1994) *Ital. J. Biochem.* **43**, 29. **22.** Shibuya-Saruta, Kasahara & Hashimoto (1996) *J. Clin. Lab. Anal.* **10**, 435. **23.** Blanco, Alana, Llama & Serra (1989) *J. Bacteriol.* **171**, 1158. **24.** Allison (1985) *Meth. Enzymol.* **113**, 419. **25.** Thompson & Meister (1977) *J. Biol. Chem.* **252**, 6792. **26.** Robichon-Szulmajster & Cherest (1967) *Biochem. Biophys. Res. Commun.* **28**, 256. **27.** Shiio & Ozaki (1981) *J. Biochem.* **89**, 1493. **28.** Jakoby & Bonner (1953) *J. Biol. Chem.* **205**, 709. **29.** Yang, Kirkpatrick, Ho, *et al.* (2001) *Biochemistry* **40**, 10645. **30.** Copik, Nocek, Swierczek, *et al.* (2005) *Biochemistry* **44**, 121. **31.** Bargoni, Fossa & Sisini (1963) *Enzymologia* **26**, 65. **32.** Tang, Hsu, Sun, *et al.* (1996) *J. Biomed. Sci.* **3**, 442. **33.** Demidkina, Zakomirdina, Kulikova, *et al.* (2003) *Biochemistry* **42**, 11161. **34.** Snell (1975) *Adv. Enzymol. Relat. Areas Mol. Biol.* **42**, 287. **35.** Gogoleva, Zakomirdina, Demidkina, Phillips & Faleev (2003) *Enzyme Microb. Technol.* **32**, 843. **36.** Lalu, Lampelo & Vanha-Perttula (1986) *Biochim. Biophys. Acta* **873**, 190. **37.** Lampelo & Vanha-Perttula (1979) *J. Reprod. Fertil.* **56**, 285. **38.** Itoh, Watanabe, Nagamatsu, *et al.* (1997) *Biol. Pharm. Bull.* **20**, 20. **39.** Bauvois & Dauzonne (2006) *Med. Res. Rev.* **26**, 88. **40.** Tokioka-Terao, Hiwada & Kokubu (1984) *Enzyme* **32**, 65. **41.** Ebner, Wolf, Gancedo, Elsässer & Holzer (1970) *Eur. J. Biochem.* **14**, 535. **42.** Heinzel & Trüper (1976) *Arch. Microbiol.* **107**, 293. **43.** Zhang, Jiang, Zhao, Yang & Chiao (2000) *Appl. Microbiol. Biotechnol.* **54**, 52. **44.** Cahyanto, Kawasaki, Nagashio, Fujiyama & Seki (2006) *Microbiology* **152**, 105. **45.** Dungan & Datta (1973) *J. Biol. Chem.* **248**, 8534. **46.** Ferreira, Meinhardt & Azevedo (2006) *Ann. Appl. Biol.* **149**, 77. **47.** Burr, Walker, Truffa-Bachi & Cohen (1976) *Eur. J. Biochem.* **62**, 519. **48.** Thoen, Rognes & Aarnes (1978) *Plant Sci. Lett.* **13**, 103. **49.** Setty & Krishnan (1972) *Biochem. J.* **126**, 313. **50.** Minatogawa, Noguchi & Kido (1977) *Hoppe-Seyler's Z. Physiol. Chem.* **358**, 59. **51.** Xing & Whitman (1992) *J. Bacteriol.* **174**, 541. **52.** Maul & Schuster (1986) *Arch. Biochem. Biophys.* **251**, 585. **53.** Cooper (1977) *J. Biol. Chem.* **252**, 2032. **54.** Yamagata,

Paszewski & Lewandowska (1990) *J. Gen. Appl. Microbiol.* **36**, 137. **55.** Shimizu, Yamagata, Masui, *et al.* (2001) *Biochim. Biophys. Acta* **1549**, 61. **56.** Yamagata (1984) *J. Biochem.* **96**, 1511. **57.** Iwama, Hosokawa, Lin, *et al.* (2004) *Biosci. Biotechnol. Biochem.* **68**, 1357. **58.** Ozaki & Shiio (1982) *J. Biochem.* **91**, 1163. **59.** Yamagata (1971) *J. Biochem.* **70**, 1035. **60.** Yeom, Hwang, Lee, Kim & Lee (2004) *J. Microbiol. Biotechnol.* **14**, 373. **61.** Kerr (1971) *J. Biol. Chem.* **246**, 95. **62.** Morinaga, Tani & Yamada (1983) *Agric. Biol. Chem.* **47**, 2855. **63.** Yamagata, Takeshima & Naiki (1975) *J. Biochem.* **77**, 1029. **64.** Ascano & Nicholas (1977) *Phytochemistry* **16**, 889. **65.** Leon & Vega (1991) *Plant Physiol. Biochem.* **29**, 595. **66.** Leon, Romero, Galvan & Vega (1987) *Plant Sci.* **53**, 93. **67.** Tamura, Iwasawa, Masada & Fukushima (1976) *Agric. Biol. Chem.* **40**, 637. **68.** Mudd (1959) *J. Biol. Chem.* **234**, 87. **69.** Garbers (1978) *Biochim. Biophys. Acta* **523**, 82. **70.** Lherbet, Gravel & Keillor (2004) *Bioorg. Med. Chem. Lett.* **14**, 3451. **71.** Lherbet & Keillor (2004) *Org. Biomol. Chem.* **2**, 238. **72.** Mares, Urbanowski & Stauffer (1992) *J. Bacteriol.* **174**, 390. **73.** Röhl, Rabenhorst & Zähner (1987) *Arch. Microbiol.* **147**, 315. **74.** Usuda & Kurahashi (2005) *Appl. Environ. Microbiol.* **71**, 3228. **75.** Wyman & Paulus (1975) *J. Biol. Chem.* **250**, 3897. **76.** Burnell & Whatley (1977) *Biochim. Biophys. Acta* **481**, 246. **77.** Smith & Thompson (1971) *Biochim. Biophys. Acta* **227**, 288. **78.** Legrain & Stalon (1976) *Eur. J. Biochem.* **63**, 289. **79.** Lusty, Jilka & Nietsch (1979) *J. Biol. Chem.* **254**, 10030. **80.** Schirch (1982) *Adv. Enzymol. Relat. Areas Mol. Biol.* **53**, 83. **81.** Waditee, Tanaka, Aoki, *et al.* (2003) *J. Biol. Chem.* **278**, 4932. **82.** Tscheme & Wainfan (1978) *Nucl. Acids Res.* **5**, 451. **83.** Goreish, Bednar, Jones, Mitchell & Steventon (2004) *Drug Metabol. Drug Interact.* **20**, 159. **84.** Carvalho, Ferreira, Coelho, *et al.* (2000) *Braz. J. Med. Biol. Res.* **33**, 355.

L-Methionine *S,S*-Dioxide

This L-methionine sulfone (FW = 181.21 g/mol; CAS 7314-32-1), also known as L-2-amino-4-(methylsulfonyl)butanoic acid, inhibits glutamine synthetase. It has also been identified as a residue in a number of proteins, including actin It is more soluble in water than methionine but not as soluble as the sulfoxide. **Target(s):** asparagine synthetase, weakly inhibited (1); glutamate synthase (2,3); glutamine synthetase (2,4-10); [glutamine synthetase] adenylyltransferase (11); methionine *S*-adenosyltransferase, weakly inhibited (12); methionyl-tRNA synthetase (13). **1.** Pike & Beevers (1982) *Biochim. Biophys. Acta* **708**, 203. **2.** Brenchley (1973) *J. Bacteriol.* **114**, 666. **3.** Meister (1985) *Meth. Enzymol.* **113**, 327. **4.** Rowe (1985) *Meth. Enzymol.* **113**, 199. **5.** Rowe, Ronzio, Wellner & Meister (1970) *Meth. Enzymol.* **17A**, 900. **6.** Meister (1974) *The Enzymes*, 3rd ed., **10**, 699. **7.** Srivastava & Amla (1997) *Indian J. Exp. Biol.* **35**, 1098. **8.** Rowe & Meister (1973) *Biochemistry* **12**, 1578. **9.** Orr & Haselkorn (1981) *J. Biol. Chem.* **256**, 13099. **10.** Ronzio, Rowe & Meister (1969) *Biochemistry* **8**, 1066. **11.** Bishop, McParland & Evans (1975) *Biochem. Biophys. Res. Commun.* **67**, 774. **12.** Berger & Knodel (2003) *BMC Microbiol.* **3**, 12. **13.** Hahn & Brown (1967) *Biochim. Biophys. Acta* **146**, 264.

Methionine *S*-Oxide

This methionine sulfoxide (FW = 165.21 g/mol; CAS 3226-65-1), also known as 2-amino-4-(methylsulfinyl)butanoic acid, inhibits glutamine synthetase. It can undergo further oxidation to form methionine sulfone. It is a substrate in the reaction catalyzed by methionine-*S*-oxide reductase and, when incorporated as a residue in a protein, it is a substrate for protein-methionine-*S*-oxide reductase. **Target(s):** asparagine synthetase (1); creatine kinase (2); glutamate synthase (3); glutamine synthetase (2,4-10); methionine *S*-adenosyltransferase, weakly inhibited (11); methionyl aminopeptidase (12). **1.** Pike & Beevers (1982) *Biochim. Biophys. Acta* **708**, 203. **2.** Haghighi & Maples (1996) *J. Neurosci. Res.* **43**, 107. **3.**

Meister (1985) *Meth. Enzymol.* **113**, 327. **4.** Elliott (1955) *Meth. Enzymol.* **2**, 337. **5.** Elliott & Gale (1948) *Nature* **161**, 129. **6.** Cohen (1952) *The Enzymes*, 1st ed., **2** (Part 2), 897. **7.** Meister (1962) *The Enzymes*, 2nd ed., **6**, 443. **8.** Rowe (1985) *Meth. Enzymol.* **113**, 199. **9.** Speck (1949) *J. Biol. Chem.* **179**, 1405. **10.** Ronzio, Rowe & Meister (1969) *Biochemistry* **8**, 1066. **11.** Berger & Knodel (2003) *BMC Microbiol.* **3**, 12. **12.** Yang, Kirkpatrick, Ho, *et al.* (2001) *Biochemistry* **40**, 10645.

L-Methionine Sulfoximine

This potent convulsant (FW = 180.23 g/mol; CAS 15985-39-4) binds at the active site of glutamine synthetase where it is phosphorylated to produce a tight-binding geometric analogue of the γ-glutamyl phosphate intermediate. Methionine sulfoximine was first isolated from nitrogen trichloride-treated protein and was later identified as the toxic factor causing convulsions in dogs fed NCl₃-treated flour (1). Nitrogen trichloride (Agene™) has been used to artificially bleach and age freshly milled wheat flour, which has a pale yellow color. **Origin of MSOX Convulsant Action:** Methionine sulfoximine inhibits brain glutamine synthetase so well that its action is essentially irreversible, with the inhibitor becoming firmly bound to the active site of the enzyme as methionine sulfoximine-phosphate. L-methionine-*S*-sulfoximine is the only one of the four methionine sulfoximine isomers to inhibit glutamine synthetase. When these isomeric forms were tested in mice for convulsant activity, only L-methionine-*S*-sulfoximine produced convulsions. The findings that only one of the four optical isomers of methionine sulfoximine induces convulsions and that same isomer inhibits brain glutamine synthetase lends strong support to the conclusion that these two effects of methionine sulfoximine are mechanistically connected (1). Recent studies suggest MSOX also increases the concentration of glutamate, which activates NMDA receptors, leading to excitotoxic cell death (2). **Target(s):** asparagine:oxo-acid aminotransferase (3); asparagine synthetase (weakly inhibited) (4); glutamate synthase (5,6); glutamin-(asparagin-)ase (7); glutamine synthetase (5,8-29); γ-glutamylcysteine synthetase (12,22,30-37); γ-glutamyl kinase (glutamate 5-kinase) (38); methionine *S*-adenosyltransferase (weakly inhibited) (39); L-methionine (*S*)-*S*-oxide reductase (methionine sulfoxide reductase) (41,42); 4-methyleneglutamine synthetase (4-methyleneglutamate:ammonia ligase; partial inhibition) (42). **1.** Rowe & Meister (1970) *Proc. Natl. Acad. Sci. U.S.A.* **66**, 500. **2.** Shaw, Bains, Pasqualotto & Curry (1999) *Can. J. Physiol. Pharmacol.* **77**, 871. **3.** Cooper (1977) *J. Biol. Chem.* **252**, 2032. **4.** Pike & Beevers (1982) *Biochim. Biophys. Acta* **708**, 203. **5.** Brenchley (1973) *J. Bacteriol.* **114**, 666. **6.** Meister (1985) *Meth. Enzymol.* **113**, 327. **7.** Steckel, Roberts, Philips & Chou (1983) *Biochem. Pharmacol.* **32**, 971. **8.** Ronzio, Rowe & Meister (1969) *Biochemistry* **8**, 1066. **9.** Rowe, Ronzio & Meister (1969) *Biochemistry* **8**, 2674. **10.** Rowe, Ronzio, Wellner & Meister (1970) *Meth. Enzymol.* **17A**, 900. **11.** Wolfenden (1977) *Meth. Enzymol.* **46**, 15. **12.** Griffith (1987) *Meth. Enzymol.* **143**, 286. **13.** Meister (1962) *The Enzymes*, 2nd ed., **6**, 443. **14.** Tate & Meister (1961) *Proc. Natl. Acad. Sci. U.S.A.* **68**, 781. **15.** Meister (1974) *The Enzymes*, 3rd ed., **10**, 699. **16.** Webb (1966) *Enzyme and Metabolic Inhibitors*, vol. **2**, p. 335, Academic Press, New York. **17.** Weisbrod & Meister (1973) *J. Biol. Chem.* **248**, 3997. **18.** Tate, Leu & Meister (1972) *J. Biol. Chem.* **247**, 5312. **19.** Ronzio & Meister (1968) *Proc. Natl. Acad. Sci. U.S.A.* **59**, 164. **20.** Wedler, Sugiyama & Fisher (1982) *Biochemistry* **21**, 2168. **21.** Rowe (1985) *Meth. Enzymol.* **113**, 199. **22.** Griffith, Anderson & Meister (1979) *J. Biol. Chem.* **254**, 1205. **23.** Orr & Haselkorn (1981) *J. Biol. Chem.* **256**, 13099. **24.** Purich (1998) *Adv. Enzymol. Relat. Areas Mol. Biol.* **72**, 9. **25.** Manning, Moore, Rowe & Meister (1969) *Biochemistry* **8**, 2681. **26.** Southern, Parker & Woods (1987) *J. Gen. Microbiol.* **133**, 2437. **27.** Blanco, Alana, Llama & Serra (1989) *J. Bacteriol.* **171**, 1158. **28.** Shatters, Liu & Kahn (1993) *J. Biol. Chem.* **268**, 469. **29.** Garcia-Dominguez, Reyes & Florencio (1997) *Eur. J. Biochem.* **244**, 258. **30.** Griffith & Meister (1979) *J. Biol. Chem.* **254**, 7558. **31.** Seelig & Meister (1985) *Meth. Enzymol.* **113**, 379. **32.** Meister (1974) *The Enzymes*, 3rd ed., **10**, 671. **33.** Richman, Orlowski & Meister (1973) *J. Biol. Chem.* **248**, 6684. **34.** Sekura & Meister (1977) *J. Biol. Chem.* **252**, 2599. **35.** Yip & Rudolph (1976) *J. Biol. Chem.* **251**, 3563. **36.** Griffith & Mulcahy (1999) *Adv. Enzymol. Relat. Areas Mol. Biol.* **73**, 209. **37.** Hell & Bergmann (1990) *Planta* **180**, 603. **38.** Krishna &

Leisinger (1979) *Biochem. J.* **181**, 215. **39**. Berger & Knodel (2003) *BMC Microbiol.* **3**, 12. **40**. Black (1962) *Meth. Enzymol.* **5**, 992. **41**. Black, Harte, Hudson & Wartofsky (1960) *J. Biol. Chem.* **235**, 2910. **42**. Winter & Dekker (1986) *J. Biol. Chem.* **261**, 11189.

Methiothepin

This piperazine derivative (FW$_{free-base}$ = 356.56 g/mol; M.P. = 88-89°C; CASs = 20229-30-5 (mesylate), 19728-88-2 (maleate)), also named Ro-8-6837, metitepine, and 1-[10,11-dihydro-8 (methylthio)dibenzo[*b,f*]thiepin-10-yl]-4-methylpiperazine, is a 5-HT1, 5-HT6, and 5-HT7 serotonin receptor antagonist. It also blocks serotonin autoreceptors. Methiothepin is typically supplied commercially as either the maleate or mesylate salt. **Target(s):** aryl-acylamidase (1); 5-HT$_1$ serotonin receptor (2-4); 5-HT$_6$ serotonin receptor (5); 5 HT$_7$ serotonin receptor (6). **1**. Hsu, Halaris & Freedman (1982) *Int. J. Biochem.* **14**, 581. **2**. Nilsson, Longmore, Shaw, *et al.* (1999) *Eur. J. Pharmacol.* **372**, 49. **3**. Granas & Larhammar (1999) *Eur. J. Pharmacol.* **380**, 171. **4**. McLoughlin & Strange (2000) *J. Neurochem.* **74**, 347. **5**. Boess, Riemer, Bos, *et al.* (1998) *Mol. Pharmacol.* **54**, 577. **6**. Schoeffter, Ullmer, Bobirnac, Gabbiani & Lubbert (1996) *Brit. J. Pharmacol.* **117**, 993.

Metoprolol

This β-blocker (FW$_{free-base}$ = 267.37 g/mol; CAS 51384-51-1; IUPAC: (*RS*)-1-(isopropylamino)-3-[4-(2-methoxyethyl)phenoxy]propan-2-ol), also named Lopressor®, selectively targets β$_1$-adrenoceptors and shows no reversal of β$_2$-mediated vasodilating effects of epinephrine. Likely modes of lopressor action include: competitive antagonism of catecholamines at peripheral (especially cardiac) adrenergic sites, attended by decreased cardiac output; CNS effects, leading to reduced sympathetic outflow to the periphery; as well as suppression of renin activity. **Pharmacokinetics:** Bioavailability 50%; Protein binding = 12%; Hepatic metabolism by CYP2D6 and CYP3A4; Biological $t_{1/2}$ = 6-7 hours; Renal excretion.

Methotrexate

This anti-rheumatic/anti-neoplastic agent (FW$_{free-acid}$ = 454.45 g/mol; CAS 59-05-2; photo- and alkali-sensitive; *Abbreviation*: MTX), also known as 4-amino-10-methylfolic acid and (+)-amethopterin as well as the trade names Rheumatrex®, Trexall®, Otrexup®, and Rasuvo®, is a potent, slow-binding inhibitor of dihydrofolate reductase (K_i = ~1 nM). Upon entry via the folate transporter, MTX becomes polyglutamylated, thereby increasing its binding

affinity for and inhibitory action on dihydrofolate reductase. (**See** *4-Amino-10-methylpteroyl-γ-glutamyl-γ-glutamyl-γ-glutamate; Methotrexate Polyglutamate; Pemetrexed*) A plasma methotrexate level of approximately 7–10 μM is cytocidal to leukemic cells. MTX is taken up by cancer cells by means of the reduced folate carrier SLC19A1, with efflux mediated by various drug-resistance-conferring ABC transporters. Once within cells, MTX is rapidly converted to active methotrexate polyglutamates (MTXPGs) by folylpolyglutamate synthetase (FPGS), the latter adding as many as six glutamate residues to MTX as isopepides. (Red blood cell uptake of MTX occurs only during the erythroblast stage of development, using the same mechanisms as folate.) The primary target of MTX and MTXPGs is dihydrofolate reductase (DHFR), which converts dihydrofolate (DHF) to tetrahydrofolate (THF), the latter required for *de novo* purine nucleotide synthesis. Inhibition also results in decreased protein and DNA methylation and impaired DNA regulation and repair. **Primary Mode of MTX Action:** Detailed analysis of the slow-binding kinetics of MTX indicates that dihydrofolate reductase rapidly binds the drug to form an Enz·NADPH·MTX ternary complex that subsequently undergoes one or more relatively slow, reversible isomerizations (1). From the apparent K_i value of 23 nM for the dissociation of methotrexate from the Enz·NADPH·MTX complex and values of 5.1 and 0.013 min^{-1} for the forward and reverse rate constants of the isomerization reaction, the overall inhibition constant for methotrexate was calculated to be 58 pM. Under conditions where reassociation of free enzyme and inhibitor is prevented by the presence of excess folate, the DHFR-bound MTX dissociated with a half-life of ~1 hour. Formation of an initial Enz·MTX complex was demonstrated by means of fluorescence quenching, and a value of 0.36 μM was determined for its dissociation constant. The same technique was used to determine dissociation constants for the reaction of MTX with the E·NADP and E·NADPH complexes. The results indicate that in the presence of either NADPH or NADP there is enancement of the binding of methotrexate to the enzyme. In this respect, that methotrexate behaves as a pseudosubstrate for dihydrofolate reductase. **Use of MTX in Cancer Chemotherapy:** By preventing cells from utilizing folate-based cofactors to make nucleotide precursors for DNA and RNA, methotrexate is especially toxic to rapidly proliferating cells. By blocking the dihydrofolate reductase reaction and depleting the one-carbon pool, methotrexate inhibits DNA biosynthesis, often leading to programmed cell death. This property and reversal by the more weak inhibitor leukovorin (**See** *N^5-Formyltetrahydrofolate*) forms the basis for the well-known "Methotrexate-Leucovorin Rescue" protocol for treating certain cancers (*e.g.,* gestational trophoblastic neoplasias). High-dose MTX (high doses 3–7.5 g/m^2) are delivered, followed by infusion of leukovorin (folinic acid). Myelotoxicity occurs in 28% of the patients, usually as a result of delayed MTX excretion. HDMTX-LCV is not recommended as conventional treatment of metastatic cancer, simply because response rates cited are not superior to those achievable by conventional MTX dosing. Given the wide spectrum of available treatment options, methotrexate is primarily used to treat choriocarcinoma, leukemia in the spinal fluid, osteosarcoma, breast cancer, lung cancer, non-Hodgkin lymphoma, and head and neck cancers. When tested at various MTX concentrations (0.5, 1, 3, and 10 μM) for 24 hours in human MCF-7 breast cancer cells, followed by rescue with labeled leucovorin (0.5 to 50 μM), total labeled intracellular folate pools increase in a log-linear fashion with respect to leucovorin exposure concentrations up to 100 μM (2). Tetrahydrofolate, 10-formyl tetrahydrofolate, 5-formyl tetrahydrofolate, 5-methyl tetrahydrofolate, and 5,10-methylene tetrahydrofolate pools reach levels present in cells not exposed to MTX at concentrations of leucovorin that are inadequate in rescuing MTX-treated cells. At rescue-concentrations of leucovorin, individual folate pool levels are up to twelve times greater than those in untreated cells, consistent with competition for catalytic activity at folate-dependent enzymes in addition to dihydrofolate reductase (2). The dihydrofolate pool also increases with increasing leucovorin concentration; however, unlike the reduced folates, this oxidized folate reaches a maximal level that is dependent on the MTX concentration to which the cells are exposed. Such findings suggest that competition between MTX and leucovorin occurs at the level of dihydrofolate reductase via a competitive interaction with dihydrofolate in this intact cell system (2). The ability of leucovorin and its metabolites to compete with direct inhibitors of dihydrofolate reductase and other metabolically important folate-dependent enzymes forms the basis of leucovorin rescue. **MTX Treatment of Rheumatoid Arthritis:** Methotrexate is generally administered to RA patients as a single dose administered either intramuscularly or orally (15 to 17.5 mg/week). Beyond its role in cancer chemotherapy, methotrexate is also the most frequently used first-line antirheumatic drug, inhibiting proliferation of the lymphocytes and other cells responsible for inflammation in joints. Mounting evidence favors the view that the endogenous anti-inflammatory autocoid adenosine mediates the anti-inflammatory effects of methotrexate. It also inhibits the formation of potentially toxic compounds (*e.g.*, the transmethylation products spermine and spermidine found in chronically inflamed tissues). Other ideas are that methotrexate inhibits proliferation of the cells responsible for synovial inflammation in RA and that methotrexate reduces intracellular glutathione levels by an oxidant-associated mechanism that reduces macrophage recruitment and function. **Use of MTX to Synchronize Cells:** Reference (3) provides detailed procedures for inducing synchrony in human cancer cell lines (Jurkat, K562, U937, SW626) and murine L1210 cells by using very low non-toxic methotrexate concentrations (0.04–0.08 µM for 13–24 hours) under standard culture conditions. This protocol offers a method for synchronization of cells at the G_1/S boundary and through the S-G_2-M phases of the cell cycle (*See also Thymidine for "Double-Thymidine Block Method for G_1/S Cell Synchnronization"*). **Mechanisms of MTX Resistance:** Resistance to can arise from decreased methotrexate uptake, induction of xenobiotic transporters, defects in folate oligoglutamylating enzyme, and increased dihydrofolate reductase (DHFR) activity. An association between lack of the retinoblastoma protein and intrinsic MTX resistance has also been reported. Methotrexate is also known to increase homocysteine, the latter associated with an elevated risk of heart disease, Alzheimer's disease, and neural tube defects. Homocysteine is formed by hydrolysis of *S*-adenosylhomocysteine, an inhibitor of SAM-dependent methyltransferases, including isoprenylcysteine carboxyl methyltransferase (ICMT), the enzyme that methylates Ras in a reaction needed for the latter's eventual plasma membrane localization and function. Upon methotrexate treatment of DKOB8 cells, Ras methylation is decreased by ~90%, resulting in mislocalization of Ras to the cytosol and a 4x decrease in the activation of p44 mitogen-activated protein kinase and Akt (4). Importantly, cells lacking ICMT are highly MTX-resistant. While cells expressing wild-type levels of ICMT are inhibited by MTX, stable expression of myristoylated H-Ras, which does not require carboxylmethylation for membrane attachment, confers resistance to methotrexate. Such findings suggest isoprenylcysteine carboxyl methyltransferase inhibition may be a critical factor influencing methotrexate effectiveness (3). **Key Pharmacokinetic Parameters:** See Appendix II in Goodman & Gilman's *THE PHARMACOLOGICAL BASIS OF THERAPEUTICS*, 12th Edition (Brunton, Chabner & Knollmann, eds.) McGraw-Hill Medical, New York (2011). **Target(s):** arylamine *N*-acetyltransferase (5-7); deoxycytidylate 5-hydroxymethyl-transferase (7); dihydrofolate reductase (8-25,54-56,58,62); dihydropterin reductase (26,27); dihydropterin reductase (NADPH) (28); formimidoyl-tetrahydrofolate cyclodeaminase, *or* formiminotetrahydrofolate cyclodeaminase (29,30); glucose-6-phosphate dehydrogenase (30); glutamate dehydrogenase (31,32); glutamate formimidoyltransferase (33); glutathione synthetase (34); glycine *N*-methyltransferase (64); glycogen phosphorylase *b* (34); lactate dehydrogenase (31); malate dehydrogenase (31); mandelate 4-monooxygenase (36); methenyltetrahydrofolate cyclohydrolase, weakly inhibited (37); methionyl-tRNA formyltransferase (38); phenylalanine 4-monooxygenase (26,39); phenylpyruvate tautomerase (40); phosphoribosyl-aminoimidazole-carboxamide formyltransferase, *or* 5-aminoimidazole-4-carboxamide ribonucleotide transformylase, weakly inhibited by the monoglutamate, but strongly inhibited by the pentaglutamate derivative (41-48); phosphoribosylglycinamide formyl-transferase, *or* glycinamide ribonucleotide transformylase (41); sepiapterin deaminase (49); serine Hydroxymethyl-transferase, *or* glycine hydroxymethyl-transferase, partial inhibitor (50,51); *Thermotoga maritima* dihydrofolate reductase (12); thymidylate synthase (52,56-62); xanthine oxidase (53). **1.** Williams, Morrison & Duggleby (1979) *Biochemistry* **18**, 2567. **2.** Boarman, Baram & Allegra (1990) *Biochem. Pharmacol.* **40**, 2651. **3.** Erba & Sen (1996) *Meth. Cell. Sci.* **18**, 149. **4.** Winter-Vann, Kamen, Bergo, *et al.* (2003) *Proc. Natl. Acad. Sci. U.S.A.* **100**, 6529. **5.** Ward, Summers & Sim (1995) *Biochem. Pharmacol.* **49**, 1759. **6.** Andres, Kolb & Weiss (1983) *Biochim. Biophys. Acta* **746**, 182. **7.** Lee, Gautam-Basak, Wooley & Sander (1988) *Biochemistry* **27**, 1367. **8.** Baker (1959) *Cancer Chemother. Rep.* **4**, 1. **9.** Roth & Burchall (1971) *Meth. Enzymol.* **18B**, 779. **10.** Matthews (1984) *Science* **197**, 452. **11.** Mathews, Scrimgeour & Huennekens (1963) *Meth. Enzymol.* **6**, 364. **12.** Scott & Tomkins (1975) *Meth. Enzymol.* **40**, 273. **14.** Dams & Jaenicke (2001) *Meth. Enzymol.* **331**, 305. **15.** Morrison (1982) *Trends Biochem. Sci.* **7**, 102. **16.** Taira, Fierke, Chen, Johnson & Benkovic (1987) *Trends Biochem. Sci.* **12**, 275. **17.** Nath & Greenberg (1962) *Biochemistry* **1**, 435. **18.** Baker (1967) *Design of Active-Site-Directed Irreversible Enzyme Inhibitors*, Wiley, New York. **19.** Osborn, Freeman & Huennekens (1958) *Proc. Soc. Exptl. Biol. Med.* **97**,

429. **20.** Kashket, Crawford, Friedkin, Humphreys & Goldin (1964) *Biochemistry* **3**, 1928. **21.** Bertino, Perkins & Johns (1965) *Biochemistry* **4**, 839. **22.** Perkins & Bertino (1966) *Biochemistry* **5**, 1005. **23.** Kumar, Kisliuk, Gaumont, Freisheim & Nair (1989) *Biochem. Pharmacol.* **38**, 541. **24.** Kumar, Kisliuk, Gaumont, *et al.* (1986) *Cancer Res.* **46**, 5020. **25.** Bertino, Gabrio & Huennekens (1960) *Biochem. Biophys. Res. Commun.* **3**, 461. **26.** Burchall & Hitchings (1965) *Mol. Pharmacol.* **1**, 126. **27.** Kaufman (1987) *Meth. Enzymol.* **142**, 97. **28.** Hasegawa & Nakanishi (1987) *Meth. Enzymol.* **142**, 103. **29.** Hasegawa & Nakanishi (1987) *Meth. Enzymol.* **142**, 111. **30.** Uyeda & Rabinowitz (1963) *Meth. Enzymol.* **6**, 380. **31.** Uyeda & Rabinowitz (1967) *J. Biol. Chem.* **242**, 24. **32.** Vogel, Snyder & Schulman (1963) *Biochem. Biophys. Res. Commun.* **10**, 97. **33.** White, Yielding & Krumdieck (1976) *Biochim. Biophys. Acta* **429**, 689. **34.** Tabor & Wyngarden (1959) *J. Biol. Chem.* **234**, 1830. **35.** Kato, Chihara, Nishioka, *et al.* (1987) *J. Biochem.* **101**, 207. **36.** Klinov, Chebotareva, Sheiman, Birinberg & Kurganov (1987) *Bioorg. Khim.* **13**, 908. **37.** Bhat & Vaidyanathan (1976) *Arch. Biochem. Biophys.* **176**, 314. **38.** Suzuki & Iwai (1973) *Plant Cell Physiol.* **14**, 319. **39.** Gambini, Crosti & Bianchetti (1980) *Biochim. Biophys. Acta* **613**, 73. **40.** Kaufman (1962) *Meth. Enzymol.* **5**, 809. **41.** Molnar & Garai (2005) *Int. Immunopharmacol.* **5**, 849. **42.** Baggott, Vaughn & Hudson (1986) *Biochem. J.* **236**, 193. **43.** Baggott, Morgan & Vaughn (1994) *Biochem. J.* **300**, 627. **44.** Allegra, Drake, Jolivet & Chabner (1985) *Proc. Natl. Acad. Sci. U.S.A.* **82**, 4881. **45.** Vergis, Bulock, Fleming & Beardsley (2001) *J. Biol. Chem.* **276**, 7727. **46.** Szabados, Hindmarsh, Phillips, Duggleby & Christopherson (1994) *Biochemistry* **33**, 14237. **47.** Rayl, Moroson & Beardsley (1996) *J. Biol. Chem.* **271**, 2225. **48.** Wolan, Greasley, Beardsley & Wilson (2002) *Biochemistry* **41**, 15505. **49.** Sugita, Aya, Ueno, Ishizuka & Kawashima (1997) *J. Biochem.* **122**, 309. **50.** Tsusue & Mazda (1977) *Experientia* **33**, 854. **51.** Ramesh & Appaji Rao (1980) *Biochem. J.* **187**, 623. **52.** Rao & Rao (1982) *Plant Physiol.* **69**, 11. **53.** Bachmann & Follmann (1987) *Arch. Biochem. Biophys.* **256**, 244. **54.** Robinson, Pilot & Meany (1990) *Physiol. Chem. Phys. Med. NMR* **22**, 95. **55.** Gangjee, Li, Yang & Kisliuk (2008) *J. Med. Chem.* **51**, 68. **56.** Gangjee, Qiu, Li & Kisliuk (2008) *J. Med. Chem.* **51**, 5789. **57.** Nakata, Tsukamoto, Miyoshi & Kojo (1987) *Biochim. Biophys. Acta* **924**, 297. **58.** McCuen & Sirotnak (1975) *Biochim. Biophys. Acta* **384**, 369. **59.** Bisson & Thorner (1981) *J. Biol. Chem.* **256**, 12456. **60.** Pattanakitsakul & Ruenwongsa (1984) *Int. J. Parasitol.* **14**, 513. **61.** So, Wong & Ko (1994) *Exp. Parasitol.* **79**, 526. **62.** Carpenter (1974) *J. Insect Physiol.* **20**, 1389. **63.** Dolnick & Cheng (1978) *J. Biol. Chem.* **253**, 3563. **64.** Aboge, Jia, Terkawi, *et al.* (2008) *Antimicrob. Agents Chemother.* **52**, 4072. **65.** Yeo & Wagner (1992) *J. Biol. Chem.* **267**, 24669.

Methotrexate Polyglutamate

These oligo(γ-glutamylated) derivatives of methotrexate (FW$_{di-glutamyl}$ = 729.69 g/mol; FW$_{tri-glutamyl}$ = 858.80 g/mol; FW$_{tetra-glutamyl}$ = 987.92 g/mol; FW$_{penta-glutamyl}$ = 1117.03 g/mol; FW$_{hexa-glutamyl}$ = 1246.15 g/mol; CAS 82334-40-5) are potent inhibitors of dihydrofolate reductase. (*See also 4-Amino-10-methylpteroyl-γ-glutamyl-γ-glutamyl-γ-glutamate; Methotrexate*) **Target(s):** amidophosphoribosyltransferase, inhibited by methotrexate pentaglutamate (1); dihydrofolate reductase (2,3); 5-formyl-tetrahydrofolate synthetase, *or* N^5,N^{10}-methenyltetrahydrofolate synthetase, inhibited by the pentaglutamate (4,5); phosphoribosylaminoimidazole carboxamide formyltransferase, *or* 5-aminoimidazole-4-carboxamide ribonucleotide transformylase, weakly inhibited by the monoglutamate but strongly inhibited by the tri- and pentaglutamate derivative (6,7); serine hydroxymethyltransferase, *or* glycine hydroxymethyl-transferase, inhibited by the triglutamate (8); thymidylate synthase, inhibited by the pentaglutamate (9). **1.** Sant, Lyons, Phillips & Christopherson (1992) *J. Biol. Chem.* **267**, 11038. **2.** Kumar, Kisliuk, Gaumont, Freisheim & Nair (1989) *Biochem. Pharmacol.* **38**, 541. **3.** Kumar, Kisliuk, Gaumont, *et al.* (1986) *Cancer Res.* **46**, 5020. **4.** Jolivet (1997) *Meth. Enzymol.* **281**, 162. **5.** Bertrand, MacKenzie & Jolivot (1987) *Biochim. Biophys. Acta* **911**, 154. **6.** Allegra, Drake, Jolivet & Chabner (1985) *Proc. Natl. Acad. Sci. U.S.A.* **82**, 4881. **7.** Baggott, Vaughn & Hudson (1986) *Biochem. J.* **236**, 193. **8.** Strong, Tendler, Seither, Goldman

& Schirch (1990) *J. Biol. Chem.* **265**, 12149. **9**. Dolnick & Cheng (1978) *J. Biol. Chem.* **253**, 3563.

Methoxetamine

This ketamine derivative and designer drug (FW = 247.33 g/mol; CAS 1239943-76-0), also named (*R*/*S*)-2-(3-methoxyphenyl)-2-(ethylamino)-cyclohexanone, is a dissociative anesthetic that exerts NMDA receptor antagonist, with appreciable affinity for serotonin transporters (1,2). Its misuse in humans has resulted in serious or even fatal outcomes. In evidence of its abuse potential, methoxetamine evokes conditioned place preference and is self-administered by rats (3). (*See Ketamine*) **1**. Ward, Rhyee, Plansky & Boyer (2011) *Clin. Toxicol.* (Philadelphia) **49**, 874. **2**. Methoxetamine Policy Paper (18 October 2012) *Advisory Council on the Misuse of Drugs*. **3**. Botanas, de la Peña, Dela Pena, *et al.* (2015) *Pharmacol. Biochem. Behav.* **133**, 31.

Methoxyamine

This highly toxic hydroxylamine derivative (FW = 47.06 g/mol; CAS 67-62-9), also called methoxylamine, is nucleophilic amine that is frequently used to derivatize aldehydes and ketones. The free base is a water-soluble liquid that boils at 49-50°C. Methoxyamine is frequently supplied commercially as the hydrochloride (FW = 85.54 g/mol). **Primary Mode of Action:** Methoxyamine reacts with aldehydes at abasic sites in DNA, making them refractory to β-elimination in deoxyribosephosphate lyase catalysis, thereby blocking single-nucleotide base-excision repair. O^6-methylguanine, 7-methylguanine, and 3-methyladenine DNA adducts are repaired by at least two mechanisms. The O^6-CH₃G DNA adduct, a cytotoxic and genotoxic lesion, is repaired by O^6-CH₃G DNA-methyltransferase (*or* MGMT), which confers resistance to anticancer methylating agents. Meanwhile, cell death caused by O^6-CH₃G adducts is promoted by mismatch repair system, such that deficiency in the latter is associated with pronounced resistance to methylating agents. N^7-CH₃G, another lesion formed by methylating agents, and N^3-CH₃A DNA adducts are removed by the base excision repair pathway. Efficient base excision repair repair minimizes the impact of these lesions in normal and tumor cells. Only when BER is disrupted, do these abundant *N*-methylated DNA adducts become highly cytotoxic. **Target(s):** carbamoyl-phosphate synthetase (1); DNA-directed DNA polymerase (2); DNA polymerase β (2); endodeoxyribonuclease (pyrimidine dimer), *or* apurinic/apyrimidinic endonuclease activity (3); hydroxymethylbilane synthase (4); serine hydroxymethyltransferase, *or* glycine hydroxymethyltransferase (5); threonine ammonia-lyase, *or* threonine dehydratase (6); thrombin (7); tyrosine decarboxylase (8); urocanate hydratase, *or* urocanase (9). **1**. Kaseman (1985) Ph.D. thesis, Dept. Biochemistry, Cornell University Medical College, New York, New York. **2**. Horton, Prasad, Hou & Wilson (2000) *J. Biol. Chem.* **275**, 2211. **3**. Liuzzi, Weinfeld & Paterson (1987) *Biochemistry* **26**, 3315. **4**. Davies & Neuberger (1973) *Biochem. J.* **133**, 471. **5**. Acharya, Prakash, Rao, Savithri & Rao (1991) *Indian J. Biochem. Biophys.* **28**, 381. **6**. Desai, Laub & Anita (1972) *Phytochemistry* **11**, 277. **7**. Longas & Finlay (1980) *Int. J. Biochem.* **11**, 565. **8**. Kezmarsky, Xu, Graham & White (2005) *Biochim. Biophys. Acta* **1722**, 175. **9**. Gerlinger & Retey (1987) *Z. Naturforsch. C* **42**, 349.

p-Methoxyamphetamine

This amphetamine and CNS stimulant (FW$_{free-base}$ = 165.24 g/mol; CAS 64-13-1; *Abbreviation*: PMA) is a potent hallucinogen that largely replaced LSD in the 1970s. When used recreationally, overdoses often prove fatal.

PMA acts as a potent selective serotonin releasing agent (SSRA) with only very weak effects on dopamine and norepinephrine. *p*-Methoxyamphetamine is also an alternative substrate of CYP2D6. **Target(s):** 5-hydroxytryptamine (serotonin) re-uptake (1); monoamine oxidase (2). **1**. Hegadoren, Greenshaw, Baker, *et al.* (1994) *J. Psychiatry Neurosci.* **19**, 57. **2**. Green & El Hait (1980) *J. Pharm. Pharmacol.* **32**, 262

2-Methoxyestradiol

This endogenous estrogen metabolite (FW = 302.41 g/mol; CAS 362-07-2) inhibits angiogenesis and mitosis, the latter accounting for its antiproliferative and antitumor effects. **Primary Mode of Action:** 2-Methoxyestradiol binds to tubulin and *in vitro* assembled tubulin sheet structures with higher affinity than to microtubules (MTs). This metabolite is readily displaced by colchicine and colchicine-like ligands. Recent studies suggest that this mammalian estradiol metabolite inhibits MT assembly and appears to be incorporated into a polymer with altered stability properties. **Target(s):** catechol *O*-methyltransferase, product inhibition (1); estrogen 2-monooxygenase (2,3); HIF-1α (hypoxia-inducible factor 1α), a key angiogenic transcription factor (4); tubulin polymerization, microtubule self-assembly, IC$_{50}$ = 5.3 μM (5-7). **1**. Ball, Knuppen, Haupt & Breuer (1972) *Eur. J. Biochem.* **26**, 560. **2**. Brueggemeier & Singh (1989) *J. Steroid Biochem.* **33**, 589. **3**. Brueggemeier (1983) *J. Steroid Biochem.* **19**, 1683. **4**. Mooberry (2003) *Drug Resist. Updat.* **6**, 355. **5**. Wang, Yang, Mohanakrishnan, *et al.* (2000) *J. Med. Chem.* **43**, 2419. **6**. D'Amato, Lin, Flynn, Folkman & Hamel (1994) *Proc. Natl. Acad. Sci. U.S.A.* **91**, 3964. **7**. Edsall, Mohanakrishnan, Yang, *et al.* (2004) *J. Med. Chem.* **47**, 5126.

N^3-(4-Methoxyfumaroyl)-(*S*)-2,3-diaminopropanoate

This amino acid (FW = 216.19 g/mol) is a specific and potent inactivator of glucosamine-6 phosphate synthase, *or* glutamine:fructose-6-phosphate aminotransferase (isomerizing). **Target(s):** glucosamine-6-phosphate deaminase (1); glutamine:fructose-6-phosphate aminotransferase (isomerizing) (2-7). **1**. Chmara, Milewski, Andruszkiewicz, Mignini & Borowski (1998) *Microbiology* **144**, 1349. **2**. Zgodka, Jedrzejczak, Milewski & Borowski (2001) *Bioorg. Med. Chem.* **9**, 931. **3**. Wojciechowski, Mazerski & Borowski (1996) *J. Enzyme Inhib.* **10**, 17. **4**. Andruszkiewicz, Chmara, Milewski & Borowski (1986) *Int. J. Pept. Protein Res.* **27**, 449. **5**. Chmara, Andruszkiewicz & Borowski (1986) *Biochim. Biophys. Acta* **870**, 357. **6**. Wojciechowski, Milewski, Mazerski & Borowski (2005) *Acta Biochim. Pol.* **52**, 647. **7**. Badet, Vermoote & Le Goffic (1988) *Biochemistry* **27**, 2282.

2-Methoxy-5-nitrotropone

This reagent (FW = 181.15 g/mol) reacts with amino groups under relatively mild conditions (pH 7.0 to 8.5). (No other functional group appears to be modified under these same conditions.) The resulting covalent adduct (λ$_{max}$ = 420 nm, ε = 20700 M^{-1}cm^{-1} for the adduct) can be removed by incubation of the modified protein with 1-2 M hydrazine at pH 8.5-9.0. *Note*: Solutions of 2-methoxy-5-nitrotropone should always be freshly prepared. It slowly decomposes at neutral pH and decomposes rapidly above pH 8.5. **Target(s):** α-amylase (1,2); CYP2B6 (*or* cytochrome P450

LM2) (3); 3-phosphoglycerate kinase4; porphobilinogen deaminase (5); ribosomal subunits (6-10); taka-amylase A (1,2); thioglucosidase, *or* myrosinase, *or* sinigrinase (11,12); uroporphyrinogen decarboxylase (13). 1. Tamaoki, Murase, Minato & Nakanishi (1967) *J. Biochem.* 62, 7. 2. Takagi, Toda & Isemura (1971) *The Enzymes*, 3rd ed., 5, 235. 3. Bernhardt, Kraft, Otto & Ruckpaul (1988) *Biomed. Biochim. Acta* 47, 581. 4. Markland, Bacharach, Weber, *et al.* (1975) *J. Biol. Chem.* 250, 1301. 5. Sancovich, Ferramola, del C. Batlle, Kivilevich & Grinstein (1976) *Acta Physiol. Lat. Amer.* 6, 379. 6. Suzuka (1971) *Eur. J. Biochem.* 23, 61. 7. Ballesta, Montejo, Hernandez & Vazquez (1974) *Eur. J. Biochem.* 42, 167. 8. Reboud, Madjar, Buisson & Reboud (1975) *Biochimie* 57, 285. 9. Chang & Craven (1977) *J. Mol. Biol.* 117, 401. 10. Reyes, Vazquez & Ballesta (1976) *Eur. J. Biochem.* 67, 267 11. Ohtsuru & Kawatani (1979) *Agric. Biol. Chem.* 43, 2249. 12. Ohtsuru & Hata (1979) *Biochim. Biophys. Acta* 567, 384. 13. Koopmann & Batlle (1987) *Int. J. Biochem.* 19, 373.

3-(4-(2-Methoxyphenoxy)piperidin-1-yl)-6-(3-methyl-1,2,4-oxadiazol-5-yl)pyridazine

This potent, selective, orally bioavailable pyridazine (FW = 367.41 g/mol) derivative inhibits stearoyl-CoA 9-desaturase (mouse, IC_{50} = 85 nM), which catalyzes the committed step in the biosynthesis of monounsaturated fatty acids from saturated, long-chain fatty acids. The rationale for targeting this enzyme is based on studies with SCD1 knockout mice that established that such animals are lean and protected from leptin deficiency-induced and diet-induced obesity, with greater whole body insulin sensitivity than wild-type animals. 1. Liu, Lynch, Freeman, *et al.* (2007) *J. Med. Chem.* 50, 3086.

4-*N*-(2-Methoxyphenyl)amino-5-(4-ethoxyphenyl)-7-(β-D-*erythro*-furanosyl)pyrrolo[2,3-*d*]pyrimidine

This tubercidin analogue (FW = 462.51 g/mol) inhibits human adenosine kinase (IC_{50} = 5.3 nM). Because the use of adenosine receptor agonists is plagued by dose-limiting cardiovascular side effects, there is great interest in developing adenosine kinase inhibitors (AKIs) as a means for raising steady-state adenosine concentrations. 1. Boyer, Ugarkar, Solbach, *et al.* (2005) *J. Med. Chem.* 48, 6430.

N-(4-(4'-Methoxyphenyl)benzoyl)-*syn*-(2-(3-amidinobenzyl)-3-*trans*-styryl)-β-alanine Methyl Ester

This 2,3-disubstituted β-alanine ($FW_{free-base}$ = 547.65 g/mol) is a first-in-class inhibitor that targets coagulation factor Xa (IC_{50} = 0.08 μM). It is a much weaker inhibitor of the related proteases, thrombin (IC_{50} = 2.7 μM) and trypsin (IC_{50} = 2.9 μM). 1. Klein, Czekaj, Gardner, *et al.* (1998) *J. Med. Chem.* 41, 437.

2'-Methoxyphenyl 4-Guanidinobenzoate

This ester ($FW_{free-base}$ = 285.30 g/mol) is a strong inhibitor of human acrosin (IC_{50} = 27 nM) and trypsin (IC_{50} = 88 nM), suggesting its potential as a contraceptive alternative to nonoxynol-9, the most commonly used active ingredient in today's vaginal contraceptive preparations. 1. Kaminski, Bauer, Mack, *et al.* (1986) *J. Med. Chem.* 29, 514.

4-Methoxyphenyl-[4-(4-methoxyphenyl)-4,5,6,7-tetrahydrothieno[3,2-*c*]pyridin-5-yl]methanone

This tetrahydrothienopyridine (FW = 363.48 g/mol) inhibits pig glucose-6-phosphatase activity (K_i = 0.61 μM) and pyrophosphatase activity (IC_{50} = 0.43-0.55 μM) of the hepatic ER-associated glucose-6-P transporter required for terminal glucose formation in the glycogenolysis and gluconeogenesis pathways. In experiments with cultured rat hepatocytes using glycerol as the substrate, this inhibitor prevented glucose production, with a concomitant 2.3-fold increase in the glucose-6-P content. 1. Westergaard, Madsen, Lundbeck, *et al.* (2002) *Diabetes Obes. Metab.* 4, 96.

[α-Methoxy Poly(ethylene glycol)]-1-(*N*-Benzyloxycarbonyl-L-leucyl)-5-(L-phenylalanyl-L-leucyl)carbohydrazide

This lysosomotropic macromolecular inhibitor (FW = indefinite), consisting of a polyethylene glycol and an a high-affinity inhibitor, targets cathepsin K ($K_{i,app}$ = 6.6 nM). It also inhibits cathepsin B and papain ($K_{i,app}$ = 1.6 and 2.4 μM, respectively). The polymer-bound inhibitor was internalized by endocytosis and was ultimately trafficked to the lysosomal compartment. 1. Wang, Pechar, Li, *et al.* (2002) *Biochemistry* 41, 8849.

5-Methoxypsoralen

This phototoxic phytoalexin (FW = 216.19 g/mol; CAS 484-20-8; *Abbreviation*: 5-MOP), found in many plants, is part of a defense scheme against fungi and insects. Upon activation by ultraviolet light, 5-

methoxypsoralen, also called bergapten, induces interstrand cross-links in DNA (1) 5-MOP intercalates between DNA base pairs and, upon uv absorption, it forms an adduct with a pyrimidine. If at an appropriate site, a second photoaddition can occur, with formation of an interstrand cross-link. 5-MOP is frequently used to investigate DNA repair and recombination. The combination of psoralens and UVA radiation (PUVA photochemotherapy) is also an established treatment for many skin disorders. Note that unsaturated fatty acids and proteins undergo a photoaddition reaction with psoralen and psoralen derivatives. (*See also other psoralen derivatives*) **Target(s):** arylamine *N*-acetyltransferase, activated at low concentrations (2); cAMP dependent protein kinase (weakly inhibited)1; CYP1A1 (3); CYP1A2 (3); CYP2A6 (4); CYP3A (4-5); CYP6D1 (6); melatonin 6-monooxygenase (7) **1**. Wang, Ternai & Polya (1997) *Phytochemistry* **44**, 787. **2**. Lee & Wu (2005) *In Vivo* **19**, 1061. **3**. Tassaneeyakul, Birkett, Veronese, *et al.* (1993) *J. Pharmacol. Exp. Ther.* **265**, 401. **4**. Koenigs & Trager (1998) *Biochemistry* **37**, 10047. **5**. Saville & Wanwimolruk (2001) *J. Pharm. Pharm. Sci.* **4**, 217. **6**. Scott (1996) *Insect Biochem. Mol. Biol.* **26**, 645. **7**. Mauviard, Raynaud, Geoffriau, Claustrat & Pevet (1995) *Biol. Signals* **4**, 32.

8-Methoxypsoralen

This phototoxic phytoalexins and minor isoprenoid product (FW = 216.19 g/mol; CAS 298-81-7; *Abbreviation*: 8-MOP), found in many plants, including false Queen Anne's Lace (*Ammi majus* in the family Apiaceae), is part of a plant defense scheme to protect against fungi and insects. Its alternative name is methoxsalen, and its IUPAC name is 9-methoxy-7*H*-furo[3,2-*g*]chromen-7-one. **Principal Mode of Inhibitory Action:** 8-Methoxypsoralen intercalates between DNA base pairs and, upon activation with UV light (A-Band: λ = 315 to 400 nm; energy = 3.94 to 3.10 eV), forms a photoadduct with a pyrimidine base. If located at a appropriate site, a second photoaddition occurs, with interstrand cross-linking. 8-MOP is frequently used to investigate DNA repair and recombination events. The combination of psoralens and UV-A radiation (PUVA) is also an established photochemotherapy for many skin disorders. **Mode of Cytochrome P450 Inhibition:** CYP2A6 is inhibited by 8-methoxypsoralen, bergapten (5-methoxypsoralen) or psoralen as a result of mechanism-based inhibition. The enzyme system initially oxidizes the psoralen to form a furanoepoxide followed by hydrolytic attack, or attack of exogenous nucleophiles, to form dihydrofuranocoumarin products. (*See also other psoralen derivatives*) **Other Actions:** Unsaturated fatty acids and proteins can also undergo a photoaddition reaction with psoralen and certain psoralen derivatives. These photobiological processes may impact cellular functions, protein immunogenicity, as well the accumulation of potentially toxic substances. **Target(s):** chemotactic activity of polymorphonuclear neutrophils (2); *trans*-cinnamate 4-monooxygenase (3,4); CYP1A1 (5); CYP1A2 (5); CYP2A6 *or* coumarin 7-hydroxylase (6-12); DNA-directed RNA polymerase (13); DNA polymerase (1,14); DNA replication (15); glutamate dehydrogenase (1); lysozyme (1); ribonuclease (1). **1**. Schmitt, Chimenti & Gasparro (1995) *J. Photochem. Photobiol. B: Biol.* **27**, 101. **2**. Esaki & Mizuno (1992) *Photochem. Photobiol.* **55**, 783. **3**. Hübner, Hehmann, Schreiner, *et al.* (2003) *Phytochemistry* **64**, 445. **4**. Gravot, Larbat, Hehn, *et al.* (2004) *Arch. Biochem. Biophys.* **422**, 71. **5**. Tassaneeyakul, Birkett, Veronese, *et al.* (1993) *J. Pharmacol. Exp. Ther.* **265**, 401. **6**. Hickman, Wang, Wang & Unadkat (1998) *Drug Metab. Dispos.* **26**, 207. **7**. Sivapathasundaram, Magnisali, Coldham, *et al.* (2003) *Toxicology* **187**, 49. **8**. Zhang, Kilicarslan, Tyndale & Sellers (2001) *Drug Metab. Dispos.* **29**, 897. **9**. Kharasch, Hankins, Fenstamaker & Cox (2000) *Eur. J. Clin. Pharmacol.* **55**, 853. **10**. Koenigs & Trager (1998) *Biochemistry* **37**, 10047. **11**. Koenigs, Peter, Thompson, Rettie & Trager (1997) *Drug Metab. Dispos.* **25**, 1407. **12**. Draper, Madan & Parkinson (1997) *Arch. Biochem. Biophys.* **341**, 47. **13**. Gniazdowski, Czyz, Wilmanska, *et al.* (1988) *Biochim. Biophys. Acta* **950**, 346. **14**. Granger, Toulme & Helene (1982) *Photochem. Photobiol.* **36**, 175. **15**. Luftl, Rocken, Plewig & Degitz (1998) *J. Invest. Dermatol.* **111**, 399.

N-(Methoxysuccinyl)-L-alanyl-L-alanyl-L-prolyl-L-valine Chloromethyl Ketone

This halomethyl ketone (FW = 503.00 g/mol) is an analogue of an acylated tetrapeptide and inhibits leukocyte elastase, chicken oviduct signal peptidase, and myeloblastin. **Target(s):** cathepsin G (1); leukocyte elastase (1-9); myeloblastin (9,10); pancreatic elastase (1); signal peptidase, chicken oviduct (11). **1**. Powers, Gupton, Harley, Nishino & Whitley (1977) *Biochim. Biophys. Acta* **485**, 156. **2**. Bieth (1998) in *Handb. Proteolytic Enzymes*, p. 54, Academic Press, San Diego. **3**. Barrett (1981) *Meth. Enzymol.* **80**, 581. **4**. Fletcher, Osinga, Hand, *et al.* (1990) *Amer. Rev. Respir. Dis.* **141**, 672. **5**. Stein & Trainor (1986) *Biochemistry* **25**, 5414. **6**. Wei, Mayr & Bode (1988) *FEBS Lett.* **234**, 367. **7**. Navia, McKeever, Springer, *et al.* (1989) *Proc. Natl. Acad. Sci. U.S.A.* **86**, 7. **8**. Jones, Elphick, Pettitt, Everard & Evans (2002) *Eur. Respir. J.* **19**, 1136. **9**. Kam, Kerrigan, Dolman, *et al.* (1992) *FEBS Lett.* **297**, 119. **10**. Hoidal, Rao & Gray (1994) *Meth. Enzymol.* **244**, 61. **11**. Walker & Lively (1998) in *Handb. Proteolytic Enzymes*, p. 455, Academic Press, San Diego.

9-Methoxythieno[2,3-*c*]isoquinolin-5(4*H*)-one

This substituted isoquinoline (FW = 231.28 g/mol), also known as 9-CH$_3$O-TIQ-A, inhibits poly(ADP-ribose) polymerase-1 (PARP-1), IC$_{50}$ = 0.21 μM. **Primary Mode of Action:** The rationale for this approach is based on the observation that excessive activation of poly(ADP-ribose) polymerase-1 (PARP-1), a nuclear enzyme catalyzing the transfer of ADP-ribose units from NAD$^+$ to acceptor proteins, induces cellular energy failure by depleting NAD$^+$ and ATP and thereby serving a causative role in ischemia/reperfusion injury. **1**. Chiarugi, Meli, Calvani, *et al.* (2003) *J. Pharmacol. Exp. Ther.* **305**, 943.

L-Methoxyvinylglycine

This amino acid (FW = 131.13 g/mol; *Abbreviation*: MVG), also known as L-2-amino-4-methoxy-3-butenoic acid and, inhibits the biosynthesis of ethylene in plants. It is produced by *Pseudomonas aeruginosa*. **Target(s):** 1-aminocyclopropane-1-carboxylate synthase (1,2); aspartate amino-transferase (3); methionine *S*-adenosyltransferase, inhibited by the *cis* isomer (4,5); tryptophan synthase (3,6). **1**. Yu, Adams & Yang (1979) *Arch. Biochem. Biophys.* **198**, 280. **2**. Jakubowicz (2002) *Acta Biochim. Pol.* **49**, 757. **3**. Rando (1977) *Meth. Enzymol.* **46**, 28. **4**. Yarlett, Garofalo, Goldberg, *et al.* (1993) *Biochim. Biophys. Acta* **1181**, 68. **5**. Sufrin, Lombardini & Alks (1993) *Biochim. Biophys. Acta* **1202**, 87. **6**. Miles, Bauerle & Ahmed (1987) *Meth. Enzymol.* **142**, 398.

Methyl Acetimidate

This imido ester (FW$_{hydrochloride}$ = 109.56 g/mol; CAS 14777-29-8) acetimidates amino groups in proteins, while allowing the protein to retain positive charges at the modified residue. **Primary Mode of Action:** All of the amino groups in pancreatic ribonuclease are modified and, although the protein is catalytically inactive, RNase A retains its physical and chemical properties (1). All fourteen ε amino groups of trypsin are modified with this reagent and the modified trypsin retains catalytic activity (2). Methyl acetimidate, which decomposes at 105°C, is an irritant and is moisture sensitive. It should be stored in a desiccator. It is slowly hydrolyzed in aqueous solutions; however, between pH values of 6.8 and 8.8, the rate of hydrolysis of methyl acetimidate decreases and the rate of amidination increases (3). **Target(s):** aspartate aminotransferase (4); chymotrypsin (5); *Eco*RI restriction endonuclease (type II site-specific deoxyribonuclease) (6,7); glyceraldehyde-3-phosphate dehydrogenase (8); inositol 1,3,4-trisphosphate 5/6-kinase (9); K^+ channel, ATP-sensitive (KATP) (10); penicillinase (11); ribonuclease A, *or* pancreatic ribonuclease (1,12); RNA polymerase (13). **1.** Reynolds (1968) *Biochemistry* **7**, 3131. **2.** Nureddin & Inagami (1975) *Biochem. J.* **147**, 71. **3.** Makoff & Malcolm (1981) *Biochem. J.* **193**, 245. **4.** Lain-Guelbenzu, Muñoz-Blanco & Cárdenas (1990) *Eur. J. Biochem.* **188**, 529. **5.** Fojo, Whitney & Awad (1983) *Arch. Biochem. Biophys.* **224**, 636. **6.** Woodhead & Malcolm (1980) *Nucl. Acids Res.* **8**, 389. **7.** Wells, Klein & Singleton (1981) *The Enzymes*, 3rd ed., **14**, 157. **8.** Lambert & Perham (1977) *Biochem. J.* **161**, 49. **9.** Hughes, Kirk & Michell (1993) *Biochem. Soc. Trans.* **21**, 365S. **10.** Lee, Ozanne, Hales & Ashford (1994) *J. Membr. Biol.* **139**, 167. **11.** Davies & Virden (1978) *Biochem. Soc. Trans.* **6**, 1222. **12.** Richards & Wyckoff (1971) *The Enzymes*, 3rd ed., **4**, 647. **13.** Makoff & Malcolm (1980) *Eur. J. Biochem.* **106**, 313.

1-Methyladenine

This modified purine (FW$_{free-base}$ = 149.16 g/mol; CAS 1670-69-5) has been identified as a component in certain RNA. For example, it is produced by the action of tRNA (adenine-N^1-) methyltransferase. It is also a product of the reaction catalyzed by 1-methyladenosine nucleosidase. In addition, 1-methyladenine is the oocyte maturation-inducing substance of starfish. 1-Methyladenine has a λ_{max} at 270 nm at pH 9 and 13 (ε = 11900 M^{-1}cm^{-1} at pH 9 which rises to 14400 M^{-1}cm^{-1} at pH 13). **Target(s):** deoxycytidine kinase, K_i = 9.4 mM (1); hypoxanthine(guanine) phosphoribosyltransferase (2); xanthine phosphoribosyltransferase (2). **1.** Krenitsky, Tuttle, Koszalka, *et al.* (1976) *J. Biol. Chem.* **251**, 4055. **2.** Naguib, Iltzsch, el Kouni, Panzica & el Kouni (1995) *Biochem. Pharmacol.* **50**, 1685.

2-Methyladenine

This modified purine (FW = 149.16 g/mol; CAS 1445-08-5; λ_{max} at pH 7 of 263 nm (ε = 12700 M^{-1}cm^{-1})) can be formed as a component of RNA via reaction with methyl radicals (1). It also protects against methotrexate inhibition (2). **Target(s):** adenine deaminase (3,4); hypoxanthine(guanine) phosphoribosyltransferase (5); inosine phosphorylase, *or* purine-nucleoside phosphorylase (3,4); xanthine phosphoribosyltransferase (5). **1.** Kang, Gallagher & Cohen (1993) *Arch. Biochem. Biophys.* **306**, 178. **2.** Loebeck (1960) *J. Bacteriol.* **79**, 384. **3.** Webb (1966) *Enzyme and Metabolic Inhibitors*, vol. **2**, Academic Press, New York. **4.** Remy (1961) *J. Biol. Chem.* **236**, 2999. **5.** Naguib, Iltzsch, el Kouni, Panzica & el Kouni (1995) *Biochem. Pharmacol.* **50**, 1685.

3-Methyladenine

This naturally occurring purine metabolite (FW = 149.16 g/mol; CAS 5142-23-4; *Symbol*: 3-MA) is released from alkylated DNA by the action of DNA-3 methyladenine glycosylases I and II. (Note bonding change in the ring relative to unmodified adenine.) 3-MA is often used as a standard in detecting methylation of nucleic acid. At 5 mM, 3-methyladenine inhibits endogenous protein degradation in isolated rat hepatocytes by ~60%, with no adverse effects on exogenous protein degradation, protein synthesis, or intracellular ATP (1). Its ability to suppress the formation of electron microscopically visible autophagosomes first suggested that 3-MA may inhibit autophagy (1). 3-MA and wortmannin are widely used as autophagy inhibitors, based on their inhibitory effect on class III phosphoinositide 3-kinase (PI3K), which is known to be essential factor in inducing autophagy (2). (*See Wortmannin*) **Target(s):** autophagosome formation (1); class III phosphoinositide 3-kinase (2); deoxycytidine kinase, K_i = 0.42 mM (3); DNA-3 methyladenine glycosylase I (4-9). **1.** Seglen & Gordon (1982) *Proc. Natl. Acad. Sci. U.S.A.* **79**, 1889. **2.** Blommaart, Krause, Schellens, Vreeling-Sindelárová & Meijer (1997) *Eur. J. Biochem.* **243**, 240. **3.** Krenitsky, Tuttle, Koszalka, *et al.* (1976) *J. Biol. Chem.* **251**, 4055. **4.** Duncan (1981) *The Enzymes*, 3rd ed., **14**, 565. **5.** Tudek, Van Zeeland, Kusmierek & Laval (1998) *Mutat. Res.* **407**, 169. **6.** Bjelland & Seeberg (1987) *Nucl. Acids Res.* **15**, 2787. **7.** Riazuddin, Athar, Ahmed, Lali & Sohail (1987) *Nucl. Acids Res.* **15**, 6607. **8.** Thomas, Yang & Goldthwait (1982) *Biochemistry* **21**, 1162. **9.** Riazuddin & Lindahl (1978) *Biochemistry* **17**, 2110.

N^6-Methyladenine

This modified purine (FW = 149.16 g/mol; CAS 443-72-1; λ_{max} = 266 nm, ε = 16200 M^{-1}cm^{-1}) at pH 7), also known as 6-(methylamino)purine, is a modified base that is often found in nucleic acids (*e.g.*, it is produced by the actions of rRNA (adenine N^6-)-methyltransferase, tRNA (adenine-N^6-)-methyltransferase, and site-specific DNA methyltransferase (adenine-specific)). N^6-methylated adenine also functions as a plant growth regulator. **Target(s):** adenine deaminase (1,2); adenosine nucleosidase (1); β-glucosidase, *Dictyostelium discoideum* (3); guanine deaminase (1); hypoxanthine(guanine) phosphoribosyltransferase (4); xanthine phosphoribosyl-transferase (4). **1.** Nolan & Kidder (1980) *Antimicrob. Agents Chemother.* **17**, 567. **2.** Jun & Sakai (1979) *J. Ferment. Technol.* **57**, 294. **3.** Parish (1977) *J. Bacteriol.* **129**, 1642. **4.** Naguib, Iltzsch, el Kouni, Panzica & el Kouni (1995) *Biochem. Pharmacol.* **50**, 1685.

7-Methyladenine

This modified purine (FW = 149.16 g/mol; CAS 935-69-3) is a minor base occasionally found in nucleic acids. (Note bonding change in the ring relative to unmodified adenine.) It is also produced nonenzymatically upon reaction of DNA with dimethyl sulfate. The methyl group is removed spontaneously from DNA *in vitro* by hydrolysis at neutral pH. Pulse-chase studies demonstrated that the half-life of 7-methyladenine *in vivo* is short (2-3 hours). At pH 1, methyladenine has a λ_{max} of 273 nm (ε = 14000 M^{-1}cm^{-1}); at pH 12, λ_{max} = 270 nm (ε = 10600 M^{-1}cm^{-1}). Adenosine levels increase at seizure foci as part of a negative feedback mechanism that controls seizure activity through adenosine receptor signaling. Agents that inhibit adenosine kinases increase/sustain site-/event-specific adenosine surges, thereby providing anti-seizure activity akin to that of adenosine receptor agonists, but with fewer dose-limiting cardiac side effects. **Target(s):** adenosine kinase (1); deoxycytidine kinase, weakly inhibited, K_i = 4.3 mM (2). **1.** Miller, Adamczyk, Miller, *et al.* (1979) *J. Biol. Chem.* **254**, 2346. **2.** Krenitsky, Tuttle, Koszalka, , *et al.* (1976) *J. Biol. Chem.* **251**, 4055.

1-Methyladenosine

This methylated purine nucleoside (FW$_{\text{free-base}}$ = 281.27 g/mol; CAS 15763-06-1) has been identified as a component in tRNA, produced by the action of tRNA (adenine-N^1-) methyltransferase. It is also a substrate of the reaction catalyzed by 1-methyladenosine nucleosidase. The λ_{max} value is 258 nm (ε = 13900 M^{-1}cm^{-1}) at pH 7 and 259 nm (ε = 14600 M^{-1}cm^{-1}) at pH 10.5. Because the use of adenosine receptor agonists is plagued by dose-limiting cardiovascular side effects, there is great interest in developing adenosine kinase inhibitors (AKIs) as a means for raising steady-state adenosine concentrations. **Target(s):** adenosine deaminase (1,2); adenosine kinase (3,4); adenosylhomocysteinase (5); cytidine deaminase (6); GMP synthetase (7). 1. Agarwal & Parks (1978) *Meth. Enzymol.* **51**, 502. 2. Challa, Johnson, Robertson & Gunasekaran (1999) *J. Basic Microbiol.* **39**, 97. 3. Long & Parker (2006) *Biochem. Pharmacol.* **71**, 1671. 4. Miller, Adamczyk, Miller, *et al.* (1979) *J. Biol. Chem.* **254**, 2346. 5. Shimizu, Shiozaki, Ohshiro & Yamada (1984) *Eur. J. Biochem.* **141**, 385. 6. Vita, Amici, Cacciamani, Lanciotti & Magni (1985) *Biochemistry* **24**, 6020. 7. Spector & Beecham (1975) *J. Biol. Chem.* **250**, 3101.

N^6-Methyladenosine

This analogue of adenosine (FW = 281.27 g/mol; at pH 7 and 13, λ_{max} = 266 nm, ε = 15900 M^{-1}cm^{-1}; at pH 7, λ_{max} = 262 nm, ε = 16600 M^{-1}cm^{-1}), also known as 6-methylaminopurine 9 ribofuranoside, inhibits calf muscle adenosine deaminase (K_i = 5.3 μM) and is also a weak alternative substrate. It has been observed as a component of certain nucleic acids (*e.g.*, it is produced via the actions of rRNA (adenine-N^6-) methyltransferase, tRNA (adenine-N^6-)-methyltransferase, and site-specific DNA methyltransferase (adenine-specific)). **Target(s):** adenosine deaminase (1-5); adenosine kinase, also as an alternative substrate (6); adenosylhomocysteinase (7,8); AMP nucleosidase (9); deoxycytidine kinase, weakly inhibited (10); GMP synthetase (11); and hypoxanthine transport (12). 1. Agarwal & Parks, (1978) *Meth. Enzymol.* **51**, 502. 2. Cory & Suhadolnik (1965) *Biochemistry* **4**, 1729 and 1733. 3. Baker (1967) *Design of Active-Site-Directed Irreversible Enzyme Inhibitors*, Wiley, New York. 4. Lindley & Pisoni (1993) *Biochem. J.* **290**, 457. 5. Daddona, Wiesmann, Lambros, Kelley & Webster (1984) *J. Biol. Chem.* **259**, 1472. 6. Kidder (1982) *Biochem. Biophys. Res. Commun.* **107**, 381. 7. Guranowski, Montgomery, Cantoni & Chiang (1981) *Biochemistry* **20**, 110. 8. Shimizu, Shiozaki, Ohshiro & Yamada (1984) *Eur. J. Biochem.* **141**, 385. 9. De Wolf, Markham & Schramm (1980) *J. Biol. Chem.* **255**, 8210. 10. Krenitsky, Tuttle, Koszalka, *et al.* (1976) *J. Biol. Chem.* **251**, 4055. 11. Spector & Beecham (1975) *J. Biol. Chem.* **250**, 3101. 12. Holland, Schein & Murphy (1983) *Res. Commun. Chem. Pathol. Pharmacol.* **41**, 111.

2'-C-Methyladenosine 5'-Diphosphate

This synthetic nucleotide (FW = 440.20 g/mol) is a mechanism-based inhibitor of ribonucleoside-diphosphate reductase (RNR). **Primary Mode of**

Action: Reaction of 2'-C-methylADP with the reduced enzyme yielded a 318-nm chromophore (2), similar that obtained upon modification of RNR by furanone (3). Incubation of radiolabed 2'-C-methylADP with reduced RNR resulted in the covalent incorporation of radiolabel into the protein and aquocobalamin, indicating it is a mechanism-based inhibitor of *C. nephridii* RNR, undergoing enzyme-catalyzed conversion to a 2'-deoxy-2'-C-methyl-3'-ketonucleotide that then collapses to a reactive furanone. 2'-C-methylADP is neither a substrate nor an inhibitor for *Escherichia coli* RNR (2). 1. Ong, McFarlan & Hogenkamp (1993) *Biochemistry* **32**, 11397. 2. McFarlan, Ong & Hogenkamp (1996) *Biochemistry* **35**, 4485. 3. Harris, Ator & Stubbe (1984) *Biochemistry* **23**, 5214.

N^6-Methyladenosine 5'-Triphosphate

This ATP analogue (FW$_{\text{free-acid}}$ = 521.21 g/mol), also known as 6-methylaminopurine riboside 5'-triphosphate, inhibits *Escherichia coli* leucyl-tRNA synthetase (1) and yeast tyrosyl-tRNA synthetase (2). 1. Marutzky, Flossdorf & Kula (1976) *Nucl. Acids Res.* **3**, 2067. 2. Freist, von der Haar, Faulhammer & Cramer (1976) *Eur. J. Biochem.* **66**, 493.

Methylamine

This simple alkylamine CH$_3$NH$_2$ (FW$_{\text{free-base}}$ = 31.06 g/mol; CAS 74-89-5; FW$_{\text{hydrochloride}}$ = 67.52 g/mol; pK_a = 10.62, dpK_a/dT = –0.032) is actuallty a stronger base than ammonia. The free base is a flammable gas at room temperature and very soluble in water. The hydrochloride is a deliquescent solid and should be stored in tightly closed containers in a desiccator. Methylamine is a substrate for the reactions catalyzed by methylamine dehydrogenase (amine dehydrogenase), methylamine: glutamate N-methyltransferase (N-methylglutamate synthase), and glutamate: methylamine ligase (γ-glutamylmethylamide synthetase). It is also an alternative substrate for many additional enzymes. It is a product of the reactions catalyzed by N methylalanine dehydrogenase, glycine oxidase (when sarcosine is the substrate), dimethylamine dehydrogenase, sarcosine reductase, N-methyl-2-oxoglutaramate hydrolase, alkylamidase, methylguanidinase, and synephrine dehydratase. **Target(s):** acetylcholinesterase, weakly inhibited by the hydrochloride (1-3); N-acetylneuraminate 7-O(or 9-O)-acetyltransferase, weakly inhibited (4); ω-amidase (5,6); anthranilate synthase (7); clostripain (8); diaminopropionate ammonia-lyase (9); dimethylargininase (10); dipeptidyl peptidase I, weakly inhibited (11); electron transport, chloroplast (12); β-glucosidase, weakly inhibited (13); glycyl-tRNA synthetase (14); hydroxymethylbilane synthase (15); lysyl endopeptidase, *or Achromobacter* proteinase I (16,17); methane monooxygenase, weakly inhibited (18); methanethiol oxidase (19); peptidyl-Lys metalloendo-peptidase (20); photophosphorylation (uncoupler) (12,21); transglutaminase, *or* protein-glutamine γ-glutamyl-transferase (22,23); trypsin, weakly inhibited, K_i = 260 mM (8,24-26); UDP-N-acetylmuramoyl-L-alanyl-D-glutamate:2,6-diaminopimelate ligase (27); UDP-N-acetylmuramoyl-tripeptide:D-alanyl-D-alanine ligase (27). 1. Krupka (1965) *Biochemistry* **4**, 429. 2. Wilson (1952) *J. Biol. Chem.* **197**, 215. 3. Wilson (1954) *J. Biol. Chem.* **208**, 123. 4. Vandamme-Feldhaus & Schauer (1998) *J. Biochem.* **124**, 111. 5. Hersh (1972) *Biochemistry* **11**, 2251. 6. Hersh (1971) *Biochemistry* **10**, 2884. 7. Zalkin & Kling (1968) *Biochemistry* **7**, 3566. 8. Cole, Murakami & Inagami (1971) *Biochemistry* **10**, 4246. 9. Nagasawa, Tanizawa, Satoda & Yamada (1988) *J. Biol. Chem.* **263**, 958. 10. Knipp, Charnock, Garner & Vasak (2001) *J. Biol. Chem.* **276**, 40449. 11. Metrione & MacGeorge (1975) *Biochemistry* **14**, 5249. 12. Izawa & Good (1972) *Meth. Enzymol.* **24**, 355. 13. Dale, Ensley, Kern, Sastry & Byers (1985) *Biochemistry* **24**, 3530. 14. Boyko & Fraser (1964) *Can. J. Biochem. Physiol.* **42**, 1677. 15. Davies & Neuberger (1973) *Biochem. J.* **133**, 471. 16. Sakiyama & Masaki (1994) *Meth. Enzymol.* **244**, 126. 17. Masaki, Fujihashi, Nakamura & Soejima (1981) *Biochim. Biophys. Acta* **660**, 51. 18. Jahng & Wood (1996) *Appl. Microbiol. Biotechnol.* **45**, 744. 19. Suylen, Large, van Dijken & Kuenen (1987) *J. Gen. Microbiol.* **133**, 2989. 20. Lewis, Basford & Walton (1978) *Biochim. Biophys. Acta* **522**, 551. 21. Cost & Frenkel (1967) *Biochemistry* **6**, 663.

22. Sener, Gomis, Lebrun, *et al.* (1984) *Mol. Cell. Endocrinol.* **36**, 175. **23.** Korner, Schneider, Purdon & Bjornsson (1989) *Biochem. J.* **262**, 633. **24.** Baker (1967) *Design of Active-Site-Directed Irreversible Enzyme Inhibitors*, Wiley, New York. **25.** Inagami (1964) *J. Biol. Chem.* **239**, 787. **26.** Inagami & York (1968) *Biochemistry* **7**, 4045. **27.** Dementin, Bouhss, Auger, *et al.* (2001) *Eur. J. Biochem.* **268**, 5800.

Methyl 3-Amino-3-deoxy-β-D-galactopyranosyl-(1→4)- 2-acetamido-2-deoxy-β-D-glucopyranoside

This glycoside, also known as methyl 3'-amino-3'-deoxy-*N*-acetyllactosaminide, has an amino group at the position that undergoes galactosylation by UDP-D-galactose:β-D-galactopyranosyl (1→4)-2-acetamido-2-deoxy-D-glucose(1→3)-α-D-galactopyranosyltransferase, strongly inhibiting that enzyme (K_i = 104 μM). **Target(s):** UDP-D-galactose:β-D-galactopyranosyl-(1→4)-2-acetamido-2-deoxy-D-glucose (1→3)-α-D-galactopyranosyltransferase. **1.** Helland, Hindsgaul, Palcic, Stults & Macher (1995) *Carbohydr. Res.* **276**, 91.

N-[2-(Methylamino)ethyl]-5-isoquinoline sulfonamide

This sulfonamide (FW = 265.34 g/mol), also known as H-8, inhibits a number of protein kinases: IC_{50} values for protein kinase G, protein kinase A, protein kinase C, and myosin light chain kinase are 0.48, 1.2, 15, and 68 μM, respectively. **Target(s):** cAMP-dependent protein kinase, *or* protein kinase A (1,2); casein kinase I, weakly inhibited (1); cGMP-dependent protein kinase, *or* protein kinase G (1,2); cyclin-dependent protein kinase 9 (3); G-protein coupled receptor kinase 2, mildly inhibited (4); myosin-light-chain kinase (1); protein kinase C (1); rhodopsin kinase, K_i = 58 μM (5); [RNA-polymerase]-subunit kinase, *or* CTD kinase (6). **1.** Hidaka, Inagaki, Kawamoto & Sasaki (1984) *Biochemistry* **23**, 5036. **2.** Hidaka, Watanabe & Kobayashi (1991) *Meth. Enzymol.* **201**, 328. **3.** Shima, Yugami, Tatsuno, *et al.* (2003) *Genes Cells* **8**, 215. **4.** Kassack, Hogger, Gschwend, *et al.* (2000) *AAPS PharmSci.* **2**, E2. **5.** Palczewski, Kahn & Hargrave (1990) *Biochemistry* **29**, 6276. **6.** Medlin, Uguen, Taylor, Bentley & Murphy (2003) *EMBO J.* **22**, 925.

O-Methyl Arachidonyl Fluorophosphonate

This arachidonate derivative (FW = 370.49 g/mol), often abbreviated MAFP and called *O*-methyl arachidonylphosphonofluoridate, is a selective irreversible inhibitor of the Group IV cytosolic phospholipase A_2 and the Ca^{2+}-independent cytosolic phospholipase A_2. *O*-methyl arachidonylphosphonofluoridate is also an cannabinoid receptor antagonist. **Target(s):** acylglycerol lipase, *or* monoacylglycerol lipase (1-6); 1-alkyl-2-acetylglycero-phosphocholine esterase, *or* platelet-activating-factor acetylhydrolase (7,8); cannabinoid receptor (9); fatty-acid amide hydrolase, *or* anandamide amidase (10-12); Lysophospholipase (13-15); phospholipase A_1, phosphatidate-specific (16,17); phospholipase A_2 (18-22). **1.** Ito, Tchoua, Okamoto & Tojo (2002) *J. Biol. Chem.* **277**, 43674. **2.** Dinh, Freund & Piomelli (2002) *Chem. Phys. Lipids* **121**, 149. **3.** Dinh, Carpenter, Leslie, *et al.* (2002) *Proc. Natl. Acad. Sci. U.S.A.* **99**, 10819. **4.** Saario, Savinainen, Laitinen, Järvinen & Niemi (2004) *Biochem.*

Pharmacol. **67**, 1381. **5.** Walter, Dinh & Stella (2004) *J. Neurosci.* **24**, 8068. **6.** Quistad, Klintenberg, Caboni, Liang & Casida (2006) *Toxicol. Appl. Pharmacol.* **211**, 78. **7.** Kell, Creer, Crown, Wirsig & McHowat (2003) *J. Pharmacol. Exp. Ther.* **307**, 1163. **8.** Vinson, Rickard, Ryerse & McHowat (2005) *J. Pharmacol. Exp. Ther.* **314**, 1241. **9.** Fernando & Pertwee (1997) *Brit. J. Pharmacol.* **121**, 1716. **10.** Deutsch, Glaser, Howell, *et al.* (2001) *J. Biol. Chem.* **276**, 6967. **11.** Deutsch, Omeir, Arreaza, *et al.* (1997) *Biochem. Pharmacol.* **53**, 255. **12.** De Petrocellis, Melck, Ueda, *et al.* (1997) *Biochem. Biophys. Res. Commun.* **231**, 82. **13.** Wang, Yang, Friedman, Johnson & Dennis (1999) *Biochim. Biophys. Acta* **1437**, 157. **14.** Tamura, Ajayi, Allmond, *et al.* (2004) *Biochem. Biophys. Res. Commun.* **316**, 323. **15.** Shanado, Kometani, Uchiyama, Koizumi & Teno (2004) *Biochem. Biophys. Res. Commun.* **325**, 1487. **16.** Miyazawa, Ikemoto, Fujii & Okuyama (2003) *Biochim. Biophys. Acta* **1631**, 17. **17.** Higgs & Glomset (1996) *J. Biol. Chem.* **271**, 10874. **18.** Lio, Reynolds, Balsinde & Dennis (1996) *Biochim. Biophys. Acta* **1302**, 55. **19.** Farooqui, Ong, Horrocks & Farooqui (2000) *Neuroscientist* **6**, 169. **20.** Sundell, Aziz, Zuzel & Theakston (2003) *Toxicon* **41**, 459. **21.** Winstead, Balsinde & Dennis (2000) *Biochim. Biophys. Acta* **1488**, 28. **22.** Stewart, Ghosh, Spencer & Leslie (2002) *J. Biol. Chem.* **277**, 29526.

*N*ω-Methyl-L-Arginine

This modified amino acid (FW = 196.26 g/mol), often referred to as *N*γ-monomethyl-L-arginine (*or* L-NMMA) inhibits nitric oxide synthase, with the IC_{50} values of 0.7, 0.65, and 3.9 μM for the endothelial, neuronal, and inducible enzymes, respectively. The D-enantiomer is not inhibitory. Methylated arginyl residues have also been identified in a number of proteins (*e.g.*, histones, cytochrome *c*, and myelin basic protein). *N*ω-Methyl-L-arginine is an alternative substrate of arginine decarboxylase. (Do not confuse this derivative, also known as *N*γ-methyl-L-arginine, with γ-methyl-L-arginine.) **Target(s):** arginine kinase, weakly inhibited (1); nitric-oxide synthase (2-23). **1.** Pereira, Alonso, Paveto, *et al.* (2000) *J. Biol. Chem.* **275**, 1495. **2.** Sakuma, Stuehr, Gross, Nathan & Levi (1988) *Proc. Natl. Acad. Sci. U.S.A.* **85**, 8664. **3.** Aisaka, Gross, Griffith & Levi (1989) *Biochem. Biophys. Res. Commun.* **160**, 881. **4.** Garvey, Furfine & Sherman (1996) *Meth. Enzymol.* **268**, 339. **5.** Griffith & Kilbourn (1996) *Meth. Enzymol.* **268**, 375. **6.** Aisaka, Gross, Griffith & Levi (1989) *Biochem. Biophys. Res. Commun.* **163**, 710. **7.** Lambert, Whitten, Baron, *et al.* (1991) *Life Sci.* **48**, 69. **8.** McCall, Feelisch, Palmer & Moncada (1991) *Brit. J. Pharmacol.* **102**, 234. **9.** Griffith & Stuehr (1995) *Ann. Rev. Physiol.* **57**, 707. **10.** Hasan, Heesen, Corbett, *et al.* (1993) *Eur. J. Pharmacol.* **249**, 101. **11.** Gross, Stuehr, Aisaka, Jaffe, Levi & Griffith (1990) *Biochem. Biophys. Res. Commun.* **170**, 96. **12.** Babu, Frey & Griffith (1999) *J. Biol. Chem.* **274**, 25218. **13.** Feldman, Griffith, Hong & Stuehr (1993) *J. Med. Chem.* **36**, 491. **14.** Hong, Kim, Choi, *et al.* 2003) *FEMS Microbiol. Lett.* **222**, 177. **15.** Chandok, Ytterberg, van Wijk & Klessig (2003) *Cell* **113**, 469. **16.** Schmidt & Murad (1991) *Biochem. Biophys. Res. Commun.* **181**, 1372. **17.** Bush, Gonzalez, Griscavage & Ignarro (1992) *Biochem. Biophys. Res. Commun.* **185**, 960. **18.** Côté & Roberge (1996) *Free Radic. Biol. Med.* **21**, 109. **19.** Knowles, Merrett, Salter & Moncada (1990) *Biochem. J.* **270**, 833. **20.** Hecker, Walsh & Vane (1991) *FEBS Lett.* **294**, 221. **21.** Knowles, Palacios, Palmer & Moncada (1990) *Biochem. J.* **269**, 207. **22.** Wolff & Datto (1992) *Biochem. J.* **285**, 201. **23.** Reif & McCreedy (1995) *Arch. Biochem. Biophys.* **320**, 170.

3-Methyl-L-aspartate (3-Methyl-L-aspartic Acid)

This methylated aspartic acid derivative (FW = 147.13 g/mol), also known as 2-amino-3-methyl-succinic acid, is an intermediate in the conversion of glutamate to citramalate in certain organisms (*e.g., Chlostridium tetanomorphum*). Both the (2*S*,3*S*)- and (2*S*,3*R*)-isomers inhibit *Escherichia coli* asparagine synthetase B (K_{is} = 0.25 mM for the (2*S*,3*R*)-isomer).

Target(s): asparagine synthetase, inhibited by both *erythro-* and *threo-*isomers (1-3); aspartate 4-decarboxylase, inhibited by the *erythro*-DL-stereoisomer, also a weak alternative substrate (4,5); aspartyl-tRNA synthetase (6); tyrosine aminotransferase (7). **1.** Lea & Fowden (1975) *Proc. R. Soc. Lond. B Biol. Sci.* **192**, 13. **2.** Parr, Boehlein, Dribben, Schuster & Richards (1996) *J. Med. Chem.* **39**, 2367. **3.** Boehlein, Stewart, Walworth, *et al.* (1998) *Biochemistry* **37**, 13230. **4.** Miles & Meister (1967) *Biochemistry* **6**, 1734. **5.** Wilson & Kornberg (1963) *Biochem. J.* **88**, 578. **6.** Lea & Fowden (1973) *Phytochemistry* **12**, 1903. **7.** Miller & Litwack (1971) *J. Biol. Chem.* **246**, 3234.

N-Methyl-D-Aspartate (*N*-Methyl-D-Aspartic Acid)

This synthetic neuroactive amino acid derivative (FW = 147.13 g/mol; CAS 6384-92-5; IUPAC Name: (2*R*)-2-(methylamino)butanedioate; water-soluble) is a specific partial agonist at the so-called NMDA receptor (NMDAR), mimicking the natural action of glutamate, its physiologic neurotransmitter agonist. **Mechanism of Action:** The NMDA receptor forms a heterotetramer between two GluN1 and two GluN2 subunits (the subunits were previously denoted as NR1 and NR2), two obligatory NR1 subunits and two regionally localized NR2 subunits. Upon specific agonist-binding to its NR2 subunits, a non-specific cation channel opens, allowing Ca^{2+} and Na^+ to pass into the cell and K^+ to pass out of the cell. The excitatory postsynaptic potential (EPSP) produced by activation of an NMDA receptor also increases the concentration of Ca^{2+} in the cell. The Ca^{2+} can in turn function as a second messenger in various signaling pathways. Activation of the *N*-methyl-D-aspartate receptor plays a role in activity-dependent synaptic plasticity, such as long-term potentiation (1), and the patterning of connections during development of the visual system (2,3). **Effect of Tau Protein Phosphorylation:** The extent of the microtubule-associated protein Tau phosphorylation at Serine 199-202 residues (but not Serine 262 and Serine 404) is decreased in NMDA-treated hippocampal slices (4). NMDA-induced reduction of Tau phosphorylation at Serine 199-202 was blocked completely by the NR2A receptor antagonist NVP-AAM077. Compared with untreated slices, NMDA receptor activation is reflected in high Serine 9 and low Tyrosine 216 phosphorylation of GSK3β, suggesting NMDA receptor activation might diminish Tau phosphorylation via a pathway involving inhibition of GSK3β. In support of this idea is the finding that PKC-mediated GSK3β is involved in the NMDA-induced reduction of Tau phosphorylation at Ser199-202. These data suggest NR2A receptor activation may be an important factor in limiting Tau phosphorylation by a PKC/GSK3β pathway. **NMDA Receptor Isoforms:** Eight NR1 subunit variants are generated by alternative splicing of GRIN1, namely NR1-1a (most abundant form) and NR1-1b; NR1-2a and NR1-2b; NR1-3a and NR1-3b; and NR1-4a and NR1-4b. In vertebrates, there are four distinct NR2 subunit isoforms: GRIN2A, GRIN2B, GRIN2C, GRIN2D. **NMDA Receptor Antagonists:** *ACBC*, acts at glycine site ligand; *D-AP5*, potent, selective NMDA antagonist; *DL-AP7*, highly sselective; *Arcaine*, competitive; *(R)-4-Carboxyphenylglycine*, moderately potent; *Cerestat*, potent and noncompetitive; *CGP-37849*, potent and selective; *CGP-39551*, potent, selective and competitive; *CGP-78608*, potent, selective, glycine-site ligand; *CGS-19755*, potent and competitive; *7-Chlorokynurenate*, glycine site ligand; *(2R,3S)-Chlorpheg*, weak antagonist; *CNQX*, glycine site ligand; *Co 101244*, NR2B-selective; *Conantokin G*, NR2B-selective; *Conantokin-R*, potent noncompetitive; *Conantokin-T*, noncompetitive; *(R)-CPP*, more active enantiomer of *(RS)*-CPP; *(RS)-CPP*, potent; *D-CPP-ene*, potent and competitive; *Dextromethorphan*; *5,7-Dichlorokynurenate*, potent, glycine site ligand; *(±)-1-(1,2-Diphenylethyl)piperidine*, acts ion channel site; *DQP-1105*, selective for NR2C/NR2D; *Eliprodil*, noncompetitive NR2B-selective; *Felbamate*, acts glycine site; *Flupirtine*, indirect antagonist, potent and selective glycine site antagonist, orally available and active *in vivo*; *(S)-(-)-HA-966*, partial agonist; *HU 211*, NMDAR antagonist and NF-κB inhibitor; *N-(4-Hydroxyphenylacetyl)spermine*, wasp toxin analogue; *Ifenprodil*, noncompetitive NMDAR antagonist, also σ ligand; *threo-Ifenprodil*, NR2B-selective NMDA antagonist; also σ agonist; *Ketamine*, noncompetitive; *L-689,560*, very potent; *L-701,252* acts glycine site; *L-701,324*, acts at glycine site; *Loperamide*, reduces Ca^{2+} flux; *LY 233053*, competitive; *LY 235959*, competitive; *Memantine*, acts at ion channel site; *(-)-MK 801*, less active

enantiomer; *(+)-MK 801*, noncompetitive, acts at ion channel site; *Norketamine*, potent and noncompetitive; *Pentamidine*, antimicrobial agent that also antagonizes NMDARs; *Phencyclidine*, noncompetitive; *PMPA*, competitive; *PPDA*, subtype-selective NR2C/NR2D antagonist; *PPPA*, competitive; *QNZ 46*, NR2C/NR2D-selective nonncompetitive antagonist; *Remacemide*, blocks ion channel and allosteric modulatory site; *Ro 04-5595*, NR2B-selective; *Ro 25-6981*, NR2B-selective; *Ro 61-8048*, increases kynurenic acid levels; *Ro 8-4304*, NR2B-selective; *SDZ 220-040*, potent and competitive; *SDZ 220-581*, competitive; *Synthalin*, noncompetitive; *TCN 201*, selective for NR1/NR2A; *TCN 213*, selective for NR2A over NR2B; *TCN 237*, highly potent and selective for NR2B; *TCS-46b*, orally active, NR1A/NR2B-subtype selective; *ZD 9379*, brain-penetrating, glycine site ligand. **1.** Bliss & Collingridge (1993) *Nature* **361**, 31. **2.** Constantine-Paton, Cline & Debski (1990) *Rev. Neurosci.* **13**, 129. **3.** Schatz (1990) *Neuron* **5**, 745. **4.** De Montigny, Elhiri, Allyson, Cyr & Massicotte (2013) *Neural Plast.* 2013, 261593 (PMID: 24349798).

(24*R*)-24-Methyl-25-azacycloartanol

This positively charged aza-analogue (FW = 433.77 g/mol), which mimics the putative high-energy C-25 carbocation intermediate, potently inhibits maize *S*-adenosyl-L-methionine-sterol-24-*C*-methyltransferase (K_i = 20 nM). Its 24*S* counterpart is a slightly weaker inhibitor (K_i = 30 nM). **1.** Rahier, Génot, Schuber, Benveniste & Narula (1984) *J. Biol. Chem.* **259**, 15215. **2.** Narula, Rahier, Benveniste & Schuber (1981) *J. Am. Chem. Soc.* **103**, 2408. **3.** Benveniste (1986) *Ann. Rev. Plant Physiol.* **37**, 275.

N-(α-Methylbenzyl)-1-aminobenzotriazole

This potent mechanism-based inhibitor (FW = 224.29 g/mol) of cytochrome P450 oxidoreductases is particularly effective with CYP2Bx. **Target(s):** CYP1A1 (1-3); CYP2B1 (4,5); CYP2B4 (1,3,5,6); CYP2B5 (6). **1.** Knickle, Philpot & Bend (1994) *J. Pharmacol. Exp. Ther.* **270**, 377. **2.** Knickle, Webb, House & Bend (1993) *J. Pharmacol. Exp. Ther.* **267**, 758. **3.** Woodcroft, Szczepan, Knickle & Bend (1990) *Drug Metab. Dispos.* **18**, 1031. **4.** Woodcroft Bend (1990) *Can. J. Physiol. Pharmacol.* **68**, 1278. **5.** Sinal, Hirst, Webb & Bend (1998) *Drug Metab. Dispos.* **26**, 681. **6.** Grimm, Bend & Halpert (1995) *Drug Metab. Dispos.* **23**, 577.

3-Methylcatechol

This substituted catechol (FW = 124.14 g/mol; M.P. = 65-68°C) inhibits 1,2-dihydroxynaphthalene dioxygenase and protocatechuate 4,5-dioxygenase. It is also an alternative substrate of 2,3-dihydroxybiphenyl 1,2-dioxygenase and catechol 1,2-dioxygenase. **Target(s):** biphenyl-2,3-diol 1,2-dioxygenase, also as a weak alternative substrate (1); catechol 1,2-dioxygenase, also as an alternative substrate (2-4); catechol 2,3-dioxygenase, also as an alternative substrate and suicide inhibitor (5); 1,2-dihydroxynaphthalene dioxygenase (6); 4,5-dihydroxyphthalate decarboxylase, weakly inhibited (7); protocatechuate 4,5-dioxygenase (8).

1. Lloyd-Jones, Ogden & Williams (1995) *Biodegradation* **6**, 11. **2**. Itoh (1981) *Agric. Biol. Chem.* **45**, 2787. **3**. Patel, Hou, Felix & Lillard (1976) *J. Bacteriol.* **127**, 536. **4**. Sauret-Ignazi, Gagnon, Béguin, *et al.* (1996) *Arch. Microbiol.* **166**, 42. **5**. Cerdan, Rekik & Harayama (1995) *Eur. J. Biochem.* **229**, 113. **6**. Patel & Barnsley (1980) *J. Bacteriol.* **143**, 668. **7**. Nakazawa & Hayashi (1978) *Appl. Environ. Microbiol.* **36**, 264. **8**. Ono, Nozaki & Hayaishi (1970) *Biochim. Biophys. Acta* **220**, 224.

4-Methylcatechol

This substituted catechol (FW = 124.14 g/mol; M.P. = 67-69°C) inhibits a number of dioxygenases as well as some metallopeptidases. It is an alternative substrate of tyrosinase, 2,3 dihydroxybiphenyl 1,2-dioxygenase, and catechol 1,2-dioxygenase. Catechol 2,3-dioxygenase is inactivated at an appreciable rate during the ring cleavage of 4-methylcatechol, a consequence of oxidation of the Fe(II) cofactor to Fe(III). **Target(s):** aminopeptidase N, *or* membrane alanyl aminopeptidase (1); biphenyl-2,3-diol 1,2 dioxygenase (2); catechol 1,2-dioxygenase, also as an altenative substrate (3,4); catechol 2,3-dioxygenase, as alternative substrate and suicide inhibitor (5-7); 1,2-dihydroxynaphthalene dioxygenase (8); 3,4 dihydroxyphenylacetate 2,3-dioxygenase (9); peptidyl-dipeptidase A, *or* angiotensin I-converting enzyme (1); protocatechuate 3,4-dioxygenase (10,11); protocatechuate 4,5-dioxygenase (12); tyrosine 3 monooxygenase (13). **1**. Bormann & Melzig (2000) *Pharmazie* **55**, 129. **2**. Lloyd-Jones, Ogden & Williams (1995) *Biodegradation* **6**, 11. **3**. Itoh (1981) *Agric. Biol. Chem.* **45**, 2787. **4**. Patel, Hou, Felix & Lillard (1976) *J. Bacteriol.* **127**, 536. **5**. Okuta, Ohnishi & Harayama (2004) *Appl. Environ. Microbiol.* **70**, 1804. **6**. Cerdan, Rekik & Harayama (1995) *Eur. J. Biochem.* **229**, 113. **7**. Tropel, Meyer, Armengaud & Jouanneau (2002) *Arch. Microbiol.* **177**, 345. **8**. Patel & Barnsley (1980) *J. Bacteriol.* **143**, 668. **9**. Que, Widom & Crawford (1981) *J. Biol. Chem.* **256**, 10941. **10**. Durham, Sterling, Ornston & Perry (1980) *Biochemistry* **19**, 149. **11**. Hou, Lillard & Schwartz (1976) *Biochemistry* **15**, 582. **12**. Ono, Nozaki & Hayaishi (1970) *Biochim. Biophys. Acta* **220**, 224. **13**. Fitzpatrick (1988) *J. Bioil. Chem.* **263**, 16058.

Methylcobalamin

This form of vitamin B_{12} (FW$_{free-acid}$ = 1344.4 g/mol), which participates in the formation of methionine from homocysteine, is rapidly oxidized in aqueous solution in the presence of dioxygen, producing aquacobalamin. This rate is significantly slower under anaerobic conditions and increased

by the presence of thiols and ethanol. **Target(s):** ethanolamine ammonia-lyase (1-6); glutamate mutase (7); propanediol dehydratase (2); HIV-1 integrase (8). **1**. Kaplan & Stadtman (1971) *Meth. Enzymol.* **17B**, 818. **2**. Abeles (1971) *The Enzymes*, 3rd ed., **5**, 481. **3**. Stadtman (1972) *The Enzymes*, 3rd ed., **6**, 539. **4**. Mauck, Hull & Babior (1975) *J. Biol. Chem.* **250**, 8997. **5**. Bradbeer (1965) *J. Biol. Chem.* **240**, 4675. **6**. Blackwell & Turner (1978) *Biochem. J.* **175**, 555. **7**. Hogenkamp & Oikawa (1964) *J. Biol. Chem.* **219**, 1911. **8**. Weinberg, Shugars, Sherman, Sauls & Fyfe (1998) *Biochem. Biophys. Res. Commun.* **246**, 393.

S-Methyl-L-cysteine

This cysteine derivative (FW = 135.19 g/mol) is found in many proteins, presumably as a consequence of nonspecific alkylation (*e.g.*, reaction with methyl mercury cation) or by methyltransfer (*e.g.*, formation via methylated-DNA:[protein]-cysteine *S*-methyltransferase.in the repair of alkylated DNA). **Target(s):** alkaline phosphatase, mildly inhibited (1); alliin lyase (2); asparagine:oxo-acid aminotransferase (3); cysteine dioxygenase (4,5); cysteine synthase, weakly inhibited (6); cysteinyl tRNA synthetase (7); glyoxalase, weakly inhibited (8); phenylalanine 4-monooxygenase (9,10); serine *O*-acetyltransferase (11,12); tryptophanase (13). **1**. Fishman & Sie (1971) *Enzymologia* **41**, 141. **2**. Schwimmer, Ryan & Wong (1964) *J. Biol. Chem.* **239**, 777. **3** . Cooper (1977) *J. Biol. Chem.* **252**, 2032. **4**. Yamaguchi & Hosokawa (1987) *Meth. Enzymol.* **143**, 395. **5**. Yamaguchi, Hosokawa, Kohashi, *et al.* (1978) *J. Biochem.* **83**, 479. **6**. Tamura, Iwasawa, Masada & Fukushima (1976) *Agric. Biol. Chem.* **40**, 637. **7**. Pan, Yu, Duh & Lee (1976) *J. Chin. Biochem. Soc.* **5**, 45. **8**. Kermack & Matheson (1957) *Biochem. J.* **65**, 48. **9**. Kaufman (1987) *Meth. Enzymol.* **142**, 3. **10**. Goreish, Bednar, Jones, Mitchell & Steventon (2004) *Drug Metabol. Drug Interact.* **20**, 159. **11**. Leu & Cook (1994) *Biochemistry* **33**, 2667. **12**. Johnson, Huang, Roderick & Cook (2004) *Arch. Biochem. Biophys.* **429**, 115. **13**. Newton, Morino & Snell (1965) *J. Biol. Chem.* **240**, 1211.

Methyl 2,5-Dihydroxycinnamate

trans-isomer　　　　*cis*-isomer

This cell-permeable erbstatin analogue (FW = 194.19 g/mol; Photosensitive) inhibits protein-tyrosine kinases (IC$_{50}$ = 0.78 μM for the epidermal growth factor-receptor (EGFR) protein-tyrosine kinase). DNA topoisomerase II is inhibited by alkylation of catalytically essential amino acyl residues. **Target(s):** *bcr-abl* protein-tyrosine kinase (1); DNA topoisomerase II (2,3); EGF-receptor protein-tyrosine kinase (4,5); Lyn protein-tyrosine kinase (6); PDGF-receptor protein-tyrosine kinase (7); protein-tyrosine kinase (1,4-6,8,9); receptor protein-tyrosine kinase (4,5,7,8); Syk protein tyrosine kinase (6). **1**. Kawada, Tawara, Tsuji, *et al.* (1993) *Drugs Exp. Clin. Res.* **19**, 235. **2**. Lassota, Singh & Kramer (1996) *J. Biol. Chem.* **271**, 26418. **3**. Azuma, Onishi, Sato & Kizaki (1995) *J. Biochem.* **118**, 312. **4**. Umezawa, Hori, Tajima, *et al.* (1990) *FEBS Lett.* **260**, 198. **5**. Umezawa, Sugata, Yamashita, Johtoh & Shibuya (1992) *FEBS Lett.* **314**, 289. **6**. Merciris, Claussen, Joiner & Giraud (2003) *Pflugers Arch.* **446**, 232. **7**. Kozawa, Suzuki, Watanabe, Shinoda & Y. Oiso (1995) *Endocrinology* **136**, 4473. **8**. Umezawa & Imoto (1991) *Meth. Enzymol.* **201**, 379. **9**. Ortega, Velasquez & Perez (2005) *Biol. Res.* **38**, 89.

α-Methyl-3,4-dihydroxyphenylalanine

This photosensitive dopa derivative (FW = 211.22 g/mol), also known as α-methyldopa and 3-(3,4-dihydroxyphenyl)-2-methylalanine, is an antihypertensive agent that is metabolized to α-methylnorepinephrine which stimulates α2 receptors. The L-enantiomer of α-methyldopa is also a potent competitive inhibitor of aromatic-L-amino-acid decarboxylase. **Target(s):** alkaline phosphatase (1); aromatic-L-amino-acid decarboxylase, or dopa decarboxylase (2-11); carnosine synthetase (12); tyrosine aminotransferase (13); tyrosine decarboxylase (11); tyrosine 3-monooxygenase (14).   **1.** Fishman & Sie (1971) *Enzymologia* **41**, 141.   **2.** Weil, Barbour & Chesne (1963) *Circulation* **28**, 165.   **3.** Lovenberg (1971) *Meth. Enzymol.* **17B**, 652.   **4.** Boeker & E. E. Snell (1972) *The Enzymes*, 3rd ed. (Boyer, ed.), **6**, 217.   **5.** Webb (1966) *Enzyme and Metabolic Inhibitors*, vol. **2**, p. 308, Academic Press, New York.   **6.** Smith (1960) *Brit. Pharm. Chemother.* **15**, 319.   **7.** Daly, Mauger, Yonemitsu, *et al.* (1967) *Biochemistry* **6**, 648.   **8.** Tocher & Tocher (1972) *Phytochemistry* **11**, 1661.   **9.** Jung (1986) *Bioorg. Chem.* **14**, 429.   **10.** Bender & Coulson (1977) *Biochem. Soc. Trans.* **5**, 1353.   **11.** Nagy & Hiripi (2002) *Neurochem. Int.* **41**, 9.   **12.** Ng & Marshall (1976) *Experientia* **32**, 839.   **13.** Jacoby & La Du (1964) *J. Biol. Chem.* **239**, 419.   **14.** Nagatsu, Levitt & Udenfriend (1964) *J. Biol. Chem.* **239**, 2910.

2-(4-Methyl-2,6-dinitroanilino)-*N*-methylpropionamide

This herbicide (FW = 282.26 g/mol), often abbreviated MDMP, is a specific inhibitor of ribosomal peptide-chain initiation, blocking the binding of the 60S subunit to the initiation conplex, the latter consisting of the 40S subunit, mRNA, and initiator RNA. MDMP also causes the diaggregation of yeast polysomes.   **1.** Fresno & Vázquez (1979) *Meth. Enzymol.* **60**, 566.   **2.** Jiménez (1976) *Trends Biochem. Sci.* **1**, 28.   **3.** Weeks & Baxter (1972) *Biochemistry* **11**, 3060.

α,β-Methyleneadenosine 5'-Diphosphate

This nonhydrolyzable ADP analogue and potent vasoconstrictor (FW_sodium-salt = 469.19 g/mol; CAS 104835-70-3; Abbreviations: AMPCP, APCP, or p(CH₂)pA) inhibits many ADP-dependent systems. With its sp^3 or tetrahedral methylene bridge, the methylene analogue is not a true ADP isostere. Its pK_a values are also 1-2 units higher than ADP, and its divalent metal ion binding is substantially weaker. Even so, many ATP-dependent enzymes bind this analogue, and its resistence to hydrolysis commends its use in many experiments. It is also a substrate for mitochondrial uptake by the ADP-ATP translocase, when assayed with isolated mitochondria. Importantly, adenosine receptor activation by adenine nucleotides requires conversion of the nucleotides to adenosine, making APCP inhibition of 5'-nucleotidase a useful way to modulate adenosine receptor activation and vascular tone. (In view of its high ionic charge, p(CH₂)pA requires intravenous administration and is unlikely to be taken up by most cells.) **Target(s):** carbamoyl-phosphate synthetase (1); 5'-nucleotidase (2-13) [*Note*: In certain MDR cell lines, ecto-5'NT serves as a required accessory molecule in resistance mediated by ATP-dependent mechanisms, and growth-sustaining nucleosides are provided by this salvage pathway (14). MDR cells with increased ecto-5'NT expression also have a lower intracellular ATP level than their parental cells.]; polynucleotide phosphorylase, *or* polyribonucleotide nucleotidyltransferase (15); thiamin-triphosphatase (16); triphosphate:protein phosphotransferase (17).   **1.** Powers & Meister (1978) *J. Biol. Chem.* **253**, 1258.   **2.** Madrid-Marina & Fox (1986) *J. Biol. Chem.* **261**, 444.   **3.** Le Hir, Gandhi & Dubach (1989) *Enzyme* **41**, 87.   **4.** Le Hir, Angielski & Dubach (1985) *Ren. Physiol.* **8**,

321.   **5.** Agnisola, Foti, Trara Genoino & Sciurba (1983) *Biochem. Biophys. Res. Commun.* **112**, 407.   **6.** Edwards, Recker, Manfredi, Rembecki & Fox (1982) *Amer. J. Physiol.* **243**, C270.   **7.** Maguire, Krishnakantha & Aronson (1984) *Placenta* **5**, 21.   **8.** Lai & Wong (1991) *Int. J. Biochem.* **23**, 1123.   **9.** Grondal & Zimmermann (1987) *Biochem. J.* **245**, 805.   **10.** García-Ayllón, Campoy, Vidal & Muñoz-Delgado (2001) *J. Neurosci. Res.* **66**, 656.   **11.** Orford & Saggerson (1996) *J. Neurochem.* **67**, 795.   **12.** Senger, Rico, Dias, Bogo & Bonan (2004) *Comp. Biochem. Physiol. B* **139**, 203.   **13.** Evans & Gurd (1973) *Biochem. J.* **133**, 189.   **14.** Ujházy, Berleth, Pietkiewicz, *et al.* (1996) *Int. J. Cancer* **68**, 493.   **15.** Godefroy-Colburn & Stetondji (1972) *Biochim. Biophys. Acta* **272**, 417.   **16.** Barchi (1979) *Meth. Enzymol.* **62**, 118.   **17.** Tsutsui (1986) *J. Biol. Chem.* **261**, 2645.

α,β-Methylene- Methyleneadenosine 5'-γ-S

This nonhydrolyzable γ-thio-ATP analogue (FW = 518.25 g/mol) protects against oxidative stress and Amyloid-β oligomers toxicity, conditions that are hallmarks in Alzheimer's Disease (1,2).. Unlike ATP-γ-S, its nonhydrolyzable analogue α,β-CH₂-ATP-γ-S resists hydrolysis. Both inhibit ROS formation in PC12 cells previously subjected to Fe(II)-oxidation, with IC₅₀ values of 0.18 μM for α,β-CH₂-ATP-γ-S and 0.20 μM for ATP-γ-S). ATP-γ-S-(α,β-CH₂) also rescues primary neurons from Aβ42 toxicity, with four times the potency if ATP-γ-S, (IC₅₀ = 0.2 and 0.8 μM, respectively). ATP-γ-S-(α,β-CH₂) also resists hydrolysis by ecto-nucleotidases such as, NPPs and TNAP, and is seven times more potent than ATP as a P2Y11 receptor agonist. *Note*: Although ATP-γ-S-(α,β-CH₂) is a promising agent for rescuing neurons from debilitating insults typical of Alzheimer's disease, its structural similarity to ATP suggests that it may likewise affect many other ATP.   **1.** Danino, Giladi, Grossman & Fischer (2014) *Biochem. Pharmacol.* **88**, 384.   **2.** Danino, Grossman & Fischer (2015) *Biochem. Biophys. Res. Commun.* **460**, 466.

α,β-Methyleneadenosine 5'-Triphosphate

This nonhydrolyzable ATP analogue (FW_lithium-salt = 511.14 g/mol; CAS 104809-20-3; Symbol: AMPCPP and pp(CH₂)pA) inhibits many ATP-dependent reactions and processes requiring scission of the P–O bond lying between the α and β phosphoryls. With its tetrahedral methylene bridge, this analogue is also not a true ATP isostere, inasmuch as the bridge oxygen is trigonal. The pK_a values of pp(CH₂)pA are also 1-2 units higher and its divalent metal ion binding substantially weaker. pp(CH₂)pA is also a P₂ purinoceptor agonist. Many ATP-hydrolyzing mechanoenzymes that employ a γ-phosphoryl sensor or molecular switch are parsimonious and will not bind pp(CH₂)pA. **Target(s):** (*N*-acetylneuraminyl)-galactosyl-glucosylceramide *N*-acetylgalactosaminyl-transferase (43); adenylate cyclase, *or* adenylyl cyclase (1,2); aminoacyl-tRNA synthetases (3); 2-amino-4-hydroxy-6-hydroxymethyl-7,8-dihydropteridine diphosphokinase (13,14,40); arginyl-tRNA synthetase (4); asparagine synthetase (5); carbamoyl phosphate synthetase (6,7); cysteinyl-tRNA synthetase (8); dTMP kinase, *or* thymidylate kinase (41); ecto-ATPase (9); endopeptidase La (31); formyltetrahydrofolate synthetase (10); glycyl-tRNA synthetase (11); GTP diphosphokinase (39); 3-hydroxymethylcephem carbamoyltransferase (12); leucyl-tRNA synthetase (15); methionine S-adenosyltransferase (16,42); mitochondrial ATP dependent protease (17); nucleoside-triphosphatase (18); 5'-nucleotidase (19); nucleotide diphosphatase (20-23); pantothenate synthetase (24); phosphodiesterase I, *or* 5'-exonuclease (32); polynucleotide adenylyltransferase, *or* poly(A) polymerase (36,37); pyruvate,orthophosphate dikinase (25); pyruvate,water dikinase (34); ribose-phosphate diphosphokinase, *or* phosphoribosyl pyrophosphate synthease (26); RNA ligase, ATP-requiring (27); selenide,water dikinase (33); streptomycin 3'' adenylyltransferase (35); sulfate adenylyltransferase38; tRNA adenylyltransferase (28); tryptophanyl tRNA synthetase (29); tyrosyl-tRNA synthetase (30). **1.** Dessauer, Scully & Gilman (1997) *J. Biol. Chem.* **272**, 22272. **2.** Dessauer & Gilman (1997) *J. Biol. Chem.* **272**, 27787. **3.** Santi, Webster & Cleland (1974) *Meth. Enzymol.* **29**, 620. **4.** Thiebe (1983) *Eur. J. Biochem.* **130**, 517. **5.** Horowitz & Meister (1972) *J. Biol. Chem.* **247**, 6708. **6.** Powers & Meister (1978) *J. Biol. Chem.* **253**, 1258. **7.** Durbecq, Legrain, Roovers, Piérard & Glansdorff (1997) *Proc. Natl. Acad. Sci. U.S.A.* **94**, 12803. **8.** Pan, Yu, Duh & Lee (1976) *J. Chin. Biochem. Soc.* **5**, 45. **9.** Dowd, Li & Zeng (1999) *Arch. Oral Biol.* **44**, 1055. **10.** Buttlaire (1980) *Meth. Enzymol.* **66**, 585. **11.** Dignam, Nada & Chaires (2003) *Biochemistry* **42**, 5333. **12.** Brewer, Taylor & Turner (1980) *Biochem. J.* **185**, 555. **13.** Yan, Blaszczyk, Xiao, Shi & Ji (2001) *J. Mol. Graph. Model.* **19**, 70. **14.** Blaszczyk, Shi &Yan (2000) *Structure Fold Des.* **8**, 1049. **15.** Marutzky, Flossdorf & Kula (1976) *Nucl. Acids Res.* **3**, 2067. **16.** H. Mudd (1973) *The Enzymes*, 3rd ed., **8**, 121. **17.** Kuzela & Goldberg (1994) *Meth. Enzymol.* **244**, 376. **18.** Zhang, Zhang & Inouye (2003) *J. Biol. Chem.* **278**, 21408. **19.** Madrid-Marina & Fox (1986) *J. Biol. Chem.* **261**, 444. **20.** Bartkiewicz, Sierakowska & Shugar (1984) *Eur. J. Biochem.* **143**, 419. **21.** Byrd, Fearney & Kim (1985) *J. Biol. Chem.* **260**, 7474. **22.** Decker & Bischoff (1972) *FEBS Lett.* **21**, 95. **23.** Bischoff, Tran-Thi & Decker (1975) *Eur. J. Biochem.* **51**, 353. **24.** Wang & D. Eisenberg (2003) *Protein Sci.* **12**, 1097. **25.** Milner, G. Michaels & H. G. Wood (1975) *Meth. Enzymol.* **42**, 199. **26.** L. Switzer & K. J. Gibson (1978) *Meth. Enzymol.* **51**, 3. **27.** Sabatini & S. L. Hajduk (1995) *J. Biol. Chem.* **270**, 7233. **28.** P. Deutscher (1982) *The Enzymes*, 3rd ed., **15**, 183. **29.** Kisselev, Favorova & Kolaleva (1979) *Meth. Enzymol.* **59**, 174. **30.** Santi & Peña (1973) *J. Med. Chem.* **16**, 273. **31.** Larimore, Waxman & Goldberg (1982) *J. Biol. Chem.* **257**, 4187. **32.** Picher & Boucher (2000) *Amer. J. Respir. Cell Mol. Biol.* **23**, 255. **33.** Veres, Kim, Scholz & Stadtman (1994) *J. Biol. Chem.* **269**, 10597. **34.** Berman & Cohn (1970) *J. Biol. Chem.* **245**, 5319. **35.** Kim, Hesek, Zajicek,Vakulenko & Mobashery (2006) *Biochemistry* **45**, 8368. **36.** Kurl, Holmes, Verney & Sidransky (1988) *Biochemistry* **27**, 8974. **37.** Balbo, Meinke & Bohm (2005) *Biochemistry* **44**, 7777. **38.** Farley, Cryns, Yang & Segel (1976) *J. Biol. Chem.* **251**, 4389. **39.** Hogg, Mechold, Malke, Cashel & Hilgenfeld (2004) *Cell* **117**, 57. **40.** Bermingham, Bottomley, Primrose & Derrick (2000) *J. Biol. Chem.* **275**, 17962. **41.** Kielley (1970) *J. Biol. Chem.* **245**, 4204. **42.** Chou & Talalay (1973) *Biochim. Biophys. Acta* **321**, 467. **43.** Senn, Cooper, Warnke, Wagner & Decker (1981) *Eur. J. Biochem.* **120**, 59.

β,γ-Methyleneadenosine 5'-Triphosphate

This nonhydrolyzable ATP analogue (FW$_{\text{free-acid}}$ = 505.21 g/mol; *Abbreviation*: AMP-PCP and p(CH$_2$)ppA) also called adenosine β,γ-methylene triphosphate and adenylyl (β,γ-methylene)-diphosphonate, is an inhibitor of many ATP-dependent enzymes that produce ADP and orthophosphate. Its P–C–P linkage has a bond angle closer to sp^3-hybridized C–C–C bonds than the P–O–P bond in ATP. p(CH$_2$)ppA can also serve as an alternative substrate for enzymes that hydrolyze ATP between the β and γ phosphorus atoms. With its tetrahedral methylene bridge, the methylene analogue is not a true ATP isostere, wherein the bridge oxygen is trigonal. Its pK_a values are also 1-2 units higher and its divalent metal ion binding substantially weaker. pp(CH$_2$)pA is also a P$_2$ purinoceptor agonist. In addition, p(CH$_2$)ppA is a selective P$_{2x}$ purinoceptor agonist that is more potent than ATP, although less potent than p(CH$_2$)ppA (*i.e.*, α,β-methyleneadenosine 5' triphosphate). *Escherichia coli* alkaline phosphatase hydrolyzes this nucleotide analogue to orthophosphate and the corresponding nucleoside diphosphate derivative. p(CH$_2$)ppA is

hygroscopic and is very soluble in water. It also binds metal ions with a great affinity. It is more stable in acidic solutions than AMP-PNP (adenosine β,γ-imidotriphosphate). Note: Many ATP-hydrolyzing mechanoenzymes that employ a γ-phosphoryl sensor or molecular switch are especially parsimonious and do not bind p(CH$_2$)ppA. **Target(s):** adenylate cyclase (1); apyrase (2); asparagine synthetase (3); ATP phosphoribosyltransferase (53); Ca^{2+}/Mg^{2+}-dependent ATPase (4); cobaltochelatase (5); diphosphomevalonate decarboxylase (6); DNA helicase (7); DNA ligase (8); DNA topoisomerase II (9); ecto-ATPase (10); endopeptidase La (46,47); endothelial growth factor receptor-2 tyrosine kinase (11); folylpolyglutamate synthetase (12); formyltetrahydrofolate synthetase (13); fructokinase (14); glucokinase (15); glutaminyl-tRNA synthetase, *or* glutamine-hydrolyzing (16); glycyl-tRNA synthetase (17); hexokinase (18); 3-hydroxymethylcephem carbamoyltransferase (19); insulysin (20); leucyl-tRNA synthetase (21); methionine *S*-adenosyltransferase (52); mitochondrial ATP-dependent protease (22); mitogen-activated protein kinase, *or* p38 mitogen-activate protein (MAP) kinase (31,50); Na$^+$/K$^+$-exchanging ATPase (23); nucleoside-triphosphatase (7,24); 5'-nucleotidase (25,26,48); nucleotide diphosphatase (27); 2',5'-oligoadenylate synthetase (28); phosphofructokinase (29); 3-phosphoglycerate kinase (30,51); polar-amino-acid-transporting ATPase, *or* histidine permease (32,33); polynucleotide phosphorylase, *or* polyribonucleotide nucleotidyltransferase (4,35); pyruvate, orthophosphate dikinase (36); RNA ligase, ATP-requiring (37); SecA ATPase (38); tetrahydrofolate synthetase (39-42); thiamin triphosphatase (43); triphosphate:protein phosphotransferase (49); UDP-*N*-acetylmuramate:L-alanine ligase (44,45). **1.** Yang & W. Epstein (1983) *J. Biol. Chem.* **258**, 3750. **2.** Kettlun, Uribe, Calvo, *et al.* (1982) *Phytochemistry* **21**, 551. **3.** Horowitz & Meister (1972) *J. Biol. Chem.* **247**, 6708. **4.** Cable, Feher & Briggs (1985) *Biochemistry* **24**, 5612. **5.** Debussche, Couder, Thibaut, Cameron, J. Crouzet & F. Blanche (1992) *J. Bacteriol.* **174**, 7445. **6.** Jabalquinto & Cardemil (1989) *Biochim. Biophys. Acta* **996**, 257. **7.** Borowski, Niebuhr, Schmitz, *et al.*(2002) *Acta Biochim. Pol.* **49**, 597. **8.** Shuman (1995) *Biochemistry* **34**, 16138. **9.** Goto, Laipis & Wang (1984) *J. Biol. Chem.* **259**, 10422. **10.** Dowd, Li & Zeng (1999) *Arch. Oral Biol.* **44**, 1055. **11.** Parast, Mroczkowski, Pinko, Misialek, Khambatta & Appelt (1998) *Biochemistry* **37**, 16788. **12.** Bognar & Shane (1986) *Meth. Enzymol.* **122**, 349. **13.** Buttlaire (1980) *Meth. Enzymol.* **66**, 585. **1L.** Copeland, Stone & Turner (1984) *Arch. Biochem. Biophys.* **233**, 748. **15.** Porter & Chassy (1982) *Meth. Enzymol.* **90**, 25. **16.** Horiuchi, Harpel, Shen, *et al.* (2001) *Biochemistry* **40**, 6450. **17.** Dignam, Nada & Chaires (2003) *Biochemistry* **42**, 5333. **18.** Ning, Purich & Fromm (1969) *J. Biol. Chem.* **244**, 3840. **19.** Brewer, Taylor & Turner (1980) *Biochem. J.* **185**, 555. **20.** Camberos, Pérez, Udrisar, Wanderley & Cresto (2001) *Exp. Biol. Med. (Maywood)* **226**, 334. **21.** Marutzky, Flossdorf & Kula (1976) *Nucl. Acids Res.* **3**, 2067. **22.** Kuzela & Goldberg (1994) *Meth. Enzymol.* **244**, 376. **23.** Askari & Huang (1984) *J. Biol. Chem.* **259**, 4169. **24.** Schröder, Rottmann, Bachmann & Müller (1986) *J. Biol. Chem.* **261**, 663. **25.** Yamazaki, Truong & Lowenstein (1991) *Biochemistry* **30**, 1503. **26.** Naito & Lowenstein (1985) *Biochem. J.* **226**, 645. **27.** Rossomando & Jahngen (1983) *J. Biol. Chem.* **258**, 7653. **28.** Ball (1982) *The Enzymes*, 3rd ed., **15**, 281. **29.** Deville-Bonne & Garel (1992) *Biochemistry* **31**, 1695. **30.** Lavoinne (1986) *Biochimie* **68**, 569. **31.** LoGrasso, Frantz, Rolandoet al. (1997) *Biochemistry* **36**, 10422. **32.** Nikaido, Liu & Ames (1997) *J. Biol. Chem.* **272**, 27745. **33.** Liu, Liu & Ames (1997) *J. Biol. Chem.* **272**, 21883. **34.** Simon & Myers (1961) *Biochim. Biopys. Acta* **51**, 178. **35.** Webb (1966) *Enzyme and Metabolic Inhibitors*, vol. 2, p. 474, Academic Press, New York. **36.** Milner, Michaels & Wood (1975) *Meth. Enzymol.* **42**, 199. **37.** Juodka & Labeikyte (1991) *Nucleosides Nucleotides* **10**, 367. **38.** Miller, Wang & Kendall (2002) *Biochemistry* **41**, 5325. **39.** Cichowicz & Shane (1987) *Biochemistry* **26**, 513. **40.** Shane & Cichowicz (1983) *Adv. Exp. Med. Biol.* **163**, 149. **41.** Shane (1982) in *Peptide Antibiotics: Biosynthesis and Functions: Enzymatic Formation of Bioactive Peptides and Related Compounds* (Kleinkauf &. von Döhren., eds.) W. de Gruyter, Berlin, pp. 353. **42.** Shane (1980) *J. Biol. Chem.* **255**, 5663. **43.** Barchi (1979) *Meth. Enzymol.* **62**, 118. **44.** Ehmann, Demeritt, Hull & Fisher (2004) *Biochim. Biophys. Acta* **1698**, 167. **45.** Deng, Gu, Marmor, *et al.* (2004) *J. Pharm. Biomed. Anal.* **35**, 817. **46.** Larimore, Waxman & Goldberg (1982) *J. Biol. Chem.* **257**, 4187. **47.** Kutejova, Durcova, Surovkova & Kuzela (1993) *FEBS Lett.* **329**, 47. **48.** Newby, Luzio & Hales (1975) *Biochem. J.* **146**, 625. **49.** Tsutsui (1986) *J. Biol. Chem.* **261**, 2645. **50.** Szafranska & Dalby (2005) *FEBS J.* **272**, 4631. **51.** Kovári, Flachner, Náray-Szabó & Vas (2002) *Biochemistry* **41**, 8796. **52.** Chou & Talalay (1973) *Biochim. Biophys. Acta* **321**, 467. **53.** Kleeman & Parsons (1976) *Arch. Biochem. Biophys.* **175**, 687.

2,2'-Methylenebis(4-chlorophenol)

This chlorinated bisphenol (FW = 269.13 g/mol; M.P. = 177-178°C; *Abbreviation*: MBC) also known as dichlorophen(e) and often abbreviated DCP and, is an antibacterial, antifungal, and anthelmintic agent that is used in agriculture as well as in soaps and shampoos. **Target(s):** chitin synthase (1); cystathionine β-synthase, *or* serine sulfhydrase (2-4); glucose-6 phosphate dehydrogenase (5); homocysteine desulfhydrase (2); hydroquinone glucosyltransferase (6); lactoylglutathione lyase, *or* glyoxalase I (7). **1.** Leighton, Marks & Leighton (1981) *Science* **213**, 905. **2.** Thong & Coombs (1987) *Exp. Parasitol.* **63**, 143. **3.** Walker & Barrett (1992) *Exp. Parasitol.* **74**, 205. **4.** Papadopoulos, Walker & Barrett (1996) *Int. J. Biochem. Cell Biol.* **28**, 543. **5.** Wang & Buhler (1981) *J. Toxicol. Environ. Health* **8**, 639. **6.** Hefner, Arend, Warzecha, Siems & Stöckigt (2002) *Bioorg. Med. Chem.* **10**, 1731. **7.** Sommer, Fischer, Krause, *et al.* (2001) *Biochem. J.* **353**, 445.

Methylene bisphosphonate

This nonhydrolyzable pyrophosphate analogue (FW$_{\text{free-acid}}$ = 176.00; CAS 194931-67-4), also named Medronate, methylenediphosphonate, and methanebisphosphonate, inhibits several pyrophosphate-dependent reactions. Methylene bisphosphonate is frequently condensed with nucleosides (N) and nucleoside 5'-monophosphates (NMP) to form the nonhydrolyzable nucleotide analogues p(CH$_2$)pN, pp(CH$_2$)pN, p(CH$_2$)ppN, *etc.* **Treating Osteoporosis:** There are many bone-stabilizing drugs that have a methylene bisphosphonate nucleus, a polydendate structure that binds metal ions and maintains bone mineralization. Bisphosphonates "home" to bone, where they concentrate. Upon acidification by bone-resorbing osteoclasts, these phosphonates are released and inhibit isopentenyl-pyrophosphate-dependent reactions, thereby reducing osteoclasts podosome. The latter are essential for bone remodeling by means of locakl acidification by podosomal V-APTase. Based on isothermal titration calorimetry (ITC) on the binding of a variety of bisphosphonates to human bone, there appear to be two binding sites: (a) a weak, highly populated site, with average ΔG of binding of −5.2 kcal/mol; and (b) stronger binding site, with a ΔG of binding of −8.5 kcal (1). Binding to both sites is entropy-driven. **Target(s):** geranyl-diphosphate cyclase, *or* bornyl-diphosphate synthase (2); GTP cyclohydrolase IIa (3); *trans*-octaprenyl*trans*transferase (21); phosphate carrier (4); phosphoenolpyruvate carboxykinase (diphosphate) (5,6); pyrophosphatase (8-14); pyrophosphate-dependent phosphofructokinase, *or* diphosphate:fructose-6 phosphate 1-phosphotransferase (7,20); pyruvate,orthophosphate dikinase (15,16); succinate dehydrogenase (17,18); vacuolar H+-translocating inorganic diphosphatase (19). **1.** Mukherjee, Huang, Guerra, *et al.* (2009) *J. Am. Chem. Soc.* **131**, 8374 **2.** Wheeler & Croteau (1988) *Arch. Biochem. Biophys.* **260**, 250. **3.** Graham, Xu & White (2002) *Biochemistry* **41**, 15074. **4.** Zackova, Kramer & Jezek (2000) *Int. J. Biochem. Cell Biol.* **32**, 499. **5.** Wood, O'Brien & Michaels (1977) *Adv. Enzymol. Relat. Areas Mol. Biol.* **45**, 85. **6.** Willard & Rose (1973) *Biochemistry* **12**, 5241. **7.** Bertagnolli & Cook (1984) *Biochemistry* **23**, 4101. **8.** Cooperman (1982) *Meth. Enzymol.* **87**, 526. **9.** Baltscheffsky & Nyrén (1986) *Meth. Enzymol.* **126**, 538. **10.** Josse & Wong (1971) *The Enzymes*, 3rd ed., **4**, 499. **11.** Butler (1971) *The Enzymes*, 3rd ed., **4**, 529. **12.** Baykov, Pavlov, Kasho & Avaeva (1989) *Arch. Biochem. Biophys.* **273**, 301. **13.** Barry & Dunaway-Mariano (1987) *Arch. Biochem. Biophys.* **259**, 196. **14.** Knight, Fitts & Dunaway-Mariano (1981) *Biochemistry* **20**, 4079. **15.** Milner, Michaels & Wood (1975) *Meth. Enzymol.* **42**, 199. **16.** Hiltpold, Thomas & Köhler (1999) *Mol. Biochem. Parasitol.* **104**, 157. **17.** Webb (1966) *Enzyme and Metabolic Inhibitors*,

vol. 2, p. 243, Academic Press, New York. **18.** Rosen & Klotz (1957) *Arch. Biochem. Biophys.* **67**, 161. **19.** Zhen, Baykov, Bakuleva & Rea (1994) *Plant Physiol.* **104**, 153. **20.** Kowalczyk, Januszewska, Cymerska & Maslowski (1984) *Physiol. Plant* **60**, 31. **21.** Gotoh, Koyama & Ogura (1992) *J. Biochem.* **112**, 20.

Methylene Blue

This widely used redox dye (FW$_{\text{anhydrous}}$ = 319.86 g/mol; CAS 61-73-4; Symbol: MB$^+$; Soluble in water and ethanol; λ$_{\text{max}}$ = 663-667 nm (undergoes concentration-dependent changes above 1 mM, with loss of 663-667 nm peak). Methylene blue ($E°'$ = 0.011 V at 30°C) is reduced to a colorless (or slightly yellow) compound in the absence, and becomes blue again, when exposed to O$_2$. This property is exploited in verifying that anaerobic bacterial culture medium is O$_2$-free. MB$^+$ has also inhibits oxidative phosphorylation and photophosphorylation. It has also inhibits tau polymerization in Alzheimer's Disease. Methylene Blue hydrate (FW = 337.87 g/mol; CAS 122965-43-9) is a frequently used nucleic acid stain in both agarose and polyacrylamide gel electrophoresis, because it is less hazardous than ethidium bromide. Moreover, DNA bands are viewable in visible light. Significantly, MB$^+$ photosensitizes DNA (occurring in both aerated or de-aerated solutions), causing guanine-specific phosphodiester-bond cleavage and making additional bonds alkali labile. **Target(s):** acetylcholinesterase (1); acetyl-CoA synthetase, *or* acetate:CoA ligase (2); adenylate kinase (3); adrenodoxin reductase, NADPH-dependent (4); alkaline mesentericopeptidase (5); alkaline phosphatase (6); amine oxidase and/or diamine oxidase (7-9); aminopeptidase B (10); arginine aminopeptidase (11,12); bromelain (13,14); carboxylesterase, *or* procaine esterase (15); cholinesterase (1,16); chymotrypsin (17); collagenase (18,19); dextransucrase (20,21); dipeptidyl-peptidase IV (70); enolase (23); β-fructofuranosidase, *or* invertase, weakly inhibited (24); β-fructosidase (25); β-galactosidase (26,27); glucose-6-phosphatase (28); glucosyltransferases (*e.g.*, levansucrase and dextransucrase) (20,21); glutaminase (29); glutamine synthetase (30); γ-glutamyl transpeptidase (31); guanylate cyclase (32); hexokinase (33); S-2-hydroxyacylglutathione hydrolase, *or* glyoxalase II (34,75,76); 3-hydroxydecanoyl-[acyl-carrier-protein] dehydratase (35,36); isocitrate dehydrogenase, NADP$^+$-dependent (37); leucyl aminopeptidase (71); levansucrase (20,21); lysozyme (17,38-41); magnesium protoporphyrin IX monomethyl ester (oxidative) cyclase (80-82); malic enzyme, NAD$^+$-dependent (42); NAD$^+$ kinase (43); NADPH peroxidase (83); Na$^+$/K$^+$-exchanging ATPase (44); nitric-oxide synthase (45,46); oligo 1,6-glucosidase (47); palmitoyl-CoA hydrolase, *or* acyl-CoA hydrolase (48,74); pancreatic ribonuclease, *or* ribonuclease A (49-53); phosphoenolpyruvate phosphatase (73); phosphofructokinase (54); phosphoglucomutase (55); 6-phosphogluconate dehydrogenase (56); phosphoglycerate mutase (57,58); phosphopantothenoyl-cysteine decarboxylase (59); progesterone monooxygenase (78); purine nucleoside phosphorylase (60,61); pyruvate kinase (62); ribonuclease T$_1$ (63,64); ribonuclease T$_2$ (72); stearoyl-CoA 9-desaturase (79); stem bromelain (13,14); steroid Δ isomerase (65); sulfate adenylyltransferase (77); sulfite oxidase (84); taurine dehydrogenase (66); transketolase (67); trypsin (17,68); urease (69). **1.** Augustinsson (1950) *The Enzymes*, 1st. ed., **1** (part 1), 443. **2.** Londesborough & Webster (1974) *The Enzymes*, 3rd ed., **10**, 469. **3.** Noda (1973) *The Enzymes*, 3rd ed., **8**, 279. **4.** Hiwatashi, Ichikawa, Yamano & Maruya (1976) *Biochemistry* **15**, 3091. **5.** Shopova & Genov (1977) *Int. J. Pept. Protein Res.* **10**, 369. **6.** Reid & Wilson (1971) *The Enzymes*, 3rd ed. (Boyer, ed.), **4**, 373. **7.** Yasunobu & Gomes (1971) *Meth. Enzymol.* **17B**, 709. **8.** Zeller (1951) *The Enzymes*, 1st ed., **2** (part 1), 536. **9.** Blaschko (1963) *The Enzymes*, 2nd ed., **8**, 337. **10.** Mäkinen & Hopsu-Havu (1967) *Enzymologia* **32**, 333. **11.** Soderling & Makinen (1983) *Arch. Biochem. Biophys.* **220**, 11. **12.** Makinen & Hopsu-Havu (1967) *Enzymologia* **32**, 333. **13.** Murachi (1976) *Meth. Enzymol.* **45**, 475. **14.** Murachi, Tsudzuki & Okumura (1975) *Biochemistry* **14**, 249. **15.** Bernheim (1952) *The Enzymes*, 1st ed., **2** (part 2), 844. **16.** Scheraga & Rupley (1962) *Adv. Enzymol.* **24**, 161. **17.** Hodgson, McVey & Spikes (1969) *Experientia* **25**, 1021. **18.** Bicsak & Harper (1984) *J. Biol. Chem.* **259**, 13145. **19.** Seifter & Harper (1971) *The Enzymes*, 3rd ed., **3**, 649. **20.** Koga & Inoue

(1981) *Carbohydr. Res.* **93**, 125. **21**. Fu & Robyt (1988) *Carbohydr. Res.* **183**, 97. **22**. Kuno, Fukui & Toraya (1990) *Arch. Biochem. Biophys.* **277**, 211. **23**. Wold (1971) *The Enzymes*, 3rd. ed., **5**, 499. **24**. Quastel & Yates (1936) *Enzymologia* **1**, 60. **25**. Korneeva, Zherebtsov, Cheremushkina & Ukhina (1998) *Biochemistry (Moscow)* **63**, 1220. **26**. Korneeva, Zherebtsov & Cheryomushkina (2001) *Biochemistry (Moscow)* **66**, 334. **27**. Wallenfels & Weil (1972) *The Enzymes*, 3rd ed. (Boyer, ed.), **7**, 617. **28**. Feldman & Butler (1972) *Biochim. Biophys. Acta* **268**, 690. **29**. Holcenberg (1985) *Meth. Enzymol.* **113**, 257. **30**. Akent'eva, Solov'eva, Pushkin, Evstigneeva & Kretovich (1983) *Biokhimiia* **48**, 833. **31**. Binkley & Olson (1951) *J. Biol. Chem.* **188**, 451. **32**. Holzmann (1983) *J. Cardiovasc. Pharmacol.* **5**, 364. **33**. Menezes, Grouselle & Pudles (1972) *Eur. J. Biochem.* **30**, 81. **34**. Ball & vander Jagt (1981) *Biochemistry* **20**, 899. **35**. Bloch (1971) *The Enzymes*, 3rd ed., **5**, 441. **36**. Helmkamp & Bloch (1969) *J. Biol. Chem.* **244**, 6014. **37**. Senkevich, Taranda, Strumilo & Vinogradov (1988) *Ukr. Biokhim. Zh.* **60**, 46. **38**. Jori, Galiazzo, Marzotto & Scoffone (1968) *J. Biol. Chem.* **243**, 4272. **39**. Silva (1979) *Radiat. Environ. Biophys.* **16**, 71. **40**. Cohen (1970) *The Enzymes*, 3rd ed., **1**, 147. **41**. Imoto, Johnson, North, Phillips Rupley (1972) *The Enzymes*, 3rd ed., **7**, 665. **42**. Saito, Yamaguchi, Charunmethee, Tokushige & Katsuki (1985) *Physiol. Chem. Phys. Med. NMR* **17**, 45. **43**. Severin, Telepneva & Tseitlin (1971) *Biokhimiia* **36**, 1014. **44**. Popova & Severin (1971) *Vopr. Med. Khim.* **17**, 575. **45**. Shimizu, Yamamoto & Momose (1993) *Res. Commun. Chem. Pathol. Pharmacol.* **82**, 35. **46**. Mayer, Brunner & Schmidt (1993) *Biochem. Pharmacol.* **45**, 367. **47**. Larner (1960) *The Enzymes*, 2nd ed., **4**, 369. **48**. Barnes (1975) *Meth. Enzymol.* **35**, 102. **49**. Weil & Seibles (1955) *Arch. Biochem. Biophys.* **54**, 368. **50**. Jori, Galiazzo, Marzotto & Scoffone (1968) *Biochim. Biophys. Acta* **154**, 1. **51**. MacKnight & Spikes (1970) *Experientia* **26**, 255. **52**. Massart (1950) *The Enzymes*, 1st ed., **1** (part 1), 307. **53**. Richards & Wyckoff (1971) *The Enzymes*, 3rd ed., **4**, 647. **54**. Ahlfors & Mansour (1969) *J. Biol. Chem.* **244**, 1247. **55**. Ray & Peck (1972) *The Enzymes*, 3rd ed., **6**, 407. **56**. Rippa & Pontremoli (1968) *Biochemistry* **7**, 1514. **57**. Grisolia & Carreras (1975) *Meth. Enzymol.* **42**, 435. **58**. Carreras, Mezquita & Pons (1982) *Comp. Biochem. Physiol. B* **72**, 401. **59**. Scandurra, Consalvi, Politi & Gallina (1988) *FEBS Lett.* **231**, 192. **60**. Lewis & Glantz (1976) *Biochemistry* **15**, 4451. **61**. Lewis & Glantz (1976) *J. Biol. Chem.* **251**, 407. **62**. Ambasht, Malhotra & Kayastha (1997) *Indian J. Biochem. Biophys.* **34**, 365. **63**. Uchida & Egami (1971) *The Enzymes*, 3rd ed., **4**, 205. **64**. Takahashi & Moore (1982) *The Enzymes*, 3rd ed., **15**, 435. **65**. Talalay & Benson (1972) *The Enzymes*, 3rd ed., **6**, 591. **66**. Kondo & Ishimoto (1987) *Meth. Enzymol.* **143**, 496. **67**. Kochetov & Kobylianskaia (1971) *Biull. Eksp. Biol. Med.* **72**, 55. **68**. Glad, Spikes & Kumagai (1968) *Experientia* **24**, 1002. **69**. Quastel (1932) *Biochem. J.* **26**, 1685. **70**. Harada, Hiraoka, Fukasawa & Fukasawa (1984) *Arch. Biochem. Biophys.* **234**, 622. **71**. Machuga & Ives (1984) *Biochim. Biophys. Acta* **789**, 26. **72**. Irie, Ohgi & Iwama (1977) *J. Biochem.* **82**, 1701. **73**. Malhotra & Kayastha (1989) *Plant Sci.* **65**, 161. **74**. Bonner & Bloch (1972) *J. Biol. Chem.* **247**, 3123. **75**. vander Jagt (1989) in *Coenzymes and Cofactors, Glutathione, Chem. Biochem. Med. Aspects Pt. A* (Dolphin, Poulson & Avromonic, eds.) **3**, 597, Wiley, New York. **76**. Principato, Rosi, Talesa, Giovannini & Uotila (1987) *Biochim. Biophys. Acta* **911**, 349. **77**. Farley, Christie, Seubert & I. H. Segel (1979) *J. Biol. Chem.* **254**, 3537. **78**. Rahm & Sih (1966) *J. Biol. Chem.* **241**, 3615. **79**. Fulco & Bloch (1964) *J. Biol. Chem.* **239**, 993. **80**. Wong & Castelfranco (1985) *Plant Physiol.* **79**, 730. **81**. Whyte & Castelfranco (1993) *Biochem. J.* **290**, 355. **82**. Wong & Castelfranco (1984) *Plant Physiol.* **75**, 658. **83**. Conn, Kraemer, Liu & Vennesland (1952) *J. Biol. Chem.* **194**, 143. **84**. Kessler & Rajagopalan (1974) *Biochim. Biophys. Acta* **370**, 399.

(Methylenecyclopropyl)acetyl-CoA

This toxic acyl-CoA is produced from hypoglycin A, isolated from unripe ackee fruit (*Blighia sapida*), and is the causative agent of Jamaican vomiting sickness. It is an irreversible inhibitor of several acyl-CoA dehydrogenases, including butyryl-CoA dehydrogenase and isovaleryl-CoA dehydrogenase, and causes severe hypoglycemia. **Target(s):** acyl-CoA dehydrogenase (1,2); butyryl-CoA dehydrogenase (2-5); glutaryl-CoA dehydrogenase (6); isovaleryl-CoA dehydrogenase (3,4); medium-chain acyl-CoA dehydrogenase (2,4); methylmalonyl-CoA mutase (7). **1**. Wenz, Thorpe & Ghisla (1981) *J. Biol. Chem.* **256**, 9809. **2**. Ator & Ortiz de

Montellano (1990) *The Enzymes*, 3rd ed., **19**, 213. **3**. Billington & Sherratt (1981) *Meth. Enzymol.* **72**, 610. **4**. Ikeda & Tanaka (1990) *Biochim. Biophys. Acta* **1038**, 216. **5**. Kean (1976) *Biochim. Biophys. Acta* **422**, 8. **6**. Ghisla, Wenz & Thorpe (1980) in *Enzyme Inhibitors*, Verlag Chemie, Weinheim. **7**. Taoka, Padmakumar, Lai, Liu & Banerjee (1994) *J. Biol. Chem.* **269**, 31630.

3,4-Methylenedioxymethamphetamine

This amphetamine derivative (FW = 193.25 g/mol), also known as MDMA and ecstasy, is a hallucinogen and potent CNS stimulator. It is a neurotoxin that promotes the release of serotonin and blocks its reuptake. MDMA induces short- and long-term neuropsychiatric behaviors and is a widely abused recreational drug. In addition to being toxic to serotonergic neurons, it induces programmed cell death in cultured human serotonergic cells and rat neocortical neurons. It also alters the release of several neurotransmitters in the brain, induces re-compartmentation of intracellular serotonin and c-fos, and modifies the expression of a number of genes. **Target(s):** monoamine oxidase (1); serotonin transport (2); tryptophan 5-monooxygenase, inhibited mainly by MDMA metabolites (3,4). **1**. Leonardi & Azmitia (1994) *Neuropsycho-pharmacol.* **10**, 231. **2**. Rudnick & Wall (1992) *Proc. Natl. Acad. Sci. U.S.A.* **89**, 1817. **3**. Schmidt & Taylor (1987) *Biochem. Pharmacol.* **36**, 4095.

α,β-Methyleneguanosine 5'-Triphosphate

This nonhydrolyzable GTP analogue (FW_{free-acid} = 521.21 g/mol; *Abbreviation*: GMPCPP and pp(CH$_2$)pG) also called guanosine α,β-methylene triphosphate and guanylyl (β,γ-methylene)-diphosphonate, is an inhibitor of many GTP-dependent enzymes that produce GMP and pyrophosphate. Its P–C–P linkage has a bond angle closer to sp^3-hybridized C–C–C bonds than the P–O–P bond in GTP. p(CH$_2$)ppA can also serve as an alternative substrate for enzymes that hydrolyze GTP between the α and β phosphorus atoms. With its tetrahedral methylene bridge, the methylene analogue is not a true GTP isostere, wherein the bridge oxygen is trigonal. Its pK_a values are also 1-2 units higher and its divalent metal ion binding substantially weaker. pp(CH$_2$)pG is also a P$_2$ purinoceptor agonist. GMPCPP inhibits GTP cyclohydrolase IIa (1). **1**. Graham, Xu & White (2002) *Biochemistry* **41**, 15074.

β,γ-Methyleneguanosine 5'-Triphosphate

This nonhydrolyzable GTP analogue (FW_{free-acid} = 521.21 g/mol; *Abbreviation*: GMP-PCP and p(CH$_2$)ppA) also called guanosine β,γ-methylene triphosphate and guanylyl (β,γ-methylene)-diphosphonate, is an inhibitor of many ATP-dependent enzymes that produce GDP and orthophosphate. Its P–C–P linkage has a bond angle closer to sp^3-hybridized C–C single-bonds than the P–O–P bond in GTP. p(CH$_2$)ppG can also serve as an alternative substrate for enzymes that hydrolyze GTP between the α and β phosphorus atoms. With its tetrahedral methylene bridge, the methylene analogue is not a true GTP isostere, wherein the bridge oxygen is trigonal. Its pK_a values are also 1-2 units higher and its divalent metal ion binding substantially weaker. pp(CH₂)pG is also a P_2 purinoceptor agonist. Guanylyl (β,γ-methylene)-diphosphonate inhibits numerous GTP-dependent enzymes that typically produce GDP and orthophosphate from GTP hydrolysis. It can also serve as a slow, alternative substrate for enzymes hydrolyzing GTP between the α and β phosphorus atoms. GMP-PCP is very soluble in water and binds metal ions with a great affinity. It is more stable in acidic solutions than GMP-PNP (guanosine β,γ-imidotriphosphate). **Target(s):** adenylosuccinate synthetase (1); dynamin GTPase (2); GTP cyclohydrolase IIa (3); phosphatidylinositol-4,5-bisphosphate phospholipase C (4); protein biosynthesis (translocation) (5,6); protein-glutamine γ-glutamyltransferase (transglutaminase) (7); releasing factor (8); microtubule depolymerization and dynamicity (9). **1.** Rudolph & Fromm (1969) *J. Biol. Chem.* **244**, 3832. **2.** Shpetner & Vallee (1992) *Nature* **355**, 733. **3.** Graham, Xu & White (2002) *Biochemistry* **41**, 15074. **4.** Gehm & McConnell (1990) *Biochemistry* **29**, 5447. **5.** Holschuh, Bonin & Gassen (1980) *Biochemistry* **19**, 5857. **6.** Lucas-Lenard & Beres (1974) *The Enzymes*, 3rd ed., **10**, 53. **7.** Siegel & Khosla (2007) *Pharmacol. Ther.* **115**, 232. **8.** Tate & Caskey (1974) *The Enzymes*, 3rd ed., **10**, 87. **9.** Terry & Purich (1980) *J. Biol. Chem.* **255**, 10532.

3,4-Methylene-β-nitrostyrene

This <u>S</u>pleen <u>ty</u>rosine <u>k</u>inase (Syk) inhibitor (FW = 193.16 g/mol; *Symbol* = MNS) represents a new class of Src and Syk tyrosine kinase inhibitors, potently preventing GPIIb/IIIa activation and platelet aggregation without directly affecting other signaling pathways required for platelet activation (1). MNS exhibits potent and broad-spectrum inhibitory effects on human platelet aggregation observed by various stimulators. Addition of MNS to already ADP-aggregated human platelets also brings about rapid disaggregation by inhibiting GPIIb/IIIa activation. MNS potently inhibits tyrosine kinases (Src and Syk) and prevented protein tyrosine phosphorylation and cytoskeletal association of GPIIb/IIIa and talin, but it has no direct effects on protein kinase C, Ca²⁺ mobilization, Ca²⁺-dependent enzymes (myosin light chain kinase and calpain), and arachidonic acid metabolism, and it does not affect the cellular levels of cyclic nucleotides. MNS also targets ATP-induced activation of the <u>NOD</u>-like <u>R</u>eceptor <u>P</u>rotein-3 (or NLRP3) inflammasome, the dysregulation of which is implicated in several inflammation-associated disorders (*e.g.*, gouty arthritis, silicosis, Type 2 diabetes, Alzheimer's disease, and Cryopyrin-associated periodic syndrome (CAPS) fever) (2). MNS does not affect the activation of the NLRC4 and AIM2 inflammasomes at concentrations that abrogated NLRP3 inflammasome activation. Moreover, deletion of Syk kinase in macrophages does not affect NLRP3 inflammasome activation-induced by ATP, nigericin and silica, a surprising finding indicating that Syk is dispensable for NLRP3-mediated inflammasome activation. MNS also inhibits NLRP3 ATPase activity *in vitro*, suggesting that MNS blocks the NLRP3 inflammasome by directly targeting NLRP3 or NLRP3-associated complexes. The site of unsaturation in 3,4-methylene-β-nitrostyrene is required for these potent pharmacologic effects. **1.** Wang, Wu & Wu (2006) *Mol. Pharmacol.* **70**, 1380. **2.** He, Varadarajan, Muñoz-Planillo, *et al.* (2013) **289**, 1142.

Methyl Gallate

This gallate ester (FW = 184.15 g/mol), also known as gallic acid methyl ester and methyl 3,4,5-trihydroxybenzoate, inhibits chitin synthase. Methyl gallate is an effective antioxidant in a variety of acellular experiments. **Target(s):** catechol *O*-methyltransferase (1); chitin synthase (2); α-glucosidase, *or* sucrase (3); 3-galactosyl-*N*-acetylglucosaminide 4-α-L-fucosyltransferase (4); glycoprotein 3-α-L-fucosyltransferase (4); 15-lipoxygenase, *or* arachidonate 15-lipoxygenase (5); ribonucleoside diphosphate reductase (6); xanthine oxidase (7). **1.** Veser (1987) *J. Bacteriol.* **169**, 3696. **2.** Hwang, Ahn, Lee, *et al.* (2001) *Planta Med.* **67**, 501. **3.** Nishioka, Kawabata & Aoyama (1998) *J. Nat. Prod.* **61**, 1413. **4.** Niu, Fan, Sun, *et al.* (2004) *Arch. Biochem. Biophys.* **425**, 51. **5.** Luther, Jordanov, Ludwig & Schewe (1991) *Pharmazie* **46**, 134. **6.** Elford, Van't Riet, Wampler, Lin & Elford (1981) *Adv. Enzyme Regul.* **19**, 151. **7.** Masuoka, Nihei & Kubo (2006) *Mol. Nutr. Food Res.* **50**, 725.

Methyl α-D-Glucopyranoside

This synthetic nonreducing sugar (FW = 194.18 g/mol), is an alternative substrate or competitive inhibitor for many β-glucosidases. **Target(s):** acetyl-CoA:α-glucosaminide *N*-acetyltransferase, *or* heparan-α-glucosaminide *N*-acetyltransferase (1); aldose 1-epimerase, *ot* mutarotase (2-7); α-amylase (8,9); β-amylase (10,11); cellobiose phosphorylase (12); concanavalin A (13); cyclomaltodextrin glucanotransferase (39); dextransucrase (40,41); exo-(1→4)-α-D-glucan lyase (14); β-fructofuranosidase, *or* invertase (8,15-20); glucan 1,4-α-glucosidase, *or* glucoamylase (21-23); glucan 1,6-α-glucosidase (24); 4-α-glucanotransferase (25,38); α glucosidase (8,26,27); β-glucosidase (28); glycogen phosphorylase (29,42); α-mannosidase (30); oligo-1,6 glucosidase (31); protein:N^π-phosphohistidine:sugar phosphotransferase (37); sucrose fructosyltransferase (8); thioglucosidase, *or* myrosinase (32,33); α,α-trehalase, weakly inhibited (34,35); *Vicia faba* lectin (36). **1.** Bame & Rome (1987) *Meth. Enzymol.* **138**, 607. **2.** Bentley (1962) *Meth. Enzymol.* **5**, 219. **3.** Bailey, Fishman, Kusiak, Mulhern & Pentchev (1975) *Meth. Enzymol.* **41**, 471. **4.** Fishman, Pentchev & Bailey (1975) *Meth. Enzymol.* **41**, 484. **5.** Bailey & Pentchev (1964) *Biochem. Biophys. Res. Commun.* **14**, 161. **6.** Fishman, Kusiak & Bailey (1973) *Biochemistry* **12**, 2540. **7.** Hucho & Wallenfels (1971) *Eur. J. Biochem.* **23**, 489. **8.** Webb (1966) *Enzyme and Metabolic Inhibitors*, vol. 2, p. 415-421, Academic Press, New York. **9.** Oosthuizen, Naude, Oelofsen, Muramoto & Kamiya (1994) *Int. J. Biochem.* **26**, 1313. **10.** Thoma & Koshland (1960) *J. Biol. Chem.* **235**, 2511. **11.** Thoma, Spradlin & Dygert (1971) *The Enzymes*, 3rd ed., **5**, 115. **12.** Sasaki (1988) *Meth. Enzymol.* **160**, 468. **13.** Goldstein, Hollerman & Smith (1965) *Biochemistry* **4**, 876. **14.** Yu, Christensen, Kragh, Bojsen & Marcussen (1997) *Biochim. Biophys. Acta* **1339**, 311. **15.** Neuberg & Mandl (1950) *The Enzymes*, 1st ed., **1** (part 1), 527. **16.** Michaelis & Pechstein (1914) *Biochem. Z.* **60**, 79. **17.** Nelson & Freeman (1925) *J. Biol. Chem.* **63**, 365. **18.** Nelson & Post (1926) *J. Biol. Chem.* **68**, 265. **19.** Papadakis (1929) *J. Biol. Chem.* **83**, 561. **20.** Isla, Vattuone, Gutierrez & Sampietro (1988) *Phytochemistry* **27**, 1993. **21.** De Mot & Verachtert (1987) *Eur. J. Biochem.* **164**, 643. **22.** De Mot, Van Oudendijck & Verachtert (1985) *Antonie Van Leeuwenhoek* **51**, 275. **23.** Fogarty & Benson (1983) *Eur. J. Appl. Microbiol. Biotechnol.* **18**, 271. **24.** Okada, Unno & Sawai (1988) *Agric. Biol. Chem.* **52**, 2169. **25.** Wiesmeyer (1962) *Meth. Enzymol.* **5**, 141. **26.** Krakenaite & Glemzha (1983) *Biokhimiia* **48**, 62. **27.** Belen'kni, Tsukerman & Rozenfel'd (1975) *Biokhimiia* **40**, 927. **28.** Dale, Ensley, Kern, Sastry & Byers (1985) *Biochemistry* **24**, 3530. **29.** Rinaudo & Bruno (1968) *Enzymologia* **34**, 45. **30.** Tulsiani, Skudlarek, Nagdas & Orgebin-Crist (1993) *Biochem. J.* **290**, 427. **31.** Plant, Parratt, Daniel & Morgan (1988) *Biochem. J.* **255**, 865. **32.** Ohtsuru, Tsuruo & Hata (1969) *Agric. Biol. Chem.* **33**, 1315. **33.** Tani, Ohtsuru & Hata (1974) *Agric. Biol. Chem.* **38**, 1623. **34.** Lúcio-Eterovic, Jorge, Polizeli & Terenzi (2005) *Biochim. Biophys. Acta* **1723**, 201. **35.** Silva, Terra & Ferreira (2004) *Insect Biochem. Mol. Biol.* **34**, 1089. **36.** Allen, Desai & Neuberger (1978) *Meth. Enzymol.* **50**, 335. **37.** García-Alles, Zahn & Erni (2002) *Biochemistry* **41**, 10077. **38.** Schmidt & John (1979) *Biochim. Biophys. Acta* **566**, 100. **39.** Bovetto, Villette, Fontaine, Sicard & Bouquelet (1992) *Biotechnol. Appl. Biochem.* **15**, 59. **40.** Kobayashi, Yokoyama & Matsuda (1986) *Agric. Biol. Chem.* **50**, 2585. **41.** Kobayashi, Mihara & Matsuda (1986) *Agric. Biol. Chem.* **50**, 551. **42.** Tanabe, Kobayashi & Matsuda (1988) *Agric. Biol. Chem.* **52**, 757.

Methyl β-D-Glucopyranoside

This synthetic nonreducing sugar (FW = 194.18 g/mol) is an alternative substrate or competitive inhibitor for many β-glucosidases. **Target(s):** aldose 1-epimerase (1); aryl-β-hexosidase (2,3); concanavalin A, poor inhibitor (4); cyclomaltodextrin glucanotransferase (5); β-fructofuranosidase, *or* invertase (6,7); 4-α-glucanotransferase (8); glucosylceramidase (9); oligo-1,6-glucosidase (10); thioglucosidase, *or* myrosinase (11,12); α,α-trehalase (13); β-xylosidase/β-glucosidase (14) **1.** Bentley (1962) *Meth. Enzymol.* **5**, 219. **2.** Distler & Jourdian (1978) *Meth. Enzymol.* **50**, 524. **3.** Distler & Jourdian (1977) *Arch. Biochem. Biophys.* **178**, 631. **4.** Goldstein, Hollerman & Smith (1965) *Biochemistry* **4**, 876. **5.** Bovetto, Villette, Fontaine, Sicard & Bouquelet (1992) *Biotechnol. Appl. Biochem.* **15**, 59. **6.** Neuberg & Mandl (1950) *The Enzymes*, 1st ed., **1** (part 1), 527. **7.** Webb (1966) *Enzyme and Metabolic Inhibitors*, vol. 2, pp. 415-421, Academic Press, New York. **8.** Wiesmeyer (1962) *Meth. Enzymol.* **5**, 141. **9.** Osiecki-Newman, Fabbro, Legler, Desnick & Grabowski (1987) *Biochim. Biophys. Acta* **915**, 87. **10.** Plant, Parratt, Daniel & Morgan (1988) *Biochem. J.* **255**, 865. **11.** Ohtsuru, Tsuruo & Hata (1969) *Agric. Biol. Chem.* **33**, 1315. **12.** Tani, Ohtsuru & Hata (1974) *Agric. Biol. Chem.* **38**, 1623. **13.** Nakano & Sacktor (1985) *J. Biochem.* **97**, 1329. **14.** Yasui & Matsuo (1988) *Meth. Enzymol.* **160**, 696.

γ-Methylglutamate (γ-Methylglutamic Acid)

This methylated amino acid (FW = 161.16 g/mol) inhibits γ-glutamylcysteine synthetase. It is an alternative substrate of glutamate dehydrogenase. The (2S,4R)-stereoisomer is a selective and potent kainate receptor antagonist. **Target(s):** [glutamine-synthetase] adenylyltransferase (1); γ-glutamylcysteine synthetase (2-4); glutamyl-tRNA synthetase, inhibited by the D-*threo*-isomer (5); kainate receptor (6); UDP-*N*-acetylmuramoylalanine:D-glutamate ligase, inhibited by the D-*erythro*-isomer, also an alternative substrate (7). **1.** Ebner, Wolf, Gancedo, Elsässer & Holzer (1970) *Eur. J. Biochem.* **14**, 535. **2.** Griffith & Meister (1977) *Proc. Natl. Acad. Sci. U.S.A.* **74**, 3330. **3.** Seelig & Meister (1985) *Meth. Enzymol.* **113**, 379. **4.** Sekura & Meister (1977) *J. Biol. Chem.* **252**, 2599. **5.** Lea & Fowden (1972) *Phytochemistry* **11**, 2129. **6.** Zhou, Gu, Costa, *et al.* (1997) *J. Pharmacol. Exp. Ther.* **280**, 422. **7.** Pratviel-Sosa, Acher, Trigalo, *et al.* (1994) *FEMS Microbiol. Lett.* **115**, 223.

(±)-*threo*-3-Methylglutamic acid

1:1 racemic mixture

This potent neurotransmitter EAAT blocker (FW = 161.16 g/mol; CAS 63088-04-0; Soluble to 100 mM, when converted to the sodium salt by addition of 1 eq. NaOH) targets the Excitatory Amino-Acid Transporters EAAT$_2$ (IC$_{50}$ = 90 µM) and EAAT$_4$ (IC$_{50}$ = 109 µM), with much weaker action against EAAT$_1$ (IC$_{50}$ = 1600 µM) and EAAT$_3$ (IC$_{50}$ = 1080 µM). As the principal excitatory neurotransmitter in the vertebrate brain, glutamate is released by a change in the action potential, whereupon EAAT transporters quickly remove glutamate from the extracellular space (synaptic groove) to terminate synaptic transmission. **1.** Vandenberg, *et al.* (1997) *Mol. Pharmacol.* **51**, 809. **2.** Mitrovic, *et al.* (1998) *J. Biol. Chem.* **273**, 14698. **3.** Eliasof, *et al.* (2001) *J. Neurochem.* **77**, 550.

S-Methylglutathione

This *S*-substituted glutathione (FW = 321.35 g/mol) inhibits γ-glutamylcysteine synthetase (1), glutathione *S*-transferase (2,3), lactoylglutathione lyase, *or* glyoxalase I (4-9), maleylacetoacetate isomerase, *or* maleylacetone isomerase (10,11), and thiopurine *S*-methyltransferase (12). **1.** Gander, Sethna & Rathbun (1983) *Eur. J. Biochem.* **133**, 635. **2.** Park, Cho & Kong (2005) *J. Biochem. Mol. Biol.* **38**, 232. **3.** Vander Jagt, Wilson & Heidrich (1981) *FEBS Lett.* **136**, 319. **4.** Kermack & Matheson (1957) *Biochem. J.* **65**, 45, 48. **5.** Vince, Daluge & Wadd (1971) *J. Med. Chem.* **14**, 402. **6.** Vince & Wadd (1969) *Biochem. Biophys. Res. Commun.* **35**, 593. **7.** Allen, C. Lo & Thornalley (1993) *Biochem. Soc. Trans.* **21**, 535. **8.** Regoli, Saccucci & Principato (1996) *Comp. Biochem. Physiol. C, Pharmacol. Toxicol. Endocrin.* **113**, 313. **9.** Iozef, Rahlfs, Chang, Schirmer & Becker (2003) *FEBS Lett.* **554**, 284. **10.** Morrison, Wong & Seltzer (1976) *Biochemistry* **15**, 4228. **11.** Seltzer (1989) in *Coenzymes and Cofactors, Glutathione, Chem. Biochem. Med. Aspects Pt. A* (Dolphin, Poulson & Avromonic, eds.), vol. **3**, p. 733, Wiley, New York. **12.** Loo & Smith (1985) *Biochem. Biophys. Res. Commun.* **126**, 1201.

Methylglyoxal

This simple α-keto aldehyde (FW = 72.06 g/mol), also known as 2-oxopropanal, pyruvaldehyde, and acetylformaldehyde, is a hygroscopic yellow liquid (boiling point = 72°C) that readily polymerizes to a glassy mass. This polymeric substance is easily dissolved into water and regenerates the monomer. Note that methylglyoxal will react with guanidinium groups and amines. Elevated levels of methylglyoxal can initiate protein cross linkage, exacerbate advanced glycation of proteins, and cause endothelial injury. It will also cross link DNA polymerase with the substrate DNA. Methylglyoxal is a product of the reactions catalyzed by D-lactaldehyde dehydrogenase, methylglyoxal synthase, and lactoylglutathione lyase (glyoxalase I) and is a substrate of aldose reductase. It is an intermediate in the metabolism of glyceraldehyde 3-phosphate in species of *Pseudomonas*. **Target(s):** aldehyde dehydrogenase, human E$_3$ isozyme (5); aminolevulinate aminotransferase (1-4,27-30); bacterial leucyl aminopeptidase (25); cathepsin (6); DNA polymerase (7); electron transport (8); formaldehyde dehydrogenase (glutathione) (9); fructose-1,6-bisphosphate aldolase (10,11); glutamine:fructose-6 phosphate aminotransferase (isomerizing) (12,31); glutamine synthetase (13); glutathione-disulfide reductase (14,15); glutathione peroxidase (16); glyceraldehyde-3-phosphate dehydrogenase (10,14,17,18); histidine ammonia-lyase (19); hydroxyacylglutathione hydrolase, *or* glyoxalase II (26); *S*-(hydroxymethyl)glutathione dehydrogenase (9); lactate dehydrogenase (14); malate synthase (32); malic enzyme, *or* malate dehydrogenase, *or* oxaloacetate decarboxylating (20); NADH dehydrogenase, *or* Complex I (8); papain (6); phosphofructokinase (10); 3-phosphoglycerate mutase (10); *Streptomyces* R61 DD-carboxypeptidase (21); succinate dehydrogenase (22); superoxide dismutase, Cu,Zn (23); L-threonine dehydrogenase (24). **1.** Sagar, Salotra, Bhatnagar & Datta (1995) *Microbiol. Res.* **150**, 419. **2.** McKinney & Ades (1991) *Int. J. Biochem.* **23**, 803. **3.** Rhee, Murata & Kimura (1988) *J. Biochem.* **103**, 1045. **4.** Shanker & Datta (1986) *Arch. Biochem. Biophys.* **248**, 652. **5.** Izaguirre, Kikonyogo & Pietruszko (1998) *Comp. Biochem. Physiol. B Biochem. Mol. Biol.* **119**, 747. **6.** Purr (1935) *Biochem. J.* **29**, 5. **7.** Murata-Kamiya & Kamiya (2001) *Nucl. Acids Res.* **29**, 3433. **8.** Ray, Dutta, Halder & Ray (1994) *Biochem. J.* **303**, 69. **9.** Uotila & Koivusalo (1974) *J. Biol. Chem.* **249**, 7653. **10.** Leoncini, Maresca & Buzzi (1989) *Cell. Biochem. Funct.* **7**, 65. **11.**

Leoncini, Ronchi, Maresca & Bonsignore (1980) *Ital. J. Biochem.* **29**, 289. **12**. Kikuchi, Ikeda & Tsuiki (1972) *Biochim. Biophys. Acta* **289**, 303. **13**. Colanduoni & Villafranca (1985) *Biochem. Biophys. Res. Commun.* **126**, 412. **14**. Morgan, Dean & Davies (2002) *Arch. Biochem. Biophys.* **403**, 259. **15**. Vander Jagt, Hunsaker, Vander Jagt, *et al.* (1997) *Biochem. Pharmacol.* **53**, 1133. **16**. Park, Koh, Takahashi, *et al.* (2003) *Free Radic. Res.* **37**, 205. **17**. Ray, Basu & Ray (1997) *Mol. Cell. Biochem.* **177**, 21. **18**. Leoncini, Maresca, Ronchi & Bonsignore (1981) *Experientia* **37**, 443. **19**. White & Kendrick (1993) *Biochim. Biophys. Acta* **1163**, 273. **20**. Chang & Huang (1981) *Biochim. Biophys. Acta* **660**, 341. **21**. Georgopapadakou, Liu, Ryono, Neubeck & Ondetti (1981) *Eur. J. Biochem.* **115**, 53. **22**. Kun (1950) *J. Biol. Chem.* **187**, 289. **23**. Kang (2003) *Mol. Cells* **15**, 194. **24**. Ray & Ray (1985) *J. Biol. Chem.* **260**, 5913. **25**. Mäkinen, Mäkinen, Wilkes, Bayliss & Prescott (1982) *J. Biol. Chem.* **257**, 1765. **26**. Oray & Norton (1980) *Biochim. Biophys. Acta* **611**, 168. **27**. Varticovski, Kushner & Burnham (1980) *J. Biol. Chem.* **255**, 3742. **28**. Bajkowski & Friedmann (1982) *J. Biol. Chem.* **257**, 2207. **29**. Sagar, Salotra, Bhatnagar & Datta (1995) *Microbiol. Res.* **150**, 419. **30**. Rhee, Murata & Kimura (1988) *J. Biochem.* **103**, 1045. **31**. Kikuchi & Tsuiki (1976) *Biochim. Biophys. Acta* **422**, 231. **32**. Miernyk & Trelease (1981) *Phytochemistry* **20**, 2657.

Methylglyoxal Bis(guanylhydrazone)

This di-cation (FW = 172.20 g/mol; *Abbreviation*: MGBG) potently inhibits *S*-adenosylmethionine decarboxylase, IC_{50} = 1.4 μM, and diamine oxidase, IC_{50} = 1.9 μM. (**See also** *1,1'-[(Methylethanediylidene)dinitrilo]bis(3-aminoguanidine)*) **Target(s):** *S*-adenosylmethionine decarboxylase (1-18); arginine decarboxylase (19); diamine oxidase (6,17). **1**. Rillema, Collins & Williams (2000) *Proc. Soc. Exp. Biol. Med.* **224**, 41. **2**. Markham, Tabor & Tabor (1983) *Meth. Enzymol.* **94**, 228. **3**. Cohn, Tabor & Tabor (1983) *Meth. Enzymol.* **94**, 231. **4**. Pegg & Pösö (1983) *Meth. Enzymol.* **94**, 234. **5**. Pegg (1983) *Meth. Enzymol.* **94**, 239. **6**. Seppänen, Alhonen-Hongisto, Käpyaho & Jänne (1983) *Meth. Enzymol.* **94**, 247. **7**. Pösö & Pegg (1982) *Biochemistry* **21**, 3116. **8**. Lu & Markham (2004) *J. Biol. Chem.* **279**, 265. **9**. Pegg (1973) *Biochem. J.* **132**, 537. **10**. Yamanoha & Cohen (1985) *Plant Physiol.* **78**, 784. **11**. Tabor & Tabor (1984) *Adv. Enzymol. Relat. Areas Mol. Biol.* **56**, 251. **12**. Pösö, Sinervirta & Jänne (1975) *Biochem. J.* **151**, 67. **13**. Ferioli, Candiani, Rocca & Scalabrino (1989) *Biogenic Amines* **6**, 513. **14**. Suzuki & Hirasawa (1980) *Plant Physiol.* **66**, 1091. **15**. Svensson, Mett & Persson (1997) *Biochem. J.* **322**, 297. **16**. Da'dara, Mett & Walter (1998) *Mol. Biochem. Parasitol.* **97**, 13. **17**. Stanek, Caravatti, Frei, *et al.* (1993) *J. Med. Chem.* **36**, 2168. **18**. Manen & Russell (1974) *Biochemistry* **13**, 4729. **19**. Choudhuri & Ghosh (1982) *Agric. Biol. Chem.* **46**, 739.

1-Methylguanine

This methylated purine (FW = 165.15 g/mol) is a component in a number of transferRNAs and ribosomalRNAs and is formed in reactions catalyzed by tRNA (guanine-N^1-)-methyltransferase and rRNA (guanine-N^1-)-methyltransferase (1). Methylguanine has λ_{max} values at 248 and 272 nm at pH 7, with respective ε values of 10000 and 7900 $M^{-1}cm^{-1}$. **Target(s):** guanosine phosphorylase, *or* purine-nucleoside phosphorylase (1); hypoxanthine(guanine) phosphoribosyltransferase (2,3); purine-nucleoside phosphorylase (4); xanthine phosphoribosyltransferase (2). **1**. Baker & Schaeffer (1971) *J. Med. Chem.* **14**, 809. **2**. Naguib, Iltzsch, el Kouni, Panzica & el Kouni (1995) *Biochem. Pharmacol.* **50**, 1685. **3**. Miller, Ramsey, Krenitsky & Elion (1972) *Biochemistry* **11**, 4723. **4**. Bzowska, Kulikowska, Darzynkiewicz & Shugar (1988) *J. Biol. Chem.* **263**, 9212.

O^6-Methylguanine

This modified guanine (FW = 165.15 g/mol), also known as 2-amino-6-methoxypurine, base pairs in DNA with thymine and is highly mutagenic. This residue in DNA is a substrate of the reaction catalyzed by methylated-DNA:[protein]-cysteine *S*-methyltransferase. **Target(s):** methionine *S*-adenosyltransferase (1); methylated-DNA:[protein]-cysteine *S* methyltransferase (O^6-alkylguanine-DNA alkyltransferase (2-5). **1**. Berger & Knodel (2003) *BMC Microbiol.* **3**, 12. **2**. Dexter, Yamashita, Donovan & Gerson (1989) *Cancer Res.* **49**, 3520. **3**. McMurry (2007) *DNA Repair* **6**, 1161. **4**. Terashima, Kawata, Sakumi, Sekiguchi & Kohda (1997) *Chem. Res. Toxicol.* **10**, 1234. **5**. Shibata, Glynn, McMurry, *et al.* (2006) *Nucl. Acids Res.* **34**, 1884.

7-Methylguanine

This methylated guanine (FW = 165.15 g/mol) is a component in eukaryotic mRNA and some tRNA molecules. It is produced in tRNA by the action of tRNA (guanine-N^7-)-methyltransferase and in mRNA by mRNA(guanine-N^7-)-methyltransferase. 7-Methylguanine has λ_{max} values at 248 and 283 nm at pH 7, with respective ε values of 5700 and 7400 $M^{-1}cm^{-1}$. **Target(s):** hypoxanthine(guanine) phosphoribosyltransferase, weakly (1); purine-nucleoside phosphorylase (2); tRNA-guanine transglycosylase (3-5); xanthine phosphoribosyltransferase, weakly (1). **1**. Naguib, Iltzsch, el Kouni, Panzica & el Kouni (1995) *Biochem. Pharmacol.* **50**, 1685. **2**. Wielgus-Kutrowska & Bzowska (2006) *Biochim. Biophys. Acta* **1764**, 887. **3**. Goodenough-Lashua & Garcia (2003) *Bioorg. Chem.* **31**, 331. **4**. Okada & Nishimura (1979) *J. Biol. Chem.* **254**, 3061. **5**. Farkas, Jacobson & Katze (1984) *Biochim. Biophys. Acta* **781**, 64.

7-Methylguanosine

This modified nucleoside (FW = 298.28 g/mol) is a component of the cap on the 5'-terminus of eukaryotic mRNA. It is also an inhibitor of pteridine biosynthesis and is a component in some tRNA molecules. It is produced in tRNA by the action of tRNA (guanine-N^7-)-methyltransferase and in mRNA by mRNA (guanine-N7-)-methyltransferase. 7-Methylguanosine has λ_{max} values at 258 and 281 nm at pH 7 (ε = 8500 and 7400 $M^{-1}cm^{-1}$, respectively). **Target(s):** $m^7G(5')$pppN diphosphatase, *or* 7-methylguanosine-5'-triphospho-5' polynucleotide 7-methylguanosine-5'-phosphohydrolase; decapase (1-3). **1**. Nuss & Furuichi (1977) *J. Biol. Chem.* **252**, 2815. **2**. Kumagi, R. Kon, T. Hoshino, T. Aramaki, Nishikawa, Hirose & Igarashi (1992) *Biochim. Biophys. Acta* **1119**, 45. **3**. Nuss, Altschuler & Peterson (1982) *J. Biol. Chem.* **257**, 6224.

7-Methylguanosine 5'-Monophosphate

This analogue (FW = 377.25 g/mol; CAS 10162-58-0) of the nucleotide cap on eukaryotic mRNAs and is an inhibitor of mRNA translation. 7-Methylguanosine 5'-monophosphate has λ_{max} values at 258 and 280 nm at pH 7.4 (ε = 10300 and 8600 $M^{-1}cm^{-1}$, respectively) and is a product of the reaction catalyzed by m7G(5')pppN diphosphatase. **Target(s):** $m^7G(5')pppN$ diphosphatase (7-methylguanosine-5'-triphospho-5' polynucleotide 7-methylguanosine-5'-phosphohydrolase; decapase; product inhibition (1,2); mRNA translation (3-12); protein biosynthesis (3-12). **1.** Nuss & Furuichi (1977) J. Biol. Chem. **252**, 2815. **2.** Kumagi, Kon, Hoshino, et al. (1992) Biochim. Biophys. Acta **1119**, 45. **3.** Seal, Schmidt, Tomaszewski & Marcus (1978) Biochem. Biophys. Res. Commun. **82**, 553. **4.** Weber, Hickey & Baglioni (1978) J. Biol. Chem. **253**, 178. **5.** Kemper & Stolarsky (1977) Biochemistry **16**, 5676. **6.** Wu, Cheung & Suhadolnik (1977) Biochem. Biophys. Res. Commun. **78**, 1079. **7.** Groner, Grosfeld & Littauer (1976) Eur. J. Biochem. **71**, 281. **8.** Sharma, Hruby & Beezley (1976) Biochem. Biophys. Res. Commun. **72**, 1392. **9.** Weber, Feman, Hickey, Williams & Baglioni (1976) J. Biol. Chem. **251**, 5657. **10.** Hickey, Weber & Baglioni (1976) Nature **261**, 71. **11.** Roman, Brooker, Seal & Marcus (1976) Nature **260**, 359. **12.** Hickey, Weber & Baglioni (1976) Proc. Natl. Acad. Sci. U.S.A. **173**, 19.

7-Methylguanosine 5'-Triphosphate

This analogue (FW = 526.21 g/mol; CAS 57718-00-0) of the nucleotide cap on eukaryotic mRNAs and is produced by the action of mRNA (guanine-N^7-)-methyltransferase. 7-Methylguanosine 5'-triphosphate has λ_{max} values at 258 and 280 nm at pH 7.4 (ε = 10300 and 8600 $M^{-1}cm^{-1}$, respectively). **Target(s):** $m^7G(5')pppN$ diphosphatase (7-methylguanosine-5'-triphospho-5'-polynucleotide 7-methylguanosine-5'-phosphohydrolase; decapase) (1,2); NMN nucleosidase (3); polypeptide biosynthesis (4-6). **1.** Kumagi, Kon, Hoshino, et al. (1992) Biochim. Biophys. Acta **1119**, 45. **2.** Nuss, Altschuler & Peterson (1982) J. Biol. Chem. **257**, 6224. **3.** Imai (1987) J. Biochem. **101**, 163. **4.** Lax, Fritz, Browning & Ravel (1985) Proc. Natl. Acad. Sci. U.S.A. **82**, 330. **5.** Chu & Rhoads (1980) Biochemistry **19**, 184. **6.** Hickey, Weber, Baglioni, Kim & Sarma (1977) J. Mol. Biol. **109**, 173.

N-Methylhistamine

This methylated histamine ($FW_{free-base}$ = 125.17 g/mol) inhibits tryptase non-linearly. Note that the name N-methylhistamine has also been used to refer to N^τ-methylhistamine, produced by histamine N-methyltransferase, and N^α-methylhistamine. **Target(s):** histamine N-methyltransferase, weakly inhibited (1); histamine uptake (2); tryptase (3). **1.** Barth, Lorenz & Niemeyer (1973) Hoppe-Seyler's Z. Physiol. Chem. **354**, 1021. **2.** Schayer & Reilly (1975) Agents Actions **5**, 231. **3.** Alter & Schwartz (1989) Biochim. Biophys. Acta **991**, 426.

Methylhydrazine

This versatile nucleophilic reagent and chemical building block (FW = 46.072 g/mol; CAS 60-34-4; M.P. = −52.4°C; B.P.= 87.5°C; clear, colorless, hygroscopic; miscible with water and ethanol), reacts with aldehydes and ketones in sugars, ketoacids, and numerous metabolites. It also reacts with the pyridoxal cofactor in vitamin B_6-dependent enzymes as well as with the pyrroloquinoline quinone (PQQ) cofactor in glucose oxidase. Methylhydrazine is also a DNA-methylating agent as a consequence of peroxidase-catalyzed oxidation, the latter generating methyl radicals. Inhalation of N-methylhydrazine induces both benign and malignant tumors in mice and hamsters. Methylhydrazine is also a skin and nasal irritant. **Target(s):** amine oxidase (1-3); 4-aminobutyrate aminotransferase (4); carbamoyl-phosphate synthetase (5); diamine oxidase (6); dopamine β-monooxygenase (7); formate dehydrogenase (8); glucose dehydrogenase (9); hydroxylamine oxidoreductase, or hydroxylamine oxidase (10); manganese peroxidase (11); methanol dehydrogenase (12); methylamine dehydrogenase (13); methylamine oxidase (14,15); methylaspartate ammonia-lyase (16); nitrite reductase, or cytochrome cd_1 (17); ornithine decarboxylase (18); peroxidase, horseradish (19); pyridoxal kinase (20). **1.** Lee, Jeon, Huang & Sayre (2001) J. Org. Chem. **66**, 1925. **2.** Holt & Callingham (1995) J. Pharm. Pharmacol. **47**, 837. **3.** Holt, Sharman & Callingham (1992) J. Pharm. Pharmacol. **44**, 494. **4.** Lightcap & Silverman (1996) J. Med. Chem. **39**, 686. **5.** Kaseman & Meister (1985) Meth. Enzymol. **113**, 305. **6.** Bieganski, Osinska & Maslinski (1982) Int. J. Biochem. **14**, 949. **7.** Fitzpatrick & Villafranca (1986) J. Biol. Chem. **261**, 4510. **8.** Kanamori & Suzuki (1968) Enzymologia **35**, 185. **9.** Oubrie, Rozeboom & Dijkstra (1999) Proc. Natl. Acad. Sci. U.S.A. **96**, 11787. **10.** Logan & Hooper (1995) Biochemistry **34**, 9257. **11.** Harris, Wariishi, Gold & Ortiz de Montellano (1991) J. Biol. Chem. **266**, 8751. **12.** Frank & Duine (1990) Meth. Enzymol. **188**, 202. **13.** Huizinga, van Zanten, Duine, et al. (1992) Biochemistry **31**, 9789. **14.** McIntire (1990) Meth. Enzymol. **188**, 227. **16.** van Iersel, van der Meer & Duine (1986) Eur. J. Biochem. **161**, 415. **16.** Pollard, Richardson, Akhtar, et al. (1999) Bioorg. Med. Chem. **7**, 949. **17.** Yap-Bondoc & Timkovich (1990) J. Biol. Chem. **265**, 4247. **18.** Klosterman (1979) Meth. Enzymol. **62**, 483. **20.** Ator, David & Ortiz de Montellano (1987) J. Biol. Chem. **262**, 14954.

(18E)- and (18Z)-29-Methylidene-2,3-oxidohexanorsqualene

This epoxysqualene analogue (FW = 356.59 g/mol), also known as (6E,10E,14E,18E)-2,3-epoxy-2,6,10,15-tetramethyl-6,10,14,18,20-heneicosapentaene, is a irreversible inhibitor of squalene:hopene cyclase, IC_{50} = 0.2 μM) (1) and lanosterol synthase, IC_{50} = 1.5 μM for Saccharomyces cerevisiae enzyme and 3.5 μM for the pig liver protein (1-4). The 18Z-isomer inhibits S. cerevisiae lanosterol synthase with IC_{50} = 15 μM. **1.** Rocco, Bosso, Viola, et al. (2003) Lipids **38**, 201. **2.** Viola, Balliano, Milla, et al. (2000) Bioorg. Med. Chem. **8**, 223. **3.** Ceruti, Rocco, Viola, et al. (1998) J. Med. Chem. **41**, 540. **4.** Ceruti, Viola, Balliano, et al. (2002) J. Chem. Soc. Perkin Trans. **1**, 1477.

1-Methylimidazole

This hygroscopic imidazole ($FW_{free-base}$ = 82.11 g/mol), also known as N-methylimidazole, forms complexes with a number of metal ions, especially Co^{2+} and Zn^{2+}. At room temperature, the free base is a liquid (M.P. = −6°C; B.P. = 198°C). **Target(s):** CYP2E1, weakly inhibited (1); estrogen 2-monooxygenase (2); glutaminyl-peptide cyclotransferase, or glutaminyl cyclase, K_i = 30 μM (3-8); histidine decarboxylase (9; thromboxane-A synthase (10). **1.** Hargreaves, Jones, Smith & Gescher (1994) *Drug Metab. Dispos.* **22**, 806. **2.** Purba, Back & Breckenridge (1986) *J. Steroid Biochem.* **24**, 1091. **3.** Schilling, Niestroj, Rahfeld, *et al.* (2003) *J. Biol. Chem.* **278**, 49773. **4.** Cynis, Rahfeld, Stephan, *et al.* (2008) *J. Mol. Biol.* **379**, 966. **5.** Schilling, Cynis, von Bohlen, *et al.* (2005) *Biochemistry* **44**, 13415. **6.** Buchholz, Heiser, Schilling, *et al* (2006) *J. Med. Chem.* **49**, 664. **7.** Schilling, Lindner, Koch, *et al.* (2007) *Biochemistry* **46**, 10921. **8.** Schilling, Appl, Hoffmann, *et al.* (2008) *J. Neurochem.* **106**, 1225. **9.** Snell (1986) *Meth. Enzymol.* **122**, 128. **10.** Tai & Yuan (1978) *Biochem. Biophys. Res. Commun.* **80**, 236.

5'-O-Methylimmucillin-A

This nucleoside analogue (FW = 277.28 g/mol), also known as (1*S*)-1-(9-deazaadenin-9-yl)-5-(methoxy)-1,4-dideoxy-1,4-imino-D-ribitol, inhibits 5'-methylthioadenosine/*S*-adenosylhomocysteine nucleosidase, K_i^* = 10 nM (1). The immucillins are transition-state mimics that are rationally designed to maximize binding affinity through their ability to closely replicate key stereoelectronic features of oxa-carbenium ion intermediates formed in ribosyltransfer reactions. (**See** *Immucillin for discussion of Primary Mode of Action*) 5'-*O*-Methyl-immucillin-A is also a slow, tight-binding inhibitor of human *S*-methyl-5' thioadenosine phosphorylase, K_i = 134 nM (2). **1.** Singh, Evans, Lenz, *et al.* (2005) *J. Biol. Chem.* **280**, 18265. **2.** Evans, Furneaux, Schramm, Singh & Tyler (2004) *J. Med. Chem.* **47**, 3275.

N-Methyl-4-isoleucine Cyclosporin

This substituted Cyclosporin A analogue (FW = 1315.88 g/mol), also known as NIM811, has the ability to bind to cyclophilin; however, the resulting Cyclophilin·NIM811 binary complex cannot bind calcineurin, and therefore lacks immunosuppressive activity of Cyclosporin A (CsA). NIM811 also blocks the mitochondrial permeability transition (MPT) that is induced by Ca^{2+} and P_i, alone or in combination with 1-methyl-4-phenyl-1,2,3,6-tetrahydro-pyridine (a dopaminergic neurotoxin), rotenone (a ETS Complex I inhibitor), and *t*-butylhydroperoxide (a pro-oxidant) (1). NIM811 was equipotent to CsA. NIM811 also blocks cell killing and prevents in situ mitochondrial inner membrane permeabilization and depolarization during tumor necrosis factor-alpha-induced apoptosis to cultured rat hepatocytes (1). NIM811 inhibition of apoptosis was equipotent with CsA except at higher concentrations: CsA lost efficacy but NIM 811 did not. We conclude that NIM811 is a useful alternative to PKF220-384 to investigate the role of the mitochondrial permeability transition in apoptotic and necrotic cell death (1). (**See** *Malonate*) Liquid chromatographic tandem mass spectrometric detection (LC-MS/MS) permits determination of NIM811 over the concentration range 1-2500 ng/mL in human whole blood, using a sample volume of 0.05 mL (2). **1.** Waldmeier, Feldtrauer, Qian & Lemasters (2002) *Mol. Pharmacol.* **62**, 22. **2.** Li, Luo, Hayes, He & Tse (2007) *Biomed. Chromatogr.* **21**, 249.

S-Methyl-L-isothiocitrulline

This citrulline derivative (FW = 205.28 g/mol), also known as *S*-methyl-L-thiocitrulline, is a potent inhibitor of nitric oxide synthase: K_i = 1.2 nM for human neuronal NOS; K_i = 11 nM for human endothelium NOS, and K_i = 40 nM for human inducible NOS. **Target(s):** arginase (1); nitric-oxide synthase (2-9). **1.** Colleluori & Ash (2001) *Biochemistry* **40**, 9356. **2.** Garvey, Furfine & Sherman (1996) *Meth. Enzymol.* **268**, 339. **3.** Ichimori, Stuehr, Atkinson & King (1999) *J. Med. Chem.* **42**, 1842. **4.** Narayanan & Griffith (1994) *J. Med. Chem.* **37**, 885. **5.** Furfine, Harmon, Paith, *et al.* (1994) *J. Biol. Chem.* **269**, 26677. **6.** Griffith & Kilbourn (1996) *Meth. Enzymol.* **268**, 375. **7.** Griffith & Stuehr (1995) *Ann. Rev. Physiol.* **57**, 707. **8.** Komers, Oyama, Chapman, Allison & Anderson (2000) *Hypertension* **35**, 655. **9.** Narayanan, Spack, McMillan, *et al.* (1995) *J. Biol. Chem.* **270**, 11103.

Methyllycaconitine

This stereochemically complex diterpenoid and insecticide (FW = 682.81 g/mol; CAS 21019-30-7; Soluble in $CHCl_3$, but not H_2O; often supplied commercially as the citrate salt; Symbol: MLA) is the principal toxin of the larkspur (*Delphinium brownie*). MLA exhibits tubocurarine-like action by blocking neuromuscular transmission in skeletal muscle, but not smooth muscle (1-4). MLA inhibits (K_i ~ 2.5 x 10^{-10} M) the binding of propionyl-α-bungarotoxin to a house-fly receptor preparation (1), competes with α-bungarotoxin (K_i ~ 1 x 10^{-9} M) and (–)-nicotine (K_i ~ 4 x 10^{-6} M) in a rat brain receptor preparation (2), and displaces α-bungarotoxin (K_i ~ 1 x 10^{-6} M) from purified Torpedo nicotinic acetylcholine receptors (2). Methyllycaconitine also attenuates methamphetimine-induced neurotoxicity in the striatum of mouse brain. **Cattle Poisoning:** This alkaloid is one of the main causes of livestock poisoning, killing some 5–15% of the cattle on North American mountain range lands. When treated with sublethal doses, test animals were reluctant to move, trembled, and developed dyspnea, muscular twitches, and convulsions (5). Within several minutes, the clinical signs abated, slowly returning to normal over ~20 min. Methyllycaconitine elimination follows a normal biphasic redistribution, with $t_{1/2}$ = 17.5 min (5). **1.** Jennings, Brown & Wright (1986). *Experientia* **42**, 611. **2.** Macallan, Lunt, Wonnacott, *et al.* (1988) *FEBS Lett.* **226**, 357. **3.** Alkondon, Pereira, Wonnacott & Albuquerque (1992) *Mol. Pharmacol.* **41**, 802. **4.** Drasdo, Caulfield, Bertrand, Bertrand & Wonnacott (1992) *Mol. Cell. Neurosci.* **3**, 237. **5.** Stegelmeier, Hall, Gardner & Panter (2003) *J. Anim. Sci.* **81**, 1237.

N-Methylmaleimide

This alkylating agent (FW = 111.10 g/mol; M.P. = 95–97°C; *Abbreviation*: NMM), also known as 1-methyl-1*H*-pyrrole-2,5-dione, reacts readily and irreversibly with sulfhydryl groups in proteins. Under alkaline conditions NMM also reacts with amino groups. Reaction with of R–SH or R–NH$_2$ occurs by addition across the double bond of NMM, yielding the corresponding thioether or secondary amine. If the modified aminoacyl residue is catalytically essential and/or is at the active site, the protein will undergo inactivation. Some enzymes are inactivated at faster rates with NMM analogues having longer alkyl chains (*e.g.*, the inactivation rate of yeast alcohol dehydrogenase with *N*-octylmaleimide is significantly faster than with *N*-ethylmaleimide (1)). (**See also** *N-Ethylmaleimide*) **Target(s):** acetyl-CoA *C*-acyltransferase (2); caffeine synthase (3); desulfo-glucosinolate sulfotransferase (4); loganate *O*-methyltransferase (5); long-chain-enoyl-CoA hydratase (6); *N*-methyl-2 oxoglutaramate hydrolase (7); mitochondrial m-AAA protease (8); 4-nitrophenol 2-monooxygenase (9); progesterone 11α-monooxygenase (10); theobromine synthase (3). **1.** Heitz, Anderson & Anderson (1968) *Arch. Biochem. Biophys.* **127**, 627. **2.** Schulz & Staack (1981) *Meth. Enzymol.* **71**, 398. **3.** Roberts & Waller (1979) *Phytochemistry* **18**, 451. **4.** Jain, Groot Wassink, Kolenovsky & Underhill (1990) *Phytochemistry* **29**, 1425. **5.** Madyastha, Guarnaccia, Baxter & Coscia (1973) *J. Biol. Chem.* **248**, 2497. **6.** Schulz (1974) *J. Biol. Chem.* **249**, 2704. **7.** Hersh (1970) *J. Biol. Chem.* **245**, 3526. **8.** Hanekamp & Thorsness (1998) in *Handb. Proteolytic Enzymes*, p. 1504, Academic Press, San Diego. **9.** Mitra & Vaidyanathan (1984) *Biochem. Int.* **8**, 609. **10.** Jayanthi, Madyastha & Madyastha (1982) *Biochem. Biophys. Res. Commun.* **106**, 1262.

Methyl malonate (Methyl Malonic Acid)

This cell-permeable pro-inhibitor (FW = 118.09 g/mol), also named propanedioic acid methyl ester and hydrogen methyl malonate, is metabolized to malonic acid, a well-known inhibitor of succinate dehydrogenase (**See** *Malonate*). Upon entry into cells, methyl malonate is hydrolyzed to malonate. Exposure of rat embryonic brain striatal and cortical neurons for 24 hours to 10 mM methylmalonate causes DNA fragmentation and reduced [ATP]/[ADP] within 3 hours, culminating in necrotic and apoptotic cell death. Cell death is attenuated in a medium containing antioxidants. (*Note:* The term methyl malonate is occasionally used for dimethylmalonate, *or* H$_3$CO–C(=O)–CH$_2$–C(=O)–OCH$_3$, whereas the single-word term methylmalonate, *or* $^-$OOC–HC(CH$_3$)–COO$^-$ (**See below**) is used to designate 2-methylpropanedioic acid.) **1.** McLaughlin, Nelson, Silver, Erecinska & Chesselet (1998) *Neuroscience* **86**, 279.

Methylmalonate (Methylmalonic Acid)

This toxic dicarboxylic acid (FW$_{free-acid}$ = 118.09 g/mol), also known as 2-methylpropanedioic acid, is very soluble in water (66 g/100 g at 20°C) and has pK_a values of 3.05 and 5.76. Methylmalonyl-CoA is converted into succinyl-CoA by the action of methylmalonyl-CoA mutase, a vitamin B$_{12}$-requiring enzyme. (*Note:* One must not confuse this compound with methyl malonate, *i.e.*, the mono-ester (H$_3$CO–C(=O)–CH$_2$–COOH). **Target(s):** cyanate hydratase, *or* cyanase (1); glycine transport (2); β-hydroxybutyrate dehydrogenase (3-5); isocitrate lyase (6); malic enzyme, *or* malate dehydrogenase (oxaloacetate decarboxylating) (7); NADH dehydrogenase, ior Complex I (8); pantothenate synthetase (9); serine 3-dehydrogenase (10); succinate dehydrogenase, Complex II (3,8,11,12); L-threonine dehydrogenase (13). **1.** Anderson, Johnson, Endrizzi, Little & Korte (1987) *Biochemistry* **26**, 3938. **2.** Benavides, Garcia, Lopez-Lahoya, Ugarte & Valdivieso (1980) *Biochim. Biophys. Acta* **598**, 588. **3.** Dutra, Dutra-Filho, Cardozo, *et al.* (1993) *J. Inherit. Metab. Dis.* **16**, 147. **4.** el Kebbaj, Gaudemer & Latruffe (1986) *Arch. Biochem. Biophys.* **244**, 671. **5.** el Kebbaj, Latruffe & Gaudemer (1984) *Biochim. Biophys. Acta* **789**, 278. **6.** Hoyt, Robertson, Berlyn & Reeves (1988) *Biochim. Biophys. Acta* **966**, 30. **7.** Sanguinetti & Rossi (1964) *Boll. Soc. Ital. Biol. Sper.* **40**, 110. **8.**

Brusque, Borba Rosa, Schuck, *et al.* (2002) *Neurochem. Int.* **40**, 593. **9.** Miyatake, Nakano & Kitaoka (1979) *Meth. Enzymol.* **62**, 215. **10.** Chowdhury, Higuchi, Nagata & Misono (1997) *Biosci. Biotechnol. Biochem.* **61**, 152. **11.** Okun, Horster, Farkas, *et al.* (2002) *J. Biol. Chem.* **277**, 14674. **12.** Kolker, Schwab, Horster, *et al.* (2003) *J. Biol. Chem.* **278**, 47388. **13.** Guerranti, Pagani, Neri, *et al.* (2001) *Biochim. Biophys. Acta* **1568**, 45.

Methylmalonyl-CoA

(S)-Methylmalonyl-CoA or, D-Methylmalonyl-CoA

(R)-Methylmalonyl-CoA or, L-Methylmalonyl-CoA

This pair of thiolester diastereoisomers (FW = 867.61 g/mol; CAS 1264-45-5) are intermediates in the conversion of propionyl-CoA to succinyl-CoA. (*Note:* These two metabolites are often incorrectly referred to as enantiomers. There are five chiral centers in the coenzyme A moiety of the molecule, and these centers remain unchanged. The thioesters thus differ only in the stereochemistry of one carbon, making them epimers, and the enzyme catalyzing their interconversion is methylmalonyl-CoA epimerase, not racemase.) The (*S*)-diastereoisomer is a product of the reaction catalyzed by propionyl-CoA carboxylase. It is a substrate for methylmalonyl-CoA carboxytransferase, (*S*)-methylmalonyl-CoA hydrolase, methylmalonyl-CoA decarboxylase, and methylmalonyl-CoA epimerase. The (*R*) diastereoisomer, produced by methylmalonyl-CoA epimerase, is a substrate in the reaction catalyzed by methylmalonyl-CoA mutase (with the (*S*)-epimer acting as an inhibitor). **Target(s):** acetoacetyl-CoA reductase (1); acetyl-CoA carboxylase (2); *N*-acetylglutamate synthase, *or* glutamate acetyltransferase (3); [acyl-carrier-protein] *S*-malonyltransferase (29); amino-acid acetyltransferase (3); [branched-chain α-keto-acid dehydrogenase] kinase, *or* 3-methyl-2-oxobutanoate dehydrogenase (acetyl-transferring) kinase] (4-6,28); carnitine *O*-palmitoyltransferase (7,8,30); fatty acid synthase complex (1,2,9-12); glycine cleavage complex, *or* glycine synthase (13); 3-hydroxyisobutyryl-CoA hydrolase, inhibited by the DL-mixture (14,15); α-ketoglutarate dehydrogenase (lipoamide) (16); malonyl-CoA decarboxylase (17-24); methylmalonyl-CoA mutase, inhibited by (*S*)-methylmalonyl-CoA (25); pyruvate carboxylase (26); pyruvate dehydrogenase (lipoamide) (27). **1.** Dodds, Guzman, Chalberg, Anderson & Kumar (1981) *J. Biol. Chem.* **256**, 6282. **2.** Wahle, Williamson, Smith & Elliot (1984) *Comp. Biochem. Physiol. B* **78**, 93. **3.** Coude, Sweetman & Nyhan (1979) *J. Clin. Invest.* **64**, 1544. **4.** Paxton (1988) *Meth. Enzymol.* **166**, 313. **5.** Randle, Patston & Espinal (1987) *The Enzymes*, 3rd ed., **18**, 97. **6.** Paxton & Harris (1984) *Arch. Biochem. Biophys.* **231**, 48. **7.** Kashfi, Mynatt & Cook (1994) *Biochim. Biophys. Acta* **1212**, 245. **8.** Brindle, Zammit & Pogson (1985) *Biochem. J.* **232**, 177. **9.** Roncari (1981) *Meth. Enzymol.* **71**, 73. **10.** Kolattukudy, Poulose & Buckner(1981) *Meth. Enzymol.* **71**, 103. **11.** Roncari & Mack (1976) *Can. J. Biochem.* **54**, 923. **12.** Forward & Gompertz (1970) *Enzymologia* **39**, 379. **13.** Hayasaka & Tada (1983) *Biochem. Int.* **6**, 225. **14.** Shimomura, Murakami, Nakai, *et al.* (2000) *Meth. Enzymol.* **324**, 229. **15.** Shimomura, Murakami, Fujitsuka, *et al.* (1994) *J. Biol. Chem.* **269**, 14248. **16.** Williamson, Corkey, Martin-Requero, Walajtys-Rode & Coll (1986) in *Problems and Potential of Branched Chain Amino Acids in Physiology and Medicine* (Odessey, ed.), pp. 135 198, Elsevier/North-Holland. **17.** Kolattukudy, Poulose & Kim (1981) *Meth. Enzymol.* **71**, 150. **18.** Buckner, Kolattukudy & Poulose (1976) *Arch. Biochem. Biophys.* **177**, 539. **19.** Kim & Kolattukudy (1978) *Arch. Biochem. Biophys.* **190**, 585. **20.** Kim, Kolattukudy & Boos (1979) *Arch. Biochem. Biophys.* **196**, 543. **21.** Koeppen, Mitzen & Ammoumi (1974) *Biochemistry* **13**, 3589. **22.** Kim & Kolattukudy (1978) *Arch. Biochem. Biophys.* **190**, 234. **23.** Kim & Kolattukudy (1978) *Biochim. Biophys. Acta* **531**, 187. **24.** Hunaiti & Kolattukudy (1984) *Arch. Biochem. Biophys.* **229**, 426. **25.** Cannata, Focesi, Mazumder, Warner & Ochoa (1965) *J. Biol. Chem.* **240**, 3249. **26.** Scrutton, Olmsted & Utter (1969) *Meth. Enzymol.* **13**, 235. **27.** Martin-Requero, Corkey, Cerdan, *et al.* (1983) *J. Biol. Chem.* **258**, 3673. **28.** Espinal, Beggs & Randle (1988) *Meth. Enzymol.* **166**, 166. **29.** Guerra & Ohlrogge (1986) *Arch. Biochem. Biophys.* **246**, 274. **30.** Brindle, Zammit & Pogson (1985) *Biochem. Soc. Trans.* **13**, 880.

6-Methylmercaptopurine 9-β-D-Ribonucleoside

This cytotoxic nucleoside with antineoplastic and anti-angiogenic properties (FW = 298.32 g/mol) is an alternative substrate of adenosine kinase, forming 6-methylmercaptopurine-9-β-D-ribonucleoside 5'-P, which then inhibits amidophosphoribosyltransferase, the first committed step in *de novo* purine nucleotide synthesis. It also inhibits fibroblast growth factor-2 (FGF2)-induced cell proliferation **Target(s):** adenosine deaminase (1,2); adenosine kinase, also as a weak alternative substrate (3-6); amidophosphoribosyltransferase (7); GMP synthetase (8); phosphoribosyl-pyrophosphate synthetase (7,9). **1.** Van Heukelom, Boom, Bartstra & Staal (1976) *Clin. Chim. Acta* **72**, 109. **2.** Daddona, Wiesmann, Lambros, Kelley & Webster (1984) *J. Biol. Chem.* **259**, 1472. **3.** Long & Parker (2006) *Biochem. Pharmacol.* **71**, 1671. **4.** Palella, Andres & Fox (1980) *J. Biol. Chem.* **255**, 5264. **5.** Datta, Bhaumik & Chatterjee (1987) *J. Biol. Chem.* **262**, 5515. **6.** Henderson, Mikoshiba, Chu & Caldwell (1972) *J. Biol. Chem.* **247**, 1972. **7.** Yen & Becker (1979) *Adv. Exp. Med. Biol.* **122B**, 137. **8.** Spector & Beecham III (1975) *J. Biol. Chem.* **250**, 3101. **9.** Planet & Fox (1976) *J. Biol. Chem.* **251**, 5839.

S-Methylmethionine Sulfonium Bromide

This methionine derivative (FW$_{bromide}$ = 244.15 g/mol), also known as (3-amino-3 carboxypropyl)dimethylsulfonium bromide, has been isolated from a large number of flowering plants and has been used in the treatment of ulcers. This photosensitive amino acid is soluble in water and slightly soluble in methanol (0.439 g/100 g at 25°C) and ethanol (0.012 g/100 g at 25°C). Note that *S*-methyl-methionine slowly decomposes at room temperature; this rate increases when exposed to light. *S*-Methylmethionine is a product of the reaction catalyzed by *S*-adenosyl-L-methionine:L methionine *S*-methyltransferase and is both an alternative substrate and an inactivator of 1 aminocyclopropane-1-carboxylate synthase. **Target(s):** *S*-adenosylmethionine cyclotransferase, weakly inhibited (1); 1-aminocyclopropane-1-carboxylate synthase (2). **1.** Mudd (1959) *J. Biol. Chem.* **234**, 87. **2.** Ko, Eliot & Kirsch (2004) *Arch. Biochem. Biophys.* **421**, 85.

N-Methyl-α-methylphenethylamine

S-enantiomer   **R-enantiomer**

This potent neurotoxin and CNS stimulant (FW = 149.23 g/mol; CAS 33817-09-3), widely known by the street names methamphetamine, "METH" and "Crystal Meth", "Crank", as well as *N*-methyl-1-phenylpropan-2-amine, has been used to treat attention deficit hyperactivity disorder (ADHD), to promote weight loss, and as a drug of abuse. There are two enantiomers: (*R*(–)-methamphetamine (the more potent) and *S*(+)-methamphetamine. Like dopamine and amphetamine, methamphetamine is a dopamine transporter (*or* DAT) substrate. It interacts with and is subsequently transported into the DAT$^+$ neurons in a Na$^+$/Cl$^-$-dependent mechanism, thereby increasing the excitability of dopaminergic neurons. **Likely Mechanisms of Action:** A stimulant that induces a cocaine-like euphoria, methamphetamine appears to operate by two mechanisms: (a) drug-induced catecholamine redistribution from synaptic vesicles to the cytosol, and (b) drug-induced induced reverse neurotransmitter transport. These processes affect extracellular catecholamine levels, uptake inhibition, exocytosis, neurotransmitter synthesis, and drug metabolism. Methamphetamine affects excitatory synaptic transmission by activating dopamine and serotonin receptor systems in the hippocampus, a modulatory effect that may contribute to synaptic maladaption during METH addiction and/or METH-mediated cognitive dysfunction (1,2). Intracellular application of methamphetamine, but not amphetamine, prevents the dopamine-induced increase in the spontaneous firing of dopaminergic neurons and the corresponding DAT-mediated inward current. **Drug Abuse:** Methamphetamine is highly addictive, and its use in humans is often associated with neurocognitive impairment (2). Ready synthesis from simple precursors also contributes to methamphetamine's illicit use. Users typically cite availability, low cost, and a longer duration of action (than cocaine) as the basis for its preferential use. METH abusers are at high risk of neurodegenerative disorders, including Parkinson disease, and there are still no effective treatments to METH-induced neurodegeneration. Phasic dopamine (DA) signaling, characterized by burst firing by DA neurons, generates short-lived elevations in extracellular DA in terminal field. These DA transients are implicated in reinforcement learning. Disrupted phasic DA signaling is thought to link DA depletions and cognitive-behavioral impairment in METH-induced neurotoxicity (3). **1.** Swant, Chirwa, Stanwood & Khoshbouei (2010) *PLoS One* **5**, e11382. **2.** North, Swant, Salvatore, *et al.* (2013) *Synapse* **67**, 245. **3.** Robinson, Howard, Pastuzyn, *et al.* (2014) *Neurotox. Res.* **26**, 152.

Methylnaltrexone

This peripheral μ-opioid receptor antagonist (FW = 356.44 g/mol; CAS 83387-25-1; *Symbol*: MNTX), also known as Relistor®, *N*-methylnaltrexone, and (5α)-17-(cyclopropylmethyl)-3,14-dihydroxy-17-methyl-4,5-epoxymorphinan-17-ium-6-one, possesses a quaternized ammonium site that prevents penetration of the blood–brain barrier. This property endows methylnaltrexone with μ-opioid receptor antagonist action throughout the body (thus counteracting such effects, as itching and constipation observed with oxycodone and naltrexone), but without affecting opioid analgesia in the brain. Notably, quaternary derivatives were ineffective as antagonists of heroin self-administration, indicating that acute reinforcing properties of intravenous opiates associated with the sensation of a "rush" involve opiate receptors located within the central nervous system and not peripheral opiate receptors (1). Intracerebroventricular administration and peripheral administration of *N*-methylnaltrexone allows one to distinguish centrally acting versus peripherally acting effects of μ-opioid receptors. **Metabolism:** Intravenous administration of [^{14}C-*N*]-methylnaltrexone, 40% to 60% of the radioactivity is excreted in urine within 24 hours, with biphasic plasma decay curves indicating a short distribution phase over 6-9 minutes, with a far longer elimination phase from 4 and 22 hours (2). MNTX undergoes sulfation of the phenolic group to MNTX-3-sulfate as well as reduction of the carbonyl group to form the epimeric alcohols, methyl-6α-naltrexol and methyl-6β-naltrexol (3). Neither naltrexone nor its metabolite 6β-naltrexol are detected in human plasma after administration of MNTX, confirming earlier observations that *N*-demethylation is not a metabolic pathway of MNTX in humans (3). **1.** Koob, Pettit, Ettenberg & Bloom (1984) *J. Pharmacol. Exp. Ther.* **229**, 481. **2.** Kotake, Kuwahara, Burton, McCoy & Goldberg (1989) *Xenobiotica* **19**, 1247. **3.** Chandrasekaran, Tong, Li, *et al.* (2010) *Drug Metab. Dispos.* **38**, 606.

1-Methylnicotinamide

This naturally occurring pyridine (FW = 137.16 g/mol), formed in the reaction catalyzed by nicotinamide *N*-methyltransferase, is an end-product of niacin metabolism. MNA also inhibited platelet-dependent thrombosis by a mechanism involving cyclooxygenase-2 and prostacyclin (1). Endogenous MNA, produced in the liver by nicotinamide *N*-methyltransferase, may well be an endogenous activator of prostacyclin production, regulating thrombotic as well as inflammatory processes in the cardiovascular system (1). **Target(s):** acetylcholinesterase (2); alcohol dehydrogenase (3-5); NAD$^+$ ADP ribosyltransferase, *or* poly(ADP-ribose) polymerase, IC$_{50}$ = 1.7 mM (6,7); NAD$^+$ nucleosidase, *or* NADase (8); nicotinate glucosyltransferase (9). **1.** Chlopicki, Swies, Mogielniki, *et al.* (2007) *Br. J. Pharmacol.* **152**, 230. **2.** Bergmann, Wilson & Nachmansohn (1950) *J. Biol. Chem.* **186**, 693. **3.** Anderson & Reynolds (1966) *Arch. Biochem. Biophys.* **114**, 299. **4.** Atkinson, Eckermann & Lilley (1967) *Biochem. J.* **104**, 872. **5.** Anderson & Anderson (1964) *Biochem. Biophys. Res. Commun.* **16**, 258. **6.** Banasik, Komura, Shimoyama & Ueda (1992) *J. Biol. Chem.* **267**, 1569. **7.** Rankin, Jacobson, Benjamin, Moss & Jacobson (1989) *J. Biol. Chem.* **264**, 4312. **8.** Yuan & Anderson (1972) *J. Biol. Chem.* **247**, 515. **9.** Taguchi, Sasatani, Nishitani & Okumura (1997) *Biosci. Biotechnol. Biochem.* **61**, 720.

5-Methylnicotinamide

This substrate analogue (FW$_{free-base}$ = 136.15 g/mol; CAS 70-57-5; *Abbreviation*: 5MN) inhibits *Micrococcus luteus* nicotinamidase (1,2). While it cannot support NAD$^+$ synthesis, 5-methylnicotinamide is an equipotent inhibitor of NAD$^+$-hydrolyzing enzymes. **Target(s):** NAD$^+$ ADP-ribosyltransferase, *or* poly(ADP-ribose) polymerase (3-5); dose-dependent increase in NAD$^+$ content in the renal cortex and specific inhibition of sodium gradient-dependent phosphate transport across the renal brush-border membrane (BBM) (6). **1.** Johnson & Gadd (1974) *Int. J. Biochem.* **5**, 633. **2.** Gadd & Johnson (1974) *Int. J. Biochem.* **5**, 397. **3.** Banasik, Komura, Shimoyama & Ueda (1992) *J. Biol. Chem.* **267**, 1569. **4.** Rankin, Jacobson, Benjamin, Moss & Jacobson (1989) *J. Biol. Chem.* **264**, 4312. **5.** Werner, Sohst, Gropp, *et al.* (1984) *Eur. J. Biochem.* **139**, 81. **6.** Kempson (1986) *Am. J. Physiol.* **251**, F520.

Methyl *p*-Nitrobenzenesulfonate

This methylating reagent (FW = 217.20 g/mol; M.P. = 93-95°C) covalently modifies a number of proteins. Note that, while methyl *p*-nitrobenzenesulfonate methylates a histidyl residue in chymotrypsin, it does not inhibit trypsin or subtilisin. **Target(s):** D-amino-acid oxidase (1); bacterial luciferase, *or* alkanal monooxygenase (2); bromelain (3); chloramphenicol acetyltransferase (4); chymotrypsin (5-9); 3-hydroxy-3-methylglutaryl CoA lyase (10); inositol 1,3,4-trisphosphate 5/6-kinase (11); poly(β-D-mannuronate) lyase (12); stem bromelain (3); system A amino acid transport (13); system ASC amino acid transport (13); L-threonine dehydrogenase (14). **1.** Swenson, Williams & Massey (1984) *J. Biol. Chem.* **259**, 5585. **2.** Paquatte & Tu (1989) *Photochem. Photobiol.* **50**, 817. **3.** Silverstein & Kezdy (1975) *Arch. Biochem. Biophys.* **167**, 678. **4.**

Lewendon & Shaw (1993) *Biochem. J.* **290**, 15. **5.** Shaw (1972) *Meth. Enzymol.* **25**, 655. **6.** Nakagawa & Bender (1969) *J. Amer. Chem. Soc.* **91**, 1566 and (1970) *Biochemistry* **9**, 259. **7.** Shaw (1970) *The Enzymes*, 3rd ed., **1**, 91. **8.** Tsai, Lu & Chuang (1991) *Biochim. Biophys. Acta* **1080**, 59. **9.** Nakagawa & Bender (1970) *Biochemistry* **9**, 259. **10.** Hruz & Miziorko (1992) *Protein Sci.* **1**, 1144. **11.** Hughes, Kirk & Michell (1993) *Biochem. Soc. Trans.* **21**, 365S. **12.** Yoon, Choi, Miyake, *et al.* (2001) *J. Microbiol. Biotechnol.* **11**, 118. **13.** Bertran, Roca, Pola, *et al.* (1991) *J. Biol. Chem.* **266**, 798. **14.** Marcus & Dekker (1995) *Arch. Biochem. Biophys.* **316**, 413.

1-Methyl-3-nitro-1-nitrosoguanidine

This water-soluble carcinogen (FW = 147.09 g/mol; decomposes at 118°C; *Abbreviation*: MNNG), also known as *N*-methyl-*N'*-nitro-*N*-nitrosoguanidine, is an extremely potent mutagen that methylates DNA. **CAUTION:** Extreme care should be exercised to avoid skin irritation and to minimize cancer-risk. MNNG also methylates cysteinyl residues in proteins. (**See also** *N-Methyl-N-nitroso-p-toluenesulfonamide*) **Target(s):** CMP-*N*-acylneuraminate phosphodiesterase (1); DNA methyltransferases (2-5); DNA polymerases (6). **1.** Corfield, Stewart, Houghton, *et al.* (1987) *Biochem. Soc. Trans.* **15**, 1052. **2.** Drahovsky & Wacker (1975) *Eur. J. Cancer* **11**, 517. **3.** Cox (1986) *Toxicol. Pathol.* **14**, 477. **4.** Boehm & Drahovsky (1981) *Int. J. Biochem.* **13**, 1225. **5.** Cox (1980) *Cancer Res.* **40**, 61. **6.** Chan & Becker (1981) *Biochem. J.* **193**, 985.

O^6-(1-Methyl-4-nitropyrrol-2-ylmethyl)guanine

This modified guanine (FW = 289.26 g/mol) inactivates methylated-DNA:[protein]-cysteine *S*-methyltransferase, *or* O^6-alkylguanine-DNA alkyltransferase (IC$_{50}$ = 0.55 µM). The enzyme reverses specific types of alkylation damage by removing the offending alkyl group, while itself becoming covalently modified and inactivated. **1.** McMurry (2007) *DNA Repair* **6**, 1161.

N^1-Methyl-N^1-nitrosourea

This carcinogenic/mutagenic alkylating reagent (FW = 103.08 g/mol; CAS 684-93-5), also known as nitrosomethylurea (NMU) modifies DNA and inhibits O^6-alkylguanine-DNA alkyltransferase (1) and DNA polymerase (2). MNU-induced rat mammary carcinogenesis is a frequently used experimental model for breast cancer. **CAUTION:** Likely carcinogen. Exercise care when handling this reagent. Acute exposure can result in eye and/or skin irritation, headache, nausea, and vomiting. Use heavy chemical rubber gloves (*not* latex hospital gloves). Sigma-Aldrich packages NMU in a 100 mL serum bottle with butyl rubber stopper and aluminum tear seal. Injecting 100 mL of any compatible solvent directly into the vial allows one to prepare a 1% (wt/vol) solution with minimal risk of exposure. **1.** Link & Tempel (1991) *J. Cancer Res. Clin. Oncol.* **117**, 549. **2.** Chan & Becker (1981) *Biochem. J.* **193**, 985.

N-Methylnorsalsolinol

This dopaminergic neurotoxin (FW$_{\text{free-base}}$ = 179.22 g/mol), also known as *N*-methyl-6,7-dihydroxy-1,2,3,4 tetrahydroisoquinoline, inhibits monoamine oxidase. It has been identified in patients with Parkinson's disease and found to be able to cross the blood-brain barrier. **Target(s):** monoamine oxidase (1,2); tyrosine 3-monooxygenase (3,4). **1.** Minami, Maruyama, Dostert, Nagatsu & Naoi (1993) *J. Neural. Transm. Gen. Sect.* **92**, 125. **2.** Moser, Scholz, Bamberg & Bohme (1996) *Neurochem. Int.* **28**, 109. **3.** Scholz, Bamberg & Moser (1997) *Neurochem. Int.* **31**, 845. **4.** Scholz, Toska, Luborzewski, *et al.* (2008) *FEBS J.* **275**, 2109.

2-Methylornithine

This amino acid (FW = 146.19 g/mol) is a potent inhibitor of ornithine decarboxylase, but activity is restored by treatment with excess pyridoxal phosphate. **Target(s):** glutamate *N*-acetyltransferase (1); ornithine decarboxylase (2-7); tyrosine:arginine ligase, *or* kyotorphin synthetase, inhibited by L-isomer (8,9). **1.** Jain & Shargool (1984) *Anal. Biochem.* **138**, 25. **2.** Tyagi, Tabor & Tabor (1983) *Meth. Enzymol.* **94**, 135. **3.** Arteaga-Nieto, Villagomez-Castro, Calvo-Mendez & Lopez-Romero (1996) I*nt. J. Parasitol.* **26**, 253. **4.** O'Leary & Herreid (1978) *Biochemistry* **17**, 1010. **5.** Bey, Danzin, Van Dorsselaer, *et al.* (1978) *J. Med. Chem.* **21**, 50. **6.** Adbel-Monem, Newton & Weeks (1974) *J. Med. Chem.* **17**, 4447. **7.** Niemann, von Besser & Walter (1996) *Biochem. J.* **317**, 135. **8.** Kawabata, Tanaka, Muguruma & Takagi (1995) *Peptides* **16**, 1317. **9.** Kawabata, Muguruma, Tanaka & Takagi (1996) *Peptides* **17**, 407.

5-Methylorotate (5-Methylorotic Acid)

This orotic acid derivative (FW$_{\text{free-acid}}$ = 170.12 g/mol) inhibits dihydroorotate dehydrogenase (1-4), orotate phosphoribosyltransferase (5), and dihydroorotase (6,7). **1.** Friedmann (1963) *Meth. Enzymol.* **6**, 197. **2.** Friedmann & Vennesland (1960) *J. Biol. Chem.* **235**, 1526. **3.** Webb (1966) *Enzyme and Metabolic Inhibitors*, vol. 2, p. 470, Academic Press, New York. **4.** Hines, Keys & Johnston (1986) *J. Biol. Chem.* **261**, 11386. **5.** Krungkrai, Aoki, Palacpac, *et al.* (2004) *Mol. Biochem. Parasitol.* **134**, 245. **6.** Christopherson & Jones (1980) *J. Biol. Chem.* **255**, 3358. **7.** Krungkrai, Krungkrai & Phakanont (1992) *Biochem. Pharmacol.* **43**, 1295.

3-(3-Methyl-1,2,4-oxadiazol-5-yl)-6-(4-(2-methylphenoxy)piperidin-1-yl)pyridazine

This potent, selective, and orally available pyridazine derivative (FW = 351.41 g/mol) inhibits stearoyl-CoA 9-desaturase (mouse, IC$_{50}$ = 5.3 nM). This DCD1 inhibitor exhibits excellent cellular activity in blocking the conversion of saturated long-chain fatty acid-CoAs (LCFA-CoAs) to mono-

unsaturated LCFA-CoAs in HepG2 cells. **1.** Liu, Lynch, Freeman, *et al.* (2007) *J. Med. Chem.* **50**, 3086.

4-Methyl-5-oxoproline

This substrate analogue (FW = 143.14 g/mol) is a weak alternative substrate (~ 4% of rate with 5-oxoproline) potently inhibits 5-oxoprolinase. **1.** Griffith & Meister (1977) *Proc. Natl. Acad. Sci. U.S.A.* **74**, 3330.

Methyl Parathion

This insecticide (FW = 263.21 g/mol), which targets acetylcholinesterase (1-4), Ca^{2+}-dependent ATPase (5,6), cholinesterase, *or* butyrylcholinesterase (3,7,8), and succinate dehydrogenase (9), must first undergo metabolic transformation to the corresponding paraoxon to inhibit AChE. Paraoxon is subsequently detoxified enzymatically by A-esterase, glutathione-*S*-aryl transferase, and glutathione-*S*-alkyl-transferase. **CAUTION:** Methyl parathion is very toxic and should be handled with care. Signs of toxicity include sweating, dizziness, vomiting, convulsions, cardiac arrest, respiratory arrest, and, in extreme cases, death. The U.S. Environmental Protection Agency now restricts the use of this insecticide. Methyl parathion is no longer be used on food crops commonly consumed by children. **1.** Schoor & Brausch (1980) *Arch. Environ. Contam. Toxicol.* **9**, 599. **2.** Lange & Wiezorek (1975) *Acta Biol. Med. Ger.* **34**, 42. **3.** Hahn, Ruhnke & Luppa (1991) *Acta Histochem.* **91**, 13. **4.** Garcia, Abu-Qare, Meeker-O'Connell, Borton & Abou-Donia (2003) *J. Toxicol. Environ. Health B Crit. Rev.* **6**, 185. **5.** Reddy & Rao (1990) *Biochem. Int.* **22**, 1053. **6.** Blasiak (1995) *Comp. Biochem. Physiol. C Pharmacol. Toxicol. Endocrinol.* **110**, 119. **7.** Benke & Murphy (1974) *Bull. Environ. Contam. Toxicol.* **12**, 117. **8.** Benke, Cheever, Mirer & Murphy (1974) *Toxicol. Appl. Pharmacol.* **28**, 97. **9.** Sreenivasa Moorthy, Kasi Reddy, Swami & Sreeramulu Chetty (1984) *Arch. Int. Physiol. Biochim.* **92**, 147.

Methylphenidate

This piperidine-class dopamine-norepinephrine reuptake inhibitor, *or* NDRI (FW$_{\text{free-base}}$ = 233.31 g/mol; CAS 113-45-1; asterisks indicate stereocenters), also known as methyl phenidylacetate, Ritalin®, Concerta®, Methylin®, Medikinet®, Equasym XL®, Quillivant XR®, and Metadate®, targets dopamine (K_i = 40 nM) and norepinephrine (K_i = 350 nM) transporters (1,2). The (2*R*,2'*R*)-(+)-*threo*-stereoisomer is more effective, with levorotary enantiomers displaying affinity toward the serotonergic 5HT$_{1A}$ and 5HT$_{2B}$ receptor subtypes, and no direct binding to the serotonin transporter (K_i > 10,000 nM). Methylphenidate is widely used in the treatment of attention-deficit hyperactivity disorder (*or* ADHD). Note that, as a dopamine reuptake inhibitor, methylphenidate differs from amphetamine, which is both a dopamine and norepinephrine releasing agent and a reuptake inhibitor. Methylphenidate has a well-know addiction liability, and ΔFosB overexpression in D1-type medium spiny neurons within the nucleus accumbens is implicated in methylphenidate addiction. **Key Pharmacokinetic Parameters:** *See* Appendix II in Goodman & Gilman's THE PHARMACOLOGICAL BASIS OF THERAPEUTICS, 12th Edition (Brunton, Chabner & Knollmann, eds.) McGraw-Hill Medical, New York (2011). **1.**

Nielsen, Chapin & Moore (1983) *Life Sci.* **33**, 1899. **2**. Schenk (2002) *Prog. Drug Res.* **59**, 111.

α-Methyl-DL-phenylalanine

This phenylalanine derivative (FW = 179.22 g/mol) is a strong inhibitor of phenylalanine monooxygenase that is used to induce experimental hyperphenylalaninemia. It is also an inhibitor of chorismate mutase. **Target(s):** chorismate mutase (1); 3-deoxy-7-phosphoheptulonate synthase, *or* 3-deoxy-D-*arabino*heptulosonate-7-phosphate synthetase (2); phenylalanine monooxygenase (3-5); phenylalanyl-tRNA synthetase, K_i = 2 mM (6); tyrosine 3-monooxygenase (7). **1**. Sugimoto & I. Shiio (1980) *J. Biochem.* **88**, 167. **2**. Simpson & Davidson (1976) *Eur. J. Biochem.* **70**, 509. **3**. Rech, Feksa, Dutra-Filho, *et al.* (2002) *Neurochem. Res.* **27**, 353. **4**. Huether & Neuhoff (1981) *J. Inherit. Metab. Dis.* **4**, 67. **5**. Greengard, Yoss & Del Valle (1976) *Science* **192**, 1007. **6**. Santi & Danenberg (1971) *Biochemistry* **10**, 4813. **7**. Moore & Dominic (1971) *Fed. Proc.* **30**, 859.

1-Methyl-4-phenyl-2,3-dihydropyridinium Ion

This photosensitive cation (FW$_{ion}$ = 172.25 g/mol; *Abbreviation*: MPDP$^+$) is a neurotoxicant and an inhibitor of monoamine oxidase B. It is an intermediate metabolite of the dopaminergic neurotoxin 1-methyl-4-phenyl-1,2,3,6-tetrahydropyridine. (*See also 1-Methyl-4-phenyl-1,2,3,6 tetrahydropyridine*) **Target(s):** acetylcholinesterase (1); cytochrome P450 (2); glutathione *S*-transferase (3); monoamine oxidase A, slowly inactivated (4-6); monoamine oxidase B (4-7). **1**. Zang & Misra (1996) *Arch. Biochem. Biophys.* **336**, 147. **2**. Das, Shahi, Moochhala, Sato & Sunamoto (1992) *Chem. Phys. Lipids* **62**, 303. **3**. Awasthi, Singh, Shen, *et al.* (1987) *Neurosci. Lett.* **81**, 159. **4**. Krueger, McKeown, Ramsay, Youngster & Singer (1990) *Biochem. J.* **268**, 219. **5**. Singer, Salach, Castagnoli & Trevor (1986) *Biochem. J.* **235**, 785. **6**. Singer, Salach & Crabtree (1985) *Biochem. Biophys. Res. Commun.* **127**, 707. **7**. Abell (1987) *Meth. Enzymol.* **142**, 638.

β-Methylphenethylamine

This amphetamine isomer (FW$_{free-base}$ = 135.21 g/mol; CAS 582-22-9; FW$_{HCl-Salt}$ = 171.67 g/mol), also known as BMPEA, β-Me-PEA, and 1-amino-2-phenylpropane, is a partial agonist for human Trace Amine-Associated Receptor-1 (TAAR1), a G$_s$- and G$_q$-coupled G protein-coupled receptor (GPCR) that is activated by endogenous monoamines (*e.g.*, *m*-tyramine, *p*-tyramine, *m*-tryptamine, tryptamine, phenylethylamine (PEA), and both *m*-octopamine and *p*-octopamine) and is located within neural presynaptic membrane and on some lymphocytes. Upon association with this receptor, BMPEA, amphetamine, methamphetamine, 3,4-methylenedioxymethamphetamine (MDMA), and 2,5-dimethoxy-4-iodoamphetamine (DOI), increase 3',5'-cyclicAMP concentration in target cells. β-methylphenethylamine has approximately 0.0013 the pressor activity of epinephrine, or about one-third that of amphetamine (1). β-methylphenethylamine also has about twice the bronchodilating power of amphetamine, when tested in rabbit lung, and an LD$_{50}$ of 50 mg/kg (rat, i.v.) (1). β-methylphenethylamine can be analyzed in urine by a two-step liquid-liquid extraction, followed by UPLC/MS/MS (2). Some over-the-counter diet supplements containing *Acacia rigidula* have β-methylphenethylamine, raising concerns over their safety (3,4). **1**. Graham, Cartland & Woodruff (1945) *Ind. Eng. Chem.* **37**, 149. **2**. Chołbiński, Wicka, Kowalczyk, *et al.* (2014) *Anal. Bioanal. Chem.* **406**, 3681. **3**. Cohen, Maller, DeSouza & Neal-Kababick (2014) *JAMA* **312**, 1691. **4**. Cohen, Bloszies, Yee & Gerona (2015) *Drug Testing & Analysis* **8**, 328.

1-Methyl-4-phenylpyridinium Cation

MPP$^+$ **MPTP**

Desmethylprodine

This photosensitive dopaminergic neurotoxin (FW$_{iodide-salt}$ = 297.14 g/mol; CAS 36913-39-0; Symbol: MPP$^+$) causes irreversible symptoms of Parkinson disorder in humans and other primates. MPP$^+$ is produced from 1-methyl-4-phenyl-1,2,3,6-tetrahydropyridine (FW = 173.26 g/mol; CAS 28289-54-5; Symbol: MPTP), the latter a byproduct in the preparation of the synthetic opioid known as desmethylprodine (FW = 247.33 g/mol; CAS 13147-09-6; Symbol: MPPP; IUPAC Name: 1-(methyl-4-phenylpiperidin-4-yl) propanoate). (*See also 1-Methyl-4-phenyl-1,2,3,6-tetrahydropyridine*) Effects are often observed in heroin users inasmuch as they also tend to abuse opioid. Like rotenone, MPP$^+$ is a potent inhibitor of Complex I in the mitochondrial electron transport system (*See Rotenone*). **Mechanism of Neurotoxicity:** Conversion of MPTP to MPP$^+$ is catalyzed by monoamine oxidase-B, an enzyme that is primarily located in serotonergic neurons and astrocytes. Neuronal transmembrane polarization is likely to contribute to MPP$^+$ transport as well as its accumulation within nerve terminals. Following concentration within dopamine neurons, MPP$^+$ is further concentrated by the electrical gradient of the mitochondrial inner membrane, attended by slow binding to NADH dehydrogenase. Both events are facilitated by tetraphenylboron anion, which forms membrane-penetrating ion-pairs with MPP$^+$ and its analogues. **Target(s):** aromatic-amino-acid decarboxylase, *or* dopa decarboxylase (1,2); dihydropteridine reductase, weakly inhibited (3;) α-ketoglutarate dehydrogenase (4); monoamine oxidase (5-7); NADH dehydrogenase, *or* Complex I (8-10); tyrosine monooxygenase (11,12). **1**. Foster, Wrona, Han & Dryhurst (2003) *Chem. Res. Toxicol.* **16**, 1372. **2**. Naoi, Takahashi, Ichinose & Nagatsu (1988) *Biochem. Biophys. Res. Commun.* **152**, 15. **3**. Bradbury, Brossi, Costall, *et al.* (1986) *Neuropharmacology* **25**, 583. **4**. Mizuno, Saitoh & Sone (1987) *Biochem. Biophys. Res. Commun.* **143**, 971. **5**. Abell (1987) *Meth. Enzymol.* **142**, 638. **6**. Singer, Salach & Crabtree (1985) *Biochem. Biophys. Res. Commun.* **127**, 707. **7**. Takamidoh, Naoi & Nagatsu (1987) *Neurosci. Lett.* **73**, 293. **8**. Singer, Trevor & Castagnoli (1987) *Trends Biochem. Sci.* **12**, 266. **9**. Ramsay, Salach, Dadgar & Singer (1986) *Biochem. Biophys. Res. Commun.* **135**, 269. **10**. Mizuno, Saitoh & Sone (1987) *Biochem. Biophys. Res. Commun.* **143**, 294. **11**. Hirata & Nagatsu (1985) *Neurosci. Lett.* **57**, 301. **12**. Nagatsu & Hirata (1987) *Eur. Neurol.* **26** Suppl. 1, 11.

N'-[(3-Methylphenyl)sulfonyl]dibenzofuran-2-carbohydrazide

This hydrazide (FW = 380.42 g/mol) inhibits human cytosolic branched-chain-amino-acid aminotransferase (IC$_{50}$ = 2.8 μM), the BCAT isoenzyme expressed specifically in neuronal tissue and which is a likely treatment target for neurodegenerative/neurological disorders in which glutamatergic mechanisms are implicated. The 2-methylphenylsulfonyl and 4-methylphenylsulfonyl analogues are weaker inhibitors (IC$_{50}$ = 3.2 amd 51.0 μM, respectively). **1**. Hu, Boxer, Kesten, *et al.* (2006) *Bioorg. Med. Chem. Lett.* **16**, 2337.

1-Methyl-4-phenyl-1,2,3,6-tetrahydropyridine

This highly neurotoxic piperidine ($FW_{free-base} = 173.26$ g/mol), often abbreviated MPTP, targets dopaminergic neurosn, devastatingly bringing about irreversible onset of the hallmarks of Parkinson's disease in humans and other primates. MPTP is a byproduct in the preparation of a synthetic opioid, and its effects as a toxic opioid impurity were first noted in drug abusers (*See also 1-Methyl-4-phenylpyridinium Cation*). MPTP preferentially destroys the *substantia nigra pars compacta*, but appears to spare the adjacent pigmented ventral tegmental areas, as well as other neuronal systems. This apparent selectivity obviously decreases with age inasmuch as more widespread damage in observed in older animals. MPTP provides an animal model for its effects in humans. In addition, it is one of the first environmental neurotoxins known to cause parkinsonism in humans. As an alternative substrate of monoamine oxidase, MPTP is metabolized in primates to the 1 methyl-4-phenylpyridinium cation (MPP+), a potent Complex I inhibitor in mitochondrial electron transport. (*See also 1-Methyl-4-phenylpyridinium Cation*) **Target(s):** benzylamine oxidase, weakly inhibited (1); dopamine uptake (2); monoamine oxidase3-5; tyrosine 3-monooxygenase (6,7). **1.** Devlin, Bhatti, Williams & Ramsden (1990) *Toxicol. Lett.* **54**, 135. **2.** Denton & Howard (1984) *Biochem. Biophys. Res. Commun.* **119**, 1186. **3.** Ator & Ortiz de Montellano (1990) *The Enzymes*, 3rd ed., **19**, 213. **4.** Singer, Salach & Crabtree (1985) *Biochem. Biophys. Res. Commun.* **127**, 707. **5.** Fuller & Hemrick-Luecke (1985) *J. Pharmacol. Exp. Ther.* **232**, 696. **6.** Nagatsu & Hirata (1987) *Eur. Neurol.* **26** Suppl. 1, 11. **7.** Scholz, Toska, Luborzewski, *et al.* (2008) *FEBS J.* **275**, 2109.

5'-(4-Methylphenylthio)immucillin-A

This nucleoside analogue (FW = 359.45 g/mol), also known as (1S)-1-(9-deazaadenin-9-yl)-1,4-dideoxy-1,4-imino-5-(4-methylphenylthio)-D-ribitol, inhibits 5'-methylthio-adenosine/S-adenosylhomocysteine nucleosidase (Ki' = 8 pM). The *meta*-methyl analogue is a weaker inhibitor (K_i' = 9 pM). The 4-methylphenyl analogue is also a tight, slow-binding inhibitor of human S-methyl-5' thioadenosine phosphorylase (K_i = 0.64 nM). The 3-methylphenyl analogue also inhibits (K_i = 0.63 nM). (**For mode of action, See** *Immucillin*) **Target(s):** adenosylhomocysteine nucleosidase (1); 5'-methylthioadenosine/S-adenosylhomocysteine nucleosidase (1); S-methyl-5'-thioadenosine phosphorylase (2). **1.** Singh, Evans, Lenz, *et al.* (2005) *J. Biol. Chem.* **280**, 18265. **2.** Evans, Furneaux, Schramm, Singh & Tyler (2004) *J. Med. Chem.* **47**, 3275.

4- Methylphenyl- 2, 4, 6- trimethylphenylsulfone

This membrane-permeant EPAC-2 selective inhibitor (FW = 274.38 g/mol; CAS 5148-64-5; Symbol: ESI-05; Spectra: λ_{max} = 244 nm; ε = 19000 M⁻¹cm⁻¹) targets the exchange protein directly activated by cAMP (low-µM IC_{50}). The major physiological effects of cAMP in mammalian cells are transduced by two ubiquitously expressed intracellular cAMP receptors: protein kinase A (PKA) and exchange protein directly activated by cAMP (EPAC), as well as cyclic nucleotide-gated ion channels in certain tissues. EPAC2-specific inhibitors exert their isoform selectivity through a unique mechanism by binding to a previously undescribed allosteric site positioned at the interface of the two cAMP binding domains, but absent in the EPAC1 isoform. **1.** Tsalkova, Mei, Li, *et al.*, *Proc. Natl. Acad. USA*, **109**, 18613 (2012).

2-Methyl-3-phytyl-1,4-naphthoquinone

This photosensitive quinone (FW = 450.71 g/mol), also known as vitamin K_1 and phylloquinone, is a photosynthetic electron carrier and a factor in the posttranslational modification of prothrombin and other coagulation factors. It is required for the formation of γ-carboxyglutamyl (*or* GLA) residues needed to bind calcium ion. Vitamin K was first identified by Henrik Dam (1,2) and the structure of K_1 was determined by Doisy *et al.* (3) and was first chemically synthesized by Fieser (1). This vitamin is a viscous yellow oil (M.P. = –20°C) that is moderately stable to air, but rapidly decomposed when exposed to UV light and alkali. The absorption spectra, in petroleum ether, has λ_{max} values (with corresponding molar extinction coefficients) of 242 nm (ε = 17850 M⁻¹cm⁻¹), 248 nm (18880 M⁻¹cm⁻¹), 260 nm (17260 M⁻¹cm⁻¹), 269 nm (17440 M⁻¹cm⁻¹), and 325 nm (3065 M⁻¹cm⁻¹). Treatment with dithionite or sodium borohydride produces the corresponding quinol. **Target(s):** arachidonate release in vascular cells (5); CYP1A1 (6); mono(ADP ribosyl)transferase (7); NAD+ ADP-ribosyltransferase, *or* poly(ADP-ribose) polymerase (8); protein glutamine γ-glutamyltransferase, *or* transglutaminase (9); retinyl-palmitate esterase, *or* retinyl ester hydrolase (10,11); testosterone 5α-reductase (12). **1.** Dam (1929) *Biochem. Z.* **215**, 475; (1930) *Biochem. Z.* **220**, 158; and (1935) *Nature* **135**, 652. **2.** Dam (1942) *Adv. Enzymol.* **2**, 285. **3.** MacCorquodale, Cheney, Binkley, *et al.* (1939) *J. Biol. Chem.* **131**, 357. **4.** Fieser (1939) *J. Amer. Chem. Soc.* **61**, 3467. **5.** Nolan & Eling (1986) *Biochem. Pharmacol.* **35**, 4273. **6.** Inouye, Mae, Kondo & Ohkawa (1999) *Biochem. Biophys. Res. Commun.* **262**, 565. **7.** Sun & Cheng (1998) *Zhongguo Yao Li Xue Bao* **19**, 104. **8.** Banasik, Komura, Shimoyama & Ueda (1992) *J. Biol. Chem.* **267**, 1569. **9.** Lai, Liu, Tucker, *et al.* (2008) *Chem. Biol.* **15**, 969. **10.** Napoli & Beck (1984) *Biochem. J.* **223**, 267. **11.** Napoli, McCormick, O'Meara & Dratz (1984) *Arch. Biochem. Biophys.* **230**, 194. **12.** Kim, Kim, Son, *et al.* (1999) *Biol. Pharm. Bull.* **22**, 1396.

N-Methyl-N-propargyl-3-(2,4-dichlorophenoxy)propylamine

This tertiary amine, also known as clorgyline (FW = 272.17 g/mol), is an antidepressant agent and a selective, mechanism-based inhibitor of monoamine oxidase A. **1.** Figueiredo, Caramona, Paiva & Guimaraes (1998) *J. Auton. Pharmacol.* **18**, 123. **2.** Coulson (1971) *Biochem. J.* **121**, 38P. **3.** Hall & Logan (1969) *Biochem. Pharmacol.* **18**, 1955. **4.** Hall, Logan & Parsons (1969) *Biochem. Pharmacol.* **18**, 1447. **5.** Mai, Artico, Esposito, *et al.* (2003) *Farmaco* **58**, 231. **6.** Kumazawa, Seno, Ishii, Suzuki & Sato (1998) *J. Enzyme Inhib.* **13**, 377. **7.** Abell (1987) *Meth. Enzymol.* **142**, 638. **8.** Ator & Ortiz de Montellano (1990) *The Enzymes*, 3rd ed., **19**, 213. **9.** J. Brush & Kozarich (1992) *The Enzymes*, 3rd ed., **20**, 317. **10.** Denney, Fritz, Patel & Abell (1982) *Science* **215**, 140. **11.** Deverina (1980) *Biokhimiia* **45**, 1897. **12.** Youdim & Holman (1975) *J. Neural Transm.* **37**, 11. **13.** Youdim (1975) *Mod. Probl. Pharmacopsychiatry* **10**, 65.

N-Methylprotoporphyrin IX

two of four possible isomeric species

These *N*-methyl regioisomers of protoporphyrin IX (FW = 576.69 g/mol; CAS 79236-56-9; *Abbreviation*: NMPP) are potent ferrochelatase inhibitors (K_i ~10 nM), resulting in accumulation of metal-free protoporphyrin IX in liver cells (1-6). **Primary Mode of Inhibitory Action:** *N*-alkylation of a pyrrole nitrogen atom introduces non-planarity into the porphyrin macrocycle, and the inhibitory potency of NMPP is thought to arise from its structural similarity to the distorted porphyrin intermediate in the ferrochelatase-catalyzed reaction. The high affinity of the enzyme towards the inhibitor is attributed to a slow, second kinetic step corresponding to the rearrangement of the initial weak enzyme–inhibitor complex into a more stable enzyme–inhibitor complex. Commercial sources typically consist of all four isomers. The *N*-ethyl derivative is also a potent inhibitor (2). **N-Methylprotoporphyrin IX Synthesis:** When administered griseofulvin or 3,5-dicarbethoxy-1,4-dihydrocollidine, porphyric patients often worsen, occasionally with fatal outcomes. While such drugs result in ferrochelatase inhibition *in vivo*, they are completely without effect when added *in vitro*. That some heme metabolite is the ferrochelatase inhibitor was indicated by the observation that, when incubated with microsomal cytochrome proteins (CYP's), these drugs give rise green pigments among them *N*-methylprotoporphyrin. **Related Aspects of Ferrochelatase Inhibition:** Manganese ion Mn^{2+} inhibits ferrochelatase competitively with respect to Fe^{2+} (K_i = 15 microM) and noncompetitively with respect to the porphyrin substrate. Heme, one of the reaction products, is a noncompetitive inhibitor with respect to iron. These findings lead to a sequential Bi Bi kinetic model for ferrochelatase with iron binding occurring prior to porphyrin binding and heme being released prior to the release of two protons (4). **Target(s):** ferrochelatase (1-6); heme oxygenase, zinc complex (6). **1.** Dailey, Fleming & Harbin (1986) *Meth. Enzymol.* **123**, 401. **2.** Ortiz de Montellano, Kunze, Cole & Marks (1980) *Biochem. Biophys. Res. Commun.* **97**, 1436. **3.** Gamble, Dailey & Marks (2000) *Drug Metab. Dispos.* **28**, 373. **4.** Dailey & Fleming (1983) *J. Biol. Chem.* **258**, 11453. **5.** Dailey, Fleming & Harbin (1986) *J. Bacteriol.* **165**, 1. **6.** De Matteis, Gibbs & Harvey (1985) *Biochem. J.* **226**, 537.

6-Methylpurine Ribonucleoside

This nucleoside analogue (FW = 266.26 g/mol) is converted metabolically to its nucleoside 5'-monophosphate which then inhibits *Escherichia coli* GMP synthetase. **Target(s):** adenosine kinase, *Mycobacterium tuberculosis*, also as alternative substrate (1); GMP synthetase (2). **1.** Long & Parker (2006) *Biochem. Pharmacol.* **71**, 1671. **2.** Spector & Beecham III (1975) *J. Biol. Chem.* **250**, 3101.

4-Methylpyrazole

This antidote (FW = 82.11 g/mol; CAS 7554-65-6; Soluble in H_2O and ethanol; M.P. = 15.5-18.5° C), also known as 4-methyl-1*H*-pyrazole and Fomepizole™, potently inhibits alcohol dehydrogenase and is indicated in cases of methanol or ethylene glycol poisoning. **Primary Mode of Action:** With rat liver alcohol dehydrogenase (ADH), 4-methylpyrazole is a potent competitive inhibitor relative to ethanol (K_{is} = 0.11 µM at pH 7.3, 37° C) and is likewise competitive *in vivo* (K_{is} = 1.4 µmol/kg) (1). Methanol is readily oxidized by ADH to formaldehyde, which is further oxidized to formic acid by formaldehyde dehydrogenase. It is formic acid that is primarily responsible for the metabolic acidosis and visual disturbances associated with methanol poisoning. Likewise, oxidation of ethylene glycol to glycolaldehyde in the first step toward forming oxalate (the toxin). By inhibiting ADH with 4-methylpyrazole, there is greater opportunity for methanol and ethylene glycol to be excreted harmlessly. Importantly, ADH

binds 4-methylpyrazole some 400-500x more tightly than glycolaldehyde. **Target(s):** alcohol dehydrogenase (1-6); CYP2E1 (7); cinnamyl-alcohol dehydrogenase (8); retinol dehydrogenase, cytosolic (9). **1.** Plapp, Leidal, Smith & Murch (1984) *Arch. Biochem. Biophys.* **230**, 30. **2.** Raia, Giordano & Rossi (2001) *Meth. Enzymol.* **331**, 176. **3.** Brändén, Jörnvall, Eklund & Furugren (1975) *The Enzymes*, 3rd ed., **11**, 103. **4.** Li & Theorell (1969) *Acta Chem. Scand.* **23**, 892. **5.** Makar & Tephly (1975) *Biochem. Med.* **13**, 334. **6.** Pietruszko (1975) *Biochem. Pharmacol.* **24**, 1603. **7.** Clejan & Cederbaum (1990) *Biochim. Biophys. Acta* **1034**, 233. **8.** Wyrambik & Grisebach (1979) *Eur. J. Biochem.* **97**, 503. **9.** Leo, Kim & Lieber (1987) *Arch. Biochem. Biophys.* **259**, 241.

{[(4-Methylpyridin-2-yl)amino]methylene}-bis(phosphonate)

This bisphosphonate (FW = 268.12 g/mol) inhibits *Trypanosoma brucei* geranyl*trans*transferase, *or* farnesyl-dipjosphate synthase (K_i = 7 nM). The 3-methyl analogue is a slightly weaker inhibitor (K_i = 10 nM). **1.** Montalvetti, Fernandez, Sanders, *et al.* (2003) *J. Biol. Chem.* **278**, 17075. **2.** Dunford, Kwaasi, Rogers, *et al.* (2008) *J. Med. Chem.* **51**, 2187.

2-({[2-({2-[1-Methyl-2-pyrrolidinyl]ethyl}amino)phenyl]sulfonyl}amino)-5,6,7,8 tetrahydro-1-naphthalenecarboxylate

This substituted anthranilic acid sulfonamide (FW = 457.58 g/mol) inhibits methionyl aminopeptidase (IC_{50} = 0.017 µM), as identified with Affinity Selection by Mass Spectrometry (ASMS) screening. **1.** Sheppard, Wang, Kawai, *et al.* (2006) *J. Med. Chem.* **49**, 3832.

3-Methylquercetin

This redox-active *O*-methylated flavonol (FW = 316.26 g/mol; CAS 480-19-3; Symbol: 3-MQ; IUPAC Name: 3,5,7-trihydroxy-2-(4-hydroxy-3-methoxyphenyl)chromen-4-one), variously known as isorhamnetol, isorhamnetin, isorhamnetine, iso-rhamnetin, 3'-methoxyquercetin, protect the host cells from a viral induced shutdown of the cellular protein synthesis. When administered intraperitoneally at a daily dose of 20 mg/kg for nine days, 3-MQ protected mice from viremia and lethal infections from coxsackie B4 virus and polio virus (IC_{50} = 6.3 x 10^{-6} M) (1). In poliovirus-infected cells, the viral protein and RNA synthesis were severely reduced, when 3-methylquercetin is present 1–2 hours post-infection (2). On the other hand, in uninfected HeLa cells, protein and RNA synthesis was inhibited only slightly by 3-methylquercetin. planar flavones and flavonols with a 7-hydroxyl group, such as 3-MQ, chrysin, luteolin, kaempferol, quercetin, and myricetin inhibited xanthine oxidase activity at low concentrations (IC_{50} range: 0.40 to 5.02 µM) in a mixed-type mode, while the nonplanar flavonoids, isoflavones and anthocyanidins were less inhibitory (3). Isorhamnetin also inhibits *Prevotella intermedia* lipopolysaccharide-induced production of interleukin-6 in murine macrophages via anti-inflammatory heme oxygenase-1 induction and inhibition of nuclear factor-κB and signal transducer and activator of transcription 1 activation (4). *P. intermedia* is a Gram-negative, obligate anaerobic pathogen involved in periodontal infections. 3-MQ also activates AMP-stimulated prorotein kinase (AMPK) to protect hepatocytes against oxidative stress and mitochondrial dysfunction (5). **1**. Van Hoof, Berghe, Hatfield, Vlietinck (1984) *Planta Med.* **50**, 513. **2**. Vrijsen, Everaert, Van Hoof, *et al.* (1987) *Antiviral Res.* **7**, 35. **3**. Nagao, Seki & Kobayashi (1999) *Biosci. Biotechnol. Biochem.* **63**, 1787. **4**. Jin, Choi, Park, *et al.* (2013) *J. Periodontal Res.* **48**, 687. **5**. Dong, Lee, Hwan, *et al.* (2014) *Eur. J. Pharmacol.* **740**, 634.

N^5-Methyltetrahydrofolate & Polyglutamylated Derivatives

This tetrahydrofolate derivative (FW = 454.42 g/mol; CAS 134-35-0; λ_{max} at pH 7.0 is 290 nm (ε = 31700 $M^{-1}cm^{-1}$), also known as N^5-methyltetrahydropteroylglutamate, is an intermediate in the formation of methylcobalamin, which is used to convert L-homocysteine to L-methionine as well as *S*-adenosyl-L-homocysteine to *S*-adenosyl-L-methionine. N^5-CH$_3$-THF is slowly oxidized in air at room temperature, forming N^5-methyldihydrofolate, a process that is accelerated by heavy metal ions. **Target(s):** folylpolyglutamate synthetase (1,2); 5-formyltetrahydro-folate cyclo-ligase, *or* N^5,N^{10}-methenyltetrahydrofolate synthetase (3,4); 5-formyltetrahydrofolate transport (5,6); glycine *N*-methyltransferase, inhibited by the mono-, tri-, penta-, and hexaglutamyl forms (7-14); glycogen phosphorylase *b* (15); methionyl-tRNA formyltransferase (16,17); 5-methyltetrahydropteroyltriglutamate: homocysteine *S*-methyltransferase, inhibited by the monoglutamate coenzyme derivative (*i.e.*, N^5-methyltetrahydrofolate) (18); serine hydroxymethyl-transferase, *or* glycine hydroxymethyltransferase, also inhibited by the polyglutamate (19-23). **1**. Masurekar & Brown (1980) *Meth. Enzymol.* **66**, 648. **2**. Cichowicz & Shane (1987) *Biochemistry* **26**, 513. **5**. Jolivet (1997) *Meth. Enzymol.* **281**, 162. **4**. Hopkins & Schirch (1984) *J. Biol. Chem.* **259**, 5618. **5**. Cai & Horne (2003) *Arch. Biochem. Biophys.* **410**, 161. **6**. Cai & Horne (2003) *Brain Res.* **962**, 151. **7**. Mudd & Wagner (2002) *Hum. Genet.* **110**, 68. **8**. Yeo, Briggs & Wagner (1999) *J. Biol. Chem.* **274**, 37559. **9**. Yeo & Wagner (1992) *J. Biol. Chem.* **267**, 24669. **10**. Luka, Pakhomova, Loukachevitch, *et al.* (2007) *J. Biol. Chem.* **282**, 4069. **11**. Luka, Loukachevitch & Wagner (2008) *Biochim. Biophys. Acta* **1784**, 1342. **12**. Wagner, Briggs & Cook (1985) *Biochim. Biophys. Res. Commun.* **127**, 746. **13**. Ogawa, Gomi, Takusagawa & Fujioka (1998) *Int. J. Biochem. Cell Biol.* **30**, 13. **14**. Wagner, Decha-Umphai & Corbin (1989) *J. Biol. Chem.* **264**, 9638. **15**. Klinov, Kurganov, Sheiman, Gorelik & Birinberg (1988) *Bioorg. Khim.* **14**, 1162. **16**. Gambini, Crosti & Bianchetti (1980) *Biochim. Biophys. Acta* **613**, 73. **17**. Dickerman & Smith (1970) *Biochemistry* **9**, 1247. **18**. Whitfield, Steers & Weisbach (1970) *J. Biol. Chem.* **245**, 390. **19**. Stover & Schirch (1991) *J. Biol. Chem.* **266**, 1543. **20**. Schirch, Tatum & Benkovic (1977) *Biochemistry* **16**, 410. **21**. Schirch & Ropp (1967) *Biochemistry* **6**, 253. **22**. Ramesh & Appaji Rao (1980) *Biochem. J.* **187**, 623. **23**. Mitchell, Reynolds & Blevins (1986) *Plant Physiol.* **81**, 553.

5'-(Methylthio)immucillin-A

This nucleoside analogue (FW = 294.36 g/mol), also known as (1*S*)-1-(9-deazaadenin-9-yl)-1,4-dideoxy-1,4-imino-5-(methylthio)-D-ribitol, inhibits 5'-methylthioadenosine/*S*-adenosylhomocysteine nucleosidase (K_i' = 77 pM) and *S*-methyl-5'-thioadenosine phosphorylase (K_i = 1 nM). (*See Immucillin for details on mode of action*) **Target(s):** adenosylhomocysteine nucleosidase (1,2); 5'-methylthioadenosine/*S* adenosylhomocysteine nucleosidase (1,2); *S*-methyl-5'-thioadenosine phosphorylase (3-5). **1**. Singh, Evans, Lenz, *et al.* (2005) *J. Biol. Chem.* **280**, 18265. **2**. Singh, Lee, Núñez, Howell & Schramm (2005) *Biochemistry* **44**, 11647. **3**. Evans, Furneaux, Schramm, Singh & Tyler (2004) *J. Med. Chem.* **47**, 3275. **4**. Singh, Shi, Evans, *et al.* (2004) *Biochemistry* **43**, 9. **5**. Evans, Furneaux, Lenz, *et al.* (2005) *J. Med. Chem.* **48**, 4679.

5'-(Methylthio)immucillin-H

This nucleoside analogue (FW = 296.35 g/mol), also known as (1*S*)-1-(9-deazahypoxanthin-9-yl)-1,4-dideoxy-1,4-imino-5-(methylthio)-D-ribitol, inhibits purine-nucleoside phosphorylase. Note that this immucillin favors inhibition of the *Plasmodium falciparum* enzyme over the human enzyme (K_d = 2.7 *versus* 303 nM). (*See Immucillin for details on mode of action*) **1**. Shi, Ting, Kicska, *et al.* (2004) *J. Biol. Chem.* **279**, 18103. **2**. Chaudhary, Ting, Kim & Roos (2006) *J. Biol. Chem.* **281**, 25652.

S-Methyl-thioinosine 5'-Monophosphate

This modified thiopurine nucleotide (FW = 377.29 g/mol) is formed by the *S*-methylation of 6-mercaptopurine (6-MP) to 6-methylmercaptopurine (6-MMP) through the action of thiopurine methyl transferase (TPMT). The latter catalyzes the main inactivation pathway of thiopurines in hematopoietic cells, and patients deficient in TPMT accumulate excessive concentrations of the active thioguanine nucleotides in blood cells when treated with conventional doses of these drugs (1-3). Delayed cytotoxicity of S-Methyl-thioinosine 5'-monophosphate is the consequence of mitotic death arising from DNA damage due to incorporation of 6-thioguanine into DN. (*See 6-Mercaptopurine*) **1**. Tay, Lilley, Murray & Atkinson (1969) *Biochem. Pharmacol.* **18**, 936. **3**. Elion (1989) *Science* **244**, 41. **3**. Sahasranaman, Howard & Roy (2008) *Eur. J. Clin. Pharmacol.* **64**, 753.

O-Methyl-DL-thyroxine

This thyroxine derivative (FW = 790.90 g/mol) inhibits thyroxine deiodinase. These data, together with the observations that 3,3',5'-triiodo-DL-thyronine blocks the metabolic effect of thyroxine *in vivo*, provided the first strong support for the concept (a) that thyroxine is converted to another compound before becoming active at the cellular level, and (b) that one step in this conversion is the removal of one iodine atom from the terminal ring of the thyronine. **1**. Larson & Albright (1961) *J. Clin. Invest.* **40**, 1132.

Methyltrienolone

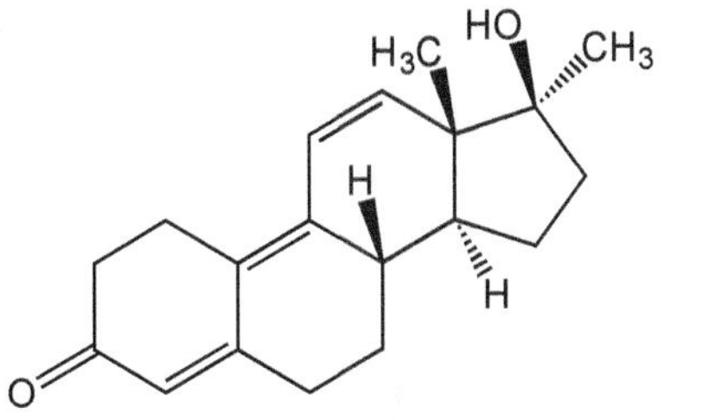

This anabolic steroid (FW = 284.40 g/mol; CAS 965-93-5; M.P. = 170°C), also known as 17β-hydroxy-17α-methylestra-4,9,11-trien-3-one and R 1881, is a non-metabolizable androgen that binds to androgen receptors. Methyltrienolone can also be used as a photoaffinity ligand for steroid binding proteins. **Target(s):** 3β-hydroxysteroid dehydrogenase (1); steroid Δ-isomerase (2); steroid 17,20 lyase (1); steroid 11β-monooxygenase, *or* CYP11B1 (3); steroid 17α-monooxygenase (1). **1**. Fanjul, Estevez, Deniz, *et al.* (1989) *Biochem. Int.* **19**, 301. **2**. Weintraub, Vincent, Baulieu & Alfsen (1977) *Biochemistry* **16**, 5045. **3**. Ohnishi, Miura & Ichikawa (1993) *Biochim. Biophys. Acta* **1161**, 257.·

β-Methyltryptophan

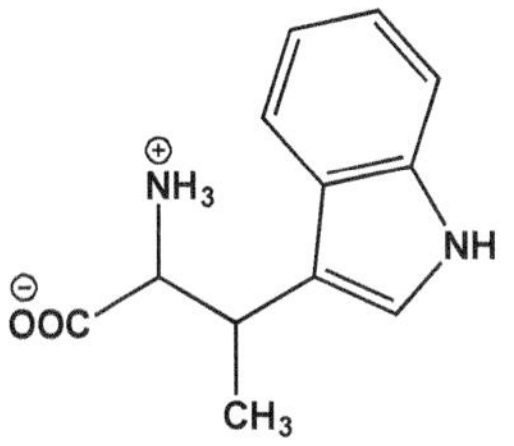

This tryptophan derivative (FW = 218.26 g/mol), also occasionally called 3-methyltryptophan, is an alternative substrate for tryptophan aminotransferase, acting competitively relative to tryptophan. **Target(s):** 3-deoxy-7-phosphoheptulonate synthase, inhibited by the DL-*erythro* isomer (1); tryptophan aminotransferase, also alternative substrate (2); tryptophan 2,3-dioxygenase (3); tryptophanyl-tRNA synthetase (4-6). **1**. Görisch & Lingens (1971) *Biochim. Biophys. Acta* **242**, 630. **2**. Speedie, Hornemann & Floss (1975) *J. Biol. Chem.* **250**, 7819. **3**. Webb (1966) *Enzyme and Metabolic Inhibitors*, vol. 2, p. 325, Academic Press, New York. **4**. Davie (1962) *Meth. Enzymol.* **5**, 718. **5**. Stulberg & Novelli (1962) *The Enzymes*, 2nd ed., **6**, 401. **6**. Sharon & Lipmann (1957) *Arch. Biochem. Biophys.* **69**, 219.

1-Methyltryptophan

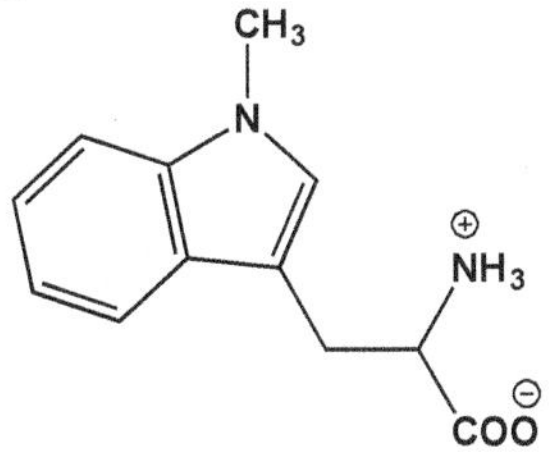

This tryptophan derivative (FW = 218.26 g/mol; CAS 21339-55-9) inhibits indoleamine 2,3-dioxygenase. Inhibition of this enzyme results in suppression of NK (natural killer) cell activity and acceleration of tumor growth. **Target(s):** indoleamine 2,3-dioxygenase (each stereoisomer selectively inhibits an isoform) (1-10); tryptophan 2,3-dioxygenase (11); tryptophan 2-C-methyltransferase (12). **1**. Cady & Sono (1991) *Arch. Biochem. Biophys.* **291**, 326. **2**. Kai, Goto, Tahara, *et al.* (2003) *J. Exp. Ther. Oncol.* **3**, 336. **3**. Southan, Truscott, Jamie, *et al.* (1996) *Med. Chem. Res.* **6**, 343. **4**. Chen, Liang, Peterson, Munn & Blazar (2008) *J. Immunol.* **181**, 5396. **5**. Ball, Yuasa, Austin, Weiser & Hunt (2009) *Int. J. Biochem. Cell Biol.* **41**, 467. **6**. Kudo & Boyd (2000) *Biochim. Biophys. Acta* **1500**, 119. **7**. Austin, Astelbauer, Kosim-Satyaputra, *et al.* (2009) *Amino Acids* **36**, 99. **8**. Macchiarulo, Camaioni, Nuti & Pellicciari (2009) *Amino Acids* **37**, 219. **9**. Ferry, Ubeaud, Lambert, *et al.* (2005) *Biochem. J.* **388**, 205. **10**. Friberg, Jennings, Alsarraj, *et al.* (2002) *Int. J. Cancer* **101**, 151. **11**. Spekker, Czesla, Ince, *et al.* (2009) *Infect. Immun.* **77**, 4496. **12**. Frenzel, Zhou & Floss (1990) *Arch. Biochem. Biophys.* **278**, 35.

1-Methyl-D-tryptophan

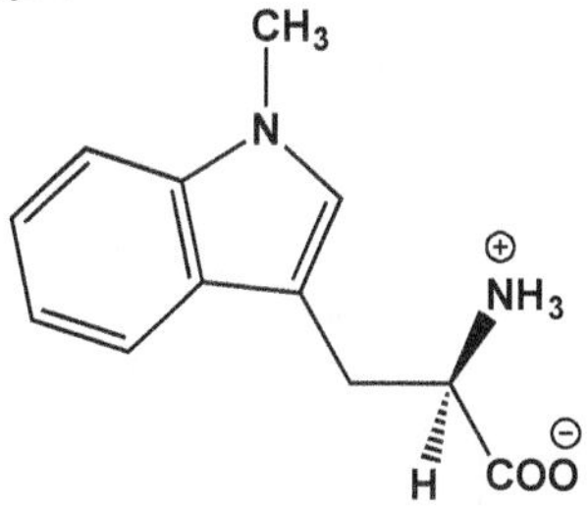

This D-tryptophan derivative (FW = 218.26 g/mol; CAS 110117-83-4; Soluble to 5 mM in water with gentle warming) inhibits indoleamine 2,3-dioxygenase (IC$_{50}$ = 7 μM), catalyzes the rate-limiting step in the metabolism of tryptophan along the kynurenine pathway. In tumors, increased IDO activity inhibits proliferation and induces apoptosis of T cells and natural killer cells. IDO is known to suppress antitumor CD8$^+$ T-cells (TCD8). 1-Methyl-D-tryptophan enhances the antitumor and antiviral immunoresponses of CD8$^+$ T-cells *in vitro*. It also reduces tumor volume in mice with xenografts overexpressing IDO, especially when used n combination with cyclophosphamide is an effective treatment for lymphoma in a mouse model (1,2). **1**. Rytelewski, Meilleur, Yekta, *et al.* (2014) *PLoS One* **9**, e90439. **2**. Nakamura, Hara, Shimizu, *et al.* (2015). *Int. J. Hematol.* **102**, 327.

2-Methyltryptophan

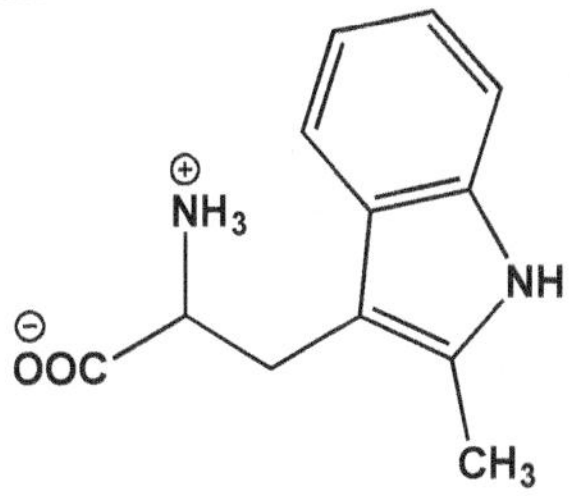

This rare tryptophan analogue (FW = 218.26 g/mol; CAS 33468-32-5) inhibits: 3-deoxy-7-phosphoheptulonate synthase, inhibited by the DL-mixture (1); phenylalanine decarboxylase, inhibited by the L-isomer (2); and tyrosine 3-monooxygenase, inhibited by the L-isomer (3). (**See also** α-Methyl L-tryptophan) **1**. Görisch & Lingens (1971) *Biochim. Biophys. Acta* **242**, 630. **2**. Lovenberg, Weissbach & Udenfriend (1962) *J. Biol. Chem.* **237**, 89. **3**. Moore & Dominic (1971) *Fed. Proc.* **30**, 859.

4-Methyltryptophan

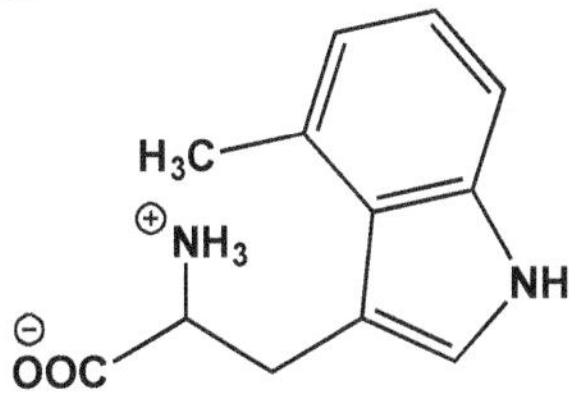

This tryptophan derivative (FW = 218.26 g/mol; CAS 1139-73-7) inhibits tryptophan synthase and is an activator of chorismate mutase. **Target(s):** alkaline phosphatase, mildly inhibited by the racemic mixture (1); anthranilate synthase (2); chorismate mutase (3); 3-deoxy-7-phosphoheptulonate synthase (4); indoleamine 2,3 dioxygenase, also weak alternative substrate (5); prephenate dehydratase6; tryptophan 2,3 dioxygenase (7); tryptophan synthase (8,9); tyrosine 3-monooxygenase, moderately inhibited (10). **1**. Fishman & Sie (1971) *Enzymologia* **41**, 141. **2**. Widholm (1972) *Biochim. Biophys. Acta* **279**, 48. **3**. Hertel, Hieke & Gröger (1991) *Acta Biotechnol.* **11**, 39. **4**. Görisch & Lingens (1971) *Biochim. Biophys. Acta* **242**, 630. **5**. Southan, Truscott, Jamie, *et al.* (1996) *Med. Chem. Res.* **6**, 343. **6**. Hagino & Nakayama (1974) *Agric. Biol. Chem.* **38**, 2367. **7**. Feigelson & Brady (1974) in *Mol. Mech. Oxygen Activ.* (Hayaishi, ed.) p. 87, Academic Press, New York. **8**. Webb (1966) *Enzyme and Metabolic Inhibitors*, vol. 2, p. 321, Academic Press, New York. **9**. Trudinger & Cohen (1956) *Biochem. J.* **62**, 488. **10**. McGeer, McGeer & Peters (1967) *Life Sci.* **6**, 2221.

5-Methyltryptophan

This methylated tryptophan (FW = 218.26 g/mol; CAS 2283-43-4) is a competitive inhibitor of tryptophanyl-tRNA synthetase and a false feedback inhibitor of anthranilate synthesis. **Target(s):** alkaline phosphatase, inhibited by the racemic mixture (1); anthranilate phosphoribosyltransferase (2); anthranilate synthase (3-7); indoleamine 2,3-dioxygenase, also alternative substrate (8); prephenate dehydratase (9); tryptophan 2,3-diooxygenase, inhibited by the DL-mixture, also weak alternative substrate (3,10,11); tryptophanyl-tRNA synthetase (12,13); tyrosine 3-monooxygenase (14,15). **1.** Fishman & Sie (1971) *Enzymologia* **41**, 141. **2.** O'Gara & Dunican (1995) *Appl. Environ. Microbiol.* **61**, 4477. **3.** Webb (1966) *Enzyme and Metabolic Inhibitors*, vol. **2**, Academic Press, New York. **4.** Bohlmann, Lins, Martin & Eilert (1996) *Plant Physiol.* **111**, 507. **5.** Moyed (1960) *J. Biol. Chem.* **235**, 1098. **6.** Matsukawa, Ishihara & Iwamura (2002) *Z. Naturforsch. [C]* **57**, 121. **7.** Widholm (1972) *Biochim. Biophys. Acta* **279**, 48. **8.** Southan, Truscott, Jamie, *et al.* (1996) *Med. Chem. Res.* **6**, 343. **9.** Hagino & Nakayama (1974) *Agric. Biol. Chem.* **38**, 2367. **10.** Hitchcock & Katz (1988) *Arch. Biochem. Biophys.* **261**, 148. **11.** Feigelson & Brady (1974) in *Mol. Mech. Oxygen Activ.* (Hayaishi, ed.) p. 87, Academic Press, New York. **12.** Davie (1962) *Meth. Enzymol.* **5**, 718. **13.** Sharon & Lipmann (1957) *Arch. Biochem. Biophys.* **69**, 219. **14.** Moore & Dominic (1971) *Fed. Proc.* **30**, 859. **15.** McGeer, McGeer & Peters (1967) *Life Sci.* **6**, 2221.

6-Methyltryptophan

This tryptophan derivative (FW = 218.26 g/mol; CAS 1991-93-1) inhibits tryptophanyl-tRNA synthetase and tryptophan 2,3-dioxygenase. **Target(s):** chorismate mutase (1); 3-deoxy-7-phosphoheptulonate synthase (2); indoleamine 2,3-dioxygenase, also alternative substrate (3); prephenate dehydratase (4); tryptophan 2,3 dioxygenase (5); tryptophanyl-tRNA synthetase (6,7); tyrosine 3-monooxygenase, moderately inhibited (8). **1.** Hertel, Hieke & Gröger (1991) *Acta Biotechnol.* **11**, 39. **2.** Görisch & Lingens (1971) *Biochim. Biophys. Acta* **242**, 630. **3.** Southan, Truscott, Jamie, *et al.* (1996) *Med. Chem. Res.* **6**, 343. **4.** Hagino & Nakayama (1974) *Agric. Biol. Chem.* **38**, 2367. **5.** Webb (1966) *Enzyme and Metabolic Inhibitors*, vol. **2**, p. 325, Academic Press, New York. **6.** Davie (1962) *Meth. Enzymol.* **5**, 718. **7.** Sharon & Lipmann (1957) *Arch. Biochem. Biophys.* **69**, 219. **8.** McGeer, McGeer & Peters (1967) *Life Sci.* **6**, 2221.

α-L-Methyltyrosine

This modified aromatic amino acid (FW = 195.22 g/mol; CAS 658-48-0), also known as metyrosine, inhibits tyrosine 3-monooxygenase and is an orally active inhibitor of catecholamine synthesis. **Target(s):** aromatic-L-amino-acid decarboxylase, *or* dopa decarboxylase (1); chorismate mutase (2); dihydropteridine reductase (3); prephenate dehydratase (4); tyrosine aminotransferase (5); tyrosine decarboxylase (1); tyrosine 3-monooxygenase (6-11). **1.** Nagy & Hiripi (2002) *Neurochem. Int.* **41**, 9. **2.**

Sugimoto & Shiio (1980) *J. Biochem.* **88**, 167. **3.** Shen (1983) *Biochim. Biophys. Acta* **743**, 129. **4.** Bode, Melo & Birnbaum (1985) *J. Basic Microbiol.* **25**, 291. **5.** Jacoby & La Du (1964) *J. Biol. Chem.* **239**, 419. **6.** Moore, Wright & Bert (1967) *J. Pharmacol. Exp. Ther.* **155**, 506. **7.** Weissman & Koe (1965) *Life Sci.* **4**, 1037. **8.** Moore & Dominic (1971) *Fed. Proc.* **30**, 859. **9.** McGeer, McGeer & Peters (1967) *Life Sci.* **6**, 2221. **10.** Nagatsu, Levitt & Udenfriend (1964) *J. Biol. Chem.* **239**, 2910. **11.** Neckameyer, Holt & Paradowski (2005) *Biochem. Genet.* **43**, 425.

α-Methyl-*m*-tyrosine

This L-phenylalanine derivative (FW = 195.22 g/mol; CAS 305-96-4), also known as α-methyl-3-hydroxy-L-phenylalanine, inhibits aromatic-L-amino-acid decarboxylase, *or* dopa decarboxylase (1-6), dopamine β-monooxygenase (1,7), phenylalanine decarboxylase (4), and tyrosine decarboxylase (5). **1.** Webb (1966) *Enzyme and Metabolic Inhibitors*, vol. **2**, Academic Press, New York. **2.** Daly, Mauger, Yonemitsu, *et al.* (1967) *Biochemistry* **6**, 648. **3.** O'Leary & Baughn (1977) *J. Biol. Chem.* **252**, 7168. **4.** Lovenberg, Weissbach & Udenfriend (1962) *J. Biol. Chem.* **237**, 89. **5.** Nagy & Hiripi (2002) *Neurochem. Int.* **41**, 9. **6.** Bender & Coulson (1977) *Biochem. Soc. Trans.* **5**, 1353. **7.** Hess, Connamacher, Ozaki & Udenfriend (1961) *J. Pharmacol. Exptl. Therap.* **134**, 129.

4-Methylumbelliferone

This fluorescent coumarin (FW = 176.17 g/mol; CAS 90-33-5), also known as hymecromone, is released by the action of specific glycosidases on synthetic glycosides containing the 4-methylumbelliferyl group. 4-Methylumbelliferyl-sulfate is also an alternative substrate for aryl sulfatases and 4-methylumbelliferyl oleate is hydrolyzed by lipases. Enzyme activity can be easily measured via fluorescence (λ_{max} = 448 nm, $\lambda_{excitation}$ = 364 nm). 4-Methylumbelliferone is also a substrate for UDP-glucuronosyltransferases and phenol sulfotransferases. **Target(s):** 17β-hydroxysteroid dehydrogenase (1); xanthine oxidase (2). **1.** Le Lain, Barrell, Saeed, *et al.* (2002) *J. Enzyme Inhib. Med. Chem.* **17**, 93. **2.** Beiler & Martin (1951) *J. Biol. Chem.* **192**, 831.

O-Methyluniconazole

This synthetic plant growth retardant (FW = 305.81 g/mol) strongly inhibits (+)-abscisate 8'-hydroxylase (K_i = 8.6 μM), a key enzyme in the catabolism of abscisic acid. The latter is a plant hormone involved in stress tolerance, stomatal closure, flowering, seed dormancy, and other physiological events. Conformational analysis and *in vitro* assays of enzyme inhibition underscore the importance of substituting a triazole in place of imidazole, the need of polar group at the 3-position to increas affinity for the active site via electrostatic or hydrogen-bonding interactions, and the conformer preference for a polar environment partially contributes to affinity for the active site. **1.** Todoroki, Kobayashi, Yoneyama, *et al.* (2008) *Bioorg. Med. Chem.* **16**, 3141.

6-Methyluracil

This pyrimidine (FW = 126.11 g/mol; CAS 626-48-2), also known as 2,4-dihydroxy-6-methylpyrimidine, inhibits dihydroorotate dehydrogenase (1,2), thymidine phosphorylase (3), thymine dioxygenase (4), and uridine phosphorylase (3). 1. Friedmann & Vennesland (1958) *J. Biol. Chem.* 233, 1398. 2. Webb (1966) *Enzyme and Metabolic Inhibitors*, vol. 2, p. 470, Academic Press, New York. 3. Avraham, Yashphe & Grossowicz (1988) *FEMS Microbiol. Lett.* 56, 29. 4. Bankel, Lindstedt & Lindstedt (1977) *Biochim. Biophys. Acta* 481, 431.

1-Methylxanthine

This substituted purine (FW = 166.14 g/mol; CAS 6136-37-4) inhibits hypoxanthine (guanine) phosphoribosyltransferase (1,2), methionine *S*-adenosyltransferase (3), xanthine phosphoribosyltransferase (1,4). 1. Naguib, Iltzsch, el Kouni, Panzica & el Kouni (1995) *Biochem. Pharmacol.* 50, 1685. 2. Miller, Ramsey, Krenitsky & Elion (1972) *Biochemistry* 11, 4723. 3. Berger & Knodel (2003) *BMC Microbiol.* 3, 12. 4. Miller, Adamczyk, Fyfe & Elion (1974) *Arch. Biochem. Biophys.* 165, 349.

Methyl β-D-Xylopyranoside

This glycoside (FW = 164.16 g/mol) inhibits acetylxylan esterase (1); aldose 1-epimerase (2), aryl-β-hexosidase (3,4), xylan 1,4-β xylosidase (5), and β-xylosidase/β-glucosidase (6). 1. Biely, Mastihubová, Côté & Greene (2003) *Biochim. Biophys. Acta* 1622, 82. 2. Bailey, Fishman, Kusiak, Mulhern & Pentchev (1975) *Meth. Enzymol.* 41, 471. 3. Distler & Jourdian (1978) *Meth. Enzymol.* 50, 524. 4. Distler & Jourdian (1977) *Arch. Biochem. Biophys.* 178, 631. 5. John & Schmidt (1988) *Meth. Enzymol.* 160, 662. 6. Yasui & Matsuo (1988) *Meth. Enzymol.* 160, 696.

Methysergide

This semisynthetic ergot alkaloid ergometrine derivative and serotonin antagonist (FW = 469.54 g/mol; CAS 129-49-7; Soluble to 10 mM in water with gentle warming), also named [8β(*S*)]-9,10-didehydro-*N*-[1-(hydroxymethyl)propyl]-1,6-dimethylergoline-8-carboxamide maleate, targets 5-HT$_1$ and 5-HT$_2$ receptors. Methysergide was synthesized from lysergic acid diethylamide (LSD) by adding a methyl group and a butanolamide group and resulting in a compound with selectivity and high potency as a serotonin (5-HT) inhibitor (1). Methysergide use in treating migraine declined considerably, when it was found to cause retroperitoneal fibrosis after chronic intake. The mechanism for its preventive effect in migraine remains elusive. 1. Koehler & Tfelt-Hansen (2008) *Cephalalgia* 28, 1126.

Metoclopramide

This antiemetic and antipsychotic agent (FW = 299.80 g/mol; CAS 364-62-5) is a dopamine D$_2$ receptor antagonist (1) as well as a 5-HT$_3$ serotonin receptor antagonist (2). It is also a reversible inhibitor of cholinesterases from human central nervous system and blood (3-5). **Key Pharmacokinetic Parameters:** *See* Appendix II in Goodman & Gilman's *THE PHARMACOLOGICAL BASIS OF THERAPEUTICS*, 12th Edition (Brunton, Chabner & Knollmann, eds.) McGraw-Hill Medical, New York (2011). 1. Kebabian & Calne (1979) *Nature* 277, 93. 2. De Winter, Boeckxstaens, De Man, *et al.* (1999) *Gut* 45, 713. 3. Kao, Tellez & Turner (1990) *Brit. J. Anaesth.* 65, 220. 4. Petroianu, Kuhn, Arafat, Zuleger & Missler (2004) *J. Appl. Toxicol.* 24, 143. 5. Graham & Crossley (1995) *Eur. J. Clin. Pharmacol.* 48, 225.

Metolachlor

This selective, pre-emergence herbicide (FW = 283.80 g/mol; CAS 51218-45-2), also known as Dual and 2-chloro-*N*-(2-ethyl-6-methylphenyl)-*N*-(2-methoxy-1-methylethyl)acetamide, is used to control certain broadleaf and annual grassy weeds. It also inhibits protein biosynthesis; thus, high-protein crops (*e.g.*, soy) can be adversely affected by excessive metolachlor applications. Additives may be included to help protect sensitive crops. It is detoxified by the formation of a glutathione adduct. **Target(s):** fatty-acid elongase; chalcone synthase; stilbene synthase. 1. Eckermann, Matthes, Nimtz, *et al.* (2003) *Phytochemistry* 64, 1045.

Metoprine

This lipid-soluble antifolate agent (FW = 269.13 g/mol; CAS 7761-45-7), also known as 2,4-diamino-5-(3,4-dichlorophenyl)-6-methylpyrimidine, reversibly inhibits dihydrofolate reductase, mouse leukemia cell line L1210, IC$_{50}$ ≈ 0.01 μM (1-3) and histamine *N*-methyltransferase (4-7). 1. Nichol, Cavillito, Woolley & Sigel (1977) *Cancer Treat Rep.* 61, 559. 2. Baker & Vermeulen (1970) *J. Med. Chem.* 13, 82. 3. Baker, Vermeulen, Ashton & Ryan (1970) *J. Med. Chem.* 13, 1130. 4. Nishibori, Oishi, Itoh & Saeki (1991) *Neurochem. Int.* 19, 135. 5. Lecklin, Tuomisto & MacDonald (1995) *Methods Find. Exp. Clin. Pharmacol.* 17, 47. 6. Hough, Khandelwal & Green (1986) *Biochem. Pharmacol.* 35, 307. 7. Horton, Sawada, Nishibori & Cheng (2005) *J. Mol. Biol.* 353, 334.

Metoprolol

This orally active, selective β$_1$ receptor blocker (FW = 267.36 g/mol; CAS 51384-51-1), also known by its trade names Lopressor™, Toprol-XL™ and by its IUPAC name (*RS*)-1-(isopropylamino)-3-[4-(2-methoxyethyl)-phenoxy]propan-2-ol, is an FDA-approved drug for treating hypertension, angina pectoris and, more generally, heart failure (1-3). While its precise antihypertensive mechanism of this β-blocker remains unknown, likely actions include: (a) competitive antagonism of catecholamines at peripheral (*i.e.,* cardiac) adrenergic neuron sites, thereby decreasing cardiac output; (b) central effects resulting in reduced sympathetic outflow to the periphery; and/or (c) suppression of renin-mediated effects. **Key Pharmacokinetic Parameters:** *See* Appendix II in Goodman & Gilman's *THE PHARMACOLOGICAL BASIS OF THERAPEUTICS,* 12th Edition (Brunton, Chabner & Knollmann, eds.) McGraw-Hill Medical, New York (2011). **Target(s):** insulin release in response to glucose, suggesting a partial dependence on intact β-adrenoceptors (4); mitochondrial coenzyme Q_{10}-enzymes of heart tissue, *i.e.,* succinoxidase and NADH-oxidase (5); hepatic 3,5,3'-triiodothyronine production (6); synaptosomal noradrenaline uptake (7); lysosomal phospholipase A_1 (8). **1.** Regårdh, Borg, Johansson, Johnsson & Palmer (1974) *J. Pharmacokinet. Biopharm.* **2**, 347. **2.** Johnsson, Nyberg & Sölvell (1975) *Acta Pharmacol. Toxicol.* (Copenhagen) **36** (Supplement 5), 69. **3.** Brogden, Heel, Speight & Avery (1977) *Drugs* **14**, 321. **4.** Ahrén & Lundquist (1981) *Eur. J. Pharmacol.* **71**, 93. **5.** Kishi, Kishi & Folkers (1975) *Res. Commun. Chem. Pathol. Pharmacol.* **12**, 533. **6.** Shulkin, Peele & Utiger (1984) *Endocrinology* **115**, 858. **7.** Street & Walsh (1984) *Eur. J. Pharmacol.* **102**, 315. **8.** Pappu, Yazaki & Hostetler (1985) *Biochem Pharmacol.* **34**, 521.

Metoxuron

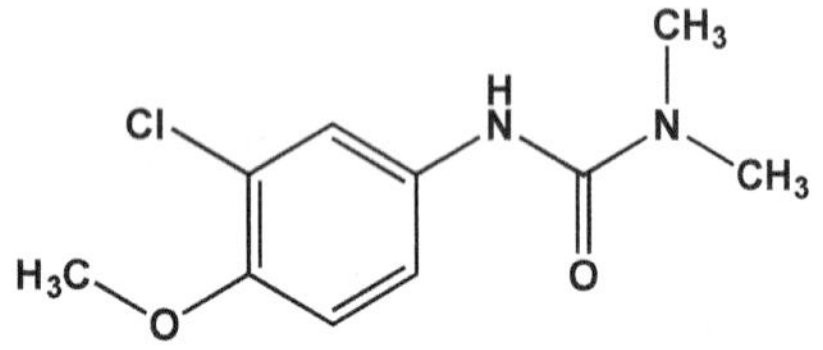

This herbicide (FW = 228.68 g/mol; CAS 19937-59-8; MP = 124-127 °C), also known as Deftor®, Pestanal®, Sulerex®, and 3-(3-chloro-4-methoxyphenyl)-1,1-dimethylurea, exerts selective pre- and post-emergence actions as a plant growth regulator and defoliant. Its primary mechanism of action is as a photosynthetic electron transport inhibitor, interrupting the flow of electrons in photosystem II and thereby blocking the chemiosmotic proton flow driving ATP synthesis. Metoxuron is effective against black grass, silky bent grass, wild oats, ryegrass, and most annual broad leaf weeds. UV irradiation of metoxuron in aerated aqueous solutions results in photohydrolysis of the C-Cl bond to form 3-(3-hydroxy-4-methoxyphenyl)-1,1 dimethylurea. **1.** Dodge (1977) *Spec. Publ., Chem. Soc.* **29**, 7.

Metronidazole

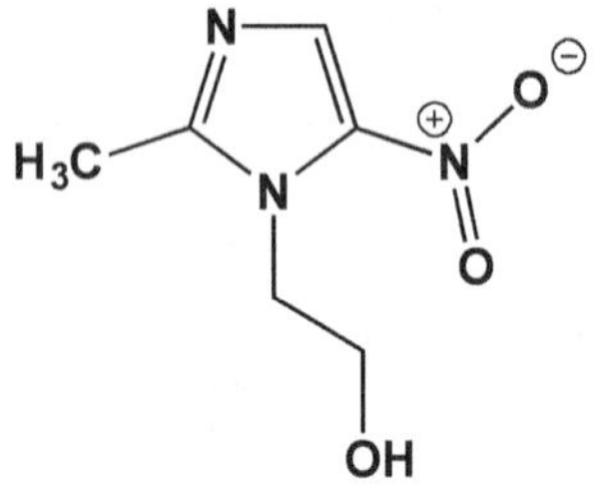

This oral antibiotic/antiprotozoal pro-drug (FW = 171.16 g/mol; CAS 443-48-1), also known by the brand name Flagyl® and as 2-methyl-5-nitroimidazole-1-ethanol and 1-(2 hydroxyethyl)-2-methyl-5-nitroimidazole, is active against anaerobic and some microaerophilic bacteria, including infections caused by *Bacteroides, Clostridium, Fusobacterium, Peptostreptococcus,* and *Prevotella* species. Metronidazole is an FDA-approved drug for the treatment of bacterial vaginosis in non-pregnant women (1). Metronidazole is also useful in treating parasitic infections, especially trichomoniasis. The U.S. Center for Disease Control and Prevention notes that trichomoniasis can be cured with a single oral dose of metronidazole or tinidazole. Given its mutagenic and potential carcinogenic properties, however, metronidazole is banned in the European Union and United States in the feed of animals and is likewise banned for use in food animals. **Mechanism of Action:** Metronidazole and related 5-nitroimidazole pro-drugs (tinidizole and ornidizole) exert their antimicrobial actions via inhibition of DNA synthesis, an action that first requires intracellular reduction of the nitro group to form a reactive radical species (likely pathway shown below).

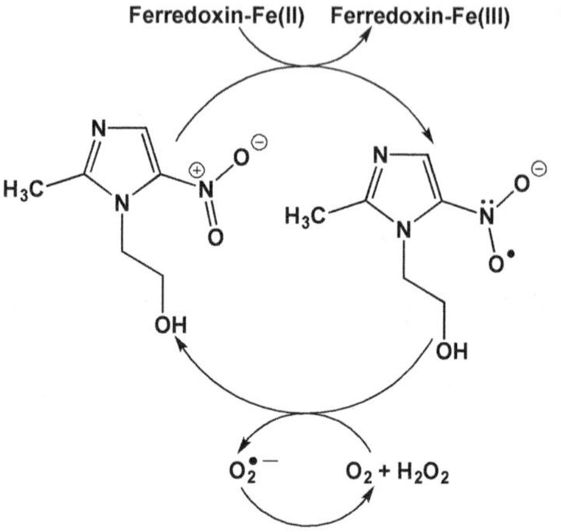

(Beyond microbial reduction, pro-drug activation can be mediated by rat liver microsomes, as demonstrated by using a mutant of of *S. typhikmurium* unable to activate metronidazole (2).) Once reduced, metronidazole binds to and lowers the melting temperature (T_m) of DNA, increasing the melting range, without affecting the cooling profile. The reduced drug also causes single- and double-strand breaks in DNA, but does not break single-stranded DNA. No such effects are observed with the unreduced drug. Upon reduction by sodium dithionite in the presence of DNA, metronidazole becomes covalently bound to DNA. Such properties are likely to explain the biological basis of Flagyl's action, not only as an antimicrobial agent, but also as a radiosensitizer (especially in hypoxic cells), and may underlie its mutagenic/carcinogenic properties. Likely radiosensitization mechanisms include enzyme-catalyzed depletion of nucleophiles, such as glutathione, fixation of organic radicals, interference with recombination reactions, and formation of toxic products (3). **Pharmacokinetics:** Oral metronidazole has a bioavailability approaching 100%. It shows limited plasma protein binding, yet attains favorable tissue distribution, even penetrating the Central Nervous System. It is oxidatively metabolized in the liver, forming two primary metabolites: the hydroxy- and acetate-containing derivatives. *See also* Appendix II in Goodman & Gilman's *THE PHARMACOLOGICAL BASIS OF THERAPEUTICS,* 12th Edition (Brunton, Chabner & Knollmann, eds.) McGraw-Hill Medical, New York (2011). **Target(s):** alcohol dehydrogenase (4-6); nitrogenase (7,8); and xanthine oxidase (9). **1.** Carey, Klebanoff, Hauth, *et al.* (2000) *New Engl. J. Med.* **342**, 534. **2.** Rosenkranz & Speck (1975) *Biochem. Biophys. Res. Comm.* **66**, 520. **3.** Wilson, Cranp & Ings (1974) *Internat. J. Rad. Biol.* **26**, 557. **4.** Brändén, Jörnvall, Eklund & Furugren (1975) *The Enzymes,* 3rd ed., **11**, 103. **5.** Paltrinieri (1967) *Farmaco* **22**, 1054. **6.** Edwards & Price (1967) *Biochem. Pharmacol.* **16**, 2026. **7.** Peterson (1988) *Biochem. Biophys Acta* **964**, 354. **8.** Kelley & Nicholas (1981) *Arch. Microbiol.* **129**, 344. **9.** Bray (1975) *The Enzymes,* 3rd ed., **12**, 299.

Metronomic Chemotherapy

This term describes a strategy for treating advanced stage cancer relies on the chronic administration of chemotherapeutic agents at relatively low, minimally toxic doses on a frequent and regular schedule of administration (hence the term "metronomic") and without prolonged drug-free breaks. Metronomic chemotherapy is believed to exert its beneficial effects by disrupting tumor-supportive angiogenesis mechanisms, thereby limiting tumor growth while significantly reducing a drug's undesirable toxic side effects. Rapidly eliminated vascular disrupting agents (VDAs), such as Combretastatin, appear to be especially well-suited for metronomic chemotherapy.

Metsulfuron-methyl

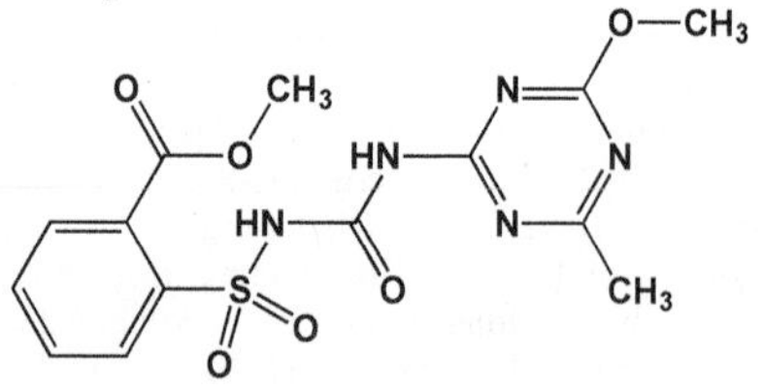

This pre- and post-emergence herbicide (FW = 381.37 g/mol; CAS 74223-64-6), also known as Ally and Allie, inhibits acetolactate synthase and branched-chain amino acid biosynthesis. **Target(s):** acetolactate synthase

(1-7); firefly luciferase, *or Photinus*-luciferin 4-monooxygenase (8). **1.** Zohar, Einav, Chipman & Barak (2003) *Biochim. Biophys. Acta* **1649**, 97. **2.** Lee, Chang & Duggleby (1999) *FEBS Lett.* **452**, 341. **3.** McCourt, Pang, Guddat & Duggleby (2005) *Biochemistry* **44**, 2330. **4.** Choi, Noh, Choi, *et al.* (2006) *Bull. Korean Chem. Soc.* **27**, 1697. **5.** Southan & Copeland (1996) *Physiol. Plant.* **98**, 824. **6.** Choi, Yu, Hahn, Choi & Yoon (2005) *FEBS Lett.* **579**, 4903. **7.** Chang, Kang, Choi & Namgoong (1997) *Biochem. Biophys. Res. Commun.* **234**, 549. **8.** Trajkovska, Tosheska, Aaron, Spirovski & Zdravkovski (2005) *Luminescence* **20**, 192.

Metyrapone

This Cytochrome P450 inhibitor and drug (FW = 226.28 g/mol; CAS 54-36-4; Soluble to 100 mM in H_2O and to 100 mM in DMSO), also known as 2-methyl-1,2-di-3-pyridyl-1 propanone, metipirone, 1,2-bis(3-pyridyl)-2-methyl-1-propanone, and SU 4885, is used to diagnose adrenal insufficiency and to treat Cushing's syndrome (hypercortisolism). Metyrapone is a nonspecific cytochrome P450 inhibitor, targeting CYP11B1 (steroid 11-β hydroxylase, IC_{50} = 7.8 μM), CYP2B, CYP 3A, and cytochrome P450-mediated prostaglandin ω/ω-1 hydroxylase. Metyrapone also induces glucocorticoid depletion, thereby modulating tyrosine hydroxylase and phenylethanolamine *N*-methyltransferase gene expression in the rat adrenal gland via noncholinergic transsynaptic activation.
Target(s): abietadiene hydroxylase (85); abietadienol hydroxylase (85); aldehyde dehydrogenase (1); aldehyde oxygenase (2); alkane-1 monooxygenase (3-6); aromatase, *or* CYP19 (7); aryl hydrocarbon monooxygenase (8); biphenyl hydroxylase (9); bufuralol 1'-hydroxylase, *or* cytochrome P450 (10,11); calcidiol 1-monooxygenase, *or* 25-hydroxyvitamin D 1α-hydroxylase (38-40,98-102); camphor 5-monooxygenase, *or* CYP101 (1); *S*-canadine synthase (77); cholestanetriol 26-monooxygenase, weakly inhibited (97); corticosterone 6β-hydroxylase (13); corticosterone 18-monooxygenase, steroid 18-monooxygenase (65); coumarin 7-hydroxylase, *or* CYP2A6 (14-19); CYP2A3 (15); CYP2A5 (16-20); CYP3A (21); CYP3A1 (21,22); CYP3A4 (21-23); CYP2B122,24-26; CYP2B224,25; CYP2B622; debrisoquine 4-monooxygenase (11); 7-deoxyloganin 7-hydroxylase (90); deoxysarpagine hydroxylase (27); dihydrochelirubine 12-monooxygenase (92); 8-dimethylallylnaringenin 2'-hydroxylase (86); ecdysone 20-monooxygenase (28,29,71-73); *ent*-isopimara 8(14),15-dien-3β-ol monooxygenase (30); estradiol 2-monooxygenase (31); estradiol 6α monooxygenase (31); estrogen-2/4-hydroxylase (32); 13-ethyl-gon-4-ene-3,17-dione 15α monooxygenase (33); glyceollin synthase (88); ω-hydroxylase, *or* alkane-1 monooxygenase (3-6); 11β hydroxysteroid dehydrogenase (34,35); 17β-hydroxysteroid dehydrogenase (36,37); 25-hydroxyvitamin D_3 24-hydroxylase (41,42); leukotriene B_4 20-monooxygenae, *or* leukotriene B_4 ω and ω-1 monooxygenase (44,96); licodione synthase (87); (*R*)-limonene 6-monooxygenase (89); (*S*)-limonene 3 monooxygenase (94); (*S*)-limonene 6-monooxygenase (94); (*S*)-limonene 7-monooxygenase (94); lipoxygenase, soybean (45); methyltetrahydroprotoberberine 14-monooxygenase (95); 11-oxo-Δ^8 tetrahydrocannabinol monooxygenase (46); progesterone 11α-monooxygenase (74-76); progesterone 14α-monooxygenase (47); prostaglandin ω and ω-1 hydroxylase (5,48,49); protopine 6 monooxygenase (93); steroid 7α-monooxygenase (50,51); steroid 9α-monooxygenase (70); steroid 11β monooxygenase, *or* CYP11B1, K_i = 0.1 μM (52-63,78-84); steroid 17α-monooxygenase (64); sterol 14-demethylase, *or* CYP51, *or* lanosterol 14α-demethylase (43,91); styrene monooxygenase (66,67); tangeretin *O*-demethylase (68); thromboxane synthase (54); vitamin D_{25}-hydroxylase, weakly inhibited (69). **1.** Martini & Murray (1996) *Biochem. Pharmacol.* **51**, 1187. **2.** Matsunaga, Iwawaki, Watanabe, *et al.* (1996) *J. Biochem.* **119**, 617. **3.** Pace-Asciak & Smith (1983) *The Enzymes*, 3rd ed. (Boyer, ed.), **16**, 543. **4.** Yamane & Abe (1995) *Neurosci. Lett.* **200**, 203. **5.** Asakura & Shichi (1992) *Exp. Eye Res.* **55**, 377. **6.** Miura (1982) *Lipids* **17**, 864. **7.** Breuer & Knuppen (1969) *Meth. Enzymol.* **15**, 691. **8.** Greiner, Malan-Shibley & Janss (1979) *Chem. Biol. Interact.* **27**, 323. **9.** Sietmann, Hammer, Moody, Cerniglia & Schauer (2000) *Arch. Microbiol* **174**, 353. **10.** Takemoto, Yamazaki, Tanaka, Nakajima & Yokoi (2003) *Xenobiotica* **33**, 43. **11.** Matsuo, Iwahashi &

Ichikawa (1992) *Biochem. Pharmacol.* **43**, 1911. **12.** Poulos & Howard (1987) *Biochemistry* **26**, 8165. **13.** Grogan, Phillips, Schuetz, Guzelian & Watlington (1990) *Amer. J. Physiol.* **258**, C480. **14.** Draper, Madan & Parkinson (1997) *Arch. Biochem. Biophys.* **341**, 47. **15.** Liu, Zhuo, Gonzalez & Ding (1996) *Mol. Pharmacol.* **50**, 781. **16.** Maenpaa, Sigusch, Raunio, *et al.* (1993) *Biochem. Pharmacol.* **45**, 1035. **17.** Goeger & Anderson (1992) *Biochem. Pharmacol.* **43**, 363. **18.** Kojo, Honkakoski, Jarvinen, Pelkonen & Lang (1989) *Pharmacol. Toxicol.* **65**, 104. **19.** Honkakoski, Kojo, Raunio, *et al.* (1988) *Arch. Biochem. Biophys.* **267**, 558. **20.** Emde, Tegtmeier, Hahnemann & Legrum (1996) *Toxicology* **108**, 73. **21.** Nebbia, Ceppa, Dacasto, Carletti & Nachtmann (1999) *Drug Metab. Dispos.* **27**, 1039. **22.** Kitamura, Shimizu, Shiraga, *et al.* (2002) *Drug Metab. Dispos.* **30**, 113. **23.** Iribarne, Berthou, Baird, *et al.* (1996) *Chem. Res. Toxicol.* **9**, 365. **24.** Dhawan, Parmar, Dayal & Seth (1999) *Mol. Cell. Biochem.* **200**, 169. **25.** Parmar, Dhawan & Seth (1998) *Mol. Cell Biochem.* **189**, 201. **26.** Goeptar, Te Koppele, *et al.* (1993) *Mol. Pharmacol.* **44**, 1267. **27.** Yu, Ruppert & Stöckigt (2002) *Bioorg. Med. Chem.* **10**, 2479. **28.** Greenwood & Rees (1984) *Biochem. J.* **223**, 837. **29.** Smith, Bollenbacher, Cooper, *et al.* (1979) *Mol. Cell. Endocrinol.* **15**, 111. **30.** Kato, Kodama & Akatsuka (1995) *Arch. Biochem. Biophys.* **316**, 707. **31.** Numazawa & Satoh (1988) *J. Steroid Biochem.* **29**, 221. **32.** Theron, Russell & Taljaard (1987) *J. Steroid Biochem.* **28**, 533. **33.** Irrgang, Schlosser & Fritsche (1997) *J. Steroid Biochem. Mol. Biol.* **60**, 339. **34.** Sampath-Kumar, Yu, Khalil & Yang (1997) *J. Steroid Biochem. Mol. Biol.* **62**, 195. **35.** Raven, Checkley & Taylor (1995) *Clin. Endocrinol.* **43**, 637. **36.** Stupans, Kong, Kirlich, *et al.* (1999) *Comp. Biochem. Physiol. C Pharmacol. Toxicol. Endocrinol.* **123**, 67. **37.** Jarabak & Sack (1969) *Biochemistry* **8**, 2203. **38.** Lobaugh, Almond & Drezner (1986) *Meth. Enzymol.* **123**, 159. **39.** Paulson, Phelps & DeLuca (1986) *Biochemistry* **25**, 6821. **40.** Paulson & DeLuca (1985) *J. Biol. Chem.* **260**, 11488. **41.** Kung, Kooh, Paek & Fraser (1988) *Can. J. Physiol. Pharmacol.* **66**, 586. **42.** Pedersen, Shobaki, Holmberg, Bergseth & Bjorkhem (1983) *J. Biol. Chem.* **258**, 742. **43.** Aoyama, Yoshida, Hata, Nishino & Katsuki (1983) *Biochem. Biophys. Res. Commun.* **115**, 642. **44.** Romano, Eckardt, Bender, *et al.* (1987) *J. Biol. Chem.* **262**, 1590. **45.** Pretus, Ignarro, Ensley & Feigen (1985) *Prostaglandins* **30**, 591. **46.** Watanabe, Hirahashi, Narimatsu, Yamamoto & Yoshimura (1991) *Drug Metab. Dispos.* **19**, 218. **47.** M. Madyastha & Joseph (1993) *J. Steroid Biochem. Mol. Biol.* **45**, 563. **48.** H. Oliw (1987) *Acta Physiol. Scand.* **129**, 259. **49.** H. Oliw & M. Hamberg (1986) *Biochim. Biophys. Acta* **879**, 113. **50.** Norlin & K. Wikvall (1998) *Biochim. Biophys. Acta* **1390**, 269. **51.** Doostzadeh, Cotillon & Morfin (1997) *J. Neuroendocrinol.* **9**, 923. **52.** Rosenthal & Narasimhulu (1969) *Meth. Enzymol.* **15**, 596. **53.** Katagiri, Takemori, Itagaki & Suhara (1978) *Meth. Enzymol.* **52**, 124. **54.** Coulson, King & Wiseman (1984) *Trends Biochem. Sci.* **9**, 446. **55.** Coon & Koop (1983) *The Enzymes*, 3rd ed. (Boyer, ed.), **16**, 645. **56.** Johansson, Sanderson & Lund (2002) *Toxicol. In Vitro* **16**, 113. **57.** Diederich, Grossmann, Hanke, *et al.* (2000) *Eur. J. Endocrinol.* **142**, 200. **58.** Yanagibashi, Shackleton & Hall (1988) *J. Steroid Biochem.* **29**, 665. **59.** Tobes, Hays, Gildersleeve, Wieland & Beierwaltes (1985) *J. Steroid Biochem.* **22**, 103. **60.** Hewick & Young (1973) *Biochem. Pharmacol.* **22**, 1653. **61.** Williamson & O'Donnell (1969) *Biochemistry* **8**, 1306. **62.** Williamson & O'Donnell (1967) *Can. J. Biochem.* **45**, 153. **63.** Wilson, Oldham & Harding (1969) *Biochemistry* **8**, 2975. **64.** Betz, Tsai & Weakley (1975) *Steroids* **25**, 791. **65.** Raman, Sharma & Dorfman (1966) *Biochemistry* **5**, 1795. **66.** Cox, Faber, Van Heiningen, *et al.* (1996) *Appl. Environ. Microbiol.* **62**, 1471. **67.** Salmona, Pachecka, Cantoni, *et al.* (1976) *Xenobiotica* **6**, 585. **68.** Canivenc-Lavier, Brunold, Siess & Suschetet (1993) *Xenobiotica* **23**, 259. **69.** Masumoto, Y. Ohyama & K. Okuda (1988) *J. Biol. Chem.* **263**, 14256. **70.** Kang & Lee (1997) *Arch. Pharm. Res.* **20**, 519. **71.** Feyereisen & Durst (1978) *Eur. J. Biochem.* **88**, 37. **72.** Johnson & Rees (1977) *Biochem. J.* **168**, 513. **73.** Horike & Sonobe (1999) *Arch. Insect Biochem. Physiol.* **41**, 9. **74.** Jayanthi, Madyastha & Madyastha (1982) *Biochem. Biophys. Res. Commun.* **106**, 1262. **75.** Kunic, Makovek & Breskvar (2000) *Pflugers Arch.* **439** (3 Suppl.), R107. **76.** Ghosh & Samanta (1981) *J. Steroid Biochem.* **14**, 1063. **77.** Rueffer & Zenk (1994) *Phytochemistry* **36**, 1219. **78.** Watanucki, Tilley & Hall (1978) *Biochemistry* **17**, 127. **79.** Yanagibashi, Haniu, Shively, Shen & Hall (1986) *J. Biol. Chem.* **261**, 3556. **80.** Denner, Vogel, Schmalix, Doehmer & Bernhardt (1995) *Pharmacogenetics* **5**, 89. **81.** Sato, Ashida, Suhara, *et al.* (1978) *Arch. Biochem. Biophys.* **190**, 307. **82.** Suzuki, Sanga, Chikaoka & Itagaki (1993) *Biochim. Biophys. Acta* **1203**, 215. **83.** Nagamine, Horisaka, Fukuyama, *et al.* (1997) *Biol. Pharm. Bull.* **20**, 188. **84.** Zuidweg (1968) *Biochim. Biophys. Acta* **152**, 144. **85.** Funk & Croteau (1994) *Arch. Biochem. Biophys.* **308**, 258. **86.** Yamamoto, Yatou & Inoue (2001) *Phytochemistry* **58**, 651. **87.** Otani, Takahashi, Furuya &

Ayabe (1994) *Plant Physiol.* **105**, 1427. **88**. Welle & Grisebach (1988) *Arch. Biochem. Biophys.* **263**, 191. **89**. Bouwmeester, Konings, Gershenzon, Karp & Croteau (1999) *Phytochemistry* **50**, 243. **90**. Katano, Yamamoto, Iio & Inoue (2001) *Phytochemistry* **58**, 53. **91**. Aoyama, Yoshida & Sato (1984) *J. Biol. Chem.* **259**, 1661. **92**. Kammerer, De-Eknamkul & Zenk (1994) *Phytochemistry* **36**, 1409. **93**. Tanahashi & Zenk (1990) *Phytochemistry* **29**, 1113. **94**. Karp, Mihaliak, Harris & Croteau (1990) *Arch. Biochem. Biophys.* **276**, 219. **95**. Rueffer & Zenk (1987) *Tetrahedron Lett.* **28**, 5307. **96**. Kikuta, Kusunose, Sumimoto, *et al.* (1998) *Arch. Biochem. Biophys.* **355**, 201. **97**. Atsuda & Okuda (1982) *J. Lipid Res.* **23**, 345. **98**. Gray, Omdahl, Ghazarian & DeLuca (1972) *J. Biol. Chem.* **247**, 7528. **99**. Ghazarian, Jefcoate, Knutson, Orme-Johnson & DeLuca (1974) *J. Biol. Chem.* **249**, 3026. **100**. Paulson & DeLuca (1985) *J. Biol. Chem.* **260**, 11488. **101**. Sakaki, Sawada, Takeyama, Kato & Inouye (1999) *Eur. J. Biochem.* **259**, 731. **102**. Paulson, Phelps & DeLuca (1986) *Biochemistry* **25**, 6821.

Mevalonate (Mevalonic Acid)

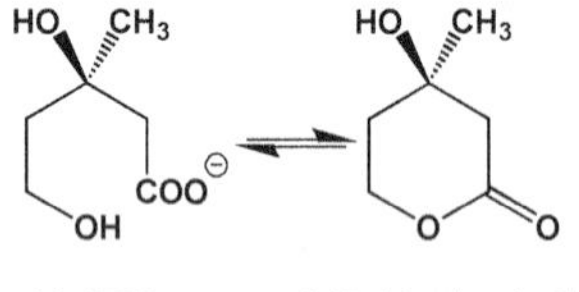

This β,δ-dihydroxy acid (FW$_{free-acid}$ = 148.16 g/mol; CAS 150-97-0; pK_a = 4.3; Very soluble in H$_2$O and many organic solvents) is a key metabolite in isoprenoid biosynthesis. The naturally occurring stereoisomer is (*R*)-mevalonate, and the (*S*)-isomer is biologically inactive. Its δ-lactone (shown above) is mevalonolactone (FW = 130.14 g/mol; CAS 674-26-0; Hygroscopic; M.P. = 28°C; rapidly hydrolyzed under alkaline conditions). It is enzymatically hydrolyzed, both in plasma and within liver cells. Mevalonolactone has no direct inhibitory effect on HMG-CoA reductase. Mevalonolactone administration to rats *in vivo* is associated with an inhibition of HMG-CoA reductase activity: (a) by increasing in the degree of phosphorylation of both HMG-CoA reductase and reductase kinase due to increased activity of reductase kinase kinase; and (b) by decreasing the dephosphorylation of both HMG-CoA reductase and reductase kinase, secondary to inhibition of phosphoprotein phosphatase activity (1). These combined effects favor an increase in the steady-state level of the phosphorylated forms of both HMG-CoA reductase and reductase kinase, resulting in a net reduction in the enzymic activity of HMG-CoA reductase and mevalonate formation. **Mevalonate's Role in Cellular Metabolism:** Mevalonate is a direct precursor of isopentenyl-pyrophosphate, which lies at a metabolic crossroad, with one path leading to isopentenyl-tRNA, and the other proceeding to farnesyl-pyrophosphate. The latter lies at a second crossroad, with paths ultimately leading to ubiquinone, cholesterol (via squalene formation), and dolichol. **Target(s):** diphosphomevalonate decarboxylase (2); phospho-mevalonate kinase (3). **1**. Beg, Stonik & Brewer (1984) *Proc. Natl. Acad. Sci. U.S.A.* **81**, 7293. **2**. Jabalquinto & Cardemil (1989) *Biochim. Biophys. Acta* **996**, 257. **3**. Pilloff, Dabovic, Romanowski, *et al.* (2003) *J. Biol. Chem.* **278**, 4510.

Mevalonic Acid 5-Diphosphate

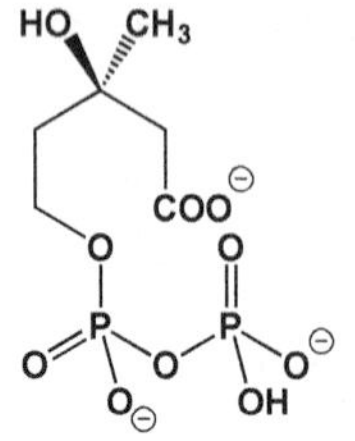

This metabolite (FW = 308.12 g/mol), also known as 5-pyrophosphomevalonate, diphosphomevalonate, and mevalonate 5-pyrophosphate, is an intermediate in the biosynthesis of sterols and isoprenoids. Note that aqueous solutions are relatively unstable: $t_{1/2}$ at 0°C and pH 6 is about three months, decomposing to mevalonate 5-phosphate and orthophosphate (the rate of decomposition is slower at alkaline pH). It is a product of the reaction catalyzed by phospho-mevalonate kinase and a substrate for diphosphomevalonate decarboxylase. **Target(s):** mevalonate kinase, IC$_{50}$ = 180 μM for the rat enzyme (1,2). **1**. Qiu & Li (2006) *Org. Lett.* **8**, 1013. **2**. Andreassi, Bilder, Vetting, Roderick & Leyh (2007) *Protein Sci.* **16**, 983.

Mevalonic Acid 5-Phosphate

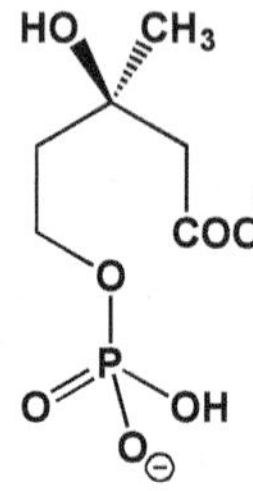

This metabolite (FW$_{trilithium-salt}$ = 245.94 g/mol (anhydrous); CAS 868553-49-5), also known as 5-phosphomevalonate, is an intermediate in isoprenoids and sterols biosynthesis. The (*R*)-enantiomer is a product of the reaction catalyzed by mevalonate kinase and is a substrate for phosphomevalonate kinase. Mevalonate 5-phosphate has pK_a values of 1.7, 4.6, and 6.9. Approximately 9% is hydrolyzed in one hour in 1 M HCl at 100°C. **Target(s):** diphosphomevalonate decarboxylase (1). **1**. Jabalquinto & Cardemil (1989) *Biochim. Biophys. Acta* **996**, 257.

Mevastatin

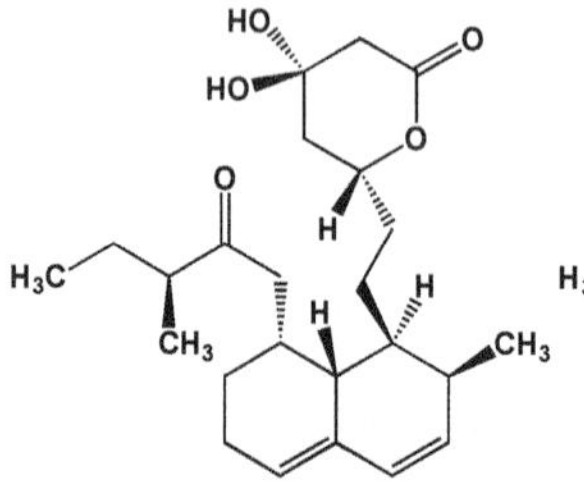

lactone 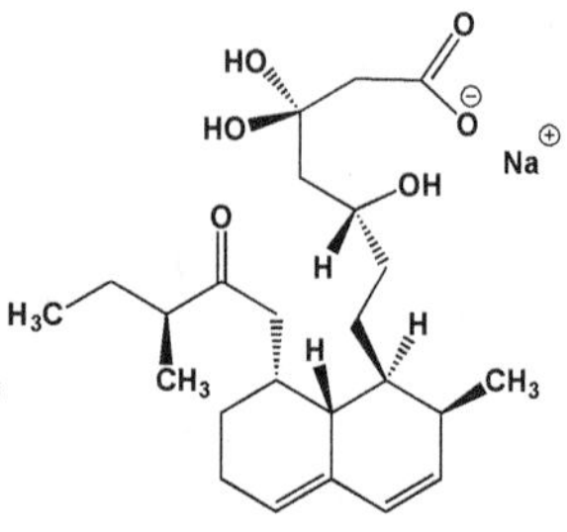   sodium salt

This fungal metabolite and hypolipemic agent (FW$_{lactone}$ = 390.52 g/mol; CAS 73573-88-3), also named ML-236B, compactin and (1*S*,7*R*,8*S*,8a*R*)-8-{2-[(2*R*,4*R*)-4-hydroxy-6-oxotetrahydro-2*H*-pyran-2-yl]ethyl}-7-methyl-1,2,3,7,8,8a-hexahydronaphthalen-1-yl(2*S*)-2-methylbutanoate, first found in *Penicillium citrinum*, is a highly effective prodrug that is metabolically hydrolyzed to the potent inhibitor (K_i = 1 nM, See structure above) of 3-hydroxy-3-methylglutaryl-CoA (*or* HMG-CoA) reductase (1-8). Although one of the first inhibitors used in the treatment of hypercholesterolemia, mevastatin was never commercialized. Mevastatin is unstable at room temperature and should be prepared immediately prior to use (1). Given mevastatin's powerful effects on the synthesis and action of other downstream isoprenoids, its's not surprising that mevastatin should affect cell cycle kinetics, often resulting in arrest in the late G$_1$ phase. Like other statins, mevastatin induces antiproliferative effects and apoptotic responses in various cancer cell types. It also sensitizes tumor cell lines of different origins to many agents. CR200 cells show > 100x higher mevastatin resistance than the parental FM3A cell line, when grown in the presence of low-density lipoprotein, *or* LDL (9). CR200 cells also have decreased levels of LDL receptor and defective mevalonate end-product regulation of HMG-CoA reductase, two properties that may underlie the over-accumulation of HMG-CoA reductase in CR200 cells and the basis of mevastatin resistance. **Prenylation Block & Release Assay:** This protocol relies upon mevastatin's ability to inhibit production of geranylgeranyl (GGPP) and farnesyl (FPP) pyrophosphates, key precursors in protein prenylation (10). Mevastatin blocks prenylation of newly synthesized substrates of GGT, RGGT and FT (*e.g.* Rho, Rab and Ras GTPases). Removal of the inhibitor releases this block, allowing GGPP and FPP biosynthesis, immediate protein pyrenylation, and subsequent membrane targeting of prenylated proteins. The process of membrane targeting from the cytosolic pool of modulator proteins like Rho, Rab and Ras to the membrane may then be observed directly using fluorescence microscopy. **1**. Endo (1981) *Meth. Enzymol.* **72**, 684. **2**. Qureshi, Nimmannit & Porter (1981) *Meth. Enzymol.* **71**, 455. **3**. Monger (1985) *Meth. Enzymol.* **110**, 51. **4**. Elbein (1987) *Meth. Enzymol.* **138**, 661. **5**. Endo (1981) *Trends Biochem. Sci.* **6**, 10. **6**. Chang (1983) *The Enzymes*, 3rd ed., **16**, 491. **7**. Gibson & Parker (1987) *The Enzymes*, 3rd ed., **18**, 179. **8**. Istvan & Deisenhofer (2001) *Science* **292**, 1160. **9**. Hasumi, Yamada, Shimizu & Endo (1987) *Eur. J. Biochem.* **164**, 547. **10**. Ali, Nouvel, Leung, Hume & Seabra (2010) *Biochem. Biophys. Res. Commun.* **397**, 34.

Mevinphos

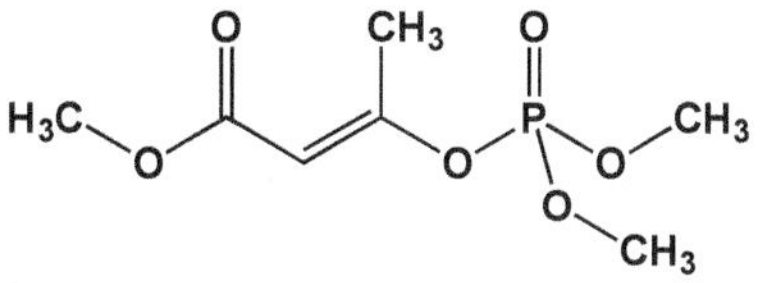

This organophosphate cholinesterase-type insecticide (FW = 224.15 g/mol; CAS 26718-65-0), readily produced by reacting trimethyl phosphite with chloroacetoacetate, is usually a mixture of the (*E*)- and (*Z*)-isomers. The (*E*)-isomer (shown above) is a hundred-times more potent. One experimental model for brain stem death employs mevinphos (10 nmol) injected bilaterally into Rostral Ventrolateral Medulla (RVLM) of Sprague-Dawley rats. **Target(s):** carboxylesterase (1); cholinesterase (2,3); Na$^+$/K$^+$-exchanging ATPase (4). **1**. Siddalinga Murthy & Veerabhadrappa (1996) *Insect Biochem. Mol. Biol.* **26**, 287. **2**. Watanabe & Sharma (1972) *Toxicol. Appl. Pharmacol.* **23**, 692. **3**. Sharma, Shupe & Potter (1973) *Toxicol. Appl. Pharmacol.* **24**, 645. **4**. Brown & Sharma (1976) *Experientia* **32**, 1540.

Mexiletine

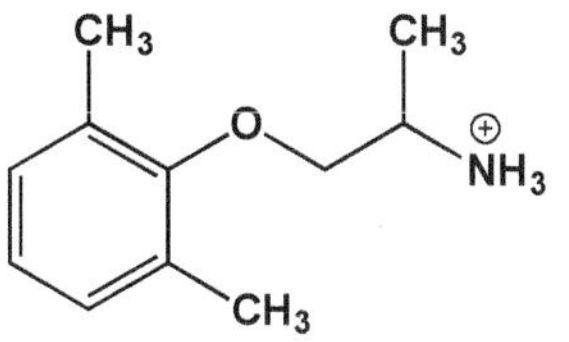

This class 1B antiarrhythmic drug (FW$_{free-base}$ = 179.26 g/mol; CAS 5370-01-4; Soluble to 100 mM in water and to 100 mM in DMSO; pK_a = 9.0; Symbol: Mex), also known by the code name Ko 1173 and the systematic name 1-(2,6-dimethylphenoxy)-2-propanamine hydrochloride, blocks the rapid inward sodium current responsible for Phase 0 of the action potential (1-4). It is primarily a use-dependent sodium channel blocker, showing IC$_{50}$ values of 75.3 and 23.6 μM for tonic and use-dependent block respectively. Mexiletine is used to treat ventricular arrhythmias. It is a neuroprotective and antimyotonic agent that has also been found to be effective for myotonia and neuropathic pain. Mexiletine, *p*-hydroxy-mexiletine (PHM), hydroxy-methyl-mexiletine (HMM), *N*-hydroxy-mexiletine (NHM) (phase I reaction products) and N-carbonyloxy-β-D-glucuronide (NMG) (phase II reaction product) gave sodium currents with depolarizing pulses from different holding potentials (HP=-140, -100, -70 mV) and stimulation frequencies (0.25, 0.5, 1, 2, 5, 10 Hz) using voltage-clamp methods. Although less potent than Mex, its phase I metabolites tested demonstrated similar pharmacological behaviour and may well contribute to its clinical profile (4). **Target(s):** Na$^+$ channel (1-4); monoamine oxidase (1-3). **1**. Clarke, Lyles & Callingham (1982) *Biochem. Pharmacol.* **31**, 27. **2**. Sicouri, Antzelevitch, Heilmann & Antzelevitch (1997) *J. Cardiovasc. Electrophysiol.* **8**, 1280. **3**. Labbe & Turgeon (1999) *Clin. Pharmacokinet.* **37**, 361. **4**. Monk & Brogden (1990) *Drugs* **40**, 374. **5**. De Bellis, De Luca, Rana, *et al.* (2006) *Brit. J Pharmacol.* **149**, 300. **6**. Eriksson & Fowler (1984) *Naunyn Schmiedebergs Arch. Pharmacol.* **327**, 273. **7**. Callingham (1977) *Brit. J. Pharmacol.* **61**, 118P.

MFZ 10-7

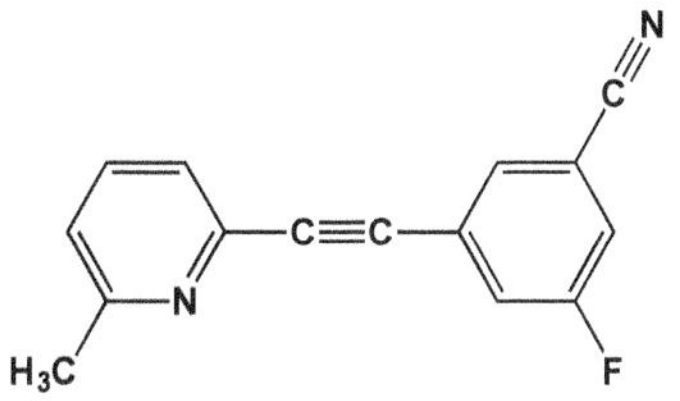

This metabotropic glutamate receptor allosteric inhibitor (FW$_{HCl-salt}$ = 272.70 g/mol; CAS 1224431-15-5; Solubility: 100 mM in DMSO or Ethanol), also named 3-fluoro-5-[2-(6-methyl-2-pyridinyl)ethynyl]-benzonitrile hydrochloride, targets the mGlu$_5$ receptor, K_i = 0.67 nM, potently inhibiting mGlu$_5$ glutamate-mediated calcium mobilization, IC$_{50}$ = 1.22 nM (1,2). MFZ 10-7 also suppresses cocaine-taking and cocaine-seeking behavior in rats (3). **1**. Alagille, *et al.* (2011) *Bioorg. Med. Chem. Lett.* **21**, 3243. **2**. Keck, *et al.* (2012) *ACS Med. Chem. Lett.* **3**, 544. **3**. Keck, *et al.* (2014) *Addict. Biol.* **19**, 195.

MG-132

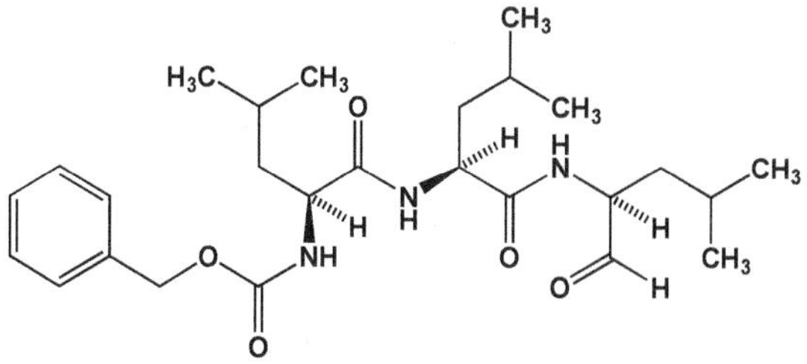

This peptidomimetic proteasome inhibitor (FW = 475.62 g/mol; CAS 133407-82-6; Solubility: 90 mg/mL DMSO; <1 mg/mL Water), also known as carbobenzoxyl-leucyl-leucyl-leucinal (or zLLLal) and systematically as benzyl (*S*)-4-methyl-1-((*S*)-4-methyl-1-((*S*)-4-methyl-1-oxopentan-2-ylamino)-1-oxopentan-2-ylamino)-1-oxopentan-2-ylcarbamate, targets the 20*S* proteasome, with IC$_{50}$ value of 100 nM. MG-132 inhibits calpain with IC$_{50}$ value of 1.2 μM and induces neurite outgrowth in PC12 cells, optimally at 20 nM, with 500x greater potency than ZLLal (1). MG-132 (10 μM) inhibits TNF-α-induced NF-κB activation, interleukin-8 (IL-8) gene transcription, and IL-8 protein release in A549 cells by inhibition of proteasome-mediated IκBα degradation (2) MG-132 treatment potently induces p53-dependent apoptosis in KIM-2 cells by 26S proteasome inhibition (3). (***See also*** *Syringolin A*) **1**. Tsubuki, *et al.* (1996) *J. Biochem.* **119**, 572. **2**. Fiedler, *et al.* (1998) *Am. J. Respir. Cell Mol. Biol.* **19**, 259. **3**. MacLaren AP, *et al.* (2001) *Cell Death Differ.* **8**, 210.

MHPT

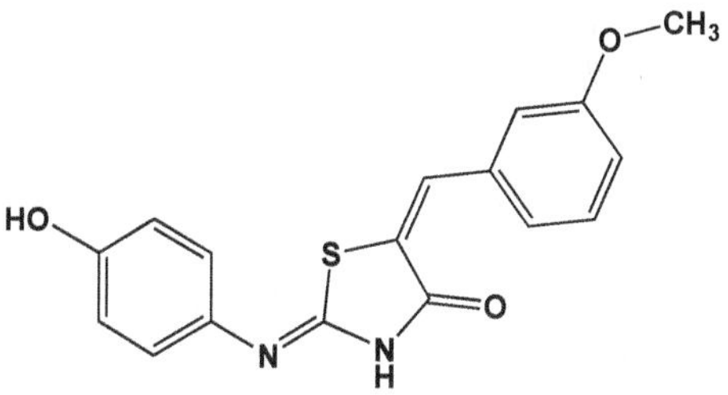

This novel synthetic tubulin polymerization inhibitor (FW = 326.37 g/mol), also named 2-((4-hydroxyphenyl)imino)-5-(3-methoxybenzylidene) thiazolidin-4-one, targets microtubule self-assembly (IC$_{50}$ = 6.3 μM), exhibiting selective cytotoxic action against rhabdomyosarcoma, both *in vitro* and *in vivo*. MHPT inhibited 50% of the growth of the rhabdomyosarcoma cell lines RD and SJ-RH30 at 0.44 μM and 1.35 μM, respectively, while displaying no obvious toxicity against normal human fibroblast cells at 100 μM. MHPT treatment leads to rhabdomyosarcoma cell cycle arrest at the G$_2$/M phase, followed by apoptosis. **1**. Mu, Liu, Li, *et al.* (2015) *PLoS One* **10**, e0121806.

MHY1485

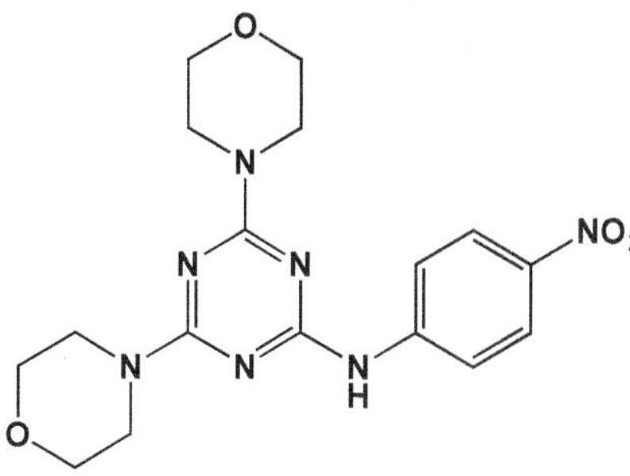

This potent, cell-permeable mTOR activator (FW = 387.39 g/mol; CAS 326914-06-1), also known as 4,6-dimorpholino-*N*-(4-nitrophenyl)-1,3,5-triazin-2-amine, potently inhibits autophagy. MHY1485 suppresses the basal autophagic flux in cells experiencing starvation, a strong physiological inducer of autophagy. The levels of p62 and beclin-1 do not change significantly after treatment with MHY1485. Decreased co-localization of autophagosomes and lysosomes in confocal microscopic images revealed MHY1485's inhibitory effect on lysosomal fusion during starvation-induced autophagy. These effects of MHY1485 led to the accumulation of LC3II and enlargement of the autophagosomes in a dose- and time-dependent manner. MHY1485 induces mTOR activation and show a higher docking score than PP242, a well-known ATP-competitive mTOR inhibitor. **1.** Choi, Park, Park, *et al.* (2012) *PLoS One* **7**, e43418.

MI-3

This protein-protein interaction inhibitor (FW = 375.55 g/mol; CAS 1271738-59-0), also named Menin-MLL Inhibitor and 4-[4-(4,5-dihydro-5,5-dimethyl-2-thiazolyl)-1-piperazinyl]-6-(1-methylethyl)thieno[2,3-*d*]pyrimidine, targets the binding of menin to Mixed Lineage Leukemia, or MLL (IC$_{50}$ = 56 nM), an interaction that plays a critical role in the viability of acute leukemias (1). This disruptor was identified by high-throughput screening, using 288,000 small molecule probes. Two aminoglycoside antibiotics (neomycin and tobramycin) are also inhibitors of menin-MLL interactions, albeit with much higher K_i values of 18.8 and 59.9 μM, respectively (2). Inhibition of this interaction therefore represents promising therapeutic strategy for combatting MLL leukemias. **1.** He, Senter, Pollock, *et al.* (2014) *J. Med. Chem.* **57**, 1543. **2.** Li, Zhou, Geng, *et al.* (2014) *Bioorg. Med. Chem. Lett.* **24**, 2090.

MI 192

This potent and selective HDAC inhibitor (FW = 383.44 g/mol; CAS 1415340-63-4; Soluble to 100 mM in DMSO), also named *N*-(2-aminophenyl)-4-[(3,4-dihydro-4-methylene-1-oxo-2(1*H*)-isoquinolinyl)-methyl]benzamide, targets histone deacetylases HDAC2 (IC$_{50}$ = 16 nM) and HDAC2 (IC$_{50}$ = 30 nM), exhibiting >250-fold selectivity over other HDAC isoforms. First-generation HDAC inhibitors (*e.g.*, butyrate and valproate) are simple short chain aliphatic compounds with a broad spectrum of activity against HDACs. MI-192 induces differentiation and was cytotoxic through promotion of apoptosis of leukemia cell lines *in vitro* (1). MI192 also inhibits TNF production at high concentrations and dose-dependently inhibited IL-6 in peripheral blood mononuclear cells (PBMCs) from RA patients, but not healthy PBMCs, when tested across a dose range of 5 nM to 10 μM (2). **1.** Boissinot, Inman, Hempshall, *et al.* (2012) *Leuk. Res.* **36**, 1304. **2.** Gillespie, Savic, Wong, *et al.* (2012) **64**, 418.

(S)-Mianserin

This tetracyclic antidepressant (FW = 264.37 g/mol; CAS 24219-97-4), also known as Depnon®, Lantanon®, Lerivon®, Lumin®, Norval®, Tolvon®, Tolmin®, and (±)-2-methyl-1,2,3,4,10,14*b*-hexahydrodibenzo[*c,f*]pyrazino[1,2-*a*]azepine, is an antagonist/inverse agonist of the H$_1$ receptors (K_i = 1 nM), 5-HT$_{1A}$ receptors (K_i ≈ 1500 nM), 5-HT$_{1F}$ receptors (K_i = 13 nM), 5-HT$_{2A}$ receptors (K_i = 3 nM), 5-HT$_{2B}$ receptors (K_i = 10 nM), 5-HT$_{2C}$ receptors (K_i = 3 nM), 5-HT$_6$ receptors (K_i = 70 nM), 5-HT$_7$ receptors (K_i = 60 nM), α$_1$-adrenoceptors (K_i = 70 nM), α$_{2A}$-adrenoceptors (K_i ≈ 5 nM), and α$_{2A}$-adrenoceptors (K_i ≈ 30 nM). Such broad action may also accounts for why mianserin also exhibits a range of anxiolytic, hypnotic, antiemetic, orexigenic, and antihistamine effects. That serotonin also plays a role in migraine, nociception, dumping syndrome, vascular disease, and hypertension, also suggests why mianserin has broad action. Used in the treatment of depressive illness and depression associated with anxiety, mianserin combines antidepressant activity with a sedative effect and has an EEG and clinical activity profile similar to that of amitriptyline (1). Its overall efficacy is comparable to amitriptyline and imipramine in depressive illness, but causes significantly fewer anticholinergic side effects and appears less likely than these drugs to cause serious cardiotoxicity on overdosage (1). Abrupt discontinuation often provokes withdrawal effects that can include depression, anxiety, panic attacks, decreased appetite, insomnia, diarrhea, nausea and vomiting. (*See also Amitriptyline; Flibanserin; Imipramin*) (Although the racemic mixture is most often used as a drug, *S*)-(+)-mianserin is some 200 to 300 times more active than (*R*)-(−)-mianserin.) **1.** Brogden, Heel, Speight & Avery (1978) *Drugs* **16**, 273.

Mibefradil

This antihypertensive agent (FW$_{free-base}$ = 495.64 g/mol; CAS 116666-63-8), also known as Ro 40-5967, is a nonselective T-type calcium channel blocker that was withdrawn from commerce amid evidence of dangerous and even fatal interactions with at least 25 other drugs, including common antibiotics, antihistamines, and cancer drugs. **Target(s):** Ca^{2+} channel (1-5); CYP3A (6,7); Na$^+$ channels (8). **1.** Mehrke, Zong, Flockerzi & Hofmann (1994) *J. Pharmacol. Exp. Ther.* **271**, 1483. **2.** Fang & Osterrieder (1991) *Eur. J. Pharmacol.* **196**, 205. **3.** Osterrieder & Holck (1989) *J. Cardiovasc. Pharmacol.* **13**, 754. **4.** Mishra & Hermsmeyer (1994) *Circ. Res.* **75**, 144. **5.** Aczel, Kurka & Hering (1998) *Brit. J. Pharmacol.* **125**, 447. **6.** Becquemont, Funck-Brentano & Jaillon (1999) *Fundam. Clin. Pharmacol.* **13**, 232. **7.** Wandel, Kim, Guengerich & Wood (2000) *Drug Metab. Dispos.* **28**, 895. **8.** McNulty & Hanck (2004) *Mol. Pharmacol.* **66**, 1652.

Micafungin

This semisynthetic echinocandin antifungal (FW$_{free-acid}$ = 1270.29 g/mol; CAS 235114-32-6; water-soluble lipopeptide), also known as Mycamine™ and FK463, is an intravenously administered drug that is effective against *Candida* and *Aspergillus* species. **Primary Mode of Action:** Micafungin inhibits β-1,3-glucan, an essential component in the formation of fungal cell walls. It also enhances the polymorphonuclear leukocytes oxidative burst in a dose-dependent manner, especially when PMN are pretreated with granulocyte-macrophage colony-stimulating factor. **Key Pharmacokinetic Parameters:** *See* Appendix II in Goodman & Gilman's *THE PHARMACOLOGICAL BASIS OF THERAPEUTICS*, 12th Edition (Brunton, Chabner & Knollmann, eds.) McGraw-Hill Medical, New York (2011). **Target(s):** CYP3A4, mildly inhibited (1,2); 1,3-β-glucan synthase (3,4). **1.** Hebert, Blough, Townsend, *et al.* (2005) *J. Clin. Pharmacol.* **45**, 1018. **2.** Sakaeda, Iwaki, Kakumoto, *et al.* (2005) *J. Pharm. Pharmacol.* **57**, 759. **3.** Maesaki,

Hossain, Miyazaki, *et al.* (2000) *Antimicrob. Agents Chemother.* **44**, 1728. **4**. Georgopapadakou (2001) *Expert Opin. Investig. Drugs* **10**, 269.

Michellamine B

This antioxidant and radical scavenger ($FW_{free\text{-}base}$ = 756.90 g/mol; Code Name: NSC 661755) from *Ancistrocladus korupensis* found in the Korup rain forest of Cameroon, inhibits HIV-1- and -2 induced cell killing. **Target(s):** bontoxilysin (1); DNA-directed DNA polymerase (2); DNA polymerases α and β (2); HIV reverse transcriptase (2); 12-lipoxygenase, *or* arachidonate 12-lipoxygenase (3); protein kinase C (4); RNA-directed DNA polymerase (2). **1**. Burnett, Schmidt, Stafford, *et al.* (2003) *Biochem. Biophys. Res. Commun.* **310**, 84. **2**. McMahon, Currens, Gulakowski, *et al.* (1995) *Antimicrob. Agents Chemother.* **39**, 484. **3**. Deschamps, Gautschi, Whitman, *et al.* (2007) *Bioorg. Med. Chem.* **15**, 6900. **4**. White, Chao, Ross, *et al.* (1999) *Arch. Biochem. Biophys.* **365**, 25.

Miconazole

This antifungal agent (FW = 416.13 g/mol; CAS 22916-47-8; Soluble to 100 mM in DMSO and to 10 mM in Ethanol), also known as Desenex™ and (*RS*)-1-(2-(2,4-dichlorobenzyloxy)-2-(2,4-dichlorophenyl)ethyl)-1*H*-imidazole, exerts its action against *Candida* and those causing athlete's foot, ringworm jock itch, oral or vaginal thrush, by inhibiting synthesis of ergosterol, a fungal membrane component critical for growth and proliferation. Miconazole and warfarin are CYP2C9 substrates and thus retard each other's hydroxylation. The net effect is that warfarin exerts a greater hypothrombinemic effect in miconazole's presence. In Ca^{2+} signaling research, miconazole has also been employed as a partial blocker for store-operated Ca^{2+} entry. **Mode of Action:** Although lanosterol 14α-demethylase (*or* CYP51) is widely distributed, it plays an essential role in mediating membrane permeability in fungi, with CYP51 catalyzing demethylation of lanosterol to create an essential precursor for ergosterol. The latter alters the permeability and rigidity of plasma membranes in a manner akin to that of cholesterol in animals. Many antifungal medications therefore target 14α-demethylase to limit fungal infections. **Target(s):** abietadiene hydroxylase (1); abietadienol hydroxylase (1); aromatase, *or* CYP19 (2,3); CYP1A1 (4); CYP3A4 (5); ATPase, plasma membrane (6); 7-deoxyloganin 7-hydroxylase (7); 1,25-dihydroxycholecalciferol metabolism (8); 8-dimethylallylnaringenin 2'-hydroxylase (9); F_1 ATPase, mitochondrial, *or* ATP synthase (10); H^+-exporting ATPase (11-13); (*R*)-limonene 6-monooxygenase (14); (*S*)-limonene 3-monooxygenase, weakly inhibited (15); (*S*) limonene 6-monooxygenase (15,16); (*S*)-limonene 7-monooxygenase, weakly inhibited (15); nitric oxide dioxygenase (17-19); steroid 11β-monooxygenase (20); sterol *O*-acyltransferase, *or* cholesterol *O*-acyltransferase, ACAT (21); sterol 14-demethylase, *or* CYP51 (22,23); tabersonine 16-hydroxylase (24); taxadiene 5α-hydroxylase (25); partial blocker for store-operated Ca^{2+} entry, such as in HL-60 cells (26), astrocytes (27), megakaryoblastic cells (28), and *Xenopus* oocytes (29). **1**. Funk & Croteau (1994) *Arch. Biochem. Biophys.* **308**, 258. **2**. Doody, Murry & Mason (1990) *J. Enzyme Inhib.* **4**, 153. **3**. Moslemi & Seralini (1997) *J. Enzyme Inhib.* **12**, 241. **4**. Petkam, Renaud & Leatherland (2003) *Comp. Biochem. Physiol. C Toxicol. Pharmacol.* **135C**, 277. **5**. Yamamoto, Suzuki

& Kohno (2004) *Xenobiotica* **34**, 87. **6**. Borst-Pauwels, Theuvenet & Stols (1983) *Biochim. Biophys. Acta* **732**, 186. **7**. Katano, Yamamoto, Iio & Inoue (2001) *Phytochemistry* **58**, 53. **8**. Napoli, Martin & Horst (1991) *Meth. Enzymol.* **206**, 491. **9**. Yamamoto, Yatou & Inoue (2001) *Phytochemistry* **58**, 651. **10**. Portillo & Gancedo (1984) *Eur. J. Biochem.* **143**, 273. **11**. Dufour, Amory & Goffeau (1988) *Meth. Enzymol.* **157**, 513. **12**. Uchida, Ohsumi & Anraku (1988) *Meth. Enzymol.* **157**, 544. **13**. Goffeau & Slayman (1981) *Biochim. Biophys. Acta* **639**, 197. **14**. Bouwmeester, Konings, Gershenzon, Karp & Croteau (1999) *Phytochemistry* **50**, 243. **15**. Karp, Mihaliak, Harris & Croteau (1990) *Arch. Biochem. Biophys.* **276**, 219. **16**. Croteau, Karp, Wagschal, *et al.* (1991) *Plant Physiol.* **96**, 744. **17**. Gardner (2005) *J. Inorg. Biochem.* **99**, 247. **18**. Helmick, Fletcher, Gardner, *et al.* (2005) *Antimicrob. Agents Chemother.* **49**, 1837. **19**. Hallstrom, Gardner & Gardner (2004) *Free Radic. Biol. Med.* **37**, 216. **20**. Denner, Vogel, Schmalix, Doehmer & Bernhardt (1995) *Pharmacogenetics* **5**, 89. **21**. Kim, Shin, Park, *et al.* (2004) *Biochem. Biophys. Res. Commun.* **319**, 911. **22**. McLean, Warman, Seward, *et al.* (2006) *Biochemistry* **45**, 8427. **23**. Trösken, Adamska, Arand, *et al.* (2006) *Toxicology* **228**, 24. **24**. St-Pierre & De Luca (1995) *Plant Physiol.* **109**, 131. **25**. Hefner, Rubenstein, Ketchum, *et al.* (1996) *Chem. Biol.* **3**, 479. **26**. Harper & Daly (2000) *Life Sci.* **67**, 651. **27**. Ju, Wang, Hung, *et al.* (2003) *Cellular Signaling* **15**, 197. **28**. Kunzelmann-Marche, Freyssinet & Martinez (2001) *J. Biol. Chem.* **276**, 5134. **29**. Lomax, Herrero, Garcia-Palomero, Garcia, & Montiel (1998) *Cell Calcium*, **23**, 229.

Microcin C₇

This toxin is a linear heptapeptide antibiotic (MW = 1171 Da; CAS 1403-96-9) is bacteriocidal by virtue of its inhibition of aspartyl-tRNA synthetase, thereby stalling the ribosomal protein translation machinery *in vivo* and *in vitro* (1-3). Microcin C_7 is effective against *Klebsiella*, *Eschericia coli*, *Shigella*, *Salmonella*, and *Yersinia*. The required aminopropylation uses *S*-adenosyl methionine to transfer 3-amino-3-carboxypropyl group onto a phosphate of an intermediate consisting of adenylated heptapeptide (4). **1**. Gonzalez-Pastor, San Millan, Castilla & Moreno (1995) *J. Bacteriol.* **177**, 7131. **2**. Guijarro, Gonzalez-Pastor, Baleux, *et al.* (1995) *J. Biol. Chem.* **270**, 23520. **3**. Kolter & Moreno (1992) *Ann. Rev. Microbiol.* **46**, 141. **4**. Kulikovsky, Serebryakova, Bantysh, *et al.* (2014) *J. Am. Chem. Soc.* **136**, 11168.

Micrococcin P₁

This protein biosynthesis inhibitor and antibiotic ($FW_{Micrococcin-P1}$ = 1151.29 g/mol; CAS 1392-45-6; Symbol: MC-LR) from *Staphylococcus equorum* Strain WS 2733 (*Micrococcus*) from French Raclette cheese species consists of two components, P_1 and P_2, with the former the major component. The antibiotic binds to the ribosomal protein L11 binding domain of 23S ribosomal RNA, preventing the binding of elongation factor G to the ribosome. **Target(s):** GTP diphosphokinase (1); protein biosynthesis (2-8). **1.** Knutsson Jenvert & Holmberg Schiavone (2005) *FEBS J.* **272**, 685. **2.** Pestka (1974) *Meth. Enzymol.* **30**, 261. **3.** Jiménez (1976) *Trends Biochem. Sci.* **1**, 28. **4.** Otaka & Kaji (1974) *Eur. J. Biochem.* **50**, 101. **5.** Cundliffe & Thompson (1981) *Eur. J. Biochem.* **118**, 47. **6.** Lentzen, Klinck, Matassova, Aboul-ela & Murchie (2003) *Chem. Biol.* **10**, 769. **7.** Cameron, Thompson, March & Dahlberg (2002) *J. Mol. Biol.* **319**, 27. **8.** Porse, Cundliffe & Garrett (1999) *J. Mol. Biol.* **287**, 33.

Microcystin LR

This cyclo-heptapeptide antibiotic (FW = 995.17 g/mol; CAS 101043-37-2), isolated form *Microcystis aeruginosa*, is a potent inhibitor of protein phosphatases types 1 and 2A (IC_{50} = 0.04 nM for phosphatase 2A and 1.7 nM for phosphatase 1). It is also a weak inhibitor of protein phosphatase 2B (IC_{50} = 200 nM). The microcystins, also known as cyanoginosins, are among the most important natural toxins of cyanobacterial origin. In cultured hepatocytes, Microcystin-LR significantly increases the expression of p53 and Bax, and decreases the expression of Bcl-2, which are involved in the regulation of Microcystin-LR-induced apoptosis (25). **Other Targets:** phosphoprotein phosphatase (1-23); [phosphorylase] phosphatase (1-7); protein phosphatase type 1 (1-7,11,14,15,19); protein phosphatase type 2A (1,3-6,12,13,15,20,22); protein phosphatase type 3 (23); protein phosphatase type 4 (8,9,15,17); protein phosphatase type 5 (10,15). MC-LR also increases Tau phosphorylation, with the dissociation of phosphorylated Tau from the cytoskeleton (24). MC-LR-induced Tau phosphorylation is suppressed by an activator of PP2A and by an inhibitor of p38 MAPK. VASP was also hyperphosphorylated upon MC-LR exposure, altering its cellular localization; its phosphorylation was unaffected by PP2A activators. Such data suggest that phosphorylated Tau is regulated by p38 MAPK, possibly as a consequence of PP2A inhibition (24). **IUPAC:** *cyclo*[2,3-didehydro-*N*-methylalanyl-D-alanyl-L-leucyl-(3*S*)-3-methyl-D-β-aspartyl-L-arginyl-(2*S*,3*S*,4*E*,6*E*,8*S*,9*S*)-3-amino-9-methoxy-2,6,8-trimethyl-10-phenyl-4,6-decadienoyl-D-γ-glutamyl]. **1.** Eriksson, Toivola, Sahlgren, Mikhailov & Härmälä-Braskén (1998) *Meth. Enzymol.* **298**, 542. **2.** Connor, Kleeman, Barik, Honkanen & Shenolikar (1999) *J. Biol. Chem.* **274**, 22366. **3.** MacKintosh, Beattie, Klumpp, Cohen & Codd (1990) *FEBS Lett.* **264**, 187. **4.** Dawson & Holmes (1999) *Front. Biosci.* **4**, D646. **5.** Lawton, Edwards, Beattie, *et al.* (1995) *Nat. Toxins* **3**, 50. **6.** Honkanen, Zwiller, Moore, *et al.* (1990) *J. Biol. Chem.* **265**, 19401. **7.** Zhang, Bai, Deans-Zirattu, Browner & Lee (1992) *J. Biol. Chem.* **267**, 1484. **8.** Huang, Cheng & Honkanen (1997) *Genomics* **44**, 336. **9.** Brewis, Street, Prescott & Cohen (1993) *EMBO J.* **12**, 987. **10.** Becker, Kentrup, Klumpp, Schultz & Joost (1994) *J. Biol. Chem.* **269**, 22586. **11.** Stubbs, Tran, Atwell, *et al.* (2001) *Biochim. Biophys. Acta* **1550**, 52. **12.** Szöör, Fehér, Bakó, *et al.* (1995) *Comp. Biochem. Physiol. B* **112**, 515. **13.** Hall, Feekes, Don, *et al.* (2006) *Biochemistry* **45**, 3448. **14.** Andrioli, Zaini, Viviani & da Silva (2003) *Biochem. J.* **373**, 703. **15.** Gallego & Virshup (2005) *Curr. Opin. Cell Biol.* **17**, 197. **16.** Solow, Young & Kennelly (1997) *J. Bacteriol.* **179**, 5072. **17.** Cohen, Philp & Vázquez-Martin (2005) *FEBS Lett.* **579**, 3278. **18.** Cohen (2004) in *Topics in Current Genetics* (Arino & Alexander, eds.) Springer **5**, 1. **19.** Hallaway & O'Kane (2000) *Meth. Enzymol.* **305**, 391. **20.** Dong, Ermolova & Chollet (2001) *Planta* **213**, 379. **21.** Douglas, Moorhead, Ye & Lees-Miller (2001) *J. Biol. Chem.* **276**, 18992. **22.** Kloeker, Reed, McConnell, *et al.* (2003) *Protein Expr. Purif.* **31**, 19. **23.**

Honkanen, Zwiller, Daily, *et al.* (1991) *J. Biol. Chem.* **266**, 6614. **24.** Sun, Liu, Huang, Guo & Xu (2013) *Environ. Toxicol.* **30**, 92. **25.** Fu, Chen, Wang & Xu (2005) *Toxicon.* **46**, 171.

MicroRNAs

These small non-coding RNAs (MW = 6-8 kDa; Length: ~22 bases), also called small interfering RNA (*or* siRNA), bind to 3'-untranslated region (3'-UTR) of a complementary mRNA sequence, regulating post-transcriptional mRNA expression and resulting in translational repression and gene silencing. Animal genomes contain numerous small genes coding for micro RNAs that are diverse, both in their sequence and expression patterns. MicroRNAs are thus widespread negative regulators of gene expression and playing pivotal roles in regulating cell proliferation, cancer, and apoptosis. **See** *PubMed listing of known microRNAs* (http://www.ncbi.nlm.nih.gov/gene/?term=microRNA).

Micrurotoxins

These first-in-class toxins, designated MmTX1 (MW = 7.2 kDa) and MmTX2 (MW = 7.2 kDa), from Costa Rican coral snake venom, bind to $GABA_A$ receptors at sub-nM concentrations, allosterically increasing $GABA_A$ receptor susceptibility to agonist, thereby potentiating receptor opening as well as desensitization, possibly by interacting at the α^+/β^- interface. Hippocampal neuron excitability measurements reveal toxin-induced transitory network inhibition, followed by an increase in spontaneous activity. MmTX1 and MmTX2 injections into mouse brain reduce basal activity between intense seizures. **1.** Rosso, Schwarz, Diaz-Bustamante, *et al.* (2015) *Proc. Natl. Acad. Sci. U.S.A.* **112**, E891.

Midazolam

This benzodiazepine-class drug (FW = 325.77 g/mol; CAS 59467-70-8) is used to treat acute seizures, moderate to severe insomnia, and to induce sedation and amnesia prior to medical procedures. Midazolam has a similar pharmacologic potency and broad therapeutic range as diazepam, exerting profound anxiolytic, amnesic, hypnotic, anticonvulsant, skeletal muscle relaxant, and sedative properties (1-3). With its fast recovery time, midazolam is often used as a premedication for sedation, and less commonly for inducing/maintaining anesthesia. Its anterograde amnesia property also commends its use as a premedication to inhibit unpleasant memories of surgery. Like classic benzodiazepines, midazolam enhances the inhibitory action of γ-aminobutyric acid by increasing the flow of chloride ions. When given intravenously, midazolam is rapidly distributed throughout the body ($t_{1/2}$ = 6–15 minutes, drowsiness commencing in 3 minutes (compared to ~15 minutes upon oral administration). Relative to other benzodiazepines, midazolam exhibits the shortest recovery time and a high metabolic clearance. When used in the U.S. as part of a three-drug cocktail: midazolam renders the condemned unconscious, at which time vecuronium bromide and potassium chloride are administered to arrest breathing and heart beat. (In the U.S., midazolam is classified as a non-narcotic agent, with low potential for abuse.) **Target(s):** Midazolam dose-dependently inhibits $K_v1.5$ current in an HEK cell line (IC_{50} = 17 μM) and in *Xenopus* oocytes (IC_{50} = 104 μM) (4). The inhibitory effects of midazolam on cloned hERG (human Ether-à-go-go-Related Gene) potassium channels have been examined using double-electrode voltage clamp on *Xenopus* oocytes heterologously expressing hERG channels, and, at 200 μM, midazolam significantly reduces the hERG activation current by 43 % and the hERG tail current amplitude by 52 % (5). Midazolam also has protective effects on the proliferation and apoptosis of astrocytes via JAK2/STAT3 signal pathway *in vitro* (6). **1.** Pieri, Schaffner, Scherschlicht, *et al.* (1981) *Arzneimittelforschung.* **31**, 2180. **2.** Carlen, Gurevich, Davies, Blaxter & O'Beirne (1985) *Can. J. Physiol. Pharmacol.* **63**, 831. **3.** Kanto & Allonen (1983) *Int. J. Clin. Pharmacol. Ther. Toxicol.* **21**, 460. **4.** Vonderlin, Fischer, Zitron, *et al.* (2015) *Drug Des. Devel. Ther.* **8**, 2263. **5.** Vonderlin, Fischer, Zitron, *et al.* (2015) *Drug Des. Devel. Ther.* **9**, 867. **6.** Liu, You, Tu, *et al.* (2015) *Cell Physiol. Biochem.* **35**, 126.

Midodrine

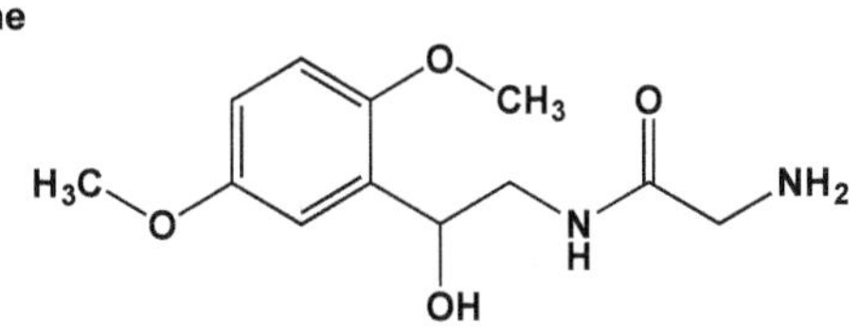

Midodrine

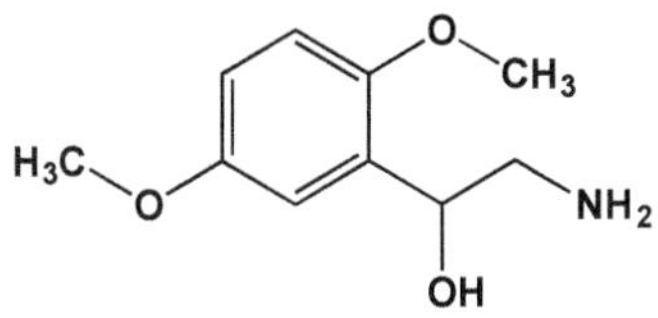

Desglymidodrine

This vasopressor/antihypotensive agent (FW = 254.28 g/mol; CAS 133163-28-7), also known as Amatine®, ProAmatine®, Gutron®, and (*RS*)- *N*-[2-(2,5-dimethoxyphenyl)-2-hydroxyethyl]glycinamide, is used to treat dysautonomia and orthostatic hypotension. Midodrine is actually a prodrug of the active metabolite, desglymidodrine (FW = 197.23 g/mol; CAS 3600-87-1; IUPAC: 1-(2',5'-dimethoxyphenyl)-2-aminoethanol), which is an α_1-receptor agonist and exerts its actions via activation of the α-adrenergic receptors of the arteriolar and venous vasculature. Accordingly, midodrine increases vascular tone and elevates blood pressure. **1**. Pittner & Turnheim (1974) *Arch. Int. Pharmacodyn. Ther.* **207**, 180. **2**. McTavish & Goa (1989) *Drugs* **38**, 757.

Mifepristone

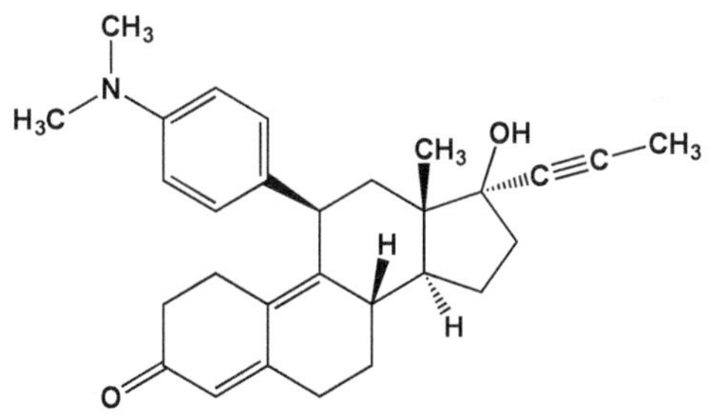

This selective progesterone receptor modulator, *or* SPRM (FW = 429.60 g/mol; CAS 84371-65-3; IUPAC: 11β-[*p*-(dimethylamino)phenyl]-17β-hydroxy-17-(1-propynyl)estra-4,9-dien-3-one), also known by the proprietary name Mifeprex® and more famously as RU-486 (previously, RU 38486), is a steroid analogue that exhibits great affinity for the progesterone receptor, but lacks progesterone activity (1). As a strong antagonist of both progesterone and glucocorticoid receptors in the ovary, RU-486 blocks progesterone effects. Indeed, oral administration interrupts the luteal phase of the menstrual cycle and early pregnancy (1-3). At all concentrations tested in castrated adult female cynomolgus monkeys, menstruation follow RU 486 injection within 48 hours and persists through 72 hours (4). A combination of the antiprogestagen mifepristone and an exogenous prostaglandin (**See** *Misoprostol*) given by intramuscular injection or intravaginal pessary is a highly effective means of inducing abortion in early pregnancy (5). Such properties commended their combined action as an oral emergency contraceptive and abortifacient, and in an FDA-approved protocol, mifeprex ends most early pregnancies within seven weeks after the start of the last menstrual period. The stereochemical antipode of RU-486 lacks antiprogesterone nor antiglucocorticoid activity (6), making it a valuable control in ligand binding experiments. RU 486 also elevates plasma ACTH, cortisol, and Plasma arginine vasopressin (AVP) concentrations in a manner that is both dose and time dependent (7) After RU 486 treatment, ACTH release occurs at an order of magnitude lower dose than did AVP release, and plasma AVP changes are glucocorticoid dependent (7). **1**. Herrmann, Wyss, Riondel, *et al.* (1982) *C. R. Seances Acad. Sci. III.* **294**, 933. **2**. Herrmann, Wyss, Riondel, *et al.* (1982) *C. R. Seances Acad. Sci. III.* **296**, 591. **3**. Schreiber, Hsueh & Baulieu (1983) *Contraception* **28**, 77. **4**. Healy, Baulieu & Hodgen (1983) *Fertil. Steril.* **40**, 253. **5**. Norman, Thong & Baird (1991) *Lancet* **338**, 1233. **6**. Ottow, Beier, Elger, *et al.* (1984) *Steroids* **44**, 519. **7**. Healy, Chrousos, Schulte, Gold & Hodgen (1985) *J. Clin. Endocrinol. Metab.* **60**, 1.

Milacemide

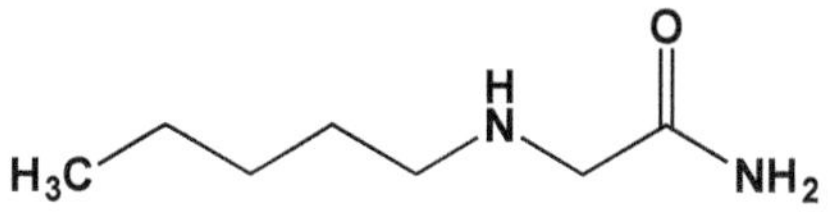

This glycinamide derivative and anti-epileptic (FW$_{\text{free-base}}$ = 144.22 g/mol; CAS 76990-56-2), also known as 2-*n*-pentylaminoacetamide, is converted to the inhibitory amino acid neurotransmitter, glycine, by monoamine oxidase B (K_m = 49 μM for the rat liver mitochondrial enzyme. A reversible inhibition of monoamine oxidase A is also observed at higher concentrations. **Target(s)**: amine oxidase (flavin-containing) (1-4). **1**. Silverman (1995) *Meth. Enzymol.* **249**, 240. **2**. O'Brien, Tipton, McCrodden & Youdim (1994) *Biochem. Pharmacol.* **47**, 617. **3**. Truong, Diamond, Pezzoli, Mena & Fahn (1989) *Life Sci.* **44**, 1059. **4**. Janssens de Varebeke, Pauwels, Buyse, *et al.* (1989) *J. Neurochem.* **53**, 1109.

Millipede Toxins

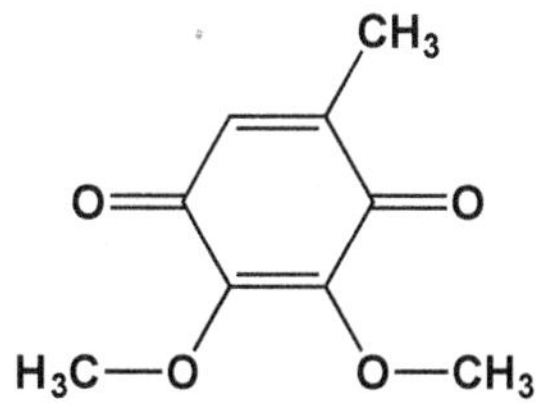

These cytotoxic substances are produced by one or more of 8,000 different species of multi-legged, worm-like arthropods (Class: *Diplopoda*) and include hydrochloric acid, hydrogen cyanide, various organic acids, phenol, cresols, benzoquinones, and hydroquinones. Some large species actually eject secretions over distances of up to 80 cm. One such substance, Ubiquinone-O (FW = 182.17 g/mol; CAS 605-94-7; IUPAC Name: 2,3-dimethoxy-5-methyl-1,4-benzoquinone) forms an unstable hydroquinone intermediate in the nonenzymatic oxidation of ascorbate and epinephrine and damages neuroblastoma cells (1,2). **1**. Sugihara, Watanabe, Kawamatsu & Morimoto (1972) *Justus Liebigs Ann, Chem.* **763**, 109. **2**. Roginsky, Bruchelt & Bartuli (1998) *Biochem. Pharmacol.* **55**, 85.

Milnacipran

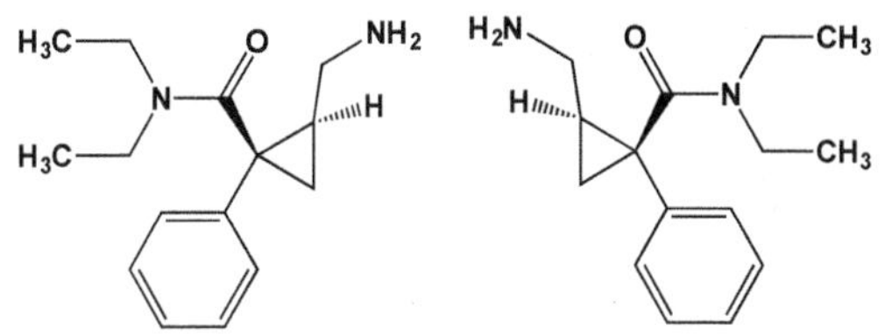

This antidepressant (FW = 246.35 g/mol; CAS 92623-85-3), also known as Ixel®, Savella®, Dalcipran®, Toledomin®, midalcipran, and (1*R**,2*S**)-2-(aminomethyl)-*N*,*N*-diethyl-1-phenylcyclopropane carboxamide, is a serotonin-norepinephrine reuptake inhibitor (*or* SNRI), with IC$_{50}$ values of 203 and 100 nM, respectively, for serotonin and norepinephrine, but without effect on dopamine reuptake (1). Milnacipran is used to treat fibromyalgia and major depressive disorder. This drug is a racemate, with one enantiomer more active (**See** *Levomilnacipran*). Midalcipran is largely eliminated in the urine as parent drug or as a glucuronide (2). **Target(s)**: Midalcipran does not inhibit monoamine oxidase, either *in vitro* or *in vivo*, and, in contrast to imipramine and desipramine, it showed no affinity for α- or β-adrenoceptors, muscarinic, histaminergic H$_1$, dopaminergic D$_2$ or serotonergic 5-HT$_2$ receptors, suggesting a general absence of anticholinergic, sedative and other side-effects (1). **1**. Moret, Charveron, Finberg, Couzinier & Briley (1985) *Neuropharmacol.* **24**, 1211. **2**. Puozzo & Leonard (1996) *Int. Clin. Psychopharmacol.* **4**, 15.

Milrinone

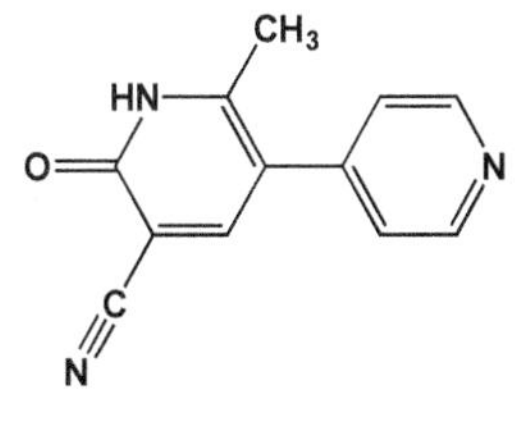

This vasodilator and cardiotonic (FW = 211.22 g/mol; CAS 78415-72-2), also known as 1,6-dihydro-2-methyl-6-oxo-(3,4' bypyridine)-5-carbonitrile, is a competitive inhibitor phosphodiesterase III (IC$_{50}$ = 0.15 μM) and is employed to treat heart failure and reduce pulmonary hypertension. At doses to 100 μM, milrinone is without effect on human cardiac cAMP-dependent protein kinase and adenylate cyclase. Moreover, cyclicAMP PDEs from human kidney, liver and lung are not readily inhibited, as IC$_{50}$ values for cAMP-PDEs from these tissues are about 7-30x higher than for the heart isozyme. Intravenous administration of milrinone produces improves hemodynamics, reducing systemic vascular resistance as well as left and right heart filling pressures (1). There is also a downward displacement of the left ventricular pressure-volume curve and acceleration of isovolumic relaxation. **Target(s):** 3',5'-cyclic-GMP phosphodiesterase (2); 3',5'-cyclic-nucleotide phosphodiesterase (2-9); phosphodiesterase 1 (2); phosphodiesterase 3A (2); phosphodiesterase 3B (2); phosphodiesterase 4 (2); phosphodiesterase 5 (2); phosphodiesterase 7 (2). 1. Colucci (1991) *Am. Heart J.* **121**, 1945. 2. Sudo, Tachibana, Toga, *et al.* (2000) *Biochem. Pharmacol.* **59**, 347. 3. Manganiello, Degerman & Elks (1988) *Meth. Enzymol.* **159**, 504. 4. Harrison, Beier, Martins & Beavo (1988) *Meth. Enzymol.* **159**, 685. 5. Houslay, Pyne & Cooper (1988) *Meth. Enzymol.* **159**, 751. 6. Matot & Gozal (2004) *Chest* **125**, 644. 7. Cheung, Yang & Boden (2003) *Metabolism* **52**, 1496. 8. Harrison, Chang & Beavo (1986) *Circulation* **73**, III109. 9. Ito, Tanaka, Saitoh, *et al.* (1988) *Biochem. Pharmacol.* **37**, 2041.

Miltefosine

This lysophospatidylcholine derivative and broad-spectrum antibiotic (FW$_{inner-salt}$ = 407.57 g/mol; CAS 58066-85-6), also known as hexadecylphosphocholine and choline hexadecyl phosphate, is a phospholipid analogue and antitumor agent that acts on cell membranes. Miltefosine also inhibits inositol phosphate formation and phosphatidylcholine biosynthesis. **Antibiotic Spectrum:** Miltefosine is primarily used in the treatment of visceral and New World cutaneous leishmaniasis, but is also available under an expanded-access IND protocol to treat infections arising from free-living amoeba, including primary amoebic meningoencephalitis caused by *Naegleria fowleri* and granulomatous amoebic encephalitis caused by *Balamuthia mandrillaris*, as well as infections by various *Acanthamoeba* species. **Target(s):** CTP:choline-phosphate cytidylyltransferase translocation (1); cytochrome-*c* oxidase (2); glyceryl-ether monooxygenase (3,4); phosphatidylcholine biosynthesis (5); phosphatidylinositol 3-kinase (6); phosphoethanolamine *N*-methyltransferase (7); phosphoinositide phospholipase C (8-10); phospholipase A$_2$ (11); phospholipase C (9,12); phosphoserine phosphatase (13); protein kinase C (8,14). 1. Geilen, Wieder & Reutter (1992) *J. Biol. Chem.* **267**, 6719. 2. Luque-Ortega & Rivas (2007) *Antimicrob. Agents Chemother.* **51**, 1327. 3. Taguchi & Armarego (1998) *Med. Res. Rev.* **18**, 43. 4. Kurisu, Taguchi, Paal & Armarego (1994) *Pteridines* **5**, 95. 5. Haase, Wieder, Geilen & Reutter (1991) *FEBS Lett.* **288**, 129. 6. Berggren, Gallegos, Dressler, Modest & Powis (1993) *Cancer Res.* **53**, 4297. 7. Pessi, Kociubinski & Mamoun (2004) *Proc. Natl. Acad. Sci. U.S.A.* **101**, 6206. 8. Uberall, Oberhuber, Maly, *et al.* (1991) *Cancer Res.* **51**, 807. 9. Berkovic, Goeckenjan, Luders, Hiddemann & Fleer (1996) *J. Exp. Ther. Oncol.* **1**, 302. 10. Ochocka & Pawelczyk (2003) *Acta Biochim. Pol.* **50**, 1097. 11. Berkovic, Luders, Goeckenjan, Hiddemann & Fleer (1997) *Biochem. Pharmacol.* **53**, 1725. 12. Pawelczyk & Lowenstein (1993) *Biochem. Pharmacol.* **45**, 493. 13. Hawkinson, Acosta-Burruel, Ta & Wood (1997) *Eur. J. Pharmacol.* **337**, 315. 14. Geilen, Haase, Buchner, *et al.* (1991) *Eur. J. Cancer* **27**, 1650.

Miltown, *See* Meprobamate

Mimosine

This amino acid (FW = 198.18 g/mol; CAS 500-44-7), also known as leucenol and 3-[3-hydroxy-4-oxo-1(4*H*)-pyridinyl]alanine, is an iron chelator that inhibits DNA replication in mammalian cells. Mimosine is also a folate antagonist and its effects are cell-specific. Large quantities of mimosine are found in species of the legume genera *Leucena* and *Mimosa*. **Target(s):** aromatic-L-amino-acid decarboxylase, *or* dopa decarboxylase, weakly inhibited (1); aspartate aminotransferase, pig heart (2); catechol oxidase (14,17,20,22); deoxyhypusine monooxygenase (3-6); L-dopachrome isomerase (7); dopamine β-hydroxylase (8); peptidylglycine α-amidating enzyme (9); polyphenol oxidase (20); procollagen-proline 4-dioxygenase (4); ribonucleotide reductase (10); serine hydroxymethyl-transferase, *or* glycine hydroxymethyl-transferase (11); tyrosinase, *or* monophenol monooxygenase, by L-isomer (8,12-17,21); tyrosine decarboxylase (18); tyrosine 3-monooxygenase (19). 1. Tocher & Tocher (1972) *Phytochemistry* **11**, 1661. 2. Lin & Tung (1964) *Kuo Li Taiwan Tah Hsueh* **10**, 69. 3. Andrus, Szabo, Grady, *et al.* (1998) *Biochem. Pharmacol.* **55**, 1807. 4. Clement, Hanauske-Abel, Wolff, Kleinman & Park (2002) *Int. J. Cancer* **100**, 491. 5. Csonga, Ettmayer, Auer, *et al.* (1996) *FEBS Lett.* **380**, 209. 6. Hanauske-Abel, Park, Hanauske, *et al.* (1994) *Biochim. Biophys. Acta* **1221**, 115. 7. Palumbo, d'Ischia, Misuraca, De Martino & Prota (1994) *Biochem. J.* **299**, 839. 8. Hashiguchi & Takahashi (1977) *Mol. Pharmacol.* **13**, 362. 9. Merkler, Kulathila & Ash (1995) *Arch. Biochem. Biophys.* **317**, 93. 10. Dai, Gold, Vishwanatha & Rhode (1994) *Virology* **205**, 210. 11. Rao, Talwar & Savithri (2000) *Int. J. Biochem. Cell Biol.* **32**, 405. 12. Lerch (1987) *Meth. Enzymol.* **142**, 165. 13. van Gastel, Bubacco, Groenen, Vijgenboom & Canters (2000) *FEBS Lett.* **474**, 228. 14. Kim & Uyama (2005) *Cell. Mol. Life Sci.* **62**, 1707. 15. Kahn & Andrawis (1985) *Phytochemistry* **24**, 905. 16. Orenes-Pinero, Garcia-Carmona & Sanchez-Ferrer (2007) *J. Mol. Catal. B* **47**, 143. 17. Khan, Mughal, Khan, *et al.* (2006) *Bioorg. Med. Chem.* **14**, 6027. 18. Crounse, Masewell & Blank (1962) *Nature* **194**, 694. 19. Yamamoto, Kobayashi, Yoshitama, Teramoto & Komamine (2001) *Plant Cell Physiol.* **42**, 969. 20. Yamamoto, Yoshitama & Teramoto (2002) *Plant Biotechnol.* **19**, 95. 21. Khan, Maharvi, Khan, *et al.* (2006) *Bioorg. Med. Chem.* **14**, 344. 22. Jaenicke & Decker (2008) *FEBS J.* **275**, 1518.

Minodronate

This bisphosphonate-based anti-osteoporosis drug (FW$_{free-acid}$ = 322.15 g/mol; CAS 180064-38-4), also named (1-hydroxy-2-imidazo[1,2-*a*]pyridin-3-yl-1-phosphonoethyl)phosphonate and YM-529, blocks osteoclast-mediated bone resorption by inhibiting *trans*-hexaprenyl*trans*-transferase and geranyl*trans*transferase. The most likely direct target is podosome assembly, a processes that assembles an actin filament-rich ring structure that localizes the proton-extruding V-ATPase for acidification and demineralization. Proetin prenylation is required for prenylation of Rac and Rho, two small G-proteins that regulate actin systems. Minodronate is also a potent inducer of apoptosis in human myeloma cells. **Target(s):** dimethylallyl*trans*transferase, *or* geranyl-diphosphate synthase (1); geranyl*trans*transferase, *or* farnesyl-diphosphate synthase (2-4); *trans*-hexaprenyl-*trans*transferase, *or* geranylgeranyl-diphosphate synthase (5). 1. Burke, Klettke & Croteau (2004) *Arch. Biochem. Biophys.* **422**, 52. 2. Dunford, Thompson, Coxon, *et al.* (2001) *J. Pharmacol. Exp. Ther.* **296**, 235. 3. Mao, Gao, Odeh, *et al.* (2004) *Acta Crystallogr. D Biol. Crystallogr.* **60**, 1863. 4. Montalvetti, Fernandez, Sanders, *et al.* (2003) *J. Biol. Chem.* **278**, 17075. 5. Guo, Cao, Liang, *et al.* (2007) *Proc. Natl. Acad. Sci. U.S.A.* **104**, 10022.

Minoxidil

This antihypertensive agent (FW = 209.25 g/mol; CAS 38304-91-5), also known as 2,4-diamino-6-piperidinopyrimidine-3-oxide, is vasodilator that activates ATP-activated potassium channels, bring about hyperpolarization. Minoxidil stops hair loss, by stimulating regrowth. Its mode of action is unclear. **Target(s):** prostaglandin-I synthase, *or* prostacyclin synthase (1); procollagen-lysine 5 dioxygenase, *or* lysyl hydroxylase (2,3); uridine kinase (4) **1**. Haurand & Ullrich (1992) *Front. Biotransform.* **6**, 183. **2**. Saika, Ooshima, Hashizume, *et al.* (1995) *Graefes Arch. Clin. Exp. Ophthalmol.* **233**, 347. **3**. Murad, Walker, Tajima & Pinnell (1994) *Arch. Biochem. Biophys.* **308**, 42. **4**. Macdonald, Walker, Assef & Duggan (1990) *Clin. Exp. Pharmacol. Physiol.* **17**, 287.

Mipomersen

This second-generation anti-sense oligonucleotide inhibitor (FW = 7595 g/mol; CAS 629167-92-6 (sodium salt); Tradename = Kynamro® Code Name = ISIS 301012) is as an adjunct to lipid-lowering medications and diet to reduce low-density lipoprotein-cholesterol (LDL-C), apolipoprotein B (apo B), total cholesterol (TC), and non-high density lipoprotein-cholesterol (non-HDL-C) in homozygous familial hypercholesterolemia (HoFH) patients. **Complete Sequence:** G*-C*-C*-U*-C*-dA-dG-dT-dC-dT-dG-dmC-dT-dT-dmC-G*-C*-A*-C*-C*, where stabilizing modifications are: d = 2'-deoxy–; m = methyl–; and * = 2'-*O*-(2-methoxyethyl)-groups, with all nucleotides joined by 3'→5' phosphorothioate linkages. The indicated chemical modifications render the drug resistant to nuclease degradation, allowing it to be administered on a once-weekly schedule. In January 2013, the FDA approved Kynamro as a once-weekly subcutaneous injection for use as an adjunct to maximally tolerated lipid-lowering medications for the treatment of patients with homozygous familial hypercholesterolemia (HoFH). **Mode of Action:** Hybridization of mipomersen by Watson-Crick base-pairing to the apolipoprotein B mRNA results in RNase H-mediated degradation of the mRNA, thereby inhibiting the biosynthesis of the apo B-100 protein (1-3). Mipomersen is a first-in-class mRNA-silencing agent that inhibits synthesis of apoB-100, the core protein for all atherogenic lipids, including low-density lipoprotein cholesterol, *or* LDL-C (1-3). The agent has been shown to produce significant reductions in LDL-C from baseline values compared with placebos. Inhibition of apolipoprotein B synthesis by antisense methodology represents a novel, effective therapy for reducing LDL cholesterol in HoFH patients who are already receiving lipid-lowering drugs, including high-dose statins. **Delivery & Lack of Pharmacokinetic Side-Effects:** The drug is administered as a single oral dose of 40 mg simvastatin or 10 mg ezetimibe, followed by four 2-hour intravenous doses of 200 mg mipomersen over an 8-day period, with ezetimibe and simvastatin again administered with the last dose of mipomersen sodium. Mipomersen exhibits no clinically relevant pharmacokinetic interactions with the disposition and clearance of simvastatin or ezetimibe, and *vice versa*. Moreover, mipomersen sodium does not inhibit CYP1A2, CYP2C9, CYP2C19, CYP2D6 or CYP3A4. The results support the use of mipomersen sodium in combination with oral lipid-lowering agents. *Note*: ApoB-100 is the principal carrier of 3-iodothyronamine, a naturally occurring derivative of thyroid hormone with biological actions that are distinct from those of T_4 and T_3 (4). **1**. Kastelein, Wedel, Baker, *et al.* (2006) *Circulation* **114**, 1729. **2**. Merki, Graham, Mullick, *et al.* (2008) Circulation **118**, 743. **3**. Yu, Geary, Flaim *et al.* (2009) *Clin. Pharmacokinet.* **48**, 39. **4**. Roy, Placzek & Scanlan (2012) *J. Biol. Chem.* **286**, 1790.

Mirtazapine

This α_2-adrenoceptor antagonist and specific serotonergic antidepressant, *or* NaSSA (FW = 265.35 g/mol; CAS 61337-67-5; Soluble to 50 mM in ethanol and to 20 mM in DMSO; IUPAC Name: (±)-2-methyl-1,2,3,4,10,14*b*-hexahydropyrazino[2,1-*a*]pyrido[2,3-*c*][2]benzazepine), known also by the trade name Remeron®, Avanza®, Axit®, Mirtazon®, an Zispin®, is used primarily to treat depression, but is also an antiemetic, anxiolytic, hypnotic, and appetite stimulant. Although mirtazapine has low affinity for 5-HT$_{1A}$ receptors, it nonetheless shows 5-HT$_{1A}$-agonistic-like

effects in a conditioned taste aversion test (1). Other studies show that both 5-HT$_2$ and 5-HT$_3$ receptors are blocked. The enhancement of both noradrenergic and serotonergic transmission is likely to underliy the therapeutic activity of mirtazapine (1). Mirtazapine also potentiates neurite outgrowth in PC12 pheochromocyte cells, indicating that this process may mediate part of its beneficial pharmacological effects (2). **Key Pharmacokinetic Parameters:** *See* Appendix II in Goodman & Gilman's *THE PHARMACOLOGICAL BASIS OF THERAPEUTICS*, 12th Edition (Brunton, Chabner & Knollmann, eds.) McGraw-Hill Medical, New York (2011). **1**. de Boer (1996) *J. Clin. Psychiatry.* **57** (Supplement 4), 19. **2**. Ishima, Fujita & Hashimoto (2004) *Eur. J. Pharmacol.* **727**, 167.

Misakinolide A

This actin filament-disrupting marine toxin (FW = 1369.80 g/mol; CAS 105304-96-9), also called bistheonellide A, was first isolated from a sponge (*Theonella* sp.) is a 40-membered dimeric lactone macrolide that differs from swinholide A with respect to the size of the macrolide ring. Misakinolide-treated cells tend to accumulate in the G$_1$ phase. **Primary Mode of Action:** Analytical ultracentrifugation and steady-state fluorescence experiments show that misakinolide A binds simultaneously to two actin subunits with virtually the same affinity (K_d ~50 nM) as swinholide A, suggesting that its ring size does not affect actin binding. Unlike swinholide A, misakinolide A does not sever actin filaments; rather, it caps actin filament barbed-ends, *or* (+)-ends. Misakinolide A has essentially no effect on the off-rate of actin subunits from the (+)-end. Energy-minimized models of misakinolide A and swinholide A are consistent with conservation of identical binding sites in both molecules, but a difference in orientation of one binding site relative to the other is likely to explain why swinholide A has severing activity, whereas misakinolide A only exhibits capping activity. **1**. Terry, Spector, Higa & Bubb (1997) *J. Biol. Chem.* **272**, 7841. **2**. Bubb & Spector (1998) *Meth. Enzymol.* **298**, 26.

Misoprostol

This synthetic prostaglandin E$_1$ (PGE$_1$) analogue (FW = 382.53; CAS 59122-46-2), also known as SC-29333, Cytotec® and its IUPAC name methyl 7-((1*R*,2*R*,3*R*)-3-hydroxy-2-((*S,E*)-4-hydroxy-4-methyloct-1-enyl)-5-oxocyclopentyl)heptanoate, is widely used to prevent gastric ulcers, to treat missed miscarriage, to induce abortion, especially when combined with RU-486 (**See Mifepristone**), and to induce labor. It inhibits secretion gastric acid secretion by G-protein coupled receptor-mediated inhibition of adenylate cyclase, leading to decreased intracellular cyclic AMP levels and decreased proton pump activity at the apical surface of the parietal cell. Other classes of drugs (especially H$_2$-receptor antagonists and proton pump inhibitors) are, however, more effective in the treatment of acute peptic ulcers, and misoprostol is only indicated for those who are taking NSAIDs and are at high risk for NSAID-induced ulcers, such as the elderly and people with ulcer complications. **1**. Akdamar, Agrawal & Ertan (1982) *Am. J. Gastroenterol.* **77**, 902. **2**. Ramage, Denton & Williams (1985) *Brit. J. Clin. Pharmacol.* **19**, 9.

Mithramycin

This chromomycin-like antibiotic (FW = 1085.16 g/mol; CAS 18378-89-7), also known as plicamycin and aureolic acid, is isolated from *Streptomyces argillaceus* and *S. tanashiensis*. Mithramycin binds reversibly to DNA with G:C base specificity and blocks protein biosynthesis via transcription inhibition. The Mg^{2+}:antibiotic complex interacts with chromatin and the complex stabilizes the DNA duplex and the histone-DNA contacts within chromatin fiber. The sugar moieties of mithramycin are thought to play a key role in the binding process. **Target(s):** DNA-directed RNA polymerase (1-4); DNA topoisomerae II (5). **1**. Kersten, Kersten, Szybalski & Fiandt (1966) *Biochemistry* **5**, 236. **2**. Mir, Majee, Das & Dasgupta (2003) *Bioorg. Med. Chem.* **11**, 2791. **3**. Chowdhury, Chowdhury & Neogy (1981) *Indian J. Med. Res.* **73**, 90. **4**. Sethi (1971) *Prog. Biophys. Mol. Biol.* **23**, 67. **5**. Christmann-Franck, Bertrand, Goupil-Lamy, *et al.* (2004) *J. Med. Chem.* **47**, 6840.

Mitochondrial ATPase Inhibitor

Mitochondrial ATP synthase (*i.e.*, F_1F_o-ATPase) is regulated by an intrinsic ATPase inhibitor protein. The bovine inhibitor protein, IF_1, contains 84 aminoacyl residues and is an α helical dimer. The proton-motive force across the inner mitochondrial membrane collapses when a cell is deprived of oxygen. If this occurs, the F_1F_o ATP synthase switches from synthesis to hydrolysis. This hydrolytic activity is prevented by a natural inhibitor protein, IF_1, and the action of this protein depends on the presence of ATP: the ATP molecule is trapped in the enzyme in the presence of IF_1. **Target(s):** F_1 ATP synthase (1-8). **1**. Monroy & Pullman (1967) *Meth. Enzymol.* **10**, 510. **2**. Ernster, Carlsson, Hundal & Nordenbrand (1979) *Meth. Enzymol.* **55**, 399. **3**. Cintrón & Pedersen (1979) *Meth. Enzymol.* **55**, 408. **4**. Satre, Klein & Vignais (1979) *Meth. Enzymol.* **55**, 421. **5**. Ichikawa & Ogura (2003) *J. Bioenerg. Biomembr.* **35**, 399. **6**. Cabezon, Montgomery, Leslie & Walker (2003) *Nat. Struct. Biol.* **10**, 744. **7**. Pullman & Monroy (1963) *J. Biol. Chem.* **238**, 3762. **8**. Green & Grover (2000) *Biochim. Biophys. Acta* **1458**, 343.

Mitomycin C

This interstrand cross-linker, *or* ICL (FW = 334.33 g/mol; CAS 50-07-7; Symbol: MMC; decomposed upon exposure to light; hence, solutions should be stored in the dark; stable in solution at pH 6–9 at 5°C for at least a week) from *Streptomyces caespitosus* chemically alters DNA, physically blocking replication, recombination, and transcription (1). In mice, the i.v. LD_{50} is 5-9 mg/kg, but there is a delayed response, with death occurring in 2-14 days following injection. **Chemistry:** Once within cells, mitomycin C undergoes rearrangement and reduction to generate a bifunctional alkylating

reagent, which then binds preferentially to the N-position of guanine in the minor groove of duplex DNA, forming monoadducts. The latter reach out to form intrastrand crosslinks with an adjacent guanine residue as well as interstrand crosslinks at opposite guanines of adjacent base-pairs, specifically in the sequence orientation of CpG, but not GpC.

Monofunctional Adduct with Guanine Residue

Bifunctional Cross-link with Guanine Residues

Mitomycin also binds to both RNA. *Note*: This dark blue solid is readily soluble in water and is Mitomycin C, the least toxic of the mitomycins, is a potent antitumor agent. **Target(s):** DNA biosynthesis (1-4); DNAse (5); DT diaphorase, *or* NAD(P)H dehydrogenase (6); indoleamide 2,3-dioxygenase (7). **1**. Ben-Yehoyada, Wang, Kozekov, *et al.* (2009) *Mol Cell* **35**, 704. **2**. Weissbach & Lisio (1965) *Biochemistry* **4**, 196. **3**. Szybalski & Iyer (1964) *Fed. Proc.* **23**, 946. **4**. Iyer & Szybalski (1964) *Science* **145**, 55. **5**. Nakata, Nakata & Sakamoto (1961) *Biochem. Biophys. Res. Commun.* **6**, 339. **6**. Siegel, Beall, Kasai, *et al.* (1993) *Mol. Pharmacol.* **44**, 1128. **7**. Lu, Lin & Yeh (2009) *J. Amer. Chem. Soc.* **131**, 12866.

Mitoxantrone

This DNA-intercalating drug ($FW_{free-base}$ = 444.49 g/mol; CAS 65271-80-9), also known by its trade name Novantrone™ and by its IUPAC name, 1,4-dihydroxy-5,8-bis[2-(2-hydroxyethylamino)ethylamino]anthracene-9,10-dione, is an immunosuppressive and cytostatic agent that exhibits antitumor activity. Mitoxantrone is a type II topoisomerase inhibitor that disrupts DNA synthesis and repair in healthy and cancer cells. Novantrone is also an FDA-approved MS drug that delays the time to first treated relapse as well as disability progression. **Key Pharmacokinetic Parameters:** *See* Appendix II in Goodman & Gilman's *THE PHARMACOLOGICAL BASIS OF THERAPEUTICS*, 12th Edition (Brunton, Chabner & Knollmann, eds.) McGraw-Hill Medical, New York (2011). **Target(s):** DNA helicase (1); DNA topoisomerase II (2-8); [myosin light-chain] kinase (9); nucleoside-triphosphatase (1). **1**. Borowski, Niebuhr, Schmitz, *et al.* (2002) *Acta Biochim. Pol.* **49**, 597. **2**. Epstein & Smith (1988) *Cancer Res.* **48**, 297. **3**. Crespi, Ivanier, Genovese & Baldi (1986) *Biochem. Biophys. Res. Commun.* **136**, 521. **4**. McDonald, Eldredge, Barrows & Ireland (1994) *J. Med. Chem.* **37**, 3819. **5**. Andersen, Bendixen & Westergaard (1996) *DNA Replication in Eucaryotic Cells, Cold Spring Harbor Laboratory Press*, p. 587. **6**. Hammonds, Maxwell &

Jenkins (1998) *Antimicrob. Agents Chemother.* **42**, 889. **7**. Insaf, Danks & Witiak (1996) *Curr. Med. Chem.* **3**, 437. **8**. Skladanowski, Come, Sabisz, Escargueil & Larsen (2005) *Mol. Pharmacol.* **68**, 625. **9**. Jinsart, Ternai & Polya (1992) *Biol. Chem. Hoppe-Seyler* **373**, 903.

Mitragynine & Mitragynine Pseudoindoxyl

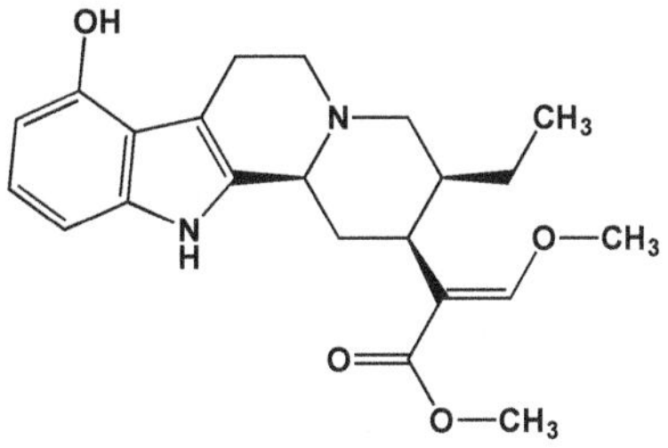

Mitragynine

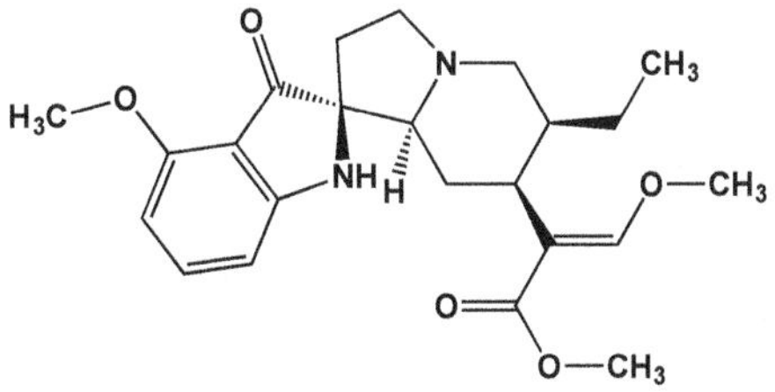

Mitragynine pseudoindoxyl

This alkaloid (FW = 398.49 g/mol; CAS 6202-22-8), from the Thai medicinal plant *Mitragyna speciosa* (kratom), is oxidized metabolically oxidized to form the μ-opioid receptor agonist, mitragynine pseudoindoxyl (FW = 414.50 g/mol), with the latter showing lower affinity for δ or κ receptors (1). Mitragynine pseudoindoxyl inhibits the electrically stimulated ileum and mouse vas deferens contractions in a concentration-dependent manner (2). In the ileum, its nM-range EC_{50} is in an nM order, being nearly equivalent to reported concentrations of the micro-opioid receptor agonist [D-Ala^2,N-MePhe^4,Gly-ol]-enkephalin (*or* DAMGO), and is 100- and 20x smaller than those of mitragynine and morphine, respectively. In the vas deferens, it is 35x smaller than that of morphine (2). The inhibitory action of mitragynine pseudoindoxyl in the ileum was antagonized by the non-selective opioid receptor antagonist naloxone and the micro-receptor antagonist naloxonazine (2). **1**. Takayama, Ishikawa, Kurihara, *et al.* (2002) *J. Med. Chem.* **45**, 1949. **2**. Yamamoto, Horie, Takayama, *et al.* (1999) *Gen. Pharmacol.* **33**, 73.

MIX, *See* *3-Isobutyl-1-methylxanthine*

Mizorbine

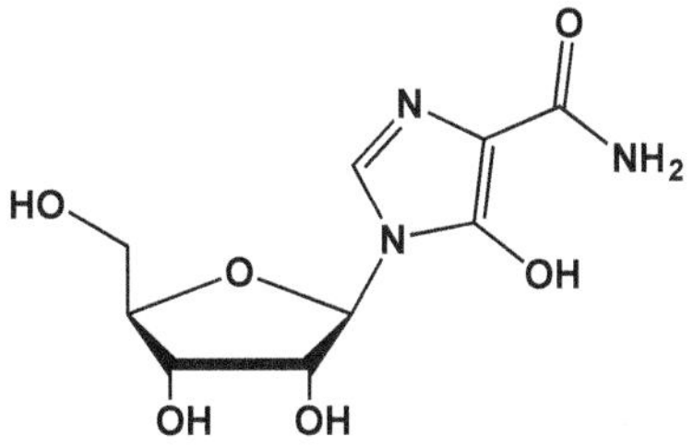

This inosine/guanosine analogue (FW = 259.22 g/mol; CAS 50924-49-7), also known as *N'*-(β-D-ribofuranosyl)-5-hydroxyimidazole-4-carboxamide, is an imidazole-based nucleoside that is metabolically phosphorylated to mizorbine 5'-monophosphate (MZP). The latter inhibits *de novo* purine nucleotide biosynthesis by targeting IMP dehydrogenase (1). X-ray crystal structure of the Enz·MZP complex reveals that IMP dehydrogenase undergoes a conformational change, wherein a flap folds into the NAD site, such that MZP, Cys319, and a water molecule are arranged in a geometry resembling the enzyme's likely transition-state intermeiate (2). Depletion of GTP and dGTP stores disrupts G-protein system and DNA replication. Indeed, measurements of purine ribonucleotide pools by HPLC show that mizoribine reduces intracellular GTP levels; repletion of GTP reverses its

antiproliferative effects (3). Mizorbine also exerts strong immuno-suppressive properties, selectively blocking T-cell proliferation in response to mitogenic and allo-antigenic stimulation. Mizoribine likewise blocks T-cell progression from G-phase to S-phase. Because the structure of MZP closely resembles the purine nucleotide biosynthetic intermediate, aminoimidazole-carboxamide ribonucleotide (AICAR), it is likely to inhibit the synthesis of *N*-formyl-aminoimidazole-carboxamide ribonucleotide. **1**. Hedstrom (1999) *Curr. Med. Chem.* **6**, 545. **2**. Gan, Seyedsayamdost, Shuto, *et al.* (2003) *Biochemistry* **42**, 857. **3**. Turka, Dayton, Sinclair, Thompson & Mitchell (1991) *J. Clin. Invest.* **87**, 940.

Mizoribine & Mizoribine 5'-Monophosphate

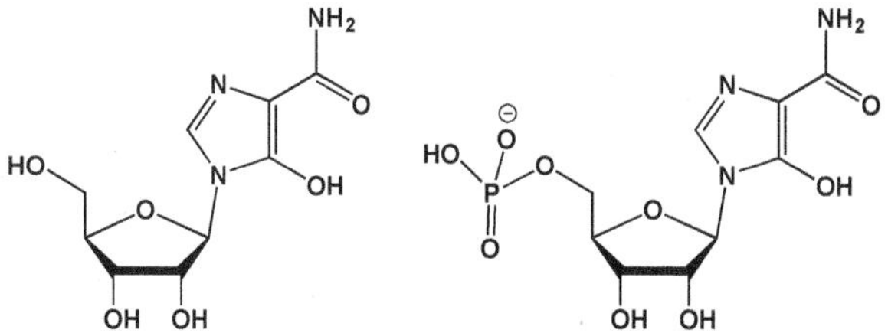

This nucleoside (FW = 259.22 g/mol), also known as bredinin, is produced by *Eupenicillium brefedianum* and is readily converted to the monophosphate, the latter a transition-state inhibitor of IMP dehydrogenase (IMPDH) with a nanomolar K_i value. The imidazole-containing molecule is a strong immunosuppressive agent that blocks T-cell proliferation. It also blocks the movement of T cells from the G_1 to the S phase. **Target(s):** DNA-directed DNA polymerase (1); DNA polymerase β (1); GMP synthetase, inhibited by the monophosphate (2); IMP dehydrogenase, inhibited by the monophosphate (2-8). **1**. Mizushina, Matsukage & Sakaguchi (1998) *Biochim. Biophys. Acta* **1403**, 5. **2**. Kusumi, Tsuda, Katsunuma & Yamamura (1989) *Cell Biochem. Funct.* **7**, 201. **3**. Koyama & Tsuji (1983) *Biochem. Pharmacol.* **32**, 3547. **4**. Hedstrom, Gan, Schlippe, Riera & Seyedsayamdost (2003) *Nucl. Acids Res. Suppl.* (3), 97. **5**. Hedstrom (1999) *Curr. Med. Chem.* **6**, 545. **6**. Goldstein & T. D. Colby (1999) *Curr. Med. Chem.* **6**, 519. **7**. Kerr & Hedstrom (1997) *Biochemistry* **36**, 13365. **8**. Hager, Collart, Huberman & Mitchell (1995) *Biochem. Pharmacol.* **49**, 1323.

MJ 15

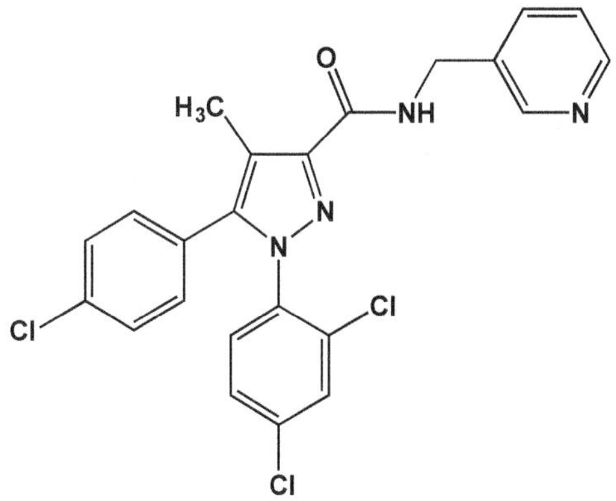

This potent cannabinoid receptor antagonist (FW = 471.77 g/mol; CAS 944154-76-1; Soluble to 100 mM in DMSO), also named 5-(4-chlorophenyl)-1-(2,4-dichlorophenyl)-4-methyl-*N*-(3-pyridinylmethyl)-1*H*-pyrazole-3-carboxamide, selectively targets CB_1 receptors (K_i = 27.2 pM, IC_{50} = 118.9 pM for rat CB_1 receptors). MJ15 dose-dependently blocks Win55212-2 mediated increase of intracellular calcium ion levels in hippocampal cells and reversed the inhibitory effects of cannabinoid CB_1 receptor agonist on forskolin-stimulated adenylyl cyclase activity in CHO cells expressing the human cannabinoid CB_1 receptor. MJ-15 also exhibits potency in obesity and hyperlipidemia models, inhibiting food intake and increasing body weight in diet-induced obese rats and mice. **1**. Chen, Tang, Liu, *et al.* (2010) *Eur. J. Pharmacol.* **637**, 178.

MK-217, *See* *Alendronate*

MK-383, *See* *Tirofiban*

MK-0431, *See* *Sitagliptin*

MK-0457

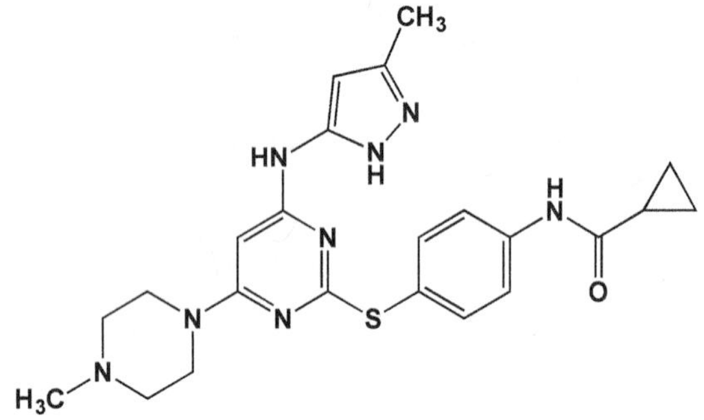

This potent membrane-permeant cell-proliferation inhibitor ($MW_{free-base}$ = 464.59 g/mol), also known as VX-680 and Tozasertib, exerts its action by inhibiting the phosphorylation of as yet unidentified protein components needed for mitosis (K_i^{app} values are 0.6, 18, and 4.6 nM for Aurora-A, Aurora-B, and Aurora-C kinase, respectively). MK-0457 also induces apoptosis in a diverse range of human tumor types. This agent inhibits all cycling cells tested with IC_{50} values ranging between 15 and 150 nM, resulting in tetraploidy. MK-0457 has been assessed in Phase II clinical trials in patients with treatment-refractory chronic myelogenous leukemia or Philadelphia chromosome-positive acute lymphoblastic leukemia containing the T315I mutation. Bebbington *et al.* (1) provides important structure-activity relationship data on this class of inhibitors. **Target(s):** Aurora kinase (1,2); [glycogen synthase III] kinase β (1); polo kinase (3); src kinase (1). **1.** Bebbington, Binch, Charrier, *et al.* (2009) *Bioorg. Med. Chem. Lett.* **19**, 3586. **2.** Harrington, Bebbington, Moore, *et al.* (2004) *Nat. Med.* **10**, 262. **3.** Johnson, Stewart, Woods, Giranda & Luo (2007) *Biochemistry* **46**, 9551.

MK-476, See *Montelukast*

MK-507, See *Dorzolamide*

MK-0518, See *Raltegravir*

MK-0571

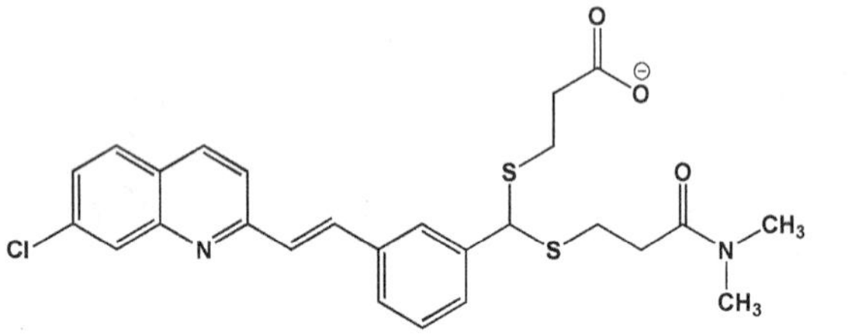

This quinolone ($FW_{free-acid}$ = 515.10 g/mol; CAS 115104-28-4), also named L-66071 and 3-[2-(7-chloro-2-quinolinyl)ethenyl]phenyl[3-(dimethylamino)-3-oxopropyl)-thio]-methyl-thiopropionic acid 1, is a leukotriene D_4 receptor antagonist. **Target(s):** leukotriene D_4 receptor (1); xenobiotic-transporting ATPase (*i.e.,* multidrug resistance protein) (2-5). **1.** Jones, Zamboni, Belley, *et al.* (1989) *Can. J. Physiol. Pharmacol.* **67**, 17. **2.** Paul, Breuninger & Kruh (1996) *Biochemistry* **35**, 14003. **3.** Gekeler, Ise, Sanders, Ulrich & Beck (1995) *Biochem. Biophys. Res. Commun.* **208**, 345. **4.** Paul, Breuninger, Tew, Shen & Kruh (1996) *Proc. Natl. Acad. Sci. U.S.A.* **93**, 6929. **5.** Bodo, Bakos, Szeri, Varadi & Sarkadi (2003) *Toxicol. Lett.* **140-141**, 133.

MK 0677

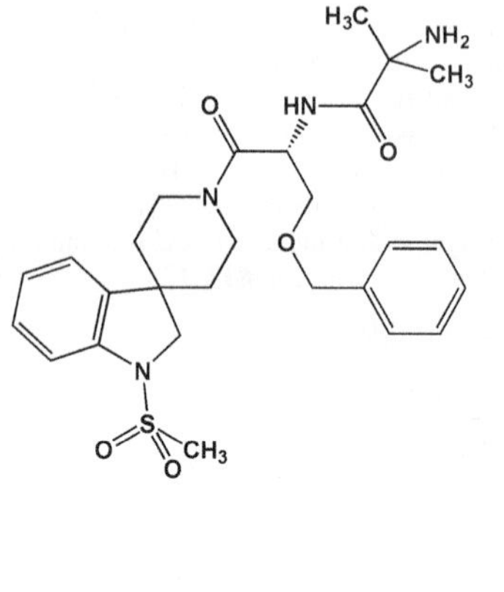

This orally available high-affinity ghrelin receptor agonist ($FW_{Mesylate-Salt}$ = 624.77 g/mol; CAS 159752-10-0; Soluble to 50 mM in water and to 100 mM in DMSO), also named 2-amino-*N*-[(1*R*)-2-[1,2-dihydro-1-(methylsulfonyl)spiro[3*H*-indole-3,4'-piperidin]-1'-yl]-2-oxo-1-[(phenylmethoxy)methyl]ethyl]-2-methylpropan-amide methanesulfonate and L-163,191, attenuates isoproterenol-induced lipolysis in rat adipocytes *in vitro* (1). MK-0677 is a growth hormone (GH) secretagogue that stimulates GH release from rat pituitary cells *in vitro* (EC_{50} = 1.3 nM) and also enhances GH plasma levels *in vivo* (2). MK-0677 also acts as orthosteric super-agonists but not allosteric regulators for activation of the G protein $G\alpha_{o1}$ by the Ghrelin receptor (3). **1.** Mucciolo, *et al.* (2004) *Eur. J. Pharmacol.* **498**, 27. **2.** Patchett, *et al.* (1995) *Proc. Natl. Acad. Sci. U.S.A.* **92**, 7001. **3.** Bennett, *et al.* (2009) *Mol. Pharmacol.* **76**, 802.

MK-0683, See *Vorinostat*

MK-0752

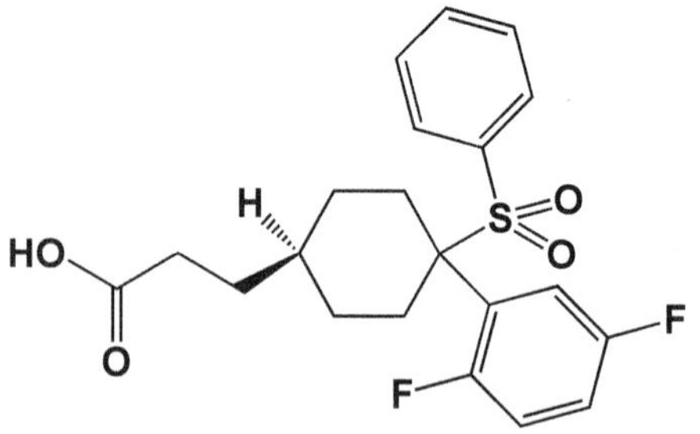

This orally bioavailable, moderately potent, protease inhibitor (FW = 442.92 g/mol; CAS 471905-41-6; Solubility: 90 mg/mL DMSO; <1 mg/mL H_2O), known systematically as 3-((1*R*,4*S*)-4-(4-chlorophenylsulfonyl)-4-(2,5-difluorophenyl)cyclohexyl)propanoic acid and MK-752, targets γ-secretase (GS), thereby inhibiting the formation of Aβ40 in human SH-SY5Y cells, with an IC_{50} value of 5 nM. **1.** Cook, *et al.* (2010) *J. Neurosci.* **30**, 6743. **2.** Harrison, *et al.* (2010) *Cancer Res.* **70**, 709.

MK-0822. See *Odanacatib*

MK 870, See *Amiloride*

MK-422, See *Enalaprilat*

MK 869 (or MK 0869), See *Aprepitant*

MK-886

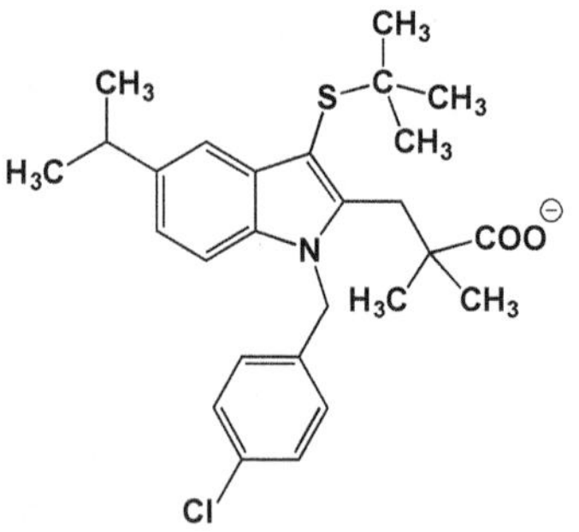

This orally active investigational drug ($FW_{free-acid}$ = 472.09 g/mol; CAS 118414-82-7; Solubility: 100 mM in DMSO; IUPAC: 1-[(4-chlorophenyl)methyl]-3-[(1,1-dimethylethyl)thio]-α,α-dimethyl-5-(1-methylethyl)-1*H*-indole-2-propanoic acid), also known as L-663,536, inhibits 5-lipoxygenase-activating protein, *or* FLAP, with IC_{50} of 30 nM (1-3). When administered *in vivo*, MK-886 is a potent inhibitor of antigen-induced dyspnea in inbred rats pretreated with methysergide (ED_{50} = 0.036 mg/kg, p.o.) and of Ascaris-induced bronchoconstriction in squirrel monkeys (ED_{50} = 1 mg/kg, p.o.) (1). Three classes of leukotriene biosynthesis inhibitors (represented by MK-886, L-674,573, and L-689,037) share a common binding site on FLAP, providing further evidence that FLAP represents a suitable target for structurally diverse classes of leukotriene biosynthesis inhibitors (3). MK-886 is also a moderately potent antagonist of Peroxisome-Proliferator-Activated Receptor, *or* PPARα (IC_{50} = 0.5-1 μM), functioning as a non-competitive inhibitor (4). The main pro-apoptotic effect of these drugs is most likely mediated through the known antiproliferative effects of the PPAR-RXR interaction, where RXR refers to the Retinoid X Receptor. **Target(s):** K+ currents, Ca^{2+}-activated ($I_{K,Ca}$) (5);

leukotriene biosynthesis (6-8); leukotriene-C4 synthase (7-9); 5-lipoxygenase, *or* arachidonate 5-lipoxygenase (10,11); 5-lipoxygenase activating protein (12); peroxisome-proliferator-activated receptor (PPAR)α (13); prostaglandin-E synthase (14,15). **1**. Gillard, Ford-Hutchinson, Chan, *et al.* (1989) *Can. J. Physiol. Pharmacol.* **67**, 456. **2**. Dixon, Diehl, Opas, *et al.* (1990) *Nature* **343**, 282. **3**. Mancini, Prasit, Coppolino, *et al.* (1992) *Mol. Pharmacol.* **41**, 267. **4**. Kehrer, Biswal, La, *et al.* (2001) *Biochem. J.* **356**, 899. **5**. Smirnov, Knock & Aaronson (1998) *Brit. J. Pharmacol.* **124**, 572. **6**. Rouzer, Ford-Hutchinson, Morton & Gillard (1990) *J. Biol. Chem.* **265**, 1436. **7**. Sala, Garcia, Zarini, *et al.* (1997) *Biochem. J.* **328**, 225. **8**. Lam, Penrose, Rokach, *et al.* (1996) *Eur. J. Biochem.* **238**, 606. **9**. Hevko & Murphy (2002) *J. Biol. Chem.* **277**, 7037. **10**. Sud'ina, Pushkareva, Shephard & Klein (2008) *Prostaglandins Leukot. Essent. Fatty Acids* **78**, 99. **11**. Ghosh & Myers (1998) *Proc. Natl. Acad. Sci. U.S.A.* **95**, 13182. **12**. Young (1999) *Eur. J. Med. Chem.* **34**, 671. **13**. Kehrer, Biswal, La, *et al.* (2001) *Biochem. J.* **356**, 899. **14**. Vazquez-Tello, Fan, Hou, *et al.* (2004) *Amer. J. Physiol.* **287**, R1155. **15**. Ouellet, Falgueyret, Ear, *et al.* (2002) *Protein Expr. Purif.* **26**, 489.

MK-906, See Finasteride

MK-0966, See Rofecoxib

MK 990, See Rafoxanide

MK-1064

This orally bioavailable 2,5-disubstituted nicotinamide and receptor antagonist (FW = 461.91 g/mol), also known as 5″-chloro-*N*-[(5,6-dimethoxypyridin-2-yl)methyl]-2,2′:5′,3″-terpyridine-3′-carboxamide, selectively targets the Orexin-2 Receptor (*or* OX$_2$R *or* Hypocretin Receptor), a G-protein coupled receptor exclusively in the brain. MK-1064 is an inveastigational drug intended for the treatment of insomnia. OX$_2$ also plays a part in a CNS feedback loop that regulates feeding behavior. **1**. Roecker, Mercer, Schreier, *et al.* (2013) *Chem. Med. Chem.* **9**, 311.

MK-1439, See Doravirine

MK-1775

This orally bioavailable non-receptor, ATP-competitive TKI (FW = 500.60; CASs = 955365-80-7, 1075739-30-8 (H$_2$O), 1075739-32-0 (xHCl); Solubility: 100 mg/mL DMSO; <1 mg/mL H$_2$O; **Formulation:** Dissolve in PBS at 0.5% methylcellulose), named systematically as 2-allyl-1-(6-(2-hydroxypropan-2-yl)pyridin-2-yl)-6-(4-(4-methylpiperazin-1-yl)phenyl-amino)-1,2-dihydropyrazolo[3,4-*d*]pyrimidin-3-one, targets Wee1 (IC$_{50}$ = 5 nM), a tyrosine kinase that phosphorylates and inactivates CDC2 at tyrosine-15, thereby abrogating the G$_2$ DNA damage checkpoint and resulting in apoptosis, when administered in combination with DNA-damaging agents, such as gemcitabine, carboplatin, and cisplatin in p53-deficient cells. **1**. Hirai, Iwasawa, Okada, *et al.* (2009) *Mol. Cancer Ther.* **8**, 2992. **2**. Davies, *et al.* (2011) *Cancer Biol. Ther.* **12**, 788.

MK-2206 Dihydrochloride

This highly selective pan-Akt inhibitor (FW = 480.39 (as dihydrochloride); CAS 1032350-13-2, 1032349-93-1 (free base), 1032349-77-1 (HCl salt); Solubility: 14 mg/mL DMSO, 1 mg/mL H$_2$O; IUPAC: 8-(4-(1-aminocyclobutyl)phenyl)-9-phenyl[1,2,4]triazolo[3,4-*f*][1,6]naphthyridin-3(2*H*)-one, is an allosteric signal-transduction pathway inhibitor targets Akt1 (IC$_{50}$ = 8 nM), Akt2 (IC$_{50}$ = 12 nM), and Akt3 (IC$_{50}$ = 65 nM), also known as the Protein Kinases B isoforms PKB-1, PKB-2, and PKB-3. MK-2206 is a tight binding, uncompetitive inhibitor of MEK1/2, binding to an allosteric binding pocket adjacent to the MgATP site. Akt's pleckstrin homology (PH) domain folds over, interacting with the bound MK-2206 and forming an auto-inhibited complex, blocking ATP binding. MK2206 also blocks influenza pH1N1 virus infection *in vitro* by inhibiting endocytic uptake of the virus (3). MK2206 does not inhibit H3N2, H7N9 and H5N1 influenza, indicating that pH1N1 apparently evolved specific requirements for efficient infection. MK2206 blocks Akt phosphorylation at Ser-473, reducing proliferation of cholangiocarcinoma (CCA) cells, an aggressive disease with limited effective treatment options (4). **1**. Hirai, Sootome, Nakatsuru, *et al.* (2010) *Mol. Cancer Ther.* **9**, 1956. **2**. Lindsley (2010) *Curr. Top. Med. Chem.* **10**, 458. **3**. Denisova, Söderholm, Virtanen, *et al.* (2014) *Antimicrob. Agents Chemother.* **58**, 3689. **4**. Wilson, Kunnimalaiyaan, Kunnimalaiyaan & Gamblin (2015) *Cancer Cell Int.* **15**, 13.

MK-2048

This second-generation integrase (*or* IN) inhibitr (FW = 435.28 g/mol; CAS 269055-15-4; Solubility: 9 mg/mL DMSO), also known as 2-[(3-chloro-4-fluorophenyl)methyl]-8-ethyl-1,2,6,7,8,9-hexahydro-10-hydroxy-*N*,6-dimethyl-1,9-dioxo-(6*S*)-pyrazino[1',2':1,5]pyrrolo[2,3-*d*]pyridazine-4-carboxamide, targets IN$^{\text{wild-type}}$ (IC$_{50}$ = 2.6 nM) and IN$^{\text{R263K}}$ (IC$_{50}$ = 1.5 nM). IC$_{50}$ values for MK-2048 against subtype B and C enzymes were 0.075 μM and 0.08 μM, respectively, and disintegration was inhibited by high concentrations of MK-2048 to a comparable extent with both subtype B and C enzymes (1). MK-2048 was specifically developed with the goal of retaining activity against viruses containing mutations associated with raltegravir- and elvitegravir-resistant HIV (2). That said, IN$^{\text{G118R}}$ and IN$^{\text{E138K}}$ confer resistance to MK-2048, but not to raltegravir or elvitegravir (2,3). **1**. Bar-Magen, Sloan, Faltenbacher, *et al.* (2009) *Retrovirology* **6**, 103. **2**. Bar-Magen, Sloan, Donahue, *et al.* (2010) *J. Virol.* **84**, 9210. **3**. Quashie, Mesplède, Han, *et al.* (2012) *J. Virol.* **86**, 2696.

MK-3102, See Omarigliptin

MK-3118

This orally bioavailable synthetic derivative (FW = 727.07 g/mol) of the novel triterpene glycoside natural product enfumafungin has improved fungal β-1,3-glucan synthase inhibitor properties. While echinocandins are also potent, naturally occurring GS inhibitors, their clinical utility is limited, because they can only be administered intravenously. MK-3118 exhibits better whole-cell antifungal activity and displays an improved spectrum of antifungal action and efficacy in animal models of invasive aspergillosis (*i.e.*, DBA/2N and neutropenic CD-1 mice after infection with *Aspergillus fumigatus*, strain MF5668; ED$_{90}$ 1-8 mg/kg). MK-3118 is also highly effective against 71 *Aspergillus* spp., including itraconazole-resistant strains (MIC ≥ 4 µg/ml). With 29 strains of *Candida albicans*, 29 strains of *Candida glabrata*, 21 strains of *Candida tropicalis*, 15 strains of *Candida parapsilosis* and 19 strains of *Candida krusei*, including azole- and echinocandin-resistant isolates, MK-3118 gave MIC values ranging from 0.06 to 16 mg/L depending on duration of treatment and endpoint criteria. (Although carrying an "MK" in a code name typical of Merck's investigational drugs, all rights were returned to Scynexis, Inc., in 2013, following Merck's review and prioritization of its infectious disease portfolio.) **1.** Pfaller, Messer, Motyl, Jones & Castanheira (2013) *Antimicrob. Agents Chemother*. **57**, 1065.

MK-3475, See *Lambrolizumab*

MK-4827, See *Niraparib*

MK-8109, See *Vintafolide*

MK-8669, See *Ridaforolimus*

MK-8742, See *Elbasvir*

MK-8745

This novel potent inhibitor (FW = 431.91 g/mol; CAS 885325-71-3; Solubility: 86 mg/mL DMSO; <1 mg/mL H$_2$O), also named (3-chloro-2-fluorophenyl)[4-[[6-(2-thiazolylamino)-2-pyridinyl]methyl]-1-piperazinyl] methanone, selectively targets Aurora A kinase (IC$_{50}$ = 0.6 nM), with more than 450x selectivity for Aurora A over Aurora B. Treatment of non-Hodgkin lymphoma (NHL) cell lines with MK-8745 results in cell cycle arrest at the G$_2$/M phase, with accumulation of tetraploid nuclei and subsequent death. Sensitivity to MK-8745 correlates with the expression level of Aurora-A activator. Indeed, siRNA knockdown of Aurora-A activator TPX2 (Targeting Protein for *Xenopus* kinase-like protein 2) increases MK-8745 sensitivity in NHL cell lines that are otherwise less sensitive to this drug; by contrast, TPX2 overexpression in NHL cell lines with high MK-8745-sensitivity increases drug resistance (1). MK8745 induces apoptotic cell death in a p53-dependent manner, when tested *in vitro* in cell lines of multiple lineages. Cells expressing wild-type p53 showed a short delay in mitosis, followed by cytokinesis and resulting in tetraploid cells, along with apoptosis. However, cells lacking or with mutant p53 resulted in a prolonged arrest in mitosis, followed by endoreduplication and polyploidy. Cytokinesis was completely inhibited in p53-deficient cells, as observed by the absence of 2N cell population. The induction of apoptosis in p53-proficient cells was associated with activation of caspase 3 and release of cytochrome c but was independent of p21 (2). **1.** Chowdhury, Chowdhury & Tsai (2012) *Leuk. Lymphoma* **53**, 462. **2.** Nair, Ho & Schwartz (2012) *Cell Cycle* **11**, 807.

MKT-077

This rhodacyanine dye and mitochondrial Hsp70 inhibitor (FW = 432.00 g/mol; CAS 147366-41-4; orange-red powder; Solubility = 200 mg/mL), also known as FJ-776 and 1-ethyl-2-[[3-ethyl-5-(3-methyl-2(3*H*)-benzothiazolylidene)-4-oxo-2-thiazolidinylidene]methyl]pyridinium chloride, is a lipophilic cation that targets mortalin, a mitochondrial Hsp70-family chaperone that palys roles in protein import and quality control, Fe-S cluster protein biogenesis, mitochondrial homeostasis, and regulation of p53 (1). Mortalin is implicated in the regulation of apoptosis, cell stress response, neurodegeneration, and cancer. Subcellular localization indicates that MKT-077 is taken up and retained by mitochondria, making it suitable for flow cytometric analysis (2). X-ray crystallography suggests how mortalin's nucleotide binding domain is likely to interact with the nucleotide exchange factor GrpEL$_1$, p53, as well as MKT-077 (3). In nude mice, MKT-077 inhibits the growth of subcutaneously implanted human renal carcinoma A498 and human prostate carcinoma DU145 and prolongs survival of mice bearing intraperitineal implanted human melanoma (2). MKT-077 treatment induces changes in mitochondrial ultrastructure in carcinoma cells, but not in similarly treated normal epithelial cells (4) MKT-077 also inhibits respiratory activity in isolated intact mitochondria and electron transport activity in freeze-thawed mitochondrial membrane fragments in a dose-dependent manner. The MKT-077 concentration needed to reach half-maximal inhibition of ADP-stimulated respiration was approximately 4-fold greater in mitochondria isolated from cells of the normal epithelial cell line, CV-1 (15 µg MKT-077/mg protein), as compared to the human colon carcinoma cell line, CX-1 (4 µg/mg protein). Treatment is attended by selective loss of mitochondrial DNA in CX-1 and CRL1420 carcinoma cells, but not CV-1 (normal epithelial) cells treated with MKT-077 at 3 µg/mL for up to 3 days. Nuclear DNA was unaffected in all three similarly treated cell lines. MKT-077sensitivity of the cell lines to mitochondrial damage correlates well with their sensitivity to MKT-077 cytotoxicity. Such results provide a basis for the selective malignant cell killing exhibited by MKT-077 (4). Significantly, the concentration of MKT-077 required to achieve half-maximal inhibition of ADP-stimulated respiration is 6x lower in the presence *versus* absence of high-intensity light, *i.e.*, 2.5 *versus* 15 µg MKT-077/mg (5). Photoirradiation also produced a 25x increase in inhibition of succinate-cytochrome c reductase activity by MKT-077 (5). **Other Target(s):** MKT-077 inhibits telomerase *in vitro* assay has no detectable effect on telomerase activity *in vivo* (6). MKT-077 is also an F-actin cross-linker (7). **1.** Wadhwa, Sugihara, Yoshida *et al.*(2000) *Cancer Res*. **60**, 6818. **2.** Koya, Li, Wang, *et al.* (1996) *Cancer Res*. **56**, 538. **3.** Amick, Schlanger, Wachnowsky, *et al.* (2014) *Protein Sci*. **23**, 833. **4.** Modica-Napolitano, Koya, Weisberg *et al.* (1996) *Cancer Res*. **56**, 544. **5.** Modica-Napolitano, Brunelli, Koya & Chen (1998) *Cancer Res*. **58**, 71. **6.** Wadhwa, Colgin, Yaguchi, *et al.* (2002) *Cancer Res*. **62**, 4434. **7.** Maruta, Tikoo, Shakri & Shishido (1999) *Ann N.Y. Acad. Sci*. **886**, 283.

ML 00253764, See *ML 253764*

ML-7

This selective, ATP-competitive protein kinase inhibitor (FW = 452.74 g/mol; CAS 110448-33-4; Solubility = 50 µM in DMSO), named systematically as 1-(5-iodonaphthalene-1-sulfonyl)-1H-hexahydro-1,4-diazepine hydrochloride, targets Ca^{2+}-calmodulin-dependent smooth muscle myosin light chain kinase (K_i = 0.3 µM), exhibiting greater potency than the parent compound ML-9. ML7 blocks or slows chromosome movement when added to anaphase crane-fly spermatocytes (2). For all chromosomes whose movements were affected by a drug, the corresponding spindle fibres of the affected chromosomes had reduced levels of 1P- and 2P-myosin. The more general phosphorylation inhibitor, staurosporine, also reduced MRLC phosphorylation, causing anaphase chromosomes to stop or slow. **1.** Dhawan & Helfman (2004) *J. Cell Sci.* **117**, 3735. **2.** Sheykhani, Shirodkar & Forer (2013) *Eur. J. Cell Biol.* **92**, 175.

ML 18

This BRS-3 antagonist (FW = 569.66 g/mol; CAS 1422269-30-4; S-enantiomer of PD176252, itself a nonpeptide antagonist for the neuromedin B receptor), also named (αS)-N-[[1-(4-methoxyphenyl)cyclohexyl]methyl-α-[[[(4-nitrophenyl)amino]carbonyl]amino]-1H-indole-3-propanamide, targets Bombesin Receptor, Subtype 3 (IC$_{50}$ = 4.8 µM), a G protein coupled receptor (GPCR) for the bombesin (BB)-family of peptides. ML-18 binds with lower affinity to the GRPR (IC$_{50}$ = 16 µM) and NMBR (with IC$_{50}$ >100µM). ML-18, but not its enantiomer EMY-98, inhibited the ability of BA1 to elevate cytosolic Ca^{2+} in a reversible manner using lung cancer cells loaded with FURA2-AM. ML18, but not EMY-98, inhibited the ability of BA1 to cause tyrosine phosphorylation of the EGFR and ERK in lung cancer cells. **1.** Moody, Mantey, Moreno, *et al.* (2015) *Peptides* **64**, 55.

ML130

This potent inhibitor (FW = 287.34 g/mol; CAS 799264-47-4; Solubility: 57 mg/mL DMSO, <1 mg/mL Water), also known systematically as 1-(4-methylphenyl)sulfonyl-1H-benzimidazol-2-amine, selectively targets NOD1, a Nucleotide-binding Oligomerization Domain protein that operates via NF-κB activation and functions as an innate-immunity sensor of microbial components derived from bacterial peptidoglycan. NOD1 participates in host defense against *Helicobacter pylori*. **Selectivity:** NOD1

(IC$_{50}$ = 0.56 µM); NOD2 (IC$_{50}$ >20 µM); and TNF-α (IC$_{50}$ >20 µM). **1.** Khan, *et al.* (2011) *ACS Med. Chem. Lett.* **2**, 780. **2.** Correa, *et al.* (2011) *Chem. Biol.* **18**, 825.

ML133

This potent and selective potassium channel blocker (FW$_{HCl-Salt}$ = 313.82 g/mol; CAS 1222781-70-5; Solubility: 100 mM in DMSO; 20 mM in H_2O, with warming), systematically named N-[(4-methoxyphenyl)methyl]-1-naphthalenemethanamine·HCL, targets inwardly rectifying K$_{ir}$2 potassium channels: mK$_{ir}$2.1 (IC$_{50}$ = 1.8 µM), hK$_{ir}$2.6 (IC$_{50}$ = 2.8 µM), hK$_{ir}$2.2 (IC$_{50}$ = 2.9 µM) and hK$_{ir}$2.3 (IC$_{50}$ = 4.0 µM). ML-1333 exhibits no effect on rK$_{ir}$1.1 (IC$_{50}$ > 300 µM) and weak activity at hK$_{ir}$7.1 (IC$_{50}$ = 33 µM) and rK$_{ir}$4.1 (IC$_{50}$ = 76 µM). **1.** Wu, *et al.* (2010) Probe Reports from the NIH Molecular Libraries Pro. **PMID**: 21433384. **2.** Wang, *et al.* (2011) *ACS Chem.Biol.* **6**, 845.

ML 141

This allosteric (noncompetitive) Cdc inhibitor (FW = 407.49 g/mol; CAS 71203-35-5; Soluble to 100 mM in DMSO), also named CID 2950007 and 4-[4,5-dihydro-5-(4-methoxyphenyl)-3-phenyl-1H-pyrazol-1-yl]benzene-sulfonamide, selectively targets Cdc42 GTPase (IC$_{50}$ ~ 200 nM); EC$_{50}$ = 2.1 µM), with weaker action against Rho GTPase family, including Rac1, Rab2 and Rab7 (all with IC$_{50}$ > 100 µM). CID2950007 decreased the B$_{max}$ of GTP-binding, while BODIPY-FL-GTP affinity to wild type Cdc42 (K_d ~4 nM) was essentially unchanged by CID2950007 (K_d ~50 nM). ML-141 also inhibits ovarian cancer cell migration *in vitro* as well as actin-based filopodia formation in 3T3 fibroblasts. **1.** Surviladze, Waller, Strouse, *et al.* (2010) *Probe Reports from the Molecular Libraries Program NBK5196.* **PMID**: 21433396. **2.** Hong, Kenney, Phillips, *et al.* (2011) *J. Biol. Chem.* **288**, 8531.

ML154

This brain-penetrant NPSR antagonist (FWHBr-salt = 545.46 g/mol; CAS 1345964-89-7; Solubility: 100 mM in DMSO; 50 mM in Ethanol), also named 3-(diphenylphosphinothioyl)-2-methyl-1-[(2E)-3-phenyl-2-propen-1-yl]imidazo[1,2-a]pyridinium bromide, targets the neuropeptide S receptor (pA$_2$ = 9.98), decreaseing operant alcohol self-administration in rats. ML 154 selectively inhibits neuropeptide S-induced ERK phosphorylation (IC$_{50}$ = 9.3 nM) over cAMP responses (IC$_{50}$ = 22 nM) and calcium responses (IC$_{50}$ = 36 nM). ML-154 also appears to modulate addictive behavior *in vivo*. This antagonist isplays no activity against vasopressin V$_{1B}$ receptors. **1.** Patnaik, *et al.* (2013) *J. Med. Chem.* **56**, 9045. **2.** Thorsell, *et al.* (2013) *J. Neurosci.* **33**, 10132.

ML188

This nonpeptidyl antiviral lead (FW = 432.56 g/mol; IUPAC Name: (R)-N-(4-(tert-butyl)phenyl)-N-(2-(tert-butylamino)-2-oxo-1-(pyridin-3-yl)ethyl)-furan-2-carboxamide) is a noncovalent inhibitor of the 3CL protease from Severe Acute Respiratory Syndrome Coronavirus (SARS-CoV), displaying good enzyme inhibition and antiviral activity. ML188 is unlike the majority of reported coronavirus 3CLpro inhibitors, which act via covalent modification of the enzyme. 1. Jacobs, Grum-Tokars, Zhou, et al. (2013) J. Med. Chem. 56, 534.

ML204

This novel antagonist (FW = 226.32 g/mol; CAS 5465-86-1; IUPAC Name: 4-methyl-2-(1-piperidinyl)quinoline) selectively modulates Transient Receptor Potential Canonical (TRPC4/C5) channels, which are nonselective Ca^{2+}-permeable channels implicated in diverse physiological functions, including smooth muscle contractility and synaptic transmission. IC_{50} values were 0.96 and 2.6 μM, respectively, based on fluorescent and electrophysiological assays. **Selectivity:** ML204 exhibits 19-fold selectivity against TRPC6, and 9-fold selectivity against TRPC5, with no significant inhibition of TRPV1, TRPV3, TRPA1 and TRPM8 channels, even when tested at concentrations up to 22 μM. 1. Miller, Shi, Zhu, et al. (2011) J. Biol. Chem. 286, 33436.

ML216

This first-in-class non-nucleoside inhibitor (FW = 383.32 g/mol; CAS 1430213-30-1), also known as 1-(4-fluoro-3-(trifluoromethyl)phenyl)-3-(5-(pyridin-4-yl)-1,3,4-thiadiazol-2-yl)urea, targets BLM helicase, a DNA unwinding enzyme that is critical in DNA repair via homologous recombination. Mutations in BLM give rise to Bloom's Syndrome and a predisposition to cancers. ML216 also sensitizes tumor cells to conventional cancer therapies, such as camptothecin. ML216 selectively inhibits cell proliferation of BLM-proficient PSNF5 fibroblasts over BLM-deficient PSNG13cells. ML216 caused a significant increase in the frequency of sister chromatid exchanges in PSNF5 cells, a cytogenetic marker for BLM – deficient cells. ML216 also exhibits excellent liver microsomal stability and plasma stability, indicating that it may retain its BLM helicase inhibitor action in vivo. ML216 shows limited oral bioavailability, additional improvement in the aqueous solubility needs to be made. **Target(s):** BLM helicase, IC_{50} = 1.2 μM; RecQ1 helicase, IC_{50} ~ 50 μM; RecQ5 helicase, IC_{50} > 50 μM; E. coli UvrD helicase, IC_{50} > 50 μM. 1. Rosenthal, Dexheimer, Nguyen, et al. (2013, updated from 2011) Probe Reports, NIH

Molecular Libraries Program, Program, Bethesda: National Center for Biotechnology Information (US), **PMID:** 24027802.

ML-236B, See Mevastatin

ML267

This novel antibiotic (FW = 431.86 g/mol) targets phosphopantetheinyl transferase, or PPTase (IC_{50} = 0.3 μM) is effective against Gram-positive bacteria, including community-acquired methicillin- resistant Staphylococcus aureus (CA MRSA). ML267 shows no cytotoxity against HepG2 cells. ML267 has promising in vitro absorption, distribution, metabolism, and excretion (ADME) profile, commending for further testing in animal models of bacterial infection and virulence. 1. Foley, Rai, Yasgar, et al. (2013) Probe Reports from the NIH Molecular Libraries Program **PMID:** 24260781.

ML295, ML296, and Analogue-12

ML 295 (CID 53364533, SID125269138)

ML 296 (CID 53364510, SID125269096)

Analogue 12 or KT182
(CID 53364491, SID125269091)

These triazole urea-based endocannabinoid inhibitors (FW_{ML-295} = 412.49 g/mol; FW_{ML-296} = 466.54 g/mol; $FW_{Analogue-12}$ = 438.53 g/mol; Solubility: ML295 (<1 μM), ML296 (12 μM) and Analogue-12 (27 μM), but higher in DMEM-FBS medium) target Abhydrolase Domain-containing protein-6 (or ABHD6), a serine hydrolase that controls the accumulation and efficacy of 2-arachidonoyl-glycerol (2-AG) at cannabinoid receptors. **Mode of Inhibitory Action:** Three enzymes (monoacylglycerol lipase (MAGL), ABHD6, and ABHD12) catalyze hydrolysis and inactivation of 2-AG, which regulates neurotransmission and neuroinflammation by activating CB_1 cannabinoid receptors on neurons and CB_2 cannabinoid receptors on microglia. Enzymes that hydrolyze 2-AG therefore regulate the accumulation and efficacy of 2-AG at these cannabinoid receptors. ML295, ML296, and Analogue-12 were examined both by means of in-gel activity assays as well as in vitro in effort to develop more potent, in vivo-active ABHD6 inhibitors to serve as either control probes for the recently developed dual DAGL-β/ABHD6 inhibitor (ML294) or for biological investigation of ABHD6 only. ML295, ML296, and Analogue-12 show IC_{50} values of 38 nM, 0.8 nM, and nM, respectively, for the target enzyme. 1. Hsu, Tsuboi, Speers, et al. (2013) Probe Reports from the NIH Molecular Libraries Program, National Center for Biotechnology Information, **PMID:** 23762934.

ML309

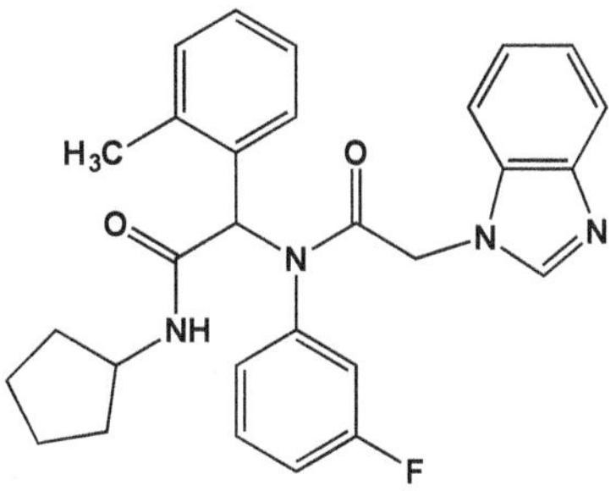

This potent, first-in-class inhibitor (FW = 484.58 g/mol; IUPAC: 2-(N-[2-(benzimidazol-1-yl)acetyl]-3-fluoroanilino)-N-cyclopentyl-2-(2-methyl-phenyl)acetamide)) targets the R132H mutant of Isocitrate Dehydrogenase I (IDH1), a common mutation in secondary gliomas and acute myeloid leukemia, generating a mutant enzyme that, while failing to catalyze its normal TCA cycle reaction, instead catalyzes the formation of the oncometabolite, 2-hydroxyglutarate. ML-309 has IC_{50} values of 96 nM and >35μM versus mutant and wild-type IDH1, respectively, as well as a 500 nM EC_{50} in assays of cell-based 2-hydroxyglutarate formation. **1.** Davis, Pragani, Popovici-Muller. *Et al.* (2012) *Probe Reports from the NIH Molecular Libraries Program*, Bethesda: National Center for Biotechnology Information (US). Updated: 2013 May 8. **PMID**: 23905201.

ML324

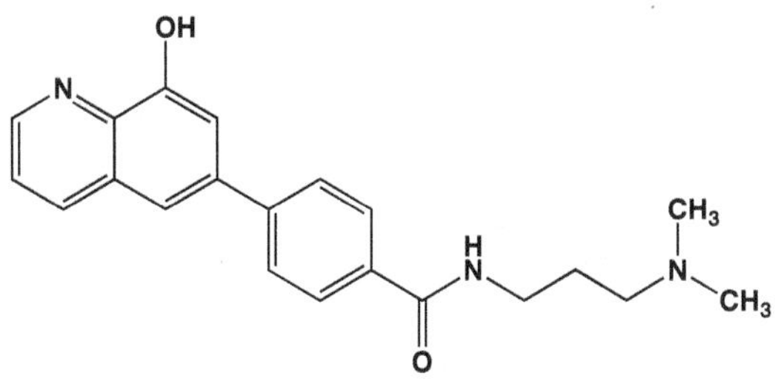

This histone demethylation inhibitor (FW = 349.43 g/mol; CAS 1432500-66-7; Soluble to 100 mM in DMSO), also named N-(3-(dimethyamino)propyl-4-(8-hydroxyquinolin-6-yl)benzamide, is cell-permeable and targets LSD1/2 and Jumonji domain (JMJD) -containing demethylase, two histone demethylases involved in the dynamic epigenetic post-translational N^{ε}-methylation/demethylation of histone lysine residues (1). ML324 possesses excellent *in vitro* absorption, distribution, metabolism, and excretion properties commending its use in cell-based studies of these enzymes. ML324 also exhibits potent anti-viral activity (IC_{50} = 10 μM) against both herpes simplex virus (HSV) and human cytomegalovirus (hCMV) infection via inhibition viral IE gene expression. It likewises suppresses the formation of HSV plaques, even at high MOI, blocking HSV-1 reactivation in a mouse ganglia explant model of latently infected mice. **1.** Rai, Kawamura, Tumber, *et al.* (2013) *Probe Reports from the NIH Molecular Libraries Program* **PMID**: 24260783.

ML345

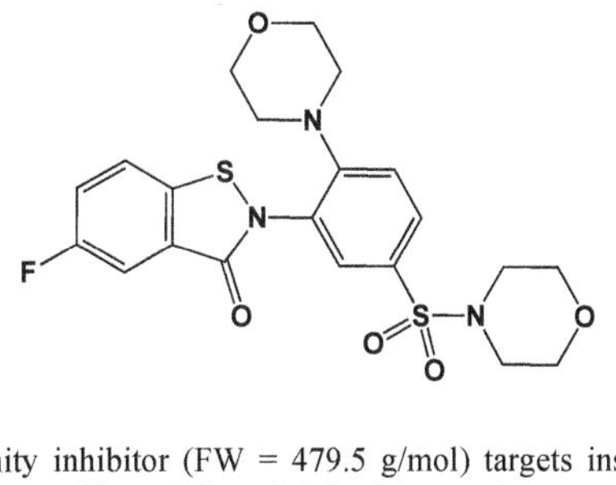

This high-affinity inhibitor (FW = 479.5 g/mol) targets insulin-degrading enzyme, *or* IDE, with an IC_{50} of 1.1 μM, promising new leads in the development of drugs that enhance/sustain insulin's role in promoting euglycemia (1). A thiol-sensitive zinc-metallopeptidase, IDE is the principal insulin-degrading enzyme *in vivo*, terminating insulin's action by modulating the hormone's off rate from its receptor (2). ML145 is superior to IDE inhibitor-1 (*or* Ii1), a peptide-derived hydroxamic acid that, while both highly potent (K_i =1.7 nM) and selective (10,000x relative to other

zinc-metalloproteases) is a peptide that us too rapidly degraded ($t_{1/2}$ ~9 min in mice) to become a useful drug. ML145 reacts with Cys819 and, to a lesser extent, Cys812, in IDE. **1.** Bannister, Wang, Abdul-Hay, *et al.* (2013) *Probe Reports from the NIH Molecular Libraries Program* **PMID**: 23833801 [PubMed]. **2.** Leissring, Malito, Hedouin, *et al.* (2010) *PLoS One* **5**, e10504.

ML352

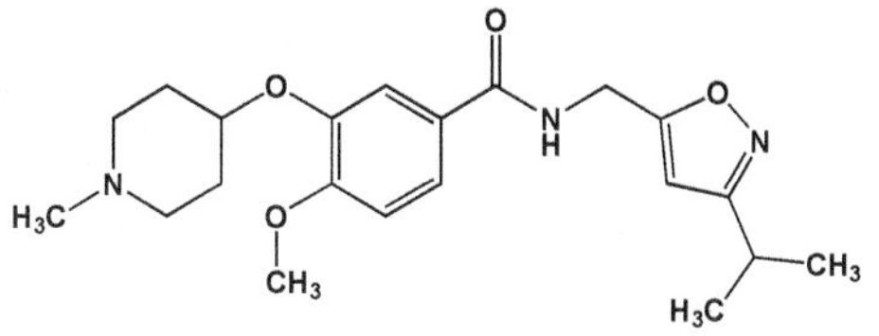

This selective CHT inhibitor (FW = 387.48 g/mol) targets the high-affinity Choline Transporter (K_i = 92 nM), the rate-limiting step in acetylcholine biosynthesis. At ML352 concentrations that fully antagonized CHT in transfected cells and nerve terminal preparations, there is no inhibition of acetylcholinesterase (AChE) or cholineacetyltransferase (ChAT). ML352 also lacks activity against many receptors and channels, including dopamine transporters, serotonin transporters, and norepinephrine transporters. ML352 noncompetitively inhibits choline uptake in intact cells and synaptosomes and reduces the apparent density of hemicholinium-3 binding sites in membrane assays. Pharmacokinetic studies reveal limited *in vitro* metabolism of ML352 as well as significant CNS penetration, with features predicting rapid clearance. **1.** Ennis, Wright, Retzlaff, *et al.* (2015) *ACS Chem. Neurosci.* **6**, 417.

ML 253764

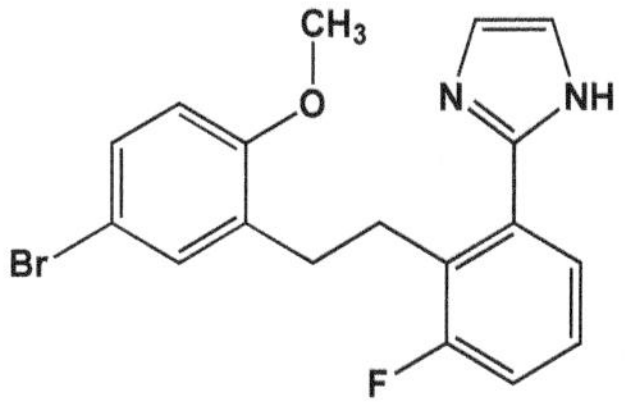

This orally available nonpeptide melanocortin receptor antagonist ($FW_{HCl\text{-}Salt}$ = 413.71 g/mol; Solubility: 10 mM in H_2O, 100 mM in DMSO), also named 2-[2-[2-(5-bromo-2-methoxyphenyl)ethyl]-3-fluorophenyl]-4,5-dihydro-2-1H-imidazole·HCl, targets MC_4 receptors (IC_{50} = 0.32 μM), MC_3 receptors (IC_{50} = 0.81 μM), and MC_5 receptors (IC_{50} = 2.12 μM) (1-3). The MC_4 receptors plays an important role in body weight regulation and energy homeostasis, and intracerebroventricular injection of peptidic antagonists (usually) helps to increase body weight and/or food intake in mouse models of tumor-induced weight loss (*i.e.,* cancer cachexia). ML 00253764 is a novel nonpeptidic antagonist that, upon peripheral administration in mice, reduces tumor-induced weight loss (1). **1.** Vos, Caracoti, Che, *et al.* (2004) *J. Med. Chem.* **47**, 1602. **2.** Tucci, White, Markison, *et al.* (2005) *Bioorg. Med. Chem. Lett.* **15**, 438.

MLN0128, See *Sapanisertib*

MLN2238, See *Ixazomib*

MLN4924, See *Pevonedistat*

MLN8237, See *Alisertib*

MLN9708, *Undergoes hydrolysis to form MLN2238*

MM-102

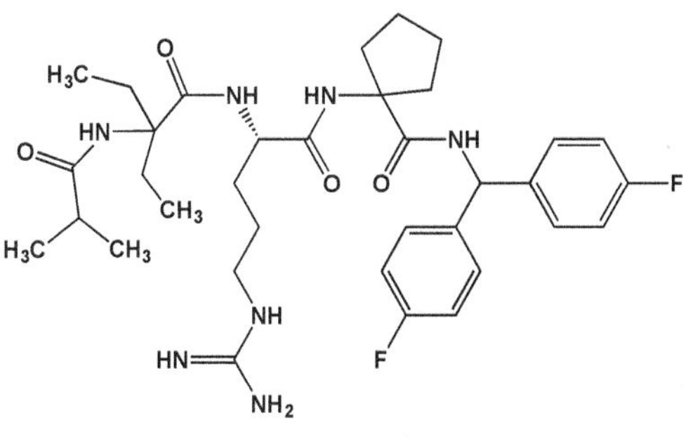

This peptidomimetic MLL1 inhibitor (FW = 669.82 g/mol; CAS 1417329-24-8; Solubility: 100 mg/mL DMSO, H_2O, or Ethanol; IUPAC Name: *See below*), targets the Mixed Lineage Leukemia 1 enzyme, *or* MLL1 (IC_{50} = 1 nM), a histone H3 lysine 4 (H3K4) methyltransferase. MLL1-WDR5 protein-protein interactions are essential for MLL1's enzymatic activity, and MM-102 binding to WDR5 potently inhibits Histone3 lysine 4 (H3K4) methylation. MM-102 inhibits expression of HoxA9 and Meis-1, two critical MLL1 target genes, in bone marrow cells transduced with an MLL1-AF9 fusion gene. Significantly, MM-102 selectively inhibits cell growth in leukemias carrying MLL1 fusion proteins. **1**. Karatas, Townsend, Cao, *et al.* (2013) *J. Am. Chem. Soc.* **135**, 669. **IUPAC Name:** 1-[[(2*S*)-5-[(aminoiminomethyl)amino]-2-[[2-ethyl-2-[(2-methyl-1-oxopropyl)amino]-1-oxobutyl]amino]-1-oxopentyl]amino]-*N*-[bis(4-fluorophenyl)methyl] cyclopentanecarboxamide.

MMI270

This orally bioavailable hydroxamate (FW = 392.48 g/mol), also named CGS 27023A and IUPAC name *N*-hydroxy-2(*R*)-[(4-methoxysulfonyl)(3-picolyl)amino]-3-methylbutaneamide, targets the active-site Zn^{2+} atom in many matrix metalloproteinases, inhibiting interstitial collagenase (K_i = 33 nM), gelatinase A (K_i = 20 nM), stromelysin 1 (Ki = 43 nM), gelatinase B (K_i = 8 nM), and neutrophil collagenase (K_i = 10 nM). MMI270 also exhibits antimetastatic effects *in vivo* as well as antiangiogenic effects *in vitro*. **Target(s):** collagenase 3 (matrix metalloproteinase 13) (1); gelatinase A (matrix metalloproteinase 2) (2,3); gelatinase B (matrix metalloproteinase 9) (2,3); interstitial collagenase (matrix metalloproteinase 1) (1-3); macrophage elastase (matrix metalloproteinase 12) (4); microbial collagenase (3); neutrophil collagenase (matrix metalloproteinase 8) (3); stromelysin 1 (matrix metalloproteinase 3) (1,2,5,6). **1**. Zhang, Gonnella, Koehn, *et al.* (2000) *J. Mol. Biol.* **301**, 513. **2**. Leung, Abbenante & Fairlie (2000) *J. Med. Chem.* **43**, 305. **3**. Ilies, Banciu, Scozzafava, *et al.* (2003) *Bioorg. Med. Chem.* **11**, 2227. **4**. Nar, Werle, Bauer, Dollinger & Jung (2001) *J. Mol. Biol.* **312**, 743. **5**. MacPherson, Bayburt, Capparelli, *et al.* (1997) *J. Med. Chem.* **40**, 2525. **6**. Li, Zhang, Melton, Ganu & Gonnella (1998) *Biochemistry* **37**, 14048.

MN-029, *See* Denibulin

MNTX, *See* Methylnaltrexone

Modafinil

(*S*)-Modanifil **(*R*)-Modanifil**

This orally available, wakefulness-promoting drug and weak, atypical dopamine reuptake inhibitor (FW = 273.36 g/mol; CAS 68693-11-8 (for racemate) and 112111-43-0 (for (−)-*R* enantiomer); off-white, crystalline powder; practically insoluble in water and cyclohexane; slightly soluble in methanol and acetone), also known as Provigil® (racemate) as well as Armodafinil® (*R*-enantiomer) and named systematically as 2-[(diphenylmethyl)sulfinyl]acetamide, exhibits robust effects on catecholamines, serotonin, glutamate, γ-amino-butyric acid, orexin, and histamine systems in the brain. Many may be secondary to catecholamine effects, with some selectivity for cortical over subcortical sites of action (1-3). At well-tolerated doses, modafinil improves cognitive function, including working memory and episodic memory. Modafinil does not bind to known sleep/wake-regulating receptors (*e.g.*, those for norepinephrine, serotonin, dopamine, GABA, adenosine, histamine-3, melatonin, or benzodiazepines), and it does not inhibit the activities of MAO-B or phosphodiesterases II-V. Given its abuse by athletes in training, the World Anti-Doping Agency (WADA) banned modafinil and its pro-drug adrafinil in sports in 2004. **Action as Weak, Atypical Dopamine Reuptake Inhibitor:** Ferraro *et al.* (2) suggested modafinil's dopamine-releasing action in the rat nucleus accumbens is secondary to its ability to reduce local GABAergic transmission, leading to reduction of $GABA_A$ receptor signaling on dopamine terminals. However, of the receptors and transporters assayed, modafinil only displays measurable potency at DA transporters (DAT), inhibiting [^{3}H]-DA uptake, with an IC_{50} of 4.0 μM (4). Moreover, modafinil pretreatment (10 μM) antagonizes methamphetamine-induced release of the DAT substrate [^{3}H]-1-methyl-4-phenylpyridinium. Intravenous modafinil (20 and 60 mg/kg) produces dose-dependent increases in motor activity and extracellular DA, without affecting serotonin (4). **Pharmacokinetics:** Modafinil is absorbed readily, reaching maximum plasma concentrations at 2-4 hours and pharmacokinetic steady-state within 2-4 days. Its elimination half-life is 12-15 hours (dominated by the longer-lived L-enantiomer). It is primarily eliminated via hepatic metabolism, with subsequent urinary excretion (<10% of the excreted unchanged). Metabolism is largely via amide hydrolysis, with lesser contributions from cytochrome P450 (CYP)-mediated oxidative pathways. Modafinil induces CYP1A2, CYP3A4, and CYP2B6, but inhibits CYP2C9 and CYP2C19 *in vitro*. *See also* Appendix II *in* Goodman & Gilman's *THE PHARMACOLOGICAL BASIS OF THERAPEUTICS*, ·12th Edition (Brunton, Chabner & Knollmann, eds.) McGraw-Hill Medical, New York (2011). **Targets:** dopamine reuptake *in vitro*, causing an increase in extracellular dopamine, but no increase in dopamine release (2); firing of dopaminergic, but not GABAergic neurons (3). **1**. Saper & Scammell (2004) *Sleep* **27**, 11. **2**. Ferraro, Tanganelli, O'Connor, *et al.* (1996) *Eur. J. Pharmacol.* **306**, 33. **3**. Korotkova, Klyuch, Ponomarenko *et al.* (2007) *Neuropharmacol.* **52**, 626. **4**. Zolkowska, Jain, Rothman, *et al.* (2009) *J. Pharmacol. Exp. Ther.* **329**, 738.

Modeccin

This protein toxin from the root of *Adenia digitata* (*Passifloraceae*), an indigenous South African plant easily mistaken for the edible root of *Coccinia* Species (*Cucurbitaceae*), is a ricin/abrin-like ribosome-inactivating agent. (**For mechanism, see** *Ricin*.) Death follows vomiting and bloody diarrhea, with the most common postmortem findings of necrosis, vascular thrombosis within the rectum and large intestine, acute nephritis, and hemorrhages within the liver. **1**. Green & Kamerman (1924) J. South Africa Chem. Inst. **7**, 3. **2**. Watt & Breyer-Brandwijk (1962) *The Medicinal and Poisonous Plants of Southern and Eastern Africa*, pp. 826-828, Livingstone, London.

Moenomycins

This antibiotic ($FW_{Moenomycin-A}$ = 1583.87 g/mol; CAS 11015-37-5; soluble in H_2O and DMF), also known as bambermycin and flavomycin, is the major component of least four liposaccharidic compounds produced by *Streptomyces bambergiensis*, *S. ghanaensis*, *S. ederensis*, and *S. geysiriensis*. **Target(s):** penicillin-binding protein 1b (1-4); penicillin-binding protein 3 (5); peptidoglycan cell wall biosynthesis (1-9); peptidoglycan glycosyltransferase (3,4,10-15); peptidoglycan polymerases (9); transglycosylase (16,17); trehalose-phosphatase (18,19); α,α-trehalose-phosphate synthase, UDP-forming (20); undecaprenol kinase, *or* C₅₅-isoprenoid-alcohol kinase (21); MurG transferase (22). **1**. Kurz, Guba & Vertesy (1998) *Eur. J. Biochem.* **252**, 500. **2**. Ruhl, Daghish, Buchynskyy, *et al.* (2003) *Bioorg. Med. Chem.* **11**, 2965. **3**. Nakagawa, Tamaki, Tomioka & Matsuhashi (1984) *J. Biol. Chem.* **259**, 13937. **4**. Chandrakala, Shandil, Mehra, *et al.* (2004) *Antimicrob. Agents Chemother.* **48**, 30. **5**. Ishino & Matsuhashi (1981) *Biochem. Biophys. Res. Commun.* **101**, 905. **6**. Elbein (1987) *Meth. Enzymol.* **138**, 661. **7**. Kusser & Ishiguro (1985) *J. Bacteriol.* **164**, 861. **8**. Huber & Nesemann (1968) *Biochem. Biophys. Res. Commun.* **30**, 7. **9**. van Heijenoort, Leduc, Singer & van Heijenoort (1987) *J. Gen. Microbiol.* **133**, 667. **10**. Terrak & Nguyen-Distèche (2006) *J. Bacteriol.* **188**, 2528. **11**. Park & Matsuhashi (1984) *J. Bacteriol.* **157**, 538. **12**. Park, Seto, Hakenbeck & Matsuhashi (1985) *FEMS Microbiol. Lett.* **27**, 45. **13**. van Heijenoort (2001) *Glycobiology* **11**, 25R. **14**. Ramachandran, Chandrakala, Kumar, *et al.* (2006) *Antimicrob. Agents Chemother.* **50**, 1425. **15**. Zawadzka-Skomial, Markiewicz, Nguyen-Distèche, *et al.* (2006) *J. Bacteriol.* **188**, 1875. **16**. Branstrom, Midha & Goldman (2000) *FEMS Microbiol. Lett.* **191**, 187. **17**. Vogel, Buchynskyy, Stembera, *et al.* (2000) *Bioorg. Med. Chem. Lett.* **10**, 1963. **18**. Klutts, Pastuszak, Edavana, *et al.* (2003) *J. Biol. Chem.* **278**, 2093. **19** Edavana, Pastuszak, Carroll, *et al.* (2004) *Arch. Biochem. Biophys.* **426**, 250. **20**. Pan & Elbein (1996) *Arch. Biochem. Biophys.* **335**, 258. **21**. Sanderman (1976) *Biochim. Biophys. Acta* **444**, 783. **22**. Liu, Ritter, Sadamoto, *et al.* (2003) *ChemBioChem* **4**, 603.

Moexipril

This potent, orally active ACE inhibitor and antihypertensive pro-drug (FW_{free-monoacid} = 498.58 g/mol; CAS 103775-10-6), also named (3*S*)-2-[(2*S*)-2-{[(2*S*)-1-ethoxy-1-oxo-4-phenylbutan-2-yl]amino}propanoyl]-6,7-dimethoxy-1,2,3,4-tetrahydroisoquinoline-3-carboxylic acid, is rapidly metabolized to the diacid, moexiprilat (*See Moexiprilat*). **Mechanism of Action:** As a long-acting, non-sulfhydryl angiotensine-converting enzyme inhibitor, moexipril is useful in treating arterial hypertension. It is a prodrug that yields moexiprilat by hydrolysis of an ethyl ester group, and moexiprilat is the pharmacologically active metabolite. The cardioprotective effects of this ACE inhibitor are mediated through decreased angiotensin II and increased bradykinin, the latter stimulating the production of prostaglandin E₂ and nitric oxide and consequentially exerting vasodilatory and anti-proliferative effects. Moexiprilat has an extended duration of action, owing to its long terminal half-life and persistent ACE inhibition. **Other Target(s):** Moexipril inhibits 3',5'-cyclicAMP phospdiesterase Type 4 (PDE4), with a micromolar range IC₅₀ (2). **1**. Chrysant & Chrysant (2004) *J. Clin. Pharmacol.* **44**, 827. **2**. Brogden & Wiseman (1998) *Drugs* **55**, 845. **3**. Cameron, Coleman, Day, *et al.* (2013) *Biochem. Pharmacol.* **85**, 1297.

Moexiprilat

This diacid (FW_{hydrochloride} = 506.98 g/mol) is a potent inhibitor of peptidyl-dipeptidase A, *or* angiotensin I-converting enzyme, IC₅₀ = 2.6 nM (1,2). Both, moexipril and moexiprilat inhibited the angiotensin I (ANG I)-induced contractions of rabbit aorta concentration-dependently, whereas contractions by other agents were not affected, indicating a high selectivity of both compounds for ACE. Moexiprilat is also the 6,7-dimethoxy analogue of quinaprilat. **1**. Klutchko, Blankley, Fleming, *et al.* (1986) *J. Med. Chem.* **29**, 1953. **2**. Friehe & Ney (1997) *Arzneimittelforschung.* **47**, 132.

Mofegiline

This orally active MAO-B inhibitor (FW_{free-base} = 197.23 g/mol; CAS 119386-96-8; Soluble as the monohydrochloride salt), also named MDL-72,974 and (2*E*)-2-(fluoromethylidene)-4-(4-fluorophenyl)butan-1-amine, selectively and irreversibly targets monoamine oxidase B (MAO-B) and vascular adhesion protein 1 (VAP-1), the latter known also as semicarbazide-sensitive amine oxidase (SSAO). Although extensively researched as a potential treatment of Parkinson's disease and Alzheimer's disease, mofegiline was never marketed. **Human Studies:** A Phase-I study showed rapid absorption and elimination (time to maximum concentration = 1 hour; half-life = 1–3 hours; maximal plasma concentration and area under the plasma concentration-time curve increased) (3). Mofegiline also rapidly and markedly inhibited platelet monoamine oxidase B (MAOB) activity, which returned to baseline within 14 days. No changes in urinary elimination of catecholamines, blood pressure, heart rate, or ECG were observed. The 48-mg single dose and the 24 mg multiple daily dose far exceeded the dose (1-mg) need to observe > 90% platelet MAOB inhibition (3). **Mode of Action:** Mofegiline competitively inhibits MAO-B (K_{i,apparent} = 28 nM), with irreversible formation of a 1:1 molar stoichiometry attended by loss of catalytic turnover (4). The absorption spectral properties of mofegiline inhibited MAO-B show features (λ_{max} ≈ 450 nm) and Visible/Near-UV CD spectra resembling *N*⁵-flavocyanine adducts. X-ray crystallography of the mofegiline-MAO-B adduct shows a covalent bond between the flavin cofactor at *N*⁵ with the distal allylamine carbon atom as well as the absence of the fluorine atom (4). **1**. Yu & Zuo (1992) *Biochem. Pharmacol.* **43**, 307. **2**. Zreika, Fozard, Dudley, *et al.* (1989) *J. Neural Transm. Park. Dis. Dement. Sect.* **1**, 243. **3**. Stoltz, Reynolds, Elkins, Salazar & Weir (1995) *Clin. Pharmacol. Ther.* **58**, 342. **4**. Milczek, Bonivento, Binda, *et al.* (2008) *J. Med. Chem.* **51**, 8019.

Molidustat

This HIF-PH inhibitor (FW = 314.00 g/mol; CAS 1154028-82-6; λ_{max} = 250 nm), also known as BAY 85-3934 and 1,2-dihydro-2-[6-(4-morpholinyl)-4-pyrimidinyl]-4-(1*H*-1,2,3-triazol-1-yl)-3*H*-pyrazol-3-one, targets Hypoxia Inducible Factor Prolyl Hydroxylase (IC₅₀ = 0.49 μM). Because oxygen-dependent degradation of the α subunit of HIF heterodimer suppresses erythropoietin expression, erythropoiesis is stimulated by agents preventing HIF. BAY 85-3934 stabilizes HIF from degradation in the proteasome by inhibiting HIF-1α PH. Molidustat increase endogenous production of erythropoietin. HIF-PH inhibitors promise to provide a novel approach for treating anemia that is based on mimicking the hypoxia-driven expression of endogenous EPO in the kidney. **1**. Flamme, Oehme, Ellinghaus, *et al.* (2014) *PLoS One* **9**, e111838.

Molybdate

This Mo(VI) oxo dianion (FW_{ion} = 159.94 g/mol; CAS 7782-91-4) is an isosteric analogue of sulfate and phosphate ions, and, as such, it inhibits many phosphatases. Molybdate dianion is also an activator of adenylate cyclase. Note that when solutions of MoO₄^{2–} are weakly acidified, oligomeric and polymeric anions form. Ammonium molybdate (NH₄)₆Mo₇O₂₄ is a redox active reagent used in spot tests and in sprays for detecting inorganic and organic phosphates in thin-layer and paper chromatography. Molybdate can also be used to test for reducing sugars. Froehde's reagent (formed by dissolving sodium molybdate (Na₂MoO₄) in concentrated sulfuric acid) is used in spot tests for alkaloids, particularly opium alkaloids. Given its promiscuity, molybdate is best used in single-enzyme experiments, where the goal is to exploit its higher affinity to conduct mechanistic studies. **Target(s):** *N*-acetylgalactosamine-4-sulfatase, *or* arylsulfatase (1); acid phosphatase (2-16,96-124); adenosine-tetraphosphatase (17); alkaline phosphatase (4,18,125-129); Arylsulfatase (1,4,55); asparagine synthetase, ADP-forming (19); cAMP phosphodiesterase (20); carbonic anhydrase, weakly inhibited (21); 2-carboxy-D-arabinitol-1-phosphatase (59); catechol *O*-methyltransferase (22); cyclomaltodextrin glucanotransferase (138); cysteine synthase (133,134); exopolyphosphatase (17); FAD diphosphatase, weakly inhibited (23); β-fructofuranosidase, *or* invertase, inhibited by Mo$_7$O$_{24}{}^{6-}$ (52-54); fructose-1,6-bisphosphate aldolase (24); glucan 1,3-β glucosidase (50,51); glucocorticoid receptor (25); glucose-1-phosphatase, inhibited by ammonium and sodium molybdate (92-94); glucose-6-phosphatase, K_i = 3.7 μM for rabbit intestinal enzyme (26-29,45); glutathione: cystine transhydrogenase (145); glyceryl-ether monooxygenase (142,143); glycogen phosphorylase (141); inositol-phosphate phosphatase (86); isoamylase, inhibited by ammonium molybdate (49); isocitrate dehydrogenase, NAD$^+$-dependent (30,31); isocitrate dehydrogenase, NADP$^+$-dependent, weakly inhibited (31); lipid-phosphate phosphatase, weakly inhibited (56); monoterpenyl diphosphatase (32); nucleotidase (79); 5'-nucleotidase (95); phloretin hydrolase (33); phosphoamidase34; *N*ω-phosphoarginine hydrolase35; phosphoenolpyruvate phosphatase (60,61); 6-phosphofructo-2 kinase, when fructose-2,6-bisphosphatase activity is activated (36); 3-phosphoglycerate phosphatase (78); phosphoprotein phosphatase (13,37-40,90,91); phosphoprotein phosphatase, CD45 (40); 3-phosphoshikimate 1-carboxyvinyltransferase, *or* 3-enolpyruvoylshikimate-5-phosphate synthase (135,136); 3-phytase (81,85); 4-phytase (80-84); phytoglycogen synthase (41); polynucleotide-5' hydroxyl-kinase (42); protein-tyrosine-phosphatase (40,63-77); purple acid phosphatase (12,114); pyridoxal phosphatase (43,57,58); pyrophosphatase (44,45); riboflavin phosphotransferase (132); ribonuclease, pea leaf (46); salicylate 1-monooxygenase, weakly inhibited (144); sorbitol-6-phosphatase (62); starch synthase (137); sucrose-phosphatase (87-89); sucrose phosphate synthase (139,140); sulfate adenylyltransferase (130,131); sulfate transport (47); tripeptide aminopeptidase (48).

1. Bleszynski & Leznicki (1967) *Enzymologia* **33**, 373. **2.** Van Etten, Waymack & Rehkop (1974) *J. Amer. Chem. Soc.* **96**, 6782. **3.** Ferreira, Taga & Aoyama (2000) *J. Enzyme Inhib.* **15**, 403. **4.** Stankiewicz & Gresser (1988) *Biochemistry* **27**, 206. **5.** Massart & Vermeyen (1942) *Naturwissenschaften* **30**, 170. **6.** Courtois & Bossard (1944) *Bull. soc. chim. biol.* **26**, 464. **7.** Massart (1950) *The Enzymes*, 1st ed., **1** (part 1), 307. **8.** Roche (1950) *The Enzymes*, 1st ed., **1** (part 1), 473. **9.** Hollander (1971) *The Enzymes*, 3rd ed., **4**, 449. **10.** Watorek, Morawiecka & Korczak (1977) *Acta Biochim. Pol.* **24**, 153. **11.** Shimada, Shinmyo & Enatsu (1977) *Biochim. Biophys. Acta* **480**, 417. **12.** Hara, Sawada, Kato, *et al.* (1984) *J. Biochem.* **95**, 67. **13.** Schlosnagle, Bazer, Tsibris & Roberts (1974) *J. Biol. Chem.* **249**, 7574. **14.** Bazer, Chen, Knight, *et al.* (1975) *J. Anim. Sci.* **41**, 1112. **15.** Ketcham, Baumbach, Bazer & Roberts (1985) *J. Biol. Chem.* **260**, 5768. **16.** Allen, Nuttleman, Ketcham & Roberts (1989) *J. Bone Miner. Res.* **4**, 47. **17.** Guranowski, Starzynska, Barnes, Robinson & Liu (1998) *Biochim. Biophys. Acta* **1380**, 232. **18.** Adler (1978) *Biochim. Biophys. Acta* **522**, 113. **19.** M. Nair (1969) *Arch. Biochem. Biophys.* **133**, 208. **20.** Ofenloch-Hahnle & Eisele (1987) *Z. Naturforsch.* **42**, 162. **21.** Innocenti, Vullo, Scozzafava & Supuran (2005) *Bioorg. Med. Chem. Lett.* **15**, 567. **22.** Beattie & Weersink (1992) *J. Inorg. Biochem.* **46**, 153. **23.** Ravindranath & Rao (1968) *Indian J. Biochem.* **5**, 137. **24.** Crans, Sudhakar & Zamborelli (1992) *Biochemistry* **31**, 6812. **25.** Murakami & Moudgil (1981) *Biochem. J.* **198**, 447. **26.** Colilla, Jorgenson & Nordlie (1975) *Biochim. Biophys. Acta* **377**, 117. **27.** Nordlie (1971) *The Enzymes*, 3rd ed., **4**, 543. **28.** Swanson (1950) *J. Biol. Chem.* **184**, 647. **29.** Lygre & Nordlie (1968) *Biochemistry* **7**, 3219. **30.** Kornberg (1955) *Meth. Enzymol.* **1**, 707. **31.** Kornberg & Pricer, Jr. (1951) *J. Biol. Chem.* **189**, 123. **32.** Croteau & Karp (1979) *Arch. Biochem. Biophys.* **198**, 523. **33.** Minamikawa, Jayasankar, Bohm, Taylor & Towers (1970) *Biochem. J.* **116**, 889. **34.** Parvin & Smith (1969) *Biochemistry* **8**, 1748. **35.** Kuba, Ohmori & Kumon (1992) *Eur. J. Biochem.* **208**, 747. **36.** Kountz, McCain, el-Maghrabi & Pilkis (1986) *Arch. Biochem. Biophys.* **251**, 104. **37.** Revel (1963) *Meth. Enzymol.* **6**, 211. **38.** Tonks, Diltz & Fischer (1991) *Meth. Enzymol.* **201**, 427. **39.** Roberts & Bazer (1976) *Biochem. Biophys. Res. Commun.* **68**, 450. **40.** Tonks, C. Diltz & Fischer (1991) *Meth. Enzymol.* **201**, 442. **41.** Cardini & Frydman (1966) *Meth. Enzymol.* **8**, 387. **42.** Zimmerman & Pheiffer (1981) *The Enzymes*, 3rd ed., **14**, 315. **43.** Fonda (1992) *J. Biol. Chem.* **267**, 15978. **44.** Felix & Fleisch

(1975) *Biochem. J.* **147**, 111. **45.** Gold & Veitch (1973) *Biochim. Biophys. Acta* **327**, 166. **46.** Anfinsen & White, Jr. (1961) *The Enzymes*, 2nd ed., **5**, 95. **47.** Cardin & Mason (1975) *Biochim. Biophys. Acta* **394**, 46. **48.** Morgan & Donlan (1985) *Eur. J. Biochem.* **146**, 429. **49.** Fujita, Kubo, Francisco, *et al.* (1999) *Planta* **208**, 283. **50.** Pitson, Seviour, McDougall, Woodward & Stone (1995) *Biochem. J.* **308**, 733. **51.** Marshall & Grand (1975) *Arch. Biochem. Biophys.* **167**, 165. **52.** Batta, Singh, Sharma & Singh (1991) *Plant Physiol. Biochem.* **29**, 415. **53.** Prado, Vattuone, Fleischmacher & Sampietro (1985) *J. Biol. Chem.* **260**, 4952. **54.** Lopez, Vattuone & Sampietro (1988) *Phytochemistry* **27**, 3077. **55.** Knoess & Glombitza (1993) *Phytochemistry* **32**, 1119. **56.** Newman, Morisseau, Harris & Hammock (2003) *Proc. Natl. Acad. Sci. U.S.A.* **100**, 1558. **57.** Fonda & Zhang (1995) *Arch. Biochem. Biophys.* **320**, 345. **58.** Tazoe, Ichikawa & Hoshino (2005) *Biosci. Biotechnol. Biochem.* **69**, 2277. **59.** Salvucci & Holbrook (1989) *Plant Physiol.* **90**, 679. **60.** Singh & Singh (1996) *Proc. Indian Acad. Sci. Sect. B* **66**, 53. **61.** Duff, Lefebvre & Plaxton (1989) *Plant Physiol.* **90**, 734. **62.** Zhou, Cheng & Wayne (2003) *Plant Sci.* **165**, 227. **63.** Aguirre-Garcia, Escalona-Montano, Bakalara, *et al.* (2006) *Parasitology* **132**, 641. **64.** Singh (1990) *Biochem. Biophys. Res. Commun.* **167**, 621. **65.** Clari, Brunati & Moret (1986) *Biochem. Biophys. Res. Commun.* **137**, 566. **66.** Harder, Owen, Wong, *et al.* (1994) *Biochem. J.* **298**, 395. **67.** Tonks, Diltz & Fischer (1988) *J. Biol. Chem.* **263**, 6731. **68.** Waheed, Laidler, Wo & Van Etten (1988) *Biochemistry* **27**, 4265. **69.** Grangeasse, Doublet, Vincent, *et al.* (1998) *J. Mol. Biol.* **278**, 339. **70.** Boivin & Galand (1986) *Biochem. Biophys. Res. Commun.* **134**, 557. **71.** Roome, O'Hare, Pilch & Brautigan (1988) *Biochem. J.* **256**, 493. **72.** Dechert, Adam, Harder, Clark-Lewis & Jirik (1994) *J. Biol. Chem.* **269**, 5602. **73.** Hoppe, Berne, Stock, *et al.* (1994) *Eur. J. Biochem.* **223**, 1069. **74.** Tonks, Diltz & Fischer (1990) *J. Biol. Chem.* **265**, 10674. **75.** Zhao, Laroque, Ho, Fischer & Shen (1994) *J. Biol. Chem.* **269**, 8780. **76.** Brunati & Pinna (1985) *Biochem. Biophys. Res. Commun.* **133**, 929. **77.** Pot, Woodford, Remboutsika, Haun & Dixon (1991) *J. Biol. Chem.* **266**, 19688. **78.** Randall & Tolbert (1971) *J. Biol. Chem.* **246**, 5510. **79.** Passariello, Schippa, Iori, *et al.* (2003) *Biochim. Biophys. Acta* **1648**, 203. **80.** Mahajan & Dua (1997) *J. Agric. Food Chem.* **45**, 2504. **81.** Greiner (2002) *J. Agric. Food Chem.* **50**, 6858. **82.** Greiner, Konietzny & Jany (1998) *J. Food Biochem.* **22**, 143. **83.** Greiner & Alminger (1999) *J. Sci. Food Agric.* **79**, 1453. **84.** Andriotis & Ross (2003) *Phytochemistry* **64**, 689. **85.** Greiner, Haller, Konietzny & Jany (1997) *Arch. Biochem. Biophys.* **341**, 201. **86.** Gumber, Loewus & Loewus (1984) *Plant Physiol.* **76**, 40. **87.** Hawker, Smith, Phillips & Wiskich (1987) *Plant Physiol.* **84**, 1281. **88.** Echeverria & Salerno (1994) *Plant Sci.* **96**, 15. **89.** Cumino, Ekeroth & Salerno (2001) *Planta* **214**, 250. **90.** Zhou, Clemens, Hakes, Barford & Dixon (1993) *J. Biol. Chem.* **268**, 17754. **91.** Polya & Haritou (1988) *Biochem. J.* **251**, 357. **92.** Joh, Yazaki, Suzuki & Hayakawa (1998) *Biosci. Biotechnol. Biochem.* **62**, 2251. **93.** Saugy, Farkas & MacLachlan (1988) *Eur. J. Biochem.* **177**, 135. **94.** Turner & Turner (1960) *Biochem. J.* **74**, 486. **95.** Reilly & Calcutt (2004) *Protein Expr. Purif.* **33**, 48. **96.** Rossi, Palma, Leone & Brigliador (1981) *Phytochemistry* **20**, 1823. **97.** Kamenan & Diopoh (1982) *Plant Sci. Lett.* **24**, 173. **98.** Hayman, Warburton, Pringle, Coles & Chambers (1989) *Biochem. J.* **261**, 601. **99.** Sugiura, Kawabe, Tanaka, Fujimoto & Ohara (1981) *J. Biol. Chem.* **256**, 10664. **100.** Lawrence & van Etten (1981) *Arch. Biochem. Biophys.* **206**, 122. **101.** Reilly, Baron, Nano & Kuhlenschmidt (1996) *J. Biol. Chem.* **271**, 10973. **102.** Panara, Pasqualini & Antonielli (1990) *Biochim. Biophys. Acta* **1037**, 73. **103.** Waymack & van Etten (1991) *Arch. Biochem. Biophys.* **288**, 621. **104.** Tejera Garcia, Olivera, Iribarne & Lluch (2004) *Plant Physiol. Biochem.* **42**, 585. **105.** Aguirre-Garcia, Cerbon & Talamas-Rohana (2000) *Int. J. Parasitol.* **30**, 585. **106.** Giordani, Nari, Noat & Sauve (1986) *Plant Sci.* **43**, 207. **107.** Turner & Plaxton (2001) *Planta* **214**, 243. **108.** Dibenedetto (1972) *Biochim. Biophys. Acta* **286**, 363. **109.** Tominaga & Mori (1974) *J. Biochem.* **76**, 397. **110.** Ching, Lin & Metzger (1987) *Plant Physiol.* **84**, 789. **111.** Lau, Farley & Baylink (1987) *Adv. Protein Phosphatases* **4**, 165. **112.** Rengasamy, Selvam & Gnanam (1981) *Arch. Biochem. Biophys.* **209**, 230. **113.** Hayman, Warburton, Pringle & Chambers (1988) *Biochem. Soc. Trans.* **16**, 895. **114.** Bozzo, Raghothama & Plaxton (2002) *Eur. J. Biochem.* **269**, 6278. **115.** Harsanyi & Dorn (1972) *J. Bacteriol.* **110**, 246. **116.** Kamenan & Diopoh (1983) *Plant Sci. Lett.* **32**, 305. **117.** Armienta-Aldana & De La Vara (2004) *Physiol. Plant.* **121**, 223. **118.** Duff, Lefebvre & Plaxton (1991) *Arch. Biochem. Biophys.* **286**, 226. **119.** Kuo & Blumenthal (1961) *Biochim. Biophys. Acta* **52**, 13. **120.** Kawabe, Sugiura, Terauchi & Tanaka (1984) *Biochim. Biophys. Acta* **784**, 81. **121.** Farooqui & Hanson (1988) *Experientia* **44**, 437. **122.** Malveaux & San Clemente (1969) *J. Bacteriol.* **97**, 1215. **123.** Coello (2002) *Physiol. Plant.* **116**, 293. **124.** Reilly, Felts, Henzl, Calcutt & Tanner (2006) *Protein Expr. Purif.* **45**,

132. **125.** Ansai, Awano, Chen, *et al.* (1998) *FEBS Lett.* **428**, 157. **126.** Ezawa, Kuwahara, Sakamoto, Yoshida & Saito (1999) *Mycologia* **91**, 636. **127.** Klumpp & Schultz (1990) *Biochim. Biophys. Acta* **1037**, 233. **128.** Al-Saleh (2002) *J. Nat. Toxins* **11**, 357. **129.** Bedu, Jeanjean & Rocca-Serra (1982) *Plant Sci. Lett.* **27**, 163. **130.** Renosto, Patel, Martin, *et al.* (1993) *Arch. Biochem. Biophys.* **307**, 272. **131.** Renosto, Martin, Wailes, Daley & Segel (1990) *J. Biol. Chem.* **265**, 10300. **132.** Katagiri, Yamada & Imai (1959) *J. Biochem.* **46**, 1119. **133.** Zhao, Kumada, Imanaka, Imamura & Nakanishi (2006) *Protein Expr. Purif.* **47**, 607. **134.** Mino, Yamanoue, Sakiyama, *et al.* (2000) *Biosci. Biotechnol. Biochem.* **64**, 1628. **135.** Steinrücken & Amrhein (1984) *Eur. J. Biochem.* **143**, 341. **136.** Ream, Steinrücken, Porter & Sikorski (1988) *Plant Physiol.* **87**, 232. **137.** Frydman & Cardini (1965) *Biochim. Biophys. Acta* **96**, 294. **138.** Fujita, Tsubouchi, Inagi, *et al.* (1990) *J. Ferment. Bioeng.* **70**, 150. **139.** Huber & Huber (1990) *Arch. Biochem. Biophys.* **282**, 421. **140.** Huber, Hite, Outlaw & Huber (1991) *Plant Physiol.* **95**, 291. **141.** Weinhausel, Griessler, Krebs, *et al.* (1997) *Biochem. J.* **326**, 773. **142.** Soodsma, Piantados & Snyder (1972) *J. Biol. Chem.* **247**, 3923. **143.** Taguchi & Armarego (1998) *Med. Res. Rev.* **18**, 43. **144.** Yamamoto, Katagiri, Maeno & Hayaishi (1965) *J. Biol. Chem.* **240**, 3408. **145.** Minoda, Kurane & Yamada (1973) *Agric. Biol. Chem.* **37**, 2511.

Molybdenum & Molybdenum Ions

This Group 6 (or Group VIA) element (Symbol = Mo; Atomic Number = 42; Atomic Weight = 95.94) , located directly beneath chromium in the Periodic Table, has the following ground state electronic configuration for the neutral atom is $1s^2 2s^2 2p^6 3s^2 3p^6 4s^2 3d^{10} 4p^6 5s^1 4d^5$. Molybdenum has valencies of 2, 3, 6, and possibly 4 and 5, with ionic radii 0.93, 0.70, and 0.62 Å of Mo^{1+}, Mo^{4+}, and Mo^{6+}, respectively. The plant *Grindelia fastigiata* may be a molybdenum accumulator. (*See also Molybdate; Permolybdate*) **Target(s):** acid phosphatase (4); alkaline phosphatase (5,6); benzoate 4-monooxygenase (7); carnosine synthetase (8); chitinase (9); chitosanase (10); cholinesterase (5); cyclomaltodextrin glucanotransferase (11,12); β-fructofuranosidase, *or* invertase, inhibited by $MoCl_2$ (13); glucose 6 phosphatase (5); indoleacetaldoxime dehydratase (14); laccase (15); phenylacetyl-CoA synthetase, *or* phenylacetate:CoA ligase (16); β-(pyradozyl-*N*)-L-alaninase (17); ribonuclease, plant (18); stizolobate synthase (19); stizolobinate synthase (19). **1.** Yenigün & Güvenilir (2003) *Appl. Biochem. Biotechnol.* **105-108**, 677. **2.** Rana & Kumar (1981) *Toxicol. Lett.* **7**, 393. **3.** Srinivas, Jayalakshmi, Sreeramulu, Sherman & Rao (2006) *Biochim. Biophys. Acta* **1760**, 310. **4.** Reddy & Vaidyanathan (1975) *Biochim. Biophys. Acta* **384**, 46. **5.** Stenesh & Winnick (1960) *Biochem. J.* **77**, 575. **6.** Park, Pan, So, *et al.* (2002) *Mol. Cells* **13**, 69. **7.** Eom & Lee (2003) *Arch. Pharm. Res.* **26**, 1036. **8.** Fujita, Tsubouchi, Inagi, *et al.* (1990) *J. Ferment. Bioeng.* **70**, 150. **9.** Tonkova (1998) *Enzyme Microb. Technol.* **22**, 678. **10.** Liebl, Brem & Gotschlich (1998) *Appl. Microbiol. Biotechnol.* **50**, 55. **11.** Shulka & Mahadevan (1968) *Arch. Biochem. Biophys.* **125**, 873. **12.** Ben Younes, Mechichi & Sayadi (2007) *J. Appl. Microbiol.* **102**, 1033. **13.** el-S. Mohamed & Fuchs (1993) *Arch. Microbiol.* **159**, 554. **14.** Nishizuka (1971) *Meth. Enzymol.* **17B**, 834. **15.** Anfinsen & White (1961) *The Enzymes*, 2nd ed., **5**, 95. **16.** Saito & Komamine (1978) *Eur. J. Biochem.* **82**, 385.

Mometasone furoate

This synthetic glucocorticoid (FW = 421.43 g/mol; CASs = 105102-22-5 and 83919-23-7), also known by the trade names Elocon® Elocom®, Elomet®, Elosalic®, Novasone®, Nasonex®, Asmanex®, Twisthaler®, Essex® and its systematic name (11β,16α)-9,21-dichloro-11-hydroxy-16-methyl-3,20-dioxopregna-1,4-dien-17-yl 2-furoate, blocks cytokine-induced surface expression of vascular cell adhesion molecule-1 (VCAM-1) and intercellular adhesion molecule-1 (ICAM-1) on BEAS-2B bronchial epithelial cells *in vitro* (IC_{50} =10 pM). Although both PDE-4 inhibitors and glucocorticoids suppress secretion of TNFα and IL-10, PDE-4 inhibitors are more potent in suppressing cytokine secretion by PBMCs from atopic individuals (2). Although less potent in inhibiting production of IL-10, PDE-4 inhibitors now have greater therapeutic potential than glucocorticoids in allergic diseases. **Formulation:** Mometasone Furoate Cream USP, 0.1%, contains mometasone furoate, USP for dermatologic use. Each gram of mometasone furoate cream (0.1% weight/weight) contains mometasone furoate (1 mg) dissolved in a cream base of hexylene glycol, phosphoric acid, propylene glycol monostearate, stearyl alcohol/ceteareth-20, aluminum starch octenylsuccinate, white wax, white petrolatum, and purified water. **1.** Atsuta, Plitt, Bochner & Schleimer (1999) *Am. J. Respir. Cell Mol. Biol.* **20**, 643. **2.** Crocker, Ohia, Church & Townley (1998) *J. Allergy Clin. Immunol.* **102**, 797.

MON-097, *See* Acetochlor

Monastrol

This cell-permeable kinesin Eg5 inhibitor (F.Wt. = 292.35; CAS 254753-54-3; Solubility = 5 mg/mL DMSO; Photosensitive), known systematically as ethyl-4-(3-hydroxyphenyl)-6-methyl-2-sulfanylidene-3,4-dihydro-1*H*-pyrimidine-5-carboxylate, arrests mammalian cells in mitosis, as characterized by monopolar spindles. Inhibition appears to be allosteric mechanisms, inasmuch as monastrol does not competitively inhibit ATP binding. [Motors actually catalyze translocation: $ATP + MT_{Position-1} \rightleftharpoons ADP + P_i + MT_{Position-2}$, where the microtubule (MT) scaffold, indicating two successive translocation states), suggesting that the rate equation has the form $v = V_m / \{1 + [ATP]/K_{ATP} + [MT_{SITES}]/K_{MT} + [ATP][MT_{SITES}]/K_{ATP}K_{MT}$, such that monastrol may compete for MT sites for kinesin docking on the microtubule.] Unlike other mitotic spindle poisons (*e.g.*, colchicine, vinblastine, and nocodazole) that target tubulin polymerization or affect microtubule stability, monastrol (IC_{50} = 14 μM) inhibits a motor protein. **1.** Mayer, Kapoor, Haggarty, King, Schreiber & Mitchison (1999) *Science* **286**, 971. **2.** Kapoor, Mayer, Coughlin & Mitchison (2000) *J. Cell Biol.* **150**, 975. **3.** Maliga, Kapoor & Mitchison (2002) *Chem. Biol.* **9**, 989.

Monensin

This monovalent ionophore ($FW_{free-acid}$ = 670.88 g/mol; CAS 17090-79-8), produced by *Streptomyces cinnamonensis*, is a powerful antibiotic that disrupts a number of cellular functions, including oxidative phosphorylation and photophosphorylation (1,2). Monensin forms a stable complex with Na^+ and exhibits high Na^+/K^+ selectivity. Monensin itself and Monensin-Na^+ complexes are very soluble in organic solvents. Monensin facilitates the transmembrane exchange of principally Na^+ ions for protons. Monensin impairs post-translational protein processing of in the Golgi complex (3), and treatment with monensin results in an accumulation of non-secreted proteins. **1.** Izawa & Good (1972) *Meth. Enzymol.* **24**, 355. **2.** Reed (1979) *Meth. Enzymol.* **55**, 435. **3.** Ledger & Tanzer (1984) *Trends Biochem. Sci.* **9**, 313.

Moniliformin

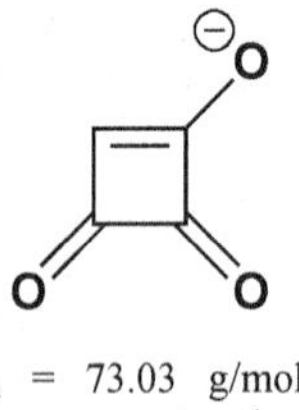

This fungal toxin (FW$_{anion}$ = 73.03 g/mol), produced by *Fusarium moniliforme* a pathogen common to stored grains, inhibits a number of thiamin-dependent enzymes by forming an adduct between the cofactor and a reactive carbonyl in moniliformin. **Target(s):** acetohydroxy acid synthase (1,2); glutathione-disulfide reductase (3); glutathione peroxidase (3); α-ketoglutarate dehydrogenase complex (1,4); pyruvate decarboxylase (1,5); pyruvate dehydrogenase complex (1,6-8); transketolase (8). **1.** Pirrung, Nauhaus & Singh (1996) *J. Org. Chem.* **61**, 2592. **2.** Singh, Stidham & Shaner (1988) *Anal. Biochem.* **171**, 173. **3.** Chen, Tian & Yang (1990) *Mycopathologia* **110**, 119. **4.** Brown & Perham (1976) *Biochem. J.* **155**, 419. **5.** Ullrich, Wittorf & Gubler (1966) *Biochim. Biophys. Acta* **113**, 595. **6.** Thiel (1978) *Biochem. Pharmacol.* **27**, 483. **7.** Gathercote, Thiel & Hofmeyer (1986) *Biochem. J.* **233**, 719. **8.** Burka, Doran & Wilson (1982) *Biochem. Pharmacol.* **31**, 79.

Monobenzone

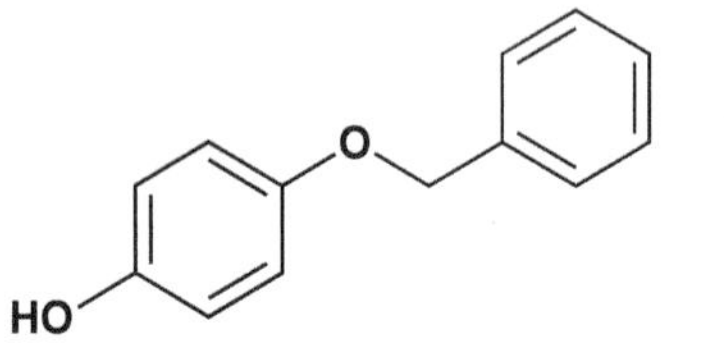

This substituted hydroquinine (FW = 200.24 g/mol; M.P. = 122.5°C), also known as 4-(phenylmethoxy)phenol, *p*-(benzyloxy)phenol, and benzylhydroquinone, is a redox-active substance that inhibits tyrosinase, *or* monophenol monooxygenase, and, as such, has been used as a depigmenting agent. **1.** Webb (1966) *Enzyme and Metabolic Inhibitors*, vol. **2**, p. 304, Academic Press, New York. **2.** Hogeboom & Adams (1942) *J. Biol. Chem.* **145**, 273. **3.** Lerner, Fitzpatrick, Calkins & Summerson (1951) *J. Biol. Chem.* **191**, 799. **4.** Parvez, Kang, Chung & Bae (2007) *Phytother. Res.* **21**, 805.

N^6-Monobutyryladenosine 3',5'-Cyclic Monophosphate

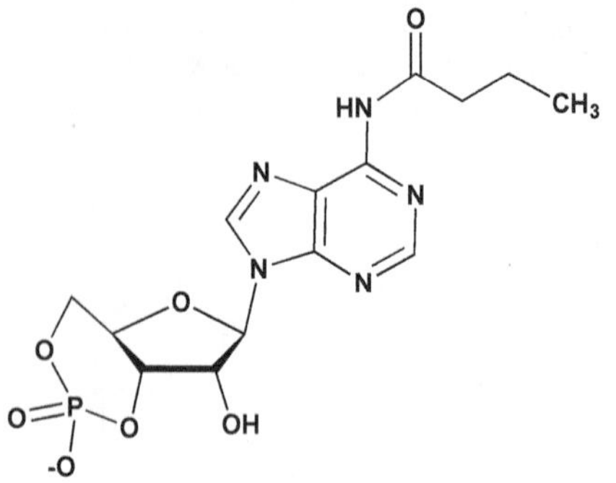

This membrane-permeable cyclic-AMP derivative (FW$_{free-acid}$ = 399.30 g/mol), often abbreviated monobutyryl-cAMP, retains the ability to activate cAMP-dependent protein kinases and the glycogen phosphorylase system. N^6-monobutyryl-cAMP is lipophilic, whereas cAMP is not. It is less susceptible to hydrolysis by phosphodiesterases. N^6-monobutyryl-cAMP is also the active form of the more popularly used cell-penetrating substance, $N^6,O^{2'}$-dibutyryladenosine 3',5'-cAMP (1). **Target(s):** poly(ADP-ribose) glycohydrolase (2). **1.** Kaukel, Mundhenk & Hilz (1972) *Eur. J. Biochem.* **27**, 197. **2.** Miwa, Tanaka, Matsushima & Sugimura (1974) *J. Biol. Chem.* **249**, 3475.

Monocrotophos

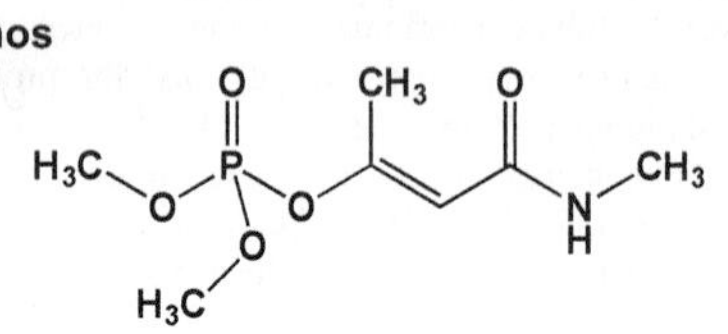

This toxic organophosphate (FW = 223.17 g/mol), itself a direct metabolite of dicrotophos, is an pesticide and cholinesterase inhibitor. **Target(s):** acetylcholinesterase (1-5); arylformamidase (6); carboxylesterase (esterase) (1); cholinesterase (7); glutaminase (8); glutathione *S*-transferase (9); monoamine oxidase (10). **1.** Siddiqui, Rahman, Mahboob, Anjum & Mustafa (1988) *J. Environ. Sci. Health B* **23**, 291. **2.** Siddiqui, Rahman, Mustafa, Rahman & Bhalerao (1991) *Ecotoxicol. Environ. Saf.* **21**, 283. **3.** Rao, Swamy & Yamin (1991) *J. Environ. Sci. Health B* **26**, 449. **4.** Qadri, Swamy & Rao (1994) *Ecotoxicol. Environ. Saf.* **28**, 91. **5.** Kato, Tanaka & Miyata (2004) *Pestic. Biochem. Physiol.* **79**, 64. **6.** Seifert & Casida (1979) *Pestic. Biochem. Physiol.* **12**, 273. **7.** Verberk (1977) *Toxicol. Appl. Pharmacol.* **42**, 345. **8.** Nag (1992) *Indian J. Exp. Biol.* **30**, 543. **9.** Siddiqui, Mahboob & Mustafa (1990) *Toxicology* **64**, 27. **10.** Nag & Nandy (2001) *Indian J. Exp. Biol.* **39**, 802.

Monomethyl Auristatin E

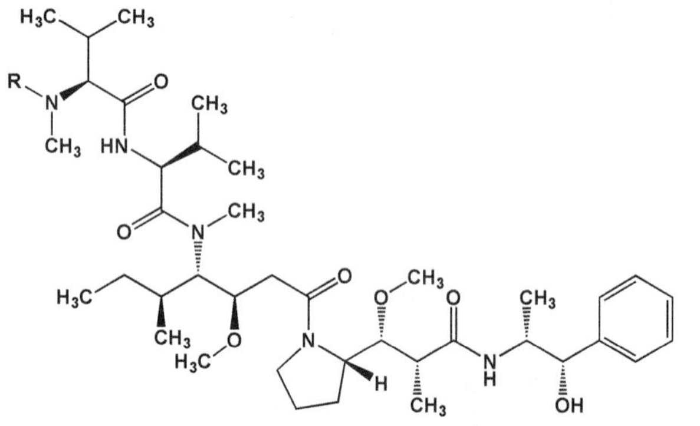

This highly toxic synthetic tubulin polymerization inhibitor and antineoplastic agent (FW = 717.99 g/mol; CAS 474645-27-7; *Symbol*: MMAE), is not employed directly as a drug, but rather as a conjugate, joined by a suitable linker (indicated as R in the above structure) to a monoclonal antibody and relying on the latter to target a specific cancer cell type (1,2). When employed as a antibody-drug conjugate (ADC), MMAE is referred to as vedotin (**See for example** *Brentuximab vedotin*). The antibody-drug linker (consisting of the amino acids valine and citrulline) is readily cleaved by cathepsin, once within the targeted cancer cell. The peptide-linked mAb-Valine-Citrulline-MMAE and mAb-phenylalanine-lysine-MMAE conjugates are much more stable in buffers and plasma than the conjugates of mAb and the hydrazone of 5-benzoylvaleric acid-AE ester (AEVB) (3). As a result, the mAb-Val-Cit-MMAE conjugates exhibited greater specificity *in vitro* and lower toxicity *in vivo* than corresponding hydrazone conjugates. *In vivo* studies demonstrated the peptide-linked conjugates induced regressions and cures of established tumor xenografts with therapeutic indices as high as 60-fold. These conjugates illustrate the importance of linker technology, drug potency and conjugation methodology in developing safe and efficacious mAb-drug conjugates for cancer therapy (3). **See also** *Dolastatin 10* **1.** Francisco, Cerveny, Meyer, *et al.* (2003) *Blood* **102**, 1458. **2.** Ma, Hopf, Malewicz, *et al.* (2006) *Clin. Cancer Res.* **12**, 2591. **3.** Doronina, Toki, Torgov *et al.* (2003) *Nature Biotechnol.* **21**, 778.

Montelukast

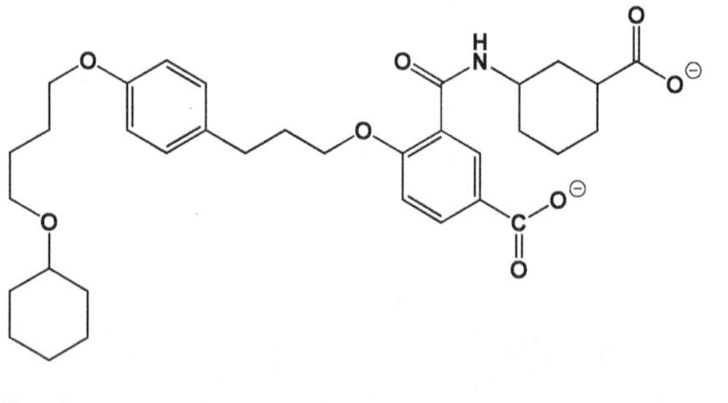

This potent, orally administered CysLT$_1$ antagonist (FW = 586.18 g/mol; CAS 158966-92-8; Code name MK-0476; Trade names: Singulair®, Montelo-10® and Monteflo®; IUPAC name (*R,E*)-2-(1-((1-(3-(2-(7-chloroquinolin-2-yl)vinyl)phenyl)-3-(2-(2-hydroxypropan-2-yl)phenyl) propylthio)methyl)cyclopropyl)acetic acid), is a selective inhibitor of cyteinyl leukotriene D_4 receptor binding in guinea pig lung (K_i = 0.18 nM), sheep lung (K_i = 4 nM), and dimethylsulfoxide-differentiated U937 cells (K_i = 0.52 nM), but it is essentially inactive against leukotriene C_4-specific binding in dimethylsulfoxide-differentiated U937 cell membranes (IC_{50} > 10 µM) and [3H]leukotriene B_4-specific binding in THP-1 cell membranes (IC_{50} ≈ 40 µM). Importantly, LTD_4 receptor blockade and LTD_4 synthesis inhibition also suppress pentylenetetrazol (PTZ)-induced kindling and pilocarpine-induced recurrent seizures (2). Montelukast (0.03 and 0.3 µM) prevented PTZ-induced BBB disruption, an effect that was reversed by LTD4 at the dose of 6 pmol, but not at the doses 0.2 and 2 pM (2) Moreover, the doses of LTD4 (0.2 and 2 pM) that reverted the effect of montelukast on seizures did not alter montelukast-induced protection of BBB, dissociating BBB protection and anticonvulsant activity (2). *See* (3) for pharmacokinetics, bioavailability, and safety data. **Key Pharmacokinetic Parameters:** *See* Appendix II in Goodman & Gilman's *THE PHARMACOLOGICAL BASIS OF THERAPEUTICS*, 12th Edition (Brunton, Chabner & Knollmann, eds.) McGraw-Hill Medical, New York (2011). **1**. Jones, Labelle, Belley, *et al.* (1995) *Can. J. Physiol. Pharmacol.* **73**, 191. **2**. Lenz, Arroyo, Temp, *et al.* (2014) *Neuroscience.* **277**, 859. **3**. Cheng, Leff, Amin, *et al.* (1996) *Pharm. Res.* **13**, 445.

Moronone

This *bis*-geranylacylphloroglucinol protonophore (FW = 502.70 g/mol), isolated from of an active *Moronobea coccinea* extract, exhibits enhanced cytotoxic and antiproliferative activity against tumor cells already subjected to rotenone-imposed metabolic stress. Surprisingly, moronone does not inhibit glycolysis, but functions instead as a protonophore, dissipating the mitochondrial proton gradient. That moronone fails to inhibit glycolysis, even in the face of likely depletion of ATP stores, suggests that glycolysis (*Overall Reaction*: Glucose + 2ATP + 4ADP + 2P$_i$ +2NAD$^+$ → 2pyruvic acid + 2ADP + 4ATP + 2NADH + 2H$^+$ +2H$_2$O) supplies sufficient ATP to sustain glucose phosphorylation. **1**. Datta, Li, Mahdi, *et al.* (2012) *J. Nat. Prod.* **75**, 2216.

Morphine

This strongly habit-forming opium alkaloid (FW$_{free-base}$ = 285.34 g/mol) is the prototypic µ-opioid receptor (µOR) agonist (as well as a κ-receptor (κOR) agonist), resulting in deranged receptor-mediated physiology that is manifested as long-lasting pain relief. (*For other alkaloids in opium poppy, see Opium*) **Primary Mode of Action:** Morphine profoundly alters receptor-mediated signal transduction throughout the nervous system. Indeed, µ-opioid receptors are prominent within the brain (concentrated mainly in the cortex, thalamus, striosomes, periaqueductal gray matter, and rostral ventromedial medulla), spinal cord, peripheral sensory neurons, and intestinal tract neurons. κ-Opioid receptors are found within the brain (mainly in the hypothalamus, periaqueductal gray matter, and claustrum), spinal cord, and peripheral sensory neurons. The robust diversity of opioid receptor pharmacology is best explained by the interplay of various receptor subtypes, thereby expanding the repertoire of responses beyond those predicted on the basis of the known opioid receptors. A likely mechanism involves the assembly of novel signaling complexes through hetero-oligomerization of receptor subunits as well as some degree of promiscuity in opioid receptor interactions with other G-protein-coupled receptors. The α_{2A}-adrenoceptor/µ-opioid receptor heterodimer complex, for example, relies on ligand binding to propagate allosteric conformational changes among its constituent subunits (*or* protomers). Upon binding to µOR, morphine triggers a conformational change in the associated norepinephrine-occupied α_{2A}-adrenoceptor, which, within milliseconds, results in reduced GTP regulatory protein activation by the α_{2A}-adrenoceptor protomer. **Uptake & Metabolism:** Morphine can be administered by many routes (*e.g.*, oral, sublingual, rectal, subcutaneous, intravenous, epidural), including inhalation of finely dispersed drug. If delivered orally, morphine's first-pass kinetics assure that less than half reaches the central nervous system. Morphine is converted to morphine-3-glucuronide and morphine-6-glucuronide by UDP-glucuronosyl transferase. Morphine's long-lasting diacetyl ester (heroin) and the even more potent semisynthetic opioids (Bentley compounds) are more active, at least in part, because they resist glucuronidation. Morphine addiction is confounded by psychological dependence, physical dependence, as well as drug tolerance. **Key Pharmacokinetic Parameters:** *See* Appendix II in Goodman & Gilman's *THE PHARMACOLOGICAL BASIS OF THERAPEUTICS*, 12th Edition (Brunton, Chabner & Knollmann, eds.) McGraw-Hill Medical, New York (2011).**Target(s):** acetylcholinesterase (1,2); cholinesterase, *or* butyrylcholinesterase (1,3,4); Ca^{2+}-dependent ATPase (5); glucuronosyl-transferase, as alternative substrate for UGT2B7 (6); phosphofructokinase (7); oxidative phosphorylation (8); oxytocin secretion (9). **1**. Eadie (1941) *J. Biol. Chem.* **138**, 597. **2**. Bernheim & M. L. C. Bernheim (1936) *J. Pharmacol. Exp. Therap.* **57**, 427. **3**. Augustinsson (1950) *The Enzymes*, 1st ed., **1** (part 1), 443. **4**. Bailey & Briggs (2005) *Amer. J. Clin. Pathol.* **124**, 226. **5**. Yamasaki & Way (1985) *Neuropeptides* **5**, 359. **6**. Watanabe, Nakajima, Ohashi, Kume & Yokoi (2003) *Drug Metab. Dispos.* **31**, 589. **7**. Deshpande & Mitchell (1980) *J. Endocrinol.* **85**, 415. **8**. Chistiakov & Gegenava (1976) *Biokhimiia* **41**, 1272. **9**. Russell, Gosden, Humphreys, *et al.* (1989) *J. Endocrinol.* **121**, 521.

3-Morpholinosynonimine

This photosensitive zwitterionic reagent (FW = 170.17 g/mol), also known as linsidomine and SIN-1, spontaneously releases nitric oxide in aqueous solution. (***See also** Nitric Oxide*) **Target(s):** ADP-ribose diphosphatase (1); caspase-3 (2); glycerophosphocholine cholinephospho-diesterase (3); heme oxygenase (4); methionine *S*-adenosyltransferase (5). **1**. Ribeiro, Cameselle, Fernandez, *et al.* (1995) *Biochem. Biophys. Res. Commun.* **213**, 1075. **2**. Mohr, Zech, Lapetina & Brune (1997) *Biochem. Biophys. Res. Commun.* **238**, 387. **3**. Sok (1998) *Neurochem. Res.* **23**, 1061. **4**. Ding, McCoubrey & Maines (1999) *Eur. J. Biochem.* **264**, 854. **5**. Avila, Corrales, Ruiz, *et al.* (1998) *Biofactors* **8**, 27.

Mosquito Oostatic Factor

This naturally occurring decapeptide (FW = 1047.18 g/mol; *Sequence*: YDPAPPPPPP; *Abbreviation*: MOOF), also known as trypsin modulating oostatic factor, inhibits egg development in mosquitos as well as trypsin- and chymotrypsin like enzyme biosynthesis (1). When microinjected into Listeria-infected cells, mosquito oostatic factor also blocks *Listeria*-induced actin rocket tail assembly as well as the intracellular locomotion of this pathogen (2). Co-administration of the actin regulatory protein profilin neutralize the inhibitory action of mosquito oostatic factor. Actin-based motility relies on two types of proline-rich sequences to assemble actin locomotory units. *Listeria monocytogenes* ActA surface protein contains a series of nearly identical EFPPPPTDE-type oligoproline sequences for binding vasodilator-stimulated phospho-protein (VASP). The latter is a tetrameric protein with numerous GPPPPP docking sites for profilin, a 15-kDa regulatory protein that promotes actin filament assembly. Analysis of known actin regulatory proteins led to the identification of the following Actin-Based Motility homology sequences: ABM-1, *Sequence* = (D/E)FPPPPX(D/E); and ABM-2, , *Sequence* = XPPPPP, where X denotes G, A, L, P and S). MOOF appears to be an ABM-2 binding antagonist (3). **1**. Borovsky, Carlson, Griffin, Shabanowitz & Hunt (1990) *FASEB J.* **4**, 3015. **2**. Southwick & Purich (1995) *Infect. Immun.* **63**, 182. **3**. Purich & Southwick (1997) *Biochem. Biophys. Res. Commun.* **231**, 686.

Motesanib

This orally bioavailable VEGF receptor kinase inhibitor (FW$_{\text{free-base}}$ = 373.46 g/mol; FW$_{\text{di-phosphate-salt}}$ = 569.44 g/mol; CAS 857876-30-3; Solubility: 100 mg/mL DMSO; 100 mg/mL H$_2$O), also known as AMG-706 and N-(2,3-dihydro-3,3-dimethyl-1H-indol-6-yl)-2-[(4-pyridinylmethyl)amino]-3-pyridinecarboxamide, VEGFR1/Flt1 (IC$_{50}$ = 2 nM); VEGFR2/Flt2 (IC$_{50}$ = 3 nM); VEGFR3/Flt4 (IC$_{50}$ = 6 nM); PDGFR (IC$_{50}$ = 84 nM); c-Kit (IC$_{50}$ = 8 nM); and Ret (IC$_{50}$ = 59 nM). AMG 706 inhibits human endothelial cell proliferation induced by VEGF, but not by basic fibroblast growth factor in vitro, as well as vascular permeability induced by VEGF in mice (1). It is also effective in combination ith radiation to treat head and neck squamous cell carcinoma (HNSCC) in mouse xenograft models (2). It is also highly effective against breast cancer xenografts (3). **1**. Polverino, Coxon, Starnes, et al. (2006) Cancer Res. **66**, 8715. **2**. Kruser, Wheeler, Armstrong, et al. (2010) Clin. Cancer Res. **16**, 3639. **3**. Coxon, Bush, Saffran, et al. (2009) Clin. Cancer Res. **15**, 110.

Motilin

This gastrointestinal peptide (FW = 2699.08 g/mol; *Human Sequence*: FVPIFTYGELQRMQEKERNKGQ) is produced by motilin-immunopositive cells (Mo cells) present in the upper small intestine. Motilin's main action is to stimulate stomach contractions, resulting in release from parietal cells in the stomach lining. Motilin also inhibits luteinizing hormone release (1). **1**. Tsukamura, Tsukahara, Maekawa, et al. (2000) J. Neuroendocrinol. **12**, 403.

Moxalactam

This cephalosporin antibiotic (FW$_{\text{free-acid}}$ = 520.48 g/mol; CAS 75007-71-5), also known as latamoxef, is a pseudosubstrate that inactivates β-lactamases via the formation of a stable covalent bond. The methylthiotetrazole side chain of moxalactam inhibits γ-carboxylation of glutamic acid residues on blood clotting factors, thereby interfering with the clot-promoting actions of vitamin K (1). Vitamin K deficiency is amplified by an administration of LMOX even in the absence of intestinal flora. **Target(s):** alcohol dehydrogenase (2); aldehyde dehydrogenase (3); β-lactamase (4-11); penicillin binding proteins (12). **1**. Shirakawa, Komai & Kimura (1990) Int. J Vitam. Nutr. Res. **60**, 245. **2**. Freundt, Schreiner & Christmann-Kleiss (1986) Arzneimittelforschung **36**, 223. **3**. Yamanaka, Yamamoto & Egashira (1983) Jpn. J. Pharmacol. **33**, 717. **4**. Zervosen, Valladares, Devreese, et al. (2001) Eur. J. Biochem. **268**, 3840. **5**. Trehan, Beadle & Shoichet (2001) Biochemistry **40**, 7992. **6**. Murata, Minami, Yasuda, et al. (1981) J. Antibiot. **34**, 1164. **7**. Labia (1982) Rev. Infect. Dis. **4** Suppl., S529. **8**. Murakami & Yoshida (1985) Antimicrob. Agents Chemother. **27**, 727. **9**. Toda, Inoue & Mitsuhashi (1981) J. Antibiot. **34**, 1469. **10**. Walsh, Gamblin, Emery, McGowan & Bennett (19960 J. Antimicrob. Chemother.

37, 423. **11**. Poirel, Brinas, Verlinde, Ide & Nordmann (2005) Antimicrob. Agents Chemother. **49**, 3743. **12**. Komatsu & Nishikawa (1980) Antimicrob. Agents Chemother. **17**, 316.

Moxifloxacin

This fourth-generation fluoroquinolone-class antibiotic (FW = 401.44 g/mol; CAS 354812-41-2), also named BAY 12-8039, Avelox®, Vigamox®, Moxeza®, and 1-cyclopropyl-7-[(1S,6S)-2,8-diazabicyclo[4.3.0]non-8-yl]-6-fluoro-8-methoxy-4-oxoquinoline-3-carboxylic acid, exhibits broad action against enteric Gram-negative rods (*Escherichia coli*, *Proteus* species, *Klebsiella* species), *Haemophilus influenzae*, as well as atypical bacteria (e.g., *Mycoplasma, Chlamydia, Legionella*), and *Streptococcus pneumoniae*, as well as many anaerobic bacteria. Moxifloxacin differs from earlier fluoroquinolone-class agents (e.g., levofloxacin and ciprofloxacin) in having greater activity against Gram-positive bacteria and anaerobes (1). MIC$_{90}$ values for moxifloxacin action against ciprofloxacin-resistant, methicillin-susceptible and methicillin-resistant *Staphylococcus aureus* are 8 µg/mL (1). Following oral administration, moxifloxacin absorption is fast, with low to medium variability, and an absolute bioavailability of 86% (2). The excretion of moxifloxacin and its metabolites was quantified in a subset of eight subjects (2). **Key Pharmacokinetic Parameters:** *See* Appendix II in Goodman & Gilman's THE PHARMACOLOGICAL BASIS OF THERAPEUTICS, 12th Edition (Brunton, Chabner & Knollmann, eds.) McGraw-Hill Medical, New York (2011). **1**. Dalhoff, Petersen & Endermann (1996) Chemotherapy **42**, 410. **2**. Stass & Kubitza (1999) J. Antimicrob. Chemother. **43** (*Supplement* B), 83.

MP-470, *See Amuvatinib*

MP A08

This first-in-class, cell-permeable SphK2/1 inhibitor (FW = 519.64 g/mol; CAS 219832-49-2; Solubility: 100 mM in DMSO), also known as MP-A08 and 4-methyl-N-[2-[[[2-[[(4-methyl-phenyl)sulfonyl]amino]phenyl]imino]methyl]phenyl]benzenesulfonamide, targets sphingosine kinase-2, K_i = 6.9 µM, and sphingosine kinase-1, K_i = 27 µM (1). MP-A08 blocks pro-proliferative signalling pathways, inducing mitochondria-associated apoptosis in a SK-dependent manner and reducing the growth of human lung adenocarcinoma tumours in a mouse xenograft model by both inducing tumor cell apoptosis and inhibiting tumour angiogenesis (1). **1**. Pitman, Powell, Coolen, et al. (2015) Oncotarget **6**, 7065.

MPC-1303, *See Aranidipine*

MPC-4326, *See Bevirimat*

MPP

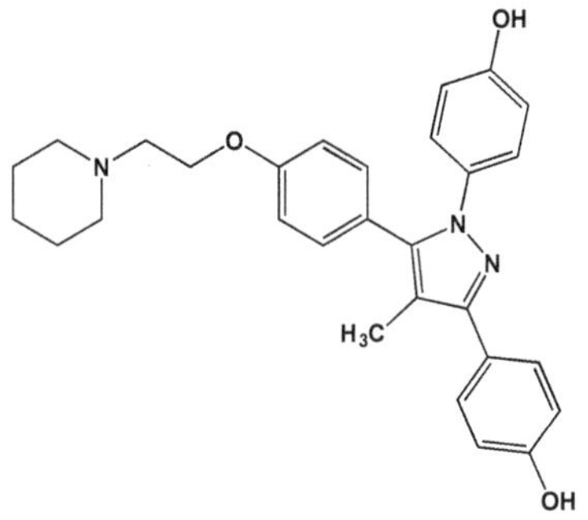

This substituted pyrazole (FW$_{free-base}$ = 469.58 g/mol), also known as methyl-piperidino-pyrazole, is an estrogen receptor α antagonist. The gene-regulating actions of estrogens are mediated through ER$_α$ and ER$_β$, two estrogen receptor subtypes receptors that function as ligand-modulated transcription factors with distinctive tissue distributions. *Note*: Do not confuse this pyrazole with the 1-methyl-4-phenylpyridinium cation, also abbreviated MPP$^+$. **1**. Sun, Huang, Harrington, *et al.* (2002) *Endocrinology* **143**, 941.

Mps1-IN-1

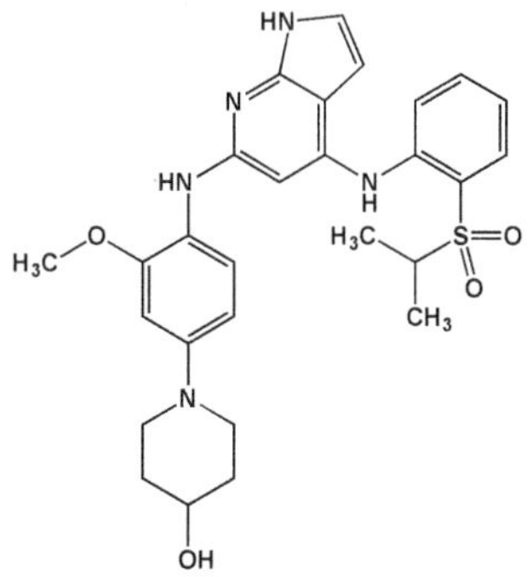

This ATP-competitive protein kinase inhibitor (FW = 535.66 g/mol) targets Mps1 (IC$_{50}$ = 367 nM), a dual-specificity kinase required for the proper spindle assembly checkpoint control and maintenance of chromosomal stability. Although spindle assembly checkpoint (SAC) is a well-conserved pathway in eukaryotes, there is good reason to believe that small-molecule inhibitors can hobble the pathway in a manner that is anti-proliferative. With the major exceptions of Alk and Ltk, this compounds demonstrated greater than 1000x selectivity relative to the 352-member kinase panel. Notably, Mps1-IN-1 treatment leads to defects in Mad1 and Mad2 establishment at unattached kinetochores, decreased Aurora B kinase activity, premature mitotic exit, and gross aneuploidy, without any evidence of centrosome duplication defects. Upon inhibitor binding, Mps1 adopts an inactive conformation as indicated by incorrect positioning of the regulatory helix αC, the lack of a salt bridge between the conserved αC glutamate (E571) and the active site lysine (K553), and an unstructured activation loop (M671-V684). (*See also MPS1-IN-2, Mps BAY 2a, and NMS-P715*) **1**. Kwiatkowski, Jelluma, Filippakopoulos, *et al.* (2010) *Nature Chem. Biol.* **6**, 359.

MPS1-IN-2

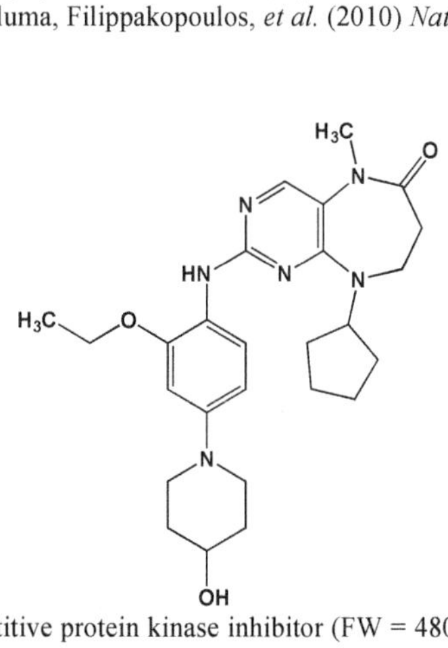

This ATP-competitive protein kinase inhibitor (FW = 480.61 g/mol) targets Mps1 (IC$_{50}$ = 145 nM), a dual-specificity kinase required for the proper

spindle assembly checkpoint control and maintenance of chromosomal stability. With the major exceptions of Gak and Plk1, this compounds demonstrated greater than 1000x selectivity relative to the 352-member kinase panel. (For other details on its mode of of inhibitor action. (*See also MPS1-IN-1, Mps BAY 2a, and NMS-P715*) **1**. Kwiatkowski, Jelluma, Filippakopoulos, *et al.* (2010) *Nature Chem. Biol.* **6**, 359.

Mps BAY 2a

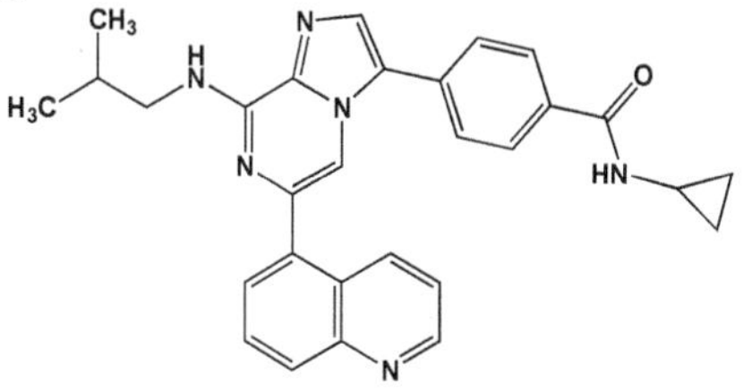

This mitosis inhibitor (FW = 476.57 g/mol; CAS 1382477-96-4; Solubiity: 100 mM in DMSO; IUPAC: *N*-Cyclopropyl-4-[8-[(2-methylpropyl)amino]-6-(5-quinolinyl)imidazo[1,2-*a*]pyrazin-3-yl]benzamide, targets monopolar spindle-1 (Mps1 *or* TTK protein kinase), an essential serine/threonine kinase in the mitotic spindle assembly checkpoint (SAC) pathway. By selectively inactivating Mps1, this inhibitor arrests cancer cell proliferation, causing polyploidy and/or cell death. Direct videomicroscopic cell-fate profiling of histone 2B-green fluorescent protein-expressing cells revealed the capacity of Mps BAY 2a to subvert the correct timing of mitosis as they induce a premature anaphase entry in the context of misaligned metaphase plates. In the presence of the inhibitor, cells either divided in a bipolar (often asymmetric) manner or entered one or more rounds of abortive mitoses, generating gross aneuploidy and polyploidy, respectively, ultimately dying as a result of mitotic catastrophe-induced activation of the mitochondrial apoptosis pathway. (*See also MPS1-IN-1, MPS1-IN-2, and NMS-P715*) **1**. Jemaà, Galluzzi, Kepp, *et al.* (2013) *Cell Death Differ.* **20**, 1532.

MPT0B098

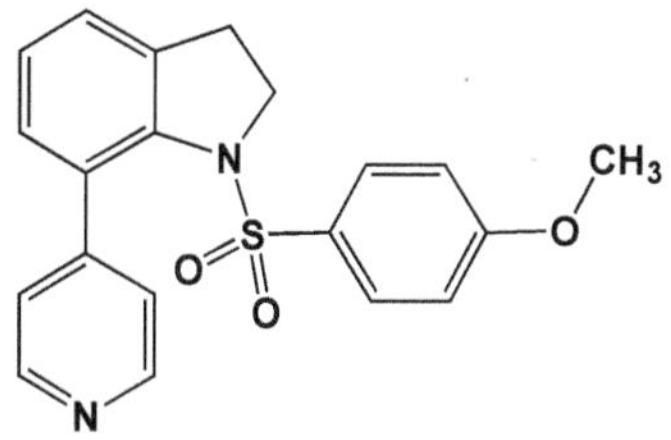

This novel vascular-disrupting agent (FW = 399.44 g/mol), also known as 7-aryl-indoline-1-benzene-sulfonamide, is a potent microtubule inhibitor that binds to the colchicine-binding site on tubulin. MPT0B098 is active against the growth of various human cancer cells, including several chemoresistant cell lines, with IC$_{50}$ values of 70-150 nM. MPT0B098 arrests cells in the G$_2$-M phase and induces subsequent apoptosis. MPT0B098 effectively suppresses VEGF-induced cell migration and disrupts capillary tube formation from human umbilical vein endothelial cells, *or* HUVECs. Vascular disrupting agents (VDAs) are compounds that block the formation of or cause the disorganization of blood vessels needed to nourish tumors, thereby starving the central tumor. The ideal VDA will bind to its target binding partner, bringing about the disruption of the capillary vasculature. It will then be released from its target binding partner, followed by timely metabolic deactivation and/orelimination. (*See Combretastatin*) **1**. Cheng, Liou, Kuo, *et al.* (2013) *Mol. Cancer Ther.* **12**, 1202.

MPTP, *See* *1-Methyl-4-phenyl-1,2,3,6-tetrahydropyridine*

MRK 560

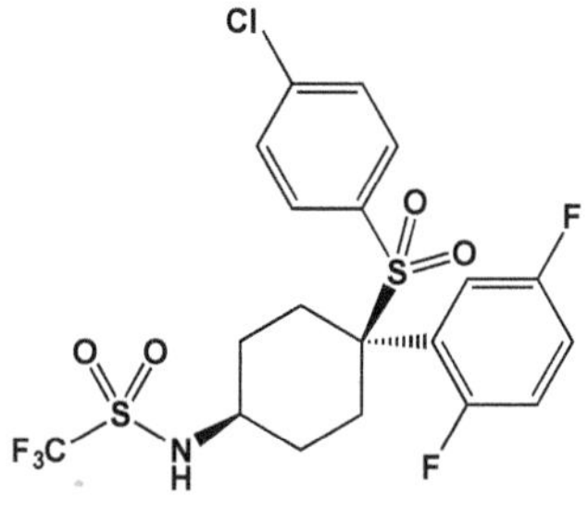

This oral γ-secretase inhibitor (FW = 517.92 g/mol; CAS 677772-84-8; Solubility: 100 mM in DMSO), also named *N*-[*cis*-4-[(4-chlorophenyl)-sulfonyl]-4-(2,5-difluorophenyl)cyclohexyl]-1,1,1-trifluoro-methanesulfon-amide, preferentially inhibits proteolytic cleavage of amyloid precursor protein (APP). MRK-560 reduces Aβ accumulation in the brain and cerebrospinal fluid, inhibiting formation of Aβ40 and Aβ42 peptides in SH-SY5Y neuroblastoma cells (IC$_{50}$ = 0.65 nM), thereby attenuating AD plaque deposition. MRK560 also reduces amyloid plaque deposition, while showing no evidence of Notch-related pathology in the Tg2576 mouse **1**. Best, *et al.* (2006) *J. Pharm. Exp. Ther.* **317**, 786. **2**. Best, *et al.* (2007) *J. Pharm. Exp. Ther.* **320**, 552.

MRS 1220

This potent and highly selective adenosine receptor antagonist (FW = 403.83 g/mol; CAS 183721-15-5; Soluble to 100 mM in DMSO with gentle warming), also named *N*-[9-chloro-2-(2-furanyl)[1,2,4]triazolo[1,5-*c*]quinazolin-5-yl]benzeneacetamide, targets the human A$_3$ adenosine receptor (K_i = 0.65 nM), rat A$_1$ (K_i = 305 nM), and rat A$_3$ (K_i = 52 nM; IC$_{50}$ > 1 μM). **1**. Kim, *et al.* (1996) *J. Med. Chem.* **39**, 4142. **2**. Jacobson, *et al.* (1997) *Neuropharmacol.* **36**, 1157. **3**. Kim, *et al.* (1998) *J. Med. Chem.* **41**, 2835.

MRS 1334

This potent and highly selective adenosine receptor antagonist (FW = 522.56 g/mol; CAS 192053-05-7; Soluble to 100 mM in DMSO), also named 1,4-dihydro-2-methyl-6-phenyl-4-(phenylethynyl)-3,5-pyridine-dicarboxylic acid 3-ethyl-5-[(3-nitrophenyl)methyl] ester, targets the human A$_3$ adenosine receptor (K_i = 2.7 nM), with K_i values in excess of 100 μM for rat A$_1$ and A$_3$ receptors. **1**. Li, *et al.* (1998) *J. Med. Chem.* **41**, 3186. **2**. Baraldi & Borea (2000) *Trends Pharmacol. Sci.* **21**, 456.

MRS 1706

This potent and highly selective adenosine receptor inverse agonist (FW = 503.56 g/mol; CAS 264622-53-9; Soluble to 5 mM in DMSO), also named *N*-(4-ccetylphenyl)-2-[4-(2,3,6,7-tetrahydro-2,6-dioxo-1,3-dipropyl-1*H*-

purin-8-yl)phenoxy]acetamide, targets the human A$_{2B}$ adenosine receptor (K_i = 1.4 nM), with much higher K_i values of 157, 112 and 230 nM for human A$_1$, A$_{2A}$ and A$_3$ receptors, respectively. **1**. Kim, *et al.* (2000) *J. Med. Chem.* **43**, 1165. **2**. Kiec-Kononowicz, *et al.* (2001) *Pure Appl. Chem.* **73**, 1411. **3**. Li, *et al.* (2007) *J. Pharmacol. Exp. Ther.* **320**, 637.

MRS 2159

This pyridoxal phosphate derivative (FW = CHNO g/mol; CAS 83654-06-2) is a P$_{2X1}$ purinoceptor antagonist. Extracellular adenine nucleosides and nucleotides act as neuromodulators and neurotransmitters by binding to P$_1$ and P$_2$ purine receptors in the central and peripheral nervous systems. Two families of P$_2$ receptors (*e.g.*, ligand-gated ion channels (the P$_{2X}$ subtype) and G protein-coupled receptors (the P$_{2Y}$ subtype) are widely studied, and it is now clear that uracil nucleotides are also active, perhaps even even preferred ligands for some P$_{2Y}$ subtypes. **1**. Kim, Camaioni, Ziganshin, *et al.* (1998) *Drug Develop. Res.* **45**, 52.

MRS 2179

This nucleotide (FW = 459.29 g/mol; IUPAC: 2'-deoxy-*N*^{6}-methyl-adenosine 3',5'-bisphosphate, is a competitive antagonist of P2Y$_1$ purinoceptor, which upon agonist binding activates phospholipase C (PLC), which generates inositol phosphates and diacylglycerol from phosphatidylinositol-4,5-bisphosphate, leading to a rise in intracellular calcium ion. The P2Y$_1$ receptor in platelets is involved in ADP-promoted aggregation, making P2Y1-selective antagonists highly promising as antithrombotic agents. Selective P2Y1 receptor agonist may also have potential as anti-hypertensive or anti-diabetic agents. **1**. Nandanan, Jang, Moro, *et al.* (2000) *J. Med. Chem.* **43**, 829. **2**. Boyer, Mohanram, Camaioni, Jacobson & Harden (1998) *Brit. J. Pharmacol.* **124**, 1.

MRS 2395

This chlorpurine diester (FW = 439.94 g/mol) is a P2Y$_{12}$ purinoceptor antagonist. Extracellular adenine nucleosides and nucleotides act as neuromodulators and neurotransmitters by binding to P$_1$ and P$_2$ purine receptors in the central and peripheral nervous systems. Two families of P$_2$ receptors (*e.g.*, ligand-gated ion channels (the P$_{2X}$ subtype) and G protein-coupled receptors (the P$_{2Y}$ subtype) are widely studied, and it is now clear that uracil nucleotides are also active, perhaps even even preferred ligands for some P$_{2Y}$ subtypes. **1**. Xu, Stephens, Kirschenheuter, *et al.* (2002) *J. Med. Chem.* **45**, 5694.

MRS 2578

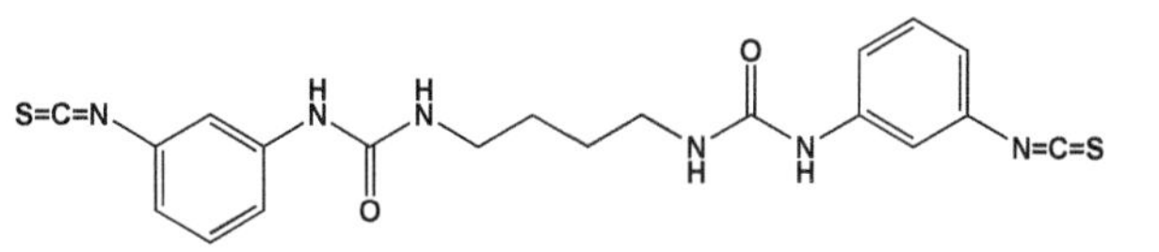

This potent, insurmountable P2Y purinoceptor-6 receptor antagonist (FW = 472.67 g/mol; CAS 711019-86-2; Solubility: 50 mg/mL DMSO; <1 mg/mL Water), named systematically as N,N'-1,4-butanediylbis[N'-(3-isothiocyanatophenyl)thiourea, targets the P2Y6 receptor (IC_{50} = 37 nM), a G-protein-coupled receptor that is activated by extracellular nucleotides. **1.** Mamedova, *et al.* (2004) *Biochem. Pharmacol.* **67**, 1763. **2.** Riegel, *et al.* (2011) *Blood* **117**, 2548. **3.** Talasila, *et al.* (2009) *Br. J. Pharmacol.* **158**, 339. **4.** Vieira, *et al.* (2011) *Am. J. Respir. Crit. Care Med.* **184**, 215.

MRT00033659

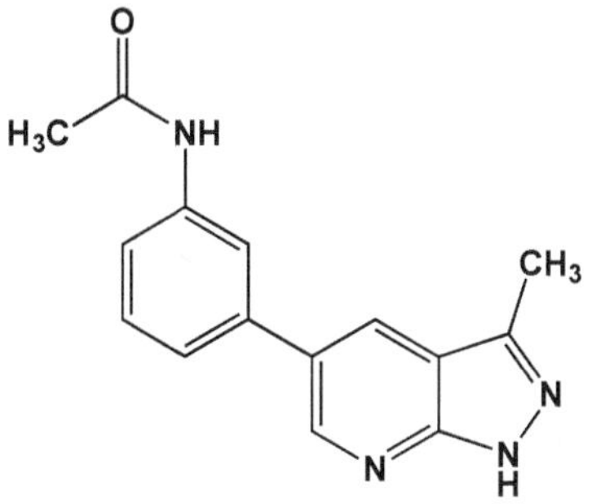

This pyrazolo-pyridine analogue (FW = 266.30 g/mol) exhibits novel multi-target inhibitory action against Checkpoint Kinase 1, *or* CHK1 (IC_{50} = 0.9 µM, *in vitro*) and Casein Kinase 1, *or* CK1δ (IC_{50} = 0.23 µM, *in vitro*). MRT00033659 shows sub-micromolar bioactivity in cells with respect to p53 protein stabilization and E2F-1 destabilization, with pharmacological similarities to the CK1-inhibitor D4476 and the MDM2-inhibitor Nutlin-3. While single-target activity has been pursued traditionally in search of selective, low-toxicity drugs, multi-target leads offer the advantage of simultaneously disrupting several signal transduction pathways, thereby placing certain cancer cell types at much higher jeopardy. **Other Target(s):** ERK8 (IC_{50} = 0.4 µM); FGF-R1 (IC_{50} = 6 µM); Nek2α (IC_{50} = 2 µM). **1.** Huart, Saxty, Merritt, *et al.* (2013) *Bioorg. Med. Chem. Lett.* **23**, 5578.

MRT67307

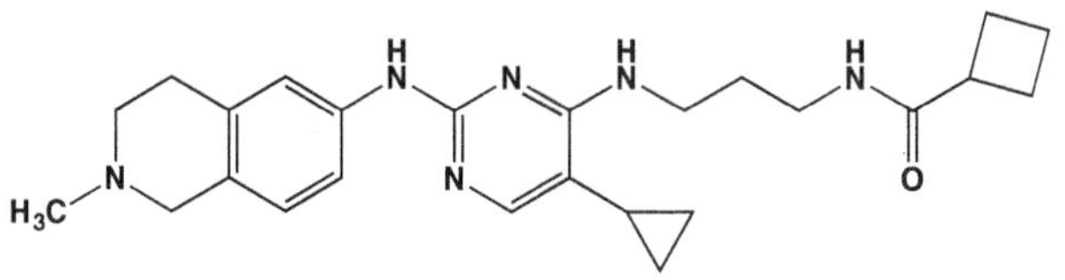

This potent and dual IKKε/TBK1 inhibitor (FW_{free-base} = 434.58 g/mol; CAS 1190379-70-4 (free base)), also named N-[3-[[5-cyclopropyl-2-[(1,2,3,4-tetrahydro-2-methyl-6-isoquinolinyl)amino]-4-pyrimidinyl]amino] propyl]cyclobutanecarboxamide hydrochloride, targets Inhibitor-κB kinase ε (IKKε), recently identified as an protein kinase with a high degree of homology to the classical I-κB kinase subunits, IKKα and IKKβ, and TANK-binding kinase-1 (TBK1), a serine/threonine kinase that plays an essential role in regulating inflammatory responses to foreign agents. Following activation of toll-like receptors by viral or bacterial components, TBK1 associates with TRAF3 and TANK and phosphorylates Interferon Regulatory Factors IRF3 and IRF7 as well as DDX3X. MRT67307 increases TLR-stimulated production of anti-inflammatory cytokines while suppressing proinflammatory cytokine secretion. MRT67307 has little effect on TLR-stimulated phosphorylation of CREB at Ser133 or the closely related ATF1 at Ser-63. Inhibition of SIKs by MRT67307, MRT199665, and HG-9-91-01 increases IL-10 production, while suppressing the secretion of IL-6, IL-12, and TNF. **1.** Clark, MacKenzie, Petkevicius, *et al.* (2012) *Proc. Natl. Acad. Sci. U.S.A.* **109**, 16986.

MS 023

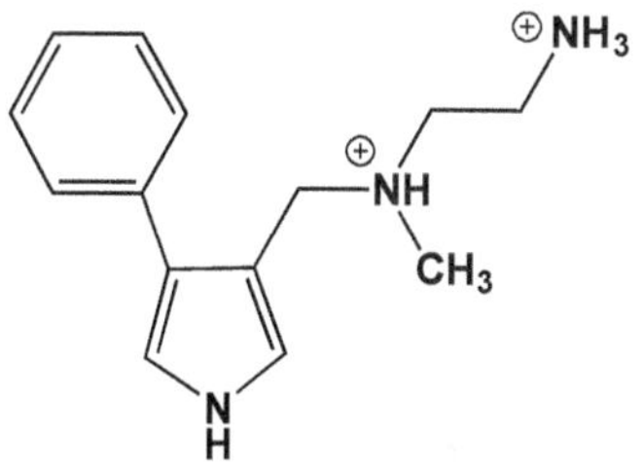

This potent and selective type I PRMT inhibitor (FW = 360.32 g/mol; Soluble to 100 mM in H_2O and to 100 mM in DMSO), also named N^1-methyl-N^1-[[4-[4-(1-methylethoxy)-phenyl]-1H-pyrrol-3-yl]methyl]-1,2-ethanediamine di-HCl, targets Protein Arginine Methyltransferase (PRMT), with respective IC_{50} values of 4, 5, 30, 83 and 119 nM for PRMT6, PRMT8, PRMT1, PRMT4 and PRMT3, while showing no significant activity against PRMT2, protein lysine methyltransferases (PKMTs), DNA mehyltransferases (DNMTs), histone lysine demethylases (or KDMs) and methyllysine reader proteins. A crystal structure of PRMT6 in complex with MS023 revealed that MS023 binds the substrate binding site. MS-023 potently decreased cellular levels of histone arginine asymmetric dimethylation. It also reduced global levels of arginine asymmetric dimethylation and concurrently increased levels of arginine monomethylation and symmetric dimethylation in cells. **1.** Eram, Shen, Szewczyk, *et al.* (2016) *ACS Chem Biol.* **11**, 772.

MS-275, See *Entinostat*

MS-752, See *Lucanthone*

MSC1992371A1

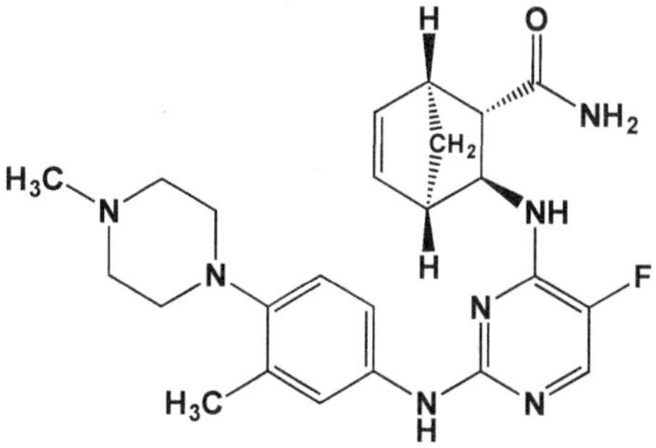

This oral, adenosine triphosphate-competitive inhibitor (FW = 451.54 g/mol; CAS 871357-89-0), also known by the code name AS703569 and systematically as (1S,2S,3R,4R)-3-((5-fluoro-2-((3-methyl-4-(4-methyl-piperazin-1-yl)phenyl)amino)pyrimidin-4-yl)amino)bicyclo-[2.2.1]hept-5-ene-2-carboxamide, targets mammalian aurora-kinase isoforms A, B, and C. MSC1992371A blocks cell separation by disrupting the mitotic spindle, leading to polyploidy and, ultimately, cell death. *In vitro*, MSC1992371A is a potent inhibitor of tumor cell growth, with IC_{50} values <100 nM [5]. *In vivo*, the compound has demonstrated significant antitumor activity in several xenograft models including those of the pancreas, breast, colon, kidney, lung, and ovary. With MiaPaca-2 pancreatic and Colo-205 colorectal xenografts, weekly dosing offered the best therapeutic window. **1.** Martinelli, Iacobucci, Papayannidis, *et al.* (2009) *Clin. Haematol.* **22**, 445. **2.** McLaughlin, Markovtsov, Li, Sacha, *et al.* (2010) *J. Cancer Res. Clin. Oncol.* **136**, 99.

MSC2032964A

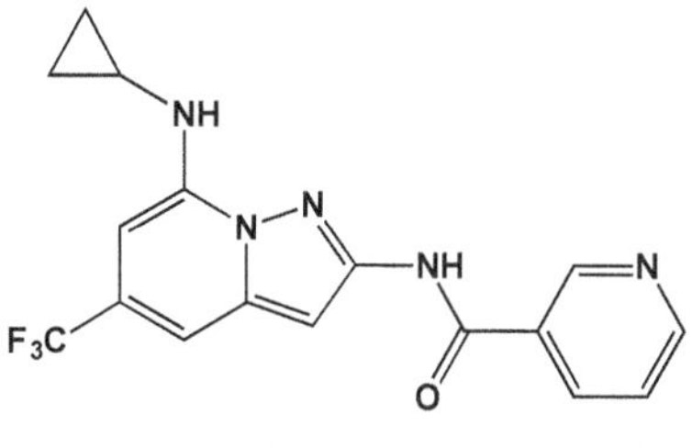

This potent and selective ASK1 inhibitor (FW = 362.31 g/mol; CAS 1124381-43-6; Solubility: 20 mM in DMSO), also named N-[5-

(cyclopropylamino)-7-(trifluoromethyl)[1,2,4]triazolo[1,5-*a*]pyridin-2-yl]-3-pyridinecarboxamide, is an orally bioavailable and brain-penetrant agent that targets Apoptosis Signal-regulating Kinase-1 (IC$_{50}$ = 93 nM), also known as mitogen-activated protein kinase kinase kinase 5 (MAP3K5), with much weaker action on 210 other kinases. (Apoptosis signal-regulating kinase 1 (ASK1) is an evolutionarily conserved mitogen-activated protein kinase (MAPK) kinase kinase which plays important roles in stress and immune responses.) MSC-2032964A blocks LPS-induced ASK1 and p38 phosphorylation in cultured mouse astrocytes. It also suppresses neuroinflammation in a mouse EAE model. **1**. Guo, Harada, Namekata, *et al.* (2010) *EMBO Mol. Med.* **2**, 504.

MTEP

This orally active anxiolytic agent (FW = 236.72 g/mol; CAS 1186195-60-7; Soluble to 100 mM in water and to 100 mM in DMSO), also named 3-((2-methyl-1,3-thiazol-4-yl)ethynyl)pyridine hydrochloride, is a potent, selective and non-competitive antagonist of the mGlu$_5$ metabolomics glutamate receptor, IC$_{50}$ = 5 nM in Ca^{2+}-flux assay, K_i = 16 nM (1-4). Metabotropic glutamate receptor mGlu$_5$ is a mediator of appetite and energy balance in rats and mice (5). **1**. Brodkin, *et al.* (2002) *Eur. J. Neurosci.* **16**, 2241. **2**. Cosford, *et al.* (2003) *J. Med. Chem.* **46**, 204. **3**. Klodzinska, *et al.* (2004) *Neuropharmacol.* **47**, 342. **4**. Roppe, *et al.* (2004) *Bioorg. Med.*

mTOR Inhibitors

These agents, which include temsirolimus (CCI-779), everolimus (RAD001) and deforolimus (AP23573), inhibit mammalian Target of Rapamycin (or mTOR), a protein kinase within the phosphatidylinositol 3-kinase (PI3K)/Akt signalling cascade. The latter plays a central role in controlling cell proliferation, survival, mobility and angiogenesis. Dysregulation of mTOR pathway has been found in many human tumors, making the mTOR pathway an important target for developing a promising new class of anticancer drugs. In tumor cells, mTOR is activated by various mechanisms, including growth factor surface receptor tyrosine kinases, oncogenes, and loss of tumor suppressor genes. Such activating factors are important in malignant transformation and progression. *See Deforolimus, Everolimus, Temsirolimus, and Zotarolimus*

MTPG

This metabotropic glutamate receptor antagonist (FW = 233.23 g/mol; CAS 169209-66-9; Soluble to 100 mM as sodium salt), also named (*RS*)-α-methyl-4-tetrazolylphenylglycine, targets mGlu$_2$ (Group II) and mGlu$_3$ (Group III) metabotropic glutamate receptor antagonist, showing selectivity for mGlu$_2$ in electrophysiological studies. **1**. Jane, *et al.* (1995) *Neuropharmacol.* **34**, 851. **2**. Bedingfield, *et al.* (1996) *Eur. J. Pharmacol.* **309**, 71. **3**. Bushell, *et al.* (1996) *Brit. J. Pharmacol.* **117**, 1457.

Mucus Proteinase Inhibitor

This 107-residue, mucus-derived, acid-stable protein (MW = 11.7 kDa), also called secretory leukocyte protease inhibitor (SLPI) and antileukoprotease, is a potent serine-protease inhibitor thought to increase in inflammation and infection and to primarily target human leukocyte elastase. **Target(s):** cathepsin G (1-3); chymase (1,3,4); chymotrypsin (1-3); elastase (1,2,5-7); leukocyte elastase (3,6,7); pancreatic elastase (5); stratum corneum chymotryptic enzyme, *or* kallikrein-7 (8); trypsin (1-3); tryptase (3,9). **1**. Sallenave (2000) *Respir. Res.* **1**, 87. **2**. Boudier & Bieth (1992) *J. Biol. Chem.* **267**, 4370. **3**. Fritz (1988) *Biol. Chem. Hoppe-Seyler* **369** Suppl., 79. **4**. Fink, Nettelbeck & Fritz (1986) *Biol. Chem. Hoppe-Seyler* **367**, 567. **5**. Bieth (1998) in *Handb. Proteolytic Enzymes*, p. 46, Academic Press, San Diego. **6**. Bieth (1998) in *Handb. Proteolytic Enzymes*, p. 54, Academic Press, San Diego. **7**. Wiesner, Litwiller, Hummel, *et al.* (2005) *FEBS Lett.* **579**, 5305. **8**. Franzke, Baici, Bartels, Christophers & Wiedow (1996) *J. Biol. Chem.* **271**, 21886. **9**. Chen, Shiota, Ohuchi, *et al.* (2000) *Eur. J. Biochem.* **267**, 3189.

μ-SLPTX-Ssm6a, See *SLPTX-Ssm6a*

Mupirocin

This isoleucyl t-RNA synthetase inhibitor (FW = 500.62 g/mol; CAS 12650-69-0; Solubility: 200 mM in DMSO, with heating), also known as pseudomonic acid A and (2*E*)-5,9-anhydro-2,3,4,8-tetradeoxy-8-[[(2*S*,3*S*)-3-[(1*S*,2*S*)-2-hydroxy-1-methylpropyl]oxiranyl]methyl]-3-methyl-L-*talo*-non-2-enonic acid 8-carboxyoctyl ester, is an antibiotic produced by *Pseudomonas fluorescens*, with a high level of activity against staphylococci and streptococci and against certain Gram-negative bacteria, including *Haemophilus influenzae* and *Neisseria gonorrhoeae*, but was much less active against most gram-negative bacilli an anaerobes (1,2). Nearly all clinical isolates of *Staphylococcus aureus* and *Staphylococcus epidermidis*, including multiply resistant strains, are susceptible to mupirocin (MIC ≤ 0.5 μg/mL). **Mechanism of Inhibitory Action:** Mupirocin is a powerful inhibitor of *Escherichia coli B* isoleucyl-tRNA synthetase (or Ile-tRNA synthetase), acting competitively (K_i = 6 nM) with respect to isoleucine (3). [9'-^{3}H]-Mupirocin forms a stable complex with Ile-tRNA synthetase, and gel-filtration and gel-electrophoresis showed that the antibiotic is fully released only upon treatment with 5 M urea treatment or upon boiling in 0.1% sodium dodecyl sulfate. The binding stoichiometry is 0.85 mol mupirocin per mol enzyme, as determined by equilibrium dialysis. Aminoacylation of yeast tRNAIle by rat liver Ile-tRNA synthetase was also competitively inhibited with respect to isoleucine (K_i = 20 μM), some 8000 times higher than for binding to the *E. coli* enzyme and providing a basis for the low toxicity of the antibiotic in mammals (3). **1**. Reilly & Spencer (1984) *J. Antimicrob. Chemother.* **13**, 295. **2**. Sutherland, Boon, Griffin, *et al.* (1985) *Antimicrob. Agents Chemother.* **27**, 495. **3**. Hughes & Mellows (1980) *Biochem. J.* **191**, 209.

Mureidomycin A

This antibiotic (FW = 840.91 g/mol; CAS 114797-04-5) and related metabolites from *Streptomyces flavidovirens* WD31235, are slow-binding inhibitors of phospho-*N*-acetylmuramoyl-pentapeptide-transferase (K_i = 36 nM and K_i* = 2 nM). **1**. Brandish, Burnham, Lonsdale, *et al.* (1996) *J. Biol. Chem.* **271**, 7609. **2**. Brandish, Kimura, Inukai, *et al.* (1996) *Antimicrob. Agents Chemother.* **40**, 1640. **3**. Stachyra, Dini, Ferrari, *et al.* (2004) *Antimicrob. Agents Chemother.* **48**, 897.

(+)-Muscarine

(+)-Muscarine   **Acetylcholine**

This toxic quaternary ammonium-containing alkaloid and acetylcholine receptor agonist (FW = 209.72 g/mol; CAS 2936-25-6; muscarine hydrolchloride is freely soluble in water and ethanol) is the prototypical muscarinic agonist of cholinergic receptors. Isolated from the mushroom *Amanita muscaria* and related gill fungi, muscarine is also a potent cholinesterase inhibitor (1). Gaskell (2) first applied galvanometry to describe the action of muscarine and pilocarpine on the dissected heart and on the electrical changes in the non-beating cardiac muscle brought about by stimulation of inhibitory and augmentor nerves. The toxic actions of muscarine include Parkinson-like CNS effects (*e.g.*, tremor, ataxia, and spasticity). The intravenous LD$_{50}$ in mice is 0.23 mg/kg, similar to pilocarpine. Treatment of muscarine poisoning involves parenteral administration of atropine, supported clinically by efforts to minimize pulmonary edema. Bebbington & Brimblecombe (3) and Kosterlitz (4) have reviewed the chemistry and pharmacology of various natural and synthetic muscarinic agents. Muscarine's pharmacologic action on mAChR's can be scored as follows: Cardiovascular, ++; Gastrointestinal, +++; Bladder, +++; Eye, ++ (applied topically). **Target(s):** acetyl-cholinesterase (1,2); calcium channels, ω-conotoxin-sensitive (5). **1**. Augustinsson (1950) *The Enzymes*, 1st. ed., **1** (part 1), 443. **2**. Gaskell (1887) *J. Physiol.* **8**, 404. (*This must-read classic is available in PubMed Central*. PMCID: PMC1485142) **3**. Bebbington & Brimblecombe (1965) *Adv. Drug Res.* **22**, 143. **4**. Kosterlitz (1967) in *Physiological Pharmacology* (Root & Hofmann, eds.) vol. **3**, p. 97, Academic Press, New York. **5**. Toselli, Perin & Taglietti (1995) *J. Neurophysiol.* **74**, 1730.

Muscarinic Acetylcholine Receptor Agonists

These agents bind to and activate muscarinic acetylcholine receptors (or mAChRs), which are receptors that form G protein-receptor complexes in the cell membranes of certain neurons and other cells. mAChRs act as the main end-receptors that are stimulated by acetylcholine, after the latter is released presynaptically from postganglionic fibers in the parasympathetic nervous system. The name "mAChRs" derives from the fact that these receptors are more sensitive to muscarine than to nicotine (*See (+)-Muscarine; Nicotine*). Muscarinic AChRs are found both in post-synaptic and pre-synaptic sites. Although this reference book focuses on enzyme inhibitors and receptor antagonists, muscarinic receptor agonists are of such widespread utility that they should be briefly described here.

Acetylcholine

mAChR Agonist: Acetylcholine (FW = 146.21 g/mol; CAS 51-84-3), which acts on both peripheral and central neurons, is the only neurotransmitter used in the motor division of the somatic nervous system and is also the principal neurotransmitter in all autonomic ganglia. Acetylcholine's pharmacologic action on mAChR's can be scored as follows: Cardiovascular, ++; Gastrointestinal, ++; Bladder, ++; Eye, + (applied topically).

Bethanechol

mAChR Agonist: Bethanechol (FW = 161.22 g/mol; CAS 674-38-4), also named Duvoid®, Myotonachol®, Urecholine®, Urocarb®, and 2-(carbamoyloxy)-*N,N,N*-trimethylpropan-1-aminium, is a parasympatho-mimetic choline carbamate that selectively stimulates muscarinic receptors without any effect on nicotinic receptors. Unlike acetylcholine, however, bethanechol is not hydrolyzed by cholinesterase and has a long duration of action. Bethanechol's pharmacologic action on mAChR's can be scored as follows: Cardiovascular, ±; Gastrointestinal, +++; Bladder, +++; Eye, ++ (applied topically).

Carbachol

mAChR Agonist: Carbachol (FW = 146.21 g/mol; CAS 51-84-3), also named Miostat® and 2-[(aminocarbonyl)oxy]-*N,N,N*-trimethylethanamine,

is a cholinomimetic drug that binds and activates acetylcholine receptors. It is used maily for ophthalmic purposes, including glaucoma and during ophthalmic surgery. Carbachol's pharmacologic action on mAChR's can be scored as follows: Cardiovascular, +; Gastrointestinal, +++; Bladder, +++; Eye, ++ (applied topically).

Methacholine

mAChR Agonist: Methacholine (FW = 160.23 g/mol; CASs = 55-92-5 (free base) and 62-51-1 (chloride)) is highly active at all muscarinic receptors, but exerts little effect on the nicotinic receptors. With its quaternary amine structure, methacholine has poor absorption from the gastrointestinal tract and does not cross the blood–brain barrier and. The additional methyl group renders methacholine highly resistant to acetylcholinesterases. Methacholine's pharmacologic action on mAChR's can be scored as follows: Cardiovascular, +++; Gastrointestinal, ++; Bladder, ++; Eye, + (applied topically).

Pilocarpine

mAChR Agonist: Pilocarpine (FW = 155.19 g/mol; CAS 63-75-2) is a parasympathomimetic alkaloid from the leaves of tropical American shrubs (genus *Pilocarpus*) that acts as a nonselective muscarinic receptor agonist in the parasympathetic nervous system, acts therapeutically at the M$_3$ muscarinic receptors and is applied topically to treat glaucoma and xerostomia. Pilocarpine's pharmacologic action on mAChR's can be scored as follows: Cardiovascular, +; Gastrointestinal, +++; Bladder, +++; Eye, ++ (applied topically). **See also** *Pilocarpine* Other mAChR agonists include: Muscarine (*See (+)-Muscarine*); Aceclidine; Alvameline; Arecoline; Cevimeline; Cl-1017; Milameline; Muscarine; Sabcomeline; Talsaclidine; Tazomeline; Vedaclidine; Xanomeline.

Muscarinic Acetylcholine Receptor Antagonists

These mAChR blockers include: Abediterol; Aclidinium bromide; AFDX-384; Atropine; Benzatropine; Benzilylcholine mustard; Bevonium; Bornaprine; Cyano-dothiepin; Darifenacin; Dexetimide; Dicycloverine; Dimenhydrinate; Diphemanil metilsulfate; Diphen-hydramine; Emepronium bromide; Etybenzatropine; Fenpiverinium; Fesoterodine; Flavoxate; Glycopyrronium bromide; Hexocyclium; Homatropine; Hydroxyzine; Hyoscyamine; Imidafenacin; Ipratropium bromide; Medrylamine; Mepenzolate; Methantheline; Methoctramine; Methylatropine; Methylhomatropine; Octatropine methylbromide; Orphenadrine; Otilonium bromide; Oxybutynin; Oxyphenonium bromide; PD-0298029; PD-102,807; Penthienate; Pipenzolate; Piperidolate; Pirenzepin; Poldine; Prifinium bromide; Procyclidine; Propantheline bromide; Propiomazine; Scopolamine; Solifenacin; Tiemonium iodide; Tiotropium bromide; Tolterodine; Trihexyphenidyl; Vedaclidine.

Muscimol

This γ-aminobutyrate analogue and CNS depressant (FW$_{free-base}$ = 114.10 g/mol) from the muishroom *Amanita muscaria* crosses the blood-brain barrier and acts as a GABA$_A$ receptor agonist (1). Muscimol injections induce an increase in serotonin in the hypothalamus and midbrain of mice and rats. (*See Ibotenic Acid*) **Target(s):** 4-aminobutyrate aminotransferase (2); 4-aminobutyrate uptake (3). **1**. DeFeudis (1982) *Rev. Pure Appl. Pharmacol. Sci.* **3**, 319. **2**. Jeffery, Rutherford, Witzman & Lunt (1988) *Biochem. J.* **249**, 795. **3**. Johnston (1971) *Psychopharm.* **22**, 230.

Mustard Gas

This membrane-penetrating chemical warfare agent (FW = 159.08 g/mol; CAS 505-60-2), also called sulfur mustard, 1,1'-thiobis(2-chloroethane), and Yperite, is a deadly vesicant. An oily liquid at room temperature (M.P.

= 13–14°C; B.P. = 215–217°C, although U.S. military reports puts that value at 227.8°C), mustard gas is readily volatilized by steam, producing a sweet, agreeable garlic-like odor. Contact with skin causes severe irritation and subepidermal blistering (1,2), with fluid accumulating in cavities created by separation of the moribund basal cell layer from the basement membrane (3). **Use in Warfare:** During World War I, Germany became the first nation to use mustard gas on July 12, 1917 as a CW agent, first delivered by artillery shells near Ypres, Belgium. Indeed, it was none other than Fritz Haber (Nobel Laureate, 1919 for chemical fixation), who put his mind to weaponizing toxic compounds (mainly chlorine, phosgene, and mustard gas) during WWI, where he personally witnessed the first large-scale CW deployment. Italy reportedly used mustard gas in Ethiopia in the 1930s, as did the Japanese against the Chinese from 1937 to 1944. **Reaction with DNA:** Mustard gas is a bifunctional alkylating agent that reacts with guanine bases, introducing intrastrand cross-links. Cells treated with mustard gas (prior to DNA synthesis in the G_1-phase) are not delayed in their progression to S-phase, but show a dose-dependent decrease in the rate and extent of DNA synthesis. Treatment during the S-phase also resulted in a dose-dependent depression in the rate of DNA synthesis, with a corresponding mitotic delay. Treatment in G_2 does not delay mitosis, but inhibits DNA synthesis in the next cell cycle. Specific endonucleases act at sites of modification to produce strand breaks. Poly(ADP-ribose) polymerase is also activated by mustard gas, resulting in a depletion of NAD^+ and inhibition of glycolysis. **Other Targets:** hexokinase (4-7); oxidative phosphorylation, as an uncoupler (8,9); pepsin (5,6,10,11); phosphoprotein phosphatase, as the mustard gas hydrolysis product, 2,2'-thiobis-ethanol, *or* thiodiglycol (1); pyrophosphatase (6); urease (12). **1.** Brimfield, Zweig, Novak & Maxwell (1998) *J. Biochem. Mol. Toxicol.* **12**, 361. **2.** Brimfield (1995) *Toxicol. Lett.* **78**, 43. **3.** Papirmeister, Gross, Meier, Petrali & Johnson (1985) *Fund. Appl. Toxicol.* **5**, S134. **4.** McDonald (1955) *Meth. Enzymol.* **1**, 269. **5.** Dixon & Needham (1946) *Nature* **158**, 432. **6.** Gilman & Phillips (1946) *Science* **103**, 409. **7.** R. Colowick (1951) *The Enzymes*, 1st ed., **2** (part 1), 114. **8.** P. Skulachev (1971) *Curr. Top. Bioenerg.* **4**, 127. **9.** G. Hanstein (1976) *Trends Biochem. Sci.* **1**, 65. **10.** A. Bovey & S. S. Yanari (1960) *The Enzymes*, 2nd ed., **4**, 63. **11.** S. Fruton (1971) *The Enzymes*, 3rd ed., **3**, 119. **12.** M. Grant & V. E. Kinsey (1946) *J. Biol. Chem.* **165**, 485.

Muzolimine

This diuretic (FW$_{free-base}$ = 272.13 g/mol), also known as Bay g 2821, inhibits human pyridoxal kinase, reportedly by forming a covalent complex with pyridoxal. The enzyme is inactivated when pyridoxal is the substrate, but not with pyridoxamine. **Target(s):** Ca^{2+} reabsorption (1); chloride transport (2,3); Na^+/K^+ cotransport, weakly inhibited (3,4); Na^+ reabsorption (1); pyridoxal kinase (5). **1.** Hanson, White & Gomoll (1982) *Miner. Electrolyte Metab.* **8**, 314. **2.** Eriksson & Wistrand (1986) *Acta Physiol. Scand.* **126**, 93. **3.** Eriksson & Wistrand (1986) *Acta Physiol. Scand.* **127**, 137. **4.** Knorr, Garthoff, Ingendoh & De Mendonça (1985) *Z. Kardiol.* **74** Suppl. 2, 175. **5.** Knorr & Garthoff (1985) *Arch. Int. Pharmacodyn. Ther.* **278**, 150. **6.** Lainé-Cessac, Cailleux & Allain (1997) *Biochem. Pharmacol.* **54**, 863.

Mycophenolate mofetil

This orally bioavailable, cell-permeant mycophenolate-based pro-drug (FW = 433.49 g/mol; CAS 128794-94-5 and 116680-01-4 (HCl); Solubility: 85 mg/mL DMSO, <1 mg/mL Water; Formulation: 65.8 mg/mL in 5% Dextrose Injection), also known as RS 61443, MMF, Myfortic, CellCept®, and named systematically as   (*E*)-2-morpholinoethyl   6-(4-hydroxy-6-methoxy-7-methyl-3-oxo-1,3-dihydroisobenzofuran-5-yl)-4-methylhex-4-

enoate, undergoes hydrolysis to the active drug mycophenolate, which targets inosine monophosphate dehydrogenase, IMPDH-1 (IC$_{50}$ = 39 nM) and IMPDH-2 (IC$_{50}$ = 10-12 nM). The IMPDH reaction is the rate-limiting reaction step of *de novo* synthesis of guanosine nucleotides. Although liver is the primary organ of purine nucleotide biosynthesis, T- and ·B-lymphocytes are highly proliferative and depend on this pathway. With IMPDH-II highly expressed in activated lymphocytes and more susceptible to MPA inhibition, this drug exerts a stronger cytostatic effect on lymphocytes than other cell types. While this is the principal mechanism by which MPA exerts its immunosuppressive effects, other mechanisms may contribute to MPA's efficacy in preventing allograft rejection: (a) induced apoptosis of activated T-lymphocytes, thereby eliminating clones that respond to antigenic stimulation; (b) induced depletion of guanosine nucleotides, suppressing expression of some adhesion molecules, thereby decreasing lymphocyte and monocyte recruitment into sites of inflammation and graft rejection; and (c) induced depletion of guanosine nucleotides, attended by diminished tetrahydrobiopterin biosynthesis and consequentially lower iNOS activity. TM-MMF suppresses T-lymphocytic responses to allogeneic cells and other antigens. **Bioactivation:** Both intestinal and hepatic microsomal carboxylesterases, and in particular microsomal carboxylesterase-1, are likely to be involved in MMF hydrolysis and play important roles in MMF bioactivation. **Inhibitory Mechanism:** The bicyclic ring system of MPA binds beneath the hypoxanthine ring of Enz-S-XMP* adduct, trapping this enzyme-bound reactant intermediate in a nonproductive Enz-S-XMP*·Mycophenolate complex. **1.** Allison & Eugui (1996) *Clin. Transplant.* **10**, 77. **2.** Sintchak & Nimmesgern (2000) *Immunopharmacology* **47**, 163. **3.** Ji, Gu, Makhov, et al. (2006) *J. Biol. Chem.* **281**, 206.

Mycophenolate (Mycophenolic Acid)

This antibiotic, immunosuppressant, and antitumor agent (FW = 320.34 g/mol; CAS 24280-93-1; *Abbreviation*: MPA; Soluble to 100 mM in DMSO), systematically known as (4*E*)-6-(1,3-duhydro-4-hydroxy 6-methoxy-7-methyl-3-oxo-5-isobenzofuranyl)-4-methyl-4-hexanoic acid, from *Penicillium brevi-compactum* and *P. stoloniferum* inhibits IMP dehydrogenase, thereby suppressing lymphocyte proliferation. MPA is a strong inhibitor of the mammalian enzyme but a much weaker inhibitor of microbial IMP dehydrogenases. It depletes guanosine nucleotides preferentially in T and B lymphocytes, thereby inhibiting their proliferation. Cell-mediated immune responses are thus suppressed, as is antibody formation. Mycophenolic acid also inhibits the glycosylation and expression of adhesion molecules, and the recruitment of lymphocytes and monocytes into sites of inflammation. In addition, it depletes tetrahydrobiopterin concentrations and decreases the production of nitric oxide by inducible nitric-oxide synthase (1). Nobelist H. W. Florey first reported mycophenolate's antimicrobial activity (2). It was the first antibiotic ever to be crystallized (3). **Key Pharmacokinetic Parameters:** *See* Appendix II in Goodman & Gilman's THE PHARMACOLOGICAL BASIS OF THERAPEUTICS, 12th Edition (Brunton, Chabner & Knollmann, eds.) McGraw-Hill Medical, New York (2011). **Target(s):** glucuronosyltransferase (4); glycosylation of adhesion molecules (5); GMP synthetase (6); IMP dehydrogenase (1,7-12); lymphocyte proliferation (1,13,14). **1.** Allison (2005) *Lupus* **14** Suppl. 1, s2. **2.** Florey, Gilliver, Jenning & Sanders (1946) *Lancet* **1**, 46. **3.** Gosio (1896) *Riv. Ig. San. publ.* **7**, 825, 869, and 961. **4.** Hara, Nakajima, Miyamoto & Yokoi (2007) *Drug Metab. Pharmacokinet.* **22**, 103. **5.** Allison, Kowalski, Muller, Waters & Eugui (1993) *Transplant. Proc.* **25** (3 Suppl. 2), 67. **6.** Spector & Beecham III (1975) *J. Biol. Chem.* **250**, 3101. **7.** Wu & Scrimgeour (1973) *Can. J. Biochem.* **51**, 1391. **8.** Smith, Fontenelle, Muzik, *et al.* (1974) *Biochem Pharmacol.* **23**, 2727. **9.** Digits & Hedstrom (1999) *Biochemistry* **38**, 15388. **10.** Nelson, Carr, Devens, *et al.* (1996) *J. Med. Chem.* **39**, 4181. **11.** Allison & Eugui (2005) *Transplantation* **80**, S181. **12.** Allison & Eugui (2000) *Immunopharmacology* **47**, 85. **13.** Eugui, Mirkovich & Allison (1991) *Scand. J. Immunol.* **33**, 175. **14.** Eugui, Almquist, Muller & Allison (1991) *Scand. J. Immunol.* **33**, 161.

Myeloid Translocation Gene-16

The 559-residue transcriptional co-repressor (MW = 61,528 Da, *Mus musculus*) inhibits glycolysis and stimulates mitochondrial respiration. Expression of certain genes involved in tumor cell metabolism (including 6-phosphofructo-2-kinase/fructose-2,6-biphosphatase 3 and 4 (PFKFB3 and PFKFB4), and pyruvate dehydrogenase kinase isoenzyme 1 (PDK1)) was rapidly diminished upon MTG16 expression. Moreover, hypoxia-stimulated production of PFKFB3, PFKFB4 and PDK1 was inhibited by MTG16 expression. Expression of MTG16 reduced glycolytic metabolism, while increasing mitochondrial respiration and formation of reactive oxygen species (ROS). These metabolic changes were paralleled by increased phosphorylation of mitogen-activated protein kinases, reduced levels of amino acids, reduced proliferation, and a decreased fraction of cells in S-phase. MTG16 may serve as a brake on glycolysis, exerting an anti-tumor effect. **1.** Kumar, Sharoyko, Spégel, *et al.* (2013) *PLoS One* **8**, e68502.

Myostatin

This secreted growth differentiation factor and myogenesis inhibitor (MW$_{Human}$ = 25.0 kDa; Accession Number = ABI48514), also known as Growth Differentiation Factor-8 (*or* GDF-8), is a TGFβ family member that binds to the activin type II receptor, triggering the recruitment of the Alk-3 or Alk-4 coreceptor. In muscle, the latter initiates a signal transduction cascade that activates SMAD2 and SMAD3, thereby inhibiting differentiation into mature muscle fibers (1). Myostatin also inhibits preadipocyte differentiation in 3T3-L1 cells, an effect that is partly mediated by altered regulation of C/EBPα and Peroxisome Proliferator-Activated Receptor-γ, or PPARγ (2). Myostatin is also an inhibitor of myoblast differentiation by Smad3-mediated down-regulation of MyoD expression via increased Smad3·MyoD complex formation (3). Myostatin induces muscle atrophy by inhibiting myoblast proliferation, with an attendant rise in ubiquitin-proteasomal activity and downregulation of the IGF-Akt pathway (4). Such effects occur in multiple atrophy-causing events, including injury, diseases such as cachexia, disuse and space flight. Significantly, myostatin knockout drives the browning of white adipose tissue by activating the AMPK-PGC1α-Fndc5 pathway in muscle (5). While myostatin deletion or loss of function induces muscle hypertrophy, its overexpression or systemic administration causes muscle atrophy. Since myostatin blockade would be effective in increasing skeletal muscle mass, myostatin inhibitors are of interest in finding treatments for muscle disorders. Indeed, myostatin blockade with a humanized monoclonal antibody induces muscle hypertrophy and reverses muscle atrophy in young and aged mice (6). By utilizing recombinant adeno-associated virus to overexpress a secretable dominant-negative myostatin exclusively in the liver of mice, systemic myostatin inhibition was found to increase skeletal muscle mass and strength in control C57 Bl/6 mice and in the dystrophin-deficient MDX model of Duchenne muscular dystrophy (7). **1.** Han, Zhou, Mitch & Goldberg (2013) *Int. J. Biochem. Cell Biol.* **45**, 2333. **2.** Kim, Liang, Dean, *et al.* (2001) *Biochem. Biophys. Res. Commun.* **281**, 902. **3.** Langley, Thomas, Bishop, *et al.* (2002) *J. Biol. Chem.* **277**, 49831. **4.** Elliott, Renshaw, Getting & Mackenzie (2012) *Acta Physiol.* (Oxford) **205**, 324. **5.** Shan, Liang, Bi & Kuang (1981) *FASEB J.* **27**, 1981. **6.** Latres, Pangilinan, Miloscio, *et al.* (2015) *Skelet Muscle* PMID: 26457176. **7.** Morine, Bish, Pendrak, *et al.* (2010) *PLoS One* **5**, e9176.

Myricetin

This DNA-intercalating bioflavinoid (FW = 318.24 g/mol; CAS 529-44-2; λ$_{max}$ = 255 and 376 nm; Solubility: 10 mg/mL DMSO), also named Cannabiscetin, NSC 407290, 3,5,7-trihydroxy-2-(3,4,5-trihydroxyphenyl)-4-chromenone and 3,5,7-trihydroxy-2-(3,4,5-trihydroxyphenyl)-4*H*-1-benzopyran-4-one, is a nutraceutical and antioxidant found in vegetables, fruits, nuts, berries, tea, and red wine. It is often marketed in doses of 100 mg, and, in some published reports, subjects consumed as much as gram-level quantities daily. Myricetin also crosses the blood-brain barrier. In rats, myricetin is metabolized to both 3-hydroxyphenylacetate and 3,5-dihydroxyphenylacetate. Myricetin is often present as glyoconjugates: myricetin 3-*O*-galactoside, myricetin 3-*O*-glucoside, myricetin 3-*O*-

rhamnoside (also known as myricitrin). **Inhibitory Targets:** Myricetin protects against some forms of cancer and cardiovascular disease, inhibiting Advanced Glycation End-product- (*or* AGE-) induced migration of retinal pericytes through enhanced phosphorylation of ERK1/2, FAK-1, and paxillin, both *in vitro* and *in vivo* (1). Myricetin also inhibits signaling by Hepatocyte Growth Factor (HGF) as well as its receptor-associated tyrosine kinase Met (*or* HGF/Met) in the medulloblastoma cell line DAOY, thereby preventing formation of actin-rich membrane ruffles and resulting in the inhibition of Met-induced cell migration in Boyden chambers (2). Myricetin is also a mixed-type noncompetitive inhibitor of both isoforms of human IMP dehydrogenase, with respective IC$_{50}$ values of 7 and 4 μM for IMPDH-I and IMPDH-II, that exerts anti-proliferative and pro-apoptotic effects on K562 human leukemia cells in a dose-dependent manner (3). This cytotoxicity is markedly attenuated by exogenous addition of guanosine *in vitro*, a finding that is fully consistent with myricetin's action as an IMPDH inhibitor. Myricetin causes toxic DNA damage by stabilizing covalent DNA-topoisomerase cleavage complexes (4). It inhibits: aldehyde oxidase (5), xanthine oxidase (5), human pancreatic α-amylase competitively (6), porcine aldose reductase, IC$_{50}$ = 40 μM (7), and glutathione-*S*-transferase (8). It also inhibits cytokine-induced cell death in RIN-m5f β-cells (9). Myricetin binds to JAK1/STAT3, inhibiting the neoplastic transformation of murine JB6 P$^+$ cells and inhibiting MEK1 kinase activity (10,11). **1.** Kim, Kim, Kim, *et al.* (2015) *Biochem. Pharmacol.* **93**, 496. **2.** Labbé, Provençal, Lamy, *et al.* (2009) *J. Nutr.* **139**, 646. **3.** Pan, Hu, Wang, *et al.* (2016) *Biochem. Biophys. Res. Commun.* **477**, 915. **4.** López-Lázaro, Martín-Cordero, Toro & Ayuso (2002) *J. Enzyme Inhib. Med. Chem.* **17**, 25. **5.** Pirouzpanah, Hanaee, Razavieh & Rashidi (2009) *J. Enzyme Inhib. Med. Chem.* **24**, 14. **6.** Tarling, Woods, Zhang, *et al.* (2008) *Chembiochem* **9**, 433. **7.** Kawanishi, Ueda & Moriyasu (2003) *Curr. Med. Chem.*, **10**, 1353. **8.** Iioa, Kawaguchi, Sakota, Otonari & Nitahara (1993) *Biosci. Biotech. & Biochem.* **57**, 1678. **9.** Ding, Zhang, Dai & Li (2012) *J. Med. Food.* **15**, 733. **10.** Kumamoto, Fujii & Hou (2009) *Cancer Lett.* **275**, 17. **11.** Lee, Kang, Rogozin, *et al.* (2007) *Carcinogenesis* **28**, 1918.

Myriocin

This sphingosine-like immunosuppressant and antibiotic, (FW = 401.54 g/mol; CAS 35891-70-4; known also as ISP-1, thermozymocidin, and 2-amino-3,4-dihydroxy-2-(hydroxymethyl)-14-oxoeicos-6-enoic acid; Soluble: 25mg/mL (dissolves best when heat briefly in boiling water bath); 5mg/mL in 50mM NaOH)), from the thermophilic fungi *Mycelia sterilia* and Chinese herb *Isaria sinclairii*, potently inhibits serine palmitoyltransferase (SPT), thereby reducing plasma sphingolipids, cholesterol and triglycerides in hyperlipidemic, apolipoprotein E knockout mice (1). Myriocin also significantly reduces the synthesis of inositol phosphorylceramide, the major sphingolipid expressed in cultured *Leishmania* (*Viannia*) *braziliensis* promastigotes (2). Log-phase promastigotes treated with 1 μM myriocin showed a 52% reduction in growth rate and morphological alterations such as more rounded shape and shorter flagellum. (**See also** Fingolimod) **1.** Miyake, Kozutsumi, Nakamura, Fujita & Kawasaki (1995) *Biochem. Biophys. Res. Commun.* **211**, 396. **2.** Castro, Yoneyama, Haapalainen, *et al.* (2013) *J. Eukaryot. Microbiol.* **60**, 377.

Myristate (Myristic Acid)

This medium chain-length saturated fatty acid (FW$_{free-acid}$ = 228.38 g/mol; CAS 208-875-2; pK_a ≈ 4.8; M.P. = 58.5°C; B.P. = 326.2°C; Solubility: 0.002 g/100 mL H$_2$O; IUPAC Name: tetradecanoic acid) is widely distributed as trimyristin (trimyristoylglcerol) in plants, paricularly in nutmeg (in which it comprises 20-25% of the dry weight of the nut and 80% of nutmeg oil) as well as cocanut oil. After conversion to myristoyl-CoA, myristoylation plays a co-/post-translational role in anchoring signal transduction enzymes and proteins to cellular membranes. (**See** *Myristoyl-CoA*) **Inhibitory Target(s):** acetyl-CoA carboxylase (1,2); alcohol dehydrogenase (3); carbonyl reductase (4); chondroitin AC lyase (5);

chondroitin B lyase, weakly inhibited (5); cytosol alanyl aminopeptidase (22); esterase (6); estrone sulfotransferase (23); ethanolamine-phospho-transferase (24); 7-ethoxycoumarin O-deethylase (P450) (7); β-galactoside α-2,6-sialyltransferase (8); glucokinase (9); glucose-6-phosphate dehydrogenase (10); glycylpeptide N-tetradecanoyltransferase (26); hexokinase (9,11,25); hyaluronate lyase, weakly inhibited (5); leukocyte elastase (12); luciferase, firefly (*Photinus*-luciferin 4-monooxygenase; IC_{50} = 0.68 μM (13,27,28); microbial collagenase (12); Na$^+$/K$^+$-exchanging ATPase (14); pectinesterase (15); peptidyl dipeptidase, *or* angiotensin I-converting enzyme (16); prostaglandin D$_2$ dehydrogenase (17); prostaglandin dehydrogenase (17,18); prostaglandin dehydrogenase (NAD$^+$) (type I) (17); prostaglandin D synthetase (17); prostaglandin E isomerase (17); pyruvate kinase (19); steroid 5α-reductase (20); L-threonine dehydrogenase (21); thromboxane synthetase (17). **1**. Miller & Levy (1975) *Meth. Enzymol.* **35**, 11. **2**. Levy (1963) *Biochem. Biophys. Res. Commun.* **13**, 267. **3**. Winer & Theorell (1960) *Acta Chem. Scand.* **14**, 1729. **4**. Imamura, Migita, Anraku & Otagiri (1999) *Biol. Pharm. Bull.* **22**, 731. **5**. Suzuki, Terasaki & Uyeda (2002) *J. Enzyme Inhib. Med. Chem.* **17**, 183. **6**. Weber & King (1935) *J. Biol. Chem.* **108**, 131. **7**. Muller-Enoch, R. Fintelmann, Nicolaev & H. Gruler (2001) *Z. Naturforsch. [C]* **56**, 1082. **8**. Bador, Morelis & Louisot (1984) *Biochimie* **66**, 223. **9**. Lea & Weber (1968) *J. Biol. Chem.* **243**, 1096. **10**. Criss & McKerns (1969) *Biochim. Biophys. Acta* **184**, 486. **11**. Stewart & Blakely (2000) *Biochim. Biophys. Acta* **1484**, 278. **12**. Rennert & Melzig (2002) *Planta Med.* **68**, 767. **13**. Ueda & Suzuki (1998) *Biophys. J.* **75**, 1052. **14**. Miller & Woodhouse (1977) *Aust. J. Exp. Biol. Med. Sci.* **55**, 741. **15**. Miller & McColloch (1959) *Biochem. Biophys. Res. Commun.* **1**, 91. **16**. Oshima & Nagasawa (1977) *J. Biochem.* **81**, 57. **17**. Osama, Narumiya, Hayaishi, *et al.* (1983) *Biochim. Biophys. Acta* **752**, 251. **18**. Mibe, Nagai, Oshige & Mori (1992) *Prostaglandins Leukot. Essent. Fatty Acids* **46**, 241. **19**. Kayne (1973) *The Enzymes*, 3rd ed., **8**, 353. **20**. Raynaud, Cousse & Martin (2002) *J. Steroid Biochem. Mol. Biol.* **82**, 233. **21**. Guerranti, Pagani, Neri, *et al.* (2001) *Biochim. Biophys. Acta* **1568**, 45. **22**. Garner & Behal (1977) *Arch. Biochem. Biophys.* **182**, 667. **23**. Adams & Ellyard (1972) *Biochim. Biophys. Acta* **260**, 724. **24**. Vecchini, Roberti, Freysz & Binaglia (1987) *Biochim. Biophys. Acta* **918**, 40. **25**. Morris, DeBruin, Yang, *et al.* (2006) *Eukaryot. Cell* **5**, 2014. **26**. Selvakumar, Lakshmikuttyamma, Shrivastav, *et al.* (2007) *Prog. Lipid Res.* **46**, 1. **27**. Matsuki, Suzuki, Kamaya & Ueda (1999) *Biochim. Biophys Acta* **1426**, 143. **28**. Inouye (2010) *Cell. Mol. Life Sci.* **67**, 387.

Myristoyl-CoA

This fatty acyl-coenzyme A derivative (FW = 991.02 g/mol), also called tetradecanoyl-CoA, is an intermediate in the β-oxidation of fatty acids and the substrate for glycylpeptide N-tetradecanoyltransferase 1 (also known as myristoysCoA:protein N-myristoyltransferase 1, *or* NMT-1). This enzyme catalyzes co-/post-translational modification of N-terminal glycine-containing proteins and enzymes destined to become bound to biomembranes. In co-translational myristoylation, which accounts for approximately 80% of known myristoylated proteins, the N-terminal glycine is modified after removal of the N-terminal methionine residue in the newly forming polypeptide. In post-translational myristoylation, caspase cleavage exposes an internal glycine residue, which then undergoes myristoylation. Myristoylate proteins tend to interact realtively weakly with biomembranes, a property suggesting that additional factors may be needed

to stabilize their docking with membrane lipids. (*See also* *Myristic Acid*) **Inhibitory Target(s):** acetyl-CoA carboxylase, K_i = 11 μM for rat liver enzyme (1-3); aralkylamine N-acetyltransferase (4); ATP:citrate lyase, *or* ATP:citrate synthase (5,6); dynein ATPase (7); fatty-acid synthase (8); glucose-6-phosphatase (9); protein kinase C (10); [pyruvate dehydrogenase, acetyl-transferring] kinase (11); RNA polymerase (12). **1**. Tanabe, Nakanishi, Hashimoto, *et al.* (1981) *Meth. Enzymol.* **71**, 5. **2**. Nikawa, Tanabe, Ogiwara, Shiba & Numa (1979) *FEBS Lett.* **102**, 223. **3**. Bortz & Lynen (1963) *Biochem. Z.* **337**, 505. **4**. Ferry, Loynel, Kucharczyk, *et al.* (2000) *J. Biol. Chem.* **275**, 8794. **5**. Shashi, Bachhawat & Joseph (1990) *Biochim. Biophys. Acta* **1033**, 23. **6**. Boulton & Ratledge (1983) *J. Gen. Microbiol.* **129**, 2863. **7**. Fujiwara, Yokokawa, Hino & Yasumasu (1982) *J. Biochem.* **92**, 441. **8**. Roncari (1975) *Can. J. Biochem.* **53**, 135. **9**. Methieux & Zitoun (1996) *Eur. J. Biochem.* **235**, 799. **10**. Yaney, Korchak &. Corkey (2000) *Endocrinology* **141**, 1989. **11**. Rahmatullah & Roche (1985) *J. Biol. Chem.* **260**, 10146. **12**. Yokokawa, Fujiwara, Shimada & Yasumasu (1983) *J. Biochem.* **94**, 415.

Myxothiazol

This bacterially derived antibiotic (FW = 487.68 g/mol; CAS 76706-55-3; IUPAC: 7-{2'-[(1*S*,2*E*,4*E*)-1,6-dimethyl-2,4-heptadienyl][2,4'-bithiazol]-4-yl}-3,5-dimethoxy-4-methyl-(2*E*,4*R*,5*S*,6*E*)-2,6-heptadienamide), isolated from *Myxobacterium fulvus*, is a strong inhibitor of Complex III of the mitochondrial electron transport chain and suppressing growth of many yeasts and fungi. Myxothiazol binds to the ubiquinol oxidation site Qo of CIII (1-13). Myxothiazol is a competitive inhibitor of ubiquinol and binds at the quinol oxidation (Qo) site of the cytochrome bc_1 complex, blocking electron transfer to the Rieske iron-sulfur protein. Binding of myxothiazol induces a red-shift to the visible absorption spectrum of reduced haem b_l. Unlike stigmatellin, myxothiazol does not form a hydrogen bond to the Rieske iron-sulfur protein, binding instead in the '*b*-proximal' region of the cytochrome *b* Quinone$_{outside}$ (*or* Qo) site. Movement of the cytoplasmic domain of the Rieske protein is therefore unaffected by the binding of this inhibitor. Myxothiazol has been utilized as a tool in studies of respiratory chain function, both in *in vitro* and with cell culture models. Davoudi *et al.* (14) recently developed a mouse model of biochemically induced and reversible CIII inhibition using myxothiazol. **Target(s):** methane monooxygenase (15); mitochondrial processing peptidase (16); ubiquinol:cytochrome-*c* reductasec, *or* Complex III. **1**. Linke & Weiss (1986) *Meth. Enzymol.* **126**, 201. **2**. von Jagow & Link (1986) *Meth. Enzymol.* **126**, 253. **3**. Thierbach & Reichenbach (1981) *Biochim. Biophys. Acta* **638**, 282. **4**. Thierbach & Reichenbach (1981) *Antimicrob. Agents Chemother.* **19**, 504. **5**. von Jagow & Engel (1981) *FEBS Lett.* **136**, 19. **6**. Trumpower (1990) *Microbiol. Rev.* **54**, 101. **7**. Hauska, Hurt, Gabellini & Lockau (1983) *Biochim. Biophys. Acta* **726**, 97. **8**. Yang & Trumpower (1986) *J. Biol. Chem.* **261**, 12282. **9**. Kriauciunas, Yu, Yu, Wynn & Knaff (1989) *Biochim. Biophys. Acta* **976**, 70. **10**. Crofts (2004) *Biochim. Biophys. Acta* **1655**, 77. **11**. Matsuno-Yagi & Hatefi (1997) *J. Biol. Chem.* **272**, 16928. **12**. Brasseur (1988) *J. Biol. Chem.* **263**, 12571. **13**. Covian & Trumpower (2006) *J. Biol. Chem.* **281**, 30925. **14**. Davoudi, Kallijärvi, Marjavaara, *et al.* (2014) *Biochem. Biophys. Res. Commun.* **446**, 1079. **15**. Zahn & DiSpirito (1996) *J. Bacteriol.* **178**, 1018. **16**. Deng, Zhang, Kachurin, *et al.* (1998) *J. Biol. Chem.* **273**, 20752.

– N –

N, See *Asparagine; Nitrogen*

N20C

This selective, non-competitive NMDA receptor antagonist (FW = 304.82 g/mol; CAS 1177583-87-7; Soluble to 100 mM in H_2O), also named 2-[(3,3-diphenylpropyl)amino]acetamide, penetrates the blood-brain barrier and binds to the receptor-associated ion channel (IC_{50} = 5 μM), thereby preventing glutamate-induced Ca^{2+} influx. Saturation of the blocker binding-site with N20C prevents dizolcipine (MK-801) blockade of the NMDA receptor, implying that both ligands bind at the same receptor site. N20C displays neuroprotective activity *in vivo*. 1. Planells-Cases, Montoliu, Humet, *et al.* (2002) *J. Pharmacol. Exp. Ther.* **302**, 163.

N6022

This potent, selective, and reversible GSNOR inhibitor (FW = 414.46 g/mol; CAS 1208315-24-5; Solubility = 82 mg/mL), also named 3-(5-(4-(1*H*-imidazol-1-yl)phenyl)-1-(4-carbamoyl-2-methylphenyl)-1*H*-pyrrol-2-yl) propanoate, targets human *S*-nitrosoglutathione reductase, (IC_{50} = 8 nM; K_i = 2.5 nM), a class III alcohol dehydrogenase. *S*-nitrosoglutathione (GSNO) serves as a reservoir of nitric oxide (NO), a key regulator of airway smooth muscle tone and inflammation (1). Decreased levels of GSNO in the lungs of asthmatics have been attributed to increased GSNO catabolism via GSNO reductase (GSNOR) leading to loss of GSNO- and NO- mediated bronchodilatory and anti-inflammatory actions (1). N6022 is well tolerated in animals. N6022 binds in the GSNO substrate binding pocket like a competitive inhibitor, although it behaves as a mixed uncompetitive inhibitor relative to GSNO substrate and a mixed competitive inhibitor relative to the formaldehyde adduct, S-hydroxymethylglutathione. N6022 is uncompetitive with cofactors NAD^+ and NADH (1). The significant bronchodilatory and anti-inflammatory actions of N6022 in the airways are consistent with restoration of GSNO levels through GSNOR inhibition (2). GSNOR inhibition may offer a therapeutic approach for the treatment of asthma and other inflammatory lung diseases. N6022 is currently being evaluated in clinical trials for the treatment of inflammatory lung disease (2). 1. Green, Chun, Patton, *et al.* (2012) *Biochemistry* **51**, 2157. 2. Blonder, Mutka, Sun, *et al.* (2014) *BMC Pulm. Med.* 10.1186/1471-2466-14-3.

NAB 2

This *N*-aryl-benzimidazole (FW = 389.88 g/mol; CAS 1504588-00-4; Solubility: 100 mM in DMSO; *N*-[(2-chlorophenyl)methyl]-1-(2,5-dimethylphenyl)-1*H*-benzimidazole-5-carboxamide) promotes endosomal transport events that depend on the E3 ubiquitin ligase Rsp5/Nedd4, strongly and selectively protecting diverse cell types from α-synuclein toxicity (1). Rsp5 is the single yeast member of the highly conserved mammalian family of HECT domain-containing Nedd4 E3 ligases, which catalyze formation of K63 linkages between ubiquitin and diverse membrane proteins, thereby regulating endosomal trafficking, *not* proteasomal degradation. NEDD4 family E3 ubiquitin ligase is involved in regulating many cellular processes including Multi-Vesicular-Body (MVB) sorting, heat shock response, transcription, endocytosis, and ribosome stability. NAB 2 reverses pathologic phenotypes in cortical neurons generated from induced pluripotent stem (iPS) cell from patients harboring critical α-synuclein mutations, placing them are at high risk of developing PD dementia (2). α-Synuclein is a small lipid-binding protein that has been implicated in several neurodegenerative diseases, including Parkinson's disease. 1. Tardiff, Jui, Khurana, *et al.* (2013) *Science* **342**, 979. 2. Chung, Khurana, Auluck *et al.* (2013) *Science* **342**, 983.

NAD^+

This centrally important redox coenzyme (FW = 663.43 g/mol; CAS 53-84-9), also known as nicotinamide adenine dinucleotide and β-NAD^+ (previously DPN^+ and diphosphopyridine nucleotide) is an essential cofactor and hydride acceptor for many oxidoreductases. Its reduced form NADH is a reducing reagent as well as the substrate for Complex I in electron transport or respiratory chain. NAD^+ also serves as ADPR donor in ADP-ribosylation reactions. It is likewise a required cofactor in Class III NAD^+-dependent histone deacetylases, *or* sirtuins, which deacetylate Lysine-9 and Lysine-14 of histone H3 and specifically Lysine-16 of histone H4. **Chemical Properties:** NAD^+ is a hygroscopic solid that is very soluble in water. NAD^+ is relatively stable in slightly acidic pH conditions (*i.e.*, pH 3–7) but will decompose in alkaline pH (the rate of decomposition increases in the presence of orthophosphate, carbonate, or maleate). The solid should be stored in a desiccator at 0°C in the dark. This photosensitive coenzyme absorbs light significantly in the ultraviolet: λ_{max} at pH 7.0 is 260 nm (ε = 18000 $M^{-1}cm^{-1}$). The reduced form, NADH, exhibits another peak at higher wavelength: λ_{max} at pH 7.0 is 339 nm (ε = 6200 $M^{-1}cm^{-1}$). **Target(s):** acetyl-CoA *C*-acyltransferase (72); *N*-acetylneuraminate 4-*O* acetyltransferase (71); acid phosphatase (1); adenylate cyclase (2); adenylate kinase (63); L-aspartate oxidase (3); CDP-diacylglycerol diphosphatase (4); choline oxidase (5); citrate (*si*)-synthase, weakly inhibited (69); creatine kinase (6,7); 3',5'-cyclic-nucleotide phosphodiesterase (47); DNA-(apurinic or apyrimidinic site) lyase (8); FAD diphosphatase (9); FAD synthetase (61); fructose-1,6 bisphosphatase (53,54); fucosterol-epoxide lyase (10); glucose-1-phosphate adenylyltransferase (58); glutamine synthetase, *Neurospora crassa* (11); γ-glutamylcysteine synthetase (12); glutathione peroxidase (13); malate oxidase (14); malic enzyme, oxaloacetate decarboxylating activity, weakly inhibited (15); methylenetetrahydrofolate dehydrogenase ($NADP^+$) (16); 6-methylsalicylic acid synthase (70); m7G(5')pppN diphosphatase, slightly inhibited (17); nicotinamidase (18-20); nicotinamide phosphoribosyltransferase (21,22,65-67); nicotinate-nucleotide adenylyltransferase (59); nicotinate-nucleotide pyrophosphorylase (carboxylating) (23); nicotinate phosphoribosyltransferase, weakly inhibited (68); nitrate reductase (NADPH) (24); NMN nucleosidase (44); nucleotide diphosphatase, alternative substrate (25-29); nucleotidases (52); 5'-nucleotidase (30,31); oligosaccharide-diphosphodolichol diphosphatase (32); oxaloacetate decarboxylase (33); palmitoyl-CoA hydrolase, *or* acyl-CoA hydrolase (55);

pectin lyase (34); phosphodiesterase I, *or* 5'-exonuclease (49-51); poly(ADP-ribose) glycohydrolase (45,46); [protein ADP-ribosylarginine] hydrolase (43); [protein-P$_{II}$] uridylyltransferase (57); [pyruvate dehydrogenase (acetyl-transferring)] kinase (56); ribose-phosphate diphosphokinase, *or* phosphoribosyl pyrophosphate synthetase (62); ribose-5-phosphate isomerase (35,36); ribosylnicotinamide kinase (64); serine hydroxymethyltransferase, *or* glycine hydroxymethyl-transferase (73); sphingomyelin phosphodiesterase, *or* sphingomyelinase (48); sterol 24-*C*-methyltransferase (74); sulfate adenylyltransferase (60); UDP-glucose 4-epimerase (37,38); UDP-glucuronate 4-epimerase (39,40); UDP sugar diphosphatase (41); xylose isomerase (42). **1.** Belfield, Ellis & Goldberg (1972) *Enzymologia* **42**, 91. **2.** Rosenberg & Pall (1983) *Arch. Biochem. Biophys.* **221**, 243. **3.** Nasu, Wicks & Gholson (1982) *J. Biol. Chem.* **257**, 626. **4.** Raetz, Hirschberg, Dowhan, Wickner & Kennedy (1972) *J. Biol. Chem.* **247**, 2245. **5.** Williams, Litwack & Elvehjem (1951) *J. Biol. Chem.* **192**, 73. **6.** Kuby, Noda & Lardy (1954) *J. Biol. Chem.* **210**, 65. **7.** Kuby & Noltmann (1962) *The Enzymes*, 2nd ed., **6**, 515. **8.** Kane & Linn (1981) *J. Biol. Chem.* **256**, 3405. **9.** Mistuda, Tsuge, Tomozawa & Kawai (1970) *J. Vitaminol.* **16**, 31. **10.** Prestwich, Angelastro, de Palma & Perino (1985) *Anal. Biochem.* **151**, 315. **11.** Kapoor & Bray (1968) *Biochemistry* **7**, 3583. **12.** Davis, Balinsky, Harington & Shepherd (1973) *Biochem. J.* **133**, 667. **13.** Little, Olinescu, Reid & O'Brien (1970) *J. Biol. Chem.* **245**, 3632. **14.** Narindrasorasak, Goldie & Sanwal (1979) *J. Biol. Chem.* **254**, 1540. **15.** Kraemer, Conn & Vennesland (1951) *J. Biol. Chem.* **188**, 583. **16.** Ljungdahl, O'Brien, Moore & Liu (1980) *Meth. Enzymol.* **66**, 599. **17.** Nuss & Furuichi (1977) *J. Biol. Chem.* **252**, 2815. **18.** Joshi, Calbreath & Handler (1971) *Meth. Enzymol.* **18B**, 180. **19.** Joshi & Handler (1962) *J. Biol. Chem.* **237**, 929. **20.** Wintzerith, Dierich & Mandel (1980) *Biochim. Biophys. Acta* **613**, 191. **21.** Ohtsu & Nishizuka (1971) *Meth. Enzymol.* **18B**, 127. **22.** Dietrich (1971) *Meth. Enzymol.* **18B**, 144. **23.** Nishizuka & Nakamura (1970) *Meth. Enzymol.* **17A**, 491. **24.** Nason & Evans (1953) *J. Biol. Chem.* **202**, 655. **25.** Bartkiewicz, Sierakowska & Shugar (1984) *Eur. J. Biochem.* **143**, 419. **26.** Byrd, Fearney & Kim (1985) *J. Biol. Chem.* **260**, 7474. **27.** Haroz, Twu & Bretthauer (1972) *J. Biol. Chem.* **247**, 1452. **28.** Jacobson & Kaplan (1957) *J. Biol. Chem.* **226**, 427. **29.** Kornberg & Pricer (1950) *J. Biol. Chem.* **182**, 763. **30.** Drummond & Yamamoto (1971) *The Enzymes*, 3rd ed., **4**, 337. **31.** Madrid-Marina & Fox (1986) *J. Biol. Chem.* **261**, 444. **32.** Belard, Cacan & Verbert (1988) *Biochem. J.* **255**, 235. **33.** Sender, Martin, Peiru & Magni (2004) *FEBS Lett.* **570**, 217. **34.** Papi & Kyriakidis (2003) *Biotechnol. Appl. Biochem.* **37**, 187. **35.** MacElroy & Middaugh (1982) *Meth. Enzymol.* **89**, 571. **36.** Rutner (1970) *Biochemistry* **9**, 178. **37.** Wilson & Hogness (1969) *J. Biol. Chem.* **244**, 2132. **38.** Maxwell (1957) *J. Biol. Chem.* **229**, 139. **39.** Gaunt, Ankel & Schutzbach (1972) *Meth. Enzymol.* **28**, 426. **40.** Gaunt, Maitra & Ankel (1974) *J. Biol. Chem.* **249**, 2366. **41.** Mauck & Glaser (1970) *Biochemistry* **9**, 1140. **42.** Vartak, Srinivasan, Powar, Rele & Khire (1984) *Biotechnol. Lett.* **6**, 493. **43.** Moss, Oppenheimer, West & Stanley (1986) *Biochemistry* **25**, 5408. **44.** Foster (1981) *J. Bacteriol.* **145**, 1002. **45.** Sugimura, Yamada, Miwa, *et. al.* (1973) *Biochem. Soc. Trans.* **1**, 642. **46.** Miwa, Tanaka, Matsushima & Sugimura (1974) *J. Biol. Chem.* **249**, 3475. **47.** Lim, Palanisamy & Ong (1986) *Arch. Microbiol.* **146**, 142. **48.** Testai, Landek, Goswami, Ahmed & Dawson (2004) *J. Neurochem.* **89**, 636. **49.** Futai & Mizuno (1967) *J. Biol. Chem.* **242**, 5301. **50.** Luthje & Ogilvie (1985) *Eur. J. Biochem.* **149**, 119. **51.** Udvardy, Marre & Farkas (1970) *Biochim. Biophys. Acta* **206**, 392. **52.** Passariello, Schippa, Iori, *et al.* (2003) *Biochim. Biophys. Acta* **1648**, 203. **53.** Fujita & Freese (1979) *J. Biol. Chem.* **254**, 5340. **54.** Amachi & Bowien (1979) *J. Gen. Microbiol.* **113**, 347. **55.** Berge & Dossland (1979) *Biochem. J.* **181**, 119. **56.** Roche, Baker, Yan, *et al.* (2001) *Prog. Nucl. Acid Res. Mol. Biol.* **70**, 33. **57.** Engleman & Francis (1978) *Arch. Biochem. Biophys.* **191**, 602. **58.** Preiss, Crawford, Downey, Lammel & Greenberg (1976) *J. Bacteriol.* **127**, 193. **59.** Dahmen, Webb & Preiss (1967) *Arch. Biochem. Biophys.* **120**, 440. **60.** Renosto, Patel, Martin, *et al.* (1993) *Arch. Biochem. Biophys.* **307**, 272. **61.** Schrecker & Kornberg (1950) *J. Biol. Chem.* **182**, 795. **62.** Fox & Kelley (1972) *J. Biol. Chem.* **247**, 2126. **63.** Storey (1976) *J. Biol. Chem.* **251**, 7810. **64.** Grose, Bergthorsson & Roth (2005) *J. Bacteriol.* **187**, 2774. **65.** Dietrich & Muniz (1972) *Biochemistry* **11**, 1691. **66.** Elliott & Rechsteiner (1982) *Biochem. Biophys. Res. Commun.* **104**, 996. **67.** Shibata, Taguchi, Nishitani, *et al.* (1989) *Agric. Biol. Chem.* **53**, 2283. **68.** Hayakawa, Shibata & Iwai (1984) *Agric. Biol. Chem.* **48**, 445. **69.** Robinson, Danson & Weitzman (1983) *Biochem. J.* **213**, 53. **70.** Lam, Neway & Gaucher (1988) *Can. J. Microbiol.* **34**, 30. **71.** Tiralongo, Schmid, Thun, Iwersen & Schauer (2001) *Glycoconjugate J.* **17**, 849. **72.** Haywood, Anderson, Chu & Dawes (1988) *FEMS Microbiol. Lett.* **52**, 91. **73.** Manohar, Ramesh & Rao (1982) *J. Biosci.* **4**, 31. **74.** Lin & Knoche (1976) *Phytochemistry* **15**, 683.

NAD 299

This selective, high-affinity serotonin antagonist (FW = 354.85 g/mol; CAS 184674-99-5; Soluble to 100 mM in H$_2$O and to 100 mM in DMSO), also named (3*R*)-3-(dicyclobutylamino)-8-fluoro-3,4-dihydro-2*H*-1-benzopyran-5-carboxamide hydrochloride, targets the 5-HT$_{1A}$ receptor antagonist (K_i = 0.6 nM, *in vitro*), with >400x selectivity over α-1 and β adrenoceptors (1-3). NAD-299 also differs in selectivity from the structurally related 5-HT$_{1A}$ receptor antagonist (*S*)-UH-301, which has considerable affinity for D$_2$ receptors (1). NAD-299 enhances the action of selective 5-HT reuptake inhibitors and reverses citalopram-induced inhibition of serotonergic cell firing. **1.** Johansson, Sohn, Thorberg, *et al.* (1997) *J. Pharmacol. Exp. Ther.* **283**, 216. **2.** Larsson, Stenfors & Ross (1998) *Eur. J. Pharmacol.* **346**, 209. **3.** Arborelius, *et al.* (1999). *Eur. J. Pharmacol.* **382**, 133.

NADH 2',3'-Dialdehyde

This periodate-modified NADH affinity-labeling reagent (FW$_{free-acid}$ = 662.85 g/mol; *Abbreviation*: oNAD$^+$) contains two aldehyde groups that facilitate imine formation with ε-amino groups of active-site lysine residues in NAD$^+$-dependent oxidoreductases. Dialdehyde-containing nucleotides are formed by periodate oxidation, usually requiring the addition of 20-40% molar excess of sodium metaperiodate (reaction period = 30 min); unreacted periodate is deactivated by the addition of 2-3x molar excess of ethylene glycol. **Target(s):** isocitrate dehydrogenase, NAD$^+$-requiring (1,2). **1.** Colman (1990) *The Enzymes*, 3rd ed., **19**, 283. **2.** Saha & Colman (1988) *Arch. Biochem. Biophys.* **264**, 665.

NADH

This vital redox coenzyme (FW$_{free-acid}$ = 664.86 g/mol), also known as reduced nicotinamide adenine dinucleotide and βNADH is a major redox energy store that is also utilized by many oxidoreductases, serving as a hydride donor and reducing agent. NADH is the substrate for the first complex in electron transport. Only the β-anomer is physiologically relevant, the α-anomer is commercially available and frequently used in specificity and inhibitor investigations. **Chemical Properties:** NADH is a photosensitive and hygroscopic solid that is very soluble in water. The solid should be stored in a desiccator at 0-4°C in the dark. It is destroyed at pH < 4 at room temperature, but is relatively stable at alkaline pH. Note that the presence of orthophosphate buffer also decreases the stability of NADH. This coenzyme absorbs UV light: λ$_{max}$ at pH 7.5 is 259 nm (ε = 16900 M^{-1}cm^{-1}) and 339 nm (ε = 6200 M^{-1}cm^{-1}. NADH is also fluorescent at 460 nm when excited at 340 nm. **Target(s):** acetyl-CoA *C*-acetyltransferase (106,107); acetyl-CoA hydrolase (70); adenylate cyclase, *or* adenylyl cyclase (1-3); adenylate kinase (73); ADP-dependent medium-chain-acyl-CoA hydrolase (68,69); ADP-dependent short-chain-acyl-CoA hydrolase (68,69); ADP-glyceromanno heptose 6-epimerase (4); ADP-ribose diphosphatase (5); alkaline phosphatase (66); arginine kinase (74); arsenate reductase, *or* glutaredoxin (110); bile-acid 7α-dehydroxylase (111); CDP-paratose 2-epimerase (6); citrate (*si*)-synthase (7-11,89-104); decylcitrate synthase, weakly inhibited (88); 3-dehydroquinate synthase (12-14); deoxyhypusine synthase (81); 3,4-dihydroxyphenylalanine oxidative deaminase (114); ferredoxin:NADP$^+$ reductase (15); fructose-1,6-bisphosphatase (64,65); GDP-mannose 3,5 epimerase (16); γ-glutamylcysteine synthetase (17,18); glyceraldehyde-3-phosphate dehydrogenase (NADP$^+$) (19); glycogen phosphorylase *b* (20); homospermidine synthase (82,83); inositol-3-phosphate synthase, *or* inositol-1-phosphate synthase (21); kynurenine 3-monooxygenase (113); 2-methylcitrate synthase (87); methylisocitrate lyase (22); 6-methylsalicylic acid synthase (105); NAD$^+$ ADP ribosyltransferase, *or* poly(ADP-ribose) synthetase (23); NAD$^+$:[dinitrogen-reductase] ADP-D-ribosyltransferase (84); NAD$^+$ kinase (25-29,76-80); NAD$^+$ nucleosidase, *or* NADase (24,58); NADPH:cytochrome c_2 reductase (30); nicotinamide phosphoribosyltransferase, weakly inhibited (85); nicotinate phosphoribosyltransferase, weakly inhibited (86); NMN nucleosidase (57); 5'-nucleotidase (31); octadecanal decarbonylase (32,33); oxaloacetate decarboxylase (34); 3-oxoadipyl CoA thiolase, *or* 3-oxoadipate CoA-transferase (35); palmitoyl-CoA hydrolase, *or* acyl-CoA hydrolase (67); pectin lyase (36); phosphate acetyltransferase (108,109); phosphodiesterase I, *or* 5'-exonuclease (60); phosphoenolpyruvate carboxykinase (37); phosphoenolpyruvate carboxykinase (ATP) (38); 6-phospho-β-glucosidase (59); phosphoketolase (39); [protein-P$_{II}$] uridylyltransferase (71); pyridoxal kinase, weakly inhibited (75); pyruvate carboxylase, *Saccharomyces cerevisiae* (40); [pyruvate dehydrogenase (acetyl-transferring)]-phosphatase (41-43,61-63); ribose-phosphate diphosphokinase, *or* phosphoribosyl-pyrophosphate synthetase (72); rifamycin-B oxidase (116); stearoyl-CoA 9-desaturase (112); succinyl-CoA synthetase (ADP-forming), *or* succinate:CoA ligase (ADP-forming) (44); thyroxine deiodinase, *or* iodothyronine deiodinase (45); tryptophan 2,3 dioxygenase (115); UDP-*N*-acetyl-D-glucosamine 4-epimerase (46); UDP-glucosamine 4-epimerase (47); UDP-glucose 4-epimerase (48-51); UDP-glucuronate decarboxylase (52-54); UDP-glucuronate 5' epimerase (55,56). **1.** Nair & Patel (1991) *Life Sci.* **49**, 915. **2.** Yang & Epstein (1983) *J. Biol. Chem.* **258**, 3750. **3.** Rosenberg & Pall (1983) *Arch. Biochem. Biophys.* **221**, 243. **4.** Ding, Seto, Ahmed & Coleman (1994) *J. Biol. Chem.* **269**, 24384. **5.** Wu, Lennon & Suhadolnik (1978) *Biochim. Biophys. Acta* **520**, 588. **6.** Matsuhashi & Strominger (1966) *Meth. Enzymol.* **8**, 310. **7.** Srere (1969) *Meth. Enzymol.* **13**, 3. **8.** Weitzman (1969) *Meth. Enzymol.* **13**, 22. **9.** Müller-Kraft & Babel (1990) *Meth. Enzymol.* **188**, 350. **10.** Spector (1972) *The Enzymes*, 3rd ed., **7**, 357. **11.** Srere & Matsuoka (1972) *Biochem. Med.* **6**, 262. **12.** Lambert, Boocock & Coggins (1985) *Biochem. J.* **226**, 817. **13.** Hasan & Nester (1978) *J. Biol. Chem.* **253**, 4999. **14.** Saijo & Kosuge (1978) *Phytochemistry* **17**, 223. **15.** Carrillo & Ceccarelli (2003) *Eur. J. Biochem.* **270**, 1900. **16.** Wolucka & Van Montagu (2003) *J. Biol. Chem.* **278**, 47483. **17.** Davis, Balinsky, Harington & Shepherd (1973) *Biochem. J.* **133**, 667. **18.** Yip & Rudolph (1976) *J. Biol. Chem.* **251**, 3563. **19.** Iglesias, Serrano & Guerrero (1987) *Biochim. Biophys. Acta* **925**, 1. **20.** Madsen & Shechosky (1967) *J. Biol. Chem.* **242**, 3301. **21.** Loewus, Bedgar & Loewus (1984) *J. Biol. Chem.* **259**, 7644. **22.** Tabuchi & Satoh (1977) *Agric. Biol. Chem.* **41**, 169. **23.** Shizuta, Ito, Nakata & Hayaishi (1980) *Meth. Enzymol.* **66**, 159. **24.** Anderson & Yuan (1980) *Meth. Enzymol.* **66**, 144. **25.** Wang (1955) *Meth. Enzymol.* **2**, 652. **26.** Blomquist (1980) *Meth. Enzymol.* **66**, 101. **27.** Anderson (1973) *The Enzymes*, 3rd ed., **9**, 49. **28.** Wang & Kaplan (1954) *J. Biol. Chem.* **206**, 311. **29.** Raffaelli, Finaurini, Mazzola, *et al.* (2004) *Biochemistry* **43**, 7610. **30.** Sabo & Orlando (1968) *J. Biol. Chem.* **243**, 3742. **31.** Madrid-Marina & Fox (1986) *J. Biol. Chem.* **261**, 444. **32.** Schneider-Belhaddad & Kolattukudy (2000) *Arch. Biochem. Biophys.* **377**, 341. **33.** Cheesbrough & Kolattukudy (1988) *J. Biol. Chem.* **263**, 2738. **34.** Sender, Martin, Peiru & Magni (2004) *FEBS Lett.* **570**, 217. **35.** Kaschabek, Kuhn, Müller, Schmidt & Reineke (2002) *J. Bacteriol.* **184**, 207. **36.** Papi & Kyriakidis (2003) *Biotechnol. Appl. Biochem.* **37**, 187. **37.** Utter & Kolenbrander (1972) *The Enzymes*, 3rd ed., **6**, 117. **38.** Krebs & Bridger (1976) *Can. J. Biochem.* **54**, 22. **39.** Whitworth & Ratledge (1977) *J. Gen. Microbiol.* **102**, 397. **40.** Scrutton & Young (1972) *The Enzymes*, 3rd ed., **6**, 1. **41.** Pettit, Teague & Reed (1982) *Meth. Enzymol.* **90**, 402. **42.** Reed & Yeaman (1987) *The Enzymes*, 3rd ed., **18**, 77. **43.** Pettit, Pelley &

Reed (1975) *Biochem. Biophys. Res. Commun.* **65**, 575. **44.** Danson, Black, Woodland & Wood (1985) *FEBS Lett.* **179**, 120. **45.** Nakagawa & Ruegamer (1967) *Biochemistry* **6**, 1249. **46.** Davidson (1966) *Meth. Enzymol.* **8**, 277. **47.** Maxwell (1957) *J. Biol. Chem.* **229**, 139. **48.** Maxwell (1961) *The Enzymes*, 2nd ed., **5**, 443. **49.** Geren, Geren & Ebner (1977) *J. Biol. Chem.* **252**, 2089. **50.** Vanhooke & Frey (1994) *J. Biol. Chem.* **269**, 31496. **51.** Dey (1984) *Phytochemistry* **23**, 729. **52.** Ankel & Feingold (1966) *Meth. Enzymol.* **8**, 287. **53.** Bar-Peled, Griffith & Doering (2001) *Proc. Natl. Acad. Sci. U.S.A.* **98**, 12003. **54.** Jacobson & Payne (1982) *J. Bacteriol.* **152**, 932. **55.** Davidson (1966) *Meth. Enzymol.* **8**, 281. **56.** Jacobson & Davidson (1962) *J. Biol. Chem.* **237**, 638. **57.** Foster (1981) *J. Bacteriol.* **145**, 1002. **58.** Schuber, Pascal & Travo (1978) *Eur. J. Biochem.* **83**, 205. **59.** Wilson & Fox (1974) *J. Biol. Chem.* **249**, 5586. **60.** Luthje & Ogilvie (1985) *Eur. J. Biochem.* **149**, 119. **61.** Rahmatullah & Roche (1988) *J. Biol. Chem.* **263**, 8106. **62.** Roche & Cate (1977) *Arch. Biochem. Biophys.* **183**, 664. **63.** Reed, Damuni & Merryfield (1985) *Curr. Top. Cell. Regul.* **27**, 41. **64.** Fujita & Freese (1979) *J. Biol. Chem.* **254**, 5340. **65.** Amachi & Bowien (1979) *J. Gen. Microbiol.* **113**, 347. **66.** Chakrabarty & Stinson (1983) *Biochim. Biophys. Acta* **839**, 174. **67.** Lee & Schulz (1979) *J. Biol. Chem.* **254**, 4516. **68.** Alexson & Nedergard (1988) *J. Biol. Chem.* **263**, 13564. **69.** Alexson, Svensson & Nedergaard (1989) *Biochim. Biophys. Acta* **1005**, 13. **70.** Bernson (1976) *Eur. J. Biochem.* **67**, 403. **71.** Engleman & Francis (1978) *Arch. Biochem. Biophys.* **191**, 602. **72.** Fox & Kelley (1972) *J. Biol. Chem.* **247**, 2126. **73.** Storey (1976) *J. Biol. Chem.* **251**, 7810. **74.** Storey (1977) *Arch. Biochem. Biophys.* **179**, 518. **75.** Takeuchi, Tsubouchi & Shibata (1985) *Biochem. J.* **227**, 537. **76.** Chung (1967) *J. Biol. Chem.* **242**, 1182. **77.** Blomquist (1973) *J. Biol. Chem.* **248**, 7044. **78.** Kawai, Mori, Mukai, Hashimoto & Murata (2001) *Eur. J. Biochem.* **268**, 4359. **79.** Muto (1983) *Z. Pflanzenphysiol.* **109**, 385. **80.** Grose, Joss, Velick & Roth (2006) *Proc. Natl. Acad. Sci. U.S.A.* **103**, 7601. **81.** Abid, Sasaki, Titani & Miyazaki (1997) *J. Biochem.* **121**, 769. **82.** Tait (1979) *Biochem. Soc. Trans.* **7**, 199. **83.** Yamamoto, Nagata & Kusaba (1993) *J. Biochem.* **114**, 45. **84.** Lowery & Ludden (1988) *J. Biol. Chem.* **263**, 16714. **85.** Dietrich & Muniz (1972) *Biochemistry* **11**, 1691. **86.** Hayakawa, Shibata & Iwai (1984) *Agric. Biol. Chem.* **48**, 445. **87.** Uchiyama & Tabuchi (1976) *Agric. Biol. Chem.* **40**, 1411. **88.** Måhlén (1971) *Eur. J. Biochem.* **22**, 104. **89.** Tabrett & Copeland (2000) *Arch. Microbiol.* **173**, 42. **90.** Robinson, Easom, Danson & Weitzman (1983) *FEBS Lett.* **154**, 51. **91.** Danson, Black, Woodland & Wood (1985) *FEBS Lett.* **179**, 120. **92.** Higa, Massarini & Cazzulo (1978) *Can. J. Microbiol.* **24**, 215. **93.** Johnson & Hanson (1974) *Biochim. Biophys. Acta* **350**, 336. **94.** Robinson, Danson & Weitzman (1983) *Biochem. J.* **213**, 53. **95.** Mitchell, Anderson & El-Mansi (1995) *Biochem. J.* **309**, 507. **96.** Kispal & Srere (1991) *Arch. Biochem. Biophys.* **286**, 132. **97.** Otto (1986) *FEMS Microbiol. Lett.* **34**, 191. **98.** Grossebuter & Görisch (1985) *Syst. Appl. Microbiol.* **6**, 119. **99.** Löhlein-Wehrhahn, Goepfert & Eggerer (1988) *Biol. Chem. Hoppe-Seyler* **369**, 109. **100.** Nguyen, Maurus, Stokell, *et al.* (2001) *Biochemistry* **40**, 13177. **101.** Mitchell & Weitzman (1986) *J. Gen. Microbiol.* **132**, 737. **102.** Molgat, Donald & Duckworth (1992) *Arch. Biochem. Biophys.* **298**, 238. **103.** Rault-Leonardon, Atkinson, Slaughter, Moomaw & Srere (1995) *Biochemistry* **34**, 257. **104.** Stokell, Donald, Maurus, *et al.* (2003) *J. Biol. Chem.* **278**, 35435. **105.** Lam, Neway & Gaucher (1988) *Can. J. Microbiol.* **34**, 30. **106.** Nishimura, Saito & Tomita (1978) *Arch. Microbiol.* **116**, 21. **107.** Suzuki, Zahler & Emerich (1987) *Arch. Biochem. Biophys.* **254**, 272. **108.** Suzuki (1969) *Biochim. Biophys. Acta* **191**, 559. **109.** Brinsmade & Escalante-Semerena (2007) *J. Biol. Chem.* **282**, 12629. **110.** Gregus & Németi (2005) *Toxicol. Sci.* **85**, 859. **111.** White, Cacciapuoti, Fricke, *et al.* (1981) *J. Lipid Res.* **22**, 891. **112.** Fulco & Bloch (1964) *J. Biol. Chem.* **239**, 993. **113.** Schott, Staudinger & Ullrich (1971) *Hoppe-Seyler's Z. Physiol. Chem.* **352**, 1654. **114.** Ranjith, Ramana &Sasikala (2008) *Can. J. Microbiol.* **54**, 829. **115.** Paglino, Lombardo, Arcà, Rizzi & Rossi (2008) *Insect Biochem. Mol. Biol.* **38**, 871. **116.** Han, Seong, Son & Mheen (1983) *FEBS Lett.* **151**, 36.

Nadifloxacin

This topical fluoroquinolone-class antibiotic (FW = 360.38 g/mol; CAS 124858-35-1), also named (*RS*)-9-Fluoro-8-(4-hydroxy-piperidin-1-yl)-5-

methyl-1-oxo-6,7-dihydro-1*H*,5*H*-pyrido[3,2,1-*ij*]quinoline-2-carboxylate, is a broad-spectrum systemic antibacterial targeting Type II DNA topoisomerases (gyrases) and thereby blocking bacterial replication and transcription. Nadifloxacin is broadly active against aerobic Gram-positive, Gram-negative and anaerobic bacteria, including *Propionibacterium acnes* and *Staphylococcus epidermidis*. For further information on mechanism of action, *See Ciprofloxacin*

Nadolol

This orally bioavailable tetrahydronaphthalene derivative and propranolol-like drug (FW = 309.41 g/mol), also known as SQ-11725 and Corgard, is a β-adrenergic blocker used in the once daily treatment of angina and hypertension. It is a preferred β-blockers for managing cardiac patients with Long QT Syndrome (LQTS) and both shortens the QT interval and prevents ventricular arrhythmia. Nadolol is more efficacious than metoprolol in the prevention of breakthrough cardiac events, while on therapy. **Target(s):** β-adrenergic receptor (1,2); ATPase, mitochondrial, $IC_{50} = 5.3$ mM (3); calmodulin-activated Ca^{2+}-dependent ATPase (4); lipoprotein lipase, weakly inhibited (5); Na^+/K^+-exchanging ATPase (6). **1.** Heel, Brogden, Pakes, Speight & Avery (1980) *Drugs* **20**, 1. **2.** Frishman (1980) *Amer. Heart J.* **99**, 124. **3.** Almotrefi & Dzimiri (1992) *Eur. J. Pharmacol.* **215**, 231. **4.** Meltzer & Kassir (1983) *Biochim. Biophys. Acta* **755**, 452. **5.** Kihara, Kubo, Ikeda, *et al.* (1989) *Biochem. Pharmacol.* **38**, 407. **6.** Almotrefi, Basco, Moorji & Dzimiri (2001) *Can. J. Physiol. Pharmacol.* **79**, 8.

NADP⁺

This important redox coenzyme ($FW_{free-acid}$ = 744.41 g/mol), also known as nicotinamide-adenine dinucleotide phosphate and β NADP⁺, is utilized by many enzymes catalyzing oxidation/reduction reactions, acting as a hydride acceptor. Only the β-anomer is physiologically relevant; however, the α-anomer is commercially available and is frequently used in specificity and inhibitor investigations. NADP⁺ is a hygroscopic solid that is very soluble in water: it is relatively stable in slightly acidic pH (*i.e.*, pH 3–7) but will decompose in alkaline or strongly acidic conditions. The solid should be stored in a desiccator at 0°C in the dark. This coenzyme absorbs light significantly in the ultraviolet: λ_{max} at pH 7.0 is 260 nm (ε = 18000 $M^{-1}cm^{-1}$). The reduced form, NADPH, exhibits another peak at higher wavelength: λ_{max} at pH 7.0 is 339 nm (ε = 6200 $M^{-1}cm^{-1}$). **Target(s):** acetolactate synthase (56); acetyl-CoA *C*-acyltransferase (55); adenylate cyclase (1); alcohol dehydrogenase (2,3); barley nuclease (4); [branched-chain α-keto-acid dehydrogenase] kinase, *or* [3-methyl-2-oxobutanoate dehydrogenase (acetyl-transferring)] kinase (5,6,40); citrate (*si*) synthase (52,53); 3′,5′-cyclic-nucleotide phosphodiesterase (33); ecdysone 20-monooxygenase (58); FAD diphosphatase (7,8); fructose-1,6-bisphosphatase (38); glucose-1-phosphate adenylyltransferase (42,43); glucose-6-phosphate 1-epimerase (9,10); γ-glutamylcysteine synthetase (11); glutathione peroxidase (12); glycogen phosphorylase (51); *myo*-inositol oxygenase (13,59); [isocitrate dehydrogenase (NADP⁺)] kinase (39); laminaribiose phosphorylase (14,50); malic enzyme, oxaloacetate decarboxylating activity (15); D-mannitol dehydrogenase, *Lactobacillus brevis* (16); methenyltetrahydrofolate cyclohydrolase (17,18); NAD⁺:[dinitrogen-reductase] ADP-D-ribosyltransferase (47); NAD⁺ nucleosidase, *or* NADase (19,20); NADH

peroxidase (21); nicotinamidase (22,23); nicotinamide-nucleotide adenylyltransferase (44); nicotinamide phosphoribosyltransferase (48); nicotinate phosphoribosyltransferase, weakly inhibited (49); NMN nucleosidase (24); nucleotidase (37); 5′-nucleotidase (25,26); phosphate butyryltransferase (54); phosphodiesterase I, *or* 5′-exonuclease (35,36); phytoene synthase (45); prenyltransferase (27); [protein P_{II}] uridylyltransferase (41); ribose-5-phosphate isomerase, weakly inhibited (28); ribozyme, group I (29); sphingomyelin phosphodiesterase, *or* sphingomyelinase (34); squalene synthase (46); sterol 24-*C* methyltransferase (57); UDP-glucuronate decarboxylase (30); UDP-glucuronate 4-epimerase (31,32). **1.** Yang & Epstein (1983) *J. Biol. Chem.* **258**, 3750. **2.** Sund & Theorell (1963) *The Enzymes*, 2nd ed., **7**, 25. **3.** Cooper & Friedman (1958) *Biochem. Pharmacol.* **1**, 76. **4.** Sasakuma & Oleson (1979) *Phytochemistry* **18**, 1873. **5.** Paxton (1988) *Meth. Enzymol.* **166**, 313. **6.** Randle, Patston & Espinal (1987) *The Enzymes*, 3rd ed., **18**, 97. **7.** Ragab, Brightwell & Tappel (1968) *Arch. Biochem. Biophys.* **123**, 179. **8.** Mistuda, Tsuge, Tomozawa & Kawai (1970) *J. Vitaminol.* **16**, 31. **9.** Wurster & Hess (1975) *Meth. Enzymol.* **41**, 488. **10.** Wurster & Hess (1974) *Hoppe-Seyler's Z. Physiol. Chem.* **355**, 255. **11.** Davis, Balinsky, Harington & Shepherd (1973) *Biochem. J.* **133**, 667. **12.** Little, Olinescu, Reid & O'Brien (1970) *J. Biol. Chem.* **245**, 3632. **13.** Charalampous (1962) *Meth. Enzymol.* **5**, 329. **14.** Goldemberg & Maréchal (1972) *Meth. Enzymol.* **28**, 953. **15.** Kraemer, Conn & Vennesland (1951) *J. Biol. Chem.* **188**, 583. **16.** Yamanaka (1975) *Meth. Enzymol.* **41**, 138. **17.** Pelletier & MacKenzie (1994) *Biochemistry* **33**, 1900. **18.** Kirk, Chen, Imeson & Cossins (1995) *Phytochemistry* **39**, 1309. **19.** Okayama, Ueda & Hayaishi (1980) *Meth. Enzymol.* **66**, 151. **20.** Balducci & Micossi (2002) *Mol. Cell. Biochem.* **233**, 127. **21.** Badwey & Karnovsky (1979) *J. Biol. Chem.* **254**, 11530. **22.** Joshi, Calbreath & Handler (1971) *Meth. Enzymol.* **18B**, 180. **23.** Joshi & Handler (1962) *J. Biol. Chem.* **237**, 929. **24.** Foster (1981) *J. Bacteriol.* **145**, 1002. **25.** Drummond & Yamamoto (1971) *The Enzymes*, 3rd ed., **4**, 337. **26.** Madrid-Marina & Fox (1986) *J. Biol. Chem.* **261**, 444. **27.** Jones & Porter (1985) *Meth. Enzymol.* **110**, 209. **28.** Rutner (1970) *Biochemistry* **9**, 178. **29.** Kim & Park (2003) *Mol. Cell Biochem.* **252**, 285. **30.** Suzuki, Watanabe, Masumura & Kitamura (2004) *Arch. Biochem. Biophys.* **431**, 169. **31.** Gaunt, Ankel & Schutzbach (1972) *Meth. Enzymol.* **28**, 426. **32.** Gaunt, Maitra & Ankel (1974) *J. Biol. Chem.* **249**, 2366. **33.** Lim, Palanisamy & Ong (1986) *Arch. Microbiol.* **146**, 142. **34.** Testai, Landek, Goswami, Ahmed & Dawson (2004) *J. Neurochem.* **89**, 636. **35.** Luthje & Ogilvie (1985) *Eur. J. Biochem.* **149**, 119. **36.** Udvardy, Marre & Farkas (1970) *Biochim. Biophys. Acta* **206**, 392. **37.** Passariello, Schippa, Iori, *et al.* (2003) *Biochim. Biophys. Acta* **1648**, 203. **38.** Fujita & Freese (1979) *J. Biol. Chem.* **254**, 5340. **39.** Nimmo & Nimmo (1984) *Eur. J. Biochem.* **141**, 409. **40.** Paxton & Harris (1984) *Arch. Biochem. Biophys.* **231**, 48. **41.** Engleman & Francis (1978) *Arch. Biochem. Biophys.* **191**, 602. **42.** Sowokinos (1981) *Plant Physiol.* **68**, 924. **43.** Sowokinos & Preiss (1982) *Plant Physiol.* **69**, 1459. **44.** Berger, Lau, Dahlmann & Ziegler (2005) *J. Biol. Chem.* **280**, 36334. **45.** Fraser, Schuch & Bramley (2000) *Planta* **211**, 361. **46.** Sasiak & Rilling (1988) *Arch. Biochem. Biophys.* **260**, 622. **47.** Lowery & Ludden (1988) *J. Biol. Chem.* **263**, 16714. **48.** Dietrich & Muniz (1972) *Biochemistry* **11**, 1691. **49.** Hayakawa, Shibata & Iwai (1984) *Agric. Biol. Chem.* **48**, 445. **50.** Goldemberg, Maréchal & De Souza (1966) *J. Biol. Chem.* **241**, 45. **51.** Robson & Morris (1974) *Biochem. J.* **144**, 513. **52.** Robinson, Danson & Weitzman (1983) *Biochem. J.* **213**, 53. **53.** Kispal & Srere (1991) *Arch. Biochem. Biophys.* **286**, 132. **54.** Wiesenborn, Rudolph & Papoutsakis (1989) *Appl. Environ. Microbiol.* **55**, 317. **55.** Haywood, Anderson, Chu & Dawes (1988) *FEMS Microbiol. Lett.* **52**, 91. **56.** Kaushal, Pabbi & Sharma (2003) *World J. Microbiol. Biotechnol.* **19**, 487. **57.** Lin & Knoche (1976) *Phytochemistry* **15**, 683. **58.** Feyereisen & Durst (1978) *Eur. J. Biochem.* **88**, 37. **59.** Charalampous (1959) *J. Biol. Chem.* **234**, 220.

NADP⁺ 2′,3′-Dialdehyde

This modified NADP$^+$ affinity-labeling reagent (FW$_{\text{free-acid}}$ = 744.40 g/mol; *Abbreviation*: oNADP$^+$) contains two aldehyde groups that facilitate imine formation with ε-amino groups of active-site lysine residues in NADP$^+$-dependent oxidoreductases. Dialdehyde-containing nucleotides are formed by periodate oxidation, usually requiring the addition of 20-40% molar excess of sodium metaperiodate (reaction period = 30 min); unreacted periodate is deactivated by the addition of 2-3x molar excess of ethylene glycol. **Target(s):** ferredoxin:NADP$^+$ reductase (1-3); glucose-6-phosphate dehydrogenase (1,4,5); isocitrate dehydrogenase (6,7); malic enzyme (1); NADPH:cytochrome P450 reductase (8); 6 phosphogluconate dehydrogenase (1,9,10); ribozyme, group I (11). **1.** Colman (1990) *The Enzymes*, 3rd ed., **19**, 283. **2.** Chan, Carrillo & Vallejos (1985) *Arch. Biochem. Biophys.* **240**, 172. **3.** Chan & Carrillo (1984) *Arch. Biochem. Biophys.* **229**, 340. **4.** Oppenheimer & Handlon (1992) *The Enzymes*, 3rd ed., **20**, 453. **5.** White & Levy (1987) *J. Biol. Chem.* **262**, 1223. **6.** Saha & Colman (1988) *Arch. Biochem. Biophys.* **264**, 665. **7.** Mas & Colman (1983) *J. Biol. Chem.* **258**, 9332. **8.** Inano, Kurihara & Tamaoki (1988) *J. Steroid Biochem.* **29**, 227. **9.** Rippa, Signorini, Signori & Dallocchio (1975) *FEBS Lett.* **51**, 281. **10.** Rippa, Bellini, Signorini & Dallocchio (1979) *Arch. Biochem. Biophys.* **196**, 619. **11.** Kim & Park (2003) *Mol. Cell Biochem.* **252**, 285.

NADPH

This important coenzyme (FW$_{\text{free-acid}}$ = 744.42 g/mol; Photosensitive; Hygroscopic and very soluble in water), also known as reduced nicotinamide-adenine dinucleotide phosphate and β-NADPH, is a hydride-donating agent in many oxidation/reduction reactions. Store in a desiccator at 0-4°C in the dark. NADPH is destroyed at pH < 4 at room temperature but is relatively stable at alkaline pH. Note that the presence of orthophosphate buffer decreases the stability of NADPH and that NADPH is slightly less stable than NADH. This coenzyme absorbs UV light, with λ_{max} at pH 7.5 is 259 nm (ε = 16900 M^{-1}cm^{-1}) and 339 nm (ε = 6200 M^{-1}cm^{-1}). The 339-nm peak is absent in the spectra of NADP+. NADPH is also fluorescent at 460 nm when excited at 340 nm. **Target(s):** flavanone 7-*O*-β-glucosyltransferase, weakly inhibited (1); fructose-1,6-bisphosphatase (2); GDP-mannose 3,5-epimerase (3); γ-glutamylcysteine synthetase (4); glutathione peroxidase (5); glutathione synthetase, weakly inhibited (6); glycogen phosphorylase (7); [3-hydroxy-3-methylglutaryl-CoA reductase (NADPH)]-phosphatase (8); isocitrate dehydrogenase (NAD$^+$) (9); [isocitrate dehydrogenase (NADP$^+$)] kinase (10,11); [isocitrate dehydrogenase] phosphatase (10-12); 15-ketoprostaglandin 13-reductase (13); methylisocitrate lyase (14); NAD$^+$ kinase (15-23); nicotinamide phosphoribosyltransferase, weakly inhibited (24); nicotinate phosphoribosyltransferase, weakly inhibited (25); 5'-nucleotidase (26); octadecanal decarbonylase (27); phosphodiesterase I, *or* 5'-exonuclease (28); phosphoketolase (29); phosphoprotein phosphatase (7); phytoene synthase (30,31); [protein-P$_{\text{II}}$] uridylyltransferase (32); ribose-phosphate diphosphokinase, *or* phosphoribosyl-pyrophosphate synthetase (33); ribozyme, group I (34); rubredoxin:NAD$^+$ reductase (35); thyroxine deiodinase, *or* iodothyronine deiodinase (36); 2-methylcitrate synthase (37); citrate (*si*)-synthase (38-41); acetylneuraminate 4-*O*-acetyltransferase (42); phosphate butyryltransferase (43); acetyl-CoA *C*-acetyltransferase (44); acetolactate synthase (45); thymidylate synthase (FAD) (46,47); *N*- bile-acid 7α-dehydroxylase (48); deoxyhypusine monooxygenase (49); NADH peroxidase (50); rifamycin-B oxidase (51). **1.** McIntosh & Mansell (1990) *Phytochemistry* **29**, 1533. **2.** Fujita & Freese (1979) *J. Biol. Chem.* **254**, 5340. **3.** Wolucka & Van Montagu (2003) *J. Biol. Chem.* **278**, 47483. **4.** Davis, Balinsky, Harington & Shepherd (1973) *Biochem. J.* **133**, 667. **5.** Little, Olinescu, Reid & O'Brien (1970) *J. Biol. Chem.* **245**, 3632. **6.** Kato, Chihara, Nishioka, *et al.* (1987) *J. Biochem.* **101**, 207. **7.** Robson & Morris (1974) *Biochem. J.* **144**, 513. **8.** Hegardt (1986) *Adv. Protein Phosphatases* **3**, 1. **9.** Stein, Kirkman & Stein (1967) *Biochemistry* **6**, 3197. **10.** Nimmo & Nimmo (1984) *Eur. J. Biochem.* **141**, 409. **11.** Miller, Chen, Karschnia, *et al.* (2000) *J. Biol. Chem.* **275**, 833. **12.** Nimmo (1984) *Trends Biochem.*

Sci. **9**, 475. **13.** Jarabak (1982) *Meth. Enzymol.* **86**, 163. **14.** Tabuchi & Satoh (1977) *Agric. Biol. Chem.* **41**, 169. **15.** Blomquist (1980) *Meth. Enzymol.* **66**, 101. **16.** Anderson (1973) *The Enzymes*, 3rd ed., **9**, 49. **17.** Raffaelli, Finaurini, Mazzola, *et al.* (2004) *Biochemistry* **43**, 7610. **18.** Blomquist (1973) *J. Biol. Chem.* **248**, 7044. **19.** Kawai, Fukuda, Mukai & Murata (2005) *J. Biol. Chem.* **280**, 39200. **20.** Kawai, Mori, Mukai, Hashimoto & Murata (2001) *Eur. J. Biochem.* **268**, 4359. **21.** Ochiai, Mori, Kawai & Murata (2004) *Protein Expr. Purif.* **36**, 124. **22.** Muto (1983) *Z. Pflanzenphysiol.* **109**, 385. **23.** Grose, Joss, Velick & Roth (2006) *Proc. Natl. Acad. Sci. U.S.A.* **103**, 7601. **24.** Dietrich & Muniz (1972) *Biochemistry* **11**, 1691. **25.** Hayakawa, Shibata & Iwai (1984) *Agric. Biol. Chem.* **48**, 445. **26.** Madrid-Marina & Fox (1986) *J. Biol. Chem.* **261**, 444. **27.** Schneider-Belhaddad & Kolattukudy (2000) *Arch. Biochem. Biophys.* **377**, 341. **28.** Luthje & Ogilvie (1985) *Eur. J. Biochem.* **149**, 119. **29.** Whitworth & Ratledge (1977) *J. Gen. Microbiol.* **102**, 397. **30.** Fraser, Schuch & Bramley (2000) *Planta* **211**, 361. **31.** Maudinas, Bucholtz, Papastephanou, *et al.* (1977) *Arch. Biochem. Biophys.* **180**, 354. **32.** Engleman & Francis (1978) *Arch. Biochem. Biophys.* **191**, 602. **33.** Fox & Kelley (1972) *J. Biol. Chem.* **247**, 2126. **34.** Kim & Park (2003) *Mol. Cell Biochem.* **252**, 285. **35.** Ueda & Coon (1972) *J. Biol. Chem.* **247**, 5010. **36.** Nakagawa & Ruegamer (1967) *Biochemistry* **6**, 1249. **37.** Uchiyama & Tabuchi (1976) *Agric. Biol. Chem.* **40**, 1411. **38.** Johnson & Hanson (1974) *Biochim. Biophys. Acta* **350**, 336. **39.** Robinson, Danson & Weitzman (1983) *Biochem. J.* **213**, 53. **40.** Kispal & Srere (1991) *Arch. Biochem. Biophys.* **286**, 132. **41.** Grossebuter & Görisch (1985) *Syst. Appl. Microbiol.* **6**, 119. **42.** Iwersen, Dora, Kohla, Gasa & Schauer (2003) *Biol. Chem.* **384**, 1035. **43.** Wiesenborn, Rudolph & Papoutsakis (1989) *Appl. Environ. Microbiol.* **55**, 317. **44.** Nishimura, Saito & Tomita (1978) *Arch. Microbiol.* **116**, 21. **45.** Kaushal, Pabbi & Sharma (2003) *World J. Microbiol. Biotechnol.* **19**, 487. **46.** Chemyshev, Fleischmann & Kohen (2007) *Appl. Microbiol Biotechnol.* **74**, 282. **47.** Agrawal, Lesley. Kuhn & Kohen (2004) *Biochemistry* **43**, 10295. **48.** White, Cacciapuoti, Fricke, *et al.* (1981) *J. Lipid Res.* **22**, 891. **49.** Abbruzzese, Park & Folk (1986) *J. Biol. Chem.* **261**, 3085. **50.** Dolin (1975) *J. Biol. Chem.* **250**, 310. **51.** Han, Seong, Son & Mheen (1983) *FEBS Lett.* **151**, 36.

NADPH 2',3'-Dialdehyde

This modified NADPH affinity-labeling reagent (FW$_{\text{free-acid}}$ = 742.40 g/mol; *Abbreviation*: oNADPH) contains two aldehyde groups that facilitate imine formation with ε-amino groups of active-site lysine residues in NADPH-dependent oxidoreductases. Dialdehyde-containing nucleotides are formed by periodate oxidation, usually requiring the addition of 20-40% molar excess of sodium metaperiodate (reaction period = 30 min); unreacted periodate is deactivated by the addition of 2-3x molar excess of ethylene glycol. **Target(s):** isocitrate dehydrogenase (1-3); NADPH-dependent enzymes (1,4); NADPH oxidase (5,6); nitric-oxide synthase (7). **1.** Colman (1990) *The Enzymes*, 3rd ed., **19**, 283. **2.** Saha & Colman (1988) *Arch. Biochem. Biophys.* **264**, 665. **3.** Mas & Colman (1983) *J. Biol. Chem.* **258**, 9332. **4.** Colman (1997) *Meth. Enzymol.* **280**, 186. **5.** Smith, Connor, Chen & Babior (1996) *J. Clin. Invest.* **98**, 977. **6.** Smith, Curnutte & Babior (1989) *J. Biol. Chem.* **264**, 1958. **7.** Schmidt, Smith, Nakane & Murad (1992) *Biochemistry* **31**, 3243.

Nafadotride

This dopamine receptor antagonist (FW = 365.47 g/mol; CAS 149649-22-9; Soluble to 100 mM in 1eq. HCl), also named N-[(1-butyl-2-pyrrolidinyl) methyl]-4-cyano-1-methoxy-2-naphthalenecarboxamide, is a centrally active agent that preferentially targets D$_3$ receptors (K_i = 0.52 nM), with lower affinity for D$_2$ (K_i = 5 nM) and D$_4$ (K_i = 269 nM) receptors (1-3). (The dextroisomer displays 20x lower affinity at the dopamine D$_3$ receptor and reduced selectivity.) Nafadotride displaces N-[^{3}H]propylnorapomorphine accumulation *in vivo* at lower dosage and for longer periods in limbic structures, containing both dopamine D$_2$ and D$_3$ receptors than in the stratum, containing dopamine D$_2$ receptor only (1). Nafadotride also inhibits augmentation of the locomotion response to repetitive amphetamine, a finding that is consistent with the proposed model of adaptive down-regulation of D$_3$ dopamine receptor function contributing to the development of behavioral sensitization (3). **1.** Sautel, Griffon, Sokoloff, *et al.* (1995) *J. Pharmacol. Exp. Ther.* **275**, 1239. **2.** Audinot, Newman-Tancredi, Gobert, *et al.* (1998) *J. Pharmacol. Exp. Ther.* **287**, 187. **3.** Richtand, Logue, Welge, *et al.* (2000) *Brain Res.* **867**, 239.

Nafamostat

This broad-spectrum nonpeptide protease inhibitor (FW$_{free-base}$ = 347.37 g/mol; CAS 82956-11-4; Soluble to 100 mM in H$_2$O and to 100 mM in DMSO), also known as 6'-amidino-2-naphthyl p-guanidinobenzoate and futhan, targets complement components C3a and C5a at micromolar concentrations. **Target(s):** alternative-complement-pathway C3/C5 convertase (1); chymotrypsin (2); classical-complement-pathway C3/C5 convertase (3,4); coagulation factor XIIa (5,6); complement subcomponent C1r (2,7,8); complement subcomponent C1s, *or* C1s esterase (2,6-8); complement factor D (1,9); hepatocyte growth factor activator (10); plasma kallikrein, IC$_{50}$ = 3 nM (2,5,6,7); plasmin, IC$_{50}$ = 2.9 µM (2,6,7,15); thrombin, IC$_{50}$ = 1.9 µM (2,6.7,15); tissue factor-factor VIIa complex (11,12); tissue kallikrein (2); trypsin (2,6,7,13,15); tryptase, K_i = 95.3 pM (14,15). **1.** Ikari, Sakai, Hitomi & Fujii (1983) *Immunology* **49**, 685. **2.** Fujii & Hitomi (1981) *Biochim. Biophys. Acta* **661**, 342. **3.** Inagi, Miyata, Maeda, *et al.* (1991) *Immunol. Lett.* **27**, 49. **4.** Fujita, Inoue, Inagi, *et al.* (1993) *Nippon Jinzo Gakkai Shi* **35**, 393. **5.** Hitomi, Ikari & Fujii (1985) *Haemostasis* **15**, 164. **6.** Hayashi, Salomon & Hugli (2002) *Int. Immunopharmacol.* **2**, 1667. **7.** Aoyama, Ino, Ozeki, *et al.* (1984) *Jpn. J. Pharmacol.* **35**, 203. **8.** Ueda, Midorikawa, Ino, *et al.* (2000) *Inflamm. Res.* **49**, 42. **9.** Volanakis (1998) in *Handb. Proteolytic Enzymes*, p. 119, Academic Press, San Diego. **10.** Kitamura (1998) in *Handb. Proteolytic Enzymes*, p. 204, Academic Press, San Diego. **11.** Uchiba, Okajima, Murakami, Okabe & Takatsuki (1995) *Thromb. Res.* **77**, 381. **12.** Uchiba, Okajima, Abe, Okabe & Takatsuki (1994) *Thromb. Res.* **74**, 155. **13.** Ramjee, Henderson, McLoughlin & Padova (2000) *Thromb. Res.* **98**, 559. **14.** Mori, Itoh, Shinohata, *et al.* (2003) *J. Pharmacol. Sci.* **92**, 420. **15.** Hijikata-Okunomiya, Tamao, Kikumoto & Okamoto (2000) *J. Biol. Chem.* **275**, 18995.

Nafcillin

This semi-synthetic penicillin (FW$_{free-acid}$ = 414.48 g/mol), also known as 6-(2-ethoxy-1-naphthamido)penicillin, is a suicide inhibitor of β-lactamase, combining with its active-site seryl residue. **Target(s):** cytosol alanyl aminopeptidase (1); β-lactamase (2-4); neprilysin, *or* enkephalinase (5). **1.** Starnes, Szechinski & Behal (1982) *Eur. J. Biochem.* **124**, 363. **2.** Tan & Fink (1992) *Biochem. J.* **281**, 191. **3.** Fink, Behner & Tan (1987) *Biochemistry* **26**, 4248. **4.** O'Callaghan & Morris (1972) *Antimicrob.*

Agents Chemother. **2**, 442. **5.** Livingston, Smith, Sewell & Ahmed (1992) *J. Enzyme Inhib.* **6**, 165.

Nafenopin

This hypolipidemic agent (FW$_{free-acid}$ = 310.39 g/mol; CAS 3771-19-5), also known as 2-methyl-2-[p-(1,2,3,4-tetrahydro-1 naphthyl)phenoxy]propionate and SU-13,437, inhibits acetyl-CoA carboxylase. Note also that administration of nafenopin can lead to hepatocellular carcinomas in mice and rats. A low incidence of pancreatic tumors was also observed in rats. **Target(s):** acetyl-CoA carboxylase (1-4); palmitate:CoA ligase (5). **1.** Tanabe, Nakanishi, Hashimoto, *et al.* (1981) *Meth. Enzymol.* **71**, 5. **2.** Scott, Reich & Goodman (1979) *J. Biol. Chem.* **254**, 4957. **3.** Lien, Goodman & Rasmussen (1975) *Biochemistry* **14**, 2749. **4.** Rainwater & Kolattukudy (1982) *Arch. Biochem. Biophys.* **213**, 372. **5.** Roberts & Knights (1992) *Biochem. Pharmacol.* **44**, 261.

Naftifine

This antifungal agent (FW$_{free-base}$ = 287.40 g/mol) inhibits cholestenol Δ-isomerase, *or* sterol Δ7,8-isomerase (1) and squalene monooxygenase, *or* squalene epoxidase (*Candida albicans* K_i = 1.1 µM, rat liver K_i = 144 µM (2-7). **1.** Moebius, Reiter, Bermoser, *et al.* (1998) *Mol. Pharmacol.* **54**, 591. **2.** Ryder, Seidl & Troke (1984) *Antimicrob. Agents Chemother.* **25**, 483. **3.** Paltauf, Daum, Zuder, *et al.* (1982) *Biochim. Biophys. Acta* **712**, 268. **4.** Petranyi, Ryder & Stutz (1984) *Science* **224**, 1239. **5.** Ryder & Dupont (1985) *Biochem. J.* **230**, 765. **6.** Favre & Ryder (1996) *Antimicrob. Agents Chemother.* **40**, 443. **7.** Ruckenstuhl, Poschenel, Possert, *et al.* (2008) *Antimicrob. Agents Chemother.* **52**, 1496.

Naftopidil

This antihypertensive agent (FW = 392.50 g/mol; CAS Registry = 1164469-60-6; Soluble to 100 mM in DMSO), also known as KT-611 and 4-(2-methoxyphenyl)-α-[(1 naphthalenyloxy)methyl]-1-piperazineethanol, is an α$_1$-adrenergic blocker, K_i = 58 nM (1-3) and a 5-HT$_{1A}$ receptor agonist (1-4). Naftopidil also inhibits the sympathetic adrenergic contraction evoked by electrical transmural stimulation in the dog mesenteric artery, and the inhibition is not relieved upon repetitive washing for 1 hour with the drug-free solution (4). **1.** Takei, Ikegaki, Shibata, Tsujimoto & Asano (1999) *Jpn. J. Pharmacol.* **79**, 447. **2.** Yamada, Suzuki, Kato, *et al.* (1992) *Life Sci.* **50**, 127. **3.** Sponer, Borbe, Muller-Beckmann, Freud & Jakob (1992) *J. Cardiovasc. Pharmacol.* **20**, 1006. **4.** Muramatsu, Yamanaka & Kigoshi (1991) *Jpn. J. Pharmacol.* **55**, 391.

Nalidixate (Nalidixic Acid)

This antibacterial agent (FW$_{free-acid}$ = 232.24 g/mol; CAS 389-08-2), systemically referred to as 1-ethyl-1,4-dihydro-7-methyl-4-oxo-1,8-naphthyridine-3-carboxylate and often abbreviated NAL, inhibits bacterial DNA biosynthesis via DNA gyrase, binding to subunit A. **Target(s):** AKR virus reverse transcriptase (1); avian myeloblastosis virus reverse transcriptase (1,2); butyryl-CoA synthetase, *or* butyrate:CoA ligase, *or* medium-chain-fatty-acyl-CoA synthetase (3,4); DNA gyrase, *or* DNA topoisomerase II (5-15); DNA topoisomerase I, *Vaccinia* (16); glycyl-tRNA synthetase (17,18); leucyl-tRNA synthetase (18); RNA-directed DNA polymerase (1,2). **1.** Sumiyoshi, Nishikawa, Watanabe & Kano (1983) *J. Gen. Virol.* **64**, 2329. **2.** Aoyama (1991) *Mol. Cell. Biochem.* **108**, 169. **3.** Kasuya, Yamaoka, Igarashi & Fukui (1998) *Biochem. Pharmacol.* **55**, 1769. **4.** Kasuya, Hiasa, Kawai, Igarashi & Fukui (2001) *Biochem. Pharmacol.* **62**, 363. **5.** Hooper (1995) *Drugs* **49**, 10. **6.** Fournier, Zhao, Lu, Drlica & Hooper (2000) *Antimicrob. Agents Chemother.* **44**, 2160. **7.** Smith, Kubo & Imamoto (1978) *Nature* **275**, 420. **8.** Gellert (1981) *The Enzymes*, 3rd ed., **14**, 345. **9.** Sugino, Peebles, Kreuzer & Cozzarelli (1977) *Proc. Natl. Acad. Sci. U.S.A.* **74**, 4767. **10.** Gellert (1981) *Ann. Rev. Biochem.* **50**, 879. **11.** Vosberg (1985) *Curr. Top. Microbiol. Immunol.* **114**, 19. **12.** Saijo, Enomoto, Hanaoka & Ui (1990) *Biochemistry* **29**, 583. **13.** Sutcliffe, Gootz & Barrett (1989) *Antimicrob. Agents Chemother.* **33**, 2027. **14.** Inoue, Sato, Fujii, *et al.* (1987) *J. Bacteriol.* **169**, 2322. **15.** Miller & Scurlock (1983) *Biochem. Biophys. Res. Commun.* **110**, 694. **16.** Shaffer & Traktman (1987) *J. Biol. Chem.* **262**, 9309. **17.** Freist, Logan & Gauss (1996) *Biol. Chem. Hoppe-Seyler* **377**, 343. **18.** Wright, Nurse & Goldstein (1981) *Science* **213**, 455.

Nalbuphine

This naltrexone analogue (FW$_{HCl-Salt}$ = 357.44 g/mol; CAS 20594-83-), also known as EN-2234A Nubain®, and (–)-17-(cyclobutylmethyl)-4,5α-epoxy-morphinan-3,6α,14-triol, is a semi-synthetic opioid used commercially as an analgesic (1). It is approximately equipotent in analgesic activity to morphine on a weight basis, with an ability to induce respiratory depression that is comparable to that of morphine. Nalbuphine binds tightly to the μ-opioid receptor (K_i = 0.89 nM) and κ-opioid receptor (K_i = 2.2 nM), but displays relatively low affinity (K_i = 240 nM) for the δ-opioid receptor (2). It is a moderate-efficacy partial agonist (or mixed agonist/antagonist) of the μ-opioid receptor (IA = 47%; EC$_{50}$ = 14 nM) and as a high-efficacy partial agonist of the κ-opioid receptor (IA = 81%; EC$_{50}$ = 27 nM) (2). **1.** Elliott, Navarro & Nomof (1970) *J. Med.* **1**, 74. **2.** Peng, Knapp, Bidlack & Neumeyer (2007) *J. Med. Chem.* **50**, 2254.

Naloxegol

This pegylated μ-opioid antagonist (FW = 651.78 g/mol; CAS 854601-70-0), also named (5α,6α)-4,5-epoxy-6-(3,6,9,12,15,18,21-heptaoxadocos-1-yloxy)-17-(2-propen-1-yl)morphinan-3,14-diol, Movantik®, and Moventig®, is a peripherally-selective opioid antagonist indicated for the treatment of opioid-induced constipation. When administered along with opioid analgesics, naloxegol reduces constipation-related side effects, while maintaining comparable levels of analgesia. Significantly, pegylation reduces naloxegol's ability to cross the blood-brain barrier. **1.** Corsetti & Tack (2015) *Expert Opin. Pharmacother.* **16**, 399.

Naloxol

α-Naloxol   β-Naloxol

These naloxone-like opioid antagonists (FW = 329.39 g/mol; CAS 20410-95-1 (α); 53154-12-4 (β); 58691-01-3 (α/β)) exist in two isomeric forms, α-naloxol and β-naloxol, the former a human metabolite of naloxone (1) and the β-naloxol prepared α-naloxol by a Mitsunobu reaction. α-Naloxol is a neutral μ-opioid receptor antagonist, regardless of the morphine pretreatment (3). **1.** Weinstein, Pfeffer, Schor, Indindoli & Mintz (1971) *J. Pharm. Sci.* **60**, 1567. **2.** Simon (1994). *Tetrahedron* **50**, 9757. **3.** Wang, Raehal, Bilsky & Sadée (2001) *J. Neurochem.* **77**, 1590.

Naloxone

This synthetic congener of oxymorphone (FW$_{free-base}$ = 327.38 g/mol; CAS 465-65-6; photosensitive), also known by the trade name Narcan™, is a tight-binding competitive antagonist for μ, κ, δ, and σ opioid receptors (1-3). It is used as an emergency medicine to counteract narcotic (most often heroin, morphine, oxycodone) overdose. Geary & Wooten (4) reported that sixty-one saturation experiments conducted in 13 regions of naive rat brains yielded monophasic Eadie-Hofstee plots for [³H]-naloxone binding, with a K_d of 1.87 nM and a B$_{max}$ of 0.1 pmol/mg protein. The sixty-one K_d values in naive rats described a normal distribution of regional binding affinities that may reflect the biological variation of a single high-affinity binding site. Similar studies in the morphine-dependent and precipitated withdrawal states showed no apparent changes in either the K_d or B$_{max}$ of regional [³H]-naloxone binding. Preincubation had no effect on the ability of either opiate agonist or antagonist to compete for [³H]-naloxone binding in the naive, morphine dependent or precipitated withdrawal states (4). *Note*: In 2014, the U.S. Food & Drug Administration approved a prescription treatment that can be used by family members or caregivers to treat a person known or suspected to have had an opioid overdose. Evzio™ (naloxone hydrochloride, injectable) rapidly delivers a single dose of naloxone via a hand-held auto-injector that can be carried in a pocket or stored in a medicine cabinet. Evzio is injected into the muscle (intramuscular) or under the skin (subcutaneous). Once actuated, the device provides verbal instructions on how to deliver the medication. Prior to the advent of this device, the CDC estimated that U.S. programs providing drug-users and their caregivers with take-home naloxone had already prevented ~10,000 deaths from opioid overdose. **Target(s):** μ, κ, δ, and σ opioid receptors (2-4); glucuronosyltranferase (1); **1.** Goodrich (1990) *AANA J.* **58**, 14. **2.** Sawynok, Pinsky & LaBella (1979) *Life Sci.* **25**, 1621. **3.** Fischer & Undem (1999) *Brit. J. Pharmacol.* **127**, 605. **4.** Geary & Wooten (1985) *Brain Res.* **360**, 214. **5.** Hara, Nakajima, Miyamoto & Yokoi (2007) *Drug Metab. Pharmacokinet.* **22**, 103.

Naloxone Benzoylhydrazone

This hydrazone (FW$_{free-base}$ = 445.52 g/mol), also known as 6-desoxy-6-benzoylhydrazido-*N*-allyl-14-hydroxydihydronormorphinone, is an antagonist at the ORL1 nociceptin receptor, *i.e.*, opioid receptor-like orphan receptor (1) as well as the μ opioid receptor (2). (*See Nociceptin*) Naloxone benzoylhydrazone is also a κ3 agonist. **1.** Chiou (2000) *J. Biomed. Sci.* **7**, 232. **2.** Gistrak, Paul, Hahn & Pasternak (1989) *J. Pharmacol. Exp. Ther.* **251**, 469.

Naloxone Methiodide

This quaternary ammonium naloxone derivative of (FW = 469.32 g/mol) is a nonselective antagonist at opioid receptors. It does not cross the blood-brain barrier and is used in the treatment of opioid overdoses. **1.** Milne, Coddington & Gamble (1990) *Neurosci. Lett.* **114**, 259.

Naltrexone

This naloxone analogue (FW$_{free-base}$ = 341.41 g/mol; CAS 16590-41-3) is a nonselective opioid receptor antagonist that acts at μ, κ, δ, and σ receptors (1-3). Like naloxone, it is used in cases of narcotic overdoses; however, it is longer acting. Naltrexone has also been used in the treatment of alcoholism. **Key Pharmacokinetic Parameters:** *See* Appendix II in Goodman & Gilman's THE PHARMACOLOGICAL BASIS OF THERAPEUTICS, 12th Edition (Brunton, Chabner & Knollmann, eds.) McGraw-Hill Medical, New York (2011). **1.** Powell & Holtzman (1999) *Eur. J. Pharmacol.* **377**, 21. **2.** Bienkowski, Kostowski & Koros (1999) *Eur. J. Pharmacol.* **374**, 321. **3.** Gonzalez & Brogden (1988) *Drugs* **35**, 192.

Naltriben

This naltrexone derivative (FW$_{free-base}$ = 415.49 g/mol; CAS 103429-31-8) is a selective δ$_2$ opioid receptor antagonist (1-3). Naltriben is also a noncompetitive antagonist for μ receptors and is an agonist for κ$_2$ receptors in rat cerebral cortex. Naltriben is often supplied commercially as the methanesulfonate salt. **1.** Takemori, Sultana, Nagase & Portoghese (1992) *Life Sci.* **50**, 1491. **2.** Miyamoto, Portoghese & Takemori (1993) *J. Pharmacol. Exp. Ther.* **264**, 1141. **3.** Kim, Son, Shin & Cho (2001) *Life Sci.* **68**, 1305.

L-NAME, *See* N^{ω}-*Nitro-L-arginine Methyl Ester*

Napabucasin

This orally bioavailable Stat3 inhibitor (FW = 240.21 g/mol; CAS 83280-65-3; Solubility: 10 mg/mL in DMSO; < 1 mg/mL in H$_2$O), also named BBI608 and 2-acetylnaphtho[2,3-*b*]furan-4,9-dione, targets cancer stemness by targeting gene transcription driven by Stat3 (IC$_{50}$ = 142 nM) and suppressing cancer relapse/metastasis. Other agents have much higher IC$_{50}$ values: Sunitinib, 9 μM; Gefitinib, 22 μM; Regorafenib, 16 μM; and Erlotinib, 12 μM). **1.** Li, Rogoff, Keates, *et al.* (2015) *Proc. Natl. Acad. Sci. USA* **112**, 1839.

Naphalecin

This novel antibiotic and likely protonophore (FW = 274.27 g/mol), also known as 1,6,8-trihydroxynaphthalen-2-yl)propan-1-one, isolated from a strictly Gram-positive anaerobe (Genus: *Sporotalea*; Strain SYB), is broadly active: *Rhodococcus rhodochrous* (Minimal Inhibitory Concentration ≈ 25 μg/mL), *Bacillus subtilis* (MIC ≈ 25 μg/mL), *Staphylococcus epidermidis* (MIC ≈ 25 μg/mL), *Micrococcus luteus* (MIC ≈ 3 μg/mL), *Pseudomonas fluorescens* (MIC > 100 μg/mL), *Pseudomonas putida* (MIC > 100 μg/mL), *Escherichia coli* (MIC > 100 μg/mL), *Saccharomyces cerevisiae* (MIC > 100 μg/mL), *Aspergillus niger* (MIC > 100 μg/mL). Toxicity to human cells was observed at concentrations exceeding 160 μg/mL. **1.** Ezaki, Muramatsu, Takase, Hashimoto & Nagai (2008) *J. Antibiot.* (Tokyo) **61**, 207.

Naphazoline

This vasoconstrictor (FW$_{free-base}$ = 210.28 g/mol; CAS 835-31-4), also called Privine and systematically named 2-(naphthalen-1-ylmethyl)-4,5-dihydro-1*H*-imidazole, is an imidazoline receptor agonist as well as an α-adrenergic receptor agonist. Naphazoline acts on α-receptors in the arterioles of the conjunctiva to produce constriction, resulting in decreased congestion. Its rapid action in reducing swelling, when applied to mucous membrane, commends its use in eye drop formulations, such as Clear Eyes® and Naphcon®. **Target(s):** cholinesterase (1); monoamine oxidase (2). **1.** Augustinsson (1950) *The Enzymes*, 1st. ed., **1** (Part 1), 443. **2.** Carpene, Collon, Remaury, *et al.* (1995) *J. Pharmacol. Exp. Ther.* **272**, 681.

1-Naphthaleneacetate

This aromatic acid (FW$_{free-acid}$ = 186.21 g/mol), also known as NAA and 1-naphthylacetic acid, is a synthetic plant growth regulator (auxin). It has been used to promote flowering in certain plants and induce root growth. The free acid has low water solubility (0.038 g/100g at 17°C) but is more soluble in ethanol and acetone. *Note*: Do not confuse naphthylacetate with the ester, naphthyl acetate (*See 1-Naphthyl Acetate; 2-Naphthyl Acetate*) **Target(s):** aryl sulfotransferase (1); β-glucosidase, maize (2); isochorismate synthase (3); tyrosine-ester sulfotransferase (4). **1.** Rao & Duffel (1991) *Drug Metab. Dispos.* **19**, 543. **2.** Feldwisch, Vente, Zettl, *et al.* (1994) *Biochem. J.* **302**, 15. **3.** Stalman, Koskamp, Luderer, *et al.* (2003) *J. Plant Physiol.* **160**, 607. **4.** Duffel (1994) *Chem. Biol. Interact.* **92**, 3.

α-Naphthol

This photosensitive naphthalene derivative (FW = 144.17 g/mol; M.P. = 96°C; pK_a = 9.34 at 25°C), also known as 1-naphthol, 1-naphthalenol, and 1-hydroxynaphthalene, is a precursor in the chemical synthesis of dyes, intermediates, inhibitors, and reagents. It is an alternative substrate for a number of UDP-glucuronosyltransferases and aryl sulfotransferases (phenol solfotransferases). **Target(s):** amine oxidase (1); catechol oxidase (2,3); chymotrypsin (4); cyclooxygenase (5); glutathione S-transferase (6); N-hydroxyarylamine O-acetyltransferase (7); lipoxygenase (8-10); nitronate monooxygenase (11); polyphenol oxidase (2,3). **1.** Yamada, Kumagai & Uwajima (1971) *Meth. Enzymol.* **17B**, 722. **2.** Richter (1934) *Biochem. J.* **28**, 901. **3.** Xu, Zheng, Meguro & Kawachi (2004) *J. Wood Sci.* **50**, 260. **4.** Wallace, Kurtz & Niemann (1963) *Biochemistry* **2**, 824. **5.** Pace-Asciak & Smith (1983) *The Enzymes*, 3rd ed., **16**, 543. **6.** Balabaskaran & Muniandy (1984) *Phytochemistry* **23**, 251. **7.** Yamamura, Sayama, Kakikawa, *et al.* (2000) *Biochim. Biophys. Acta* **1475**, 10. **8.** Schewe, Wiesner & Rapoport (1981) *Meth. Enzymol.* **71**, 430. **9.** Holman & Bergström (1951) *The Enzymes*, 1st ed., **2** (part 1), 559. **10.** Tappel (1963) *The Enzymes*, 2nd ed., **8**, 275. **11.** Kido, Soda & Asada (1978) *J. Biol. Chem.* **253**, 226.

β-Naphthol

This photosensitive naphthalene derivative (FW = 144.17 g/mol; CAS 90-15-3; M.P. = 121-123°C; pK_a = 9.51 at 25°C), also known as 2-naphthol, 2-naphthalenol, and 2-hydroxynaphthalene, is a precursor in the chemical synthesis of dyes, intermediates, inhibitors, and reagents. It is an alternative substrate for a number of UDP-glucuronosyltransferases and aryl sulfotransferases. **Target(s):** N-hydroxyarylamine O-acetyltransferase (1); D-3-hydroxybutyrate dehydrogenase (2); monoamine oxidase (3,4); nitronate monooxygenase (5); tyrosine-ester sulfotransferase (6). **1.** Yamamura, Sayama, Kakikawa, *et al.* (2000) *Biochim. Biophys. Acta* **1475**, 10. **2.** Gotterer (1969) *Biochemistry* **8**, 641. **3.** Yamada, Kumagai & Uwajima (1971) *Meth. Enzymol.* **17B**, 722. **4.** McEwan, Sasaki & Jones (1969) *Biochemistry* **8**, 3952. **,5.** Kido, Soda & Asada (1978) *J. Biol. Chem.* **253**, 226. **6.** Duffel & Jakoby (1981) *J. Biol. Chem.* **256**, 11123.

1,2-Naphthoquinone

This quinone, also known as β-naphthoquinone (FW = 158.16 g/mol; CAS 524-42-5; M.P. = 145 147°C), is a yellow solid that is practically insoluble in water, but dissolves upon addition of NaOH or concentrated sulfuric acid. **Target(s):** α-amylase (1); β-amylase (1); catalase (2); catechol 1,2-dioxygenase (3); ζ-crystallin, *or* NADPH:quinone oxidoreductase (4); glutathione S-transferase (5-7); α-glycerophosphatase (8); 15-hydroxyprostaglandin dehydrogenase (NAD$^+$) (9); 3α-hydroxysteroid transhydrogenase (10); indoleamine 2,3-dioxygenase (11); NAD$^+$:protein-arginine ADP-ribosyltrabsferase, *or* arginine-specific mono(ADP-ribosyl)transferase (12); nicotinate oxidase (13); papain (14); pyruvate decarboxylase (15). **1.** Owens (1953) *Contrib. Boyce Thompson Inst.* **17**, 273. **2.** Hoffmann-Ostenhof & E. Biach (1947) *Monatsh. Chem.* **76**, 319. **3.** Itoh (1981) *Agric. Biol. Chem.* **45**, 2787. **4.** Bazzi (2001) *Arch. Biochem. Biophys.* **395**, 185. **5.** Nishinaka, Yasunari, Abe, *et al.* (1993) *Curr. Eye Res.* **12**, 333. **6.** Nishinaka, Terada, Nanjo, Mizoguchi & Nishihara (1993) *Exp. Eye Res.* **56**, 299. **7.** Kulkarni, Nelson & Radulovic (1987) *Comp. Biochem. Physiol. B* **87**, 1005. **8.** Hoffmann-Ostenhof & Putz (1948) *Monatsh. Chem.* **79**, 421. **9.** Jarabak (1992) *Arch. Biochem. Biophys.* **292**, 239. **10.** Koide (1962) *Biochim. Biophys. Acta* **59**, 708. **11.** Kumar,

Malachowski, DuHadaway, *et al.* (2008) *J. Med. Chem.* **51**, 1706. **12.** Davis, Sabir, Tavassoli & Shall (1997) *Adv. Exp. Med. Biol.* **419**, 145. **13.** Pinsky & Michaelis (1952) *Biochem. J.* **52**, 33. **14.** Hoffmann-Ostenhof & Biach (1946) *Experentia* **2**, 405. **15.** Kuhn & Beinert (1947) *Ber.* **80**, 101.

1,4-Naphthoquinone

This quinone (FW = 158.16 g/mol; CAS 130-15-4; M.P. = 126°C), also called α-naphthoquinone, is a yellow solid that is sparingly soluble in water. When dissolved in aqueous NaOH, a reddish brown solution is obtained. Among the many naturally occurring 1,4-naphthoquinone derivatives are juglone, lapachol, alkannin, *etc.* (**See** *specific compound*). **Target(s):** alkaline phosphatase (1); α-amylase (2); β-amylase (2); carbonic anhydrase (4); catalase (5); catechol 2,3-dioxygenase (6); *trans*-cinnamate 4-monooxygenase (7); ζ-crystallin, *or* NADPH:quinone oxidoreductase (8); deoxyribonuclease (9); glutamate dehydrogenase (10); 3α-hydroxysteroid transhydrogenase (11); indoleamine 2,3-dioxygenase (12); NAD$^+$ ADP-ribosyltransferase, *or* poly(ADP-ribose) polymerase (13); NADH:tetrazolium oxidoreductase, *or* NADH dehydrogenase (14); NAD$^+$:protein-arginine ADP-ribosyltransferase, *or* arginine-specific mono(ADP-ribosyl)transferase (3); nicotinate oxidase (15); nitrite reductase (16); papain, weakly inhibited (17); polyphenol oxidase (18); pyruvate decarboxylase (19,20); ribonuclease H activity associated with HIV-1 reverse transcriptase (21); succinate dehydrogenase (22,23); succinate oxidase (22); urease (24); vitamin K-dependent carboxylase, *or* protein-glutamate carboxylase (25). **1.** Anderson (1961) *Biochim. Biophys. Acta* **54**, 110. **2.** Owens (1953) *Contrib. Boyce Thompson Inst.* **17**, 273. **3.** Davis, Sabir, Tavassoli & Shall (1997) *Adv. Exp. Med. Biol.* **419**, 145. **4.** Chiba, Kawai & Kondo (1954) *Bull. Res. Inst. Food Sci., Kyoto Univ.* **13**, 12. **5.** Hoffmann-Ostenhof & Biach (1947) *Monatsh. Chem.* **76**, 319. **6.** Hayaishi, Nozaki & Abbott (1975) *The Enzymes*, 3rd ed., **12**, 119. **7.** Potts, Weklych & Conn (1974) *J. Biol. Chem.* **249**, 5019. **8.** Bazzi (2001) *Arch. Biochem. Biophys.* **395**, 185. **9.** Hoffmann-Ostenhof & Frisch-Niggemeyer (1952) *Monatsh. Chem.* **83**, 1175. **10.** Ernster, Dallner & Azzone (1963) *J. Biol. Chem.* **238**, 1124. **11.** Koide (1962) *Biochim. Biophys. Acta* **59**, 708. **12.** Kumar, Malachowski, DuHadaway, *et al.* (2008) *J. Med. Chem.* **51**, 1706. **13.** Banasik, Komura, Shimoyama & Ueda (1992) *J. Biol. Chem.* **267**, 1569. **14.** Slater (1959) *Nature* **183**, 1659. **15.** Pinsky & Michaelis (1952) *Biochem. J.* **52**, 33. **16.** Medina & Nicholas (1957) *Biochim. Biophys. Acta* **25**, 138. **17.** Hoffmann-Ostenhof & Biach (1946) *Experentia* **2**, 405. **18.** Owens (1953) *Contrib. Boyce Thompson Inst.* **17**, 221. **19.** Kuhn & Beinert (1947) *Ber.* **80**, 101. **20.** Foote, Little & Sproston (1949) *J. Biol. Chem.* **181**, 481. **21.** Min, Miyashiro & Hattori (2002) *Phytother. Res.* **16** Suppl. 1, S57. **22.** Herz (1954) *Biochem. Z.* **325**, 83. **23.** Potter & DuBois (1943) *J. Gen. Physiol.* **26**, 391. **24.** Schopfer & Grob (1949) *Helv. Chim. Acta* **32**, 829. **25.** Johnson (1980) *Meth. Enzymol.* **67**, 165.

1-Naphthylacetylspermine

This synthetic spermine derivative and selective AMPAR antagonist (FW$_{free-base}$ = 370.54 g/mol; FW$_{triHCl-salt}$ = 479.91 g/mol; CAS 1049731-36-3; Tri-HCl salt is soluble to 100 mM in H$_2$O), also named Naspm and N-[3-[[4-[(3-aminopropyl)amino]butyl]amino]propyl]-1-naphthaleneacetamide, is a Joro spider toxin, JSTX (**See** *Joro Spider Toxin*) analogue that targets Ca^{2+}-permeable α-amino-3-hydroxy-5-methyl-4-isoxazolepropionic acid (*or* AMP$_A$) receptors that lack the GluR2 subunit. (AMP$_A$R's are glutamate-gated ion channels that play central roles in the mammalian brain, mediating fast excitatory synaptic transmission and facilitating several forms of synaptic plasticity.) Naspm depresses the excitatory postsynaptic currents (EPSCs) mediated by non-NMDA receptor channels (1). EPSCs in CA1 neurons after ischemia are mediated by Ca^{2+}-permeable, non-NMDA receptor-mediated conductances, and Naspm and JSTX are effective at blocking abnormal EPSCs that induce Ca^{2+} accumulation leading to delayed neuronal death after transient ischemic insult (1). In cultured rat hippocampal neurons probed by whole-cell patch clamp techniques, NASPM selectively suppressed the inwardly rectifying and Ca^{2+}-permeable AMP_A receptors expressed in type II neurons (2). When the effect of NASPM reaches a steady state, current responses induced by ionophoretic applications of kainate, a non-desensitizing agonist of AMP_A receptors, in type II neurons are suppressed by NASPM in a dose-dependent manner at -60 mV, with $IC_{50} = 0.33$ μM and a Hill coefficient of 0.94 (2). Naspm was also used to implicate GluR2-lacking AMPA receptors in the ischemia-induced rise in free Zn^{2+} and death of CA1 neurons, although a direct action at the time of the rise in Zn^{2+} remains unproven (3). **1**. Tsubokawa, Oguro, Masuzawa, Nakaima & Kawai (1995) *J. Neurophysiol.* **74**, 218. **2**. Koike, Iino & Ozawa (1997) *Neurosci. Res.* 29, 27. **3**. Noh, Yokota, Mashiko, *et al.* (2005) *Proc. Natl. Acad. Sci. U.S.A.* 102, 12230.

N^{α}-(2-Naphthylsulfonylglycyl)-4-amidino-DL-phenylalanylpiperidine

This naphthyl sulfonamine ($FW_{acetate-salt} = 581.69$ g/mol; *Abbreviation*: NAPAP), also known as 4-amidino-N^{α}-(naphthalene-2-sulfonylglycyl)-DL-phenylalanine piperidide, N^{α}-(2-naphthalenesulfonylglycyl)-4-amidino-DL-phenylalanine piperidide, and by its proprietary name Pefabloc TH, is a potent inhibitor of thrombin ($K_i = 6.5$ nM). It is a slightly weaker inhibitor of trypsin ($K_i = 330$ nM). **Target(s)**: coagulation factor Xa, weakly inhibited (1); plasmin, weakly inhibited (1,2); complement factor I (3); platelet-aggregating endopeptidase of *Bothrops jararaca* venom (4); thrombin (1,2,4-13); trypsin (1,2). **1**. Stürzebecher, Markwardt, Voigt, Wagner & Walsmann (1983) *Thromb. Res.* **29**, 635. **2**. Tapparelli, Matternich, Ehrardt, *et al.* (1993) *J. Biol. Chem.* **268**, 4734. **3**. Tsiftsoglou & Sim (2004) *J. Immunol.* **173**, 367. **4**. Serrano (1998) in *Handb. Proteolytic Enzymes*, p. 229, Academic Press, San Diego. **5**. Svendsen, Brogli, Lindeberg & Stocker (1984) *Thromb. Res.* **34**, 457. **6**. Nilsson, Sjoling-Ericksson & Deinum (1998) *J. Enzyme Inhib.* **13**, 11. **7**. Markwardt, Nowak & Hoffmann (1983) *Thromb. Haemost.* **49**, 235. **8**. Kaiser & Markwardt (1986) *Thromb. Haemost.* **55**, 194. **9**. Banner & Hadvary (1991) *J. Biol. Chem.* **266**, 20085. **10**. Reers, Koschinsky, Dickneite, *et al.* (1995) *J. Enzyme Inhib.* **9**, 61. **11**. Stürzebecher, Prasa, Wikstrom & Vieweg (1995) *J. Enzyme Inhib.* **9**, 87. **12**. De Cristofaro, Akhavan, Altomare, *et al.* (2004) *J. Biol. Chem.* **279**, 13035. **13**. Engh, Brandstetter, Sucher, *et al.* (1996) *Structure* **4**, 1353.

5'-(1-Naphthylthio)-immucillin-A

This nucleoside analogue ($FW_{free-base} = 408.50$ g/mol), also known as (1*S*)-1-(9-deazaadenin-9-yl)-5-(1-naphthylthio)-1,4-dideoxy-1,4-imino-D-ribitol,

inhibits 5'-methylthioadenosine/*S*-adenosylhomocysteine nucleosidase (K_i' = 750 pM). It is also a tight, slow-binding inhibitor of human *S*-methyl-5' thioadenosine phosphorylase ($K_i = 90$ nM). **Target(s)**: adenosyl-homocysteine nucleosidase (1); 5'-methylthioadenosine/*S*-adenosyl-homocysteine nucleosidase (1); *S*-methyl-5'-thioadenosine phosphorylase (2). **1**. Singh, Evans, Lenz, *et al.* (2005) *J. Biol. Chem.* **280**, 18265. **2**. Evans, Furneaux, Schramm, Singh & Tyler (2004) *J. Med. Chem.* **47**, 3275.

α-Naphthylthiourea

This substituted thiourea (FW = 202.28 g/mol; *Abbreviation*: ANTU) inhibits flavin-containing monooxygenases and is frequently used as a rodenticide. Although highly toxic for the adult Norway rat, this thiourea is less toxic to most other rat strains. **Target(s)**: flavin-containing monooxygenase (1-3); succinate dehydrogenase (4); tyrosinase, *or* monophenol monooxygenase, weakly inhibited (4); unspecific monooxygenase (1,2,5). **1**. Itoh, Kimura, Yokoi, Itoh & Kamataki (1993) *Biochim. Biophys. Acta* **1173**, 165. **2**. Ziegler & Mitchell (1972) *Arch. Biochem. Biophys.* **160**, 116. **3**. Chiba, Kobayashi, Itoh, *et al.* (1995) *Eur. J. Pharmacol.* **293**, 97. **4**. DuBois & Erway (1946) *J. Biol. Chem.* **165**, 711. **5**. Diaz Gomez & Castro (1986) *Arch. Toxicol.* **59**, 64.

Naproxen

This anti-inflammatory agent ($FW_{free-acid} = 230.26$ g/mol; CAS 22204-53-1), also known as ($α_S$)-6-methoxy-α-methyl-2-naphthaleneacetic acid, inhibits the cyclooxygenase activity of prostaglandin endoperoxide synthases 1 and 2. Naproxen photolysis generates radicals that result in single-strand DNA cleavage. **Primary Mode of Inhibitory Action:** Based on their kinetic behavior, nonsteroidal anti-inflammatory drugs (NSAIDs) exhibit four modes of COX inhibition: Type-1, simple competitive inhibition (*example*: ibuprofen); Type-2, weak, time-dependent binding (*example*: oxicam); Type-3, tight binding, time-dependent (*example*: indomethacin); Type-4, site-directed covalent modification (*example*: aspirin) (1). Naproxen's action resembles that of oxicam, but with mixed inhibition ($K_i \approx 10$-15 μM), with time-dependence toward COX-2. **Pharmacokinetic Properties:** When administered orally, naproxen is absorbed rapidly and virtually completely, remaining protein-bound (mainly to serum albumin) during its distribution. Its metabolism relatively simple, as is its renal excretion. The mean half-life in humans is 13 hours, nearly ideal for twice-daily administration. In man, naproxen and its only detected metabolite (demethylated at position-6) are excreted in the urine, mainly as conjugates. **Key Pharmacokinetic Parameters:** *See* Appendix II in Goodman & Gilman's *THE PHARMACOLOGICAL BASIS OF THERAPEUTICS*, 12th Edition (Brunton, Chabner & Knollmann, eds.) McGraw-Hill Medical, New York (2011). (**See also** *ATB-346, an H_2S-producing naproxen derivative*) **Target(s)**: cyclooxygenase, *or* prostaglandin-endoperoside synthase 1 and 2 (2,3); estrone sulfotransferase (4); long-chain-fatty-acyl-CoA synthetase, (5); phenol sulfotransferase (4); phosphoribosyl-aminoimidazolecarboxamide formyltransferase (6); thiopurine *S*-methyltransferase (7); tyrosine-ester sulfotransferase (8). **1**. Gierse, Koboldt, Walker, Seibert & Isakson (1999) *Biochem. J.* **339** (Part 3), 607. **2**. Barnett, Chow, Ives, *et al.* (1994) *Biochim. Biophys. Acta* **1209**, 130. **3**. Miller, Munster, Wasvary, *et al.* (1994) *Biochem. Biophys. Res. Commun.* **201**, 356. **4**. King, Ghosh & Wu (2006) *Curr. Drug Metab.* **7**, 745. **5**. Knights & Jones (1992) *Biochem. Pharmacol.* **43**, 1465. **6**. Ha, Morgan, Vaughn, Eto & Baggott (1990)

Biochem. J. **272**, 339. **7**. Oselin & Anier (2007) *Drug Metab. Dispos.* **35**, 1452. **8**. Duffel (1994) *Chem. Biol. Interact.* **92**, 3.

Napsagatran

This antithrombotic agent (FW = 558.66 g/mol; CAS 159668-20-9), also known as Ro 46-6240, is a potent inhibitor of thrombin, K_i = 0.27 nM (1-3) and a weaker inhibitor of trypsin, K_i = 1.9 μM (1,2). The ethyl ester is a weaker inhibitor. **1**. Hilpert, Ackermann, Banner, *et al.* (1994) *J. Med. Chem.* **37**, 3889. **2**. Dullweber, Stubbs, Musil, Stürzebecher & Klebe (2001) *J. Mol. Biol.* **313**, 593. **3**. Russo, Gast, Schlaeger, Angiolillo & Pietropaolo (1997) *Protein Expr. Purif.* **10**, 214.

Naptalam

This aromatic amide (FW = 291.31 g/mol; CAS 132-66-1), also known as N^1-naphthylphthalamic acid (NPA) and Alanap™, is a selective pre-emergence herbicide and anti-auxin. The application of this phytotropin to various plant tissues results in inhibition of auxin efflux carrier activity and, as a consequence, increases auxin accumulation in cells. Although the mechanism of its inhibitory action on polar auxin transport remains unclear, it appears to be mediated by a specific, naptalam binding protein. Naptalam has a low solubility in water and is hydrolyzed by strong acids and bases. **Target(s):** auxin transport (1-4); indole-3-acetate β-glucosyltransferase (5). **1**. Morris (2000) *Plant Growth Regul.* **32**, 161. **2**. Katekar & Geissler (1980) *Plant Physiol.* **66**, 1190. **3**. Rubery (1990) *Symp. Soc. Exp. Biol.* **44**, 119. **4**. Petráek, Cerná, Schwarzerová, *et al.* (2003) *Plant Physiol.* **131**, 254. **5**. Leznicki & Bandurski (1988) *Plant Physiol.* **88**, 1481.

Naringin

This flavanone glycoside (FW = 580.54 g/mol; CAS 10236-47-2), also known as naringenin 7-rhamnoglucoside, is found in a number of plants, and is the primary bitter component of the grapefruit. (**See also** *Naringenin*) **Target(s):** CYP3A4, weakly inhibited (1); flavanone 7-*O*-glucoside 2"-*O*-β-L-rhamnosyltransferase, weakly inhibited (2); β-glucuronidase, weakly inhibited (3;) [3-hydroxy-3 methylglutaryl-CoA reductase (NADPH)] kinase (4); iodide peroxidase, *or* thyroid peroxidase (5); phenol sulfotransferase, *or* aryl sulfotransferase (6,7); tyrosinase, weakly inhibited

(3); xanthine oxidase (8). **1**. Ho, Saville & Wanwimolruk (2001) *J. Pharm. Pharm. Sci.* **4**, 217. **2**. Bar-Peled, Lewinsohn, Fluhr & Gressel (1991) *J. Biol. Chem.* **266**, 20953. **3**. Rodney, Swanson, Wheeler, Smith & Worrel (1950) *J. Biol. Chem.* **183**, 739. **4**. Samari, Møller, Asmyhr & Seglen (2005) *Biochem. J.* **386**, 237. **5**. Divi & Doerge (1996) *Chem. Res. Toxicol.* **9**, 16. **6**. Walle, Eaton & Walle (1995) *Biochem. Pharmacol.* **50**, 731. **7**. Nishimuta, Ohtani, Tsujimoto, *et al.* (2007) *Biopharm. Drug Dispos.* **28**, 491. **8**. Dew, Day & Morgan (2005) *J. Agric. Food Chem.* **53**, 6510.

Navitoclax

This apoptosis-promoting antineoplastic agent (FW = 974.61 g/mol; CAS 923564-51-6; Solubility: 190 mg/mL DMSO, < 1mg/mL H_2O), also known by the developmental code name ABT-263 and structurally related to ABT-737, is a potent, orally bioavailable disruptor of Bcl-2/Bcl-X_L interactions with anti-apoptotic proteins. IC_{50} values for Bcl-X_L, Bcl-2, and Bcl-w interactions are < 0.5 nM, < 1 nM, and < 1 nM, respectively (1). Oral administration of ABT-263 alone induces complete tumor regression in xenograft models of small-cell lung cancer and acute lymphoblastic leukemia (1). Evasion of apoptosis can be achieved by overexpression of anti-apoptotic BCL2 family members. Navitoclax is a first-in-class BCL2 family inhibitor that restores the ability of cancer cells to undergo apoptosis by inhibiting protein-protein interactions. Cells that successfully resist ABT-263 possess high levels of MCL1, aBCL2 family member that is not targeted by the drug. MCL1 expression is regulated transcriptionally, translationally, and through proteasome-mediated degradation. Significantly, several recently screened micro-RNA mimics were also found to induce apoptosis, but only in the presence of ABT-263 (2). Whereas all 12 of these micro-RNAs reduced MCL1 protein expression, only 10 of them targeted MCL1 through direct binding to the 3'-untranslated region of the gene, raising the possibility that other resistance regulators of MCL1 expression may be identified using this approach (2). The pro-apoptotic action of Navitoclax is also potentiated by 2-deoxyglucose, which is transported into normal and cancer cells and subsequently undergoes phosphorylation by hexokinase (3). With malignant (*i.e.*, rapidly growing tumor cells) typically displaying glycolytic rates up to 200x faster than their normal tissues of origin, inhibition of glycolysis by 2-deoxyglucose-6-P apparently stresses cancer cells, making them more susceptible to Navitoclax. Another important finding is that Navitoclax has little effect on cell death during interphase, but strongly accelerates apoptosis during mitotic arrest (4). With reduced glycolysis in the presence of 2-deoxyglucose likely to reduce ATP stores needed for cell division and with cancer cell aneuploidy delaying mitosis by checkpoint control, Navitoclax is likely to act on an increased fraction of poorly dividing cells. Such findings also suggest Navitoclax will prove to be more effective as an antineoplastic agent when administered along with cytostatic agent(s). **1**. Tse, Shoemaker, Adickes, *et al.* (2008) *Cancer Res.* **68**, 3421. **2**. Lam, Lu, Zhang, *et al.* (2010) *Mol. Cancer Ther.* **9**, 2943. **3**. Yamaguchi, Janssen, Perkins, *et al.* (20 11) *PLoS One* **6**, e24102. **4**. Shi, Zhou, Huang & Mitchison (2011) *Cancer Res.* **71**, 4518.

NBD 556

This small-molecule CD4 mimetic (FW = 337.84 g/mol; CAS 333353-44-9; Soluble to 100 mg/mL in DMSO), also named N^1-(4-chlorophenyl)-N^2-(2,2,6,6-tetramethyl-4-piperidinyl)ethanediamide, induces structural changes in the HIV-1 envelope protein gp120 in a manner analogous to CD4, thereby blocking virus-cell and cell-cell fusion as well as inhibiting HIV-1 entry (IC$_{50}$ ≈ 2-5 mM, depending on cell type and HIV strain) in CXCR4- and CCR5-expressing cell lines (1). To initiate HIV entry into host cells, the HIV envelope protein gp120 must engage its primary receptor CD4 as well as a coreceptor (*i.e.*, CCR5 or CXCR4). Systematic studies showed this compound targets viral entry by inhibiting the binding of HIV-1 envelope glycoprotein gp120 to the cellular receptor CD4 but did not inhibit reverse transcriptase, integrase, or protease, indicating that they do not target the later stages of the HIV-1 life cycle to inhibit HIV-1 infection. This compound was equally effective as inhibitors of both X4 and R5 viruses tested in CXCR4- and CCR5-expressing cell lines, respectively, indicating that their anti-HIV-1 activity is not dependent on the coreceptor tropism of the virus. Surface plasmon resonance demonstrates that NBD 556 bind to unliganded HIV-1 gp120, but not to the cellular receptor CD4. **1**. Zhao, Ma, *et al.* (2005) *Virology* **339**, 213.

NB-598

This benzylamine derivative (FW$_{free-base}$ = 449.68 g/mol) is a strong competitive inhibitor of cholesterol biosynthesis and squalene monooxygenase, *or* squalene epoxidase: K_i = 0.68 nM for the HepG2 epoxidase. NB-598 also suppresses triacylglycerol biosynthesis. **1**. Horie, Tsuchiya, Hayashi, *et al.* (1990) *J. Biol. Chem.* **265**, 18075. **2**. Nagumo, Kamei, Sakakibara & Ono (1995) *J. Lipid Res.* **36**, 1489. **3**. Satoh, Horie, Watanabe, Tsuchiya & Kamei (1993) *Biol. Pharm. Bull.* **16**, 349.

NC-1400, *See Bevantolol*

NCI 47729-M

This non-folate dye (FW = 563.00 g/mol) inhibits (IC$_{50}$ = 3.3 μM) phosphoribosylaminoimidazolecarboxamide formyltransferase, one of the two folate-dependent enzymes in the *de novo* purine biosynthesis pathway. This formyltransferase is a promising target for anti-neoplastic chemotherapy. **1**. Xu, Li, Olson & Wilson (2004) *J. Biol. Chem.* **279**, 50555.

NCX 429

This nitric oxide- and naproxen-releasing drug (FW = 361.45 g/mol), also named (*S*)-6-(nitrooxy)hexyl-2-(6-methoxynaphthalen-2-yl)propanoate, is a new class of cyclooxygenase (COX)-inhibiting nitric oxide donors (CINODs) that combine a COX inhibitor with a nitric oxide- (NO-) donating moiety, reducing the toxicity of the non-steroidal anti-inflammatory drugs (NSAIDs), while exerting their analgesic and anti-

inflammatory effects. NSAID-induced gastroenteropathy is a significant limitation to the use of this class of drugs, and nitric oxide is an important mediator of gastric mucosal defense. (*See also Naproxen; NCX 429*) **1**. Amoruso, Fresu, Dalli, *et al.* (2015) *Life Sci.* **126**, 28.

NE-10064, *See Azimilide*

Nebivolol

This presumptive third-generation β$_1$-adrenoceptor blocker and anti-hypertensive drug (FW = 405.43 g/mol (free base) = 441.9 g/mol (as hydrochloride); CAS 99200-09-6; Solubility: 88 mg/mL DMSO; <1 mg/mL Water), also known as R-67145, Bystolic, and named systematically as 1-(6-fluorochroman-2-yl)-{[2-(6-fluorochroman-2-yl)-2-hydroxyethyl]-amino}ethanol, is a selective antagonist (IC$_{50}$ = 0.8 nM), displaying nitric oxide-potentiating vasodilatory effects (1-3). Significantly, nebivolol treatment (intravenously within 10 min of reperfusion and continued orally) reduces the myocardial apoptosis after myocardial infarction (4). **Primary Mode of Action:** Although the mechanism of nebivolol-induced NO production remains controversial, there are parallels with carvedilol, another β-blocker that displays biased agonism that is independent of G$_{αs}$ and involves G-protein-coupled receptor kinase (GRK)/β-arrestin signaling, with downstream activation of EGFR and ERK (5). In experiments with mouse embryonic fibroblasts (MEFs), a cell line solely expressing β$_2$-ARs, and HL-1 cardiac myocytes, a cell line expressing both β$_1$- and β$_2$-ARs (but no detectable β$_3$-ARs), nebivolol did not significantly alter cAMP levels, demonstrating that it cannot be a classical agonist. In both cell types, nebivolol induced rapid internalization of β-ARs, indicating that it is also not a classical β-blocker (5). Furthermore, nebivolol treatment resulted in a time-dependent phosphorylation of ERK that was indistinguishable from the action of carvedilol and similar in duration, but not amplitude, to isoproterenol. Nebivolol-mediated ERK phosphorylation was sensitive to propranolol (a non-selective β-AR-blocker) and AG1478 (an EGFR inhibitor), indicating that the signaling emanates from β-ARs and involves the EGFR. Moreover, nebivolol-mediated phosphorylation of ERK in MEFs was sensitive to pharmacological inhibition of GRK2 as well as siRNA knockdown of β-arrestin 1/2. Such findings indicate that nebivolol is a β$_2$-AR, and likely a β$_1$-AR, GRK/β-arrestin-biased agonist (5). **1**. Pauwels, *et al.* (1988) *Mol. Pharmacol.* **34**, 843. **2**. Brixius, Bundkirchen, Bölck, *et al.* (2001) *Brit. J. Pharmacol.* **133**, 1330. **3**. Brehm, Wolf, Bertsch, *et al.* (2001) *Cardiovasc. Res.* **49**, 430. **4**. Mercanoglu, Safran, Gungor, *et al.* (2008) *Circulation* **72**, 660. **5**. Erickson, Gul, Blessing, *et al.* (2013) *PLoS One* **8**, e71980.

Nebularine

This toxic and water-soluble nucleoside (FW = 252.23 g/mol; CAS 118457-14-0), also called 9-(β-D-ribofuranosyl)purine, 9-(β-D-ribosyl)purine, 9-purine ribonucleoside, and purine riboside, is a structural analogue of adenosine and inhibitor of adenosine deaminase (K_i = 8.8 μM). It has been isolated from the mushroom *Clitocybe nebularis* as well as species of *Streptomyces*. **Target(s):** adenosine deaminase (1-11); adenosine kinase, also weak alternative substrate (12); adenosine-phosphate deaminase (13); adenosylhomocysteinase, *or* S-adenosylhomocysteine hydrolase, weakly inhibited (14-17). **1**. Baker (1967) *Design of Active-Site-Directed Irreversible Enzyme Inhibitors*, Wiley, New York. **2**. Wolfenden, Sharpless & Allan (1967) *J. Biol. Chem.* **242**, 977. **3**. Cory & Suhadolnik (1965) *Biochemistry* **4**, 1729 and 1733. **4**. Evans & Wolfenden (1970) *J. Amer. Chem. Soc.* **92**, 4751. **5**. Peterson & Koch (1966) *Biochim. Biophys. Acta* **126**, 129. **6**. Wolfenden, Kaufman & Macon (1969) *Biochemistry* **8**, 2412. **7**. Philips, Robbins & Coleman (1987) *Biochemistry* **26**, 2893. **8**. Singh &

Sharma (2000) *Mol. Cell. Biochem.* **204**, 127. **9**. Harbison & Fisher (1973) *Arch. Biochem. Biophys.* **154**, 84. **10**. Rossi & Lucacchini (1974) *Biochem. Soc. Trans.* **2**, 1313. **11**. Alrokayan (2002) *Int. J. Biochem. Cell Biol.* **34**, 1608. **12**. Long & Parker (2006) *Biochem. Pharmacol.* **71**, 1671. **13**. Uchida, Narita, Masuda, Chen & Nomura (1995) *Biosci. Biotechnol. Biochem.* **59**, 2120. **14**. Corbo, Ingianna, Scacchi & Bozzi (1992) *Ann. Hum. Genet.* **56**, 35. **15**. Guranowski, Montgomery, Cantoni & Chiang (1981) *Biochemistry* **20**, 110. **16**. Shimizu, Shiozaki, Ohshiro & Yamada (1984) *Eur. J. Biochem.* **141**, 385. **17**. Bozzi, Parisi & Martini (1993) *J. Enzyme Inhib.* **7**, 159.

Nebularine 5'-Triphosphate

This nucleotide (FW$_{\text{free-acid}}$ = 492.17 g/mol), also known as 9-purine riboside 5'-triphosphate, inhibits several aminoacyl-tRNA synthetases. **Target(s):** arginyl-tRNA synthetase (1); isoleucyl-tRNA synthetase (2); leucyl-tRNA synthetase (3); lysyl-tRNA synthetase (1); threonyl-tRNA synthetase (1). **1**. Freist, Sternbach, von der Haar & Cramer (1978) *Eur. J. Biochem.* **84**, 499. **2**. Freist, von der Haar & Cramer (1981) *Eur. J. Biochem.* **119**, 151. **3**. Marutzky, Flossdorf & Kula (1976) *Nucl. Acids Res.* **3**, 2067.

Nefazodone

This antidepressant (FW$_{\text{free-base}}$ = 470.01 g/mol; CAS 83366-66-9), also known as 2-[3-[4-(3-chlorophenyl)-1-piperazinyl]propyl]-5-ethyl-2,4-dihydro-4-(2-phenoxyethyl)-3*H*-1,2,4-triazol-3-one, is a mixed serotonin 5-HT$_{2A}$ receptor antagonist and serotonin uptake inhibitor. **Target(s):** CYP1A2, moderately inhibited (1,2); CYP3A3/4, strongly inhibited (2,3); CYP2B6 (4); CYP2D6 (2,3); serotonin 5-HT$_{2A}$ receptor (5,6); serotonin uptake (5,7,8). **1**. von Moltke, Greenblatt, Duan, *et al.* (1996) *Psychopharmacology (Berlin)* **128**, 398. **2**. Owen & Nemeroff (1998) *Depress. Anxiety* **7** Suppl. 1, 24. **3**. Schmider, Greenblatt, von Moltke, Harmatz & Shader (1996) *Brit. J. Clin. Pharmacol.* **41**, 339. **4**. Hesse, Venkatakrishnan, Court, *et al.* (2000) *Drug Metab. Dispos.* **28**, 1176. **5**. Hemrick-Luecke, Snoddy & Fuller (1994) *Life Sci.* **55**, 479. **6**. Eison, Eison, Torrente, Wright & Yocca (1990) *Psychopharmacol. Bull.* **26**, 311. **7**. Bolden-Watson & Richelson (1993) *Life Sci.* **52**, 1023. **8**. Owens, Ieni, Knight, Winders & Nemeroff (1995) *Life Sci.* **57**, PL373.

Nefopam

This benzoxazocine-class analgesic and muscle relaxant (FW = 253.34 g/mol; CAS 13669-70-0), also known as Acupan®, Ajan®, Nefadol®, and Silentan®, is a centrally acting, non-opioid drug for managing moderate to severe pain, with analgesic potency similar to that of codeine phosphate (1). Nefopam is a "triple reuptake inhibitor" (*or* TRI), exhibiting the capacity to inhibit transporters responsible for the reuptake for serotonin, norepinephrine, and dopamine. Nefopam also inhibits calcium influx, cGMP formation, and NMDA receptor-dependent neurotoxicity following activation of voltage sensitive calcium channels (2). (+)-Nefopam is

substantially more potent than (–)-nefopam. **1**. Tobin & Gold (1972) *J. Clin. Pharmacol. New Drugs* **12**, 230. **2**. Novelli, Díaz-Trelles, Groppetti & Fernández-Sánchez (2005) *Amino Acids* **28**, 183.

Negamycin

This antibiotic (FW$_{\text{free-base}}$ = 248.28 g/mol; CAS 83366-66-9; Insoluble in ethanol; Soluble in water; pK_a values of 3.55, 8.10, and 9.75), isolated from strains of *Streptomyces purpeofuscus*, inhibits bacterial protein biosynthesis. **1**. Uehara, Hori & Umezawa (1976) *Biochim. Biophys. Acta* **442**, 251. **2**. Uehara, Hori & Umezawa (1974) *Biochim. Biophys. Acta* **374**, 82. **3**. Mizuno, Nitta & Umezawa (1970) *J. Antibiot.* **23**, 581. **4**. Uehara, Hori & Umezawa (1976) *Biochim. Biophys. Acta* **447**, 406.

Nelfinavir

This peptide analogue, also known as AG-1346 (FW$_{\text{free-base}}$ = 567.79 g/mol; CAS 159989-64-7), is an oral antiviral agent that inhibits human immunodeficiency virus-1 protease (K_i = 1.5 nM). The methane sulfonate salt (AG-1343) is known commercially as Viracept or Agouron. It was approved for human use in 1997. **Target(s):** candidapepsin (1); glucuronosyltransferase, *or* UDP-glucuronosyltransferase, UGT1A (1), UGT1A (3), and UGT1A4 (2); HIV-1 retropepsin (1,3-8). **1**. Tossi, Benedetti, Norbedo, *et al.* (2003) *Bioorg. Med. Chem.* **11**, 4719. **2**. Zhang, Chando, Everett, *et al.* (2005) *Drug Metab. Dispos.* **33**, 1729. **3**. Patick, Mo, Markowitz, *et al.* (1996) *Antimicrob. Agents Chemother.* **40**, 292. **4**. Clemente, Moose, Hemrajani, *et al.* (2004) *Biochemistry* **43**, 12141. **5**. Wilson, Phylip, Mills, *et al.* (1997) *Biochim. Biophys. Acta* **1339**, 113. **6**. Clemente, Hemrajani, Blum, Goodenow & Dunn (2003) *Biochemistry* **42**, 15029. **7**. Clemente, Coman, Thiaville, *et al.* (2006) *Biochemistry* **45**, 5468. **8**. Surleraux, de Kock, Verschueren, *et al.* (2005) *J. Med. Chem.* **48**, 1965.

Nematode Anticoagulant Protein c$_2$

This 85-residue polypeptide, originally isolated from the hematophagous hookworm *Ancylostoma caninum*, is a tight binding, competitive inhibitor of coagulation factor Xa ($K_d \approx$ 1 nM) and coagulation factor VIIa ($K_i \approx$ 10 pM). **1**. Stassens, Bergum, Gansemans, *et al.* (1996) *Proc. Natl. Acad. Sci. U.S.A.* **93**, 2149. **2**. Buddai, Toulokhonova, Bergum, Vlasuk & Krishnaswamy (2002) *J. Biol. Chem.* **277**, 26689.

Nemonapride

This benzamide (FW$_{\text{free-base}}$ = 387.91 g/mol; CAS 75272-39-8), also known as YM-09151-2 and *rel*-5-chloro-2-methoxy-4-(methylamino)-*N*-[(2*R*,3*R*)-2-methyl-2-(phenylmethyl)-3-pyrrolidinyl]benzamide, is a selective D$_2$ dopamine receptor antagonist (K_i = 0.1 nM), with much les affinity for D$_1$ receptors) (K_i = 740 nM). Nemonapride is also a potent SR-1A agonist with an IC$_{50}$ value of 34 nM (1,2). With potassium phosphate buffer with 25 nM spiperone, [^{3}H]-nemonapride bound with high affinity to σ-receptors, showing no affinity for D_2 dopamine or 5-HT$_{1A}$ receptors in rat brain (3). The order of pK_i values of various σ-compounds at [³H]-nemonapride binding sites and stereoisomer selectivity were consistent with previous studies using other σ-ligands. Although Scatchard analysis fitted a one-site model, competition between [³H]-nemonapride and (+)-pentazocine revealed two sites, σ-1 and σ-2 receptors, at which the K_i values of nemonapride were 8.4 nM and 9.6nM, respectively (3). Unlike clozapine and bromerguride, nemonapride exhibited marked 5-HT$_{1A}$ receptor agonist properties both *in vitro* and *in vivo* (4) **1.** Grewe, Frey, Cote & Kebabian (1982) *Eur. J. Pharmacol.* **81**, 149. **2.** Terai, Usuda, Kuroiwa, Noshiro & Maeno (1983) *Jpn. J. Pharmacol.* **33**, 749. **3.** Ujike, Akiyama & Kuroda (1996) *Neuroreport* **7**, 1057. **4.** Assié, Cosi & Koek (1997) *Eur. J. Pharmacol.* **334**, 141.

Neocitreamicin I & II

These citreamicin-class antibiotics (FW$_I$ = 825.82 g/mol; FW$_{II}$ = 653.54 g/mol), from a novel *Nocardia* strain G0655 found in a sandy soil sample collected in Falmouth, Massachusetts (USA), show good antibacterial activity against Gram-positive bacteria, including methicillin-resistant *Staphylococcus aureus* and vancomycin-resistant *Enterococcus faecalis*. Both show little or no activity against Gram-negative bacteria. All previously reported citreamicins bear a methoxy group at Position-17, where the neocitreamicins lack any substitution at the 17-position. While Position-18 is occupied by a free hydroxyl group in Neocitreamicin I, it was glycosylated in Neocitreamicin II. The fact that both show similar antibacterial activity suggests that the ring end of the structure can be modified without loss of activity. The free hydroxyl group at Position-18 makes Neocitreamicin I more amenable to medicinal chemistry efforts for the improvement of drug properties. **1.** Peoples, Zhang, Millett, Rothfeder, *et al.* (2008) *J. Antibiot.* (Tokyo) **61**, 457.

Neocuproine

This metal ion chelator (FW = 208.26 g/mol; CAS 484-11-7), also known as 2,9-dimethyl-1,10-phenanthroline, is frequently used in the determination of copper(I). **Target(s):** alcohol dehydrogenase, yeast (1,2); amine oxidase (copper-containing) (3-6); catechol oxidase (7); cysteamine dioxygenase (8); cysteine dioxygenase (9); cytosol alanyl aminopeptidase (10); 2,3-dihydroxybenzoate 2,3-dioxygenase (11,12); fructose-2,6-bisphosphate 6-phosphatase (13); histone deacetylase (14); human immunodeficiency virus type 1 (HIV-1) integrase, inhibited by the cuprous-neocuproine complex (15); indole 2,3-dioxygenase (16); laccase (22); porphobilinogen synthase, *or* δ-aminolevulinate dehydratase (17); RNA polymerase and transcription, inhibited by the cuprous-neocuproine complex (18); stizolobate synthase (19); stizolobinate synthase (19); taurine dehydrogenase (20); vellosimine reductase (21). **1.** Anderson & Reynolds (1966) *Arch. Biochem. Biophys.* **114**, 299. **2.** Brändén, Jörnvall, Eklund & Furugren (1975) *The Enzymes*, 3rd ed., **11**, 103. **3.** Yasunobu & Smith (1971) *Meth. Enzymol.* **17B**, 698. **4.** Yasunobu & Gomes (1971) *Meth. Enzymol.* **17B**, 709. **5.** Malmström, Andréasson & Reinhammar (1975) *The Enzymes*, 3rd ed., **12**, 507. **6.** Yamada & Yasunobu (1963) *J. Biol. Chem.* **238**, 2669. **7.** Pomerantz (1963) *J. Biol. Chem.* **238**, 2351. **8.** Cavallini, Duprè, Scandurra, Graziani & Cotta-Rasusino (1968) *Eur. J. Biochem.* **4**, 209. **9.** Yamaguchi, Hosokawa, Kohashi, Y. Kori, S. Sakakibara & I. Ueda (1978) *J. Biochem.*

83, 479. **10.** Garner & Behal (1974) *Biochemistry* **13**, 3227. **11.** Sharma & Vaidyanathan (1975) *Eur. J. Biochem.* **56**, 163. **12.** Sharma & Vaidyanathan (1975) *Phytochemistry* **14**, 2135. **13.** Plankert, Purwin & Holzer (1991) *Eur. J. Biochem.* **196**, 191. **14.** Hay & Candido (1983) *Biochemistry* **22**, 6175. **15.** Mazumder, Gupta, Perrin, *et al.* (1995) *AIDS Res. Hum. Retroviruses* **11**, 115. **16.** Divakar, Subramanian, Sugumaran & Vaidyanathan (1979) *Plant Sci. Lett.* **15**, 177. **17.** Komai & Neilands (1969) *Biochim. Biophys. Acta* **171**, 311. **18.** Mazumder, Perrin, McMillin & Sigman (1994) *Biochemistry* **33**, 2262. **19.** Saito & Komamine (1978) *Eur. J. Biochem.* **82**, 385. **20.** Kondo & Ishimoto (1987) *Meth. Enzymol.* **143**, 496. **21.** Pfitzner, Krausch & Stoeckigt (1984) *Tetrahedron* **40**, 1691. **22.** Lehman, Harel & Mayer (1974) *Phytochemistry* **13**, 1713.

Neo-Kyotorphin

This basic pentapeptide (FW = 653.74 g/mol; *Sequence*: TSKYR), corresponding to residues 137-141 in the α-chain of hemoglobin, was first isolated as an analgesic peptide from bovine brain. Hibernating ground squirrels (*Citellus undulatus*) produce this bioactive peptide. ITSKYR also stimulates proliferation of normal cells and tumor cells in the absence of fetal bovine serum. **Target(s):** aminopeptidase, IC$_{50}$ = 131 μM; angiotensin-converting enzyme, *or* peptidyl dipeptidase A, IC$_{50}$ = 200 μM; dipeptidyl aminopeptidase, IC$_{50}$ = 306 μM. **1.** Hazato, Kase, Ueda, Takagi & Katayama (1986) *Biochem. Int.* **12**, 379.

Neomycins

These aminoglycoside antibiotics (FW$_{free-base}$ = 908.88 g/mol; CAS 1404-04-2), first isolated from *Streptomyces fradiae* by Salman Waksman in 1949, was later shown 16S and 23S ribosomal RNA. Neomycin decreases translational fidelity, altering the error rate from its normal value of ~1:1000 to ~1:100. This effect arises because neomycin binds to the decoding region (interacting with Helix-45) of 16S rRNA, changing the conformation near nucleotide residues A1492 and A1493, thereby reducing the ability to discriminate between cognate and noncognate anticodons. Neomycin also binds to Helix-69 of the 23S rRNA in the 50S ribosome sununit. It is typically used as a topical preparation, such as Neosporin. Neomycin can also be given orally, where it is usually combined with other antibiotics. It is not absorbed from the gastrointestinal tract and has been used as a preventive measure for hepatic encephalopathy and hypercholesterolemia. By killing bacteria in the intestinal tract, it keeps ammonia levels low to prevent hepatic encephalopathy prior to GI surgery. Neomycin is not administered via injection, because it is extremely nephrotoxic (causes kidney damage). In experimental molecular biology, the cDNA for a neomycin-resistance is commonly incorporated into plasmids used to establish stable mammalian cell lines expressing a desired recombinant protein (the cDNA for which is likewise in the plasmid). Non-transfected cells die off when the culture is treated with neomycin. Many commercially available protein expression plasmids use the neomycin resistance-conferring gene as a selectable marker. **Target(s):** catalase (1); cellulose 1,4-β-cellobiosidase (2); DNA-directed DNA polymerase (Klenow polymerase) (3); DNA-directed RNA polymerase (4); DNA topoisomerase I (5,6); exoribonuclease H (7); gentamicin 3'-N-acetyltransferase, K_i = 0.4 μM (8); hammerhead ribozyme (9); ornithine decarboxylase, K_i = 1.3 mM (10); phosphoinositide 5-phosphatase (11); phosphoinositide phospholipase C, *or* phosphatidylinositol-specific phospholipase C, *or* 1-phosphatidylinositol-4,5 bisphosphate phosphodiesterase (12-16); phospholipase D (17-20); poly(A)-specific ribonuclease (3); protein biosynthesis (4,21-23); protein kinase C (24); protein-synthesizing GTPase (elongation factor) (23); ribonuclease H (7); ribonuclease P (25,26); ribozymes, group I (27); streptomycin 3''-adenylyltransferase (28). **1.** Ghatak, Saxena & Agarwala (1958) *Enzymologia* **19**, 261. **2.** Singh, Agrawal, Abidi & Darmwal (1990) *J. Gen. Appl. Microbiol.* **36**, 245. **3.** Ren, Martínez, Kirsebom & Virtanen (2002) *RNA* **8**, 1393. **4.** Chamberlin (1974) *The Enzymes*, 3rd ed., **10**, 333. **5.** Kohsaka (1989) *Agric. Biol. Chem.* **53**, 3357. **6.** Vosberg (1985) *Curr. Top. Microbiol. Immunol.* **114**, 19. **7.** Li, Barbieri, Lin, et al. (2004) *Biochemistry* **43**, 9732. **8.** Williams & Northrop (1978) *J. Biol. Chem.* **253**, 5902 and 5908. **9.** Stage, Hertel & Uhlenbeck (1995) *RNA* **1**, 95. **10.** Henley, Mahran & Schacht (1988) *Biochem. Pharmacol.* **37**, 1679. **11.** Roach & Palmer (1981) *Biochim. Biophys. Acta* **661**, 323. **12.** Lipsky & Lietman (1982) *J. Pharmacol. Exp. Ther.* **220**, 287. **13.** Schwertz, Kreisberg & Venkatachalam (1984) *J. Pharmacol. Exp. Ther.* **231**, 48. **14.** Downes & Michell (1981) *Biochem. J.* **198**, 133. **15.** Piacentini, Piatti, Fraternale, et al. (2004) *Biochimie* **86**, 343. **16.** Fang, Marchesini & Moreno (2006) *Biochem. J.* **394**, 417. **17.** Huang, Qureshi & Chen (1999) *Mol. Cell Biochem.* **197**, 195. **18.** Liscovitch, Chalifa, Danin & Eli (1991) *Biochem. J.* **279**, 319. **19.** Huang, Zhang, Liu & Chen (2000) *Mol. Cell Biochem.* **207**, 3. **20.** Kurz, Kemken, Mier, Weber & Richardt (2004) *J. Mol. Cell. Cardiol.* **36**, 225. **21.** Pestka (1974) *Meth. Enzymol.* **30**, 261. **22.** Jiménez (1976) *Trends Biochem. Sci.* **1**, 28. **23.** Campuzano & Modolell (1981) *Eur. J. Biochem.* **117**, 27. **24.** Hagiwara, Inagaki, Kanamura, Ohta & H. Hidaka (1988) *J. Pharmacol. Exp. Ther.* **244**, 355. **25.** Tekos, Tsagla, Stathopoulos & Drainas (2000) *FEBS Lett.* **485**, 71. **26.** Mikkelsen, Brannvall, Virtanen & Kirsebom (1999) *Proc. Natl. Acad. Sci. U.S.A.* **96**, 6155. **27.** von Ahsen, Davies & Schroeder (1991) *Nature* **353**, 368. **28.** Jana & Deb (2005) *Biotechnol. Lett.* **27**, 519.

Neopterin

D-*erythro*-neopterin

This fluorescent pteridinone (FW = 253.22 g/mol; CAS 670-65-5), also known as 2-amino-6-(1,2,3-trihydroxypropyl)-4(3H)-pteridinone, can exist as four possible stereoisomers, of which two have been identified in nature. The D-*erythro* isomer is a precursor to biopterin while the L-*threo* form is a growth factor for the protozoan *Crithidia fasciculata* and has also been identified in *Serratia indica*. **Target(s):** GTP cyclohydrolase I (1); NADPH-dependent superoxide-generating oxidase, *or* NADPH oxidase (2); ricin (3); xanthine oxidase (4,5). **1.** Shen, Alam & Zhang (1988) *Biochim. Biophys. Acta* **965**, 9. **2.** Kojima, Nomura, Icho, et al. (1993) *FEBS Lett.* **329**, 125. **3.** Yan, Hollis, Svinth, et al. (1997) *J. Mol. Biol.* **266**, 1043. **4.** Wede, Altindag, Widner, Wachter & Fuchs (1998) *Free Radic. Res.* **29**, 331. **5.** Watanabe, Arai, Mori, et al. (1997) *Biochem. Biophys. Res. Commun.* **233**, 447.

Neostigmine

This N,N,N-trimethylaniline derivative (FW$_{bromide}$ = 303.20 g/mol; CAS 114-80-7; photosensitive), also known as prostigmine and proserine, is a strong reversible acetyl-/butyryl- cholinesterase inhibitor that has been used as an antidote for curare and to treat myasthenia gravis. **Target(s):** acetylcholinesterase, IC$_{50}$ = 11.3 nM for electric eel enzyme (1-12,31-35); acid phosphatase (24); aryl-acylamidase (12-16); carboxylesterase, esterase (17-19,36,37); carnitine acetyltransferase (39); cholinesterase, *or* butyrylcholinesterase (1,7,12,20,21,26-30); choline sulfotransferase (38); 1,4-lactonase (25); nicotinic acetylcholine receptors (22); sodium pump, electrogenic (23). **1.** Augustinsson (1950) *The Enzymes*, 1st. ed., **1** (part 1), 443. **2.** Wilson (1960) *The Enzymes*, 2nd ed., **4**, 501. **3.** Shaw (1970) *The Enzymes*, 3rd ed., **1**, 91. **4.** Iwanaga, Kimura, Miyashita, et al. (1994) *Jpn. J. Pharmacol.* **66**, 317. **5.** Pauling & Petcher (1971) *J. Med. Chem.* **14**, 1. **6.** Froede & Wilson (1971) *The Enzymes*, 3rd ed., **5**, 87. **7.** Eadie (1942) *J. Biol. Chem.* **146**, 85. **8.** Nachmansohn, Rothenberg & Feld (1948) *J. Biol. Chem.* **174**, 247. **9.** Augustinsson & Nachmansohn (1949) *J. Biol. Chem.* **179**, 543. **10.** Wilson & Bergmann (1950) *J. Biol. Chem.* **185**, 479. **11.** Iverson & Main (1969) *Biochemistry* **8**, 1889. **12.** Costagli & Galli (1998) *Biochem. Pharmacol.* **55**, 1733. **13.** Hsu, Halaris & Freedman (1982) *Int. J. Biochem.* **14**, 581. **14.** Oommen & Balasubramanian (1979) *Eur. J. Biochem.* **94**, 135. **15.** George & Balasubramanian (1980) *Eur. J. Biochem.* **111**, 511. **16.** Hsu (1982) *Int. J. Biochem.* **14**, 1037. **17.** Bernheim (1952) *The Enzymes*, 1st ed., **2** (part 2), 844. **18.** Hofstee (1960) *The Enzymes*, 2nd ed., **4**, 485. **19.** Muftic (1954) *Enzymologia* **17**, 123. **20.** Massart (1950) *The Enzymes*, 1st ed., **1** (part 1), 307. **21.** Fitch (1963) *Biochemistry* **2**, 1221. **22.** Zheng, He, Yang & Liu (1998) *Life Sci.* **62**, 1171. **23.** Arvanov, Liou, Chang, et al. (1994) *Neuroscience* **62**, 581. **24.** Lisa, Garrido & Domenech (1984) *Mol. Cell. Biochem.* **63**, 113. **25.** Fishbein & Bessman (1966) *J. Biol. Chem.* **241**, 4835. **26.** Riov & Jaffe (1973) *Plant Physiol.* **51**, 520. **27.** Fluck & Jaffe (1975) *Biochim. Biophys. Acta* **410**, 130. **28.** Brown, Kalow, Pilz, Whittaker & Woronick (1981) *Adv. Clin. Chem.* **22**, 1. **29.** Kovalev, Rozengart & Chepkasova (2004) *J. Evol. Biochem. Physiol.* **40**, 258. **30.** Voigtmann & Uhlenbruck (1971) *Z. Naturforsch. B* **26**, 1374. **31.** Geyer, Muralidharan, Cherni, et al. (2005) *Chem. Biol. Interact.* **157-158**, 331. **32.** Joshi & Singh (2000) *Indian J. Biochem. Biophys.* **37**, 192. **33.** Kasturi & Vasantharajan (1976) *Phytochemistry* **15**, 1345. **34.** Muralidharan, Soreq & Mor (2005) *Chem. Biol. Interact.* **157-158**, 406. **35.** Sagane, Nakagawa, Yamamoto, et al. (2005) *Plant Physiol.* **138**, 1359. **36.** McGhee (1987) *Biochemistry* **26**, 4101. **37.** Tsujita & Okuda (1983) *J. Biochem.* **94**, 793. **38.** Orsi & Spencer (1964) *J. Biochem.* **56**, 81. **39.** Fritz & Schultz (1965) *J. Biol. Chem.* **240**, 2188.

Nepicastat

This potent and selective dopamine β-hydroxylase inhibitor (FW = 295.35 g/mol; CAS 173997-05-2), also known by the code names SYN117 and RS-25560-197 and by its systematic name, 5-(aminomethyl)-1-[(2S)-5,7-difluoro-1,2,3,4-tetrahydronaphthalen-2-yl]-1,3-dihydro-2H-imidazole-2-thione, inhibits the bovine and human enzymes, both with K_i values ~9 nM (1). The R-enantiomer (RS-25560-198) was approximately 2-3 times less potent than nepicastat. When tested in the artery, left ventricle and cerebral cortex of spontaneously hypertensive rats, nepicastat reduces noradrenaline content, increases dopamine content, and increases the Dopamine-to-Noradrenaline concentration ratio in a dose-dependent manner (1). Nepicastat modulates sympathetic drive to cardiovascular tissues, indicating its potential value in treating cardiovascular disorders associated with overactivation of the sympathetic nervous system, including hypertension and congestive heart failure (2). **1.** Stanley, Li, Bonhaus, et al. (1997) *Br. J. Pharmacol.* **121**, 1803. **2.** Stanley, Lee, Johnson, et al. (1998) *J. Cardiovasc. Pharmacol.* **31**, 963.

Neplanocin A

This cyclopentenyl adenosine analogue (FW = 263.26 g/mol; CAS 72877-50-0) is a mechanism-based inhibitor of *S*-adenosylhomocysteine hydrolase (1-10) and also inhibits tRNA-guanine transglycosylase (11) and vaccinia virus multiplication (10,12). **1.** Chiang (1987) *Meth. Enzymol.* **143**, 377. **2.** Ator & Ortiz de Montellano (1990) *The Enzymes*, 3rd ed., **19**, 213. **3.** Gordon, Ginalski, Rudnicki, *et al.* (2003) *Eur. J. Biochem.* **270**, 3507. **4.** Bujnicki, Prigge, Caridha & Chiang (2003) *Proteins* **52**, 624. **5.** Yang, Hu, Yin, *et al.* (2003) *Biochemistry* **42**, 1900. **6.** Wolfson, Chisholm, Tashjian, Fish & Abeles (1986) *J. Biol. Chem.* **261**, 4492. **7.** de Clercq (2005) *Nucleosides Nucleotides Nucleic Acids* **24**, 1395. **8.** Nakanishi, Iwata, Yatome & Kitade (2001) *J. Biochem.* **129**, 101. **9.** Fabianowska-Majewska, Duley & Simmonds (1994) *Biochem. Pharmacol.* **48**, 897. **10.** Borchardt, Keller & Patel-Thombre (1984) *J. Biol. Chem.* **259**, 4353. **11.** Farkas, Jacobson & Katze (1984) *Biochim. Biophys. Acta* **781**, 64. **12.** De Clercq (2001) *Clin. Microbiol. Rev.* **14**, 382.

Neral

This monoterpene aldehyde (FW = 152.24 g/mol; CAS 5392-40-5), also known as *cis*-citral (it is one of the two components of citral, is a geometric isomer of geranial and is found in many natural oils, including lemon and orange oils. Neral has also been observed in a number of insect pheromone mixtures. Upon heating, neral converts to isocitral. (*See also* Geranial) **Target(s):** aldehyde dehydrogenase (1-3); CYP1A1, inhibited by citral (4); CYP1A2, inhibited by citral (5); CYP2B4, inhibited by citral (5); retinal dehydrogenase, inhibited by citral (6,7); retinol dehydrogenase, inhibited by citral (8,9). **1.** Kikonyogo, Abriola, Dryjanski & Pietruszko (1999) *Eur. J. Biochem.* **262**, 704. **2.** Pietruszko, Abriola, Izaguirre, *et al.* (1999) *Adv. Exp. Med. Biol.* **463**, 79. **3.** Boyer & Petersen (1991) *Drug Metab. Dispos.* **19**, 81. **4.** Tomita, Okuyama, Ohnishi & Ichikawa (1996) *Biochim. Biophys. Acta* **1290**, 273. **5.** Raner, Vaz & Coon (1996) *Mol. Pharmacol.* **49**, 515. **6.** Gagnon, Duester & Bhat (2002) *Biochim. Biophys. Acta* **1596**, 156. **7.** Penzes, Wang & Napoli (1997) *Biochim. Biophys. Acta* **1342**, 175. **8.** Chen, Namkung & Juchau (1995) *Biochem. Pharmacol.* **50**, 1257. **9.** Connor & Smit (1987) *Biochem. J.* **244**, 489.

Neratinib

This potent irreversible inhibitor (FW = 557.04 g/mol; CAS 698387-09-6 and 915942-22-2 (Maleate Salt); Solubility (25°C): 2 mg/mL DMSO, <1 mg/mL Water), also known as HKI-272 and by its IUPAC name (2*E*)-*N*-[4-[[3-chloro-4-[(pyridin-2-yl)methoxy]phenyl]amino]-3-cyano-7-ethoxyquinolin-6-yl]-4-(dimethylamino)but-2-enamide, targets HER-2 (IC$_{50}$ = 59 nM)

in vitro, while showing weaker inhibition (IC$_{50}$ = 92 nM) of epidermal growth factor receptor autophos-phorylation (1). Unlike structurally related EKB-569, which potently inhibits EGFR autophosphorylation (IC$_{50}$ = 39 nM), HKI-272 possesses an anilino group that fits into a long lipophilic pocket on Her-2. A lipophilic 2-pyridinylmethyl moiety, itself placed at the para-position of the aniline, and a lipophilic chlorine atom at the meta-position allow HKI-272 to interact directly with its target, with which its form a covalent adduct between its Michael acceptor (located at the 6-position) and a nearby sulfhydryl group. This mode of inhibition, which accounts for the sustained inhibition of phosphorylation after withdrawal of drug from the medium, was confirmed by direct labeling of HER-2 by [^{14}C]HKI-272 (1). Neratinib is in development for treating early- and late-stage HER2-positive breast cancers. **1.** Rabindran, Discafani, Rosfjord (2004) *Cancer Res.* **64**, 3958. Burstein, Sun, Dirix, *et al.* (2010) *J. Clin. Oncol.* **28**, 1301.

Neridronate

This alkyl-bisphosphonate (FW = 277.15 g/mol; CAS 79778-41-9) is a slow, tight-binding inhibitor of geranyl*trans*transferase, *or* farnesyl-diphosphate synthase, suppressing bone turnover in postmenopausal women. Nitrogen-containing bisphosphonates (N-BPs) inhibit by a mechanism involving time-dependent isomerization of the enzyme. **1.** Dunford, Kwaasi, Rogers, *et al.* (2008) *J. Med. Chem.* **51**, 2187.

Nervonate (Nervonic Acid)

This very-long-chain monounsaturated acid (FW$_{free-acid}$ = 366.63 g/mol; CAS 506-37-6), also called *cis*-15-tetracosenoic acid and *cis*-15-tetraeicosenoic acid, is present in marine oils and brain lipids (*e.g.*, the cerebroside known as nervon). Nervonate is also present in milk phosphatidylcholines as well as in sphingolipids. The naturally occurring *cis*-isomer is soluble in ethanol and chloroform and has a melting point of 39-43°C. The *trans*-isomer has a melting point of 61°C. **Target(s):** chondroitin AC lyase (1); DNA polymerase β (2,3); DNA topoisomerase II (4); HIV 1 reverse transcriptase (2); hyaluronate lyase (1). **1.** Suzuki, Terasaki & Uyeda (2002) *J. Enzyme Inhib. Med. Chem.* **17**, 183. **2.** Mizushina, Yoshida, Matsukage & Sakaguchi (1997) *Biochim. Biophys. Acta* **1336**, 509. **3.** Mizushina, Ohkubo, Date, *et al.* (1999) *J. Biol. Chem.* **274**, 25599. **4.** Mizushina, Sugawara, Iida & Sakaguchi (2000) *J. Mol. Biol.* **304**, 385.

Neryl Diphosphate

This monoterpene phosphoester and *cis* isomer of geranyl diphosphate (FW$_{free-acid}$ = 314.21 g/mol), also known as neryl pyrophosphate, is a precursor for the biosynthesis of a number of cyclic monoterpenes. Neryl diphosphate is also an alternative substrate for *Z*-farnesyl diphosphate synthase and geranyl pyrophosphate:sabinene hydrate cyclase. Carbocylase can utilize either geranyl or neryl pyrophosphate as a substrate; however, the ability to utilize neryl pyrophosphate is reportedly almost completely lost with aging (1). **Target(s):** isopentenyl-diphosphate Δ-isomerase (2). 1. Rojas, Chayet, Portilla & Cori (1983) *Arch. Biochem. Biophys.* **222**, 389. 2. Banthorpe, Doonan & Gutowski (1977) *Arch. Biochem. Biophys.* **184**, 381.

Netilmicin

This semi-synthetic aminoglycoside antibiotic ($FW_{free-base}$ = 475.59 g/mol; CAS 56391-56-1) binds to the 30S ribosomal subunit and causes mRNA misreading. Except for an ethyl group on the 2-deoxystreptamine residue, netilmicin is identical to sisomicin. **Target(s):** glucose-6-phosphate dehydrogenase (1); ornithine decarboxylase, K_i = 1.7 mM (2); 6-phosphogluconate dehydrogenase (3); phospholipase A_1 (4); phospholipase A_2 (4); protein biosynthesis (5,6). 1. Ciftci, Kufrevioglu, Gundogdu & Ozmen (2000) *Pharmacol. Res.* **41**, 109. 2. Henley, Mahran & Schacht (1988) *Biochem. Pharmacol.* **37**, 1679. 3. Ciftci, Beydemir, Yilmaz & Bakan (2002) *Pol. J. Pharmacol.* **54**, 275. 4. Carlier, Laurent & Tulkens (1984) *Arch. Toxicol. Suppl.* **7**, 282. 5. Buss, Piatt & Kauten (1984) *J. Antimicrob. Chemother.* **14**, 231. 6. Loveless, Kohlhepp & Gilbert (1984) *J. Lab. Clin. Med.* **103**, 294.

Netropsin

This non-intercalating, photosensitive peptide antibiotic ($FW_{dihydrochloride}$ = 503.39 g/mol; CAS 1438-30-8), produced by *Streptomyces netropsis*, selectively binds to Adenine:Thymine base pairs within the minor groove of double-stranded B-DNA, thereby inhibiting deoxyribonuclease I. By binding in the minor groove, netropsin causes under-condensation of heterochromatic regions of metaphase chromosomes, disrupting the cell cycle, prolonging G_1, and arresting G_2. Netropsin is typically supplied as either the disulfate or dihydrochloride salt. **Target(s):** deoxyribonuclease I (1); DNA-adenine methyltransferase (2); DNA-dependent RNA polymerase (3); DNA helicase (4); DNA polymerase (5-7); DNA topoisomerase II (8,9); *Bal*I restriction endonuclease (10); *Eco*RI restriction endonuclease (10-12); *Eco*RV restriction endonuclease (10); *Hinc*II restriction endonuclease (13); *Hpa*I restriction endonuclease (13); phage T7 RNA polymerase (14); *Sgr*AI restriction endonuclease (15); reverse transcriptase, *or* RNA-directed DNA polymerase (16). 1. Zimmer, Luck & Nuske (1980) *Nucl. Acids Res.* **8**, 2999. 2. Schmidt, Reinert, Venner & Bieber (1979) *Z. Allg. Mikrobiol.* **19**, 489. 3. Puschendorf, Petersen, Wolf, Werchau & Grunicke (1971) *Biochem. Biophys. Res. Commun.* **43**, 617. 4. Brosh, Karow, White, *et al.* (2000) *Nucl. Acids Res.* **28**, 2420. 5. Philippe & Chevaillier (1976) *Biochim. Biophys. Acta* **447**, 188. 6. Fox, Popanda, Edler & Thielmann (1996) *J. Cancer Res. Clin. Oncol.* **122**, 78. 7. Zimmer,

Puschendorf, Grunicke, Chandra & Venner (1971) *Eur. J. Biochem.* **21**, 269. 8. Beerman, Woynarowski, Sigmund, *et al.* (1991) *Biochim. Biophys. Acta* **1090**, 52. 9. Poot, Hiller, Heimpel & Hoehn (1995) *Exp. Cell Res.* **218**, 326. 10. Forrow, Lee, Souhami & Hartley (1995) *Chem. Biol. Interact.* **96**, 125. 11. Goppelt, Langowski, Pingoud, *et al.* (1981) *Nucl. Acids Res.* **9**, 6115. 12. Sidorova, Gazoni & Rau (1995) *J. Biomol. Struct. Dyn.* **13**, 367. 13. Malcolm & Moffatt (1981) *Biochim. Biophys. Acta* **655**, 128. 14. Piestrzeniewicz, Studzian, Wilmanska, Plucienniczak & Gniazdowski (1998) *Acta Biochim. Pol.* **45**, 127. 15. Laue, Ankenbauer, Schmitz & Kessler (1990) *Nucl. Acids Res.* **18**, 3421. 16. Juca & Aoyama (1995) *J. Enzyme Inhib.* **9**, 171.

Neuraminate (Neuraminic Acid)

α-Neuraminate

This nine-carbon amino sugar acid (FW = 295.29 g/mol; CAS 497-43-8) is the parent compound of a family of closely related sugar acids. Note that, although free neuraminate is not typically observed in nature, many *N*- and *O*-substituted derivatives have been isolated. *N*-acetylneuraminate is the most notable example. Indeed, the influenza C virus uses 9-*O*-acetyl-*N*-acetylneuraminic acid as a high-affinity receptor determinant for attachment to cells. *N*-Acylation increases the stability of the metabolite (2). **Target(s):** exo-α-sialidase, *or* neuraminidase, *or* sialidase. 1. Martinez-Zorzano, Feijoo, de la Cadena, *et al.* (1996) *Enzyme Protein* **48**, 282. 2. Rogers, Herrler, Paulson & Klenck (1986) *J. Biol. Chem.* **261**, 5947.

Neuropeptide S

This endogenous NPSR agonist (FW_{human} = 2187.50 g/mol; Sequence: SFR NGVGTGMKKTSFQRAKS; CAS 412938-67-1; Solubility: 1 mg/mL H_2O; FW_{mouse} = 2182.47 g/mol; Sequence: SFRNGVGSGAKKTSFRRA KQ; CAS 412938-74-0; Solubility: 2 mg/mL H_2O), binds to the Neuropeptide S Receptor ($EC_{50,hNSP}$ = 9.4 nM; $EC_{50,mNSP}$ = 3 nM), mobilizing intracellular Ca^{2+}, increasing locomotor activity and wakefulness in mice, and reducing anxiety-like behavior in mice. 1. Xu *et al.* (2004) *Neuron* **43**, 487.

Neuropeptide Y

This vasoconstricting brain peptide amide (FW = 4272 g/mol; *Sequence*: YPSKPDNPGEDAPAEDMARYYSALR-HYINLITRQRY-NH_2), formerly known as melanostatin, inhibits Ca^{2+}-activated K^+ channels (1,2) and is a potent stimulator of feeding (3). Neuropeptide Y also inhibits cholecystokinin- and secretin-stimulated pancreatic secretions (1-3). The porcine sequence is identical, except for Leu-17, and the ovine peptide is identical, albeit for Asp-10 and Leu17. 1. Beech (1997) *Pharmacol. Ther.* **73**, 91. 2. Sun, Philipson & Miller (1998) *J. Pharmacol. Exp. Ther.* **284**, 625. 3. Clark, Kalra, Crowley & Kalra (1984) *Endocrinology* **115**, 427.

Neuroserpin

This 410-residue human serpin (MW = 46427 g; GenBank: CAB03626.1) inhibits tissue-type plasminogen activator and is located primarily in neurons within the central nervous system. Neuroserpin plays a role in neuronal survival following strokes by regulating the morphology of neuroendocrine cells in culture by modulating the proteolytic degradation of the extracellular matrix. Although neuroserpin binds to and inhibits tPA, uPA, and plasmin, its CNS target remains uncertain. Familial encephalopathy, a neurodegenerative disease caused by point mutations within the neuroserpin gene, is characterized by the intracellular accumulation and polymerization of mutated neuroserpin, eventually leading to neuronal death and dementia. (*For a discussion of the likely mechanism of inhibitory action,* **See** *Serpins; also α₁-Antichymotrypsin*) **Target(s):** nerve growth factor-γ (1); plasmin, weakly inhibited (1,2); thrombin, weakly inhibited (1,2); t-plasminogen activator, *or* tissue-type plasminogen activator (1-5); trypsin (1); u-plasminogen activator, *or* urokinase (1,2). 1. Hastings, Coleman, Haudenschild, *et al.* (1997) *J. Biol. Chem.* **272**, 33062. 2. Osterwalder, Cinelli, Baici, *et al.* (1998) *J. Biol. Chem.* **273**, 2312. 3. Yepes, Sandkvist, Wong, *et al.* (2000) *Blood* **96**, 569. 4. Osterwalder, Contartese, Stoeckli, Kuhn & Sonderegger (1996) *EMBO J.* **15**, 2944. 5. Yepes & Lawrence (2004) *Thromb. Haemost.* **91**, 457.

Nevirapine

This dipyridodiazepinone-based, non-nucleoside reverse transcriptase inhibitor (*or* NNRTI) and first-line drug for HIV infection (FW = 266.30 g/mol; CAS 129618-40-2; *Symbol*: NVP), also known as BI-RG-587, inhibits human immunodeficiency virus type 1 (HIV-1) reverse transcriptase, IC_{50} = 80-90 nM (1-10). Nevirapine rapidly selects for mutant human immunodeficiency virus (HIV) *in vivo*, with the most common mutation occurring at residue-181 in patients on NVP monotherapy. The X-ray structure of HIV-1 RT heterodimer with bound nevirapine reveals that NVP is almost completely buried in a pocket near, but not overlapping with, the polymerase active site (11). NVP and MgATP show strong negative cooperativity, such that binding of MgATP reduces the fraction of reverse transcriptase (RT) bound to NVP and enhances the dissociation rate constant of the NVP, resulting in an increase of the open/closed conformation of the RT fingers/thumb subdomains (12). Comparisons of the NMR spectra of three conservative mutations, Ile-63-Met, Leu-74-Met, and Leu-289-Met, indicated that Met-63 showed the greatest shift sensitivity to the addition of NVP. Protein-mediated interactions thus appear to explain most of the affinity variation of NVP for RT. Sex-dependent conversion to 12-hydroxy-nevirapine and 3-hydroxy-nevirapine may contribute to the sexual dimorphism in nevirapine toxicity (13). **1.** Merluzzi, Hargrave, Labadia, *et al.* (1990) *Science* **250**, 1411. **2.** Mui, Jacober, Hargrave & Adams (1992) *J. Med. Chem.* **35**, 201. **3.** De Clercq (1992) *AIDS Res. Human Retrovir.* **8**, 119. **4.** El Safadi, Vivet-Boudou & Marquet (2007) *Appl. Microbiol. Biotechnol.* **75**, 723. **5.** Nissley, Radzio, Ambrose, *et al.* (2007) *Biochem. J.* **404**, 151. **6.** Herschhorn, Lerman, Weitman, *et al.* (2007) *J. Med. Chem.* **50**, 2370. **7.** Quinones-Mateu, Soriano, Domingo & Menendez-Arias (1997) *Virology* **236**, 364. **8.** Hang, Li, Yang, *et al.* (2007) *Biochem. Biophys. Res. Commun.* **352**, 341. **9.** Tucker, Lumma & Culberson (1996) *Meth. Enzymol.* **275**, 440. **10.** Balzarini & De Clercq (1996) *Meth. Enzymol.* **275**, 472. **11.** Kohlstaedt, Wang, Friedman, Rice & Steitz (1992) *Science* **256**, 1783. **12.** Zheng, Mueller, DeRose & London (2013) *Biophys J.* **104**, 2695. **13.** Marinho, Rodrigues, Caixas, *et al.* (2013) *J. Antimicrob. Chemother.* **69**, 476.

NF-κB Activation Inhibitor VIII

This cell-permeable peptide (FW = 2781.5 g/mol; CAS CAS 213546-43-3; Sequence: AAVALLPAVLLALLAPVQRKRQKLMP), also known as NF-κB SN50, contains the nuclear localization sequence (NLS) of the transcription factor NF-κB p50 linked to the hydrophobic region (h-region) of the signal peptide of Kaposi fibroblast growth factor (K-FGF). SN50 inhibits NF-κB translocation (IC_{90} = 18 μM) into the nucleus in cultured endothelial and monocytic cells stimulated with lipopolysaccharide or Tumor Necrosis Factor-α, *or* TNFα). SN50 attenuates inflammation and acute respiratory distress syndrome, but the mechanisms associated with the effects of SN50 in VILI have not been fully elucidated (2). **1.** Lin, Yao, Veach, Torgerson & Hawiger (1995) *J. Biol. Chem.* **270**, 14255. **2.** Chiang, Chiang, Chuang, Liu & Tsai (2016) *Shock* **46**, 194.

NG2 Proteoglycan

This 2300-residue major chondroitin sulfate proteoglycan (MW = ~280 kDa) is produced upon spinal cord injury and expressed by macrophages and oligodendrocyte progenitors (1). It inhibits axon regeneration within the injured central nervous system, presumably by binding to growth cone receptors (Rc), via inhibition of Rho family GTP-regulatory proteins, and disruption of the actin cytoskeleton. NG proteoglycan plays a role in regulating endothelial cell proliferation and migration during motility during microvascular morphogenesis. It may also inhibit neurite outgrowth and growth cone collapse during axon regeneration. The net effect of NG2 proteoglycan is growth cone collapse/repulsion (2). **1.** Dou & Levine (1994) *J. Neurosci.* **14**, 7616. **2.** Sandvig, Berry, Barrett, *et al.* (2004) *Glia* **46**, 225.

NH125

This imidazolium salt ($FW_{iodide\text{-}salt}$ = 524.57 g/mol), also known as 1-benzyl-3-cetyl-2-methylimidazolium cation, inhibits bacterial protein-histidine kinase and eukaryotic [elongation factor 2] kinase. **Target(s):** [elongation factor-2] kinase, eukaryotic, *or* eEF-2 kinase, IC_{50} = 60 nM (1); protein-histidine kinase, IC_{50} = 6.6 μM for the *Bacillus subtilis* system (2,3); protein kinase A, *or* PKA, IC_{50} = 80 μM (1); protein kinase C, *or* PKC, IC_{50} = 7.5 μM (1). **1.** Arora, Yang, Kinzy, *et al.* (2003) *Cancer Res.* **63**, 6894. **2.** Yamamoto, Kitayama, Ishida, *et al.* (2000) *Biosci. Biotechnol. Biochem.* **64**, 919. **3.** Yamamoto, Kitayama, Minagawa, *et al.* (2001) *Biosci. Biotechnol. Biochem.* **65**, 2306.

NI 42

This orally bioavailable, high-affinity BRPF inhibitor (FW = 353.39 g/mol; CAS 1884640-99-6), also named 4-cyano-*N*-(1,2-dihydro-1,3-dimethyl-2-oxo-6-quinolinyl)benzenesulfonamide, targets bromodomain-containing proteins BRPF1B (K_d = 40 nM), BRPF2 (K_d = 0.21 μM), BRPF1B (K_d = 0.94 μM), and BRD9 (K_d = 1.13 μM), with much weaker action against forty-four other bromodomain-containing proteins. Bromodomains (the 110-residue domains that recognize acetylated lysine residues, such as those on the N-terminal tails of histones.) are epigenetic readers of histone acetylation, which factors into chromatin remodeling and transcriptional regulation. The human proteome comprises forty-six bromodomain-containing proteins with a total of sixty-one bromodomains, which, despite highly conserved structural features, recognize a wide array of natural peptide ligands. *See NI 57* Some enzymes (*e.g.*, histone acetyltransferases, HATs, *etc.*) "write" transcription markers, other enzymes (*e.g.*, histone deacetylases, HDACs, *etc.*) "erase" these signals. Still other enzymes and regulatory factors (*e.g.*, bromodomain-containing protein) "read" the state of histone acetylation.

NI 57

This potent and selective BRPF inhibitor (FW = 383.42 g/mol; λ_{max} at 210, 236, 279, 310, and 334 nm; Soluble to 100 mM in DMSO; 4-cyano-*N*-(1,3-dimethyl-2-oxo-1,2-dihydroquinolin-6-yl)-2-methoxybenzene-1-sulfon-amide, targets Bromodomain/PHD finger-containing proteins (BRPFs), binding to BRPF1B (K_d = 31 nM), BRPF2 (K_d = 108 nM), and BRPF3 (K_d = 408 nM), as determined by isothermal calorimetry. NI-57 is very selective against other non-Class IV bromodomains, including the BETs, as measured by both biophysical and biochemical methods. The bromodomain and PHD finger-containing (BRPF) proteins are scaffolding components of chromatin-binding MOZ/MORF histone acetyltransferase complexes, which are transcriptional regulators. Closest off-target effects are against BRD9 (32-fold less effective). **See:** http://www.thesgc.org/chemical-probes/NI-57/

Niacin, *See Nicotinate (Nicotinic Acid)*

Niacinamide, *See Nicotinamide*

NIBR 189

This potent and selective EBI2 receptor antagonist (FW = 429.31 g/mol; CAS 1599432-08-2; Solubility: 100 mM in DMSO), also named (2*E*)-3-(4-bromophenyl)-1-[4-4-methoxybenzoyl)-1-piperazinyl]-2-propene-1-one, targets (Binding IC_{50} = 16 nM; Functional IC_{50} = 11 nM) the G protein-coupled receptor known as EBI2 (*or* Epstein–Barr Virus-Induced Gene 2, also known as GPR183). Activation of this receptor has been shown to be critical for the adaptive immune response and has been genetically linked to autoimmune diseases, including type I diabetes. NIBR-189 also blocks the migration of the monocyte cell line U937, suggesting a functional role of the oxysterol/EBI2 pathway in these immune cells. **1.** Gessier, Preuss, Yin, *et al.* (2014) *J. Med. Chem.* **57**, 3358.

Nicardipine

This antihypertensive agent ($FW_{free-base}$ = 479.53 g/mol; CAS 55985-32-5), originally referred to as YC-93, is the first intravenously administered dihydropyridine calcium channel blocker and has been used in the treatment of angina, hypertension, and cardiovascular disorders. **Target(s):** Ca^{2+} channels, L-type voltage-dependent (1-3); 3',5'-cyclic-nucleotide phosphodiesterase, cAMP phosphodiesterase (3-5); fatty acid Δ^5-desaturase (6); phosphodiesterase 7, weakly inhibited (5). **1.** Zangar, Okita, Kim, *et al.* (1999) *J. Pharmacol. Exp. Ther.* **290**, 1436. **2.** Thellung, Florio, Villa, *et al.* (2000) *Neurobiol. Dis.* **7**, 299. **3.** Satoh, Yanagisawa & Taira (1980) *Clin. Exp. Pharmacol. Physiol.* **7**, 249. **4.** Sharma (2003) *Ind. J. Biochem. Biophys.* **40**, 77. **5.** Mukai, Asai, Naka & Tanaka (1994) *Brit. J. Pharmacol.* **111**, 389. **6.** Kawashima, Akimoto, Jareonkitmongkol, Shirasaka & Shimizu (1996) *Biosci. Biotechnol. Biochem.* **60**, 1672.

Nicergoline

This ergoline derivative (FW = 484.39 g/mol; CAS 27848-84-6), also known as 10-methoxy-1,6-dimethylergoline-8β-methanol-(5 bromonicotinate) and Sermion™, is both a cerebral and peripheral vasodilator. **Primary Mode of Action:** This alkaloid is a potent and selective α_{1A} adrenergic receptor antagonist (*in vitro* IC_{50} = 0.2 nM), with primary action of increasing arterial blood flow by vasodilation and secondarily by blocking platelet aggregation. **Target(s):** α_1-adrenergic receptor (1,2); CYP1A2, weakly inhibited (3); phospholipase A_2 (4). **1.** Alvarez-Guerra, Bertholom

& Garay (1999) *Fundam. Clin. Pharmacol.* **13**, 50. **2.** Heitz, Descombes, Miller & Stoclet (1986) *Eur. J. Pharmacol.* **123**, 279. **3.** Sanderink, Bournique, Stevens, Petry & Martinet (1997) *J. Pharmacol. Exp. Ther.* **282**, 1465. **4.** Nikolov & Koburova (1984) *Methods Find. Exp. Clin. Pharmacol.* **6**, 429.

Nickel Ions

These Group 10 (or Group VIIIA) cations (Atomic Symbol = Ni; Atomic Number = 28; Atomic Weight = 58.6934) have valencies of 0, 1, 2, or 3. The ionic radius of Ni^{2+} is 0.69 Å (compare to 0.86 Å for Mg^{2+} and 1.14 Å for Ca^{2+}). There are five stable isotopes: ^{58}Ni (natural abundance = 68.077%), ^{60}Ni (28.223%), ^{61}Ni (1.140%), ^{62}Ni (3.634%), and ^{64}Ni (0.926%). The radioisotope with the longest half-life is 7.6 x 10^4 years for ^{59}Ni. Nickel is an essential trace element for a number of plants and microorganisms as well as birds and mammals. Nickel is an important component for a number of macromolecules. Enzymes requiring nickel ions include carbon-monoxide dehydrogenase (EC 1.2.99.2), hydrogen dehydrogenase (EC 1.12.1.2), coenzyme F420 hydrogenase (EC 1.12.99.1), hydrogenase (EC 1.18.99.1), and Jack bean urease (EC 3.5.1.5). Perhaps the best known of these is urease, an enzyme that contains two atoms of Ni(II) and catalyzes the hydrolysis of urea to ammonium ion and carbamate, the latter spontaneously undergoing nonenzymatic hydrolysis to produce bicarbonate and a second molecule of ammonia. Nickel ions have also been demonstrated to stabilize ribosomes. Nickel ions inhibit numerous enzymes, but, given the low concentration of uncomplexed nickel ion in most living organisms, these effects are unlikely to be physiologically relevant. Ni^{2+} is also a contact allergen, giving rise to allergic dermatitis. In humans, nickel's toxic range begins at ~40 ppm. In the 1990s, American astrophysicist Thomas Wdowiak suggested that the dinosaurs became extinct after a nickel-laden asteroid impacted Earth, briefly raising environmental nickel to around 150 ppm.

Niclosamide

This classical anthelmintic (FW = 327.12 g/mol; CAS 50-65-7), also named 5-chloro-*N*-(2-chloro-4-nitrophenyl)-2-hydroxybenzamide and Niclocide®, inhibits glucose uptake, oxidative phosphorylation, and anaerobic metabolism in tapeworms, but is not cytotoxic other worms, such as pinworms or roundworms (1). Niclosamide undergoes metabolic hydroxylation (forming 3-hydroxyniclosamide) and glucuronidation (forming niclosamide-2-*O*-glucuronide), with CYP1A2 and UGT1A1 mainly responsible (2). Adult T-cell leukemia and lymphoma is caused by chronic infection by HTLV-1 (Human T cell Leukemia Virus Type 1), and the oncogenic viral protein, known as Tax, plays a role in trans-activating viral gene transcription and deregulating oncogenic signaling, thereby promoting survival, proliferation and transformation of virally infected T cells (4). Niclosamide induces apoptosis of HTLV-1-transformed T cells by facilitating Tax degradation in proteasomes, a process linked to MAPK/ERK1/2 and IκB inhibition, and down-regulation of Stat3 and Mcl-1, the latter a pro-survival Bcl-2 family member (3). Niclosamide also suppresses cell migration and invasion in enzalutamide-resistant prostate cancer cells via Stat3-AR axis inhibition (4). **1.** Katz (1977) *Drugs* **13**, 124. **2.** Lu, Ma, Zhang, Zhang & Wu (2015) *Xenobiotica* **11**, 1. **3.** Xiang, Yuan, Chen, *et al.* (2015) *Biochem. Biophys. Res. Commun.* **464**, 221. **4.** Liu, Lou, Armstrong, *et al.* (2015) *Prostate* **75**, 1341.

Nicotianamine

This naturally occurring azetidine, found in many plants, is a metal ion chelator and it inhibits peptidyl-dipeptidase A (IC_{50} = 0.085 μM). **Target(s):** peptidyl-dipeptidase A, *or* angiotensin I-converting enzyme (1-3); gelatinase (4). **1.** Aoyagi (2006) *Phytochemistry* **67**, 618. **2.** Kinoshita,

Yamakoshi & Kikuchi (1993) *Biosci. Biotechnol. Biochem.* **57**, 1107. **3.** Hayashi & Kimoto (2007) *J. Nutr. Sci. Vitaminol.* **53**, 331. **4.** Suzuki, Shimada, Nozoe, Tanzawa & Ogita (1996) *J. Antibiot.* **49**, 1284.

Nicotinamide

This essential redox-active constituent of pyridine dinucleotide coenzymes (FW = 122.13 g/mol; CAS 98-92-0), also known as niacinamide and 3-pyridinecarboxamide, functions as a vitamin (*i.e.*, as vitamin B₃). Nicotinamide is nutritionally equivalent to nicotinate; however, elevated nicotinamide concentrations have a different effect upon lipoprotein concentrations compared to nicotinate. The pyridinium nitrogen has a pK_a of 3.3 at 20°C. Nicotinamide is water-soluble (~1 g/mL) and stable at neutral pH, but is hydrolyzed slowly to nicotinic acid under acidic or alkaline conditions. Nicotinamide has a λ_{max} at 261.5 nm, and the molar extinction coefficient varies with pH. **Inhibition of Histone Deacetylation:** Silent information regulator 2 (Sir2) enzymes catalyze NAD⁺-dependent protein/histone deacetylation, where the acetyl group from the lysine epsilon-amino group is transferred to the ADP-ribose moiety of NAD⁺, producing nicotinamide and the novel metabolite *O*-acetyl-ADP-ribose (1). Nicotinamide inhibition is the result of nicotinamide intercepting an ADP-ribosyl-Sir2-acetyl peptide intermediate, with regeneration of NAD⁺ via transglycosidation. Nicotinamide inhibits both Sir2 and SIRT1 noncompetitively *in vitro*. Its IC₅₀ value of <50 µM is equivalent to, or better than the most effective known synthetic inhibitors of Sir2 and SIRT1. **Target(s):** acetylcholinesterase (2); alcohol dehydrogenase (3); alkane 1-monooxygenase, *or* ω-hydroxylase (4); cAMP phosphodiesterase (5); choline acetyltransferase (6); choline oxidase (7); cytochrome P450 (8,9); ecdysone 20-monooxygenase (47); glucose dehydrogenase (10-12); glucose-6 phosphate dehydrogenase (12); L-lactate dehydrogenase (12); malate dehydrogenase (13); NAD⁺ ADP ribosyltransferase, *or* poly(ADP-ribose) polymerase, product inhibition (14-17,40-46); NAD⁺ nucleosidase, *or* NADase, product inhibition (9,18-32); NAD+:protein-arginine ADP ribosyltransferase, product inhibition (33,34); nicotinate phosphoribosyltransferase (35); nucleotide diphosphatase, *or* nucleotide pyrophosphatase (36); 6-phosphogluconate dehydrogenase (17,37); tRNA methyltransferases (38); xanthine oxidase (39). acetate incorporation into free fatty acids and lipids of rat tissues (48); DNA methylation in bacteriophage *ΦX*174 (49); Aryl hydrocarbon hydroxylase, both constitutive and induced forms (50); *in vitro* production of interleukin-12 and TNF-a in peripheral whole blood of people at high risk of developing type-1 diabetes and those with newly diagnosed type 1 diabetes (51); β-NAD⁺-induced, but not cADPR-elicited, elevation of cytoplasmic Ca²⁺_free concentration and calcium ion influx through voltage-activated Ca²⁺ channels, *or* VACCs (52); sodium-dependent phosphate co-transport activity in rat small intestine (53); macrophage NO synthase, noncompetitive relative to L-arginine (K_i = 13 mM) and uncompetitive versus NADPH and tetrahydrobiopterin (54); alternative vascularization pathway observed in highly aggressive melanoma (55).

1. Jackson, Schmidt, Oppenheimer & Denu (2003) *J. Biol. Chem.* **278**, 50985. **2.** Bergmann, Wilson & Nachmansohn (1950) *J. Biol. Chem.* **186**, 693. **3.** Sund & Theorell (1963) *The Enzymes*, 2nd ed., 7, 25. **4.** Pace-Asciak & Smith (1983) *The Enzymes*, 3rd ed.; **16**, 543. **5.** Shimoyama, Kawai, Nasu, Shioji & Hoshi (1975) *Physiol. Chem. Phys.* **7**, 125. **6.** Korey, de Braganza & Nachmansohn (1951) *J. Biol. Chem.* **189**, 705. **7.** Williams, Litwack & Elvehjem (1951) *J. Biol. Chem.* **192**, 73. **8.** Hageman & Stierum (2001) *Mutat. Res.* **475**, 45. **9.** Theoharides & Kupfer (1981) *J. Biol. Chem.* **256**, 2168. **10.** Webb (1966) *Enzyme and Metabolic Inhibitors*, vol. **2**, Academic Press, New York. **11.** von Euler (1942) *Ber.* **75**, 1876. **11.** Brink (1953) *Acta Chem. Scand.* **7**, 1090. **12.** Alivisatos & Denstedt (1952) *J. Biol. Chem.* **199**, 493. **13.** Feigelson, Williams & Elvehjem (1951) *J. Biol. Chem.* **189**, 361. **14.** Nishizuka, Ueda & Hayaishi (1971) *Meth. Enzymol.* **18B**, 230. **15.** Shizuta, Ito, Nakata & Hayaishi (1980) *Meth. Enzymol.* **66**, 159. **16.** Villamil, Podesta, Molina Portela & Stoppani (2001) *Mol. Biochem. Parasitol.* **115**, 249. **17.** Dickens & Glock (1951) *Biochem. J.* **50**, 81. **18.** Okayama, Ueda & Hayaishi (1980) *Meth. Enzymol.* **66**, 151. **19.** Mann & Quastel (1941) *Biochem. J.* **35**, 502. **20.** Alivisatos, Kashket & Denstedt (1956) *Can. J. Biochem. Physiol.* **34**, 46. **21.** Hofmann & Rapaport (1957) *Biochem. Z.* **329**, 437. **22.** Hofmann (1955) *Biochem. Z.*

327, 273. **23.** Branster & Morton (1956) *Biochem. J.* **63**, 640. **24.** Kaplan, Colowick & Nason (1951) *J. Biol. Chem.* **191**, 473. **25.** Zatman, Kaplan & Colowick (1953) *J. Biol. Chem.* **200**, 197. **26.** Nakazawa, Ueda, Honjo, *et al.* (1968) *Biochem. Biophys. Res. Commun.* **32**, 143. **27.** De Wolf, van Dessel, Lagrou, Hilderson & Dierick (1985) *Biochem. J.* **226**, 415. **28.** Kim, Kim, Kim, *et al.* (1993) *Arch. Biochem. Biophys.* **305**, 147. **29.** Moser, Winterhalter & Richter (1983) *Arch. Biochem. Biophys.* **224**, 358. **30.** Zielinska, Barata & Chini (2004) *Life Sci.* **74**, 1781. **31.** Mellors, Lun & Peled (1975) *Can. J. Biochem.* **53**, 143. **32.** Green & Dobrjansky (1972) *Biochemistry* **11**, 4108. **33.** Moss, Stanley & Watkins (1980) *J. Biol. Chem.* **255**, 5838. **34.** Osborne, Stanley & Moss (1985) *Biochemistry* **24**, 5235. **35.** Blum & Kahn (1971) *Meth. Enzymol.* **18B**, 138. **36.** Kornberg & Pricer (1950) *J. Biol. Chem.* **182**, 763. **37.** Noltmann & Kuby (1963) *The Enzymes*, 2nd ed., 7, 223. **38.** Kerr & Borek (1973) *The Enzymes*, 3rd ed., **9**, 167. **39.** Murray (1971) *Meth. Enzymol.* **18B**, 210. **40.** Banasik, Komura, Shimoyama & Ueda (1992) *J. Biol. Chem.* **267**, 1569. **41.** Rankin, Jacobson, Benjamin, Moss & Jacobson (1989) *J. Biol. Chem.* **264**, 4312. **42.** Kofler, Wallraff, Herzog, *et al.* (1993) *Biochem. J.* **293**, 275. **43.** Burtscher, Klocker, Schneider, *et al.* (1987) *Biochem. J.* **248**, 859. **44.** Ushiro, Yokoyama & Shizuta (1987) *J. Biol. Chem.* **262**, 2352. **45.** Werner, Sohst, Gropp, *et al.* (1984) *Eur. J. Biochem.* **139**, 81. **46.** Furneaux & Pearson (1980) *Biochem. J.* **187**, 91. **47.** Feyereisen & Durst (1978) *Eur. J. Biochem.* **88**, 37. **48.** Velikii, Mogilevich & Simonova (1978) *Ukr. Biokhim. Zh.* **50**, 368. **49.** Razin, Goren & Friedman (1975) *Nucleic Acids Res.* **2**, 1967. **50.** Watanabe, Ariji, Takusagawa &, Konno (1975) *Gann.* **66**, 399. **51.** Kretowski, Myśliwiec, Szelachowska, Kinalski & Kinalska (2000) *Diabetes Res. Clin. Pract.* **47**, 81. **52.** Hashii, Minabe & Higashida (2000) *Biochem. J.* **345** (Part 2), 207. **53.** Katai, Tanaka, Tatsumi, *et al.* (1999) *Nephrol. Dial. Transplant.* **14**, 1195. **54.** Andrade, Ramírez, Conde, Sobrino & Bedoya (1997) *Life Sci.* **61**, 1843. **55.** Itzhaki, Greenberg, Shalmon, *et al.* (2013) *PLoS One* **8**, e57160.

Nicotinamide 2'-Deoxy-2'-fluoroarabinoside Adenine Dinucleotide

This NAD⁺ analogue (FW = 681.42 g/mol), also known as araF-NAD⁺, is a slow-binding inhibitor of NAD⁺ nucleosidase (*or* NAD⁺ glycohydrolase), K_i = 32 nM (**See *Nicotinamide Arabinoside Adenine Dinucleotide***). The kinetic data are consistent with both slow k_{on} and slow k_{off} in the formation of an enzyme-inhibitor complex (*i.e.*, the association rate constants are about 10⁴ and 10⁶ times slower than diffusion rates, respectively, for araF-NAD⁺ and ara-NAD⁺, and the half-life of the complex is about 3-10 min for both analogues. The kinetic model does not account for a slow turnover of an ADP-ribosyl-enzyme intermediary complex. **1.** Muller-Steffner, Malver, Hosie, Oppenheimer & Schuber (1992) *J. Biol. Chem.* **267**, 9606.

Nicotinamide Mononucleotide

This naturally occurring redox nucleotide (FW = 334.22 g/mol; CAS 1094-61-7; *Abbreviation*: NMN⁺ (or NMN)), is a product of the reaction catalyzed by NAD⁺ diphosphatase, ribosylnicotinamide kinase, nicotinamide phosphoribosyltransferase, and the NAD⁺-dependent DNA ligases. NMN is a substrate for nicotinamine-nucleotide adenylyltransferase (NMN adenylyltransferase), NMN nucleosidase, and nicotinamide-nucleotide amidase. It is typically stored frozen in a solution rather than as a solid. The λ_{max} value at pH 7.0 is 266 nm (ε = 4600 M⁻¹cm⁻¹). **Target(s):** ADP-ribose diphosphatase (1); DNA ligase (NAD⁺) (2); DNA polymerase I with RecBC, *or* exodeoxyribonuclease V (3); isocitrate dehydrogenase (NADP$^+$) (4,5); glutamate dehydrogenase (6,7); 3α,20β-hydroxysteroid dehydrogenase, in the presence of AMP (8); L-lactate dehydrogenase, in the presence of AMP (9); NADase (10-12); NAD$^+$ kinase, weakly inhibited (13); NAD$^+$:protein-arginine ADP-ribosyltransferase (14); nicotinamide phosphoribosyl-transferase (15); nicotinate-nucleotide adenylyltransferase (16,17); 5′-nucleotidase (18); nucleotide diphosphatase, nucleotide pyrophosphatase, weakly inhibited (19). **1.** Wu, Lennon & Suhadolnik (1978) *Biochim. Biophys. Acta* **520**, 588. **2.** Olivera (1971) *Meth. Enzymol.* **21**, 311. **3.** Pyhtila & Syvaoja (1980) *Eur. J. Biochem.* **112**, 125. **4.** Seelig & Colman (1977) *J. Biol. Chem.* **252**, 3671. **5.** Seelig & Colman (1978) *Arch. Biochem. Biophys.* **188**, 394. **6.** Smith, Austen, Blumenthal & Nyc (1975) *The Enzymes*, 3rd ed., **11**, 293. **7.** Blumenthal & Smith (1975) *J. Biol. Chem.* **250**, 6560. **8.** Sweet & Samant (1981) *Biochemistry* **20**, 5170. **9.** Holbrook, Liljas, Steindel & Rossmann (1975) *The Enzymes*, 3rd ed., **11**, 191. **10.** Okayama, Ueda & Hayaishi (1980) *Meth. Enzymol.* **66**, 151. **11.** Hofmann & Rapaport (1957) *Biochem. Z.* **329**, 437. **12.** Snell, Snell & Richards (1984) *J. Neurochem.* **43**, 1610. **13.** Chung (1967) *J. Biol. Chem.* **242**, 1182. **14.** Terashima, Osago, Hara, *et al.* (2005) *Biochem. J.* **389**, 853. **15.** Dietrich & Muniz (1972) *Biochemistry* **11**, 1691. **16.** Sestini, Ricci, Micheli & Pompucci (1993) *Arch. Biochem. Biophys.* **302**, 206. **17.** Olland, Underwood, Czerwinski, *et al.* (2002) *J. Biol. Chem.* **277**, 3698. **18.** Drummond & Yamamoto (1971) *The Enzymes*, 3rd ed., **4**, 337. **19.** Kornberg & Pricer (1950) *J. Biol. Chem.* **182**, 763.

Nicotinamide-Mononucleotide Adenylyltransferase

This enzyme, also known as NAD+ pyrophosphorylase, catalyzes the reaction of ATP with nicotinamide ribonucleotide (NMN) to produce NAD$^+$ and pyrophosphate. When pure enzyme is incubated in the presence of the poly(ADP-ribose) polymerase, the polymerase activity is inhibited, both in the presence and in the absence of histones. Elevated concentrations of DNA, NAD$^+$, or Mg^{2+} did not affect the inhibition. **1.** Ruggieri, Gregori, Natalini, *et al.* (1990) *Biochemistry* **29**, 2501.

Nicotinate (Nicotinic Acid)

This vitamin (FW$_{free-acid}$ = 123.11 g/mol; CAS 59-67-6; λ$_{max}$ at 261.5 nm), also commonly known as niacin, but also named 3-picolinic acid, vitamin B$_3$, and 3-pyridinecarboxylic acid, is a precursor of the pyridine coenzymes (1). Niacin is soluble in water (1 g/60 mL at 20°C) and is stable in both acidic and alkaline conditions. This vitamin is anionic under most physiological conditions. **Pellagra:** Niacin deficiency occurs only rarely in developed countries, and is associated with poverty, malnutrition, and chronic alcohol use. Corn is low in digestible niacin, giving rise to niacin insufficiency in places where carn is the main starch-containing food. Severe niacin deficiency of in causes pellagra, a disorder characterized by diarrhea, dermatitis, and dementia. **Anti-atherogenic Properties:** Elevated niacin levels are known to increase plasma HDL concentrations, but the mechanism is uncertain. Recent work (2) suggests that, through ROS-mediated signaling, niacin decreases neutrophil myeloperoxidase (MPO) release as well as its catalytic activity, protecting apolipoprotein-AI (apo-AI) modification and improving HDL function. Beyond its potent bacteriocidal action against invading pathogens by generating HOCl from H$_2$O$_2$ and Cl$^-$, MPO-derived oxidants also play critical roles in tissue injury and such chronic diseases/conditions as cancer, renal disease, lung injury, atherosclerosis, multiple sclerosis, Alzheimer's Disease and Parkinson's Disease. Specifically, 0.25-0.5 mM niacin decreases phorbol myristic acid- (*or* PMA-) induced MPO activity in cells as well as that released into the medium. Niacin also reduces MPO protein mass in the medium without affecting its mRNA expression. Increased NADPH oxidase and ROS production by phorbol myristic acid (PMA) are significantly inhibited by niacin. Studies with specific inhibitors suggest that ROS-dependent Src and p38MAP kinase mediate decreased MPO activity by niacin. Niacin blocked apo-AI degradation, and apo-AI from niacin treated cells decreased monocyte adhesion to aortic endothelial cells (2). **Target(s):** acetylcholinesterase (3); acetyl-CoA carboxylase (4,5); adenylate cyclase (6); D-amino-acid oxidase, K_i = 0.11 mM (7,8); γ-butyrobetaine dioxygenase, *or* γ-butyrobetaine hydroxylase, K_i = 0.41 mM (9); choline acetyltransferase (10); diacylglycerol *O*-acyltransferase (11,12); glucose dehydrogenase (7,13,14); histidine kinase (15); imidazoleacetate 4-monooxygenase (16); kynurenine:oxoglutarate aminotransferase, weakly inhibited (17); lactate dehydrogenase, weakly inhibited (7,13); monophenol monooxygenase (7); NAD$^+$ nucleosidase, *or* NADase (7,18-22); nicotinamide nucleotide amidase (23); nitrilase (24); orotate phosphoribosyltransferase, weakly inhibited (25); palmitoyl-CoA hydrolase, *or* acyl-CoA hydrolase (26); procollagen-lysine 5-dioxygenase (27); procollagen-proline 3-dioxygenase (28); procollagen-proline 4-dioxygenase (27,28). **1.** Elvehjem, Madden, Strong & Woolley (1938) *J. Biol. Chem.* **123**, 137. **2.** Ganji, Kamanna & Kashyap (2014) *Atherosclerosis*. **235**, 554. **3.** Bergmann, Wilson & Nachmansohn (1950) *J. Biol. Chem.* **186**, 693. **4.** Fomenko, Pozharun, Shushevich & Khalmuradov (1983) *Biokhimya* **48**, 714. **5.** Khalmuradov & Fomenko (1984) *Ukr. Biokhim. Zh.* **56**, 363. **6.** Aktories, Schultz & Jakobs (1983) *Arzneimittelforschung* **33**, 1525. **7.** Webb (1966) *Enzyme and Metabolic Inhibitors*, vol. **2**, Academic Press, New York. **8.** Klein & Austin (1953) *J. Biol. Chem.* **205**, 725. **9.** Ng, Hanauske-Abel & England (1991) *J. Biol. Chem.* **266**, 1526. **10.** Korey, de Braganza & Nachmansohn (1951) *J. Biol. Chem.* **189**, 705. **11.** Lung & Weselake (2006) *Lipids* **41**, 1073. **12.** Hobbs & Hills (2000) *Biochem. Soc. Trans.* **28**, 687. **13.** von Euler (1942) *Ber.* **75**, 1876. **14.** Brink (1953) *Acta Chem. Scand.* **7**, 1090. **15.** Bock & Gross (2002) *Eur. J. Biochem.* **269**, 3479. **16.** Maki, Yamamoto, Nozaki & Hayaishi (1969) *J. Biol. Chem.* **244**, 2942. **17.** Takeuchi, Otsuka & Shibata (1983) *Biochim. Biophys. Acta* **743**, 323. **18.** Gopinathan, Sirsi & Vaidyanathan (1964) *Biochem. J.* **91**, 277. **19.** Zatman, Kaplan, Colowick & Ciotti (1954) *J. Biol. Chem.* **209**, 453. **20.** Hofmann & Rapoport (1957) *Biochem. Z.* **329**, 437. **21.** Balducci & Micossi (2002) *Mol. Cell. Biochem.* **233**, 127. **22.** Zielinska, Barata & Chini (2004) *Life Sci.* **74**, 1781. **23.** Imai (1973) *J. Biochem.* **73**, 139. **24.** Almatawah & Cowan (1999) *Enzyme Microb. Technol.* **25**, 718. **25.** Victor, Greenberg & Sloan (1979) *J. Biol. Chem.* **254**, 2647. **26.** Berge & Dossland (1979) *Biochem. J.* **181**, 119. **27.** Majamaa, Turpeenniemi-Hujanen, Latipää, *et al.* (1985) *Biochem. J.* **229**, 127. **28.** Majamaa, Hanauske-Abel, Günzler & Kivirikko (1984) *Eur. J. Biochem.* **138**, 239.

Nicotine

This toxic alkaloid (FW$_{free-base}$ = 162.23 g/mol; CAS 54-11-5), abundantly present in the leaves of the tobacco plant (*Nicotiana tabacum* and *N. rustica*), is the prototypic nicotinic acetylcholine receptor agonist. While known as early as 1571, the earliest report on nicotine was included in *Esperienze inforno a diverse Cose Naturli* (1686) by Francesco Redi (the very same Redi who convincingly demonstrated that insects (and hence all life forms) are not the product of spontaneous generation). **Chemical Properties:** In tobacco, nicotine is largely (>99%) the (*S*)-isomer. It is weak base with a pK_a of 8.0. The free base is a hygroscopic liquid that turns brown upon exposure to air or light. It is miscible with water below 60°C., The free base is itself well absorbed through the skin, a property that explains occupational risks of nicotine poisoning (or Green Tobacco Sickness) in those harvesting wet tobacco leaves with bare hands. **Pharmacokinetics:** The disposition of nicotine in humans is complex, showing multi-compartment kinetics for its metabolic and physiologic effects (1). Briefly, much of nicotine is protonated in physiologic fluids and does not cross membranes rapidly. That said, the huge surface area of the alveoli and small airways favors efficient nicotine absorption from cigarette smoke in the lungs. Indded, the lag-time between inhaling cigarette smoke and nicotine arrival at the brain is only 10–20 seconds. Once in circulation, nicotine shows highest affinity for liver, kidney, spleen, and lung. It likewise concentrates in the brain, where high-affinity binding sites are plentiful. Nicotine is metabolized primarily by the liver enzymes UDPglucuronosyltransferase (*or* UGT), CYP2A6, and flavin-containing monooxygenase (FMO). **Primary Mode of Action:** Nicotine binds to nicotinic cholinergic receptors, facilitating the release of neurotransmitters (particularly dopamine, glutamate, and γ-aminobutyric acid) that mediate a manifold of psychophysiologic actions, leading to and maintaining tobacco (nicotine) dependence. With regular use, one develops nicotine dependence, a condition that is manifestly an addiction that is characterized by nicotinic receptor activation, desensitization, up-regulation, and subsequent modulation of the mesocorticolimbic dopaminergic system. The complexity of nicotine dependence is indicated in functional magnetic resonance images (fMRI) demonstrating involvement of the anterior and posterior cingulate cortex, medial and lateral orbitofrontal cortex, ventral striatum, amygdala, thalamus and insula in the maintenance of smoking and nicotine withdrawal. **Effect on Chromatin:** Nicotine enhances the extent of histone acetylation by inhibiting histone deacetylases (HDACs), the action of the latter favoring gene silencing. This nicotine-induced epigenetic modification is likely give rise to an open chromatin structure that primes FosB for enhanced transcriptional activation after chronic cocaine use, thus facilitating or promoting cocaine-use behaviors (2). (**See also the naturally occurring structural analogues** Anabasine & Anabaseine) **Target(s):** antheraxanthin κ-cyclase (3); aromatase, or CYP19 (3,5); azaheterocycle N-methyltransferase, inhibited by S-(–)-nicotine (6); choline acetylase, or choline acetyltransferase (7); CYP2E1 (8); β-fructofuranosidase, or invertase (9); 3α-hydroxysteroid dehydrogenase (10); lycopene β cyclase (3); NADH dehydrogenase, or Complex I (11); sea urchin cortical granule protease (12); sodium channels, voltage-dependent (13); steroid 17,20-lyase (14); steroid 11-β-monooxygenase (15,16); steroid 17α-monooxygenase (14); steroid 21-monooxygenase (15,16); thromboxane synthase (17). **1.** Benowitz, Hukkanen & Jacob (2009) Nicotine chemistry, metabolism, kinetics and biomarkers, in *Handbook of Experimental Pharmacology* 192, 29–60. **2.** Volkow (2011) *Sci. Transl. Med.* **3**, 107, ps43. **3.** Bouvier, d'Harlingue & Camara (1997) *Arch. Biochem. Biophys.* **346**, 53. **4.** Barbieri, Gochberg & Ryan (1986) *J. Clin. Invest.* **77**, 1727. **5.** Barbieri, McShane & Ryan (1986) *Fertil. Steril.* **46**, 232. **6.** Cundy & Crooks (1985) *Drug Metab. Dispos.* **13**, 658. **7.** Fahmy, Ryman & Walsh (1954) *J. Pharm. Pharmacol.* **6**, 607. **8.** Van Vleet, Bombick & Coulombe (2001) *Toxicol. Sci.* **64**, 185. **9.** Rojo, Quiroga, Vattuone & Sampietro (1998) *Phytochemistry* **49**, 965. **10.** Meikle, Liu, Taylor & Stringham (1988) *Life Sci.* **43**, 1845. **11.** Cormier, Morin, Zini, Tillement & Lagrue (2001) *Brain Res.* **900**, 72. **12.** Carroll (1976) *Meth. Enzymol.* **45**, 343. **13.** Liu, Zhu, Zhang, *et al.* (2004) *J. Neurophysiol.* **91**, 1482. **14.** Yeh, Barbieri & Friedman (1989) *J. Steroid Biochem.* **33**, 627. **15.** Barbieri, Friedman & Osathanondh (1989) *J. Clin. Endocrinol. Metab.* **69**, 1221. **16.** Barbieri, York, Cherry & Ryan (1987) *J. Steroid Biochem.* **28**, 25. **17.** Goerig, Ullrich, Schettler, Foltis & Habenicht (1992) *Clin. Investig.* **70**, 239.

Nicotinic Acid Adenine Dinucleotide

This NAD⁺ analogue (FW = 686.39 g/mol; CAS 6450-77-7), also known as deamido-NAD⁺, is the product of the reaction catalyzed by nicotinate-nucleotide adenylyltransferase and is the naturally occurring substrate of NAD⁺ synthetase and NAD⁺ synthetase (glutamine-hydrolyzing). **Target(s):** alcohol dehydrogenase, yeast (1); lactate dehydrogenase (2); NADase (3,4); NAD⁺ kinase (5); nicotinamide-nucleotide amidase (6); nicotinamide phosphoribosyltransferase (7); nicotinate phosphoribosyl-transferase (8); NMN nucleosidase (9). **1.** Sund & Theorell (1963) *The Enzymes*, 2nd ed., 7, 25. **2.** Chang, Huang & Chiou (1991) *Arch. Biochem. Biophys.* **284**, 285. **3.** Okayama, Ueda & Hayaishi (1980) *Meth. Enzymol.* **66**, 151. **4.** Ueda, Fukushima, Okayama & Hayaishi (1975) *J. Biol. Chem.* **250**, 7541. **5.** Lerner, Niere, Ludwig & Ziegler (2001) *Biochem. Biophys. Res. Commun.* **288**, 69. **6.** Imai (1973) *J. Biochem.* **73**, 139. **7.** Dietrich & Muniz (1972) *Biochemistry* **11**, 1691. **8.** Blum & Kahn (1971) *Meth. Enzymol.* **18B**, 138. **9.** Foster (1981) *J. Bacteriol.* **145**, 1002.

Nicotinic Acid Hydrazide

This toxic hydrazide (FW$_{free-base}$ = 137.14 g/mol; CAS 553-53-7), also known as 3-pyridine carboxylic acid hydrazide, inhibits diamine oxidase. Note that this hydrazide is the *meta*-isomer of isonicotinic acid hydrazide (*i.e.*, isoniazid), a well-known tuberculostatic agent. **Target(s):** diamine oxidase, or amine oxidase, copper-containing (1); NADase, weakly inhibited (2); nicotinamidase (3,4). **1.** Pec, Haviger, Kopecna & Frebort (1992) *J. Enzyme Inhib.* **6**, 243. **2.** Hofmann & Rapaport (1957) *Biochem. Z.* **329**, 437. **3.** Tanigawa, Shimiyama, Dohi & Ueda (1972) *J. Biol. Chem.* **247**, 8036. **4.** Johnson & Gadd (1974) *Int. J. Biochem.* **5**, 633.

Nifedipine

This photosensitive pyridine derivative (FW = 346.34 g/mol; CAS 21829-25-4), also known as Procardia™ and 2,6-dimethyl-3,5-dicarbomethoxy-4-(2-nitrophenyl)-1,4-dihydropyridine, is a dihydropyridine-based calcium channel blocker that acts as a vasodilator, reduces blood pressure. Procardia has been used to treat angina. It also induces apoptosis in human glioblastoma cells. **Primary Mode of Inhibitory Action:** By virtue of its ability to inhibit calcium ion influx via L-type calcium channels, nifedipine decreases arterial smooth muscle contractility and subsequent vasoconstriction Calcium ions normally enter these cells and bind to calmodulin, thereby activating myosin light chain kinase (MLCK). The latter catalyzes myosin light chain phosphorylation. Signal amplification is achieved by calcium-induced calcium release from the sarcoplasmic reticulum through ryanodine receptors. By inhibiting the contractile processes of smooth muscle cells, nifedipine administration causes of coronary and systemic artery dilation, increasing oxygen delivery to the myocardial tissue, decreasing peripheral resistance, lowering systemic blood pressure, and decreasing afterload. **Key Pharmacokinetic Parameters:** *See* Appendix II in Goodman & Gilman's *THE PHARMACOLOGICAL BASIS OF THERAPEUTICS*, 12th Edition (Brunton, Chabner & Knollmann, eds.) McGraw-Hill Medical, New York (2011). **Target(s):** acetylcholinesterase (1); Ca²⁺ channel (2-4); cAMP-dependent protein kinase, or Protein Kinase A (5); cholinesterase, or butyrylcholinesterase (1); fatty acid Δ⁶-desaturase (6); orotidine 5'-phosphate decarboxylase (7); uridine kinase (7). **1.** Chiou, Lai, Tsai, *et al.* (2005) *J. Chin. Chem. Soc.* **52**, 843. **2.** Fleckenstein, Tritthart, Doring & Byon (1972) *Arzneimittelforschung* **22**, 22. **3.** Vater, Kroneberg, Hoffmeister, *et al.* (1972) *Arzneimittelforschung* **22**, 1. **4.** Triggle, Shefter & Triggle (1980) *J. Med. Chem.* **23**, 1442. **5.** Su, Chen, Zhou & Vacquier (2005) *Dev. Biol.* **285**, 116. **6.** Kawashima, Akimoto, Jareonkitmongkol, Shirasaka & Shimizu (1996) *Biosci. Biotechnol. Biochem.* **60**, 1672. **7.** Najarian & Traut (2000) *Neurorehabil. Neural Repair* **14**, 237.

Niflumate (Niflumic Acid)

This anti-inflammatory drug and chelating agent (FW$_{free-acid}$ = 282.22 g/mol; CAS 4394-00-7) inhibits prostaglandin H synthase 2 (IC$_{50}$ = 20 nM) and blocks calcium-activated chloride channels. **Primary Mode of Action:** By promoting a depolarizing influence on the resting membrane potential of vascular smooth muscle cells, calcium-activated chloride channels are crucially important in regulating vascular tone. A potent blocker of calcium-activated chloride channels in vascular myocytes, NFA's action results from sustained calcium-activated chloride currents. Niflumic acid also forms complexes copper ion, significantly inhibiting polymorphonuclear leukocyte oxidative metabolism. In addition to these actions, it appears to affect brain GABA$_A$ receptors (1). **Target(s):** Al^{3+}-activated malate channel (2); anion channel (3,4); catechol sulfotransferase (5); channel-conductance-controlling ATPase, *or* cystic-fibrosis membrane-conductance-regulating protein, *or* CFTR (6); chloride channels, Ca^{2+}-activated (7-11); cyclooxygenase 2, *or* prostaglandin-endoperoxide synthase (12,13); fatty acid amide hydrolase (14,15); gap junction intercellular communication (16); glucuronosyltransferase, substrate for certain transferases (17,18); nonselective cation channels (19); 15-oxoprostaglandin 13-reductase/leukotriene B$_4$ 12-hydroxydehydrogenase/eicosanoid oxido-reductase (20); phenol sulfotransferase (5); phosphate transport (21); sulfate transport (22). **1.** Sinkkonen, Mansikkamaki, Moykkynen, *et al.* (2003) *Mol. Pharmacol.* **64**, 753. **2.** Zhang, Ryan & Tyerman (2001) *Plant Physiol.* **125**, 1459. **3.** Hall, Kirk, Potts, Rae & Kirk (1996) *Amer. J. Physiol.* **271**, C579. **4.** Simchowitz (1988) *Soc. Gen. Physiol. Ser.* **43**, 193. **5.** Vietri, De Santi, Pietrabissa, Mosca & Pacifici (2000) *Xenobiotica* **30**, 111. **6.** van Kuijck, van Aubel, Busch, *et al.* (1996) *Proc. Natl. Acad. Sci. U.S.A.* **93**, 5401. **7.** Coelho, Souza, Soares, *et al.* (2004) *Brit. J. Pharmacol.* **141**, 367. **8.** Parai & Tabrizchi (2002) *Eur. J. Pharmacol.* **448**, 59. **9.** Fuller, Ji, Tousson, *et al.* (2001) *Pflugers Arch.* **443** Suppl. 1, S107. **10.** Gandhi, Elble, Gruber, *et al.* (1998) *J. Biol. Chem.* **273**, 32096. **11.** White & Aylwin (1990) *Mol. Pharmacol.* **37**, 720. **12.** Barnett, Chow, Ives, *et al.* (1994) *Biochim. Biophys. Acta* **1209**, 130. **13.** Johnson, Wimsatt, Buckel, Dyer & Maddipati (1995) *Arch. Biochem. Biophys.* **324**, 26. **14.** Fowler, Holt & Tiger (2003) *J. Enzyme Inhib. Med. Chem.* **18**, 55. **15.** Fowler, Borjesson & Tiger (2000) *Brit. J. Pharmacol.* **131**, 498. **16.** Harks, de Roos, Peters, *et al.* (2001) *J. Pharmacol. Exp. Ther.* **298**, 1033. **17.** Gaganis, Miners & Knights (2007) *Biochem. Pharmacol.* **73**, 1683. **18.** Vietri, Pietrabissa, Mosca & Pacifici (2002) *Eur. J. Clin. Pharmacol.* **58**, 93. **19.** Gogelein, Dahlem, Englert & Lang (1990) *FEBS Lett.* **268**, 79. **20.** Clish, Sun & Serhan (2001) *Biochem. Biophys. Res. Commun.* **288**, 868. **21.** Meng, Timmer, Gunn & Abercrombie (2000) *Amer. J. Physiol. Cell Physiol.* **278**, C1183. **22.** Lopez, Bracht, Yamamoto, *et al.* (1999) *Gen. Pharmacol.* **32**, 713.

Nifurtimox

This orally available 5-nitrofuran and anti-trypanosomal agent (FW = 287.30 g/mol; CAS 23256-30-6), also known by its code name Bayer-2502, trade name Lampit™, and systematic name (*RS*)-3-methyl-*N*-[(1*E*)-(5-nitro-2-furyl)methylene]thiomorpholin-4-amine 1,1-dioxide, inhibits is used to treat of Chagas' disease and African Sleeping Sickness (1-5). **Likely Mechanisms of Trypanocidal Action:** Although the precise biochemical basis for nifurtimox poisoning of trypanosomes has yet to be established, several ideas have gained traction. The first relates to nifurtimox 's ability to generate a nitro-anion radical metabolite that reacts with parasite's DNA, inducing strand breakage and ensuing apotosis. The second relates the production of superoxide anion, which also generates hydrogen peroxide through the action of superoxide dismutase. Both oxidants might then inhibit trypanothione reductase (a parasite-specific antioxidant defense enzyme) and ascorbate-linked peroxidase (another likely nifurtimox target). A third model (5) views nifurtimox as a prodrug that is activated by trypanosomal type I nitroreductases, which catalyze its reduction to an unsaturated open-chain nitrile derivative that has growth inhibitory properties against parasite and mammalian cells. In principle, nitrofuran activation would occur by either of two routes: (a) by a two-electron reduction mediated by type I nitroreductases, or (b) by a one-electron pathway catalyzed by type II enzymes. **Target(s):** DNA topoisomerase I (1); DNA topoisomerase II (1); glutathione-disulfide reductase, IC$_{50}$ = 40 μM (2,3); oxidative phosphorylation (4); trypanothione reductase, IC$_{50}$ = 200 μM (3). **1.** Riou, Douc-Rasy & Kayser (1986) *Biochem. Soc. Trans.* **14**, 496. **2.** Grinblat, Sreider & Stoppani (1989) *Biochem. Pharmacol.* **38**, 767. **3.** Jockers-Scherubl, Schirmer & Krauth-Siegel (1989) *Eur. J. Biochem.* **180**, 267. **4.** Biscardi, Fernandez Villamil & Stoppani (1994) *Rev. Argent. Microbiol.* **26**, 72. **5.** Hall, Bot & Wilkinson (2011) *J. Biol. Chem.* **286**, 13088.

Nigericin

This polyether K$^+$/H$^+$ ionophore and antibiotic (FW$_{free-acid}$ = 724.97 g/mol; CAS 28380-24-7) from *Streptomyces hygroscopicus*, inhibits both oxidative phosphorylation and photophosphorylation. Its metal ion specificity is: K$^+$ > Rb$^+$ ≥ Cs$^+$ >> Na$^+$ (1). **Primary Mechanism of Inhibitory Action:** As a potassium ionophore, nigericin induces caspase-1-mediated pro-protein processing and release of IL-1β, an important 17.5-kDa protein mediator of the inflammatory response, and one that is involved in regulating cell proliferation, differentiation, and apoptosis. **Target(s):** cadmium-exporting ATPase (2); F$_1$Fo ATP synthase (3,4); insulin-stimulated glucose transport (5); oxidative and photophosphorylation (1,3,4,6,7). **1.** Reed (1979) *Meth. Enzymol.* **55**, 435. **2.** Tsai, Yoon & Lynn (1992) *J. Bacteriol.* **174**, 116. **3.** Stiggall, Galante & Hatefi (1979) *Meth. Enzymol.* **55**, 308. **4.** Gautheron, Penin, Deléage & Godinot (1986) *Meth. Enzymol.* **126**, 417. **5.** Chu, Kao & Fong (2002) *J. Cell Biochem.* **85**, 83. **6.** Izawa & Good (1972) *Meth. Enzymol.* **24**, 355. **7.** Shavit, Dilley & San Pietro (1968) *Biochemistry* **7**, 2356.

Nigerose

α-anomer   β-anomer

This α$_{1,3}$-linked glucosyl disaccharide (FW = 342.30 g/mol; CAS 497-48-3) inhibits glucosidase II. **Target(s):** concanavalin A (1); glucan 1,3-α-glucosidase, *or* glucosidase II, *or* mannosyl oligosaccharide glucosidase II (2-6); mycodextranase (7); *Streptococcus sobrinus* GTF-I glucosyltransferase (8). **1.** Goldstein, Hollerman & Smith (1965) *Biochemistry* **4**, 876. **2.** Kaushal & Elbein (1989) *Meth. Enzymol.* **179**, 452. **3.** Presper & Heath (1983) *The Enzymes*, 3rd ed. (Boyer, ed.), **16**, 449. **4.** Kaushal, Pastuszak, Hatanaka & Elbein (1990) *J. Biol. Chem.* **265**, 16271. **5.** Ugalde, Staneloni & Leloir (1980) *Eur. J. Biochem.* **113**, 97. **6.** Takeuchi, Kamata, Yoshida, Kameda & Matsui (1990) *J. Biochem.* **108**, 42. **7.** Tung, Rosenthal & Nordin (1971) *J. Biol. Chem.* **246**, 6722. **8.** McAlister, Doyle & Taylor (1989) *Carbohydr. Res.* **187**, 131.

Nikkomycin Wx

This nucleoside peptide antibiotic (FW = 450.41 g/mol; CAS 59456-70-1), from *Streptomyces tendae* is a conjugate of L-tyrosine, 5-amino-5-deoxy-D-*allo*-furanuronic acid, and 4-formyl-4 imidazolin-2-one. Although nikkomycin Wx is an effective *in vitro* inhibitor (K_i = 12 μM) of chitin synthase from *Coprinus cinereus*, it is rapidly hydrolyzed *in vivo*, thereby negating any inhibitory effect on the growth of this multicellular basidiomycete. **1**. Decker, Walz, Bormann, *et al.* (1990) *J. Antibiot.* **43**, 43.

Nikkomycin Z

This nucleoside peptide antibiotic (FW = 495.45 g/mol; CAS 59456-70-1), first isolated from *Streptomyces tendae*, is an UDP-*N*-acetylglucosamine analogue that inhibits chitin synthase, showing potent antifungal, insecticidal, and acaridicial properties. **1**. Cabib, Kang & Au-Young (1987) *Meth. Enzymol.* **138**, 643. **2**. Dahn, Hagenmaier, Hohne, *et al.* (1976) *Arch. Microbiol.* **107**, 143. **3**. Yadan, Gonneau, Sarthou & Le Goffic (1984) *J. Bacteriol.* **160**, 884. **4**. McCarthy, Troke & Gull (1985) *J. Gen. Microbiol.* **131**, 775. **5**. Hector & Braun (1986) *Antimicrob. Agents Chemother.* **29**, 389. **6**. Cabib (1991) *Antimicrob. Agents Chemother.* **35**, 170. **7**. Kang, Hwang, Yun, Shin & Kim (2008) *Biol. Pharm. Bull.* **31**, 755. **8**. Cohen, Elster & Chet (1986) *Pestic. Sci.* **17**, 175. **9**. Haenseler, Nyhlen & Rast (1983) *Exp. Mycol.* **7**, 17. **10**. Wang & Szaniszlo (2002) *Med. Mycol.* **40**, 283. **11**. Hardy & Gooday (1983) *Curr. Microbiol.* **9**, 51. **12**. Uchida, Shimmi, Sudoh, Arisawa & Yamada-Okabe (1996) *J. Biochem.* **119**, 659. **13**. Plant, Thompson & Williams (2008) *J. Org. Chem.* **73**, 3714.

Nilotinib

This TKI (FW = 529.52 g/mol; CAS 641571-10-0), also known as AMN107 and Tasigna™, specifically targets Bcr-Abl protein-tyrosine kinase (1-4). In 2007, nilotinib was approved by the U.S. Food and Drug Administration for the treatment of adult patients with chronic-phase and accelerated-phase Philadelphia chromosome-positive (Ph⁺) chronic myelogenous leukemia resistant to or intolerant of prior treatment that included imatinib. Parkinson Disease (PD) -associated dyskinesia is striatal dysfunction, arising from an imbalance of dopamine and glutamate transmissions that are integrated by the 32-kDa Dopamine- and cAMP-Regulated Phosphoprotein (DARPP-32), the phosphorylation of which requires c-Abl activation of Cdk5 by phosphorylation in the striatum. Systemic administration of nilotinib, which inhibits both Cdk5 and DARPP-32 phosphorylation, normalizes striatal motor behaviors in a MPTP mouse model for PD (5). c-Abl is activated in the brains of PD patients and in 1-methyl-4-phenyl-1,2,3,6-tetrahydropyridine- (MPTP-) poisoned mice, where it inhibits parkin through tyrosine phosphorylation, leading to the accumulation of parkin substrates and neuronal cell death. Administration of nilotinib reduces c-Abl activation and the levels of the parkin substrate, PARIS, resulting in prevention of dopamine (DA) neuron loss and behavioral deficits following MPTP intoxication (6). **Target(s):** Bcr-Abl protein-tyrosine kinase (1-4); protein-tyrosine kinase (1-4); riboksyldihydronicotinamide dehydrogenase (quinone) (7). **1**. Tauchi & Ohyashiki (2006) *Int. J. Hematol.* **83**, 294. **2**. O'Hare, Walters, Deininger & Druker (2005) *Cancer Cell* **7**, 11. **3**. Weisberg, Manley, Breitenstein, *et al.* (2005) *Cancer Cell* **7**, 129. **4**. Cortes, Jabbour, Kantarjian, *et al.* (2007) *Blood* **110**, 4005. **5**. Tanabe A1, Yamamura Y1, Kasahara, *et al.* (2014) *Front Cell Neurosci.* **8**, 50. **6**. Karuppagounder, Brahmachari, Lee, *et al.*

(2014) *Sci. Rep.* **4**, 4874. **7**. Winger, Hantschel, Superti-Furga & Kuriyan (2009) *BMC Struct. Biol.* **9**, 7.

Nilutamide

This antiandrogen (FW = 317.22 g/mol; CAS 63612-50-0), also known as RU23908 and Anandron™, is used for the combined antihormonal treatment of prostatic cancer. **Target(s):** benzo[*a*]pyrene hydroxylase, cytochrome P450 (1); benzphetamine *N*-demethylase, cytochrome P450 (1); 7-ethoxycoumarin *O*-deethylase, cytochrome P450 (1); hexobarbital hydroxylase, cytochrome P450 (1); mephenytoin 4-hydroxylase (2); NADH dehydrogenase, mitochondrial Complex I (3); steroid 17,20-lyase (4); steroid 17α-monooxygenase (4). **1**. Babany, Tinel, Letteron, *et al.* (1989) *Biochem. Pharmacol.* **38**, 941. **2**. Horsmans, Lannes, Larrey, *et al.* (1991) *Xenobiotica* **21**, 1559. **3**. Berson, Schmets, Fisch, *et al.* (1994) *J. Pharmacol. Exp. Ther.* **270**, 167. **4**. Ayub & Levell (1987) *J. Steroid Biochem.* **28**, 43.

(+)-Nilvadipine

This L-type calcium channel antagonist, antihypertensive agent and Syk inhibitor (FW = 385.37 g/mol; CAS 75530-68-6), also named 2-cyano-1,4-dihydro-6-methyl-4-(3-nitrophenyl)-3,5-pyridinedicarboxylic acid 3-methyl 5-(1-methylethyl) ester, exerts anti-hypertensive activity, showing good antiatherogenic properties, high vascular selectivity, an elimination half-life of 15-20 hours that commends once daily dosing, and good tolerance (1,2). The (−)-enantiomer is around 1000x less potent than its antipode in opposing vasoconstriction induced by FPL64176, a selective L-type calcium channel agonist (3). **Alzheimer Disease Treatment:** Syk regulates NFκB activity, which beyond its role in neuroinflammatory processes also regulates BACE-1 expression, the rate-limiting enzyme in Aβ production. Aβ has been shown to stimulate Syk activity resulting in microglial activation and neurotoxicity Both nilvadipine enantiomers are equipotent in reducing Aβ production in a cell line that overexpresses Aβ, showing that its Aβ-lowering properties are not related to L-type calcium channel inhibition (3). Notably, both (+)- and (−)-nilvadipine target Spleen tyrosine kinase, a non-receptor cytoplasmic enzyme that encoded by the *SYK* gene in humans (3). Both enantiomers also inhibit Syk in a cell-free assay, and binding kinetics using biolayer interferometry show that both nilvadipine enantiomers bind to full-length human Syk with relatively high affinity (K_d = 2.1 μm). Such findings identify Syk as the likely nilvadipine target responsible for its Aβ-lowering properties. This property was demonstrated by showing that pharmacological inhibition of Syk *in vitro* and *in vivo* as well as down-regulation of *syk* expression result in a reduction of Aβ production (3). Syk inhibition also reduces sAPPβ production and BACE-1 transcription, mimicking the observed effect of (−)-nilvadipine. Syk inhibition also increases the clearance of Aβ across the BBB recapitulating to the full extent the Aβ-lowering properties of (−)-nilvadipine. *Note*: The related dihydropyridine- (DHP-) class antitensive Nimodipine (Nimotop) passes through the blood-brain barrier and is currently used to prevent cerebral vasospasm. **1**. Cho, Ueda, Shima, *et al.* (1989) *J. Med. Chem.* **32**, 2399. **2**. Rosenthal (1994) *J Cardiovasc Pharmacol.* **24** (Suppl 2), S92-107 (a review). **3**. Paris, Ait-Ghezala, Bachmeier *et al.* (2014) *J. Biol. Chem.* **289**, 33927.

NIM811

This orally available cyclophilin inhibitor (MW = 1183.67 g/mol), also known as N-methyl-4-isoleucine cyclosporine, has a higher affinity than cyclosporine, but is non-immunosuppressive that exhibits potent anti-human immunodeficiency virus type 1 (HIV-1) activity, selectively inhibiting HIV-1 replication in T4 lymphocyte cell lines, in a monocytic cell line, and in HeLa T4 cells (1). NIM811 also induces a concentration-dependent reduction of hepatitis C virus (HCV) RNA in the replicon cells (IC_{50} of 0.66 μM at 48 hours), with >1000x reduction after treating the cells with as little as 1 μM of NIM811 for 9 days (2). In addition, the combination of NIM811 with α-interferon significantly enhances anti-HCV activities without causing any increase of cytotoxicity. **Mechanism of Action:** NIM811 differs from other anti-HIV agents that inhibit reverse transcriptases, proteases, and integrases or interfere with Rev or Tat function. Instead, NIM 811 inhibits (IC_{50} = 0.17 μg/mL) formation of two-long terminal repeat (2-LTR) circles in a concentration-dependent manner that is indicative of nuclear localization of preintegration complexes, with IC_{50} of 0.01-0.2 μg/mL for viral growth inhibition. Integration of proviral DNA into cellular DNA was likewise inhibited by NIM811. Treatment of chronically infected cells reveals the presence capsid proteins, reverse transcriptase activity, and viral RNA comparable to those of the untreated control. However, these particles showed a dose-dependent reduction in infectivity (50% inhibitory concentration of 0.03 μg/mL), indicating that viral assembly is also impaired. Given the central involvement of Gag proteins are in virus assembly and also in nuclear localization of the preintegration complex, Gag-cyclophilin interactions are clearly responsible for the antiviral action of NIM811. Indeed, Cyclophilin A binds to HIV-1 p24^{gag}, and this cyclophilin–Gag interaction leads to the incorporation of cyclophilin A into HIV-1 virions. NIM811 inhibits this protein interaction is likely to be the molecular basis for its antiviral activity (3). In activated primary T cells, NIM811 interferes with two stages of the virus replication cycle: (a) translocation of pre-integration complexes into the nucleus and (b) production of infectious virus particles. NIM811 not only inhibits translocation of HIV-1 pre-integration complexes in primary T cells, but also in a growth-arrested T cell line. **1.** Rosenwirth, Billich, Datema, *et al.* (1994) *Antimicrob. Agents Chemother.* **38**, 1763. **2.** Ma, Boerner, TiongYip, *et al.* (2006) *Antimicrob. Agents Chemother.* **50**, 2976. **3.** Mlynar, Blevec, Billich, Rosenwirth & Steinkasserer (1997) *J. Gen. Virol.* **78**, 825

Nimesulide

This anti-inflammatory agent (FW = 308.31 g/mol; CAS 51803-78-2), also known as N-(4-nitro-2 phenoxyphenyl)methanesulfonamide, and nemesulide, inhibits prostaglandin biosynthesis and platelet aggregation. **Target(s):** collagenase (1); cyclic-nucleotide phosphodiesterase type IV (2); cyclooxygenase-2, *or* prostaglandin-endoperoxide synthase-2 (3-12); estrone sulfotransferase (13); oxidative phosphorylation, mitochondrial (uncoupling) (14); phenol sulfotransferase, *or* aryl sulfotransferase (13,15). **1.** Barracchini, Franceschini, Amicosante, *et al.* (1998) *J. Pharm. Pharmacol.* **50**, 1417. **2.** Bevilacqua, Vago, Baldi, *et al.* (1994) *Eur.*

Pharmacol. **268**, 415. **3.** van Ryn & Pairet (1999) *Inflamm. Res.* **48**, 247. **4.** Bjarnason (1999) *Ital. J. Gastroenterol. Hepatol.* **31** Suppl. 1, S27. **5.** Tavares, Bishai & Bennett (1995) *Arzneimittelforschung* **45**, 1093. **6.** Vago, Bevilacqua & Norbiato (1995) *Arzneimittelforschung* **45**, 1096. **7.** Taniguchi, Yokoyama, Inui, *et al.* (1997) *Eur. J. Pharmacol.* **330**, 221. **8.** Reininger & Bauer (2006) *Phytomedicine* **13**, 164. **9.** Barnett, Chow, Ives, *et al.* (1994) *Biochim. Biophys. Acta* **1209**, 130. **10.** Rogge, Ho, Liu, Kulmacz & Tsai (2006) *Biochemistry* **45**, 523. **11.** Patrignani, Panara, Sciulli, *et al.* (1997) *J. Physiol. Pharmacol.* **48**, 623. **12.** Sud'ina, Pushkareva, Shephard & Klein (2008) *Prostaglandins Leukot. Essent. Fatty Acids* **78**, 99. **13.** King, Ghosh & Wu (2006) *Curr. Drug Metab.* 7, 745. **14.** Moreno-Sanchez, Bravo, Vasquez, *et al.* (1999) *Biochem. Pharmacol.* **57**, 743. **15.** Vietri, De Santi, Pietrabissa, Mosca & Pacifici (2000) *Eur. J. Clin. Pharmacol.* **56**, 81.

Nimodipine

This dihydropyridine calcium channel blocker (FW = 418.44 g/mol; CAS 66085-59-4), also known as BAY e 9736, Nimotop™, Nymalize™ and 3-(2-methoxyethyl) 5-propan-2-yl 2,6-dimethyl-4-(3-nitrophenyl)-1,4-dihydropyridine-3,5-dicarboxylate, is an FDA-approved treatment for reducing the incidence and severity of ischemic deficits in adult patients with subarachnoid hemorrhage (SAH) from ruptured intracranial berry aneurysms, mainly acting to reduce blood pressure (1-3). **Primary Mechanism of Inhibitory Action:** Nimodipine inhibits calcium ion transfer into vascular smooth muscle cells, thereby inhibiting contractions and reducing the incidence and severity of ischemic deficits in adult patients with subarachnoid hemorrhage from ruptured intracranial berry aneurysms is unknown. Significantly, nimodipine has no effect on cerebral metabolism but increases CBF, particularly after disruption of the blood-brain barrier. The apparent K_d is 0.3-0.4 nM for binding sites in brain membranes, with maximum number of binding sites (B_{max}) of 300-350 fmol/mg protein, roughly in the range commonly observed for neurotransmitters, hormones or channel toxins (4). **Target(s):** Ca^{2+} channel, L-type (5,6); calmodulin-dependent cAMP phosphodiesterase (7); nitric-oxide synthase (8); orotidine-5'-phosphate decarboxylase (9); uridine kinase (9). **1.** Haws, Gourley & Heistad (1983) *J. Pharmacol. Exp. Ther.* **225**, 24. **2.** Van Meel, Wilffert, De Zoeten, Timmermans & Van Zwieten (1982) *Arch. Int. Pharmacodyn. Ther.* **260**, 206. **3.** Harris, Branston, Symon, Bayhan & Watson (1982) *Stroke* **13**, 759. **4.** Bellemann, Ferry, Lübbecke & Glossmann (1982) *Arzneimittelforschung* **32**, 361. **5.** McCarthy & TanPiengco (1992) *J. Neurosci.* **12**, 2225. **6.** Dolin, Hunter, Halsey & Little (1988) *Eur. J. Pharmacol.* **152**, 19. **7.** Schachtele, Wagner & Marme (1987) *Naunyn Schmiedebergs Arch. Pharmacol.* **335**, 340. **8.** Zhu, Li, Liu & Hua (1999) *Life Sci.* **65**, PL221. **9.** Najarian & Traut (2000) *Neurorehabil. Neural Repair* **14**, 237.

Nimustine

This nitrosourea (FW = 272.69 g/mol; CAS 42471-28-3), also known as N'-[(4-amino-2-methyl-5-pyrimidinyl)methyl]-N-(2-chloroethyl)-N-nitrosourea, exhibits antitumor activity by alkylating DNA, which alters DNA replication and repair and triggering apoptosis. (*See Nitrosoureas*) **Target(s):** acetylcholinesterase (1); butyrylcholinesterase (1); glutathione-disulfide reductase (2); ribonucleotide reductase (2); thioredoxin reductase (2). **1.** Fujii, Ohnoshi, Namba & Kimura (1982) *Gan To Kagaku Ryoho* 9, 831. **2.** Schallreuter, Gleason & Wood (1990) *Biochim. Biophys. Acta* **1054**, 14.

Nintedanib

This oral angiokinase inhibitor and antineoplastic agent (FW = 539.64 g/mol; CAS 656247-17-5), also known by its code name BIBF 1120, its trade name Vargatef®, and its systematic name methyl (3Z)-3-{[(4-{methyl[(4-methylpiperazin-1-yl)acetyl]amino}phenyl)amino](phenyl)methylidene}-2-oxo-2,3-dihydro-1H-indole-6-carboxylate, inhibits vascular endothelial growth factor receptor (VEGFR1, IC_{50} = 34 nM; VEGFR2, IC_{50} = 13 nM; VEGFR3, IC_{50} = 13 nM), fibroblast growth factor receptor (FGFR1, IC_{50} = 69 nM; FGFR2, IC_{50} = 37 nM; FGFR3, IC_{50} = 108 nM), and platelet-derived growth factor receptor (PDGFRα, IC_{50} = 59 nM; PDGFRβ, IC_{50} = 65 nM), thus providing sustained receptor blockade and good antitumor efficacy. In tested tumor models that included human tumor xenografts growing in nude mice and a syngeneic rat tumor model, BIBF 1120 is highly active at well-tolerated doses (25-100 mg/kg daily p.o.), as evaluated by (a) MRI of tumor perfusion after three days, (b) reduction in vessel density and vessel integrity after five days, and (c) profound tumor growth inhibition (1). So designated as a triple angiokinase inhibitor, Nintedanib binds to the ATP-binding site in the cleft between the N- and C-terminal lobes of the kinase domain. Significantly, Nintedanib induces hypoxia, but not epithelial-to-mesenchymal transition (EMT), and blocks progression of preclinical models of lung and pancreatic cancer (3). 1. Hilberg, Roth, Krssak, et al. (2008) Cancer Res. 68, 4774. 2. Roth, Heckel, Colbatzky, et al. (2009) J. Med. Chem. 52, 4466. 3. Kutluk Cenik, Ostapoff, Gerber & Brekken (2013) Mol. Cancer Ther. 12, 992.

Nipecotate (Nipecotic Acid)

This water-soluble piperidine derivative (FW = 129.16 g/mol; CAS 498-95-3), also known as 3-piperidinecarboxylic acid, is a structural analogue of γ-aminobutyric acid (GABA) and a strong inhibitor of GABA uptake. **Target(s):** γ-aminobutyrate (GABA) transport (1-3); L-carnitine uptake (4); carnosine synthetase (5). 1. Balcar, Mark, Borg & Mandel (1979) Neurochem. Res. 4, 339. 2. Johnston, Krogsgaard-Larsen, Stephanson & Twitchin (1976) J. Neurochem. 26, 1029. 3. Krogsgaard-Larsen & Johnston (1975) J. Neurochem. 25, 797. 4. Hannuniemi & Kontro (1988) Neurochem. Res. 13, 317. 5. Seely & Marshall (1982) Life Sci. 30, 1763.

Niraparib

This selective PARP inhibitor (FW = 320.39 g/mol; CAS 1038915-60-4), also named (S)-2-(4-(piperidin-3-yl)phenyl)-2H-indazole-7-carboxamide and MK-4827, targets the poly(D-phosphoribosyl)polymerases PARP1 (IC_{50} = 3.8 nM) and PARP2 (IC_{50} = 2.1 nM), with >330x selectivity against PARP3, V-PARP and Tank1 (1). Poly(ADP-ribose) polymerases are involved in DNA repair following damage by endogenous or exogenous processes, and it has become clear that, in the context of defects in other DNA repair mechanisms, PARP inhibition can provide for tumor specific chemotherapy. Niraparib is highly active against cancer cells with BRCA-1 and BRCA-2 mutations. MK-4827 shows a high potential for improving radiotherapy efficacy, based on results of a study of fractionated radiotherapy (e.g., 2 Gy per fraction, given once or twice a day for 1-2 weeks) combined with MK-4827 (e.g., 50 mg/kg once daily or 25 mg/kg twice daily), using a variety of human tumor xenografts of differing p53

status (e.g., Calu-6 (p53null), A549 (p53wild-type) and H-460 (p53wild-type) lung cancers as well as triple-negative MDA-MB-231 human breast carcinoma (2). Combined use of MK-4827 and radiation may also provide effective therapy for children with high-risk neuroblastoma (3). Significantly, niraparib radiosensitizes human lung and breast cancer cells (4). 1. Jones, Altamura, Boueres, et al. (2009) J. Med. Chem. 52, 7170. 2. Wang, Mason, Ang, et al. (2012) Invest. New Drugs 30, 2113. 3. Mueller, Bhargava, Molinaro, et al. (2013) Anticancer Res. 33, 755. 4. Bridges, Toniatti, Buser, et al. (2014) Oncotarget 5, 5076.

Nisin A

This polypeptide antibiotic (FW = 3354.07 g/mol; CAS 1414-45-5; where Abu = α-aminobutyric acid, Dha = dehydroalanine, and Dhb = dehydrbutyrine) from Streptococcus lactis (Lactococcus lactis) binds specifically to the peptidoglycan precursor Lipid II, thereby combining the formation of bacteriocidal membrane pores with inhibition of cell wall biosynthesis. The presence of dehydrated residues and thioether bonds constrain the peptide's conformation. It is bacteriocidal at nM concentrations. Note that Nisin Z is identical toNnisin A, except for an asparaginyl residue at position 27 in place of a histidyl. **Pore Formation:** Nisin first associates tightly with the anionic membrane surface leading to a high local concentration, disturbing the lipid dynamics near the interface of the solvent and phospholipid polar head groups, thereby immobilizing nearby lipids. Then, in the presence of a transmembrane electrical potential, the molecules reorient, giving rise to a wedge-like, nonspecific, water-filled pore (1). **Target(s):** peptidoglycan biosynthesis (1-4); peptidoglycan glycosyltransferase (5); undecaprenyl-diphospho-muramoylpentapeptide β-N-acetylglucosaminyl-transferase, or MurG transferase (6). 1. Moll, Roberts, Konings & Driessen (1996) Antonie Van Leeuwenhoek 69, 185 2. Reisinger, Seidel, Tschesche & Hammes (1980) Arch. Microbiol. 127, 187. 3. Wiedemann, Breukink, van Kraaij, et al. (2001) J. Biol. Chem. 276, 1772. 4. Gross & Morell (1971) J. Amer. Chem. Soc. 93, 4634. 5. Ramachandran, Chandrakala, Kumar, et al. (2006) Antimicrob. Agents Chemother. 50, 1425. 6. Ravishankar, Kumar, Chandrakala, et al. (2005) Antimicrob. Agents Chemother. 49, 1410.

NITD008

This nucleoside analogue and pro-drug (FW = 290.28 g/mol; CAS 1044589-82-3), also named 7-deaza-2'-C-acetylene-adenosine, potently inhibits the four serotypes of the mosquito-borne Dengue virus (DENV), both in vitro and in vivo (1). Its triphosphate form directly inhibits the RNA-dependent RNA polymerase activity of DENV, indicating that NITD008 must first be phosphorylated to the triphosphate, which then functions as a polymeras chain terminator during viral RNA synthesis. NITD008 also inhibits other flaviviruses, including West Nile virus, yellow fever virus, and Powassan virus. While this agent also suppresses hepatitis C virus, it does not inhibit nonflaviviruses, such as Western equine encephalitis virus and vesicular stomatitis virus (1). Treatment of DENV-infected mice with NITD008 suppressed peak viremia, reduced cytokine elevation, and completely prevented the infected mice from death (1). It is also effective against Enterovirus 71 (EV71), a major viral pathogen in China and Southeast Asia (2). Combined treatment with NITD008 and the histone deacetylase inhibitor vorinostat is a useful immunotherapy for ameliorating West Nile virus infection (3). 1. Yin, Chen, Schul, et al. (2009) Proc. Natl. Acad. Sci. U.S.A. 106, 20435. 2. Deng, Yeo, Ye, et al. (2014) J. Virol. 88, 11915. 3. Nelson, Roe, Orillo, Shi & Verma (2015) Antiviral Res. 122, 39.

Nitecapone

This antiulcer drug (FW = 265.22 g/mol; CAS 116313-94-1), also known as 3-(3,4-dihydroxy-5-nitrobenzylidene)-2,4-pentanedione and OR-462, targets catechol O-methyltransferase (COMT) and was originally intended to be a treatment for Parkinson Disease. Later work, however, showed that low COMT is associated with increased sensitivity to acute pre-/post-operative pain. In fact, a single nucleotide polymorphism (Val-158-Met) results in the low-activity variant of COMT that is associated with migraine and/or increased risk for fibromyalgia. Nitecapone is a water-soluble compound with antioxidative properties and inhibits oxidation of low-density lipoproteins promoted by copper ions. **Target(s):** catechol O-methyltransferase (1-6); glycosulfatase (7,8); NF-κB activation induced by tumor necrosis factor-α (9); xanthine oxidase (1,2). **1.** Marcocci, Suzuki, Tsuchiya & Packer (1994) *Meth. Enzymol.* **234**, 526. **2.** Suzuki, Tsuchiya, Safadi, Kagan & Packer (1992) *Free Radic. Biol. Med.* **13**, 517. **3.** Kaakkola, Gordin & Mannisto (1994) *Gen. Pharmacol.* **25**, 813. **4.** Nissinen, Linden, Schultz, *et al.* (1988) *Eur. J. Pharmacol.* **153**, 263. **5.** Bonifácio, Palma, Almeida & Soares-da-Silva (2007) *CNS Drug Reviews* **13**, 352. **6.** Schultz & Nissinen (1989) *Biochem. Pharmacol.* **38**, 3953. **7.** Slomiany, Murty, Piotrowski, Morita & Slomiany (1993) *J. Physiol. Pharmacol.* **44**, 7. **8.** Murty, Piotrowski, Morita, Slomiany & Slomiany (1992) *Biochem. Int.* **26**, 1091. **9.** Suzuki & Packer (1994) *Biochem. Mol. Biol. Int.* **32**, 299.

Nitenpyram

This neonicotinoid (FW = 270.72 g/mol; CAS 150824-47-8), also known as Capstar® and (E)-N-(6-chloro-3-pyridylmethyl)-N-ethyl-N'-methyl-2-nitro-vinylidenediamine, is an insecticide/pesticide used to exterminate external parasites of livestock and pets. Nitenpyram is a short-acting agent that is without effect on insect eggs and pupa. Capstar is therefore especially useful in treating flea infestations in household pets. (**See also** *Imidacloprid*) Nitenpyram exhibits much lower toxicity to nicotinic acetylcholine receptors in mammals.

Nitidine

This DNA-intercalating benzo[c]phenanthridinium alkaloid (FW$_{chloride}$ = 383.83 g/mol; CAS 13063-04-2), isolated from trees belonging to the genus *Zanthoxylum*, inhibits a number of enzymes with nucleotide or DNA substrates or cofactors. **Target(s):** catechol O-methyltransferase (1); DNA-directed DNA polymerase (2,3); DNA ligase (ATP), human, IC$_{50}$ = 69 μM (4); DNA topoisomerase I (5-7); DNA topoisomerase II (5,7); polyadenylic acid polymerase (3); RNA-directed DNA polymerase, *or* reverse transcriptase (2,3,8-10); RNA polymerase (2,3); tRNA methyltransferase (1). **1.** Lee, MacFarlane, Zee-Cheng & Cheng (1977) *J. Pharm. Sci.* **66**, 839. **2.** Sethi (1977) *Ann. N. Y. Acad. Sci.* **284**, 508. **3.** Sethi (1976) *Cancer Res.* **36**, 2390. **4.** Tan, Lee, Lee, *et al.* (1996) *Biochem. J.* **314**, 993. **5.** Prado, Michel, Tillequin, *et al.* (2004) *Bioorg. Med. Chem.* **12**, 3943. **6.** Wang, Johnson & Hecht (1993) *Chem. Res. Toxicol.* **6**, 813. **7.**

Meng, Liao & Pommier (2003) *Curr. Top. Med. Chem.* **3**, 305. **8.** Tan, Pezzuto, Kinghorn & Hughes (1991) *J. Nat. Prod.* **54**, 143. **9.** Kakiuchi, Hattori, Namba, *et al.* (1985) *J. Nat. Prod.* **48**, 614. **10.** Sethi (1979) *J. Nat. Prod.* **42**, 187.

Nitisinone

This herbicidal triketone (FW = 329.23 g/mol), also known as 2-[2-nitro-4-(trifluoromethyl)benzoyl]-1,3 cyclohexane-dione and NTBC, inhibits 4-hydroxyphenylpyruvate dioxygenase in plants and animals. It has been used in the treatment of type I tyrosinemia. **Target(s):** 4-hydroxymandelate synthase (1); 4-hydroxyphenylpyruvate dioxygenase (2-7). **1.** Conrad & Moran (2008) *Inorganica Chim. Acta* **361**, 1197. **2.** Kavana & Moran (2003) *Biochemistry* **42**, 10238. **3.** Mitchell (1996) *Hum. Exp. Toxicol.* **15**, 179. **4.** Ellis, Whitfield, Gowans, *et al.* (1995) *Toxicol. Appl. Pharmacol.* **133**, 12. **5.** Lindstedt, Holme, Lock, Hjalmarson & Strandvik (1992) *Lancet* **340**, 813. **6.** Purpero & Moran (2006) *Biochemistry* **45**, 6044. **7.** Brownlee, Johnson-Winters, Harrison & Moran (2004) *Biochemistry* **43**, 6370.

Nitrazepam

This depressant and hypnosis-inducing agent (FW = 281.27 g/mol; CAS 146-22-5), also known as 1,3-dihydro-7-nitro-5-phenyl-2H-1,4-benzodiazepin-2-one, is practically insoluble in water, but very soluble in ethanol and n-octanol. Note: Nitrazepam is a controlled substance. **Target(s):** alcohol dehydrogenase (1); glycine receptors (2); 3(20)α-hydroxysteroid dehydrogenase (3). **1.** Roig, Bello, Burguillo, Cachaza & Kennedy (1991) *J. Pharm. Sci.* **80**, 267. **2.** Thio, Shanmugam, Isenberg & Yamada (2003) *J. Neurophysiol.* **90**, 89. **3.** Usami, Yamamoto, Shintani, *et al.* (2002) *Biol. Pharm. Bull.* **25**, 441.

Nitric Oxide

$$:\dot{N}=\ddot{O}:$$

This key signaling gas (FW = 30.01 g/mol; CAS 10102-43-9) has an unpaired electron in an antibonding orbital (configuration = $\sigma_1^2\sigma_2^2\sigma_3^2\pi^4\pi^*$; bond-length = 115.06 pm), making it paramagnetic and highly reactive. Nitric oxide can be generated nonenzymatically. Peroxynitrite (ONOO⁻) reacts with reduced glutathione (GSH) to form GS–NO$_2$, which decomposes to NO. (**See** *Nitroprusside; S-Nitroso-N-acetylpenicillamine; 3,3-Bis(aminoethyl)-1-hydroxy-2-oxo-1-triazine; S-Nitrosoglutathione*) In physiological solutions, ·NO has a short half-life (~5 seconds), reacting rapidly with dissolved dioxygen to form the nitrogen dioxide radical (*Reaction*: 2 ·NO + O$_2$ ⇌ 2 ·NO$_2$). In cellular processes, nitric oxide is formed by various constitutive and inducible nitric oxide synthases (*Reaction*: L-arginine + 3/2 NADPH + H⁺ + 2 O$_2$ ⇌ Citrulline + ·NO + 3/2 NADP⁺). Previously called endothelium-derived relaxation factor (EDRF), nitric oxide has many diverse biochemical and physiological roles as a neurotransmitter, vasodilator, nitrogen fixation, and cytostatic agent. Binding of nitric oxide to the heme iron of soluble guanylyl cyclase results in activation of the enzyme. Nitric oxide also activates cyclooxygenase and p21ras. Note that many reported cases of enzyme inhibition by nitric oxide results from the presence of peroxynitrite (**See** *Peroxynitrite*). Although ethylene is a long-known signaling gas in plants, NO, CO, and H$_2$S are the three gasotransmitters known to be active humans. Nitrate, the main nitric oxide metabolite, is excreted in the urine. **Target(s):** aconitase (1-5); adenain, *or* adenovirus proteinase (6); ADP-ribose diphosphatase (7); [β-adrenergic-receptor] kinase (76); alcohol dehydrogenase (8); aldehyde dehydrogenase (9); O^6-

alkylguanine-DNA alkyltransferase (10,11); L-ascorbate peroxidase (12); caspase-3 (13,73,74); catalase (12,93); catechol 2,3-dioxygenase (89); Ca^{2+}-transporting ATPase (72); CF_1 ATP synthase, or chloroplast H^+-transporting two-sector ATPase (14); coagulation factor XIII (15); creatine kinase (16,17); cruzipain (18,19); cystathionine β-synthase (20); cytochrome *c* oxidase, or Complex IV (21-25,96-105); cytochrome *c* peroxidase (94,95); cytochrome P450 (1); dimethylargininase (26,27); DNA ligase (28); dopamine β-monooxygenase (29); epidermal growth factor receptor protein-tyrosine kinase (30); falcipain (19,31); flavin-containing monooxygenase (83); formamidopyrimidine-DNA glycolyase (32); glutamate decarboxylase (33); γ-glutamylcysteine synthetase (43); glutathione peroxidase (92); glyceraldehyde-3-phosphate dehydrogenase (34-36); heme oxygenase (82); HIV-1 protease (37); HIV-1 reverse transcriptase (38); hydrogenase (39,40); hydrogenase (acceptor) (90); hydrogen dehydrogenase (91); indoleamine 2,3-dioxygenase (41,84-88); insulin receptor kinase (42); K^+ current, glutamate-induced (44); α-ketoglutarate dehydrogenase (4); lactoylglutathione lyase, or glyoxalase I (45,46); *Leishmania infantum* cysteine proteinase (19,47); 12-lipoxygenase, or arachidonate 12-lipoxygenase (48); methionine *S*-adenosyltransferase (77-80); methionine synthase, or homocysteine methyltransferase (49); methylmalonyl-CoA mutase (50); Na^+/K^+-exchanging sATPase (51,52); nitrile hydratase (53); nitrogenase (54-56); ornithine decarboxylase (57,58); papain (59,60); phospholipase D (61); poly(ADPribose) polymerase (62); prolyl monooxygenase (63); prostaglandin-I synthase, inhibited at elevated concentrations (64); protein-glutamine γ-glutamyltransferase, or transglutaminase (81); protein-tyrosine-phosphatase (75); protocatechuate 4,5-dioxygenase (89); ribonucleotide reductase (66); sulfinoalanine decarboxylase, or cysteinesulfinate decarboxylase (33); thiosulfate sulfurtransferase, or rhodanese (65); thromboxane-A synthase (64); tryptophan monooxygenase (67,68); xanthine dehydrogenase (69,70); xanthine oxidase (69-71). **1**. Wink, Grisham, Mitchell & Ford (1996) *Meth. Enzymol.* **268**, 12. **2**. Drapier & Hibbs (1996) *Meth. Enzymol.* **269**, 26. **3**. Hausladen & Fridovich (1996) *Meth. Enzymol.* **269**, 37. **4**. Andersson, Leighton, Young, Blomstrand & Newsholme (1998) *Biochem. Biophys. Res. Commun.* **249**, 512. **5**. Gardner, Costantino, Szabo & Salzman (1997) *J. Biol. Chem.* **272**, 25071. **6**. Cao, Baniecki, McGrath, et al. (2003) *FASEB J.* **17**, 2345. **7**. Ribeiro, Cameselle, Fernandez, et al. (1995) *Biochem. Biophys. Res. Commun.* **213**, 1075. **8**. Gergel & Cederbaum (1996) *Biochemistry* **35**, 16186. **9**. DeMaster, Redfern, Quast, Dahlseid & Nagasawa (1997) *Alcohol* **14**, 181. **10**. Liu, Xu-Welliver, Kanugula & Pegg (2002) *Cancer Res.* **62**, 3037. **11**. Laval & Wink (1994) *Carcinogenesis* **15**, 443. **12**. Clark, Durner, Navarre & Klessig (2000) *Mol. Plant Microbe Interact.* **13**. 1380. **13**. Mohr, Zech, Lapetina & Brune (1997) *Biochem. Biophys. Res. Commun.* **238**, 387. **14**. Satoh, Moritani, Ohhashi, Konishi & Ikeda (1994) *Biosci. Biotechnol. Biochem.* **58**, 521. **15**. Catani, Bernassola, Rossi & Melino (1998) *Biochem. Biophys. Res. Commun.* **249**, 275. **16**. Kaasik, Minajeva, De Sousa, Ventura-Clapier & Veksler (1999) *FEBS Lett.* **444**, 75. **17**. Wolosker, Panizzutti & Engelender (1996) *FEBS Lett.* **392**, 274. **18**. Venturini, Salvati, Muolo, et al. (2000) *Biochem. Biophys. Res. Commun.* **270**, 437. **19**. Bocedi, Gradoni, Menegatti & Ascenzi (2004) *Biochem. Biophys. Res. Commun.* **315**, 710. **20**. Taoka & Banerjee (2001) *J. Inorg. Biochem.* **87**, 245. **21**. Wilson & Erecinska (1978) *Meth. Enzymol.* **53**, 191. **22**. Torres & Wilson (1996) *Meth. Enzymol.* **269**, 3. **23**. Torres, Davies, Darley-Usmar & Wilson (1997) *J. Inorg. Biochem.* **66**, 207. **24**. Cassina & Radi (1996) *Arch. Biochem. Biophys.* **328**, 309. **25**. Cleeter, Cooper, Darley-Usmar, Moncada & Schapira (1994) *FEBS Lett.* **345**, 50. **26**. Knipp, Braun, Gehrig, Sack & Vasak (2003) *J. Biol. Chem.* **278**, 3410. **27**. Leiper, Murray-Rust, McDonald & Vallance (2002) *Proc. Natl. Acad. Sci. U.S.A.* **99**, 13527. **28**. Graziewicz, Wink & Laval (1996) *Carcinogenesis* **17**, 2501. **29**. Zhou, Espey, Chen, et al. (2000) *J. Biol. Chem.* **275**, 21241. **30**. Estrada, Gomez, Martin-Nieto, et al. (1997) *Biochem. J.* **326**, 369. **31**. Venturini, Colasanti, Salvati, Gradoni & Ascenzi (2000) *Biochem. Biophys. Res. Commun.* **267**, 190. **32**. Wink & Laval (1994) *Carcinogenesis* **15**, 2125. **33**. Davis, Foos, Wu & Schloss (2001) *J. Biomed. Sci.* **8**, 359. **34**. Albina, Mastrofrancesco & Reichner (1999) *Biochem. J.* **341**, 5. **35**. Mohr, Hallak, de Boitte, Lapetina & Brune (1999) *J. Biol. Chem.* **274**, 9427. **36**. Padgett & Whorton (1995) *Amer. J. Physiol.* **269**, C739. **37**. Persichini, Colasanti, Lauro & Ascenzi (1998) *Biochem. Biophys. Res. Commun.* **250**, 575. **38**. Persichini, Colasanti, Fraziano, et al. (1999) *Biochem. Biophys. Res. Commun.* **258**, 624. **39**. San Pietro (1955) *Meth. Enzymol.* **2**, 861. **40**. Tibelius & Knowles (1984) *J. Bacteriol.* **160**, 103. **41**. Thomas, Mohr & Stocker (1994) *J. Biol. Chem.* **269**, 14457. **42**. Schmid, Hotz-Wagenblatt & Droge (1999) *Antioxid. Redox Signal.* **1**, 45. **43**. Griffith & Mulcahy (1999) *Adv. Enzymol. Relat. Areas Mol. Biol.* **73**,

209. **44**. Sawada, Ichinose & Anraku (2000) *J. Neurosci. Res.* **60**, 642. **45**. Mitsumoto, Kim, Oshima, et al. (2000) *J. Biochem.* **128**, 647. **46**. Mitsumoto, Kim, Oshima, et al. (1999) *Biochem. J.* **344**, 837. **47**. Salvati, Mattu, Colasanti, et al. (2001) *Biochim. Biophys. Acta* **1545**, 357. **48**. Nakatsuka & Osawa (1994) *Biochem. Biophys. Res. Commun.* **200**, 1630. **49**. Nicolaou, Waterfield, Kenyon & Gibbons (1997) *Eur. J. Biochem.* **244**, 876. **50**. Kambo, Sharma, Casteel, et al. (2005) *J. Biol. Chem.* **280**, 10073. **51**. Tipsmark & Madsen (2003) *J. Exp. Biol.* **206**, 1503. **52**. Sato, Kamata, Irifune & Nishikawa (1997) *J. Neurochem.* **68**, 1312. **53**. Nojiri, Yohda, Odaka, et al. (1999) *J. Biochem.* **125**, 696. **54**. Burns & Hardy (1972) *Meth. Enzymol.* **24**, 480. **55**. Webb (1966) *Enzyme and Metabolic Inhibitors*, vol. **2**, p. 292, Academic Press, New York. **56**. Meyer (1981) *Arch. Biochem. Biophys.* **210**, 246. **57**. Bauer, Buga, Fukuto, Pegg & Ignarro (2001) *J. Biol. Chem.* **276**, 34458. **58**. Bauer, Fukuto, Buga, Pegg & Ignarro (1999) *Biochem. Biophys. Res. Commun.* **262**, 355. **59**. Xian, Chen, Liu, Wang & Wang (2000) *J. Biol. Chem.* **275**, 20467. **60**. Venturini, Fioravanti, Colasanti, Persichini & Ascenzi (1998) *Biochem. Mol. Biol. Int.* **46**, 425. **61**. Madesh & Balasubramanian (1997) *FEBS Lett.* **413**, 269. **62**. Sidorkina, Espey, Miranda, Wink & Laval (2003) *Free Radic. Biol. Med.* **35**, 1431. **63**. Metzen, Zhou, Jelkmann, Fandrey & Brune (2003) *Mol. Biol. Cell* **14**, 3470. **64**. Wade & Fitzpatrick (1997) *Arch. Biochem. Biophys.* **347**, 174. **65**. Spallarossa, Forlani, Pagani, et al. (2003) *Biochem. Biophys. Res. Commun.* **306**, 1002. **66**. Lepoivre, Fieschi, Coves, Thelander & Fontecave (1991) *Biochem. Biophys. Res. Commun.* **179**, 442. **67**. Kuhn & Arthur (1996) *J. Neurochem.* **67**, 1072. **68**. Kuhn & Arthur (1997) *J. Neurochem.* **68**, 1495. **69**. Ichimori, Fukahori, Nakazawa, Okamoto & Nishino (1999) *J. Biol. Chem.* **274**, 7763. **70**. Cote, Yu, Zulueta, Vosatka & Hassoun (1996) *Amer. J. Physiol.* **271**, L869. **71**. Houston, Chumley, Radi, Rubbo & Freeman (1998) *Arch. Biochem. Biophys.* **355**, 1. **72**. Ishii, Sunami, Saitoh, et al. (1998) *FEBS Lett.* **440**, 218. **73**. Guo, Xian, Zhang, McGill & Wang (2001) *Bioorg. Med. Chem.* **9**, 99. **74**. Rössig, Fichtlscherer, Breitschopf, et al. (1999) *J. Biol. Chem.* **274**, 6823. **75**. Raugei, Ramponi & Chiarugi (2002) *Cell. Mol. Life Sci.* **59**, 941. **76**. Whalen, Foster, Matsumoto, et al. (2007) *Cell* **129**, 511. **77**. Avila, Corrales, Ruiz, et al. (1998) *Biofactors* **8**, 27. **78**. del Pino, Corrales & Mato (2002) *J. Biol. Chem.* **275**, 23476. **79**. Corrales, Pérez-Mato, Sánchez del Pino, et al. (2002) *J. Nutr.* **132**, 2377S. **80**. Lindermayr, Saalbach, Bahnweg & Durner (2006) *J. Biol. Chem.* **281**, 4285. **81**. Hitomi, Yamagiwa, Ikura, Yamanishi & Maki (2000) *Biosci. Biotechnol. Biochem.* **64**, 2128. **82**. Ding, McCoubrey & Maines (1999) *Eur. J. Biochem.* **264**, 854. **83**. Ryu, Yi, Cha, et al. (2004) *Life Sci.* **75**, 2559. **84**. Fallarino, Vacca, Orabona, et al. (2002) *Int. Immunol.* **14**, 65. **85**. Thomas & Stocker (1999) *Redox Rep.* **4**, 199. **86**. Sono (1989) *Biochemistry* **28**, 5400. **87**. Samelson-Jones & Yeh (2006) *Biochemistry* **45**, 8527. **88**. Thomas, Terentis, Cai, et al. (2007) *J. Biol. Chem.* **282**, 23778. **89**. Arciero, Orville & Lipscomb (1985) *J. Biol. Chem.* **260**, 14035. **90**. Fauque, Czechowski, Berlier, et al. (1992) *Biochem. Biophys. Res. Commun.* **184**, 1256. **91**. Hyman & Arp (1988) *Biochem. J.* **254**, 469. **92**. Asahi, Fujii, Suzuki, et al. (1995) *J. Biol. Chem.* **270**, 21035. **93**. Titov, Petrenko & Vanin (2008) *Biochemistry (Moscow)* **73**, 92. **94**. Osipov, Stepanov, Vladimirov, Kozlov & Kagan (2006) *Biochemistry (Moscow)* **71**, 1128. **95**. Edwards, Kraut & Poulos (1988) *Biochemistry* **27**, 8074. **96**. Nichols & Chance (1974) in *Mol. Mech. Oxygen Activ.* (O. Hayaishi, ed.), p. 479, Academic Press, New York. **97**. Sharpe & Cooper (1998) *J. Biol. Chem.* **273**, 30961. **98**. Collman, Dey, Barile, Ghosh & Decréau (2009) *Inorg. Chem.* **48**, 10528. **99**. Jekabsone, Neher, Borutaite & Brown (2007) *J. Neurochem.* **103**, 346. **100**. Hüttemann, Lee, Kreipke & Petrov (2008) *Neuroscience* **151**, 148. **101**. Parihar, Vaccaro & Ghafourifar (2008) *IUBMB Life* **60**, 64. **102**. Cooper, Mason & Nicholls (2008) *Biochim. Biophys. Acta* **1777**, 867. **103**. Cooper, Davies, Psychoulis, et al. (2003) *Biochim. Biophys. Acta* **1607**, 27. **104**. Mason, Nicholls, Wilson & Cooper (2006) *Proc. Natl. Acad. Sci. U.S.A.* **103**, 708. **105**. Palacios-Callender, Hollis, Frakich, Mateo & Moncada (2007) *J. Cell Sci.* **120**, 160.

4-Nitro-5-aminoimidazole Ribonucleoside 5'-Monophosphate

This synthetic nucleotide (FW$_{free-acid}$ = 340.19 g/mol; Symbol: NAIR), also known as 5-amino-1-(5'-phospho-β-D-ribofuranosyl)-4-nitroimidazole, is a tight-binding inhibitor of *Gallus gallus* phosphoribosylaminoimidazole carboxylase (K_i = 0.34 nM), but a weaker inhibitor (K_i = 0.5 μM) of the *Escherichia coli* enzyme. While the inhibition kinetics suggested this compound as a transition state or reactive intermediate mimic, it is unclear what molecular features of NAIR contribute to the mimetic properties for either of the two proposed mechanisms of AIR carboxylase (3). Significantly, a variety of AIR, CAIR and NAIR analogues were found to exhibit less binding affinity than NAIR, despite only modest structural modifications. Such investigations are significant in asmuch as vertebrates synthesize purines in ten steps (as originally proposed by Buchanan), whereas bacteria, yeast and fungi make an additional pathway intermediate (*i.e.*, 4-carboxy-5-aminoimidazole ribonucleotide). 1. Firestine, Poon, Mueller, Stubbe & Davisson (1994) *Biochemistry* 33, 11927. 2. Firestine & Davisson (1993) *J. Med. Chem.* 36, 3484. 3. Firestine, Wu, Youn & Davisson (2009) *Bioorg. Med. Chem.* 17, 794.

Nitroarachidonate (Nitroarachidonic Acid)

12-Nitroarachidonate

This isomeric mixture of mono-nitrated arachidonic acids (FW = 351.48 g/mol) includes 9-nitroeicosa-5,8,11,14-tetraenoic acid (9-AANO$_2$), 12-nitroeicosa-5,8,11,14-tetraenoic acid (12-AANO$_2$), 14-nitroeicosa-5,8,11,14-tetraenoic acid (14-AANO$_2$), and 15-nitroeicosa-5,8,11,14-tetraenoic acid (15-AANO$_2$) (1). They inhibit both the COX activity (*i.e.*, conversion of AA to prostaglandin G$_2$, *or* PGG$_2$) and POX activity (*i.e.*, reduction of PGG$_2$ to PGH$_2$) of Prostaglandin Endoperoxide H Synthase-1, but *only* the POX activity of PGHS-2. Kinetic analysis suggests that inactivation of PGHS by AANO$_2$ is a two-step process: first, reversible binding (E + I ⇌ E·I), described by K_i, followed by a practically irreversible event (E + I ⇌ E*–I), described by K_i^{*app} and leading to an inactivated enzyme E*–I (2). The plot of k_{obs} *versus* [AANO$_2$] showed a hyperbolic function with k_{inact} = 0.045 s^{-1} and K_i^{*app} = 0.019 μM for PGHS-1 and k_{inact} = 0.057 s^{-1} and K_i^{*app} = 0.020 μM for PGHS-2. Nitroarachidonic acids are likely immunomodulators. 1. Trostchansky, Souza, Ferreira, *et al.* (2007) *Biochemistry* 46, 4645. 2. Trostchansky, Bonilla, Thomas, *et al.* (2011) *J. Biol. Chem.* 286, 12891.

N^ω-Nitro-L-arginine

This arginine derivative (FW$_{zwitterion}$ = 219.20 g/mol; FW$_{HCl-Salt}$ = 269.70 g/mol) is an irreversible inhibitor of constitutive nitric oxide synthase (nNOS) and a reversible inhibitor of inducible nitric oxide synthase (iNOS). N^ω-Nitro-L-arginine inhibits brain nitric-oxide synthase and endothelial nitric-oxide synthase with K_i values of 25 nM and 90 nM, respectively, while the inducible enzyme has a K_i value of 8.1 μM. N^ω-Nitro-L-arginine reduces cardiac output and causes significant vasoconstriction. A good control is the use of N^ω-nitro-D-arginine, an agent that is likewise commercially available. **Target(s):** arginase (1,2); arginine kinase (3,4); catalase (5); nitric-oxide synthase (6-21). 1. Colleluori & Ash (2001) *Biochemistry* 40, 9356. 2. Robertson, Green, Niedzwiecki, Harrison & Grant (1993) *Biochem. Biophys. Res. Commun.* 197, 523. 3. Pereira, Alonso, Ivaldi, *et al.* (2003) *J. Eukaryot. Microbiol.* 50, 132. 4. Pereira, Alonso, Paveto, *et al.* (2000) *J. Biol. Chem.* 275, 1495. 5. Titov, Petrenko & Vanin (2008) *Biochemistry (Moscow)* 73, 92. 6. Garvey, Furfine & Sherman (1996) *Meth. Enzymol.* 268, 339. 7. Griffith & Kilbourn (1996)

Meth. Enzymol. 268, 375. 8. Reif & McCreedy (1995) *Arch. Biochem. Biophys.* 320, 170. 9. Modin, Weitzberg, Hokfelt & Lundberg (1994) *Neuroscience* 62, 189. 10. Furfine, Harmon, Paith & Garvey (1993) *Biochemistry* 32, 8512. 11. Michel, Phul, Stewart & Humphrey (1993) *Brit. J. Pharmacol.* 109, 287. 12. Lambert, Whitten, Baron, *et al.* (1991) *Life Sci.* 48, 69. 13. McCall, Feelisch, Palmer & Moncada (1991) *Brit. J. Pharmacol.* 102, 234. 14. Griffith & Stuehr (1995) *Ann. Rev. Physiol.* 57, 707. 15. Gross, Stuehr, Aisaka, *et al.* (1990) *Biochem. Biophys. Res. Commun.* 170, 96. 16. García-Pascual, Costa, Labadia, Persson & Triguero (1996) *Brit. J. Pharmacol.* 118, 905. 17. Schmidt & Murad (1991) *Biochem. Biophys. Res. Commun.* 181, 1372. 18. Côté & Roberge (1996) *Free Radic. Biol. Med.* 21, 109. 19. Hecker, Walsh & Vane (1991) *FEBS Lett.* 294, 221. 20. Knowles, Palacios, Palmer & Moncada (1990) *Biochem. J.* 269, 207. 21. Wolff & Datto (1992) *Biochem. J.* 285, 201.

N^ω-Nitro-L-arginine Methyl Ester

This arginine derivative (FW = 233.23 g/mol; CAS 51298-62-5; Soluble to 100 mM in water and to 100 mM in DMSO), often abbreviated L-NAME, is much more water soluble than N^ω-nitro-L-arginine and is also an inhibitor of nitric-oxide synthase, albeit not as potent as N^ω-nitro-L-arginine. Note that L-NAME elicits a variety of physiological effects (*e.g.*, induces leukocyte adhesion, inhibits relaxation induced by acetylcholine, impairs reference memory formation, and modulates microvascular permeability). L-NAME inhibits cGMP formation in endothelial cells with an IC$_{50}$ of 3.1 μM (in the presence of 30 μM arginine) and reverses the vasodilatory effects of acetylcholine in rat aorta rings (EC$_{50}$ = 0.54 μM). **Target(s):** arginase (1); nitric-oxide synthase (2-11); protein-arginine deaminase (12); tyrosine:arginine ligase, *or* kyotorphin synthetase (13,14). A good control is the use of the same concentration N^ω-nitro-D-arginine methy ester, an agent that is likewise commercially available. 1. Reisser, Onier-Cherix & Jeannin (2002) *J. Enzyme Inhib. Med. Chem.* 17, 267. 2. Griffith & Kilbourn (1996) *Meth. Enzymol.* 268, 375. 3. Rees, Palmer, Schulz, Hodson & Moncada (1990) *Brit. J. Pharmacol.* 101, 746. 4. Babbedge, Wallace, Gaffen, Hart & Moore (1993) *Neuroreport* 4, 307. 5. Griffith & Stuehr (1995) *Ann. Rev. Physiol.* 57, 707. 6. García-Pascual, Costa, Labadia, Persson & Triguero (1996) *Brit. J. Pharmacol.* 118, 905. 7. Hong, Kim, Choi, *et al.* (2003) *FEMS Microbiol. Lett.* 222, 177. 8. Bodnárová, Martásek & Moroz (2005) *J. Inorg. Biochem.* 99, 922. 9. Xia, Tsai, Berka & Zweier (1998) *J. Biol. Chem.* 273, 25804. 10. Presta, Liu, Sessa & Stuehr (1997) *Nitric Oxide* 1, 74. 11. Genestra, Guedes-Silva, Souza, *et al.* (2006) *Arch. Med. Res.* 37, 328. 12. McGraw, Potempa, Farley & Travis (1999) *Infect. Immun.* 67, 3248. 13. Kawabata, Muguruma, Tanaka & Takagi (1996) *Peptides* 17, 407. 14. Kawabata, Tanaka, Muguruma & Takagi (1995) *Peptides* 16, 1317.

2-Nitrobenzenesulfenyl Chloride

This moisture-sensitive reagent (FW = 189.6 g/mol; CAS 7669-54-7; M.P. = 74–76°C), also called 2-nitrophenylsulfenylchloride, reacts with tryptophanyl and cysteinyl residues in peptides and proteins, in the presence of 20–60% acetic acid (producing a 2-thioether tryptophanyl derivitive and a mixed disulfide, respectively). Reaction conditions for 2-nitrobenzenesulfenyl chloride are often too harsh to observe any effects on enzyme activity. Many investigators argue for the importance of an active-site tryptophanyl residue, when a substrate or active-site metal ion inhibits the labeling of an enzyme. **Target(s):** Brazilian scorpion (*Tityus serrulatus*) toxin VII (Ts γ) (1); cholera toxin (toxicity and ADP-ribosylating activity are lost) (2); king cobra (*Ophiophagus hannah*) venom α-neurotoxin Oh-4 (3); king cobra (*Ophiophagus hannah*) venom α-neurotoxin Oh-7 (3); lactose synthase (4); lysophospholipase, Australian elapid snake (*Pseudechis australis*) (5); lysozyme (6,7); phospholipase A$_2$ (8,9); *Rhizopus* ribonuclease (10). 1. Hassani, Mansuelle, Cestele, *et al.* (1999) *Eur. J. Biochem.* 260, 76. 2. De Wolf, Fridkin, Epstein & Kohn

(1981) *J. Biol. Chem.* **256**, 5481. **3**. Chang, Lin, Chang & Kuo (1995) *J. Protein Chem.* **14**, 89. **4**. Hill & Brew (1975) *Adv. Enzymol. Relat. Areas Mol. Biol.* **43**, 411. **5**. Takasaki & Tamiya (1982) *Biochem. J.* **203**, 269. **6**. Habeeb & Atassi (1969) *Immunochemistry* **6**, 555. **7**. Perraudin, Looze, Vincentelli & Leonis (1979) *Arch. Int. Physiol. Biochim.* **87**, 203. **8**. Chang, Kuo & Chang (1993) *Biochim. Biophys. Acta* **1202**, 216. **9**. Andrião-Escarso, Soares, Rodrigues, *et al* (2000) *Biochimie* **82**, 755. **10**. Sanda & Irie (1980) *J. Biochem.* **87**, 1079.

p-Nitrobenzoate (*p*-Nitrobenzoic Acid)

This aromatic acid (FW$_{free-acid}$ = 167.12 g/mol; CAS 62-23-7; pK_a value of 3.41 at 25°C), also known as 4-nitrobenzoic acid, inhibits propionyl-CoA synthetase, monoamine oxidase, and D-amino-acid oxidase. The free acid (M.P. = 242.4°C) is very slightly soluble (1 g/2380 mL H$_2$O), and the sodium salt is much more soluble). **Target(s):** D-amino-acid oxidase (1-3); benzoate 4-monooxygenase (4); monoamine oxidase (5); propionyl-CoA synthetase (6); thiopurine *S*-methyltransferase, IC$_{50}$ = 0.5 mM (7). **1**. Webb (1966) *Enzyme and Metabolic Inhibitors*, vol. 2, p. 341, Academic Press, New York. **2**. Bartlett (1948) *J. Amer. Chem. Soc.* **70**, 1010. **3**. Yagi, Osawa & Okada (1959) *Biochim. Biophys. Acta* **35**, 102. **4**. Reddy & Vaidyanathan (1976) *Arch. Biochem. Biophys.* **177**, 488. **5**. Kimes & Carr (1982) *Biochem. Pharmacol.* **31**, 2639. **6**. Krahenbuhl & Brass (1991) *Biochem. Pharmacol.* **41**, 1015. **7**. Ames, Selassie, Woodson, *et al.* (1986) *J. Med. Chem.* **29**, 354.

3-(*m*-Nitrobenzoyl)alanine

This unusual aromatic amino acid (FW = 238.20 g/mol) inhibits kynureninase, IC$_{50}$ = 100 μM (1-3) and kynurenine 3-monooxygenase, IC$_{50}$ = 077 μM (1,2,4,5). **1**. Pellicciari, Natalini, Costantino, *et al.* (1994) *J. Med. Chem.* **37**, 647. **2**. Chiarugi, Carpenedo & Moroni (1996) *J. Neurochem.* **67**, 692. **3**. Moroni, Carpenedo & Chiarugi (1996) *Adv. Exp. Med. Biol.* **398**, 203. **4**. Cozzi, Carpenedo & Moroni (1999) *J. Cereb. Blood Flow Metab.* **19**, 771. **5**. Röver, Cesura, Huguenin, Kettler & Szente (1997) *J. Med. Chem.* **40**, 4378.

N-(2-Nitrobenzyl)-*N*-(4-nitrophenylsulfenyl)-β-alanine Hydroxamate

This hydroxamate (FW = 420.41 g/mol) inhibits interstitial collagenase, *or* MMP-1 (K_i = 28 nM), gelatinase A, *or* MMP-2 (K_i = 15 nM), neutrophil collagenase, *or* MMP-8 (K_i = 14 nM), gelatinase B, *or* MMP-9 (K_i = 19 nM), and microbial collagenase (K_i = 30 nM). **1**. Scozzafava, Ilies, Manole & Supuran (2000) *Eur. J. Pharm. Sci.* **11**, 69.

S-(*p*-Nitrobenzyloxycarbonyl)glutathione

This glutathione derivative (FW = 486.46 g/mol) inhibits hydroxyacyl-glutathione hydrolase, *or* glyoxalase II (1-4) and glyoxalase I (5-7). The *o*- and *m*-nitro analogues also inhibit glyoxalase II. **1**. Uotila (1989) in *Coenzymes and Cofactors, Glutathione, Chem. Biochem. Med. Aspects Pt. A* (Dolphin, Poulson & Avromonic, eds.) **3**, 767, Wiley, New York. **2**. Norton, Principato, Talesa, Lupattelli & Rosi (1989) *Enzyme* **42**, 189. **3**. Norton, Elia, Chyan, *et al.* (1993) *Biochem. Soc. Trans.* **21**, 545. **4**. Allen, Lo & Thornalley (1993) *Eur. J. Biochem.* **213**, 1261. **5**. Allen, Lo & Thornalley (1993) *Biochem. Soc. Trans.* **21**, 535 **6**. Allen, Lo & Thornalley (1993) *J. Protein Chem.* **12**, 111. **7**. Cameron, Ridderstrom, Olin, *et al.* (1999) *Biochemistry* **38**, 13480.

3-[*N*$^\epsilon$-(4-Nitrobenzyloxycarbonyl)-L-lysyl]thiazolidine

This dipeptide analogue (FW = 396.47 g/mol), also known as 3-[*N*$^\epsilon$-(4-nitrobenzyloxycarbonyl)-L-lysyl]thiazolidide, is a strong inhibitor of dipeptidyl-peptidase IV. **1**. Schön, Born, Demuth, *et al.* (1991) *Biol. Chem. Hoppe-Seyler* **372**, 305. **2**. Faust, Fuchs, Wrenger, *et al.* (2003) *Adv. Exp. Med. Biol.* **524**, 175. **3**. Stöckel-Maschek, Stiebitz, Faust, *et al.* (2003) *Adv. Exp. Med. Biol.* **524**, 69. **4**. Kähne, Reinhold, Neubert, *et al.* (2000) *Adv. Exp. Med. Biol.* **477**, 131.

S-(*p*-Nitrobenzyl)-6-thioinosine

This potent nucleoside transport inhibitor (FW = 419.41 g/mol; CAS 38048-32-7; *Abbreviation*: NBMPR) binds tightly and reversibly to HeLa cell membrane sites associated with the nucleoside transport. Site-specific binding was assayed with [^{33}S]NBMPR and a competing, unlabeled congener. HeLa cells possess ~10^3 binding sites of a single class that bound NBMPR tightly (K_i ≈ 0.1 nM. Occupancy of these binding sites by NBMPR inhibits uridine and thymidine uptake, but the relationship between these parameters is complex; as binding saturation is approached (at ~5 nM NBMPR), a substantial fraction (25-30%) of the transport capability remained active. By blocking adenosine re-uptake, NBMPR increases the extracellular adenosine concentration, thereby increasing adenosinergic neurotransmission. Indeed, the adenosine transporter (mENT1) is a target for adenosine receptor signaling (2). **Target(s):** adenosine kinase, also alternative substrate (The *meta*-analogue also inhibits.) (3,4); adenosine transport (5-8); thymidine transport (9); uridine transport (10,11). **1**. Lauzon & Paterson (1977) *Mol. Pharmacol.* **13**, 88. **2**. Chaudary, Naydenova, Shuralyova & Coe (2004) *J. Pharmacol. Exp. Ther.* **310**, 1190. **3**. Rais, Al Safarjalani, Yadav, *et al.* (2005) *Biochem. Pharmacol.* **69**, 1409. **4**. Galazka, Striepen & Ullman (2006) *Mol. Biochem. Parasitol.* **149**, 223. **5**. Pearson, Carleton, Hutchings & Gordon (1978) *Biochem. J.* **170**, 265. **6**. Poucher, Brooks, Pleeth, Conant & Collis (1994) *Eur. J. Pharmacol.* **252**, 19. **7**. Jarvis & Young (1980) *Biochem. J.* **190**, 377. **8**. Dhalla, Dodam, Jones & Rubin (2001) *J. Mol. Cell. Cardiol.* **33**, 1143. **9**. Wohlhueter, Marz & Plagemann (1978) *J. Membr. Biol.* **42**, 247. **10**. Heichal, Bibi, Katz & Cabantchik (1978) *J. Membr. Biol.* **39**, 133. **11**. Choi, Shin, Choi, *et al.* (2008) *J. Neurotrauma* **25**, 695.

2-Nitro-4-carboxyphenyl-*N,N*-Diphenylcarbamate

This carbamate (FW = 378.34 g/mol; *Abbreviation*: NCDC) is an inactivator of chymotrypsin and has been used to quantify that enzyme. This reagent produces a modified protein (*i.e.*, diphenylcarbamoyl-chymotrypsin) and releases an equimolar amount of 3-nitro-4 hydroxybenzoate. The latter has a molar extinction coefficient at 410 nm of 3910 $M^{-1}cm^{-1}$, where as NCDC has negligible absorbance at this wavelength. **Target(s):** chymotrypsin (1); envelysin, *or* sea urchin hatching enzyme (2); phospholipase C (3,4); pancreatic elastase II (5). **1.** Erlanger & Edel (1964) *Biochemistry* **3**, 346. **2.** Post, Schuel & Schuel (1988) *Biochem. Cell. Biol.* **66**, 1200. **3.** Saetrum Opgaard, Nothacker, Ehlert & Krause (2000) *Eur. J. Pharmacol.* **406**, 265. **4.** Fink, Finiasz, Sterin-Borda, Borda & de Bracco (1988) *Immunol. Lett.* **17**, 183. **5.** Azuma, Banshou & Suzuki (2001) *J. Protein Chem.* **20**, 577.

4-Nitrocatechol

This aromatic diol (FW = 155.11 g/mol; CAS 3316-09-4; M.P. = 174-176°C), also known as 1,2-dihydroxy-4-nitrobenzene, is an alternative substrate for catechol *O*-methyltransferase and CYP2E1 and inhibits a numberof non-heme iron dioxygenases. Interestingly, Fe(III)-soybean lipoxygenase-1 binds 4-nitrocatechol and produces a green-colored complex, which exhibits λ_{max} values at 385 and 650 nm at pH 7.0. **Target(s):** catechol 1,2-dioxygenase, *or* pyrocatechase (1-3); catechol *O*-methyltransferase, weak alternative substrate (4); catechol oxidase (5,6); 3,4-dihydroxyphenylacetate 2,3-dioxygenase (7,8); 15-lipoxygenase, *or* arachidonate 15-lipoxygenase, IC_{50} = 4.6 µM (9,10); lipoxygenase, soybean (11-14); protocatechuate 2,3-dioxygenase and homoprotocatechuate 2,3-dioxygenase, *or* 3,4-dihydroxyphenylacetate 2,3-dioxygenase (15,16); protocatechuate 3,4 dioxygenase (1,17-22); protocatechuate 4,5-dioxygenase (23); tyrosinase, *or* monophenol monooxygenase (6,24). **1.** Tyson (1975) *J. Biol. Chem.* **250**, 1765. **2.** Itoh (1981) *Agric. Biol. Chem.* **45**, 2787. **3.** Patel, Hou, Felix & Lillard (1976) *J. Bacteriol.* **127**, 536. **4.** Veser (1987) *J. Bacteriol.* **169**, 3696. **5.** Dawson & Magee (1955) *Meth. Enzymol.* **2**, 817. **6.** Dawson & Tarpley (1951) *The Enzymes*, 1st ed., **2** (part 1), 454. **7.** Que, Widom & Crawford (1981) *J. Biol. Chem.* **256**, 10941. **8.** Emerson, Wagner, Reynolds, *et al.* (2005) *J. Biol. Inorg. Chem.* **10**, 751. **9.** Walther, Holzhütter, Kuban, *et al.* (1999) *Mol. Pharmacol.* **56**, 196. **10.** Schewe, Kühn & Rapaport (1986) *Prostaglandins Leukotr. Med.* **23**, 155. **11.** Spaapen, Verhagen, Veldink & Vliegenthart (1980) *Biochim. Biophys. Acta* **617**, 132. **12.** Sadik, Sies & Schewe (2003) *Biochem. Pharmacol.* **65**, 773. **13.** Galliard & Chan (1980) in *The Biochemistry of Plants* (Stumpf, ed.) **4**, 131. **14.** Skrzypczak-Jankun, Borbulevych, Zavodszky, *et al.* (2006) *Acta Crystallogr D Biol. Crystallog.* **62**, 766. **15.** Wolgel & Lipscomb (1990) *Meth. Enzymol.* **188**, 95. **16.** Miller & Lipscomb (1996) *J. Biol. Chem.* **271**, 5524. **17.** May, Phillips & Oldham (1978) *Biochemistry* **17**, 1853. **18.** Vetting, D'Argenio, Ornston & Ohlendorf (2000) *Biochemistry* **39**, 7943. **19.** Que & Epstein (1981) *Biochemistry* **20**, 2545. **20.** Hou (1975) *Biochemistry* **14**, 3899. **21.** Durham, Sterling, Ornston & Perry (1980) *Biochemistry* **19**, 149. **22.** Hou, Lillard & Schwartz (1976) *Biochemistry* **15**, 582. **23.** Zabinski, Münck, Champion & Wood (1972) *Biochemistry* **11**, 3212. **24.** Ullrich & Duppel (1975) *The Enzymes*, 3rd ed., **12**, 253.

Nitroethane & Nitroethane Anion

This nitroalkane (FW = 75.067 g/mol; CAS 79-24-3), which generates a carbanion at the α-carbon in the presence of excess base (pK_a = 8.6), is an alternative substrate for D-amino-acid oxidase, yielding acetaldehyde and nitrite stoichiometrically. When present simultaneously, nitroethane anion and cyanide inhibit this oxidase irreversibly, and spectral properties of the enzyme-bound FAD cofactor indicate that the N^5-adduct is formed. Nitroethane is also a substrate for the nitroalkane-oxidizing enzyme 2-nitropropane dioxygenase. **Target(s):** D-amino acid oxidase, in the presence of cyanide (1,2); flavocytochrome b_2, *or* L-lactate dehydrogenase (cytochrome) (3). **1.** Walsh, Cromartie, Marcotte & Spencer (1978) *Meth. Enzymol.* **53**, 437. **2.** Porter, Voet & Bright (1973) *J. Biol. Chem.* **248**, 4400. **3.** Genet & Lederer (1990) *Biochem. J.* **266**, 301.

Nitroglycerin

This oral vasodilator (FW = 227.09 g/mol; CAS 55-63-0; Abbreviation: GTN; sweet tasting, clear viscous liquid; odorless), also called glycerine trinitrate, glycerol trinitrate, 1,3-dinitrooxypropan-2-yl nitrate, and propane-1,2,3-triyl trinitrate, has been employed as a fast-acting antianginal agent for well over a century. **Mechanism of Nitric Oxide Formation:** When taken sublingually, nitroglycerin relieves angina (acute chest pain) by forming nitric oxide *(See Nitric Oxide)* upon reaction with active-site cysteine thiol groups in aldehyde dehydrogenase-2 (ALDH2). The precise mechanism by which this mitochondrial matrix enzyme catalyzes NO formation remains unknown. **Mechanism of Vasodilatiory Action:** Nitroglycerin is a direct precursor of nitric oxide and consequently a guanylate cyclase activator that promotes vasodilation therapeutically relevant concentrations (≤1 µM) of nitroglycerin. Nitroglycerin relaxes blood vessels mainly by way of activating soluble guanylyl cyclase (sGC) and the cGMP-dependent protein kinase (cGK-I) signal transduction pathway (1-3). Vasodilation strongly correlates with cGK-I–dependent phosphorylation of the vasodilator-stimulated phosphoprotein (VASP), a key actin-filament (+)-end-tracking motor (actoclampin) protein. Consistent with ALDH2's role in nitroglycerin bioactivation was suggested in earlier studies showing that the East Asian variant of ALDH2 exhibits significantly reduced dehydrogenase, esterase, and nitroglycerin denitrating activities (4). **Dynamite:** Nitroglycerin is the famously shock-sensitive explosive that was first prepared by dissolving glycerol in concentrated nitric acid in the presence of sulfuric acid (5). Seeking to increase its safety and reliability, Alfred Nobel invented Dynamite by combining nitroglycerin, sorbents (e.g., powdered shells or clay), and stabilizers. His formulation proved so stable that a blasting cap is required for detonation. **Key Pharmacokinetic Parameters:** *See* Appendix II in Goodman & Gilman's *THE PHARMACOLOGICAL BASIS OF THERAPEUTICS*, 12th Edition (Brunton, Chabner & Knollmann, eds.) McGraw-Hill Medical, New York (2011). **1.** Opelt, Eroglu, Waldeck-Weiermair, *et al.* (2016) *J. Biol. Chem.* **261**, 24076. **2.** Kleschyov, Oelze, Daiber, *et al.* (2003) *Circ. Res.* **93**, e104. **3.** Agvald, *et al.* (2002) *Brit. J. Pharmacol.* **135**, 373. **4.** Beretta, Gorren, Wenzl, *et al.* (2010) *J. Biol. Chem.* **285**, 943. **5.** Sobrero (1847) *Ann.* **64**, 398.

7-Nitroindazole

This NOS inhibitor (FW = 163.13 g/mol; CAS 2942-42-9; Symbol = 7-NI), also named 7-nitro-1*H*-indazole, selectively targets neuronal nitric oxide synthase, IC_{50} = 0.5 µM (1,2). 7-NI also attenuates cocaine kindling and partially blocked the effects produced by activation of the NMDA receptor, but not the effects induced by acute cocaine administration (3). 7-NI treatment provides full protection against dopamine depletion and loss of dopamine transporter binding sites (4), indicating a role for NO in methamphetamine-induced neurotoxicity and suggesting that NOS blockade may be used to manage Parkinson's disease. When delivered at 40 or 80 mg/kg (i.p.), 7-NI inhibits the development of tolerance to the antinociceptive action of morphine (μ-opioid agonist) and U-50,488H (κ-opioid agonist), but lower doses are ineffective (5). **1**. Moore, Babbedge, Wallace, Gaffen & Hart (1993) *Brit. J. Pharmacol.* **108**, 296. **2**. Moore & Bland-Ward (1996) *Meth. Enzymol.* **268**, 393. **3**. Itzhak (1996) *Neuropharmacology* **35**, 1065. **4**. Itzhak & Ali (1996) *J. Neurochem.* **67**, 1770. **5**. Bhargava, Cao & Zhao (1997) *Peptides* **18**, 797.

Nitromethane

This powerful methylating agent (FW = 61.040 g/mol; M.P = –29°C; B.P. = 101.2°C). It is slightly soluble in water (9.5% by volume at 20°C) and will react with dehydroalanyl residues in proteins. Nitromethane's acidity allows it to undergo deprotonation, enabling condensation reactions analogous to those of carbonyl compounds. Under conditions of base catalysis, nitromethane adds to aldehydes in 1,2-addition in the nitroaldol reaction Care should be exercised when handling sodium salts of nitromethane, which are explosive and will burst into flame upon contact with water. **Target(s):** D-amino-acid oxidase (1); glyceraldehyde-3-phosphate dehydrogenase (sulfenic acid form), acyl phosphatase activity (2); histidine ammonia-lyase (3-5); horseradish peroxidase (6); nitronate monooxygenase (7); peroxidase (6); phenylalanine ammonia-lyase (3). **1**. Walsh, Cromartie, Marcotte & Spencer (1978) *Meth. Enzymol.* **53**, 437. **2**. Benitez & Allison (1974) *J. Biol. Chem.* **249**, 6234. **3**. Hanson & Havir (1972) *The Enzymes*, 3rd ed., **7**, 75. **4**. Okamura, Nishida & Nakagawa (1974) *J. Biochem.* **75**, 139. **5**. Brand & Harper (1976) *Arch. Biochem. Biophys.* **177**, 123. **6**. Ator & Ortiz de Montellano (1990) *The Enzymes*, 3rd ed., **19**, 213. **7**. Kido, Soda, Suzuki & Asada (1976) *J. Biol. Chem.* **251**, 6994.

N-(5-(5-Nitro-2-oxo-1,2-dihydro-3H-indol-3-ylidene)4-oxo-2-thioxo-1,3-thiazolidin-3-yl)nicotinamide

This bacterial MurD ligase inhibitor (FW = 427.41 g/mol) targets UDP-*N*-acetylmuramyl-L-alanine:D-glutamate ligase (MurD) belongs to a family of Mur ligases that are essential for the synthesis of bacterial peptidoglycan. Indeed, it is a competitive inhibitor (K_i = 115 μM) with respect to the substrate UDP-MurNAc-l-Ala (UMA). NMR and molecular dynamics data of this new MurD inhibitor reveal a different binding mode compared to the naphthalene-*N*-sulfonyl-D-glutamic acid derivatives. The pattern of the CSP of MurD methyl groups upon binding of this compound clearly indicates that it is located within the UMA binding site and interacts only with the *N*-terminal and central MurD domains. The 2-oxoindolinylidene ring is located withhin the uracil binding site, while the 4-oxo-2-thioxo-1,3-thiazolidine ring and the pyridine ring are close to the part of the binding site occupied by the UMA phosphate groups and *N*-acetylmuramic acid ring respectively. **1**. Simčič, Pureber, Kristan, *et al.* (2014) *Eur. J. Med. Chem.* **83**, 92.

m-Nitrophenol

This protonophore (FW = 139.11 g/mol; CAS 554-84-7) is soluble in water (3.02 g/100 g at 40°C) and ethanol (116.9 g/100 g at 1°C) and has a melting point of 97°C. With a pK_a value of 8.34 at 18°C, *m*-nitrophenol has been used as a pH indicator: a 0.3% solution in 50% ethanol is colorless below

pH 6.8 and yellow above pH 8.6. **Target(s):** catechol 2,3-dioxygenase (1); Ca^{2+}-transporting ATPase (2); β-fructofuranosidase, *or* invertase (3); glycogen phosphorylase (4); haloalkane dehalogenases (5); (*R*)-2-hydroxyglutaryl-CoA dehydratase (6); protocatechuate 3,4-dioxygenase (7). **1**. Kobayashi, Ishida, Horiike, *et al.* (1995) *J. Biochem.* **117**, 614. **2**. Wolosker, Petretski & De Meis (1990) *Eur. J. Biochem.* **193**, 873. **3**. Neuberg & Mandl (1950) *The Enzymes*, 1st ed., **1** (part 1), 527. **4**. Soman & Philip (1974) *Biochim. Biophys. Acta* **358**, 359. **5**. Schindler, Naranjo, Honaberger, *et al.* (1999) *Biochemistry* **38**, 5772. **6**. Muller & Buckel (1995) *Eur. J. Biochem.* **230**, 698. **7**. Hou, Lillard & Schwartz (1976) *Biochemistry* **15**, 582.

o-Nitrophenol

This phenol (FW = 139.11 g/mol; CAS 88-75-5), also known as 2-nitrophenol, is slightly soluble in cold water, soluble in ethanol and hot water, and has a melting point of 44-45°C. Its pK_a value of 7.17 at 25°C and has been used as a pH indicator (a 2% aqueous solution is colorless below pH 5 and yellow above pH 7). **Target(s):** amidase, *or* acetanilide hydrolase (1); *o*-aminophenol oxidase (2); carbonic anhydrase, carbonate dehydratase (3,4); catechol 2,3-dioxygenase (5-7); catechol oxidase (8,9); (*R*)-2 hydroxyglutaryl-CoA dehydratase (10); malate dehydrogenase, weakly inhibited (11); monophenol monooxygenase (9); NADH oxidase (12); NAD(P)H dehydrogenase (quinone) (13-15); quinone reductase (14,15). **1**. Alt, Heymann & Krisch (1975) *Eur. J. Biochem.* **53**, 357. **2**. Suzuki, Furusho, Higashi, Ohnishi & Horinouchi (2006) *J. Biol. Chem.* **281**, 824. **3**. Tibell, Forsman, Simonsson & Lindskog (1985) *Biochim. Biophys. Acta* **829**, 202. **4**. Elleby, Sjoeblom & Lindskog (1999) *Eur. J. Biochem.* **262**, 516. **5**. Hayaishi, Nozaki & Abbott (1975) *The Enzymes*, 3rd ed., **12**, 119. **6**. Kobayashi, Ishida, Horiike, *et al.* (1995) *J. Biochem.* **117**, 614. **7**. Hori, Hashimoto & Nozaki (1973) *J. Biochem.* **74**, 375. **8**. Dawson & Magee (1955) *Meth. Enzymol.* **2**, 817. **9**. Webb (1966) *Enzyme and Metabolic Inhibitors*, vol. 2, p. 298, Academic Press, New York. **10**. Muller & Buckel (1995) *Eur. J. Biochem.* **230**, 698. **11**. Wedding, Hansch & Fukuto (1967) *Arch. Biochem. Biophys.* **121**, 9. **12**. Suzuki (1966) *Enzymologia* **30**, 215. **13**. Wosilait & Nason (1955) *Meth. Enzymol.* **2**, 725. **14**. Wosilait, Nason & Terrell (1954) *J. Biol. Chem.* **206**, 271. **15**. Wosilait & Nason (1954) *J. Biol. Chem.* **208**, 785.

p-Nitrophenol

This highly effective oxidative phosphorylation uncoupler (FW = 139.11 g/mol; CAS 100-02-7; Moderately soluble in cold water; Soluble in ethanol; M.P. = 113-114°C; λ_{max} = 400 nm, ε = 18300 $M^{-1}cm^{-1}$ at pH 10.2), also known as 4-nitrophenol, is a protonophore that dissipates transmembrane proton gradients generated by the electron transport chain. It is also a substrate for a number of UDP-glucuronosyltransferases, phenol sulfotransferases, and cytochrome P450 systems (particularly CYP2E1). *p*-Nitrophenol (pK_a = 7.15 at 25°C) is also a pH indicator (*e.g.*, a 0.1% solution in ethanol is colorless below pH 5.6 and yellow above pH 7.6). 4-Nitrophenol absorbs only weakly at 405 nm (ε = 0.2 $mM^{-1}cm^{-1}$), whereas its nitrophenolate anion absorbs more intensely at 405 nm (ε = 18.1 $mM^{-1}cm^{-1}$). The isosbestic point occurs at 348 nm for 4-nitrophenol/4-nitrophenoxide (ε = 5.4 $mM^{-1}cm^{-1}$). **Action in Dissipating pH-Gradients and Uncoupling Oxidative Phosphorylation:** Facilitated transport of protons across the biological membrane by protonophore is achieved in a multi-step process: (a) an electron-rich, anionic protonophore species (P⁻) is adsorbed onto the net-positively charged face of a biological membrane; (b) protons (H⁺) leave the aqueous solution phase and combine with the anionic protonophore to produce the neutral from (H·P); (c) the latter enters and diffuses across the biological membrane, whereupon H·P dissociates into H⁺ and P⁻ on the opposite face; (d) protons are released from the biological membrane into the other aqueous phase; and (e) the reformed P⁻ diffuses through the membrane, returning to the initial face by electrostatic attraction. Because ATP synthase catalysis is driven by proton gradients, and *p*-nitrophenol prevents proton gradient-driven ATP formation from

ADP and P_i. **Target(s):** β-*N*-acetylhexosaminidase (1); D-amino-acid oxidase (2); β-apiosyl-β-glucosidase (3); L-ascorbate oxidase (5); carbonic anhydrase (6); carboxylesterase (7); catalase (41); catechol 2,3-dioxygenase (40); catechol oxidase (8,9,44); Ca^{2+}-transporting ATPase (10); choline sulfotransferase (11); galactosylgalactosylglucosylceramidase (12); β-glucosidase (13,14); β-glucuronidase (15); glucuronosyltransferase, substrate for UGT1A6 and UGT1A9 (30,34,35); glycogen phosphorylase (36); 4-hydroxybenzoate 3-monooxygenase, weakly inhibited (39); hydroxyglutaryl-CoA dehydratase (16); laccase (42,43); malate dehydrogenase (17); monophenol monooxygenase (18); NADH oxidase (19); NAD(P)H dehydrogenase (quinone) (20-22); neuraminidase (23); nicotinamidase (24); oligo-1,6-glucosidase (25,26); phenol sulfotransferase, *or* aryl sulfotransferase, also alternative substrate (4,32,33); 4-phytase (27); polyphenol oxidase (44); quinine 3-monooxygenase (38); quinone reductase (22); tyrosinase, *or* monophenol monooxygenase (18,28,29,37); (*R*)-2-UDP glucuronosyltransferase (30); xylan 1,4-β-xylosidase (31). **1.** Johnson, Mook & Brady (1972) *Meth. Enzymol.* **28**, 857. **2.** Yagi, Osawa & Okada (1959) *Biochim. Biophys. Acta* **35**, 102. **3.** Hósel & Barz (1975) *Eur. J. Biochem.* **57**, 607. **4.** Kwon, Yun & Choi (2001) *Biochem. Biophys. Res. Commun.* **285**, 526. **5.** Gaspard, Monzani, Casella, *et al.* (1997) *Biochemistry* **36**, 4852. **6.** Sarraf, Saboury & Moosavi-Movahedi (2002) *J. Enzyme Inhib. Med. Chem.* **17**, 203. **7.** Veerabhadrappa (1996) *Insect Biochem. Mol. Biol.* **26**, 287. **8.** Mayer & Harel (1979) *Phytochemistry* **18**, 193. **9.** Richter (1934) *Biochem. J.* **28**, 901. **10.** Petretski, Wolosker & de Meis (1989) *J. Biol. Chem.* **264**, 20339. **11.** Orsi & Spencer (1964) *J. Biochem.* **56**, 81. **12.** Johnson & Brady (1972) *Meth. Enzymol.* **28**, 849. **13.** Dale, Ensley, Kern, Sastry & Byers (1985) *Biochemistry* **24**, 3530. **14.** Nagatomo, Matsushita, Sugamoto & Matsui (2005) *Biosci. Biotechnol. Biochem.* **69**, 128. **15.** Kim, Jin, Jung, Han & Kobashi (1995) *Biol. Pharm. Bull.* **18**, 1184. **16.** Muller & Buckel (1995) *Eur. J. Biochem.* **230**, 698. **17.** Wedding, Hansch & Fukuto (1967) *Arch. Biochem. Biophys.* **121**, 9. **18.** Webb (1966) *Enzyme and Metabolic Inhibitors*, vol. 2, p. 298, Academic Press, New York. **19.** Suzuki (1966) *Enzymologia* **30**, 215. **20.** Wosilait & A. Nason (1955) *Meth. Enzymol.* **2**, 725. **21.** Martius (1963) *The Enzymes*, 2nd ed., **7**, 517. **22.** Wosilait, Nason & Terrell (1954) *J. Biol. Chem.* **206**, 271. **23.** Brossmer, Keilich & Ziegler (1977) *Hoppe-Seyler's Z. Physiol. Chem.* **358**, 391. **24.** Albizati & J. L. Hedrick (1972) *Biochemistry* **11**, 1508. **25.** Suzuki & Tomura (1986) *Eur. J. Biochem.* **158**, 77. **26.** Suzuki, Aoki & Hayashi (1982) *Biochim. Biophys. Acta* **704**, 476. **27.** Ramakrishnan & Bhandari (1979) *Experientia* **35**, 994. **28.** Dawson & Tarpley (1951) *The Enzymes*, 1st ed., **2** (part 1), 454. **29.** van Gastel, Bubacco, Groenen, Vijgenboom & Canters (2000) *FEBS Lett.* **474**, 228. **30.** Watanabe, Nakajima, Ohashi, Kume & Yokoi (2003) *Drug Metab. Dispos.* **31**, 589. **31.** Gomez, Isorna, Rojo & Estrada (2001) *Biochimie* **83**, 961. **32.** Baranczyk-Kuzma, Borchardt, Schasteen & Pinnick (1981) in *Phenol Sulfotransferase, Ment. Health Res.*. p. 55, Macmillan, London. **33.** Seah & Wong (1994) *Biochem. Pharmacol.* **47**, 1743. **34.** Watanabe, Nakajima, Ohashi, Kume & Yokoi (2003) *Drug Metab. Dispos.* **31**, 589. **35.** Matern, Matern, Schelzig & Gerok (1980) *FEBS Lett.* **118**, 251. **36.** Soman & Philip (1974) *Biochim. Biophys. Acta* **358**, 359. **37.** Anosike & Ayaebene (1982) *Phytochemistry* **21**, 1889. **38.** Zhao & Ishizaki (1997) *J. Pharmacol. Exp. Ther.* **283**, 1168. **39.** Sterjiades (1993) *Biotechnol. Appl. Biochem.* **17**, 77. **40.** Kobayashi, Ishida, Horiike, *et al.* (1995) *J. Biochem.* **117**, 614. **41.** Titov, Petrenko & Vanin (2008) *Biochemistry (Moscow)* **73**, 92. **42.** Lehman, Harel & Mayer (1974) *Phytochemistry* **13**, 1713. **43.** Casella, Gullotti, Monzani, *et al.* (2006) *J. Inorg. Biochem.* **100**, 2127. **44.** Xu, Zheng, Meguro & Kawachi (2004) *J. Wood Sci.* **50**, 260.

p-Nitrophenylacetate (*p*-Nitrophenylacetic Acid)

This substituted acetic acid ($FW_{\text{free-acid}}$ = 181.15 g/mol; CAS 830-03-5), also known as 4-nitrobenzeneacetic acid, is a pale yellow solid with a melting point of 153°C and a pK_a of 3.98 at 25°C. The free acid is sparingly soluble in water, but the sodium salt is water-soluble. **Target(s):** carboxypeptidase A (1-4); chymotrypsin (5-7); creatine kinase (8-10); glyceraldehyde 3-phosphate dehydrogenase (11-13); lipase, *or* triacylglycerol lipase (14). **1.** Pétra (1970) *Meth. Enzymol.* **19**, 460. **2.** Smith (1951) *The Enzymes*, 1st ed., **1** (part 2), 793. **3.** Webb (1966) *Enzyme and Metabolic Inhibitors*, vol.

2, p. 365, Academic Press, New York. **4.** Elkins-Kaufman, Neurath & De Maria (1949) *J. Biol. Chem.* **178**, 645. **5.** Hartley & Kilby (1954) *Biochem. J.* **56**, 288. **6.** Balls & Wood (1956) *J. Biol. Chem.* **219**, 245. **7.** Oosterbaan, van Adrichem & Cohen (1962) *Biochim. Biophys. Acta* **63**, 204. **8.** Watts (1973) *The Enzymes*, 3rd ed., **8**, 383. **9.** Clark & Cunningham (1965) *Biochemistry* **4**, 2637. **10.** Lui & Cunningham (1966) *Biochemistry* **5**, 144. **11.** Harris & Waters (1976) *The Enzymes*, 3rd ed., **13**, 1. **12.** Park, Meriwether, Clodfelder & Cunningham (1961) *J. Biol. Chem.* **236**, 136. **13.** Cunningham & Schepman (1963) *Biochim. Biophys. Acta* **73**, 406. **14.** Sigman & Mooser (1975) *Ann. Rev. Biochem.* **44**, 889.

p-Nitrophenyl Phosphate

This organophosphate ($FW_{\text{free-acid}}$ = 219.09 g/mol; λ_{max} = 290 nm at pH 2.3 and 310 nm at pH 9.2, where ε = 9500 $M^{-1}cm^{-1}$ at both pH values) is a chromogenic substrate for many phosphatases, including alkaline phosphatase. The second pK_a value is about 5.7. Note that the λ_{max} value for *p*-nitrophenol product is 400 nm (ε = 18300 $M^{-1}cm^{-1}$ in alkaline pH); however, the product is typically measured at 405-420 nm, because *p*-nitrophenyl phosphate has a slight absorbance at 400 nm. Solutions of *p*-nitrophenyl phosphate decompose slowly and must be stored in the dark at about 4°C. The solid should also be stored in the dark. **Target(s):** arylsulfatase (1); bis(2-ethylhexyl)phthalate esterase (2); glycerol-1,2-cyclic phosphate phosphodiesterase (3); glycerophosphocholine cholinephosphodiesterase (4); nonspecific lipase5; 5'-nucleotidase (6); phosphatidate phosphatase, alternative substrate (7); phosphoprotein phosphatase (8); protein-tyrosine-phosphatase (9,10); pyridoxal phosphatase (11); [pyruvate dehydrogenase (acetyl-transferring)]-phosphatase (12); *Xenopus* ribosomal protein S6 kinase II (13). **1.** Sampson, Vergara, Fedor, Funk & Benkovic (1975) *Arch. Biochem. Biophys.* **169**, 372. **2.** Krell & Sandermann (1984) *Eur. J. Biochem.* **143**, 57. **3.** Clarke & Dawson (1978) *Biochem. J.* **173**, 579. **4.** Kanfer & McCartney (1989) *J. Neurosci. Res.* **24**, 231. **5.** Wells & DiRenzo (1983) *The Enzymes*, 3rd ed., **16**, 113. **6.** Skladanowski & Newby (1990) *Biochem. J.* **268**, 117. **7.** Berg & Wieslander (1997) *Biochim. Biophys. Acta* **1330**, 225. **8.** Polya & Haritou (1988) *Biochem. J.* **251**, 357. **9.** Ballou & Fischer (1986) *The Enzymes*, 3rd ed., **17**, 311. **10.** Hörlein, Gallis, Brautigan & Bornstein (1982) *Biochemistry* **21**, 5577. **11.** Fonda (1992) *J. Biol. Chem.* **267**, 15978. **12.** Reed, Damuni & Merryfield (1985) *Curr. Top. Cell. Regul.* **27**, 41. **13.** Erikson, Maller & Erikson (1991) *Meth. Enzymol.* **200**, 252.

5-Nitro-2-(3-phenylpropylamino)benzoic acid

This chloride current blocker and putative GPR35 agonist (FW = 300.31 g/mol; CAS 107254-86-4; Symbol: NPPB; Solubility: 100 mM in DMSO) inhibits Ca^{2+}-sensitive chloride currents (ICl_{Ca}) evoked by caffeine (10 mM) and by noradrenaline (10 μM) by 58% and 96%, respectively, at a holding potential of –60 mV in K^+-free, caesium-containing solutions, NPPB (10 μM) (1). At a holding potential of –2 mV in K^+-containing solutions, NPPB (10 μM) inhibited charybdotoxin-sensitive K-currents (IBK_{Ca}) induced by noradrenaline (10 μM) and acetylcholine (10 μM) by ~90% (1). In contrast, IBK(Ca) induced by caffeine (10 mM) was unaffected in the presence of NPPB (10 μM). Conversely, IBK_{Ca} elicited by caffeine (2 mM) was reduced by ~50% whereas IBK_{Ca} evoked by noradrenaline (50 μM) was not significantly inhibited by NPPB (1) In K-containing solutions, NPPB (10 μM) abolished spontaneous transient outward currents (STOCs) and induced a slowly-developing outward K^+-current (1). NPPB is also a GPR35 agonist. In human embryonic kidney 293 (HEK293) cells, NPPB activates the GPR35-$G_{i/o}$ and GPR35-G16 pathways (2). It also mobilizes intracellular calcium in a concentration-dependent manner in HEK293 cells coexpressing human, rat or mouse GPR35 and the chimeric G protein G_{qi5}. Kynurenate, an ionotropic glutamate receptor antagonist, and zaprinast, a phosphodiesterase inhibitor, are also GPR35 agonists. **1**. Kirkup, Edwards & Weston (1996) *Br. J. Pharmacol.* **117**, 175. **2**. Taniguchi, Tonai-Kachi & Shinjo (2008) *Pharmacology* **82**, 245.

p-Nitrophenyl riboside 5'-phosphate

This synthetic substrate analogue (FW = g/mol) targets orotate phosphoribosyltransferase (OPRT), an indispensible enzyme for the malaria-causative parasite, *Plasmodium falciparum*. Human and parasite OPRTs are characterized by highly dissociative transition states, showing ribocation character. *p*-Nitrophenyl riboside 5'-phosphate binds surprisingly well to OPRTs with K_d values near 40 nM. Nitrophenyl β-D-ribose 5'-phosphate is a mechanistic tool used to distinguish relative forces for leaving group activation and for ribocation formation (2). Presence of the 5'-phosphate causes only 3-4x increased binding affinity, indicating a relatively weak role of the phosphate region for inhibitor binding. Notably, iminoribitol mimics of the ribocation transition state give K_d values as low as 80 nM. Inhibitors with pyrrolidine groups as ribocation mimics display slightly weaker binding affinities to OPRTs. Analogues with a C^5 pyrimidine carbon-carbon bond as ribocation mimics give K_d values in the range of 80-500 nM. Acyclic inhibitors with achiral serinol groups as the ribocation mimics also display nanomolar inhibition against OPRTs. Despite their tight-binding to the targets, none of these inhibitors is effective in killing *P. falciparum* in culture. **1**. Zhang, Evans, Clinch (2013) *J. Biol. Chem.* **288**, 34746. **2**. Mazzella, Parkin, Tyler, Furneaux & Schramm (1996) *J. Am. Chem. Soc.* **118**, 211.

3-Nitropropionate (3-Nitropropionic Acid)

This naturally occurring excitotoxin (FW = 119.08 g/mol; *Symbol*: 3-NP) is a succinate analogue and the intermediate in the reaction catalyzed by isocitrate lyase, K_i = 1.5 nM (1-4). Inhibition of isocitrate lyase by 3-NP proceeds through a "double slow-onset" process, in which an initial complex (K_i = 3.3 μM) forms during the first minute, followed by formation of a final complex (K_i^* = 44 nM) over the next minutes to hours (5). 3-NP inhibition involves an unfavorable isomerization of active-site Cys191 and His193 to its thiolate-imidazolium pair prior to inhibitor binding, followed by proton transfer from 3-NP to Cys191. Propionate-3-nitronate, the conjugate base of 3-NP, is thus the "true inhibitor", one that does not bind directly, but instead must be generated enzymatically. 3-NP is also a suicide inactivator of succinate dehydrogenase, producing 3-nitroacrylate (O_2N–CH=CH–COO$^-$), which inactivates the enzyme rapidly and irreversibly (6,7). **Other Target(s):** aspartate ammonia-lyase, *or* aspartase (8-11); aspartate carbamoyltransferase (12); cyanate hydratase, *or* cyanase (13); fumarase, by the C3 carbanion (7); nitronate monooxygenase (14). **1**. Schloss & Cleland (1982) *Biochemistry* **21**, 4420. **2**. Sharma, Sharma, Höner zu Bentrup, *et al.* (2000) *Nat. Struct. Biol.* **7**, 663. **3**. Hoyt, Robertson, Berlyn & Reeves (1988) *Biochim. Biophys. Acta* **966**, 30. **4**. Höner zu Bentrup, Miczak, Swenson & Russell (1999) *J. Bacteriol.* **181**, 7161. **5**. Moynihan & Murkin (2013) *Biochemistry* **53**, 178. **6**. Coles, Edmondson & Singer (1979) *J. Biol. Chem.* **254**, 5161. **7**. Alston, Mela & Bright (1977) *Proc. Natl. Acad. Sci. U.S.A.* **74**, 3767. **8**. Porter & Bright (1980) *J. Biol. Chem.* **255**, 4772. **9**. Falzone, Karsten, Conley & Viola

(1988) *Biochemistry* **27**, 9089. **10**. Viola (2000) *Adv. Enzymol. Relat. Areas Mol. Biol.* **74**, 295. **11**. Karsten & Viola (1991) *Arch. Biochem. Biophys.* **287**, 60. **12**. Foote, Lauritzen & Lipscomb (1985) *J. Biol. Chem.* **260**, 9624. **13**. Anderson, Johnson, Endrizzi, Little & Korte (1987) *Biochemistry* **26**, 3938. **14**. Gadda & Fitzpatrick (1999) *Arch. Biochem. Biophys.* **363**, 309.

3-Nitrosobenzamide

This light-yellow *C*-nitroso compound (FW = 150.14 g/mol; CAS 144189-66-2; *Abbreviation*: NOBA) inactivates caspase 3 by *S*-nitrosylation of cysteinyl residues. It will also induce DNA fragmentation and has been shown to remove zinc from the retroviral-type zinc finger of HIV-1 nucleocapsid proteins. **Target(s):** caspase-3 (1); HIV-1 infectivity (2); poly(ADP-ribose) polymerase, *or* NAD$^+$:poly(adenosine diphosphate-D-ribose) ADP-ribosyltransferase (3). **1**. Mihalik, Bauer, Petak, *et al.* (1999) *Int. J. Cancer* **82**, 875. **2**. Rice, Schaeffer, Harten, *et al.* (1993) *Nature* **361**, 473. **3**. Rice, Hillyer, Harten, *et al.* (1992) *Proc. Natl. Acad. Sci. U.S.A.* **89**, 7703.

S-Nitroso-L-cysteine

This modified cysteine (FW = 136.15 g/mol; CAS 51209-75-7) inactivates mammalian γ-glutamylcysteine synthetase. **Target(s):** [β-adrenergic-receptor] kinase (1); catalase (2); γ-glutamylcysteine synthetase (3); lactoylglutathione lyase, *or* glyoxalase I (4). **1**. Whalen, Foster, Matsumoto, *et al.* (2007) *Cell* **129**, 511. **2**. Titov, Petrenko & Vanin (2008) *Biochemistry (Moscow)* **73**, 92. **3**. Griffith & Mulcahy (1999) *Adv. Enzymol. Relat. Areas Mol. Biol.* **73**, 209. **4**. Mitsumoto, Kim, Oshima, *et al.* (1999) *Biochem. J.* **344**, 837.

S-Nitrosoglutathione

This vasodilator and smooth muscle relaxant (FW = 336.32 g/mol; CAS 57564-91-7) is a nitric oxide donor and inhibitor of platelet activation. Ornithine decarboxylase is inhibited by via *S*-transnitrosation. (**See also Nitric Oxide**) **Target(s):** alcohol dehydrogenase-2, *Entamoeba histolytica* (1); aldose reductase (2); annexin II (3); arylamine *N*-acetyltransferase (34); Ca^{2+} channels, L-type (4); caspase-3 (5); catalase (35); cathepsin K (6); coagulation factor XIII (7); creatine kinase (8,9); *Entamoeba histolytica* cysteine proteinases (1); falcipain (10); glutathione-disulfide reductase (11-13); glutathione *S*-transferase (14); glyceraldehyde-3 phosphate dehydrogenase (15,16); hexokinase (17); lactoylglutathione lyase, *or* glyoxalase I (15,18-20); methionine *S*-adenosyltransferase (21-24); ornithine decarboxylase (25,26); papain (27); platelet aggregation (28); prolyl monooxygenase (29); thiosulfate sulfurtransferase, *or* rhodanese (30,31); vacuolar H$^+$-ATPase (32); xanthine dehydrogenase (33); xanthine oxidase (33). **1**. Siman-Tov & Ankri (2003) *Parasitol. Res.* **89**, 146. **2**. Chandra, Srivastava, Petrash, Bhatnagar & Srivastava (1997) *Biochemistry* **36**, 15801. **3**. Liu, Enright, Sun, *et al.* (2002) *Eur. J. Biochem.* **269**, 4277. **4**. Hu, Chiamvimonvat, Yamagishi & Marban (1997) *Circ. Res.* **81**, 742. **5**. Mohr, Zech, Lapetina & Brune (1997) *Biochem. Biophys. Res. Commun.* **238**, 387. **6**. Percival, Ouellet, Campagnolo, Claveau & Li (1999) *Biochemistry* **38**, 13574. **7**. Catani, Bernassola, Rossi & Melino (1998) *Biochem. Biophys. Res. Commun.* **249**, 275. **8**. Konorev, Kalyanaraman & Hogg (2000) *Free Radic. Biol. Med.* **28**, 1671. **9**. Wolosker, Panizzutti &

Engelender (1996) *FEBS Lett.* **392**, 274. **10**. Venturini, Colasanti, Salvati, Gradoni & Ascenzi (2000) *Biochem. Biophys. Res. Commun.* **267**, 190. **11**. Fujii, Hamaoka, Fujii & Taniguchi (2000) *Arch. Biochem. Biophys.* **378**, 123. **12**. Becker, Savvides, Keese, Schirmer & Karplus (1998) *Nat. Struct. Biol.* **5**, 267. **13**. Becker, Gui & Schirmer (1995) *Eur. J. Biochem.* **234**, 472. **14**. Clark & Debnam (1988) *Biochem. Pharmacol.* **37**, 3199. **15**. Sahoo, Sengupta & Ghosh (2003) *Biochem. Biophys. Res. Commun.* **302**, 665. **16**. Mohr, Hallak, de Boitte, Lapetina & Brune (1999) *J. Biol. Chem.* **274**, 9427. **17**. Miller, Ross-Inta & Giulivi (2007) *Amino Acids* **32**, 593. **18**. Mitsumoto, Kim, Oshima, *et al.* (2000) *J. Biochem.* **128**, 647. **19**. Iozef, Rahlfs, Chang, Schirmer & Becker (2003) *FEBS Lett.* **554**, 284. **20**. Mitsumoto, Kim, Oshima, *et al.* (1999) *Biochem.* **344**, 837. **21**. Avila, Corrales, Ruiz, *et al.* (1998) *Biofactors* **8**, 27. **22**. del Pino, Corrales & Mato (2000) *J. Biol. Chem.* **275**, 23476. **23**. Corrales, Pérez-Mato, Sánchez del Pino, *et al.* (2002) *J. Nutr.* **132**, 2377S. **24**. Lindermayr, Saalbach, Bahnweg & Durner (2006) *J. Biol. Chem.* **281**, 4285. **25**. Bauer, Buga, Fukuto, Pegg & Ignarro (2001) *J. Biol. Chem.* **276**, 34458. **26**. Bauer, Fukuto, Buga, Pegg & Ignarro (1999) *Biochem. Biophys. Res. Commun.* **262**, 355. **27**. Xian, Chen, Liu, Wang & Wang (2000) *J. Biol. Chem.* **275**, 20467. **28**. Radomski, Rees, Dutra & Moncada (1992) *Brit. J. Pharmacol.* **107**, 745. **29**. Metzen, Zhou, Jelkmann, Fandrey & Brune (2003) *Mol. Biol. Cell* **14**, 3470. **30**. Kwiecien, Sokolowska, Luchter-Wasylewska & Wlodek (2003) *Int. J. Biochem. Cell Biol.* **35**, 1645. **31**. Spallarossa, Forlani, Pagani, *et al.* (2003) *Biochem. Biophys. Res. Commun.* **306**, 1002. **32**. Forgac (1999) *J. Biol. Chem.* **274**, 1301. **33**. Cote, Yu, Zulueta, Vosatka & Hassoun (1996) *Amer. J. Physiol.* **271**, L869. **34**. Dairou, Atmane, Dupret & Rodrigues-Lima (2003) *Biochem. Biophys. Res. Commun.* **307**, 1059. **35**. Titov, Petrenko & Vanin (2008) *Biochemistry (Moscow)* **73**, 92.

2-Nitro-5-thiocyanatobenzoate

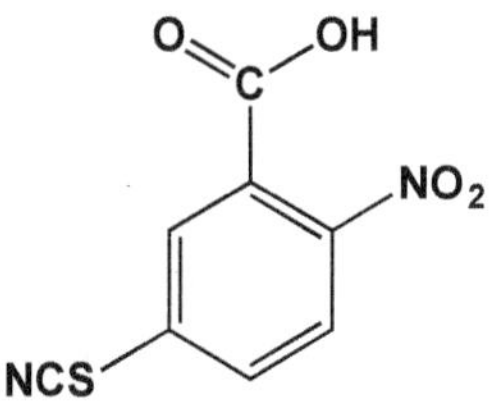

This moisture-sensitive reagent (FW$_{\text{free-acid}}$ = 224.20 g/mol; M.P. = 156-157°C) selectively cyanylates cysteinyl thiols in peptides and proteins, inhibiting enzymes with catalytically essential thiols. Note that these modified cysteinyl residues can be cleaved on the N-terminal side after incubation for a few hours at pH 9. Interestingly, although pig heart 3-oxoacid CoA-transferase is inactivated with this reagent, the thiol group is not catalytically essential. **Target(s):** acetolactate synthase (1); aspartate racemase (2); bifunctional regulatory protein of maize leaf pyruvate,orthophosphate dikinase (3); deoxyribonuclease I (4-6); glucosamine-1-phosphate *N*-acetyltransferase (7,8); glucose-6-phosphatase (9); glutamate racemase (10-12); glutathione synthetase (13); glyceraldehyde-3-phosphate dehydrogenase (14); NADPH:ferrihemoprotein reductase, *or* NADPH:cytochrome P450 reductase (15); nitrile hydratase (16); 2,3-oxidosqualene cyclase (17); 3-oxoacid CoA-transferase, *or* 3-ketoacid CoA-transferase, *or* succinyl-CoA:3-ketoacid CoA-transferase (18); pantetheine hydrolase (19); phosphoribulokinase (20); tryptophanase (21); UDP-*N* acetylmuramoylalanine:D-glutamate ligase (22). **1**. Shin, Chong, Chang & Choi (2000) *Biochem. Biophys. Res. Commun.* **271**, 801. **2**. Yamaguchi, Choi, Okada, *et al.* (1992) *J. Biol. Chem.* **267**, 18361. **3**. Roeske & Chollet (1987) *J. Biol. Chem.* **262**, 12575. **4**. Moore (1981) *The Enzymes*, 3rd ed., **14**, 281. **5**. Liao & Wadano (1979) *J. Biol. Chem.* **254**, 9602. **6**. Kreuder, Dieckhoff, Sittig & Mannherz (1984) *Eur. J. Biochem.* **139**, 389. **7**. Mengin-Lecreulx & van Heijenoort (1994) *J. Bacteriol.* **176**, 5788. **8**. Pompeo, van Heijenoort & Mengin-Lecreulx (1998) *J. Bacteriol.* **180**, 4799. **9**. Clottes, Middleditch & Burchell (2002) *Arch. Biochem. Biophys.* **408**, 33. **10**. Ashiuchi, Tani, Soda & Misono (1998) *J. Biochem.* **123**, 1156. **11**. Nakajima, Tanizawa, Tanaka & Soda (1986) *Agric. Biol. Chem.* **50**, 2823. **12**. Nakajima, Tanizawa, Tanaka & Soda (1988) *Agric. Biol. Chem.* **52**, 3099. **13**. Kato, Tanaka, Nishioka, Kimura & Oda (1988) *J. Biol. Chem.* **263**, 11646. **14**. Byers & Koshland (1975) *Biochemistry* **14**, 3661. **15**. Strobel & Dignam (1978) *Meth. Enzymol.* **52**, 89. **16**. Bonnet, Stevens, de Sousa, *et al.* (2001) *J. Biochem.* **130**, 227. **17**. Balliano, Grosa, Milla, Viola & Cattel (1993) *Lipids* **28**, 903. **18**. Kindman & Jencks (1981) *Biochemistry* **20**, 5183. **19**. Ricci, Nardini, Chiaraluce, Dupre & Cavallini (1986) *Biochim. Biophys. Acta* **870**, 82. **20**. Porter & Hartman (1990) *Arch.*

Biochem. Biophys. **281**, 330. **21**. Nihira, Yasuda, Kakizono, *et al.* (1985) *Eur. J. Biochem.* **149**, 129. **22**. Vaganay, Tanner, van Heijenoort & Blanot (1996) *Microb. Drug Resist.* **2**, 51.

Nitrosoureas

These agents (**See** *Nimustine, Carmustine (BCNU), Lomustine, Semustine, Ethylnitrosourea*) are among the more active, if not the most active, anticancer drugs – both quantitatively (*e.g.*, the *in vivo* log[Kill] of sensitive tumor cells) and qualitatively (*e.g.*, spectrum of tumors responding to treatment), whether administered orally or parenterally. **Primary Mode of Inhibitory Action:** These agents are highly active against experimental meningeal leukemias and intracerebrally implanted transplantable primary tumors of CNS origin, including astrocytomas, ependymoblastomas, and gliomas. By their action on DNA, nitrosoureas often impart additive lethal toxicity when combined with other classes of anticancer agents, such as folate antagonists, purine antagonists, pyrimidine antagonists, DNA polymerase inhibitors, ribonucleotide reductase inhibitors, mitotic inhibitors, DNA intercalators, topoisomerase inhibitors, as well as other types of alkylating agents. The demonstrated lack of cross-resistance of several cancer types for resistance to nitrosoureas and other alkylating suggests these agents need not share a similar mode of inhibitory action, such that resistance to one alkylating is not predictive for cross-resistance to all alkylating agents (1). **DNA Alkylation Mechanism:** Alkyl-nitrosoureas are prodrugs that first must be hydroxylated by cytochrome P450 hydroxylases, followed by chemical rearrangement to generate the active methylating agent.

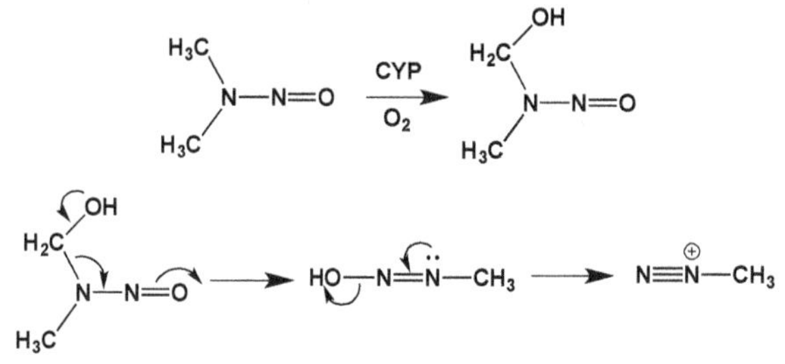

Only then may a base undergo *N*-alkylation or *O*-alkylation, as outlined in the following reaction scheme.

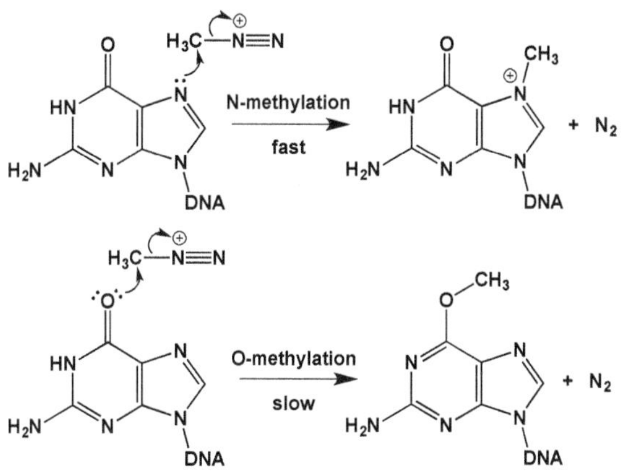

Specificity/Selectivity of Nitrosourea Toxicity: Although DNA modification appears to be the primary mode of nitrosourea cellular toxicity, the tissue- and cancer-type selectivity of drugs like Nimustine, Carmustine, Lomustine, Semustine, and Ethylnitrosourea, appear to be determined by other factors. Thioredoxin reductase and glutathione reductase, for example, function as alternative electron donors for ribonucleotide reductase, and inhibition of these thiol-containing electron transfer proteins by nitrosoureas determines the availability of deoxynucleotide precursors needed for DNA synthesis. Calcium ion is also known to shield thioredoxin reductase from nitrosourea-mediated deactivation, and reduced thioredoxin is known to reactivate thioredoxin reductase and glutathione reductase by reducing and displacing nitrosoureas from their active sites. Celluar levels of glutathione are likewise a factor. Differences in the availability of calcium ion and reduced thioredoxin are likely to explain why certain cell types are more or less susceptible to nitrosoureas. The lipophilicity of some nitrosoureas also accounts for their ability to penetrate the blood-brain barrier, rendering them especially effective against the devastating cancers such as glioblastoma multiforme. Moreover, the unique killing action of streptozotocin (another nitrosourea) on pancreatic β-cells is related to its structural resemblance to glucose, allowing it to gain access to these cells via the GLUT2 transporter. **Nitrosourea Resistance:** As is also true for the action of other alkylating agents (including busulfan, procarbazine, temozolomide, mercaptopurine, 6-thioguanine, Cisplatin, and carboplatin), mounting evidence indicates that tumors acquire resistance to nitrosoureas by loss of DNA-mismatch repair (MMR) activity and MMR-associated signal-transduction pathways causing programmed cell death (3). Other mutations in MMR-deficient tumors provide additional secondary phenotypes for drug resistance to other antineoplastic agents. The antitumor activity of bifunctional nitrosoureas, like BCNU, is based on their ability to form cytosine-guanine crosslinks, and resistance occurs when the enzyme, O^6-alkylguanine-DNA alkyltransferase, repairs an intermediate in crosslink formation (4). Inhibition of this alkyltransferase often restores BCNU sensitivity, although other repair mechanisms may contribute to resistance. **1.** Schabel (1976) *Cancer Treat Rep.* **60**, 665. **2.** Bono (1976) *Cancer Treat Rep.* 60, 699. **3.** Lage & Dietel (1999) *J. Cancer Res. Clin. Oncol.* **125**, 156. **4.** Ludlum (1997) *Cancer Invest.* **15**, 588.

5-Nitrouracil

This uracil derivative (FW = 157.09 g/mol; CAS 611-08-5) is a strong inhibitor of ribonuclease and thymidine phosphorylase. **Target(s):** 4-aminobutyrate aminotransferase, weakly inhibited (1); (*R*)-3-amino-2 methylpropionate:pyruvate aminotransferase (2); isoorotate decarboxylase (3); NAD^+ ADP ribosyltransferase, *or* poly(ADP-ribose) polymerase, IC_{50} = 0.43 mM (4); orotate phosphoribosyltransferase (5); pancreatic ribonuclease, *or* ribonuclease A (6); thymidine phosphorylase (7-10); thymine dioxygenase (11); UDP-glucuronosyltransferase (12); uracil phosphoribosyltransferase (13); uridine phosphorylase, K_i = 56 μM (14,15). **1.** Tamaki, Kubo, Aoyama & Funatsuka (1983) *J. Biochem.* **93**, 955. **2.** Kaneko, Kontani, Kikugawa & Tamaki (1992) *Biochim. Biophys. Acta* **1122**, 45. **3.** Smiley, Angelot, Cannon, Marshall & Asch (1999) *Anal. Biochem.* **266**, 85. **4.** Banasik, Komura, Shimoyama & Ueda (1992) *J. Biol. Chem.* **267**, 1569. **5.** Javaid, el Kouni & Iltzsch (1999) *Biochem. Pharmacol.* **58**, 1457. **6.** Richards & Wyckoff (1971) *The Enzymes*, 3rd ed. (Boyer, ed.), **4**, 647. **7.** Baker (1967) *Design of Active-Site-Directed Irreversible Enzyme Inhibitors*, Wiley, New York. **8.** Kita, Takahashi & Hashimoto (2001) *Biol. Pharm. Bull.* **24**, 860. **9.** Miszczak-Zaborska & Wozniak (1997) *Z. Naturforsch. [C]* **52**, 670. **10.** Baker & Kelley (1971) *J. Med. Chem.* **14**, 812. **11.** Bankel, Lindstedt & Lindstedt (1977) *Biochim. Biophys. Acta* **481**, 431. **12.** Naydenova, Grancharov, Shopova & Golovinsky (1995) *Comp. Biochem. Physiol. C Pharmacol. Toxicol. Endocrinol.* **112**, 321. **13.** Dai, Lee & O'Sullivan (1995) *Int. J. Parasitol.* **25**, 207. **14.** Jimenez, Kranz, Lee, Gero & O'Sullivan (1989) *Biochem. Pharmacol.* **38**, 3785. **15.** Baker & Kelley (1970) *J. Med. Chem.* **13**, 461.

Nitrous Acid

This weak acid (FW = 47.01 g/mol; CAS 7782-77-6; pK_a = 3.35 at 25°C) is known only in solution and gas phases, where it is produced by the action of strong acids on inorganic nitrites. (For example, nitrous acid can be formed by the reaction of sulfuric acid on barium nitrite, with the insoluble barium sulfate coproduct removed by precipitation.) Freshly prepared solutions of nitrous acid are pale blue in color. Nitrous acid solutions are unstable, disproportionating into nitric acid and nitric oxide. **Role in Diazotization:** Reaction of primary aromatic amines (Ar–NH₂) with nitrous acid (the latter generated *in situ* by combining sodium nitrite with a strong acid, such as HCl, H_2SO_4, or HBF_4) leads to the formation diazonium salts (Ar-N⁺=N). Although diazonium salt formation occurs reasonably rapidly (on the 5–15 minute time-scale), the likely stepwise chemical scheme is surprisingly

belabored, detailing all of the rearrangements, with protonation/ deprotonation playing central roles.

The latter is a versatile synthetic intermediate for preparation of halides, azo compounds, and phenols. A classic example is the nitrous acid-mediated conversion of adenine to hypoxanthine. Note that intermediates resulting from the diazotization of primary, aliphatic amines are unstable, converting rapidly to carbocations, attended by loss of dinitrogen (N_2). Nitrous acid also reacts with α-amino groups in free amino acids, peptides, and proteins, resulting in deamination with release of dinitrogen (1). Note that α-amino groups are more reactive than ε amino groups. Other modifications include reaction with tryptophanyl and tyrosyl residues as well as reacting with sulfhydryl groups. Note that modification by nitrous acid does not necessarily result in enzyme inhibition. For example, asparaginase undergoes deamination yet retains its activity (2). (***See also*** *Nitrite; Nitrate; Nitric Oxide*) **Target(s):** alkaline phosphatase (3,4); β-amylase (5-7); chymotrypsin (8); elastase (9); β-fructofuranosidase, *or* invertase (10,11); hurain, *or* cucumisin (12); papain (13); pepsin (14); thrombin (15-17); trypsin (18). **1.** Van Slyke (1929) *J. Biol. Chem.* **83**, 425. **2.** Wagner, Irion, Arens & Bauer (1969) *Biochim. Biophys. Res. Commun.* **37**, 383. **3.** Fernley (1971) *The Enzymes*, 3rd ed., **4**, 417. **4.** Gould (1944) *J. Biol. Chem.* **156**, 365. **5.** Myrbäck & Neumüller (1950) *The Enzymes*, 1st ed., **1** (part 1), 653. **6.** Little & Caldwell (1942) *J. Biol. Chem.* **142**, 585. **7.** Little & Caldwell (1943) *J. Biol. Chem.* **147**, 229. **8.** Sizer (1945) *J. Biol. Chem.* **160**, 547. **9.** Gertler & Hofmann (1967) *J. Biol. Chem.* **242**, 2522. **10.** Sumner & Myrbäck (1930) *Z. physiol. Chem.* **189**, 2. **11.** Neuberg & Mandl (1950) *The Enzymes*, 1st ed., **1** (part 1), 527. **12.** Jaffé (1943) *J. Biol. Chem.* **149**, 1. **13.** Smith & Kimmel (1960) *The Enzymes*, 2nd ed., **4**, 133. **14.** Philpot & Small (1938) *Biochem. J.* **32**, 542. **15.** Lundblad, Kingdon & Mann (1976) *Meth. Enzymol.* **45**, 156. **16.** Seegers (1951) *The Enzymes*, 1st ed., **1** (part 2), 1106. **17.** Seegers (1940) *J. Biol. Chem.* **136**, 103. **18.** Scrimger & Hofmann (1967) *J. Biol. Chem.* **242**, 2528.

Nitrous Oxide

This colorless gas (FW = 44.013 g/mol; CAS 10024-97-2; M.P. = – 90.91°C; B.P. = –88.46°C; Soluble in water, ethanol, and diethyl ether), also known as dinitrogen monoxide and laughing gas, is a linear molecule with two major resonance hybrids (see above). The dipole moment is 0.161 D. Cobalamin-dependent methionine synthases from both *Escherichia coli* and pig liver are irreversibly inactivated by N_2O, with the latter generating hydroxyl radical. **Target(s):** Ca^{2+}-transporting ATPase (1); CO-methylating acetyl-CoA synthase, *or* carbon monoxide dehydrogenase (2); homocysteine methyltransferase, *or* methionine synthase (3); 5-

mcthyltctrahydrofolate: homocysteine methyltransferase (3); nitrogenase (4); tetrachloroethene reductive dehalogenase (5,6). **1**. Franks, Horn, Janicki & Singh (1995) *Anesthesiology* **82**, 108. **2**. Lu & Ragsdale (1991) *J. Biol. Chem.* **266**, 3554. **3**. Frasca, Riazzi & Matthews (1986) *J. Biol. Chem.* **261**, 15823. **4**. Webb (1966) *Enzyme and Metabolic Inhibitors*, vol. **2**, p. 292, Academic Press, New York. **5**. Neumann, Wohlfarth & Diekert (1995) *Arch. Microbiol.* **163**, 276. **6**. van de Pas, Gerritse, De Vos, Schraa & Stams (2001) *Arch. Microbiol.* **176**, 165.

Nitroxyl

This triatomic molecule (FW = 31.014 g/mol; CAS 14332-28-6), also known as nitrosyl hydride, is a degradation product of a number of metabolites and synthetic donors (*e.g.*, *N*-hydroxycyanamide, HONHCN). It can also be generated nonenzymatically from Angeli's salt ($Na_2N_2O_3$) at physiological pH. **Chemical Reactivity:** Deprotonation of HNO is a slow spin-forbidden process that controls the observed chemistry in alkaline solutions. The simplest molecule with nitrogen in the +1 oxidation state, nitroxyl readily dimerizes and then disproportionates to form nitrous oxide and water. A frequently used source for aqueous HNO is sodium trioxodinitrate, also known as Angeli's salt, which is an HNO-releasing reagent (1). It is widely accepted that slow decomposition of the monoprotonated anion occurs through heterolytic N–N bond cleavage (Reaction: $HN_2O_3^- \rightleftharpoons HNO + NO_2^-$). Nitroxyl is also oxidized by O_2 to form peroxynitrite. Superoxide dismutase also produces nitric oxide from HNO, with the latter more cytotoxic than NO. (**See also** *Nitrosobenzene*) **Aldehyde Dehydrogenase Inhibition:** Nitroxyl is a potent inhibitor of aldehyde dehydrogenase, *or* AIDH in a time- and concentration-dependent manner (2-4). IC_{50} values are 1.3 and 1.8 µM under anaerobic conditions, with and without NAD^+, respectively (4). Benzaldehyde, a well-known AlDH substrate, competitively blocks nitroxyl inhibition in the presence of NAD^+, but not in its absence, a finding that fits with the ordered kinetic mechanism employed by this enzyme (4). Dithiothreitol (5 mM) and reduced glutathione (10 mM) also block inhibition of AlDH by Angeli's salt (1). They can also restore activity partially, with the extent of reactivation depending on the pH at which the inactivation occurs. The reversible form of the inhibited enzyme is thought to be an intra-subunit disulfide, and the irreversible form a sulfinamide. Both forms of the inhibited enzyme are derived via a common *N*-hydroxysulfenamide intermediate produced by the addition of nitroxyl to active site thiol(s) (4). **Target(s):** aldehyde dehydrogenase, irreversibly (2-4); aldehyde succinate dehydrogenase (5); nitric oxide synthase (6); papain (7); poly(ADPribose) polymerase (8). **1**. Shafirovich & Lymar (2002) *Proc. Natl. Acad. Sci.* **99**, 7340. **2**. Nagasawa, DeMaster, Goon, Kawle & Shirota (1990) *J. Med. Chem.* **38**, 1872. **3**. Nagasawa, DeMaster, Redfern, Shirota & Goon (1990) *J. Med. Chem.* **33**, 3120. **4**. DeMaster, Redfern & Nagasawa (1998) *Biochem. Pharmacol.* **55**, 2007. **5**. Shiva, Crawford, Ramachandran, *et al.* (2004) *Biochem. J.* **379**, 359. **6**. Kotsonis, Frey, Frohlich, *et al.* (1999) *Biochem. J.* **340**, 745. **7**. Vaananen, Kankuri & Rauhala (2005) *Free Radic. Biol. Med.* **38**, 1102. **8**. Sidorkina, Espey, Miranda, Wink & Laval (2003) *Free Radic. Biol. Med.* **35**, 1431.

Nivalenol

This mycotoxin and protein biosynthesis inhibitor (FW = 312.32 g/mol; CAS 23282-20-4), the toxic principle isolated from *Fusarium nivale*, is structurally related to T-2 toxin and should be handled with extreme care. **1**. Ueno & Fukushima (1968) *Experientia* **24**, 1032. **2**. Ueno, Hosoya, Morita, Ueno & Tatsuno (1968) *J. Biochem.* **64**, 479. **3**. Ohtsubo, Yamada & Saito (1968) *Jpn. J. Med. Sci. Biol.* **21**, 185.

Nivolumab

This fully humanized IgG4 antibody and PD-1 checkpoint inhibitor (MW = 143.6 kDa; CAS 946414-94-4), also known as BMS-936558, MDX1106 and Opdivo®, blocks ligand binding to Programmed Death-1 (PD-1), an immunologic checkpoint receptor on the surface of activated T-cells. When bound by Programmed Death Ligand-1 (PD-L1), T cells quiesce/die, limiting bystander tissue damage and preventing development of autoimmunity during an inflammatory response. Many cancer cell types make PD-L1, preventing T-cells from mounting an immune attack on tumors. Nivolumab binds to PD-1 with high affinity and specificity, and *in vitro* assays demonstrate its ability to enhance T-cell responses and cytokine production in mixed lymphocyte reaction and superantigen or cytomegalovirus-stimulation assays. No antibody-dependent cell-mediated or complement-dependent cytotoxicity is observed *in vitro*, when nivolumab and activated T cells are used as targets. Moreover, nivolumab treatment does not induce adverse events when given to cynomolgus macaques, even at high concentrations. Among previously untreated patients with metastatic melanoma, nivolumab alone or combined with ipilimumab resulted in significantly longer progression-free survival than ipilimumab alone (2). **See** *Ipilimumab* **1**. Wang, Thudium, Han, *et al.* (2014) *Cancer Immunol. Res.* **2**, 846. **2**. Larkin, Chiarion-Sileni, Gonzalez, *et al.* (2015) *New Engl. J. Med.* **373**, 23.

Nizatidine

This anti-ulcerative agent (FW = 331.46 g/mol; CAS 76963-41-2) is a histamine H_2 receptor antagonist that has been used to treat gastroesophogeal reflux disease. **Target(s):** acetylcholinesterase, weakly inhibited (1,2); CYP3A (3); histamine H_2 receptor (4). **1**. Laine-Cessac, Turcant, Premel-Cabic, Boyer & Allain (1993) *Res. Commun. Chem. Pathol. Pharmacol.* **79**, 185. **2**. Kounenis, Voutsas, Koutsoviti-Papadopoulou & Elezoglou (1988) *J. Pharmacobiodyn.* **11**, 767. **3**. Wrighton & Ring (1994) *Pharm. Res.* **11**, 921. **4**. Hill, Ganellin, Timmerman, *et al.* (1997) *Pharmacol. Rev.* **49**, 253.

NLG919

This potent, orally active IDO inhibitor (FW = 282.39 g/mol; CAS 1402836-58-1; Solubility: 56 mg/mL DMSO or Ethanol; <1 mg/mL H_2O), also named α-cyclohexyl-5*H*-imidazo[5,1-*a*]isoindole-5-ethanol, targets indoleamine-(2,3)-dioxygenase, with a K_i of 7 nM and EC_{50} of 75 nM (). The IDO pathway mediates immunosuppressive effects through the metabolic conversion of tryptophan (Trp) to kynurenine (Kyn), triggering downstream signaling through GCN2, mTOR and AHR and affecting T cell differentiation and proliferation. A single oral dose of NLG919 reduces the concentration of plasma and tissue Kyn by ~50%. Similarly, using IDO-expressing mouse DCs from tumor-draining lymph nodes, NLG919 abrogated IDO-induced suppression of antigen-specific T cells (OT-I) *in vitro*, with ED_{50} = 120 nM. **1**. Mautino, Jaipuri, Waldo, *et al.* (2013) *Cancer Res.* **73**, Suppl. 1: Abstract 491.

NLX104, *See Avibactam*

NMS-873

This is an allosteric and specific ERAD inhibitor (FW = 520.67 g/mol; CAS 1418013-75-8), also known as 3-[3-(cyclopentylthio)-5-[[[2-methyl-4'-(methylsulfonyl)[1,1'-biphenyl]-4-yl]oxy]methyl]-4H-1,2,4-triazol-4-yl] pyridine, targets p97 (IC_{50} = 30 nM), reducing its sensitivity to trypsin digestion. p97 (also known as VCP, or Cdc48p in yeast) is an AAA+ ATPase that regulates endoplasmic reticulum-associated degradation (or ERAD). NMS-873 modulates ERAD, resulting in the formation of mislocalized polypeptides that are ubiquitinated and subsequently degraded as part of the Unfolded Protein Response (UPR). NMS-873 is an allosteric inhibitor that binds to a region spanning the D_1 and D_2 domains of adjacent p97 protomers, encompassing elements important for nucleotide-state sensing and ATP hydrolysis. **1.** Magnaghi, D'Alessio, Valsasina, et al. (2013) Nature Chem. Biol. **9**, 548.

NMS-P715

This oral ATP-competitive inhibitor (FW = 675.76 g/mol; CAS 1202055-34-2; IUPAC = N-(2,6-diethylphenyl)-1-methyl-8-({4-[(1-methylpiperidin-4-yl)carbamoyl]-2-(trifluoromethoxy)phenyl}amino)-4,5-dihydro-1H-pyrazolo[4,3-h]quinazoline-3-carboxamide) targets the mitosis-regulating protein kinase MPS1 (IC_{50} = 180 nM, K_i = 1 nM; reversible/time-dependent), causing cell death associated with impaired spindle assembly checkpoint (SAC) function and increasing chromosome mis-segregation (1). MPS1 kinases target chromosome attachment and spindle checkpoint, and, beyond mitosis, they have been implicated in development and cytokinesis (2). In vitro kinase assays of 60 kinases revealed that NMS-P715 is selective for MPS1, displaying activity against only three kinases (CK2, MELK, and NEK6), with IC_{50} <10 µM), but not against other mitotic kinases (e.g., PLK1, CDK1, Aurora A, and Aurora B) or the SAC kinase BUB1. NMS-P715 promotes massive SAC override, with an EC_{50} of 65 nM, in nocodazole-arrested U2OS cells, reducing the G_1 and G_2/M phase of the cell cycle in A2780 ovarian cancer cells. NMS-P715 also inhibits the proliferation of select cancer cell lines (IC_{50} ~ 1 µM), without noticeable effects on 127 normal cell lines. NMS-P715 (90 mg/kg/day) also inhibits growth of A2780 tumor xenograft in mice by 53% without marked loss of body weight. In vitro growth of human and murine pancreatic ductal adenocarcinoma (PDAC) cells is also inhibited by NMS-P715, whereas adipose-derived human mesenchymal stem cells are relatively resistant and maintain chromosome stability upon exposure to NMS-P715 (3). (**See also** MPS1-IN-1 and MPS1-IN-2) **1.** Colombo, Caldarelli, Mennecozzi, et al. (2010) Cancer Res. **70**, 10255. **2.** Liu & Winey (2012) Annu. Rev. Biochem. **81**, 561. **3.** Slee, Grimes, Bansal, et al. (2013) Mol. Cancer Ther. **13**, 3071.

NNRTI, See Etravirine

NNVP-BEZ235, See Dactolisib

NO-328, See Tiagabine

Nobiletin

This citrus-derived hexamethoxyflavone (FW = 402.39 g/mol; CAS 478-01-3), also known as 2-(3,4-dimethoxyphenyl)-5,6,7,8-tetramethoxy-chromen-4-one, suppresses phorbol ester-induced expression of multiple scavenger receptor genes in THP-1 human monocytic cells. Nobiletin also induces G_1 cell cycle arrest, but not apoptosis, in human breast and colon cancer cells. **Target(s):** cAMP phosphodiesterase (1); 15-lipoxygenase (2); mitogen-activated protein/extracellular signal-regulated kinase (or MEK) (3); P-glycoprotein (4,5); phenol sulfotransferase, or aryl sulfotransferase (6); tyrosinase, or monophenol monooxygenase (7,8). **1.** Nikaido, Ohmoto, Sankawa, Hamanaka & Totsuka (1982) Planta Med. **46**, 162. **2.** Malterud & Rydland (2000) J. Agric. Food Chem. **48**, 5576. **3.** Miyata, Sato, Imada, et al. (2008) Biochem. Biophys. Res. Commun. **366**, 168. **4.** Takanaga, Ohnishi, Yamada, et al. (2000) J. Pharmacol. Exp. Ther. **293**, 230. **5.** Ikegawa, Ushigome, Koyabu, et al. (2000) Cancer Lett. **160**, 21. **6.** Nishimuta, Ohtani, Tsujimoto, et al. (2007) Biopharm. Drug Dispos. **28**, 491. **7.** Sasaki & Yoshizaki (2002) Biol. Pharm. Bull. **25**, 806. **8.** Zhang, Lu, Tao, et al. (2007) J. Enzyme Inhib. Med. Chem. **22**, 83 and 91.

Nociceptin

This 17-residue opioid-related neuropeptide and potent anti-analgesic (FW = 1809.04 g/mol; CAS 170713-75-4; Sequence: FGGFTGARKSARKLA NQ), also known as Orphanin FQ, is the endogenous ligand for the nociceptin receptor NOP, or ORL-1. Nociceptin is derived from the pre-pro-nociceptin, the human gene for which is located on Ch8p21. While displaying some opioid-related peptide properties, nociceptin does not act at µ ℜ, κ ℜ, and δ ℜ opioid receptors; moreover, its actions are not antagonized by the opioid antagonist naloxone. (See Naloxone) Nociceptin is widely distributed in the hypothalamus, brainstem, and forebrain, and both ventral and dorsal horns of the spinal cord.

Nocodazole

This cytotoxic carbamate (FW = 300.33 g/mol; CAS 31430-18-9), also known as methyl N-(5-[2-thienylcarbonyl]-1H-benzimidazol 2-yl) carbamate, reversibly inhibits tubulin polymerization and microtubule assembly. Nocodazole binds to the β-subunit ,of the αβ-heterodimer, promoting rapid microtubule disassembly, while also preventing self-assembly of nascent microtubules. Microtubules in interphase as well as early mitosis are especially susceptible to nocodazole. Cells treated with 30-40 µM nocodazole undergo cell-cycle arrest, accumulating in G_2- and/or M-phase. Depending on the cell type, Golgi complexes are also dispersed when cells are similarly treated for several hours. Unlike high-affinity tubulin ligands (such as colchicine, vinblastine, or taxol), the action of nocodazole is reversible, and cells washed several times with nocodazole-free buffer will soon recover, displaying apparently normal microtubule arrays and Golgi location. When used in conjunction with thymidine, colcemid can be used to synchronize cells in a single division cycle. N-(7-

Nitrobenz-2-oxa-1,3-diazol-4-yl)-colcemid is a useful fluorescent probe for colcemid binding to tubulin (1). **Target(s):** glucose transport by GLUT4 (2); tubulin polymerization (3-6). **1.** Hiratsuka & Kato (1987) *J. Biol. Chem.* **262**, 6318. **2.** Molero, Whitehead, Meerloo & James (2001) *J. Biol. Chem.* **276**, 43829. **3.** Rosenshine, Ruschkowski & Finlay (1994) *Meth. Enzymol.* **236**, 467. **4.** Jordan & Wilson (1998) *Meth. Enzymol.* **298**, 252. **5.** Lee, Field & Lee (1980) *Biochemistry* **19**, 6209. **6.** Head, Lee, Field & Lee (1985) *J. Biol. Chem.* **260**, 11060.

Nodularin

This toxic protein phosphatase 2A inhibitor (FW = 824.96 g/mol; CAS 118399-22-7) from the cyanobacterium *Nodularia spumigena* is a cyclic pentapeptide that contains D-glutamic acid (linked via an isopeptide bond), D-*erythro*-β-methylaspartic acid, L-arginine, *N*-methyldehydrobutyrine (*i.e.*, 2-(methylamino)-2(*Z*) dehydrobutyric acid), and (2*S*,3*S*,8*S*,9*S*)-3-amino-9-methoxy-2,6,8-trimethyl-10-phenyldeca 4,6,-dienoic acid. This last amino acid (often referred to as Adda) is unique to nodularin, motuporin, and microcystins. Different forms of nodularin have been identified, differing in what is known as the variable residue (*i.e.*, the L-arginyl residue in nodularin R) as well as the degree of methylation and the stereochemistry. Motuporin, also known as nodularin-V, has L-valyl residue instead of an arginyl residue. There is a non-toxic nodularin isomer, usually occurring as minor contaminant in nodularin samples, in which the double bond C6-C7 of Adda is in the *cis* configuration. Nodularin is a water-soluble hepatotoxin, which can persist in seawater well after the decay of the cyanobacteria. The process of nodularin degradation takes days or weeks, depending on conditions. On acute exposure, nodularin causes vomiting, nausea, sensation of cold-warm extremities, and anorexia. Liver damage follows if exposure to nodularin continues; in extreme cases, liver necrosis, coma, and death can occur. **Target(s):** phosphoprotein phosphatase (1-8); protein phosphatase 1 (1,2,5); protein phosphatase 2A (1-4,6,7); protein phosphatase 4 (7); protein phosphatase 5 (7). **1.** Honkanen, Dukelow, Zwiller, *et al.* (1991) *Mol. Pharmacol.* **40**, 577. **2.** Gulledgea, Aggena, Huangb, Nairnc & Chamberlin (2002) *Curr. Med. Chem.* **9**, 1991. **3.** Matsushima, Yoshizawa, Watanabe, *et al.* (1990) *Biochem. Biophys. Res. Commun.* **171**, 867. **4.** Yoshizawa, Matsushima, Watanabe, *et al.* (1990) *J. Cancer Res. Clin. Oncol.* **116**, 609. **5.** Stubbs, Tran, Atwell, *et al.* (2001) *Biochim. Biophys. Acta* **1550**, 52. **6.** Zabrocki, Swiatek, Sugajska, *et al.* (2002) *Eur. J. Biochem.* **269**, 3372. **7.** Gallego & Virshup (2005) *Curr. Opin. Cell Biol.* **17**, 197. **8.** Pilecki, Grzyb, Zien, Sekula & Szyszka (2000) *J. Basic Microbiol.* **40**, 251.

Nogalamycin

This intercalating anthracycline-based antibiotic (FW$_{free-base}$ = 787.82 g/mol; CAS 1404-15-5), produced by *Streptomyces nogalater*, inhibits DNA-directed RNA polymerase. **Target(s):** deoxyribonuclease (1); DNA-directed RNA polymerase (2-5); DNA helicase/ATPase (6-12); DNA topoisomerase I (13,14); nucleoside-triphosphatase (12); ribonuclease (1). **1.** Zeleznick & Sweeney (1967) *Arch. Biochem. Biophys.* **120**, 292. **2.** Kersten, Kersten, Szybalski & Fiandt (1966) *Biochemistry* **5**, 236. **3.** Chowdhury, Chowdhury & Neogy (1981) *Indian J. Med. Res.* **73**, 90. **4.** Ellem & Rhode (1970) *Biochim. Biophys. Acta* **209**, 415. **5.** Sethi (1971) *Prog. Biophys. Mol. Biol.* **23**, 67. **6.** Phan, Ehtesham, Tuteja & Tuteja (2003) *Eur. J. Biochem.* **270**, 1735. **7.** Pham & Tuteja (2002) *Biochem. Biophys. Res. Commun.* **294**, 334. **8.** Tuteja & Phan (1998) *Plant Physiol.* **118**, 1029. **9.** Tuteja & Phan (1998) *Biochem. Biophys. Res. Commun.* **244**, 861. **10.** Bachur, Yu, Johnson, *et al.* (1992) *Mol. Pharmacol.* **41**, 993. **11.** George, Ghate, Matson & Besterman (1992) *J. Biol. Chem.* **267**, 10683. **12.** Borowski, Niebuhr, Schmitz, *et al.* (2002) *Acta Biochim. Pol.* **49**, 597. **13.** Tudan, Jackson, Higo & Burt (2003) *Inflamm. Res.* **52**, 8. **14.** van Dross & Sanders (2002) *Antimicrob. Agents Chemother.* **46**, 2145. **15.** Biswas, Chakraborty, Choudhury & Neogy (1983) *Indian J. Exp. Biol.* **21**, 517.

Nojirimycin

This sugar derivative (FW$_{free-base}$ = 179.17 g/mol; pK_a = 5.3; CAS 15218-38-9), also called 5-amino-5-deoxy-D-glucopyranose, was isolated in 1966 from the culture broth of *Streptomyces roseochromogenes* R-468 and *S. lavendulae* SF-425, becoming the first 5-aminosugar found in nature. It was was soon found to be a potent inhibitor of a number of α- and β-glucosidases, presumably by mimicking the oxacarbenium ion intermediate formed during catalysis. This inhibitor is relatively unstable, owing to the presence of the hydroxyl group at the C_1 position, and should be stored as the bisulfite adduct. The reduced form, 1-deoxynojirimycin, is more stable and can be prepared from nojirimycin by reaction with sodium borohydride or by catalytic hydrogenation with platinum. 1-Deoxynojirimycin has also been found naturally. (***See also*** *1-Deoxynojirimycin; N-Methyl-1-deoxynojirimycin; N-Nonyl-1-deoxynojirimycin*) **Target(s):** α-amylase (1); amylo-α-1,6-glucosidase/4-α-glucanotransferase, *or* glycogen debranching enzyme (2,3); cellobiose phosphorylase (4,5); dextransucrase (19); glucan 1,3-β glucosidase (6); α-glucosidase (1,7,8,18); β-glucosidase (2,8-14); glycoprotein processing (1,7,15); mannosyl oligosaccharide glucosidase, *or* glucosidase I (15,16); sucrose α-glucosidase, *or* sucrase (17); β xylosidase/β-glucosidase (4). **1.** Elbein (1987) *Meth. Enzymol.* **138**, 661. **2.** Gillard & Nelson (1977) *Biochemistry* **16**, 3978. **3.** Takrama & Madsen (1988) *Biochemistry* **27**, 3308. **4.** Sasaki (1988) *Meth. Enzymol.* **160**, 468. **5.** Sasaki, Tanaka, Nakagawa & Kainuma (1983) *Biochem. J.* **209**, 803. **6.** Bucheli, Durr, Buchala & Meier (1985) *Planta* **166**, 530. **7.** Schwarz & Datema (1984) *Trends Biochem. Sci.* **9**, 32. **8.** Mooser (1992) *The Enzymes*, 3rd ed., **20**, 187. **9.** Niwa, Inoue, Tsuruoka, Koaze & Niida (1970) *Agric. Biol. Chem.* **34**, 966. **10.** Sadana, Patil & Shewale (1988) *Meth. Enzymol.* **160**, 424. **11.** Donsimoni, Legler, Bourbouze & Lalegerie (1988) *Enzyme* **39**, 78. **12.** Yang, Akao & Kobashi (1995) *Biol. Pharm. Bull.* **18**, 1175. **13.** Sanyal, Kundu, Dube & Dube (1988) *Enzyme Microb. Technol.* **10**, 91. **14.** Shewale & Sadana (1981) *Arch. Biochem. Biophys.* **207**, 185. **15.** Hettkamp, Legler & Bause (1984) *Eur. J. Biochem.* **142**, 85. **16.** Hettkamp, Bause & Legler (1982) *Biosci. Rep.* **2**, 899. **17.** Hanozet, Pircher, Vanni, Oesch & Semenza (1981) *J. Biol. Chem.* **256**, 3703. **18.** Minamiura, Matoba, Nishinaka & Yamamoto (1982) *J. Biochem.* **91**, 809. **19.** Kobayashi, Yokoyama & Matsuda (1986) *Agric. Biol. Chem.* **50**, 2585.

Nojirimycin 6-Phosphate

This nojirimycin derivative (FW = 179.17 g/mol; CAS 116026-31-4) is a mechanism-based inactivator of rabbit muscle phosphoglucomutase. The phosphorylated form of the enzyme catalyzes the phosphorylation of nojirimycin 6-phosphate at the C_1 hydroxyl, and this intermediate rapidly eliminates orthophosphate to form an imine and the dephosphorylated enzyme. The dephosphorylated enzyme is then rephosphorylated by the D-glucose 1,6-bisphosphate cofactor, concomitantly forming D-glucose 6-phosphate. The imine is nonenzymatically hydrated back to nojirimycin 6-phosphate. Occasionally (*i.e.*, ~5% of the times), the imine isomerizes to a compound that is not processed by PGM. **1.** Kim & Raushel (1988) *Biochemistry* **27**, 7328.

Nolatrexed

This novel lipophilic and orally active antineoplastic antifolate (FW = 284.34 g/mol; CAS 147149-76-6), also known as AG-337, Thymitaq™, and 2-amino-6-methyl-5-(4-pyridylthio)-1*H*-quinazolin-4-one, inhibits thymidylate synthase (1,2), inducing by S-phase arrest, biphasic mitochondrial alterations, and caspase-dependent apoptosis in leukemia cells (3). Lipophilic inhibitors should gain entry in ways that are independent of the reduced folate carrier transporter (RFC) for uptake, allowing them to avoid resistance arising from impaired RFC. **1.** Tong, Liu-Chen, Ercikan-Abali, *et al.* (1998) *J. Biol. Chem.* **273**, 11611. **2.** Webber, Bleckman, Attard, *et al.* (1993) *J. Med. Chem.* **36**, 733. **3.** Sakoff & Ackland (2000) *Cancer Chemother. Pharmacol.* **46**, 477.

Nomifensine

This antidepressant (FW$_{free-base}$ = 238.33 g/mol; CAS 89664-20-0 (free base) and 32795-47-4 (maleate salt)), also known as 1,2,3,4-tetrahydro-2-methyl-4-phenyl-8-isoquinolinamine, binds to the dopamine transporter at a site that is topologically distinct from that for cocaine. Nomifensine has an intravenous LD_{50} of 72 mg/kg in rats. **Target(s):** amine oxidase (1-3); dihydropteridine reductase (4); dopamine transporter (5-7). **1.** Obata & Yamanaka (2000) *Neurosci. Lett.* **286**, 131. **2.** Egashira, Takayama & Yamanaka (1999) *Jpn. J. Pharmacol.* **81**, 115. **3.** Izumi, Togashi, Hayakari, Hayashi & Ozawa (1976) *Tohoku J. Exp. Med.* **118**, 223. **4.** Shen, Sheng & Abell (1984) *J. Pharm. Pharmacol.* **36**, 411. **5.** Dembiec-Cohen (1998) *J. Neural. Transm.* **105**, 735. **6.** Meiergerd & Schenk (1994) *J. Neurochem.* **63**, 1683. **7.** Ehsanullah & Turner (1977) *Brit. J. Clin. Pharmacol.* **4** Suppl. **2**, 159S.

Nonactin

This ionophoric antibiotic (FW = 736.94 g/mol; CAS 6833-84-7), produced by certain strains of *Streptomyces*, is a neutral ionophoric homologue of monactin. When present at or above 10^{-7} M, nonactin binds to membranes and transports monovalent cations, particularly potassium and ammonium ions. Nonactin's selectivity is: $NH_4^+ > K^+ \approx Rb^+ > Cs^+ > Na^+$ (1-3).

Target(s): adenylate cyclase (4); oxidative phosphorylation, as an ionophoric uncoupler (1,3,5); photophosphorylation, as an ionophoric uncoupler (3). **1.** Pressman (1976) *Ann. Rev. Biochem.* **45**, 501. **2.** Haynes & Pressman (1974) *J. Membr. Biol.* **18**, 1. **3.** Izawa & Good (1972) *Meth. Enzymol.* **24**, 355. **4.** Kuo & Dill (1968) *Biochem. Biophys. Res. Commun.* **32**, 333. **5.** Graven, Lardy, Johnson & Rutter (1966) *Biochemistry* **5**, 1729.

Nonoxynols

These nonionic detergents (CAS 26027-38-3), which differ only in their degree of polymerization, are frequently employed to solubilize membrane-bound or membrane-associated enzymes. Nonoxynol-9 (CAS 20427-84-3), which has an average of nine repeating ethoxy groups, is an especially effective spermatocide solubilizes acrosin, a protease released from the sperm acrosome to cleave egg surface glycoproteins, thereby facilitating fertilization. (*See also* Nonadet P-40) **1.** Muller-Esterl & Schill (1982) *Andrologia* **14**, 309.

N-Nonyl-1-deoxynojirimycin

This sugar analogue (FW = 289.42 g/mol; CAS 81117-35-3) inhibits processing α-glucosidases and is also highly effective in inhibiting hepatitis B virus and bovine viral diarrhea virus in cell-based assays. Viral DNA replication is also inhibited. It is believed that *N*-nonyl-1-deoxynojirimycin prevents the proper encapsidation of the viral pregenomic RNA. In addition, *N*-nonyl-1-deoxynojirimycin can act as a chemical chaperone. Gaucher disease is a lysosomal storage disorder caused by deficient lysosomal β-glucosidase. The presence of this inhibitor in fibroblast culture medium leads to a two-fold increase in the activity of N370S β-glucosidase, the most common mutation causing Gaucher disease. It has been proposed that this inhibitor chaperones β-glucosidase folding, stabilizes the enzyme in transit from the endoplasmic reticulum to the Golgi, and enables proper trafficking to the lysosome. **Target(s):** α-glucosidase (1,2); α-glucosidase I, *or* mannosyl-oligosaccharide α-glucosidase I (2); α-glucosidase II, mannosyl-oligosaccharide α-glucosidase II (2); α-1,6-glucosidase (1); β-glucosidase (3); glucosylceramidase, *or* ceramide glucosyltransferase (4,5). **1.** Andersson, Reinkensmeier, Butters, Dwek & Platt (2004) *Biochem. Pharmacol.* **67**, 697. **2.** Mellor, Neville, Harvey, *et al.* (2004) *Biochem. J.* **381**, 867. **3.** Sawkar, Cheng, Beutler, *et al.* (2002) *Proc. Natl. Acad. Sci. U.S.A.* **99**, 15428. **4.** Legler & Liedtke (1985) *Biol. Chem. Hoppe Seyler* **366**, 1113. **5.** Mellor, Neville, Harvey, *et al.* (2004) *Biochem. J.* **381**, 861.

2-*n*-Nonyl-4-hydroxyquinoline-*N*-oxide

This quinoline derivative (FW = 287.40 g/mol; *Abbreviations*: NOQNO or NQNO) inhibits mitochondrial and chloroplast electron transport. It is more effective than 2-*n*-heptyl-4-hydroxyquinoline-*N*-oxide, albeit ten-times less soluble. **Target(s):** cytochrome *bo₃* ubiquinol oxidase complex, *Escherichia coli* (1); electron transport, chloroplast (2); electron transport, mitochondrial (3-8); sulfide:quinone reductase (9-11); ubiquinol: cytochrome-c reductase, or Complex III (3-6,12,13). **1**. Musser, Stowell, Lee, Rumbley & Chan (1997) *Biochemistry* **36**, 894. **2**. Izawa & Good (1972) *Meth. Enzymol.* **24**, 355. **3**. Rieske (1967) *Meth. Enzymol.* **10**, 239. **4**. Hatefi (1978) *Meth. Enzymol.* **53**, 35. **5**. von Jagow & Link (1986) *Meth. Enzymol.* **126**, 253. **6**. Gao, Wen, Esser, *et al.* (2003) *Biochemistry* **42**, 9067. **7**. Bermudez, Dagger, D'Aquino, Benaim & Dawidowicz (1997) *Mol. Biochem. Parasitol.* **90**, 43. **8**. Riccio, Passarella, Simone & Quagliariello (1979) *Boll. Soc. Ital. Biol. Sper.* **55**, 2406. **9**. Bronstein, Schutz, Hauska, Padan & Shahak (2000) *J. Bacteriol.* **182**, 3336. **10**. Shahak, Arieli, Padan & Hauska (1992) *FEBS Lett.* **299**, 127. **11**. Arieli, Padan & Shahak (1991) *J. Biol. Chem.* **266**, 104. **12**. Kamensky, Konstantinov, Kunz & Surkov (1985) *FEBS Lett.* **181**, 95. **13**. Alric, Pierre, Picot, Lavergne & Rappaport (2005) *Proc. Natl. Acad. Sci. U.S.A.* **102**, 15860.

S-(3-Nonynoyl)-N-acetylcysteamine

This acylated acyl-carrier protein analogue (FW = 255.38 g/mol), also known as *N*-acetylcysteamine 3 nonynoate thioester, inhibits 3-hydroxydecanoyl-[acyl-carrier-protein] dehydratase within the fatty acid synthase complex. **1**. Bloch (1971) *The Enzymes*, 3rd ed., **5**, 441. **2**. Helmkamp, Rando, Brock & Bloch (1968) *J. Biol. Chem.* **243**, 3229.

Norartocarpetin

This flavone (FW = 286.24 g/mol; CAS 520-30-9), isolated from the wood of *Artocarpus heterophyllus*, inhibits tyrosinase, *or* monophenool monooxygenase (IC$_{50}$ = 0.46 µM). **1**. Zheng, Cheng, To, Li & Wang (2008) *Mol. Nutr. Food Res.* **52**, 1530. **2**. Likhitwitayawuld, Sritularak & De-Eknamkul (2000) *Planta Med.* **66**, 275. **3**. Ryu, Ha, Curtis-Long, *et al.* (2008) *J. Enzyme Inhib. Med. Chem.* **23**, 922.

Norbuprenorphine, See Buprenorphine

Norcantharidin

This orally bioavailable cantharidin analogue (FW = 168.15 g/mol; CAS 29745-04-8; Symbol: NCTD) also known as 7-oxabicyclo[2.2.1]heptane-2,3-dicarboxylic anhydride, is the demethylated derivative of cantharidin. (As a cyclic carboxylic anhydride, one would anticipate NCTD would indiscriminantly modify many enzymes and proteins via nucleophilic displacement.) NCTD inhibits phosphoprotein phosphatase-2A, *or* PP2A, an enzyme that regulates the restriction point and G$_2$ and M stages of the cell cycle (1). Daily intraperitoneal injection induces dose- and circadian time-dependent transient leukocytosis in normal mice, without accelerating bone marrow (BM) regeneration. NCTD stimulates the cell cycle progression of granulocyte-macrophage colony-forming cells (GM-CFC), increasing DNA synthesis and the frequency of mitosis in short-term human BM cultures. NCTD also stimulates the production of interleukin IL-1β, colony-stimulating activity (CSA) and tumor necrosis factor TNFα (1). Norcantharidin inhibits HeLa cell growth in a time- and dose-dependent manner, showing characteristic apoptotic morphological changes and DNA

fragmentation (2). Caspase family inhibitor (z-VAD-fmk), caspase-8 inhibitor (z-IETD-fmk), caspase-9 inhibitor (Ac-LEHD-CHO), and caspase-3 inhibitor (z-DEVD-fmk) partially prevent norcantharidin-induced apoptosis, but initiator caspase-1 inhibitor (Ac-YVAD-fmk) will not (2). In the DU145 prostate cancer cell line, NCTD induces the degradation of Cdc6, an essential initiation protein for DNA replication, impairing pre-replication complex assembly and inhibiting DNA replication (3). **Target(s):** arylamine *N*-acetyltransferase (4); protein-serine/threonine phosphatases 1, 2A, and 2B (5-7). **1**. Liu, Blazsek, Comisso, *et al.* (1995) *Eur. J. Cancer* **31A**, 953. **2**. An, Gong, Wang, *et al.* (2004) *Acta Pharmacol. Sin.* **25**, 1502. **3**. Chen, Wan, Ding, *et al.* (2013) *Int. J. Mol. Med.* **32**, 43. **4**. Wu, Chung, Chen & Tsauer (2001) *Am. J. Chin. Med.* **29**, 161. **5**. Baba, Hirukawa, Tanohira & Sodeoka (2003) *J. Amer. Chem. Soc.* **125**, 9740. **6**. McCluskey, Ackland, Gardiner, Walkom & Sakoff (2001) *Anticancer Drug Des.* **16**, 291. **7**. McCluskey, Walkom, Bowyer, *et al.* (2001) *Bioorg. Med. Chem. Lett.* **11**, 2941.

Nordihydroguaiaretic Acid

This bis(benzenediol) (FW = 302.37 g/mol; CAS 500-38-9) is a strong lipoxygenase inhibitor that is often used as an antioxidant in foods. The *meso*-form occurs naturally in the creosote bush (*Larrea divaricata*). Nordihydroguaiaretic acid is soluble in ethanol and diethyl ether and slightly soluble in hot water. (*See also Heminordihydroguaiaretic Acid*) **Target(s):** 1-alkylglycerophosphocholine *O*-acetyltransferase (26,27); aryl hydrocarbon monooxygenase, *or* cytochrome P450 (1); ascorbate oxidase (2); carboxylesterase (3); β-carotene (15,16); monooxygenase (29); carotene oxidase (4); catechol *O*-methyltransferase (28); electron transport (5); 9-*cis*-epoxycarotenoid dioxygenase (30,31); 7-ethoxyresorufin *O*-deethylase activity, *or* cytochrome P450 (1); fatty-acid synthase (25); formyltetrahydrofolate synthetase (3); glucuronosyltransferase (6); hyaluronoglucosaminidase, *or* hyaluronidase (7); hydroperoxide dehydratase (8,9); K$^+$ and Ca^{2+} channels, voltage-sensitive (10,11); linoleate diol synthase (32); lipoxygenase (12-14,48-54); 5-lipoxygenase, *or* arachidonate 5-lipoxygenase (15,34-37); 8-lipoxygenase, arachidonate 8-lipoxygenase (33); 12-lipoxygenase, arachidonate 12-lipoxygenase (36,39,40,45-47); 15-lipoxygenase, arachidonate 15-lipoxygenase (38-44); myosin ATPase (16); Na$^+$/K$^+$-exchanging ATPase (17); palmitoyl-CoA hydrolase, *or* acyl-CoA hydrolase (18); phospholipase A$_2$ (19,20); respiratory burst (21); steroid 5α-reductase (22); taurine channel, swelling-activated, ATP-sensitive (23); vesicle-mediated protein transport (24). **1**. Agarwal, Wang, Bik & Mukhtar (1991) *Drug Metab. Dispos.* **19**, 620. **2**. Stark & Dawson (1963) *The Enzymes*, 2nd ed., **8**, 297. **3**. Schegg & Welch (1984) *Biochim. Biophys. Acta* **788**, 167. **4**. Jüttner (1988) *Meth. Enzymol.* **167**, 336. **5**. Bhuvaneswaran & Dakshinamurti (1972) *Biochemistry* **11**, 85. **6**. Basu, Ciotti, Hwang, *et al.* (2004) *J. Biol. Chem.* **279**, 1429. **7**. Girish & Kemparaju (2005) *Biochemistry (Moscow)* **70**, 948. **8**. Grossman, Bergman & Sofer (1983) *Biochim. Biophys. Acta* **752**, 65. **9**. Hamberg & Gerwick (1993) *Arch. Biochem. Biophys.* **305**, 115. **10**. Hatton & Peers (1997) *Brit. J. Pharmacol.* **122**, 923. **11**. Korn & Horn (1990) *Mol. Pharmacol.* **38**, 524. **12**. Tappel (1962) *Meth. Enzymol.* **5**, 539. **13**. Tappel (1963) *The Enzymes*, 2nd ed., **8**, 275. **14**. Sircar, Schwender & Johnson (1983) *Prostaglandins* **25**, 393. **15**. Pace-Asciak & Smith (1983) *The Enzymes*, 3rd ed., **16**, 543. **16**. Hirata, Inamitsu, Hashimoto & Koga (1984) *J. Biochem.* **95**, 891. **17**. Kellet, Barker, Beach & Dempster (1993) *Biochem. Pharmacol.* **45**, 1932. **18**. Maloberti, Lozano, Mele, *et al.* (2002) *Eur. J. Biochem.* **269**, 5599. **19**. Lanni & Becker (1985) *Int. Arch. Allergy Appl. Immunol.* **76**, 214. **20**. Holk, Rietz, Zahn, Quader & Scherer (2002) *Plant Physiol.* **130**, 90. **21**. Rossi, Della Bianca & Bellavite (1981) *FEBS Lett.* **127**, 183. **22**. Hiipakka, Zhang, Dai, Dai & Liao (2002) *Biochem. Pharmacol.* **63**, 1165. **23**. Ballatori & Wang (1997) *Amer. J. Physiol.* **272**, C1429. **24**. Tagaya, Henomatsu, Yoshimori, *et al.* (1996) *J. Biochem.* **119**, 863. **25**. Li, Ma, Wang & Tian (2005) *J. Biochem.* **138**, 679. **26**. Hurst & Bazan (1997) *J. Ocul. Pharmacol. Ther.* **13**, 415. **27**. Yamazaki, Sugatani, Fujii, *et al.* (1994) *Biochem. Pharmacol.* **47**, 995. **28**. Jeffery & Roth (1984) *J. Neurochem.* **42**, 826. **29**. Nagao, Maeda, Lim, Kobayashi & Terao (2000) *Nutr. Biochem.* **11**, 348. **30**. Han, Kitahata, Saito, *et al.* (2004) *Bioorg. Med. Chem. Lett.* **14**, 3033. **31**. Han, Kitahata, Sekimata, *et al.* (2004) *Plant*

Physiol. **135**, 1574. **32**. Brodowsky, Hamberg & Oliw (1994) *Eur. J. Pharmacol.* **254**, 43. **33**. Fürstenberger, Hagedorn, Jacobi, *et al.* (1991) *J. Biol. Chem.* **266**, 15738. **34**. Furakawa, Yoshimoto, Ochi & Yamamoto (1984) *Biochim. Biophys. Acta* **795**, 458. **35**. Pufahl, Kasten, Hills, *et al.* (2007) *Anal. Biochem.* **364**, 2004. **36**. Chi, Jong, Son, *et al.* (2001) *Biochem. Pharmacol.* **62**, 1185. **37**. Mulliez, Leblanc, Girard, Rigaud & Chottard (1987) *Biochim. Biophys. Acta* **916**, 13. **38**. Weinstein, Liu, Gu, *et al.* (2005) *Bioorg. Med. Chem. Lett.* **15**, 1435. **39**. Segraves, Shah, Segraves, *et al.* (2004) *J. Med. Chem.* **47**, 4060. **40**. Vasquez-Martinez, Ohri, Kenyon, Holman & Sepúlveda-Boza (2007) *Bioorg. Med. Chem.* **15**, 7408. **41**. Walther, Holzhütter, Kuban, *et al.* (1999) *Mol. Pharmacol.* **56**, 196. **42**. Schewe, Kühn & Rapaport (1986) *Prostaglandins Leukotr. Med.* **23**, 155. **43**. Burrall, Cheung, Chiu & Goetzl (1988) *J. Invest. Dermatol.* **91**, 294. **44**. Kühn, Barnett, Grunberger, *et al.* (1993) *Biochim. Biophys. Acta* **1169**, 80. **45**. Yokoyama, Mizuno, Mitachi, *et al.* (1983) *Biochim. Biophys. Acta* **750**, 237. **46**. Yoshimoto, Miyamoto, Ochi & Yamamoto (1982) *Biochim. Biophys. Acta* **713**, 638. **47**. Flatman, Hurst, McDonald-Gibson, Jonas & Slater (1986) *Biochim. Biophys. Acta* **883**, 7. **48**. Iny, Grossman & Pinsky (1993) *Int. J. Biochem.* **25**, 1325. **49**. Pinto, Tejeda, Duque & Macías (2007) *J. Agric. Food Chem.* **55**, 5956. **50**. Beneytout, Andrianarison, Rakotoarisoa & Tixier (1989) *Plant Physiol.* **91**, 367. **51**. Fornaroli, Petrussa, Braidot, Vianello & Macri (1999) *Plant Sci.* **145**, 1. **52**. Macias, Pinto & Campillo (1987) *Z. Naturforsch. B* **42**, 1343. **53**. Galliard & Phillips (1971) *Biochem. J.* **124**, 431. **54**. Lorenzi, Maury, Casanova & Berti (2006) *Plant Physiol. Biochem.* **44**, 450.

Norepinephrine

This photosensitive adrenergic neurotransmitter and vasodilator (FW$_{\text{free-base}}$ = 169.18 g/mol; CAS 51-41-2), also known as noradrenaline, L-(−)-norepinephrine, and 4 [(1*R*)-2-amino-1-hydroxyethyl]-1,2-benzenediol, is an agonist for adrenoceptors and the immediate precursor to epinephrine. Norepinephrine is a product of the dopamine β monooxygenase reaction. Note that the amino group has a pK_a value of 8.64, while one of the phenolic groups has a pK_a value of 9.70. **Target(s):** alanine aminotransferase, weakly inhibited (1); aldehyde oxidase (2); aromatic-L amino-acid decarboxylase, *or* dopa decarboxylase (3,4); carnosine synthetase (5); catechol *O*-methyltransferase, product inhibition (6); diacylglycerol cholinephosphotransferase (7); dihydropteridine reductase (8,9); ethanolamine phosphotransferase (7); glutamate decarboxylase, weakly inhibited (1); glutamate dehydrogenase (10); γ-glutamylhistamine synthetase (11); 5-lipoxygenase, *or* arachidonate 5-lipoxygenase (30); phenol sulfotransferase, *or* aryl sulfotransferase (12); phenylalanine 4-monooxygenase (13-15); phenylethanolamine *N*-methyltransferase (16); procollagen lysine 5-dioxygenase (17); pyridoxal kinase (18); L-serine dehydratase (1); sulfinoalanine decarboxylase, *or* cysteinesulfinate decarboxylase (19); tyrosine aminotransferase (1,20,21); tyrosine 3-monooxygenase (22-29); xanthine oxidase (2). **1**. Pestaña, Sandoval & Sols (1971) *Arch. Biochem. Biophys.* **146**, 373. **2**. Panoutsopoulos & Beedham (2004) *Acta Biochim. Pol.* **51**, 649. **3**. Fellman (1959) *Enzymologia* **20**, 366. **4**. Nakazawa, Kumagai & Yamada (1987) *Agric. Biol. Chem.* **51**, 2531. **5**. Ng & Marshall (1976) *Experientia* **32**, 839. **6**. Zaccharia, Dubay, Mi & Jackson (2003) *Hypertension* **42**, 82. **7**. Strosznajder, Radominska-Pyrek & Horrocks (1979) *Biochim. Biophys. Acta* **574**, 48. **8**. Shen (1985) *J. Enzyme Inhib.* **1**, 61. **9**. Shen (1983) *Biochim. Biophys. Acta* **743**, 129. **10**. Smith, Austen, Blumenthal & Nyc (1975) *The Enzymes*, 3rd ed., **11**, 293. **11**. Stein & Weinreich (1982) *J. Neurochem.* **38**, 204. **12**. Pennings, Vrielink, Wolters & van Kempen (1976) *J. Neurochem.* **27**, 915. **13**. Kaufman (1978) *Meth. Enzymol.* **53**, 278. **14**. Kaufman (1987) *Meth. Enzymol.* **142**, 3. **15**. Martínez, Andersson, Haavik & Flatmark (1991) *Eur. J. Biochem.* **198**, 675. **16**. Pohorecky & Baliga (1973) *Arch. Biochem. Biophys.* **156**, 703. **17**. Murray, Cassell & Pinnell (1977) *Biochim. Biophys. Acta* **481**, 63. **18**. Neary, Meneely, Grever & Diven (1972) *Arch. Biochem. Biophys.* **151**, 42. **19**. Guion-Rain, Portemer & Chatagner (1975) *Biochim. Biophys. Acta* **384**, 265. **20**. Granner & Tomkins (1970) *Meth. Enzymol.* **17A**, 633. **21**. Jacoby & La Du (1964) *J. Biol. Chem.* **239**, 419. **22**. Fujisawa & Okuno (1987) *Meth. Enzymol.* **142**, 63. **23**. Udenfriend, Zaltzman-Nirenberg & Nagatsu (1965)

Biochem. Pharmacol. **14**, 837. **24**. Moore & Dominic (1971) *Fed. Proc.* **30**, 859. **25**. Haavik (1997) *J. Neurochem.* **69**, 1720. **26**. Nagatsu, Levitt & Udenfriend (1964) *J. Biol. Chem.* **239**, 2910. **27**. Chaube & Joy (2003) *J. Neuroendocrinol.* **15**, 273. **28**. Neckameyer, Holt & Paradowski (2005) *Biochem. Genet.* **43**, 425. **29**. Gordon, Quinsey, Dunkley & Dickson (2008) *J. Neurochem.* **106**, 1614. **30**. Furakawa, Yoshimoto, Ochi & Yamamoto (1984) *Biochim. Biophys. Acta* **795**, 458.

Norfloxacin

This fluoroquinolone (FW$_{\text{free-acid}}$ = 319.34 g/mol; CAS 70458-96-7) is an orally absorbed antibacterial agent that is structurally related to nalidixic acid, but with fluorine at position-6 and a piperazine ring at position-7. These two changes markedly enhance the *in vitro* antibacterial activity. The antibacterial spectrum includes *Pseudomonas aeruginosa* as well as enteric pathogens. Norfloxacin is also active against both penicillin-susceptible and penicillin resistant strains of *Neisseria gonorrhoeae*. **Primary Mode of Action:** The mechanism of action of norfloxacin involves inhibition of the A subunit of DNA gyrase, which is essential for bacterial DNA replication. The inhibition of DNA gyrase reportedly occurs as a result of binding to a site, which appears after the formation of a gyrase-DNA complex. **Target(s):** CYP1A2 (1); DNA gyrase, *or* DNA topoisomerase II (2-7). **1**. Kinzig-Schippers, Fuhr, Zaigler, *et al.* (1999) *Clin. Pharmacol. Ther.* **65**, 262. **2**. Landini, Pagani, Debiaggi, Cereda & Romero (1989) *Microbiologica* **12**, 247. **3**. Shen, Kohlbrenner, Weigl & Baranowski (1989) *J. Biol. Chem.* **264**, 2973. **4**. Ferrazzi, Peracchi, Biasolo, *et al.* (1988) *Biochem. Pharmacol.* **37**, 188. **5**. Zweerink & Edison (1986) *Antimicrob. Agents Chemother.* **29**, 598. **6**. Sutcliffe, Gootz & Barrett (1989) *Antimicrob. Agents Chemother.* **33**, 2027. **7**. Inoue, Sato, Fujii, *et al.* (1987) *J. Bacteriol.* **169**, 2322.

Norharmane

This indole derivative (FW = 168.20 g/mol; CAS 244-63-3), also known as 9*H*-pyrido[3,4-*b*]indole and β-carboline, is an alkaloid that modulates the activity of a number of oxidases. **Target(s):** *S*-adenosylmethionine decarboxylase (1); aldehyde oxidase (2,3); benzodiazepine receptor (4); cholesterol-5,6-oxide hydrolase (5); CYP1A (6); CYP2E1 (7); DNA topoisomerase I (6,8); IκB kinase (9); indoleamine 2,3-dioxygenase (10-14); monoamine oxidase A15; monoamine oxidase B16; NAD+ ADP-ribosyltransferase, *or* poly(ADP-ribose) polymerase, weakly inhibited (17); steroid 11β-monooxygenase (18); steroid 17α-monooxygenase (18). **1**. Lapinjoki & Gynther (1986) *Biochem. Int.* **12**, 847. **2**. Vila, Kurosaki, Barzago, *et al.* (2004) *J. Biol. Chem.* **279**, 8668. **3**. Yoshihara & Tatsumi (1997) *Arch. Biochem. Biophys.* **338**, 29. **4**. Rommelspacher, Nanz, Borbe, *et al.* (1980) *Naunyn Schmiedebergs Arch. Pharmacol.* **314**, 97. **5**. Palakodety, Vaz & Griffin (1987) *Biochem. Pharmacol.* **36**, 2424. **6**. Nii (2003) *Mutat. Res.* **541**, 123. **7**. Stawowy, Bonnet & Rommelspacher (1999) *Biochem. Pharmacol.* **57**, 511. **8**. Funayama, Nishio, Wakabayashi, *et al.* (1996) *Mutat. Res.* **349**, 183. **9**. Castro, Dang, Soucy, *et al.* (2003) *Bioorg. Med. Chem. Lett.* **13**, 2419. **10**. Chiarugi, Dello Sbarba, Paccagnini, *et al.* (2000) *J. Leukoc. Biol.* **68**, 260. **11**. Sono & Cady (1989) *Biochemistry* **28**, 5392. **12**. Sono (1989) *Biochemistry* **28**, 5400. **13**. Kudo & Boyd (2000) *Biochim. Biophys. Acta* **1500**, 119. **14**. Austin, Astelbauer, Kosim-Satyaputra, *et al.* (2009) *Amino Acids* **36**, 99. **15**. Rommelspacher, May & Salewski (1994) *Eur. J. Pharmacol.* **252**, 51. **16**. Rommelspacher, Meier-Henco, Smolka & Kloft (2002) *Eur. J. Pharmacol.* **441**, 115. **17**. Banasik, Komura, Shimoyama & Ueda (1992) *J. Biol. Chem.* **267**, 1569. **18**. Kühn-Velten (1993) *Eur. J. Pharmacol.* **250**, R1.

Norisoboldine

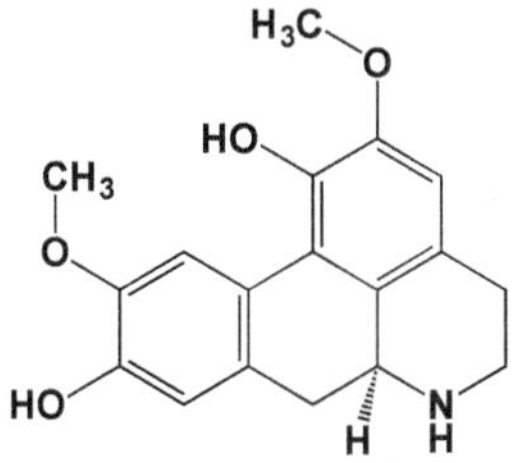

This alkaloid (FW = 313.35 g/mol; CAS 23599-69-1; *Symbol*: NOR) from *Radix linderae* significantly suppresses synovial angiogenesis by selectively inhibiting endothelial cell migration via a cAMP-PKA-NF-κB/Notch1 signaling pathway (1). NOR markedly suppresses VEGF-induced intra-cytoplasmic cAMP production and PKA activation and thereby down-regulating the activation of downstream components of the PKA pathway, including src tyrosine kinase, the actin filament motor component VASP, and eNOS, as well as the transcription factor NF-κB. The transcription activation potential of NF-κB is reduced by NOR. NOR selectively inhibits expression of p-p65 (ser276), but not p-p65 (ser536) or PKAc, indicating that PKAc participates in the regulation of NF-κB by NOR. Simulation of ligand docking suggests that NOR affinity for PKA is likely to be higher than that of the original PKA ligand (1). Norisoboldine also suppresses osteoclast differentiation, most likely by preventing the accumulation of TRAF6-TAK1 complexes and the activation of MAPKs/NF-κB/c-Fos/NFATc1 pathways (2). 1. Lu, Tong, Luo, *et al.* (2013) *PLoS One* 8, e81220. 2. Wei, Tong, Xia, *et al.* (2013) *PLoS One* 8, e59171.

Norleucine

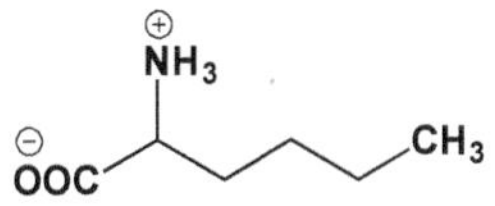

This leucine/isoleucine/methionine analogue (FW = 131.17 g/mol; CAS 327-56-0; pK_a values of 2.34 and 9.81; Solubility: 1.5 g/100 mL H$_2$O at 25°C; insoluble in ethanol; *Abbreviation*: Ahx or Nle), also known as 2-aminohexanoic acid and α-aminocaproic acid, is a lysine deamination product and, as such, is found in collagen. Norvaline substitutes for methionine in protein biosynthesis. **Target(s):** alanine aminotransferase, by L-enantiomer (1); arginase, weakly inhibited (2); homocitrate synthase, by L-enantiomer (3); insulin release (4); lipid methylation (5); membrane alanyl aminopeptidase, microsomal aminopeptidase (6); methionine γ-lyase, by L-enantiomer (7,8); methionyl aminopeptidase, by L-enantiomer (9); methionyl-tRNA synthetase, also alternative substrate (10); ornithine carbamoyltransferase (11,12); saccharopine dehydrogenase (13); tRNA (uracil-5-)-methyltransferase, weakly inhibited (14). 1. Saier & Jenkins (1967) *J. Biol. Chem.* 242, 101. 2. Hunter & Downs (1945) *J. Biol. Chem.* 157, 427. 3. Gray & Bhattacharjee (1976) *J. Gen. Microbiol.* 97, 117. 4. Sener & Malaisse (1984) *Biochimie* 66, 353. 5. David, Clavel-Seres, Clement, Lazlo & Rastogi (1989) *Acta Leprol.* 7 Suppl 1, 77. 6. Wachsmuth, Fritze & Pfleiderer (1966) *Biochemistry* 5, 175. 7. Lockwood & Coombs (1991) *Biochem. J.* 279, 675. 8. Nakayama, Esaki, Tanaka & Soda (1988) *Agric. Biol. Chem.* 52, 177. 9. Yang, Kirkpatrick, Ho, *et al.* (2001) *Biochemistry* 40, 10645. 10. Kiick & Tirrell (2000) *Tetrahedron* 56, 9487. 11. Nakamura & Jones (1970) *Meth. Enzymol.* 17A, 286. 12. Marshall & Cohen (1972) *J. Biol. Chem.* 247, 1654. 13. Fujioka & Nakatani (1972) *Eur. J. Biochem.* 25, 301. 14. Tscheme & Wainfan (1978) *Nucl. Acids Res.* 5, 451.

19-Nortestosterone

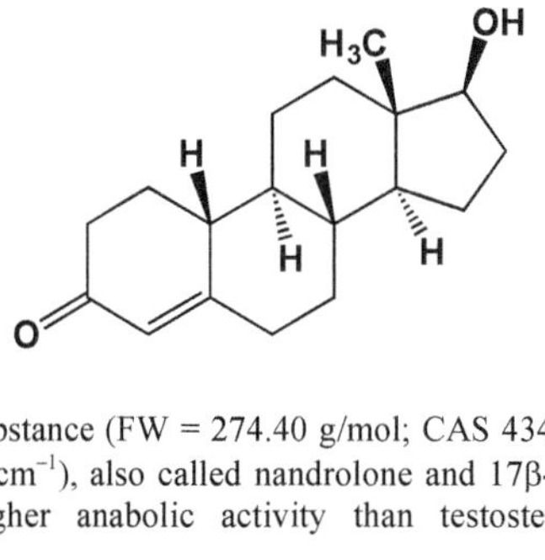

This controlled substance (FW = 274.40 g/mol; CAS 434-22-0; λ_{max} at 241 nm; ε = 1700 M^{-1}cm^{-1}), also called nandrolone and 17β-hydroxyestr-4-en-3-one, has a higher anabolic activity than testosterone. **Target(s):**

[pyruvate dehydrogenase (acetyl-transferring)] kinase, IC$_{50}$ = 120 μM (1); steroid Δ-isomerase (K_i = 13 μM (2-7); steroid 5α-reductase (8). 1. Aicher, Damon, Koletar, *et al.* (1999) *Bioorgan. Med. Chem. Lett.* 9, 2223. 2. Jarabak, Colvin, Moolgavkar & Talalay (1969) *Meth. Enzymol.* 15, 642. 3. Benisek (1977) *Meth. Enzymol.* 46, 469. 4. Talalay & Benson (1972) *The Enzymes*, 3rd ed., 6, 591. 5. Falcoz-Kelly, Baulieu & Alfsen (1968) *Biochemistry* 7, 4119. 6. Choi, Ha, Kim, *et al.* (2000) *Biochemistry* 39, 903. 7. Weintraub, Vincent, Baulieu & Alfsen (1977) *Biochemistry* 16, 5045. 8. Faredin, Toth, Oszlanczy & Scultety (1992) *Int. Urol. Nephrol.* 24, 145.

11-Nor-Δ⁹-tetrahydrocannabinol-9-carboxylate

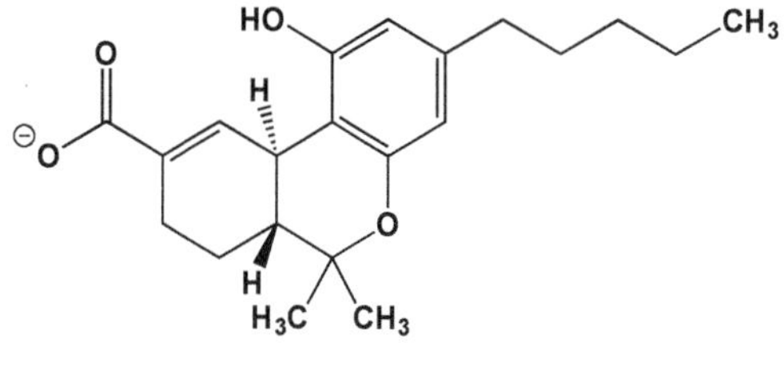

This nonpsychoactive plant isoprenoid (FW$_{free-acid}$ = 344.45 g/mol), also called Δ¹-tetrahydrocannabinol-7-oic acid, is a major Δ⁹-tetrahydrocannabinol (THC) metabolite in most species, including humans and is likely responsible for the well-known analgesic properties of THC. It also exhibits antinociceptive properties. **Target(s):** cyclooxygenase; 5-lipoxygenase; platelet activating factor. 1. Burstein, Audette, Doyle, *et al.* (1989) *J. Pharmacol. Exp. Ther.* 251, 531.

Nortriptyline

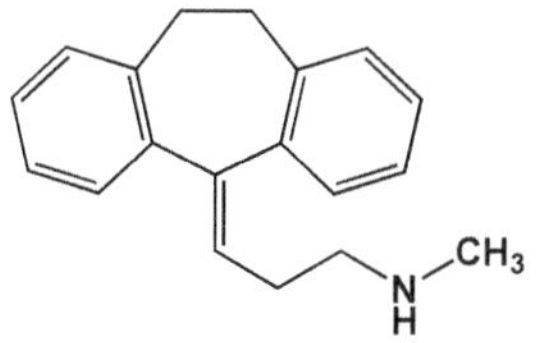

This antidepressant (FW$_{free-base}$ = 263.38 g/mol; CAS 72-69-5) is a serotonin receptor antagonist that preferentially inhibits norepinephrine uptake. **Key Pharmacokinetic Parameters:** *See* Appendix II in Goodman & Gilman's *THE PHARMACOLOGICAL BASIS OF THERAPEUTICS*, 12th Edition (Brunton, Chabner & Knollmann, eds.) McGraw-Hill Medical, New York (2011). **Target(s):** acetylcholinesterase, *Electrophorus electricus* (1); γ-aminobutyrate transport (2); bufuralol L'-hydroxylase (3); CYP2C11 (4,5); CYP2D6 (6); norepinephrine uptake (7-10); serotonin, *or* 5-hydroxytryptamine, receptor (11). 1. Nunes-Tavares, Nery da Matta, Batista e Silva, *et al.* (2002) *Int. J. Biochem. Cell Biol.* 34, 1071. 2. Nakashita, Sasaki, Sakai & Saito (1997) *Neurosci. Res.* 29, 87. 3. Tu, Seddon, Boobis & Davies (1989) *Zhongguo Yao Li Xue Bao* 10, 465. 4. Murray & Murray (2003) *Xenobiotica* 33, 973. 5. Murray (1992) *Mol. Pharmacol.* 42, 931. 6. Shin, Park, Kim, *et al.* (2002) *Drug Metab. Dispos.* 30, 1102. 7. Maxwell, Keenan, Chaplin, Roth & Eckhardt (1969) *J. Pharmacol. Exp. Ther.* 166, 320. 8. Borga, Hamberger, Malmfors & Sjoqvist (1970) *Clin. Pharmacol. Ther.* 11, 581. 9. Maxwell, Chaplin, Eckhardt, Soares & Hite (1970) *J. Pharmacol. Exp. Ther.* 173, 158. 10. Maxwell, Eckhardt & Hite (1970) *J. Pharmacol. Exp. Ther.* 171, 62. 11. Sanchez & Hyttel (1999) *Cell Mol. Neurobiol.* 19, 467.

L-Norvaline

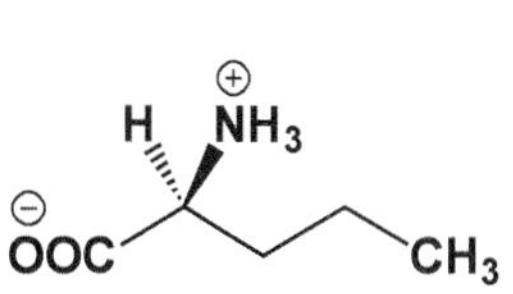

This metabolite (FW = 117.17 g/mol; CAS 6600-40-4; Solubility = 10.7 g/100 g H$_2$O at 15°C; pK_a values of 2.36 and 9.72), also known as L-2-aminovalerate and L-2-aminopentanoate, is a naturally occurring, albeit rare, amino acid. **Target(s):** alanine aminotransferase (1); alkaline phosphatase, mildly inhibited by the racemic mixture (2); D-amino-acid oxidase (3); 2-aminohexanoate transaminase (4); arginase (5,6); arginine deaminase (7); argininosuccinate synthetase (8); asparagine:oxo-acid aminotransferase (9); aspartate kinase (10); cytosol alanyl aminopeptidase (11); glycine amidinotransferase (12); homoserine kinase (13); methionyl-tRNA synthetase (14); ornithine aminotransferase (15-18); ornithine carbamoyltransferase (19-31); putrescine carbamoyltransferase (32); saccharopine dehydrogenase (33). **1**. Saier & Jenkins (1967) *J. Biol. Chem.* **242**, 101. **2**. Fishman & Sie (1971) *Enzymologia* **41**, 141. **3**. Dixon & Kleppe (1965) *Biochim. Biophys. Acta* **96**, 368. **4**. Der Garabedian & Vermeersch (1987) *Eur. J. Biochem.* **167**, 141. **5**. Hunter & Downs (1945) *J. Biol. Chem.* **157**, 427. **6**. Webb (1966) *Enzyme and Metabolic Inhibitors*, vol. **2**, p. 337, Academic Press, New York. **7**. Manca de Nadra, Pesce de Ruiz Holgado & Oliver (1984) *J. Appl. Biochem.* **6**, 184. **8**. Takada, Saheki, Igarashi & Katsunuma (1979) *J. Biochem.* **85**, 1309. **9**. Maul & Schuster (1986) *Arch. Biochem. Biophys.* **251**, 585. **10**. Keng & Viola (1996) *Arch. Biochem. Biophys.* **335**, 73. **11**. Garner & Behal (1977) *Arch. Biochem. Biophys.* **182**, 667. **12**. Walker (1973) *The Enzymes*, 3rd ed., **9**, 497. **13**. Huo & Viola (1996) *Biochemistry* **35**, 16180. **14**. Kiick & Tirrell (2000) *Tetrahedron* **56**, 9487. **15**. Jenkins & Tsai (1970) *Meth. Enzymol.* **17A**, 281. **16**. Strecker (1965) *J. Biol. Chem.* **240**, 1225. **17**. Kalita, Kerman & Strecker (1976) *Biochim. Biophys. Acta* **429**, 780. **18**. Matsuzawa (1974) *J. Biochem.* **75**, 601. **19**. Nakamura & Jones (1970) *Meth. Enzymol.* **17A**, 286. **20**. Ahmad, Bhatnagar & Venkitasubramanian (1986) *Biochem. Cell Biol.* **64**, 1349. **21**. Shi, H. Morizono, Aoyagi, Tuchman & Allewell (2000) *Proteins* **39**, 271. **22**. Marshall & Cohen (1972) *J. Biol. Chem.* **247**, 1654. **23**. Kuo, Herzberg & Lipscomb (1985) *Biochemistry* **24**, 4754. **24**. Baker & Yon (1983) *Phytochemistry* **22**, 2171. **25**. Kurtin, Bishop & Himoe (1971) *Biochem. Biophys. Res. Commun.* **45**, 551. **26**. Legrain & Stalon (1976) *Eur. J. Biochem.* **63**, 289. **27**. Templeton, Reinhardt, Collyer, Mitchell & Cleland (2005) *Biochemistry* **44**, 4408. **28**. Xiong & Anderson (1989) *Arch. Biochem. Biophys.* **270**, 198. **29**. Lusty, Jilka & Nietsch (1979) *J. Biol. Chem.* **254**, 10030. **30**. Pierson, Cox & Gilbert (1977) *J. Biol. Chem.* **252**, 6464. **31**. Sankaranarayanan, Cherney, Cherney, *et al.* (2008) *J. Mol. Biol.* **375**, 1052. **32**. Stalon (1983) *Meth. Enzymol.* **94**, 339. **33**. Fujioka & Y. Nakatani (1972) *Eur. J. Biochem.* **25**, 301.

Noscapine

This opium alkaloid (FW$_{free-base}$ = 413.43 g/mol; CAS 128-62-1; triboluminescent solid that forms unstable salts with acids and strong bases), known also as narcosine, narcotine and gnoscopine (*dl*-noscapine) from the plant *Papaver somniferum* is reduces microtubule assembly/disassembly dynamics, thereby arresting mitosis, inducing apoptosis, and exerting potent antitumor activity. **1**. Landen, Lang, McMahon, *et al.* (2002) *Cancer Res.* **62**, 4109.

Nosiheptide

This branched polypeptide antibiotic (FW = 1222.38 g/mol; CAS 56377-79-8), also known as multhiomycin, from *Streptomyces actuosus* binds to bacterial ribosome-associated elongation factors Tu and G$_1$, thereby inhibiting protein biosynthesis. **1**. Cundliffe & Thompson (1981) *J. Gen. Microbiol.* **126**, 185. **2**. Lentzen, Klinck, Matassova, Aboul-ela & Murchie (2003) *Chem. Biol.* **10**, 769.

Novobiocin

This photosensitive aminocoumarin antibiotic (FW = 612.63 g/mol; CAS 303-81-1; λ_{max} = 307 nm (0.1N NaOH), λ_{max} = 324 nm (0.1N CH$_3$OH/HCl; E$_{1\%,1-cm}$ = 390); λ_{max} = 390 nm (pH 7); E$_{1\%,1-cm}$ = 350), from *Streptomyces spheroides* and *S. niveus*, also named 4-hydroxy-3-[4-hydroxy-3-(3-methylbut-2-enyl)benzamido]-8-methylcoumarin-7-yl 3-*O*-carbamoyl-5,5-di-*C*-methyl-α-L-lyxofuranoside, is a competitive inhibitor of ATP binding to topoisomerase II, thereby blocking DNA synthesis. Aminocoumarins are potent inhibitors of bacterial DNA gyrase, targeting the GyrB subunit. **Target(s):** CDP-glycerol glycerophosphotransferase (3,35); CDP-ribitol ribitolphosphotransferase, *or* teichoic acid synthase (4); CYP3A (5); DNA ligase (ATP), *Vaccinia* virus (6); DNA-directed DNA polymerase (7); DNA polymerase α (7); DNA topoisomerase I (8-11); DNA topoisomerase II, *or* DNA gyrase (9,12-23); F$_1$ ATP synthase, yeast (24); glycyl-tRNA synthetase (25); leucyl-tRNA synthetase (25); NAD$^+$ ADP-ribosyl-transferase, *or* poly(ADP-ribose) polymerase (37); NAD$^+$:protein-arginine ADP-ribosyltransferase, *or* arginine-specific mono-ADP ribosyltransferase (1,2,36); nucleolar ATPase (26); oxidative phosphorylation, uncoupling (27); RNA directed DNA polymerase, *or* reverse transcriptase (7,28,29); RNA polymerase (30-32); self-splicing of transcripts of the T$_4$ phage thymidylate synthase gene (39); α,α-trehalose-phosphate synthase, UDP-forming (38); UDP-glucuronosyltransferase (5,33); undecaprenyldiphospho muramoyl-pentapeptide β-*N*-acetylglucosaminyltransferase, weakly inhibited (34). **1**. Lodhi, Clift, Omann, *et al.* (2001) *Arch. Biochem. Biophys.* **387**, 66. **2**. Yamada, Tsuchiya, Nishikori & Shimoyama (1994) *Arch. Biochem. Biophys.* **308**, 31. **3**. Burger & Glaser (1966) *Meth. Enzymol.* **8**, 430. **4**. Ishimoto & Strominger (1966) *J. Biol. Chem.* **241**, 639. **5**. Villar, Furasawa, Monshouwer & Van Miert (1998) *Vet. Res. Commun.* **22**, 405. **6**. Shuman (1995) *Biochemistry* **34**, 16138. **7**. Sarih, Garret, Aoyama, *et al.* (1983) *Biochem. Int.* **7**, 79. **8**. Sekiguchi, Stivers, Mildvan & Shuman (1996) *J. Biol. Chem.* **271**, 2313. **9**. Goto, Laipis & Wang (1984) *J. Biol. Chem.* **259**, 10422. **10**. Lazarus, Henrich, Kelly, Schmitz & Castora (1987) *Biochemistry* **26**, 6195. **11**. Shaffer & Traktman (1987) *J. Biol. Chem.* **262**, 9309. **12**. Gellert (1981) *The Enzymes*, 3rd ed., **14**, 345. **13**. Larsen, Escargueil & Skladanowski (2003) *Pharmacol. Ther.* **99**, 167. **14**. Assairi (1994) *Biochim. Biophys. Acta* **1219**, 107. **15**. Gellert, O'Dea, Itoh & Tomizawa (1976) *Proc. Natl. Acad. Sci. U.S.A.* **73**, 4474. **16**. Gellert (1981) *Ann. Rev. Biochem.* **50**, 879. **17**. Low, Orton & Friedman (2003) *Eur. J. Biochem.* **270**, 4173. **18**. Vosberg (1985) *Curr. Top. Microbiol. Immunol.* **114**, 19. **19**. Saijo, Enomoto, Hanaoka & Ui (1990) *Biochemistry* **29**, 583. **20**. Assairi (1994) *Biochim. Biophys. Acta* **1219**, 107. **21**. Jain & Nagaraja (2005) *Mol. Microbiol.* **58**, 1392. **22**. Riou, Douc-Rasy & Kayser (1986) *Biochem. Soc. Trans.* **14**, 496. **23**. Miller & Scurlock (1983) *Biochem. Biophys. Res. Commun.* **110**, 694. **24**. Jenkins, Pocklington & Orr (1990) *J. Cell Sci.* **96**, 675. **25**. Wright, Nurse & Goldstein (1981) *Science* **213**, 455. **26**. Fox & Studzinski (1982) *J. Histochem. Cytochem.* **30**, 364. **27**. Gallagher, Weinberg & Simpson (1986) *J. Biol. Chem.* **261**, 8604. **28**. Juca & Aoyama (1995) *J. Enzyme Inhib.* **9**, 171. **29**. Sumiyoshi, Nishikawa, Watanabe & Kano (1983) *J. Gen. Virol.* **64**, 2329. **30**. Webb & Jacob (1988) *J. Biol. Chem.* **263**, 4745. **31**. Webb, Maguire & Jacob (1987) *Nucl. Acids Res.* **15**, 8547. **32**. Gottesfeld (1986) *Nucl. Acids Res.* **14**, 2075. **33**. Duvaldestin, Mahu, Preaux & Berthelot (1976) *Biochem. Pharmacol.* **25**, 2587. **34**. Meadow, Anderson & Strominger (1964) *Biochem. Biophys. Res. Commun.* **14**, 382. **35**. Burger & Glaser (1964) *J. Biol. Chem.* **239**, 3168. **36**. Taniguchi, Tsuchiya & Shimoyama (1993) *Biochim. Biophys. Acta* **1161**, 265. **37**. Banasik, Komura, Shimoyama & Ueda (1992) *J. Biol. Chem.* **267**, 1569. **38**. Pan & Elbein (1996) *Arch. Biochem. Biophys.* **335**, 258. **39**. Jung, Shin & Park (2008) *Mol. Cell. Biochem.* **314**, 143.

Noxiustoxin

This arachnid toxin (MW = 4201.95 g/mol; *Sequence*: TIINVKCTSPK QCSKPCKELYGSSAGAKCMNGKCKCYNN-NH$_2$) from the predatory Mexican scorpion *Centruoides noxius* is a powerful blocker of lymphocyte voltage-dependent K$^+$ channels ($K_i \approx$ 2 nM), but is without effect on Ca^{2+}-activated K$^+$ channels. Synthetic peptide NTX10-20 (corresponding to residues 10-20 of noxiustox) showed remarkable preference for the IA K$^+$ current of cerebellum granular cells, using the patch-clamp technique in whole-cell assay (3). **1**. Grissmer, Nguyen, Aiyar, *et al.* (1994) *Mol. Pharmacol.* **45**, 1227. **2**. Drakopoulou, Cotton, Virelizier, *et al.* (1995) *Biochem. Biophys. Res. Commun.* **213**, 901. **3**. Frau, Pisciotta, Gurrola, Possani & Prestipino (2001) *Eur. Biophys. J.* **29**, 569.

NP-12, *See Tideglusib*

NP031112, *See Tideglusib*

NPC-15437

This short-chain fatty acyl amide (FW$_{free-base}$ = 438.70 g/mol), also known as (*S*)-2,6-diamino-*N*-([1-oxotridecyl)-2-piperidinyl]methyl)hexanamide, is a selective inhibitor of protein kinase C (IC$_{50}$ = 19 μM), binding at the regulatory domain. No inhibition of cAMP-dependent or calcium/calmodulin-dependent protein kinases was observed at concentrations of the inhibitor up to 300 μM. **1**. Sullivan, Connor, Shearer & Burch (1991) *Agents Actions* **34**, 142. **2**. Sullivan, Connor, Shearer & Burch (1992) *Mol. Pharmacol.* **41**, 38. **3**. Sullivan, Connor, Tiffany, Shearer & Burch (1991) *FEBS Lett.* **285**, 120. **4**. Saraiva, Fresco, Pinto & Goncalves (2003) *J. Enzyme Inhib. Med. Chem.* **18**, 475.

NS 102

This oxime (FW = 261.24 g/mol), systematically referred to as 6,7,8,9-tetrahydro-5-nitro-1*H*-benz[*g*]indole-2,3-dione 3-oxime, is an ionotropic glutamate receptor antagonist, showing selective displacement of low-affinity kainate binding. **1**. Johansen, Drejer, Watjen & Nielsen (1993) *Eur. J. Pharmacol.* **246**, 195. **2**. Verdoorn, Johansen, Drejer & Nielsen (1994) *Eur. J. Pharmacol.* **269**, 43.

NS 187

This imatinib analogue (FW = 630.60 g/mol; CAS 859212-16-1, 859212-17-2 (HCl); Solubility: < 1 mg/mL H$_2$O, 115 mg/mL DMSO), also named INNO-406, its trade name Bafetinib™, and its IUPAC name (*S*)-4-((3-(dimethylamino)pyrrolidin-1-yl)methyl)-*N*-(4-methyl-3-(4-(pyrimidin-5-yl)pyrimidin-2-ylamino)phenyl)-3-(trifluoromethyl)benzamide, inhibits both Bcr-Abl (IC$_{50}$ = 6 nM) and Lyn (IC$_{50}$ = 19 nM) protein tyrosine kinases (1,2) and suppresses the growth of Philadelphia chromosome-positive (*Ph$^+$*) acute lymphoblastic *leukemia* (ALL) cells. **1**. Tauchi & Ohyashiki (2006) *Int. J. Hematol.* **83**, 294. **2**. Kimura, Naito, Segawa, *et al.* (2005) *Blood* **106**, 3948.

NS-398

This anti-inflammatory agent (FW = 314.36 g/mol), also known as *N*-(2-[cyclohexyloxy]-4 nitrophenyl)methanesulfonamide, selectively inhibits cyclooxygenase-2 (COX-2), with an IC$_{50}$ value of 3.8 μM, but with an IC$_{50}$ value >100 μM for COX-1. **1**. Futaki, Takahashi, Yokoyama, *et al.* (1994) *Prostaglandins* **47**, 55. **2**. Forghani, Ouellet, Keen, Percival & Tagari (1998) *Anal. Biochem.* **264**, 216. **3**. Kargman, Wong, Greig, *et al.* (1996) *Biochem. Pharmacol.* **52**, 1113. **4**. Reininger & Bauer (2006) *Phytomedicine* **13**, 164. **5**. Barnett, Chow, Ives, *et al.* (1994) *Biochim. Biophys. Acta* **1209**, 130. **6**. Patrignani, Panara, Sciulli, *et al.* (1997) *J. Physiol. Pharmacol.* **48**, 623. **7**. Winnall, Ali, O'Bryan, *et al.* (2007) *Biol. Reprod.* **76**, 759. **8**. Chi, Jong, Son, *et al.* (2001) *Biochem. Pharmacol.* **62**, 1185. **9**. Sud'ina, Pushkareva, Shephard & Klein (2008) *Prostaglandins Leukot. Essent. Fatty Acids* **78**, 99.

NS 1619

This benzimidazolone (FW = 362.23 g/mol), also named 1,3-dihydro-1-[2-hydroxy-5-(trifluoromethyl)phenyl]-5-(trifluoromethyl)-2*H*-benzimidazol-2-one, is a selective large conductance Ca^{2+}-activated K$^+$-channel activator and an inhibitor of mitochondrial electron transport. **1**. Debska, Kicinska, Dobrucki, *et al.* (2003) *Biochem. Pharmacol.* **65**, 1827.

NSC 23766

This Rac GTPase inhibitor (FW$_{free-base}$ = 427.62 g/mol; FW$_{tri-HCl}$ = 530.96 g/mol; CAS 1177865-17-6; Solubility: 100 mg/mL DMSO, 100 mg/mL H$_2$O), also known as N^6-[2-[[4-(diethylamino)-1-methylbutyl]amino]-6-methyl-4-pyrimidinyl]-2-methyl-4,6-quinolinediamine, targets activation of Rac by guanine nucleotide exchange factors (*or* GEFs), displaying an IC$_{50}$ of ~50 μM, with no inhibition of such structurally related targets as Cdc42 or RhoA (**See also** *EHop-016*). NSC23766 was identified by a structure-based virtual screening of compounds that fit into a surface groove of Rac1 known to be critical for GEF specification (1,2). *In vitro*, it effectively inhibits Rac1 binding and activation by the Rac-specific GEF Trio or Tiam1 in a dose-dependent manner, without interfering with the closely related Cdc42 or RhoA binding or activation by their respective GEFs or with Rac1 interaction with BcrGAP or effector PAK1 (1). In cells, NSC23766 potently blocked serum or platelet-derived growth factor-induced Rac1 activation and lamellipodia formation without affecting the activity of endogenous Cdc42 or RhoA (1). Blockade of Rac1 activity by NSC23766 induces G$_1$ cell-cycle arrest or apoptosis in breast cancer cells through downregulation of cyclin D$_1$, survivin, and X-linked inhibitor of apoptosis protein (3). Although LPS challenge leads to increases of both Rac1 and Rac2, (but not

CDC42 or RhoA) activities in lungs, intraperitoneal administration with NSC23766 inhibited both Rac1 and Rac2, but not CDC42 or RhoA activities (4). Treatment with NSC23766 at 1 or 3mg/kg not only reduced the inflammatory cells infiltration and MPO activities, but also inhibited pro-inflammatory mediators, tumor necrosis factor-α and interleukin-1β, mRNA expression. Moreover, *in vitro* neutrophil migration assay and in vivo microvascular permeability assay indicates that NSC23766 not only inhibits neutrophil transwell migration toward the chemoattractant fMLP, but also reduces Evans Blue and albumin accumulation in LPS-challenged lungs (4). NSC23766 inhibition of Rac1 also impairs replication of a wide variety of influenza viruses, including a human virus strain of the pandemic from 2009 as well as highly pathogenic avian virus strains (5). These and other findings identified a crucial role of Rac1 for the activity of the viral polymerase complex, and the antiviral potential of NSC23766 was confirmed in mouse experiments, identifying Rac1 as a new cellular target for therapeutic treatment of influenza virus infections (5).   **1**. Gao, Dickerson, Gao, Zheng & Zheng (2004) *Proc. Natl. Acad. Sci. U.S.A.* **101**, 7618.   **2**. Akbar, Cancelas, Williams, Zheng & Zheng (2006) *Meth. Enzymol.* **406**, 554.   **3**. Yoshida, et al. (2010) *Mol. Cancer Ther.* **9**, 1657.   **4**. Yao, et al. (2011) *Biochim. Biophys. Acta* **1810**, 666.   **5**. Dierkes, Warnkin, Liedmann, *et al.* (2014) *PLoS One* **9**, e88520.

NSC 30552, See *Mangostin*

NSC 52947, See *Pactamycin*

NSC 66811

This p53/MDM2 antagonist/inhibitor (FW = 340.42 g/mol; CAS 6964-62-1; Solubility: 25 mg/mL DMSO; <1 mg/mL H_2O), also named 2-methyl-7-[phenyl(phenylamino)methyl]-8-quinolinol, targets Mouse Double Minute-2 (MDM2), also known as E_3 ubiquitin-protein ligase that recognizes the N-terminal trans-activation domain of the p53 tumor suppressor, inhibiting p53 transcriptional activation (IC_{50} = 120 μM). NSC 66811 binds MDM2 in the p53-binding pocket, leading to cell cycle arrest, apoptosis, and growth inhibition of human tumor xenografts in nude mice (1,2). NSC 66811 mimics three p53 residues critical in the binding to MDM2.   **1**. Lu, Nikolovska-Coleska, Fang, *et al.* (2006) *Med. Chem.* **49**, 3759.   **2**. Shangary & Wang (2009) *Annu. Rev. Pharmacol. Toxicol.* **49**, 223.

NSC-101088, See *Withaferin A*

NSC113939, See *5-Iodotubercidin*

NSC127716, See *Decitabine*

NSC136476, See *GANT61*

NSC-165301, See *Acivicin*

NSC-241240, See *Carboplatin*

NSC 336628, See *Merbarone*

NSC 405020

This cell-permeable MT1-MMP inhibitor (FW = 260.16 g/mol; CAS 7497-07-6; Solubility: 52 mg/mL DMSO; <1 mg/mL H_2O), also named 3,4-dichloro-N-(1-methylbutyl)benzamide, does not directly bind to the active site to alter catalysis, but instead binds to a subunit-subunit interaction site within the PEX domain of MT1-MMP, thereby blocking PEX homo-

dimerization. **Mode of Action:** Most matrix metalloproteinases (MMPs inhibitors) target the conserved active site and consequently inhibit multiple MMPs instead. The MT1-MMP, however, contains a hemopexin domain (PEX) that is distinct from and additional to the catalytic domain, making it an ideal druggable target. NSC 405020 only weakly inhibits MT1-MMP (IC_{50} = 100 μM), suggesting it is best suited for *in vitro* experiments or as a lead molecule for developing higher affinity agents.   **1**. Remacle, Golubkov, Shiryaev, *et al.* (2012) *Cancer Res.* **72**, 2339.

NSC 407290, See *Myricetin*

NSC 609699, See *Topotecan*

NSC624206

This disulfide-containing ubiquitin pathway inhibitor (FW = 389.08 g/mol) specifically blocks (IC_{50} = 9 μM) ubiquitin-thiolester formation in the E_1 activation reaction, without any effect on ubiquitin adenylylation, thus defining a promising new way for limiting excessive ubiquitin-mediated proteolysis *in vivo*. NSC524206 is believed to react with ubiquitin ligase E_1-SH to form E_1-S–S-$(CH_2)_2$-NH-$(CH_2)_9$-CH_3 with the release of *p*-chlorobenzyl thiol. Treatment of Kip16 cells with 3 μM of NSC524206 for 24 h results in a uniform stabilization of the tumor suppressor protein p27. (**See also** *PYR-41*)   **1**. Ungermannova, Parker, Nasveschuk, *et al.* (2012) *J. Biomol. Screen.* **17**, 421.

NSC 630176, See *Romidepsin*

NSC 632839

This deubiquitylase (DUB) and deSUMOylase inhibitor (FW = 339.86 g/mol; CAS 157654-67-6), also named 3,5-bis[(4-methylphenyl)-methylene]-4-piperidinone, targets USP2 (IC_{50} = 45 μM), USP7 (IC_{50} = 37 μM), and SENP2 (IC_{50} = 10 μM). Deconjugation of ubiquitin (Ub) or ubiquitin-like proteins (UBL) from proteins is performed by a specific class of isopeptidases. Small ubiquitin-like modifier (SUMO) are removed by sentrin/SUMO-specific proteases (SENPs), the first known class of deSUMOylase. NSC 632839 was identified on the basis of its ability to initiate a Bcl-2-dependent but apoptosome-independent pathway of caspase activation. DUBs have been divided into five distinct subfamilies based on structural similarities and mechanisms of action and include a large group of enzymes that are mainly cysteine proteases. The pharmacophore NSC 632839 should inhibit cysteine proteases, but all of the targets are still unknown (2). Interestingly, prostaglandin Δ^{12}-PGJ2 holds the same pharmacophore and similarly induces apoptosis *via* accumulation of polyubiquitinated proteins (3, 4), and Δ^{12}-PGJ2) has been suggested to inhibit UCH-L1 and UCH-L3 at micromolar concentrations (5).   **1**. Nicholson, Leach, Goldenberg, *et al.* (2008) *Protein Sci.* **17**, 1035.   **2**. Aleo, Henderson, Fontanini, Solazzo & Brancolini (2006) *Cancer Res.* **66**, 9235.   **3**. Mullally & Fitzpatrick (2002) *Mol. Pharmacol.* **62**, 351.   **4**. Mullally, Moos, Edes & Fitzpatrick (2001) *J. Biol. Chem.* **276**, 30366.   **5**. Zongmin, Melandri, Berdo, *et al.* (2004) *Biochem. Biophys. Res. Commun.* **319**, 1171.

NSC 648766, See *Ecteinascidin 743*

NSC 652287, See *RITA*

NSC 642492, See *TNP-470*

NSC 718781, See *Erlotinib*

NSC-732208, See *Cediranib*

NSD 1055, See *4-Bromo-3-hydroxybenzyloxyamine*

NT219

This novel, long-lasting and highly efficient inhibitor (FW = 412.25 g/mol) targets the insulin/IGF signaling pathway and protects against age-onset, neurodegeneration-linked proteotoxicity of the sort observed in Alzheimer's and Huntington's Diseases. Prior studies suggested that reducing the activity of the insulin/IGF signaling cascade (IIS), a highly conserved aging-regulating pathway, elevates stress resistance, thereby extending the lifespans of worms, flies, and mice. NT219 efficiently blocks IGF1R to AKT signaling in mammalian cells by (a) reducing IGF1-induced phosphorylation and by activating AKT, directing IRS1/2 for inhibitory serine-phosphorylation and degradation and (b) by increasing expression of FOXO target genes in mammals. 1. El-Ami, Moll, Carvalhal Marques, et al. (2013) Aging Cell 13, 165.

NU6027

This potent ATR/CDK inhibitor (FW = 251.29 g/mol; CAS 220036-08-8; Solubility: 50 mg/mL DMSO; <1 mg/mL H_2O), also known as 6-(cyclohexylmethoxy)-5-nitroso-2,4-pyrimidinediamine, targets Cyclin-Dependent Kinases CDK1 (K_i = 2.5 μM) and CDK2 (K_i = 1.3 μM) as well as ATR (K_i = 0.4 μM) and DNA-PK (K_i = 2.2 μM), entering cells more readily than well-known 6-aminopurine-based inhibitors (1.2). NU6027 exerts its cytotoxicity by interfering with the DNA damage response (DDR), for which ATR is a major regulator. This inhibitor shows >600-fold selectivity over related kinases ATM or DNA-PK, and markedly enhances death induced by DNA-damaging agents in certain cancers, but not in normal cells. This differential response highlights the great potential for ATR inhibition as a druggable target to increase dramatically the efficacy of many established DNA-damaging drugs and ionizing radiation (2). NU6027 inhibits ATR, resulting in G_2/M arrest, inhibition of homologous recombination, and increasing sensitivity to DNA-damaging agents and PARP inhibitors (3). 1. Arris, Boyle, Calvert, et al. (2000) J. Med. Chem. 43, 2797. 2. Charrier, Durrant, Golec, et al. (2011) J. Med. Chem. 54, 2320. 3. Peasland, Wang, Rowling, et al. (201) Br. J. Cancer. 105, 372.

NU7441

This selective, ATP-competitive, DNA-PK inhibitor (F.Wt. = 413.49; CAS 503468-95-9; Solubility (25°C): 3 mg/mL DMSO, <1 mg/mL Water), also known as KU-57788 and 8-(4-dibenzothienyl)-2-(4-morpholinyl)-4H-1-benzopyran-4-one, inhibits DNA-dependent protein kinase, a trimeric nuclear serine/threonine kinase composed of a large catalytic subunit and two DNA-targeting proteins (1, 2). NU7441 potentiates the apoptotic and/or toxic effects of radiation, doxorubicin, and etoposide in human tumor cell lines in vitro and etoposide in a human tumor xenograft model in vivo (3). **Targets:** DNA-PK (IC_{50} = 14 nM); mTOR (IC_{50} = 1.7 μM); PI$_3$-K (IC_{50} = 5 μM); ATM (IC_{50} = 100 μM); and ATR (IC_{50} = 100 μM). 1. Leahy, et al. (2004) Bioorg. Med. Chem. Lett. 14, 6083. 2. Zhao, et al. (2006) Cancer Res. 66, 5354. 3. Willmore, et al. (2008) Clin. Cancer Res. 14, 3984.

Nucleocidin

This purine antibiotic (FW = 364.31 g/mol; CAS 24751-69-7), produced by Streptomyces calvus, has been used as an antitrypanosomal agent. The first naturally occurring fluoro-sugar identified, nucleocidin has a potency that exceeds puromycin, making it one of the most potent inhibitors of translaion. **Target(s):** adenosylhomocysteinase (1); protein biosynthesis (2). 1. Chiang (1987) Meth. Enzymol. 143, 377. 2. Florini, Bird & Bell (1966) J. Biol. Chem. 241, 1091.

Nucleoside diphosphate kinase

This ubiquitous transphosphorylase (EC 2.7.4.6) is a hexameric enzyme that catalyzes the interconversion nucleoside 5'-di- and tri-phosphates (Reaction: NDP + YTP $\rightleftharpoons$ NTP + YDP), possesses a single nucleotide binding site and employs an active-site histidinyl residue to preserve group transfer potential as a phosphoryl-enzyme intermediate. Beyond its role in the balanced synthesis of nucleotide triphosphates as well as nucleic acids, proteins and nucleotidylated sugars and lipids, NDPK isoforms play regulatory functions in diverse cellular processes, including proliferation and endocytosis. Although it plays a positive regulatory role in clathrin- and dynamin-mediated micropinocytosis, NDPK also alters growth, endocytosis and exocytosis. In Dictyostelium, NDPK negatively regulates endocytosis, contrasting with its positive regulatory role in higher eukaryotes. micropinocytosis in mammalian cells. 1. Annesley, Bago, Bosnar et al. (2011) PLoS One 6, e26024.

Nutlin-3b

This p53/MDM2 antagonist/inhibitor (FW = 581.49 g/mol; CAS 675576-97-3) targets Mouse Double Minute-2 (MDM2), also known as E$_3$ ubiquitin-protein ligase that recognizes the N-terminal trans-activation domain of the p53 tumor suppressor, inhibiting p53 transcriptional activation (IC_{50} = 13.6 μM). Nutlin-3b binds MDM2 in the p53-binding pocket and activates the p53 pathway in cancer cells, leading to cell cycle arrest, apoptosis, and growth inhibition of human tumor xenografts in nude mice. 1. Vassilev, Vu, Graves, et al. (2004) Science 303, 844.

NVP 231

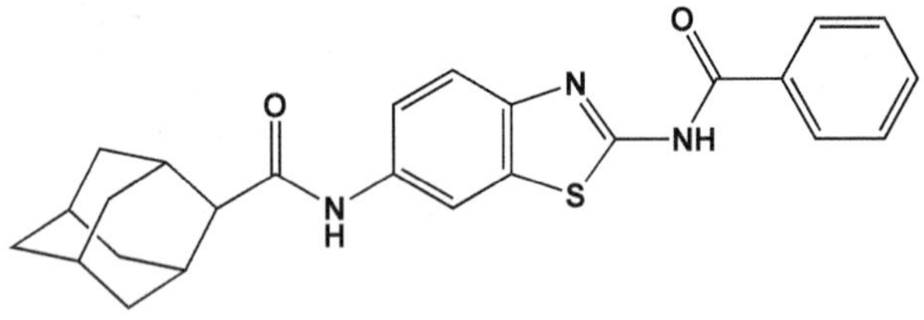

This CerK inhibitor (FW = 431.55 g/mol; CAS 362003-83-6; Soluble: 100 mM in DMSO and to 5 mM in ethanol), also named N-[2-(benzoylamino)-6-benzothiazolyl]tricyclo[3.3.1.1^{3,7}]-decane-1-carboxamide, selectively and reversibly targets ceramide kinase (IC$_{50}$ = 12 nM, *in vitro*), displaying 8000x selectivity versus sphingosine kinases (IC$_{50}$ ≥ 100 μM) and other lipid kinases. **1**. Graf, *et al.* (2009) *Mol. Pharmacol.* **74**, 925. **2**. Niwa, *et al.* (2009) *Microvasc. Res.* **77**, 389.

NVP CXCR2 20

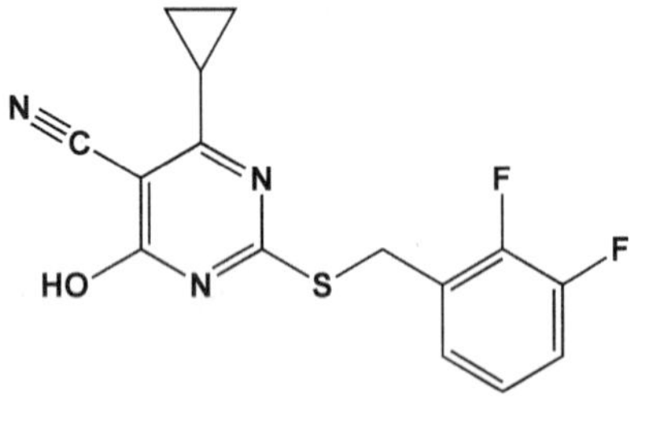

This rrally bioavailable and potent CXCR2 antagonist (FW = 319.33 g/mol; CAS 1029521-30-9; Soluble to 50 mM in DMSO), systematically named 4-cyclopropyl-2-[[(2,3-difluorophenyl)methyl]thio]-1,6-dihydro-6-oxo-5-pyrimidinecarbonitrile, selectively targets the chemokine CXCR2 receptors (IC$_{50}$ = 40 nM), which are G-protein-coupled seven-transmembrane domain receptors that are found on the surface of leukocytes.. NVP CXCR2 20 exhibits selectivity over a panel of 49 other GPCRs. **1**. Porter, Bradley, Brown, *et al.* (2014) *Bioorg. Med. Chem. Lett.* 24, 3285.

N-WASP Inhibitor

This cyclic tetradecapeptide inhibitor (MW = 1784.13; Sequence: *cyclo*-(L-Lys-D-Phe-D-Pro-D-Phe-L-Phe-D-Pro-L-Gln)$_2$; CAS 380488-27-7; Store dessicated at –20°C) binds to <u>N</u>eural <u>W</u>iskott-<u>A</u>ldrich <u>S</u>yndrome <u>P</u>rotein (N-WASP), thereby stabilizing its autoinhibited and blocking (IC$_{50}$ ~ 2 μM) phosphatidylinositol 4,5-bisphosphate- (PIP$_2$-) stimulated actin cytoskeletal assembly. Notably, N-WASP inhibitor is without any direct inhibitory effect on the polymerization of pure G-actin. **1**. Peterson *et al* (2001) *Proc. Natl. Acad. Sci. USA* **98**, 10624. **2**. Suetsugu *et al* (2001) *J. Biol. Chem.* **276**, 33175. **3**. Prehoda & Lim (2002) *Curr. Opin. Cell Biol.* **14**, 149.

NVP-AUY922, See *Luminespib*

NVP-BEZ235, See *ETP-46464*

NVP-LAF237, See *Vildagliptin*

NVP-TAE684, See *TAE684*

Nystatin

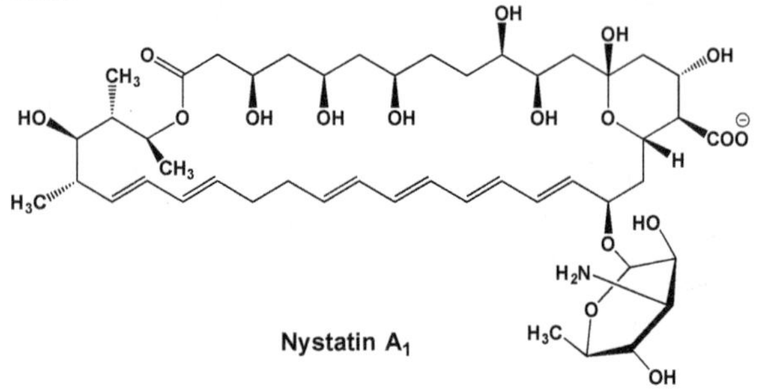

Nystatin A$_1$

This *Streptomyces*-derived antifungal antibiotic (FW$_{free-base}$ = 926.11 g/mol for nystatin A$_3$; CAS 1400-61-9), most often composed of three active components (designated A$_1$, A$_2$, and A$_3$) are polyene derivatives that bind to ergosterol, an essential fungal cell component. By forming transmembrane ion pores, treatment results in fungal death, making nystatin an effective fungicide. Nystatin is used prophylactically in patients at risk for fungal infections, especially those with HIV AIDS who have a low CD4$^+$. Patients undergoing chemotherapy similarly require nystatin. For nystatin-sensitive fungi, the minimal inhibitory concentration is typically 1.5-7 μg/mL, often attended by resultant loss of cations (*e.g.* K$^+$, Na$^+$, and H$^+$) as well as other low-molecular-weight substances, including sugars, amino acids and nucleotides. **Applications in Neurophysiologic Experiments:** Nystatin increases membrane conductance of *Aplysia* neurons in a dose-dependent manner, with little or no effect <10 μM, but with effects too large to measure at concentrations >20 μM (1). Upon return to antibiotic-free solution, membrane conductance returned to pre-treatment levels within 30 minutes. When ion-specific microelectrodes are used, it is possible to record changes of intracellular univalent ion activities that attend nystatin-induced permeability. Changes in permeability mainly involve univalent cations, but chloride ion permeability also increases. Nystatin may therefore be used to selectively rearrange the cell's internal ionic milieu to study the effect of such changes on membrane transport, electrical properties, or metabolic processes. For example, while conventional recording methods, using either KCl or CsCl in the patch pipet, reveal *N*-methyl-D-aspartate (NMDA) currents of neocortical neurons to be biphasic (consisting of peak and steady-state currents with similar reversal potentials), recordings with the nystatin-perforated patch method and KCl as the principal intracellular cation disclosed an NMDA-evoked outward current (2). **Target(s):** alkaline phosphatase, weakly inhibited (3); chitin synthase (4); dimethylpropiothetin dethiomethylase (5); sterol 24-*C*-methyltransferase, slightly inhibited, activates at high concentrations (6). **1**. Russell, Eaton & Brodwick (1977) *J. Membr. Biol.* **37**, 137. **2**. Mistry & Hablitz (1990) *Brain Res.* **535**, 318. **3**. Solov'ena, Belousova & Tereshin (1977) *Biokhimiia* **42**, 277. **4**. Rast & Bartnicki-Garcia (1981) *Proc. Natl. Acad. Sci. U.S.A.* **78**, 1233. **5**. Bacic & Yoch (1998) *Appl. Environ. Microbiol.* **64**, 106. **6**. Mukhtar, Hakkou & Bonaly (1994) *Mycopathologia* **126**, 75.

– O –

O, *See* Ornithine; Orotidine; Oxygen

O-587, *See* Altropane

O-1918

This endocannabinoid antagonist (FW = 286.42 g/mol; CAS 536697-79-7), systematically named 1,3-dimethoxy-5-methyl-2-[(1R,6R)-3-methyl-6-(1-methylethenyl)-2-cyclohexen-1-yl]benzene, is a cannabidinodiol analogue that selectively targets a putative G-coupled endothelial anandamide receptor that is distinct from CB_1 or CB_2 endocannabinoid receptors. O-1918 does not bind to CB_1 or CB_2 receptors and does not cause vasorelaxation at concentrations up to 30 μM, but inhibits the vasorelaxant effects of abn-cbd and anandamide in a concentration-dependent manner (1-30 μM). (*See* abn-cbd; Anandamide; Oleamide). While the atypical cannabinoids O-1602 and abn-cbd (*or* abnormal cannabidiol) stimulate GPR55-dependent GTPγS activity (EC_{50} ~ 2 nM), O-1918 antagonizes such effects (1). O-1918 is involved in the delayed hypotension induced by anandamide in anaesthetized rats (2). **1**. Johns, Behm, Walker, *et al.* (2007) *Brit. J. Pharmacol.* **152**, 825. **2**. Zakrzeska, Schlicker, Baranowska, *et al.* (2010) *Brit. J. Pharmacol.* **160**, 574.

O-2093

This anandamide uptake inhibitor (FW = 584.62g/mol; CAS 439080-01-0), also named *N*-(*bis*-3-chloro-4-hydroxybenzyl)-5*Z*,8*Z*,11*Z*,14*Z*-eicosatetra-enamide, exhibits an IC_{50} of 17.3 μM, with little or no activity detected at cannabinoid receptors CB_1 and CB_2, Transient receptor potential subfamily V member-1 cation channel (TrpV1), and Fatty Acid Amide Hydrolase, *or* FAAH (also known as oleamide hydrolase and anandamide amidohydrolase). It shows anti-spastic activity in a mouse model of multiple sclerosis. **1**. Di Marzo, *et al.* (2002) *J. Pharmacol. Exp. Ther.* **300**, 984. **2**. Ligresti, *et al.* (2006) *Brit. J. Pharmacol.* **147**, 83.

OAA, *See* Oxaloacetate (Oxaloacetic Acid)

OBAA

This potent PLA inhibitor (FW = 428.65 g/mol; CAS 221632-26-4; Soluble to 10 mM in DMSO with gentle warming), also named 4-(4-octadecylphenyl)-4-oxobutenoic acid, targets phospholipase A_2 (IC_{50} = 70 nM). OBAA is a mechanism-based inhibitor, showing a typical time-dependence as well as a molar inactivation ratio of twenty OBAA hydrolyzed per molecule of irreversibly inactivated enzyme (1). Double-reciprocal replots of the apparent inactivation constants versus OBAA concentration gave a (pseudo) first-order rate constant of inactivation of 2.3 min^{-1} and a dissociation constant of 6 x 10^{-6} M^{-1} for the enzyme-inhibitor intermediate (1). OBAA reduces bronchospasm in guinea pigs *in vivo* (2). **1**. Köhler, Heinisch, Kirchner, *et al.* (1992) *Biochem. Pharmacol.* **44**, 805. **2**.

Madi, Giessler, Hirschelmann, Friedrich, Braquet (1991) *Agents Actions* **32**, 144.

Obatoclax

This cell-permeable antiapoptotic agent ($FW_{free-base}$ = 317.39 g/mol; $FW_{mesylate-salt}$ = 413.49 g/mol; CAS 803712-79-0; Solubility: 83 mg/mL DMSO, also known as GX15-070 and (*Z*)-2-(5-((3,5-dimethyl-1*H*-pyrrol-2-yl)methylene)-4-methoxy-5*H*-pyrrol-2-yl)-1*H*-indole mesylate, is a pan-Bcl-2 antagonist (K_i = 0.22 μM), binding to the BH3-binding groove of the B-cell lymphoma-2 (i.e., Bcl) family of proteins (1). Bcl-2 proteins confer a protective effect on malignant cells against death signals of apoptosis, and cancer cells that are resistant to various anti-cancer drugs and treatment regimen are often found to overexpress these Bcl-2, Bcl-X_L, Mcl-1, Bcl-w, and A1/Bfl1. Obatoclax inhibits the binding of Bak to Mcl-1, up-regulating Bim, inducing cytochrome *c* release, and activating capase-3 in human myeloma cell lines (1). Acting as a single agent, it also induces potent cytotoxic responses against diverse patient-derived neoplasias, including multiple myeloma (1). Obatoclax retains its efficacy in cells overexpressing the P-glycoprotein multidrug-resistance transporter (P-gp), multidrug resistance-associated protein 2 (MRP2), or breast cancer resistance protein (BCRP) and might also act as a perpetrator drug in interactions with drugs, for example being substrates of CYP1A2 or BCRP (2). In human pharmacokinetic experiments, Obatoclax has C_{max} of 10 to 80 ng/mL, below the level needed for effective *in vitro* activity against most tumor targets, a finding that is consistent with its failure to reach levels in a mouse xenograft tumor model sufficient to disrupt Mcl-1/Bax protein-protein interaction (3). **1**. Trudel, Li, Rauw, *et al.* (2007) *Blood* **109**, 5430. **2**. Theile, Allendorf, Köhler, Jassowicz & Weiss (2015) *J. Pharm. Pharmacol.* **67**, 1575. **3**. Azmi & Mohammad (2009) *J. Cell Physiol.* **218**, 13.

Obeticholate (Obeticholic Acid)

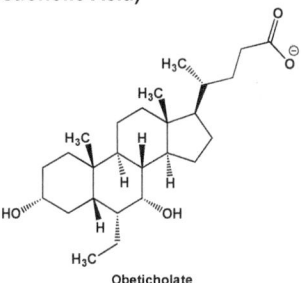

Obeticholate

This semisynthetic bile acid analogue (FW = 420.63 g/mol; CAS 459789-99-2), also named INT-747, 6α-ethyl-chenodeoxycholate, and (3α,5β,6α,7α)-6-ethyl-3,7-dihydroxycholan-24-oic acid, is an analogue (1) of the naturally occurring bile acid (FW = 392.57 g/mol; CAS 474-25-9). The latter is synthesized in the liver, where it conjugated to form tauro-chenodeoxycholate and glycol-chenodeoxycholate, reducing its pK_a and increasing retention in the gastrointestinal tract, until reabsorption by the ileum. Chenodeoxycholate is the most active physiological ligand known for the farnesoid X receptor, *or* FXR (encoded by the *NR1H4* gene in humans) that translocates to the nucleus, dimerizes, and binds to hormone response elements (2). Obeticholic acid reduces bacterial translocation and invasion in cirrhotic rats by restoring intestinal barrier integrity (through increased expression of tight junction proteins) and by inhibiting inflammation (3). Obeticholate likewise up-regulated expression of the FXR-associated gene small heterodimer partner (SHP). **1**. Gioiello, Macchiarulo, Carotti, *et al.* (2011) *Bioorg. Med. Chem.* **19**, 2650. **2**. Forman, Goode, Chen, *et al.* (1995) *Cell* **81**, 687. **3**. Úbeda, Lario, Muñoz, *et al.* (2015) *J. Hepatol.* **64**, 1049.

Obiltoxaximab

This anti-anthrax therapeutic (MW = 145.5 kDa; CAS 1351337-07-9; Trade Name = Anthim®) is an FDA-approved mouse/human chimeric monoclonal antibody that binds to protective antigen (K_d = 0.3 nM), blocking the latter's interaction with host cell anthrax receptor, which mediates the translocation of lethal factor [LF] and edema factor into the host cell cytoplasm (**See Anthrax Lethal Toxin**). Anthim is recommended for the treatment (when combined with antibacterial drugs, such as levofloxacin, ciprofloxacin and doxycycline) and prophylactically (when used alone) treat inhalational anthrax (*i.e.*, exposure to *Bacillus anthracis* spores. Obiltoxaximab was been designed to neutralize the free protective antigen of *B. anthracis*, thereby inhibiting the lethal effects of anthrax toxins. Inasmuch as it would be unethical to conduct clinical trials in humans with clinically inducced inhalational anthrax, Anthim's efficacy was evaluated in NZW rabbits and cynomolgus macaques. **See also** *JMV 390-1* **1**. Greig (2016) *Drugs* **76**, 823.

Obtustatin

This 41-residue disintegrin (FW = 4401 g/mol; *Sequence* CTTGPCCRQ CKLKPAGTTCWKTSLTSHYCTGKSCDCPLYPG, with disulfide bonds at Cys1-Cys10, Cys6-Cys29, Cys7-Cys34, and Cys19-Cys36) from the venom of the blunt-nosed viper *Vipera lebetina obtusa* exhibits tight binding to integrins (1), inhibiting platelet aggregation, IC_{50} = 2 nM (2). It is the shortest polypeptide to contain an integrin-recognition loop. **Target(s):** Obtustatin is a highly potent integrin $\alpha_1\beta_1$ inhibitor (IC_{50} = 0.8 nM for $\alpha_1\beta_1$ binding to type IV collagen), displaying specificity for $\alpha_1\beta_1$ over $\alpha_2\beta_1$, $\alpha_{IIb}\beta_3$, $\alpha_v\beta_3$, $\alpha_4\beta_1$, $\alpha_5\beta_6$, $\alpha_9\beta_1$ and $\alpha_4\beta_7$. **1**. Marcinkiewicz, Weinreb, Calvete, *et al.* (2003) *Cancer Res.* **63**, 2020. **2**. Moreno-Murciano, Monleon, Calvete, Celda & Marcinkiewicz (2003) *Protein Sci.* **12**, 366.

OC252-324

This highly constrained pseudo-tetrapeptide ($FW_{inner-salt}$ = 541.65 g/mol) defines a new allosteric binding site (n_{Hill} = 2) that is located near the center of *Escherichia coli* fructose-1,6-bisphosphatase, showing uncompetitive inhibition. Binding is synergistic with both AMP and fructose 2,6-bisphosphate. OC252-324 contact with three of four subunits of the tetramer, hydrogen bonding with the side chain of Asp-187 and the backbone carbonyl of Residue-71, and electrostatically interacting with the backbone carbonyl of Residue-51. **1**. Choe, Nelson, Arienti, *et al.* (2003) *J. Biol. Chem.* **278**, 51176.

Ochratoxins

Ochratoxin A

Ochratoxin B

Ochratoxin C

These mycotoxins, produced by *Aspergillus ochraceous* and certain other fungi, are calcium ion modulators frequently found in decaying vegetables and grain. The major component is ochratoxin A ($FW_{free-acid}$ = 403.82 g/mol;

CAS 303-47-9), also known as (*R*)-*N*-[(5-chloro-3,4-dihydro-8-hydroxy-3-methyl-1-oxo-1*H*-2-benzopyran-7-yl)carbonyl]-L-phenylalanine. It has a green fluorescence and induces differentiation in cloned renal cell lines. It also promotes free radical formation, lipid peroxidation, and stimulates Ca^{2+}-exporting ATPase activity in the endoplasmic reticulum. Ochratoxin B is less toxic than Ochratoxin A and has a blue fluorescence. Ochratoxin C is the ethyl ester of A. **Target(s):** electron transport, mitochondrial (1-3); mitochondrial respiration (1-3); phenylalanine monooxygenase (4); phenylalanyl-tRNA synthetase (1,5-9); protein biosynthesis (10-12); succinate: cytochrome *c* reductase (2); succinate dehydrogenase (2); succinate oxidase (2). **1**. Hohler (1998) *Z. Ernahrungswiss.* **37**, 2. **2**. Wei, Lu, Lin & Wei (1985) *Toxicology* **36**, 119. **3**. Moore & Truelove (1970) *Science* **168**, 1102. **4**. Zanic-Grubisic, Zrinski, Cepelak, *et al.* (2000) *Toxicol. Appl. Pharmacol.* **167**, 132. **5**. Vedani & Arend (1996) *ALTEX* **13**, 124. **6**. McMasters & Vedani (1999) *J. Med. Chem.* **42**, 3075. 7. Creppy, Lugnier, Fasiolo, *et al.* (1979) *Chem. Biol. Interact.* **24**, 257. **8**. Konrad & Röschenthaler (1977) *FEBS Lett.* **83**, 341. **9**. Roth, Eriani, Dirheimer & Gangloff (1993) *FEBS Lett.* **326**, 87. **10**. Creppy, Röschenthaler & Dirheimer (1984) *Food Chem. Toxicol.* **22**, 883. **11**. Bunge, Dirheimer & Röschenthaler (1978) *Biochem. Biophys. Res. Commun.* **83**, 398. **12**. Heller & Röschenthaler (1978) *Can. J. Microbiol.* **24**, 467.

Oclacitinib

This JAK family inhibitor (FW = 453.51 g/mol; CAS 1640292-55-2; Solubility: 90 mg/mL DMSO; 18 mg/mL H_2O), also named Apoquel and *N*-methyl-4-(methyl-7*H*-pyrrolo[2,3-*d*]pyrimidin-4-ylamino)cyclohexane-methane sulfonamide and, targets Janus Kinases (JAK1, 10 nM; JAK2, 18 nM; JAK3, 99 nM; TYK2, 84 nM), inhibiting the function of JAK1-dependent cytokines involved in allergy and inflammation (IL-2, IL-4, IL-6, and IL-13) as well as pruritus (IL-31). Oclacitinib had minimal effects on cytokines that did not activate the JAK1 enzyme in cells (erythropoietin, granulocyte/macrophage colony-stimulating factor, IL-12, IL-23; IC_{50} values greater than 1000 nM). These results demonstrate that oclacitinib is a targeted therapy that selectively inhibits JAK1-dependent cytokines involved in allergy, inflammation, and pruritus and suggests these are the mechanisms by which oclacitinib effectively controls clinical signs associated with allergic skin disease in dogs. Apoquel is FDA-approved for canine allergic dermatitis, including flea allergy dermatitis, food allergy, atopic dermatitis, and contact dermatitis. **Other Targets:** (% Inhibition at 1 μM): ABL tyrosine kinase, 13%; RAC-α serine/threonine kinase (Protein kinase B), 3%; ZAP-70 tyrosine kinase, 10%; Angiopoietin 1 receptor (TIE-2) (TEK), 3%; Aurora-related kinase 1 (ARK1), 40%; Ste20-like kinase MST2, 18%; SRC kinase, 17%; Serum glucocorticoid-regulated kinase (SGK1), 14%; p160ROCK protein kinase, 33%; Protein kinase C β II Isoform, 7%; Protein kinase A α, 17%; Serine/threonine protein kinase PIM-2, 9%; PAK-4 (p21 activated kinase 4), 13%; High-affinity nerve growth factor receptor (TRK-A), 31%; Serine/threonine protein kinase NEK2, 3%; Myosin light chain kinase (skeletal), 6%; Serine/threonine protein kinase MASK, 4%; Hepatocyte growth factor receptor (MET Proto-oncogene tyrosine kinase), 15%; Serine/threonine protein kinase MARK, 42%; MAP kinase activated protein kinase 2 (MAPKAPK-2), 3%; Mitogen-activated protein kinase 1 (MAPK1/ERK2), 8%; Mitogen-activated protein kinase kinase kinase kinase 4 (MAP4K4/HGK), 38%; Proto-oncogene LCK tyrosine kinase, 22%; Vascular Endothelial Growth Factor Receptor 2 (VEGFR-2/FLK1), 39%; Tyrosine-protein kinase JAK3 (JANUS KINASE 3), 95%; Insulin receptor, 5%; Glycogen Synthase Kinase-3 β (GSK-3β), 8%; FGF Receptor-1 (FGFR1), 10%; Ephrin type-A Receptor 2 (ECK), 5%; Epidermal growth factor receptor, 8%; Casein Kinase II α prime chain (CK II), 5%; Casein Kinase-1, α, 1%; Checkpoint kinase (CHK2), 7%; Checkpoint kinase CHK1, 14%; CDK2/Cyclin-A, 10%; Calcium/calmodulin-dependent protein kinase II α-B subunit, 18%; Bruton's tyrosine kinase (BTK), 7%; TAO Kinase-2, 10%; Mitogen-Activated Protein Kinase-14 (MAPK14/P38-α), 8%. **1**. Gonzales, Bowman, Fici, *et al.* (2014) *J. Vet. Pharmacol. Ther.* **37**, 317.

Ocrelizumab

This second-generation humanized anti-CD20 monoclonal antibody and CD20 antagonist (MW = 148 kDa; CAS 637334-45-3), also named Ocrevus®, targets mature B lymphocytes, making it a highly effective, FDA-approved immunosuppressive drug for the treatment of Primary Progressive Multiple Sclerosis (PPMS). **Mode of Action:** B cells are known to play central roles in MS pathogenesis (*e.g.*, activation of proinflammatory T cells, secretion of proinflammatory cytokines, as well as production of anti-myelin autoantibodies). CD20 (also called B-lymphocyte antigen CD20) is an activated-glycosylated phosphoprotein of unknown function that is expressed on the surface of all B-cells, beginning at the pro-B phase (CD45R⁺, CD117⁺), with its concentration increasing progressively until maturity. Compared with placebo, two doses of Ocrevus (600 and 2000 mg on days 1 and 15) show a pronounced effect on disease activity seen in MRI as gadolinium-enhanced lesions and also has a significant effect on relapses. Other CD20-targeting drugs include ofatumumab, hA20, and TRU-015, again mainly acting to deplete B cells. 1. Sorensen & Blinkenberg (2016) *Ther. Adv. Neurol. Disord.* **9**, 44.

9Z,11E-Octadecadienoate

This unsaturated fatty acid ($FW_{free\text{-}acid}$ = 280.45 g/mol), also known as *cis⁹,trans¹¹*-conjugated linoleic acid, inhibits linoleoyl-CoA Desaturase (*Reaction*: Linoleoyl-CoA + AH_2 + O_2 ⇌ γ-Linolenoyl-CoA + A + 2 H_2O, where AH_2 is the electron acceptor) (1,2) and stearoyl-CoA 9-desaturase (*Reaction*: Stearoyl-CoA + 2 Ferrocytochrome b_5 + O_2 + 2 H^+ ⇌ Oleoyl-CoA + 2 Ferricytochrome b_5 + 2 H_2O) (3). 1. Chuang, Leonard, Liu, *et al.* (2001) *Lipids* **36**, 1099. 2. Chuang, Thurmond, Liu, *et al.* (2001) *Lipids* **36**, 139. 3. Choi, Park, Storkson, Pariza & Ntambi (2002) *Biochem. Biophys. Res. Commun.* **294**, 785.

9-Octadecanyl Sulfate

This lipid sulfate ($FW_{free\text{-}acid}$ = 350.56 g/mol) is allosteric competitive inhibitor of the intrinsic phosphatase activity of soluble epoxide hydrolase (IC_{50} = 4.7 μM). Intriguingly, soluble epoxide hydrolase (sEH) is a homodimeric enzyme. Each monomer contains a C-terminal domain with epoxide hydrolase activity that is involved in the metabolism of arachidonic acid epoxides, endogenous chemical mediators that play important roles in blood pressure regulation, cell growth, and inflammation. The *N*-terminal domain, however, contains a Mg^{2+}-dependent lipid phosphate phosphatase activity. 1. Tran, Aronov, Tanaka, *et al.* (2005) *Biochemistry* **44**, 12179.

9Z-Octadecen-12-ynoate

This linoleic acid analogue ($FW_{free\text{-}acid}$ = 278.44 g/mol), also known as 9*Z*-octadecen-12-ynoic acid, inhibits linoleate isomerase (1) and inactivates soybean lipoxygenase (2). 1. Kepler, Tucker & Tove (1970) *J. Biol. Chem.* **245**, 3612. 2. Nieuwenhuizen, Van der Kerk-Van Hoof, van Lenthe, *et al.* (1997) *Biochemistry* **36**, 4480.

Octadecylamine

This long-chain aliphatic amine ($FW_{free\text{-}base}$ = 269.51 g/mol; CAS 124-30-1), also known as stearylamine and 1-aminooctadecane, is a cationic detergent

that inhibits numerous enzymes at relatively low concentrations. Stearylamine can be considered a structural analogue of sphingosine. **Target(s):** 2-acylglycerol *O*-acyltransferase, *or* monoacyl-glycerol *O*-acyltransferase (1); glucosylceramidase (2); [myosin light-chain] kinase (3); NADPH oxidase (4); phosphatidyl-glycerophosphatase (5); phosphatidyl-inositol diacylglycerol-lyase, *or* phosphatidylinositol-specific phospholipase C (6); phospholipase A_2 (7); phospholipase D (8); protein kinase C (8); protein-tyrosine sulfotransferase (9). 1. Bhat, Wang & Coleman (1995) *Biochemistry* **34**, 11237. 2. Osiecki-Newman, Fabbro, Legler, Desnick & Grabowski (1987) *Biochim. Biophys. Acta* **915**, 87. 3. Jinsart, Ternai & Polya (1991) *Plant Sci.* **78**, 165. 4. Sawai, Asada, Nishizawa, Nunoi & Katayama (1999) *Jpn. J. Pharmacol.* **80**, 237. 5. Chang & Kennedy (1967) *J. Lipid Res.* **8**, 456. 6. Hirasawa, Irvine & Dawson (1981) *Biochem. J.* **193**, 607. 7. Franson, Harris, Ghosh & Rosenthal (1992) *Biochim. Biophys. Acta* **1136**, 169. 8. Elstad, McIntyre, Prescott & Zimmerman (1991) *Amer. J. Respir. Cell. Mol. Biol.* **4**, 148. 9. Kasinathan, Sundaram, Slomiany & Slomiany (1993) *Biochemistry* **32**, 1194.

1-*O*-Octadecyl-2-*O*-methyl-*sn*-glycero-3-phosphocholine

This phosphatidylcholine/lysophosphatidylcholine analogue (FW = 523.73 g/mol), also known as edelfosine and ET-18-OCH₃, is an antiproliferative agent that is cytotoxic to transformed cells both *in vitro* and *in vivo*. It induces apoptosis in HeLa and HL-60 cells and is a noncompetitive inhibitor of phosphatidylinositol 3-kinase (IC_{50} = 35 μM). **Target(s):** arachidonoyl-CoA:1-acyl-*sn*-glycero-3-phosphocholine acyltransferase (1); asialomucin-sialyltransferase (2); choline-phosphate cytidylyltransferase, *or* CTP: phosphocholine cytidylyltransferase; inhibited by the racemic compound)3-5; glycerophospholipid arachidonoyltransferase (CoA-independent), IC_{50} = 0.5 μM (6); phosphoinositide phospholipase C, *or* phosphatidylinositol-dependent phospholipase C (7,8); phosphatidylinositol 3-kinase (9); phospholipase C (7,10); protein kinase C (11). 1. Herrmann, Ferber & Munder (1986) *Biochim. Biophys. Acta* **876**, 28. 2. Bador, Morelis & Louisot (1983) *Int. J. Biochem.* **15**, 1137. 3. Baburina & Jackowski (1998) *J. Biol. Chem.* **273**, 2169. 4. Boggs, Rock & Jackowski (1995) *J. Biol. Chem.* **270**, 7757. 5. Yang, Boggs & Jackowski (1995) *J. Biol. Chem.* **270**, 23951. 6. Winkler, Eris, Sung, *et al.* (1996) *J. Pharmacol. Exp. Ther.* **279**, 956. 7. Powis, Seewald, Gratas, *et al.* (1992) *Cancer Res.* **52**, 2835. 8. Piacentini, Piatti, Fraternale, *et al.* (2004) *Biochimie* **86**, 343. 9. Berggren, Gallegos, Dressler, Modest & Powis (1993) *Cancer Res.* **53**, 4297. 10. Llansola, Monfort & Felipo (2000) *J. Pharmacol. Exp. Ther.* **292**, 870. 11. Zhou & Arthur (1997) *Biochem. J.* **324**, 897.

R-3,4-Octadienoyl-CoA

This unusual thiolester ($FW_{free\text{-}acid}$ = 889.71 g/mol) inhibits medium-chain acyl-Coenzyme A dehydrogenase, rapidly undergoing inhibition, with a stoichiometry of two mol racemate per mol enzyme-bound flavin. Synthesis of *R*- and *S*-3,4-decadienoyl-CoA shows that the R-enantiomer is the inhibitor. α-Proton abstraction yields an enolate to oxidized flavin charge-transfer intermediate prior to adduct formation. Crystal structure data on the reduced, inactive enzyme shows a single covalent bond linking the C-4 carbon of the 2,4-dienoyl-CoA moiety and the N5 locus of reduced flavin. The kinetics of reversible adduct formation, with release of the conjugated 2,4-diene, were evaluated as a function of both acyl chain length and truncation of the CoA moiety. The adduct is most stable with medium chain length allenic inhibitors. However, the adducts with *R*-3,4-decadienoyl-pantetheine and *N*-acetylcysteamine are some 9- and >100-fold more kinetically stable than the full-length CoA thioester. Crystal structures of these reduced enzyme species, determined to 2.4 Å, suggesting the placement of H-bonds to the inhibitor carbonyl oxygen and the positioning of the catalytic base are important determinants of adduct stability. While not an inhibitor, the *S*-isomer is rapidly isomerized to the *trans-trans*-2,4-conjugated diene. Protein modeling suggests the *S*-enantiomer cannot approach close enough to the isoalloxazine ring to form a flavin adduct, but can be easily reprotonated by the catalytic base. **1**. Wang, Fu, Zhou, Kim & Thorpe (2001) *Biochemistry* **40**, 12266.

Octahydropiericidin A

Octahydropiericidin A₁

This reduced antibiotic (FW = 423.64 g/mol; CAS 2738-64-9; Soluble in ethanol, methanol, DMF or DMSO) is a structural analogue of ubiquinone and a partially competitive inhibitor of bacterial and mitochondrial Type-I NADH-ubiquinone oxidoreductases (*or* Complex I). Octahydropiericidin A also inhibits glucose dehydrogenase (3). **1**. Jeng, Hall, Crane, *et al.* (1968) *Biochemistry* **7**, 1311. **2**. Ramsay, Krueger, Youngster & Singer (1991*) Biochem. J.* **273**, 481. **3**. Friedrich, van Heek, Leif, *et al.* (1994) *Eur. J. Biochem.* **219**, 691.

1-Octanethiol

This foul-smelling thiol (FW = 146.30 g/mol; B.P. = 197-200°C) is an irreversible inhibitor of soybean liopxygenase-1, which catalyzes an oxygenation reaction, converting 1-octanethiol to 1-octanesulfonic acid, with 1 in 48 turnovers resulting in enzyme inactivation. No inactivation was observed with several more polar thiols, such as mercaptoethanol, dithiothreitol, L-cysteine, glutathione, *N*-acetylcysteamine, or captopril, suggesting that a hydrophobic thiol is required to irreversibly inactivate soybean lipoxygenase. **1**. Clapp, Grandizio, Yang, *et al.* (2002) *Biochemistry* **41**, 11504.

Octanoate (Octanoic Acid)

This medium-chain fatty acid (FW$_{free-acid}$ = 144.21 g/mol), also called *n*-octanoic acid and caprylic acid, has a pK_a value of 4.85 at 25°C. It is an oily liquid (M.P. = 16.7°C; B.P. = 239.7°C) with a slightly rancid taste. Octanoic acid is slightly soluble in water (0.068 g/100 g H$_2$O at 20°C) and soluble in ethanol, diethyl ether, benzene, and glacial acetic acid. **Target(s)**: alanyl aminopeptidase (1,30); alcohol dehydrogenase (2); aldehyde reductase, aldose reductase (3); D-amino-acid oxidase, weakly inhibited; K_i = 6.3 mM (4,5); [branched-chain α-keto-acid dehydrogenase] kinase, *or* 3-methyl-2-oxobutanoate dehydrogenase (acetyl-transferring)] kinase (6-8,32,33); carbonyl reductase (9); choline acetyltransferase (10); cutinase (11); cytosol alanyl aminopeptidase (1,30); esterase (12,13); firefly luciferase, IC$_{50}$ = 2.9 mM (35); β-galactoside α-2,6 sialyltransferase, weakly inhibited (34); glucokinase (14,15); glucose-6-phosphate dehydrogenase (15); glutathione *S*-transferase (16); hexokinase (14,15,17); kynurenine aminotransferase, weakly inhibited (18); L-lactate dehydrogenase (19-21); lipoxygenase (22); monoamine oxidase (23); NAD$^+$ nucleosidase, NADase, K_i = 56 mM (31); penicillin acylase (24); penicillocarboxypeptidases (25); phenylacetyl-CoA synthetase, *or* phenylacetate:CoA ligase (26); phosphofructokinase (14,15,17); propionyl-CoA synthetase, K_i = 58 μM (27,28); pyruvate kinase (15,29). **1**. Garner & Behal (1975) *Biochemistry* **14**, 5084. **2**. Winer & Theorell (1960) *Acta Chem. Scand.* **14**, 1729. **3**. Hayman & Kinoshita (1965) *J. Biol. Chem.* **240**, 877. **4**. Brown & Scholefield (1953) *Proc. Soc. Exp. Biol. Med.* **82**, 34. **5**. Klein & Austin (1953) *J. Biol. Chem.* **205**, 725. **6**. Paxton (1988) *Meth. Enzymol.* **166**, 313. **7**. Harris, Paxton, Powell, *et al.* (1986) *Adv. Enzyme Regul.* **25**, 219. **8**. Randle, Patston & Espinal (1987) *The Enzymes*, 3rd ed., **18**, 97. **9**. Imamura, Migita, Anraku & Otagiri (1999) *Biol. Pharm. Bull.* **22**, 731. **10**. Ninomiya & Kayama (1998) *Neurochem. Res.* **23**, 1303. **11**. Maeda, Yamagata, Abe, *et al.* (2005) *Appl. Microbiol. Biotechnol.* **67**, 778. **12**. Horsted, Dey, Holmberg & Kielland-Brandt (1998) *Yeast* **14**, 793. **13**. Weber & King (1935) *J. Biol. Chem.* **108**, 131. **14**. Lea & Weber

(1968) *J. Biol. Chem.* **243**, 1096. **15**. Weber, Lea, Convery & Stamm (1967) *Adv. Enzyme Reg.* **5**, 257. **16**. Young & Briedis (1989) *Biochem. J.* **257**, 541. **17**. Lea, Weber & Morris (1976) *Oncology* **33**, 205. **18**. Mason (1959) *J. Biol. Chem.* **234**, 2770. **19**. Hinkson & Mahler (1963) *Biochemistry* **2**, 216. **20**. Webb (1966) *Enzyme and Metabolic Inhibitors*, vol. **2**, p. 436, Academic Press, New York. **21**. Dikstein (1959) *Biochem. Biopys. Acta* **36**, 397. **22**. Holman & Bergström (1951) *The Enzymes*, 1st ed., **2** (part 1), 559. **23**. McEwan, Sasaki & Jones (1969) *Biochemistry* **8**, 3952. **24**. Torres-Guzman, de la Mata, Torres-Bacete, *et al.* (2002) *Biochem. Biophys. Res. Commun.* **291**, 593. **25**. Hofmann (1976) *Meth. Enzymol.* **45**, 587. **26**. Martinez-Blanco, Reglero, Rodriguez-Aparicio & Luengo (1990) *J. Biol. Chem.* **265**, 7084. **27**. Krahenbuhl & Brass (1991) *Biochem. Pharmacol.* **41**, 1015. **28**. Groot (1975) *Biochim. Biophys. Acta* **380**, 12. **29**. Kayne (1973) *The Enzymes*, 3rd ed., **8**, 353. **30**. Garner & Behal (1977) *Arch. Biochem. Biophys.* **182**, 667. **31**. Pace, Agnellini, Lippoli & Berger (1997) *Adv. Exp. Med. Biol.* **419**, 389. **32**. Espinal, Beggs & Randle (1988) *Meth. Enzymol.* **166**, 166. **33**. Paxton & Harris (1984) *Arch. Biochem. Biophys.* **231**, 48. **34**. Hickman, Ashwell, Morell, van den Hamer & Scheinberg (1970) *J. Biol. Chem.* **245**, 759. **35**. Matsuki, Suzuki, Kamaya & Ueda (1999) *Biochim. Biophys Acta* **1426**, 143.

Octanoyl-CoA

This coenzyme thiolester (FW$_{free-acid}$ = 893.72 g/mol), also known as caproyl-CoA and capryl-CoA (note that this term has also been used for decanoyl-CoA), is an inhibitor of enoyl-CoA hydratase in β-oxidation. **Target(s)**: acetyl-CoA carboxylase (1); 1-alkyl-*sn*-glycero-3-phosphate acetyltransferase (2); aralkylamine *N*-acetyltransferase (3); arylamine *N*-acetyltransferase (4); citrate synthase (5); enoyl-CoA hydratase (6,7); fatty-acid synthase (8); glutamate dehydrogenase (5); 3-hydroxy-3-methylglutaryl-CoA synthase (9); isocitrate dehydrogenase (NADP$^+$) (10); α-ketoglutarate dehydrogenase (10); lipoyl(octanoyl) transferase (11); methylmalonyl-CoA epimerase (12); (*S*)-methylmalonyl-CoA hydrolase (13); pantothenate kinase (14); [pyruvate dehydrogenase] kinase (15). **1**. Bortz & Lynen (1963) *Biochem. Z.* **337**, 505. **2**. Baker & Chang (1995) *J. Neurochem.* **64**, 364. **3**. Ferry, Loynel, Kucharczyk, *et al.* (2000) *J. Biol. Chem.* **275**, 8794. **4**. Kawamura, Graham, Mushtaq, *et al.* (2005) *Biochem. Pharmacol.* **69**, 347. **5**. Lai, Liang, Zhai, Jarvi & Lu (1994) *Metab. Brain Dis.* **9**, 143. **6**. Engel, Kiema, Hiltunen & Wierenga (1998) *J. Mol. Biol.* **275**, 859. **7**. Lau, Powell, Buettner, Ghisla & Thorpe (1986) *Biochemistry* **25**, 4184. **8**. Podack, Saathoff & Seubert (1974) *Eur. J. Biochem.* **50**, 237. **9**. Middleton (1972) *Biochem. J.* **126**, 35. **10**. Lai & Cooper (1991) *Neurochem. Res.* **16**, 795. **11**. Nesbitt, Baleanu-Gogonea, Cicchillo, *et al.* (2005) *Protein Expr. Purif.* **39**, 269. **12**. Stabler, Marcell & Allen (1985) *Arch. Biochem. Biophys.* **241**, 252. **13**. Kovachy, Copley & Allen (1983) *J. Biol. Chem.* **258**, 11415. **14**. Halvorsen & Skrede (1982) *Eur. J. Biochem.* **124**, 211. **15**. Batenburg & Olson (1976) *J. Biol. Chem.* **251**, 1364.

2-Octenoyl-CoA

trans-2-Octenoyl-CoA

cis-2-Octenoyl-CoA

This unsaturated fatty-acyl-CoA metabolite (FW$_{free\ acid}$ = 891.72 g/mol) is a product of the reaction catalyzed by medium-chain fatty-acyl-CoA dehydrogenase and will act as a product inhibitor (the *trans*-isomer is an intermediate in the β-oxidation of saturated fatty acids). **Target(s)**: acyl-CoA dehydrogenases (1,2); fatty-acid synthase (3). **1**. Beinert (1963) *The Enzymes*, 2nd ed., **7**, 447. **2**. Peterson, Sergienko, Wu, *et al.* (1995) *Biochemistry* **34**, 14942. **3**. Podack, Saathoff & Seubert (1974) *Eur. J. Biochem.* **50**, 237.

Octopamine

This substituted tyramine (FW = 153.18 g/mol; CAS 104-14-3), also called β-hydroxytyramine, is an α-adrenergic receptor agonist. The D-enantiomer was first isolated from the salivary glands of a species of *Octopus* in 1951, hence the name, and has now been found in many invertebrates, in which it acts as neurohormone, a neuromodulator, and a neurotransmitter. It will inhibit adenylate cyclase via the $α_{2A}$-adrenoreceptor. **Target(s):** arylsulfatase (1,2); CYPc1 (13); dihydropteridine reductase (4); γ-glutamylhistamine synthetase (5). **1.** Okamura, Yamada, Murooka & Harada (1976) *Agric. Biol. Chem.* **40**, 2071. **2.** Murooka, Yim & Harada (1980) *Appl. Environ. Microbiol.* **39**, 81 2. **3.** Louw, Allie, Swart & Swart (2000) *Endocr. Res.* **26**, 729. **4.** Shen (1983) *Biochim. Biophys. Acta* **743**, 129. **5.** Stein & Weinreich (1982) *J. Neurochem.* **38**, 204.

D-Octopin

This arginine derivative (FW$_{hydrochloride}$ = 282.73 g/mol; CAS 34522-32-2), also known as N^2-([R]-1-carboxyethyl)-L-arginine (note that it can also be regarded as an *N*-substituted derivative of D-alanine), is found in the muscles of various invertebrates (first isolated from octopus). In addition, it is found in plant tumors of crown gall disease. Note that the term has also been used to describe similar *N*-substituted alanines. Octopine is a substrate for D-octopine dehydrogenase. **Target(s):** histidine ammonia-lyase. **1.** Leuthardt (1951) *The Enzymes*, 1st ed., **1** (part 2), 1156.

Octreotide

This orally bioavailable octapeptide somatostatin analogue (FW = 1019.24 g/mol; CAS 83150-76-9; 79517-01-4 (acetate salt); 135467-16-2 (pamoate salt); D-Phe-Cys-Phe-D-Trp-Lys-Thr-Cys-NH-CH(CH$_2$OH)CHOHCH$_3$), also known as SMS 201-995 and Sandostatin®, is a long-lasting and more potent inhibitor of growth hormone, glucagon, and insulin than is somatostatin. Its synthesis was enabled by stepwise modification of a conformationally stabilized analogue of the fragment of somatostatin that had been thought to be a biologically active moiety (1). Indeed, octreotide displaces [³H[naloxone from its binding sites (IC$_{50}$ = 38 nM), or >200x more potent than somatostatin (2). As measured by the difference between the binding of [³H]-dihydromorphine, [³H]-[D-Ala$_2$,D-Leu$_5$]enkephalin, and (–)-[³H]-bremazocine, SMS 201-995 appears to be highly specific for the opiate μ binding site (2). Octreotide immediate-release injection is used to decrease the amount of growth hormone produced in acromegaly. **1.** Bauer, Briner, Doepfner, *et al.* (1982) *Life Sci.* **31**, 1133. **2.** Maurer, Gaehwiler, Buescher, Hill & Roemer (1982) *Proc. Natl. Acad. Sci. U.S.A.* **79**, 4815. **3.**

Octylamine

This toxic alkylamine (FW$_{free-base}$ = 129.25 g/mol; M.P. = –5 to –1°C; B.P. = 175-176°C; 111-86-4), also known as 1-aminooctane and caprylamine, inhibits many enzyme activities. **Target(s):** alanyl aminopeptidase (1,2); benzo(*a*)pyrene hydroxylase (3); carnosine synthetase (4); cysteine-conjugate *S*-oxidase activity (5); cytosol alanyl aminopeptidase (1,2); deoxyhypusine synthase (6); flavin-containing monooxygenase (7); β-glucosidase (8); glucosylceramidase (9); guanidinoacetate (10); guanylate cyclase (11); NAD$^+$ nucleosidase, *or* NADase, K_i = 21 mM (12); NADPH oxidase (13); phospho-*N*-acetylmuramoyl pentapeptide-transferase (14). **1.** Garner & Behal (1975) *Biochemistry* **14**, 5084. **2.** Garner & Behal (1977) *Arch. Biochem. Biophys.* **182**, 667. **3.** Ahokas, Pelkonen & Karki (1977) *Cancer Res.* **37**, 3737. **4.** Seely & Marshall (1982) in *Peptide Antibiotics: Biosynthesis and Functions: Enzymatic Formation of Bioactive Peptides and Related Compounds* (Kleinkauf & von Döhren, eds.) W. de Gruyter, Berlin, pp. 347. **5.** Sausen & Elfarra (1990) *J. Biol. Chem.* **265**, 6139. **6.** Jakus, Wolff, Park & Folk (1993) *J. Biol. Chem.* **268**, 13151. **7.** Demirdoegen & Adall (2005) *Cell Biochem. Funct.* **23**, 245. **8.** Grabowski & Dagan (1984) *Anal. Biochem.* **141**, 267. **9.** Osiecki-Newman, Fabbro, Legler, Desnick & Grabowski (1987) *Biochim. Biophys. Acta* **915**, 87. **10.** Shirokane & Nakayima (1986) *J. Ferment. Technol.* **64**, 29. **11.** Gorczyca, Van Hooser & Palczewski (1994) *Biochemistry* **33**, 3217. **12.** Pace, Agnellini, Lippoli & Berger (1997) *Adv. Exp. Med. Biol.* **419**, 389. **13.** Sawai, Asada, Nishizawa, Nunoi & Katayama (1999) *Jpn. J. Pharmacol.* **80**, 237. **14.** Heydanek, Struve & Neuhaus (1969) *Biochemistry* **8**, 1214.

Octyl-CoA Disulfide

This synthetic disulfide (FW$_{free-acid}$ = 943.89 g/mol) inhibits *Mycobacterium tuberculosis* β-ketoacyl [acyl-carrier-protein] synthase III, IC$_{50}$ = 30 μM. The kinetics of MeSSCoA indicate rapid inhibition of one monomer of ecFabH by forming a methyl disulfide conjugate with this cysteine. Reaction of the second subunit with either MeSSCoA or acetyl-CoA is much slower. In the presence of malonyl-ACP, the acylation rate of the second subunit is restored to that of the native ecFabH. Such findings suggest a catalytic model, in which a structurally disordered apo-ecFabH dimer orders on binding either the first substrate, acetyl-CoA, or the inhibitor MeSSCoA, and is restored to a disordered state on binding of malonyl-ACP. **1.** Alhamadsheh, Musayev, Komissarov, *et al.* (2007) *Chem. Biol.* **14**, 513.

S-Octylglutathione

This *S*-alkylated glutathione (FW$_{hydrochloride}$ = 456.00 g/mol) inhibits glutathione *S*-transferase (1,2); hydroxyacyl-glutathione hydrolase, *or* glyoxalase II (7); lactoylglutathione lyase, *or* glyoxalase I (3-8); xenobiotic-transporting ATPase (multidrug resistance protein) (9). **1.** Pace-Asciak & Smith (1983) *The Enzymes*, 3rd ed. (Boyer, ed.), **16**, 543. **2.** Ortiz-Salmeron, Yassin, Clemente-Jimenez, *et al.* (2001) *Biochim. Biophys. Acta* **1548**, 106. **3.** Kraus, Pernice & Ponce (1988) *Res. Commun. Chem. Pathol. Pharmacol.* **59**, 419. **4.** Kraus & Castaing (1989) *Res. Commun. Chem. Pathol. Pharmacol.* **65**, 105. **5.** Vince & Wadd (1969) *Biochem. Biophys. Res. Commun.* **35**, 593. **6.** Allen, Lo & Thornalley (1993) *Biochem. Soc. Trans.* **21**, 535. **7.** Norton, Talesa, Yuan & Principato (1990) *Biochem. Int.* **22**, 411. **8.** Regoli, Saccucci & Principato (1996) *Comp. Biochem. Physiol.*

C, Pharmacol. Toxicol. Endocrin. **113**, 313. **9**. Loe, Stewart, Massey, Deeley & Cole (1997) Mol. Pharm. **51**, 1034.

Octylguanidine

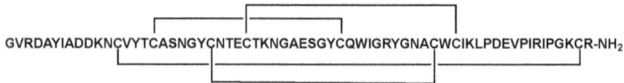

This hygroscopic guanidine derivative ($FW_{hydrochloride}$ = 217.82 g/mol; CAS 3038-42-4) will inactivate the mitochondrial permeability transition pore. It is also an inhibitor of oxidative and photophosphorylation. **Target(s):** ATPase, plasma membrane (1); cation translocation (2); chloroplast uncoupler (3); F_1 ATPase, or H^+-transporting two-sector ATPase (1,4-6); oxidative phosphorylation (6). **1**. Bowman, Mainzer, Allen & Slayman (1978) Biochim. Biophys. Acta **512**, 13. **2**. Beaty, Gutierrez, Lopez-Vancell & Estrada (1986) Acta Physiol. Pharmacol. Latinoam. **36**, 217. **3**. Izawa & Good (1972) Meth. Enzymol. **24**, 355. **4**. Galante, Wong & Hatefi (1982) Biochemistry **21**, 680. **5**. Tuena de Gomez Puyou, Gomez Puyou & Salmom (1977) Biochim. Biophys. Acta **461**, 101. **6**. Papa, Tuena de Gomez-Puyou & Gomez-Puyou (1975) Eur. J. Biochem. **55**, 1.

N-Octylindomethacin Amide

This anti-inflammatory (FW = 469.02 g/mol), also known as 1-(p-chlorobenzoyl)-5-methoxy-2-methyl-1H indole-3-octylacetamide, is a selective inhibitor of cyclooxygenase-2. The IC_{50} value for COX-2 is 40 nM, whereas the value for COX-1 is 66 μM. **1**. Kalgutkar, Marnett, Crews, Remmel & Marnett (2000) J. Med. Chem. **43**, 2860. **2**. Kalgutkar, Crews, Rowlinson, et al. (2000) Proc. Natl Acad. Sci. U.S.A. **97**, 925.

3-Octylthio-1,1,1-trifluoropropan-2-one

This halomethyl ketone (FW = 256.33 g/mol), also known as 1,1,1-trifluoro-(3-octylsulfanyl)-2-propanone, inhibits Tenebrio molitor juvenile-hormone esterase (IC_{50} = 4.6 nM). **Target(s):** acetylcholinesterase (1); carboxylesterase (1); cutinase, IC_{50} = 1.6 μM (2); juvenile hormone epoxide hydrolase (3); juvenile-hormone esterase (1,4-13); microsomal epoxide hydrolase (3). **1**. Hammock, Abdel-Aal, Mullin, Hanzlik & Roe (1984) Pest. Biochem. Physiol. **22**, 209. **2**. Wang, Michailides, Hammock, Lee & Bostock (2000) Arch. Biochem. Biophys. **382**, 31. **3**. Touhara & Prestwich (1993) J. Biol. Chem. **268**, 19604. **4**. Hinton & Hammock (2003) Insect Biochem. Mol. Biol. **33**, 317 and 477. **5**. Kamita, Hinton, Wheelock, et al. (2003) Insect Biochem. Mol. Biol. **33**, 1261. **6**. Abdel-Aal & Hammock (1985) Arch. Biochem. Biophys. **243**, 206. **7**. Zera & Zeisset (1996) Biochem. Genet. **34**, 421. **8**. Stauffer, Shiotsuki, Chan & Hammock (1997) Arch. Insect Biochem. Physiol. **34**, 203. **9**. Venkatesh, Abdel-Aal, Armstrong & Roe (1990) J. Biol. Chem. **265**, 21727. **10**. Wogulis, Wheelock, Kamita, et al. (2006) Biochemistry **45**, 4045. **11**. Abdel-Aal & Hammock (1985) Insect Biochem. **15**, 111. **12**. Abdel-Aal & Hammock (1986) Science **233**, 1073. **13**. Abdel-Aal, Hanzlik, Hammock, Harshman & Prestwich (1988) Comp. Biochem. Physiol. B Comp. Biochem. **90**, 117.

Odanacatib

This investigational osteoporosis/bone-metastasis drug (FW = 525.56 g/mol; CAS 603139-19-1), also known by the code name MK-0822 and its systematic name N-(1-cyanocyclopropyl)-4-fluoro-N^2-{(1S)-2,2,2-trifluoro-1-[4'-(methylsulfonyl)biphenyl-4-yl]ethyl}-L-leucinamide, targets the cathepsin K, a unique collagenolytic lysosomal cysteine protease that is secreted, thereby facilitating bone resorption by cleaving type-1 collagen under acidic pH (1,2). Cathepsin K exhibits a unique cleavage pattern of type I collagen molecules that is fundamentally different from that of other endogenous collagenases. Women receiving ODN (10-50 mg) for 5 years had gains in spine and hip bone mineral density (BMD), showing larger reductions in bone resorption than bone formation markers (4). Significantly, discontinuation of ODN resulted in reversal of treatment effects. An odanacatib synthesis, which describes a novel stereospecific S_N2 triflate displacement of a chiral α-trifluoromethylbenzyl triflate with (S)-γ-fluoroleucine ethyl ester, achieves a 61% overall yield in 6 steps (4). **1**. Gauthier, Chauret, Cromlish, et al. (2008) Bioorg. Med. Chem. Lett. **18**, 923. **2**. Le Gall, Bonnelye & Clézardin (2008) Curr. Opin. Support Palliat. Care. **2**, 218. **3**. Langdahl, Binkley, Bone, et al. (2012) J. Bone Miner. Res. **27**, 2251. **4**. O'Shea, Chen, Gauvreau, et al. (2009) J. Org. Chem. **74**, 1605.

OD1

GVRDAYIADDKNCVYTCASNGYCNTECTKNGAESGYCQWIGRYGNACWCIKLPDEVPIRIPGKCR-NH₂

This scorpion toxin (FW = 7206.10 g/mol; Disulfide bridge: Cys13-Cys64, Cys17-Cys37, Cys23-Cys47, Cys27-Cys49), from the venom of the scorpion Odonthobuthus doriae, modulates $Na_V1.7$ (1), dramatically impairing fast inactivation (EC_{50} = 4.5 nM), substantially increasing the peak current at all voltages, and inducing a substantial persistent current (2). $Na_V1.8$ is unaffected up to 2 μM OD1, whereas $Na_V1.3$ is sensitive only to concentrations >100 nM. OD1 impairs the inactivation process of $Na_V1.3$ (EC_{50} = 1.3 μM. The effects of OD1 were compared with a classic α-toxin, AahII from Androctonus australis Hector and a classic α-like toxin, BmK M1 from Buthus martensii Karsch. At a concentration of 50 nM, both toxins affected $Na_V1.7$. $Na_V1.3$ was sensitive to AahII but not to BmK M1, whereas $Na_V1.8$ was affected by neither toxin. The scorpion toxin OD1 is therefore a potent $Na_V1.7$ modulator, with a unique selectivity pattern. A high-resolution X-ray structure (1.8 Å) of synthetic OD1 showed the typical βαββ α-toxin fold and revealed important conformational differences in the pharmacophore region when compared with other α-toxin structures (3). Investigation of nine OD1 mutants revealed that three residues in the reverse turn contributed significantly to selectivity, with the triple OD1 mutant (D9K, D10P, K11H) being 40-fold more selective for $Na_V1.7$ over $Na_V1.6$, while OD1 K11V was 5-fold more selective for $Na_V1.6$ than $Na_V1.7$ (3). **1**. Jalali, Bosmans, Amininasab, et al. (2005) FEBS Lett. **579**, 4181. **2**. Maertens, Cuypers, Amininasab, et al. (2006) Mol. Pharmacol. **70**, 405. **3**. Durek, Vetter, Wang, et al. (2013) ACS Chem. Biol. **8**, 1215.

Oenothein B

This macrocyclic ellagitannin (FW = 1569.09 g/mol; CAS 104987-36-2) inhibits neprilysin, peptidyl-dipeptidase A, and poly(ADP-ribose) glycohydrolase. It also suppresses glucocorticoid-sensitive mousemammary tumor virus transcription. **Target(s):** aromatase, or CYP 19 (1); DNA polymerase, Epstein-Barr virus (2); neprilysin (3); peptidyl-dipeptidase A, or angiotensin I-converting enzyme (3); poly(ADP-ribose) glycohydrolase (4); steroid 5α-reductase (1,5). **1.** Ducrey, Marston, Göhring, Hartmann & Hostettmann (1997) *Planta Med.* **63**, 111. **2.** Lee, Chiou, Yen & Yang (2000) *Cancer Lett.* **154**, 131. **3.** Kiss, Kowalski & Melzig (2004) *Planta Med.* **70**, 919. **4.** Aoki, Maruta, Uchiumi, *et al.* (1995) *Biochem. Biophys. Res. Commun.* **210**, 329. **5.** Lesuisse, Berjonneau, Ciot, *et al.* (1996) *J. Nat. Prod.* **59**, 490.

Ofloxacin

This potent fluoroquinolone antibacterial ($FW_{free-acid}$ = 361.37 g/mol; CAS 82419-36-1) inhibits DNA gyrase and is active against both Gram-positive and Gram-negative bacteria (1-7). (For the prototypical member of this antibiotic class, **See** *Ciprofloxacin*) **Mode of Inhibitory Action:** The drug's bactericidal effect is mediated through the stabilization of a cleavable complex via a cooperative drug binding to a partially denatured DNA pocket created by DNA gyrase. The drug binds to supercoiled DNA in a manner similar to that to which it binds to the enzyme-DNA complex. The bacteria chromosome is composed mainly of double-stranded DNA, with 60-70 domains of supercoiling, each attached to an RNA core and organized by its own second echelon of supercoiling. DNA supercoiling is determined by DNA gyrase, an enzyme that introduces transient breaks into both DNA strands of each domain, removing ~400 helical turns before resealing the DNA and locking in the supercoiling. Ofloxacin blocks gyrase action by preferential binding to supercoiled DNA. The (*S*)-isomer, levofloxacin, has twice the antimicrobial activity of the (*R*)-isomer. Uptake of ofloxacin in moderately resistant strains is almost the same in the presence or absence of carbonyl cyanide m-chlorophenylhydrazone (CCCP), whereas in highly resistant strains, uptake increased when CCCP is added (8). Such findings suggest that resistance may depend, at least in part, to an energy-requiring drug export system. **Target(s):** DNA gyrase, or DNA topoisomerase II (1-7); glutathione-disulfide reductase, inhibited by levofloxacin (9); medium-chain-fatty-acyl-CoA synthetase, or butyryl-CoA synthetase, or butyrate:CoA ligase (10). **1.** Morrissey, Hoshino, Sato, *et al.* (1996) *Antimicrob. Agents Chemother.* **40**, 1775. **2.** Mitscher, Sharma, Chu, Shen & Pernet (1987) *J. Med. Chem.* **30**, 2283. **3.** Wolfson, Hooper, Ng, *et al.* (1987) *Antimicrob. Agents Chemother.* **31**, 1861. **4.** Ikeda, Yazawa & Nishimura (1987) *Antiviral Res.* **8**, 103. **5.** Imamura, Shibamura, Hayakawa & Osada (1987) *Antimicrob. Agents Chemother.* **31**, 325. **6.** Sato, Inoue, Fujii, Aoyama & Mitsuhashi (1986) *Infection* **14** Suppl. 4, S226. **7.** Inoue, Sato, Fujii, *et al.* (1987) *J. Bacteriol.* **169**, 2322. **8.** Tanaka, Zhang, Ishida, *et al.* (1995) *J. Med. Microbiol.* **42**, 214. **9.** Erat & Ciftci (2003) *J. Enzyme Inhib. Med. Chem.* **18**, 545. **10.** Kasuya, Hiasa, Kawai, Igarashi & Fukui (2001) *Biochem. Pharmacol.* **62**, 363.

OG-L002

This histone H3K9 demethylase inhibitor (FW = 669.82 g/mol; CAS 1357302-64-7; Solubility: 45 mg/mL DMSO; <1 mg/mL H_2O) potently and specifically inhibits the histone demethylase LSD1 (IC_{50} = 20 nM), resulting in increased repressive chromatin assembly and suppression of gene expression during the lytic phase of herpes simplex virus and varicella-zoster virus infection. This LSD1 inhibitor also blocks initial gene expression of the human cytomegalovirus and adenovirus type 5, attesting to the critical role of LSD1 in regulating DNA virus infection. Because this small-molecule inhibitor was originally designed to inhibit monoamine oxidases, it is noteworthy that its affinity for LSD1 is 36x and 69x higher than MAO-B and MAO-A, respectively. **1.** Liang, Quenelle, Vogel, *et al.* (2013) *mBio* **4**, e00558.

Ogerin

This GPR68 activator (FW = 307.35 g/mol; CAS 1309198-71-7), also named 2-[4-amino-6-[(phenylmethyl)amino]-1,3,5-triazin-2-yl]benzene-methanol, selectively and allosterically targets G-Protein-coupled Receptor 68, one of at least 120 non-olfactory G-protein-coupled receptors in the human genome that remain 'orphans' (*i.e.*, they are pharmacologically dark, with no known endogenous ligands). Ogerin suppresses recall in fear conditioning in wild-type, but not in GPR68-knockout mice. GPR68 is also a newly recognized metastasis suppressor gene in certain cancers. **1.** Huang, Karpiak, Kroeze, (2015) *Nature* **527**, 477.

Okadaic Acid

This marine sponge toxin and potent cell-permeable ionophoric polyether (FW = 824.42 g/mol; CAS 78111-17-8; *Symbol*: OA), also known as 9,10-deepithio-9,10-didehydroacanthifolicin, is a widely distributed marine toxin that accumulates by filtering shellfish (mainly bivalve mollusks) as well as fish, and is then consumed by humans, giving rise to intestinal poisoning. **Primary Mode of Action:** Kinetic studies showed that okadaic acid acts as a noncompetitive (mixed-type) inhibitor of phosphoprotein phosphatase types PP-1 and PP-2A. (1). The Type-2A protein phosphatases in mammalian tissue extracts are inhibited completely and specifically by 1–2 nM okadaic acid (1). In contrast, Type-1 protein phosphatases are hardly affected at these concentrations, complete inhibition requiring 1 μM okadaic acid (IC_{50} ≈ 100 nM). These observations provide a means for identifying and quantifying type 1, type 2A, and type 2C protein phosphatases in tissue extracts. Bialojan & Takai (2) also investigated the inhibitory action of this dinoflagellate natural product. Of the protein phosphatases examined, the catalytic subunit of rabbit skeletal muscle 2A phosphatase was most potently inhibited. For the phosphorylated myosin light-chain (PMLC) phosphatase activity of the enzyme, the okadaic acid concentration required to obtain 50% inhibition was about 1 nM. The PMLC phosphatase activities of type 1 and PCM phosphatase were also strongly inhibited (IC_{50} ≈ 0.1–0.5 μM). The PMCL phosphatase activity of type 2B phosphatase (calcineurin) was inhibited to a lesser extent (IC_{50} ≈ 4–5 μM). These investigators obtained similar results for the phosphorylase *a* phosphatase activity of type 1 and PCM phosphatases and for calcineurin's *p* nitrophenyl phosphate phosphatase activity. The following phosphatases were not affected by up to 10 μM okadaic acid: type 2C phosphatase, phosphotyrosyl phosphatase, inositol-1,4,5 trisphosphate phosphatase, acid phosphatases, and alkaline phosphatases. These results suggest that okadaic acid is relatively specific for type 2A, type 1, and PCM phosphatases. **Other Actions:** This light-sensitive substance also disrupts Golgi, arrests transport on the rough endoplasmic reticulum, alters nitric oxide metabolism, and markedly influences calcium ion signalling. **CAUTION:** In view of its *documented capacity to promote tumorigenesis*, investigators should carefully read precautions and recommended procedures in commercial product insert data pertaining to okadaic acid. Avoid direct contact with or inhalation of the dry powder. **Target(s):** [acetyl-CoA carboxylase]-phosphatase (3); [myosin-light-chain] phosphatase (4-8); phosphoprotein phosphatase (1,2,9-18,21-40); [phosphorylase] phosphatase (9-11,15); protein phosphatase 1 (9-11,15,16,22,23,32,37,38); protein phosphatase 2A (1-3,9-14,17,18,21,23,28,30,31,34,35); protein phosphatase 3 (36); protein phosphatase 4 (23,26); protein phosphatase 5 (23); protein-tyrosine phosphatase (19); rod outer segment protein phosphatase 2A (opsin phosphatase) (13,14); soluble epoxide hydrolase, weakly inhibited (20). **1.**

Cohen, Klumpp & Schelling (1989) *FEBS Lett.* **250**, 596. **2.** Bialojan & Takai (1988) *Biochem. J.* **256**, 283. **3.** Palanivel, Veluthakal, McDonald & Kowluru (2005) *Endocrine* **26**, 71. **4.** Ito, Feng, Tsujino, *et al.* (1997) *Biochemistry* **36**, 7607. **5.** Parizi, Howard & Tomasek (2000) *Exp. Cell Res.* **254**, 210. **6.** Schmidt, Troschka & Pfitzer (1995) *Pfluger's Arch.* **429**, 708. **7.** Murányi, Erdodi, Ito, Gergely & Hartshorne (1998) *Biochem. J.* **330**, 225. **8.** Etter, Eto, Wardle, Brautigan & Murphy (2001) *J. Biol. Chem.* **276**, 34681. **9.** Cohen (1991) *Meth. Enzymol.* **201**, 389. **10.** Hardie, Haystead & Sim (1991) *Meth. Enzymol.* **201**, 469. **11.** Eriksson, Toivola, Sahlgren, Mikhailov & Härmälä-Braskén (1998) *Meth. Enzymol.* **298**, 542. **12.** Dounay & Forsyth (2002) *Curr. Med. Chem.* **9**, 1939. **13.** Akhtar, King & McCarthy (2000) *Meth. Enzymol.* **315**, 557. **14.** Palczewski, McDowell, Jakes, Ingebritsen & Hargraves (1989) *J. Biol. Chem.* **264**, 15770. **15.** Zhang, Bai, Deans-Zirattu, Browner & Lee (1992) *J. Biol. Chem.* **267**, 1484. **16.** Stubbs, Tran, Atwell, *et al.* (2001) *Biochim. Biophys. Acta* **1550**, 52. **17.** Szöör, Fehér, Bakó, *et al.* (1995) *Comp. Biochem. Physiol. B* **112**, 515. **18.** Zabrocki, Swiatek, Sugajska, *et al.* (2002) *Eur. J. Biochem.* **269**, 3372. **19.** Aguirre-Garcia, Escalona-Montano, Bakalara, *et al.* (2006) *Parasitology* **132**, 641. **20.** Newman, Morisseau, Harris & Hammock (2003) *Proc. Natl. Acad. Sci. U.S.A.* **100**, 1558. **21.** Hall, Feekes, Don, *et al.* (2006) *Biochemistry* **45**, 3448. **22.** Andrioli, Zaini, Viviani & da Silva (2003) *Biochem. J.* **373**, 703. **23.** Gallego & Virshup (2005) *Curr. Opin. Cell Biol.* **17**, 197. **24.** Solow, Young & Kennelly (1997) *J. Bacteriol.* **179**, 5072. **25.** Sayed, Whitehouse & Jones (1997) *J. Endocrinol.* **154**, 449. **26.** Cohen, Philp & Vázquez-Martin (2005) *FEBS Lett.* **579**, 3278. **27.** Pilecki, Grzyb, Zien, Sekula & Szyszka (2000) *J. Basic Microbiol.* **40**, 251. **28.** Swiatek, Sugajska, Lankiewicz, Hemmings & Zolnierowicz (2000) *Eur. J. Biochem.* **267**, 5209. **29.** Cohen (2004) in *Topics in Current Genetics* (Arino & Alexander, eds.) Springer **5**, 1. **30.** Hallaway & O'Kane (2000) *Meth. Enzymol.* **305**, 391. **31.** Dong, Ermolova & Chollet (2001) *Planta* **213**, 379. **32.** Szöör, Gross & Alphey (2001) *Arch. Biochem. Biophys.* **396**, 213. **33.** Douglas, Moorhead, Ye & Lees-Miller (2001) *J. Biol. Chem.* **276**, 18992. **34.** Kray, Carter, Pennington, *et al.* (2005) *J. Biol. Chem.* **280**, 35974. **35.** Kloeker, Reed, McConnell, *et al.* (2003) *Protein Expr. Purif.* **31**, 19. **36.** Honkanen, Zwiller, Daily, *et al.* (1991) *J. Biol. Chem.* **266**, 6614. **37.** Gibbons, Weiser & Shenolikar (2005) *J. Biol. Chem.* **280**, 15903. **38.** Haneda, Kojima, Nishikimi, *et al.* (2004) *FEBS Lett.* **567**, 171.**39.** Ganguly & Singh (1999) *Phytochemistry* **52**, 239. **40.** Yokoyama, Kobayashi, Tamura & Sugiya (1996) *Arch. Biochem. Biophys.* **331**, 1.

Olanzapine

This widely used atypical antipsychotic drug, *or AAPD* (FW$_{free-base}$ = 312.44 g/mol; CAS 132539-06-1), also known as Zyprexa® and 2-methyl-4-(4-methyl-1-piperazinyl)-10*H*-thieno[2,3-*b*][1,5]benzo-diazepine, is a dual serotonin (5-HT$_2$) and dopamine (D$_1$/D$_2$) receptor antagonist used to treat bipolar disorders (1). Olanzapine is metabolized by the CYP1A2 and to a lesser extent by CYP2D6, resulting in removal of more than 40% of the oral dose by the liver on first pass. Olanzapine pharmacokinetics display large inter-individual variation, with multiple-fold differences in drug exposure between patients given the same dose. Formation of the various olanzapine metabolites is influenced by polymorphisms in the genes coding for CYP1A2, CYP1A expression regulator AHR, UGT1A4 and UGT2B10, as well as FMO3 (2). *Note*: Olanzapine is FDA-approved for treating symptoms of schizophrenia and bipolar disorder in adults and teenagers age thirteen and older, but is not approved for treating behavior disorders in older adults with dementia. **Key Pharmacokinetic Parameters:** *See* Appendix II in Goodman & Gilman's THE PHARMACOLOGICAL BASIS OF THERAPEUTICS, 12th Edition (Brunton, Chabner & Knollmann, eds.) McGraw-Hill Medical, New York (2011). **Target(s):** Dopamine D$_1$ receptor, K_i = 70 nM; dopamine D$_2$ receptor, K_i = 31 nM; dopamine D$_3$ receptor, K_i = 11 nM; dopamine D$_4$ receptor, K_i = 18 nM; serotonin 5-HT$_{1A}$ receptor, K_i = 2300 nM; serotonin 5-HT$_{2A}$ receptor, K_i = 3.7 nM; serotonin 5-HT$_{2B}$ receptor, K_i = 8.2 nM; serotonin 5-HT$_{2C}$ receptor, K_i = 10 nM; serotonin 5-HT$_3$ receptor, K_i = 57 nM; serotonin 5-HT$_6$ receptor, K_i = 5 nM; serotonin 5-HT$_7$ receptor, K_i = 7.1 nM; muscarinic M$_1$ receptor, K_i = 2.5 nM; muscarinic M$_2$ receptor, K_i = 18-96 nM; muscarinic M$_3$ receptor, K_i = 25-132 nM; muscarinic M$_4$ receptor, K_i = 13-32 nM; muscarinic M$_5$ receptor, K_i = 48 nM; adrenergic α_{1A} receptor, K_i = 110 nM; adrenergic α_{2A} receptor, K_i = 310 nM; and histamine H$_1$ receptor, K_i = 2.2 nM; CYP3A (2); CYP2C9 (3) CYP2C19 (3); CYP2D6 (3); dopamine receptor (1,4); glucuronosyltransferase (5). (*See* **LLY-2707**) **1.** Bymaster, Nelson, DeLapp, *et al.* (1999) *Schizophr. Res.* **37**, 107. **2.** Söderberg & Dahl (2013) *Pharmacogenomics* **14**, 1319. **3.** Ring, Binkley, Vandenbranden & Wrighton (1996) *Brit. J. Clin. Pharmacol.* **41**, 181. **4.** Fuller & Snoddy (1992) *Res. Commun. Chem. Pathol. Pharmacol.* **77**, 87. **5.** Hara, Nakajima, Miyamoto & Yokoi (2007) *Drug Metab. Pharmacokinet.* **22**, 103.

Olaparib

This orally active poly-ADP ribose polymerase inhibitor (FW = 435.08 g/mol; CAS 763113-22-0; Solubility: 80 mg/mL; << 1 mg/mL H$_2$O), also known as AZD-2281, Lynparza®, and 4-[(3-[(4-cyclopropyl-carbonyl) piperazin-4-yl]carbonyl)4-fluorophenyl]methyl-(2*H*)-phthalazin-1-one, targets PARP1 (IC$_{50}$ = 5 nM) and PARP 2 (IC$_{50}$ = 1 nM). Clinical evaluation of olaparib action against cancers lacking the BRCA1 or BRCA2 genes exemplifies the promise of targeted therapies exploiting specific molecular defects in cancer cells. Since <5 % of patients are BRCA1 or BRCA2 mutation carriers, small molecules that functionally mimic BRCA1 or BRCA2 mutations promise to extend the synthetic lethal therapeutic option for non-mutation carriers (2). A deficiency of RAD51C, a key participant in recombinational repair of damaged DNA, is also a biomarker for predicting the patients most likely to respond to olaparib (2). In patients with prostate cancers that no longer responded to standard treatments and who had defects in DNA-repair genes, olaparib treatment led to a high response rate (3). **1.** Fong, Boss, Yap, *et al.* (2009) *New Engl. J. Med.* **361**, 123. **2.** Pessetto, Yan, Bessho & Natarajan A. (2012) *Breast Cancer Res Treat.* **134**, 511. **3.** Min, Im, Yoon *et al.* (2013) *Molec. Cancer Ther.* **12**, 865. **3.** Mateo, Carreira, Sandhu, *et al.* (2015) *New Engl. J. Med.* **373**, 1697.

Oleamide

This fatty acid amide (FW = 281.48 g/mol; CAS 301-02-0), also known as *cis*-9-octadecenamide and oleic acid amide, accumulates in the cerebral spinal fluid on sleep deprivation. It also induces sleep in animals, reversibly enhances GABA$_A$ currents, and depresses the frequency of spontaneous excitatory and inhibitory synaptic activity in cultured networks. The *trans* form (known as elaidamide) is inactive. Oleamide is a substrate of fatty acid amide hydrolase. Originally referred to as cerebrodiene, oleamide was originally isolated from the brain of sleep-deprived cats (1,2). **Target(s):** chymase (3); 5-HT$_7$ serotonin receptors modulation (4); microsomal epoxide hydrolase (5); Na$^+$ channel-dependent rises in intrasynaptosomal calcium ions (6). **1.** Lerner, Siuzdak, Prospero-Garcia, *et al.* (1994) *Proc. Natl. Acad. Sci. U.S.A.* **91**, 9505. **2.** Cravatt, Prospero-Garcia, Siuzdak, *et al.* (1995) *Science* **268**, 1506. **3.** Kido, Fukusen & Katunuma (1985) *Arch. Biochem. Biophys.* **239**, 436. **4.** Hedlund, Carson, Sutcliffe & Thomas (1999) *Biochem. Pharmacol.* **58**, 1807. **5.** Morisseau, Newman, Dowdy, Goodrow & Hammock (2001) *Chem. Res. Toxicol.* **14**, 409. **6.** Verdon, Zheng, Nicholson, Ganelli & Lees (2000) *Brit. J. Pharmacol.* **129**, 283.

Oleanolate

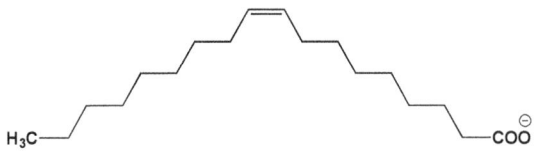

This triterpene alcohol (FW$_{\text{free-acid}}$ = 456.71 g/mol; CAS 508-02-1), also known as oleanolic acid, is present in certain plants (*e.g.*, *Olea europeaea*, *Viscum album*, mistletoe, cloves, cacti, sugar beets, *etc.*), where it is found in free form, esterified with acetic acid, and as a glycoside. It has a pK_a value of 2.52 and is practically insoluble in water. Barton and Holness (1) were able to use conformational analysis to assign a conformational structure to oleanolic acid in the early 1950s. X-ray crystallographic studies demonstrated that one of their two possible conformations was correct (2). **Target(s):** α-amylase (3); cAMP-dependent protein kinase (4); chitin synthase II (5); complement C3 convertase (6); cyclooxygenase-2 (COX-2) (7,8); CYP1A2 (9); CYP3A4 (9); diacylglycerol *O*-acyltransferase (10); DNA-directed DNA polymerase (12,13); DNA ligase (ATP), human, IC$_{50}$ = 216 μM (11); DNA polymerase α (12); DNA polymerase β (12,13); DNA topoisomerase I (14); DNA topoisomerase II (12,14); α-glucosidase (15,16); glucosyltransferase (17); glycogen phosphorylase, IC$_{50}$ = 14 μM (16,18,19,22); 11β-hydroxysteroid dehydrogenase (20); leucocyte elastase (21); 5-lipoxygenase, *or* arachidonate 5-lipoxygenase (24); sterol *O*-acyltransferase, *or* cholesterol *O*-acyltransferase, ACAT (23). **1.** Barton & Holness (1952) *J. Chem. Soc.*, **78**. **2.** Abel el Rehim & Carlisle (1954) *Chem. Ind. (London)*, 279. **3.** Ali, Houghton & Soumyanath (2006) *J. Ethnopharmacol.* **107**, 449. **4.** Wang & Polya (1996) *Phytochemistry* **41**, 55. **5.** Jeong, Hwang, Lee, *et al.* (1999) *Planta Med.* **65**, 261. **6.** Kapil & Sharma (1994) *J. Pharm. Pharmacol.* **46**, 922. **7.** Huss, Ringbom, Perera, Bohlin & Vasange (2002) *J. Nat. Prod.* **65**, 1517. **8.** Ringbom, Segura, Noreen, Perera & Bohlin (1998) *J. Nat. Prod.* **61**, 1212. **9.** Kim, Lee, Park, *et al.* (2004) *Life Sci.* **74**, 2769. **10.** Dat, Cai, Rho, *et al.* (2005) *Arch. Pharm. Res.* **28**, 164. **11.** Tan, Lee, Lee, *et al.* (1996) *Biochem. J.* **314**, 993. **12.** Mizushina, Ikuta, Endoh, *et al.* (2003) *Biochem. Biophys. Res. Commun.* **305**, 365. **13.** Deng, Starck & Hecht (1999) *J. Nat. Prod.* **62**, 1624. **14.** Syrovets, Buchele, Gedig, Slupsky & Simmet (2000) *Mol. Pharmacol.* **58**, 71. **15.** Ali, Jahangir, Hussan & Choudhary (2002) *Phytochemistry* **60**, 295. **16.** Li, Lu, Su, *et al.* (2008) *Planta Med.* **74**, 287. **17.** Kozai, Miyake, Kohda, *et al.* (1987) *Caries Res.* **21**, 104. **18.** Chen, Liu, Zhang, *et al.* (2006) *Bioorg. Med. Chem. Lett.* **16**, 2915. **19.** Chen, Gong, Liu, *et al.* (2008) *Chem. Biodivers.* **5**, 1304. **20.** Zhang & Wang (1997) *Zhongguo Yao Li Xue Bao* **18**, 240. **21.** Ying, Rinehart, Simon & Cheronis (1991) *Biochem. J.* **277**, 521. **22.** Wen, Sun, Liu, *et al.* (2008) *J. Med. Chem.* **51**, 3540. **23.** Kim, Han, Chung, *et al.* (2005) *Arch. Pharm. Res.* **28**, 550. **24.** Schneider & Bucar (2005) *Phytother. Res.* **19**, 81.

Oleate (Oleic Acid)

This unsaturated fatty acid (FW$_{\text{free-acid}}$ = 282.47 g/mol; CAS 112-80-1), also known as oleic acid and (*Z*)-octadec-9-enoic acid, is found in most organisms, often as a constituent of phospholipids and triacylglycerols. It is insoluble in water and the sodium salt, sodium oleate, has a low solubility. The free acid is a colorless liquid at room temperature and readily oxidizes in air, acquiring rancid odor. Oleic acid should be stored in a tight container and protected from light. Note that the *trans*-isomer is elaidic acid (**See also** *Elaidate & Elaidic Acid*). Elevated concentrations of oleate are toxic and will uncouple oxidative phosphorylation. **Target(s):** 2-acylglycerol *O*-acyltransferase, *or* monoacylglycerol *O*-acyltransferase (57-59); 1-acylglycerophosphocholine *O*-acyltransferase, *or* lysophosphaitidylcholine *O*-acyltransferase (56); adenine nucleotide translocase (1); adenylate kinase (2); 1-alkylglycerophosphocholine *O* acetyltransferase (55); arachidonyl-CoA synthetase, *or* arachidonate:CoA ligase (3); ATPase (4); choloyl CoA

synthetase, *or* cholate:CoA ligase (6,7); chondroitin AC lyase (8); chondroitin B lyase (8); chymase (9,36); diacylglycerol *O*-acyltransferase (60); DNA (adenine-*N*^{6}-)-methyltransferase (10); DNA directed DNA polymerase (11,12); DNA polymerases α, β, γ, δ, and ε (11,12); DNA topoisomerase I (12); DNA topoisomerase II (12); electron transport and uncoupling, mitochondrial (13,14); ethanolamine-phosphate cytidylyltransferase (49); 7-ethoxycoumarin *O*-deethylase (15); β-galactoside α-2,6-sialyltransferase (50); gelatinase A (16); 1,3-β-glucan synthase (53,54); β-glucuronidase (17); glycerol 3-phosphate *O*-acyltransferase (61-63); hexokinase (18); hormone-sensitive lipase (19,20); hyaluronate lyase (8); leukocyte elastase, *or* neutrophil elastase (25,35); linoleate isomerase (21,22); lipase, *or* triacylglycerol lipase (46); lipoxygenase (23); 5-lipoxygenase, *or* arachidonate 5-lipoxygenase (66); *N*-(long-chain-acyl)ethanolamine deacylase (24); lysophospholipase (45); microbial collagenase (35); monogalactosyldiacylglycerol synthase (52); [myosin light-chain] kinase (47); NAD$^+$ ADP ribosyltransferase, *or* poly(ADP-ribose) polymerase (51); NADH dehydrogenase (13); palmitoyl-CoA hydrolase, *or* acyl-CoA hydrolase (41,42); peroxidase (68); peroxidase activity of metmyoglobin (26); phosphatidate cytidylyltransferase (48); phosphatidate phosphatase (27,28); phosphatidylglycerophosphatase, inhibited at elevated concentrations (*i.e.*, >1.5 mM) (40); phospholipase A$_1$ (43); phospholipase D (38,39); phospholipid-hydroperoxide glutathione peroxidase (67); plasmanylethanolamine desaturase, weakly inhibited (64); prostaglandin D$_2$ dehydrogenase (29); prostaglandin D synthetase (29); prostaglandin-endoperoxide synthase, *or* cyclooxygenase (65); pyruvate kinase (30); sphingomyelin phosphodiesterase (37); steroid 5α-reductase (31,32); sterol *O*-acyltransferase, *or* cholesterol *O*-acyltransferase, ACAT (5); sterol esterase, *or* cholesterol esterase (44); tyrosine kinase (33); uroporphyrinogen decarboxylase (34). **1.** Bell (1980) *Atherosclerosis* **37**, 21. **2.** Criss & Pradhan (1978) *Meth. Enzymol.* **51**, 459. **3.** Reddy & Bazan (1983) *Arch. Biochem. Biophys.* **226**, 125. **4.** Davis, Davis, Blas & Schoenl (1987) *Biochem. J.* **248**, 511. **5.** Goodman (1969) *Meth. Enzymol.* **15**, 522. **6.** Vessey & Zakim (1977) *Biochem. J.* **163**, 357. **7.** Falany, Xie, Wheeler, *et al.* (2002) *J. Lipid Res.* **43**, 2062. **8.** Suzuki, Terasaki & Uyeda (2002) *J. Enzyme Inhib. Med. Chem.* **17**, 183. **9.** Kido, Fukusen & Katunuma (1984) *Arch. Biochem. Biophys.* **230**, 610. **10.** Suzuki, Nagao, Tokunaga & Katayama & Uyeda (1996) *J. Enzyme Inhib.* **10**, 271. **11.** Mizushina, Tanaka, Yagi, *et al.* (1996) *Biochim. Biophys. Acta* **1308**, 256. **12.** Mizushina, Tsuzuki, Eitsuka, *et al.* (2004) *Lipids* **39**, 977. **13.** Rapoport & Schewe (1977) *Trends Biochem. Sci.* **2**, 186. **14.** Vázquez-Colón, Ziegler & Elliott (1966) *Biochemistry* **5**, 1134. **15.** Muller-Enoch, Fintelmann, Nicolaev & Gruler (2001) *Z. Naturforsch. [C]* **56**, 1082. **16.** Emonard, Marcq, Mirand & Hornebeck (1999) *Ann. N.Y. Acad. Sci.* **878**, 647. **17.** Tappel & Dillard (1967) *J. Biol. Chem.* **242**, 2463. **18.** Stewart & Blakely (2000) *Biochim. Biophys. Acta* **1484**, 278. **19.** Jepson & Yeaman (1992) *FEBS Lett.* **310**, 197. **20.** Severson & Hurley (1984) *Lipids* **19**, 134. **21.** Kepler & Tove (1969) *Meth. Enzymol.* **14**, 105. **22.** Kepler, Tucker & Tove (1970) *J. Biol. Chem.* **245**, 3612. **23.** Holman & Bergström (1951) *The Enzymes*, 1st ed., **2** (part 1), 559. **24.** Schmid, Zuzarte-Augustin & Schmid (1985) *J. Biol. Chem.* **260**, 14145. **25.** Tyagi & Simon (1990) *Biochemistry* **29**, 9970. **26.** Stewart (1990) *Biochem. Cell Biol.* **68**, 1096. **27.** Elabbadi, Day, Virden & Yeaman (2002) *Lipids* **37**, 69. **28.** Moller, Green & Harkness (1977) *Biochim. Biophys. Acta* **486**, 359. **29.** Osama, Narumiya, Hayaishi, *et al.* (1983) *Biochim. Biophys. Acta* **752**, 251. **30.** Kayne (1973) *The Enzymes*, 3rd ed., **8**, 353. **31.** Liang & Liao (1992) *Biochem. J.* **285**, 557. **32.** Raynaud, Cousse & Martin (2002) *J. Steroid Biochem. Mol. Biol.* **82**, 233. **33.** Tomaska & Resnick (1993) *J. Biol. Chem.* **268**, 5317. **34.** Seki, Kawanishi & Sano (1986) *Meth. Enzymol.* **123**, 415. **35.** Rennert & Melzig (2002) *Planta Med.* **68**, 767. **36.** Kido, Fukusen & Katunuma (1985) *Arch. Biochem. Biophys.* **239**, 436. **37.** Liu, Nilsson & Duan (2002) *Lipids* **37**, 469. **38.** Min, Park & Exton (1998) *J. Biol. Chem.* **273**, 7044. **39.** Vinggaard & Hansen (1995) *Biochim. Biophys. Acta* **1258**, 169. **40.** Cao & Hatch (1994) *Lipids* **29**, 475. **41.** Ohkawa, Shiga & Kageyama (1979) *J. Biochem.* **86**, 643. **42.** Joyard & Stumpf (1980) *Plant Physiol.* **65**, 1039. **43.** Nishijima, Akamatsu & Nojima (1974) *J. Biol. Chem.* **249**, 5658. **44.** Wee & Grogan (1993) *J. Biol. Chem.* **268**, 8158. **45.** Zhang & Dennis (1988) *J. Biol. Chem.* **263**, 9965. **46.** Sharma, Chisti & Banerjee (2001) *Biotechnol. Adv.* **19**, 627. **47.** Kigoshi, Uchida, Kaneko, *et al.* (1990) *Biochem. Biophys. Res. Commun.* **171**, 369. **48.** Sribney & Hegadorn (1982) *Can. J. Biochem.* **60**, 668. **49.** Vermeulen, Tijburg, Geelen & van Golde (1993) *J. Biol. Chem.* **268**, 7458. **50.** Hickman, Ashwell, Morell, van den Hamer & Scheinberg (1970) *J. Biol. Chem.* **245**, 759. **51.** Banasik, Komura, Shimoyama & Ueda (1992) *J. Biol. Chem.* **267**, 1569. **52.** Heemskerk, Jacobs, Scheijen, Helsper & Wintermans (1987) *Biochim.*

Biophys. Acta **918**, 189. **53**. Kauss & Jeblick (1986) *Plant Physiol.* **80**, 7. **54**. Ko, Frost, Ho, Ludescher & Wasserman (1994) *Biochim. Biophys. Acta* **1193**, 31. **55**. Baker & Chang (1998) *Biochim. Biophys. Acta* **1390**, 215. **56**. Kerkhoff, Habben, Gehring, Resch & Kaever (1998) *Arch. Biochem. Biophys.* **351**, 220. **57**. Coleman, Wang & Bhat (1998) *Biochemistry* **37**, 5916. **58**. Bhat, Wang & Coleman (1994) *J. Biol. Chem.* **269**, 13172. **59**. Coleman, Wang & Bhat (1996) *Biochemistry* **35**, 9576. **60**. Chung, Rho, Lee, *et al.* (2006) *Planta Med.* **72**, 267. **61**. Kito, Aibara, Hasegawa & Hata (1972) *J. Biochem.* **71**, 99. **62**. Douady & Dubacq (1987) *Biochim. Biophys. Acta* **921**, 615. **63**. Frentzen, Heinz, McKeon & Stumpf (1983) *Eur. J. Biochem.* **129**, 629. **64**. Paltauf & Holasek (1973) *J. Biol. Chem.* **248**, 1609. **65**. Reininger & Bauer (2006) *Phytomedicine* **13**, 164. **66**. Schneider & Bucar (2005) *Phytother. Res.* **19**, 81. **67**. Maiorino, Roveri, Gregolin & Ursini (1986) *Arch. Biochem. Biophys.* **251**, 600. **68**. Suzuki, Honda, Mukasa & Kim (2006) *Phytochemistry* **67**, 219.

1-Oleoyl-2-acetyl-*sn*-glycerol

This 1,2-diacylglycerol (FW = 398.58 g/mol), often abbreviated OAG, is an activator of protein kinase C and a cell-permeable analogue of larger 1,2-diacylglycerols. Commercial preparations often contain 1-elaidoyl-2-acetyl *sn*-glycerol (*i.e.*, the *trans* isomer), which accumulates upon storage. **Target(s):** 1-alkyl-2-acetylglycerol *O*-acyltransferase (1); nonspecific serine/threonine protein kinase (2); phospholipase C (3); protein kinase D (2). **1**. Kawasaki & Snyder (1988) *J. Biol. Chem.* **263**, 2593. **2**. Valverde, Sinnett-Smith, Van Lint & Rozengurt (1994) *Proc. Natl. Acad. Sci. U.S.A.* **91**, 8572. **3**. Haeffner & Wittmann (1992) *J. Lipid Mediat.* **5**, 237.

Oleoylcarnitine

This acylated carnitine (FW = 425.65 g/mol; CAS 13962-05-5) is transported across the mitochondrial membrane prior to β-oxidation of the unsaturated fatty acid. **Target(s):** adenine nucleotide translocase (1); phosphatidylcholine: cholesterol acyltransferase, LCAT of rat and rabbit (2); Na$^+$/Cl$^-$-dependent glycine transporter GlyT$_2$ (IC$_{50}$ = 340 nM, ~15x lower than *N*-arachidonyl-glycine). **1**. Bell (1980) *Atherosclerosis* **37**, 21. **2**. Bell (1983) *Int. J. Biochem.* **15**, 133. **3**. Carland, Mansfield, Ryan & Vandenberg (2013) *Br. J. Pharmacol.* **168**, 891.

Oleoyl-CoA

This fatty acyl-CoA thiolester (FW$_{\text{free-acid}}$ = 1031.99 g/mol; CAS 1716-06-9; Critical Micelle Concentration ≈ 4.7 µM) is a common intermediate in fatty acid metabolism and is a substrate of many acyltransferases. *Note*: When oleoyl-CoA is present at concentrations exceeding its critical micelle concentration, micelles will form, giving rise to the likelihood that the observed inhibition is merely the nonphysiologic consequence of clustered surface charge. Experiments with mixed micelles may be required to reach

an unambiguous conclusion. **Target(s):** acetyl-CoA carboxylase, K_i = 1.3 µM for the rat liver enzyme (1-5); acetyl CoA synthetase, *or* acetate:CoA ligase (6,7); adenine nucleotide translocase (8,9); alcohol *O*-acetyltransferase (41); 1-alkyl-*sn*-glycero-3-phosphate *O*-acetyltransferase (10,12); 1-alkylglycerophosphocholine *O*-acetyltransferase (11,12,42); ATP:citrate lyase, *or* ATP:citrate synthase (13,39,40); citrate synthase (14); cyclooxygenase (15); fatty-acid synthase (16); glucose-6 phosphatase (35,36); glucose-6-phosphate dehydrogenase (17,18); glucose-6-phosphate translocase (19); glycerol-3-phosphate dehydrogenase (20); hexokinase (21); hormone-sensitive lipase (22,23); 3-hydroxydecanoyl-[acyl-carrier-protein] dehydratase (24); lipoxygenase (15); lyso-[platelet-activating factor] acetyltransferase (12); nicotinamide nucleotide transhydrogenase (25); *p*-nitroanisole *O*-demethylase (25); nuclear thyroid hormone receptor (26); pantothenate kinase (38); phosphatidate phosphatase (27,28); phosphoglyceride:lysophosphatidylglycerol *O*-acyltransferase (29); phospholipase A$_2$ (30); protein kinase C (31); [pyruvate dehydrogenase (acetyl-transferring)] kinase (37); RNA polymerase (32); stearoyl-CoA 9-desaturase, as product inhibitor (43); UDPglucuronosyl-transferase (33,34). **1**. Inoue & Lowenstein (1975) *Meth. Enzymol.* **35**, 3. **2**. Tanabe, Nakanishi, Hashimoto, *et al.* (1981) *Meth. Enzymol.* **71**, 5. **3**. Nikawa, Tanabe, Ogiwara, Shiba & Numa (1979) *FEBS Lett.* **102**, 223. **4**. McGee & Spector (1975) *J. Biol. Chem.* **250**, 5419. **5**. Bortz & Lynen (1963) *Biochem. Z.* **337**, 505. **6**. Preston, Wall & Emerich (1990) *Biochem. J.* **267**, 179. **7**. Satyanarayana & Klein (1973) *J. Bacteriol.* **115**, 600. **8**. Shug, Lerner, Elson & Shrago (1971) *Biochem. Biophys. Res. Commun.* **43**, 557. **9**. Bell (1980) *Atherosclerosis* **37**, 21. **10**. Baker & Chang (1995) *J. Neurochem.* **64**, 364. **11**. Lee, Vallari & Snyder (1992) *Meth. Enzymol.* **209**, 396. **12**. Baker & Chang (1996) *Biochim. Biophys. Acta* **1302**, 257. **13**. Shashi, Bachhawat & Joseph (1990) *Biochim. Biophys. Acta* **1033**, 23. **14**. Hsu & Powell (1975) *Proc. Natl. Acad. Sci. U.S.A.* **72**, 4729. **15**. Fujimoto, Tsunomori, Sumiya, *et al.* (1995) *Prostaglandins Leukot. Essent. Fatty Acids* **52**, 255. **16**. Roncari (1975) *Can. J. Biochem.* **53**, 135. **17**. Ros, Cubero, Lobato, Garcia-Ruiz & Moreno (1984) *Mol. Cell Biochem.* **63**, 119. **18**. Cacciapuoti & Morse (1980) *Can. J. Microbiol.* **26**, 863. **19**. Fox, Hill, Rawsthorne & Hills (2000) *Biochem. J.* **352**, 525. **20**. Edgar & Bell (1979) *J. Biol. Chem.* **254**, 1016. **21**. Thompson & Cooney (2000) *Diabetes* **49**, 1761. **22**. Jepson & Yeaman (1992) *FEBS Lett.* **310**, 197. **23**. Severson & Hurley (1984) *Lipids* **19**, 134. **24**. Brock, Kass & Bloch (1967) *J. Biol. Chem.* **242**, 4432. **25**. Danis, Kauffman, Evans, *et al.* (1985) *Biochem. Pharmacol.* **34**, 609. **26**. Li, Yamamoto, Inoue & Morisawa (1990) *J. Biochem.* **107**, 699. **27**. Elabbadi, Day, Virden & Yeaman (2002) *Lipids* **37**, 69. **28**. Moller, Green & Harkness (1977) *Biochim. Biophys. Acta* **486**, 359. **29**. Hostetler, Huterer & Wherrett (1992) *Meth. Enzymol.* **209**, 104. **30**. Marki & Franson (1986) *Biochim. Biophys. Acta* **879**, 149. **31**. Yaney, Korchak & Corkey (2000) *Endocrinology* **141**, 1989. **32**. Yokokawa, Fujiwara, Shimada & Yasumasu (1983) *J. Biochem.* **94**, 415. **33**. Zhong, Kauffman & Thurman (1991) *Cancer Res.* **51**, 4511. **34**. Krcmery & Zakim (1993) *Biochem. Pharmacol.* **46**, 897. **35**. Fulceri, Gamberucci, Scott, *et al.* (1995) *Biochem. J.* **307**, 391. **36**. Methieux & Zitoun (1996) *Eur. J. Biochem.* **235**, 799. **37**. Rahmatullah & Roche (1985) *J. Biol. Chem.* **260**, 10146. **38**. Halvorsen & Skrede (1982) *Eur. J. Biochem.* **124**, 211. **39**. Evans & Ratledge (1985) *Can. J. Microbiol.* **31**, 1000. **40**. Boulton & Ratledge (1983) *J. Gen. Microbiol.* **129**, 2863. **41**. Yoshioka & Hashimoto (1981) *Agric. Biol. Chem.* **45**, 2183. **42**. Baker & Chang (1998) *Biochim. Biophys. Acta* **1390**, 215. **43**. McDonald & Kinsella (1973) *Arch. Biochem. Biophys.* **156**, 223.

N-Oleoylethanolamide

This endocannabinoid-related metabolite (FW = 325.54 g/mol; CAS 111-58-0; *Symbol*: NOE and OEA; Soluble in Ethanol, Chloroform, or Methanol) is widely employed to inhibit acid and neutral ceramidases (IC$_{50}$ ~ 500 µM) that cleave fatty acids from ceramide, producing sphingosine (SPH), which is then enzymatically phosphorylated to form the receptor-sensed metabolite, sphingosine-1-phosphate, *or* S1P (1-4). NOE also inhibits the glucosylation of naturally occurring ceramides (5). In CHP-100 neuroepithelioma cells treated with N-hexanoylsphingosine (C_6-Cer; 30 µM), NOE affected only marginally short-chain glucocerebroside accumulation, but markedly decreased accumulation of glucocerebrosides originating from glucosylation of a long-chain ceramide (Lc-Cer) produced upon C_6-Cer treatment (5). NOE also inhibits fatty acid hydrolase (or FAAH), an integral membrane hydrolase possessing an unusual catalytic triad Ser-Ser-Lys (6). FAAH catalysis is based on the formation of a tetrahedral intermediate, which is a product of the nucleophilic attack of Ser241 on the carbonyl group of the substrate. In the presence of NOE, TNFα leads to a sustained accumulation of ceramide and induces DNA fragmentation in renal mesangial cells (7). **Membrane & Receptor Interactions:** NOE forms stable complexes with phospholipid vesicles, lowering diphenylhexatriene polarization ratios in dimyristoyl-phosphatidylcholine and dipalmitoylphosphatidylcholine uni- and multi-lamellar bilayer vesicles (8). Once incorporated into the lipid bilayer of phospholipid vesicles, N-oleoylethanolamine partitions preferentially into more fluid areas, producing a concentration-dependent decrease in their phase transitions (9). N-Oleoylethanolamine is also an effective inhibitor of mitochondrial swelling, but does not inhibit phospholipase A_2 or ruthenium red-induced Ca^{2+} release (10). Moreover, among naturally occurring N-acylethanolamines, only NOE (at 10 µM) inhibits the accumulation of N-arachidonoylethanolamine (or Anandamide, AEA), a putative endogenous ligand of the cannabinoid receptor, into cerebellar granule cells occurs by means of facilitated diffusion (10). Ethanolamides of palmitic acid and linolenic acid are inactive at 10 µM. **Role as an Anorexic Lipid Mediator:** N-Oleoylethanolamine is a structural analogue of the endogenous cannabinoid anandamide, but does not activate cannabinoid receptors. NOE biosynthesis in rat small intestine is increased by feeding and reduced by fasting. NOE decreases food intake in food-deprived rats via a mechanism that requires intact sensory fibers. Such results suggest that OEA may contribute to the peripheral regulation of feeding. NOE may contribute to regulation of satiety, providing a chemical scaffold for the design of novel appetite-suppressing medications (11). Despite its designation as endocannabinoid-like compound, NOE is unable to bind with significant affinity to either CB_1 or CB_2 cannabinoid receptors (12). **SIRT6 Activation:** N-Oleoylethanolamine binds to and activates <u>Sirtuin 6</u> (EC_{50} = 3.1 µM), an NAD^+-dependent histone deacetylase that selectively deacylates H3K9Ac and H3K56Ac, but acts preferentially on long-chain fatty acyl groups over acetyl groups *in vitro* (13). (**See also** *Oleamide; N-Oleoyl-2-amino-1,3-propanediol; Oleoyl-N-hydroxylamide*) 1. Nikolova-Karakashian & Merrill (2000) *Meth. Enzymol.* **311**, 194. 2. Shiraishi, Imai & Uda (2003) *Biol. Pharm. Bull.* **26**, 775. 3. Tani, Okino, Mitsutake, *et al.* (2000) *J. Biol. Chem.* **275**, 3462. 4. Okino, Tani, Imayama & Ito (1998) *J. Biol. Chem.* **273**, 14368. 5. Spinedi, Di Bartolomeo & Piacentini (1999) *Biochem Biophys Res Commun.* **255**, 456. 6. Sałaga, Sobczak & Fichna (2014) *Eur. J. Pharm. Sci.* **52**, 173. 7. Huwiler, Pfeilschifter & van den Bosch (1999) *J. Biol. Chem.* **274**, 7190. 8. Epps & Cardin (1987) *Biochim. Biophys. Acta* **903**, 533. 9. Broekemeier, Schmid, Schmid & Pfeiffer (1985) *J. Biol. Chem.* **260**, 105. 10. Hillard, Edgemond, Jarrahian & Campbell (1987) *J. Neurochem.* **69**, 631. 11. Gaetani, Oveisi & Piomelli (2003) *Neuropsychopharm.* **28**, 1311. 12. Di Sabatino, Battista, Biancheri, *et al.* (2011) *Mucosal Immunol.* **4**, 574. 13. Rahnasto-Rilla, Kokkola, Jarho, Lahtela-Kakkonen & Moaddel (2015) *Chembiochem.* **17**, 77.

1-Oleoylglycerol

This monoacylglycerol (FW = 356.55 g/mol; M.P. = 35°C; CAS 111-03-5), also known as glycerol 1-oleate, 1-monoolein, and 1 monooleoylglycerol, inhibits electron transport and diacylglycerol kinase. Note that commercial sources are often the racemic mixture (1-oleoyl *rac*-glycerol), consisting of a mixture of 1-oleoyl-*sn*-glycerol and 3-oleoyl-*sn*-glycerol. In addition, acyl migration readily occurs. **Target(s):** alcohol O-acetyltransferase (1); diacylglycerol kinase, K_i = 91 µM (2,3); electron transport, mitochondrial (4); 7-ethoxycoumarin O-deethylase (cytochrome P450) (5); 1,3-β-glucan

synthase (6); glycerol-3-phosphate O-acyltransferase (7); NADH dihydro-genase (4); sphingomyelin phosphodiesterase, inhibited by the racemic mixture (8); triacylglycerol lipase, also a substrate (9). 1. Yoshioka & Hashimoto (1981) *Agric. Biol. Chem.* **45**, 2183. 2. Zawalich & Zawalich (1990) *Mol. Cell. Endocrinol.* **68**, 129. 3. Bishop, Ganong & Bell (1986) *J. Biol. Chem.* **261**, 6993. 4. Rapoport & Schewe (1977) *Trends Biochem. Sci.* **2**, 186. 5. Muller-Enoch, Fintelmann, Nicolaev & Gruler (2001) *Z. Naturforsch. [C]* **56**, 1082. 6. Ko, Frost, Ho, Ludescher & Wasserman (1994) *Biochim. Biophys. Acta* **1193**, 31. 7. Coleman (1988) *Biochim. Biophys. Acta* **963**, 367. 8. Liu, Nilsson & Duan (2002) *Lipids* **37**, 469. 9. Twu, Garfinkel & Schotz (1984) *Biochim. Biophys. Acta* **792**, 330.

2-Oleoylglycerol

This 2-monoacylglycerol (FW = 356.55 g/mol; CAS 3443-84-3) is not optically active and inhibits fatty acid amide hydrolase. Acyl migration readily occurs. **Target(s):** fatty acid amide hydrolase, or anandamide amidohydrolase (1); glycerol-3 phosphate O-acyltransferase (2). 1. Ghafouri, Tiger, Razdan, *et al.* (2004) *Brit. J. Pharmacol.* **143**, 774. 2. Coleman (1988) *Biochim. Biophys. Acta* **963**, 367.

1-Oleoyl-*sn*-glycerol 3-Phosphate

This lysophosphatidate ($FW_{free-acid}$ = 436.53 g/mol; CAS 22556-62-3; Hygroscopic, Photosensitive), also called monooleoylphosphatidate, is prepared from 1,2-dioleoyl-*sn*-glycerol 3-phosphate by the action of phospholipase A_2. It is the endogenous agonist for the LPA receptor. (**See also** *Lysophosphatidate and Lysophosphatidic Acid*) **Target(s):** adenylyl cyclase, mediated by the inhibitory coupling protein, Ni (1); alkylglycerophosphoethanolamine phosphodiesterase, or lysophospholipase D (2); chymase (3). 1. Proll, Clark & Butcher (1985) *Mol. Pharmacol.* **28**, 331. 2. van Meeteren, Ruurs, Christodoulou, *et al.* (2005) *J. Biol. Chem.* **280**, 21155. 3. Kido, Fukusen & Katunuma (1985) *Arch. Biochem. Biophys.* **239**, 436.

1-Oleoyl-*sn*-glycero-3-phosphocholine

This lysophospholipid (FW = 521.68 g/mol; CAS 19420-56-5), also known as oleoyl lysolecithin and 1-oleoyllysophosphatidylcholine, inhibits choline-phosphate cytidylyltransferase (1), 1,3-β-glucan synthase (2), and Na^+/K^+ exchanging ATPase (3-5). 1. Vance, Pelech & Choy (1981) *Meth. Enzymol.* **71**, 576. 2. Ko, Frost, Ho, Ludescher & Wasserman (1994) *Biochim. Biophys. Acta* **1193**, 31. 3. Lijnen, Huysecom, Fagard, Staessen & Amery (1990) *Methods Find. Exp. Clin. Pharmacol.* **12**, 281. 4. Tamura, Harris, Higashimori, *et al.* (1987) *Biochemistry* **26**, 2797. 5. Tamura, Inagami, Kinoshita & Kuwano (1987) *J. Hypertens.* **5**, 219.

Oligomycin

Oligomycin A (R = OH)
Oligomycin A (R = H)

Oligomycin B

These macrolide antibiotics from *Streptomyces diastatochromogenes* have three major members: Oligomycin A (FW = 791.08 g/mol; CAS 14104-19-9), Oligomycin B (FW = 805.06 g/mol; CAS 11050-94-5), and Oligomycin C (FW = 775.08 g/mol; CAS 11052-72-5). Others include Oligomycin D (also known as rutamycin; FW = 777.05 g/mol; CAS 1404-59-7), Rutamycin B (isolated from *Streptomyces aureofaciens* (**See also** *Rutamycin*)), and Oligomycin E (FW = 822.07 g/mol; CAS 110231-34-0; isolated from *Streptomyces* sp. MCI-2225). **Primary Mode of Inhibitory Action:** The oligomycins inhibit the ATP synthesis activity of the mitochondrial F_oF_1 ATP synthase, defining the "o" (*not* 0) in the first subscript. The ATPase activity of the isolated F_1 complex is unaffected by oligomycins, unlike the F_o portion. Oligomycin-sensitivity conferring ptrotein (OSCP) interacts with other components in the ATP synthase complex and facilitates oligomycin's binding to the F_o unit. The F_o complex is, in fact, an integral membrane protein that functions as a proton channel. **Other Actions:** Oligomycins also inhibit Na^+/K^+-exchanging ATPases that transport sodium ions from the inside to the outside of animal cells while translocating potassium ions in the reverse direction with the concomitant hydrolysis of ATP. Oligomycin inhibits Na^+ translocation *in vivo* (*i.e.*, the enzyme activity is inhibited by stabilizing the Na^+ occlusion but not the K^+ occlusion). **Other Properties:** The oligomycins are slightly soluble in water (*e.g.*, oligomycin A has a solubility of 2 μg per 100 mL at 25°C) and are considerable more soluble in water-miscible solvents such as ethanol, acetone, and glacial acetic acid. **Target(s):** aldehyde oxidase (1); ammonia kinase (2); apyrase, membrane-bound (3); ATP synthase, *or* F_oF_1 ATP synthase (4-19); calcidiol 1-monooxygenase (20); copper-exporting ATPase, partially inhibited (21); endopeptidase La, weakly inhibited (22); H^+-transporting ATPase, lysosomal (23); Na^+/K^+-exchanging ATPase (24-32); nucleoside triphosphatase (33); photophosphorylation in *Rhodospirillum rubrum* (34); xenobiotic-transporting ATPases (35,36). **1.** Rajagopalan & Handler (1966) *Meth. Enzymol.* **9**, 364. **2.** Dowler & Nakada (1968) *J. Biol. Chem.* **243**, 1434. **3.** Curdova, Jechova & Hostalek (1982) *Folia Microbiol.* **27**, 159. **4.** Slater (1967) *Meth. Enzymol.* **10**, 48. **5.** Kagawa (1967) *Meth. Enzymol.* **10**, 505. **6.** Penefsky (1979) *Meth. Enzymol.* **55**, 297. **7.** Soper & Pedersen (1979) *Meth. Enzymol.* **55**, 328. **8.** Tzagoloff (1979) *Meth. Enzymol.* **55**, 351. **9.** Linnett & Beechey (1979) *Meth. Enzymol.* **55**, 472. **10.** Gautheron, Penin, Déléage & Godinot (1986) *Meth. Enzymol.* **126**, 417. **11.** McEnery & Pedersen (1986) *Meth. Enzymol.* **126**, 470. **12.** Williams & Pedersen (1986) *Meth. Enzymol.* **126**, 477. **13.** Penefsky (1974) *The Enzymes*, 3rd ed., **10**, 375. **14.** Lardy, Witonsky & Johnson (1965) *Biochemistry* **4**, 552. **15.** Lardy, Johnson & McMurray (1958) *Arch. Biochem. Biophys.* **78**, 587. **16.** Sebald & Hoppe (1981) *Curr. Top. Bioenerg.* **12**, 1. **17.** Fillingame (1981) *Curr. Top. Bioenerg.* **11**, 35. **18.** Criddle, Johnston & Stack (1979) *Curr. Top. Bioenerg.* **9**, 89. **19.** Sebald (1977) *Biochim. Biophys. Acta* **463**, 1. **20.** Gray, Omdahl, Ghazarian & DeLuca (1972) *J. Biol. Chem.* **247**, 7528. **21.** Mandal, Cheung & Arguello (2002) *J. Biol. Chem.* **277**, 7201. **22.** Desautels & Goldberg (1982) *J. Biol. Chem.* **257**, 11673. **23.** Jonas, Smith, Allison, *et al.* (1983) *J. Biol. Chem.* **258**, 11727. **24.** Post & Sen (1967) *Meth. Enzymol.* **10**, 762. **25.** Skou (1988) *Meth. Enzymol.* **156**, 1. **26.** Askari & Koyal (1968) *Biochem. Biophys. Res. Commun.* **32**, 227. **27.** Homareda & Matsui (1982) *J. Biochem.* **92**, 219. **28.** Arato-Oshima, Matsui, Wakizaka & Homareda (1996) *J. Biol. Chem.* **271**, 25604. **29.** Cavieres (1977) in *Membrane Translocation in Red Cells* (Ellory & Lew, eds.), pp. 1 37, Academic Press, New York. **30.** Robinson & Flashner (1979) *Biochim. Biophys. Acta* **549**, 145. **31.** Homareda, Ishii & Takeyasu (2000) *Eur. J. Pharmacol.* **400**, 177. **32.** Robinson & Flashner (1979) *Biochim. Biophys. Acta* **549**, 145. **33.** Schröder, Rottmann, Bachmann & Müller (1986) *J. Biol. Chem.* **261**, 663. **34.** Izawa & Good (1972) *Meth. Enzymol.* **24**, 355. **35.** Eytan, Borgnia, Regev & Y. G. Assaraf (1994) *J. Biol. Chem.* **269**, 26058. **36.** Decottignies, Lambert, Catty, *et al.* (1995) *J. Biol. Chem.* **270**, 18150.

Olivanates

Olivanate MM 4550

Olivanate MM 13902

These Δ^2-penem-type antibiotics ($FW_{free-acid}$ = 394.40 g/mol for MM4550; CAS 64761-66-6) from *Streptomyces olivaceus* are β-lactamase inhibitors, consisting of seven structurally related members (of which three are sulfated). Certain ampicillin-resistant bacteria are rendered sensitive to ampicillin in the presence of olivanic acid at concentrations that did not inhibit growth when administered separately. Note that these antibiotics are also unstable in the presence of hydroxylamine and cysteine. **1.** Brown, Corbett, Eglington & Howarth (1977) *J. Chem. Soc., Chem. Commun.* (15), 523. **2.** Matagne, Ghuysen & Frere (1993) *Biochem. J.* **295**, 705. **3.** Reading & Cole (1986) *J. Enzyme Inhib.* **1**, 83. **4.** Hood, Box & Verrall (1979) *J. Antibiot. (Tokyo)* **32**, 295. **5.** Butterworth, Cole, Hanscomb & Rolinson (1979) *J. Antibiot. (Tokyo)* **32**, 287.

Olivomycins

Olivomycin A

These chromomycin-like antibiotics ($FW_{Olivomycin-A}$ = 1197.3 g/mol; CAS 6988-58-5) produced by *Streptomyces olivoreticuli* anthat bind to DNA in a base pair-specific manner, preferring Guanine:Cytosine pairs. The major component, olivimycin A, is a yellow solid that is insoluble in water and soluble in ethanol and chloroform. **Target(s):** DNA-directed DNA polymerase (1-3); DNA-directed RNA polymerase (4-7); *Eco*RI restriction endonuclease, *or* Type II site-specific deoxyribonuclease (8). **1.** Skripal', Bezuglyi & Babichev (1993) *Mikrobiol. Zh.* **55**, 24. **2.** Müller, Zahn & Seidel (1971) *Nature New Biol.* **232**, 143. **3.** Müller, Yamazaki & Zahn (1972) *Enzymologia* **43**, 1. **4.** Kersten, Kersten, Szybalski & Fiandt (1966) *Biochemistry* **5**, 236. **5.** Kullyev & Gauze (1975) *Sov. J. Dev. Biol.* **5**, 350. **6.** Goldberg & Friedman (1971) *Ann. Rev. Biochem.* **40**, 775. **7.** Sethi (1971) *Prog. Biophys. Mol. Biol.* **23**, 67. **8.** Wells, Klein & Singleton (1981) *The Enzymes*, 3rd ed., **14**, 157.

Olmesartan

Olmesartan

Olmesartan Medoxomil

This antihypertensive agent and ARB (FW = 445.50 g/mol; Code Name CS-866) is the pharmacologically active angiotensin receptor target formed by enzymatic hydrolysis of its prodrug olmesartan medoxomil (FW =

458.60 g/mol; CAS 144689-63-4; Code Name: CS-866; IUPAC: (5-methyl-2-oxo-2H-1,3-dioxolen-4-yl)methyl-4-(2-hydroxypropan-2-yl)-2-propyl-1-({4-[2-(2H-1,2,3,4-tetrazol-5-yl)phenyl]phenyl}methyl)-1H-imidazole-5-carboxylate). Olmesartan blocks [^{125}I]angiotensin II binding to bovine adrenal cortical membranes (containing angiotensin AT$_1$ receptors) with an IC$_{50}$ of 7.7 nM, but not [^{125}I]angiotensin II binding to bovine cerebellar membranes (containing angiotensin AT$_2$ receptors), thereby indicating its selectivity for AT$_1$ receptors (1). Unlike other ACE inhibitors, however, its actions are therefore independent of angiotensin II synthesis pathways. (**See also** Candesartan; Eprosartan; Irbesartan; Losartan; Telmisartan; Valsartan) Although SKF-525A, a potent P-450 inhibitor, suppresses the angiotensin II inhibitory effect of losartan, the same was not true for olmesartan medoxomil (1). Plasma levels of angiotensin II, renin, angiotensin converting enzyme and chymase are unchanged by the high-cholesterol diet, whereas vascular angiotensin converting enzyme, but not chymase, is significantly increased (2). Serum levels of macrophage-colony stimulating factor, transforming growth factor-β_1 and intracellular adhesion molecule-1 are significantly increased in monkeys fed a high-cholesterol diet but they were suppressed by olmesartan medoxomil (2). The relaxation response of isolated carotid arteries to acetylcholine is suppressed in the high-cholesterol group, whereas it was improved by CS-866 (2). Olmesartan-associated sprue-like enteropathy was first described in 2012, and the syndrome is characterized by severe diarrhea and sprue-like histopathologic findings in the intestine, often leading to increased sub-epithelial collagen. **Key Pharmacological Parameters:** Bioavailability = 26 %; Food Effect? None; Drug $t_{1/2}$ = 3.5–4 hours; Metabolite $t_{1/2}$ = 3–11; Drug's Protein Binding = 99 %; Metabolite's Protein Binding = not determined; Route of Elimination = 35–50 % Renal, 50–65 % Hepatic. **1**. Mizuno, Sada, Ikeda, *et al.* (1995) *Eur. J. Pharmacol.* **285**, 181. **2**. Takai, Kim, Sakonjo & Miyazaki (2003) *J. Hypertens.* **21**, 361.

Olodaterol

This long-acting β-adrenoceptor agonist (FW = 386.44 g/mol; CAS 868049-49-4), also known as Striverdi$^®$ and 6-hydroxy-8-{(1R)-1-hydroxy-2-{[1-(4-methoxyphenyl)-2-methylpropan-2-yl]amino}ethyl}-4H-1,4-benzoxazin-3-one, is used to treat patients with chronic obstructive pulmonary disease (COPD) by causing the bronchi to relax and reducing resistance to airflow. Olodaterol is nearly full agonist (EC$_{50}$ = 0.1 nM), displaying 88% intrinsic activity of isoprenaline. Once olodaterol is bound to a β_2-receptor, the dissociation half-life is 17.8 hours, permitting once-a-day application. Olodaterol has a higher *in vitro* selectivity for β_2-receptors, as compared, with 241-fold *versus* β_1-receptors and 2300-fold *versus* β_3-receptors. Stiolto Respimat$^®$ is a once-daily fixed-dose drug consisting of olodaterol and tiotropium. **1**. Bouyssou, Hoenke, Rudolf, *et al.* (2010). *Bioorg. Med. Chem. Lett.* **20**, 1410. **2**. Bouyssou, Casarosa, Naline, *et al.* (2010) *J. Pharmacol. Exper. Ther.* **334**, 53. **3**. Casarosa, Kollak, Kiechle, *et al.* (2011) *J. Pharmacol. Exper. Ther.* **337**, 600.

Olomoucine

This ATP site-competitive purine derivative (FW = 298.35 g/mol; CAS 101622-51-9), systematically named 6-benzylamino-2-[2-hydroxyethyl-amino]-9-methylpurine, is a potent inhibitor of cyclin-dependent kinases, with an IC$_{50}$ value of 7 µM for both cdk1 and cdk2 and an IC$_{50}$ value of 3 µM for cdk5, arresting cells in G$_1$ (1-6). Olomoucine also inhibits cytokinin 7β-glucosyltransferase (7). **1**. Meijer & Kim (1997) *Meth. Enzymol.* **283**, 113. **2**. Vesely, Havlicek, Strnad, *et al.* (1994) *Eur. J. Biochem.* **224**, 771. **3**. Braña, Cacho, Garcia, *et al.* (2005) *J. Med. Chem.* **48**, 6843. **4**. VanderWel, Harvey, McNamara, *et al.* (2005) *J. Med. Chem.* **48**, 2371. **5**. MacKintosh & MacKintosh (1994) *Trends Biochem. Sci.* **19**, 444. **6**.

Corellou, Camasses, Ligat, Peaucellier & Bouget (2005) *Plant Physiol.* **138**, 1627. **7**. Parker, Entsch & Letham (1986) *Phytochemistry* **25**, 303.

Olsalazine

This anti-inflammatory agent (FW$_{free-acid}$ = 302.24 g/mol; CAS 15722-48-2) inhibits myeloperoxidase myeloperoxidase (1), Na$^+$/K$^+$-exchanging ATPase (2), and thiopurine S methyltransferase (3,4). **1**. Kettle & C. C. Winterbourn (1991) *Biochem. Pharmacol.* **41**, 1485. **2**. Scheurlen, Allgayer, Kruis, Erdmann & Sauerbruch (1993) *Clin. Investig.* **71**, 286. **3**. Zhou & Chowbay (2007) *Curr. Pharmacogenomics* **5**, 103. **4**. Oselin & Anier (2007) *Drug Metab. Dispos.* **35**, 1452.

Oltipraz

This schistosomicide (FW = 226.35 g/mol; CAS 64224-21-1), also known as 5-(2-pyrazinyl)-4-methyl-1,2-dithiole-3-thione, is used to treat disease caused by parasitic flatworms and is also a potent cancer chemopreventive agent. Oltipraz was found to protect against chemically induced carcinogens within the lung, stomach, colon, and urinary bladder. The anti-HIV activity of the pharmaceutical is believed to be due to its ability to irreversibly inhibit HIV reverse transcriptase. **Target(s):** CYP1A2 (1-3); CYP3A4 (3); HIV I reverse transcriptase (4); RNA-directed DNA polymerase (4). **1**. Sofowora, Choo, Mayo, Shyr & Wilkinson (2001) *Cancer Chemother. Pharmacol.* **47**, 505. **2**. Langouet, Furge, Kerriguy, *et al.* (2000) *Chem. Res. Toxicol.* **13**, 245. **3**. Langouet, Coles, Morel, *et al.* (1995) *Cancer Res.* **55**, 5574. **4**. Chavan, Bornmann, Flexner & Prochaska (1995) *Arch. Biochem. Biophys.* **324**, 143.

Omacetaxine

This alkaloid and protein synthesis inhibitor (FW = 545.63 g/mol; CAS 26833-87-4), also known as homoharringtonine and Synribo$^®$, from *Cephalotaxus harringtonia* is indicated for treatment of Chronic Myelogenous Leukemia (CML). Omacetaxine binds to the ribosomal A-site, thereby preventing proper positioning of amino acid side-chains of incoming aminoacyl-tRNAs. With mounting evidence of resistance to imatinib and other tyrosine kinase inhibitors (TKIs), particularly as a consequence of an ABL mutation at position-315), omacetaxine offers hope for CML patients failing to respond to TKIs (1). Indeed, Synribo is a FDA-approved treatment for Philadelphia-positive CML either in chronic or accelerated phase whose disease failed two prior TKIs. **1**. Wetzler & Segal (2011) *Curr. Pharm. Des.* **17**, 59.

Omarigliptin

This oral DPP-4 inhibitor (MW = 398.43 g/mol; CAS 1226781-44-7), also known MK-3102 and systemically as (2R,3S,5R)-2-(2,5-difluorophenyl)-5-[2-(methylsulfonyl)-2,6-dihydropyrrolo[3,4-c]pyrazol-5(4H)-yl]tetrahydro-2H-pyran-3-amine, is a potent and selective reversible competitive inhibitor of Dipeptidyl Peptidase-4 (IC_{50} = 1.6 nM, K_i = 0.8 nM). It is highly selective over all proteases tested (IC_{50} > 67 µM), including QPP (or quiescent cell proline dipeptidase), FAP (or fibroblast activation protein, or seprase), DPP8 and DPP9 (or dipeptidyl peptidase 8 and 9). Omarigliptin has weak ion channel activity (IC_{50} > 30 nM at IKr, Ca_v1.2, and Na_v1.5). **Mode of Inhibitory Action:** Glucagon-like peptide 1 (GLP-1), which is released by the gut in response to food intake, stimulates insulin secretion in a glucose-dependent manner. Inhibition of DPP-4 prolongs the circulating half-life of glucagon-like peptide 1 and glucose-dependent insulinotropic polypeptide. By increasing the circulating concentration of GLP-1 and GIP, omarigliptin improves glucose control in Type-2 diabetes. Omarigliptin has a long half-life (rat, $t_{1/2}$ = 11 hours; dog, $t_{1/2}$ = 22 hours) and lower clearance (rat, 1.1 mL/min/kg; dog, 0.9 mL/min/kg). Based on the human pharmacokinetics prediction, omarigliptin is projected to be amenable for once-weekly dosing. This conclusion is recapitulated in the clinical studies, where omarigliptin is shown to have a biphasic PK profile with a terminal $t_{1/2}$ of 120 hours. Indeed, the average renal clearance of omarigliptin was ~2 Liters/hour (2). Over the dose range studied, DPP-4 inhibition ranged from ~77-°89% at one week following the last of three once-weekly doses Omarigliptin resulted in ~ 2-fold increases in weighted average postprandial active GLP-1 (2). **1.** Biftu, Sinha-Roy, Chen, et al. (2014) J. Med. Chem. **57**, 3205. **2.** Krishna, Addy, Tatosia, et al. (2016) J. Clin. Pharmacol. **56**, 1528.

Omaveloxolone

This antioxidant inflammation modulator, or AIM (FW = 554.71 g/mol; CAS 1474034-05-3; Solubility: 100 mg/mL DMSO; < 1 mg/mL H_2O; topical formulation vehicle: Sesame oil), also known as RTA-408 and N-((4aS,6aR,6bS,8aR,12aS)-11-cyano-2,2,6a,6b,9,9,12a-heptamethyl-10,14-dioxo-1,2,3,4,4a,5,6,6a,6b,7,8,8a,9,10,12a,14,14a,14b-octadecahydropicen-4a-yl)-2,2-difluoropropanamide, is a synthetic triterpenoid that activates nuclear factor erythroid 2-related factor 2, or Nrf2 (1). A powerfully cytoprotective transcription factor, Nrf2 elicits the coordinated induction of cytoprotective genes in response to oxidative and electrophilic stress. It rapidly translocates to the nucleus, recognizes and binds to antioxidant response elements in the upstream promoter regions of its target genes, thereby facilitating coordinated responses that serve to alleviate oxidative trauma (2,3). It also inhibits signaling by the proinflammatory transcription factor NF-κB. Topical application of RTA-408 protects mice from radiation-induced dermatitis, with significant increases in Nrf2 target genes and significant decreases in NF-κB target genes in mouse skin tissue. **1.** Probst, Trevino, McCauley, et al. (2015) PLoS One **10**, e0122942. **2.** Klaassen & Reisman (2010) Toxicol. Appl. Pharmacol. **244**, 57. **3.** Taguchi, Motohashi & Yamamoto (2011) Genes Cells **16**, 123. **3.**

Ombitasvir

This N-phenylpyrrolidine-based, pan-genotypic HCV NS5A protease inhibitor (FW = 893.14 g/mol), also named ABT-267, targets Nonstructural protein 5A (NS5A), a zinc-binding and proline-rich hydrophilic phosphoprotein that plays a key role in Hepatitis C virus RNA replication.

Ombitasvir exhibits low-picomolar EC_{50} values against Hepatitis C virus, with superior pharmacokinetics (1). Although ombitasvir shows a low barrier to resistance, when given as monotherapy, co-administration with other antivirals enhances its barrier to resistance. Indeed, a 12-week, Phase-3 study demonstrated that a multi-targeted regimen consisting of ombitasvir, dasabuvir and ribavirin is highly effective and showed a low rate of treatment discontinuation (2). **1.** DeGoey, Randolph, Liu, et al. (2014) J. Med. Chem. **57**, 2047. **2.** Feld, Kowdley, Coakley, et al. (2014) New Engl. J. Med. **370**, 1594.

Omeprazole

This proton pump inhibitor and antacid drug (FW = 345.42 g/mol; CASs = 73590-58-6 and 119141-88-7; Soluble in ethanol and methanol and only very slightly soluble in H_2O), also known as Nexium®, Essocam®, Esomeprazole®, and Esomezol®, and (S)-5-methoxy-2-[(4-methoxy-3,5-dimethylpyridin-2-yl)methylsulfinyl]-3H-benzoimidazole, is used to treat dyspepsia, peptic ulcer disease, gastroesophageal reflux disease, and Zollinger-Ellison syndrome. **Primary Mode of Inhibitory Action:** Omeprazole is a prodrug which is converted to its active form only at the site of action, namely the parietal cell. It then binds and reacts irreversibly with an active-site thiol on H^+/K^+-ATPase, causing effective and long-lasting inhibition (IC_{50} = 3.9 µM) of gastric acid secretion.

Omeprazole

Inhibition is selective and irreversible, suppressing gastric acid secretion by targeting the H^+/K^+-ATPase. Gastro-esophageal reflux disease arises from the interaction of several anatomical and physiological factors, with a striking correlation between symptoms of heartburn with acid regurgitation and reflux esophagitis first noted in 1934. Allison (1,2) first described hiatus hernia as a factor in the development of the reflux disease. Episodes of reflux were related to transient relaxations of lower oesophageal sphincter. Omeprazole was introduced in 1989 as the first proton pump inhibitor for treating gastro-esophageal reflux disease. **Target(s):** F_1 ATPase, or H^+-transporting two-sector ATPase (3); H^+/K^+-exchanging ATPase (4-11); pyruvate decarboxylase (12); Na^+/K^+-ATPase activity, IC_{50} = 186 µM (13). **Inhibition of Melanogenesis:** There is great interest in topical products capable of reducing undesired hyperpigmentation, such as irregular hyperpigmentation of the skin, leading to melasma, solar lentigens, age spots, and uneven skin color. Omeprazole and closely related congeners inhibit melanogenesis at µM concentrations in B16 mouse melanoma cells, normal human epidermal melanocytes, and in a reconstructed human skin model (14). When omeprazole is applied topically to the skin of UV-irradiated human subjects, pigment levels after 3 weeks are significantly reduced, when compared with untreated controls. Omeprazole has no significant inhibitory effect on purified human tyrosinase, a copper-containing glycoprotein that catalyzes the hydroxylation of tyrosine to dihydroxyphenylalanine (DOPA); nor does omeprazole affect levels of mRNAs for tyrosinase, dopachrome tautomerase, Pmel17, or MITF. Although melanocytes do not express ATP$_4$A, they do express ATP7A, a copper transporting P-type ATPase in the *trans*-Golgi network required for copper acquisition by tyrosinase. Omeprazole inhibits ATP7A relocalization from the *trans*-Golgi network to the plasma membrane in response to elevated copper concentrations in melanocytes. Omeprazole treatment increased the proportion of EndoH-sensitive tyrosinase, indicating that tyrosinase maturation was impaired. In addition, omeprazole reduced tyrosinase protein abundance in the presence of cycloheximide, a finding that suggests increased degradation (14). **1.** Allison (1946) *J. Thorac. Surg.* **15**, 308. **2.** Allison (1973) *Ann. Surg.* **178**, 273. **3.** Beil, Birkholz, Wagner & Sewing (1995) *Pharmacology* **50**, 333. **4.** Morii & Takeguchi (1993) *J. Biol. Chem.* **268**, 21553. **5.** Wallmark, Brandstrom & Larsson (1984) *Biochim. Biophys. Acta* **778**, 549. **6.** Hango, Nojima & Setaka (1990) *Jpn. J. Pharmacol.* **52**, 295. **7.** Chatterjee & Das (1995) *Mol. Cell. Biochem.* **148**, 95. **8.** Planelles, Anagnostopoulos, Cheval & Doucet (1991) *Amer. J. Physiol.* **260**, F806. **9.** Hersey, Perez, Matheravidathu & Sachs (1989) *Amer. J. Physiol.* **257**, G539. **10.** Shin & Sachs (2004) *Biochem. Pharmacol.* **68**, 2117. **11.** Lorentzon, Jackson, Wallmark & Sachs (1987) *Biochim. Biophys. Acta* **897**, 41. **12.** Sutak, Tachezy, Kulda & Hrdy (2004) *Antimicrob. Agents Chemother.* **48**, 2185. **13.** Keeling, Fallowfield, Milliner, *et al.* (1985) *Biochem. Pharmacol.* **34**, 2967. **14.** Matsui, Petris, Niki, *et al.* (2015) *J. Invest. Dermatol.* **135**, 834.

ON-01910, *See* Rigoserib

Ondansetron

This serotonin 5-HT$_3$ receptor antagonist and anti-emetic (FW = 293.40 g/mol; CAS 99614-02-5), also known by its code name GR38032F, its trade names Zofran® and Emetron®, as well as by its systematic name (*RS*)-9-methyl-3-[(2-methyl-1*H*-imidazol-1-yl)methyl]-2,3-dihydro-1*H*-carbazol-4(9*H*)-one, is used to treat episodes of nausea and vomiting following chemotherapy. Ondansetron affects both peripheral and central nervous systems, reducing vagus nerve activity and thereby deactivating the 5-HT$_3$ receptor-dense vomiting center located in the medulla oblongata. Ondansetron is without effect on dopamine receptors or muscarinic receptors. **Primary Mode of Inhibitor Action:** On the isolated vagus nerve and superior cervical ganglion of the rat, racemic GR38032F behaved as a reversible competitive antagonist of 5-HT-induced depolarization with pK_B values of 8.61 and 8.13, respectively (1). The resolved *R*- and *S*-isomers were equipotent 5-HT antagonists on the rat vagus nerve, with pK_B values were 8.95 and 8.63, respectively. (Similar 5-HT$_3$ antagonists include: tropisetron, granisetron, dolasetron, and palonosetron. The effectiveness of each depends on particular variants of 5-HT$_3$ receptors expressed by the patient, including changes in promoters for the receptor genes.) Ondansetron had a terminal plasma half-life of 3.0-3.5 h and plasma clearance of 600 mL/min, with no evidence of accumulation at steady state (2). Its absolute oral bioavailability was 59%, and its metabolism involves CYP3A4, CYP1A2, and CYP2D6. **Key Pharmacokinetic Parameters:** *See* Appendix II in Goodman & Gilman's THE PHARMACOLOGICAL BASIS OF THERAPEUTICS, 12th Edition (Brunton, Chabner & Knollmann, eds.) McGraw-Hill Medical, New York (2011). **1.** Butler, Hill, Ireland, Jordan & Tyers (1988) *Br. J. Pharmacol.* **94**, 397. **2.** Blackwell & Harding (1989) *Eur. J. Cancer Clin. Oncol.* **Suppl 1**, S21.

ONO-2235, *See* Epalrestat

ONO-5046, *See* Sivelestat

ONX-0801

This antifolate (FW = 647.63 g/mol; CAS 501332-69-0), also known as BGC 945, CB 300945 as well as *N*-[4-[2-propyn-1-yl][(6*S*)-4,6,7,8-tetrahydro-2-(hydroxymethyl)-4-oxo-3*H*-cyclopenta[*g*]quinazolin-6-yl]amino]benzoyl]-L-γ-glutamyl-D-glutamic acid, is an anticancer drug that combines the virtues of thymidylate synthase transition-state inhibition with lower toxicity due to selective intracellular accumulation through α-folate receptor (α-FR) transport in tumor cells. Indeed, α-FR is overexpressed in many carcinomas, particularly those of the ovary and uterus, and this property has been exploited for folate-mediated targeting of macromolecules, anticancer drugs, imaging agents, and nucleic acids to cancer cells. ONX 0801 forms a high-affinity complex, when incubated with thymidylate synthase in the presence of 2'-dUMP. The K_i for ONX 0801 with isolated thymidylate synthase is 1.2 nM, whereas the IC$_{50}$ for growth inhibition of α-FR-negative mouse L1210 or human A431 cells is approximately 7 μM. In contrast, BGC 945 is highly potent (IC$_{50}$ ≈ 1-300 nM) in α-FR-overexpressing human tumor cell lines (1). **1.** Gibbs, Theti, Wood, *et al.* (2005) *Cancer Res.* **65**, 11721.

ONX 0912, *See* Oprozomib

ONX-0914

This potent protease inhibitor (FW = 580.67 g/mol; CAS 960374-59-8; Solubility: 100 mg/mL DMSO or H$_2$O), also known as PR-957 and *N*-[2-(4-morpholinyl)acetyl]-L-alanyl-*O*-methyl-*N*-[(1*S*)-2-[(2*R*)-2-methyl-2-oxiranyl]-2-oxo-1-(phenylmethyl)ethyl]-L-tyrosinamide, targets (IC$_{50}$ = 10 nM) the chymotrypsin-like protease activity of immunoproteasomes, a distinct class of proteasome found mainly in monocytes and lymphocytes. These specialized proteasomes shape the antigenic repertoire presented on class I major histocompatibility complexes (MHC-I). ONX-0914 shows minimal cross-reactivity for the constitutive proteasome. Selective inhibition blocks production of Interleukin-23 (IL-23) by activated monocytes and interferon-γ and IL-2 by T cells. **1.** Muchamuel, Basler, Aujay, *et al.* (2009) *Nature Med.* **15**, 781.

OP-1068, *See* Solithromycin

OPC 3689, *See* Cilostamide

OPC-34712, *See* Brexpiprazole

OPC-145977, *See* Aripiprazole

OPC-41061, *See* Tolvaptan

OPC-67683, *See* Delamanid

Opium

While not a drug *per se*, opium refers to the latex-like substance exuded by the seedpod of the opium poppy *Papaver somniferum*. Because it contains mainly morphine (~12% by dry weight), the extract is a powerful analgesic. Even so, opium is so bitter that the only convenient route for consuming this plant extract is through smoking it, and that process undoubtedly alters the composition, if not the chemical nature, of opium ingredients. The following excerpt from the *Pharmaceutical Journal & Transactions* of the Pharmaceutical Society of Great Britain (Third Series, Volume XIV, p. 28,

July 14, 1883) is illuminating: "It has not as yet been found practicable to take, with any approximation to accuracy, the temperature at which the prepared opium is smoked, but it would appear to be a higher temperature than that at which the opium alkaloids are decomposed. Should this prove to be the case, it will be of itself sufficient to show that none of the active constituents exist in the smoke in the same condition as they do in opium. It appears perfectly clear from the general composition of opium and the method of manufacturing the extract, that morphine, codeine, and narceine are the only bodies of importance. Codiene and narceine give no sublimate. Moreover, opium smoke does not possess the well-marked bitterness of morphine. The variety of opium most affected by smokers is characterized by the small percentage of morphine which it contains." We now know that temperature, moisture, and even construction of the opium pipe itself can influence the composition of opium smoke, and, even though the resulting smoke is rich in only the most volatile (free-base) form of morphine, other agents are present in lesser abundance. Collectively, these substances are apt to exert a manifold of pharmacologic actions within the lungs and beyond. Although the composition of opium smoke is highly variable, the raw opium extract is the repository of many constituents, and among them: α-*Allocryptopine* (CAS 24240-04-8, an anti-arrhythmia drug); *Berberine* (CAS 2086-83-1, an anti-pyretic); *Canadine* (CAS 522-97-4, a mild sedative); *Coptisine* (CAS 3486-66-6, an antibacterial and anti-inflammatory agent); *Corytuberine* (CAS 517-56-6, a toxin causing asphyxia and convulsive seizures); *Cryptopine* (CAS 482-74-6, anti-arrhythmia drug); *Dihydrosanguinarine* (CAS 3606-45-9, a sanguinarine mimic); *Isoboldine* (CAS 3019-51-0, an antimicrobial agent); *Isocorypalmine* (CAS 483-34-1; a platelet aggregation inhibitor); *Laudanine* (CAS 85-64-3, an accelerator of respiration); *Laudanosine* (CAS 1699-51-0, a hypotensive angent); *Magnofluorine* (CAS 2141-09-5, an anti-inflammatory agent); *Narceine* (CAS 131-28-2, a mild codeine mimic); *Narcotoline* (CAS 521-40-4, a cough suppressant); *Neopine* (CAS 32404-16-3, a mild codeine mimic); *Oripavine* (CAS 467-04-9, an analgesic); *Protopine* (CAS 130-86-9, an anti-arrhythmia drug); *Pseudomorphine* (CAS 57-27-2, an antihypertensive agent); *Reticuline* (CAS 485-19-8, a mild DOPA antagonist), *Rhoeadine* (CAS 2718-25-4, a mild sedative); *Salutaridine* (CAS 1936-18-1, a mild GABA receptor agonist); *Sanguinarine* (CAS 2447-54-3, a hypertensive agent); *Scoulerine* (CAS 6451-73-6, a mild sedative); and *Stepholidine* (CAS 16562-13-3, an antihypertensive agent), and *Thebaine*, or *Paramorphine* (CAS 115-37-7, exerts stimulatory rather than depressant effects).

Ophthalmate (Ophthalmic Acid)

This tripeptide (FW$_{hydrochloride}$ = 325.75 g/mol; CAS 495-27-2), also known as L-γ-glutamyl-L-α-aminobutyrylglycine, is a glutathione analogue first found in lens of the eye. **Target(s):** formaldehyde dehydrogenase, glutathione-dependent (1); γ-glutamylcysteine synthetase (2-6); lactoylglutathione lyase, *or* glyoxalase I (7). 1. Valiakina, Reznikova, Novikova, Gorbunov & Zhuze (1972) *Biokhimiia* 37, 757. 2. Seelig & Meister (1985) *Meth. Enzymol.* 113, 379. 3. Meister (1995) *Meth. Enzymol.* 252, 26. 4. Richman & Meister (1975) *J. Biol. Chem.* 250, 1422. 5. Griffith & Mulcahy (1999) *Adv. Enzymol. Relat. Areas Mol. Biol.* 73, 209. 6. Huang, Chang, Anderson & Meister (1993) *J. Biol. Chem.* 268, 19675. 7. Cliffe & Waley (1961) *Biochem. J.* 79, 475.

Oprozomib

This orally active inhibitor (FW = 532.61 g/mol; CAS 935888-69-0; Solubility: 105 mg/mL DMSO, <1 mg/mL H$_2$O), also known as ONX 0912 and *O*-methyl-*N*-[(2-methyl-5-thiazolyl)carbonyl]-L-seryl-*O*-methyl-*N*-

[(1*S*)-2-[(2*R*)-2-methyl-2-oxiranyl]-2-oxo-1-(phenylmethyl)-ethyl]-L-serinamide, selectively targets the chymotrypsin-like (CT-L) activity of 20S proteasome β5 (IC$_{50}$ = 36 nM) and 20S proteasome LMP7 (IC$_{50}$ = 82 nM). In animal tumor model studies, ONX 0912 significantly reduced tumor progression and prolonged survival. Immunostaining of multiple myeloma tumors from ONX 0912-treated mice showed growth inhibition, apoptosis, and a decrease in associated angiogenesis (2). Oprozomib is distinct from carfilzomib, even though the same chemistry was employed to selectively target the proteasome. Oprozomib is under development as an oral therapy for hematologic malignancies, including multiple myeloma, and for patients with recurrent or refractory solid tumors (*See also Carfilzomib*). 1. Zhou, Aujay, Bennett, *et al.* (2009) *J. Med. Chem.* 52, 3028. 2. Chauhan, Singh, Aujay, *et al.* (2010) *Blood* 116, 4906.

Opsopyrroledicarboxylic Acid

This porphobilinogen analogue (FW$_{free-acid}$ = 197.19 g/mol), also known as pyrrole-3-(acetic acid)-4-(propionic acid) and 4-(carboxymethyl)-1*H*-pyrrole-3-propionic acid, competitively inhibits hydroxymethylbilane synthase, *or* porphobilinogen deaminase. 1. Bogorad (1962) *Meth. Enzymol.* 5, 885. 2. Carpenter & Scott (1961) *Biochim. Biophys. Acta* 52, 195. 3. Bogorad & Pluscec (1970) *Biochemistry* 9, 4736. 4. Davies & Neuberger (1973) *Biochem. J.* 133, 471. 5. Russell & Rockwell (1980) *FEBS Lett.* 116, 199.

OPT-80, *See Fidaxomicin*

OR-462, *See Nitecapone*

OR-611, *See Entacapone*

Orbifloxacin

This fluoroquinolone-class antibiotic (FW = 395.37 g/mol; CAS 113617-63-3), also named 1-cyclopropyl-7-[(3*S*,5*R*)-3,5-dimethylpiperazin-1-yl]-5,6,8-trifluoro-4-oxo-1,4-dihydroquinoline-3-carboxylate, is a broad-spectrum systemic antibacterial that targets Type II DNA topoisomerases (gyrases), thereby blocking bacterial replication and transcription. For the prototypical member of this antibiotic class, *See Ciprofloxacin*

Orexin A & B

These 33- and 28-residue appetite- and sleep-regulating neuropeptide hormones (FW$_{Orexin-A}$ = 3561.12 g/mol; CAS 205640-90-0; *Sequence:* XPLPDCCRQKTCSCRLYELLHGAGNHAAGILTL-NH$_2$, where X = Glp, Disulfides join residues 6 to 12 and 7 to 14; FW$_{Orexin-B}$ = 2899.36 g/mol; CAS 205640-91-1; *Sequence:* RSGPPGLQGRLQRLLQASGNHAAGIL TM-NH$_2$), also called hypocretins, are produced from a common 131-residue precursor named prepro-orexin. The latter is exclusively expressed by orexin-producing neurons in the perifornical nucleus, the dorsomedial hypothalamic nucleus, and the dorsal and lateral hypothalamic areas. Although rodents have only a few thousand orexin-producing neurons, humans have more (~30–70,000). Intracerebroventricular injections of orexins A and B stimulate feeding in a dose-related manner, with orexin A significantly more effective than orexin B. Orexins are strongly linked to the regulation of sleep-activity cycles, the dysfunction of which results in some forms of narcolepsy. Both receptors can couple to G$_q$ and mobilize intracellular Ca^{2+} via activation of phospholipase C, whereas OX$_2$R can also couple G_i/G_o and inhibit cAMP production via inhibition of adenylate cyclase. In non-neuronal cells OX_2R mediates extracellular signal-regulated kinase activation via G_s, G_q, and G_i. In competition radioligand binding OX_1R has a 10–100x higher affinity for orexin A (20 nM) than for orexin B (250 nM), whereas OX_2R binds both orexin peptides with similar affinity. **1.** de Lecea, Kilduff, Peyron, *et al.* (1998) *Proc. Natl. Acad. Sci. U.S.A.* **95**, 322. **2.** Sakurai, Amemiya, Ishii, *et al.* (1998) *Cell* **92**, 573.

Org 5222, *See* Asenapine

Org 25935

This synthetic drug (FW = 339.43 g/mol; CAS 949588-40-3), known systematically as 2-([(1R,2S)-6-methoxy-1-phenyl-1,2,3,4-tetrahydronaph-thalen-2-yl]methylmethylamino)acetate, is a selective inhibitor of the sodium/chloride-dependent glycine transporter $GlyT_1$, counteracting the effects of the dissociative drug ketamine (1). **Rationale:** Elevation of extracellular synaptic concentration of glycine by blockade of GlyT1 is believed to potentiate NMDA receptor function *in vivo*, thereby suggesting a rational approach for the treatment of schizophrenia and cognitive disorders. Moreover, accumbal glycine receptors (GlyRs) are involved in mediating the dopamine activating effects of ethanol, and that administration of glycine locally into the nucleus accumbens (nAc) reduces ethanol consumption in ethanol high-preferring rats (1). Org-25935 decreased ethanol intake and ethanol preference, as compared with vehicle, whereas water intake was unaffected. This effect was dose-dependent, developed gradually and was sustained for up to 40 days (1) *Note*: Other GlyT1 inhibitors include GSK931145, Lu AA21279, NFPS, Org 25935, SB-710622, SSR 504734, whereas D-serine is an agonist. All GlyT1 inhibitors increased seizure thresholds dose-dependently, indicative of anticonvulsant activity (2). **1.** Molander, Lidö, Löf, Ericson & Söderpalm (2007) *Alcohol Alcohol.* **42**, 11. **2.** Kalinichev, Starr, Teague, *et al.* (2010) *Brain Res.* **1331**, 105.

Oritavancin

This novel, semi-synthetic glycopeptide antibiotic (FW = 1793.10 g/mol; CAS 171099-57-3), also known as LY333328, Orbactiv®, and (4R)-22-*O*-(3-amino-2,3,6-trideoxy-3-*C*-methyl-α-L-arabinohexopyranosyl)-N^3-(p-(p-chlorophenyl)benzyl)vancomycin, disrupts the cell membrane of Gram-positive bacteria (1-3) and inhibits both transglycosylation and transpeptidation (2). Oritavancin differs from vancomycin by the presence of a hydrophobic *N*-4-(4-chlorophenyl)benzyl substituent on the disaccharide, the addition of a 4-epi-vancosamine monosaccharide to the amino acid residue in Ring-6, and replacement of the vancosamine moiety by 4-epi-vancosamine (3). When compared vancomycin, teicoplanin, and quinupristin-dalfopristin (Synercid) against 219 strains of enterococci and staphylococci, including vancomycin-resistant enterococci and methicillin-

resistant *Staphylococcus aureus*, LY333328 demonstrated superior activity against vancomycin-resistant enterococci and was the only antibiotic which was bactericidal. Indeed, a single infusion of this antibiotic (1200 mg) can clear serious bacterial skin infections, including methicillin-resistant *Staphylococcus aureus*, or MRSA, as effectively as the usual 7-10 day, twice-daily regimen of vancomycin now needed to treat patients (5). **1.** Cooper, Snyder, Zweifel, *et al.* (1996) *J. Antibiot.* (Tokyo) **49**, 575. **2.** Schwalbe, McIntosh, Qaiyumi, *et al.* (1996) *Antimicrob. Agents Chemother.* **40**, 2416. **3.** Belley, McKay, Arhin, *et al.* (2010) *Antimicrob. Agents & Chemother.* **54**, 5369. **4.** Zhanel, Schweizer & Karlowsky (2012) *Clin. Infect. Dis.* **54** (Suppl 3), S214. **5.** Corey, Kabler, Mehra, *et al.* (2014) *New Engl. J. Med.* **370**, 2180.

Orlistat

This potent pancreatic lipase inhibitor (FW = 509.77 g/mol; CAS 96829-58-2), also known as tetrahydrolipstatin and Ro 18-0647 as well as by its proprietary name Xenical™ and Alli™, is the saturated derivative of lipstatin and reacts irreversibly with an essential serine (1,2). An amphiphile, orlistat efficiently inactivates enzymes catalyzing lipolytic reactions at the lipid/water interface. One lipase molecule among 10^5 substrate molecules in a membrane is sufficient to reduce lipase activity to half its initial velocity (1). Lipase inactivation is surface pressure-independent. (**See also *Lipstatin*) Use in Weight Control:** Orlistat acts locally within the gastrointestinal tract, and systemic absorption is not required for efficacy (3). Serum total cholesterol and low-density lipoprotein-cholesterol levels are reduced in obese, but not in healthy, patients. Obese patients who maintained a hypocaloric diet and received orlistat (360 mg/day for 12 weeks) lost 5% of bodyweight, significantly greater than placebo recipients (3.5%) (3). In 2-year studies, weight loss was significantly greater in orlistat than in placebo recipients. Ro 18-0647 also reliably increases fecal fat excretion, commending its use in treating obesity and hyperlipidemia by limiting the hydrolysis of triglycerides. **Other Target(s):** Orlistat also inhibits human gastric lipase, pancreatic carboxyl ester lipase (*or* cholesterol esterase), and the closely related bile-salt-stimulated lipase from human milk (1). It does not inhibit the exocellular lipase from *Rhizopus arrhizus* or a lipase recently isolated from *Staphylococcus aureus* (1). **1.** Borgström (1988) *Biochim. Biophys. Acta* **962**, 308. **2.** Ransac, Gargouri, Moreau & Verger (1991) *Eur. J. Biochem.* **202**, 395. **3.** McNeely & Benfield (1998) *Drugs* **56**, 241.

ORM-10103

This novel and specific sodium/calcium exchanger inhibitor (FW = 348.36 g/mol), also known as 5-nitro-2-(2-phenychroman-6-yloxy)pyridine, significantly reduces both the inward (EC_{50} = 780 nM) and outward (EC_{50} = 960 nM) currents of the electrogenic sodium-calcium exchanger (*or* NCX), without significantly change in the L-type Ca^{2+} current (*or* I_{CaL}) or the maximum rate of depolarization (*or* dV/dt_{max}), indicative of the fast inward Na^+ current. The NCX is widely expressed in different tissues. In heart, NCX plays an important role in calcium ion homeostasis, in excitation-contraction coupling, and in the electrophysiological properties of cardiac myocytes. ORM-10103 does not influence Na^+/K^+ pump or the main K^+ currents of canine ventricular myocytes, but at 3 μM slightly diminishes the rapid delayed rectifier K^+ current. Amplitudes of early and delayed after-depolarizations were significantly decreased by ORM-10103 in a concentration-dependent manner. Such findings suggest that ORM-10103 can abolish triggered arrhythmias and may have antiarrhythmic electrophysiological properties. Other NCX inhibitors include: KB-R7943, SEA0400, SN-6, and YM-244769, but KB-R7943 (2) and SEA-0400 (3) also inhibit the I_{CaL}, leaving the interpretation of their antiarrhythmic effects less certain. **1.** Jost, Nagy, Kohajda, *et al.* (2013) *Br. J. Pharmacol.* **170**, 768. **2.** Tanaka, Nishimaru, Aikawa, *et al.* (2002) *Br. J. Pharmacol.* **135**, 1096. **3.** Birinyi, Acsai, Banyasz, *et al.* (2005) *Naunyn. Schmiedebergs. Arch. Pharmacol.* **372**, 63.

D-Ornithine

This dibasic amino acid ($FW_{dihydrochloride}$ = 205.08 g/mol; CAS 410523-47-6) is the unnatural enantiomer of the common intermediate of the urea cycle and can be found in a number of antibiotics, including bacitracin A. D-Ornithine is a product of the reaction catalyzed by D-arginase and ornithine racemase and is a substrate for D-ornithine 4,5-aminomutase. (*See also L-Ornithine*) **Target(s):** 2-aminohexano-6-lactam racemase (1); L-arginine deaminase, weakly inhibited (2); arginine racemase (3); hydroxylysine kinase (4,5); L-ornithine decarboxylase (6-10); pyrroline 5-carboxylate synthase, *or* glutamate-5-semialdehyde dehydrogenase (11). **1.** Ahmed, Esaki, Tanaka & Soda (1985) *Agric. Biol. Chem.* **49**, 2991. **2.** Eichler (1989) *Biol. Chem. Hoppe Seyler* **370**, 1127. **3.** Yorifuji (1971) *J. Biol. Chem.* **246**, 5093. **4.** Hiles & Henderson (1972) *J. Biol. Chem.* **247**, 646. **5.** Chang (1977) *Enzyme* **22**, 230. **6.** Eichler (1989) *J. Protozool.* **36**, 577. **7.** Kitani & Fujisawa (1988) *J. Biochem.* **103**, 547. **8.** Qu, Ignatenko, Yamauchi, *et al.* (2003) *Biochem. J.* **375**, 465. **9.** Ono, Inoue, Suzuki & Takeda (1972) *Biochim. Biophys. Acta* **284**, 285. **10.** Jackson, Goldsmith & Phillips (2003) *J. Biol. Chem.* **278**, 22037. **11.** Henslee, Wakabayashi, Small & Jones (1983) *Arch. Biochem. Biophys.* **226**, 693.

L-Ornithine

This dibasic amino acid ($FW_{dihydrochloride}$ = 205.08 g/mol; CAS 70-26-8; *Symbol:* Orn; pK_a values of 1.94, 8.65, and 10.76), also known as 2,5-diaminovaleric acid and 5 aminonorvaline, is a key intermediate in the urea cycle (*and* arginine biosynthesis). It is also a precursor for the tropane alkaloids and is found in a number antibiotics and naturally occurring peptides. The monohydrochloride is readily soluble in water and practically insoluble in methanol, ethanol, and diethyl ether (*See also D-Ornithine*). L-Ornithine is a substrate of reactions catalyzed by N^δ-(carboxyethyl)ornithine synthase, ornithine carbamoyl-transferase, ornithine *N*-benzoyltransferase, ornithine:oxo-acid aminotransferase, ornithine(lysine) aminotransferase, ornithine decarboxylase, ornithine cyclodeaminase, and ornithine racemase. It is a product in reactions catalyzed by glycine amidinotransferase, *scyllo*-inosamine-4-phosphate amidinotransferase, glutamate *N*-acetyltransferase, acetylornithine deacetylase, citrullinase, and arginase. **Target(s):** acetylornithine aminotransferase, also alternative substrate (47); agmatinase (2); aminoacyl-lysine dipeptidase, *or* Xaa-Arg dipeptidase (3); 2-aminohexano-6-lactam racemase (4,5); aminolevulinate aminotransferase, weakly inhibited (46); aminopeptidase B, weakly inhibited (40); 5-aminovalerate transaminase (45); arginase (6); D-arginase (33); arginine decarboxylase (7-12); arginine deaminase (13,34-37); arginine racemase (14,15); arginyl-tRNA synthetase (16); carboxypeptidase B (18); diaminopimelate decarboxylase (19); γ-glutamyl kinase, *or* glutamate 5-kinase, weakly inhibited (41); glycine amidinotransferase (49); histidyl-tRNA synthetase, weakly inhibited (20); homoserine kinase (44); hydroxylysine kinase (42,43); kynureninase (21-23); D-lysine 5,6-aminomutase (24,25); lysine 2-monooxygenase (50); lysyl endopeptidase, *or Achromobacter* proteinase I (26); lysyl-tRNA synthetase (27); methionine *S*-adenosyltransferase (48); D-ornithine 4,5-aminomutase (28); ornithine carbamoyltransferase, inhibited by the zwitterionic neutral species (the monocationic form is the true substrate) (29); peptidyl-Lys metalloendopeptidase (17,38,39); Δ¹-pyrroline 5-carboxylate synthase, *or* glutamate-5-semialdehyde dehydrogenase (30-32); Xaa-Arg dipeptidase (3). **1.** Rivard (1953) *Biochem. Prep.* **3**, 97. **2.** Satishchandran & Boyle (1986) *J. Bacteriol.* **165**, 843. **3.** Kumon, Matsuoka, Kakimoto, Nakajima & Sano (1970) *Biochim. Biophys. Acta* **200**, 466. **4.** Ahmed, Esaki, Tanaka

& Soda (1983) *Agric. Biol. Chem.* **47**, 1887. **5.** Ahmed, Esaki, Tanaka & Soda (1985) *Agric. Biol. Chem.* **49**, 2991. **6.** Hunter & Downs (1945) *J. Biol. Chem.* **157**, 427. **7.** Boeker & Snell (1971) *Meth. Enzymol.* **17B**, 657. **8.** Ramakrishna & Adiga (1975) *Eur. J. Biochem.* **59**, 377. **9.** Blethen, Boeker & Snell (1968) *J. Biol. Chem.* **243**, 1671. **10.** Smith (1979) *Phytochemistry* **18**, 1447. **11.** Rosenfeld & Roberts (1976) *J. Bacteriol.* **125**, 601. **12.** Winer, Vinkler & Apelbaum (1984) *Plant Physiol.* **76**, 233. **13.** Eichler (1989) *Biol. Chem. Hoppe Seyler* **370**, 1127. **14.** Soda, Yorifuji & Ogata (1970) *Meth. Enzymol.* **17A**, 341. **15.** Yorifuji (1971) *J. Biol. Chem.* **246**, 5093. **16.** Williams, Yem, McGinnis & Williams (1973) *J. Bacteriol.* **115**, 228. **17.** Takio (1998) in *Handb. Proteolytic Enzymes*, p. 1538, Academic Press, San Diego. **18.** Wolff, Schirmer & Folk (1962) *J. Biol. Chem.* **237**, 3094. **19.** Rosner (1975) *J. Bacteriol.* **121**, 20. **20.** Chen & Somberg (1980) *Biochim. Biophys. Acta* **613**, 514. **21.** Moriguchi & Soda (1973) *Biochemistry* **12**, 2974. **22.** Tanizawa & Soda (1979) *J. Biochem.* **86**, 499. **23.** Soda & Tanizawa (1979) *Adv. Enzymol. Relat. Areas Mol. Biol.* **49**, 1. **24.** Morley & Stadtman (1970) *Biochemistry* **9**, 4890. **25.** Stadtman (1972) *The Enzymes*, 3rd ed., **6**, 539. **26.** Masaki, Fujihashi, Nakamura & Soejima (1981) *Biochim. Biophys. Acta* **660**, 51. **27.** Levengood, Ataide, Roy & Ibba (2004) *J. Biol. Chem.* **279**, 17707. **28.** Somack & Costilow (1973) *Biochemistry* **12**, 2597. **29.** Nakamura & Jones (1970) *Meth. Enzymol.* **17A**, 286. **30.** Kramer, Henslee, Wakabashi & Jones (1985) *Meth. Enzymol.* **113**, 113. **31.** Lodato, Smith, Valle, Phang & Aoki (1981) *Metabolism* **30**, 908. **32.** Henslee, Wakabayashi, Small & Jones (1983) *Arch. Biochem. Biophys.* **226**, 693. **33.** Nadai (1958) *J. Biochem.* **45**, 1011. **34.** Park, Hirotani, Nakano & Kitaoka (1984) *Agric. Biol. Chem.* **48**, 483. **35.** Knodler, Sekyere, Stewart, Schofield & Edwards (1998) *J. Biol. Chem.* **273**, 4470. **36.** Manca de Nadra, Pesce de Ruiz Holgado & Oliver (1984) *J. Appl. Biochem.* **6**, 184. **37.** Monstadt & Holldorf (1990) *Biochem. J.* **273**, 739. **38.** Lewis, Basford & Walton (1978) *Biochim. Biophys. Acta* **522**, 551. **39.** Nonaka, Ishikawa, Tsumuraya, *et al.* (1995) *J. Biochem.* **118**, 1014. **40.** Kawata, Takayama, Ninomiya & Makisumi (1980) *J. Biochem.* **88**, 1601. **41.** Krishna & Leisinger (1979) *Biochem. J.* **181**, 215. **42.** Hiles & Henderson (1972) *J. Biol. Chem.* **247**, 646. **43.** Chang (1977) *Enzyme* **22**, 230. **44.** Thoen, Rognes & Aarnes (1978) *Plant Sci. Lett.* **13**, 103. **45.** Der Garabedian (1986) *Biochemistry* **25**, 5507. **46.** Neuberger & Turner (1963) *Biochim. Biophys. Acta* **67**, 342. **47.** Billheimer, Carnevale, Leisinger, Eckardt & Jones (1976) *J. Bacteriol.* **127**, 1315. **48.** Schröder, Eichel, Breinig & Schröder (1997) *Plant Mol. Biol.* **33**, 211. **49.** Walker (1957) *J. Biol. Chem.* **224**, 57. **50.** Nakazawa, Hori & Hayaishi (1972) *J. Biol. Chem.* **247**, 3439.

Ornithodorin

This toxic protein (MW = 12.6 kDa) from the blood-sucking tick *Ornithodoros moubata*, is a potent inhibitor (K_i = 1-3 pM) of thrombin, binding near the catalytic site (1,2). The N-terminal portion and the C-terminal helix of each domain are structurally very similar to BPTI, whereas the regions corresponding to the binding loop of BPTI adopt different conformations (3). Neither of the two 'reactive site loops' of ornithodorin contacts the protease in the ornithodorin-thrombin complex. Instead, the N-terminal residues of ornithodorin bind to the active site of thrombin, reminiscent of the thrombin-hirudin interaction. The C-terminal domain binds at the fibrinogen recognition exosite (3). **1.** Van de Locht, Stubbs, Bode, *et al.* (1996) *EMBO J.* **15**, 6011. **2.** Klinger & Friedrich (1997) *Biophys. J.* **73**, 2195. **3.** van de Locht, Stubbs, Bode, *et al.* (1996) *EMBO J.* **15**, 6011.

Orotate (Orotic Acid)

This pyrimidine biosynthesis intermediate ($FW_{free-acid}$ = 156.10 g/mol; Solubility = 0.2 g/100 mL H_2O at 18°C; λ_{max} at 279 nm at pH 7 (ε = 7700 $M^{-1}cm^{-1}$); pK_a values of 2.4, 9.5, and >13.0), also known as uracil-4-carboxylic acid and uracil-6-carboxylic acid, is the product of dihydroorotate oxidase and dihydroorotate dehydrogenase reactions. It is also a substrate in the reactions catalyzed by the orotate reductases and orotate phosphoribosyltransferase. **Target(s):** acetyl-CoA synthetase (1); acylphosphatase (2); dihydroorotase (3-9); ornithine carbamoyltransferase (with D-ornithine concentrations above 5 mM) (10); orotidine-5'-phosphate decarboxylase (11); urate-ribonucleotide phosphorylase, weakly inhibited

(12). **1**. Bernstein, Richardson & Amundson (1977) *J. Dairy Sci.* **60**, 1846. **2**. Ramponi (1975) *Meth. Enzymol.* **42**, 409. **3**. Smith & Sullivan (1960) *Biochim. Biophys. Acta* **39**, 554. **4**. Webb (1966) *Enzyme and Metabolic Inhibitors*, vol. **2**, p. 470, Academic Press, New York. **5**. Ogawa & Shimizu (1995) *Arch. Microbiol.* **164**, 353. **6**. Chen & Jones (1979) *J. Biol. Chem.* **254**, 4908. **7**. Jones (1980) *Ann. Rev. Biochem.* **49**, 253. **8**. Christopherson & Jones (1980) *J. Biol. Chem.* **255**, 3358. **9**. Krungkrai, Krungkrai & Phakanont (1992) *Biochem. Pharmacol.* **43**, 1295. **10**. Knight & Jones (1977) *J. Biol. Chem.* **252**, 5928. **11**. Miller & Wolfenden (2002) *Ann. Rev. Biochem.* **71**, 847. **12**. Laster & Blair (1963) *J. Biol. Chem.* **238**, 3348.

Orotidine

This water-soluble ribonucleoside (FW$_{free-acid}$ = 288.21 g/mol; CAS 314-50-1; λ_{max} of 267 nm in 0.1 N HCl (ε = 9570 cm^{-1}M^{-1}; another report lists 9800) and 265 nm in 0.1 N methanolic NaOH (ε = 8960 cm^{-1}M^{-1}; another report lists 7800)), first observed in *Neurospora crassa*, is also foound in the urine of cancer patients treated with 6-azauridine, Orotidine is most likely formed from OMP, its 5'-monophosphate ester and the direct precursor of UMP. Orotidine is readily hydrolyzed in dilute mineral acids. **Target(s):** orotate phosphoribosyltransferase (1); orotidine-5'-phosphate decarboxylase (2,3). **1**. Javaid, el Kouni & Iltzsch (1999) *Biochem. Pharmacol.* **58**, 1457. **2**. Miller, Butterfoss, Short & Wolfenden (2001) *Biochemistry* **40**, 6227. **3**. Sievers & Wolfenden (2005) *Bioorg. Chem.* **33**, 45.

Orotidine 5'-Monophosphate

This pyrimidine nucleotide (FW$_{free-acid}$ = 368.19 g/mol; CAS 2149-82-8; *Abbreviation*: OMP; λ_{max} at 266 nm at pH 7), also called orotidylic acid, is an intermediate in the biosynthesis of the urridine monophosphate. OMP is a product of the reaction catalyzed by orotate phosphoribosyltransferase and is a substrate of orotidine-5'-phosphate decarboxylase. **Target(s):** adenylosuccinate synthetase (1); amidophospho-ribosyltransferase, weakly inhibited (2); cytosine deaminase (3,4). **1**. Van der Weyden & Kelly (1974) *J. Biol. Chem.* **249**, 7282. **2**. Holmes, McDonald, McCord, Wyngaarden & Kelley (1973) *J. Biol. Chem.* **248**, 144. **3**. West, Shanley & O'Donovan (1982) *Biochim. Biophys. Acta* **719**, 251. **4**. West (1988) *Experientia* **41**, 1563.

Orphenadrine

This muscle relaxant and antihistaminic agent (FW$_{free-base}$ = 269.39 g/mol; CAS 83-98-7) is both a muscarinic and histamine H$_1$ receptor antagonist that also inhibits some cytochrome P450 systems. Classified as an anticholinergic drug (with 58% as potent as atropine) and an ethanolamine-class antihistamine, it also inhibits the noradrenergic transporter and NMDA receptor ion channel ($K_i \approx 6$ μM). Commercially available as either the hydrochloride or citrate salt, orphenadrines used to relieve pain of muscle injuries, such as strains and sprains. **Target(s):** aldehyde oxidase (1); CYP1A2 (2); CYP2A6 (2); CYP3A4 (2); CYP2B (3,4); CYP2B6 (2,5,6); CYP2C9 (2); CYP2C19 (2); CYP2D6 (2); histamine H$_1$ receptor (7); *N*-methyl-D-aspartate (NMDA) receptor (8); muscarinic receptor (9,10); noradrenaline reuptake (11). **1**. Robertson & Gamage (1994) *Biochem. Pharmacol.* **47**, 584. **2**. Guo, Raeissi, White & Stevens (1997) *Drug Metab. Dispos.* **25**, 390. **3**. Skaanild & Friis (2002) *Pharmacol. Toxicol.* **91**, 198. **4**. Murray, Fiala-Beer & Sutton (2003) *Brit. J. Pharmacol.* **139**, 787. **5**. Hijazi & Boulieu (2002) *Drug Metab. Dispos.* **30**, 853. **6**. Ekins, VandenBranden, Ring & Wrighton (1997) *Pharmacogenetics* **7**, 165. **7**. Gibson, Roques & Young (1994) *Brit. J. Pharmacol.* **111**, 1262. **8**. Kornhuber, Parsons, Hartmann, *et al.* (1995) *J. Neural. Transm. Gen. Sect.* **102**, 237. **9**. Syvalahti, Kunelius & Lauren (1988) *Pharmacol. Toxicol.* **62**, 90. **10**. Fernando, Hoskins & Ho (1985) *Life Sci.* **37**, 883. **11**. Pubill, Canudas, Pallas, *et al.* (1999) *J. Pharm. Pharmacol.* **51**, 307.

Orthophosphate (or Inorganic Phosphate)

monobasic　dibasic　tribasic

These conjugate bases (Symbol: P$_i$) of phosphoric acid H$_3$PO$_4$ (pK_a values of 2.15, 7.20, and 12.33 at 25°C, with corresponding dpK_a/dT values are 0.0044, –0.0028, and –0.026) are commonly used as buffers in biochemical experiments. Monobasic orthophosphate is a major intracellular buffer, second only to bicarbonate for pH control in humans. Monobasic phosphate (NaH$_2$PO$_4$; FW = 119.98 g/mol; CAS 7558-79-4; KH$_2$PO$_4$; FW = 136.09 g/mol; CAS 7778-77-0) is typically found as the mono- or dehydrate, and both are freely soluble in water. **Use as Buffers:** Dibasic phosphate (Na$_2$HPO$_4$; FW = 141.96 g/mol; CAS 7782-85-6; K$_2$HPO$_4$; FW = 174.20 g/mol; CAS 7758-11-4) is found as the anhydrous powder as well as the di-, hepta-, and dodecahydrate; all are soluble in water. The anhydrous material will readily absorb atmospheric water. Tribasic phosphate (Na$_3$PO$_4$; FW = 163.94 g/mol; CAS 7601-54-9; K$_3$PO$_4$; FW = 212.27 g/mol; CAS 7778-53-2) is typically supplied as the dodecahydrate, which is soluble in about 3.5 parts water. NaH$_2$PO$_4$ and Na$_2$HPO$_4$ salts are solids, and buffer solutions are easily and reproducibly prepared. (Even so, care must be exercised because these salts are hygroscopic). Orthophosphate is also transparent to ultraviolet light. On the other hand, orthophosphate ions bind metal ions (*e.g.*, calcium phosphate has a low solubility in water and precipitates out of solution, depending on the conditions). Phosphate-containing solutions also tend to support fungal and algal growth. Most importantly, orthophosphate is a substrate, product, and/or effector of a wide number of enzymes, and, when preset at elevated concentrations in buffer, phosphate can mask a kinetic or regulatory event that would be evident had a different buffer had been used. **Phosphate-Buffered Saline (PBS):** A 1.0-liter stock of 10x stock solution may be prepared by combining 80.00 g NaCl, 2.00 g KCl, 14.40 g Na$_2$HPO$_4$ · 2H$_2$O, 2.40 g KH$_2$PO$_4$ in 800 mL distilled water. Bring the final solution to 1.00 L. The pH of the 10X stock is will be approximately 6.8, but changes to 7.4, when diluted to 10x to obtain PBS. It is wise to measure the pH with a pH meter, adjusting it to 7.4 by the addition of either 1 M hydrochloric acid or 1 M sodium hydroxide (The latter should be freshly prepared to prevent introduction of atmospheric carbon dioxide.) **Targets:** Orthophosphate is a substrate/product in over 2000 enzyme-catalyzed reactions, and its inhibitory effects, while too numerous to list here, are generally mild, with K_i values typically in the 0.1–2 mM range.

Orthovanadate Ion

This vanadium(V) anion VO$_4^{3-}$ (FW$_{trisodium-orthovanadate}$ = 183.91 g/mol) inhibits many phosphotransferases and phosphohydrolases, often adopting a structure that mimics the geometry of intermediates in associative (in-line) S$_N$2 nucleophilic displacement reactions. Because tungstate and molybdate ions are isoelectronic with orthovanadate and exert similar effects, they are likely to interact with enzyme targets in a similar manner, albeit with different affinities. (***See also** Decavanadate; Pervanadate; Metavanadate*) **Solution Behavior:** At physiological pH values, orthovanadate tends to form larger aggregations (*e.g.*, V$_3$O$_9^{3-}$ and V$_4$O$_{10}^{4-}$). At lower pH, the predominant species is decavanadate (V$_{10}$O$_{28}$H$_5^-$). After making a vanadate solution, the latter should stand for several days to ensure that equilibrium has been reached among its various forms. Solutions of orthovanadate may be prepared as follows: dissolve Vanadium(V) oxide in 2.1 molar equivalents of 1 N NaOH, stir 2-3 days until the yellow color becomes faint, and dilute with water to a final concentration of 100 mM. Vanadate-inhibited enzymes often reactivate on addition of EDTA, which forms high-affinity Vanadate:EDTA complexes. When divalent cations must be chelated, one may use EGTA. Vanadate also interacts with many other organic compounds and buffer salts; HEPES is a useful exception. Some commercial preparations of nucleoside triphosphates and carbamoyl phosphate also contain trace amounts of vanadate, and one should ascertain that reagents are vanadate-free before using them in kinetic and mechanistic studies. Orthovanadate also complexes with peroxide anion to produce pervanadate, a reagent that can oxidatively inactivate enzymes with active-site thiols. **Does Vanadate Form True Transition State-like Complexes?** Vanadate often binds to phosphotransferases to form a trigonal-bipyramidal structure at the active site, with the enzyme-vanadate K_d values that are much lower than those for phosphate. Although enzyme-bound vanadate moieties are often touted as transition state analogues, one must learn whether the bond orders of the VO bonds in the complex approach those of the PO bonds in an authentic transition state (1). In steady-state kinetic measurements and difference Raman experiments on vanadate binding to *Yersinia* protein tyrosine phosphatase (PTPase) and its T410A, D356N, W354A, R409K, and D356A mutants, there is clear no correlation between K_i and k_{cat} or k_{cat}/K_m. The bond order change of the nonbridging V–O bonds in the vanadate complexes does not correlate with the kinetic parameters in a number of PTPase variants, as required by the transition state binding assumption. Moreover, the ionization state of bound vanadate is not invariant across the PTPase variants studied, and the average bond order of the nonbridging VO bonds decreased by 0.06−0.07 valence unit in wild type and all the mutant PTPases, either in dianionic or in monoanionic form. In this respect, the complex would resemble an associative transition state, contrary to the previously determined dissociative structure of the transition state, suggesting that vanadate cannot be a true transition-state analogue for the PTPase reactions (1). Such findings also point to the overly simplistic view that vanadate combines with target enzymes to form a true transition-state analogue. **Other Effects:** Vanadate also exerts insulin-like effects on glucose metabolism by activating glucose transport and oxidation. It also raises fructose 2,6-bisphosphate (*or* Fru(2,6)P$_2$) levels, activating glycolysis in hepatocytes from normal or diabetic rats. In other reports, oral administration of vanadate to diabetic rats normalizes hepatic glucose metabolism and glucose blood levels and restores insulin-responsiveness of target tissues. **Target(s):** *N*-acetylgalactosamine-4-sulfatase, *or* arylsulfatase B (141); acid phosphatase (2-3,216-224); adenosine-tetraphosphatase (4); Ag$^+$-exporting ATPase (5); alkaline phosphatase (2,3,6,225-228); 1-alkyl-2-acetylglycerophosphatase (148); alkylglycerol-phosphate 2-*O*-acetyltransferase (239); apyrase (other apyrases report no inhibition by orthovanadate or different degrees of inhibition of ADP *vs.* ATP hydrolysis) (7-10); arylsulfatase (2,142); ATPase, slow-binding inhibition (3,11); cadmium-transporting ATPase, glutathione-conjugate-transporting ATPase (12-14); Ca^{2+} transporting ATPase (3,15-23); CD45 protein phosphatase (24); centromere binding protein E (25); channel-conductance-controlling ATPase, *or* cystic-fibrosis membrane-conductance-regulating protein (26); chloroplast protein-transporting ATPase (27); Cl$^-$-transporting ATPase (28-33); copper-exporting ATPase (34-37); DNA topoisomerase II (38,39); dolichyl-phosphatase (149); dual-specificity protein phosphatase (208,210); dynein ATPase (3,40-54); endopeptidase La, inhibition attributed to decavanadate (55,134-140); exopolyphosphatase (4,56); Fe^{3+}-transporting ATPase (57); fructose-1,6 bisphosphate aldolase (58); fructose-2,6-bisphosphate 2-phosphatase (185,186); FtsH protease (59); glucose-1-phosphatase (211); glucose-6-phosphatase (212,213); glyceraldehyde-3-phosphate dehydrogenase (3); glycerol-1-phosphatase (202); glycogen phosphorylase (238); glycosylphosphatidyl-inositol diacylglycerol-lyase (60); H$^+$-exporting ATPase (61-64); H$^+$/K$^+$ exchanging ATPase (3,65-69,188); inositol-trisphosphate 3-kinase (233); kinesin (25,70-72); K$^+$ transporting ATPase (73-81); β-lactamase, in the presence of hydroxamates (82); lipid-phosphate phosphatase (144); maltose-transporting ATPase (83); minus-end-directed kinesin ATPase (25,48,70); mitochondrial ATP-dependent protease (84); monoterpenyl-diphosphatase (85); MutS ABC ATPase (86); myosin ATPase (3,87-90); NADH:cytochrome *c* reductase, *or* NADH dehydrogenase (91); Na$^+$/K$^+$-exchanging ATPase (3,92-97,187); 4-nitrophenylphosphatase activity of H$^+$/K$^+$-exchanging ATPase (188); 4-nitrophenylphosphatase activity of Na$^+$/K$^+$-exchanging ATPase (187); nucleotidases (189); 5'-nucleotidase (214); P-glycoprotein98-103; phosphatidylinositol diacylglycerol lyase, *or* phosphatidylinositol-specific phospholipase C (104); phosphatidylinositol-3 phosphatase (147); phosphoenolpyruvate mutase (105); 6-phosphofructo-2-kinase (106); phosphoglucomutase (107); phospholipase A$_1$, phosphatidate-specific (229); phospholipase C (108); phospholipid-translocating ATPase, *or* flippase (103,109-113); phosphomannomutase (114); phosphoprotein phosphatase (24,115-118,203-210); phosphoserine phosphatase (215); 2-phosphosulfolactate phosphatase (146); 3-phytase (119,191,197,198); 3- or 4-phytase (194); 4-phytase (190- 193,195,196); 5-phytase (145); plus-end-directed kinesin ATPase (71,72); polar-amino-acid-transporting ATPase, *or* histidine permease (120,121); polynucleotide adenylyltransferase, *or* poly(A) polymerase (231); protein-synthesizing GTPase (elongation factor) (122); protein-tyrosine kinase (230); protein-tyrosine-phosphatase (117,118,123,151-184); ribonuclease (3); soluble epoxide hydrolase, weakly inhibited (143); sorbitol-6-phosphatase (150); sucrose-phosphatase (199-201); sucrose phosphate synthase (237); sulfate adenylyltransferase (ADP) (232); thiamin-triphosphatase (124); thymidine-triphosphatase (125); α,α-trehalose phosphorylase (configuration-retaining) (234-236); xenobiotic-transporting ATPase, multidrug-resistance protein (98-103,126-131); zinc-exporting ATPase (132,133).

1. Deng, Callender, Huang & Zhang (2002) *Biochemistry* **41**, 5865. **2.** Stankiewicz & Gresser (1988) *Biochemistry* **27**, 206. **3.** Macara (1980) *Trends Biochem. Sci.* **5**, 92. **4.** Guranowski, Starzynska, Barnes, Robinson & Liu (1998) *Biochim. Biophys. Acta* **1380**, 232. **5.** Bury, Grosell, Grover & Wood (1999) *Toxicol. Appl. Pharmacol.* **159**, 1. **6.** Lopez, Stevens & Lindquist (1976) *Arch. Biochem. Biophys.* **175**, 31. **7.** Kettlun, Alvarez, Quintar, *et al.* (1994) *Int. J. Biochem.* **26**, 437. **8.** Knowles, Nagy, Strobel & Wu-Weis (2002) *Eur. J. Biochem.* **269**, 2373. **9.** Fietto, DeMarco, Nascimento, *et al.* (2004) *Biochem. Biophys. Res. Commun.* **316**, 454. **10.** Oses, Cardoso, Germano, *et al.* (2004) *Life Sci.* **74**, 3275. **11.** Morrison (1982) *Trends Biochem. Sci.* **7**, 102. **12.** Rebbeor, Connolly, Dumont & Ballatori (1998) *J. Biol. Chem.* **273**, 33449. **13.** Li, Szczypka, Lu, Thiele & Rea (1996) *J. Biol. Chem.* **271**, 6509. **14.** Petrovic, Pascolo, Gallo, *et al.* (2000) *Yeast* **16**, 561. **15.** Csermely, Varga & Martonosi (1985) *Eur. J. Biochem.* **150**, 455. **16.** Carafoli & Zurini (1982) *Biochim. Biophys. Acta* **683**, 279. **17.** Wright & van Houten (1990) *Biochim. Biophys. Acta* **1029**, 241. **18.** Sorin, Rosas & Rao (1997) *J. Biol. Chem.* **272**, 9895. **19.** Lenoir, Picard, Moller, *et al.* (2004) *J. Biol. Chem.* **279**, 32125. **20.** Almeida, Benchimol, De Souza & Okorokov (2003) *Biochim. Biophys. Acta* **1615**, 60. **21.** Carafoli (1992) *J. Biol. Chem.* **267**, 2115. **22.** Caroni & Carafoli (1981) *J. Biol. Chem.* **256**, 3263. **23.** Vicenzi & Ashleman (1980) in *Calcium-binding Proteins, Structure and Function*, 173. **24.** Tonks, Diltz & Fischer (1991) *Meth. Enzymol.* **201**, 442. **25.** Thrower, Jordan, Schaar, Yen & Wilson (1995) *EMBO J.* **14**, 918. **26.** Annereau, Ko & Pedersen (2003) *Biochem. J.* **371**, 451. **27.** Pain & Blobel (1987) *Proc. Natl. Acad. Sci. U.S.A.* **84**, 3288. **28.** Gerencser (1996) *CRC Crit. Rev. Biochem.* **31**, 303. **29.** Moritani, Ohhashi, Satoh, Oesterhelt & Ikeda (1994) *Biosci. Biotechnol. Biochem.* **58**, 2087. **30.** Zeng, Hara & Inagaki (1994) *Brain Res.* **641**, 167. **31.** Gerencser & Purushotham (1996) *J. Bioenerg. Biomembr.* **28**, 459. **32.** Shiroya, Fukunaga, Akashi, *et al.* (1989) *J. Biol. Chem.* **264**, 17416. **33.** Gerencser & Zhang (2001) *Can. J. Physiol. Pharmacol.* **79**, 367. **34.** Voskoboinik, Mar, Strausak & Camakaris (2001) *J. Biol. Chem.* **276**, 28620. **35.** Fan & Rosen (2002) *J. Biol. Chem.* **277**, 46987. **36.** Mandal, Cheung & Arguello (2002) *J. Biol. Chem.* **277**, 7201. **37.** Takeda, Ushimaru, Fukushima & M. Kawamura (1999) *J. Membr. Biol.* **170**, 13. **38.** Vaughn, Huang, Wessel, *et al.* (2005) *J. Biol. Chem.* **280**, 11920. **39.** Sorensen, Grauslund, Jensen, Sehested & Jensen (2005) *Biochem. Biophys. Res. Commun.* **334**, 853. **40.** Bell, Fraser, Sale, Tang & Gibbons (1982) *Meth. Enzymol.* **85**, 450. **41.** Hisanaga & Sakai (1986) *Meth. Enzymol.* **134**, 337. **42.** Paschal, Shpetner & Vallee (1991) *Meth. Enzymol.* **196**, 181. **43.** Moss, Gatti, King & Witman (1991) *Meth. Enzymol.* **196**, 201. **44.** Gibbons & Mocz (1991) *Meth. Enzymol.* **196**, 438. **45.** Shimizu (1981) *Biochemistry* **20**, 4347. **46.** Anderson & Purich (1982) *J. Biol. Chem.* **257**, 6656. **47.** Terry & Purich (1982) *Adv. Enzymol.* **53**, 113. **48.** Shimizu, Toyoshima, Edamatsu & Vale (1995) *Biochemistry* **34**, 1575. **49.** Romac, Zanic-Grubisic, Culic, Cvitkovic & Flogel (1994) *Hum. Reprod.* **9**, 1474. **50.** Belles-Isles, Chapeau, White & Gagnon (1986) *Biochem. J.* **240**, 863. **51.** Kobayashi, Martensen, Nath & Flavin (1978) *Biochem. Biophys. Res. Commun.* **81**, 1313. **52.** Shimizu & Johnson (1983) *J. Biol. Chem.* **258**, 13833. **53.** Mocz & Gibbons (1990) *J. Biol. Chem.* **265**, 2917. **54.** Gibbons, Tang & Gibbons (1985) *J. Cell Biol.* **101**, 1281. **55.** Goldberg, Sreedhara Swamy, Chung & Larimore (1981) *Meth. Enzymol.* **80**, 680. **56.** Lorenz, Müller, Kualev & Schröder (1994) *J. Biol. Chem.* **269**, 22198. **57.** Anderson, Adhikari, Nowalk, Chen & Mietzner (2004) *J. Bacteriol.* **186**, 6220. **58.** Crans, Sudhakar & Zamborelli (1992) *Biochemistry* **31**, 6812. **59.** Akiyama & Ito (1998) in *Handb. Proteolytic Enzymes*, p. 1502, Academic Press, San Diego. **60.** Bütikofer & Brodbeck (1993) *J. Biol. Chem.* **268**, 17794. **61.** Dufour, Amory & Goffeau (1988) *Meth. Enzymol.* **157**, 513. **62.** Serrano (1988) *Meth. Enzymol.* **157**, 533. **63.** Huang & Berry (1990) *Biochim.*

Biophys. Acta 1039, 241. 64. Goffeau & Slayman (1981) Biochim. Biophys. Acta 639, 197. 65. Siebers, Wieczorek & Altendorf (1988) Meth. Enzymol. 157, 668. 66. De Pont & Bonting (1981) New Compr. Biochem. 2, 209. 67. Chatterjee & Das (1995) Mol. Cell. Biochem. 148, 95. 68. Planelles, Anagnostopoulos, Cheval & Doucet (1991) Amer. J. Physiol. 260, F806. 69. Faller, Rabon & Sachs (1983) Biochemistry 22, 4676. 70. Walker, Salmon & Endow (1990) Nature 347, 780. 71. Vale, Reese & Sheetz (1985) Cell 42, 39. 72. Mitchison (1986) J. Cell Sci. Suppl. 5, 121. 73. Epstein, Wieczorek, Siebers & Altendorf (1984) Biochem. Soc. Trans. 12, 235. 74. Hafer, Siebers & Bakker (1989) Mol. Microbiol. 3, 487. 75. Siebers & Altendorf (1989) J. Biol. Chem. 264, 5831. 76. Kollmann & Altendorf (1993) Biochim. Biophys. Acta 1143, 62. 77. Abee, Siebers, Altendorf & Konings (1992) J. Bacteriol. 174, 6911. 78. Siebers & Altendorf (1988) Eur. J. Biochem. 178, 131. 79. Fendler, Dröse, Epstein, Bamberg & Altendorf (1999) Biochemistry 38, 1850. 80. Sebestian, Petrmichlova, Sebestianova, Naprstek & Svobodova (2001) Can. J. Microbiol. 47, 1116. 81. Fendler, Dröse, Altendorf & Bamberg (1996) Biochemistry 35, 8009. 82. Bell & Pratt (2002) Biochemistry 41, 4329. 83. Landmesser, Stein, Bluschke, et al. (2002) Biochim. Biophys. Acta 1565, 64. 84. Kuzela & Goldberg (1994) Meth. Enzymol. 244, 376. 85. Croteau & Karp (1979) Arch. Biochem. Biophys. 198, 523. 86. Pezza, Villarreal, Montich & Argarana (2002) Nucl. Acids Res. 30, 4700. 87. Goodno (1982) Meth. Enzymol. 85, 116. 88. Cremo, Grammer & Yount (1991) Meth. Enzymol. 196, 442. 89. Tiago, Aureliano & Gutierrez-Merino (2004) Biochemistry 43, 5551. 90. Park, Ajtai & Burghardt (1999) Biochim. Biophys. Acta 1430, 127. 91. Vernon, Mahler & Sarkar (1952) J. Biol. Chem. 199, 599. 92. Skou (1988) Meth. Enzymol. 156, 1. 93. Beaugé (1988) Meth. Enzymol. 156, 251. 94. Furriel, McNamara & Leone (2001) Comp. Biochem. Physiol. A 130, 665. 95. Almansa, Sanchez, Cozzi, et al. (2001) J. Comp. Physiol. B 171, 557. 96. Segall, Daly & Blostein (2001) J. Biol. Chem. 276, 31535. 97. de Lima Santos & Ciancaglini (2003) Comp. Biochem. Physiol. B 135, 539. 98. Sharom, Liu, Romsicki & Lu (1999) Biochim. Biophys. Acta 1461, 327. 99. Lu, Liu & Sharom (2001) Eur. J. Biochem. 268, 1687. 100. Wang, Casciano, Clement & Johnson (2001) Drug Metab. Dispos. 29, 1080. 101. Omote & Al-Shawi (2002) J. Biol. Chem. 277, 45688. 102. Qu & Sharom (2002) Biochemistry 41, 4744. 103. Romsicki & Sharom (2001) Biochemistry 40, 6937. 104. Vizitiu, Kriste, Campbell & Thatcher (1996) J. Mol. Recognit. 9, 197. 105. Seidel & Knowles (1994) Biochemistry 33, 5641. 106. Kountz, McCain, el-Maghrabi & Pilkis (1986) Arch. Biochem. Biophys. 251, 104. 107. Percival, Doherty & Gresser (1990) Biochemistry 29, 2764. 108. Tan & Roberts (1996) Biochim. Biophys. Acta 1298, 58. 109. Auland, Morris & Roufogalis (1994) Arch. Biochem. Biophys. 312, 272. 110. Daleke & Lyles (2000) Biochim. Biophys. Acta 1486, 108. 111. Devaux, Zachowski, Morrot, et al. (1990) Biotechnol. Appl. Biochem. 12, 517. 112. Ding, Wu, Crider, et al. (2000) J. Biol. Chem. 275, 23378. 113. Doerrler & Raetz (2002) J. Biol. Chem. 277, 36697. 114. Pirard, Achouri, Collet, et al. (1999) Biochem. J. 339, 201. 115. Tonks, Diltz & Fischer (1991) Meth. Enzymol. 201, 427. 116. Ingebritsen (1991) Meth. Enzymol. 201, 451. 117. Gordon (1991) Meth. Enzymol. 201, 477. 118. Ballou & Fischer (1986) The Enzymes, 3rd ed., 17, 311. 119. Gibson & Ullah (1988) Arch. Biochem. Biophys. 260, 503. 120. Nikaido, Liu & Ames (1997) J. Biol. Chem. 272, 27745. 121. Liu, Liu & Ames (1997) J. Biol. Chem. 272, 21883. 122. Uritani & Miyazaki (1988) J. Biochem. 103, 522. 123. Huyer, Liu, Kelly, et al. (1997) J. Biol. Chem. 272, 843. 124. Matsuda, Tonomura, Baba & Iwata (1991) Int. J. Biochem. 23, 1111. 125. Schultes, Fischbach & Dahlmann (1992) Biol. Chem. Hoppe-Seyler 373, 237. 126. Bodo, Bakos, Szeri, Varadi & Sarkadi (2003) Toxicol. Lett. 140-141, 133. 127. Van Veen, Venema, Bolhuis, et al. (1996) Proc. Natl. Acad. Sci. U.S.A. 93, 10668. 128. Kern, Szentpetery, Liliom, et al. (2004) Biochem. J. 380, 549. 129. Orelle, Dalmas, Gros, Di Pietro & Jault (2003) J. Biol. Chem. 278, 47002. 130. Steinfels, Orelle, Dalmas, et al. (2002) Biochim. Biophys. Acta 1565, 1. 131. Steinfels, Orelle, Fantino, et al. (2004) Biochemistry 43, 7491. 132. Okkeri & Haltia (1999) Biochemistry 38, 14109. 133. Hou & Mitra (2003) J. Biol. Chem. 278, 28455. 134. Larimore, Waxman & Goldberg (1982) J. Biol. Chem. 257, 4187. 135. Waxman & Goldberg (1982) Proc. Natl. Acad. Sci. U.S.A. 79, 4883. 136. Kutejova, Durcova, Surovkova & Kuzela (1993) FEBS Lett. 329, 47. 137. Goldberg (1992) Eur. J. Biochem. 203, 9. 138. Goldberg & Waxman (1985) J. Biol. Chem. 260, 12029. 139. Menon & Goldberg (1987) J. Biol. Chem. 262, 14921. 140. Desautels & Goldberg (1982) J. Biol. Chem. 257, 11673. 141. Bond, Clements, Ashby, et al. (1997) Structure 5, 277. 142. Knoess & Glombitza (1993) Phytochemistry 32, 1119. 143. Newman, Morisseau, Harris & Hammock (2003) Proc. Natl. Acad. Sci. U.S.A. 100, 1558. 144. Enayetallah & Grant (2006) Biochem. Biophys. Res. Commun. 341, 254. 145. Jog, Garchow, Mehta & Murthy (2005) Arch. Biochem. Biophys. 440, 133. 146. Graham, Graupner, Xu & White (2001) Eur. J. Biochem. 268, 5176. 147. Lips & Majerus (1989) J. Biol. Chem. 264, 19911. 148. Lee, Malone & Snyder (1986) J. Biol. Chem. 261, 5373. 149. Frank & Waechter (1998) J. Biol. Chem. 273, 11791. 150. Zhou, Cheng & Wayne (2003) Plant Sci. 165, 227. 151. Aguirre-Garcia, Escalona-Montano, Bakalara, et al. (2006) Parasitology 132, 641. 152. Zhou, Bhattacharjee & Mukhopadhyay (2006) Mol. Biochem. Parasitol. 148, 161. 153. Singh (1990) Biochem. Biophys. Res. Commun. 167, 621. 154. Harder, Owen, Wong, et al. (1994) Biochem. J. 298, 395. 155. Li & Strohl (1996) J. Bacteriol. 178, 136. 156. Mijakovic, Musumeci, Tautz, et al. (2005) J. Bacteriol. 187, 3384. 157. Tonks, Diltz & Fischer (1988) J. Biol. Chem. 263, 6731. 158. Waheed, Laidler, Wo & Van Etten (1988) Biochemistry 27, 4265. 159. Kim, Jung & Kang (1996) Exp. Mol. Med. 28, 207. 160. Grangeasse, Doublet, Vincent, et al. (1998) J. Mol. Biol. 278, 339. 161. Tung & Reed (1987) Anal. Biochem. 161, 412. 162. Chernoff & Li (1983) Arch. Biochem. Biophys. 226, 517. 163. Foulkes (1983) Curr. Top. Microbiol. Immunol. 107, 163. 164. Lau, Farley & Baylink (1989) Biochem. J. 257, 23. 165. Boivin & Galand (1986) Biochem. Biophys. Res. Commun. 134, 557. 166. Roome, O'Hare, Pilch & Brautigan (1988) Biochem. J. 256, 493. 167. Cui, Yu, DeAizpurua, Schmidli & Pallen (1996) J. Biol. Chem. 271, 24817. 168. Sorio, Mendrola, Lou, et al. (1995) Cancer Res. 55, 4855. 169. Dechert, Adam, Harder, Clark-Lewis & Jirik (1994) J. Biol. Chem. 269, 5602. 170. Hoppe, Berne, Stock, et al. (1994) Eur. J. Biochem. 223, 1069. 171. Tonks, Diltz & Fischer (1990) J. Biol. Chem. 265, 10674. 172. Zhao, Laroque, Ho, Fischer & Shen (1994) J. Biol. Chem. 269, 8780. 173. Aoyama, Matsuda & Aoki (1999) Biochem. Biophys. Res. Commun. 266, 523. 174. Ohsugi, Kuramochi, Matsuda & Yamamoto (1997) J. Biol. Chem. 272, 33092. 175. Rotenberg & Brautigan (1987) Biochem. J. 243, 747. 176. Tamura, Suzuki, Kikuchi & Tsuiki (1986) Biochem. Biophys. Res. Commun. 140, 212. 177. Okada, Owada & Nakagawa (1986) Biochem. J. 239, 155. 178. Brunati & Pinna (1985) Biochem. Biophys. Res. Commun. 133, 929. 179. Bakalara, Seyfang, Davis & Baltz (1995) Eur. J. Biochem. 234, 871. 180. Koul, Choidas, Treder, et al. (2000) J. Bacteriol. 182, 5425. 181. Fukaya & Yamaguchi (2004) Int. J. Mol. Med. 14, 427. 182. Cuevas, Rohloff, Sánchez & Docampo (2005) Eukaryot. Cell 4, 1550. 183. Soulat, Vaganay, Duclos, et al. (2002) J. Bacteriol. 184, 5194. 184. Qi, Zhao, Cao, et al. (2002) J. Cell. Biochem. 86, 79. 185. Rider, Bartrons & Hue (1990) Eur. J. Biochem. 190, 53. 186. Brauer & Stitt (1990) Physiol. Plant. 78, 568. 187. Mendonca, Masui, McNamara, Leone & Furriel (2007) Comp. Biochem. Physiol. A 146, 534. 188. Bramkamp, Gassel & Altendorf (2004) Biochemistry 43, 4559. 189. Passariello, Schippa, Iori, et al. (2003) Biochim. Biophys. Acta 1648, 203. 190. Greiner, Konietzny & Jany (1993) Arch. Biochem. Biophys. 303, 107. 191. Greiner (2002) J. Agric. Food Chem. 50, 6858. 192. Greiner, Konietzny & Jany (1998) J. Food Biochem. 22, 143. 193. Greiner & Alminger (1999) J. Sci. Food Agric. 79, 1453. 194. Ullah & Sethumadhavan (2003) Biochem. Biophys. Res. Commun. 303, 463. 195. Gibson & Ullah (1988) Arch. Biochem. Biophys. 260, 503. 196. Andriotis & Ross (2003) Phytochemistry 64, 689. 197. Greiner, Haller, Konietzny & Jany (1997) Arch. Biochem. Biophys. 341, 201. 198. Kerovuo, Rouvinen & Hatzack (2000) Biochem. J. 352, 623. 199. Hawker, Smith, Phillips & Wiskich (1987) Plant Physiol. 84, 1281. 200. Echeverria & Salerno (1994) Plant Sci. 96, 15. 201. Cumino, Ekeroth & Salerno (2001) Planta 214, 250. 202. Paltauf, Zinser & Daum (1985) Biochim. Biophys. Acta 835, 322. 203. Zhou, Clemens, Hakes, Barford & Dixon (1993) J. Biol. Chem. 268, 17754. 204. Polya & Haritou (1988) Biochem. J. 251, 357. 205. Cheng, Wang, Gong, et al. (2001) Neurochem. Res. 26, 425. 206. Szöor, Gross & Alphey (2001) Arch. Biochem. Biophys. 396, 213. 207. Barik (1993) Proc. Natl. Acad. Sci. U.S.A. 90, 10633. 208. Peters, Jackson, Crabb & Browning (2002) J. Biol. Chem. 277, 39566. 209. Ganguly & Singh (1999) Phytochemistry 52, 239. 210. Zhao, Qi & Zhao (2000) Biochem. Biophys. Res. Commun. 270, 222. 211. Saugy, Farkas & MacLachlan (1988) Eur. J. Biochem. 177, 135. 212. Van Schaftingen & Gerin (2002) Biochem. J. 362, 513. 213. Ghosh, Cheung, Mansfield & Chou (2005) J. Biol. Chem. 280, 11114. 214. Reilly & Calcutt (2004) Protein Expr. Purif. 33, 48. 215. Shetty & Shetty (1991) Neurochem. Res. 16, 1203. 216. Lawrence & van Etten (1981) Arch. Biochem. Biophys. 206, 122. 217. Reilly, Baron, Nano & Kuhlenschmidt (1996) J. Biol. Chem. 271, 10973. 218. Aguirre-Garcia, Cerbon & Talamas-Rohana (2000) Int. J. Parasitol. 30, 585. 219. Turner & Plaxton (2001) Planta 214, 243. 220. Lau, Farley & Baylink (1987) Adv. Protein Phosphatases 4, 165. 221. Armienta-Aldana & De La Vara (2004) Physiol. Plant. 121, 223. 222. Tanaka, Kishi, Takanezawa, et al. (2004) FEBS Lett. 571, 197. 223. Coello (2002) Physiol. Plant. 116, 293. 224. Reilly, Felts, Henzl, Calcutt & Tanner (2006) Protein Expr. Purif. 45, 132. 225. Chakrabartty & Stinson (1983) Biochim. Biophys. Acta 839, 174. 226.

Ansai, Awano, Chen, *et al.* (1998) *FEBS Lett.* **428**, 157. **227**. Klumpp & Schultz (1990) *Biochim. Biophys. Acta* **1037**, 233. **228**. Bedu, Jeanjean & Rocca-Serra (1982) *Plant Sci. Lett.* **27**, 163. **229**. Hiramatsu, Sonoda, Takanezawa, *et al.* (2003) *J. Biol. Chem.* **278**, 49438. **230**. Hunter (1995) *Cell* **80**, 225. **231**. Kurl, Holmes, Verney & Sidransky (1988) *Biochemistry* **27**, 8974. **232**. Grunberg-Manago, Del Campillo-Campbell, Dondon & Michelson (1966) *Biochim. Biophys. Acta* **123**, 1. **233**. Carrasco & Figueroa (1995) *Comp. Biochem. Physiol. B* **110**, 747. **234**. Nidetzky & Eis (2001) *Biochem. J.* **360**, 727. **235**. Eis & Nidetzky (2002) *Biochem. J.* **363**, 335. **236**. Schwarz, Goedl, Minani & Nidetzky (2007) *J. Biotechnol.* **129**, 140. **237**. Huber & Huber (1990) *Arch. Biochem. Biophys.* **282**, 421. **238**. Weinhausel, Griessler, Krebs, *et al.* (1997) *Biochem. J.* **326**, 773. **239**. Lee, Malone & Snyder (1986) *J. Biol. Chem.* **261**, 5373.

Orvepitant

This selective NK$_1$ receptor antagonist (FW = 628.62 g/mol; CAS 579475-18-6; Soluble in DMSO, but not H$_2$O), also named (2R,4S)-N-((R)-1-(3,5-bis(trifluoromethyl)phenyl)-ethyl)-2-(4-fluoro-2-methylphenyl)-N-methyl-4-((S)-6-oxohexahydropyrrolo[1,2-a]pyrazin-2(1H)-yl)piperidine-1-carbox-amide and GW823296, provides full and long-lasting blockade of central Neurokinin-1 receptors and may represent an efficacious mechanism for treating Major Depressive Disorder (MDD), anxiety, and insomnia. (**See also the NK$_1$ antagonist,** *Aprepitant*) 1. Ratti, Bettica, Alexander, *et al.* (2013) *J. Psychopharmacol.* **27**, 424.

Oseltamivir (Oseltamivir Carboxylate)

Oseltamivir

Oseltamivir Carboxylate

This zanamivir analogue (FW = 284.36 g/mol), also known by its code name GS 4071, and systematic name (3R,4R,5S)-5-amino-4-acetamido-3-(pentan-3-yloxy)cyclohex-1-ene-1-carboxylate, is a potent inhibitor of influenza A and B neuraminidase (IC$_{50}$ = 1 nM). The pro-drug ester (FW = 312.40 g/mol; CAS 196618-13-0; Generic Name: Oseltamivir; Systematic Name: ethyl (3R,4R,5S)-5-amino-4-acetamido-3-(pentan-3-yloxy)-cyclohex-1-ene-1-carboxylate; tradename: Tamiflu®) is known as oseltamivir. **Pharmacokinetics:** Oseltamivir is readily absorbed from the gastrointestinal tract and is subsequently converted to its pharmacologically active form (oseltamivir carboxylate). It is widely distributed in the body, showing a half-life is 6 to 10 hours. The drug is excreted primarily through the kidneys, and dosing must be adjusted for patients with renal insufficiency. Oseltamivir achieves high plasma levels and thus can be effective beyond the respiratory tract. (**See** *Zanamivir*) **Potential Use as an Antipandemic Agent:** Analysis data for patients (all ages) admitted to hospital worldwide with laboratory-confirmed or clinically diagnosed infection with the pandemic influenza A H1N1pdm09 virus indicates that early treatment in adults admitted to hospital with suspected/proven influenza infection can be effective in reducing mortality

(3). In the United States, tamiflu has been stockpiled as the front line of its national pandemic defense strategy. Indeed oseltamivir's producer, Roche, in coordination with external partners now possesses the capacity to produce more than 400 million courses of Tamiflu yearly. 1. Kim, Lew, Williams, *et al.* (1997) *J. Amer. Chem. Soc.* **119**, 681. 2. Doucette & Aoki (2001) *Expert Opin. Pharmacother.* **2**, 1671. 3. Muthuri, Venkatesan, Myles, *et al.* (2014) *Lancet Respir. Med.* **2**, 395.

OSI-774, See *Erlotinib*
Osimertinib, See *AZD9291*

OSU-03012

This PDK1 inhibitor (FW = 460.45 g/mol; CAS 742112-33-0; Soluble to 100 mM in DMSO), systematically named 2-Amino-N-[4-[5-(2-phenanthrenyl)-3-(trifluoromethyl)-1H-pyrazol-1-yl]phenyl]acetamide, blocks Akt signaling by targeting 3-Phosphoinositide-Dependent Protein Kinase-1 (IC50 = 2-3 μM) (1). Exposure of PC-3 cells to this agent leads to Akt dephosphorylation and inhibition of p70 S6 kinase activity. OSU-03012 promotes glioma cell killing that is dependent on endoplasmic reticulum stress, lysosomal dysfunction, and BID-dependent release of AIF from mitochondria, and whose lethality is enhanced by irradiation or by inhibition of protective signaling pathways (2). OSU03012 induced mitochondrial-dependent apoptosis of medulloblastoma cells and enhanced the cytotoxic effects of chemotherapeutic drugs in a synergistic or additive manner (3) When tested *in vivo*, OSU03012 inhibits the growth of established medulloblastoma xenograft tumors in a dose-dependent manner and augmented the antitumor effects of mammalian target of rapamycin inhibitor CCI-779 (3). 1. Zhu, Huang, Tseng, *et al.* (2004) *Cancer Res.* **64**, 4309. 2. Yacoub, Park, Hanna, *et al.* (2006) *Mol. Pharmacol.* **70**, 589. 3. Baryawno, Sveinbjörnsson, Eksborg, *et al.* (2010) *Cancer Res.* **70**, 266.

Otlertuzumab

This intravenously administered anti-CD37 IgG fusion protein (MW = 107.5 kDa; CAS 1372645-37-8), also known by its code name TRU-016, is intended for the treatment of B-cell malignancies, including chronic lymphocytic leukemia (CLL) and non-Hodgkin's lymphoma (NHL), as well as for autoimmune and inflammatory diseases. Otlertuzumab was created by humanizing SMIP-016, a mouse/human chimeric protein that demonstrates antitumor activity against lymphoid malignancies in preclinical studies, including in human B-cell tumor mouse xenograft models. TRU-016 demonstrates synergistic or additive activity in NHL cells in combination with rituximab, rapamycin, doxorubicin and bendamustine. In a phase I/II clinical trial in refractory or relapsed patients with CLL or small lymphocytic lymphoma, Otlertuzumab is well tolerated, with clinical benefit and a reduced absolute lymphocyte count observed in all cohorts dosed at > 0.1 mg/kg. TRU-016 is a promising therapeutic agent for patients with B-cell lymphoid malignancies, especially patients refractory to standard treatment.

OTX015

This bromodomain inhibitor (FW = 491.99 g/mol; CAS 202590-98-5; Solubility: 98 mg/mL DMSO; <1mg/mL H$_2$O), also named (6S)-4-(4-

chlorophenyl)-*N*-(4-hydroxyphenyl)-2,3,9-trimethyl-6*H*-thieno[3,2-*f*][1,2,4]triazolo[4,3-*a*][1,4]diazepine-6-acetamide, targets the Bromodomain and Extra Terminal (BET) domain proteins BRD2, BRD3, and BRD4, with EC_{50} values in the 10-19 nM range, exhibiting cytostatic anti-proliferative activity in several anaplastic large cell lymphoma (ALCL) cell lines. Treatment leads to rapid down-regulation of c-MYC expression followed by the cell cycle G_1 arrest. The synergistic effects of OTX-015 with anti-ALK inhibitors provide the rational for lower-toxicity protocols for the treatment of ALKpos ALCL patients. **1**. Noel, *et al.* (2013) *Mol. Cancer Ther.* **12**, C244. **2**. Boi, *et al.* (2013) *Mol Cancer Ther.* **12**, A219.

Ouabain

This cardiac glycoside (FW = 584.66 g/mol) from *Strophanthus gratus* and *Acokanthera ouabaio*, also called γ-strophanthin and *gratus* strophanthin, selectively inhibits Na^+/K^+ exchanging ATPases by binding to the α-subunit near its amino terminal and to the H_5-H_6 hairpin region, blocking the active efflux of sodium ion as well as the reuptake of potassium ion by blocking movement of the H_5 and H6 transmembrane domains (1,2). Ouabain increases natriuresis (3), slows heart rate with its positive ionotropic effect (4), increases intracellular Ca^{2+} (4,5), and induces hypertension (6). Recently, endogenous inhibitors of the Na^+/K^+-exchanging ATPase have been identified in adrenal cortex and brain that are stereoisomers of ouabain and possess inotropic and vasopressor activities (7-9). Hydrolysis of ouabain will yield ouabagenin and L-rhamnose. Ouabain octahydrate (FW = 728.78 g/mol) is soluble in boiling water (200 mg/mL), cold water (10 mg/mL), and ethanol (10 mg/mL at room temperature). Ouabain should be stored in the dark. **Target(s):** adenylate cyclase, *Phycomyces blakesleeanus* (10); H^+/K^+-exchanging ATPase, moderately sensitive (11); Na^+/K^+-exchanging ATPase (12-32); 4-nitrophenylphosphatase activity of Na^+/K^+-exchanging ATPase (31,32). **1**. Lingrel, Croyle, Woo & Arguello (1998) *Acta Physiol. Scand. Suppl.* **643**, 69. **2**. Juhaszova & Blaustein (1997) *Proc. Natl. Acad. Sci. U.S.A.* **94**, 1800. **3**. McDougall & Yates (1998) *Clin. Exp. Pharmacol. Physiol. Suppl.* **25**, S57. **4**. Blaustein, Juhaszova & Golovina (1998) *Clin. Exp. Hypertens.* **20**, 691. **5**. Peng, Huang, Xie, Huang & Askari (1996) *J. Biol. Chem.* **271**, 10372. **6**. Veerasingham & Leenen (1999) *Amer. J. Physiol.* **276**, H63. **7**. Hamlyn, Lu, Manunta, *et al.* (1998) *Clin. Exp. Hypertens.* **20**, 523. **8**. van Huysse & Leenen (1998) *Clin. Exp. Hypertens.* **20**, 657. **9**. Doris & Bagrov (1998) *Proc. Soc. Exp. Biol. Med.* **218**, 156. **10**. Cohen, Ness & Whiddon (1980) *Phytochemistry* **19**, 1913. **11**. Adams, Tillekeratne, Yu, Pestov & Modyanov (2001) *Biochemistry* **40**, 5765. **12**. Post & Sen (1967) *Meth. Enzymol.* **10**, 762. **13**. Jørgensen (1974) *Meth. Enzymol.* **32**, 277. **14**. Jørgensen (1975) *Meth. Enzymol.* **36**, 434. **15**. Ruoho & Kyte (1977) *Meth. Enzymol.* **46**, 523. **16**. Skou (1988) *Meth. Enzymol.* **156**, 1. **17**. Wallick & Schwartz (1988) *Meth. Enzymol.* **156**, 201. **18**. Hootman & Ernst (1988) *Meth. Enzymol.* **156**, 213. **19**. English & Schulz (1989) *Meth. Enzymol.* **173**, 676. **20**. Sjodin (1989) *Meth. Enzymol.* **173**, 695. **21**. Furriel, McNamara & Leone (2001) *Comp. Biochem. Physiol. A* **130**, 665. **22**. Almansa, Sanchez, Cozzi, *et al.* (2001) *J. Comp. Physiol. B* **171**, 557. **23**. Schuurman, Stekhoven & Bonting (1981) *Physiol. Rev.* **61**, 1. **24**. Robinson & Flashner (1979) *Biochim. Biophys. Acta* **549**, 145. **25**. de Lima Santos & Ciancaglini (2003) *Comp. Biochem. Physiol. B* **135**, 539. **26**. Taniguchi & Iida (1972) *Biochim. Biophys. Acta* **288**, 98. **27**. Cortas & Walser (1971) *Biochim. Biophys. Acta* **249**, 181. **28**. Post, Sen & Rosenthal (1965) *J. Biol. Chem.* **240**, 1437. **29**. Su & Scheiner-Bobis (2004) *Biochemistry* **43**, 4731. **30**. Akera (1971) *Biochim. Biophys. Acta* **249**, 53. **31**. Homareda &

Ushimaru (2005) *FEBS J.* **272**, 673. **32**. Mendonca, Masui, McNamara, Leone & Furriel (2007) *Comp. Biochem. Physiol. A* **146**, 534.

Ovomucoid

This egg white glycoprotein (MW ≈ 28 kDa) is a multidomain Kazal-type protease inhibitor. The chicken (*Gallus gallus*) protein contains 210 aminoacyl residues, and its carbohydrate chains are N-linked at asparaginyl residues. (*See also Ovoinhibitors*). **Target(s):** acrosin (1,21-23); archaebacterial serine proteinase of *Halobacterium mediterranei* (2); brachyuran (3); chymotrypsin (4-6,26); envelysin, *Xenopus* (7); glutamyl endopeptidase II, inhibited notably by [Leu18→Glu]-modified ovomucoid (8,9); leukocyte elastase (10-12); pancreatic elastase, inhibited by third domain (6,12,13); peptidase K, *or* proteinase K, inhibited by third domain (6); peptidyl-glycinamidase, *or* carboxamidopeptidase (14); spermosin (1); streptogrisin A, inhibited by third domain (6); streptogrisin B, inhibited by third domain (6,15); *Streptomyces erythraeus* trypsin (16); subtilisin (6,17); subtilisin BPN' (6); subtilisin Carlsberg (6); trypsin (4,5,18-20,24-26). **1**. Sawada, Yokosawa & Ishii (1984) *J. Biol. Chem.* **259**, 2900. **2**. Vanoni & Tortora (1998) in *Handb. Proteolytic Enzymes*, p. 334, Academic Press, San Diego. **3**. Kristjansson, Gudmundsdottir, Fox & Bjarnason (1995) *Comp. Biochem. Physiol. B* **110**, 707. **4**. Kassell (1970) *Meth. Enzymol.* **19**, 890. **5**. Feinstein & Feeney (1967) *Biochemistry* **6**, 749. **6**. Ardelt & Laskowski, Jr. (1985) *Biochemistry* **24**, 5313. **7**. Fan & Katagiri (2001) *Eur. J. Biochem.* **268**, 4892. **8**. Stennicke & Breddam (1998) in *Handb. Proteolytic Enzymes*, p. 246, Academic Press, San Diego. **9**. Komiyama, Bigler, Yoshida, Noda & Laskowski (1991) *J. Biol. Chem.* **266**, 10727. **10**. Bieth (1998) in *Handb. Proteolytic Enzymes*, p. 54, Academic Press, San Diego. **11**. Barrett (1981) *Meth. Enzymol.* **80**, 581. **12**. Starkey & Barrett (1976) *Biochem. J.* **155**, 265. **13**. Bieth (1998) in *Handb. Proteolytic Enzymes*, p. 42, Academic Press, San Diego. **14**. Simmons & Walter (1980) *Biochemistry* **19**, 39. **15**. Fujinaga, Read, Sielecki, *et al.* (1982) *Proc. Natl. Acad. Sci. U.S.A.* **79**, 4868. **16**. Norioka & Sakiyama (1998) in *Handb. Proteolytic Enzymes*, p. 22, Academic Press, San Diego. **17**. Ballinger & Wells (1998) in *Handb. Proteolytic Enzymes*, p. 289, Academic Press, San Diego. **18**. Lineweaver & Murray (1947) *J. Biol. Chem.* **171**, 565. **19**. Fraenkel-Conrat, Bean & Lineweaver (1949) *J. Biol. Chem.* **177**, 385. **20**. Zahnley & Davis (1970) *Biochemistry* **9**, 1428. **21**. Hardy, Schoots & Hedrick (1989) *Biochem. J.* **257**, 447. **22**. Richardson, Korn, Bodine & Thurston (1992) *Poult. Sci.* **71**, 1789. **23**. Gilboa, Elkana & Rigbi (1973) *Eur. J. Biochem.* **39**, 85. **24**. Ohlsson & Tegner (1973) *Biochim. Biophys. Acta* **317**, 328. **25**. Asgeirsson, Fox & Bjarnason (1989) *Eur. J. Biochem.* **180**, 85. **26**. Ryan, Clary & Tomimatsu (1965) *Arch. Biochem. Biophys.* **110**, 175.

Ovostatin

This tetrameric protein, also called ovomacroglobulin, inhibits proteinases of all four mechanistic classes. Ovostatin inhibits proteinases in a manner similar to α₂-macroglobulin. Protease-catalyzed hydrolysis of a peptide bond in the susceptible region of the ovostatin polypeptide chain triggers a conformational change that hinders binding of macromolecular substrates, while remaining accessible to small synthetic substrates. **Target(s):** chymotrypsin (1,2); cortical granule protease (1); interstitial collagenase, matrix metalloproteinase 1 (3-5); leukocyte elastase (6-8); matrilysin, matrix metalloproteinase 7 (9-11); papain (2,12); procollagen C-endopeptidase (13); *Pseudomonas aeruginosa* alkaline protease (7); serralysin (14); stromelysin 1, matrix metalloproteinase 3 (3,15,16); thermolysin (2,4,5,12,17,18); trypsin (1,2,12). **1**. Yamada & Aketa (1988) *Gamete Res.* **19**, 265. **2**. Osada, Sasaki & Ikai (1988) *J. Biochem.* **103**, 212. **3**. Enghild, Salvesen, Brew & Nagase (1989) *J. Biol. Chem.* **264**, 8779. **4**. Nagase, Harris, Woessner & Brew (1983) *J. Biol. Chem.* **258**, 7481. **5**. Nagase & Harris (1983) *J. Biol. Chem.* **258**, 7490. **6**. Miyagawa, Kamata, Matsumoto, Okamura & Maeda (1994) *Graefes Arch. Clin. Exp. Ophthalmol.* **232**, 488. **7**. Miyagawa, Kamata, Matsumoto, Okamura & Maeda (1991) *Graefes Arch. Clin. Exp. Ophthalmol.* **229**, 281. **8**. Ikai, Nakashima & Aoki (1989) *Biochem. Biophys. Res. Commun.* **158**, 831. **9**. Woessner (1998) in *Handb. Proteolytic Enzymes*, p. 1183, Academic Press, San Diego. **10**. Woessner (1995) *Meth. Enzymol.* **248**, 485. **11**. Woessner & Taplin (1988) *J. Biol. Chem.* **263**, 16918. **12**. Kitamoto, Nakashima & Ikai (1982) *J. Biochem.* **92**, 1679. **13**. Hojima, van der Rest & Prockop (1985) *J. Biol. Chem.* **260**, 15996. **14**. Molla, Oda & Maeda (1987) *J. Biochem.* **101**, 199. **15**. Nagase (1998) in *Handb. Proteolytic Enzymes*, p. 1172, Academic Press, San Diego. **16**. Nagase (1995) *Meth. Enzymol.* **248**, 449. **17**. Ruben, Harris & Nagase (1988) *J. Biol. Chem.* **263**, 2861. **18**. Molla, Matsumura, Yamamoto, Okamura & Maeda (1987) *Infect. Immun.* **55**, 2509.

Oxacillin

This semisynthetic antibiotic (FW$_{\text{free-acid}}$ = 401.44 g/mol) is structurally related to the penicillins and is resistant to hydrolysis by a number of β-lactamases. **Target(s):** aminopeptidase D (1); cytosol alanyl aminopeptidase (2); β-lactamase, *or* penicillinase (3-5); serine-type D-Ala-D-Ala carboxypeptidase (6). **1.** Starnes, Desmond & Behal (1980) *Biochim. Biophys. Acta* **616**, 290. **2.** Starnes, Szechinski & Behal (1982) *Eur. J. Biochem.* **124**, 363. **3.** Hamilton-Miller & Smith (1964) *Nature* **201**, 999. **4.** Fujii-Kuriyama, Yamamoto & Sugawara (1977) *J. Bacteriol.* **131**, 726. **5.** Feller, Zekhnini, Lamotte-Brasseur & Gerday (1997) *Eur. J. Biochem.* **244**, 186. **6.** Coyette, Ghuysen & Fontana (1978) *Eur. J. Biochem.* **88**, 297.

1*H*-[1,2,4]Oxadiazolo[4,3-*a*]quinoxalin-1-one

This hygroscopic quinoxaline (FW = 187.16 g/mol; *Abbreviation*: ODQ), is an irreversible inhibitor of soluble guanylate cyclase, whereas particulate guanylate cyclase is unaffected. It is insoluble in water and soluble in dimethyl sulfoxide (5 mg/mL) and acetonitrile (20 mg/mL). **1.** Schrammel, Behrends, Schmidt, Koesling & Mayer (1996) *Mol. Pharmacol.* **50**, 1. **2.** Garthwaite, Southam, Boulton, *et al.* (1995) *Mol. Pharmacol.* **48**, 184. **3.** Schmidt, Schramm, Schroder & Stasch (2003) *Eur. J. Pharmacol.* **468**, 167. **4.** Li, Jin & Campbell (1998) *Hypertension* **31**, 303. **5.** Hoenicka, Becker, Apeler, *et al.* (1999) *J. Mol. Med.* **77**, 14. **6.** Lee, Martin & Murad (2000) *Proc. Natl. Acad. Sci. U.S.A.* **97**, 10763.

Oxalate (Oxalic Acid)

This dicarboxylic acid (FW$_{\text{free-acid}}$ = 90.04 g/mol; CAS 144-62-7), is the simplest dicarboxylic acid (the pKa values are 1.27 and 4.29 at 25°C). Oxalate is produced by a number of plants (*e.g.*, species of *Oxalis* and *Rumex*) and is toxic to mammals. Small amounts are formed in humans, particularly in individuals with primary hyperoxaluria. Life-threatening oxalate poisoning occurs whenever humans mistakenly consume ethylene glycol. Oxalate is also a natural product of ascorbate and proline degradation. The most common renal calculi are composed of calcium oxalate monohydrate, which is virtually insoluble in water and forms opportunistically when oxalate and calcium ions combine within renal tubules, eventually filling and occluding the renal pelvis and ureters, thereby obstructing urine flow. The physicochemical basis oxalate biomineralization and stone formation is best understood as a condensation-equilibrium process (1). Oxalate also inhibits many enzymes, mainly as a consequence of its metal-ion chelating properties. **Target(s):** acetylpyruvate hydrolase (2); acid phosphatase (3-9,158,159); ε-acyl-lysine deacylase (10); acylphosphatase (11); alanine aminotransferase (178); alcohol dehydrogenase (12); alkaline phosphatase (9,13); allantoicase (14); arylsulfatase A, *or* cerebroside-sulfatase, mildly inhibited (15); arylsulfatase B, *or* N-acetylgalactosamine-4-sulfatase (15); aspartate ammonia lyase (16); D-aspartate oxidase (17); atrazine chlorohydrolase (18); benzoate 4-monooxygenase, weakly inhibited (187); bothropasin (146); carbamoyl-phosphate synthetase (19); carboxy-*cis,cis*-muconate cyclase (20); carboxypeptidase A (21-23); catalase (192); catechol oxidase, *or* polyphenol oxidase, *or* tyrosinase (5,24); cholinesterase (25); cyanate

hydratase, cyanase (26-28); cytosol alanyl aminopeptidase (148); dehydrogluconate dehydrogenase (29); esterase, *or* lipase (30); 5-formyltetrahydrofolate cyclo-ligase, *or* 5,10-methenyltetrahydrofilate synthetase (31); fumarylacetoacetase (32); gluconate 2-dehydrogenase (33); glucose-6-phosphatase (34); glutamate decarboxylase, (35); glycogen synthase (36); homocitrate synthase (179); D-2 hydroxy-acid dehydrogenase (5,37-41); (*S*)-2-hydroxy-acid oxidase (42); hydroxyacylglutathione hydrolase (glyoxalase II)160; [3-hydroxy-3-methylglutaryl-CoA reductase (NADPH)] phosphatase (154); inositol oxygenase (188); isocitrate lyase (43-57); L-kynurenine aminotransferase (58,59); laccase (193); lactase-phlorizin hydrolase, *or* glycosylceramidase (152); D-lactate dehydrogenase (5,38,60-64); L-lactate dehydrogenase (5,60,62,63,65-79); D-lactate dehydrogenase (cytochrome) (5,38,63); L-lactate dehydrogenase (cytochrome) (63); lactate 2-monooxygenase, slow-binding inhibition (80,81,189-191); levanase (150,151); malate dehydrogenase (82); malate synthase (83-86,180-185); malic enzyme, *or* malate dehydrogenase (decarboxylating) (87); malyl-CoA lyase (88); D-mandelate dehydrogenase (64); L-mandelate dehydrogenase (89); membrane alanyl aminopeptidase, *or* aminopeptidase N, inhibited by ammonium salt (149); *N*-methylglutamate dehydrogenase, mildly inhibited (/90); methylmalonyl-CoA carboxyltransferase, *or* transcarboxylase (68,91-93); nucleotidases (94,155); oxaloacetate decarboxylase (68,95-103); oxaloacetate tautomerase (104,105); 3-oxoacid CoA-transferase (106); 2 oxopent-4-enoate hydratase (107); pantothenase (108,143-145); peptidyl-dipeptidase A, *or* angiotensin I converting enzyme (147); peroxidase (109); phosphoenolpyruvate carboxykinase (ATP) (110); phosphoenolpyruvate carboxykinase (GTP) (111-114); phosphoenolpyruvate carboxylase (115); phosphoenolpyruvate mutase (K_i = 32 μM (116-118); phosphoenolpyruvate phosphatase (153); phosphoenolpyruvate:protein phosphotransferase , (164,165); phosphoglycerate mutase (119); phosphonopyruvate hydrolase (120); phytase (121); 4-phytase (156,157); procollagen-proline 4-dioxygenase (122); pyrophosphatase (123,124); pyruvate carboxylase (125-129); pyruvate kinase (68,130-132,166-176); pyruvate,orthophosphate dikinase (133,161-163); pyruvate,water dikinase (134); serine:glyoxylate aminotransferase (177); succinate dehydrogenase (135-138); tartronate semialdehyde reductase, *or* 2-hydroxy-3-oxopropionate reductase (139); thermolysin (140); triose phosphate isomerase (141); tyrosinase, *or* monophenol monooxygenase, *or* polyphenol oxidase (186); urease (142).

1. Brown & Purich (1992) Physical-Chemical Processes in Kidney Stone Formation, in *Disorders of Bone and Mineral Metabolism* (Coe and Favus, eds.), Raven Press, New York, Chapter 29, pp. 613-624. **2.** Davey & Ribbons (1975) *J. Biol. Chem.* **250**, 3826. **3.** Roche (1950) *The Enzymes*, 1st ed., **1** (part 1), 473. **4.** Hollander (1971) *The Enzymes*, 3rd ed., **4**, 449. **5.** Webb (1966) *Enzyme and Metabolic Inhibitors*, vol. 2, Academic Press, New York. **6.** Nigam, Davidson & Fishman (1959) *J. Biol. Chem.* **234**, 1550. **7.** Kilsheimer & Axelrod (1957) *J. Biol. Chem.* **227**, 879. **8.** Albers, Büsing & Schudt (1966) *Enzymologia* **30**, 149. **9.** Belfanti, Contardi & Ercoli (1935) *Biochem. J.* **29**, 517 and 842. **10.** Chibata, Ishikawa & Tosa (1970) *Meth. Enzymol.* **19**, 756. **11.** Koshland (1955) *Meth. Enzymol.* **2**, 555. **12.** Sund & Theorell (1963) *The Enzymes*, 2nd ed., **7**, 25. **13.** Belfanti, Contardi & Ercoli (1935) *Biochem. J.* **29**, 1491. **14.** van der Drift & Vogels (1970) *Biochim. Biophys. Acta* **198**, 339. **15.** Bleszynski & Leznicki (1967) *Enzymologia* **33**, 373. **16.** Virtanen & Ellfolk (1955) *Meth. Enzymol.* **2**, 386. **17.** Dixon & Kenworthy (1967) *Biochim. Biophys. Acta* **146**, 54. **18.** Seffernick, McTavish, Osborne, *et al.* (2002) *Biochemistry* **41**, 14430. **19.** Jones (1962) *Meth. Enzymol.* **5**, 903. **20.** Thatcher & Cain (1975) *Eur. J. Biochem.* **56**, 193. **21.** Pétra (1970) *Meth. Enzymol.* **19**, 460. **22.** Smith, Hanson & Wendelboe (1949) *J. Biol. Chem.* **179**, 803. **23.** Smith (1951) *The Enzymes*, 1st ed., **1** (part 2), 793. **24.** Malkin, Thickman, Markworth, Wilcox & Kull (2001) *J. Enzyme Inhib.* **16**, 135. **25.** Massart & Dufait (1939) *Enzymologia* **6**, 282. **26.** Anderson, Johnson, Endrizzi, Little & Korte (1987) *Biochemistry* **26**, 3938. **27.** Anderson & Little (1986) *Biochemistry* **25**, 1621. **28.** Walsh, Otwinowski, Perrakis, Anderson & Joachimiak (2000) *Structure Fold Des.* **8**, 505. **29.** Shinagawa & Ameyama (1982) *Meth. Enzymol.* **89**, 194. **30.** Falk (1913) *J. Amer. Chem. Soc.* **35**, 601. **31.** Grimshaw, Henderson, Soppe, *et al.* (1984) *J. Biol. Chem.* **259**, 2728. **32.** Connors & Stotz (1949) *J. Biol. Chem.* **178**, 881. **33.** Matsushita, Shinagawa & Ameyama (1982) *Meth. Enzymol.* **89**, 187. **34.** Nordlie (1971) *The Enzymes*, 3rd ed., **4**, 543. **35.** Fonda (1972) *Biochemistry* **11**, 1304. **36.** Rothman & Cabib (1967) *Biochemistry* **6**, 2098. **37.** Cammack (1975) *Meth. Enzymol.* **41**, 323. **38.** Hatefi & Stiggall (1976) *The Enzymes*, 3rd ed., **13**, 175. **39.** Boeri, Cremona & Singer (1960) *Biochem. Biophys. Res. Commun.* **2**, 298. **40.** Cremona (1964) *J. Biol. Chem.* **239**, 1457. **41.** Tubbs (1960) *Biochem. Biophys. Res. Commun.* **3**, 513. **42.** Jorns (1975) *Meth. Enzymol.* **41**, 337. **43.** Daron & Gunsalus (1963) *Meth. Enzymol.* **6**,

622. **44.** McFadden (1969) *Meth. Enzymol.* **13**, 163. **45.** Munir, Hattori & Shimada (2002) *Arch. Biochem. Biophys.* **399**, 225. **46.** Spector (1972) *The Enzymes*, 3rd ed., **7**, 357. **47.** Daron, Rutter & Gunsalus (1966) *Biochemistry* **5**, 895. **48.** Jameel, El-Gul & McFadden (1984) *Phytochemistry* **23**, 2753. **49.** Tsukamoto, Ejiri & Katsumata (1986) *Agric. Biol. Chem.* **50**, 409. **50.** Hoyt, Robertson, Berlyn & Reeves (1988) *Biochim. Biophys. Acta* **966**, 30. **51.** Höner zu Bentrup, Miczak, Swenson & Russell (1999) *J. Bacteriol.* **181**, 7161. **52.** Tanaka, Yoshida, Watanabe, Izumi & Mitsunaga (1997) *Eur. J. Biochem.* **249**, 820. **53.** Hoyt, Johnson & Reeves (1991) *J. Bacteriol.* **173**, 6844. **54.** Giachetti, Pinzauti, Bonaccorsi, Vincenzini & Vanni (1987) *Phytochemistry* **26**, 2439. **55.** DeLucas, Amor, Diaz, Turner & Laborda (1997) *Mycol. Res.* **101**, 410. **56.** Vicenzini, Nerozzi, Vincieri & Vanni (1980) *Phytochemistry* **19**, 769. **57.** Vanni, Vicenzini, Nerozzi & Sinha (1979) *Can. J. Biochem.* **57**, 1131. **58.** Tanizawa, Asada & Soda (1985) *Meth. Enzymol.* **113**, 90. **59.** Tobes (1987) *Meth. Enzymol.* **142**, 217. **60.** Snoswell (1966) *Meth. Enzymol.* **9**, 321. **61.** Kaczorowski, Kohn & Kaback (1978) *Meth. Enzymol.* **53**, 519. **62.** Jasso-Chavez, Torres-Marquez & Moreno-Sanchez (2001) *Arch. Biochem. Biophys.* **390**, 295. **63.** Nygaard (1963) *The Enzymes*, 2nd ed., **7**, 557. **64.** Allison, O'Donnell & Fewson (1985) *Biochem. J.* **231**, 407. **65.** Zewe & Fromm (1965) *Biochemistry* **4**, 782. **66.** Goldberg (1975) *Meth. Enzymol.* **41**, 318. **67.** Javed & Waqar (1996) *J. Enzyme Inhib.* **10**, 187. **68.** Wolfenden (1977) *Meth. Enzymol.* **46**, 15. **69.** Schwert & Winer (1963) *The Enzymes*, 2nd ed., **7**, 142. **70.** Novoa, Winer, Glaid & Schwert (1959) *J. Biol. Chem.* **234**, 1143. **71.** Dalziel (1975) *The Enzymes*, 3rd ed., **11**, 1. **72.** Holbrook, Liljas, Steindel & Rossmann (1975) *The Enzymes*, 3rd ed., **11**, 191. **73.** Fondy, Pesce, Freedberg, Stolzenbach & Kaplan (1964) *Biochemistry* **3**, 522. **74.** Okabe, Hayakawa, Hamada & Koike (1968) *Biochemistry* **7**, 79. **75.** Leathwood, Gilford & Plummer (1972) *Enzymologia* **42**, 285. **76.** Bernheim (1928) *Biochem. J.* **22**, 1178. **77.** Pirie (1934) *Biochem. J.* **28**, 411. **78.** Neilands (1954) *J. Biol. Chem.* **208**, 225. **79.** Hakala, Glaid & Schwert (1953) *Fed. Proc.* **12**, 213. **80.** Ghisla & Massey (1975) *J. Biol. Chem.* **250**, 577. **81.** Morrison (1982) *Trends Biochem. Sci.* **7**, 102. **82.** Green (1936) *Biochem. J.* **30**, 2095. **83.** Dixon & Kornberg (1962) *Meth. Enzymol.* **5**, 633. **84.** Munir, Hattori & Shimada (2002) *Biosci. Biotechnol. Biochem.* **66**, 576. **85.** Higgins, Kornblatt & Rudney (1972) *The Enzymes*, 3rd ed., **7**, 407. **86.** Dixon, Kornberg & Lund (1960) *Biochim. Biophys. Acta* **41**, 217. **87.** Kun (1963) *The Enzymes*, 2nd ed., **7**, 149. **88.** Hersh (1974) *J. Biol. Chem.* **249**, 5208. **89.** Hoey, Allison, Scott & Fewson (1987) *Biochem. J.* **248**, 871. **90.** Hersh, Stark, Worthen & Fiero (1972) *Arch. Biochem. Biophys.* **150**, 219. **91.** Northrop & Wood (1969) *J. Biol. Chem.* **244**, 5820. **92.** Wood, Jacobson, Gerwin & Northrop (1969) *Meth. Enzymol.* **13**, 215. **93.** Wood (1972) *The Enzymes*, 3rd ed., **6**, 83. **94.** Arsenis & Touster (1978) *Meth. Enzymol.* **51**, 271. **95.** Schmitt, Bottke & Siebert (1966) *Hoppe-Seyler's Z. Physiol. Chem.* **347**, 18. **96.** Dimroth (1986) *Meth. Enzymol.* **125**, 530. **97.** Kosicki & Westheimer (1968) *Biochemistry* **7**, 4303. **98.** Sender, Martin, Peiru & Magni (2004) *FEBS Lett.* **570**, 217. **99.** Dimroth (1981) *Eur. J. Biochem.* **115**, 353. **100.** Dimroth & Thomer (1986) *Eur. J. Biochem.* **156**, 157. **101.** Labrou & Clonis (1999) *Arch. Biochem. Biophys.* **365**, 17. **102.** Morinaga & Shirakawa (1971) *Agric. Biol. Chem.* **35**, 1166. **103.** Dahinden, Auchli, Granjon, *et al.* (2005) *Arch. Microbiol.* **183**, 121. **104.** Vinogradov, Kotlyar, Burov & Belikova (1989) *Adv. Enzyme Regul.* **28**, 271. **105.** Belikova, Burov & Vinogradov (1988) *Biochim. Biophys. Acta* **936**, 10. **106.** Fenselau & Wallis (1974) *Biochemistry* **13**, 3884. **107.** Pollard & Bugg (1998) *Eur. J. Biochem.* **251**, 98. **108.** Airas (1979) *Meth. Enzymol.* **62**, 267. **109.** Lück (1958) *Enzymologia* **19**, 227. **110.** Delbaere, Sudom, Prasad, Leduc & Goldie (2004) *Biochim. Biophys. Acta* **1697**, 271. **111.** Hebda & Nowak (1982) *J. Biol. Chem.* **257**, 5503. **112.** Guidinger & Nowak (1990) *Arch. Biochem. Biophys.* **178**, 131. **113.** Mukhopadhyay, Concar & Wolfe (2001) *J. Biol. Chem.* **276**, 16137. **114.** Ash, Emig, Chowdhury, Satoh & Schramm (1990) *J. Biol. Chem.* **265**, 7377. **115.** Miziorko, Nowak & Mildvan (1974) *Arch. Biochem. Biophys.* **163**, 378. **116.** Seidel & Knowles (1994) *Biochemistry* **33**, 5641. **117.** Huang, Li, Jia, Dunaway-Mariano & Herzberg (1999) *Structure Fold. Des.* **7**, 539. **118.** Kim & Dunaway-Mariano (1996) *Biochemistry* **35**, 4628. **119.** Ray & Peck (1972) *The Enzymes*, 3rd ed., **6**, 407. **120.** Chen, Han, Niu, *et al.* (2006) *Biochemistry* **45**, 11491. **121.** Rapoport, Leva & Guest (1941) *J. Biol. Chem.* **139**, 621. **122.** Hutton, Tappel & Udenfriend (1967) *Arch. Biochem. Biophys.* **118**, 231. **123.** Naganna & Menon (1948) *J. Biol. Chem.* **174**, 501. **124.** Naganna (1950) *J. Biol. Chem.* **183**, 693. **125.** Mildvan, Scrutton & Utter (1966) *J. Biol. Chem.* **241**, 3488. **126.** Scrutton, Olmsted & Utter (1969) *Meth. Enzymol.* **13**, 235. **127.** Young, Tolbert & Utter (1969) *Meth. Enzymol.* **13**, 250. **128.** Seubert & Weicker (1969) *Meth. Enzymol.* **13**, 258. **129.** Scrutton & Mildvan (1968) *Biochemistry* **7**, 1490. **130.** Reed &

Morgan (1974) *Biochemistry* **13**, 3537. **131.** Jursinic & Robinson (1978) *Biochim. Biophys. Acta* **523**, 358. **132.** Ambasht, Malhotra & Kayastha (1997) *Indian J. Biochem. Biophys.* **34**, 365. **133.** South & Reeves (1975) *Meth. Enzymol.* **42**, 187. **134.** Narindrasorasak & Bridger (1978) *Can. J. Biochem.* **56**, 816. **135.** Potter & Elvehjem (1937) *J. Biol. Chem.* **117**, 341. **136.** Quastel & Wooldridge (1928) *Biochem. J.* **22**, 689. **137.** Rosen & Klotz (1957) *Arch. Biochem. Biophys.* **67**, 161. **138.** Potter & Elvehjem (1937) *J. Biol. Chem.* **117**, 341. **139.** Gotto & Kornberg (1961) *Biochem. J.* **81**, 273. **140.** Matsubara (1970) *Meth. Enzymol.* **19**, 642. **141.** Noltmann (1972) *The Enzymes*, 3rd ed., **6**, 271. **142.** Onodera (1915) *Biochem. J.* **9**, 544. **143.** Airas (1976) *Biochim. Biophys. Acta* **452**, 193 and 201. **144.** Airas (1988) *Biochem. J.* **250**, 447. **145.** Airas (1976) *Biochem. J.* **157**, 415. **146.** Mandelbaum, Reichel & Assakura (1982) *Toxicon* **20**, 955. **147.** Oshima & Nagasawa (1977) *J. Biochem.* **81**, 57. **148.** Garner & Behal (1974) *Biochemistry* **13**, 3227. **149.** Bauvois & Dauzonne (2006) *Med. Res. Rev.* **26**, 88. **150.** Kang, Lee, Lee & Lee (1999) *Biotechnol. Appl. Biochem.* **29**, 263. **151.** Chaudhary, Gupta, Gupta & Banerjee (1996) *J. Biotechnol.* **46**, 55. **152.** Hamaswamy & Radhakrishnan (1973) *Biochem. Biophys. Res. Commun.* **54**, 197. **153.** Malhotra & Kayastha (1990) *Plant Physiol.* **93**, 194. **154.** Hegardt, Gil & Calvet (1983) *J. Lipid Res.* **24**, 821. **155.** Tjernshaugen (1978) *Biochem. J.* **169**, 597. **156.** Sutardi & Buckle (1986) *J. Food Biochem.* **10**, 197. **157.** Gibson & Ullah (1988) *Arch. Biochem. Biophys.* **260**, 503. **158.** Kuo & Blumenthal (1961) *Biochim. Biophys. Acta* **52**, 13. **159.** Igarashi & Hollander (1968) *J. Biol. Chem.* **243**, 6084. **160.** Uotila (1973) *Biochemistry* **12**, 3944. **161.** Hiltpold, Thomas & Köhler (1999) *Mol. Biochem. Parasitol.* **104**, 157. **162.** Michaels, Milner & Reed (1975) *Biochemistry* **14**, 3213. **163.** Meyer, Kelly & Latzko (1978) *Plant Sci. Lett.* **12**, 35. **164.** Saier, Schmidt & Lin (1980) *J. Biol. Chem.* **255**, 8579. **165.** Teplyakov, Lim, Zhu, *et al.* (2006) *Proc. Natl. Acad. Sci. U.S.A.* **103**, 16218. **166.** Wu & Turpin (1992) *J. Phycol.* **28**, 472. **167.** Lin, Turpin & Plaxton (1989) *Arch. Biochem. Biophys.* **269**, 228. **168.** Plaxton, Smith & Knowles (2002) *Arch. Biochem. Biophys.* **400**, 54. **169.** Baysdorfer & Bassham (1984) *Plant Physiol.* **74**, 374. **170.** Oria-Hernández, Cabrera, Pérez-Montfort & Ramírez-Silva (2005) *J. Biol. Chem.* **280**, 37924. **171.** Andre, Froehlich, Moll & Benning (2007) *Plant Cell* **19**, 2006. **172.** Gupta & Singh (1989) *Plant Physiol. Biochem.* **27**, 703. **173.** Smith, Knowles & Plaxton (2000) *Eur. J. Biochem.* **267**, 4477. **174.** Hu & Plaxton (1996) *Arch. Biochem. Biophys.* **333**, 298. **175.** Vetter & Buchholz (1997) *Comp. Biochem. Physiol. A* **116A**, 1. **176.** Singh, Malhotra & Singh (2000) *Indian J. Biochem. Biophys.* **37**, 51. **177.** Karsten, Ohshiro, Izumi & Cook (2001) *Arch. Biochem. Biophys.* **388**, 267. **178.** Agarwal (1985) *Indian J. Biochem. Biophys.* **22**, 102. **179.** Qian, West & Cook (2006) *Biochemistry* **45**, 12136. **180.** Miernyk & Trelease (1981) *Phytochemistry* **20**, 2657. **181.** Reinscheid, Eikmanns & Sahm (1994) *Microbiology* **140**, 3099. **182.** Fukawa, Ejiri & Katsumata (1987) *Agric. Biol. Chem.* **51**, 1553. **183.** Smith, Huang, Miczak, *et al.* (2003) *J. Biol. Chem.* **278**, 1735. **184.** Sundaram, Chell & Wilkinson (1980) *Arch. Biochem. Biophys.* **199**, 515. **185.** Durchschlag, Biedermann & Eggerer (1981) *Eur. J. Biochem.* **114**, 255. **186.** Malkin, Thickman, Markworth, Wilcox & Kull (2001) *J. Enzyme Inhib.* **16**, 135. **187.** Reddy & Vaidyanathan (1975) *Biochim. Biophys. Acta* **384**, 46. **188.** Reddy, Pierzchala & Hamilton (1981) *J. Biol. Chem.* **256**, 8519. **189.** Durfor & Cromartie (1981) *Arch. Biochem. Biophys.* **210**, 710. **190.** Massey, Ghisla & Kieschke (1980) *J. Biol. Chem.* **255**, 2796. **191.** Ghisla & Massey (1977) *J. Biol. Chem.* **252**, 6729. **192.** Chatterjee & Sanwal (1993) *Mol. Cell. Biochem.* **126**, 125. **193.** Lorenzo, Moldes, Rodríguez Couto & Sanromán (2005) *Chemosphere* **60**, 1124.

Oxaliplatin

This potent anti-cancer drug (FW = 397.29 g/mol; CAS 63121-00-6; IUPAC Name: [(1*R*,2*R*)-cyclohexane-1,2-diamine](ethanedioato-*O,O'*)platinum(II)), also known as Eloxatin®, is a intravenously administered DNA crosslinker and mutagen. The combination of amine group- and carboxyl group-donating bidentate ligands results in significantly lower pharmacologic inactivation as a consequence of nonenzymatic hydrolysis. (**See also** *Cisplatin (for mechanism of action)* and *Carbonatoplatin*) Unlike cisplatin, oxaliplatin forms both interstrand and intrastrand DNA cross links that prevent DNA replication and transcription, thereby promoting programmed cell death (2). Oxaliplatin also crosses the blood-brain barrier. **Key Pharmacokinetic Parameters:** *See* Appendix II in Goodman & Gilman's THE PHARMACOLOGICAL BASIS OF THERAPEUTICS, 12th Edition (Brunton, Chabner & Knollmann, eds.) McGraw-Hill Medical, New York (2011). **1.** Kidani, Noji & Tashiro (1980) *Gann.* **71**, 637. **2.** Graham, Mushin & Kirkpatrick (2004) *Nature Rev. Drug Discovery* **3**, 11.

Oxaloacetate (Oxaloacetic Acid)

This tricarboxylic acid cycle intermediate (FW$_{\text{free-acid}}$ = 132.07 g/mol; CAS 328-42-7; *Abbreviation*: OAA), also known as 2-oxosuccinic acid, 2-oxobutanedioic acid, and oxalacetic acid, is the α-keto acid transamination product of aspartate. **Solution Chemistry:** OAA has pK_a values of 2.22 and 3.67, and the OH group in enol form has a pK_a value of 13.03. Note that it is relatively unstable in aqueous solutions and undergoes β-decarboxylation. The half-life of oxaloacetate at pH 7 and 25°C is between 30 and 60 minutes. Equilibrium exchange studies of malate dehydrogenase were made at 1°C to minimize the decarboxylation rate of oxaloacetate. While the solid dicarboxylic acid exists primarily as the enol, aqueous oxaloacetate, at pH 6 to 10, is roughly 82-88% keto, 7-10% enol, and 5-8% *gem*-diol. At pH 1.3, most of the oxaloacetic acid is present as the *gem*-diol (~81% vs. 13% keto and 6% enol). In diethyl ether, more than 90% is found as the enol. The rate of decarboxylation is maximum at pH 3–4; this rate is accelerated in the presence of metal ions, with the pH optimum shifting to 5–6.5. **Metabolism:** Oxaloacetate is a product of the reactions catalyzed by the malate dehydrogenases, malate oxidase, D- and L-aspartate oxidases, methylmalonyl-CoA carboxyltransferase, aspartate aminotransferase, aspartate:phenylalanine aminotransferase, 4-methyloxaloacetate esterase, citrate lyase, oxalomalate lyase, citryl-CoA lyase, L(+)- and D(−)-tartrate dehydratases, *threo-* and *erythro*-3-hydroxyaspartate ammonia-lyases, and pyruvate carboxylase. It is a substrate for lactate:malate transhydrogenase, citrate (*si*)- and (*re*)-synthases, decylcitrate synthase, 2-methylcitrate synthase, ATP citrate synthase pyridoxamine:oxaloacetate aminotransferase, glycine:oxaloacetate aminotransferase, oxaloacetase, oxaloacetate decarboxylase, phosphoenol-pyruvate carboxylase, phosphoenolpyruvate carboxy-kinases, and oxaloacetate tautomerase. It is also an alternative substrate/product for many other aminotransferases, homocitrate synthase, and opine dehydrogenase. **Target(s):** 2-(acetamidomethylene)-succinate hydrolase (78); acetoacetate decarboxylase (1); acetylpyruvate hydrolase (2); aconitase (3); adenylate cyclase (4-6); adenylosuccinate synthetase (7); aldehyde reductase, aldose reductase (8); allophanate hydrolase (76); amino-acid *N* acetyltransferase (100); 4-aminobutyrate aminotransferase (93); aminolevulinate aminotransferase (91,92); asparagine synthetase (9); aspartate 1-decarboxylase (10,11); γ-butyrobetaine dioxygenase, γ-butyrobetaine hydroxylase, K_i = 0.12 mM (12); citrate lyase (13-15); cyanate hydratase, *or* cyanase (16,17); decylhomocitrate synthase18; fructose-1,6-bisphosphate aldolase (19); fumarylacetoacetase (20); glucose-6-phosphatase (84); glucose-6-phosphate isomerase, weakly inhibited (21); glutamate decarboxylase, K_i = 2.9 mM (22-24); glutamate synthase (25); glycogen synthase (26); guanylate cyclase (27-29); histidine decarboxylase, weakly inhibited (30); homocitrate synthase (97); 3-hydroxybutyrate dehydrogenase (31); 4-hydroxy-4-methyl-2-oxoglutarate aldolase (32); 4-hydroxy-2-oxoglutarate aldolase (33,34); inositol oxygenase (105); isocitrate dehydrogenase (NADP$^+$), particularly in the presence of glyoxylate (36-40); [isocitrate dehydrogenase (NADP$^+$)] kinase (35,85,86); isocitrate lyase (41-45); α-ketoglutarate dehydrogenase (46,47); lactase-phlorizin hydrolase, *or* glycosylceramidase (83); D-lactate dehydrogenase, cytochrome (48,49); lysine 2-monooxygenase (106); malate synthase, weakly inhibited (98,99); maleate hydratase (50); malic enzyme (51); *N*-methylglutamate dehydrogenase (52); *N*-methyl-2-oxoglutaramate hydrolase (77); nicotinate-nucleotide diphosphorylase, carboxylating (95,96); ornithine aminotransferase (94); pantothenase (53,79-82); phospho-gluconate dehydrogenase, decarboxylating (54); procollagen-lysine 5-dioxygenase (101); procollagen-proline 3-dioxygenase (101); procollagen-proline 4-dioxygenase (101-104); pyridoxamine-5-phosphate oxidase (55); pyruvate kinase, weakly inhibited (89,90); pyruvate,water dikinase (87,88); saccharopine dehydrogenase (57); succinate dehydrogenase (49,58-73); succinyl-CoA synthetase (74); theanine hydrolase (75); thiosulfate

sulfurtransferase, *or* rhodanese (56). **1.** Davies (1943) *Biochem. J.* **37**, 230. **2.** Davey & Ribbons (1975) *J. Biol. Chem.* **250**, 3826. **3.** Glusker (1971) *The Enzymes*, 3rd ed., **5**, 413. **4.** Takai, Kurashina & Hayaishi (1974) *Meth. Enzymol.* **38**, 160. **5.** Ide (1971) *Arch. Biochem. Biophys.* **144**, 262. **6.** Yang & Epstein (1983) *J. Biol. Chem.* **258**, 3750. **7.** Markham & Reed (1977) *Arch. Biochem. Biophys.* **184**, 24. **8.** Hayman & Kinoshita (1965) *J. Biol. Chem.* **240**, 877. **9.** Rognes (1980) *Phytochemistry* **19**, 2287. **10.** Williamson (1985) *Meth. Enzymol.* **113**, 589. **11.** Williamson & Brown (1979) *J. Biol. Chem.* **254**, 8074. **12.** Ng, Hanauske-Abel & Englard (1991) *J. Biol. Chem.* **266**, 1526. **13.** Bowen & Rogers (1963) *Biochim. Biophys. Acta* **67**, 633. **14.** Tate & Datta (1965) *Biochem. J.* **94**, 470. **15.** Spector (1972) *The Enzymes*, 3rd ed., **7**, 378. **16.** Anderson, Johnson, Endrizzi, R. M. Little & J. J. Korte (1987) *Biochemistry* **26**, 3938. **17.** Anderson & Little (1986) *Biochemistry* **25**, 1621. **18.** Måhlén (1973) *Eur. J. Biochem.* **38**, 32. **19.** Ujita & Kimura (1982) *Meth. Enzymol.* **90**, 235. **20.** Braun & Schmidt (1973) *Biochemistry* **12**, 4878. **21.** Sangwan & Singh (1989) *J. Biosci.* **14**, 47. **22.** Fonda (1972) *Biochemistry* **11**, 1304. **23.** Gerig & Kwock (1979) *FEBS Lett.* **105**, 155. **24.** Prabhakaran, Harris & Kirchheimer (1983) *Arch. Microbiol.* **134**, 320. **25.** Boland & Benny (1977) *Eur. J. Biochem.* **79**, 355. **26.** Rothman & Cabib (1967) *Biochemistry* **6**, 2098. **27.** White & Zenser (1974) *Meth. Enzymol.* **38**, 192. **28.** Hardman & Sutherland (1969) *J. Biol. Chem.* **244**, 6363. **29.** de Jonge (1975) *FEBS Lett.* **55**, 143. **30.** Sakamoto, Watanabe, Hayashi, Taguchi & Wada (1985) *Agents Actions* **17**, 32. **31.** Green, Dewan & Leloir (1937) *Biochem. J.* **31**, 934. **32.** Tack, Chapman & Dagley (1972) *J. Biol. Chem.* **247**, 6444. **33.** Scholtz & Schuster (1986) *Biochim. Biophys. Acta* **869**, 192. **34.** Anderson, Scholtz & Schuster (1985) *Arch. Biochem. Biophys.* **236**, 82. **35.** Nimmo (1984) *Trends Biochem. Sci.* **9**, 475. **36.** Ochoa & Weisz-Tabori (1948) *J. Biol. Chem.* **174**, 123. **37.** Yoon, Hattori & Shimada (2003) *Biosci. Biotechnol. Biochem.* **67**, 114. **38.** Dhariwal & Venkitasubramanian (1987) *J. Gen. Microbiol.* **133**, 2457. **39.** Nimmo (1986) *Biochem. J.* **234**, 317. **40.** Saiki & Arima (1975) *J. Biochem.* **77**, 233. **41.** McFadden (1969) *Meth. Enzymol.* **13**, 163. **42.** Tanaka, Nabeshima, Tokuda & Fukui (1977) *Agric. Biol. Chem.* **41**, 795. **43.** Takao, Takahashi, Tanida & Takahashi (1984) *J. Ferment. Technol.* **62**, 577. **44.** Munir, Hattori & Shimada (2002) *Arch. Biochem. Biophys.* **399**, 225. **45.** Hönes, Simon & Weber (1991) *J. Basic Microbiol.* **31**, 251. **46.** Bunik & Pavlova (1997) *Biochemistry (Moscow)* **62**, 1012. **47.** Meixner-Monori, Kubicek, Habison, Kubicek-Pranz & Röhr (1985) *J. Bacteriol.* **161**, 265. **48.** Nygaard (1963) *The Enzymes*, 2nd ed., **7**, 557. **49.** Hatefi & Stiggall (1976) *The Enzymes*, 3rd ed., **13**, 175. **50.** Ueda, Yamada & Asano (1994) *Appl. Microbiol. Biotechnol.* **41**, 215. **51.** Kun (1963) *The Enzymes*, 2nd ed., **7**, 149. **52.** Hersh, Stark, Worthen & Fiero (1972) *Arch. Biochem. Biophys.* **150**, 219. **53.** Airas (1979) *Meth. Enzymol.* **62**, 267. **54.** Silverberg & Dalziel (1975) *Meth. Enzymol.* **41**, 214. **55.** Pogell (1975) *Meth. Enzymol.* **6**, 331. **56.** Oi (1975) *J. Biochem.* **78**, 825. **57.** Fujioka & Nakatani (1972) *Eur. J. Biochem.* **25**, 301. **58.** Bonner (1955) *Meth. Enzymol.* **1**, 722. **59.** Bernath & Singer (1962) *Meth. Enzymol.* **5**, 597. **60.** Veeger, DerVartanian & Zeylemaker (1969) *Meth. Enzymol.* **13**, 81. **61.** Hatefi (1978) *Meth. Enzymol.* **53**, 27. **62.** Schäfer, Moll & Schmidt (2001) *Meth. Enzymol.* **331**, 369. **63.** Schlenk (1951) *The Enzymes*, 1st ed., **2** (part 1), 316. **64.** Singer & Kearney (1963) *The Enzymes*, 2nd ed., **7**, 383. **65.** Kusunose, Nagai, Kusunose & Yamamura (1956) *J. Bacteriol.* **72**, 754. **66.** Godzeski & Stone (1955) *Arch. Biochem. Biophys.* **59**, 132. **67.** Swingle, Axelrod & Elvehjem (1942) *J. Biol. Chem.* **145**, 581. **68.** Dickens (1946) *Biochem. J.* **40**, 171. **69.** Pardee & Potter (1948) *J. Biol. Chem.* **176**, 1085. **70.** Mandrik, Vonsovich & Vinogradov (1983) *Ukr. Biokhim. Zh.* **55**, 503. **71.** Suraveratum, Krungkrai, Leangaramgul, Prapunwattana & Krungkrai (2000) *Mol. Biochem. Parasitol.* **105**, 215. **72.** Pardee, Potter & Lyle (1948) *J. Biol. Chem.* **176**, 1085. **73.** Das (1937) *Biochem. J.* **31**, 1124. **74.** Leitzmann, Wu & Boyer (1970) *Biochemistry* **9**, 2338. **75.** Tsushida & Takeo (1985) *Agric. Biol. Chem.* **49**, 2913. **76.** Maitz, Haas & Castric (1982) *Biochim. Biophys. Acta* **714**, 486. **77.** Hersh (1970) *J. Biol. Chem.* **245**, 3526. **78.** Huynh & Snell (1985) *J. Biol. Chem.* **260**, 2379. **79.** Airas (1976) *Biochim. Biophys. Acta* **452**, 193 and 201. **80.** Airas (1983) *Anal. Biochem.* **134**, 122. **81.** Airas (1988) *Biochem. J.* **250**, 447. **82.** Airas (1976) *Biochem. J.* **157**, 415. **83.** Hamaswamy & Radhakrishnan (1973) *Biochem. Biophys. Res. Commun.* **54**, 197. **84.** Mithieux, Vega & Riou (1990) *J. Biol. Chem.* **265**, 20364. **85.** Nimmo & Nimmo (1984) *Eur. J. Biochem.* **141**, 409. **86.** Wang & Koshland (1982) *Arch. Biochem. Biophys.* **218**, 59. **87.** Cooper & Kornberg (1974) *The Enzymes*, 3rd ed. (Boyer, ed.) **10**, 631. **88.** Chulavatnatol & Atkinson (1973) *J. Biol. Chem.* **248**, 2712. **89.** Andre, Froehlich, Moll & Benning (2007) *Plant Cell* **19**, 2006. **90.** Gupta & Singh (1989) *Plant Physiol. Biochem.* **27**, 703. **91.** Bajkowski & Friedmann (1982) *J. Biol. Chem.* **257**, 2207. **92.** Neuberger & Turner

(1963) *Biochim. Biophys. Acta* **67**, 342. **93**. Tamaki, Kubo, Aoyama & Funatsuka (1983) *J. Biochem.* **93**, 955. **94**. Goldberg, Flescher & Lengy (1979) *Exp. Parasitol.* **47**, 333. **95**. Shibata & Iwai (1980) *Biochim. Biophys. Acta* **611**, 280. **96**. Iwai, Shibata & Taguchi (1979) *Agric. Biol. Chem.* **43**, 351. **97**. Qian, West & Cook (2006) *Biochemistry* **45**, 12136. **98**. Munir, Hattori & Shimada (2002) *Biosci. Biotechnol. Biochem.* **66**, 576. **99**. Durchschlag, Biedermann & Eggerer (1981) *Eur. J. Biochem.* **114**, 255. **100**. Hinde, Jacobson, Weiss & Davis (1986) *J. Biol. Chem.* **261**, 5848. **101**. Majamaa, Turpeenniemi-Hujanen, Latipää, *et al.* (1985) *Biochem. J.* **229**, 127. **102**. Majamaa, Hanauske-Abel, Günzler & Kivirikko (1984) *Eur. J. Biochem.* **138**, 239. **103**. Kaska, Günzler, Kivirikko & Myllylä (1987) *Biochem. J.* **241**, 483. **104**. Tuderman, Myllylä & Kivirikko (1977) *Eur. J. Biochem.* **80**, 341. **105**. Reddy, Pierzchala & Hamilton (1981) *J. Biol. Chem.* **256**, 8519. **106**. Vandecasteele & Hermann (1972) *Eur. J. Biochem.* **31**, 80.

N-Oxaloglycine

N-Oxaloglycine

Dimethyloxalylglycine

This iron-binding α-ketoglutarate (2-oxoglutarate) analogue (FW = 147.09 g/mol; also named *N*-oxaloglycine) competitively inhibits prolyl-4-hydroxylase (*Reaction*: Procollagen(L-proline) + α-ketoglutarate + O_2 → Procollagen((2S,4R)-4-hydroxyproline) + succinate + CO_2) (1). During catalysis, prolyl-4-hydroxylase forms Fe(III), and the latter most likely makes an extremely stable metal ion complex with *N*-oxaloglycine. Substitution on the glycine moiety alters inhibitor activity stereoselectively and that, if the ω-carboxylate is homologated or replaced, either by acylsulfonamides or anilide, activity is likewise sharply reduced (1). **Prolyl-4-Hydroxylase Catalysis:** Each catalytic round of this posttranslational modifying enzyme reaction occurs in two stages. O_2 is bound end-on in an axial position, producing a dioxygen unit. Nucleophilic attack at C-2 generates a tetrahedral intermediate, with loss of the double bond in dioxygen and bonds to iron and the α-carbon of α-ketoglutarate. Elimination of CO_2 coincides with formation of the Fe(IV)=O species. The second stage involves the abstraction of the pro-*R* hydrogen atom from C-4 of proline, followed by radical combination, yielding hydroxyproline (2). In the presence of α-ketoglutarate, enzyme-bound Fe^{2+} is rapidly converted to Fe^{3+}, resulting in inactivation of the enzyme (3) Ascorbate is utilized as a cofactor to reduce Fe(III) back to Fe(II) (4). **Cell-Permeable Analogue:** Dimethyloxalylglycine (FW = 175.14 g/mol; CAS 89464-63-1; Symbol: DMOG, also named *N*-(methoxyoxoacetyl)-glycine methyl ester) is metabolicaly demethylated to form *N*-oxaloglycine upon entry to many cells. **1**. Cunliffe, Franklin, Hales & Hill (1992) *J. Med. Chem.* **35**, 2652. **2**. Fujita, Gottlieb, Peterkofsky, Udenfriend, & Witkop (1964) *J. Amer. Chem. Soc.* **86**, 4709. **3**. De Jong, Albracht & Kemp, A (1982) *Biochim. Biophys. Acta* **704**, 326. **4**. De Jong & Kemp (1984) *Biochim. Biophys. Acta* **787**, 105.

Oxalomalate (Oxalomalic Acid)

This α-hydroxy-β-oxalosuccinate (FW$_{free-acid}$ = 206.11 g/mol), which is formed by the combination of glyoxylate and oxaloacetate, spontaneously loses carbon dioxide to form γ-hydroxy-α-ketoglutarate ($^-$OOC-CH(OH)-CH$_2$-COCOO$^-$). Oxalomalate is a competitive inhibitor of aconitase (1,3) and NADP$^+$-dependent isocitrate dehydrogenase (3-7). It also modulates the RNA-binding activity of iron-regulatory proteins. **1**. Glusker (1971) *The Enzymes*, 3rd ed., **5**, 413. **2**. Adinolfi, Guarriera-Bobyleva, Olezza & Ruffo (1971) *Biochem. J.* **125**, 557. **3**. Dhariwal & Venkitasubramanian (1987) *J.*

Gen. Microbiol. **133**, 2457. **4**. Nimmo (1986) *Biochem. J.* **234**, 317. **5**. Johanson & Reeves (1977) *Biochim. Biophys. Acta* **483**, 24. **6**. Ingebretsen (1976) *Biochim. Biophys. Acta* **452**, 302. **7**. Ruffo, Moratti, Montani & d'Eril (1974) *Ital. J. Biochem.* **23**, 357.

β-*N*-Oxalyl-L-α,β-diaminopropionate

This excitatory neurotoxin (FW$_{hydrochloride}$ = 212.59 g/mol; water-soluble; pK_a values of 1.95, 2.95, and 9.25), also known as β-*N*-oxalyl-amino-L-alanine, was first isolated from the seeds of *Lathyrus sativus* (1) results in the progressive neurodegenerative condition known as neurolathyrism. Ingestion of this amino acid results in a paralytic condition characterized by lack of strength in or inability to move the lower limbs. Neurolathyrism is a motor neuron disease affecting upper motor neurons and anterior horn cells of the lumbar spinal cord. Note that β-*N*-oxalyl-L-α,β-diaminopropionate is a glutamate analogue and acts as an agonist at the AMPA receptor. **Mechanism of Action:** Although the receptor antagonist 2-amino-5-phosphonovaleric acid and divalent cadmium ion did not alter the conductance increase evoked by β-ODAP, they markedly depressed responses to L-aspartate. Such differences suggest that β-ODAP's mechanism of excitatory action is similar to that of kainic and quisqualic acids (3). The membrane conductance effects of α-ODAP, the 2-oxalylamino isomer, are about 10x weaker (1). With voltage-clamp recordings of an AMPA receptor, β-ODAP was a strong agonist, the potency being almost the same as L-glutamate. On the other hand, β-ODAP had little effect on the glutamate-evoked currents through the expressed NMDA receptor, showing only a weak inhibitory effect on the glycine-modulatory site (4). **Target(s):** NADH dehydrogenase, *or* Complex I (2); tyrosine aminotransferase (3). **1**. Rao, Adiga & Sarma (1964) *Biochemistry* **3**, 432. **2**. Ravindranath (2002) *Neurochem. Int.* **40**, 505. **3**. Shasi Vardhan, Pratap Rudra & Rao (1997) *J. Neurochem.* **68**, 2477. **3**. MacDonald & Morris (1984) *Exp. Brain Res.* **57**, 158. **4**. Kusama-Eguchi, Ikegami, Kusama, *et al.* (1996) *Environ. Toxicol. Pharmacol.* **2**, 339.

N-(Oxalyl)glycine

This acylated glycine (FW$_{free-acid}$ = 147.09 g/mol), also called oxalglycine and oxalylglycine, is a structural analogue of α-ketoglutarate. **Target(s):** glutamate dehydrogenase (1-3); [histone-H$_3$]-lysine-36 demethylase (4); homocitrate synthase (5); isocitrate dehydrogenase (6-8); peptide-aspartate β-dioxygenase (9); procollagen-proline 4-dioxygenase (9-14); proline 4-dioxygenase (15); taurine dioxygenase (16). **1**. Rife & Cleland (1980) *Biochemistry* **19**, 2238. **2**. Fisher, Pazhanisamy & Medary (1987) *J. Biol. Chem.* **262**, 11684. **3**. Fisher, Medary, Wykes & Wolfe (1984) *J. Biol. Chem.* **259**, 4105. **4**. Marmorstein & Trievel (2009) *Biochim. Biophys. Acta* **1789**, 58. **5**. Qian, West & Cook (2006) *Biochemistry* **45**, 12136. **6**. Levy & Villafranca (1977) *Biochemistry* **16**, 3301. **7**. Grissom & Cleland (1988) *Biochemistry* **27**, 2934. **8**. Northrop & Cleland (1974) *J. Biol. Chem.* **249**, 2928. **9**. Koivunen, Hirsilä, Günzler, Kivirikko & Myllyharju (2004) *J. Biol. Chem.* **279**, 9899. **10**. Baader, Tschank, Baringhaus, Burghard & Günzler (1994) *Biochem. J.* **300**, 525. **11**. Cunliffe, Franklin, Hales & Hill (1992) *J. Med. Chem.* **35**, 2652. **12**. Wojtaszek, Smith & Bolwell (1999) *Int. J. Biochem. Cell Biol.* **31**, 463. **13**. Myllyharju (2008) *Ann. Med.* **40**, 402. **14**. Hirsilä, Koivunen, Günzler, Kivirikko & Myllyhatju (2003) *J. Biol. Chem.* **278**, 30772. **15**. Lawrence, Sobey, Field, Baldwin & Schofield (1996) *Biochem. J.* **313**, 185. **16**. Kalliri, Grzyska & Hausinger (2005) *Biochem. Biophys. Res. Commun.* **338**, 191.

Oxamate (Oxamic Acid)

This dicarboxylic monoamide (FW$_{\text{free-acid}}$ = 89.05 g/mol), also known as aminooxoacetic acid and oxalic acid monoamide, is an isoelectronic and isosteric analogue of pyruvic acid. The free acid is sparingly soluble in water whereas the ammonium and sodium salts are very soluble. **Target(s):** acetolactate synthase (1); alanopine dehydrogenase (2); aspartate aminotransferase (3); carbonic anhydrase (4); ε-crystallin L-lactate dehydrogenase (5); dehydrogluconate dehydrogenase (6); gluconate 2-dehydrogenase (7); glutamate dehydrogenase, weakly inhibited (8); glyoxylate reductase (9,10); D-lactate dehydrogenase (11,12); L-lactate dehydrogenase (5,8,9,13-24); lactate 2-monooxygenase (25); lactate racemase (26); malate synthase (27); pyruvate carboxylase (26,28-30). **1.** Huseby & Stormer (1971) *Eur. J. Biochem.* **20**, 215. **2.** Fields & Hochachka (1981) *Eur. J. Biochem.* **114**, 615. **3.** Rej (1979) *Clin. Chem.* **25**, 555. **4.** Rogers, Mukherjee & Khalifah (1987) *Biochemistry* **26**, 5672. **5.** Chang, Huang & Chiou (1991) *Arch. Biochem. Biophys.* **284**, 285. **6.** Shinagawa & Ameyama (1982) *Meth. Enzymol.* **89**, 194. **7.** Matsushita, Shinagawa & Ameyama (1982) *Meth. Enzymol.* **89**, 187. **8.** Baker (1967) *Design of Active-Site-Directed Irreversible Enzyme Inhibitors*, Wiley, New York. **9.** Webb (1966) *Enzyme and Metabolic Inhibitors*, vol. **2**, Academic Press, New York. **10.** Zelitch (1955) *J. Biol. Chem.* **216**, 553. **11.** Long (1975) *Meth. Enzymol.* **41**, 313. **12.** Kaczorowski, Kohn & Kaback (1978) *Meth. Enzymol.* **53**, 519. **13.** Schwert & Winer (1963) *The Enzymes*, 3rd. ed., **7**, 127. **14.** Neilands (1955) *Meth. Enzymol.* **1**, 449. **15.** Dennis (1962) *Meth. Enzymol.* **5**, 426. **16.** Snoswell (1966) *Meth. Enzymol.* **9**, 321. **17.** Goldberg (1975) *Meth. Enzymol.* **41**, 318. **18.** Hillier & Jago (1982) *Meth. Enzymol.* **89**, 362. **19.** Schwert & Winer (1963) *The Enzymes*, 2nd ed., **7**, 142. **20.** Novoa, Winer, Glaid & Schwert (1959) *J. Biol. Chem.* **234**, 1143. **21.** Dalziel (1975) *The Enzymes*, 3rd ed., **11**, 1. **22.** Holbrook, Liljas, Steindel & Rossmann (1975) *The Enzymes*, 3rd ed., **11**, 191. **23.** Hakala, Glaid & Schwert (1953) *Fed. Proc.* **12**, 213. **24.** Okabe, Hayakawa, Hamada & Koike (1968) *Biochemistry* **7**, 79. **25.** Ghisla & Massey (1975) *J. Biol. Chem.* **250**, 577. **26.** Hiyama, Fukui & Kitahara (1968) *J. Biochem.* **64**, 99. **27.** Miernyk & Trelease (1981) *Phytochemistry* **20**, 2657. **28.** Martin-Requero, Ayuso & Parrilla (1986) *J. Biol. Chem.* **261**, 13973. **29.** Scrutton, Olmsted & Utter (1969) *Meth. Enzymol.* **13**, 235. **30.** Seubert & Weicker (1969) *Meth. Enzymol.* **13**, 258.

Oxamflatin

This anti-tumor agent (FW = 342.38 g/mol; CAS 151720-43-3; Solubility: 13 mg/mL DMSO, methanol, acetonitrile, 10 mg/mL), also known as (2*E*)-5-[3-[(phenylsulfonyl)amino]phenyl]-pent-2-en-4-ynohydroxamate, inhibits histone deacetylase. Oxamflatin shows antiproliferative activity *in vitro* against various mouse and human tumor cell lines, attended by drastic changes in the cell morphology and *in vivo* antitumor activity against B16 melanoma. Oxamflatin causes an elongated cell shape with filamentous protrusions as well as arrest of the cell cycle at the G$_1$ phase in HeLa cells. These phenotypic changes in HeLa cells were similar to those induced by trichostatin A (TSA), a specific inhibitor of histone deacetylase (HDAC). **1.** Kim, Lee, Sugita, Yoshida & Horinouchi (1999) *Oncogene* **18**, 2461. **2.** Monneret (2005) *Eur. J. Med. Chem.* **40**, 1.

Oxanilate (Oxanilic Acid)

This acid (FW$_{\text{free acid}}$ = 165.15 g/mol), also known as *N*-phenyloxamate and oxanilic acid, is a structural analogue of phenylpyruvate. **Target(s):** glutamate dehydrogenase, weakly inhibited (1); lactate dehydrogenase (1,2); neuraminidase, *or* sialidase (3); pyruvate decarboxylase, weakly inhibited (4). **1.** Baker, Lee, Tong, Ross & Martinez (1962) *J. Theoret. Biol.* **3**, 446. **2.** Yu, Deck, Hunsaker, Deck, Royer, Goldberg & Vander Jagt (2001) *Biochem. Pharmacol.* **62**, 81. **3.** Gottschalk & Bhargava (1971) *The Enzymes*, 3rd ed., **5**, 321. **4.** Gale (1961) *Arch. Biochem. Biophys.* **94**, 236.

4-(2-Oxapentadeca-4-yne)phenylpropanoate

This isoform-specific inhibitor (FW$_{\text{free-acid}}$ = 358.52 g/mol) binds to both the ferrous- and ferric-forms of 12-lipoxygenase (arachidonate 12-lipoxygenase), binding to the Fe^{2+} form with K_i of 70 nM. **1.** Moody & Marnett (2002) *Biochemistry* **41**, 10297. **2.** Richards, Moody & Marnett (1999) *Biochemistry* **38**, 16529.

Oxatomide

This anti-allergic agent (FW = 426.56 g/mol), initially characterized as an H$_1$ antagonist, inhibits the secretion of a number of inflammation mediators in human basophils and mast cells. *In vitro*, oxatimide inhibits the release of both preformed (*e.g.*, histamine and tryptase) and *de novo* synthesized mediators (*e.g.*, leukotriene C$_4$ and prostaglandin D$_2$). **Target(s):** dopamine uptake (1); inflammation mediator release (2,3); NADPH oxidase, human neutrophil (4) **1.** Matsunaga, Sato, Shuto, *et al.* (1998) *Eur. J. Pharmacol.* **350**, 165. **2.** Marone, Granata, Spadaro, Onorati & Triggiani (1999) *J. Investig. Allergol. Clin. Immunol.* **9**, 207. **3.** Awouters, Niemegeers, Van den Berk, *et al.* (1977) *Experientia* **33**, 165. **4.** Umeki (1992) *Biochem. Pharmacol.* **43**, 1109.

Oxazepam

This benzodiazepinone (FW = 286.72 g/mol) is an anxiolytic agent that inhibits a number of UDP-glucuronosyltransferases (1-8) and nitric-oxide synthase (9). **1.** Kirkwood, Nation & Somogyi (1998) *Clin. Exp. Pharmacol. Physiol.* **25**, 266. **2.** Sim, Back & Breckenridge (1991) *Brit. J. Clin. Pharmacol.* **32**, 17. **3.** Rajaonarison, Lacarelle, De Sousa, Catalin & Rahmani (1991) *Drug Metab. Dispos.* **19**, 809. **4.** Wahlstrom, Pacifici, Lindstrom, Hammar & Rane (1988) *Brit. J. Pharmacol.* **94**, 864. **5.** Meacham, Sisenwine, Liu, *et al.* (1986) *Drug Metab. Dispos.* **14**, 430. **6.** Pacifici & Rane (1981) *Drug Metab. Dispos.* **9**, 569. **7.** Dybing (1976) *Biochem. Pharmacol.* **25**, 1421. **8.** Hara, Nakajima, Miyamoto & Yokoi (2007) *Drug Metab. Pharmacokinet.* **22**, 103. **9.** Fernandez-Cancio, Fernandez-Vitos, Imperial & Centelles (2001) *Brain Res. Mol. Brain Res.* **96**, 87.

Oxindolyl-L-alanine

This tryptophan analogue (FW = 220.23 g/mol) is a tight-binding competitive inhibitor of both and tryptophanase (*or* tryptophan indole-lyase) (1,2) and tryptophan synthase (3,4), with Ki values in the 10 μM range, corresponding to an affinity that is 10-100-fold higher than the corresponding K_m or K_i values for the substrate L-tryptophan. Its mode of inhibition suggests that substrate analogues with tetrahedral geometry at C-3 induce new enzyme-substrate contacts, resulting in the accumulation of a putative a *gem*-diamine intermediate (4). **1**. Kiick & Phillips (1988) *Biochemistry* **27**, 7339. **2**. Faleev, Dementieva, Zakomirdina, Gogoleva & Belikov (1994) *Biochem. Mol. Biol. Int.* **34**, 209. **3**. Roy & Miles (1999) *J. Biol. Chem.* **274**, 31189. **4**. Roy, Miles, Phillips & Dunn (1988) *Biochemistry* **27**, 8661.

Oxipurinol

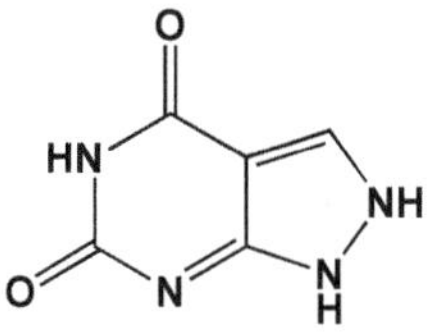

This xanthine analogue and potent xanthine oxidase inhibitor (FW = 152.11 g/mol; CAAS Registry Number = 2465-59-0), also known as alloxanthine, oxoallopurinol, and 1*H*-pyrazolo[3,4-*d*]pyrimidine-4,6-diol; is formed metabolically from allopurinol, thereby accounting for allopurinol hypouricemic efficacy. (**See Allopurinol for oxypurinol pharmacokinetics.**) **Target(s):** hypoxanthine (guanine) phosphoribosyltransferase (1); orotate phosphoribosyltransferase (2); purine-nucleoside phosphorylase (3); salicylhydroxamic acid reductase (4); xanthine dehydrogenase (5-8); xanthine oxidase (9-14); xanthine phosphoribosyltransferase, also an alternative substrate (15). **1**. Miller, Ramsey, Krenitsky & Elion (1972) *Biochemistry* **11**, 4723. **2**. Jones, Kavipurapu & Traut (1978) *Meth. Enzymol.* **51**, 155. **3**. Parks & Agarwal (1972) *The Enzymes*, 3rd ed., 7, 483. **4**. Katsura, Kitamura & Tatsumi (1993) *Arch. Biochem. Biophys.* **302**, 356. **5**. Truglio, Theis, Leimkühler, *et al.* (2002) *Structure* **10**, 115. **6**. Fhaolain & Coughlan (1978) *FEBS Lett.* **90**, 305. **7**. Atmani, Baghiani, Harrison & Benboubetra (2005) *Int. Dairy J.* **15**, 1113. **8**. Okamoto, Eger, Nishino, Pai & Nishino (2008) *Nucleosides Nucleotides Nucleic Acids* **27**, 888. **9**. Massey, Komai, Palmer & Elion (1970) *J. Biol. Chem.* **245**, 2837. **10**. Bray (1975) *The Enzymes*, 3rd ed., **12**, 299. **11**. Spector, Hall & Krenitsky (1986) *Biochem. Pharmacol.* **35**, 3109. **12**. Hawkes, George & Bray (1984) *Biochem. J.* **218**, 961. **13**. Chalmers, Kromer, Scott & Watts (1968) *Clin. Sci.* **35**, 353. **14**. Terada, Leff & Repine (1990) *Meth. Enzymol.* **186**, 651. **15**. Miller, Adamczyk, Fyfe & Elion (1974) *Arch. Biochem. Biophys.* **165**, 349.

1-Oxipurinol Ribonucleoside 5'-Monophosphate

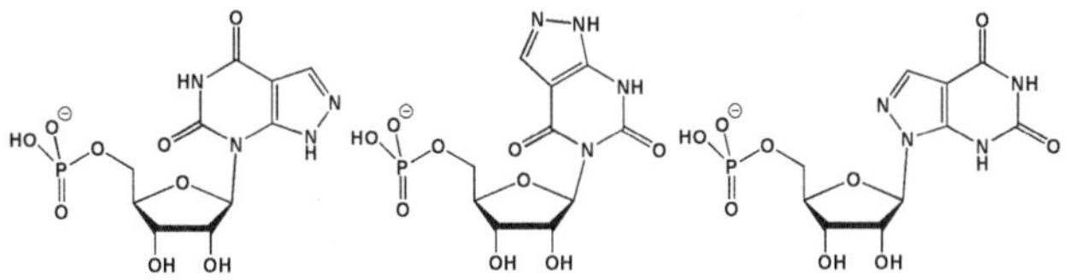

This ribonucleotide (FW$_{free-acid}$ = 364.21 g/mol), also known as 1-ribosyloxipurinol 5'-phosphate, inhibits orotidylate decarboxylase. Three separate numbering schemes have been used with respect to allopurinol and oxipurinol; hence, the term could be applied to each of the three nucleotides shown above. Orotidine-5'-phosphate decarboxylase is inhibited by the first and third nucleotides depicted above, and the third is the one originally called the 1-oxipurinol nucleotide. **1**. Silva & Hatfield (1978) *Meth. Enzymol.* **51**, 143. **2**. Brown & O'Sullivan (1977) *Biochem. Pharmacol.* **26**, 1947. **3**. Porter & Short (2000) *Biochemistry* **39**, 11788. **4**. Acheson, Bell, Jones & Wolfenden (1990) *Biochemistry* **29**, 3198. **5**. Fyfe, Miller & Krenitsky (1973) *J. Biol. Chem.* **248**, 3801.

7-Oxipurinol Ribonucleoside 5'-Monophosphate

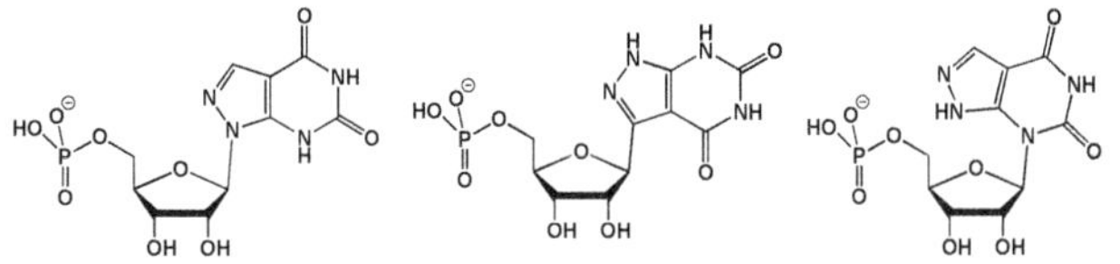

This ribonucleotide (FW$_{free-acid}$ = 364.21 g/mol) inhibits orotidylate decarboxylase. Care should be exercised when this name appears in the literature. Three separate numbering schemes have been used with respect to allopurinol and oxipurinol; hence, this name can apply to each of the three depicted nucleotides depicted. Orotidine-5'-phosphate decarboxylase

is inhibited by the left and right structures shown above. (The rightmost derivative was originally called the 7-oxipurinol nucleotide.) **1**. Silva & Hatfield (1978) *Meth. Enzymol.* **51**, 143. **2**. Brown & O'Sullivan (1977) *Biochem. Pharmacol.* **26**, 1947. **3**. Porter & Short (2000) *Biochemistry* **39**, 11788. **4**. Fyfe, Miller & Krenitsky (1973) *J. Biol. Chem.* **248**, 3801.

Oxo acid, *See the corresponding* keto acid

cis-3-Oxo-8-*trans*-(*N*[1]-acrylamidospermidyl)-9*b*-*trans*-ethylacrylyl-1,2,3,4,4*a*,9*b*-hexahydrodibenzofuran

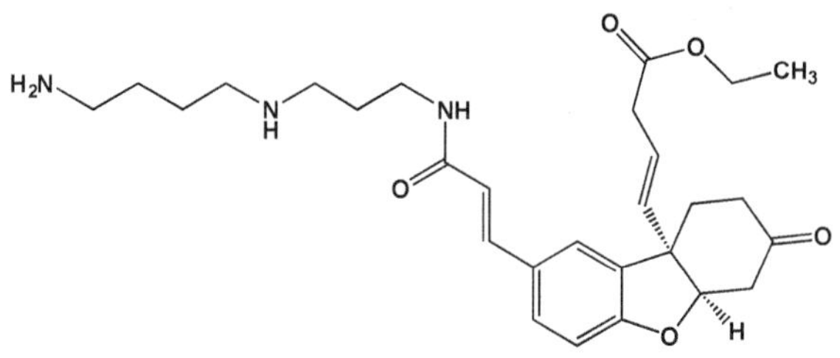

This lunarine analogue (FW$_{free-base}$ = 497.63 g/mol) is a time-dependent inhibitor of trypanothione-disulfide reductase. Kinetic data are consistent with an inactivation mechanism involving a conjugate addition of an active site cysteine residue onto the C^{24}-C^{25} double bond of its tricyclic nucleus, confirming the importance of the unique structure of the tricyclic core as a motif for inhibitor design. **1**. Hamilton, Saravanamuthu, Poupat, Fairlamb & Eggleston (2005) *Bioorg. Med. Chem.* **14**, 2266.

24-Oxo-25-azacycloartanol

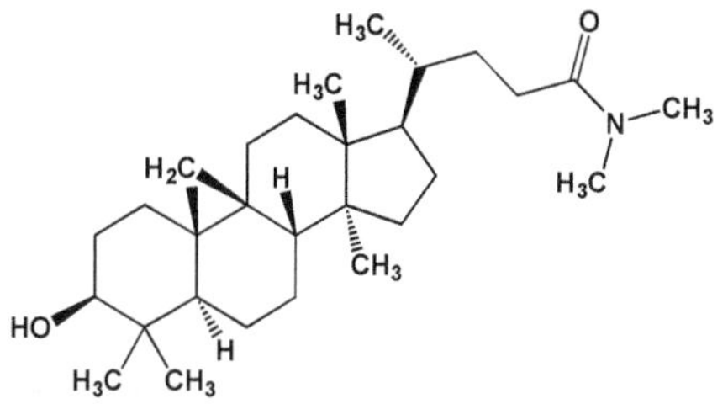

This sterol analogue (FW = 443.71 g/mol) inhibits sterol 24-*C*-methyltransferase (maize, K_i = 170 nM). Microsomes from maize seedlings catalyze C-24 alkylation of 4,4,14 α-trimethyl-9-β,19-cyclo-5-α-cholest-24-en-3-β-ol (*i.e.*, cycloartenol) by (*S*)-adenosyl-L-methionine leading to 24-methylene cycloartanol. Derivatives of cycloartenol bearing a nitrogen atom at C-25 have been previously shown to be potent inhibitors of the AdoMet-cycloartenol-C-24-methyltransferase (2). The presence of a positive charge at Position-25 is the major cause of inhibition, because electrostatically neutral isosteric compounds (with a carbon in place of the nitrogen atom) do not inhibit (1). The positive charge leading to inhibition may be conferred by a protonated amine, a quaternary ammonium group, as well as by a sulfonium or an arsonium group (1). A steroid-like structure of the inhibitor is also important. Finally, the presence of a free 3β-hydroxy group and the bent conformation of cycloartenol, which are essential molecular features of the substrate for the methylation reaction, are no longer required for inhibition (1). **1**. Rahier, Génot, Schuber, Benveniste & Narula (1984) *J. Biol. Chem.* **259**, 15215. **2**. Narula, Rahier, Benveniste & Schuber (1981) *J. Am. Chem. Soc.* **103**, 2408.

Oxodipine

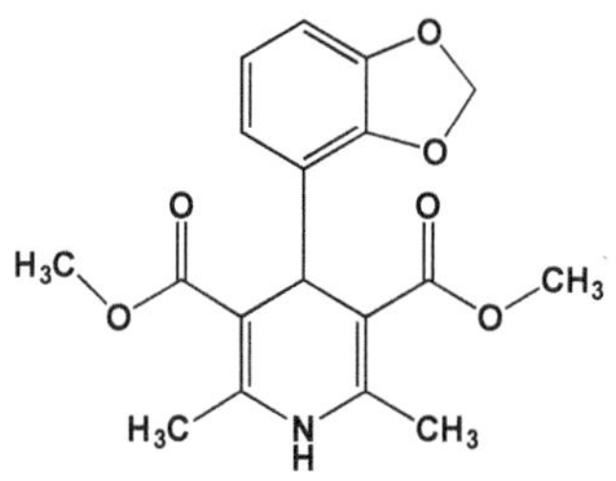

This dihydropyridine-class calcium ion blocker (FW = 359.37 g/mol; CAS 90729-41-2), also known as 4-(1,3-benzodioxol-4-yl)-2,6-dimethyl-1,4-dihydropyridine-3,5-dicarboxylic acid O^3-ethyl,O^5-methyl ester, greatly depresses KCl-induced contraction of rabbit aorta and also decreases the contractile force of rat ventricular strips, the latter with lower potency. Oxodipine markedly shortens cardiac action potentials, and, in rat cultured neonatal ventricular myocytes, oxodipine decreases the L-type Ca^{2+} current (I_{CaL}) with IC_{50} of 0.24 μM and is also an effective blocker of T-type Ca^{2+} current (I_{CaT}) than elgodipine ($IC_{50} = 0.41$ μM). 1. Galán, Talavera, Vassort & Alvarez (1998) *Eur. J. Pharmacol.* **357**, 93.

2-Oxoglutaconic Acid & Dimethyl Esters

trans-isomer   **cis-isomer**

trans-isomer   **cis-isomer**

A *cis/trans*-mixture of these α-ketoglutarate analogues (FW$_{acid}$ = 142.07 g/mol; FW$_{ester}$ = 155.13 g/mol), also known as α-ketoglutaconic acid, irreversibly inhibits aspartate aminotransferase. It is an alternative substrate for glutamate dehydrogenase. Note that the dimethyl ester also inactivates. 1. Kato, Asano, Makar & Cooper (1996) *J. Biochem.* **120**, 531.

Oxolinate (Oxolinic Acid)

This quinolone-type antibiotic (FW$_{free-acid}$ = 261.23 g/mol), also known as 5-ethyl-5,8-dihydro-8-oxo-1,3-dioxolo[4,5-g]quinoline-7-carboxylate, inhibits bacterial DNA gyrase, binding to subunit A and arresting DNA replication (1-7). Certain mutant DNA gyrases fail to bind oxolinic acid, accounting for oxolinate-resistant bacteria. In male rats, oxolinate raises the serum levels of luteinizing hormone (LH), increasing the incidence of testicular Leydig cell tumors (8). It also inhibits (IC_{50} = 4.3 μM) neuronal dopamine uptake (9). **Target(s):** DNA topoisomerase IV (2,10); glycyl tRNA synthetase (11,12); and leucyl-tRNA synthetase (11). 1. Gellert (1981) *The Enzymes*, 3rd ed., **14**, 345. 2. Chen, Malik, Snyder & Drlica (1996) *J. Mol. Biol.* **258**, 627. 3. Thielmann, Popanda & Edler (1991) *J. Cancer Res. Clin. Oncol.* **117**, 19. 4. Huff & Kreuzer (1990) *J. Biol. Chem.* **265**, 20496. 5. Sugino, Peebles, Kreuzer & Cozzarelli (1977) *Proc. Natl. Acad. Sci. U.S.A.* **74**, 4767. 6. Gellert (1981) *Ann. Rev. Biochem.* **50**, 879. 7. Vosberg (1985) *Curr. Top. Microbiol. Immunol.* **114**, 19. 8. Yamada, Nakamura, Okuno, et al. (1995) *Toxicol. Appl. Pharmacol.* **134**, 35. 9. Garcia de Mateos-Verchere, Vaugeois, Naudin & Costentin (1998) *Eur. Neuropsychopharmacol.* **8**, 255. 10. Saiki, Shen, Chen, Baranowski & Lerner (1999) *Antimicrob. Agents Chemother.* **43**, 1574. 11. Wright, Nurse & Goldstein (1981) *Science* **213**, 455. 12. Freist, Logan & Gauss (1996) *Biol. Chem. Hoppe-Seyler* **377**, 343.

Oxonate (Oxonic Acid)

This orotic acid analogue (FW$_{free-acid}$ = 157.09 g/mol), also known as 5-azaorotic acid and allantoxanic acid, is often used in cancer chemotherapy, because it suppresses the gastrointestinal toxicity of 5-flurouracil and its derivatives without altering their antitumor activity. Note that the free acid is very unstable (pK_a = 6.85-6.90), and it is frequently prepared as the monopotassium salt. **Target(s):** dihydroorotate dehydrogenase (1); orotate

phosphoribosyl-transferase (1-4); orotidylate decarboxylase (1,3); urate oxidase, uricase; K_i = 0.1 μM (5-9); urate transport (9-11). 1. Gero, O'Sullivan & Brown (1985) *Biochem. Med.* **34**, 60. 2. Iltzsch, Niedzwicki, Senft, Cha & el Kouni (1984) *Mol. Biochem. Parasitol.* **12**, 153. 3. O'Sullivan & Ketley (1980) *Ann. Trop. Med. Parasitol.* **74**, 109. 4. Javaid, el Kouni & Iltzsch (1999) *Biochem. Pharmacol.* **58**, 1457. 5. Fridovich (1965) *J. Biol. Chem.* **240**, 2491. 6. Cartier & Thuillier (1974) *Anal. Biochem.* **61**, 416. 7. Truscoe (1968) *Enzymologia* **34**, 337. 8. Bongaerts, Uitzetter, Brouns & Vogels (1978) *Biochim. Biophys. Acta* **527**, 348. 9. Leal-Pinto, London, Knorr & Abramson (1995) *J. Membr. Biol.* **146**, 123. 10. Nies, Kinne, Kinne-Saffran & Grieshaber (1995) *Amer. J. Physiol.* **269**, R339. 11. Knorr, Beck & Abramson (1994) *Kidney Int.* **45**, 727.

Oxophenarsine

This sulfhydryl-reactive arsenical (FW = 199.04 g/mol; CAS 306-12-7), variously named Mapharsen, Ehrlich-5, oxyphenarsine, oxarsan, arsinoxide, 3-amino-4-hydroxybenzenearsenoxide, and 2-amino-4-arsenosophenol, was originally developed by Nobelist Paul Ehrlich as an anti-syphilis drug. (Note: do not confuse this compound with oxophenylarsine, also known phenylarsine oxide). It is now used as an antiprotozomal agent, particularly against *Trypanosoma*. Oxophenarsine is typically supplied as the hydrochloride (FW = 235.50 g/mol) and is readily soluble in water. (*See also Dichlorophenarsine*) **Target(s):** acetyl-CoA acyltransferase (1); actomyosin ATPase (2); aldehyde dehydrogenase (3-5); alkaline phosphatase (6); asparagine synthetase (7); ATPase, rat brain (8); ATPase, trypanosomal (9); cathepsin (10); cholinesterase (11,12); deoxyribonuclease (13); diisopropylfluorophosphatase (14); γ-glutamyl hydrolase, *or* pteroylglutamate conjugase (15); glyceraldehyde-3-phosphate dehydrogenase (9); α-glycerophosphate oxidase (16); hexokinase (9); myosin ATPase (2,17,18); pyrophosphatase (19); pyruvate decarboxylase (20,21); pyruvate oxidase (22); ribose-5-phosphate isomerase (23); succinate dehydrogenase (12,24,25); succinate oxidase (26); urease (12,27). 1. Stern (1955) *Meth. Enzymol.* **1**, 573. 2. Mugikura, Miyazaki & Nagai (1956) *Enzymologia* **17**, 321. 3. Stoppani & Milstein (1957) *Biochim. Biophys. Acta* **24**, 655. 4. Bradbury, Clark, Steinman & Jakoby (1975) *Meth. Enzymol.* **41**, 354. 5. Jakoby (1963) *The Enzymes*, 2nd ed., **7**, 203. 6. Lazdunski & Ouellet (1962) *Can.J. Biochem. Physiol.* **40**, 1619. 7. Milman & Cooney (1979) *Biochem. J.* **181**, 51. 8. Gordon (1953) *Biochem. J.* **55**, 812. 9. Chen (1948) *J. Infect. Diseases* **82**, 226. 10. Maschmann (1935) *Biochem. Z.* **280**, 204. 11. Markwardt (1953) *Naturwissenschaften* **40**, 341. 12. Gordon & Quastel (1948) *Biochem. J.* **42**, 337. 13. Maver & Voegtlin (1937) *Amer. J. Cancer* **29**, 333. 14. Mounter, Floyd & Chanutin (1953) *J. Biol. Chem.* **204**, 221. 15. Mims, Swendseid & Bird (1947) *J. Biol. Chem.* **170**, 367. 16. Grant & Sargent (1960) *Biochem. J.* **76**, 229. 17. Turba & Kuschinsky (1952) *Biochim. Biophys. Acta* **8**, 76. 18. Bárány & Bárány (1959) *Biochim. Biophys. Acta* **35**, 293. 19. Gordon (1950) *Biochem. J.* **46**, 96. 20. Stoppani, Actis, Deferrari & Gonzalez (1952) *Nature* **170**, 842. 21. Stoppani, Actis, Deferrari & Gonzalez (1953) *Biochem. J.* **54**, 378. 22. Stocken, Thompson & Whittaker (1947) *Biochem. J.* **41**, 47. 23. Bruns, Noltmann & Vahlhaus (1958) *Biochem. Z.* **330**, 483. 24. Stoppani & Brignone (1956) *Biochem. J.* **64**, 196. 25. Stoppani & Brignone (1957) *Arch. Biochem. Biophys.* **68**, 432. 26. Barron & Singer (1945) *J. Biol. Chem.* **157**, 221. 27. Yall & Green (1952) *Proc. Soc. Exptl. Biol. Med.* **79**, 306.

N-[2-(4-oxo-1-phenyl-1,3,8-triazaspiro[4,5]dec-8-yl)ethyl]-2-naphthalene-carboxamide

This phospholipase D inhibitor (FW = 428.54 g/mol; CAS 1130067-34-3) targets phospholipase D isoforms PLD$_1$ and PLD$_2$, which catalyze phospholipid hydrolysis and release phosphatidic acid. The latter is a second messenger that promotes Ras activation, cell spreading, stress fiber formation, chemotaxis, and membrane vesicle trafficking. Indeed, the machinery of PLD-induced cell invasion is mediated by PA and involves Wiscott-Aldrich Syndrome protein, Growth Receptor-bound Protein-2 and Rac2 signaling events that ultimately affect actin polymerization and cell invasion. The use of this inhibitor helped to demonstrate that PLD2 has a central role in the development, metastasis, and level of aggressiveness of breast cancer, suggesting that PLD_2 could become a useful therapeutic target. (*See also 5-Fluoro-2-Indolyl des-Chlorohalopemide*) **1.** Henkels, Boivin, Dudley, Berberich & Gomez-Cambronero (2013) *Oncogene* **32**, 5551.

5-Oxoprolinal

This 5-oxoproline derivative (FW = 113.12 g/mol) is a transition-state aldehyde inhibitor of pyroglutamyl-peptide hydrolase. It is a competitive inhibitor with a K_i value of 26 nM. **1.** Friedman, Kline & Wilk (1985) *Biochemistry* **24**, 3907. **2.** Robert-Baudouy, Clauziat & Thierry (1998) in *Handb. Proteolytic Enzymes*, p. 791, Academic Press, San Diego. **3.** McKeon & O'Connor (1998) in *Handb. Proteolytic Enzymes*, p. 796, Academic Press, San Diego.

L-5-Oxoproline

This pyrrolidone-carboxylic acid (FW = 129.12 g/mol; pK_a = 3.32; Specific rotation at the D-line = –11.3° at 20°C for 2 g/100 mL), also known as pyroglutamic acid, is an amino acid that is found in many peptides and proteins (often at the N-terminus) as well as in physiological fluids. Although oxoproline can form nonenzymatically from L-glutamine, L-glutamate, and γ-glutamylated peptides, it is commonly the product of the reaction catalyzed by γ-glutamylcyclotransferase. 5-Oxoproline is converted to glutamic acid upon boiling for 1-2 hours in 2.0 M HCl or 0.5 M NaOH. **Target(s):** CTP synthetase, moderately inhibited (1); glutamate carboxypeptidase II (2); γ-glutamyl kinase, *or* glutamate 5-kinase (3,4); Na^+/K^+-exchanging ATPase (5); prolactin release (6); pyroglutamyl-peptidase II (7); 1-pyrroline-5-carboxylate reductase (8). **1.** Bearne, Hekmat & MacDonnell (2001) *Biochem. J.* **356**, 223. **2.** Robinson, Blakely, Couto & Coyle (1987) *J. Biol. Chem.* **262**, 14498. **3.** Krishna & Leisinger (1979) *Biochem. J.* **181**, 215. **4.** Marco-Marín, Gil-Ortiz, Pérez-Arellano, *et al.* (2007) *J. Mol. Biol.* **367**, 1431. **5.** Escobedo & Cravioto (1973) *Clin. Chim. Acta* **49**, 147. **6.** Lam, Knudsen & Folkers (1978) *Biochem. Biophys. Res. Commun.* **81**, 680. **7.** Bauer (1994) *Eur. J. Biochem.* **224**, 387. **8.** Krishna, Beilstein & Leisinger (1979) *Biochem. J.* **181**, 233.

4-Oxo-4-[(pyrazolo[4,3-*b*]olean-12-en-28-yl)oxy]butanoic Acid

This oleanolic acid derivative (FW = 564.81 g/mol) inhibits rabbit muscle glycogen phosphorylase (IC_{50} = 24.7 μM). **1.** Chen, Gong, Liu, *et al.* (2008) *Chem. Biodivers.* **5**, 1304.

(*E*,*E*)-2-[2-Oxo-2-[[(3,7,11-trimethyl-2,6,10-dodecatrienyl)oxy]amino]ethyl]phosphonate

This long-chain phosphonate ($FW_{free-acid}$ = 359.40 g/mol; *Abbreviation*: FPT inhibitor II) is an unreactive farnesyl pyrophosphate analogue that inhibits Ras protein farnesyltransferase, IC_{50} = 75 nM (1). It also inhibits protein geranylgeranyltransferases I and II at elevated concentrations, IC_{50} = 24 μM (2). **1.** Terry, Long & Beese (2001) *The Enzymes*, 3rd ed., **21**, 19. **2.** Manne, Ricca, Brown, *et al.* (1995) *Drug Development Res.* **34**, 121.

Oxycodone

This thebaine derivative (FW = 315.36 g/mol; CAS 76-42-6), also known as Roxicodone®, OxyContin®, Oxecta®, OxyIR®, and systematically as (5*R*,9*R*,13*S*,14*S*)-4,5α-epoxy-14-hydroxy-3-methoxy-17-methylmorphinan-6-one, is a widely used (*and* abused) narcotic analgesic that is indicated for the relief of moderate-to-severe pain. In use since its successful semi-synthesis in 1917, oxycodone carries with it a significant risk for dependence, especially when administered for chronic pain. Athough the precise receptor that binds oxycodone has been disputed, it now appears that oxycodone binds to μ, κ and δ receptors, inhibiting adenylyl cyclase, and resulting in neuron hyperpolarization and decreased excitability. **Pharmacokinetics:** Oxycodone's analgesic effects commence ~1 hour after administration and last for ~12 hours, when administered in the controlled-release form (1). These properties distinguish oxycodone from its rapid-onset and longer-lasting opiods, morphine and heroin. Electrophysiological measurements reveal low-affinity oxycodone inhibition (IC_{50} = 170 μM) of the potassium current through hERG channels expressed in HEK293 cells (2). Oxycodone plasma half-life is 3-5 hours (roughly half that of morphine), with stable plasma levels reached within 24 hours (compare to 2-7 days for morphine) (1). Oral bioavailability ranges from 60 to 87%, and plasma protein binding is 45%. Most of the drug is metabolized in the liver, with the remainder excreted by the kidney (1). Its major metabolites are oxymorphone (a highly potent analgesic) and noroxycodone (weaker analgesic). (*See Tapentadol*) **Key Pharmacokinetic Parameters:** *See* Appendix II in Goodman & Gilman's THE PHARMACOLOGICAL BASIS OF THERAPEUTICS, 12ᵗʰ Edition (Brunton, Chabner & Knollmann, eds.) McGraw-Hill Medical, New York (2011). **1.** Ordóñez Gallego, González Barón & Espinosa Arranz (2007) *Clin. Transl. Oncol.* **9**, 298. **2.** Fanoe, Jensen, Sjøgren, Korsgaard & Grunnet (2009) *Br. J. Clin. Pharmacol.* **67**, 172.

Oxy-coenzyme A

This biochemically inert coenzyme A analogue ($FW_{free-acid}$ = 751.46 g/mol), which contains a hydroxyl group in place of its thiol group, binds to and inhibits most Coenzyme A-dependent enzymes. **Target(s):** 3-ketoacid

CoA-transferase (1); phosphate acetyltransferase (2,3). **1**. Jencks (1973) *The Enzymes*, 3rd ed., **9**, 483. **2**. Shimizu (1970) *Meth. Enzymol.* **18A**, 322. **3**. Stewart & Miller (1965) *Biochem. Biophys. Res. Commun.* **20**, 433.

Oxygen

This diatomic gas (FW = 31.99 g/mol) of the Group 16 (or Group VIB) element oxygen (Symbol: O; Atomic Number: 8; Atomic Weight: 15.9994; Valence: 2; with ionic radii for O^{2-}, O^-, O^+, and O^{6+} are 1.32, 1.76, 0.22, 0.09 Å, respectively), also known as dioxygen, is a colorless and odorless gas that constitutes 20.946% of Earth's dry atmosphere. At saturation, aqueous O_2 is about 1.4 mM. Its structure and reactivity are predicted by valence bond theory to be O=O, having one σ and one π bond, with each oxygen having two pairs of unshared electrons. Such a structure does not account for the observation that O_2 is paramagnetic in all phases, but molecular orbital theory correctly predicts the two unpaired electrons in dioxygen both in π* orbitals (the bond length ois 0.1208 nm). The singlet state of oxygen (the $^1\Delta_g$ state), in which the two π electrons are paired, is actually an excited state and is about 92 kJ·mol^{-1} above the triplet ground state (the $^3\Sigma_{g-}$ state). Singlet dioxygen can react with many unsaturated compounds and can be readily generated by numerous photochemical reactions. Singlet oxygen is generated and/or utilized in many biochemical processes, particularly in the presence of light. Many enzymes are inhibited by dioxygen, among them a significant contain active-site sulfhydryl groups that participate in substrate binding and/or catalysis. Enzymes requiring pantothenate and Coenzyme A are also inhibited by oxygen, which most often converts the thiol-containing metabolites to their inactive disulfide forms. Dioxygen also inhibits photophosphorylation and is a competitive inhibitor of the carboxylase activity of ribulose bisphosphate carboxylase/oxygenase. **See also** *Ozone*

Oxymatrine

This quinolizidine alkaloid (FW = 264.37 g/mol), also known as matrine oxide, matrine *N*-oxide, matrine 1-oxide, and (7*aS*,13*aR*,13*bR*,3*cS*) dodecahydro-1*H*,5*H*,10*H*-dipyrido[2,1-*f*:3',2',1'-*ij*][1,6]naphthyridin-10-one 4-oxide, from the root of the traditional Chinese herb *Sophora flavescens* inhibits/suppresses ulceration (*e.g.,* experimental gastric ulcer, pylorus ligation ulcer and indomethacin), most likely by decreasing acid secretion and inhibiting gastric motility. Oxymatrine induces mitochondria dependent apoptosis in human osteosarcoma MNNG and HOS cells through the inhibition of PI3K/Akt pathway (2). Oxymatrine also prevents NF-κB nuclear translocation, thereby ameliorating acute intestinal inflammation (3). **1**. Yamazaki (2000) *Yakugaku Zasshi.* **120**, 1025. **2**. Zhang, Sun, Chen, *et al.* (2013) *Tumor Biol.* . (2014) *Tumor Biol.* **35**, 1619. **3**. Guzman, Koo, Goldsmith, *et al.* (2013) *Sci. Rep.* **3**, 1629.

Oxymetazoline

This vasoconstrictor and nasal decongestant (FW = 260.38 g/mol; CAS 1491-59-4), systematically named as 6-*t*-butyl-3-(4,5-dihydro-1*H*-imidazol-2-ylmethyl)-2,4-dimethylphenol, is a partial α_{2A}-adrenergic agonist as well as a 5-HT$_{1A}$, 5-HT$_{1B}$, and 5-HT$_{1D}$ serotonin receptor agonist. It is an over-the-counter drug that is most often used topically to treat rhinitis and sinusitis. Oxymetazoline exerts a dose-dependent inhibitory effect on total iNOS activity, as indicated by nitrite/nitrate formation, and this effect is attributable to inhibition of enzyme induction rather than direct inhibition of the enzyme itself. **1**. Westerveld, Voss, van der Hee, *et al.* (2000) *Eur. Respir. J.* **16**, 437.

Oxymorphone

This semi-synthetic opioid analgesic and oxycodone metabolite (FW = 301.34 g/mol; CAS 76-41-5), also named 4,5α-epoxy-3,14-dihydroxy-17-methylmorphinan-6-one and marketed under the trade names Numorphan® (suppository and injectable solution), Opana ER® (extended-release tablet), and Opana IR® (immediate-release tablet), is used to relieve moderate to severe pain and preoperatively to alleviate apprehension and to maintain anaesthesia, especially as an obstetric analgesic. Opana IR is an oral drug that is indicated for the management of chronic pain; however, when injected recreationally, Opana can be deadly. Overdose is attended by respiratory depression, somnolence (progressing to stupor and/or coma), skeletal muscle flaccidity, bradycardia, and hypotension. In severe cases, cardiac arrest and death can occur. **1**. Vadivelu, Maria, Jolly, *et al.* (2013) *J. Opioid Manag.* **9**, 439.

Oxytetracycline

This tetracycline analogue (FW = 460.44 g/mol; CAS 79-57-2), also called terramycin, inhibits protein biosynthesis by blocking aminoacyl-tRNA binding to the acceptor site (A-site) on ribosomes. The half-life of oxytetracycline at 37°C and pH 7.0 is 26 hours. Oxytetracycline also binds metal ions tightly. **Target(s):** anhydrotetracycline monooxygenase (1); catalase (2); cytochrome *c* oxidase (3); cystathionine γ-lyase, *or* cysteine desulfhydrase (4); fructose-1,6-bisphosphate aldolase (5); glyoxalase I, *or* lactoylglutathione lyase (6); lactate dehydrogenase (7); malate dehydrogenase (7); phosphatase (8); protein biosynthesis (6-11); urease (12). **1**. Behal, Neuzil & Hostalek (1983) *Biotechnol. Lett.* **5**, 537. **2**. Ghatak, Saxena & Agarwala (1958) *Enzymologia* **19**, 261. **3**. Krcmery & Kellen (1966) *J. Hyg. Epidemiol. Microbiol. Immunol.* **10**, 339. **4**. Bernheim & Deturk (1953) *Enzymologia* **16**, 69. **5**. Ghatak & Shrivastava (1958) *Enzymologia* **19**, 237. **6**. Van Brummelen, Myburgh & Bissbort (1993) *Res. Commun. Chem. Pathol. Pharmacol.* **82**, 339. **7**. Kellen, Krcmery & Hustavova (1967) *Folia Microbiol. (Praha)* **12**, 362. **8**. Krcmery (1966) *Cesk. Epidemiol. Mikrobiol. Imunol.* **15**, 28. **9**. van den Bogert, Holtrop, Melis, Roefsema & Kroon (1987) *Biochem. Pharmacol.* **36**, 1555. **10**. de Jonge & Hulsmann (1973) *Eur. J. Biochem.* **32**, 356. **11**. Laskin & Chan (1964) *Biochem. Biophys. Res. Commun.* **14**, 137. **12**. Reithel (1971) *The Enzymes*, 3rd ed., **4**, 1.

Oxytocin

This peptide hormone (MW = 1007.19 g/mol; CAS: 50-56-6; *Sequence*: CYIQNCPLG-NH$_2$, with an intramolecular disulfide bond) induces uterine contraction and lactation, increases sodium ion secretion, and stimulates myometrial GTPase and phospholipase C. Oxytocin is produced via proteolysis of a precursor protein and released by the posterior pituitary gland. (*See also* *Isotocin; Glumitocin; Mesotocin; Vasopressin*) Oxytocin was one of the first naturally occurring polypeptides to be synthesized in the laboratory. **Target(s):** carboxypeptidase E (1); hydroxyindole *O*-

methyltransferase, *or* acetylserotonin *O*-methyltransferase (2); protein-disulfide reductase (glutathione), $K_i = 20$ μM, also possible alternative substrate (3); serotonin *N*-acetyltransferase (4). **1**. Hook & LaGamma (1987) *J. Biol. Chem.* **262**, 12583. **2**. Sugden & Klein (1987) *J. Biol. Chem.* **262**, 6489. **3**. Varandani, Nafz & Chandler (1975) *Biochemistry* **14**, 2115. **4**. Klein & Namboodiri (1982) *Trends Biochem. Sci.* **7**, 98.

Ozanimod

This orally available $S1P_{R1}$ modulator (FW = 404.46 g/mol; CAS 1306760-87-1; Solubility: 81 mg/mL DMSO; < 1 mg/mL H_2O), also named RPC1063 and 5-[3-[(1*S*)-2,3-dihydro-1-[(2-hydroxyethyl)amino]-1*H*-inden-4-yl]-1,2,4-oxadiazol-5-yl]-2-(1-methylethoxy)benzonitrile, selectively targets sphingosine 1-Phosphate (S1P) receptors, which mediate multiple processes, including lymphocyte trafficking, cardiac function, and endothelial barrier integrity (1). Ozanimod is specific for $S1P_{1R}$ ($IC_{50} = 0.4$ nM) and $S1P_{5R}$, inducing $S1P_{1R}$ internalization and reversibly reducing the number of circulating B and CCR7+ T lymphocytes *in vivo*. It shows high oral bioavailability and volume of distribution, and a circulatory half-life that supports once daily dosing (1). Oral ozanimod reduced inflammation and disease parameters in three autoimmune disease models, commending its use in the treatment of relapsing multiple sclerosis (RMS) and inflammatory bowel disease (IBD). Indeed, the safety and efficacy of this modulator ozanimod in relapsing multiple sclerosis is indicated on the basis of a randomized, placebo-controlled, Phase 2 trial (2). Other $S1P_{R1}$-directed immunomodulators include laquinimod, ponesimod, and siponimod. **1**. Scott, Clemons, Brooks, *et al.* (2016) *Br. J. Pharmacol.* **173**, 1778. **2**. Cohen, Arnold, Comi, *et al.* (2016) *Lancet Neurol.* **15**, 373.

Ozone

This molecular oxygen allotrope and powerful oxidant (FW = 48.00 g/mol; M.P. = –192.7°C and B.P. = –111.9°C) is a colorless gas with a distinctive odor. Whereas O_2 is paramagnetic, O_3 is diamagnetic. Ozone is formed by electrical discharge or UV irradiation in an O_2-rich atmosphere. Even the Greek Poet Homer (800 BC) noted the odor often associated with thunderstorms. The molecule is symmetrical and bent, having a bond angle of 117° and an oxygen–oxygen bond length of 0.127 nm, showing considerable double-bond character. Ozone is highly toxic, and inhalation is lethal to the ciliated cells of the upper airways and centriacinar region. Ozone exposure leads to free-radical-mediated or lipid peroxide-mediated toxicity. **Target(s):** catalase (1); β-glucosidase (2-4); glyceraldehyde-3-phosphate dehydrogenase (5); lysozyme (6,7); Na^+/K^+-exchanging ATPase (8); ribonuclease T_1 (9); superoxide dismutase (1). **1**. Whiteside & Hassan (1987) *Arch. Biochem. Biophys.* **257**, 464. **2**. Hestrin, Feingold & Scramm (1955) *Meth. Enzymol.* **1**, 231. **3**. Helferich & Petersen (1935) *Hoppe Seyler's Z. Physiol. Chem.* **233**, 75. **4**. Veibel (1950) *The Enzymes*, 1st ed, **1** (part 1), 583. **5**. Knight & Mudd (1984) *Arch. Biochem. Biophys.* **229**, 259. **6**. Imoto, Johnson, North, Phillips & Rupley (1972) *The Enzymes*, 3rd ed., **7**, 665. **7**. Holzman, Gardner & Coffin (1968) *J. Bacteriol.* **96**, 1562. **8**. Chan, Kindya & Kesner (1977) *J. Biol. Chem.* **252**, 8537. **9**. Tamaoki, Sakiyama & Narita (1978) *J. Biochem.* **83**, 771.

– P –

P

Symbol for Phosphorus in its elemental or combined form; Proline, especially to denote prolyl residues within a peptide or protein.

P1

This cell-permeable PDI inhibitor (FW = 583.70 g/mol; CAS 1461648-55-4; Soluble to 100 mM in DMSO; N-[(1,1-dimethylethoxy)carbonyl]-L-phenylalanyl-O-(ethenyl-sulfonyl)-N-4-pentyn-1-yl-L-tyrosinamide) targets protein disulfide isomerase (IC$_{50}$ = 1.7 μM), an enzyme that assists in the folding of cellular proteins within the endoplasmic reticulum of mammalian cells. P1 reduces the proliferation of numerous cancer cell lines, with nominal GI$_{50}$ values of ~4 μM. PDI binds to the axonal microtubule-associated protein known as Tau, mainly through its thioredoxin-like catalytic domain, by forming a one-to-one complex that prevents Tau misfolding. **1.** Ge, Zhang, Li, *et al.* (2013) *ACS Chem. Biol.* **8**, 2577. **2.** Xu, Liu, Chen & Liang (2013) *PLoS One* **8**, e76657.

P$_{II}$ (or P$_2$)

This 44-kDa homotetrameric regulatory protein operates both as an inhibitor and activator in the glutamine synthetase (GS) adenylylation/deadenylylation cascade in *Escherichia coli* and related Gram-negative bacteria (1-3). Glutamine synthetase plays the crucial role of providing sufficient glutamine to satisfy biosynthetic needs as a source of nascent ammonia in those enzymes possessing an ammonia-transfer tunnel connecting glutaminase and biosynthetic active sites. At low ammonia concentration, dodecameric GS is enzymatically deadenylylated (*Reaction*: GS-$(O$-pA)$_m$ + mP$_i$ $\rightleftharpoons$ GS$_0$ + mADP, where the O-pA represents the adenylyl group attached by a phosphodiester linkage to Tyrosine-397, and m indicates the degree of dodecamer adenylylation (*i.e.*, $0 < m < 12$)). At high intracellular concentrations, ammonia replaces glutamine as a nitrogen donor in various biosynthetic reactions, thereby limiting cellular glutamine needs to that mainly required for protein synthesis. Under such conditions, GS is enzymatically adenylylated (*Reaction*: GS$_0$ + mATP $\rightleftharpoons$ GS-$(O$-pA)$_m$ + PP$_i$), with commensurate loss of catalytic activity. Unmodified protein P$_{IIA}$ promotes adenylylation and GS inhibition, whereas P$_{IID}$ (containing an enzymatically introduced Tyr-O-pU linkage) promotes deadenylylation and GS activation (1-3). P$_{II}$ is also an important effector in the broader context of microbial nitrogen metabolism. **1.** Brown, Segal & Stadtman (1971) *Proc. Natl. Acad. Sci. U.S.A.* **68**, 2949. **2.** Mangum, Magni & Stadtman (1973) *Arch. Biochem. Biophys.* **158**, 514. **3.** Adler, Purich & Stadtman (1975) *J. Biol. Chem.* **250**, 6264.

P11

This potent PDZ motif-containing anti-angiogenic agent (FW = 720.78 g/mol; CAS 848644-86-0; Sequence: HSDVHK-NH$_2$) targets the integrin $\alpha_v\beta_3$-vitronectin interaction, blocking proliferation and inducing apoptosis in human vascular endothelial cells (HUVECs) (1,2). Arg-Gly-Asp (or RGD) binding-site recognition by P11 is site-specific, showing a strong antagonism against $\alpha_v\beta_3$-vitronectin interaction, IC$_{50}$ = 26 nM (2). The binding orientation of docked P11 in $\alpha_v\beta_3$ is similar to that for RGD, suggesting a divalent metal-ion coordination is a common driving force for the formation of both SDV·$\alpha_v\beta_3$ and RGD·$\alpha_v\beta_3$ complexes. P11 appears to inhibit β-Fibroblast Growth Factor (or βFGF) -induced HUVEC proliferation via mitogen-activated protein kinase kinase and extracellular-signal regulated kinase inhibition as well as p53-mediated apoptosis related with activation of caspases (3). (*Note*: Consisting of 80-90 amino-acids in far-flung signaling proteins and binding to short C-terminal regions in protein interaction partners, "PDZ" is an acronym that includes the first letters of Post-synaptic density protein (PSD95), *Drosophila* Dlg1 tumor suppressor, and Zonula occludens-1 protein (zo-1), the first proteins

discovered to share this domain.) **1.** Lee, Kang, Chang, Han & Kang (2004) *J. Biomol. Screen.* **9**, 687. **2.** Choi, Kim, Lee, Han & Kang (2010) *Proteomics* **10**, 72. **3.** Bang, Kim, Kang, *et al.* (2011) *Molec. Cell. Proteom.* **10**, M110.005264.

p35

This single-chain baculovirus protein (MW = 35 kDa) inhibits apoptosis in virally infected host insect cells and also potently inhibits (K_i < 9 nM) human caspase-1 (1,2), caspase-1, *Spodoptera frugiperda* (1,3), caspase-2 (1), caspase-3 (1,2), caspase-4 (1), caspase-6 (1,2), caspase-8 (1,3,4), caspase-10 (1,2), and Ced-3 (1,5). This protein belongs to MEROPS proteinase inhibitor family I50, clan IQ. Purified recombinant p35 does not significantly inhibit unrelated serine or cysteine proteases, a property implying p35 is a potent caspase-specific inhibitor. The interaction of p35 with caspase-3, as a model of the inhibitory mechanism, revealed classic slow-binding inhibition, with both active-sites of the caspase-3 dimer acting equally and independently. **1.** Zhou & Salvesen (2000) *Meth. Enzymol.* **322**, 143. **2.** Zhou, Krebs, Snipas, *et al.* (1998) *Biochemistry* **37**, 10757. **3.** Ahmad, Srinivasula, Wang, *et al.* (1997) *J. Biol. Chem.* **272**, 1421. **4.** Xu, Cirilli, Huang, *et al.* (2001) *Nature* **410**, 494. **5.** Bertin, Mendrysa, LaCount, *et al.* (1996) *J. Virol.* **70**, 6251.

p38 Mitogen-Acivated Protein Kinase Inhibitor (p39 MAP Kinase Inhibitor), *See* specific inhibitor; e.g., *PD 169316; 2-(4-Chloro-phenyl)-4-(4-fluorophenyl)-5-pyridin-4-yl-1,2-dihydropyrazol-3-one*

p425

This C$_2$-rotationally symmetric, synthetic allosteric inhibitor (FW = 876.83 g/mol), identified by high throughput screening, targets the cytokine known as Macrophage Migration Inhibitory Factor (MIF), strongly inhibiting MIF-mediated catalysis of 4-hydroxyphenyl pyruvate tautomerization. p425 also blocks the interaction of MIF with its receptor, CD74, and likewise interfere with the pro-inflammatory activities of the cytokine. Structural studies revealed a unique mode of allosteric binding, wherein a single p425 molecule occupies the interface of two MIF trimers. The inhibitor binds mainly on MIF's surface by means of hydrophobic interactions that are stabilized by hydrogen bonding with four highly specific residues contributed by three different monomers. The mode of p425 binding reveals a unique way to block the activity of the cytokine for potential therapeutic benefit in MIF-associated diseases. (*See Macrophage Migration Inhibitory Factor*) **1.** Bai, Asojo, Cirillo, *et al.* (2012) *J. Biol. Chem.* **287**, 30653.

p276-00

This cell cycle-inhibiting flavone and anticancer agent (FW = 438.30 g/mol; CAS 920113-03-7), also named 2-(2-chlorophenyl)-5,7-dihydroxy-8-[(2R,3S)-2-(hydroxymethyl)-1-methyl-3-pyrrolidinyl]-4H-1-benzopyran-4-one, targets the Cyclin-Dependent Kinases CDK1 (IC$_{50}$ = 79 nM), CDK4 (IC$_{50}$ = 63 nM) and CDK9 (IC$_{50}$ = 20 nM). p276-00 showed potent antiproliferative effects against various human cancer cell lines (IC$_{50}$ values ranging from 300 to 800 nmol/L), but has little effect on cultured fibroblasts (1). A significant down-regulation of cyclin D1 and Cdk4 and a decrease in Cdk4-specific pRb Ser(780) phosphorylation is observed. P276-00 produces potent inhibition of Cdk4-D1 activity that is competitive with ATP, and not with retinoblastoma protein. The compound also induced apoptosis in human promyelocytic leukemia (HL-60) cells, as evidenced by the induction of caspase-3 and DNA ladder studies (1). In 22 human cancer xenografts, P276-00 is approximately 26x more potent than cisplatin, and is also active against cisplatin-resistant tumors of central nervous system, melanoma, prostate, and renal cancers (2). Synchronized human non-small cell lung carcinoma (H-460) and human normal lung fibroblast (WI-38) cells are arrested in G_1 (2).   **1**. Joshi, Rathos, Joshi, *et al.* (2007) *Mol. Cancer Ther.* **6**, 918.   **2**. Joshi, Rathos, Mahajan, *et al.* (2007) *Mol. Cancer Ther.* **6**, 926.

P107

This potent potassium ion channel opener, *or* KCO (FW = 231.30 g/mol; CAS 60559-98-0; Solubility: 50 mM in Ethanol; 100 mM in DMSO), also known as *N*-cyano-*N*'-(1,1-dimethylpropyl)-*N*"-3-pyridyl-guanidine, targets $K_{ir}6$ (*or* K_{ATP}) channels (EC_{50} = 7.5 nM; for relaxation of rat aorta). In studies on rats, rabbit and dogs, P1075 effects only correlate with the presence of specific receptor binding sites in rat vascular preparations, showing there are significant species-specific and raising questions about the pharmacological significance of this K_{ATP} opener binding site (1). These channels are heteromultimers composed with a 4:4 stoichiometry of an inwardly rectifying K^+ channel subunit plus a regulatory subunit comprising the receptor sites for hypoglycemic sulfonylureas and KCOs (a sulfonylurea receptor) (2). P1075-induced channel activation IS mediated by drug interaction with a single binding site per tetradimeric complex (2). P-1075 also binds to sulfonylurea receptors SUR2A (K_d = 17 nM) and SUR2B (K_d = 3 nM) (3).   **1**. Higdon, Khan, Buchanan & Meisheri (1997) *J. Pharmacol. Exp. Ther.* **280**, 255.   **2**. Gross, Toman, Uhde, Schwanstecher & Schwanstecher (1999) *Mol. Pharmacol.* **56**, 1370.   **3**. Buckner, Milicic, Daza, *et al.* (2000) *Eur. J.* **Pharmacol. 400, 287.**

P22077

This ubiquitin cycle inhibitor (FW = 315.33 g/mol; CAS 1247819-59-5; Solubility: 60 mg/mL DMSO), also known as 1-[5-[(2,4-difluorophenyl)thio]-4-nitro-2-thienyl]ethanone, targets Ubiquitin-Specific Peptidase-7 (*or* USP-7) with an EC_{50} value of 8.6 µM. P22077 also inhibits the structurally related peptidase USP-47.   **1**. Altun, Kramer, Willems, *et al.* (2011) *Chem. Biol.* **18**, 1401.   **2**. Tian, Isamiddinova, Peroutka, *et al.* (2011) *Assay Drug Dev. Technol.* **9**, 165.

17-PA, *See* *17-Phenyl-(3α,5α)-androst-16-en-3-ol*

PA-452

This RXR antagonist (FW = 439.59 g/mol; CAS 457657-34-0; Solubility: 100 mM in DMSO; 10 mM in Ethanol), also named 2-[[3-(hexyloxy)-5,6,7,8-tetrahydro-5,5,8,8-tetramethyl-2-naphthalenyl]methylamino]-5-pyrimidinecarboxylic acid, blocks RXR-RAR heterodimer interactions (IC_{50} = 90 nM). By inducing dissociation of nuclear RXR tetramers into dimers,

RXR antagonists can display potent anticarcinogenic activities that alter gene expression by modulating the chromatin architecture.   **1**. Takahashi, *et al.* (2002) *J. Med. Chem.* **45**, 3327.   **2**. Yasmin, *et al.* (2010) *J. Mol. Biol.* **397**, 1121.   **3**. Nakayama, *et al.* (2011) *J. Med. Chem. Lett.* **2**, 896.

PA-457, *See* *Bevirimat*

PA 824

This first-in-class tuberculosis pro-drug (FW = 359.26 g/mol; CAS 187235-37-6), also named (6*S*)-2-nitro-6-{[4-(trifluoromethoxy)-benzyl]oxy}-6,7-dihydro-5*H*-imidazo[2,1-*b*][1,3]oxazine, is active against both replicating and hypoxic, non-replicating forms of *Mycobacterium tuberculosis*. After activation by a process that depends on *M. tuberculosis* F420 cofactor, PA-824 exerts its bacteriocidal action by inhibiting protein biosynthesis and cell wall lipid synthesis (1). Its reduction potential does not appear to be the main driver for inhibitor efficacy (2). Significantly, PA-824 generates a des-nitro metabolite, along with a reactive nitrogen species (including nitric oxide), that may be effectors of the anaerobic activity of these compounds (3). Furthermore, nitric oxide scavengers protected *M. tuberculosis* from the lethal effects of the drug. This finding suggests PA-824 releases a lethal nitrogen oxide-like metabolite that factors into its bacteriocidal action. That PA-824 is without effect on *M. leprae* is consistent with the absence of a required activating enzyme, namely nitroimidazo-oxazine-specific nitroreductase (encoded by Rv3547), in the *M. leprae* genome (4).   **1**. Stover, Warrener, VanDevanter, *et al.* (2000) *Nature* **405**, 962.   **2**. Thompson, Blaser, Anderson, *et al.* (2009) *J. Med. Chem.* **52**, 637.   **3**. Singh, Manjunatha, Boshoff, *et al.* (2008) *Science* **322**, 1392.   **4**. Manjunatha, Lahiri, Randhawa, *et al.* (2006) *Antimicrob. Agents Chemother.* **50**, 3350.

PACAP$_{6-38}$

This potent and competitive PAC antagonist (FW = 4024.78 g/mol; CAS 143748-18-9; Sequence: FTDSYSRYRKQMAVKKYLAAVLGKRYKQR VKNK-NH$_2$; Solubility: 2 mg/mL in H$_2$O) targets Pituitary Adenylate Cyclase-Activating Polypeptide receptor *or* PAC$_1$ (IC_{50} = 2 nM), blocking PACAP$_{1-27}$-induced stimulation of adenylate cyclase (K_i = 1.5 nM) (1). PACAP$_{6-38}$ also inhibits growth of prostate cancer cells (2). PACAP$_{6-38}$ inhibition aided in demonstrating that the neuropeptide PACAP promotes α-secretase pathway for processing Alzheimer amyloid precursor protein (3). Note that the longer peptide, PACAP$_{1-38}$ (FW = 4535 g/mol; CAS 137061-48-4), Sequence: HSDGIFTDSYSRYRKQMAVKKYLAAVLGK RYKQRVKNK-NH$_2$) is an endogenous neuropeptide agonist, showing considerable homology with vasoactive intestinal peptide (VIP), but with more potent stimulation of adenylyl cyclase (4).   **1**. Robberecht, *et al.* (1992) *Eur. J. Biochem.* **207**, 239.   **2**. Leyton, *et al.* (1998) *Cancer Lett.* **125**, 131.   **3**. Kojro, *et al.* (2006) *FASEB J.* **20**, 512.   **4**. Rawlings, *et al.* (1994) *J. Biol. Chem.* **269**, 5680.

Paclitaxel

This potent anticancer agent (FW = 853.92 g/mol; CAS 33069-62-4; Solubility = 170 mg/mL DMSO, <<1 mg/mL H$_2$O), more widely known as Taxol® and systematically as (2α,4α,5β,7β,10β,13α)-4,10-bis(acetyloxy)-13-{[(2*R*,3*S*)-3-(benzoylamino)-2-hydroxy-3-phenylpropanoyl]oxy}-1,7-

dihydroxy-9-oxo-5,20-epoxytax-11-en-2-yl benzoate, from the bark of the Pacific yew *Taxus brevifolia* inhibits microtubule disassembly and induces microtubule bundling (1-4). Taxol greatly reduces the normal turnover of cellular microtubules, perturbs kinetochore-to-microtubule attachments, and disrupts chromosome segregation, thereby activating checkpoint pathway(s) that delay cell cycle progression and induce programmed cell death. Paclitaxel thus potently inhibits replication and proliferation, arresting cells at the G_2-M stage, with an IC_{50} value of 0.1 pM in human endothelial cells. **Paclitaxel-Induced Tubulin Polymerization:** With purified tubulin (*i.e.*, containing no microtubule-associated proteins, *or* MAPs), taxol induces polymerization only if GTP is present when the reaction warmed to 30-37° C (5). Under such conditions, GTP is stoichiometrically hydrolyzed relative to tubulin in tandem with polymerization, whereas GDP inhibits both polymerization and hydrolysis. The polymerized protein shows only a few microtubules and instead consists mainly of sheets of protofilaments that undergo cold-induced disassembly. To prevent gross morphologic changes in microtubules in *in vitro* experiments, one should limit the amount of taxol added to 10-20% of the total tubulin. **Mode of Action:** Taxol alters the steady-state tubulin flux and the apparent molecular rate constants for tubulin addition and loss at the two ends of bovine brain microtubules *in vitro* (4-6). Reassembled bovine brain microtubules (consisting of 70% tubulin and 30% microtubule-associated proteins, *or* MAPs), undergo subunit exchange reactions at both ends with a net steady state flux. Once treated with taxol, these microtubules lose their dynamicity, showing little evidence of dynamic instability. Taxol (5-7 µM) slows the flux by 50% inhibition. To a first approximation, taxol decreases the magnitudes of the dissociation rate constants at the two ends to similar extents, while exerting little effect on the tubulin association rate constants. **Key Pharmacokinetic Parameters:** *See* Appendix II in Goodman & Gilman's THE PHARMACOLOGICAL BASIS OF THERAPEUTICS, 12th Edition (Brunton, Chabner & Knollmann, eds.) McGraw-Hill Medical, New York (2011). **Paclitaxel Resistance:** Although paclitaxel has improved both duration and quality of life for some cancer patients, the majority eventually develop progressive disease, representing a major obstacle to improving overall response and survival. Resistance arises mainly from tubulin mutations, tubulin isotype selection and post-translational modifications (7). Overexpression of Class III β-tubulin (TUBB3), for example, has emerged as a biomarker in a number of cancers, a prognostic indicator of more aggressive disease, and a predictor of resistance to taxanes and vinca alkaloids (8). The semisynthetic taxane Yg-3-46a effectively evades P-glycoprotein-mediated drug efflux and β-III tubulin-mediated drug resistance *in vitro* (9). **Formulations:** Paclitaxel is highly insoluble in water, necessitating initial dissolution in DMSO for biochemical and cell experiments. The resulting solution may then be diluted 20x with aqueous solutions of tubulin or cell suspensions. In the clinic, Taxol is principally used to combat ovarian and breast carcinomas. Taxol Injection® is a clear, colorless to slightly yellow viscous nonaqueous solution. (Each mL of this sterile nonpyrogenic solution contains 6 mg paclitaxel, 527 mg purified Cremophor® (*i.e.*, polyoxyethylated castor oil) and 49.7% (vol/vol) dehydrated alcohol) that is diluted with a suitable parenteral fluid prior to intravenous infusion.) Abraxane® is a nanoparticle-based formulation that exploits the ability of human albumin to bind paclitaxel, thereby avoiding disadvantages of cremaphor. **Effect of Microtubule End-Binding Proteins:** Analysis of microtubule (MT) growth times supports the view that catastrophic MT disassembly depends on the age of MTs (10). Colchicine and Vinblastine accelerate aging in a manner that depends on the presence of end-binding proteins. On the other hand, while Paclitaxel and Peloruside A induce catastrophes, they compensate by promoting MT rescues and reversing the MT aging. **Low-Temperature Tubulin Polymerization in the Presence of Paclitaxel:** Depolymerized bovine brain microtubule protein reassembles at 0° C in the presence of paclitaxel (11). Initial polymerization does not form microtubules, resulting instead in the formation of protofilamentous ribbons. Microtubules are first noted after 1 hour of incubation with the drug. After 20 hours with taxol at 0° C, the bulk of the polymerized tubulin is present as microtubules. **1**. Schiff & Horwitz (1980) *Proc. Natl. Acad. Sci. U. S.A.* **77**, 1561. **2**. Schiff, Fant & Horwitz (1979) *Nature* **277**, 665. **3**. Purich & Kristofferson (1984) *Adv. Protein Chem.* **36**, 133. **4**. Wilson, Miller, Ferrell, Snyder, Thompson & Purich (1985) *Biochemistry* **24**, 5254. **5**. Hamel, del Campo, Lowe & Lin (1981) *J. Biol. Chem.* **256**, 11887. **6**. Jordan & Wilson (1998) *Meth. Enzymol.* **298**, 252.**6**. **7**. Orr, Verdier-Pinard, McDaid & Horwitz (2003) *Oncogene* **22**, 7280. **8**. Powell, Kaizer, Koopmeiners, Iwamoto & Klein (2014) *Oncol Lett.* **7**, 405. **9**. Cai, Sharom & Fang (2013) *Cancer. Lett.* **341**, 214. **10**. Mohan, Katrukha, Doodhi, *et al.* (2013) *Proc. Natl. Acad. Sci. U.S.A.* **110**, 8900. **11**. Thompson, Wilson & Purich (1981) *Cell Motil.* **1**, 445.

Paclobutrazol

This synthetic plant growth regulator (FW = 293.80 g/mol; CAS 76738-62-0; white solid; solubility = 35 mg/L water; (2*RS*,3*RS*)-1-(4-chlorophenyl)-4,4-dimethyl-2-([1*H*]-1,2,4-triazol-1-yl)pentan-3-ol) inhibits three steps in the conversion of *ent*-kaurene to *ent*-kaurenoate, exerting the overall effect of blocking gibberellin biosynthesis. **Target(s):** cytochrome P450 (1-3); *ent*-kaurene oxidase, inhibited by both 2*R*,3*R*- and 2*S*,3*S*-isomers (2,3); and phytosterol biosynthesis (4). **1**. Coulson, King & Wiseman (1984) *Trends Biochem. Sci.* **9**, 446. **2**. Ashman, Mackenzie & Bramley (1990) *Biochim. Biophys. Acta* **1036**, 151. **3**. Swain, Singh, Helliwell & Poole (2005) *Plant Cell Physiol.* **46**, 284. **4**. Haughan, Lenton & Goad (1987) *Biochem. Biophys. Res. Commun.* **146**, 510.

Pacritinib

This pyrimidine-based macrocycle and potent protein kinase inhibitor (FW = 472.59 g/mol; CAS 937272-79-2; Solubility: 11 mg/mL DMSO), also known as SB1518, selectively targets Janus Kinase 2 (*or* JAK2; IC_{50} = 23 nM) and Fms-Like Tyrosine Kinase-3 (*or* FLT3; IC_{50} = 22 nM), with values of 1280 and 522 nM for JAK1 and JAK3. FLT3 is genetically altered in up to 35% of acute myeloblastic leukemias, making it an attractive target for AML patients. Pacritinib has potent anti-proliferative effects on myeloid and lymphoid cell lines driven by mutant or wild-type JAK2 or FLT3, resulting from cell cycle arrest and induction of apoptosis (1). SB1518 has favorable pharmacokinetic properties after oral dosing in mice, is well tolerated and significantly reduces splenomegaly and hepatomegaly in a JAK2(Val-617-Phe)-driven disease model (1). **Primary Mode of Inhibitory Action:** Upregulation of JAK2 in FLT3-TKI-resistant AML cells was identified as a potential mechanism of resistance to selective FLT3 inhibition (2). This resistance could be overcome by pacritinib's combined inhibition of FLT3 and JAK2 in this cellular model. Pacritinib potently inhibits FLT3 auto-phosphorylation and downstream STAT5, MAPK and PI3K signaling pathways in AML cell lines with highest potency against cells harboring FLT3-ITD mutations (2). Blockade of FLT3 signaling was also demonstrated in primary AML blasts treated *ex vivo* with pacritinib. In both cell lines and primary blasts, pacritinib treatment led to the induction of G_1 arrest, inhibition of cell proliferation, as well as caspase-dependent apoptosis. The anti-proliferative effects of pacritinib on the FLT3-ITD harboring cell lines MV4-11 (IC_{50} = 47 nM) and MOLM-13 (IC_{50} = 67 nM), which have been reported previously,16 are in the same range as the inhibition of intracellular FLT3 signalling (3). **1**. Hart, Goh, Novotny-Diermayr, *et al.* (2011) *Leukemia* **25**, 1751. **2**. Hart, Goh, Novotny-Diermayr, *et al.* (2011) *Blood Cancer J.* **1**, e44.

Pactamycin

This universal translation inhibitor (FW = 558.63 g/mol; CAS 23668-11-3; soluble in ethanol and benzene; *Abbreviation*: Pct) from *Streptomyces pactum*, previously known as NSC 52947 and named systematically as benzoic acid, 2-hydroxy-6-methyl-[(1*S*,2*R*,3*R*,4*S*,5*S*)-5-[(3-acetylphenyl) amino]-4-amino-3-[[(dimethylamino)carbonyl]amino]-1,2-dihydroxy-3-[(1*S*)-1-hydroxyethyl]-2-methylcyclopentyl]methyl ester, inhibits protein synthesis and exhibits antitumor activity (1-10). Because the optical rotation of dissolved pactamycin changes upon standing, some as-yet undefined chemical rearrangement must take place. Solutions should be freshly prepared. **Primary Mode of Inhibitory Action:** Pct restricts structural transitions in ribosomal RNA, preventing the ribosome blocks the binding of initiator tRNA to the initiator complex, preventing formation of the 80S ribosomal complex and arresting protein biosynthesis immediately after the initial dipeptide is formed. **Binding Interactions:** Pct binds in a single site on the 30S ribosome subunit in the upper part of the platform, very close to the cleft in the subunit that is responsible for binding of the three tRNA molecules (1). It interacts primarily with residues at the tips of the stem loops H23b and H24a in the central domain of 16S RNA, where it folds up to mimic an RNA dinucleotide. The interaction between N6 of A694 and Pct is crucial for binding of the drug since this is the only interaction with that particular base. *N*-methylation of this residue causes resistance, most likely from distortions of the local structure. **1.** Brodersen, Clemons, Carter, *et al.* (2000) *Cell* **103**, 1143. **2.** Pestka (1974) *Meth. Enzymol.* **30**, 261. **3.** Fresno & Vázquez (1979) *Meth. Enzymol.* **60**, 566. **4.** Nash & Tate (1984) *J. Biol. Chem.* **259**, 678. **5.** Jiménez (1976) *Trends Biochem. Sci.* **1**, 28. **6.** Ochoa & Mazumder (1974) *The Enzymes*, 3rd ed., **10**, 1. **7.** Mankin (1997) *J. Mol. Biol.* **274**, 8. **8.** Kappen & Goldberg (1976) *Biochemistry* **15**, 811. **9.** Cohen, Herner & Goldberg (1969) *Biochemistry* **8**, 1312. **10.** Cohen, Goldberg & Herner (1969) *Biochemistry* **8**, 1327.

Pactimibe

This indoline (FW = 416.61 g/mol; CAS 189198-30-9; liquid; Density = 1.071), also known as CS-505 and systematically as 7-[(2,2-dimethyl-1-oxopropyl)amino]-2,3-dihydro-4,6-dimethyl-1-octyl-1*H*-indole-5-acetic acid, inhibits cholesterol *O*-acyltransferase, or cholesterol *O*-acyltransferase (ACAT), K_i = 5.6 μM, and reduces plasma cholesterol concentrations (1-3). Although this ACAT-inhibiting drug appears to be of limited value in the management of atherosclerosis (4), CS-505 is being investigated with respect to Alzheimer's disease. **1.** Terasaka, Miyazaki, Kasanuki, *et al.* (2007) *Atherosclerosis* **190**, 239. **2.** Kitayama, Tanimoto, Koga, *et al.* (2006) *Eur. J. Pharmacol.* **540**, 121. **3.** Takahashi, Kasai, Ohta, *et al.* (2008) *J. Med. Chem.* **51**, 4823. **4.** Nissen, Tuzcu, Brewer, *et al.* (2006) *New Engl. J. Med.* **354**, 1253 (Erratum: *NEJM* **355**, 638).

Paecilopeptin, See *N-Acetyl-L-leucyl-L-valinal*

PAF, See *Platelet-Activating Factor*

PAH, See *4-Aminohippurate*

PAI-1 and PAI-2, See *Plasminogen Activator Inhibitor Types 1 and 2*

PAI-749

This potent and selective synthetic antagonist of plasminogen activator inhibitor 1, *or* PAI-1 (FW = 501.63 g/mol; CAS 481631-45-2), systematically named as 3-pentyl-1-(phenylmethyl)-2-[6-(2*H*-tetrazol-5-ylmethoxy)naphthalen-2-yl]indole, maintains the activity of tissue-type plasminogen activator (tPA) and urokinase-type plasminogen activator (uPA) activities in the presence of PAI-1, with IC_{50} values of 157 and 87 nM, respectively (1). PAI-749 preserved tPA and uPA activity in the face of neutralizing amounts of PAI-1. That PAI-1 is the target is strongly suggested by observation of the same rank order for PAI inhibition and its binding to PAI, the latter deduced from fluorescence binding experiments. Even so, one unresolved point is the observation that the apparent IC_{50} values of PAI-749 with tPA and uPA are not identical. That PAI-749 induces PAI-1 polymerization may explain this discrepancy. PAI-749 does not affect thrombus formation or fibrinolysis in a range of established human plasma and whole blood-based systems (2). **See also** *Plasminogen Activator Inhibitor* **1.** Gardell, Krueger, Antrilli, *et al.* (2007) *Mol. Pharmacol.* **72**, 897. **2.** Lucking, Visvanathan, Philippou, *et al.* (2010) *J. Thromb. Haemost.* **8**, 1333.

PAK-104P

This MDR inhibitor (FW = 698.76 g/mol; CAS 131356-86-0), also known as 2-[4-(diphenylmethyl)-1-piperazinyl]ethyl 5-(*trans*-4,6-dimethyl-1,3,2-dioxaphosphorinan-2-yl)-2,6-dimethyl-4-(3-nitrophenyl)-3-pyridine-carboxylate *P*-oxide, reverses multidrug resistance and lowers calcium channel blocking activity (1). PAK-104P also inhibits cadmium-transporting ATPase (2). **1.** Shudo, Mizoguchi, Kiyosue, *et al.* (1990) *Cancer Res.* **50**, 3055. **2.** Ren, Furukawa, Chen, *et al.* (2000) *Biochem. Biophys. Res. Commun.* **270**, 608.

PALA, See *N-(Phosphonoacetyl)-L-aspartate*

Palatinose, See *Isomaltulose*

Palbociclib

This orally active, non-ATP-competitive cyclin kinase-directed inhibitor (FW = 483.99 g/mol (mono-HCl); CASs = 827022-32-2 (mono-hydrochloride, 571190-30-2 (free base); Solubility: 10 mg/mL DMSO; 30 mg/mL Water; Formulation: Dissolved in sodium lactate buffer (50 mM, pH 4.0)), also known as PD-0332991, Ibrance®, and 6-acetyl-8-cyclopentyl-5-methyl-2-(5-(piperazin-1-yl)pyridin-2-ylamino)pyrido[2,3-*d*]pyrimidin-7(8*H*)-one hydrochloride, targets Cdk-4 (Cyclin D_1) and Cdk-6 (Cyclin D_2), enzymes that participate in the so-called CDK4/6-retinoblastoma signaling pathway governing the cell-cycle restriction point. Palbociclib induces rapid G_1 cell-cycle arrest in primary human myeloma cells (1-4). This agent also shows significant efficacy in a broad spectrum of human tumor xenografts *in vivo*, resulting in complete regression in some tumors with no evidence of acquired resistance or ability to circumvent the growth inhibitory properties of this agent (1). Ibrance received FDA approval in 2015 for combined use with letrozole to treat postmenopausal women with estrogen receptor-positive, (HER2)-negative advanced breast cancer as an initial endocrine-based therapy for metastatic disease. **Cyclin Target Selectivity:** Cdk1 (weak, if any), Cdk2 (weak, if any), Cdk3 (weak, if any), Cdk4 (IC_{50} = 11 nM), Cdk5 (weak, if any), Cdk6 (IC_{50} = 16 nM), Cdk7 (weak, if any), Cdk8 (weak, if any), Cdk9 (weak, if any), Cdk10 (weak, if any). **1.** Fry, *et al.* (2004) *Mol. Cancer Ther.* **3**, 1427. **2.** Menu, *et al.* (2008) *Cancer Res.* **68**,

5519. **3**. Finn, *et al.* (2009) *Breast Cancer Res.* **11**, R77. **4**. Katsumi, *et al.* (2011) *Biochem. Biophys. Res. Commun.* **413**, 62.

Palinavir

This peptide analogue (FW$_{free-base}$ = 708.90 g/mol; CAS 154612-39-2) is a potent inhbitor of HIV-1, K_i = 30 nM, and HIV-2 proteases, K_i = 130 pM. **1**. Lamarre, Croteau, Wardrop, *et al.* (1997) *Antimicrob. Agents Chemother.* **41**, 965.

Paliperidone

Risperidone

Paliperidone

Paliperidone Pamityl Ester

This long-acting atypical antipsychotic drug (FW = 426.48 g/mol; CAS 144598-75-4), also named (*RS*)-3-[2-[4-(6-fluorobenzo[*d*]isoxazol-3-yl)-1-piperidyl]-ethyl]-7-hydroxy-4-methyl-1,5-diazabicyclo[4.4.0]deca-3,5-dien-2-one, is an α_1 and α_2 adrenergic receptor antagonist and H$_1$ histamine receptor antagonist. Routes of delivery include long-acting oral form (Chemical Form: Paliperidone; Trade Name: Invega™) and an even longer-acting intramuscular form (Chemical Form: Paliperidone palmityl ester; Trade Name(s): Invega Sustenna or Xeplion). Paliperidone is also the primary metabolite of another antipsychotic drug, Resperidone. The half-life of risperidone conversion to paliperidone is about 3 hours for typical metabolizing individuals, compared with about 19 hours in slow-metabolizing subjects. **Primary Mode of Action:** Risperidone and 9-hydroxyrisperidone had similar *in vitro* binding profiles, with highest affinity was for 5-HT$_{2A}$ receptors (K_i = 0.4 nM); K_i values for other 5-HT-receptor subtypes were at least 100-times higher. The binding affinity of risperidone and 9-hydroxyrisperidone for the D$_2$ family of receptors (D$_{2L}$, D$_{2S}$, D$_3$, D$_4$) was one order of magnitude lower than their affinity for 5-HT$_{2A}$ receptors. **Key Pharmacokinetic Parameters:** *See* Appendix II in Goodman & Gilman's THE PHARMACOLOGICAL BASIS OF THERAPEUTICS, 12th Edition (Brunton, Chabner & Knollmann, eds.) McGraw-Hill Medical, New York (2011). **1**. Leysen, Janssen, Megens & Schotte (1994) *J. Clin. Psychiatry* **55**, Supplement 5-12.

Paliperidone palmityl ester, *See Paliperidone*

Palmitate

This saturated fatty acid (FW$_{free-acid}$ = 256.43 g/mol; CAS 57-10-3; pK_a ≈ 4.8; M.P. = 63-64°C; B.P. = 351.5°C), also called palmitic acid, hexadecanoic acid and cetylic acid, is found throughout nature. Palmitic acid is insoluble in water but is soluble in ethanol (7.21 g per 100 mL), chloroform (15.1 g per 100 mL), as well as benzene, acetone, and ethyl acetate. **Micelle Formation:** Dissolve 1 mg into 1 mL of a chloroform:methanol solution (1:4, vol/vol), add to aqueous buffer, and disperse with the aid of a sonicator (5-6 30-sec bursts) while maintained at 24°C in a water bath. Resulting lipid solutions/dispersions can be stored temporarily at 4°C. Before use, solutions are warmed to RT and vortexed three times for a few seconds with a Vortex Genie (at maximum setting) to obtain a homogenous preparation. *Note*: Because this substance readily forms micelles, one must take care to discriminate whether the monomeric and/or micelle form(s) is(are) inhibitory, when present at concentrations above the critical micelle concentration (*cmc*). **Target(s):** acetyl-CoA carboxylase (1,2); 1-acylglycerophosphocholine *O*-acyltransferase, *or* lysophosphatidylcholine *O*-acyltransferase (39); alcohol dehydrogenase (3); alkyldihydroxyacetone-phosphate synthase (4); arachidonyl-CoA synthetase, *or* arachidonate:CoA ligase (23); aspartate aminotransferase (35); branched-chain α-keto-acid dehydrogenase (5); ceramidase (26); choloyl-CoA synthetase, *or* cholate:CoA ligase (24); chondroitin AC lyase (25); chondroitin B lyase, weakly inhibited (25); DNA-directed DNA polymerase α (6); DNA polymerase α (6); 7-ethoxycoumarin *O*-deethylase (7); fatty-acid synthase (1,8); firefly luciferase, *or Photinus*-luciferin 4-monooxygenase, IC$_{50}$ = 0.67 μM (9,40); β-galactoside α-2,6-sialyltransferase (10,36); 1,3-β-glucan synthase (37); glucokinase (11); glucose-6-phosphate dehydrogenase (12); glutathione *S*-transferase (13); glycylpeptide *N*-tetradecanoyltransferase (38); hexokinase (14,34); hyaluronate lyase, weakly inhibited (25); leukocyte elastase, *or* neutrophil elastase (27); Lysophospholipase (30,31); microbial collagenases (27); Na$^+$/K$^+$-exchanging ATPase (15); palmitoyl-CoA hydrolase, *or* acyl-CoA hydrolase (29); phosphatidate cytidylyltransferase (32); phosphatidate phosphatase (16); phosphatidylinositol 3-kinase (33); phospholipase D (28); photophosphorylation (17); prostaglandin dehydrogenase (18); protein kinase B (19); pyrophosphatase, *or* inorganic diphosphatase (20); pyruvate kinase (21); and L-threonine dehydrogenase (22). **1**. Ascenzi & Vestal (1979) *J. Bacteriol.* **137**, 384. **2**. Roessler (1990) *Plant Physiol.* **92**, 73. **3**. Winer & Theorell (1960) *Acta Chem. Scand.* **14**, 1729. **4**. Brown & Snyder (1982) *J. Biol. Chem.* **257**, 8835. **5**. Buxton, Barron, Taylor & Olson (1984) *Biochem. J.* **221**, 593. **6**. Mizushina, Tanaka, Yagi, *et al.* (1996) *Biochim. Biophys. Acta* **1308**, 256 **7**. Muller-Enoch, Fintelmann, Nicolaev & Gruler (2001) *Z. Naturforsch. [C]* **56**, 1082. **8**. Wright, Cant & McBride (2002) *J. Dairy Sci.* **85**, 642. **9**. Ueda & Suzuki (1998) *Biophys. J.* **75**, 1052. **10**. Bador, Morelis & Louisot (1984) *Biochimie* **66**, 223. **11**. Lea & Weber (1968) *J. Biol. Chem.* **243**, 1096. **12**. Criss & McKerns (1969) *Biochim. Biophys. Acta* **184**, 486. **13**. Mitra, Govindwar, Joseph & Kulkarni (1992) *Toxicol. Lett.* **60**, 281. **14**. Stewart & Blakely (2000) *Biochim. Biophys. Acta* **1484**, 278. **15**. Smith, Solar, Paulson, Hill & Broderick (1999) *Mol. Cell. Biochem.* **194**, 125. **16**. Elabbadi, Day, Virden & Yeaman (2002) *Lipids* **37**, 69. **17**. Pick, Weiss & Rottenberg (1987) *Biochemistry* **26**, 8295. **18**. Mibe, Nagai, Oshige & Mori (1992) *Prostaglandins Leukot. Essent. Fatty Acids* **46**, 241. **19**. Cazzolli, Carpenter, Biden & Schmitz-Peiffer (2001) *Diabetes* **50**, 2210. **20**. Balocco (1962) *Enzymologia* **24**, 275. **21**. Kayne (1973) *The Enzymes*, 3rd ed. (Boyer, ed.), **8**, 353. **22**. Guerranti, Pagani, Neri, *et al.* (2001) *Biochim. Biophys. Acta* **1568**, 45. **23**. Reddy & Bazan (1983) *Arch. Biochem. Biophys.* **226**, 125. **24**. Schepers, Casteels, Verheyden, *et al.* (1989) *Biochem. J.* **257**, 221. **25**. Suzuki, Terasaki & Uyeda (2002) *J. Enzyme Inhib. Med. Chem.* **17**, 183. **26**. Gatt (1966) *J. Biol. Chem.* **241**, 3724. **27**. Rennert & Melzig (2002) *Planta Med.* **68**, 767. **28**. Okamura & Yamashita (1994) *J. Biol. Chem.* **269**, 31207. **29**. Joyard & Stumpf (1980) *Plant Physiol.* **65**, 1039. **30**. Sugimoto & Yamashita (1994) *J. Biol. Chem.* **269**, 6252. **31**. Zhang & Dennis (1988) *J. Biol. Chem.* **263**, 9965. **32**. Sribney & Hegadorn (1982) *Can. J. Biochem.* **60**, 668. **33**. Hardy, Langelier & Prentki (2000) *Cancer Res.* **60**, 6353. **34**. Morris, DeBruin, Yang, *et al.* (2006) *Eukaryot. Cell* **5**, 2014. **35**. McKenna, Hopkins, Lindauer & Bamford (2006) *Neurochem. Int.* **48**, 629. **36**. Hickman, Ashwell, Morell, van den Hamer & Scheinberg (1970) *J. Biol. Chem.* **245**, 759. **37**. Ko, Frost, Ho, Ludescher & Wasserman (1994) *Biochim. Biophys. Acta* **1193**, 31. **38**. Selvakumar, Lakshmikuttyamma, Shrivastav, *et al.* (2007) *Prog. Lipid Res.* **46**, 1. **39**. Kerkhoff, Habben, Gehring, Resch & Kaever (1998) *Arch. Biochem. Biophys.* **351**, 220. **40**. Matsuki, Suzuki, Kamaya & Ueda (1999) *Biochim. Biophys Acta* **1426**, 143.

1-Palmitin, *See Palmitoylglycerol*

Palmitoleate

This common unsaturated fatty acid ($FW_{\text{free-acid}}$ = 254.41 g/mol; CAS 373-49-9; $pK_a \approx 4.8$; M.P. = −0.5-0.5°C) is known commonly as palmitoleic acid, less frequently as zoomaric acid, and systematically as (Z)-Δ^9-hexadecenoic acid. Palmitoleic acid is the *cis*-isomer of palmitelaidic acid. At elevated concentrations, palmitoleate is a protein denaturant. *Note*: Because this substance readily forms micelles, one must take care to discriminate whether the monomeric and/or micelle form(s) is(are) inhibitory, when present at concentrations above the critical micelle concentration (*cmc*). **Target(s):** ammonia-dependent O_2 uptake activity and ammonia monooxygenase (1); Ca^{2+} uptake, mitochondrial (2); cGMP-stimulated cAMP phosphodiesterase (3,4); choloyl-CoA synthetase, cholate:CoA ligase (5); chondroitin AC lyase (17); chondroitin B lyase (17); DNA polymerase (6); epidermal-growth-factor-induced increases in cytosolic $[Ca^{2+}]$, membrane potential, and inositol-1,4,5-trisphosphate generation (7); 1,3-β-glucan synthase (21,22); glucose-6-phosphate dehydrogenase (8); glycerol-3-phosphate *O*-acyltransferase (23); hexokinase (9); hyaluronate lyase (17); isocitrate dehydrogenase (8); KinA signal-transducing protein kinase (10); lactate dehydrogenase (8); leukocyte elastase, *or* neutrophil elastase (18); malate dehydrogenase (8); microbial collagenases (18); NAD^+ ADP-ribosyltransferase, *or* poly(ADP-ribose) polymerase (20); NAD^+ kinase (11); Na^+/K^+-exchanging ATPase (12); palmitoyl-CoA hydrolase, *or* acyl-CoA hydrolase (19); phospholipase A_2 (13); renin (14); steroid Δ^5-reductase (15); and valyl-tRNA synthetase (16). **1**. Juliette, Hyman & Arp (1995) *J. Bacteriol.* **177**, 4908. **2**. Rustenbeck & Lenzen (1989) *Biochim. Biophys. Acta* **982**, 147. **3**. Pyne, Cooper & Houslay (1986) *Biochem. J.* **234**, 325. **4**. Yamamoto, Yamamoto, Manganiello & Vaughan (1984) *Arch. Biochem. Biophys.* **229**, 81. **5**. Falany, Xie, Wheeler, *et al.* (2002) *J. Lipid Res.* **43**, 2062. **6**. Mizushina, Tanaka, Yagi, *et al.* (1996) *Biochim. Biophys. Acta* **1308**, 256. **7**. Casabiell, Pandiella & Casanueva (1991) *Biochem. J.* **278**, 679. **8**. Vincenzini, Favilli, Treves, Vanni & Baccari (1983) *Int. J. Biochem.* **15**, 1283. **9**. Stewart & Blakely (2000) *Biochim. Biophys. Acta* **1484**, 278. **10**. Strauch, de Mendoza & Hoch (1992) *Mol. Microbiol.* **6**, 2909. **11**. Harmon, Jarrett & Cormier (1984) *Anal. Biochem.* **141**, 168. **12**. Burth, Younes-Ibrahim, Goncalez, Costa & Faria (1997) *Infect. Immun.* **65**, 1557. **13**. Ballou & Cheung (1985) *Proc. Natl. Acad. Sci. U.S.A.* **82**, 371. **14**. Kotchen, Welch & Talwalkar (1978) *Amer. J. Physiol.* **234**, E593. **15**. Liang & Liao (1992) *Biochem. J.* **285**, 557. **16**. Black (1985) *J. Biol. Chem.* **260**, 433. **17**. Suzuki, Terasaki & Uyeda (2002) *J. Enzyme Inhib. Med. Chem.* **17**, 183. **18**. Rennert & Melzig (2002) *Planta Med.* **68**, 767. **19**. Joyard & Stumpf (1980) *Plant Physiol.* **65**, 1039. **20**. Banasik, Komura, Shimoyama & Ueda (1992) *J. Biol. Chem.* **267**, 1569. **21**. Kauss & Jeblick (1986) *Plant Physiol.* **80**, 7. **22**. Ko, Frost, Ho, Ludescher & Wasserman (1994) *Biochim. Biophys. Acta* **1193**, 31. **23**. Kito, Aibara, Hasegawa & Hata (1972) *J. Biochem.* **71**, 99.

Palmitoleoyl-Coenzyme A

This fatty acyl-CoA metabolite (FW = 1003.94 g/mol), also known as *cis*-9-hexadecenoyl-CoA, which is an alternative substrate for several acyltransferases as well as in β-oxidation, inhibits acetyl-CoA carboxylase (1). *Note*: Because this substance readily forms micelles, one must take care to discriminate whether the monomeric and/or micelle form(s) is(are) inhibitory, when present at concentrations above the critical micelle concentration (*cmc*). **1**. Tanabe, Nakanishi, Hashimoto, *et al.* (1981) *Meth. Enzymol.* **71**, 5.

Palmitoyl-D-carnitine

This acylated carnitine derivative (FW = 399.62 g/mol) is not the naturally occurring ester; it is the enantiomer of the natural intermediate of the mitochondrial degradation of palmitate. *Note*: Because this substance readily forms micelles, one must take care to discriminate whether the monomeric and/or micelle form(s) is(are) inhibitory, when present at concentrations above the critical micelle concentration (*cmc*). **Target(s):** carnitine:acylcarnitine translocase (1); and carnitine palmitoyltransferase (2,3), however, see also (1). **1**. Baillet, Mullur, Esser & McGarry (2000) *J. Biol. Chem.* **275**, 36766. **2**. Bieber & Farrell (1983) *The Enzymes*, 3rd ed. (Boyer, ed.), **16**, 627. **3**. Arduini, Mancinelli, Radatti, *et al.* (1992) *J. Biol. Chem.* **267**, 12673.

Palmitoyl-DL-carnitine

This acylcarnitine racemate (FW = 436.07 g/mol) is significantly cheaper source than the pure naturally occurring L-enantiomer and is frequently used instead in investigations into lipid metabolism and regulation. *Note*: Because this substance readily forms micelles, one must take care to discriminate whether the monomeric and/or micelle form(s) is(are) inhibitory, when present at concentrations above the critical micelle concentration (*cmc*). **See also** *Palmitoyl-L-carnitine; Palmitoyl-D-carnitine* **Target(s):** 1,3-β-glucan synthase (1); NADH oxidation, mitochondrial (2); and protein kinase C (3-7). **1**. Kauss & Jeblick (1986) *Plant Physiol.* **80**, 7. **2**. Dargel & Strack (1966) *Acta Biol. Med. Ger.* **16**, 271. **3**. Katoh, Wrenn, Wise, Shoji & Kuo (1981) *Proc. Natl. Acad. Sci. U.S.A.* **78**, 4813. **4**. Kikkawa & Nishizuka (1986) *The Enzymes*, 3rd ed. (Boyer & Krebs, eds.), **17**, 167. **5**. Wise & Kuo (1983) *Biochem. Pharmacol.* **32**, 1259. **6**. Pelaez, de Herreros & de Haro (1987) *Arch. Biochem. Biophys.* **257**, 328. **7**. Takahashi, Kamimura, Shirai & Yokoo (2000) *Skin Pharmacol. Appl. Skin Physiol.* **13**, 133.

Palmitoyl-L-carnitine

This naturally occurring acylated derivative (FW = 436.07 g/mol; CAS 18877-64-0) is an intermediate in the mitochondrial palmitic acid oxidation. Palmitoyl-L-carnitine reduces negative charge of membrane surfaces, when incorporated into the plasma membrane in erythrocytes and myocytes (1). *Note*: Because this substance readily forms micelles, one must take care to discriminate whether the monomeric and/or micelle form(s) is(are) inhibitory, when present at concentrations above the critical micelle concentration (*cmc*). **Target(s):** 7-ethoxycoumarin *O*-deethylase, cytochrome P450 (2); lysophospholipase (3-9); [myosin light-chain] kinase (10); Na^+/K^+-exchanging ATPase (11-13); palmitoyl-CoA hydrolase, *or* acyl-CoA hydrolase (14-16); phosphatidylcholine:cholesterol acyltransferase, LCAT (17); and protein kinase C (18). **1**. Gruver & Pappano (1993) *J. Mol. Cell Cardiol.* **25**, 1275. **2**. Muller-Enoch, Fintelmann, Nicolaev & Gruler (2001) *Z. Naturforsch. [C]* **56**, 1082. **3**. Gross & Sobel (1983) *J. Biol. Chem.* **258**, 5221. **4**. Chen, Wright, Golding & Sorrell (2000) *Biochem. J.* **347**, 431. **5**. Gross, Ahumada & Sobel (1984) *Amer. J. Physiol.* **246**, C266. **6**. Mirbod, Banno, Ghannoum, *et al.* (1995) *Biochim. Biophys. Acta* **1257**, 181. **7**. Sugimoto & Yamashita (1994) *J. Biol. Chem.* **269**, 6252. **8**. Wright, Payne, Santangelo, *et al.* (2004) *Biochem. J.* **384**, 377. **9**. Takahashi, Banno, Shikano, Mori & Nozawa (1991) *Biochim. Biophys. Acta* **1082**, 161. **10**. Jinsart, Ternai & Polya (1991) *Plant Sci.* **78**, 165. **11**. Wood, Bush, Pitts & Schwartz (1977) *Biochem. Biophys. Res. Commun.* **74**, 677. **12**. Shen & Pappano (1995) *Amer. J. Physiol.* **268**, H1027. **13**. Kramer & Weglicki (1985) *Amer. J. Physiol.* **248**, H75. **14**. Broustas & Hajra (1995) *J. Neurochem.* **64**, 2345. **15**. Gross (1983) *Biochemistry* **22**, 5641. **16**. Gross (1984) *Biochim. Biophys. Acta* **802**, 197. **17**. Bell (1983) *Int. J. Biochem.* **15**, 133. **18**. Nakadate & Blumberg (1987) *Cancer Res.* **47**, 6537.

Palmitoylcholine

This acylated choline (FW$_{bromide}$ = 422.47 g/mol; CAS 2932-74-3 for the chloride salt) inhibits carnitine O-palmitoyltransferase (1), [myosin light-chain] kinase (2), and phosphatidylinositol diacylglycerol-lyase, *or* phosphatidylinositol-specific phospholipase C (3). *Nte*: Because this substance readily forms micelles, one must take care to discriminate whether the monomeric and/or micelle form(s) is(are) inhibitory, when present at concentrations above the critical micelle concentration (*cmc*). **1.** Gandour, Colucci, Stelly, Brady & Brady (1988) *Arch. Biochem. Biophys.* **267**, 515. **2.** Jinsart, Ternai & Polya (1991) *Plant Sci.* **78**, 165. **3.** Hirasawa, Irvine & Dawson (1981) *Biochem. J.* **193**, 607.

Palmitoyl-Coenzyme A

This thiolester (FW = 1005.95 g/mol), consisting of palmitic acid and Coenzyme A, often abreviated as Palmitoyl-S-CoA and Palmitoyl-CoA, is a key metabolite in the degradation of longer chain fatty acids and in the posttranslational palmitoylation of membrane-targeted proteins. Palmitoyl-CoA readily forms micelles, with critical micelle concentration (*cmc*) of 30-60 μM (1); one must take care to discriminate whether the monomeric and/or micelle form(s) is(are) inhibitory, when palmitoyl-CoA is present at concentrations above the *cmc*. **Target(s):** acetoacetyl-CoA synthetase (2); acetyl-CoA carboxylase, K_i = 7.2 μM for the rat liver enzyme (3-16), acetyl-CoA hydrolase (119,120); acetyl-CoA synthetase, *or* acetate:CoA ligase (108-111); acylglycerol lipase (monoacylglycerol lipase; inhibited by micellar concentrations of propionyl-CoA) (121); acylglycerone-phosphate reductase, *or* acyl/alkyl dihydroxyacetone-phosphate reductase (17,18); adenine-nucleotide translocase (19-23); 1-alkyl-n-glycerol-3-phosphate O-acetyltransferase (25); 1-alkylglycero-phosphocholine O-acetyltransferase (24,144); amine acetyltransferase (26); AMP deaminase (27); anion-conducting channel (28); arylalkylamine N-acetyltransferase (143); arylamine N-acetyltransferase (29,148); ATP:citrate lyase, *or* ATP:citrate synthase (30,31,136,137); carbamoyl-phosphate synthetase I, ammonia-requiring (107); carnitine O-acetyltransferase (32-34,147); carnitine O-octanoyltransferase, also alternative substrate (139,140); citrate synthase (35-43); cyclooxygenase (44); decylcitrate synthase (138); decylhomocitrate synthase (45); diacylglycerol cholinephosphotransferase (122); dicarboxylic acid carrier (46); dihydrofolate reductase (47); dihydrolipoyllysine-residue acetyltransferase (145); dihydroxyacetone phosphate O-acyltransferase, *or* glycerone-3-phosphate O-acyltransferase (48); DNA polymerases α and γ (49); dopamine sulfotransferase (50); dynein ATPase (51); enoyl-[acyl-carrier protein] reductase (52); *trans*-2-enoyl-CoA reductase, NADPH-dependent *or* crotonyl-CoA reductase (53); ethanolamine phosphotransferase (122,123); fatty-acid synthase (15,54-61); glucokinase (62-65,133); glucose-6-phosphatase (66,67,117); glucose-6-phosphate dehydrogenase (68-76); glutamate dehydrogenase (40,77); glutaminase, phosphate-activated (78); glycerol-3-phosphate dehydrogenase (79-82); glycogen synthase (83); hexokinase (62-65,84,133); hormone-sensitive lipase (85); 4-hydroxybenzoate nonaprenyltransferase (134); 3-hydroxydecanoyl-[acyl-carrier-protein] dehydratase (113); 3-hydroxy-3-methylglutaryl-CoA synthase (135); isocitrate dehydrogenase, NADP$^+$-dependent (86,87); K$_{ATP}$ channel, mitochondrial (88); α-ketoglutarate dehydrigenase complex (87); lipase (triacylglycerol lipase (85,89,90); lipoxygenase (42); long-chain-acyl-CoA synthetase (137); malate dehydrogenase (77); malonyl-CoA decarboxylase (114,115); meromyosin ATPase (92); α-methylacyl-CoA epimerase, inhibition due to micelle formation (112); 2-methyl branched-chain acyl-CoA dehydrogenase (93); 3-methylcrotonyl-CoA carboxylase (94,95); (S)-methylmalonyl-CoA hydrolase (118); NADH dehydrogenase transhydrogenase activity (96); NAD(P)$^+$ transhydrogenase (97-99); Na$^+$/K$^+$-exchanging ATPase (100); palmitoyl-CoA hydrolase, by micellar form (101); pantothenate kinase (126-132); phenol sulfotransferase (50); phosphate acetyltransferase (146);

phosphatidate cytidylyltransferase (124,125); phosphatidate phosphatase (102); phosphatidylcholine:retinol O-acyltransferase (141); 6-phosphogluconate dehydrogenase (74); phytanoyl-CoA dioxygenase (150); propionyl-CoA carboxylase (103); protein kinase C (104,105); RNA polymerase (106); stearoyl-CoA 9-desaturase (149); and 3α,7α,12α-trihydroxy-5β-cholestanoyl-CoA 24-hydroxylase (116). **1.** Powell, Grothusen, Zimmerman, Evans & Fish (1981) *J. Biol. Chem.* **256**, 12740. **2.** Ito, Fukui, Saito & Tomita (1987) *Biochim. Biophys. Acta* **922**, 287. **3.** Ogiwara, Tanabe, Nikawa & Numa (1978) *Eur. J. Biochem.* **89**, 33. **4.** Numa (1969) *Meth. Enzymol.* **14**, 9. **5.** Inoue & Lowenstein (1975) *Meth. Enzymol.* **35**, 3. **6.** Miller & Levy (1975) *Meth. Enzymol.* **35**, 11. **7.** Tanabe, SNakanishi, Hashimoto, *et al.* (1981) *Meth. Enzymol.* **71**, 5. **8.** Alberts & Vagelos (1972) *The Enzymes*, 3rd ed. (Boyer, ed.), **6**, 37. **9.** Herbert, Price, Alban, *et al.* (1996) *Biochem. J.* **318**, 997. **10.** Brownsey & Denton (1987) *The Enzymes*, 3rd ed. (Boyer & Krebs, eds.), **18**, 123. **11.** Nikawa, Tanabe, Ogiwara, Shiba & Numa (1979) *FEBS Lett.* **102**, 223. **12.** Trumble, Smith & Winder (1995) *Eur. J. Biochem.* **231**, 192. **13.** Thampy & Koshy (1991) *J. Lipid Res.* **32**, 1667 and Thampy (1989) *J. Biol. Chem.* **264**, 17631. **14.** Ogiwara, Tanabe, Nikawa & Numa (1978) *Eur. J. Biochem.* **89**, 33. **15.** Ascenzi & Vestal (1979) *J. Bacteriol.* **137**, 384. **16.** Bortz & Lynen (1963) *Biochem. Z.* **337**, 505. **17.** Hajra, Datta & Ghosh (1992) *Meth. Enzymol.* **209**, 402. **18.** Datta, Ghosh & Hajra (1990) *J. Biol. Chem.* **265**, 8268. **19.** Ho & Pande (1974) *Biochim. Biophys. Acta* **369**, 86. **20.** Morel, Lauquin, Lunardi, Duszynski & Vignais (1974) *FEBS Lett.* **39**, 133. **21.** Ciapaite, van Eikenhorst & Krab (2002) *Mol. Biol. Rep.* **29**, 13. **22.** Paulson & Shug (1984) *Biochim. Biophys. Acta* **766**, 70. **23.** Shrago, Ball, Sul, Baquer & McLean (1977) *Eur. J. Biochem.* **75**, 83. **24.** Lee, Vallari & Snyder (1992) *Meth. Enzymol.* **209**, 396. **25.** Baker & Chang (1995) *J. Neurochem.* **64**, 364. **26.** Barenholz & Gatt (1975) *Meth. Enzymol.* **35**, 247. **27.** Skladanowski, Kaletha & Zydowo (1978) *Int. J. Biochem.* **9**, 43. **28.** Halle-Smith, Murray & Selwyn (1988) *FEBS Lett.* **236**, 155. **29.** Tabor (1955) *Meth. Enzymol.* **1**, 608. **30.** Pfitzner, Kubicek & Rohr (1987) *Arch. Microbiol.* **147**, 88. **31.** Shashi, Bachhawat & Joseph (1990) *Biochim. Biophys. Acta* **1033**, 23. **32.** Chase (1969) *Meth. Enzymol.* **13**, 387. **33.** Chase (1965) *Biochem. J.* **95**, 50. **34.** Mittal & Kurup (1980) *Biochim. Biophys. Acta* **619**, 90. **35.** Srere (1969) *Meth. Enzymol.* **13**, 3. **36.** Shepherd & Garland (1969) *Meth. Enzymol.* **13**, 11. **37.** Weitzman (1969) *Meth. Enzymol.* **13**, 22. **38.** Spector (1972) *The Enzymes*, 3rd ed. (Boyer, ed.), **7**, 357. **39.** Fritz (1966) *Biochim. Biophys. Res. Commun.* **22**, 744. **40.** Lai, Liang, Zhai, Jarvi & Lu (1994) *Metab. Brain Dis.* **9**, 143. **41.** Else, Barnes, Danson & Weitzman (1988) *Biochem. J.* **251**, 803. **42.** Srere (1965) *Biochim. Biophys. Acta* **106**, 445. **43.** Wieland & Weiss (1963) *Biochem. Biophys. Res. Commun.* **13**, 26. **44.** Fujimoto, Tsunomori, Sumiya, *et al.* (1995) *Prostaglandins Leukot. Essent. Fatty Acids* **52**, 255. **45.** Måhlén (1973) *Eur. J. Biochem.* **38**, 32. **46.** Beatrice & Pfeiffer (1981) *Biochem. J.* **194**, 71. **47.** Mita & Yasumasu (1981) *Int. J. Biochem.* **13**, 229. **48.** Webber & Hajra (1992) *Meth. Enzymol.* **209**, 92. **49.** Shimada, Haraguchi, Nagano, Fujiwara & Yasumasu (1983) *Biochim. Biophys. Res. Commun.* **110**, 902. **50.** Tulik, Chodavarapu, Edgar, *et al.* (2002) *J. Biol. Chem.* **277**, 39296. **51.** Fujiwara, Yokokawa, Hino & Yasumasu (1982) *J. Biochem.* **92**, 441. **52.** Bergler, Fuchsbichler, Hogenauer & Turnowsky (1996) *Eur. J. Biochem.* **242**, 689. **53.** Wallace, Bao, Dai, *et al.* (1995) *Eur. J. Biochem.* **233**, 954. **54.** Hsu, Butterworth & Porter (1969) *Meth. Enzymol.* **14**, 33. **55.** Muesing & Porter (1975) *Meth. Enzymol.* **35**, 45. **56.** Bloch (1975) *Meth. Enzymol.* **35**, 84. **57.** Flick & Bloch (1975) *J. Biol. Chem.* **250**, 3348. **58.** Knoche, Esders, Koths & Bloch (1973) *J. Biol. Chem.* **248**, 2317. **59.** Mahmoud, Abu el Souod & Niehaus (1996-1997) *Mycopathologia* **136**, 75. **60.** Gavilanes, Lizarbe, Municio & Onaderra (1982) *Int. J. Biochem.* **14**, 1061. **61.** Roncari (1975) *Can. J. Biochem.* **53**, 135. **62.** Vandercammen & Van Schaftingen (1991) *Eur. J. Biochem.* **200**, 545. **63.** Lin, Vogel & Neet (1989) *Int. J. Pept. Protein Res.* **34**, 333. **64.** Vogel, Keenan, Gelev & Neet (1989) *Mol. Cell Biochem.* **86**, 171. **65.** Dawson & Hales (1969) *Biochim. Biophys. Acta* **176**, 657. **66.** Nordlie (1971) *The Enzymes*, 3rd ed. (Boyer, ed.), **4**, 543. **67.** Fulceri, Gamberucci, Scott, *et al.* (1995) *Biochem. J.* **307**, 391. **68.** Kawaguchi & Bloch (1974) *J. Biol. Chem.* **249**, 5793. **69.** Anderson, Wise & Anderson (1997) *Biochim. Biophys. Acta* **1340**, 268. **70.** Young, Schmotzer & Swanson (1990) *Proc. Soc. Exp. Biol. Med.* **193**, 274. **71.** Cho & Joshi (1990) *Neuroscience* **38**, 819. **72.** Rodriguez-Torres, Villamarin, Carballal, Vazquez-Illanez & Ramos-Martinez (1987) *Rev. Esp. Fisiol.* **43**, 7. **73.** Rodriguez-Torres & Ramos-Martinez (1987) *Rev. Esp. Fisiol.* **43**, 119. **74.** Mita & Yasumasu (1980) *Arch. Biochem. Biophys.* **201**, 322. **75.** Olive & Levy (1967) *Biochemistry* **6**, 730. **76.** Eger-Neufeldt, Teinzer, Weiss & Wieland (1965) *Biochem. Biophys. Res. Commun.* **19**, 43. **77.** Kawaguchi & Bloch (1976) *J. Biol. Chem.* **251**, 1406. **78.** Kvamme, Torgner & Svenneby (1985) *Meth.*

Enzymol. **113**, 241. **79**. Swierczynski, Scislowski & Aleksandrowicz (1976) *Biochim. Biophys. Acta* **452**, 310. **80**. Pieringer (1983) *The Enzymes*, 3rd ed. (Boyer, ed.), **16**, 255. **81**. Kawaguchi & Bloch (1974) *J. Biol. Chem.* **249**, 5793. **82**. Edgar & Bell (1979) *J. Biol. Chem.* **254**, 1016. **83**. Wititsuwannakul & Kim (1977) *J. Biol. Chem.* **252**, 7812. **84**. Thompson & Cooney (2000) *Diabetes* **49**, 1761. **85**. Severson & Hurley (1984) *Lipids* **19**, 134. **86**. Farrell, Jr., Wickham & Reeves (1995) *Arch. Biochem. Biophys.* **321**, 199. **87**. Lai & Cooper (1991) *Neurochem. Res.* **16**, 795. **88**. Paucek, Yarov-Yarovoy, Sun & Garlid (1996) *J. Biol. Chem.* **271**, 32084. **89**. McDonough & Neely (1988) *J. Mol. Cell Cardiol.* **20** Suppl. 2, 31. **90**. Severson & Hurley (1984) *Lipids* **19**, 134. **91**. Pande (1973) *Biochim. Biophys. Acta* **306**, 1. **92**. Fujiwara, Fujisaki, Asai & Yasumasu (1981) *J. Biochem.* **90**, 757. **93**. Komuniecki, Fekete & Thissen-Parra (1985) *J. Biol. Chem.* **260**, 4770. **94**. Chen, Wurtele, Wang & Nikolau (1993) *Arch. Biochem. Biophys.* **305**, 103. **95**. Maier & Lichtenthaler (1998) *J. Plant Physiol.* **152**, 213. **96**. Hatefi & Stiggall (1976) *The Enzymes*, 3rd ed. (Boyer, ed.), **13**, 175. **97**. Rydström, Hoek & Ernster (1976) *The Enzymes*, 3rd ed. (Boyer, ed.), **13**, 51. **98**. Danis, Kauffman, Evans, *et al.* (1985) *Biochem. Pharmacol.* **34**, 609. **99**. Fristedt, Rydstrom & Persson (1994) *Biochim. Biophys. Res. Commun.* **198**, 928. **100**. Wood, Bush, Pitts & Schwartz (1977) *Biochim. Biophys. Res. Commun.* **74**, 677. **101**. Berge & Farstad (1981) *Meth. Enzymol.* **71**, 234. **102**. Brandes & Shapiro (1967) *Biochim. Biophys. Acta* **137**, 202. **103**. Erfle (1973) *Biochim. Biophys. Acta* **316**, 143. **104**. Wise & Kuo (1983) *Biochem. Pharmacol.* **32**, 1259. **105**. Stasia, Dianoux & Vignais (1987) *Biochem. Biophys. Res. Commun.* **147**, 428. **106**. Yokokawa, Fujiwara, Shimada & Yasumasu (1983) *J. Biochem.* **94**, 415. **107**. Corvi, Soltys & Berthiaume (2001) *J. Biol. Chem.* **276**, 45704. **108**. Preston, Wall & Emerich (1990) *Biochem. J.* **267**, 179. **109**. Satyanarayana & Klein (1976) *Arch. Biochem. Biophys.* **174**, 480. **110**. Satyanarayana & Klein (1973) *J. Bacteriol.* **115**, 600. **111**. Guranowski, Gunther Sillero & Sillero (1994) *J. Bacteriol.* **176**, 2986. **112**. Schmitz, Albers, Fingerhut & Conzelmann (1995) *Eur. J. Biochem.* **231**, 815. **113**. Brock, Kass & Bloch (1967) *J. Biol. Chem.* **242**, 4432. **114**. Scholte (1973) *Biochim. Biophys. Acta* **309**, 457. **115**. Landriscina, Gnoni & Quagliariello (1971) *Eur. J. Biochem.* **19**, 573. **116**. Casteels, Schepers, Van Eldere, Eyssen & Mannaerts (1988) *J. Biol. Chem.* **263**, 4654. **117**. Methieux & Zitoun (1996) *Eur. J. Biochem.* **235**, 799. **118**. Kovachy, Copley & Allen (1983) *J. Biol. Chem.* **258**, 11415. **119**. Prass, Isohashi & Utter (1980) *J. Biol. Chem.* **255**, 5215. **120**. Bernson (1976) *Eur. J. Biochem.* **67**, 403. **121**. De Jong, Kalkman & Hülsmann (1978) *Biochim. Biophys. Acta* **530**, 56. **122**. Coleman & Bell (1977) *J. Biol. Chem.* **252**, 3050. **123**. Sparace, Wagner & Moore (1981) *Plant Physiol.* **67**, 922. **124**. Sribney & Hegadorn (1982) *Can. J. Biochem.* **60**, 668. **125**. Petzold & Agranoff (1967) *J. Biol. Chem.* **242**, 1187. **126**. Vallari, Jackowski & Rock (1987) *J. Biol. Chem.* **262**, 2468. **127**. Leonardi, Rock, Jackowski & Zhang (2007) *Proc. Natl. Acad. Sci. U.S.A.* **104**, 1494. **128**. Zhang, Rock & Jackowski (2005) *J. Biol. Chem.* **280**, 32594. **129**. Halvorsen & Skrede (1982) *Eur. J. Biochem.* **124**, 211. **130**. Fisher, Robishaw & Neely (1985) *J. Biol. Chem.* **260**, 15745. **131**. Zhang, Rock & Jackowski (2006) *J. Biol. Chem.* **281**, 107. **132**. Kotzbauer, Truax, Trojanowski & Lee (2005) *J. Neurosci.* **25**, 689. **133**. Kim, Kalinowski & Marcinkeviciene (2007) *Biochemistry* **46**, 1423. **134**. Kawahara, Koizumi, Kawaji, *et al.* (1991) *Agric. Biol. Chem.* **55**, 2307. **135**. Reed, Clinkenbeard & Lane (1975) *J. Biol. Chem.* **250**, 3117. **136**. Evans & Ratledge (1985) *Can. J. Microbiol.* **31**, 1000. **137**. Boulton & Ratledge (1983) *J. Gen. Microbiol.* **129**, 2863. **138**. Måhlén (1971) *Eur. J. Biochem.* **22**, 104. **139**. Solberg (1974) *Biochim. Biophys. Acta* **360**, 101. **140**. Lilly, Bugaisky, Umeda & Bieber (1990) *Arch. Biochem. Biophys.* **280**, 167. **141**. Kaschula, Jin, Desmond-Smith & Travis (2005) *Exp. Eye Res.* **82**, 111. **142**. Glover, Goddard & Felsted (1988) *Biochem. J.* **250**, 485. **143**. Ferry, Loynel, Kucharczyk, *et al.* (2000) *J. Biol. Chem.* **275**, 8794. **144**. Wykle, Malone & Snyder (1980) *J. Biol. Chem.* **255**, 10256. **145**. Butterworth, Tsai, Eley, Roche & Reed (1975) *J. Biol. Chem.* **250**, 1921. **146**. Rado & Hoch (1973) *Biochim. Biophys. Acta* **321**, 114. **147**. Huckle & Tamblyn (1983) *Arch. Biochem. Biophys.* **226**, 94. **148**. Barenholz, Edelman & Gatt (1974) *Biochim. Biophys. Acta* **358**, 262. **149**. McDonald & Kinsella (1973) *Arch. Biochem. Biophys.* **156**, 223. **150**. Croes, Foulon, Casteels, Van Veldhoven & Mannaerts (2000) *J. Lipid Res.* **41**, 629.

2-Palmitoylglycerol

This monoacylglyceride (FW = 330.51 g/mol; CAS 23470-00-0; M.P. = 68.5°C), also known as 2-monopalmitoylglycerol and 2-monopalmitin, inhibits hormone-sensitive lipase and cytochrome P450-dependent 7-ethoxycoumarin *O*-deethylase. Acyl group migration can readily occur. *Note*: Because this substance readily forms micelles, one must take care to discriminate whether the monomeric and/or micelle form(s) is(are) inhibitory, when present at concentrations above the critical micelle concentration (*cmc*). **Target(s):** 7-ethoxycoumarin *O*-deethylase, *or* cytochrome P450 (1); 1,3-β-glucan synthase (2); and hormone-sensitive lipase (3). **1**. Muller-Enoch, Fintelmann, Nicolaev & Gruler (2001) *Z. Naturforsch. [C]* **56**, 1082. **2**. Ko, Frost, Ho, Ludescher & Wasserman (1994) *Biochim. Biophys. Acta* **1193**, 31. **3**. Jepson & Yeaman (1992) *FEBS Lett.* **310**, 197.

1-Palmitoyl-*sn*-glycero-3-phosphate

This monoacylglycerol phosphate (FW = 407.46 g/mol), also known as 1-palmitoyllysophosphatidate, inhibits 1-alkyl-2-acetylglycerophosphatase (1), alkylglycerophospho-ethanolamine phosphodiesterase, *or* lysophospholipase D (2), and chymase (3). *Note*: Because palmitoylglycero-P readily forms micelles, one must take care to discriminate whether the monomeric and/or micelle form(s) is(are) inhibitory, when palmitoylglycero-P is present at concentrations above the critical micelle concentration (*cmc*). **1**. Tellis & Lekka (2000) *J. Eukaryot. Microbiol.* **47**, 122. **2**. van Meeteren, Ruurs, Christodoulou, *et al.* (2005) *J. Biol. Chem.* **280**, 21155. **3**. Kido, Fukusen & Katunuma (1985) *Arch. Biochem. Biophys.* **239**, 436.

1-Palmitoyl-*sn*-glycero-3-phosphocholine

This lysolecithin (FW = 495.64 g/mol; CAS 17364-16-8), also known as 1-palmitoylglycerophosphocholine or lysophosphatidylcholine (palmitate-containing), is the hydrolysis reaction product formed by phospholipase A₂ and is a substrate for 1-acylglycerophosphocholine *O*-acyltransferase, lysophospholipase, and lysolecithin acylmutase. *Note*: Because palmitoylglycero-phosphocholine readily forms micelles, one must take care to discriminate whether the monomeric and/or micelle form(s) is(are) inhibitory, when palmitoylglycero-phosphoryl choline is present above the critical micelle concentration (*cmc*). *See also 1-Acylglycero-phosphocholine* **Target(s):** 1-alkyl-2-acetylglycerophosphocholine esterase, *or* platelet-activating-factor acetylhydrolase (1,2); 1-alkylglycero-phosphocholine *O*-acetyltransferase (3,4); 1,3-β-glucan synthase (5); glycerophosphatidylinositol phospholipase D (6); glyceryl-ether monooxygenase (7); lipase, *or* triacylglycerol lipase (8); and Na⁺/K⁺-exchanging ATPase (9,10). **1**. Tsoukatos, Liapikos, Tselepis, Chapman & Ninio (2001) *Biochem. J.* **357**, 457. **2**. Chroni & Mavri-Vavayanni (2000) *Life Sci.* **67**, 2807. **3**. Lee, Vallari & Snyder (1992) *Meth. Enzymol.* **209**, 396. **4**. Lee (1985) *J. Biol. Chem.* **260**, 10952. **5**. Ko, Frost, Ho, Ludescher & Wasserman (1994) *Biochim. Biophys. Acta* **1193**, 31. **6**. Lee, Lee, Kim, Myung & Sok (1999) *Neurochem. Res.* **24**, 1577. **7**. Kurisu, Taguchi, Imai & Armarego (1994) *Pteridines* **5**, 95. **8**. W. Tsuzuki, A. Ue, A. Nagao, M. Endo & M. Abe (2004) *Biochim. Biophys. Acta* **1684**, 1. **9**. Lijnen, Huysecom, Fagard, Staessen & Amery (1990) *Methods Find. Exp. Clin. Pharmacol.* **12**, 281. **10**. Tamura, Inagami, Kinoshita & Kuwano (1987) *J. Hypertens.* **5**, 219.

3-Palmitoyl-5-hydroxymethyltetronic Acid, *See* RK-682

Palmitoyl-inosino-Coenzyme A

This inosine-containing acyl-CoA analogue (FW = 1005.93 g/mol) inhibits acetyl-CoA carboxylase (1,2). *Note*: Because palmitoyl-inosino-CoA readily forms micelles, one must take care to discriminate whether the monomeric and/or micelle form(s) is(are) inhibitory, when palmitoyl-inosino-CoA is present at concentrations above the critical micelle concentration (*cmc*). **1**. Tanabe, Nakanishi, Hashimoto, *et al.* (1981) *Meth. Enzymol.* **71**, 5. **2**. Nikawa, Tanabe, Ogiwara, Shiba & Numa (1979) *FEBS Lett.* **102**, 223.

Palmitoyl-keto-Coenzyme A

This thiol ester (FW = 1021.85 g/mol), consisting of a hexadecanoyl moiety linked to an oxidized form of natural coenzyme A, inhibits acetyl-CoA carboxylase (1,2). *Note*: Because palmitoyl-keto-CoA readily forms micelles, one must take care to discriminate whether the monomeric and/or micelle form(s) is(are) inhibitory, when palmitoyl-keto-CoA is present at concentrations above the critical micelle concentration (*cmc*). **1**. Tanabe, Nakanishi, Hashimoto, *et al.* (1981) *Meth. Enzymol.* **71**, 5. **2**. Nikawa, Tanabe, Ogiwara, Shiba & Numa (1979) *FEBS Lett.* **102**, 223.

1-Palmitoyl-2-palmitoylamino-2-deoxy-*sn*-glycero-3-phosphorylcholine

This amide-contaning phosphatidylcholine analogue (FW = 733.07 g/mol) inhibits cobra venom (*Naja naja naja*) phospholipase A$_2$, IC$_{50}$ = 156 μM (1). *Note*: Because this substance readily forms micelles, one must take care to discriminate whether the monomeric and/or micelle form(s) is(are) inhibitory, monomeric and/or micelle form(s) is(are) inhibitory, when present at concentrations above the critical micelle concentration (*cmc*). **1**. Yu, Deems, Hajdu & Dennis (1990) *J. Biol. Chem.* **265**, 2657.

N-Palmitoyl-D-sphingomyelin, *See* *N-Palmitoyl-D-sphingosine-1-phosphocholine*

N-Palmitoyl-D-sphingosine-1-phosphocholine

This sphingomyelin, frequently misnamed *N*-palmitoyl-D-sphingomyelin (FW = 703.04 g/mol), inhibits phospholipase C δ (1). *Note*: Because this substance readily forms micelles, one must take care to discriminate whether the monomeric and/or micelle form(s) is(are) inhibitory, when present at concentrations above the critical micelle concentration (*cmc*). **1**. Pawelczyk & Lowenstein (1997) *Biochimie* **79**, 741.

Palmityl Alcohol, *See* *Cetyl Alcohol*

Palmityl-CoA, *See* *Palmitoyl-CoA; S-Cetyl-CoA*

Palmityl Gallate, *See* *Cetyl Gallate*

O-(Palmityloxyethyl)phosphocholine

This alkylated phosphocholine (FW = 451.63 g/mol) inhibits phospholipase A$_2$ (1). *Note*: Because this substance readily forms micelles, one must take care to discriminate whether monomeric and/or micelle form(s) is(are) inhibitory, when present at concentrations above the critical micelle concentration (*cmc*). **1**. Magolda, Ripka, Galbraith, Johnson & Rudnick

(1985) in *Prostaglandins, Leukotrienes, and Lipoxins* (Bailey, ed.), p. 669, Plenum Press, New York.

1-Palmityl-2-palmitoylamino-2-deoxy-*sn*-glycero-3-phosphorylcholine

This ether amide-phosphocholine (FW = 719.08 g/mol) is an analogue of phosphatidylcholine and inhibits cobra venom (*Naja naja naja*) phospholipase A$_2$, IC$_{50}$ = 38 μM (1). A six-step synthesis of this compound is presented. *Note*: Because this substance readily forms micelles, one must take care to discriminate whether monomeric and/or micelle form(s) is(are) inhibitory, when present at concentrations above the critical micelle concentration (*cmc*). **1**. Yu, Deems, Hajdu & Dennis (1990) *J. Biol. Chem.* **265**, 2657.

Palmitylsulfonate, *See* *Cetyl Sulfonate, Sodium*

1-Palmitylthio-2-palmitoylamino-1,2-dideoxy-*sn*-glycero-3-phosphorylcholine

This thioether amide-containing phosphatidylcholine analogue (FW = 735.15 g/mol) strongly inhibits cobra venom (*Naja naja naja*) phospholipase A$_2$, IC$_{50}$ = 2 μM (1). *Note*: Because this substance readily forms micelles, one must take care to discriminate whether the monomeric and/or micelle form(s) is(are) inhibitory, when present at concentrations above the critical micelle concentration (*cmc*). **1**. Yu, Deems, Hajdu & Dennis (1990) *J. Biol. Chem.* **265**, 2657.

1-Palmitylthio-2-palmitoylamino-1,2-dideoxy-*sn*-glycero-3-phosphoryl-ethanolamine

This thioether amide-containing phosphatidylethanolamine analogue (FW = 693.07 g/mol) potently inhibits cobra venom (*Naja naja naja*) phospholipase A$_2$, IC$_{50}$ = 0.45 μM (1). *Note*: Because this substance readily forms micelles, one must take care to discriminate whether the monomeric and/or micelle form(s) is(are) inhibitory, when present at concentrations above the critical micelle concentration (*cmc*). **1**. Yu, Deems, Hajdu & Dennis (1990) *J. Biol. Chem.* **265**, 2657.

O-(Palmitylthiopropyl)phosphocholine

This alkylated phosphocholine (FW = 481.72 g/mol) inhibits phospholipase A$_2$ (1). *Note*: Because this substance readily forms micelles, one must take care to discriminate whether the monomeric and/or micelle form(s) is(are) inhibitory, when present at concentrations above the critical micelle concentration (*cmc*). **1**. Magolda, Ripka, Galbraith, Johnson & Rudnick (1985) in *Prostaglandins, Leukotrienes, and Lipoxins* (Bailey, ed.), p. 669, Plenum Press, New York.

Palmityl Trifluoromethyl Ketone

Palmityl Trifluoromethyl Ketone

Palmityl Trifluoromethyl Ketone Hydrate

This palmityl trifluoromethyl ketone (FW = 308.43 g/mol), also named 1,1,1-trifluoro-2-heptadecanone, is a potent reversible inhibitor of acylglycerol lipase, *or* monoacylglycerol lipase (1), 1-acylglycerophosphocholine *O*-acyltransferase, *or* lysophosphatidylcholine *O*-acyltransferase (2), fatty acid amide hydrolase, *or* anandamide amidohydrolase (1), and phospholipase A$_2$ (3,4). Inhibition is doubtlessly the consequence of hydration of the ketone (strongly favored by the presence of its trifluoromethyl group.) and attack of an active-site nucleophile. The hydrate then forms a tight-binding analogue of presumed tetrahedral and/or acyl-enzyme intermediates in the hydrolase-type reaction. **1.** Ghafouri, Tiger, Razdan, *et al.* (2004) *Brit. J. Pharmacol.* **143**, 774. **2.** Kerkhoff, Habben, Gehring, Resch & Kaever (1998) *Arch. Biochem. Biophys.* **351**, 220. **3.** Ackermann, Conde-Frieboes & Dennis (1995) *J. Biol. Chem.* **270**, 445. **4.** Holk, Rietz, Zahn, Quader & Scherer (2002) *Plant Physiol.* **130**, 90.

PALO, See *N-(Phosphonacetyl)-L-ornithine*

Palomid 529

This novel Akt inhibitor (FW = 406.43 g/mol; CAS 914913-88-5; Solubility: 80 mg/mL DMSO; <1 mg/mL H$_2$O or Ethanol; Abbreviation: P529), also named 8-(1-hydroxyethyl)-2-methoxy-3-(4-methoxybenzyloxy) benzo[*c*]chromen-6-one, inhibits tumor angiogenesis, vascular permeability, and tumour growth (1). P529 has the additional benefit of blocking pAktS473 signaling consistent with blocking TORC2 in all cells and thus bypassing feedback loops that lead to increased Akt signaling in some tumor cells (1). P529 significantly enhance the antiproliferative effect of radiation in prostate cancer PC-3 cells (2). Analysis of signal transduction pathways targeted by P529 exhibited a decrease in p-Akt, VEGF, MMP-2, MMP-9, and Id-1 levels after radiation treatment. Palomid 529 is an effective suppressor of Müller cell proliferation, glial scar formation, and photoreceptor cell death in a rabbit model of retinal detachment, *or* RD (3). Palomid 529 significantly suppresses breast cancer susceptibility gene (Brca1) -deficient tumor growth in mice through inhibition of both Akt and mTOR signaling (4). **1.** Xue, Hopkins, Perruzz, *et al.* (2008) *Cancer Res.* **68**, 9551. **2.** Diaz, Nguewa, Diaz-Gonzalez, *et al.* (2009) *Brit. J. Cancer* **100**, 932. **3.** Lewis, Chapin, Byun, *et al.* (2009) *Invest. Ophthalmol. Vis. Sci.* **50**, 4429. **4.** Xiang, Jia, Sherris, *et al.* (2011) *Oncogene* **30**, 2443.

Palonosetron

This serotonin 5-HT$_3$ receptor antagonist and longlasting anti-emetic (FW = 296.41 g/mol; CAS 135729-61-2), known by code name RS 25259, RS 25259-197, trade name Aloxi®, and systematic name (3a*S*)-2-[(3*S*)-1-azabicyclo[2.2.2]oct-3-yl]-2,3,3a,4,5,6-hexahydro-1*H*-benz[*de*]isoquinolin-1-one, is used to treat chemotherapy-induced nausea/vomiting (1-3). Polonosetron affects both peripheral and central nervous systems, reducing vagus nerve activity, thereby deactivating the 5-HT$_3$ receptor-dense vomiting center located in the medulla oblongata. It is without effect on dopamine receptors or muscarinic receptors. **Primary Mode of Inhibitor Action:** Intravenously administered palonosetron has a linear pharmacokinetic profile, with a long terminal elimination half-life of ~40 hours and moderate (62%) plasma protein binding (4). Its high affinity and long half-life most likely explains its persistent antiemetic effect. Similar 5-HT$_3$ antagonists include tropisetron, ondansetron, dolasetron, and granisetron. The effectiveness of each depends on particular variants of 5-HT$_3$ receptors expressed by the patient, including changes in promoters for the receptor genes. **1.** Clark, Miller, Berger, *et al.* (1993) *J. Med. Chem.* **36**,

2645. **2.** Wong, Clark, Leung, *et al.* (1995) *Brit. J. Pharmacol.* **114**, 851. **3.** Eglen, Lee, Smith, *et al.* (1995) *Brit. J. Pharmacol.* **114**, 860. **4.** Siddiqui & Scott (2004) *Drugs* **64**, 1125. **5.** Grunberg & Koeller (2003) *Expert Opin. Pharmacother.* **4**, 2297.

Palytoxin

This structurally complex neurotoxin (FW = 2680.17 g/mol; CAS 11077-03-5; Abbreviation: PLTX), from the Pacific soft coral *Palythoa toxica*, binds slowly, but with extreme affinity, to Na$^+$/K$^+$-exchanging ATPase, converting the latter into an open channel, allowing unhindered passage of potassium and sodium ions. Less than 1 pM palytoxin is required for rapid potassium ion outflow from erythrocytes, thereby dissipating its transmembrane electric potential (1-7). Palytoxin is lethal in mice, exhibiting an impressive LD$_{50}$ of 15 ng/kg. The pore-forming action of palytoxin is not restricted to Na$^+$/K$^+$-ATPase, but is also observed with the colonic H$^+$/K$^+$-ATPase (8). This observation raises the likelihood that the toxin reacts at functionally and/or structurally similar manner with these pumps. PLTX brings about depolarization of the cellular membrane, disrupting intracellular calcium concentration ([Ca^{2+}]$_i$) and leading to smooth and cardiac muscle contraction, cytoskeletal dysregulation, and neurotransmitter release (9). Second only to maitotoxin as the most toxic natural product known, palytoxin evokes prompt onset of severe agina and asthma-like airway effects, followed quickly by tachycardia and death, often within minutes. SKF-96365 and Gd^{3+} are widely used to block store-operated and stretch-activated Ca^{2+} channels, respectively. SKF-96365 fails to affect the long-lasting phase elicited by PLTX, excluding the activation of the store-operated channels during this phase (9). In contrast, Gd^{3+} abolishes the long-lasting phase of the [Ca^{2+}]$_i$ increase, suggesting a possible role for stretch-activated channels in PLTX action. While Gd^{3+} is also known to inhibit voltage-dependent Ca^{2+} conductance in neural cells, voltage-dependent Ca^{2+} blockade seems unlikely in muscle cells. Even at a level (100 μM) commonly used to investigate stretch-activated channels activity, Gd^{3+} does not affect the transient phase as Verapamil and La^{3+}/Cd^{2+} does. In addition to the selective inhibitory action of Gd^{3+} on the long-lasting phase, Gd^{3+} also significantly reduces, albeit incompletely, the toxic effects of PLTX on skeletal muscle cells, suggesting a role for the stretch-activated channels in the chain of events culminating in death of cultured muscle cells (9). Palytoxin also binds to erythrocyte Band-3 protein (B3 or AE1), altering the anionic flux and seriously compromising not only CO$_2$ (bicarbonate) transport, thereby also affecting hemoglobin oxygenation/deoxygenation (10). The stereocontrolled synthesis of palytoxin (11) was heralded as one of the most complicated syntheses ever undertaken. (*See Maitotoxin*) **1.** Habermann (1989) *Toxicon* **27**, 1171. **2.** Ito, Toyoda, Higashiyama, *et al.* (2003) *FEBS Lett.* **543**, 108. **3.** Wang & Horisberger (1997) *FEBS Lett.* **409**, 391. **4.** Kim, Marx & Wu (1995) *Naunyn Schmiedebergs Arch. Pharmacol.* **351**, 542. **5.** Bottinger & Habermann (1984) *Naunyn Schmiedebergs Arch. Pharmacol.* **325**, 85. **6.** Ishida, Takagi, Takahashi, Satake & Shibata (1983) *J. Biol. Chem.* **258**, 7900. **7.** Su & Scheiner-Bobis (2004) *Biochemistry* **43**, 4731. **8.** Scheiner-Bobis, Hübschle, Diener (2002) *Eur. J Biochem.* **269**, 3905. **9.** Del Favero, Florio, Codan, *et al.* (2012) *Chem. Res Toxicol.* **25**, 1912. **10.** Ficarra,

Russo, Stefanizzi, *et al.* (2011) *J. Membr. Biol.* **242**, 31. **11.** Armstrong, Cheon, Christ *et al.* (1994) *J. Am. Chem. Soc.* **111**, 7530.

Pancreastatin

This polypeptide (MW = 5103 g/mol; CAS 106477-83-2; Porcine Sequence: GWPQAPAMDGAGKTGAEEAQPPEGKGAREHSRQEEEEE TAGAPQGLFRG-NH₂; pI = 4.6), formed by processing of the glycoprotein Chromogranin A in neuroendocrine cells, inhibits glucose-induced insulin release from the pancreas. Human, porcine, and bovine pancreastatins have 52, 49, and 47 residues, respectively. Pancreastatin reportedly inhibits glucose uptake in rat adipocytes by signaling through the Gαq-subunit of heterotrimeric G proteins in the insulin signaling pathway, thus preventing GLUT4 translocation to the plasma membrane. Pancreastatin also inhibits glycogen synthesis in rat adipocytes by activating [glycogen synthase] kinase-3 activity through the activation of protein kinase C (1). *Note:* Porcine pancreastatin is commercially available (Gene-Script). *Storage:* Store in dry form in a dessicator at 0–5°C. For most reliable results, rehydrate the peptide immediately prior to use, and do not re-freeze any unused portion of the dissolved peptide. **Target(s):** glucose-induced insulin release (2-4); and GLUT-4-mediated glucose transport (5). **1.** Gonzalez-Yanes & Sanchez-Margalet (2001) *Biochem. Biophys. Res. Commun.* **289**, 282. **2.** Schmidt & Creutzfeldt (1991) *Acta Oncol.* **30**, 441. **3.** Ahren, Bertrand, Roye & Ribes (1996) *Acta Physiol. Scand.* **158**, 63. **4.** Tatemoto, Efendic, Mutt, *et al.* (1986) *Nature* **324**, 476. **5.** Sanchez-Margalet & Gonzalez-Yanes (1998) *Amer. J. Physiol.* **275**, E1055.

Pancreatic Secretory Trypsin Inhibitor

This polypeptide (FW = 6247 g/mol; Abbreviation: PSTI), also known as pancreatic trypsin inhibitor and Kunitz & Northrop inhibitor, forms a high-affinity complex with trypsin. Synthesized in the gut mucosa and secreted from pancreatic acinar cells into pancreatic juice, PSTI prevents premature trypsin-catalyzed zymogen activation within the pancreas and/or pancreatic duct. Mutations in its gene are associated with hereditary pancreatitis and tropical calcific pancreatitis. **See Trypsin Inhibitors** **1.** Kunitz & Northrop (1936) *J. Gen. Physiol.* **19**, 991. **2.** Li & Chung (1983) *Proc. Natl. Acad. Sci. U.S.A.* **80**, 1204. **3.** Chauvet, Nouvel & Acher (1966) *Biochim. Biophys. Acta* **115**, 121 and 130. **4.** Chauvet & Acher (1966) *Bull. Soc. Chim. Biol.* **48**, 1284. **5.** Kassell, Radicevic, Ansfield & Laskowski, Sr. (1965) *Biochem. Biophys. Res. Commun.* **18**, 255.

Pancreozymin, *See Cholecystokinin*

Pancuronium Bromide

This cationic aminosteroid (FW = 732.68 g/mol; CAS 15500-66-0; Soluble to 1 g/mL H₂O at 20°C), is a long-acting, nondepolarizing relaxant used to induce skeletal muscle relaxation during anesthesia. It binds to post-synaptic nicotinic acetylcholine receptors, displacing acetylcholine competitively (1-4). Infusion of the reversible competitive inhibitor pancuronium for up to 12 hrs does not reduce acetylcholine receptor number, an action that contrasts with that of the irreversible acetylcholine receptor blocker α-bungarotoxin. Because equilibrium conditions of neurotransmitter concentration and receptor binding are rarely, if ever, achieved during synaptic transmission at the neuromuscular junction, one must consider the binding kinetics of drugs acting at this synapse. Based on a single high-affinity site model of competitive inhibition, rate constants for pancuronium (k_{on} is 2.7 x 10⁸ M⁻¹s⁻¹; k_{off} is 2.1 s⁻¹) are in reasonably good agreement with an IC₅₀ value of 5.5 nM (4). Pancuronium bromide has been used in lethal injection protocols for capital punishment. Sodium pentathol is administered first, followed by pancuronium bromide, with potassium chloride given last. **Target(s):** acetylcholine receptors (1-3); acetylcholine esterase, weakly inhibited (5); choline acetyltransferase (6); cholinesterase (5); diamine oxidase (6); histamine *N*-methyltransferase (8,9); and Na⁺ currents (10). **1.** Wenningmann & Dilger (2001) *Mol. Pharmacol.* **60**, 790.

2. Kerr, Stevenson & Mitchelson (1995) *J. Pharm. Pharmacol.* **47**, 1002. **3.** Nedoma, Dorofeeva, Tucek, Shelkovnikov & Danilov (1985) *Naunyn Schmiedebergs Arch. Pharmacol.* **329**, 176. **4.** Wenningmann & Dilger (2001) *Molec. Pharmacol.* **60**, 790. **5.** Stovner, Oftedal & Holmboe (1975) *Brit. J. Anaesth.* **47**, 949. **6.** Kambam, Janson, Day & Sastry (1990) *Can. J. Anaesth.* **37**, 690. **7.** Sattler, Hesterberg, Lorenz, *et al.* (1985) *Agents Actions* **16**, 91. **8.** Futo, Kupferberg & Moss (1990) *Biochem. Pharmacol.* **39**, 415. **9.** Harle, Baldo & Fisher (1985) *Agents Actions* **17**, 27. **10.** Maestrone, Magnelli, Nobile & Usai (1994) *Brit. J. Pharmacol.* **111**, 283.

Pandinotoxin-Kα

This internally disulfide cross-linked, 35-residue oligopeptide (MW = 4010 g/mol; CAS 185529-64-0) from the venom of the scorpion *Pandinus imperator*, inhibits K⁺ currents, voltage-gated, rapidly inactivating A-type K⁺ channels, IC₅₀ = 6 nM (1-3) as well as K$_v$1.2 potassium channels, IC₅₀ = 32 pM (2,3). **1.** Matteson & Blaustein (1997) *J. Physiol.* **503**, 285. **2.** Tenenholz, Rogowski, Collins, Blaustein & Weber (1997) *Biochemistry* **36**, 2763. **3.** Rogowski, Collins, O'Neil, *et al.* (1996) *Mol. Pharmacol.* **50**, 1167.

Pantherine, *See Muscimol*

Panobinostat

This pan-HDAC inhibitor (FW = 349.43 g/mol; CASs = 404950-80-7, 960055-57-6 (Maleic acid), 960055-60-1 (methanesulfonate); Solubility: 70 mg/mL DMSO; <1 mg/mL Water), also known as LBH589 and named systematically as (*E*)-*N*-hydroxy-3-(4-((2-(2-methyl-1*H*-indol-3-yl)ethyl amino)-methyl)phenyl)acrylamide, exhibits a broad spectrum of action, targeting histone acetylases with IC₅₀ values of 5-20 nM and resulting in histone hyperacetylation (1,2). Low nanomolar concentrations of LBH589 inhibit the growth of all melanoma cell lines tested, but not normal melanocytes. Inhibition is characterized by increased apoptosis as well as a G₁ cell-cycle arrest (3). Panobinostat induces E-cadherin expression on cell membranes of MDA-MB-231 cells (as a model for triple-negative breast cancer, *or* TNBC), reducing cell invasion and migration (4). When LBH589 is combined with the mTORC1 inhibitor RAD001, there is strong synergy in inducing cytotoxicity in multiple myeloma cells (5). **1.** Scuto A, *et al.* (2008) *Blood* **111**, 5093. **2.** Crisanti, *et al.* (2009) *Mol. Cancer Ther.* **8**, 2221. **3.** Woods, Woan, Cheng, *et al.* (2013) *Melanoma Res.* **23**, 341. **4.** Fortunati, Marano, Bandino, *et al.* (2013) *Int. J. Oncol.* **44**, 700. **5.** Ramakrishnan, Kimlinger, Timm, *et al.* (2014) *Leuk. Res.* **38**, 1358.

Pantoprazole

This racemic benzimidazole (FW = 383.38 g/mol; CAS 102625-70-7), also known as BY-1023, SKF 96022 and marketed under the names Protium™ and Protonix™, inhibits the gastric proton pump, *or* H⁺/K⁺-exchanging ATPase (1,2). **Primary Mode of Action:** Like other proton pump inhibitors, *or* PPIs (such as omeprazole, lansoprazole and rabeprazole), pantoprazole is a prodrug that undergoes acid-catalyzed conversion to its sulfhydryl-reactive form. The activated PPI then binds to and covalently modifies gastric H⁺/K⁺-ATPase, forming a disulfide bond with Cys813 and inhibiting the pump. Pantoprazole also inhibits the exchange between ¹⁸O-labeled orthophosphate and HOH, a well-known feature of H⁺/K⁺-ATPase catalysis. **Metabolism:** Pantoprazole degradation proceeds mainly by CYP2C19-catalyzed demethylation, followed by sulfation. Another pathway involves CYP3A4-catalyzed oxidation. In either case, pantoprazole metabolites are not believed to be pharmacologically significant. **Key Pharmacokinetic Parameters:** *See* Appendix II in Goodman & Gilman's *THE PHARMACOLOGICAL BASIS OF THERAPEUTICS*, 12ᵗʰ Edition (Brunton, Chabner

& Knollmann, eds.) McGraw-Hill Medical, New York (2011). **1**. Shin & Sachs (2004) *Biochem. Pharmacol.* **68**, 2117. **2**. Simon, Keeling, Laing, Fallowfield & Taylor (1990) *Biochem. Pharmacol.* **39**, 1799.

PAP or pAp. *See Adenosine 3',5'-Bisphosphate*

Papaverine

This opium metabolite (FW$_{\text{free-base}}$ = 339.39 g/mol; CAS 61-25-6; triboluminescent solid; pK_a = 8.07 at 25°C), also known as 1-[(3,4-dimethoxy-phenyl)methyl]-6,7-dimethoxyisoquinoline, is a smooth muscle relaxant and cerebral vasodilator. It is most stable when stored as the hydrochloride or in aqueous solutions at pH 2.0-2.8. **Target(s):** acetylcholinesterase (1,2); Ca^{2+} current, voltage-dependent (3); 3',5'-cyclic-GMP phosphodiesterase, *or* cGMP phosphodiesterase (5,7,9, 12,24,26); 3',5'-cyclic-nucleotide phosphodiesterase, *or* cAMP phosphodiesterase (4-11,24-27); guanylate cyclase (13); HIV replication (14,15); HVJ (Sendai virus) replication (16); 15-hydroxyprostaglandin dehydrogenase (17); K$^+$ current, Ca^{2+}-activated oscillatory (3); K$^+$ current, voltage-dependent transient outward (3); lactate dehydrogenase (18); malate dehydrogenase (18); monoamine oxidase (19); NADH dehydrogenase, *or* complex I (20); NAD(P)H dehydrogenase, *or* DT diaphorase (21); oxidative phosphorylation (22); purine transport (23); and rhodopsin kinase, weakly inhibited (28). .1. Whiteley & Daya (1995) *J. Enzyme Inhib.* **9**, 285. **2**. Augustinsson (1950) *The Enzymes*, 1st ed. (Sumner & Myrbäck, eds.), **1** (part 1), 443. **3**. Iguchi, Nakajima, Hisada, Sugimoto & Kurachi (1992) *J. Pharmacol. Exp. Ther.* **263**, 194. **4**. Hosono (1988) *Meth. Enzymol.* **159**, 497. **5**. Harrison, Beier, Martins & Beavo (1988) *Meth. Enzymol.* **159**, 685. **6**. Poch & Kukovetz (1971) *Life Sci. I* **10**, 133. **7**. Bergstrand, Kristoffersson, Lundquist & Schurmann (1977) *Mol. Pharmacol.* **13**, 38. **8**. Mannhold (1988) *Arzneimittelforschung* **38**, 1806. **9**. Lugnier & Stoclet (1974) *Biochem. Pharmacol.* **23**, 3071. **10**. Rufeger, Tellhelm & Frimmer (1971) *Naunyn Schmiedebergs Arch. Pharmakol.* **270**, 428. **11**. Kukovetz & Poch (1970) *Naunyn Schmiedebergs Arch. Pharmakol.* **267**, 189. **12**. Liebman & Evanczuk (1982) *Meth. Enzymol.* **81**, 532. **13**. Strinden & Stellwagen (1984) *Biochem. Biophys. Res. Commun.* **123**, 1194. **14**. Nokta, Albrecht & Pollard (1993) *Immuno-pharmacology* **26**, 181. **15**. Turano, Scura, Caruso, *et al.* (1989) *AIDS Res. Hum. Retroviruses* **5**, 183. **16**. Ogura, Sato & Hatano (1987) *J. Gen. Virol.* **68**, 1143. **17**. Iijima, Ueno, Sasagawa & Yamazaki (1978) *Biochem. Biophys. Res. Commun.* **80**, 484. **18**. Kapp & Whiteley (1991) *J. Enzyme Inhib.* **4**, 233. **19**. Lee, Lee, Cheong, *et al.* (2001) *Biol. Pharm. Bull.* **24**, 838. **20**. Morikawa, Nakagawa-Hattori & Mizuno (1996) *J. Neurochem.* **66**, 1174. **21**. Tsuda, Urakawa & Fukami (1977) *Jpn. J. Pharmacol.* **27**, 855. **22**. Santi, Contessa & Ferrari (1963) *Biochem. Biophys. Res. Commun.* **11**, 156. **23**. Kraupp, Paskutti, Schon & Marz (1994) *Biochem. Pharmacol.* **48**, 41. **24**. Methven, Lemon & Bhoola (1980) *Biochem. J.* **186**, 491. **25**. Grant & Colman (1984) *Biochemistry* **23**, 1801. **26**. Pannbacker, Fleischman & Reed (1972) *Science* **175**, 757. **27**. Kunz, Oberholzer & Seebeck (2005) *FEBS J.* **272**, 6412. **28**. Weller, Virmaux & Mandel (1975) *Proc. Natl. Acad. Sci. U.S.A.* **72**, 381.

PAQ-22

This isoquinoline derivative (FW = 294.35 g/mol), also known as 3-(2,6-diethylphenyl)-2,4-(1*H*,3*H*)-quinazolinedione, inhibits cytosol (1) and membrane (2) alanine aminopeptidase. **1**. Thielitz, Bukowska, Wolke, *et al.* 2004) *Biochem. Biophys. Res. Commun.* **321**, 795. **2**. Kakuta, Koiso, Takahashi, Nagasawa & Hashimoto (2001) *Heterocycles* **55**, 1433.

PAR-101, *See Fidaxomicin*

Paracetamol, *See Acetaminophen*

Paradaxin, *See Pardaxin*

Paradoxin

This presynaptic neurotoxin (MW = 17.5 kDa; CAS 70699-67-1; Symbol: PDX) from the venom of the inland taipan snake (*Oxyuranus microlepidotus*) is a heterotrimeric phospholipase A$_2$ that catalyzes the hydrolysis of 2-acyl groups within 3-*sn*-phosphoglycerides (1). Paradoxin abolishes indirect twitches of the chick biventer cervicis and mouse phrenic nerve diaphragm preparations, when present at 65 nM (2). The time to 90% inhibition by PDX was significantly increases by replacing 2.5 mM Ca^{2+} in the physiological solution with 10 mM Sr^{2+} (2). In the mouse diaphragm (at low Ca^{2+} and at room temperature), the inhibitory effect of PDX (6.5 nM) is delayed, attended by a transient increase of contractions. In intracellular recording experiments using the mouse hemidiaphragm, PDX (6.5-65 nM) significantly increases quantal content and miniature endplate potential frequency prior to blocking evoked release of acetylcholine. In extracellular recording experiments using the mouse triangularis sterni, PDX (2.2-65 nM) significantly inhibits the voltage-dependent potassium ion, but not sodium ion, waveform (2). In patch clamp experiments on B82 mouse fibroblasts that had been stably transfected with rKv 1.2, PDX (22 nM) is without significant effect on currents evoked by 10mV step depolarizations from -60 to +20mV. PDX exhibits the pharmacological properties of a β-neurotoxin, and is likely to be one of the most potent, if not the most potent β-neurotoxin yet discovered (2). The toxic effects of paradoxin are blocked upon incubation with 4-bromophenacyl bromide (1.8 mM). **1**. Bell, Sutherland & Hodgson (1998) *Toxicon* **36**, 63. **2**. Hodgson, Dal Belo & Rowan (2007) *Neuropharmacol.* **52**, 1229.

Parahematin, *See Ferrihemochrome*

Paramomycin, *See Paromomycin*

Paraoxon, *See Diethyl p-Nitrophenyl Phosphate*

Parathion

This insecticide (FW = 291.26 g/mol; CAS 56-38-2), also known as E605, phosphostigmine, and phosphorothioic acid *O,O*-diethyl *O*-(4-nitrophenyl) ester, is a cholinesterase inhibitor. **Primary Mode of Action:** Like other phosphorothioate organo-phosphorus pesticides, parathion is metabolized by cytochrome P450s, either through a dearylation reaction to form an inactive metabolite, or through a desulfuration to form the active oxon metabolite, paraoxon-ethyl (*e.g.*, (CH$_3$CH$_2$–O–)$_2$P(=O)–O–C$_6$H$_4$–NO$_2$). The latter is a highly effective mechanism-based acetylcholinesterase inhibitor. The primary enzymes involved in parathion bioactivation are CYP2B6, CYP2C19, and CYP1A2. **CAUTION:** Parathion is highly toxic by all routes of exposure. Its use and that of related organophosphorus compounds annually accounts for 200,000 deaths worldwide, mainly farmworkers. The lowest dosage with detectible toxic effects in humans is 240 µg/kg (< 0.1 ounce). The most common cause of death is failure to maintain adequate oxygenation. Clinical signs of organophosphate poisoning include excess salivation, lacrimation, abdominal pain, vomiting, intestinal hypermotility, and diarrhea. The consensus treatment is early resuscitation with atropine and use of a mechanical respirator, supplemented with oxygen, to provide adequate oxygenation of vital tissues. **Target(s):** acetylcholinesterase (1-5); (Ca^{2+}/Mg^{2+})-ATPase (6); carboxylesterase (7-9); cholinesterase, *or* butyrylcholinesterase (1-3,10,11); CYP1A1 (12); CYP3A2 (13); CYP3A4 (14); CYP3A27 (12); CYP2B1 (15); CYP2C6 (13); CYP2C11 (13); cytochrome P450 (12-17); nicotinic acetylcholine receptor (18); and photosynthetic system (19). **1**. Gibson & Ludke (1971) *Bull. Environ. Contam. Toxicol.* **6**, 97. **2**. Namba (1971) *Bull. World Health Organ.* **44**, 289. **3**. Roex, Keijzers & van Gestel (2003) *Aquat. Toxicol.* **64**, 451. **4**. Kasturi & Vasantharajan (1976) *Phytochemistry* **15**, 1345. **5**. Pralovorio & Fournier (1991) *Biochem. Genet.* **30**, 77. **6**. Blasiak (1995) *Comp. Biochem. Physiol. C Pharmacol. Toxicol. Endocrinol.* **110**, 119. **7**. Lee & Waters, III (1977) *Blood* **50**, 947. **8**. Jansen, Nutting, Jang & Balls (1949) *J. Biol.*

Chem. **179**, 189. **9**. Zhang, Qiao & Lan (2005) *Enzyme Microb. Technol.* **36**, 648. **10**. Reiner, Simeon-Rudolf & Skrinjaric-Spoljar (1995) *Toxicol. Lett.* **82/83**, 447. **11**. Mehrani (2004) *Proc. Biochem.* **39**, 877. **12**. Miranda, Henderson & Buhler (1998) *Toxicol. Appl. Pharmacol.* **148**, 237. **13**. Butler & Murray (1993) *Mol. Pharmacol.* **43**, 902. **14**. Butler & Murray (1997) *J. Pharmacol. Exp. Ther.* **280**, 966. **15**. Murray & Butler (1995) *J. Pharmacol. Exp. Ther.* **272**, 639. **16**. Ator & Ortiz de Montellano (1990) *The Enzymes*, 3rd ed. (Sigman & Boyer, eds.), **19**, 213. **17**. Halpert & Neal (1981) *Drug Metab. Rev.* **12**, 239. **18**. Katz, Cortes, Eldefrawi & Eldefrawi (1997) *Toxicol. Appl. Pharmacol.* **146**, 227. **19**. Suzuki & Uchiyama (1977) *Ecotoxicol. Environ. Saf.* **1**, 263.

Pargyline

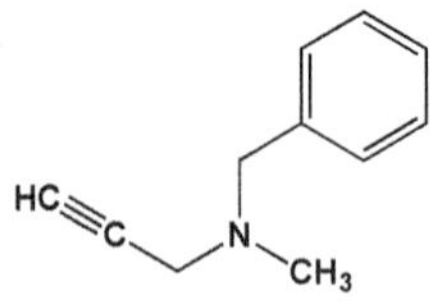

This antihypertensive agent (FW$_{\text{free-base}}$ = 159.23 g/mol; CAS 555-57-7), also known as Eutonyl, *N*-benzyl-*N*-methyl-2-propynylamine, is a mechanism-based inhibitor that targets monoamine oxidase, beginning with an initial reversible interaction (K_i = 4-5 µM) and proceeding to irreversible loss of enzyme activity. **Target(s):** aldehyde dehydrogenase, pargyline is the precursor of the actual inhibitor (1-5); bacterial bioluminescence (6); and monoamine oxidase (7-18). **1**. Shirota, DeMaster & Nagasawa (1979) *J. Med. Chem.* **22**, 463. **2**. DeMaster & Stevens (1988) *Biochem. Pharmacol.* **37**, 229. **3**. DeMaster, Shirota & Nagasawa (1980) *Adv. Exp. Med. Biol.* **132**, 219. **4**. Lebsack & Anderson (1979) *Res. Commun. Chem. Pathol. Pharmacol.* **26**, 263. **5**. Lebsack, Petersen, Collins & Anderson (1977) *Biochem. Pharmacol.* **26**, 1151. **6**. Makemson & Hastings (1979) *Arch. Biochem. Biophys.* **196**, 396. **7**. Sigman & Mooser (1975) *Ann. Rev. Biochem.* **44**, 889. **8**. Rando (1977) *Meth. Enzymol.* **46**, 158. **9**. Walsh, Cromartie, Marcotte & Spencer (1978) *Meth. Enzymol.* **53**, 437. **10**. Maycock (1980) *Meth. Enzymol.* **66**, 294. **11**. Abell (1987) *Meth. Enzymol.* **142**, 638. **12**. Bright & Porter (1975) *The Enzymes*, 3rd ed., **12**, 421. **13**. Ator & Ortiz de Montellano (1990) *The Enzymes*, 3rd ed., **19**, 213. **14**. Brush & Kozarich (1992) *The Enzymes*, 3rd ed., **20**, 317. **15**. Schraven & Reibert (1984) *Arzneimittelforschung* **34**, 1258. **16**. Fowler, Mantle & Tipton (1982) *Biochem. Pharmacol.* **31**, 3555. **17**. Green (1981) *J. Pharm. Pharmacol.* **33**, 798. **18**. McEwan, Sasaki & Jones (1969) *Biochemistry* **8**, 3963.

Paricalcitol

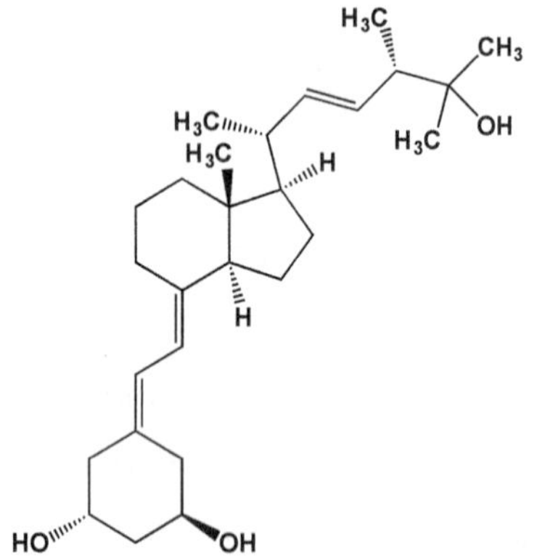

This vitamin D receptor agonist (FW = 416.64 g/mol; CAS 131918-61-1), also named (1*R*,3*R*,5*Z*)-5-[(2*E*)-2-[(1*R*,3a*S*,7a*R*)-octahydro-1-[(1*R*,2*E*,4*S*)-5-hydroxy-1,4,5-trimethyl-2-hexen-1-yl]-7a-methyl-4*H*-inden-4-ylidene] ethylidene]-1,3-cyclohexanediol, is a vitamin D analogue that suppresses elevated parathyroid hormone (PTH) concentrations in patients with hyperparathyroidism, especially those maintained on chronic hemodialysis (1). The vitamin D receptor is a ligand-induced transcription factor that regulates the expression of thye genes involved in controlling calcium homeostasis, bone remodeling, hormone secretion, as well as the growth and differentiation of target cells (1). Paricalcitol also inhibits tumor formation in a murine model of uterine fibroids, suggesting its use as a noninvasive medical treatment option for uterine fibroids (2). **1**. Goldenberg (1999) *Clin. Ther.* **21**, 432. **2**. Halder, Sharan, Al-Hendy & Al-Hendy (2014) *Reprod Sci.* **21**, 1108.

Parinaric Acids

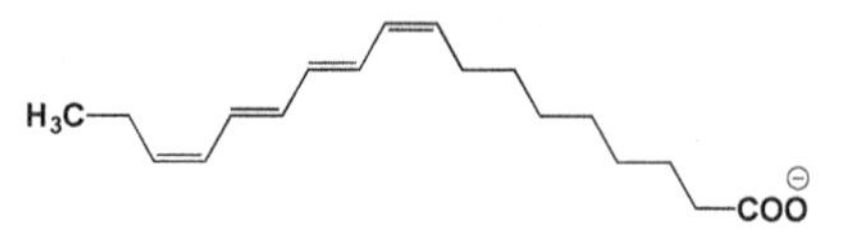

9Z,11E,13E,15Z)-isomer

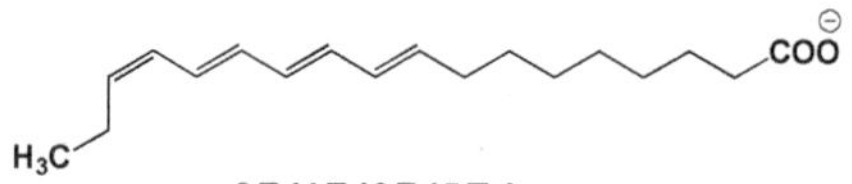

9E,11E,13E,15Z)-isomer

This term refers to any of octadeca-9,11,13,15-tetraenoic acid's sixteen different *cis/trans*-isomers (FW$_{\text{free-acid}}$ = 276.42 g/mol), all possessing a conjugated double-bond network first found in the seeds of *Parinarium laurinum*. The (9*Z*,11*E*,13*E*,15*Z*)-isomer is frequently referred to as α- or *cis*-parinaric acid, while the (9*E*,11*E*,13*E*,15*Z*)-isomer is usually referred to as β- or *trans*-parinaric acid (however, note that the all-*E*-isomer has also been called β-parinaric acid). *Note*: The common terms *cis*- or *trans*-parinaric acid do not refer to either the all-*cis*- or all-*trans*-isomers. **Target(s):** DNA-directed DNA polymerase (1); DNA polymerases α, β, and δ, inhibited by α-parinarate (1); DNA topoisomerase I, inhibited by α-parinarate (1); DNA topoisomerase II, inhibited by α-parinarate (1); gelatinase A, inhibited by α- and β-parinarate (2); gelatinase B, inhibited by α- and β-parinarate (2); hexokinase, inhibited by α-parinarate (3); neutophil elastase, inhibited by α- and β-parinarate (4). **1**. Mizushina, Tsuzuki, Eitsuka, *et al.* (2004) *Lipids* **39**, 977. **2**. Berton, Rigot, Huet, *et al.* 2001) *J. Biol. Chem.* **276**, 20458. **3**. Stewart & Blakely (2000) *Biochim. Biophys. Acta* **1484**, 278. **4**. Tyagi & Simon (1991) *J. Biol. Chem.* **266**, 15185.

trans-Parinaroyl Ganglioside G$_{M3}$

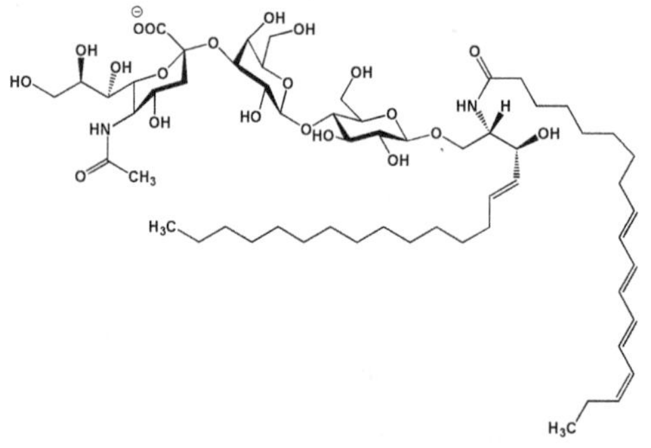

This fluorescent ganglioside G$_{M3}$ analogue (FW = 1173.44 g/mol), with the fatty acyl component is replaced with *trans*-9,11,13,15-octadecatetraenoic acid, inhibits the EGF receptor protein-tyrosine kinase of the epidermal growth factor receptor. The deacetylated analogue is not inhibitory. *Note*: Because this substance readily forms micelles, one must take care to discriminate whether the monomeric and/or micelle form(s) is(are) inhibitory, when present at concentrations above the critical micelle concentration (*cmc*). **1**. Song, Welti, Hafner-Strauss & Rintoul (1993) *Biochemistry* **32**, 8602.

Paritaprevir

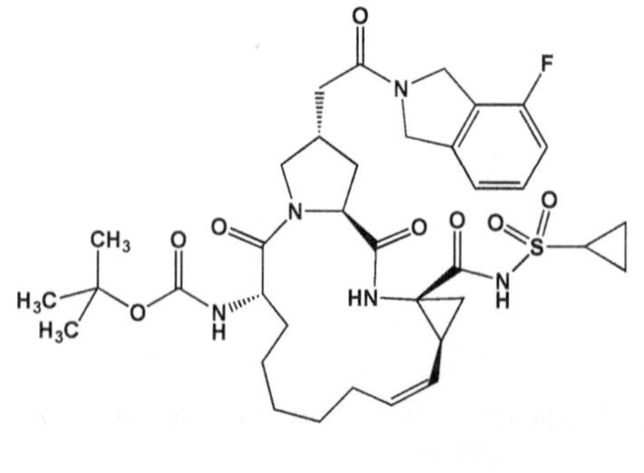

This potent macrocyclic HCV NS3 protease inhibitor (FW = 729.87 g/mol; CAS 1216941-48-8), also known as ABT-450, causes a mean maximum viral decline of about 4 logs, when given in monotherapy to Hepatitis C

Virus-infected patients for 3 days at different doses (1). ABT-450 has been combined with ritonavir (designated ABT-450/r) to increase the ABT-450 plasma concentration and half-life, permitting once-daily dosing for interferon-free treatment of chronic HCV infection (2). The macrocycle-based structure gives ABT-450 higher efficacy and better safety than Telaprevir or Boceprevir, first-generation HCV drugs that possess so-called linear structures. Several second-generation HCV NS3 protease inhibitors (*e.g.*, ABT-450 from AbbVie, RG-7227 from Roche, and MK-7009 from Merck) have emerged from clinical trials, and the U.S. Food and Drug Administration approved Viekira Pak® (tablets consting of ombitasvir, paritaprevir and ritonavir that co-packaged with dasabuvir tablets) to treat patients with chronic hepatitis C virus (HCV) genotype 1 infection, including those with a type of advanced liver disease called cirrhosis. **1**. Gentile, Borgia, Buonomo, *et al.* (2014) *Curr. Med. Chem.* **21**, 3261. **2**. Poordad, Lawitz, Kowdley, *et al.* (2013) *New Engl. J. Med.* **368**, 45.

Park Nucleotide, See *UDP-N-acetylmuramyl-L-alanyl-γ-(D-glutamyl)-(meso-2,6-diaminopimeloyl)-D-alanyl-D-alanine; UDP-N-acetylmuramyl-L-alanyl-γ-(D-glutamyl)-L-lysyl-D-alanyl-D-alanine*

Paromomycin

This water-soluble aminoglucoside antibiotic (FW$_{free-base}$ = 615.64 g/mol; CAS 7542-37-2) from a number of *Streptomyces* species inhibits protein biosynthesis in a streptomycin-like manner (1,2) by binding to 16S ribosomal RNA at the aminoacyl-tRNA site, resulting in misreading and inhibition of translocation. Note the structural similarity to neomycin B. The numbers adjacent to the amino groups correspond to the pK_a values. **Target(s):** aminoglycoside $N^{6'}$-acetyltransferase, *or* kanamycin 6'-acetyltransferase (1-5); DNA-directed DNA polymerase, Klenow polymerase (6); exoribonuclease H (7,8); gentamicin 3'-N-acetyltransferase, K_i = 8.3 μM (9); poly(A)-specific ribonuclease (6); protein biosynthesis, initiation and translocation (8,10-13); ribonuclease A (7); ribonuclease H (7,8); and ribonuclease P (14-16). **1**. Haas & Dowding (1975) *Meth. Enzymol.* **43**, 611. **2**. Boehr, Jenkins & Wright (2003) *J. Biol. Chem.* **278**, 12873. **3**. Martel, Masson, Moreau & Le Goffic (1983) *Eur. J. Biochem.* **133**, 515. **4**. Benveniste & Davies (1971) *Biochemistry* **10**, 1787. **5**. Kim, Villegas-Estrada, Hesek & Mobashery (2007) *Biochemistry* **46**, 5270. **6**. Ren, Martínez, Kirsebom & Virtanen (2002) *RNA* **8**, 1393. **7**. Barbieri, Li, Guo, *et al.* (2003) *J. Amer. Chem. Soc.* **125**, 6469. **8**. Li, Barbieri, Lin, *et al.* (2004) *Biochemistry* **43**, 9732. **9**. Williams & Northrop (1978) *J. Biol. Chem.* **253**, 5908. **10**. Pestka (1974) *Meth. Enzymol.* **30**, 261. **11**. Jiménez (1976) *Trends Biochem. Sci.* **1**, 28. **12**. Lynch, Gonzalez & Puglisi (2003) *Structure (Cambridge)* **11**, 43. **13**. Fourmy, Yoshizawa & Puglisi (1998) *J. Mol. Biol.* **277**, 333. **14**. Tekos, Prodromaki, Papadimou, *et al.* (2003) *Skin Pharmacol. Appl. Skin Physiol.* **16**, 252. **15**. Tekos, Tsagla, Stathopoulos & Drainas (2000) *FEBS Lett.* **485**, 71. **16**. Mikkelsen, Brannvall, Virtanen & Kirsebom (1999) *Proc. Natl. Acad. Sci. U.S.A.* **96**, 6155.

Paroxetine

This SSRI antidepressant (FW = 260.25 g/mol; CAS 22994-85-0; *Abbreviation*: PXT), known by its trade name Aropax®, Paxil®, Pexeva®, Seroxat®, Sereupin® and Brisdelle® as well as its systematic name (3S,4R)-3-[(2H-1,3-benzodioxol-5-yloxy)methyl]-4-(4-fluorophenyl)piperidine, is a selective serotonin reuptake inhibitor (IC$_{50}$ = 33 μM) that is indicated for the treatment of major depression, obsessive-compulsive disorder, panic disorder, posttraumatic stress disorder, generalized anxiety disorder and vasomotor symptoms, such as hot flashes and night sweats, that are associated with menopause (1-4). Significantly, Antidepressant responses to selective serotonin reuptake inhibitors (citalopram or paroxetine) are abolished in mice unable to synthesize histamine due to either targeted disruption of histidine decarboxylase gene (*or* HDC$^{-/-}$) or injection of α-fluoromethylhistidine, a suicide inhibitor of this enzyme (5). Such findings demonstrate that SSRIs selectively require the integrity of the brain histamine system to exert their preclinical responses (5). (*See also Citalopram; α-Fluoromethyl histidine*) **Pharmacokinetic Properties:** Plasma concentration and time-curves fit a two-compartment open model (6), with the oral route giving a longer $t_{1/2}$ (30 hours) than by the intravenous route (12 hours). Deviations probably reflect saturated elimination kinetics during first-pass metabolism (6). Co-administration of lipoic acid and PXT may improve anxiolytic and antidepressant responses (7), suggesting that PXT may deplete lipoic acid stores or that PXT interferes with some lipoic acid-requiring metabolic pathway. Somewhat surprisingly, Paroxetine (20-40 μM) induces growth inhibition and apoptosis in prostate cancer cells *in vitro* (8). Paroxetine is metabolized by CYP2D6 *via* demethylenation of the methylenedioxy group, yielding a catechol metabolite and formic acid. Paroxetine is also a potent inhibitor of cytochrome P450 2D6 (CYP2D6). Time- dependent inhibition was demonstrated with an apparent K_i of 4.8 μM and an apparent k_{inact} value of 0.17 min^{-1} (9). Paroxetine has critical but differential effects on IL-6 and TNFα production in macrophages and likely regulates their formation by different mechanisms (10). **Key Pharmacokinetic Parameters:** *See* Appendix II in Goodman & Gilman's *THE PHARMACOLOGICAL BASIS OF THERAPEUTICS*, 12th Edition (Brunton, Chabner & Knollmann, eds.) McGraw-Hill Medical, New York (2011). **1**. Heydorn (1999) *Expert Opin. Investig. Drugs* **8**, 417. **2**. Habert, Graham, Tahraoui, Claustre & Langer (1985) *Eur. J. Pharmacol.* **118**, 107. **3**. Lassen (1978) *Psychopharm-acology (Berlin)* **57**, 151. **4**. Lund, Thayssen, Menge, *et al.* (1982) *Acta Pharmacol. Toxicol.* (Copenhagen) **51**, 351. **5**. **2**. Munari, Provensi, Passani, *et al.* (2015) *Int. J. Neuropsychopharmacol.* **18**, pyv045. **6**. Beyazyüz, Albayrak, Eğilmez, Albayrak & Beyazyüz (2013) *Psychiatry Investig.* **10**, 148. **7**. Silva, Sampaio, de Araújo, *et al.* (2013) *Am. J. Ther.* **21**, 85. **8**. Shibuya (2011) *Cancer Res.* **71**, Supplement 1: Proceedings: AACR 102nd Annual Meeting, Apr 2-6, Orlando, FL. **9**. Bertelsen, Venkatakrishnan, Von Moltke, Obach & Greenblatt (2003) *Drug Metab. Dispos.* **31**, 289. **10**. Durairaj, Steury & Parameswaran (2015) *Int. Immunopharmacol.* **25**, 485..

Parthenolide

This sesquiterpene lactone (FW = 248.32 g/mol; CAS 70-70-2), also named [1aR-(1aR*,4E,7aS*,10aS*,-10bR*)]-2,3,6,7,7a,8,10a,10b-octahydro-1a,5-dimethyl-8-methylene-oxireno[9,10]cyclodeca[1,2-b]furan-9(1aH)-one, is an active component of the herb feverfew (*Tanacetum parthenium*) and a number of Mexican-Indian medicinal plants. Parthenolide is a potent anti-inflammatory agent that suppresses expression of certain interleukins and inhibits the activation of p42/44 mitogen-activated protein kinase. It also inhibits the release of 5-HT from blood platelets. Also inhibits generation of leukotriene B$_4$ and thromboxane B$_2$. **Target(s):** IκB kinase (1-3); NF-κB (1,3-5); phenylpyruvate tautomerase, weakly inhibited (6); prostaglandin synthase (7); and eicosanoid generation (appears to be irreversible but not time-dependent) (8). **1**. Zingarelli, Hake, Denenberg & Wong (2002) *Shock* **17**, 127. **2**. Kwok, Koh, Ndubuisi, Elofsson & Crews (2001) *Chem. Biol.* **8**, 759. **3**. Hehner, Hofmann, Droge & Schmitz (1999) *J. Immunol.* **163**, 5617. **4**. Pozarowski, Halicka & Darzynkiewicz (2003) *Cell Cycle* **2**, 377. **5**. Garcia-Pineres, Castro, Mora, *et al.* (2001) *J. Biol. Chem.* **276**, 39713. **6**. Molnar & Garai (2005) *Int. Immunopharmacol.* **5**, 849. **7**. Pugh & Sambo (1988) *J. Pharm. Pharmacol.* **40**, 743. **8**. Sumner, Salan, Knight & Hoult (1992) *Biochem. Pharmacol.* **43**, 2313.

Parvaquone

This hydroxynaphthylquinone antiprotozoal agent (FW = 256.30 g/mol; CAS 4042-30-2), first synthesized by Fieser in 1948 (U.S. Patent Number 2553648), is a yellow solid that resembles ubiquinone and acts as an electron transport inhibitor and uncoupler, with some 20 times greater toxicity to *Toxoplasma* than to humans (1,2). **1**. Tappel (1960) *Biochem. Pharmacol.* **3**, 289. **2**. Howland (1967) *Biochim. Biophys. Acta* **131**, 247.

PAS, See *4-Aminosalicylate and 4-Aminosalicylic Acid*

Pathocidin, See *8-Azaguanine*

Patulin

This antibiotic (FW = 154.12 g/mol; CAS 149-29-1), known also as clavacin and systematically as 4-hydroxy-4*H*-furo[3,2-*c*]pyran-2(6*H*)-one, from *Aspergillus*, *Gymnoascus*, and *Penicillium* inhibits K^+ uptake. Patulin also causes breaks single-stranded DNA and even double-stranded DNA at higher concentrations. *Note*: This water-soluble mycotoxin is unstable in alkaline solutions, with consequential loss of biological activity. **Target(s):** alcohol dehydrogenase (1); aldolase (2); aminoacyl-tRNA synthetases (3); carboxylase (4,5); DNA-directed RNA polymerase (6); K^+ uptake (7); lactate dehydrogenase (1); 6-methylsalicylate synthetase (8); Na^+/K^+-exchanging ATPase (9); protein farnesyltransferase (10); protein prenylation (10); ribonuclease H (6); and urease (11). **1**. Ashoor & Chu (1973) *Food Cosmet. Toxicol.* **11**, 617. **2**. Ashoor & Chu (1973) *Food Cosmet. Toxicol.* **11**, 995. **3**. Arafat, Kern & Dirheimer (1985) *Chem. Biol. Interact.* **56**, 333. **4**. Karrer & Visconti (1947) *Helv. Chim. Acta* **30**, 268. **5**. Massart (1950) *The Enzymes*, 1st ed. (Sumner & Myrbäck, eds.), **1** (part 1), 307. **6**. Tashiro, Hiral & Ueno (1979) *Appl. Environ. Microbiol.* **38**, 191. **7**. Kahn (1957) *J. Pharmacol. Exper. Therap.* **121**, 234. **8**. Malik (1980) *Trends Biochem. Sci.* **5**, 68. **9**. Phillips & Hayes (1979) *Toxicology* **13**, 17. **10**. Miura, Hasumi & Endo (1993) *FEBS Lett.* **318**, 88. **11**. Reiss (1977) *Naturwissenschaften* **64**, 97.

Paxilline

This $BK_{Ca}/K_{Ca}1.1$) channel blocker (FW = 435.56 g/mol; CAS 57186-25-1; Solubility: 100 mM in DMSO), also named (2*R*,4*bS*,6*aS*,12*bS*,12*cR*,14*aS*)-5,6,6*a*,7,12,12*b*,12*c*,13,14,14*a*-decahydro-4*b*-hydroxy-2-(1-hydroxy-1-methylethyl)-12*b*,12*c*-dimethyl-2*H*-pyrano[2",3": 5',6']benz[1',2':6,7]indeno[1,2-*b*]indol-3(4*bH*)-one, binds to the α-subunit of BK_{Ca}, exhibiting a K_i value of 1.9 nM in blocking currents in α-subunit-expressing oocytes and enhancing binding of to BK_{Ca} channels in vascular smooth muscle (1,2). (**See** *Charybdotoxin*) Paxilline also inhibits sarco/endoplasmic reticulum Ca^{2+}-ATPase, IC$_{50}$ = 5 - 50 μM (3). **1**. Knaus, *et al.* (1994) *Biochemistry* **33**, 5819. **2**. Sanchez & McManus (1996) *Neuropharmacol.* **35**, 963. **3**. Bilmen, *et al.* (2002) *Arch. Biochem. Biophys.* **406**, 55.

Pazopanib

This novel multi-kinase inhibitor (FW = 437.52 g/mol; CAS 444731-52-6; Solubility: 17 mg/mL DMSO, <1 mg/mL H_2O), also known by its trade name Votrient®, and its systematic name 5-[[4-[(2,3-dimethyl-2*H*-indazol-6-yl)methylamino]-2-pyrimidinyl]amino]-2-methyl-benzolsulfonamide, targets VEGFR1 (IC$_{50}$ = 10 nM), VEGFR2 (IC$_{50}$ = 30 nM), VEGFR3 (IC$_{50}$ = 47 nM), PDGFR (IC$_{50}$ = 84 nM), FGFR (IC$_{50}$ = 74 nM), c-Kit (IC$_{50}$ = 140 nM), and c-Fms (IC$_{50}$ = 146 nM). *In vitro*, pazopanib inhibits the growth of synovial sarcoma cells, inducing G_1 arrest, with high suppression of the PI3K-AKT pathway (2). Moreover, pazopanib administration suppresses the tumor growth in a xenograft model. Pazopanib also reduces light-induced overexpression and secretion of VEGF and platelet-derived growth factor in human retinal pigment epithelial cells (3). **1**. Harris, Boloor, Cheun, *et al.* (2008) *J. Med. Chem.* **51**, 4632. **2**. Hosaka, Horiuchi, Yoda, *et al.* (2012) *J. Orthop. Res.* **30**, 1493. **3**. Kernt, *et al.* (2012) *Retina* **32**, 1652.

Pazufloxacin

This fluoroquinolone-class antibiotic (FW = 318.29 g/mol; CAS 127046-18-8; IUPAC Name: (3*R*)-10-(1-aminocyclopropyl)-9-fluoro-3-methyl-7-oxo-1*H*,7*H*-[1,3]oxazino[5,4,3-*ij*]quinoline-carboxylate) is a broad-spectrum systemic antibacterial agent that targets Type II DNA topoisomerases (gyrases), which are required for bacterial replication and transcription. For the prototypical member of this antibiotic class, **See** *Ciprofloxacin*

PBA, See *N⁶-Benzyl-N⁹-(2-tetrahydropyranyl)adenine*

PBD, See *5-(4-Biphenylyl)-2-phenyl-1,3,4-oxadiazole*

PBDA, See *4,4'-Phosphonicobis(butane-1,3-dicarboxylate); Phenylbutyldopamine*

PBG, See *Porphobilinogen*

PBIT

This JARID1 Histone Demethylase inhibitor (FW = 241.31 g/mol; CAS 2514-30-9; Soluble to 100 mM in DMSO), also named 2-(4-methylphenyl)-1,2-benzisothiazol-3(2*H*)-one, targets the Jumonji AT-Rich Interactive Domain 1, with respective IC$_{50}$ values of 3, 4.9 and 6 μM for JARID1B, JARID1A and JARID1C. The enzymes responsible for the demethylation of trimethylated lysine 4 in histone H3 (H3K4me3) are the Jumonji AT-rich interactive domain 1 (JARID1) or lysine demethylase-5 (KDM5) family of lysine demethylases. This family consists of JARID1A (also known as KDM5A or RBP2), JARID1B (also known as KDM5B or PLU1), JARID1C (also known as KDM5C or SMCX), and JARID1D (also known as KDM5D or SMCY) in mammals. Like other JmjC domain-containing demethylases, JARID1 enzymes catalyze histon demethylation in a Fe(II) and α-ketoglutarate (α-KG)-dependent reaction. Oxidative decarboxylation of α-KG results in an unstable hydroxylated methyl-lysine intermediate. Release of the hydroxyl and methyl groups as formaldehyde from this intermediate results in demethylation. JARID1 demethylases have been linked to human diseases such as cancer and X-linked mental retardation. Both JARID1A and JARID1B are potential oncoproteins, and both are overexpressed in a variety of cancers. Increased expression of JARID1A promotes a more stem-like phenotype and enhanced resistance to anticancer agents. PBIT was identified by high-throughput screening (using biotinylated H3K4me3 peptide substrate) of agents that are selective for JARID1 over UTX and JMJD3. PBIT increases levels of methylated H3K4 in JARIDB1-transfected HeLa cells and MCF7 cells. (JARID1A and JARID1B knock-out mice are viable, suggesting that inhibition of JARID1A or JARID1B has minimal effects on normal cells *in vivo.*) PBIT also inhibits proliferation of breast cancer cell lines expressing high levels of JARIDB1. **1**. Sayegh, Cao, Zou, *et al.* (2013) *J. Biol. Chem.* **288**, 9408.

1,3-PBIT

This symmetrical arginine analogue ($FW_{free-base}$ = 282.43 g/mol), systematically known as phenylene-1,3-bis(ethane-2-isothiourea), exhibits a K_i value of 47 nM for the inducible nitric-oxide synthase, whereas the values for the neuronal and endothelial enzymes are 0.25 and 9 μM, respectively (1,2). As a selective iNOS blocker, cells may be incubated with 1,3-PBIT (50-70 nM) in modified Krebs solution. **1**. Garvey, Oplinger, Tanoury, *et al.* (1994) *J. Biol. Chem.* **269**, 26669. **2**. Raman, Li, Martásek, Babu, Griffith, *et al.* (2001) *J. Biol. Chem.* **276**, 26486.

1,4-PBIT

This bisisothioureas nitric oxide synthase inhibitor ($FW_{free-base}$ = 282.43 g/mol), also known as 1,4-phenylene-bis(1,2-ethanediyl)bisisothiourea, targets inducible NO synthase (K_i = 7.4 nM) and endothelial NO synthase (K_i = 0.36 nM) (1,2). Crystal structures of the heme domain of the three NOS isoforms show a very high degree of similarity in the immediate vicinity of the heme active site illustrating the challenge of isoform-selective inhibitor design. Isothioureas are potent NOS inhibitors, and the structures of the endothelial NOS heme domain complexed with isothioureas bearing small *S*-alkyl substituents have been determined (3). **1**. Garvey, Oplinger, Tanoury, *et al.* (1994) *J. Biol. Chem.* **269**, 26669. **2**. Raman, Li, Martásek, Babu, Griffith, *et al.* (2001) *J. Biol. Chem.* **276**, 26486. **3**. Li, Raman, Martásek, *et al.* (2000) *J. Inorg. Biochem.* **81**, 133

1,3-PBITU, See 1,3-PBIT

PBN, See N-(t-Butyl)-a-phenylnitrone

PC, See Phosphatidylcholine; Phosphocreatine

PC 15

This thiol (FW = 152.31 g/mol), also known as 2-amino-4-methylthiobutane-1-thiol and methionine thiol, inhibits aminopeptidase N, *or* APN/CD13 (IC_{50} = 11 nM), a transmembrane protease that us present in endothelial, epithelial, fibroblast, leukocyte cell types (1,2). APN/CD13 expression is dysregulated in inflammatory diseases and in both solid and hematologic tumors. APN/CD13 is also a receptor for coronaviruses. The corresponding sulfoxide also inhibits (IC_{50} = 20 nM). **1**. Fournié-Zaluski, Coric, Turcaud, *et al.* (1992) *J. Med. Chem.* **35**, 1259. **2**. Bauvois & Dauzonne (2006) *Med. Res. Rev.* **26**, 88.

PC 57

This tripeptide analogue (FW = 409.53 g/mol) exerts its anti-hypertensive effects by inhibiting neprilysin, K_i = 1.4 nM, and dipeptidyl-peptidase A, *or* angiotensin I-converting enzyme, with a K_i value of 0.2 nM. **1**. Roques, Noble, Crine & Fournié-Zaluski (1995) *Meth. Enzymol.* **248**, 263.

PC190723

This bacterial cytoskeletal inhibitor (FW = 355.74 g/mol) inhibits the assembly of the tubulin-like protein FtsZ, arresting its polymerization in *Staphylococcus aureus* and blocking cell division, showing strong bacteriocidal action (1,2). The likely interaction site is analogous to the paclitaxel (taxol) binding site on tubulin. In keeping with the microtubule-stabilizing effects of paclitaxel, the crystal structure of the *S. aureus* FtsZ-PC190723 complex suggests that a domain movement is likely to stabilize the FtsZ protofilant rather than the monomer, with the conformational change transmitted from the GTP binding site to the C-terminal domain *via* FtsZ Helix-7 (2). (*See SB-RA-2001*) **1**. Haydon, Stokes, Ure, *et al.* (2008) *Science* **321**, 1644. **2**. Andreu, Schaffner-Barbero, Huecas, *et al.* ()2010) *J. Biol. Chem.* **285**, 14239. **3**. Elsen, Parthsarathy, Reid, *et al.* (2012) *J. Am. Chem. Soc.* **134**, 12342.

PCA, See Perchlorate (Perchloric Acid)

p(CH₂)ppA, See β,γ-Methyleneadenosine 5'-Triphosphate

p(CH₂)ppG, See β,γ-Methyleneguanosine 5'-Triphosphate

PCI, See Protein C Inhibitor

PCI-24781

This broad-spectrum histone deacetylase inhibitor (FW = 394.48 g/mol; CASs = 783355-60-2, 783356-67-2 (HCl)); Solubility: 80 mg/mL DMSO; <1 mg/mL Water; Formulated in 30% HP-cyclodextrin in water), also known as CRA-024781 and systematically as 3-((dimethylamino)methyl)-*N*-(2-(4-(hydroxycarbamoyl)phenyl)ethyl)benzofuran-2-carboxamide, is a novel broad-spectrum HDAC inhibitor targeting HDAC1, HDAC2, HDAC3, HDAC6, HDAC8 and HDAC10 with K_i of 7 nM, 19 nM, 8.2 nM, 17 nM, 280 nM, 24 nM, respectively. PCI-24781 treatment causes dose-dependent accumulation of both acetylated histones and acetylated tubulin in HCT116 or DLD-1 cells (1). It also induces expression of p21 and leads to PARP cleavage and accumulation of the γH2AX. **1**. Buggy, *et al.* (2006) *Mol. Cancer Ther.* **5**, 1309. **2**. Adimoolam, *et al.* (2007) *Proc. Natl. Acad.*

Sci. USA, **104**, 19482. **3**. Lopez, *et al.* (2009) *Clin. Cancer Res.* **15**, 3472. **4**. Bhalla, *et al.* (2009) *Clin. Cancer Res.* **15**, 3354.

PCI-32765, *See Ibrutinib*

PCI-34051

This selective histone deacetylase (HDAC) inhibitor (FW = 296.32 g/mol; CAS 950762-95-5, 1072027-64-5); Solubility: 60 mg/mL DMSO; <1 mg/mL H_2O), also known as 1-(4-methoxybenzyl)-*N*-hydroxy-1*H*-indole-6-carboxamide, selectively targets HDAC8 (K_i = 10 nM), about five times lower than the K_i for HDAC1, 200-times lower than the K_i for HDAC1 and HDAC6, and 1000-times lower than the K_i values for HDAC2, HDAC3 and HDAC10. **1**. Balasubramanian, *et al.* (2008) *Leukemia* **22**, 1026.

PCMB, *See p-Chloromercuribenzoate*

PCMBS, *See p-Chloromercuribenzenesulfonate*

PCP, *See Phencyclidine; Pentachlorophenol; Methylenebisphosphonate*

PCS1055

This novel M_4 subtype-preferring acetylcholine receptor antagonist (FW = 384.53 g/mol) inhibits [^{3}H]-*N*-methylscopolamine binding to the muscarinic M_4 receptor (K_i = 6.5 nM). Although of lower potency than the pan-muscarinic antagonist atropine, PCS1055 exhibits better subtype selectivity over previously reported M_4-selective reagents, such as the muscarinic-peptide toxins and PD102807. In GTP-γ-[^{35}S] binding studies, PCS1055 exhibits 255-, 69-, 342- and >1000-fold greater inhibition of Oxo-M activity at the M_4 versus the M_1-, M_2-, M_3- or M_5-receptor subtypes, respectively. Schild analysis indicates PCS1055 is a competitive antagonist to muscarinic M_4 receptor (K_b = 5.7 nM). Identification of receptor subtype-selective ligands is challenging in view of the high sequence identity and structural homology of muscarinic acetylcholine receptors. · **1**. Croy, Chan, Castetter, *et al.* (2016) *Eur. J. Pharmacol.* **782**, 70.

PD 083176

This peptide (FW$_{free-base}$ = 976.10 g/mol), also known as N^α-Cbz-L-His-L-Tyr(*O*-Bn)-L-Ser(*O*-Bn)-L-Trp-D-Ala-NH$_2$, is an early inhibitor of protein farnesyltransferase, IC$_{50}$ = 76 nM (1,2) and protein geranylgeranyl-transferase type I (3). **1**. Gibbs (2001) *The Enzymes*, 3rd ed., **21**, 81. **2**. Leonard, Shuler, Poulter, *et al.* (1997) *J. Med. Chem.* **40**, 192. **3**. Yokoyama, McGeady & Gelb (1995) *Biochemistry* **34**, 1344.

PD 98059

This cell-permeable agent (FW = 267.28 g/mol; IUPAC Name: 2-(2'-amino-3'-methoxyphenyl)oxanaphthalen-4-one and 2'-amino-3'-methoxyflavone) inhibits [mitogen-activated protein kinase] kinase (MAP kinase kinase), thus blocking the activation of MAP kinase and subsequent phosphorylation of MAP kinase substrates. PD98059 also inhibits cell growth and reverses the phenotype of *Ras*-transformed BALB3T3 mouse fibroblasts and rat kidney cells. **Target(s):** cyclooxygenase-1 and -2, via MAPK; prostaglandin-endoperoxide synthase (1,2); mitogen-activated protein kinase (3-5); [mitogen-activated protein kinase] kinase (6-13); and Raf activation of [mitogen-activated protein kinase] kinase (3,9). **1**. Borsch-Haubold, Pasquet & Watson (1998) *J. Biol. Chem.* **273**, 28766. **2**. Zhang & Wood (2005) *Brain Res.* **1060**, 100. **3**. Eriksson, Toivola, Sahlgren, Mikhailov & Härmälä-Braskén (1998) *Meth. Enzymol.* **298**, 542. **4**. Ratner, Bryan, Weber, *et al.* (2001) *J. Biol. Chem.* **276**, 19267. **5**. Lee, Yu & Chung (2005) *Free Radic. Res.* **39**, 399. **6**. Ahn, Nahreini, Tolwinski & Resing (2001) *Meth. Enzymol.* **332**, 417. **7**. Pang, Sawada, Decker & Saltiel (1995) *J. Biol. Chem.* **270**, 13585. **8**. Dudley, Pang, Decker, Bridges & Saltiel (1995) *Proc. Natl. Acad. Sci. U.S.A.* **92**, 7686. **9**. Alessi, Cuenda, Cohen, Dudley & Saltiel (1995) *J. Biol. Chem.* **270**, 27489. **10**. Du, Cai, Suzuki, *et al.* (2003) *J. Cell. Biochem.* **88**, 1235. **11**. Ross, Corey, Dunn & Kelley (2007) *Cell. Signal.* **19**, 923. **12**. Chen, Lin & Jeng (2008) *Plant Cell Environ.* **31**, 62. **13**. Kojima, Konopleva, Samudio, Ruvolo & Andreeff (2007) *Cancer Res.* **67**, 3210.

PD-116948, *See Dipropylcyclopentylxanthine*

PD 125754

This peptide analogue (FW$_{free-base}$ = 752.01 g/mol) inhibits a number of aspartic proteinases: *e.g.*, endothiapepsin, K_i = 16.2 μM (1,2) and renin, IC$_{50}$ = 22 nM (1,3). **1**. Cooper, Quail, Frazao, *et al.* (1992) *Biochemistry* **31**, 8142. **2**. Bailey & Cooper (1994) *Protein Sci.* **3**, 2129. **3**. Kaltenbronn, Hudspeth, Lunney, *et al.* (1990) *J. Med. Chem.* **33**, 838.

PD 128042, *See CI-976*

PD 129541

This peptide analogue (FW = 840.08 g/mol) is a strong inhibitor of saccharopepsin, K_i = 4 nM (1), and a weak inhibitor of human renin (2), and saccharopepsin, *or* yeast proteinase A (2,3). **1**. Coates, Erskine, Crump, Wood & Cooper (2002) *J. Mol. Biol.* **318**, 1405. **2**. Cronin, Badasso, Tickle, *et al.* (2000) *J. Mol. Biol.* **303**, 745. **3**. Dunn (2002) *Chem. Rev.* **102**, 4431.

PD 123319

This potent non-peptide receptor antagonist (FW = 508.61 g/mol; CAS 130663-39-7), also named 1-[[4-(dimethylamino)-3-methylphenyl]methyl]-5-(2,2-diphenylacetyl)-4,5,6,7-tetrahydro-(6S)-1H-imidazo[4,5-c]pyridine-6-carboxylate, selectively targets AT_2 angiotensin II receptor (IC_{50} = 34 nM). This class of 4,5,6,7-tetrahydro-1H-imidazo[4,5-c]pyridine-6-carboxylic acid derivatives displace ^{125}I-labeled angiotensin II from a specific subset of angiotensin II (Ang II) binding sites (that are different from those mediating vascular contraction or aldosterone release) that are likely to have novel properties (1). PD-123319 and another nonpeptide antagonist (DuP-753) discriminate between two subclasses of AII receptors in many different tissues (2). PD 123319 does not influence baseline cerebral blood flow, with minor drop in blood pressure (3). **1**. Blankley, Hodges, Klutchko, et al. (1991) *J. Med. Chem.* **34**, 3248. **2**. Boulay, Servant, Luong, Escher & Guillemette (1992) *Mol. Pharmacol.* **41**, 809. **3**. Estrup, Paulson & Strandgaard (2001) *J. Renin Angiotensin Aldosterone Syst.* **2**, 188.

PD 150606, See (Z)-3-(4-Iodophenyl)-2-mercapto-2-propenoate

PD 151746

This cell-permeable thiol-reactive agent (FW = 237.25 g/mol; CAS 179461-52-0), also named (Z)-3-(5-fluoroindol-3-yl)-2-mercapto-2-propenoate, inhibits μ-calpain (K_i = 0.21 μM) decreases oxLDL-induced cytotoxicity, whereas the general caspase inhibitor BAF (t-butoxycarbonyl-Asp-methoxyfluoromethylketone) is without effect. Note that the structurally related inhibitor (Z)-3-(4-iodophenyl)-2-mercapto-2-propenoate does not appear to react at the active site, instead targeting the calcium ion binding site. (*See also* (Z)-3-(4-Iodophenyl)-2-mercapto-2-propenoate) **1**. Pörn-Ares, Saido, Andersson & Ares (2003) *Biochem. J.* **374**, 403.

PD 152247, See PNQX

PD153035

This extraordinarily potent EGFR inhibitor (F.Wt. = 396.67 (as hydrochloride); CAS 183322-45-4 (hydrochloride), 153436-54-5 (free base); Solubility (25°C): <1 mg/mL DMSO, <1 mg/mL Water), also known systematically as N-(3-bromophenyl)-6,7-dimethoxyquinazoline-4-amine, rapidly suppresses in vitro EGFR autophosphorylation (K_i = 5.2 pM; IC_{50} = 29 pM) and selectively blocks EGF-mediated cellular processes including mitogenesis, early gene expression, and oncogenic transformation (1). PD153035 exposure does not affect the expression of either EGF receptors or HER2/neu (2). PD153035 causes dose-dependent growth inhibition of EGF receptor-overexpressing cell lines at low micromolar concentrations, and the IC_{50} values in monolayer cultures at less than 1 μM in most cell lines tested (2). **1**. Fry, Kraker, McMichael, et al. (1994) *Science* **265**, 1093. **2**. Bos, Mendelsohn, Kim, et al. (1997) *Clin. Cancer Res.* **3**, 2099.

PD 156273

This photosensitive EPGFR inhibitor (FW = 344.21 g/mol), also named 6-amino-4-[(3-bromophenyl)amino]-7-(methylamino)quinazoline, targets epidermal growth factor receptor protein-tyrosine kinase, IC_{50} = 0.69 nM

(1,2). **1**. Baguley, Marshall, Holdaway, Rewcastle & Denny (1998) *Eur. J. Cancer* **34**, 1086. **2**. Bridges, Zhou, Cody, et al. (1996) *J. Med. Chem.* **39**, 267.

PD 157432, See 2'-Thioadenoside

PD 158780

This photosensitive pyridopyrimidine (FW = 330.19 g/mol), known systematically as 4-[(3-bromophenyl)amino]-6-(methylamino)pyrido[3,4-d]pyrimidine, inhibits EPGFR (epidermal growth factor receptor) protein-tyrosine kinase (IC_{50} = 8 pM). In addition, PD 158780 inhibits heregulin-stimulated autophosphorylation in SK-BR-3 (IC_{50} = 49 nM) and MDA-MB-453 (IC_{50} = 52 nM) breast carcinomas. **1**. Rewcastle, Murray, Elliott, et al. (1998) *J. Med. Chem.* **41**, 742.

PD0166285

This potent G_2 checkpoint abrogator (FW = 512.43 g/mol; CAS 185039-89-8; Solubility = 100 mg/mL DMSO), also named 6-(2,6-dichlorophenyl)-2-[[4-2-(diethylamino)ethoxy]phenyl] amino]-8-methylpyrido[2,3-d] pyrimidin-7(8H)-one, targets Wee1 kinase (IC_{50} = 24 nM), a enzyme that is crucial for maintaining G_2 cell-cycle arrest through its inhibitory phosphorylation of Cdc2 (1). PD-166285 was identified in a screening campaign that was premised on the idea that cells that lack p53 would lack the capacity to engage effective G_1 checkpoint regulation, such that they would depend on the G_2 checkpoint to permit DNA repair prior to mitosis. This logic led to the hypothesis that a G_2 checkpoint abrogator would preferentially kill p53-inactive cancer cells by removing the only checkpoint protecting such cells from premature mitosis in response to DNA damage (1). At an intracellular concentration of 0.5 μM, PD0166285 potently inhibits irradiation-induced Cdc2 phosphorylation at Tyr-15 and Thr-14 in seven of seven cancer cell lines tested, showing that this G_2 checkpoint abrogator can kill cancer cells (1). Notably, PD0166285 is a radiosensitizer, enhancing cell sensitivity to radiation-induced cell death, showing a sensitivity enhancement ratio of 1.23 in a standard clonogenic assay (1). Its radiosensitizing activity is p53-dependent, showing a higher efficacy in p53-inactive cells (1). Treatment of B16 mouse melanoma cells with the inhibitor B16 cells also dramatically abrogates the G_2 checkpoint, with arrest in the early G_1 phase at 0.5 muM for 4 hours observed by flow cytometry. Cyclin D mRNA decreased within 4 hours observed by Real-time PCR. Rb was dephosphrylated for 24 hours. However, B16 cells did not undergo cell death after treatment with 0.5 μM PD0166285 for 24 hours. Immnofluoscence microscopy also showed that the cells become round and small in the morphogenesis, suggesting that microtubule stabilization is blocked and that Wee1 distribution was restricted after treatment for 4 hours (2). PD0166285 also abrogates the G_2 checkpoint in osteosarcoma (OS) cells, pushing them into mitotic catastrophe and sensitizing them to irradiation-induced cell death (3). Other agents, like caffeine (4) and UCN-01 (5), can also abrogate the G_2 checkpoint, thereby sensitizing p53 inactive cells to apoptosis. **Other Targets:** Myelin transcription factor-1, *or* Myt1 (IC_{50} = 72 nM); checkpoint kinase Chk1 IC_{50} = 3.4 μM). **1**. Wang, Li, Booher, et al. (2001) *Cancer Res.* **61**, 8211. **2**. Hashimoto, Shinkawa, Torimura, et al. (2006) *BMC Cancer* **6**, 292. **3**. PosthumaDeBoer, Würdinger, Graat, et al. (2011) *BMC Cancer* **11**, 156. **4**. Yao, Akhtar, McKenna, et al. (1996) *Nature Med.* **2**, 1140. **5**. Yu, Orlandi, Wang, et al. (1998) *J. Biol. Chem.* **273**, 33455.

PD 166793

This diphenylsulfonamide (FW = 412.30 g/mol) inhibits matrix metalloproteinases, or MMP (1,2), including collagenase 3 (IC_{50} = 8 nM), gelatinase A or MMP 2 (IC_{50} = 4 nM), gelatinase B or MMP 9 (IC_{50} = 7.9 μM), interstitial collagenase or MMP 1 (IC_{50} = 6 μM), matrilysin or MMP 7 (IC_{50} = 7.2 μM), and stromelysin 1 or MMP 3 (IC_{50} = 7 nM). 1. O'Brien, Ortwine, Pavlovsky, et al. (2000) J. Med. Chem. 43, 156. 2. Parker, Lunney, Ortwine, et al. (1999) Biochemistry 38, 13592.

PD 168393

This photosensitive quinazoline (FW = 369.22 g/mol), also known as 4-[(3-bromophenyl)amino]-6-acrylamido-quinazoline, irreversibly inhibits epidermal growth factor receptor (EGFR) protein-tyrosine kinase, IC_{50} = 0.7 nM (1,2). See also PD 174265 1. Fry, Bridges, Denny, et al. (1998) Proc. Natl. Acad. Sci. U.S.A. 95, 12022. 2. Hubbard & Miller (2007) Curr. Opin. Cell Biol. 19, 117.

PD 173074

This pyrido[2,3-d]pyrimidine ($FW_{free-base}$ = 523.68 g/mol; CAS 219580-11-7; Solubility: 100 mM in DMSO, 100 mM in ethanol), also known as N-[2-[[4-(diethylamino)butyl]amino]-6-(3,5-dimethoxyphenyl)pyrido[2,3-d]pyrimidin-7-yl]-N'-(1,1-dimethylethyl)urea), inhibits the fibroblast growth factor receptor protein-tyrosine kinase and vascular endothelial growth factor receptor protein-tyrosine kinase (1). PD173074 inhibits FGF-2-mediated effects on proliferation, differentiation, as well as MAPK activation in oligodendrocyte-lineage cells (2). It likewise inhibits fibroblast growth factor receptor-3, inducing differentiation and apoptosis in t(4;14) myeloma (3). PD173074 also selectively inhibits fibroblast growth factor receptor-3 (FGFR3) tyrosine kinase. It inhibits proliferation of bladder cancer cell carrying the FGFR3 gene mutation along with up-regulation of p27/Kip1 and G_1/G_0 arrest (4). PD173074 blocks small cell lung cancer growth in vitro and in vivo (5). 1. Mohammadi, Froum, Hamby, et al. (1998) EMBO J. 17, 5896. 2. Bansal, et al. (2003) J. Neurosci. Res. 74, 486. 3. Trudel, et al. (2004) Neoplasia 103, 3521. 4. Miyake, et al. (2010) J. Pharmacol. Exp. Ther. 332, 795. 5. Pardo, et al. (2010) Cancer Res. 69, 8645.

PD173955

This tyrosine kinase inhibitor (FW = 443.35 g/mol; CAS 260415-63-2; Solubility: <1 mg/mL DMSO, Ethanol, H_2O), 6-(2,6-dichlorophenyl)-8-methyl-2-((3-(methylthio)phenyl)amino)pyrido[2,3-d]pyrimidin-7(8H)-one targets Bcr-Abl (IC_{50} = 1-2 nM, in vitro), the constitutively active tyrosine kinase that results from inadvertent fusion of bcr and abl genes and that causes oncogenic transformation in chronic myelogenous leukemia, or CML, inhibiting Bcr-Abl-dependent cell growth with an IC_{50} of 2-35 nM in different cell lines (1,2). PD173955 also inhibits kit ligand-dependent c-kit autophosphorylation (IC_{50} = ~25 nM) and kit ligand-dependent proliferation of M07e cells (IC_{50} = 40 nM), but had a lesser effect on interleukin 3-dependent (IC_{50} = 250 nM) or granulocyte macrophage colony-stimulating factor (IC_{50} = 1 μM)-dependent cell growth. PD173955 also increases the susceptibility of HT29 cells to detachment-induced apoptosis (anoikis) in a dose- and time-dependent manner (3). **Structural Features Distinguishing Imatinib & PD173955 Binding:** Crystal structures of Abl kinase domain complexes with imatinib (Gleevec) and PD173955 show that both bind to the canonical ATP-binding site, but in distinctive ways. Imatinib captures a specific inactive conformation of Abl's activation loop, mimicking the bound peptide substrate. In contrast, PD173955 binds to the Abl activation loop in a way that resembles the active kinase conformation. The 10x greater potency of PD173955 over imatinib is attributed to its ability to target multiple active and inactive forms of Abl, whereas imatinib binds only to a specific catalytically inactive conformation (4). 1. Moasser, Srethapakdi, Sachar, Kraker & Rosen (1999) Cancer Res. 59, 6145. 2. Wisniewski, Lambek, Liu, et al. (1999) Cancer Res. 59, 4244. 3. Windham, Parikh, Siwak, et al. (2002) Oncogene 21, 7797. 4. Nagar, Bornmann, Pellicena, et al. (2002) Cancer Res. 62, 4236.

PD 174265

This substituted quinazoline ($FW_{free-base}$ = 371.24 g/mol), also known as 4-[(3-bromophenyl)amino]-6-propionylamidoquinazoline, inhibits epidermal growth factor receptor (EGFR) protein-tyrosine kinase, IC_{50} = 0.45 nM (1). PD 174265 selectively targets and irreversibly inactivates the epidermal growth factor receptor tyrosine kinase through specific, covalent modification of a Cys-773. See also PD 168393 1. Fry, Bridges, Denny, et al. (1998) Proc. Natl. Acad. Sci. U.S.A. 95, 12022.

PD 180557

This diphenylsulfonamide ($FW_{free-base}$ = 551.46 g/mol) inhibits various matrix metalloproteinases, or MMPs (1,2), including collagenase 3, or MMP 13 (IC_{50} = 12 nM), gelatinase A, or MMP 2 (IC_{50} = 2 nM), gelatinase B, or MMP 9 (IC_{50} = 2.3 μM), interstitial collagenase, or MMP 1 (IC_{50} = 27 μM), matrilysin, or MMP 7 (IC_{50} = 15 μM), and stromelysin 1, or MMP 3 (IC_{50} = 7 nM). Systematic use of isothermal calorimetry provided estimates of the solvation, translational, and conformational components of the entropy term, and this analysis suggests: (a) that a polar group at the P_1 position contributes the observed, large and favorable enthalpy, (b) that a hydrophobic group at $P_{2'}$ contributes favorably to the entropy of desolvation, and (3) that $P_{1'}$ substituents may trigger an entropically unfavorable conformational change in the enzyme upon inhibitor binding. 1. O'Brien, Ortwine, Pavlovsky, et al. (2000) J. Med. Chem. 43, 156. 2. Parker, Lunney, Ortwine, et al. (1999) Biochemistry 38, 13592.

PD 180970

This pyridopyrimidine (FW$_{free-base}$ = 429.28 g/mol) inhibits Bcr-Abl protein-tyrosine kinase and induces apoptosis of K562 leukemic cells (1-3). *In vitro*, PD180970 potently inhibited p210Bcr-Abl autophosphorylation (IC$_{50}$ = 5 nM) as well as the kinase activity of purified Abl tyrosine kinase (IC$_{50}$ = 2.2 nM). Incubation of K562 cells with PD180970 resulted in cell death, with results of nuclear staining, apoptotic-specific poly(ADP-ribose) polymerase cleavage, and annexin V binding assays confirming as much. Significantly, PD180970 is without effect on the growth and viability of p210Bcr-Abl-negative HL60 human leukemic cells. Such findings demonstrate that PD180970 is among the most potent known inhibitors of the p210Bcr-Abl tyrosine kinase and that it is a promising candidate as a novel therapeutic agent for Bcr-Abl-positive leukemia. **1**. Tauchi & Ohyashiki (2006) *Int. J. Hematol.* **83**, 294. **2**. Dorsey, Jove, Kraker & Wu (2000) *Cancer Res.* **60**, 3127. **3**. Wisniewski, Lambek, Liu, *et al.* (2002) *Cancer Res.* **62**, 4244.

PD 183805, See *CI-1033*

PD 184352

This substituted benzamide (FW = 478.66 g/mol), also known as 2-(chloro-4-iodophenylamino)-*N*-cyclopropylmethoxy-3,4-difluorobenzamide and CI-1040, inhibits ERK activation, IC$_{50}$ = 16 nM (1-6). **1**. Ahn, Nahreini, Tolwinski & Resing (2001) *Meth. Enzymol.* **332**, 417. **2**. Yu, Wang, Dent & Grant (2001) *Mol. Pharmacol.* **60**, 143. **3**. Davies, Reddy, Caivano & Cohen (2000) *Biochem. J.* **351**, 95. **4**. Sebolt-Leopold (2000) *Oncogene* **19**, 6594. **5**. Sebolt-Leopold, Dudley, Herrera, *et al.* (1999) *Nat. Med.* **5**, 810. **6**. Spicer, Rewcastle, Kaufman, *et al.* (2007) *J. Med. Chem.* **50**, 5090.

PD 0313052

This substituted benzamidine (FW$_{free-base}$ = 448.59 g/mol) is a slow, tight-binding inhibitor of coagulation factor Xa. Analysis of the association and dissociation kinetics of PD0313052 with human factor Xa demonstrated a reversible, slow-onset mechanism of inhibition, with a simple, single-step bimolecular association between factor Xa and inhibitor. This interaction was governed by association k_{on} and dissociation k_{off} rate constants of 10^7 M^{-1}s^{-1} and 1.9 x 10^{-3} s^{-1}. The inhibition of human factor Xa by PD0313052 displayed tight-binding behavior, as indicated by a K_i* value of 0.3 nM. **1**. Gould, Cladera, Harris, *et al.* (2005) *Biochemistry* **44**, 9280.

PD 404182

This KDO-8-P synthase inhibitor and potential antibiotic (FW = 217.29 g/mol), also named 6*H*-6-imino(2,3,4,5-tetrahydropyrimido)[1,2-*c*][1,3]benzothiazine, is a slow, tight-binding agent that targets 3-deoxy-D-*manno*-octulosonate-8-phosphate synthase, an enzyme participating in the synthesis of 3-deoxy-D-manno-2-octulosonic acid (KDO). The latter is an essential component of LPS (lipopolysaccharide) in the outer membrane of Gram-negative bacteria. PD 404182 is an antiretroviral agent with submicromolar

inhibitory activity against human immunodeficiency virus-1 (HIV-1) and HIV-2 infection (2). PD-404182 also potently inhibits dimethylarginine dimethylaminohydrolase isoform-1 (DDAH1), the human enzyme that degrades Asymmetric Dimethylarginine, *or* ADMA (**See** *Asymmetric Dimethylarginine*) (3). Because ADMA inhibits nitric oxide synthesis and because ADMA levels are predictive for future cardiovascular disease, chronic use of PD-404182 is apt to have side-effects. **1**. Birck, Holler & Woodard (2000) *J. Am. Chem. Soc.* **122**, 9334. **2**. Mizuhara, Oishi, Ohno, *et al.* (2012) *Bioorg. Med. Chem.* **20**, 6434. **3**. Ghebremariam, Erlanson & Cooke (2013) *J. Pharmacol. Exp. Ther.* **348**, 69.

PDBu, See *Phorbol 12,13-Dibutyrate*

PDD, See *Phorbol 12,13-Didecanoate*

DL-*threo*-PDMP, See *DL-threo-1-Phenyl-2-decanoyl-amino-3-morpholino-1-propanol*

α$_1$-PDX, See *α$_1$-Antitrypsin Portland*

PDR-192, See *Enavatuzumab*

PE, See *Phosphatidylethanolamine*

PE-104, See *2-[4-(4'-Chlorophenoxy)phenoxy-acetylamino]-ethylphosphorylethanolamine*

PE859

This orally bioavailable tau aggregation inhibitor (FW = 448.53 g/mol; IUPAC: 3-[(1*E*)-2-(1*H*-indol-6-yl)ethenyl]-5-[(1*E*)-2-[2-methoxy-4-(2-pyridylmethoxy)phenyl]ethenyl]-1*H*-pyrazole) significantly reduces formation of sarkosyl-insoluble aggregates of Tau, the principal axonal microtubule-associated protein, and prevents onset and progression of the motor dysfunction in JNPL3 Pro-301-Leu-mutated human tau transgenic mice. PE859 also inhibits aggregation of three-repeat Tau, *or* 3RMBD-Tau (IC$_{50}$ = 0.81 µM), and four-repeat Tau, *or* 2N4R-Tau (IC$_{50}$ = 2.23 µM). **1**. Okuda, Hijikuro, Fujita, *et al.* (2015) *PLoS One* **10**, e0117511.

PEAQX

This potent NMDA receptor antagonist and experimental anticonvulsant (FW = 542.14 g/mol), also known as NVP-AAM077 and tetrasodium [[[(1*S*)-1-(4-bromophenyl)ethyl]amino](1,2,3,4-tetrahydro-2,3-dioxo-5-quinoxalinyl)methyl]] phosphonate, is a competitive antagonist for ionotropic glutamate [*N*-methyl-D-aspartate] receptors (*or* GluN receptors). Although originally described as being 100x more active against GluN1/GluN2A receptors *versus* GluN1/GluN2B receptors (1), later studies revealed only a 5x difference in affinity (2). There is no difference in the steady-state levels of inhibition produced by NVP-AAM077 when it was either preapplied or coapplied with glutamate. **1**. Auberson, Allgeier, Bischoff, *et al.* (2002) *Bioorg. Med. Chem. Letts.* **12**, 1099. **2**. Frizelle, Chen, Wyllie, *et al.* (2006) *Molec. Pharmacol.* **70**, 1022.

PEC-60

This 60-residue intestine and brain polypeptide (MW = 6835 g/mol) inhibits glucose-induced insulin secretion from perfused pancreas, activates Na$^+$/K$^+$-exchanging ATPase and reduces cAMP production. **Target(s):** adenylate cyclase (1); and insulin secretion, glucose-induced (2-4). **1**. Laasik & Sillard (1993) *Biochem. Biophys. Res. Commun.* **197**, 849. **2**. Norberg, Gruber, Angelucei, *et al.* (2003) *Cell Mol. Life Sci.* **60**, 378. **3**. Ahren,

Ostenson & Efendic (1992) *Pancreas* **7**, 443.   4. Agerberth, Soderling-Barros, Jornvall, *et al.* (1989) *Proc. Natl. Acad. Sci. U.S.A.* **86**, 8590.

Pectin Methylesterase Inhibitors (PMEI's)

These proteins assist in regulating the methylation of pectin, a plant cell wall component that is secreted in highly methyl-esterified form and is subsequently dimethylated *in muro* by pectin methylesterase (PME). Pectin is a structurally complex polysaccharide that, by weight, comprises slightly more than one-third of the primary cell wall of dicots and nongraminaceous monocots. Demethylated pectin is susceptible to hydrolysis and fragmentation by host- and pathogen-derived endopolygalacturonases (1,2). In *Arabidopsis thaliana*, the only two PMEI's are the 151-residue AtPMEI-1 (MW = 16,266 Da; pI = 7.7) and the 148-residue AtPMEI-2 (MW = 15,615 Da; pI = 9.0), consisting of an α-helix up-and-down four-helical bundle fold and five strictly conserved Cys residues, with four forming two structure-maintaining disulfide bridges. These proteins inhibit PMEs of plant origin by forming a noncovalent stoichiometric 1:1 complex with the latter. That they do not typically inhibit PMEs produced by plant pathogenic microorganisms suggests roles in modulating endogenous PME activity during development and growth. Overexpression of a pectin methylesterase inhibitor in *Arabidopsis* also alters growth morphology of the stem and defective organ separation. **1.** Lionetti, Raiola, Camardella, *et al.* (2007) *Plant Physiol.* **143**, 1871. **2.** Müller, Levesque-Tremblay, Fernandes, *et al.* (2013) *Plant Signal Behav.* **8**, e26464.

Pectinose, *See L-Arabinose*

Pectin Sugar, *See L-Arabinose*

Pederin

This toxic agent (FW = 505.29 g/mo; CAS 27973-72-4l), also spelled pederine and paederine, from the blister beetle *Paederus fuscipes* inhibits eukaryotic protein biosynthesis and mitosis, even at concentrations of 2–3 nM (1-6). **1.** Carrasco, Battaner & Vazquez (1974) *Meth. Enzymol.* **30**, 282. **2.** Jiménez (1976) *Trends Biochem. Sci.* **1**, 28. **3.** Richter, Kocienski, Raubo & Davies (1997) *Anticancer Drug Des.* **12**, 217. **4.** Barbacid, Fresno & Vazquez (1975) *J. Antibiot.* **28**, 453. **5.** Jacobs-Lorena, Brega & Baglioni (1971) *Biochim. Biophys. Acta* **240**, 263. **6.** Brega, Falaschi, De Carli & Pavan (1968) *J. Cell Biol.* **36**, 485.

Pefabloc SC Hydrochloride, *See 4-(2-Aminoethyl)-benzenesulfonyl Fluoride Hydrochloride*

Pefabloc TH, *See N^α-(2-Naphthalenesulfonylglycyl)-4-amidino-DL-phenylalaninepiperidide*

Pefabloc Xa, *See N^ε-Tosylglycyl-3-amidino-DL-phenylalanine Methyl Ester*

Peficitinib

This orally bioavailable JAK inhibitor and anti-rheumatoid arthritis drug (FW = 326.39 g/mol; CAS 944118-01-8; Solubility: 200 mM in DMSO), also named ASP015K, JNJ-54781532, and *trans*-4-[[5-hydroxy-2-adamantyl]amino]-1*H*-pyrrolo[2,3-*b*]pyridine-5-carboxamide, targets Janus Kinase JAK1, JAK2, JAK3 and the nonreceptor tyrosine kinase Tyk2 enzyme with IC$_{50}$ values of 3.9, 5.0, 0.71 and 4.8 nM, respectively (1). Its

milder inhibition of JAK2, compared tofacitinib and baricitinib, which selectively suppress JAK3 or JAK1/2, respectively, may explain the weaker effects on red blood cells and platelets reported to be caused by JAK2 inhibition (2). Moreover, peficitinib improves symptoms in RA animal models after once-daily oral administration (3) and demonstrates dose-dependent improvement in psoriatic disease activities in a 6-week phase IIa study (4). The terminal mean half-life of peficitinib is 7–13 hours in pharmacological studies with healthy subjects (5). **1.** Takeuchi, Tanaka, Iwasaki, *et al.* (2015) *Ann. Rheum. Dis.* E-document 208279. **2.** Parganas, Wang, Stravopodis, *et al.* (1998) *Cell* **93**, 385. **3.** Yamazaki, Inami, Ito, *et al.* (2012) *Arthritis Rheum.* **64**, 2084. **4.** Papp, Pariser, Catlin, *et al.* (2015) *Br. J. Dermatol.* **173**, 767. **5.** Zhu, Sawamoto, Valluri, *et al.* (2013) *Rheum. Dis.* **72**, 898.

Pefloxacin

This antibiotic (FW = 333.36 g/mol; CAS 70458-95-6), an analogue of norfloxacin, derives its antibacterial effects by inhibiting DNA topoisomerase. (For the prototypical member of this antibiotic class, *See Ciprofloxacin*) **Target(s):** CYP1A2 (1); DNA-directed DNA polymerase (2); DNA polymerase α (2); DNA polymerase β (2); DNA topoisomerase I (3); DNA topoisomerase II (4,5); and terminal deoxynucleotidyltransferase (2). **1.** Kinzig-Schippers, Fuhr, Zaigler, *et al.* (1999) *Clin. Pharmacol. Ther.* **65**, 262. **2.** Rusquet, Bonhommet & David (1984) *Biochem. Biophys. Res. Commun.* **121**, 762. **3.** Tabary, Moreau, Dureuil & Le Goffic (1987) *Antimicrob. Agents Chemother.* **31**, 1925. **4.** Riou, Douc-Rasy & Kayser (1986) *Biochem. Soc. Trans.* **14**, 496. **5.** Borner & Lode (1986) *Infection* **14** Suppl. 1, S54.

Peganine, *See Vasicine*

PEG Phenol, *See specific polyethyleneglycol alcohol; e.g., Brij 35; Lubrol PX*

PEG Sorbitol Esters, *See specific polyethyleneglycol sorbitol ester; e.g., Tween 20*

Pelargonate (Pelargonic Acid)

This medium-chain fatty acid (FW = 158.24 g/mol; CAS 112-05-0; pK_a ≈ 4.8 at 25°C) also called nonanoic acid, is an oily liquid (M.P. = 12.5°C; B.P. = 252-253°C). Pelargonic acid has a low solublity in water (0.026 g/100 mL at 20°C), but is soluble in ethanol, diethyl ether, benzene, and chloroform. **Target(s):** alcohol dehydrogenase (1); *p*-aminohippurate transport (2); kynurenine aminotransferase, weakly inhibited (3); and phenylacetyl-CoA synthetase, *or* phenylacetate:CoA ligase (4). **1.** Winer & Theorell (1960) *Acta Chem. Scand.* **14**, 1729. **2.** Ullrich, Rumrich & Kloss (1987) *Pflugers Arch.* **409**, 547. **3.** Mason (1959) *J. Biol. Chem.* **234**, 2770. **4.** Martinez-Blanco, Reglero, Rodriguez-Aparicio & Luengo (1990) *J. Biol. Chem.* **265**, 7084.

Pelargonidin

This flavylium cation (FW$_{chloride}$ = 306.79 g/mol; CAS 134-04-3), also spelled pelargonidine chloride and named systematically as 3,5,7,4'-tetrahydroxyflavylium cation, is the aglycon of many plant anthocyanins (*e.g.*, pelargonin, callistephin, and fragarin). The chloride salt is a reddish-brown solid that has a λ_{max} of 530 nm ($\varepsilon = 32000$ $M^{-1}cm^{-1}$) in ethanol and 0.01% HCl. **Target(s):** aflatoxin B_1 biosynthesis (1); K^+ channel, ATP-dependent (2,3); DNA topoisomerase II decatenation activity (4); glycogen phosphorylase (5); inositol-trisphosphate 3-kinase (6); leucoanthocyanidin reductase (7); and nitric oxide production (8). **1.** Norton (1999) *J. Agric. Food Chem.* **47**, 1230. **2.** Marinov, Grigoriev, Skarga, Olovjanishnikova & Mironova (2001) *Membr. Cell Biol.* **14**, 663. **3.** Grigoriev, Skarga, Mironova & Marinov (1999) *Biochim. Biophys. Acta* **1410**, 91. **4.** Habermeyer, Fritz, Barthelmes, *et al.* (2005) *Chem. Res. Toxicol.* **18**, 1395. **5.** Jakobs, Fridrich, Hofem, Pahlke & Eisenbrand (2006) *Mol. Nutr. Food Res.* **50**, 52. **6.** Mayr, Windhorst & Hillemeier (2005) *J. Biol. Chem.* **280**, 13229. **7.** Tanner, Francki, Abrahams, *et al.* (2003) *J. Biol. Chem.* **278**, 31647. **8.** Wang & Mazza (2002) *J. Agric. Food Chem.* **50**, 850.

Pelargonidin 3-(β-Galactoside), *See Fragarin*

Pelargonidin 3-(β-Glucoside), *See Callistephin*

Peldesine

This guanosine nucleoside analogue (FW = 241.25 g/mol), also known as BCX-34 and 9-deaza-9-(3-pyridinyl-methyl)guanine, inhibits purine-nucleoside phosphorylase, human K_i = 13.5 nM (1-3). The pyridin-2-ylmethyl and pyridin-4-ylmethyl analogues also inhibit (calf spleen IC_{50} = 15 and 64 nM, respectively). **1.** Bzowska, Kulikowska & Shugar (2000) *Pharmacol. Ther.* **88**, 349. **2.** Castilho, Postigo, de Paula, *et al.* (2006) *Bioorg. Med. Chem.* **14**, 516. **3.** Wada, Yagihashi, Terasawa, *et al.* (1996) *Artif. Organs* **20**, 849.

Pellagra-Preventative Factor, *See Nicotinate and Nicotinic Acid; Nicotinamide*

Peloruside A

This novel marine metabolite and microtubule-stabilizing agent (FW = 548.67 g/mol; IUPAC: (1S,3S,4R,7R,9R,11R,13S,14S,15S)-4,11,13,14-tetrahydroxy-7-[(2Z,4S)-4-(hydroxymethyl)hex-2-en-2-yl]-3,9,15-trimethoxy-12,12-dimethyl-6,17-dioxabicyclo[11.3.1]heptadecan-5-one) from the New Zealand sponge *Mycale hentscheli* has potent paclitaxel-like activity and is cytotoxic at low-nM concentrations. Its 16-membered macrolide ring resembles that found in the anticancer agent epothilone (1,2). Peloruside A arrests cells in the G_2-M phase of the cell cycle and induces apoptosis (2). The peloruside A site on the α,β-tubulin heterodimer is topologically distinct from the taxoid site (which is used by paclitaxel, docetaxel, epothilone A, and discodermolide) and synergizes with these agents, enhancing their antimitotic action (3). Based on an earlier model for peloruside A binding to β-tubulin (4), a later and more persuasive model (5) shows more extensive desolvation and a greater array of favorable hydrophobic and electrostatic interactions for peloruside A binding. The latter model is also suitable for laulimalide binding. Peloruside A-resistant lines contain single-base mutations in β-tubulin that result in the following substitutions: Arg-306-His, Tyr-340-Ser, Asn-337-Asp, and Ala-296-Ser in various combinations (6). These mutations are localized to peptides previously identified by Hydrogen-Deuterium exchange mapping, and center on a cleft in which the drug side chain appears to dock (6). Significantly, β_{II}-tubulin and β_{III}-tubulin mediate sensitivity to peloruside A and laulimalide, but not paclitaxel or vinblastine, in human ovarian

carcinoma cells (7). Analysis of microtubule (MT) growth times supports the view that catastrophic MT disassembly depends on the age of MTs (8). Colchicine and vinblastine accelerate aging in a manner that depends on the presence of end-binding proteins. On the other hand, while Paclitaxel and Peloruside A induce catastrophes, they compensate by promoting MT rescues and reversing the MT aging. **1.** West, Northcote, Battershill, *et al.* (2000) *J. Org. Chem.* **65**, 445. **2.** Hood, West, Rouwé, *et al.* (2002) *Cancer Res.* **62**, 3356. **3.** Wilmes, Bargh, Kelly, Northcote & Miller (2007) *Mol. Pharm.* **4**, 269. **4.** Huzil et al. (2008) *J. Mol. Biol.* **378**, 1016. **5.** Nguyen, Xu, Gussio, Ghosh & Hamel (2010) *Chem. Inf. Model.* **50**, 2019. **6.** Begaye, Trostel, Zhao, *et al.* (2011) *Cell Cycle* **10**, 3387. **5.** Kanakkanthara, Northcote & Miller (2012) *Mol. Cancer Ther.* **11**, 393. **6.** Mohan, Katrukha, Doodhi, *et al.* (2013) *Proc. Natl. Acad. Sci. U.S.A.* **110**, 8900.

Pembrolizumab

This humanized receptor-directed monoclonal antibody and anticancer agent (MW = 146.3 kDa; CAS 1374853-91-4), also known by its code name MK-3475, its former name lambrolizumab, and its trade name Keytruda®, targets the Programmed Death-1 (PD-1 *or* Pdcd1) receptor, a well-known negative regulator of T-cell effector mechanisms that limits immune responses against cancer. Having received "Breakthrough Therapy" designation in April, 2013 to expedite its development as a melanoma therapy, pembrolizumab won final FDA approval in September, 2014. **Mechanism of Inhibitor Action:** An immunoreceptor belonging to the CD28/CTLA-4 family, PD-1 negatively regulates antigen receptor signaling by recruiting the protein tyrosine phosphatase SHP-2, after interacting with either of two ligands, PD-L1 or PD-L2. With its wide range of ligand distribution in the body, PD-1 is significant in nearly every aspect of immune responses (*e.g.* autoimmunity, tumor immunity, infectious immunity, transplantation immunity, allergy and immunological privilege). Pembrolizumab blocks the interaction between PD-1 and its ligands, PD-L1 and PD-L2. PD-1 and related target PD-ligand 1 (PD-L1) are normally expressed on the surface of activated T cells, and formation of the PD-L1·PD-1 complex inhibits immune activation and reduces T-cell cytotoxic activity when bound. This negative feedback loop maintains normal immune responses by limiting T-cell activity and thereby protecting normal cells during chronic inflammation. Tumor cells can circumvent T-cell–mediated cytotoxicity by expressing PD-L1 on their outer surface or on tumor-infiltrating immune cells to inhibit immune-mediated tumor cell killing. Use of pembrolizumab is indicated in the treatment of patients with unresectable or metastatic melanoma and disease progression following ipilimumab, especially if positive for the $BRAF^{V600}$ mutation. MK3475 is also highly selective, humanized monoclonal IgG4-κ isotype antibody against PD-1 that is designed to block the negative immune regulatory signaling of the PD-1 receptor on T cells. **Inhibitor Design:** To prepare pembrolizumab, the variable region sequences of a very high affinity (K_d = 28 pM) mouse anti-human PD-1 antibody were grafted into a human IgG4 immunoglobulin with a stabilizing Ser-228-Pro alteration in Fc receptors. The IgG4 subtype does not engage Fc receptors and does not activate complement, thereby avoiding cytotoxic effects of the antibody when it binds to the T cells that it is intended to activate. **Pharmacokinetics:** The recommended dose (2 mg/kg) is administered as an intravenous infusion over 30 minutes every 3 weeks, or until disease progression resumes or toxicity becomes unacceptable. Its steady-state concentration is reached by 18 weeks. (Half-life: 26 days; Clearance: 0.22 L/day) **1.** Hamid, Robert, Daud, et al. (2013) *New Engl. J. Med.* **369**, 134. **2.** Ascierto, Kalos, Schaer, Callahan & Wolchok (2013) *Clin. Cancer Res.* **19**, 1009. **2.** Sheridan (2014) *Nature Biotechnol.* **32**, 847.

Pemetrexed

This antifolate prodrug ($FW_{free-base}$ = 425.40 g/mol; CAS 357166-30-4), known by the code name LY 231514, the trade name Alimta™ and systematically as *N*-[4-[2-(2-amino-3,4-dihydro-4-oxo-7*H*-pyrrolo[2,3-*d*]pyrimidin-5-yl)ethyl]benzoyl]-L-glutamic acid, undergoes enzyme-catalyzed polyglutamylation after cellular uptake, whereupon it inhibits dihydrofolate reductase (1-4), 10-formyltetrahydrofolate synthase (2); 5,10-

methylene-tetrahydrofolate dehydrogenase (2); phosphoribosyl-amino-imidazolecarboxamide formyltransferase (1,2); phosphoribosylglycinamide formyltransferase (1,2,5); and thymidylate synthase (1-4,6,7). By inhibiting the formation of purine and pyrimidine nucleotides, pemetrexed blocks DNA and RNA synthesis needed for the growth and survival of proliferating cells. Pemetrexed is recommended for the treatment of malignant pleural mesothelioma. When used in combination with carboplatin or cisplatin, it is also effective in the treatment of non-small cell lung cancer. (**See** *Methotrexate*) **1**. Shih, Habeck, Mendelsohn, Chen & Schultz (1998) *Adv. Enzyme Regul.* **38**, 135. **2**. Shih, Chen, Gossett, *et al.* (1997) *Cancer Res.* **57**, 1116. **3**. Gangjee, Li, Yang & Kisliuk (2008) *J. Med. Chem.* **51**, 68. **4**. Gangjee, Qiu, Li & Kisliuk (2008) *J. Med. Chem.* **51**, 5789. **5**. Deng, Wang, Cherian, *et al.* (2008) *J. Med. Chem.* **51**, 5052. **6**. Taylor, Kuhnt, Shih, *et al.* (1992) *J. Med. Chem.* **35**, 4450. **7**. Bijnsdorp, Comijn, Padron, Gmeiner & Peters (2007) *Oncol. Rep.* **18**, 287.

Penbutolol

This non-selective antihypertensive β-blocker (FW = 291.43 g/mol; CAS 36507-48-9; IUPAC: (*S*)-1-(*tert*-butylamino)-3-(2-cyclopentylphenoxy)-propan-2-ol), known variously as Levatol®, Levatolol®, Lobeta®, Paginol®, Hostablock®, and Betapressin), binds to both β₁- and β₂-adrenergic receptors, thus exhibiting the sympathomimetic effects of a partial β-adrenergic agonist, while also acting as a serotonin (5-HT₁ₐ) receptor antagonist (1,2). By blocking β-adrenergic receptors and the sympathetic nervous system, penbutolol decreases heart rate and cardiac output, thereby lowering arterial blood pressure. **1**. Katzung, Bertram G. (1998). *Basic & Clinical Pharmacology* (7th ed.). London: Appleton & Lange. **2**. Frishman & Covey (1990). *J. Clin. Pharmacol.* **30**, 412.

Penciclovir

This ganciclovir analogue (FW = 253.26 g/mol; CAS 39809-25-1), also known as 9-[4-hydroxy-3-(hydroxymethyl)but-1-yl]guanine, is (upon phosphorylation to tis triphosphate form) a potent inhibitor of viral DNA polymerases and is also an analogue of. The (*S*)-enantiomer of penciclovir triphosphate is preferentially formed in herpes virus-infected cells and is a more active agent against the herpes simplex virus. However, the (*R*)-enantiomer of penciclovir triphosphate is the more potent inhibitor of hepatitis B virus DNA polymerase-reverse transcriptase. Penciclovir is soluble in water (1.7 mg/mL at 20°C; the sodium salt has a solubility of > 200 mg/mL) and has λ_max values (in 0.01 N NaOH) of 215 and 268 nm (ε = 18140 and 10710 M⁻¹cm⁻¹, respectively). The prodrug is called famciclovir. (**See also prodrug** *Famciclovir*) **Target(s):** DNA polymerases α, δ, and ε (1); DNA polymerases, viral (2); DNA primase (1); Epstein-Barr virus replication (2,3); hepatitis B virus DNA polymerase (4-10); herpes simplex virus replication (2,11-13); and varicella zoster virus replication (2). **1**. Ilsley, Lee, Miller & Kuchta (1995) *Biochemistry* **34**, 2504. **2**. Prisbe & Chen (1996) *Meth. Enzymol.* **275**, 425. **3**. Bacon & Boyd (1995) *Antimicrob. Agents Chemother.* **39**, 1599. **4**. Offensperger, Offensperger, Keppler-Hafkemeyer, Hafkemeyer & Blum (1996) *Antivir. Ther.* **1**, 141. **5**. Das, Xiong, Yang, *et al.* (2001) *J. Virol.* **75**, 4771. **6**. Ying, De Clercq, Nicholson, Furman & Neyts (2000) *J. Viral Hepat.* **7**, 161. **7**. De Clercq (1999) *Int. J. Antimicrob. Agents* **12**, 81. **8**. Shaw, Mok & Locarnini (1996) *Hepatology* **24**, 996. **9**. Lin, Luscombe, Wang, Shaw & Locarnini (1996) *Antimicrob. Agents Chemother.* **40**, 413. **10**. Shaw, Amor, Civitico, Boyd & Locarnini (1994) *Antimicrob. Agents Chemother.* **38**, 719. **11**. Bacon, Howard, Spender & Boyd (1996) *J. Antimicrob. Chemother.* **37**, 303. **12.**

Weinberg, Bate, Masters, *et al.* (1992) *Antimicrob. Agents Chemother.* **36**, 2037. **13**. Hodge & Perkins (1989) *Antimicrob. Agents Chemother.* **33**, 223.

Penciclovir Triphosphate, *See* Penciclovir

D-Penicillamine

This water-soluble amino acid (FW = 149.21 g/mol; CAS 52-67-5), also known as D-2-amino-3-mercapto-3-methylbutyrate, 3-mercapto-D-valine, and 3,3-dimethyl-D-cysteine, is a penicillin degradation. Penicillamine is a strong chelator of metal ions as well as a radical and H_2O_2 scavenger. The reported pK_a values for DL-penicillamine are 1.8, 7.9, and 10.5. **Target(s):** alanine racemase, *Acidiphilium organovorum* (1); alkaline phosphatase (2-4); D-amino-acid aminotransferase (36); amino-acid racemase (23); 4-aminobutyrate transaminase (37,38); 8-amino-7-oxononanoate synthase (5,39); aminopeptidase B (6); arginine racemase, inhibited by DL-penicillamine (24); aspartate racemase (7); carboxy-peptidase A (8); 3-chloro-D-alanine dehydrochlorinase (25); collagenases (9); complement component C4 (10); cystathionine γ-lyase (11,27-29); D-cysteine desulfhydrase (26); cysteinylglycine dipeptidase (12,13); diaminopimelate decarboxylase (14); diaminopropionate ammonia-lyase (30); gelatinase B (15); glucose-6-phosphate dehydrogenase (16); glutamate decarboxylase, weakly inhibited (17,32); glutamate-1-semialdehyde 2,1-aminomutase, *or* glutamate-1-semialdehyde aminotransferase (18); glutathione peroxidase (19); glyceraldehyde-3-phosphate dehydrogenase (3); kynureninase (33,34); β-lactamase, *or* penicillinase (20); leukotriene D₄ dipeptidase (13); matrilysin (8); nitrile hydratase (21); phosphatidylserine decarboxylase (31); pyridoxal kinase (35); L-serine dehydratase (17); serine hydroxymethyltransferase, *or* glycine hydroxymethyltransferase (22); thermolysin (8); tyrosinase, *or* monophenol monooxygenase (40); and tyrosine aminotransferase (17). **1**. Seow, Inagaki, Tamura, Soda & Tanaka (1998) *Biosci. Biotechnol. Biochem.* **62**, 242. **2**. Agus, Cox & Griffin (1966) *Biochim. Biophys. Acta* **118**, 363. **3**. Raab & Gmeiner (1975) *Arch. Dermatol. Res.* **254**, 87. **4**. Raab & Morth (1974) *Z. Klin. Chem. Klin. Biochem.* **12**, 309. **5**. Izumi, Tani & Ogata (1979) *Meth. Enzymol.* **62**, 326. **6**. Herranz, Garcia-Lopez & Perez (1991) *Arch. Pharm. (Weinheim)* **324**, 239. **7**. Shibata, Watanabe, Yoshikawa, *et al.* (2003) *Comp. Biochem. Physiol. B* **134**, 307. **8**. Chong & Auld (2000) *Biochemistry* **39**, 7580. **9**. Francois, Cambie & Feher (1973) *Ophthalmologica* **166**, 222. **10**. Sim, Dodds & Goldin (1989) *Biochem. J.* **259**, 415. **11**. Nagasawa, Kanzaki & Yamada (1987) *Meth. Enzymol.* **143**, 486. **12**. Tate (1985) *Meth. Enzymol.* **113**, 471. **13**. Huber & Keppler (1987) *Eur. J. Biochem.* **167**, 73. **14**. White (1971) *Meth. Enzymol.* **17B**, 140. **15**. Norga, Grillet, Masure, Paemen & Opdenakker (1996) *Clin. Rheumatol.* **15**, 31. **16**. Raab & Gmeiner (1976) *J. Clin. Chem. Clin. Biochem.* **14**, 173. **17**. Pestaña, Sandoval & Sols (1971) *Arch. Biochem. Biophys.* **146**, 373. **18**. Kannangara & Gough (1978) *Carlsberg Res. Commun.* **43**, 185. **19**. Tappel (1984) *Curr. Top. Cell. Regul.* **24**, 87. **20**. Pollock (1960) *The Enzymes*, 2nd ed. (Boyer, Lardy & Myrbäck, eds.), **4**, 269. **21**. Nagasawa, Nanba, Ryuno, Takeuchi & Yamada (1987) *Eur. J. Biochem.* **162**, 691. **22**. Miyazaki, Toki, Izumi & Yamada (1987) *Eur. J. Biochem.* **162**, 533. **23**. Asano & Endo (1988) *Appl. Microbiol. Biotechnol.* **29**, 523. **24**. Yorifuji & Ogata (1971) *J. Biol. Chem.* **246**, 5085. **25**. Nagasawa, Ohkishi, Kavakami, *et al.* (1982) *J. Biol. Chem.* **257**, 13749. **26**. Nagasawa, Ishii, Kumagai & Yamada (1985) *Eur. J. Biochem.* **153**, 541. **27**. De Angelis, Curtin, McSweeney, Faccia & Gobbetti (2002) *J. Dairy Res.* **69**, 255. **28**. Bruinenberg, de Roo & Limsowtin (1997) *Appl. Environ. Microbiol.* **63**, 561. **29**. Nagasawa, Kanzaki & Yamada (1984) *J. Biol. Chem.* **259**, 10393. **30**. Nagasawa, Tanizawa, Satoda & Yamada (1988) *J. Biol. Chem.* **263**, 958. **31**. Suda & Matsuda (1974) *Biochim. Biophys. Acta* **369**, 331. **32**. Tunnicliff (1990) *Int. J. Biochem.* **27**, 1235. **33**. Moriguchi, Yamamoto & Soda (1973) *Biochemistry* **12**, 2969. **34**. Tanizawa & Soda (1979) *J. Biochem.* **85**, 901. **35**. Lainé-Cessac, Cailleux & Allain (1997) *Biochem. Pharmacol.* **54**, 863. **36**. Yonaha, Misono, Yamamoto & Soda (1975) *J. Biol. Chem.* **250**, 6983. **37**. Yonaha & Toyama (1980) *Arch. Biochem. Biophys.* **200**, 156. **38**. Yonaha, Suzuki & Toyama (1985) *Eur. J. Biochem.* **146**, 101. **39**. Izumi, Morita, Tani & Ogata (1973) *Agric. Biol. Chem.* **37**, 1327. **40**. Lovstad (1976) *Biochem. Pharmacol.* **25**, 533.

L-Penicillamine

This water-soluble amino acid (FW = 149.21 g/mol;), also known as L-2-amino-3-mercapto-3-methylbutyrate, 3-mercapto-L-valine, and 3,3-dimethyl-L-cysteine, is a strong chelator of metal ions as well as a radical and H_2O_2 scavenger. L-Penicillamine also inhibits several several pyridoxal-phosphate-dependent enzymes. The reported pK_a values for DL-penicillamine are 1.8, 7.9, and 10.5. *Note*: Both D- and L-forms are readily oxidized to the corresponding disulfides. **Target(s):** *O*-acetylhomoserine aminocarboxy-propyltransferase (48,49); adenosyl-methionine: 8-amino-7-oxononanoate transaminase, *or* diaminopelargonate aminotransferase, inhibited by DL-penicillamine (42); alanine aminotransferase (2); alanine:glyoxylate aminotransferase (44); alanine racemase, mildly inhibited (3); alkaline phosphatase (4); D-amino-acid aminotransferase (46); 5-aminolevulinate aminotransferase (5,6,45); 5-aminolevulinate synthase (7); 8-amino-7-oxononanoate synthase (8); aminopeptidase B (9); arginine racemase, inhibited by DL-penicillamine (24); aspartate racemase (10); carboxypeptidase A (41); 3-chloro-D-alanine dehydrochlorinase (25); cystathionine β-lyase, inhibited by DL-penicillamine (29); cystathionine γ-lyase (11,31-33); D-cysteine desulfhydrase (26); cysteinylglycine dipeptidase (12); diaminopimelate decarboxylase (13); diamino-propionate ammonia-lyase (34); D-glucosaminate dehydratase (14); glutamate decarboxylase, inhibited by DL-mixture (38); homocysteine desulfhydrase, inhibited by DL-penicillamine (30); homoserine *O*-acetyltransferase, inhibited by DL-penicillamine (51); isocitrate lyase, inhibited by DL-penicillamine (36); kynureninase (40); kynurenine:oxoglutarate amino-transferase (47); β-lactamase, *or* penicillinase (15,16); leukotriene D₄ dipeptidase (17); matrilysin (41); methionine *S*-adenosyltransferase, weakly inhibited (50); methionine γ-lyase, inhibited by DL- and L-penicillamine (18,19,27,28); nitrile hydratase (20,35); ornithine:α-keto-acid amino-transferase (21); phosphatidylserine decarboxylase (37); serine:glyoxylate aminotransferase, weakly inhibited (43); serine Hydroxymethyltransferase, *or* glycine hydroxymethyltransferase (22); superoxide dismutase (52); thermolysin (41); valine decarboxylase, stereochemistry not reported (39); and valyl-tRNA synthetase (23). **1.** Wilson & du Vigneaud (1950) *J. Biol. Chem.* **184**, 63. **2.** Evered & Hargreaves & Verjee (1969) *Biochem. J.* **111**, 15P. **3.** Seow, Inagaki, Tamura, Soda & Tanaka (1998) *Biosci. Biotechnol. Biochem.* **62**, 242. **4.** Agus, Cox & Griffin (1966) *Biochim. Biophys. Acta* **118**, 363. **5.** Turner & Neuberger (1970) *Meth. Enzymol.* **17A**, 188. **6.** Braunstein (1973) *The Enzymes*, 3rd ed. (Boyer, ed.), **9**, 379. **7.** Jordan & Shemin (1972) *The Enzymes*, 3rd ed. (Boyer, ed.), **7**, 339. **8.** Izumi, Tani & Ogata (1979) *Meth. Enzymol.* **62**, 326. **9.** Herranz, Garcia-Lopez & Perez (1991) *Arch. Pharm. (Weinheim)* **324**, 239. **10.** Shibata, Watanabe, Yoshikawa, *et al.* (2003) *Comp. Biochem. Physiol. B* **134**, 307. **11.** Nagasawa, Kanzaki & Yamada (1987) *Meth. Enzymol.* **143**, 486. **12.** Tate (1985) *Meth. Enzymol.* **113**, 471. **13.** White (1971) *Meth. Enzymol.* **17B**, 140. **14.** Iwamoto, Imanaga & Soda (1982) *J. Biochem.* **91**, 283. **15.** Abraham (1955) *Meth. Enzymol.* **2**, 120. **16.** Pollock (1960) *The Enzymes*, 2nd ed. (Boyer, Lardy & Myrbäck, eds.), **4**, 269. **17.** Huber & Keppler (1987) *Eur. J. Biochem.* **167**, 73. **18.** Esaki & Soda (1987) *Meth. Enzymol.* **143**, 459. **19.** Lockwood & Coombs (1991) *Biochem. J.* **279**, 675. **20.** Nagasawa, Nanba, Ryuno, Takeuchi & Yamada (1987) *Eur. J. Biochem.* **162**, 691. **21.** Jenkins & Tsai (1970) *Meth. Enzymol.* **17A**, 281. **22.** Miyazaki, Toki, Izumi & Yamada (1987) *Eur. J. Biochem.* **162**, 533. **23.** Natarajan & Gopinathan (1979) *Biochim. Biophys. Acta* **568**, 253. **24.** Yorifuji & Ogata (1971) *J. Biol. Chem.* **246**, 5085. **25.** Nagasawa, Ohkishi, Kavakami, *et al.* (1982) *J. Biol. Chem.* **257**, 13749. **26.** Nagasawa, Ishii, Kumagai & Yamada (1985) *Eur. J. Biochem.* **153**, 541. **27.** Tanaka, Esaki & Soda (1977) *Biochemistry* **16**, 100. **28.** Dias & Weimer (1998) *Appl. Environ. Microbiol.* **64**, 3327. **29.** Alting, Engels, van Schalkwijk & Exterkate (1995) *Appl. Environ. Microbiol.* **61**, 4037. **30.** Thong & Coombs (1985) *IRCS Med. Sci. Libr. Compend.* **13**, 493. **31.** De Angelis, Curtin, McSweeney, Faccia & Gobbetti (2002) *J. Dairy Res.* **69**, 255. **32.** Bruinenberg, de Roo & Limsowtin (1997) *Appl. Environ. Microbiol.* **63**, 561. **33.** Nagasawa, Kanzaki & Yamada (1984) *J. Biol. Chem.* **259**, 10393. **34.** Nagasawa, Tanizawa, Satoda & Yamada (1988) *J. Biol. Chem.* **263**, 958. **35.** Nagasawa & Yamada (1987) *Biochem. Biophys. Res. Commun.* **147**, 701. **36.** Tanaka, Yoshida, Watanabe, Izumi & Mitsunaga (1997) *Eur.*

J. Biochem. **249**, 820. **37.** Suda & Matsuda (1974) *Biochim. Biophys. Acta* **369**, 331. **38.** Blindermann, Maitre, Ossola & Mandel (1978) *Eur. J. Biochem.* **86**, 143. **39.** Bast, Hartmann & Steiner (1971) *Arch. Mikrobiol.* **79**, 12. **40.** Tanizawa & Soda (1979) *J. Biochem.* **85**, 901. **41.** Chong & Auld (2000) *Biochemistry* **39**, 7580. **42.** Izumi, Sato, Tani & Ogata (1975) *Agric. Biol. Chem.* **39**, 175. **43.** Izumi, Yoshida & Yamada (1990) *Eur. J. Biochem.* **190**, 285. **44.** Sakuraba, Kawakami, Takahashi & Ohshima (2004) *J. Bacteriol.* **186**, 5513. **45.** Neuberger & Turner (1963) *Biochim. Biophys. Acta* **67**, 342. **46.** Yonaha, Misono, Yamamoto & Soda (1975) *J. Biol. Chem.* **250**, 6983. **47.** Asada, Sawa, Tanizawa & Soda (1986) *J. Biochem.* **99**, 1101. **48.** Yamagata (1984) *J. Biochem.* **96**, 1511. **49.** Yamagata, Isaji, Nakamura, *et al.* (1994) *Appl. Microbiol. Biotechnol.* **42**, 92. **50.** Berger & Knodel (2003) *BMC Microbiol.* **3**, 12. **51.** Yamagata (1987) *J. Bacteriol.* **169**, 3458. **52.** Kim, Eum & Kang (1998) *Mol. Cells* **8**, 478.

Penicillanate

This penicillin building block (FW_{free-acid} = 201.25 g/mol), also known as penicillanic acid and (2S,5R)-3,3-dimethyl-7-oxo-4-thia-1-azabicyclo [3.2.0]heptane-2-carboxylic acid, weakly inhibits serine-type D-Ala-D-Ala carboxypeptidase but is devoid of significant antibacterial activity. **1.** Frère, Kelly, Klein & Ghuysen (1982) *Biochem. J.* **203**, 223.

Penicillanic Acid Sulfone, *See* Sulbactam

Penicillate (Penicillic Acid) & Lactone

This water-soluble toxic metabolite (FW = 170.17 g/mol), also known 3-methoxy-5-methyl-4-oxo-2,5-hexadienoic acid and produced by a number of fungi (*e.g.*, certain species of *Penicillium* and *Aspergillus*) exists as an equilibrium mixture of uncyclized and lactone forms. Note that this mycotoxin is not related to the penicillins nor to penicillanic acid. **Target(s):** alcohol dehydrogenase (1); Ca^{2+} channels (2); carboxypeptidase A (3); caspase-8 self-processing (4); fructose-1,6-bisphosphate aldolase (5); glutathione *S*-transferase (6); K^+ channels (2); lactate dehydrogenase (1); and Na^+ channels (2). **1.** Ashoor & Chu (1973) *Food Cosmet. Toxicol.* **11**, 617. **2.** Pandiyan, Nayeem, Nanjappan & Ramamurti (1990) *Indian J. Exp. Biol.* **28**, 295. **3.** Parker, Phillips, Kubena, Russell & Heidelbaugh (1982) *J. Environ. Sci. Health B* **17**, 77. 4. Bando, Hasegawa, Tsuboi, *et al.* (2003) *J. Biol. Chem.* **278**, 5786. **5.** Ashoor & Chu (1973) *Food Cosmet. Toxicol.* **11**, 995. **6.** Dierickx & De Beer (1984) *Mycopathologia* **86**, 137.

Penicillins

penicillin core structure

These structurally related bacterial cell wall biosynthesis inhibitors, which possess a distictive β-lactam/ thiazolidine nucleus (*or* drug warhead), are used to treat a wide range of infections caused by susceptible bacteria, including *Streptococci*, *Staphylococci*, *Clostridium*, and *Listeria* genera. Others include *Listeria monocytogenes* (MIC = 0.06–0.25 μg/mL), *Neisseria meningitides* (MIC = 0.03–0.5 μg/mL), and *Staphylococcus aureus* (MIC − 0.015 0.3 μg/mL). Although superceded by the action of later-generation derivatives, pencillin G (*see entry below*)) is still highly effective in controlling *Treptonema palladum*, the causative agent for syphilis. Naturally occurring penicillins are produced by species of *Penicillium*, *Aspergillus*, *Cephalosporium*, *Trichophyton*, *Epidermophyton*;

others are produced by semisynthesis. **Core Sructure:** The term *penam* refers to the common core skeleton of the penicillins, consisting of a β-lactam ring fused to a five-membered thiazolidine ring. The β-lactam ring is highly strained, with its four atoms joined at nearly 90° angles. Adding to the ring's instability is the geometrically imposed lack of amide-bond resonance. **Classification:** The R-group in the structure above is the variable component. *Naturally Occurring* – Penicillin G and Penicillin V; *β-Lactamase-Resistant* – Methicillin, Nafcillin, Oxacillin, Cloxacillin, and Dicloxacillin; *Aminopenicillins* – Ampicillin, Amoxicillin, Bacampicillin, Hetacillin, Metampicillin, Pivampicillin, Talampicillin, and Epicillin; *Carboxypenicillins* – Carbenicillin, Ticarcillin; *Ureidopenicillins* – Mezlocillin and Piperacillin. **Mode of Inhibitor Action:** That penicillin is a highly specific inhibitor of bacterial cell wall biosynthesis was first indicated by the formation of penicillin-induced *Escherichia coli* spheroplasts, the latter membrane-bounded cells lacking an intact peptidoglycan cell walls. In penicillin-treated bacteria, the enzymes that hydrolyze the peptidoglycan cross-links continue to function, while those that form such cross-links cannot. This weakens the cell wall of the bacterium, and uncompensated osmotic pressure eventually causes cytolysis and death. Penicillin also leads to the accumulation of a uridine nucleotide in penicillin-treated *Staphylococcus aureus*, which, in view of its structural similarity to the cell wall, appeared to be a biosynthetic precursor of the cell wall. This idea was amply documented by direct isotopic measurements of cell wall synthesis carried out in the presence of several penicillins and in both Gram-negative and Gram-positive bacteria. A linear cell wall glycopeptide is synthesized from UDP-acetylmuramyl-L-Ala-D-Glu-L-Lys-D-Ala-D-Ala (*or* UDP-MurNAc-pentapeptide) and is cross-linked by peptide bridges. From studies of the structure of the cell wall glycopeptide in normal cells and in penicillin-induced L-forms of *Proteus mirabilis*, it was evident that penicillin blocks formation of essential peptide crosslinks (**See** *Amoxicillin*). Penicillin thus kills susceptible bacteria by specifically inhibiting the transpeptidase that catalyzes the cross-linking of peptidoglycan, the final step in cell wall biosynthesis. As originally hypothesized (1), penicillin is a structural analogue of the acyl-D-alanyl-D-alanine terminus of the pentapeptide side chains of nascent cell wall peptidoglycan, such that, by virtue of its highly reactive β-lactam ring, penicillin irreversibly acylates the active site of the transpeptidase. Indeed, when the cell wall transpeptidase was first purified from membranes of *Bacillus stearothermophilus* using a penicillin affinity chromatography, this enzyme was shown to be covalently labeled with either [^{14}C]penicillin G or the substrate, [^{14}C]diacetyl-L-lysyl-D-alanyl-D-lactate, with both bound as esters to [Serine-36] (2). The remarkable effectiveness of penicillins in inactivating the cell wall transpeptidase has been attributed to their ability to mimic a catalytic transition-state in the naturally occurring transpeptidase reaction, a property allowing them to bind tightly and afford a more efficient path to acylation. A more likely explanation is that they are slow alternative substrates, the bactericidal action of which arises from their ability to form long-lived acyl-enzyme intermediates, thereby blocking the peptidoglycan-forming transpeptidase reaction. In some cases, the penicillanoyl-ester requires hours to days to undergo hydrolysis (3), wheras penicillin-inactivating β-lactamases transiently form acyl-enzyme intermediates, which are hydrolyzed in only a few microseconds.

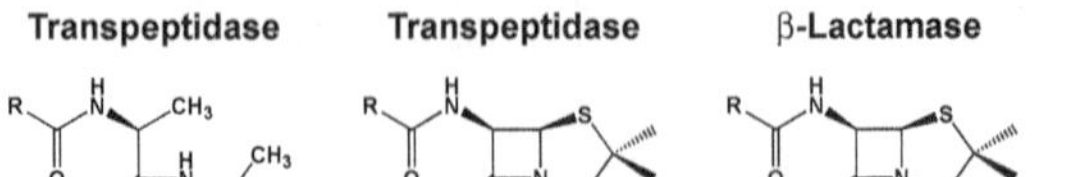

As noted by Walsh & Wencewicz (3), the penicillanoyl scaffold imposes an orientation on the transpeptidase that prevents displacement by the normal cosubstrate (*i.e.*, a nearby ε-NH$_2$ on an adjacent pentapeptidyl arm of another Lipid II molecule). Water access to the transpeptidase active site

must likewise be greatly reduced to prevent hydrolysis. **Penicillin Resistance:** Organism-to-organism transmission, *or* horizontal gene transfer, plays the predominant role in the acquisition and evolution of antibiotic resistance, both in community- and hospital-based infections (4). The catchy term "superbugs" refers to microbes exhibiting enhanced morbidity and mortality due to multiple mutations endowing microorganisms with a high-level of antibiotic resistance. Gram-negative pathogens, such as *Escherichia coli*, *Salmonella enterica*, and *Klebsiella pneumoniae*, show a strong correlation between antibiotic use in the treatment and development of β-lactam antibiotic resistance attributable to the emergence of numerous antibiotic-inactivating β-lactamases that bind and hydrolyze penicillin and many of its derivatives. Super-resistant strains also often acquire increased virulence and transmissibility. The emergence of widespread acquired resistance underscores the potential for antibiotic failure. **1.** Tipper & Strominger (1965) *Proc. Natl. Acad. Sci. U.S.A.* **54**, 1133. **2.** Yocum, Rasmussen & Strominger (1980) *J. Biol. Chem.* **255**, 3977. **3.** Walsh & Wencewicz (2016) *Actiobiotics: Challenges, Mechanisms, Opportunities*, ASM Press, Washington, D.C. **4.** Davies & Davies (2010) *Microbiol. Mol. Biol. Rev.* **74**, 417.

Penicillin G

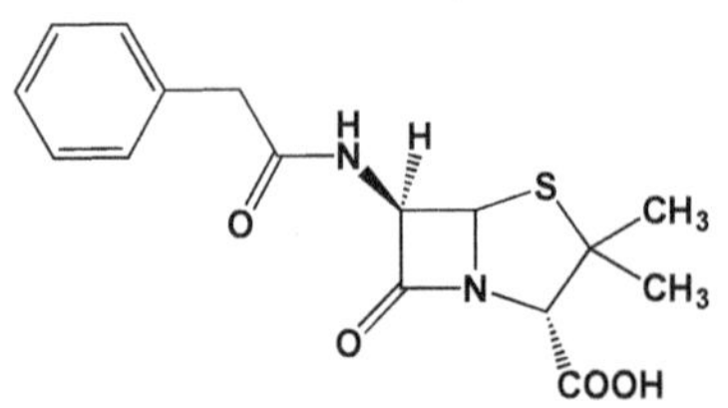

This naturally-occurring antibiotic (FW$_{free-acid}$ = 334.39 g/mol), also known as benzylpenicillin and penicillin II, from *Penicillium* species was the first penicillin, discovered by Alexander Fleming in 1928. Penicillin G's bacteriostatic action stems from its inhibition of peptidoglycan biosynthesis, a key requirement for bacterial proliferation. While the free acid is sparingly soluble in water, the sodium and potassium salts are very soluble. **As Treatment for Syphilis:** A chronic multi-stage infectious disease, syphilis is transmitted sexually by contact with an active lesion of a partner or congenitally from an infected woman to her fetus. Although endemic in developing countries, syphilis remains a problem in even developed countries. Its resurgence is a global public health issue, because the lesions of early syphilis increase the risk of acquiring and transmitting infection, along with HIV-AIDS. Absent a vaccine, syphilis control requires treatment of infected individuals and their sexual partners with penicillin G (less so sith azithromycin). Penicillin G is thus the first-line drug for syphilis in all its stages. *For details on mode of action, See Penicillins* **Target(s):** acylaminoacyl-peptidase (1); aminopeptidase D, *or* D-stereospecific aminopeptidase (2,3); fructose-1,6-bisphosphatase (4); glucose-6-phosphate dehydrogenase (5); γ-glutamyl transpeptidase (6,7); glycopeptide transpeptidase (8); β-lactamase, penicillinase, also substrate (9); lecithin:cholesterol acyltransferase, *or* LCAT (10); muramoylpentapeptide carboxypeptidase (8,11-13); muramoyltetrapeptide carboxypeptidase (14); neprilysin, *or* enkephalinase (15); penicillin amidase, also alternative substrate (16-18); penicillin-binding proteins (19); penicillin-binding protein 1b transpeptidase (20); penicillin-binding protein 2a (21); penicillin-binding protein 4 (22); phospholipase (23); serine-type D-Ala-D-Ala carboxypeptidase (24-30); D-stereospecific aminopeptidase (2,3); u-plasminogen activator, *or* urokinase (31). **1.** Sharma & Orthwerth (1993) *Eur. J. Biochem.* **216**, 631. **2.** Starnes, Desmond & Behal (1980) *Biochim. Biophys. Acta* **616**, 290. **3.** Asano (1998) in *Handb. Proteolytic Enzymes* (Barrett, Rawlings & Woessner, eds.), p. 430, Academic Press, San Diego. **4.** Han, Pollock, Matthews & Johnson (1979) *Experientia* **35**, 586. **5.** Beydemir, Ciftci & Kufrevioglu (2002) *J. Enzyme Inhib. Med. Chem.* **17**, 271. **6.** Hanes, Hird & Isherwood (1950) *Nature* **166**, 288. **7.** Binkley & Olson (1951) *J. Biol. Chem.* **188**, 451. **8.** Izaki, Matsuhashi & Strominger (1966) *Meth. Enzymol.* **8**, 487. **9.** Pollock (1960) *The Enzymes*, 2nd ed., **4**, 269. **10.** Bojesen (1978) *Scand. J. Clin. Lab. Invest. Suppl.* **150**, 26. **11.** Diaz-Mauriño, Nieto & Perkins (1974) *Biochem. J.* **143**, 391. **12.** Martin, Maskos & Burger (1975) *Eur. J. Biochem.* **55**, 465. **13.** Leyh-Bouille, Ghuysen, Nieto, *et al.* (1970) *Biochemistry* **9**, 2971. **14.** Hammes & Seidel (1978) *Eur. J. Biochem.* **91**, 509. **15.** Livingston, Smith, Sewell & Ahmed (1992) *J. Enzyme Inhib.* **6**, 165. **16.** Savidge & Cole (1975) *Meth. Enzymol.* **43**, 705. **17.** Warburton, Balasingham, Dunnil & Lilly (1972) *Biochim. Biophys. Acta* **284**, 278. **18.** Balasingham, Warburton, Dunnil & Lilly (1972) *Biochim. Biophys. Acta* **276**, 250. **19.** Dougherty, Kennedy, Kessler

& Pucci (1996) *J. Bacteriol.* **178**, 6110. **20**. Guilmi, Dessen, Dideberg & Vernet (2003) *J. Bacteriol.* **185**, 1650. **21**. Zhao, Meier, Hoskins & Jaskunas (1999) *Protein Expr. Purif.* **16**, 331. **22**. Wilkin (1998) in *Handb. Proteolytic Enzymes* (Barrett, Rawlings & Woessner, eds.), p. 435, Academic Press, San Diego. **23**. Matsumoto & Saito (1984) *J. Antibiot.* **37**, 910. **24**. Izaki & Strominger (1970) *Meth. Enzymol.* **17A**, 182. **25**. Umbreit & Strominger (1972) *Meth. Enzymol.* **28**, 692. **26**. Wilkin (1998) in *Handb. Proteolytic Enzymes* (Barrett, Rawlings & Woessner, eds.), p. 418, Academic Press, San Diego. **27**. Blumberg & Strominger (1971) *Proc. Natl. Acad. Sci. U.S.A.* **68**, 2814. **28**. Leyh-Bouille, Coyette, Ghuysen, *et al.* (1971) *Biochemistry* **10**, 2163. **29**. Ghuysen (1998) in *Handb. Proteolytic Enzymes* (Barrett, Rawlings & Woessner, eds.), p. 424, Academic Press, San Diego. **30**. Reynolds, Ambur, Casadewall & Courvalin (2001) *Microbiology* **147**, 2571. **31**. Dotzlaf & Yeh (1989) *J. Biol. Chem.* **264**, 10219.

Penicillin N

This antibiotic (FW$_{free-acid}$ = 359.40 g/mol; CAS 525-94-0; Water-soluble), also known as (2*S*,5*R*,6*R*)-6-[[(5*R*)-5-amino-5-carboxy-1-oxopentyl]amino]-3,3-dimethyl-7-oxo-4-thia-1-azabicyclo[3.2.0]heptane-2-carboxylic acid, (D-4-amino-4-carboxybutyl)-penicillinic acid, Adicillin®, Synnematin B®, as well as cephalosporin N, is produced by species of *Cephalosporium* as well as by *Paecilomyces percicimus* and *Penicillium chrysogenum*. (*For details on mode of action, See* Penicillins) **Spectrum of Antibiotic Action:** Penicillin N bacteriocida against *Diplococcus pneumoniae*, *Proteus vulgaris*, *Salmonella typhimurium*, *Sarcina lutea*, but shows little no activity against *Bacillus subtilis* and *Staphylococcus aureus*. **Target(s):** deacetoxycephalosporin-C hydroxylase (1); and deacetylcephalosporin-C acetyltransferase (2). **1**. Baker, Dotzlaf & Yeh (1991) *J. Biol. Chem.* **266**, 5087. **2**. Scheidegger, Gutzwiller, Kuenzi, Fiechter & Nuesch (1985) *J. Biotechnol.* **3**, 109.

Penicillin V

This peptidoglycan biosynthesis inhibitor and antibiotic (FW$_{free-acid}$ = 350.40 g/mol), also known as phenoxymethylpenicillin, is produced by species of *Penicillium* grown on yeast broth supplemented with 2-phenoxyethanol. While the free acid is sparingly soluble in water, the sodium and potassium salts are very soluble. Penicillin V is relatively stable is not hydrolyzed in the stomach. (*For details on mode of action, See* Penicillins) **Target(s):** aminopeptidase D (1); cytosol alanyl aminopeptidase (2); deacetoxycephalosporin-C synthase (3); penicillin-binding protein 4 (4); serine-type D-Ala-D-Ala carboxypeptidase, *or* penicillin-binding protein 5 (5-9). **1**. Starnes, Desmond & Behal (1980) *Biochim. Biophys. Acta* **616**, 290. **2**. Starnes, Szechinski & Behal (1982) *Eur. J. Biochem.* **124**, 363. **3**. Dotzlaf & Yeh (1989) *J. Biol. Chem.* **264**, 10219. **4**. Wilkin (1998) in *Handb. Proteolytic Enzymes* (Barrett, Rawlings & Woessner, eds.), p. 435, Academic Press, San Diego. **5**. Wilkin (1998) in *Handb. Proteolytic Enzymes* (Barrett, Rawlings & Woessner, eds.), p. 418, Academic Press, San Diego. **6**. Blumberg & Strominger (1971) *Proc. Natl. Acad. Sci. U.S.A.* **68**, 2814. **7**. Frère, Ghuysen & Perkins (1975) *Eur. J. Biochem.* **57**, 353. **8**.

Coyette, Ghuysen & Fontana (1978) *Eur. J. Biochem.* **88**, 297. **9**. Coyette, Perkins, Polacheck, Shockman & Ghuysen (1974) *Eur. J. Biochem.* **44**, 459.

Penitrem A

This neurotoxin (FW = 634.21 g/mol; CAS 12627-35-9), also called tremortin A, from a common fungus (*Penicillium palitans*) associated with rye grass, inhibits Ca^{2+}-activated K^{+} channels (EC$_{50}$ = 10 nM), which are found at highest density within cerebellar Purkinje cells. **1**. Knaus, McManus, Lee, *et al.* (1994) *Biochemistry* **33**, 5819.

PENTA, *See* Fondaparinux

Penta(L-arginine), *See* L-Arginyl-L-arginyl-L-arginyl-L-arginyl-L-arginine

2,4,5,3',4'-Pentabromobiphenyl

This persistent environmental pollutant (FW = 548.69 g/mol; CAS 67888-97-5), often abbreviated 2,4,5,3',4'-PBB and 2,4,5,3',4'-PBBP, inhibits uroporphyrinogen decarboxylase. **1**. Sinclair, Bement, Bonkovsky & Sinclair (1984) *Biochem. J.* **222**, 737.

Pentachlorophenol

This electron transport uncoupler (FW = 266.34 g/mol; abbreviated PCP; pK_a = 4.89) is often employed as an insecticide, fungicide, herbicide, and a pre-emergence defoliant that acts by inhibiting both mitochondrial oxidative phosphorylation and photophosphorylation. Pentachlorophenol is slightly soluble in water (8 mg/100 mL), but the commercially available sodium salt is much more soluble). **Target(s):** acetylcholinesterase (1,2); adenosine deaminase (3); arylamine *N*-acetyltransferase (4); arylhydroxamic acid *N,O*-acetyltransferase (4); ATP phosphoribosyltransferase (5); ATP synthase (6); bioluminescence, bacterial (7); chitin synthase (26); electron transport (8); estrone sulfotransferase (9); fumarate reductase, *or* menaquinol:fumarate oxidoreductase, *Escherichia coli* (10); *N*-hydroxyarylamine *O*-acetyltransferase (4,27); iodothyronine sulfotransferase, *or* thyroid hormone sulfotransferase (11); lipase, *or* triacylglycerol lipase (12); luciferase, bacterial (7); NAD(P)H dehydrogenase, *or* diaphorase (13); oxidative phosphorylation, uncoupler (6,14-16); phenol sulfotransferase, *or* aryl sulfotransferase (11,17-23); photophosphorylation, uncoupler (8); porphyrinogen carboxy-lyase (24); steroid 5α-reductase, NADH-dependent (25); succinate dehydrogenase, *or* succinate:ubiquinone oxidoreductase, *Escherichia coli* (10). **1**. Matsumura, Matsuoka, Igisu & Ikeda (1997) *Arch. Toxicol.* **71**, 151. **2**. Igisu, Hamasaki & Ikeda (1993) *Biochem. Pharmacol.* **46**, 175. **3**. Jun, Kim & Yeeh (1994) *Biotechnol. Appl. Biochem.* **20**, 265. **4**. Saito, Shinohara, Kamataki & Kato (1986) *J. Biochem.* **99**, 1689. **5**. Dall-Larsen, Kryvi & Klungsoyr (1976) *Eur. J. Biochem.* **66**, 443. **6**. Stiggall, Galante & Hatefi (1979) *Meth. Enzymol.* **55**, 308. **7**. Ismailov, Pogosian, Mitrofanova, *et al.* (2000) *Prikl. Biokhim. Mikrobiol.* **36**, 469. **8**.

Izawa & Good (1972) *Meth. Enzymol.* **24**, 355. **9.** Kester, Bulduk, Tibboel, *et al.* (2000) *Endocrinology* **141**, 1897. **10.** Maklashina & Cecchini (1999) *Arch. Biochem. Biophys.* **369**, 223. **11.** Visser, Kaptein, Glatt, *et al.* (1998) *Chem. Biol. Interact.* **109**, 279. **12.** Christensen & Riedel (1981) *Arch. Environ. Contam. Toxicol.* **10**, 357. **13.** Raw, Nogueira & Filho (1961) *Enzymologia* **23**, 123. **14.** Hanstein (1976) *Trends Biochem. Sci.* **1**, 65. **15.** Stockdale & Selwyn (1971) *Eur. J. Biochem.* **21**, 565. **16.** Reddy & Peck, Jr. (1978) *J. Bacteriol.* **134**, 982. **17.** Konishi-Imamura, Kim, Koizumi & Kobashi (1995) *J. Enzyme Inhib.* **8**, 233. **18.** Bostrom, Becedas & DePierre (2000) *Chem. Biol. Interact.* **124**, 103. **19.** Beckmann, Henry, Ulphani & Lee (1998) *Chem. Biol. Interact.* **109**, 93. **20.** Rein, Glover & Sandler (1982) *Biochem. Pharmacol.* **31**, 1893. **21.** Duffel, Marshall, McPhie, Sharma & Jakoby (2001) *Drug Metab. Rev.* **33**, 369. **22.** Sekura, Duffel & Jakoby (1981) *Meth. Enzymol.* **77**, 197. **23.** Meinl, Pabel, Osterloh-Quiroz, Hengstler & Glatt (2006) *Int. J. Cancer* **118**, 1090. **24.** Billi, Koss & San Martin de Viale (1986) *IARC Sci. Publ.* (77), 471. **25.** Golf, Graef, Rempeters & Mersdorf (1985) *Biol. Chem. Hoppe-Seyler* **366**, 647. **26.** Pfefferle, Anke, Bross & Steglich (1990) *Agric. Biol. Chem.* **54**, 1381. **27.** Saito, Shinohara, Kamataki & Kato (1985) *Arch. Biochem. Biophys.* **239**, 286.

Pentadecane-2-one

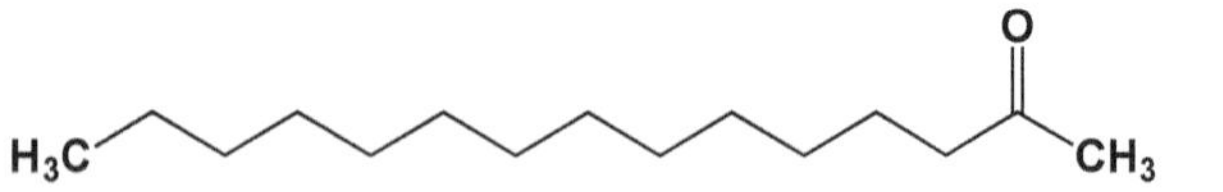

This long-chain ketone (FW = 226.40 g/mol), also called 2-pentadecanone, from the essential oils of the leaves and stems of *Pilocarpus microphyllus* inhibits lanosterol demethylase and lathosterol isomerase. Cholesterol biosynthesis, which is stimulated by cell culture in lipid-depleted medium, is inhibited by pentadecane-2-one, the latter acting mainly at two post-HMG-CoA steps: lanosterol demethylation and lathosterol isomerisation to cholesterol. A parallel pentadecane-2-one inhibition of cell growth is also observed, even when cells were cultured in the presence of whole serum, a finding suggesting a causal relationship between endogenous cholesterol synthesis and cell growth. **1.** Tabacik, Aliau & Sultan (1985) *Biochim. Biophys. Acta* **837**, 152.

1-*N*-Pentadecanoyl-3"-*N*-trifluoroacetyl-kanamycin A

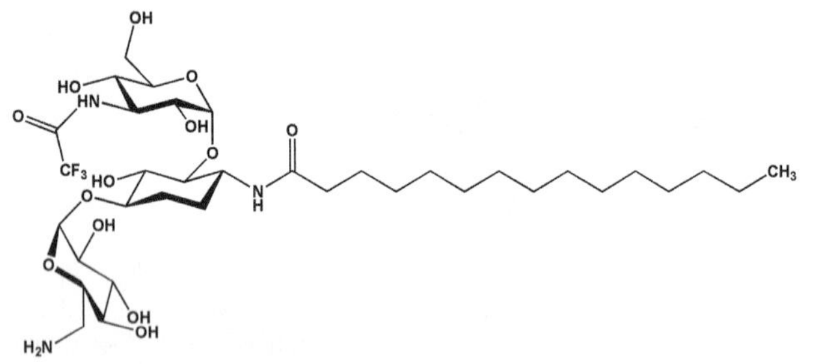

This antiviral agent (FW$_{free-base}$ = 804.90 g/mol), often abbreviated PTKA, inhibits herpes simplex virus type 2 replication *in vitro* (1). *Note*: Because this substance readily forms micelles, one must take care to discriminate whether the monomeric or micelle is inhibitory, when present at concentrations above the critical micelle concentration (*cmc*). **1.** Rahman, Yamamoto, Morishima, Maeno & Nishiyama (1988) *Antiviral Res.* **9**, 11.

Pentadecylate, *See* Pentadecanoate

2,3-Pentadienoyl-*S*-pantetheine 11-Pivalate

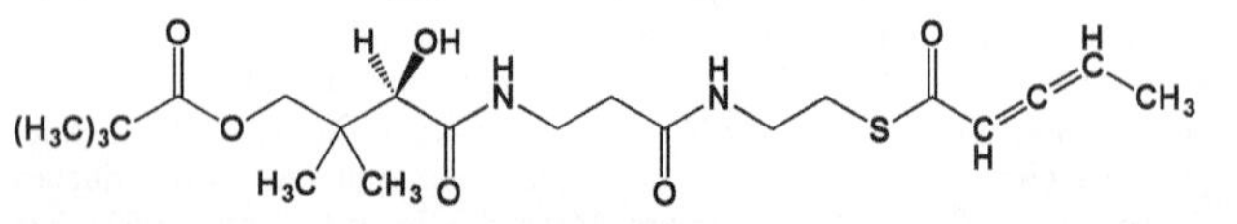

This Michael-type mechanism-based inhibitor (FW = 442.58 g/mol) irreversibly inhibits acetyl-CoA *C*-acetyltransferase (K_i = 1.25 mM; k_{inact} = 0.26 min^{-1}), trapping Cys-378, the active site cysteine residue. **1.** Palmer, Differding, Gamboni, *et al.* (1991) *J. Biol. Chem.* **266**, 8369.

Pentaeicosanoate, *See* Pentacosanoate

2-Pentafluoropropionamidobenzenesulfonyl Fluoride

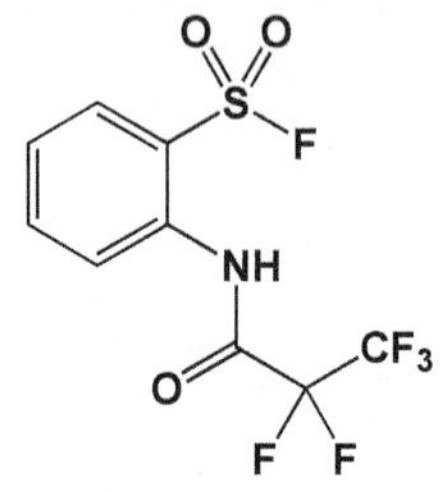

This aryl sulfonyl fluoride (FW = 321.20 g/mol) is an irreversible inhibitor of pancreatic and leukocyte elastases, but a weak inhibitor of cathepsin G and chymotrypsin. A model for the elastase inhibition reaction is proposed which involves interaction of the fluroacyl group of the inhibitor with the primary substrate recognition site S$_1$ of the enzyme. Hydrogen bonding also occurs between the inhibitor NH and a backbone peptide carbonyl group, probably from residue-214. The 2-fluoroacyl group plays the dual role of binding in the hydrophobic S$_1$ pocket and through electronic effects, increasing the strength f the hydrogen bond. **1.** Yoshimura, Barker & Powers (1982) *J. Biol. Chem.* **257**, 5077.

Pentagastrin

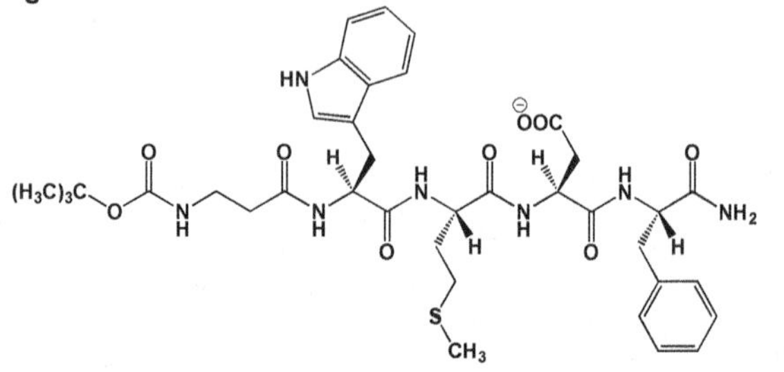

This synthetic peptide (FW$_{free-acid}$ = 767.90 g/ml), also named *N*-[(1,1-dimethylethoxy)carbonyl]-β-alanyl-L-tryptophanyl-L-methionyl-L-aspartyl-L-phenylalaninamide, is an analogue of gastrin I (the so-called 'big gastrin'), except that the C-terminal tetrapeptidamide Trp-Met-Asp-Phe-NH$_2$ of gastrin I has an additional N-terminal *N*-(*t*-butyloxycarbonyl)-β-alanyl residue. Pentagastrin is a cholecystokinin B receptor agonist and stimulates gastric acid secretion. **Target(s):** Na$^+$/K$^+$-exchanging ATPase (1). **1.** Mozsik, Kutas, Nagy & Tarnok (1979) *Acta Med. Acad. Sci. Hung.* **36**, 459.

Pentamethylenediamine, *See* Cadaverine

1,1'-Pentamethylenepyridinium Iodide, 3-Hydroxypyridinium Iodide Methanesulfonate

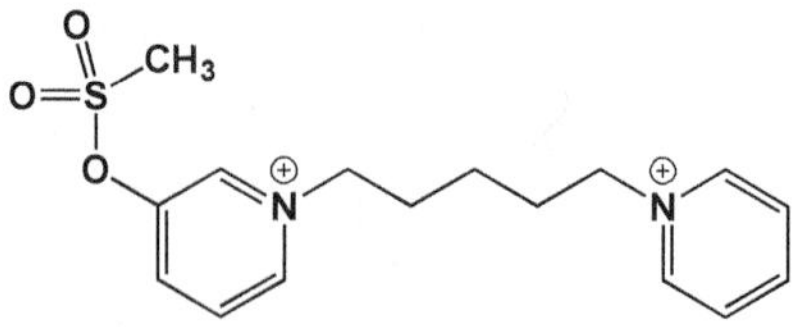

This pyridinium salt (FW = 576.24 g/mol) irreversibly inhibits acetylcholinesterase with a K_i value of 20 μM and k_{inact} = 5.5 x 10^{-4} sec^{-1}. *Note*: The authors also developed a graphical method (now called the Kitz-Wilson Plot) for analyzing irreversible enzyme inhibition kinetics (2). **1.** Kitz & Wilson (1962) *J. Biol. Chem.* **237**, 3245. **2.** Purich & Allison (1995) *Handbook of Biochemical Kinetics*, Academic Press, New York.

Pentamethylenetetrazole, *See* Pentylenetetrazole

Pentamidine & Pentamidine Methylene Analogues

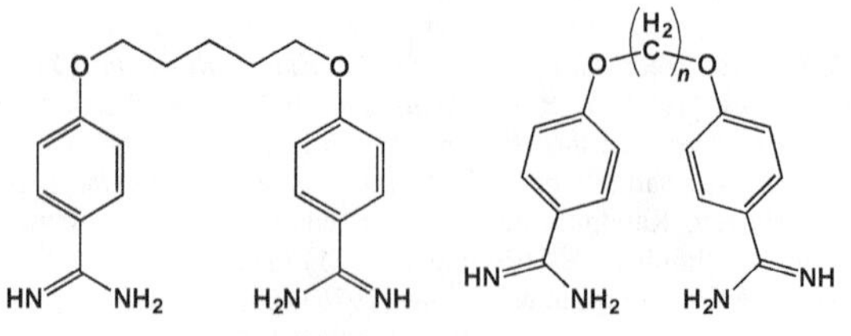

This aromatic diamidine-class antiprotozoal (FW = 340.42 g/mol; CAS 100-33-4; $FW_{isethionate}$ = 592.68 g/mol; CAS 140-64-7 (isethionate salt)), known systematically as 4,4'-[pentane-1,5-diylbis(oxy)]dibenzenecarbox imidamide, targets an as-yet unspecified vital process in parasite mitochondria. Although readily metabolized by human P450 systems, pentamidine is not processed likewise in trypanosomes, suggesting an explanation for its differential action (1). Pentamidine typically requires intracellular concentrations of 1 mM, or greater. Its uptake by *T. brucei* in the bloodstream involves the *T. brucei* P_2 aminopurine permease (*or* $TbAT_1$), a transporter that likewise mediates the uptake of melarsoprol. Because this parasite cannot synthesize adenine, purine uptake is a druggable target (1). (*See also Suramin, Melarsoprol, and Eflornithine*) Pentamidine and related homologues also bind the CTG*CAG repeat in DNA to inhibit transcription, and analysis of methylene linker analogues, *n*,*n*'-[$(CH_2)_n$-1,5-diylbis(oxy)]dibenzene-carboximidamide, where the integer *n* varies from 3 to 9, reveals that heptamidine (*n* = 7) is exceptionally effective at reversing splicing defects and rescues myotonic dystonia in a mouse model (2). *Note*: Amidines are reasonably good bases, mainly because the protonation sp^2-hybridized nitrogen can be stabilized by delocalization of the resulting positive charge on both nitrogen atoms. **1.** Wilkinson & Kelly (2009) *Expert Rev. Molec. Med.* **1**, e31. **2.** Coonrod, Nakamori, Wang, *et al.* (2013) *ACS Chem Biol.* **8**, 2528.

Pentamidine Isethionate, *See Pentamidine*

Pentanal, *See n-Valeraldehyde*

1,5-Pentanediamine, *See Cadaverine*

Pentanedioate, *See Glutarate*

1,5-Pentanediol 1,5-Bisphosphate

This fructose 1,6-bisphosphate analogue (FW = 260.08 g/mol) inhibits fructose-1,6-bisphosphatase, $K_{0.5}$ = 5 mM (1), fructose-1,6-bisphosphate aldolase, K_i = 29 μM (2), and pyrophosphate-dependent phosphofructokinase, *or* diphosphate:fructose-6-phosphate 1-phosphotransferase (3). **1.** Marcus (1976) *J. Biol. Chem.* **251**, 2963. **2.** Hartman & Barker (1965) *Biochemistry* **4**, 1068. **3.** Bertagnolli, Younathan, Voll, Pittman & Cook (1986) *Biochemistry* **25**, 4674.

2,4-Pentanedione, *See Acetylacetone*

Pentanoate, *See Valerate*

Pentanochlor

This photosynthetic electron transport inhibitor (FW = 239.74 g/mol; CAS 2307-68-8), also known as Solan®, (*RS*)-3'-chloro-2-methylvaler-*p*-toluidine, and *N*-(3-chloro-4-methylphenyl)-2-methylpentanamide, is a selective pre- and post-emergence herbicide used to control annual weeds in fields producing carrots, celery, fennel, and parsley. Pentanochlor is also used in contact sprays with flowers and ornamental plants. **1.** Dodge (1977) *Spec. Publ., Chem. Soc.* **29**, 7.

Pentanoyl-CoA, *See Valeryl-CoA*

4-Pentenoate (4-Pentenoic Acid)

This unsaturated carboxylic acid ($FW_{free-acid}$ = 100.12 g/mol; CAS 591-80-0; M.P. = –22.5°C), also known as 4-pentenoic acid, reversibly inhibits carbamoyl-phosphate synthetase (1) and is also metabolized to form an irreversible inhibitor of 3-ketoacyl-CoA thiolase, *or* acetyl-CoA acyltransferase, and acetoacetyl-CoA thiolase, *or* acetyl-CoA acetyltransferase (2,3), and palmitoyl-CoA hydrolase, *or* acyl-CoA hydrolase (4) **1.** Pausch, Rasenack, Haussinger & Gerok (1985) *Eur. J. Biochem.* **150**, 189. **2.** Schulz & Fong (1981) *Meth. Enzymol.* **72**, 604. **3.** Schulz (1987) *Life Sci.* **40**, 1443. **4.** Dixon, Osterloh & Becker (1990) *J. Pharm. Sci.* **79**, 103.

Pentenylguanidine Sulfate, *See Galegine Sulfate*

Pentetate (Pentetic Acid)

This EDTA-like chelator, antimicrobial, and food stabilizer (FW = 393.35 g/mol; CAS 67-43-6; Symbol: DTPA), systematically named 2-[bis[2-[bis(carboxymethyl)amino]ethyl]amino]acetic acid, inhibits many biochemical processes by binding divalent and trivalent metal ions, especially Fe^{3+}. DTPA inhibits bacterial growth by sequestering Fe(III)), a nutrient metal ion required by aconitase (also known as iron-responsive element-binding protein), cytochromes, other redox enzymes, and ribonucleotide reductase (needed for DNA synthesis). The US Food and Drug Administration has also determined that Zn^{2+}-DTPA and Ca^{2+}-DTPA are safe and effective oral treatments for individuals who inhaled or ingested Plutonium, Americium, or Curium.

Pentifylline

This xanthine-based vasodilator (FW = 264.33 g/mol; CAS 6493-05-6; M.P. = 82-83°C), also named Cosaldon®, 1-hexyl-3,7-dimethylxanthine, and 1-hexyl-theobromine, exhibits many of the biological effects of theobromine, but is more lipid-soluble. **Target(s):** cAMP phosphodiesterase (1,2); and Mg^{2+}-ATPase (3). **1.** Stefanovich (1975) *Arzneimittelforschung* **25**, 740. **2.** Stefanovich, von Polnitz & Reiser (1974) *Arzneimittelforschung* **24**, 1747. **3.** Porsche & Stefanovich (1979) *Arzneimittelforschung* **29**, 1089.

Pentobarbital

This barbiturate (FW = 226.28 g/mol; CAS 76-74-4), known systematically as 5-ethyl-(1-methylbutyl)-2,4,6(1*H*,3*H*,5*H*)-pyrimidinetrione, is an anesthetic and sedative. The sodium salt, also called pentobarbitone sodium and Nembutal, is freely water soluble, but unstable, in water. The (+)-isomer is predominantly excitatory, while the (–)-enantiomer is predominantly inhibitory. The latter is considerably more effective in potentiating inhibitory responses to the transmitter γ-aminobutyrate, *or* GABA (1). **Target(s):** acetylcholine release (2); adenylate cyclase (3,4); alcohol dehydrogenase, $NADP^+$-dependent, *or* aldehyde reductase (5); α-amino-3-hydroxy-5-methyl-4-isoxazolepropionic acid (AMPA) receptor (6,7); Ca^{2+} current, voltage-dependent (8); GABA receptor, both stimulation and inhibition (9); GABA uptake (10); glucose-6-phosphatase (11); D-glucose transport (12); inositol phospholipid hydrolysis (13); K^+ currents (14); monoamine oxidase (15); norepinephrin release (16); phospholipase C (17); and substance P receptor (18). **1.** Huang & Barker (1980) *Science* **207**, 195. **2.** Lindmar, Loffelholz & Weide (1979) *J. Pharmacol. Exp. Ther.* **210**, 166. **3.** Dan'ura, Kurokawa, Yamashita & Ishibashi (1989) *Chem. Pharm. Bull. (Tokyo)* **37**, 3142. **4.** Dan'ura, Kurokawa, Yamashita, Yanagiuchi & Ishibashi (1986) *Biochem. Biophys. Res. Commun.* **140**, 237. **5.** Erwin & Deitrich (1973) *Biochem. Pharmacol.* **22**, 2615. **6.** Jackson,

Joo, Al-Mahrouki, Orser & Macdonald (2003) *Mol. Pharmacol.* **64**, 395. **7.** Yamakura, Sakimura, Mishina & Shimoji (1995) *FEBS Lett.* **374**, 412. **8.** Ikemoto, Mitsuiye & Ishizuka (1986) *Cell. Mol. Neurobiol.* **6**, 293. **9.** Cash & Subbarao (1988) *Biochemistry* **27**, 4580. **10.** Hertz & Sastry (1978) *Can. J. Physiol. Pharmacol.* **56**, 1083. **11.** Anchors & Karnovsky (1975) *J. Biol. Chem.* **250**, 6408. **12.** Naftalin & Arain (1999) *Biochim. Biophys. Acta* **1419**, 78. **13.** Robinson-White, Muldoon & Robinson (1989) *Eur. J. Pharmacol.* **172**, 291. **14.** Bachmann, Mueller, Kopp, *et al.* (2002) *Naunyn Schmiedebergs Arch. Pharmacol.* **365**, 29. **15.** Quevedo & D'Iorio (1970) *Can. J. Biochem.* **48**, 187. **16.** Gothert & Rieckesmann (1978) *Experientia* **34**, 282. **17.** Hattori, Nishimura, Sakai, *et al.* (1987) *Neurol. Res.* **9**, 164. **18.** Okamoto, Minami, Uezono, *et al.* (2003) *Anesth. Analg.* **97**, 104.

D-*glycero*-2,3-Pentodiulose 1,5-Bisphosphate

This bisphospho-diketo sugar (FW$_{free-acid}$ = 308.08 g/mol) is a potent slow, tight-binding inhibitor of ribulose-bisphosphate carboxylase/ oxygenase (1,2). H$_2$O$_2$ and D-*glycero*-2,3-pentodiulose-1,5-bisphosphate are predicted products arising from elimination of H$_2$O$_2$ from a peroxyketone intermediate, a reaction that is specific to Rubisco-catalyzed oxidation of D-ribulose-1,5-bisphosphate during synthesis and/or storage. **1.** Kane, Wilkin, Portis, Jr., & Andrews (1998) *Plant Physiol.* **117**, 1059. **2.** Kim & Portis (2004) *FEBS Lett.* **571**, 124.

Pentopril (and Pentoprilat)

Pentopril   **Pentoprilat**

This indole derivative (FW$_{pentopril}$ = 318.37 g/mol; CAS 82924-03-6), also known as CGS 13945, is an angiotensin I-converting enzyme inhibitor. The pharmacologically active metabolite is the diacid, known as pentoprilat and CGS 13934. **Target(s):** peptidyl-dipeptidase A, *or* angiotensin I-converting enzyme (1,2); and Xaa-Trp aminopeptidase, *or* aminopeptidase W (3). **1.** Jacot des Combes, Turini, Brunner, *et al.* (1983) *J. Cardiovasc. Pharmacol.* **5**, 511. **2.** Rakhit, Hurley, Tipnis, *et al.* (1986) *J. Clin. Pharmacol.* **26**, 156. **3.** Tieku & Hooper (1982) *Biochem. Pharmacol.* **44**, 1725.

Pentosan Polysulfate

This heparin-like polysulfated polysaccharide anticoagulant (MW = indefinite: CAS 37300-21-3), also called xylan polysulfate, xylan hydrogen sulfate, and Elmiron™, is a semi-synthetic inhibitor of the cell surface protease, often referred to as guanidinobenzoatase, associated with colonic carcinoma tissues (1). The polymer does not, however, bind to the guanidinobenzoatase isoenzyme of normal colon epithelial cell surfaces. Pentosan polysulfates electrostatically block the action of antithrombin III and heparin cofactor II inhibition (2). Xylan polysulfate also binds to cationic proteins, such as ribonuclease H, inhibiting the latter (3). The sodium salt of this polymer is a slightly hygroscopic white solid that is freely soluble in water. Note that the name is ambiguous, also referring to the sulfated polymer of any pentose. ***See also Xylans* Target(s):** alcohol dehydrogenase (4); calcium oxalate crystal growth (5,6); cAMP-dependent protein kinase (7); cathepsin B1, *or* cathepsin B (8); cathepsin G9; coagulation factor VIIIa (10); guanidinobenzoatase (1); hyaluronidase (11);

protein kinase C (7,12,13); protein-serine/threonine kinases (7); protein-tyrosine kinase (7); and ribonuclease H (3). **1.** Anees (1996) *J. Enzyme Inhib.* **10**, 203. **2.** Carson & Doctor (1990) *Thromb. Res.* **58**, 367. **3.** Moelling, Schulze & Diringer (1989) *J. Virol.* **63**, 5489. **4.** Paulikova, Molnarova & Podhradsky (1998) *Biochem. Mol. Biol. Int.* **46**, 887. **5.** Senthil, Subha, Saravanan & Varalakshmi (1996) *Mol. Cell. Biochem.* **156**, 31. **6.** Martin, Werness, Bergert & Smith (1984) *J. Urol.* **132**, 786. **7.** Srivastava, Sekaly & Chiasson (1993) *Mol. Cell. Biochem.* **120**, 127. **8.** Kruze, Fehr & Boni (1976) *Z. Rheumatol.* **35**, 95. **9.** Steinmeyer & Kalbhen (1991) *Arzneimittelforschung* **41**, 77. **10.** Wagenvoord, Hendrix, Soria & Hemker (1988) *Thromb. Haemost.* **60**, 220. **11.** Meyer, Hoffman & Linker (1960) *The Enzymes*, 2nd ed. (Boyer, Lardy & Myrbäck, eds.), **4**, 447. **12.** Khaled, Rideout, O'Driscoll, *et al.* (1995) *Clin. Cancer Res.* **1**, 113. **13.** Herbert & Maffrand (1991) *Biochim. Biophys. Acta* **1091**, 432.

Pentostatin, *See 2'-Deoxycoformycin*

Pentostatin 5'-Monophosphate, *See 2'-Deoxy-coformycin 5'-Monophosphate*

Pentothal, *See Thiopental*

Pentoxifylline

This methylxanthine (FW = 278.31 g/mol; CAS 6493-05-6; Solubility = 77 mg/mL H$_2$O at 25°C), known also as Trental™ and systematically as 1-(5-oxohexyl)-3,7-dimethylxanthine, is a theobromine derivative that improves blood flow by decreasing blood viscosity. Pentoxifylline also inhibits the synthesis of tumor necrosis factor α. Major metabolites are 1-[5-hydroxy-hexyl]-3,7-dimethylxanthine and 1-[3-carboxypropyl]-3,7-dimethylxan-thine, accumulating to plasma levels that are 5 and 8 times higher, respectively, than pentoxifylline. (***See Torbafylline*) Target(s):** alkaline phosphatase (1); and cAMP phosphodiesterase (2-4). **1.** Glogowski, Danforth & Ciereszko (2002) *J. Androl.* **23**, 783. **2.** Reuter & Wallace (1999) *Amer. J. Physiol.* **277**, G847. **3.** Nandi, Nair & Deo (1980) *Adv. Myocardiol.* **1**, 359. **4.** Stefanovich (1974) *Res. Commun. Chem. Pathol. Pharmacol.* **8**, 673.

9-(*n*-Pentyl)adenine

This alkylated adenine analogue (FW = 205.26 g/mol), known also as 9-amyladenine, is a moderately potent inhibitor of calf muscle adenosine deaminase, K_i = 50 μM (1) Its inhibitory behavior is consistent with Baker's concept of "exo"-inhibition, whereby incorporation of a bulky side-chain favors tight binding of a reversible enzyme inhibitor by means of interactions at sites immediately adjacent to the substrate binding pocket (2). **1.** Cory & Suhadolnik (1965) *Biochemistry* **4**, 1729 and 1733. **2.** Baker (1967) *Design of Active-Site-Directed Irreversible Enzyme Inhibitors*, Wiley, New York.

tert-Pentyl Alcohol, *See 2-Methyl-2-butanol*

1-Pentyl-3-(4'-aminophenyl)pyrrolidine-2,5-dione, *See 3-(4'-Aminophenyl)-1-pentylpyrrolidine-2,5-dione*

Pentylenetetrazole

This stable tetrazole (FW = 138.17 g/mol; CAS 54-95-5; M.P. = 57-60°C; soluble in water and many polar organic solvents), also known as 6,7,8,9-tetrahydro-5*H*-tetrazolo[1,5-*a*)azepine and metrazol and 1,5-

pentamethylenetetrazole, is a non-specific central-nervous-system stimulant and a convulsant. It is used experimentally to induce epileptic activity. **Target(s):** acetylcholinesterase (1); cholesterol biosynthesis (2); dihydropyrimidinase, *or* hydantoinase, *or* dihydroorotase (3,4); and Na⁺/K⁺-exchanging ATPase (5). **1.** Mahon & Brink (1970) *J. Neurochem.* **17**, 949. **2.** Alexander & Alexander (1962) *Biochemistry* **1**, 783. **3.** Bernheim (1952) *The Enzymes*, 1st ed. (Sumner & Myrbäck, eds.), **2** (part 2), 844. **4.** Eadie, Bernheim & Bernheim (1949) *J. Biol. Chem.* **181**, 449. **5.** Dubberke, Vasilets & Schwarz (1998) *Pflugers Arch.* **437**, 79.

S-Pentylglutathione

This *S*-alkylated glutathione derivative (FW = 378.46 g/mol), also known as *S*-amylglutathione, inhibits glutathione *S*-transferase (1) and lactoylglutathione lyase, *or* glyoxalase I (2-4). **1.** Board & Mannervik (1991) *Biochem. J.* **275**, 171. **2.** Vince & Wadd (1969) *Biochem. Biophys. Res. Commun.* **35**, 593. **3.** Allen, Lo & Thornalley (1993) *Biochem. Soc. Trans.* **21**, 535. **4.** Regoli, Saccucci & Principato (1996) *Comp. Biochem. Physiol. C, Pharmacol. Toxicol. Endocrin.* **113**, 313.

(3-Pentynoyl)pantetheine

This mechanism-based agent (FW = 358.46 g/mol) inactivates acetyl-CoA *C*-acetyltransferase (1) and butyryl-CoA dehydrogenase from *Megasphaera elsdenii* (2,3). For the latter, the inactivator becomes covalently attached, with a stoichiometry of 0.61 mol ¹⁴C-inactivator/mol per enzyme-bound flavin, but with no modification of the flavin itself. The covalent adduct is labile toward base and neutral NH₂OH, releasing 85% of incorporated ¹⁴C-label from the protein. Base-catalyzed hydrolysis of the adduct releases 3-oxopentanoic acid (0.6 mol/mol of incorporated inactivator). Moreover, NH₂OH treatment forms an enzyme-bound hydroxamate (0.64 mol/mol of incorporated inactivator). The covalent adduct is reduced with NaBH₄, liberating 1,3-pentanediol. Hydrolysis of the protein with 6 N HCl after NaBH₄ reduction yields 2-amino-5-hydroxyvaleric acid and proline, indicating that the inactivator reactes with the γ-carboxyl group of a glutamyl residue in the active site. Inactivation most likely proceeds through enzyme-catalyzed rearrangement of the acetylene to an allene, followed by nucleophilic addition of the carboxyl group to the allene. Similarly, (3-chloro-3-butenoyl)pantetheine inactivates the enzyme and modifies a glutamyl residue. Butyryl-CoA dehydrogenase also catalyzes the isomerization of (3-butenoyl)pantetheine to (2-butenoyl)pantetheine. The enzyme catalyzes the elimination of HF from 3-fluoropropionyl-CoA and (3,3-difluorobutyryl)pantetheine, again supporting an oxidative mechanism that is initiated by α-proton abstraction. **1.** Holland, Clark & Bloxham (1973) *Biochemistry* **12**, 3309. **2.** Ator & Ortiz de Montellano (1990) *The Enzymes*, 3rd ed. (Sigman & Boyer, eds.), **19**, 213. **3.** Fendrich & Abeles (1982) *Biochemistry* **21**, 6685.

Peonidin

This inhibitor of nitric oxide formation (FW_chloride = 336.73 g/mol; reddish-brown anthocyanidin flavylium cation), also spelled paeonidin, is the aglucon of peonin, a major pigment in peonies. Glycosides of peonidin are also found in red wine and exhibits antioxidant activity. **Target(s):** 3',5'-cyclic-GMP phosphodiesterase, *or* cGMP phosphodiesterase, isozyme-5 (1); DNA topoisomerase II decatenation activity (2); glycogen phosphorylase (3); and nitric oxide production (4). **1.** Dell'Agli, Galli, Vrhovsek, Mattivi & Bosisio (2005) *J. Agric. Food Chem.* **53**, 1960. **2.** Habermeyer, Fritz, Barthelmes, *et al.* (2005) *Chem. Res. Toxicol.* **18**, 1395. **3.** Jakobs, Fridrich, Hofem, Pahlke & Eisenbrand (2006) *Mol. Nutr. Food Res.* **50**, 52. **4.** Wang & Mazza (2002) *J. Agric. Food Chem.* **50**, 850.

Peonidin 3-[6'-Caffeoyl-6"-feruloylsophoroside] 5-Glucoside

This natural acylated anthocyanin (FW_chloride = 1149.46 g/mol), known systematically as 3-*O*-[2-*O*-(6-*O*-*E*-feruloyl-β-D-glucopyranosyl)-6-*O*-*E*-caffeoyl-β-D-glucopyranoside]-5-*O*-β-D-glucopyranoside, from the the red flowers of the morning glory (*Pharbitis nil*) and the storage roots of purple sweet potato (*Ipomoea batatas*) strongly inhibits α-glucosidase (1,2). **1.** Matsui, Ueda, Oki, *et al.* (2001) *J. Agric. Food Chem.* **49**, 1952. **2.** Matsui, Ebuchi, Kobayashi, *et al.* (2002) *J. Agric. Food Chem.* **50**, 7244.

Pepsinostreptin

This statin-containing peptidomimetic (FW = 670.87 g/mol)) Structure: Isobutyryl-L-Val-L-Val-Sta-L-Ala-Sta, where Sta is statine, *or* (3S,4S)-4-amino-3-hydroxy-6-methylheptanoic acid, from a stain of *Streptomyces* forms a high-affinity complex with pepsin, inhibiting its activity. Pepsinostreptin, which is identical to pepstatin A, except for the N-terminal isobutyryl group in place of pepstatin's isovaleryl group, probably inhibits other aspartate proteinases. **1.** Kakinuma & Kanamaru (1976) *J. Takeda Res. Lab.* **35**, 123, 128, and 136.

Pepstatin

This naturally occurring statine-containing peptidomimetic (FW_free-acid = 685.90 g/mol; CAS 26305-03-3), also called pepstatin A and isovaleryl-pepstatin and named systematically as *N*-(isovaleryl)-L-valyl-L-valyl-statyl-L-alanyl-statine (where the statyl residue is a (3S,4S)-4-amino-3-hydroxy-6-methylheptanoyl residue and the C-terminal statine is the corresponding acid), is a presumptive transition-state analogue for prototypical aspartate proteinase pepsin (K_i = 46 pM). Pepstatin will also inhibit other carboxy proteinases. The inhibitory effectiveness of statine-containing peptides has been widely exploited by incorporating the statine residue into peptides that otherwise match sequence preferences of the target enzyme's sub-sites. Because pepstatin has a low solubility in water, it is often dissolved in dimethyl sulfoxide, methanol, or ethanol. Note that many derivatives of pepstatin are available, each displaying a different spectrum of inhibitory effects. Pepstatin B and C are the *N*-(*n*-caproyl)- and *N*-(*iso*-caproyl)-derivatives, respectively. *Note*: Above a critical concentration of 0.1 mM in low ionic-strength and neutral buffers, pepstatin often polymerizes into filaments that be several micrometers in length and have characteristic diameters ranging between 6 and 12 nm. **1.** Umezawa, Aoyagi, Morishima, Matsuzaki & Hamada (1970) *J. Antibiot.* **23**, 259. **2.** Umezawa, Aoyagi, Morishima, *et al.* (1970) *J. Antibiot.* **23**, 259. **3.** Seglen (1983) *Meth. Enzymol.* **96**, 737. 2. Knight (1995) *Meth. Enzymol.* **248**, 85. **4.** Kageyama & Takahashi (1976) *J. Biochem.* **79**, 455. **5.** Kageyama & Takahashi (1980) *J. Biochem.* **88**, 635. **6.** Morrison (1982) *Trends Biochem. Sci.* **7**, 102. **7.** Kageyama, Moriyama & Takahashi (1983) *J. Biochem.* **94**, 1557. **8.** Boger, Lohr, Ulm, *et al.* (1983) *Nature* **303**, 81. **9.** Kay, Valler & Dunn (1983) in *Proteinase Inhibitors: Medical and Biological Aspects* (Katunema, Holzer & Umezawa, eds.), pp. 201-210, Springer-Verlag, Berlin. **10.** Valler, Kay, Aoyagi & Dunn (1985) *J. Enzyme Inhib.* **1**, 77. 4. Umezawa (1976) *Meth. Enzymol.* **45**, 678. 5. Seglen (1983) *Meth. Enzymol.* **96**, 737. **11.** Rich, Bernatowicz, Agarwal, *et al.* (1985) *Biochemistry* **24**, 3165. **12.** Tanji, Kageyama & Takahashi (1988) *Eur. J. Biochem.* **177**, 251. **13.** Sarubbi, Seneci, Angelastro, *et al.* (1993) *FEBS Lett.* **319**, 253.

Peptide 6

This twenty-residue peptide (*Sequence*: DAAREGFLDTLVVLHRAGAR; MW = 2167.46 g/mol), derived from the third ankyrin-like repeat (residues 84-103) of human p16CDKN2/INK4a (p16) tumor suppressor protein, binds to cyclin-dependent kinases cdk4 and cdk6 and inhibits cdk4-cyclin D1 kinase activity *in vitro* and also blocks cell cycle progression through G1 (1). [Ala92]-Peptide 6 (*Sequence*: DAAREGFLATLVVLHRAGAR; MW = 2123.45 g/mol) binds to these targets with hig82.74her affinity (1). 1. Fahraeus, Paramio, Ball, Lain & Lane (1996) *Curr. Biol.* **6**, 84.

Peptide A

This 21-residue peptide (*Sequence*: VAPSDSIQAEEWYFGKITRRE; MW = 2482.74 g/mol), corresponding to a part of the noncatalytic domain of p60v-src, inhibit the protein-tyrosine kinase activity of p60v-src. It is a noncompetitive inhibitor of Src/EGF (epidermal growth factor) receptor. Peptide A is not a substrate for tyrosyl residue phosphorylation. 1. Sato, Miki, Tachibana, *et al.* (1990) *Biochem. Biophys. Res. Commun.* **171**, 1152.

Peramivir

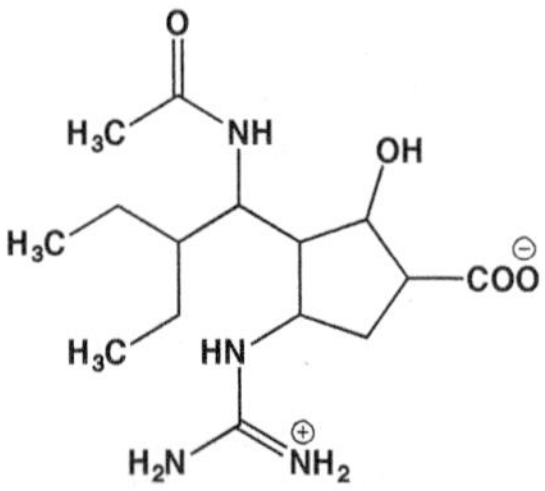

This substituted cyclopentane (FW = 345.42 g/mol; CAS 330600-85-6), also known as BCX 1812 and RWJ 270201, inhibits influenza A and B neuraminidase (sialidase), IC$_{50}$ values of < 1 and 10 nM, respectively (1,2). 1. Chand, Kotian, Dehghani, *et al.* (2001) *J. Med. Chem.* **44**, 4379. **2.** Sidwell & Smee (2002) *Expert Opin. Investig. Drugs* **11**, 859.

Perampanel

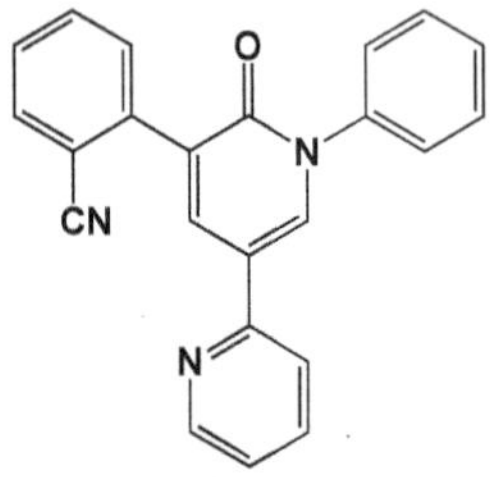

This highly potent, first-in-class antiepileptic drug (FW = 349.39 g/mol; CAS 380917-97-5), also known by its trade name Fycampa® and its systematic name 2-(2-oxo-1-phenyl-5-pyridin-2-yl-1,2-dihydropyridin-3-yl)benzonitrile, is a selective noncompetitive antagonist (IC$_{50}$ = 180 nM) of α-amino-3-hydroxy-5-methyl-4-isoxazolepropionic acid (*or* AMPA) receptors, the major subtype of ionotropic glutamate receptors. (*Note*: AMPA receptors are designated GluR1 (*or* GRIA1), GluR2 (*or* GRIA2),

GluR3 (*or* GRIA3), and GluR4 (*or* GRIA4), the latter also called GluRA-D2.) Perampanel increases intracellular Ca^{2+} and selectively blocks AMPA receptor-mediated synaptic transmission, thereby reducing neuronal excitation. In hippocampal slices, perampanel reduces AMPA receptor-mediated excitatory postsynaptic field potentials (f-EPSPs) with an IC$_{50}$ of 0.23 μM, with full block at 3 μM. This behavior compares with an IC$_{50}$ of 7.8 μM for GYKI52466. Perampanel (10 μM) was without effect on NMDA or kainate receptors (1). As an AMPA-receptor antagonist, perampanel may possess antiepileptogenic properties in addition to its demonstrated antiseizure properties (2,3). **1.** Ceolin, Bortolotto, Bannister, *et al.* (2012) *Neurochem Int.* **61**, 517. **2.** Shih, Tatum & Rudzinski (2013) *Ther. Clin. Risk Manag.* **9**, 285. **3.** Krauss (2013) *Epilepsy Curr.* **13**, 269.

Percaine, *See* Dibucaine Hydrochloride

Perchloric Acid

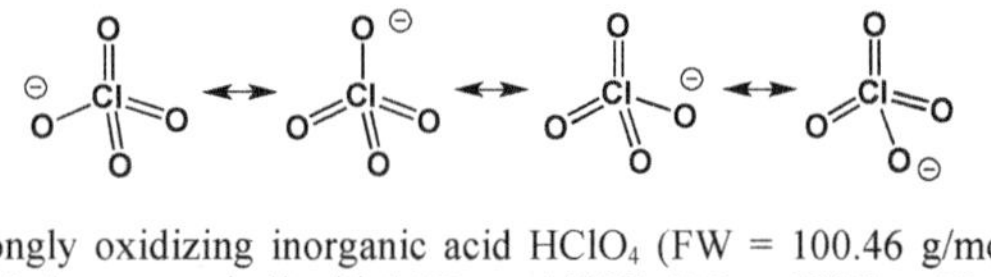

This strongly oxidizing inorganic acid HClO$_4$ (FW = 100.46 g/mol; CAS 7601-90-3; hygroscopic liquid; M.P. = –112°C; B.P. = 39°C at 56 mm Hg) readily dissociates. (This anion has tetrahedral geometry with a Cl–O bond distance of 1.42 Å. Shown above are the chlorate monanion's four resonance species, all obeying the Octet Rule.) **Caution:** This agent often explodes upon distilled at atmospheric pressure (typically at about 90°C) and will also deflagrate upon contact with oxidizable substances. Commercial perchloric acid is often supplied as a 60-72% solution in water (note that water and perchloric acid form a binary azeotrope: 72.4% perchloric acid which boils at 203°C). The pK_a value is 1.77 at 20°C. The perchlorate anion (ClO$_4^-$), also known as chlorate(VII), inhibits many enzymes. Perchlorate anion is a strong chaotropic reagent that is capable of acting as a protein denaturant (it is a destabilizing anion in the Hofmeister series). **Target(s):** acetoacetate decarboxylase (1,36); acylase (2); acylphosphatase (37); alcohol dehydrogenase (3,4); alkaline phosphatase (5); β-amylase (2); ATPase, bicarbonate-stimulated (6,7); carbonic anhydrase (8-14); catalase (15); chymotrypsin (2); cytidylate kinase (16); dihydroorotate dehydrogenase, at elevated concentrations (17); enolase, *or* phosphopyruvate hydratase (18,19); estradiol-17β dehydrogenase (2); fumarylacetoacetase (20); 3-β-galactosidase2; glutaminase (41); glycogen phosphorylase *b* (21); guanine deaminase (40); iodide uptake (22,23); 3-ketoacid CoA-transferase, *or* 3-oxoacid CoA-transferase (24,39); lactate dehydrogenase (25-27); lipase (2); luciferase, firefly (28); NADH:cytochrome *c* reductase, *or* NADH dehydrogenase (29); NAD(P)$^+$ nucleosidase (42); Na$^+$/K$^+$-exchanging ATPase (30); peptide α-*N*-acetyltransferase (48); phosphoglucomutase (31,32); succinate dehydrogenase (33); succinate:3-hydroxy-3-methylglutarate CoA-transferase (38); sulfate adenylyltransferase (43-47); superoxide dismutase (49); tRNA methyltransferases (34); and vacuolar ATPase (35). **1.** Westheimer (1969) *Meth. Enzymol.* **14**, 231. **2.** Warren & Cheatum (1966) *Biochemistry* **5**, 1702. **3.** Sund & Theorell (1963) *The Enzymes*, 2nd ed. (Boyer, Lardy & Myrbäck, eds.), **7**, 25. **4.** Bränden, Jörnvall, Eklund & Furugren (1975) *The Enzymes*, 3rd ed. (Boyer, ed.), **11**, 103. **5.** Sizer (1942) *J. Biol. Chem.* **145**, 405. **6.** Ivashchenko (1978) *Tsitologiia* **20**, 113. **7.** Ivashchenko & Zhubanova (1976) *Vopr. Med. Khim.* **22**, 258. **8.** Lindskog, Henderson, Kannan, *et al.* (1971) *The Enzymes*, 3rd. ed. (Boyer, ed.), **5**, 587. **9.** Verpoorte, Mehta & Edsall (1967) *J. Biol. Chem.* **242**, 4221. **10.** Lindskog (1966) *Biochemistry* **5**, 2641. **11.** Kiese and Hastings (1940) *J. Biol. Chem.* **132**, 281. **12.** Innocenti, Vullo, Scozzafava & Supuran (2005) *Bioorg. Med. Chem. Lett.* **15**, 567. **13.** van Goor (1948) *Enzymologia* **13**, 73. **14.** Pocker & Stone (1968) *Biochemistry* **7**, 2936. **15.** Blaschko (1935) *Biochem. J.* **29**, 2303. **16.** Orengo & Maness (1978) *Meth. Enzymol.* **51**, 321. **17.** Forman & Kennedy (1977) *J. Biol. Chem.* **252**, 3379. **18.** Kornblatt, Zheng, Lamande & Lazar (2002) *Biochim. Biophys. Acta* **1597**, 311. **19.** Lin & Kornblatt (2000) *Biochim. Biophys. Acta* **1476**, 279. **20.** Braun & Schmidt, Jr. (1973) *Biochemistry* **12**, 4878. **21.** Sealock & Graves (1967) *Biochemistry* **6**, 201. **22.** Yoshida, Sasaki, A. Mori, *et al.* (1997) *Biochem. Biophys. Res. Commun.* **231**, 731. **23.** Saito, Yamamoto, Takai & Yoshida (1983) *Acta Endocrinol. (Copenhagen)* **104**, 456. **24.** Hersh & Jencks (1969) *Meth. Enzymol.* **13**, 75. **25.** Panteghini, Bonora & Pagani (1991) *Clin. Chem.* **37**, 1300. **26.** Sanders, van der Neut & van Straalen (1990) *Clin. Chem.* **36**, 1964. **27.** Paz, Garcia, Gonzales, *et al.* (1990) *Clin. Chem.* **36**, 355. **28.** Ugarova, Filippova & Berezin (1981) *Biokhimiia* **46**, 851. **29.** Vernon, Mahler & Sarkar (1952) *J. Biol. Chem.*

199, 599. **30**. Ganea, Babes, Lupfert, *et al.* (1999) *Biophys. J.* **77**, 267. **31**. Ray, Jr., & Peck, Jr. (1972) *The Enzymes*, 3rd ed. (Boyer, ed.), **6**, 407. **32**. Ray, Jr., & Roscelli (1966) *J. Biol. Chem.* **241**, 2596. **33**. Hatefi & Stiggall (1976) *The Enzymes*, 3rd ed. (Boyer, ed.), **13**, 175. **34**. Kerr & Borek (1973) *The Enzymes*, 3rd ed. (Boyer, ed.), **9**, 167. **35**. Dschida & Bowman (1995) *J. Biol. Chem.* **270**, 1557. **36**. Kimura, Yasuda, Tanigaki-Nagae, *et al.* (1986) *Agric. Biol. Chem.* **50**, 2509. **37**. Satchell, Spencer & White (1972) *Biochim. Biophys. Acta* **268**, 233. **38**. Francesconi, Donella-Deana, Furlanetto, *et al.* (1989) *Biochim. Biophys. Acta* **999**, 163. **39**. Hersh & Jencks (1967) *J. Biol. Chem.* **242**, 3468. **40**. Kumar, Sitaramayya & Krishnan (1972) *Biochem. J.* **128**, 1079. **41**. Nandakumar, Wakayama, Nagano, *et al.* (1999) *Protein Expr. Purif.* **15**, 155. **42**. Broeker, Schindelmeiser & Pape (1979) *FEMS Microbiol. Lett.* **6**, 245. **43**. Renosto, Patel, Martin, *et al.* (1993) *Arch. Biochem. Biophys.* **307**, 272. **44**. Seubert, Hoang, Renosto & Segel (1983) *Arch. Biochem. Biophys.* **225**, 679. **45**. Burnell & Roy (1978) *Biochim. Biophys. Acta* **527**, 239. **46**. Hanna, MacRae, Medina, Fisher & Segel (2002) *Arch. Biochem. Biophys.* **406**, 275. **47**. Yu, Martin, Jain, Chen & Segel (1989) *Arch. Biochem. Biophys.* **269**, 156. **48**. Yamada & Bradshaw (1991) *Biochemistry* **30**, 1010. **49**. Cass (1985) *Top. Mol. Struct. Biol.* **6**, 121.

Permethrin

This synthetic pyrethroid and insect neurotoxin (FW = 391.21 g/mol; CAS 52645-53-1), also known as 3-phenoxybenzyl (1*RS*)-*cis*,*trans*-3-(2,2-dichloro-vinyl)-2,2-dimethylcyclopropanecarboxylate, is a widely used insecticide, acaricide, and repellent that causes prolonged activation of voltage-gated sodium channel (*See Pyrethrins*). Although permethrin finds wide use in agriculture to rid cotton, wheat, maize, and alfalfa crops of insect infestations, its broad spectrum action also kills many beneficial insects. **Treatment of Scabies:** Scabies is a skin disorder caused by the mite *Sarcoptes scabiei var hominis*. About 300 million cases of scabies occur annually worldwide, aided by several predisposing factors (*e.g.,* overcrowding, immigration, poor hygiene, poor nutritional status, homelessness, dementia, and sexual contact). Direct skin-to-skin contact for 15-20 minutes is needed to transfer the mites from one person to another. The first-line treatment permethrin (5% weight/weight), marketed as Lyclear® and Nix®. The degree of permethrin absorption is assessed by determining conjugated and unconjugated *cis*/*trans* 3-(2,2-dichlorovinyl)-2,2-dimethylcyclopropane-carboxylic acid, a permethrin metabolite that is excreted in the urine. The major pyrethroid-hydrolyzing esterase is located in mammalian liver microsomes. **Pyrethroid Resistance in *Aedes aegypti*:** Mutations in hydrophobic segment-6 of Domain II of para in *Ae. Aegypti* (the most prevalent vector of Dengue and Yellow Fever viruses.) reduce nerve sensitivity to permethrin inhibition (3). Two occurr at codons Iso1011 and Val1016 in Exons 20 and 21, respectively. A transition in the third position of Iso1011 encoded a Met1011 replacement and a transversion in the second position of Val1016 encoded a Gly1016 replacement. **Health Effects of Permethrin-Impregnated Army Battle-Dress Uniforms:** Because the U.S. Environmental Protection Agency classifies permethrin as a potential carcinogen, concerns arose about health risks to soldiers wearing permethrin-impregnated battle dress uniforms. A full account of the current status was prepared by a National Research Council Subcommittee (4). **1**. Miyamoto (1976) *Environ Health Perspect.* **14**, 15. **2**. Taplin & Meinking (1996) *Curr. Probl. Dermatol.* **24**, 255. **3**. Saavedra-Rodriguez, Urdaneta-Marquez, Rajatileka, *et al.* (2007) *Insect Mol Biol.* **16**, 785. **4**. *National Research Council (US) Subcommittee to Review Permethrin Toxicity from Military Uniforms* (1994) National Academies Press, Washington, DC. **PMID**: 25101418

Performate (Performic Acid)

This strong oxidizing agent (FW = 62.03 g/mol; CAS 107-32-4), also called peroxyformic acid, will oxidize cysteinyl and cystinyl residues in proteins to cysteic acid residues. Indeed, judicious use of performic acid enabled

Sanger to separate the two chains of insulin, an essential simplifying step in his successful Nobel Prize-winning determination of the primary structure of insulin A and B chains. Methionine is also converted to methionine sulfone. Degradation of tryptophanyl residues is typically minimized by reacting the protein at low temperature (–10°C) in the presence of a cryoagent, such as methanol. Harsher conditions result in the oxidation of tyrosyl, seryl, and threonyl residues. A 90% solution of performic acid is produced by reacting formic acid with hydrogen peroxide in the presence of H_2SO_4. **Caution:** Concentrated performic acid solutions often explode upon contact with readily oxidized metals, metal oxides, reducing agents, or even during distillation. *Note*: This reagent is likely to be nonselective in its oxidative action, and the fact that only two targets are listed below is probably misleading. **Target(s):** pancreatic ribonuclease, *or* ribonuclease A (1,2); and papain (3). **1**. Scheraga & Rupley (1962) *Adv. Enzymol.* **24**, 161. **2**. Richards & Wyckoff (1971) *The Enzymes*, 3rd ed., **4**, 647. **3**. Smith & Kimmel (1960) *The Enzymes*, 2nd ed. (Boyer, Lardy & Myrbäck, eds.), **4**, 133.

Perhexiline

This calcium-blocking vasodilator (FW = 277.49 g/mol; CAS 6621-47-2; typically supplied as the maleate or HCl salt), also known as Pexid™ and 2-(2,2-dicyclohexylethyl)piperidine, is used to treat angina. It does so by targetting carnitine palmitoyltransferase-1 (CPT-1), an enzyme that controls the access of long chain fatty acids to the mitochondrial site of β-oxidation, showing concentration-dependent inhibition *in vitro*, using rat cardiac mitochondria, $IC_{50} = 77$ μM, and hepatic mitochondria, $IC_{50} = 148$ μM (1). Amiodarone, another drug with anti-anginal properties, likewise inhibits cardiac CPT-1 ($IC_{50} = 228$ μM). The rank order of potency for inhibition was malonyl-CoA > 4-hydroxyphenylglyoxylate = perhexiline > amiodarone = monohydroxy-perhexiline. Inhibition was competitive with respect to palmitoyl-CoA but non-competitive inhibition with respect to carnitine. This shifts myocardial metabolism from fatty acid to glucose utilisation which results in increased ATP production for the same O_2 consumption and consequently increases myocardial efficiency (1). (*See also Meldonium*) **Other Target(s):** CYP2D6 (3); glutathione *S*-transferase (4); Mg^{2+}-ATPase (5); NADH dehydrogenase (6); Na^+/K^+-exchanging ATPase (5); β-oxidation (6,7); oxidative phosphorylation (6,7); and succinate dehydrogenase (6). **1**. Kennedy, Unger & Horowitz (1996) *Biochem. Pharmacol.* **52**, 273. **2**. Kennedy & Horowitz (1998) *Cardiovasc. Drugs Ther.* **12**, 359. **3**. Kerry, Somogyi, Bochner & Mikus (1994) *Brit. J. Clin. Pharmacol.* **38**, 243. **4**. Mariscal, Munoz, Collado, Esteller & Gonzalez (1988) *Biochem. Pharmacol.* **37**, 3461. **5**. Shacoori, Leray, Guenet, *et al.* (1988) *Res. Commun. Chem. Pathol. Pharmacol.* **59**, 161. **6**. Deschamps, DeBeco, Fisch, *et al.* (1994) *Hepatology* **19**, 948. **7**. Fromenty & Pessayre (1995) *Pharmacol. Ther.* **67**, 101.

Perifosine

This novel Akt signal-transduction pathway inhibitor (F.Wt. = 461.66; CAS 157716-52-4; Solubility (25°C): <1 mg/mL DMSO, 8 mg/mL Water), also known as KRX-0401 and 1,1-dimethylpiperidinium-4-yl octadecyl phosphate (zwitterion), inhibits Akt protein kinases, $IC_{50} = 5$ μM. Perifosine strongly inhibits Akt and extracellular signal-regulated kinase (Erk) 1/2, inducing cell cycle arrest in G_1 and G_2, as well as dose-dependent growth inhibition of mouse glial progenitors (2). Perifosine (10 μM) completely inhibits Akt in MM.1S cells (3) and induces cell cycle arrest and apoptosis in human hepatocellular carcinoma cell lines, again by inhibiting Akt phosphorylation (4). *Note*: Perifosine is related structurally to miltefosine. **1**. Vyomesh Patel, *et al.* (2002) Cancer Res. **62**, 1401. **2**. Momota, *et al.* (2005) Cancer Res. **65**, 7429. **3**. Hideshima, *et al.* (2006) Blood **107**, 4053. **4**. Fei, *et al.* (2010) *Cytotechnology* **62**, 449.

Perillyl Alcohol

This limonene metabolite (FW = 152.24 g/mol; CAS 536-59-4), which has been used to treat breast cancer, inhibits post-translational isoprenylation of cysteinyl residues of susceptible proteins. **Target(s):** protein farnesyltransferase (1,2); and protein geranylgeranyltransferase (1-4). **1.** Gelb, Tamanoi, Yokoyama, *et al.* (1995) *Cancer Lett.* **91**, 169. **2.** Hardcastle, Rowlands, Barber, *et al.* (1999) *Biochem. Pharmacol.* **57**, 801. **3**. Unlu, Mason, Schachter & Hughes (2000) *J. Cardiovasc. Pharmacol.* **35**, 341. **4.** Ren, Elson & Gould (1997) *Biochem. Pharmacol.* **54**, 113.

Periodate Ions

These strongly oxidizing inorganic ions (Sodium Hydrogen Orthoperiodate: $Na_2H_3IO_6$; $FW_{anhydrous}$ = 273.91; CAS 15599-97-0; Sodium Metaperiodate: $NaIO_4$; FW = 213.89 g/mol; CAS 7790-28-5) is a hygroscopic solid that is freely soluble in water and has pK_a values of 1.6 and 5.7 at 20°C. H_5IO_6 is the "ortho" form of periodic acid (*i.e.*, *ortho*-periodic acid or *o*-periodic acid). When orthoperiodate is heated at 100°C and 12 mm Hg, the *meta*-periodic acid is produced. Both forms are strong oxidizing agents. *meta*-Periodic acid will bind to and oxidatively cleave the carbon-carbon bond of susceptible organic compounds, namely those containing (a) two hydroxyl groups, each attached to adjacent carbon atoms (*i.e.*, 1,2-glycols), (b) an alcohol and an aldehyde/ketone, or (c) one hydroxyl group and one primary or secondary amino group, each again attached to adjacent carbon atoms (*i.e.*, β-amino alcohols). If neither carbon atom is a terminal carbon, then the resulting product will contain aldehyde or ketone moieties, depending on what other groups are attached to those carbon atoms (*e.g.*, $RCH(OH)CH(OH)R' + HIO_4 \rightarrow RCHO + R'CHO + H_2O + HIO_3$ and $RCH(OH)CH(NH_2)R' + HIO_4 \rightarrow RCHO + R'CHO + NH_3 + HIO_3$). If there are three or more hydroxyl groups on three or more adjacent carbon atoms, the middle carbon atoms are converted to formic acid. Periodic acid oxidation will also work on α-hydroxy aldehydes and ketones, α-diketones, α-keto aldehydes, and glyoxal. If the two neighboring hydroxyl groups are on an aromatic ring, the carbon-carbon bond is not cleave but the reactant is still oxidized. Thus, catechol is oxidized to the corresponding quinone. Periodic acid is a highly effective oxidizing agent in aqueous solution. If the reaction must be carried out in organic solvents, lead tetraacetate is often included. *Note*: Periodic acid will not act on α-keto acids or α-hydroxy acids whereas lead tetraacetate does. The corresponding reactions are actually oxidative decarboxylations. **Preparation of Enzyme Inhibitors:** Periodic acid oxidation has also proved to be a very useful tool in enzymology, simply because a wide variety of biochemicals contain hydroxyl groups on adjacent carbon atoms. For example, periodate-oxidized ATP (also called adenosine 5'-triphosphate 2',3'-dialdehyde) has often been used as an alternative substrate or irreversible inhibitor for a wide variety of ATP-utilizing enzymes. This and similar compounds are now commercially available, even though they are readily synthesized: *e.g.*, periodic acid-oxidized ADP, AMP, adenosine, P^1,P^5-di(adenosine-5')pentaphosphate, P^1,P^4-di(adenosine-5')tetraphosphate, GTP, GDP, GMP, guanosine, CTP, CDP, CMP, *etc.* In the case of nucleosides, commercial sources also can supply the dialcohol form of the nucleoside: *i.e.*, the nucleoside has first been oxidized with periodic acid and then reduced to the dialcohol with borohydride. **Target(s):** *N*-acetylglucosamine kinase (1); acid phosphatase (2); *N*-acylmannosamine kinase (1); alkaline phosphatase (3); carbonic anhydrase, *or* carbonate dehydratase (4-6); chymotrypsin, partial inhibition (7); dextranase (8); β-fructofuranosidase (9); γ-glutamyl transpeptidase (10); lysozyme (11); maleate isomerase (12,13); pancreatic ribonuclease (14,15); 6-phosphofructo-2-kinase (16); and thymidylate synthase (17,18). **1.** Hinderlich, Nöhring, Weise, *et al.* (1998) *Eur. J. Biochem.* **252**, 133. **2.** Andrews & Pallavicini (1973) *Biochim. Biophys. Acta* **321**, 197. **3.** Ohlsson & Wilson (1974) *Biochim. Biophys. Acta* **350**, 48. **4.** Kiese and Hastings (1940) *J. Biol. Chem.* **132**, 281. **5**. van Goor (1948) *Enzymologia* **13**, 73. **6.** Innocenti, Vullo, Scozzafava & Supuran (2005) *Bioorg. Med. Chem. Lett.* **15**, 567. **7**. Jansen, Curl & Balls (1951) *J. Biol. Chem.* **189**, 671. **8.** Koenig & Day (1989) *Eur. J. Biochem.* **183**, 161. **9.** Myrbäck (1957) *Festschr. Arthur Stoll*, p. 551.

10. Szewczuk & Baranowski (1963) *Biochem. Zeits.* **338**, 317. **11.** Imoto, Johnson, North, Phillips & Rupley (1972) *The Enzymes*, 3rd ed., **7**, 665. **12.** Seltzer (1972) *The Enzymes*, 3rd ed., **6**, 381. **13**. Takamura, Takamura, Soejima & Uemura (1969) *Agric. Biol. Chem.* **33**, 718. **14**. McDonald (1955) *Meth. Enzymol.* **2**, 427. **15**. Richards & Wyckoff (1971) *The Enzymes*, 3rd ed. (Boyer, ed.), **4**, 647. **16.** Kountz, McCain, el-Maghrabi & Pilkis (1986) *Arch. Biochem. Biophys.* **251**, 104. **17**. Ivery, Daron & Aull (1984) *J. Inorg. Biochem.* **22**, 259. **16.** Aull, Ivery & Daron (1984) *J. Inorg. Biochem.* **22**, 119.

Periodate-Oxidized Nucleosides and Nucleotides, *See specific compound; e.g., Guanosine 5'-Monophosphate 2',3'-Dialdehyde*

PERK Inhibitor I

This first-in-class, orally bioavailable pyrrolopyrimidamine and small-molecule protein kinase inhibitor (FW = 449.42 g/mol; CAS 1337531-89-1; Soluble in DMSO; Cell Permeable), also known by its code name GSK-2606414 (or GSK2606414), its systematic name 7-methyl-5-(1-((3-trifluoromethyl)phenyl)acetyl)-2,3-dihydro-1*H*-indol-5-yl)-7*H*-pyrrolo[2,3-*d*]pyrimidin-4-amine, and by many aliases (1), targets the protein kinase R (PKR)-like endoplasmic reticulum kinase, *or* PERK, IC_{50} = 0.4 nM (2), a key transducer of the unfolded protein response, *or* UPR. **Primary Mode of Inhibition:** Accumulation of unfolded proteins within the ER elicits release of ER chaperones from the stress-sensing domain of PERK. Once activated by oligomerization and autophosphorylation, PERK phosphorylates Ser-51 of eukaryotic initiation factor 2α (eIF2α), thereby inhibiting overall protein synthesis and providing time for the ER to clear itself of accumulated unfolded proteins. PERK Inhibitor I blocks ER stress-induced PERK autophosphorylation (nearly 100% inhibition at ≤30 nM after 30 min) after thapsigargin treatment (*See Thapsigargin*) of A549 cultures *in vitro* and effectively retards PxBC-3 tumor growth in mice *in vivo* (2). **Other Target(s):** c-Kit (IC_{50} = 150 nM), Aurora B (IC_{50} = 410 nM), BRK (IC_{50} = 410 nM), HRI/EIF2AK1 (IC_{50} = 420 nM), MLK2/MAP3K10 (IC_{50} = 450 nM), c-MER (IC_{50} = 470 nM), DDR2 (IC_{50} = 520 nM), PKR/EIF2AK2 (IC_{50} = 700 nM), and MLCK2/MYLK2 (IC_{50} = 700 nM), with little activity against more than 280 other kinases (IC_{50} >1 μM). **1.** *Aliases:* 1-(5-(4-Amino-7-methyl-7*H*-pyrrolo[2,3-*d*]pyrimidin-5-yl)indolin-1-yl)-2-(3-(trifluoromethyl)phenyl)ethanone; DDR2 Inhibitor I, EIF2AK1 Inhibitor I, EIF2AK2 Inhibitor II, HRI Inhibitor I, MLCK Inhibitor VI, MLK2/MAP3K10 Inhibitor I, PKR-like ER Kinase Inhibitor I, Aurora Kinase Inhibitor IX, BRK Inhibitor I, c-Kit Inhibitor III, Mer RTK Inhibitor II, EIF2AK3 Inhibitor I, PKR Inhibitor II, MERTK Inhibitor II, RP38 Inhibitor II, TAM Family RTK Inhibitor II, Double-stranded RNA-activated Protein Kinase Inhibitor II, Double-stranded RNA-dependent Protein Kinase Inhibitor II. **2.** Axten, Medina, Feng, *et al.* (2012) *J. Med. Chem.* **55**, 7193.

Permanganate

This extremely reactive oxidizing Mn(VII) oxide ion inhibits many enzymes, often nonspecifically, by modifying susceptible functional groups (*e.g.*, cysteine (–SH), methionine (–CH_2–S–CH_3), tryptophan (indole), and tyrosine (phenol). Potassium permanganate (FW = 158.03 g/mol; CAS 7722-64-7) is a purple solid that is soluble in water (14.2 parts cold water) and will decompose in ethanol. It readily reacts with oxidizable compounds: *e.g.*, alcohols, sulfites, organohalides, arsenites, *etc.* **Target(s):** acid phosphatase (1); adenosine deaminase, weakly inhibited (35); alkaline phosphatase (1-4); α-amylase (5); β-amylase (6); arginine decarboxylase (7); aspartate carbamoyltransferase (8,9); bromelain (10); Ca^{2+}-dependent ATPase (11); calotropain, *or Calotropis gigantea* latex protease (12); carbonic anhydrase, *or* carbonate dehydratase (13-16); cathepsin D (41); cellulase (60); CF_1 ATP synthase, *or* chloroplast H^+-transporting two-sector ATPase (33); chymotrypsin (17); dextranase (56); dextrin dextranase (63,64); endo-1,3(4)-β-glucanase (59); endo-1,4-β-xylanase (57,58); filamentous actin (18); β-fructofuranosidase (19); glucan endo-1,3-β-D-glucosidase (52-54); glucan endo-1,6-β-glucosidase (45); glucan 1,3-β-glucosidase (49); glucan 1,4-α-glucosidase, *or* glucoamylase (61); glucan 1,6-α-glucosidase (46,47); glucan 1,6-α-isomaltosidase (43,44); β-glucosidase (55); glutamate decarboxylase (7,20); β-glycerophosphatase (21); histidine decarboxylase (7,34); hydrogenlyase (22); isopullulanase (50); levanase (23,48); levan fructotransferase, DFA-IV forming (31); lipoxygenase (24); lysine decarboxylase (7); mucorpepsin (39); nucleotide diphosphokinase (62); ornithine decarboxylase (7); oryzin (42); pectin lyase (32); penicillopepsin (40); pullulanase (51); riboflavin kinase (25,26); scytalisopepsin A, weakly inhibited (38); serralysin (36,37); succinyl-CoA synthetase, ADP-forming (27); thrombin (28); thymidylate synthase (29); tyrosine decarboxylase (7); and urocanase, *or* urocanate hydratase (30). **1.** Sizer (1942) *J. Biol. Chem.* **145**, 405. **2.** Ohlsson & Wilson (1974) *Biochim. Biophys. Acta* **350**, 48. **3.** Al-Saleh (2002) *J. Nat. Toxins* **11**, 357. **4.** Thomas & Kirsch (1980) *Biochemistry* **19**, 5328. **5.** Di Carlo & Redfern (1947) *Arch. Biochem.* **15**, 343. **6.** Ghosh (1958) *Proc. Intern. Symp. Enzyme Chem., Tokyo Kyoto 1957*, p. 532, Maruzen, Tokyo. **7.** Schales (1951) *The Enzymes*, 1st ed., **2** (part 1), 216. **8.** Sigman & Mooser (1975) *Ann. Rev. Biochem.* **44**, 889. **9.** Jacobson & Stark (1973) *The Enzymes*, 3rd ed. (Boyer, ed.), **9**, 225. **10.** Greenberg & Winnick (1940) *J. Biol. Chem.* **135**, 761. **11.** Ariki & Shamoo (1983) *Biochim. Biophys. Acta* **734**, 83. **12.** Bose & Krishna (1958) *Enzymologia* **19**, 186. **13.** Roughton & Clark (1951) *The Enzymes*, 1st ed., **1** (part 2), 1250. **14.** Davis (1961) *The Enzymes*, 2nd ed., **5**, 545. **15.** Kiese and Hastings (1940) *J. Biol. Chem.* **132**, 281. **16.** van Goor (1948) *Enzymologia* **13**, 73. **17.** Sizer (1945) *J. Biol. Chem.* **160**, 547. **18.** Gicquaud, Gruda & Pollender (1980) *Eur. J. Cell Biol.* **20**, 234. **19.** Sizer (1942) *J. Gen. Physiol.* **25**, 399. **20.** Ambe & Sohonie (1963) *Enzymologia* **26**, 98. **21.** Rao, Cama, Kumar & Vaidyanathan (1960) *J. Biol. Chem.* **235**, 3353. **22.** Crewther (1953) *Australian J. Biol. Sci.* **6**, 205. **23.** Kang, Lee, Lee & Lee (1999) *Biotechnol. Appl. Biochem.* **29**, 263. **24.** Holman & Bergström (1951) *The Enzymes*, 1st ed., **2** (part 1), 559. **25.** Caputto (1962) *The Enzymes*, 2nd ed., **6**, 133. **26.** Giri, Krishnaswamy & Rao (1958) *Biochem. J.* **70**, 66. **27.** Nishimura, Mitchell & Matula (1976) *Biochem. Biophys. Res. Commun.* **69**, 1057. **28.** Seegers (1951) *The Enzymes*, 1st ed., **1** (part 2), 1106. **29.** Aull, Ivery & Daron (1984) *J. Inorg. Biochem.* **22**, 119. **30.** Feinberg & Greenberg (1959) *J. Biol. Chem.* **234**, 2670. **31.** Cha, Park, Yang & Lee (2001) *J. Biotechnol.* **91**, 49. **32.** Afifi, Fawzi & Foaad (2002) *Ann. Microbiol.* **52**, 287. **33.** Nelson (1976) *Biochim. Biophys. Acta* **456**, 314. **34.** Gale (1941) *Biochem. J.* **35**, 66. **35.** Brady (1942) *Biochem. J.* **36**, 478. **36.** Morihara (1963) *Biochim. Biophys. Acta* **73**, 113. **37.** Lyerly & Kreger (1979) *Infect. Immun.* **24**, 411. **38.** Oda & Murao (1974) *Agric. Biol. Chem.* **38**, 2435. **39.** Arima, Yu, Iwasaki & Tamura (1968) *Appl. Microbiol.* **16**, 1727. **40.** Hashimoto, Iwaasa & Yokotsuka (1973) *Appl. Microbiol.* **25**, 578. **41.** Fukushima, Gnoh & Shinano (1971) *Agric. Biol. Chem.* **35**, 1495. **42.** Kundu & Manna (1975) *Appl. Microbiol.* **30**, 507. **43.** Okada, Takayanagi, Miyahara & Sawai (1988) *Agric. Biol. Chem.* **52**, 829. **44.** Okada, Takayanagi & Sawai (1988) *Agric. Biol. Chem.* **52**, 495. **45.** Pitson, Seviour, McDougall, Stone & Sadek (1996) *Biochem. J.* **316**, 841. **46.** Okada & Unno (1989) *Agric. Biol. Chem.* **53**, 223. **47.** Okada, Unno & Sawai (1988) *Agric. Biol. Chem.* **52**, 2169. **48.** Chaudhary, Gupta, Gupta & Banerjee (1996) *J. Biotechnol.* **46**, 55. **49.** Pitson, Seviour, McDougall, Woodward & Stone (1995) *Biochem. J.* **308**, 733. **50.** Okada, Takayanagi, Miyahara & Sawai (1988) *Agric. Biol. Chem.* **52**, 829. **51.** Odibo & Obi (1988) *J. Ind. Microbiol.* **3**, 343. **52.** Nagasaki, Nishioka, Mori & Yamamoto (1976) *Agric. Biol. Chem.* **40**, 1059. **53.** Nagasaki, Fukuyama, Yamamoto & Kobayashi (1974) *Agric. Biol. Chem.* **38**, 349. **54.** Galan, Mendoza, Calonje & Novaes-Ledieu (1999) *Curr. Microbiol.* **38**, 190. **55.** Pitson, Seviour & McDougall (1997) *Enzyme Microb. Technol.* **21**, 182. **56.** Das & Dutta (1996) *Int. J. Biochem. Cell Biol.* **28**, 107. **57.** Nihira, Kansarn, Kono & Okada (2001) *J. Appl. Glycosci.* **48**, 45. **58.** Yoshioka, Nagato, Chavanich, Nilubol & Hayashida (1981) *Agric. Biol. Chem.* **45**, 2425. **59.** Sharma & Nakas (1987) *Enzyme Microb. Technol.* **9**, 89. **60.** Okada (1975) *J. Biochem.* **77**, 33. **61.** Yasuda, Kuwae & Matsushita (1989) *Agric. Biol. Chem.* **53**, 247. **62.** Nishino & Murao (1975) *Agric. Biol. Chem.* **39**, 1827. **63.** Naessens, Cerdobbel, Soetaert & Vandamme (2005) *J.* *Ind. Microbiol. Biotechnol.* **32**, 323. **64.** Suzuki, Unno & Okada (1999) *J. Appl. Glycosci.* **46**, 469.

Permethrin

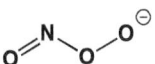

This synthetic insecticide and acaricide (FW = 391.29 g/mol; CAS 52645-53-1; M.P. ≈ 35°C) is insoluble in water but readily soluble in many organic solvents. *Note*: There are four possible stereoisomers, and commercial sources typically supply their own proprietary mixtures of these isomers. As a consequence, the reproducibility of inhibitory effects may depend on its source. **Target(s):** acetylcholinesterase (1,2); adenylate cyclase (3); Ca^{2+}-stimulated ATPase (4); glutamate uptake (5); glutathione *S*-transferase (6-8); monoamine oxidase A (9); NADH dehydrogenase, *or* complex I (10); and tryptophan 2,3-dioxygenase (11). **1.** Rao & Rao (1995) *J. Neurochem.* **65**, 2259. **2.** Bandyopadhyay (1982) *Indian J. Exp. Biol.* **20**, 488. **3.** Sahib, Prasada Rao & Desaiah (1987) *J. Appl. Toxicol.* **7**, 75. **4.** Clark & Matsumura (1987) *Comp. Biochem. Physiol. C* **86**, 135. **5.** Wu, Xia, Shi & Liu (1999) *Wei Sheng Yan Jiu* **28**, 261. **6.** Jirajaroenrat, Pongjaroenkit, Krittanai, Prapanthadara & Ketterman (2001) *Insect Biochem. Mol. Biol.* **31**, 867. **7.** Wongtrakul, Sramala, Prapanthadara & Ketterman (2005) *Insect Biochem. Mol. Biol.* **35**, 197. **8.** Yamamoto, Fujii, Aso, Banno & Koga (2007) *Biosci. Biotechnol. Biochem.* **71**, 553. **9.** Rao & Rao (1993) *Mol. Cell. Biochem.* **124**, 107. **10.** Gassner, Wuthrich, Scholtysik & Solioz (1997) *J. Pharmacol. Exp. Ther.* **281**, 855. **11.** el-Toukhy, Ebied, Hassan & el-Sewedy (1989) *J. Environ. Sci. Health B* **24**, 265.

Peroxynitrite

This highly reactive, naturally occurring inorganic anion (FW_{anion} = 62 g/mol; CAS 19059-14-4) is responsible for lipid peroxidation, protein modification, as well as DNA damage, the latter forming 8-nitroguanine and 8-oxoguanine and also inducing single-strand breaks. **Chemical Properties:** Peroxynitrite is a relatively long-lived oxidant that is produced *in vivo* from the diffusion-controlled ($k = 10^{10}$ $M^{-1}sec^{-1}$) reaction of superoxide anion radical and nitric oxide (itself a radical). This reaction occurs in neutrophils, activated macrophages, and endothelial cells, where mitochondria are a primary site of formation. because peroxynitrite crosses cell membranes, immediately neighboring cells are also targets during its 10-millisecond lifetime. Because NO is constitutively produced *in vivo* and is uncharged and freely diffusible, production of peroxynitrite is more likely governed by local production of superoxide. Under physiological conditions the major reactive form of peroxynitrite (pK_a = 6.8) is peroxynitrous acid (HOONO), a reactive species that itself decomposes on the one-second timescale. **Preparation:** Sodium peroxynitrite (FW = 85.02 g/mol; CAS 14042-01-4; λ_{max} = 302 nm; Formulated as a solution in 0.3 M sodium hydroxide;) is commercially available (Cayman Chemicals, Inc.; Purity ≥ 90%, with balance as nitrite and nitrate), but is expensive. Peroxynitrite is prepared by treating acidified hydrogen peroxide with a solution of sodium nitrite, followed by rapid addition of NaOH. 3-Morpholino-sydnonimine hydrochloride (also called SIN-1·HCl), may also be used to generate both the superoxide anion and nitric oxide, which spontaneously react to produce peroxynitrite. **Detection:** Peroxynitrite concentration is determined by its absorbance at 302 nm (pH 12, ε = 1670 $M^{-1}cm^{-1}$). **Lipid Peroxidation:** Addition of peroxynitrite to liposomes results in the formation of malondialdehyde and conjugated dienes, consuming O_2 in the process. Lipid peroxidation is greater at acidic and neutral pH, with little lipid peroxidation occurring above pH 9.5. **Protein Modification:** Peroxynitrite promotes the nitration of protein hydroxyls, thiols, and tyrosines. It also activates src protein-tyrosine kinase, while inhibiting protein tyrosine phosphatases (*e.g.*, PTP 1B) by reacting rapidly ($k = 2.2 \times 10^8$ $M^{-1}sec^{-1}$) with the active-site cysteine thiolate. Such reactivity is most likely due to the fact that the active-site thiol is ionized at pH 7.4, with peroxynitrite resembling the natural phosphate in terms of molecular radii and charge. Glutathione also undergoes *S*-nitrosation in the presence of this anion. **Metalloporphyrin-mediated Decomposition of Peroxynitrite:** In the presence of Fe(III)-porphyrins, peroxynitrite rapidly isomerizes to nitrate; however, the same cannot be said for Mn(III)-porphyrins. Unlike Fe(III)-porphyrins,

Mn(III)-porphyrins are readily reduced by mitochondrial complexes I and II, yielding Mn(II). Once in this reduced state, Mn-porphyrins promote the two-electron reduction of peroxynitrite to nitrite and Mn(IV) complex, a reaction that can be readily reversed to Mn(III) by endogenous reductants, thereby completing a cycle for peroxynitrite decomposition. Mn- and Fe-porphyrins may also promote the one-electron oxidation of peroxynitrite to NO_2 and Mn/Fe(IV) state, which is then reduced back to the (III) redox state. **Target(s):** aconitase, *or* aconitate hydratase (1-4); actin polymerization (5); arylamine *N*-acetyltransferase (50-52); Ca^{2+}-dependent ATPase (6-8); caspase-3 (9); catalase (10); choline acetyltransferase (11,12); creatine kinase (13-15); cyclooxygenase, *or* prostaglandin-endoperoxide synthase (16,55,56); cytochrome *c* oxidase (17,18); electron transport (19); glutaredoxin (20); glyceraldehyde-3-phosphate dehydrogenase (21,22); glycerophosphocholine cholinephosphodiesterase (48); glycogen phosphorylase (49); K^+ channels, Ca^{2+}-activated (23); NADH dehydrogenase, *or* complex I (24,25); NADH oxidase (26); Na^+/K^+-exchanging ATPase (27,28); neutral sphingomyelinase 1 (29); nitric-oxide synthase (30.31); ornithine decarboxylase (32); prostaglandin-I synthase, *or* prostacyclin synthase (33-36); protein kinase C (37); protein-tyrosine phosphatase (38); ribonucleotide reductase (39); superoxide dismutase (40-43,53,54); tryptophan monooxygenase (44); tubulin polymerization, *or* microtubule self-assembly (45); tyrosine monooxygenase (46); and xanthine oxidase (47) **1.** Grune, Blasig, Sitte, Kuwae (1998) *J. Biol. Chem.* **273**, 10857. **2.** Cheung, Danial, Jong & Schulz (1998) *Arch. Biochem. Biophys.* **350**, 104. **3.** Castro, Rodriguez & Radi (1994) *J. Biol. Chem.* **269**, 29409. **4.** Hausladen & Fridovich (1994) *J. Biol. Chem.* **269**, 29405. **5.** Clements, Siemsen, Swain, *et al.* (2003) *J. Leukoc. Biol.* **73**, 344. **6.** Grover, Samson, Robinson & Kwan (2003) *Amer. J. Physiol. Cell Physiol.* **284**, C294. **7.** Muriel & Sandoval (2000) *J. Appl. Toxicol.* **20**, 435. **8.** Stepien, Zajdel, Wilczok, *et al.* (2000) *Biochim. Biophys. Acta* **1523**, 189. **9.** Mohr, Zech, Lapetina & Brune (1997) *Biochem. Biophys. Res. Commun.* **238**, 387. **10.** Kocis, Kuo, Liu, Guruvadoo & Langat (2002) *Front. Biosci.* **7**, a175. **11.** Guermonprez, Ducrocq & Gaudry-Talarmain (2001) *Mol. Pharmacol.* **60**, 838. **12.** Morot Gaudry-Talarmain, Moulian, Meunier, *et al.* (1997) *Nitric Oxide* **1**, 330. **13.** Mihm & Bauer (2002) *Biochimie* **84**, 1013. **14.** Mihm, Coyle, Schanbacher, Weinstein & Bauer (2001) *Cardiovasc Res.* **49**, 798. **15.** Konorev, Hogg & Kalyanaraman (1998) *FEBS Lett.* **427**, 171. **16.** Boulos, Jiang & Balazy (2000) *J. Pharmacol. Exp. Ther.* **293**, 222. **17.** Barone, Darley-Usmar & Brookes (2003) *J. Biol. Chem.* **278**, 27520. **18.** Sharpe & Cooper (1998) *J. Biol. Chem.* **273**, 30961. **19.** Radi, Rodriguez, Castro & Telleri (1994) *Arch. Biochem. Biophys.* **308**, 89. **20.** Aykac-Toker, Bulgurcuoglu & Kocak-Toker (2001) *Hum. Exp. Toxicol.* **20**, 373. **21.** Buchczyk, Grune, Sies & Klotz (2003) *Biol. Chem.* **384**, 237. **22.** Buchczyk, Briviba, Hartl & Sies (2000) *Biol. Chem.* **381**, 121. **23.** Brzezinska, Gebremedhin, Chilian, Kalyanaraman & Elliott (2000) *Amer. J. Physiol. Heart Circ. Physiol.* **278**, H1883. **24.** Murray, Taylor, Zhang, Ghosh & Capaldi (2003) *J. Biol. Chem.* **278**, 37223. **25.** Riobo, Clementi, Melani, *et al.* (2001) *Biochem J.* **359**, 139. **26.** Martin-Romero, Gutierrez-Martin, Henao & Gutierrez-Merino (2002) *J. Neurochem.* **82**, 604. **27.** Muriel, Castaneda, Ortega & Noel (2003) *J. Appl. Toxicol.* **23**, 275. **28.** Muriel & Sandoval (2000) *Nitric Oxide* **4**, 333. **29.** Josephs, Katan & Rodrigues-Lima (2002) *FEBS Lett.* **531**, 329. **30.** Robinson, Sato, Nelson, *et al.* (2001) *Free Radic. Biol. Med.* **30**, 986. **31.** Pasquet, Zou & Ullrich (1996) *Biochimie* **78**, 785. **32.** Seidel, Ragan & Liu (2001) *Life Sci.* **68**, 1477. **33.** Zou & Ullrich (1996) *FEBS Lett.* **382**, 101. **34.** Zou, Yesilkaya & Ullrich (1999) *Drug Metab. Rev.* **31**, 343. **35.** Zou, Martin & Ullrich (1997) *Biol. Chem.* **378**, 707. **36.** Wu & Liou (2005) *Biochem. Biophys. Res. Commun.* **338**, 45. **37.** Knapp, Kanterewicz, Hayes & Klann (2001) *Biochem. Biophys. Res. Commun.* **286**, 764. **38.** Takakura, Beckman, MacMillan-Crow & Crow (1999) *Arch. Biochem. Biophys.* **369**, 197. **39.** Guittet, Decottignies, Serani, *et al.* (2000) *Biochemistry* **39**, 4640. **40.** MacMillan-Crow, Crow & Thompson (1998) *Biochemistry* **37**, 1613. **41.** Yamakura, Taka, Fujimura & Murayama (1998) *J. Biol. Chem.* **273**, 14085. **42.** MacMillan-Crow & Thompson (1999) *Arch. Biochem. Biophys.* **366**, 82. **43.** Nilakantan, Halligan, Nguyen, *et al.* (2005) *J. Heart Lung Transplant.* **24**, 1591. **44.** Kuhn & Geddes (1999) *J. Biol. Chem.* **274**, 29726. **45.** Landino, Hasan, McGaw, *et al.* (2002) *Arch. Biochem. Biophys.* **398**, 213. **46.** Park, Geddes, Javitch & Kuhn (2003) *J. Biol. Chem.* **278**, 28736. **47.** Lee, Liu & Zweier (2000) *J. Biol. Chem.* **275**, 9369. **48.** Sok (1998) *Neurochem. Res.* **23**, 1061. **49.** Dairou, Pluvinage, Noiran, *et al.* (2007) *J. Mol. Biol.* **372**, 1009. **50.** Dupret, Dairou, Atmane & Rodrigues-Lima (2005) *Meth. Enzymol.* **400**, 215. **51.** Sim, Westwood & Fullam (2007) *Expert. Opin. Drug Metab. Toxicol.* **3**, 169. **52.** Dairou, Atmane, Rodrigues-Lima & Dupret (2004) *J. Biol. Chem.* **279**, 7708. **53.** Quint, Reutzel, Mikulski, McKenna & Silverman (2006) *Free Radic. Biol. Med.*

40, 453. **54.** Castellano, Di Maro, Ruocco, *et al.* (2006) *Biochimie* **88**, 1377. **55.** Deeb, Hao, Gross, *et al.* (2006) *J. Lipid Res.* **47**, 898. **56.** Trostchansky, O'Donnell, Goodwin, *et al.* (2007) *Free Radic. Biol. Med.* **42**, 1049.

Perphenazine

This photosensitive phenothiazine ($FW_{free-base}$ = 403.98 g/mol; CAS 58-39-9) is a D_2 dopamine receptor antagonist and a σ receptor agonist. Perphenazine is insoluble in water and soluble in ethanol (153 mg/mL). **Target(s):** acetylcholinesterase (1); Ca^{2+}-dependent ATPase (2); CYP2D6 (3,4); D_2 dopamine receptor (5); glutamate dehydrogenase (6-8); monoamine oxidase (9); trypanothione reductase, inhibited by cationic free radicals (10,11). **1.** Spinedi, Pacini, Limatola, Luly & Farias (1992) *Biochem. Pharmacol.* **44**, 1511. **2.** Karelson, Tarve & Tiakhepyl'd (1981) *Biull. Eksp. Biol. Med.* **91**, 576. **3.** Shin, Soukhova & Flockhart (1999) *Drug Metab. Dispos.* **27**, 1078. **4.** Hamelin, Bouayad, Drolet, Gravel & Turgeon (1998) *Drug Metab. Dispos.* **26**, 536. **5.** Christensen, Arnt & Svendsen (1985) *Psychopharmacology* Suppl. **2**, 182. **6.** Yoon, Hwang, Lee, *et al.* (2001) *Biochimie* **83**, 907. **7.** Couee & Tipton (1991) *Neurochem. Res.* **16**, 773. **8.** Couee & Tipton (1990) *Biochem. Pharmacol.* **39**, 1167. **9.** Suzuki, Seno & Kumazawa (1988) *Life Sci.* **42**, 2131. **10.** Gutierrez-Correa, Fairlamb & Stoppani (2001) *Rev. Argent. Microbiol.* **33**, 36. **11.** Gutierrez-Correa, Fairlamb & Stoppani (2001) *Free Radic. Res.* **34**, 363.

Perseitol

This seven-carbon sugar polyol (FW = 212,20 g/mol; CAS 527-06-0), also known as D-*glycero*-D-*galacto*-heptitol and α-mannoheptitol, from the avocado (*Persea americana*) and leaves of *Scurrula fusca*. Perseitol inhibits EIImtl, the membrane-bound mannitol-specific enzyme II (EIIMtl) of the phosphoenolpyruvate-dependent phosphotransferase of *Escherichia coli*. **1.** Lolkema, Wartna & Robillard (1993) *Biochemistry* **32**, 5848.

Persulfate

This free-radical initiator (Ammonium Persulfate: $(NH_4)_2S_2O_8$, FW = 228.20, CAS 7727-54-0; Potassium Persulfate: $K_2S_2O_8$, FW = 270.32, CAS 7727-21-1; Sodium Persulfate: $Na_2S_2O_8$, FW = 238.11 g/mol, CAS 7775-27-1; Redox Potential = 2.1 V), perhaps more rationally described as peroxydisulfate, is strongly oxidizing and generates free radicals in the presence of a catalyst. For example, in the presence of tetramethylethylenediamine (TEMED), persulfate forms sulfate radicals ($SO_4^{\cdot-}$) that initiate the free radical polymerization of alkenes, including the famously important polyacrylamide gels formed from acrylamide and bis-acrylamide. Indeed, ammonium persulfate is the reagent of choice for making polyacrylamide gels, but potassium persulfate is recommended when one must work with weakly buffered basic systems (pH ≈ 9). Because some persulfate inevitably persists after polyacrylamide gel formation, pre-electrophoresis is advised, when analyzing persulfate-sensitive enzymes in active-enzyme electrophoresis experiments. Persulfate salts, particularly ammonium persulfate, should be stored in air-tight containers in a cool environment. In aqueous solution, ammonium persulfate slowly decomposes with liberation of O_2 to form NH_4HSO_4. *Note:* Beyond those specifically listed below, many other enzymes, especially those employing thiol groups in catalysis, are likely to be inhibited by persulfate. **Target(s):** acid phosphatase (1); carbonic anhydrase, weakly inhibited (2,3); deoxyribopyrimidine photo-lyase (4); fumarase, *or* fumarate hydratase, inhibited by ammonium persulfate (5); gallate decarboxylase, inhibited by ammonium persulfate (6); glycogen phosphorylase (7); levanase, inhibited by ammonium persulfate (8); lipase, *or* triacylglycerol lipase, inhibited by ammonium persulfate (9); procollagen N-endopeptidase, inhibited by ammonium persulfate (10); and tannase (11). **1.** Andrews & Pallavicini (1973) *Biochim. Biophys. Acta* **321**, 197. **2.** Kiese & Hastings (1940) *J. Biol. Chem.* **132**, 281. **3.** van Goor (1948) *Enzymologia* **13**, 73. **4.** Saito & Werbin (1970) *Biochemistry* **9**, 2610. **5.** Ueda, Yumoto, Tokushige, Fukui & Ohya-Nishiguchi (1991) *J. Biochem.* **109**, 728. **6.** Zeida, Wieser, Yoshida, Sugio & Nagasawa (1998) *Appl. Environ. Microbiol.* **64**, 4743. **7.** Chen & Segel (1968) *Arch. Biochem. Biophys.* **127**, 175. **8.** Chaudhary, Gupta, Gupta & Banerjee (1996) *J. Biotechnol.* **46**, 55. **9.** Sharma, Soni, Vohra, Gupta & Gupta (2002) *Process Biochem.* **37**, 1075. **10.** Hojima, Morgelin, Engel, *et al.* (1994) *J. Biol. Chem.* **269**, 11381. **11.** Kar, Banerjee & Bhattacharyya (2003) *Proc. Biochem.* **38**, 1285.

Pertussis Toxin

This enterotoxin (EC 2.4.2.31; Symbol: PT) produced by the Gram-negative pathogenic aerobe *Bordetella pertussis* is an oligomeric enzyme catalyzing the NAD^+-dependent ADP-ribosylation of α-subunits of the GTP-regulatory protein G_i, other trimeric cytosolic G-proteins, as well as transducin (1,2). Modification prevents the G proteins from binding to their respective receptors, and the associated adenylate cyclase remains inactive. A cysteinyl residue within the α-subunit of G_i, G_o, and G_{gust} undergoes ADPribosylation, but G_s is not a substrate. PT is known to dissociate into two components: the enzymatically active A subunit (S_1) and the cell-binding B subunit. S_1 is thus a major virulence factor of *Bordetella pertussis*, with ADP-ribosylation of host-cell G-proteins a way to inhibit GPCR signaling. The intracellular pathway of PT action includes endocytosis and retrograde transport to the trans-Golgi network (TGN) and endoplasmic reticulum (ER). Subsequent translocation of S_1 to the cytosol is presumably preceded by dissociation from the holotoxin. PT dissociation is stimulated by ATP *in vitro*. **1.** Sprang (1997) *Ann. Rev. Biochem.* **66**, 639. **2.** Wise, Watson-Koken, Rees, Lee & Milligan (1997) *Biochem. J.* **321**, 721.

Pertuzumab

This breast cancer drug (MW = 148 kDa; CAS 380610-27-5), also known by the trade name Perjeta®, is a humanized, first-in-class recombinant IgG1 monoclonal antibody that inhibits dimerization of HER/erbB tyrosine kinases. **Mechanism of Inhibitory Action:** The EGF receptor family consists of four transmembrane receptors: EGFR (HER1/erbB-1), HER2 (erbB-2/neu), HER3 (erbB-3), and HER4 (erbB-4) (1-3). HER1, HER3, and HER4 each contain three major functional domains: an extracellular ligand-binding domain, a hydrophobic transmembrane domain, and a cytoplasmic tyrosine kinase domain. No ligand has been clearly identified for HER2; however, HER2 can be activated as a result of ligand binding to other HER receptors with the formation of receptor homodimers and/or heterodimers. Pertuxumab prevents ERBB2 receptor protein from binding (dimerizing) to other HER family receptors. Recent work also suggests that the HER-2-targeting antibodies Pertuzumab and Trastuzumab synergistically inhibit the survival of breast cancer cells growing *in vitro* (4). Because HER1 and HER2 are overexpressed on many solid tumors (*e.g.*, breast, non-small cell lung, head and neck, and colon), pertuzumab prevents tumor cell growth by blocking the pairing of the most potent signaling heterodimer: HER2/HER3. **Use in Combination with Trastuzumab:** Because pertuzamab and trastuzumab (**See also** *Trastuzumab*) bind to unique antigenic regions on HER2, use of both provides a more complete blockade of HER2-mediated signal transduction than either agent alone. Clinical studies on their combined action suggest their likely utility as a first-line treatment for HER2-positive metastatic breast cancer treatment (5). The initial dose of 840 mg pertuzamab (60-minute I.V. infusion) is followed every 3 weeks thereafter by a 420-mg I.V. dose administered over 30-60 minutes (6). When administered with pertuzamab, the recommended initial dose of trastuzumab is 8 mg/kg administered as a 90-minute I.V. infusion, followed every 3 weeks thereafter by a dose of 6 mg/kg administered as an I.V. infusion over 30 to 90 minutes (6). **1.** Adams, Allison, Flagella, *et al.* (2006) *Cancer Immunol. Immunother.* **55**, 717. **2.** Agus, Gordon, Taylor, *et al.* (2005) *J. Clin. Oncol.* **23**, 2534. **3.** Ng, Lum, Gimenez, Kelsey &

Allison (2006) *Pharm. Res.* **23**, 1275. **4.** Nahta, Hung & Esteva (2004) *Cancer. Res.* **64**, 2343. **5.** Harbeck, Beckmann, Rody, *et al.* (2013) *Breast Care* (Basel). **8**, 49. **6.** Genentech Inc., Reference ID: 3143182.

Peruvoside

This cardiotonic glycoside or cardenoide (FW = 548.67 g/mol; CAS 1182-87-2), named systematically as (3β,5β)-3-(6-deoxy-3-O-methyl-α-L-glucopyranosyl)oxyl-14-hydroxy-19-oxocard-20(22)-enolide, a component of neriperside from *Thevetia neriifolia* (yellow oleander) resembles oubain in its inotropic (force of contraction) and chronotropic (rate of contraction) effects on heart function. Orally administered peruvoside is effective in treating congestive heart failure. Note the structural similarities of the aglycon to digitoxigenin. **Target(s):** Na^+/K^+-exchanging ATPase (1). **1.** Ye & Yang (1990) *Zhongguo Yao Li Xue Bao* **11**, 491.

5-(Perylen-3-yl)ethynyl-arabino-uridine

This bulky aryl-alkyne-substituted arabinonucleoside (FW = 516.55 g/mol), also known as aUY11, is a rigid amphipathic fusion inhibitor that targets virion envelope lipids, blocking fusion of influenza virus, hepatitis C virus (HCV), vesicular stomatitis virus (VSV), and even protein-free liposomes to cells (1). Infection by viruses having lipid-bilayer envelopes proceeds through fusion of the viral membrane with a target cell membrane, a process facilitated by viral fusion proteins that vary greatly in structure, but have a common mechanism of action, *i.e.*, ligand-triggered, large-scale conformational change in the fusion protein, coupled to merger of suitably apposed viral and host bilayers (2). Significantly, aUY11 inhibits formation of negative curvature in model lipid bilayers, such that its effect results from a purely biophysical phenomenon relating to its shape and amphipathicity, and not by direct binding to an enzyme active site or a receptor binding site (1) **1.** Colpitts, Ustinov, Epand, *et al.* (2013) *J. Virol.* **87**, 3640. **2.** Harrison (2008) *Nature Struct. Mol. Biol.* **15**, 690.

Petroselaidate

This unsaturated fatty acid ($FW_{free-acid}$ = 282.47 g/mol; M.P. = 53°C), also known as petroselaidic acid and named systematically as (*E*)-octadec-6-enoic acid, inhibits DNA-directed DNA polymerase and DNA polymerase α non-competitively with respect to the DNA template and substrate (dTTP), while DNA polymerase β inhibited competitively with both DNA and substrate. **1.** Mizushina, Tanaka, Yagi, *et al.* (1996) *Biochim. Biophys. Acta* **1308**, 256.

Petroselenate, *See Petroselinate*

Petroselinate

This unsaturated fatty acid ($FW_{free-acid}$ = 282.47 g/mol; M.P. = 29.5–30.1°C; CAS 4712-34-9; (*Z*)-isomer, CAS 593-39-5; (*E*)-isomer, CAS 593-40-8), also known as petroselinic acid and named systematically (*Z*)-octadec-6-

enoic acid, is found many plants (especially celery, parsley, and caraway seeds). **Target(s):** chondroitin AC lyase (1); chondroitin B lyase (1); DNA polymerase (2); DNA topoisom\erase I (3); DNA topoisomerase II (3); hyaluronate lyase (1); and linoleate isomerase (4). **1**. Suzuki, Terasaki & Uyeda (2002) *J. Enzyme Inhib. Med. Chem.* **17**, 183. **2**. Mizushina, Tanaka, Yagi, *et al.* (1996) *Biochim. Biophys. Acta* **1308**, 256. **3**. Suzuki, Shono, Kai, Uno & Uyeda (2000) *J. Enzyme Inhib.* **15**, 357. **4**. Kepler & Tove (1969) *Meth. Enzymol.* **14**, 105.

Petunidin 3-(β-Glucoside), *similar in action to:* Petunidin

Pevonedistat

This first-in-class small molecule inhibitor (FW = 443.52 g/mol; CAS 905579-51-3; Solubility = 10 mg/mL DMSO), also named MLN4924 and [(1S,2S,4R)-4-[4-[[(1S)-2,3-dihydro-1H-inden-1-yl]amino]-7H-pyrrolo[2,3-d]pyrimidin-7-yl]-2-hydroxycyclopentyl]sulfamate methyl ester, targets Nedd8 activating enzyme, or NAE (IC$_{50}$ = 4.7 nM), with much weaker action against UAE (IC$_{50}$ = 1.5 μM), SAE (IC$_{50}$ = 8.2 μM), and UBA6 (IC$_{50}$ = 1.8 μM). In most cancer cells, pevonedistat treatment results in the induction of DNA re-replication, resulting in DNA damage and cell death (1). MLN4924 also exhibits an alternative mechanism of action. Treatment of activated B cell-like (ABC) diffuse large B-cell lymphoma (DLBCL) cells with pevonedistat resulted in rapid accumulation of pIκBα, decrease in nuclear p65 content, reduction of transcriptional activity of NF-κB (or nuclear factor κ-light-chain-enhancer of activated B cells), and G$_1$ arrest, ultimately resulting in apoptosis induction, events consistent with potent NF-κB pathway inhibition. Treatment of germinal-center B cell-like (GCB) DLBCL cells resulted in an increase in cellular Cdt-1 and accumulation of cells in S-phase, consistent with cells undergoing DNA rereplication (1). Pevonedistat also inhibits Vpx/Vpr-induced SAMHD1 degradation by inhibiting the neddylation of E$_3$ ubiquitin-ligase and blocking SIVmac replication in myeloid cells, therebyindicating the potential efficacy of inhibiting neddylation as an antiretroviral strategy (2). **1**. Milhollen, Traore, Adams-Duffy, *et al.* (2010) *Blood* **116**, 1515. **2**. Wei W, Guo, Liu, *et al.* (2013) *J. Virol.* **88**, 745.

PF 543

This potent SphK1 inhibitor (FW$_{HCl-Salt}$ = 502.07 g/mol; CAS 1706522-79-3; Solubility: 10 mM in H$_2$O, with gentle warming; 100 mM in DMSO), also named (2R)-1-[[(4-[[3-methyl-5-[(phenylsulfonyl)methyl]phenoxy] methyl]phenyl]methyl]-2-pyrrolidinemethanol hydrochloride, selectively targets sphingosine kinase 1 (IC$_{50}$ = 2 nM; K$_i$ = 3.6 nM) (1). The latter phosphorylates sphingosine to form sphingosine-1-phosphate (S1P), a lipid messenger with both intracellular functions (regulating cell proliferation and survival) and extracellular functions (as a ligand for EDG1, or sphingosine-1-phosphate receptor 1). PF-543 also exhibits >100-fold selectivity for Sphk1 over Sphk2 as well as >5000 fold selectivity over S1P$_{1-5}$ receptors and 48 protein and lipid kinases. P 543 attenuates proliferation and induces necrosis in human colorectal cancer cells *in vitro*. It also suppresses human colorectal cancer cell line HCT-116 as a tumor xenograft growth in mice (2). **1**. Schnute, McReynolds, Kasten, *et al.* (2012) *Biochem J.* **444**, 79. **2**. Ju, Gao & Fang (2016) *Biochem. Biophys. Res. Commun.* **470**, 728.

PF-3845

This irreversible inhibitor (FW = 456.47 g/mol), also known as N-arachidonoylethanolamine, or AEA, selectively targets Fatty Acid Amide Hydrolase (or FAAH), an integral membrane enzyme principally responsible for degrading anandamide, the lipid-signaling endocannabinoid that regulates a wide range of mammalian behaviors, including pain, inflammation, and cognitive/emotional state (**See** *Anandamide*). As a covalent inhibitor, PF-3845 binds to FAAH and carbamylates Ser-241 (k$_{inactivation}$ = 0.0033 s^{-1}; K$_i$ = 0.23 μM; k$_{inactivation}$/K$_i$ = 14,300 M^{-1}s^{-1}). PF-3845 also inhibits FAAH *in vivo*, raising brain anandamide levels for up to 24 hours and producing a significant cannabinoid receptor-dependent reduction in inflammatory pain. **1**. Ahn, Johnson, Mileni, *et al.* (2009) *Chem. & Biol.* **16**, 411.

PF 9184

This potent mPGES-1 inhibitor (FW = 461.32 g/mol; CAS 1221971-47-6; Solubility: 100 mM in DMSO), also named N-[3',4'-dichloro(1,1'-biphenyl)yl]-4-hydroxy-2H-1,2-benzothiazine-3-carboxamide 1,1-dioxide), targets human and rat microsomal prostaglandin E synthase 1, with respective IC$_{50}$ values of 16.5 nM and 1.08 μM, exhibiting >6500x selectivity toward mPGES-1 *versus* COX1 and COX2. PF-9184 inhibits IL-1β-induced PGE$_2$ synthesis *in vitro*. In inflammation and clinically relevant biological systems, mPGES-1 expression, like COX-2 expression is induced in cell context- and time-dependent manner, a behavior that is fully consistent with the kinetics of PGE$_2$ synthesis. **1**. Mbalaviele, Pauley, Shaffer, *et al.* (2010) *Biochem. Pharmacol.* **79**, 1445.

PF-562271

This potent, yet reversible ATP-competitive protein kinase inhibitor (FW$_{free-base}$ = 507.49 g/mol; CAS 717907-75-0; FW$_{HCl-Salt}$ = 543.95 g/mol; FW = 665.66 g/mol (benzenesulfonate salt); CAS 939791-38-5), also named, N-[3-[[[2-[(2,3-dihydro-2-oxo-1H-indol-5-yl)amino]-5-(trifluoromethyl)-4-pyrimidinyl]amino]methyl]-2-pyridinyl]-N-methylmethanesulfonamide, targets Focal Adhesion Kinase, or FAK, (IC$_{50}$ = 1.5 nM), with about 10x greater potency than for Pyk2 (IC$_{50}$ = 14 nM) and >100x more than for other protein kinases. PF-562271 binds within the ATP-binding cleft of FAK, where it forms two of the three "canonical" H-bonds between the inhibitor and main-chain atoms in the protein kinase's hinge region. **Mode of Inhibitory Action:** Many cancer cells grow in an anchorage-independent manner, and the nonreceptor tyrosine kinase FAK is thought to contribute to this phenotype by localizing in focal adhesion plaques and has playing roles as a scaffolding and signaling component. Integrin signals within the tumor microenvironment also affect cancer cell survival and invasion during tumor progression, and FAK is activated by β-integrins in both normal and transformed cells. PF-562271 inhibits FAK phosphorylation in vivo in a dose-dependent fashion (calculated EC$_{50}$ = 93 ng/mL) after p.o. administration to tumor-bearing mice (1,2). PF-562271 also has potent effects on metastatic prostate cancer growth *in vivo* (3). Oral administration of PF-562271 at a dose of 5 mg/kg suppressed the growth and local spread of intratibial tumors and restored tumor-induced bone loss. The unique ability of PF-562271 to both curb tumor growth and safely increase bone formation may be an effective therapy for many cancer patients with bone metastases and cancer-associated osteoporosis (4). **1**. Roberts, Ung, Whalen, *et al.* (2008) *Cancer Res.* **68**, 1935. **2**. Canel, Serrels, Miller, *et al.*

(2010) *Cancer Res.* **70**, 9413. **3**. Sun, Pisle, Gardner & Figg (2010) *Cancer Biol. Ther.* **10**, 38. **4**. Bagi, Roberts & Andresen (2008) *Cancer* **112**, 2313.

PF-868554, *See Filibuvir*

PF-915275

This orally bioavailable 11βHSD1 inhibitor (FW = 350.39 g/mol; CAS 857290-04-1; Soluble to 100 mM in DMSO), also named *N*-(6-amino-2-pyridinyl)-4'-cyano-[1,1'-biphenyl]-4-sulfonamide, potent and selective targets 1β-hydroxysteroid dehydrogenase type 1 (11βHSD1) inhibitor (K_i = 2.3 nM) that displays little activity at 11βHSD2 (1.5% inhibition at 10 μM). PF-915275 inhibits the conversion of prednisone to prednisolone in human hepatocytes *in vitro* (EC_{50} = 15 nM) and has antidiabetic activity *in vivo*. **1**. Bhat, *et al.* (2008) *J. Pharmacol. Exp. Ther.* **324**, 299. **2**. Courtney, *et al.* (2008) *J. Clin. Endocrinol. Metab.* **93**, 550. **3**. Fotsch & Wang (2008) *J. Med. Chem.* **51**, 4851.

PF-1247324

This novel oral Na$_V$1.8 blocker (FW = 330.59 g/mol) attenuates nociception and neuronal excitability by selectively targeting voltage-gated sodium transporter Na$_V$1.8, with much weaker action against Na$_V$1.1, Na$_V$1.2, Na$_V$1.4, Na$_V$1.5, Na$_V$1.6, and Na$_V$1.7 transporters. PF-01247324 inhibited native tetrodotoxin-resistant (TTX-R) currents in human dorsal root ganglion (DRG) neurons (IC_{50} = 331 nM) and in recombinantly expressed hNa$_V$ 1.8 (IC_{50} = 196 nM), with 50-fold selectivity over recombinantly expressed TTX-R hNa$_V$ 1.5 channels (IC_{50} ~ 10 μM) and 65-100 greater selectivity over TTX-sensitive (TTX-S) channels (IC_{50} ~ 10-18 μM). Native TTX-R currents in small diameter rodent DRG neurons were inhibited with an IC_{50} of 448 nM, and the block of both human recombinant Na$_V$1.8 and TTX-R from rat DRG neurons was both frequency and state-dependent. Unlike previously published Na$_V$1.8 blockers, PF-01247324 demonstrates frequency-dependence, and off-target frequency-dependence at other sodium channel subtypes may reduce its selectivity window. The majority of small molecule sodium channel blockers interact at the local anesthetic binding site, which due to a high level of sequence homology across voltage-gated sodium channel subtypes seems an unlikely site for interaction of selective agents such as PF-01247324. The majority of small molecule sodium channel blockers interact at the local anesthetic binding site, which due to a high level of sequence homology across voltage-gated sodium channel subtypes seems an unlikely site for interaction of selective agents such as PF-01247324. **1**. Payne, Brown, Theile, *et al.* (2015) *Brit. J. Pharmacol.* **172**, 1258.

PF-1367338, *See Rucaparib*

PF-2341066

This antitumor agent (FW = 449.34 g/mol), also known as [(*R*)-3-[1-(2,6-dichloro-3-fluorophenyl)ethoxy]-5-(1-piperidin-4-yl-1*H*-pyrazol-4-yl)pyridin-2-ylamine, inhibits c-Met and RON receptor protein-tyrosine kinases, thereby exhibiting antiproliferative and antiangiogenic mechanisms (1).

Aberrant expression of RON (*Recepteur d'Origine Nantais*) receptor tyrosine kinase and the generation of multiple-splicing/truncation variants contributes to pathogenesis of epithelial cancers. The six variants include RON-Δ[55], RON-Δ[110], RON-Δ[155], RON-Δ[160], RON-Δ[165], and RON-Δ[170], and deletions/truncations in the extracellular or intracellular regions have been identified. The extracellular sequences contain functional structures such as a SEMA domain, PSI motif, and IPT units. Deletion or truncation results in constitutive phosphorylation and increased kinase activities, attended by or associated with epithelial cell cancer. **1**. Lu, Yao & Wang (2007) *Cancer Lett.* **257**, 157. **2**. Zou, Li, Lee, *et al.* (2007) *Cancer Res.* **67**, 4408.

PF-299804, *See Dacomitib*

PF-3084014

This novel γ-secretase inhibitor (FW = 489.66 g/mol; CAS 1290543-63-3; soluble in DMSO), also code-named HY-15185 and systematically named [(*S*)-2-((*S*)-5,7-difluoro-1,2,3,4-tetrahydronaphthalen-3-ylamino)-*N*-(1-(2-methyl-1-(neopentylamino)propan-2-yl)-1*H*-imidazol-4-yl)pentanamide], inhibits γ-secretase reversibly, noncompetitively, and selectively, thereby reducing amyloid-β (Aβ) production, with an *in vitro* IC_{50} of 1.2 nM in a whole-cell assay and 6.2 nM in cell-free assay (1). PF-03084014 inhibits Notch-related T- and B-cell maturation in an *in vitro* thymocyte assay with an EC_{50} of 2.1 μM. PF-03084014 had an IC_{50} on B- and T-cell reductions of 1.3 to 3 μM with a mean EC_{50} of 2.1 μM. This represents >300-fold separation from the broken-cell Aβ IC_{50} and >1500x separation from the whole-cell IC_{50}. A single acute dose showed dose-dependent reduction in brain, cerebrospinal fluid (CSF), and plasma Aβ. When dosed with PF-3084014 for 5 days using an osmotic minipump (0.03 to 3 mg/kg/day), Guinea pigs exhibited dose-dependent Aβ reduction in brain, CSF, and plasma. While other γ-secretase inhibitors show high potency at elevating Aβ in the conditioned media of whole cells and the plasma of multiple animal models and humans, such potentiation is not observed with PF-3084014 (1). By evoking antitumor and antimetastatic properties via pleiotropic mechanisms, experiments with PF-03084014 offer hope that Notch pathway downstream genes may be used to predict the antitumor activity of PF-03084014 in breast cancer patients (2). **1**. Lanz, Wood, Richter, *et al.* (2010) *J. Pharmacol. Exp. Ther.* **334**, 269. **2**. Zhang, Pavlicek, Zhang, *et al.* (2012) *Clin. Cancer Res.* **18**, 5008.

PF 3491390, *See Emricasan*

PF-4691502

This ATP-competitive PI3K/mTOR dual inhibitor (FW = 325.49 g/mol; CAS 1013101-36-4; Solubility = 14 mg/mL DMSO, < 1 mg/ML H₂O), also systematically named 2-amino-8-((1*R*,4*R*)-4-(2-hydroxyethoxy)cyclohexyl)-6-(6-methoxypyridin-3-yl)-4-methylpyrido[2,3-*d*]pyrimidin-7(8*H*)-one, potently inhibits recombinant Class-I PI3Kα (K_i = 1.8 nM), PI3Kβ (K_i = 2.1 nM), PI3Kδ (K_i = 1.6 nM), PI3Kγ (K_i = 1.9 nM), and mTOR (K_i = 16 nM) in biochemical assays, with little activity against either Vps34, AKT, PDK1, p70S6K, MEK, ERK, p38, or JNK. PF-04691502 also suppresses avian fibroblast transformation mediated by wild-type PI3K γ, δ, or mutant PI3Kα (1). In PIK3CA-mutant and PTEN-deleted cancer cell lines, PF-04691502 reduces phosphorylation of AKT T308 (IC_{50} = 7.5–47 nM) and AKT S473 (IC_{50} = 3.8–20 nM) and inhibits cell proliferation (IC_{50} = 180–310 nM). PF-04691502 also inhibite mTORC1 activity within cells, as measured by PI3K-independent, nutrient-stimulated assay (IC$_{50}$ = 32 nM) and inhibited the activation of PI3K and mTOR downstream effectors, including AKT, FKHRL1, PRAS40, p70S6K, 4EBP1, and S6RP. Short-term exposure to PF-04691502 predominantly inhibits PI3K, whereas mTOR inhibition persists for 24 to 48 hours. PF-04691502 also induces cell cycle G$_1$ arrest, concomitant with upregulation of p27 Kip1 and reduction of retinblastoma protein, or Rb (1). At doses below the maximal tolerable dose, PD-0325901 potently inhibits tumor growth, when Kras and/or PI3K are drivers of tumor growth and progression (2). **1**. Yuan, Mehta, Yin, *et al.* (2011) *Mol. Cancer Ther.* **10**, 2189. **2**. Simmons, Lee, Lalwani, *et al.* (2012) *Cancer Chemother. Pharmacol.* **70**, 213.

PF-4708671

This cell-permeable protein kinase inhibitor (FW = 390.41 g/mol; CAS 1255517-76-0; Solubility = 30 mg/mL DMSO, <1 mg/mL H$_2$O) targets p70 ribosomal S6 kinase (S6K$_1$ isoform) with K_i of 20 nM and IC$_{50}$ of 160 nM, without significant inhibition of S6K$_2$ isoform or Akt1, Akt2, PKA, PKCα, PKCε, PRK2, ROCK2, RSK1, RSK2, or SGK1. PF-4708671 prevents the S6K1-mediated phosphorylation of S6 protein in response to IGF-1 (insulin-like growth factor 1), while having no effect upon the PMA-induced phosphorylation of substrates of the highly related RSK (p90 ribosomal S6 kinase) and MSK (mitogen- and stress-activated kinase) kinases. **1**. Pearce, Alton, Richter, *et al.* (2010) *Biochem. J.* **431**, 245.

PF-4859989

This systemically available KAT II inhibitor (FW = 178.19 g/mol; CAS 34783-48-7; IUPAC: (3*S*)-3-amino-1-hydroxy-3,4-dihydroquinolin-2(1*H*)-one) irreversibly inhibits kynurenine amino transferase II, or KAT II, with K_i = 23 nM, k_{inact}/K_i 112,000 M^{-1}s^{-1}, IC$_{50}$ = 263 nM, and Partition-Coefficient$_{n\text{-octanol/H2O}}$ = 3 x 10^8 (1). This inhibitor most likely makes an enamine adduct with enzyme-bound pyridoxal-P. **Rationale:** Kynurenate is a tryptophan-derived endogenous antagonist for glutamate and α$_7$-nicotinic acetylcholine receptors in the brain. Increased levels are observed postmortem in the brains of patients with a range of neurocognitive disorders, including schizophrenia. Kynurenate may well contribute to the cognitive symptoms of these conditions. **Pharmacology:** PF-04859989 restores glutamate release "transients" evoked by ejections of nicotine into the prefrontal cortex of rats exhibiting elevated kynurenate levels. Systemic administration of PF-04859989 30 min prior to administration of L-kynurenine (but not when administered 30 min after L-kynurenine) restores glutamatergic transients recorded up to 120 min after the administration of the KAT II inhibitor. Furthermore, the KAT-II inhibitor significantly reverses L-kynurenine-induced elevations of brain KYNA levels. The KAT-II inhibitor does not affect nicotine-evoked glutamatergic transients in rats not pre-treated with L-kynurenine. **1**. Dounay, Anderson, Bechle, *et al.* (2013) *Bioorg. Med. Chem. Lett.* **23**, 1961. **2**. Cherian, Gritton, Johnson, *et al.* (2014) *Neuropharmacol.* **82**, 41.

PF-4929113

This potent and selective HSP90 inhibitor (FW = 521.54 g/mol; CAS 908115-27-5; Solubility: 100 mg/mL DMSO, <1mg/mL H$_2$O), also known

as SNX-5422 and (1*R*,4*R*)-4-(2-carbamoyl-5-(6,6-dimethyl-4-oxo-3-(trifluoromethyl)-4,5,6,7-tetrahydroindazol-1-yl)phenylamino)cyclohexyl 2-aminoacetate, targets heat shock protein-90 (K_d = 41 nM) and also inhibits Her-2 degradation (IC$_{50}$ = 37 nM). PF-04929113 also shows potent anti-proliferative activity against a broad range of cancer cell types, e.g. MCF-7 (IC$_{50}$ =16 nM), SW620 (IC$_{50}$ =19 nM), K562 (IC$_{50}$ =23 nM), SK-MEL-5 (IC$_{50}$ =25 nM), and A375 (IC$_{50}$ =51 nM) (1). SNX2112 inhibits *in vitro* proliferation, inducing G$_2$/M arrest, and enhanced cytotoxicity, chemosensitivity, and radiosensitivity between 25 and 250 nM (2), decreasing expression and/or phosphorylation of EGFR, c-MET, AKT, ERK-1 and -2, IκB kinase, and STAT3, corresponding downstream NFκB, AP-1, and STAT3 reporter genes, and target oncogenes and angiogenic cytokines (2). **1**. Huang, Veal, Fadden, *et al.* (2009) *J. Med. Chem.* **52**, 4288. **2**. Friedman, Wise, Hu, *et al.* (2013) *Transl. Oncol.* **6**, 429.

PF-05089771

This potent voltage-gated sodium ion channel blocker (FW = 500.38 g/mol; CAS 1430806-04-4; Solubility: 100 mM in DMSO), also known as 4-[2-(3-Amino-1*H*-pyrazol-4-yl)-4-chlorophenoxy]-5-chloro-2-fluoro-*N*-4-thiazol-ylbenzenesulfonamide tosylate, targets the Na$_V$1.7 channel, with IC$_{50}$ values of 8, 11 and 171 nM for mouse, human and rat. Nav1.7 is the predominant functional tetrodotoxin-sensitive Na$_V$ in mouse and human nociceptors and contributes to the initiation and the upstroke phase of the nociceptor action potential. Na$_V$1.7 also influences synaptic transmission in the dorsal horn of the spinal cord as well as peripheral neuropeptide release in the skin. PF-05089771 also exhibits selectivity for Na$_V$1.7 over Na$_V$1.2 (IC$_{50}$ = 0.11 μM), Na$_V$1.6 (IC$_{50}$ = 0.16 μM), Na$_V$1.1 (IC$_{50}$ = 0.85 μM), Na$_V$1.4 (IC$_{50}$ = 10 μM), Na$_V$1.3 (IC$_{50}$ = 11 μM) and Na$_V$1.5 (IC$_{50}$ = 25 μM). It likewise exhibits selectivity over a panel of 81 other ion channels, receptors, enzymes, and transporters. PF-05089771 blocks spontaneous firing of inherited erythromelalgia (IEM)-derived iPSC sensory neurons *in vitro*. **1**. Alexandrou, Brown, Chapman, *et al.* (2016) *PLoS One* **11**, e0152405. **2**. Cao, McDonnell, Nitzsche, *et al.* (2016) *Sci. Transl. Med.* **8**, 335ra56.

PF-5175157

This orally bioavailable ACC inhibitor (FW = 405.49 g/mol; CAS 1301214-47-0; Solubility: 50 mM in DMSO), also named 1,4-dihydro-1'-[2-methyl-1*H*-benzimidazol-6-yl)carbonyl]-1-(1-methylethyl)-spiro[5*H*-indazole-5,4'-piperidin]-7(6*H*)-one, targets Acetyl-CoA Carboxylase-1 (IC$_{50}$ = 98 nM) and ACC-2 (IC$_{50}$ = 45 nM). Reduces hepatic and skeletal muscle malonyl-CoA levels in rats. PF-05175157 also inhibits hepatic *de novo* lipogenesis and reduces whole body respiratory exchange ratio in rats. **1**. Griffith, Kung, Esler, *et al.* (2014) *J. Med. Chem.* **57**, 10512.

PF-5274857

This brain-penetrating Smoothened (Smo) antagonist (FW = 436.96 g/mol; CAS 1373615-35-0), also known as 1-(4-(5-chloro-4-(3,5-dimethylpyridin-2-yl)pyridin-2-yl)piperazin-1-yl)-3-(methylsulfonyl)propan-1-one, potently and selectively inhibits Hedgehog (Hh) signaling (IC_{50} = 5.8 nM; K_i = 4.6 nM). PF-5274857 elicits robust antitumor activity in a mouse model of medulloblastoma (IC_{50} = 8.9 nmol/L), with down-regulation of Gli1, a marker closely linked to the tumor growth inhibition in patched (+/-) medulloblastoma mice. PF-5274857 also inhibits Smo activity within the brain of primary medulloblastoma mice, improving survival. PF-5274857 is orally available and metabolically stable *in vivo*. **Other Hedgehog inhibitors include:** GANT61, BMS-833923, Purmorphamine, SANT-1, LY2940680, Cyclopamine, LDE225 (or NVP-LDE225, Erismodegib), Vismodegib (GDC-0449) **1.** Rohner, Spilker, Lam, *et al.* (2012) *Mol. Cancer Ther.* **11**, 57.

PF-6274484

Michael Adduct

This EGFR kinase inactivator (FW = 372.78 g/mol; CAS 1035638-91-5; Soluble to 100 mM in DMSO), also named *N*-[4-[(3-chloro-4-fluorophenyl)amino]-7-methoxy-6-quinazolinyl]-2-propenamide, targets Epidermal Growth Factor Receptor (K_i = 0.14 nM), inhibiting autophosphorylation of wild-type EGFR (IC_{50} = 5.8 nM; k_{inact} = 11 ms^{-1}; k_{inact}/K_i = 100 $\mu M^{-1}s^{-1}$) and mutant EGFR$^{L858R/T790M}$ (IC_{50} = 6.6 nM). Although this covalent inhibitor targets a cysteinyl residue in a Michael-type addition reaction, oxidation of the EGFR cysteine nucleophile does not alter catalysis, but has widely varied effects on inhibitor potency, depending on the EGFR context (*e.g.*, oncogenic mutations), type of oxidation (sulfinylation or glutathiolation), and inhibitor architecture. **1.** Schwartz, Kuzmic, Solowiej, *et al.* (2014) *Proc. Natl. Acad. Sci. U.S.A.* **111**, 173.

PF-6447475

This potent, brain penetrant and selective LRRK2 inhibitor (FW = 305.44 g/mol; CAS 1527473-33-1), also named 3-[4-(4-morpholinyl)-7*H*-pyrrolo[2,3-*d*]pyrimidin-5-yl]benzonitrile, targets Leucine-Rich Repeat Kinase 2, IC_{50} = 3 nM, (encoded by the *PARK8* gene), which phosphorylates Akt1 (Ser-473), suggesting that Akt1 is a convincing candidate for the physiological substrate of LRRK2 (1,2). Disease-associated mutations forms of LRRK2 (including Arg-1441-Cys, Gly-2019-Ser, and Ile-2020-Thr) exhibit reduced interaction with, and phosphorylation of, Akt1, a finding that suggests a possible mechanism for the neurodegeneration caused by LRRK2 mutations. Therapeutic approaches to slow or block the progression of Parkinson disease (PD) do not exist. Given that genetic and biochemical studies implicate α-synuclein and leucine-rich repeat kinase 2 (LRRK2) in late-onset PD. In wild-type rats as well as transgenic [Gly-2019-Ser]-LRRK2 rats that were injected intracranially with adeno-associated viral vectors expressing human α-synuclein in the substantia nigra, those expressing [Gly-2019-Ser]-LRRK2 show exacerbated dopaminergic neurodegeneration and inflammation in response to the overexpression of α-synuclein (2). Both neurodegeneration and neuroinflammation associated with [Gly-2019-Ser]-LRRK2 expression were mitigated by PF-06447475, which provided neuroprotection in wild-

type rats. There are no adverse pathological indications in the lung, kidney, or liver of rats treated with PF-06447475. Pharmacological inhibition of LRRK2 is well tolerated for a 4-week period of time in rats and can counteract dopaminergic neurodegeneration caused by acute α-synuclein overexpression. **1.** Henderson, Kormos, Hayward, *et al.* (2015) *J. Med. Chem.* **58**, 419. **2.** Daher, Abdelmotilib, Hu, *et al.* (2015) *J. Biol. Chem.* **290**, 19433.

PF-6463922

This potent, dual ALK/ROS1 inhibitor (FW = 406.41 g/mol; CAS 1454846-35-5), also named (10R)-7-amino-12-fluoro-10,15,16,17-tetrahydro-2,10,16-trimethyl-15-oxo-2*H*-4,8-methenopyrazolo[4,3-*h*][2,5,11]benzoxadiazacyclotetradecine-3-carbonitrile, targets the proto-oncogene tyrosine-protein kinase ROS1 (K_i < 0.02 nM) as well as wild-type anaplastic lymphoma kinase ALK^{WT} (K_i < 0.07 nM) and its Leu-to-Met mutant ALK^{L1196M} (K_i of = 0.7 nM) (1). PF-06463922 significantly inhibits cell proliferation and induces cell apoptosis in the HCC78 human NSCLC cells harboring SLC34A2-ROS1 fusions and the BaF3-CD74-ROS1 cells expressing human CD74-ROS1 (2). **1.** Johnson, Richardson, Bailey, *et al.* (2014) *J. Med. Chem.* **57**, 4720. **2.** Zou, Engstrom, Li, *et al.* (2013) "PF-06463922, a novel ROS1/ALK inhibitor, demonstrates sub-nanomolar potency against oncogenic ROS1 fusions and capable of blocking the resistant ROS1^{G2032R} mutant in preclinical tumor models" *Mol. Cancer Ther.* **12**, Abstract 277.

(R)-PFI

This SETD7 inhibitor (FW = 499.52 g/mol; CAS 1627676-59-8; IUPAC: 8-fluoro-1,2,3,4-tetrahydro-*N*-[(1*R*)-2-oxo-2-(1-pyrrolidinyl)-1-[[3-(trifluoromethyl)phenyl]methyl]ethyl]-6-isoquinolinesulfonamide, targets the SET domain containing lysine methyltransferase, *or* SETD7 (K_i = 0.33 nM; IC_{50} = 2 nM), showing 1000-times selectivity versus other methyl-transferases and other non-epigenetic targets. (*R*)-PFI-2 exhibits an unusual cofactor-dependent and substrate-competitive inhibitory mechanism by occupying the substrate peptide binding groove of SETD7, including the catalytic lysine-binding channel, and by making direct contact with the donor methyl group of the cofactor, S-adenosylmethionine. In murine embryonic fibroblasts, (*R*)-PFI-2 treatment phenocopies the effects of Setd7 deficiency on Hippo pathway signaling, via modulation of the transcriptional coactivator Yes-associated protein (YAP) and regulation of YAP target genes. In confluent MCF7 cells, (*R*)-PFI-2 rapidly alters YAP localization, suggesting continuous and dynamic regulation of YAP by the methyltransferase activity of SETD7. **1.** Barsyte-Lovejoy, Li, Oudhoff, *et al.* (2014) *Proc. Natl. Acad. Sci. U.S.A.* **111**, 12853.

PGE$_2$, PGF$_{2a}$, etc., See *specific prostaglandin*

[D-pGlu^1,D-Phe^2,D-Trp^{3,6}]-Luteinizing Hormone-Releasing Hormone

This decapeptide amide (Sequence: D-pGlu-D-Phe-L-Trp-L-Ser-L-Tyr-D-Ala-L-Leu-L-Arg-L-Pro-Gly-NH$_2$, where pG is the cyclic pyroglutamate; MW = 1207.35 g/mol), an analogue of the luteinizing hormone-releasing hormone (*i.e.*, gonadotropin releasing hormone), is a substance P antagonist and inhibits sympathetic vasomotor outflow. **1.** Takano, Sawyer, Sanders & Loewy (1985) *Brain Res.* **337**, 357.

PH 797804

This orally active, potent, and selective p38α/β inhibitor (FW = 477.31 g/mol; CAS 1358027-80-1; Solubility: 50 mM in DMSO), also named (-)-3-[3-bromo-4-[(2,4-difluorophenyl)methoxy]-6-methyl-2-oxo-1(2*H*)-pyridin-yl]-*N*,4-dimethylbenzamide, targets p38 mitogen-activated protein kinases, with IC$_{50}$ values of 26 and 102 nM for p38α (K_i = 5.8 nM; ATP-competitive and readily reversible) and p38β, respectively (1). The latter are signal-transducing enzymes that respond to stress stimuli (*e.g.*, cytokines, ultraviolet irradiation, heat shock, and osmotic shock) and are involved in regulating cell differentiation, apoptosis and autophagy. PH-797804 blocks inflammation-induced production of cytokines and proinflammatory mediators, such as prostaglandin E$_2$, at concentrations that parallel inhibition of cell-associated p38 MAP kinase (2). When administered orally, PH-797804 inhibits acute inflammatory responses induced by systemically administered endotoxin in both rat and cynomolgus monkeys (2). It also demonstrates robust anti-inflammatory activity in chronic disease models, significantly reducing both joint inflammation and associated bone loss in streptococcal cell wall-induced arthritis in rats and mouse collagen-induced arthritis. PH-797804 reduces Tumor Necrosis Factor-α (TNFα) and interleukin-6 production in clinical studies after endotoxin administration in a dose-dependent manner, paralleling inhibition of the target enzyme (2). Low-nanomolar biochemical enzyme inhibition potency correlated with p38 MAP kinase inhibition in human cells and in vivo studies. 1. Selness, Devraj, Devadas, *et al.* (2011) *Bioorg. Med. Chem. Lett.* 21, 4066. 2. Hope, Anderson, Burnette, *et al.* (2009) *J. Pharmacol. Exp. Ther.* 331, 882.

PHA 408

This potent and highly selective, ATP-competitive IKK-2 inhibitor (FW = 560.02 g/mol; CAS 503555-55-3; Soluble in DMSO), also named 8-(5-chloro-2-(4-methylpiperazin-1-yl)isonicotinamido)-1-(4-fluorophenyl)-4,5-dihydro-1*H*-benzo[*g*]indazole-3-carboxamide, targets IKB kinase-2 (IC$_{50}$ = 40 nM). PHA-408 binds tightly to IKK-2, showing a relatively slow off rate (1). In arthritis-relevant cells and animal models, PHA-408 suppresses inflammation-induced cellular events, including IκBα phosphorylation and degradation, p65 phosphorylation and DNA binding activity, the expression of inflammatory mediators, and joint pathology (1). It selectively inhibits IL-1β- and LPS-induced expression of various inflammatory mediators. PHA-408 inhibits LPS-induced TNF-α production *in vivo* in a rat acute model of inflammation. Significantly, PHA-408 is efficacious in a chronic model of arthritis with no adverse effects at maximally efficacious doses. 1. Mbalaviele, Sommers, Bonar, *et al.* (2009) *J. Pharmacol. Exp. Ther.* 329, 14.

PHA-767491

This cyclin kinase-directed inhibitor (FW = 213.24 g/mol; CAS 845714-00-3, 942425-68-5 (HCl); Solubility: 24 mg/mL DMSO; <1 mg/mL Water;

Formulation: Dissolve in DMSO, and then dilute in saline), named systematically as 2-(pyridin-4-yl)-6,7-dihydro-1*H*-pyrrolo[3,2-*c*]pyridin-4(5*H*)-one, targets Cdc7 and CDK9, with IC$_{50}$ values of 10 nM and 34 nM, respectively. PHA-767491 displays approximately 20-times lower inhibition of Cdk1, Cdk2 and GSK3-β, 50-times lower for MK2 and Cdk5 and 100-times lower for PLK1 and Chk2. **Cyclin Target Selectivity:** Cdk1 (weak, if any), Cdk2 (weak, if any), Cdk3 (weak, if any), Cdk4 (weak, if any), Cdk5 (weak, if any), Cdk6 (weak, if any), Cdk7 (+++), Cdk8 (weak, if any), Cdk9 (++), Cdk10 (weak, if any). 1. Montagnoli, *et al.* (2008) *Nat. Chem. Biol.* 4, 357. 2, Yecies, *et al.* (2010) *Blood* 115, 3304. 3. Natoni, *et al.* (2011) *Mol. Cancer Ther.* 10, 1624.

PHA-793887

This cyclin kinase-directed inhibitor (FW = 361.48 g/mol; CAS 718630-59-2, 718630-60-5 (HCl); Solubility: 70 mg/mL DMSO; <1 mg/mL Water; Formulation: Dissolve in 5% dextrose solution), also known as *N*-(6,6-dimethyl-5-(1-methylpiperidine-4-carbonyl)-1,4,5,6-tetrahydropyrrolo[3,4-*c*]pyrazol-3-yl)-3-methylbutanamide, targets CDK2, CDK5 and CDK7 with IC$_{50}$ values of 8 nM, 5 nM and 10 nM, respectively. **Alternative target**: glycogen synthase-3 kinase-β (GSK-3β). **Cyclin Target Selectivity:** Cdk1 (weak, if any), Cdk2 (+++), Cdk3 (weak, if any), Cdk4 (weak, if any), Cdk5 (+++), Cdk6 (weak, if any), Cdk7 (+++), Cdk8 (weak, if any), Cdk9 (weak, if any), Cdk10 (weak, if any). 1. Brasca, *et al.* (2010) *Bioorg. Med. Chem.* 18, 1844. 2. Alzani, *et al.* (2010) *Exp. Hematol.* 38, 259. 3. Pevarello, et al. (2004) *J. Med. Chem.* 47, 3367.

PHA-848125

This cyclin kinase-directed inhibitor (FW = 460.57 g/mol; CAS 802539-81-7, 802540-32-5 (3HCl), 1253645-38-3 (Maleic acid); Solubility: 90 mg/mL DMSO; <1 mg/mL Water), systematically named *N*,1,4,4-tetramethyl-8-(4-(4-methylpiperazin-1-yl)phenylamino)-4,5-dihydro-1*H*-pyrazolo[4,3-*h*]quinazoline-3-carboxamide, targets cyclin A/CDK2 inhibitor with IC$_{50}$ of 45 nM. PHA-848125-treated cells show cell cycle arrest in G$_1$ and reduced DNA synthesis, accompanied by inhibition of pRb phosphorylation and modulation of other CDK-dependent markers. **Cyclin Target Selectivity:** Cdk1 (weak, if any), Cdk2 (++), Cdk3 (weak, if any), Cdk4 (weak, if any), Cdk5 (weak, if any), Cdk6 (weak, if any), Cdk7 (weak, if any), Cdk8 (weak, if any), Cdk9 (weak, if any), Cdk10 (weak, if any). 1. Brasca, *et al.* (2009) *J. Med. Chem.* 52, 5152. 2. Albanese, *et al.* (2010) *Mol. Cancer Ther.* 9, 2243.

Phaclofen

This (–)-(*R*)-phosphonic acid (FW = 249.63 g/mol; CAS 108351-35-5), systematically named as 3-amino-2-(4-chlorophenyl)propanephosphonic acid, is a GABA$_B$ γ-aminobutyric acid receptor antagonist (1-3). Although synaptic action of GABA is mediated by bicuculline-sensitive GABA$_A$ receptors that selectively increase chloride conductance, GABA has a presynaptic inhibitory action that is bicuculline-insensitive and is mimicked by baclofen. This distinctive behavior defines the GABA$_B$ receptor, allowing one to distinguish them from classic bicuculline-sensitive GABA$_A$

receptors. In hippocampal slices, phaclofen (baclofen's phosphonate derivative) is a remarkably selective antagonist of both the postsynaptic action of baclofen and GABA's bicuculline-resistant action (1). Phaclofen selectively abolishes the slow inhibitory postsynaptic potential in pyramidal cells (1). It is ineffective in antagonizing competitively the synaptically-mediated Late Hyperpolarizing Response (LHP) recorded from the same or similar Dorsolateral Septal Nucleus (DLSN) neurons, from which baclofen responses were recorded (3). Such findings support the usefulness of phaclofen as a competitive antagonist of baclofen, and suggest that when larger stimulus intensities are applied, the LHP in the rat DLS may be mediated by a transmitter in addition to GABA (3). Moreover, (-)-(R)-phaclofen inhibits the binding of [^{3}H]-(R)-baclofen to GABA$_B$ receptor sites on rat cerebellar membranes (IC$_{50}$ = 76 μM), whereas (+)-(S)-phaclofen is inactive (IC$_{50}$ > 1000 μM) (4). **1.** Dutar & Nicoll (1988) *Nature* **332**, 156. **2.** Kerr, Ong, Prager, Gynther & Curtis (1987) *Brain Res.* **405**, 150. **3.** Hasuo & Gallagher (1988) *Neurosci. Lett.* **86**, 77. **4.** Frydenvang, Hansen, Krogsgaard-Larsen, *et al.* (1994) *Chirality* **6**, 583.

Phallacidin

This cyclic heptapeptide toxin (FW$_{free-acid}$ = 846.92 g/mol; CAS 26645-35-2) from the poisonous green fungus *Amanita phalloides* stabilizes actin filaments and inhibits actin filament depolymerization. NBD phallacidin is a high-affinity, filamentous actin probe consisting of rabbit actin that is conjugated by reacting its thiol group with 4-chloro-7- nitrobenzofurazan, a yellow-green fluorescent dye. NBD phallacidin (Excitation at 465 nm; Emission at 536 nm) selectively stains F-actin un a manner that is superior to antibody staining, preferably in fixed and permeabilized samples. **See** *Phalloidin* **1.** Fechheimer & Zigmond (1983) *Cell Motil.* **3**, 349.

Phalloidin

This bicyclic heptapeptide toxin (FW = 788.88 g/mol; CAS 17466-45-4) is was first isolated from the poisonous green fungus *Amanita phalloides*. **Primary Mode of Action:** Phalloidin binds preferentially to filamentous actin; little or no binding to globular actin has been detected. such preferential action stimulates actin polymerization, and phalloidin lowers the actin monomer critical concentration by 30x, from 50-100 nM down to 2-3 nM. When present at 1 to 10 concentration ratio of phalloidin to total actin subunits, actin filaments are also greatly stabilized toward depolymerization. **Effects on Actin Filaments:** Cellular processes requiring filament disassembly are likewise inhibited. Depolymerization of F-actin by cytochalasins, potassium iodide, and elevated temperatures are inhibited by phalloidin binding. Because the toxin and its fluorescent derivatives are relatively small, a wide variety of actin-binding proteins can still bind to phalloidin-labeled filamentts. Perhaps more significantly, phalloidin-labeled actin filaments retain many of their functional properties, such that phalloidin-labeled, glycerinated muscle fibers can still contract, and labeled actin filaments still move on myosin that has been tethered to solid-phase substrates. Phalloidin can be also be used to assess the relative

concentrations of these two forms of actin as well as a means to label actin filaments in a cell (**See** *Phallacidin*). **1.** Cooper (1987) *J. Cell Biol.* **105**, 1473. **2.** Wieland & Faulstich (1978) *CRC Crit. Rev. Biochem.* **5**, 185. **3.** De La Cruz & Pollard (1996) *Biochemistry* **35**, 14054. **4.** Dancker, Low, Hasselbach & Wieland (1975) *Biochim. Biophys. Acta* **400**, 407.

Pharbitate, See Gibberethione

PhcrTx1 Toxin

This marine peptide toxin (MW = 3476.03 g/mol; *Sequence:* CASQGQ KCKTKSDCCNGMWCAGTRGHTCYKPK) from the sea anemone *Phymanthus crucifer* is a novel Acid-Sensing Ion Channel (ASIC) inhibitor that partially blocks ASIC currents (IC$_{50}$ ~100 nM), as well as voltage-gated K$^+$ currents (or K$_v$ channels), the latter with much lower potency. PhcrTx1's effects on the peak and steady-state currents were lower than 20% in dorsal root ganglion (DRG) neurons at micromolar concentrations. No significant effect was observed on Na$^+$ voltage-gated currents in these neurons. Analysis of the cysteine pattern and predicted secondary structure suggest this peptide is likely to have an Inhibitor Cystine Knot scaffold already observed in other spider, scorpions and *Conus* toxins. **1.** Rodríguez, Salceda, Garateix, *et al.* (2013) *Peptides* **53**, 3.

Phebestin

This bestatin analogue (FW = 441.53 g/mol), known systematically as (2S,3R)-3-amino-2-hydroxy-4-phenylbutanoyl-L-valyl-L-phenylalanine, isolated from *Streptomyces* inhibits a number of aminopeptidases. **Target(s):** aminopeptidase I1; Jc-peptidase, or Japanese cedar pollen aminopeptidase (1); and membrane alanyl aminopeptidase, or aminopeptidase N (2-4). **1.** Noguchi, Nagata, Koganei, *et al.* (2002) *J. Agric. Food Chem.* **50**, 3540. **2.** Nagai, Kojima, Naganawa, *et al.* (1997) *J. Antibiot.* **50**, 82. **3.** Xu & Li (2005) *Curr. Med. Chem. Anticancer Agents* **5**, 281. **4.** Bauvois & Dauzonne (2006) *Med. Res. Rev.* **26**, 88.

Phenacetin

This first-in-class, synthetic antipyretic (FW = 179.22 g/mol; CAS 62-44-2), also known as *N*-(4-ethoxyphenyl)acetamido, was synthesized by American chemist Harmon Northrop Morse in 1897. Phenacetin was used as an analgesic/antipyretic, but was withdrawn from the market in view of adverse effects, including methemoglobinemia and renal failure (1,2). Phenacetin is metabolized to acetaminophen (**See** *Acetaminophen*). Hydrolysis by arylacetamide deacetylase (*or* AADAC) and subsequent metabolism by CYP1A2 and CYP2E1 play predominant roles in phenacetin-induced methemoglobinemia (3). Indeed, treatment with eserine, a potent AADAC inhibitor, greatly reduces metabolic hydrolysis of phenacetin as well as the severity of methemoglobinemia (3). **1.** Boyd (1970) *Appl. Ther.* **12**, 9. **2.** Hinson (1983) *Environ. Health Perspect.* **49**, 71. **3.** Kobayashi, Fukami, Higuchi, Nakajima & Yokoi (2012) *Biochem. Pharmacol.* **84**, 1196.

1-Phenacyl-2-methylpyridinium Chloride

This cationic pyridinium salt (FW = 247.72 g/mol) inhibits human blood acetylcholinesterase, binding to both the catalytic and the allosteric sites of the enzyme, with K_i and K_d values of 6.9 μM and 27 μM, respectively. **1.** Skrinjaric-Spoljar, Burger & Lovric (1999) *J. Enzyme Inhib.* **14**, 331.

Phenamil

This amiloride derivative (FW = 306.73 g/mol; CAS 2038-35-9), also known as 3,5-diamino-6-chloro-N-[imino(phenylamino)methyl]pyrazine carboxamide, irreversibly inhibits amiloride-sensitive Na$^+$ channels. It is often supplied commercially as the methanesulfonate salt. Na$^+$ channel blockers, such as phenamil and benzamil, are be alsoing explored as a way to counteract the hyperabsorption of NaCl in cystic fibrosis airways. **Target(s):** diamine oxidase (1); Na$^+$ channels (2,3); and Na$^+$-driven flagellar motility (4-9). **1.** Novotny, Chassande, Baker, Lazdunski & Barbry (1994) *J. Biol. Chem.* **269**, 9921. **2.** Garvin, Simon, Cragoe, Jr., & Mandel (1985) *J. Membr. Biol.* **87**, 45. **3.** AGuia, Chau, Bose & Bose (1996) *J. Pharmacol. Exp. Ther.* **277**, 163. **4.** Gosink & Hase (2000) *J. Bacteriol.* **182**, 4234. **5.** Sato & Homma (2000) *J. Biol. Chem.* **275**, 20223. **6.** Jaques, Kim & McCarter (1999) *Proc. Natl. Acad. Sci. U.S.A.* **96**, 5740. **7.** Kojima, Asai, Atsumi, Kawagishi & Homma (1999) *J. Mol. Biol.* **285**, 1537. **8.** Kawagishi, Imagawa, Imae, McCarter & Homma (1996) *Mol. Microbiol.* **20**, 693. **9.** Atsumi, Sugiyama, Cragoe, Jr., & Imae (1990) *J. Bacteriol.* **172**, 1634.

9,10-Phenanthrenequinone

This orange-red diketone (FW = 208.22 g/mol; M.P. = 206–207°C; CAS 84-11-7), also called 9,10-phenanthrenedione and phenanthraquinone, reacts with the guanidinium group of arginyl residues in peptides and proteins (1). 9,10-Phenathrenequinone is also an alternative substrate for a number of oxidoreductases (*e.g.*, 20α-hydroxysteroid dehydrogenase, ζ-crystallin, and prostaglandin D$_2$ 11-ketoreductase). **Target(s):** cAMP-dependent protein kinase (2); Deoxyribonuclease, weakly inhibited (3); and nitric-oxide synthase4; protein-tyrosine phosphatase (5). **1.** Yamada & Itano (1966) *Biochim. Biophys. Acta* **130**, 538. **2.** Wang, Ternai & Polya (1994) *Biol. Chem. Hoppe Seyler* **375**, 527. **3.** Hoffmann-Ostenhof & Frisch-Niggemeyer (1952) *Monatsh. Chem.* **83**, 1175. **4.** Kumagai, Nakajima, Midorikawa, Homma-Takeda & Shimojo (1998) *Chem. Res. Toxicol.* **11**, 608. **5.** Urbanek, Suchard, Steelman, *et al.* (2001) *J. Med. Chem.* **44**, 1777.

Phenanthriplatin

This potent Pt(II) complex (FW = 505.82 g/mol; CAS 1416900-51-0), also known as *cis*-diamminephenanthridinechloroplatinum(II), penetrates cell membranes either by passive diffusion or carrier-mediated active transport, with the hydrophobic ligand maximizing cellular uptake and making it cytotoxic compared with cisplatin and carbinatoplatin (1). Its molecular structure was designed with based the X-ray structure of RNA polymerase II (Pol II), when stalled at a monofunctional pyriplatin-DNA adduct (1). Phenanthriplatin interacts with guanine residues, forming highly potent, monofunctional DNA adducts. While many Pol II elongation complexes stall after successful addition of CTP opposite the phenanthriplatin-dG adduct in an error-free manner, a few slowly undergo error-prone bypass of the phenanthriplatin-dG lesion, a behavior that resembles DNA polymerases that similarly switch from high-fidelity replicative DNA processing (error-free) to low-fidelity translesion DNA synthesis (error-prone) at DNA damage sites (2). **1.** Park, Wilson, Song & Lippard (2012) *Proc. Natl. Acad. Sci. U.S.A.* **109**, 11987. **2.** Kellinger, Park, Chong, Lippard & Wang (2013) *J. Am. Chem. Soc.* **135**, 13054.

1,10-Phenanthroline

This metal ion chelator (FW$_{free-base}$ = 180.21 g/mol; CAS 66-71-7), also called *o*-phenanthroline, is especially effective as a zinc ion chelator. It is a relatively low-affinity reversible inhibitor that acts by chelating active-site metal ions in certain metal-activated and metalloenzymes. Given the fact that active-site metal ions are most often bound by multiple amino acid side-chains, one must consider the likelihood that 1,10-phenanthrolein will act slowly in a manner akin to most slow-binding enzyme inhibitors. Another consideration is that the observed inhibitory effect of 1,10-phenanthrolene has nothing to do with chelation but is simply a consequence ofd its hydrophobicity. (*See comment under 4,7-Phenanthroline*) The inhibitory range is typically 1–5 mM; however, some iron-dependent proteins are reportedly inhibited at 10^{-8} M. The reported K_i values for angiotensin I-converting-enzyme (peptidyl-dipeptidase A), thermolysin, and astacin are 6, 50, and 50 mM for, respectively. In addition to iron and zinc, it will also bind Mn, Co, Ni, Ag, Ru, and Cu. This chelator will also inhibit dioxygen production in photosynthesis and has been reported to exhibit nuclease activity when complexed with copper. The ferrous complex, frequently referred to as Ferroin and available commercially, is used as an indicator in oxidation/reduction systems. 1,10-Phenanthroline is hygroscopic and is usually supplied as the monohydrate (FW = 198.22 g/mol). It has a melting point of 93-94°C (anhydrous melting point = 117°C) and is soluble in about 300 parts water. It is typically prepared as 100-200 mM stock solution using dimethyl sulfoxide (preferred), methanol, or ethanol. *Note*: When testing for the inhibitory effects of 1,10-phenanthroline, a good practice is to also test 1,7-phenanthroline or 4,7-phenanthroline as controls for nonchelating binding effects. Another distinguishing feature is that metal ion resolution by 1,10-phenanthroline is typically time-dependent, whereas binding of non-chelating isomers is rarely time-dependent.

4,7-Phenanthroline

This hygroscopic reagent (FW$_{free-base}$ = 180.21 g/mol; M.P. = 172-174°C; CAS 230-07-9), is a non-chelating isomer of 1,10-phenanthroline. Before inferring that the action of 1,10-phenanthroline is metal ion chelation, one should also assess the inhibitory effect of this non-chelating isomer. If a similar degree of inhibition is observed with both, one may assume that the effect has nothing to do with chelation an essential metal ion activator. **See also** *1,10-Phenanthroline; 1,7-Phenanthroline* **Target(s):** aldehyde dehydrogenase (1); leucyl aminopeptidase, weakly inhibited (2); nitrilase (3); pyroglutamyl-peptidase II, *or* thyrotropin-releasing-hormone-degrading ectoenzyme (4-6). **1.** Hoe & Crabbe (1987) *Exp. Eye Res.* **44**, 663. **2.** Cahan, Axelrad, Safrin, Ohmann & Kessler (2001) *J. Biol. Chem.* **276**, 43645. **3.** Almatawah, Cramp & Cowan (1999) *Extremophiles* **3**, 283. **4.** Czekay & Bauer (1993) *Biochem. J.* **290**, 921. **5.** Gallagher & O'Connor (1998) *Int. J. Biochem. Cell Biol.* **30**, 115. **6.** Bauer (1994) *Eur. J. Biochem.* **224**, 387.

m-Phenanthroline, *See 1,5-Phenanthroline; 1,7-Phenanthroline*

o-Phenanthroline, *See 1,10-Phenanthroline*

p-Phenanthroline, *See 1,8-Phenanthroline*

1,6-Phenazinediol 5,10-Dioxide, *See Iodinin*

Phenazine Methosulfate

This redox-active reagent (FW = 306.34 g/mol; CAS 299-11-6; E_o' = +0.080 V, pH = 7 and T = 30°C), also known as *N*-methylphenazonium methosulfat, is frequently used as an artificial electron acceptor and carrier in studies of redox reactions. The reduced semiquinone, which may be prepared nonenzymatically from NADH or NADPH, is a colorless product (occasionally, a green color is reported) and can be used as an electron donor. This reduced compound is rapidly oxidized by dioxygen and will reduce cytochrome c, indophenol dyes, and many other electron acceptors. It is often used with ascorbic acid to determine nitric oxide reductase activity. **Action as a Redox Substrate:** Phenazine methosulfate (MTT) is a synthetic electron acceptor substttate for many enzymes (*e.g.,* succinate dehydrogenase, holine dehydrogenase, glycolate dehydrogenase, polyvinyl-alcohol dehydrogenase, (*R*)-pantolactone dehydrogenase, formate dihydro-genase, isoquinoline 1-oxidoreductase, quinaldate 4-oxidoreductase, aralkylamine dehydrogenase, glycine dehydrogenase (cyanide-forming), trimethylamine dehydrogenase, cytokinin dehydrogenase, and 4-cresol dehydrogenase (hydroxylating). When used in enzyme assays, MTT is converted to formazin, an intensely purple-colored product. To achieve high-sensitivity and a linear dependence, one must use a solubilization solution (usually either dimethyl sulfoxide, an acidified ethanol solution, or a solution of the detergent sodium dodecyl sulfate in diluted hydrochloric acid) to disperse/dissolve the otherwise insoluble formazan to obtain colored suspension/solution. The absorbance of this solution may then be quantified by measuring the wavelength between 500 and 600 nm in a spectrophotometer. The degree of light absorption depends on the solvent. **Cell Viability Asaays:** MTT has been widely employed in colorimetric assays for assessing the viability of cells. When tested under defined conditions, NAD(P)H-dependent cellular oxidoreductase enzymes catalyze the conversion of MTT to formazan, the intensity of which indicates cell viability. Other related tetrazolium dyes (including XTT, MTS and the WSTs) are used in conjunction with the intermediate electron acceptor, 1-methoxy phenazine methosulfate (PMS). **Target(s):** ferredoxin:NADP$^+$ reductase (1); magnesium-protoporphyrin IX monomethyl ester (oxidative) cyclase (2); photophosphorylation (3); progesterone monooxygenase (4); protein-N^π-phosphohistidine:sugar phosphotransferase (5); stearoyl-CoA 9-desaturase (6,7); steroid 9α-monooxygenase (8); steroid 11β-monooxygenase (9); and testosterone 5α-reductase (10). **1.** San Pietro (1963) *Meth. Enzymol.* **6**, 439. **2.** Wong & Castelfranco (1984) *Plant Physiol.* **75**, 658. **3.** Lynn (1967) *J. Biol. Chem.* **242**, 2186. **4.** Rahm & Sih (1966) *J. Biol. Chem.* **241**, 3615. **5.** Grenier, Waygood & Saier, Jr. (1985) *Biochemistry* **24**, 47 and 4872. **6.** Fulco & Bloch (1964) *J. Biol. Chem.* **239**, 993. **7.** Wilson & Miller (1978) *Can. J. Biochem.* **56**, 1109. **8.** Chang & Sih (1964) *Biochemistry* **3**, 1551. **9.** Zuidweg (1968) *Biochim. Biophys. Acta* **152**, 144. **10.** Kim, Kim, Son, *et al.* (1999) *Biol. Pharm. Bull.* **22**, 1396.

Phenazone

This antipyretic agent (FW = 188.23 g/mol; CAS 60-80-0; pK_a = 1.4; IUPAC Name: 1,2-dihydro-1,5-dimethyl-2-phenyl-3*H*-pyrazol-3-one), also known as antipyrine, nonselectively inhibits COX-1 and COX-2, showing little anti-inflammatory activity. It was first synthesized in 1884, more than a decade before the much-heralded synthesis of acetylsalicylic acid. *See Antipyretic Agents*

Phenbenicillin, *See Fenbenicillin*

Phencyclidine

This piperidine (FW$_{\text{free-base}}$ = 243.39 g/mol; CAS 77-10-1: *Symbol*: PCP), commonly known as Angel Dust (1) and systematically as 1-(1-phenylcyclohexyl)piperidine, is a NMDA (*N*-methyl-D-aspartate) antagonist, psychostimulant, σ receptor agonist, and frequent drug of abuse (2-5). Formerly used as a veterinary anesthetic and briefly as a general

anesthetic in humans, phencyclidine is also a powerful hallucinogen. **Mode of Action:** NMDA receptors play important roles in mediating excitatory neurotransmission and are preferentially inhibited by some general anesthetics. Phencyclidine is similar to ketamine in structure and in many of its effects; both produce a dissociative state (*See Ketamine*). Phencyclidine inhibits activation of NMDA receptors (3-5), inducing schizophrenia-like symptoms in healthy individuals and exacerbating pre-existing symptoms in patients with schizophrenia (6). PCP behavioral effects are strongly dose-dependent: low doses (3-5 mg) produce intoxication (characteristics: numbness in the extremities, staggering or unsteady gait, slurred speech, and bloodshot eyes); moderate doses (*e.g.,* 5–10 mg intranasal or 0.01–0.02 mg/kg IM or IV) produce analgesia and anesthesia; and high doses often lead to convulsions. **Pharmacokinetics & Metabolism:** PCP is well absorbed by all routes of administration, but, in cigarette smoke, about half is converted to an inactive thermal degradation product. PCP is highly lipid-soluble, accouting for its tendency to concentrate in fat and brain tissue. The plasma binding of PCP is 65%, and its $t_{1/2}$ ranges from 7-46 hours, with a 21-hour average. PCP is extensively metabolized to inactive metabolites by a variety of metabolic routes. Benzodiazepines decrease PCP's hypertensive effects and reverse seizure activity. PCP is also an inhibitor of a number of cytochrome P450 systems and a suicide inhibitor of nitric-oxide synthase (7). However, the inhibition of nitric-oxide synthase is not related to the psychotomimetic action of phencyclidine. **Target(s):** Ca^{2+}-dependent ATPase (8,9); CYP2B1 (10); K$^+$ channels, ATP-sensitive (11); nitric-oxide synthase (7); and NMDA-receptor channel (4-6,12). Phencyclidine is a powerful, noncompetitive inhibitor of the nicotinic acetylcholine receptor in a sympathetic nerve cell line, PC-12. In the presence of 1 mM carbamoylcholine, the rate of the receptor-controlled influx of ^{22}Na$^+$ is reduced by a factor of 2 by 0.7 μM phencyclidine (13,14). **1. Alternative Street Names:** Dust, Cadillac, Stardust, Supergrass, Crystal, Surfer, Tranks. **2.** Sharp (1997) *Brain Res.* **758**, 51. **3.** Porter & Greenamyre (1995) *J. Neurochem.* **64**, 614. **4.** O'Donnell & Grace (1998) *Neuroscience* **87**, 823. **5.** Johnson & Jones (1990) *Annu. Rev. Pharmacol. Toxicol.* **30**, 707. **6.** Jodo (2013) *J. Physiol. Paris* **107**, 434. **7.** Osawa & Davila (1993) *Biochem. Biophys. Res. Commun.* **194**, 1435. **8.** Pande, Cameron, Vig, Ali & Desaiah (1999) *Mol. Cell Biochem.* **194**, 173. **9.** Pande, Cameron, Vig & Desaiah (1998) *Toxicology* **129**, 95. **10.** Crowley & Hollenberg (1995) *Drug Metab. Dispos.* **23**, 786. **11.** Kokoz, Alekseev, Povzun, Korystova & Peres-Saad (1994) *FEBS Lett.* **337**, 277. **12.** Yi, Snell & Johnson (1988) *Brain Res.* **445**, 147. **13.** Sachs, Leprince, Karpen, *et al.* (1983) *Arch. Biochem. Biophys.* **225**, 500. **14.** Karpen, Aoshima, Abood & Hess (1982) *Proc. Natl. Acad. Sci. U.S.A.* **79**, 2509.

Phenelzine

This hydrazine (FW = 136.20 g/mol; CAS 51-71-8 and 156-51-4), also known as (2-phenethyl)hydrazine and (2-phenylethyl)hydrazine, is an antidepressant and nonselective monoamine oxidase inhibitor. Phenelzzine is often supplied commercially as the sulfate or hydrochloride salt. **Target(s):** alanine aminotransferase (1); 4-aminobutyrate aminotransferase (1); aromatic-L-amino acid decarboxylase (2); cytochrome P450 (3-5); dopamine β-monooxygenase (6); histidine ammonia-lyase (7); monoamine oxidase (2,8-12); taurine dehydrogenase (13); and tyrosine aminotransferase (2,14). **1.** Tanay, Parent, Wong, *et al.* (2001) *Cell. Mol. Neurobiol.* **21**, 325. **2.** Dyck & Dewar (1986) *J. Neurochem.* **46**, 1899. **3.** Belanger & Atitse-Gbeassor (1983) *Can. J. Physiol. Pharmacol.* **61**, 524. **4.** Clark (1967) *Biochem. Pharmacol.* **16**, 2369. **5.** Ator & Ortiz de Montellano (1990) *The Enzymes*, 3rd ed. (Sigman & Boyer, eds.), **19**, 213. **6.** Fitzpatrick & Villafranca (1986) *J. Biol. Chem.* **261**, 4510. **7.** Brand & Harper (1976) *Biochemistry* **15**, 1814. **8.** Neff & Yang (1974) *Life Sci.* **14**, 2061. **9.** Baker, Coutts, McKenna & Sherry-McKenna (1992) *J. Psychiatry Neurosci.* **17**, 206. **10.** Bright & Porter (1975) *The Enzymes*, 3rd ed. (Boyer, ed.), **12**, 421. **11.** Yu & Tipton (1989) *Biochem. Pharmacol.* **38**, 4245. **12.** Urichuk, Allison, Holt, Greenshaw & Baker (2000) *J. Affect. Disord.* **58**, 135. **13.** Kondo & Ishimoto (1987) *Meth. Enzymol.* **143**, 496. **14.** Dyck (1987) *Biochem. Pharmacol.* **36**, 1373.

Phenethyl Alcohol, *See 2-Phenylethanol*

β-Phenethylamine, *See β-Phenylethylamine*

Phenethylbiguanide, *See Phenformin*

N^1-Phenethylcymserine

This cymserine analogue (FW = 457.62 g/mol) inhibits butyrylcholinesterase, IC$_{50}$ = 6 nM (1-3). **1**. Giacobini (2003) *Neurochem. Res.* **28**, 515. **2**. Yu, Holloway, Utsuki, Brossi & Greig (1999) *J. Med. Chem.* **42**, 1855. **3**. Cerasoli, Griffiths, Doctor, *et al.* (2005) *Chem. Biol. Interact.* **157-158**, 363.

N'-β-Phenethylformamidinyliminourea, *See* *Phenformin*

9-(2'-Phenethyl)guanine, *See* *9-(2'-Phenylethyl)guanine*

Phenethylhydrazine, *See* *Phenelzine*

Phenformin

This oral, biguanide-class antidiabetic (FW$_{hydrochloride}$ = 241.72 g/mol; M.P. = 175-178°C; typically supplied as the water-soluble hydrochloride salt), also known as phenethylbiguanide and phenylethylbiguanide, and named systematically as N'-β-phenethylformamidinyliminourea, was withdrawn from the U.S. market in 1977 due to high risk of inducing often fatal lactic acidosis. **Target(s):** cholesterol biosynthesis (1,2); 7-dehydrocholesterol reductase (3); diamine oxidase (4,5); glucose transport (6); insulysin (7); pyruvate kinase (8-10); and ubiquinol:cytochrome *c* reductase (11). Phenformin is an activator of AMP-stimulated protein kinase, and like ther AMPK activators also induce nucleoli re-organization, with attendant changes in cell proliferation. Among the compounds tested, phenformin and resveratrol had the most pronounced impact on nucleolar organization (12). **1**. McDonald & Dalidowicz (1962) *Biochemistry* **1**, 1187. **2**. Dalidowicz & McDonald (1965) *Biochemistry* **4**, 1138. **3**. Dempsey (1969) *Meth. Enzymol.* **15**, 501. **4**. Cubria, Ordonez, Alvarez-Bujidos, Negro & Ortiz (1991) *Comp. Biochem. Physiol. B* **100**, 543. **5**. Finazzi-Agro, Floris, Fadda & Crifo (1979) *Agents Actions* **9**, 244. **6**. Kruger, Altschuld, Hollobaugh & Jewett (1970) *Diabetes* **19**, 50. **7**. Burghen, Kitabchi & Brush (1972) *Endocrinology* **91**, 633. **8**. Davidoff & Reid (1977) *Meth. Enzymol.* **46**, 548. **9**. Davidoff & Carr (1972) *Proc. Natl. Acad. Sci. U.S.A.* **69**, 1957. **10**. Kayne (1973) *The Enzymes*, 3rd ed. (Boyer, ed.), **8**, 353. **11**. Slater (1967) *Meth. Enzymol.* **10**, 48. **12**. Kodiha, Salimi, Wang & Stochaj (2014) *PLoS One* **9**, e88087.

Phenidone, *See* *1-Phenyl-3-pyrazolidone*

Pheniprazine, *See* *β-Phenylisopropylhydrazine*

Pheniramine

This antihistamine (FW$_{free-base}$ = 240.35 g/mol; CAS 86-21-5; insoluble in water; maleate and *p*-aminosalicylate salts are water-soluble), also known as *N,N*-dimethyl-γ-phenyl-2-pyridinepropanamide, is more frequently abused than other antihistamines relative to its market share. One of the histamine H$_1$ antagonists with little sedative action, pheniramine is often used in treatment of hay fever, rhinitis, allergic dermatoses, and pruritus. **Target(s):** H$_1$ histamine receptor (1,2); and histamine *N*-methyltransferase, mildly inhibited (3). **1**. Karadag, Ulugol, Dokmeci & Dokmeci (1996) *Jpn. J. Pharmacol.* **71**, 109. **2**. Paton & Webster (1985) *Clin. Pharmacokinet.* **10**, 477. **3**. Thithapandha & Cohn (1978) *Biochem. Pharmacol.* **27**, 263.

Phenobarbital

This barbiturate (FW = 232.24 g/mol; CAS 50-06-6; pK_1 = 7.3 and pK_2 = 11.8; water-soluble), known also as phenobarbitone, luminal, and 5-ethyl-5-phenyl-2,4,6(1H,3H,5H)-pyrimidinetrione, is a sedative and anticonvulsant that induces the biosynthesis of several proteins. **Key Pharmacokinetic Parameters:** *See* Appendix II in Goodman & Gilman's *THE PHARMACOLOGICAL BASIS OF THERAPEUTICS*, 12th Edition (Brunton, Chabner & Knollmann, eds.) McGraw-Hill Medical, New York (2011). **Target(s):** alcohol dehydrogenase, NADP$^+$-dependent, *or* aldehyde reductase (1-5); alkaline phosphatase (6); 1,6-anhydro-D-fructose reductase (7); γ-aminobutyrate aminotransferase (8); cholesterol-5,6-oxide hydrolase (9); glucose-6-phosphatase (10); glucuronate reductase (11); γ-glutamyl transpeptidase, *or* γ-glutamyltransferase (12,13); glycerol dehydrogenase (14); inositol oxygenase (15); pyrophosphatase, *or* inorganic diphosphatase (16); and succinate-semialdehyde dehydrogenase (8). **1**. Erwin & Deitrich (1973) *Biochem. Pharmacol.* **22**, 2615. **2**. Hara, Seiriki, Nakayama & Sawada (1985) *Prog. Clin. Biol. Res.* **174**, 291. **3**. Davidson & Murphy (1985) *Prog. Clin. Biol. Res.* **174**, 251. **4**. Baumann (1982) *Arch. Toxicol. Suppl.* **5**, 136. **5**. Branlant (1982) *Eur. J. Biochem.* **129**, 99. **6**. Tardivel, Banide, Porembska, *et al.* (1992) *Enzyme* **46**, 276. **7**. Sakuma, Kametani & Akanuma (1998) *J. Biochem.* **123**, 189. **8**. Whittle & Turner (1978) *J. Neurochem.* **31**, 1453. **9**. Levin, Michaud, Thomas & Jerina (1983) *Arch. Biochem. Biophys.* **220**, 485. **10**. Plewka, Kaminski, Plewka & Nowaczyk (2000) *Mech. Ageing Dev.* **113**, 49. **11**. Flynn, Cromlish & Davidson (1982) *Meth. Enzymol.* **89**, 501. **12**. Sachdev, Leahy & Chace (1983) *Biochim. Biophys. Acta* **749**, 125. **13**. Allison (1985) *Meth. Enzymol.* **113**, 419. **14**. Flynn & Cromlish (1982) *Meth. Enzymol.* **89**, 237. **15**. Reddy, Pierzchala & Hamilton (1981) *J. Biol. Chem.* **256**, 8519. **16**. Naganna (1950) *J. Biol. Chem.* **183**, 693.

Phenobarbitol, *See* *Phenobarbital*

Phenobarbitone, *See* *Phenobarbital*

[*N*-Phe^1]-Nociceptin Fragment 1-13 Amide, *See* *[N-Phe^1]-Nociceptin Fragment 1-13 Amide* under *Nociceptin*

Phenolphthalein

This common pH-indicating dye and laxative (FW = 318.33 g/mol; CAS 77-09-8; pK_a = 9.7 at 25°C), named systematically as 3,3-bis(4-hydroxyphenyl)-1(3H)-isobenzofuranone, is soluble in ethanol (~1 g/12 mL) and nearly insoluble in water. Phenolphthalein is colorless at pH ≤ 8.2, but turns red at pH ≥ 10.

A typical stock indicator solution is 0.05 g phenolphthalein in 100 mL of 50% v/v ethanol/water. A drop added per 50 mL is sufficient for use as a visual indicator in acid/base titrimetry. **Use as a Laxative:** Phenolphthalein was for many years a key component of several commercial laxatives. Perfusion of phenolphthalein results in significantly reduced water absorption in both ileum and colon. In humans, 30 mg phenolphthalein is the optimum laxative dose. Even so, the typical single oral dose was 30–200 mg for adults and children 12 years of age and older. (It was often also present as an undeclared additive in weight-loss products sold as food supplements.) Although laxatives are infrequently used, some 3-4% of the U.S. population had a lifetime use of a laxative product that exceeded 350 times. Concerns about its potential as a cancer-causing agent and other questions about its effectiveness eventually resulted in its removal from many over-the-counter laxative formulations. **Carcinogenic Potential:** Phenolphthalein is *reasonably anticipated to be a human carcinogen*, based on sufficient evidence of carcinogenicity in experimental animals (1). In humans, it is absorbed from the gastrointestinal tract and undergoes extensive first-pass metabolism in the intestinal epithelium and liver, resulting in almost complete conversion to its glucuronide and consequential elimination in the bile. Aneuploidy is a major cause of human reproductive failure and plays a large role in cancer. Phenolphthalein induces tumors in rodents, but its primary mechanism does not seem to be DNA damage. In heterozygous TSG-p53 mice, PHT induces lymphomas and also micronuclei, many containing kinetochores indicative of chromosome loss (2). The induction of aneuploidy would be compatible with the loss of the normal p53 gene seen in the lymphomas. Phenolphthalein also induces abnormalities in tubulin polymerization and deregulates the centrosome duplication cycle, causing centrosome amplification, often attended by apoptosis (2). **Target(s):** aldo-keto reductases 1C1, 1C2, and 1C4, *or* dihydrodiol dehydrogenases DD1, DD2, and DD4 (3,4); cytochrome P450 (5); glucose transport (6,7); glucuronosyltransferase (8,9); glutathione *S*-transferase (10); iodothyronine deiodinase (11); and thymidylate synthase (11-14). **1.** *National Toxicology Program Report on Carcinogens Background Document for Phenolphthalein* (1999) Research Triangle Park, 92 pp. **2.** Heard, Rubitski, Spellman & Schuler (2013) *Environ. Mol. Mutagen.* 54, 308. **3.** Atalla, Breyer-Pfaff & Maser (2000) *Xenobiotica* 30, 755. **4.** Matsuura, Hara, Deyashiki, *et al.* (1998) *Biochem. J.* 336, 429. **5.** Beyhl (1981) *Experientia* 37, 943. **6.** Adamic & Bihler (1967) *Mol. Pharmacol.* 3, 188. **7.** Forsling & Widdas (1968) *J. Physiol.* 194, 545. **8.** Rao, Rao & Breuer (1976) *Biochim. Biophys. Acta* 452, 89. **9.** Mulder (1974) *Biochem. Soc. Trans.* 2, 1172. **10.** Balabaskaran & Muniandy (1984) *Phytochemistry* 23, 251. **11.** Fekkes, Hennemann & Visser (1982) *FEBS Lett.* 137, 40. **12.** Stout, Tondi, Rinaldi, *et al.* (1999) *Biochemistry* 38, 1607. **13.** Shoichet, Stroud, Santi, Kuntz & Perry (1993) *Science* 259, 1445. **14.** Costi, Rinaldi, Tondi, *et al.* (1999) *J. Med. Chem.* 42, 2112.

Phenolphthalein Bisphosphate

This water-soluble reagent (FW = 491.26 g/mol), also known as phenolphthalein diphosphate and phenolphthalein phosphate, is an alternative substrate for phosphomonoesterases (*e.g.*, alkaline phosphatase), as is the monophosphate. Hydrolysis of the phosphate group generates free phenolphthalein which is red in alkaline conditions (1). **Target(s):** inositol 1,4,5-trisphosphate receptors, weakly inhibited (2); phosphoprotein phosphatase (3). **1.** Manning, Steinetz, Babson & Butler (1966) *Enzymologia* 31, 309. **2.** Richard, Bernier, Boulay & Guillemette (1994) *Can. J. Physiol. Pharmacol.* 72, 174. **3.** Polya & Haritou (1988) *Biochem. J.* 251, 357.

Phenolphthalein Bissulfate

This water-soluble reagent (FW = 493.45 g/mol), also known as phenolphthalein disulfate, inhibits iduronate 2-sulfatase. It is a weak alternative substrate for steryl-sulfatase. **Target(s):** arylsulfatase (1); iduronate-2-sulfatase (2;) and inositol 1,4,5-trisphosphate receptors, weakly inhibited (3). **1.** Sampson, Vergara, Fedor, Funk & Benkovic (1975) *Arch. Biochem. Biophys.* 169, 372. **2.** Yutaka, Fluharty, Stevens & Kihara (1982) *J. Biochem.* 91, 433. **3.** Richard, Bernier, Boulay & Guillemette (1994) *Can. J. Physiol. Pharmacol.* 72, 174.

Phenolphthalein Monophosphate, See *Phenolphthalein Bisphosphate*

Phenolphthalein Phosphate, See *Phenolphthalei Bisphosphate*

Phenolphthalin, See *Kastle-Meyer Reagent*

Phenol Red

This pH-indicating dye (FW = 354.38 g/mol; CAS 143-74-8; pK_a = 7.9; yellow < pH 6.8; red > pH 8.4), also known as phenolsulfonphthalein and sulfophthalein, is a diagnostic aid for evaluating renal function. A typical stock indicator solution is 0.1 g phenol red in 28.2 mL 10 mM NaOH and 221.8 mL water. **Target(s):** aldehyde reductase (1); glutaminase (2,3); γ-glutamyl transpeptidase (4); glutathione *S*-transferase (5-7); 4-methyleneglutaminase (8); phosphate uptake (9); thromboxane A_2/prostaglandin H_2 receptor (10); and thyroglobulin iodination (11). **1.** Shinoda, Mori, Shintani, Ishikura & Hara (1999) *Biol. Pharm. Bull.* 22, 741. **2.** Richterich-Van Baerle, Goldstein & Dearborn (1957) *Enzymologia* 18, 190. **3.** Goldstein, Richterich-Van Baerle & Dearborn (1957) *Enzymologia* 18, 355. **4.** Binkley & Olson (1951) *J. Biol. Chem.* 188, 451. **5.** Balabaskaran & Muniandy (1984) *Phytochemistry* 23, 251. **6.** Warholm, Jensson, Tahir & Mannervik (1986) *Biochemistry* 25, 4119 **7.** Alin, Jensson, Guthenberg, *et al.* (1985) *Anal. Biochem.* 146, 313. **8.** Ibrahim, Lea & Fowden (1984) *Phytochemistry* 23, 1545. **9.** Mary & Rao (2002) *Kobe J. Med. Sci.* 48, 59. **10.** Greenberg, Johns, Kleha, *et al.* (1994) *J. Pharmacol. Exp. Ther.* 268, 1352. **11.** Gruffat, Gonzalvez, Mauchamp & Chabaud (1991) *Mol. Cell. Endocrinol.* 81, 195.

Phenol *N*-(5-Salicylato)carbamate, See *5-(Carbophenoxyamino)salicylate*

Phenolsulfonphthalein, See *Phenol Red*

Phenothiazine

This antiprotozoal agent (FW = 199.28 g/mol; CAS 92-84-2), also known as dibenzothiazine, is a yellow solid that is used as an insecticide and in the chemical synthesis of many pharmaceuticals (*e.g.*, chlorpromazine and thioridazine). Phenothiazine is soluble in benzene and insoluble in chloroform and water. It is also readily oxidized in air. **Target(s):** cGMP phosphodiesterase (1); cholinesterase, *or* butyrylcholinesterase, K_i = 31 μM (2); 3',5'-cyclic-nucleotide phosphodiesterase (3); pyruvate dehydrogenase complex (4). **1.** Filburn, Colpo & Sacktor (1979) *Mol. Pharmacol.* 15, 257. **2.** Darvesh, McDonald, Penwell, *et al.* (2005) *Bioorg. Med. Chem.* 13, 211. **3.** Thompson & Strada (1984) in *Methods in Enzymatic Analysis* (Bergmeyer, ed.) 4, 127, Verlag Chemie, Weinheim. **4.** Miernyk, Fang & Randall (1987) *J. Biol. Chem.* 262, 15338.

Phenothiazone

This phenthiazine metabolite (FW = 213.26 g/mol), found in urine and feces, inhibits creatine kinase (1,2) and succinate dehydrogenase (3). **1.** Kuby & Noltmann (1962) *The Enzymes*, 2nd ed. (Boyer, Lardy & Myrbäck, eds.), **6**, 515. **2**. Solvonuk, McRae & Collier (1956) *Can. J. Biochem. Physiol.* **34**, 481. **3**. Collier & Allen (1941) *J. Biol. Chem.* **140**, 675.

Phenoxyacetamide

This amide (FW = 151.17 g/mol; CAS 129689-30-1; M.P. = 101-103°C) inhibits soluble epoxide hydrolase, K_i = 0.6 mM (1) and weakly inhibits chymotrypsin, IC$_{50}$ ≈ 25 mM (2). Notably, phenoxyacetamide has been used as a pharmacophore to produce small, drug-like molecules that are low-μM inhibitors of Type III Secretion System (T3SS) in *Pseudomonas aeruginosa* in assays of T3SS-mediated secretion and translocation (3). **1**. Rink, Kingma, Lutje Spelberg & Janssen (2000) *Biochemistry* **39**, 5600. **2**. Baker & Hurlbut (1967) *J. Med. Chem.* **10**, 1129. **3**. Williams, Torhan, Neelagiri, *et al.* (2015) *Bioorg. Med. Chem.* **23**, 1027.

Phenoxyacetic Acid

This acid (FW = 152.15 g/mol), also known as *O*-phenylglycolic acid and phenyloxyacetic acid (FW$_{free-acid}$ = 152.15 g/mol; melting point = 98°C) has a pK_a value of 3.12 at 25°C and exhibits a fungicide activity. **Target(s):** benzoylformate decarboxylase (1); lactate dehydrogenase (2,3); mandelate racemase (4); penicillin amidase, *or* penicillin V acylase, product inhibition (5). **1**. Weiss, Garcia, Kenyon, Cleland & Cook (1988) *Biochemistry* **27**, 2197. **2**. Baker (1967) *Design of Active-Site-Directed Irreversible Enzyme Inhibitors*, Wiley, New York. **3**. Ottolenghi & Denstedt (1958) *Can. J. Biochem. Physiol.* **36**, 1075. **4**. Hegeman, Rosenberg & Kenyon (1970) *Biochemistry* **9**, 4029. **5**. Sudhakaran & Shewale (1995) *Hindustan Antibiot. Bull.* **37**, 9.

***N*-(Phenoxyacetyl)aniline,** *See Phenoxyacetanilide*

***N*-[(*N*-Phenoxyacetyl-L-histidyl)-(5-amino-4-hydroxy-2-isopropyl-7-methyloctanoyl)]-L-isoleucine 2-Pyridylmethylamide,** *See U-71017*

N-(Phenoxyacetyl)-L-valyl-L-prolyl-DL-valine trifluoromethyl ketone

This halomethyl ketone (FW = 485.50 g/mol) inhibits leukocyte elastase, K_i = 17 nM, which, in addition to catalyzing elastin proteolysis, also disrupts tight junctions, damages certain tissues, breaks down cytokines and alpha proteinase inhibitor, cleaves IgA and IgG, cleave complement component C3bi, a component and the CR1, receptor that triggers phagocytosis in neutrophils. **1**. Veale, Bernstein, Bohnert, *et al.* (1997) *J. Med. Chem.* **40**, 3173.

4-(4'-Phenoxyanilino)-6,7-dimethoxyquinazoline, *See 6,7-Dimethoxy-4-(4'-phenoxyanilino)quinazoline*

Phenoxybenzamine

This tertiary amine (FW$_{free-base}$ = 303.83 g/mol; CAS 59-96-1; M.P. = 38-40°C), known systematically as *N*-(2-chloroethyl)-*N*-(1-methyl-2-phenoxyethyl)benzenemethanamine, is a calmodulin antagonist and a nonselective α-adrenergic antagonist. The free base is soluble in benzene

and the hydrochloride salt is soluble in ethanol and chloroform and sparingly soluble in water. **Target(s):** adenylate cyclase (1); α-adrenergic receptor (2,3); Ca^{2+}-dependent cyclic-nucleotide phosphodiesterase (4); calmodulin (5-8); prostaglandin synthase, weakly inhibited (9); and retinyl ester hydrolase (10). **1**. Walton, Liepmann & Baldessarini (1978) *Eur. J. Pharmacol.* **52**, 231. **2**. Thoenen, Hurlimann & Haefely (1964) *Experientia* **20**, 272. **3**. Yu & Koss (2003) *J. Ocul. Pharmacol. Ther.* **19**, 255. **4**. Hidaka, Inagaki, Nishikawa & Tanaka (1988) *Meth. Enzymol.* **159**, 652. **5**. Weiss (1983) *Meth. Enzymol.* **102**, 171. **6**. Ning & Sanchez (1996) *Mol. Endocrinol.* **10**, 14. **7**. Zhang, Prozialeck & Weiss (1990) *Mol. Pharmacol.* **38**, 698. **8**. Cimino & Weiss (1988) *Biochem. Pharmacol.* **37**, 2739. **9**. Kandasamy (1977) *Clin. Exp. Pharmacol. Physiol.* **4**, 585. **10**. Schindler (2001) *Lipids* **36**, 543.

2-(3-Phenoxybenzoylamino)benzoic Acid

This benzoic acid derivative (FW = 333.34 g/mol) inhibits *Enterococcus faecalis* β-ketoacyl-[acyl-carrier-protein] synthase III, IC$_{50}$ = 2.7 μM (1,2). **1**. Nie, Perretta, Lu, *et al.* (2005) *J. Med. Chem.* **48**, 1596. **2**. Ashek, San Juan & Cho (2007) *J. Enzyme Inhib. Med. Chem.* **22**, 7.

3-(3-Phenoxybenzoylamino)thiophene-2-carboxylic Acid

This benzoic acid derivative (FW = 338.36 g/mol) inhibits *Enterococcus faecalis* β-ketoacyl-[acyl-carrier-protein] synthase III, IC$_{50}$ = 10 μM (1,2). Note that the 2-(3-phenoxybenzoylamino)thiophene-3-carboxylic acid analogue is an even weaker inhibitor, IC$_{50}$ = 43 μM. **1**. Nie, Perretta, Lu, *et al.* (2005) *J. Med. Chem.* **48**, 1596. **2**. Ashek, San Juan & Cho (2007) *J. Enzyme Inhib. Med. Chem.* **22**, 7.

3-(4-Phenoxybenzyl)thiazolidin-2-one

This thiazolidin-2-thione (FW = 285.37 g/mol) inhibits *Escherichia coli* β-ketoacyl-[acyl-carrier-protein] synthase III. The 1,1-dioxide also inhibits. **1**. Alhamadsheh, Waters, Huddler, *et al.* (2007) *Bioorg. Med. Chem. Lett.* **17**, 879.

2-[(2-Phenoxybiphenyl-4-ylcarbonyl)amino]benzoic Acid

This aminobenzoic acid derivative (FW = 408.43 g/mol) inhibits *Enterococcus faecalis* β-ketoacyl-[acyl-carrier-protein] synthase III, IC$_{50}$ = 56 nM (1,2). **1**. Nie, Perretta, Lu, *et al.* (2005) *J. Med. Chem.* **48**, 1596. **2**. Ashek, San Juan & Cho (2007) *J. Enzyme Inhib. Med. Chem.* **22**, 7.

Phenoxymethylpenicillin, *See Penicillin V*

2(R)-[2-[4-Phenoxyphenyl]ethyl]-4(S)-butylpentanedioic Acid 1-((S)-tert-Butylglycine Methylamide) Amide

This peptidomimetic (FW = 509.67 g/mol), also known as N-[4(S)-carboxy-2(R)-[2-[4-phenoxyphenyl]ethyl]octanoyl]-α(S)-(tert-butyl)glycine methylamide, inhibits gelatinase A, or matrix metalloproteinase 2 and stromelysin 1 or matrix metalloproteinase 3, with K_i values of 3000 and 200 nM, respectively. 1. Esser, Bugianesi, Caldwell, et al. (1997) J. Med. Chem. **40**, 1026.

4-(4-Phenoxyphenylsulfonyl)butane-1,2-dithiol

This dithiol (FW = 354.50 g/mol) is a slow-binding inhibitor of gelatinase A, or matrux metalloproteinase 2, K_i = 0.046 μM (1,2), gelatinase B, or matrix metalloproteinase 9, K_i = 0.1 μM (2), membrane-type matrix metalloproteinase 1, matrix metalloproteinase 14 (2), and stromelysin, or matrix metalloproteinase 3 (2). 1. Rosenblum, Meroueh, Kleifeld, et al. (2003) J. Biol. Chem. **278**, 27009. 2. Bernardo, Brown, Li, Fridman & Mobashery (2002) J. Biol. Chem. **277**, 11201.

(4-Phenoxyphenylsulfonyl)methylthiirane

This thiirane (FW = 306.41 g/mol), also known as matrix metalloproteinase-2/matrix metalloproteinase-9 inhibitor IV (MMP-2/MMP-9 inhibitor IV) and SB-3CT, is a mechanism-based inhibitor of gelatinase A, or matrux metalloproteinase 2, K_i = 0.014 μM (1,2), gelatinase B, or matrix metalloproteinase 9, K_i = 0.6 μM (2), membrane-type matrix metalloproteinase 1, matrix metalloproteinase 14 (2), matrilysin, K_i = 96 μM, stromelysin 1, or matrix metalloproteinase 3, K_i = 15 μM (2). 1. Rosenblum, Meroueh, Kleifeld, et al. (2003) J. Biol. Chem. **278**, 27009. 2. Bernardo, Brown, Li, Fridman & Mobashery (2002) J. Biol. Chem. **277**, 11201.

2-(3-Phenoxy-4-piperidin-1-ylbenzoylamino)benzoic Acid

This ACPS inhibitor (FW = 415.47 g/mol) targets Enterococcus faecalis β-ketoacyl-[acyl-carrier-protein] synthase III, IC$_{50}$ = 0.29 μM (1-3). A related analogue with piperazine in place of piperidine is a weaker inhibitor, IC$_{50}$ = 185 μM. 1. Castillo & Perez (2008) Mini Rev. Med. Chem. **8**, 36. 2. Nie, Perretta, Lu, et al. (2005) J. Med. Chem. **48**, 1596. 3. Ashek, San Juan & Cho (2007) J. Enzyme Inhib. Med. Chem. **22**, 7.

6-[4-(Phenoxy)piperidin-1-yl]-N-(3-methylbutyl)pyridazine-3-carboxamide

This pyridazine derivative (FW = 365.19 g/mol) inhibits stearoyl-CoA 9-desaturase (mouse IC$_{50}$ = 63 nM), which catalyzes the rate-limiting step in the cellular synthesis of monounsaturated fatty acids from saturated fatty acids (1). A proper ratio of saturated to monounsaturated fatty acids contributes to membrane fluidity. Alterations in this ratio have been implicated in various disease states including cardiovascular disease, obesity, non-insulin-dependent diabetes mellitus, hypertension, neurological diseases, immune disorders, and cancer. Studies with SCD-1 knockout mice established that these animals are lean and protected from leptin deficiency-induced and diet-induced obesity, with greater whole body insulin sensitivity than wild-type animals. 1. Liu, Lynch, Freeman, et al. (2007) J. Med. Chem. **50**, 3086.

Phenoxy-2-propanone, See Phenoxyacetone

(–)-Phenserine

This long-acting AChE inhibitor and Aß$_{42}$-lowering agent (FW = 337,42 g/mol; CAS 101246-66-6), also known as (–)-N-phenylcarbamoyl eseroline, is a carbamate analogue of eserine (or physostigmine) that inhibits acetylcholinesterase and is under active investigation for its potential in cholinomimetic therapy to reduce cognitive impairments associated with aging and Alzheimer's Disease. Phenserine is a potent and highly selective AChE inhibitor (IC$_{50}$ = 22 nM), displaying 70x greater inhibitory action than observed with butyrylcholinesterase, or BChE (IC$_{50}$ = 1560 nM), both in vitro and in clinical trials for treatment of Alzheimer's disease (1-5). Compared to physostigmine and tacrine, phenserine appears to be less toxic and robustly enhances cognition in animal models. In rats, phenserine achieves maximum acetylcholinesterase inhibition of 73.5% at 5 min, maintaining high and relatively constant inhibition for >8 hours (6). Phenserine decreased the levels of secreted β-amyloid precursor protein (β-APP) in the cerebrospinal fluid (CSF) of forebrain cholinergic system-lesioned rats, whereas DFP, a relatively non-specific cholinesterase inhibitor, failed to affect CSF levels of secreted β-APP. Such findings suggest that phenserine alters the induction of cortical β-APP mRNAs and increased levels of secreted β-APP in the CSF (7). Phenserine reduces Aβ levels by regulating β-APP translation via a recently described iron regulatory element in the 5'-untranslated region of β-APP mRNA that was previously shown to be up-regulated in the presence of interleukin-1 (8). Other dual AChE/Aß$_{42}$ inhibitors include: rivastigmine, ladostigil, asenapine, phenserine, amitriptyline, clomipramine, doxepin and desipramine. Posiphen, the better-tolerated (+)-enantiomer of phenserine, is devoid of anticholinesterase action, but represses translation of neural α-synuclein, a druggable target in the treatment of Parkinson Disease (9). 1. Luo, Yu, Zhan, et al. (2005) J. Med. Chem. **48**, 986. 2. Iijima, Greig, Garofalo, et al. (1993) Psychopharmacology (Berlin) **112**, 415. 3. al-Jafari, Kamal, Greig, Alhomida & Perry (1998) Biochem. Biophys. Res. Commun. **248**, 180. 4. Giacobini (2003) Neurochem. Res. **28**, 515. 5. Cerasoli, Griffiths, Doctor, et al. (2005) Chem. Biol. Interact. **157-158**, 363. 6. Iijima, Greig, Garofalo, et al. (1993) Psychopharmacology (Berlin) **112**, 415. 7. Haroutunian, Greig, Pei, et al. (1997) Brain Res. Mol. Brain Res. **46**, 161. 8. Shaw, Utsuki, Rogers, et al. (2001) Proc. Natl. Acad. Sci U.S.A. **98**,7605. 9. Mikkilineni, Cantuti-Castelvetri, Cahill, et al. (2012) Parkinson's Disease Article ID: 142372.

Phentermine

This central nervous system stimulant and sympathomimetic agent ($FW_{free-base}$ = 149.24 g/mol; CAS 122-09-8 (free-base) and 1197-21-3 (HCl salt); B.P. = 205°C at 750 mmHg), also known as α,α-dimethylbenzene ethanamine, is a tertiary butylamine that exhibits pharmacologic actions and uses similar to those of dextroamphetamine, has been used most frequently in the treatment of obesity and as an anorexic agent. One example is Qsymia®, an extended release formulation of phentermine and topiramate used in weight control. *Note*: Phentermine is classified as a controlled substance. **Target(s):** K$^+$ channel, voltage-gated, hK$_v$1.5 (1); monoamine oxidase, weakly inhibited (2-4). 1. Perchenet, Hilfiger, Mizrahi & Clement-Chomienne (2001) *J. Pharmacol. Exp. Ther.* **298**, 1108. 2. Nandigama, Newton-Vinson & Edmondson (2002) *Biochem. Pharmacol.* **63**, 865. 3. Kilpatrick, Traut & Heal (2001) *Int. J. Obes. Relat. Metab. Disord.* **25**, 1454. 4. Ulus, Maher & Wurtman (2000) *Biochem. Pharmacol.* **59**, 1611.

Phentolamine

This versatile nonselective α-adrenergic receptor antagonist ($FW_{free-base}$ = 281.36 g/mol; CAS 50-60-2 (free base), 65-28-1 (mesylate salt), and 73-05-2 (mono-HCl salt, which is slightly soluble in water (1 g/ 50 mL water)), known systematically as 3-[4,5-dihydro-1*H*-imidazol-2-ylmethyl-(4-methylphenyl)amino]phenol, has been used to treat hypertension, hypertensive emergencies, pheochromocytoma, vasospasm associated with Reynaud's Disorder and frostbite. The methanesulfonate salt is also known as Regitine™. *Note*: Aqueous solutions are unstable and must be freshly prepared and cannot be stored. **Target(s):** α-adrenergic receptor (1); K$^+$-channels, ATP-sensitive, *or* K$_{ATP}$ channels (2-4); and protein kinase C (5,6). 1. Langer (1998) *Eur. Urol.* **33** Suppl. 2, 2. 2. Proks & Ashcroft (1997) *Proc. Natl. Acad. Sci. U.S.A.* **94**, 11716. 3. Dunne (1991) *Brit. J. Pharmacol.* **103**, 1847. 4. Plant & Henquin (1990) *Brit. J. Pharmacol.* **101**, 115. 5. Kikkawa, Minakuchi, Takai & Nishizuka (1983) *Meth. Enzymol.* **99**, 288. 6. Kikkawa & Nishizuka (1986) *The Enzymes*, 3rd ed. (Boyer & Krebs, eds.), **17**, 167.

Phenylacetaldehyde

This aldehyde (FW = 120.15 g/mol; CAS 122-78-1; M.P. = 33-34°C; B.P. = 195°C), also known as α-tolualdehyde and Hyacinthin, is an oily liquid with an odor that is reminiscent of hyacinth. Phenylacetaldehyde is produced in the degradation of L-phenylalanine and is a mechanism-based inhibitor of dopamine β-monooxygenase (1,2). *Note*: Phenylacetaldehyde polymerizes upon standing. 1. Ator & Ortiz de Montellano (1990) *The Enzymes*, 3rd ed., **19**, 213. 2. Bossard & Klinman (1986) *J. Biol. Chem.* **261**, 16421.

2-Phenylacetamidine

This amidine ($FW_{free-base}$ = 134.18 g/mol) inhibits trypsin (K_i = 15.1 mM) and phenylalanyl-tRNA synthetase (K_i = 0.23 mM). 2-Phenylacetamidine should not be confused with *N*-phenylacetamidine (CH$_3$C(NH$_2^+$)NHC$_6$H$_5$). **Target(s):** complement system, guinea pig, weakly inhibited (1); phenylalanyl-tRNA synthetase (2); tryptophan 2-monooxygenase (3); trypsin, weakly inhibited (4). 1. Baker & Erickson (1969) *J. Med. Chem.* **12**, 408. 2. Danenberg & Santi (1975) *J. Med. Chem.* **18**, 528. 3. Emanuele, Heasley & Fitzpatrick (1995) *Arch. Biochem. Biophys.* **316**, 241. 4. Mares-Guia & Shaw (1965) *J. Biol. Chem.* **240**, 1579.

1-Phenyl-2-acetamido-3-butanone, *See* 2-Acetamido-1-phenyl-3-butanone

Phenylacetic Acid

This phenylalanine-derived metabolite ($FW_{free-acid}$ = 136.15 g/mol; CAS 103-82-2 (acid) and 114-70-5 (Na salt); pK_a = 4.25 at 25°C) has a melting point of 76.5°C, a boiling point of 265.5°C, and is slightly soluble in water (1.6 g per 100 mL at 20°C). It is more soluble in ethanol, chloroform, and diethyl ether. *Note*: Because phenylacetic acid is a precursor of phenylacetone, which in turn is used in the illicit production of methamphetamine, its purchase is regulated in the U.S. **Target(s):** D-amino-acid oxidase (1); decarboxylase (2); [branched-chain α-keto-acid dehydrogenase] kinase, *or* [3-methyl-2-oxobutanoate dehydrogenase (acetyl-transferring)] kinase (3,4); carboxypeptidase A (5-11); catechol oxidase (12); Ca^{2+}-transporting ATPase (13); choline *O*-acetyltransferase (14,15); chymotrypsin, weakly inhibited (16,17); glutamate decarboxylase (9); haloacetate dehalogenase (18); (*S*)-2-hydroxy-acid oxidase (19); hydroxylaminolysis (nonenzymatic) of amino acid esters (20); imidazoleacetate 4-monooxygenase (21,22); (*S*)-mandelate dehydrogenase (23); mandelate racemase, K_i = 0.2 mM (24,25); monophenol monooxygenase (9); Na$^+$/K$^+$-exchanging ATPase (26); penicillin amidase (27-32); phenylalanine 2-monooxygenase (33); pyruvate carboxylase (34,35); stearoyl-CoA 9-desaturase (36); succinate dehydrogenase (37,38); tyrosine aminotransferase (39); tyrosine-ester sulfotransferase (40); and zeatin 9-aminocarboxyethyltransferase (41). 1. Klein & Austin (1953) *J. Biol. Chem.* **205**, 725. 2. Weiss, Garcia, Kenyon, Cleland & Cook (1988) *Biochemistry* **27**, 2197. 3. Paxton (1988) *Meth. Enzymol.* **166**, 313. 4. Paxton & Harris (1984) *Arch. Biochem. Biophys.* **231**, 58. 5. Pétra (1970) *Meth. Enzymol.* **19**, 460. 6. Elkins-Kaufman, Neurath & De Maria (1949) *J. Biol. Chem.* **178**, 645. 7. Smith (1951) *The Enzymes*, 1st ed. (Sumner & Myrbäck, eds.), **1** (part 2), 793. 8. Neurath (1960) *The Enzymes*, 2nd ed. (Boyer, Lardy & Myrbäck, eds.), **4**, 11. 9. Webb (1966) *Enzyme and Metabolic Inhibitors*, vol. 2, Academic Press, New York. 10. Coleman & Vallee (1964) *Biochemistry* **3**, 1874. 11. Auld & Vallee (1970) *Biochemistry* **9**, 602. 12. Aerts & Vercauteren (1964) *Enzymologia* **28**, 1. 13. Jankowski, Luftmann, Tepel, *et al.* (1998) *J. Amer. Soc. Nephrol.* **9**, 1249. 14. Potempska, Loo & Wisniewski (1984) *J. Neurochem.* **42**, 1499. 15. Korey, de Braganza & Nachmansohn (1951) *J. Biol. Chem.* **189**, 705. 16. Huang & Niemann (1952) *J. Amer. Chem. Soc.* **74**, 5963. 17. Neurath & Gladner (1951) *J. Biol. Chem.* **188**, 407. 18. Goldman (1965) *J. Biol. Chem.* **240**, 3434. 19. Robinson, Keay, Molinari & Sizer (1962) *J. Biol. Chem.* **327**, 2001. 20. Grant & Alburn (1965) *Biochemistry* **4**, 1913. 21. Watanabe, Kambe, Imamura, *et al.* (1983) *Anal. Biochem.* **130**, 321. 22. Maki, Yamamoto, Nozaki & Hayaishi (1969) *J. Biol. Chem.* **244**, 2942. 23. Lehoux & Mitra (1999) *Biochemistry* **38**, 5836. 24. St. Maurice & Bearne (2000) *Biochemistry* **39**, 13324. 25. Hegeman, Rosenberg & Kenyon (1970) *Biochemistry* **9**, 4029. 26. Dwivedy & Shah (1982) *Neurochem. Res.* **7**, 717. 27. Savidge & Cole (1975) *Meth. Enzymol.* **43**, 705. 28. Alkema, Floris & Janssen (1999) *Anal. Biochem.* **275**, 47. 29. Niersbach, Kuhne, Tischer, *et al.* (1995) *Appl. Microbiol. Biotechnol.* **43**, 679. 30. Mandel, Kostner, Sijmer, *et al.* (1975) *Prikl. Biokhim. Mikrobiol.* **11**, 219. 31. Warburton, Balasingham, Dunnil & Lilly (1972) *Biochim. Biophys. Acta* **284**, 278. 32. Balasingham, Warburton, Dunnil & Lilly (1972) *Biochim. Biophys. Acta* **276**, 250. 33. Koyama (1984) *J. Biochem.* **96**, 421. 34. Bahl, Matsuda, DeFronzo & Bressler (1997) *Biochem. Pharmacol.* **53**, 67. 35. Liu, Jetton & Leahy (2002) *J. Biol. Chem.* **277**, 39163. 36. Scott & Foote (1979) *Biochim. Biophys. Acta* **573**, 197. 37. Quastel & Woolbridge (1928) *Biochem. J.* **22**, 689. 38. Massart (1950) *The Enzymes*, 1st ed. (Sumner & Myrbäck, eds.), **1** (part 1), 307. 39. Jacoby & La Du (1964) *J. Biol. Chem.* **239**, 419. 40. Duffel (1994) *Chem. Biol. Interact.* **92**, 3. 41. Parker, Entsch & Letham (1986) *Phytochemistry* **25**, 303.

Phenylacetone

This plant auxin (FW = 134.18 g/mol; oily liquid. M.P. = −16 to −15°C; B.P. = 214°C), also known as 1-phenyl-2-propanone, It is a competitive inhibitor of acetylcholinesterase (1) and a weak inhibitor of chymotrypsin,

IC$_{50}$ ≈ 19 mM (2). *Note*: Phenylacetone is photosensitive and liberates highly reactive benzyl radicals upon photolysis. Moreover, because phenylacetone is used in the illicit production of methamphetamine, its purchase is regulated in the U.S. **1.** Dafforn, Anderson, Ash, *et al.* (1977) *Biochim. Biophys. Acta* **484**, 375. **2.** Baker & Hurlbut (1967) *J. Med. Chem.* **10**, 1129.

Phenylacetonitrile, *See* Benzyl Cyanide

Phenylacetyl-CoA

This arylacyl coenzyme A derivative (FW = 785.67 g/mol) is a product of phenylacetyl-CoA synthetase and a substrate for glutamine *N*-phenylacetyltransferase and isopenicillin N *N*-acyltransferase. **Target(s):** choline *O*-acetyltransferase (1); glycine acyltransferase (2); and glycine *N*-benzoyltransferase (3). **1.** Potempska, Loo & Wisniewski (1984) *J. Neurochem.* **42**, 1499. **2.** Webster, Jr. (1981) *Meth. Enzymol.* **77**, 301. **3.** Nandi, Lucas & Webster, Jr. (1979) *J. Biol. Chem.* **254**, 7230.

Phenylacetylene

This alkyne (FW = 102.14 g/mol; CAS 536-74-3; M.P. = –44.8°C; B.P. = 142.4°C)), known systematically as ethynylbenzene, inhibits ammonia monooxygenase (1), CYP2B1, *or* cytochrome P450 2B1 (2), methane monooxygenase, soluble (1), phenol 2-monooxygenase (3), and toluene 4-monooxygenase (4). **1.** Lontoh, DiSpirito, Krema, *et al.* (2000) *Environ. Microbiol.* **2**, 485. **2.** Ortiz de Montellano (1991) *Meth. Enzymol.* **206**, 533. **3.** Kagle & Hay (2006) *Appl. Microbiol. Biotechnol.* **72**, 306. **4.** Keener, Watwood, Schaller, *et al.* (2001) *J. Microbiol. Methods* **46**, 171.

D-Phenylalanine

This aromatic amino acid (FW = 165.19 g/mol; CAS 673-06-3) is the enantiomer of the proteogenic amino acid, L-phenylalanine. D-Phenylalanine occurs in a number of antibiotics, including gramicidin S, tyrocidin A, and bacitracin A. *See L-Phenylalanine* **Target(s):** alkaline phosphatase, weakly inhibited (1); aminoacylase (2); aminopeptidase B, weakly inhibited (3); carboxypeptidase A (4-10); carboxypeptidase C, *or* carboxypeptidase Y (11); neprilysin, *or* enkephalinase (12-14); phenylalanine ammonia-lyase (15-20); phenylalanine dehydrogenase (21); phenylalanine 2-monooxygenase (22); phenylalanyl-tRNA synthetase (23); pyruvate kinase, Co(II)-substituted (24); tryptophan 2-monooxygenase (25); tyrosine phenol-lyase (26); and urease (27,28). **1.** Hoylaerts, Manes & Millán (1992) *Biochem. J.* **286**, 23. **2.** Park & Fox (1960) *J. Biol. Chem.* **235**, 3193. **3.** Kawata, Takayama, Ninomiya & Makisumi (1980) *J. Biochem.* **88**, 1601. **4.** Pétra (1970) *Meth. Enzymol.* **19**, 460. **5.** Elkins-Kaufman & Neurath (1948) *J. Biol. Chem.* **175**, 893. **6.** Elkins-Kaufman, Neurath & De Maria (1949) *J. Biol. Chem.* **178**, 645. **7.** Hartsuck & Lipscomb (1971) *The Enzymes*, 3rd ed. (Boyer, ed.), **3**, 1. **8.** Webb (1966) *Enzyme and Metabolic Inhibitors*, vol. 2, p. 365, Academic Press, New York. **9.** Coleman & Vallee (1964) *Biochemistry* **3**, 1874. **10.** Auld, Bertini, Donaire, Messori & Moratal (1992) *Biochemistry* **31**, 3840. **11.** Bai, Hayashi & Hata (1975) *J. Biochem.* **77**, 81. **12.** Halpern & Dong (1986) *Pain* **24**, 223. **13.** Blum, Briggs, Trachtenberg, Delallo & Wallace (1987) *Alcohol* **4**, 449. **14.** Marcello, Grazia, Sergio & Federigo (1986) *Adv. Exp. Med. Biol.* **198**, Pt. B, 153. **15.** Havir & Hanson (1970) *Meth. Enzymol.* **17A**, 575. **16.** Kalghatgi & Subba Rao (1975) *Biochem. J.* **149**, 65. **17.** Abell & Shen (1987) *Meth. Enzymol.* **142**, 242. **18.** Hanson & Havir (1972) *The Enzymes*,

3rd ed. (Boyer, ed.), **7**, 75. **19.** Camm & Towers (1973) *Phytochemistry* **12**, 961. **20.** Jorrin, Lopez-Valbuena & Tena (1988) *Biochim. Biophys. Acta* **964**, 73. **21.** Misono, Yonezawa, Nagata & Nagasaki (1989) *J. Bacteriol.* **171**, 30. **22.** Koyama (1984) *J. Biochem.* **96**, 421. **23.** Santi & Danenberg (1971) *Biochemistry* **10**, 4813. **24.** Kwan & Davis (1982) *Can. J. Biochem.* **60**, 86. **25.** Emanuele, Heasley & Fitzpatrick (1995) *Arch. Biochem. Biophys.* **316**, 241. **26.** Faleev, Ruvinov, Demidkina, *et al.* (1988) *Eur. J. Biochem.* **177**, 395. **27.** Kleczkowski & Dabrowska (1968) *Bull. Acad. Pol. Sci. Biol.* **16**, 267. **28.** Reithel (1971) *The Enzymes*, 3rd ed. (Boyer, ed.), **4**, 1.

L-Phenylalanine

This aromatic amino acid (FW = 165.19 g/mol; CAS 63-91-2; pK_1 = 2.20 and pK_2 = 9.31 at 25°C; decomposes at 283-284°C), named systematically as (*S*)-2-amino-3-phenylpropanoic acid and symbolized by Phe and F) is one of the twenty proteogenic amino acids. Phenylalanine is nutritionally essential in the diets of mammals.The L-stereoisomer has a specific rotation of –4.47° at 25°C of the D line for 1.0–2.0 g/100 mL in 5 M HCl (–34.5° in distilled water). It is soluble in water (2.96 g/100 mL at 25°C; 4.46 g/100 mL at 50°C; and 6.62 g/100 mL at 75°C). L-Phenylalanine is very slightly soluble in ethanol and methanol and insoluble in diethyl ether. The solubility of the racemic mixture in water is 1.411 g/100 mL at 25°C (3.708 g/100 mL at 75°C). This aromatic amino acid has both useful absorbance and fluorescent properties. It absorbs light in the ultraviolet region of the spectrum (λ_{max} = 257.5 nm with ε = 195 M^{-1}cm^{-1} in 0.1 M HCl). DL-Phenylalanine, which has a sweetish taste, is unstable in alkaline solutions. Approximately three-fourths of the amino acid will decompose in five hours at 110–115°C in a sealed ampule in 5 M NaOH. **Role in Phenylketonuria:** The autosomal recessive disorder Phenylketonuria (PKU) is caused by mutations in the phenylalanine hydroxylase (*PAH*) gene, leading to childhood mental retardation by exposing neurons to cytotoxic phenylalanine (Phe) levels. Although many of the effects of PKU-associated Phe accumulation are indirect (*e.g.*, decreased creatine kinase activity, deficient myelin production, and reduced dopamine synthesis), high Phe concentrations also trigger apoptosis by activating the mitochondria-initiated intrinsic pathway through the RhoA/Rho-associated kinase pathway (140) and through the Fas receptor (FasR)-mediated cell death receptor pathway (141). **Target(s):** acetylcholinesterase, inhibited by DL- and L-phenylalanine (1,100); agaritine γ-glutamyltransferase (132); alkaline phosphatase, also inhibited by DL-phenylalanine (3-7,86-91); D-amino-acid oxidase (8); 2-aminohexano-6-lactam racemase, weakly inhibited (48); aminopeptidase B, weakly inhibited (77,78); apyrase (9,10); arginase, weakly inhibited (11); arogenate dehydrogenase, *or* cyclohexadienyl dehydrogenase (12); arylformamidase, weakly inhibited (69); asparagine:oxo-acid aminotransferase (103,104); aspartokinase III, weakly inhibited (13); benzoate 4-monooxygenase (136); carboxypeptidase A (72,73); carboxypeptidase C, *or* carboxypeptidase Y (74); catechol oxidase (14,15); chorismate mutase (16-22); cystathionine γ-lyase, *or* cysteine desulfhydrase, mildly inhibited by racemic mixture (23); cystinyl aminopeptidase, *or* oxytocinase (79); cytosol alanyl aminopeptidase (2,75,76); 3-deoxy-7-phosphoheptulonate synthase, *or* 3-deoxy-D-*arabino*-heptulosonate-7-phosphate synthetase (24,25,108-121); dimethylglycine *N*-methyltransferase, weakly inhibited (125); β-glucosidase (83); hemoglobin S polymerization (26,27); (*S*)-2-hydroxy-acid oxidase (28); 2-isopropylmalate synthase (29); kynurenine:oxoglutarate aminotransferase (105-108); lactate 2-monooxygenase (138); membrane alanyl aminopeptidase, *or* aminopeptidase N (80-82); Na$^+$/K$^+$-exchanging ATPase (68); ornithine (lysine) transaminase, weakly inhibited (102); pepsin, weakly inhibited (30); 4-phytase (85); polyphenol oxidase (127,128); prephenate dehydratase (16,17,22,31-33,49-57); prephenate dehydrogenase (34); procollagen-lysine 5-dioxygenase, weakly inhibited (137); propionyl-CoA carboxylase (35); pyruvate kinase (36-39,93-101); sarcosine/ dimethylglycine *N*-methyltransferase, weakly inhibited (125); shikimate kinase (40); starch phosphorylase (122,123); thermolysin, K_i = 15.6 mM

(70,71); tryptophanase, *or* tryptophan indole-lyase (41,42,63-67); tryptophan 5-monooxygenase (43,44,130-133); tyrosinase, *or* monophenol monooxygenase (45,126-129); tyrosine 3-monooxygenase (46,47,132,134, 135); tyrosine phenol-lyase (58-62); and tyrosine-phosphatase (84); 3-deoxy-D-arabino-heptulosonate 7-phosphate synthase (DAH7PS), *Neisseria meningitidis*, allosteric (139). **1.** Bergmann, Wilson & Nachmansohn (1950) *J. Biol. Chem.* **186**, 693. **2.** Garner & Behal (1975) *Biochemistry* **14**, 3208. **3.** Ghosh & Fishman (1966) *J. Biol. Chem.* **241**, 2516. **4.** Fernley (1971) *The Enzymes*, 3rd ed., **4**, 417. **5.** Hoylaerts, Manes & Millán (1992) *Biochem. J.* **286**, 23. **6.** Ghosh, Goldman & Fishman (1967) *Enzymologia* **33**, 113. **7.** Manning, Inglis, Green & Fishman (1969) *Enzymologia* **37**, 251. **8.** Burton (1955) *Meth. Enzymol.* **2**, 199. **9.** Berti, Bonan, da Silva, *et al.* (2001) *Int. J. Dev. Neurosci.* **19**, 649. **10.** Wyse, Sarkis, Cunha-Filho, *et al.* (1994) *Neurochem. Res.* **19**, 1175. **11.** Hunter & Downs (1945) *J. Biol. Chem.* **157**, 427. **12.** Zamir, Tiberio, Fiske, Berry & Jensen (1985) *Biochemistry* **24**, 1607. **13.** Truffa-Bachi (1973) *The Enzymes*, 3rd ed., **8**, 509. **14.** Pomerantz (1963) *J. Biol. Chem.* **238**, 2351. **15.** Aerts & Vercauteren (1964) *Enzymologia* **28**, 1. **16.** Cotton & Gibson (1970) *Meth. Enzymol.* **17A**, 564 and (1965) *Biochim. Biophys. Acta* **100**, 76. **17.** Davidson (1987) *Meth. Enzymol.* **142**, 432. **18.** Gilchrist & Connelly (1987) *Meth. Enzymol.* **142**, 450. **19.** Zhang, Wilson & Ganem(2003) *Bioorg. Med. Chem.* **11**, 3109. **20.** Baker (1966) *Biochemistry* **5**, 2655. **21.** Sugimoto & Shiio (1980) *J. Biochem.* **88**, 167. **22.** Schmit & Zalkin (1969) *Biochemistry* **8**, 174. **23.** Fromageot & Grand (1942) *Enzymologia* **11**, 81. **24.** Simpson & Davidson (1976) *Eur. J. Biochem.* **70**, 509. **25.** Previc & Binkley (1964) *Biochem. Biophys. Res. Commun.* **16**, 162. **26.** Noguchi & Schechter (1978) *Biochemistry* **17**, 5455. **27.** Noguchi & Schechter (1977) *Biochem. Biophys. Res. Commun.* **74**, 637. **28.** Robinson, Keay, Molinari & Sizer (1962) *J. Biol. Chem.* **237**, 2001. **29.** Gross (1970) *Meth. Enzymol.* **17A**, 777. **30.** Inouye & Fruton (1968) *Biochemistry* **7**, 1611. **31.** Fischer & Jensen (1987) *Meth. Enzymol.* **142**, 507. **32.** Malik (1980) *Trends Biochem. Sci.* **5**, 68. **33.** Euverink, Wolters & Dijkhuizen (1995) *Biochem. J.* **308**, 313. **34.** Fischer & Jensen (1987) *Meth. Enzymol.* **142**, 503. **35.** Kimura, Kojyo, Kimura & Sato (1998) *Arch. Microbiol.* **170**, 179. **36.** Imamura & Tanaka (1982) *Meth. Enzymol.* **90**, 150. **37.** Kayne (1973) *The Enzymes*, 3rd ed., **8**, 353. **38.** Engström, Ekman, Humble & Zetterqvist (1987) *The Enzymes*, 3rd ed., **18**, 47. **39.** Izbicka-Dimitrijevic, Dimitrijevic & Kochman (1982) *Biochim. Biophys. Acta* **703**, 101. **40.** De Feyter (1987) *Meth. Enzymol.* **142**, 355. **41.** Hoch, Simpson & DeMoss (1966) *Biochemistry* **5**, 2229. **42.** Demidkina, Zakomirdina, Kulikova, *et al.* (2003) *Biochemistry* **42**, 11161. **43.** Ichiyama & Nakamura (1970) *Meth. Enzymol.* **17A**, 449. **44.** Freedland, Wadzinski & Waisman (1961) *Biochem. Biophys. Res. Commun.* **6**, 227. **45.** Webb (1966) *Enzyme and Metabolic Inhibitors*, vol. **2**, p. 305, Academic Press, New York. **46.** Nagatsu, Levitt & Udenfriend (1964) *J. Biol. Chem.* **239**, 2910. **47.** Moore & Dominic (1971) *Fed. Proc.* **30**, 859. **48.** Ahmed, Esaki, Tanaka & Soda (1985) *Agric. Biol. Chem.* **49**, 2991. **49.** Dopheide, Crewther & Davidson (1972) *J. Biol. Chem.* **247**, 4447. **50.** Hagino & Nakayama (1974) *Agric. Biol. Chem.* **38**, 2367. **51.** Gething, Davidson & Dopheide (1976) *Eur. J. Biochem.* **71**, 317. **52.** Ahmad, Wilson & Jensen (1988) *Eur. J. Biochem.* **176**, 69. **53.** Riepl & Glover (1978) *Arch. Biochem. Biophys.* **191**, 192. **54.** Hsu, Lin, Lo & Hsu (2004) *Arch. Microbiol.* **181**, 237. **55.** Cerutti & Guroff (1965) *J. Biol. Chem.* **240**, 3034. **56.** Friedrich, Friedrich & Schlegel (1976) *J. Bacteriol.* **126**, 723. **57.** Schmit & Zalkin (1969) *Biochemistry* **8**, 174. **58.** Faleev, Zhukov, Khurs, *et al.* (2000) *Eur. J. Biochem.* **267**, 6897. **59.** Faleev, Sadovnikova, Martinkova & Belikov (1981) *Enzyme Microb. Technol.* **3**, 219. **60.** Demidkina, Myagkikh & Azhayev (1987) *Eur. J. Biochem.* **170**, 311. **61.** Kumagai, Kashima, Torii, *et al.* (1972) *Agric. Biol. Chem.* **36**, 472. **62.** Demidkina, Barbolina, Faleev, *et al.* (2002) *Biochem. J.* **363**, 745. **63.** Snell (1975) *Adv. Enzymol. Relat. Areas Mol. Biol.* **42**, 287. **64.** Cowell, Moser & DeMoss (1973) *Biochim. Biophys. Acta* **315**, 449. **65.** Yoshida, Utagawa, Kumagai & Yamada (1974) *Agric. Biol. Chem.* **38**, 2065. **66.** Yoshida, Kumagai & Yamada (1974) *Agric. Biol. Chem.* **38**, 463. **67.** Gogoleva, Zakomirdina, Demidkina, Phillips & Faleev (2003) *Enzyme Microb. Technol.* **32**, 843. **68.** Streck, Zugno, Tagliari, *et al.* (2002) *Int. J. Dev. Neurosci.* **20**, 77. **69.** Serrano & Nagayama (1991) *Comp. Biochem. Physiol. B Comp. Biochem.* **99**, 281. **70.** Kester & Matthews (1977) *Biochemistry* **16**, 2506. **71.** Feder, Aufheide & Wildi (1976) in *Enzymes and Proteins from Thermophilic Microorganisms* (Zuber, ed.), Birkhäuser Verlag, Basel, p. 31. **72.** Ludwig (1981) *Inorg. Biochem.* **1**, 438. **73.** Auld, Bertini, Donaire, Messori & Moratal (1992) *Biochemistry* **31**, 3840. **74.** Bai, Hayashi & Hata (1975) *J. Biochem.* **77**, 81. **75.** Garner & Behal (1977) *Arch. Biochem. Biophys.* **182**, 667. **76.** Garner & Behal (1975) *Biochemistry* **14**, 5084. **77.** Kawata, Takayama, Ninomiya & Makisumi (1980) *J. Biochem.* **88**, 1601. **78.** Lauffart, McDermott, Jones & Mantle

(1988) *Biochem. Soc. Trans.* **16**, 849. **79.** Krishna & Kanagasabapathy (1989) *J. Endocrinol.* **121**, 537. **80.** McCaman & Villarejo (1982) *Arch. Biochem. Biophys.* **213**, 384. **81.** McClellan, Jr, & Garner (1980) *Biochim. Biophys. Acta* **613**, 160. **82.** Tokioka-Terao, Hiwada & Kokubu (1984) *Enzyme* **32**, 65. **83.** Dale, Ensley, Kern, Sastry & Byers (1985) *Biochemistry* **24**, 3530. **84.** Fukami & Lipmann (1982) *Proc. Natl. Acad. Sci. U.S.A.* **79**, 4275. **85.** Bitar & Reinhold (1972) *Biochim. Biophys. Acta* **268**, 442. **86.** Srinivas, Jayalakshmi, Sreeramulu, Sherman & Rao (2006) *Biochim. Biophys. Acta* **1760**, 310. **87.** Al-Saleh (2002) *J. Nat. Toxins* **11**, 357. **88.** Le Du, Stigbrand, Taussig, Menez & Stura (2001) *J. Biol. Chem.* **276**, 9158. **89.** Belland, Visser, Poppema & Stinson (1993) *Enzyme Protein* **47**, 73. **90.** Goldstein & Harris (1979) *Nature* **280**, 602. **91.** Yedlin, Young, Seetharam, Seetharam & Alpers (1981) *J. Biol. Chem.* **256**, 5620. **92.** Tsakiris, Giannoulia-Karantana, Simintzi & Schulpis (2006) *Pharmacol. Res.* **53**, 1. **93.** Plaxton & Storey (1984) *Eur. J. Biochem.* **143**, 257. **94.** Ohta, Nishikawa & Imamura (2003) *Comp. Biochem. Physiol. B* **135**, 397. **95.** Srivastava & Baquer (1985) *Arch. Biochem. Biophys.* **236**, 703. **96.** Schering, Eigenbrodt, Linder & Schoner (1982) *Biochim. Biophys. Acta* **717**, 337. **97.** Feksa, Cornelio, Dutra-Filho, *et al.* (2005) *Int. J. Dev. Neurosci.* **23**, 509. **98.** Feksa, Cornelio, Dutra-Filho, *et al.* (2003) *Brain Res.* **968**, 546. **99.** Roberts & Anderson (1985) *Comp. Biochem. Physiol. B Comp. Biochem.* **80**, 51. **100.** Terlecki (1989) *Int. J. Biochem.* **21**, 1053. **101.** Leon, Moran & Gonzalez (1982) *Comp. Biochem. Physiol. B* **72**, 65. **102.** Lowe & Rowe (1986) *Mol. Biochem. Parasitol.* **21**, 65. **103.** Maul & Schuster (1986) *Arch. Biochem. Biophys.* **251**, 585. **104.** Hongo, Ito, Takeda & Sato (1986) *Enzyme* **36**, 232. **105.** Guidetti, Okuno & Schwarcz (1997) *J. Neurosci. Res.* **50**, 457. **106.** Okuno, Nishikawa & Nakamura (1996) *Adv. Exp. Med. Biol.* **398**, 455. **107.** Milart, Urbanska, Turski, Paszkowski & Sikorski (2001) *Placenta* **22**, 259. **108.** Ray & Bauerle (1991) *J. Bacteriol.* **173**, 1894. **109.** McCandliss, Poling & Herrmann (1978) *J. Biol. Chem.* **253**, 4259. **110.** Wu, Sheflyan & Woodard (2005) *Biochem. J.* **390**, 583. **111.** Friedrich & Schlegel (1975) *Arch. Microbiol.* **103**, 133. **112.** Park & Bauerle (1999) *J. Bacteriol.* **181**, 1636. **113.** Parker, Bulloch, Jameson & Abell (2001) *Biochemistry* **40**, 14821. **114.** Stephens & Bauerle (1991) *J. Biol. Chem.* **266**, 20810. **115.** Paravicini, Schmidheini & Braus (1989) *Eur. J. Biochem.* **186**, 361. **116.** Schnappauf, Hartmann, Künzler & Braus (1998) *Arch. Microbiol.* **169**, 517. **117.** Wu, Howe & Woodard (2003) *J. Biol. Chem.* **278**, 27525. **118.** Euverink, Hessels, Franke & Dijkhuizen (1995) *Appl. Environ. Microbiol.* **61**, 3796. **119.** Liao, Lin, Chien & Hsu (2001) *FEMS Microbiol. Lett.* **194**, 59. **120.** Shumilin, Kretsinger & Bauerle (1996) *Proteins Struct. Funct. Genet.* **24**, 404. **121.** Hu, Jiang, Xu, Wu & Huang (2003) *J. Basic Microbiol.* **43**, 399. **122.** Kumar & Sanwal (1982) *Biochemistry* **21**, 4152. **123.** Singh & Sanwal (1976) *Phytochemistry* **15**, 1447. **124.** Gigliotti & Levenberg (1964) *J. Biol. Chem.* **239**, 2274. **125.** Waditee, Tanaka, Aoki, *et al.* (2003) *J. Biol. Chem.* **278**, 4932. **126.** Kim & Uyama (2005) *Cell. Mol. Life Sci.* **62**, 1707. **127.** Anosike & Ayaebene (1981) *Phytochemistry* **20**, 2625. **128.** Anosike & Ayaebene (1982) *Phytochemistry* **21**, 1889. **129.** Kong, Park, Hong & Cho (2000) *Comp. Biochem. Physiol. B Biochem. Mol. Biol.* **125**, 563. **130.** Kowlessur & Kaufman (1999) *Biochim. Biophys. Acta* **1434**, 317. **131.** Nakata & Fujisawa (1982) *Eur. J. Biochem.* **124**, 595. **132.** Ogawa & Ichinose (2006) *Neurosci. Lett.* **401**, 261. **133.** Tong & Kaufman (1975) *J. Biol. Chem.* **250**, 4152. **134.** Neckameyer, Holt & Paradowski (2005) *Biochem. Genet.* **43**, 425. **135.** Ikeda, Levitt & Udenfriend (1967) *Arch. Biochem. Biophys.* **120**, 420. **136.** McNamee & Durham (1985) *Biochem. Biophys. Res. Commun.* **129**, 485. **137.** Murray, Cassell & Pinnell (1977) *Biochim. Biophys. Acta* **481**, 63. **138.** Xu, Yano, Yamamoto, Suzuki & Kumagai (1996) *Appl. Biochem. Biotechnol.* **56**, 277. **139.** Cross, Pietersma, Allison, *et al.* (2013) *Protein Sci.* **22**, 1087. **140.** Zhang, Gu & Yuan (2007) *Eur. J. Neurosci.* **25**, 1341. **141.** Huang, Lu, Lv, *et al.* (2013) *PLoS One* **8**, e71553.

β-Phenyl-β-alanine

This β-amino acid (FW = 165.19 g/mol), also known as 3-amino-3-phenylpropionic acid, inhibits alkaline phosphatase (1,2), GABA_B receptor (3,4), and gramicidin S biosynthesis (5-7). **1.** Fernley (1971) *The Enzymes*,

3rd ed. (Boyer, ed.), **4**, 417. **2**. Fishman & Sie (1971) *Enzymologia* **41**, 141. **3**. Humeniuk, White & Ong (1993) *Psychopharmacology (Berlin)* **111**, 219. **4**. Yamasaki & Goto (1992) *Jpn. J. Pharmacol.* **60**, 55. **5**. Egorov, Vypiiach, Zharikova & Maksimov (1975) *Mikrobiologiia* **44**, 237. **6**. Egorov, Vypiiach, Zarubina & Markelova (1979) *Antibiotiki* **24**, 906. **7**. Vostroknutova, Simakova, Kharat'ian, Bulgakova & Udalova (1983) *Biokhimiia* **48**, 818.

L-Phenylalanine Chloromethyl Ketone

This halomethyl ketone, also known as (*S*)-3-amino-1-chloro-4-phenyl-2-butanone (FW$_{free-base}$ = 197.66 g/mol), inhibits leucyl aminopeptidase. (*See also N-(p-Tosyl-L-phenylalanine Chloromethyl Ketone)* **Target(s):** leucyl aminopeptidase (1,2); phenylalanine racemase, ATP-hydrolyzing (3); tripeptidyl-peptidase I (4); and Tyr/Phe transport protein (5). **1**. Hanson & Frohne (1976) *Meth. Enzymol.* **45**, 504. **2**. Fittkau (1972) *Acta Biol. Med. Ger.* **28**, 259. **3**. Nguyen Huu, von Dungen & Kleinkauf (1976) *FEBS Lett.* **67**, 75. **4**. Vines & Warburton (1998) *Biochim. Biophys. Acta* **1384**, 233. **5**. Hartman (1977) *Meth. Enzymol.* **46**, 130.

Phenylalanine Hydroxamate

This amino acid hydroxamate (FW$_{free-base}$ = 180.21 g/mol) inhibits atrolysin A (1), bacterial leucyl aminopeptidase, inhibited by both L- and DL-phenylalanine hydroxamate (2,3), chorismate mutase (4), and vibriolysin (5). **1**. Fox, Campbell, Beggerly & Bjarnason (1986) *Eur. J. Biochem.* **156**, 65. **2**. Baker, Wilkes, Bayliss & Prescott (1983) *Biochemistry* **22**, 2098. **3**. Wilkes & Prescott (1983) *J. Biol. Chem.* **258**, 13517. **4**. Sugimoto & Shiio (1980) *J. Biochem.* **88**, 167. **5**. Wilkes, Bayliss & Prescott (1988) *J. Biol. Chem.* **263**, 1821.

L-Phenylalanine β-Naphthylamide

This amino acid amide (FW$_{free-base}$ = 290.36 g/mol) inhibits *Escherichia coli* phenylalanyl-tRNA synthetase: K_i = 0.41 mM (1) and tyramine *N*-feruloyltransferase (2). **1**. Santi & Danenberg (1971) *Biochemistry* **10**, 4813. **2**. Negrel & Javelle (2001) *Phytochemistry* **56**, 523.

L-Phenylalaninol Adenylate

This aminoacyl adenylate analogue (FW = 480.42 g/mol), also known as adenosine 5'-*O*-(2(*S*)-amino-3-phenylpropyl)phosphate, has a methylene group substituted for the carbonyl group and is a competitive inhibitor of phenylalanyl-tRNA synthetase (1-3). **1**. Moor, Lavrik, Favre & Safro (2003) *Biochemistry* **42**, 10697. **2**. Lavrik, Moor & Nevinsky (1978) *Bioorg. Khim.* **4**, 1480. **3**. Hertz & Zachau (1973) *Eur. J. Biochem.* **37**, 203.

L-Phenylalanyl-L-alanine

This dipeptide (FW = 236.27 g/mol; Sequence: FA), inhibits alanine carboxypeptidase, also alternative substrate (1) and dipeptidyl-peptidase IV2; neprilysin, K_i = 5.5 μM (3,4), and vibriolysin (5). **1**. Levy & Goldman (1969) *J. Biol. Chem.* **244**, 4467. **2**. Bella, Jr., Erickson & Kim (1982) *Arch. Biochem. Biophys.* **218**, 156. **3**. Marie-Claire, Ruffet, Antonczak, *et al.* (1997) *Biochemistry* **36**, 13938. **4**. Dion, Cohen, Crine & Boileau (1997) *FEBS Lett.* **411**, 140. **5**. Wilkes, Bayliss & Prescott (1988) *J. Biol. Chem.* **263**, 1821.

L-Phenylalanyl-L-alanine Methyl Ester

This dipeptide ester (FW$_{free-base}$ = 250.30 g/mol), abbreviated by FA-OMe, inhibits rabbit muscle pyruvate kinase (ATP:pyruvate 2-*O*-phosphotransferase, EC 2.7.1.40), as analyzed with the Yagi-Ozawa equation. Straight-line plots obtained for L-phenylalanine + L-alanine and L-phenylalanine + L-phenylalanyl methyl ester (PheAlaOMe), indicative of competition for common sites. Parabolic plots for 2-phosphoglycerate and L-phenylalanine or PheAlaOMe suggest a lack of competition between these compounds. Biphasic concave plots were obtained for 2-phosphoglycerate and L-1-amino-2-phenylethyl phosphonic acid (PnPhe), and mixtures of inhibitory (L-phenylalanine, PheAlaOMe) and activatory (PnPhe, L-alanine and L-1-aminoethyl phosphonic acid (PnAla)) amino acid derivatives. Such non-classical Yagi-Ozawa plots suggest the existence of distinct regulatory sites for inhibitory and activatory ligands. 1. Izbicka-Dimitrijevic, Dimitrijevic & Kochman (1982) *Biochim. Biophys. Acta* **703**, 101.

L-Phenylalanyl-L-alanyl-L-alanine Chloromethyl Ketone

This halomethyl ketone (FW = 340.83 g/mol), abbreviated by FAA-cmk, inhibits tripeptidyl-peptidase I, which cleaves tripeptides from synthetic substrates provided that the N-terminus is unsubstituted and the amino acid in the P$_1$ position is not charged. The enzyme also cleaves small peptides (angiotensin II and glucagon) releasing tripeptides but does not appear to demonstrate any preference for amino acids on either side of the cleavage site. The peptidase was potently inhibited by a series of substrate-based tripeptidyl chloromethyl ketones (K_i values ranging from 10-100 nM). Inhibition was rapid and reversible, a mode of inhibition that differs from chloromethyl ketones interactions with cysteine or serine peptidases. These tripeptidyl chloromethyl ketones were also inhibitors of bone resorption using an in vitro assay suggesting that a tripeptidyl peptidase is involved in the degradation of bone matrix proteins. **1**. Vines & Warburton (1998) *Biochim. Biophys. Acta* **1384**, 233.

D-Phenylalanyl-3,4-dehydro-L-prolyl-L-arginine Chloromethyl Ketone

This halomethyl ketone (FW = 417.53 g/mol) inhibits activated protein C, or blood coagulation factor XIV, a serine protease that plays an pivotal role in regulating anticoagulation, inflammation, and cell death, as wellas maintaining the permeability of blood vessel walls in humans. Activated protein C (APC) proteolytically inactivates Factor Va and Factor VIIIa. Unliike thrombin, APC accommodates large hydrophobic residues such as phenylalanine and leucine in the P^2 position. In the P^3 position, the enzyme prefers an apolar D-amino acid residue. **1**. Stone & Hofsteenge (1985) *Biochem. J.* **230**, 497.

L-Phenylalanylglycine

This dipeptide (FW = 206.24 g/mol; *Sequence*: FG) inhibits cytochrome *c* aggregation (1), γ-glutamyl transpeptidase (2,3), neprilysin, IC_{50} = 30.2 μM (4), and peptidyl-dipeptidase A (5). **1**. La Rosa, Milardi, Amato, Pappalardo & Grasso (2005) *Arch. Biochem. Biophys.* **435**, 182. **2**. Allison (1985) *Meth. Enzymol.* **113**, 419. **3**. Thompson & Meister (1977) *J. Biol. Chem.* **252**, 6792. **4**. Dion, Cohen, Crine & Boileau (1997) *FEBS Lett.* **411**, 140. **5**. Kawamura, Kikuno, Oda & Muramatsu (2000) *Biosci. Biotechnol. Biochem.* **64**, 2193.

L-Phenylalanyl-L-norleucyl-statinyl-L-alanyl-statine Methyl Ester

This pepstatin-like peptidomimetic (FW$_{\text{free-base}}$ = 677.88 g/mol; Sequence: Phe-Nor-(Sta)-Ala-(Sta)-CH$_3$, where Sta = (*S,S*)-statine), also known as (3*S*,4*S*)-4-amino-3-hydroxy-6-methylheptanoate, inhibits renin, IC_{50} = 7.7 μM. **1**. Guégan, Diaz, Cazaubon, *et al.* (1986) *J. Med. Chem.* **29**, 1152.

D-Phenylalanyl-L-phenylalanyl-L-arginine Chloromethyl Ketone

This tripeptide halomethyl ketone (FW$_{\text{free-base}}$ = 501.03 g/mol; Sequence: FFR-cmk) inhibits a number of proteinases, including coagulation factor VIIa (1,2), coagulation factor XIIa (3), gingipain R (4,5), mouse γ-nerve growth factor (6), plasma kallikrein (3,6), plasmin (3), activated protein C (7), thrombin (3,6), tissue kallikrein (6), u-plasminogen activator, *or* urokinase (8). **1**. Persson, Bak & Olsen (2001) *J. Biol. Chem.* **276**, 29195. **2**. Pike, Brzozowski, Roberts, Olsen & Persson (1999) *Proc. Natl. Acad. Sci. U.S.A.* **96**, 8925. **3**. Hayashi, Salomon & Hugli (2002) *Int. Immunopharmacol.* **2**, 1667. **4**. Eichinger, Beisel, Jacob, *et al.* (1999) *EMBO J.* **18**, 5453. **5**. Banbula, Potempa, Travis, Bode & Medrano (1998) *Protein Sci.* **7**, 1259. **6**. Kettner & Shaw (1981) *Meth. Enzymol.* **80**, 826. **7**. Stone & Hofsteenge (1985) *Biochem. J.* **230**, 497. **8**. Chao (1983) *J. Biol. Chem.* **258**, 4434.

L-Phenylalanyl-L-phenylalanyl-L-arginine Chloromethyl Ketone

This tripeptide halomethyl ketone FFR-cmk (FW$_{\text{free-base}}$ = 501.03 g/mol), irreversibly inhibits a number of proteinases, including coagulation factor VIIa (1), coagulation factor XIIa (2), mouse γ-nerve growth factor (3), plasma kallikrein (3,4), activated protein C (5), and tissue kallikrein (6). **1**. Morrissey (1998) in *Handb. Proteolytic Enzymes* (Barrett, Rawlings & Woessner, eds.), p. 161, Academic Press, San Diego. **2**. Silverberg & Kaplan (1988) *Meth. Enzymol.* **163**, 68. **3**. Kettner & Shaw (1981) *Meth. Enzymol.* **80**, 826. **4**. Silverberg & Kaplan (1988) *Meth. Enzymol.* **163**, 85. **5**. Stone & Hofsteenge (1985) *Biochem. J.* **230**, 497. **6**. Chao & Chao (1988) *Meth. Enzymol.* **163**, 128.

D-Phenylalanyl-L-prolyl-L-arginal

This tripeptide aldehyde (FW$_{\text{free-base}}$ = 402.50 g/mol; unstable in neutral aqueous solutions) inhibits a number of proteinases, including plasmin (1), protein-glutamate methylesterase (2), thrombin (1,3), and urokinase, *or* u-plasminogen activator (1). **1**. Bajusz, Széll, Barabás & Bagdy (1982) *Folia Haematol. Int. Mag. Klin. Morphol. Blutforsch.* **109**, 16. **2**. Veeraragavan & Gagnon (1987) *Biochem. Biophys. Res. Commun.* **142**, 603. **3**. Bajusz, Szell, Bagdy, *et al.* (1990) *J. Med. Chem.* **33**, 1729.

D-Phenylalanyl-L-prolyl-L-arginine Chloromethyl Ketone

This halomethyl ketone containing tripeptide (FW$_{\text{free-base}}$ = 450.97 g/mol) inhibits a number of proteinases, including batroxobin (1,2), cancer procoagulant (3), clostripain (4), coagulation factor IXa (5), coagulation factor XIIa (6), limulus clotting factor C (7,8) plasmin (9), activated protein C (10), snake venom factor V activator, *or* thrombocytin (11), thrombin (12-15), and t-plasminogen activator (7,16). **1**. Markland, Jr. & Pirkle (1998) in *Handb. Proteolytic Enzymes* (Barrett, Rawlings & Woessner, eds.), p. 216, Academic Press, San Diego. **2**. Sturzebecher, Sturzebecher & Markwardt (1986) *Toxicon* **24**, 585. **3**. Moore (1992) *Biochem. Biophys. Res. Commun.* **184**, 819. **4**. Kembhavi, Buttle, Rauber & Barrett (1991) *FEBS Lett.* **283**, 277. **5**. Bajaj (1998) in *Handb. Proteolytic Enzymes* (Barrett, Rawlings & Woessner, eds.), p. 157, Academic Press, San Diego. **6**. Silverberg & Kaplan (1988) *Meth. Enzymol.* **163**, 68. **7**. Muta, Tokunaga, Nakamura, Morita & Iwanaga (1993) *Meth. Enzymol.* **223**, 336. **8**. Tokunaga, Nakajima & Iwanaga (1991) *J. Biochem.* **109**, 150. **9**. Gilboa, Villannueva & Fenton II (1988) *Enzyme* **40**, 144. **10**. Stone & Hofsteenge (1985) *Biochem. J.* **230**, 497. **11**. Kisiel & Canfield (1981) *Meth. Enzymol.* **80**, 275. **12**. Odake, Kam & Powers (1995) *J. Enzyme Inhib.* **9**, 17. **13**. Kettner & Shaw (1981) *Meth. Enzymol.* **80**, 826. **14**. Backes, Harris, Leonctti, Craik & Ellman (2000) *Nat. Biotechnol.* **18**, 187. **15**. Engh,

Brandstetter, Sucher, *et al.* (1996) *Structure* **4**, 1353 **16**. Lijnen & Collen (1998) in *Handb. Proteolytic Enzymes* (Barrett, Rawlings & Woessner, eds.), p. 184, Academic Press, San Diego.

L-Phenylalanyl-L-prolyl-L-arginine Chloromethyl Ketone

This tripeptide halomethyl ketone FPR-cmk (FW$_{free-base}$ = 450.97 g/mol), inhibits gingipains K and R. The arginine-specific enzyme was found to be a high-molecular-mass gingipain, formed by 50-kDa gingipain complexed to a 44-kDa hemagglutinin. This complex is believed to be play a role in the uptake of hemin, a vital metabolite for *Pseudomonas gingivalis*. **1**. Pike, McGraw, Potempa & Travis (1994) *J. Biol. Chem.* **269**, 406.

D-Phenylalanyl-L-prolyl-L-boroarginine

This boroarginine-containing tripeptide (FW$_{dihydrochloride}$ = 491.23 g/mol) is a potent inhibitor of thrombin, K_i < 1 pM, and a weaker inhibitor of plasma kallikrein, K_i = 0.6 nM, coagulation factor Xa, K_i = 8.2 nM, and t-plasminogen activator, K_i = 2.3 nM. **1**. Kettner, Mersinger & Knabb (1990) *J. Biol. Chem.* **265**, 18289.

L-Phenylalanyl-L-seryl-L-arginyl-L-phenylalanyl-L-leucyl-L-asparaginyl-L-lysyl-L-glutaminyl-L-proline

This nonapeptide (*Sequence*: FSRFLNKQP; MW = 1136.32 g/mol), corresponding to a sequence in the phosphotransferase recognition site in cathepsin A, inhibits UDP-*N*-acetylglucosamine:lysosomal-enzyme *N*-acetylglucosaminephosphotransferase, K_i = 0.45 mM. **1**. Lukong, Elsliger, Mort, Potier & Pshezhetsky (1999) *Biochemistry* **38**, 73.

4-*N*-Phenylamino-5-phenyl-7-(5'-deoxy-4'-*C*-methyl-β-D-*ribo*-furanosyl)pyrrolo[2,3-*d*]pyrimidine

This tubercidin analogue (FW = 416.48 g/mol) inhibits human adenosine kinase, IC$_{50}$ = 4 µM. Because the use of adenosine receptor agonists is plagued by dose-limiting cardiovascular side effects, there is great interest in developing adenosine kinase inhibitors (AKIs) as a means for raising

steady-state adenosine concentrations. **1**. Boyer, Ugarkar, Solbach, *et al.* (2005) *J. Med. Chem.* **48**, 6430.

4-*N*-Phenylamino-5-phenyl-7-(β-D-*erythro*-furanosyl)pyrrolo[2,3-*d*]pyrimidine

This tubercidin analogue (FW = 388.43 g/mol) inhibits human adenosine kinase, IC$_{50}$ = 4 nM. An endogenous neuromodulator when produced in the central and peripheral nervous systems, adenosine has anticonvulsant, anti-inflammatory, and analgesic properties. Even so, efforts to use adenosine receptor agonists have been limited by dose-limiting cardiovascular side effects. The use of adenosine kinase inhibitors (AKIs) is yet another potential route to prevent seizures in the absence of adenosine's overt cardiovascular side effects. **1**. Boyer, Ugarkar, Solbach, *et al.* (2005) *J. Med. Chem.* **48**, 6430.

17-Phenyl-(3α,5α)-androst-16-en-3-ol

This selective neurosteroid antagonist (FW = 350.54 g/mol; CAS 694438-95-4; Soluble to 25 mM in DMSO and to 50 mM in ethanol), also known as 17-PA, targets neurosteroid potentiation (1) and directly gates GABA$_A$ receptors (2). 17-PA inhibition was also useful in demonstrating that ethanol modulates the interaction of the endogenous neurosteroid allopregnanolone with the $\alpha_1\beta_2\gamma_2$L GABA$_A$ receptor. 17-PA selectively reduces the effects of 5α-reduced steroids compared to 5β-reduced steroids and displays no effect on potentiation evoked by barbiturates and benzodiazepines (2). It also attenuates 3α,5α-THP-induced loss of righting reflex and total sleep time, following intracerebroventricular administration in rats (3). **1**. Mennerick, *et al.* (2004) *Mol. Pharmacol.* **65**, 1191. PMID: 15102947. **2**. Akk, *et al.* (2007) *Mol. Pharmacol.* **71**, 461. **3**. Kelley, *et al.* (2007) Eur.J.Pharmacol. 572 94.

N-Phenylanthranilate (Phenylanthranilic Acid)

This acid (FW$_{free-acid}$ = 213.24 g/mol), also known as diphenylamine-2-carboxylic acid, targets the channel-conductance-controlling ATPase, *or* cystic-fibrosis membrane-conductance-regulating protein (1,2). **1**. Kogan, Ramjeesingh, Huan, Wang & Bear (2001) *J. Biol. Chem.* **276**, 11575. **2**. McDonough, Davidson, Lester & McCarty (1994) *Neuron* **13**, 623.

Phenylarsine Oxide

This arsenical (FW = 168.03 g/mol; CAS 8052-79-7; M.P. = 144-146°C), also known as phenylarsenoxide and oxophenylarsine, reacts with neighboring cysteinyl residues within peptides and proteins. Although enzyme inhibition by phenylarsinoxides is taken to indicate the presence of functionally important neighboring thiols, such is not always the case. For example, arginyl-tRNA:protein transferase (or arginyltransferase) is inhibited by phenylarsine oxide by a thiol-independent mechanism (1). Phenylarsine oxide is a cell-permeant protein-tyrosine phosphatase inhibitor. Phenylarsine oxide stimulates 2-deoxyglucose transport in insulin-resistant human skeletal muscle and activates $p56^{lck}$ protein-tyrosine kinase. In addition, it blocks TNF-α-dependent activation of NF-κB in human myeloid ML-1a cells. **Target(s):** acetate kinase (2,3); alcohol dehydrogenase, yeast (4,5); arginyltransferase, or arginyl-tRNA:protein transferase (1); aspartate ammonia-lyase (6,7); Ca^{2+}-dependent ATPase (8); calcineurin (9,10); carboxylesterase (68); caspases (11); choline oxidase (12); dipeptide/proton cotransport system (13); diphospho-mevalonate decarboxylase (63); glucose transporter trafficking (14); glucose transport, insulin-stimulated (15-17); glyceraldehyde-3-phosphate dehydrogenase, or acetylphosphatase activity (18); D-β-hydroxybutyrate dehydrogenase (19); 3-hydroxy-3-methylglutaryl-CoA synthase, weakly inhibited (74); inositol 1,2,3,5,6-pentakisphosphate 5-phosphatase (20); insulin activation of phosphatidylinositol 3'-kinase (21); α-ketoglutarate oxidase (22); lactose carrier (23); malate dehydrogenase, or malate oxidase (22); morphine 6-dehydrogenase (24); NADH oxidase (25,26); NADPH oxidase (27-31); Na^+/H^+ exchanger (32); NF-κB activation (33-35); nicotinamide-nucleotide transhydrogenase (36,37); nucleoside-triphosphatase (64); oxidative phosphorylation (38,39); pantetheine hydrolase (65); phosphate uptake (40); phosphatidylcholine:sterol O-acyltransferase, or LCAT (75); 1-phosphatidylinositol 4-kinase (41-44,70-73); phospholipase Cγ (45); protein-tyrosine-phosphatase (46,47,66,67); pyruvate dehydrogenase (48); pyruvate oxidase (49,50); retinol dehydrogenase (51-53); RhoA GTPase (54); rhodanese, or thiosulfate sulfurtransferase (55); squalene monooxygenase, or squalene epoxidase (56); succinate oxidase, or succinate dehydrogenase (22,57); system A amino acid transporter (58); tubulin polymerization, or microtubule assembly (59); tyrosine-ester sulfotransferase, or aryl sulfotransferase IV (89); ubiquitin-conjugating enzyme E2-230K (60); and urease (61,62). 1. Li & Pickart (1995) *Biochemistry* **34**, 139 and 15829. 2. Rose (1955) *Meth. Enzymol.* **1**, 591. 3. Rose, Grunberg-Manago, Korey & Ochoa (1954) *J. Biol. Chem.* **211**, 737. 4. Sund & Theorell (1963) *The Enzymes*, 2nd ed., 7, 25. 5. Barron & Levine (1952) *Arch. Biochem. Biophys.* **41**, 175. 6. Virtanen & Ellfolk (1955) *Meth. Enzymol.* **2**, 386. 7. Ellfolk (1953) *Acta Chem. Scand.* 7, 1155. 8. Hmadcha, Carballo, Conde, *et al.* (1999) *Mol. Genet. Metab.* **68**, 363. 9. Bogumil & Ullrich (2002) *Meth. Enzymol.* **348**, 271. 10. Bogumil, Namgaladze, Schaarschmidt, *et al.* (2000) *Eur. J. Biochem.* **267**, 1407. 11. Takahashi, Goldschmidt-Clermont, Alnemri, *et al.* (1997) *Exp. Cell Res.* **231**, 123. 12. Rothschild, Cori & Barron (1954) *J. Biol. Chem.* **208**, 41. 13. Miyamoto, Tiruppathi, Ganapathy & Leibach (1989) *Biochim. Biophys. Acta* **978**, 25. 14. Yang, Clark, Harrison, Kozka & Holman (1992) *Biochem. J.* **281**, 809. 15. Frost & Lane (1985) *J. Biol. Chem.* **260**, 2646. 16. Wang, Hsieh & Wu (1991) *Biochem. Biophys. Res. Commun.* **176**, 201. 17. Frost, Kohanski & Lane (1987) *J. Biol. Chem.* **262**, 9872. 18. Allison & Conners (1970) *Arch. Biochem. Biophys.* **136**, 383. 19. Phelps & Hatefi (1981) *Biochemistry* **20**, 453. 20. Bandyopadhyay, Kaiser, Rudolf, *et al.* (1997) *Biochem. Biophys. Res. Commun.* **240**, 146. 21. Han & Kohanski (1997) *Biochem. Biophys. Res. Commun.* **239**, 316. 22. Barron & Singer (1945) *J. Biol. Chem.* **157**, 221. 23. Neuhaus & Wright (1983) *Eur. J. Biochem.* **137**, 615. 24. Yamano, Kageura, Ishida & Toki (1985) *J. Biol. Chem.* **260**, 5259. 25. Eichel (1956) *J. Biol. Chem.* **222**, 137. 26. Janiszewski, Pedro, Scheffer, *et al.* (2000) *Free Radic. Biol. Med.* **29**, 889. 27. Le Cabec & Maridonneau-Parini (1995) *J. Biol. Chem.* **270**, 2067. 28. O'Kelly, Peers & Kemp (2001) *Biochem. Biophys. Res. Commun.* **283**, 1131. 29. Doussiere, Bouzidi, Poinas, Gaillard & Vignais (1999) *Biochemistry* **38**, 16394. 30. Obeso, Gomez-Nino & Gonzalez (1999) *Amer. J. Physiol.* **276**, C593. 31. Doussiere, Poinas, Blais & Vignais (1998) *Eur. J. Biochem.* **251**, 649. 32. Kulanthaivel, Simon, Leibach, Mahesh & Ganapathy (1990) *Biochim. Biophys. Acta* **1024**, 385. 33. Estrov, Manna, Harris, *et al.* (1999) *Blood* **94**, 2844. 34. Ponnappan, Trebilcock & Zheng (1999) *Exp. Gerontol.* **34**, 95. 35. Mahboubi, Young & Ferreri (1998) *J. Pharmacol. Exp. Ther.* **285**, 862. 36. Persson, Berden, Rydstrom & van Dam (1987) *Biochim. Biophys. Acta* **894**, 239. 37. Fristedt, Rydstrom & Persson (1994) *Biochem. Biophys. Res. Commun.* **198**, 928. 38. Joshi & Hughes (1981) *J. Biol. Chem.* **256**, 11112. 39. Yagi & Hatefi (1984) *Biochemistry* **23**, 2449. 40. Dibas, Prasanna & Yorio (1999) *J. Ocul.*

Pharmacol. Ther. **15**, 241. 41. Searl & Silinsky (2000) *Brit. J. Pharmacol.* **130**, 418. 42. Bartlett, Reynolds, Weible II, & Hendry (2002) *J. Neurosci. Res.* **68**, 169. 43. Yue, Liu & Shen (2001) *J. Biol. Chem.* **276**, 49093. 44. Searl & Silinsky (2000) *Brit. J. Pharmacol.* **130**, 418. 45. Yanaga, Asselin, Schieven & Watson (1995) *FEBS Lett.* **368**, 377. 46. Garcia-Morales, Minami, Luong, Klausner & Samelson (1990) *Proc. Natl. Acad. Sci. U.S.A.* **87**, 9255. 47. Liao, Hoffman & Lane (1991) *J. Biol. Chem.* **266**, 6544. 48. Stocken & Thompson (1946) *Biochem. J.* **40**, 535. 49. Samikkannu, Chen, Yih, *et al.* (2003) *Chem. Res. Toxicol.* **16**, 409. 50. Peters, Sinclair & Thompson (1946) *Biochem. J.* **40**, 516. 51. Chai, Zhai, Popescu & Napoli (1995) *J. Biol. Chem.* **270**, 28408. 52. Boerman & Napoli (1995) *Arch. Biochem. Biophys.* **321**, 434. 53. Boerman & Napoli (1995) *Biochemistry* **34**, 7027. 54. Gerhard, John, Aktories & Just (2003) *Mol. Pharmacol.* **63**, 1349. 55. Prasad & Horowitz (1987) *Biochim. Biophys. Acta* **911**, 102. 56. Laden & Porter (2001) *J. Lipid Res.* **42**, 235. 57. Aldridge & Cremer (1955) *Biochem. J.* **61**, 406. 58. Henriksen (1991) *Amer. J. Physiol.* **261**, C608. 59. Werbovetz, Brendle & Sackett (1999) *Mol. Biochem. Parasitol.* **98**, 53. 60. Berleth & Pickart (1996) *Biochemistry* **35**, 1664. 61. Rona & György (1920) *Biochem. Z.* **111**, 115. 62. Gordon & Quastel (1948) *Biochem. J.* **42**, 337. 63. Alvear, Jabalquinto & Cardemil (1989) *Biochim. Biophys. Acta* **994**, 7. 64. Schröder, Rottmann, Bachmann & Müller (1986) *J. Biol. Chem.* **261**, 663. 65. Ricci, Nardini, Chiaraluce, Dupre & Cavallini (1986) *Biochim. Biophys. Acta* **870**, 82. 66. Li & Strohl (1996) *J. Bacteriol.* **178**, 136. 67. Dechert, Adam, Harder, Clark-Lewis & Jirik (1994) *J. Biol. Chem.* **269**, 5602. 68. Choi, Miguez & Lee (2004) *Appl. Environ. Microbiol.* **70**, 3213. 69. Marshall, Darbyshire, McPhie & Jakoby (1998) *Chem. Biol. Interact.* **109**, 107. 70. Krinke, Ruelland, Valentová, *et al.* (2007) *Plant Physiol.* **144**, 1347. 71. Tóth, Balla, Ma, *et al.* (2006) *J. Biol. Chem.* **281**, 36369. 72. Balla & Balla (2006) *Trends Cell Biol.* **16**, 351. 73. Minogue, Waugh, De Matteis, *et al.* (2006) *J. Cell Sci.* **119**, 571. 74. Middleton & Tubbs (1972) *Biochem. J.* **126**, 27. 75. Zhou, Jauhiainen, Stevenson & Dolphin (1991) *J. Chromatogr.* **568**, 69.

N-(4-Phenylbenzoyl)-*syn*-(2-(3-amidinobenzyl)-3-*trans*-styryl)-β-alanine

This substituted β-alanine (FW = 503.60 g/mol) inhibits coagulation factor Xa, IC_{50} = 2.7 μM, thrombin, IC_{50} = 7.7 μM, and trypsin, IC_{50} = 22 μM, and its methyl ester also inhibits these three enzymes, with IC_{50} values of 0.11, 18, and 5.3 μM, respectively. 1. Klein, Czekaj, Gardner, *et al.* (1998) *J. Med. Chem.* **41**, 437.

N$^{\alpha}$-(3-Phenylbenzoyl)-L-boroarginine *O*,*O*-(+)-pinanediol Ester

This boronate ester ($FW_{hydrochloride}$ = 524.90 g/mol) inhibits thrombin and trypsin, with K_i values of 110 and 7.4 nM, respectively. The corresponding 4-phenylbenzoyl analogue inhibits thrombin and trypsin, with K_i values of 0.94 and 33 nM, respectively. 1. Quan, Wityak, Dominguez, *et al.* (1997) *Bioorg. Med. Chem. Lett.* **7**, 1595.

2(R)-[2-(4'-Phenylbiphenyl-4-yl)ethyl]-4(S)-(n-butyl)-1,5-pentanedioic Acid 1-(α(S)-tert-Butylglycine Methylamide) Amide

This protease inhibitor (FW = 570.77 g/mol), known systematically as 2(R)-[2-[4-(4-phenylphenyl)phenyl]ethyl]-4(S)-butylpentanedioic acid 1-((S)-tert-butylglycine methylamide) amide, inhibits gelatinase A, or matrix metalloproteinase 2, and stromelysin 1, or matrix metalloproteinase 3, with K_i values of 84 and 39 nM, respectively. **1.** Esser, Bugianesi, Caldwell, et al. (1997) J. Med. Chem. **40**, 1026.

2-Phenyl-5-(4-biphenylyl)-1,3,4-oxadiazole, See 5-(4-Biphenylyl)-2-phenyl-1,3,4-oxadiazole

Phenylboronic Acid

This arylboronic acid (FW = 121.93 g/mol; CAS 98-80-6), also known as benzeneboronic acid, phenylboric acid, and phenylboron dihydroxide, inhibits a number of serine-dependent enzymes. Note: Commercial sources typically contain varying amounts of phenylboronic anhydride, and prolonged storage results in the formation of significant amounts of phenylboroxide. **Target(s):** aminoacylase (1); Aspergillus melleus semi-alkaline protease, identical or closely related to oryzin (2); bacterial leucyl aminopeptidase (21); cholesterol esterase, or sterol esterase, K_i = 36 μM (3-5,23); chymotrypsin (19); cutinase (6,22); diacylglycerol lipase (7); β-lactamase (8); lecithin:cholesterol acyltransferase, or LCAT (9); lipase, or triacylglycerol lipase (10-13); lipoprotein lipas (5,14); oryzin (2); pancreatic elastase (19); peptidyltransferase (24); serine-type D-Ala-D-Ala carboxypeptidase (20); subtilisin (15,16); Treponema denticola protease (17); and urease, weakly inhibited (18). **1.** Wakayama, Shiiba, Sakai & Moriguchi (1998) J. Ferment. Bioeng. **85**, 278. **2.** Nakatani, Fujiwake & Hiromi (1977) J. Biochem. **81**, 1269. **3.** Feaster & Quinn (1997) Meth. Enzymol. **286**, 231. **4.** Hosie, Sutton & Quinn (1987) J. Biol. Chem. **262**, 260. **5.** Sutton, Stout, Hosie, Spencer & Quinn (1986) Biochem. Biophys. Res. Commun. **134**, 386. **6.** Koller & Kolattukudy (1982) Biochemistry **21**, 3083. **7.** Lee, Kraemer & Severson (1995) Biochim. Biophys. Acta **1254**, 311. **8.** Beesley, Gascoyne, Knott-Hunziker, et al. (1983) Biochem. J. **209**, 229. **9.** Jauhiainen, Ridgway & Dolphin (1987) Biochim. Biophys. Acta **918**, 175. **10.** Abouakil & Lombardo (1989) Biochim. Biophys. Acta **1004**, 215. **11.** Raghavendra & Prakash (2002) J. Agric. Food Chem. **50**, 6037. **12.** Gjellesvik, Lombardo & Walther (1992) Biochim. Biophys. Acta **1124**, 123. **13.** Garner (1980) J. Biol. Chem. **255**, 5064. **14.** Vainio, Virtanen & Kinnunen (1982) Biochim. Biophys. Acta **711**, 386. **15.** Lindquist & Terry (1974) Arch. Biochem. Biophys. **160**, 135. **16.** Matthews, Alden, Birktoft, Freer & Kraut (1975) J. Biol. Chem. **250**, 7120. **17.** Rosen, Naor, Rahamim, Yishai & Sela (1995) Infect. Immun. **63**, 3973. **18.** Todd & Hausinger (1989) J. Biol. Chem. **264**, 15835. **19.** Smoum, Rubinstein & Srebnik (2003) Bioorg. Chem. **31**, 464. **20.** Stefanova, Davies, Nicholas & Gutheil (2002) Biochim. Biophys. Acta **1597**, 292. **21.** Baker, Wilkes, Bayliss & Prescott (1983) Biochemistry **22**, 2098. **22.** Sebastian & Kolattukudy (1988) Arch. Biochem. Biophys. **263**, 77. **23.** Sando & Rosenbaum (1985) J. Biol. Chem. **260**, 15186. **24.** Cerná & Rychlík (1980) FEBS Lett. **119**, 342.

(N-[N-(4-Phenylbutanoyl)-L-prolyl]pyrrolidine

This peptide analogue (FW = 314.43 g/mol), also known as SUAM-1221, inhibits prolyl oligopeptidase, IC_{50} = 2.2 nM (1-5). **See also** N-[N-(4-Phenylbutanoyl)-5(R)-methyl-L-prolyl]-2(S)-(hydroxyacetyl)pyrrolidine **1.** Wallén, Christiaans, Forsberg, et al. (2002) J. Med. Chem. **45**, 4581. **2.** Arai, Nishioka, Niwa, et al. (1993) Chem. Pharm. Bull. (Tokyo) **41**, 1583. **3.** Saito, Hashimoto, Kawaguchi, et al. (1991) J. Enzyme Inhib. **5**, 51. **4.** Wallén, Christiaans, Saarinen, et al. (2003) Bioorg. Med. Chem. **11**, 3611. **5.** Wallén, Christiaans, Saario, et al. (2002) Bioorg. Med. Chem. **10**, 2199.

Phenylbutazone

This anti-inflammatory agent (FW = 308.38 g/mol), also known as 4-butyl-1,2-diphenyl-3,5-pyrazolidinedione, butazolidine, and 1,2-diphenyl-3,5-dioxo-4-n-butylpyrazolidine, is an alternative substrate for prostaglandin peroxidase and glucuronosyltransferase. Phenylbutazone also forms carbon-centered radicals in a reaction catalyzed horseradish peroxidase and hydrogen peroxide. **Target(s):** alcohol dehydrogenase (1,2); ATPase, mitochondrial (3); carbonyl reductase (4); cathepsin L (5); creatine kinase, in the presence of hydrogen peroxide and horseradish peroxidase (2); cyclooxygenase (6-8); CYP2C, probable alternative substrate (9); glutamate dehydrogenase (10); L-glutamine:D-fructose-6-phosphate aminotransferase (11,12); glyceraldehyde-3-phosphate dehydrogenase, in the presence of hydrogen peroxide and horseradish peroxidase (2); hyaluronidase (13); K+-ATPase (14); phenylpyruvate tautomerase, weakly inhibited (15); urate-ribonucleotide phosphorylase (16); and xanthine dehydrogenase (17). **1.** Sund & Theorell (1963) The Enzymes, 2nd ed. (Boyer, Lardy & Myrbäck, eds.), **7**, 25. **2.** Miura, Muraoka & Fujimoto (2001) Free Radic. Res. **34**, 167. **3.** Chatterjee & Stefanovich (1976) Arzneimittelforschung **26**, 499. **4.** Higuchi, Imamura & Otagiri (1995) Biol. Pharm. Bull. **18**, 618. **5.** Raghav & Singh (1993) Indian J. Med. Res. **98**, 188. **6.** Hughes, Mason & Eling (1988) Mol. Pharmacol. **34**, 186. **7.** Bryant, Farnfield & Janicke (2003) Amer. J. Vet. Res. **64**, 211. **8.** Cheng, Nolan & McKellar (1998) Inflammation **22**, 353. **9.** Zweers-Zeilmaker, Horbach & Witkamp (1997) Xenobiotica **27**, 769. **10.** Frieden (1963) The Enzymes, 2nd ed. (Boyer, Lardy & Myrbäck, eds.), **7**, 3. **11.** Schonhofer & Anspach (1967) Arch. Int. Pharmacodyn. Ther. **166**, 382. **12.** Chatterjee & Stefanovich (1976) Arzneimittelforschung **26**, 502. **13.** Lotmar (1960) Z. Rheumaforsch. **19**, 1. **14.** Spenney & Mize (1977) Biochem. Pharmacol. **26**, 1241. **15.** Molnar & Garai (2005) Int. Immunopharmacol. **5**, 849. **16.** Laster & Blair (1963) J. Biol. Chem. **238**, 3348. **17.** Martin-Esteve, Bozal & Calvet (1964) Arch. Interam. Rheumatol. **15**, 462.

4-Phenyl-3-butenoate

This acid ($FW_{free-acid}$ = 162.19 g/mol) is a mechanism-based inhibitor of peptidylglycine monooxygenase (1-4). The methyl ester also inhibits. **1.** Katopodis & May (1990) Biochemistry **29**, 4541. **2.** Sunman, Foster, Folse, May & Matesic (2004) Mol. Carcinog. **41**, 231. **3.** Bauer, Sunman, Foster, et al. (2007) J. Pharmacol. Exp. Ther. **320**, 1171. **4.** Labrador, Brun, König, Roatti & Baertschi (2004) Circ. Res. **95**, e98.

Phenyl-*n*-butylborinic Acid

This putative transition-state analogue (FW = 162.04 g/mol) has a K_i value for cholesterol esterase, *or* sterol esterase, of 2.9 nM (1-3) and lipoprotein lipase, K_i = 1.7 μM (2). **1.** Feaster & Quinn (1997) *Meth. Enzymol.* **286**, 231. **2.** Sutton, Stout, Hosie, Spencer & Quinn (1986) *Biochem. Biophys. Res. Commun.* **134**, 386. **3.** Sutton, Froelich, Hendrickson & Quinn (1991) *Biochemistry* **30**, 5888.

4-Phenyl-1-butyne

This arylalkyne (FW = 130.19 g/mol) is a mechanism-based (or suicide) inhibitor of the mouse cytochrome P450 CYP2b (*i.e.*, dealkylation of benzyloxyresorufin) and rat CYP2B1. **1.** Beebe, Roberts, Fornwald, Hollenberg & Alworth (1996) *Biochem. Pharmacol.* **52**, 1507.

2-Phenylbutyrate (2-Phenylbutyric Acid)

This carboxylic acid (FW$_{free-acid}$ = 164.20 g/mol; CAS 1009-67-2; M.P. = 42-44°C) inhibits acetyl-CoA synthetase (1,2), histone deacetylase (3), and kynurenine aminotransferase, weakly inhibited (4). The (*R*)-(−)- and (*S*)-(+)-enantiomers are frequently used as chiral building blocks in organic syntheses and as resolving agents. *Note*: 2-Phenylbutyric acid should not be confused with the ester phenyl butyrate (*i.e.*, $CH_3CH_2CH_2COOC_6H_5$). **1.** Webster, Jr., (1969) *Meth. Enzymol.* **13**, 375. **2.** Jencks (1962) *The Enzymes*, 2nd ed., **6**, 373. **3.** Lea, Randolph & Hodge (1999) *Anticancer Res.* **19**, 1971. 4. Mason (1959) *J. Biol. Chem.* **234**, 2770.

3-Phenylbutyrate (3-Phenylbutyric Acid)

This carboxylic acid (FW$_{free-acid}$ = 164.20 g/mol; CAS 4593-90-2; M.P. = 37-39°C) inhibits carboxypeptidase A (1) and histone deacetylase (2). The (*R*)- and (*S*)-enantiomers are frequently used as chiral building blocks in organic syntheses. *Note*: 2-Phenylbutyric acid should not be confused with the ester phenyl butyrate (*i.e.*, $CH_3CH_2CH_2COOC_6H_5$). **1.** Elkins-Kaufman, Neurath & De Maria (1949) *J. Biol. Chem.* **178**, 645. **2.** Lea, Randolph & Hodge (1999) *Anticancer Res.* **19**, 1971.

4-Phenylbutyrate (4-Phenylbutyric Acid)

This short-chain fatty acid and ammonia-scavenging pro-drug (FW$_{free-acid}$ = 164.20 g/mol; CAS 1821-12-1; *Abbreviation*: PBA), which is administered orally as sodium phenylbutyrate (FW$_{sodium-salt}$ = 186.18 g/mol; CAS 1716-12-7; Trade Name = Buphenyl®; *Abbreviation*: NaPBA), is an FDA-approved treatment for hyperammonia as a result of urea cycle disorders. **Primary Mode of Action:** PBA undergoes two-carbon loss to form the active drug, phenylacetate, which is then enzymatically conjugated to glutamine to form phenylacetylglutamine, with the latter excreted in urine (1). Because glutamine contains two equivalents of ammonia, PBA is especially effective in its action. Twenty-four-hour urinary phenylacetylglutamine levels correlate strongly with dose, making it a clinically useful non-invasive biomarker for compliance and therapeutic monitoring (2). Glycerol phenylbutyrate (FW = 530.67 g/mol; CAS; 611168-24-2; IUPAC Name: propane-1,2,3-triyl-*tris*(4-phenylbutanoate; Other Names: HPN-100 and Ravicti®) is a non-sodium-releasing FDA-approved pro-pro-drug that releases PBA and glycerol upon metabolic hydrolysis (3). **Other PBA Actions:** Sodium PBA also acts as a chaperone, reducing the load of mutant or unfolded proteins retained in the ER during cellular stress. For example, PBA protects against liver ischemia re-perfusion injury by inhibiting endoplasmic reticulum stress-mediated apoptosis (4). By stimulating fetal hemoglobin synthesis, PBA is useful in treating sickle cell disease (5). PBA also induces Arginase I in many tissues (by as much as 4x in brain) (6), a finding that suggests reduced arginine may be manifested as reduced nitric oxide synthesis. **Historical Note:** In his classic studies elucidating the β-oxidation pathway, Knoop (7) used phenylbutyrate to demonstrate that, after oral administration, phenylacetylglycine (phenaceturic acid) appears in the urine. This observation suggested phenylbutyrate underwent one cycle of two-carbon chain shortening, forming phenylacetate, which was then coupled to glycine. **1.** Walker (2009) *Diabetes Obes Metab.* **11**, 823. **2.** Mokhtarani, Diaz, Rhead, *et al.* (2012) *Mol. Genet. Metab.* **107**, 308. **3.** Lichter-Konecki, Diaz, Merritt, *et al.* (2011) *Mol. Genet. Metab.* **103**, 323. **4.** Vilatoba, Eckstein, Bilbao, *et al.* (2005) *Surgery* **138**, 342. **5.** Dover, Brusilow & Charache (1994) *Blood* **84**, 339. **6.** Kern, Yang & Kim (2007) *Mol. Genet. Metab.* **90**, 37. **7.** Knoop (1904) *Der Abbau aromatischer Fettsauren im Tierkörper*, Ernst Kuttruff, Freiburg, Germany.

N-[*N*-(4-Phenylbutyryl)-L-prolyl]pyrrolidine, *See* *N*-[*N*-(4-Phenylbutanoyl)-L-prolyl]pyrrolidine

4-Phenylchalcone Oxide

This chalcone oxide derivative (FW = 300.36 g/mol), also known as [3-(4-phenylphenyl)oxiran-2-yl](phenyl)methanone and 2,3-epoxy-3-(4-biphenyl)-1-phenyl-1-propanone, inhibits soluble epoxide hydrolase, IC$_{50}$ = 0.20 μM for the human enzyme (1-7). The 4'-phenyl analogue also inhibits (IC$_{50}$ = 0.28 μM). **1.** Debernard, Morisseau, Severson, *et al.* (1998) *Insect Biochem. Mol. Biol.* **28**, 409. 2. Newman, Morisseau, Harris & Hammock (2003) *Proc. Natl. Acad. Sci. U.S.A.* **100**, 1558. **3.** Moody & Hammock (1987) *Arch. Biochem. Biophys.* **258**, 156. **4.** Morisseau, Du, Newman & Hammock (1998) *Arch. Biochem. Biophys.* **356**, 214. **5.** Morisseau, Beetham, Pinot, *et al.* (2000) *Arch. Biochem. Biophys.* **378**, 321. **6.** Hammock, Prestwich, Loury, *et al.* (1986) *Arch. Biochem. Biophys.* **244**, 292. **7.** Meijer & Depierre (1985) *Eur. J. Biochem.* **150**, 7.

1-Phenylcyclopropylamine

This amine (FW$_{free-base}$ = 133.19 g/mol) is a mechanism-based (or suicide) inactivator of both monoamine oxidases A and B, with preference for the latter (1-7). Monoamine oxidase B is inactivated via attachment to both a cysteinyl residue and the flavin, while monoamine oxidase A is inhibited only via a covalent adduct formed with the flavin cofactor. Another target is trimethylamine dehydrogenase (8). **1.** Silverman (1995) *Meth. Enzymol.* **249**, 240. **2.** Brush & Kozarich (1992) *The Enzymes*, 3rd ed. (Sigman, ed.), **20**, 317. **3.** Mitchell, Nikolic, van Breemen & Silverman (2001) *Bioorg. Med. Chem. Lett.* **11**, 1757. **4.** Mitchell, Nikolic, Rivera, *et al.* (2001) *Biochemistry* **40**, 5447. **5.** Silverman & Hiebert (1988) *Biochemistry* **27**, 8448. **6.** Silverman & Zieske (1986) *Biochem. Biophys. Res. Commun.* **135**, 154. **7.** Silverman & Zieske (1985) *Biochemistry* **24**, 2128. **8.** Mitchell, Nikolic, Jang, *et al.* (2001) *Biochemistry* **40**, 8523.

***trans*-2-Phenylcyclopropylamine, *See* Tranylcypromine**

(+)-(*E*)-7-[4-[4-[[[2-(*trans*)-Phenylcyclopropyl]amino]carbonyl]-2-oxazolyl]phenyl]-7-(3-pyridyl)hept-6-enoate

This substituted oxazolecarboxamide ($FW_{free-acid}$ = 508.60 g/mol) inhibits human thromboxane-A synthase (IC_{50} = 8.5 nM) and is a thromboxane receptor antagonist (K_d = 62.6 nM). With the (–)-isomer, IC_{50} = 5.9 nM and K_d = 104.7 nM, respectively. **1**. Takeuchi, Kohn, True, *et al.* (1998) *J. Med. Chem.* **41**, 5362.

D-*threo*-1-Phenyl-2-decanoylamino-3-morpholino-1-propanol

This structural analogue ($FW_{free-base}$ = 390.57 g/mol) of the natural sphingolipid substrate of ceramide glucosyltransferase inhibits that enzyme (the actual active agent is the D-*threo*-isomer [*i.e.*, the (*R,R*)-isomer]) and reversibly depletes cellular glycosphingolipids. Note that intracellular ceramide concentrations also increase cytotoxic levels. **Target(s):** 1-*O*-acylceramide synthase (1); ceramide glucosyltransferase, *or* glucosylceramide synthase (2-13); ceramide glycanase (2); glucosylceramide galactosyl-transferase, *or* lactosylceramide synthase (14). **1**. Abe & Shayman (1998) *J. Biol. Chem.* **273**, 8467. **2**. Basu, Kelly, Girzadas, Li & Basu (2000) *Meth. Enzymol.* **311**, 287. **3**. Shayman & Abe (2000) *Meth. Enzymol.* **311**, 42. **4**. Shayman, Lee, Lee & Shu (2000) *Meth. Enzymol.* **311**, 373. **5**. Inokuchi, Momosaki, Shimeno, Nagamatsu & Radin (1989) *J. Cell Physiol.* **141**, 573. **6**. Radin, Shayman & Inokuchi (1993) *Adv. Lipid Res.* **26**, 183. **7**. Inokuchi & Radin (1987) *J. Lipid Res.* **28**, 565. **8**. Vunnam & Radin (1980) *Chem. Phys. Lipids* **26**, 265. **9**. Di Sano, Fazi, Citro, *et al.* (2003) *Cancer Res.* **63**, 3860. **10**. Saito, Fukushima, Tatsumi, *et al.* (2002) *Arch. Biochem. Biophys.* **403**, 171. **11**. Ichikawa & Hirabayashi (1998) *Trends Cell Biol.* **8**, 198. **12**. Wu, Marks, Watanabe, *et al.* (1999) *Biochem. J.* **341**, 395. **13**. Chujor, Feingold, Elias & Holleran (1998) *J. Lipid Res.* **39**, 277. **14**. Chatterjee (2000) *Meth. Enzymol.* **311**, 73.

Phenyldichloroarsine

This sulfhydryl group-directed organoarsenical (FW = 222.93 g/mol) reacts with many proteins, inactivating those enzymes with essential thiols. *Caution*: A strong vesicant that is readily hydrolyzed in the lung to form hydrogen chloride and phenylarsenious oxide, phenyldichloroarsine should be handled only in chemical fume hoods and with protective equipment. It also has an immediate effect on eyes and will cause blisters on skin and can induce vomiting. First synthesized by Fritz Haber, and briefly used as chemical warfare agent by Germany in World War I in 1917-18, phenyldichloroarsine has been given the chemical warfare designation PD. When administered in a timely manner, 2,3-Dimercapto-1-propanol (British Anti-Lewisite, *or* BAL) and 2,3-dithioerythritol (DTE) can be effective antidotes (1). Phenyldichloroarsine readily enters cells and forms adducts with glutathione (2). **Target(s):** SH-dependent enzymes. **1**. Boyd, Harbell, O'Connor & McGown (1989) *Chem. Res. Toxicol.* **2**, 301. **2**. Dill,

Adams, O'Connor, Chong & McGown (1987) *Arch. Biochem. Biophys.* **257**, 293.

Phenyl Dichlorophosphate, *See* Phenylphosphoro-dichloridate

Phenyl Disulfide, *See* Diphenyl Disulfide

***o*-Phenylenediamine**

This aromatic amine ($FW_{free-base}$ = 108.14 g/mol; M.P. = 103-104°C), also known as 1,2-diaminobenzene, inhibits catalase (1) and succinate oxidase (2). *o*-Phenylenediamine is a precursor in the chemical synthesis of dyes and inhibitors as well as a reagent in the qualitative detection of mono- and disaccharides and uronic acids. **1**. Ceriotti, Spandrio & Agradi (1966) *Enzymologia* **30**, 290. **2**. Potter & Dubois (1943) *J. Gen. Physiol.* **26**, 391.

***p*-Phenylenediamine**

This aromatic amine ($FW_{free-base}$ = 108.14 g/mol; M.P. = 145-147 °C; pK_1 = 2.67 and pK_2 = 6.16.), also known as 1,3-diaminobenzene, inhibits aspartate aminotransferase (1), catalase (2), ornithine decarboxylase (3), pyruvate decarboxylase (4,5), succinate dehydrogenase (6), succinate oxidase (7), and urease (8). *m*-Phenylenediamine is also an artificial electron donor that can reduce cytochrome *c* and is frequently employed in qualitative assays for sugar acids. *Note*: *p*-Phenylenediamine darkens when exposed to air and should be tightly closed and protected from light. **1**. Cohen (1951) *The Enzymes*, 1st ed. (Sumner & Myrbäck, eds.), **1** (part 2), 1040. **2**. Ceriotti, Spandrio & Agradi (1966) *Enzymologia* **30**, 290. **3**. Solano, Penafiel, Solano & Lozano (1988) *Int. J. Biochem.* **20**, 463. **4**. Kensler, Young & Rhoads (1942) *J. Biol. Chem.* **143**, 465. **5**. Kensler, Dexter & Rhoads (1942) *Cancer Res.* **2**, 1. **6**. Schlenk (1951) *The Enzymes*, 1st ed. (Sumner & Myrbäck, eds.), **2** (part 1), 316. **7**. Potter & Dubois (1943) *J. Gen. Physiol.* **26**, 391. **8**. Potter (1942) *Cancer Res.* **2**, 688.

***N,N'*-1,2-Phenylenedimaleimide, *See* N,N'-(1,2-Phenylene)bismaleimide**

2-Phenylethaneboronic Acid, *See* 2-Phenylethylboronic Acid

1-Phenylethanol

This alcohol (FW = 122.17 g/mol; M.P. = 19-20°C) inhibits β-alanine:pyruvate aminotransferase, weakly inhibited by the *R*-enantiomer (1) and β-glucosidase (2). **1**. Yun, Yang, Cho, Hwang & Kim (2003) *Biotechnol. Lett.* **25**, 809. **2**. Dale, Ensley, Kern, Sastry & Byers (1985) *Biochemistry* **24**, 3530.

2-Phenylethanol

This alcohol (FW = 122.17 g/mol; M.P. = –27°C; B.P. = 219-221°C at 750 mm Hg; slightly soluble in water (1.6 g/100 mL at 20°C); miscible in ethanol), also known as phenethyl alcohol, β-phenylethyl alcohol, and benzyl carbinol, is found in a number of essential oils. 2-Phenylethanol has a distinct floral odor, reminiscent of roses, and is often used in formulating perfumes and flavors. At a concentration of 0.25%, 2-phenylethanol will inhibit DNA biosynthesis without affecting RNA or protein biosynthesis. At higher concentrations, RNA and protein biosynthesis are also inhibited. **Target(s):** choline linase (1); chymotrypsin (2); DNA biosynthesis (3,4);

DNA-dependent DNA polymerase (5); excision repair of DNA (5); β-glucosidase (6); *sn*-glycerol 3-phosphate acyltransferase (7); signal peptidase II (8); and uridine kinase (1). **1.** Plagermann, Roth & Erbe (1969) *Biochemistry* **8**, 4782. **2.** Wallace, Kurtz & Niemann (1963) *Biochemistry* **2**, 824. **3.** Konetzka & Berrah (1962) *Biochem. Biophys. Res. Commun.* **8**, 407. **4.** Müller, Yamazaki & Zahn (1972) *Enzymologia* **43**, 1. **5.** Tachibana & Yonei (1985) *Int. J. Radiat. Biol. Relat. Stud. Phys. Chem. Med.* **47**, 663. **6.** Dale, Ensley, Kern, Sastry & Byers (1985) *Biochemistry* **24**, 3530. **7.** Nunn, Cheng, Deutsch, Tang & Tropp (1977) *J. Bacteriol.* **130**, 620. **8.** Sankaran & Wu (1995) *Meth. Enzymol.* **248**, 169.

2-Phenylethylamine

This phenylalanine-derived monoamine (FW$_{free-base}$ = 121.18 g/mol; M.P. < –20°C; B.P. = 194.5°C), also known as β-phenethylamine, which is structurally related to amphetamine, is normally present in human urine at ~30 μg/L. 2-Phenylethylamine is also used as an organic solvent in chromatography, spectrophotometry, and in liquid scintillation counting. It will absorb carbon dioxide from air and has a low solubility in water (3.4% at 20°C). **Target(s):** aldehyde oxidase, weakly inhibited (1); aromatic-L-amino-acid decarboxylase (2); chymotrypsin (3); dopamine β-monooxygenase (4-6); β-glucosidase (7); glutamate:ethylamine ligase, *or* theanine synthetase (8); hemoglobin S polymerization (9); histamine *N*-methyltransferase (10); monoamine oxidase (11); phenylalanine ammonia-lyase (12); phenylalanine racemase, ATP-hydrolyzing (13); phenylalanyl-tRNA synthetase, K_i = 93 μM (14); renin (15); trypsin, weakly inhibited (16,17); tryptophan 2-monooxygenase (18); and tyrosine aminotransferase (19,20). **1.** Panoutsopoulos & Beedham (2004) *Acta Biochim. Pol.* **51**, 649. **2.** Nakazawa, Kumagai & Yamada (1987) *Agric. Biol. Chem.* **51**, 2531. **3.** Kaufman & Neurath (1949) *J. Biol. Chem.* **181**, 623. **4.** Webb (1966) *Enzyme and Metabolic Inhibitors*, vol. 2, pp. 320, 600, Academic Press, New York. **5.** Goldstein & Contrera (1961) *Experentia* **17**, 267. **6.** Goldstein & Contrera (1962) *J. Biol. Chem.* **327**, 1898. **7.** Dale, Ensley, Kern, Sastry & Byers (1985) *Biochemistry* **24**, 3530. **8.** Sasaoka, Kito & Onishi (1965) *Agric. Biol. Chem.* **29**, 984. **9.** Noguchi & Schechter (1978) *Biochemistry* **17**, 5455. **10.** Thithapandha & Cohn (1978) *Biochem. Pharmacol.* **27**, 263. **11.** Kinemuchi, Arai, Oreland, Tipton & Fowler (1982) *Biochem. Pharmacol.* **31**, 959. **12.** Dubery & Smit (1994) *Biochim. Biophys. Acta* **1207**, 24. **13.** Vater & Kleinkauf (1976) *Biochim. Biophys. Acta* **429**, 1062. **14.** Santi & Danenberg (1971) *Biochemistry* **10**, 41913. **15.** Dworschack & Printz (1978) *Biochemistry* **17**, 2484. **16.** Baker & Erickson (1967) *J. Med. Chem.* **10**, 11. **17.** Leiros, Brandsdal, Andersen, *et al.* (2004) *Protein Sci.* **13**, 1056. **18.** Emanuele, Heasley & Fitzpatrick (1995) *Arch. Biochem. Biophys.* **316**, 241. **19.** Granner & Tomkins (1970) *Meth. Enzymol.* **17A**, 633. **20.** Jacoby & La Du (1964) *J. Biol. Chem.* **239**, 419.

(2-Phenyl-1-ethyl)guanidine, *See* Phenethyl- guanidine

N-(Phenylethyl)oxamate, *See* N-(Phenethyl)oxamate

N-(Phenylethylphosphonyl)glycyl-L-prolyl-L-norleucine

This phosphonamide-based peptidomimetic (FW$_{free-acid}$ = 453.48 g/mol), also called *N*-(phenylethylphosphonyl)-glycyl-L-prolyl-L-aminohexanoate and phosphodiepryl 03, is a potent inhibitor of microbial collagenases, K_i = 60 nM (1), neurolysin, K_i = 0.3 nM (2-6), and thimet oligopeptidase, K_i = 0.9 nM (2-4,7). **1.** Dive, Yiotakis, Nicolaou & Toma (1990) *Eur. J. Biochem.* **191**, 685. **2.** Barrett, Brown, Dando, *et al.* (1995) *Meth. Enzymol.* **248**, 529. **3.** Checler, Barelli, Dauch, *et al.* (1995) *Meth. Enzymol.* **248**, 593. **4.** Barelli, Dive, Yiotakis, Vincent & Checler (1992) *Biochem. J.* **287**,

621. **5.** Checler, Barelli, Dauch, *et al.* (1993) *Biochem. Soc. Trans.* **21**, 692. **6.** Barelli, Fox-Threlkeld, Dive, *et al.* (1994) *Brit. J. Pharmacol.* **112**, 127. **7.** Akopyan, Couedel, Orlowski, Fournie-Zaluski & Roques (1994) *Biochem. Biophys. Res. Commun.* **198**, 787.

2-Phenylethynesulfonamide

This synthetic Hsp70 protein substrate-binding inhibitor (FW = 181.21 g/mol; CAS 64984-31-2; *Abbreviation*: PES) selectively interacts with the multifunctional, stress-inducible molecular chaperone HSP70, disrupting its association with several of co-chaperones and client proteins (1). Treatment of cultured tumor cells with PES promotes cell death as a consequence of protein aggregation, impaired autophagy, and inhibition of lysosomal function. (Typical does are 1–10 μg/g injected intraperitoneally in mice, and 1–10 μM *in vitro*.) PES also exerts profound effects both *in vivo* and *in vitro* by: (a) preventing LPS-induced increase in serum alanine aminotransferase (ALT) and aspartate aminotransferase (AST) activity, reducing infiltration of inflammatory cells, and promoting liver cell apoptosis; (b) reducing inducible nitric oxide synthase (iNOS) protein expression as well as that of serum nitric oxide (NO), tumor necrosis factor-α (TNF-α), and interleukin-6 (IL-6) content in LPS-stimulated mice; (c) decreasing the mRNA levels of iNOS, TNF-α, and IL-6 in LPS-stimulated liver; (d) attenuating the degradation of inhibitor of κB-α (IκB-α) as well as the phosphorylation and nuclear translocation of nuclear factor-κB (NF-κB) in LPS-stimulated liver (1). PES also remarkably reduces typically observed increases in intracellular [Ca^{2+}] and intracellular pH value in LPS-stimulated RAW 264.7 cells. Moreover, PES significantly reduces the increase in Na$^+$/H$^+$ exchanger 1 (NHE1) association to Hsp70 in LPS-stimulated macrophages and liver, suggesting that NHE1-Hsp70 interaction is required for the involvement of NHE1 in the inflammation response. By disrupting HSP70 actions of in multiple cell signaling pathways, PES offers a way to probe the diverse functions of this molecular. **1.** Leu, Pimkina, Frank, Murphy & George (2009) *Mol. Cell.* **36**, 15. **2.** Huang, Wang, Chen, Wang & Zhang (2013) *PLoS One* **8**, e67582.

DL-α-Phenylglycidate (DL-α-Phenylglycidic Acid)

R-isomer   *S*-isomer

Thisoxirane-containing agent (FW$_{free-acid}$ = 164.16 g/mol), also known as phenyloxiranecarboxylate, is an active-site-directed, irreversible inhibitor of mandelate racemase, with inhibition only expressed by the (*R*)-enantiomer. The enzyme is inactivated following nucleophilic attack of the ε-amino group of a lysyl residue on the distal carbon on the epoxide ring of (*R*)-isomer (1-6). The (*S*)-isomer also binds to the active site, albeit with a weaker affinity. **1.** Mitra, Kallarakal, Kozarich, *et al.* (1995) *Biochemistry* **34**, 2777. **2.** Kenyon & Hegeman (1977) *Meth. Enzymol.* **46**, 541. **3.** Fee, Hegeman & Kenyon (1974) *Biochemistry* **13**, 2533. **4.** Landro, Gerlt, Kozarich, *et al.* (1994) *Biochemistry* **33**, 635. **5.** Whitman, Hegeman, Cleland & Kenyon (1985) *Biochemistry* **24**, 3936. **6.** Kenyon & Hegeman (1979) *Adv. Enzymol. Relat. Areas Mol. Biol.* **50**, 325.

(2R,3R)- and (2S,3S)-3-Phenylglycidol

2R,3R-isomer

These two stereoisomers (FW = 150.18 g/mol) inhibit soluble epoxide hydrolase, with respective I$_{50}$ values of 430 and 370 μM for the (2*R*,3*R*)- and (2*S*,3*S*)-forms (1-3). **1.** Dietze, Kuwano, Casas & Hammock (1991) *Biochem. Pharmacol.* **42**, 1163. **2.** Dietze, Kuwano & Hammock (1993) *Int. J. Biochem.* **25**, 43. **3.** Stapleton, Beetham, Pinot, *et al.* (1994) *Plant J.* **6**, 251.

Phenylglyoxal

This commercially available peptide/protein-modifying reagent (FW = 134.13 g/mol; CAS 1074-12-0), occasionally called benzoylformaldehyde, is a mild oxidant that reacts covalently with guanidinium groups at pH ~ 7, consuming two moles of reagent per mole of reactive arginyl residue. Modified guanidinium groups are relatively stable below pH 4; however, the original guanidinium group is slowly regenerated at neutral or alkaline pH. Upon prolonged incubation with phenylglyoxal, α-amino groups may also be modified. *See also Glyoxal; 4-Hydroxyphenylglyoxal; etc.* **Target(s):** To date, over 240 enzymes are inhibited by phenylglyoxal.

Phenylguanidine

This guanidine salt (FW = 136.18 g/mol; CAS 2002-16-6) inhibits trypsin, K_i = 72.5 µM (1-7) and urokinase, *or* u-plasminogen activator (8). **1**. Walsh (1970) *Meth. Enzymol.* **19**, 41. **2**. Shaw (1970) *The Enzymes*, 3rd ed. (Boyer, ed.), **1**, 91. **3**. Baker (1967) *Design of Active-Site-Directed Irreversible Enzyme Inhibitors*, Wiley, New York. **4**. Mares-Guia & Shaw (1965) *J. Biol. Chem.* **240**, 1579. **5**. Baker & Erickson (1967) *J. Med. Chem.* **10**, 1123. **6**. Sanborn & Hein (1968) *Biochemistry* **7**, 3616. **7**. Sanborn & Bryan (1968) *Biochemistry* **7**, 3624. **8**. Nienaber, Wang, Davidson & Henkin (2000) *J. Biol. Chem.* **275**, 7239.

8-Phenylguanine

This substituted purine (FW = 227.23 g/mol) inhibits adenylosuccinate synthetase, $I_{50} \approx 0.23$ mM (1), guanine deaminase (2,3), guanosine phosphorylase, *or* purine-nucleoside phosphorylase, $I_{50} \approx 0.25$ mM (4), and xanthine oxidase (5). **1**. Baker & Erickson (1967) *J. Pharmaceut. Sci.* **56**, 1075. **2**. Baker (1967) *Design of Active-Site-Directed Irreversible Enzyme Inhibitors*, Wiley, New York. **3**. Baker & Santi (1967) *J. Med. Chem.* **10**, 62. **4**. Baker & Schaeffer (1971) *J. Med. Chem.* **14**, 809. **5**. Baker (1967) *J. Pharmaceut. Sci.* **56**, 959.

DL-*threo*-1-Phenyl-2-hexadecanoylamino-3-morpholino-1-propanol

D-enantiomer

This inhibitor (FW = 474.73 g/mol), also known as DL-*threo*-1-phenyl-2-palmitoylamino-3-morpholino-1-propanol and often abbreviated DL-*threo*-PHMP or PPMP, is a structural analogue of the natural sphingolipid substrate of ceramide glucosyltransferase. *See also DL-threo-1-Phenyl-2-decanoylamino-3-morpholino-1-propanol* **Target(s):** ceramide glucosyltransferase (1-5); and ceramide glycanase (6,7). **1**. Nakamura, Kuroiwa, Kono & Takatsuki (2001) *Biosci. Biotechnol. Biochem.* **65**, 1369. **2**. di Bartolomeo & Spinedi (2001) *Biochem. Biophys. Res. Commun.* **288**, 269. **3**. Chujor,

Feingold, Elias & Holleran (1998) *J. Lipid Res.* **39**, 277. **4**. Abe, Inokuchi, Jimbo, *et al.* (1992) *J. Biochem.* **111**, 191. **5**. Couto, Caffaro, Uhrig, *et al.* (2004) *Eur. J. Biochem.* **271**, 2204. **6**. Basu, Kelly, Girzadas, Li & Basu (2000) *Meth. Enzymol.* **311**, 287. **7**. Basu, Girzadas, Dastgheib, *et al.* (1997) *Indian J. Biochem. Biophys.* **34**, 142.

Phenylhydrazine

This toxic hydrazine (FW$_{free-base}$ = 108.14 g/mol; CAS 100-63-0; M.P. = 19.5°C) inhibits many pyridoxal 5'-phosphate-dependent enzymes and also reacts with aldehyde- and ketone-containing substances, especially reducing sugars. *Caution*: This substance is highly toxic, giving rise to oxidative hemolysis, the latter accounting for its use in stimulating reticulocyte formation in experimental animals. In the 1870s, phenylhydrazine proved invaluable in the structural characterization of monosaccharides, for which Fischer was awarded the Nobel Prize in Chemistry (1902). However, Fischer's overexposure to phenylhydrazine was responsible in part for his failing health and contributed to his decision to end his own life with cyanide in 1919. **Target(s):** Literature survey indicates >120 enzyme targets, too many to list here.

Phenylhydrazone Malononitrile, *See Mesoxalonitrile Phenylhydrazone*

2-Phenyl-1,3-indandione, *See Phenindione*

Phenylisocyanate

This toxic liquid, also known as isocyanatobenzene and phenyl isocyanate (FW = 119.12 g/mol; melting point = –30 °C boiling point = 158-168°C), can react with the free amino group at the N-terminus of a peptide or protein, transiently producing the hydantoic acid derivative. This product subsequently generates the phenyl hydantoin derivative upon cleavage of the N-terminal aminoacyl residue. **Target(s):** alkaline phosphatase (1,2); amylase, pancreatic (3); ribonuclease, pancreatic (4); and urease (5). **1**. Fernley (1971) *The Enzymes*, 3rd ed. (Boyer, ed.), **4**, 417. **2**. Gould (1944) *J. Biol. Chem.* **156**, 365. **3**. Little & Caldwell (1942) *J. Biol. Chem.* **142**, 585. **4**. McDonald (1955) *Meth. Enzymol.* **2**, 427. **5**. Varner (1960) *The Enzymes*, 2nd ed. (Boyer, Lardy & Myrbäck, eds.), **4**, 247.

β-Phenylisopropylamine, *See Amphetamine*

N-Phenyl Isopropyl Carbamate, *See Propham*

β-Phenylisopropylhydrazine

This substituted hydrazine (FW$_{free-base}$ = 150.22 g/mol), also known as pheniprazine and (1-methyl-2-phenylethyl)hydrazine, is an antihypertensive agent that also irreversibly inhibits monoamine oxidase. **Target(s):** alcohol dehydrogenase, particularly the yeast enzyme (1); and monoamine oxidase (2-10). **1**. Brändén, Jörnvall, Eklund & Furugren (1975) *The Enzymes*, 3rd ed. (Boyer, ed.), **11**, 103. **2**. Blaschko (1963) *The Enzymes*, 2nd ed. (Boyer, Lardy & Myrbäck, eds.), **8**, 337. **3**. Smith, Weissbach & Udenfriend (1963) *Biochemistry* **2**, 746. **4**. Wong & Tyce (1979) *Neurochem. Res.* **4**, 821. **5**. Bouchaud (1969) *C. R. Acad. Sci. Hebd. Seances Acad. Sci. D* **268**, 531. **6**. Bouchaud, Couteaux & Gautron (1965) *C. R. Hebd. Seances Acad. Sci.* **260**, 348. **7**. Eberson & Persson (1962) *J. Med. Pharm. Chem.* **91**, 738. **8**. Horita (1958) *J. Pharmacol. Exp. Ther.* **122**, 176. **9**. Udenfriend, Witkop, Redfield & Weissbach (1958) *Biochem. Pharmacol.* **1**, 160. **10**. Nagatsu, Yamamoto & Harada (1970) *Enzymologia* **39**, 15.

3-Phenyllactic Acid

L-3-phenyllactate

This α-hydroxy acid (FW$_{free-acid}$ = 166.18 g/mol; CAS 156-05-8), also known as α-hydroxyhydrocinnamic acid and β-phenyllactic acid, is found naturally as the L-enantiomer and is a degradation product of L-phenylalanine *via* phenylpyruvate. The free acid of the L-enantiomer is a solid (melting point = 124°C) that is soluble in hot water. **Target(s):** aminoacylase I, by the L-enantiomer (1); [branched-chain α-keto-acid dehydrogenase] kinase, *or* [3-methyl-2-oxobutanoate dehydrogenase (acetyl-transferring)] kinase (2,3); carboxypeptidase A, by the L-enantiomer (4-8); carboxypeptidase C, *or* carboxypeptidase Y (9,10); crayfish carboxypeptidase (11,12); diphosphomevalonate decarboxylase (27); fumarase, *or* fumarate hydratase, inhibited by DL-mixture (13); histidine decarboxylase, weakly inhibited (14); hydroxylaminolysis of amino acid esters at −18°C, nonenzymatic (15); mandelate racemase (16); Na$^+$/K$^+$-exchanging ATPase (17); phenylalanine ammonia-lyase, by the L-enantiomer (18,19); phenylalanine dehydrogenase, by the L-enantiomer (20); phenylalanyl-tRNA synthetase, weakly inhibited, K_i = 2.2 mM (21); stearoyl-CoA desaturase (22); thyroid-hormone aminotransferase (23); tryptophanase, tryptophan indole-lyase (24); tryptophan 2-monooxygenase (25); and tryptophan 5-monooxygenase (26). **1.** Tamura, Oki, Yoshida, *et al.* (2000) *Arch. Biochem. Biophys.* **379**, 261. **2.** Paxton (1988) *Meth. Enzymol.* **166**, 313. **3.** Paxton & Harris (1984) *Arch. Biochem. Biophys.* **231**, 58. **4.** Hartsuck & Lipscomb (1971) *The Enzymes*, 3rd ed. (Boyer, ed.), **3**, 1. **5.** Martinelli, Hanson, Thompson, *et al.* (1989) *Biochemistry* **28**, 2251. **6.** Johansen, Klyosov & Vallee (1976) *Biochemistry* **15**, 296. **7.** Hall, Kaiser & Kaiser (1969) *J. Amer. Chem. Soc.* **91**, 485. **8.** Ludwig (1973) *Inorg. Biochem.* **1**, 438. **9.** Hayashi (1976) *Meth. Enzymol.* **45**, 568. **10.** Jonson & Aswad (1990) *Biochemistry* **29**, 4373. **11.** Zwilling & Neurath (1981) *Meth. Enzymol.* **80**, 633. **12.** Zwilling, Jakob, Bauer, Neurath & Enfield (1979) *Eur. J. Biochem.* **94**, 223. **13.** Flint (1994) *Arch. Biochem. Biophys.* **311**, 509. **14.** Sakamoto, Watanabe, Hayashi, Taguchi & Wada (1985) *Agents Actions* **17**, 32. **15.** Grant & Alburn (1965) *Biochemistry* **4**, 1913. **16.** Hegeman, Rosenberg & Kenyon (1970) *Biochemistry* **9**, 4029. **17.** Dwivedy & Shah (1982) *Neurochem. Res.* **7**, 717. **18.** Hanson & Havir (1972) *The Enzymes*, 3rd ed. (Boyer, ed.), **7**, 75. **19.** Parkhurst & Hodgins (1972) *Arch. Biochem. Biophys.* **152**, 597. **20.** Brunhuber, Thoden, Blanchard & Vanhooke (2000) *Biochemistry* **39**, 9174. **21.** Santi & Danenberg (1971) *Biochemistry* **10**, 4813. **22.** Scott & Foote (1979) *Biochim. Biophys. Acta* **573**, 197. **23.** Nakano (1970) *Meth. Enzymol.* **17A**, 660. **24.** Snell (1975) *Adv. Enzymol. Relat. Areas Mol. Biol.* **42**, 287. **25.** Emanuele, Heasley & Fitzpatrick (1995) *Arch. Biochem. Biophys.* **316**, 241. **26.** Freedland, Wadzinski & Waisman (1961) *Biochem. Biophys. Res. Commun.* **6**, 227. **27.** Shama Bhat & Ramasarma (1979) *Biochem. J.* **181**, 143.

Phenylmercuric Acetate

This reagent (FW = 336.74 g/mol), which is often used as a fungicide and herbicide, inhibits enzymes with a catalytically essential sulfhydryl group or when a thiol is critical for maintaining a state of oligomerization required for enzymatic activity. The reactive species in aqueous solutions is the phenylmercuric cation, C$_6$H$_5$–Hg$^+$. Phenylmercuric acetate is soluble in about 600 parts water. ***See also*** *p-Hydroxymercuribenzoate* **Target(s):** adenosine deaminase (1-3); adenylyl-sulfate kinase (37); alliin lyase (4); D-amino-acid oxidase (5,6); α-amylase (31); asclepain (7); bromelain (8,9); bromoxyniol nitrilase (10); carboxypeptidase C, carboxypeptidase Y (28); desulfoglucosinolate sulfotransferase (11); dextransucrase (45-47); electron transport, bacterial photosynthetic (12); flavonol 3-O-glucosyltransferase (42,43); flavonol 7-O-β-glucosyltransferase (38,39); formate dehydrogenase (13); fruit bromelain (8); β-glucosidase (30); glutamate racemase (14); glutathione-disulfide reductase (15); glycerol dehydratase (16); glycogen synthase (44); haloalkane dehalogenase (17); [heparan sulfate]-glucosamine

N-sulfotransferase (35); hydroxyanthraquinone glucosyltransferase (41); imidazoleacetate 4-monooxygenase (52); isoamylase (29); isobutyraldoxime O-methyltransferase (50); kaempferol 3-O-galactosyltransferase (40); kaempferol 4'-O-methyltransferase (48); myosin ATPase (18); NADH oxidase (19); nitrilase, *or* nitrile hydratase (20-22); nucleoside-diphosphate kinase (23); phosphoinositide 5-phosphatase (34); phosphoinositide phospholipase C (32,33); photophosphorylation (12); pyrophosphatase (24); pyruvate carboxylase (25); ribonuclease (26); ribose-5-phosphate isomerase (27); serine-phosphoethanolamine synthase (36); stem bromelain (8,9); sterigmatocystin 8-O-methyltransferase (49); superoxide dismutase (51). **1.** Ronca, Lucacchini & (1977) *Meth. Enzymol.* **46**, 327. **2.** Ronca, Bauer & Rossi (1967) *Eur. J. Biochem.* **1**, 434. **3.** Zielke & Suelter (1971) *The Enzymes*, 3rd ed., **4**, 47. **4.** Jansen, Muller & Knobloch (1989) *Planta Med.* **55**, 440. **5.** Frisell & Hellerman (1957) *J. Biol. Chem.* **225**, 53. **6.** Boyer (1959) *The Enzymes*, 2nd ed., **1**, 511. **7.** Rodriguez-Romero & Hernández-Arana (1998) in *Handb. Proteolytic Enzymes*, p. 576, Academic Press, San Diego. . Ota, Moore & Stein (1964) *Biochemistry* **3**, 180. **9.** Silverstein & Kezdy (1975) *Arch. Biochem. Biophys.* **167**, 678. **10.** Stalker, Malyj & McBride (1988) *J. Biol. Chem.* **263**, 6310. **11.** Jain, Groot Wassink, Kolenovsky & Underhill (1990) *Phytochemistry* **29**, 1425. **12.** Izawa & Good (1972) *Meth. Enzymol.* **24**, 355. **13.** Kanamori & Suzuki (1968) *Enzymologia* **35**, 185. **14.** Tanaka, Kato & Kinoshita (1961) *Biochem. Biophys. Res. Commun.* **4**, 114. **15.** Williams, Jr. (1976) *The Enzymes*, 3rd ed. (Boyer, ed.), **13**, 89. **16.** Smiley & Sobolov (1962) *Arch. Biochem. Biophys.* **97**, 538. **17.** Scholtz, Leisinger, Suter & Cook (1987) *J. Bacteriol.* **169**, 5016. **18.** Kielley (1961) *The Enzymes*, 2nd ed., **5**, 159. **19.** Suzuki (1966) *Enzymologia* **30**, 215. **20.** Li, Zhang & Yang (1992) *Appl. Biochem. Biotechnol.* **36**, 171. **21.** Almatawah, Cramp & Cowan (1999) *Extremophiles* **3**, 283. **22.** Harper (1977) *Biochem. J.* **165**, 309. **23.** Glaze & Wadkins (1967) *J. Biol. Chem.* **242**, 2139. **24.** Schwarm, Vigenschow & Knobloch (1986) *Biol. Chem. Hoppe-Seyler* **367**, 119. **25.** Tu & Hagedorn (1992) *Arch. Insect Biochem. Physiol.* **19**, 53. **26.** Zittle (1946) *J. Biol. Chem.* **163**, 111. **27.** Middaugh & MacElroy (1976) *J. Biochem.* **79**, 1331. **28.** Bai & Hayashi (1979) *J. Biol. Chem.* **254**, 8473. **29.** Kitagawa, Amemura & Harada (1975) *Agric. Biol. Chem.* **39**, 989. **30.** Hornberger, Bottigheimer, Hillier-Kaiser & Kreis (2000) *Plant Physiol. Biochem.* **38**, 929. **31.** Bealin-Kelly, Kelly & Fogarty (1990) *Biochem. Soc. Trans.* **18**, 310. **32.** Palmer (1973) *Biochim. Biophys. Acta* **326**, 194. **33.** Thompson & Dawson (1964) *Biochem. J.* **91**, 237. **34.** Dawson & Thompson (1964) *Biochem. J.* **91**, 244. **35.** Eisenman, Balasubramanian & Marx (1967) *Arch. Biochem. Biophys.* **119**, 387. **36.** Allen & Rosenberg (1968) *Biochim. Biophys. Acta* **151**, 504. **37.** Schriek & Schwenn (1986) *Arch. Microbiol.* **145**, 32. **38.** Ishikura & Yang (1994) *Phytochemistry* **36**, 1139. **39.** Ishikura, Yang & Teramoto (1993) *Z. Naturforsch. C* **48**, 563. **40.** Ishikura & Mato (1993) *Plant Cell Physiol.* **34**, 329. **41.** Khouri & Ibrahim (1987) *Phytochemistry* **26**, 2531. **42.** Larson & Lonergan (1972) *Planta* **103**, 361. **43.** Ishikura & Mato (1993) *Plant Cell Physiol.* **34**, 329. **44.** Takahara & Matsuda (1978) *Biochim. Biophys. Acta* **522**, 363. **45.** Kobayashi & Matsuda (1976) *J. Biochem.* **79**, 1301. **46.** Kobayashi & Matsuda (1980) *Biochim. Biophys. Acta* **614**, 46. **47.** Kobayashi & Matsuda (1975) *Biochim. Biophys. Acta* **397**, 69. **48.** Curir, Lanzotti, Dolci, *et al.* (2003) *Eur. J. Biochem.* **270**, 3422. **49.** Liu, Bhatnagar & Chu (1999) *Nat. Toxins* **7**, 63. **50.** Harper & Kennedy (1985) *Biochem. J.* **226**, 147. **51.** Briggs & Fee (1978) *Biochim. Biophys. Acta* **537**, 86. **52.** Okamoto, Nozaki & Hayaishi (1968) *Biochem. Biophys. Res. Commun.* **32**, 30.

Phenylmethanesulfonyl Fluoride

This toxic sulfonylating agent, abbreviated as PMSF (FW = 174.20 g/mol; CAS 329-98-6), also known as phenylmethylsulfonyl fluoride and α-tolylsulfonyl fluoride, irreversibly inhibits many enzymes by reacting with an active-site hydroxyl group. Our literature survey indicates that more than 260 enzymes are inactivated by PMSF. Reactivity is enhanced by proton transfer from the nucleophilic serine hydroxyl group to a nearby imidazole nitrogen, such as that within the so-called Ser/His/Asp catalytic triad of many serine proteinases and serine esterases (1,2). Inactivation requires 1-3 mM concentrations (higher concentrations precipitate spontaneously). Typically, a 20-50 mM solution is freshly prepared in 2-propanol or dioxane, followed by dilution with water or directly into the enzyme-containing solution, with the aid of a vortex mixer. The half-life of an aqueous 0.1–1.0 mM PMSF solution is about one hour at pH 7.5. *Note:*

PMSF is the recommended alternative to the neurotoxic fluorophosphates and fluoro-phosphonates. (Even so, PMSF should still be handled with care to avoid inhalation of the powdered reagent.) PMSF also inactivates papain, and reactivation occurs upon treating the modified protein with a thiol reagent (3-5). **1**. Gold (1967) *Meth. Enzymol.* **11**, 706. **2**. Halfon & Craik (1998) in *Handb. Proteolytic Enzymes*, p. 12, Academic Press, San Diego. **3**. Whitaker & Perez-Villasenor (1968) *Arch. Biochem. Biophys.* **124**, 70. **4**. Cohen (1970) *The Enzymes*, 3rd ed., **1**, 147. **5**. Glazer & Smith (1971) *The Enzymes*, 3rd ed., **3**, 501.

5-Phenyl-1-pentyne

This alkyne (FW = 144.22 g/mol) is a mechanism-based inhibitor of certain cytochrome P450 systems, including CYP2b (1), CYP2B1 (1,2), CYP2E1 (2), and CYP2F2 (3-6). **1**. Beebe, Roberts, Fornwald, Hollenberg & Alworth (1996) *Biochem. Pharmacol.* **52**, 1507. **2**. Roberts, Alworth & Hollenberg (1998) *Arch. Biochem. Biophys.* **354**, 295. **3**. Carlson (2003) *J. Toxicol. Environ. Health A* **66**, 861. **4**. Carlson (2002) *Toxicology* **179**, 129. **5**. Born, Caudill, Fliter & Purdon (2002) *Drug Metab. Dispos.* **30**, 483. **6**. Carlson (1997) *J. Toxicol. Environ. Health* **51**, 477.

2-Phenylphenol

This biphenyl fungicide (FW = 170.21 g/mol; CAS 90-43-7; M.P. = 55.5-57.5°C), also known as *o*-phenylphenol and 2-hydroxybiphenyl, is metabolized to phenylhydroquinone. 2-Phenylphenol used in the fluorometric determination of trioses (1). **Target(s):** deoxyribonuclease I (2); 2'-hydroxybiphenyl-2-sulfinate desulfinase, product inhibition (3,4); and prostaglandin-H synthase (5). **1**. Asabe, Momose, Suzuki & Takitani (1979) *Anal. Biochem.* **94**, 1. **2**. Gottesfeld, Adams, el-Badry, Moses & Calvin (1971) *Biochim. Biophys. Acta* **228**, 365. **3**. Nakayama, Matsubara, Ohshiro, *et al.* (2002) *Biochim. Biophys. Acta* **1598**, 122. **4**. Konishi & Maruhashi (2003) *Appl. Microbiol. Biotechnol.* **62**, 356. **5**. Freyberger & Degen (1998) *Arch. Toxicol.* **72**, 637.

Phenyl Phosphate

This phosphorylated phenol (FW_{disodium salt} = 218.06 g/mol; soluble in water and methanol) is an chromogenic alternative substrate for phosphomonoesterases (*i.e.*, activity can be spectrophotometrically assayed by the release of phenolate ion). Phenol phosphate should be stored a refrigerated, light-tight desiccator. Once prepared, aqueous solutions hydrolyze slowly. **Target(s):** Arylsulfatase (1); glycerol-1,2-cyclic-phosphate phosphodiesterase (2); pancreatic ribonuclease, *or* ribonuclease A (3); and pyridoxal phosphatase (4). **1**. Webb (1966) *Enzyme and Metabolic Inhibitors*, vol. **2**, p. 443, Academic Press, New York. **2**. Clarke & Dawson (1978) *Biochem. J.* **173**, 579. **3**. Richards & Wyckoff (1971) *The Enzymes*, 3rd ed., **4**, 647. **4**. Fonda (1992) *J. Biol. Chem.* **267**, 15978.

Phenylphosphonate (Phenylphosphonic Acid)

This corrosive aryl phosphonic acid (FW_{free-acid} = 158.09 g/mol; CAS 1571-33-1; M.P. = 163-166 °C) inhibits alkaline phosphatase (1-4). **1**. Reid & Wilson (1971) *The Enzymes*, 3rd ed. (Boyer, ed.), **4**, 373. **2**. Farley, Puzas

& Baylink (1982) *Miner. Electrolyte Metab.* **7**, 316. **3**. Chakrabartty & Stinson (1983) *Biochim. Biophys. Acta* **839**, 174. **4**. Stinson & Chan (1987) *Adv. Protein Phosphatases* **4**, 127.

Phenylphosphorodiamidate

This diphosphoramidate (FW = 172.12 g/mol; M.P. = 185-190°C) is a slow-binding competitive inhibitor of *Klebsiella aerogenes* urease, K_i = 94 pM (1-3) and urethanase (4). **1**. Todd & Hausinger (1989) *J. Biol. Chem.* **264**, 15835. **2**. Clemens, Lee & Horwitz (1995) *J. Bacteriol.* **177**, 5644. **3**. Benini, Ciurli, Nolting & Mangani (1996) *Eur. J. Biochem.* **239**, 61. **4**. Kobashi, Takebe & Sakai (1990) *Chem. Pharm. Bull.* **38**, 1326.

N-Phenylphosphoryl-L-phenylalanyl-L-alanine

This phosphoramidate (FW_{free-acid} = 392.35 g/mol) inhibits a number of metalloproteinases, including bacillolysin, peptidyl-dipeptidase A. *or* angiotensin-I-converting enzyme, and thermolysin. **1**. Holmquist & Vallee (1979) *Proc. Natl. Acad. Sci. U.S.A.* **76**, 6216.

(20*R*)-20-Phenyl-5-pregnene-3β,20-diol

This sterol derivative (FW = 394.60 g/mol) is a potent competitive inhibitor of CYP11A, *or* cholesterol side-chain-cleaving enzyme, $K_{i,app}$ = 30 nM); it is not an alternative substrate (1-4). **1**. Vickery (1991) *Meth. Enzymol.* **206**, 548. **2**. Jacobs, Singh & Vickery (1987) *Biochemistry* **26**, 4541. **3**. Vickery & Kellis (1983) *J. Biol. Chem.* **258**, 3832. **4**. Weeks, Strong, Daux & Vickery (1982) *Steroids* **40**, 359.

4*S*-(3-Phenyl)propargyl-D-glutamic Acid, *See* *(2R,4S)-2-Amino-4-(3-phenyl)propargylpentanedioic Acid*

trans-Phenylpropenoic Acid, *See* Cinnamic Acid

2-Phenylprop-2-enylamine, *See* 1-Phenyl-1-(aminomethyl)ethane
4*S*-(3-Phenylprop-2-enyl)-D-glutamic Acid, *See* *(2R,4S)-2-Amino-4-(3-phenylprop-2-enyl)pentanedioic Acid*

Phenylpropiolic Acid

This mechanism-based (suicide) inhibitor (FW_{free-acid} = 146.15 g/mol; CAS 637-44-5) inactivates Fe(II)- and α-ketoglutarate-dependent dioxygenase (the latter catalyzing the hydroxylation of the herbicide 2,4-dichlorophenoxyacetic acid), with K_i = 38.1 μM and k_{inact} = 2.3 min⁻¹, covalently modifiing a side-chain functional group in the proteolytic fragment (166-AEHYALNSR-174) forming one side of the substrate-binding pocket. Phenylpropioloc acid also inhibits esterase (2) and phenylalanine ammonia-lyase (3). **1**. Hotopp & Hausinger (2002) *Biochemistry* **41**, 9787. **2**. Weber & King (1935) *J. Biol. Chem.* **108**, 131. **3**. Dubery & Smit (1994) *Biochim. Biophys. Acta* **1207**, 24.

4S-(3-Phenyl)propyl-D-glutamic Acid, See *(2R,4S)-2-Amino-4-(3-phenyl)propylpentanedioic Acid*

1-Phenyl-3-pyrazolidinone, See *1-Phenyl-3-pyrazolidone*

4-Phenylpyridine

This substituted pyridine (FW = 155.20 g/mol; CAS 939-23-1; M.P. = 69-73°C; B.P. = 274-275°C), which occurs naturally and is likewise a manmade pollutant, inhibits human placental aromatase, K_i = 0.36 µM (1), monoamine oxidase (2), and NADH dehydrogenase (3,4). 4-Phenylpyridine is also an effective inhibitor of mitochondrial complex I. Note that it both occurs naturally and is an environmental pollutant. **1.** Vaz, Coon, H. Peegel & Menon (1992) *Drug Metab. Dispos.* **20**, 108. **2.** Arai, Kinemuchi, Hamamichi, *et al.* (1986) *Neurosci. Lett.* **66**, 43. **3.** Ramsay, McKeown, Johnson, Booth & Singer (1987) *Biochem. Biophys. Res. Commun.* **146**, 53. **4.** Ramsay, Salach, Dadgar & Singer (1986) *Biochem. Biophys. Res. Commun.* **135**, 269.

Phenylpyruvate (Phenylpyruvic Acid)

This aromatic α-keto acid and PKU biomarker (FW$_{free-acid}$ = 164.16 g/mol; CAS 156-06-9 for acid; CAS 114-76-1 for sodium salt; CAS 51828-93-4 for calcium salt); λ_{max} = 320 nm, ε = 17500 M^{-1}cm^{-1} in 0.7 M NaOH), also known as 2-oxo-3-phenylpropanoic acid and α-oxohydrocinnamic acid, is readily formed by the action of a number of aminotransferases on phenylalanine. Phenylpyruvate is only slightly soluble in boiling water but is soluble in ethanol and diethyl ether. The solid oxidizes in air and decomposes on storage, if not dry. Sodium phenylpyruvate is a water-soluble monohydrate salt. **Phenylketonuria:** Phenylpyruvate is present at high urinary concentrations in phenylalanine monooxygenase deficiency, better known as Phenylketonuria. Indeed, individuals suffering PKU tend to excrete urine containing large quantities of phenyl-pyruvate, phenyl-acetate and phenyl-lactate, along with phenylalanine. If untreated, mental retardation and microcephaly are evident by the first year, along with irritability, epileptic seizures, and skin lesions. **Target(s):** acetolactate synthase (1); aldehyde reductase (2); apyrase (3); branched-chain α-keto acid dehydrogenase (4); [branched-chain α-keto-acid dehydrogenase] kinase, *or* [3-methyl-2-oxobutanoate dehydrogenase (acetyl-transferring)] kinase (5-8); carnitine-acylcarnitine translocase (9); cathepsin D (10); choline acetyltransferase (11); 3-deoxy-7-phosphoheptulonate synthase, *or* 3-deoxy-D-*arabino*-heptulosonate 7-phosphate synthetase (12-14); dihydropteridine reductase (15); diphosphomevalonate decarboxylase (43); glutamate decarboxylase (16); glutamine:pyruvate aminotransferase (17); glycerate dehydrogenase, weakly inhibited (16,18); hexokinase (human brain), K_i = 2 mM (19); histidine decarboxylase (20); 4-hydroxyphenylpyruvate dioxygenase (16); 3α-hydroxysteroid dihydrogenase (2); 3β-hydroxysteroid dehydrogenase (2); hypothalamus acid proteinase (21); indolepyruvate decarboxylase, K_i = 50 µM (22); kynureninase (23); L-lactate dehydrogenase (16,24); L-lactate dehydrogenase (cytochrome) (24,25); mercaptopyruvate sulfurtransferase (26); mevalonate-5-pyrophosphate decarboxylase (27); oxaloacetate tautomerase (28); phenylalanine ammonia-lyase (29-31); phenylalanyl-tRNA synthetase, weakly inhibited, K_i = 7.6 mM (32); pyruvate carboxylase (33,34); pyruvate decarboxylase (16,35,36); pyruvate dehydrogenase (37); pyruvate kinase (human), K_i = 8.5 mM (19,38); pyruvate oxidase (16); pyruvate transport (39); stearoyl-CoA 9-desaturase (40); tryptophan 2-monooxygenase (44); tryptophan 5-monooxygenase (41); tyrosinase (16); and tyrosine 2,3-aminomutase (42). **1.** Stormer (1975) *Meth. Enzymol.* **41**, 518. **2.** Sawada, Hara, Nakayama & Hayashibara (1982) *J. Biochem.* **92**, 185. **3.** Berti, Bonan, da Silva, *et al.* (2001) *Int. J. Dev. Neurosci.* **19**, 649. **4.** Danner, Sewell & Elsas (1982) *J. Biol. Chem.* **257**, 659. **5.** Paxton (1988) *Meth. Enzymol.* **166**, 313. **6.** Randle, Patston & Espinal (1987) *The Enzymes*, 3rd ed., **18**, 97. **7.** Paxton & Harris (1984) *Arch. Biochem. Biophys.* **231**, 58. **8.** Harris, Paxton, Powell, *et al.* (1986) *Adv. Enzyme Regul.* **25**, 219. **9.** Parvin & Pande (1978) *J. Biol. Chem.* **253**, 1944. **10.** Mycek (1970) *Meth. Enzymol.* **19**, 285. **11.** Nachmansohn & John (1945) *J. Biol. Chem.* **158**, 157. **12.** Jensen, Calhoun & Stenmark (1973) *Biochim.*

Biophys. Acta **293**, 256. **13.** Olekhnovich, Maksimova & Fomichev (1986) *Mol. Gen. Mikrobiol. Virusol.* **1986(12)**, 34. **14.** Whitaker, Fiske & Jensen (1982) *J. Biol. Chem.* **257**, 12789. **15.** Shen, Smith, Davis, Brubaker & Abell (1982) *J. Biol. Chem.* **257**, 7294. **16.** Webb (1966) *Enzyme and Metabolic Inhibitors*, vol. **2**, Academic Press, New York. **17.** Braunstein (1973) *The Enzymes*, 3rd ed. (Boyer, ed.), **9**, 379. **18.** Holzer & Holldorf (1957) *Biochem. Z.* **329**, 292. **19.** Weber (1969) *Proc. Natl. Acad. Sci. U.S.A.* **63**, 1365. **20.** Sakamoto, Watanabe, Hayashi, Taguchi & Wada (1985) *Agents Actions* **17**, 32. **21.** Akopyan, Arutunyan, Lajtha & Galoyan (1978) *Neurochem. Res.* **3**, 89. **22.** Koga, Adachi & Hidaka (1992) *J. Biol. Chem.* **267**, 15823. **23.** Shibata, Takeuchi, Tsubouchi, *et al.* (1991) *Adv. Exp. Med. Biol.* **294**, 523. **24.** Dikstein (1959) *Biochim. Biophys. Acta* **36**, 397. **25.** Nygaard (1963) *The Enzymes*, 2nd ed., **7**, 557. **26.** Wing & Baskin (1992) *J. Biochem. Toxicol.* **7**, 65. **27.** Castillo, Martinez-Cayuela, Zafra & Garcia-Peregrin (1991) *Mol. Cell. Biochem.* **105**, 21. **28.** Belikova, Burov & Vinogradov (1988) *Biochim. Biophys. Acta* **936**, 10. **29.** Kalghatgi & Subba Rao (1975) *Biochem. J.* **149**, 65. **30.** Dahiya (1993) *Indian J. Exp. Biol.* **31**, 874. **31.** Jorrin, Lopez-Valbuena & Tena (1988) *Biochim. Biophys. Acta* **964**, 73. **32.** Santi & Danenberg (1971) *Biochemistry* **10**, 4813. **33.** Scrutton, Olmsted & Utter (1969) *Meth. Enzymol.* **13**, 235. **34.** Scrutton & Mildvan (1968) *Biochemistry* **7**, 1490. **35.** Gale (1961) *Arch. Biochem. Biophys.* **94**, 236. **36.** Neuser, Zorn, Richter & Berger (2000) *Biol. Chem.* **381**, 349. **37.** Swierczynski, Aleksandrowicz & Zydowo (1976) *Acta Biochim. Pol.* **23**, 85. **38.** Feksa, Cornelio, Dutra-Filho, *et al.* (2003) *Brain Res.* **968**, 199. **39.** Halestrap, Brand & Denton (1974) *Biochim. Biophys. Acta* **367**, 102. **40.** Scott & Foote (1979) *Biochim. Biophys. Acta* **573**, 197. **41.** Freedland, Wadzinski & Waisman (1961) *Biochim. Biophys. Res. Commun.* **6**, 227. **42.** Kurylo-Borowska & Abramsky (1972) *Biochim. Biophys. Acta* **264**, 1. **43.** Shama Bhat & Ramasarma (1979) *Biochem. J.* **181**, 143. **44.** Emanuele, Heasley & Fitzpatrick (1995) *Arch. Biochem. Biophys.* **316**, 241.

Phenyl-p-quinone, See *Phenyl-p-benzoquinone*

5-(Phenylselenenyl)acyclouridine

This nucleoside analogue (FW = 341.23 g/mol) inhibits uridine phosphorylase (1,2). **1.** Ashour, Naguib, Goudgaon, Schinazi & el Kouni (2000) *Cancer Chemother. Pharmacol.* **46**, 235. **2.** Ashour, Al Safarjalani, Naguib, *et al.* (2000) *Cancer Chemother. Pharmacol.* **45**, 351.

Phenyl Selenoacetylene

This selenide (FW = 257.19 g/mol) inhibits rat porphobilinogen synthase, *or* (δ-aminolevulinate dehydratase, IC$_{50}$ = 0.25 mM, with inhibition requiring formation diphenyl diselenide which mediates oxidation of a catalytically essential sulfhydryl group (1,2). **1.** Bolzan, Folmer, Farina, *et al.* (2002) *Pharmacol. Toxicol.* **90**, 214. **2.** Folmer, Bolzan, Farina, *et al.* (2005) *Toxicology* **206**, 403.

Se-Phenyl-L-selenocysteine

This S-phenyl-L-cysteine and homophenylalanine analogue (FW = 244.15 g/mol) inhibits a number of cytochrome P450 systems, including CYP1A1,

CYP1A2, CYP3A4, CYP2C9, CYP2D6, and CYP2E1 (1,2). **1.** Rooseboom, Vermeulen, Durgut & Commandeur (2002) *Chem. Res. Toxicol.* **15**, 1610. **2.** Venhorst, Rooseboom, Vermeulen & Commandeur (2003) *Xenobiotica* **33**, 57.

β-Phenylserine

This serine derivative (FW = 169.18 g/mol), which is an alternative substrate for serine hydroxymethyltransferase, phenylserine aldolase (L-*threo*-isomer), and threonine dehydratase, inhibits alkaline phosphatase (1), 3-deoxy-7-phosphoheptulonate synthase (2), nonenzymatic hydroxyl-aminolysis of amino acid esters at –18°C (3), phenylalanine ammonia-lyase (4,5) and tryptophanase, *or* tryptophan indole-lyase (6). **1.** Fishman & Sie (1971) *Enzymologia* **41**, 141. **2.** Simpson & Davidson (1976) *Eur. J. Biochem.* **70**, 509. **3.** Grant & Alburn (1965) *Biochemistry* **4**, 1913. **4.** Koukol & Conn (1961) *J. Biol. Chem.* **236**, 2692. **5.** Jorrin, Lopez-Valbuena & Tena (1988) *Biochim. Biophys. Acta* **964**, 73. **6.** Snell (1975) *Adv. Enzymol. Relat. Areas Mol. Biol.* **42**, 287.

Phenylsuccinic Acid

This dicarboxylic acid (FW$_{free-acid}$ = 194.19 g/mol; M.P. = 167-169°C) inhibits the α-ketoglutarate transporter. *Note:* Phenylsuccinic acid should not be confused with phenyl succinate monoester ($^-$OOCCH$_2$CH$_2$COOC$_6$H$_5$). **Target(s):** dicarboxylic acid transporter (1); glutamate decarboxylase, K_i = 55 mM (2); hydroxyproline transport (3); α-ketoglutarate transporter (4); and oxalate uptake (5). **1.** Quagliariello, Passarella & Palmieri (1974) in *Dynamics of Energy-Transducing Membranes* (Ernster, Estabrook & Slater, eds.), p. 483, Elsevier, Amsterdam. **2.** Fonda (1972) *Biochemistry* **11**, 1304. **3.** Atlante, Passarella & Quagliariello (1994) *Biochem. Biophys. Res. Commun.* **202**, 58. **4.** Coll, Colell, Garcia-Ruiz, Kaplowitz & Fernandez-Checa (2003) *Hepatology* **38**, 692. **5.** Strzelecki & Menon (1986) *J. Biol. Chem.* **261**, 12197.

7-Phenylsulfonyl-1,2,3,4-tetrahydroisoquinoline

This substrate analogue (FW = 272.35 g/mol) inhibits phenylethanolamine *N*-methyltransferase, K_i = 14 μM (1,2). **1.** Grunewald, Dahanukar, Jalluri & Criscione (1999) *J. Med. Chem.* **42**, 118. **2.** Grunewald, Seim, Regier, *et al.* (2006) *J. Med. Chem.* **49**, 5424.

4-Phenyl-1,2,3,4-tetrahydroisoquinoline

This dopamine release inhibitor (FW = 245.75 g/mol; CAS 6109-35-9; Soluble in DMSO and to 25 mM in H$_2$O; Symbol: 4-PTIQ) 10(-6 M) inhibited the dopamine-releasing effect of methamphetamine when present

at 10^{-6} M. 4-PTIQ did not affect the elevation of the extracellular dopamine level induced by high concentrations of nomifensine (10^{-5} M) and cocaine (3 x 10^{-5} M). 4-PTIQ was the weakest inhibitor of dopamine uptake by rat striatal synaptosomes. Such results suggest that 4-PTIQ is a selective antagonist against the dopamine-releasing effect of methamphetamine in the nucleus accumbens. **1.** Tateyama, Nagao, Ohta, Hirobe & Ono (1993) *Eur. J. Pharmacol.* **240**, 51.

2-Phenyl-4,4,5,5-tetramethylimidazoline-1-oxyl 3-oxide, *See PTIO*

O^6-(4-Phenylthenyl)guanine

This modified guanine inactivates methylated-DNA:[protein]-cysteine *S*-methyltransferase, *or* O^6-alkylguanine-DNA alkyltransferase, I$_{50}$ = 15 nM). The 5-phenyl analogue is a weaker inhibitor, I$_{50}$ = 0.75 μM. **1.** McMurry (2007) *DNA Repair* **6**, 1161.

Phenylthiocarbamate, *See Phenylthiourea*

2-Phenylthiochromen-4-one 1-Oxide

This substituted 2-phenyl-4-quinolone (FW = 254.31 g/mol) targets DNA topoisomerase I, displaying significant growth inhibitory action against a panel of tumor cell lines, including human ileocecal carcinoma (HCT-8), murine leukemia (P-388), human melanoma (RPMI), and human central nervous system tumor (TE671) cells. **1.** Wang, Bastow, Cosentino & Lee (1996) *J. Med. Chem.* **39**, 1975.

5'-(Phenylthio)-immucillin-A

This nucleoside-based transition-state analogue (FW = 355.42 g/mol), also known as (1*S*)-1-(9-deazaadenin-9-yl)-1,4-dideoxy-1,4-imino-5-(phenyl-thio)-D-ritiol, inhibits 5'-methylthioadenosine/*S*-adenosylhomocysteine nucleosidase, K_i' = 32 pM (1) and *S*-methyl-5'-thioadenosine phosphoryl-ase, K_i = 1 nM (1), and *S*-methyl-5'-thioadenosine phosphorylase (2,3). **1.** Singh, Evans, Lenz, *et al.* (2005) *J. Biol. Chem.* **280**, 18265. **2.** Evans, Furneaux, Schramm, Singh & Tyler (2004) *J. Med. Chem.* **47**, 3275. **3.** Singh, Shi, Evans, *et al.* (2004) *Biochemistry* **43**, 9.

3-(Phenylthio)-1,1,1-trifluoropropan-2-one

This halomethyl ketone (FW = 220.21 g/mol), also known as 3-phenylsulfanyl-1,1,1-trifluoro-2-propanone, inhibits juvenile-hormone esterase, IC$_{50}$ = 8.2 μM, and acetylcholinesterase, IC$_{50}$ = 3.7 μM. **1.** Hammock, Abdel-Aal, Mullin, Hanzlik & Roe (1984) *Pest. Biochem. Physiol.* **22**, 209.

Phenylthiourea

This thiourea (FW = 152.22 g/mol; M.P. = 154°C; = 0.25 g/100 g cold H_2O) either tastes bitter or is tasteless, depending on the heredity of the taster. **Target(s):** N-acetyl-6-hydroxytryptophan oxidase (1); aromatic-L-amino-acid decarboxylase, or dopa decarboxylase (2); aureusidin synthase (3); bacterial leucyl aminopeptidase (4); catechol oxidase (5); cytochome c oxidase (6;) dynein ATPase (7); succinate dehydrogenase (6); tyrosinase, or monophenol monooxygenase, or polyphenol oxidase (1,6,8-17); and xanthine oxidase (18). **1.** Birse & Clutterbuck (1990) *J. Gen. Microbiol.* **136**, 1725. **2.** Tocher & Tocher (1972) *Phytochemistry* **11**, 1661. **3.** Nakayama, Sato, Fukui, et al. (2001) *FEBS Lett.* **499**, 107. **4.** Bienvenue, Gilner & Holz (2002) *Biochemistry* **41**, 3712. **5.** Rompel, Fischer, Meiwes, et al. (1999) *J. Biol. Inorg. Chem.* **4**, 56. **6.** DuBois & Erway (1946) *J. Biol. Chem.* **165**, 711. **7.** Blum & Hayes (1979) *J. Supramol. Struct.* **12**, 23. **8.** Horowitz, Fling & Horn (1970) *Meth. Enzymol.* **17A**, 615. **9.** Lerch (1987) *Meth. Enzymol.* **142**, 165. **10.** Bernheim & Bernheim (1942) *J. Biol. Chem.* **145**, 213. **11.** Zarivi, Bonfigli, Cesare, et al. (2003) *FEMS Microbiol. Lett.* **220**, 81. **12.** Lerner, Fitzpatrick, Calkins & Summerson (1950) *J. Biol. Chem.* **187**, 793. **13.** Criton & Le Mellay-Hamon (2008) *Bioorg. Med. Chem. Lett.* **18**, 3607. **14.** Hata, Azumi & Yokosawa (1998) *Comp. Biochem. Physiol. B Biochem. Mol. Biol.* **119**, 769. **15.** Jimenez-Cervantes, Garcia-Borron, Valverde, Solano & Lozano (1993) *Eur. J. Biochem.* **217**, 549. **16.** Liu, Zhang, Wang, et al. (2004) *Lett. Appl. Microbiol.* **39**, 407. **17.** Benjakul, Visessanguan & Tanaka (2005) *J. Food Biochem.* **29**, 470. **18.** Roussos (1967) *Meth. Enzymol.* **12A**, 5.

Phenytoin, *See* 5,5-Diphenylhydantoin

Pheophorbide *a*

This chlorophyll *a* degradation intermediate (MW = 579.68 g/mol; CAS 15664-29-6), which is identical to chlorophyll *a*, except for the absence of the latter's magnesium ion and ester at the 17-position, is obtained by acid hydrolysis of chlorophyll *a*. Pheophorbide *a* inhibits acyl-CoA:cholesterol acyltransferase (1) and magnesium protoporphyrin IX methyltransferase (2), acts as an endothelin receptor antagonist (3), and also induces apoptosis. Pheophorbide *a* is also as an anti-tumor promoter. **1.** Song, Rho, Lee, et al. (2002) *Planta Med.* **68**, 845. **2.** Ellsworth & St. Pierre (1976) *Photosynthetica* **10**, 291. **3.** Ohshima, Hirata, Oda, Sasaki & Shiratsuchi (1994) *Chem. Pharm. Bull.* **42**, 2174.

Phethiol

This thiol (FW$_{free-base}$ = 167.27 g/mol), also known as 2-amino-3-phenylpropane-1-thiol, inhibits membrane alanyl aminopeptidase, or aminopeptidase N, K_i = 39 nM (1-4) and *Clostridium botulinum* bontoxilysin, K_i = 100 μM (5). **1.** Gros, Giros, Schwartz, et al. (1988) *Neuropeptides* **12**, 111. **2.** Xu & Li (2005) *Curr. Med. Chem. Anticancer Agents* **5**, 281. **3.** Bauvois & Dauzonne (2006) *Med. Res. Rev.* **26**, 88. **3.** Noble, Luciani, Da Nacimento, et al. (2000) *FEBS Lett.* **467**, 81. **4.** Luciani, Marie-Claire, Ruffet, et al. (1998) *Biochemistry* **37**, 686. **5.** Anne, Turcaud, Quancard, et al. (2003) *J. Med. Chem.* **46**, 4648.

Philanthotoxin 343

This hygroscopic polyamine derivative (FW$_{free-base}$ = 435.61 g/mol; CAS 115976-93-7), often abbreviated PhTX-343, is a synthetic analogue of the wasp toxin philanthotoxin 433 that blocks NMDA-gated ion channels (1-3) and nicotinic acetylcholine receptors (4). The designation 343 refers to the number of methylene groups between the nitrogen atoms in the polyamine portion of the molecule (*i.e.*, 3, 4, and 3). With whole-cell ion transport recordings and outside-out patch clamp recordings, Brier *et al.* (4) investigated antagonism of human muscle nicotinic acetylcholine receptors (nAChR) by PhTX-343. When co-applied with AChR, PhTX-343 caused activation-dependent, noncompetitive inhibition (IC$_{50}$ = 17 μM at −100 mV) of whole-cell currents that was strongly voltage-dependent. Preapplication of PhTX-343 unveiled a voltage-independent antagonism that also required receptor activation, which is suggestive of desensitization enhancement. In single-channel studies, 10 μM PhTX-343 significantly reduced the mean open time of channel openings evoked by 1 μM acetylcholine, indicating that PhTX-343 predominantly blocks the opening of channels by acetylcholine. **Target(s):** acetylcholine receptor (4,5); GluR6(Q) glutamate receptor channel (6); IRK1 K^+ channel (7); NMDA-gated (N-methyl-D-aspartate-gated) channels (1-3). **1.** Jones & Lodge (1991) *Eur. J. Pharmacol.* **204**, 203. **2.** Fedorov, Srebitsky & Reymann (1992) *Eur. J. Pharmacol.* **228**, 201. **3.** Brackley, Goodnow, Jr., Nakanishi, Sudan & Usherwood (1990) *Neurosci. Lett.* **114**, 51. **4.** Brier, Mellor, Tikhonov, et al. (2003) *Mol. Pharmacol.* **64**, 954. **5.** Jayaraman, Usherwood & Hess (1999) *Biochemistry* **38**, 11406. **6.** Bahring & Mayer (1998) *J. Physiol.* **509**, 635. **7.** Guo & Lu (2000) *J. Gen. Physiol.* **115**, 799.

Philanthotoxin 433

This hygroscopic polyamine toxin (FW$_{free-base}$ = 435.61 g/mol; CAS 77108-00-0), abbreviated PhTX-443, from the Digger wasp *Philanthus triangulum* blocks NMDA-gated ion channels (1) and nicotinic acetylcholine receptors (2). The designation 433 refers to the number of methylene groups between the nitrogen atoms in the polyamine portion of the molecule (*i.e.*, 4, 3, and 3). **Target(s):** acetylcholine receptor (2); NMDA-gated (N-methyl-D-aspartate-gated) channels (1,3); and porin OmpF-mediated current, *Escherichia coli* (4). **1.** Jones & Lodge (1991) *Eur. J. Pharmacol.* **204**, 203. **2.** Rozental, Scoble, Albuquerque, et al. (1989) *J. Pharmacol. Exp. Ther.* **249**, 123. **3.** Ragsdale, Gant, Anis, et al. (1989) *J. Pharmacol. Exp. Ther.* **251**, 156. **4.** Basle & Delcour (2001) *Biochem. Biophys. Res. Commun.* **285**, 550.

Phleomycins

Pheomycin D1

These glycopeptide antibiotics (FW$_{Pheomycin-D1}$ = 1427.60 g/mol; CAS 11031-11-1), like the closely related bleomycins, inhibit DNA biosynthesis, and are active *in vivo* against most bacteria (including *Escherichia coli*), filamentous fungi, yeasts, plant, and animal cells. Phleomycins are bleomycin analogues in which the penultimate thiazolium ring of the bithiazole tail is reduced. Phleomycins partially intercalate into DNA and participate in the iron- and oxygen-dependent cleavage of the DNA duplex. **1**. Tanaka, Yamaguchi & Umezawa (1963) *Biochem. Biophys. Res. Commun.* **10**, 171.

Phloretin

This naturally occurring aglucon (FW = 274.27 g/mol; CAS 60-82-2) of phlorizin (or phloridzin) is generated from the latter by the action of β-glucosidases, the phlorizin hydrolase activity of lactase, and by dilute mineral acids. Phloretin is soluble in ethanol and acetone. It is almost insoluble in water; however, the solubility is increased upon addition of ethanol. It adsorbs to lipid surfaces and alters the dipole potential of lipid monolayers and bilayers. Adsorption to biological and artificial membranes results in a change of the membrane permeability for a variety of charged and neutral compounds. **Target(s):** aldose 1-epimerase, *or* mutarotase (1,27,28); ATPases, chloroplast (29); Ca^{2+} channels, L-type (2); Ca^{2+}-dependent ATPase (29); cAMP phosphodiesterase (3); β-carotene 15,15'-monooxygenase (4); chalcone isomerase (5); electron transport, chloroplast (29); fatty-acid synthase (32); glucose transport (6-10); glycogen phosphorylase (11,12); inositol-trisphosphate 3-kinase (31); iodothyronine deiodinase (13); lactate transport (14); L-leucine transport (15); Na^+ and K^+ uptake (16); oxidative phosphorylation (mitochondrial), uncoupler (17); phenylpyruvate tautomerase (18); photophosphorylation (29); prostaglandin $F_{2\alpha}$ receptor (19); protein kinase C (20,21); pyruvate transport (14); starch phosphorylase (22); α,α-trehalase (23,30); tyrosinase, *or* monophenol monooxygenase (33); urea transport (24,25); and water permeability (25,26). **1**. Bailey, Fishman, Kusiak, Mulhern & Pentchev (1975) *Meth. Enzymol.* **41**, 471. **2**. Prevarskaya, Skryma, Vacher, *et al.* (1994) *Mol. Cell Neurosci.* **5**, 699. **3**. Kuppusamy & Das (1992) *Biochem. Pharmacol.* **44**, 1307. **4**. Nagao, Maeda, Lim, Kobayashi & Terao (2000) *J. Nutr. Biochem.* **11**, 348. **5**. Herles, Braune & Blaut (2004) *Arch. Microbiol.* **181**, 428. **6**. Sahagian (1965) *Can. J. Biochem.* **43**, 851. **7**. Yokota, Nishi & Takesue (1983) *Biochem. Pharmacol.* **32**, 3453. **8**. Seyfang & Duszenko (1991) *Eur. J. Biochem.* **202**, 191. **9**. Betz, Drewes & Gilboe (1975) *Biochim. Biophys. Acta* **406**, 505. **10**. Martin, Kornmann & Fuhrmann (2003) *Chem. Biol. Interact.* **146**, 225. **11**. Cori, Illingworth & Keller (1955) *Meth. Enzymol.* **1**, 200. **12**. Brown & Cori (1961) *The Enzymes*, 2nd ed., **5**, 207. **13**. Aufmkolk, Koehrle, Hesch, Ingbar & Cody (1986) *Biochem. Pharmacol.* **35**, 2221. **14**. Wang, Poole, Halestrap & Levi (1993) *Biochem. J.* **290**, 249. **15**. Rosenberg (1979) *J. Physiol.* **291**, 53P. **16**. Skriabin, Orlov, Masse & Berthiaume (2000) *Exp. Lung Res.* **26**, 319. **17**. De Jonge, Wieringa, Van Putten, Krans & Van Dam (1983) *Biochim. Biophys. Acta* **722**, 21. **18**. Molnar & Garai (2005) *Int. Immunopharmacol.* **5**, 849. **19**. Kitanaka, Ishibashi & Baba (1993) *J. Neurochem.* **60**, 704. **20**. Gschwendt, Horn, Kittstein, *et al.* (1984) *Biochem. Biophys. Res. Commun.* **124**, 63. **21**. Horn, Gschwendt & Marks (1985) *Eur. J. Biochem.* **148**, 533. **22**. Whelan (1955) *Meth. Enzymol.* **1**, 192. **23**. Nakano & Sacktor (1984) *Biochim. Biophys. Acta* **791**, 45. **24**. Chou & Knepper (1989) *Amer. J. Physiol.* **257**, F359. **25**. Toon & Solomon (1987) *J. Membr. Biol.* **99**, 157. **26**. Benga, Popescu, Pop, Hodor & Borza (1992) *Eur. J. Cell Biol.* **59**, 219. **27**. Fishman, Pentchev & Bailey (1973) *Biochemistry* **12**, 2490. **28**. Mulhern, Fishman, Kusiak & Bailey (1973) *J. Biol. Chem.* **248**, 4163. **29**. Uribe (1970) *Biochemistry* **9**, 2100. **30**. Lúcio-Eterovic, Jorge, Polizeli & Terenzi (2005) *Biochim. Biophys. Acta* **1723**, 201. **31**. Mayr, Windhorst & Hillemeier (2005) *J. Biol. Chem.* **280**, 13229. **32**. Li, Ma, Wang & Tian (2005) *J. Biochem.* **138**, 679. **33**. Lin, Hsu, Chen, Chern & Lee (2007) *Phytochemistry* **68**, 1189.

Phlorizin

This glucosuria-inducing agent (FW = 436.42 g/mol, CAS 60-81-1), also known as phlorhizin, phloridzin, phlorrhizin, and 4,4',6'-trihydroxy-2'-glucosidodihydrochalcone, is found in all parts of the apple tree except for the mature fruit. Phlorizin blocks glucose reabsorption in the kidney and intestine by inhibiting glucose/Na^+ cotransporters as well as glucose transport into red blood cells and hepatocytes. Commercial phlorizin, typically supplied as the dihydrate, should be treated with activated charcoal before use and recrystallized from hot water. Note that muscle phosphorylase is much more inhibited than potato phosphorylase. **Target(s):** aldose 1-epimerase, *or* mutarotase (1,2,40,41); ATPase, chloroplast, and photophosphorylation (3-7); ATPase, mitochondrial (8); bicarbonate-chloride exchange (9); complement component C'3, guinea pig (10); galactosyltramsferase (11); glucose-6-phosphatase (12-15); glucose transport (16-23); glucoside 3-dehydrogenase (24); glycogen phosphorylase (25-29); glycogen synthase (30); lactase, *or* lactase/phlorizin hydrolase, *or* glycosylceramidase (42-46); laminaribiose phosphorylase (31,57); Na^+/*myo*-inositol cotransporter (32,33); Na^+/K^+-exchanging ATPase (34); phloretin hydrolase (42); 3-phytase (55,56); starch phosphorylase, weakly inhibited (25,35); sucrose phosphorylase (36); α,α-trehalase (37-39,50-54); tyrosinase, monophenol monooxygenase, weakly inhibited (58). **1**. Bentley (1962) *Meth. Enzymol.* **5**, 219. **2**. Bailey, Fishman, Kusiak, Mulhern & Pentchev (1975) *Meth. Enzymol.* **41**, 471. **3**. Carmeli & Avron (1972) *Meth. Enzymol.* **24**, 92. **4**. Izawa & Good (1972) *Meth. Enzymol.* **24**, 355. **5**. McCarty (1980) *Meth. Enzymol.* **69**, 719. **6**. Izawa, Connolly, Winget & Good (1966) *Brookhaven Symp. Biol.* **19**, 169. **7**. Winget, Izawa & Good (1969) *Biochemistry* **8**, 2067. **8**. Tellez de Inon (1968) *Acta Physiol. Lat. Amer.* **18**, 268. **9**. Chow & Chen (1982) *Biochim. Biophys. Acta* **685**, 196. **10**. Shin & Mayer (1968) *Biochemistry* **7**, 3003. **11**. Kuhn & White (1980) *Biochem. J.* **188**, 503. **12**. Nordlie (1971) *The Enzymes*, 3rd ed. (Boyer, ed.), **4**, 543. **13**. Lygre & Nordlie (1969) *Biochim. Biophys. Acta* **185**, 360. **14**. Zerr & Novoa (1968) *Biochem. Biophys. Res. Commun.* **32**, 129. **15**. Soodsma, Legler & Nordlie (1967) *J. Biol. Chem.* **242**, 1955. **16**. Sahagian (1965) *Can. J. Biochem.* **43**, 851. **17**. Diedrich (1966) *Arch. Biochem. Biophys.* **117**, 248. **18**. Kotyk, Kolinska, Veres & Szammer (1965) *Biochem. Z.* **342**, 129. **19**. McCracken & Lumsden (1975) *Comp. Biochem. Physiol. B* **50**, 153. **20**. Lauterbach (1975) *Naunyn Schmiedebergs Arch. Pharmacol.* **287** Suppl., R69. **21**. Vick, Diedrich & Baumann (1973) *Amer. J. Physiol.* **224**, 552. **22**. Betz, Drewes & Gilboe (1975) *Biochim. Biophys. Acta* **406**, 505. **23**. Alvarado (1978) *J. Physiol. (Paris)* **74**, 633. **24**. Fukui & Hochster (1963) *Biochem. Biophys. Res. Commun.* **11**, 50. **25**. Hassid, Doudoroff & Barker (1951) *The Enzymes*, 1st ed. (Sumner & Myrbäck, eds.), **1** (part 2), 1014. **26**. Cori, Illingworth & Keller (1955) *Meth. Enzymol.* **1**, 200. **27**. Brown & Cori (1961) *The Enzymes*, 2nd ed. (Boyer, Lardy & Myrbäck, eds.), **5**, 207. **28**. Cori, Cori & Green (1943) *J. Biol. Chem.* **151**, 39. **29**. Rinaudo & Bruno (1968) *Enzymologia* **34**, 45. **30**. Leloir & Cardini (1962) *The Enzymes*, 2nd ed. (Boyer, Lardy & Myrbäck, eds.), **6**, 317. **31**. Goldemberg & Maréchal (1972) *Meth. Enzymol.* **28**, 953. **32**. Coady, Wallendorff, Gagnon & Lapointe (2002) *J. Biol. Chem.* **277**, 35219. **33**. Diecke, Beyer-Mears & Mistry (1995) *J. Cell Physiol.* **162**, 290. **34**. Nakagawa & Nakao (1977) *J. Biochem.* **81**, 1511. **35**. Green & Stumpf (1942) *J. Biol. Chem.* **142**, 355. **36**. Doudoroff (1943) *J. Biol. Chem.* **151**, 351. **37**. Semenza & Rihova (1969) *Biochim. Biophys. Acta* **178**, 393. **38**. Nakano & Sacktor (1984) *Biochim. Biophys. Acta* **791**, 45. **39**. Yoneyama (1987) *Arch. Biochem. Biophys.* **255**, 168. **40**. Hucho & Wallenfels (1971) *Eur. J. Biochem.* **23**, 489. **41**. Mulhern, Fishman, Kusiak & Bailey (1973) *J. Biol. Chem.* **248**, 4163. **42**. Ramaswamy & Radhakrishnan (1975) *Biochim. Biophys. Acta* **403**, 446. **43**. Skovbjerg, Sjöström & Noren (1981) *Eur. J. Biochem.* **114**, 653. **44**. Skovbjerg, Noren, Sjöström, Danielsen & Enevoldsen (1982) *Biochim. Biophys. Acta* **707**, 89. **45**. Leese & Semenza (1973) *J. Biol. Chem.* **248**, 8170. **46**. Kraml, Kolínská, Ellederová & Hirsová (1972) *Biochim. Biophys. Acta* **258**, 520. **50**. Silva, Terra & Ferreira (2006) *Comp. Biochem. Physiol. B Biochem. Mol. Biol.* **143**, 367. **51**. Silva, Terra & Ferreira (2004) *Insect Biochem. Mol. Biol.* **34**, 1089. **52**. Nakano & Sacktor (1985) *J. Biochem.* **97**, 1329. **53**. Nakano, Sumi & Miyakawa (1977) *J. Biochem.* **81**, 1041. **54**. Galand (1984) *Biochim. Biophys. Acta* **789**, 10. **55**. Youssef, Ghareib & Nour el Dein (1987) *Zentralbl. Mikrobiol.* **142**, 397. **56**. Ghareib, Youssef & Nour el Dein (1988) *Zentralbl. Mikrobiol.* **143**, 397. **57**. Goldemberg, Maréchal & De Souza (1966) *J. Biol. Chem.* **241**, 45. **58**. Lin, Hsu, Chen, Chern & Lee (2007) *Phytochemistry* **68**, 1189.

Phloroglucinol

This reagent (FW = 126.11 g/mol; CAS 108-73-6), also known as 1,3,5-trihydroxybenzene and phloroglucin, is a white solid, soluble in ten parts ethanol and in 100 parts water at 25°C. Resorcinol is a component of many naturally-occurring compounds (*e.g.*, phloretin) and is often used in the qualitative detection of carbohydrates, aldehydes, and lignin. A solution of phloroglucinol acidified with HCl turns red, when boiled in the presence of pentoses. Phloroglucinol is a product of the reaction catalyzed by phloretin hydrolase and is a substrate of phloroglucinol reductase. **Target(s):** carboxylase (1,2); catechol oxidase (3,4); β-glucosidase (5); lipoxygenase (6); nitronate monooxygenase (7); phloretin hydrolase, product inhibition (8); 3-phytase (9); 4-phytase (9); aand tyrosinase (10). **1.** Karrer & Visconti (1947) *Helv. Chim. Acta* **30**, 268. **2.** Massart (1950) *The Enzymes*, 1st ed. (Sumner & Myrbäck, eds.), **1** (part 1), 307. **3.** Webb (1966) *Enzyme and Metabolic Inhibitors*, vol. 2, p. 297, Academic Press, New York. **4.** Richter (1934) *Biochem. J.* **28**, 901. **5.** Dale, Ensley, Kern, Sastry & Byers (1985) *Biochemistry* **24**, 3530. **6.** Holman & Bergström (1951) *The Enzymes*, 1st ed. (Sumner & Myrbäck, eds.), **2** (part 1), 559. **7.** Kido, Soda & Asada (1978) *J. Biol. Chem.* **253**, 226. **8.** Schoefer, Braune & Blaut (2004) *Appl. Environ. Microbiol.* **70**, 6131. **9.** Goel & Sharma (1979) *Phytochemistry* **18**, 939. **10.** Gortner (1911) *J. Biol. Chem.* **10**, 113.

Phlorrhizin, *See* Phlorizin

PHM-27, *See* PHI-27

DL-*threo*-PHMP, *See* DL-threo-1-Phenyl-2-hexadecanoylamino-3-morpholino-1-propanol

Phomopsin A

This mycotoxin and antibiotic (FW = 789.24 g/mol; CAS 64925-80-0) is obtained from *Diaporthe toxica* (formerly *Phomopsis leptostromiformis*), a fungus that grows mainly within lupin stems, the consumption of which leads to lupinosis in sheep grazing on lupin stubble. Intoxication results in liver damage, disorientation, blindness, lethargy and death in severe cases. Phomopsin A is a 13-membered ether oxygen-containing macrolide that blocks tubulin polymerization (K_i <1 μM). It also inhibits vinblastine binding to tubulin and, in common with vinblastine and maytansine, enhances colchicine binding (1). Phomopsin A and the depsipeptide, Dolastatin 10, bind to a site adjacent to the vinca alkaloid and nucleotide sites. This mycotoxin induces tubulin oligomerization into ring structures that cannot form microtubules (2). Scatchard analysis suggests two classes of binding sites: a high-affinity site (K_1 = 1 x 10^{-8} M) and a low-affinity site (K_2 = 3 x 10^{-7} M) (3). Phomopsin A also inhibits rhizoxin binding, with a K_i of 0.8 x 10^{-8} M, suggesting that the high-affinity site of phomopsin A is identical to the rhizoxin binding site (3). The development and validation of an LC-MS/MS method for detecting Phomopsin A in lupin and lupin-containing retail food has been reported (4). **1.** Lacey, Edgar & Culvenor (1987) *Biochem. Pharmacol.* 36, 2133. **2.** Correia (1991) *Pharmacol. Ther.* **52**, 127. **3.** Li, Kobayashi, Tokiwa, Hashimoto & Iwasaki (1992) *Biochem. Pharmacol.* **43**, 219. **4.** de Nijs, Pereboom-de Fauw, van Dam, *et al.* (2013) *Food Addit. Contam. (Part A: Chem. Anal. Control Expo. Risk Assess.)* **30**, 1819.

Phorbol 12-Myristate 13-Acetate

This photosensitive diester (FW = 616.84 g/mol; Symbol: PMA), also known as 12-*O*-tetradecanoylphorbol 13-acetate, found naturally in oil expressed from *Croton tiglium* seeds, is an activator of protein kinase C, a ligand for histone H$_1$, and a potent tumor promoter. It also induces nitric oxide production. Phorbol 12-myristate 13-acetate will inhibit apoptosis induced by the Fas antigen but will induce apoptosis in HL-60 leukemia cells. *Note*: The 4α-epimer (4α-phorbol 12-myristate 13-acetate) is not biologically active and can be used as a negative control. **Target(s):** [elongation factor 2] kinase (1); and protein-disulfide isomerase (2). **1.** Sans, Xie & Williams (2004) *Biochem. Biophys. Res. Commun.* **319**, 144. **2.** Mayumi, Azuma, Kobayashi, *et al.* (2000) *Biol. Pharm. Bull.* **23**, 1111.

Phosdrin, *See* Mevinphos

Phosfon, *See* Chlorphonium Chloride

Phosfon D, *See* Chlorphonium Chloride

Phosfon S, *See* Tributyl-2,4-dichlorobenzylammonium Chloride

Phosphamidon

This insecticide (FW = 299.69 g/mol; CAS 13171-21-6; 297-99-4 for (*E*)-isomer; 23783-98-4 for (*Z*)-isomer; Density = 1.2132 g/cm^3; M.P. = 120-123 °C; B.P. = 162 °C at 1.5 mm Hg; miscible in water; LD$_{50}$ = 13 mg/kg (mouse, oral) = 6 mg/kg (mouse, IV); LD$_{50}$ = 20 mg/kg (rat, oral) = 26 mg/kg (rat, subcutaneous)), known systematically as (*E/Z*)-[3-chloro-4-(diethylamino)-4-oxobut-2-en-2-yl] dimethyl phosphate and phosphoric acid, 2-chloro-3-(diethylamino)-1-methyl-3-oxo-1-propenyl dimethyl ester, inhibits cholinesterases. Commercial sources typically are a mixture of *cis* and *trans* isomers. Phosphamidon is membrane-permeant, and contact with the skin should be avoided. Prolonged treatment with vitamin C partially corrects the oxidative stress observed in phosphamidon-treated animals. **Target(s):** acetylcholinesterase (1-5); carboxylesterase (5,6); cholinesterase (7,8); monoamine oxidase (9); and neuropathy target esterase (5). **1.** Braid & Nix (1969) *Can. J. Biochem.* **47**, 1. **2.** Rose & Voss (1971) *Bull. Environ. Contam. Toxicol.* **6**, 205. **3.** Chattopadhyay, Choudhuri & Maity (1986) *Indian J. Med. Res.* **83**, 435. **4.** Datta, Gupta & Sengupta (1994) *Indian J. Med. Res.* **100**, 87. **5.** Mokanovic', Maksimovic' & Stepanovic' (1995) *Arch. Toxicol.* **69**, 425. **6.** Siddalinga Murthy & Veerabhadrappa (1996) *Insect Biochem. Mol. Biol.* **26**, 287. **7.** Anliker, Beriger, Geiger & Schmid (1961) *Helv. Chim. Acta* **44**, 1622. **8.** Awal, Malik & Sandhu (1988) *Vet. Hum. Toxicol.* **30**, 444. **9.** Nag & Nandy (2001) *Indian J. Exp. Biol.* **39**, 802.

Phosphapyrimidine Nucleoside 1

This tetrahedral transition-state analogue (FW = g/mol) resembles the putative tetrahedral intermediate in the associative (S_N2) reaction catalyzed by cytidine deaminase. As such, it is a slow-binding inhibitor at pH 6, with the following parameters: K_i = 0.9 nM; k_{on} = 8300 M^{-1}s^{-1}, k_{off} = 7.8 x 10^{-6} s^{-1} (1,2). With as tetrahydrouridine as an inhibitor of this enzyme, the initial interaction is a loose complex (K_D = 1.2 μM), attended by slower isomerization to give a more tightly bound complex (K_i = 0.24 μM, with forward and reverse rate constants of 3.81 min^{-1} and 0.95 min^{-1}, respectively. **1.** Ashley & Bartlett (1984) *J. Biol. Chem.* **259**, 13621. **2.** Ashley & Bartlett (1982) *Biochem. Biophys. Res. Commun.* **108**, 1467.

Phosphatidate (Phosphatidic Acid)

CH$_2$OCOR'

RCOO—C—H

CH$_2$OPO$_3$H

Generic Structural Formula

1-Palmitoyl-2-stearoyl-*sn*-glycerol 3-phosphate

These fatty acyl glycerophosphate derivatives are intermediates in the biosynthesis of phosphoglyceride and complex phospholipids. Although the term does not stipulate the hydroxyl group(s) that is(are) acylated, nor the location of the phosphoryl group, phosphatidate refers to 1,2-diacyl-*sn*-glycerol 3-phosphate. *Note*: Acid-catalyzed hydrolysis yields both α- and β-glycerophosphate, whereas base-catalyzed hydrolysis is unattended by phosphoryl group migration. (*See also specific phosphatidate*) **Target(s):** β-*N*-acetylglucosaminyl-glycopeptide β-1,4-galactosyl-transferase (47,48); *N*-acetyllactosamine synthase (47,48); adenylyl cyclase (1); AMP deaminase, inhibited by 1,2-dioleoyl-*sn*-glycerol 3-phosphate (2); Ca^{2+}/calmodulin-dependent protein kinase, activated at low concentrations (40); cardiolipin synthase (3); Ca^{2+}-transporting ATPase, inhibited by 1,2-dioleoyl-*sn*-glycerol 3-phosphate (4); ceramidase (5); chymase (6,24); CMP-*N*-acetylneuraminate monooxygenase (53); cytidylate cyclase (7); diacylglycerol kinase, product inhibition (44); dolichyldiphosphatase (8); dolichyl-phosphatase (8,9,31-33); dopamine β-monooxygenase (10); ethanolaminephosphotransferase (42); glycerol-1,2-cyclic-phosphate phosphodiesterase (27); glycerol-3-phosphate *O*-acyltransferase (11,51,52); glycerone-phosphate *O*-acyltransferase, dihydroxyacetone-phosphate *O*-acyltransferase (50); glycosylphosphatidylinositol phospholipase D (12,25,26); G-protein signaling regulator (13); hydroxyacylglutathione hydrolase, glyoxalase II (36); indole-3-acetate β-glucosyltransferase (46); insulin receptor protein-tyrosine kinase (14); Lysophospholipase (15,38,39); lysophospholipase/transacylase (15,38); [myosin-light-chain] phosphatase (30); nitric-oxide synthase, neuronal (16); phosphatidylglycerophosphatase (34); 1-phosphatidylinositol 3-kinase (17,18,43); 1-phosphatidylinositol 4-kinase, weakly inhibited (45); phospholipase C (also weak substrate (29); phospholipase D (28); phosphoprotein phosphatase (20,21,35); polar-amino-acid-transporting ATPase, *or* histidine permease (19); protein phosphatase-1 (20,21,35); tubulin polymerization, *or* microtubule assembly, as stimulated by microtubule-associated proteins (22,23); UDP-*N*-acetylglucosamine: lysosomal-enzyme *N*-acetylglucosamine phosphotransferase (41). **1.** Proll, Clark & Butcher (1985) *Mol. Pharmacol.* **28**, 331. **2.** Tanfani, Kulawiak, Kossowska, *et al.* (1998) *Mol. Genet. Metab.* **65**, 51. **3.** Tropp (1997) *Biochim. Biophys. Acta* **1348**, 192. **4.** Dalton, East, Mall, *et al.* (1998) *Biochem. J.* **329**, 637. **5.** El Bawab, Birbes, Roddy, *et al.* (2001) *J. Biol. Chem.* **276**, 16758. **6.** Fukusen, Kido & Katunuma (1985) *Arch. Biochem. Biophys.* **237**, 118. **7.** Mori, Muto & Yamamoto (1989) *Biochem. Biophys. Res. Commun.* **162**, 1502. **8.** Belocopitow & Boscoboinik (1982) *Eur. J. Biochem.* **125**, 167. **9.** DeRosa & Lucas (1984) *Arch. Biochem. Biophys.* **234**, 537. **10.** Boyadzhyan & Karagezyan (1990) *Biomed. Sci.* **1**, 379. **11.** Haldar & Vancura (1992) *Meth. Enzymol.* **209**, 64. **12.** Low & Huang (1993) *J. Biol. Chem.* **268**, 8480. **13.** Ouyang, Tu, Barker & Yang (2003) *J. Biol. Chem.* **278**, 11115. **14.** Arnold & Newton (1996) *J. Cell Biochem.* **62**, 516. **15.** Sugimoto & Yamashita (1994) *J. Biol. Chem.* **269**, 6252. **16.** Hori, Iwasaki, Hayashi, *et al.* (1999) *J. Biochem.* **126**, 829. **17.** Lavie & Agranoff (1996) *J. Neurochem.* **66**, 811. **18.** Lauener, Shen, Duronio & Salari (1995) *Biochem. Biophys. Res. Commun.* **215**, 8. **19.** Nikaido, Liu & Ames (1997) *J. Biol. Chem.* **272**, 27745. **20.** Jones & Hannun (2000) *J. Biol. Chem.* **277**, 15530. **21.** Kishikawa, Chalfant, Perry, Bielawska & Hannun (1999) *J. Biol. Chem.* **274**, 21335. **22.** Lee, Flynn, Yamauchi & Purich (1986) *Ann. N. Y. Acad. Sci.* **466**, 519. **23.** Yamauchi & Purich (1987) *J. Biol. Chem.* **262**, 3369. **24.** Kido, Fukusen & Katunuma (1985) *Arch. Biochem. Biophys.* **239**, 436. **25.** Rhode, Schulze, Cumme, *et al.* (2000) *Biol. Chem.* **381**, 471. **26.** Lee, Lee, Kim, Myung & Sok (1999) *Neurochem. Res.* **24**, 1577. **27.** Clarke & Dawson (1978) *Biochem. J.* **173**, 579. **28.** Okamura & Yamashita (1994) *J. Biol. Chem.* **269**, 31207. **29.** Tan & Roberts (1998) *Biochemistry* **37**, 4275. **30.** Ito, Feng, Tsujino, *et al.* (1997) *Biochemistry* **36**, 7607. **31.** Ravi, Rip & Carroll (1983) *Biochem. J.* **213**, 513. **32.** Rip, Rupar, Chaudhary & Carroll (1981) *J. Biol. Chem.* **256**, 1929. **33.** Boscoboinik, Morera & Belocopitow

(1984) *Biochim. Biophys. Acta* **794**, 41. **34.** Icho & Raetz (1983) *J. Bacteriol.* **153**, 722. **35.** Plummer, Perreault, Holmes & Posse de Chaves (2005) *Biochem. J.* **385**, 685. **36.** Scirè, Tanfani, Saccucci, Bertoli & Principato (2000) *Proteins* **41**, 33. **37.** Tsujita & Okuda (1993) *J. Biochem.* **113**, 264. **38.** Mirbod, Banno, Ghannoum, *et al.* (1995) *Biochim. Biophys. Acta* **1257**, 181. **39.** Wright, Payne, Santangelo, *et al.* (2004) *Biochem. J.* **384**, 377. **40.** Baudier & Cole (1987) *J. Biol. Chem.* **262**, 17577. **41.** Zhao, Yeh & Miller (1992) *Glycobiology* **2**, 119. **42.** Vecchini, Roberti, Freysz & Binaglia (1987) *Biochim. Biophys. Acta* **918**, 40. **43.** Carpenter, Duckworth, Auger, *et al.* (1990) *J. Biol. Chem.* **265**, 19704. **44.** Kanoh, Kondoh & Ono (1983) *J. Biol. Chem.* **258**, 1767. **45.** Collins & Wells (1983) *J. Biol. Chem.* **258**, 2130. **46.** Leznicki & Bandurski (1988) *Plant Physiol.* **88**, 1481. **47.** Mitranic & Moscarello (1980) *Can. J. Biochem.* **58**, 809. **48.** Clark & Moscarello (1986) *Biochim. Biophys. Acta* **859**, 143. **49.** Garcia, Montero, Alvarez & Sanchez Crespo (1993) *J. Biol. Chem.* **268**, 4001. **50.** Jones & Hajra (1983) *Arch. Biochem. Biophys.* **226**, 155. **51.** Monroy, Kelker & Pullman (1973) *J. Biol. Chem.* **248**, 2845. **52.** Vancura & Haldar (1994) *J. Biol. Chem.* **269**, 27209. **53.** Shaw, Schneckenburger, Carlsen, Christiansen & Schauer (1992) *Eur. J. Biochem.* **206**, 269.

Phosphatidylcholine

CH$_2$OCOR'

RCOO—C—H

CH$_2$OP(O$_2$$^-$)OCH$_2CH_2N^+$(CH$_3$)$_3$

Generic Structural Formula

1-Palmitoyl-2-stearoyl-*sn*-glycerol 3-phosphocholine

This phospholipid, also called lecithin and 1,2-diacyl-*sn*-glycero-3-phosphocholine, is an important constituent of biomembranes and lipoproteins. The physical properties depend on the nature of the acyl groups and on the nature of other lipids. Freshly prepared phosphatidylcholines are white solids that turn yellow and brown on exposure to air. They typically sinter between 75 and 120°C and melt in the neighborhood of 230°C. *See also specific compound* **Target(s):** β-*N*-acetylhexosaminidase (1); alcohol *O*-acetyltransferase (47); 1-alkyl-2-acetylglycerophospho-choline esterase, weakly inhibited (33); acylglycerol kinase, *or* monoacylglycerol kinase (43); aryldialkylphosphatase, *or* paraoxonase (37); arylesterase (37); carnitine *O*-palmitoyltransferase (48); ceramidase (2,3); cerebroside-sulfatase, *or* arylsulfatase A (4); cholesterol monooxygenase, side-chain-cleaving, *or* CYP11A1 (50,51); choline-phosphate cytidylyltransferase (5); cyclopropane-fatty-acyl-phospholipid synthase (6); diacylglycerol kinase (42); dolichyldiphosphatase (7); dolichyl-phosphatase (7-9,31); α-glucosidase (10); β-glucosidase (11,12); glucuronosyltransferase (46); glutamate dehydrogenase (16); glycosylphosphatidylinositol diacylglycerol-lyase, weakly inhibited (13); hormone-sensitive lipase (35); indole-3-acetate β-glucosyltransferase, weakly inhibited (45); lecithin:cholesterol acyltransferase, *or* LCAT, inhibited by the diphytanoyl derivative (14); linoleate diol synthase (52); lipoxygenase (15); lysophospholipase36; microsomal epoxide hydrolase, inhibited when lipid is above the critical micelle concentration (17); phosphatidate phosphatase, weakly inhibited (32); phosphatidyl-ethanolamine methyltransferase (18); phosphatidylinositol diacylglycerol-lyase, phosphatidylinositol-specific phospho-lipase C (19-22); 1-phosphatidylinositol 3-kinase (40,41); 1-phosphatidylinositol 4-kinase (44); phosphatidyl-*N*-methylethanolamine *N*-methyltransferase (49); phosphoinositide phospholipase C (23); protein-tyrosine sulfotransferase, weakly inhibited (38); sphingomyelin phosphodiesterase, *or* sphingomyelinase (24-29); sterol esterase, *or* cholesterol esterase (34,35); steryl-β-glucosidase (30); and UDP-*N*-acetylglucosamine:dolichyl-phosphate *N*-acetylglucosaminephosphotransferase (39). **1.** Frohwein & Gatt (1967) *Biochemistry* **6**, 2783. **2.** El Bawab, Birbes, Roddy, *et al.* (2001) *J. Biol. Chem.* **276**, 16758. **3.** Yada, Higuchi & Imokawa (1995) *J. Biol. Chem.* **270**, 12677. **4.** Stinshoff & Jatzkewitz (1975) *Biochim. Biophys. Acta* **377**, 126. **5.** Vance, Pelech & Choy (1981) *Meth. Enzymol.* **71**, 576. **6.** Chung & Law (1964) *Biochemistry* **3**, 967. **7.** Belocopitow & Boscoboinik (1982) *Eur. J. Biochem.* **125**, 167. **8.** Rip, Rupar, Chaudhary & Carroll (1981) *J. Biol. Chem.* **256**, 1929. **9.** Adrian & Keenan (1981) *Biochem. J.* **197**, 233. **10.** Bravo-Torres, Villagómez-Castro, Calvo-

Méndez, Flores-Carreón & López-Romero (2004) *Int. J. Parasitol.* **34**, 455. **11.** Daniels, Coyle, Chiao, Glew & Labow (1981) *J. Biol. Chem.* **256**, 13004. **12.** Glew, Peters & Christopher (1976) *Biochim. Biophys. Acta* **422**, 179. **13.** Morris, Ping-Sheng, Shen & Mensa-Wilmot (1995) *J. Biol. Chem.* **270**, 2517. **14.** Pownall, Pao, Brockman & Massey (1987) *J. Biol. Chem.* **262**, 9033. **15.** Holman & Bergström (1951) *The Enzymes*, 1st ed. (Sumner & Myrbäck, eds.), **2** (part 1), 559. **16.** Smith, Austen, Blumenthal & Nyc (1975) *The Enzymes*, 3rd ed. (Boyer, ed.), **11**, 293. **17.** Lu & Miwa (1980) *Ann. Rev. Pharmacol. Toxicol.* **20**, 513. **18.** Schneider & Vance (1979) *J. Biol. Chem.* **254**, 3886. **19.** Dawson, Hemington & Irvine (1980) *Eur. J. Biochem.* **112**, 33. **20.** Dawson, Hemington & Irvine (1985) *Biochem. J.* **230**, 61. **21.** Hirasawa, Irvine & Dawson (1981) *Biochem. J.* **193**, 607. **22.** Irvine, Letcher & Dawson (1980) *Biochem. J.* **192**, 279. **23.** Ochocka & Pawelczyk (2003) *Acta Biochim. Pol.* **50**, 1097. **24.** Kanfer & Brady (1969) *Meth. Enzymol.* **14**, 131. **25.** Gatt & Barenholz (1969) *Meth. Enzymol.* **14**, 144. **26.** Sloan (1972) *Meth. Enzymol.* **28**, 874. **27.** Duan & Nilsson (2000) *Meth. Enzymol.* **311**, 276. **28.** Testai, Landek, Goswami, Ahmed & Dawson (2004) *J. Neurochem.* **89**, 636. **29.** Barnholz, Roitman & Gatt (1966) *J. Biol. Chem.* **241**, 3731. **30.** Kalinowska & Wojciechowski (1986) *Phytochemistry* **25**, 45. **31.** Ravi, Rip & Carroll (1983) *Biochem. J.* **213**, 513. **32.** Casola & Possmayer (1979) *Biochim. Biophys. Acta* **574**, 212. **33.** Blank, Lee, Fitzgerald & Snyder (1981) *J. Biol. Chem.* **256**, 175. **34.** Okawa & Yamaguchi (1977) *J. Biochem.* **81**, 1209. **35.** Tsujita & Okuda (1993) *J. Biochem.* **113**, 264. **36.** Wright, Payne, Santangelo, *et al.* (2004) *Biochem. J.* **384**, 377. **37.** Billecke, Draganov, Counsell, *et al.* (2000) *Drug Metab. Dispos.* **28**, 1335. **38.** Kasinathan, Sundaram, Slomiany & Slomiany (1993) *Biochemistry* **32**, 1194. **39.** Kaushal & Elbein (1986) *Plant Physiol.* **82**, 748. **40.** Shibasaki, Fukui & Takenawa (1993) *Biochem. J.* **289**, 227. **41.** Carpenter, Duckworth, Auger, *et al.* (1990) *J. Biol. Chem.* **265**, 19704. **42.** Yada, Ozeki, Kanoh & Nozawa (1990) *J. Biol. Chem.* **265**, 19237. **43.** Shim, Lin & Strickland (1989) *Biochem. Cell Biol.* **67**, 233. **44.** Endemann, Dunn & Cantley (1987) *Biochemistry* **26**, 6845. **45.** Leznicki & Bandurski (1988) *Plant Physiol.* **88**, 1481. **46.** Yokota, Fukuda & Yuasa (1991) *J. Biochem.* **110**, 50. **47.** Furukawa, Yamada, Mizoguchi & Hara (2003) *J. Biosci. Bioeng.* **96**, 380. **48.** Pande, Murthy & Noel (1986) *Biochim. Biophys. Acta* **877**, 223. **49.** Schneider & Vance (1979) *J. Biol. Chem.* **254**, 3886. **50.** Wang & Kimura (1976) *J. Biol. Chem.* **251**, 6068. **51.** Wang, Pfeiffer, Kimura & Tchen (1974) *Biochem. Biophys. Res. Commun.* **57**, 93. **52.** Su, Sahlin & Oliw (1998) *J. Biol. Chem.* **273**, 20744.

Phosphatidylethanolamine

CH_2OCOR'
$RCOO-C-H$
$CH_2OP(O_2^-)OCH_2CH_2NH_3^+$

Generic Structural Formula

1-Palmitoyl-2-stearoyl-*sn*-glycerol 3-phosphoethanolamine

This hygroscopic phospholipid, also called cephalin and 1,2-diacyl-*sn*-glycero-3-phosphoethanolamine, is an important constituent of biomembranes and lipoproteins. It is the major structural phospholipid in brain tissue. The physical properties depend on the nature of the acyl groups and on the presence of other lipids. Freshly prepared phosphatidylethanolamines are white solids that turn yellow and brown on exposure to air. They typically sinter between 80 and 90°C and melt in the neighborhood of 175°C. Solutions of phosphatidylethanolamine slowly decompose at room temperature (roughly 0.3-0.5% per day); hence, solutions should always be freshly prepared and kept cold prior to use. It is labile in alkaline conditions. *See also specific compound* **Target(s):** β-*N*-acetylglucosaminyl-glycopeptide β-1,4-galactosyltransferase (23); *N*-acetyllactosamine synthase (23); acylglycerol kinase, *or* monoacylglycerol kinase (1); cholesterol monooxygenase, side-chain-cleaving, *or* CYP11A1 (26,27); chymase (3); cytidylate cyclase (3); dolichyldiphosphatase (4); dolichyl-phosphatase (4,5,16); dopamine β-monooxygenase (6); ethanolaminephosphotransferase (18); glutamate dehydrogenase (7); γ-glutamyl transpeptidase, mildly inhibited (8); glycerol-3-phosphate *O*-acyltransferase (25); glycerone-phosphate *O*-acyltransferase, *or* dihydroxyacetone-phosphate *O*-acyltransferase (24); hormone-sensitive lipase (9); 3-hydroxybutyrate dehydrogenase (10); indole-3-acetate β-

glucosyltransferase (22); phosphatidate cytidylyltransferase (19); 1-phosphatidylinositol 4-kinase (20,21); phosphoinositide phospholipase C (11,12); phospholipase D, phosphatidylcholine-specific enzyme (13); sphingomyelin phosphodiesterase (14); steryl-β-glucosidase (15); UDP-*N*-acetylglucosamine:dolichyl-phosphate *N*-acetylglucosaminephosphotransferase (17). **1.** Shim, Lin & Strickland (1989) *Biochem. Cell Biol.* **67**, 233. **2.** Kido, Fukusen & Katunuma (1985) *Arch. Biochem. Biophys.* **239**, 436. **3.** Mori, Muto & Yamamoto (1989) *Biochem. Biophys. Res. Commun.* **162**, 1502. **4.** Belocopitow & Boscoboinik (1982) *Eur. J. Biochem.* **125**, 167. **5.** Rip, Rupar, Chaudhary & Carroll (1981) *J. Biol. Chem.* **256**, 1929. **6.** A. S. Boyadzhyan & K. G. Karagezyan (1990) *Biomed. Sci.* **1**, 379. **7.** Smith, Austen, Blumenthal & Nyc (1975) *The Enzymes*, 3rd ed. (Boyer, ed.), **11**, 293. **8.** Butler & Spielberg (1979) *J. Biol. Chem.* **254**, 3152. **9.** Tsujita & Okuda (1993) *J. Biochem.* **113**, 264. **10.** Gotterer (1967) *Biochemistry* **6**, 2147. **11.** Ochocka & Pawelczyk (2003) *Acta Biochim. Pol.* **50**, 1097. **12.** Ginger & Parker (1992) *Eur. J. Biochem.* **210**, 155. **13.** Okamura & Yamashita (1994) *J. Biol. Chem.* **269**, 31207. **14.** Gatt & Barenholz (1969) *Meth. Enzymol.* **14**, 144. **15.** Kalinowska & Wojciechowski (1986) *Phytochemistry* **25**, 45. **16.** Ravi, Rip & Carroll (1983) *Biochem. J.* **213**, 513. **17.** Kaushal & Elbein (1986) *Plant Physiol.* **82**, 748. **18.** Vecchini, Roberti, Freysz & Binaglia (1987) *Biochim. Biophys. Acta* **918**, 40. **19.** Sribney & Hegadorn (1982) *Can. J. Biochem.* **60**, 668. **20.** Steinert, Wissing & Wagner (1994) *Plant Sci.* **101**, 105. **21.** Hou, Zhang & Tai (1988) *Biochim. Biophys. Acta* **959**, 67. **22.** Leznicki & Bandurski (1988) *Plant Physiol.* **88**, 1481. **23.** Clark & Moscarello (1986) *Biochim. Biophys. Acta* **859**, 143. **24.** Jones & Hajra (1983) *Arch. Biochem. Biophys.* **226**, 155. **25.** Kito, Ishinaga & Nishihara (1978) *Biochim. Biophys. Acta* **529**, 237. **26.** Wang & Kimura (1976) *J. Biol. Chem.* **251**, 6068. **27.** Wang, Pfeiffer, Kimura & Tchen (1974) *Biochem. Biophys. Res. Commun.* **57**, 93.

Phosphatidylglycerol

CH_2OCOR'
$RCOO-C-H$
$CH_2OP(O_2^-)OCH_2CH_2CH(OH)CH_2OH$

Generic Structural Formula

1-Palmitoyl-2-stearoyl-*sn*-glycerol 3-phospho-(1-glycerol)

These phospholipids, also called 1,2-diacyl-*sn*-glycero-3-phospho-(1-glycerol), is an important constituent of some biomembranes and a structural component of more complex phospholipids and is found more frequently in plants. The physical properties depend on the nature of the acyl groups and on the presence of other lipids. *See specific compound* **Target(s):** β-*N*-acetylglucos-aminylglycopeptide β-1,4-galactosyl-transferase (1); *N*-acetyllactosamine synthase (1); *N*-acetylmuramoyl-L-alanine amidase, at concentrations above the 0.3 mM critical micelle concentration (2,3); alcohol *O*-acetyltransferase (20); carbamoyl-phosphate synthetase I, ammonia-utilizing (4); ceramidase (5); chlorophyllase (6); diacylglycerol kinase (7); glycerol-3-phosphate *O*-acyltransferase, activated at low concentrations (21); α-glucosidase (8); glycosylphosphatidylinositol diacylglycerol-lyase (9); glycosylphosphatidylinositol phospholipase D (10); α-mannosidase (11); peptidase Do (HtrA heat shock protease) (12); phosphatidate phosphatase (13); 1-phosphatidylinositol 4-kinase (14); polar-amino-acid-transporting ATPase, *or* histidine permease (15); tubulin polymerization, *or* microtubule assembly, as stimulated by microtubule-associated proteins, weakly inhibited (16,17); UDP-*N*-acetylglucos-amine:lysosomal-enzyme *N*-acetylglucosamine phosphotransferase (18); and undecaprenyl-phosphate mannosyltransferase (19). **1.** Clark & Moscarello (1986) *Biochim. Biophys. Acta* **859**, 143. **2.** Vanderwinkel & De Vlieghere (1985) *Biochim. Biophys. Acta* **838**, 54. **3.** Vanderwinkel, De Vlieghere & De Volcsey (1981) *Biochim. Biophys. Acta* **663**, 46. **4.** Brandt & Powers-Lee (1991) *Arch. Biochem. Biophys.* **290**, 14. **5.** El Bawab, Birbes, Roddy, *et al.* (2001) *J. Biol. Chem.* **276**, 16758. **6.** Lambers & Terpstra (1985) *Biochim. Biophys. Acta* **831**, 225. **7.** Kanoh, Kondoh & Ono (1983) *J. Biol. Chem.* **258**, 1767. **8.** Bravo-Torres, Villagómez-Castro, Calvo-Méndez, Flores-Carreón & López-Romero (2004) *Int. J. Parasitol.* **34**, 455. **9.** Morris, Ping-Sheng, Shen & Mensa-Wilmot (1995) *J. Biol. Chem.* **270**, 2517. **10.** Rhode, Schulze, Cumme, *et al.* (2000) *Biol. Chem.* **381**, 471. **11.** Forsee & Schutzbach (1981) *J. Biol. Chem.* **256**, 6577. **12.**

Skorko-Glonek, Lipinska, Krzewski, *et al.* (1997) *J. Biol. Chem.* **272**, 8974. **13**. Humble & Berglund (1991) *J. Lipid Res.* **32**, 1869. **14**. Hou, Zhang & Tai (1988) *Biochim. Biophys. Acta* **959**, 67. **15**. Nikaido, Liu & Ames (1997) *J. Biol. Chem.* **272**, 27745. **16**. Lee, Flynn, Yamauchi & Purich (1986) *Ann. N. Y. Acad. Sci.* **466**, 519. **17**. Yamauchi & Purich (1987) *J. Biol. Chem.* **262**, 3369. **18**. Zhao, Yeh & Miller (1992) *Glycobiology* **2**, 119. **19**. Lahav, Chiu & Lennarz (1969) *J. Biol. Chem.* **244**, 5890. **20**. Furukawa, Yamada, Mizoguchi & Hara (2003) *J. Biosci. Bioeng.* **96**, 380. **21**. Kito, Ishinaga & Nishihara (1978) *Biochim. Biophys. Acta* **529**, 237.

Phosphatidylinositol

CH₂OCOR'
RCOO—C—H
CH₂OP(O₂–)-1D-*myo*-inositol

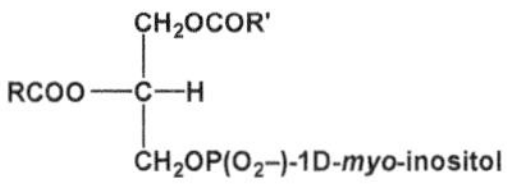

Generic Structural Formula

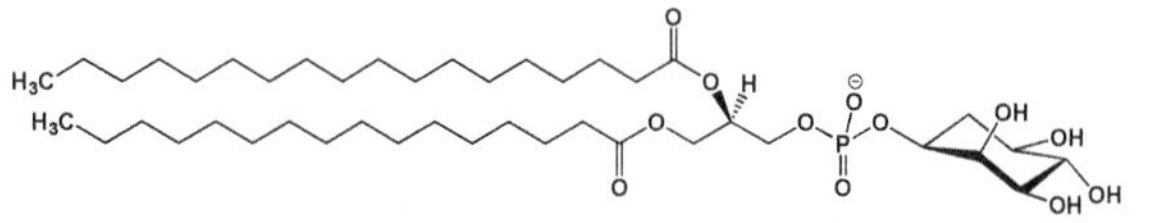

1-Palmitoyl-2-stearoyl-*sn*-glycerol 3-phospho-1D-*myo*-inositol

These phospholipid, often abbreviated PI or PtdIns, containing *sn*-glycerol 3-phosphate, two fatty acyl residues, and *myo*-inositol, is a common constituent of the membranes of eukaryotes and eubacteria and a key precursor in an intracellular signaling pathway. In addition, a derivative of phosphatidylinositol, namely glycosylphosphatidylinositol, assists in anchoring a number of proteins to membranes. *See specific compound* **Target(s):** actin filament bundling in the presence of microtubule-associate protein (21); alcohol *O*-acetyltransferase (30,31); Ca²⁺/calmodulin-dependent protein kinase (24); carbamoyl-phosphate synthetase I, ammonia-requiring (2); chymase (3,4); CMP-*N*-acetylneuraminate monooxygenase (35); diacylglycerol kinase (26); dolichyl-phosphatase (20); dolichyl-phosphate-mannose:protein mannosyltransferase (29); glucose-6-phosphatase (21); β-glucosidase (5); glycerol-3-phosphate *O*-acyltransferase (6,7,34); glycerone-phosphate *O*-acyltransferase, *or* dihydroxyacetone-phosphate *O*-acyltransferase (32); glycosylphosphatidylinositol diacylglycerol-lyase (8); glycosylphosphatidylinositol phospholipase D (9); hexokinase (27); 3-hydroxybutyrate dehydrogenase (10); indole-3-acetate β-glucosyltransferase (28); α-mannosidase (11); mannosyl-oligosaccharide 1,2-α-mannosidase, *or* mannosidase I (11,12); [myosin-light-chain] phosphatase (13); phosphatidate cytidylyltransferase (25); phosphatidate phosphatase (22); phospholipase D (14); sphingomyelin phosphodiesterase (15-17); sterol *O*-acyltransferase, *or* cholesterol *O*-acyltransferase, *or* ACAT (33); sterol esterase, *or* cholesterol esterase (23); tubulin polymerization, *or* microtubule assembly, as stimulated by microtubule-associated proteins (18,19). **1**. Yamauchi & Purich (1993) *Biochem. Biophys. Res. Commun.* **190**, 710. **2**. Brandt & Powers-Lee (1991) *Arch. Biochem. Biophys.* **290**, 14. **3**. Fukusen, Kido & Katunuma (1985) *Arch. Biochem. Biophys.* **237**, 118. **4**. Kido, Fukusen & Katunuma (1985) *Arch. Biochem. Biophys.* **239**, 436. **5**. Daniels, Coyle, Chiao, Glew & Labow (1981) *J. Biol. Chem.* **256**, 13004. **6**. Yamashita & Numa (1981) *Meth. Enzymol.* **71**, 550. **7**. Haldar & Vancura (1992) *Meth. Enzymol.* **209**, 64. **8**. Morris, Ping-Sheng, Shen & Mensa-Wilmot (1995) *J. Biol. Chem.* **270**, 2517. **9**. Rhode, Schulze, Cumme, *et al.* (2000) *Biol. Chem.* **381**, 471. **10**. Gotterer (1967) *Biochemistry* **6**, 2147. **11**. Forsee & Schutzbach (1981) *J. Biol. Chem.* **256**, 6577. **12**. Forsee, Palmer & Schutzbach (1989) *J. Biol. Chem.* **264**, 3869. **13**. Ito, Feng, Tsujino, *et al.* (1997) *Biochemistry* **36**, 7607. **14**. Okamura & Yamashita (1994) *J. Biol. Chem.* **269**, 31207. **15**. Gatt & Barenholz (1969) *Meth. Enzymol.* **14**, 144. **16**. Quintern, Weitz, Nehrkorn, *et al.* (1987) *Biochim. Biophys. Acta* **922**, 323. **17**. Barnholz, Roitman & Gatt (1966) *J. Biol. Chem.* **241**, 3731. **18**. Lee, Flynn, Yamauchi & Purich (1986) *Ann. N. Y. Acad. Sci.* **466**, 519. **19**. Yamauchi & Purich (1987) *J. Biol. Chem.* **262**, 3369. **20**. Boscoboinik, Morera & Belocopitow (1984) *Biochim. Biophys. Acta* **794**, 41. **21**. Mithieux, Daniele, Payrastre & Zitoun (1998) *J. Biol. Chem.* **273**, 17. **22**. Humble & Berglund (1991) *J. Lipid Res.* **32**, 1869. **23**. Sando & Rosenbaum (1985) *J. Biol. Chem.* **260**, 15186. **24**. Baudier & Cole (1987) *J. Biol. Chem.* **262**, 17577. **25**. Sribney & Hegadorn (1982) *Can. J. Biochem.* **60**, 668. **26**. Kanoh, Kondoh & Ono (1983) *J. Biol. Chem.* **258**, 1767. **27**. Nemat-Gorgani & Wilson (1985) *Arch. Biochem. Biophys.* **236**, 220. **28**. Leznicki & Bandurski (1988) *Plant Physiol.* **88**, 1481. **29**. Weston, Nassau, Henley & Marriott (1993) *Eur. J. Biochem.* **215**,

845. **30**. Yoshioka & Hashimoto (1981) *Agric. Biol. Chem.* **45**, 2183. **31**. Furukawa, Yamada, Mizoguchi & Hara (2003) *J. Biosci. Bioeng.* **96**, 380. **32**. Jones & Hajra (1983) *Arch. Biochem. Biophys.* **226**, 155. **33**. Doolittle & Chang (1982) *Biochemistry* **21**, 674. **34**. Yamashita & Numa (1972) *Eur. J. Biochem.* **31**, 565. **35**. Shaw, Schneckenburger, Carlsen, Christiansen & Schauer (1992) *Eur. J. Biochem.* **206**, 269.

Phosphatidylinositol 3,4-Bisphosphate

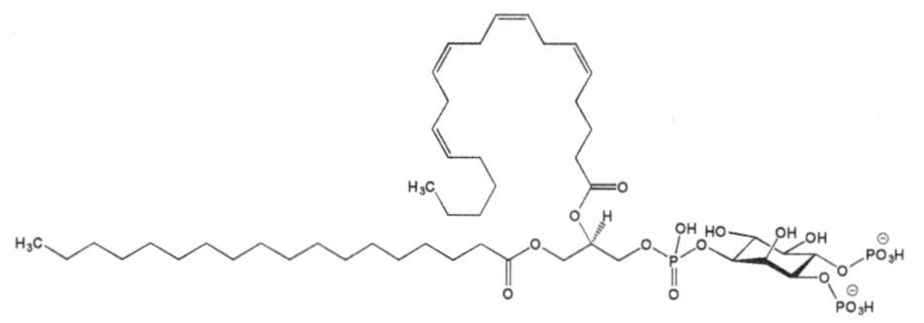

These low-abundance membrane phospholipids (FW = 1011.07 g/mol for 1'-stearoyl-2'-arachidonoyl-*sn*-glycerol-3-phospho-1D-*myo*-inositol 3,4-bisphosphate (shown above)), also called triphosphoinositides and abbreviated PIP₂ or PI-3,4-P₂, are important intermediates in signal transduction pathways that generate regulatory effectors of a number of metabolic processes. Each member of this class differs only in the fatty acyl components linked to C¹ and C² on the glycerol moiety. **Target(s):** glucose-6-phosphatase. **1**. Mithieux, Daniele, Payrastre & Zitoun (1998) *J. Biol. Chem.* **273**, 17.

Phosphatidylinositol 4,5-Bisphosphate

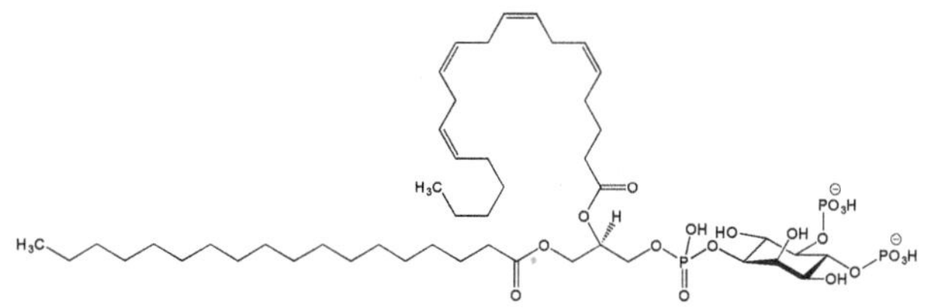

Thes membrane phospholipids, differing only in the fatty acyl components linked to C1 and C2 of the glycerol moiety, also called triphosphoinositides and abbreviated PIP₂ or PI-4,5-P₂, are important intermediates in a metabolic scheme that generates effectors for a number of metabolic processes. Phosphatidylinositol 4,5-bisphosphate is a precursor to the intracellular messengers inositol 1,4,5-trisphosphate and *sn*-1,2-diacylglycerol. Phosphatidylinositol 4,5-bisphosphate is an essential cofactor and stimulator of phospholipase D and is a substrate for specific isoforms of phospholipase C. *See specific compound* *Note*: The free acid slowly decomposes whereas the sodium salt can be stored cold in chloroform for extended periods. The stearoyl arachidonyl form (*i.e.*, 1'-stearoyl-2'-arachidonoyl-*sn*-glycerol-3-phospho-1D-*myo*-inositol 4,5-bisphosphate, shown above) is a form commonly found in mammals. **Target(s):** [β-adrenergic-receptor] kinase (1); AMP deaminase (2); arachidonoyl-diacylglycerol kinase, a specific diacylglycerol kinase (3); casein kinase I (4-7); caspase 8 (8); caspase 9 (8); ER-60 protease (9); glucose-6-phosphatase (10,11); nitric-oxide synthase (12); phosphatidate cytidylyltransferase (13); 1-phosphatidylinositol 4-kinase (20,25,26); 1-phosphatidylinositol-4-phosphate 5-kinase, product inhibition (14-19); protein kinase C (20); protein-tyrosine kinase (21); and sphingomyelin phosphodiesterase, *or* sphingomyelinase (22-24). **1**. Onorato, Gillis, Liu, Benovic & Ruoho (1995) *J. Biol. Chem.* **270**, 21346. **2**. Sims, Mahnke-Zizelman, Profit, *et al.* (1999) *J. Biol. Chem.* **274**, 25701. **3**. Walsh, Suen & Glomset (1995) *J. Biol. Chem.* **270**, 28647. **4**. Gross, Hoffman, Fisette, Baas & Anderson (1995) *J. Cell Biol.* **130**, 711. **5**. Chauhan, Singh, Chauhan & Brockerhoff (1993) *Biochim. Biophys. Acta* **1177**, 318. **6**. Bazenet, Brockman, Lewis, Chan & Anderson (1990) *J. Biol. Chem.* **265**, 7369. **7**. Brockman & Anderson (1991) *J. Biol. Chem.* **266**, 2508. **8**. Mejillano, Yamamoto, Rozelle, *et al.* (2001) *J. Biol. Chem.* **276**, 1865. **9**. Urade & Kito (1992) *FEBS Lett.* **312**, 83. **10**. Van Schaftingen & Gerin (2002) *Biochem. J.* **362**, 513. **11**. Mithieux, Daniele, Payrastre & Zitoun (1998) *J. Biol. Chem.* **273**, 17. **12**. Hori, Iwasaki, Hayashi, *et al.* (1999) *J. Biochem.* **126**, 829. **13**. Saito, Goto, Tonosaki & Kondo (1997) *J. Biol. Chem.* **272**, 9503. **14**. Van Rooijen, Rossowska & Bazan (1985) *Biochem. Biophys. Res. Commun.* **126**, 150. **15**. Smith, Wreggett & Irvine (1986) *Biochem. Soc. Trans.* **14**, 1145. **16**. Ling, Schulz & Cantley (1989) *J. Biol.*

Chem. **264**, 5080. **17**. Lundberg, Jergil & Sundler (1986) *Eur. J. Biochem.* **161**, 257. **18**. Kai, Salway & Hawthorne (1968) *Biochem. J.* **106**, 791. **19**. Vancurova, Choi, Lin, Kuret & Vancura (1999) *J. Biol. Chem.* **274**, 1147. **20**. Yao, Suzuki, Ozawa, *et al.* (1997) *J. Biol. Chem.* **272**, 13033. **21**. Bordin, Clari, Baggio & Moret (1992) *Biochem. Biophys. Res. Commun.* **187**, 853. **22**. Quintern & Sandhoff (1991) *Meth. Enzymol.* **197**, 536. **23**. Quintern, Weitz, Nehrkorn, *et al.* (1987) *Biochim. Biophys. Acta* **922**, 323. **24**. Testai, Landek, Goswami, Ahmed & Dawson (2004) *J. Neurochem.* **89**, 636. **25**. Graziani, Ling, Endemann, Carpenter & Cantley (1992) *Biochem. J.* **284**, 39. **26**. Porter, Li & Deuel (1988) *J. Biol. Chem.* **263**, 8989.

Phosphatidylinositol 3-Phosphate

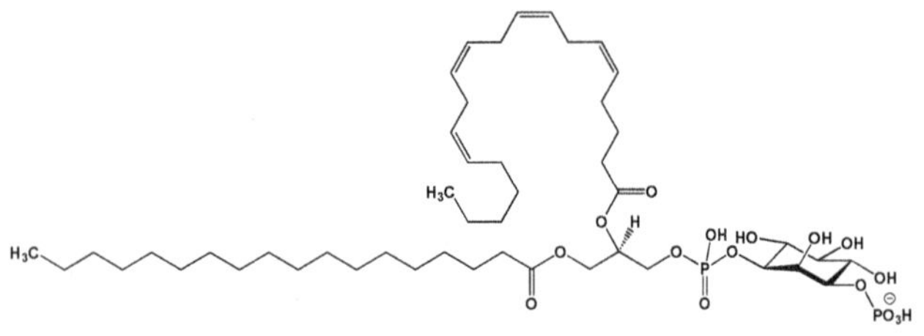

These membrane phospholipids, also called phosphoinositides and abbreviated PIP or PI-3-P, are important intermediates in a metabolic scheme that generates effectors of a number of metabolic processes (each member of this class differ only in the fatty acyl components linked to C1 and C2 of the glycerol moiety: 1'-stearoyl-2'-arachidonoyl-*sn*-glycerol-3-phospho-1D-*myo*-inositol 3-phosphate is shown above). **Target(s):** glucose-6-phosphatase (1). **1**. Mithieux, Daniele, Payrastre & Zitoun (1998) *J. Biol. Chem.* **273**, 17.

Phosphatidylinositol 4-Phosphate

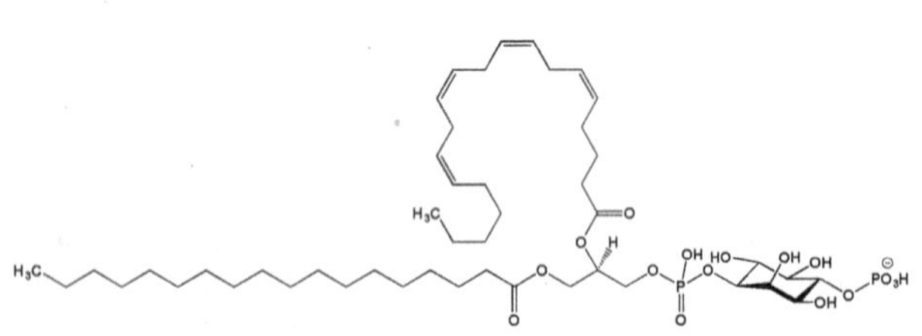

These membrane phospholipids, also called phosphoinositides and abbreviated PIP or PI-4-P, are important intermediates in a metabolic scheme that generates effectors of a number of metabolic processes (each member of this class differ only in the fatty acyl components linked to C1 and C2 of the glycerol moiety: 1'-stearoyl-2'-arachidonoyl-*sn*-glycerol-3-phospho-1D-*myo*-inositol 4-phosphate is shown above). It is a component of the lipid signaling pathway and promotes β-adrenergic receptor kinase phosphorylation. **Target(s):** glucose-6-phosphatase (1); and phosphatidate cytidylyltransferase, weakly inhibited (2). **1**. Mithieux, Daniele, Payrastre & Zitoun (1998) *J. Biol. Chem.* **273**, 17. **2**. Saito, Goto, Tonosaki & Kondo (1997) *J. Biol. Chem.* **272**, 9503.

Phosphatidylinositol 3,4,5-Trisphosphate

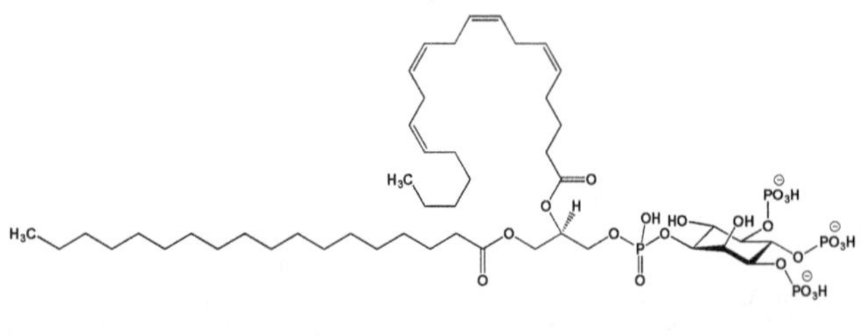

These membrane phospholipids, differing only in the fatty acyl components linked to C1 and C2 of the glycerol moiety, are important intermediates in a signal transduction cascades that generate a number of key metabolic effectors. **Target(s):** glucose-6-phosphatase (1); and sphingomyelin phosphodiesterase (2). **1**. Mithieux, Daniele, Payrastre & Zitoun (1998) *J. Biol. Chem.* **273**, 17. **2**. Testai, Landek, Goswami, Ahmed & Dawson (2004) *J. Neurochem.* **89**, 636.

Phosphatidyl-L-serine

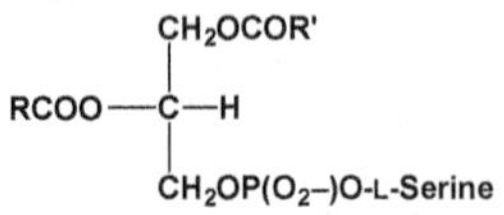

Generic Structural Formula

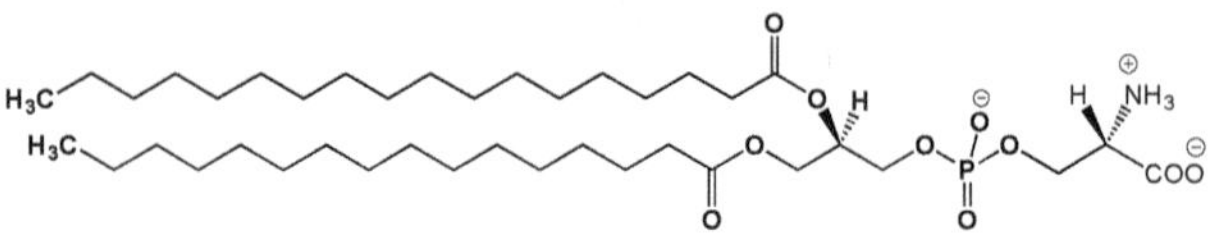

1-Palmitoyl-2-stearoyl-*sn*-glycerol 3-phospho-L-Serine

These phospholipids, also called cephalin and 1,2-diacyl-*sn*-glycero-3-phospho-L-serine, is an important constituent of biomembranes and lipoproteins and the major structural phospholipid in brain tissue. The physical properties depend on the nature of the acyl groups and on the presence of other lipids. Freshly prepared phosphatidylserines are white solids that turn yellow and brown on exposure to air. They typically sinter about 120°C and melt in the neighborhood of 159°C. Solutions will typically undergo decomposition at room temperature at a rate of 0.5% per day. (**See** *specific compound*) **Target(s):** acetylcholinesterase (37); β-*N*-acetylglucosaminylglycopeptide β-1,4-galactosyltransferase (44); *N*-acetyllactosamine synthase (44); alcohol *O*-acetyltransferase (45); ATP diphosphatase1; Ca²⁺-exporting ATPase (2,3); carbamoyl-phosphate synthetase I, ammonia-requiring) (4); ceramidase (5); chymase (6,7); cytidylate cyclase (8); dopamine β-monooxygenase (9); elastase (6); fructose-1,6-bisphosphate aldolase, inhibited by phosphatidylserine liposomes (10); β-glucosidase (11); glutamate dehydrogenase (12); γ-glutamyl transpeptidase (13); glycerol-3-phosphate *O*-acyltransferase (14,15,48); glycerone-phosphate *O*-acyltransferase, *or* dihydroxyacetone-phosphate *O*-acyltransferase, weakly inhibited (46); glycosylphosphatidyl-inositol diacylglycerol-lyase (16); glycosylphosphatidylinositol phospholipase D (17); hexokinase (43); hormone-sensitive lipase (36); hydroxyacyl-glutathione hydrolase, *or* glyoxalase II (34); 3-hydroxybutyrate dehydrogenase (18); lactate dehydrogenase (19); monoamine oxidase (20,21); [myosin-light-chain] phosphatase (22); phosphatidate cytidylyltransferase (40); phosphatidate phosphatase (23); 1-phosphatidylinositol 4-kinase (41,42); phosphatidylinositol-3,4,5-trisphosphate 3-phosphatase (24); phosphoinositide phospholipase C (25-27); polar-amino-acid-transporting ATPase, histidine permease (28); sphingomyelin phosphodiesterase, *or* sphingomyelinase (29,30); sterol *O*-acyltransferase, *or* cholesterol *O*-acyltransferase, *or* ACAT (47); sterol esterase, *or* cholesterol esterase (35,36); tubulin polymerization, *or* microtubule assembly, as stimulated by microtubule-associated proteins, weakly inhibited (31,32); tumor necrosis factor (33); UDP-*N*-acetylglucosamine: dolichyl-phosphate *N*-acetylglucosamine-phospho-transferase (38); and UDP-*N*-acetylglucosamine:lysosomal-enzyme *N*-acetylglucosaminephospho-transferase (39). **1**. Capito, Hansen, Hedeskov & Thams (1986) *Diabetes* **35**, 1096. **2**. Dalton, East, Mall, *et al.* (1998) *Biochem. J.* **329**, 637. **3**. Tsakiris & Deliconstantinos (1985) *Int. J. Biochem.* **17**, 1117. **4**. Brandt & Powers-Lee (1991) *Arch. Biochem. Biophys.* **290**, 14. **5**. El Bawab, Birbes, Roddy, *et al.* (2001) *J. Biol. Chem.* **276**, 16758. **6**. Fukusen, Kido & Katunuma (1985) *Arch. Biochem. Biophys.* **237**, 118. **7**. Kido, Fukusen & Katunuma (1985) *Arch. Biochem. Biophys.* **239**, 436. **8**. Mori, Muto & Yamamoto (1989) *Biochem. Biophys. Res. Commun.* **162**, 1502. **9**. Boyadzhyan & Karagezyan (1990) *Biomed. Sci.* **1**, 379. **10**. Kwiatkowska, Modrzycka & Sidorowicz (1994) *Gen. Physiol. Biophys.* **13**, 425. **11**. Daniels, Coyle, Chiao, Glew & Labow (1981) *J. Biol. Chem.* **256**, 13004. **12**. Smith, Austen, Blumenthal & Nyc (1975) *The Enzymes*, 3rd ed. (Boyer, ed.), **11**, 293. **13**. Butler & Spielberg (1979) *J. Biol. Chem.* **254**, 3152. **14**. Yamashita & Numa (1981) *Meth. Enzymol.* **71**, 550. **15**. Haldar & Vancura (1992) *Meth. Enzymol.* **209**, 64. **16**. Morris, Ping-Sheng, Shen & Mensa-Wilmot (1995) *J. Biol. Chem.* **270**, 2517. **17**. Rhode, Schulze, Cumme, *et al.* (2000) *Biol. Chem.* **381**, 471. **18**. Gotterer (1967) *Biochemistry* **6**, 2147. **19**. Terlecki, Czapinska & Gutowicz (2002) *Cell Mol. Biol. Lett.* **7**, 895. **20**. Tachiki, Buckman, Eiduson, Kling & Hullett (1986) *Biol. Psychiatry* **21**, 59. **21**. Buckman, Eiduson & Boscia (1983) *Biochem. Pharmacol.* **32**, 3639. **22**. Ito, Feng, Tsujino, *et al.* (1997) *Biochemistry* **36**, 7607. **23**. Humble & Berglund (1991) *J. Lipid Res.* **32**, 1869. **24**. Kabuyama, Nakatsu, Homma & Fukui (1996) *Eur. J. Biochem.* **238**, 350. **25**. Litosch (2000) *Biochemistry* **39**, 7736. **26**. Ochocka & Pawelczyk (2003) *Acta Biochim. Pol.* **50**, 1097. **27**. Ginger & Parker

(1992) *Eur. J. Biochem.* **210**, 155. **28**. Nikaido, Liu & Ames (1997) *J. Biol. Chem.* **272**, 27745. **29**. Gatt & Barenholz (1969) *Meth. Enzymol.* **14**, 144. **30**. Barnholz, Roitman & Gatt (1966) *J. Biol. Chem.* **241**, 3731. **31**. Lee, Flynn, Yamauchi & Purich (1986) *Ann. N. Y. Acad. Sci.* **466**, 519. **32**. Yamauchi & Purich (1987) *J. Biol. Chem.* **262**, 3369. **33**. Monastra, Cross, Bruni & Raine (1993) *Neurology* **43**, 153. **34**. Scirè, Tanfani, Saccucci, Bertoli & Principato (2000) *Proteins* **41**, 33. **35**. Sando & Rosenbaum (1985) *J. Biol. Chem.* **260**, 15186. **36**. Tsujita & Okuda (1993) *J. Biochem.* **113**, 264. **37**. Webb (1978) *Can. J. Biochem.* **56**, 1124. **38**. Kaushal & Elbein (1986) *Plant Physiol.* **82**, 748. **39**. Zhao, Yeh & Miller (1992) *Glycobiology* **2**, 119. **40**. Sribney & Hegadorn (1982) *Can. J. Biochem.* **60**, 668. **41**. Hou, Zhang & Tai (1988) *Biochim. Biophys. Acta* **959**, 67. **42**. Kanoh, Banno, Hirata & Nozawa (1990) *Biochim. Biophys. Acta* **1046**, 120. **43**. Nemat-Gorgani & Wilson (1985) *Arch. Biochem. Biophys.* **236**, 220. **44**. Mitranic & Moscarello (1980) *Can. J. Biochem.* **58**, 809. **45**. Yoshioka & Hashimoto (1981) *Agric. Biol. Chem.* **45**, 2183. **46**. Jones & Hajra (1983) *Arch. Biochem. Biophys.* **226**, 155. **47**. Doolittle & Chang (1982) *Biochemistry* **21**, 674. **48**. Yamashita & Numa (1972) *Eur. J. Biochem.* **31**, 565.

4,4'-Phosphinicobis(butane-1,3-dicarboxylic Acid)

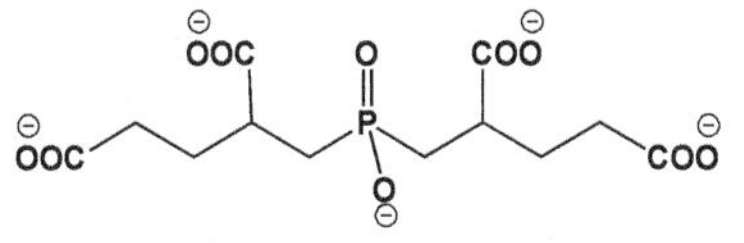

This tetracarboxylic acid ($FW_{free-acid}$ = 354.25 g/mol) inhibits glutamate carboxypeptidase II (K_i = 21.7 nM) and is a mGluR3 receptor agonist (1,2). **1**. Kozikowski, Zhang, Nan, *et al.* (2004) *J. Med. Chem.* **47**, 1729. **2**. Kozikowski, Nan, Conti, *et al.* (2001) *J. Med. Chem.* **44**, 298.

Phosphinothricin

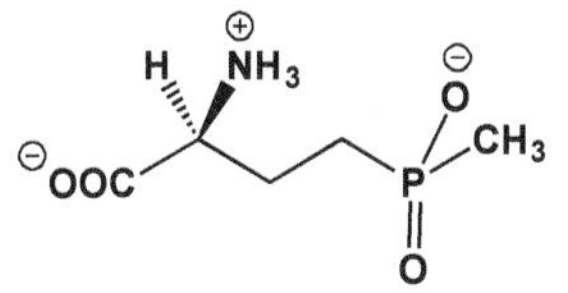

This dephospho transition-state analogue (FW = 177.10 g/mol; CAS 51276-47-2) is naturally occurring glutamate analogue, first isolated from *Streptomyces viridochromogenes*. Phosphonothricin is a slow, tight-binding inhibitor of glutamine synthetase (1-9) that binds at the glutamate binding site and stabilizes the flap of a glutamyl residue in a position blocking glutamate entry into the active site, thereby trapping the inhibitor on the enzyme. Phosphinothricin undergoes an ATP-dependent phosphorylation. *Note*: Phosphinothricin is a component of the antibiotics bialaphos and phosalacine. The monoammonium salt of the racemic compound, also called glufosinate-ammonium, is a post-emergent herbicide. **1**. Logusch, Walker, McDonald & Franz (1989) *Biochemistry* **28**, 3043. **2**. Gill & Eisenberg (2001) *Biochemistry* **40**, 1903. **3**. Abell & Villafranca (1991) *Biochemistry* **30**, 6135. **4**. Robertson & Alberte (1996) *Plant Physiol.* **111**, 1169. **5**. Garcia-Dominguez, Reyes & Florencio (1997) *Eur. J. Biochem.* **244**, 258. **6**. Montanini, Betti, Marquez, *et al.* (2003) *Biochem. J.* **373**, 357. **7**. Crespo, Guerrero & Florencio (1999) *Eur. J. Biochem.* **266**, 1202. **8**. Ericson (1985) *Plant Physiol.* **79**, 923. **9**. Colanduoni & Villafranca (1986) *Bioorg. Chem.* **14**, 163.

2'-Phospho-ADP-ribose

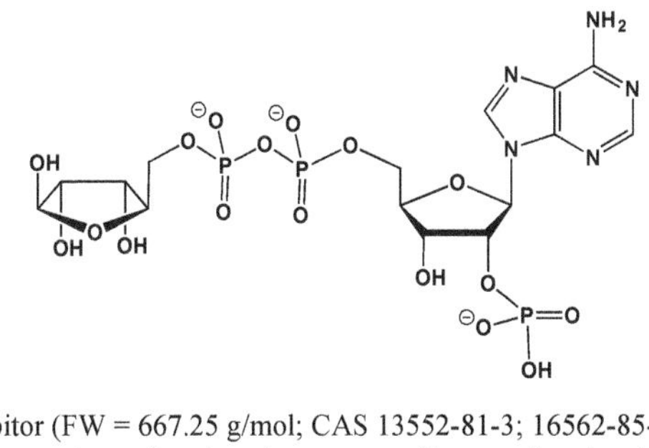

This inhibitor (FW = 667.25 g/mol; CAS 13552-81-3; 16562-85-9 for the β-D-ribofuranose-isomer), also known as 2'-monophosphoadenosine 5'-diphosphoribose and adenosine-2'-monophospho-5'-diphosphoribose, is structurally equivalent to pyridine-less $NADP^+$ or NADPH. **Target(s):**

glucose-6-phosphate dehydrogenase (1); glutamate dehydrogenase (2); glutathione-disulfide reductase, NADPH-dependent (3); 3-hydroxy-3-methylglutaryl-CoA reductase (4-7); isocitrate dehydrogenase, $NADP^+$-dependent (1,8); and 6-phosphogluconate dehydrogenase (1). **1**. Grafe, Bormann & Truckenbrodt (1980) *Z. Allg. Mikrobiol.* **20**, 607. **2**. Smith, Austen, Blumenthal & Nyc (1975) *The Enzymes*, 3rd ed., **11**, 293. **3**. Llobell, Fernandez & Lopez-Barea (1986) *Arch. Biochem. Biophys.* **250**, 373. **4**. Chang (1983) *The Enzymes*, 3rd ed., **16**, 491. **5**. Feingold & Moser (1986) *Arch. Biochem. Biophys.* **249**, 46. **6**. Tanazawa & Endo (1979) *Eur. J. Biochem.* **98**, 195. **7**. Roitelman & Shechter (1986) *J. Lipid Res.* **27**, 828. **8**. Seelig & Colman (1978) *Arch. Biochem. Biophys.* **188**, 394.

5-Phospho-D-arabinonamide

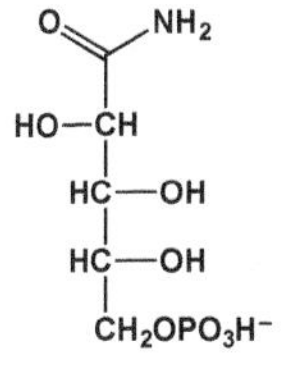

This glucosamine 6-phosphate analogue (FW = 258.12 g/mol) inhibits glutamine:fructose-6-phosphate aminotransferase (isomerizing) and glucose-6-phosphate isomerase, with respective K_i values of 6.2 mM and 4.6 μM. **1**. Milewski, Janiak & Wojciechowski (2006) *Arch. Biochem. Biophys.* **450**, 39.

5-Phospho-D-arabinonate, *See* D-Arabinonic Acid 5-Phosphate

5-Phosphoarabonate, *See* D-Arabinonic Acid 5-Phosphate

N^{ω}-Phospho-L-arginine

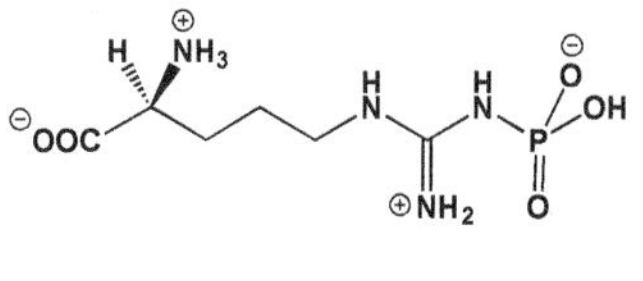

This metabolite (FW = 254.18 g/mol; CAS 1189-11-3; pK_1 = 2.0, pK_2 = 4.5, pK_3 = 9.4, and pK_4 = 11.2), formerly called arginine phosphate, is the most common phosphagen in invertebrates. It is produced by the action of arginine kinase and is a substrate of phosphoamidase and alkaline phosphatase. Complete hydrolysis occurs after one minute at 100°C in 0.1 M HCl. **Target(s):** adenylate kinase, weakly inhibited (1); creatine kinase (2); 6-phosphofructokinase (3); and pyruvate kinase (4-7). **1**. Storey (1976) *J. Biol. Chem.* **251**, 7810. **2**. Watts (1973) *The Enzymes*, 3rd ed. (Boyer, ed.), **8**, 383. **3**. Storey (1982) *Meth. Enzymol.* **90**, 39. **4**. Giles, Poat & Munday (1980) *Biochem. Soc. Trans.* **8**, 143. **5**. Wu, Wong & Yeung (1979) *Comp. Biochem. Physiol. B* **63**, 29. **6**. Plaxton & Storey (1984) *Eur. J. Biochem.* **143**, 257. **7**. Storey (1985) *J. Comp. Physiol. B* **155**, 339.

Phosphoaspartate, *See* β-Aspartyl Phosphate

Phosphocholine

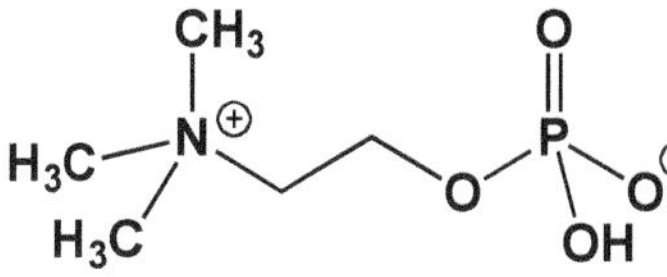

This choline phosphomonoester ($FW_{zwitterion}$ = 183.14 g/mol; CAS 107-73-3), also known as choline phosphate and phosphorylcholine, is a key intermediate in the biosynthesis of many phospholipids. Both the free acid and the calcium salt are very soluble in water. The free acid is relatively stable in acidic conditions: only 15% is hydrolyzed in five hours in 1 M HCl at 100°C. Under alkaline conditions, there is complete hydrolysis after four hours by refluxing with barium hydroxide. The phosphocholine chloride calcium salt is stable for months at room temperature. **Target(s):** 1-alkyl-2-acetylglycerophosphocholine esterase, *or* platelet-activating-factor acetylhydrolase (1); choline-sulfatase (2,3); choline sulfotransferase (4); ethanolamine kinase (5-7); ethanolamine-phosphate cytidylyltransferase (8); glutamate decarboxylase (9); glycerophosphocholine cholinephospho-diesterase (10); ornithine decarboxylase (9); phosphoethanolamine *N*-methyltransferase (11-14); phosphoserine phosphatase (15); and sphingomyelin phosphodiesterase, *or* sphingomyelinase (16-18). **1**. Chroni & Mavri-Vavayanni (2000) *Life Sci.* **67**, 2807. **2**. Lucas, Burchiel & Segel (1972) *Arch. Biochem. Biophys.* **153**, 664. **3**. Osteras, Boncompagni, Vincent, Poggi & Le Rudulier (1998) *Proc. Natl. Acad. Sci. U.S.A.* **95**, 11394. **4**. Renosto & Segel (1977) *Arch. Biochem. Biophys.* **180**, 416. **5**. Weinhold & Rethy (1972) *Biochim. Biophys. Acta* **276**, 143. **6**. Sung & Johnstone (1967) *Biochem. J.* **105**, 497. **7**. Brophy, Choy, Toone & Vance (1977) *Eur. J. Biochem.* **78**, 491. **8**. Vermeulen, Geelen & van Golde (1994) *Biochim. Biophys. Acta* **1211**, 343. **9**. Gilad & Gilad (1984) *Biochem. Biophys. Res. Commun.* **122**, 277. **10**. Sok (1998) *Neurochem. Res.* **23**, 1061. **11**. Smith, Summers & Weretilnyk (2000) *Physiol. Plant.* **108**, 286. **12**. Brendza, Haakenson, Cahoon, *et al.* (2007) *Biochem. J.* **404**, 439. **13**. Pessi, Kociubinski & Mamoun (2004) *Proc. Natl. Acad. Sci. U.S.A.* **101**, 6206. **14**. Nuccio, Ziemak, Henry, Weretilnyk & Hanson (2000) *J. Biol. Chem.* **275**, 14095. **15**. Hawkinson, Acosta-Burruel, Ta & Wood (1997) *Eur. J. Pharmacol.* **337**, 315. **16**. Callahan, Jones, Davidson & Shankaran (1983) *J. Neurosci. Res.* **10**, 151. **17**. Duan & Nilsson (1997) *Hepatology* **26**, 823. **18**. Heller & Shapiro (1966) *Biochem. J.* **98**, 763.

Phospho-CPI-17, *See CPI-17*

Phosphocreatine

This phosphagen (FW = 211.11 g/mol; CAS 67-07-2; 922-32-7 for disodium salt), known almost as commonly as creatine phosphate, is the primary cytosolic phosphoryl donor for the resynthesis of MgATP from MgADP through the action of creatine kinase in vertebrates and some invertebrates. While abundant in skeletal and cardiac muscle (40-60 mM), nonmuscle cells contain 3-5 mM, and some fungi (*e.g.*, *Candida*) have intermediate levels of this metabolite. *Note*: The reported inhibitory action by creatine kinase is often due to impurities (*e.g.*, the inhibition of 5′-nucleotidase is due to an impurity in the commercial phosphocreatine). The reported inhibition of AMP deaminase is due mainly to the presence of pyrophosphate (1). Phosphocreatine is a hygroscopic solid and very soluble in water. Unstable in acid, all is hydrolyzed after one minute at 100°C in 1.0 M HCl, and half is hydrolyzed in 4 min at 25°C in 0.5 M HCl. Molybdate and certain buffers reportedly accelerate hydrolysis. Hydrolysis is slow under slightly alkaline conditions (*e.g.*, pH 7.8). Nevertheless, solutions should always be freshly prepared and stored frozen. **Target(s)**: adenylate cyclase (1); AMP deaminase (2-4); glutamine synthetase (5); H⁺-extrusion by plasma membrane ATPase (6); NADH peroxidase (7); phosphofructokinase (8,9); and thiamin-phosphate diphosphorylase (10-12). **1**. Khan, Quemener & Moulinoux (1990) *Life Sci.* **46**, 43. **2**. Wheeler & Lowenstein (1979) *J. Biol. Chem.* **254**, 1484. **3**. Coffee (1978) *Meth. Enzymol.* **51**, 490. **4**. Zielke & Suelter (1971) *The Enzymes*, 3rd ed., **4**, 47. **5**. Wu (1977) *Can. J. Biochem.* **55**, 332. **6**. Manzoor, Amin & Khan (2002) *Indian J. Exp. Biol.* **40**, 785. **7**. Badwey & Karnovsky (1979) *J. Biol. Chem.* **254**, 11530. **8**. Kemp (1975) *Meth. Enzymol.* **42**, 67. **9**. Bloxham & Lardy (1973) *The Enzymes*, 3rd ed., **8**, 239. **10**. Penttinen (1979) *Meth. Enzymol.* **62**, 68. **11**. Kawasaki & Esaki (1970) *Biochem. Biophys. Res. Commun.* **40**, 1468. **12**. Kayama & Kawasaki (1973) *Arch. Biochem. Biophys.* **158**, 242.

Phosphoenol-3-fluoropyruvic Acid, *See 3-Fluoro-phosphoenolpyruvic Acid*

Phosphoenol-α-ketobutyric Acid

(E)-isomer   **(Z)-isomer**

Both the (*Z*)- and (*E*)-isomers of this phosphoenolpyruvate homologue (FW$_{\text{free-acid}}$ = 182.07 g/mol) are alternative substrates for pyruvate kinase. There is less than a three-fold difference between the two $V_{\text{max}}/K_{\text{m}}$ ratios for

the two isomers (note that the V_{max} values are substantially smaller than that of phosphoenolpyruvate (PEP), but the K_{m} value of (*Z*)-phosphoenol-α-ketobutyrate is close to that of PEP (1). Both isomers inhibit phosphoenolpyruvate carboxykinase, with the (*E*)-isomer being the more potent inhibitor, and neither isomer is a substrate for this enzyme (2,3). The (*Z*)-isomer is the more effective inhibitor of enolase, with the (*E*)-isomer becoming a better inhibitor in the presence of Mg²⁺ (3). Both isomers also inhibited plant phosphoenolpyruvate carboxylase in the presence of 5 mM Mg²⁺, with the (*Z*)-isomer having the smaller K_{i} value. In the presence of 0.5 mM Mn²⁺, the (*Z*)-isomer is still effective as inhibitor while the (*E*)-isomer produced activation by binding at an allosteric site. Both compounds were alternative substrates of the enzyme with similar $V_{\text{max}}/K_{\text{m}}$ values. **Target(s)**: 3-deoxy-7-phosphoheptulonate synthase (4,5); enolase (3); phosphoenolpyruvate carboxykinase (2,3); phosphoenolpyruvate carboxylase (6); pyruvate kinase, K_{i} = 65 µM (3,7-9). **1**. Adlersberg, Dayan, Bondinell & Sprinson (1977) *Biochemistry* **16**, 4382. **2**. Duffy, Markovitz, Chuang, Utter & Nowak (1981) *Proc. Natl. Acad. Sci. U.S.A.* **78**, 6680. **3**. Duffy, Saz & Nowak (1982) *Biochemistry* **21**, 132. **4**. Hu & Sprinson (1977) *J. Bacteriol.* **129**, 177. **5**. Simpson & Davidson (1976) *Eur. J. Biochem.* **70**, 501. **6**. Gonzalez & Andreo (1988) *Eur. J. Biochem.* **173**, 339. **7**. Jursinic & Robinson (1978) *Biochim. Biophys. Acta* **523**, 358. **8**. James, Reuben & Cohn (1973) *J. Biol. Chem.* **248**, 6443. **9**. Woods, O'Bryan, Mui & Crowder (1970) *Biochemistry* **9**, 2334.

Phosphoenolpyruvate

This key intermediate in glycolysis and gluconeogenesis (FW$_{\text{free-acid}}$ = 168.04 g/mol; CAS 73-89-2), frequently abbreviated as PEP, is the phosphate ester formed from the hydroxyl group of the enol tautomer of pyruvate. The $\Delta G°$ value for hydrolysis is –62 kJ/mol at pH 7. The half-life of PEP is 8.3 min in 1.0 N HCl at 100°C. PEP is also readily hydrolyzed by base. Solutions should be immediately neutralized after dissolution of the vacuum-dessicated powder and never stored beyond a few days. **Target(s)**: acetolactate synthase (131); acylphosphatase (57,58); adenylate kinase (78,79); adenylosuccinate synthetase (1); cAMP-dependent protein kinase phosphorylation of pyruvate kinase (2); diglucosyl diacylglycerol synthase (124); farnesyl-diphosphate kinase, weakly inhibited (77); fructose-1,6-bisphosphatase (Note: PEP is an activator of many FBPases.) (66-68); fructose-1,6-bisphosphate aldolase (51,52); 1,3-β-glucan synthase (125); glucose-1,6-bisphosphate synthase (80-82); glucose-1-phosphate adenylyltransferase (3,74-76); glucose-6-phosphate isomerase (4); glycerol-3-phosphate dehydrogenase (5); glycogen synthase (6); guanylate cyclase (7-9); hexokinase (122); [isocitrate dehydrogenase (NADP⁺)] kinase (10,69); isocitrate lyase (11,12,42-50); L-lactate dehydrogenase (13); malate synthase (128-130); methylglyoxal synthase (14,36-41); NADH kinase (106,107); NADH peroxidase (15); NAD⁺ kinase, weakly inhibited (109); oxaloacetate decarboxylase (54,55); oxaloacetate tautomerase (16); 6-phosphofructokinase (17-24,115-121); 6-phosphofructo-2-kinase (83-100); 6-phospho-β-galactosidase (62,63); phosphoglucomutase (25); 6-phospho-β-glucosidase (61); phosphoketolase (53); phosphonopyruvate hydrolase (56); phosphoprotein phosphatase (64,65); phosphoribulokinase (110-114); protein phosphatase 2A (65); [protein-PII] uridylyltransferase (73); pyrophosphate-dependent phosphofructokinase, *or* diphosphate: fructose-6-phosphate 1-phosphotransferase (101-105); pyruvate carboxylase (26); pyruvate kinase (2,27,108); pyruvate,orthophosphate dikinase (71,72); pyruvate,water dikinase (70); ribulose-5-phosphate kinase (28); starch phosphorylase (127); sucrose-phosphate synthase, weakly inhibited (126); thiamin-phosphate diphosphorylase (29,30,123); threonine ammonia-lyase, *or* threonine dehydratase (31); triose-phosphate isomerase (4,32-35); and Xaa-Pro Dipeptidase, *or* prolidase, K_{i} = 8.5 nM (59,60). **1**. Spector, Jones & Elion (1979) *J. Biol. Chem.* **254**, 8422. **2**. El-Maghrabi, Haston, Flockhart, Claus & Pilkis (1980) *J. Biol. Chem.* **255**, 668. **3**. Preiss (1973) *The Enzymes*, 3rd ed., **8**, 73. **4**. Noltmann (1972) *The Enzymes*, 3rd ed. (Boyer, ed.), **6**, 271. **5**. Schryvers, Lohmeier & Weiner (1978) *J. Biol. Chem.* **253**, 783. **6**. Rothman & Cabib (1967) *Biochemistry* **6**, 2107. **7**. White & Zenser (1974) *Meth. Enzymol.* **38**, 192. **8**. Hardman & Sutherland (1969) *J. Biol. Chem.* **244**, 6363. **9**. de Jonge (1975) *FEBS Lett.* **55**, 143. **10**. Nimmo (1984) *Trends Biochem. Sci.* **9**, 475. **11**. McFadden (1969)

Meth. Enzymol. **13**, 163. **12**. Spector (1972) *The Enzymes*, 3rd ed., 7, 357. **13**. Oba & Uritani (1982) *Meth. Enzymol.* **89**, 345. **14**. Cooper (1975) *Meth. Enzymol.* **41**, 502. **15**. Badwey & Karnovsky (1979) *J. Biol. Chem.* **254**, 11530. **16**. Belikova, Burov & Vinogradov (1988) *Biochim. Biophys. Acta* **936**, 10. **17**. Kemp (1975) *Meth. Enzymol.* **42**, 67. **18**. Kemerer, Griffin & Brand (1975) *Meth. Enzymol.* **42**, 91. **19**. Kotlarz & Buc (1982) *Meth. Enzymol.* **90**, 60. **20**. Marschke & Bernlohr (1982) *Meth. Enzymol.* **90**, 70. **21**. Fordyce, Moore & Pritchard (1982) *Meth. Enzymol.* **90**, 77. **22**. Bloxham & Lardy (1973) *The Enzymes*, 3rd ed., 8, 239. **23**. Tlapak-Simmons & Reinhart (1994) *Arch. Biochem. Biophys.* **308**, 226. **24**. Ortigosa, Kimmel & Reinhart (2004) *Biochemistry* **43**, 57. **25**. Chiba, Ueda & Hirose (1976) *Agric. Biol. Chem.* **40**, 2423. **26**. Scrutton & Young (1972) *The Enzymes*, 3rd ed., 6, 1. **27**. Del Valle, Busto, De Arriaga & Soler (1990) *J. Enzyme Inhib.* **3**, 21. **28**. Hart & Gibson (1975) *Meth. Enzymol.* **42**, 115. **29**. Penttinen (1979) *Meth. Enzymol.* **62**, 68. **30**. Kawasaki & Esaki (1970) *Biochem. Biophys. Res. Commun.* **40**, 1468. **31**. Choi & Kim (1995) *J. Biochem. Mol. Biol.* **28**, 118. **32**. Krietsch (1975) *Meth. Enzymol.* **41**, 434. **33**. Krietsch (1975) *Meth. Enzymol.* **41**, 438. **34**. Tomlinson & Turner (1979) *Phytochemistry* **18**, 1959. **35**. Lambeir, Opperdoes & Wierenga (1987) *Eur. J. Biochem.* **168**, 69. **36**. Murata, Fukuda, Watanabe, *et al.* (1985) *Biochem. Biophys. Res. Commun.* **131**, 190. **37**. Hopper & Cooper (1971) *FEBS Lett.* **13**, 213. **38**. Hopper & Cooper (1972) *Biochem. J.* **128**, 321. **39**. Cooper (1974) *Eur. J. Biochem.* **44**, 81. **40**. Ray & Ray (1981) *J. Biol. Chem.* **256**, 6230. **41**. Huang, Rudolph & Bennett (1999) *Appl. Environ. Microbiol.* **65**, 3244. **42**. Tsukamoto, Ejiri & Katsumata (1986) *Agric. Biol. Chem.* **50**, 409. **43**. Tanaka, Nabeshima, Tokuda & Fukui (1977) *Agric. Biol. Chem.* **41**, 795. **44**. MacKintosh & Nimmo (1988) *Biochem. J.* **250**, 25. **45**. Takao, Takahashi, Tanida & Takahashi (1984) *J. Ferment. Technol.* **62**, 577. **46**. Giachetti, Pinzauti, Bonaccorsi, Vincenzini & Vanni (1987) *Phytochemistry* **26**, 2439. **47**. Johanson, Hill & McFadden (1974) *Biochim. Biophys. Acta* **364**, 327. **48**. Chell, Sundaram & Wilkinson (1978) *Biochem. J.* **173**, 165. **49**. Vanni, Vicenzini, Nerozzi & Sinha (1979) *Can. J. Biochem.* **57**, 1131. **50**. Hönes, Simon & Weber (1991) *J. Basic Microbiol.* **31**, 251. **51**. Moorhead & Plaxton (1990) *Biochem. J.* **269**, 133. **52**. Botha & O'Kennedy (1989) *J. Plant Physiol.* **135**, 433. **53**. Whitworth & Ratledge (1977) *J. Gen. Microbiol.* **102**, 397. **54**. Labrou & Clonis (1999) *Arch. Biochem. Biophys.* **365**, 17. **55**. Dean & Bartley (1973) *Biochem. J.* **135**, 667. **56**. Chen, Han, Niu, *et al.* (2006) *Biochemistry* **45**, 11491. **57**. Pazzagli, Ikram, Liguri, *et al.* (1993) *Ital. J. Biochem.* **42**, 233. **58**. Liguri, Camici, Manao, *et al.* (1986) *Biochemistry* **25**, 8089. **59**. Radzicka & Wolfenden (1990) *J. Amer. Chem. Soc.* **112**, 1248. **60**. Mock & Liu (1995) *Bioorg. Med. Chem. Lett.* **5**, 627. **61**. Wilson & Fox (1974) *J. Biol. Chem.* **249**, 5586. **62**. Calmes & Brown (1979) *Infect. Immun.* **23**, 68. **63**. Thompson (2002) *FEMS Microbiol. Lett.* **214**, 183. **64**. Zhou, Clemens, Hakes, Barford & Dixon (1993) *J. Biol. Chem.* **268**, 17754. **65**. Dong, Ermolova & Chollet (2001) *Planta* **213**, 379. **66**. Kruger & Beevers (1984) *Plant Physiol.* **76**, 49. **67**. Rittmann, Schaffer, Wendisch & Sahm (2003) *Arch. Microbiol.* **180**, 285. **68**. Babul & Guixe (1983) *Arch. Biochem. Biophys.* **225**, 944. **69**. Nimmo & Nimmo (1984) *Eur. J. Biochem.* **141**, 409. **70**. Chulavatnatol & Atkinson (1973) *J. Biol. Chem.* **248**, 2712. **71**. Schwitzguebel & Ettlinger (1979) *Arch. Microbiol.* **122**, 103. **72**. Jenkins & Hatch (1985) *Arch. Biochem. Biophys.* **239**, 53. **73**. Engleman & Francis (1978) *Arch. Biochem. Biophys.* **191**, 602. **74**. Preiss (1978) *Adv. Enzymol. Relat. Areas Mol. Biol.* **46**, 317. **75**. Meyer, Borra, Igarashi, Lin & Springsteel (1999) *Arch. Biochem. Biophys.* **372**, 179. **76**. Amir & Cherry (1972) *Plant Physiol.* **49**, 893. **77**. Shechter (1974) *Biochim. Biophys. Acta* **362**, 233. **78**. Storey (1976) *J. Biol. Chem.* **251**, 7810. **79**. McKellar, Charles & Butler (1980) *Arch. Microbiol.* **124**, 275. **80**. Rose, Warms & Kaklij (1975) *J. Biol. Chem.* **250**, 3466. **81**. Rose, Warms & Wong (1977) *J. Biol. Chem.* **252**, 4262. **82**. Ueda, Hirose, Sasaki & Chiba (1978) *J. Biochem.* **83**, 1721. **83**. Markham & Kruger (2002) *Eur. J. Biochem.* **269**, 1267. **84**. Villadsen & Nielsen (2001) *Biochem. J.* **359**, 591. **85**. Kretschmer & Hofmann (1984) *Biochem. Biophys. Res. Commun.* **124**, 793. **86**. García de Frutos & Baanante (1994) *Arch. Biochem. Biophys.* **308**, 461. **87**. Harmsen, Kubicek-Pranz, Röhr, Visser & Kubicek (1992) *Appl. Microbiol. Biotechnol.* **37**, 784. **88**. Aragón, Gómez & Ganccdo (1987) *FEBS Lett.* **226**, 121. **89**. Bedri, Kretschmer, Schellenberger & Hofmann (1989) *Biomed. Biochim. Acta* **48**, 403. **90**. Laloux, Van Schaftingen, Francois & Hers (1985) *Eur. J. Biochem.* **148**, 155. **91**. Manes & el-Maghrabi (2005) *Arch. Biochem. Biophys.* **438**, 125. **92**. Sakurai, Johnson & Uyeda (1996) *Biochem. Biophys. Res. Commun.* **218**, 159. **93**. Rider (1987) *Biochem. Soc. Trans.* **15**, 988. **94**. Vásquez-Illanes, Barcia, Ibarguren, Villamarin & Ramos-Martinez (1992) *Mar. Biol.* **112**, 277. **95**. Larondelle, Mertens, Van Schaftingen & Hers (1986) *Eur. J. Biochem.* **161**, 351. **96**. Loiseau, Rider & Hue (1987) *Biochem. Soc. Trans.* **15**, 384. **97**. Pyko, Rider, Hue & Wegener (1993) *J. Comp. Physiol. B* **163**, 89. **98**. Chevalier, Bertrand, Rider, *et al.* (2005) *FEBS J.* **272**, 3542. **99**. Francois, Van Schaftingen & Hers (1988) *Eur. J. Biochem.* **171**, 599. **100**. Vásquez-Illanes & Ramos-Martinez (1991) *FEBS Lett.* **295**, 176. **101**. Theodorou & Plaxton (1996) *Plant Physiol.* **112**, 343. **102**. Kombrink & Kruger (1984) *Z. Pflanzenphysiol.* **114**, 443. **103**. Theodorou & Kruger (2001) *Planta* **213**, 147. **104**. Pfleiderer & Klemme (1980) *Z. Naturforsch. C* **35c**, 229. **105**. Kowalczyk, Januszewska, Cymerska & Maslowski (1984) *Physiol. Plant* **60**, 31. **106**. Iwahashi, Hitoshio, Tajima & Nakamura (1989) *J. Biochem.* **105**, 588. **107**. Iwahashi & Nakamura (1989) *J. Biochem.* **105**, 922. **108**. Saavedra, Olivos, Encalada & Moreno-Sanchez (2004) *Exp. Parasitol.* **106**, 11. **109**. Blomquist (1973) *J. Biol. Chem.* **248**, 7044. **110**. Marsden & Codd (1984) *J. Gen. Microbiol.* **130**, 999. **111**. Miziorko (2000) *Adv. Enzymol. Relat. Areas Mol. Biol.* **74**, 95. **112**. Abdelal & Schlegel (1974) *Biochem. J.* **139**, 481. **113**. Rippel & Bowien (1984) *Arch. Microbiol.* **139**, 207. **114**. MacElroy, Mack & Johnson (1972) *J. Bacteriol.* **112**, 532. **115**. Turner & Plaxton (2003) *Planta* **217**, 113. **116**. Botha & Turpin (1990) *Plant Physiol.* **93**, 871. **117**. Cawood, Botha & Small (1988) *J. Plant Physiol.* **132**, 204. **118**. Kombrink & Wöber (1982) *Arch. Biochem. Biophys.* **213**, 602. **119**. Häusler, Holtum & Latzko (1989) *Plant Physiol.* **90**, 1498. **120**. Knowles, Greyson & Dennis (1990) *Plant Physiol.* **92**, 155. **121**. Mediavilla, Meton & Baanante (2007) *Biochim. Biophys. Acta* **1770**, 706. **122**. Magnani, Stocchi, Serafini, *et al.* (1983) *Arch. Biochem. Biophys.* **226**, 377. **123**. Kayama & Kawasaki (1973) *Arch. Biochem. Biophys.* **158**, 242. **124**. Vikström, Li & Wieslander (2000) *J. Biol. Chem.* **275**, 9296. **125**. Marechal & Goldemberg (1964) *J. Biol. Chem.* **239**, 3163. **126**. Doehlert & Huber (1983) *Plant Physiol.* **73**, 989. **127**. Sutton (1975) *Aust. J. Plant Physiol.* **2**, 403. **128**. Munir, Hattori & Shimada (2002) *Biosci. Biotechnol. Biochem.* **66**, 576. **129**. Fukawa, Ejiri & Katsumata (1987) *Agric. Biol. Chem.* **51**, 1553. **130**. Smith, Huang, Miczak, *et al.* (2003) *J. Biol. Chem.* **278**, 1735. **131**. Phalip, Schmitt & Divies (1995) *Curr. Microbiol.* **31**, 316.

4-Phosphoerythronate

4-phospho-
D-erythronate   4-phospho-
L-erythronate

This phosphorylated aldonic acid (FW$_{\text{free-acid}}$ = 216.08 g/mol), also known as erythronic acid 4-phosphate, inhibits arabinose-5-phosphate isomerase (1), glucose-6-phosphate isomerase (2), ribose-5-phosphate isomerase (3-5), and ribulose-bisphosphate carboxylase/oxygenase (6). **1**. Bigham, Gragg, Hall, *et al.* (1984) *J. Med. Chem.* **27**, 717. **2**. Chirgwin & Noltmann (1975) *J. Biol. Chem.* **250**, 7272. **3**. Woodruff III & Wolfenden (1979) *J. Biol. Chem.* **254**, 5866. **4**. Roos, Burgos, Ericsson, Salmon & Mowbray (2005) *J. Biol. Chem.* **280**, 6416. **5**. Burgos & Salmon (2004) *Tetrahedron Lett.* **45**, 753. **6**. Purohit, Becker & Evans (1982) *Biochim. Biophys. Acta* **715**, 230.

O-Phosphoethanolamine

This phosphorylated metabolite (FW = 141.06 g/mol; CAS 1071-23-4; 10389-08-9 (Ca[1:1] salt); 24027-75-6 (^{3}H-labeled); 37785-54-9 (Zn salt); 34851-96-2 (Mg salt)), also called ethanolamine phosphate, phosphoryl-ethanolamine, and 2-aminoethanol O-phosphate, is a precursor in the biosynthesis of phosphatidylethanolamine. Both the zwitterion and the sodium salt are very soluble in water (20 g/100 mL and 75 g/100 mL at 20°C, respectively): the amino group has a pK_a value of 10.20. O-Phosphoethanolamine is relatively stable in both acidic and basic conditions. After five hours at 100°C in 1.0 M HCl, 5% has hydrolyzed. **Target(s):** D-alanyl-D-alanine synthetase, *or* D-alanine:D-alanine ligase (1); carnosine synthetase, weakly inhibited (2); choline-phosphate cytidylyltransferase, weak alternative substrate (3); glutamate decarboxylase (4); ornithine decarboxylase (4); and sphingomyelinase (5). **1**. Lacoste, Poulsen, Cassaigne & Neuzil (1979) *Curr. Microbiol.* **2**, 113. **2**.

Seely & Marshall (1982) *Life Sci.* **30**, 1763. **3**. Jamil & Vance (1991) *Biochim. Blophys. Acta* **1086**, 335. **4**. Gilad & Gilad (1984) *Biochem. Biophys. Res. Commun.* **122**, 277. **5**. Callahan, Jones, Davidson & Shankaran (1983) *J. Neurosci. Res.* **10**, 151.

6-Phospho-D-gluconate

$$
\begin{array}{c}
\text{COO}^{\ominus} \\
\text{HC} - \text{OH} \\
\text{HO} - \text{CH} \\
\text{HC} - \text{OH} \\
\text{HC} - \text{OH} \\
\text{CH}_2\text{OPO}_3\text{H}^{\ominus}
\end{array}
$$

This water-soluble metabolite (FW$_{\text{free-acid}}$ = 276.14 g/mol), also called D-gluconic acid 6-phosphate, is a key intermediate in the pentose phosphate and Entner-Doudoroff pathways. It lactonizes upon heating to 70°C. **Target(s):** cellobiose phosphorylase (1); fructose-1,6-bisphosphate aldolase (46); fructose-2,6-bisphosphate 2-phosphatase (51,52); glucose-6-phosphate dehydrogenase (2); glucose-6-phosphate isomerase (3-20,37-41,60); glyceraldehyde-3-phosphate dehydrogenase (21); 1L-*myo*-inositol-1-phosphatase (22); inositol-3-phosphate synthase, *or* *myo*-inositol-1-phosphate synthase (22-25); isocitrate lyase (26); L-lactate dehydrogenase (27); levansucrase (61); mannose-6-phosphate isomerase (6,19,20); phosphoenolpyruvate carboxykinase (28); phosphofructokinase (29); phosphoglucomutase (30); phosphoglycolate phosphatase (53); phosphoribulokinase (55-59); pyruvate kinase (54); ribose-5-phosphate isomerase (31,42-44); ribulose-1,5-bisphosphate carboxylase/oxygenase (32-36,47-50);and triose-phosphate isomerase, weakly inhibited (45). **1**. Sasaki (1988) *Meth. Enzymol.* **160**, 468. **2**. Anderson, Wise & Anderson (1997) *Biochim. Biophys. Acta* **1340**, 268. **3**. Nosoh (1975) *Meth. Enzymol.* **41**, 383. **4**. Hizukuri, Takeda & Nikuni (1975) *Meth. Enzymol.* **41**, 388. **5**. Gracy & Tilley (1975) *Meth. Enzymol.* **41**, 392. **6**. Noltmann (1972) *The Enzymes*, 3rd ed., **6**, 271. **7**. Webb (1966) *Enzyme and Metabolic Inhibitors*, vol. **2**, p. 406, Academic Press, New York. **8**. Ruijter & Visser (1999) *Biochimie* **81**, 267. **9**. Sangwan & Singh (1990) *Indian J. Biochem. Biophys.* **27**, 23 and (1989) *J. Biosci.* **14**, 47. **10**. Gaitonde, Murray & Cunningham (1989) *J. Neurochem.* **52**, 1348. **11**. Zera (1987) *Biochem. Genet.* **25**, 205. **12**. Van Beneden & Powers (1985) *J. Biol. Chem.* **260**, 14596. **13**. Takama & Nosoh (1982) *Biochim. Biophys. Acta* **705**, 127. **14**. Howell & Schray (1981) *Mol. Cell. Biochem.* **37**, 101. **15**. Takama & Nosoh (1980) *J. Biochem.* **87**, 1821. **16**. Chirgwin & Noltmann (1975) *J. Biol. Chem.* **250**, 7272. **17**. Parr (1956) *Nature* **178**, 1401. **18**. Backhausen, Jöstingmeyer & Scheibe (1997) *Plant Sci.* **130**, 121. **19**. Lee & Matheson (1984) *Phytochemistry* **23**, 983. **20**. Hansen, Wendorff & Schonheit (2004) *J. Biol. Chem.* **279**, 2262. **21**. Rivers & Blevins (1987) *J. Gen. Microbiol.* **133**, 3159. **22**. Charalampous & Chen (1966) *Meth. Enzymol.* **9**, 698. **23**. Wong, Mauck & Sherman (1982) *Meth. Enzymol.* **90**, 309. **24**. Barnett, Rasheed & Corina (1973) *Biochem. Soc. Trans.* **1**, 1267. **25**. Naccarato, Ray & Wells (1974) *Arch. Biochem. Biophys.* **164**, 194. **26**. Reinscheid, Eikmanns & Sahm (1994) *J. Bacteriol.* **176**, 3474. **27**. Gordon & Doelle (1976) *Eur. J. Biochem.* **67**, 543. **28**. Kaloyianni (1991) *Experientia* **47**, 248. **29**. Bloxham & Lardy (1973) *The Enzymes*, 3rd ed. (Boyer, ed.), **8**, 239. **30**. Ray & Peck (1972) *The Enzymes*, 3rd ed. (Boyer, ed.), **6**, 407. **31**. MacElroy & Middaugh (1982) *Meth. Enzymol.* **89**, 571. **32**. Schloss, Phares, Long, *et al.* (1982) *Meth. Enzymol.* **90**, 522. **33**. Lee, Read & Tabita (1991) *Arch. Biochem. Biophys.* **291**, 263. **34**. Shively, Saluja & McFadden (1978) *J. Bacteriol.* **134**, 1123. **35**. Gibson & Tabita (1977) *J. Biol. Chem.* **252**, 943. **36**. Tabita & McFadden (1976) *J. Bacteriol.* **126**, 1271. **37**. Jeong, Fushinobu, Ito, *et al.* (2003) *FEBS Lett.* **535**, 200. **38**. Marchand, Kooystra, Wierenga, *et al.* (1989) *Eur. J. Biochem.* **184**, 455. **39**. Hansen, Schlichting, Felgendreher & Schonheit (2005) *J. Bacteriol.* **187**, 1621. **40**. Thomas (1981) *J. Gen. Microbiol.* **124**, 403. **41**. Mathur, Ahsan, Tiwari & Garg (2005) *Biochem. Biophys. Res. Commun.* **337**, 626. **42**. Horitsu, Sasaki, Kikuchi, *et al.* (1976) *Agric. Biol. Chem.* **40**, 257. **43**. Skrukrud, Gordon, Dorwin, *et al.* (1991) *Plant Physiol.* **97**, 730. **44**. Middaugh & MacElroy (1976) *J. Biochem.* **79**, 1331. **45**. Tomlinson & Turner (1979) *Phytochemistry* **18**, 1959. **46**. Moorhead & Plaxton (1990) *Biochem. J.* **269**, 133. **47**. Heda & Madigan (1989) *Eur. J. Biochem.* **184**, 313. **48**. Bowien (1977) *FEMS Microbiol. Lett.* **2**, 263. **49**. Taylor & Dow (1980) *J. Gen. Microbiol.* **116**, 81. **50**. Holuigue, Herrera, Phillips, Young & Alende (1987) *Biotechnol. Appl. Biochem.* **9**, 497. **51**. Sakurai, Johnson & Uyeda (1996) *Biochem. Biophys. Res. Commun.* **218**, 159. **52**. Villadsen &

Nielsen (2001) *Biochem. J.* **359**, 591. **53**. Verin-Vergeau, Baldy & Cavalie (1980) *Phytochemistry* **19**, 763. **54**. De Médicis, Laliberté & Vass-Marengo (1982) *Biochim. Biophys. Acta* **708**, 57. **55**. Shabnam, Saharan & Singh (1993) *J. Plant Biochem. Biotechnol.* **2**, 121. **56**. Runquist, Ríos, Vinarov & Miziorko (2001) *Biochemistry* **40**, 14530. **57**. Runquist, Harrison & Miziorko (1999) *Biochemistry* **38**, 13999. **58**. Wadano, Nishikawa, Hirashi, Satoh & Kwaki (1998) *Photosynth. Res.* **56**, 27. **59**. Runquist & Miziorko (2006) *Protein Sci.* **15**, 837. **60**. Milewski, Janiak & Wojciechowski (2006) *Arch. Biochem. Biophys.* **450**, 39. **61**. Lyness & Doelle (1983) *Biotechnol. Lett.* **5**, 345.

D-2-Phosphoglycerate (D-2-Phosphoglyceric Acid)

$$
\begin{array}{c}
\text{COO}^{\ominus} \\
\text{HC} - \text{OPO}_3\text{H}^{\ominus} \\
\text{CH}_2\text{OH}
\end{array}
$$

This metabolite (FW$_{\text{free-acid}}$ = 186.06 g/mol; CAS 2553-59-5; 23295-92-3 for (*S*)-isomer; 3443-57-0 for (*R*)-isomer), also called 2-phospho-D-glycerate, D-glycerate 2-phosphate, and (*R*)-2,3-dihydroxypropionate 2-(dihydrogen phosphate), is an important intermediate in carbohydrate metabolism. D-2-Phosphoglycerate is slowly hydrolyzed by acids and salts are readily soluble in water and. It is a substrate or product in the reactions catalyzed by phosphoglycerate phosphatase, enolase, and phosphoglycerate mutase. D-2-Phosphoglycerate is an ADPglucose synthetase activator (1). **Target(s):** bisphosphoglycerate mutase (2); bisphosphoglycerate phosphatase (2-4); 3-deoxy-7-phosphoheptulonate synthase, *or* 3-deoxy-D-*arabino*heptulosonate-7-phosphate synthetase (5); glycerol-3-phosphate dehydrogenase (6); hexokinase (7,8); methylglyoxal synthase (9); NADH peroxidase (10); phosphoenolpyruvate carboxylase (11); phosphoenol-pyruvate:protein phosphotransferase (12); 6-phosphofructokinase (13-18); 6-phosphofructo-2-kinase (19-21); 6-phospho-β-galactosidase (22); phosphoglycerate kinase (23); pyrophosphatase, *or* inorganic diphosphatase (24); pyruvate kinase (25-27); and triose-phosphate isomerase (28). **1**. Gardiol & Preiss (1990) *Arch. Biochem. Biophys.* **280**, 175. **2**. Harkness, Isaacks & Roth (1977) *Eur. J. Biochem.* **78**, 343. **3**. Webb (1966) *Enzyme and Metabolic Inhibitors*, vol. **2**, p. 413, Academic Press, New York. **4**. Sasaki, Hirose, Sugimoto & Chiba (1971) *Biochim. Biophys. Acta* **227**, 595. **5**. Simpson & Davidson (1976) *Eur. J. Biochem.* **70**, 501. **6**. Schryvers, Lohmeier & Weiner (1978) *J. Biol. Chem.* **253**, 783. **7**. Cesar, Colepicolo, Rosa & Rosa (1997) *Comp. Biochem. Physiol. B* **118**, 395. **8**. Magnani, Stocchi, Serafini, *et al.* (1983) *Arch. Biochem. Biophys.* **226**, 377. **9**. Murata, Fukuda, Watanabe, *et al.* (1985) *Biochem. Biophys. Res. Commun.* **131**, 190. **10**. Badwey & Karnovsky (1979) *J. Biol. Chem.* **254**, 11530. **11**. Schuller, Turpin & Plaxton (1990) *Plant Physiol.* **94**, 1429. **12**. Saier, Jr., Schmidt & Lin (1980) *J. Biol. Chem.* **255**, 8579. **13**. Shimizu, Kono, Mineo, *et al.* (1984) *J. Inherit. Metab. Dis.* **7**, 107. **14**. Bloxham & Lardy (1973) *The Enzymes*, 3rd ed. (Boyer, ed.), **8**, 239. **15**. Turner & Plaxton (2003) *Planta* **217**, 113. **16**. Cawood, Botha & Small (1988) *J. Plant Physiol.* **132**, 204. **17**. Kombrink & Wöber (1982) *Arch. Biochem. Biophys.* **213**, 602. **18**. Häusler, Holtum & Latzko (1989) *Plant Physiol.* **90**, 1498. **19**. Markham & Kruger (2002) *Eur. J. Biochem.* **269**, 1267. **20**. Villadsen & Nielsen (2001) *Biochem. J.* **359**, 591. **21**. Larondelle, Mertens, Van Schaftingen & Hers (1986) *Eur. J. Biochem.* **161**, 351. **22**. Calmes & Brown (1979) *Infect. Immun.* **23**, 68. **23**. Lavoinne (1986) *Biochimie* **68**, 569. **24**. Klemme, Klemme & Gest (1971) *J. Bacteriol.* **108**, 1122. **25**. Podesta & Plaxton (1991) *Biochem. J.* **279**, 495. **26**. Lin, Turpin & Plaxton (1989) *Arch. Biochem. Biophys.* **269**, 228. **27**. Singh, Malhotra & Singh (2000) *Indian J. Biochem. Biophys.* **37**, 51. **28**. Tomlinson & Turner (1979) *Phytochemistry* **18**, 1959.

D-3-Phosphoglycerate (D-3-Phosphoglyceric Acid)

$$
\begin{array}{c}
\text{COO}^{\ominus} \\
\text{HC} - \text{OH} \\
\text{CH}_2\text{OPO}_3\text{H}^{\ominus}
\end{array}
$$

This metabolite (FW$_{\text{free-acid}}$ = 186.06 g/mol; CAS 820-11-1; 3443-58-1 for (*R*)-isomer; 61546-67-6 for tri-Na salt; 927-10-6 for mono-Na salt), also called 3-phospho-D-glycerate, D-glycerate 3-phosphate, and (*R*)-2,3-dihydroxypropionate 3-(dihydrogen phosphate), is an important intermediate in carbohydrate metabolism. It is also a precursor in the biosynthesis of L-serine and L-cysteine. In addition, D-3-phosphoglycerate was the first intermediate identified in the photosynthetic carbon reduction cycle by Melvin Calvin and coworkers1. D-3-Phosphoglycerate was the first example of a feedback inhibitor (2,3). The alkaline metal salts of this metabolite are readily soluble in water; however, the barium salt is sparingly soluble (albeit, soluble in dilute mineral acids). D-3-Phosphoglycerate is very slowly hydrolyzed by acids. Half is hydrolyzed by 1.0 M HCl at 100°C in 35.7 hours (and about 3 hours at 125°C). Note that solutions of D-3-phosphoglycerate can quickly become contaminated with microorganisms: hence, solutions should be stored frozen. **Target(s):** acetolactate synthase (80); acylphosphatase (50); ADP-sugar diphosphatase (49); bisphosphoglycerate mutase (4); bisphosphoglycerate phosphatase (4-6,57-60); 1-deoxy-D-xylulose-5-phosphate synthase (79); enolase (7-10); fructose-1,6-bisphosphate aldolase (46); glucose-1,6-bisphosphate synthase (64,65); glucose-6-phosphate dehydrogenase (11); glucose-6-phosphate isomerase (12,13); [glutamine-synthetase] adenylyltransferase (62); glyceraldehyde-3-phosphate dehydrogenase (8); glycerol-3-phosphate dehydrogenase (14,15); hexokinase (16,77,78); inositol-phosphate phosphatase, weakly inhibited (54); [isocitrate dehydrogenase] kinase/phosphatase (17,18); isocitrate dehydrogenase, NADP$^+$-dependent (19); isocitrate lyase (20,44,45); methylglyoxal synthase (21,40-43); NADH peroxidase (22); phosphoenolpyruvate carboxykinase, ATP-dependent (47); phosphoenolpyruvate carboxylase (48); phosphoenolpyruvate phosphatase (51-53); phosphoenolpyruvate:protein phosphotransferase (63); 6-phosphofructokinase (23-26,71-76); 6-phosphofructo-2-kinase (27,28); phosphoglucomutase (29,30); phosphoglycerate phosphatase55; phospho-protein phosphatase (weakly inhibited)56; phosphoribulokinase (70); [protein-PII] uridylyltransferase (61); pyrophosphate-dependent phosphofructokinase, or diphosphate:fructose-6-phosphate 1-phospho-transferase (66,67); pyruvate kinase (31,32,68,69); ribulose-bisphosphate carboxylase (33-35); ribose-5-phosphate isomerase (36,37); triose-phosphate isomerase (38,39). **1.** Calvin & Benson (1948) *Science* **107**, 476. **2.** Dische (1976) *Trends Biochem. Sci.* **1**, N269. **3.** Dische (1940) *Bull. Soc. Chim. Biol.* **23**, 1140. **4.** Harkness, Isaacks & Roth (1977) *Eur. J. Biochem.* **78**, 343. **5.** Sasaki, Ikura, Narita, Yanagawa & Chiba (1982) *Trends Biochem. Sci.* **7**, 140. **6.** Rapoport & Luebering (1951) *J. Biol. Chem.* **189**, 683.7. Spring (1970) Ph.D. Thesis, Univ. of Minnesota, Minneapolis. **8.** Wold & Ballou (1957) *J. Biol. Chem.* **227**, 313. **9.** Wold (1971) *The Enzymes*, 3rd. ed. (Boyer, ed.), **5**, 499. **10.** Webb (1966) *Enzyme and Metabolic Inhibitors*, vol. 2, p. 410, Academic Press, New York. **11.** Sokolov & Trotsenko (1990) *Meth. Enzymol.* **188**, 339. **12.** Takama & Nosoh (1982) *Biochim. Biophys. Acta* **705**, 127. **13.** Backhausen, Jöstingmeyer & Scheibe (1997) *Plant Sci.* **130**, 121. **14.** Dawson & Thorne (1975) *Meth. Enzymol.* **41**, 254. **15.** Schryvers, Lohmeier & Weiner (1978) *J. Biol. Chem.* **253**, 783. **16.** Ning, Purich & Fromm (1969) *J. Biol. Chem.* **244**, 3840. **17.** Nimmo (1984) *Trends Biochem. Sci.* **9**, 475. **18.** Miller, Chen, Karschnia, *et al.* (2000) *J. Biol. Chem.* **275**, 833. **19.** Shikata, Ozaki, Kawai, Ito & Okamoto (1988) *Biochim. Biophys. Acta* **952**, 282. **20.** McFadden (1969) *Meth. Enzymol.* **13**, 163. **21.** Cooper (1975) *Meth. Enzymol.* **41**, 502. **22.** Badwey & Karnovsky (1979) *J. Biol. Chem.* **254**, 11530. **23.** Kemp (1975) *Meth. Enzymol.* **42**, 67. **24.** Shimizu, Kono, Mineo, *et al.* (1984) *J. Inherit. Metab. Dis.* **7**, 107. **25.** Li, Rivera, Ru, Gunasekera & Kemp (1999) *Biochemistry* **38**, 16407. **26.** Bloxham & Lardy (1973) *The Enzymes*, 3rd ed. (Boyer, ed.), **8**, 239. **27.** Cseke, Balogh, Wong, *et al.* (1984) *Trends Biochem. Sci.* **9**, 533. **28.** Markham & Kruger (2002) *Eur. J. Biochem.* **269**, 1267. **29.** Popova, Matasova & Lapot'ko (1998) *Biochemistry (Moscow)* **63**, 697. **30.** Ray, Jr., & Peck, Jr. (1972) *The Enzymes*, 3rd ed. (Boyer, ed.), **6**, 407. **31.** Podesta & Plaxton (1991) *Biochem. J.* **279**, 495. **32.** Ng & Hamilton (1975) *J. Bacteriol.* **122**, 1274. **33.** Wishnick & Lane (1971) *Meth. Enzymol.* **23**, 570. **34.** Paulsen & Lane (1966) *Biochemistry* **5**, 2350. **35.** Kuehn & McFadden (1969) *Biochemistry* **8**, 2394. **36.** Skrukrud, Gordon, Dorwin, *et al.* (1991) *Plant Physiol.* **97**, 730. **37.** Woodruff III & Wolfenden (1979) *J. Biol. Chem.* **254**, 5866. **38.** Noble, Wierenga, Lambeir, *et al.* (1991) *Proteins* **10**, 50. **39.** Tomlinson & Turner (1979) *Phytochemistry* **18**, 1959. **40.** Murata, Fukuda, Watanabe, *et al.* (1985) *Biochem. Biophys. Res. Commun.* **131**, 190. **41.** Hopper & Cooper (1971) *FEBS Lett.* **13**, 213. **42.** Hopper & Cooper (1972) *Biochem. J.* **128**, 321. **43.** Cooper (1974) *Eur. J. Biochem.* **44**, 81. **44.** Tsukamoto, Ejiri & Katsumata (1986) *Agric. Biol. Chem.* **50**, 409. **45.** MacKintosh & Nimmo (1988) *Biochem. J.* **250**, 25. **46.** Szwergold, Ugurbil & Brown (1995) *Arch. Biochem. Biophys.* **317**, 244. **47.** Lea, Chen, Leegood & Walker (2001) *Amino Acids* **20**, 225. **48.** Schuller, Turpin & Plaxton (1990) *Plant Physiol.* **94**, 1429. **49.** Moreno-Bruna, Baroja-Fernandez, Munoz, *et al.* (2001) *Proc. Natl. Acad. Sci. U.S.A.* **98**, 8128. **50.** Pazzagli, Ikram, Liguri, *et al.* (1993) *Ital. J. Biochem.* **42**, 233. **51.** Singh & Singh (1996) *Proc. Indian Acad. Sci. Sect. B* **66**, 53. **52.** Duff, Lefebvre & Plaxton (1989) *Plant Physiol.* **90**, 734. **53.** Malhotra & Kayastha (1990) *Plant Physiol.* **93**, 194. **54.** Loewus & Loewus (1982) *Plant Physiol.* **70**, 765. **55.** Fallon & Byrne (1965) *Biochim. Biophys. Acta* **105**, 43. **56.** Zhou, Clemens, Hakes, Barford & Dixon (1993) *J. Biol. Chem.* **268**, 17754. **57.** Sasaki, Hirose, Sugimoto & Chiba (1971) *Biochim. Biophys. Acta* **227**, 595. **58.** Sasaki, Ikura, Sugimoto & Chiba (1975) *Eur. J. Biochem.* **50**, 581. **59.** Tauler, Pons & Carreras (1986) *Biochim. Biophys. Acta* **872**, 201. **60.** Pons & Carreras (1982) *Biochim. Biophys. Acta* **842**, 56. **61.** Engleman & Francis (1978) *Arch. Biochem. Biophys.* **191**, 602. **62.** Wolf & Ebner (1972) *J. Biol. Chem.* **247**, 4208. **63.** Saier, Jr., Schmidt & Lin (1980) *J. Biol. Chem.* **255**, 8579. **64.** Rose, Warms & Kaklij (1975) *J. Biol. Chem.* **250**, 3466. **65.** Rose, Warms & Wong (1977) *J. Biol. Chem.* **252**, 4262. **66.** Kombrink & Kruger (1984) *Z. Pflanzenphysiol.* **114**, 443. **67.** Theodorou & Kruger (2001) *Planta* **213**, 147. **68.** Lin, Turpin & Plaxton (1989) *Arch. Biochem. Biophys.* **269**, 228. **69.** Turner, Knowles & Plaxton (2005) *Planta* **222**, 1051. **70.** Surek, Heilbronn, Austen & Latzko (1985) *Planta* **165**, 507. **71.** Botha & Turpin (1990) *Plant Physiol.* **93**, 871. **72.** Cawood, Botha & Small (1988) *J. Plant Physiol.* **132**, 204. **73.** Kombrink & Wöber (1982) *Arch. Biochem. Biophys.* **213**, 602. **74.** Li, Rivera, Wu, Gunasekera & Kemp (1999) *Biochemistry* **38**, 16407. **75.** Knowles, Greyson & Dennis (1990) *Plant Physiol.* **92**, 155. **76.** Mediavilla, Meton & Baanante (2007) *Biochim. Biophys. Acta* **1770**, 706. **77.** Cesar, Colepicolo, Rosa & Rosa (1997) *Comp. Biochem. Physiol. B* **118**, 395. **78.** Magnani, Stocchi, Serafini, *et al.* (1983) *Arch. Biochem. Biophys.* **226**, 377. **79.** Eubanks & Poulter (2003) *Biochemistry* **42**, 1140. **80.** Phalip, Schmitt & Divies (1995) *Curr. Microbiol.* **31**, 316.

3-Phospho-D-glyceroyl Phosphate

This key glycolytic pathway intermediate (FW$_{free-acid}$ = 266.04 g/mol), more commonly known as 1,3-bisphospho-D-glycerate (albeit incorrectly), is quite unstable: 50% is hydrolyzed in 27 minutes at pH 7.2 and 38°C, producing D-glycerate 3-phosphate. **Target(s):** ATPase (1); methylglyoxal synthase (2); 6-phosphofructokinase (3); and phosphoglucomutase (4,5). **1.** Ells & Faulkner (1961) *Biochem. Biophys. Res. Commun.* **5**, 255. **2.** Murata, Fukuda, Watanabe, *et al.* (1985) *Biochem. Biophys. Res. Commun.* **131**, 190. **3.** Turner & Plaxton (2003) *Planta* **217**, 113. **4.** Hirose, Ueda & Chiba (1976) *Agric. Biol. Chem.* **40**, 2433. **5.** Chiba, Ueda & Hirose (1976) *Agric. Biol. Chem.* **40**, 2423.

2-Phosphoglycolate (2-Phosphoglycolic Acid)

This metabolite (FW$_{free-acid}$ = 156.03 g/mol), also known as phospho-glycollate and glycollic acid phosphate, is a putative transition-state analogue of the reaction catalyzed by triose-phosphate isomerase. **Target(s):** bisphosphoglycerate mutase (1); enolase (2-4); glycerol-3-phosphate dehydrogenase (5); methylglyoxal synthase (6-8); phosphoenolpyruvate carboxylase (9); phosphoenolpyruvate mutase (10); 6-phosphofructokinase (11-15); 6-phosphofructo-2-kinase (16); 3-phosphoglycerate kinase (17); phosphoprotein phosphatase (18); pyrophosphate-dependent phosphofructokinase, or diphosphate:fructose-6-phosphate 1-phosphotransferase (19); pyruvate kinase (20-22); pyruvate,orthophosphate dikinase (23); and triose-phosphate isomerase (24-36). **1.** Sasaki, Ikura, Sugimoto & Chiba (1975) *Eur. J. Biochem.* **50**, 581. **2.** Wold (1971) *The Enzymes*, 3rd. ed. (Boyer, ed.), **5**, 499. **3.** Nazarian, Vais, Egorin, Kustzheba-Vuitsntska & Petkevich (1992) *Biokhimiia* **57**, 1827. **4.** Lebioda, Stec, Brewer & Tykarska (1991) *Biochemistry* **30**, 2823. **5.** Schryvers, Lohmeier & Weiner (1978) *J. Biol. Chem.* **253**, 783. **6.** Saadat & Harrison (1998) *Biochemistry* **37**, 10074. **7.** Saadat & Harrison (2000) *Biochemistry* **39**, 2950.8. Tsai & Gracy (1976) *J. Biol. Chem.* **251**, 364. **9.** Miziorko, Nowak & Mildvan (1974) *Arch. Biochem. Biophys.* **163**, 378. **10.** Seidel & Knowles (1994) *Biochemistry* **33**, 5641. **11.** Kimmel & Reinhart (2000) *Proc. Natl. Acad. Sci. U.S.A.* **97**, 3844. **12.** Tlapak-Simmons & Reinhart (1994) *Arch. Biochem. Biophys.* **308**, 226. **13.** Turner & Plaxton (2003) *Planta* **217**, 113. **14.** Kombrink & Wöber (1982) *Arch.*

Biochem. Biophys. **213**, 602. **15.** Knowles, Greyson & Dennis (1990) *Plant Physiol.* **92**, 155. **16.** Larondelle, Mertens, Van Schaftingen & Hers (1986) *Eur. J. Biochem.* **161**, 351. **17.** Szilágyi & Vas (1998) *Biochemistry* **37**, 8551. **18.** Polya & Haritou (1988) *Biochem. J.* **251**, 357. **19.** Kombrink & Kruger (1984) *Z. Pflanzenphysiol.* **114**, 443. **20.** Podesta & Plaxton (1991) *Biochem. J.* **279**, 495. **21.** Wu & Turpin (1992) *J. Phycol.* **28**, 472. **22.** Lin, Turpin & Plaxton (1989) *Arch. Biochem. Biophys.* **269**, 228. **23.** Milner, Michaels & Wood (1975) *Meth. Enzymol.* **42**, 199. **24.** Wolfenden (1969) *Nature* **223**, 704. **25.** Wolfenden (1970) *Biochemistry* **9**, 3404. **26.** Krietsch (1975) *Meth. Enzymol.* **41**, 434. **27.** Krietsch (1975) *Meth. Enzymol.* **41**, 438. **28.** Gracy (1975) *Meth. Enzymol.* **41**, 442. **29.** Hartman & Norton (1975) *Meth. Enzymol.* **41**, 447. **30.** R. Wolfenden (1977) *Meth. Enzymol.* **46**, 15. **31.** Noltmann (1972) *The Enzymes*, 3rd ed. (Boyer, ed.), **6**, 271. **32.** Lambeir, Opperdoes & Wierenga (1987) *Eur. J. Biochem.* **168**, 69. **33.** Lolis & Petsko (1990) *Biochemistry* **29**, 6619. **34.** Bell, Russell, Kohlhoff, *et al.* (1998) *Acta Crystallogr. Sect. D* **54**, 1419. **35.** Eber & Krietsch (1980) *Biochim. Biophys. Acta* **614**, 173. **36.** Nickbarg & Knowles (1988) *Biochemistry* **27**, 5939.

Phosphoglycolohydrazide

This hydrazide (FW$_{free-acid}$ = 169.05 g/mol) is a strong reversible inhibitor of class II fructose-1,6-bisphosphate aldolase, K_i = 0.34 μM (1,2) and rabbit muscle triose-phosphate isomerase, K_i = 111 μM (1). **1.** Fonvielle, Mariano & Therisod (2005) *Bioorg. Med. Chem. Lett.* **15**, 2906. **2.** Fonvielle, Weber, Dabkowska & Therisod (2004) *Bioorg. Med. Chem. Lett.* **14**, 2923.

Phosphoglycolohydroxamate

This hydroxamate (FW$_{free-acid}$ = 171.05 g/mol) is a mimic of dihydroxyacetone phosphate and an enediolate transition-state analogue for triose-phosphate isomerase, fructose-1,6-bisphosphate aldolase, and other aldolases. **Target(s):** fructose 1,6-bisphosphate aldolase (1-7); L-fuculose-phosphate aldolase (8-10); methylglyoxal synthase, K_i = 39 nM (11); L-rhamnulose-1-phosphate aldolase (8); tagatose-1,6-bisphosphate aldolase (12); and triose-phosphate isomerase (1,5,7,13-21). **1.** Wolfenden (1969) *Nature* **223**, 704. **2.** Wolfenden (1977) *Meth. Enzymol.* **46**, 15. **3.** Sigman & Mooser (1975) *Ann. Rev. Biochem.* **44**, 889. **4.** Zgiby, Plater, Bates, Thomson & Berry (2002) *J. Mol. Biol.* **315**, 131. **5.** Hall, Leonard, Reed, *et al.* (1999) *J. Mol. Biol.* **287**, 383. **6.** Fonvielle, Mariano & Therisod (2005) *Bioorg. Med. Chem. Lett.* **15**, 2906. **7.** Fonvielle, Weber, Dabkowska & Therisod (2004) *Bioorg. Med. Chem. Lett.* **14**, 2923. **8.** Kroemer, Merkel & Schulz (2003) *Biochemistry* **42**, 10560. **9.** Fessner, Schneider, Held, *et al.* (1996) *Angew. Chem., Int. Ed. Engl.* **35**, 2219. **10.** Dreyer & Schulz (1996) *J. Mol. Biol.* **259**, 458. **11.** Marks, Harris, Massiah, Mildvan & Harrison (2001) *Biochemistry* **40**, 6805. **12.** Hall, Bond, Leonard, *et al.* (2002) *J. Biol. Chem.* **277**, 22018. **13.** Collins (1974) *J. Biol. Chem.* **249**, 136. **14.** Hartman & Norton (1975) *Meth. Enzymol.* **41**, 447. **15.** Kursula, Partanen, Lambeir, *et al.* (2001) *Eur. J. Biochem.* **268**, 5189. **16.** Zhang, Komives, Sugio, *et al.* (1999) *Biochemistry* **38**, 4389. **17.** Harris, Abeygunawardana & Mildvan (1997) *Biochemistry* **36**, 14661. **18.** Komives, Lougheed, Zhang, *et al.* (1996) *Biochemistry* **35**, 15474. **19.** Davenport, Bash, Seaton, *et al.* (1991) *Biochemistry* **30**, 5821. **20.** Nickbarg & Knowles (1988) *Biochemistry* **27**, 5939. **21.** Fonvielle, Mariano & Therisod (2005) *Bioorg. Med. Chem. Lett.* **15**, 2906.

O-Phosphohomocholine

This homologue of phosphocholine inhibits 1-acylglycerophosphocholine *O*-acyltransferase, *or* lysophosphatidylcholine *O*-acyltransferase. **1.** Lands & Hart (1965) *J. Biol. Chem.* **240**, 1905.

Phospho-3-hydroxypropionate, *See* 3-Hydroxypropionate Phosphate

Phosphohydroxypyruvate, *See* Hydroxypyruvate 3-phosphate

N-Phospho-L-isoleucyl(*O*-ethyl)-L-tyrosyl(*O*-benzyl)glycine, *See* N^{α}-(*O*-Ethylphospho)-L-isoleucyl-L-(*O*-benzyl)tyrosylglycine

7-Phospho-2-keto-3-deoxy-D-*arabino*heptonate, *See* 3-Deoxy-D-arabino-heptulosonate 7-Phosphate

D-*O*-Phospholactate

This phosphoenolpyruvate analogue (FW$_{free-acid}$ = 170.06 g/mol) inhibits enolase, K_i = 0.35 mM (1-4) and phosphoenolpyruvate carboxylase, also alternative substrate (5,6). **1.** Wold (1971) *The Enzymes*, 3rd. ed. (Boyer, ed.), **5**, 499. **2.** Webb (1966) *Enzyme and Metabolic Inhibitors*, vol. **2**, p. 410, Academic Press, New York. **3.** Wold & Ballou (1957) *J. Biol. Chem.* **227**, 313. **4.** Spring (1970) Ph.D. Thesis, Univ. of Minnesota, Minneapolis. **5.** Izui, Matsuda, Kameshita, Katsuki & Woods (1983) *J. Biochem.* **94**, 1789. **6.** Miziorko, Nowak & Mildvan (1974) *Arch. Biochem. Biophys.* **163**, 378.

DL-*O*-Phospholactate (DL-*O*-Phospholactic Acid)

This phosphoenolpyruvate analogue (FW$_{free-acid}$ = 170.06 g/mol) inhibits phosphoenolpyruvate carboxykinase, GTP-dependent (1), phosphoenolpyruvate carboxylase (2-4), and phosphoglycerate mutase (5). **1.** Lane, Chang & Miller (1969) *Meth. Enzymol.* **13**, 270. **2.** Lane, Maruyama & Easterday (1969) *Meth. Enzymol.* **13**, 277. **3.** Kodaki, Fujita, Kameshita, Izui & Katsuki (1984) *J. Biochem.* **95**, 637. **4.** Naide, Izui, Yoshinaga & Katsuki (1979) *J. Biochem.* **85**, 423. **5.** Ray & Peck (1972) *The Enzymes*, 3rd ed. (Boyer, ed.), **6**, 407.

Phospholipase C-γ1 D-domain Hexadecapeptide

This hexadecapeptide (*Sequence*: RRKKIALELSELVVYC; FW = 1920.34 g/mol) inhibits mitogen-activated protein kinase. **1.** Buckley, Sekiya, Kim, Rhee & Caldwell (2004) *J. Biol. Chem.* **279**, 41807.

Phosphonoacetate (Phosphonoacetic Acid)

This pyrophosphate and carbamoyl phosphate analogue (FW$_{free-acid}$ = 140.03 g/mol; CAS 4408-78-0) exhibits antiherpetic activity. **See also** *Phosphonoformate* **Target(s):** adenosine kinase (25,26); African swine fever virus-induced DNA polymerase (1,23); aspartate carbamoyltransferase (2); bacteriophage T$_4$ DNA polymerase (3,22); bacteriophage λ protein phosphatase (4); biotin carboxylase, K_{is} = 1.8 mM for the *Escherichia coli* enzyme (5); carbamate kinase (24); DNA-directed DNA polymerase (1,3,6-8,22,23); DNA polymerases α, β, and γ (6-8); DNA polymerase, cytomegalovirus (8); DNA polymerase, viral (*e.g.*, Herpes simplex-induced DNA polymerase and Vaccinia-induced DNA polymerase) (1,8-16); Epstein-Barr virus induced DNA polymerase (12,16); formyltetrahydrofolate synthetase (17); phosphate transport (18,19); phosphoenolpyruvate mutase (20); phosphonoacetaldehyde hydrolase (21); ribokinase (27); and RNA-directed DNA polymerase (6)/ **1.** Marques & Costa (1992) *Virology* **191**, 498. **2.** Burns, Mendz & Hazell (1997) *Arch. Biochem. Biophys.* **347**, 119. **3.** Reha-Krantz & Wong (1996) *Mutat. Res.* **350**, 9. **4.** Reiter, White

& Rusnak (2002) *Biochemistry* **41**, 1051.**5**. Blanchard, Amspacher, Strongin & Waldrop (1999) *Biochem. Biophys. Res. Commun.* **266**, 466. **6**. Allaudeen & Bertino (1978) *Biochim. Biophys. Acta* **520**, 490. **7**. Sabourin, Reno & Boezi (1978) *Arch. Biochem. Biophys.* **187**, 96. **8**. Eriksson, Oberg & Wahren (1982) *Biochim. Biophys. Acta* **696**, 115. **9**. Prisbe & Chen (1996) *Meth. Enzymol.* **275**, 425. **10**. Weissbach (1981) *The Enzymes*, 3rd ed., **14**, 67. **11**. Crumpacker (1992) *Amer. J. Med.* **92**, 3S-7S. **12**. Grossberger & Clough (1981) *Biochemistry* **20**, 4049. **13**. Eriksson, Larsson, Helgstrand, Johansson & Oberg (1980) *Biochim. Biophys. Acta* **607**, 53. **14**. Leinbach, Reno, Lee, Isbell & Boezi (1976) *Biochemistry* **15**, 426. **15**. Mao & Robishaw (1975) *Biochemistry* **14**, 5475. **16**. Allaudeen (1985) *Antiviral Res.* **5**, 1. **17**. Buttlaire, Balfe, Wendland & Himes (1979) *Biochim. Biophys. Acta* **567**, 453. **18**. Loghman-Adham, Szczepanska-Konkel, Yusufi, Van Scoy & Dousa (1987) *Amer. J. Physiol.* **252**, G244. **19**. Szczepanska-Konkel, Yusufi, VanScoy, Webster & Dousa (1986) *J. Biol. Chem.* **261**, 6375. **20**. Seidel & Knowles (1994) *Biochemistry* **33**, 5641. **21**. Olsen, Hepburn, Lee, *et al.* (1992) *Arch. Biochem. Biophys.* **296**, 144. **22**. Khan, Reha-Krantz & Wright (1994) *Nucl. Acids Res.* **22**, 232. **23**. Oliveros, Yanez, Salas, *et al.* (1997) *J. Biol. Chem.* **272**, 30899. **24**. Marina, Uriarte, Barcelona, *et al.* (1998) *Eur. J. Biochem.* **253**, 280. **25**. Park, Singh, Maj & Gupta (2004) *Protein J.* **23**, 167. **26**. Park, Singh & Gupta (2006) *Mol. Cell. Biochem.* **283**, 11. **27**. Park, van Koeverden, Singh & Gupta (2007) *FEBS Lett.* **581**, 3211.

Phosphonoacetohydroxamate

This hydroxamate (FW$_{\text{free-acid}}$ = 155.05 g/mol) inhibits 1-deoxy-D-xylulose-5-phosphate synthase (1) and enolase, *or* phosphopyruvate hydratase, the latter with a K_i value of 15 pM in the presence of optimal Mg^{2+} conecentration (1,2). **1**. Eubanks & Poulter (2003) *Biochemistry* **42**, 1140. **2**. Brewer, Glover, Holland & Lebioda (1998) *Biochim. Biophys. Acta* **1383**, 351. **3**. Anderson, Weiss & Cleland (1984) *Biochemistry* **23**, 2779.

3-(Phosphonoacetylamido)-L-alanine

This unreactive structural analogue (FW$_{\text{ion}}$ = 224.11 g/mol) of the γ-glutamyl phosphate intermediate formed during glutamine synthetase catalysis is initially a low-affinity competitive inhibitor of the synthetase, which upon binding of PA$_2$LA, undergoes subsequent conformational changes. It remains to be determined whether its weak binding relative to the γ-glutamyl-phosphate intermediate results from restricted rotation of the internal amide group, differences in its acid/base properties, or some obligatory conformational changes during ATP-dependent phosphorylation of the γ-carboxyl group. **Target(s):** glutamate-semialdehyde dehydrogenase, *or* γ-glutamylphosphate reductase (1,2); and glutamine synthetase, slow-binding inhibition (3-5). **1**. Hayzer & Leisinger (1983) *Biochim. Biophys. Acta* **742**, 391. **2**. Krishna, Beilstein & Leisinger (1979) *Biochem. J.* **181**, 223. **3**. Morrison (1982) *Trends Biochem. Sci.* **7**, 102. **4**. Wedler & Horn (1976) *J. Biol. Chem.* **251**, 7530. **5**. Purich (1982) *Meth. Enzymol.* **87**, 3.

N$^\alpha$-(Phosphonoacetyl)-L-asparagine

This substrate analogue (FW$_{\text{ion}}$ = 251.11 g/mol) inhibits *Escherichia coli* aspartate carbamoyltransferase, IC$_{50}$ = 225 nM, albeit not as potently as *N*-(phosphonoacetyl)-L-aspartate. **1**. Eldo, Heng & Kantrowitz (2007) *Bioorg.*

Med. Chem. Lett. **17**, 2086. **2**. Cardia, Eldo, Xia, *et al.* (2008) *Proteins Struct. Funct. Bioinform.* **71**, 1088.

N$^\alpha$-(Phosphonoacetyl)-α-aspartamide

This substrate analogue (FW$_{\text{ion}}$ = 251.11 g/mol) inhibits aspartate carbamoyltransferase (*Escherichia coli* IC$_{50}$ = 87 nM), albeit not a potently as *N*-(phosphonoacetyl)-L-aspartate (1). **1**. Eldo, Heng & Kantrowitz (2007) *Bioorg. Med. Chem. Lett.* **17**, 2086.

N-(Phosphonoacetyl)-L-aspartate

This tight-binding inhibitor (FW$_{\text{free-acid}}$ = 255.12 g/mol), often abbreviated as PALA, is a geometric analogue that collectively incorporates key features of the two natural substrates of aspartate carbamoyltransferase to achieve nanomolar-level dissociation constants. While PALA potently inhibits the enzyme, some cells compensate for loss of enzymatic activity by synthesizing elevated levels of the enzyme. **1**. Collins & Stark (1971) *J. Biol. Chem.* **246**, 6599. **2**. Wolfenden (1977) *Meth. Enzymol.* **46**, 15. **3**. Sigman & Mooser (1975) *Ann. Rev. Biochem.* **44**, 889. **4**. Purcarea (2001) *Meth. Enzymol.* **331**, 248. **5**. Kantrowitz, Pastra-Landis & Lipscomb (1980) *Trends Biochem. Sci.* **5**, 124. **6**. Jacobson & Stark (1973) *The Enzymes*, 3rd ed. (Boyer, ed.), **9**, 225. **7**. Hoogenraad (1974) *Arch. Biochem. Biophys.* **161**, 76. **8**. Phillips, Bordas, Foote, Koch & Moody (1982) *Biochemistry* **21**, 830. **9**. Parmentier, O'Leary, Schachman & Cleland (1992) *Biochemistry* **31**, 6598. **10**. Wang, Yang, Hu & Schachman (1981) *J. Biol. Chem.* **256**, 7028. **11**. Burns, Mendz & Hazell (1997) *Arch. Biochem. Biophys.* **347**, 119. **12**. Savithri, Vaidyanathan & Rao (1978) *Proc. Indian Acad. Sci. Sect. B* **87B**, 81. **13**. Vickrey, Herve & Evans (2002) *J. Biol. Chem.* **277**, 24490. **14**. Laing, Chan, Hutchinson & Íberg (1990) *FEBS Lett.* **260**, 206. **15**. Purcarea, Erauso, Prieur & Herve (1994) *Microbiology* **140**, 1967. **16**. Chen & Slocum (2008) *Plant Physiol. Biochem.* **46**, 150. **17**. Brabson, Maurizi & Switzer (1985) *Meth. Enzymol.* **113**, 627. **18**. Swyryd, Seaver & Stark (1974) *J. Biol. Chem.* **249**, 6945. **19**. Van Boxstael, Cunin, Khan & Maes (2003) *J. Mol. Biol.* **326**, 203.

N$^\delta$-(Phosphonoacetyl)-L-ornithine

This ornithine derivative (FW = 252.16 g/mol), often abbreviated PALO, is a bisubstrate or geometric analogue of the two natural substrates of ornithine carbamoyltransferase (1-11). **1**. Mori, Aoyagi, Tatibana Ishikawa & Ishii (1977) *Biochem. Biophys. Res. Commun.* **76**, 900. **2**. Hoogenraad (1978) *Arch. Biochem. Biopys.* **188**, 137. **3**. Ha, McCann, Tuchman & Allewell (1997) *Proc. Natl. Acad. Sci. U.S.A.* **94**, 9550. **4**. Legrain, Villeret, Roovers, *et al.* (2001) *Meth. Enzymol.* **331**, 227. **5**. Shi, Morizono, Ha, *et al.* (1998) *J. Biol. Chem.* **273**, 34247. **6**. Neway & Switzer (1983) *J. Bacteriol.* **155**, 512. **7**. Xu, Feller, Gerday & Glansdorff (2003) *J. Bacteriol.* **185**, 2161. **8**. Legrain, Villeret, Roovers, *et al.* (1997) *Eur. J. Biochem.* **247**, 1046. **9**. Cohen, Cheung, Sijuwade & Raijman (1992) *Biochem. J.* **282**, 173. **10**. Koger, Howell, Kelly & Jones (1994) *Arch. Biochem. Biophys.* **309**, 293. **11**. Sanchez, Baetens, Van De Casteele, *et al.* (1997) *Eur. J. Biochem.* **248**, 466.

3-Phosphono-DL-alanine, *See* 2-Amino-3- phosphonopropionate

3-Phosphono-2-aminopropionate, *See* 2-Amino-3- phosphonopropionate

N-Phosphonoarginine, See *N-Phospho-L-arginine*

N-Phosphonocreatine, See *N-Phosphocreatine*

(Z)-3-(2-Phosphonoethen-1-yl)pyridine-2-carboxylic Acid

This homoserine phosphate analogue (FW_{ion} = 226.10 g/mol) inhibits wheat cystathionine γ-synthase, K_i = 40 μM. The same enzyme is irreversibly inhibited by DL-propargylglycine (Ki = 45 μM, k_{inact} = 0.16 min^{-1}). Other homoserine phosphate analogues, such as 4-(phosphonomethyl)-pyridine-2-carboxylic acid and DL-*E*-2-amino-5-phosphono-3-pentenoic acid, are also reversible competitive inhibitors of wheat cystathionine γ-synthase, with respective K_i values of 40 and 1.1 μM. **1.** Kreft, Townsend, Pohlenz & Laber (1994) *Plant Physiol.* **104**, 1215.

Phosphonoformate (Phosphonoformic Acid)

This pyrophosphate analogue ($FW_{trisodium\ salt}$ = 191.95 g/mol; CAS 4428-95-9), also called foscarnet, exhibits antiherpetic activity. It is a strong inhibitor of viral reverse transcriptases and viral replication as well as Na$^+$/phosphate cotransporters. **Target(s):** adenylyl cyclase (1); avian myeloblastosis virus reverse transcriptase (2,3); cytomegalovirus DNA polymerase (4-6); DNA-directed DNA polymerase (7,46); DNA polymerase β (7); DNA polymerases, viral (2-6,8-29); duck hepatitis B virus reverse transcriptase (13,17); Epstein-Barr virus DNA polymerase (16,21); equine infectious anemia virus reverse transcriptase (12); exoribonuclease H (44); feline immunodeficiency virus reverse transcriptase (11,29); geranyl-diphosphate cyclase, *or* bornyl-diphosphate synthase (30); guanylyl cyclase (1); HIV-1 reverse transcriptase (4,8,10,11,14, 18,31,44,45); human T-lymphotropic virus type III reverse transcriptase (19); herpes virus reverse transcriptase (6,20,21,25,28); human hepatitis B virus reverse transcriptase (22,26); Moloney murine leukemia virus reverse transcriptase (3); Na$^+$-dependent phosphate transport (32-38); phosphoenolpyruvate carboxylase (39); phosphoenolpyruvate mutase (40); phosphonoacetate hydrolase (41,43); phosphonopyruvate hydrolase (42); ribonuclease H (44); RNA-directed DNA polymerase (2-6,8-29,31,44,45); simian immunodeficiency virus reverse transcriptase (15); and woodchuck hepatitis virus reverse transcriptase (22,24). **1.** Kudlacek, Mitterauer, Nanoff, *et al.* (2001) *J. Biol. Chem.* **276**, 3010. **2.** Eriksson, Stening & Oberg (1982) *Antiviral Res.* **2**, 81. **3.** Margalith, Falk & Panet (1982) *Mol. Cell Biochem.* **43**, 97. **4.** Prisbe & Chen (1996) *Meth. Enzymol.* **275**, 425. **5.** Wahren & Eriksson (1985) *Adv. Enzyme Regul.* **23**, 263. **6.** Eriksson, Oberg & Wahren (1982) *Biochim. Biophys. Acta* **696**, 115. **7.** Rozovskaya, Tarussova, Minassian, *et al.* (1989) *FEBS Lett.* **247**, 289. **8.** Balzarini & De Clercq (1996) *Meth. Enzymol.* **275**, 472. **9.** Weissbach (1981) *The Enzymes*, 3rd ed., **14**, 67. **10.** Shaw-Reid, Munshi, Graham, *et al.* (2003) *J. Biol. Chem.* **278**, 2777. **11.** Auwerx, North, Preston, *et al.* (2002) *Mol. Pharmacol.* **61**, 400. **12.** Tao, Zhang & Quan (1990) *Proc. Chin. Acad. Med. Sci. Peking Union Med. Coll.* **5**, 130. **13.** Lofgren, Nordenfelt & Oberg (1989) *Antiviral Res.* **12**, 301. **14.** Mizrahi, Lazarus, Miles, Meyers & Debouck (1989) *Arch. Biochem. Biophys.* **273**, 347. **15.** Vrang, Oberg, Lower & Kurth (1988) *Antimicrob. Agents Chemother.* **32**, 1733. **16.** Li & Cheng (1988) *Virus Genes* **1**, 369. **17.** Offensperger, Walter, Offensperger, *et al.* (1988) *Virology* **164**, 48. **18.** Wondrak, Lower & Kurth (1988) *J. Antimicrob. Chemother.* **21**, 151. **19.** Vrang & Oberg (1986) *Antimicrob. Agents Chemother.* **29**, 867. **20.** Frank & Cheng (1985) *Antimicrob. Agents Chemother.* **27**, 445. **21.** Allaudeen (1985) *Antiviral Res.* **5**, 1. **22.** Hantz, Ooka, Vitvitski, Pichoud & Trepo (1984) *Antimicrob. Agents Chemother.* **25**, 242. **23.** Modak, Srivastava & Gillerman (1980) *Biochem. Biophys. Res. Commun.* **96**, 931. **24.** Nordenfelt & Werner (1980) *Acta Pathol. Microbiol. Scand. [B]* **88**, 183. **25.** Eriksson, Larsson, Helgstrand, Johansson & Oberg (1980) *Biochim. Biophys. Acta* **607**, 53. **26.** Hess, Arnold & Meyer zum Buschenfelde (1980) *J. Med. Virol.* **5**, 309. **27.** Sundquist & Oberg (1979) *J. Gen. Virol.* **45**, 273. **28.** Reno, Lee & Boezi (1978) *Antimicrob. Agents Chemother.* **13**, 188. **29.**

North, Cronn, Remington & Tandberg (1990) *Antimicrob. Agents Chemother.* **34**, 1505. **30.** Wheeler & Croteau (1988) *Arch. Biochem. Biophys.* **260**, 250. **31.** Tan, Zhang, Li, *et al.* (1991) *Biochemistry* **30**, 2651. **32.** Timmer & Gunn (2000) *J. Gen. Physiol.* **116**, 363. **33.** Timmer & Gunn (1998) *Amer. J. Physiol.* **274**, C757. **34.** Loghman-Adham (1996) *Gen. Pharmacol.* **27**, 305. **35.** Loghman-Adham & Dousa (1992) *Amer. J. Physiol.* **263**, F301. **36.** Loghman-Adham, Szczepanska-Konkel, Yusufi, Van Scoy & Dousa (1987) *Amer. J. Physiol.* **252**, G244. **37.** Yusufi, Szczepanska-Konkel, Kempson, McAteer & Dousa (1986) *Biochem. Biophys. Res. Commun.* **139**, 679. **38.** Szczepanska-Konkel, Yusufi, VanScoy, Webster & Dousa (1986) *J. Biol. Chem.* **261**, 6375. **39.** Janc, O'Leary & Cleland (1992) *Biochemistry* **31**, 6421. **40.** Seidel & Knowles (1994) *Biochemistry* **33**, 5641. **41.** McGrath, Wisdom, McMullan, Larkin & Quinn (1995) *Eur. J. Biochem.* **234**, 225. **42.** Kulakova, Wisdom, Kulakov & Quinn (2003) *J. Biol. Chem.* **278**, 23426. **43.** McGrath, Kulakova & Quinn (1999) *J. Appl. Microbiol.* **86**, 834. **44.** Shaw-Reid, Feuston, Munshi, *et al.* (2005) *Biochemistry* **44**, 1595. **45.** El Safadi, Vivet-Boudou & Marquet (2007) *Appl. Microbiol. Biotechnol.* **75**, 723. **46.** Khan, Reha-Krantz & Wright (1994) *Nucl. Acids Res.* **22**, 232.

***rac*-1-Phosphono-*all-trans*-geranylgeraniol**

This geranylgeranyl diphosphate analogue ($FW_{free-acid}$ = 370.47 g/mol) inhibits protein geranylgeranyltransferase type I: K_i = 0.72 μM (1-3). **1.** Yokoyama & Gelb (2001) *The Enzymes*, 3rd. ed. (Tamanoi & Sigman, eds.), **21**, 105. **2.** Yokoyama, McGeady & Gelb (1995) *Biochemistry* **34**, 1344. **3.** Stirtan & Poulter (1997) *Biochemistry* **36**, 4552.

9-[2-(Phosphonomethoxy)ethyl]adenine

This acyclic AMP analogue ($FW_{free-acid}$ = 273.19 g/mol), also known as adefovir and 9-(2-phosphonylmethoxyethyl)adenine (abbreviated PMEA), exhibits a broad spectrum of antiviral activity. The diphosphate (*i.e.*, an analogue of ATP) inhibits avian myeloblastosis virus reverse transcriptase and herpes simplex virus DNA polymerase via chain termination after its incorporation into the growing chain. Interestingly, the diphosphate, PMEApp, is produced by the action of 5-phosphoribosyl-1-pyrophosphate synthetase, in which the pyrophosphate group of PRPP is directly transferred to PMEA (1). PMEA has pK_a values of 2.0 and 6.8. **Target(s):** adenylyl cyclase, inhibited by the diphosphate (2); avian myeloblastosis virus reverse transcriptase, inhibited by the diphosphate (3); DNA-directed DNA polymerase (4,5); DNA polymerase α, inhibited by the diphosphate, IC_{50} = 3.7 μM (4); DNA polymerase δ, moderately inhibited by the diphosphate (5); DNA polymerases, viral (6-11); feline immunodeficiency virus reverse transcriptase, inhibited by the diphosphate (9); herpes simplex virus DNA polymerase, inhibited by the diphosphate (8,10); herpes viruses replication (6); HIV-1 and HIV-2 (4,6,7,11); human immunodeficiency virus reverse transcriptase, inhibited by the diphosphate (4,11); purine-nucleoside phosphorylase, inhibited by the monophosphate (12); Rauscher murine leukemia virus reverse transcriptase (11); ribose-phosphate diphosphokinase (13); and RNA-directed DNA polymerase (3,4,6-11). **1.** Balzarini & De Clercq (1991) *J. Biol. Chem.* **266**, 8686. **2.** Shoshani, Laux, Perigaud, Gosselin & Johnson (1999) *J. Biol. Chem.* **274**, 34742. **3.** Votruba, Travnicek, Rosenberg, *et al.* (1990) *Antiviral Res.* **13**, 287. **4.** De Clercq (1992) *AIDS Res. Human Retrovir.* **8**, 119. **5.** Kramata, Votruba, Otova & Holy (1996) *Mol. Pharmacol.* **49**, 1005. **6.** Prisbe & Chen (1996) *Meth. Enzymol.* **275**, 425. **7.** Balzarini & De Clercq (1996) *Meth. Enzymol.* **275**, 472. **8.** Merta, Votruba, Rosenberg, *et al.* (1990) *Antiviral Res.* **13**, 209. **9.** Cronn, Remington, Preston & North (1992) *Biochem. Pharmacol.* **44**, 1375. **10.** Foster, Cerny & Cheng (1991) *J. Biol. Chem.* **266**, 238. **11.** Cherrington, Fuller, Mulato, *et al.* (1996) *Antimicrob. Agents Chemother.* **40**, 1270. **12.** Bzowska, Kulikowska & Shugar (2000) *Pharmacol. Ther.*

88, 349. **13**. Balzarini, Nave, Becker, Tatibana & De Clercq (1995) *Nucleosides Nucleotides* **14**, 1861.

9-[2-(Phosphonomethoxy)ethyl]guanine

This acyclic GMP analogue (FW$_{free-acid}$ = 289.19 g/mol), often abbreviated PMEG, exhibits a broad spectrum of antiviral activity. The diphosphate (*i.e.*, an analogue of GTP) inhibits avian myeloblastosis virus reverse transcriptase and herpes simplex virus DNA polymerase. **Target(s):** avian myeloblastosis virus reverse transcriptase, inhibited by the diphosphate (1); DNA-directed DNA polymerase (2); DNA polymerases α and ε, inhibited by the diphosphate (2); GMP kinase, also alternative substrate (3); herpes simplex virus DNA polymerase, inhibited by the diphosphate (4). **1**. Votruba, Travnicek, Rosenberg, *et al.* (1990) *Antiviral Res.* **13**, 287. **2**. Kramata, Votruba, Otova & Holy (1996) *Mol. Pharmacol.* **49**, 1005. **3**. Krejcova, Horska, Votruba & Holy (2000) *Biochem. Pharmacol.* **60**, 1907. **4**. Merta, Votruba, Rosenberg, *et al.* (1990) *Antiviral Res.* **13**, 209.

9-[2-(Phosphonomethoxy)ethyl]guanine Diphosphate, *See 9-[2-(Phosphonomethoxy)-ethyl]guanine*

9-[2-(Phosphonomethoxy)ethyl]-6-thioguanine

This acyclic GMP analogue (FW$_{free-acid}$ = 304.24 g/mol), also known as 2-amino-9-[2-(phosphonomethoxy)ethyl]-6-sulfanylpurine, inhibits purine-nucleoside phosphorylase (1-3). **1**. Wielgus-Kutrowska & Bzowska (2006) *Biochim. Biophys. Acta* **1764**, 887. **2**. Wielgus-Kutrowska, Antosiewicz, Dlogosz, Holy & Bzowska (2007) *Biophys. Chem.* **125**, 260. **3**. Antosiewicz, Wielgus-Kutrowska, Dlugosz, Holy & Bzowska (2007) *Nucleosides Nucleotides Nucleic Acids* **26**, 969.

1-[2-(Phosphonomethoxy)ethyl]thymine

This thymine derivative (FW$_{ion}$ = 262.16 g/mol) inhibits rat T-cell lymphoma thymidine phosphorylase, K_i = 1.5 μM (1,2). **1**. Votruba, Pomeisl, Tloust'ová, Holy & Otová (2005) *Biochem. Pharmacol.* **69**, 1517. **2**. Pomeisl, Holy, Votruba & Pohl (2008) *Bioorg. Med. Chem. Lett.* **18**, 1364.

(R)-9-[2-(Phosphonomethoxy)propyl]adenine

This acyclic AMP analogue (FW$_{free-acid}$ = 287.22 g/mol), also known as (*R*)-9-(2-phosphonylmethoxypropyl)adenine and tenofovir and abbreviated PMPA, exhibits a broad spectrum of antiviral activity. The diphosphate (*i.e.*, an analogue of ATP) inhibits reverse transcriptases. **Target(s):** HIV-1 reverse transcriptase, inhibited by the diphosphate (1-4); purine-nucleoside phosphorylase (5); and RNA-directed DNA polymerase (1-4). **1**. Balzarini & De Clercq (1996) *Meth. Enzymol.* **275**, 472. **2**. Suo & Johnson (1998) *J. Biol. Chem.* **273**, 27250. **3**. Cihlar, Birkus, Greenwalt & Hitchcock (2002) *Antiviral Res.* **54**, 37. **4**. El Safadi, Vivet-Boudou & Marquet (2007) *Appl. Microbiol. Biotechnol.* **75**, 723. **5**. Bzowska, Kulikowska & Shugar (2000) *Pharmacol. Ther.* **88**, 349.

(S)-1-[2-(Phosphonomethoxy)propyl]thymine

This thymine derivative (FW$_{ion}$ = 263.17 g/mol) inhibits rat T-cell lymphoma thymidine phosphorylase, K_i = 0.68 μM (1). The (*R*)-enantiomer is a slightly weaker inhibitor (K_i = 1.90 μM). **1**. Votruba, Pomeisl, Tloust'ová, Holy & Otová (2005) *Biochem. Pharmacol.* **69**, 1517.

N-(Phosphonomethyl)iminodiacetic Acid

This modified triacid (FW = 224.08 g/mol) inhibits adenosine kinase (1,2) and ribokinase (3). Because the use of adenosine receptor agonists is plagued by dose-limiting cardiovascular side effects, there is great interest in developing adenosine kinase inhibitors (AKIs) as a means for raising steady-state adenosine concentrations. **1**. Park, Singh, Maj & Gupta (2004) *Protein J.* **23**, 167. **2**. Park, Singh & Gupta (2006) *Mol. Cell. Biochem.* **283**, 11. **3**. Park, van Koeverden, Singh & Gupta (2007) *FEBS Lett.* **581**, 3211.

2-(Phosphonomethyl)pentanedioate

This hygroscopic phosphono-dicarboxylic acid (FW$_{free-acid}$ = 226.12 g/mol), often abbreviated 2-PMPA, is a competitive inhibitor of glutamate carboxypeptidase II, *or* N-acetylated-α-linked-acidic Dipeptidase, K_i = 0.275 nM (1-9). **1**. Jackson, Cole, Slusher, *et al.* (1996) *J. Med. Chem.* **39**, 619. **2**. Luthi-Carter, Barczak, Speno & Coyle (1998) *Brain Res.* **795**, 341. **3**. Rojas, Frazier, Flanary & Slusher (2002) *Anal. Biochem.* **310**, 50. **4**. Kozikowski, Zhang, Nan, *et al.* (2004) *J. Med. Chem.* **47**, 1729. **5**. Kozikowski, Nan, Conti, *et al.* (2001) *J. Med. Chem.* **44**, 298. **6**. Majer, Jackson, Delahanty, *et al.* (2003) *J. Med. Chem.* **46**, 1989. **7**. Ghadge, Slusher, Bodner, *et al.* (2003) *Proc. Natl. Acad. Sci. U.S.A.* **100**, 9554. **8**. Jackson, Tays, Maclin, *et al.* (2001) *J. Med. Chem.* **44**, 4170. **9**. Ajit, Corse, Coccia, *et al.* (2003) *Eur. J. Pharmacol.* **471**, 177.

4-(Phosphonomethyl)pyridine-2-carboxylic Acid

This homoserine phosphate analogue (FW$_{ion}$ = 214.09 g/mol) inhibits cystathionine γ-synthase, with K_i values of 45 and 300 μM for the wheat and tobacco enzymes, respectively (1). **1**. Kreft, Townsend, Pohlenz &

Laber (1994) *Plant Physiol.* **104**, 1215. **2.** Clausen, Wahl, Messerschmidt, *et al.* (1999) *Biol. Chem.* **380**, 1237.

3-Phosphonopropionate (3-Phosphonopropionic Acid)

This succinate analogue (FW$_{free-acid}$ = 154.06 g/mol), also known as 2-carboxyethylphosphonic acid,. Inhibits adenosine kinase, thereby raising steady-state adenosine concentrations and consequentially stimulating adenosine receptors. **Target(s):** adenosine kinase (1,2); aspartate carbamoyltransferase (3); carboxypeptidase A (4); dicarboxylate transporter (5); herpes simplex virus type-1 DNA polymerase (6); phosphonoacetate hydrolase (7,8); phosphonopyruvate hydrolase (9); 2-phosphosulfolactate phosphatase, weakly inhibited (10); ribokinase (11); succinate dehydrogenase (12,13); and triose-phosphate isomerase (14,15). **1.** Park, Singh, Maj & Gupta (2004) *Protein J.* **23**, 167. **2.** Park, Singh & Gupta (2006) *Mol. Cell. Biochem.* **283**, 11. **3.** Laing, Chan, Hutchinson & Íberg (1990) *FEBS Lett.* **260**, 206. **4.** Grobelny, Goli & Galardy (1985) *Biochem. J.* **232**, 15. **5.** Ullrich, Rumrich, Burke, Shirazi-Beechey & Lang (1997) *J. Pharmacol. Exp. Ther.* **283**, 1223. **6.** Eriksson, Larsson, Helgstrand, Johansson & Oberg (1980) *Biochim. Biophys. Acta* **607**, 53. **7.** McGrath, Wisdom, McMullan, Larkin & Quinn (1995) *Eur. J. Biochem.* **234**, 225. **8.** McGrath, Kulakova & Quinn (1999) *J. Appl. Microbiol.* **86**, 834. **9.** Kulakova, Wisdom, Kulakov & Quinn (2003) *J. Biol. Chem.* **278**, 23426. **10.** Graham, Graupner, Xu & White (2001) *Eur. J. Biochem.* **268**, 5176. **11.** Park, van Koeverden, Singh & Gupta (2007) *FEBS Lett.* **581**, 3211. **12.** Webb (1966) *Enzyme and Metabolic Inhibitors*, vol. **2**, p. 242, Academic Press, New York. **13.** Seaman (1952) *Arch. Biochem. Biophys.* **39**, 241. **14.** Noble, Wierenga, Lambeir, *et al.* (1991) *Proteins* **10**, 50. **15.** Bell, Russell, Kohlhoff, *et al.* (1998) *Acta Crystallogr. Sect. D* **54**, 1419.

9-[3-(3-Phosphonoprop-1-yl)benzyl]guanine

This nucleoside analogue (FW$_{ion}$ = 344.25 g/mol) inhibits purine-nucleoside phosphorylase (K_i = 1.3 nM), a ubiquitous enzyme that plays a key role in the purine salvage pathway. In humans PNP deficiency impairs T-cell function, usually with no effect on B-cell function. This property has been exploited in the synthesis of much higher affinity transition-state analogues (*See Immucillins*) to treat T-cell lymphomas. **1.** Bzowska, Kulikowska & Shugar (2000) *Pharmacol. Ther.* **88**, 349.

Phosphon S, See *Tributyl-2,4-dichlorobenzyl-ammonium Chloride*

Phosphoramidon

This glycosyl phosphoramidopeptide (FW$_{free-acid}$ = 543.51 g/mol; CAS 36357-77-4), also systematically named as *N*-(α-rhamnopyranosyl-oxyhydroxyphosphinyl)-L-leucyl-L-tryptophan, is a slow-binding thermolysin inhibitor (K_i = 30 nM). *Note:* Phosphoramidon is an epimer of

talopeptin. Stock 1 mg/mL aqueous solutions of phosphoramidon are stable for one month at –20°C. **Target(s):** atrolysin A (56); atrolysin C (1-3); bacillolysin (4); carboxypeptidase C, weakly inhibited (59); coccolysin (5); cytosol alanyl aminopeptidase, weakly inhibited (62); dactylysin (6); endothelin-converting enzyme (17-13); endothelin-converting enzyme 2 (14,15); fungalysin (16); gametolysin (17-19); granzyme B (57); *Legionella* metalloendopeptidase (20,21); leucyl aminopeptidase, weakly inhibited (64); *Listeria* metalloprotease Mpl (22); lysine carboxypeptidase, *or* lysine(arginine) carboxypeptidase; carboxypeptidase N (58); mycolysin (23); neprilysin, *or* enkephalinase, K_i = 0.8 nM (14,24-34); oligopeptidase O (35); peptidyl-dipeptidase A, *or* angiotensin I-converting enzyme (60,61); pitrilysin, *or* protease Pi (36); pseudolysin, *or Pseudomonas aeruginosa* elastase (37-41); pyruvate decarboxylase (42); thermolysin (27,28,41,43-53); urethanase (54); vibriolysin (55); and Xaa-Pro aminopeptidase, *or* aminopeptidase P (63). **1.** Fox & Bjarnason (1998) in *Handb. Proteolytic Enzymes*, p. 1266, Academic Press, San Diego. **2.** Fox, Campbell, Beggerly & Bjarnason (1986) *Eur. J. Biochem.* **156**, 65. **3.** Shannon, Baramova, Bjarnason & Fox (1989) *J. Biol. Chem.* **264**, 11575. **4.** Van den Burg, Eijsink, Vriend, Veltman & Venema (1998) *Biotechnol. Appl. Biochem.* **27**, 125. **5.** Mäkinen, Clewell, An & Mäkinen (1989) *J. Biol. Chem.* **264**, 3325. **6.** Carvalho, Joudiou, Boussetta, Leseney & Cohen (1992) *Proc. Natl. Acad. Sci. U.S.A.* **89**, 84. **7.** Ahn (1998) in *Handb. Proteolytic Enzymes*, p. 1085, Academic Press, San Diego. **8.** Roden, Prskavec, Furnsinn, *et al.* (1994) *Regul. Pept.* **51**, 207. **9.** Xu, Emoto, Giaid, *et al.* (1994) *Cell* **78**, 473. **10.** Ahn, Herman & Fahnoe (1998) *Arch. Biochem. Biophys.* **359**, 258. **11.** Korth, Egidy, Parnot, *et al.* (1997) *FEBS Lett.* **417**, 365. **12.** Takahashi, Matsushita, Iijima & Tanzawa (1993) *J. Biol. Chem.* **268**, 21394. **13.** Hasegawa, Hiki, Sawamura, *et al.* (1998) *FEBS Lett.* **428**, 304. **14.** Ahn (1998) in *Handb. Proteolytic Enzymes*, p. 1090, Academic Press, San Diego. **15.** Emoto & Yanagisawa (1995) *J. Biol. Chem.* **270**, 15262. **16.** Kolattukudy & Sirakova (1998) in *Handb. Proteolytic Enzymes*, p. 1498, Academic Press, San Diego. **17.** Matsuda (1998) in *Handb. Proteolytic Enzymes*, p. 1140, Academic Press, San Diego. **18.** Jaenicke, Kuhne, Spessert, Wahle & Waffenschmidt (1987) *Eur. J. Biochem.* **170**, 485. **19.** Matsuda, Saito, Yamaguchi & Kawase (1985) *J. Biol. Chem.* **260**, 6373. **20.** Woessner (1998) in *Handb. Proteolytic Enzymes*, p. 1064, Academic Press, San Diego. **21.** Black, Quinn & Tompkins (1990) *J. Bacteriol.* **172**, 2608. **22.** Broer, Wehland & Chakraborty (1998) in *Handb. Proteolytic Enzymes*, p. 1050, Academic Press, San Diego. **23.** Ishii & Kumazaki (1998) in *Handb. Proteolytic Enzymes*, p. 1078, Academic Press, San Diego. **24.** Schulz, Sakane, Berry & Ghai (1991) *J. Enzyme Inhib.* **4**, 347. **25.** Deddish, Marcic, Tan, *et al.* (2002) *Hypertension* **39**, 619. **26.** Jeng, Ansell & Erion (1989) *Life Sci.* **45**, 2109. **27.** Marie-Claire, Ruffet, Tiraboschi & Fournié-Zaluski (1998) *FEBS Lett.* **438**, 215. **28.** Marie-Claire, Ruffet, Antonczak, *et al.* (1997) *Biochemistry* **36**, 13938. **29.** Isaac (1988) *Biochem. J.* **255**, 843. **30.** Fulcher & Kenny (1983) *Biochem. J.* **211**, 743. **31.** Landry, Santagata, Bawab, *et al.* (1993) *Biochem. J.* **291**, 773. **32.** Lian, Wu, Konings, Mierau & Hersh (1996) *Arch. Biochem. Biophys.* **333**, 121. **33.** Welches, Broshihan & Ferrario (1993) *Life Sci.* **52**, 1461. **34.** Malfroy & Schwartz (1982) *Biochem. Biophys. Res. Commun.* **106**, 276. **35.** Mierau & Kok (1998) in *Handb. Proteolytic Enzymes*, p. 1091, Academic Press, San Diego. **36.** Fricke, Betz & Friebe (1995) *J. Basic Microbiol.* **35**, 21. **37.** Kessler & Ohman (1998) in *Handb. Proteolytic Enzymes*, p. 1058, Academic Press, San Diego. **38.** Morihara (1995) *Meth. Enzymol.* **248**, 242. **39.** Kessler & Spierer (1984) *Curr. Eye Res.* **3**, 1075. **40.** McKay, Thayer, Flaherty, Pley & Benvegnu (1992) *Matrix Suppl.* **1**, 112. **41.** Nishino & Powers (1980) *J. Biol. Chem.* **255**, 3482. **42.** de la Plaza, Fernandez de Palencia, Pelaez & Requena (2004) *FEMS Microbiol. Lett.* **238**, 367. **43.** Beynon & Beaumont (1998) in *Handb. Proteolytic Enzymes*, p. 1037, Academic Press, San Diego. **44.** Umezawa (1976) *Meth. Enzymol.* **45**, 678. **45.** Knight (1995) *Meth. Enzymol.* **248**, 85. **46.** Komiyama, Suda, Aoyagi, Takeuchi & Umezawa (1975) *Arch. Biochem. Biophys.* **171**, 727. **47.** Suda, Aoyagi, Takeuchi & Umezawa (1973) *J. Antibiot.* **26**, 621. **48.** Durrant, Beynon & Rodgers (1986) *Biochem. Soc. Trans.* **14**, 143. **49.** Kam, Nishino & Powers (1979) *Biochemistry* **18**, 3032. **50.** Gettins (1988) *J. Biol. Chem.* **263**, 10208. **51.** Kitagishi & Hiromi (1986) *J. Biochem.* **99**, 191. **52.** Kitagishi & Hiromi (1984) *J. Biochem.* **95**, 529. **53.** Weaver, Lester & Matthews (1977) *J. Mol. Biol.* **114**, 119. **54.** Zhao & Kobashi (1994) *Biol. Pharm. Bull.* **17**, 773. **55.** Durham (1998) in *Handb. Proteolytic Enzymes*, p. 1053, Academic Press, San Diego. **56.** Fox, Campbell, Beggerly & Bjarnason (1986) *Eur. J. Biochem.* **156**, 65. **57.** Poe, Blake, Boulton, *et al.* (1991) *J. Biol. Chem.* **266**, 98. **58.** Barabé & Huberdeau (1991) *Biochem. Pharmacol.* **41**, 821. **59.** Liu, Tachibana, Taira, *et al.* (2004) *J. Ind. Microbiol. Biotechnol.* **31**, 572. **60.** Kase, Hazato, Shimamura, Kiuchi & Katayama (1985) *Arch. Biochem. Biophys.*

240, 330. **61**. Garats, Nikolskaya, Binevski, Pozdnev & Kost (2001) *Biochemistry (Moscow)* **66**, 429. **62**. Yamamoto, Li, Huang, Ohkubo & Nishi (1998) *Biol. Chem.* **379**, 711. **63**. Ryan, Valido, Berryer, Chung & Ripka (1992) *Biochim. Biophys. Acta* **1119**, 140. **64**. Morty & Morehead (2002) *J. Biol. Chem.* **277**, 26057.

N-(5'-Phosphoribityl)anthranilate

This inhibitory deoxyphosphoribityl-anthranilate (FW$_{\text{free-acid}}$ = 351.25 g/mol) targets N-(5-phosphoribosyl)anthranilate isomerase/indoleglycerol-P synthase (1-6). **1**. Kirschner, Szadkowski, Jardetzky & Hager (1987) *Meth. Enzymol.* **142**, 386. **2**. Bisswanger, Kirschner, Cohn, Hager & Hansson (1979) *Biochemistry* **18**, 5946. **3**. Eberhard, Tsai-Pflugfelder, Bolewska, Hommel & Kirschner (1995) *Biochemistry* **34**, 5419. **4**. Hommel, Eberhard & Kirschner (1995) *Biochemistry* **34**, 5429. **5**. Priestle, Gruetter, White, *et al.* (1987) *Proc. Natl. Acad. Sci. U.S.A.* **84**, 5690. **6**. Sterner, Kleemann, Szadkowski, *et al.* (1996) *Protein Sci.* **5**, 2000.

5-Phosphoribonic Acid

D-isomer   L-isomer

This phosphorylated aldonic acid (FW$_{\text{free-acid}}$ = 246.11 g/mol), also known as ribonic acid 5-phosphate, inhibits arabinose-5-phosphate isomerase (1), 6-phosphogluconate dehydrogenase (2), phosphoribulokinase (3), and ribose-5-phosphate isomerase (3-6). **1**. Bigham, Gragg, Hall, *et al.* (1984) *J. Med. Chem.* **27**, 717. **2**. Price & Cook (1996) *Arch. Biochem. Biophys.* **336**, 215. **3**. Webb (1966) *Enzyme and Metabolic Inhibitors*, vol. 2, pp. 411 & 413, Academic Press, New York. **4**. Axelrod (1955) *Meth. Enzymol.* **1**, 363. **5**. Topper (1961) *The Enzymes*, 2nd ed., **5**, 429. **6**. Woodruff III & Wolfenden (1979) *J. Biol. Chem.* **254**, 5866.

Phosphoribosylaminoimidazole-succinocarboxamide

This purine nucleotide intermediate (FW$_{\text{free-acid}}$ = 451.26 g/mol), abbreviated SAICAR and also known as 4-(N-succinocarboxamide)-5-aminoimidazole ribotide, inhibits adenylosuccinate synthetase. **1**. Ryzhova, Andreichuk & Domkin (1998) *Biochemistry (Moscow)* **63**, 650.

1-(5'-Phospho-D-ribosyl)barbituric Acid

This UMP and OMP analogue (FW$_{\text{free-acid}}$ = 340.17 g/mol), also known as 6-hydroxyuridine 5'-phosphate and barbiturate ribofuranoside 5'-monophosphate, inhibits orotidine 5'-phosphate decarboxylase (1-10). **1**.

Potvin, Stern, May, Lam & Krooth (1978) *Biochem. Pharmacol.* **27**, 655. **2**. Bell & Jones (1991) *J. Biol. Chem.* **266**, 12662. **3**. Levine, Brody & Westheimer (1980) *Biochemistry* **19**, 4993. **4**. Harris, Poulsen, Jensen & Larsen (2002) *J. Mol. Biol.* **318**, 1019. **5**. Harris, Poulsen, Jensen & Larsen (2000) *Biochemistry* **39**, 4217. **6**. Porter & Short (2000) *Biochemistry* **39**, 11788. **7**. Miller, Butterfoss, Short & Wolfenden (2001) *Biochemistry* **40**, 6227. **8**. Wu, Gillon & Pai (2002) *Biochemistry* **41**, 4002. **9**. Miller, Snider, Wolfenden & Short (2001) *J. Biol. Chem.* **276**, 15174. **10**. Sievers & Wolfenden (2005) *Bioorg. Chem.* **33**, 45.

5-Phospho-α-D-ribosyl 1-Pyrophosphate

This key metabolite (FW$_{\text{free-acid}}$ = 390.07 g/mol), abbreviated PRPP and also known as 5-phospho-α-D-ribosyl diphosphate and 5-phosphoribose 1-diphosphate, is an intermediate in the biosynthesis of the purines, pyrimidines, and histidine. It is also an essential co-substrate in purine salvage (*e.g.*, Hypoxanthine:Guanine Phosphoribosyltransferase). PRPP is completely hydrolyzed to ribose 5-phosphate and pyrophosphate in 20 min at pH 3.0 and 65°C, whereas half is hydrolyzed at pH 6.7. Aqueous solutions are stable, if kept frozen at pH 8.0. The solid should be stored in a desiccator and protected from light. **Target(s):** NMN nucleosidase (1); purine-nucleoside phosphorylase (2); ribose-phosphate diphosphokinase, *or* phosphorinosyl-pyrophosphate synthetase (3-8); and transketolase (9). **1**. Imai (1987) *J. Biochem.* **101**, 163. **2**. Ropp & Traut (1991) *Arch. Biochem. Biophys.* **288**, 614. **3**. Becker, Kostel & Meyer (1975) *J. Biol. Chem.* **250**, 6822. **4**. Fox & Kelley (1972) *J. Biol. Chem.* **247**, 2126. **5**. Hove-Jensen, Bentsen & Harlow (2005) *FEBS J.* **272**, 3631. **6**. Gallois, Prevot, Clement & Jacob (1997) *Plant Physiol.* **115**, 847. **7**. Krath & Hove-Jensen (2001) *Protein Sci.* **10**, 2317. **8**. Krath & Hove-Jensen (2001) *J. Biol. Chem.* **276**, 17851. **9**. Hosomi, Tara, Terada & Mizoguchi (1989) *Biochem. Med. Metab. Biol.* **42**, 52.

N$^{\alpha}$-Phosphoryl-L-alanyl-L-proline

This phosphorylated dipeptide (FW$_{\text{free-acid}}$ = 266.19 g/mol) is a strong inhibitor of angiotensin I-converting enzyme, *or* peptidyl-dipeptidase A, K_i = 9 nM (pH 8.3) and 1.4 nM (pH 7.5) (1-3). **1**. Galardy (1980) *Biochem. Biophys. Res. Commun.* **97**, 94. **2**. Galardy (1982) *Biochemistry* **21**, 5777. **3**. Ondetti & Cushman (1982) *Ann. Rev. Biochem.* **51**, 283.

N-Phosphoryl-L-leucinamide

This phosphorylated amino acid amide (FW$_{\text{free-acid}}$ = 210.17 g/mol) inhibits thermolysin, K_i = 1.3 μM (1). Upon binding to the enzyme, two of the phosphoramidate oxygens of the inhibitor interact with the protein's zinc ion to form a complex that approximates a pentacoordinate system and mimics the transition-state. The O-methyl phosphoester is a weaker thermolysin inhibitor (K_i = 150 μM). **Target(s):** carboxypeptidase A, K_i = 160 μM (1); and thermolysin, K_i = 1.3 μM (1-4). **1**. Kam, Nishino & Powers (1979) *Biochemistry* **18**, 3032. **2**. Beynon & Beaumont (1998) in *Handb. Proteolytic Enzymes* (Barrett, Rawlings & Woessner, eds.), p. 1037,

Academic Press, San Diego. **3**. Tronrud, Monzingo & Matthews (1986) *Eur. J. Biochem.* **157**, 261. **4**. Gettins (1988) *J. Biol. Chem.* **263**, 10208.

N^{α}-Phosphoryl-L-leucyl-L-phenylalanine

This phosphorylated dipeptide ($FW_{\text{free-acid}}$ = 358.33 g/mol) inhibits neprilysin, *or* enkephalinase (1), pseudolysin, K_i = 0.2 µM (2-4), and thermolysin, K_i = 19 nM (2,5). **1**. Altstein, Blumberg & Vogel (1981) *Eur. J. Pharmacol.* **76**, 299. **2**. Kessler & Ohman (1998) in *Handb. Proteolytic Enzymes*, p. 1058, Academic Press, San Diego. **3**. Kessler, Spierer & Blumberg (1983) *Invest. Ophthalmol. Vis. Sci.* **24**, 1093. **4**. Kessler, Israel, Landshman, Chechick & Blumberg (1982) *Infect. Immun.* **38**, 716. **5**. Kam, Nishino & Powers (1979) *Biochemistry* **18**, 3032.

N^{α}-Phosphoryl-L-leucyl-L-tryptophan

This phosphorylated dipeptide ($FW_{\text{free-acid}}$ = 397.37 g/mol) inhibits carboxypeptidase A, K_i = 54 µM (1), pseudolysin (2,3), and thermolysin, K_i = 15 nM (1,2,4). **1**. Kam, Nishino & Powers (1979) *Biochemistry* **18**, 3032. **2**. Kessler & Ohman (1998) in *Handb. Proteolytic Enzymes* (Barrett, Rawlings & Woessner, eds.), p. 1058, Academic Press, San Diego. **3**. Poncz, Gerken, Dearborn, Grobelny & Galardy (1984) *Biochemistry* **23**, 2766. **4**. Kitagishi & Hiromi (1986) *J. Biochem.* **99**, 191.

O-Phospho-L-serine

This phosphomonoester of L-serine (FW = 183.06 g/mol), also called serine phosphate, is an intermediate in L-serine biosynthesis and an agonist at metabotropic the AP4 receptors $GluR_4$, $GluR_8$, and $GluR_7$. O-Phospho-L-serine concentrations are elevated in Alzheimer's Disease brains. In addition, O-phospho-L-seryl residues are generated as intermediates in enzyme-catalyzed reactions and specific proteins are phosphorylated at seryl residues by the action of a number of protein kinases. O-Phospho-L-serine is soluble in water (2.8 g/100 mL at 20°C for the free acid and 25 g/100 mL for the sodium salt) and is hydrolyzed in acidic or basic conditions (61% hydrolysis in 24 h in 1 M HCl and 51% hydrolysis in 24 h in 0.5 M NaOH). **Target(s):** O-acetylhomoserine aminocarboxypropyl-transferase, *or* O-acetylhomoserine sulfhydrylase (1,2); arylsulfatase B (3); aspartate ammonia-lyase (4); choline acetyltransferase (5); γ-glutamylcysteine synthetase (6); γ-glutamyl transpeptidase (7); glycine transport (8); homoserine kinase (9); kynurenine:oxoglutarate aminotransferase (10); ornithine decarboxylase (11); phosphoglycerate phosphatase (12); phosphoprotein phosphatase (13); serine hydroxymethyl-transferase, *or* glycine hydroxymethyltransferase (14); serine C-

palmitoyltransferase (15); serine-sulfate ammonia-lyase (16-19); and threonine synthase (20). **1**. Yamagata (1987) *Meth. Enzymol.* **143**, 465. **2**. Yamagata (1984) *J. Biochem.* **96**, 1511. **3**. Rao & Christe (1984) *Biochim. Biophys. Acta* **788**, 58. **4**. Giorgianni, Beranova, Wesdemiotis & Viola (1997) *Arch. Biochem. Biophys.* **341**, 329. **5**. Andriamampandry & Kanfer (1993) *Neurobiol. Aging* **14**, 367. **6**. Yip & Rudolph (1976) *J. Biol. Chem.* **251**, 3563. **7**. Butler & Spielberg (1979) *J. Biol. Chem.* **254**, 3152. **8**. Benavides, Garcia, Lopez-Lahoya, Ugarte & Valdivieso (1980) *Biochim. Biophys. Acta* **598**, 588. **9**. Huo & Viola (1996) *Biochemistry* **35**, 16180. **10**. Battaglia, Rassoulpour, Wu, *et al.* (2000) *J. Neurochem.* **75**, 2051. **11**. Tsirka, Sklaviadis & Kyriakidis (1986) *Biochim. Biophys. Acta* **884**, 482. **12**. Fallon & Byrne (1965) *Biochim. Biophys. Acta* **105**, 43. **13**. Polya & Haritou (1988) *Biochem. J.* **251**, 357. **14**. Rao & Rao (1982) *Plant Physiol.* **69**, 11. **15**. Ikushiro, Hayashi & Kagamiyama (2001) *J. Biol. Chem.* **276**, 18249. **16**. Tudball & Thomas (1971) *Meth. Enzymol.* **17B**, 361. **17**. Tudball & Thomas (1973) *Eur. J. Biochem.* **40**, 25. **18**. Murakoshi, Sanda & Haginiwa (1977) *Chem. Pharm. Bull.* **25**, 1829. **19**. Thomas & Tudball (1967) *Biochem. J.* **105**, 467. **20**. Giovanelli, Veluthambi, Thompson, Mudd & Datko (1984) *Plant Physiol.* **76**, 285.

Phosphostim

This isopentenyl pyrophosphate bromohydrin (FW = 340.99 g/mol) inhibits geranyl*trans*transferase, *or* farnesyl-diphosphate synthase (1) and isopentenyl-diphosphate Δ-isomerase, *or* isopentenyl pyrophosphate/dimethylallyl pyrophosphate isomerase (2). **1**. Montalvetti, Fernandez, Sanders, *et al.* (2003) *J. Biol. Chem.* **278**, 17075. **2**. Wouters, Yin, Song, *et al.* (2005) *J. Amer. Chem. Soc.* **127**, 536.

Phosphosulfate, See *Thiophosphate, Sodium*

O-Phospho-L-threonine

This phosphorylated amino acid (FW_{ion} = 198.09 g/mol), also called threonine phosphate, is found as a residue in the amino acid sequence of a number of proteins. These residues are generated as intermediates in enzyme-catalyzed reactions and specific proteins are phosphorylated at threonyl residues by the action of a number of protein kinases. **Target(s):** phosphoprotein phosphatase (1); phosphoserine aminotransferase (2); and threonine synthase (3-5). **1**. Polya & Haritou (1988) *Biochem. J.* **251**, 357. **2**. Ali & Nozaki (2006) *Mol. Biochem. Parasitol.* **145**, 71. **3**. Webb (1966) *Enzyme and Metabolic Inhibitors*, vol. 2, p. 357, Academic Press, New York. **4**. Flavin & Slaughter (1960) *J. Biol. Chem.* **235**, 1103. **5**. Giovanelli, Veluthambi, Thompson, Mudd & Datko (1984) *Plant Physiol.* **76**, 285.

5'-Phosphothymidine 3'-Phosphate, See *Thymidine 3',5'-Bisphosphate*

Phosphotrienin, See *Fostriecin*

O-Phospho-L-tyrosine

This phosphorylated amino acid (FW = 260.16 g/mol), also called tyrosine phosphate, is found as a residue in the amino acid sequence of a number of proteins. Specific proteins are phosphorylated at tyrosyl residues by the action of a number of protein kinases. **Target(s):** ornithine decarboxylase (1); phosphoprotein phosphatase (2); and protein-tyrosine phosphatase (3,4). **1**. Tsirka, Sklaviadis & Kyriakidis (1986) *Biochim. Biophys. Acta*

884, 482. **2**. Polya & Haritou (1988) *Biochem. J.* **251**, 357. **3**. Ballou & Fischer (1986) *The Enzymes*, 3rd ed. (Boyer & Krebs, eds.), **17**, 311. **4**. Hörlein, Gallis, Brautigan & Bornstein (1982) *Biochemistry* **21**, 5577.

Phrixotoxin-1 and -2

These spider venom oligopeptides (Phrixotoxin-1 *Sequence*: TCQKWMK TCDSARKCCEGLVCRLWCKKII; MW = 3435 g/mol; Phrixotoxin-2 *Sequence*: YCQKWMWTCEERDKCCEGLVCRLWCKRIINM; MW = 3929 g/mol) from the tarantula *Phrixotrichus auratus*, are high-affinity,reversible voltage-gated $K_V4.2$ and $K_V4.3$ channel blockers (1,2). Phrixotoxin-1 and -2 each have three intrachain disulfide bonds. **1**. Diochot, Drici, Moinier, Fink & Lazdunski (1999) *Brit. J. Pharmacol.* **126**, 251. **2**. Chagot, Escoubas, Villegas, *et al.* (2004) *Protein Sci.* **13**, 1197.

PHT-427

This orally bioavailable, dual Akt/PDPK1 inhibitor (F.Wt. = 409.61; CAS 1191951-57-1 and 1178893-77-0; Solubility (25°C): 80 mg/mL DMSO, <1 mg/mL Water), known systematically as 4-dodecyl-*N*-(1,3,4-thiadiazol-2-yl)benzenesulfonamide, targets Akt (also known as Protein Kinase B, *or* PKB) and 3-phosphoinositide-dependent protein kinase-1, with K_i of 2.7 μM and 5.2 μM, respectively (1). PHT-427 was designed to bind to the pleckstrin homology (PH) auto-inhibitory domains of the signal cascade protein kinase Akt (2). **1**. Meuillet, *et al.* (2010) *Mol. Cancer Ther.* **9**, 706. **2**. Moses, *et al.* (2009) *Cancer Res.* **69**, 5073. **2**. Mahadevan, Powis, Mash, *et al.* (2008) *Mol. Cancer Ther.* **7**, 2621.

o-Phthalaldehyde

This dialdehyde (FW = 134.13 g/mol; M.P. = 55-58°C), also known as *o*-phthaldialdehyde and benzene-1,2-dicarboxaldehyde reacts readily with primary amines (*e.g.*, of amino acids) to produce a fluorescent derivative. Hence, this derivatization reagent is frequently used in chromatography and amino acid analysis. It has also been used as a bifunctional cross-linking reagent. In octopine dehydrogenase, *o*-phthalaldehyde binds to proximal cysteinyl and lysyl residues leading to the formation of an isoindole derivative (1). **Target(s):** alcohol dehydrogenase (2); aminolevulinate aminotransferase (31); α-amylase (3); *Aspergillus* nuclease S (14); biliverdin reductase (5); Ca^{2+}-dependent ATPase (6); creatine kinase (7,8); ζ-crystalline9; cytochrome b_6f complex (10); dextransucrase (11-13); electron transport (14); fatty-acid synthase (15-17); glucan 1,4-β-glucosidase (18); glutamate dehydrogenase (19); glutathione-disulfide reductase (20); glycerol dehydrogenase (21); glyoxylate synthetase (22); 3-ketoacyl-[acyl-carrier protein] synthase (16,17); malate dehydrogenase (23); nitrogenase (24); octopine dehydrogenase (1,25); oxidative phosphorylation (14); 6-phosphofructo-2-kinase (26); porphobilinogen synthase, *or* 5-aminolevulinate dehydratase (27); pyruvate kinase (28); pyruvate carboxylase (29); and succinate-semialdehyde dehydrogenase (30). **1**. Sheikh & Katiyar (1994) *J. Enzyme Inhib.* **8**, 39. **2**. Le, Yan, Huang, Zhang & Zhou (1994-1995) *Enzyme Protein* **48**, 183. **3**. Ueyama, Chiba & Kobayashi (1995) *Biosci. Biotechnol. Biochem.* **59**, 864. **4**. Gite & Shankar (1995) *J. Mol. Recognit.* **8**, 281. **5**. Frydman, Tomaro, Rosenfeld & Frydman (1990) *Biochim. Biophys. Acta* **1040**, 119. **6**. Khan, Starling, East & Lee (1996) *Biochem. J.* **317**, 439. **7**. Wang, Xu & Zhou (1994-1995) *Enzyme Protein* **48**, 1. **8**. Sheikh, Mukunda & Katiyar (1993) *Biochim. Biophys. Acta* **1203**, 276. **9**. Bazzi, Rabbani & Duhaiman (2002) *Biochim. Biophys. Acta* **1597**, 67. **10**. Bhagwat, Blokesch, Irrgang, Salnikow & Vater (1993) *Arch. Biochem. Biophys.* **304**, 38. **11**. Goyal & Katiyar (1998) *J. Enzyme Inhib.* **13**, 147. **12**. Goyal & Katiyar (1995) *Biochem. Mol. Biol. Int.* **36**, 579. **13**. Majumder, Purama & Goyal (2007) *Indian J. Microbiol.* **47**, 197. **14**. White & Elliott (1980) *Can. J. Biochem.* **58**, 9. **15**. Mukherjee & Katiyar (1999) *Indian J. Biochem. Biophys.* **36**, 63. **16**. Wakil & Stoops (1983) *The Enzymes*, 3rd ed. (Boyer, ed.), **16**, 3. **17**. Stoops, Henry &

Wakil (1983) *J. Biol. Chem.* **258**, 12482. **18**. Jagtap & Rao (2005) *Biochem. Biophys. Res. Commun.* **329**, 111. **19**. Ahn, Lee, Choi & Cho (2000) *Mol. Cells* **10**, 25. **20**. Pandey & Katiyar (1996) *J. Enzyme Inhib.* **11**, 141. **21**. Pandey & Iyengar (2002) *J. Enzyme Inhib. Med. Chem.* **17**, 49. **22**. Janave, Ramaswamy & Nair (1993) *Eur. J. Biochem.* **214**, 889. **23**. Sheikh & Katiyar (1995) *J. Enzyme Inhib.* **9**, 235. **24**. Miller, Eady, Gormal, Fairhurst & Smith (1997) *Biochem. J.* **322**, 737. **25**. Sheikh & Katiyar (1993) *Biochem. Mol. Biol. Int.* **29**, 719. **26**. Rider & Hue (1989) *Biochem. J.* **262**, 97. **27**. Block, Lohmann & Beyersmann (1990) *Biol. Chem. Hoppe-Seyler* **371**, 1145. **28**. Yilmaz & Ozer (1990) *Arch. Biochem. Biophys.* **279**, 32. **29**. Werneburg & Ash (1993) *Arch. Biochem. Biophys.* **303**, 214. **30**. Blaner & Churchich (1979) *J. Biol. Chem.* **254**, 1794. **31**. Neuberger & Turner (1963) *Biochim. Biophys. Acta* **67**, 342.

m-Phthalamide

This amide (FW = 140.14 g/mol) inhibits bovine poly(ADP-ribose) polymerase, $IC_{50} = 0.05$ mM (1). The *o*-analogue is a weaker inhibitor (IC_{50} = 1 mM). **1**. Banasik, Komura, Shimoyama & Ueda (1992) *J. Biol. Chem.* **267**, 1569.

Phthalascidin

This semisynthetic antitumor agent ($FW_{free-base}$ = 650.69 g/mol), also known as Pt 650, binds to the minor groove of DNA and inhibits DNA topoisomerase I (1,2). Phthalascidin is prepared from ecteinascidin 743. **1**. Meng, Liao & Pommier (2003) *Curr. Top. Med. Chem.* **3**, 305. **2**. Martinez, Owa, Schreiber & Corey (1999) *Proc. Natl. Acad. Sci. U.S.A.* **96**, 3496.

Phthalate (Phthalic Acid)

This aromatic acid ($FW_{free-acid}$ = 166.13 g/mol; M.P. = ~230°C), also known as 1,2-benzenedicarboxylic acid, is readily converted to the anhydride when heated. The free acid is soluble in water (1 g/160 mL) and has pK_a values (at 25°C) of 2.95 ($dpK_a/dT = 0.0020$) and 5.41 ($dpK_a/dT = -0.001$ at T < 25°C and +0.001 at T > 25°C). **Target(s):** aconitase (1); aspartate aminotransferase (2); aspartate 4-decarboxylase (3); benzoylformate decarboxylase (4); γ-butyrobetaine dioxygenase, *or* γ-butyrobetaine hydroxylase, K_i = 17 mM (5); dehydro-L-gulonate decarboxylase (6); 4,5-dihydroxyphthalate decarboxylase (7); diphospho-mevalonate decarboxylase (8,9); glucose oxidase (10); glutamate decarboxylase (11); kynurenine aminotransferase (12); nicotinate-nucleotide diphosphorylase, carboxylating, *or* quinolinate phosphoribosyltransferase (13-18); peptidyl-dipeptidase A, *or* angiotensin I-converting enzyme (19); procollagen-proline 3-dioxygenase (20); succinate dehydrogenase (21); and tyrosine decarboxylase (22). **1**. Glusker (1971) *The Enzymes*, 3rd ed. (Boyer, ed.), **5**, 413. **2**. Rej & Horder (1984) in *Methods in Enzymatic Analysis* (Bergmeyer, ed.) **3**, 416, Verlag Chemie, Weinheim. **3**. Soda, Novogrodsky & Meister (1964) *Biochemistry* **3**, 1450. **4**. Weiss, Garcia, Kenyon, Cleland & Cook (1988) *Biochemistry* **27**, 2197. **5**. Ng, Hanauske-Abel & Englard (1991) *J. Biol. Chem.* **266**, 1526. **6**. Kagawa & Shimazono (1970) *Meth.*

Enzymol. **18A**, 46.   **7.** Nakazawa & Hayashi (1978) *Appl. Environ. Microbiol.* **36**, 264.   **8.** Cardemil & Jabalquinto (1985) *Meth. Enzymol.* **110**, 86.   **9.** Alvear, Jabalquinto, Eyzaguirre & Cardemil (1982) *Biochemistry* **21**, 4646.   **10.** Kelley & Reddy (1988) *Meth. Enzymol.* **161**, 307.   **11.** Fonda (1972) *Biochemistry* **11**, 1304.   **12.** Mason (1959) *J. Biol. Chem.* **234**, 2770.   **13.** Iwai & Taguchi (1980) *Meth. Enzymol.* **66**, 96.   **14.** Cao, Pietrak & Grubmeyer (2002) *Biochemistry* **41**, 3520.   **15.** Bhatia & Calvo (1996) *Arch. Biochem. Biophys.* **325**, 270.   **16.** Okuno & Schwarcz (1985) *Biochim. Biophys. Acta* **841**, 112.   **17.** Taguchi & Iwai (1976) *Agric. Biol. Chem.* **40**, 385.   **18.** Chang & Zylstra (1999) *J. Bacteriol.* **181**, 3069.   **19.** Oshima & Nagasawa (1977) *J. Biochem.* **81**, 57.   **20.** Majamaa, Turpeenniemi-Hujanen, Latipää, *et al.* (1985) *Biochem. J.* **229**, 127.   **21.** Tietze & Klotz (1952) *Arch. Biochem. Biophys.* **35**, 355.   **22.** Sundaresan & Coursin (1970) *Meth. Enzymol.* **18A**, 509.

Phthalide

This lactone on the above left (FW = 134.13 g/mol; CAS 87-41-2; M.P. = 72-74°C) reportedly inhibits chymotrypsin. Phthalide is a fungicide that inhibits melanin biosynthesis and has been used in the treatment of rice blast disease. *Note:* The term phthalide has also been used to refer to the tetrachloro derivative (*i.e.*, 4,5,6,7-tetrachlorophthalide (on the right above); FW = 271.91 g/mol; CAS: 27355-22-2; also known as BAYER 96610, KF-32, Rabcide, and TCP). **Target(s):** chymotrypsin (1); scytalone dehydratase, inhibited by the tetrachloride derivative (2); and trihydroxynaphthalene reductase, inhibited by the tetrachloride derivative (3,4).   **1.** Wallace, Kurtz & Niemann (1963) *Biochemistry* **2**, 824.   **2.** Wheeler & Greenblatt (1988) *Exp. Mycol.* **12**, 151.   **3.** Liao, Thompson, Fahnestock, Valent & Jordan (2001) *Biochemistry* **40**, 8696.   **4.** Liao, Basarab, Gatenby, Valent & Jordan (2001) *Structure* **9**, 19.

Phthiocol

This quinone pigment (FW = 188.18 g/mol; yellow solid; slightly soluble in water; freely soluble in most organic solvents), also called 3-hydroxymenadione, from *Mycobacterium tuberculosis* possesses some vitamin K activity. **Target(s):** lactoylglutathione lyase, *or* glyoxalase I (1); pyruvate decarboxylase (2); succinate dehydrogenase, *or* succinate oxidase (3,4).   **1.** Allen, Lo & Thornalley (1993) *Biochem. Soc. Trans.* **21**, 535.   **2.** Kuhn & Beinert (1947) *Ber.* **80**, 101.   **3.** Herz (1954) *Biochem. Z.* **325**, 83.   **4.** Ball, Anfinsen & Cooper (1947) *J. Biol. Chem.* **168**, 257.

Phycocyanin, *See* Phycocyanobilin

Phycocyanobilin

This linear tetrapyrrole (FW = 559.64 g/mol) is a prosthetic group of blue, highly fluorescent photoreceptor pigments (phycocyanins) found in certain algae. It is soluble in water and polar organic solvents. Phycocyanins, found in Cyanophyta, are proteins, which typically contain three molecules of covalently bound phycocyanobilin per subunit. One molecule of phycocyanobilin is bound per molecule of allophycocyanin, found in Cyanophyta and Rhodophyta. Phycocyanins and their chromophore phycocyanobilin are peroxy radical scavengers and inhibits ONOO⁻-mediated biological effects (1,2). **Target(s):** cyclooxygenase-2, inhibited by C-phycocyanin and poorly by phycocyanobilin (3).   **1.** Bhat & Madyastha (2001) *Biochem. Biophys. Res. Commun.* **285**, 262.   **2.** Bhat & Madyastha (2000) *Biochem. Biophys. Res. Commun.* **275**, 20.   **3.** Reddy, Bhat, Kiranmai, *et al.* (2000) *Biochem. Biophys. Res. Commun.* **277**, 599.

Phylloquinone, *See* 2-Methyl-3-phytyl-1,4-naphthoquinone

Phylloquinone 2,3-Epoxide, *See* 2,3-Epoxy-2,3-dihydro-2-methyl-3-phytyl-1,4-naphthoquinone

PHYLPA

This novel lysophosphatidic acid (FW = 403.48 g/mol), named systematically as 1-*O*-(9',10'-methanohexadecanoyl)-*sn*-glycero-2,3-cyclic phosphate, from the slime mold *Physarum polycephalum*, inhibits DNA-directed DNA polymerase and DNA polymerase α (1,2).   **1.** Murakami-Murofushi, Shioda, Kaji, Yoshida & Murofushi (1992) *J. Biol. Chem.* **267**, 21512.   **2.** Takahashi, Shimada, Shioda, *et al.* (1993) *Cell Struct. Funct.* **18**, 135.

Physostigmine, *See* Eserine

Physostigmine *N*-Oxide, *See* Eseridine˙

Physovenine

This physostigmine analogue (FW = 250.30 g/mol; CAS 6091-05-0) is a Calabar bean alkaloid that inhibits acetylcholinesterase and butyrylcholinesterase, with IC$_{50}$ values of 27 and 4 nM, respectively (1,2).   **1.** Luo, Yu, Zhan, *et al.* (2005) *J. Med. Chem.* **48**, 986.   **2.** Dale & Robinson (1970) *J. Pharm. Pharmacol.* **22**, 889.

Phytanate (Phytanic Acid)

This branched-chain fatty acid (FW = 312.54 g/mol; CAS 14721-66-5; M.P. = −65°C), also called 3,7,11,15-tetramethylhexadecanoic acid, is produced from the degradation of phytol, a constituent of chlorophyll. **Target(s):** cholesterol biosynthesis (1); and long-chain fatty acid activation (2).   **1.** Kuroda & Endo (1976) *Biochim. Biophys. Acta* **486**, 70.   **2.** Pande, Siddiqui & Gattereau (1971) *Biochim. Biophys. Acta* **248**, 156.

Phytate (Phytic Acid)

This water-soluble polyanion (FW$_{free-acid}$ = 660.04 g/mol; CAS 83-86-3), also known as *myo*-inositol hexakisphosphate, is an important phosphate storage molecule in plants. It is found in elevated concentrations in seeds, legumes, and grains. Phytic acid binds di- and trivalent cations tightly and is known to inhibit a number of metalloenzymes. It has also been identified in insulin-secreting pancreatic β-cells. It has been reported to modulate the binding of oxygen to avian and reptilian hemoglobins. **Target(s):** acid phosphatase (31); [β-adrenergic-receptor] kinase (1,32); alkaline phosphatase (2,3); AMP deaminase (4); arrestin binding to rhodopsin kinase (5); bisphosphoglycerate mutase (6); bisphospho-glycerate phosphatase (6); carboxypeptidase A (7,8); cytochrome *c* oxidase (8); diphosphoinositol-polyphosphate diphosphatase (9); diphtheria toxin binding (10); fructose-1,6-bisphosphate aldolase (11); guanylate cyclase (12); inositol-polyphosphate 5-phosphatase, or inositol-1,4,5-trisphosphate/1,3,4,5-tetrakisphosphate 5-phosphatase (13); inositol-tetrakisphosphate 1-kinase (33); iron uptake (14); multiple inositol-polyphosphate phosphatase, or inositol 1,3,4,5-tetrakisphosphate 3-phosphatase (13,15-18); phosphatidyl-inositol-3,4-bisphosphate 4-phosphatase (19); phosphatidylinositol 3'-kinase (20); phosphatidylinositol-specific phospholipase C (21); phosphoglycerate mutase (22); phospholipase A_2 (23); phosphoprotein phosphatase (30); polyphenol oxidase (8); pteroylpolyglutamate hydrolase (24); L- and P-selectin (25); serine/threonine protein phosphatases type 1, type 2A, and type 3 (26); ultraviolet B-induced signal transduction (27); and zinc absorption (28,29). **1.** Benovic (1991) *Meth. Enzymol.* **200**, 351. **2.** Martin & Evans (1991) *J. Inorg. Biochem.* **42**, 161 and 177. **3.** Martin & Evans (1989) *Res. Commun. Chem. Pathol. Pharmacol.* **65**, 289. **4.** Yoshino, Kawamura, Fujisawa & Ogasawara (1976) *J. Biochem.* **80**, 309. **5.** Palczewski, Rispoli & Detwiler (1992) *Neuron* **8**, 117. **6.** Harkness, Isaacks & Roth (1977) *Eur. J. Biochem.* **78**, 343. **7.** Friedman, Grosjean & Zahnley (1986) *Food Chem. Toxicol.* **24**, 897. **8.** Friedman, Grosjean & Zahnley (1986) *Adv. Exp. Med. Biol.* **199**, 531. **9.** Safrany, Caffrey, Yang, *et al.* (1998) *EMBO J.* **17**, 6599. **10.** Proia, Hart & Eidels (1979) *Infect. Immun.* **26**, 942. **11.** Koppitz, Vogel & Mayr (1986) *Eur. J. Biochem.* **161**, 421. **12.** Suzuki, Suematsu & Makino (2001) *FEBS Lett.* **507**, 49. **13.** Höer & Oberdisse (1991) *Biochem. J.* **278**, 219. **14.** Glahn, Wortley, South & Miller (2002) *J. Agric. Food Chem.* **50**, 390. **15.** Estrada-Garcia, Craxton, Kirk & Michell (1991) *Proc. R. Soc. Lond. B Biol. Sci.* **244**, 63. **16.** Hughes & Shears (1990) *J. Biol. Chem.* **265**, 9869. **17.** Nogimori, Hughes, Glennon, *et al.* (1991) *J. Biol. Chem.* **266**, 16499. **18.** Craxton, Ali & Shears (1995) *Biochem. J.* **305**, 491. **19.** Norris, Atkins & Majerus (1997) *J. Biol. Chem.* **272**, 23859. **20.** Huang, Ma, Hecht & Dong (1997) *Cancer Res.* **57**, 2873. **21.** Vizitiu, Kriste, Campbell & Thatcher (1996) *J. Mol. Recognit.* **9**, 197. **22.** Rigden, Walter, Phillips & Fothergill-Gilmore (1999) *J. Mol. Biol.* **289**, 691. **23.** Tykka, Mahlberg, Pantzar & Tallberg (1980) *Scand. J. Gastroenterol.* **15**, 519. **24.** Wei & Gregory III (1998) *J. Agric. Food Chem.* **46**, 211. **25.** Cecconi, Nelson, Roberts, *et al.* (1994) *J. Biol. Chem.* **269**, 15060. **26.** Larsson, Barker, Sj-oholm, *et al.* (1997) *Science* **278**, 471. **27.** Chen, Ma & Dong (2001) *Mol. Carcinog.* **31**, 139. **28.** Solomons & Cousins (1984) in *Absorption and Malabsorption of Mineral Nutrients* (Solomon & Rosenberg, eds.), pp. 125-197, Alan R. Liss, Inc., New York. **29.** O'Dell & Savage (1960) *Proc. Soc. Exp. Biol. Med.* **103**, 304. **30.** Polya & Haritou (1988) *Biochem. J.* **251**, 357. **31.** Turner & Plaxton (2001) *Planta* **214**, 243. **32.** Benovic, Stone, Caron & Lefkowitz (1989) *J. Biol. Chem.* **264**, 6707. **33.** Phillippy (1998) *Plant Physiol.* **116**, 291.

Phytohemagglutinins

This class of toxic plant lectins (MW depends on species; CAS 9008-97-3) is a carbohydrate-binding protein, consisting mainly of the leukoagglutinins PHA-L and PHA-E, indicating their ability to agglutinate leukocytes and erythrocytes. Phytohemagglutinins occur at the highest concentrations in uncooked red and white kidney beans, green beans, broad beans, and fava beans. *Phaseolus vulgaris* phytohemagglutinin consists of seven different glycosylated polypeptides. Phytohemagglutinins induce mitosis, alter membrane transport, and increase membrane permeability to proteins. In humans, poisoning can be induced from as few as five raw beans, with symptoms occurring within three hours, beginning with nausea, then vomiting, followed by diarrhea. Medical intervention is rarely required.

Phytol

This long-chain alcohol (FW = 296.54 g/mol; CAS 150-86-7), also known as (2E,7R,11R)-3,7,11,15-tetramethyl-2-hexadecen-1-ol, is a component of chlorophyll and vitamin K_1. It is an oily liquid with a very low solubility in water that is always release in the degradation of chlorophyll (it is a product in the reaction catalyzed by chlorophyllase). **Target(s):** Ca^{2+}-induced cytosolic enzyme efflux from skeletal muscle (1); and chlorophyllase (2-4). **1.** Phoenix, Edwards & Jackson (1989) *Biochem. J.* **257**, 207. **2.** Khalyfa, Kermasha, Marsot & Goetghebeur (1995) *Appl. Biochem. Biotechnol.* **53**, 11. **3.** Samaha & Kermasha (1997) *J. Biotechnol.* **55**, 181. **4.** Moll & Stegwee (1978) *Planta* **140**, 75.

Phytol Pyrophosphate, *See* Phytyl Pyrophosphate

Phytosphingosine

This triol (FW$_{\text{free-base}}$ = 317.51 g/mol; CAS 554-62-1), also known as 2S,3S,4R-2-amino-1,3,4-octadecanetriol and D-ribo-1,3,4-trihydroxy-2-aminooctadecane, is a component of plant and fungal sphingolipids. **Target(s):** 2-acylglycerol *O*-acyltransferase, or monoacylglycerol *O*-acyltransferase (1); Ca^{2+} release channel, or ryanodine receptor (2); CDP-diacylglycerol:serine *O*-phosphatidyltransferase, or phosphatidylserine synthase (3); ceramidase (4,5); DNA primase (6); 1,3-β-glucan synthase (7); histidine uptake (8); leucine uptake (8); phosphatidate phosphatase (9,10); phospholipase C-δl, weakly inhibited (11); sphingomyelin phosphodiesterase, or sphingomyelinase (12); tryptophan uptake (8); uracil uptake (8). **1.** Bhat, Wang & Coleman (1995) *Biochemistry* **34**, 11237. **2.** Sharma, Smith, Li & Schroepfer, Jr., & Needleman (2000) *Chem. Phys. Lipids* **104**, 1. **3.** Yamashita & Nikawa (1997) *Biochim. Biophys. Acta* **1348**, 228. **4.** Okino, Tani, Imayama & Ito (1998) *J. Biol. Chem.* **273**, 14368. **5.** Nieuwenhuizen, van Leeuwen, Jack, Egmond & Gotz (2003) *Protein Expr. Purif.* **30**, 94. **6.** Tamiya-Koizumi, Murate, Suzuki, *et al.* (1997) *Biochem. Mol. Biol. Int.* **41**, 1179. **7.** Abe, Nishida, Minemura, *et al.* (2001) *J. Biol. Chem.* **276**, 26923. **8.** Chung, Mao, Heitman, Hannun & Obeid (2001) *J. Biol. Chem.* **276**, 35614. **9.** Wu & Carman (2000) *Meth. Enzymol.* **312**, 373. **10.** Wu, Lin, Wang, Merrill, Jr., & Carman (1993) *J. Biol. Chem.* **268**, 13830. **11.** Matecki & Pawelczyk (1997) *Biochim. Biophys. Acta* **1325**, 287. **12.** Ella, Qi, Dolan, Thompson & Meier (1997) *Arch. Biochem. Biophys.* **340**, 101.

Phytyl Pyrophosphate

2E-isomer

2Z-isomer

This phytol derivative (FW$_{\text{free-acid}}$ = 456.50 g/mol), also called phytyl diphosphate, is a precursor in the biosynthesis of the tocopherols and related molecules. A synthetic preparation of a mixture of the 2E and 2Z isomers was found to inhibit squalene synthase. **Target(s):** mevalonate kinase (1,2); and squalene synthase (3). **1.** Gray & Kekwick (1973) *Biochem. J.* **133**, 335. **2.** Gray & Kekwick (1972) *Biochim. Biophys. Acta* **279**, 290. **3.** de Montellano, Wei, Castillo, Hsu & Boparai (1977) *J. Med. Chem.* **20**, 243.

PI-103

This potent, ATP-competitive, phosphoinositide-3-kinase (PI3K) inhibitor (FW = 348.4; CAS = 371935-74-9; Solubility: 0.25 mg/mL in 1 part PBS and 2 parts DMSO), named systematically as 3-[4-(4-morpholinyl)pyrido[3',2':4,5]furo[3,2-d]pyrimidin-2-yl]phenol and also known as S1038, exhibits selectivity toward: DNA-PK, IC$_{50}$ = 2 nM; p110α, IC$_{50}$ = 8 nM; mTORC1, IC$_{50}$ = 20 nM; PI3KC2β, IC$_{50}$ = 26 nM; p110δ, IC$_{50}$ = 48 nM; mTORC2, IC$_{50}$ = 83 nM; p110β, IC$_{50}$ = 88 nM; and p110γ, IC$_{50}$ = 150 nM (1). **1.** Knight, Gonzalez, Feldman, *et al.* (2006) *Cell* **125**, 733. **2.** Rückel, Scwartz & Rommel (2006) *Drug Discovery* **5**, 903. **3.** Fan, Knight, Goldenberg, *et al.* (2006) *Cancer Cell* **9**, 341.

PI-1840

This potent proteasome inhibitor (FW = 394.38 g/mol) selectively targets its chymotrypsin-like activity, *or CT-L* (IC_{50} = 27 nM), while showing much weaker action (IC_{50} values >100 μM) against its Trypsin-Like (T-L) and peptidylglutamyl peptide hydrolyzing, *or PGPH* activities. Mass spectrometry and equilibrium dialysis show no evidence of covalent linkage of PI-1840 with any proteasomal protein component. In intact cancer cells, PI-1840 inhibits CT-L activity, inducing the accumulation of proteasome substrates p27, Bax and IκB-α, thereby inhibiting survival pathways and viability, and inducing apoptosis. PI=1840 is also >100x more active against the constitutive proteasome, compared to immunoproteasome. 1. Kazi, Ozcan, Tecleab, *et al.* (2014) *J. Biol. Chem.* **289**, 11906.

PI3Kα inhibitor-2

This PI3Kα-/KDR-selective inhibitor (F.Wt. = 313.4 g/mol; CAS = 371943-05-4; λ_{max} = 255, 299 nm), also known as 3-[4-(4-morpholinyl)thieno[3,2-*d*]pyrimidin-2-yl]phenol, inhibits phosphatidylinositol 3-kinase (PI3K), which catalyzes the phosphorylation of the 3'-hydroxyl position of phosphatidylinositols to produce the second messengers PtdIns-(3,4)-P_2 and PtdIns-(3,4,5)-P_3 (1-4). **Targets:** PI3Kα, IC_{50} = 2 nM; KDR (Kinase-insert Domain-containing Receptor, alternatively referred to as VEGFR-2), IC_{50} = 3.4 nM; PI3Kβ, IC_{50} = 16 nM; cyclin E/CDK2, IC_{50} = 28 nM; PKA, IC_{50} = 91 nM; PI3K2Cβ, IC_{50} = 220 nM; PKCα, IC_{50} = 466 nM; and PI3Kγ, IC_{50} = 660 nM. 1. Rameh & Cantley (1999) *J. Biol. Chem.* **274**, 8347. 2. Vivanco & Sawyers (2002) *Nature Rev. Cancer* **2**, 489. 3. Hennessy, Smith & Ram (2005) *Nature Rev. Drug Disc.* **4**, 988. 4. Hayakawa, Kaizawa & Moritomo (2006) *Bioorg. Med. Chem.* **14**, 3847.

Piceatannol

This photosensitive *trans*-stilbene phytopolyphenol and F_1 ATPase inhibitor (FW = 244.25 g/mol; CAS 10083-24-6), also named 4-[(*E*)-2-(3,5-Dihydroxyphenyl)ethenyl]benzene-1,2-diol or 3,4,3',5'-tetrahydroxy-*trans*-stilbene, from the mycorrhizal and non-mycorrhizal roots of Norway spruces (*Picea abies*), binds to a pocket on ATP synthase that is formed by contributions from α and β stator subunits and the carboxyl-terminal region of the rotor γ subunit (1-3). Piceatannol also exerts anti-leukemic and immunomodulatory actions by interfering with the cytokine signaling pathway by inhibiting certain protein-tyrosine kinases. It inhibits myosin light-chain kinase (IC_{50} = 12 μM), protein kinase A (IC_{50} = 3 μM), and protein kinase C (IC_{50} = 8 μM). **Mode of F_1 ATPase Inhibition:** Single-molecule observations, made under V_{max} conditions and minimal load, show that the uninhibited F_1 sector of the ATP synthase rotates through cycles, with catalytic dwells of ~0.2 msec and 120° rotational steps in ~0.6 msec time-frame (3). The rate-limiting transition step occurs during the catalytic dwell at the initiation of the 120° rotation. Piceatannol does not interfere with the movement through the 120° rotation step, but causes both an increase in the duration of the catalytic dwell as well as a significantly longer duration time for the intrinsic inhibited state of F_1 (3) All of the beads rotated at a lower rate in the presence of saturating piceatannol, indicating that the inhibitor stays bound throughout the rotational catalytic cycle. An Arrhenius plot for the reciprocal of catalytic dwell duration (*i.e.,* the catalytic rate) shows a significantly higher activation energy for the rate-limiting step to trigger the 120° rotation (3). Because the activation energy is further increased by combination of piceatannol and a M23K substitution in the γ subunit, it appears that the inhibitor and the β/γ interface mutation affect the same transition step, even though they perturb topologically distinct rotor-stator interactions (3). **Target(s):** F_oF_1-ATPase, *or* ATP synthase; H^+-transporting two-sector ATPase (1-3); IκBα kinase (4); protein-tyrosine kinase, *or* non-specific protein-tyrosine kinase (5-10); serine/threonine-specific protein kinases, *or* protein kinase A, protein kinase C, and Ca^{2+}-dependent protein kinase (11); and Syk protein-tyrosine kinase, IC_{50} = 10 μM (7,10). 1. Zheng & Ramirez (2000) *Brit. J. Pharmacol.* **130**, 1115. 2. Zheng & Ramirez (1999) *Biochem. Biophys. Res. Commun.* **261**, 499. 3. Sekiya, Nakamoto, Nakanishi-Matsui & Futai (2012) *J. Biol. Chem.* **287**, 22771. 4. Ashikawa, Majumdar, Banerjee, *et al.* (2002) *J. Immunol.* **169**, 6490. 5. Oliver, Burg, Wilson, McLaughlin & Geahlen (1994) *J. Biol. Chem.* **269**, 29697. 6. Geahlen & McLaughlin (1989) *Biochem. Biophys. Res. Commun.* **165**, 241. 7. Mahabeleshwar & Kundu (2003) *J. Biol. Chem.* **278**, 6209. 8. Thakkar, Geahlen & Cushman (1993) *J. Med. Chem.* **36**, 2950. 9. Hollósy & Kéri (2004) *Curr. Med. Chem. Anticancer Agents* **4**, 173. 10. Gevrey, Isaac & Cox (2005) *J. Immunol.* **175**, 3737. 11. Wang, Lu & Polya (1998) *Planta Med.* **64**, 195.

Piclamilast

This benzamide (FW = 381.25 g/mol; CAS 144035-83-6), also known as RP-73401 and named systemtically as 3-(cyclopentyloxy)-*N*-(3,5-dichloro-4-pyridyl)-4-methoxybenzamide, is a selective cyclic-nucleotide phosphodiesterase 4 inhibitor, IC_{50} = 1 nM (1,2). 1. O'Grady, Jiang, Maniak, *et al.* (2002) *J. Membr. Biol.* **185**, 137. 2. Ashton, Cook, Fenton, *et al.* (1994) *J. Med. Chem.* **37**, 1696.

2-Picolinate (2-Picolinic Acid)

This acid ($FW_{free-acid}$ = 123.11 g/mol; CAS 98-98-6; pK_1 = 1.07 and pK_2 = 5.25), also known as 2-pyridinecarboxylic acid, is soluble in water and very soluble in glacial acetic acid. **Target(s):** acetylcholinesterase (1); aconitase, *or* aconitate hydratase (2); altronate dehydratase (3); D-amino-acid oxidase (4); γ-butyrobetaine dioxygenase, *or* γ-butyrobetaine hydroxylase, K_i = 39 μM (5); deoxyhypusine monooxygenase (6-9); homocitrate synthase (10); 3-hydroxyanthranilate 3,4-dioxygenase (11); mannonate dehydratase (3); nicotinamide-nucleotide amidase (12); nicotinate *N*-methyltransferase (13); nicotinate-nucleotide diphosphorylase, carboxylating (14); nicotinate phosphoenolpyruvate carboxylase (15); nicotinate phosphoribosyltransferase (16); phosphoribosyltransferase (17); phenol oxidase (18); procollagen-lysine 5-dioxygenase (19); procollagen-proline 3-dioxygenase (19); procollagen-proline 4-dioxygenase, *or* prolyl 4-hydroxylase (19-24); pyridoxal oxidase (25); and the zinc transporter (26). 1. Bergmann, Wilson & Nachmansohn (1950) *J. Biol. Chem.* **186**, 693. 2. Glusker (1971) *The Enzymes*, 3rd ed. (Boyer, ed.), **5**, 413. 3. Dreyer (1987) *Eur. J. Biochem.* **166**, 623. 4. Fitzpatrick & Massey (1982) *J. Biol. Chem.* **257**, 9958. 5. Ng, Hanauske-Abel & England (1991) *J. Biol. Chem.* **266**, 1526. 6. Abbruzzese, Park & Folk (1986) *J. Biol. Chem.* **261**, 3085. 7. Abbruzzese, Liguori & Folk (1988) *Adv. Exp. Med. Biol.* **250**, 459. 8. Park, Cooper & Folk (1982) *J. Biol. Chem.* **257**, 7217. 9. Beninati, Ferraro & Abbruzzese (1990) *Ital. J. Biochem.* **39**, 183A. 10. Qian, West & Cook (2006) *Biochemistry* **45**, 12136. 11. Hayaishi, Nozaki & Abbott (1975) *The*

Enzymes, 3rd ed. (Boyer, ed.), **12**, 119.   **12**. Imai (1973) *J. Biochem.* **73**, 139.   **13**. Taguchi, Nishitani, Okumura, Shimabayashi & Iwai (1989) *Agri. Biol. Chem.* **53**, 2867.   **14**. Mann & Byerrum (1974) *J. Biol. Chem.* **249**, 6817.   **15**. Owttrim & Colman (1986) *J. Bacteriol.* **168**, 207.   **16**. Gaut & Solomon (1971) *Biochem. Pharmacol.* **20**, 2903.   **17**. Blum & Kahn (1971) *Meth. Enzymol.* **18B**, 138.   **18**. Dowd (1999) *Nat. Toxins* **7**, 337.   **19**. Majamaa, Turpeenniemi-Hujanen, Latipää, *et al.* (1985) *Biochem. J.* **229**, 127.   **20**. Tschank, Hanauske-Abel & Peterkofsky (1988) *Arch. Biochem. Biophys.* **261**, 312.   **21**. Majamaa, Hanauske-Abel, Günzler & Kivirikko (1984) *Eur. J. Biochem.* **138**, 239.   **22**. Majamaa, Günzler, Hanauske-Abel, Myllylä & Kivirikko (1986) *J. Biol. Chem.* **261**, 7819.   **23**. Kaska, Günzler, Kivirikko & Myllylä (1987) *Biochem. J.* **241**, 483.   **24**. de Waal, Hartog & de Jong (1988) *Biochim. Biophys. Acta* **953**, 20.   **25**. Takeuchi, Tsubouchi & Shibata (1985) *Biochem. J.* **227**, 537.   **26**. Menard & Cousins (1983) *J. Nutr.* **113**, 1653.

3-Picolinic Acid, *See* Nicotinic Acid

β-Picoline, *See* 3-Methylpyridine

γ-Picoline, *See* 4-Methylpyridine

Picolinohydroxamate (Picolinohydroxamic Acid)

This iron-binding hydroxamic acid (FW = 138.13 g/mol; CAS 31888-72-9), also named *N*-hydroxy-2-pyridinecarboxamide, 2-pyridinecarboxamide, and *N*-hydroxy-2-pyridinehydroxamic acid, inhibits ribonucleoside diphosphate reductase, most likely by chelating one or both of the active-site iron ions. **1**. Elford, Van't Riet, Wampler, Lin & Elford (1981) *Adv. Enzyme Regul.* **19**, 151.

Picotamide

This antithrombotic agent (FW = 376.41 g/mol; CAS 32828-81-2), also known as 4-methoxy-*N*,*N*'-bis(3-pyridinylmethyl)-1,3-benzenedicarbox-amide, is a dual eicosanoid receptor antagonist (1-4) and a thromboxane A$_2$ synthase inhibitor (2,5). **1**. Modesti, Colella, Abbate, Gensini & Neri Serneri (1989) *Eur. J. Pharmacol.* **169**, 85.   **2**. Berrettini, De Cunto, Parise, Grasselli & Nenci (1990) *Eur. J. Clin. Pharmacol.* **39**, 495.   **3**. Pulcinelli, Pignatelli, Pesciotti, *et al.* (1997) *Thromb. Res.* **85**, 207.   **4**. Modesti, Cecioni, Colella, *et al.* (1994) *Brit. J. Pharmacol.* **112**, 81.   **5**. Matera, Chisari, Altavilla, Foca & Cook (1988) *Proc. Soc. Exp. Biol. Med.* **187**, 58.

Picrate (Picric Acid)

This nitrated phenol (FW = 229.11 g/mol; CAS 88-89-1; M.P. = 122-123°C; pK_a = 0.38 at 25°C), also called 2,4,6-trinitrophenol and nitroxanthic acid, is a pale yellow solid that explodes when heated above 300°C. Caution: Picric acid also explodes by percussion, and serious laboratory injuries and death have occurred with this common laboratory reagent. **Target(s):** D-amino-acid oxidase (1); ATP synthase (2); β-fructofuranosidase, *or* invertase (3); malate dehydrogenase (4); NAD(P)H dehydrogenase, quinone (5-8); oxidative phosphorylation, protonophoric uncoupler in suitably prepared mitochondria (9); photosystem II, inhibited by picrate and 18-crown-6 potassium picrate (10,11); and quinone reductase

(7,8). **1**. Yagi, Osawa & Okada (1959) *Biochim. Biophys. Acta* **35**, 102.   **2**. Stiggall, Galante & Hatefi (1979) *Meth. Enzymol.* **55**, 308.   **3**. Neuberg & Mandl (1950) *The Enzymes*, 1st ed. (Sumner & Myrbäck, eds.), **1** (part 1), 527.   **4**. Wedding, Hansch & Fukuto (1967) *Arch. Biochem. Biophys.* **121**, 9.   **5**. Wosilait & Nason (1955) *Meth. Enzymol.* **2**, 725.   **6**. Martius (1963) *The Enzymes*, 2nd ed. (Boyer, Lardy & Myrbäck, eds.), **7**, 517.   **7**. Wosilait, Nason & Terrell (1954) *J. Biol. Chem.* **206**, 271.   **8**. Wosilait & Nason (1954) *J. Biol. Chem.* **208**, 785.   **9**. Hanstein (1976) *Trends Biochem. Sci.* **1**, 65.   **10**. Oettmeier & Masson (1982) *Eur. J. Biochem.* **122**, 163.   **11**. Mohanty & Seibert (1997) *Indian J. Biochem. Biophys.* **34**, 241.

Picromycin

This polyketide (FW = 525.68 g/mol; CAS 19721-56-3; bitter-tasting solid; low water solubility; soluble in benzene and acetone) from *Actinomyces* spp., and originally spelled pikromycin, was the first identified macrolide antibiotic (1-3). Picromycin is most likely produced as the inactive diglucoside and liberated extracellularly as the active antibiotic through the action of a β-glucosidase. **1**. Maezawa, Hori, Kinumaki & Suzuki (1973) *J. Antibiot.* **26**, 771.   **2**. Brockmann & Henkel (1951) *Chem. Ber.* **84**, 284.   **3**. Brockmann & Henkel (1950) *Naturwissenschaften* **37**, 138.

Picrotoxin

Picrotoxinin	**Picrotin**

These neurotoxic GABA$_A$ receptor antagonists, consisting of a nearly 1:1 mixture of picrotoxinin, (FW = 292.29 g/mol; CAS 17617-45-7) and picrotin (FW = 310.30 g/mol; CAS 124-87-8), is a bitter-tasting (hence the Greek prefix *picros*) principle, also known as cocculin, from fishberry seeds (*Anamirta cocculus*) and the Philippine bayating (*Tinomiscium philippinense*). Picrotoxin binds to the chloride channel of the receptor without displacing γ-aminobutyric acid (1-3). Picrotoxin is highly toxic, exerting both stimulant and convulsant effects, and care must be exercised when handling this agent. **1**. Chapouthier & Venault (2002) *Curr. Top. Med. Chem.* **2**, 841.   **2**. Yoon, Covey & Rothman (1993) *J. Physiol.* **464**, 423.   **3**. Gottesfeld & Elliott (1971) *J. Neurochem.* **18**, 683.

Picrylsulfonate (Picrylsulfonic Acid)

This yellow-colored protein-modifying reagent (FW$_{free-acid}$ = 293.17 g/mol; CAS 2508-19-2; 5400-70-4 for sodium salt), abbreviated TNBS and also known as 2,4,6-trinitrobenzenesulfonic acid, reacts with amino groups in peptides and proteins under mildly alkaline conditions (pH ~ 9), with the release of a sulfite ion. The absorption spectrum of the product has a λ_{max} of about 345 nm. (*Note*: Sulfite ion affects the absorption spectra by forming an aromatic ring adduct with the modified amino acid). Picrylsulfonic acid also reacts with sulfhydryl groups to form an unstable product that decomposes to regenerate the sulfhydryl group. **Target(s):** acetate kinase (72); acetolactate decarboxylase, weakly inhibited (46); acetylcholinesterase (54); *N*-acyl-D-amino-acid deacylase (52); 2-acylglycerol *O*-acyltransferase, *or* monoacylglycerol *O*-acyltransferase (23,82,83); adenylyl-sulfate kinase (73,74); alcohol *O*-acetyltransferase (80,81); alkaline phosphatase (62); alliin lyase (37); AMP deaminase (51); aryl-acylamidase (53,54); *Aspergillus* nuclease S1 (59); bisphosphoglycerate mutase (3-5); bisphosphoglycerate phosphatase (4,61); butyrylcholinesterase (53); cellulase (58); chorismate mutase (6); collagenase, neutrophil (7); 3-deoxy-7-phosphoheptulonate synthase (75); dextransucrase (79); 4,5-dihydrophthalate decarboxylase (45); elastase (8); exo-(1→4)-α-D-glucan lyase (41); F$_1$ ATPase, *or* H$^+$-transporting two-sector ATPase (9); fatty-acid synthase (10); *S*-formylglutathione hydrolase (11,64); fumarylacetoacetase (47); 3-galactosyl-*N*-acetylglucosaminide 4-α-L-fucosyltransferase (77); glutamate dehydrogenase (12); γ-glutamylcysteine synthetase (13,14); glutathione-disulfide reductase (15,16); glutathione *S*-transferase (17,76); glyceraldehyde-3-phosphate dehydrogenase (18); glycerophosphocholine cholinephosphodiesterase (60); hydroxyacylglutathione hydrolase, glyoxalase II (65-69); inositol-3-phosphate synthase, *or* inositol-1-phosphate synthase (19); 3-ketovalidoxylamine C-N-lyase (38); lactose synthase (78); lactoylglutathione lyase, *or* glyoxalase I (20,21); leukotriene-A$_4$ hydrolase (55); lima bean trypsin inhibitor (22); lipase, *or* triacylglycerol lipase (71); myosin ATPase (48); Na$^+$/K$^+$-exchanging ATPase (24-26); neutrophil collagenases (7); ornithine carbamoyltransferase (84); ovomucoids (22); pectate lyase (44); pepsin (27); phenylalanine ammonia-lyase (39,40); phosphoglycerate mutase (3-5,28); poly(α-L-guluronate) lyase (42); poly(β-D-mannuronate) lyase (43); protein-glutamate methylesterase (70); pyrophosphatase, *or* inorganic diphosphatase (29,49,50); pyruvate carboxylase (30,31); pyruvate kinase (32,33); rhodanese, *or* thiosulfate sulfurtransferase (34,35); salicylate 1-monooxygenase (86); *S*-succinylglutathione hydrolase (63); superoxide dismutase (85); thioglucosidase, *or* myrosinase, *or* sinigrinase (56,57); and UDP-*N*-acetylmuramoylalanine:D-glutamate ligase (36). **1.** Ajtai, Peyser, Park, Burghardt & Muhlrad (1999) *Biochemistry* **38**, 6428. **2.** Chapman-Smith, Booker, Clements, Wallace & Keech (1991) *Biochem. J.* **276**, 759. **3.** Ikura, Narita, Sasaki & Chiba (1978) *Eur. J. Biochem.* **89**, 23. **4.** Ikura, Sasaki, Narita, Sugimoto & Chiba (1976) *Eur. J. Biochem.* **66**, 515. **5.** Sasaki, Utsumi, Sugimoto & Chiba (1976) *Eur. J. Biochem.* **66**, 523. **6.** Davidson (1987) *Meth. Enzymol.* **142**, 432. **7.** Mookhtiar, Wang & Van Wart (1986) *Arch. Biochem. Biophys.* **246**, 645. **8.** Shotton (1970) *Meth. Enzymol.* **19**, 113. **9.** Satre, Lunardi, Dianoux, *et al.* (1986) *Meth. Enzymol.* **126**, 712. **10.** Mukherjee & Katiyar (2000) *J. Enzyme Inhib.* **15**, 421. **11.** Uotila & Koivusalo (1981) *Meth. Enzymol.* **77**, 320. **12.** Freedman & Radda (1968) *Biochem. J.* **108**, 383. **13.** Griffith & Mulcahy (1999) *Adv. Enzymol. Relat. Areas Mol. Biol.* **73**, 209. **14.** Chang (1996) *J. Protein Chem.* **15**, 321. **15.** Carlberg, Sahlman & Mannervik (1985) *FEBS Lett.* **180**, 102. **16.** Carlberg & Mannervik (1979) *FEBS Lett.* **98**, 263. **17.** Weinander, Ekstrom, Andersson, *et al.* (1997) *J. Biol. Chem.* **272**, 8871. **18.** Ghosh, Mukherjee, Ray & Ray (2001) *Eur. J. Biochem.* **268**, 6037. **19.** Naccarato, Ray & Wells (1974) *Arch. Biochem. Biophys.* **164**, 194. **20.** Lupidi, Bollettini, Venardi, Marmocchi & Rotilio (2001) *Prep. Biochem. Biotechnol.* **31**, 317. **21.** Baskaran & Balasubramanian (1987) *Biochim. Biophys. Acta* **913**, 377. **22.** Haynes, Osuga & Feeney (1967) *Biochemistry* **6**, 541. **23.** Coleman (1992) *Meth. Enzymol.* **209**, 98. **24.** Breier, Monosikova, Ziegelhoffer & Dzurba (1986) *Gen. Physiol. Biophys.* **5**, 537. **25.** Breier, Ziegelhoffer, Stankovicova, *et al.* (1995) *Mol. Cell Biochem.* **147**, 187. **26.** De Pont, Van Emst-De Vries & Bonting (1984) *J. Bioenerg. Biomembr.* **16**, 263. **27.** Bohak (1970) *Meth. Enzymol.* **19**, 347. **28.** Ray, Jr., & Peck, Jr. (1972) *The Enzymes*, 3rd ed. (Boyer, ed.), **6**, 407. **29.** Josse & Wong (1971) *The Enzymes*, 3rd ed. (Boyer, ed.), **4**, 499. **30.** Libor, Sundaram & Scrutton (1978) *Biochem. J.* **169**, 543. **31.** Chapman-Smith, Booker, Clements, Wallace & Keech (1991) *Biochem. J.* **276**, 759. **32.** Jursinic & Robinson (1978) *Biochim. Biophys. Acta* **523**, 358. **33.** Kayne (1973) *The Enzymes*, 3rd ed. (Boyer, ed.), **8**, 353. **34.** Malliopoulou & Rakitzis (1988) *J. Enzyme Inhib.* **2**, 99. **35.** Malliopoulou, Rakitzis & Malliopoulou (1990) *J. Enzyme Inhib.* **4**, 27. **36.** Vaganay, Tanner, van Heijenoort & Blanot (1996) *Microbiol. Drug Resist.* **2**, 51. **37.** Won & Mazelis (1989) *Physiol. Plant* **77**, 87. **38.** Takeuchi, Neyazaki & Matsui (1990) *Chem. Pharm. Bull.* **38**, 1419. **39.** Parkhurst & Hodgins (1972) *Arch. Biochem. Biophys.* **152**, 597. **40.** Hodgins (1972) *Arch. Biochem.*

Biophys. **149**, 91. **41.** Yoshinaga, Fujisue, Abe, *et al.* (1999) *Biochim. Biophys. Acta* **1472**, 447. **42.** Takeuchi, Nibu, Murata & Kusakabe (1997) *Food Sci. Technol. Int.* **3**, 22. **43.** Iwamoto, Iriyama, Osatomi, Oda & Muramatsu (2002) *J. Protein Chem.* **21**, 455. **44.** Kobayashi, Koike, Yoshimatsu, *et al.* (1999) *Biosci. Biotechnol. Biochem.* **63**, 65. **45.** Nakazawa & Hayashi (1978) *Appl. Environ. Microbiol.* **36**, 264. **46.** Oshiro, Aisaka & Uwajima (1989) *Agric. Biol. Chem.* **53**, 1913. **47.** Hsiang, Sim, Mahuran & Schmidt, Jr., (1972) *Biochemistry* **11**, 2098. **48.** Kameyama, Ichikawa, Sunaga, *et al.* (1985) *J. Biochem.* **97**, 625. **49.** Lathi & Raudaskoski (1983) *Folia Microbiol.* **28**, 371. **50.** Lahti & Niemi (1981) *J. Biochem.* **90**, 79. **51.** Martini, Ranieri-Raggi, Sabbatini & Raggi (2001) *Biochim. Biophys. Acta* **1544**, 123. **52.** Wakayama, Yada, Kanda, *et al.* (2000) *Biosci. Biotechnol. Biochem.* **64**, 1. **53.** Boopathy & Balasubramanian (1985) *Eur. J. Biochem.* **151**, 351. **54.** Majumdar & Balasubramanian (1984) *Biochemistry* **23**, 4088. **55.** Mueller, Samuelsson & Haeggström (1995) *Biochemistry* **34**, 3536. **56.** Ohtsuru & Kawatani (1979) *Agric. Biol. Chem.* **43**, 2249. **57.** Ohtsuru & Hata (1973) *Agric. Biol. Chem.* **37**, 269. **58.** Matsumoto, Endo, Tamiya, *et al.* (1974) *J. Biochem.* **76**, 563. **59.** Gite, Reddy & Shankar (1992) *Biochem. J.* **285**, 489. **60.** Sok (1998) *Neurochem. Res.* **23**, 1061. **61.** Sasaki, Hirose, Sugimoto & Chiba (1971) *Biochim. Biophys. Acta* **227**, 595. **62.** Chen, Xie, Yu, Xu & Zhang (2005) *Int. J. Biochem. Cell Biol.* **37**, 1446. **63.** Uotila (1979) *J. Biol. Chem.* **254**, 7024. **64.** Uotila & Koivusalo (1974) *J. Biol. Chem.* **249**, 7664. **65.** vander Jagt (1989) in *Coenzymes and Cofactors, Glutathione, Chem. Biochem. Med. Aspects Pt. A* (Dolphin, Poulson & Avromonic, eds.) **3**, 597, Wiley, New York. **66.** Uotila (1973) *Biochemistry* **12**, 3944. **67.** vander Jagt (1993) *Biochem. Soc. Trans.* **21**, 522. **68.** Principato, Rosi, Talesa, Giovannini & Uotila (1987) *Biochim. Biophys. Acta* **911**, 349. **69.** Ball & vander Jagt (1981) *Biochemistry* **20**, 899. **70.** Snyder, Stock & Koshland (1984) *Meth. Enzymol.* **106**, 321. **71.** Kundu, Basu, Guchhait & Chakrabarti (1987) *J. Gen. Microbiol.* **133**, 149. **72.** Nakajima, Suzuki & Imahori (1978) *J. Biochem.* **84**, 193. **73.** Schriek & Schwenn (1986) *Arch. Microbiol.* **145**, 32. **74.** Renosto, Seubert, Knudson & Segel (1985) *J. Biol. Chem.* **260**, 1535. **75.** Simpson & Davidson (1976) *Eur. J. Biochem.* **70**, 509. **76.** Asaoka & Takahashi (1977) *J. Biochem.* **82**, 1313. **77.** Léonard, Lhernould, Carlué, *et al.* (2005) *Glycoconj. J.* **22**, 71. **78.** Hill & Brew (1975) *Adv. Enzymol. Relat. Areas Mol. Biol.* **43**, 411. **79.** Majumder, Purama & Goyal (2007) *Indian J. Microbiol.* **47**, 197. **80.** Akita, Suzuki & Obata (1990) *Agric. Biol. Chem.* **54**, 1485. **81.** Yoshioka & Hashimoto (1981) *Agric. Biol. Chem.* **45**, 2183. **82.** Mostafa, Bhat & Coleman (1993) *Biochim. Biophys. Acta* **1169**, 189. **83.** Coleman & Haynes (1986) *J. Biol. Chem.* **261**, 224. **84.** Marshall & Cohen (1972) *J. Biol. Chem.* **247**, 1669. **85.** Borders, Jr., Bjerrum, Schirmer & Oliver (1998) *Biochemistry* **37**, 11323. **86.** Suzuki, Mizuguchi, Gomi & Itagaki (1995) *J. Biochem.* **117**, 579.

Pictilisib

This potent PI3K inhibitor (FW = 513.64 g/mol; CAS 957054-30-7; Solubility: 44 mg/mL DMSO), also known as GDC-0941 and 2-(1*H*-indazol-4-yl)-6-((4-(methylsulfonyl)piperazin-1-yl)methyl)-4-morpholino-thieno[3,2-*d*]pyrimidine, targets phosphatidyl-inositol-4,5-bisphosphate 3-kinase, catalytic subunit α, *or* PIK3CA and p110α (IC$_{50}$ = 3 nM); PI3Kδ, *or* p110δ (IC$_{50}$ = 3 nM); PI3Kβ, *or* p110β (IC$_{50}$ = 33 nM); and PI3Kγ, *or* p110γ (IC$_{50}$ = 75 nM), important targets in cancers exploiting the dysregulation of PI3K/Akt signaling (1). Ellagic acid enhances the efficacy GDC-0941 in breast cancer cells (2). **Binding Mode:** An X-ray crystal structure of GDC-0941 bound to human p110γ shows that it is very efficiently anchored in the ATP-binding site (1). Similar to LY294002, GDC-0941 uses an essential morpholino group to form the key hydrogen bond with the hinge. Its indazole moiety also fits snugly deep in the affinity pocket with the two indazole nitrogen atoms forming hydrogen bonds with the hydroxyl group of Tyr 867 and the carboxylate group of Asp 841, further strengthening the interactions in this pocket (1). **1**. Folkes, Ahmadi, Alderton, *et al.* (2008) *J. Med. Chem.* **51**, 5522. **2**. Shi, Gao, Li, *et al.* (2015) *Curr. Mol. Med.* **15**, 478.

Piericidin A₁

Piericidin A₁

This antibiotic (FW = 415.57 g/mol; CAS 2738-64-9; IUPAC: 2-(10-hydroxy-3,7,9,11-tetramethyl-2,4,7,11-tridecatetraenyl)-5,6-dimethoxy-3-methyl-4-pyridinol, from species of *Streptomyces*, is a structural analogue of ubiquinone and inhibits NADH dehydrogenase and electron transport by binding to the ubiquinone binding site. **Target(s):** cytochrome *d* complex (1); cytochrome *o* complex (1); electron transport (2-8); glucose dehydrogenase, ubiquinone (9); NADH dehydrogenase, complex I (2-7,9-21); photosystem II (22); succinate dehydrogenase, inhibited at high concentrations (7); ubiquinol:cytochrome *c* reductase activity of complex III (23); and ubiquinol oxidase, *Escherichia coli* (1,24). **1**. Anraku & Gennis (1987) *Trends Biochem. Sci.* **12**, 262. **2**. Izawa & Good (1972) *Meth. Enzymol.* **24**, 355. **3**. Hatefi & Stiggall (1978) *Meth. Enzymol.* **53**, 5. **4**. Hatefi (1978) *Meth. Enzymol.* **53**, 11. **5**. Singer (1979) *Meth. Enzymol.* **55**, 454. **6**. Rapoport & Schewe (1977) *Trends Biochem. Sci.* **2**, 186. **7**. Jeng, Hall, Crane, *et al.* (1968) *Biochemistry* **7**, 1311. **8**. Hall, Wu, Crane, *et al.* (1966) *Biochem. Biophys. Res. Commun.* **25**, 373. **9**. Friedrich, van Heek, Leif, *et al.* (1994) *Eur. J. Biochem.* **219**, 691. **10**. Hatefi & Stiggall (1976) *The Enzymes*, 3rd ed. (Boyer, ed.), **13**, 175. **11**. Magnitsky, Toulokhonova, Yano, *et al.* (2002) *J. Bioenerg. Biomembr.* **34**, 193. **12**. Scheide, Huber & Friedrich (2002) *FEBS Lett.* **512**, 80. **13**. Galkin, Grivennikova & Vinogradov (2001) *Biochemistry (Moscow)* **66**, 435. **14**. Fang, Wang & Beattie (2001) *Eur. J. Biochem.* **268**, 3075. **15**. Kerscher, Okun & Brandt (1999) *J. Cell Sci.* **112**, 2347. **16**. Singer & Ramsay (1994) *Biochim. Biophys. Acta* **1187**, 198. **17**. Ramsay & Singer (1992) *Biochem. Biophys. Res. Commun.* **189**, 47. **18**. Ramsay, Krueger, Youngster & Singer (1991) *Biochem. J.* **273**, 481. **19**. Palmer, Horgan, Tisdale, Singer & Beinert (1968) *J. Biol. Chem.* **243**, 844. **20**. Darrouzet, Issartel, Lunardi & Dupuis (1998) *FEBS Lett.* **431**, 34. **21**. Prieur, Lunardi & Dupuis (2001) *Biochim. Biophys. Acta* **1504**, 173. **22**. Ikezawa, Ifuku, Endo & Sato (2002) *Biosci. Biotechnol. Biochem.* **66**, 1925. **23**. Degli Esposti, Ghelli, Crimi, *et al.* (1993) *Biochem. Biophys. Res. Commun.* **190**, 1090. **24**. Kita, Konishi & Anraku (1986) *Meth. Enzymol.* **126**, 94.

Piericidin A, Reduced, *See* Octahydropiericidin A

Piericidin B, Reduced, *See* Octahydropiericidin B

Pifithrin-α

This p53 inhibitor (FW_hydrobromide = 367.31 g/mol), also known as 2-(2-imino-4,5,6,7-tetrahydrobenzothiazol-3-yl)-1-*p*-tolylethanone, protects wild type mice against lethal doses of radiation, but is without effect on p53-deficient animals. **Target(s):** caspase-2 activation (1); luciferase, firefly, *or Photinus*-luciferin 4-monooxygenase, ATP-hydrolyzing (2-4); p53 (5,6); p53-mediated apoptosis (5,6); and p53-dependent gene transcription (5,6). **1**. Tyagi, Singh, Agarwal & Agarwal (2006) *Carcinogenesis* **27**, 2269. **2**. Rocha, Campbell, Roche & Perkins (2003) *BMC Mol. Biol.* **4**, 9. **3**. Auld, Southall, Jadhav, *et al.* (2008) *J. Med. Chem.* **51**, 2372. **4**. Inouye (2010) *Cell. Mol. Life Sci.* **67**, 387. **5**. Komarova & Gudkov (2000) *Biochemistry (Moscow)* **65**, 41. **6**. Komarov, Komarova, Kondratov, *et al.* (1999) *Science* **285**, 1733.

Pifithrin-μ

This cell-permeable sulfonamide-based inhibitor and anti-apoptotic factor (FW = 181.20 g/mol; CAS 64984-31-2; Solubility: >10 mg/mL DMSO, <2 mg/mL H₂O; pK_a = 8; *Symbol* = PFTμ and PAS), also known as 2-phenylethynesulfonamide, targets p53 and Heat Shock Protein-70, *or* HSP 70 (1). Because it only targets the mitochondrial branch of the p53 pathway without affecting the important transcriptional functions of p53, Pifithrin-μ is recommended over Pifithrin-α for *in vivo* studies. PFTμ exhibits high specificity for p53 and does not protect cells from apoptosis induced by overexpression of the proapoptotic protein Bax or by treatment with dexamethasone. With B-chronic lymphocytic leukemia (CLL) cells, Pifithrin-μ (5–20 μM) initiated apoptosis within 24 hours, with maximal death at 48 hours, as assessed by cell morphology, cleavage of poly(ADP-ribose) polymerase (PARP), caspase-3 activation, and annexin V staining (2). **1**. Strom, Sathe, Komarov, *et al.* (2006) *Nature Chem Biol.* **2**, 474. **2**. Steele, Prentice, Hoffbrand, *et al.* (2009) *Blood* **114**, 1217.

Pilaralisib

This selective, reversible, and ATP-competitive Class I PI3K inhibitor (FW = 541.02 g/mol; CAS 934526-89-3; Solubility: 100 mg/mL DMSO, when warmed; < 1 mg/mL H₂O), also known as XL147, SAR245408, and *N*-(3-{[(3-{[2-chloro-5-(methoxy)phenyl]-amino}quinoxalin-2-yl)amino]sulfonyl}phenyl)-2-methylalaninamide, targets PI3Kα (IC₅₀ = 39 nM), PI3Kδ (IC₅₀ = 36 nM), and PI3Kγ (IC₅₀ = 23 nM), but only weakly for PI3Kβ. Pilaralisib is active against human breast cancer cell lines with constitutive PI3K activation (1). PI3K inhibitors reduce AKT activity and relieves suppression of receptor tyrosine kinase expression and activity. XL147 shows dose-dependent inhibition of cell growth and levels of pAKT and pS6, signal transducers in the PI3K/AKT/TOR pathway (1). In HER2-overexpressing cells, pilaralisib inhibition of PI3K is attended by up-regulation of expression and phosphorylation of multiple receptor tyrosine kinases, including HER3. Knockdown of FoxO1 and FoxO3a transcription factors suppressed the induction of HER3, InsR, IGF1R, and FGFR2 mRNAs upon inhibition of PI3K. In HER2(+) cells, knockdown of HER3 with siRNA or cotreatment with the HER2 inhibitors trastuzumab or lapatinib enhance XL147-induced cell death and inhibition of pAKT and pS6 (1). When tested separately, trastuzumab and lapatinib synergized with XL147 for inhibition of pAKT and growth of established BT474 xenografts. Compared with XL147 alone, the combination exhibited a superior antitumor effect against trastuzumab-resistant tumor xenografts. **1**. Chakrabarty, Sánchez, Kuba, Rinehart & Arteaga (2012) *Proc. Natl. Acad. Sci. U.S.A.* **109**, 2718.

PIK-75

This novel phosphoinositide-3 kinase p110α inhibitor (F.Wt. = 488.74; CAS 372196-67-3; Solubility (25°C): 98 mg/mL DMSO, <1 mg/mL Water), also known as (*E*)-*N*'-((6-bromoH-imidazo[1,2-*a*]pyridin-3-yl)methylene)-*N*,2-dimethyl-5-nitrobenzenesulfonohydrazide-HCl, competes with respect to a substrate (phosphatidylinositol, PI), unlike most other PI3K inhibitors, which bind at or near the ATP site. PIK-75 exhibits impressive isoform selectivity, with an IC₅₀ value of 5.8 nM for p110α, while the corresponding IC₅₀ values are 1300 nM, 76 nM and 510 nM for p110β, p110γ, and p110δ, respectively. **1**. Zheng, Amran, Zhu, *et. al.* (2012) *Biochem. J.* **444**, 529.

PIK-90

This potent, cell-permeable p110 inhibitor (F.Wt. = 351.36; CAS 1349796-36-6 and 1123889-87-1; Solubility (25°C): <1 mg/mL DMSO, <1 mg/mL Water), is also known as N-(2,3-dihydro-7,8-dimethoxyimidazo[1,2-c]quinazolin-5-yl)-3-pyridinecarboxamide, has the following targets: p110α, IC_{50} = 11 nM; p110β, IC_{50} = 350 nM; p110γ, IC_{50} = 18 nM; and p110δ, IC_{50} = 58 nM (1). PIK-90 also exhibits significant antiproliferative activity by blocking Akt phosphorylation. 1. [1] Van Keymeulen, et al. (2006) J. Cell Biol. 174, 437.

PIK-93

This potent and selective PI3Kγ/PI₄KIIIβ inhibitor (FW = 389.88; CAS 593960-11-3; Solubility (25°C): 78 mg/mL DMSO, <1 mg/mL Water), also known as N-[5-[4-chloro-3-[(2-hydroxyethyl)sulfamoyl]phenyl]-4-methylthiazol-2-yl]acetamide, impairs actin filament consolidation and stability at the leading edge in cells treated with N-formyl-Met-Leu-Phe (2) as well as ceramide transport between ER and Golgi compartments (3). The inhibition of PI3K by PIK-93, LY294002, or wortmannin decreased carbachol-induced translocation of TRPC6 to the plasma membrane and carbachol-induced net Ca^{2+} entry into T6.11 cells (4). PIK-93 inhibits both poliovirus (PV) and hepatitis C virus (HCV) replication, with EC_{50} values of 0.14 μM and 1.9 μM, respectively (5). Targets: PI3Kγ, IC_{50} = 16 nM; PI₄KIIIβ, IC_{50} = 19 nM; PI3Kα, IC_{50} = 39 nM; PI3Kδ, IC_{50} = 0.12 μM; and PI3Kβ, IC_{50} = 0.59 μM (1). 1. Knight (2006) Cell 125, 733. 2. Van Keymeulen, et al. (2006) Cell Biol. 174, 437. 3. Tóth B, et al. (2006) J. Biol. Chem. 281, 36369. 4. Monet, Francoeur & Boulay (2012) J. Biol. Chem. 287, 17672. 5. Arita, et al. (2011) J. Virol. 85, 2364.

PIK-294

This pseudopurine-containing phosphatidylinositol kinase inhibitor (FW = 489.53 g/mol; CAS = 900185-02-6; Solubility: 98 mg/mL DMSO; <1 mg/mL H₂O), also named 2-((4-amino-3-(3-hydroxyphenyl)-1H-pyrazolo[3,4-d]pyrimidin-1-yl)methyl)-5-methyl-3-ortho-tolylquinazolin-4(3H)-one, selectively targets p110δ (IC_{50} = 10 nM), with much weaker inhibition of PI3Kα (IC_{50} = 10,000 nM), PI3Kβ (IC_{50} = 490 nM), and PI3Kγ (IC_{50} = 160 nM), respectively. 1. Knight, Gonzalez, Feldman, et al. (2006) Cell 125, 733. 2. Bobrovnikova-Marjon, Pytel, Riese, et al. (2012) Mol. Cell Biol. 32, 2268.

Pilocarpine

This photosensitive alkaloid (FW_free-base = 208.26 g/mol; CAS 92-13-7; 148-72-1 for the (3S-cis)-isomer, mononitrate salt; 54-71-7 for the (3S-cis)-isomer, mono-HCl salt; M.P. = 34°C; water-soluble; imidazole pK_a = 7.15 at 20°C) from the leaves of Pilocarpus jaborandi and other P. species is a nonselective muscarinic acetylcholine receptor agonist used in the treatment of glaucoma, as a diaphoretic, and as an agent to produce experimental epilepsy. Target(s): CYP2A5 (1,2); CYP2A6 (1-4); CYP3A, moderately inhibited (2); CYP2B (1,2,5); β-fructofuranosidase, or invertase (6); and urease, inhibited by free base (7). 1. Kinonen, Pasanen, Gynther, et al. (1995) Brit. J. Pharmacol. 116, 2625. 2. Kimonen, Juvonen, Alhava & Pasanen (1995) Brit. J. Pharmacol. 114, 832. 3. Xia, Peng, Yu, Wang & Wang (2002) Acta Pharmacol. Sin. 23, 471. 4. Li, Li & Sellers (1997) Eur. J. Drug Metab. Pharmacokinet. 22, 295. 5. Skaanild & Friis (2002) Pharmacol. Toxicol. 91, 198. 6. Neuberg & Mandl (1950) The Enzymes, 1st ed. (Sumner & Myrbäck, eds.), 1 (part 1), 527. 7. Onodera (1915) Biochem. J. 9, 544.

Pimasertib

This selective, orally bioavailable MEK inhibitor (FW = 431.20 g/mol; CAS 1236699-92-5; Solubility: 86 mg/mL DMSO), also known as AS-703026 and (S)-N-(2,3-dihydroxypropyl)-3-(2-fluoro-4-iodophenylamino)-isonicotinamide, targets MEK1 (IC_{50} = 5 nM) and MEK2 (IC_{50} = 2 μM) in multiple myeloma (MM) cell lines. MEK is an alternate name for Mitogen-Activated Protein Kinase Kinase (MAP2K or MAPKK), a signal transduction kinase that phosphorylates Mitogen-Activated Protein Kinase (MAPK). AS703026 is an ATP-noncompetitive (allosteric) inhibitor that exhibits exquisite kinase selectivity. AS703026 sensitizes MM cells to a broad spectrum of conventional (e.g., dexamethasone, melphalan) and novel/emerging (e.g., lenalidomide, perifosine, bortezomib, rapamycin) anti-MM therapies (1). Significant tumor growth reduction is observed in AS703026–treated versus vehicle-treated mice bearing H929 MM xenograft tumors, which correlated with downregulation of pERK1/2, induction of PARP cleavage, and decreased microvessel formation in vivo (1). AS703026 circumvents resistance to the BRAF inhibitor PLX4032 in human malignant melanoma cells (e.g., RPMI-7951 and SK-MEL5) harboring BRAF^V600E (2). Combination of a PI3K/mTOR inhibitor (SAR245409) with a MEK inhibitor (AS703026) induces a synergistic antitumor effect in certain endometrial cancer cells, underscores the utility of optimized doses of antitumor agents in treating endometrial cancer (3). A kinome siRNA library assay targeting 790 kinases, plus or minus AS703026, indicated that SAR245409 and AS703026 acting in combination is likely to be an effective treatment of Triple-Negative Breast Cancer, or TNBC (4). 1. Kim, Kong, Fulciniti, et al. (2010) Brit. J. Haematol. 149, 537. 2. Park, Hong, Moon, et al. (2013) Am. J. Med. Sci. 346, 494. 3. Inaba, Oda, Ikeda, et al. (2015) Gynecol. Oncol. 138, 323. 4. Lee, Galloway, Grandjean, et al. (2015) J. Cancer. 6, 1306.

Pimavanserin

This orally active atypical non-dopaminergic antipsychotic[(FW_free-base = 427.56 g/mol; CAS 706779-91-1 (free base) and 706782-28-7 (tartrate salt)), also known as ACP-103, Nuplazid® and N-(4-fluorophenylmethyl)-N-(1-methyl-piperidin-4-yl)-N′-(4-(2-methylpropyloxy)phenyl-methyl)carbamide, is a potent inverse agonist and antagonist at serotonin 5-HT_2A receptors (K_i = 0.087 nM) and less so at serotonin 5-HT_2C receptors (K_i = 0.44 nM) (1-3). Pimavanserin shows low binding to σ_i receptors (K_i = 120 nM), with no appreciable affinity (K_i > 300 nM) to serotonin 5-HT_2B, dopaminergic (including D₂), muscarinic, histaminergic, or adrenergic receptors, or to calcium channels. 5-HT$_{2A}$ receptor dysregulation is implicated in both the etiology and treatment of schizophrenia, and the same receptor plays an active role in the regulation of sleep architecture. Nuplazid is specifically indicated in its FDA approval (April, 2016) for the treatment of hallucinations and delusions associated with Parkinson disease. Significantly, 5-HT$_{2A}$ serotonin receptor antagonists differentially regulate 5-HT$_{2A}$ receptor protein level in vivo (4). *See also Ketanserin; RH-34; Risperidone; Ritanserin; Setoperone; Volinanserin.* **1.** Vanover, Weiner, Makhay, *et al.* (2006) *J. Pharmacol. Exp. Ther.* **317**, 910. **2.** Meltzer, Mills, Revell, *et al.* (2010) *Neuropsychopharmacol.* **35**, 881. **3.** *Nuplazid (pimavanserin) Tablets, for Oral Use. U.S. Full Prescribing Information* (PDF). ACADIA Pharmaceuticals, Inc. (Retrieved May, 2016). **4.** Yadav, Kroeze, Farrell & Roth (2011) *J. Pharmacol. Exp. Ther.* **339**, 99.

Pimecrolimus

This immunophilin ligand (FW = 810.46 g/mol; CAS 137071-32-0; Solubility = 100 mg/mL DMSO), also known as ASM 981 and SDZ ASM 981, is a ascomycin macrolactam that binds to macrophilin-12 (also known as FKBP-12) and inhibits phosphoprotein phosphatase calcineurin, thereby blocking T-cell activation by inhibiting cytokine synthesis and release from T-cells. Pimecrolimus downregulates production of Th1 [interleukin (IL)-2, interferon-gamma] and Th2 (IL-4, IL-10) type cytokines after antigen-specific stimulation of a human T-helper cell clone isolated from the skin of an atopic dermatitis patient (1). SDZ ASM 981 inhibits phorbol myristate/phytohemagglutinin-stimulated transcription of a reporter gene coupled to the human IL-2 promoter in the human T-cell line Jurkat and the IgE/antigen-mediated transcription of a reporter gene coupled to the human tumour necrosis factor (TNF)-alpha promoter in the murine mast-cell line CPII (1). Pharmacokinetic studies demonstrated very low blood levels of pimecrolimus following topical application, with no accumulation after repeated applications. Pimecrolimus is thus a versatile treatment for inflammatory skin diseases. **1.** Grassberger, Baumruker, Enz, *et al.* (1999) *Brit. J. Dermatol.* **141**, 264.

Pimelate (Pimelic Acid)

This dicarboxylic acid (FW$_{free-acid}$ = 160.17 g/mol; CAS 111-16-0; pK_1 = 4.46 and pK_2 = 5.58), also known as heptanedioic acid, is soluble in water (5 g/100 mL at 20°C). Pimelate is excreted in elevated amounts in urine in individuals with disorders in mitochondrial and peroxisomal β-oxidation, for which it is of significant diagnostic value. **Target(s):** 2-aminoadipate aminotransferase (1); 4-aminobutyrate aminotransferase (2); carbamoyl-phosphate synthetase (3); carboxy-*cis,cis*-muconate cyclase (4); glutamate decarboxylase, K_i = 0.81 mM (5,6); kynurenine:oxoglutarate aminotransferase (7-11); and 2,3,4,5-tetrahydropyridine-2,6-dicarboxylate *N*-succinyltransferase (12). **1.** Deshmukh & Mungre (1989) *Biochem. J.* **261**, 761. **2.** Schousboe, Wu & Roberts (1974) *J. Neurochem.* **23**, 1189. **3.** Jones (1962) *Meth. Enzymol.* **5**, 903. **4.** Thatcher & Cain (1975) *Eur. J. Biochem.* **56**, 193. **5.** Fonda (1972) *Biochemistry* **11**, 1304. **6.** Webb (1966) *Enzyme and Metabolic Inhibitors*, vol. **2**, p. 328, Academic Press, New York. **7.** Tanizawa, Asada & Soda (1985) *Meth. Enzymol.* **113**, 90. **8.** Tobes (1987) *Meth. Enzymol.* **142**, 217. **9.** Mason (1959) *J. Biol. Chem.*

234, 2770. **10.** Asada, Sawa, Tanizawa & Soda (1986) *J. Biochem.* **99**, 1101. **11.** Mawal, Mukhopadhyay & Deshmukh (1991) *Biochem. J.* **279**, 595. **12.** Berges, DeWolf, Dunn, *et al.* (1986) *J. Biol. Chem.* **261**, 6160.

Pimozide

This oral antipsychotic drug (FW$_{free-base}$ = 461.55 g/mol; CAS 2062-78-4) is a D$_2$ receptor antagonist, Ca^{2+} channel antagonist, and high-affinity antagonist for the 5-hydroxytryptamine-7 receptor. Pimozide is used to control motor or verbal tics (*i.e.*, an uncontrollable need to repeat certain movements or sounds) caused by Tourette's disorder. Pimozide likewise inhibits the dopamine transporter (DAT), a property that accounts for its stimulant properties. Inhibition of dopamine-reuptake may explain the synergistic effects of pimozide in the treatment of ADHD, when administered with a stimulant. *Caution*: Pimozide can have severe, even potentially fatal side-effects. **Target(s):** adenylate cyclase, dopamine-stimulated (1,2); and CYP2D6 (3). **1.** Iversen (1975) *Science* **188**, 1084. **2.** Iversen (1976) *Trends Biochem. Sci.* **1**, 121. **3.** Desta, Kerbusch, Soukhova, *et al.* (1998) *J. Pharmacol. Exp. Ther.* **285**, 428.

Pinacoyl Methylphosphofluorodate, *See Soman*

Pinane Thromboxane A$_2$

This stable thromboxane A$_2$ analogue (FW$_{free-acid}$ = 376.58 g/mol) is a partial thromboxane A$_2$ receptor agonist. **Target(s):** GABA-gated chloride channels (1); 15-hydroxyprostaglandin dehydrogenase (2); thromboxane receptor (3); and thromboxane synthase (4). **1.** Schwartz-Bloom, Cook & Yu (1996) *Neuropharmacology* **35**, 1347. **2.** Robinson & Hoult (1983) *Biochim. Biophys. Acta* **754**, 190. **3.** Armstrong, Jones, Peesapati, Will & Wilson (1985) *Brit. J. Pharmacol.* **84**, 595. **4.** Nicolaou, Magolda, Smith, *et al.* (1979) *Proc. Natl. Acad. Sci. U.S.A.* **76**, 2566.

Pinan-3-ol

This bicyclic terpenol (FW = 154.25 g/mol; CAS 25465-65-0) inhibits α-pinene-oxide decyclase by 88% when present at 2.8 mM. **1.** Griffiths, Harries, Jeffcoat & Trudgill (1987) *J. Bacteriol.* **169**, 4980.

Pindobind

This α-haloacetamido pindolol analogue (FW = 480.45 g/mol; hygroscopic), also known as N^8-(bromoacetyl)-N^1-[3-(4-indolyloxy)-2'-hydroxypropyl]-(Z)-1,8-diamino-p-menthane, is an affinity label for β-adrenoceptors (1). Pindobind does not stimulate adenylate cyclase activity, but does inhibit (−)-isoproterenol-stimulated adenylate cyclase activity. **1.** Pitha, Buchowiecki, Milecki & Kusiak (1987) *J. Med. Chem.* **30**, 612.

Pindolol

This antihypertensive agent (FW = 248.33 g/mol; CAS 13523-86-9), also known as 1-(1H-indol-4-yloxy)-3-[(1-methylethyl)amino]-2-propanol and Visken, is a moderately lipophilic β$_1$-adrenergic receptor antagonist and a 5-HT$_{1A}$ serotonin receptor partial agonist, K_i = 30 nM (1-3). *Note:* The pharmacologically active stereomer is the (−)-S-enantiomer. *See also Iodocyanopindolol* **Target(s):** glycogenolysis in rat liver (4); endogenous prostaglandin synthesis (5); hepatic 3,5,3'-triiodothyronine production (6); inhibition of thromboxane and prostacyclin production in whole blood (7); rat brain serotonin (5-HT) synthesis (8). **1.** Watkins, Lawrence, Lewis & Jarrott (1996) *J. Auton. Nerv. Syst.* **60**, 12. **2.** Salimi (1975) *Pharmacology* **13**, 441. **3.** Frishman (1983) *N. Engl. J. Med.* **308**, 940. **4.** Pogátsa, Káldor & Vizi (1973) *Arzneimittelforschung.* **23**, 1085. **5.** Durão, Prata & Gonçalves (1977) *Lancet* **2**, 1005. **6.** Shulkin BL, Peele ME, Utiger (1984) *Endocrinology* **115**, 858. **7.** Greer, Walker, McLaren, Calder & Forbes (1985) *Prostaglandins Leukot. Med.* **19**, 209. **8.** Hjorth & Carlsson (1985) *Neuropharmacology* **24**, 1143.

α-Pinene

(1S)-(−)-α-pinene   (1R)-(−)-α-pinene

This cyclic monoterpene antifungal agent (FW = 136.24 g/mol; CAS 80-56-8; 7785-70-8 for (+)-α-pinene; 7785-26-4 for (−)-α-pinene), named systematically as (1S,5S)-2,6,6-trimethyl bicyclo[3.1.1]hept-2-ene, is found in turpentine oil. (Note: α-Pinene from North American turpentine is primarily dextrorotatory, whereas European turpentine α-pinene is levorotatory.) The d-form is found in Port Oxford cedar wood oil (*Chamaecyparis lawsoniana*) and the l-stereoisomer is in mandarin peel oil. The boiling point of each enantiomer is 155-156°C. **Target(s):** acetylcholinesterase, both (+)- and (−)-isomers inhibit, with respective K_i values of 0.15 and 0.17 mM for the bovine enzyme (1,2); CYP3A (3); CYP2B1 (4); and methane monooxygenase (5). **1.** Perry, Houghton, Theobald, Jenner & Perry (2000) *J. Pharm. Pharmacol.* **52**, 895. **2.** Miyazawa & Yamafuji (2005) *J. Agric. Food Chem.* **53**, 1765. **3.** Pass & McLean (2002) *Xenobiotica* **32**, 1109. **4.** De-Oliveira, Ribeiro-Pinto & Paumgartten (1997) *Toxicol. Lett.* **92**, 39. **5.** Amaral, Ekins, Richards & Knowles (1998) *Appl. Environ. Microbiol.* **64**, 520.

Pinosylvin

This *trans*-stilbene derivative (FW = 212.25 g/mol; CAS 102-61-4; M.P. = 155.5-156°C; low solubility in water), also known as (E)-3,5-stilbenediol and *trans*-3,5-dihydroxystilbene and named systematically as 5-[(E)-2-phenylethenyl]benzene-1,3-diol, occurs naturally in the hardwood of pine

and other woody plants. Pinosylvin exxhibits micromolar K_i values for specific isozymes of stilbene synthase and chalcone synthase (1). **Target(s):** chalcone synthase (1); stilbene synthase (1); tyrosinase, *or* monophenol monooxygenase (2). **1.** Kodan, Kuroda & Sakai (2002) *Proc. Natl. Acad. Sci. U.S.A.* **99**, 3335. **2.** Lin, Hsu, Chen, Chern & Lee (2007) *Phytochemistry* **68**, 1189.

Pioglitazone

This thiazolidinedione (FW = 356.45 g/mol; CAS 11025-46-8), also known by the trade names Actos®, Glustin®, Glizone®, and Pioz® as well as (*RS*)-5-(4-[2-(5-ethylpyridin-2-yl)ethoxy]benzyl)thiazolidine-2,4-dione, is an insulin sensitizer and a specific Peroxisome Proliferator-Activated Receptor-γ, *or* PPAR-γ agonist. **Mode of Action:** Pioglitazone selectively stimulates the nuclear receptor PPAR-γ (and PPAR-α to a lesser degree), thereby modulating insulin-sensitive gene transcription, reducing insulin resistance in the liver and peripheral tissues, increasing the expense of insulin-dependent glucose, decreasing circulating levels of glucose, insulin and glycated hemoglobin. Pioglitazone likewise decreases the level of triglycerides and increases that of high-density lipoproteins (HDL) without altering low-density lipoproteins and total cholesterol in patients with lipid metabolism disorders. **Key Pharmacokinetic Parameters:** *See* Appendix II in Goodman & Gilman's *THE PHARMACOLOGICAL BASIS OF THERAPEUTICS*, 12th Edition (Brunton, Chabner & Knollmann, eds.) McGraw-Hill Medical, New York (2011). **Target(s):** CYP2C8, weakly inhibited (1); CYP2C9, weakly inhibited (1); CYPc17 (2); 3β-hydroxysteroid dehydrogenase type II (2); long-chain-fatty-acyl-CoA synthetase 5, *or* long-chain-fatty-acid:CoA ligase 5 (3). **1.** Yamazaki, Suzuki, Tane, *et al.* (2000) *Xenobiotica* **30**, 61. **2.** Arlt, Auchus & Miller (2001) *J. Biol. Chem.* **276**, 16767. **3.** Kim, Lewin & Coleman (2001) *J. Biol. Chem.* **276**, 24667.

PIP, *See* *Phosphatidylinositol 4-Phosphate*

PIP$_2$, *See* *Phosphatidylinositol 4,5-Bisphosphate*

Pipamperone

This orally bioavailable antipsychotic agent (FW$_{free-base}$ = 375.49 g/mol; CAS 1892-33-0), marketed as Dipiperon® and known systematically as 1'-[4-(4-fluorophenyl)-4-oxobutyl]-1,4'-bipiperidine-4'-carboxamide, is a D$_2$ dopamine receptor and 5-HT$_2$ serotonin receptor antagonist (1-4). **1.** Vanhauwe, Ercken, van de Wiel, Jurzak & Leysen (2000) *Psychopharmacology (Berlin)* **150**, 383. **2.** Van Oekelen, Megens, Meert, Luyten & Leysen (2002) *Behav. Pharmacol.* **13**, 313. **3.** Nisijima, Yoshino, Yui & Katoh (2001) *Brain Res.* **890**, 23. **4.** Ilien, Gorissen & Laduron (1982) *Mol. Pharmacol.* **22**, 243.

Pipecolic Acid

This amino acid (FW = 129.16 g/mol; CAS 3105-95-1), also called 2-piperidinecarboxylic acid and pipecolinic acid, is readily prepared from lysine and can be regarded as a proline homologue. The L-enantiomer has been identified in mushrooms, fruits, malt, and seeds. Elevated levels of pipecolate (*i.e.*, hyperpipecolic acidemia) have been observed in certain inborn errors in metabolism. Note that the pK_a values of pipecolic acid are 2.28 and 10.72; hence, under most aqueous conditions, this molecule exists as a zwitterion. The L-isomer is a product of the reaction catalyzed by Δ^1-piperideine-2-carboxylate reductase and is a substrate for L-pipecolate oxidase and L-pipecolate dehydrogenase. **Target(s):** bovine leukemia virus retropepsin (1); and proline racemase, weakly inhibited (2). **1.** Ménard & Guillemain (1998) in *Handb. Proteolytic Enzymes* (Barrett, Rawlings & Woessner, eds.), p. 940, Academic Press, San Diego. **2.** Cardinale & Abeles (1968) *Biochemistry* **7**, 3970.

Pipecolinic Acid, *See Pipecolic Acid*

Piperacillin

This semisynthetic broad-spectrum antibiotic (FW$_{\text{free-acid}}$ = 517.56 g/mol; CAS 61477-96-1; 59703-84-3 for monosodium salt), known systematically as (2S,5R,6R)-6-{[(2R)-2-[(4-ethyl-2,3-dioxo-piperazine-1-carbonyl)-amino]-2-phenylacetyl]amino}-3,3-dimethyl-7-oxo-4-thia-1-azabicyclo[3.2.0]heptane-2-carboxylic acid, inhibits proteoglycan biosynthesis. Piperacillin is often used in combination with tazobactam and marketed as Tazocin®, Zosyn®, Brodactam® and Trezora®. **Target(s):** penicillin-binding protein-3 (1); serine-type D-Ala-D-Ala carboxypeptidase, *or* D-alanine carboxypeptidase (2). **1.** Botta & Park (1981) *J. Bacteriol.* **145**, 333. **2.** Suginaka, Shimatani, Ogawa & Kotani (1979) *J. Gen. Microbiol.* **112**, 181.

Piperine

This traditional medicine and insecticide (FW = 285.34 g/mol; CAS 94-62-2; Solubility; IUPAC Name: 1-[5-(1,3-benzodioxol-5-yl)-1-oxo-2,4-pentadienyl]piperidine), which gives black pepper (*Piper nigrum*) and long peppers their pungency, enhances the potency of various drugs by inhibiting enzymes catalyzing arylhydrocarbon hydroxylation, ethylmorphine-*N*-demethylation, 7-ethoxycoumarin-*O*-deethylation and 3-hydroxy-benzo(*a*)pyrene glucuronidation (1). Piperine also inhibits the drug transporter P-glycoprotein and CYP3A4, the latter a major drug-metabolizing enzyme (2). Because both enzymes are expressed in enterocytes and hepatocytes, where they play major roles in first-pass elimination of many orally administered drugs, dietary piperine is likely to affect plasma concentrations of drugs in humans. In Ca^{2+}-free medium, piperine (1-30 µM) exhibits vasoconstrictor effects. In rat aorta, piperine demonstrates endothelium-independent vasodilator effect and is more potent against high K^+ pre-contractions than phenylephrine (3). In bovine coronary artery preparations, piperine likewise inhibits high K^+-stimulated pre-contractions completely. Piperine also suppresses phorbol-12-myristate-13-acetate-induced tumor cell invasion by inhibiting PKCα/ERK1/2-dependent matrix metalloproteinase-9 expression (4). At 75-150 µM, piperine inhibits the growth of several colon cancer cell lines, with little effect on the growth of normal fibroblasts and epithelial cells (5). Notably, piperine inhibits HT-29 colon carcinoma cell proliferation by causing G_1 phase cell cycle arrest that was associated with decreased expression of cyclins D1 and D3 and their activating partner CDK-4 and -6, as well as reduced phosphorylation of the retinoblastoma protein and up-regulation of p21/WAF1 and p27/KIP1 expression. In addition, piperine causes hydroxyl radical production and apoptosis. Piperine-treated HT-29 cells show loss of mitochondrial membrane integrity and cleavage of poly(ADP-ribose) polymerase-1, as well as caspase activation and reduced apoptosis in the presence of the pan-caspase inhibitor zVAD-FMK. Increased expression of the endoplasmic reticulum stress-associated proteins inositol-requiring 1α protein, C/EBP homologous protein, and binding immunoglobulin protein, and activation of c-Jun N-terminal kinase and p38 mitogen-activated protein kinase, as well as decreased phosphorylation of Akt and reduced survivin expression are also observed in piperine-treated HT-29 cells (5). **1.** Atal, Dubey & Singh (1985) *J. Pharmacol. Exp. Ther.* **232**, 258. **2.** Bhardwaj,

Glaeser, Becquemont, *et al.* (2002) *J. Pharmacol. Exp. Ther.* **302**, 645. **3.** Taqvi, Shah & Gilani (2008) *J. Cardiovasc. Pharmacol.* **52**, 452. **4.** Hwang, Yun, Kim, *et al.* (2011) *Toxicol. Lett.* **203**, 9. **5.** Yaffe, Power Coombs, Doucette, Walsh & Hoskin (2014) *Mol. Carcinog.* **54**, 1070.

PI3K/HDAC Inhibitor I

This orally available, dual-PI3K/HDAC inhibitor (FW = 508.55 g/mol; CAS 1339928-25-4); Solubility: 100 mg/mL DMSO; <1 mg/mL Water), known as *N*-hydroxy-2-(((2-(6-methoxypyridin-3-yl)-4-morpholino-thieno[3,2-*d*]pyrimidin-6-yl)methyl)(methyl)amino)-pyrimidine-5-carbox-amide, targets the phosphatidylinositol 3-kinase PI3Kα, and histone deacetylases HDAC1, HDAC2, HDAC3 and HDAC10 with IC_{50} values of 19 nM, 1.7 nM, 5 nM, 1.8 nM and 2.8 nM, respectively. PI3K/HDAC Inhibitor I inhibits other PI3K isoforms such as PI3Kβ, PI3Kγ, PI3Kδ, PI3KαH1047R and PI3KαE545K with IC50 of 54 nM, 311 nM, 39 nM, 73 nM and 62 nM, respectively. [1] Bao, Wang, Qu, *et al.* (2012), *American Association for Cancer Research Meeting*, Chicago, Illinois, Poster # 3744.

Piperastatin A

This piperazate-containing peptidomimetic (FW = 809.1 g/mol; NLM Unique Identifier C103102), corresponding to *N*-formyl-*allo*-Ile-Thr-Leu-Val-Pip-Leu-Pip, where Pip is piperazate (*or* hexahydropyridadine-3-carboxylate), is obtained from *Streptomyces lavendofoliae* MJ908-WF13. *Note*: This inhibitor is chemically unstable, and its inhibitory activity diminishes in a time-dependent manner. **Target(s):** carboxypeptidase C, *or* carboxypeptidase Y (1-5). **1.** Murakami, Harada, Yamazaki, *et al.* (1996) *J. Enzyme Inhib.* **10**, 93. **2.** Murakami, Takahashi, Naganawa, Takeuchi & Aoyagi (1996) *J. Enzyme Inhib.* **10**, 105. **3.** Liu, Tachibana, Taira, *et al.* (2004) *J. Ind. Microbiol. Biotechnol.* **31**, 572. **4.** Liu, Tachibana, Taira, Ishihara & Yasuda (2004) *J. Ind. Microbiol. Biotechnol.* **31**, 23. **5.** Satoh, Kadota, Oheda, *et al.* (2004) *J. Antibiot.* **57**, 316.

2-(2-Piperidinyl)ethylidene-1,1-bisphosphonate

This bisphosphonate (FW$_{\text{ion}}$ = 271.15 g/mol), also known as NE58034, inhibits geranyl*trans*transferase, *or* farnesyl-diphosphate synthase. The main drugs currently used to treat diseases characterized by excessive bone resorption, <u>n</u>itrogen-containing <u>b</u>isphosphonates (N-BPs) The major molecular target of N-BPs inhibit farnesylpyrophosphate synthase. by a mechanism that involves time-dependent isomerization of the enzyme. **1.** Dunford, Kwaasi, Rogers, *et al.* (2008) *J. Med. Chem.* **51**, 2187.

20-Piperidin-2-yl-pregnan-20(*R*),3β-diol, *See 22,26- Azasterol*

3-[5-(2-(Piperidin-1-ylsulfonylamino)ethyl)-5-(pyrid-3-ylmethyl)-1-benzene-propionate

This sulfonamide (FW$_{\text{free-acid}}$ = 431.56 g/mol) inhibits human thromboxane-A synthase (IC$_{50}$ = 43 nM) and also is a thromboxane receptor antagonist. Introduction of a 5-(1*H*-imidazol-1-ylmethyl), a 5-(3-pyridinyl-methyl), or a 5-(3-pyridinyloxy) substituent leads to dual agents with thromboxane synthase inhibitory activity comparable with that of dazmegrel (7). In addition, 3-pyridinylalkyl substituents also make a significant contribution to thromboxane receptor binding. **1**. Dickinson, Dack, Long & Steele (1997) *J. Med. Chem.* **40**, 3442.

2-Piperidone, *See* δ-Valerolactam

Piperine

This peppery-tasting alkaloid (FW = 285.34 g/mol; CAS 7780-20-3; 30511-76-3 for (*Z,E*)-isomer; 30511-77-4 for (*E,Z*)-isomer; 495-91-0 (*Z,Z*)-isomer; 94-62-2 (*E,E*)-isomer), also known as piperic acid piperidide and (*E,E*)-1-piperoylpiperidine, from black pepper (*Piper nigrum*) inhibits arylhydrocarbon hydroxylase (1), CYP1A1 (2), CYP3A4 (3), electron transport (4), malate dehydrogenase (5), P-glycoprotein (3), UDP-glucose dehydrogenase (6), and UDP-glucuronosyltransferase (6). **1**. Reen & Singh (1991) *Indian J. Exp. Biol.* **29**, 568. **2**. Reen, Roesch, Kiefer, Wiebel & Singh (1996) *Biochem. Biophys. Res. Commun.* **218**, 562. **3**. Bhardwaj, Glaeser, Becquemont, *et al.* (2002) *J. Pharmacol. Exp. Ther.* **302**, 645. **4**. Reanmongkol, Janthasoot, Wattanatorn, Dhumma-Upakorn & Chudapongse (1988) *Biochem. Pharmacol.* **37**, 753. **5**. Jamwal & Singh (1993) *J. Biochem. Toxicol.* **8**, 167. **6**. Reen, Jamwal, Taneja, *et al.* (1993) *Biochem. Pharmacol.* **46**, 229.

Piperonyl Butoxide

This insecticide synergist (FW = 338.44 g/mol), which is frequently used in mixtures with rotenone or pyrethroids, is a substrate for cytochrome P450 systems and competes with other xenobiotics for available enzymes. In addition, an intermediate produced during its metabolism binds cytochrome P450, forming a catalytically inactive complex. **Target(s):** arylesterase (1); cytochrome P450 (2,3); and ecdysone 20-monooxygenase (4). **1**. Young, Gunning & Moores (2005) *Pest Manag. Sci.* **61**, 397. **2**. Friedman & Woods (1977) *Res. Commun. Chem. Pathol. Pharmacol.* **17**, 623. **3**. Franklin (1976) *Environ. Health Perspect.* **14**, 29. **4**. Feyereisen & Durst (1978) *Eur. J. Biochem.* **88**, 37.

Pipsyl Fluoride, *See* p-Iodobenzenesulfonyl Fluoride

Pirarubicin

This doxorubicin-derived antineoplastic agent (FW$_{free-base}$ = 627.65 g/mol; CAS 72496-41-4, 95343-20-7 (HCl); Solubility: 125 mg/mL DMSO; <1 mg/mL Water), also known as THP-adraimycin and systematically as (7*S*,9*S*)-7-((2*R*,4*S*,5*S*,6*S*)-4-amino-6-methyl-5-((*R*)-tetrahydro-2*H*-pyran-2-yloxy)tetrahydro-2*H*-pyran-2-yloxy)-6,9,11-trihydroxy-9-(2-hydroxyacetyl)-4-methoxy-7,8,9,10-tetrahydrotetracene-5,12-dione, inhibits DNA polymerases and DNA topoisomerase II. Pirarubicin intercalates into DNA, while also interacting with topoisomerase II, thereby inhibiting DNA replication repair, and transcription. This agent is less cardiotoxic than doxorubicin and exhibits activity against some doxorubicin-resistant cell lines. **Target(s):** cytochrome *c* oxidase (1); DNA-directed DNA polymerase (2); DNA polymerase I, *Escherichia coli* (2); DNA polymerase α, calf thymus (2); DNA polymerase β, calf thymus (2); DNA topoisomerase II (3); and oxidative phosphorylation (4). **1**. Del Tacca, Danesi, Solaini, Bernardini & Bertelli (1987) *Anticancer Res.* **7**, 803. **2**. Tanaka, Yoshida & Kimura (1983) *Gann* **74**, 829. **3**. Insaf, Danks & Witiak (1996) *Curr. Med. Chem.* **3**, 437. **4**. Solaini, Ronca & Bertelli (1985) *Drugs Exp. Clin. Res.* **11**, 115 and 533.

Pirenperone

This agent (FW$_{free-base}$ = 393.46 g/mol; CAS 75444-65-4), also known as 3-[2-[4-(4-fluorobenzoyl)-1-piperidinyl]ethyl-2-methyl-4*H*-pyrido[1,2-*a*]pyrimidin-4-one, is a 5-HT$_2$ serotonin receptor antagonist (1-4) and an LSD (lysergic acid diethylamine) antagonist. **1**. Walker, Poulos & Le (1994) *Psychopharmacology (Berlin)* **113**, 527. **2**. Fiorella, Rabin & Winter (1995) *Psychopharmacology (Berlin)* **119**, 222. **3**. Valentin, Bessac, Colpaert & John (1995) *Methods Find. Exp. Clin. Pharmacol.* **17**, 267. **4**. Colpaert, Niemegeers & Janssen (1982) *J. Pharmacol. Exp. Ther.* **221**, 206.

Pirenzepine

This hygroscopic antiulcerative (FW$_{free-base}$ = 351.41 g/mol; CAS 28797-61-7) is a gastric acid inhibitor and selective M$_1$ muscarinic receptor antagonist (1,4), and the antisecretory properties of pirenzepine on gastric acid and pepsin secretion may be attributed to this antagonistic activity on muscarinic M$_1$ receptors. **1**. Pelat, Lazartigues, Tran, *et al.* (1999) *Eur. J. Pharmacol.* **379**, 117. **2**. Eberlein, Schmidt, Reuter & Kutter (1977) *Arzneimittelforschung* **27**, 356. **3**. Heller, Karn, Neubauer, Althoff & Schoffling (1978) *Verh. Dtsch. Ges. Inn. Med.* **84**, 991. **4**. Del Tacca, Danesi, Blandizzi & Bernardini (1989) *Minerva Dietol. Gastroenterol.* **35**, 175.

Pirfenidone

This anti-fibrotic/anti-inflammatory drug (FW = 185.23 g/mol; CAS 53179-13-8), also known as 5-methyl-1-phenylpyridin-2-one, reduces fibroblast proliferation (1-4), inhibits TGF-β stimulated collagen production (1,2,5-7), and reduces the production of fibrogenic mediators such as TGF-β1 (3,6). The latter is secreted by aberrantly activated alveolar epithelial cells (AEC), and, along with other pro-fibrotic factors, promotes fibroblast and AEC differentiation into myofibroblasts, resulting in overproduction of ECM within the lung. Pirfenidone also reduces the production of TNF-α and IL-1β in both cultured cells and isolated human peripheral mononuclear cells (7,8). Pirfenidone attenuates the IL-1β-induced hyaluronic acid production in orbital fibroblasts from patients with thyroid-associated ophthalmopathy (TAO), at least in part, through suppression of the MAPK-mediated hyaluronic acid synthase (HAS) expression (9). Pirfedone is indicated for the treatment of idiopathic pulmonary fibrosis (IPF), the most common form of the interstitial lung diseases, marked by chronic, progressive fibrosing interstitial pneumonia, primarily affecting older adults. is approved for use in Europe and U.S. (trade name: Esbriet®), Japan (trade name: Pirespa®), and India (trade name: Pirfenex®). With IPF remaining a ill-defined collection of disorders, it seems unlikely that any single drug will be an effective treatment for all forms of IPF. **1.** Di Sario, Bendia, Svegliati, *et al.* (2002) *J. Hepatol.* **37**, 584. **2.** Hewitson, Kelynack, Tait, *et al.* (2001) *J. Nephrol.* **14**, 453. **3.** Lin, Yu, Wu, Yuan & Zhong (2009) *Invest. Ophthalmol. Vis. Sci.* **50**, 3763. **4.** Lee, Margolin & Nowak (1998) *J. Clin. Endocrinol. Metab.* **83**, 219. **5.** Ozes & Blatt LM (2006) *Chest* **130**, 230S. **6.** Sulfab (2007) *Am. J. Respir. Crit. Care Med.* **175**, A730. **7.** Nakayama, Mukae, Sakamoto, *et al.* (2008) *Life Sci.* **82**, 210. **8.** Grattendick, Nakashima, Feng, Giri & Margolin (2008) *Int. Immunopharmacol.* **8**, 679. **9.** Chung, Jeon, Choi, *et al.* (2014) *Invest. Ophthalmol. Vis. Sci.* **55**, 2276,

Pirimicarb

This pyrimidine carbamate-based insecticide (FW = 238.29 g/mol; CAS 23103-98-2), systematically named 2-Dimethylamino-5,6-dimethylpyrimidin-4-yl dimethylcarbamate, is a frequently used miticide and aphicide, inhibits acetylcholinesterase (1-3). **Toxicity:** LD_{50} (Dermal) > 500 mg/kg in rats; LD_{50} (Oral) = 147 mg/kg in rats. **1.** Li & Han (2002) *Arch. Insect Biochem. Physiol.* **51**, 37. **2.** Belzunces & Colin (1991) *Neuroreport* **2**, 265. **3.** Baranyovits & Ghosh (1969) *Chem Ind.* **30**, 1018.

Piriprost

This lipoxygenase-inhibiting pyrrole analogue of prostacyclin (FW_potassium salt = 463.66 g/mol; CAS 79672-88-1), also known as U60257B and 6,9-deepoxy-6,9-(phenylimino)-$\Delta^{6,8}$-prostaglandin I1, [4R-[4α(1E,3S*),5β]]-1,4,5,6-tetrahydro-5-hydroxy-4-(3-hydroxy-1-octenyl)-1-phenyl-cyclopenta[b]pyrrole-2-pentanoate, inhibits leukotriene C and D formation.1-5 Piriprost also inhibits the release of histamine and leukotrienes from isolated porcine lung cells.6 Note that both priprost and its methyl ester (U-56467) inhibit leukotriene biosynthesis: ID_{50} = 4.6 and 0.31 μM, respectively. **Target(s):** glutathione *S*-transferase (4,5); and 5-lipoxygenase, *or* arachidonate 5-lipoxygenase (1-3). **1.** Cejic & Kennedy (1991) *Prostaglandins* **42**, 179. **2.** Lawson, Smith & Fitzpatrick (1986) *Wien Klin. Wochenschr.* **98**, 110. **3.** Sun & McGuire (1983) *Prostaglandins* **26**, 211. **4.** Bach, O'Brien, Brashler, Johnson & Morton, Jr. (1985) *Res. Commun. Chem. Pathol. Pharmacol.* **49**, 361. **5.** Bach, Brashler, Smith, *et al.* (1982) *Prostaglandins* **23**, 759. **6.** McCormack & Peterson (1989) *Amer. Rev. Respir. Dis.* **139**, 100.

Piritrexim

This antifolate (FW = 325.37 g/mol; CAS 72732-56-0; 72676-60-9 for the mono-HCl), also known as 2,4-diamino-6-(2,5-dimethoxybenzyl)-5-methylpyrido[2,3-d]pyrimidine and BW301U, is an antineoplastic agent and potent inhibitor of dihydrofolate reductase. **Target(s):** amidophospho-ribosyl-transferase (1,2); dihydrofolate reductase (3-6); and histamine *N*-methyltransferase (7). **1.** Schoettle, Crisp, Szabados & Christopherson (1997) *Biochemistry* **36**, 6377. **2.** Sant, Lyons, Phillips & Christopherson (1992) *J. Biol. Chem.* **267**, 11038. **3.** Grivsky, Lee, Sigel, Duch & Nichol (1980) *J. Med. Chem.* **23**, 327. **4.** Duch, Edelstein, Bowers & Nichol (1982) *Cancer Res.* **42**, 3987. **5.** Fry & Jackson (1987) *Cancer Metastasis Rev.* **5**, 251. **6.** Champness, Achari, Ballantine, *et al.* (1994) *Structure* **2**, 915. **7.** Malmberg-Aiello, Lamberti, Ipponi, *et al.* (1997) *Naunyn Schmiedebergs Arch. Pharmacol.* **355**, 354.

Piroxicam

This long-lasting anti-inflammatory agent (FW = 331.35 g/mol; CAS 36322-90-4), also known as 4-hydroxy-2-methyl-*N*-(2-pyridyl)-2*H*-1,2-benzothiazine-3-carboxamide 1,1-dioxide, is a potent inhibitor of prostaglandin production (1). **Target(s):** chloride/ bicarbonate antiports (2); collagenase (3); cyclooxygenase, *or* prostaglandin-endoperoxide synthase, selective for COX-1 (4-7); dihydrofolate reductase (8); elastase, granulocyte (9); estrone sulfotransferase (10); leukocyte thioltransferase (11); myeloperoxidase (12); ornithine decarboxylase (13); oxidative phosphorylation, as an uncoupler (14); phenol sulfotransferase, *or* aryl sulfotransferase (10); phenylpyruvate tautomerase (15); phosphoribosylamino-imidazolecarboxamide formyltransferase, *or* AICAR transformylase (8); and thiopurine *S*-methyltransferase (16). **1.** Carty, Eskra, Lombardino & Hoffman (1980) *Prostaglandins* **19**, 51. **2.** Tonnessen, Aas, Sandvig & Olsnes (1989) *Biochem. Pharmacol.* **38**, 3583. **3.** Barracchini, Franceschini, Amicosante, *et al.* (1998) *J. Pharm. Pharmacol.* **50**, 1417. **4.** Rowlinson, Kiefer, Prusakiewicz, *et al.* (2003) *J. Biol. Chem.* **278**, 45763. **5.** Carty, Stevens, Lombardino, Parry & Randall (1980) *Prostaglandins* **19**, 671. **6.** Meade, Smith & DeWitt (1993) *J. Biol. Chem.* **268**, 6610. **7.** Patrignani, Panara, Sciulli, *et al.* (1997) *J. Physiol. Pharmacol.* **48**, 623. **8.** Baggott, Morgan, Ha, Vaughn & Hine (1992) *Biochem. J.* **282**, 197. **9.** Lentini, Ternai & Ghosh (1987) *Biochem. Int.* **15**, 1069. **10.** King, Ghosh & Wu (2006) *Curr. Drug Metab.* **7**, 745. **11.** Mizoguchi, Nishnaka, Uchida, *et al.* (1993) *Biol. Pharm. Bull.* **16**, 840. **12.** Kettle & Winterbourn (1991) *Biochem. Pharmacol.* **41**, 1485. **13.** Bruni, Dal Pra & Segre (1984) *Int. J. Tissue React.* **6**, 463. **14.** Moreno-Sanchez, Bravo, Vasquez, *et al.* (1999) *Biochem. Pharmacol.* **57**, 743. **15.** Molnar & Garai (2005) *Int. Immunopharmacol.* **5**, 849. **16.** Oselin & Anier (2007) *Drug Metab. Dispos.* **35**, 1452.

Pitavastatin

This HMG-CoA reductase inhibitor (FW = 421.46 g/mol; CAS 147511-69-1; IUPAC Name: (3R,5S,6E)-7-[2-cyclopropyl-4-(4-fluorophenyl)quinolin-3-yl]-3,5-dihydroxyhept-6-enoic acid), also known as itavastatin, itabavastin, nisvastatin, NK-104, NKS-104, and the trade name Livalo®, is indicated for ameliorating hypercholesterolemia and preventing cardiovascular disease (1,2). NK-104 potency is dose-dependent and is roughly equivalent to that of atorvastatin. It is well-tolerated in the treatment of patients with hypercholesterolemia. Pitavastatin uptake is carrier-mediated (3). **Target(s):** Reduces inflammatory cytokine production from human bronchial epithelial cells (4); Decreases microtubule tau protein levels via the inactivation of Rho/ROCK (5); Inhibits hepatic steatosis and fibrosis in non-alcoholic steatohepatitis model (6); Suppresses theroslerosis induced by chronic inhibition of the synthesis of nitric oxide in moderately hypercholesterolemic rabbits (7); Decreases the expression of endothelial lipase both *in vitro* and *in vivo* (8); Inhibits NFκB pathway in brain (9); Inactivates NFκB and decreases IL-6 production through Rho kinase pathway in MCF-7 human breast cancer cells (10); Reduces C-reactive-protein-induced interleukin-8 production in human aortic endothelial cells (11); Inhibits lysophosphatidic acid-induced proliferation and monocyte chemoattractant protein-1 expression in aortic smooth muscle cells by suppressing Rac-1-mediated reactive oxygen species generation (12); Inhibits upregulation of intermediate conductance calcium-activated potassium channels and coronary arteriolar remodeling induced by long-term blockade of nitric oxide synthesis (13); Inhibits migration and proliferation of rat vascular smooth muscle cells (14). **1.** Aoki, Nishimura, Nakagawa *et al.* (1997) *Arzneimittelforschung.* **47**, 904. **2.** Kitahara, Kanaki, Toyoda *et al.* (1998) *Japan. J. Pharmacol.* **77**, 117. **3.** Shimada, Fujino, Morikawa, Moriyasu & Kojima (2003) *Drug Metab. Pharmacokinet.* **18**, 245. **4.** Iwata, Shirai, Ishii, *et al.* (2012) *Clin. Exp. Immunol.* **168**, 234. **5.** Hamano, Yen, Gendron, *et al.* (2012) *Aging* **33**, 2306. **6.** Miyaki, Nojiri, Shinkai, *et al.* (2011) *Hepatol Res.* **41**, 375. **7.** Kitahara, Kanaki, Ishii & Saito (2010) *Brit. J. Pharmacol.* **159**, 1418. **8.** Kojima, Ishida, Sun, *et al* (2010) *Cardiovasc. Res.* **87**, 385. **9.** Aoki, Kataoka, Ishibashi R *et al* (2009) *Neurosurgery* 64, 357. **10.** Wang & Kitajima (2007) *Oncol. Rep.* **17**, 1149. **11.** Kibayashi, Urakaze, Kobashi, *et al.* (2005) Clin. Sci. (London). **108**, 515. **12.** Kaneyuki, Ueda, Yamagishi (2007) *Vascul. Pharmacol.* **46**, 286. **13.** Terata, Saito, Fujiwara *et al.* (2003) *Pharmacology* **68**, 169. **14.** Kohno, Shinomiya, Abe, *et al.* (2002) *Hypertens. Res.* **25**, 279.

PITC, *See* Phenyl Isothiocyanate

Pitocin, *See* Oxytocin

Pitressin, *See* Vasopressin

Pixantrone

This DNA-intercalator and mitoxantrone analogue (FW = 325.37 g/mol; CAS 144510-96-3), also known as Pixuvri® and 6,9-bis[(2-aminoethyl)amino]benzo[g]isoquinoline-5,10-dione, is a second-generation anthracenedione that selectively targets topoisomerase II and is an experimental anti-cancer agent (1). (Note: The code name BBR-2778 is assigned to pixantrone dimaleate, the salt most often used in clinical trials.) In vitro pixantrone showed a median relative IC_{50} value of 54 nM (range <3 nM to 1.03 μM) against Wilms tumor and other cancers (2). **1.** Krapcho, Petry, Getahun, *et al.* (1994) *J. Med. Chem.* **37**, 828. **2.** Kurmasheva, Reynolds, Kang, *et al.* (2014) *Pediatr. Blood Cancer* **62**, 922.

2-Pivaloyl-1,3-indandione, *See* Pindone

PJ-34

This PARP inhibitor ($FW_{free-base}$ = 295.34 g/mol; $FW_{hydrochloride}$ = 331.80 g/mol), systematically named *N*-(6-oxo-5,6-dihydro-phenanthridin-2-yl)-*N,N*-dimethylacetamide and known for its activity in neuroprotection under stress conditions, exclusively eradicates multi-centrosomal human mammary, colon, lung, pancreas, ovarian cancer cells, by acting as an extra-centrosome(s) de-clustering agent in mitosis (1). When applied at 20-30 μM, PJ-34 caused G_2/M-arrest and a massive cell death Normal human proliferating endothelial, epithelial and mesenchymal cells were unaffected. PJ-34's cytotoxicity on cancer cells is not attributable to PARP inhibition alone. PJ-34 was originally designed to protect neuronal cells in the central nervous system from cell death evoked by high activity of PARP-1 in response to DNA damage caused by brain injury, stroke or inflammation (2,3). **Targets:** extra-centrosome de-clustering (1); polyADP-ribose polymerase-1, EC_{50} = 20 nM (2-6) **1.** Castiel, Visochek, Mittelman, *et al.* (2011) *BMC Cancer* **11**, 412. **2.** Chiarugi, Meli, Calvani, *et al.* (2003) *J. Pharmacol. Exp. Ther.* **305**, 943. **3.** Jagtap & Szabo (2005) *Nat. Rev. Drug Discov.* **4**, 421. **4.** Abdelkarim, Gertz, Harms, *et al.* (2001) *Int. J. Mol. Med.* **7**, 255. **5.** Garcia Soriano, Virag, Jagtap, *et al.* (2001) *Nat. Med.* **7**, 108. **6.** Pagano, Métrailler-Ruchonnet, Aurrand-Lions, *et al.* (2007) *Amer. J. Physiol. Lung Cell Mol. Physiol.* **293**, L619.

PK13, *See* Proteinase K/α-Amylase Inhibitor

PKAI, *See* Protein Kinase A Inhibitors

PKC412

This broad-spectrum serine/threonine/tyrosine protein kinase inhibitor (PKI) and radiooncology sensitizing agent (FW = 570.65 g/mol; CAS 120685-11-2), also known as CGP 41251 and [9S-(9α,10β,11β,13α)]-*N*-(2,3,10,11,12,13-hexahydro-10-methoxy-9-methyl-1-oxo-9,13-epoxy-1*H*,9*H*-diindolo[1,2,3-*gh*:3',2',1'-*lm*]pyrrolo[3,4-*j*][1,7]benzodiazonin-11-yl)-*N*-methylbenzamide targets conventional PKCα, PKCβ, PKCγ, PDFRβ, VEGFR2, Syk, PKCη, Flk-1, Flt3, Cdk1/B, PKA, c-Kit, c-Fgr, c-Src, VEGFR1 and EGFR, displaying potent antitumor activity (*See also AC220 & G-749*). **1.** Fabbro, Ruetz S, Bodis, *et al.* (2000) *Anticancer Drug Des.* **15**, 17. **2.** Tenzer, Zingg, Rocha, *et al.* (2001) *Cancer Res.* **61**, 8203. **3.** Nakazono-Kusaba, Takahashi-Yanaga, Miwa, *et al.* (2004) *Eur. J. Pharmacol.* **497**, 155.

PKI-587

This dual PI3K/mTOR (phosphoinositide-3-kinase and mammalian target of rapamycin) signal-transduction pathway inhibitor and antineoplastic agent (F.Wt. = 615.73; CAS 1197160-78-3); Solubility (25°C): 2 mg/mL DMSO, <1 mg/mL Water), also known as PF-05212384 and systematically as 1-(4-(4-(dimethylamino)piperidine-1-carbonyl)phenyl)-3-(4-(4,6-dimorpholino-1,3,5-triazin-2-yl)phenyl)urea, inhibits PI3K-α, PI3K-γ and mTOR with IC_{50} of 0.4 nM, 5.4 nM and 1.6 nM, respectively. PKI-587 also inhibits mutant forms of PI3Kα, including the PI3Kα-H1047R and PI3Kα-E545K with IC_{50} of 0.6 nM and 0.6 nM, respectively. **1.** Venkatesan, Dehnhardt, Delos Santos (2010) *J. Med. Chem.* **53**, 2636.

Plasdone, *See* Polyvinylpyrrolidone

α2-Plasmin Inhibitor, *See* α2-Antiplasmin

Plasmin Inhibitor, Leech, *See* Bdellin

Plasminogen Activator Inhibitor Type 1

This single-chain glycoprotein (MW = 42.5 kDa), often abbreviated PAI-1, consisting of 379 residues and produced by the vascular endothelium and various other cells, is a serpin family member that controls fibrinolysis. PAI-1 also interacts with vitronectin and with the urokinase receptor and its co-receptors. Defects in the PAI-1 gene are the cause of plasminogen activator inhibitor-1 deficiency, whereas high concentrations of the gene product are associated with thrombophilia. Alternatively spliced transcript variants encoding different isoforms have been found for this gene. Plasma levels of PAI-1 also increase in several chronic inflammatory states that are associated with chronic kidney disease. (*For a discussion of the likely mechanism of inhibitory action, See Serpins; also α_1-Antichymotrypsin*) PAI-1 has many targets (1-24). The novel and specific small-molecule PAI-1 inhibitor IMD-4690 reduces allergic airway remodeling in a mouse model of chronic asthma via regulating angiogenesis and remodeling-related mediators (25). **Target(s):** acrosin (1); bat salivary plasminogen activator (2); matriptase (3); pancreatic elastase (24); plasminogen activator Pla4, protein C activated (5); t-plasminogen activator (2,4-15,24); u-plasminogen activator, *or* urokinase (2,4,5,7,16-24). **1.** Zheng, Geiger, Ecke, *et al.* (1994) *Fibrinolysis* **8**, 364. **2.** Gardell & Friedman (1993) *Meth. Enzymol.* **223**, 233. **3.** Szabo, Netzel-Arnett, Hobson, Antalis & Bugge (2005) *Biochem. J.* **390**, 231. **4.** Lahteenmaki, Kuusela & Korhonen (2001) *FEMS Microbiol. Rev.* **25**, 531. **5.** Fay & Owen (1989) *Biochemistry* **28**, 5773. **6.** Lijnen & Collen (1998) in *Handb. Proteolytic Enzymes* (Barrett, Rawlings & Woessner, eds.), p. 184, Academic Press, San Diego. **7.** Loskutoff & Schleef (1988) *Meth. Enzymol.* **163**, 293. **8.** Verheijen (1988) *Meth. Enzymol.* **163**, 302. **9.** Saksela & Rifkin (1988) *Ann. Rev. Cell Biol.* **4**, 93. **10.** Johnsen, Ravn, Berglund, *et al.* (1998) *Biochemistry* **37**, 12631. **11.** Urano, Strandberg, Johansson & Ny (1992) *Eur. J. Biochem.* **209**, 985. **12.** Sherman, Lawrence, Verhamme, *et al.* (1995) *J. Biol. Chem.* **270**, 9301. **13.** Kaneko, Sakata, Matsuda & Mimuro (1992) *J. Biochem.* **111**, 244. **14.** Declerck, De Mol, Vaughan & Collen (1992) *J. Biol. Chem.* **267**, 11693. **15.** Hofmann, Mayer, Schultz, Socher & Reilley (1992) *Fibrinolysis* **6**, 263. **16.** Loskutoff & Edgington (1981) *J. Biol. Chem.* **256**, 4142. **17.** Eddy & Fogo (2006) *J. Amer. Soc. Nephrol.* **17**, 2999. **18.** Ellis & Danø (1998) in *Handb. Proteolytic Enzymes* (Barrett, Rawlings & Woessner, eds.), p. 177, Academic Press, San Diego. **19.** Saksela (1985) *Biochim. Biophys. Acta* **823**, 35. **20.** Takahashi, Kwaan, Koh & Tanabe (1992) *Biochem. Biophys. Res. Commun.* **182**, 1473. **21.** Schmitt, Jänicke, Moniwa, *et al.* (1992) *Biol. Chem. Hoppe-Seyler* **373**, 611. **22.** Lijnen, De Cock & Collen (1994) *Eur. J. Biochem.* **224**, 567. **23.** Franco, Mastronicola, De Cesare, *et al.* (1992) *J. Biol. Chem.* **267**, 19369. **24.** Komissarov, Declerck & Shore (2004) *J. Biol. Chem.* **279**, 23007. **25.** Tezuka, Ogawa, Azuma, *et al.* (2015) *PLoS One* **10**, e0121615.

Plasminogen Activator Inhibitor Type 2

This protein inhibitor (MW = 44.5 kDa), often abbreviated PAI-2, inhibits plasminogen activators and is a member of the serpin family. It has been isolated from placental extracts and it controls fibrinolysis. **Target(s):** acrosin (1); t-plasminogen activator (2-5); u-plasminogen activator, *or* urokinase (4-8). (*For a discussion of the likely mechanism of inhibitory action, See Serpins; also α_1-Antichymotrypsin*) **1.** Zheng, Geiger, Ecke, *et al.* (1994) *Fibrinolysis* **8**, 364. **2.** Johnsen, Ravn, Berglund, *et al.* (1998) *Biochemistry* **37**, 12631. **3.** Mikus, Urano, Liljeström & Ny (1993) *Eur. J. Biochem.* **218**, 1071. **4.** Lahteenmaki, Kuusela & Korhonen (2001) *FEMS Microbiol. Rev.* **25**, 531. **5.** Saksela & Rifkin (1988) *Ann. Rev. Cell Biol.* **4**, 93. **6.** Ellis & Danø (1998) in *Handb. Proteolytic Enzymes* (Barrett, Rawlings & Woessner, eds.), p. 177, Academic Press, San Diego. **7.** Takahashi, Kwaan, Koh & Tanabe (1992) *Biochem. Biophys. Res. Commun.* **182**, 1473. **8.** Schmitt, Jänicke, Moniwa, *et al.* (1992) *Biol. Chem. Hoppe-Seyler* **373**, 611.

Plasminogen Activator Inhibitor Type 3, *See* Protein C Inhibitor

Plasminostreptin

This 109-residue protease inhibitor (MW = 11402 g/mol; CAS 50864-63-6) from *Streptomyces antifibrinolyticus*, inhibits selected serine proteases, but not chymotrypsin, thrombin, kallikrein, or elastase (1). Plasminostreptin reportedly also inhibits certain metalloendoproteinases, such as mycolysin. **Target(s):** mycolysin (1,2); plasmin (3,4); *Streptomyces griseus* metalloendopeptidases (12); *Streptomyces griseus* metalloendopeptidases II (2,5); subtilisin (3,4); and trypsin (3,4,6). **1.** Ishii & Kumazaki (1998) in *Handb. Proteolytic Enzymes* (Barrett, Rawlings & Woessner, eds.), p. 1078, Academic Press, San Diego. **2.** Tsuyuki, Kajiwara, Fujita, Kumazaki & Ishii (1991) *J. Biochem.* **110**, 339. **3.** Kakinuma, Sugino, Moriya & Isono

(1978) *J. Biol. Chem.* **253**, 1529. **4.** Sugino, Nakagawa & Kakinuma (1978) *J. Biol. Chem.* **253**, 1538. **5.** Kajiwara, Fujita, Tsuyuki, Kumazaki & Ishii (1991) *J. Biochem.* **110**, 350. **6.** Sugino, Kakinuma & Iwanaga (1978) *J. Biol. Chem.* **253**, 1546.

Plasmochin, *See* Pamaquine Pamoate

Plastoquinones

Plastoquinone 9

This family of polyisoprenoid quinones participates in photosynthetic electron transport and vary only in the length of the isoprenoid side chain. The most common member is plastoquinone 9 (FW = 749.22 g/mol, M.P. = 48-49°C; PQ-10 at 50-51°C) consisting of nine isoprene units. PQ-9 is a yellow solid with λ_{max} values (in isooctane) of 254 and 262 nm. Plastoquinones are unstable in oxygen and light, dictating storage in a well-sealed and light-tight container. **Target(s):** 5-lipoxygenase, *or* arachidomate 5-lipoxygenase, inhibited by PQ-7 (1); vitamin K-dependent carboxylase, inhibited by PQ-9 (2); and vitamin K-epoxide reductase, inhibited by PQ-9 (2). **1.** Odukoya, Houghton & Raman (1999) *Phytomedicine* **6**, 251. **2.** Ronden, Soute, Thijssen, Saupe & Vermeer (1996) *Biochim. Biophys. Acta* **1298**, 87.

Platelet Activating Factor

1-O-hexadecyl-2-acetyl-sn-glycero-3-phosphocholine

This phospholipid (abbreviated PAF), typically consisting of either 1-*O*-hexadecyl-2-acetyl-*sn*-glycerol 3-phosphocholine or 1-*O*-octadecyl-2-acetyl-*sn*-glycerol 3-phosphocholine, is a key mediator of platelet aggregation, inflammation, and anaphylaxis. PAF also stimulates uterine contractions, decrease cardiac output, and mediate hypertension. **Target(s):** 1,3-β-glucan synthase (1,2); sphingomyelin phosphodiesterase, *or* sphingomyelinase, also an alternative substrate for the rat alkaline enzyme (3). **1.** Kauss & Jeblick (1986) *Plant Physiol.* **80**, 7. **2.** Ko, Frost, Ho, Ludescher & Wasserman (1994) *Biochim. Biophys. Acta* **1193**, 31. **3.** Wu, Nilsson, Jönsson, *et al.* (2006) *Biochem. J.* **394**, 299.

Platelet Aggregation Inhibitors

These agents antagonize or impair mechanisms leading to blood platelet aggregation during the phases of platelet activation and shape change, or after the dense-granule release reaction and stimulation of the prostaglandin-thromboxane system. Platelet aggregation inhibitor types include: *cyclooxygenase inhibitors*, such as acetylsalicylic acid (aspirin); *adenosine 5'-diphosphate receptor inhibitors*, such as clopidogrel (Plavix™), prasugrel (Effient™), ticlopidine (Ticlid™); *phosphodiesterase inhibitors*, such as cilostazol (Pletal™); *glycoprotein IIB/IIIA inhibitors*, including abciximab (ReoPro™), eptifibatide (Integrilin™), tirofiban (Aggrastat™); and *adenosine reuptake inhibitors*, such as dipyridamole (Persantine™) and picotamide. *See also Abciximab; Acetylsalicylic Acid; Cilostazol; Clopidogrel; Dipyridamole; Eptifibatide; Picotamide; Prasugrel; Ticlopidine; Tirofiban; Vorapaxar*

Platelet-Derived Growth Factor Receptor β Fragment 751-755, Phosphorylated

This phosphorylated pentapeptide (MW = 700.76 g/mol; *Sequence*: Y(PO$_3^{2-}$)VPML), which contains a phosphorylated tyrosyl residue, inhibits binding of the 85-kDa subunit of phosphatidylinositol 3-kinase to the PDGF β-receptor. It was observed (a) that neutralization of the amine and carboxy terminus led to analogues with enhanced activity and (b) that only minimal modifications were allowed for pTyr and Met, while the other positions were quite tolerant of modification. **1.** Ramalingam, Eaton, Cody, *et al.* (1995) *Bioorg. Med. Chem.* **3**, 126.

Placotylene A

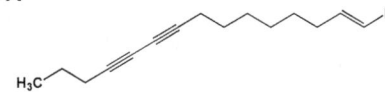

This novel bone resorption inhibitor (FW = 328.24 g/mol), an iodinated polyacetylene class of natural products from a Korean marine sponge (*Placospongia* sp.), targets the Receptor Activator of NF-κB Ligand (RANKL)-induced osteoclast differentiation. The chemical structure was elucidated on the basis of 1D and 2D NMR, supplemented by MS data. Placotylene A (10 μM) effectively inhibits RANKL-induced osteoclast differentiation, but the structurally placotylene B does not show significant activity up to 100 μM. **1.** Kim, Kim, Yeon, *et al.* (2014) *Marine Drugs* **12**, 2054.

Pladienolide B

This *Streptomyces platensis* macrocyclic lactone (FW = 636.71 g/mol; CAS 445493-23-2; Soluble in DMSO) exhibits strong antitumor activities, both *in vitro* and in xenograft models (1,2). Pladienolide binds to the SF3b complex and inhibits mRNA splicing and changes splicing patterns (1). Pladienolide-resistant clones from several colorectal cancer cells are insensitive to pladienolide's inhibitory action on cell proliferation and splicing (3). Moreover, an mRNA-Seq differential analysis revealed that these cell lines have an identical mutation at Arg1074 in the gene for SF3B1, which encodes a subunit of the SF3b complex. Reverse expression of the mutant protein transferred pladienolide resistance to WiDr cells. Furthermore, immunoprecipitation analysis using a radiolabeled probe showed that the mutation impaired the binding affinity of paldienolide to its target (3). Significantly, pladienolide-B induce an early pattern of mRNA intron retention, or spliceosome modulation. This process was associated with apoptosis preferentially in cancer cells as compared to normal lymphocytes. Its pro-apoptotic activity is observed regardless of poor prognostic factors such as Δ17p, TP53 or SF3B1 mutations and can overcome the protective effect of culture conditions that resemble the tumor microenvironment (4). **1.** Mizui, Sakai, Iwata, *et al.* (2004) *J. Antibiot.* (Tokyo) **57**, 188. **2.** Kanada, Itoh, Nagai, *et al.* (2007) *Angew, Chem. Int. Ed. Engl.* **46**, 4350. **3.** Yokoi, Kotake, Takahashi, *et al.* (2011) *FEBS J.* **278**, 4870. **4.** Kashyap, Kumar, Villa, *et al.* (2015) *Haematologica* **100**, 945.

Plasminogen Activator Inhibitor-1

This 402-residue serine protease inhibitor, *or* serpin (MW = 45074 g/mol; Abbreviation: PAI-1) binds to and inhibits plasminogen activators-tissue-type plasminogen activator (tPA) and urokinase-type plasminogen activator (uPA). This inhibition reduces plasmin production and suppresses dissolution of fibrin clots. Elevated PAI-1 levels correlate with an increased cardiovascular disease risk, a behavior that has also been linked to obesity and metabolic syndrome. Pharmacological suppression of PAI-1 is a likely way to prevent or treat vascular disease. Reduced PAI-1 levels may result in increased fibrinolysis and an associated bleeding diathesis. PAI-1 was initially identified in the 1980s, and the first reported case of PAI-1 deficiency appeared in 1989. Unambiguous proof that PAI-1 deficiency as a cause of a bleeding disorder has been rare, but use of selective PAI inhibitors (**See** PAI-749) may clarify this point. Because of lack of standardized commercially available PAI-1 activity assay sensitive in the lowest range, the true prevalence of this rare condition has yet to be established. **1.** Vaughan (2005) *J. Thromb. Haemost.* **3**, 1879. **2.** Fortenberry (2013) *Expert Opin. Ther. Pat.* **23**, 801.

Platencin

This antibiotic (FW = 425.2 g/mol; CAS 869898-86-2; Soluble in ethanol, methanol, DMF or DMSO) inhibits both *Staphylococcus aureus* β-ketoacyl-[acyl-carrier-protein] synthases II and III, with IC₅₀ values of 1.95 and 3.91 μg/mL, respectively (1). Platencin exhibits strong, broad-spectrum, Gram-positive antibacterial activity to key antibiotic resistant strains, including methicillin-resistant *Staphylococcus aureus*, vancomycin-intermediate *S. aureus*, and vancomycin-resistant *Enterococcus faecium*, without evidence of toxicity in humans. **1.** Wang, Kodali, Lee, *et al.* (2007) *Proc. Natl. Acad. Sci. U.S.A.* **104**, 7612.

Platensimycin

This thiol-reactive *Streptomyces platensis*-derived antibiotic (FW = 438.50 g/mol; CAS 835876-32-9), also named 3-[[3-[(1R,3R,4R,5aR,9R,9aS)-1,4,5,8,9,9a-hexahydro-3,9-dimethyl-8-oxo-3H-1,4:3,5a-dimethano-2-benzoxepin-9-yl]-1-oxopropyl]amino]-2,4-dihydroxybenzoic acid, inhibits *Staphylococcus aureus* β-ketoacyl-[acyl-carrier-protein] synthase II, *or* FabF (IC₅₀ = 290 nM) (1-3). This enzyme, which allows bacteria to produce the fatty acids needed for making cell membranes, is thus a uniquely druggable prokaryotic target. Platensimycin has potent, broad-spectrum Gram-positive activity *in vitro* and exhibits no cross-resistance to other key antibiotic-resistant bacteria including Methicillin-resistant *Staphylococcus aureus* (MRSA), vancomycin-intermediate *S. aureus*, vancomycin-resistant *Enterococci*, as well as linezolid-resistant and macrolide-resistant pathogens. Platencin is a more potent analogue of Platensimycin. While effective *in vivo* when continuously administered, its efficacy is reduced when administered by more conventional means. **Mechanism of Action:** The Cys163 within the FabF active site is activated through the dipole moment of helix N-α-3, lowering its pK_a, also increasing its nucleophilicity by the stabilizing effects of FabF's catalytically required oxyanion hole. Interestingly, the crystal structure complex with platensimycin employed a C163Q mutant which gave a 50-fold increase in apparent binding. **See** *Platencin* **1.** Wang, Kodali, Lee, *et al.* (2007) *Proc. Natl. Acad. Sci. U.S.A.* **104**, 7612. **2.** Wang, Soisson, Young, *et al.* (2006) *Nature* **441**, 358. **3.** Häbich, F. Von Nussbaum (2006) *ChemMedChem* **1**, 951.

Platinic Chloride, *See Chloroplatinic Acid*

Platinum

Platinum (Pt; atomic number 78; atomic weight 195.084) is a group 10 (or group VIIIA) element, located directly beneath palladium in the Periodic Table. The ground state electronic configuration of the neutral atom is $1s^2 2s^2 2p^6 3s^2\ 3p^6 4s^2 3d^{10} 4p^6 5s^2 4d^{10} 5p^6 6s^1 4f^{14} 5d^9$. The principal oxidation states are II and IV, with respective ionic radii of 0.80 and 0.65 Å, respectively); however, compounds of ixidation states III, IV, V, and VI have also been identified. In addition to its numerous uses in laboratory equipment and instruments, platinum derivatives have been found to have important pharmacological properties. Perhaps the best known is cisplatin (or, *cis*-diamminedichloroplatinum), noted for its antitumor properties. Glutathione reportedly contributes to tumor cell resistance toward cisplatin. There are also numerous reports on the interaction of platinum derivatives with DNA to form platinum-adducts. When the chlorides of these complexes are replaced with adenine, cytosine, or other purine/pyrimidine, the resulting complexes have been used as models in the investigation of DNA interactions. **See** *specific platinum complex*

cis-Platinum(II) Diammine Dichloride, *See Cisplatin*

trans-Platinum(II) Diammine Dichloride, *See Transplatin*

Plecanatide

Plecanitide:　NDECELCVNVACTGCL

Uroguanylin:　NDDCELCVNVACTGCL

This uroguanylin analogue (FW = 1681.90 g/mol; CAS 467426-54-6), also known by the codename SP-304, binds to and activates guanylate cyclase-C (GC-C) receptors (EC₅₀ ≈ 0.3 μM) expressed on the epithelial cells lining the gastrointestinal mucosa, stimulating cyclic GMP production, which in turn sequentially activates protein kinase G-II and the chloride channel known as the Cystic Fibrosis Transmembrane-conductance Regulator (CFTR) to regulate ion and fluid transport, to promote epithelial cell homeostasis, and to maintain barrier function in the GI mucosa. Plecanatide is a hexadecapeptide that is structurally identical to uroguanylin, except for an glutamate substitution at position-3. Oral treatment with plecanatide at a dose range between 0.05-2.5 mg/kg per day is as effective as once-daily treatment with 5-amino salicylic acid (100 mg/kg) or sulfasalazine (80 mg/kg). Amelioration of colitis by the treatment with plecanatide or dolcanatide was not dose-dependent, most likely due to saturation of available GC-C receptors. **1**. Shailubhai, Palejwala, Arjunan, *et al.* (2015) *World J. Gastrointest. Pharmacol. Therapeut.* **6**, 213

Plerixafor

This metal ion-chelating immunostimulant (FW = 502.80 g/mol; CAS 155148-31-5), also known by its code names AMD3100 and JM3100, its trade name Mozobil™ as well as its systematic name 1,1'-[1,4-phenylenebis(methylene)]*bis*-[1,4,8,11-tetraazacyclotetradecane], is an α-C-X-C chemokine receptor CXCR₄ antagonist (*or* partial agonist) and an allosteric CXCR7 agonist (1-4). The active chemical form is likely to contain divalent zinc ion. (*See also TC 14012*) **Likely Mode of Action:** The CXCR4 α-chemokine is stimulated by Stromal cell-Derived Factor-1 (*or* SDF-1) that plays roles in hematopoietic stem cell homing to the bone marrow and in hematopoietic stem cell (HSC) quiescence. CXCR7) is a G-protein-coupled receptor (GPCR) that binds the chemokines CXCL12/SDF-1 and CXCL11 and serves as a coreceptor for human immunodeficiency viruses. Plerixafor inhibits the replication of various HIV-1 and HIV-2 strains in various cell lines (EC₅₀ = 1-10 ng/mL, *or* >100,000x lower than its cytotoxic concentration of 500 µg/mL) (1). When combined with either 3'-azido-2',3'-dideoxythymidine or 2',3'-dideoxyinosine, JM3100 achieved a additive inhibition of HIV replication, and when repeatedly subcultivated in the presence of JM3100, with the virus remaining sensitive to the compound for at least 30 passages in culture (1). AMD3100 blocks HIV-1 entry and membrane fusion via the CXCR4 co-receptor, blocking the latter's role as both a HIV-1 co-receptor and a CXC-chemokine receptor (2). **Use as an Investigational Drug:** Plerixafor's discovery provided a new way to mobilize HSC for autologous transplantation (4). In 2008, plerixafor was approved by the FDA for the mobilization of hematopoietic stem cells. Plerixafor is also an investigational drug for the treatment of WHIM Syndrome (standing for Warts, Hypogammaglobulinemia, Infection & Myelokathexis Syndrome), a rare congenital immunodeficiency disorder characterized by chronic noncyclic neutropenia and arising from gain-of-function deletion mutations in CXCR₄. In a Phase-I study, circulating leukocytes were durably increased throughout the trial in all patients, and this was associated with fewer infections and improvement in warts in combination with imiquimod; however immunoglobulin levels and specific vaccine responses were not fully restored (5). **Metal Ion Binding:** Plerixafor binds zinc, copper nickel, cobalt and rhodium ions, and the biologically active form is the 1:2 Plerixafor-Zn²⁺ ternary complex (FW = 627.53 g/mol). Other CXCR4 antagonists include: AMD070, T140, FC131, FC122, with several being evident metal ion-binding chelators. The diversity of their structures suggests the possibility that they may act as divalent metallophores, a unique class of organic molecules that facilitate metal ion transfer across membrane bilayers in a manner akin to the valinomycin-facilitated transport of potassium ion or by means of a vesicle pathway, such as receptor-mediated endocytosis. **1**. De Clercq, Yamamoto, Pauwels, *et al.* (1994) *Antimicrob. Agents Chemother.* **38**, 668. **2**. Donzella, Schols, Lin, *et al.* (1998) *Nature Med.* **4**, 72. **3**. Kalatskaya, Berchiche, Gravel, *et al.* (2009) *Molec. Pharmacol.* **75**: 1240. **4**. Fricker (2013) *Transfus. Med. Hemother.* **40**, 237. **5**. McDermott, Liu, Velez, *et al.* (2014) *Blood* **123**, 2308.

Plevitrexed, *See ZD 9331*

Plicacetin, *See Amicetins*

Plinabulin

This vascular disrupting agent and antineoplastic drug (FW = 336.39 g/mol; CAS 714272-27-2 and 1367709-16-7 (TFA); Solubility = 67 g/mol DMSO; <1 g/mol Water; <1 g/mol Ethanol), also known as NPI-2358 and systematically as (3*Z*,6*Z*)-3-benzylidene-6-((5-*tert*-butyl-1*H*-imidazol-4-yl)methylene)piperazine-2,5-dione, binds reversibly with tubulin (IC₅₀ = 9-18 nM) to prevent microtubule assembly (1-3), selectively disrupting the abnormal vasculature associated with disease processes such as cancer and macular degeneration. Such action has been attributed to the ability of these agents to selectively destroy the central regions of tumors, areas widely believed to contain cell populations resistant to cytotoxic therapies (4). **1**. Nicholson, *et al.* (2006) *Anticancer Drugs* **17**, 25. **2**. Singh, *et al.* (2011) *Blood* **117**, 5692. **3**. Bertelsen, *et al.* (2011) *Int. J. Radiat. Biol.* **87**, 1126. **4**. Chaplin, Horsman & Siemann (2006) *Curr. Opin. Investig. Drugs* **7**, 522.

Plicamycin, *See Mithramycin*

Pleuromutilin

This antibiotic (FW = 378.51 g/mol; CAS 125-65-5) from *Pleurotus mutilus* and *P. passeckerianus* is a derivative of cyclooctane that inhibits protein biosynthesis at the ribosomal elongation step (1-3). **1**. Jiménez (1976) *Trends Biochem. Sci.* **1**, 28. **2**. Hogenauer (1975) *Eur. J. Biochem.* **52**, 93. **3**. Riedl (1976) *J. Antibiot.* **29**, 132.

PLP, *See Pyridoxal 5'-Phosphate*

PLT 98625, *See SKF 98625*

Plumbagin

This yellow-colored naphthoquinone (FW = 188.18 g/mol; CAS 481-42-5; M.P. = 78-79°C; Soluble to 100 mM in DMSO and to 50 mM in Ethanol), also known as 5-hydroxymenadione and 5-hydroxy-2-methyl-1,4-naphthoquinone, from the roots of *Plumbago* species is a membrane-permeant, sulfhydryl-modifying reagent. At 50 µM, plumbagin induces oxidative stress in treated cells by depleting intracellular stores of reduced glutathione and inducing reactive oxygen species (ROS). It is also an anticancer agent that induces cell-cycle arrest at G₂/M and apoptosis in A549 cells via JNK-dependent p53 Ser-15 phosphorylation. inhibits A549 and MDA-MD-231 tumour xenograft growth in nude mice (1). Plumbagin promotes autophagic cell death in MDA-MB-231 and MCF-7 cells and inhibits Akt/mTOR signaling (2). Plumbagin inhibits bacterial proliferation by inhibiting FtsZ assembly (3). **Target(s):** (*S*)-canadine synthase (4); chitin synthase (5); cytochrome P450-linked monooxygenase (6); dihydroxy-acid dehydratase, *or* α,β-dihydroxyisovalerate dehydratase (7,8); ecdysone 20-monooxygenase (2); lactose transporter (9); NAD⁺ ADP-ribosyltransferse, *or* poly(ADP-ribose) polymerase (10); NADH dehydrogenase (11); prostaglandin-endoperoxide synthase (12); protein-*N*ᵖ-phosphohistidine:sugar phosphotransferase, *or* glucose enzyme II and mannose enzyme II of bacterial phosphotransferase system (13); ribonuclease H (14); succinate dehydrogenase (15); vitamin K-dependent carboxylase, *or* protein-glutamate carboxylase (16). **1**. Hsu, *et al.* (2006) *J. Pharmacol. Exp. Ther.* **318**, 484. **2**. Kuo, *et al.* (2006) *Mol. Cancer Ther.* **5**, 3209. **3**. Bhattacharya, Jindal, Singh, Datta & Panda **(2013)** *FEBS J.* **280**, 4585. **4**. Rueffer & Zenk (1994) *Phytochemistry* **36**, 1219. **5**. Mitchell & Smith (1988) *Experientia* **44**, 990. **6**. Muto, Inouye, Inada, Nakanishi & Tan (1987) *Biochem. Biophys. Res. Commun.* **146**, 487. **7**. Babu & Brown (1995) *Microbios* **82**, 157. **8**. Kuo, Mashino & Fridovich (1987) *J. Biol. Chem.* **262**, 4724. **9**. Neuhaus & Wright (1983) *Eur. J. Biochem.* **137**, 615. **10**. Banasik, Komura, Shimoyama & Ueda (1992) *J. Biol. Chem.* **267**, 1569. **11**. Imlay & Fridovich (1992) *Arch. Biochem. Biophys.* **296**, 337. **12**. Wurm, Baumann, Geres & Schmidt (1984) *Arzneimittelforschung* **34**, 652. **13**. Grenier, Waygood & Saier, Jr. (1985) *Biochemistry* **24**, 47 and 4872. **14**. Min, Miyashiro & Hattori (2002) *Phytother. Res.* **16** Suppl. 1, S57. **15**. Suraveratum, Krungkrai, Leangaramgul, Prapunwattana & Krungkrai (2000) *Mol. Biochem. Parasitol.* **105**, 215. **16**. Johnson (1980) *Meth. Enzymol.* **67**, 165.

PluriSIn #1

This stearoyl-CoA sesaturase 1 inhibitor, *or* SCD1 (FW = 213.24 g/mol; CAS 91396-88-2; Solubility: 43 mg/mL DMSO; <1 mg/mL H$_2$O), also known as NSC 14613 an 4-pyridinecarboxylic acid, 2-phenylhydrazide, selectively eliminates <u>h</u>uman <u>P</u>luripotent <u>S</u>tem <u>C</u>ells (hPSCs), while sparing a large array of progenitor and differentiated cells, thus providing a promising approach for minimizing the tumorigenic risk from residual undifferentiated cells. Closer examination identified PluriSIn #1 as an SCD1, thereby indicating a role for lipid metabolism in hPSC survival. PluriSIn #1 is likewise cytotoxic to mouse blastocysts, again suggesting a dependence on oleate in cells in a pluripotent state. SCD is key and highly regulated enzyme which catalyzes the Δ^9-*cis* desaturation of a range of palmitoyl- and stearoyl-CoA to palmitoleoyl- and oleoyl-CoA respectively. **1**. Ben-David, Gan, Golan-Lev, *et al.* (2013) *Cell Stem Cell* **12**, 167.

PLX-4032, *See* **Vemurafenib**

PMA, *See* *Phorbol 12-Myristate 13-Acetate*

PMP-C

This insect peptide (*Sequence*: EISCEPGKTFKDKCNTCRCGADGKSA ACTLKACPNQ; FW = 3779.32 g/mol), also known as *pars intercerebralis* major peptide C, from *Locusta migratoria* is a potent inhibitor of chymotrypsin (K_i = 0.13 nM) and a weaker inhibitor of leukocyte elastase (K_i = 180 nM). PMP-C contains three intrachain disulfide bonds: Cys4-Cys19, Cys17-Cys28, and Cys14-Cys33 (1). **1**. Roussel, Mathieu, Dobbs, *et al.* (2001) *J. Biol. Chem.* **276**, 38893.

PMP-D2

This insect peptide (*Sequence*: EEKCTPGQVKQQDCNTCTCTPYGVW GCTRKGCQPA; MW = 3821.32 g/mol), also known as *pars intercerebralis* major peptide D2, is isolated from *Locusta migratoria* and inhibits trypsin and chymotrypsin (K_i = 100 and 1500 nM for the bovine enzymes, and is a much stronger inhibitor of *Locusta* trypsin. PMP-D2 consists of thirty-five aminoacyl residues with three disulfide bonds: Cys4-Cys19, Cys17-Cys27, and Cys14-Cys32 (1). **1**. Roussel, Mathieu, Dobbs, *et al.* (2001) *J. Biol. Chem.* **276**, 38893.

PMX 205

This potent C5aR antagonist (FW = 839.05 g/mol; CAS 514814-49-4; Soluble to 1 mg/mL in 20% Ethanol/H$_2$O) is a cyclic peptide analogue (Sequence: X$_1$PX$_3$WR, where X$_1$ = N^2-(1-cxo-3-phenylpropyl)-ornithine, X$_3$

= D-Cha (*or* β-<u>c</u>yclohexylala<u>n</u>ine), with a lactam bridge between Orn-1 and Arg-5) that targets human Complement 5a (IC$_{50}$ = 31 nM), a plasma protein with both potent chemoattractant and pro-inflammatory properties (1). C5a binds to its G protein-coupled receptor (C5aR) on polymorphonuclear leukocytes (PMNLs) through a high-affinity helical bundle and a low-affinity C-terminus, the latter solely responsible for receptor activation (1). C5aR overexpression correlates with severity of inflammatory diseases, suggesting a rationale for it serving as a druggable target. Computer modeling suggests that potent antagonists have a cyclic backbone shape, with affinity-determining side-chains of defined volume projecting from the cyclic scaffold (1). PMX-205 was helpful in adducing first evidence that C5aR inhibition can interfere with neuroinflammation and neurodegeneration in AD rodent models, making C5aR a novel therapeutic target for reducing pathology and improving cognitive function in human AD patients (2). **1**. March, Proctor, Stoermer, *et al.* (2004) *Mol. Pharmacol.* **65**, 868. **2**. Fonseca, Ager, Chu, *et al.* (2009) *J. Immunol.* **183**, 1375.

PNA, *See* *Peptide Nucleic Acids; p-Nitroaniline*

p(NH)ppA, *See* *Adenosine 5'-[β,γ-Imido]triphosphate*

p(NH)ppG, *See* *Guanosine 5'-[β,γ-Imido]triphosphate*

PNP, *See* *p-Nitrophenol*

PNU-183792

This 4-oxo-dihydroquinoline (FW = 425.91 g/mol) inhibits DNA polymerase of human and animal herpesviruses (1,2). PNU-183792 exhibits a novel inhibition mechanism in which the inhibitor binds at the polymerase active site interacting non-covalently with both the polymerase and the DNA duplex (3). **1**. Thomsen, Oien, Hopkins, *et al.* (2003) *J. Virol.* **77**, 1868. **2**. Brideau, Knechtel, Huang, *et al.* (2002) *Antiviral Res.* **54**, 19. **3**. Liu, Knafels, Chang, *et al.* (2006) *J. Biol. Chem.* **281**, 18193.

POB, *See* *p-Hydroxybenzoic Acid*

Podophyllotoxin

This naturally occurring microtubule-directed antineoplastic agent (FW = 414.41 g/mol; CAS 518-28-5; soluble in ethanol; slightly soluble in water (120 mg/L at 23°C), first isolated from the rhizomes of the common mayapple (*Podophyllum peltatum*) inhibits microtubule self-assembly as well as mitotic spindle formation, the latter arresting nuclear division at metaphase (1-8). Several podohyllotoxin derivatives are inhibitors of DNA topoisomerase II. **1**. Podwyssotzki (1880) *Arch. Exp. Path. Pharmakol.* **13**, 29. **2**. Purich, Terry, MacNeal & Karr (1982) *Meth. Enzymol.* **85**, 416. **3**. Jordan & Wilson (1998) *Meth. Enzymol.* **298**, 252. **4**. Desbene & Giorgi-Renault (2002) *Curr. Med. Chem. Anti-Canc. Agents* **2**, 71. **5**. Sackett (1993) *Pharmacol. Ther.* **59**, 163. **6**. Karr, White & Purich (1979) *J. Biol. Chem.* **254**, 6107. **7**. Wilson, Anderson & Chin (1976) in *Cold Spring Harbor Conference on Cell Proliferation III. Cell Motility* (Goldman, Pollard & Rosenbaum, eds.), pp. 1051-1064, Cold Spring Harbor Laboratory, New York. **9**. Karr & Purich (1979) *J. Biol. Chem.* **254**, 10885. **8**. Karr, White, Coughlin & Purich (1982) *Meth. Cell Biol.* **24**, 51.

Poison Ivy, *See* *Urushiols*

Poison Oak, *See* *Urushiols*

Poison Sumac, *See* *Urushiols*

Pol 647, *See* *N-Furoyl-L-leucyl-L-tryptophan*

Polychlorinated Biphenyls (or, PCBs), *See specific compound; e.g., 2,2',4,4',5,5'-Hexachlorobiphenyl; Aroclor 1254*

Poly(L-glutamate)

This L-glutamate homopolymer (MW = indefinite, depending on the degree of polymerization; CAS 25513-46-6 (free acid) and 26247-79-0 (sodium salt)) possesses little α-helical content at neutral pH, but its helicity increases dramatically with decreasing pH. Titration of the γ-carboxyl group yields a pK_a value of 4.94 (1). **Target(s):** β-*N*-acetylglucosaminylglycopeptide β-1,4-galactosyltransferase (2); *N*-acetyllactosamine synthase (2); [β-adrenergic-receptor] kinase (3,4); cartilage metalloproteoglycanases (5); casein kinase type 2 (6,7); DNA polymerase-α-primase (8); elongation factor-3 ATPase (9); histone acetyltranferase (10); lysozyme, weakly inhibited (11-13); microtubule assembly, *or* MAP-stimulated tubulin polymerization (14); polygalacturonidase (15); procollagen N-endopeptidase (16,17); rhodopsin kinase (18,19); stromelysin (5); trypsin (12); tubulin-tyrosine carboxypeptidase (20); UDPgalactose:glycoprotein galactosyltransferase (2); vaccinia RNA polymerase (21); vesicular stomatitis virus-associated RNA-directed RNA polymerase (22). **1.** Sage & Fasman (1966) *Biochemistry* **5**, 286. **2.** Rao, Chyatte & Nadler (1978) *Biochim. Biophys. Acta* **541**, 435. **3.** Benovic, Stone, Caron & Lefkowitz (1989) *J. Biol. Chem.* **264**, 6707. **4.** Benovic (1991) *Meth. Enzymol.* **200**, 351. **5.** Sheff & Sapolsky (1994) *Ann. N. Y. Acad. Sci.* **732**, 398. **6.** Meggio, Pinna, Marchiori & Borin (1983) *FEBS Lett.* **162**, 235. **7.** Meggio, Grankowski, Kudlicki, *et al.* (1986) *Eur. J. Biochem.* **159**, 31. **8.** Holler, Achhammer, Angerer, *et al.* (1992) *Eur. J. Biochem.* **206**, 1. **9.** Uritani, Nakano, Aoki, Shimada & Arisawa (1994) *J. Biochem.* **115**, 820. **10.** Wiegand & Brutlag (1981) *J. Biol. Chem.* **256**, 4578. **11.** Sela & Steiner (1963) *Biochemistry* **2**, 416. **12.** Webb (1966) *Enzyme and Metabolic Inhibitors*, vol. **2**, pp. 456, 459, Academic Press, New York. **13.** Skarnes & Watson (1955) *J. Bacteriol.* **70**, 110. **14.** Fujii, Suzuki, Fujii, *et al.* (1986) *Biochem. Cell Biol.* **64**, 615. **15.** Schejter & Marcus (1988) *Meth. Enzymol.* **161**, 366. **16.** Hojima, McKenzie, van der Rest & Prockop (1980) *J. Biol. Chem.* **264**, 11336. **17.** Hojima, Morgelin, Engel, *et al.* (1994) *J. Biol. Chem.* **269**, 11381. **18.** Palczewski, Arendt, McDowell & Hargrave (1989) *Biochemistry* **28**, 8764. **19.** Kikkawa, Yoshida, Nakagawa, Iwasa & Tsuda (1998) *J. Biol. Chem.* **273**, 7441. **20.** Lopez, Arce & Barra (1990) *Biochim. Biophys. Acta* **1039**, 209. **21.** Gershon & Moss (1996) *Meth. Enzymol.* **275**, 208. **22.** Carroll & Wagner (1978) *J. Biol. Chem.* **253**, 3361.

Polygodial

This sesquiterpene (FW = 424.85 g/mol; CAS 6754-20-7), isolated from the leaves of *Polygonum hydropiper* and *Drymis winteri,* inhibits 5-lipoxygenase (IC$_{50}$ = 8.6 μM). **Target(s):** ATPase, mitochondrial (1); glutamate transport (2); 5-lipoxygenase, *or* arachidonate 5-lipoxygenase (3); and phosphorylation, oxidative, as an uncoupler (4). **1.** Lunde & Kubo (2000) *Antimicrob. Agents Chemother.* **44**, 1943. **2.** Martini, Cereser, Junior, *et al.* (2006) *Neurochem. Res.* **31**, 431. **3.** Abe, Ozawa, Uda, *et al.* (2006) *Biosci. Biotechnol. Biochem.* **70**, 2494. **4.** Castelli, Lodeyro, Malheiros, Zacchino & Roveri (2005) *Biochem. Pharmacol.* **70**, 82.

Poly(L-lysine)

This water-soluble L-lysine homopolymer (MW = 30–50 kDa; indefinite polymer, depending on the degree of polymerization) is an inhibitor of several enzymes as well as a stabilizer/activator of other proteins. High-molecular-weight poly(L-lysine) (MW = 100–300 kDa) promotes the adhesion of cell to solid substrates, most often without blocking their proliferation. This property commending its use as a coating on plastic cell culture dishes, serving in place of more costly collagen and fibronectin. **Target(s):** adenylate cyclase (1); [β-adrenergic-receptor] kinase (2,3); alkaline phosphatase (47); Ca^{2+}-dependent calmodulin-like domain protein kinase, *Arabidopsis* (4); carboxylester lipase (5); carboxypeptidase C, *or* carboxypeptidase Y (6); casein kinase I, *Trypanosoma cruzi* (7); chloroplast electron transport, as an uncoupler (8); cholinesterase (9); cytochrome *c* oxidase, complex IV (10-12); cytochrome *c* peroxidase (13); elongation factor-3 ATPase (14); [glycogen-synthase] kinase 3, *or* [tau protein] kinase (15); GroEL16; guanylate cyclase (17); histone acetyltransferase (18); inositol-1,3,4-trisphosphate 5/6-kinase (19); iodide uptake (20); lipase, *or* pancreatic lipase, *or* triacylglycerol lipase (5); lipoprotein lipase (21); lipoxygenase, soybean, inhibited by the octamer (53); [low-density-lipoprotein receptor] kinase (48); lysozyme (22); [myosin heavy-chain] kinase (51); ornithine decarboxylase (23); papain (24); pepsin A (21,25-30); peroxidase (31); phospholipase A$_2$ (32); phosphoprotein phosphatase (33-35); phosphorylase kinase (49); [phosphorylase] phosphatase (33,36); photosystem I (37-39); poly(ADP-ribose) glycohydrolase (40-42); protein kinase C (43); protein phosphatase 1 (33); protein phosphatase 7 (35); rhodopsin kinase (50); ribonuclease U2, *or* *Ustilago sphaerogena* ribonuclease (44); RNA-directed RNA polymerase (45,52); trypsin, immobilized trypsin inhibited by polylysine of MW < 13000 (46). **1.** Raptis, Knipling, Gentile & Wolff (1989) *Infect. Immun.* **57**, 1066. **2.** Benovic (1991) *Meth. Enzymol.* **200**, 351. **3.** Benovic, Stone, Caron & Lefkowitz (1989) *J. Biol. Chem.* **264**, 6707. **4.** Binder, Harper & Sussman (1994) *Biochemistry* **33**, 2033. **5.** Tsujita, Sumiyoshi, Takaku, *et al.* (2003) *J. Lipid Res.* **44**, 2278. **6.** Hayashi (1976) *Meth. Enzymol.* **45**, 568. **7.** Calabokis, Kurz, Wilkesman, *et al.* (2002) *Parasitol. Int.* **51**, 25. **8.** Izawa & Good (1972) *Meth. Enzymol.* **24**, 355. **9.** Lin, Liao & Lee (1977) *Biochem. J.* **161**, 229. **10.** Errede, Kamen & Hatefi (1978) *Meth. Enzymol.* **53**, 40. **11.** Dickerson & Timkovich (1975) *The Enzymes*, 3rd ed. (Boyer, ed.), **11**, 397. **12.** Person, Zipper, Fine & Mora (1964) *J. Biol. Chem.* **239**, 4159. **13.** Hirota, Tsukazaki & Yamauchi (2000) *Biochem. Biophys. Res. Commun.* **268**, 395. **14.** Uritani, Nakano, Aoki, Shimada & Arisawa (1994) *J. Biochem.* **115**, 820. **15.** Hegazy, Schlender, Wilson & Reimann (1989) *Biochim. Biophys. Acta* **1011**, 198. **16.** Lau & Churchich (1999) *Biochim. Biophys. Acta* **1431**, 282. **17.** Gorczyca, Van Hooser & Palczewski (1994) *Biochemistry* **33**, 3217. **18.** Wiegand & Brutlag (1981) *J. Biol. Chem.* **256**, 578. **19.** Hughes, Kirk & Michell (1994) *Biochim. Biophys. Acta* **1223**, 57. **20.** Kawada, Yoshimura & Minami (1976) *Endocrinol. Jpn.* **23**, 221. **21.** Webb (1966) *Enzyme and Metabolic Inhibitors*, vol. **2**, Academic Press, New York. **22.** Audy, Trudel & Asselin (1988) *Plant Sci.* **58**, 43. **23.** Kashiwagi & Igarashi (1987) *Biochim. Biophys. Acta* **911**, 180. **24.** Mekras, Lawton & Washington (1989) *J. Pharm. Pharmacol.* **41**, 22. **25.** Bovey & Yanari (1960) *The Enzymes*, 2nd ed. (Boyer, Lardy & Myrbäck, eds.), **4**, 63. **26.** Katchalski, Berger & Neumann (1954) *Nature* **173**, 998. **27.** Dellert & Stahmann (1955) *Nature* **176**, 1028. **28.** Lawton & Mekas (1985) *J. Pharm. Pharmacol.* **37**, 396. **29.** Anderson, Harthill & Rahmatalla (1980) *J. Pharm. Pharmacol.* **32**, 248. **30.** Rao & Dunn (1981) *Arch. Biochem. Biophys.* **207**, 135. **31.** Puliti, Giovannini, Megazzini, *et al.* (1980) *Boll. Soc. Ital. Biol. Sper.* **56**, 2530. **32.** Emadi, Elalamy, Vargaftig & Hatmi (1996) *Biochim. Biophys. Acta* **1300**, 226. **33.** Ballou & Fischer (1986) *The Enzymes*, 3rd ed. (Boyer & Krebs, eds.), **17**, 311. **34.** Schlender, Wilson, Thysseril & Mellgren (1985) *Adv. Protein Phosphatases* **1**, 311. **35.** Kutuzov, Evans & Andreeva (1998) *FEBS Lett.* **440**, 147. **36.** Gratecos,

Detwiler, Hurd & Fischer (1977) *Biochemistry* **16**, 4812. **37**. Trebst (1980) *Meth. Enzymol.* **69**, 675. **38**. Davis, Krogmann & San Pietro (1979) *Biochem. Biophys. Res. Commun.* **90**, 110. **39**. Brand, San Pietro & Mayne (1972) *Arch. Biochem. Biophys.* **152**, 426. **40**. Tavassoli, Tavassoli & Shall (1983) *Eur. J. Biochem.* **135**, 449. **41**. Niedergang, Okazaki & Mandel (1979) *Eur. J. Biochem.* **102**, 43. **42**. Miwa, Tanaka, Matsushima & Sugimura (1974) *J. Biol. Chem.* **249**, 3475. **43**. Leventhal & Bertics (1991) *Biochemistry* **30**, 1385. **44**. Glitz & Dekker (1964) *Biochemistry* **3**, 1391. **45**. Lazarus & Itin (1973) *Arch. Biochem. Biophys.* **156**, 154. **46**. Shtelzer, Rappoport, Avnir, Ottolenghi & Braun (1992) *Biotechnol. Appl. Biochem.* **15**, 227. **47**. Stinson & Chan (1987) *Adv. Protein Phosphatases* **4**, 127. **48**. Kishimoto, Brown, Slaughter & Goldstein (1987) *J. Biol. Chem.* **262**, 1344. **49**. Negami, Sakai, Kobayashi, *et al.* (1984) *FEBS Lett.* **166**, 335. **50**. Kikkawa, Yoshida, Nakagawa, Iwasa & Tsuda (1998) *J. Biol. Chem.* **273**, 7441. **51**. Medley, Bagshaw, Truong & Côté (1992) *Biochim. Biophys. Acta* **1175**, 7. **52**. Johnson, Sun, Hockman, *et al.* (2000) *Arch. Biochem. Biophys.* **377**, 129. **53**. Schurink, van Berkel, Wichers & Boeriu (2007) *Peptides* **28**, 2268.

Polymethacrylate (Poly(methacrylic acid))

This homopolymeric polyelectrolyte (MW = Indefinite Polymer) inhibits and/or inactivates many enzymes. Polymethacrylate should not be confused with the following polyesters: poly(methyl methacrylate) (*i.e.*, ···–CH₂C(CH₃)(CO₂CH₃)–···) or poly(methyl acrylate) (*i.e.*, ···–CH₂CH(CO₂CH₃)–···). **Target(s):** complement component C1q (1); hyaluronidase (2); glutamate dehydrogenase (3); and lactate dehydrogenase (3). **1**. Burdelev, Kaplun, Kozlov, *et al.* (2003) *Bioorg. Khim.* **29**, 159. **2**. Webb (1966) *Enzyme and Metabolic Inhibitors*, vol. **2**, p. 459, Academic Press, New York. **3**. Saburova, Bobreshova, Elphimova & Sukhorukov (2000) *Biochemistry (Moscow)* **65**, 976.

Polymin P, *See* Poly(ethyleneimine)

Polymyxins

polymyxin B₁

These decapeptide antibiotics (MW ≈ 1300 g/mol; NLM Unique Identifier D011113), containing a heterodetic ring, are produced by *Bacillus polymyxa*, and contain five or six residues of L-2,4-diaminobutyrate (L-DAB). The N-terminal residue is typically acylated with a 6-methyloctanoyl or 6-methylheptanoyl group (*e.g.*, polymyxin B₁ and B₂, respectively; note that the term polymyxin B refers to a mixture of both B₁ and B₂). Polymyxins are effective against Gram-negative bacteria and typically bind to and interfere with the permeability of the cytoplasmic membrane, resulting in loss of cytosolic components. The membranes of eukaryotic cells can also undergo attack. Circulins are structurally related polypeptide antibiotics isolated from *Bacillus circulans*. Circulin A is identical to polymyxin B₁, albeit with a D-leucyl residue instead of D-phenylalanine and an L-isoleucyl residue instead of L-leucine in the cycloheptapeptide component of the antibiotic. (*See also* Colistins) **Target(s):** Ca²⁺/calmodulin-dependent protein kinase II, weakly inhibited (1); calmodulin (2); glucose transporter, insulin-induced, inhibited by polymyxin B (3,4); glyce rol-3-phosphate *O*-acyltransferase, microsomal, inhibited by polymyxin B; however, the mitochondrial enzyme is stimulated (5-8); IGF II receptor translocation, inhibited by polymyxin B (3); K⁺ channel, ATP-sensitive, inhibited by polymyxin B (9); K⁺ channel, Ca²⁺-activated (10); Na⁺/K⁺-exchanging ATPase, weakly inhibited (1); protein kinase C, inhibited by polymyxin B (1,11); ribonuclease I (12); and α,α-trehalose-phosphate synthase (UDP-forming), inhibited by circulins (13). **1**. Raynor, Zheng & Kuo (1991) *J. Biol. Chem.* **266**, 2753. **2**. Hegemann, van Rooijen, Traber & Schmidt (1991) *Eur. J. Pharmacol.* **207**, 17. **3**. Cormont, Gremeaux, Tanti, Van Obberghen & Le Marchand-Brustel (1992) *Eur. J.*

Biochem. **207**, 185. **4**. Goto, Kida, Ikeuchi, Kaino & Matsuda (1991) *Biochem. Pharmacol.* **42**, 1399. **5**. Haldar & Vancura (1992) *Meth. Enzymol.* **209**, 64. **6**. Bell & Coleman (1983) *The Enzymes*, 3rd ed. (Boyer, ed.), **16**, 87. **7**. Carroll, Morris, Grosjean, Anzalone & Haldar (1982) *Arch. Biochem. Biophys.* **214**, 17. **8**. Lewin, Schwerbrock, Lee & Coleman (2004) *J. Biol. Chem.* **279**, 13488. **9**. Harding, Jaggar, Squires & Dunne (1994) *Pflugers Arch.* **426**, 31. **10**. Varecka, Peterajova & Pogady (1987) *FEBS Lett.* **225**, 173. **11**. Kikkawa & Nishizuka (1986) *The Enzymes*, 3rd ed. (Boyer & Krebs, eds.), **17**, 167. **12**. Teuber (1967) *Z. Naturforsch. B* **22**, 562. **13**. Pan & Elbein (1996) *Arch. Biochem. Biophys.* **335**, 258.

Poly(L-ornithine)

This positively charged homopolymer (MW = Indefinite Polymer; CAS 25104-12-5) of indefinite length is commercially available in many molecular weight ranges. Large polymers have been used to transfect DNA into mammalian cells. **Target(s):** angiogenin binding to ribonuclease inhibitor (1); foot-and-mouth disease virus RNA-directed RNA polymerase (2); iodide uptake (3); pepsin (4); poly(ADP-ribose) glycohydrolase (5); protein biosynthesis (6); α,α-trehalose-phosphate synthase, UDP-forming (7). **1**. Moenner, Chauviere, Chevaillier & Badet (1999) *FEBS Lett.* **443**, 303. **2**. Lazarus & Itin (1973) *Arch. Biochem. Biophys.* **156**, 154. **3**. Kawada, Yoshimura & Minami (1976) *Endocrinol. Jpn.* **23**, 221. **4**. Webb (1966) *Enzyme and Metabolic Inhibitors*, vol. **2**, p. 457, Academic Press, New York. **5**. Tavassoli, Tavassoli & Shall (1983) *Eur. J. Biochem.* **135**, 449. **6**. Ogawa & Ichihara (1978) *J. Biochem.* **83**, 519. **7**. Elbein & Mitchell (1974) *Carbohyd. Res.* **37**, 223.

Polyoxin A

This antifungal agent (FW_free-acid = 616.54 g/mol; CAS 19396-03-3; IUPAC: (S-(E))-1-(5-((2-amino-5-*O*-(aminocarbonyl)-2-deoxy-L-xylonoyl)amino)-1,5-dideoxy-1-(3,4-dihydro-5-(hydroxymethyl)-2,4-dioxo-1(2*H*)-pyrimidin-yl)-β-D-allofuranuronoyl)-3-ethylidene-2- azetidinecarboxylic acid and produced by *Streptomyces cacaoi* and *S. piomogenus*, inhibits chitin biosynthesis (1-4). **1**. Cabib (1972) *Meth. Enzymol.* **28**, 572. **2**. Cabib, Kang & Au-Young (1987) *Meth. Enzymol.* **138**, 643. **3**. Elbein (1987) *Meth. Enzymol.* **138**, 661. **4**. Ruiz-Herrera & San-Blas (2003) *Curr. Drug Targets Infect. Disord.* **3**, 77.

Polyoxin B

This antifungal antibiotic (FW_free-acid = 507.14 g/mol), produced by *Streptomyces cacaoi* and *S. piomogenus*, inhibits chitin synthase (1-3) and is frequently used to control many plant fungal pathogens. **1**. Cann, Kobayashi, Onoda, Wakita & Hoshino (1993) *J. Appl. Bacteriol.* **74**, 127. **2**. Cohen & Casida (1982) *Pestic. Biochem. Physiol.* **17**, 301. **3**. Gow & Selitrennikoff (1984) *Curr. Microbiol.* **11**, 211.

Polyoxin D

This pyrimidin nucleoside analogue and antifungal antibiotic ($FW_{free-acid}$ = 521.39 g/mol; CAS 22976-86-9; 33401-46-6 for the zinc salt), produced by soil organisms *Streptomyces cacaoi* and *S. piomogenus*, inhibits chitin biosynthesis, K_i = 6 μM, thereby blocking septum formation. The zinc salt of polyyoxin is used to control Rice Sheath Blight, caused by the fungus *Rhizoctonia solani*, as well as the turfgrass disease Brown Patch, caused by other *Rhizoctonia* species. Polyoxin D also inhibits *Neurospora crassa*, but is nontoxic to land mammals, insects, or birds, when applied to plants as the wettable powder form known as Endorse™. Target(s): chitin synthase (1-24); chitin synthase B (5); chitin synthase, canChs1A (6); WdChs4p, a chitin synthase 3 homologue (25). **1.** Cann, Kobayashi, Onoda, Wakita & Hoshino (1993) *J. Appl. Bacteriol.* **74**, 127. **2.** Cabib, Kang & Au-Young (1987) *Meth. Enzymol.* **138**, 643. **3.** Elbein (1987) *Meth. Enzymol.* **138**, 661. **4.** Leighton, Marks & Leighton (1981) *Science* **213**, 905. **5.** Tatsuno, Yamada-Okabe, Takagi, Arisawa & Sudoh (1997) *FEMS Microbiol. Lett.* **149**, 279. **6.** Sudoh, Watanabe, Mio, *et al.* (1995) *Microbiology* **141**, 2673. **7.** Causier, Milling, Foster & Adams (1994) *Microbiology* **140**, 2199. **8.** Machida & Saito (1993) *J. Biol. Chem.* **268**, 1702. **9.** Cabib (1991) *Antimicrob. Agents Chemother.* **35**, 170. **10.** Lyr & Seyd (1978) *Z. Allg. Mikrobiol.* **18**, 721. **11.** Cohen, Elster & Chet (1986) *Pestic. Sci.* **17**, 175. **12.** Cohen & Casida (1980) *Pestic. Biochem. Physiol.* **13**, 121. **13.** Kang, Hwang, Yun, Shin & Kim (2008) *Biol. Pharm. Bull.* **31**, 755. **14.** Ruiz-Herrera, Lopez-Romero & Bartnicki-Garcia (1977) *J. Biol. Chem.* **252**, 3338. **15.** Jan (1974) *J. Biol. Chem.* **249**, 1973. **16.** Cohen & Casida (1982) *Pestic. Biochem. Physiol.* **17**, 301. **17.** Sburlati & Cabib (1986) *J. Biol. Chem.* **261**, 15147. **18.** Orlean (1987) *J. Biol. Chem.* **262**, 5732. **19.** Huizar & Aronson (1985) *Exp. Mycol.* **9**, 302. **20.** Wang & Szaniszlo (2002) *Med. Mycol.* **40**, 283. **21.** Manocha & Begum (1985) *Can. J. Microbiol.* **31**, 6. **22.** Mayer, Chen & DeLoach (1980) *Insect Biochem.* **10**, 5422. **23.** Gow & Selitrennikoff (1984) *Curr. Microbiol.* **11**, 211. **24.** Plant, Thompson & Williams (2008) *J. Org. Chem.* **73**, 3714. **25.** Wang, Zheng, Hauser, Becker & Szaniszlo (1999) *Infect. Immun.* **67**, 6619.

Polyphloretin Phosphate

This water-soluble polyanion (MW = Indefinite Polymer; *Abbreviation*: PPP) is a mixture of polyesters of phosphoric acid and phloretin. PPP is a light-green or brown powder that inhibits alkaline phosphatase and is a prostaglandin antagonist. PPP is a non-specific and reversible antagonist of the response of non-gravid rat uterine smooth muscle to oxytocics *in vitro*, and its specificity of action as a prostaglandin antagonist is likely to be species/tissue-dependent. Target(s): acid phosphatase (1); adenylate cyclase (2); alkaline phosphatase (3,4); cystinyl aminopeptidase, *or* oxytocinase (5,6); hyaluronidase (4); 15-hydroxyprostaglandin dehydrogenase (7); phospholipase A (8); and ribonuclease (9). **1.** Webb (1966) *Enzyme and Metabolic Inhibitors*, vol. 2, p. 464, Academic Press, New York. **2.** Hynie & Klenerova (1979) *Prostaglandins Med.* **2**, 11. **3.** Stadtman (1961) *The Enzymes*, 2nd ed. (Boyer, Lardy & Myrbäck, eds.), **5**, 55. **4.** Diczfalusy, Ferno, Fex, *et al.* (1953) *Acta Chem. Scand.* **7**, 913. **5.** Roy, Yeang & Karim (1982) *Prostaglandins Leukot. Med.* **8**, 173. **6.** Roy, Yeang & Karim (1981) *Prostaglandins Med.* **6**, 577. **7.** Pace-Asciak & Smith (1983) *The Enzymes*, 3rd ed. (Boyer, ed.), **16**, 543. **8.** Arnesjo, Ihse & Qvist (1973) *Acta Chem. Scand.* **27**, 2225. **9.** Stockx & Dierick (1959) *Enzymologia* **21**, 189.

Polyphosphate

This homopolymeric anhydride, also called polymeta-phosphate and hexametaphosphate (with sodium hexametaphosphate known also as Calgon®), is a linear sequence of phosphoric acid (or, polyphosphoric acid) in any of its many ionization states. Polyphosphate is produced in a number of microorganisms and accumulates in highly refractile subcellular regions known as metachromatic granules. Oligo- and poly-phosphates are thought to be phosphate reserves. Cyclic metaphosphates are known to have 3–10 metaphosphate subunits, and ultraphosphate, containing variously branched units, are also known to exist. Polyphosphate dissolves slowly in water and slowly hydrolyzes to produce trimetaphosphate and orthophosphate. *See also Tripolyphosphate; Tetrapolyphosphate* Target(s): acid phosphatase (1); adenylate kinase (2); AMP deaminase (3); α-amylase (4); cAMP phosphodiesterase (5); DNA ligase6; DNA polymerase (6); fructose-1,6-bisphosphate aldolase (7); glutamate dehydrogenase (8); lactate dehydrogenase (8); NMN nucleosidase (9); phosphofructokinase (10); 3-phosphoglycerate kinase (16); phosphorylase kinase (11); 3- or 4-phytase (12); polynucleotide adenylyltransferase, *or* poly(A) polymerase (15); pyrophosphatase, *or* inorganic diphosphatase (13); restriction endonucleases (6); RNA polymerase (14); *Taq*I DNA polymerase (6); and T4 DNA ligase (6). **1.** Magboul & McSweeney (2000) *Int. Dairy J.* **9**, 849. **2.** Colowick (1955) *Meth. Enzymol.* **2**, 598. **3.** Yoshino & Murakami (1988) *Biochim. Biophys. Acta* **954**, 271. **4.** Schwimmer & Balls (1949) *J. Biol. Chem.* **179**, 1063. **5.** Speziali & Van Wijk (1971) *Biochim. Biophys. Acta* **235**, 466. **6.** Rodriguez (1993) *Anal. Biochem.* **209**, 291. **7.** Szwergold, Ugurbil & Brown (1995) *Arch. Biochem. Biophys.* **317**, 244. **8.** Saburova, Bobreshova, Elphimova & Sukhorukov (2000) *Biochemistry (Moscow)* **65**, 976. **9.** Imai (1987) *J. Biochem.* **101**, 163. **10.** Lardy (1962) *The Enzymes*, 2nd ed. (Boyer, Lardy & Myrbäck, eds.), **6**, 67. **11.** Krebs, Love, Bratvold, *et al.* (1964) *Biochemistry* **3**, 1022. **12.** Hegeman & Grabau (2001) *Plant Physiol.* **126**, 1598. **13.** Heppel & Hilmoe (1951) *J. Biol. Chem.* **192**, 87. **14.** Kusano & Ishihama (1997) *Genes Cells* **2**, 433. **15.** Sillero, de Diego, Silles, Osorio & Sillero (2003) *FEBS Lett.* **550**, 41. **16.** Boyle, Fairbrother & Williams (1989) *Eur. J. Biochem.* **184**, 535.

3-Polyprenyl-4-hydroxybenzoate and 3-Polyprenyl-4-hydroxybenzoic Acid, *See specific compound; e.g., 3-Nonaprenyl-4-hydroxybenzoate*

Poly(L-proline)

This indefinite L-proline homopolymer (CAS 25191-13-3) of indefinite length, is frequently used in structural and conformational investigations, adopting one of two major conformations: Poly(L-Pro) I helix (with φ and ψ angles of –83° and 158°, respectively, with peptide bonds that are all *cis*) and poly(L-Pro) II helix (with φ and ψ angles of –78° and 149°, and containing peptide bonds that are all *trans*). Collagen strands adopt a conformation similar to that of poly(L-Pro) II helix. Proline-rich regions in other proteins also have secondary structures similar to poly(L-Pro) II. Poly(L-proline) also binds tightly (K_i = 10 μM) to the actin regulatory protein profilin, which within all eukaryotic cells interacts with corresponding Actin-Based Motility-2 sequences (*e.g.,* Xaa-Pro-Pro-Pro-Pro-Pro sequences) found as multiple registers within the actin-filament end-tracking Ena/VASP proteins, WASP, N-WASP, and formins. When microinjected into cells at 10-30 μM final concentrations, poly(L-proline) and analogous GPPPPPGPPPPPGPPPPP sequences in VASP immediately disperse profilin·Actin·ATP from active polymerization sites in lamellipodia and filopodia, thereby arresting all actin-based motility (2). *See also GPPPPPGPPPPPGPPPPP* Target(s): actin-based motility (1,2); procollagen-proline 3-dioxygenase (3); and procollagen-proline 4-dioxygenase, *or* prolyl hydroxylase (4-11). **1.** Purich & Southwick (1997) *Biochem. Biophys. Res. Commun.* **231**, 686. **2.** Kang, Laine, Bubb, Southwick & Purich (1997) *Biochemistry* **36**, 8384-92. **3.** Tiainen, Pasanen, Sormunen & Myllyharju (2008) *J. Biol. Chem.* **283**, 19432. **4.** Hayaishi, Nozaki & Abbott (1975) *The Enzymes*, 3rd ed. (Boyer, ed.), **12**, 119. **5.** Prockop & Kivirikko (1969) *J. Biol. Chem.* **244**, 4838. **6.** Kivirikko & Myllylä (1980) *Enzymol. Post-transl. Modif. Proteins* (Freedman & Hawkins, eds.) **1**, 53, Academic Press, New York. **7.** Laushner & Pasternak (1978) *Can. J. Zool.* **56**, 159. **8.** Marumo & Waite (1987) *J. Exp. Zool.* **244**, 365. **9.** Tuderman, Myllylä & Kivirikko (1977) *Eur. J. Biochem.* **80**, 341. **10.** Myllylä, Tuderman & Kivirikko (1977) *Eur. J. Biochem.* **80**, 349. **11.** Kivirikko, Myllylä & Pihlajaniemi (1989) *FASEB J.* **3**, 1609.

Pomalidomide

This orally active, immunomodulatory antineoplastic agent (FW = 273.24 g/mol; CAS 19171-19-8; Solubility = ≥14 mg/mL DMSO), also known as Pomalyst™, CC-4047, 4-amino-thalidomide, and systematically as 4-amino-2-(2,6-dioxopiperidin-3-yl)isoindole-1,3-dione, inhibits proliferation of hematopoietic tumor cells (attended by induced apoptosis) as well as monocyte production of pro-inflammatory cytokines, while potentiating T cell- and natural killer cell-mediated immunity. Receiving FDA approval in 2013 for the treatment for relapsed/refractory multiple myeloma, pomalidomide is indicated for patients, who were unsuccessfully treated with bortezomib and lenalidomide (1). When tested in fifteen patients with relapsed/refractory myeloma, pomalidomide gave clear evidence of activation of endothelium, coagulation, and fibrinolysis, in the absence of either platelet activation or endothelial cell damage (2). Although pomalidomide treatment frequently results in deep vein thrombosis, alternate-day pomalidomide administration was associated with a marked reduction in the incidence of DVT, while retaining excellent anti-myeloma properties (3). Co-administration of pomalidomide plus low-dose dexamethasone is therapeutically active and well tolerated in lenalidomide-refractory multiple myeloma (4). Repression of TNF-α expression is the crucial factor of many of pomalidomide's anti-inflammatory properties. Pomalidomide also regulates the activity of Rho GTPases, altering the formation of actin filaments in primary human T cells and monocytes, showing that RhoA activation was essential for pomalidomide-induced interleukin-2 expression in T cells (5). 1. Koh, Janz, Mapara, *et al.* (2005) *Blood* **105**, 3833. 2. Streetly, Hunt, Parmar, *et al.* (2005) *Eur. J. Haematol.* **74**, 293. 3. Streetly, Gyertson, Daniel, *et al.* (2008) *Br. J. Haematol.* **141**, 41. 4. Lacy, Hayman, Gertz, *et al.* (2010) *Leukemia* **24**, 1934. 5. Xu, Li, Ferguson, *et al.* (2009) *Blood* **114**, 338.

α-Pompilidotoxin & β-Pompilidotoxin

α-PMTX: RIKIGLFDQLSKL-NH$_2$
β-PMTX: RIKIGLFDQLSRL-NH$_2$

These structurally related 13-residue neurotoxins (FW$_{\alpha\text{-PMTX}}$ = 1530.87 g/mol; Isoelectric Point = 9.99; FW$_{\beta\text{-PMTX}}$ = 1558.89 g/mol; Isoelectric Point = 10.84) are novel neurotoxins from the venom of a solitary wasp *Anoplius safnariensis* (1). With neuromuscular synapses in lobster walking leg and the rat trigeminal ganglion (TG) neurons, paired intracellular recordings from the presynaptic axon terminals and the innervating lobster leg muscles revealed that α-PMTX (at 10-100 μM) induces long bursts of action potentials in the presynaptic axon, resulting in both facilitated excitatory and inhibitory synaptic transmission (2). α-PMTX's action is distinct from that of other known facilitatory presynaptic toxins, including sea anemone toxins (**See** *AETX*) and α-scorpion toxins, which modify the fast inactivation of Na$^+$ current. Using whole-cell recordings from rat trigeminal neurons, α-PMTX slowed the Na$^+$ channels inactivation process without changing the peak current-voltage relationship or the activation time course of tetrodotoxin-sensitive Na$^+$ currents (**See** *Tetrodotoxin*) and that α-PMTX had voltage-dependent effects on the rate of recovery from Na$^+$ current inactivation and deactivating tail currents (2). Such results suggest that α-PMTX slows or blocks conformational changes required for fast inactivation of the Na$^+$ channels on the extracellular surface (2). Voltage-gated tetrodotoxin-sensitive sodium channels of Purkinje neurons produce "resurgent" current with repolarization, resulting from relief of an open-channel block that terminates current flow at positive potentials (3). The associated recovery of sodium channels from inactivation is thought to facilitate the rapid firing patterns characteristic of Purkinje neurons. Resurgent current appears to depend primarily on Na$_V$1.6 α-subunits, because it is greatly reduced in "med" mutant mice that lack Na$_V$1.6. β-PMTX increases resurgent current in wild-type neurons and induces resurgent current in med neurons (3). In med cells, the resurgent component of β-PMTX-modified sodium currents cannot be selectively abolished by application of intracellular alkaline phosphatase, suggesting that, like in Na$_V$1.6-expressing cells, the open-channel block of Na$_V$1.1 and Na$_V$1.2 subunits is regulated by constitutive phosphorylation (3). Such results

indicate the endogenous blocker exists independently of Na$_V$1.6 expression, and conventional inactivation regulates resurgent current by controlling the extent of open-channel block. In Purkinje cells, therefore, the relatively slow conventional inactivation kinetics of Na$_V$1.6 appears to be well adapted to carry resurgent current (3). 1. Konno, Miwa, Takayama, *et al.* (1997) *Neurosci. Lett.* **238**, 99. 2. Sahara, Gotoh, Konno, *et al.* (2000) *Eur. J. Neurosci.* **12**, 1961. 3. Grieco & Raman (2004) *J. Neurosci.* **24**, 35.

Ponatinib

This orally bioavailable tyrosine kinase inhibitor (FW = 532.56 g/mol; CAS 943319-70-8), also known as AP24534, Iclusig®, and systematically named as 3-2-(imidazo1,2-bipyridazin-3-yl)ethynyl-4-methyl-*N*-{4-(4-methylpi-perazin-1-yl)methyl-3-trifluoromethyl)phenyl}benzamide, targets the Breakpoint Cluster Region-Abelson (BCR-ABL) kinase. Ponatinib was developed by a structure-guided design for use against chronic myeloid leukemia (CML) and Philadelphia chromosome positive (Ph$^+$) acute lymphoblastic leukemia (ALL), including the so-called Bcr-Abl^{T315I} gatekeeper mutant that is resistant to Gleevec (Imatinib), Tasigna and Sprycel. On October 31, 2013, the U.S. Food and Drug Administration announced that it had asked the manufacturer, Ariad Pharmaceuticals, to suspend marketing and sales of Iclusig in view of the risk of life-threatening blood clots and severe narrowing of blood vessels. 1. Huang, Metcalf, Sundaramoorthi, *et al.* (2010) *J. Med. Chem.* **53**, 4701. 2. O'Hare, Shakespeare, Zhu, *et al.* (2009) *Cancer Cell* **16**, 401.

POPSO

This widely used Good's Buffer (FW$_{\text{free-acid}}$ = 362.43 g/mol; pK_a = 7.8 (25°C); ΔpK_a/Δ°C = −0.013), also known as piperazine-*N,N'*-bis(2-hydroxypropanesulfonic acid), inhibits the anion uniport, a member of the solute carrier (SLC) family (1). *Note*: Chloroethane sulfonate, which is an essential ingredient in the synthesis of Good's Buffers that contain ethane sulfonic acid, gives rise to variable amounts of oligovinylsulfonate (OVS). The latter strongly inhibits ribonuclease (2) and is likely to inhibit other enzymes that bind polyanions. 1. Ng, Selwyn & Choo (1993) *Biochim. Biophys. Acta* **1143**, 29. 2. Smith, Soellner & Raines (203) *J. Biol. Chem.* **278**, 20934.

Porfiromycin

This bioreductive antibiotic (FW = 349.36 g/mol; CAS 801-52-5; E^0 = −0.4 V) from *Streptomyces ardus* alkylates DNA and RNA, producing inter-strand cross-links and single-strand breaks, with selective toxicity to cells operating at low oxygen, making it an effective agent in solid tumor radiotherapy. Target(s): DNA biosynthesis (2); NAD(P)H dehydrogenase, *or* DT diaphorase (3). 1. Keyes, Rockwell & Sartorelli (1984) *Cancer Research* **45**, 3642. 2. Weissbach & Lisio (1965) *Biochemistry* **4**, 196. 3. Siegel, Beall, Kasai, *et al.* (1993) *Mol. Pharmacol.* **44**, 1128.

Porphycenes

These synthetic photosensitizers (FW = 310.36 g/mol; CAS 100572-96-1) are Fe(II)- and Fe(III)-binding structural isomers of porphyrin that promote singlet oxygen-mediated photodamage and are used in photodynamic therapy, *or* PTD (1-3). Their unique solubility and photochemical properties depend on the nature of ring substituents and bound metal ion. Photosensitizers with specific targets (*e.g.*, mitochondria, lysosomes, and plasma membrane) may be used to delineate the mechanism of PDT-induced apoptosis in cancer cells (3). One example is 9-acetoxy-2,7,12,17-tetrakis-(β-methoxyethyl)-porphycene (ATMPn), which enters cells and concentrates within mitochondria (4). **1**. Aramendia, Redmond, Nonell, *et al.* (1986) *Photochem. Photobiol.* **44**, 555. **2**. Richert, Wessels, Müller, *et al.* (1994) *J. Med. Chem.* **37**, 2797. **3**. Kessel & Luo (1998) *J. Photochem. Photobiol. B.* **42**, 89. **4**. Szeimies, Karrer, Abels, *et al.* (1996) *J. Photochem. Photobiol. B.* **34**, 67.

Porphyrexide

This paramagnetic sulfhydryl-modifying reagent (FW = 129.14 g/mol; CAS 15622-62-5; λ_{max} = 460 nm), also known systematically as 4-amino-2,5-dihydro-2-imino-5,5-dimethyl-1*H*-imidazol-1-yloxy, reacts with many sulfhydryl-containing substances, with loss of its characteristic red color (1). Porphyrexide is also be used as an oxidant substrate by a number of enzymes, including galactose oxidase (2). The *Escherichia coli* sulfite reductase heme protein subunit reacts at nearly unit stoichiometry with porphyrexide to produce a ferrisiroheme π-cation radical (3). **Target(s)**: alcohol dehydrogenase (4); glyceraldehyde-3-phosphate dehydrogenase (4); α-glycerophosphate dehydrogenase (4); lactate dehydrogenase (4); and pyruvate decarboxylase (5). **1**. Webb (1966) *Enzyme and Metabolic Inhibitors*, vol. 2, pp. 664-670, Academic Press, New York. **2**. Kwiatkowski, Adelman, Pennelly & Kosman (1981) *J. Inorg. Biochem.* **14**, 209. **3**. Young & Siegel (1988) *Biochemistry* **27**, 5984. **4**. Cedrangolo & Adler (1939) *Enzymologia* **7**, 297. **5**. Kuhn & Beinert (1943) *Ber.* **76**, 904.

Potato Protease Inhibitors

These proteolysis inhibitors are abundant in tubers and plant seeds, where they are frequently regarded as storage and/or defense proteins. They accumulate in potato leaves in response to wounding or ultraviolet irradiation. Potato inhibitors have been classified into seven different families: potato inhibitor I, potato inhibitor II, potato cysteine protease inhibitor, potato aspartate protease inhibitor, potato Kunitz-type protease inhibitor, potato carboxypeptidase inhibitor, and other serine protease inhibitors (1) **See also** *Trypsin Inhibitors* The potato carboxypeptidase inhibitor (MW ~4 kDa), often abbreviated CPI, inhibits carboxypeptidases from numerous sources. Three isoforms CPI-I, -II, and –III have been reported with pI values of 4.6, 5.6, and 6.5. Carboxipeptidase A is inhibited with K_i = 1.5–3.5 nM and carboxypeptidase B is inhibited with K_i = 1–6 nM. **Target(s)**: actinidain, inhibited by potato cysteine proteinase inhibitor (2); ananain, inhibited by potato cysteine proteinase inhibitor (2); carboxypeptidase A (21-30); carboxypeptidase B (16,20,21,28,30,34); carboxypeptidase U (15-18,31); caricain, *or* papaya proteinase III, inhibited by potato cysteine proteinase inhibitor (2); cathepsin B, inhibited by potato cysteine proteinase inhibitor (2,3); cathepsin D (4-6); cathepsin H, inhibited by potato cysteine proteinase inhibitor (2); cathepsin L, inhibited by potato cysteine proteinase inhibitor (2,3); cerevisin (7); chymopapain, inhibited by potato cysteine proteinase inhibitor (2,8); chymotrypsin (9); ficain, *or* ficin,

inhibited by potato cysteine proteinase inhibitor (2,8); fruit bromelain, inhibited by potato cysteine proteinase inhibitor (2); lysine carboxypeptidase, *or* lysine(arginine) carboxypeptidase, *or* carboxypeptidase N (19); mast cell carboxypeptidase (29,32,33); oryzin (1,11-13); papain, inhibited by potato cysteine proteinase inhibitor (2,3,8); pinguinain, inhibited by potato cysteine proteinase inhibitor (2); saccharopepsin (4); stem bromelain, inhibited by potato cysteine proteinase inhibitor (2); streptogrisin B (10); and trypsin (9,14). **1**. Kolattukudy & Sirakova (1998) in *Handb. Proteolytic Enzymes* (Barrett, Rawlings & Woessner, eds.), p. 320, Academic Press, San Diego. **2**. Rowan, Brzin, Buttle & Barrett (1990) *FEBS Lett.* **269**, 328. **3**. Brzin, Popovic, Drobnic-Kosorok, Kotnik & Turk (1988) *Biol. Chem. Hoppe-Seyler* **369**, 233. **4**. Cater, Lees, Hill, *et al.* (2002) *Biochim. Biophys. Acta* **1596**, 76. **5**. Pohl, Bures & Slavik (1981) *Collect. Czech. Chem. Commun.* **46**, 3302. **6**. Rupova, Keilova & Tomasek (1977) *Collect. Czech. Chem. Commun.* **42**, 2279. **7**. Nowak & Tsai (1989) *Can. J. Microbiol.* **35**, 295. **8**. Rodis & Hoff (1984) *Plant Physiol.* **74**, 907. **9**. Pouvreau, Gruppen, Piersma, *et al.* (2001) *J. Agric. Food Chem.* **49**, 2864. **10**. Qasim (1998) in *Handb. Proteolytic Enzymes* (Barrett, Rawlings & Woessner, eds.), p. 236, Academic Press, San Diego. **11**. Ohara & Nasuno (1972) *Agric. Biol. Chem.* **36**, 1797. **12**. Nakagawa (1970) *Meth. Enzymol.* **19**, 581. **13**. Nakadai, Nasuno & Iguchi (1973) *Agric. Biol. Chem.* **37**, 2685. **14**. Volpicella, Ceci, Cordewener, *et al.* (2003) *Eur. J. Biochem.* **270**, 10. **15**. Bouma, Marx, Mosnier & Meijers (2001) *Thromb. Res.* **101**, 329. **16**. Mao, Colussi, Bailey, *et al.* (2003) *Anal. Biochem.* **319**, 159. **17**. Boffa, Wang, Bajzar & Nesheim (1998) *J. Biol. Chem.* **273**, 2127. **18**. Lazoura, Campbell, Yamaguchi, *et al.* (2002) *Chem. Biol.* **9**, 1129. **19**. Sato, Miwa, Akatsu, *et al.* (2000) *J. Immunol.* **165**, 1053. **20**. Bradley, Naudé, Muramoto, Yamauchi & Oelofsen (1996) *Int. J. Biochem. Cell Biol.* **28**, 521. **21**. Hass, Derr, Makus & Ryan (1979) *Plant Physiol.* **64**, 1022. **22**. Hass (1979) *Arch. Biochem. Biophys.* **198**, 247. **23**. Everitt & Neurath (1980) *FEBS Lett.* **110**, 292. **24**. Bradley, Naudé, Muramoto, Yamauchi & Oelofsen (1994) *Int. J. Biochem.* **26**, 555. **25**. Gettins (1986) *J. Biol. Chem.* **261**, 15513. **26**. Marino-Buslje, Venhudová, Molina, *et al.* (2000) *Eur. J. Biochem.* **267**, 1502. **27**. Auld (1998) in *Handb. Proteolytic Enzymes* (Barrett, Rawlings & Woessner, eds.), p. 1321, Academic Press, San Diego. **28**. Birk (1976) *Meth. Enzymol.* **45**, 728. **29**. Hass & Ryan (1981) *Meth. Enzymol.* **80**, 778. **30**. Ryan, Hass & Kuhn (1974) *J. Biol. Chem.* **249**, 5495. **31**. Hendriks (1998) in *Handb. Proteolytic Enzymes* (Barrett, Rawlings & Woessner, eds.), p. 1328, Academic Press, San Diego. **32**. Springman (1998) in *Handb. Proteolytic Enzymes* (Barrett, Rawlings & Woessner, eds.), p. 1330, Academic Press, San Diego. **33**. Woodbury, Everitt & Neurath (1981) *Meth. Enzymol.* **80**, 588. **34**. Avilés & Vendrell (1998) in *Handb. Proteolytic Enzymes* (Barrett, Rawlings & Woessner, eds.), p. 1333, Academic Press, San Diego.

Poziotinib

This orally administered, irreversible pan-Human Epidermal growth factor Receptor (HER) inhibitor (FW$_{free-base}$ = 441.28 g/mol), also known by the code name HM781-36B and systematically as 1-[4-[4-(3,4-dichloro-2-fluorophenylamino)-7-methoxyquinolin-6-yloxy]-piperidin-1-yl]prop-2-en-1-one hydrochloride, is a new anticancer drug intended to treat advanced solid tumors in clinical trial. HER2-amplified cells were sensitive to HM781-36B (IC$_{50}$=1-10 nM), arresting in G$_1$ arrest and undergoing subsequent apoptosis. HM781-36B shows promise in the treatment for HER2-amplified breast cancer, especially as a chemosensitizer when used in combination with other cytotoxic agents (*e.g.*, 5-fluorouracil, cisplatin, paclitaxel, or gemcitabine). **1**. Cha, Lee, Kim, *et al.* (2012) *Int. J. Cancer* **130**, 2445. **2**. Kim, Kim, Yoon, *et al.* (2012) *Anticancer Drugs* **23**, 288.

PP1

This cell-permeable ATP site-directed pyrazolopyrimidine (FW = 281.36 g/mol), also known as 4-amino-5-(4-methylphenyl)-7-(*t*-butyl)pyrazolo[3,4-*d*]pyrimidine, inhibits a number of protein kinases, including the non-receptor, proto-oncogenic Src protein-tyrosine kinase (1-4). Target(s): Bcr-Abl protein-tyrosine kinase (5); casein kinase 1δ , IC50 = 1.06 μM (1); Csk protein-tyrosine kinase, IC50 = 0.52 μM (2); lymphocyte kinase, IC50 = 0.05 μM (1,2); polo kinase (7); protein-tyrosine kinase (1-6); Src protein-tyrosine kinase (1-4,6); and stress-activated protein kinase 2a/p38, IC50 = 0.64 μM (1). **1.** Bain, McLauchlan, Elliott & Cohen (2003) *Biochem. J.* **371**, 199. **2.** Hanke, Gardner, Dow, *et al.* (1996) *J. Biol. Chem.* **271**, 695. **3.** Backlund & Ingelman-Sundberg (2005) *Cell. Signal.* **17**, 39. **4.** Chong, Ia, Mulhern & Cheng (2005) *Biochim. Biophys. Acta* **1754**, 210. **5.** Tauchi & Ohyashiki (2006) *Int. J. Hematol.* **83**, 294. **6.** Crosby & Poole (2003) *J. Biol. Chem.* **278**, 24533. **7.** Johnson, Stewart, Woods, Giranda & Luo (2007) *Biochemistry* **46**, 9551.

PPI-0903, *See Ceftaroline Fosamil*

PP1 Analogue

This cell-permeable ATP site-directed pyrazolopyrimidine, also known as 4-amino-5-(1-naphthyl)-7-(*t*-butyl)pyrazolo[3,4-*d*]pyrimidine (FW = 317.39 g/mol), inhibits Src-family protein-tyrosine kinases, including Ile-338-Gly v-Src, IC50 = 1.5 nM, wild-type v-Src, IC50 = 1.0 μM, and Fyn, IC50 = 0.6 μM (1-3). **1.** Bishop, Ubersax, Petsch, *et al.* (2000) *Nature* **407**, 395. **2.** Bishop, Kung, Shah, *et al.* (1999) *J. Amer. Chem. Soc.* **121**, 627. **3.** Merciris, Claussen, Joiner & Giraud (2003) *Pflugers Arch.* **446**, 232.

PP2

This cell-permeable and photosensitive ATP site-directed pyrazole-pyrimidine (FW = 301.78 g/mol), also known as 4-amino-5-(4-chlorophenyl)-7-(*t*-butyl)pyrazolo[3,4-*d*]pyrimidine and tyrphostin AG 1879, potently inhibits Lck and Fyn protein kinases, with IC50 values of 4 nM and 5 nM. PP2 is ~100x less potent toward EGFR and is inactive for ZAP-70, JAK2 and PKA (1-3). PP2 prevented serum-independent growth of RET/PTC1-transformed NIH3T3 fibroblasts and of TPC1 and FB2, two human papillary thyroid carcinoma cell lines that carry spontaneous RET/PTC1 rearrangements (1). PP2 activates the E-cadherin-mediated cell adhesion system, suppressing metastasis in cancer cells (2). In cervical cancer cells (HeLa and SiHa), 10 μM PP2 down-regulates pSrc-Y416, pEGFR-Y845, and pEGFR-Y1173 expression levels, while down-regulatING pSrc-Y416 and pEGFR-Y845, but not pEGFR-Y1173 (3). PP2 is also a potential neuroprotective agent in cerebral ischemia-reperfusion (4). Target(s): casein kinase 1δ, IC50 = 1.3 μM (5); Csk protein-tyrosine kinase (6); focal adhesion kinase (7); lymphocyte kinase, IC50 = 0.06 μM (5,8); p56*lck*, IC50 = 4 nM (8,9); p59*fyn*T, IC50 = 5 nM (8); receptor protein-tyrosine kinase (10); src protein-tyrosine kinase, IC50 = 0.7 μM (5,11,12-23); stress-activated protein kinase 2a/p38, IC50 = 1.4 μM (5); and Yes kinase (13). **1.** Carlomagno, Vitagliano, Guida, *et al.* (2003) *J. Clin. Endocrinol. Metab.* **88**, 1897. **2.** Nam, Ino, Sakamoto & Hirohashi (2002) *Clin. Cancer Res.* **8**, 2430. **3.** Kong, Deng, Shen & Zhang (2011) *Mol. Cell Biochem.* **348**, 11. **4.** Lennmyr, Ericsson, Gerwins, *et al.* (2004) *Acta Neurol. Scand.* **110**, 175. **5.** Bain, McLauchlan, Elliott & Cohen (2003) *Biochem. J.* **371**, 199. **6.** Ayrapetov, Lee & Sun (2003) *Protein Expr. Purif.* **29**, 1486. **7.** Browe & Baumgarten (2003) *J. Gen. Physiol.* **122**, 689. **8.** Hanke, Gardner, Dow, *et al.* (1996) *J. Biol. Chem.* **271**, 695. **9.** Mahabeleshwar & Kundu (2003) *J. Biol. Chem.* **278**, 52598. **10.** Mehdi, Azar & Srivastava (2007) *Cell Biochem. Biophys.* **47**, 1. **11.** Okutani, Lodyga, Han & Liu (2006) *Amer. J. Physiol. Lung Cell Mol. Physiol.* **291**, L129. **12.** Tsai, Zhang, Sharma, Wu & Kinsey (2005) *Dev. Biol.* **277**, 129. **13.** Szigligeti, Neumeier, Duke, *et al.* (2006) *J. Physiol.* **573**, 357. **14.** Øvrevik, Låg, Schwarze & Refsnes (2004) *Toxicol. Sci.* **81**, 480. **15.** Piccardoni, Manarini, Federico, *et al.* (2004) *Biochem. J.* **380**, 57. **16.** Backlund & Ingelman-Sundberg (2005) *Cell. Signal.* **17**, 39. **17.** Chong, Ia, Mulhern & Cheng (2005) *Biochim. Biophys. Acta* **1754**, 210. **18.** Meyn III, Schreiner, Dumitrescu, Nau & Smithgall (2005) *Mol. Pharmacol.* **68**, 1320. **19.** Arnaud, Ballif & Cooper (2003) *Mol. Cell. Biol.* **23**, 9293. **20.** Ren & Baumgarten (2005) *Amer. J. Physiol.* **288**, H2628. **21.** Gao, Lau, Wong & Li (2004) *Cell. Signal.* **16**, 333. **22.** Merciris, Claussen, Joiner & Giraud (2003) *Pflugers Arch.* **446**, 232. **23.** Browe & Baumgarten (2003) *J. Gen. Physiol.* **122**, 689.

PP3

This ATP site-directed pyrazolopyrimidine (FW = 211.23 g/mol), also known as 4-amino-7-phenylpyrazolo[3,4-*d*]pyrimidine, inhibits casein kinase and EGFR (epidermal growth factor receptor) protein tyrosine kinase. Target(s): casein kinase 1δ, IC50 = 9.9 μM (1,2); EGFR (epidermal growth factor receptor) protein-tyrosine kinase, IC50 = 2.7 μM (3); protein-tyrosine kinase, *or* non-specific protein-tyrosine kinase (4); and Yes kinase (4). **1.** Bain, McLauchlan, Elliott & Cohen (2003) *Biochem. J.* **371**, 199. **2.** Liu, Bishop, Witucki, *et al.* (1999) *Chem. Biol.* **6**, 671. **3.** Traxler, Bold, Frei, *et al.* (1997) *J. Med. Chem.* **40**, 3601. **4.** Tsai, Zhang, Sharma, Wu & Kinsey (2005) *Dev. Biol.* **277**, 129.

PP 242

This ATP-competitive mTORC1/mTORC2 inhibitor (FW = 308.34 g/mol; CAS 1092351-67-1; Solubility: 25 mM in DMSO), also named 2-[4-amino-1-(1-methylethyl)-1*H*-pyrazolo[3,4-*d*]pyrimidin-3-yl]-1*H*-indol-5-ol, has a IC50 = 8 nM for both isoforms of mammalian target of rapamycin (mTOR), also known as FK506-binding protein 12-rapamycin-associated protein 1, *or* FRAP1, a serine/threonine protein kinase that regulates cell growth, cell proliferation, cell motility, cell survival, protein synthesis, autophagy, and transcription. PP-242 shows considerable selectivity for mTOR versus

PI3K family kinases (1,2). In models of acute leukemia harboring the Philadelphia chromosome (Ph) translocation, PP242, but not rapamycin, causes death of mouse and human leukemia cells (3). Indeed, PP242 completely inhibits TORC1 and TORC2 signaling in BCR-ABL$^+$ cells, whereas rapamycin partially suppresses TORC1 and drives a PI3K/AKT surge. Moreover, PP242 augments TNF-related apoptosis-inducing ligand- (TRAIL-) induced apoptosis in NSCLC cells, indicating that mTORC2 regulates Cbl-dependent FLIP$_S$ degradation and TRAIL-induced apoptosis (4) **Other Targets:** p110γ (IC$_{50}$ = 0.102 μM); DNA-PK ((IC$_{50}$ = 0.408 μM); p110δ ((IC$_{50}$ = 1.27 μM); p110α ((IC$_{50}$ = 1.96 μM), and p110β ((IC$_{50}$ = 2.2 μM), with much weaker inhibition of 215 other kinases. **1**. Apsel *et al.* (2008) *Nature Chem. Biol.* **4**, 691. **3**. Feldman, et al. (2009) *PLoS Biol.* **7**, 371. **3**. Janes, Limon, So, *et al.* (2010) *Nature Med.* **16**, 205. **4**. Zhao, Yue, Khuri & Sun (2013) *Cancer Res.* **73**, 1946.

PPI-2458

This orally bioavailable, selectively cytotoxic agent (FW$_{free-acid}$ = 424.54 g/mol; Photosensitive; Store in dark: IUPAC: [(3R,4S,5S,6R)-5-methoxy-4-[(2R,3R)-2-methyl-3-(3-methylbut-2-enyl)oxiran-2-yl]-1-oxaspiro2.5octan-6-yl] *N*-[(2R)-1-amino-3-methyl-1-oxobutan-2-yl]carbamate), a synthetic analogue of an *Aspergillus fumigatus* secondary metabolite (**See** *Fumagillin*), suppresses the formation of new blood vessels and inhibits endothelial cell proliferation and angiogenesis. Its epoxide group reacts with an active-site histidyl residue in methionyl aminopeptidase type II. Because removal of the *N*-terminal methionine from many proteins is required for their biologic activity, subcellular localization, and stability, methionyl-aminopeptidase plays a pivotal co-regulatory role in translation. PPI-2458 potently inhibits the proliferation of human fibroblast-like synoviocytes, *or* HFLS-RA (GI$_{50}$ = 0.04 nM) derived from RA patients, showing >95% inhibition at 1 nM. Proliferation of human umbilical vein endothelial cells (HUVEC) is similarly inhibited (GI$_{50}$ = 0.2 nM) by PPI-2458. Moreover, PPI-2458 inhibition of MetAP-2 catalysis (IC$_{50}$ = 0.2 nM) in HFLS-RA is directly correlated with cell growth inhibition and a decrease in the DNA polymerase processivity factor PCNA. PPP-2458 also protects the α-subunit of eukaryotic initiation factor 2 from inhibitory phosphorylation. Based on the structure of TNP-470, a fumagillin analogue that exhibits dose-limiting CNS toxicity, PPI-2458 was designed to retain antiproliferative activity while improving its CNS toxicity profile. PPI-2458 also inhibits proliferation of B16F10 melanoma cells *in vitro*, (GI$_{50}$ = 0.2 nM). This property, coupled with the absence of detectable resistance to PPI-2458 and the induction of morphological features of differentiated melanocytes, commends this agent for melanoma chemotherapy (2). PPI-2458 also inhibits non-Hodgkin's lymphoma cell proliferation *in vitro* and *in vivo* (3). Metabolic data demonstrate the participation of active metabolites in the *in vivo* efficacy of PPI-2458 (4). **1**. Bernier, Lazarus, Clark, *et al.* (2004) *Proc. Natl. Acad. Sci. U.S.A.* **101**, 10768. **2**. Hannig, Lazarus, Bernier, *et al.* (2006) *Int. J. Oncol.* **28**, 955. **3**. Cooper, Karp, Clark, *et al.* (2006) *Clin. Cancer Res.* **12**, 2583. **4**. Arico-Muendel, Belanger, Benjamin, *et al.* (2013) *Drug Metab Dispos.* **41**, 814.

PPTN

This highly potent and receptor type-selective P2Y$_{14}$ antagonist (FW = 511.96 g/mol; Solubility: 100 mM in DMSO), also named 4-[4-(4-piperidinyl)phenyl]-7-[4-(trifluoromethyl)phenyl]-2-naphthalenecarboxylic acid hydrochloride, targets the UDP-glucose–specific G$_i$ protein–coupled P2Y purinoceptor-14, *or* named previously as UDP-glucose receptor (K$_b$ =

0.43 nM), which binds uridine nucleotides and UDP-glucose that are released into the extracellular environment upon cell lysis (1). Two UDP analogues selectively activate the P2Y14 receptor over the UDP-activated P2Y6 receptor, and these molecules stimulate phosphorylation of ERK1/2 in differentiated human HL-60 promyeloleukemia cells (2). PPTN exhibits >10,000x selectivity toward P2Y14 relative other P2Y receptors and inhibits UDP-glucose and MRS 2690 (*or* 2-thiouridine-5'-diphosphoglucose) -induced porcine pancreatic artery contraction *ex vivo*. Indeed, PPTN also shows no agonist or antagonist affinity at P2Y$_1$, P2Y$_2$, P2Y$_4$, P2Y$_6$, P2Y$_{11}$, P2Y$_{12}$ or P2Y$_{13}$ receptors. It also blocks UDP-glucose-induced chemotaxis of HL-60 leukemia cells *in vitro* (3). The P2Y14 receptor (P2Y14-R) is highly expressed in hematopoietic cells, and, while its physiologic functions remain undefined, it has been linked to immune and inflammatory responses. **1**. Barrett, Sesma, Ball, *et al.* (2013) *Mol. Pharmacol.* **84**, 41. **2**. Carter, Fricks, Barrett, *et al.* (2009) *Mol. Pharmacol.* **76**, 1341. **3**. Alsaqati, Latif, Chan & Ralevic (2014) *Brit. J. Pharmacol.* **171**, 701.

PQ 69

This potent and selective hA$_1$ receptor antagonist (FW = 359.39 g/mol; CAS 910045-32-8; Soluble to 100 mM in DMSO), also named 4-(butylamino)-2-(3-fluorophenyl)-1,2-dihydro-3*H*-pyrazolo4,3-*c*quinolin-3-one, targets adenosine A$_1$ receptors, with an *in vitro* K$_i$ of 0.96 nM for the cloned receptor, with 217-fold lower activity against hA$_{2A}$ receptors and >1,000-fold lower action against hA$_3$ receptors. [^{35}S]-GTPγS binding and cAMP assays indicates that PQ-69 is an A1AR inverse agonist activity. PQ-69 displays highly inhibitory activities on isolated guinea pig contraction (pA$_2$ = 9), as induced by an A1AR agonist, 2-chloro-N^6-cyclopentyl adenosine. Systemic administration of PQ-69 increases urine flow and sodium excretion in normal rats. It has better metabolic stability thab 1,3-dipropyl-8-cyclopentylxanthine, in terms of *in vitro* properties and longer terminal elimination t$_{1/2}$ *in vivo*. **1**. Lu, Wang, Zhang, *et al.* (2014) *Purinergic Signal.* **10**, 619.

PR-619

This non-selective, cell-permeable pyridinamine (FW = 223.28 g/mol; CAS 2645-32-1) reversibly inhibits deubiquitinylating enzymes, *or* DUBs, with EC$_{50}$ values typically in the 1 to 20 μM range. **Targets:** ATXN3, B̲R̲C̲A̲1̲ A̲ssociated P̲rotein-1 (*or* BAP1), J̲osephin d̲omain c̲ontaining-2 (*or* JOSD2), O̲varian T̲umor D̲omain-containing-5 (*or* OTUD5), U̲biquitin C̲arboxyl-Terminal H̲ydrolase Isozyme-1 (*or* UCH-L1, UCH-L3, UCH-L5, UCH37, U̲biquitin-S̲pecific P̲eptidase-1 (*or* USP-1), USP-2, USP-4, USP-5, USP-7, USP-8, USP-9X, USP-10, USP-14, USP15, USP-16, USP-19, USP-20, USP-22, USP-24, USP-28, USP-47, USP-48, V̲alosin-c̲ontaining protein (*or* VCIP135), YOD1, as well as deISGylase PLpro, deNEDDylase DEN1, and deSUMOlyase SENP6. **1**. Altun, Kramer, Willems, *et al.* (2011) *Chem. Biol.* **18**, 1401.

PRA, See *Phosphoribosylamine*

Pralidoxime

This water-soluble, oxime-containing 2-formyl-1-methylpyrinidinium ion analogue (FW = 172.61 g/mol; CAS 6735-59-7; 1200-55-1 for the methyl sulfate salt; 154-97-2 for mesylate salt; 21239-05-4 for the monolactate; 3687-33-0 for trichloroacetate salt; 51-15-0 for the chloride salt), also called pyridine-2-aldoxime methochloride and 2-PAM, effectively reverses the inhibition of acetylcholinesterase and cholinesterase (1-3) by organophosphate-containing insecticide, thereby serving as an antidote. 2-PAM has even been observed to reactivate blood and diaphragm cholinesterase inhibited by the biowarfare agent VX. **1**. Kovarik, Ciban, Radic, Simeon-Rudolf & Taylor (2006) *Biochem. Biophys. Res. Commun.* **342**, 973. **2**. Reiner, Simeon-Rudolf & Skrinjaric-Spoljar (1995) *Toxicol. Lett.* **82/83**, 447. **3**. Petroianu, Nurulain, Arafat, Rajan & Hasan (2006) *Arch. Toxicol.* **80**, 777.

Pramipexole

This non-ergoline receptor agonist ($FW_{free-base}$ = 211.33 g/mol; CAS 104632-26-0; $FW_{diHCl-monohydrate}$ = 302.26 g/mol; CAS = 191217-81-9; Solubility = 42 mg/mL DMSO), also known as (S)-N^6-propyl-4,5,6,7-tetrahydrobenzo[d]thiazole-2,6-diamine, Mirapexin, Sifrol, and SND-919, exerts partial/full antagonist action on dopamine D_{2S} (K_i = 3.9 nM), D_{2L} (K_i = 2.2 nM), D_3 (K_i = 0.5 nM), and D_4 (K_i = 5.1 nM) receptors (1). D_{2S} and D_{2L} are G protein-coupled receptors that respectively participate in presynaptic and postsynaptic dopaminergic transmission; D_3 subtype receptors inhibit adenylyl cyclase through inhibitory G-proteins; as does the D_4 subtype. Although L-DOPA is the gold-standard for treating Parkinson Disease, its effectiveness fades rapidly and its continued use can result in serious motor fluctuations. Pramipexole selectively acts at dopamine receptors in the D_2 subfamily an possesses full activity that is comparable to dopamine (2). Its preferential affinity for the D_3 receptor subtype may well account for its efficacy in the treatment of both the motor and psychiatric symptoms of Parkinson disease (2). Unlike the long-term use of the ergot-derived dopamine agonists, pramipexole does not carry the risk to induce valvular heart disease or pulmonary and retroperitoneal fibrosis. **Key Pharmacokinetic Parameters:** *See* Appendix II in Goodman & Gilman's *THE PHARMACOLOGICAL BASIS OF THERAPEUTICS*, 12th Edition (Brunton, Chabner & Knollmann, eds.) McGraw-Hill Medical, New York (2011). **1**. Mierau & Schingnitz (1992) *Eur. J. Pharmacol.* **215**, 161. **2**. Bennett & Piercey (1999) *J. Neurol. Sci.* **163**, 25.

Pramiracetam

This memory-enhancing (*or* nootropic) drug (FW = 269.39 g/mol; CAS 68497-62-1) also known by the trade names Remen®, Neupramir®, and Pramistar®, as well as by its systematic name *N*-[2-(diisopropylamino)ethyl]-2-(2-oxopyrrolidin-1-yl)acetamide, reverses electroconvulsive shock- (ECS-) induced amnesia in mice, when administered after ECS treatment, but is inactive in a general observational test for CNS activity (1). The anti-amnesic effect of pramiracetam and related nootropic drugs parallel their *in vitro* inhibitory activities on prolyl endopeptidase, suggesting that the inhibitors exhibit their anti-amnesic effects by reducing this enzyme's activity in the brain. For pharmacokinetics of oral pramiracetam in normal volunteers, see (3). Prolyl oligopeptidase a cytosolic serine peptidase that hydrolyzes proline-containing peptides at the carboxy terminus of proline residues. It has been associated with schizophrenia, bipolar affective disorder, and related neuropsychiatric disorders. (**See** *Baicalin*) **1**. Butler, Nordin, L'Italien, *et al.* (1984) *J. Med. Chem.* **27**, 684. **2**. Yoshimoto T, Kado K, Matsubara, *et al.* (1987) *J. Pharmacobiodyn.* **10**, 730. **3**. Chang, Young, Goulet & Yakatan (1985) *J. Clin. Pharmacol.* **25**, 291.

Prasugrel

This oral thienopyridine-containing antithrombotic prodrug (FW = 373.44 g/mol; CAS 150322-43-3), known also by its codename KS-747, its trade names Effient™ (U.S.), Efient™ (E.U.), and Prasita™ (India), and its systematic name 5-[2-cyclopropyl-1-(2-fluorophenyl)-2-oxoethyl]-4,5,6,7-

tetrahydrothieno[3,2-c]pyridin-2-yl acetate, is metabolized to the thiol-containing active metabolite R-138727. In the following scheme, esterase-catalyzed hydrolysis of the acetate function results in a thiolactone species that then undergoes cytochrome P450-dependent oxidative opening of the thiolactone ring to form a reactive sulfenic acid metabolite that is eventually reduced to the active thiol-containing active drug (2).

R-138727 binds to and irreversibly inhibits the $P2Y_{12}$-subtype of the adenosine 5'-diphosphate (ADP) receptor, blocking platelet aggregation in the presence of fibrin (1). R-138727 is actually a mixture of (R, S), (R, R), (S, S), and (S, R) stereoisomers, where the first letter indicates the configuration of the sulfur-bearing carbon and the second indicates the configuration of the benzylic carbon. As compared with clopidogrel, lower levels of prasugrel consistently resulted in a higher degree of inhibition of platelet aggregation (3). There was no relevant effect of genetic variation in CYP2B6, CYP2C9, CYP2C19, or CYP3A5 on the pharmacokinetics of prasugrel's active metabolite or its inhibition of platelet aggregation. **1**. Hasegawa, Sugidachi, Ogawa *et al.* (2007) *Thromb. Hemost.* **94**, 593. **2**. Dansette, Rosi, Debernardi, Bertho & Mansuy (2012) *Chem. Res. Toxicol.* **25**, 1058. **3**. Jakubowski, Matsushima, Asai *et al.* (2007) *Brit. J. Clin. Pharmacol.* **63**, 421.

Pravastatin

This therapeutic anticholesterogenic HMG-CoA inhibitor ($FW_{free-acid}$ = 424.53 g/mol; CAS 81093-37-0), also known as CS-514 and Pravachol®, inhibits 3-hydroxy-3-methylglutaryl-CoA reductase, $K_i \approx 1$ nM (1-3). Orally administered pravastatin exhibits one-pass absorption by the liver. Moreover, because there is no uptake of pravastatin by most extrahepatic cells, this statin will be unable to mitigate the increase in mevalonate synthesis in extrahepatic tissues accompanying the decrease in circulating cholesterol caused by its inhibition of hepatic HMG-CoA reductase. This inhibitor produced dose-related reduction in total cholesterol and low-density lipoprotein cholesterol. Pravastatin reportedly modifies cholesterol exchange between plasma and the erythrocyte membrane. **Key Pharmacokinetic Parameters:** *See* Appendix II in Goodman & Gilman's *THE PHARMACOLOGICAL BASIS OF THERAPEUTICS*, 12th Edition (Brunton, Chabner & Knollmann, eds.) McGraw-Hill Medical, New York (2011). **1**. Nakaya, Homma, Tamachi & Goto (1986) *Atherosclerosis* **61**, 125. **2**. Yoshino, Kazumi, Kasama, *et al.* (1986) *Diabetes Res. Clin. Pract.* **2**, 179. **3**. Rabini, Polenta, Staffolani, *et al.* (1993) *Exp. Mol. Pathol.* **59**, 51.

Praziquantel

This antihelmintic agent (FW = 312.41 g/mol; CAS 55268-74-1; 57452-97-8 for the (S)-isomer; 57452-98-9 for the (R)-isomer), also known as 2-(cyclohexylcarbonyl)-1,2,3,6,7,11*b*-hexahydro-4*H*-pyrazino[2,1-*a*]isoquinolin-4-one, is a *Schistosoma japonicum* glutathione *S*-transferase inhibitor used in the treatment of schistosomiasis. It reportedly has a selective effect on the tegument of trematodes and increases permeability of calcium. Praziquantel is soluble in ethanol (9.7 g/100 g) and has a low solubility in water (0.04 g/100 g). **Key Pharmacokinetic Parameters:** *See* Appendix II in Goodman & Gilman's THE PHARMACOLOGICAL BASIS OF THERAPEUTICS, 12th Edition (Brunton, Chabner & Knollmann, eds.) McGraw-Hill Medical, New York (2011). **Target(s):** *N*-acetylgalactosamine-4-sulfatase, *or* arylsulfatase B (1); γ-glutamyl transpeptidase, *or* γ-glutamyltransferase (2); glutathione *S*-transferase (3,4); and lactate dehydrogenase (5). **1**. Balbaa & Bassiouny (2006) *J. Enzyme Inhib. Med. Chem.* **21**, 81. **2**. Wang, Cheng, Huang & Zheng (1987) *Zhongguo Ji Sheng Chong Xue Yu Ji Sheng Chong Bing Za Zhi* **5**, 32. **3**. McTigue, Williams & Tainer (1995) *J. Mol. Biol.* **246**, 21. **4**. Rufer, Thiebach, Baer, Klein & Hennig (2005) *Acta Crystallogr. Sect. F* **61**, 263. **5**. Veerakumari & Munuswamy (2000) *Vet. Parasitol.* **91**, 129.

Prazobind

This alkylating agent and antihypertensive (FW = 422.49 g/mol; CAS 107021-36-3; Soluble in Chloroform, DMSO, Ethanol and Methanol), also known as SZL-49, irreversibly antagonizes α_1 adrenoceptors (1-4). **1**. Helman, Kusiak, Pitha & Baum (1987) *Biochem. Biophys. Res. Commun.* **142**, 403. **2**. Mante & Minneman (1991) *Eur. J. Pharmacol.* **208**, 113. **3**. Pitha, Szabo, Szurmai, Buchowiecki & Kusiak (1989) *J. Med. Chem.* **32**, 96. **4**. Piascik, Butler, Kusiak, Pitha & Holtman (1989) *J. Pharmacol. Exp. Ther.* **251**, 878.

Prednisolone

This semisynthetic corticosteroid (FW = 360.45 g/mol; CAS 50-24-8), also known as (11β)-11,17,21-trihydroxypregna-1,14-diene-3,20-dione, is a potent antiinflammatory and antiallergic agent. Note that prednisolone has a short biological half-life and is interconverted to prednisone (via the action of 11β-hydroxysteroid dehydrogenase). Prednisolone binds irreversibly to the glucocorticoid receptors αGR and βGR, of which 3-10,000 are found in virtually all cell types, depending on the tissue involved. The steroid/receptor complexes dimerize, whereupon they interact with DNA in the nucleus, binding to steroid-response elements and modifying the transcription of target genes. **Key Pharmacokinetic Parameters:** *See* Appendix II in Goodman & Gilman's THE PHARMACOLOGICAL BASIS OF THERAPEUTICS, 12th Edition (Brunton, Chabner & Knollmann, eds.) McGraw-Hill Medical, New York (2011). **Target(s):** catalase (1); fumarase, *or* fumarate hydratase (1); interleukin-1β release (2); and peptidyl-dipeptidase A, *or* angiotensin-converting enzyme (3). **1**. Shyadehi & Harding (2002) *Biochim. Biophys. Acta* **1587**, 31. **2**. Uehara, Kohda, Sekiya, Takasugi & Namiki (1989) *Experientia* **45**, 166. **3**. Roulston & Galloway (1986) *Clin. Chem.* **32**, 697.

Pregabalin

This novel Ca^{2+} channel $\alpha_2\delta$-subunit antagonist and anticonvulsant drug (FW = 159.23 g/mol; CAS 148553-50-8), known by its trade name, Lyrica™, and its IUPAC name, (*S*)-3-(aminomethyl)-5-methylhexanoic

acid, targets γ-aminobutyric acid (GABA) aminotransferase (*Reaction*: GABA + α-Ketoglutarate ⇌ Succinate Semialdehyde + L-Glutamate). This pyridoxal 5'-dependent enzyme is responsible for the degradation of the inhibitory neurotransmitter GABA and stoichiometric formation of the excitatory neurotransmitter L-glutamate (1,2). Notably, compounds inhibiting this enzyme often exert anticonvulsant actions and may also affect other disorders, such as Alzheimer's Disease, Huntington's disease, Parkinson Disease, as well as drug addiction. With its low potential for abuse and limited dependence liability, pregabalin is classified by the FDA as a Schedule V drug. **Primary Mode of Action:** Discovery of this alternative substrate inhibitor was premised on the need to explore how the introduction of a lipophilic side-chain might allow a GABA analogue to cross the blood-brain barrier as a prelude to designing mechanism-based irreversible inhibitors. Ironically, the need for developing a suicide inhibitor was deemed unnecessary, inasmuch as this substituted GABA analogue proved to be a poor alternative substrate that effectively ties up GABA-AT, reducing the enzyme's ability to form L-Glutamate. Beyond blocking L-Glutamate formation, pregabalin has the added benefit of activating glutamate decarboxylase (*Reaction*: L-Glutamate ⇌ GABA + CO₂), the latter raising the concentration of the inhibitory neurotransmitter. The mechanism by which pregabalin inhibits pain is by suppressing Ca^{2+} influx through $\alpha_2\delta$ subunit-containing voltage-gated calcium channel (3,4). **Key Pharmacokinetic Parameters:** *See* Appendix II in Goodman & Gilman's THE PHARMACOLOGICAL BASIS OF THERAPEUTICS, 12th Edition (Brunton, Chabner & Knollmann, eds.) McGraw-Hill Medical, New York (2011). **1**. Andruszkiewicz & Silverman (1990) *J. Biol. Chem.* **265**, 22288. **2**. Silverman (2008) *Angew. Chem. Int. Ed.* **47**, 3500. **3**. Bian, Li, Offord *et al.* (2006) *Brain Res.* **1075**, 68. **4**. Taylor, Angelotti & Fauman (2007) *Epilepsy Res.* **73**, 137.

4,9-Pregnadiene-3,20-dione

This steroid (FW = 298.43 g/mol) competitively inhibits *Pseudomonas testosteroni* 3-oxosteroid Δ^5-Δ^4-isomerase (K_i = 5.3 μM), helping to define the structural features of sterols acting as competitive or noncompetitive inhibitors of the interconversion of 3-oxo-Δ^5-steroids and 3-oxo-Δ^4-steroids. **1**. Weintraub, Vincent, Baulieu & Alfsen (1977) *Biochemistry* **16**, 5045.

5α-Pregnane-3β,20α-diol

This isomer of pregnanediol (FW = 318.50 g/mol) a mixed-type noncompetitive inhibitor ($K_{i,app}$ = 0.7 μM) of 3β-hydroxysteroid dehydrogenase. **1**. Chavatte, Rossdale & Tait (1995) *Equine Vet. J.* **27**, 342.

5α-Pregnane-3β,17α-diol-20-one

This steroid (FW = 334.50 g/mol) is one of 14 steroids that inhibit rat mammary glucose-6-phosphate dehydrogenase *in vitro*. Inhibition requires the presence of a keto group at position C-17 for androstanes and estanes or at position C-20 for pregnanes. Binding occurs through hydrophobic interactions, and all modifications that diminish ligand planarity likewise reduce binding affinity. Note also that other hydrophobic compounds also inhibit G6PDH, including various antimalarials. **1**. Raineri & Levy (1970) *Biochemistry* **9**, 2233.

5α-Pregnane-3α-ol-20-one, See *5α-Pregnan-3α-ol-20-one*

5α-Pregnane-3β-ol-20-one, See *5α-Pregnan-3β-ol-20-one*

5β-Pregnane-3β-ol-20-one, See *5β-Pregnan-3β-ol-20-one*

5α-Pregnane-3α,11β,17α,21-tetrol-20-one, See *Allotetrahydrocortisol*

Pregnane-3α,11β,17α,21-tetrol-20-one, See *Tetrahydrocortisol*

Pregnane-3α,17α,21-triol-11,20-dione, See *Tetrahydrocortisone*

5α-Pregnan-3β-ol-20-one

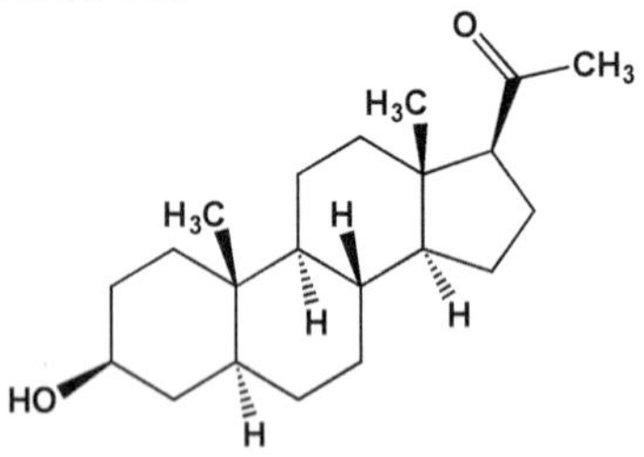

This neurosteroid (FW = 318.50 g/mol; CAS 128-20-1), also known as allopregnanolone, inhibits 3β-hydroxysteroid dehydrogenase and glucose-6-phosphate dehydrogenase (K_i = 0.57 μM) (1.2) and also exhibits anxiolytic and anticonvulsant activities through potentiation of the GABA$_A$ receptor. (**For additional comment, see** *5α-Pregnane-3β,17α-diol-20-one*) **1**. Raineri & Levy (1970) *Biochemistry* **9**, 2233. **2**. Chavatte, Rossdale & Tait (1995) *Equine Vet. J.* **27**, 342.

4-Pregnene-11β,21-diol-3,20-dione, See *Corticosterone*

4-Pregnene-17α,21-diol-3,11,20-trione, See *Cortisone*

4-Pregnene-3,20-dione, See *Progesterone*

4-Pregnene-21-ol-3,20-dione, See *Deoxycorticosterone*

4-Pregnene-3β-ol-20-one, See *4-Pregnen-3β-ol-20-one*

5-Pregnene-3β-ol-20-one, See *Pregnenolone*

4-Pregnene-20α-ol-3-one, See *20α-Dihydroprogesterone*

4-Pregnene-11β,17α,21-triol-3,20-dione, See *Cortisol*

Pregneninolone, See *Ethisterone*

4-Pregnen-21-ol-3,20-dione, See *11-Deoxycorticosterone*

Pregnenolone

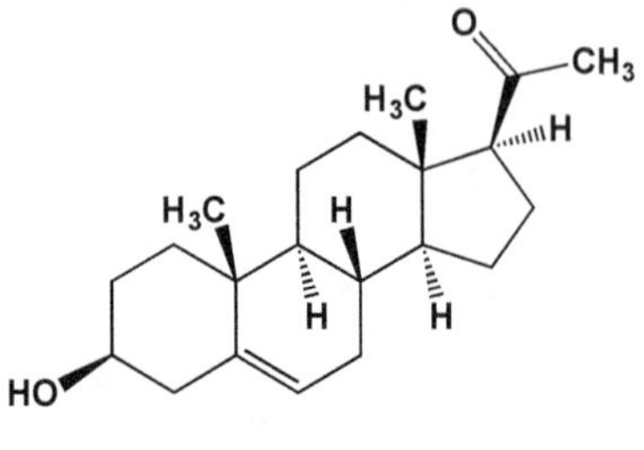

This steroid (FW = 316.48 g/mol; CAS 145-13-1), also known as (3β)-3-hydroxypregn-5-en-20-one and Δ^5-pregnen-3β-ol-20-one, is a key intermediate in the biosynthesis of progesterone and has glucocorticoid activity. Pregnenolone has a low solubility in water and is used in the treatment of rheumatoid arthritis. Target(s): arylamine *N*-acetyltransferase (1); cholesterol 7α-monooxygenase (2); glucose-6-phosphate dehydrogenase (3-7); ketosteroid monooxygenase, *or* androst-4-ene-3,17-dione monooxygenase, inhibits lactonization of androstenedione (8); malic

enzyme (9); steroid 11β-monooxygenase (10); steroid 16α-monooxygenase (11); steroid 20-monooxygenase (12); steroid sulfatase (13); steroid sulfotransferase, also alternative substrate (14); sterol *O*-acyltransferase, *or* cholesterol *O*-acyltransferase, *or* ACAT (15,16); sterol demethylase (17); steryl-sulfatase, weakly inhibited (18). **1**. Kawamura, Westwood, Wakefield, *et al.* (2008) *Biochem. Pharmacol.* **75**, 1550. **2**. Van Cantfort & Gielen (1975) *Eur. J. Biochem.* **55**, 33. **3**. Langdon (1966) *Meth. Enzymol.* **9**, 126. **4**. Noltmann & Kuby (1963) *The Enzymes*, 2nd ed. (Boyer, Lardy & Myrbäck, eds.), **7**, 223. **5**. McKerns & Kaleita (1960) *Biochem. Biophys. Res. Commun.* **2**, 344. **6**. Criss & McKerns (1969) *Biochim. Biophys. Acta* **184**, 486. **7**. Raineri & Levy (1970) *Biochemistry* **9**, 2233. **8**. Itagaki (1986) *J. Biochem.* **99**, 825. **9**. Reddy (1974) *Biochem. Biophys. Res. Commun.* **59**, 1005. **10**. Ullrich & Duppel (1975) *The Enzymes*, 3rd ed. (Boyer, ed.), **12**, 253. **11**. Sano, Shibusawa, Yoshida, *et al.* (1980) *Acta Obstet. Gynecol. Scand.* **59**, 245. **12**. Koritz & Hall (1964) *Biochemistry* **3**, 1298. **13**. Notation & Ungar (1969) *Biochemistry* **8**, 501. **14**. Falany, Vasquez & Kalb (1989) *Biochem. J.* **260**, 641. **15**. Billheimer (1985) *Meth. Enzymol.* **111**, 286. **16**. Simpson & Burkhart (1980) *Arch. Biochem. Biophys.* **200**, 79. **17**. Gaylor, Chang, Nightingale, Recio & Ying (1965) *Biochemistry* **4**, 1144. **18**. Norkowska & Gniot-Szulzycka (2002) *J. Steroid Biochem. Mol. Biol.* **81**, 263.

5-Pregnen-3β-ol-20-one, See *Pregnenolone*

4-Pregnen-20α-ol-3-one, See *20α-Dihydroprogesterone*

5-Pregnen-3β-ol-20-one-16α-carbonitrile, See *Pregnenolone 16α-Carbonitrile*

Pregnenolone Sulfate

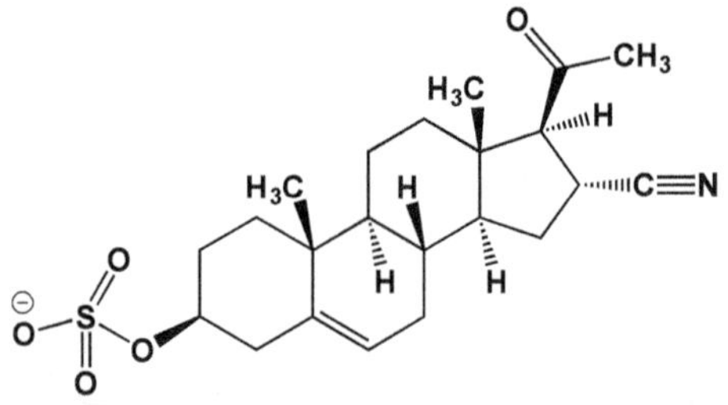

This neurosteroid (FW$_{ion}$ = 333.52 g/mol), also known as 3-hydroxypregn-5-en-20-one sulfate and 3β-hydroxy-5-pregnen-20-one 3-sulfate, is an inhibitor of estrone sulfatase (1,2), GABA$_A$ receptor chloride channel (3), and steroid 16α-monooxygenase (4). **1**. Santner & Santen (1993) *J. Steroid Biochem. Mol. Biol.* **45**, 383. **2**. MacIndoe (1988) *Endocrinology* **123**, 1281. **3**. Mienville & Vicini (1989) *Brain Res.* **489**, 190. **4**. Sano, Shibusawa, Yoshida, *et al.* (1980) *Acta Obstet. Gynecol. Scand.* **59**, 245.

Prenylamine

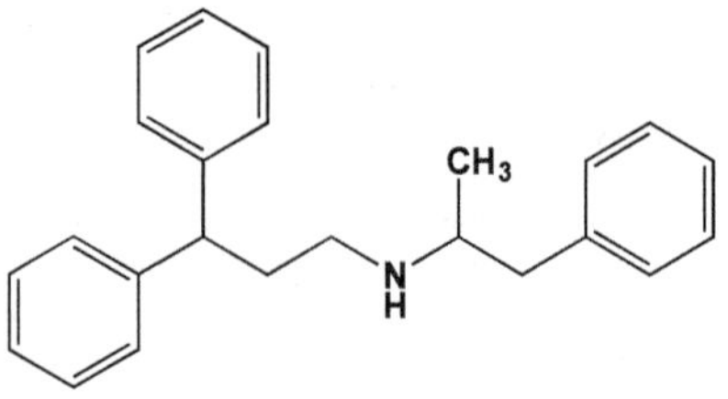

This vasodilator and a Ca^{2+} channel blocker (FW$_{free-base}$ = 329.49 g/mol; CAS 390-64-7), also named *N*-(1-methyl-2-phenylethyl)-γ-phenylbenzene propanamine, is often supplied as the lactate salt (FW = 419.56 g/mol), which is slightly soluble in water (approximately 0.5%) and melts at 140-142°C. Target(s): Ca^{2+}-dependent cyclic-nucleotide phosphodiesterase (1); Ca^{2+}-transporting ATPase (2-5); and glibenclamide-sensitive K$^+$ channels (6,7). **1**. Hidaka, Inagaki, Nishikawa & Tanaka (1988) *Meth. Enzymol.* **159**, 652. **2**. Hasselbach (1974) *The Enzymes*, 3rd ed. (Boyer, ed.), **10**, 431. **3**. Balzer, Makinose & Hasselbach (1968) *Naunyn Schmiedebergs Arch. Exp. Pathol. Pharmakol.* **260**, 444. **4**. Balzer & Makinose (1968) *Naunyn Schmiedebergs Arch. Exp. Pathol. Pharmakol.* **259**, 151. **5**. Balzer, Makinose & Hasselbach (1967) *Naunyn Schmiedebergs Arch. Exp. Pathol. Pharmakol.* **257**, 7. **6**. Sakuta, Okamoto & Watanabe (1992) *Brit. J. Pharmacol.* **107**, 1061. **7**. Sakuta, Sekiguchi, Okamoto & Sakai (1992) *Eur. J. Pharmacol.* **226**, 199.

8-Prenylnaringenin

This prenylated phytoestrogenic flavonoid (FW = 340.38 g/mol), isolated from hops (*Humulus lupulus*), inhibits cytochrome P450 systems CYP1A2 (1,2) and CYP19, *or* aromatase (3). 8-Prenylnaringenin also inhibits angiogenesis *in vitro* and *in vivo*. **1**. Henderson, Miranda, Stevens, Deinzer & Buhler (2000) *Xenobiotica* **30**, 235. **2**. Miranda, Yang, Henderson, *et al.* (2000) *Drug Metab. Dispos.* **28**, 1297. **3**. Monteiro, Hecker, Azevedo & Calhau (2006) *J. Agric. Food Chem.* **54**, 2938.

Prenyl Pyrophosphate, See *specific metabolite*

Prephenic Acid

This dicarboxylic acid (FW$_{free-acid}$ = 226.19 g/mol; CAS 126-49-8; IUPAC: 1-carboxy-4-hydroxy-2,5-cyclohexadiene-1-pyruvate), an intermediate in the biosynthesis of L-phenylalanine and L-tyrosine, inhibits arogenate dehydratase (1,2), arogenate dehydrogenase, *or* cyclohexadienyl dehydrogenase (3), 3-deoxy-7-phosphoheptulonate synthase (4-7), and shikimate kinase (8). Under strongly acidic conditions (0.5 M HCl), this metabolite is unstable, converting to phenylpyruvic acid and liberating carbon dioxide and reaching 100% completion at 37° C in 10 min. The half-life at pH 4 and 0° C is seven hours. It is more stable at neutral or alkaline pH, but still decomposes upon heating to form *p*-hydroxyphenyllactate. **1**. Xia, Ahmad, Zhao & Jensen (1991) *Arch. Biochem. Biophys.* **286**, 461. **2**. Zhao, Xia, Fischer & Jensen (1992) *J. Biol. Chem.* **267**, 2487. **3**. Lingens, Keller & Keller (1987) *Meth. Enzymol.* **142**, 513. **4**. Rubin & Jensen (1985) *Plant Physiol.* **79**, 711. **5**. Kim (2001) *J. Biochem. Mol. Biol.* **34**, 299. **6**. Wu, Sheflyan & Woodard (2005) *Biochem. J.* **390**, 583. **7**. Wu & Woodard (2006) *J. Biol. Chem.* **281**, 4042. **8**. Huang, Montoya & Nester (1975) *J. Biol. Chem.* **250**, 7675.

Pressinoic Acid

C-Y-F-Q-N-C

This cyclic hexapeptide (FW = 774.86 g/mol; CAS 35748-51-7) corresponds to the N-terminal six amino acyl residues of vasopressin. If Phe-3 is replaced with Ile, the resulting cyclic hexapeptide, known as tocinoic acid. is identical to the N-terminal six amino acid residues of oxytocin. Pressinoic acid has no known pressor activity, perhaps as result of the loss of the carboxy-terminal tripeptide and/or the incorrect orientation of the Asn-5 side-chain (1) This hexapeptide also activates pineal acetyl-CoA hydrolase (2) and has a potent corticotrophin-releasing activity (3). Target(s): serotonin *N*-acetyltransferase (4). **1**. Langs, Smith, Stezowski & Hughes (1986) *Science* **232**, 1240. **2**. Namboodiri, Favilla & Klein (1982) *J. Biol. Chem.* **257**, 10030. **3**. Saffran, Pearlmutter, Rapino & Upton (1972) *Biochem. Biophys. Res. Commun.* **49**, 748. **4**. Klein & Namboodiri (1982) *Trends Biochem. Sci.* **7**, 98.

P-RibPP, See *5-Phospho-α-D-ribosyl 1-Pyrophosphate*

Prilocaine

This powerful Na$^+$ channel-blocking anesthetic (FW$_{free-base}$ = 220.31 g/mol; CAS 721-50-6; HCl salt freely soluble in water) with properties resembling

those of lidocaine, is an antagonist of β$_2$-adrenergic receptors (1) as well as an inhibitor of monoamine oxidase (2), Na$^+$ channels (3), and Na$^+$/K$^+$-exchanging ATPase (4). **1**. Butterworth, James & Grimes (1997) *Anesth. Analg.* **85**, 336. **2**. Yasuhara, Wada, Sakamoto & Kamijo (1982) *Jpn. J. Pharmacol.* **32**, 213. **3**. Tella & Goldberg (1998) *Pharmacol. Biochem. Behav.* **59**, 305. **4**. Kutchai, Geddis & Farley (2000) *Pharmacol. Res.* **41**, 1.

Primaquine

This antimalarial (FW$_{free-base}$ = 259.35 g/mol; CAS 90-34-6), also known systematically as N^4-(6-methoxy-8-quinolinyl)-1,4-pentanediamine, was used widely during World War II, until it was discovered that individuals with glucose-6-phosphate dehydrogenase deficiency are especially sensitive to this and related antimalarial drugs. Primaquine retards nonenzymatic formation of hemozoin, the malaria parasite pigment that is a paracrystalline aggregate of ferriprotoporphyrin IX (FP-Fe(III)) produced during hemoglobin digestion by the *Plasmodium* enzyme plasmepsin. Because hemozoin formation is essential for FP-Fe(III) detoxification within the parasite, quinolines are highly effective antimalarials. Target(s): acetylcholinesterase (1); alcohol dehydrogenase (2,3); aminopyrine *N*-demethylase (4); calcium-release-activated current (5); CYP3A4 (6,7,19); estrogen 2-hydroxylase (8); glutathione *S*-transferase (9); histamine *N*-methyltransferase, weakly inhibited (10); HIV integrase (11); monoamine oxidases A and B (12); myeloperoxidase (13); Na$^+$ channels (14); Na$^+$/K$^+$-exchanging ATPase (15); peroxidase (13); phospholipase A$_1$ (16); phospholipase A$_2$ (16); protein biosynthesis (17); protein kinase A (18); quinine 3-monooxygenase (19); tolbutamide hydroxylase (20); and vesicular transport (5). **1**. Katewa & Katyare (2005) *Drug Chem. Toxicol.* **28**, 467. **2**. Bränden, Jörnvall, Eklund & Furugren (1975) *The Enzymes*, 3rd ed. (Boyer, ed.), **11**, 103. **3**. Li & Magnes (1972) *Biochem. Pharmacol.* **21**, 17. **4**. Murray (1984) *Biochem. Pharmacol.* **33**, 3277. **5**. Somasundaram, Norman & Mahaut-Smith (1995) *Biochem. J.* **309**, 725. **6**. Baune, Furlan, Taburet & Farinotti (1999) *Drug Metab. Dispos.* **27**, 565. **7**. Guengerich (1990) *Amer. J. Obstet. Gynecol.* **163**, 2159. **8**. Purba, Back & Breckenridge (1986) *J. Steroid Biochem.* **24**, 1091. **9**. Srivastava, Puri, Kamboj & Pandey (1999) *Trop. Med. Int. Health* **4**, 251. **10**. Thithapandha & Cohn (1978) *Biochem. Pharmacol.* **27**, 263. **11**. Fesen, Kohn, Leteurtre & Pommier (1993) *Proc. Natl. Acad. Sci. U.S.A.* **90**, 2399. **12**. Brossi, Millet, Landau, Bembenek & Abell (1987) *FEBS Lett.* **214**, 291. **13**. Kettle & Winterbourn (1991) *Biochem. Pharmacol.* **41**, 1485. **14**. Orta-Salazar, Bouchard, Morales-Salgado & Salinas-Stefanon (2002) *Brit. J. Pharmacol.* **135**, 751. **15**. Sides & Wittels (1975) *Biochem. Pharmacol.* **24**, 1246. **16**. Löffler & Kunze (1987) *FEBS Lett.* **216**, 51. **17**. Olenick & Hahn (1972) *Antimicrob. Agents Chemother.* **1**, 259. **18**. Wang, Ternai & Polya (1994) *Biol. Chem. Hoppe Seyler* **375**, 527. **19**. Zhao & Ishizaki (1997) *J. Pharmacol. Exp. Ther.* **283**, 1168. **20**. Back, Tjia, Karbwang & Colbert (1988) *Brit. J. Clin. Pharmacol.* **26**, 23.

Primisulfuron-methyl

This post-emergence herbicide (FW = 468.34 g/mol) inhibits acetolactate synthase and branched-chain amino acid biosynthesis (1-4). **1**. Zohar, Einav, Chipman & Barak (2003) *Biochim. Biophys. Acta* **1649**, 97. **2**. Choi, Noh, Choi, *et al.* (2006) *Bull. Korean Chem. Soc.* **27**, 1697. **3**. Yang & Kim (1997) *J. Biochem. Mol. Biol.* **30**, 13. **4**. Choi, Yu, Hahn, Choi & Yoon (2005) *FEBS Lett.* **579**, 4903.

Prinomastat

This antineoplastic agent (FW = 423.51 g/mol; NLM Unique Identifier C113282), also known as AG3340, inhibits a number of matrix metalloproteinases, including collagenase-3 (K_i = 38 pM), gelatinase A (K_i = 83 pM), stromelysin 1 (K_i = 0.27 nM), and interstitial collagenase (K_i = 8.2 nM). Note that both the 3- and 4-pyridyl structures have been reported for prinomastat; however, the 4-analogue is the original structure reported for this inhibitor. Target(s): ADAM 17 endopeptidase, or tumor necrosis factor-α converting enzyme, or TACE (1,2); ADAMTS-4 endopeptidase, or aggrecanase (3); collagenase-3 (2,4,5); gelatinase A (1,2,4,5); gelatinase B (4,5); interstitial collagenase (1,2,4,5); matrilysin (4,5); membrane-type matrix metalloproteinase-1, or matrix metalloproteinase 14 (4-6); and stromelysin 1 (2,4,5). **1.** Leung, Abbenante & Fairlie (2000) *J. Med. Chem.* **43**, 305. **2.** Holms, Mast, Marcotte, *et al.* (2001) *Bioorg. Med. Chem. Lett.* **11**, 2907. **3.** Sugimoto, Takahashi, Yamamoto, Shimada & Tanzawa (1999) *J. Biochem.* **126**, 449. **4.** Shalinsky, Brekken, Zou, *et al.* (1999) *Clin. Cancer Res.* **5**, 1905. **5.** Shalinsky, Brekken, Zou, *et al.* (1999) *Ann. N. Y. Acad. Sci.* **878**, 236. **6.** Rozanov, Ghebrehiwet, Postnova, *et al.* (2002) *J. Biol. Chem.* **277**, 9318.

Pristinamycin IA

This antibacterial agent (FW = 866.97 g/mol; CAS 3131-03-1), also known as streptogramin B, mikamycin IA, ostreogrycin B, synergistin B, and vernamycin Bα, is one of two components produced by *Streptomyces pristinaespiralis*, the other component being pristinamycin IIA, or virginiamycin M (1). This white powder inhibits protein biosynthesis by binding to the 50S ribosomal subunit. Greater inhibition is observed with ribosomes from Gram-positive microorganisms compared to those from Gram-negative bacteria. Target(s): peptidyltransferase (2); and protein biosynthesis (1-3). **1.** Jiménez (1976) *Trends Biochem. Sci.* **1**, 28. **2.** Porse & Garrett (1999) *J. Mol. Biol.* **286**, 375. **3.** Pestka (1974) *Meth. Enzymol.* **30**, 261.

Pritelivir

This antiviral (FW = 402.49 g/mol; CAS 348086-71-5), also named BAY 57-1293 and *N*-methyl-*N*-(4-methyl-5-sulfamoyl-1,3-thiazol-2-yl)-2-[4-(pyridin-2-yl)phenyl]acetamide, inhibits replication of herpes simplex virus (HSV) types 1 and 2 by blocking the enzymatic activity of the viral primase-helicase complex ($IC_{50} \approx 20$ nM)). Although pritelivir is highly active against HSV-1 and superior to acyclovir in Vero cells, drug resistance has been detected in two different strains of HSV at 10^{-4} to 10^{-5}, and resistant variants are present, even before the selection was applied (2). Resistant variants did not readily revert to a drug-sensitive phenotype in the absence of the inhibitor, and representative pritelivir-resistant variants were cross-resistant to BILS-22-BS, another helicase-primase inhibitor (HPI). Notably, variants resistant to BAY 57-1293 retained sensitivity to the nucleoside analogue, ACV (2). Two HSV-1 mutants, that are 100x and >3000x resistant to ritelivir, have substitutions in the UL5 helicase protein (*e.g.*, one with A4V and K356Q mutations, and the other with a single

G352R mutation) (3). Marker transfer experiments confirmed that K356Q and G352R (near the predicted functional domain of HSV-1 helicase)) are the drug-resistance mutations that are directly associated with differences in virus growth in tissue culture (3). Moreover, to achieve maximum potency, pritelivir interacts with both components (UL5 and UL52) of the helicase-primase complex, but BILS-22-BS does not (4). Herpes simplex type 1 (HSV1) infected cells form β-amyloid (Aβ) and abnormally phosphorylated Tau (P-Tau) *in vitro*, and pritelivir is more effective that acyclovir in preventing these Alzheimer-associated proteins (5). **1.** Betz, Fischer, Kleymann, Hendrix & Rübsamen-Waigmann (2002) *Antimicrob. Agents Chemother.* **46**, 1766. **2.** Biswas, Swift & Field (2007) *Antivir. Chem. Chemother.* **18**, 13. **3.** Biswas, Jennens & Field (2007) *Arch. Virol.* **152**, 1489. **4.** Biswas, Kleymann, Swift, *et al.* (2008) *J. Antimicrob. Chemother.* **61**, 1044. **5.** Wozniak, Frost & Itzhaki (2013) *Antiviral Res.* **99**, 401.

Privine Hydrochloride, *See Naphazoline*

[Pro3]-GIP (Mouse)

This GIPR antagonist (FW = 4971.62 g/mol; Sequence: YAPGTFISDY SIAMDKIRQQDFVNWLLAQRGKKSDWKHNITQ, noting Prolyl for Glutamyl substitution at Position-3; Solubility: 2 mg/mL in H_2O) inhibits Glucose-dependent Insulinotropic Polypeptide Receptor-mediated release of insulin from pancreatic β-cells *in vitro* (1). Although native GIP (also previously termed Gastric Inhibitory Polypeptide) is rapidly degraded by human plasma (~40 % remains after 8 hours), (Pro³)GIP is completely stable, even after 24 h. In CHL cells expressing the human GIP receptor, (Pro(3))GIP antagonizes the 3',5'-cyclicAMP stimulatory effects of 10^{-7} M native GIP (IC_{50} = 2.6 µM). In the clonal pancreatic β-cell line (BRIN-BD11), (Pro(3))GIP dose dependently inhibits GIP-stimulated insulin release by 1.2x to 1.7x over the 10^{-13}–10^{-8} M concentration range (1). In *ob/ob* mice, it blocks the effects of GIP on insulin release and plasma glucose levels. Also improves intraperitoneal glucose tolerance, insulin sensitivity, and glucose response to feeding in *ob/ob* mice (1). Early administration of GIP^{P3E} prevents development of diabetes and related metabolic abnormalities associated with genetically inherited obesity in *ob/ob* mice (2). **1.** Gault, O'Harte, Harriott & Flatt (2002) *Biochem. Biophys. Res. Commun.* **290**, 1420. **2.** Irwin, McClean, O'Harte, *et al.* (2007) *Diabetologia* **50**, 1532.

PRO 140

This first-in-class, humanized anti-CCR5 monoclonal antibody (MW ≈ 150,000) potently inhibits HIV-1 entry and replication in both primary peripheral blood mononuclear cells and primary macrophages at concentrations that do not affect CCR5's chemokine receptor activity (1). CCR5 is a member of G-protein-coupled, seven transmembrane segment receptors, also known as C-C motif chemokine receptor-5. Acute HIV infection almost always involves virus binding to the CCR5 co-receptor, and, during advanced HIV disease, the virus may shift to the alternate CXCR4 receptor, but half of the virus still binds to the CCR5 co-receptor (2). When tested as single intravenous doses (ranging up to 5 mg/kg of body weight or three subcutaneous doses of up to 324 mg), viral loads reached their nadir by post-treatment day-12 and remained significantly reduced through day-29, with no dose-limiting toxicity (3). Although intravenously administered HIV drugs are unlikely to rival their orally available, small-molecule counterparts (*e.g.*, aplaviroc, vivriviroc, maraviroc, and SCH-C, or SCH 351125), PRO 140 nonetheless promises to be of great prophylactic value in lowering the risk of HIV infection in rape victims as well in other clinical settings. PRO 140 will also have enduring value as a tool for investigating the cellular mechanisms uderlying HIV infection. **1.** Trkola, Ketas, Nagashima, *et al.* (2001) *J. Virol.* **75**, 579. **2.** Cilliers, Nhlapo, Coetzer, Orlovic *et al.* (2003) *J. Virol.* **77**, 4449. **3.** Jacobson, Lalezari, Thompson, *et al.* (2010) *Antimicrob Agents Chemother.* **54**, 4137.

Proadifen

This nonselective P450 inhibitor ($FW_{free-base}$ = 353.50 g/mol; CAS 302-33-0; 62-68-0 for the HCl salt), also known as SKF-525A, β-diethylaminoethyl 2,2-diphenylpropylacetate, and *N,N*-diethylaminoethyl 2,2-diphenylvalerate, blocks glibenclamide-sensitive K^+ channels, including ATP-sensitive inward rectifier potassium channel 8 (KIR6.1). As a nonselective inhibitor of cytochrome P450 systems, proadifen also inhibits drug metabolism, thereby potentiating the action of many pharmaceuticals. For example, proadifen inhibits the demethylation of 2-methoxyestrogens (1). **Target(s)**: aldehyde oxidase (2,3); alkanal monooxygenase, FMN-linked (4,5); aromatase, *or* CYP19 (6); bacterial luciferase, *or* alkanal monooxygenase, FMN-linked (4,5); benzoate 4-monooxygenase (45); bromomethane monooxygenase7; calcium influx8; *ent*-copalyl-diphosphate synthase (9); CYPC211 (10); cytochrome P450 (1,6,10-23); diamine *N*-acetyltransferase (34); *N,N*-dimethylaniline demethylase (24); ecdysone 20-monooxygenase (37); ethoxyresorufin *O*-deethylase (14); flavin-containing monooxygenase, *or* dimethylaniline-*N*-oxide aldolase (17,18); glyceollin synthase (42); 2-hydroxyisoflavanone synthase (41); 25-hydroxyvitamin D₃ 24-monooxygenase (24); isopentenyl-diphosphate Δ-isomerase, *or* isopentenyl-pyrophosphate Δ-isomerase (25,26); *ent*-kaurene synthase (9); leukotriene B₄ 20-hydroxylase (12); (*S*)-limonene 3-monooxygenase (44); (*S*)-limonene 6-monooxygenase (44); (*S*)-limonene 7-monooxygenase (44); methane monooxygenase (27); monoamine oxidase (28,29); NADPH:cytochrome P450 reductase (16); nitric-oxide synthase, neuronal (30); pentoxyresorufin *O*-depentylase (16); phenol *O*-methyltransferase (35); progesterone 11α-monooxygenase (38); steroid 11β-monooxygenase (40); steroid 21-monooxygenase, *or* CYP21A1, weakly inhibited (39); sterol 14-demethylase, *or* CYP51 (43); styrene monooxygenase (31); thiol *S*-methyltransferase (36,37); thromboxane production (32); and trimethylamine-oxide aldolase (33). **1.** Breuer & Knuppen (1969) *Meth. Enzymol.* **15**, 691. **2.** Yoshihara & Tatsumi (1985) *Arch. Biochem. Biophys.* **242**, 213. **3.** Robertson & Bland (1993) *Biochem. Pharmacol.* **45**, 2159. **4.** Hastings, Baldwin & Nicoli (1978) *Meth. Enzymol.* **57**, 135. **5.** Nealson & Hastings (1972) *J. Biol. Chem.* **247**, 888. **6.** Coulson, King & Wiseman (1984) *Trends Biochem. Sci.* **9**, 446. **7.** Colby, Dalton & Whittenbury (1975) *Biochem. J.* **151**, 459. **8.** Lee & Berkowitz (1975) *J. Pharmacol. Exp. Ther.* **198**, 347. **9.** Frost & West (1977) *Plant Physiol.* **59**, 22. **10.** Chang, Chen & Waxman (1995) *J. Pharmacol. Exp. Ther.* **274**, 270. **11.** Khan, Sood & O'Brien (1993) *Biochem. Pharmacol.* **45**, 439. **12.** Soberman & Okita (1988) *Meth. Enzymol.* **163**, 349. **13.** Grasdalen, Backstrom, Eriksson & Ehrenberg (1975) *FEBS Lett.* **60**, 294. **14.** Pennanen, Kojo, Pasanen, *et al.* (1996) *Hum. Exp. Toxicol.* **15**, 435. **15.** Chung & Cha (1997) *Biochem. Biophys. Res. Commun.* **235**, 685. **16.** Can-Eke, Puskullu, Buyukbingol & Iscan (1998) *Chem. Biol. Interact.* **113**, 65. **17.** Ziegler & Pettit (1966) *Biochemistry* **5**, 2932. **18.** Machinist, Orme-Johnson & Ziegler (1966) *Biochemistry* **5**, 2939. **19.** Stralka & Strobel (1989) *Cancer* **64**, 2111. **20.** Kamataki, Ando, Ishii & Kato (1980) *Jpn. J. Pharmacol.* **30**, 841. **21.** Freedman, Parker, Marinello, Gurtoo & Minowada (1979) *Cancer Res.* **39**, 4612. **22.** Anders & Mannering (1966) *Mol. Pharmacol.* **2**, 319. **23.** Abou-Donia & Menzel (1968) *Biochemistry* **7**, 3788. **24.** Bikle (1980) *Biochim. Biophys. Acta* **615**, 208. **25.** Zabkiewicz, Keates & Brooks (1969) *Phytochemistry* **8**, 2087. **26.** Banthorpe, Doonan & Gutowski (1977) *Arch. Biochem. Biophys.* **184**, 381. **27.** Tonge, Harrison & Higgins (1977) *Biochem. J.* **161**, 333. **28.** Pfeffer, Semmel & Schor (1968) *Enzymologia* **34**, 299. **29.** Carrano & Malone (1966) *J. Pharm. Sci.* **55**, 563. **30.** Hecker, Mulsch & Busse (1994) *J. Neurochem.* **62**, 1524. **31.** Salmona, Pachecka, Cantoni, *et al.* (1976) *Xenobiotica* **6**, 585. **32.** Boeynaems, Demolle & Van Coevorden (1986) *Prostaglandins* **32**, 145. **33.** Myers & Zatman (1971) *Biochem. J.* **121**, 10P. **34.** Seiler & al-Therib (1974) *Biochim. Biophys. Acta* **354**, 206. **35.** Axelrod & Daly (1968) *Biochim. Biophys. Acta* **159**, 472. **36.** Weinshilboum, Sladek & Klumpp (1979) *Clin. Chim. Acta* **97**, 59. **36.** Otterness, Keith, Kerremans & Weinshilboum (1986) *Drug Metab. Dispos.* **14**, 680. **37.** Feyereisen & Durst (1978) *Eur. J. Biochem.* **88**, 37. **38.** Jayanthi, Madyastha & Madyastha (1982) *Biochem. Biophys. Res. Commun.* **106**, 1262. **39.** Ryan & Engel (1957) *J. Biol. Chem.* **225**, 103. **40.** Yanagibashi, Haniu, Shively, Shen & Hall (1986) *J. Biol. Chem.* **261**, 3556. **41.** Kochs & Grisebach (1986) *Eur. J. Biochem.* **115**, 311. **42.** Welle & Grisebach (1988) *Arch. Biochem. Biophys.* **263**, 191. **43.** Aoyama, Yoshida & Sato (1984) *J. Biol. Chem.* **259**, 1661. **44.** Karp, Mihaliak, Harris & Croteau (1990) *Arch. Biochem. Biophys.* **276**, 219. **45.** McNamee & Durham (1985) *Biochem. Biophys. Res. Commun.* **129**, 485.

Probenecid

This prototypical uricosuric agent (FW$_{free-acid}$ = 285.36 g/mol; CAS 57-66-9; pK_a = 5.8), also known as 4-[(dipropylamino)sulfonyl]benzoic acid, inhibits the renal excretion of organic anions and reduces tubular reabsorption of urate. Probenecid has also been used to treat patients with renal impairment, and, because it reduces the renal tubular excretion of other drugs, has been used as an adjunct to antibacterial therapy. **Target(s)**: cadmium-transporting ATPase (1), cysteine-*S*-conjugate *N*-acetyltransferase, weakly inhibited (2), glucuronosyltransferase, *or* UDP-glucuronosyltransferase (3-7), glutathione *S*-transferase (8,9), glutathione-transporting ATPase (1), leukotriene-C₄ synthase (9), and organic anion transporter (10,11). **1.** Rebbeor, Connolly, Dumont & Ballatori (1998) *J. Biol. Chem.* **273**, 33449. **2.** Aigner, Jaeger, Pasternack, *et al.* (1996) *Biochem. J.* **317**, 213. **3.** Haumont, Magdalou, Lafaurie, *et al.* (1990) *Arch. Biochem. Biophys.* **281**, 264. **4.** Kirkwood, Nation & Somogyi (1998) *Clin. Exp. Pharmacol. Physiol.* **25**, 266. **5.** Sim, Back & Breckenridge (1991) *Brit. J. Clin. Pharmacol.* **32**, 17. **6.** Yue, von Bahr, Odar-Cederlof & Sawe (1990) *Pharmacol. Toxicol.* **66**, 221. **7.** Uchaipichat, Mackenzie, Guo, *et al.* (2004) *Drug Metab. Dispos.* **32**, 413. **8.** Kaplowitz, Clifton, Kuhlenkamp & Wallin (1976) *Biochem. J.* **158**, 243. **9.** Bach, Brashler & Morton (1984) *Arch. Biochem. Biophys.* **230**, 455. **10.** Burckhardt & Burckhardt (2003) *Rev. Physiol. Biochem. Pharmacol.* **146**, 95. **11.** Cunningham, Israili & Dayton (1981) *Clin. Pharmacokinet.* **6**, 135.

Probestin

This aminopeptidase inhibitor (FW = 502.61 g/mol; CAS 123652-87-9), also known as *N*-[(2*S*,3*R*)-3-amino-2-hydroxy-4-phenylbutanoyl]-L-leucyl-L-prolyl-L-proline, from a strain of *Streptomyces azureus* inhibits dipeptidyl-peptidase III (1), glutamyl aminopeptidase, *or* aminopeptidase A (2), membrane alanyl aminopeptidase, aminopeptidase M, *or* aminopeptidase N, K_i = 19 nM (2-9); Xaa-Trp aminopeptidase, *or* aminopeptidase W (2,10,11). **1.** Abramic, Schleuder, Dolovcak, *et al.* (2000) *Biol. Chem.* **381**, 1233. **2.** Tieku & Hooper (1992) *Biochem. Pharmacol.* **44**, 1725. **3.** Turner (1998) in *Handb. Proteolytic Enzymes* (Barrett, Rawlings & Woessner, eds.), p. 996, Academic Press, San Diego. **4.** Lendeckel, Kahne, Arndt, Frank & Ansorge (1998) *Biochem. Biophys. Res. Commun.* **252**, 5. **5.** Hua, Tsukamoto, Taguchi, *et al.* (1998) *Biochim. Biophys. Acta* **1383**, 301. **6.** Yoshida, Nakamura, Naganawa, Aoyagi & Takeuchi (1990) *J. Antibiot.* **43**, 149. **7.** Aoyagi, Yoshida, Nakamura, *et al.* (1990) *J. Antibiot.* **43**, 143. **8.** Xu & Li (2005) *Curr. Med. Chem. Anticancer Agents* **5**, 281. **9.** Bauvois & Dauzonne (2006) *Med. Res. Rev.* **26**, 88. **10.** Hooper (1998) in *Handb. Proteolytic Enzymes* (Barrett, Rawlings & Woessner, eds.), p. 1513, Academic Press, San Diego. **11.** Tieku & Hooper (1993) *Biochem. Soc. Trans.* **21**, 250S.

Probucol

This is an orally active cholesterol-reducing agent (FW = 516.85 g/mol; CAS 23288-49-5), also known as Lorelco, reduces total cholesterol and low-density-lipoprotein cholesterol levels and also serves as an free radical-scavenging antioxidant. **Target(s)**: Ca²⁺-sensitive K⁺ channel (1); lipid efflux (2); low density lipoprotein oxidation (3,4); and retinyl ester hydrolase (5). **1.** Howland, Daughtey, Donatelli & Theofrastous (1984) *Pharmacol. Res. Commun.* **16**, 1057. **2.** Tsujita & Yokoyama (1996) *Biochemistry* **35**, 13011. **3.** Hanna, Feller, Witiak & Newman (1993) *Biochem. Pharmacol.* **45**, 753. **4.** Parthasarathy, Young, Witztum, Pittman & Steinberg (1986) *J. Clin. Invest.* **77**, 641. **5.** Schindler (2001) *Lipids* **36**, 543.

Procainamide

This local anesthetic and cardioactive drug (FW$_{\text{free-base}}$ = 235.33 g/mol; CAS 51-06-9; 614-39-1 for HCl salt), also known as 4-amino-N-[2-(diethylamino)ethyl]benzamide and pronestyl, is a Na$^+$ channel blocker that is a widely used antiarrhythmic agent. Along with hydralazine, RG108, procaine, IM25, and disulfiram, it is also a potent nonnucleoside DNA methylase inhibitor. **Key Pharmacokinetic Parameters:** *See* Appendix II in Goodman & Gilman's THE PHARMACOLOGICAL BASIS OF THERAPEUTICS, 12th Edition (Brunton, Chabner & Knollmann, eds.) McGraw-Hill Medical, New York (2011). **Target(s):** acetylcholinesterase (1); ATPase, mitochondrial (2); cocaine degradation (3); DNA methyltransferase (4,5); F$_1$ ATPase, H$^+$-transporting two-sector ATPase (6); K$^+$ channel (7,8); lactate dehydrogenase, weakly inhibited (9); Na$^+$ channel (7,8); and Na$^+$/K$^+$-exchanging ATPase (10-13). **1.** Talesa, Romani, Rosi & Giovanni (1996) *Eur. J. Biochem.* **238**, 538. **2.** Dzimiri (1993) *Res. Commun. Chem. Pathol. Pharmacol.* **80**, 121. **3.** Bailey (1999) *J. Anal. Toxicol.* **23**, 173. **4.** Quddus, Johnson, Gavalchin, *et al.* (1993) *J. Clin. Invest.* **92**, 38. **5.** Scheinbart, Johnson, Gross, Edelstein & Richardson (1991) *J. Rheumatol.* **18**, 530. **6.** Chazotte, Vanderkooi & Chignell (1982) *Biochim. Biophys. Acta* **680**, 310. **7.** Ridley, Milnes, Benest, *et al.* (2003) *Biochem. Biophys. Res. Commun.* **306**, 388. **8.** Sakuta, Okamoto & Watanabe (1992) *Brit. J. Pharmacol.* **107**, 1061. **9.** Dzimiri & Almotrefi (1993) *Clin. Exp. Pharmacol. Physiol.* **20**, 201. **10.** Kutchai, Geddis & Farley (2000) *Pharmacol. Res.* **41**, 1. **11.** Almotrefi, Basco, Moorji & Dzimiri (1999) *Can. J. Physiol. Pharmacol.* **77**, 866. **12.** Dzimiri & Almotrefi (1991) *Arch. Int. Pharmacodyn. Ther.* **314**, 34. **13.** Dzimiri & Almotrefi (1991) *Gen. Pharmacol.* **22**, 403.

Procaine

This local anesthetic and vasodilator (FW$_{\text{free-base}}$ = 236.31 g/mol; CAS 59-46-1; 51-05-8 for the HCl salt), also known as 4-aminobenzoic acid 2-(diethylamino)ethyl ester, is a Na$^+$ channel blocker and calmodulin antagonist. It also constricts blood vessels, reducing bleeding. The hydrochloride is Novocain, which is highly water-soluble (~1 g/mL). Procaine is hydrolyzed by esterases to *p*-aminobenzoate and diethylaminoethanol, and is somewhat unstable in solution and is slowly oxidized in the presence of heavy metal ions and alkali. Procaine was first synthesized in 1905 by German chemist Alfred Einhorn, who named it Novocain (<u>nov</u>el + <u>cocaine</u>). Prior to its discovery, cocaine was the most commonly used local anesthetic. Procaine became the first injectable man-made local anesthetic; epinephrine was often co-administered to counteract its vasodilator effects. Along with hydralazine, RG108, procainamide, IM25, and disulfiram, procaine is also a potent nonnucleoside DNA methylase inhibitor. **Target(s):** acetylcholine receptor (1-3); acetylcholinesterase (4,5,28); acid phosphatase (26); 1-acylglycerophosphocholine *O*-acyltransferase, *or* lysophosphatidylcholine *O*-acyltransferase, *or* alkanal monooxygenase (7); bacterial luciferase, *or* alkanal monooxygenase (7); calmodulin (8); calmodulin stimulation of Ca^{2+}-dependent ATPase (8); calmodulin stimulation of cyclic nucleotide phosphodiesterase (8); cholesterol esterase, weakly inhibited (9); cholinesterase, *or* butyrylcholinesterase (10,27); electron transport, mitochondrial, *or* cytochrome *c* oxidase, *or* durohydroquinone oxidase, *or* succinate oxidase, *or* NADH oxidase, *or* succinate dehydrogenase, *or* succinate: cytochrome *c* oxidoreductase, *or* NADH:cytochrome *c* oxidoreductase (11); F$_1$ ATPase, *or* H$^+$-trransporting two-sector ATPase (12); histamine *N*-methyltransferase, mildly inhibited (30); lecithin:cholesterol acyltransferase, *or* LCAT, weakly inhibited (13); *N*-methyl-D-aspartate (NMDA) receptor (14,15); monoamine oxidase, weakly inhibited (16-18); Na$^+$ channel (19,20); Na$^+$/K$^+$-exchanging ATPase, weakly inhibited (21-23); phospholipase A$_2$, weakly inhibited (24,25); sterol *O*-acyltransferase, *or* cholesterol *O*-acyltransferase, *or* ACAT, weakly inhibited (6). **1.** Yost & Dodson (1993) *Cell. Mol. Neurobiol.* **13**, 159. **2.** Forman & Miller (1989) *Biochemistry* **28**, 1678. **3.** Shiono, Takeyasu, Udgaonkar, *et al.* (1984) *Biochemistry* **23**, 6889. **4.** al-Jafari, Kamal,

Duhaiman & Alhomida (1996) *J. Enzyme Inhib.* **11**, 123. **5.** Marquis (1983) *Comp. Biochem. Physiol. C* **74**, 119. **6.** Bell & Hubert (1980) *Biochim. Biophys. Acta* **619**, 302. **7.** Harvey (1951) *The Enzymes*, 1st ed. (Sumner & Myrbäck, eds.), **2** (part 1), 581. **8.** Volpi, Sha'afi, Epstein, Andrenyak & Feinstein (1981) *Proc. Natl. Acad. Sci. U.S.A.* **78**, 795. **9.** Traynor & Kunze (1975) *Biochim. Biophys. Acta* **409**, 68. **10.** Augustinsson (1950) *The Enzymes*, 1st. ed. (Sumner & Myrbäck, eds.), **1** (part 1), 443. **11.** Chazotte & Vanderkooi (1981) *Biochim. Biophys. Acta* **636**, 153. **12.** Chazotte, Vanderkooi & Chignell (1982) *Biochim. Biophys. Acta* **680**, 310. **13.** Bell & Hubert (1980) *Lipids* **15**, 811. **14.** Sugimoto, Uchida & Mashimo (2003) *Brit. J. Pharmacol.* **138**, 876. **15.** Nishizawa, Shirasaki, Nakao, Matsuda & Shingu (2002) *Anesth. Analg.* **94**, 325. **16.** Fuller & Roush (1977) *J. Amer. Geriatr. Soc.* **25**, 90. **17.** MacFarlane (1975) *Fed. Proc.* **34**, 108. **18.** Yasuhara, Wada, Sakamoto & Kamijo (1982) *Jpn. J. Pharmacol.* **32**, 213. **19.** Hille (1976) *Annu. Rev. Physiol.* **38**, 139. **20.** Berde & Strichartz (1999) in *Anesthesia* (Miller, ed.), 491, Churchill Livingstone, Philadelphia. **21.** Kutchai & Geddis (2001) *Pharmacol. Res.* **43**, 399. **22.** Kutchai, Geddis & Farley (2000) *Pharmacol. Res.* **41**, 1. **23.** Hudgins & Bond (1984) *Biochem. Pharmacol.* **33**, 1789. **24.** Hendrickson & van Dam-Mieras (1976) *J. Lipid Res.* **17**, 399. **25.** Blackwell, Flower, Nijkamp & Vane (1978) *Brit. J. Pharmacol.* **62**, 79. **26.** Lisa, Garrido & Domenech (1984) *Mol. Cell. Biochem.* **63**, 113. **27.** Nagasawa, Sugisaki, Tani & Ogata (1976) *Biochim. Biophys. Acta* **429**, 817. **28.** Duhaiman, Alhomida, Rabbani, Kamal & al-Jafari (1996) *Biochimie* **77**, 46. **29.** Sanjanwala, Sun & MacQuarrie (1989) *Arch. Biochem. Biophys.* **271**, 407. **30.** Thithapandha & Cohn (1978) *Biochem. Pharmacol.* **27**, 263.

Prochloraz

This imidazole fungicide (FW = 376.67 g/mol; CAS 67747-09-5; M.P. = 38.5-41°C) is widely applied to packaging used for cereals, fruits, and vegetables. Prochloraz exerts multiple mechanisms of action *in vitro*, acting to antagonize androgen and oestrogen receptors, to agonize aryl hydrocarbon receptors, and to inhibit aromatase activity. **Target(s):** aldrin epoxidase (1); aromatase, *or* CYP19 (2,3); (*S*)-canadine synthase (4); cytochrome P450 (2,3,5-8); 7-ethoxyresorufin-*O*-deethylase (1,7); methyltetrahydro-protoberberine 14-monooxygenase (9); protopine 6-monooxygenase (10); salutaridine synthase (11); steroid 16α-monooxygenase (2); steroid 17α-monooxygenase (2); steroid 21-monooxygenase (2); steroid 5α-reductase (8); sterol 14-demethylase, *or* CYP51 (12,13). **1.** Snegaroff & Bach (1989) *Xenobiotica* **19**, 255. **2.** Mason, Carr & Murry (1987) *Steroids* **50**, 179. **3.** Vinggaard, Hnida, Breinholt & Larsen (2000) *Toxicol. In Vitro* **14**, 227. **4.** Rueffer & Zenk (1994) *Phytochemistry* **36**, 1219. **5.** Riviere, Leroux, Bach & Gredt (1984) *Pestic. Sci.* **15**, 317. **6.** Antignac, Koch, Delaforge & Narbonne (1991) *Xenobiotica* **21**, 669. **7.** Laignelet, Narbonne, Lhuguenot & Riviere (1989) *Toxicology* **59**, 271. **8.** Lo, King, Alléra & Klingmüller (2007) *Toxicol. In Vitro* **21**, 502. **9.** Rueffer & Zenk (1987) *Tetrahedron Lett.* **28**, 5307. **10.** Tanahashi & Zenk (1990) *Phytochemistry* **29**, 1113. **11.** Gerardy & Zenk (1993) *Phytochemistry* **32**, 79. **12.** Dyer, Hansen, Delaney & Lucas (2000) *Appl. Environ. Microbiol.* **66**, 4599. **13.** Trösken, Adamska, Arand, *et al.* (2006) *Toxicology* **228**, 24.

Prochlorperazine

This antipsychotic drug (FW$_{\text{free-base}}$ = 373.95 g/mol; CAS 58-38-8), systematically named as 2-chloro-10-(3-(4-methyl-1-piperazinyl)propyl)-10*H*-phenothiazine, is a D$_2$ antagonist that is used in the treatment of dizziness and vertigo. Prochlorperazine inhibits adenylate cyclase (1-3). Prochlorperazine also exhibits antiemetic properties. **1.** Iversen (1975) *Science* **188**, 1084. **2.** Iversen (1976) *Trends Biochem. Sci.* **1**, 121. **3.** Wolff & Jones (1970) *Proc. Natl. Acad. Sci. U.S.A.* **65**, 454.

Proscillaridin A

This cardiac glycoside (FW = 530.65 g/mol; CAS 466-06-8) was identified as an inhibitor of HIF-1α protein translation and HIF-2α mRNA expression in a cell-based screen of a library of drugs (at 50 nM in the case of Proscillaridin A) currently in clinical trials (*See also Digoxin; Oabain*). 1. Zhang, Qian, Tan, *et al.* (2008) *Proc. Natl. Acad. Sci. USA* **105**, 19579.

Procion Blue H-B

This reactive dye (FW$_{trisodium-salt}$ = 840.11 g/mol; λ$_{max}$ = 607 nm), also known as Reactive $_{Blue}$ 2 and Cibacron Blue F3G-A, is a sulfonated triazine often used to probe nucleotide binding sites. Procion Blue H-B is also a P2Y purinoceptor antagonist and is one of the most potent antagonists for ATP-activated channels (*See also Cibacron Blue*). When immobilized on a support matrix, this triazine can be used to purify proteins that bind nucleotides or nucleotide-containing coenzymes. *Note*: Given its tendency to bind in place of nucleotides, this agent is apt to have inhibitory effects beyond those listed below. Target(s): aldehyde reductase (1); alkaline phosphatase (2); ATP-activated channels (3); carbonyl reductase (4); choline *O*-acetyltransferase (5); choline-phosphate cytidylyltransferase (6); coenzyme F$_{420}$ hydrogenase (7); ferredoxin:NADP$^+$ oxidoreductase (8); glucose-6-phosphate dehydrogenase (9); glutathione *S*-transferase (10); hexokinase (11); NAD$^+$ nucleosidase, *or* NADase, *or* NAD$^+$ glycohydrolase (12); nucleotide diphosphatase (13); phosphoglycerate kinase (14); P2Y purinoceptor (15,16); thymidylate synthase (17); and UDP-glucose 4-epimerase (18). 1. Turner & Hryszko (1980) *Biochim. Biophys. Acta* **613**, 256. 2. Kirchberger, Seidel & Kopperschlager (1987) *Biomed. Biochim. Acta* **46**, 653. 3. Bean (1992) *Trends Pharmacol. Sci.* **13**, 87. 4. Wermuth (1981) *J. Biol. Chem.* **256**, 1206. 5. Polsky & Shuster (1976) *Biochim. Biophys. Acta* **445**, 25. 6. Hunt & Postle (1986) *Biochem. Soc. Trans.* **14**, 1279. 7. Livingston, Fox, Orme-Johnson & Walsh (1987) *Biochemistry* **26**, 4228. 8. Carrillo & Vallejos (1983) *Biochim. Biophys. Acta* **742**, 285. 9. Reuter, Metz, Lorenz & Kopperschlager (1990) *Biomed. Biochim. Acta* **49**, 151. 10. Ajele & Afolayan (1992) *Comp. Biochem. Physiol. B* **103**, 47. 11. Clonis, Goldfinch & Lowe (1981) *Biochem. J.* **197**, 203. 12. Yost & Anderson (1981) *J. Biol. Chem.* **256**, 3647. 13. Grobben, Claes, Roymans, *et al.* (2000) *Brit. J. Pharmacol.* **130**, 139. 14. Beissner & Rudolph (1979) *J. Biol. Chem.* **254**, 6273. 15. Inoue, Nakazawa, Ohara-Imaizumi, *et al.* (1991) *Brit. J. Pharmacol.* **102**, 851. 16. Nakazawa, Inoue, Fujimori & Takanaka (1991) *Pflugers Arch.* **418**, 214. 17. Nakata, Tsukamoto, Miyoshi & Kojo (1987) *Biochim. Biophys. Acta* **924**, 297. 18. Samanta & Bhaduri (1982) *Biochim. Biophys. Acta* **707**, 129.

Procion Red HE-3B

This triazine dye (FW$_{free-acid}$ = 1338.11 g/mol), which inhibits several enzymes, is frequently immobilized on a polymeric matrix and used in dye-affinity chromatography in protein purification. *Note*: Given its tendency to bind in place of nucleotides, this agent is apt to have inhibitory effects beyond those listed below. Target(s): aldehyde reductase, *or* alcohol dehydrogenase, NADP$^+$-dependent (1); alkaline phosphatase (2); ferredoxin hydrogenase (3); ferredoxin:NADP$^+$ reductase (4,5); glucose-6-phosphate dehydrogenase (6); hexokinase (7); and 6-phosphogluconate dehydrogenase (8). 1. Turner & Hryszko (1980) *Biochim. Biophys. Acta* **613**, 256. 2. Kirchberger, Seidel & Kopperschlager (1987) *Biomed. Biochim. Acta* **46**, 653. 3. Schneider, Pinkwart & Jochim (1983) *Biochem. J.* **213**, 391. 4. Carrillo & Vallejos (1983) *Biochim. Biophys. Acta* **742**, 285. 5. Levy & Bets (1988) *Biochim. Biophys. Acta* **955**, 236. 6. Reuter, Metz, Lorenz & Kopperschlager (1990) *Biomed. Biochim. Acta* **49**, 151. 7. Clonis, Goldfinch & Lowe (1981) *Biochem. J.* **197**, 203. 8. Qadri & Dean (1980) *Biochem. J.* **191**, 53.

Procion Red MX-2B

This dye (FW$_{free-acid}$ = 648.39 g/mol), also known as Reactive Red 1 and Procion Brilliant Red 2B, inhibits several enzymes. *Note*: Given its tendency to bind in place of nucleotides, this agent is apt to have inhibitory effects beyond those listed below. Target(s): 2,4-dichlorophenol hydroxylase (1); ferredoxin:NADP$^+$ oxidoreductase (2); glutamate carboxypeptidase, *or* carboxypeptidase G-2 (3); and ribulose-5-phosphate kinase (4). 1. Radjendirane, Bhat & Vaidyanathan (1991) *Arch. Biochem. Biophys.* **288**, 169. 2. Levy & Betts (1988) *Biochim. Biophys. Acta* **955**, 236. 3. Hughes, Sherwood & Lowe (1984) *Eur. J. Biochem.* **144**, 135. 4. Ashton (1984) *Biochem. J.* **217**, 79.

Proctolin

This insect neuromodulator/neurotransmitter/ neurohormone (FW = 648.76 g/mol; CAS : 100930-02-7; *Sequence*: L-Arg-L-Tyr-L-Leu-L-Pro-L-Thr) exerts its proctolinergic action by changing the way impulses are transmitted across a synapse, often assisting a more common neurotransmitter, such as glutamate. **Mode of Action:** Proctolin potently stimulates visceral and skeletal muscles contraction of a number of in insects. The *Drosophila melanogaster* proctolin receptor (CG6986) is an orphan G-protein coupled receptor and is strongly expressed in the head, the larval hindgut, the aorta and on neuronal endings in adult hearts. Unlike classical neurotransmitters, proctolin does not change the postsynaptic conductance. It also modulates reproductive tissue, stimulating contractions of the oviducts in the American cockroach *Periplaneta americana*, as well as *Leucophaea maderae*, *L. migratoria*, and spermathecae in *L. migratoria* and *Rhodnius prolixus*. Proctolin also speeds up heart rate in some insects. **Target(s);** dipeptidyl-peptidase III, *or* aminoenkephalinase (1,2). 1. Hui, Hui, Ling & Lajtha (1985) *Life Sci.* **36**, 2309. 2. Abramic, Schleuder, Dolovcak, *et al.* (2000) *Biol. Chem.* **381**, 1233.

Procyanidins

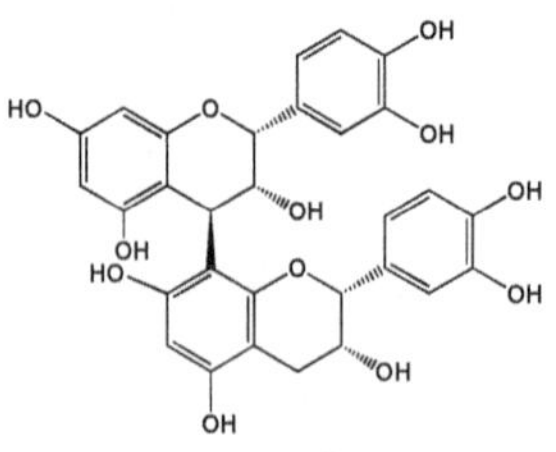

Procyanidin B-2

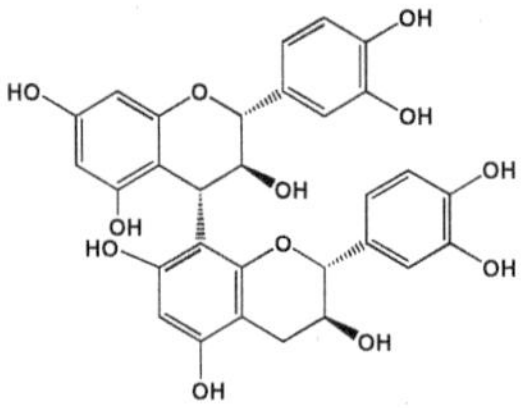

Procyanidin B-3

These redox-active catechin oligomers (procyanidin B-2; FW = 575.50 g/mol; procyanidin B-3; FW = 575.50 g/mol) exhibit potent antioxidant activity. After lignin, they are among the most abundant natural phenolics. Fourteen dimeric, eleven trimeric, and one tetrameric procyanidins alreadyt identified within grape seeds. Target(s): 1-alkylglycerophosphocholine O-acetyltransferase, inhibited by procyanidin B-2 3,3'-di-O-gallate and procyanidin B-5 3,3'-di-O-gallate (1); 5-lipoxygenase, or arachidonate 5-lipoxygenase, inhibited by di- and trimeric procyanidins (2); 12-lipoxygenase, or arachidonate 12-lipoxygenase, inhibited most by decameric procyanidins (3); 15-lipoxygenase, or arachidonate 15-lipoxygenase, inhibited most by decameric procyanidins (3); myosin-light-chain kinase (4); peptidyl-dipeptidase A, or angiotensin I-converting enzyme (5,6); protein kinase A (4); protein kinase C, inhibited by procyanidin B-2 and procyanidin C-1 (4,7); squalene monooxygenase, or squalene epoxidase, inhibited by procyanidin B-2 3,3'-di-O-gallate and procyanidin B-5 3,3'-di-O-gallate (8); tyrosinase, or monophenol monooxygenase, weakly inhibited (9). 1. Sugatani, Fukazawa, Ujihara, et al. (2004) Int. Arch. Allergy Immunol. 134, 17. 2. Schewe, Kuhn & Sies (2002) J. Nutr. 132, 1825. 3. Schewe, Sadik, Klotz, et al. (2001) Biol. Chem. 382, 1687. 4. Wang, Foo & Polya (1996) Phytochemistry 43, 359. 5. Uchida, Ikari, Ohta, et al. (1987) Jpn. J. Pharmacol. 43, 242. 6. Actis-Goretta, Ottaviani, Keen & Fraga (2003) FEBS Lett. 555, 597. 7. Takahashi, Kamimura, Shirai & Yokoo (2000) Skin Pharmacol. Appl. Skin Physiol. 13, 133. 8. Abe, Seki, Noguchi & Kashiwada (2000) Planta Med. 66, 753. 9. Momtaz, Mapunya, Houghton, et al. (2008) J. Ethnopharmacol. 119, 507.

Prodigiosin

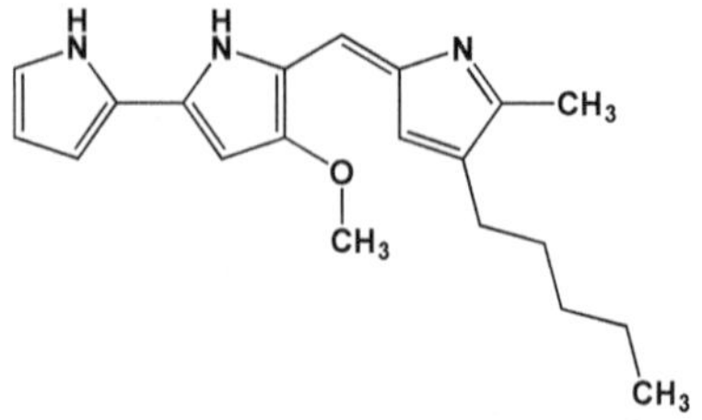

This red bacterial pigment (FW$_{free-base}$ = 323.44 g/mol; melting point = 151-152°C) is an antibiotic produced by Serratia marcescens (formerly Chromobacterium prodigiosum). Prodigiosin is the parent member of a class of pyrrole-containing natural products that exhibit immunosuppressive, cytotoxic, and apoptotic activity and can facilitate copper-promoted oxidative double-strand DNA cleavage through reductive activation of Cu^{2+}. They also uncouple lysosomal vacuolar-type ATPases by promoting H^{+}/Cl^{-} symport activity. The solid pigment is dark red and displays a green reflex. Acidic aqueous solutions are also red, but neutral or alkaline solutions are orange-yellow. Target(s): DNA topoisomerase I (1); DNA topoisomerase II (1); F$_{o}$F$_{1}$ ATPase, or H^{+}-transporting two-sector ATPase, as an uncoupler (2); H^{+}/K^{+}-ATPase, as an uncoupler (3); proton translocation, as an uncoupler (4); and Serratia marcescens nuclease (5). 1. Montaner, Castillo-Avila, Martinell, et al. (2005) Toxicol. Sci. 85, 870. 2.

Konno, Matsuya, Okamoto, et al. (1998) J. Biochem. 124, 547. 3. Matsuya, Okamoto, Ochi, et al. (2000) Biochem. Pharmacol. 60, 1855. 4. Sato, Konno, Tanaka, et al. (1998) J. Biol. Chem. 273, 21455. 5. Insupova, Kireeva, Beliaeva, Vinogradova & Gareishina (1977) Mikrobiologiia 46, 245.

Prodipine

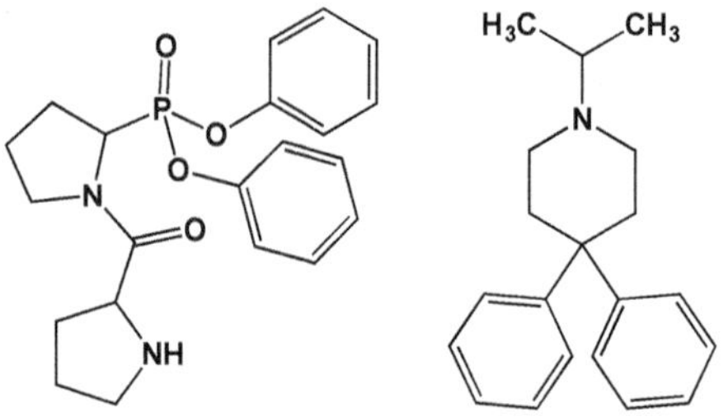

This name refers to two distinctly different molecular entities: the phosphonate (above left; FW$_{free-base}$ = 400.41 g/mol), also known as diphenyl 1-(S)-prolylpyrrolidine-2-(R,S)-phosphonate and Pro-Pro-diphenyl-phosphonate, that inhibits CYP2D6 (1) and dipeptidyl peptidase IV (2); and the antiparkinson drug (above right, FW$_{free-base}$ = 279.43 g/mol), also known as 1-(1-methylethyl)-4,4-diphenylpiperidine, that inhibits CYP2D6 and monoamine oxidase (3). 1. de Groot, Bijloo, van Acker, et al. (1997) Xenobiotica 27, 357. 2. De Meester, Belyaev, Lambeir, et al. (1997) Biochem. Pharmacol. 54, 173. 3. Planz, Palm & Quiring (1973) Arzneimittelforschung 23, 281.

Profenofos

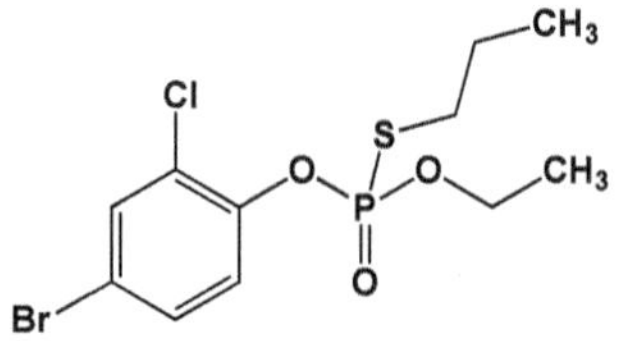

This broad-spectrum thiophosphorotriester insecticide (FW = 373.64 g/mol; CAS 41198-08-7), also known as Curacron™ and Selecron™™ and named systematically as O-(4-bromo-2-chlorophenyl) O-ethyl S-propyl phosphorothioate, inhibits cholinesterase's and other hydrolases. After feeding on a treated plant or crawling over a treated leaf, the pest is first paralyzed and then quickly succumbs. In vitro, the (–)-isomer is 34x more active as a cholinesterase inhibitor. Chromosome aberrations in spermatogonia and sperm abnormalities have been reported in Curacron-treated mice. Target(s): acetylcholinesterase (1-4); acylglycerol lipase, or monoacylglycerol lipase (4); 1-alkyl-2-acetylglycerophosphocholine esterase, or platelet-activating-factor acetylhydrolase (5); butyrylcholinesterase (1); carboxylesterase (1,4); fatty acid amide hydrolase, or anandamide amidohydrolase (4); lysophospholipase (4); and neuropathy target esterase (4). 1. Leader & Casida (1982) J. Agric. Food Chem. 30, 546. 2. El-Sebae, Enan, Soliman, El-Fiki & Khamees (1981) J. Environ. Sci. Health B 16, 475. 3. Nillos, Rodriguez-Fuentes, Gan & Schlenk (2007) Environ. Toxicol. Chem. 26, 1949. 4. Quistad, Klintenberg, Caboni, Liang & Casida (2006) Toxicol. Appl. Pharmacol. 211, 78. 5. Quistad, Fisher, Owen, Klintenberg & Casida (2005) Toxicol. Appl. Pharmacol. 205, 149.

Proflavine

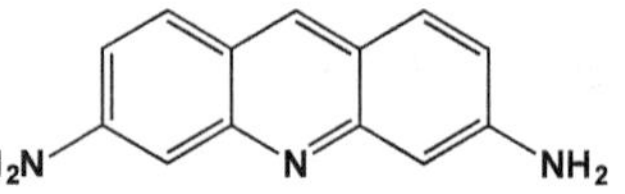

This dye (FW$_{free-base}$ = 209.25 g/mol; CAS 92-62-6; orange-red solid; M.P. = 281 or 288°C; soluble in water; pK_a = 9.65; solutions are light sensitive; insoluble in organic solvents such as benzene, diethyl ether, and chloroform), also known as proflavin and 3,6-acridinediamine, is a mutagen that intercalates in double-stranded DNA, inhibiting both DNA and RNA biosynthesis. Proflavine can also participate in the photooxidation of proteins (e.g., dopa decarboxylase (1)). In addition, proflavine can inhibit a number of enzymes directly. For xample scriflavine, a mixture of proflavine with 3,6-diamino-10-methylacridinium chloride, inhibits protein kinase C. Target(s): L-amino-acid oxidase (2); carbonic anhydrase, or carbonate dehydratase, $K_i \approx$ 10 mM (3); chymotrypsin (4-7,25); DNA (cytosine-5-)-

methyltransferase (8); DNA-directed RNA polymerase (9,29); dopa decarboxylase (1); F_1F_o ATPase, *or* F_1F_o ATP synthase, *or* H$^+$-transporting two-sector ATPase (10); ficain, *or* ficin (11); glutamate decarboxylase (12); lysozyme (13); monoamine oxidase (14); NADH:cytochrome b_5 reductase (15); nucleoside-triphosphatase, nuclear-envelope (16,26); papain (11); poly(ADP-ribose) glycohydrolase (17); polynucleotide adenylyltransferase, *or* poly(A) polymerase (27,28); pyruvate kinase (18); RNA-directed DNA polymerase (19); thrombin (20); tRNA adenylytltransferase (22); tRNA (guanine-N^2-)-methyltransferase (30); tRNA cytidylyltransferase (22); tRNA methyltransferases (8,21); tRNA nucleotidyltransferase (22); and trypsin (4,7,23,24). **1**. Pasqua, Dominici, Murgia, Poletti & Borri Voltattorni (1984) *Biochem. Int.* **9**, 437. **2**. Singer & Kearney (1950) *Arch. Biochem.* **27**, 348. **3**. Pocker & Stone (1968) *Biochemistry* **7**, 2936. **4**. Markland, Jr., & Smith (1971) *The Enzymes*, 3rd ed. (Boyer, ed.), **3**, 561. **5**. Wallace, Kurtz & Niemann (1963) *Biochemistry* **2**, 824. **6**. Glazer (1968) *J. Biol. Chem.* **243**, 3693. **7**. Feinstein & Feeney (1967) *Biochemistry* **6**, 749. **8**. Kerr & Borek (1973) *The Enzymes*, 3rd ed. (Boyer, ed.), **9**, 167. **9**. Chamberlin (1974) *The Enzymes*, 3rd ed. (Boyer, ed.), **10**, 333. **10**. Mai & Allison (1983) *Arch. Biochem. Biophys.* **221**, 467. **11**. Hall & Anderson (1974) *Biochemistry* **13**, 2082 and 2087. **12**. Cozzani & Jori (1980) *Biochim. Biophys. Acta* **623**, 84. **13**. Imoto, Johnson, North, Phillips & Rupley (1972) *The Enzymes*, 3rd ed. (Boyer, ed.), **7**, 665. **14**. Gorkin, Komisarova, Lerman & Veryovkina (1964) *Biochem. Biophys. Res. Commun.* **15**, 383. **15**. Yubisui & Takeshita (1980) *J. Biol. Chem.* **255**, 2454. **16**. Agutter, Cockrill, Lavine, McCaldin & Sim (1979) *Biochem. J.* **181**, 647. **17**. Tavassoli, Tavassoli & Shall (1985) *Biochim. Biophys. Acta* **827**, 228. **18**. McKellar & Kushner (1977) *Can. J. Biochem.* **55**, 618. **19**. Hirschman (1971) *Trans. N. Y. Acad. Sci.* **33**, 595. **20**. Lundblad, Kingdon & Mann (1976) *Meth. Enzymol.* **45**, 156. **21**. Kerr (1974) *Meth. Enzymol.* **29**, 716. **22**. Girgenti, Whitford, Jr., & Cory (1976) *Enzyme* **21**, 225. **23**. Walsh (1970) *Meth. Enzymol.* **19**, 41. **24**. Antonini, Ascenzi, Bolognesi, Menegatti & Guarneri (1983) *J. Biol. Chem.* **258**, 4676. **25**. Shiao & Sturtevant (1969) *Biochemistry* **8**, 4910. **26**. Schröder, Rottmann, Bachmann & Müller (1986) *J. Biol. Chem.* **261**, 663. **27**. Kurl, Holmes, Verney & Sidransky (1988) *Biochemistry* **27**, 8974. **28**. Tsiapalis, Dorson & Bollum (1975) *J. Biol. Chem.* **250**, 4486. **29**. Sethi (1971) *Prog. Biophys. Mol. Biol.* **23**, 67. **30**. Taylor & Gantt (1979) *Biochemistry* **18**, 5253.

Profluralin

This pre-planting herbicide and anti-mitotic agent (FW = 347.29 g/mol; CAS 26399-36-0; M.P. = 33-36°C), also known as Tolban® and named systematically as *N*-(cyclopropylmethyl)-2,6-dinitro-*N*-propyl-4-(trifluoromethyl)benzamine aids in controlling such weeds as barnyard grass, panicums, foxtails, goosegrass, crabgrass, witchgrass, pigweed, Florida pursley, kochia, pigweed, purslane, stinkgrass, and sprangletop. **Target(s):** tubulin polymerization (microtubule self-assembly (1). **1**. Schibler & Huang (1991) *J. Cell Biol.* **113**, 605.

Progabide

This anticonvulsant, antidepressive, and anti-parkinson disease drug (FW = 334.78 g/mol; CAS 62666-20-0), which displaces γ-aminobutyrate from its binding sites in membranes and behaves as a GABA agonist, inhibits γ-aminobutyrate receptor (1,2), microsomal epoxide hydrolase (3), and pyridoxal kinase (4). **1**. Kaplan, Raizon, Desarmenien, *et al.* (1980) *J. Med. Chem.* **23**, 702. **2**. Lloyd, Arbilla, Beaumont, *et al.* (1982) *J. Pharmacol. Exp. Ther.* **220**, 672. **3**. Kroetz, Loiseau, Guyot & Levy (1993) *Clin. Pharmacol. Ther.* **54**, 485. **4**. Lainé-Cessac, Cailleux & Allain (1997) *Biochem. Pharmacol.* **54**, 863.

Progesterone

This steroid hormone (FW = 314.47 g/mol; CAS 57-83-0), also known as progestin (although the term progestin historically refers to a mixture of steroids, with progesterone is the major component), pregn-4-ene-3,20-dione, and 4-pregnene-3,20-dione, is synthesized in the corpus luteum and is secreted during the latter half of the menstrual cycle. Progesterone induces the biosynthesis of uterus-specific proteins. Progesterone is also present in the testis and the adrenal cortex. It modulates the acrosome reaction in human sperm cells and also exhibits antiglucocorticoid and antimineralcorticoid activity (1). This hormone has also been identified in certain plants (*e.g.*, *Holarrhena floribunda*). **Target(s):** aldehyde dehydrogenase (2); aldehyde oxidase (3); aldose 1-epimerase, *or* mutarotase, weakly inhibited (4); cortisol sulfotransferase (29-31); F_1F_o ATPase, *or* H$^+$-transporting two-sector ATPase (5); glucose-6-phosphate dehydrogenase (however, note that (6) reports no inhibition (2,7,8); glutamate dehydrogenase (2,9,11); hydroxyindole *O*-methyltransferase (11); 15-hydroxyprostaglandin dehydrogenase (12); 3β-hydroxysteroid dehydrogenase (13,14); 3β-hydroxy-5α-steroid dehydrogenase (15); 21-hydroxysteroid dehydrogenase, NAD$^+$-dependent (16); isocitrate dehydrogenase, weakly inhibited (6); ketosteroid monooxygenase, *or* androst-4-ene-3,17-dione monooxygenase, inhibits lactonization of 17-ketosteroids (37); Na$^+$/K$^+$-exchanging ATPase (5); 6-phosphogluconate dehydrogenase8; retinol *O*-fatty-acyltransferase19,34,35; steroid Δ-isomerase20,21; steroid 21-monooxygenase (20,38); steroid sulfatase (21); steroid sulfotransferase (30); sterol *O*-acyltransferase, *or* cholesterol *O*-acyltransferase, *or* ACAT (22,23,34-36); sterol demethylase (24); sterol esterase, *or* cholesterol esterase (27,28); tryptophan 2,3-dioxygenase (29); and xanthine oxidase (30). **1**. Scott & Tomkins (1975) *Meth. Enzymol.* **40**, 273. **2**. Douville & Warren (1968) *Biochemistry* **7**, 4052. **3**. Rajagopalan & Handler (1966) *Meth. Enzymol.* **9**, 364. **4**. Mulhern, Fishman, Kusiak & Bailey (1973) *J. Biol. Chem.* **248**, 4163. **5**. Zheng & Ramirez (1999) *Eur. J. Pharmacol.* **368**, 95. **6**. McKerns (1963) *Biochim. Biophys. Acta* **73**, 507. **7**. Noltmann & Kuby (1963) *The Enzymes*, 2nd ed. (Boyer, Lardy & Myrbäck, eds.), **7**, 223. **8**. Criss & McKerns (1969) *Biochim. Biophys. Acta* **184**, 486. **9**. Tappel & Dillard (1967) *J. Biol. Chem.* **242**, 2463. **10**. Frieden (1963) *The Enzymes*, 2nd ed. (Boyer, Lardy & Myrbäck, eds.), **7**, 3. **11**. Morton & Forbes (1989) *J. Pineal Res.* **6**, 259. **12**. Pace-Asciak & Smith (1983) *The Enzymes*, 3rd ed. (Boyer, ed.), **16**, 543. **13**. Townsley (1975) *Acta Endocrinol. (Copenh.)* **79**, 740. **14**. Goldman (1967) *J. Clin. Endocrinol. Metab.* **27**, 320. **15**. Giacomini & Wright (1980) *J. Steroid Biochem.* **13**, 645. **16**. Monder & Furfine (1969) *Meth. Enzymol.* **15**, 667. **17**. Ross (1982) *J. Biol. Chem.* **257**, 2453. **18**. Talalay & Benson (1972) *The Enzymes*, 3rd ed. (Boyer, ed.), **6**, 591. **19**. Weintraub, Vincent, Baulieu & Alfsen (1977) *Biochemistry* **16**, 5045. **20**. Yoshida, Sekiba, Yanaihara, *et al.* (1978) *Endocrinol. Jpn.* **25**, 349. **21**. Notation & Ungar (1969) *Biochemistry* **8**, 501. **22**. Billheimer (1985) *Meth. Enzymol.* **111**, 286. **23**. Chang & Doolittle (1983) *The Enzymes*, 3rd ed. (Boyer, ed.), **16**, 523. **24**. Gaylor, Chang, Nightingale, Recio & Ying (1965) *Biochemistry* **4**, 1144. **25**. Slotte & Ekman (1986) *Biochim. Biophys. Acta* **879**, 221. **26**. Tomita, Sawamura, Uetsuka, *et al.* (1996) *Biochim. Biophys. Acta* **1300**, 210. **27**. Braidman & Rose (1970) *Biochem. J.* **118**, 7P. **28**. Roussos (1967) *Meth. Enzymol.* **12A**, 5. **29**. Singer, Gebhart & Hess (1978) *Can. J. Biochem.* **56**, 1028. **30**. Singer (1979) *Arch. Biochem. Biophys.* **196**, 340. **31**. Singer & Bruns (1980) *Can. J. Biochem.* **58**, 660. **32**. Kaschula, Jin, Desmond-Smith & Travis (2005) *Exp. Eye Res.* **82**, 111. **33**. Muniz, Villazana-Espinoza, Thackeray & Tsin (2006) *Biochemistry* **45**, 12265. **34**. Simpson & Burkhart (1980) *Arch. Biochem. Biophys.* **200**, 79. **35**. Erickson, Shrewsbury, Brooks & Meyer (1980) *J. Lipid Res.* **21**, 930. **36**. Tabas, Chen, Clader, *et al.* (1990) *J. Biol. Chem.* **265**, 8042. **37**. Itagaki (1986) *J. Biochem.* **99**, 825. **38**. Blom, Förlin & Andersson (2001) *Fish Physiol. Biochem.* **24**, 1.

Progestin, *See* Progesterone

Proglumide

This anticholinergic agent (FW$_{free-acid}$ = 334.42 g/mol; CAS 6620-60-6), also known as DL-4-benzamido-*N*,*N*-dipropylglutaramic acid and 4-(benzoylamino)-5-(dipropylamino)-5-oxopentanoic acid, is a cholecystokinin receptor antagonist. Proglumide A also exerts an inhibitory effect on gastric secretion and reduces gastrointestinal motility. It finds use clinically as a therapy for gastrointestinal ulcers. **Target(s):** cholecystokinin receptor (1); gastrin receptor (2). **1**. Hahne, Jensen, Lemp & Gardner (1981) *Proc. Natl. Acad. Sci. U.S.A.* **78**, 6304. **2**. Johnson & Guthrie (1984) *Amer. J. Physiol.* **246**, G62.

Progranulin

This 93-residue, cysteine-rich granulin precursor (MW = 88 kDa; Symbol: PGRN), also called proepithelin and PC cell-derived growth factor, is a pleiotrophic growth factor with important roles in maintaining and regulating the homeostatic dynamics of normal tissue development, proliferation, regeneration, and host-defense. PGRN also has potent anti-inflammatory properties, and its deregulation is associated with rheumatoid arthritis and inflammatory bowel disease. Progranulin plays crucial roles in preserving bone mass by inhibiting TNF-α-induced osteoclastogenesis and promoting osteoblastic differentiation in mice (1). Recombinant PGRN protein strongly inhibits TNF and the expression of interferon-γ-induced chemokines CXCL9 and CXCL10 (2). Moreover, in PGRN^{null} mice, many chemokines, including CXCL9 and CXCL10, are significantly induced (2). **1**. Noguchi, Ebina, Hirao, *et al.* (2015) *Biochem. Biophys. Res. Commun.* **465**, 638. **2**. Mundra, Jian, Bhagat & Liu (2016) *Sci. Rep.* **6**, 21115.

Proguanil, *See* Chlorguanide

Prohexadione

This plant growth regulator (FW$_{free-acid}$ = 178.19 g/mol), also named 3-hydroxy-5-oxo-4-propionylcyclohex-3-ene-1-carboxylic acid, inhibits the biosynthesis of growth active gibberellins (GAs), thereby reducing longitudinal shoot growth. Prohexadione resembles 2-oxoglutaric acid, the co-substrate the dioxygenases that catalyze hydroxylation steps in late stages of GA biosynthesis. Its primary target of is 3β hydroxylation, thereby reducing levels of GA$_1$ (the highly active form) and bringing about theaccumulation of its immediate precursor (inactive). **Targets:** flavanone 3-dioxygenase (1) flavonol synthase (1); and gibberellin biosynthesis (1). **1**. Halbwirth, Fischer, Schlangen, *et al.* (2006) *Plant Sci.* **171**, 194.

L-Prolinal

This proline derivative (FW = 100.14 g/mol), with an aldehyde that undoubtedly hydrates to a tetrahedral species, inhibits tripeptide aminopeptidase. Aldehyde hydrates and bestatin resemble intermediates formed during direct attack by water on peptide substrates. Alternatively, inhibitors of both kinds might form derivatives of an active site nucleophile, resembling intermediates in a double-displacement mechanism. Exchange experiments with H$_2$^{18}O suggest that bestatin is bound intact by leucine aminopeptidase, lending support to the first mechanism. **1**. Frick & Wolfenden (1985) *Biochim. Biophys. Acta* **829**, 311.

L-Proline

This pyrrolidine (FW = 115.13 g/mol; pK_1 = 1.95; pK_2 = 10.64), symbolized as eithet Pro or P, is one of the twenty proteogenic amino acids. Its five-membered ring is usually puckered, with the γ carbon (position C4 of the ring) out of the plane by about 0.5 Å. There are six codons that code for L-proline: CCC, CCU, CCA, and CCG. This amino acid is unique among the twenty proteogenic amino acids in that the "side chain" is covalently linked to the nitrogen. Proline is regarded as a nonpolar amino acid; interestingly, however, most prolyl residue in proteins are located on the surface of proteins, since the steric constraints of the pyrrolidine ring results in reverse turns. Such constraints make prolyl residues incompatible with α-helical structures and β-pleated sheets. Even so, prolyl residues can be found at N-termini of α-helices. In addition, prolyl residues can be found in long helices; the residue cause a small distortion of the helical structure. Prolyl residues are also abundant and structural important in the collagen triple helix. In homologous proteins, if a prolyl residue is interchanged, it is usually with either an alanyl or a seryl residue. The L-stereoisomer has a specific rotation of –60.4° at 25°C of the D line for a 1.0–2.0 g/100 mL solution in 5 M HCl (–86.2° in distilled water). It is soluble in water (162.3 g/100 mL at 25°C; 206.7 g/100 mL at 50°C; and 239 g/100 mL at 65 °C) and is one of the few common amino acids that is soluble in ethanol (67 g/100 mL at 19°C). It is insoluble in diethyl ether, butanol, and isopropanol. While the proline "side chain" is relatively inert, prolyl residues in certain proteins, including collagen, undergo enzyme-catalyzed hydroxylation reactions at the C3 or C4 positions. Free proline also react swith periodate to produce 2-pyrrolidone. A number of enzymes use D- or L-proline as an alternative substrate (and competitive inhibitor). These include the amino-acid dehydrogenase, the amino acid oxidases, alanine:oxo-acid aminotransferase, ornithine:oxo-acid aminotransferase, asparagine:oxo-acid aminotransferase, and amino-acid racemase. **Target(s):** acetyl-cholinesterase, weakly inhibited (1); 2-aminohexano-6-lactam racemase, weakly inhibited (2); arginase, weakly inhibited (3-5); arginine deiminase (6); aspartate aminotransferase (7); [citrate lyase] deacetylase, weakly inhibited (8); CTP synthetase, weakly inhibited (9); cysteinyl-tRNA synthetase activity of *Methanococcus jannaschii* prolyl-cysteinyl-tRNA synthetase (10); glutamine synthetase, weakly inhibited (11); γ-glutamyl kinase, *or* glutamate 5-kinase (12-19); glycine/sarcosine *N*-methyltransferase (30); hydroxy-L-proline oxidase (20); membrane alanyl aminopeptidase, *or* aminopeptidase N (21); phosphorylase kinase (22); polar-amino-acid-transporting ATPase, histidine permease, also alternative substrate (23); pyrroline-5-carboxylate dehydrogenase (24,25); 1-pyrroline-5-carboxylate reductase (14); pyruvate kinase (26); α,α-trehalose-phosphate synthase, *or* UDP-forming (27); and Xaa-Pro dipeptidase, *or* prolidase, *or* imidodipeptidase (28,29). **1**. Bergmann, Wilson & Nachmansohn (1950) *J. Biol. Chem.* **186**, 693. **2**. Ahmed, Esaki, Tanaka & Soda (1983) *Agric. Biol. Chem.* **47**, 1887. **3**. Hunter & Downs (1945) *J. Biol. Chem.* **157**, 427. **4**. Patchett, Daniel & Morgan (1991) *Biochim. Biophys. Acta* **1077**, 291. **5**. Kaysen & Strecker (1973) *Biochem. J.* **133**, 779. **6**. Park, Hirotani, Nakano & Kitaoka (1984) *Agric. Biol. Chem.* **48**, 483. **7**. Shanti, Shashikumar & Desai (2004) *Neurochem. Res.* **29**, 2197. **8**. Giffhorn & Gottschalk (1975) *J. Bacteriol.* **124**, 1052. **9**. Bearne, Hekmat & MacDonnell (2001) *Biochem. J.* **356**, 223. **10**. Stathopoulos, Jacquin-Becker, Becker, *et al.* (2001) *Biochemistry* **40**, 46. **11**. Southern, Parker & Woods (1987) *J. Gen. Microbiol.* **133**, 2437. **12**. Kramer, Henslee, Wakabashi & Jones (1985) *Meth. Enzymol.* **113**, 113. **13**. Baich (1969) *Biochim. Biophys. Acta* **192**, 462. **14**. Krishna & Leisinger (1979) *Biochem. J.* **181**, 215. **15**. Hayzer & Moses (1978) *Biochem. J.* **173**, 219. **16**. Vasáková & Stefl (1982) *Coll. Czech. Chem. Commun.*, No. 42, 349. **17**. Marco-Marín, Gil-Ortiz, Pérez-Arellano, *et al.* (2007) *J. Mol. Biol.* **367**, 1431. **18**. Fujita, Maggio, Garcia-Rios, *et al.* (2003) *J. Biol. Chem.* **278**, 14203. **19**. Stefl & Vasakova (1982) *Collect. Czech. Chem. Commun.* **47**, 360 and (1984) **49**, 2698. **20**. Heacock & Adams (1975) *J. Biol. Chem.* **250**, 2599. **21**. Bauvois & Dauzonne (2006) *Med. Res. Rev.* **26**, 88. **22**. Chebotareva, Andreeva, Makeeva, Livanova & Kurganov (2004) *J. Mol. Recognit.* **17**, 426. **23**. Boncompagni, Dupont, Mignot, *et al.* (2000) *J. Bacteriol.* **182**, 3717. **24**. Strecker (1971) *Meth. Enzymol.* **17B**, 262. **25**. Webb (1966) *Enzyme and Metabolic Inhibitors*, vol. **2**, p. 355, Academic Press, New York. **26**. Plaxton & Storey (1984) *Eur. J. Biochem.* **143**, 257. **27**. Valenzuela-Soto, Márquez-Escalante, Iturriaga & Figueroa-Soto (2004) *Biochem. Biophys. Res. Commun.* **313**, 314. **28**. Sjöström (1974) *Acta Chem. Scand. Ser. B* **28**, 802. **29**. Mock & Green (1990) *J. Biol. Chem.* **265**, 19606. **30**. Waditee, Tanaka, Aoki, *et al.* (2003) *J. Biol. Chem.* **278**, 4932.

Prolixin-S

This 20-kDa hemoprotein from the blood-sucking insect *Rhodnius prolixus* (a vector for American trypanosomiasis) binds and tansports nitric oxide. Prolixin-S also binds directly to coagulation factors IX and IXa (1). *Note:*

This term is also a proprietary name for fluphenazine dihydrochloride. **1.** Isawa, Yuda, Yoneda & Chinzei (2000) *J. Biol. Chem.* **275**, 6636.

Proluton, See *Progesterone*

L-Prolyl-L-alanine

This dipeptide (FW = 186.21 g/mol) inhibits dipeptidyl-peptidase IV (1) and Xaa-Pro dipeptidyl-peptidase (2). **1.** Yoshimoto, Fischl, Orlowski & Walter (1978) *J. Biol. Chem.* **253**, 3708. **2.** Yan, Ho & Hou (1992) *Biosci. Biotechnol. Biochem.* **56**, 704.

N-[L-Prolyl-L-histidyl-L-prolyl-L-phenylalanyl-L-histidyl]-(2(S)-amino-1-hydroxy-4-methylpentyl)-L-valyl-L-isoleucyl-L-histidyl-L-lysine, See *Renin Inhibitor H142*

L-Prolyl-L-leucylglycinamide, See *Melanostatin*

L-Prolyl-L-leucylglycine Hydroxamate

This tripeptide hydroxamate (*Sequence*: PLG-(NHOH); $FW_{free-base}$ = 300.36 g/mol), which bears notable structural similarity to actinonin, inhibits a number of metalloendoproteinases, including: astacin (1-3); collagenase (4); meprin A (3); and meprin B (3). **1.** Stöcker (1998) in *Handb. Proteolytic Enzymes* (Barrett, Rawlings & Woessner, eds.), p. 1219, Academic Press, San Diego. **2.** Reyda, Jacob, Zwilling & Stöcker (1999) *Biochem. J.* **344**, 851. **3.** Bertenshaw, Turk, Hubbard, *et al.* (2001) *J. Biol. Chem.* **276**, 13248. **4.** Moore & Spilburg (1986) *Biochemistry* **25**, 5189.

L-Prolyl-L-leucyl-L-phosphonotryptophan

This phosphonate (FW = 437.46 g/mol) inhibits gelatinase B, *or* matrix metalloproteinase 9 (1) and neutrophil collagenase, *or* matrix metalloproteinase 8 (1,2). **1.** Gallina, Gavuzzo, Giordano, *et al.* (1999) *Ann. N. Y. Acad. Sci.* **878**, 700. **2.** Gavuzzo, Pochetti, Mazza, *et al.* (2000) *J. Med. Chem.* **43**, 3377.

L-Prolyl-L-lysyl-L-methionyl-L-cysteinylglycyl-L-valinamide

This hexapeptide amide (*Sequence*: PKMCGV-NH$_2$; $FW_{free-base}$ = 632.85 g/mol) inhibits *Crotalus adamanteus* adamalysin (IC$_{50}$ = 3.2 μM, the corresponding D-cysteinyl residue-containing hexapeptide amide has a IC$_{50}$ of 74 μM (1,2). **1.** Kress & Catanese (1998) in *Handb. Proteolytic Enzymes*

(Barrett, Rawlings & Woessner, eds.), p. 1262, Academic Press, San Diego. **2.** Grams, Huber, Kress, Moroder & Bode (1993) *FEBS Lett.* **335**, 76.

D-Prolyl-L-phenylalanyl-L-arginal

This tripeptide aldehyde (FW = 302.48 g/mol) inhibits plasma kallikrein, and its sequence is based on cleavage sites in the natural Kex2 substrate pro-alpha-factor. This inhibitor exhibited a K_i of 3.7 nM and a second-order inactivation rate constant (k_2/K_i) of 1.3 x 10^7 M^{-1}s^{-1} and is comparable with the value of k_{cat} and the K_m obtained with Kex2 for the corresponding peptidyl methylcoumarinyl-amide substrate. The enzyme is also sensitive to the other peptidyl chloromethanes over a range of concentrations, depending on peptide sequence and α-amino decanoylation, but was completely resistant to peptidyl sulphonium salts. **1.** Evans, Jones, Pitt, *et al.* (1996) *Immunopharmacology* **32**, 117.

L-Prolyl-L-phenylalanyl-L-arginine Chloromethyl Ketone

This halomethyl ketone ($FW_{free-base}$ = 450.97 g/mol) typically inhibits amidohydrolase targets irreversibly by virtue of its susceptibility to attack by an active-site nucleophile (see exception below). **Target(s):** coagulation factor XIIa (1,2); crotalase, *or* fibrinogen-clotting snake venom enzyme (3,4); histolysain, *or Entamoeba histolytica* cysteine endopeptidase, *or* histolysin, inhibited reversibly (5,6); mouse γ-nerve growth factor (7); plasma kallikrein (8.9); tissue kallikrein (7,8); and venombin A (3,4). **1.** Silverberg & Kaplan (1988) *Meth. Enzymol.* **163**, 68. **2.** Silverberg & Kaplan (1982) *Blood* **60**, 64. **3.** Markland, Jr. & Pirkle (1998) in *Handb. Proteolytic Enzymes* (Barrett, Rawlings & Woessner, eds.), p. 216, Academic Press, San Diego. **4.** Markland, Kettner, Schiffman, *et al.* (1982) *Proc. Natl. Acad. Sci. U.S.A.* **79**, 1688. **5.** Scholze & Tannich (1994) *Meth. Enzymol.* **244**, 512. **6.** Luaces & Barrett (1988) *Biochem. J.* **250**, 903. **7.** Kettner & Shaw (1981) *Meth. Enzymol.* **80**, 826. **8.** Geiger & Fritz (1981) *Meth. Enzymol.* **80**, 466. **9.** Nagase & Barrett (1981) *Biochem. J.* **193**, 187.

L-Prolyl-L-prolylglycyl-L-phenylalanyl-L-seryl-L-proline
This hexapeptide PPGFSP (FW = 600.67 g/mol), corresponding to residues 2–7 within bradykinin and hence known as bradykinin$_{2-7}$, is a strong inhibitor of Xaa-Pro aminopeptidase, *or* aminopeptidase P (1-3). **1.** Matos & Monnet (1998) in *Handb. Proteolytic Enzymes* (Barrett, Rawlings & Woessner, eds.), p. 1410, Academic Press, San Diego. **2.** Orawski & Simmons (1995) *Biochemistry* **34**, 11227. **3.** Simmons & Orawski (1992) *J. Biol. Chem.* **267**, 4897.

L-Prolyl-L-prolyl-L-prolyl-L-prolyl-L-prolyl-L-proline, See *Hexaproline*

L-Prolyl-L-threonyl-L-glutamyl-L-phenylalanyl-Ψ[CH$_2$NH]-L-norleucyl-L-arginyl-L-leucine, See *H-297*

L-Prolyl-L-threonyl-L-glutamyl-L-phenylalanyl-Ψ[CH₂NH]-L-phenylalanyl-L-arginyl-L-glutamate, *See* H-256

Promazine

This pheniazine antipsychotic (FW_free-base = 284.43 g/mol; CAS 58-40-2; free base is oily liquid; hydrochloride is very soluble in water) is a dopamine receptor antagonist and tranquilizer. Although similar in action to chlorpromazine, promazine exhibits less antipsychotic activity, and is used primarily for short-term treatment of disturbed behavior and as an antiemetic. **Target(s):** CYP2B1 (1); D₂ dopamine receptor (2,3); monoamine oxidase (4); and Na⁺/K⁺-exchanging ATPase (5,6). **1.** Murray (1992) *Biochem. Pharmacol.* **44**, 121. **2.** Daniel, Syrek & Wojcikowski (1999) *Eur. Neuropsychopharmacol.* **9**, 337. **3.** Horn, Post & Kennard (1975) *J. Pharm. Pharmacol.* **27**, 553. **4.** Suzuki, Seno & Kumazawa (1988) *Life Sci.* **42**, 2131. **5.** Hackenberg & Krieglstein (1972) *Naunyn Schmiedebergs Arch. Pharmacol.* **274**, 63. **6.** Davis & Brody (1966) *Biochem. Pharmacol.* **15**, 703.

Promegestone

This synthetic progestin (FW = 326.48 g/mol; CAS 34184-77-5), also known as (17β)-17-methyl-17-(1-oxopropyl)estra-4,9-dien-3-one, is a tight-binding ligand of the progesterone/progestin receptor. It is not bound by transcortin and binds to progesterone receptors with higher affinity than progesterone. Promegestone is also a noncompetitive acetylcholine receptor antagonist that may alter the receptor's function by interacting at the lipid-protein interface (1). **Target(s):** acetylcholine receptor, Torpedo nicotinic (1); CYP17 (2;) estrone sulfatase (3-5); estrone sulfotransferase, but activated at low concentrations (6); gonadotropin secretion (7); 17α-hydroxyprogesterone aldolase (8); steroid Δ-isomerase, *or* 3β-hydroxysteroid dehydrogenase/Δ⁵,Δ⁴-isomerase (9). **1.** Blanton, Xie, Dangott & Cohen (1999) *Mol. Pharmacol.* **55**, 269. **2.** Kuhn-Velten, Bunse & Forster (1991) *J. Biol. Chem.* **266**, 6291. **3.** Chetrite, Varin, Delalonde & Pasqualini (1993) *Anticancer Res.* **13**, 931. **4.** Pasqualini, Schatz, Varin & Nguyen (1992) *J. Steroid Biochem. Mol. Biol.* **41**, 323. **5.** Pasqualini, Ebert & Chetrite (2001) *Gynecol. Endocrinol.* **15** Suppl. 6, 44. **6.** Pasqualini & Chetrite (2007) *Anticancer Res.* **27**, 3219. **7.** Labrie, Ferland, Lagace, *et al.* (1977) *Fertil. Steril.* **28**, 1104. **8.** Dalla Valle, Ramina, Vianello, Belvedere & Colombo (1996) *J. Steroid Biochem. Mol. Biol.* **58**, 577. **9.** Takahashi, Luu-The & Labrie (1990) *J. Steroid Biochem. Mol. Biol.* **37**, 231.

Promethazine

This antihistaminic drug (FW_free-base = 284.43 g/mol; CAS 60-87-7), also known as 10-[(2-dimethylamino)propyl]phenothiazine, is a histamine H₁ receptor antagonist and CNS depressant. **Key Pharmacokinetic Parameters:** *See* Appendix II in Goodman & Gilman's THE PHARMACOLOGICAL BASIS OF THERAPEUTICS, 12th Edition (Brunton, Chabner & Knollmann, eds.) McGraw-Hill Medical, New York (2011). **Target(s):** *Bacillus subtilis*

sporulation (1); Ca²⁺-dependent ATPase (2,3); cAMP phosphodiesterase (4,5); CYP2D6 (6,7); H₁ histamine receptor (8-11); hydroxysteroid sulfotransferase (12); monoamine oxidase (13); Na⁺/K⁺-exchanging ATPase (3,14); nucleoside-diphosphatase (15); osteoclastic bone resorption (16); phospholipase A₂ (17); UDP-glucuronosyltransferase (18). **1.** Burke, Jr., & Spizizen (1977) *J. Bacteriol.* **129**, 1215. **2.** Agarwal, Shukla & Tekwani (1992) *Int. J. Biochem.* **24**, 1447. **3.** Agarwal, Tekwani, Shukla & Ghatak (1990) *Indian J. Exp. Biol.* **28**, 245. **4.** Curtis-Prior, Jenner, Chan & McColl (1979) *Experientia* **35**, 1430. **5.** Uzan & Le Fur (1976) *Arch. Int. Pharmacodyn. Ther.* **219**, 160. **6.** Hamelin, Bouayad, Drolet, Gravel & Turgeon (1998) *Drug Metab. Dispos.* **26**, 536. **7.** Nakamura, Yokoi, Inoue, *et al.* (1996) *Pharmacogenetics* **6**, 449. **8.** Li & Hatton (1996) *Neuroscience* **70**, 145. **9.** Levi, Capurro & Lee (1975) *Eur. J. Pharmacol.* **30**, 328. **10.** Sharma & Hamelin (2003) *Curr. Drug Metab.* **4**, 105. **11.** Hill, Ganellin, Timmerman, *et al.* (1997) *Pharmacol. Rev.* **49**, 253. **12.** Matsui, Takahashi, Miwa, Motoyoshi & Homma (1995) *Biochem. Pharmacol.* **49**, 739. **13.** Suzuki, Seno & Kumazawa (1988) *Life Sci.* **42**, 2131. **14.** Davis & Brody (1966) *Biochem. Pharmacol.* **15**, 703. **15.** Sano, Matsuda & Nakagawa (1988) *Eur. J. Biochem.* **171**, 231. **16.** Hall, Nyugen, Schaeublin, Michalsky & Missbach (1996) *Gen. Pharmacol.* **27**, 845. **17.** Vadas, Stefanski & Pruzanski (1986) *Agents Actions* **19**, 194. **18.** Sharp, Mak, Smith & Coughtrie (1992) *Xenobiotica* **22**, 13.

Prometryn

This selective post-emergence triazine herbicide (FW = 241.36 g/mol; CAS 7287-19-6; M.P. = 118-120°C), also known as Gesagard, Caparol, and systematically as *N,N'*-bis(1-methylethyl)-6-methylthio-1,3,5-triazine-2,4-diamine, is used to control annual grasses and broadleaf weeds in the cultivation of a variety of crops, including cotton and celery. *Note*: Prometryn is toxic to a number of organisms, including humans and fish, but is not toxic to birds, bees, and earthworms. **Target(s):** electron transport, photosynthetic1 **1.** Dodge (1977) *Spec. Publ., Chem. Soc.* **29**, 7.

Prontosil, *See* Sulfamidochrysoidine

Propachlor

This herbicide (FW = 211.69 g/mol; CAS 1918-16-7; M.P. = 67-76°C), also known as Bexton, Ramrod, and systematically as α-chloro-*N*-isopropylacetanilide, is widely used to protect corn, maize, onion, cotton, sugarcane, sorghum, cabbage, rose bushes, and ornamental plants. Propachlor is effective against annual grasses and certain broad-leaved weeds. More than one degradative pathway for propachlor has been identified. **Target(s):** protein biosynthesis (1,2). **1.** Jaworski (1969) *J. Agric. Food Chem.* **17**, 165. **2.** Duke, Slife, Hanson & Butler (1975) *Weed Sci.* **29**, 142.

Propafenone

This class 1C antiarrhythmic agent (FW_free-base = 341.45 g/mol; CAS 54063-53-5), also known as 1-[2-[2-hydroxy-3-(propylamino)propoxy]-phenyl]-3-phenyl-1-propanone, is a β-adrenergic receptor antagonist that also blocks a number of K⁺ channels. The (*S*)-(–)-enantiomer is the more potent blocker of cardiac β-adrenoceptors. **Target(s):** β-adrenoceptor (1); aldehyde oxidase (2); calcium ion current (3); CYP1A2 (4); CYP2D1 (5); HERG K$^+$ channel (6,7); hK$_{v1.5}$ channels (8); IK$_r$ potassium current (9); K$^+$ channels, ATP-sensitive (10); and the TWIK-related spinal cord K$^+$ channel, *or* TRESK (11). **1.** Groschner, Lindner, Schnedl & Kukovetz (1991) *Brit. J. Pharmacol.* **102**, 669. **2.** Obach, Huynh, Allen & Beedham (2004) *J. Clin. Pharmacol.* **44**, 7. **3.** Fei, Gill, McKenna & Camm (1993) *Brit. J. Pharmacol.* **109**, 178. **4.** Kobayashi, Nakajima, Chiba, *et al.* (1998) *Brit. J. Clin. Pharmacol.* **45**, 361. **5.** Xu, Aasmundstad, Christophersen, Morland & Bjorneboe (1997) *Biochem. Pharmacol.* **53**, 603. **6.** Arias, Gonzalez, Moreno, *et al.* (2003) *Cardiovasc. Res.* **57**, 660. **7.** Paul, Witchel & Hancox (2002) *Brit. J. Pharmacol.* **136**, 717. **8.** Franqueza, Valenzuela, Delpon, *et al.* (1998) *Brit. J. Pharmacol.* **125**, 969. **9.** Cahill & Gross (2004) *J. Pharmacol. Exp. Ther.* **308**, 59. **10.** Christe, Tebbakh, Simurdova, Forrat & Simurda (1999) *Eur. J. Pharmacol.* **373**, 223. **11.** Sano, Inamura, Miyake, *et al.* (2003) *J. Biol. Chem.* **278**, 27406.

Propamidine

This trypanocidal agent (FW$_{free-base}$ = 312.37 g/mol; CAS 104-32-5) is a DNA intercalator that binds within the minor groove. **Target(s):** *S*-adenosylmethionine decarboxylase (1); aldehyde oxidase (2); complement component C5 binding to C3b (3); deoxyribonuclease I (4); and diamine oxidase (5). **1.** Hugo & Byers (1993) *Biochem. J.* **295**, 203. **2.** Knox (1946) *J. Biol. Chem.* **163**, 699. **3.** Vogt, Schmidt & Hinsch (1979) *Immunology* **36**, 139. **4.** Sutton, Conn, Brown & Lane (1997) *Biochem. J.* **321**, 481. **5.** Cubria, Balana Fouce, Alvarez-Bujidos, *et al.* (1993) *Biochem. Pharmacol.* **45**, 1355.

Propane-1,3-bisphosphonate, *See 1,3-Bis(phosphono)propane*

Propanedioic Acid, *See Malonic Acid*

1,2-Propanediol

This vicinal diol (FW = 76.095 g/mol; CAS 57-55-6; 4254-14-2 for the (*R*)-isomer; 4254-15-3 for the (*S*)-isomer); 4254-16-4 for the (+/−)-isomer), also known as propylene glycol and propane-1,2-diol, is a product of the reaction catalyzed by lactaldehyde reductase. 1,2-Propanediol is a substrate of propanediol dehydratase and an alternative substrate for (*R*,*R*)-butanediol dehydrogenase and glycerol dehydrogenase. Both enantiomers are hygroscopic and miscible with water. The racemic mixture melts at −59°C and boils at 188.2°C. Note that the (+)-enantiomer corresponds to 3-deoxy-*sn*-glycerol. **Target(s):** actin polymerization (1); bacterial leucyl aminopeptidase (2); CYP2E1 (3); glucan 1,4-α-glucosidase, *or* glucoamylase (4); glycerol kinase (5-7); oxidative phosphorylation (8); and propanediol dehydratase, as substrate and inactivator (9). **1.** Vincent, Pruliere, Pajot-Augy, *et al.* (1990) *Cryobiology* **27**, 9. **2.** Dick, Matheson & Wang (1970) *Can. J. Biochem.* **48**, 1181. **3.** Thomsen, Loft, Roberts & Poulsen (1995) *Pharmacol. Toxicol.* **76**, 395. **4.** Kumar & Satyanarayana (2003) *Biotechnol. Prog.* **19**, 936. **5.** Eisenthal, Harrison, Lloyd & Taylor (1972) *Biochem. J.* **130**, 199. **6.** Schneider (1975) *Biochim. Biophys. Acta* **397**, 110. **7.** Eisenthal, Harrison & Lloyd (1974) *Biochem. J.* **141**, 305. **8.** Lovett & Sweetman (1983) *Methods Find. Exp. Clin. Pharmacol.* **5**, 695. **9.** Bachovchin, Eagar, Jr., Moore & Richards (1977) *Biochemistry* **16**, 1082.

1,3-Propanediol

HO—____—OH

This diol (FW = 76.095 g/mol; CAS 504-63-2; B.P. = 214.22°C; miscible with water), also known as trimethylene glycol, is a viscous liquid with a sweet taste. It is a glycerol fermentation product and substrate for 1,3-propanediol dehydrogenase. **Target(s):** diol dehydratase (1); glycerol dehydratase (1); and *N*-methyl-D-aspartate receptor, *or* NMDA receptor (2). **1.** Knietsch, Bowien, Whited, Gottschalk & Daniel (2003) *Appl. Environ. Microbiol.* **69**, 3048. **2.** Peoples & Ren (2002) *Mol. Pharmacol.* **61**, 169.

n-Propanethiol

This foul-smelling thiol CH$_3$CH$_2$CH$_2$SH (FW = 76.163 g/mol; CAS 107-03-9 as well as 6898-84-6 for sodium salt; M.P. = −113°C; B.P. = 67-68°C), also known as propyl mercaptan, inhibits

Propane-1,2,3-tricarboxylic Acid, *See Tricarballylic Acid*

1,2,3-Propanetriol, *See Glycerol*

Propanil

This contact herbicide and nematocide (FW = 218.08 g/mol; CAS 709-98-8), also known as Stam, Stampede, and Rogue, is effective against numerous grasses and broad-leaved weeds in crops of rice, which possesses an aryl acylamidase that inactivates propanil. Weeds that do not have this activity are killed by propanil. Propanil also inhibits tumor necrosis factor-α production by reducing nuclear levels of the transcription factor nuclear factor-κB. **Target(s):** electron transport, photosynthetic (1). **1.** Dodge (1977) *Spec. Publ., Chem. Soc.* **29**, 7.

Propanoic Acid, *See Propionic Acid*

3-Propanolamine, *See 3-Aminopropanol*

(+)-8'-Propargylabscisic Acid

This substrate analogue (FW = 272.34 g/mol) is a competitive inhibitor of (+)-abscisate 8'-hydroxylase, K_i = 1.1 μM. The (−)-stereoisomer is a significantly weaker inhibitor, K_i = 56 μM. **1.** Cutler, Rose, Squires, *et al.* (2000) *Biochemistry* **39**, 13614.

(+)-9'-Propargylabscisic Acid

This seed germination inhibitor (FW$_{free-acid}$ = 288.34 g/mol) is mechanism-based inactivator of (+)-abscisate 8'-hydroxylase, K_i = 0.27 μM, k_{inact} = 0.11 min^{-1} (1). The (−)-stereoisomer is a weaker inhibitor. **1.** Cutler, Rose, Squires, *et al.* (2000) *Biochemistry* **39**, 13614.

Propargyl Amine

This alkynyl amine HC≡C–CH$_2$–NH$_2$ (FW$_{free-base}$ = 55.08 g/mol), also known as propargylamine and 3-amino-1-propyne, is a toxic liquid that boils at 83°C. The water-soluble hydrochloride melts at 179-182°C. **Target(s):** aldehyde dehydrogenase (1); diamine oxidase, *or* amine oxidase [copper-containing] (2,3); glycine cleavage complex, *or* glycine synthase (4); and monoamine oxidase, *or* amine oxidase [flavin-containing] (2,5-9). **1.** DeMaster, Shirota & Nagasawa (1986) *Biochem. Pharmacol.* **35**, 1481. **2.** Rando (1977) *Meth. Enzymol.* **46**, 158. **3.** Jeon & Sayre (2003) *Biochem. Biophys. Res. Commun.* **304**, 788. **4.** Benavides, Croci & Strolin Benedetti (1983) *Biochem. Pharmacol.* **32**, 287. **5.** Salach, Jr. (1978) *Meth. Enzymol.* **53**, 495. **6.** Brush & Kozarich (1992) *The Enzymes*, 3rd ed. (Sigman, ed.), **20**, 317. **7.** Gartner & Hemmerich (1975) *Angew. Chem. Int. Ed. Engl.* **14**, 110. **8.** Abeles & Tashjian (1974) *Biochem. Pharmacol.* **23**, 2205. **9.** McEwan, Jr., Sasaki & Jones (1969) *Biochemistry* **8**, 3963.

Propargylate

This alkynyl carboxylic acid (FW = 70.05 g/mol; M.P. = 9°C; B.P. = 144°C), also known as propiolic acid and 2-propynoic acid, inhibits acyl-CoA dehydrogenase (1) and pyruvate formate-lyase, *or* formate acetyltransferase (2). **1.** Freund, Mizzer, Dick & Thorpe (1985) *Biochemistry* **24**, 5996. **2.** Brush & Kozarich (1992) *The Enzymes*, 3rd ed. (Sigman, ed.), **20**, 317.

10-Propargyl-5,8-dideazafolic Acid

This folate analogue (FW$_{free-acid}$ = 477.47 g/mol), often abbreviated PDDF and also known as CB3717, is a potent inhibitor of thymidylate synthase, human IC$_{50}$ = 85 nM (1-5). It is a weaker inhibitor of dihydrofolate reductase (human IC$_{50}$ = 1.9 µM (4,5). **1.** Spencer, Villafranca & Appleman (1997) *Biochemistry* **36**, 4212. **2.** Anderson, O'Neil, DeLano & Stroud (1999) *Biochemistry* **38**, 13829. **3.** Arvizu-Flores, Sugich-Miranda, Arreola, *et al.* (2008) *Int. J. Biochem Cell Biol.* **40**, 2206. **4.** Gangjee, Li, Yang & Kisliuk (2008) *J. Med. Chem.* **51**, 68. **5.** Gangjee, Qiu, Li & Kisliuk (2008) *J. Med. Chem.* **51**, 5789.

10-Propargylestr-4-ene-3,17-dione

This 10β-propynyl-substituted steroid (FW = 310.44 g/mol), often abbreviated PED, is a Michael addition-type suicide inhibitor of aromatase, K_i = 23 nM and k_{inact} = 1.1 x 10^{-3} s^{-1} (1-3). **1.** Doody, Murry & Mason (1990) *J. Enzyme Inhib.* **4**, 153. **2.** Brandt, Covey & Zimniski (1990) *J. Enzyme Inhib.* **4**, 143. **3.** Covey, Hood & Parikh (1981) *J. Biol. Chem.* **256**, 1076.

D-Propargylglycine

This acetylenic amino acid (FW = 113.12 g/mol; CAS 23235-03-2), also known as D-2-amino-4-pentynoate and D-2-amino-4-pentynoic acid, is a Michael addition-type suicide inhibitor of D-amino-acid oxidase (1-5). (*See also DL-Propargylglycine; L-Propargylglycine*) **1.** Marcotte & Walsh (1978) *Biochemistry* **17**, 2864. **2.** Walsh, Cromartie, Marcotte & Spencer (1978) *Meth. Enzymol.* **53**, 437. **3.** Horiike, Nishina, Miake & Yamano (1975) *J. Biochem.* **78**, 57. **4.** Miyano, Fukui, Watanabe, *et al.* (1991) *J. Biochem.* **109**, 171. **5.** Marcotte & Walsh (1976) *Biochemistry* **15**, 3070.

DL-Propargylglycine

This naturally occurring δ-acetylenic α-amino acid (FW = 113.12 g/mol; CAS 64165-64-6; 16900-57-5 for HCl salt; 23235-03-2 for the (*R*)-isomer), also known as DL-2-amino-4-pentynoate, *C*-propargylglycine, and systematically as DL-2-amino-4-pentynoic acid, is a racemic mixture of a

potent inhibitor of cystathionine γ-lyase, cystathionine γ-synthase, and methionine γ-lyase. **See also** *D-Propargylglycine; L-Propargylglycine* **Target(s):** *O*-acetylhomoserine aminocarboxypropylt-ransferase (1-3); alanine aminotransferase (4,5); cystathionine β-lyase (6); cystathionine γ-lyase (7-13); cystathionine γ-synthase, *or* *O*-succinylhomoserine (thiol)-lyase) (5,13-17); cysteine dioxygenase (18); homocysteine desulfhydrase (19,20); and methionine γ-lyase (21-23). **1.** Yamagata, Paszewski & Lewandowska (1990) *J. Gen. Appl. Microbiol.* **36**, 137. **2.** Shimizu, Yamagata, Masui, *et al.* (2001) *Biochim. Biophys. Acta* **1549**, 61. **3.** Iwama, Hosokawa, Lin, *et al.* (2004) *Biosci. Biotechnol. Biochem.* **68**, 1357. **4.** Cornell, Zuurendonk, J. Kerich & Straight (1984) *Biochem. J.* **220**, 707. **5.** Marcotte & Walsh (1975) *Biochem. Biophys. Res. Commun.* **62**, 677. **6.** Alting, Engels, van Schalkwijk & Exterkate (1995) *Appl. Environ. Microbiol.* **61**, 4037. **7.** Nagasawa, Kanzaki & Yamada (1987) *Meth. Enzymol.* **143**, 486. **8.** De Angelis, Curtin, McSweeney, Faccia & Gobbetti (2002) *J. Dairy Res.* **69**, 255. **9.** Bruinenberg, de Roo & Limsowtin (1997) *Appl. Environ. Microbiol.* **63**, 561. **10.** Nagasawa, Kanzaki & Yamada (1984) *J. Biol. Chem.* **259**, 10393. **11.** Steegborn, Clausen, Sondermann, *et al.* (1999) *J. Biol. Chem.* **274**, 12675. **12.** Silverman & Abeles (1977) *Biochemistry* **16**, 5515. **13.** Ator & Ortiz de Montellano (1990) *The Enzymes*, 3rd ed. (Sigman & Boyer, eds.), **19**, 213. **14.** Kreft, Townsend, Pohlenz & Laber (1994) *Plant Physiol.* **104**, 1215. **15.** Clausen, Wahl, Messerschmidt, *et al.* (1999) *Biol. Chem.* **380**, 1237. **16.** Ravanel, Droux & Douce (1995) *Arch. Biochem. Biophys.* **316**, 572. **17.** Thompson, Datko & Mudd (1982) *Plant Physiol.* **70**, 1347. **18.** Bagley & Stipanuk (1994) *J. Nutr.* **124**, 2410. **19.** Thong & Coombs (1987) *Exp. Parasitol.* **63**, 143. **20.** Thong & Coombs (1985) *IRCS Med. Sci. Libr. Compend.* **13**, 493. **21.** Tokoro, Asai, Kobayashi, Takeuchi & Nozaki (2003) *J. Biol. Chem.* **278**, 42717. **22.** Dias & Weimer (1998) *Appl. Environ. Microbiol.* **64**, 3327. **23.** Lockwood & Coombs (1991) *Biochem. J.* **279**, 675.

L-Propargylglycine

This alkynyl amino acid and mechanism-based inhibitor (FW = 113.11 g/mol; CAS 23235-01-0), also known as (*S*)-2-amino-4-pentynoic acid, irreversibly inactivates γ-cystathionase (an enzyme that also catalyzes the synthesis of the metabolic signaling gas, H$_2$S, **See also** *Hydrogen Sulfide*), acting as a Michael addition type suicide inhibitor of cystathionine γ-lyase, cystathionine γ-synthase, alanine aminotransferase, and methionine γ-lyase. Propargylglycine inactivates γ-cystathionase with pseudo-first-order kinetics and incorporation of 1 mol inhibitor per 80 kDa enzyme (1,2). When studied *in vivo*, inactivation of cystathionine γ-lyase in rat kidney was less than that in the liver, owing to the presence of a higher cysteine concentration in kidney (3). L-Propargylglycine was found to inactivate pig heart L-alanine transaminase (EC 2.6.1.2) at 37° C with a K_i = 3.9 mM, an observed maximal first order rate constant, k_{inact} = 0.26 min^{-1}, a minimal stoichiometric ratio necessary for inactivation of 2.7 L-propargylglycine molecules/enzyme subunit, with 2.2 molecules/subunit undergoing transamination before inactivation ensues (4). Experimental cystathioninuria, induced in rats by administration of D,L-propargylglycine, results in the formation of the cystathionine metabolites, cystathionine ketimine and perhydro-1,4-thiazepine-3,5-dicarboxylic acid, in various regions of the brain (5). L-Propargylglycine irreversibly inactivates proline dehydrogenase, which catalyzes the first step of proline catabolism, oxidizing proline to pyrroline-5-carboxylate (6). The 1.9-Å resolution structure of the inactivated *Thermus thermophilus* enzyme shows that N^5 of the flavin cofactor is covalently connected to the ε-amino group of Lys-99 via a three-carbon linkage, consistent with the mass spectral analysis of the inactivated enzyme. The isoalloxazine ring has a butterfly angle of 25°, suggesting the cofactor is reduced. Two mechanisms, both involving oxidation to *N*-propargyliminoglycine, can account for these properties (6). L-Propargylglycine irreversibly inhibits L-amino acid oxidase from the venom of *Crotalus adamanteus* (Eastern diamondback rattlesnake) and *Crotalus atrox* (Western diamondback rattlesnake) in a dose- and time-dependent manner that was blocked by the substrate L-phenylalanine (7). Other targets include methionine γ-lyase (8,9) and UDP-*N*-acetylmuramate:L-alanine ligase, *or* L-alanine-adding enzyme, *or* UDP-*N*-acetylmuramoyl-L-alanine synthetase (7,8). **1.** Abeles & Walsh (1973) *J. Am. Chem. Soc.* **95**, 6124. **2.** Washtien & Abeles (1977) *Biochemistry* **16**,

2485. **3**. Awata, Nakayama, Suzuki & Kodama (1989) *Acta Med. Okayama.* **43**, 329. **4**. Burnett, Marcotte & Walsh (1980) *J. Biol. Chem.* **255**, 3487. **5**. Yu, Sugahara, Nakayama, Awata & Kodama (2000) *Metabolism* **49**, 1025. **6**. White, Johnson, Whitman & Tanner (2008) *Biochemistry* **47**, 5573. **7**. Mitra & Bhattacharyya (2013) *FEBS Open Bio.* **3**, 135. **8**. Ator & Ortiz de Montellano (1990) *The Enzymes*, 3rd ed. (Sigman & Boyer, eds.), **19**, 213. **9**. Johnston, Jankowski, Marcotte, *et al.* (1979) *Biochemistry* **18**, 4690. **10**. Liger, Blanot & van Heijenoort (1991) *FEMS Microbiol. Lett.* **64**, 111. **11**. Liger, Masson, Blanot, van Heijenoort & Parquet (1995) *Eur. J. Biochem.* **230**, 80.

Propavane, *See* Propiomazine
2-Propenoic Acid, *See* Acrylic Acid

Propentofylline

This peripheral xanthine-based vasodilator and neuroprotector (FW = 306.36 g/mol; CAS 55242-55-2), known systematically as 3,7-dihydro-3-methyl-1-(5-oxohexyl)-7-propyl-1*H*-purine-2,6-dione, is a nerve growth factor stimulator that is also a non-selective adenosine receptor antagonist. Propentofylline used to treat or ameliorate peptic ulcer or irritation of the gastrointestinal tract. **Target(s):** adenosine receptor (1); adenosine transport (1,2); cAMP phosphodiesterase (1,3-5); and cGMP-stimulated phosphodiesterase (5). **1**. Borgland, Castanon, Spevak & Parkinson (1998) *Can. J. Physiol. Pharmacol.* **76**, 1132. **2**. Peralta, Hotter, Closa, *et al.* (1999) *Hepatology* **29**, 126. **3**. Yamagishi, Homma, Haruta, Iwatsuki & Chiba (1985) *Jpn. J. Pharmacol.* **39**, 31. **4**. Nagata, Ogawa, Omosu, Fujimoto & Hayashi (1985) *Arzneimittelforschung* **35**, 1034. **5**. Meskini, Nemoz, Okyayuz-Baklouti, Lagarde & Prigent (1994) *Biochem. Pharmacol.* **47**, 781.

Propiconazole

This fungicide (FW = 342.22 g/mol; CAS 60207-90-1), named systematically as 1-((2-(2,4-dichlorophenyl)-4-propyl-1,3-dioxolan-2-yl)methyl)-1*H*-1,2,4-triazole, inhibits aromatase, *or* CYP19 (1), (*S*)-canadine synthase (2), laurate ω-hydroxylase (3), sterol 14-demethylase, *or* CYP5 (1), and lanosterol 14α-demethylase (3,4). **1**. Sanderson, Boerma, Lansbergen & van den Berg (2002) *Toxicol. Appl. Pharmacol.* **182**, 44. **2**. Rueffer & Zenk (1994) *Phytochemistry* **36**, 1219. **3**. Sanglard, Käppeli & Fiechter (1986) *Arch. Biochem. Biophys.* **251**, 276. **4**. Trösken, Adamska, Arand, *et al.* (2006) *Toxicology* **228**, 24.

Propidium Diiodide

This fluorescent DNA- and RNA-intercalating dye (FW = 668.40 g/mol; CAS 36015-30-2 and 25535-16-4) is often used as a stain for nucleic acids and as a probe of chromatin structure and exhibits a ten-fold enhancement of fluorescence upon binding to acetylcholinesterase (1). **Target(s):** acetylcholinesterase (1-6); cholinesterase, *or* butyrylcholinesterase (7); protein kinase C, IC$_{50}$ = 16 μM (8); and restriction endonuclease (9). **1**. Taylor & Lappi (1975) *Biochemistry* **14**, 1989. **2**. Costagli & Galli (1998) *Biochem. Pharmacol.* **55**, 1733. **3**. Cohen, Kronman, Chitlaru, *et al.* (2001) *Biochem. J.* **357**, 795. **4**. Sanders, Mathews, Sutherland, *et al.* (1996) *Comp. Biochem. Physiol. B* **115**, 97. **5**. Hussein, Chacón, Smith, Tosado-Acevedo & Selkirk (1999) *J. Biol. Chem.* **274**, 9312. **6**. Bentley, Jones & Agnew (2005) *Mol. Biochem. Parasitol.* **141**, 119. **7**. McClellan, Coblentz, Sapp, *et al.* (1998) *Eur. J. Biochem.* **258**, 419. **8**. Abeywickrama, Rotenberg & Baker (2006) *Bioorg. Med. Chem.* **14**, 7796. **9**. Soslau & Pirollo (1983) *Biochem. Biophys. Res. Commun.* **115**, 484.

Propiolic Acid, *See* Propargylic Acid

Propiolaldehyde

This aldehyde (FW = 54.048 g/mol; CAS 624-67-9), also known as propynal, synthesized from propargyl alcohol, is an irreversible inhibitor of aldehyde dehydrogenases (1-6). **1**. DeMaster, Shirota & Nagasawa (1986) *Biochem. Pharmacol.* **35**, 1481. **2**. Shirota, DeMaster & Nagasawa (1979) *J. Med. Chem.* **22**, 463. **3**. Pietruszko, MacKerell, Jr., & Ferencz-Biro (1985) *Prog. Clin. Biol. Res.* **183**, 67. **4**. Ferencz-Biro & Pietruszko (1984) *Alcohol Clin. Exp. Res.* **8**, 302. **5**. Shirota, DeMaster, Elberling & Nagasawa (1980) *J. Med. Chem.* **23**, 669. **6**. DeMaster, Shirota & Nagasawa (1980) *Adv. Exp. Med. Biol.* **132**, 219.

Propiolyl-CoA

This fatty-acyl-CoA analogue (FW = 820.58 g/mol), also known as 2-propynoyl-CoA, inactivates general acyl-CoA dehydrogenase. **1**. Freund, Mizzer, Dick & Thorpe (1985) *Biochemistry* **24**, 5996.

Propiomazine

This substituted phenothiazine (FW$_{free-base}$ = 340.49 g/mol; CAS 362-29-8), known systematically as 1-(10-(2-(dimethylamino)propyl)-10*H*-phenothiazin-2-yl)-1-propanone and often used clinically as a sedative, inhibits oxidative phosphorylation. The maleate salt is frequently referred to as Propavane or Propavan. **1**. Rasmussen (1964) *Biochem. Biophys. Res. Commun.* **16**, 19.

Propionaldehyde, *See* Propanal

Propionamide

This simple alkylamide CH$_3$CH$_2$CONH$_2$ (FW = 73.095 g/mol; CAS 79-05-0; M.P. = 79°C; B.P. = 222.2°C) is freely soluble in water, ethanol, and chloroform. Propionaldehyde is an alternative substrate for a many amidases. **Target(s):** alcohol dehydrogenase (1,2); dihydropyrimidinase, *or* hydantoinase (3); and nitrile hydratase, weakly inhibited (4). **1**. Woronick (1961) *Acta Chem. Scand.* **15**, 2062. **2**. Winer & Theorell (1960) *Acta Chem. Scand.* **14**, 1729. **3**. Jahnke, Podschun, Schnackerz, Kautz & Cook (1993) *Biochemistry* **32**, 5160. **4**. Hjort, Godtfredsen & Emborg (1990) *J. Chem. Technol. Biotechnol.* **48**, 217.

Propionic Acid *N*-Hydroxysuccinimide Ester, *See* N-Succinimidyl Propionate

Propionitrile

This toxic nitrile CH_3CH_2CN (FW = 55.079 g/mol; CAS 107-12-0; M.P. = –91.8°C; B.P. = 97.2°C), itself a substrate for nitrile hydratase, inhibits amidase (1) and succinate dehydrogenase, or succinate oxidase (2,3). **1.** Maestracci, Bui, Thiery, Arnaud & Galzy (1984) *Biotechnol. Lett.* **6**, 149. **2.** Stoppani (1948) *Enzymologia* **13**, 165. **3.** Sen (1931) *Biochem. J.* **25**, 849.

Propionohydroxamic Acid

This metal ion-chelating hydroxamic acid (FW = 89.094 g/mol) inhibits ribonucleoside-diphosphate reductase (1) and urease (2,3), most likely by forming coordination complexes with their respective active-site iron and nickel ions. **1.** Larsen, Sjöberg & Thelander (1982) *Eur. J. Biochem.* **125**, 75. **2.** Fishbein & Daly (1970) *Proc. Soc. Exp. Biol. Med.* **134**, 1083. **3.** Martelli, Buli, Cortecchia, *et al.* (1987) *Contrib. Nephrol.* **58**, 196.

Propionylcholine Chloride

This hygroscopic acetylcholine analogue (FW = 195.69 g/mol) is a substrate of cholinesterase (or butyrylcholinesterase) and a weak substrate for acetylcholinesterase. Solutions should always be freshly prepared as propionylcholine chloride hydolyzes nonenzymatically. **See also Acetylcholine Target(s):** aryl-acylamidase (1); choline kinase (2); ethanolamine kinase (3); and Na^+/K^+-exchanging ATPase (4). **1.** Oommen & Balasubramanian (1979) *Eur. J. Biochem.* **94**, 135. **2.** Ishidate & Nakazawa (1992) *Meth. Enzymol.* **209**, 121. **3.** Sung & Johnstone (1967) *Biochem. J.* **105**, 497. **4.** Takachuk, Lopina & Boldyrev (1975) *Biokhimiia* **40**, 1032.

Propionyl-CoA

This thio ester ($FW_{free-acid}$ = 823.61 g/mol), also known as propanoyl-CoA, is a key intermediate in the degradation of odd-chain fatty acids, L-methionine, L-valine, L-threonine, and L-isoleucine. **Target(s):** acetyl-CoA synthetase, or acetate:CoA ligase (1); *N*-acetylglutamate synthase, or glutamate *N*-acetyltransferase, or amino-acid *N*-acetyltransferase (2); [acyl-carrier-protein] *S*-malonyltransferase (32); 1-alkyl-*sn*-glycero-3-phosphate acetyltransferase, weakly inhibited (3); amino-acid acetyltransferase, or glutamate acetyltransferase, or *N*-acetylglutamate synthase (2); arylamine *N*-acetyltransferase, as alternative substrate for the enzyme from certain sources (34); carbamoyl-phosphate synthetase, ammonia-requiring (4); carnitine *O*-palmitoyltransferase (5); citrate (*si*)-synthase, as alternative substrate for the enzyme from several sources (6); glycine cleavage complex, or glycine synthase (7); 3-hydroxyisobutyryl-CoA hydrolase (8,9); 3-hydroxy-3-methylglutaryl-CoA reductase, weakly inhibited (10); 3-hydroxy-3-methylglutaryl-CoA synthase (11,12); α-ketoglutarate dehydrogenase (13); malonyl-CoA decarboxylase (14-20); malyl-CoA lyase (21); 3-methylcrotonoyl-CoA carboxylase, K_i = 1.78 mM (22); 3-methylcrotonoyl-CoA hydratase, or enoyl-CoA hydratase (23); methylmalonyl-CoA epimerase (24); (*S*)-methylmalonyl-CoA hydrolase (25); [3-methyl-2-oxobutanoate dehydrogenase, or 2-methylpropanoyl-transferring] phosphatase, or [branched-chain-α-keto-acid dehydrogenase] phosphatase (26,27); oxaloacetate decarboxylase (28); pantothenate kinase (29); phosphoprotein phosphatase (27); pyruvate dehydrogenase complex (30); and succinyl-CoA synthetase (31). **1.** Preston, Wall & Emerich (1990) *Biochem. J.* **267**, 179. **2.** Coude, Sweetman & Nyhan (1979) *J. Clin. Invest.* **64**, 1544. **3.** Baker & Chang (1995) *J. Neurochem.* **64**, 364. **4.** Martin-Requero, Corkey, Cerdan, *et al.* (1983) *J. Biol. Chem.* **258**, 3673. **5.** Kashfi, Mynatt & Cook (1994) *Biochim. Biophys. Acta* **1212**, 245. **6.** Lee, Park &

Yim (1997) *Mol. Cells* **7**, 599. **7.** Hayasaka & Tada (1983) *Biochem. Int.* **6**, 225. **8.** Shimomura, Murakami, Nakai, *et al.* (2000) *Meth. Enzymol.* **324**, 229. **9.** Shimomura, Murakami, Fujitsuka, *et al.* (1994) *J. Biol. Chem.* **269**, 14248. **10.** Kirtley & Rudney (1967) *Biochemistry* **6**, 230. **11.** Higgins, Kornblatt & Rudney (1972) *The Enzymes*, 3rd ed. (Boyer, ed.), **7**, 407. **12.** Middleton (1972) *Biochem. J.* **126**, 35. **13.** Dynnik, Maevskii, Grigorenko & Kim (1984) *Biofizika* **29**, 954. **14.** Kolattukudy, Poulose & Kim (1981) *Meth. Enzymol.* **71**, 150. **15.** Scholte (1973) *Biochim. Biophys. Acta* **309**, 457. **16.** Kim, Kolattukudy & Boos (1979) *Arch. Biochem. Biophys.* **196**, 543. **17.** Koeppen, Mitzen & Ammoumi (1974) *Biochemistry* **13**, 3589. **18.** Kim & Kolattukudy (1978) *Arch. Biochem. Biophys.* **190**, 234. **19.** Kim & Kolattukudy (1978) *Biochim. Biophys. Acta* **531**, 187. **20.** Hunaiti & Kolattukudy (1984) *Arch. Biochem. Biophys.* **229**, 426. **21.** Hersh (1974) *J. Biol. Chem.* **249**, 5208. **22.** Lau, Cochran & Fall (1980) *Arch. Biochem. Biophys.* **205**, 352. **23.** Dhar, Dhar & Rosazza (2002) *J. Ind. Microbiol. Biotechnol.* **28**, 81. **24.** Stabler, Marcell & Allen (1985) *Arch. Biochem. Biophys.* **241**, 252. **25.** Kovachy, Copley & Allen (1983) *J. Biol. Chem.* **258**, 11415. **26.** Damuni, Merryfield, Humphreys & Reed (1984) *Proc. Natl. Acad. Sci. U.S.A.* **81**, 4335. **27.** Reed & Damuni (1987) *Adv. Protein Phosphatases* **4**, 59. **28.** Horton & Kornberg (1964) *Biochim. Biophys. Acta* **89**, 381. **29.** Halvorsen & Skrede (1982) *Eur. J. Biochem.* **124**, 211. **30.** Gregersen (1981) *Biochem. Med.* **26**, 20. **31.** Stumpf, McAfee, Parks & Eguren (1980) *Pediatr. Res.* **14**, 1127. **32.** Guerra & Ohlrogge (1986) *Arch. Biochem. Biophys.* **246**, 274. **33.** Goldstein (1959) *J. Biol. Chem.* **234**, 2702. **34.** Paul & Ratledge (1973) *Biochim. Biophys. Acta* **320**, 9.

Propionyl Pantetheine

This acylated pantetheine (FW = 334.44 g/mol) inhibits methylmalonyl-CoA carboxyltransferase, or transcarboxylase, K_i = 4 mM (1,2). **1.** Wood, Jacobson, Gerwin & Northrop (1969) *Meth. Enzymol.* **13**, 215. **2.** Wood (1972) *The Enzymes*, 3rd ed. (PBoyer, ed.), **6**, 83.

Propioxatin A and B

These hydroxamic acids (MW = 360.43g/mol for proploxatin A (R = H); CAS 102962-94-7; MW = 374.46 g/mol for proploxatin B (R = CH_3; CAS 102962-95-8) from *Kitasatosporia setae*, inhibit dipeptidyl-peptidase III with respective K_i values of 13 and 110 nM (1-3) as well as *Serratia* metalloproteinase (4). **1.** Inaoka & Tamaoki (1987) *Biochim. Biophys. Acta* **925**, 27. **2.** Inaoka, Tamaoki, Takahashi, Enokita & Okazaki (1986) *J. Antibiot.* **39**, 1368. **3.** Inaoka & Naruto (1988) *J. Biochem.* **104**, 706. **4.** Murao, Imafuku & Oyama (1997) *Biosci. Biotechnol. Biochem.* **61**, 561.

Propofol

This intravenously administered, nonanalgesic anesthetic (FW = 178.27 g/mol; CAS 2078-54-8; pK_a = 11; *n*-Octanol:H_2O Partition Coefficient = 6761:1, over pH 6–8.5 range), known also by its brand name Diprivan® and systematically as 2,6-diisopropylphenol, induces and maintains general anesthesia, as required for surgery. Propofol has a predictable duration of action, arising from its rapid penetration of the blood-brain barrier, followed by redistribution to inactive tissue depots, such as muscle and fat. **Likely Mode(s) of Action:** Anesthetics typically exert wide-ranging effects, targeting ligand-gated ion channels or receptors for GABA, glycine, and NMDA as well as channels for K^+, Na^+, and Ca^{2+}. Propofol enhances two types of GABAergic inhibition: (a) a synaptic form (referred to as phasic inhibition) that regulates neural excitability by means of activating postsynaptic $GABA_A$ receptors through intermittent GABA release from presynaptic terminals, and (b) a persistent form (known as tonic inhibition) that is generated through continuous activation of extrasynaptic $GABA_A$ receptors by low concentrations of ambient GABA (1). **Pharmacokinetics:** In humans, a two-compartment analysis (2) yields the central volume of distribution (V_C = 24.7 liters), the peripheral volume of distribution (V_T = 112 liters), the metabolic clearance (Cl = 2.64 l/min), and the inter-compartmental clearance (Q = 0.989 l/min). The delay of the anesthetic effect (expressed with respect to plasma concentrations) is described by the effect compartment with the rate constant for the distribution to the effector compartment equal to 0.24 min^{-1} (2). Despite its 2–24 hour biological half-life, propofol's sedative effect is much shorter, allowing for rapid recovery of consciousness. In a three-compartment model (3), the first exponential ($t_{1/2}$ = 2–3 minutes) describes its rapid onset of action, the second exponential ($t_{1/2}$ = 0.5–1 hour) defines its high metabolic clearance, and the long third exponential phase ($t_{1/2}$ = 3–6 hours) describes the slow elimination of a small proportion of the drug remaining in poorly perfused tissues. Importantly, propofol sedation should only occur while patients are under mechanical ventilation, a precaution relying on its fast metabolism and clearance to assure full regain of consciousness. (The notorious death of Singer Superstar Michael Jackson is believed to have occurred as the result of the tragic failure to heed such precautions.) Reference-(4) provides information on inter-racial variability, defined the difference between races in the median dose of propofol required for loss of consciousness within a race. Propofol is cleared reasonably quickly after metabolic glucuronidation, forming propofol-glucuronide, as well as sulfo- and glucuro-conjugation of its hydroxylated metabolite (6-diisopropyl-1,4-quinol) mainly by CYP2C9, but also CYP2A6, CYP2C8, CYP2C18, CYP2C19 and CYP1A2 (5). *See also:* Appendix II in Goodman & Gilman's THE PHARMACOLOGICAL BASIS OF THERAPEUTICS, 12th Edition (Brunton, Chabner & Knollmann, eds.) McGraw-Hill Medical, New York (2011). **Other Effects:** Rho-kinase is essential for propofol-induced neurite retraction in cortical neurons, and activation of PKC inhibits neurite retraction caused by propofol (5). Orexin A blocks propofol-induced neurite retraction by a PLD/PKCε-mediated pathway, and PKCε maybe the key enzyme, where wakefulness and anaesthesia signal pathways converge (6). **Target(s):** Cytochrome P450, as an alternative hydroxylation substrate (7); glucuronosyltransferase, as an alternative bioconjugation substrate (8); glutamate-dependent Ca^{2+} entry (8); glutamate transport, ATP-dependent (9); and ouabain-insensitive synaptosomal ATPase (9). **Formulation:** A typical neutral-pH formulation contains: the active ingredient propofol at 1% wt/vol (10 mg/mL); emulsifiers like soybean oil at 10% wt/vol (100 mg/mL) and egg phospholipid at 1.2% wt/vol (12 mg/mL); glycerol at 2% wt/vol (20 mg/mL); as well as EDTA to suppress microbial growth during storage. **1.** Nishikawa (2011) *Masui.* **60**, 534. **2.** Wiczling, Bienert, Sobczyński *et al.* (2012) *Pharmacol. Rep.* **64**, 113. **3.** Kanto & Gepts (1989) *Clin. Pharmacokinet.* **17**, 308. **4.** Lampotang, *et al.* (2016) *J. Clin. Pharmacol.* **56**, 1141. **5.** Favetta, Degoute, Perdrix, Dufresne, Boulieu & Guitton (2002) *British J. Anaesthes.* **88**, 653. **6.** Björnström, Turina, Strid, Sundqvist & Eintrei (2014) *PLoS One* **9**, e97. **7.** Baker, Chadam & Ronnenberg, Jr. (1993) *Anesth. Analg.* **76**, 817. **8.** Zhang, Chando, Everett, *et al.* (2005) *Drug Metab. Dispos.* **33**, 1729. **9.** Bianchi, Battistin & Galzigna (1991) *Neurochem Res.* **16**, 443.

Propoxur

This methylcarbamate insecticide (FW = 209.25 g/mol; CAS 114-26-1), known also as Aprocarb, Baygon, Sendran, and Unden and systematically as *o*-isopropoxyphenoyl *N*-methylcarbamate, inhibits acetylcholinesterase (1-4), carboxylesterase (5), cholinesterase, *or* butyrylcholineesterase (6), and nicotinic acetylcholine receptor (4). **1.** O'Brien, Hilton & Gilmour (1966) *Mol. Pharmacol.* **2**, 593. **2.** Kato, Tanaka & Miyata (2004) *Pestic. Biochem. Physiol.* **79**, 64. **3.** Frasco, Fournier, Carvalho & Guilhermino (2006) *Aquat. Toxicol.* **77**, 412. **4.** Smulders, Bueters, Van Kleef & Vijverberg (2003) *Toxicol. Appl. Pharmacol.* **193**, 139. **5.** Valles, Oi &

Strong (2001) *Insect Biochem. Mol. Biol.* **31**, 715. **6.** Iverson (1975) *Biochem. Pharmacol.* **24**, 1537.

D-Propoxyphene

This widely used analgesic ($FW_{free-base}$ = 325.45 g/mol; CAS 469-62-5), also known as (αS)-α-[(1R)-2-(dimethylamino)-1-methylethyl]-α-phenyl-benzeneethanol propionate and Darvon® for its hydrochloride salt), is sufficiently habit-forming that its use is controlled. First marketed as an alternative to codeine by Eli Lilly and Co. in 1957, propoxyphene is an opioid receptor agonist. In addition, it is an NMDA (*N*-methyl-D-aspartate) receptor antagonist. Only the dextro-isomer has an analgesic effect, and the levo-isomer appears to exert an antitussive effect. **Target(s):** aldehyde oxidase (1); Ca^{2+} channels (2); carboxylesterase (3); CYP2C9 (4); and NMDA receptor (5,6). **1.** Robertson & Gamage (1994) *Biochem. Pharmacol.* **47**, 584. **2.** Seyler, Borowitz & Maickel (1983) *Fundam. Appl. Toxicol.* **3**, 536. **3.** Zhang, Burnell, Dumaual & Bosron (1999) *J. Pharmacol. Exp. Ther.* **290**, 314. **4.** Levy (1995) *Epilepsia* **36** Suppl. 5, S8. **5.** Ebert, Thorkildsen, Andersen, Christrup & Hjeds (1998) *Biochem. Pharmacol.* **56**, 553. **6.** Ebert, Andersen, Hjeds & Dickenson (1998) *J. Pain Symptom Manage.* **15**, 269.

Propranolol

R-Propanolol   *S*-Propanolol

This β-blocking antihypertensive ($FW_{free-base}$ = 259.35 g/mol; CAS 525-66-6; stable at acidic/neutral pH, but not at alkaline pH), widely known as Inderal® and systematically named as 1-[(1-methylethyl)amino]-3-(1-naphthalenyloxy)-2-propanol, is widely used to treat hypertension, cardiac arrhythmia, and migraines. Propranolol is both a β-adrenergic receptor antagonist and a 5-HT$_1$/5-HT$_2$ serotonin receptor antagonist. While the (S)-enantiomer is the active β-antagonist, racemic propranolol is most often dispensed. Propranolol is also effective for controlling intentional tremor, but with a requirement for escalating drug levels, when used chronically. For his germinal work on β-antagonists, including propranolol, the Scottish pharmacologist James Whyte Black was awarded the Nobel Prize in Physiology or Medicine in 1988. **Key Pharmacokinetic Parameters:** *See* Appendix II in Goodman & Gilman's THE PHARMACOLOGICAL BASIS OF THERAPEUTICS, 12th Edition (Brunton, Chabner & Knollmann, eds.) McGraw-Hill Medical, New York (2011). **Target(s):** acetylcholine esterase (1); β-adrenergic receptor (2-4); [β-adrenergic-receptor] kinase (5); alcohol dehydrogenase (6,7); aldehyde dehydrogenase (7); 1-alkylglycerophosphocholine *O*-acetyltransferase (38); 5-aminolevulinate synthase (39); Ca^{2+}-dependent ATPase (8-10); cellobiohydrolase (11); choline-phosphate cytidylyltransferase (37); cholinesterase (1,12); CYP2D (6) and a few other cytochrome P450, or CYP, family members (13); glycosylphosphatidylinositol diacylglycerol-lyase (14); hexokinase (15); K^+ channels, ATP-sensitive (16); lecithin:cholesterol acyltransferase, *or* LCAT (17); lipoprotein lipase (18); Mg^{2+}-dependent ATPase (1,8,19); Na^+/K^+-exchanging ATPase (1,8); phosphatidate phosphatase (20-29); phospholipase A (30); prolyl hydroxylase, *or* procollagen-proline monooxygenase (31); prostaglandin-endoperoxide synthase, *or* cyclooxygenase (40); protein kinase C (22); serotonin 5-HT$_1$/5-HT$_2$ receptor (32-34); thyroxine 5'-deiodinase (35); and xanthine oxidase (36). **1.** Whittaker, Wicks & Britten (1982) *Clin. Chim. Acta* **119**, 107. **2.** Whitehurst, Vick, Alleva, *et al.* (1999) *Proc. Soc. Exp. Biol. Med.* **221**, 382. **3.** Tan & Summers (1995) *J. Auton. Pharmacol.* **15**, 421. **4.** Black, Duncan

& Shanks (1965) *Brit. J. Pharmacol. Chemother.* **25**, 577. **5**. Müller, Straub & Lohse (1997) *FEBS Lett.* **401**, 25. **6**. Brändén, Jörnvall, Eklund & Furugren (1975) *The Enzymes*, 3rd ed. (Boyer, ed.), **11**, 103. **7**. Julian & Duncan (1973) *Mol. Pharmacol.* **9**, 191. **8**. Gopalaswamy, Satav, Katyare & Bhattacharya (1997) *Chem. Biol. Interact.* **103**, 51. **9**. Meltzer & Kassir (1983) *Biochim. Biophys. Acta* **755**, 452. **10**. Hasselbach (1974) *The Enzymes*, 3rd ed. (Boyer, ed.), **10**, 431. **11**. Stahlberg, Henriksson, Divne, et al. (2001) *J. Mol. Biol.* **305**, 79. **12**. Alkondon, Ray & Sen (1986) *J. Pharm. Pharmacol.* **38**, 848. **13**. Kastelova & Yanev (2002) *Methods Find. Exp. Clin. Pharmacol.* **24**, 189. **14**. Butikofer & Brodbeck (1993) *J. Biol. Chem.* **268**, 17794. **15**. Dovrat, Horwitz, Sivak, et al. (1993) *Exp. Eye Res.* **57**, 747. **16**. Xie, Takano & Noma (1998) *Brit. J. Pharmacol.* **123**, 599. **17**. Schauer, Schauer & Thielmann (1984) *Int. J. Clin. Pharmacol. Ther. Toxicol.* **22**, 608. **18**. Kubo & Hostetler (1987) *Biochim. Biophys. Acta* **918**, 168. **19**. Wei, Lin, Hong & Chiang (1985) *Biochem. Pharmacol.* **34**, 911. **20**. Carman & Quinlan (1992) *Meth. Enzymol.* **209**, 219. **21**. Meier, Gause, Wisehart-Johnson, et al. (1998) *Cell Signal* **10**, 415. **22**. Sozzani, Agwu, McCall, et al. (1992) *J. Biol. Chem.* **267**, 20481. **23**. Abdel-Latif & Smith (1984) *Can. J. Biochem. Cell Biol.* **62**, 170. **24**. Fleming & Yeaman (1995) *Biochem. J.* **308**, 983. **25**. Kanoh, Imai, Yamada & Sakane (1992) *J. Biol. Chem.* **267**, 25309. **26**. Furneisen & Carman (2000) *Biochim. Biophys. Acta* **1484**, 71. **27**. English, Martin, Harvey, et al. (1997) *Biochem. J.* **324**, 941. **28**. Morlock, McLaughlin, Lin & Carman (1991) *J. Biol. Chem.* **266**, 3586. **29**. Roberts, Sciorra & Morris (1998) *J. Biol. Chem.* **273**, 22059. **30**. Trotz, Jellison & Hostetler (1987) *Biochem. Pharmacol.* **36**, 425. **31**. Iwatsuki (1980) *Tohoku J. Exp. Med.* **132**, 365. **32**. Tinajero, Fabbri & Dufau (1993) *Endocrinology* **133**, 257. **33**. Alexander & Wood (1987) *J. Pharm. Pharmacol.* **39**, 664. **34**. Goodwin & Green (1985) *Brit. J. Pharmacol.* **84**, 743. **35**. Garcia-Macias, Molinero, Guerrero & Osuna (1994) *FEBS Lett.* **354**, 110. **36**. Janero, Lopez, Pittman & Burghardt (1989) *Life Sci.* **44**, 1579. **37**. Pelech, Jetha & Vance (1983) *FEBS Lett.* **158**, 89. **38**. Garcia, Montero, Alvarez & Sanchez Crespo (1993) *J. Biol. Chem.* **268**, 4001. **39**. Kaliman & Barannik (1999) *Biochemistry* **64**, 699. **40**. Pace-Asciak (1972) *Biochim. Biophys. Acta* **280**, 161.

N-Propylamido-epoxysuccinyl-L-isoleucyl-L-proline, *See N-[L-3-trans-(Propyl-carbomoyl)oxirane-2-carbonyl]-L-isoleucyl-L-proline*

N^ω-Propyl-L-arginine

This alkylated arginine/mol derivative (FW = 216.28 g/mol; CAS 137361-05-8) inhibits neuronal nitric-oxide synthase (IC$_{50}$ = 57 nM). It is a weaker inhibitor of the endothelial and inducible forms of the enzyme, which have IC$_{50}$ values of 8.5 and 180 μM, respectively (1-4). **1**. Griffith & Kilbourn (1996) *Meth. Enzymol.* **268**, 375. **2**. Cooper, Mialkowski & Wolff (2000) *Arch. Biochem. Biophys.* **375**, 183. **3**. Zhang, Fast, Marletta, Martasek & Silverman (1997) *J. Med. Chem.* **40**, 3869. **4**. Babu, Frey & Griffith (1999) *J. Biol. Chem.* **274**, 25218.

N-(α(R)-Propylbenzylaminocarbonyl) 4(S)-(4-carboxymethyl-phenoxy)-3,3-diethylazetidin-2-one

This nonpeptidic alkylazetidinone (FW$_{free-acid}$ = 452.55 g/mol), also known as L-680414, is stable, potent, highly specific, and time-dependent monocyclic β-lactam inhibitor of human leucocyte elastase (HLE). The specificity of this compound toward HLE *versus* porcine pancreatic elastase (PPE) is consistent with the differences in substrate specificity reported for

these enzymes. **1**. Knight, Green, Chabin, et al. (1992) *Biochemistry* **31**, 8160.

N-L-3-trans-(Propylcarbamoyl)oxirane-2-carbonyl-L-isoleucyl-L-proline

This oxirane-containing peptide (FW$_{free-acid}$ = 383.44 g/mol), known also as cathepsin B inhibitor III and CA-074 and named systematically as *N*-propylamido-epoxysuccinyl-L-isoleucyl-L-proline, is a potent, irreversible, and specific inhibitor of cathepsin B, with a IC$_{50}$ value of 2.2 nM for the rat liver enzyme (1-8), but weakly inhibits cathepsin K (9,10) cathepsin X (11), and miltpain (12). CA-074 also reduces ischemia-induced neural death. **1**. Buttle, Murata, Knight & Barrett (1992) *Arch. Biochem. Biophys.* **299**, 377. **2**. Mort (1998) in *Handb. Proteolytic Enzymes* (Barrett, Rawlings & Woessner, eds.), p. 609, Academic Press, San Diego. **3**. Katunuma & Kominami (1995) *Meth. Enzymol.* **251**, 382. **4**. Chan, Selzer, McKerrow & Sakanari (1999) *Biochem. J.* **340**, 113. **5**. Murata, Miyashita, Yokoo, et al. (1991) *FEBS Lett.* **280**, 307. **6**. Towatari, Nikawa, Murata, et al. (1991) *FEBS Lett.* **280**, 311. **7**. Ghoneim & Klinkert (1995) *Int. J. Parasitol.* **25**, 1515. **8**. Yamamoto, Tomoo, Hara, et al. (2000) *J. Biochem.* **127**, 635. **9**. Menard, Therrien, Lachance, et al. (2001) *Biol. Chem.* **382**, 839. **10**. Therrien, Lachance, Sulea, et al. (2001) *Biochemistry* **40**, 2702. **11**. Klemencic, Carmona, Cezari, et al. (2000) *Eur. J. Biochem.* **267**, 5404. **12**. Kawabata, Doi & Ichishima (2000) *Comp. Biochem. Physiol. B Biochem. Mol. Biol.* **125**, 533.

N-L-3-trans-(Propylcarbamoyl)oxirane-2-carbonyl-L-isoleucyl-L-proline Methyl Ester

This methyl ester (FW = 397.47 g/mol), also known as CA-074 methyl ester, is a prodrug for *N*-L-3-*trans*-(propylcarbamoyl)oxirane-2-carbonyl-L-isoleucyl-L-proline, *or* CA-074, a potent inhibitor of cathepsin B (1), with weaker action on miltpain (2). **1**. Buttle, Murata, Knight & Barrett (1992) *Arch. Biochem. Biophys.* **299**, 377. **2**. Kawabata & Ichishima (1997) *Comp. Biochem. Physiol. B Biochem. Mol. Biol.* **117**, 445.

α-Propyl-3,4-dihydroxyphenylacetamine, *See 2-(3,4-Dihydroxyphenyl)-2-propylacetamide*

Propyl-2,2-diphenyl Diselenide

This poorly named organoselenium compound (FW = 354.21 g/mol) inhibits 5-aminolevulinate dehydratase, *or* porphobilinogen synthase (1). The effect may be mediated by diphenyl diselenide, because propyl-2,2-diphenyl diselenide is known to decomposes rapidly, forming diphenyl diselenide in aqueous media. **1**. Barbosa, Rocha, Zeni, et al. (1998) *Toxicol. Appl. Pharmacol.* **149**, 243.

Propylene Dichloride, *See 1,2-Dichloropropane*

Propylene Glycol, *See 1,2-Propanediol*

Propylene Oxide

This oxirane (FW = 58.080 g/mol; CAS 75-56-9; M.P. = –112.13°C; B.P. = 34.23°C; very soluble in water), also known as 1,2-epoxypropane, is produced enzymatically from propene by methane monooxygenase. *Note:* There are two stereoisomers of propylene oxide. **Target(s):** alkene monooxygenase (1); glutathione *S*-transferase (2); and methane monooxygenase, as alternative product inhibition (3). **1**. Habets-Cruetzen & de Bont (1985) *Appl. Microbiol. Biotechnol.* **22**, 428. **2**. Ansari, Singh, Gan & Awasthi (1987) *Toxicol. Lett.* **37**, 57. **3**. Vasil'ev, Tikhonova, Gvozdev, Tukhvatullin & Popov (2006) *Biochemistry (Moscow)* **71**, 1329.

n-Propyl Gallate

This antioxidant (FW = 212.20 g/mol; CAS 121-79-9; B.P. = 150°C, Solubility: 0.35 g/100 mL water at 25°C; Decomposes in the presence of iron ions), also known as Tenox PG and 3,4,5-trihydroxybenzoic acid propyl ester, exhibits antimicrobial activity and is frequently used as an additive in food products. *n*-Propyl gallate also inhibits the production of ethylene (1,2). In the presence of copper ions, elevated concentrations of *n*-propyl gallate, induce single-strand DNA breaks. *n*-Propyl gallate also inhibits the superoxide-generating activity of xanthine oxidase. Note that this reagent is also reported to prevent neuronal apoptosis. In addition, this reagent has proved useful in fluorescence microscopy, reducing photobleaching. **Target(s):** alkylglycerophosphoethanolamine phosphodiesterase, *or* lysophospholipase D (17,27); alternative oxidase (3-6); aminocyclopropane-carboxylate oxidase (36); 1-aminocyclopropane-1-carboxylate synthase (7); ascorbate oxidase (8); β-carotene 15,15'-monooxygenase (30); carotene oxidase (9); cyclooxygenase (10); CYP1A2 (11); ethylene production (1,2,7); fatty-acid Δ^5-desaturase (21); fatty-acid Δ^6-desaturase (21); fatty-acid synthase (29); 3-galactosyl-*N*-acetylglucosaminide 4-α-L-fucosyltransferase (28); glucan 1,3-α-glucosidase, *or* glucosidase II, *or* mannosyl-oligosaccharide glucosidase II (25); glucose-6-phosphatase (12); glycoprotein 3-α-L-fucosyltransferase (28); horseradish peroxidase (13); hyaluronoglucosaminidase, *or* hyaluronidase (26); iodide peroxidase, *or* thyroid peroxidase (13); lipoxygenase (10,14-16,34,35,38-40); 15-lipoxygenase, *or* arachidonate 15-lipoxygenase (33,34); NADPH:cytochrome *c* reductase (18); peroxidase (13,41); phytanoyl-CoA dioxygenase (31); procollagen-proline 4-dioxygenase (32); retinal oxidase (19); staphylococcal thermonuclease (20); tyrosine 3-monooxygenase (22,23); and xanthine oxidase (24,37). **1**. Weinke, Kahl & Kappus (1987) *Toxicol. Lett.* **35**, 247. **2**. Mattoo & Lieberman (1977) *Plant Physiol.* **60**, 794. **3**. Murphy & Lang-Unnasch (1999) *Antimicrob Agents Chemother.* **43**, 651. **4**. Popov, Simonian, Skulachev & Starkov (1997) *FEBS Lett.* **415**, 87. **5**. Kay & Palmer (1985) *Biochem. J.* **228**, 309. **6**. Parrish & Leopold (1978) *Plant Physiol.* **62**, 470. **7**. Boller, Herner & Kende (1979) *Planta* **145**, 293. **8**. Stark & Dawson (1963) *The Enzymes*, 2nd ed. (Boyer, Lardy & Myrbäck, eds.), **8**, 297. **9**. Jüttner (1988) *Meth. Enzymol.* **167**, 336. **10**. Van Wauwe & Goossens (1983) *Prostaglandins* **26**, 725. **11**. Baer-Dubowska, Szaefer & Krajka-Kuzniak (1998) *Xenobiotica* **28**, 735. **12**. Paradisi, Negro, Panagini & Torrielli (1979) *Boll. Soc. Ital. Biol. Sper.* **55**, 1877. **13**. Grintsevich, Senchuk, Puchkaev & Metelitza (2000) *Biochemistry (Moscow)* **65**, 924. **14**. Tappel (1962) *Meth. Enzymol.* **5**, 539. **15**. Schewe, Wiesner & Rapoport (1981) *Meth. Enzymol.* **71**, 430. **16**. Tappel (1963) *The Enzymes*, 2nd ed. (Boyer, Lardy & Myrbäck, eds.), **8**, 275. **17**. Tokumura, Miyake, Yoshimoto, Shimizu & Fukuzawa (1998) *Lipids* **33**, 1009. **18**. Torrielli & Slater (1971) *Biochem. Pharmacol.* **20**, 2027. **19**. Bhat, Poissant & Lacroix (1988) *Biochim. Biophys. Acta* **967**, 211. **20**. Kumar, Sharma & Kulkarni (2000) *Nahrung* **44**, 272. **21**. Kawashima, Akimoto, Shirasaka & Shimizu (1996) *Biochim. Biophys. Acta* **1299**, 34. **22**. Shiman & Kaufman (1970) *Meth. Enzymol.* **17A**, 609. **23**. Moore & Dominic (1971) *Fed. Proc.* **30**, 859. **24**. Lin, Chen, Ho & Lin-Shiau (2000) *J. Agric. Food Chem.* **48**, 2736. **25**. Gamberucci, Konta, Colucci, *et al.* (2006) *Biochem. Pharmacol.* **72**, 640. **26**. Girish & Kemparaju (2005) *Biochemistry (Moscow)* **70**, 948. **27**. Xie & Meier (2004) *Cell. Signal.* **16**, 975. **28**. Niu, Fan, Sun, *et al.* (2004) *Arch. Biochem. Biophys.* **425**, 51. **29**. Wang, Song, Guo & Tian (2003) *Biochem. Pharmacol.* **66**, 2039. **30**. Nagao, Maeda, Lim, Kobayashi & Terao (2000) *Nutr. Biochem.* **11**, 348. **31**. Jansen, Mihalik, Watkins, *et al.* (1998) *Clin. Chim. Acta* **271**, 203. **32**. Hutton, Jr., Tappel & Udenfriend (1967) *Arch. Biochem. Biophys.* **118**, 231. **33**. Luther, Jordanov, Ludwig & Schewe (1991) *Pharmazie* **46**, 134. **34**. Schurink, van Berkel, Wichers & Boeriu (2007) *Peptides* **28**, 2268. **35**. Iny, Grossman & Pinsky (1993) *Int. J. Biochem.* **25**, 1325. **36**. Bidonde, Ferrer, Zegzouti, *et al.* (1998) *Eur. J. Biochem.* **253**, 20. **37**. Masuoka, Nihei & Kubo (2006) *Mol. Nutr. Food*

Res. **50**, 725. **38**. Fornaroli, Petrussa, Braidot, Vianello & Macri (1999) *Plant Sci.* **145**, 1. **39**. Todd, Paliyath & Thompson (1990) *Plant Physiol.* **94**, 1225. **40**. Lorenzi, Maury, Casanova & Berti (2006) *Plant Physiol. Biochem.* **44**, 450. **41**. Martinez, Civello, Chaves & Anon (2001) *Phytochemistry* **58**, 379.

S-Propylglutathione

This *S*-alkylated glutathione (FW = 337.40 g/mol) inhibits glutathione *S*-transferase (1), hydroxyacylglutathione hydrolase, *or* glyoxalase II (2,3), and lactoylglutathione lyase, *or* glyoxalase I (4-6). **1**. Pace-Asciak & Smith (1983) *The Enzymes*, 3rd ed. (Boyer, ed.), **16**, 543. **2**. Norton, Principato, Talesa, Lupattelli & Rosi (1989) *Enzyme* **42**, 189. **3**. Akoachere, Iozef, Rahlfs, *et al.* (2005) *Biol. Chem.* **386**, 41. **4**. Vince & Wadd (1969) *Biochem. Biophys. Res. Commun.* **35**, 593. **5**. Allen, Lo & Thornalley (1993) *Biochem. Soc. Trans.* **21**, 535. **6**. Regoli, Saccucci & Principato (1996) *Comp. Biochem. Physiol. C, Pharmacol. Toxicol. Endocrin.* **113**, 313.

1-Propylguanidinium Ion

This substituted guanidine (FW$_{\text{free-base}}$ = 101.15 g/mol) is a simple structural analogue of arginine and, as expected, inhibits trypsin. **Target(s):** arginine decarboxylase (1), clostripain (2,3), guanidinoacetase (4), Na$^+$ channels (5), and trypsin, K_i = 0.9 mM at pH 6 and 0.53 mM at pH 8 (3,6-8). **1**. Balbo, Patel, Sell, *et al.* (2003) *Biochemistry* **42**, 15189. **2**. Mitchell & Harrington (1971) *The Enzymes*, 3rd ed. (Boyer, ed.), **3**, 699. **3**. Cole, Murakami & Inagami (1971) *Biochemistry* **10**, 4246. **4**. Shirokane & Nakayima (1986) *J. Ferment. Technol.* **64**, 29. **5**. Lo & Shrager (1981) *Biophys J.* **35**, 31 and 45. **6**. Walsh (1970) *Meth. Enzymol.* **19**, 41. **7**. Shaw (1970) *The Enzymes*, 3rd ed. (Boyer, ed.), **1**, 91. **8**. Inagami & York (1968) *Biochemistry* **7**, 4045.

N-(*n*-Propyl)-3-(3-hydroxyphenyl)piperidine, *See* 3-(3-Hydroxyphenyl)-N-propylpiperidine

n-Propyl Iodide, *See* 1-Iodopropane

Propyl Isocyanate

This alkyl isocyanate (FW = 85.106 g/mol; CAS 110-78-1; B.P. = 83-84°C) is a lachrymator that inhibits aldehyde dehydrogenase (1) and protein-glutamine γ-glutamyltransferase, *or* transglutaminase (2). **1**. Nagasawa, Elberling, Goon & Shirota (1994) *J. Med. Chem.* **37**, 4222. **2**. Gross, Whetzel & Folk (1975) *J. Biol. Chem.* **250**, 7693.

N-Propylmaleimide

This thiol-reactive reagent (FW = 139.15 g/mol; CAS 21746-40-7), also known as NPM and 1-propyl-1*H*-pyrrole-2,5-dione, readily alkylates protein sulfhydryl groups. Under alkaline conditions it can also react with amino groups. Some enzymes are inactivated at faster rates with analogues having longer alkyl chains (*e.g.*, the inactivation rate of yeast alcohol dehydrogenase with *N*-octylmaleimide is significantly faster than with *N*-

ethylmaleimide1). *See also* N-Ethylmaleimide **Target(s):** acylglycerol lipase, *or* monoacylglycerol lipase (2). **1.** Heitz, Anderson & Anderson (1968) *Arch. Biochem. Biophys.* **127**, 627. **2.** Saario, Salo, Nevalainen, *et al.* (2005) *Chem. Biol.* **12**, 649.

N,N-Propyl-D-mannoamidine

This mannose (FW = 217.26 g/mol) analogue inhibits several glycosidases: α-mannosidase (K_i = 0.11 μM), β-mannosidase (K_i = 0.19 μM), α-glucosidase (K_i = 125 μM), β-glucosidase (K_i = 111 μM), α-galactosidase (K_i = 4700 μM), β-galactosidase (K_i = 2200 μM), α-fucosidase (K_i = 34 μM), and β-N-acetylhexosaminidase (K_i = 900 μM). **1.** Heck, Vincent, Murray, *et al.* (2004) *J. Amer. Chem. Soc.* **126**, 1971.

Propyl-2-methoxy-2-phenyl Selenide

This selenide (FW = 229.18 g/mol) inhibits 5-aminolevulinate dehydratase, *or* porphobilinogen synthase (1). This inhibitory effect may be mediated by diphenyl diselenide, because propyl-2-methoxy-2-phenyl selenide is known to decompose rapidly, forming diphenyl diselenide in aqueous media. **1.** Barbosa, Rocha, Zeni, *et al.* (1998) *Toxicol. Appl. Pharmacol.* **149**, 243.

1-Propyl-3-methylimidazolium Tetrafluoroborate, *See* 1-Methyl-3-propylimidazolium Tetrafluoroborate

2-Propyl Methylphosphonochloridate

This phosphonate ester (FW = 156.55 g/mol; CAS 1445-76-7) inhibits chymotrypsin, forming diastereomeric phosphonate ester adducts with chymotrypsin's active-site serine hydroxyl(1). *Caution:* Because this agent is an obvious analogue of di-isopropylfluorophosphate and other skin-penetrating cholinesterase-directed chemical warfare agents, caution should be be exercised. Use of heavy rubber (not latex!) gloves and a fume hood is strongly advised. **1.** Kovach, McKay & Vander Velde (1993) *Chirality* **5**, 143.

N-n-Propylnorapomorphine

This apomorphine analogue (FW = 295.38 g/mol; CAS 58479-52-0 (for racemate); 18426-20-5 (for (R)-isomer); 20382-71-2 ((R)-isomer hydrochloride)) interacts with dopamine receptor, with the R(–)-enantiomer acting as a strong agonist and the S(+)-isomer acting as a weak antagonist (1), as well as an inhibitor of dihydropteridine reductase (2). **1.** Neumeyer, Reischig, Arana, *et al.* (1983) *J. Med. Chem.* **26**, 516. **2.** Shen, Smith, Davis & Abell (1984) *J. Biol. Chem.* **259**, 8994.

2(R)-[2-[4-(n-Propyl)phenyl]ethyl]-4(S)-butylpentanedioic Acid 1-((S)-tert-Butylglycine Methylamide) Amide

This peptidomimetic (FW$_{free-acid}$ = 447.64 g/mol), also known as N-[4(S)-carboxy-2(R)-[2-[4-(n-propyl)phenyl]ethyl]-octanoyl]-α(S)-(tert-butyl)glycine methylamide, inhibits stromelysin 2 (matrix metalloproteinase 3) and gelatinase A (matrix metalloproteinase 2), with K_i values of 140 and 47 nM, respectively. **1.** Esser, Bugianesi, Caldwell, *et al.* (1997) *J. Med. Chem.* **40**, 1026.

2(R)-[2-[4-(n-Propyl)phenyl]ethyl]-4(S)-butylpentanedioic Acid 1-((S)-tert-Butylglycine 4-Pyridylamide) Amide

This peptidomimetic (FW$_{free-acid}$ = 512.72 g/mol) inhibits stromelysin 2 (matrix metalloproteinase 3) and gelatinase A (matrix metalloproteinase 2), with K_i values of 11 and 22 nM, respectively). **1.** Esser, Bugianesi, Caldwell, *et al.* (1997) *J. Med. Chem.* **40**, 1026.

3-[5-(2-(n-Propylsulfonylamino)ethyl)-5-(pyrid-3-ylmethyl)-1-benzenepropionate

This sulfonamide (FW$_{free-acid}$ = 390.50 g/mol) inhibits human thromboxane-A synthase (IC_{50} = 29 nM) and is a thromboxane receptor antagonist (1). **1.** Dickinson, Dack, Long & Steele (1997) *J. Med. Chem.* **40**, 3442.

S-Propyl-L-thiocitrulline

This thiocitrulline derivative (FW = 234.34 g/mol) inhibits neuronal and inducible nitric-oxide synthases, with IC_{50} values of 3.75 and 2.55 μM, respectively (1,2). **1.** Griffith & Kilbourn (1996) *Meth. Enzymol.* **268**, 375. **2.** Narayanan, Spack, McMillan, *et al.* (1995) *J. Biol. Chem.* **270**, 11103.

6-(n-Propyl)-2-thiouracil

This uracil derivative (FW = 170.24 g/mol), also known as 2,3-dihydro-6-propyl-2-thioxo-4(1H)-pyrimidinone, inhibits the production of thyroxine and triiodothyronine and is used in the treatment of hyperthyroidism. It is slightly soluble in water (approximately 1 g/900 g H_2O) and more soluble in ethanol (1 g/60 g). 6-(n-Propyl)-2-thiouracil is also a mechanism-based inactivator of an isoform of nitric-oxide synthase (1). **Target(s):** dopamine β-hydroxylase (2); glutathione S-transferase (3); glyceraldehyde-3-phosphate dehydrogenase, acetylphosphatase activity thereof (4); iodide peroxidase, thyroid peroxidase (8,12,13,27,28); iodothyronine secretion (7); lactoperoxidase (8,29); myeloperoxidase (9.10); NADH:cytochrome b_5 reductase (11); nitric-oxide synthase (1); peroxidase (5,6,8-10,12,13,29); peroxidase, horseradish (5,6); thyroxine 5-deiodinase (14,15); thyroxine 5'-deiodinase, *or* iodothyronine deiodinase (14,16-26). **1.** Wolff & Marks (2002) *Arch. Biochem. Biophys.* **407**, 83. **2.** Hidaka & Nagasaka (1977)

Biochem. Pharmacol. **26**, 1092. **3**. Kariya, Sawahata, Okuno & Lee (1986) *Biochem. Pharmacol.* **35**, 1475. **4**. Allison & Conners (1970) *Arch. Biochem. Biophys.* **136**, 383. **5**. Zaton & Ochoa de Aspuru (1988) *Biochem. Biophys. Res. Commun.* **153**, 904. **6**. Zaton & Ochoa de Aspuru (1995) *FEBS Lett.* **374**, 192.**7**. Laurberg (1978) *Endocrinology* **103**, 900. **8**. Ator & Ortiz de Montellano (1990) *The Enzymes*, 3rd ed. (Sigman & Boyer, eds.), **19**, 213. **9**. Lee, Miki, Katsura & Kariya (1990) *Biochem. Pharmacol.* **39**, 1467. **10**. Taurog & Dorris (1992) *Arch. Biochem. Biophys.* **296**, 239. **11**. Lee & Kariya (1986) *FEBS Lett.* **209**, 49. **12**. Taurog (1976) *Endocrinology* **98**, 1031. **13**. Davidson, Soodak, Neary, *et al.* (1978) *Endocrinology* **103**, 871. **14**. Köhrle (2002) *Meth. Enzymol.* **347**, 125. **15**. Sanders, Van der Geyten, Kaptein, *et al.* (1999) *Endocrinology* **140**, 3666. **16**. Visser (1980) *Trends Biochem. Sci.* **5**, 222. **17**. Fenton & Valverde-R (2000) *Gen. Comp. Endocrinol.* **117**, 77. **18**. Leonard & Rosenberg (1978) *Endocrinology* **103**, 2137. **19**. Visser, Leonard, Kaplan & Larsen (1982) *Proc. Natl. Acad. Sci. U.S.A.* **79**, 5080. **20**. Orozco, Silva & Valverde-R (1997) *Endocrinology* **138**, 254. **21**. Berry, Kieffer, Harney & Larsen (1991) *J. Biol. Chem.* **266**, 14155. **22**. Sharifi & St. Germain (1992) *J. Biol. Chem.* **267**, 12539. **23**. Mori, Nishikawa, Toyoda, *et al.* (1991) *Endocrinology* **128**, 3105. **24**. Molinero, Osuna & Guerrero (1995) *J. Endocrinol.* **146**, 105. **25**. Kuiper, Wassen, Klootwijk, *et al.* (2003) *Endocrinology* **144**, 5411. **26**. Garcia-Macias, Molinero, Guerrero & Osuna (1994) *FEBS Lett.* **354**, 110. **27**. Schmutzler, Bacinski, Gotthardt, *et al.* (2007) *Endocrinology* **148**, 2835. **28**. Carvalho, Ferreira, Coelho, *et al.* (2000) *Braz. J. Med. Biol. Res.* **33**, 355. **29**. Bhuyan & Mugesh (2008) *Inorg. Chem.* **47**, 6569.

Propyl Xanthate

This xanthic acid salt (FW = 146.21 g/mol) inhibits mushroom tyrosinase, *or* monophenol monooxygenase, with a K_i value of 5 ‡M for the catecholase activity. The length of the hydrophobic tail of the xanthates has a stronger effect on the Ki values for catecholase inhibition than for cresolase inhibition. Increasing the length of the hydrophobic tail leads to a decrease of the K_i values for cresolase inhibition and an increase of the K_i values for catecholase inhibition. **1**. Saboury, Alijanianzadeh & Mansoori-Torshizi (2007) *Acta Biochim. Pol.* **54**, 183.

Propyne

$$H_3C — C \equiv CH$$

This simple alkyne (FW = 40.065 g/mol; CAS 74-99-7; M.P. = –102.7°C; B.P. = –23.2°C) inhibits the growth of *Burkholderia cepacia* G4 on toluene. Propyne is a suicide inhibitor of alkene monooxygenase. **Target(s):** alkene monooxygenase (1-3); and toluene 2-monooxygenase (4). **1**. Fosdike, Smith & Dalton (2005) *FEBS J.* **272**, 2661. **2**. Small & Ensign (1997) *J. Biol. Chem.* **272**, 24913. **3**. Gallagher, Cammack & Dalton (1997) *Eur. J. Biochem.* **247**, 635. **4**. Yeager, Bottomley, Arp & Hyman (1999) *Appl. Environ. Microbiol.* **65**, 632.

1-Propynyl Diethyl Phosphate

This phosphate ester (FW = 192.15 g/mol) is an irreversible inhibitor of a number of enzymes, with the enzyme-activated reagent converting to a highly reactive ketene intermediate that reacts with an active-site nucleophile. **Target(s):** acetylcholinesterase (1); aryldialkyl-phosphatase, *or* phosphotriesterase (2); butyryl-cholinesterase (1); cholesterol esterase (1); chymotrypsin (1); elastase (1); esterase (1); kallikrein (1); lipase (1); proteinase K (1); subtilisin BPN (1); subtilisin Carlsberg (1); and thrombin (1). **1**. Segal, Shalitin, Shalitin, Fischer & Stang (1996) *FEBS Lett.* **392**, 117. **2**. Banzon, Kuo, Miles, *et al.* (1995) *Biochemistry* **34**, 743.

4S-(Prop-2-ynyl)-D-glutamate, *See (2R,4S)-2-Amino-4-(prop-2-ynyl)pentanedioic Acid*

2-Propynyl-17α-methylestradiol

This estradiol derivative (FW = 324.46 g/mol) inhibits tubulin polymerization, or microtubule sel-assembly (IC_{50} = 11 μM) and possesses moieties that are expected to inhibit/resist deactivating enzymes. **1**. Edsall, Mohanakrishnan, Yang, *et al.* (2004) *J. Med. Chem.* **47**, 5126.

Propyzamide

This herbicide (FW = 256.13 g/mol; CAS 23950-58-5; M.P. = 155-156°C), also known as Pronamide and Kerb, is a preemergence agent (1) that also inhibits microtubule formation (2,3). Propyzamide is often used in fields of alfalfa, bird's-foot trefoil, woody nursery stock, lettuce, apples, pears, and lowbush blueberries. Perennial grasses including quack grass, annual grasses, volunteer cereals, and common chickweed are sensitive to propyzamide. **1**. Swithenbank, McNulty & Viste (1971) *J. Agric. Food Chem.* **19**, 417. **2**. Hirai, Sonobe & Hayashi (1998) *Proc. Natl. Acad. Sci. U.S.A.* **95**, 15102. **3**. Schibler & Huang (1991) *J. Cell Biol.* **113**, 605.

Prorenin Fragment 10P-20P, Human

This undecapeptide (*Sequence:* RIFLKRMPSIR; FW = 1416.79 g/mol), corresponding to residues 10P to 20P of the propart region of human prorenin, is a mild competitive inhibitor of endothiapepsin and calf chymosin, with respective K_i values of 11 and 37 μM (1). **1**. Richards, Kay, Dunn, Bessant & Charlton (1992) *Int. J. Biochem.* **24**, 297.

Prorenin Fragment 32P-43P, Human

This dodecapeptide (*Sequence:* RLGPEWSQPMKR; FW = 1484.74 g/mol), corresponding to residues 32P to 43P of the propart region of human prorenin, is a competitive inhibitor of several proteases: the apparent K_i values for endothiapepsin, human pepsin, human gastricsin, and human renin are 7, 15, and 18 and 115 μM, respectively (1). **1**. Richards, Kay, Dunn, Bessant & Charlton (1992) *Int. J. Biochem.* **24**, 297.

pro-SAAS Fragment 235-246

This dodecapeptide (*Sequence:* VLGALLRVKRLE; FW = 1366.71 g/mol) corresponds to residues 235-246 of the neuroendocrine secretory protein pro-SAAS, inhibits proprotein convertase 1, K_i = 51 nM (1), and is a significantly weaker inhibitor of furin, proprotein convertase 5, and proprotein convertase 7 (K_i = 39.4, 100, and 4.75 μM, respectively). Twelve variants of this peptide containing an alanyl residue substitution at each position were reported to be weaker inhibitors (1). **1**. Basak, Koch, Dupelle, *et al.* (2001) *J. Biol. Chem.* **276**, 32720.

Proscillaridin

This cardiac glycoside (FW = 530.66 g/mol), also known as desglucoproscillaridin A and scillarenin 3β-rhamnoside, is almost insoluble in water and soluble in ethanol and dioxane. Proscillaridin is obtained from scillarin A by partial enzymatic degradation. **Target(s):** acetyl-cholinesterase (1); ATPase (2); H^+/K^+-exchanging ATPase (3); and Na^+/K^+-

exchanging ATPase (4,5). **1.** Hanke, Nelson & Baskin (1991) *J. Appl. Toxicol.* **11**, 119. **2.** Kleeberg & Belz (1974) *Naunyn Schmiedebergs Arch. Pharmacol.* **282**, 433. **3.** Modyanov, Pestov, Adams, *et al.* 2003) *Ann. N.Y. Acad. Sci.* **986**, 183. **4.** Balzan, D'Urso, Ghione, Martinelli & Montali (2000) *Life Sci.* **67**, 1921. **5.** Erdmann (1978) *Arzneimittelforschung* **28**, 531.

Proserine, See Neostigmine

[D-Pro⁹-(spiro-g-lactam)⁹,¹⁰,Trp¹¹]-Substance P

This substance P analogue (FW_free-acid 1444.38 g/mol; *Sequence*: L-Arg-L-Pro-L-Lys-L-Pro-L-Gln-L-Gln-L-Phe-L-Phe-D-Pro-L-Leu-L-Trp-NH₂, with an ethylene moiety bridging the α-position of Pro⁹ and the nitrogen of Leu¹⁰) is a strong NK-1 receptor antagonist. **1.** Ward, Ewan, Jordan, *et al.* (1990) *J. Med. Chem.* **33**, 1848.

Prostaglandin A₁

This eicosanoid (FW_free-acid = 336.47 g/mol; CAS 14152-28-4; Symbol: PGA₁) induces apoptosis and stimulates the expression of stress genes. The structure and role of the prostaglandin A₁ was investigated by Bergström and Samuelsson, who shared the 1982 Nobel Prize in Physiology or Medicine for their pioneering work on the eicosanoids. **Target(s):** CYP4F2 (1); DNA topoisomerase II (2); HIV-1 transcription (3); 3α-hydroxysteroid dehydrogenase, NADP⁺-dependent (4); IκB kinase (5,6); Mayaro virus replication (7); stress-induced NF-κB activation (8); and virus protein biosynthesis, *or* translation (9). **1.** Jin, Koop, Raucy & Lasker (1998) *Arch. Biochem. Biophys.* **359**, 89. **2.** Suzuki & Uyeda (2002) *Biosci. Biotechnol. Biochem.* **66**, 1706. **3.** Carattoli, Fortini, Rozera & Giorgi (2000) *J. Biol. Regul. Homeost. Agents* **14**, 209. **4.** Hara, Inoue, Nakagawa, Naganeo & Sawada (1988) *J. Biochem.* **103**, 1027. **5.** Rossi, Kapahi, Natoli, *et al.* (2000) *Nature* **403**, 103. **6.** Amici, Belardo, Rozera, Bernasconi & Santoro (2004) *AIDS* **18**, 1271. **7.** Burlandy & Rebello (2001) *Intervirology* **44**, 344. **8.** Boller, Brandes, Russell, *et al.* (2000) *Oncol. Res.* **12**, 383. **9.** Amici, Giorgi, Rossi & Santoro (1994) *J. Virol.* **68**, 6890.

Prostaglandin A₂

This eicosanoid (FW_free-acid = 347.48 g/mol; CAS 13345-50-1; Symbol: PGA₂) induces apoptosis in tumor cells, inhibits viral replication, and blocks the cell cycle at G2/M. **Target(s):** DNA topoisomerase II (1); glutathione *S*-transferase P1-1 (2); G₁ phase cyclin-dependent kinases (3,4); IκB kinase (5); multidrug resistance-associated protein (2); Na⁺/K⁺-exchanging ATPase (6,7); and prostaglandin E₂ 9-reductase (8,9). **1.** Suzuki & Uyeda (2002) *Biosci. Biotechnol. Biochem.* **66**, 1706. **2.** van

Iersel, Cnubben, Smink, Koeman & van Bladeren (1999) *Biochem. Pharmacol.* **57**, 1383. **3.** Hitomi, Shu, Agarwal, Agarwal & Stacey (1998) *Oncogene* **17**, 959. **4.** Gorospe, Liu, Xu, Chrest & Holbrook (1996) *Mol. Cell Biol.* **16**, 762. **5.** Amici, Belardo, Rozera, Bernasconi & Santoro (2004) *AIDS* **18**, 1271. **6.** Matsukawa, Terao, Hayakawa & Takiguchi (1983) *Int. J. Biochem.* **15**, 739. **7.** Shibata, Ohzeki, Sato, Suzuki & Takiguchi (1982) *Int. J. Biochem.* **14**, 347. **8.** Tai & Yuan (1982) *Meth. Enzymol.* **86**, 113. **9.** Pace-Asciak & Smith (1983) *The Enzymes*, 3rd ed. (Boyer, ed.), **16**, 543.

Prostaglandin B₁ and Prostaglandin Bₓ

This eicosanoid (FW_B1(free-acid) = 336.47 g/mol; CAS 13345-51-2, for PGB₁, and 39306-29-1, for Bₓ) is a metabolite of prostaglandin E₁: prostaglandin Bₓ is an oligomer of prostaglandin B₁. The X-ray crystal structure of prostaglandin B₁ suggests this prostaglandin has an unusual L-shaped configuration, with the α and ω side chains roughly perpendicular to one another (1). The 15-hydroxyl group, which is normally directed away from the centroid of the prostaglandin in the standard hairpin model, is turned inward in prostaglandin B₁. **Target(s):** F₀F₁ ATPase, *or* H⁺-transporting two-sector ATPase, inhibited by prostaglandin Bₓ (2,3); 15-hydroxyprostaglandin dehydrogenase (4); phospholipase A₂, inhibited by prostaglandin Bₓ (5,6). **1.** DeTitta (1976) *Science* **191**, 1271. **2.** Devlin, Krupinski-Olsen, Uribe & Nelson (1986) *Biochem. Biophys. Res. Commun.* **137**, 215. **3.** Kreutter & Devlin (1983) *Arch. Biochem. Biophys.* **221**, 216. **4.** Jarabak (1982) *Meth. Enzymol.* **86**, 126. **5.** Franson, Raghupathi, Fry, *et al.* (1990) *Adv. Exp. Med. Biol.* **279**, 219. **6.** Franson & Rosenthal (1989) *Biochim. Biophys. Acta* **1006**, 272.

Prostaglandin B₂

This eicosanoid (FW_free-acid = 334.46 g/mol; CAS 13367-85-6; Symbol: PGB₂), the major prostaglandin of osteoblasts, induces pulmonary hypertension. **Target(s):** 15-hydroxyprostaglandin dihydrogenase (1); and prostaglandin E₂ 9-reductase (2,3). **1.** Lee & Levine (1975) *J. Biol. Chem.* **250**, 548. **2.** Tai & Yuan (1982) *Meth. Enzymol.* **86**, 113. **3.** Pace-Asciak & Smith (1983) *The Enzymes*, 3rd ed. (Boyer, ed.), **16**, 543.

Prostaglandin Bₓ, See Prostaglandin B₁ and Prostaglandin Bₓ

Prostaglandin D₂

This eicosanoid (FW_free-acid = 352.47 g/mol; CAS 41598-07-6; Symbol: PGD₂) is the principal cyclooxygenase metabolite of arachidonic acid.the primary prostaglandin in brain is a stimulator of adenylate cyclase and can induce inflammation. Prostagladin D₂ is released upon activation of mast cells and is also synthesized by alveolar macrophages. Among its many biological actions, the most important are its role as a bronchoconstrictor, acting as a platelet-activating-factor inhibitory. **Target(s):** platelet aggregation (1); and prostaglandin E_2 9-reductase (2,3). **1.** Nishizawa, Miller, Gorman, *et al.* (1975) *Prostaglandins* **9**, 109. **2.** Tai & Yuan (1982) *Meth. Enzymol.* **86**, 113. **3.** Pace-Asciak & Smith (1983) *The Enzymes*, 3rd ed. (Boyer, ed.), **16**, 543.

Prostaglandin E₁

This bioactive prostaglandin ($FW_{free-acid}$ = 354.49 g/mol; CAS 745-65-3; Symbol: PGE_1) is a vasodilator that activates adenylyl cyclase through a G-protein-coupled receptor. **Target(s):** adenylate cyclase (1); cystinyl aminopeptidase, *or* oxytocinase (2-6); and IκB kinase (7). **1.** Reddy, Oliver, Festoff & Engel (1978) *Biochim. Biophys. Acta* **540**, 371. **2.** Roy, Yeang & Karim (1982) *Prostaglandins Leukot. Med.* **8**, 173. **3.** Roy, Yeang, Tan, Kottegoda & Ratnam (1985) *Prostaglandins* **30**, 255. **4.** Roy, Yeang, Kottegoda & Ratnam (1984) *Prostaglandins Leukot. Med.* **14**, 105. **5.** Roy, Yeang & Karim (1981) *Prostaglandins Med.* **6**, 577. **6.** Roy, Sen & Ratnam (1989) *J. Reprod. Fertil.* **87**, 163. **7.** Amici, Belardo, Rozera, Bernasconi & Santoro (2004) *AIDS* **18**, 1271.

Prostaglandin E₂

This bioactive prostaglandin ($FW_{free-acid}$ = 352.47 g/mol; CAS 363-24-6; Symbol: PGE_2) is a vasodilator. It can be used to induce cervical ripening and parturition as well as regulate the sleep/wake cycle. In addition, it modulates adenylate cyclase via G protein receptors. **Target(s):** cystinyl aminopeptidase, *or* oxytocinase (1-5); IκB kinase (6); MAP kinase (7); Na^+/K^+-exchanging ATPase (8); sterol acyltransferase, *or* cholesterol acyltransferase, *or* ACAT (9). **1.** Roy, Yeang & Karim (1982) *Prostaglandins Leukot. Med.* **8**, 173. **2.** Roy, Yeang, Tan, Kottegoda & Ratnam (1985) *Prostaglandins* **30**, 255. **3.** Roy, Yeang, Kottegoda & Ratnam (1984) *Prostaglandins Leukot. Med.* **14**, 105. **4.** Roy, Yeang & Karim (1981) *Prostaglandins Med.* **6**, 577. **5.** Roy, Sen & Ratnam (1989) *J. Reprod. Fertil.* **87**, 163. **6.** Amici, Belardo, Rozera, Bernasconi & Santoro (2004) *AIDS* **18**, 1271. **7.** Li, Zarinetchi, Schrier & Nemenoff (1995) *Amer. J. Physiol.* **269**, C986. **8.** Rubinger, Wald, Scherzer & Popovtzer (1990) *Prostaglandins* **39**, 179. **9.** Billheimer (1985) *Meth. Enzymol.* **111**, 286.

Prostaglandin F₂α

This bioactive prostaglandin ($FW_{free-acid}$ = 354.49 g/mol; Symbol: $PGF_{2\alpha}$) is a vasoconstrictor and can induce parturition. **Target(s):** cystinyl aminopeptidase, *or* oxytocinase (1-5); and sterol acyltransferase, *or* cholesterol acyltransferase, *or* ACAT (6). **1.** Roy, Yeang & Karim (1982) *Prostaglandins Leukot. Med.* **8**, 173. **2.** Roy, Yeang, Tan, Kottegoda & Ratnam (1985) *Prostaglandins* **30**, 255. **3.** Roy, Yeang, Kottegoda &

Ratnam (1984) *Prostaglandins Leukot. Med.* **14**, 105. **4.** Roy, Yeang & Karim (1981) *Prostaglandins Med.* **6**, 577. **5.** Roy, Sen & Ratnam (1989) *J. Reprod. Fertil.* **87**, 163. **6.** Billheimer (1985) *Meth. Enzymol.* **111**, 286.

Prostaglandin G₁

This 9,11-endoperoxide 15-hydroperoxide ($FW_{free-acid}$ = 370.49 g/mol; Symbol: PGG_1), itself an intermediate in the biosynthesis of biologically active series-1 prostaglandins, inhibits prostacyclin synthase. **1.** Ham, Egan, Soderman, Gale & Kuehl, Jr. (1979) *J. Biol. Chem.* **254**, 2191.

Prostaglandin I₂

This (FW = 374.45 g/mol; CAS 61849-14-7; Symbol: PGI_1), also called prostacyclin and epoprostenol, is produced in endothelial cells from prostaglandin H_2 by the action of the enzyme prostacyclin synthase. Prostacyclin chiefly prevents formation of the platelet plug involved in primary hemostasis (a part of blood clot formation). It does this by inhibiting platelet activation.[5] It is also an effective vasodilator.

Prostigmine, *See Neostigmine*

Prostratin

This orally active, nontumorigenic phorbol ester (FW = 390.47 g/mol; CAS 60857-08-1; Soluble to 75 mM in DMSO; Code Name: P-4462), isolated from the groundcover plant *Pimelea prostrate* as well as the bark of the Samoan mamala tree, *Homalanthus nutans* (*Euphorbiaceae*) (1), is a selective Protein Kinase C activator (*See Phorbol Myristate Acetate*), but exhibits profound growth-suppressing action against certain tumors. Prostratin inhibits the growth of myeloid leukemia cells by a predominant G_1 arrest and variable induction of apoptosis (2). It also induces significant differentiation of AML cell lines and primary AML blasts by reprogramming transcriptional factor expression. Ectopic expression of c-Myc in HL-60 cells eliminates prostratin-mediated cellular differentiation and cell cycle arrest, indicating an essential role for c-Myc suppression in the differentiation-inducing effects of prostratin (1). Prostratin also potentiates induction of cellular differentiation by Ara-C (2). Prostratin also represses tumorigenesis in K-Ras mutant pancreatic cancer cells (3). Prostratin also shows promise for eradicating the latent HIV-1 provirus by inducing HIV-1 transcriptional activation, most likely through effects on protein kinase D_3 (3). Silencing PKD3, but not the other PKD family members, blocked prostratin-induced transcription of HIV-1. Over-expression of the constitutively active form of PKD3, but not the wild-type or kinase-dead form of PKD3, augments the expression of HIV-1. Moreover, prostratin triggers PKD3 activation by inducing the phosphorylation of its activation loop by PKCε. The activating effect of PKD3 on HIV-1 transcription depends on the presence of κB element and the prostratin-induced activation of NF-κB, and PKD3 silencing blocks prostratin-induced NF-κB activation and NF-κB-dependent HIV-1 transcription (4). **Use in HIV Therapy:** Prostratin is of interest because of its ability to activate otherwise latent viral reservoirs. Prostratin also inhibits *de novo* infection by Human Immunodeficiency Virus type-1 (HIV-1) (5), while up-regulating viral expression from latent proviruses, commending it for use as an inductive adjuvant therapy for patients treated with highly active antiretroviral therapy (HAART) (4). Prostratin blocks HIV-1 infection by inducing the down-regulation of CD4 receptor expression, with similar down-regulation of the HIV-1 co-receptors, CXCR4 and CCR5, actions that may also reduce viral infectivity of treated host cells. Prostratin also up-regulates HIV-1 expression from CD8$^+$ T lymphocyte–depleted peripheral blood mononuclear cells of patients undergoing HAART (5). Such observations suggest Prostratin is a candidate for augmenting HAART by inducing expression of latent HIV-1, and when combined with other agents, helping to eliminate persistent viral reservoirs in certain individuals infected with HIV-1 (5). **1.** Cashmore, *et al.* (1976) *Tetrahedron Lett.* **20**, 1737. **2.** Shen, Xiong, Jing, *et al.* (2015) *Cancer Lett.* **356**, 686. **2.** Wang, Holderfield, Galeas, *et al.* (2015) *Cell* 163, 1237. **3.** Wang, Zhu, Zhu, *et al.* (2014) *Biomed. Res. Int.* **2014**, 968027. **4.** Gustafson, Cardellina, McMahon, *et al.* (1992) *J. Med. Chem.* **35**, 1978. **5.** Kulkosky, Culnan, Roman, *et al.* (2001) *Blood* **98**, 3006. **IUPAC Name:** (1*aR*,1*bS*,4*aR*,7*aS*,7*bR*,8*R*,9*aS*)-9*a*-(acetyloxy)-1,1*a*,1*b*,4,4*a*,7*a*,7*b*,8,9,9*a*-decahydro-4*a*,7*b*-dihydroxy-3-(hydroxymethyl)-1,1,6,8-tetramethyl-5*H*-cyclopropa[3,4]benz[1,2-*e*]azulen-5-one

pro-TAME

This cell-permeable TAME analogue (FW$_{hydrochloride}$ = 670.23 g/mol) hydrolyzes upon entering cells, yielding a synthetic substrate that is useful for quantifying the esterase activity of many neutral proteinases, including trypsin, plasmin, and papain (1-10). Depending on the target enzyme, TAME may act as a slow alternative substrate that competitively inhibits the peptide-hydrolyzing activities of proteases (*See TAME*). **1.** Zeng, Sigoillot, Gaur, *et al.* (2010) *Cancer Cell* **18**, 382.

Protamine & Protamine Sulfate

Protamines are small basic proteins and oligopeptides, typically rich in arginyl residues, that are found in the sperm of all animals. They are often associated with polynucleic acids such as double-stranded DNA (they compact DNA into a more condensed structure). Protamine sulfate is sulfated salt of a protamine. It binds to and inactivates heparin; hence, it is often used in cases of heparin overdose. As mentioned above, it also binds to and precipitates DNA and can be used to remove DNA from a protein sample or in the purification of DNA-binding proteins. It has also been used in retroviral-mediated gene transfers as a substitute for polybrene. Note that salmine is the protamine sulfate salt from salmon and clupeine is the salt from herring. *See also specific protamine; e.g., Salmine; Clupeine* **Target(s):** acetylcholinesterase (1); acylglycerol lipase, *or* monoacylglycerol lipase (34); adenylate cyclase (2); alkaline phosphatase (28); AMP deaminase (3); carbamoyl-phosphate synthetase (4); cholinesterase (1); cytochrome *c* oxidase (6,7); diacylglycerol lipase (38); electron transport (6,7); glycoprotein Ib-von Willebrand factor activity (8); histone acetyltransferase (42); IgA-specific serine endopeptidase (9); lipase, *or* triacylglycerol lipase (8,10,37-39); lipoprotein lipase (11-13,29-33); lysine carboxypeptidase, *or*

carboxypeptidase N, *or* lysine(arginine)carboxypeptidase (5,24,25); oxidative phosphorylation (7); pantothenoylcysteine decarboxylase (14); phospholipase A$_2$ (15); phospholipase D (16); phosphoprotein phosphatase-I, *or* [phosphorylase] phosphatase (17,18); phosphorylase kinase (19,40); poly(ADP-ribose) glycohydrolase (26,27); protein xylosyltransferase (41); sterol esterase, *or* cholesterol esterase (35,36); thrombin (20); tryptase (21); and tryptophan synthase (22,23). **1.** Lin, Liao & Lee (1977) *Biochem. J.* **161**, 229. **2.** Kiss & Zamfirova (1983) *Experientia* **39**, 1381. **3.** Lee (1957) *J. Biol. Chem.* **227**, 999. **4.** Guthöhrlein & Knappe (1969) *Eur. J. Biochem.* **8**, 207. **5.** Tan, Jackman, Skidgel, Zsigmond & Erdos (1989) *Anesthesiology* **70**, 267. **6.** Konstantinov (1975) *Biokhimiia* **40**, 401. **7.** Kossekova, Mitovska & Dancheva (1975) *Acta Biol. Med. Ger.* **34**, 539. **8.** Barstad, Stephens, Hamers & Sakariassen (2000) *Thromb. Haemost.* **83**, 334. **9.** Bleeg, Reinholdt & Kilian (1985) *FEBS Lett.* **188**, 357. **10.** Tsujita, Matsuura & Okuda (1996) *J. Lipid Res.* **37**, 1481. **11.** Hofstee (1960) *The Enzymes*, 2nd ed. (Boyer, Lardy & Myrbäck, eds.), **4**, 485. **12.** Korn (1962) *Meth. Enzymol.* **5**, 542. **13.** Posner & Morrison (1979) *Acta Cient. Venez.* **30**, 152. **14.** Brown (1957) *J. Biol. Chem.* **226**, 651. **15.** Emadi, Elalamy, Vargaftig & Hatmi (1996) *Biochim. Biophys. Acta* **1300**, 226. **16.** Kates & Sastry (1969) *Meth. Enzymol.* **14**, 197. **17.** Cohen, Alemany, Hemmings, *et al.* (1988) *Meth. Enzymol.* **159**, 390. **18.** Ballou & Fischer (1986) *The Enzymes*, 3rd ed. (Boyer & Krebs, eds.), **17**, 311. **19.** Krebs, Love, Bratvold, *et al.* (1964) *Biochemistry* **3**, 1022. **20.** Cobel-Geard & Hassouna (1983) *Amer. J. Hematol.* **14**, 227. **21.** Hallgren, Estrada, Karlson, Alving & Pejler (2001) *Biochemistry* **40**, 7342. **22.** Meyer, Germershausen & Suskind (1970) *Meth. Enzymol.* **17A**, 406. **23.** Tsai & Suskind (1972) *Biochim. Biophys. Acta* **284**, 324. **24.** Oshima, Kato & Erdös (1975) *Arch. Biochem. Biophys.* **170**, 132. **25.** Mao, Colussi, Bailey, *et al.* (2003) *Anal. Biochem.* **319**, 159. **26.** Tavassoli, Tavassoli & Shall (1983) *Eur. J. Biochem.* **135**, 449. **27.** Miwa, Tanaka, Matsushima & Sugimura (1974) *J. Biol. Chem.* **249**, 3475. **28.** Stinson & Chan (1987) *Adv. Protein Phosphatases* **4**, 127. **29.** Sato, Akiba & Horiguchi (1997) *Comp. Biochem. Physiol. A* **118**, 855. **30.** LaDu, Schultz, Essig & Palmer (1991) *Int. J. Biochem.* **23**, 405. **31.** Aisaka & Terada (1980) *Agric. Biol. Chem.* **44**, 799. **32.** Skinner & Youssef (1982) *Biochem. J.* **203**, 727. **33.** Augustin, Freeze, Tejada & Brown (1978) *J. Biol. Chem.* **253**, 2912. **34.** Sakurada & Noma (1981) *J. Biochem.* **90**, 1413. **35.** Brown & Sgoutas (1980) *Biochim. Biophys. Acta* **617**, 305. **36.** van Berkel, Vaandrager, Kruijt & Koster (1980) *Biochim. Biophys. Acta* **617**, 446. **37.** Finkelstein, Strawich & Sonnino (1970) *Biochim. Biophys. Acta* **206**, 380. **38.** Stam, Broekhoven-Schokker & Hülsmann (1986) *Biochim. Biophys. Acta* **875**, 76. **39.** Palmer & Oscai (1990) *Can. J. Physiol. Pharmacol.* **68**, 689. **40.** Carlson, Bechtel & Graves (1979) *Adv. Enzymol. Relat. Areas Mol. Biol.* **50**, 41. **41.** Casanova, Kuhn, Kleesiek & Götting (2008) *Biochem. Biophys. Res. Commun.* **365**, 678. **42.** Wiegand & Brutlag (1981) *J. Biol. Chem.* **256**, 4578.

Protease Inhibitor 4, *See Kallistatin*

Protease Inhibitor "Cocktails"

Enzymatic proteolysis can be arrested or minimized through the use of so-called cocktails containing combinations of serine protease, cysteine protease, aspartic protease, and metalloprotease inhibitors. The following combinations (or cocktails) have proved to be highly effective in the isolation of intact proteins, and some have are so popular that they are commercially available. **General Protease Cocktails** typically contain AEBSF, E-64, bestatin, leupeptin, aprotinin, and sodium EDTA. The latter is included only if the protein or enzyme being purified does not require a divalent metal cofactor for structural integrity. **Fungal and Yeast Protease Cocktails** typically contain AEBSF, pepstatin A, E-64, and 1,10-phenanthroline. **Mammalian Cell Protease Inhibitor Cocktails** typically contain AEBSF, pepstatin A, E-64, bestatin, leupeptin, and aprotinin. (Other metal ion chelators, such as 0.5 mM EGTA, may be added to suppress the activity of calcium ion-dependent proteases such as calpain. Again, one must determine whether the protein or enzyme being purified does not require a divalent metal cofactor for structural integrity or biological activity. **Bacterial Extract Protease Cocktails** typically contain AEBSF), pepstatin A, E-64, bestatin, and sodium EDTA. For convenience, the inhibitory properties of these reagents are summarized as follows. **AEBSF:** This broad-spectrum serine protease inhibitor, which is stable in water for one month when stored at –20 °C, has an inhibitory range 0.1-1.0 mM. AEBSF also inhibits cysteine proteases, such as papain. **Antipain:** This inhibitor is effective against serine proteases (*e.g.*, plasmin, thrombin and trypsin, but not chymotrypsin, as well as some cysteine proteases, such as calpain and papain. Its inhibitory range is 1-100 μM, with IC$_{50}$ = 0.15 mg/mL for papain, 0.25 mg/mL for trypsin, 1.2 mg/mL for cathepsin A). Prepare at 10 mM in water. **Aprotinin:** This water-soluble inhibitor, which is stable for months, when stored at 4 °C, blocks and many serine proteases at 0.3 μM (or equimolar with proteinase), but is without effect on thrombin or blood-clotting factor Xa. **Bestatin:** This aminopeptidase inhibitor, which has an inhibitory range of 1-10 μM, is prepared as a 1 mM stock solution in methanol and stable for 1 month, when stored at –20 °C. **Chymostatin:** This chymotrypsin-like serine protease inhibitor blocks chymase and cathepsins A,B,D and G, as well as some cysteine proteases (including papain). Its inhibitory range is 10–100 μM; IC_{50} = 0.1 mg/mL for chymotrypsin). Stock solutions are prepared at 10 mM in DMSO. **DFP:** This highly effective broad-spectrum serine-protease inhibitor is also a highly toxic nerve poison, and a well-ventilated fume hood and heavy rubber gloves should be used to avoid inhalation or contact with skin. For safety, DFP should be prepared as a 200 mM solution in anhydrous isopropanol or propylene glycol and then diluted 60-100x into cell extract. In aqueous solution, DFP rapidly decomposes ($t_{1/2}$ = 20 min in water). **E-64:** This thiol protease inhibitor, which is effective in the 1-10 μM concentration range, is prepared as a 1 mM stock solution in water and is stable 1-2 weeks at –20 °C. **EDTA:** This broad-spectrum metalloproteinase inhibitor (effective range = 1-10 mM) nonspecifically inhibits many metalloenzymes, especially those with metal ion K_d values above 1 mM. **N-Ethylmaleimide:** This chemical modifying agent inhibits most thiol-proteases by reacting irreversibly with essential active-site thiol groups. NEM is water-soluble at >10mg/ml, and must be prepared freshly. **Leupeptin:** This inhibitor blocks trypsin-like serine proteases, such as trypsin, chymotrypsin, chymase, pepsin and thrombin. Leupeptin inhibits selected cysteine proteases (calpain, cathepsin B, H & L and papain). The inhibitory range is 10-100 μM, with the following IC_{50} values: 2 mg/mL for trypsin; 8 mg/mL for plasmin; and 0.5 mg/mL for cathepsin B. When prepared at 10 mM in water, leupeptin is stable 6 months when stored at –20 °C. **Pepstatin A:** This inhibitor blocks aspartic proteases, such as renin, chymosin and pepsin (inhibits at 1 μM). Prepare as 1mM stock solution in methanol or DMSO. **1,10-Phenanthroline:** This broad-spectrum metallo-proteinase inhibitor (with an inhibitory range of 1–10 mM) is prepared at 200mM in methanol or DMSO. This reagent nonspecifically inhibits many metalloenzymes, especially zinc- and manganese-dependent enzymes and those with metal ion K_d values above 1 mM. **PMSF:** This broad-spectrum serine protease inhibitor, which is soluble in DMSO and stable for 1 month at –20 °C, has an inhibitory range of 0.1–1.0 mM. PMSF also inhibits cysteine proteases (including papain), which are reactivated by 1 mM 2-mercaptoethanol or dithiothreitol. **Phosphoramidon:** This agent potently inhibits metalloendoproteinases, including thermolysin and elastases at 1–10 μM, but only weakly inhibits collagenase. When prepared at 1 mM in water, phosphoramidon solutions are stable for 1 month at –20 °C. **TLCK:** This haloketone-containing reagent inhibits trypsin-like serine proteases at 10-100 μM. TLCK solutions must be freshly prepared at 10 mM in 1 mM HCl and then diluted 100x into cell extracts. **TPCK:** This haloketone-containing reagent inhibits chymotrypsin-like serine proteases at 10-100 μM. TPCK solutions must be freshly prepared at 10 mM in 1 mM HCl and then diluted 100x into cell extracts. **See also** AEBSF; Antipain; Aprotinin; Bestatin; Chymostatin; E-64; N-Ethylmaleimide; Leupeptin; Pepstatin A; 1,10-Phenanthroline; PMSF; Phosphoramidon; TPCK

Protease Inhibitor Nexin I

This 43.9-kDa glycoprotein serpin, also called protease nexin-1, inhibits thrombin-like proteases. It is found in several tissues throughout the body including a platelet-bound form. Protease nexin-1 forms tight complexes with the target proteases, acting as a suicide inhibitor, and is presumably cleaved during the inhibitory reaction. It also promotes neurite extension. (For a discussion of the likely mechanism of inhibitory action, See Serpins; also α_1-Antichymotrypsin) **Target(s):** acrosin (1,2); coagulation factor Xa (3); coagulation factor Xia (4); granzyme A (5,6); plasmin (3,7); protein C, activated (8); thrombin (3,7,9,10); trypsin (3); and u-plasminogen activator, or urokinase (3,7,11-16). **1.** Hermans, Monard, Jones & Stone (1995) Biochemistry **34**, 3678. **2.** Zheng, Geiger, Ecke, et al. (1994) Fibrinolysis **8**, 364. **3.** Scott, Bergman, Bajpai, et al. (1985) J. Biol. Chem. **260**, 7029. **4.** Knauer, Majumdar, Fong & Knauer (2000) J. Biol. Chem. **275**, 37340. **5.** Froelich & Salvesen (1998) in Handb. Proteolytic Enzymes (Barrett, Rawlings & Woessner, eds.), p. 77, Academic Press, San Diego. **6.** Gurwitz, Simon, Fruth & Cunningham (1989) Biochem. Biophys. Res. Commun. **161**, 300. **7.** Howard & Knauer (1987) J. Cell Physiol. **131**, 276. **8.** Hermans & Stone (1993) Biochem. J. **295**, 239. **9.** Festoff, Smirnova, Ma & Citron (1996) Semin. Thromb. Hemost. **22**, 267. **10.** Stone & Hermans (1995) Biochemistry **34**, 5164. **11.** Ellis & Danø (1998) in Handb.

Proteolytic Enzymes (Barrett, Rawlings & Woessner, eds.), p. 177, Academic Press, San Diego. **12.** Kruithof (1988) Enzyme **40**, 113. **13.** Crisp, Knauer & Knauer (2002) J. Biol. Chem. **277**, 47285. **14.** Saksela (1985) Biochim. Biophys. Acta **823**, 35. **15.** Schmitt, Jänicke, Moniwa, et al. (1992) Biol. Chem. Hoppe-Seyler **373**, 611. **16.** Saksela & Rifkin (1988) Ann. Rev. Cell Biol. **4**, 93.

Protease Inhibitor Nexin II

This term, also called protease nexin-2 (PN-2), refers to the subset of the several differentially spliced forms (including the 751-aminoacyl residue isoform) of the amyloid precursor protein that contains a Kunitz-type inhibitor domain. PN-2 reversibly inhibits a number of coagulation factors. **Target(s):** chymotrypsin (1); coagulation factor VIIa (2); coagulation factor IXa (3); coagulation factor Xa (4); coagulation factor Xia, K_i = ~400 pM, heparin improves the inhibition (5-14). **1.** Van Nostrand, Wagner, Suzuki, et al. (1989) Nature **341**, 546. **2.** Mahdi, Rehemtulla, Van Nostrand, Bajaj & Schmaier (2000) Thromb. Res. **99**, 267. **3.** Schmaier, Dahl, Hasan, et al. (1995) Biochemistry **34**, 1171. **4.** Mahdi, Van Nostrand & Schmaier (1995) J. Biol. Chem. **270**, 23468. **5.** Walsh (1998) in Handb. Proteolytic Enzymes (Barrett, Rawlings & Woessner, eds.), p. 153, Academic Press, San Diego. **6.** Walsh, Baglia & Jameson (1993) Meth. Enzymol. **222**, 65. **7.** Van Nostrand, Schmaier & Wagner (1992) Ann. N. Y. Acad. Sci. **674**, 243. **8.** Van Nostrand, Schmaier, Farrow & Cunningham (1991) Ann. N. Y. Acad. Sci. **640**, 140. **9.** Van Nostrand, Wagner, Farrow & Cunningham (1990) J. Biol. Chem. **265**, 9591. **10.** Van Nostrand (1995) Thromb. Res. **78**, 43. **11.** Wagner, Keane, Melchor, Auspaker & Van Nostrand (2000) Biochemistry **39**, 7420. **12.** Badellino & Walsh (2000) Biochemistry **39**, 4769. **13.** Scandura, Zhang, Van Nostrand & Walsh (1997) Biochemistry **36**, 412. **14.** Zhang, Scandura, Van Nostrand & Walsh (1997) J. Biol. Chem. **272**, 26139.

Protease Nexin I, See Protease Inhibitor Nexin I

Protease Nexin II, See Protease Inhibitor Nexin II

Proteasome Inhibitor I, See N^α-Benzyloxycarbonyl-L-isoleucyl-L-glutamyl(O'-t-butyl ester)-L-alanyl-L-leucinal

Proteasome Inhibitor II, See N^α-Benzyloxycarbonyl-L-leucyl-L-leucyl-L-phenylalaninal

Proteasome Inhibitor III, See N^α-Benzyloxycarbonyl-L-leucyl-L-leucyl-L-leucineboronic Acid

Proteasome Inhibitor IV, See N^α-Benzyloxycarbonyl-glycyl-L-prolyl-L-phenylalanyl-L-leucinal

Proteinase B Inhibitors

These two 8.5 kDa proteins (each consisting of 74 aminoacyl residues) have been isolated from yeast that inhibit cerevisin, yeast proteinase B (1,2). **1.** Maier, Müller & Holzer (1979) J. Biol. Chem. **254**, 8491. **2.** Holzer (1975) Adv. Enzyme Regul. **13**, 125.

α_1-**Proteinase Inhibitor,** See α_1-Antitrypsin

Proteinase Inhibitor, Japanese Horseshoe Crab (Tachypleus tridentatus)

This proteinase inhibitor from the Japanese horseshoe crab is a 61-aminoacyl residue protein is a Kunitz-type protease inhibitor, with a 0.5-nM K_i value for trypsin. **Target(s):** chymotrypsin (1); elastase (1); kallikrein (1); plasmin (1); and trypsin (1). **1.** Nakamura, Hirai, Tokunaga, Kawabata & Iwanaga (1987) J. Biochem. **101**, 1297.

Proteinase Inhibitor 8, See PI8

Proteinase Inhibitor 9, See PI9

Protein C Inhibitor (Activated Protein C Inhibitor)

This secreted heparin-binding glycoprotein (MW = 57000 g; Symbol: PCI), commonly abbreviated PCI and also called plasminogen activator inhibitor 3 (PAI-3), is a member of the plasma serine protease inhibitor (serpin) family. It was initially identified as an inhibitor of activated Protein C, the main proteolytic enzyme in the protein C anticoagulant pathway (K_i = 58 nM). This inhibitor forms an acyl-bond complex with activated protein C, a reaction that is enhanced by heparin and dextran sulfate. (For a discussion of the likely mechanism of inhibitory action, See Serpins; also α_1-Antichymotrypsin) Givern its cationic nature, PCI likewise binds negatively charged phospholipids (e.g., unsaturated phosphatidylserine (PS), oxidised phosphatidyl-ethanolamine, phosphatidic acid (PA), cardiolipin (CL), and phosphoinositides (PIPs)), thereby stimulating PCI's inhibitory action on different proteases (1). The interaction of phospholipids with PCI might also alter the lipid distribution pattern of blood cells and influence the remodeling of cellular membranes. **Modulation of the Phagocytosis of PS-Presenting Cells:** PCI binds to PS exposed on apoptotic cells, thereby regulating their removal by phagocytosis (2). With Jurkat T-lymphocytes and U937 myeloid cells, PCI binds to apoptotic cells to a similar extent at the same sites as Annexin V, but in a different manner compared to live cells (defined as spots of concentration on ~10-30% of cells). PCI dose dependently decreases phagocytosis of apoptotic Jurkat cells by U937 macrophages. Phagocytosis of PS exposing, activated platelets by human blood derived monocytes likewise declines in the presence of PCI. In U937 cells, PCI expression and cell surface binding increase with time after phorbol ester treatment/macrophage differentiation. Such results suggest a role of PCI as a negative regulator of apoptotic cell and activated platelets removal (2). **Target(s):** acrosin (3-6); chymotrypsin (7); coagulation factor Xa (6-8); coagulation factor Xia (6,8,9); matriptase (10); plasmin, with the bovine inhibitor transiently inhibiting bovine plasmin, but not human plasmin (11); protein C (activated) (6-8,12,13,23-26); semenogelase (14); thrombin (6-8,15,25); thrombin-thrombomodulin complex (15); tissue kallikrein (6,8,9,16-19); tissue kallikrein hK (4), human (9,17-19); t-plasminogen activator (6,20); trypsin (7); u-plasminogen activator, *or* urokinase (6,20-22,25). **1.** Wahlmüller, Sokolikova, Rieger & Geiger (2013) *Thromb. Haemost.* **111**, 41. **2.** Rieger, Assinger, Einfinger, Sokolikova & Geiger (2014) *PLoS One* **9**, e101794. **3.** Zheng, Geiger, Ecke, *et al.* (1994) *Amer. J. Physiol.* **267**, C466. **4.** Zheng, Geiger, Ecke, *et al.* (1994) *Fibrinolysis* **8**, 364. **5.** Hermans, Jones & Stone (1994) *Biochemistry* **33**, 5440. **6.** Suzuki (1993) *Meth. Enzymol.* **222**, 385. **7.** Suzuki, Nishioka, Kusumoto & Hashimoto (1984) *J. Biochem.* **95**, 187. **8.** Suzuki, Kusumoto, Nishioka & Komiyama (1990) *J. Biochem.* **107**, 381. **9.** Meijers, Kanters, Vlooswijk, *et al.* (1988) *Biochemistry* **27**, 4231. **10.** Szabo, Netzel-Arnett, Hobson, Antalis & Bugge (2005) *Biochem. J.* **390**, 231. **11.** Yuasa, Tanaka, Hayashi, *et al.* (2000) *Thromb. Haemost.* **83**, 262. **12.** Shen & Dahlbäck (1998) in *Handb. Proteolytic Enzymes* (Barrett, Rawlings & Woessner, eds.), p. 174, Academic Press, San Diego. **13.** Nishioka, Ning, Hayashi & Suzuki (1998) *J. Biol. Chem.* **273**, 11281. **14.** Chao (1998) in *Handb. Proteolytic Enzymes* (Barrett, Rawlings & Woessner, eds.), p. 102, Academic Press, San Diego. **15.** Rezaie, Cooper, Church & Esmon (1995) *J. Biol. Chem.* **270**, 25336. **16.** Chao (1998) in *Handb. Proteolytic Enzymes* (Barrett, Rawlings & Woessner, eds.), p. 97, Academic Press, San Diego. **17.** Chao (1998) in *Handb. Proteolytic Enzymes* (Barrett, Rawlings & Woessner, eds.), p. 100, Academic Press, San Diego. **18.** Cao, Becker, Lundwall, *et al.* (2003) *Prostate* **57**, 196. **19.** Mikolajczyk, Millar, Kumar & Saedi (1999) *Int. J. Cancer* **81**, 438. **20.** Heeb, España, Geiger, *et al.* (1987) *J. Biol. Chem.* **262**, 15813. **21.** Ellis & Danø (1998) in *Handb. Proteolytic Enzymes* (Barrett, Rawlings & Woessner, eds.), p. 177, Academic Press, San Diego. **22.** Schmitt, Jänicke, Moniwa, *et al.* (1992) *Biol. Chem. Hoppe-Seyler* **373**, 611. **23.** Hermans & Stone (1993) *Biochem. J.* **295**, 239. **24.** Suzuki, Nishioka & Hashimoto (1983) *J. Biol. Chem.* **258**, 163. **25.** Pratt, Macik & Church (1989) *Thromb. Res.* **53**, 595. **26.** Friedrich, Blom, Dahlbaeck & Villoutreix (2001) *J. Biol. Chem.* **276**, 24122.

Protein Kinase A Inhibitors

These small, heat-stable proteins, also known as adenosine-3',5'-cyclic-monophosphate-dependent protein kinase inhibitor proteins and abbreviated PKIs or PKAIs, bind to the catalytic subunit of protein kinase A and inhibit the activity (1-10). The K_i value for the α isoform is 98 pM, and many fragments have 1-3 nM K_i values. This inhibitor was first identified in the laboratories of Edmund H. Fischer and Edwin G. Krebs, who shared the 1992 Nobel Prize in Physiology or Medicine for their discovery on protein kinases. *Note:* The TTTADFIASGRTGRRNAIHD corresponds to the active site of the rabbit skeletal muscle inhibitor protein and also inhibits the catalytic subunit of protein kinase A. **1.** Whitehouse & Walsh (1983) *Meth. Enzymol.* **99**, 80. **2.** Walsh & Glass (1991) *Meth. Enzymol.* **201**, 304. **3.** Beebe & Corbin (1986) *The Enzymes*, 3rd ed. (Boyer & Krebs, eds.), **17**, 43. **4.** Walsh, Ashby, Gonzalez, *et al.* (1971) *J. Biol. Chem.* **246**, 1977. **5.** Demaille, Peters & Fischer (1977) *Biochemistry* **16**, 3080. **6.** Whitehouse & Walsh (1983) *J. Biol. Chem.* **258**, 3682. **7.** Scott, Fischer, Takio, Demaille & Krebs (1985) *Proc. Natl. Acad. Sci. U.S.A.* **82**, 5732. **8.** Thomas, Van Patten, Howard, *et al.* (1991) *J. Biol. Chem.* **266**, 10906. **9.** Baude, Dignam, Olsen, Reimann & Uhler (1994) *J. Biol. Chem.* **269**, 2316. **10.** Adams (2001) *Chem. Rev.* **101**, 2271.

Protein Kinase A Inhibitor Fragment 5-24

This eicosapeptide fragment from the rabbit skeletal muscle protein kinase A inhibitor, consisting of residues TTYADFIASGRTGRRNAIHD (also known as IP20) has a K_i value of 2.3 nM for the catalytic subunit of protein kinase A (1). Kinases other than the catalytic subunit of protein kinase A are also inhibited; hence, this peptide is also simply called protein kinase inhibitor. Note also that the *N*-myristoylated peptide is also commercially available. **Target(s):** Ca^{2+}/calmodulin-dependent protein kinase (2); cAMP-dependent protein kinase, *or* protein kinase A (1-13); cGMP-dependent protein kinase, *or* protein kinase G (2); and [myosin light-chain] (2). **1.** Kemp, Pearson & House (1991) *Meth. Enzymol.* **201**, 287. **2.** Kemp, Cheng & Walsh (1988) *Meth. Enzymol.* **159**, 173. **3.** Cheng, Kemp, Pearson, *et al.* (1986) *J. Biol. Chem.* **261**, 989. **4.** Walsh & Glass (1991) *Meth. Enzymol.* **201**, 304. **5.** Adams (2001) *Chem. Rev.* **101**, 2271. **6.** Patel, Soulages, Wells & Arrese (2004) *Insect Biochem. Mol. Biol.* **34**, 1269. **7.** Gardner, Delos Santos, Matta, Whitt & Bahouth (2004) *J. Biol. Chem.* **279**, 21135. **8.** Bejar & Villamarin (2006) *Arch. Biochem. Biophys.* **450**, 133. **9.** Wu, Yang, Kannan, *et al.* (2005) *Protein Sci.* **14**, 2871. **10.** Yang, Ten Eyck, Xuong & Taylor (2004) *J. Mol. Biol.* **336**, 473. **11.** Akamine, Madhusudan, Wu, *et al.* (2003) *J. Mol. Biol.* **327**, 159. **12.** Kim, Xuong & Taylor (2005) *Science* **307**, 690. **13.** Zhang, Morris & Beebe (2004) *Protein Expr. Purif.* **35**, 156.

Protein Kinase A Inhibitor Fragment 6-22 Amide

This octadecapeptide amide (*Sequence:* TYADFIASGR-TGRRNAI-NH$_2$) corresponds to residues 6-22 of the rabbit skeletal muscle protein inhibitor (1,2). The peptide amide inhibits the catalytic subunit of the kinase with a K_i value of 1.7-1.9 nM. Note that the phenylalanyl residue is a major factor in the inhibitory activity of this peptide. **1.** Glass, Lundquist, Katz & Walsh (1989) *J. Biol. Chem.* **264**, 14579. **2.** Walsh & Glass (1991) *Meth. Enzymol.* **201**, 304.

Protein Kinase A Inhibitor Fragment 14-24 Amide

This undecapeptide amide (*Sequence:* GRTGRRNAIHD-NH$_2$) inhibits protein kinase A. **1.** Norenberg, Wirkner, Assmann, Richter & Illes (1998) *Amino Acids* **14**, 33.

Protein Kinase A Inhibitor Fragment 14-22 Amide, Myristoylated, *See N-(Myristoyl)-Protein Kinase A Inhibitor Fragment 14-22 Amide*
Protein Kinase A Inhibitor Fragment 17-22, Myristoylated, *See N-(Myristoyl)-Protein Kinase A Inhibitor Fragment 17-22*

Protein Kinase A Regulatory Subunit

Type I protein kinase A (*i.e.,* cAMP-dependent protein kinase) consists of two catalytic and two regulatory subunits. The regulatory subunits inhibit the activity of the catalytic subunits. Upon binding of cyclic AMP to the regulatory subunits, the catalytic subunits dissociate from the complex and become catalytically active. **Target(s):** cAMP-dependent protein kinase, *or* protein kinase A (1-5); and phosphorylase phosphatase (6). **1.** Traugh, Ashby & Walsh (1974) *Meth. Enzymol.* **38**, 290. **2.** Beavo, Bechtel & Krebs (1974) *Meth. Enzymol.* **38**, 299. **3.** Lohmann, De Camilli & Walter (1988) *Meth. Enzymol.* **159**, 183. **4.** Seville & Holbrook (1988) *Meth. Enzymol.* **159**, 208. **5.** Fletcher, Ishida, Van Patten & Walsh (1988) *Meth. Enzymol.* **159**, 255. **6.** McNall, Ballou, Villa-Moruzzi & Fischer (1988) *Meth. Enzymol.* **159**, 377.

Protein Kinase C Inhibitor Fragment 19-31

This tridecapeptide (FW = 1543.84 g/mol; *Sequence:* RFARKGALRQ KNV), corresponding to a sequence of residues in the regulatory domain of protein kinase C, inhibits protein kinase C with a IC$_{50}$ value of about 0.1 μM. **1.** House & Kemp (1987) *Science* **238**, 1726.

Protein Kinase C Inhibitor Fragment 19-36

This fragment (FW = 2151.51 g/mol; *Sequence:* RFARKGALRQKN VHEVKN) of the regulatory domain of protein kinase C resembles a substrate phosphorylation site and acts as a potent inhibitor of protein kinase C autophosphorylation and protein substrate phosphorylation, K_i = of 147 nM (1,2). **1.** House & Kemp (1987) *Science* **238**, 1726. **2.** Kemp, Pearson & House (1991) *Meth. Enzymol.* **201**, 287.

Protein Kinase C Inhibitor Fragment 20-28, Myristoylated, *See N-(Myristoyl)-Protein Kinase C Inhibitor Fragment 20-28*

Protein Kinase C-ζ Pseudosubstrate Inhibitor, 113-125

This tridecapeptide (FW = 1718.04 g/mol; *Sequence*: SIYRRGARRWRKL, matching residues 113-125 of the autoinhibitory domain) is a selective inhibitor of protein kinase C-ζ. **1**. Laudanna, Mochly-Rosen, Liron, Constantin & Butcher (1998) *J. Biol. Chem.* **273**, 30306.

Protein Kinase C-ζ Pseudosubstrate Inhibitor, 113-125, Myristoylated, See *N-(Myristoyl)-Protein Kinase C-ζ Pseudosubstrate Inhibitor, 113-125*

Protein Kinase C-η Pseudosubstrate Inhibitor, Myristoylated, See *N-(Myristoyl)-Protein Kinase C-η Pseudosubstrate Inhibitor*

Protein Kinase C-θ Pseudosubstrate Inhibitor, Myristoylated, See *N-(Myristoyl)-Protein Kinase C-θ Pseudosubstrate Inhibitor*

Protein Kinase G Inhibitor, See *L-Arginyl-L-lysyl-L-arginyl-L-alanyl-L-arginyl-L-lysyl-L-glutamate*

Protein Kinase Inhibitor, See *Protein Kinase A Inhibitor Fragment 5-24; Protein Kinase A Inhibitors; Protein Kinase C Inhibitor Peptide 19-31; Protein Kinase C Inhibitor Peptide 19-36; Protein Kinase G Inhibitor*

Protein Phosphatase Inhibitor-1

This endogenous regulatory protein (MW$_{human}$ = 28 kDa; 171 residues) is a specific inhibitor of type 1 serine/threonine protein phosphatases. Its phosphorylation at Threonine-35 (T35) by PKA greatly enhances its ability to bind and inhibit protein phosphatase-1 (PP1), a serine/threonine phosphatase that plays a critical role in glycogen metabolism, muscle contraction, cell progression, neurotransmission, RNA splicing, mitosis, cell division, apoptosis, protein synthesis, and regulation of membrane receptors and channels. A hallmark of heart failure, for example, is deficient sarcoplasmic reticulum (SR) Ca^{2+} uptake, which impairs contractility. This disorder results, at least partially, from dephosphorylation of phospholamban by a disease-associated increase in protein phosphatase 1 (PP1) activity. Decreased expression and phosphorylation of inhibitor-1 at Thr35 contribute to observed increases in PP1 activity and dysregulated Ca^{2+} homeostasis. **Target(s)**: phosphoprotein phosphatase (1-13); [phosphorylase] phosphatase (1-10); and protein phosphatase-1 (1-11,13). **1**. McNall, Ballou, Villa-Moruzzi & Fischer (1988) *Meth. Enzymol.* **159**, 377. **2**. Ballou & Fischer (1986) *The Enzymes*, 3rd ed. (Boyer & Krebs, eds.), **17**, 311. **3**. Cohen, Alemany, Hemmings, *et al.* (1988) *Meth. Enzymol.* **159**, 390. **4**. Cohen, Foulkes, Holmes, Nimmo & Tonks (1988) *Meth. Enzymol.* **159**, 427. **5**. Cohen (1991) *Meth. Enzymol.* **201**, 389. **6**. Connor, Kleeman, Barik, Honkanen & Shenolikar (1999) *J. Biol. Chem.* **274**, 22366. **7**. Connor, Quan, Ramaswamy, *et al.* (1998) *J. Biol. Chem.* **273**, 27716. **8**. Cohen (1989) *Ann. Rev. Biochem.* **58**, 453. **9**. Khatra (1988) *Meth. Enzymol.* **159**, 368. **10**. Bollen, Vandenheede, Goris & Stalmans (1988) *Biochim. Biophys. Acta* **969**, 66. **11**. Gallego & Virshup (2005) *Curr. Opin. Cell Biol.* **17**, 197. **12**. Cohen (2004) in *Topics in Current Genetics* (Arino & Alexander, eds.) Springer **5**, 1. **13**. Gibbons, Weiser & Shenolikar (2005) *J. Biol. Chem.* **280**, 15903.

Protein Phosphatase Inhibitor 2

This 204-residue, heat-stable protein (MW = 31000 for rabbit skeletal muscle) is a specific inhibitor of protein phosphatase-1 (IC$_{50}$ = 1-2 nM) that is active in its dephosphorylated form. This protein is the regulatory subunit of the cytosolic form of protein phosphatase-1. **Target(s)**: phosphoprotein phosphatase (1-18); [phosphorylase] phosphatase (1-13); and protein phosphatase-1 (1-15,17,18). **1**. McNall, Ballou, Villa-Moruzzi & Fischer (1988) *Meth. Enzymol.* **159**, 377. **2**. Ballou & Fischer (1986) *The Enzymes*, 3rd ed. (Boyer & Krebs, eds.), **17**, 311. **3**. Cohen, Alemany, Hemmings, *et al.* (1988) *Meth. Enzymol.* **159**, 390. **4**. Cohen, Foulkes, Holmes, Nimmo & Tonks (1988) *Meth. Enzymol.* **159**, 427. **5**. Cohen (1991) *Meth. Enzymol.* **201**, 389. **6**. Eriksson, Toivola, Sahlgren, Mikhailov & Härmälä-Braskén (1998) *Meth. Enzymol.* **298**, 542. **7**. Park, Roach, Bondor, Fox & DePaoli-Roach (1994) *J. Biol. Chem.* **269**, 944. **8**. Huang & Glinsmann (1976) *Eur. J. Biochem.* **70**, 419. **9**. Ingebritsen, Foulkes & Cohen (1980) *FEBS Lett.* **119**, 9. **10**. Cohen (1989) *Ann. Rev. Biochem.* **58**, 453. **11**. Khatra (1988) *Meth. Enzymol.* **159**, 368. **12**. Villa-Moruzzi (1986) *Arch. Biochem. Biophys.* **247**, 155. **13**. Zhang, Bai, Deans-Zirattu, Browner & Lee (1992) *J. Biol. Chem.* **267**, 1484. **14**. Stubbs, Tran, Atwell, *et al.* (2001) *Biochim. Biophys. Acta* **1550**, 52. **15**. Gallego & Virshup (2005) *Curr. Opin. Cell Biol.* **17**, 197. **16**. Cohen (2004) in *Topics in Current Genetics* (Arino & Alexander, eds.) Springer **5**, 1. **17**. Szöor, Gross & Alphey (2001) *Arch.*

Biochem. Biophys. **396**, 213. **18**. Gibbons, Weiser & Shenolikar (2005) *J. Biol. Chem.* **280**, 15903.

Protein Phosphatase 2A Inhibitor I$_1$^{PP2A}

This heat-stable protein (MW = 39 kDa) acts as a potent, non-competitive inhibitor of protein phosphatase 2A, K_i = 30 nM with myelin basic protein, histone H$_1$, or phosphorylase as substrate (1,2). Protein Phosphatase 2A Inhibitor I$_1$^{PP2A} is highly expressed in some cancers and may trigger apoptosis. **1**. Li, Guo & Damuni (1995) *Biochemistry* **34**, 1988. **2**. Li, Makkinje & Damuni (1996) *Biochemistry* **35**, 6998.

Protein Phosphatase 2A Inhibitor I$_2$^{PP2A}

This heat-stable protein (MW = 39 kDa) inhibits protein phosphatase 2A, K_i ≈ 0.1 nM with myelin basic protein, histone H$_1$, and other protein substrates (1,2). Protein Phosphatase 2A Inhibitor I$_2$^{PP2A} also stimulates protein phosphatase 1 and regulates histone binding to DNA. I$_2$^{PP2A} is a truncated form of SET, a largely nuclear protein that is fused to nucleoporin Nup214 in acute non-lymphocytic myeloid leukemia. **1**. Li, Guo & Damuni (1995) *Biochemistry* **34**, 1988. **2**. Li, Makkinje & Damuni (1996) *Biochemistry* **35**, 6998.

14-3-3 Protein-Protein Interaction Inhibitors 9 & 10

14-3-3 PPI 9   14-3-3 PPI 10

These small-molecule PPIs (FW$_{Compound-9}$ = 395.37g/mol; FW$_{Compound-10}$ = 438.44 g/mol) target are 14-3-3 Protein–Protein Interaction Inhibitors that sensitize multidrug-resistant (MDR) cancer cells to doxorubicin and the Akt inhibitor GSK690693 (1). The 14-3-3 family of highly conserved adapter proteins adapter proteins play roles in eukaryotic cellular signaling, cell cycle regulation, protein trafficking, metabolism, and control of apoptosis. Based on bioinformatic and proteomic studies, 14-3-3 isoforms carry out their functions by interacting with >500 protein partners, including c-Abl and Raf kinases, p53, MLF1, FOXO transcription factors, Cdc25, Bad, and Tau. This fact has motivated the search for small-molecule protein-protein interaction inhibitors (PPIs) that can disrupt interactions of 14-3-3 isoforms with their partners. The most active of these PPIs, designated Compounds, 9 and 10 promote translocation of c-Abl into the nucleus, suggesting that they interact with 14-3-3σ. Notably, Compound 9 udergoes spontaneous dehydration to form a phthalimide form, a property that was eliminated through the synthesis of Compound 10. These 14-3-3 PPIs provide anticancer effects in cells by significantly enhancing the activity of the doxorubicin and the pan-Akt inhibitor GSK690693 in MDR cancer cells by decreasing P-glycoprotein expression and inducing apoptosis. **1**. Mori, Vignaroli, Cau, *et al.* (2014) *Chem. Med. Chem.* **9**, 973.

Protein Synthesis Inhibitors

These agents retard or arrest protein synthesis by interacting with essential reaction components in multistep ribosome-catalyzed processes. Cycloheximide effectively inhibits eukaryotic protein synthesis on cytoplasmic ribosomes, at 1 mg/mL resulting in 70–75% inhibition for cell-free translation, whereas 30 mg/mL almost totally blocks intracellular protein synthesis. Chloramphenicol blocks the peptidyltransferase reaction, but DNA synthesis can also be reduced by this inhibitor. Other protein synthesis inhibitors include: α-amino-β-butyric acid (an analogue of valine); aurintricarboxylic acid (an inhibitor of initiation at 50 mM); 7-azatryptophan (a tryptophan analogue that blocks bacteriophage synthesis); azetidine-2-carboxylic acid (a four-membered ring analogue of proline that can be incorporated into bacterial proteins); canavanine (an arginine analogue that inhibits bacterial protein synthesis); emetine (binds to the 40S eukaryotic ribosome and inhibits translocation); erythromycin (bacterial protein synthetase inhibitor); ethionine (a methionine analogue that inhibits bacterial protein synthesis); 5-fluorotryptophan; 6-fluorotryptophan; methyltryptophan; fusidic acid; kasugamycin; 7-methylguanosine 5′-

phosphate; ω-methyllysine; O-methylthreonine; norleucine; pactamycin; puromycin (a potent inhibitor of prokaryotic and eukaryotic protein synthesis); selenomethionine; sparsomycin; streptomycin; tetracycline antibiotics; and tryptazan. **See** specific inhibitor; α-Amino-β-Butyric Acid; 7-Azatryptophan; Azetidine-2-Carboxylic Acid; Canavanine; Chloramphenicol; Cycloheximide; Emetine; Erythromycin; Ethionine; 5-Fluorotryptophan; 6-Fluorotryptophan; Fusidic acid; Kasugamycin; 7-Methylguanosine 5'-phosphate; w-Methyllysine; O-Methyl-threonine; Methyltryptophan Norleucine; Pactamycin; Puromycin; Selenomethionine; Sparsomycin; Streptomycin; Tetracycline; and Tryptazan.

Protein-Tyrosine Phosphatase Inhibitor I, See α-Bromo-4-hydroxyacetophenone

Protein-Tyrosine Phosphatase Inhibitor II, See α-Bromo-4-methoxyacetophenone

Protein-Tyrosine Phosphatase Inhibitor III, See α-Bromo-4-(carboxymethoxy)-acetophenone

Prothionine & Prothionine Sulfoximine Phosphate

These tetrahedral transition-state analogues PSOX (FW_{ion} = 207.27 g/mol) and PSOX-P (FW_{ion} = 285.24 g/mol) potently inhibit γ-glutamylcysteine synthetase in a manner similar to the inhibition of glutamine synthetase by methionine sulfoximine. The phosphorylated analogue mimics the acyl-P reaction intermediate. The propyl group presumably binds at the enzyme site pocket normally occupied by L-cysteine. **Target(s):** glutamine synthetase, mildly inhibited (1); and γ-glutamylcysteine synthetase (1-4). **1.** Griffith, Anderson & Meister (1979) J. Biol. Chem. **254**, 1205. **2.** Seelig & Meister (1985) Meth. Enzymol. **113**, 379. **3.** Griffith (1987) Meth. Enzymol. **143**, 286. **4.** Griffith & Meister (1979) J. Biol. Chem. **254**, 7558.

Protoanemonin

This antibacterial and antifungal lactone (FW = 96.09 g/mol), also called 5-methylene-2(5H)-furanone, 4-methylenebut-2-en-4-olide, and γ-hydroxy-vinylacrylic acid lactone, is produced as a glucoside by Anemone pulsatilla. Protoanemonin is also formed in the degradation of some chloroaromatic compounds, such as polychlorinated biphenyls, by natural microbial action. It not only inhibits the growth of bacteria, it will also inhibit elongation of oat coleoptile segments and growth of roots. Protoanemonin reportedly reacts with sulfhydryl groups in proteins. *Note:* Protoanemonin can also be produced from 2-chloro-cis,cis-muconate by the combined action of muconate cycloisomerase and muconolactone isomerase (1). **Target(s):** carboxymethylenebutenolidase, or dienelactone hydrolase, also weak alternative substrtate (2); and pyruvate oxidase (3). **1.** Skiba, Hecht & Pieper (2002) J. Bacteriol. **184**, 5402. **2.** Brückmann, Blasco, Timmis & Pieper (1998) J. Bacteriol. **180**, 400. **3.** Baer (1948) J. Biol. Chem. **173**, 211.

Protocatechualdehyde

This aldehyde (FW = 138.12 g/mol), also known as 3,4-dihydroxybenzaldehyde and 4-formylcatechol, is soluble in water (5 g/100 mL at 20°C) and has a pK_a value of 7.55 at 25°C. Protocatechualdehyde induces apoptosis in cytotoxic T cells. **Target(s):** aldehyde oxidase (1); aldose reductase (2); β-carotene 15,15'-monooxygenase (3); mandelonitrile lyase (4); protocatechuate 3,4-dioxygenase (5-11); protocatechuate 4,5-dioxygenase (12,13); tyrosinase, weakly inhibited, IC_{50} = 620 μM (14); and xanthine oxidase (15,16). **1.** Panoutsopoulos & Beedham (2004) Acta Biochim. Pol. **51**, 649. **2.** Lee, Shim, Kim, Shin & Kang (2005) Biol. Pharm. Bull. **28**, 1103. **3.** Nagao, Maeda, Lim, Kobayashi & Terao (2000) Nutr. Biochem. **11**, 348. **4.** Xu, Singh & Conn (1986) Arch. Biochem. Biophys. **250**, 322. **5.** Fujisawa (1970) Meth. Enzymol. **17A**, 526. **6.** Hayaishi, Nozaki & Abbott (1975) The Enzymes, 3rd ed. (Boyer, ed.), **12**, 119. **7.** Hou, Lillard & Schwartz (1976) Biochemistry **15**, 582. **8.** Durham, Sterling, Ornston & Perry (1980) Biochemistry **19**, 149. **9.** Fujisawa & Hayaishi (1968) J. Biol. Chem. **243**, 2673. **10.** Hou, Lillard & Schwartz (1976) Biochemistry **15**, 582. **11.** Que, Lipscomb, Münck & Wood (1977) Biochim. Biophys. Acta **485**, 60. **12.** Zabinski, Münck, Champion & Wood (1972) Biochemistry **11**, 3212. **13.** Arciero, Orville & Lipscomb (1985) J. Biol. Chem. **260**, 14035. **14.** Ley & Bertram (2001) Bioorg. Med. Chem. **9**, 1879. **15.** Beiler & Martin (1951) J. Biol. Chem. **192**, 831. **16.** Nguyen, Awale, Tezuka, et al. (2006) Planta Med. **72**, 46.

Protocatechuate (Protocatechuic Acid)

This aromatic acid ($FW_{free-acid}$ = 154.12 g/mol), also known as 3,4-dihydroxybenzoic acid, is a white-to-brownish solid that discolors in air. The free acid is slightly soluble in water (1.8 g/100 g H_2O at 14°C) and has pK_a values of 4.49, 8.83, and 12.6. Store in a well sealed container. Protocatechuateis produced in reactions catalyzed by 4-hydroxybenzoate 3-monooxygenase, 4droxybenzoate 3-monooxygenase (NAD(P)H); vanillate demethylase, (3S,4R)-3,4-dihydroxy-cyclohexa-1,5-diene-1,4-dicarboxylate dehydrogenase, terephthalate 1,2-cis-dihydrodiol dehydrogenase, 4-sulfobenzoate 3,4-dioxygenase, 3-hydroxybenzoate 4-monooxygenase, 4,5-dihydroxyphthalate decarboxylase, and 3,4-dihydroxyphthalate decarboxylase. It is a substrate for protocatechuate 4,5-dioxygenase, protocatechuate 3,4-dioxygenase, and protocatechuate decarboxylase. **Target(s):** γ-butyrobetaine dioxygenase, or γ-butyrobetaine hydroxylase, K_i = 0.6 μM (1); CYP1A2 (2); CYP2B (2); dehydroquinate synthase (3); 5-demethylubiquinone-9 3-O-methyltransferase (4;) deoxyhypusine monooxygenase (26); dihydroorotate dehydrogenase (5); 1,2-dihydroxynaphthalene dioxygenase (40); 3,4-dihydroxyphenylacetate 2,3-dioxygenase (43,44); 4,5-dihydroxyphthalate decarboxylase (6); diphosphomevalonate decarboxylase (7,8); DNA topoisomerase I (9); fatty-acid synthase, weakly inhibited (10); 3-galactosyl-N-acetylglucosaminide 4-α-L-fucosyltransferase (11); glutamate decarboxylase (12,13); glycoprotein 3-α-L-fucosyltransferase (11); 4-hydroxybenzoate 3-monooxygenase, product inhibition (30-32); 4-hydroxyphenylpyruvate dioxygenase (42); hyoscyamine (6S)-dioxygenase (14); lipase, or triacylglycerol lipase (15); 15-lipoxygenase,or arachidonate 15-lipoxygenase (41); pancreatic lipase (15); peptide-aspartate β-dioxygenase (33); phenol sulfotransferase, or aryl sulfotransferase (16); phenylalanine ammonia-lyase (17); procollagen-proline 4-dioxygenase, or prolyl 4-hydroxylase (18,33-35,37-39); proline 4-dioxygenase (36); prolyl hydroxylase, Skp1 (19); o-pyrocatechuate decarboxylase, weakly inhibited (20,21); shikimate dehydrogenase (22,23); tannase (24); tyrosinase, or monophenol monooxygenase (27,28); tyrosine 3-monooxygenase (29); and urease (25). **1.** Ng, Hanauske-Abel & Englard (1991) J. Biol. Chem. **266**, 1526. **2.** Baer-Dubowska, Szaefer & Krajka-Kuzniak (1998) Xenobiotica **28**, 735. **3.** Chandran & Frost (2001) Bioorg. Med. Chem. Lett. **11**, 1493. **4.** Houser & Olson (1977) J. Biol. Chem. **252**, 4017. **5.** Palfey, Bjornberg & Jensen (2001) J. Med. Chem. **44**, 2861. **6.** Nakazawa & Hayashi (1978) Appl. Environ. Microbiol. **36**, 264. **7.** Lalitha, George & Ramasarma (1985) Phytochemistry **24**, 2569. **8.** Shama Bhat & Ramasarma (1979) Biochem. J. **181**, 143. **9.** Stagos, Kazantzoglou, Magiatis, et al. (2005) Int. J. Mol. Med. **15**, 1013. **10.** Wang, Song, Guo & Tian (2003) Biochem. Pharmacol. **66**, 2039. **11.** Niu, Fan, Sun, et al. (2004) Arch. Biochem. Biophys. **425**, 51. **12.** Tunnicliff (1990) Int. J. Biochem. **27**, 1235. **13.** Youngs & Tunnicliff (1991) Biochem. Int. **23**, 915.

14. Hashimoto & Yamada (1987) *Eur. J. Biochem.* **164**, 277.   **15.** Karamac & Amarowicz (1996) *Z. Naturforsch. [C]* **51**, 903.   **16.** Yeh & Yen (2003) *J. Agric. Food Chem.* **51**, 1474.   **17.** Sarma & Sharma (1999) *Phytochemistry* **50**, 729.   **18.** Tschank, Hanauske-Abel & Peterkofsky (1988) *Arch. Biochem. Biophys.* **261**, 312.   **19.** van der Wel, Ercan & West (2005) *J. Biol. Chem.* **280**, 14645.   **20.** Santha, Rao & Vaidyanathan (1996) *Biochim. Biophys. Acta* **1293**, 191.   **21.** Kamath, Dasgupta & Vaidyanathan (1987) *Biochem. Biophys. Res. Commun.* **145**, 586.   **22.** Balinsky & Dennis (1970) *Meth. Enzymol.* **17A**, 354.   **23.** Balinsky & Davies (1961) *Biochem. J.* **80**, 296.   **24.** Iibuchi, Minoda & Yamada (1972) *Agric. Biol. Chem.* **36**, 1553.   **25.** Quastel (1933) *Biochem. J.* **27**, 1116.   **26.** Abbruzzese, Hanauske-Abel, Park, Henke & Folk (1991) *Biochim. Biophys. Acta* **1077**, 159.   **27.** McIntyre & Vaughan (1975) *Biochem. J.* **149**, 447.   **28.** Lin, Hsu, Chen, Chern & Lee (2007) *Phytochemistry* **68**, 1189.   **29.** Nagatsu, Levitt & Udenfriend (1964) *J. Biol. Chem.* **239**, 2910.   **30.** Fernandez, Dimarco, Ornston & Harayama (1995) *J. Biochem.* **117**, 1261.   **31.** Spector & Massey (1972) *J. Biol. Chem.* **247**, 4679.   **32.** Hosokawa & Stanier (1966) *J. Biol. Chem.* **241**, 2453.   **33.** Koivunen, Hirsilä, Günzler, Kivirikko & Myllyharju (2004) *J. Biol. Chem.* **279**, 9899.   **34.** Majamaa, Günzler, Hanauske-Abel, Myllylä & Kivirikko (1986) *J. Biol. Chem.* **261**, 7819.   **35.** Kaska, Günzler, Kivirikko & Myllylä (1987) *Biochem. J.* **241**, 483.   **36.** Lawrence, Sobey, Field, Baldwin & Schofield (1996) *Biochem. J.* **313**, 185.   **37.** Kivirikko, Myllylä & Pihlajaniemi (1989) *FASEB J.* **3**, 1609.   **38.** Günzler, Hanauske-Abel, Myllylä, Mohr, Kivirikko (1987) *Biochem. J.* **242**, 163.   **39.** Myllyharju (2008) *Ann. Med.* **40**, 402.   **40.** Patel & Barnsley (1980) *J. Bacteriol.* **143**, 668.   **41.** Russell, Scobbie, Duthie & Chesson (2008) *Bioorg. Med. Chem.* **16**, 4589.   **42.** Lindstedt & Rundgren (1982) *J. Biol. Chem.* **257**, 11922.   **43.** Que, Jr., Widom & Crawford (1981) *J. Biol. Chem.* **256**, 10941.   **44.** Gabello, Ferrer, Martín & Garrido-Pertierra (1994) *Biochem. J.* **301**, 145.

Protocatechuic Acid Methyl Ester

This aromatic ester (FW = 168.15 g/mol), also known as methyl 3,4-dihydroxybenzoate, inhibits protocatechiate 3,4-dioxygenase.   **1.** Que, Lipscomb, Münck & Wood (1977) *Biochim. Biophys. Acta* **485**, 60.

Protohemin IX, *See Hemin*

Protolichesterinic Acid

This aliphatic α-methylene-γ-lactone (FW = 324.46 g/mol; CAS 493-46-9) from the lichen *Cetraria islandica* inhibits human DNA ligase (ATP-dependent), $IC_{50} = 20$ μM (1). Protolichesterinate is a strong inhibitor of the DNA polymerase activity associated with the HIV-1 reverse transcriptase (2) and 5-lipoxygenase, *or* arachidonate 5-lipoxygenase (3,4). It also exhibits both antibiotic and antitumor properties.   **1.** Tan, Lee, Lee, *et al.* (1996) *Biochem. J.* **314**, 993.   **2.** Pengsuparp, Cai, Constant, *et al.* (1995) *J. Nat. Prod.-Lloydia* **58**, 1024.   **3.** Ogmundsdottir, Zoega, Gissurarson & Ingolfsdottir (1998) *J. Pharm. Pharmacol.* **50**, 107.   **4.** Schneider & Bucar (2005) *Phytother. Res.* **19**, 81.

Protonophores, *See Uncouplers (Gradient-Dissipating Proton Carriers)*

Protopine

This photosensitive antiplatelet agent (FW_{free-base} = 353.37 g/mol; CAS 130-86-9) from a number of plants, including the herb *Fumaria officinalis* and opium, is a Ca^{2+} channel blocker. Protopine is commonly supplied commercially as the hydrochloride or the methiodide. **Target(s):** acetylcholinesterase (1); Ca^{2+} channels (2); cation channel currents, including $I_{Ca,L}$, I_K, I_{K1}, as well as I_{Na} (3); certain cytochrome P450 systems (4); phospholipase (5); and thromboxane synthase (5,6).   **1.** Kim, Hwang, Jang, *et al.* (1999) *Planta Med.* **65**, 218.   **2.** Ko, Wu, Lu, *et al.* (1992) *Jpn. J. Pharmacol.* **58**, 1.   **3.** Song, Ren, Chen, *et al.* (2000) *Brit. J. Pharmacol.* **129**, 893.   **4.** Janbaz, Saeed & Gilani (1998) *Pharmacol. Res.* **38**, 215.   **5.** Shiomoto, Matsuda & Kubo (1991) *Chem. Pharm. Bull. (Tokyo)* **39**, 474.   **6.** Saeed, Gilani, Majoo & Shah (1997) *Pharmacol. Res.* **36**, 1.

Protoporphyrin IX

This iiron-free, immediate precursor of heme (FW_{free-acid} = 562.67 g/mol; CAS 553-12-8; brownish-yellow solid; soluble in a number of organic solvents; disodium and dipotassium salts solublized in the presence of Tween 80) has λ_{max} values in 25% HCl are 602.4, 582.2, and 557.2 nm. Protoporphryin IX is also an activator of guanylate cyclase. *See also Iron Protoporphyrin IX; Heme; Hemin* **Target(s):** aminolevulinate aminotransferase (1); 5-aminolevulinate synthase (2-5); glutamate: glyoxylate aminotransferase (6); glutathione *S*-transferase (7-10); glyoxalase I, *or* lactoylglutathione lyase (11); guanylate cyclase (12); heme oxygenase (13); hydroxymethylbilane synthase, *or* porphobilinogen deaminase (14,15); nitric-oxide synthase (16); porphobilinogen synthase, *or* 5-aminolevulinate dehydratase (17,18); succinyl-CoA synthetase (19,20); tryptophan pyrrolase, *or* tryptophan 2,3-dioxygenase (21); and uroporphyrinogen decarboxylase (18,22).   **1.** Singh & Datta (1985) *Biochim. Biophys. Acta* **827**, 305.   **2.** Burnham (1970) *Meth. Enzymol.* **17A**, 195.   **3.** Yubisui & Yoneyama (1972) *Arch. Biochem. Biophys.* **150**, 77.   **4.** Scholnick, Hammaker & Marver (1972) *J. Biol. Chem.* **247**, 4132.   **5.** Tait (1973) *Biochem. J.* **131**, 389.   **6.** Paszkowski (1992) *Acta Biochim. Pol.* **39**, 345.   **7.** Lederer & Boger (2003) *Biochim. Biophys. Acta* **1621**, 226.   **8.** Harwaldt, Rahlfs & Becker (2002) *Biol. Chem.* **383**, 821.   **9.** Smith, Nuiry & Awasthi (1985) *Biochem. J.* **229**, 823.   **10.** Liebau, Eckelt, Wildenburg, *et al.* (1997) *Biochem. J.* **324**, 659.   **11.** Douglas & Sharif (1983) *Biochim. Biophys. Acta* **748**, 184.   **12.** El Deib, Parker & White (1987) *Biochim. Biophys. Acta* **928**, 83.   **13.** Frydman, Tomaro, Buldain, *et al.* (1981) *Biochemistry* **20**, 5177.   **14.** Wolff, Naddelman, Lubeskie & Saks (1996) *Arch. Biochem. Biophys.* **333**, 27.   **15.** Meissner, Adams & Kirsch (1993) *J. Clin. Invest.* **91**, 1436.   **16.** Cardalda, Juknat, Princ & Batlle (1997) *Arch. Biochem. Biophys.* **347**, 69.   **17.** Coleman (1970) *Meth. Enzymol.* **17A**, 211.   **18.** Afonso, Chinarro, Munoz, de Salamanca & Batlle (1990) *J. Enzyme Inhib.* **3**, 303.   **19.** Bridger (1974) *The Enzymes*, 3rd ed. (Boyer, ed.), **10**, 581.   **20.** Wider & Tigier (1971) *Enzymologia* **41**, 217.   **21.** Greengard & Feigelson (1962) *J. Biol. Chem.* **237**, 1903.   **22.** Jones & Jordan (1993) *Biochem. J.* **293**, 703.

Protoporphyrinogen IX

This heme biosynthetic intermediate (FW = 566.70 g/mol; CAS 7412-77-3) is formed in the reaction catalyzed by coproporphyrinogen oxidase and is a substrate of protoporphyrinogen oxidase. **Target(s):** coproporphyrinogen oxidase (1,2); glutathione *S*-transferase (3); porphobilinogen deaminase, *or* hydroxymethylbilane synthase (4). **1.** Jones, He & Lash (2002) *J. Biochem.* **131**, 201. **2.** Breckau, Mahlitz, Sauerwald, Layer & Jahn (2003) *J. Biol. Chem.* **278**, 46625. **3.** Lederer & Boger (2003) *Biochim. Biophys. Acta* **1621**, 226. **4.** Meissner, Adams & Kirsch (1993) *J. Clin. Invest.* **91**, 1436.

Protothebaine, *See* Reticuline

Provitamin A, *See* β-Carotene

Provitamin D₂, *See* Ergosterol

Provitamin D₃, *See* 7-Dehydrocholesterol

Proxyphylline, *See* 7-(β -Hydroxypropyl)theophylline

Prozac, *See* Fluoxetine

PRPP, *See* 5-Phospho- α-D-ribosyl 1-Pyrophosphate

PRT062607

This novel orally available Syk inhibitor ($FW_{free-base}$ = 393.46 g/mol; FW_{HCl} = 429.91 g/mol; CAS 1370261-97-4), also known as P505-15 and BIIB057 and systematically named 2-[[(1*R*,2*S*)-2-aminocyclohexyl]amino]-4-[[3-(2*H*-1,2,3-triazol-2-yl)phenyl]amino]-5-pyrimidinecarboxamide, exhibits greater than 80x selectivity for Spleen Tyrosine Kinase Syk (IC_{50} = 1 nM) over Fgr, Lyn, FAK, Pyk2 and Zap70.Syk is a B-cell signal-transduction kinase, suggesting that its inhibition should have implications in immune processes. Mice lacking Syk die during embryonic development around midgestation. Abnormal Syk activity is observed in several hematopoeitic malignancies, including those having translocations in Itk and Tel. In human whole blood, PRT062607 potently inhibits B cell antigen receptor-mediated B cell signaling and activation (IC_{50} = 0.27 and 0.28 μM, respectively) and Fcε receptor 1-mediated basophil degranulation (IC_{50} = 0.15 μM) (1). Similar levels of *ex vivo* inhibition are measured after dosing in mice (Syk signaling IC_{50} = 0.32 μM). Syk-independent signaling and activation are unaffected at much higher PRT062607 concentrations, demonstrating the specificity of kinase inhibition in cellular systems (1). In assays that model chronic lymphocytic leukemia (CLL) interactions with the microenvironment, PRT062607 effectively antagonizes CLL cell survival after BCR triggering and in nurse-like cell-co-cultures (2). Moreover, it inhibits BCR-dependent secretion of the chemokines CCL3 and CCL4 by CLL cells, and leukemia cell migration toward the tissue homing chemokines CXCL12, CXCL13, and beneath stromal cells. PRT062607 also inhibits Syk and extracellular signal-regulated kinase phosphorylation after BCR triggering (2). Such findings demonstrate that selective Syk inhibitors inhibition CLL survival and tissue homing circuits, supporting therapeutic development of such agents in patients with CLL, other B cell malignancies, and autoimmune disorders (2). **Target(s):** Syk (IC_{50} = 1 nM); Fgr (IC_{50} = 81 nM); Mlk1 (IC_{50} = 88 nM). **1.** Coffey, DeGuzman, Inagaki, *et al.* (2012) *J. Pharmacol. Exp. Ther.* **340**, 350. **2.** Hoellenriegel, Coffey, Sinha, *et al.* (2012) *Leukemia* **26**, 1576.

PRX-08066

This selective 5-hydroxytryptamine (serotonin) receptor antagonist (FW = 517.96 g/mol; CAS = 866206-55-5), also named 5-((4-(6-chlorothieno[2,3-*d*]pyrimidin-4-ylamino)piperidin-1-yl)methyl)-2-fluorobenzonitrile, targets 5-HT₂B (IC_{50} of 3.4 nM), showing high selectivity over the closely related 5-

HT₂A, 5-HT₂C and other receptors and preventing the severity of pulmonary arterial hypertension in the monocrotaline (MCT)-induced rat pulmonary artery hypertension model. PRX-08066 significantly attenuates the elevation in pulmonary artery pressure and right ventricular hypertrophy, thereby improving cardiac function (1). Targeting the 5-5-HT₂B receptor also appears be an effective antiproliferative and antifibrotic strategy for small intestinal neuroendocrine tumors (SI-NETs), because it inhibits tumor microenvironment fibroblasts as well as NET cells (2). **1.** Porvasnik, Germain, Embury, *et al.* (2010) *J. Pharmacol. Exp. Ther.* **334**, 364. **2.** Svejda, Kidd, Giovinazzo, *et al.* (2010) *Cancer* **116**, 2902.

Prucalopride

This novel enterokinetic drug (FW = 367.87 g/mol; CAS 179474-81-8), also known by the tradename Resolor® and its systematic name, 4-amino-5-chloro-*N*-[1-(3-methoxypropyl)piperidin-4-yl]-2,3-dihydro-1-benzofuran-7-carboxamide, is a selective, high affinity serotonin 5-HT₄ receptor agonist with that stimulates colonic mass movements, which provide the main propulsive force for defecation. It exhibits high affinity to both 5-HT₄ receptor isoforms, with respective pK_i values of 8.60 and 8.10 for the human 5-HT₄A and 5-HT₄B receptors. Based on 50 other binding assays,tonly the human D₄ receptor (pK_i = 5.63), the mouse 5-HT₃ receptor (pK_i = 5.41) and the human σ₁ (pK_i = 5.43) shown measurable affinity, resulting in >290x selectivity for 5-HT₄ receptors. Prucalopride also differs from other 5-HT₄ agonists, such as tegaserod and cisapride, that interact with other receptors (5-HT₁B/D and cardiac human ether-a-go-go K⁺ or hERG channel, respectively). **1.** Briejer, Bosmans, Van Daele, *et al.* (2001) *Eur. J. Pharmacol.* **423**, 71.

Prulifloxacin

This fluoroquinolone-class antibiotic (FW = 461.46 g/mol; CAS 123447-62-1; IUPAC: (*RS*)-6-fluoro-1-methyl-7-[4-(5-methyl-2-oxo-1,3-dioxolen-4-yl)methyl-1-piperazinyl]-4-oxo-4*H*-[1,3]thiazeto[3,2-*a*]quinoline-3-carboxylate), marketed in Japan, Italy and Austria under the tradenames Quisnon®, Unidrox®, Prixina®, and Glimbax®, is metabolically de-esterified to ulifloxacin, a broad-spectrum systemic antibacterial agent that targets Type IV DNA topoisomerases (gyrases), thereby impairing replication and transcription. Prulifloxacin is used orally to treat urinary tract infections, community-acquired respiratory tract infections, gastroenteritis, and infectious diarrheas. For the prototypical member of this antibiotic class, *See* Ciprofloxacin

Prunasin

This cyanogenic glucoside (FW = 295.29 g/mol; CAS 99-18-3), also known as D(*R*)-mandelonitrile-β-D-glucoside, is readily obtained from amygdalin and is found in the pits of *Prunus serotina*. Prunasin is soluble in water and will give rise to prulaurasin (*i.e.*, the diastereoisomeric mixture of DL(*RS*)-mandelonitrile-β-D-glucoside [*i.e.*, prunasin and sambunigrin]). **Target(s):** DNA-directed DNA polymerase (1); DNA polymerase β (1); and α,α-trehalase (2,3). **1**. Mizushina, Takahashi, Ogawa, *et al.* (1999) *J. Biochem.* **126**, 430. **2**. Silva, Terra & Ferreira (2006) *Comp. Biochem. Physiol. B Biochem. Mol. Biol.* **143**, 367. **3**. Silva, Terra & Ferreira (2004) *Insect Biochem. Mol. Biol.* **34**, 1089.

Prunetin

This *Prunus* species isoflavone (FW = 284.27 g/mol; CAS 552-59-0; melting point = 240°C), also known as 4',5-dihydroxy-7-methoxyisoflavone and 5-hydroxy-3-(4-hydroxyphenyl)-7-methoxy-4*H*-1-benzopyran-4-one, inhibits alcohol dehydrogenase (1), aldehyde dehydrogenase (2-4), 3',5'-cyclic-nucleotide phosphodiesterase (5), phosphodiesterase 4 (5), and protein-tyrosine kinase (6). Its 4'-glucoside is prunitrin. Methylated isoflavones, such as prunetin, are rapidly converted by CYP1A2 to more active genistein and daidzein, which inhibit.formation of acetaminophen from phenacetin with an IC$_{50}$ value of 16 µM. **1**. Keung (1993) *Alcohol Clin. Exp. Res.* **17**, 1254. **2**. Shen, Benson, Johnson, Lipsky & Naylor (2001) *J. Amer. Soc. Mass Spectrom.* **12**, 97. **3**. Sheikh & Weiner (1997) *Biochem. Pharmacol.* **53**, 471. **4**. Keung & Vallee (1993) *Proc. Natl. Acad. Sci. U.S.A.* **90**, 1247. **5**. Ko, Shih, Lai, Chen & Huang (2004) *Biochem. Pharmacol.* **68**, 2087. **6**. Akiyama & Ogawara (1991) *Meth. Enzymol.* **201**, 362.

Prussic Acid, *See Hydrogen Cyanide; Cyanide*

PS, *See Phosphatidyl-L-serine*

PS10

This novel ATP binding pocket-directed inhibitor (FW = 323.32 g/mol), systematically named 2-[(2,4-dihydroxyphenyl)sulfonyl]isoindoline-4,6-diol, targets all four PDK isoforms, with an IC$_{50}$ = 0.8 µM and K_i = 0.24 µM for PDK2. Its design is based on structure-guided design, converting a known Hsp90 inhibitor into a series of highly specific PDK inhibitors, substituting a sulfonyl group in place of a carbonyl in the parent compound. This modification results in weak binding to Hsp90 (K_i = 47 µM). PS10 administration (70 mg/kg) to diet-induced obese mice significantly augments PDC activity with reduced phosphorylation in different tissues. Prolonged treatment improves glucose tolerance, with notably lessened hepatic steatosis in a mouse model. Such findings suggest pharmacological targeting of PDK may be useful in controlling glucose and fat levels in obesity and type 2 diabetes. **1**. Tso, Qi, Gui, *et al.* (2014) *J. Biol. Chem.* **289**, 4432.

PS-341, *See Bortezomib*

PS 1145, *See 6-Chloro-8-nicotinamido-β-carboline*

Psalmotoxin-1
This 40-residue spider toxin (MW = 4695 Da; Smbol = PcTx1) from the venom of the Trinidad tarantula *Psalmopoeus cambridgei* binds to cysteine-rich domains (IC$_{50}$ = 0.9 nM) within the extracellular loop of the pain-sensing Acid-Sensing Ion Channel (*or* ASIC), rapidly desensitizing these proton-gated sodium channels. **1**. Escoubas, De Weille, Lecoq, *et al.* (2000) *J. Biol. Chem.* **275**, 25116.

Psammaplin A

This marine natural product (FW = 640.37 g/mol), first isolated from the sponges *Poecillastra* sp. and *Jaspis* sp., is cytotoxic against several cancer cell lines. **Target(s):** chitinase (1,2); DNA methyltransferase (3); DNA

topoisomerase II (4); histone deacetylase (3,5); membrane alanyl aminopeptidase, *or* aminopeptidase N (6-8); and mycothiol-*S*-conjugate amidase (9). **1**. Tabudravu, Eijsink, Gooday, *et al.* (2002) *Bioorg. Med. Chem.* **10**, 112. **2**. Andersen, Dixon, Eggleston & van Aalten (2005) *Nat. Prod. Rep.* **22**, 563. **3**. Piña, Gautschi, Wang, *et al.* (2003) *J. Org. Chem.* **68**, 3866. **4**. Kim, Lee, Jung, Lee & Choi (1999) *Anticancer Res.* **19**, 4085. **5**. Kim, Shin & Kwon (2007) *Exp. Mol. Med.* **39**, 47. **6**. Shim, Lee, Shin & Kwon (2004) *Cancer Lett.* **203**, 163. **7**. Xu & Li (2005) *Curr. Med. Chem. Anticancer Agents* **5**, 281. **8**. Bauvois & Dauzonne (2006) *Med. Res. Rev.* **26**, 88. **9**. Nicholas, Eckman, Ray, *et al.* (2002) *Bioorg. Med. Chem. Lett.* **12**, 2487.

PSC-833, *See Valspodar*

Pseudocarba-NAD⁺

This NAD⁺ analogue (FW = 925.39 g/mol), also known as carbanicotin-amide adenine dinucleotide, has an L-2,3-dihydroxycyclopentyl moiety in place of the D-ribosyl residue of the nicotinamide nucleotide. It is a potent inhibitor of *Bungarus fasciatus* venom NAD⁺ nucleosidase as well as the bovine and ovine brain enzymes. It is not a substrate of alcohol dehydrogenase. **Target(s):** NAD⁺ nucleosidase, *or* NADase, *or* NAD⁺ glycohydrolase (1-4). **1**. Oppenheimer & Handlon (1992) *The Enzymes*, 3rd ed. (Sigman, ed.), **20**, 453. **2**. Slama & Simmons (1989) *Biochemistry* **28**, 7688. **3**. Wall, Klis, Kornet, *et al.* (1998) *Biochem J.* **335**, 631. **4**. Muller-Steffner, Slama & Schuber (1996) *Biochem. Biophys. Res. Commun.* **228**, 128.

Pseudocarbanicotinamide Adenine Dinucleotide, *See Pseudocarba-NAD⁺*

Pseudoephedrine

This common decongestant and monoamine oxidase inhibitor (FW = 165.24 g/mol) is a nonselective adrenergic agonist and is the (+)-*threo*-isomer of ephedrine. Pseudophedrine is the sympathomimetic amine found in species of *Ephedra*. Its hydrochloride salt, also known by the proprietary names Novafed and Sudafed, is soluble in water (2 g/mL at 20°C). **Key Pharmacokinetic Parameters:** *See* Appendix II in Goodman & Gilman's *THE PHARMACOLOGICAL BASIS OF THERAPEUTICS*, 12th Edition (Brunton, Chabner & Knollmann, eds.) McGraw-Hill Medical, New York (2011). **1**. Ulus, Maher & Wurtman (2000) *Biochem. Pharmacol.* **59**, 1611.

Pseudohypericin

This anti-retroviral agent (FW = 520.45 g/mol) inhibits protein kinase C with an IC$_{50}$ value of 28.7 µM and dopamine β-hydroxylase with IC$_{50}$ = 3 µM. Pseudohypericin is a component of St. John's Wort (*Hypericum perforatum*). **Target(s):** dopamine β-hydroxylase (1); inositol-trisphosphate 3-kinase (2); and protein kinase C (3). **1**. Denke, Schempp, Weiser & Elstner (2000) *Arzneimittelforschung* **50**, 415. **2**. Mayr, Windhorst & Hillemeier (2005) *J. Biol. Chem.* **280**, 13229. **3**. Takahashi, Nakanishi, Kobayashi, *et al.* (1989) *Biochem. Biophys. Res. Commun.* **165**, 1207.

Pseudo-iodotyrostatin

This tyrostatin analogue (FW = 522.38 g/mol) inhibits the pepstatin-insensitive carboxyl protease from *Pseudomonas* sp.: K_i = 14 nM (1). **Target(s):** pseudomonalisin, *or* pseudomonapepsin (1-3); and sedolisin (3). **1**. Wlodawer, Li, Dauter, *et al.* (2001) *Nature Struct. Biol.* **8**, 442. **2**. Wlodawer, Li, Gustchina, *et al.* (2001) *Biochemistry* **40**, 15602. **3**. Wlodawer, Li, Gustchina, *et al.* (2004) *Biochem. Biophys. Res. Commun.* **314**, 638.

Pseudomauveine, *See* Mauve

Pseudomonas Exotoxin A

This 613-residue toxin (Molecular Mass = 71.5 kDa) excreted by *Pseudomonas aeruginosa* catalyzes the NAD$^+$-dependent ADPribosylation of elongation factor 2 (EF-2), inhibiting protein biosynthesis (1-4). **1**. Iglewski & Sadoff (1979) *Meth. Enzymol.* **60**, 780. **2**. Saelinger (1988) *Meth. Enzymol.* **165**, 226. **3**. Passador & Iglewski (1994) *Meth. Enzymol.* **235**, 617. **4**. Lai (1986) *Adv. Enzymol.* **58**, 99.

Pseudomonic Acids

These carboxylic acids (FW$_{Pseudomonate-A}$ = 499.62 g/mol; FW$_{Pseudomonate-B}$ = 483.62 g/mol; FW$_{Pseudomonate-C}$ = 497.61 g/mol;) are antibiotics produced by a strain of *Pseudomonas fluorescens*. Pseudomonic acid A, also known as mupriocin, is the major component of the isolated mixture of pseudomonic acids. Pseudomonic acid A inhibits isoleucyl-tRNA synthetase via a slow-tight binding competitive mechanism. **1**. Jenal, Rechsteiner, Tan, *et al.* (1991) *J. Biol. Chem.* **266**, 10570. **2**. Brown, Mensah, Doyle, *et al.* (2000) *Biochemistry* **39**, 6003. **3**. Pope, Moore, McVey, *et al.* (1998) *J. Biol. Chem.* **273**, 31691. **4**. Pope, McVey, Fantom & Moore (1998) *J. Biol. Chem.* **273**, 31702. **5**. Hughes & Mellows (1980) *Biochem. J.* **191**, 209. **6**. Hughes & Mellows (1978) *Biochem. J.* **176**, 305. **7**. Beyer, Kroll, Endermann, *et al.* (2004) *Antimicrob. Agents Chemother.* **48**, 525. **8**. Nakama, Nureki & Yokoyama (2001) *J. Biol. Chem.* **276**, 47387. **9**. Yanagisawa & Kawakami (2003) *J. Biol. Chem.* **278**, 25887. **10**. Silvian, Wang & Steitz (1999) *Science* **285**, 1074.

Pseudo-tyrostatin

This tyrostatin analogue (FW = 412.49 g/mol) inhibits the pepstatin-insensitive carboxyl protease from *Pseudomonas* sp., K_i = 56 nM (1) and kumamolysin (K_i = 17 µM at 22.4°C (2). **1**. Wlodawer, Li, Gustchina, *et al.* (2001) *Biochemistry* **40**, 15602. **2**. Oyama, Hamada, Ogasawara, *et al.* (2002) *J. Biochem.* **131**, 757.

Pseudouridine

This pseudonucleoside (FW = 244.20 g/mol), also known as 5-D-β-ribofuranosyluracil and 5-D-β-ribosyluracil and symbolized by ψ, is a nonglycosidic constituent of tRNA and snRNA. The λ_{max} value at pH 7 is 263 nm (ε = 8100 M^{-1}cm^{-1}); the value at pH 12 is 286 nm (ε = 7700 M^{-1}cm^{-1}). It has pK_a values of 9.0 and > 13. **Target(s):** cytidine deaminase (1); and UDP-*N*-acetylglucosamine diphosphorylase, *or* *N*-acetylglucosamine-1-phosphate uridylyltransferase (2). **1**. Cacciamani, Vita, Cristalli, *et al.* (1991) *Arch. Biochem. Biophys.* **290**, 285. **2**. Yamamoto, Moriguchi, Kawai & Tochikura (1980) *Biochim. Biophys. Acta* **614**, 367.

PSI, *See* N-(Benzyloxycarbonyl)-L-isoleucyl-L-(glutamyl-O^g-t-butyl ester)-L-alanyl-L-leucinal

PSI-7977

This prodrug of 2'-deoxy-2'-α-F-2'- β-C-methyluridine 5'-monophosphate (FW = 541.47 g/mol; CAS 1190307-88-0), known systematically as isopropyl (2S)-2-[[[(2R,3R,4R,5R)-5-(2,4-dioxopyrimidin-1-yl)-4-fluoro-3-hydroxy-4-methyl-tetrahydrofuran-2-yl]methoxyphenoxy-phosphoryl]-amino]propanoate, potently inhibits chronic hepatitis C virus. Once within liver cells, PSI-7977 is converted to 2'-deoxy-2'-α-F-2'- β-C-methyluridine 5'-triphosphate in a multi-step process: (a) stereospecific hydrolysis of the carboxyl ester by human cathepsin A and/or carboxylesterase; (b) nucleophilic attack on the phosphorus by the carboxyl group, resulting in the spontaneous elimination of phenol and the production of an alaninyl phosphate metabolite, PSI-352707; (c) removal of the amino acid moiety by histidine triad nucleotide-binding protein 1; and (d) phosphorylation of the 5'-monophosphate to the active triphosphate by sequential action by

UMP/CMP kinase and nucleoside diphosphate kinase, respectively (1). **1.** Murakami , Tolstykh, Bao, *et al.* (2011) *J. Biol. Chem.* **285**, 34337.

Psicofuranine

This adenosine analogue (FW = 297.27 g/mol; CAS 1874-54-0), also known as 6-amino-9-D-psicofuranosylpurine, 9β-D-(psicofuranosyl)-adenine, and angustmycin C, is a nucleoside antibiotic isolated from *Streptomyces hygroscopicus* that also exhibits antitumor activity. Psicofuranine is soluble in water (8 mg/mL at 25°C) and has a λ_{max} value of 261 nm in 10 mM base or 259 nm in 10 mM acid. Psicofuranine is an irreversible inhibitor of GMP synthetase that acts by preventing the release (or further reaction) of adenyl-XMP with water or ammonia; however, it does not suppress exchange reactions (1). **Target(s):** deoxycytidine kinase, weakly inhibited, K_i = 12 mM (1); GMP synthetase, *or* XMP aminase (2-11); and NAD$^+$ synthetase (12,13). **1.** Krenitsky, Tuttle, Koszalka, *et al.* (1976) *J. Biol. Chem.* **251**, 4055. **2.** Fukuyama (1966) *J. Biol. Chem.* **241**, 4745. **3.** Sakamoto (1978) *Meth. Enzymol.* **51**, 213. **4.** Slechta (1960) *Biochem. Biophys. Res. Commun.* **3**, 596. **5.** Fukuyama & Moyed (1964) *Biochemistry* **3**, 1488. **6.** Patel, Moyed & Kane (1975) *J. Biol. Chem.* **250**, 2609. **7.** von der Saal, Crysler & Villafranca (1985) *Biochemistry* **24**, 5343. **8.** Hanka (1960) *J. Bacteriol.* **80**, 30. **9.** Nakamura & Lou (1995) *J. Biol. Chem.* **270**, 7347. **10.** Matsui, Sato, Enei & Hirose (1979) *Agric. Biol. Chem.* **43**, 1739. **11.** Lou, Nakamura, Tsing, *et al.* (1995) *Protein Expr. Purif.* **6**, 487. **12.** Zalkin (1985) *Meth. Enzymol.* **113**, 297. **13.** Spencer & Preiss (1967) *J. Biol. Chem.* **242**, 385.

D-Psicose

This ketohexose (FW = 180.16 g/mol), also known as D-*ribo*-2-hexulose and reportedly fpound in cane molasses, has epimers of D-psicose are D-fructose, D-sorbose, and L-tagatose. D-Psicose is an alternative substrate for fructokinase. The structural composition of D-psicose at 27°C in D$_2$O is 22% α-pyranose, 24% β-pyranose, 39% α-furanose, and 15% β-furanose. **Target(s):** ketohexokinase, *or* hepatic fructokinase; also alternative substrate (1). **1.** Raushel & Cleland (1977) *Biochemistry* **16**, 2169.

D-Psicose 1,6-Bisphosphate

This phosphorylated ketohexose (FW$_{free-acid}$ = 340.12 g/mol) is an epimer of D-fructose 1,6-bisphosphate. It inhibits bacterial and plant pyrophosphate-dependent phospho-fructokinase (K_i = 0.88 mM for the bacterial enzyme), *or* diphosphate:fructose-6-phosphate 1-phosphotransferase. **1.** Bertagnolli, Younathan, Voll, Pittman & Cook (1986) *Biochemistry* **25**, 4674.

D-Psicose 6-Phosphate

This phosphorylated ketose (FW$_{free-acid}$ = 260.14 g/mol), an alternative substrate for 6-phosphofructo-2-kinase/fructose-2,6-bisphosphatase, inhibits fructose-2,6-bisphosphate 2-phosphatase, weakly inhibited (1,2) and pyrophosphate-dependent phosphofructokinase, *or* diphosphate:fructose-6-phosphate 1-phosphotransferase, K_i = 3.78 mM (3). **1.** Pilkis, Pilkis, El-Maghrabi & Claus (1985) *J. Biol. Chem.* **260**, 7551. **2.** Pilkis, Claus, Kountz & El-Maghrabi (1987) *The Enzymes*, 3rd ed. (Boyer & Krebs, eds.), **18**, 3. **3.** Bertagnolli, Younathan, Voll, Pittman & Cook (1986) *Biochemistry* **25**, 4674.

Psilocybin

Psilocybin   **Psilocin**

This serotonergic psychedelic (mind-altering) agent (FW = 284.25 g/mol; CAS 520-52-5); IUPAC Name: 4-hydroxy-*N,N*-dimethyltryptamine), also known as 3-[2-(dimethyl-amino)ethyl]-4-indolol, is found in psilocybin mushrooms (*e.g.*, *Psilocybe azurescens*, *P. semilanceata*, and *P. cyanescens*), but is rapidly dephosphorylated in humans to form psilocin (FW = 204.27 g/mol; CAS 520-53-6; IUPAC: 3-[2-(dimethylamino)ethyl]-4-indolol), which is a serotonin 5-HT$_{2A}$, 5-HT$_{2C}$, and 5-HT$_{1A}$ receptor agonist (or partial agonist). Psilocin is a phospholipase A$_2$ activator, unlike the endogenous ligand serotonin, which activates phospholipase C. Psilocin is structurally similar to serotonin, differing only by the position of its hydroxyl group and the dimethylated nitrogen. Psilocin is thought to be a partial agonist activity at 5-HT$_{2A}$ serotonin receptors in the prefrontal cortex. Ingestion of psilocybin or psilocybin mushrooms leads dilated pupils, restlessness, euphoria, open- and closed-eye visualizations, synesthesia, tachycardia, elevated temperature, headache, sweating, chills, and nausea. **1.** Passie, Seifert, Schneider & Emrich (2002) *Addict. Biol.* **7**, 357.

Psoralen

This phototoxic phytoalexin (FW = 186.17 g/mol), found in many plants, is part of a plant defense scheme to protect against fungi and insects. Upon activation by ultraviolet light, psoralen induces interstrand cross-links in DNA.1 Psoralen intercalates between DNA base pairs and, upon UV irradiation, forms covalent bonds with a pyrimidine. If posed at a susceptible site, a second photoaddition can occur, forming interstrand cross-links. The combination of psoralens and UVA radiation (PUVA photochemotherapy) is an established treatment for many skin disorders. Note that unsaturated fatty acids and proteins can also undergo a photoaddition reaction with psoralen and psoralen derivatives. *See also psoralen derivatives (e.g., 8-Methoxypsoralen)* **Target(s):** *trans*-cinnamate 4-monooxygenase (2,3); CYP2A6 (4); Cyp2a-5 (5); CYP6D1 (6); DNA replication (7); lysozyme (1); and monoamine oxidase (8). **1.** Schmitt, Chimenti & Gasparro (1995) *J. Photochem. Photobiol. B: Biol.* **27**, 101. **2.** Hübner, Hehmann, Schreiner, *et al.* (2003) *Phytochemistry* **64**, 445. **3.** Gravot, Larbat, Hehn, *et al.* (2004) *Arch. Biochem. Biophys.* **422**, 71. **4.** Koenigs & Trager (1998) *Biochemistry* **37**, 10047. **5.** Maenpaa, Sigusch,

Raunio, *et al.* (1993) *Biochem. Pharmacol.* **45**, 1035. **6**. Scott (1996) *Insect Biochem. Mol. Biol.* **26**, 645. **7**. Luftl, Rocken, Plewig & Degitz (1998) *J. Invest. Dermatol.* **111**, 399. **8**. Kong, Tan, Woo & Cheng (2001) *Pharmacol. Toxicol.* **88**, 75.

Psychosine

This cationic brain lysosphingolipid (FW$_{free-base}$ = 461.64 g/mol; CAS 2238-90-6), also known as 1-*O*-β-D-galactosylsphingosine, inhibits metabotropic α$_1$-adrenergic receptor signaling-induced phospholipase C activation in rat brain astrocytes. Psychosine also exerts protective effects against quisqualate, suggesting it may be useful in prophylaxis of neurodegenerative disorders arising from over-stimulation of hippocampal Group 1 metabolotropic glutamate receptor, *or* mGluR (1). Psychosine is the product of the reaction catalyzed by sphingosine β-galactosyltransferase and is the substrate for psychosine sulfotransferase and galactosylceramidase. **Target(s):** cerebroside-sulfatase, *or* arylsulfatase A (2); cytochrome *c* oxidase (3-5); cytokinesis (6); electron transport (7); galactosylgalactosylglucosylceramidase (8); glucocerebrosidase, lysosome (9); β-glucosidase (10); phosphorylase kinase (10); protein kinase C (11-14); protein-tyrosine sulfotransferase (15); and sphingomyelinase (16). **1**. Hodgson, Taylor, Zhang & Rosenberg (1998) Brain Res. **802**, 1. **2**. Stinshoff & Jatzkewitz (1975) *Biochim. Biophys. Acta* **377**, 126. **3**. Cooper, Markus, Seetulsingh & Wrigglesworth (1993) *Biochem J.* **290**, 139. **4**. Igisu, Hamasaki, Ito & Ou (1988) *Lipids* **23**, 345. **5**. Igisu & Nakamura (1986) *Biochem. Biophys. Res. Commun.* **197**, 323. **6**. Kanazawa, Nakamura, Momoi, *et al.* (2000) *J. Cell Biol.* **149**, 943. **7**. Tapasi, Padma & Setty (1998) *Indian. J. Biochem. Biophys.* **35**, 161. **8**. Brady, Gal, Bradley & Martensson (1967) *J. Biol. Chem.* **242**, 1021. **9**. LaMarco & Glew (1985) *Arch. Biochem. Biophys.* **236**, 669. **10**. Baltas, Zevgolis, Kyriakidis, Sotiroudis & Evangelopoulos (1989) *Biochem. Int.* **19**, 99. **11**. Yoshimura, Kobayashi & Goto (1992) *Neurochem. Res.* **17**, 1021. **12**. Dawson & McAtee (1989) *J. Cell Biochem.* **40**, 261. **13**. Vartanian, Dawson, Soliven, Nelson & Szuchet (1989) *Glia* **2**, 370. **14**. Hannun & Bell (1987) *Science* **235**, 670. **15**. Kasinathan, Sundaram, Slomiany & Slomiany (1993) *Biochemistry* **32**, 1194. **16**. Brady (1983) *The Enzymes*, 3rd ed. (Boyer, ed.), **16**, 409.

PT1

This AMPK activator (FW = 497.91 g/mol; CAS 331002-70-1), systematically named 2-chloro-5-[[5-[[5-(4,5-dimethyl-2-nitrophenyl)-2-furanyl]methylene]-4,5-dihydro-4-oxo-2-thiazolyl]amino]benzoic acid, targets AMP-activated protein kinase (EC$_{50}$ = 0.3 μM) by antagonizing its built-in autoinhibition mechanism (1). PT1 dose-dependently activates AMPK α1$_{394}$, α1$_{335}$, α2$_{398}$, and even the heterotrimer α1β1γ1 form. Based on the structure of PT1 docked to AMPK α1 subunit, it appears that PT1 interacts with Glu-96 and Lys-156 near the autoinhibitory domain, directly relieving autoinhibition. In studies using L6 myotubes, the phosphorylation of AMPK and its downstream substrate, acetyl-CoA carboxylase, were dose-dependently and time-dependently increased by PT1 without any change in cellular AMP:ATP concentration ratio. **1**. Pang, Zhang, Gu, *et al.* (2008) *J.Biol. Chem.* **283**, 16051.

PT70

This slow, tight-binding enzyme inhibitor (FW = 284.40 g/mol; IUPAC: 2-(*o*-tolyloxy)-5-hexylphenol) targets InhA, the enoyl-ACP reductase (K$_i$ = 22 pM) in *Mycobacterium tuberculosis*, a pathogen that kills more than two

million people annually (1). PT70 binds preferentially to the InhA·NAD$^+$ complex, after which it has a residence time of 24 min, or ~14,000 times longer than that of the rapidly reversible inhibitor from which it is derived. The off-rate constant for PT70 is 0.043 min^{-1}. Parameters for related inhibitors are: PT10 (K$_i$ = 129 pM, k$_{off}$ = 0.037 min^{-1}); PT91 (K$_i$ = 0.96 pM, k$_{off}$ = 0.048 min^{-1}); PT92 (K$_i$ = 0.2 pM, k$_{off}$ = 0.033 min^{-1}); PT155 (K$_i$ = 12 pM, k$_{off}$ = rapid). A large-scale conformational change accounts for the slow conversion of EI to EI* in a two-step, induced-fit scheme: *Step*-1, E·NAD$^+$ + I $\rightleftharpoons$ E·NAD$^+$·I; *Step*-2, E·NAD$^+$·I $\rightleftharpoons$ E·NAD$^+$·I* (2). Analogues of the slow-onset inhibitor PT70 permitted identification of the structure of the initial reversible E·I complex, which resembled the E·S complex. Energy profiles were then calculated, using the structures of E·I and E·I* complexes as the respective initial and final points on the reaction coordinate, leading from the "open" to the "closed" state (2). Crystal structures of the slow-onset and rapidly-reversible ternary complexes suggest that there exist two major enzyme conformations from which the diphenyl ether may select, thus influencing the binding kinetics. Since slow inhibition lags about 4 orders of magnitude behind the rate of catalysis, structural conversions between these two major enzyme conformers when bound to inhibitor must be separated by a much larger energy barrier than that between any conformers on the catalytic reaction coordinate (2). The energy profiles rationalize the observed inhibition kinetics by rapid reversible and slow-onset inhibitors, providing a framework for rationally modulating residence time for InhA inhibitors (2). **1**. Luckner, Liu, am Ende, Tonge & Kisker (2010) *J. Biol. Chem.* **285**, 14330. **2**. Li, Lai, Pan, *et al.* (2014) *ACS Chem Biol.* **9**, 986.

PTC124, *See Ataluren*

PTEN

This tumor-suppressing, Mg^{2+}-dependent dual-specificity protein phosphatase (MW = 47166 g/mol; GenBank: AAD13528.1), which catalyzes the dephosphorylation of phosphotyrosine-, phosphoserine- and phosphothreonine-containing proteins, is also a lipid phosphatase (*Reaction*: Phosphatidylinositol 3,4,5-trisphosphate + H$_2$O $\rightleftharpoons$ Phosphatidylinositol 4,5-bisphosphate + P$_i$), with the following *in vitro* substrate preference: PtdIns(3,4,5)P$_3$ > PtdIns(3,4)P$_2$ > PtdIns3P > Ins(1,3,4,5)P$_4$. PTEN antagonizes PI3K-AKT/PKB signaling by dephosphorylating phosphoinositides and reduces cell cycle progression and survival. The unphosphorylated form cooperates with AIP1 to suppress AKT1 activation. PTEN catalyzes the dephosphorylation of tyrosine-phosphorylated focal adhesion kinase, inhibiting cell migration and integrin-mediated cell spreading. PTEN also modulates AKT-mTOR signaling. PTEN plays an essential role in the maintenance of genome stability (1,2). PTEN also dephosphorylates MCM2 and inhibits replication fork progression under replication stress, suggesting that PTEN is involved in an S-phase checkpoint (3). DNA molecular combing analysis revealed that replication forks in PTEN-proficient cells are subject to more efficient braking in response to fork stalling than the forks in PTEN-deficient cells. **1**. Puc, Keniry, Li, *et al.* (2005) *Cancer Cell* **7**, 193. **2**. Shen, Balajee, Wang, *et al.* (2007) *Cell* **128**, 157. **3**. Feng, Liang, Li, *et al.* (2015) *Cell Rep.* **64**, 156.

Pterin

This aromatic reagent (FW = 163.14 g/mol), also known as 2-amino-4-hydroxypteridine and 2-amino-4-oxopteridine, is the structural unit for folic acid, folic acid antagonists, plant and animal pigments, enzyme cofactors, *etc.* **Target(s):** adenosine deaminase (1); γ-glutamyl hydrolase, *or* pteroylpolyglutamate hydrolase (2); hypoxanthine(guanine) phosphoribosyltransferase (3); NAD(P)H dehydrogenase, quinone-dependent (4); sepiapterin deaminase (5-7); and tRNA-guanine transglycosylase (8). **1**. Harbison & Fisher (1973) *Arch. Biochem. Biophys.* **154**, 84. **2**. Wang, Chandler & Halsted (1986) *J. Biol. Chem.* **261**, 13551. **3**. Miller, Ramsey, Krenitsky & Elion (1972) *Biochemistry* **11**, 4723. **4**. Koli, Yearby, Scott & Donaldson (1969) *J. Biol. Chem.* **244**, 621. **5**. Tsusue (1971) *J. Biochem.* **69**, 781. **6**. Tsusue & Mazda (1977) *Experientia* **33**, 854. **7**. Tsusue, Kuroda & Sawada (1990) *Pteridines* **2**, 175. **8**. Farkas, Jacobson & Katze (1984) *Biochim. Biophys. Acta* **781**, 64.

Pterin-6-aldehyde, *See 2-Amino-4-hydroxypteridine-6-aldehyde*

Pteroic Acid

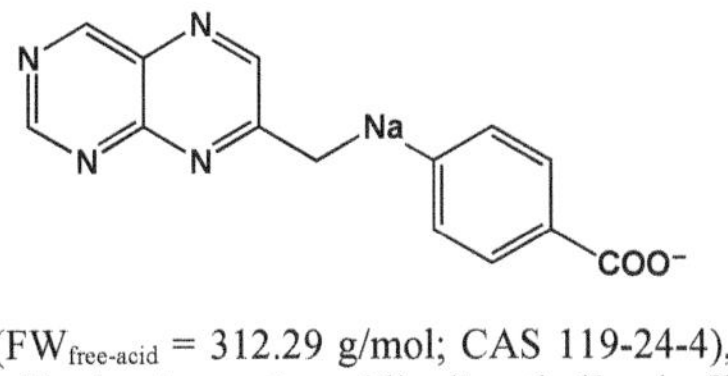

This metabolite (FW$_{\text{free-acid}}$ = 312.29 g/mol; CAS 119-24-4), also known as 4-[[(2-amino-1,4-dihydro-4-oxo-6-pteridinyl)methyl]amino]benzoic acid, is a structural component of folates. Pteroic acid is a growth factor for a number of *Enterococci*. Pteroic acid is a yellow powder that is slightly soluble in aqueous NaOH. The λ_{max} values in 0.1 M NaOH are 255, 275, and 365 nm (ε = 26300, 23400, and 8900 M^{-1}cm^{-1}, respectively). **Target(s):** arylamine *N*-acetyltransferase (1); dihydrofolate reductase (2,3); dihydrofolate synthetase, weakly inhibited (4); dihydropteroate synthase (5); glutamate carboxypeptidase (6); and xanthine oxidase (7-9). **1**. Ward, Summers & Sim (1995) *Biochem. Pharmacol.* **49**, 1759. **2**. Baker (1967) *Design of Active-Site-Directed Irreversible Enzyme Inhibitors*, Wiley, New York. **3**. Bertino, Perkins & Johns (1965) *Biochemistry* **4**, 839. **4**. Webb & Ferone (1976) *Biochim. Biophys. Acta* **422**, 419. **5**. Babaoglu, Qi, Lee & White (2004) *Structure* **12**, 1705. **6**. Albrecht, Boldizsar & Hutchinson (1978) *J. Bacteriol.* **134**, 506. **7**. Webb (1966) *Enzyme and Metabolic Inhibitors*, vol. **2**, p. 289, Academic Press, New York. **8**. Kalckar & Klenow (1948) *J. Biol. Chem.* **172**, 34. **9**. Hofstee (1949) *J. Biol. Chem.* **179**, 633.

Pteropterin, See Pteroyltriglutamate

Pteroyldiglutamate, See Pteroyl-(γ-glutamyl)-glutamate; Diopterin

Pteroylglutamate, See Folic Acid

Pteroyl-(γ-glutamyl)glutamate

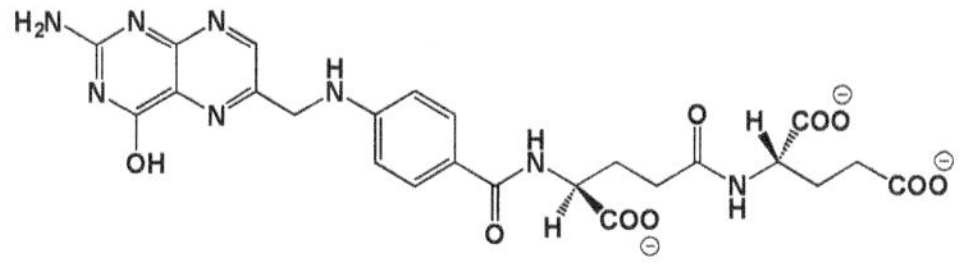

This form of folic acid, also called pteroyldiglutamate (FW$_{\text{free-acid}}$ = 569.51 g/mol), a precursor in the formation of longer polyglutamate forms, is reportedly the only folate form in the *Halobacterium* strain GN-1. Note: This precursor should not be confused with diopterin, also known as pteroyl-α-glutamylglutamate. **Target(s):** N^5-methyltetrahydrofolate: homocysteine *S*-methyltransferase (1); and thymidylate synthase (2,3). **1**. Whitfield, Steers & Weisbach (1970) *J. Biol. Chem.* **245**, 390. **2**. Danenberg (1977) *Biochim. Biophys. Acta* **473**, 73. **3**. Radparvar, Houghton & Houghton (1988) *Arch. Biochem. Biophys.* **260**, 342.

Pteroyl-γ-glutamyl-γ-glutamylglutamate, See Pteroyltriglutamate

Pteroylhexaglutamate

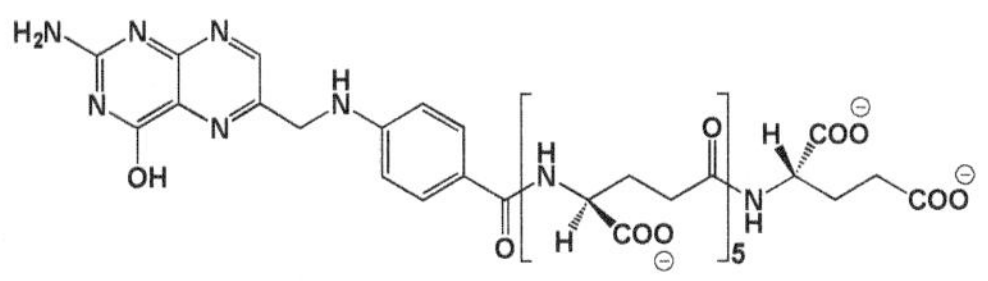

This folic acid polyglutamate (FW = 1091.1 g/mol) inhibits thymidylate synthase (1-3). **1**. Kisliuk, Gaumont & Baugh (1974) *J. Biol. Chem.* **249**, 4100. **2**. Radparvar, Houghton & Houghton (1988) *Arch. Biochem. Biophys.* **260**, 342. **3**. Danenberg (1977) *Biochim. Biophys. Acta* **473**, 73.

Pteroylpentaglutamate

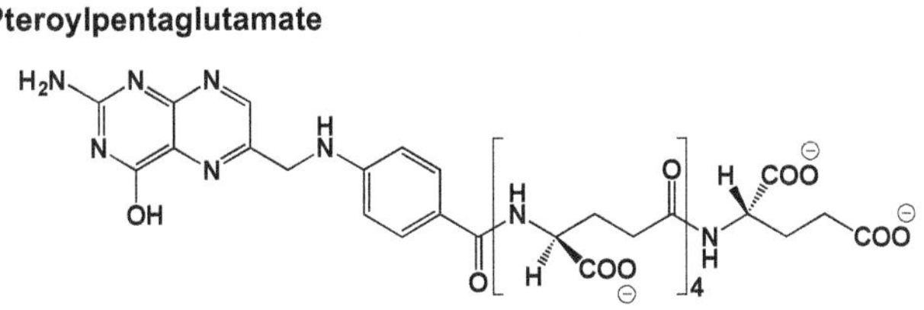

This folic acid polyglutamate (FW$_{\text{free-acid}}$ = 957.87 g/mol), the major form in liver and murine erythroleukemia cells, inhibits amidophosphoribosyl-transferase (1), N^5,N^{10}-methenyltetrahydrofolate synthetase (2), phosphoribosyl-aminoimidazolecarboxamide formyltransferase, *or* 5-aminoimidazole-4-carboxamide ribotide transformylase (3), and thymidylate synthase (4,5). **1**. Santi, Lyons, Phillips & Christopherson (1992) *J. Biol. Chem.* **267**, 11038. **2**. Jolivet (1997) *Meth. Enzymol.* **281**, 162. **3**. Baggott, Vaughn & Hudson (1986) *Biochem. J.* **236**, 193. **4**. Dolnick & Cheng (1978) *J. Biol. Chem.* **253**, 3563. **5**. Radparvar, Houghton & Houghton (1988) *Arch. Biochem. Biophys.* **260**, 342.

Pteroyltriglutamic Acid

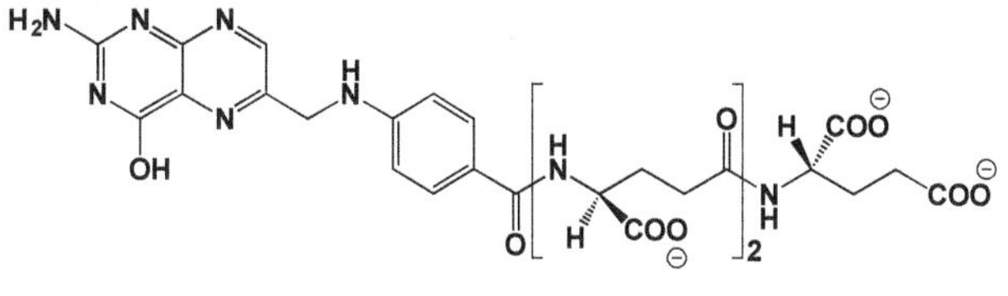

This folate (FW$_{\text{free-acid}}$ = 699.63 g/mol), also known as pteropterin and pteroyl-γ-glutamyl-γ-glutamylglutamate, is a precursor to polyglutamated forms of folic acid. It is the major form of folate in certain microorganisms. **Target(s):** 5-methyl-tetrahydro-pteroyl-triglutamate:homocysteine *S*-methyltransferase (1); and thymidylate synthase (2-5). **1**. Whitfield, Steers & Weisbach (1970) *J. Biol. Chem.* **245**, 390. **2**. Lockshin & Danenberg (1979) *J. Biol. Chem.* **254**, 12285. **3**. Kisliuk, Gaumont & Baugh (1974) *J. Biol. Chem.* **249**, 4100. **4**. Dolnick & Cheng (1978) *J. Biol. Chem.* **253**, 3563. **5**. Radparvar, Houghton & Houghton (1988) *Arch. Biochem. Biophys.* **260**, 342.

Pterulinic Acid

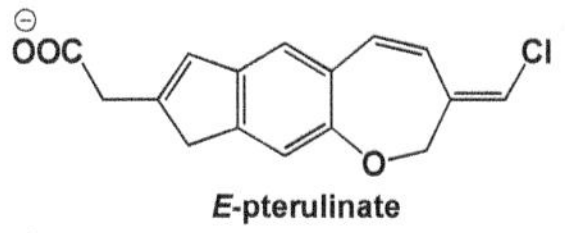

E-pterulinate

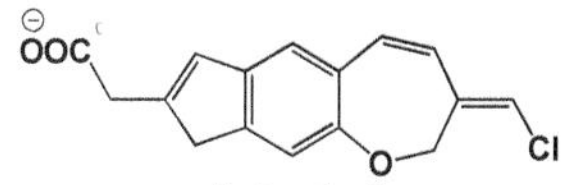

Z-pterulinate

This novel acid (FW$_{\text{free-acid}}$ = 290.70 g/mol), produced by species of *Pterula*, inhibits complex I of mitochondrial electron transport, also known as NADH dehydrogenase and NADH:ubiquinone oxidoreductase (1,2). **1**. Engler, Anke, Sterner & Brandt (1997) *J. Antibiot.* **50**, 325. **2**. Engler, Anke & Sterner (1997) *J. Antibiot.* **50**, 330.

PTP Inhibitor

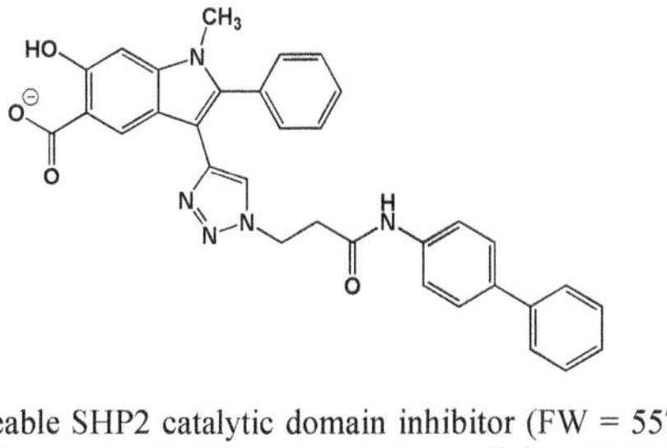

This cell-permeable SHP2 catalytic domain inhibitor (FW = 557.61 g/mol; Solubility: 100 mg/mL DMSO; Protect from light: IUPAC: 3-(1-(3-(biphenyl-4-ylamino)-3-oxopropyl)-1*H*-1,2,3-triazol-4-yl)-6-hydroxy-1-methyl-2-phenyl-1*H*-indole-5-carboxylate), also variously known as SHP2 Inhibitor VI, CD45 Inhibitor V, FAP1 Inhibitor, LMWPTP Inhibitor, Lyp Inhibitor III, PTP1B Inhibitor IX, SHP1 Inhibitor IX, and code-named II-B08, targets SHP2-catalyzed pNPP hydrolysis (IC$_{50}$ = 5.5 μM; reversible and noncompetitive manner, K_i = 5.2 μM), exhibiting selectivity over PTP1B (IC$_{50}$ = 14 μM), SHP1 (IC$_{50}$ = 16 μM), FAP1 (IC$_{50}$ = 20 μM), Lyp (IC$_{50}$ = 25 μM), CD45 (IC$_{50}$ = 30 μM), and LMWPTP (IC$_{50}$ = 31 μM, respectively), with >9-fold selectivity over Cdc14A, HePTP, LAR, PTPα, and VHR, all having IC$_{50}$ values in excess of 50 μM. Gain-of-function KIT receptor mutation (*e.g.*, KIT$^{\text{D814V}}$ in mice and KIT$^{\text{D816V}}$ in humans) results in altered substrate recognition and constitutive tyrosine autophosphorylation leading to promiscuous signaling that is associated with gastrointestinal stromal tumors (GIST), systemic mastocytosis (SM), and acute myelogenous leukemia (AML). Likewise, the Src Homology-2 domain containing protein tyrosine Phosphatase-2 (SHP2) plays a pivotal role in growth factor and cytokine signaling, and gain-of-function SHP2 mutations are associated with various leukemias and solid tumors. Screening of a salicylate-based combinatorial library led to the discover of this inhibitor's highly efficacious cellular activity, blocking growth factor stimulated ERK1/2 activation and hematopoietic progenitor proliferation, thereby showing SHP2 inhibition may be exploited to combat cancer (1). X-ray crystallographic analysis of the structure of SHP2 in complex with this inhibitor reveals molecular determinants that can be exploited for the acquisition of other SHP2 inhibitors (1). Indeed, pharmacologic inhibition of SHP2 phosphatase by II-B08 results in reduced SHP2 constitutive phosphorylation in KIT^{D814V}-bearing cells (2). Inhibition of SHP2 phosphorylation by II-B08 also results in significant repression in ligand-independent growth and survival of cells bearing KIT^{D814V} and primary human KIT^{D816V}-bearing $CD34^+$ cells (2). Consistent with such findings, deficiency of SHP2 in HSC/Ps expressing KIT^{D814V} likewise inhibits ligand-independent growth and survival (2). Inhibition of protein tyrosine phosphatase 1B (PTP1B) by II-B08 also promotes endothelial cell migration and angiogenesis by enhancing VEGFR2 signaling and increases p130Cas phosphorylation as well as the interactions among p130Cas, Crk and DOCK180 (3). (Phosphorylation levels of focal adhesion kinase, Src, paxillin, or Vav2 remain unchanged.) Gene silencing of DOCK180, but not Vav2, was found to abrogate the effects of PTP1B inhibitor on EC motility (3). The effects of PTP1B inhibitor on EC motility and p130Cas/DOCK180 activation persisted in the presence of the VEGFR2 antagonist, suggesting stimulation of the DOCK180 pathway represents an alternative mechanism of PTP1B inhibitor-stimulated EC motility, one that does not require concomitant VEGFR2 activation as a prerequisite (3). **1.** Zhang, He, Liu, *et al.* (2010) *J. Med. Chem.* **53**, 2482. **2.** Mali, Ma, Zeng, *et al.* (2012) *Blood* **120**, 2669. **3.** Wang, Yan, Ye, Wu & Jiang (2016) *Sci. Rep.* **6**, 24111.

PTX042695

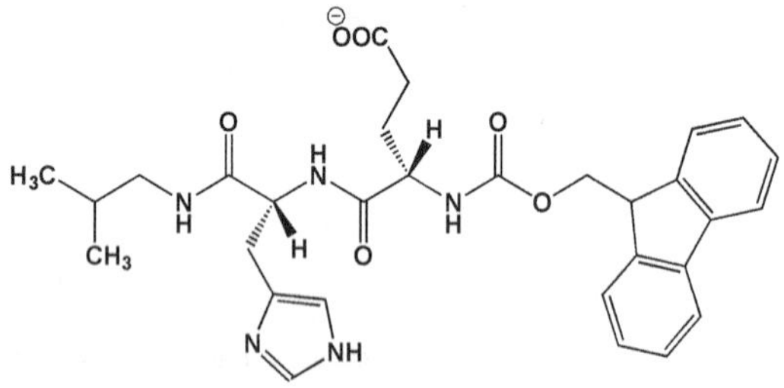

This peptide analogue (FW = 561.63 g/mol) is a strong inhibitor of *Escherichia coli* pantetheine-phosphate adenylyltransferase (IC_{50} = 6 nM) but does not inhibit the porcine enzyme (1). Replacing the D-histidyl residue with D-serine weakens the inhibition (IC_{50} = 0.12 μM). Replacing the L-glutamyl residue with *S*-carboxymethyl-D-cysteine or the corresponding sulfone slightly weakens the inhibition (IC_{50} = 20 and 10 nM, respectively). **1.** Zhao, Allanson, Thomson, *et al.* (2003) *Eur. J. Med. Chem.* **38**, 345.

Puerarin

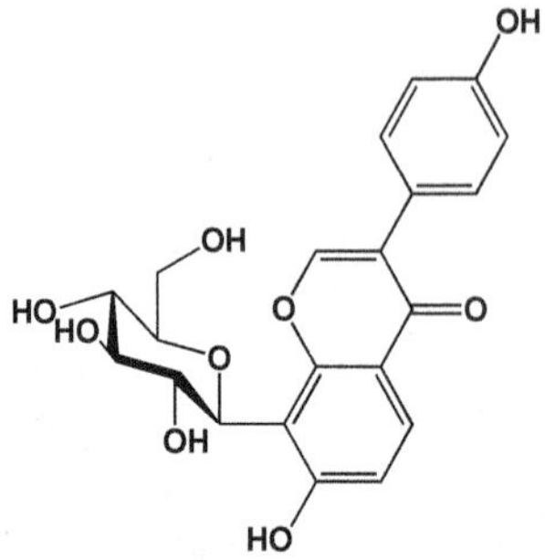

This C-glycoside isoflavone (FW = 416.38 g/mol; M.P. = 190°C), isolated from the root of a wild leguminous creeper, *Pueraria lobata*, is a major component of ge-gen, which is used in traditional Asian medicine. **Target(s):** aldose reductase (1); Ca^{2+} channel, L-type (2); K^+ currents, transient outward and delayed rectifier currents (3); and Na^+ current, tetrodotoxin-resistant (4). **1.** Zhang & Zhou (1989) *Zhongguo Zhong Yao Za Zhi* **14**, 557. **2.** Qian, Li, Huang, *et al.* (1999) *Chin. Med. J. (Engl.)* **112**, 787. **3.** Zhang, Hao, Zhou, Wu & Dai (2001) *Acta Pharmacol. Sin.* **22**, 253. **4.** Ji & Wang (1996) *Zhongguo Yao Li Xue Bao* **17**, 115.

Pullulan

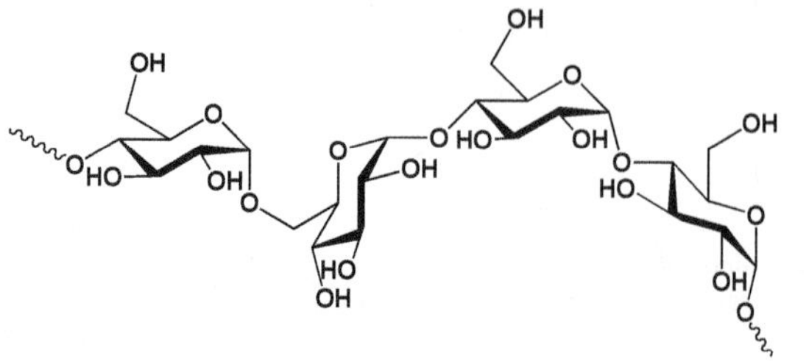

This glucan is a branched polymer (M.W. ranging between 10^5 and 10^6 Da) of D-glucose linked $\alpha(1\rightarrow4)$ and $\alpha(1\rightarrow6)$, first isolated from the fungi *Aureobasidium pullulans* and later observed in other fungi. Pullulan is much more highly branched than either amylopectin or glycogen and can be regarded as a polymer of α-maltotriosyl units joined by $\alpha(1\rightarrow6)$ linkages with a few α-maltotetraosyl units also present. **Target(s):** cyclomaltodextrin glucanotransferase, alkaliphilic bacteria, slight inhibition (1); and leukocyte elastase, inhibited by sulfated pullulans (2). **1.** Zhang, Yan & Zhang (1998) *Wei Sheng Wu Xue Bao* **38**, 98. **2.** Becker, Franz & Alban (2003) *Thromb. Haemost.* **89**, 915.

Pulvomycin

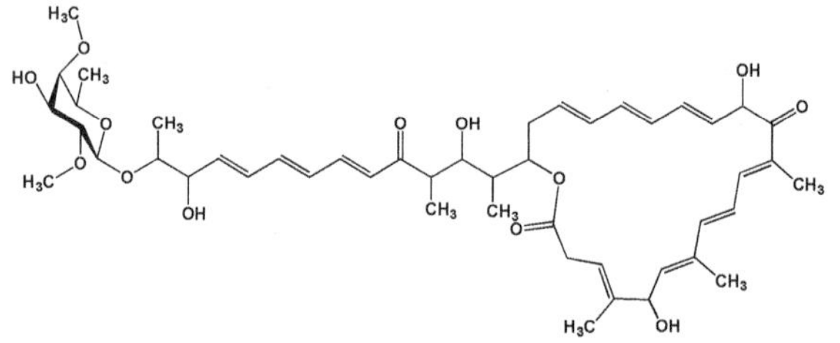

This antibiotic (FW = 839.03 g/mol), also known as antibiotic 1063-Z, inhibits protein biosynthesis by blocking ternary complex formation between elongation factor Tu, GTP, and aminoacyl-tRNA (1-6). **1.** Mesters, Potapov, de Graaf & Kraal (1994) *J. Mol. Biol.* **242**, 644. **2.** Wolf, Assmann & Fischer (1978) *Proc. Natl. Acad. Sci. U.S.A.* **75**, 5324. **3.** Pingoud, Block, Urbanke & Wolf (1982) *Eur. J. Biochem.* **123**, 261. **4.** Assmann & Wolf (1979) *Arch. Microbiol.* **120**, 297. **5.** Pingoud, Block, Wittinghofer, Wolf & Fischer (1982) *J. Biol. Chem.* **257**, 11261. **6.** Parmeggiani, Krab, Okamura, *et al.* (2006) *Biochemistry* **45**, 6846.

Pumiliotoxins

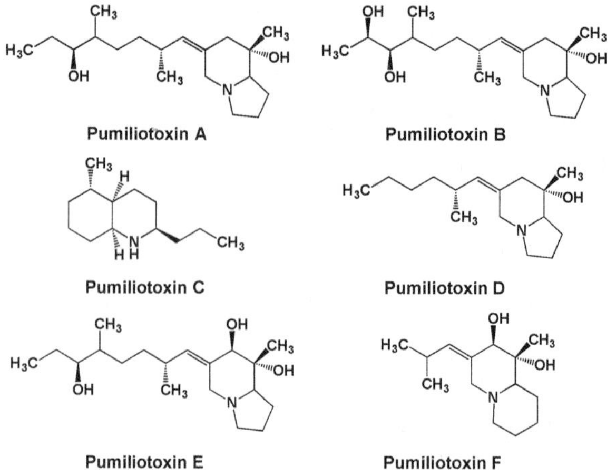

Pumiliotoxin A

Pumiliotoxin B

Pumiliotoxin C

Pumiliotoxin D

Pumiliotoxin E

Pumiliotoxin F

These neurotoxins are found in the skin secretions of neotropical poison arrow frogs, occurring in all species groups of *Dendrobates* and *Phyllobates*. There are more than one hundred toxins designated as pumiliotoxins, and they are currently divided into major groups referred to as Pumiliotoxins A, B, and C. Pumiliotoxins A and B, also known as pumiliotoxin 307A and pumiliotoxin 323A (with the number refers to the nominal mass: *i.e.*, 307.48 and 323.48 g/mol) are far more toxic than the pumiliotoxins C, which is virtually nontoxic. Subcutaneous injection of pumiliotoxin A or B in mice causes locomotor difficulties, partial paralysis of hind limbs, salivation, extensor movements, and finally, clonic convulsions and death, all in less than ten minutes. Even at lower doses, one member of the B group can cause death in less than twenty minures. Pumiliotoxin C, now referred to as decahydroquinoline cis-195A, interacts with acetylcholine receptor-regulated ion channels (2,3), whereas pumiliotoxins A and B stimulate Na⁺ channels. **Target(s):** Ca²⁺-transporting ATPase, inhibited by pumiliotoxins A, B, and 251D (4); nicotinic acetylcholine receptor-channels, by pumiliotoxin C (2,3). **1.** Daly & Myers (1967) *Science* **156**, 970. **2.** Conn (1983) *Meth. Enzymol.* **103**, 401. **3.** Daly, Nishizawa, Padgett, *et al.* (1991) *Neurochem. Res.* **16**, 1207. **4.** Tamburini, Albuquerque, Daly & Kauffman (1981) *J. Neurochem.* **37**, 775.

Punicic Acid

This conjugated unsaturated fatty acid ($FW_{free-acid}$ = 278.44 g/mol), also known as 9Z,11E,13Z-octadecatrienoic acid, from *Punica granatum* inhibits DNA-directed DNA polymerase, DNA polymerase α, and DNA topoisomerase II (1). **1.** Mizushina, Tsuzuki, Eitsuka, *et al.* (2004) *Lipids* **39**, 977.

Purine

This heterocyclic base (FW = 120.11 g/mol; CAS 120-73-0) is the structural foundation for a wide variety of naturally occurring compounds. It is soluble in water and absorbs light in the ultraviolet. Purine was first synthesized, characterized, and named in the late nineteenth century by Emil Fischer, the 1902 Nobel Laureate in Chemistry. **Target(s):** adenine deaminase (1-3); adenosine deaminase (4-6); 5-phosphoribosylamine synthetase, *or* ribose-5-phosphate:ammonia ligase (7); xanthine dehydrogenase, weakly inhibited (8); xanthine oxidase (9); and xanthine phosphoribosyltransferase (10). **1.** Hartenstein & Fridovich (1967) *J. Biol. Chem.* **242**, 740. **2.** Jun & Sakai (1979) *J. Ferment. Technol.* **57**, 294. **3.** Kidder, Dewey & Nolan (1977) *Arch. Biochem. Biophys.* **183**, 7. **4.** Wolfenden, Sharpless & Allan (1967) *J. Biol. Chem.* **242**, 977. **5.** Wolfenden, Kaufman & Macon (1969) *Biochemistry* **8**, 2412. **6.** Singh & Sharma (2000) *Mol. Cell. Biochem.* **204**, 127. **7.** Reem (1968) *J. Biol. Chem.* **243**, 5695. **8.** Sin (1975) *Biochim. Biophys. Acta* **410**, 12. **9.** Sheu, Lin & Chiang (1996) *Anticancer Res.* **16**, 3571. **10.** Miller, Adamczyk, Fyfe & Elion (1974) *Arch. Biochem. Biophys.* **165**, 349.

Purmorphamine

This Smo activator (FW = 520.62 g/mol; CAS 483367-10-8; Solubility: 4 mg/mL DMSO; <1 mg/mL H₂O), also named 9-cyclohexyl-N-[4-(4-morpholinyl)phenyl]-2-(1-naphthalenyloxy)-9H-purin-6-amine, targets Smoothened and blocks BODIPY-cyclopamine binding (IC₅₀ ~ 1.5 μM). Purmorphamine is also an inducer of osteoblast differentiation (EC₅₀ = 1 μM). **1.** Sinha, *et al.* (2006) *Nature Chem. Biol.* **2**, 29. **2.** Wu, *et al.* (2002) *J. Am. Chem. Soc.* **124**, 14520. **3.** Wu, *et al.* (2004) *Chem. Biol.* **11**, 1229.

Puromycin

Puromycin

3'-O-Tyrosyl-adenosine

This tyrosylated aminonucleoside antibiotic ($FW_{free-base}$ = 471.52 g/mol) from *Streptomyces alboniger*, named systematically as 3'-((2-amino-3-(4-methoxyphenyl)-1-oxopropyl)amino)-3'-deoxy-N,N-dimethyl-(S)-adenosine, is a potent protein biosynthesis inhibitor (1,2). It is highly cytotoxic and is unselective in its action against both prokaryotes or eukaryotes. Puromycin resistance is conferred by the Pac gene, which encodes a puromycin N-acetyl-transferase (PAC), in its *Streptomyces* producer strain. The multi-paper description of its synthesis was reported in 1954 by B. R. Baker and coworkers in Issue 4 of Vol 19 of *The Journal of Organic Chemistry*. **Mechanism of Action:** Puromycin binds at the A-site on the ribosome and is incorporated into the growing polypeptide chain. (Note the similarity in structure to the aminoacyl-adenylyl terminus of aminoacyl-tRNAs.) The carboxyl group of the growing peptide is linked by a peptide bond to the amino group of the puromycin side chain. Protein synthesis is then terminated and the puromycinyl peptide is released from the ribosome. The free base form is sparingly soluble in water whereas the dihydrochloride and monosulfate are readily soluble. **Key Role in mRNA Display Technique:** This peptide- and protein-selection approach, which provides a way to carry out both ligand discovery and interaction analysis, utilizes encoded peptide and protein libraries that are covalently fused to their own mRNA (3). Indeed, the mRNA display technique has been used to discover novel peptide and protein ligands for RNA, small molecules and proteins, as well as to define cellular interaction partners of proteins and drugs. In this method, puromycin serves as a chemically stable, small-molecule mimic of aminoacyl tRNA. Synthetic oligonucleotide containing a 3'-puromycin molecule is ligated to the 3'-end of an mRNA, and the product is translated in rabbit reticulocyte lysate. The sequence present in the peptide is therefore encoded in the covalently attached mRNA, allowing the sequence information in the protein to be read and recovered after selection via RT-PCR, such that exceedingly small amounts of material can be amplified and analyzed. **Properties:** The pK_a values are 6.8 and 7.2. The nucleoside also absorbs light significantly in the ultraviolet: λ_{max} in 0.1 N HCl is 267.5 nm (ε = 19500 M⁻¹cm⁻¹) whereas the value in 0.13 M NaOH is 275 nm (ε = 20300 M⁻¹cm⁻¹). **Target(s):** acetylcholinesterase (4,5); aminopeptidase B (41,42); aminopeptidase PS (6); aminopeptidase, rat brain (5); aromatic-hydroxylamine O-acetyltransferase (60); aryl-acylamidase (7,8); biotinidase (10); bleomycin hydrolase (11); cathepsin H, weakly inhibited (12,13); 3',5'-cyclic-nucleotide phosphodiesterase (14); [cytochrome c]-lysine N-methyltransferase (61); cytosol alanyl aminopeptidase (39); O-demethylpuromycin O-methyltransferase, product inhibition (62); dipeptidyl-peptidase I, *or* cathepsin C (15); dipeptidyl-peptidase II (16-23); dipeptidyl-peptidase III (24-26); dipeptidyl-peptidase IV, *or* post-proline dipeptidyl aminopeptidase (27); glutamate carboxypeptidase II (28); glutamyl aminopeptidase, *or* aminopeptidase A (40); leucyl aminopeptidase (43,51-53); leucyl aminopeptidase/cystinyl aminopeptidase (43); leucyltransferase, *or* leucyl-tRNA:protein leucyltransferase (29,57-59); membrane alanyl aminopeptidase, *or* aminopeptidase N (30-33,40,44-50); neprilysin (34); protein biosynthesis, *or* elongation (1-3,35-37); pyroglutamyl-peptidase I (9); ribonuclease P (54-56); subtilopeptidase A, *or Bacillus subtilis* protease, type 8 (38); and tripeptidyl-peptidase (15). **1.** Pestka (1974) *Meth. Enzymol.* **30**, 261. **2.** Jiménez (1976) *Trends Biochem. Sci.* **1**, 28. **3.** Takahash, Austin & Roberts (2003) Trends in Biochem. Sci. **28**, 159. **4.** Moss, Moss & Fahrney (1974) *Biochim. Biophys. Acta* **350**, 95. **5.** Hersh (1981) *J. Neurochem.* **36**, 1594. **6.** Dando & Barrett (1998) in *Handb. Proteolytic Enzymes* (Barrett, Rawlings & Woessner, eds.), p. 1013, Academic Press, San Diego. **7.** Bury, Coolbear & Savery (1977) *Biochem. J.* **163**, 565. **8.** Engelhardt, Wallnöfer & Plapp (1973) *Appl. Microbiol.* **26**, 709. **9.** Robert-Baudouy, Clauziat & Thierry (1998) in *Handb. Proteolytic Enzymes* (Barrett, Rawlings & Woessner, eds.), p. 791, Academic Press, San Diego. **10.** Oizumi & Hayakawa (1991) *Biochim. Biophys. Acta* **1074**, 433. **11.** Sebti, Celeon & Lazo (1987) *Biochemistry* **26**, 4213. **12.** Raghav, Kamboj, Parnami & Singh (1995) *Indian J. Biochem. Biophys.* **32**, 279. **13.** Schwartz & Barrett (1980) *Biochem. J.* **191**, 487. **14.** Drummond & Yamamoto (1971) *The Enzymes*, 3rd ed. (Boyer, ed.), **4**, 355. **15.** Mantle (1991) *Clin. Chim. Acta* **196**, 135. **16.** McDonald (1998) in *Handb. Proteolytic Enzymes* (Barrett, Rawlings & Woessner, eds.), p. 408, Academic Press, San Diego. **17.** Stevens, Raizada, Sumners & Fernandez (1987) *Brain Res.* **406**, 113. **18.** McDonald, Leibach, Grindeland & Ellis (1968) *J. Biol. Chem.* **243**, 4143. **19.** McDonald & Schwabe (1980) *Biochim. Biophys. Acta* **616**, 68. **20.** Eisenhauer & McDonald (1986) *J. Biol. Chem.* **261**, 8859. **21.** Fukasawa, Fukasawa, Hiraoka & Harada (1983) *Biochim. Biophys. Acta* **745**, 6. **22.** Gossrau & Lojda (1980) *Histochemistry* **70**, 53. **23.** McDonald, Reilly, Zeitman & Ellis (1968) *J. Biol. Chem.* **243**, 2028. **24.** Smyth & O'Cuinn (1994) *J. Neurochem.* **63**, 1439. **25.** Swanson, Albers-Jackson & McDonald (1978) *Biochem. Biophys. Res. Commun.* **84**, 1151. **26.** Abramic, Schleuder, Dolovcak, *et al.* (2000) *Biol. Chem.* **381**, 1233. **27.** Emerson (1989) *Meth. Enzymol.* **168**, 365. **28.** Robinson, Blakely, Couto & Coyle (1987) *J. Biol. Chem.* **262**, 14498. **29.** Soffer (1974) *Adv. Enzymol. Relat. Areas Mol. Biol.* **40**, 91. **30.** Jahreis & Aurich (1990) *Biomed. Biochim. Acta* **49**, 339. **31.** Mikhailova, Vorotyntseva, Bessmertnaia & Antonov (1984) *Biokhimiia* **49**, 1733. **32.** McClellan, Jr, & Garner (1980) *Biochim. Biophys. Acta* **613**, 160. **33.** Vanha-Perttula (1988) *Clin. Chim. Acta* **177**, 179. **34.** Fulcher & Kenny (1983) *Biochem. J.* **211**, 743. **35.** Downey, So & Davie (1965) *Biochemistry* **4**, 1702. **36.** Coutsogeorgopoulos (1967) *Biochemistry* **6**, 1704. **37.** Morris, Favelukes, Arlinghaus & Schweet (1962) *Biochem. Biophys. Res. Commun.* **7**, 326. **38.** Kuo (1968) *Biochim. Biophys. Acta* **165**, 208. **39.** Yamamoto, Li, Huang, Ohkubo & Nishi (1998) *Biol. Chem.* **379**, 711. **40.** Lalu, Lampelo & Vanha-Perttula (1986) *Biochim. Biophys. Acta* **873**, 190. **41.** Sharma, Padwal-Desai & Ninjoor (1989) *Biochem. Biophys. Res. Commun.* **159**, 464. **42.** Tanioka, Hattori, Masuda, *et al.* (2003) *J. Biol. Chem.* **278**, 32275. **43.** Matsumoto, Rogi, Yamashiro, *et al.* (2000) *Eur. J. Biochem.* **267**, 46. **44.** Xu & Li (2005) *Curr. Med. Chem. Anticancer Agents* **5**, 281. **45.** Bauvois & Dauzonne (2006) *Med. Res. Rev.* **26**, 88. **46.** Jamadar, Jamdar, Dandekar & Harikumar (2003) *J. Food Sci.* **68**, 438. **47.** Yoshimoto, Tamesa, Gushi, Murayama & Tsuru (1988) *Agric. Biol. Chem.* **52**, 217. **48.** Huang, Takahara, Kinouchi, *et al.* (1997) *J. Biochem.* **122**, 779. **49.** McClellan, Jr, & Garner (1980) *Biochim. Biophys. Acta* **613**, 160. **50.** Gros, Giros & Schwartz (1985) *Biochemistry* **24**, 2179. **51.** Ichishima, Yamagata, Chiba, Sawaguchi & Tanaka (1989) *Agric. Biol. Chem.* **53**, 1867. **52.** Morty & Morehead (2002) *J. Biol. Chem.* **277**, 26057. **53.** Kumagai, Watanabe & Fujimoto (1991) *Biochem. Med. Metab. Biol.* **46**, 110. **54.** Kirsebom (1999) *Nucleic Acids Symp. Ser.* **41**, 17. **55.** Salavati, Panigrahi & Stuart (2001) *Mol. Biochem. Parasitol.* **115**, 109. **56.** Rossmanith & Karwan (1998) *Biochem. Biophys. Res. Commun.* **247**, 234. **57.** Abramochkin & Shrader (1996) *J. Biol. Chem.* **271**, 22901. **58.** Kaji, Kaji & Novelli (1965) *J. Biol. Chem.* **240**, 1185. **59.** Momose & Kaji (1966) *J. Biol. Chem.* **241**, 3294. **60.** Bartsch, Dworkin, Miller & Miller (1973) *Biochim. Biophys. Acta* **304**, 42. **61.** Park, Frost, Tuck, *et al.* (1987) *J. Biol. Chem.* **262**, 14702. **62.** Sankaran & Pogell (1975) *Antimicrob. Agents Chemother.* **8**, 721.

Puromycin Aminonucleoside

This aminoribose-containing nucleoside (FW = 294.31 g/mol), also known as 3'-amino-3'-deoxy-N^6,N^6-dimethyladenosine, inhibits RNA biosynthesis. Because the use of adenosine receptor agonists is plagued by dose-limiting cardiovascular side effects, there is great interest in developing adenosine kinase inhibitors (AKIs) as a means for raising steady-state adenosine concentrations. **Target(s):** adenosine deaminase (1); adenosine kinase (2); aminopeptidase PS (3); dipeptidyl-peptidase II (4); protein biosynthesis (5); protein degradation (5); and RNA biosynthesis (6,7). **1.** Dickie, Norton, Derr, Alexander & Nagasawa (1966) *Proc. Soc. Exp. Biol. Med.* **123**, 421. **2.** Miller, Adamczyk, Miller, *et al.* (1979) *J. Biol. Chem.* **254**, 2346. **3.** Dando & Barrett (1998) in *Handb. Proteolytic Enzymes* (Barrett, Rawlings & Woessner, eds.), p. 1013, Academic Press, San Diego. **4.** McDonald, Reilly, Zeitman & Ellis (1968) *J. Biol. Chem.* **243**, 2028. **5.** Kovacs & Seglen (1981) *Biochim. Biophys. Acta* **676**, 213. **6.** Tunkel & Studzinski (1981) *J. Cell Physiol.* **108**, 239. **7.** Albanese & Studzinski (1979) *J. Cell Physiol.* **99**, 55.

Purpurin

This red/orange pigment (FW = 256.21 g/mol), also known as 1,2,4-trihydroxyanthraquinone, is found with alizarin in madder root (*Rubia tinctorum*) and related plants. Purpurin produces a yellow solution in water and diethyl ether and a red solution in ethanol. It forms colored chelates with metal ions and is used in the detection of boron. *Note:* The term purpurin has also been used to refer to a lectin from *Dictyostelium purpureum*. In addition, a 20-kDa protein from neural retina adherons is referred to as retinal purpurin. **Target(s):** CYP1A1 (1); CYP1A2 (1); CYP2A6, weakly inhibited (1); CYP1B1 (1); CYP2E1, weakly inhibited (1); inositol-trisphosphate 3-kinase (2); [myosin light-chain] kinase (3); protein kinase A (4); protein-tyrosine kinase, *or* non-specific protein-tyrosine kinase (5); and xanthine oxidase (6). **1.** Takahashi, Fujita, Kamataki, *et al.* (2002) *Mutat. Res.* **508**, 147. **2.** Mayr, Windhorst & Hillemeier (2005) *J. Biol. Chem.* **280**, 13229. **3.** Jinsart, Ternai & Polya (1992) *Biol. Chem. Hoppe-Seyler* **373**, 903. **4.** Zhao, Polya, Wang, *et al.* (1995) *Neurobiol. Learn Mem.* **64**, 106. **5.** Hollósy & Kéri (2004) *Curr. Med. Chem. Anticancer Agents* **4**, 173. **6.** Sheu & Chiang (1997) *Anticancer Res.* **17**, 3293.

Purpurogallin

This aglycone (FW = 220.18 g/mol) of a number of glycosides (*e.g.*, dryophantin) from several nutgalls. Purpurogallin is a scavenger of polymorphonuclear leukocyte-derived oxyradicals and acts as a cardioprotector. **Target(s):** catechol *O*-methyltransferase (1,2); cystathionine β-synthase (3); glutathione-disulfide reductase (4); glutathione *S*-transferase (4-7); HIV-1 integrase (8); lactoylglutathione lyase, *or* glyoxalase I (9,10); 3-phosphoglycerate kinase (11); prolyl endopeptidase (12); protein-tyrosine kinase (13); and xanthine oxidase (14). **1.** Veser (1987) *J. Bacteriol.* **169**, 3696. **2.** Lambert, Chen, Wang, *et al.* (2005) *Bioorg. Med. Chem.* **13**, 2501. **3.** Walker & Barrett (1992) *Exp. Parasitol.* **74**, 205. **4.** Kurata, Suzuki & Takeda (1992) *Comp. Biochem. Physiol. B* **103**, 863. **5.** Das, Singh, Mukhtar & Awasthi (1986) *Biochem. Biophys. Res. Commun.* **141**, 1170. **6.** Das, Bickers & Mukhtar (1984) *Biochem. Biophys. Res. Commun.* **120**, 427. **7.** Liebau, Eckelt, Wildenburg, *et al.* (1997) *Biochem. J.* **324**, 659. **8.** Farnet, Wang, Hansen, *et al.* (1998) *Antimicrob. Agents Chemother.* **42**, 2245. **9.** Allen, Lo & Thornalley (1993) *Biochem. Soc. Trans.* **21**, 535. **10.** Sommer, Fischer, Krause, *et al.* (2001) *Biochem. J.* **353**, 445. **11.** Hickey, Coutts, Tsang-Tan & Pogson (1995) *Biochem. Soc. Trans.* **23**, 607S. **12.** Inamori, Muro, Sajima, *et al.* (1997) *Biosci. Biotechnol. Biochem.* **61**, 890. **13.** Abou-Karam & Shier (1999) *Phytother. Res.* **13**, 337. **14.** Sheu, Lai & Chiang (1998) *Anticancer Res.* **18**, 263.

Purvalanol A and B

Purvalanol A   Purvalinol B

This highly substituted adenine/valanol conjugate and cell-permeable inhibitor (FW$_A$ = 388.90 g/mol), also known as (2R)-2-[[6-[(3-chlorophenyl)amino]-9-(1-methylethyl)-9H-purin-2-yl]amino]-3-methyl-1-butanol, targets cyclin-dependent kinases and is ATP-competitive. IC$_{50}$ values are 4, 70, 35, 850 and 75 nM for Cdc2/Cyclin B, Cdk2/Cyclin A, Cdk2/Cyclin E, Cdk4/Cyclin D1 and Cdk5-p35 respectively. Purvalanol reversibly arrests synchronised cells in G$_1$ and G$_2$, inhibiting cell proliferation and inducing cell death. Purvalanol B (FW$_B$ = 431.91) is more water-soluble and therefore less cell-permeant. **Target(s):** ABC transporter ABCG2 (1); cAMP-dependent protein kinase, or protein kinase A, weakly inhibited (2); cyclin-dependent kinase, or cdk2, IC$_{50}$ = 0.1 μM (2-5); dual-specificity kinase, or dual-specificity, thyrosine-phosphorylated and regulated kinase 1A, abbreviated DYRK1A, IC$_{50}$ = 0.3 μM (2); Lck protein-tyrosine kinase (2); mitogen-activated protein kinase, or p42/p44 MAPK (2,6); non-specific serine/threonine protein kinase (2); protein-tyrosine kinase, or non-specific protein-tyrosine kinase (2); and ribosomal S6 kinase 1, abbreviated RSK1, IC$_{50}$ = 1.5 μM (2). **1.** An, Hagiya, Tamura, et al. (2009) Pharm. Res. **26**, 449. **2.** Bain, McLauchlan, Elliott & Cohen (2003) Biochem. J. **371**, 199. **3.** Gray, Wodicka, Thunnissen, et al. (1998) Science **281**, 533. **4.** Woodard, Li, Kathcart, et al. (2003) J. Med. Chem. **46**, 3877. **5.** Braña, Cacho, Garcia, et al. (2005) J. Med. Chem. **48**, 6843. **6.** Knockaert, Lenormand, Gray, et al. (2002) Oncogene **21**, 6413.

Putrescine

This diamine (FW$_{free-base}$ = 88.15 g/mol; CAS 110-60-1; colorless oil; M.P. = 23-24°C; B.P. = 158-160°C), also known as tetramethylenediamine, 1,4-diaminobutane, and 1,4-butanediamine, is produced by the decarboxylation of ornithine and is a precursor for the polyamines spermidine and spermine. Purescine has a strong odor similar to that of piperidine. It is very soluble in water and ethanol. The dihydrochloride (FW = 161.07 g/mol) has a melting point of about 280°C, at which it decomposes. The pK_a values at 20°C are 9.35 and 10.80 (at 30°C are 9.04 and 10.50). **Target(s):** acetylspermidine deacetylase (1,44,45); adenosylmethionine decarboxylase (32); adenylate cyclase (2); agmatinase (3); agmatine deiminase (40-42); Agropyron elongatum nuclease (48); amino-acid N-acetyltransferase (63); 1-amino-cyclopropane-1-carboxylate synthase (31); amylo-α-1,6-glucosidase/4-α-glucanotransferase, or glycogen debranching enzyme (47); arginine decarboxylase (4,33-39); arginine deiminase (5,43); arginyltransferase, weakly inhibited (62); carbamoyl-phosphate synthetase, glutamine-hydrolyzing (6,7); carbonic anhydrase, or carbonate dehydratase (8); deoxyhypusine monooxygenase, weakly inhibited (69); deoxyhypusine synthase (54); DNA (cytosine-5-)-methyltranferase (66); DNA-directed RNA polymerase (9); glucose-6-phosphate dehydrogenase (8); glutamate dehydrogenase (10,11); glutamate:ethylamine ligase, or theanine synthetase, weakly inhibited (12); indolethylamine-N-methyltransferase (13); kynureninase (14); lipoxygenase-1 (15); lysine 5,6-aminomutase (16,17); methionine S-adenosyltransferase (18,57); nitric-oxide synthase (19,20); ornithine aminotransferase (52,53); ornithine carbamoyltransferase (64,65); pancreatic ribonuclease, ribonuclease A (21); phosphatidylinositol 3-kinase (22); phospholipase C (23); phosphoprotein phosphatase I, e.g., phosphorylase phosphatase (24); polynucleotide adenylyltransferase, or poly(A) polymerase (50); protein-arginine N-methyltransferase (25); protein-glutamine γ-glutamyltransferase, or transglutaminase; also alternative substrate (58-61); protein kinase C (26); pyridoxal kinase (27,51); ribonuclease (21); RNA-directed RNA polymerase (49); spermine synthase, or spermidine aminopropyltransferase (28,29,55,56); tRNA (adenine-N1-)-methyltransferase (67); tRNA adenylyltransferase (30); tRNA (cytosine-5-)-methyltransferase (67); tRNA (uracil-5-)-methyltransferase (68); and tubulinyl-tyrosine carboxypeptidase, inhibited

above 6 mM (46). **1.** Libby (1983) Meth. Enzymol. **94**, 329. **2.** Khan, Quemener & Moulinoux (1990) Life Sci. **46**, 43. **3.** Salas, Rodriguez, Lopez, et al. (2002) Eur. J. Biochem. **269**, 5522. **4.** Rosenfeld & Roberts (1976) J. Bacteriol. **125**, 601. **5.** Eichler (1989) Biol. Chem. Hoppe Seyler **370**, 1127. **6.** Mori & Tatibana (1978) Meth. Enzymol. **51**, 111. **7.** Szondy, Matyasi & Elodi (1989) Acta Biochim. Biophys. Hung. **24**, 107. **8.** Ciftci, Demir, Ozmen & Atici (2003) J. Enzyme Inhib. Med. Chem. **18**, 71. **9.** Krakow & Ochoa (1963) Meth. Enzymol. **6**, 11. **10.** Kuo, Michalik & Erecinska (1994) J. Neurochem. **63**, 751. **11.** Jarzyna, Lietz & Bryla (1994) Biochem. Pharmacol. **47**, 1387. **12.** Sasaoka, Kito & Onishi (1965) Agric. Biol. Chem. **29**, 984. **13.** Porta, Camardella, Esposito & Della Pietra (1977) Biochem. Biophys. Res. Commun. **77**, 1196. **14.** Jakoby & Bonner (1953) J. Biol. Chem. **205**, 709. **15.** Maccarrone, Baroni & Finazzi-Agro (1998) Arch. Biochem. Biophys. **356**, 35. **16.** Morley & Stadtman (1970) Biochemistry **9**, 4890. **17.** Stadtman (1973) Adv. Enzymol. **38**, 413. **18.** Geller, Legros, Wherry & Kotb (1997) Arch. Biochem. Biophys. **345**, 97. **19.** Blachier, Mignon & Soubrane (1997) Nitric Oxide **1**, 268. **20.** Hu, Mahmoud & el-Fakahany (1994) Neurosci. Lett. **175**, 41. **21.** Richards & Wyckoff (1971) The Enzymes, 3rd ed. (Boyer, ed.), **4**, 647. **22.** Singh, Chauhan, Brockerhoff & Chauhan (1995) Life Sci. **57**, 685. **23.** Takahashi, Sugahara & Ohsaka (1981) Meth. Enzymol. **71**, 710. **24.** Ballou & Fischer (1986) The Enzymes, 3rd ed. (Boyer & Krebs, eds.), **17**, 311. **25.** Yoo, Park, Okuda, et al. (1999) Amino Acids **17**, 391. **26.** Kikkawa & Nishizuka (1986) The Enzymes, 3rd ed. (Boyer & Krebs, eds.), **17**, 167. **27.** Kerry & Kwok (1986) Prep. Biochem. **16**, 199. **28.** Rauna, Pajula & Eloranta (1983) Meth. Enzymol. **94**, 276. **29.** Pegg (1983) Meth. Enzymol. **94**, 294. **30.** Deutscher (1974) Meth. Enzymol. **29**, 706. **31.** Miyazaki & Yang (1987) Phytochemistry **26**, 2655. **32.** Yamanoha & Cohen (1985) Plant Physiol. **78**, 784. **33.** Smith (1979) Phytochemistry **18**, 1447. **34.** Ramakrishna & Adiga (1975) Eur. J. Biochem. **59**, 377. **35.** Das, Bhaduri, Bose & Ghosh (1996) J. Plant Biochem. Biotechnol. **5**, 123. **36.** Balasundaram & Tyagi (1989) Eur. J. Biochem. **183**, 339. **37.** Nam, Lee & Lee (1997) Plant Cell Physiol. **38**, 1150. **38.** Winer, Vinkler & Apelbaum (1984) Plant Physiol. **76**, 233. **39.** Li, Regunathan & Reis (1995) Ann. N. Y. Acad. Sci. **763**, 325. **40.** Chaudhuri & Ghosh (1985) Phytochemistry **24**, 2433. **41.** Yanagisawa & Suzuki (1981) Plant Physiol. **67**, 697. **42.** Sindhu & Desai (1979) Phytochemistry **18**, 1937. **43.** Knodler, Sekyere, Stewart, Schofield & Edwards (1998) J. Biol. Chem. **273**, 4470. **44.** Libby (1978) Arch. Biochem. Biophys. **188**, 360. **45.** Santacroce & Blankenship (1982) Proc. West. Pharmacol. Soc. **25**, 113. **46.** Barra & Argarana (1982) Biochem. Biophys. Res. Commun. **108**, 654. **47.** Gillard & Nelson (1977) Biochemistry **16**, 3978. **48.** Yupsanis, Symeonidis, Kalemi, Moustaka & Yupsani (2004) Plant Physiol. Biochem. **42**, 795. **49.** Lazarus & Itin (1973) Arch. Biochem. Biophys. **156**, 154. **50.** Rose & Jacob (1976) Arch. Biochem. Biophys. **175**, 748. **51.** Gäng & von Collins (1973) Int. J. Vitam. Nutr. Res. **43**, 318. **52.** Strecker (1965) J. Biol. Chem. **240**, 1225. **53.** Yasuda, Tanizawa, Misono, Toyama & Soda (1981) J. Bacteriol. **148**, 43. **54.** Jakus, Wolff, Park & Folk (1993) J. Biol. Chem. **268**, 13151. **55.** Pajula, Raina & Eloranta (1979) Eur. J. Biochem. **101**, 619. **56.** Hannonen, Jänne & Raina (1972) Biochim. Biophys. Acta **289**, 225. **57.** Schröder, Eichel, Breinig & Schröder (1997) Plant Mol. Biol. **33**, 211. **58.** Harrison, Layton, Hau, et al. (2007) Brit. J. Dermatol. **156**, 247. **59.** Siegel & Khosla (2007) Pharmacol. Ther. **115**, 232. **60.** Wu, Lai & Tsai (2005) Int. J. Biochem. Cell Biol. **37**, 386. **61.** Jeon, Lee, Jang, et al. (2004) Exp. Mol. Med. **36**, 576. **62.** Kato (1983) J. Biochem. **94**, 2015. **63.** Haas & Leisinger (1974) Biochem. Biophys. Res. Commun. **60**, 42. **64.** Baker & Yon (1983) Phytochemistry **22**, 2171. **65.** Tricot, Schmid, Baur, et al. (1994) Eur. J. Biochem. **221**, 555. **66.** Cox (1979) Biochem. Biophys. Res. Commun. **86**, 594. **67.** Nau, Pham-Coeur-Joly & Dubert (1983) Eur. J. Biochem. **130**, 261. **68.** Ny, Lindström, Hagervall & Björk (1988) Eur. J. Biochem. **177**, 467. **69.** Abbruzzese, Park, Beninati & Folk (1989) Biochim. Biophys. Acta **997**, 248.

PX-866

This orally active antifungal and antineoplastic agent (FW = 541.60 g/mol; CAS 502632-66-8; Solubility: 40 mg/mL DMSO, water-insoluble; stable for six months, when stored in a refrigerated desicator) is a chemically stable wortmannin derivative that potently and irreversibly inhibits phosphoinositide 3-kinase (IC$_{50}$ = 0.1 nM) and phosphoinositide-3-kinase-dependent signaling (1). PX-866 blocks PtdIns-3-kinase signaling, as measured by phospho-Ser^{473}-Akt levels in HT-29 colon cancer cells (IC$_{50}$ = 20 nM). PX-866 also overcomes resistance to the EGFR inhibitor gefitinib in A-549 human non-small cell lung cancer xenografts (2). **1**. Ihle, Williams, Chow, *et al.* (2004) *Mol. Cancer Ther.* **3**, 763. **2**. Ihle, Paine-Murrieta, Berggren *et al.* (2005) *Mol Cancer Ther.* **4**, 1349.

PXD101, See Belinostat

PXS-4681A

This SSAO/VAP-1-directed anti-inflammatory agent (FW = 342.84 g/mol) is a potent and selective mechanism-based inhibitor (K_i = 37 nM; k_{inact} = 0.26 min^{-1}) of Semicarbazide-Sensitive Amine Oxidase (SSAO), *or* Vascular Adhesion Protein-1 (VAP-1), a copper-dependent amine oxidase associated with various forms of inflammation and fibrosis (1). SSAO/VAP-1 catalyzes the oxidation of primary amine substrates (including benzylamine, tyramine, methylamine, *n*-decylamine, histamine, tryptamine or β-phenylethylamine) to aldehydes, releasing ammonia and hydrogen peroxide upon regeneration of its 6-hydroxy-dopa-quinone (TPQ) co-factor. PXS-4681A is highly selective for SSAO/VAP-1, when profiled against related amine oxidases, ion channels and 7-TM receptors, superior to inhibitors reported previously. While the exact physiological role of this enzyme is presently not well understood, PXS-4681A (at 2 mg/kg) attenuates neutrophil migration, TNF-α and IL-6 levels in mouse models of lung inflammation and localized inflammation. Such findings suggest SSAO inhibition leads to decreased neutrophil rolling/extravasation, resulting in a reduction in inflammation. *See also* PXS-4728A **1**. Foot, Yow, Schilter, *et al.* (2013) *J. Pharmacol. Exp. Ther.* **347**, 365.

PXS-4728A

This potent and highly selective mechanism-based VAP-1 inhibitor (FW$_{HCl-salt}$ = 316.80 g/mol; Solubility: >10 mg/mL H$_2$O) targets Semicarbazide-Sensitive Amine Oxidase, *or* Vascular Adhesion Protein-1 (K_i = 175 nM; k_{inact} of 0.68 min^{-1}). PXS-4728A provides >90% inhibition for 24 hours after a single oral dose. PXS-4728A in development for the treatment of cardiometabolic diseases like the liver-related disease Nonalcoholic Steatohepatitis (NASH), which is common to people who are overweight or obese. Given the increasing rate of obesity around the world, made worse by sedentary lifestyles and poor food choices, the condition is likely to become a major cause of liver disease and transplantations in coming years. PXS-4728A exhibits an IC$_{50}$ of <10 nM against the human VAP-1/SSAO, and this activity was maintained across all the mammalian species tested. It is 500x selective for VAP-1/SSAO *versus* related human amine oxidases. It shares the excellent potency, selectivity with PXS-4681A, but has a more balanced blood/plasma ratio (1.2 for human and dog, 1.45 for rat) due to substitution of the sulfonamide group with an amide, thereby removing all inhibition of carbonic anhydrase II and thus avoiding the potential for sequestration by erythrocytes in whole blood. Like PXS-4681A, PXS-4728A is stable in human, rat and dog plasma (>90% remaining after 1 hour incubation) and has high water solubility >10 mg/mL at pH 7.4. *See also* PXS-4681A **1**. Schilter, Collison, Russo, *et al.* (2015) *Respir. Res.* **16**, 42.

Pygenic Acid A

This triterpene (FW$_{free-acid}$ = 492.12 g/mol) inhibits rabbit muscle glycogen phosphorylase, IC$_{50}$ = 213 μM. X-ray analysis indicates that the inhibitor binds at the allosteric activator site, where the physiological activator AMP binds. **1**. Wen, Sun, Liu, *et al.* (2008) *J. Med. Chem.* **51**, 3540.

Pyocyanine

This natural pigment (FW = 210.24 g/mol), which has an antifungal activity, is a secretory agent produced by *Pseudomonas aeruginosa* (formerly *P. pyocyanea* and *Bacillus pyocyaneus*). It is a blue solid that is soluble in chloroform and pyridine generates reactive oxygen species, such as superoxide and hydrogen peroxide. Pyocyanine is slightly soluble in cold water and the aqueous blue solution can be decolorized under alkaline conditions upon the addition of glucose or sodium hydrosulfite. **Target(s):** catalase (1); cytochrome P450 leukotriene B$_4$ ω-oxidation (2); and diamine oxidase (3,4). **1**. O'Malley, Reszka, Rasmussen, *et al.* (2003) *Amer. J. Physiol. Lung Cell Mol. Physiol.* **285**, L1077. **2**. Muller & Sorrell (1992) *Infect. Immun.* **60**, 2536. **3**. Zeller (1951) *The Enzymes*, 1st ed. (Sumner & Myrbäck, eds.), **2** (part 1), 536. **4**. Zeller (1963) *The Enzymes*, 2nd ed. (Boyer, Lardy & Myrbäck, eds.), **8**, 313.

PYR-41

This cell-permeable pyrazone (FW = 425.40 g/mol; CAS 418805-02-4), also known as 4-(4-(5-nitrofuran-2-ylmethylene)-3,5-dioxo-pyrazolidin-1-yl)benzoic acid ethyl ester, irreversibly inhibits the ubiquitin-activating enzyme E$_1$ activity (IC$_{50}$ <10 μM in cell-free E$_1$ ubiquitination assays) but exhibits little or no activity against E$_3$, E$_2$, or caspase enzymatic activity. PYR-41 blocks ubiquitination-dependent protein degradation and other ubiquitination-mediated cellular activities. Unexpectedly, in addition to blocking ubiquitylation, PYR-41 increases total sumoylation in cells. Although the molecular basis for this effect is unknown, increased sumoylation is also observed in cells harboring a temperature-sensitive form of E$_1$. PYR-41 attenuates cytokine-mediated NFκβ activation. This behavior correlates with inhibition of nonproteasomal (Lys-63) ubiquitylation of TRAF6, a protein that is essential to IκB kinase activation. PYR-41 also prevents the downstream ubiquitylation and proteasomal degradation of IκβB. (*See also* NSC624206) Later studies demonstrated not only PYR-41 inhibited UBE$_1$ but also had equal or greater inhibitory activity against several deubiquitinases (DUBs) within intact cells and purified Ubiquitin-Specific Peptidase 5 (USP5) *in vitro* (2). Both UBE$_1$ and DUB inhibition are mediated through covalent protein cross-linking, which parallels the inhibition of the target proteins enzymatic activity. PYR-41 also mediates cross-linking of specific protein kinases (Bcr-Abl, Jak2) to inhibit their signaling activity (2). Degradation of Repressor Element 1-Silencing transcription factor (REST), itself is a major neuronal and tumor suppressor, is blocked by both PYR-41 and the proteasome inhibitor MG-132, but not by the nuclear export inhibitor leptomycin B (3). **1**. Yang, Kitagaki, Dai, *et al.* (2007) *Cancer Res.* **67**, 9472. **2**. Kapuria, Peterson, Showalter, *et al.* (2011) *Biochem. Pharmacol.* **82**, 341. **3**. Guan & Ricciardi (2012) *J. Virol.* **86**, 5594.

Pyrazinamide

This pyrazinecarboxamide (FW = 123.11 g/mol), is an important tuberculosis-sterilizing drug that helps to shorten the duration of current chemotherapy regimens for tuberculosis. Pyrazinamide, which is hydrolyzed by an amidase to pyrazinoic acid, has a pK_a value of 0.5 and is soluble in water (15 mg/mL). **Key Pharmacokinetic Parameters:** *See* Appendix II in Goodman & Gilman's *THE PHARMACOLOGICAL BASIS OF THERAPEUTICS*, 12th Edition (Brunton, Chabner & Knollmann, eds.) McGraw-Hill Medical, New York (2011). **Target(s):** fatty-acid synthase (1-4); NAD$^+$ ADP-ribosyltransferase, *or* poly(ADP-ribose) polymerase, IC$_{50}$ = 0.13 mM (5); nicotinamidase, also alternative substrate (6,7); and nicotinate *N*-methyltransferase (8). **1**. Schroeder, de Souza, Santos, Blanchard & Basso (2002) *Curr. Pharm. Biotechnol.* **3**, 197. **2**. Zimhony, Cox, Welch, Vilcheze & Jacobs, Jr. (2000) *Nature Med.* **6**, 1043. **3**. Boshoff, Mizrahi & Barry III (2002) *J. Bacteriol.* **184**, 2167. **4**. Ngo, Zimhony, Chung, *et al.* (2007) *Antimicrob. Agents Chemother.* **51**, 2430. **5**. Rankin, Jacobson, Benjamin, Moss & Jacobson (1989) *J. Biol. Chem.* **264**, 4312. **6**. Johnson & Gadd (1974) *Int. J. Biochem.* **5**, 633. **7**. Sun & Zhang (1999) *Antimicrob. Agents Chemother.* **43**, 537. **8**. Taguchi, Nishitani, Okumura, Shimabayashi & Iwai (1989) *Agri. Biol. Chem.* **53**, 2867.

Pyrazinecarboxamide, See *Pyrazinamide*

2-(Pyrazin-2-yl)-1*H*-benzimidazole

This substituted benzimidazole (FW = 196.21 g/mol) inhibits *Escherichia coli* methionyl aminopeptidase *in vitro* (IC$_{50}$ = 4.6 μM) by binding to an additional Co(II) ion situated at the entrance of the active site. This unexpected finding explains the inactivity of this and other compounds under *in vivo* conditions.. **1**. Schiffmann, Neugebauer & Klein (2006) *J. Med. Chem.* **49**, 511.

Pyrazofurin

This pyrimidine nucleoside (FW = 261.23 g/mol), also known as pyrazomycin, exhibits antineoplastic activity and inhibits cell proliferation and DNA synthesis in cells by inhibiting orotidylate decarboxylase (uridine 5'-phosphate synthase). Pyrazofurin, also known as 3,β-D-ribofuranosyl-4-hydroxypyrazole-5-carboxamide, is the prodrug of the 5'-phosphate derivative. **Target(s):** adenosylhomocysteinase (1); 5-aminoimidazole-4-carboxamide-1-β-D-ribofuranosyl 5'-monophosphate formyltransferase (2); orotate phosphoribosyltransferase (3); orotidylate decarboxylase, *or* UMP synthase (4). **1**. Guranowski, Montgomery, Cantoni & Chiang (1981) *Biochemistry* **20**, 110. **2**. Worzalla & Sweeney (1980) *Cancer Res.* **40**, 1482. **3**. Krungkrai, Aoki, Palacpac, *et al.* (2004) *Mol. Biochem. Parasitol.* **134**, 245. **4**. Ringer, Howell & Etheredge (1991) *J. Biochem. Toxicol.* **6**, 19.

Pyrazofurin 5'-Phosphate

This nucleotide (FW$_{free-acid}$ = 341.21 g/mol), itself a five-membered ring analogue of orotidine 5'-phosphate (OMP), inhibits orotidylate decarboxylase, *or* UMP synthase (1-3). **1**. Jones, Kavipurapu & Traut (1978) *Meth. Enzymol.* **51**, 155. **2**. Ringer, Howell & Etheredge (1991) *J. Biochem. Toxicol.* **6**, 19. **3**. Suttle & Stark (1979) *J. Biol. Chem.* **254**, 4602.

1*H*-Pyrazole-1-carboxamidine

This pyrazole derivative (FW$_{free-base}$ = 110.12 g/mol) inhibits nitric-oxide synthase, IC$_{50}$ = 0.2 μM (1,2). **1**. Southan, Gauld, Lubeskie, *et al.* (1997) *Biochem. Pharmacol.* **54**, 409. **2**. Lee, Martasek, Roman & Silverman (2000) *Bioorg. Med. Chem. Lett.* **10**, 2771.

1,9-Pyrazoloanthrone, See *SP 600125*

Pyrazoloisoguanine

This allopurinol-like, guanine analogue (FW = 151.13 g/mol), also known as 4-amino-6-hydroxypyraxolo[3,4-*d*]pyrimidine, inhibits xanthine oxidase (1,2). Pyrazoloisoguanine can be produced by the action of xanthine oxidase on pyrazoloadenine. **1**. Webb (1966) *Enzyme and Metabolic Inhibitors*, vol. 2, p. 279, Academic Press, New York. **2**. Feigelson, Davidson & Robins (1957) *J. Biol. Chem.* **226**, 993.

Pyrazolo[4,3-*b*]olean-12-en-28-oate

This oleanolic acid derivative (FW = 466.71 g/mol), *or* pyrazolo[4,3-*b*]olean-12-en-28-oic acid, inhibits rabbit muscle glycogen phosphorylase, IC$_{50}$ = 9.9 μM. **1**. Chen, Gong, Liu, *et al.* (2008) *Chem. Biodivers.* **5**, 1304.

17-(3'-Pyrazolyl)androsta-5,16-dien-3β-ol

This synthetic steroid (FW = 334.46 g/mol) inhibits steroid 17α-monooxygenase, *or* CYP17, human IC_{50} = 42 nM (1-3). The 17-(1*H*-pyrazol-1-yl) analogue is a weaker inhibitor. **1**. Ling, Li, Liu, *et al.* (1997) *J. Med. Chem.* **40**, 3297. **2**. Njar, Kato, Nnane, *et al.* (1998) *J. Med. Chem.* **41**, 902. **3**. Njar & Brodie (1999) *Curr. Pharm. Des.* **5**, 163.

Pyrazon, See *Chloridazon*

1-Pyrenylglyoxal

This pyrenylated derivative of glyoxal (FW = 258.28 g/mol) is an irreversible inhibitor of E_1 of the pyruvate dehydrogenase complex; an arginyl residue at the active site is modified. **See also** *Phenylglyoxal* **Target(s):** phosphoenolpyruvate carboxykinase (1); and pyruvate dehydrogenase (2). **1**. Bazaes, Silva, Goldie, Cardemil & Jabalquinto (1993) *J. Protein Chem.* **12**, 571. **2**. Eswaran, Ali, Shenoy, *et al.* (1995) *Biochim. Biophys. Acta* **1252**, 203.

N-(1-Pyrenyl)maleimide

This fluorescent labeling reagent (FW = 297.31 g/mol) is a *N*-arylmaleimide derivative that reacts with thiol groups to produce fluorescently tagged natural products. Pyrenyl-actin has proven to be a remarkably valuable tool in filament assembly studies, because the filamentous form is some 20-25 times mor fluorescent than unpolymerized actin. Nonetheles,, modification of actin by this reagent greatly reduces monomeric actin's affinity for profilin and thymosin β4 (1). **Target(s):** acylglycerol lipase, *or* monoacylglycerol lipase (2); desulfoglucosinolate sulfotransferase (3); H⁺-translocating pyridine nucleotide transhydrogenase (4); H⁺-translocating pyrophosphatase, vacuolar (5); and 6-phosphofructo-2-kinase (6,7). **1**. Kang, Purich & Southwick (1999) *J. Biol. Chem.* **274**, 36963. **2**. Saario, Salo, Nevalainen, *et al.* (2005) *Chem. Biol.* **12**, 649. **3**. Jain, Groot Wassink, Kolenovsky & Underhill (1990) *Phytochemistry* **29**, 1425. **4**. Sedgwick, Meuller, Hou, Rydstrom & Bragg (1997) *Biochemistry* **36**, 15285. **5**. Maruyama, Tanaka, Takeyasu, Yoshida & Sato (1998) *Plant Cell Physiol.* **39**, 1045. **6**. Baez, Rodríguez, Babul & Guixé (2003) *Biochem. J.* **376**, 277. **7**. Guixé (2000) *Arch. Biochem. Biophys.* **376**, 313.

Pyrethrins

These plant-derived insecticides are the active constituents of *Chrysanthemum cinerariaefolium* (or *Pyrethrum cinerariaefolium*). Pyrethrin I (FW = 328.45 g/mol), systematically named as (2*Z*)-(1*S*)-2-methyl-4-oxo-3-(penta-2,4-dienyl)cyclopent-2-enyl (+)-*trans*-chrysanthemate, is a viscous liquid that oxidizes readily and is quickly inactivated in air and must be stored in the dark and cold. Pyrethrin II (FW = 372.46 g/mol), systematically named as (2*Z*)-(1*S*)-2-methyl-4-oxo-3-(penta-2,4-dienyl)cyclopent-2-enyl pyrethrate, is also a viscous liquid that is likewise oxidized in air. Both pyrethrins are practically insoluble in water. These esters also inhibit multidrug-resistant *Mycobacterium tuberculosis*. Formed by the action of chrysanthemyl diphosphate synthase on two molecules of dimethylallyl pyrophosphate, pyrethrins concentrate mainly in the seed cases, where they ward of invasive insects. *Note*: The term pyrethrins also refers to a mixture of natural insecticides that contain various amounts of cinerin I, cinerin II, jasmolin I, jasmolin II, pyrethrin I, and pyrethrin II. **See also** synthetic pyrethroids (e.g., *Permethrin; Cypermethrin; Deltamethrin*) **1**. Kakko, Toimela & Tahti (2000) *Chemosphere* **40**, 301.

Pyridine

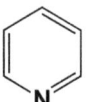

This aromatic nitrogen heterocycle (FW = 79.101 g/mol; CAS 110-86-1; colorless; M.P. = –41.5° C; B.P. = 115.25° C; unpleasant odor; weak base; pK_a = 5.23˙ at 25° C, with dpK_a/dT = –0.014) has many uses in organic synthesis and analytical chemistry. Pyridine should only be used in a fume hood or where there is sufficient ventilation. Miscible with water, pyridine has a larger dipole moment (2.37 *versus* 1.87 for water), and a dielectric constant of 12.4 (a sixth that of water). **Target(s):** acid phosphatase (1); adenain (2); alcohol dehydrogenase (3-7); D-amino-acid oxidase, weakly inhibited (8); aromatase, *or* CYP19 (9); carbonic anhydrase, *or* carbonate dehydratase, $K_i \approx 0.5$ M (10); catechol 2,3-dioxygenase (11,12); chymotrypsin (13); dimethylaniline-*N*-oxide aldolase (14); glucose dehydrogenase, weakly inhibited (15); histidine decarboxylase, $K_i = 1.4$ mM (16-18); lipase, *or* triacylglycerol lipase (19,20); nicotinate phosphoribosyltransferase (21); and xyloglucan:xyloglucosyl transferase (22). **1**. Sugiura, Kawabe, Tanaka, Fujimoto & Ohara (1981) *J. Biol. Chem.* **256**, 10664. **2**. McGrath, Abola, Toledo, Brown & Mangel (1996) *Virology* **217**, 131. **3**. Atkinson, Eckermann & Lilley (1967) *Biochem. J.* **104**, 872. **4**. Sund & Theorell (1963) *The Enzymes*, 2nd ed. (Boyer, Lardy & Myrbäck, eds.), **7**, 25. **5**. Shaw (1970) *The Enzymes*, 3rd ed. (Boyer, ed.), **1**, 91. **6**. van Eys & Kaplan (1957) *Biochim. Biophys. Acta* **23**, 574. **7**. Webb (1966) *Enzyme and Metabolic Inhibitors*, vol. **2**, p. 498, Academic Press, New York. **8**. Hellerman, Lindsay & Bovarnick (1946) *J. Biol. Chem.* **163**, 553. **9**. Vaz, Coon, Peegel & Menon (1992) *Drug Metab. Dispos.* **20**, 108. **10**. Pocker & Stone (1968) *Biochemistry* **7**, 2936. **11**. Nozaki (1970) *Meth. Enzymol.* **17A**, 522. **12**. Hayaishi, Nozaki & Abbott (1975) *The Enzymes*, 3rd ed. (Boyer, ed.), **12**, 119. **13**. Wallace, Kurtz & Niemann (1963) *Biochemistry* **2**, 824. **14**. Machinist, Orme-Johnson & Ziegler (1966) *Biochemistry* **5**, 2939. **15**. Brink (1953) *Acta Chem. Scand.* **7**, 1090. **16**. Chang & Snell (1971) *Meth. Enzymol.* **17B**, 663. **17**. Snell (1986) *Meth. Enzymol.* **122**, 128. **18**. Chang & Snell (1968) *Biochemistry* **7**, 2005. **19**. E. Böer, H. P. Mock, R. Bode, G. Gellissen & G. Kunze (2005) *Yeast* **22**, 523. **20**. Snellman, Sullivan & Colwell (2002) *Eur. J. Biochem.* **269**, 5771. **21**. Gaut & Solomon (1971) *Biochem. Pharmacol.* **20**, 2903. **22**. Fry (1997) *Plant J.* **11**, 1141.

Pyridine-3-aldehyde, See *3-Pyridinecarboxaldehyde*

3-Pyridinealdehyde Adenine Dinucleotide

This NAD⁺ analogue (FW_{ion} = 646.40 g/mol) is an alternative substrate for many oxidoreductases and an inhibitor of others (1-3). The aldehyde moiety forms a thiohemiacetal linkage with an active-site cysteinyl residue of glyceraldehyde-3-phosphate dehydrogenase and inhibits the dehydrogenase and esterase activities of that enzyme (4). **Target(s):** aldehyde dehydrogenase (5); dihydrolipoamide dehydrogenase (1); glyceraldehyde-3-phosphate dehydrogenase (4,6,7); lactate dehydrogenase (8); NAD^+ kinase (9,10); nicotinamide phosphoribosyl-transferase (2); and nitrate reductase (3). **1.** Koike & Hayakawa (1970) *Meth. Enzymol.* **18A**, 298. **2.** Dietrich (1971) *Meth. Enzymol.* **18B**, 144. **3.** Trimboli & Barber (1994) *Arch. Biochem. Biophys.* **315**, 48. **4.** Hill, Chou, Shih & Park (1975) *J. Biol. Chem.* **250**, 1734. **5.** Schwarcz & Stoppani (1970) *Enzymologia* **38**, 269. **6.** Kaplan & Ciotti (1961) *Ann. N. Y. Acad. Sci.* **94**, 701. **7.** Yang & Deal, Jr. (1969) *Biochemistry* **8**, 2806. **8.** Okabe, Hayakawa, Hamada & Koike (1968) *Biochemistry* **7**, 79. **9.** Chung (1967) *J. Biol. Chem.* **242**, 1182. **10.** Lerner, Niere, Ludwig & Ziegler (2001) *Biochem. Biophys. Res. Commun.* **288**, 69.

3-Pyridinealdehyde-NAD⁺, *See* 3-Pyridinealdehyde Adenine Dinucleotide

Pyridine-2-aldoxime Methochloride, *See* Pralidoxime Chloride

3-Pyridinecarboxaldehyde

This photosensitive aldehyde (FW = 107.11 g/mol; CAS 500-22-1), also known as pyridine-3-aldehyde, 3-pyridylcarboxaldehyde, and nicotin-aldehyde, inhibits xanthine oxidase. **Target(s):** nicotinamidase (1,2); nicotinate phosphoribosyltransferase (3); starch phosphorylase (4); and xanthine oxidase (5). **1.** Johnson & Gadd (1974) *Int. J. Biochem.* **5**, 633. **2.** Yan & Sloan (1987) *J. Biol. Chem.* **262**, 9082. **3.** Imsande (1964) *Biochim. Biophys. Acta* **85**, 255. **4.** Matheson & Richardson (1973) *Phytochemistry* **17**, 195. **5.** Morpeth & Bray (1984) *Biochemistry* **23**, 1332.

3-Pyridinecarboxaldehyde N',N'-(Dibenzyl)hydrazone

This substituted hydrazine (FW = 301.39 g/mol) inhibits glucose-6-phosphatase, IC_{50} = 3.7 μM (1). The 2- and 4-pyridinecarboxaldehyde analogues are weaker inhibitors, with IC_{50} values of 10 and 17 μM, respectively. **1.** Madsen, Jakobsen & Westergaard (2001) *Bioorg. Med. Chem. Lett.* **11**, 2165.

Pyridine-2-carboxaldehyde Thiosemicarbazone

This thiosemicarbazone (FW = 166.23 g/mol) is a strong inhibitor of ribonucleoside-diphosphate reductase. Cell cycle arrest and cell death produced by simple chemicals like N-hydroxyurea and hydroxamates can be traced back to their interaction with metal ions within the ribonucleotide reductase active site. **1.** Lammers & Follmann (1983) *Struct. Bonding* **54**, 27.

2-Pyridinecarboxylic Acid, *See* 2-Picolinic Acid

3-Pyridinecarboxylic Acid, *See* Nicotinic Acid

4-Pyridinecarboxylic Acid, *See* Isonicotinic Acid

Pyridine-2-carboxylic Acid Thiazol-2-ylamide

This amide (FW = 205.24 g/mol), also known as N-1,3-thiazol-2-ylpyridine-2-carboxamide, inhibits *Escherichia coli*, *Saccharomyces cerevisiae*, and human methionyl aminopeptidase, with an IC_{50} value of 1.3 μM for the human enzyme (1-5). **1.** Luo, Li, Liu, *et al.* (2003) *J. Med. Chem.* **46**, 2631. **2.** Li, Chen, Cui, *et al.* (2003) *Biochem. Biophys. Res. Commun.* **307**, 172. **3.** Schiffmann, Neugebauer & Klein (2006) *J. Med. Chem.* **49**, 511. **4.** Chen, Li, Li, *et al.* (2004) *Acta Pharmacol. Sin.* **25**, 907. **5.** Li, Chen, Cui, *et al.* (2004) *Biochemistry* **43**, 7892.

Pyridine-2,3-dicarboxylic Acid, *See* Quinolinic Acid

Pyridine-2,4-dicarboxylic Acid

This dicarboxylic acid ($FW_{free-acid}$ = 167.12 g/mol), also known as lutidinic acid, is a structural analogue of α-ketoglutarate and accordingly inhibits a number of α-ketoglutarate-dependent enzymes. **Target(s):** γ-butyrobetaine dioxygenase, *or* γ-butyrobetaine hydroxylase, K_i = 0.2 μM (1); deoxyhypusine monooxygenase (2,3); flavanone 3-dioxygenase, *or* flavone 3-dioxygenase (4-6); flavone synthase (6); flavonol synthase (5); homocitrate synthase (7); 6β-hydroxyhyoscyamine epoxidase (8); hyoscyamine (6S)-dioxygenase (9); α-ketoglutarate dehydrogenase (10); nicotinate-nucleotide diphosphorylase (carboxylating), weakly inhibited (11); peptide-aspartate β-dioxygenase (12,13); procollagen-lysine 5-dioxygenase (10); procollagen-proline 3-dioxygenase (10,14); procollagen-proline 4-dioxygenase, *or* prolyl 4-hydroxylase (10,13,15-25); and proline 4-dioxygenase (26). **1.** Ng, Hanauske-Abel & England (1991) *J. Biol. Chem.* **266**, 1526. **2.** Abbruzzese, Park & Folk (1986) *J. Biol. Chem.* **261**, 3085. **3.** Beninati, Ferraro & Abbruzzese (1990) *Ital. J. Biochem.* **39**, 183A. **4.** Britsch & Grisebach (1986) *Eur. J. Biochem.* **156**, 569. **5.** Halbwirth, Fischer, Schlangen, *et al.* (2006) *Plant Sci.* **171**, 194. **6.** Britsch (1990) *Arch. Biochem. Biophys.* **282**, 152. **7.** Qian, West & Cook (2006) *Biochemistry* **45**, 12136. **8.** Hashimoto, Kohno & Yamada (1989) *Phytochemistry* **28**, 1077. **9.** Hashimoto & Yamada (1987) *Eur. J. Biochem.* **164**, 277. **10.** Majamaa, Turpeenniemi-Hujanen, Latipää, *et al.* (1985) *Biochem. J.* **229**, 127. **11.** Taguchi & Iwai (1976) *Agric. Biol. Chem.* **40**, 385. **12.** Derian, VanDusen, Przysiecki, *et al.* (1989) *J. Biol. Chem.* **264**, 6615. **13.** Koivunen, Hirsilä, Günzler, Kivirikko & Myllyharju (2004) *J. Biol. Chem.* **279**, 9899. **14.** Tiainen, Pasanen, Sormunen & Myllyharju (2008) *J. Biol. Chem.* **283**, 19432. **15.** Tschank, Hanauske-Abel & Peterkofsky (1988) *Arch. Biochem. Biophys.* **261**, 312. **16.** Hanauske-Abel (1991) *J. Hepatol.* **13** Suppl. 3, S8. **17.** Majamaa, Hanauske-Abel, Günzler & Kivirikko (1984) *Eur. J. Biochem.* **138**, 239. **18.** Majamaa, Günzler, Hanauske-Abel, Myllylä & Kivirikko (1986) *J. Biol. Chem.* **261**, 7819. **19.** Wojtaszek, Smith & Bolwell (1999) *Int. J. Biochem. Cell Biol.* **31**, 463. **20.** Kaska, Günzler, Kivirikko & Myllylä (1987) *Biochem. J.* **241**, 483. **21.** Kivirikko, Myllylä & Pihlajaniemi (1989) *FASEB J.* **3**, 1609. **22.** Günzler, Hanauske-Abel, Myllylä, Mohr & Kivirikko (1987) *Biochem. J.* **242**, 163. **23.** Myllyharju (2008) *Ann. Med.* **40**, 402. **24.** Kaska, Myllylä, Günzler, Gibor & Kivirikko (1988) *Biochem. J.* **256**, 257. **25.** Hirsilä, Koivunen, Günzler, Kivirikko & Myllyhatju (2003) *J. Biol. Chem.* **278**, 30772. **26.** Lawrence, Sobey, Field, Baldwin & Schofield (1996) *Biochem. J.* **313**, 185.

Pyridine-2,5-dicarboxylic Acid

This dicarboxylic acid ($FW_{free-acid}$ = 167.12 g/mol), also known as isocinchomeronic acid, is a structural analogue of α-ketoglutarate and will inhibit a number of α-ketoglutarate-dependent enzymes. When heated above its melting point of 254°C, this dicarboxylic acid sublimes as nicotinic acid. **Target(s):** γ-butyrobetaine dioxygenase, *or* γ-butyrobetaine hydroxylase, K_i = 14 μM (1); deoxyhypusine monooxygenase (2,3); flavanone 3-dioxygenase, *or* flavone 3-dioxygenase, *or* naringenin 3-dioxygenase (4); flavonol synthase (5); homocitrate synthase (6); α-ketoglutarate dehydrogenase (7); nicotinate-nucleotide diphosphorylase (carboxylating), weakly inhibited (8); peptide-aspartate β-dioxygenase (9); procollagen-lysine 5-dioxygenase (7); procollagen-proline 3-dioxygenase (7,10); procollagen-proline 4-dioxygenase, *or* prolyl 4-hydroxylase (7,9,11-18); and proline 4-dioxygenase (19). **1**. Ng, Hanauske-Abel & England (1991) *J. Biol. Chem.* **266**, 1526. **2**. Abbruzzese, Park & Folk (1986) *J. Biol. Chem.* **261**, 3085. **3**. Beninati, Ferraro & Abbruzzese (1990) *Ital. J. Biochem.* **39**, 183A. **4**. Britsch & Grisebach (1986) *Eur. J. Biochem.* **156**, 569. **5**. Halbwirth, Fischer, Schlangen, *et al.* (2006) *Plant Sci.* **171**, 194. **6**. Qian, West & Cook (2006) *Biochemistry* **45**, 12136. **7**. Majamaa, Turpeenniemi-Hujanen, Latipää, *et al.* (1985) *Biochem. J.* **229**, 127. **8**. Taguchi & Iwai (1976) *Agric. Biol. Chem.* **40**, 385. **9**. Koivunen, Hirsilä, Günzler, Kivirikko & Myllyharju (2004) *J. Biol. Chem.* **279**, 9899. **10**. Tiainen, Pasanen, Sormunen & Myllyharju (2008) *J. Biol. Chem.* **283**, 194320 **11**. Tschank, Hanauske-Abel & Peterkofsky (1988) *Arch. Biochem. Biophys.* **261**, 312. **12**. Majamaa, Hanauske-Abel, Günzler & Kivirikko (1984) *Eur. J. Biochem.* **138**, 239. **13**. Majamaa, Günzler, Hanauske-Abel, Myllylä & Kivirikko (1986) *J. Biol. Chem.* **261**, 7819. **14**. Kaska, Günzler, Kivirikko & Myllylä (1987) *Biochem. J.* **241**, 483. **15**. Kivirikko, Myllylä & Pihlajaniemi (1989) *FASEB J.* **3**, 1609. **16**. Günzler, Hanauske-Abel, Myllylä, Mohr & Kivirikko (1987) *Biochem. J.* **242**, 163. **17**. Myllyharju (2008) *Ann. Med.* **40**, 402. **18**. Kaska, Myllylä, Günzler, Gibor & Kivirikko (1988) *Biochem. J.* **256**, 257. **19**. Lawrence, Sobey, Field, Baldwin & Schofield (1996) *Biochem. J.* **313**, 185.

Pyridine-2,6-dicarboxylic Acid, *See Dipicolinic Acid*

Pyridine-3,4-dicarboxylic Acid

This dicarboxylic acid ($FW_{\text{free-acid}}$ = 167.12 g/mol), also known as cinchomeronic acid, is a structural analogue of maleate and succinate. **Target(s):** aspartate 4-decarboxylase (1); γ-butyrobetaine hydroxylase, *or* γ-butyrobetaine:2-oxoglutarate dioxygenase, K_i = 0.74 mM (2); deoxyhypusine monooxygenase (3); nicotinate-nucleotide diphosphorylase, carboxylating (4); procollagen-lysine 5-dioxygenase (5); procollagen-proline 3-dioxygenase (5); and procollagen-proline 4-dioxygenase (5,6). **1**. Soda, Novogrodsky & Meister (1964) *Biochemistry* **3**, 1450. **2**. Ng, Hanauske-Abel & England (1991) *J. Biol. Chem.* **266**, 1526. **3**. Beninati, Ferraro & Abbruzzese (1990) *Ital. J. Biochem.* **39**, 183A. **4**. Mann & Byerrum (1974) *J. Biol. Chem.* **249**, 6817. **5**. Majamaa, Turpeenniemi-Hujanen, Latipää, *et al.* (1985) *Biochem. J.* **229**, 127. **6**. Majamaa, Hanauske-Abel, Günzler & Kivirikko (1984) *Eur. J. Biochem.* **138**, 239.

Pyridine-3,5-dicarboxylic Acid

This dicarboxylic acid ($FW_{\text{free-acid}}$ = 167.12 g/mol), also known as dinicotinic acid, is a structural analogue of glutaric acid. **Target(s):** aspartate 4-decarboxylase (1); γ-butyrobetaine hydroxylase, *or* γ-butyrobetaine:2-oxoglutarate dioxygenase, K_i = 1.03 mM (2); deoxyhypusine monooxygenase (3); glutamate dehydrogenase (4); procollagen-lysine 5-dioxygenase (5); procollagen-proline 3-dioxygenase (5); and procollagen-proline 4-dioxygenase (5,6). **1**. Soda, Novogrodsky & Meister (1964) *Biochemistry* **3**, 1450. **2**. Ng, Hanauske-Abel & England (1991) *J. Biol. Chem.* **266**, 1526. **3**. Beninati, Ferraro & Abbruzzese (1990) *Ital. J. Biochem.* **39**, 183A. **4**. Rogers, Boots & Boots (1972) *Biochim.*

Biophys. Acta **258**, 343. **5**. Majamaa, Turpeenniemi-Hujanen, Latipää, *et al.* (1985) *Biochem. J.* **229**, 127. **6**. Majamaa, Hanauske-Abel, Günzler & Kivirikko (1984) *Eur. J. Biochem.* **138**, 239.

Pyridine-2,3-diol, *See 2,3-Dihydroxypyridine*

Pyridine-2,4-diol, *See 3-Deazauracil*

Pyridine-3,4-diol, *See 3,4-Dihydroxypyridine*

4-Pyridine Disulfide, *See 4,4'-Dithiopyridine*

4,4'-[2,6-Pyridinediylbis(oxy)]bis(benzamidine)

This substituted benzamidine (FW = 349.39 g/mol) inhibits coagulation factor Xa, thrombin, and trypsin, with K_i values of 400, 550, and 2800 nM, respectively (1,2). **1**. Phillips, Davey, Eagen, *et al.* (1999) *J. Med. Chem.* **42**, 1749. **2**. Liang, Light, Kochanny, *et al.* (2003) *Biochem. Pharmacol.* **65**, 1407.

3-Pyridinepropanol, *See 3-(3-Pyridyl)-1-propanol*

Pyridine-3-sulfonamide

This sulfonamide (FW = 158.18 g/mol) inhibits carbonic anhydrase, *or* carbonate dehydratase (1,2). **1**. van Goor (1948) *Enzymologia* **13**, 73. **2**. Taylor, King & Burgen (1970) *Biochemistry* **9**, 2638.

Pyridine-3-sulfonic Acid

This water-soluble analogue of nicotinic acid (FW = 159.17 g/mol), also called 3-pyridylsulfonic acid, is a nicotinamide antagonist in some organisms. Pyridine-3-sulfonic acid is corrosive and decomposes at about 357°C. **Target(s):** alcohol dehydrogenase (1,2); glucose (3); L-lactate dehydrogenase (3); NADase, weakly inhibited (4); and nicotinate phosphoribosyltransferase (5). **1**. van Eys & Kaplan (1957) *Biochim. Biophys. Acta* **23**, 574. **2**. Webb (1966) *Enzyme and Metabolic Inhibitors*, vol. 2, p. 498, Academic Press, New York. **3**. von Euler (1942) *Ber.* **75**, 1876. **4**. Zatman, Kaplan, Colowick & Ciotti (1954) *J. Biol. Chem.* **209**, 453. **5**. Gaut & Solomon (1971) *Biochem. Pharmacol.* **20**, 2903.

2-[(Pyridin-2-yl)amino]ethylidene-1,1-bis(phosphonate)

This bisphosphonate (FW_{ion} = 280.11 g/mol), also known as NE 11808, inhibits geranyl*trans*transferase, *or* farnesyl-diphosphate synthase (1,2). **1**. Dunford, Thompson, Coxon, *et al.* (2001) *J. Pharmacol. Exp. Ther.* **296**, 235. **2**. Dunford, Kwaasi, Rogers, *et al.* (2008) *J. Med. Chem.* **51**, 2187.

2-{[(2-{[2-(4-Pyridinyl)ethyl]amino}phenyl)sulfonyl]amino}-5,6,7,8-tetrahydro-1-naphthalenecarboxylic Acid

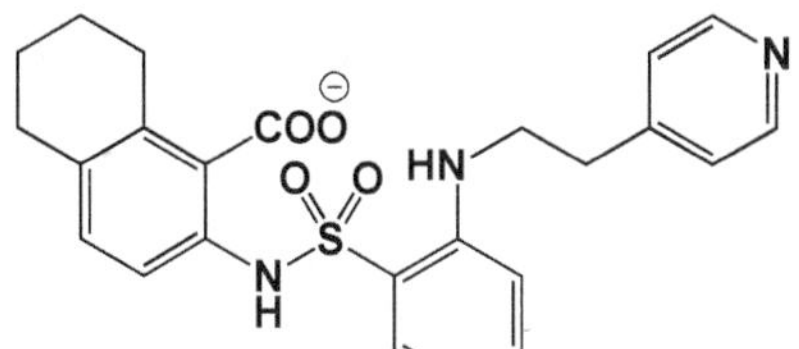

This substituted anthranilic acid sulfonamide (FW$_{ion}$ = 451.54 g/mol) inhibits methionyl aminopeptidase, IC$_{50}$ = 0.016 μM (1). The 2-(2-pyridinyl)ethyl analogue is a slightly weaker inhibitor (IC$_{50}$ = 0.053 μM). 1. Sheppard, Wang, Kawai, et al. (2006) J. Med. Chem. **49**, 3832.

2-(3-Pyridinyl)ethylidene-1,1-phosphonocarboxylic Acid

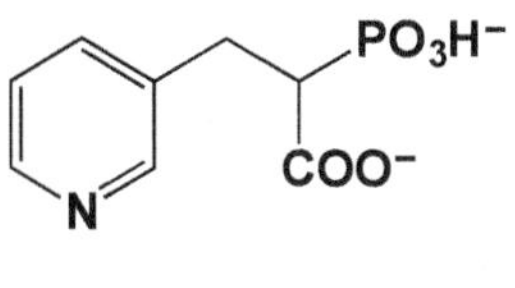

This risedronate analogue (FW$_{ion}$ = 228.12 g/mol), in which a second phosphonate is replaced by a carboxyl group, inhibits protein geranylgeranyltransferase type II. 1. Coxon, Ebetino, Mules, et al. (2005) Bone **37**, 349.

2-(Pyridin-4-yl)-1-hydroxyethane-1,1-bis(phosphonic Acid), See 1-Hydroxy-2-(pyridin-4-yl)ethylidene-1,1-bis(phosphonate

2-(3-Pyridinyl)-1-hydroxyethylidene-1,1-phosphonocarboxyliate

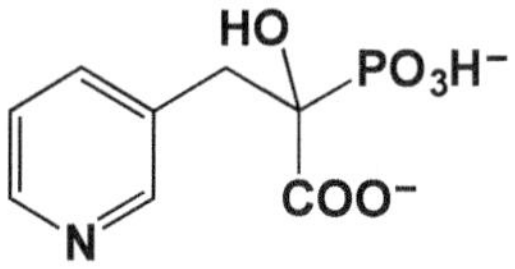

This risedronate analogue (FW$_{ION}$ = 245.13 g/mol), also known as NE10790, inhibits geranyl*trans*transferase, or farnesyl-diphosphate synthase (1) and protein geranylgeranyltransferase type II (2,3). 1. Dunford, Kwaasi, Rogers, et al. (2008) J. Med. Chem. **51**, 2187. 2. Coxon, Ebetino, Mules, et al. (2005) Bone **37**, 349. 3. Wu, Waldmann, Reents, et al. (2006) Chembiochem **7**, 1859.

8-[2-(3-Pyridinyl)-1H-indol-1-yl]octanoic Acid 1-(Carboxymethyl)-3-[(1-carboxy-3-phenylpropyl)-amino]-2,3,4,5-tetrahydro-2-oxo-1H-1-benzazepin-5-yl Ester

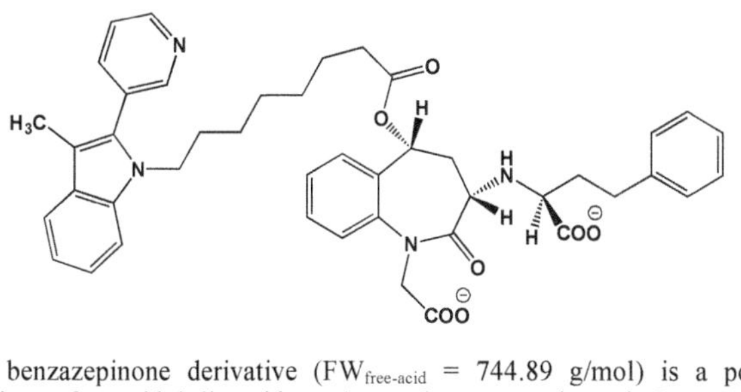

This benzazepinone derivative (FW$_{free-acid}$ = 744.89 g/mol) is a potent inhibitor of peptidyl-dipeptidase A, or human angiotensin-I-converting enzyme (IC$_{50}$ = 90 nM) and thromboxane-A synthase (IC$_{50}$ = 20 nM). 1. Ksander, Erion, Yuan, et al. (1994) J. Med. Chem. **37**, 1823.

5-(3'-Pyridinylmethyl)benzofuran-2-carboxylic Acid

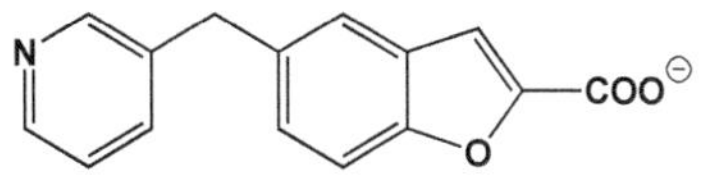

This benzofuran (FW$_{free-acid}$ = 253.26 g/mol), also known as U-63557A, is a potent human thromboxane-A synthase inhibitor in platelets (1-3). A single oral dose of 3.0 mg/kg U-63557A inhibits the platelet thromboxane

synthase in macaques approximately 80% for at least twelve hours. 1. Jones & Fitzpatrick (1991) J. Biol. Chem. **266**, 23510. 2. Gorman, Johnson, Spilman & Aiken (1983) Prostaglandins **26**, 325. 3. Hall, Tuan & Venton (1986) Biochem. J. **233**, 637.

N-(Pyridin-2-ylmethyl)-N'-(1,3-thiazol-2-yl)oxamide

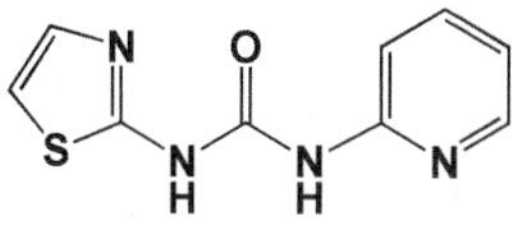

This diamide (FW = 250.28 g/mol), also known as N-(pyridin-2-ylmethyl)-N'-1,3-thiazol-2-ylethanediamide, inhibits the Co(II) and Mn(II) forms of methionyl aminopeptidase, particularly the former. Intrguingly, although methionine aminopeptidases require a divalent metal ion, such as Mn(II), Fe(II), Co(II), Ni(II), or Zn(II), to catalyze the removal of the N-terminal methionine from newly synthesized proteins, it is uncertain which of these ions is most important *in vivo*. 1. Ye, Xie, Huang, et al. (2004) J. Amer. Chem. Soc. **126**, 13940.

2-[3-(Pyridin-4-yloxy)benzoylamino)benzoic Acid

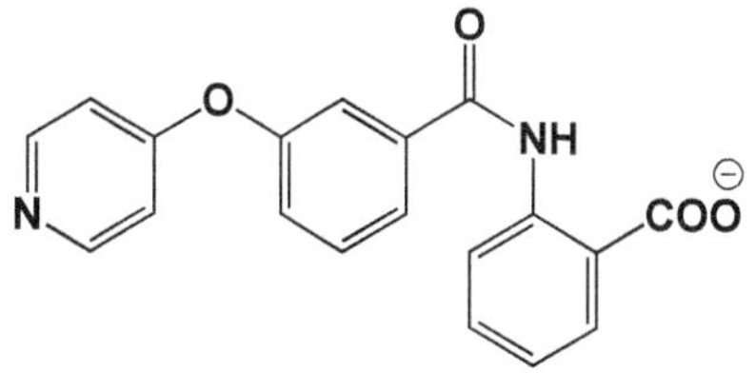

This benzoic acid derivative (FW$_{ion}$ = 333.32 g/mol) inhibits Enterococcus faecalis β-ketoacyl-[acyl-carrier-protein] synthase III, IC$_{50}$ = 10 μM (1,2). 1. Nie, Perretta, Lu, et al. (2005) J. Med. Chem. **48**, 1596. 2. Ashek, San Juan & Cho (2007) J. Enzyme Inhib. Med. Chem. **22**, 7.

Pyridin-3-ylpropyl α-Cyano-3,4-dihydroxycinnamate

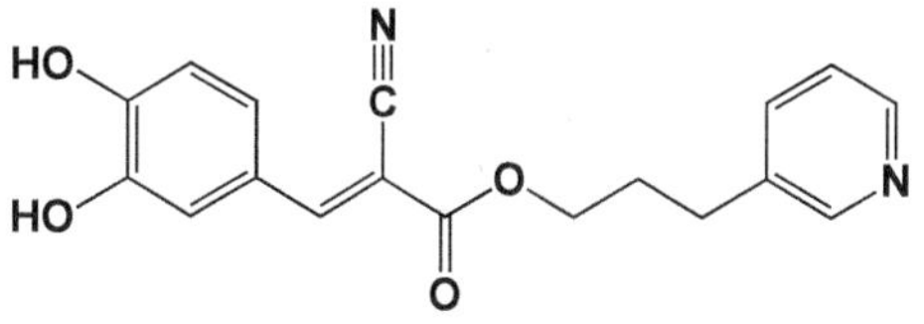

This caffeic acid derivative (FW = 425.14 g/mol) inhibits 12-lipoxygenase (or arachidonate 12-lipoxygenase) and 15-lipoxygenase (or arachidonate 15-lipoxygenase), with respective IC$_{50}$ values of 0.47 and 3.93 μM. 1. Cho, Ueda, Tamaoka, et al. (1991) J. Med. Chem. **34**, 1503.

5-(Pyridin-2-yl)-1H-1,2,3-triazole

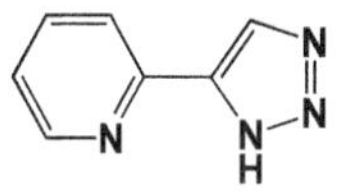

This aryl-1,2,3-triazole (FW = 145.15 g/mol) inhibits human methionyl aminopeptidase type 2, K$_{i,app}$ = 41 nM for the Co(II)-substituted protein (1). The 5-(pyridin-3-yl) analogue is a weaker inhibitor (K$_{i,app}$ = 260 nM). 1. Kallander, Lu, Chen, et al. (2005) J. Med. Chem. **48**, 5644.

Pyridochromanones

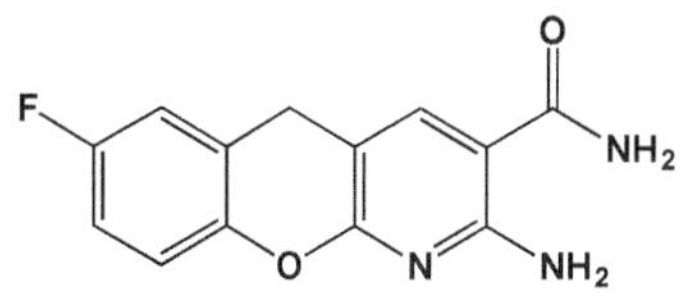

These so-caled chromones are strong inhibitors of Escherichia coli and Mycobacterium tuberculosis NAD$^+$-dependent DNA ligases. Shown above is the most active known chromone (FW = 273.22 g/mol), which exhibits an IC$_{50}$ value of 0.04 μM for the E. coli enzyme (1,2). 1. Gong, Martins, Bongiorno, Glickman & Shuman (2004) J. Biol. Chem. **279**, 20594. 2. Brötz-Oesterhelt, Knezevic, Bartel, et al. (2003) J. Biol. Chem. **278**, 39435.

Pyridoglutethimide, *See Rogletimide*

α-Pyridone, *See 2-Hydroxypyridine*

Pyridostigmine Bromide

This hygroscopic pyridinium salt (FW = 261.12 g/mol) is a reversible cholinesterases inhibitor, with a K_d value of 5.5 μM for acetylcholinesterase (1-5). It is soluble in water and ethanol and almost insoluble in acetone and diethyl ether. Pyridostigmine is effective in protecting against irreversible inhibitors of acetylcholinesterase such as organophosphates and chemical warfare agents (*e.g.*, soman).   **Target(s):** acetylcholinesterase (1-5); catalytic antibody with acetylcholinesterase activity (5); cholinesterase, *or* butyrylcholinesterase (2,6,7); and microtubule assembly, *or* tubulin polymerization (8). **1.** Forsberg & Puu (1984) *Eur. J. Biochem.* **140**, 153. **2.** Xia, Wang & Pei (1981) *Fundam. Appl. Toxicol.* **1**, 217. **3.** Laine-Cessac, Turcant, Premel-Cabic, Boyer & Allain (1993) *Res. Commun. Chem. Pathol. Pharmacol.* **79**, 185. **4.** Froede & Wilson (1971) *The Enzymes*, 3rd ed. (Boyer, ed.), **5**, 87. **5.** Johnson & Moore (2002) *J. Immunol. Methods* **269**, 13. **6.** Cerasoli, Griffiths, Doctor, *et al.* (2005) *Chem. Biol. Interact.* **157-158**, 363. **7.** Petroianu, Nurulain, Arafat, Rajan & Hasan (2006) *Arch. Toxicol.* **80**, 777. **8.** Prasad, Scotch, Chaudhuri, *et al.* (2000) *Neurochem. Res.* **25**, 19.

Pyridoxal

This photosensitive aldehyde form of vitamin B_6 (FW = 167.16 g/mol), systematically referred to as 3-hydroxy-5-(hydroxymethyl)-2-methyl-4-pyridinecarboxaldehyde, is the immediate precursor to the coenzyme pyridoxal 5'-phosphate (PLP) and is often a weaker competitive inhibitor of PLP binding to PLP-dependent enzymes). Pyridoxal is soluble in water (1 g/2 mL) and sensitive to heat, particularly at alkaline pH. The pK_a values are 4.23 (phenol OH), 8.70 (pyridinium NH$^+$), and 13.0. Pyridoxal has a λ_{max} value of 252 nm at pH 7.0 ($\varepsilon = 8200$ M^{-1}cm^{-1}). It is typically supplied as the hydrochloride. **Target(s):** *O*-acetylhomoserine aminocarboxy-propyltransferase, *or O*-acetylhomoserine (thiol)-lyase (1); alanine racemase (2); aldehyde dehydrogenase (3); arginine decarboxylase (4,5,25); cystathionine β-lyase, *or* cystine lyase (6); cysteine synthase (34); dextransucrase (7); β-fructofuranosidase, *or* invertase (8-10,27-32); glutamate dehydrogenase (11-13); hemoglobin S polymerization (14,15); hydroxyacyl-glutathione hydrolase, *or* glyoxalase II (16); kynurenine 3-monooxygenase (37); lactoylglutathione lyase, *or* glyoxalase I (16); NMN nucleosidase (26); phosphopantothenoylcysteine decarboxylase (17-19); porphobilinogen synthase, *or* δ-aminolevulinate dehydratase (20); pyridoxal kinase (21); pyridoxal phosphatase, weakly inhibited (33); starch phosphorylase (35); thiamin phosphatase (22); tyrosinase (23); and xanthine dehydrogenase (24,36). **1.** Yamagata & Takeshima (1976) *J. Biochem.* **80**, 777. **2.** Olivard & Snell (1955) *J. Biol. Chem.* **213**, 203. **3.** Lee, Manthey & Sladek (1991) *Biochem. Pharmacol.* **42**, 1279. **4.** Ramakrishna & Adiga (1975) *Eur. J. Biochem.* **59**, 377. **5.** Blethen, Boeker & Snell (1968) *J. Biol. Chem.* **243**, 1671. **6.** Ramirez & Whitaker (1999) *J. Agric. Food Chem.* **47**, 2218. **7.** Thaniyavarn, Taylor, Singh & Doyle (1982) *Infect. Immun.* **37**, 1101. **8.** Pressey (1968) *Biochim. Biophys. Acta* **159**, 414. **9.** Lampen (1971) *The Enzymes*, 3rd ed. (Boyer, ed.), **5**, 291. **10.** Obenland, Simmen, Boller & Wiemken (1993) *Plant Physiol.* **101**, 1331. **11.** Smith, Austen, Blumenthal & Nyc (1975) *The Enzymes*, 3rd ed. (Boyer, ed.), **11**, 293. **12.** Brink (1953) *Acta Chem. Scand.* **7**, 1090. **13.** Brown, Culver & Fisher

(1973) *Biochemistry* **12**, 4367. **14.** Zaugg, Walder & Klotz (1977) *J. Biol. Chem.* **252**, 8542. **15.** Kark, Kale, Tarassoff, Woods & Lessin (1978) *J. Clin. Invest.* **62**, 888. **16.** Oray & Norton (1980) *Biochem. Biophys. Res. Commun.* **95**, 624. **17.** Abiko (1970) *Meth. Enzymol.* **18A**, 354. **18.** Abiko (1967) *J. Biochem.* **61**, 300. **19.** Yang & Abeles (1987) *Biochemistry* **26**, 4076. **20.** Van Heyningen & Shemin (1971) *Biochem. J.* **124**, 68P. **21.** Furukawa, Yamada & Iwashima (1981) *Acta Vitaminol. Enzymol.* **3**, 145. **22.** Lewin (1967) *Proc. Soc. Exp. Biol. Med.* **124**, 39. **23.** Yokochi, Morita & Yagi (2003) *J. Agric. Food Chem.* **51**, 2733. **24.** Bray (1975) *The Enzymes*, 3rd ed. (Boyer, ed.), **12**, 299. **25.** Smith (1979) *Phytochemistry* **18**, 1447. **26.** Foster (1981) *J. Bacteriol.* **145**, 1002. **27.** Pressey & Avants (1980) *Plant Physiol.* **65**, 135. **28.** Krishnan, Blanchette & Okita (1985) *Plant Physiol.* **78**, 241. **29.** Rojo, Quiroga, Vattuone & Sampietro (1998) *Phytochemistry* **49**, 965. **30.** Prado, Vattuone, Fleischmacher & Sampietro (1985) *J. Biol. Chem.* **260**, 4952. **31.** Lopez, Vattuone & Sampietro (1988) *Phytochemistry* **27**, 3077. **32.** Obenland, Simmen, Boller & Wiemken (1993) *Plant Physiol.* **101**, 1331. **33.** Fonda (1992) *J. Biol. Chem.* **267**, 15978. **34.** Yamagata & Takeshima (1976) *J. Biochem.* **80**, 777. **35.** Matheson & Richardson (1973) *Phytochemistry* **17**, 195. **36.** Yen & Glassman (1967) *Biochim. Biophys. Acta* **146**, 35. **37.** Breton, Avanzi, Magagnin, *et al.* (2000) *Eur. J. Biochem.* **267**, 1092.

Pyridoxal-Alanine, *See N-(4-Pyridoxyl)-L-alanine*

Pyridoxal 5'-Diphospho-5'-adenosine, *See ADP-pyridoxal*

Pyridoxal 5'-Diphospho-1-α-D-glucose

Pyridoxal 5'-diphospho-1-α-D-glucose (PLDP-Glc; FW$_{free-acid}$ = 488.26 g/mol) is an active-site probe that inhibits glycogen phosphorylase *a*, AMP-activated (1), glycogen phosphorylase *b* (1,2), and UDP-glucose pyrophosphorylase (glucose-1-phosphate uridylyltransferase, in the latter case reacting with with five lysyl residues. **1.** Withers (1985) *J. Biol. Chem.* **260**, 841. **2.** Takagi, Fukui & Shimomura (1982) *Proc. Natl. Acad. Sci. U.S.A.* **79**, 3716. **3.** Fukui & Tanizawa (1997) *Meth. Enzymol.* **280**, 41. **4.** Kazuta, Tanizawa & Fukui (1991) *J. Biochem.* **110**, 708.

Pyridoxal 5'-Diphospho-5'-guanosine, *See GDP-Pyridoxal*

Pyridoxal 5'-Diphospho-5'-uridine, *See UDP-Pyridoxal*

Pyridoxal-Isoleucine, *See N-(4-Pyridoxyl)-L-isoleucine*

4-Pyridoxal Lactone, *See 4-Pyridoxolactone*

Pyridoxal 5'-Phosphate

This vitamin B_6-derived coenzyme (FW$_{free-acid}$ = 247.14 g/mol; CAS 853645-22-4), often abbreviated PLP, plays a central role in metabolism. In addition to facilitating aminotransfer reactions, PLP is a coenzyme in reactions involving: (a) loss of the α-proton, resulting in racemization, cyclization, or β-elimination/replacement (*e.g.*, alanine racemase, 1-aminocyclopropane-1-carboxylate synthase, and serine dehydratase, respectively); (b) loss of the α-carboxylate as carbon dioxide (*e.g.*, glutamate decarboxylase); (c) removal/replacement of a group by aldol cleavage (*e.g.*, threonine aldolase); and (d) catalysis via ketimine intermediates (*e.g.*, selenocysteine lyase). PLP is also often employed as an inhibitor and site-directed, conformationally sensitive reporter group for studying the environment of reactive amino groups within proteins. PLP forms aldimine adducts that can be reduced with nucleophilic reducing reagents such as sodium borohydride or sodium cyanoborohydride. Solid

PLP is light-sensitive, and solutions should be prepared fresh. There are four pK_a values associated with this coenzyme: a pK_a < 2.5 and another at 6.20 for the phosphate group, a pK_a of 4.14 for the hydroxyl group, and a pK_a of 8.69 for the pyridinium group. The coenzyme has an ultraviolet/visible spectrum: λ_{max} = 330 nm and 388 nm (ϵ = 2500 and 4900 $M^{-1}cm^{-1}$, respectively) in 50 mM phosphate buffer at pH 7.0; λ_{max} = 305 nm and 388 nm (ϵ = 1100 and 6550 $M^{-1}cm^{-1}$, respectively) in 100 mM NaOH. Pyridoxal 5'-phosphate may also be used as a photosensitizing agent to modify nearby aminoacyl residues (e.g., histidyl residues in 6-phosphogluconate dehydrogenase and tryptophanase). **Target(s):** acetylcholinesterase (143,160); acetyl-CoA carboxylase (2); acetyl-CoA synthetase, or acetate:CoA ligase (3); N-acetylneuraminate 4-O-acetyltransferase (190); acyl-[acyl-carrier-protein]:UDP-N-acetylglucosamine O-acyltransferase, followed by $NABH_4$ reduction (188); acylphosphatase (4,5); adenylate cyclase (6-8); [β-adrenergic-receptor] kinase (9,171); alanine racemase (10,11); alcohol dehydrogenase, followed by $NABH_4$ reduction (12); alcohol sulfotransferase (170; aldose reductase (13); allantoate deiminase (159); N-(5-amino-5-carboxypentanoyl)-L-cysteinyl-D-valine synthetase (14); 5-aminoleulinate synthase, in the presence of borohydride (191); aryl-acylamidase (160); aspartate carbamoyltransferase (15-17); aspartyl aminopeptidase (162); ATP-dependent deoxyribonuclease (18); BglI restriction endonuclease, type II site-specific deoxyribonuclease (19); biliverdin reductase (20); carbonic anhydrase, or carbonate dehydratase (151); casein kinase II (21,22); catechol O-methyltransferase (23); cathepsin B (24,161); cathepsin K (24,161); cathepsin L (24,161); cathepsin S (24,161); Ca^{2+}-transporting ATPase (25); cerebroside sulfotransferase (26); cholesterol monooxygenase, side-chain cleaving, or cytochrome P450^{scc} (27); ζ-crystallin/NADPH:quinone oxidoreductase (28); cytosine deaminase (157,158); dextransucrase (29,30,185); dipeptidyl-peptidase I, or cathepsin C (161); DNA-directed DNA polymerase (31-36,173); φ29 DNA polymerase (37); DNA topoisomerase I (36); DNA topoisomerase II (36); dopa decarboxylase (38); estrone sulfotransferase (170); exodeoxyribonuclease V (39); F_1 ATPase, or H^+-transporting two-sector ATPase (40); fatty-acid synthase, at enoyl-[acyl-carrier protein] reductase (41-43,189); fatty-acyl-CoA synthase, yeast, at 3-ketoacyl-[acyl-carrier protein] reductase step (43); flavin-containing monooxygenase (197); β-fructofuranosidase, or invertase (44,164); fructose-1,6-bisphosphate aldolase (15,45-47,152); fructose-6-phosphate 2-kinase/fructose-2,6-bisphosphatase (48); α1→3 fucosyltransferase (49); galactosylceramide sulfotransferase (169); glucan 1,4-α-glucosidase, or glucoamylase, weakly inhibited (166); glucose-6-phosphatase (50); glucose-1-phosphate adenylyltransferase (172); glucose-6-phosphate dehydrogenase (51-53); α-glucosidase (165); glucuronate reductase (54); glutamate decarboxylase (55,56); glutamate dehydrogenase (57-61); glutathione-disulfide reductase (62,63); glutathione synthetase, slowly inhibited (64); glyceraldehyde-3-phosphate dehydrogenase (65-67); glycerol-1,2-cyclic-phosphate phosphodiesterase (167); glycerol dehydrogenase (68); glycogen phosphorylase (69); glycolipid sulfotransferase (70); glycoprotein N-acetylglucosaminyltransferase (71); glycoprotein 6-α-L-fucosyltransferase (183); glycoprotein galactosyltransferase (71); glycoprotein sialyltransferase (71); hemoglobin S polymerization (72); hexokinase (73); HIV (human immunodeficiency virus) reverse transcriptase (74,75); H^+/K^+-exchanging ATPase (76); homoserine kinase (176); H^+-transporting ATPase, chloroplast (77); 2'-hydroxybiphenyl-2-sulfinate desulfinase (156); ω-hydroxy-fatty-acid:NADP$^+$ oxidoreductase (78); hydroxyindole-O-methyltransferase (acetylserotonin O-methyltransferase (79,80); hydroxymethylbilane synthase (180); 4-hydroxyphenylpyruvate dioxygenase (19,80); inositol-phosphate phosphatase (168); inositol-3-phosphate synthase, or inositol-1-phosphate synthase (81); integrase, HIV (173); α-ketoglutarate carrier (82); α-ketoglutarate dehydrogenase (83); kynurenine 3-monooxygenase (195,196); lactate dehydrogenase (84); lactoylglutathione lyase, or glyoxalase I (144); malate dehydrogenase (85); malate synthase (187); methanol:5-hydroxybenzimidazolylcobamide Co-methyltransferase (192); mevalonate kinase (86); murine leukemia virus reverse transcriptase (87); myosin ATPase (88); NAD(P)$^+$ transhydrogenase (89); Na^+/K^+-exchanging ATPase (90); neuraminidase (91); NMN nucleosidase, weakly inhibited (163); nucleoside-diphosphatase (92); pancreatic ribonuclease, or ribonuclease A (93-101); phenol 2-monooxygenase (102); phenol sulfotransferase, or aryl sulfotransferase (103); phenylalanyl-tRNA synthetase (104); phosphoenolpyruvate carboxylase (105); phosphofructokinase (106-108); phosphogluconate dehydrogenase (109-111); phosphoglucose isomerase (112-114); 3-phosphoglycerate kinase (115); phospholipase C (116); phosphomevalonate

kinase (174); phosphopantothenoylcysteine decarboxylase (117,118,153-155); phosphoribulokinase (119,177); porphobilinogen synthase, or δ-aminolevulinate dehydratase (149,150); pyridoxamine:oxaloacetate aminotransferase (178,179); pyridoxine 4-oxidase (120); pyruvate dehydrogenase (121); pyruvate kinase (122,175); pyruvate,orthophosphate dikinase (123.124); retinal oxidase (125); ribonuclease A (93-101); ribonucleoside-diphosphate reductase (126,193,194); ribosomal proteins (127); ribulose-1,5-bisphosphate carboxylase (128,129); RNA-directed DNA polymerase, or reverse transcriptase (130,131); RNA polymerase (132-134); serine-sulfate ammonia-lyase (145); starch phosphorylase (186); starch synthase (184); succinic-semialdehyde dehydrogenase (135); thiamin-phosphate pyrophosphorylase (136); thymidylate synthase (137,138); tissue-specific transcription factor HNF1 (139); tryptophanase (140); tryptophan synthase (15); uracilylalanine synthase (181,182); urocanase (141); uroporphyrinogen-III synthase (146-148); xanthine dehydrogenase, Drosophila melanogaster (142). **1.** Peterson, Sober & Meister (1953) Biochem. Prep. **3**, 34. **2.** Rainwater & Kolattukudy (1982) Arch. Biochem. Biophys. **213**, 372. **3.** Preston, Wall & Emerich (1990) Biochem. J. **267**, 179. **4.** Ramponi (1975) Meth. Enzymol. **42**, 409. **5.** Ramponi, Manao, Camici & White (1975) Biochim. Biophys. Acta **391**, 486. **6.** Tao (1974) Meth. Enzymol. **38**, 155. **7.** Ide (1971) Arch. Biochem. Biophys. **144**, 262. **8.** Yang & Epstein (1983) J. Biol. Chem. **258**, 3750. **9.** Benovic (1991) Meth. Enzymol. **200**, 351. **10.** Shibata, Shirasuna, Motegi, et al. (2000) Comp. Biochem. Physiol. B **126**, 599. **11.** Fujita, Okuma & Abe (1997) Fish. Sci. **63**, 440. **12.** Brändén, Jörnvall, Eklund & Furugren (1975) The Enzymes, 3rd ed. (Boyer, ed.), **11**, 103. **13.** Morjana, Lyons & Flynn (1989) J. Biol. Chem. **264**, 2912. **14.** Zhang, Wolfe & Demain (1992) Biochem. J. **283**, 691. **15.** Sigman & Mooser (1975) Ann. Rev. Biochem. **44**, 889. **16.** Jacobson & Stark (1973) The Enzymes, 3rd ed. (Boyer, ed.), **9**, 225. **17.** Cole & Yon (1987) Biochem. J. **248**, 403. **18.** Fujiyoshi, Nakayama & Anai (1981) J. Biochem. **89**, 1137. **19.** Wells, Klein & Singleton (1981) The Enzymes, 3rd ed. (Boyer, ed.), **14**, 157. **20.** Frydman, Tomaro, Rosenfeld & Frydman (1990) Biochim. Biophys. Acta **1040**, 119.**21.** Hathaway, Tuazon & Traugh (1983) Meth. Enzymol. **99**, 308. **22.** Hathaway & Traugh (1983) Meth. Enzymol. **99**, 317. **23.** Borchardt (1973) J. Med. Chem. **16**, 387. **24.** Matsui, Tsuzuki, Murata, et al. (2000) Biofactors **11**, 117. **25.** Yamagata, Daiho & Kanazawa (1993) J. Biol. Chem. **268**, 20930. **26.** Jungalwala, Natowicz, Chaturvedi & Newburg (2000) Meth. Enzymol. **311**, 94. **27.** Tsubaki, Iwamoto, Hiwatashi & Ichikawa (1989) Biochemistry **28**, 689. **28.** Rabbani & Duhaiman (1998) Biochim. Biophys. Acta **1388**, 175. **29.** Goyal & Katiyar (1995) J. Enzyme Inhib. **8**, 291. **30.** Thaniyavarn, Taylor, Singh & Doyle (1982) Infect. Immun. **37**, 1101. **31.** Modak (1976) Biochem. Biophys. Res. Commun. **71**, 180. **32.** Monaghan & Hay (1996) J. Biol. Chem. **271**, 24242. **33.** Modak & Dumaswala (1981) Biochim. Biophys. Acta **654**, 227. **34.** Oguro, Nagano & Mano (1979) Nucl. Acids Res. **7**, 727. **35.** Modak (1976) Biochemistry **15**, 3620. **36.** Matsubara, Matsumoto, Mizushina, Lee & Kato (2003) Int. J. Mol. Med. **12**, 51. **37.** Lázaro, Blanco & Salas (1995) Meth. Enzymol. **262**, 42. **38.** Pasqua, Dominici, Murgia, Poletti & Borri Voltattorni (1984) Biochem. Int. **9**, 437. **39.** Telander, Muskavitch & Linn (1981) The Enzymes, 3rd ed. (Boyer, ed.), **14**, 233. **40.** Satre, Lunardi, Dianoux, et al. (1986) Meth. Enzymol. **126**, 712. **41.** Mukherjee & Katiyar (1998) J. Enzyme Inhib. **13**, 217. **42.** Kolattukudy, Poulose & Buckner (1981) Meth. Enzymol. **71**, 103. **43.** Wakil & Stoops (1983) The Enzymes, 3rd ed. (Boyer, ed.), **16**, 3. **44.** Lampen (1971) The Enzymes, 3rd ed. (Boyer, ed.), **5**, 291. **45.** Shapiro, Enser, Pugh & Horecker (1968) Arch. Biochem. Biophys. **128**, 554. **46.** Davis, Ribereau-Gayon & Horecker (1971) Proc. Natl. Acad. Sci. U.S.A. **68**, 416. **47.** Horecker, Tsolas & Lai (1972) The Enzymes, 3rd ed. (Boyer, ed.), **7**, 213. **48.** Kitajima, Thomas & Uyeda (1985) J. Biol. Chem. **260**, 13995. **49.** Holmes (1992) Arch. Biochem. Biophys. **296**, 562. **50.** Gold & Widnell (1976) J. Biol. Chem. **251**, 1035. **51.** Domagk & Chilla (1975) Meth. Enzymol. **41**, 205. **52.** Milhausen & Levy (1975) Eur. J. Biochem. **50**, 453. **53.** Domagk, Chilla, Domschke, Engel & Sörensen (1969) Hoppe-Seyler's Z. physiol. Chem. **350**, 626. **54.** Flynn, Cromlish & Davidson (1982) Meth. Enzymol. **89**, 501. **55.** Cozzani & Jori (1980) Biochim. Biophys. Acta **623**, 84. **56.** Cozzani, Santoni, Jori, Gennari & Tamburro (1974) Biochem. J. **141**, 463. **57.** Anderson, Anderson & Churchich (1966) Biochemistry **5**, 2893. **58.** Bedino (1976) Ital. J. Biochem. **25**, 304. **59.** Chen & Engel (1975) Biochem. J. **149**, 619. **60.** Smith, Austen, Blumenthal & Nyc (1975) The Enzymes, 3rd ed. (Boyer, ed.), **11**, 293. **61.** Brown, Culver & Fisher (1973) Biochemistry **12**, 4367. **62.** Pandey, Iyengar & Katiyar (1997) J. Enzyme Inhib. **12**, 143. **63.** Pandey & Katiyar (1995) Biochem. Mol. Biol. Int. **36**, 347. **64.** Hibi, Kato, Nishioka, et al. (1993) Biochemistry **32**, 1548. **65.** Ghosh, Mukherjee, Ray & Ray (2001) Eur. J. Biochem. **268**, 6037. **66.** Ronchi, Zapponi & Ferri

(1969) *Eur. J. Biochem.* **8**, 325. **67**. Harris & Waters (1976) *The Enzymes*, 3rd ed. (Boyer, ed.), **13**, 1. **68**. Pandey & Iyengar (2002) *J. Enzyme Inhib. Med. Chem.* **17**, 49. **69**. Avramovic-Zikic & Madsen (1972) *J. Biol. Chem.* **247**, 6999 and 7005. **70**. Kamio, Honke & Makita (1995) *Glycoconj. J.* **12**, 762. **71**. Stibler & Borg (1991) *Scand. J. Clin. Lab. Invest.* **51**, 43. **72**. Kark, Tarassoff & Bongiovanni (1983) *J. Clin. Invest.* **71**, 1224. **73**. Grillo (1968) *Enzymologia* **34**, 7. **74**. Basu, Tirumalai & Modak (1989) *J. Biol. Chem.* **264**, 8746. **75**. Mitchell & Cooperman (1992) *Biochemistry* **31**, 7707. **76**. Maeda, Tagaya & Futai (1988) *J. Biol. Chem.* **263**, 3652. **77**. Horbach, Meyer & Bickel-Sandkotter (1991) *Eur. J. Biochem.* **200**, 449. **78**. Kolattukudy & Agarwal (1981) *Meth. Enzymol.* **71**, 411. **79**. Borchardt, Wu & Wu (1978) *Biochem. Pharmacol.* **27**, 120. **80**. Nir, Hirschmann & Sulman (1976) *Biochem. Pharmacol.* **25**, 581. **81**. Naccarato, Ray & Wells (1974) *Arch. Biochem. Biophys.* **164**, 194. **82**. Natuzzi, Daddabbo, Stipani, *et al.* (1999) *J. Bioenerg. Biomembr.* **31**, 535. **83**. Ostrovtsova (1998) *Acta Biochim. Pol.* **45**, 1031. **84**. Gould & Engel (1980) *Biochem. J.* **191**, 365. **85**. Banaszak & Bradshaw (1975) *The Enzymes*, 3rd ed. (Boyer, ed.), **11**, 369. **86**. Soler, Jabalquinto & Beytía (1979) *Int. J. Biochem.* **10**, 931. **87**. Basu, Nanduri, Gerard & Modak (1988) *J. Biol. Chem.* **263**, 1648. **88**. Sarkozi & Szilagyi (1989) *Acta Biochim. Biophys. Hung.* **24**, 317. **89**. Yamaguchi & Hatefi (1985) *Arch. Biochem. Biophys.* **243**, 20. **90**. Hinz & Kirley (1990) *J. Biol. Chem.* **265**, 10260. **91**. Bellini, Tomasi & Dallocchio (1993) *Biochim. Biophys. Acta* **1161**, 323. **92**. Sano, Matsuda & (1988) *Eur. J. Biochem.* **171**, 231. **93**. Means & Feeney (1971) *J. Biol. Chem.* **246**, 5532. **94**. Raetz & Auld (1972) *Biochemistry* **11**, 2229. **95**. Stewart & Stevenson (1973) *Biochem. J.* **135**, 427. **96**. Borisova, Matrosov, Shlyapnikov & Karpeiskii (1974) *Mol. Biol.* **8**, 228. **97**. Dudkin, Karabachyan, Borisova, *et al.* (1975) *Biochim. Biophys. Acta* **386**, 275. **98**. Riquelme, Brown & Marcus (1975) *Int. J. Pept. Protein Res.* **7**, 37. **99**. Moroz, Kondakov, Stepuro & Iaroshevich (1987) *Biokhimiia* **52**, 550. **100**. Xiao & Zhou (1996) *Biochim. Biophys. Acta* **1294**, 1. **101**. Blackburn & Moore (1982) *The Enzymes*, 3rd ed. (Boyer, ed.), **15**, 317. **102**. Neujahr & Kjellén (1980) *Biochemistry* **19**, 4967. **103**. Bartzatt & Beckmann (1994) *Biochem. Pharmacol.* **47**, 2087. **104**. Baltzinger, Fasiolo & Remy (1979) *Eur. J. Biochem.* **97**, 481. **105**. Podesta, Iglesias & Andreo (1986) *Arch. Biochem. Biophys.* **246**, 546. **106**. Reuter, Sellschopp, Xylander & Hofmann (1983) *Biomed. Biochim. Acta* **42**, 1067. **107**. Uyeda (1969) *Biochemistry* **8**, 2366. **108**. Bloxham & Lardy (1973) *The Enzymes*, 3rd ed. (Boyer, ed.), **8**, 239. **109**. Rippa, Spanio & Pontremoli (1967) *Arch. Biochem. Biophys.* **118**, 48. **110**. Rippa, Spanio & Pontremoli (1966) *Boll. Soc. Ital. Biol. Sper.* **42**, 748. **111**. Rippa & Pontremoli (1969) *Arch. Biochem. Biophys.* **133**, 112. **112**. Howell & Schray (1981) *Mol. Cell Biochem.* **37**, 101. **113**. Schnackerz & Noltmann (1971) *Biochemistry* **10**, 4837. **114**. Hathaway & Noltmann (1977) *Arch. Biochem. Biophys.* **179**, 24. **115**. Markland, Bacharach, Weber, *et al.* (1975) *J. Biol. Chem.* **250**, 1301. **116**. Dennis (1983) *The Enzymes*, 3rd ed. (Boyer, ed.), **16**, 307. **117**. Abiko (1970) *Meth. Enzymol.* **18A**, 354. **118**. Scandurra, Moriggi, Consalvi & Politi (1979) *Meth. Enzymol.* **62**, 245. **119**. Kiesow, Lindsley & Bless (1977) *J. Bacteriol.* **130**, 20. **120**. Choi, Churchich, Zaiden & Kwok (1987) *J. Biol. Chem.* **262**, 12013. **121**. Stepp & Reed (1985) *Biochemistry* **24**, 7187. **122**. Kayne (1973) *The Enzymes*, 3rd ed. (Boyer, ed.), **8**, 353. **123**. Roeske & Chollet (1987) *J. Biol. Chem.* **262**, 12575. **124**. Phillips, Goss & Wood (1983) *Biochemistry* **22**, 2518. **125**. Huang & Ichikawa (1995) *Biochim. Biophys. Acta* **1243**, 431. **126**. Cory & Mansell (1975) *Cancer Res.* **35**, 390. **127**. Ohsawa & Gualerzi (1983) *J. Biol. Chem.* **258**, 150. **128**. Schloss, Phares, Long, *et al.* (1982) *Meth. Enzymol.* **90**, 522. **129**. Bhagwat & McFadden (1983) *Arch. Biochem. Biophys.* **223**, 610. **130**. Verma (1981) *The Enzymes*, 3rd ed. (Boyer, ed.), **14**, 87. **131**. Venegas, Martial & Valenzuela (1973) *Biochem. Biophys. Res. Commun.* **55**, 1053. **132**. Martial, Zaldivar, Bull, Venegas & Valenzuela (1975) *Biochemistry* **14**, 4907. **133**. Venegas, Martial & Valenzuela (1973) *Biochem. Biophys. Res. Commun.* **55**, 1053. **134**. Lewis & Burgess (1982) *The Enzymes*, 3rd ed. (Boyer, ed.), **15**, 109. **135**. Blaner & Churchich (1979) *J. Biol. Chem.* **254**, 1794. **136**. Leder (1970) *Meth. Enzymol.* **18A**, 207. **137**. Rao (1998) *Indian J. Biochem. Biophys.* **35**, 229. **138**. Chen, Daron & Aull (1989) *Int. J. Biochem.* **21**, 1217. **139**. Oka, Sugitatsu, Nordin, *et al.* (2001) *Biochim. Biophys. Acta* **1568**, 189. **140**. Nihira, Toraya & Fukui (1979) *Eur. J. Biochem.* **101**, 341. **141**. Hug, Hunter & O'Donnell (1977) *Photochem. Photobiol.* **25**, 175. **142**. Bray (1975) *The Enzymes*, 3rd ed. (Boyer, ed.), **12**, 299. **143**. Ghag, Wright & Moudgil (1986) *Biochim. Biophys. Acta* **881**, 30. **144**. Baskaran & Balasubramanian (1987) *Biochim. Biophys. Acta* **913**, 377. **145**. Murakoshi, Sanda & Haginiwa (1977) *Chem. Pharm. Bull.* **25**, 1829. **146**. Stamford, Capretta & Battersby (1995) *Eur. J. Biochem.* **231**, 236. **147**. Hart & Battersby (1985) *Biochem. J.* **232**, 151. **148**. Smythe & Williams (1988) *Biochem. J.* **253**, 275. **149**. Anderson & Desnick (1979) *J.*

Biol. Chem. **254**, 6924. **150**. Van Heyningen & Shemin (1971) *Biochem. J.* **124**, 68P. **151**. Gill, Fedorka-Cray, Tweten & Sleeper (1984) *Arch. Microbiol.* **138**, 113. **152**. Pasha & Salahuddin (1977) *Biochim. Biophys. Acta* **483**, 435. **153**. Scandurra, Barboni, Granata, Pensa & Costa (1974) *Eur. J. Biochem.* **49**, 1. **154**. Abiko (1967) *J. Biochem.* **61**, 300. **155**. Yang & Abeles (1987) *Biochemistry* **26**, 4076. **156**. Watkins, Rodriguez, Schneider, *et al.* (2003) *Arch. Biochem. Biophys.* **415**, 14. **157**. Kim & Yu (1998) *J. Microbiol. Biotechnol.* **8**, 581. **158**. Yu, Kim & Kim (1998) *J. Microbiol.* **36**, 39. **159**. Xu, Zhou & Huang (2004) *Zhi Wu Sheng Li Yu Fen Zi Sheng Wu Xue Xue Bao* **30**, 460. **160**. Majumdar & Balasubramanian (1984) *Biochemistry* **23**, 4088. **161**. Katunuma, Matsui, Inubushi, *et al.* (2000) *Biochem. Biophys. Res. Commun.* **267**, 850. **162**. Zhang, Zhang, Wang, Yang & Lu (1998) *Ann. N. Y. Acad. Sci.* **864**, 621. **163**. Foster (1981) *J. Bacteriol.* **145**, 1002. **164**. de la Vega, Cejudo & Paneque (1991) *Enzyme Microb. Technol.* **13**, 267. **165**. Im & Henson (1995) *Carbohydr. Res.* **277**, 145. **166**. Han & Yu (2005) *J. Microbiol. Biotechnol.* **15**, 239. **167**. Clarke & Dawson (1978) *Biochem. J.* **173**, 579. **168**. Kim, Hong, Eum, *et al.* (2005) *J. Biochem. Mol. Biol.* **38**, 58. **169**. Kamio, Honke & Makita (1995) *Glycoconjugate J.* **12**, 762. **170**. Allali-Hassani, Pan, Dombrovski, *et al.* (2007) *PLoS Biol.* **5**, e97. **171**. Benovic, Stone, Caron & Lefkowitz (1989) *J. Biol. Chem.* **264**, 6707. **172**. Meyer, Borra, Igarashi, Lin & Springsteel (1999) *Arch. Biochem. Biophys.* **372**, 179. **173**. Acel, Udashkin, Wainberg & Faust (1998) *J. Virol.* **72**, 2062. **174**. Bazaes, Beytía, Jabalquinto, *et al.* (1980) *Biochemistry* **19**, 2305. **175**. Chan & Sim (2005) *Biochem. Biophys. Res. Commun.* **326**, 188. **176**. Huo & Viola (1996) *Arch. Biochem. Biophys.* **330**, 373. **177**. Runquist, Harrison & Miziorko (1999) *Biochemistry* **38**, 13999. **178**. Wu & Mason (1964) *J. Biol. Chem.* **239**, 1492. **179**. Wada & Snell (1962) *J. Biol. Chem.* **237**, 127. **180**. Hädener, Alefounder, Hart, Abell & Battersby (1990) *Biochem. J.* **271**, 487. **181**. Ahmmad, Maskall & Brown (1984) *Phytochemistry* **23**, 265. **182**. Murakoshi, Ikegami, Ookawa, *et al.* (1978) *Phytochemistry* **17**, 1571. **183**. Kaminska, Wisniewska & Koscielak (2003) *Biochimie* **85**, 303. **184**. Gao, Keeling, Shibles & Guan (2004) *Arch. Biochem. Biophys.* **427**, 1. **185**. Majumder, Purama & Goyal (2007) *Indian J. Microbiol.* **47**, 197. **186**. Matheson & Richardson (1973) *Phytochemistry* **17**, 195. **187**. Miernyk & Trelease (1981) *Phytochemistry* **20**, 2657. **188**. Wyckoff & Raetz (1999) *J. Biol. Chem.* **274**, 27047. **189**. Wakil, Stoops & Joshi (1983) *Ann. Rev. Biochem.* **52**, 537. **190**. Iwersen, Dora, Kohla, Gasa & Schauer (2003) *Biol. Chem.* **384**, 1035. **191**. Nandi (1978) *J. Biol. Chem.* **253**, 8872. **192**. van der Meijden, te Brömmelstroet, Poirot, van der Drift & Vogels (1984) *J. Bacteriol.* **160**, 629. **193**. Lammers & Follmann (1983) *Struct. Bonding* **54**, 27. **194**. Huszar & Bacchetti (1981) *J. Virol.* **37**, 580. **195**. Han, Calvo, Marinotti, *et al.* (2003) *Insect Mol. Biol.* **12**, 483. **196**. Breton, Avanzi, Magagnin, *et al.* (2000) *Eur. J. Biochem.* **267**, 1092. **197**. Wu & Ichikawa (1995) *Eur. J. Biochem.* **229**, 749. **198**. Lindstedt & Rundgren (1982) *J. Biol. Chem.* **257**, 11922.

Pyridoxal-5'-phosphate-6-azo(benzene-2,4-disulfonic Acid)

This reagent (FW_{ion} = 330.35 g/mol), often abbreviated PPADS and also known as 4-[[4-formyl-5-hydroxy-6-methyl-3-[(phosphonooxy)methyl]-2-pyridinyl]-azo]-1,3-benzenedisulfonic acid, is a functionally selective antagonist of P2 purinoceptors. **Target(s):** apyrase (1); ecto-ATPase (1-5); nucleotide diphosphatase, ecto-nucleotide pyrophosphatase (5-7); and P2 purinoceptor (8). **1**. Ziganshin, Zigashina, Bodin, Bailey & Burnstock (1995) *Biochem. Mol. Biol. Int.* **36**, 863. **2**. Yegutkin & Burnstock (2000) *Biochim. Biophys. Acta* **1466**, 234. **3**. Chen & Lin (1997) *Biochem. Biophys. Res. Commun.* **233**, 442. **4**. Chen, Lee & Lin (1996) *Brit. J. Pharmacol.* **119**, 1628. **5**. Grobben, Claes, Roymans, *et al.* (2000) *Brit. J. Pharmacol.* **130**, 139. **6**. Grobben, Anciaux, Roymans, *et al.* (1999) *J. Neurochem.* **72**, 826. **7**. Vollmayer, Clair, Goding, *et al.* (2003) *Eur. J. Biochem.* **270**, 2971. **8**. Lambrecht, Friebe, Grimm, *et al.* (1992) *Eur. J. Pharmacol.* **217**, 217.

Pyridoxal-5'-phosphate-γ-L-Glutamyl Hydrazone

This amino acyl hydrazone (FW$_{ion}$ = 386.26 g/mol) inhibits glutamate decarboxylase (1). **1**. Tapia & Salazar (1991) *Neurochem. Res.* **16**, 263.

Pyridoxal-5'-phosphate-6-(2'-naphthylazo-6'-nitro-4',8'-disulfonate, See PPNDS

Pyridoxal 5'-Phosphate Oxime

This oxime of pyridoxal 5'-phosphate (FW = 259.13 g/mol), which is practically insoluble in water, ethanol, and diethyl ether, inhibits aromatic-L-amino-acid decarboxylase (1), *or* DOPA decarboxylase (1), pyridoxamine-phosphate oxidase (2), and pyridoxine 4-oxidase (3). **1**. Gonnard, Camier,Boigne, Duhault & Nguyen-Philippon (1965) *Enzymologia* **28**, 211. **2**. Merrill, Kazarinoff, Tsuge, Horiike & McCormick (1979) *Meth. Enzymol.* **62**, 568. **3**. Korytnyk & Ikawa (1970) *Meth. Enzymol.* **18A**, 524.

Pyridoxal 5'-Phosphate Oxime-O-acetic Acid

This pyridoxal 5'-phosphate derivative (FW$_{ion}$ = 317.17 g/mol), which is readily prepared by the reaction of pyridoxal 5'-phosphate with (aminooxy)acetic acid (*i.e.*, carboxymethoxylamine), inhibits glutamate decarboxylase and is believed to be an analogue of the aldimine intermediate. **1**. Tapia & Sandoval (1971) *J. Neurochem.* **18**, 2051.

Pyridoxal Semicarbazone

This semicarbazone (FW = 224.22 g/mol) inhibits pyridoxal kinase (1-3). **1**. Korytnyk & Ikawa (1970) *Meth. Enzymol.* **18A**, 524. **2**. Webb (1966)

Enzyme and Metabolic Inhibitors, vol. **2**, p. 564, Academic Press, New York. **3**. McCormick & Snell (1961) *J. Biol. Chem.* **236**, 2085.

Pyridoxal 5'-Sulfate

This pyridoxal 5'-phosphate analogue (FW$_{ion}$ = 246.22 g/mol) inhibits glutamate decarboxylase (1) and hemoglobin S polymerization (2). **1**. Korytnyk & Ikawa (1970) *Meth. Enzymol.* **18A**, 524. **2**. Benesch, Benesch & Yung (1974) *Proc. Natl. Acad. Sci. U.S.A.* **71**, 1504.

Pyridoxal 5'-Triphospho-5'-adenosine, See ATP-Pyridoxal

Pyridoxal 5'-Triphospho-5'-guanosine, See GTP-pyridoxal

Pyridoxamine

This photosensitive vitamer (FW$_{dihydrochloride}$ = 241.12 g/mol; CAS 524-36-7; Solubility: ~ 0.5 g/mL H$_2$O; pK_a values: 3.54 (the unusually acidic phenolic OH group), 8.21 (pyridinium NH), and 10.63 (primary amine)), also known as 4-(aminomethyl)-5-(hydroxymethyl)-2-methylpyridin-3-ol, is formed during digestion by enzymatic hydrolysis of dietary pyridoxamine-5-P and absorbed within the jejunum. The ionized phenolic hydroxyl and the methylamino groups probably account for the ability of pyridoxamine to coordinate with Cu(II) and Fe(II) ions. The 3'-hydroxyl group of pyridoxamine is thought to play an important role in scavenging free radicals. Pyridoxamine also blocks the formation of advanced glycation end-products formed by the Maillard reaction and associated with medical complications in diabetess. Pyridoxamine is believed to trap intermediates in the formation of Amadori products released from glycated proteins. Pyridoxamine is commonly supplied as the dihydrochloride and should be stored in tightly closed containers in desiccator. **Target(s):** alanine racemase (1); amine oxidase (2,3); dextransucrase (4); β-fructofuranosidase, *or* invertase (5-8); hydroxyacylglutathione hydrolase, *or* glyoxalase II (9); lactoylglutathione lyase, *or* glyoxalase I (9); pyridoxal dehydrogenase (10,11); pyridoxamine-5-phosphate oxidase (12); starch phosphorylase (13); and tyrosinase (14). **1**. Olivard & Snell (1955) *J. Biol. Chem.* **213**, 203. **2**. Tabor, Tabor & Rosenthal (1955) *Meth. Enzymol.* **2**, 390. **3**. Zeller (1951) *The Enzymes*, 1st ed. (Sumner & Myrbäck, eds.), **2** (part 1), 536. **4**. Thaniyavarn, Taylor, Singh & Doyle (1982) *Infect. Immun.* **37**, 1101. **5**. Pressey & Avants (1980) *Plant Physiol.* **65**, 135. **6**. Krishnan, Blanchette & Okita (1985) *Plant Physiol.* **78**, 241. **7**. De la Vega, Cejudo & Paneque (1991) *Enzyme Microb. Technol.* **13**, 267. **8**. Lopez, Vattuone & Sampietro (1988) *Phytochemistry* **27**, 3077. **9**. Oray & Norton (1980) *Biochem. Biophys. Res. Commun.* **95**, 624. **10**. Burg (1970) *Meth. Enzymol.* **18A**, 634. **11**. Burg & Snell (1969) *J. Biol. Chem.* **244**, 2585. **12**. Pogell (1963) *Meth. Enzymol.* **6**, 331. **13**. Matheson & Richardson (1973) *Phytochemistry* **17**, 195. **14**. Yokochi, Morita & Yagi (2003) *J. Agric. Food Chem.* **51**, 2733.

Pyridoxamine 5-Phosphate

This vitamin B_6 metabolite (FW = 248.18 g/mol) is derived biosynthetically from pyridoxamine and by transamination from pyridoxal 5'-phosphate. The hygroscopic solid is light-sensitive, and solutions should be prepared freshly, despite their stability when stored cold and in the absence of light. Solutions are more labile under alkaline conditions. There are four pK_a values associated with this coenzyme: a value less than 2.5 and a value of 5.76 for the phosphate group, 3.69 for the hydroxyl group, 8.61 for the pyridinium group, and 10.92 for the primary amine. The coenzyme has an ultraviolet/visible spectrum: λ_{max} = 254 nm and 327 nm (ε = 5200 and 9400 $M^{-1}cm^{-1}$, respectively) at pH 7.2; λ_{max} = 244 nm and 312 nm (ε = 7500 and 8300 $M^{-1}cm^{-1}$, respectively) at pH 10. (*See also Pyridoxal 5'-Phosphate*) **Target(s):** dextransucrase (1); glutamine:fructose-6-phosphate aminotransferase, isomerizing (2,3); phenol 2-monooxygenase, weakly inhibited (4); phosphomevalonate kinase (5); phosphoprotein phosphatase, weakly inhibited (6); porphobilinogen synthase, *or* δ-aminolevulinate dehydratase (7); pyridoxal kinase (8); pyridoxal phosphatase, also as an alternative substrate (9); pyridoxamine:oxaloacetate aminotransferase (10,11); starch phosphorylase (12); tyrosinase (13); and tyrosine aminotransferase (14). **1.** Thaniyavarn, Taylor, Singh & Doyle (1982) *Infect. Immun.* **37**, 1101. **2.** Pogell (1962) *Meth. Enzymol.* **5**, 408. **3.** Gryder & Pogell (1960) *J. Biol. Chem.* **235**, 558. **4.** Neujahr & Kjellén (1980) *Biochemistry* **19**, 4967. **5.** Bazaes, Beytía, Jabalquinto, *et al.* (1980) *Biochemistry* **19**, 2305. **6.** Zhou, Clemens, Hakes, Barford & Dixon (1993) *J. Biol. Chem.* **268**, 17754. **7.** Van Heyningen & Shemin (1971) *Biochem. J.* **124**, 68P. **8.** White & Dempsey (1970) *Biochemistry* **9**, 4057. **9.** Fonda (1992) *J. Biol. Chem.* **267**, 15978. **10.** Wu & Mason (1964) *J. Biol. Chem.* **239**, 1492. **11.** Wada & Snell (1962) *J. Biol. Chem.* **237**, 127. **12.** Matheson & Richardson (1973) *Phytochemistry* **17**, 195. **13.** Yokochi, Morita & Yagi (2003) *J. Agric. Food Chem.* **51**, 2733. **14.** Hayashi, Granner & Tomkins (1967) *J. Biol. Chem.* **242**, 3998.

4-Pyridoxic Acid

This oxidized derivative (FW = 171.15 g/mol) of pyridoxal is the chief catabolic product of pyridoxine, pyridoxal, and pyridoxamine degradation; it can be isolated from human urine. The carboxyl group has a pK_a value of 5.50 while the phenolic OH has a value of 9.75. 4-Pyridoxic acid has a blue fluorescence which is maximal between pH 3 and 4. Heating 4-pyridoxic acid in 0.5 N acid will convert the metabolite to its lactone, which has an even stronger fluorescence. **Target(s):** dextransucrase (1); glucose dehydrogenase (2,3); pyridoxal dehydrogenase (4,5); and pyridoxamine-phosphate oxidase (6). **1.** Thaniyavarn, Taylor, Singh & Doyle (1982) *Infect. Immun.* **37**, 1101. **2.** Strecker (1955) *Meth. Enzymol.* **1**, 335. **3.** Brink (1953) *Acta Chem. Scand.* **7**, 1090. **4.** Burg (1970) *Meth. Enzymol.* **18A**, 634. **5.** Burg & Snell (1969) *J. Biol. Chem.* **244**, 2585. **6.** Wada & Snell (1961) *J. Biol. Chem.* **236**, 2089.

5-Pyridoxic Acid

This pyridoxine metabolite (FW = 171.15 g/mol), also known as 3-hydroxy-4-hydroxymethyl-2-methylpyridine-5-carboxylate and isopyridoxic acid, inhibits 3-hydroxy-2-methylpyridinecarboxylate dioxygenase (1-3). **1.** Burg (1970) *Meth. Enzymol.* **18A**, 634. **2.** Sparrow, Ho, Sundaram, *et al.* (1969) *J. Biol. Chem.* **244**, 2590. **3.** Kishore & Snell (1981) *J. Biol. Chem.* **256**, 4234.

Pyridoxine

This essential nutritional factor vitamin B_6 (FW$_{free-base}$ = 169.18 g/mol), also called pyridoxol, is a white solid that is typically supplied as the hydrochloride (the two pK_a values are 4.94 and 8.89). (*Note*: The term vitamin B_6 actually refers to a mixture of pyridoxine, pyridoxal, and pyridoxamine.) Pyridoxine is stable in light and air, but is somewhat photolabile in neutral and alkaline solutions. It is soluble in water (1 g/4.5 mL) and has λ_{max} values at 253 and 325 nm at pH 7 in phosphate buffer (ε = 3700 and 7100 $M^{-1}cm^{-1}$, respectively). Pyridoxine is a substrate of the reactions catalyzed by pyridoxine 4-dehydrogenase, pyridoxine 4-oxidase, pyridoxine 5-dehydrogenase, and pyridoxine 5'-O-β-D-glucosyltransferase. **Target(s):** alanine racemase (1); arginine decarboxylase (2,3); dextransucrase (4); β-fructofuranosidase, *or* invertase (5-10); glycogen phosphorylase (11); hydroxymethylpyrimidine kinase, also alternative substrate, K_i = 2.7 μM (12); pyridoxal dehydrogenase (13,14); pyridoxamine-5-phosphate oxidase (15); pyridoxamine: pyruvate aminotransferase (13,16); tyrosinase (17). **1.** Olivard & Snell (1955) *J. Biol. Chem.* **213**, 203. **2.** Das, Bhaduri, Bose & Ghosh (1996) *J. Plant Biochem. Biotechnol.* **5**, 123. **3.** Choudhuri & Ghosh (1982) *Agric. Biol. Chem.* **46**, 739. **4.** Thaniyavarn, Taylor, Singh & Doyle (1982) *Infect. Immun.* **37**, 1101. **5.** Liu, Huang, Chang & Sung (2006) *Food Chem.* **96**, 62. **6.** Pressey & Avants (1980) *Plant Physiol.* **65**, 135. **7.** Krishnan, Blanchette & Okita (1985) *Plant Physiol.* **78**, 241. **8.** Rojo, Quiroga, Vattuone & Sampietro (1998) *Phytochemistry* **49**, 965. **9.** Prado, Vattuone, Fleischmacher & Sampietro (1985) *J. Biol. Chem.* **260**, 4952. **10.** Lopez, Vattuone & Sampietro (1988) *Phytochemistry* **27**, 3077. **11.** Klinova, Klinov, Kurganov, Mikhno & Baliakina (1988) *Bioorg. Khim.* **14**, 1520. **12.** Mizote & Nakayama (1989) *Biochim. Biophys. Acta* **991**, 109. **13.** Burg (1970) *Meth. Enzymol.* **18A**, 634. **14.** Burg & Snell (1969) *J. Biol. Chem.* **244**, 2585. **15.** Pogell (1963) *Meth. Enzymol.* **6**, 331. **16.** Dempsey & Snell (1963) *Biochemistry* **2**, 1414. **17.** Yokochi, Morita & Yagi (2003) *J. Agric. Food Chem.* **51**, 2733.

N-(4-Pyridoxyl)-L-alanine

This amino acid-vitamin B_6 conjugate (FW$_{ion}$ = 240.26 g/mol) str0ngly inhibits pyridoxamine:pyruvate aminotransferase, K_i = 0.18 μM (1-4). **1.** Dempsey & Snell (1963) *Biochemistry* **2**, 1414. **2.** Korytnyk & Ikawa (1970) *Meth. Enzymol.* **18A**, 524. **3.** Burg (1970) *Meth. Enzymol.* **18A**, 634. **4.** Wolfenden (1977) *Meth. Enzymol.* **46**, 15.

Pyridoxal-5'-phosphate-6-azo(benzene-2,4-disulfonic Acid)

This PLP adduct (FW = 507.35 g/mol), often abbreviated PPADS and also known as 4-[[4-formyl-5-hydroxy-6-methyl-3-[(phosphonooxy)methyl]-2-pyridinyl]azo]-1,3-benzenedisulfonic acid, is a functionally selective antagonist of P2 purinoceptors. **Target(s):** apyrase (1); ecto-ATPase (1-5); nucleotide diphosphatase, ecto-nucleotide pyrophosphatase (5-7); and P2 purinoceptor (8). **1**. Ziganshin, Zigashina, Bodin, Bailey & Burnstock (1995) *Biochem. Mol. Biol. Int.* **36**, 863. **2**. Yegutkin & Burnstock (2000) *Biochim. Biophys. Acta* **1466**, 234. **3**. Chen & Lin (1997) *Biochem. Biophys. Res. Commun.* **233**, 442. **4**. Chen, Lee & Lin (1996) *Brit. J. Pharmacol.* **119**, 1628. **5**. Grobben, Claes, Roymans, *et al.* (2000) *Brit. J. Pharmacol.* **130**, 139. **6**. Grobben, Anciaux, Roymans, *et al.* (1999) *J. Neurochem.* **72**, 826. **7**. Vollmayer, Clair, Goding, *et al.* (2003) *Eur. J. Biochem.* **270**, 2971. **8**. Lambrecht, Friebe, Grimm, *et al.* (1992) *Eur. J. Pharmacol.* **217**, 217.

Pyridoxal-5'-phosphate-6-(2'-naphthylazo-6'-nitro-4',8'-disulfonate, See PPNDS

Pyridoxal 5'-Phosphate Oxime

This oxime of pyridoxal 5'-phosphate (FW = 260.14 g/mol), which is practically insoluble in water, ethanol, and diethyl ether, inhibits aromatic-L-amino-acid decarboxylase, *or* DOPA decarboxylase (1), pyridoxamine-phosphate oxidase (2), and pyridoxine 4-oxidase (3). **1**. Gonnard, Camier,Boigne, Duhault & Nguyen-Philippon (1965) *Enzymologia* **28**, 211. **2**. Merrill, Kazarinoff, Tsuge, Horiike & McCormick (1979) *Meth. Enzymol.* **62**, 568. **3**. Korytnyk & Ikawa (1970) *Meth. Enzymol.* **18A**, 524.

Pyridoxal Semicarbazone

This semicarbazone (FW = 224.22 g/mol), formed by the combination of pyridoxal and semicarbazide, inhibits pyridoxal kinase (1-3). **1**. Korytnyk & Ikawa (1970) *Meth. Enzymol.* **18A**, 524. **2**. Webb (1966) *Enzyme and Metabolic Inhibitors*, vol. 2, p. 564, Academic Press, New York. **3**. McCormick & Snell (1961) *J. Biol. Chem.* **236**, 2085.

Pyridoxamine

This key metabolite (FW$_{dihydrochloride}$ = 241.12 g/mol; CAS 524-36-7; photosensitive; deliquescent, Solubility ~0.5 g/mL H$_2$O, pK_a values of 3.54 (the unusually acidic phenolic OH group), 8.21 (pyridinium NH), and 10.63 (primary amine)) is formed from vitamin B$_6$. The ionized phenolic hydroxyl and the methylamino groups probably account for its ability to coordinate with Cu(II) and Fe(II) ions. The 3'-hydroxyl group is thought to play an important role in scavenging free radicals. Pyridoxamine also blocks formation of advanced glycation end-products that are formed by the Maillard reaction. Pyridoxamine is believed to trap intermediates in the formation of Amadori products released from glycated proteins. Pyridoxamine is commonly supplied as the dihydrochloride and should be stored in tightly closed containers in desiccator. **Target(s):** alanine racemase (1); amine oxidase (2,3); dextransucrase (4); β-fructofuranosidase, *or* invertase (5-8); hydroxyacylglutathione hydrolase, *or* glyoxalase II (9);

lactoylglutathione lyase, *or* glyoxalase I (9); pyridoxal dehydrogenase (10,11); pyridoxamine-5-phosphate oxidase (12); starch phosphorylase (13); and tyrosinase (14). **1**. Olivard & Snell (1955) *J. Biol. Chem.* **213**, 203. **2**. Tabor, Tabor & Rosenthal (1955) *Meth. Enzymol.* **2**, 390. **3**. Zeller (1951) *The Enzymes*, 1st ed. (Sumner & Myrbäck, eds.), **2** (part 1), 536. **4**. Thaniyavarn, Taylor, Singh & Doyle (1982) *Infect. Immun.* **37**, 1101. **5**. Pressey & Avants (1980) *Plant Physiol.* **65**, 135. **6**. Krishnan, Blanchette & Okita (1985) *Plant Physiol.* **78**, 241. **7**. De la Vega, Cejudo & Paneque (1991) *Enzyme Microb. Technol.* **13**, 267. **8**. Lopez, Vattuone & Sampietro (1988) *Phytochemistry* **27**, 3077. **9**. Oray & Norton (1980) *Biochem. Biophys. Res. Commun.* **95**, 624. **10**. Burg (1970) *Meth. Enzymol.* **18A**, 634. **11**. Burg & Snell (1969) *J. Biol. Chem.* **244**, 2585. **12**. Pogell (1963) *Meth. Enzymol.* **6**, 331. **13**. Matheson & Richardson (1973) *Phytochemistry* **17**, 195. **14**. Yokochi, Morita & Yagi (2003) *J. Agric. Food Chem.* **51**, 2733.

Pyridoxamine 5'-Phosphate

This vitamin B$_6$ metabolite (FW = 248.18 g/mol; CAS 529-96-4) is derived biosynthetically from pyridoxamine and by transamination from pyridoxal 5'-phosphate. The hygroscopic solid is light-sensitive, and solutions should be prepared freshly, despite their stability when stored cold and in the absence of light. Solutions are more labile under alkaline conditions. There are four pK_a values associated with this coenzyme: a value less than 2.5 and a value of 5.76 for the phosphate group, 3.69 for the hydroxyl group, 8.61 for the pyridinium group, and 10.92 for the primary amine. The coenzyme has an ultraviolet/visible spectrum: λ_{max} = 254 nm and 327 nm (ε = 5200 and 9400 M^{-1}cm^{-1}, respectively) at pH 7.2; λ_{max} = 244 nm and 312 nm (ε = 7500 and 8300 M^{-1}cm^{-1}, respectively) at pH 10. ***See also*** *Pyridoxal 5'-Phosphate* **Target(s):** dextransucrase (1); glutamine:fructose-6-phosphate aminotransferase, isomerizing (2,3); phenol 2-monooxygenase, weakly inhibited (4); phosphomevalonate kinase (5); phosphoprotein phosphatase, weakly inhibited (6); porphobilinogen synthase, *or* δ-aminolevulinate dehydratase (7); pyridoxal kinase (8); pyridoxal phosphatase, also as an alternative substrate (9); pyridoxamine:oxaloacetate aminotransferase (10,11); starch phosphorylase (12); tyrosinase (13); and tyrosine aminotransferase (14). **1**. Thaniyavarn, Taylor, Singh & Doyle (1982) *Infect. Immun.* **37**, 1101. **2**. Pogell (1962) *Meth. Enzymol.* **5**, 408. **3**. Gryder & Pogell (1960) *J. Biol. Chem.* **235**, 558. **4**. Neujahr & Kjellén (1980) *Biochemistry* **19**, 4967. **5**. Bazaes, Beytía, Jabalquinto, *et al.* (1980) *Biochemistry* **19**, 2305. **6**. Zhou, Clemens, Hakes, Barford & Dixon (1993) *J. Biol. Chem.* **268**, 17754. **7**. Van Heyningen & Shemin (1971) *Biochem. J.* **124**, 68P. **8**. White & Dempsey (1970) *Biochemistry* **9**, 4057. **9**. Fonda (1992) *J. Biol. Chem.* **267**, 15978. **10**. Wu & Mason (1964) *J. Biol. Chem.* **239**, 1492. **11**. Wada & Snell (1962) *J. Biol. Chem.* **237**, 127. **12**. Matheson & Richardson (1973) *Phytochemistry* **17**, 195. **13**. Yokochi, Morita & Yagi (2003) *J. Agric. Food Chem.* **51**, 2733. **14**. Hayashi, Granner & Tomkins (1967) *J. Biol. Chem.* **242**, 3998.

4-Pyridoxic Acid

This oxidized derivative (FW$_{ion}$ = 180.14 g/mol) of pyridoxal is the chief catabolic product of pyridoxine, pyridoxal, and pyridoxamine degradation; it can be isolated from human urine. The carboxyl group has a pK_a value of 5.50 while the phenolic OH has a value of 9.75. 4-Pyridoxic acid has a blue fluorescence which is maximum between pH 3 and 4. Heating 4-pyridoxic acid in 0.5 N acid will convert the metabolite to the lactone, which has a stronger fluorescence. **Target(s):** dextransucrase (1); glucose dehydrogenase (2,3); pyridoxal dehydrogenase (4,5); and pyridoxamine-phosphate oxidase (6). **1**. Thaniyavarn, Taylor, Singh & Doyle (1982) *Infect. Immun.* **37**, 1101. **2**. Strecker (1955) *Meth. Enzymol.* **1**, 335. **3**.

Brink (1953) *Acta Chem. Scand.* **7**, 1090.   **4**. Burg (1970) *Meth. Enzymol.* **18A**, 634.   **5**. Burg & Snell (1969) *J. Biol. Chem.* **244**, 2585.   **6**. Wada & Snell (1961) *J. Biol. Chem.* **236**, 2089.

5-Pyridoxic Acid

This pyridoxine metabolite (FW = 183.16 g/mol), also known as 3-hydroxy-4-hydroxymethyl-2-methylpyridine-5-carboxylate and isopyridoxic acid, inhibits 3-hydroxy-2-methylpyridinecarboxylate dioxygenase. It is a substrate for 5-pyridoxate dioxygenase. **Target(s):** 3-hydroxy-2-methylpyridinecarboxylate dioxygenase (1-3).   **1**. Burg (1970) *Meth. Enzymol.* **18A**, 634.   **2**. Sparrow, Ho, Sundaram, *et al.* (1969) *J. Biol. Chem.* **244**, 2590.   **3**. Kishore & Snell (1981) *J. Biol. Chem.* **256**, 4234.

Pyridoxine

This nutritionally essential factor (FW$_{free\ base}$ = 169.18 g/mol; CAS 65-23-6), also called vitamin B$_6$ and pyridoxol, is a white solid that is typically supplied as the hydrochloride (the two pK_a values are 4.94 and 8.89). (*Note:* The term vitamin B$_6$ actually refers to a mixture of pyridoxine, pyridoxal, and pyridoxamine.) Pyridoxine is stable in light and air, but is somewhat photolabile in neutral and alkaline solutions. It is soluble in water (1 g/4.5 mL) and has λ_{max} values at 253 and 325 nm at pH 7 in phosphate buffer (ε = 3700 and 7100 M^{-1}cm^{-1}, respectively). Pyridoxine is a substrate of the reactions catalyzed by pyridoxine 4-dehydrogenase, pyridoxine 4-oxidase, pyridoxine 5-dehydrogenase, and pyridoxine 5'-*O*-β-D-glucosyltransferase. **Target(s):** alanine racemase (1); arginine decarboxylase (2,3); dextransucrase (4); β-fructofuranosidase, *or* invertase (5-10); glycogen phosphorylase *b* (11); hydroxymethylpyrimidine kinase, also alternative substrate; K_i = 2.7 μM (12); pyridoxal dehydrogenase (13,14); pyridoxamine-5-phosphate oxidase (15); pyridoxamine:pyruvate aminotransferase (13,16); and tyrosinase (17).   **1**. Olivard & Snell (1955) *J. Biol. Chem.* **213**, 203.   **2**. Das, Bhaduri, Bose & Ghosh (1996) *J. Plant Biochem. Biotechnol.* **5**, 123.   **3**. Choudhuri & Ghosh (1982) *Agric. Biol. Chem.* **46**, 739.   **4**. Thaniyavarn, Taylor, Singh & Doyle (1982) *Infect. Immun.* **37**, 1101.   **5**. Liu, Huang, Chang & Sung (2006) *Food Chem.* **96**, 62.   **6**. Pressey & Avants (1980) *Plant Physiol.* **65**, 135.   **7**. Krishnan, Blanchette & Okita (1985) *Plant Physiol.* **78**, 241.   **8**. Rojo, Quiroga, Vattuone & Sampietro (1998) *Phytochemistry* **49**, 965.   **9**. Prado, Vattuone, Fleischmacher & Sampietro (1985) *J. Biol. Chem.* **260**, 4952.   **10**. Lopez, Vattuone & Sampietro (1988) *Phytochemistry* **27**, 3077.   **11**. Klinova, Klinov, Kurganov, Mikhno & Baliakina (1988) *Bioorg. Khim.* **14**, 1520.   **12**. Mizote & Nakayama (1989) *Biochim. Biophys. Acta* **991**, 109.   **13**. Burg (1970) *Meth. Enzymol.* **18A**, 634.   **14**. Burg & Snell (1969) *J. Biol. Chem.* **244**, 2585.   **15**. Pogell (1963) *Meth. Enzymol.* **6**, 331.   **16**. Dempsey & Snell (1963) *Biochemistry* **2**, 1414.   **17**. Yokochi, Morita & Yagi (2003) *J. Agric. Food Chem.* **51**, 2733.

N-(4-Pyridoxyl)-L-alanine

This amino acid-vitamin B$_6$ conjugate (FW = 224.26 g/mol) strongly inhibits pyridoxamine:pyruvate aminotransferase, K_i = 0.18 μM (1-4).   **1**.

Dempsey & Snell (1963) *Biochemistry* **2**, 1414.   **2**. Korytnyk & Ikawa (1970) *Meth. Enzymol.* **18A**, 524.   **3**. Burg (1970) *Meth. Enzymol.* **18A**, 634.   **4**. Wolfenden (1977) *Meth. Enzymol.* **46**, 15.

N-(4-Pyridoxylidene)phenethylamine

This pyridoxal-phenethylamine adduct (FW = 270.33 g/mol), which has been identified in the urine of individuals with phenylketonuria, inhibits pyridoxal kinase (1-2).   **1**. Korytnyk & Ikawa (1970) *Meth. Enzymol.* **18A**, 524.   **2**. Loo & Whittaker (1967) *J. Neurochem.* **14**, 997.

N-(4-Pyridoxyl-5'-phosphate)-L-dopa

This pyridoxal 5'-phosphate conjugate (FW = 426.32 g/mol), also known as *N*-(5'-phospho-4-pyridoxyl)-L-dopa and *N*-(4-pyridoxyl-5'-phosphate)-L-3,4-dihydroxy-phenylalanine, inhibits aromatic-L-amino-acid decarboxylase, *or* dopa decarboxylase.   **1**. Jung (1986) *Bioorg. Chem.* **14**, 429.

(*R*)-(+)-*N*-(4-Pyridyl)-4-(1-aminoethyl)benzamide, *See* (*R*)-(+)-4-(1-Aminoethyl)-*N*-(4-pyridyl)benzamide

17-(3-Pyridyl)androsta-5,16-dien-3β-ol

This synthetic steroid (FW = 351.26 g/mol) inhibits steroid 17α-monooxygenase, human IC$_{50}$ = 4 nM (1,2). The 2-pyridyl and 4-pyridyl analogues are weaker inhibitors (human IC$_{50}$ = 270 and 4000 nM, respectively).   **1**. Njar & Brodie (1999) *Curr. Pharm. Des.* **5**, 163.   **2**. Potter, Barrie, Jarman & Rowlands (1995) *J. Med. Chem.* **38**, 2463.

4-(2-Pyridylazo)resorcinol

This metal ion chelator (FW$_{free\text{-}acid}$ = 215.21 g/mol), which has been used in the determination of trace amounts of Cr^{3+}, inhibits many metalloenzymes, including: alcohol dehydrogenase (1); collagenase (2); 4-methoxybenzoate monooxygenase (*O*-demethylating), weakly inhibited (3); and protein-*S*-prenylcysteine *O*-methyltransferase (4).   **1**. Zhang & Zhou (1996) *J. Enzyme Inhib.* **10**, 239.   **2**. Springman, Nagase, Birkedal-Hansen & Van Wart (1995) *Biochemistry* **34**, 15713.   **3**. Bernhardt, Nastainczyk & Seydewitz (1977) *Eur. J. Biochem.* **72**, 107.   **4**. Anderson, Frase, Michaelis & Hrycyna (2005) *J. Biol. Chem.* **280**, 7336.

1-[Nα-(4-Pyridylcarboxy)-L-leucylamino]-3-(phenoxyphenylsulfonamido)-2-propanone

This substituted diaminopropanone (FW = 538.62 g/mol) is a strong reversible inhibitor of cathepsin K, apparent K_i = 13 nM. 1. Yamashita, Smith, Zhao, *et al.* (1997) *J. Amer. Chem. Soc.* **119**, 11351.

3-(2-Pyridyl)-5,6-diphenyl-1,2,4-triazine-4′,4″-disulfonate

This chelator, also known as ferrozine (FW$_{\text{free-acid}}$ = 470.49 g/mol), binds cobalt, silver, copper, nickel, and iron ions, the latter with particularly high affinity. **Target(s):** CMP-*N*-acetylneuraminate monooxygenase (1-5). 1. Schlenzka, Shaw, Schneckenburger & Schauer (1994) *Glycobiology* **4**, 675. 2. Schlenzka, Shaw & Schauer (1993) *Biochim. Biophys. Acta* **1161**, 131. 3. Gollub & Shaw (2003) *Comp. Biochem. Physiol. B Biochem. Mol. Biol.* **134**, 89. 4. Gollub, Schauer & Shaw (1998) *Comp. Biochem. Physiol. B Biochem. Mol. Biol.* **120**, 605. 5. Shaw, Schneckenburger, Carlsen, Christiansen & Schauer (1992) *Eur. J. Biochem.* **206**, 269.

Pyridyldithioethylamine

This sulfhydryl-reactive reagent (FW$_{\text{free-base}}$ = 186.30 g/mol) inhibits the phospholipid-translocating ATPase, also known as flippase (1,2) and inhibits taurine transport (3). In general, the pyridine-thiolate is the leaving group in a reaction that transfers a thioethylamine to a reactive thiol on susceptible enzyme targets. 1. Connor & Schroit (1991) *Biochim. Biophys. Acta* **1066**, 37. 2. Connor & Schroit (1988) *Biochemistry* **27**, 848. 3. Dumaswala & Brown (1996) *Placenta* **17**, 329.

(S)-1-(4-Pyridyl)ethyl 1-Adamantanecarboxylate

This ester (FW = 270.35 g/mol) inhibits human steroid 17α-monooxygenase, IC$_{50}$ = 3.3 nM, and aromatase, *or* CYP19, IC$_{50}$ = 5.6 μM (1). The (*R*)-1-(4-pyridyl)ethyl epimer is a weaker inhibitor (IC$_{50}$ = 340 nM and 16 μM, respectively) In addition, the corresponding 3-pyridylethyl analogues are weaker inhibitor (IC$_{50}$ = 840 nM and 4.6 μM, respectively, for the (*S*)-epimer and 580 nM and 36 μM for the (*R*)-epimer). 1. Chan, Potter, Barrie, *et al.* (1996) *J. Med. Chem.* **39**, 3319.

2-Pyridylferrocene

This ferrocene derivative (FW = 293.19 g/mol) inhibits rat brain DNA topoisomerase II, IC$_{50}$ = 0.1 mM. Thiomorpholide amido methyl ferrocene shows higher inhibition of catalytic activity (IC$_{50}$ = 50 μM) against topoisomerase IIβ, when compared to azalactone ferrocene. Analysis of protein-DNA intermediates formed in the presence of these two compounds suggests that azalactone ferrocene readily induces formation of cleavable complex in a dose-dependent manner, in comparison with thiomorpholide amido methyl ferrocene. Both the compounds show significant inhibition of DNA-dependent ATPase activity of enzyme 1. Sai Krishna, Panda & Kondapi (2005) *Arch. Biochem. Biophys.* **438**, 206.

N-(3-Pyridyl)indomethacinamide

This indomethacin derivative (FW = 433.89 g/mol; Symbol = N-3PylA), also named *N*-(3-pyridyl)-1-(4-chlorobenzoyl)-5-methoxy-2-methyl-1*H*-indole-3-acetamide, potently inhibits human cyclooxygenase-2, IC$_{50}$ = 52 nM, but is more than 300-fold less effective on human cyclooxygenase-1 (1). *N*-(3-Pyridyl)indomethacinamide could be a useful tool for determining whether a COX metabolite plays a downstream signaling role in mediating a cellular process of interest. It has, for example been employed to probe whether bradykinin-mediated contraction of the pig iris sphincter muscle is mediated by cyclooxygenase-2 metabolites (2). Another example is the demonstration that α-linolenic acid inhibits human renal cell carcinoma cell proliferation in part through COX-2 inhibition (3). There is less selectivity with respect to the ovine cyclooxygenases (IC$_{50}$ values of 50 and 75 μM for COX-2 and COX-1, respectively). 1. Kalgutkar, Marnett, Crews, Remmel & Marnett (2000) *J. Med. Chem.* **43**, 2860. 2. El Sayah & Calixto (2003) *Eur. J. Pharmacol.* **458**, 175. 3. Yang, Yuan, Liu, *et al.* (2013) *Oncol. Lett.* **6**, 197.

(2R,3S)-N-{(1S)-4-(2-Pyridylsulfonyl)guanyl-1-[(1,3-thiazol-2-ylamino)-carbonyl]-butyl}-3-[formyl(hydroxy)-amino]-2-isobutylhexanamide

This metalloproteinase inhibitor (FW = 610.76 g/mol) inhibits: ADAM 17 endopeptidase, *or* tumor necrosis factor-α converting enzyme, *or* TACE, IC$_{50}$ = 6 nM; interstitial collagenase, *or* matrix metalloproteinase 1, IC$_{50}$ = 76 nM; gelatinase B, *or* matrix metalloproteinase 9, IC$_{50}$ = 43 nM; and stromelysin 1, *or* matrix metalloproteinase 9, IC$_{50}$ = 84 nM (1). 1. Rabinowitz, Andrews, Becherer, *et al.* (2001) *J. Med. Chem.* **44**, 4252.

Pyrilamine

This antihistamine ($FW_{free\text{-}base}$ = 285.39 g/mol), first prepared and used as an antihistamine in the mid-1940s in the laboratory of Nobelist Daniel Bovet, is a potent H_1 histamine receptor antagonist. The free base is an oily liquid, but the commercially available hydrochloride and maleate salts are highly water-soluble. **Target(s):** H_1 histamine receptor (1); histamine *N*-methyltransferase (2). **1.** Haley (1983) *J. Pharm. Sci.* **72**, 3. **2.** Thithapandha & Cohn (1978) *Biochem. Pharmacol.* **27**, 263.

Pyrimethamine

This folate antagonist (FW = 248.71 g/mol; slightly soluble in ethanol (~ 9 g/L) and dilute HCl (~ 5 g/L)), also known as 2,4-diamino-5-(*p*-chlorophenyl)-6-ethylpyrimidine, 5-(*p*-chlorophenyl)-2,4-diamino-6-ethylpyrimidine, as well as Daraprim®, is an antiprotozoal and antimalarial agent first synthesized in the early 1950s by Nobelist George Hitchings and coworkers (1). While primarily active against *Plasmodium falciparum*, it is also active against *P. vivax*. Other antifolates (*e.g.,* methotrexate and trimethoprim) potentiate the action of pyrimethamine and can lead to folate deficiency anemia and other blood dyscrasias. Resistance to pyrimethamine is widespread, mainly resulting from mutations in the malarial dihydrofolate reductase that decrease binding affinity via loss of required hydrogen bonds and steric interactions **Target(s):** dihydrofolate reductase, K_i = 17 nM (1-10); glutathione *S*-transferase (11); histamine *N*-methyltransferase (12). **1.** Russell & Hitchings (1951) *J. Amer. Chem. Soc.* **73**, 3763. **2.** Roth & Burchall (1971) *Meth. Enzymol.* **18B**, 779. **3.** Baker (1967) *Design of Active-Site-Directed Irreversible Enzyme Inhibitors*, Wiley, New York. **4.** Bertino, Perkins & Johns (1965) *Biochemistry* **4**, 839. **5.** Tahar, de Pecoulas, Basco, Chiadmi & Mazabraud (2001) *Mol. Biochem. Parasitol.* **113**, 241. **6.** Rastelli, Sirawaraporn, Sompornpisut, *et al.* (2000) *Bioorg. Med. Chem.* **8**, 1117. **7.** Kompis, Then, Wick & Montavon (1980) in *Enzyme Inhibitors* (Brodbeck, ed.), pp. 177-189, Verlag Chemie, Weinheim. **8.** Burchall & Hitchings (1965) *Mol. Pharmacol.* **1**, 126. **9.** Pattanakitsakul & Ruenwongsa (1984) *Int. J. Parasitol.* **14**, 513. **10.** Aboge, Jia, Terkawi, *et al.* (2008) *Antimicrob. Agents Chemother.* **52**, 4072. **11.** Mukanganyama, Widersten, Naik, Mannervik & Hasler (2002) *Int. J. Cancer* **97**, 700. **12.** Thithapandha & Cohn (1978) *Biochem. Pharmacol.* **27**, 263.

2-Pyrimidinone

This hygroscopic pyrimidine (FW = 96.09 g/mol), also known as 2-hydroxypyrimidine and 2-pyrimidinol, inhibits cytosine deaminase (1-3) and pancreatic ribonuclease (4), the latter weakly. **1.** Kornblatt & Tee (1986) *Eur. J. Biochem.* **756**, 297. **2.** Ko, Lin, Hu, *et al.* (2003) *J. Biol. Chem.* **278**, 19111. **3.** Hsu, Hu, Lin & Liaw (2003) *Acta Crystallogr. Sect. D* **59**, 950. **4.** Richards & Wyckoff (1971) *The Enzymes*, 3rd ed. (Boyer, ed.), **4**, 647.

2-Pyrimidinone Nucleosides

R Group	Abbreviation
– CH₃	MPdR
– I	IPdR
– C ≡ CH	EPdR
– C ≡ CCH₃	PPDR

These 5-substituted deoxythymidine analogues (including 5-methyl-2-pyrimidinone, FW_{MPdR} = 226.23 g/mol; 5-iodo-2-pyrimidinone, FW_{IPdR} = 338.10 g/mol; 5-ethynyl-2-pyrimidinone, FW_{PdR} = 336.23 g/mol; and 5-propynyl-2-pyrimidinone, FW_{PPdR} = 250.25 g/mol) target Herpes Simplex Virus Types 1 and 2 (HSV-1 and HSV-2), viruses that induce type-specific thymidine kinase (TK) in virus-infected cells. HSV TKs have a broader substrate specificity than host TK, suggesting that unusual 2-pyriminone

nucleosides can be activated to their nucleotide derivatives in HSV-laden cells and selectively exert their cytotoxic effects. **1.** Lewandowski, Grill, Fisher, *et al.* (1989) *Antimicrob. Agents Chemother.* **33**, 340.

4-Pyrimidinone

This hygroscopic pyrimidine, also known as 4-hydroxypyrimidine and 4-pyrimidinol (FW = 96.09 g/mol), inhibits both hypoxanthine(guanine) and xanthine phosphoribosyltransferases (1). **1.** Naguib, Iltzsch, el Kouni, Panzica & el Kouni (1995) *Biochem. Pharmacol.* **50**, 1685.

N^1-(2-Pyrimidyl)sulfanilamide, *See* Sulfadiazine

Pyriminil

This substituted urea-based rodenticide (FW = 272.26 g/mol), also known as Vacor and named systematically as *N*-(4-nitrophenyl)-*N'*-(3-pyridinylmethyl)urea, inhibits, acetylcholinesterase (1)), insulin release (2). and NADH:ubiquinone reductase activity of mitochondrial complex I, *or* NADH dehydrogenase (3). **1.** Devi & Krishnamoorthy (1979) *Indian J. Physiol. Pharmacol.* **23**, 285. **2.** Taniguchi, Yamashiro, Chung, *et al.* (1989) *J. Endocrinol. Invest.* **12**, 273. **3.** Esposti, Ngo & Myers (1996) *Diabetes* **45**, 1531.

Pyripyropene A

This fungal sesquiterpene (FW = 580.65 g/mol), first isolated from *Aspergillus fumigatus*, inhibits sterol *O*-acyltransferase, *or* cholesterol *O*-acyltransferase, *abbreviated* ACAT (1-3). **1.** Ohshiro, Rudel, Ômura & Tomoda (2007) *J. Antibiot.* **60**, 43. **2.** Ômura, Tomoda, Kim & Nishida (1993) *J. Antibiot.* **46**, 1168. **3.** Das, Davis, Tomoda, Ômura & Rudel (2008) *J. Biol. Chem.* **283**, 10453.

Pyrithiamin

This thiamin analogue ($FW_{bromide/hydrobromide\ salt}$ = 420.15 g/mol; hygroscopic solid; M.P. = 205-210°C) soluble in water, chloroform, and ethanol), also known as pyrithiamine and neopyrithiamin, is a thiamin antagonist in many organisms. **Target(s):** esterase, weakly inhibited (1); Na^+/K^+-exchanging ATPase (2-4); thiaminase (5); thiamin kinase (6); thiamin diphosphokinase (7-14); thiamin oxidase (15); thiamin-phosphate kinase (16); thiamin transport (17-19); and thiamin-triphosphatase (20). **1.** Muftic (1954) *Enzymologia* **17**, 123. **2.** Matsuda, Iwata & Cooper (1985) *Biochim. Biophys. Acta* **817**, 17. **3.** Matsuda, Iwata & Cooper (1984) *J. Biol. Chem.* **259**, 3858. **4.** Matsuda & Cooper (1983) *Biochemistry* **22**, 2209. **5.** Sealock & White (1949) *J. Biol. Chem.* **181**, 393. **6.** Rogers (1970) *Meth. Enzymol.* **18A**, 245. **7.** Gubler (1970) *Meth. Enzymol.* **18A**, 219. **8.** Sanemori & Kawasaki (1980) *J. Biochem.* **88**, 223. **9.** Artsukevich, Voskoboev & Ostrovskii (1977) *Vopr. Med. Khim.* **23**, 203. **10.** Eich & Cerecedo (1954) *J. Biol. Chem.* **207**, 295. **11.** Molin & Fites (1980) *Plant Physiol.* **66**, 308. **12.** Peterson, Gubler & Kuby (1975) *Biochim. Biophys. Acta* **397**, 377. **13.** Sanemori & Kawasaki (1980) *J. Biochem.* **88**, 223. **14.** Liu, Timm & Hurley (2006) *J. Biol. Chem.* **281**, 6601. **15.** Gomez-Moreno & Edmondson (1985) *Arch. Biochem. Biophys.* **239**, 46. **16.** Nishino (1972) *J. Biochem.* **72**, 1093. **17.** Casirola, Patrini, Ferrari & Rindi (1990) *J. Membr. Biol.* **118**, 11. **18.** Toburen-Bots & Hagedorn (1977) *Arch. Microbiol.* **113**, 23. **19.** Hoyumpa, Jr., Middleton III, Wilson & Schenker (1975) *Gastroenterology* **68**, 1218. **20.** Nishimune & Hayashi (1987) *J. Nutr. Sci. Vitaminol.* **33**, 113.

Pyrithiamin Pyrophosphate

This thiamin pyrophosphate analogue (FW = 417.28 g/mol) inhibits pyruvate decarboxylase (1) and pyruvate oxidase (2,3). **1.** Rogers (1970) *Meth. Enzymol.* **18A**, 245. **2.** Woolley (1951) *J. Biol. Chem.* **191**, 43. **3.** Onrust, van der Linden & Jansen (1954) *Enzymologia* **16**, 289.

Pyrithione, *See* 2-Mercaptopyridine N-Oxide; Bis[1-hydroxypyridine-2(1H)-thionato-S,O]zinc(II)

Pyrizinostatin

This antibiotic (FW = 281.27 g/mol; melting point = 188-190°C) inhibits pyroglutamyl-peptidase, IC_{50} = 1 μM (1-4). **1.** Robert-Baudouy, Clauziat & Thierry (1998) in *Handb. Proteolytic Enzymes* (Barrett, Rawlings & Woessner, eds.), p. 791, Academic Press, San Diego. **2.** Tatsuta & Kitagawa (1994) *J. Antibiot.* **47**, 389. **3.** Hatsu, Naganawa, Aoyagi, Takeuchi & Kodama (1992) *J. Antibiot.* **45**, 1961. **4.** Aoyagi, Hatsu, Imada, *et al.* (1992) *J. Antibiot.* **45**, 1795.

Pyrogallol

This poisonous substance (FW = 126.11 g/mol; M.P. = 131-133°C; solubility, or 1 g per 1.7 mL water), also called 1,2,3-trihydroxybenzene and pyrogallic acid, is a white solid that turns gray upon exposure to light and air. Aqueous solutions also darken when exposed to air and light, becoming alkaline in pH. Autooxidation of pyrogallol also generates several reactive oxygen species, including superoxide anion. **Target(s):** aldehyde dehydrogenase (1); catechol 1,2-dioxygenase (2); catechol *O*-methyltransferase (3-7); 3-demethylubiquinone-9 3-*O*-methyltransferase, weakly inhibited (8); dihydroceramide desaturase (9); 1,2-dihydroxynaphthalene dioxygenase (10); histidine decarboxylase (11); hydroxyquinol 1,2-dioxygenase (12); lipoxygenase (13); nitronate monooxygenase (14); procollagen-lysine 5-dioxygenase (15); protocatechuate 4,5-dioxygenase (16); RNA-directed RNA polymerase (17); shikimate dehydrogenase (18); tannase (19); and urease (20). **1.** Rubenstein, Collins & Tabakoff (1975) *Experientia* **31**, 414. **2.** Itoh (1981) *Agric. Biol. Chem.* **45**, 2787. **3.** Axelrod & Larouche (1959) *Science* **130**, 800. **4.** Crout (1961) *Biochem. Pharmacol.* **6**, 47. **5.** Bonifácio, Palma, Almeida & Soares-da-Silva (2007) *CNS Drug Reviews* **13**, 352. **6.** Veser (1987) *J. Bacteriol.* **169**, 3696. **7.** Tong & D'Iorio (1977) *Can. J. Biochem.* **55**, 1108. **8.** Houser & Olson (1977) *J. Biol. Chem.* **252**, 4017. **9.** Schulze, Michel & van Echten-Deckert (2000) *Meth. Enzymol.* **311**, 22. **10.** Patel & Barnsley (1980) *J. Bacteriol.* **143**, 668. **11.** Webb (1966) *Enzyme and Metabolic Inhibitors*, vol. 2, p. 352, Academic Press, New York. **12.** Zaborina, Seitz, Sidorov, *et al.* (1999) *J. Basic Microbiol.* **39**, 61. **13.** Holman & Bergström (1951) *The Enzymes*, 1st ed. (Sumner & Myrbäck, eds.), **2** (part 1), 559. **14.** Kido, Soda & Asada (1978) *J. Biol. Chem.* **253**, 226. **15.** Murray, Cassell & Pinnell (1977) *Biochim. Biophys. Acta* **481**, 63. **16.** Ono, Nozaki & Hayaishi (1970) *Biochim. Biophys. Acta* **220**, 224. **17.** Kozlov, Polyakov, Ivanov, *et al.* (2006) *Biochemistry (Moscow)* **71**, 1021. **18.** Balinsky & Davies (1961) *Biochem. J.* **80**, 296. **19.** Iibuchi, Minoda & Yamada (1972) *Agric. Biol. Chem.* **36**, 1553. **20.** Quastel (1933) *Biochem. J.* **27**, 1116.

Pyroglutamic Acid, *See* 5-Oxoproline

L-Pyroglutamyl-L-asparaginyl-L-prolinamide

This thyrotropin-releasing factor analogue (FW = 339.35 g/mol) is not hydrolyzed by pyroglutamyl-peptidase II (thyrotropin-releasing hormone-degrading ectoenzyme) but is an inhibitor, K_i = 17.5 μM (1). The free acid (*i.e.*, L-pyroglutamyl-L-asparaginyl-L-proline) is a weaker inhibitor, K_i = 750 μM (2). **1.** Kelly, Slator, Tipton, Williams & Bauer (2000) *J. Biol. Chem.* **275**, 16746. **2.** O'Connor & O'Cuinn (1985) *Eur. J. Biochem.* **150**, 47.

L-Pyroglutamyl-L-asparaginyl-L-proline 7-Amido-4-methylcoumarin

This thyrotropin-releasing factor analogue (FW = 497.19 g/mol) is not hydrolyzed by pyroglutamyl-peptidase II, *or* thyrotropin-releasing hormone-degrading ectoenzyme, but is an inhibitor, K_i = 0.97 μM (1). **1.** Kelly, Slator, Tipton, Williams & Bauer (2000) *J. Biol. Chem.* **275**, 16746.

L-Pyroglutamyl-L-asparaginyl-L-tryptophan

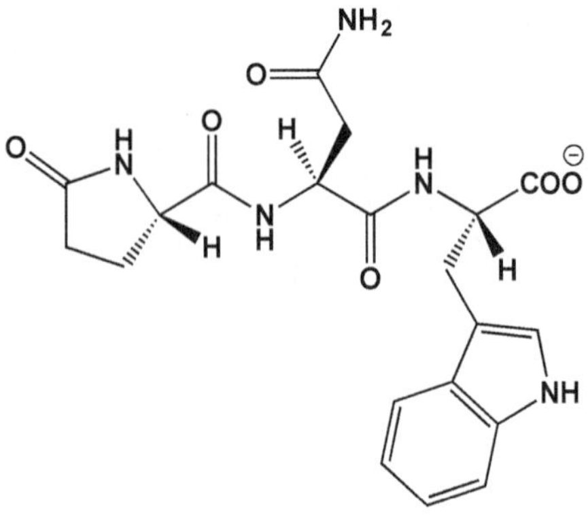

This naturally occurring tripeptide ($FW_{\text{free-acid}}$ = 429.43 g/mol) inhibits adamalysin (1), atrolysin E, K_i = 11 μM (3), and *Trimeresurus mucrosquamatus* (Taiwan habu) venom metalloproteinase, K_i = 0.16 μM (4,5). **1**. Gomis-Rüth, Meyer, Kress & Politi (1998) *Protein Sci.* **7**, 283. **2**. Zhang, Botos, Gomis-Rüth, *et al.* (1994) *Proc. Natl. Acad. Sci. U.S.A.* **91**, 8447. **3**. Fox & Bjarnason (1995) *Meth. Enzymol.* **248**, 368. **4**. Huang, Hung, Wu & Chiou (1998) *Biochem. Biophys. Res. Commun.* **248**, 562. **5**. Huang, Chiou, Ko & Wang (2002) *Eur. J. Biochem.* **269**, 3047.

L-Pyroglutamyl-L-glutaminyl-L-arginyl-L-leucylglycyl-L-asparaginyl-L-glutaminyl-L-tryptophanyl-L-alanyl-L-valylglycyl-L-histidyl-L-leucyl-L-methioninamide, *See Bombesin*

L-Pyroglutamylglycyl-L-valyl-L-asparaginyl-L-aspartyl-L-asparaginyl-L-glutamyl-L-glutamylglycyl-L-phenylalanyl-L-phenylalanyl-L-seryl-L-alanyl-7 -arginine, *See Fibrinopeptide B*

L-Pyroglutamyl-L-histidyl-L-prolinamide, *See Thyrotropin-Releasing Hormone*

L-Pyroglutamyl-L-histidyl-L-tryptophanyl-L-seryl-L-tyrosylglycyl-L-leucyl-L-arginyl-L-prolylglycinamide, *See Gonadotropin-Releasing Hormone*

L-Pyroglutamyl-L-histidyl-L-tryptophanyl-L-seryl-L-tyrosylglycyl-L-tryptophan

This heptapeptide (*Sequence*: pGHTSTGW (where pG = pyroglutamyl); FW = 726.75 g/mol), corresponding to the N-terminus of salmon gonadotropin-releasing hormone and also known as luteinizing hormone releasing hormone (LHRH), is not hydrolyzed by pyroglutamyl-peptidase II but does inhibit (K_i = 60.3 μM). **1**. Gallagher & O'Connor (1998) *Int. J. Biochem. Cell Biol.* **30**, 115.

L-Pyroglutamyl-L-lysyl-L-tryptophan

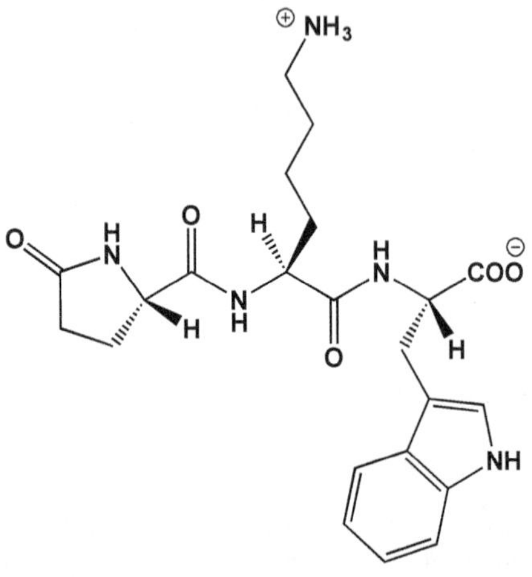

This naturally occurring tripeptide (FW = 443.50 g/mol) inhibits *Trimeresurus mucrosquamatus* (Taiwan habu) venom metalloproteinase, K_i = 0.124 μM (1,2). **1**. Huang, Hung, Wu & Chiou (1998) *Biochem. Biophys.*

Res. Commun. **248**, 562. **2**. Huang, Chiou, Ko & Wang (2002) *Eur. J. Biochem.* **269**, 3047.

L-Pyroglutamyl-L-lysyl-L-tryptophanyl-L-alanyl-L-proline, *See Bradykinin Potentiating Peptide 5a*

D-Pyroglutamyl-D-phenylalanyl-L-tryptophanyl-L-seryl-L-tyrosyl-D-alanyl-L-leucyl-L-arginyl-L-prolylglycinamide, *See [D-pGlu¹,D-Phe²,D-Trp³,⁶]-Luteinizing Hormone-Releasing Hormone*

L-Pyroglutamyl-D-phenylalanyl-L-tryptophanyl-L-seryl-L-tyrosyl-D-alanyl-L-leucyl-L-arginyl-L-prolylglycinamide, *See [D-Phe²,D-Ala⁶]-Luteinizing Hormone-Releasing Hormone*

L-Pyroglutamyl-L-tryptophanyl-L-prolyl-L-arginyl-L-prolyl-L-glutaminyl-L-isoleucyl-L-prolyl-L-proline, *See Bradykinin Potentiating Peptide 9a*

Pyromellitic Acid

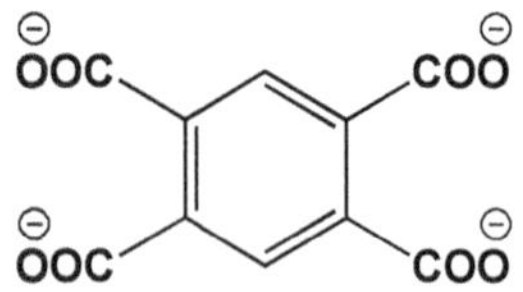

This aromatic tetracarboxylic acid ($FW_{\text{free-acid}}$ = 254.15 g/mol), also known as 1,2,4,5-benzenetetracarboxylic acid, is converted to the dianhydride upon heating to its melting point (281-281.5°C). **Target(s):** aconitase, *or* aconitate hydratase (1); fumarase, *or* fumarate hydratase (2,3); phosphofructokinase (4); and phosphoglycerate mutase (5). **1**. Glusker (1971) *The Enzymes*, 3rd ed. (Boyer, ed.), **5**, 413. **2**. Ueda, Yumoto, Tokushige, Fukui & Ohya-Nishiguchi (1991) *J. Biochem.* **109**, 728. **3**. Beeckmans & Kanarek (1977) *Eur. J. Biochem.* **78**, 437. **4**. Passonneau & Lowry (1963) *Biochem. Biophys. Res. Commun.* **13**, 372. **5**. Rigden, Walter, Phillips & Fothergill-Gilmore (1999) *J. Mol. Biol.* **289**, 691.

1,2-Pyrone-5-carboxylic Acid, *See Coumalic Acid*

1,4-Pyrone-2,6-dicarboxylic Acid, *See Chelidonic Acid*

Pyronin Y

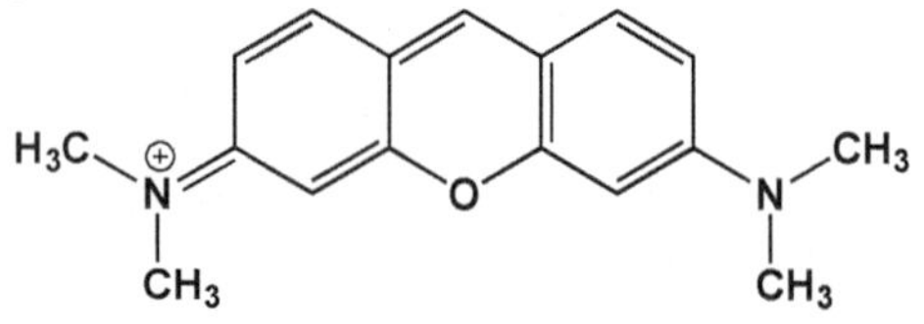

This cationic dye (FW_{ion} = 267.35 g/mol), also known as puronine Y and pyronine G, is used to selectively demonstrate RNA (red) in contrast to DNA (green) in the Unna-Pappenheim stain. **Target(s):** F_1 ATPase, *or* H^+-transporting two-sector ATPase (1,2). **1**. Bullough, Ceccarelli, Roise & Allison (1989) *Biochim. Biophys. Acta* **975**, 377. **2**. Mai & Allison (1983) *Arch. Biochem. Biophys.* **221**, 467.

Pyrophosphate

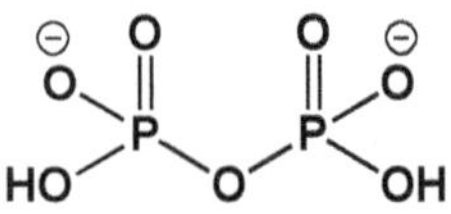

This conjugate base (also called diphosphate and previously inorganic pyrophosphate; abbreviated PP_i) of pyrophosphoric acid (FW = 177.98 g/mol; pK_1 = 0.85, pK_2 = 1.96, pK_3 = 6.60, and pK_4 = 9.41 (at 25°C); pK_a values sensitive to changes in ionic strength) is commonly supplied as its crystalline sodium salts that are very soluble in water. Pyrophosphate anion hydrolyzes to orthophosphate, but much more slowly pyrophosphoric acid. Pyrophosphate binds many metal ions quite tightly. This property alone often accounts for its inhibition of numerous enzymes. The Mg^{2+}-$HP_2O_7^{3-}$ stability constant is reported to be 1200 M^{-1}.1 Values for Ni^{2+}, Cu^{2+}, Co^{2+}, and Zn^{2+} are higher.

Pyrophosphomevalonic Acid, *See Mevalonic Acid 5-Diphosphate*
1α-Pyrophosphoryl-2α,3α-dihydroxy-4β-cyclo-pentanemethanol 5-Phosphate, *See 1α-Diphosphoryl-2α,3α-dihydroxy-4β-cyclopentanemethanol 5-Phosphate*

Pyroquilon

This fungicide (FW = 173.21 g/mol), also known as 1,2,5,6-tetrahydro-4*H*-pyrrolo[3,2,1-*i,j*]quinolin-4-one, inhibits melanin biosynthesis in *Pyricularia oryzae* and has been used in the treatment of rice blast disease. **Target(s):** scytalone dehydratase (1); tetrahydroxynaphthalene reductase (2); and trihydroxynaphthalene reductase (2,3). **1**. Wheeler & Greenblatt (1988) *Exp. Mycol.* **12**, 151. **2**. Liao, Thompson, Fahnestock, Valent & Jordan (2001) *Biochemistry* **40**, 8696. **3**. Liao, Basarab, Gatenby, Valent & Jordan (2001) *Structure* **9**, 19.

Pyrostatins

Pyrostatin A   **Pyrostatin B**

These *N*-acetyl-β-D-glucosaminidase inhibitors (Pyrostatin A, FW_{ion} = 159.17 g/mol; Pyrostatin B, FW_{ion} = 143.17 g/mol), also named systematically as 4-hydroxy-2-imino-1-methylpyrrolidine-5-carboxylate) and (2-imino-1-methylpyrrolidine-5-carboxylate, respectively, were isolated from a species of *Streptomyces*. Their respective K_i values are 1.7 and 2.0 μM (1). **1**. Aoyama, Kojima, Imada, *et al.* (1995) *J. Enzyme Inhib.* **8**, 223.

Pyrrobutamine

This antihistamine (FW = 311.86 g/mol; CAS 91-82-7), also known as 1-[(2*E*)-4-(4-chlorophenyl)-3-phenylbut-2-en-1-yl]pyrrolidine, inhibits histamine *N*-methyltransferase. **1**. Thithapandha & Cohn (1978) *Biochem. Pharmacol.* **27**, 263.

Pyrrole-2-carboxylic Acid

This derivative of pyrole ($FW_{free-acid}$ = 111.10 g/mol; melting point = 204-208°C [decomposes]) can be regarded as an unsaturated, planar analogue of proline, albeit with substantial π-orbital electron density. **Target(s):** D-amino-acid oxidase (1-3); CTP synthase, weakly inhibited (4); 4-hydroxyproline epimerase (5,6); procollagen N-endopeptidase (7); and proline racemase (8-11). **1**. Yagi (1971) *Meth. Enzymol.* **17B**, 608. **2**. Webb (1966) *Enzyme and Metabolic Inhibitors*, vol. **2**, p. 342, Academic Press, New York. **3**. Klein & Austin (1953) *J. Biol. Chem.* **205**, 725. **4**. Bearne, Hekmat & Macdonnell (2001) *Biochem. J.* **356**, 223. **5**. Adams (1971) *Meth. Enzymol.* **17B**, 266. **6**. Finlay & Adams (1970) *J. Biol. Chem.* **245**, 5248. **7**. Dombrowski, Sheats & Prockop (1986) *Biochemistry* **25**, 4302. **8**. Wolfenden (1977) *Meth. Enzymol.* **46**, 15. **9**. Cardinale & Abeles (1968) *Biochemistry* **7**, 3970. **10**. Reina-San-Martin, Degrave, Rougeot, *et al.* (2000) *Nature Med.* **6**, 890. **11**. Chamond, Gregoire, Coatnoan, *et al.* (2003) *J. Biol. Chem.* **278**, 15484.

Pyrrolidine

This colorless liquid ($FW_{free-base}$ = 71.122 g/mol; B.P. = 88.5-89°C), also known as tetrahydropyrrole, is the structural foundation of many biomolecules, including proline. Pyrrolidine has also been reported in tobacco and carrot leaves. Pyrrolidine is miscible in water and is a strong base (pK_a = 11.27 at 25°C). **Target(s):** D-amino-acid oxidase, weakly inhibited (1); and prolyl-tRNA synthetase (2). **1**. Dixon & Kleppe (1965) *Biochim. Biophys. Acta* **96**, 368. **2**. Norris & Fowden (1972) *Phytochemistry* **11**, 2921.

Pyrrolidine-2-carboxylic Acid, *See* L-Proline; D-Proline

L-*trans*-Pyrrolidine-2,4-dicarboxylic Acid

This L-proline derivative (FW = 145.11 g/mol), also known as *trans*-4-carboxy-L-proline, inhibits glutamate transport (1,2) and kynurenine:oxoglutarate aminotransferase (3). **1**. Bridges, Stanley, Anderson, Cotman & Chamberlin (1991) *J. Med. Chem.* **34**, 717. **2**. Balcar (1992) *FEBS Lett.* **300**, 203. **3**. Battaglia, Rassoulpour, Wu, *et al.* (2000) *J. Neurochem.* **75**, 2051.

2-Pyrrolidinemethanol, *See* Prolinol

3-Pyrrolidinol

This alcohol (FW = 87.122 g/mol) inhibits 4-hydroxyproline epimerase (1,2). *Note*: 3-Pyrrolodinol is the decarboxylated analogue of both 3-hydroxy- and 4-hydroxyproline. **1**. Adams (1971) *Meth. Enzymol.* **17B**, 266. **2**. Finlay & Adams (1970) *J. Biol. Chem.* **245**, 5248.

7-(4'-(Pyrrolidin-1-yl)but-2-ynyloxy)chromen-2-one

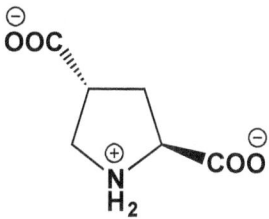

This chromenone, also known as 7-(4'-(*N*-pyrrolidynyl)but-2-ynyloxy)chromen-2-one ($FW_{free-base}$ = 283.33 g/mol), inhibits *Alicyclobacillus acidocaldarius* squalene:hopene cyclase, IC_{50} = 5 μM (1,2) and *Saccharomyces cerevisiae* lanosterol synthase, IC_{50} = 10.3 μM (1). **1**. Oliaro-Bosso, Viola, Matsuda, *et al.* (2004) *Lipids* **39**, 1007. **2**. Cravotto, Balliano, Tagliapietra, Palmisano & Penoni (2004) *Eur. J. Med. Chem.* **39**, 917.

2-Pyrrolidone

This toxic lactam (FW = 85.106 g/mol), also called 2-pyrrolidinone, is soluble in water and many organic solvents. **Target(s):** cholinesterase, weakly inhibited (1); prolyl-tRNA synthetase (2); and pyroglutamyl-peptidase I (3-7). **1**. Fitch (1963) *Biochemistry* **2**, 1221. **2**. Norris & Fowden (1972) *Phytochemistry* **11**, 2921. **3**. McKeon & O'Connor (1998) in *Handb. Proteolytic Enzymes* (Barrett, Rawlings & Woessner, eds.), p. 796, Academic Press, San Diego. **4**. Cummins & O'Connor (1996) *Int. J. Biochem. Cell Biol.* **28**, 883. **5**. Dando, Fortunato, Strand, Smith & Barrett (2003) *Protein Expr. Purif.* **28**, 111. **6**. Armentrout & Doolittle (1969) *Arch. Biochem. Biophys.* **132**, 80. **7**. Awadé, Gonzàles, Cleuziat & Robert-Baudouy (1992) *FEBS Lett.* **308**, 70.

Δ¹-Pyrroline-5-carboxylic Acid

This metabolite (FW$_{free-acid}$ = 113.12 g/mol) is an intermediate in L-proline biosynthesis, is rapidly formed in aqueous solutions from glutamate γ-semialdehyde, the equilibrium favoring the cyclic structure by roughly 20-to-1. ***See also*** *L-Glutamate γ-Semialdehyde* **Target(s):** CTP synthetase, moderately inhibited (1); glutamine:fructose-6-phosphate aminotransferase, isomerizing, *or* glucosamine-6-phosphate synthase (2,3). **1.** Bearne, Hekmat & MacDonnell (2001) *Biochem. J.* **356**, 223. **2.** Bearne & Wolfenden (1995) *Biochemistry* **34**, 11515. **3.** Badet-Denisot, Leriche, Massière & Badet (1995) *Bioorg. Med. Chem. Lett.* **5**, 815.

Pyrrolnitrin

This photosensitive chloropyrrole (FW = 257.08 g/mol), also known as 3-chloro-4-(2'-nitro-3'-chlorophenyl)pyrrole, is an antifungal reagent that inhibits electron transport. **Target(s):** complex I, *or* NADH dehydrogenase (1); complex II, *or* succinate dehydrogenase (1); and electron transport (1-4). **1.** el-Banna & Winkelmann (1998) *J. Appl. Microbiol.* **85**, 69. **2.** Rapoport & Schewe (1977) *Trends Biochem. Sci.* **2**, 186. **3.** Lambowitz & Slayman (1972) *J. Bacteriol.* **112**, 1020. **4.** Warden & Edwards (1976) *Eur. J. Biochem.* **71**, 411.

3-(1H-Pyrrol-2-ylmethylene)-1,3-dihydroindol-2-one

This indole derivative (FW = 210.24 g/mol), also known as oxindole I, is a potent and selective inhibitor of the vascular endothelial growth factor (VEGF) receptor protein-tyrosine kinase Flk-1 (*i.e.*, fetal liver kinase-1), IC$_{50}$ = 390 nM (1,2), but a weaker inhibitor of platelet-derived growth factor receptor (PDGFR) protein-tyrosine kinase, IC$_{50}$ = 12 μM. In addition, it also inhibits certain cyclin-dependent kinases: cyclin D1/cdk4 enzyme (IC$_{50}$ = 4.9 μM) > cyclin E/cdk2 (IC$_{50}$ = 10 μM) and cyclin B/cdk1 (IC$_{50}$ = 10.2 μM). **1.** Kent, Hull-Campbell, Lau, *et al.* (1999) *Biochem. Biophys. Res. Commun.* **260**, 768. **2.** Sun, Tran, Tang, *et al.* (1998) *J. Med. Chem.* **41**, 2588.

Pyruvaldehyde, *See Methylglyoxal*

Pyruvamide

This synthetic activator (FW= 87.08 g/mol) of *Kluyveromyces lactis* and *Saccharomyces cerevisiae* pyruvate decarboxylase actually becomes a mixed-type inhibitor at elevated concentrations. **Target(s):** pyruvate decarboxylase (1,2); pyruvate dehydrogenase (acetyl-transferring) complex (3); and [pyruvate dehydrogenase (acetyl-transferring)] kinase (4). **1.** Krieger, Spinka, Golbik, Hubner & Konig (2002) *Eur. J. Biochem.* **269**,

3256. **2.** Wei, Liu & Jordan (2002) *Biochemistry* **41**, 451. **3.** Liu & Bisswanger (2005) *Biol. Chem.* **386**, 11. **4.** Korotchkina, Sidhu & Patel (2004) *Free Radic. Res.* **38**, 1083.

Pyruvate (Pyruvic Acid)

This pivotal α-keto acid metabolite (FW$_{free-acid}$ = 88.06 g/mol) is found as its conjugate, pyruvate, in almost every cell as an important intermediate in both the catabolism and biosynthesis of carbohydrates and amino acids. Pyruvic acid was the first α-keto acid synthesized in the laboratory (by Berzelius in 1835). The free acid is a liquid at room temperature (M.P. = 11.8°C; B.P. (with decomposition) = 165°C), has a pK_a value of 2.39 at 25°C, and is very soluble in water and ethanol. In concentrated solutions, the free acid polymerizes. Aqueous solutions of the free acid exist primarily in the *gem*-diol form (CH$_3$C(OH)$_2$COOH), typically at 54–71%. By contrast, aqueous solutions of the anion form consists of only 3–5% *gem*-diol. Aqueous solutions polymerize on storage to DL-γ-hydroxy-γ-methyl-α-ketoglutaric acid and higher molecular weight compounds. The aldol dimer and aldol trimer of pyruvic acid have also been called meta- and parapyruvic acid, respectively. ***See also*** *Parapyruvic Acid* **Target(s):** acetoacetate decarboxylase (2); acetyl-CoA synthetase, *or* acetate:CoA ligase (3); *N*-acylglucosamine 2-epimerase (4); alanine racemase (5); 4-aminobutyrate aminotransferase (101); δ-aminolevulinate synthase (6); 8-amino-7-oxononanoate synthase (109); asparaginase, weakly inhibited (69); asparagine synthetase, weakly inhibited (7); [branched-chain α-keto-acid dehydrogenase] kinase, *or* [3-methyl-2-oxobutanoate dehydrogenase (acetyl-transferring)] kinase (8,80,81); γ-butyrobetaine dioxygenase, *or* γ-butyrobetaine hydroxylase, K_i = 1.4 mM (9); catalase (1); choline acetyltransferase (10,11); cystathionine γ-lyase, *or* cysteine desulfhydrase (12); deacetylcephalosporin-C acetyltransferase (108); 2-dehydro-3-deoxy-L-arabinonate dehydratase (13); exo-α-sialidase, *or* neuraminidase, *or* sialidase (71); fructose-1,6-bisphosphate aldolase (14); fructose-2,6-bisphosphate 2-phosphatase (72); fumarase, *or* fumarate hydratase (54); fumarylacetoacetase (64); gluconate 2-dehydrogenase (15); glucose-1-phosphate adenylyltransferase (95); glutaminase (16,17,68); glycerate dehydrogenase (18); glycerate kinase (97,98); glycogen synthase (19); glyoxylate reductase (20); guanylate cyclase (21); hexose oxidase (22); histidine ammonia-lyase (23); 3-hydroxybutyrate dehydrogenase (24); 4-hydroxy-2-ketoglutarate aldolase (25); indolepyruvate decarboxylase, also alternative substrate (59); inositol oxygenase (114); [isocitrate dehydrogenase (NADP$^+$)] kinase (26,77-79); isocitrate lyase (27,55-57); isopenicillin-N synthase, weakly inhibited (65); kynurenine 3-monooxygenase (28); lactase-phlorizin hydrolase, *or* glycosylceramidase (70); malate dehydrogenase (29,30); malate synthase (102-107); *N*-methylglutamate dehydrogenase (31); 3-methyl-2-ketobutanoate hydroxymethyltransferase, *or* 3-methyl-2-oxobutanoate hydroxymethyltransferase (32); 3-mercaptopyruvate sulfurtransferase (33,73-76); NAD$^+$ kinase, weakly inhibited (99); oxaloacetate tautomerase (34); pantothenase (66,67); phenylalanine monooxygenase (35); phosphoenolpyruvate carboxylase (61-63); phosphoenolpyruvate:protein phosphotransferase (96); phosphonopyruvate decarboxylase, pyruvate is also a slow alternative substrate (58); procollagen-lysine 5-dioxygenase (110); procollagen-proline 3-dioxygenase (110); procollagen-proline 4-dioxygenase (110-113); proline dehydrogenase (36); pyrophosphatase, inorganic diphosphatase (37); [pyruvate dehydrogenase, acetyl-transferring)] kinase (1,38-44,82-94); ribulose-bisphosphate carboxylase/oxygenase (60); saccharopine dehydrogenase (46); L-serine ammonia-lyase, *or* L-serine dehydratase (47); serine:glyoxylate aminotransferase (100); tartronate-semialdehyde reductase, *or* 2-hydroxy-3-oxopropionate reductase, weakly inhibited (48); thiosulfate sulfurtransferase, rhodanese (45); threonine ammonia-lyase, threonine dehydratase (49-51); D-threonine dehydrogenase (52); and tyrosine 2,3-aminomutase (53). **1.** Chatterjee & Sanwal (1993) *Mol. Cell. Biochem.* **126**, 125. **2.** Utter (1961) *The Enzymes*, 2nd ed. (Boyer, Lardy & Myrbäck, eds.), **5**, 319. **3.** Preston, Wall & Emerich (1990) *Biochem. J.* **267**, 179. **4.** Lee, Yi, Lee, Takahashi & Kim (2004) *Enzyme Microb. Technol.* **35**, 121. **5.** Fujita, Okuma & Abe (1997) *Fish. Sci.* **63**, 440. **6.** Jordan & Shemin (1972) *The Enzymes*, 3rd ed. (Boyer, ed.), **7**, 339. **7.** Rognes (1980) *Phytochemistry* **19**, 2287. **8.** Paxton (1988) *Meth. Enzymol.* **166**, 313. **9.** Ng, Hanauske-Abel & England (1991) *J. Biol. Chem.* **266**, 1526. **10.** Nachmansohn & John (1945) *J. Biol. Chem.* **158**, 157. **11.**

Korey, de Braganza & Nachmansohn (1951) *J. Biol. Chem.* **189**, 705. **12.** Bernheim & Deturk (1953) *Enzymologia* **16**, 69. **13.** Stoolmiller (1975) *Meth. Enzymol.* **42**, 308. **14.** Ujita & Kimura (1982) *Meth. Enzymol.* **90**, 235. **15.** Matsushita, Shinagawa & Ameyama (1982) *Meth. Enzymol.* **89**, 187. **16.** Goldstein, Richterich-Van Baerle & Dearborn (1957) *Enzymologia* **18**, 355. **17.** Guha (1962) *Enzymologia* **24**, 310. **18.** Holzer & Holldorf (1957) *Biochem. Z.* **329**, 292. **19.** Rothman & Cabib (1967) *Biochemistry* **6**, 2098. **20.** Zelitch (1955) *J. Biol. Chem.* **216**, 553. **21.** Sun, Shapiro & Rosen (1974) *Biochem. Biophys. Res. Commun.* **61**, 193. **22.** Ikawa (1982) *Meth. Enzymol.* **89**, 145. **23.** Leuthardt (1951) *The Enzymes*, 1st ed. (Sumner & Myrbäck, eds.), **1** (part 2), 1156. **24.** Delafield & Doudoroff (1969) *Meth. Enzymol.* **14**, 227. **25.** Scholtz & Schuster (1986) *Biochim. Biophys. Acta* **869**, 192. **26.** Nimmo (1984) *Trends Biochem. Sci.* **9**, 475. **27.** McFadden (1969) *Meth. Enzymol.* **13**, 163. **28.** Shin, Sano & Umezawa (1982) *J. Nutr. Sci. Vitaminol. (Tokyo)* **28**, 191. **29.** Berger & Avery, Jr. (1943) *Amer. J. Botany* **30**, 297. **30.** Green (1936) *Biochem. J.* **30**, 2095. **31.** Hersh, Stark, Worthen & Fiero (1972) *Arch. Biochem. Biophys.* **150**, 219. **32.** Powers & Snell (1976) *J. Biol. Chem.* **251**, 3786. **33.** Meister, Fraser & Tice (1954) *J. Biol. Chem.* **206**, 561. **34.** Annett & Kosicki (1969) *J. Biol. Chem.* **244**, 2059. **35.** Udenfriend & Cooper (1952) *J. Biol. Chem.* **194**, 503. **36.** Scarpulla & Soffer (1978) *J. Biol. Chem.* **253**, 5997. **37.** Naganna (1950) *J. Biol. Chem.* **183**, 693. **38.** Priestman, Orfali & Sugden (1996) *FEBS Lett.* **393**, 174. **39.** Pettit, Yeaman & Reed (1982) *Meth. Enzymol.* **90**, 195. **40.** Pettit, Yeaman & Reed (1983) *Meth. Enzymol.* **99**, 331. **41.** Roach (1984) *Meth. Enzymol.* **107**, 81. **42.** Harris, Kuntz & Simpson (1988) *Meth. Enzymol.* **166**, 114. **43.** Randle (1978) *Trends Biochem. Sci.* **3**, 217. **44.** Reed & Yeaman (1987) *The Enzymes*, 3rd ed. (Boyer & Krebs, eds.), **18**, 77. **45.** Oi (1975) *J. Biochem.* **78**, 825. **46.** Fujioka & Nakatani (1972) *Eur. J. Biochem.* **25**, 301. **47.** Farias, Strasser de Saad, Pesce de Ruiz Holgado & Oliver (1985) *J. Gen. Appl. Microbiol.* **31**, 563. **48.** Gotto & Kornberg (1961) *Biochem. J.* **81**, 273. **49.** Choi & Kim (1995) *J. Biochem. Mol. Biol.* **28**, 118. **50.** Park & Datta (1979) *J. Bacteriol.* **138**, 1026. **51.** Kim & Datta (1982) *Biochim. Biophys. Acta* **706**, 27. **52.** Misono, Kato, Packdibamrung, Nagata & Nagasaki (1993) *Appl. Environ. Microbiol.* **59**, 2963. **53.** Kurylo-Borowska & Abramsky (1972) *Biochim. Biophys. Acta* **264**, 1. **54.** Behal & Oliver (1997) *Arch. Biochem. Biophys.* **348**, 65. **55.** Tanaka, Nabeshima, Tokuda & Fukui (1977) *Agric. Biol. Chem.* **41**, 795. **56.** Takao, Takahashi, Tanida & Takahashi (1984) *J. Ferment. Technol.* **62**, 577. **57.** Munir, Hattori & Shimada (2002) *Arch. Biochem. Biophys.* **399**, 225. **58.** Zhang, Dai, Lu & Dunaway-Mariano (2003) *J. Biol. Chem.* **278**, 41302. **59.** Koga, Adachi & Hidaka (1992) *J. Biol. Chem.* **267**, 15823. **60.** Wang & Tabita (1992) *J. Bacteriol.* **174**, 3593. **61.** Marczewski (1989) *Physiol. Plant* **76**, 539. **62.** Mares, Barthova & Leblova (1979) *Collect. Czech. Chem. Commun.* **44**, 1835. **63.** Munoz, Escribano & Merodio (2001) *Phytochemistry* **58**, 1007. **64.** Braun & Schmidt, Jr., (1973) *Biochemistry* **12**, 4878. **65.** Castro, Liras, Laiz, Cortes & Martin (1988) *J. Gen. Microbiol.* **134**, 133. **66.** Airas (1976) *Biochim. Biophys. Acta* **452**, 201. **67.** Airas (1976) *Biochem. J.* **157**, 415. **68.** Huerta-Saquero, Calderon-Flores, Diaz-Villasenor, Du Pont & Duran (2004) *Biochim. Biophys. Acta* **1673**, 201. **69.** Sakamoto, Araki, Beppu &

Arima (1977) *Agric. Biol. Chem.* **41**, 1359. **70.** Hamaswamy & Radhakrishnan (1973) *Biochem. Biophys. Res. Commun.* **54**, 197. **71.** Engstler, Reuter & Schauer (1992) *Mol. Biochem. Parasitol.* **54**, 21. **72.** Villadsen & Nielsen (2001) *Biochem. J.* **359**, 591. **73.** Vachek & Wood (1972) *Biochim. Biophys. Acta* **258**, 133. **74.** Porter & Baskin (1996) *J. Biochem. Toxicol.* **11**, 45. **75.** Jarabak (1981) *Meth. Enzymol.* **77**, 291. **76.** Jarabak & Westley (1978) *Arch. Biochem. Biophys.* **185**, 458. **77.** Nimmo & Nimmo (1984) *Eur. J. Biochem.* **141**, 409. **78.** Miller, Chen, Karschnia, et al. (2000) *J. Biol. Chem.* **275**, 833. **79.** Wang & Koshland (1982) *Arch. Biochem. Biophys.* **218**, 59. **80.** Harris, Paxton & DePaoli-Roach (1982) *J. Biol. Chem.* **257**, 139 15. **81.** Paxton & Harris (1984) *Arch. Biochem. Biophys.* **231**, 48. **82.** Hucho, Randall, Roche, et al. (1972) *Arch. Biochem. Biophys.* **151**, 328. **83.** Bao, Kasten, Yan & Roche (2004) *Biochemistry* **43**, 13432. **84.** Roche, Baker, Yan, et al. (2001) *Prog. Nucl. Acid Res. Mol. Biol.* **70**, 33. **85.** Reed, Damuni & Merryfield (1985) *Curr. Top. Cell. Regul.* **27**, 41. **86.** Roche & Hiromasa (2007) *Cell. Mol. Life Sci.* **64**, 830. **87.** Schuller & Randall (1990) *Arch. Biochem. Biophys.* **278**, 211. **88.** Korotchkina, Sidhu & Patel (2004) *Free Radic. Res.* **38**, 1083. **89.** Kerbey & Randle (1985) *Biochem. J.* **231**, 523. **90.** Roche, Hiromasa, Turkan, et al. (2003) *Eur. J. Biochem.* **270**, 1050. **91.** Sugden, Langdowsn, Harris & Holness (2000) *Biochem. J.* **352**, 731. **92.** Sheu, Lai & Blass (1984) *J. Neurochem.* **42**, 230. **93.** Hiromasa, Hu & Roche (2006) *J. Biol. Chem.* **281**, 12568. **94.** Pratt & Roche (1979) *J. Biol. Chem.* **254**, 7191. **95.** Amir & Cherry (1972) *Plant Physiol.* **49**, 893. **96.** Dimitrova, Peterkofsky & Ginsburg (2003) *Protein Sci.* **12**, 2047. **97.** Kleczkowski & Randall (1988) *Planta* **173**, 221. **98.** Saharan & Singh (1993) *Plant Physiol. Biochem.* **31**, 559. **99.** Blomquist (1973) *J. Biol. Chem.* **248**, 7044. **100.** Ireland & Joy (1983) *Arch. Biochem. Biophys.* **223**, 291. **101.** Tamaki, Kubo, Aoyama & Funatsuka (1983) *J. Biochem.* **93**, 955. **102.** Miernyk & Trelease (1981) *Phytochemistry* **20**, 2657. **103.** Munir, Hattori & Shimada (2002) *Biosci. Biotechnol. Biochem.* **66**, 576. **104.** Beeckmans, Khan, Kanarek & Van Driessche (1994) *Biochem. J.* **303**, 413. **105.** Durchschlag, Biedermann & Eggerer (1981) *Eur. J. Biochem.* **114**, 255. **106.** Chell & Sundaram (1978) *J. Bacteriol.* **135**, 334. **107.** Anstrom, Kallio & Remington (2003) *Protein Sci.* **12**, 1822. **108.** Scheidegger, Gutzwiller, Kuenzi, Fiechter & Nuesch (1985) *J. Biotechnol.* **3**, 109. **109.** Izumi, Sato, Tani & Ogata (1973) *Agric. Biol. Chem.* **37**, 1335. **110.** Majamaa, Turpeenniemi-Hujanen, Latipää, et al. (1985) *Biochem. J.* **229**, 127. **111.** Majamaa, Hanauske-Abel, Günzler & Kivirikko (1984) *Eur. J. Biochem.* **138**, 239. **112.** Kaska, Günzler, Kivirikko & Myllylä (1987) *Biochem. J.* **241**, 483. **113.** Tuderman, Myllylä & Kivirikko (1977) *Eur. J. Biochem.* **80**, 341. **114.** Reddy, Pierzchala & Hamilton (1981) *J. Biol. Chem.* **256**, 8519.

Pyruvate,Orthophosphate Dikinase Regulatory Protein

This maize protein is a bifunctional protein-serine/threonine kinase/protein phosphatase (1-3). **1.** Burnell & Chastain (2006) *Biochem. Biophys. Res. Commun.* **345**, 675. **2.** Burnell & Hatch (1985) *Arch. Biochem. Biophys.* **237**, 490. **3.** Burnell & Hatch (1983) *Biochem. Biophys. Res. Commun.* **111**, 288.

– Q –

Q, Symbol for Glutamine; Quinone

Q-58, *See* 2-(N,N-Dimethyl-N-heptylammonium Bromide)-p-methan-1-ol

Q-64, *See* 2-(N,N-Dimethyl-N-octylammonium Bromide)-p-methan-1-ol

Qc 1

This embryonic stem cell-selective inhibitor (FW = 455.46 g/mol; CAS 403718-45-6; Soluble: 50 mM in DMSO; 1,2,3,4-tetrahydro-4-oxo-*N*-(phenylmethyl)-2-thioxo-3-[3-(trifluoromethyl)phenyl]-7-quinazoline-carboxamide, selectively targets threonine dehydrogenase (IC$_{50}$ ~ 0.5 µM, reversible noncompetitive), inducing autophagy and inhibiting cell proliferation in mouse embryonic stem (ES) cells. Significantly, ES cells use a mitochondrial threonine dehydrogenase to catabolize threonine into glycine and acetyl-CoA, expressing >1,000x higher levels of TDH mRNA than any of seven other mouse tissues tested. Moreover, when cell culture medium is deprived of threonine, ES cells rapidly discontinue DNA synthesis, arrest cell division, and eventually die. Because glycine is required in purine nucleotide metabolism, TDH inhibition is likely to result in a shortage of nucleotides required for DNA replication, energy metabolism, and signal transduction. **1.** Alexander, Wang & McKnight (2011) *Proc. Natl. Acad. Sci. U.S.A.* **108**, 15828.

QNZ

This signal transduction inhibitor (FW = 356.42 g/mol; CAS 545380-34-5; Solubility: 5 mg/mL DMSO; <1 mg/mL H$_2$O), also known as EVP4593 and N^4-[2-(4-phenoxyphenyl)ethyl]-4,6-quinazolinediamine, potently inhibits both NF-κB activation and TNF-α production with IC$_{50}$ values of 11 nM and 7 nM, respectively (1-3). QNZ was useful in demonstrating that NF-κB enhances the potassium currents elicited by treatment with the growth related oncogene (CXCL1) chemokine in nociceptive neurons (3). **1.** Tobe, Isobe, Tomizawa, *et al.* (2003) *Bioorg. Med. Chem.* **11**, 383. **2.** Li, Whiteman & Moore (2009) *J. Cell Mol. Med.* **13**, 2684. **3.** Yang, Strong & Zhang (2009) *Mol. Pain* **28**, 26.

QO 58

This potassium channel opener (FW = 443.18 g/mol; CAS 1259536-62-3; Solubility: Soluble to 50 mM in DMSO), also named 5-(2,6-dichloro-5-fluoro-3-pyridinyl)-3-phenyl-2-(trifluoromethyl)pyrazolo[1,5-*a*]pyrimidin-

7(4*H*)-one, targets K$_v$7.4 (EC$_{50}$ = 0.6 µM), K$_v$7.2 (EC$_{50}$ = 1.0 µM), K$_v$7.3/7.5 (EC$_{50}$ = 5.2 µM), and K$_v$7.1 (EC$_{50}$ = 7.0 µM) channels, thereby increasing the threshold for neuropathic pain in a sciatic nerve CCI *in vivo* model. Voltage-gated K$_v$7/KCNQ/M-potassium channels play a pivotal role in controlling neuronal excitability and are involved in epilepsy and arrhythmia. A chain of amino acids (Val^{224}Val^{225}Tyr^{226}) in K$_v$7.2 appears to be important for QO-58 activation of this channel (2). **1.** Qi, Zhang, Mi, *et al.* (2011) *Eur. J. Med. Chem.* **46**, 934. **2.** Zhang, Mi, Qi, *et al.* (2013) *Br. J. Pharmacol.* **168**, 1030.

Quaternary Amiton, *See* O,O-Diethyl S-[2-(N,N,N-Triethylammonio)-ethyl]thiophosphate Cation

Quazinone

This cardiotonic agent (FW = 235.67 g/mol; CAS 70018-51-8), known also as Dozonone and (*R*)-6-chloro-1,5-dihydro-3-methylimidazo[2,1-*b*]quinazolin-2[3*H*]-one, exerts its vasodilative properties by inhibiting cyclic-GMP phosphodiesterase, acting selectively on the type-III isozyme, IC$_{50}$ = 0.6 µM (1,2). **1.** Holck, Thorens, Muggli & Eigenmann (1984) *J. Cardiovasc. Pharmacol.* **6**, 520. **2.** Osinski & Schror (2000) *Biochem. Pharmacol.* **60**, 381.

Queen Substance

This honeybee pheromone (FW$_{free-acid}$ = 184.24 g/mol; M.P. = 54.5-55.5°C), also known as 9-oxo-*trans*-2-decenoic acid and Queen Mandibular Pheromone, is secreted by the queen bee to block ovary formation in worker bees and to attract male suitors (1,2). Synthetic and natural 9-oxodec-*trans*-2-enoic acid have the same biological activity (3). The methyl ester of 9-oxodecenoic acid is also active. As pointed out by Velthuis (4), Queen Substance accounts, at least in part, for the dominant status of a queen within the hive, as indicated by the attention paid her by worker bees and by the reactions of the colony to her departure as a consequence of swarming or regicide. By secreting this volatile agent, the queen bee identifies herself in the hive, thereby maintaining the social hierarchy and order. More recent work suggests that queen-worker pheromone communication is a multi-component, labile dialogue between the castes, rather than a simple, fixed signal-response system (5). QS also inhibits tRNA methyltransferases (6). The other inhibitory factor, known simply as 'Queen Scent', does not completely inhibit queen rearing; however, when combined with 9-oxodecenoic acid, the two cause complete inhibition (3). **1.** Moritz, Simon & Crewe (2000) *Naturwissenschaften* **87**, 395. **2.** Johnston, Law & Weaver (1965) *Biochemistry* **4**, 1615. **3.** Butler, Callow & Johnston (1962) *Proc. Royal Soc. Series B* **155**, 960. **4.** Velthuis (1970) *Z. vergl. Physiologie* **70**, 210. **5.** Kocher & Grozinger (2011) *J. Chem. Ecol.* **37**, 1263. **6.** Rojas, Pain, Tekitek, *et al.* (1979) *C. R. Seances Acad. Sci. D* **288**, 1335.

Quercetagetin

This flavone (FW = 318.24 g/mol; CAS 90-18-6), the yellow pigment (as the glycoside) of the French and African marigold, exhibits anti-inflammatory and anti-microbial properties. In the marigold, quercitrin is concentrated in its UV-absorbing, insect-attracting petal zones (known also as "nectar guides"). The latter are visible to certain insects (not humans), providing them pollination-facilitating orientation cues. (*See Quercitin*) **Inhibitory Target(s):** aldose reductase (1); DNA polymerase (2,3); glycogen phosphorylase (4); lipoxygenase (5); [myosin light-chain] kinase (6); phospholipase A$_2$ (7); reverse transcriptase, HIV (2,3); reverse transcriptase, Rauscher murine leukemia virus (3); RNA-directed DNA polymerase (2,3); and RNA polymerase (2,3). 1. Li, Mao, Cao, *et al.* (1991) *Yan Ke Xue Bao* 7, 29. 2. Ono & Nakane (1990) *J. Biochem.* 108, 609. 3. Ono, Nakane, Fukushima, Chermann & Barre-Sinoussi (1990) *Eur. J. Biochem.* 190, 469. 4. Kato, Nasu, Takebayashi, *et al.* (2008) *J. Agric. Food Chem.* 56, 4469. 5. Robak, Shridi, Wolbis & Krolikowska (1988) *Pol. J. Pharmacol. Pharm.* 40, 451. 6. Jinsart, Ternai & Polya (1991) *Biol. Chem. Hoppe-Seyler* 372, 819. 7. Gil, Sanz, Terencio, *et al.* (1994) *Life Sci.* 54, PL333.

Quercetin

This plant flavonoid (FW = 302.24 g/mol; CAS 117-39-5; λ_{max} at 258 nm and 375 nm), also known as quercitin, quercetol, and quertine, asnd named systematically as 2-(3,4-dihydroxyphenyl)-3,5,7-trihydroxy-4*H*-chromen-4-one, exhibits anti-thrombotic and anti-inflammatory properties, and also inhibits numerous enzymes. Quercetin inhibits phosphatidylinositol (PtdIns) 3-kinase (IC$_{50}$ ≈ 4 μM), an enzyme implicated in growth factor signal transduction by associating with receptor and nonreceptor tyrosine kinases, including the platelet-derived growth factor receptor. Given the nonspecificity of quercitin, it seems that other, more specific quercetin-derived PtdIns 3-kinase inhibitors (*e.g.*, LY294002) will prove more useful in exploring the function and regulatory mechanisms of the enzyme. *See also Quercitrin; Quercitin* **Target(s):** adenosine deaminase (1); alcohol dehydrogenase, NADP$^+$-dependent (2); aldose reductase, *or* aldehyde reductase (3-7); 1-alkylglycerophosphocholine *O*-acetyltransferase (150,151); amine sulfotransferase (122); AMP-activated protein kinase (8); aromatase, *or* CYP19 (9,10); arylamine *N*-acetyltransferase (152,153); ATPase (11,101); ATP synthase (12,13); avian myeloblastosis reverse transcriptase (14); Ca^{2+}/calmodulin-dependent protein kinase (129); caffeate *O*-methyltransferase (154); calmodulin (15); carbonyl reductase (16-19); β-carotene 15,15'-monooxygenase (163); casein kinase (28); casein kinase type G (20); catechol *O*-methyltransferase (155-160); CF$_1$ ATP synthase, *or* chloroplast H$^+$-transporting two-sector ATPase (21); chalcone isomerase (103); chitin synthase II (23); chymotrypsin (24); creatine kinase (133); 3',5'-cyclic-GMP phosphodiesterase, *or* cGMP phosphodiesterase (22), although (27) reports little inhibition; 3',5'-cyclic-nucleotide phosphodiesterase (22,25-27,142); cyclin-dependent kinase, *e.g.*, Cdk1/B, Cdk5, and Cdk6 (126); CYP1A2 (28); CYP3A4 (28); CYP2C9 (28); cystathionine β-synthase (109); diacylglycerol kinase (137); DNA-dependent protein kinase (29); DNA-directed DNA polymerase (132); DNA topoisomerase I (30); DNA topoisomerase II (30,102,136); dopamine β-hydroxylase (31); ecdysone 20-monooxygenase (32); endopeptidase La (33,116,117); epidermal growth-factor receptor protein-tyrosine kinase (131); estrone sulfatase (34); estrone sulfotransferase (123,124); 7-ethoxycoumarin *O*-deethylase activity, cytochrome P450 (35,36); ethoxyresorufin deethylase, cytochrome P450 (37,38); eukaryotic initiation factor 2α kinases, *or* interferon-inducible double-stranded RNA-dependent kinase (39); fatty-acid synthase (40,149); flavanone 7-*O*-glucoside 2"-*O*-β-L-rhamnosyltransferase (145); β-D-glucuronidase (42,43); glutathione-disulfide reductase (44,45); glutathione *S*-transferase (45,143,144); glycogen phosphorylase (146-148); [glycogen synthase] kinase-3, *or* [tau protein] kinase (126); hexokinase (46); HIV-1 retropepsin (47); H$^+$-transporting ATPase (41,100); hyaluronoglucosaminidase, *or* hyaluronidase (43,120); 17β-hydroxysteroid dehydrogenase (48); IκB kinases α and β (49); inositol-trisphosphate 3-kinase (136); integrase, HIV (50); iodothyronine deiodinase, type I (51); β-lactamase (24,114); lactate dehydrogenase (52); lactoylglutathione lyase, *or* glyoxalase I (104-106);

leucoanthocyanidin reductase (161,162); lipoxygenase (53-55); 5-lipoxygenase (56,57); 15-lipoxygenase (58); lysozyme (43); malate dehydrogenase (24); malic enzyme, NADP$^+$-dependent (59); 3-methylbutanal reductase113; mitogen-activated protein kinase60; [mitogen-activated protein kinase]-activated protein kinase (18); Moloney murine leukemia virus reverse transcriptase (14,61); [myosin light-chain] kinase (128); NADH oxidase (101); NADPH diaphorase (62); Na$^+$/K$^+$-exchanging ATPase (63-65); nitric-oxide dioxygenase (167); nucleoside diphosphate kinase (66); nucleoside-triphosphatase (112); 5'-nucleotidase (67); ornithine decarboxylase (55); peptidyl-dipeptidase A, *or* angiotensin I-converting enzyme; weakly inhibited (118); P-glycoprotein (110,111); phenol sulfotransferase, *or* aryl sulfotransferase (122,125); phenylalanine ammonia-lyase (68,107,108); 1-phosphatidylinositol 3-kinase, *or* (PtdIns) 3-kinase, IC$_{50}$ = 3.8 μM (8,69,70,134,135); 1-phosphatidylinositol 4-kinase (13, 9-141); 1-phosphatidyl-inositol-4-phosphate 5-kinase (71,138); phosphodiesterase (127); phosphodiesterase (227); phosphodiesterase 3 (27); phosphodiesterase 4 (27); phosphoenolpyruvate carboxylase (72,73); phospholipase A$_2$ (74-76); phosphorylase kinase (77,127); p70 ribosomal protein S6 kinase (8); prostaglandin endoperoxide synthase, *or* cyclooxygenase (56,78); prostaglandin E$_2$ 9-reductase (79); prostaglandin F synthase (80); proteasome endopeptidase complex (115); protein kinase (8,20,29,69,81-84); protein kinase C (69,81,84,85); protein-tyrosine kinase, *or* non-specific protein-tyrosine kinase (69,77,82,86-89,130); pyruvate kinase (52,90); receptor protein-tyrosine kinase (131); RNA-directed DNA polymerase (14,61); RNA polymerase II (91); Rous-associated virus-2 reverse transcriptase (14); Rous sarcoma virus src gene product pp60src protein kinase (83); soluble epoxide hydrolase (119); sorbitol dehydrogenase, *or* L-iditol dehydrogenase (59); steroid sulfotransferase (121); steroid 5α-reductase (92); succinoxidase (101); sulfotransferase (93-95); tyrosinase, *or* monophenol monooxygenase (43,96,164-166); xanthine oxidase (97-99); and xenobiotic-transporting ATPase, *or* multidrug-resistance protein (110,111). 1. Melzig (1996) *Planta Med.* 62, 20. 2. Flynn & Cromlish (1982) *Prog. Clin. Biol. Res.* 114, 209. 3. Wermuth & von Wartburg (1982) *Meth. Enzymol.* 89, 181. 4. Chaudhry, Cabrera, Juliani & Varma (1983) *Biochem. Pharmacol.* 32, 1995. 5. Wermuth, Burgisser, Bohren & von Wartburg (1982) *Eur. J. Biochem.* 127, 279. 6. Varma, Mikumi & Kinoshita (1975) *Science* 188, 1215. 7. Branlant (1982) *Eur. J. Biochem.* 129, 99. 8. Davies, Reddy, Caivano & Cohen (2000) *Biochem. J.* 351, 95. 9. Pelissero, Lenczowski, Chinzi, *et al.* (1996) *J. Steroid Biochem. Mol. Biol.* 57, 215. 10. Kellis, Jr., & Vickery (1984) *Science* 225, 1032. 11. Yarlett & Lloyd (1981) *Mol. Biochem. Parasitol.* 3, 13. 12. Lang & Racker (1974) *Biochim. Biophys. Acta* 333, 180. 13. Linnett & Beechey (1979) *Meth. Enzymol.* 55, 472. 14. Spedding, Ratty & Middleton, Jr. (1989) *Antiviral Res.* 12, 99. 15. Nishino, Naitoh, Iwashima & Umezawa (1984) *Experientia* 40, 184. 16. Kataoka, Doi, Sim, Shimizu & Yamada (1992) *Arch. Biochem. Biophys.* 294, 469. 17. Shimizu, Hattori, Hata & Yamada (1988) *Eur. J. Biochem.* 174, 37. 18. Wermuth (1981) *J. Biol. Chem.* 256, 1206. 19. Imamura, Migita, Uriu, Otagiri & Okawara (2000) *J. Biochem.* 127, 653. 20. Cochet, Feige, Pirollet, Keramidas & Chambaz (1982) *Biochem. Pharmacol.* 31, 1357. 21. Shoshan, Shahak & Shavit (1980) *Biochim. Biophys. Acta* 591, 421. 22. Ruckstuhl, Beretz, Anton & Landry (1979) *Biochem. Pharmacol.* 28, 535. 23. Hwang, Ahn, Lee, *et al.* (2001) *Planta Med.* 67, 501. 24. McGovern & Shoichet (2003) *J. Med. Chem.* 46, 1478. 25. Lanza, Beretz, Stierle, Corre & Cazenave (1987) *Thromb. Res.* 45, 477. 26. Beretz, Anton & Stoclet (1978) *Experientia* 34, 1054. 27. Ko, Shih, Lai, Chen & Huang (2004) *Biochem. Pharmacol.* 68, 2087. 28. Obach (2000) *J. Pharmacol. Exp. Ther.* 294, 88. 29. Izzard, Jackson & Smith (1999) *Cancer Res.* 59, 2581. 30. Constantinou, Mehta, Runyan, *et al.* (1995) *J. Nat. Prod.* 58, 217. 31. Denke, Schempp, Weiser & Elstner (2000) *Arzneimittelforschung* 50, 415. 32. Mitchell, Keogh, Crooks & Smith (1993) *Insect Biochem. Mol. Biol.* 23, 65. 33. Goldberg, Sreedhara Swamy, Chung & Larimore (1981) *Meth. Enzymol.* 80, 680. 34. Huang, Fasco & Kaminsky (1997) *J. Steroid Biochem. Mol. Biol.* 63, 9. 35. Moon, Lee & Park (1998) *Xenobiotica* 28, 117. 36. Musonda, Helsby & Chipman (1997) *Hum. Exp. Toxicol.* 16, 700. 37. Siess, Le Bon, Suschetet & Rat (1990) *Food Addit. Contam.* 7 Suppl. 1, S178. 38. Sousa & Marletta (1985) *Arch. Biochem. Biophys.* 240, 345. 39. Ito, Warnken & May (1999) *Biochem. Biophys. Res. Commun.* 265, 589. 40. Li & Tian (2003) *J. Enzyme Inhib. Med. Chem.* 18, 349. 41. Uchida, Ohsumi & Anraku (1988) *Meth. Enzymol.* 157, 544. 42. Mariscal, Gomez-Aracena, Varo & Fernandez-Crehuet (1998) *Arch. Environ. Contam. Toxicol.* 35, 588. 43. Rodney, Swanson, Wheeler, Smith & Worrel (1950) *J. Biol. Chem.* 183, 739. 44. Elliott, Scheiber, Thomas & Pardini (1992) *Biochem. Pharmacol.* 44. 1603. 45. Kurata, Suzuki & Takeda (1992) *Comp. Biochem. Physiol. B* 103, 86. 46. Graziani (1977) *Biochim. Biophys. Acta* 460, 364. 47. Xu,

Wan, Dong, But & Foo (2000) *Biol. Pharm. Bull.* **23**, 1072. **48.** Krazeisen, Breitling, Moller & Adamski (2001) *Mol. Cell. Endocrinol.* **171**, 151 and (2002) *Adv. Exp. Med. Biol.* **505**, 151. **49.** Peet & Li (1999) *J. Biol. Chem.* **274**, 32655. **50.** Fesen, Kohn, Leteurtre & Pommier (1993) *Proc. Natl. Acad. Sci. U.S.A.* **90**, 2399. **51.** Ferreira, Lisboa, Oliveira, *et al.* (2002) *Food Chem. Toxicol.* **40**, 913. **52.** Grisolia, Rubio, Feijoo & Mendelson (1975) *Physiol. Chem. Phys.* **7**, 473. **53.** Forstermann, Alheid, Frolich & Mulsch (1988) *Brit. J. Pharmacol.* **93**, 569. **54.** Wheeler & Berry (1986) *Carcinogenesis* **7**, 33. **55.** Kato, Nakadate, Yamamoto & Sugimura (1983) *Carcinogenesis* **4**, 1301. **56.** Hoult, Moroney & Payá (1994) *Meth. Enzymol.* **234**, 443. **57.** Laughton, Evans, Moroney, Hoult & Halliwell (1991) *Biochem. Pharmacol.* **42**, 1673. **58.** Luiz da Silva, Tsushida & Terao (1998) *Arch. Biochem. Biophys.* **349**, 313. **59.** O'Brien, Schofield & Edwards (1983) *Biochem. J.* **211**, 81. **60.** Bird, Schule, Delaney, *et al.* (1992) *Cytokine* **4**, 429. **61.** Chu, Hsieh & Lin (1992) *J. Nat. Prod.* **55**, 179. **62.** Tamura, Kagawa, Tsuruo, Ishimura & Morita (1994) *Jpn. J. Pharmacol.* **65**, 371. **63.** Umarova, Khushbactova, Batirov & Mekler (1998) *Membr. Cell. Biol.* **12**, 27. **64.** Robinson, Robinson & Martin (1984) *Biochim. Biophys. Acta* **772**, 295. **65.** Nakamura & Racker (1984) *Biochemistry* **23**, 385. **66.** Martin, O'Sullivan & Gomperts (1995) *Brit. J. Pharmacol.* **115**, 1080. **67.** Kavutcu & Melzig (1999) *Pharmazie* **54**, 457. **68.** Hanson & Havir (1972) *The Enzymes*, 3rd ed. (Boyer, ed.), **7**, 75. **69.** Agullo, Gamet-Payrastre, Manenti, *et al.* (1997) *Biochem. Pharmacol.* **53**, 1649. **70.** Matter, Brown & Vlahos (1992) *Biochem. Biophys. Res. Commun.* **186**, 624. **71.** Cochet & Chambaz (1986) *Biochem. J.* **237**, 25. **72.** Moraes & Plaxton (2000) *Eur. J. Biochem.* **267**, 4465. **73.** Pairoba, Colombo & Andreo (1996) *Biosci. Biotechnol. Biochem.* **60**, 779. **74.** Lindahl & Tagesson (1993) *Inflammation* **17**, 573. **75.** Fawzy, Vishwanath & Franson (1988) *Agents Actions* **25**, 394. **76.** Lanni & Becker (1985) *Int. Arch. Allergy Appl. Immunol.* **76**, 214. **77.** Srivastava (1985) *Biochem. Biophys. Res. Commun.* **131**, 1. **78.** Kalkbrenner, Wurm & von Bruchhausen (1992) *Pharmacology* **44**, 1. **79.** Pace-Asciak & Smith (1983) *The Enzymes*, 3rd ed. (Boyer, ed.), **16**, 543. **80.** Chen, Watanabe & Hayaishi (1992) *Arch. Biochem. Biophys.* **296**, 17. **81.** Davies, Reddy, Caivano & Cohen (2000) *Biochem. J.* **351**, 95. **82.** Hagiwara, Inoue, Tanaka, *et al.* (1988) *Biochem. Pharmacol.* **37**, 2987. **83.** Graziani, Erikson & Erikson (1983) *Eur. J. Biochem.* **135**, 583. **84.** Gschwendt, Horn, Kittstein & Marks (1983) *Biochem. Biophys. Res. Commun.* **117**, 444. **85.** Varadkar, Dubey, Krishna & Verma (2001) *J. Radiol. Prot.* **21**, 361. **86.** Wenzel, Kuntz & Daniel (2001) *J. Pharmacol. Exp. Ther.* **299**, 351. **87.** Levy, Teuerstein, Marbach, Radian & Sharoni (1984) *Biochem. Biophys. Res. Commun.* **123**, 1227. **88.** Akiyama & Ogawara (1991) *Meth. Enzymol.* **201**, 362. **89.** Brunati & Pinna (1988) *Eur. J. Biochem.* **172**, 451. **90.** Smith, Knowles & Plaxton (2000) *Eur. J. Biochem.* **267**, 4477. **91.** Nose (1984) *Biochem. Pharmacol.* **33**, 3823. **92.** Hiipakka, Zhang, Dai, Dai & Liao (2002) *Biochem. Pharmacol.* **63**, 1165. **93.** Marchetti, De Santi, Vietri, *et al.* (2001) *Xenobiotica* **31**, 841. **94.** Eaton, Walle, Lewis, *et al.* (1996) *Drug Metab. Dispos.* **24**, 232. **95.** Mesia-Vela & Kauffman (2003) *Xenobiotica* **33**, 1211. **96.** Kubo & Kinst-Hori (1999) *J. Agric. Food Chem.* **47**, 4121. **97.** Iio, Ono, Kai & Fukumoto (1986) *J. Nutr. Sci. Vitaminol.* **32**, 635. **98.** Selloum, Reichl, Muller, Sebihi & Arnhold (2001) *Arch. Biochem. Biophys.* **395**, 49. **99.** Beiler & Martin (1951) *J. Biol. Chem.* **192**, 831. **100.** Jonas, Smith, Allison, *et al.* (1983) *J. Biol. Chem.* **258**, 11727. **101.** Bohmont, Aaronson, Mann & Pardini (1987) *J. Nat. Prod.* **50**, 427. **102.** Jo, Gonzalez de Mejia & Lila (2005) *J. Agric. Food Chem.* **53**, 2489. **103.** Dixon, Dey & Whitehead (1982) *Biochim. Biophys. Acta* **715**, 25. **104.** Allen, Lo & Thornalley (1993) *Biochem. Soc. Trans.* **21**, 535. **105.** Ito, Sadakane, Shiotsuki & Eto (1992) *Biosci. Biotechnol. Biochem.* **56**, 1461. **106.** Sommer, Fischer, Krause, *et al.* (2001) *Biochem. J.* **353**, 445. **107.** Sarma & Sharma (1999) *Phytochemistry* **50**, 729. **108.** Iredale & Smith (1974) *Phytochemistry* **13**, 575. **109.** Walker & Barrett (2002) *Exp. Parasitol.* **74**, 205. **110.** Sharom, Liu, Romsicki & Lu (1999) *Biochim. Biophys. Acta* **1461**, 327. **111.** Shapiro & Ling (1997) *Eur. J. Biochem.* **250**, 130. **112.** Schröder, Rottmann, Bachmann & Müller (1986) *J. Biol. Chem.* **261**, 663. **113.** Van Nedervelde, Verlingen, Philipp & Debourg (1997) *Proc. Congr. Eur. Brew. Conv.* **26**, 447. **114.** Denny, Lambert & West (2002) *FEMS Microbiol. Lett.* **208**, 21. **115.** Akaishi, Sawada & Yokosawa (1996) *Biochem. Mol. Biol. Int.* **39**, 1017. **116.** Larimore, Waxman & Goldberg (1982) *J. Biol. Chem.* **257**, 4187. **117.** Desautels & Goldberg (1982) *J. Biol. Chem.* **257**, 11673. **118.** Actis-Goretta, Ottaviani & Fraga (2006) *J. Agric. Food Chem.* **54**, 229. **119.** Bellevik, Zhang & Meijer (2002) *Eur. J. Biochem.* **269**, 5295. **120.** Girish & Kemparaju (2005) *Biochemistry (Moscow)* **70**, 948. **121.** Ohkimoto, Liu, Suiko, Sakakibara & Liu (2004) *Chem. Biol. Interact.* **147**, 1. **122.** Harris, Wood, Bottomley, *et al.* (2004) *J. Clin. Endocrinol. Metab.* **89**, 1779. **123.** Allali-Hassani, Pan, Dombrovski, *et al.* (2007) *PLoS Biol.* **5**, e97. **124.** Otake, Nolan, Walle & Walle (2000) *J. Steroid Biochem. Mol. Biol.* **73**, 265. **125.** Nishimuta, Ohtani, Tsujimoto, *et al.* (2007) *Biopharm. Drug Dispos.* **28**, 491. **126.** Lu, Chang, Baratte, Meijer & Schulze-Gahmen (2005) *J. Med. Chem.* **48**, 737. **127.** Nikolaropoulos & Sotiroudis (1985) *Eur. J. Biochem.* **151**, 467. **128.** Jinsart, Ternai & Polya (1991) *Biol. Chem. Hoppe-Seyler* **372**, 819. **129.** Boulton, Gregory & Cobb (1991) *Biochemistry* **30**, 278. **130.** F. Hollósy & G. Kéri (2004) *Curr. Med. Chem. Anticancer Agents* **4**, 173. **131.** Lee, Huang, Hwang, *et al.* (2004) *Biochem. Pharmacol.* **67**, 2103. **132.** Spampinato, Pairoba, Colombo, Benediktsson & Andreo (1994) *Biosci. Biotechnol. Biochem.* **58**, 822. **133.** Miura, Muraoka & Fujimoto (2003) *Food Chem. Toxicol.* **41**, 759. **134.** Shibasaki, Fukui & Takenawa (1993) *Biochem. J.* **289**, 227. **135.** Baxter, Bramham, Thomason, Downes & Carter (1993) *Biochem. Soc. Trans.* **21**, 359S. **136.** Mayr, Windhorst & Hillemeier (2005) *J. Biol. Chem.* **280**, 13229. **137.** Kato & Takenawa (1990) *J. Biol. Chem.* **265**, 794. **138.** Urumow & Wieland (1990) *Biochim. Biophys. Acta* **1052**, 152. **139.** Hou, Zhang & Tai (1988) *Biochim. Biophys. Acta* **959**, 67. **140.** Yamakawa & Takenawa (1988) *J. Biol. Chem.* **263**, 17555. **141.** Prajda, Singhal, Yeh, *et al.* (1995) *Life Sci.* **56**, 1587. **142.** Mata, Gamboa, Macias, *et al.* (2003) *J. Agric. Food Chem.* **51**, 4559. **143.** Hayeshi, Mutingwende, Mavengere, Masiyanise & Mukanganyama (2007) *Food Chem. Toxicol.* **45**, 286. **144.** Liebau, Eckelt, Wildenburg, *et al.* (1997) *Biochem. J.* **324**, 659. **145.** Bar-Peled, Lewinsohn, Fluhr & Gressel (1991) *J. Biol. Chem.* **266**, 20953. **146.** Gregus & Németi (2007) *Toxicol. Sci.* **100**, 44. **147.** Kato, Nasu, Takebayashi, *et al.* (2008) *J. Agric. Food Chem.* **56**, 4469. **148.** Jakobs, Fridrich, Hofem, Pahlke & Eisenbrand (2006) *Mol. Nutr. Food Res.* **50**, 52. **149.** Li & Tian (2004) *J. Biochem.* **135**, 85. **150.** Yanoshita, Chang, Son, Kudo & Samejina (1996) *Inflamm. Res.* **45**, 546. **151.** Hurst & Bazan (1997) *J. Ocul. Pharmacol. Ther.* **13**, 415. **152.** Kukongviriyapan, Phromsopha, Tassaneeyakul, *et al.* (2006) *Xenobiotica* **36**, 15. **153.** Makarova (2008) *Curr. Drug Metab.* **9**, 538. **154.** Poulton, Hahlbrock & Grisebach (1976) *Arch. Biochem. Biophys.* **176**, 449. **155.** Chen, Wang, Lambert, *et al.* (2005) *Biochem. Pharmacol.* **69**, 1523. **156.** Nagai, Conney & Zhu (2004) *Drug Metab. Dispos.* **32**, 497. **157.** Rutherford, Le Trong, Stenkamp & Parson (2008) *J. Mol. Biol.* **380**, 120. **158.** Bonifácio, Palma, Almeida & Soares-da-Silva (2007) *CNS Drug Reviews* **13**, 352. **159.** van Duursen, Sanderson, de Jong, Kraaij & van den Berg (2004) *Toxicol. Sci.* **81**, 316. **160.** L. C. Zaccharia, R. K. Dubay, Z. Mi & E. K. Jackson (2003) *Hypertension* **42**, 82. **161.** D. Y. Xie, S. B. Sharma & R. A. Dixon (2004) *Arch. Biochem. Biophys.* **422**, 91. **162.** Tanner, Francki, Abrahams, *et al.* (2003) *J. Biol. Chem.* **278**, 31647. **163.** Nagao, Maeda, Lim, Kobayashi & Terao (2000) *Nutr. Biochem.* **11**, 348. **164.** Kim & Uyama (2005) *Cell. Mol. Life Sci.* **62**, 1707. **165.** Lin, Hsu, Chen, Chern & Lee (2007) *Phytochemistry* **68**, 1189. **166.** Karioti, Protopappa, Magoulas & Skaltsa (2007) *Bioorg. Med. Chem.* **15**, 2708. **167.** Hallstrom, Gardner & Gardner (2004) *Free Radic. Biol. Med.* **37**, 216.

Quercetin-3-*O*-rutinoside, See *Rutin*

Quercitrin

This plant arylglycoside and antioxidant (FW = 448.38 g/mol; CAS 522-12-3; λ_{max} at 258 and 350 nm), also known as 2-(3,4-dihydroxyphenyl)-5,7-dihydroxy-3-[[(2*S*,3*R*,4*R*,5*R*,6*S*)-3,4,5-trihydroxy-6-methyl-2-tetrahydro-pyranyl]oxy]-4-chromenone, inhibits many enzymes: acetoin dehydrogenase (1); NADP$^+$-dependent alcohol dehydrogenase (1); aldose reductase, *or* aldehyde reductase (2-4); ATP synthase (5); carbonyl reductase (1,6,7); catechol *O*-methyltransferase (8,9); chitin synthase II (10); DNA-directed DNA polymerase (11); dopamine β-hydroxylase (12); fatty-acid synthase (13); [myosin light-chain] kinase (14); peptidyl-dipeptidase A, *or* angiotensin I-converting enzyme (15); phosphoenolpyruvate carboxylase (16); and xanthine oxidase (17). The *in vivo* hydrolysis of quercitrin liberates quercitin a potent anti-inflammatory agent (**See** *Quercitin*). **1.** Hara, Seirik, Nakayama & Sawada (1985) *Prog. Clin. Biol. Res.* **174**, 291. **2.** Wermuth & von Wartburg (1982) *Meth. Enzymol.* **89**, 181. **3.** Chaudhry, Cabrera, Juliani & Varma (1983) *Biochem. Pharmacol.* **32**, 1995. **4.** Ahmed, Felsted & Bachur (1979) *J. Pharmacol. Exp. Ther.* **209**, 12. **5.** Linnett & Beechey (1979) *Meth Enzymol.* **55**, 472. **6.** Wermuth (1981) *J. Biol. Chem.* **256**, 1206. **7.** Hara, Deyashiki, Nakagawa, Nakayama & Sawada (1982) *J. Biochem.* **92**, 1753. **8.** Axelrod (1962) *Meth. Enzymol.* **5**, 748. **9.** Axelrod & Laroche (1959) *Science* **130**, 800. **10.** Hwang, Ahn, Lee, *et al.* (2001) *Planta Med.* **67**, 501. **11.** Spampinato, Pairoba, Colombo, Benediktsson & Andreo (1994) *Biosci. Biotechnol. Biochem.* **58**, 822. **12.** Denke, Schempp, Weiser & Elstner (2000) *Arzneimittelforschung* **50**, 415. **13.** Wang, Zhang, Ma & Tian (2006) *J. Enzyme Inhib. Med. Chem.* **21**, 87. **14.** Jinsart, Ternai & Polya (1991) *Biol. Chem. Hoppe-Seyler* **372**, 819. **15.** Kiss, Kowalski & Melzig (2004) *Planta Med.* **70**, 919. **16.** Pairoba, Colombo & Andreo (1996) *Biosci. Biotechnol. Biochem.* **60**, 779. **17.** Iio, Ono, Kai & Fukumoto (1986) *J. Nutr. Sci. Vitaminol.* **32**, 635.

Quetiapine

This atypical antipsychotic agent (FW = 383.51 g/mol; CAS 111974-69-7; formulated as the fumarate salt as well as in once-daily, sustained release forms), known commonly as ICI 204,636, Seroquel®, Seroquel-XR®, and Ketipinor®, and systematically named 2-(2-(4-dibenzo[*b,f*][1,4]thiazepin-11-yl-1-piperazinyl)ethoxy)ethanol, is a dopamine, serotonin, and andrenergic antagonist, with antihistamine effects and little anticholinergic action. Consistent with many studies indicating the efficacy of dopamine D_2 receptor antagonists in treating psychotic symptoms, this agent has been used to manage schizophrenia, depressive episodes associated with bipolar disorder, as well as acute manic episodes associated with Type 1 bipolar disorder (1-3). Quetiapine has the following pharmacologic actions: (a) as an α_1-adrenergic receptor antagonist, IC_{50} = 90 nM, and as an α_2 antagonist, IC_{50} = 0.33 µM; (b) as a dopamine D_1 receptor antagonist, IC_{50} = 1.3 µM; as D_2 antagonist, IC_{50} = 0.3 µM, and more weakly as D_3 and D_4 antagonists; (c) as a histamine H_1 receptor antagonist, IC_{50} = 30 nM; and (d) as a serotonin 5-HT_{1A} receptor antagonist, IC_{50} = 0.7 µM, and more weakly as 5-T_{2A}, 5-T_{2C} and 5-T_7 antagonists. The efficacy of Seroquel in schizophrenia is thought to be mediated through a combination of dopamine D_2 and serotonin 5-HT_{2A} receptor antagonism. The active metabolite *N*-desalkyl quetiapine (norquetiapine) has similar activity at D_2, but greater activity at 5-HT_{2A} receptors, than quetiapine. While both exhibit affinity for multiple neurotransmitter receptors, quetiapine differs from norquetiapine in having no appreciable affinity for muscarinic M_1 receptors whereas norquetiapine has high affinity. Quetiapine and norquetiapine also lack appreciable affinity for benzodiazepine receptors. HPLC and GC-MS detection of seroquel and its 7-hydroxy metabolite within human plasma is also well defined (4). PK profiles of the immediate-release formulation (quetiapine IR) show a rapid peak in plasma level and striatal dopamine D_2 receptor occupancy, followed by a rapid decrease to baseline levels, necessitating twice-daily dosing (5). Its externded release (ER) formulation does not change the overall absorption or elimination profile, a finding that supports the use of quetiapine XR as a once daily treatment in patients initiating therapy or those established on the immediate release drug (quetiapine IR) (6). **Key Pharmacokinetic Parameters:** *See* Appendix II in Goodman & Gilman's THE PHARMACOLOGICAL BASIS OF THERAPEUTICS, 12th Edition

(Brunton, Chabner & Knollmann, eds.) McGraw-Hill Medical, New York (2011). **1.** Campbell, Yeghiayan, Baldessarini & Neumeyer (1991) *Psychopharmacology* (Berlin) **103**, 323. **2.** Goren & Levin (1998) *Pharmacotherapy* **18**, 1183. **3.** Green (1999) *Curr. Med. Res. Opin.* **15**, 145. **4.** Pullen Palermo & Curtis (1992) *J. Chromatogr.* **573**, 49. **5.** Mamo, Uchida, Vitcu, *et al.* (2008) *J. Clin. Psychiatry* **69**, 81. **6.** Figueroa, Brecher, Hamer-Maansson & Winter (2009) *Prog. Neuropsychopharmacol. Biol. Psychiatry* **33**, 199.

Quinacillin

This β-lactamase inhibitor (FW$_{free-acid}$ = 416.41 g/mol; CAS 1596-63-0; photosensitive), systematically named (2*S*,5*R*,6*R*)-6-[[(3-carboxy-2-quinoxalinyl)carbonyl]amino]-3,3-dimethyl-7-oxo-4-thia-1-azabicyclo[3.2.0]heptane-2-carboxylic acid, is an antimicrobial agent (1,2). CD measurements of the deactivated protein, separated from excess quinacillin, showed that the quinacillin side-chain chromophore was bound in an asymmetric environment (2). **1.** Persaud, Pain & Virden (1986) *Biochem. J.* **237**, 723. **2.** Carrey, Virden & Pain (1984) *Biochim. Biophys. Acta* **785**, 104.

Quinacrine

This acridine-based antimalarial (FW$_{free-base}$ = 399.97 g/mol; CAS 83-89-6 (free base), 130-42-7 9 (HCl salt); and 6151-30-0 (di-HCl, dehydrate); pK_1 = 7.7; pK_2 = 10), also called Acrichine, Mepacrine, Atebrin, and Atabrine, and named systematically as N^4-(6-chloro-2-methoxy-9-acridinyl)-N^1,N^1-diethyl-1,4-pentanediamine, is a DNA intercalator used to fluorescently label and/or detect DNA. In human cells, quinacrine treatment results in enhanced fluorescence of the Y-chromosome, a property that is not due to increased binding of the dye. First introduced in 1932 as a treatment for malaria, quinacrine suppresses the growth of the malaria parasite and many other microorganisms, including *Lactobacillus casei*, as well as spore germination of *Bacillus subtilis* and *B. coagulans*. Although formerly used as an antimalarial, quincrine has been superseded by chloroquine. Even so, both enantiomers of quinacrine display equal activity *in vitro* against chloroquine-sensitive and -resistant strains of *Plasmodium falciparum*. Quinacrine has also been used as an anthelmintic and in the treatment of giardiasis and malignant effusions. Its use in cell biology relates mainly to its potent inhibition of phospholipase A_2, a rate-limiting step in the synthesis of anti-inflammatory eicosanoids. Quinacrine inhibits the transcription and activity of both basal and inducible NF-κB, inducing tumor suppressor p53 transcription, restoring p53-dependent apoptotic pathways, as well as inducing tumor cell apoptosis. This pluripotent agent also inhibits oxidative phosphorylation, uncouples photophosphorylation, prevents formation of prion aggregates *in vitro* by binding to prion monomers, and acts as an effective antifertility agent, when delivered by a drug-releasing IUD. **Target(s):** acetylcholinesterase (1,2);

acetylspermidine deacetylase (111); alcohol dehydrogenase (3); aldehyde oxidase (4-7); allohydroxy-D-proline oxidase (8); amine oxidase (9-11,105); D-amino-acid oxidase (12-16); L-amino-acid oxidase (17); asparaginase (112); ATPase (18,19); benzoate 4-monooxygenase (122); biotin sulfoxide reductase (20); bontoxilysin (113); γ-butyrobetaine:2-oxoglutarate dioxygenase (21); Ca^{2+}-dependent ATPase (22); catechol oxidase (23); cGMP phosphodiesterase (24); cholest-5-ene-3β,7α-diol 3β-dehydrogenase (25); choline dehydrogenase (26); cholinesterase1,2; cyclohexanone monooxygenase (121); cyclooxygenase (27); cytochrome *c* oxidase (28); cytochrome *c* reductase (28); cytochrome *c*-554 reductase (29); diamine oxidase (11,30); diferric-transferrin reductase (119); DNA ligase, NAD⁺-dependent (31); DNA methylases (32); DNA polymerase (33,34); ethanolamine oxidase (35,36); F_1 ATPase, *or* F_1F_o ATP synthase, *or* H⁺-transporting two-sector ATPase (37); ferrooxidase (38); galactonolactone dehydrogenase (39); glucose dehydrogenase (40,41); glucose-6-phosphate dehydrogenase (28,42,43); glutaminase (44-49); hexokinase (50), however (19) lists quinacrine as a stimulator; histamine *N*-methyltransferase (51,52,116); hydrogenase (53); L-2-hydroxyacid oxidase (54); hydroxylamine oxidase (55); hydroxylamine reductase, NADH-dependent (56); 4-hydroxymandelate oxidase (57); 4-hydroxyproline epimerase (108); 2-hydroxyquinoline 5,6-dioxygenase, weakly inhibited (123); ITPase (58); lactate dehydrogenase (59,60); L-lactate oxidase (61,62); lactate racemase (107); lipase, *or* aliesterase (43); 5-lipoxygenase (63); malate dehydrogenase (3,64); malate dehydrogenase (acceptor), *or* malate:vitamin K reductase (65); L-mandelate dehydrogenase (104); 4-methoxybenzoate monooxygenase, *O*-demethylating (120); myeloperoxidase (66); myosin ATPase (67); NADH dehydrogenase, *or* NADH:cytochrome *c* reductase (3,28,68-72); NAD(P)H dehydrogenase (quinone), *or* diaphorase (73-76,106); nicotinate dehydrogenase (117); nitrate reductase (29,77-79); nitrite reductase (56,80); nitroaryl nitroreductase (81); oxidative (mitochondrial) and photophosphorylation uncoupler (18); phospholipase A_1 (114); phospholipase A_2 (82-87,114); phospholipase C (24); L-pipecolate dehydrogenase (88); *Plasmodium falciparum* aminopeptidase (90); platelet-activating factor acetyltransferase (116); polyamine oxidase (90); protein-disulfide reductase (91); retinal oxidase, *or* retinene oxidase (92); riboflavin transglucosidase (93); RNA nucleotidyltransferase (94); rubredoxin:NAD⁺ reductase (110); spermidine dehydrogenase (95); succinate dehydrogenase (96); sulfite reductase (53,97); sulfur reductase (109); taurine dehydrogenase (98); thiosulfate reductase (53,97); thyroxine deiodinase (99); trypanothione reductase (100); xanthine dehydrogenase (118); and xanthine oxidase (101-103). **1.** Augustinsson (1950) *The Enzymes*, 1st ed. (Sumner & Myrbäck, eds.), **1** (part 1), 443. **2.** Wright & Sabine (1948) *J. Pharmacol. Exptl. Therap.* **93**, 230. **3.** Frimmer (1961) *Arch. Exptl. Pathol. Pharmakol.* **242**, 96. **4.** Mahler (1955) *Meth. Enzymol.* **1**, 523. **5.** Mahler, Mackler, Green & Bock (1954) *J. Biol. Chem.* **210**, 465. **6.** Rajagopalan, Fridovich & Handler (1962) *J. Biol. Chem.* **237**, 922. **7.** Obach, Huynh, Allen & Beedham (2004) *J. Clin. Pharmacol.* **44**, 7. **8.** Yoneya & Adams (1961) *J. Biol. Chem.* **236**, 3272. **9.** Tabor, Tabor & Rosenthal (1955) *Meth. Enzymol.* **2**, 390. **10.** Ma & Sourkes (1980) *Agents Actions* **10**, 395. **11.** Zeller (1963) *The Enzymes*, 2nd ed. (Boyer, Lardy & Myrbäck, eds.), **8**, 313. **12.** Burton (1955) *Meth. Enzymol.* **2**, 199. **13.** Hellerman, Lindsay & Bovarnick (1946) *J. Biol. Chem.* **163**, 553. **14.** Massart (1950) *The Enzymes*, 1st ed. (Sumner & Myrbäck, eds.), **1** (part 1), 307. **15.** Krebs (1951) *The Enzymes*, 1st ed. (Sumner & Myrbäck, eds.), **2** (part 1), 499. **16.** Meister & Wellner (1963) *The Enzymes*, 2nd ed. (Boyer, Lardy & Myrbäck, eds.), **7**, 609. **17.** Singer & Kearney (1950) *Arch. Biochem.* **27**, 348. **18.** Izawa & Good (1972) *Meth. Enzymol.* **24**, 355. **19.** Löw (1959) *Exptl. Cell Res.* **16**, 456. **20.** del Campillo-Campbell, Dykhuizen & Cleary (1979) *Meth. Enzymol.* **62**, 379. **21.** Lindstedt (1967) *Biochemistry* **6**, 1271. **22.** Garcia-Martin & Gutierrez-Merino (1986) *J. Neurochem.* **47**, 668. **23.** Nair & Vining (1964) *Arch. Biochem. Biophys.* **106**, 422. **24.** Yamakado, Tanaka & Hidaka (1984) *Biochim. Biophys. Acta* **801**, 111. **25.** Wikvall (1981) *J. Biol. Chem.* **256**, 3376. **26.** Bargoni (1963) *Z. Physiol. Chem.* **333**, 242. **27.** Raz (1983) *Thromb. Haemost.* **50**, 784. **28.** Haas (1944) *J. Biol. Chem.* **155**, 321. **29.** Hori (1963) *J. Biochem.* **53**, 354. **30.** Suzuki & Yamasaki (1968) *Enzymologia* **35**, 198. **31.** Ciarrocchi, MacPhee, Deady & Tilley (1999) *Antimicrob. Agents Chemother.* **43**, 2766. **32.** Tanaka, Hibasami, Nagai & Ikeda (1982) *Aust. J. Exp. Biol. Med. Sci.* **60**, 223. **33.** Fox, Popanda, Edler & Thielmann (1996) *J. Cancer Res. Clin. Oncol.* **122**, 78. **34.** Bohner & Hagen (1977) *Biochim. Biophys. Acta* **479**, 300. **35.** Narrod & Jakoby (1966) *Meth. Enzymol.* **9**, 354. **36.** Narrod & Jakoby (1964) *J. Biol. Chem.* **239**, 2189. **37.** Laikind & Allison (1983) *J. Biol. Chem.* **258**, 11700. **38.** Blaylock & Nason (1963) *J. Biol. Chem.* **238**, 3453. **39.** Mapson & Breslow (1958) *Biochem. J.* **68**, 395. **40.** Hauge (1966) *Meth. Enzymol.* **9**, 92 and 107. **41.** Eichel & Wainio (1948) *J. Biol. Chem.*

175, 155. **42.** Noltmann & Kuby (1963) *The Enzymes*, 2nd ed. (Boyer, Lardy & Myrbäck, eds.), **7**, 223. **43.** Hemker & Hülsmann (1960) *Biochim. Biophys. Acta* **44**, 175. **44.** Zittle (1951) *The Enzymes*, 1st ed. (Sumner & Myrbäck, eds.), **1** (part 2), 922. **45.** Archibald (1944) *J. Biol. Chem.* **154**, 657. **46.** Guha (1962) *Enzymologia* **24**, 310. **47.** Richterich-Van Baerle, Goldstein & Dearborn (1957) *Enzymologia* **18**, 190. **48.** Goldstein, Richterich-Van Baerle & Dearborn (1957) *Enzymologia* **18**, 261. **49.** Guha (1962) *Enzymologia* **24**, 310. **50.** Fraser & Kermack (1957) *Brit. J. Pharmacol.* **12**, 16. **51.** Harle & Baldo (1988) *Biochem. Pharmacol.* **37**, 385. **52.** Zawilska & Nowak (1985) *Pol. J. Pharmacol. Pharm.* **37**, 821. **53.** Ishimoto, Kondo, Kameyama, Yagi & Shiraki (1958) *Proc. Intern. Symp. Enzyme Chem., Tokyo Kyoto 1957*, p. 229, Maruzen, Tokyo. **54.** Robinson, Keay, Molinari & Sizer (1962) *J. Biol. Chem.* **237**, 2001. **55.** Aleem & Lees (1963) *Can. J. Biochem. Physiol.* **41**, 763. **56.** Roussos & Nason (1960) *J. Biol. Chem.* **235**, 2997. **57.** Bhat & Vaidyanathan (1976) *Eur. J. Biochem.* **68**, 323. **58.** Kaldor & Gitlin (1963) *Arch. Biochem. Biophys.* **102**, 216. **59.** Snoswell (1959) *Australian J. Exptl. Biol. Med. Sci.* **37**, 49. **60.** Molinari & Lara (1960) *Biochem. J.* **75**, 57. **61.** Boeri, Cutolo, Luzzati & Tosi (1955) *Arch. Biochem. Biophys.* **56**, 487. **62.** Eichel & Rem (1959) *Biochim. Biophys. Acta* **35**, 571. **63.** Stewart, White, Jobling, *et al.* (2001) *J. Neurosci. Res.* **65**, 565. **64.** Benziman (1969) *Meth. Enzymol.* **13**, 129. **65.** Asano & Brodie (1963) *Biochem.Biophys. Res. Commun.* **13**, 423. **66.** Kettle & Winterbourn (1991) *Biochem. Pharmacol.* **41**, 1485. **67.** Grillo, Rinaudo & Vergani (1964) *Enzymologia* **27**, 41. **68.** Brodie (1952) *J. Biol. Chem.* **199**, 835. **69.** Doi & Halvorson (1961) *J. Bacteriol.* **81**, 51. **70.** Ryan & King (1962) *Arch. Biochem. Biophys.* **85**, 450. **71.** Dolin (1959) *J. Bacteriol.* **77**, 383. **72.** Vernon, Mahler & Sarkar (1952) *J. Biol. Chem.* **199**, 599. **73.** Ernster (1967) *Meth. Enzymol.* **10**, 309. **74.** Martius (1963) *The Enzymes*, 2nd ed. (Boyer, Lardy & Myrbäck, eds.), **7**, 517. **75.** Harper & Strecker (1962) *J. Neurochem.* **9**, 125. **76.** Raw, Nogueira & Filho (1961) *Enzymologia* **23**, 123. **77.** Hageman & Hucklesby (1971) *Meth. Enzymol.* **23**, 491. **78.** Hageman & Reed (1980) *Meth. Enzymol.* **69**, 270. **79.** Heredia & Medina (1960) *Biochem. J.* **77**, 24. **80.** Nicholas, Medina & Jones (1960) *Biochim. Biophys. Acta* **37**, 468. **81.** Higgins (1961) *Enzymologia* **23**, 176. **82.** Al Moutaery & Tariq (1997) *Digestion* **58**, 129. **83.** Sargent, Vesterqvist, McCullough, Ogletree & Grover (1992) *J. Pharmacol. Exp. Ther.* **262**, 1161. **84.** Jain, Yu, Rogers, Ranadive & Berg (1991) *Biochemistry* **30**, 7306. **85.** Kench, Seale, Temple & Tennant (1985) *Prostaglandins* **30**, 199. **86.** Schoene (1978) *Adv. Prostaglandin Thromboxane Res.* **3**, 121. **87.** Magolda, Ripka, Galbraith, Johnson & Rudnick (1985) in *Prostaglandins, Leukotrienes, and Lipoxins* (Bailey, ed.), p. 669, Plenum Press, New York. **88.** Rodwell (1971) *Meth. Enzymol.* **17B**, 174. **89.** Vander Jagt, Baack & Hunsaker (1984) *Mol. Biochem. Parasitol.* **10**, 45. **90.** Hölttä (1983) *Meth. Enzymol.* **94**, 306. **91.** Asahi, Bandurski & Wilson (1961) *J. Biol. Chem.* **236**, 1830. **92.** Mahadevan, Murthy & Ganguly (1962) *Biochem. J.* **85**, 326. **93.** Tachibana, Katagiri & Yamada (1958) *Proc. Intern. Symp. Enzyme Chem., Tokyo Kyoto 1957*, p. 154, Maruzen, Tokyo. **94.** Hochster & Chang (1963) *Can. J. Biochem. Physiol.* **41**, 1503. **95.** Tabor & Kellogg (1971) *Meth. Enzymol.* **17B**, 746. **96.** Wadkins & Mills (1956) *Fed. Proc.* **15**, 377. **97.** Ishimoto & Yagi (1961) *J. Biochem.* **49**, 103. **98.** Kondo & Ishimoto (1987) *Meth. Enzymol.* **143**, 496. **99.** Tata (1959) *Biochim. Biophys. Acta* **35**, 567. **100.** Bonse, Santelli-Rouvier, Barbe & Krauth-Siegel (1999) *J. Med. Chem.* **42**, 5448. **101.** Roussos (1967) *Meth. Enzymol.* **12A**, 5. **102.** Mackler, Mahler & Gleen (1954) *J. Biol. Chem.* **210**, 149. **103.** Fridovich & Handler (1957) *J. Biol. Chem.* **228**, 67. **104.** Hoey, Allison, Scott & Fewson (1987) *Biochem. J.* **248**, 871. **105.** Tabor, Tabor & Rosenthal (1954) *J. Biol. Chem.* **208**, 645. **106.** Wosilait & Nason (1954) *J. Biol. Chem.* **208**, 785. **107.** Hiyama, Fukui & Kitahara (1968) *J. Biochem.* **64**, 99. **108.** Adams & Norton (1964) *J. Biol. Chem.* **239**, 1525. **109.** Zöphel, Kennedy, Beinert & Kroneck (1988) *Arch. Microbiol.* **150**, 72. **110.** Petitdemange, Marczak, Blusson & Gay (1979) *Biochem. Biophys. Res. Commun.* **91**, 1258. **111.** Suzuke, Kumazawa, Seno & Matsumoto (1987) *Med. Sci. Res.* **15**, 675. **112.** Gaffar & Shetna (1977) *Appl. Environ. Microbiol.* **33**, 508. **113.** Burnett, Schmidt, Stafford, *et al.* (2003) *Biochem. Biophys. Res. Commun.* **310**, 84. **114.** Löffler & Kunze (1987) *FEBS Lett.* **216**, 51. **115.** Lee, Uemura & Snyder (1992) *J. Biol. Chem.* **267**, 19992. **116.** Thithapandha & Cohn (1978) *Biochem. Pharmacol.* **27**, 263. **117.** Hirschberg & Ensign (1971) *J. Bacteriol.* **108**, 751. **118.** Ohe & Watanabe (1979) *J. Biochem.* **86**, 45. **119.** Sun, Navas, Crane, Morré & Löw (1987) *J. Biol. Chem.* **262**, 15915. **120.** Bernhardt, Pachowsky & Staudinger (1975) *Eur. J. Biochem.* **57**, 241. **121.** Hasegawa, Nakai, Tokuyama & Iwaki (2000) *Biosci. Biotechnol. Biochem.* **64**, 2696. **122.** McNamee & Durham (1985) *Biochem. Biophys. Res. Commun.* **129**, 485. **123.** Schach, Tshisuaka, Fetzner & Lingens (1995) *Eur. J. Biochem.* **232**, 536.

Quinacrine Mustard

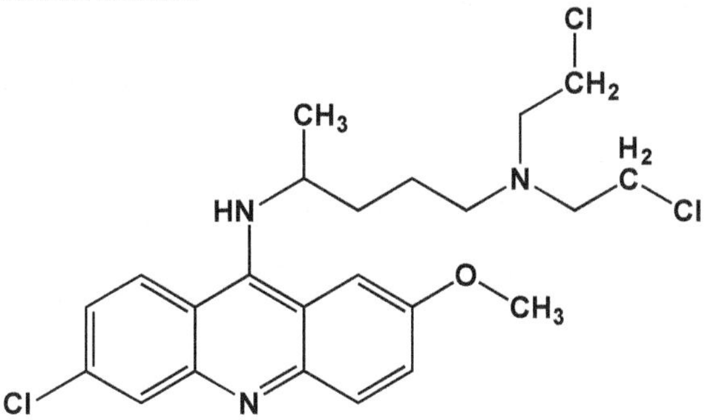

This chloro-substituted quinacrine ($FW_{free-base}$ = 468.85 g/mol; CAS 4213-45-0) is an alkylating agent that inactivates bovine heart F_1 ATPase, labeling the β-subunit in the process. Quinacrine mustard exhibits cytostatic and cytotoxic effects and is used as a DNA-labeling reagent. **Target(s):** F_1 ATPase, *or* F_1F_o ATP synthase, *or* H^+-transporting two-sector ATPase (1-6); histamine *N*-methyltransferase (7); TF_1 ATPase (4,5); vacuolar H^+-translocating ATPase (1); and ATPase, *Escherichia coli* (5). **1**. Moriyama, Patel & Futai (1995) *FEBS Lett.* **359**, 69. **2**. Bullough, Ceccarelli, Verburg & Allison (1989) *J. Biol. Chem.* **264**, 9155. **3**. Laikind & Allison (1983) *J. Biol. Chem.* **258**, 11700. **4**. Yoshida & Allison (1986) *J. Biol. Chem.* **261**, 5714. **5**. Bullough, Kwan, Laikind, Yoshida & Allison (1985) *Arch. Biochem. Biophys.* **236**, 567. **6**. Kasho, Allison & Boyer (1993) *Arch. Biochem. Biophys.* **300**, 293. **7**. Thithapandha & Cohn (1978) *Biochem. Pharmacol.* **27**, 263.

Quinaldate

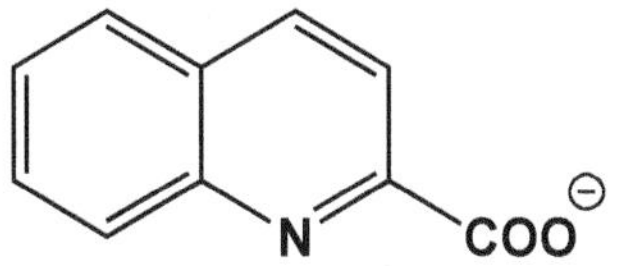

This tryptophan catabolite ($FW_{free-acid}$ = 173.127 g/mol; CAS 93-10-7), also known as quinaldic acid, quinaldinic acid and 2-quinolinecarboxylic acid, forms complexes with a number of metal ions, especially zinc and copper ions. Target(s): aconitase, or aconitate hydratase (1,2); alcohol dehydrogenase (3); dopamine β-hydroxylase (4); D-glyceraldehyde-3-phosphate dehydrogenase (5-7); kynureninase (8); lactate dehydrogenase (3,6); and pyruvate oxidase (9). **1**. Glusker (1971) *The Enzymes*, 3rd ed. (Boyer, ed.), **5**, 413. **2**. Treton & Heslot (1978) *Agric. Biol. Chem.* **42**, 1201. **3**. Konig, Kocsis & Pocskay (1975) *Acta Biochim. Biophys. Acad. Sci. Hung.* **10**, 171. **4**. Townes, Titone & Rosenberg (1990) *Biochim. Biophys. Acta* **1037**, 240. **5**. Lien & Keleti (1979) *Acta Biochim. Biophys. Acad. Sci. Hung.* **14**, 1. **6**. Lien, Ecsedi & Keleti (1979) *Acta Biochim. Biophys. Acad. Sci. Hung.* **14**, 11. **7**. Lien, Koubakouenda & Keleti (1979) *Acta Biochim. Biophys. Acad. Sci. Hung.* **14**, 19. **8**. Tanizawa & Soda (1979) *J. Biochem.* **86**, 499. **9**. Lardy (1969) in *Inhibitors: Tools in Cell Research* (Bücher & Siess, eds.), p. 374, Springer, New York.

Quinalizarin

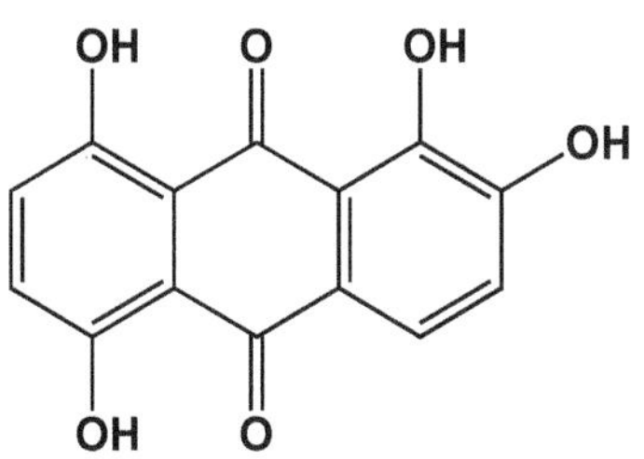

This red dye (FW = 272.21 g/mol; CAS 81-61-8; M.P. > 275°C), named systematically as 1,2,5,8-tetrahydroxyanthraquinone, inhibits HIV-1 integrase (1), inositol-trisphosphate 3-kinase (2), [myosin light-chain] kinase (3), and protein kinase C (3). With a panel of 140 kinases, quinalizarin is one of the most selective inhibitors of CK2, superior to silmitasertib (CX-4945), a first-in-class CK2 inhibitor in clinical trials for cancer treatment (4). Quinalizarin discriminates between the isolated CK2 catalytic subunit (CK2α) and CK2 holoenzyme (CK2$α_2β_2$). **1**. Hong, Neamati, Winslow, *et al.* (1998) *Antivir. Chem. Chemother.* **9**, 461. **2**.

Mayr, Windhorst & Hillemeier (2005) *J. Biol. Chem.* **280**, 13229. **3**. Jinsart, Ternai & Polya (1992) *Biol. Chem. Hoppe Seyler* **373**, 903. **4**. Cozza, Venerando, Sarno & Pinna (2015) *Biomed. Res. Int.* **2015**, 734127.

Quinapril & Quinaprilat

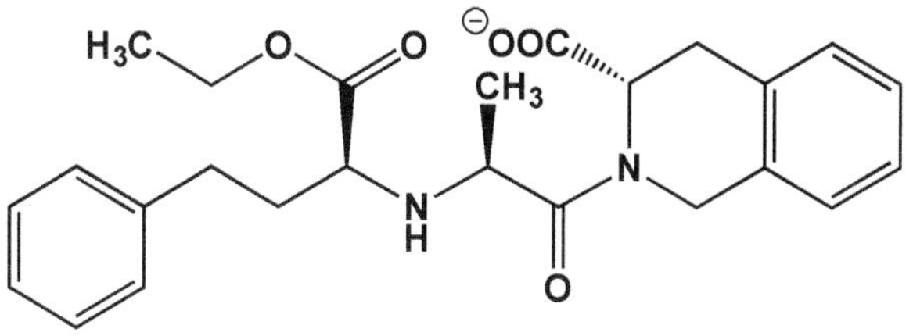

Quinapril

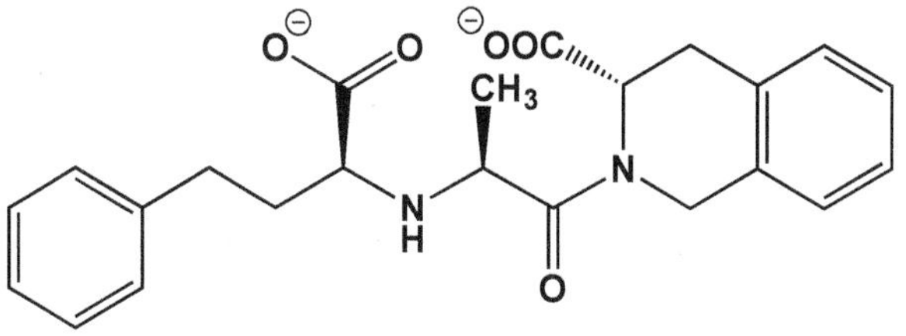

Quinaprilat

This substituted isoquinoline ($FW_{free-acid}$ = 438.52 g/mol; CAS 85441-61-8), also known as (3S)-2-[(2S)-2-[[(1S)-1-(ethoxycarbonyl)-3-phenylpropyl]amino]-1-oxopropyl]-1,2,3,4-tetrahydro-3-isoquinolinecarboxylic acid, is an antihypertensive prodrug that, like other ACE inhibitors, decreases systemic vascular resistance without increasing heart rate. The actual active agent is the diacid metabolite, quinaprilat ($FW_{free-acid}$ = 410.46 g/mol; CAS 82768-85-2), targets peptidyl-dipeptidase A, *or* angiotensin I-converting enzyme, IC_{50} = 8.3 nM (1-3). Due to reduced angiotensin production, plasma concentrations of aldosterone are likewise reduced, resulting in increased sodium ion excretion in the urine and an increased blood potassium ion level. With different chemical structures and susceptibilities to activation in the liver, ACE prodrugs differ in their bioavailability, plasma half-life, route of elimination, and volume of distribution. This is true for moexipril, which is the 6,7-dimethoxy analogue of quinapril. **Key Pharmacokinetic Parameters:** *See* Appendix II in Goodman & Gilman's *THE PHARMACOLOGICAL BASIS OF THERAPEUTICS*, 12th Edition (Brunton, Chabner & Knollmann, eds.) McGraw-Hill Medical, New York (2011). **1**. McAreavey & Robertson (1990) *Drugs* **40**, 326. **2**. Klutchko, Blankley, Fleming, *et al.* (1986) *J. Med. Chem.* **29**, 1953. **3**. Van Dyck, Novakova, Van Schepdael & Hoogmartens (2003) *J. Chromatogr. A* **1013**, 149.

Quinate

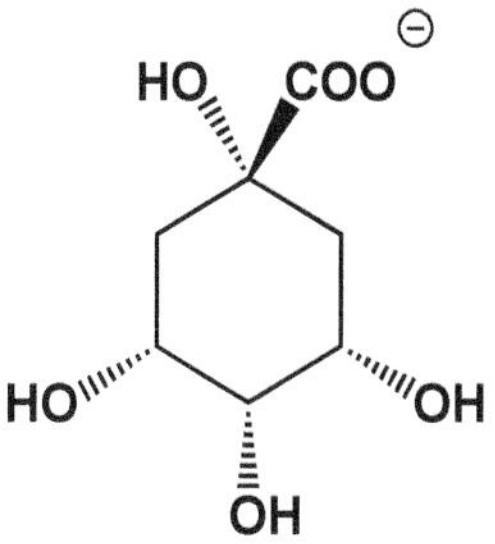

This cyclic polyol carboxylic acid ($FW_{free-acid}$ = 192.17 g/mol; CAS 60321-02-0; M.P. = 162-163°C), also known as quinic acid and named systematically as (1R,3R,4S,5R)-1,3,4,5-tetrahydroxycyclohexanecarboxylic acid, is found in many plants, especially cinchona bark, tobacco leaves, carrot leaves, and some fruits. While initially thought to be pharmacologically inert, quinic acid is now believed to displace binding of the mu opioid receptor antagonists. **Target(s):** α-amylase (1,2); chitin synthase (3); kynurenine aminotransferase, weakly inhibited (4); lysozyme (1); and trypsin (1). **1**. Rohn, Rawel & Kroll (2002) *J. Agric. Food Chem.* **50**, 3566. **2**. Funke & Melzig (2005) *Pharmazie* **60**, 796. **3**. Hwang, Ahn, Lee, *et al.* (2001) *Planta Med.* **67**, 501. **4**. Mason (1959) *J. Biol. Chem.* **234**, 2770.

Quinidine

This triboluminescent quinine diastereoisomer (FW$_{free-base}$ = 324.42 g/mol; CAS 56-54-2; M.P. = 174-175°C), found in a number of plants (often with quinine), exhibits cardiac anti-arrhythmic effects, mainly by blocking the fast inward Na$^+$ current (I$_{Na}$), but also blocking the slow-inactivating, tetrodotoxin-sensitive Na$^+$ current, the slow inward Ca^{2+} current (I$_{Ca}$), the rapid (I$_{Kr}$) and slow (I$_{Ks}$) components of the delayed K$^+$ rectifier current, the inward K$^+$ rectifier current (I$_{KI}$), the ATP-sensitive K$^+$ channel (I$_{KATP}$) and I$_{to}$. **Key Pharmaco-kinetic Parameters:** *See* Appendix II in Goodman & Gilman's *THE PHARMACOLOGICAL BASIS OF THERAPEUTICS*, 12th Edition (Brunton, Chabner & Knollmann, eds.) McGraw-Hill Medical, New York (2011). **Target(s):** acetylcholinesterase (1-3); adenylate cyclase (4); ATPase (5); cholesterol 24-hydroxylase, moderately inhibited (6); cholinesterase, *or* butyrylcholinesterase (1,2,7,8); CYP2C3 (9); CYP2D6 (10-14); debrisoquine 4-hydroxylase (15,16); ecdysone 20-monooxygenase (17); 7-esterase, nonspecific (2); ethoxyresorufin *O*-deethylase (18); K$^+$ channel blocker (19); 4-nitrophenol 2-monooxygenase, *or* CYP2E1, weakly inhibited (20); steroid 21-monooxygenase (21); and strictosidine synthase (22). **1**. Augustinsson (1950) *The Enzymes*, 1st. ed. (Sumner & Myrbäck, eds.), **1** (part 1), 443. **2**. Nachmansohn & Schneemann (1945) *J. Biol. Chem.* **159**, 239. **3**. Sihotang (1974) *J. Biochem.* **75**, 939. **4**. Needham, Dodd & Houslay (1987) *Biochim. Biophys. Acta* **899**, 44. **5**. Almotrefi (1993) *Gen. Pharmacol.* **24**, 233. **6**. Mast, White, Bjorkhem, *et al.* (2008) *Proc. Natl. Acad. Sci. U.S.A.* **105**, 9546. **7**. Nagasawa, Sugisaki, Tani & Ogata (1976) *Biochim. Biophys. Acta* **429**, 817. **8**. Bailey & Briggs (2005) *Amer. J. Clin. Pathol.* **124**, 226. **9**. Shiiyama, Soejima-Ohkuma, Honda, *et al.* (1997) *Xenobiotica* **27**, 379. **10**. Palamanda, Casciano, Norton, *et al.* (2001) *Drug Metab. Dispos.* **29**, 863. **11**. Hosseinpour & Wikvall (2000) *J. Biol. Chem.* **275**, 34650. **12**. von Moltke, Greenblatt, Duan, *et al.* (1998) *J. Pharm. Sci.* **87**, 1184. **13**. Ching, Blake, Ghabrial, *et al.* (1995) *Biochem. Pharmacol.* **50**, 833. **14**. Rodrigues (1996) *Meth. Enzymol.* **272**, 186. **15**. Kobayashi, Murray, Watson, *et al.* (1989) *Biochem. Pharmacol.* **38**, 2795. **16**. von Bahr, Spina, Birgersson, *et al.* (1985) *Biochem. Pharmacol.* **34**, 2501. **17**. Mitchell, Keogh, Crooks & Smith (1993) *Insect Biochem. Mol. Biol.* **23**, 65. **18**. Pasanen, Taskinen, Sotaniemi, Kairaluoma & Pelkonen (1988) *Pharmacol. Toxicol.* **62**, 311. **19**. Pelassy & Aussel (1993) *Pharmacology* **47**, 28. **20**. Tassaneeyakui, Veronese, Birkett, Gonzalez & Miners (1993) *Biochem. Pharmacol.* **46**, 1975. **21**. Kishimoto, Hiroi, Sharaishi, *et al.* (2004) *Endocrinology* **145**, 699. **22**. Stevens, Giroud, Pennings & Verpoorte (1993) *Phytochemistry* **33**, 99.

Quinimine Form of *p*-Aminophenol, *See* p-Aminophenol

Quinine

This quinidine diastereoisomer (FW$_{free-base}$ = 324.42 g/mol; CAS 130-95-0; M.P. = 174-177 °C; triboluminescent crystals; pK_1 = 4.13 and pK_2 = 8.3)

from the bark of the cinchona is a potassium channel blocker. Quinine has been widely used in the treatment and prevention of malaria. Quinine is believed to be useful in treating some muscular disorders, especially nocturnal leg cramps and myotonia congenital, affecting muscle membrane depolarization by acting directly on sodium channels. The refreshing quality of quinine-containing drinks are may have a similar explanation. **Key Pharmacokinetic Parameters:** *See* Appendix II in Goodman & Gilman's *THE PHARMACOLOGICAL BASIS OF THERAPEUTICS*, 12th Edition (Brunton, Chabner & Knollmann, eds.) McGraw-Hill Medical, New York (2011). **Target(s):** acetylcholinesterase (1,2); D-amino-acid oxidase (3-6); aminopyrine *N*-demethylase (7); benzoate 4-monooxygenase (8); bontoxilysin, weakly inhibited (9); carboxylesterase, esterase (10,11); cholinesterase (11,12); choline exchanger (13); choline oxidase (14); cyclohexanone monooxygenase (15); CYP3A4 (16,17); CYP2C8 (18); CYP2D1 (19,20); debrisoquine 4-hydroxylase (21,22); ecdysone 20-monooxygenase (23); 7-ethoxyresorufin *O*-deethylase (24); F$_O$F$_1$ ATP synthase, *or* H$^+$-transporting two-sector ATPase, *Streptococcus pneumoniae* (25); β-fructofuranosidase, *or* saccharase, *or* invertase (26); α-glucosidase (27); heme polymerase (28); D-hydroxybutyrate dehydrogenase (29); K$^+$ channels (30); lipase (31-35); phenylpyruvate tautomerase (36); sorbose reductase (37); strictosidine synthase (38); and urease (39). **1**. Richter & Croft (1942) *Biochem. J.* **36**, 746. **2**. Katewa & Katyare (2005) *Drug Chem. Toxicol.* **28**, 467. **3**. Hellerman, Lindsay & Bovarnick (1946) *J. Biol. Chem.* **163**, 553. **4**. Massart (1950) *The Enzymes*, 1st ed. (Sumner & Myrbäck, eds.), **1** (part 1), 307. **5**. Krebs (1951) *The Enzymes*, 1st ed. (Sumner & Myrbäck, eds.), **2** (part 1), 499. **6**. Meister & Wellner (1963) *The Enzymes*, 2nd ed. (Boyer, Lardy & Myrbäck, eds.), **7**, 609. **7**. Murray (1984) *Biochem. Pharmacol.* **33**, 3277. **8**. McNamee & Durham (1985) *Biochem. Biophys. Res. Commun.* **129**, 485. **9**. Burnett, Schmidt, Stafford, *et al.* (2003) *Biochem. Biophys. Res. Commun.* **310**, 84. **10**. Krisch (1971) *The Enzymes*, 3rd ed. (Boyer, ed.), **5**, 43. **11**. Nachmansohn & Schneemann (1945) *J. Biol. Chem.* **159**, 239. **12**. Augustinsson (1950) *The Enzymes*, 1st. ed. (Sumner & Myrbäck, eds.), **1** (part 1), 443. **13**. Ebel, Hollstein & Gunther (2002) *Biochim. Biophys. Acta* **1559**, 135. **14**. Ebisuzaki & Williams, Jr. (1953) *J. Biol. Chem.* **200**, 297. **15**. Hasegawa, Nakai, Tokuyama & Iwaki (2000) *Biosci. Biotechnol. Biochem.* **64**, 2696. **16**. Baune, Furlan, Taburet & Farinotti (1999) *Drug Metab. Dispos.* **27**, 565. **17**. Zhao, Kawashiro & Ishizaki (1998) *Drug Metab. Dispos.* **26**, 188. **18**. Ong, Coulter, Birkett, Bhasker & Miners (2000) *Brit. J. Clin. Pharmacol.* **50**, 573. **19**. Daniel, Syrek & Haduch (1999) *Pol. J. Pharmacol.* **51**, 435. **20**. Xu, Aasmundstad, Christophersen, Morland & Bjorneboe (1997) *Biochem. Pharmacol.* **53**, 603. **21**. Kobayashi, Murray, Watson, *et al.* (1989) *Biochem. Pharmacol.* **38**, 2795. **22**. von Bahr, Spina, Birgersson, *et al.* (1985) *Biochem. Pharmacol.* **34**, 2501. **23**. Mitchell, Keogh, Crooks & Smith (1993) *Insect Biochem. Mol. Biol.* **23**, 65. **24**. Pasanen, Taskinen, Sotaniemi, Kairaluoma & Pelkonen (1988) *Pharmacol. Toxicol.* **62**, 311. **25**. Munoz, Garcia & De la Campa (1996) *J. Bacteriol.* **178**, 2455. **26**. Rona & Block (1921) *Biochem. Z.* **118**, 185. **27**. Halvorson & Ellias (1958) *Biochem. Biophys. Acta* **30**, 28. **28**. Slater & Cerami (1992) *Nature* **355**, 167. **29**. Gotterer (1969) *Biochemistry* **8**, 641. **30**. Pelassy & Aussel (1993) *Pharmacology* **47**, 28. **31**. Bier (1955) *Meth. Enzymol.* **1**, 627. **32**. Rona & Reinecke (1921) *Biochem. Z.* **118**, 213. **33**. Rona & Takata (1923) *Biochem. Z.* **134**, 118. **34**. Rona & Pavlovic (1923) *Biochem. Z.* **134**, 108. **35**. Nachlas & Seligman (1949) *J. Biol. Chem.* **181**, 343. **36**. Molnar & Garai (2005) *Int. Immunopharmacol.* **5**, 849. **37**. Sugisawa, Hoshino & Fujiwara (1991) *Agric. Biol. Chem.* **55**, 2043. **38**. Stevens, Giroud, Pennings & Verpoorte (1993) *Phytochemistry* **33**, 99. **39**. Onodera (1915) *Biochem. J.* **9**, 544.

Quinlobelane

This water-soluble lobelane analogue (FW = 445.03 g/mol), also named *N*-methyl-*cis*-2,6-diphenethylpiperidine, inhibits vesicular monamine transporter-2, *or* VMAT2 (K_i = 51 nM), the membrane carrier responsible for sequestering cytosolic dopamine (DA) into synaptic vesicles (*See Dopamine*), thereby preventing dopamine degradation by monoamine oxidase. Quinlobelane is therefore some 1000-tmes more potent than lobelane (1). Interesting, structure-activity studies with quinlobelane and structurally related molecules showed that that their affinity for the VMAT2 binding site does not necessarily predict their ability to inhibit dopamine transport by VMAT2 (1). Such findings also suggest that a ligand-induced conformational change, mimicking that of dopamine and other transport substrates, is a requirement for effective transport inhibition. **1**. Vartak, Deaciuc, Dwoskin, & Crooks (2010) *Bioorg. Med. Chem. Lett.* **20**, 3584.

Quinoid 6(*S*)-Methyldihydropterin

This product analogue (FW = 179.18 g/mol) is a strong competitive inhibitor of rat liver 4a-hydroxytetrahydro-biopterin dehydratase: K_i = 1.5 µM. **1**. Rebrin, Bailey, Boerth, Ardell & Ayling (1995) *Biochemistry* **34**, 5801.

m-**Quinol**, *See Resorcinol*

o-**Quinol**, *See Catechol*

p-**Quinol**, *See Hydroquinone*

Quinolinate

This dicarboxylic acid ($FW_{free-acid}$ = 167.12 g/mol; CAS 89-00-9; λ_{max} = 268 nm; ε = 4000 $M^{-1}cm^{-1}$ in acid), also known as quinolinic acid and pyridine-2,3-dicarboxylic acid, is an intermediate in the formation of NAD^+ from tryptophan. Quinolinate is also an NMDA (*N*-methyl-D-aspartate) agonist and differentiates between cerebellar and forebrain NMDA receptors. Spectral analysis also confirms that quinolinate complexes with Fe^{2+}. **Target(s):** aspartate 4-decarboxylase (1); γ-butyrobetaine hydroxylase, *or* γ-butyrobetaine:2-oxoglutarate dioxygenase, K_i = 0.15 mM (2); deoxyhypusine monooxygenase (3); glutamate carboxypeptidase II (4); homocitrate synthase (5); 3-hydroxyanthranilate 3,4-dioxygenase (6); kynurenine:oxoglutarate aminotransferase, weakly inhibited (7,8); monoamine oxidase (9); nicotinamide-nucleotide amidase (10); nicotinate phosphoribosyltransferase (11); phosphoenolpyruvate carboxykinase, ATP-dependent (12); PEP carboxykinase, GTP-dependent (13-15); phosphoenolpyruvate carboxylase (16); and pyridoxal kinase (17,18). **1**. Soda, Novogrodsky & Meister (1964) *Biochemistry* **3**, 1450. **2**. Ng, Hanauske-Abel & England (1991) *J. Biol. Chem.* **266**, 1526. **3**. Beninati, Ferraro & Abbruzzese (1990) *Ital. J. Biochem.* **39**, 183A. **4**. Robinson, Blakely, Couto & Coyle (1987) *J. Biol. Chem.* **262**, 14498. **5**. Qian, West & Cook (2006) *Biochemistry* **45**, 12136. **6**. Hayaishi, Nozaki & Abbott (1975) *The Enzymes*, 3rd ed. (Boyer, ed.), **12**, 119. **7**. Mason (1959) *J. Biol. Chem.* **234**, 2770. **8**. Mason (1974) *Biochem. Biophys. Res. Commun.* **60**, 64. **9**. Naoi, Ishiki, Nomura, Hasegawa & Nagatsu (1987) *Neurosci. Lett.* **74**, 232. **10**. Imai (1973) *J. Biochem.* **73**, 139. **11**. Blum & Kahn (1971) *Meth. Enzymol.* **18B**, 138. **12**. Hunt & Köhler (1995) *Biochim. Biophys. Acta* **1249**, 15. **13**. Maxwell & Ray (1980) *Biochim. Biophys. Acta* **614**, 163. **14**. Utter & Kolenbrander (1972) *The Enzymes*, 3rd ed. (Boyer, ed.), **6**, 117. **15**. Bentle & Lardy (1977) *J. Biol. Chem.* **252**, 1431. **16**. Owttrim & Colman (1986) *J. Bacteriol.* **168**, 207. **17**. Karawya, Mostafa & Osman (1981) *Biochim. Biophys. Acta* **657**, 153. **18**. Takeuchi, Tsubouchi & Shibata (1985) *Biochem. J.* **227**, 537.

Quinolinic Acid, *See Quinolinate*

2-Quinolinol, *See 2-Hydroxyquinolinek*

8-Quinolinol, *See 8-Hydroxyquinoline*

N-[*N*-(2-Quinolinylcarbonyl)-L-valyl]-(2*S*,3*R*)-2-amino-4-((2*S*,4*R*)-2-(*tert*-butylcarbamoyl)-4-(pyridin-2-ylthio)piperidin-1-yl)-3-hydroxy-1-phenylbutane

This pseudopeptide linkage-containing peptide analogue (FW = 699.94 g/mol) inhibits HIV-1 and HIV-2 retropepsins, with respective K_i values of 1.9 and 1.8 nM. X-ray crystallography reveals that the substrate flap and residues 79-81 in the S_1 substrate-binding pocket undergo conformational changes upon inhibitor binding. Residues 29 and 30 also adapt their conformation to accommodate certain inhibitors, showing that HIV protease flexibility plays an important role in inhibitor binding. **1**. Tong, Pav, Mui, *et al.* (1995) *Structure* **3**, 33.

N-Quinolin-4-yl-3-(1*H*-imidazol-4-yl)propanamine

This imidazole derivative (FW = 252.31 g/mol) is a histamine H_3 receptor antagonist (K_i = 4.1 nM) and inhibits histamine *N*-methyltransferase (IC_{50} = 24 nM), making this derivative a valuable pharmacological tool for further development of dual-action drugs that inhibit histamine N^τ-methyltransferase (HMT) and antagonizing histamine H_3 receptors. The *N*-quinolin-6-yl analogue is a weaker effector (K_i = 71 nM and IC_{50} = 6.1 µM for the receptor and transferase, respectively. **1**. Grabmann, Apelt, Sippl, *et al.* (2003) *Bioorg. Med. Chem.* **11**, 2163.

Quinomycin A, *See Echinomycin*

p-Quinone

This quinone (FW = 108.10 g/mol; CAS 106-51-4; M.P = 115.7°C), also known as *p*-benzoquinone and 1,4-benzoquinone, is a common laboratory reagent used as an oxidizing agent and in organic syntheses. Care should be exercised in its handling, as its vapors can cause eye and skin irritation. *p*-Quinone also reacts with amino groups in proteins to produce 1,4-disubstituted adducts. If quinone-reactive amino group(s) of an enzyme is(are) catalytically essential, its biological activity may be inhibited. *See also Hydroquinone* **Target(s):** alanine:oxomalonate aminotransferase (62); aldehyde:nitrate oxidoreductase (1); aldehyde oxidase (2); alkaline phosphatase (3,4); amidase (5); amine oxidase (6); D-amino-acid oxidase (7); α-amylase (8-10); β-amylase (10,11); aryl-acylamidase (59,60); arylamine glucosyltransferase (64); aspartate aminotransferase (12); carbonic anhydrase (13); catalase (7,10,14); cholinesterase (15); *trans*-cinnamate 4-monooxygenase (66); complement system, guinea pig (16); cytochrome oxidase (17); dihydroxy-acid dehydratase (57); DNA topoisomerase II (18,55,56); fructose-1,6-bisphosphatase (19); glucose dehydrogenase (20,21); glutaminase (22-25); glyceraldehyde-3-phosphate dehydrogenase (26); α-glycerophosphatase (27); homogentisate oxidase, *or* homogentisate 1,2-dioxygenase (28); hyaluronoglucosaminidase, *or* hyaluronidase (61); lactate dehydrogenase (29); lipoxygenase (7,30,31); lysozyme (8); magnesium-protoporphyrin IX monomethyl ester (oxidative) cyclase (65); NAD^+ ADP-ribosyltransferase, *or* poly(ADP-ribose) polymerase (63); nicotinate oxidase (32); oxalate oxidase (5); papain (33-35); peroxidase, horseradish (36); polyphenol oxidase (10); pyruvate decarboxylase (37-39); pyruvate oxidase (40); succinate dehydrogenase (41-45); succinate oxidase (7,41,45); triose-phosphate isomerase (46); trypsin (8); tryptophan pyrrolase, *or* tryptophan 2,3-dioxygenase (47); urease (48-51,58); xanthine oxidase (52); and xylose isomerase (53,54).**1.** Franke & Schumann (1942) *Ann.* **552**, 243. **2.** Rajagopalan, Fridovich & Handler (1962) *J. Biol. Chem.* **237**, 922. **3.** Anderson (1961) *Biochim. Biophys. Acta* **54**, 110. **4.** Sizer (1942) *J. Biol. Chem.* **145**, 405. **5.** Nordwig & Strauch (1963) *Z. Physiol. Chem.* **330**, 145. **6.** Kenten & Mann (1952) *Biochem. J.* **50**, 360. **7.** Franke (1944) *Z. Physiol. Chem.* **281**, 162. **8.** Rohn, Rawel & Kroll (2002) *J. Agric. Food Chem.* **50**, 3566. **9.** Di Carlo & Redfern (1947) *Arch. Biochem.* **15**, 343. **10.** Owens (1953) *Contrib. Boyce Thompson Inst.* **17**, 221. **11.** Owens (1953) *Contrib. Boyce Thompson Inst.* **17**, 273. **12.** Cohen (1951) *The Enzymes*, 1st ed. (Sumner & Myrbäck, eds.), **1** (part 2), 1040. **13.** Chiba, Kawai & Kondo (1953) *Bull. Res. Inst. Food Sci., Kyoto Univ.* **13**, 12. **14.** Hoffmann-Ostenhof & Biach (1947) *Monatsh. Chem.* **76**, 319. **15.** Chaudhuri (1950) *Ann. Biochem. Exptl. Med. (Calcutta)* **10**, 71. **16.** Ecker, Pillemer, Martiensen & Wertheimer (1938) *J. Biol. Chem.* **123**, 351. **17.** Sisakyan & Filippovich (1956) *Biochemistry (U.S.S.R.)* [Engl. trans.] **21**, 159. **18.** Hutt & Kalf (1996) *Environ. Health Perspect.* **104** Suppl. 6, 1265. **19.** Stadtman (1961) *The Enzymes*, 2nd ed. (Boyer, Lardy & Myrbäck, eds.), **5**, 55. **20.** Matsushita & Ameyama (1982) *Meth. Enzymol.* **89**, 149. **21.** Nakamura (1954) *J. Biochem.* **41**, 67. **22.** Zittle (1951) *The Enzymes*, 1st ed. (Sumner & Myrbäck, eds.), **1** (part 2), 922. **23.** Roberts (1960) *The Enzymes*, 2nd ed. (Boyer, Lardy & Myrbäck, eds.), **4**, 285. **24.** Archibald (1944) *J. Biol. Chem.* **154**, 657. **25.** Sayre & Roberts (1958) *J. Biol. Chem.* **233**, 1128. **26.** Holzer (1956) *Medizinische* No. 15, 576. **27.** Hoffmann-Ostenhof & Putz (1948) *Monatsh. Chem.* **79**, 421. **28.** Schepartz (1953) *J. Biol. Chem.* **205**, 185. **29.** Pfleiderer, Jeckel & Wieland (1959) *Arch. Biochem. Biophys.* **83**, 275. **30.** Holman & Bergström (1951) *The Enzymes*, 1st ed. (Sumner & Myrbäck, eds.), **2** (part 1), 559. **31.** Tappel (1963) *The Enzymes*, 2nd ed. (Boyer, Lardy & Myrbäck, eds.), **8**, 275. **32.** Pinsky & Michaelis (1952) *Biochem. J.* **52**, 33. **33.** Hoffmann-Ostenhof & Biach (1946) *Experentia* **2**, 405. **34.** Bersin & Logemann (1933) *Z. Physiol. Chem.* **220**, 209. **35.** Bahadur & Atreya (1960) *Enzymologia* **21**, 238. **36.** Klapper & Hackett (1963) *J. Biol. Chem.* **238**, 3736. **37.** Kuhn & Beinert (1947) *Ber.* **80**, 101. **38.** Kuhn & Beinert (1943) *Ber.* **76**, 904. **39.** Schales & Schales (1957) *Arch. Biochem. Biophys.* **69**, 378. **40.** Barron (1936) *J. Biol. Chem.* **113**, 695. **41.** Bergstermann & Stein (1944) *Biochem. Z.* **317**, 217. **42.** Schlenk (1951) *The Enzymes*, 1st ed. (Sumner & Myrbäck, eds.), **2** (part 1), 316. **43.** Potter & DuBois (1943) *J. Gen. Physiol.* **26**, 391. **44.** Potter (1942) *Cancer Res.* **2**, 688. **45.** Herz (1954) *Biochem. Z.* **325**, 83. **46.** Krietsch (1975) *Meth. Enzymol.* **41**, 438. **47.** Frieden, Westmark & Schor (1961) *Arch. Biochem. Biophys.* **92**, 176. **48.** Sumner (1951) *The Enzymes*, 1st ed. (Sumner & Myrbäck, eds.), **1** (part 2), 873. **49.** Quastel (1933) *Biochem. J.* **27**, 1116. **50.** Grant & Kinsey (1946) *J. Biol. Chem.* **165**, 485. **51.** Zaborska, Kot & Superata (2002) *J. Enzyme Inhib. Med. Chem.* **17**, 247. **52.** Beiler & Martin (1951) *J. Biol. Chem.* **192**, 831. **53.** Slein (1962) *Meth. Enzymol.* **5**, 347. **54.** Hochster (1955) *Can. J. Microbiol.* **1**, 589. **55.** Lindsey, Jr., Bender & Osheroff (2005) *Chem. Res. Toxicol.* **18**, 761. **56.** Lindsey, Jr., Bromberg, Felix & Osheroff (2004) *Biochemistry* **43**, 7563. **57.** Babu & Brown (1995) *Microbios* **82**, 157. **58.** Pearson & Smith (1943) *Biochem. J.* **37**, 148. **59.** Hoagland & Graf (1974) *Can. J. Biochem.* **52**, 903. **60.** Hoagland (1975) *Phytochemistry* **14**, 383. **61.** Meyer & Rapport (1952) *Adv. Enzymol.* **13**, 199. **62.** Nagayama, Muramatsu & Shimura (1958) *Nature* **181**, 417. **63.** Banasik, Komura, Shimoyama & Ueda (1992) *J. Biol. Chem.* **267**, 1569. **64.** Frear (1968) *Phytochemistry* **7**, 381. **65.** Whyte & Castelfranco (1993) *Biochem. J.* **290**, 355. **66.** Potts, Weklych & Conn (1974) *J. Biol. Chem.* **249**, 5019.

Quinovose

This deoxy hexose (FW = 164.16 g/mol; CAS 7658-08-4), also known as chinovose, D-*epi*rhamnose, and D-*iso*rhamnose, and 6-deoxy-D-glucose, is a structural component of chloroplast glycolipids and certain cardiac glycosides. Quinovose is soluble in water and ethanol and melts at 146°C. In D_2O at 44°C, quinovose exists primarily in the β-pyranose form (64% *vs.* 36% α-pyranose and 0.002% aldehyde). **Target(s):** aldose 1-epimerase, *or*

mutarotase (1); α-glucosidase, *Entamoeba histolytica* (2,3); β-glucosidase, weakly inhibited (4); hexokinase (5); and α-L-rhamnosidase (6,7). **1.** Keston (1964) *J. Biol. Chem.* **239**, 3241. **2.** Bravo-Torres, Villagómez-Castro, Calvo-Méndez, Flores-Carreón & López-Romero (2004) *Int. J. Parasitol.* **34**, 455. **3.** Bravo-Torres, Calvo-Méndez, Flores-Carreón & López-Romero (2003) *Antonie Leeuwenhoek* **84**, 169. **4.** Dale, Ensley, Kern, Sastry & Byers (1985) *Biochemistry* **24**, 3530. **5.** Machado de Domenech & Sols (1980) *FEBS Lett.* **119**, 174. **6.** Jang & Kim (1996) *Biol. Pharm. Bull.* **19**, 1546. **7.** Bourbouze, Percheron & Courtois (1976) *Eur. J. Biochem.* **63**, 331.

2-Quinoxalinecarboxamide-*N*-adamantan-1-yl, *See NPS 2390*

(±)-3-Quinuclidinyl Benzilate

This powerfully incapacitating chemical warfare agent (FW$_{free-base}$ = 337.42 g/mol; CAS 6581-06-2; Symbol: QNB; M.P. = 164-165°C; λ_{max} = 207.4 nm in methanol (ε = 13500 $M^{-1}cm^{-1}$)), commonly referred to as BZ, is a non-selective, competitive antagonist of muscarinic cholinergic receptors, that has been used to investigate acetylcholine receptors (1). One target, the M_2 muscarinic acetylcholine receptor, is essential for controlling cardiovascular function by activating G-protein-coupled inwardly rectifying potassium channels. In M_2 receptors, QNB binds in the orthosteric binding pocket, in the middle of a long aqueous channel extending approximately two-thirds of the way through the membrane (2). A layer of tyrosine residues forms an aromatic cap that restricts dissociation of the bound ligand. This channel is formed by amino acids found to be identical in all five muscarinic receptor subtypes. This binding site also shares structural homology with other functionally unrelated acetylcholine binding proteins from different species (2). When given to rats in the passive avoidance task, either before training (a memory acquisition test), immediately post-training (a memory consolidation test) or 24-hours pre-retention (a memory retrieval test), QNB greatly impairs acquisition in the water maze at doses 0.5-5.0 mg/kg as well as the acquisition of passive avoidance task (3). In contrast, consolidation and retrieval were unaffected by QNB, indicating that QNB specifically affects the stage of acquisition (3). **CAUTION:** Symptoms of QNB overexposure include dizziness, tachycardia, headaches, disorientation, and hallucinations. Decontamination can be accomplished by washing the skin vigorously with soap and water. **1.** Saunders, Hough & Chuang (1996) *Brain Res.* **713**, 29. **2.** Haga, Kruse, Asada, *et al.* (2012) *Nature* **482**, 547. **3.** Misik, Vanek, Musilek & Kassa (2014) *Behav. Brain Res.* **266**, 193.

Quinupristin

This antibiotic (FW = 1022.23 g/mol; CAS 120138-50-3), also named *N*-{(6*R*,9*S*,10*R*,13*S*,15*aS*,18*R*,22*S*,24*aS*)-18-{[(3*S*)-1-azabicyclo[2.2.2]oct-3-ylthio]methyl}-22-[4-(dimethylamino)benzyl]-6-ethyl-10,23-dimethyl-5,8, 12,15,17,21,24-heptaoxo-13-phenyl-docosahydro-12*H*-pyrido[2,1-*f*]pyr-rolo[2,1-*l*][1,4,7,10,13,16]oxapentaazacyclononadecin-9-yl}-3-hydroxypyridine-2-carboxamide, is used in combination with dalfopristin (marketed as Synercid®), which shows a selective spectrum of antibacterial activity, mainly against gram-positive aerobic bacteria. Synercid has been assessed primarily in emergency-use protocols, in hospitalized patients with skin and skin-structure infections, and in patients with vancomycin-resistant *Enterococcus faecium* (VREF) bacteremia. **1**. Allington & Rivey (2001) *Clin. Ther.* **23**, 24.

Quisinostat

This broad-spectrum histone deacetylase inhibitor (FW = 394.48 g/mol; CAS 875320-29-9, 875320-30-2 (TFA), 875320-31-3 (2HCl); Solubility: 80 mg/mL DMSO; <1 mg/mL H_2O; Formulation for Animal Studies: 2 mg/mL in 20% hydroxypropyl-β-cyclodextrin (final pH 8.7)), also known as JNJ-26481585 and systematically as *N*-hydroxy-2-(4-(((1-methyl-1*H*-indol-3-yl)methylamino)methyl)piperidin-1-yl)pyrimidine-5-carboxamide, targets HDAC1, HDAC2, HDAC4, HDAC10 and HDAC11 with IC_{50} values of 0.11 nM, 0.33 nM, 0.64 nM, 0.46 nM and 0.37 nM, respectively. JNJ-26481585 exhibits broad-spectrum anti-proliferative activity in solid and hematologic cancer cell lines (*e.g.*, lung, breast, colon, prostate, brain, and ovarian tumor cell lines), with IC_{50} values ranging from 3 to 250 nM. **1**. Arts, *et al.* (2009) *Clin. Cancer Res.* **15**, 6841. **2**. Stühmer T, *et al.* (2010) *Brit. J. Haematol.* **149**, 529.

L-Quisqualate

This glutamate analogue (FW = 189.13 g/mol), also known as L-quisqualic acid and 3-(3,5-dioxo-1,2,4-oxadiazolidin-2-yl)-L-alanine; is an excitatory amino acid used to identify certain receptors. **Target(s):** aspartate aminotransferase (1); choline *O*-acetyltransferase (2); *N*-formylglutamate deformylase, *or* *N*-β-citrylglutamate deacylase (3); glutamate carboxypeptidase II (4-10); and kynurenine:oxoglutarate aminotransferase (11-13). **1**. Guidetti, Amori, Sapko, Okuno & Schwarcz (2007) *J. Neurochem.* **102**, 103. **2**. Loureiro-Dos-Santos, Reis, Kubrusly, *et al.* (2001) *J. Neurochem.* **77**, 1136. **3**. Asakura, Nagahashi, Hamada, *et al.* (1995) *Biochim. Biophys. Acta* **1250**, 35. **4**. Carter & Coyle (1998) in *Handb. Proteolytic Enzymes* (Barrett, Rawlings & Woessner, eds.), p. 1434, Academic Press, San Diego. **5**. Robinson, Blakely, Couto & Coyle (1987) *J. Biol. Chem.* **262**, 14498. **6**. Slusher, Robinson, Tsai, *et al.* (1990) *J. Biol. Chem.* **265**, 21297. **7**. Serval, Galli, Glowinski & Lavielle (1992) *J. Pharmacol. Exp. Ther.* **260**, 1093. **8**. Serval, Barbeito, Pittaluga, *et al.* (1990) *J. Neurochem.* **55**, 39. **9**. Carter, Feldman & Coyle (1996) *Proc. Natl. Acad. Sci. U.S.A.* **93**, 749. **10**. Bzdega, Turi, Wroblewska, *et al.* (1997) *J. Neurochem.* **69**, 2270. **11**. Guidetti, Okuno & Schwarcz (1997) *J. Neurosci. Res.* **50**, 457. **12**. Battaglia, Rassoulpour, Wu, *et al.* (2000) *J. Neurochem.* **75**, 2051. **13**. Zarnowski, Rejdak, Zagorski, *et al.* (2004) *Ophthalmic Res.* **36**, 124.

Quizalofop

This herbicide ($FW_{\text{free-acid}}$ = 344.75 g/mol; CAS 76578-14-8) and several of its esters are inhibitors of plant acetyl-CoA carboxylase (1-4). *Note*: Its ethyl ester is known as Targa™, and its 2-isopropylideneamino-oxyethylester is known as Agil™. **1**. Dehaye, Alban, Job, Douce & Job (1994) *Eur. J. Biochem.* **225**, 1113. **2**. Herbert, Cole, Pallett & Harwood (1996) *Pestic. Biochem. Physiol.* **55**, 129. **3**. Zuther, Johnson, Haselkorn, McLeod & Gornicki (1999) *Proc. Natl. Acad. Sci. U.S.A.* **96**, 13387. **4**. Price, Herbert, Moss, Cole & Harwood (2003) *Biochem. J.* **375**, 415.

Quizartinib

This second-generation FLT3 inhibitor (FW = 560.67 g/mol; CAS 950769-58-1), also known as AC220 and 1-(5-(*tert*-butyl)isoxazol-3-yl)-3-(4-(7-(2-morpholinoethoxy)benzo[*d*]imidazo[2,1-*b*]thiazol-2-yl)phenyl)urea, exerts low-nM potency in biochemical and cellular assays aw well as exceptional kinase selectivity, (*See also G749*). In animal models, AC220 is efficacious at doses as low as 1 mg/kg given orally once daily. **Rationale:** Activating mutations in the FLT3 receptor tyrosine kinase occur ~30% of acute myeloid leukemia (AML) patients, with the most common class of internal tandem duplications (ITDs) in the juxtamembrane domain, resulting in lead constitutive, ligand-independent activation of the kinase (1). FLT3 has thus become an attractive druggable target. First-generation FLT3 inhibitors include: CEP-701, MLN-518, PKC-412, sorafenib, and sunitinib. **1**. Zarrinkar, Gunawardane, Cramer, *et al.* (2009) *Blood* **114**, 2984. **2**. Chao, Sprankle, Grotzfeld, *et al.* (2009) *J. Med.Chem.* **52**, 7808.

Q-VD-Oph

This cell-permeable, pan-caspase inhibitor (FW = 513.49 g/mol; CAS 1135695-98-5; Solubility = 10 mM in DMSO; IUPAC: *N*-(2-quinolyl)valyl-*O*-methylaspartyl-(2,6-difluorophenoxy)methyl ketone; *Symbol:* QVD;, selectively protects cells from capsase-dependent apoptosis. The difluoro-phenoxymethyl ketone moiety serves as the warhead for reaction with the catalytic nucleophile, displacing the *o*-difluorophenol leaving group (pK_a = 6.3) in the process. Compared to ZVA-D-fmk or Boc-D-fmk, Q-VD-OPh is more effective in preventing apoptosis mediated by caspase 9/3, caspase 8/10, and caspase 12, with the following inhibitory action: caspase-3 (IC_{50} = 25 nM), caspase-1 (IC_{50} = 50 nM), caspase-8 (IC_{50} = 100 nM) and caspase-9 (IC_{50} = 430 nM). **1**. Caserta, Smith, Gultice, Reedy, Brown (2003) *Apoptosis* **8**, 345.

– R –

R, *See* Arginine

R16 Peptide
This high-affinity peptide antagonist and apoptosis inducer (FW = 2309.69 g/mol; Sequence = PHCVPRDLSWLDLEANMCLP; CAS 211364-78-2; Soluble to 1 mg/mL in H_2O), first identified through the use of phage display techniques, targets the binding of 14-3-3 protein to signal transduction proteins that control key aspects of cell proliferation, transformation, and apoptosis. The ability of the R16 peptide to inhibit Bad and other proapoptotic proteins suggests that 14-3-3 promotes/maintains cell survival. R-16 peptide blocks the ability of 14.3.3 proteins to bind to Raf-1, Bad, ASK1 and exoenzyme S, thereby inducing apoptosis. **1.** Wang, *et al.* (1999) *Biochemistry* **38**, 12499. **2.** Masters & Fu (2001) *J. Biol. Chem.* **276**, 45193. **3.** Masters, *et al.* (2002) *Biochem. Soc. Transact.* **30**, 360.

R106, *See* Aureobasidins

R428

This Axl inhibitor (FW = 506.64 g/mol; CAS 1037624-75-1; Solubility: 24 mg/mL DMSO, with warming), also known as BGB324 and 1-(6,7-dihydro-5H-benzo[6,7]cyclohepta[1,2-c]pyridazin-3-yl)-N^3-[(7S)-6,7,8,9-tetrahydro-7-(1-pyrrolidinyl)-5H-benzocyclohepten-2-yl]-1H-1,2,4-triazole-3,5-diamine, targets tyrosine-protein kinase receptor UFO (IC_{50} = 14 nM), an enzyme that transduces extracellular matrix signals into the cytoplasm by binding <u>G</u>rowth <u>A</u>rrest-<u>S</u>pecific <u>6</u> (*or* GAS6) and regulating many physiological processes, including cell survival, cell proliferation, migration, and differentiation (1,2). Axl is a novel regulator of endothelial cell haptotactic migration towards the matrix factor vitronectin (3). R428 shows >100-fold selective toward Axl *versus* Abl. Selectivty for Axl is also greater than Mer and Tyro3 (50x to 100x more selective) and InsR, EGFR, HER2, and PDGFRβ (100x more selective). **1.** Holland, Pan, Franci, *et al.* (2010) *Cancer Res.* **70**, 1544. **2.** Ghosh, Secreto, Boysen, *et al. Blood* **117**, 1928. **3.** Holland, Powell, Franci, *et al.* (2005) *Cancer Res.* **65**, 9294.

R547

This cyclin kinase-directed inhibitor (FW = 441.46 g/mol; CAS 741713-40-6), named systematically as 4-amino-2-(1-methanesulfonylpiperidin-4-ylamino)pyrimidin-5-yl](2,3-difluoro-6-methoxyphenyl)methanone, targets CDK1/cyclin B, CDK2/cyclin E, and CDK4/cyclin D1 (K_i = 1–3 nM) and was inactive (K_i > 5,000 nM) against a panel of >120 unrelated kinases in cell-free assays. The growth-inhibitory activity is characterized by a cell cycle block at G_1 and G_2 phases and induction of apoptosis. **Cyclin Target Selectivity:** Cdk1 (++++), Cdk2 (++++), Cdk3 (weak, if any), Cdk4 (++++), Cdk5 (weak, if any), Cdk6 (weak, if any), Cdk7 (weak, if any), Cdk8 (weak, if any), Cdk9 (weak, if any), Cdk10 (weak, if any). **1.** DePinto, *et al.* (2006) *Mol. Cancer Ther.* **5**, 2644.

R 715
This potent, selective and metabolically stable bradykinin antagonist (FW = 1140.35 g/mol; CAS 185052-09-9; Sequence = Acetyl-KRPPGFSXI, where X = D-β-Nalidixic Acid; Solubility: 5 mg/mL H_2O) targets B_1 receptors ($IC_{50} \approx$ 3 nM), with no activity toward B_2 receptors. R715 reduces mechanical hyper-nociception in a mouse model of neuropathic pain. **1.** Gobeil, *et al.* (1996) *Hypertension* **28**, 833. **2.** Abdouh, *et al.* (2008) *Brit. J. Pharmacol.* **154**, 136. **3.** Quintao, *et al.* (2008) *J. Neurosci.* **28**, 2856.

R 830, *See* 2,6-Di(t-butyl)-4-(2'-thenoyl)phenol

R-848, *See* Resiquimod

R 1485

This selective, high-affinity and brain-penetrant serotonin antagonist ($FW_{free-base}$ = 377.48 g/mol; FW_{di-HCl} = 450.35 g/mol; Solubility: 100 mM dihydrochloride in H_2O; 100 mM free base in DMSO), also named 4-[(2-fluorophenyl)sulfonyl]-3,4-dihydro-8-(1-piperazinyl)-2H-1,4-benzoxazine dihydrochloride, targets $5-HT_6$ receptors (pK_i = 8.9), with low hERG inhibition and >100x selectivity against a panel of 50 targets, including other 5-HT receptor subtypes. **1.** Zhao, Berger, Clark, *et al.* (2007) *Bioorg. Med. Chem. Lett.* **17**, 3504. **2.** Liu, *et al.* (2009) *Drug Des. Rev.* **70**, 145.

R 1530

This potent and orally available multi-kinase inhibitor (FW = 356.79 g/mol; CAS 882531-87-5; Solubility: 100 mM in DMSO), also named 5-(2-chlorophenyl)-7-fluoro-1,2-dihydro-8-methoxy-3-methylpyrazolo[3,4-b][1,4]benzodiazepine, targets Checkpoint Kinase-2, *or* Chk2 (IC_{50} = 24 nM), Kinase insert Domain Receptor kinase, *or* KDR (IC_{50} = 34 nM), fibroblast growth factor receptor kinase, *or* FGFR (IC_{50} = 50 nM), Aurora A kinase (IC_{50} = 58 nM), and Cyclin-dependent kinase Cdk2 (IC_{50} = 88 nM), with weaker action against VEGFR-2 (K_d = 15 nM), FGFR1 (K_d = 61 nM) and PDGFRβ (K_d = 88 nM). R1530 induces polyploidy, interfering with tubulin polymerization as well as mitotic checkpoint function in cancer cells, leading to abortive mitosis, endo-reduplication and polyploidy (1). R1530-induced polyploid cancer cells undergo apoptosis or become senescent, whereas normally proliferating cells are resistant to R1530-induced polyploidy, a finding favoring its use in treating cancers (1). R-1530 also displays antiproliferative activity *in vitro*, inhibiting both mitosis and angiogenesis (2). **1.** Tovar, Higgins, Deo, *et al.* (2010) *Cell Cycle* **9**, 3364. **2.** Liu, *et al.* (2013) *ACS Med. Chem. Lett.* **4**, 259.

R 7050

This cell-permeable TNFα receptor antagonist (FW = 380.77 g/mol; CAS 303997-35-5), also named 8-chloro-4-(phenylthio)-1-(trifluoromethyl)[1,2,4]triazolo[4,3-a]quinoxaline, targets TNFα receptor 1, blocking its association with with <u>T</u>NFα <u>R</u>eceptor-<u>A</u>ssociated <u>D</u>eath <u>D</u>omain protein (*or* TRADD) and <u>R</u>eceptor <u>I</u>nteracting <u>P</u>rotein <u>1</u> (*or* RIP1) in the initial intracellular signaling event following TNFα stimulation (1). It also inhibits TNFα-induced NF-κB and MAPK signaling pathway activation. R-7050 attenuates intracerebral hemorrhage-induced neurovascular injury in mice (2). **1**. Gururaja, Yung, Ding, *et al.* (2007) *Chem. Biol.* **14**, 1105. **2**. King, Alleyne & Dhandapani (2013) *Neurosci. Lett.* 2013 May 10;542:92

R 24571, *See* Calmidazolium Chloride

R-39209, *See* Alfentanil

R43512, *See* Astemizole

R 59022, *See* 6-(2-[4-([4-Fluorophenyl]phenyl-methylene)-1-piperidinyl]ethyl)-7-methyl-5H-thiazolo(3,2-a)pyrimidin-5-one

R-64766, *See* Risperidone

R-67145, *See* Nebivolol

R 68070, *See* Ridogrel

R 82150, *See* S-4,5,6,7-Tetrahydro-5-methyl-6-(3-methyl-2-butenyl)-imidazo[4,5,1-jk][1,4]-benzo-diazepin-2(1H)-thione

R 82913. *See* (+)-S-4,5,6,7-Tetrahydro-9-chloro-5-methyl-6-(3-methyl-2-butenyl)-imidazo[4,5,1-jk][1,4]-benzodiazepin-2(1H)-thione

R83842, *See* Vorozole

R 89439, *See* Loviride

R115777, *See* Tipifarnib

R207910, *See* Bedaquiline

RA 839

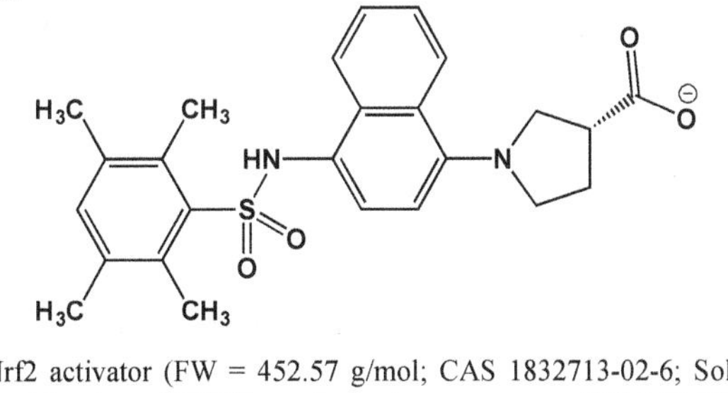

This Nrf2 activator (FW = 452.57 g/mol; CAS 1832713-02-6; Solubility: 100 mM in DMSO), also known as (3S)-1-[4-[[(2,3,5,6-tetramethylphenyl)sulfonyl]amino]-1-naphthalenyl]-3-pyrrolidinecarboxyl-ate, is an anti-inflammatory agent that inhibits Nrf2/Keap1 interaction (K_d = 6 µM), suppressing lipopolysaccharide (LPS) induction of iNOS expression and nhence nitric oxide formation in macrophages. The the transcription factor NF-E2-related factor 2 (Nrf2) is a basic leucine zipper (bZIP) protein that serves as athe a master regulator of the antioxidant response by regulating the expression of antioxidant proteins that protect against oxidative damage triggered by injury and inflammation. The multi-domain Keap1 (<u>Kel</u>ch-like <u>ECH</u>-<u>a</u>ssociated <u>p</u>rotein <u>1</u>) contains a stress-sensing cysteinyl residue (C151). Absent stressors, the activity of Nrf2 is inhibited by its interaction with the Keap1. Similar to the activation of Nrf2 by either silencing of Keap1 expression or by the reactive compound CDDO-Me (2-cyano-3,12-dioxooleana-1,9-dien-28-oic acid methyl ester), RA839 prevents induction of both inducible nitric-oxide synthase expression and nitric oxide release in response to lipopolysaccharides in macrophages. In mice, RA839 acutely induced Nrf2 target gene expression in liver. RA839 is a selective inhibitor of the Keap1/Nrf2 interaction and a useful tool compound to study the biology of Nrf2. **1**. Winkel, Engel, Margerie, *et al.* (2015) *J. Biol. Chem.* **290**, 28446.

Rab GDP Dissociation Inhibitor

This monomeric cytosolic protein ($MW_{isoform-α}$ = 50,500; $MW_{isoform-β}$ = 50,700), often abbreviated Rab GDI, inhibits exchange of GTP with Rab·GDP complex by preventing GDP dissociation from small G proteins and subsequent the binding of GTP (1-7). GDI is required to maintain a cytosolic pool of Rab proteins. Rab GDI also inhibits intra-cisternal

transport between the Golgi stacks and participates in the removal of Rab proteins from the Golgi. Rab GDI only interacts with lipid-modified GDP-Rab3A. Other Rab family members that can be acted on include Rab1, Rab2, and Rab4. ***See also Rho GDP Dissociation Inhibitor*** **1**. Takai, Kaibuchi, Kikuchi & Sasaki (1995) *Meth. Enzymol.* **250**, 122. **2**. Sasaki & Takai (1995) *Meth. Enzymol.* **257**, 70. **3**. Shisheva (2001) *Meth. Enzymol.* **329**, 39. **4**. Sasaki, Kikuchi, Araki, *et al.* (1990) *J. Biol. Chem.* **265**, 2333. **5**. Wu, Zeng, Wilson & Balch (1996) *Trends Biochem. Sci.* **21**, 472. **6**. Elazar, Mayer & Rothman (1994) *J. Biol. Chem.* **269**, 794. **7**. Dirac-Svejstrup, Soldati, Shapiro & Pfeffer (1994) *J. Biol. Chem.* **269**, 15427.

Racecadotril

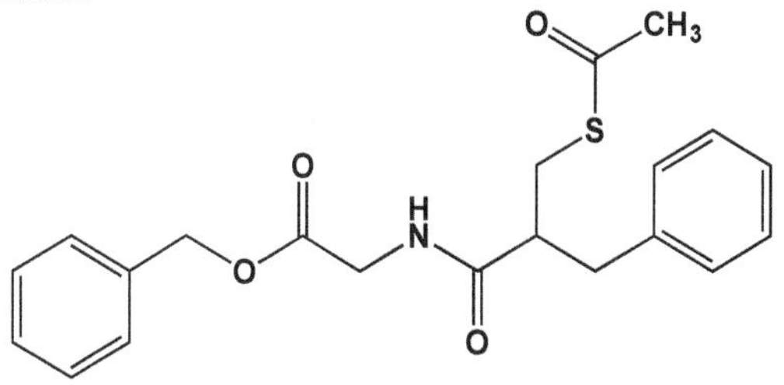

This lipophilic thiorphan derivative and antidiarrheal agent (FW = 385.48 g/mol; CAS 81110-73-8), also known as acetorphan and *N*-[(*R,S*)-3-acetylmercapto-2-benzylpropanoyl]glycine benzyl ester, inhibits dactylysin (1) and neprilysin, *or* enkephalinase (2-4). The *S*-isomer is called ecadotril. **1**. Carvalho, Joudiou, Boussetta, Leseney & Cohen (1992) *Proc. Natl. Acad. Sci. U.S.A.* **89**, 84. **2**. Lecomte, Costentin, Vlaiculescu, *et al.* (1986) *J. Pharmacol. Exp. Ther.* **237**, 937. **3**. Gros, Souque, Schwartz, *et al.* (1989) *Proc. Natl. Acad. Sci. U.S.A.* **86**, 7580. **4**. Erdös & Skidgel (1989) *FASEB J.* **3**, 145.

Racemethorphan, *See* Dextromethorphan; Levomethorphan

Racivir

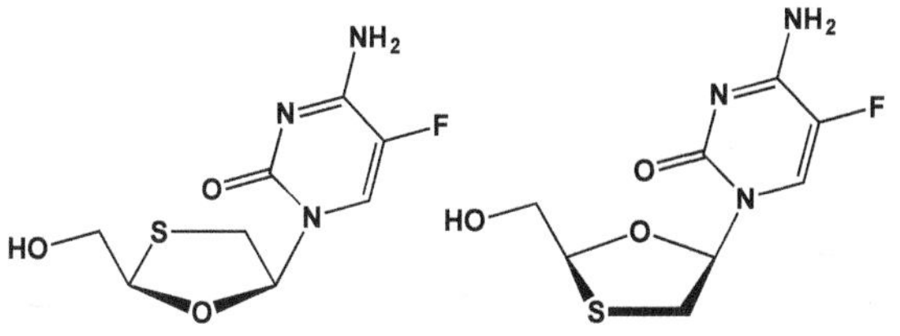

This proprietary product (CAS 143491-54-7) consists of a racemic mixture of the two β-enantiomers of emtricitabine which inhibits HIV-1 reverse transcriptase via the 5'-triphosphates. The L-enantiomer is more potent than the D-isomer. ***See also Emtricitabine*** **Target(s):** HIV-1 reverse transcriptase (1-4); and RNA-directed DNA polymerase (1-4). **1**. El Safadi, Vivet-Boudou & Marquet (2007) *Appl. Microbiol. Biotechnol.* **75**, 723. **2**. Wilson, Martin, Borroto-Esoda, *et al.* (1993) *Antimicrob. Agents Chemother.* **37**, 1720. **3**. Faraj, Agrofoglio, Wakefield, *et al.* (1994) *Antimicrob. Agents Chemother.* **38**, 2300. **4**. Feng, Shi, Schinazi & Anderson (1999) *FASEB J.* **13**, 1511.

(*S*)-Raclopride

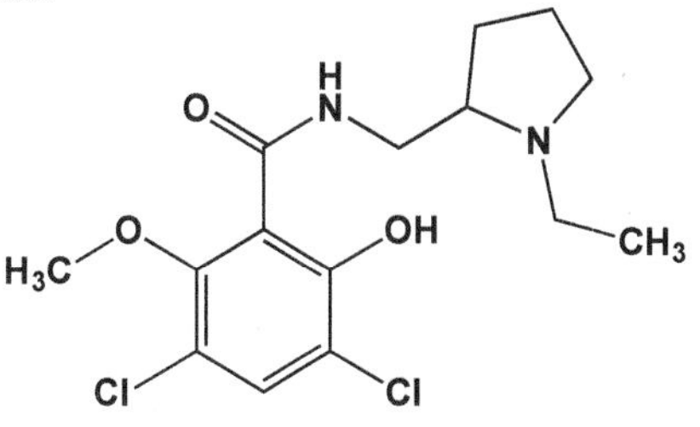

This CNS-acting dopamine receptor antagonist ($FW_{free-base}$ = 347.24 g/mol; CAS 84225-95-6; Solubility: 100 mM in DMSO), also known as 3,5-dichloro-*N*-(1-ethylpyrrolidin-2-ylmethyl)-2-hydroxy-6-methoxybenz-amide, selectively targets D_2 (K_i = 1.8 nM) and D_3 (K_i = 3.5 nM) receptors, reacting much more weakly with D_4 (K_i = 2.4 µM) and D_1 (K_i = 1.8 µM) receptors. (***See also Remoxipride***) **1**. Protais, Chagraoui, Arbaoui & Mocaer (1994) *Eur. J. Pharmacol.* **271**, 167. **2**. Kohler, Hall, Ogren & Gawell (1985) *Biochem. Pharmacol.* **34**, 2251.

RAD001, See *Everlimus*

Radicicol

This photosensitive antifungal macrolide (FW = 364.78 g/mol; CAS 12772-57-5), also called monorden, from *Diheterospora chlamydosporia* and *Monosporium bonorden* inhibits p60v-*scr* protein-tyrosine kinase activity, IC_{50} = 0.27 nM. Radicicol also acts as a cell differentiation modulator, suppresses the expression of mitogen-inducible cyclooxygenase, and inhibits receptor-mediated activation of p85 phosphatidylinositol 3-kinase. Radicicol inhibits [pyruvate dehydrogenase] kinase (PDK) activity by binding directly to the ATP-binding pocket of PDK3, similar to Hsp90 and Topo VI from the same ATPase/kinase superfamily. **Target(s):** ATP:citrate lyase, *or* ATP:citrate synthase, showing noncompetitive inhibitor of ATP citrate lyase with K_i values for citrate and ATP of 13 and 7 μM (1); [branched-chain α-keto acid dehydrogenase] kinase, *or* [3-methyl-2-oxobutyrate dehydrogenase (acetyl-transferring)] kinase (2); heat shock protein 90-kDa ATPase (2-4); nonchaperonin molecular chaperone ATPase (4); protein-histidine kinase, Sln1 yeast (2); protein-tyrosine kinase (5,6); and [pyruvate dehydrogenase (acetyl-transferring)] kinase (7). **1.** Ki, Ishigami, Kitahara, *et al.* (2000) *J. Biol. Chem.* **275**, 39231. **2.** Besant, Lasker, Bui & Turck (2002) *Mol. Pharmacol.* **62**, 289. **3.** Roe, Prodromou, O'Brien, *et al.* (1999) *J. Med. Chem.* **42**, 260. **4.** Rowlands, Newbatt, Prodromou, *et al.* (2004) *Anal. Biochem.* **327**, 176. **5.** Pillay, Nakano & Sharma (1996) *Cell Growth Differ.* **7**, 1487. **6.** Kwon, Yoshida, Fukui, Horinouchi & Beppu (1992) *Cancer Res.* **52**, 6926. **7.** Kato, Li, Chuang & Chuang (2007) *Structure* **15**, 992.

Raffinose

This nonreducing trisaccharide (FW = 504.44 g/mol; CAS 17629-30-0), also known as β-D-fructofuranosyl-*O*-α-D-galactopyranosyl-(1→6)-α-D-glucopyranoside, melitose, and melitriose, occurs widely in higher plants, often functioning as a transport carbohydrate. **Raffinose Space:** Raffinose is not taken up by most animal cells, a property that makes radiolabeled raffinose an especially useful tool in investigations of ion and metabolite transport investigations. One may estimate the volume of extracellular space (V_e) by determining the dilution (in cpm/mL) after a given volume V_0 (units mL) of radiolabel of known concentration (say C_0 = $cpm_{initial}$/mL) is mixed with a tissue, followed by centrifugation to recover some of the fluid. All dilutions follow the equation: $C_0V_0 = C_fV_f$, where V_f = $(V_0 + V_e)$. Therefore, measurement of C_f (*or* cpm_{final}/mL), allows one to correct for the amount of ion or metabolite present extracellularly (*i.e.*, in the interstitial space), which is often called the "raffinose space". Such corrections can be substantial (*e.g.*, rabbit muscle has an interstitial space of 23.5 mL per 100 g tissue). This method for the estimating the raffinose space is essentially the same as the time-honored inulin method (1). **Target(s):** α-galactosidase (2,3); β-galactosidase, weakly inhibited (2,4,5); α-glucosidase (2,6); inulinase (7); and phosphoenolpyruvate-dependent sucrose phosphotransferase system (8). **1.** Fisher & Lindsay (1956) *J.*

Physiol. **131**, 536. **2.** Webb (1966) *Enzyme and Metabolic Inhibitors*, vol. 2, pp. 416-418, Academic Press, New York. **3.** Sheinin & Crocker (1961) *Can. J. Biochem. Physiol.* **39**, 55. **4.** Lester & Bonner (1952) *J. Bacteriol.* **63**, 759. **5.** Levin & Mahoney (1981) *Antonie Leeuwenhoek* **47**, 53. **6.** Kato, Matsushima & Akabori (1960) *J. Biochem.* **48**, 199. **7.** Avigad & Bauer (1966) *Meth. Enzymol.* **8**, 621. **8.** Slee & Tanzer (1979) *Infect. Immun.* **24**, 821.

Rafoxanide

This anthelminthic (FW = 626.02 g/mol; CAS 22662-39-1), known as Disalan™ Flukanide™, MK 990, Rafoxanid™, and Ranide™, and named systematically as *N*-(3-chloro-4-(4-chlorophenoxy)phenyl)-2-hydroxy-3,5-diiodobenzamide, is often used to treat grazing livestock, This fasciolicidic agent inhibits cystathionine β-synthase and is a proton ionophore. Rafoxanide is also used to treat fluke, hookworm and other infestations. **Target(s):** cystathionine β-synthase (1); fumarate reductase (2); oxidative phosphorylation (3); and succinate dehydrogenase (2). **1.** Walker & Barrett (1992) *Exp. Parasitol.* **74**, 205. **2.** Kaur & Sood (1983) *Vet. Parasitol.* **13**, 333. **3.** Martin (1997) *Vet. J.* **154**, 11.

Raloxifene

This second-generation selective estrogen receptor modulator (FW = 473.59 g/mol; CAS 84449-90-1) is effective in the treatment of osteoporosis in postmenopausal women. While exerting estrogen agonist effects on bone and cholesterol metabolism, raloxifene behaves as a complete estrogen antagonist on mammary gland and uterine tissue. It is also a potent inhibitor of aldehyde oxidase, which catalyzes the oxidation of aldehydes (*Reaction*: Aldehyde + H_2O + O_2 ⇌ Carboxylate + H_2O_2 + H^+) as well as the hydroxylation of some heterocycles. Aldehyde oxidase also catalyzes the oxidation of intermediate products in both CYP450 and monoamine oxidase (MAO) reactions. **Key Pharmacokinetic Parameters:** *See* Appendix II in Goodman & Gilman's THE *PHARMACOLOGICAL BASIS OF THERAPEUTICS*, 12th Edition (Brunton, Chabner & Knollmann, eds.) McGraw-Hill Medical, New York (2011). **Target(s):** aldehyde oxidase, IC_{50} = 2.9 nM (1); CYP3A4 (2); estrogen receptor (3,4); LDL oxidation (5); and myeloperoxidase (5); anoxic xanthine oxidoreductase-catalyzed reduction of NO_2^- to ·NO, EC_{50} = 64 μM (6). In the latter case, Exposure of purified XO to raloxifene (PBS, pH 7.4) resulted in a dose-dependent (12.5-100 μM) inhibition of xanthine oxidation to uric acid. Although Dixon plot analysis suggested a competitive inhibition, K_i =13 μM (6), Dixon Plots cannot distinguish between competitive inhibition and certain kinds of mixed-type competitive inhibition (7). **1.** Obach, Huynh, Allen & Beedham (2004) *J. Clin. Pharmacol.* **44**, 7. **2.** Chen, Ngui, Doss, *et al.* (2002) *Chem. Res. Toxicol.* **15**, 907. **3.** Kellen (2001) *Curr. Drug Targets* **2**, 423. **4.** Kellen (2001) *In Vivo* **15**, 459. **5.** Zuckerman & Bryan (1996) *Atherosclerosis* **126**, 65. **6.** Weidert, Schoenborn, Cantu-Medline, *et al.* (2014) *Nitric Oxide* **37**, 41. **7.** Purich & Fromm (1972) *Biochim. Biophys. Acta.* **268**, 1.

Raltegravir

This potent integrase inhibitor (FW = 444.42 g/mol; CAS 518048-05-0; Solubility: 89 mg/mL DMSO; <1 mg/mL H_2O; pK_a= 6.7), also known as MK-0518, Isentress, and N-(4-fluorobenzyl)-5-hydroxy-1-methyl-2-(2-(2-methyl-1,3,4-oxadiazole-5-carboxamido)propan-2-yl)-6-oxo-1,6-dihydro-pyrimidine-4-carboxamide, targets wild-type HIV integrase (IC_{50} = 90 nM) and Prototype Foamy Virus (PFV) intasome IN^{S217Q} (IC_{50} = 40 nM) (1-3). Targeting the strand transfer step of viral integration, efficiently blocking viral replication *in vitro*, and suppressing viremia in patients, raltegravir is the first FDA-approved human HIV-1 integrase drug. Notably, HIV-1 $IN^{Q148H/G140S}$ and HIV-1 IN^{N155H} and the analogous PFV IN^{S217H} and PFV IN^{N224H} display reduced sensitivity to raltegravir *in vitro* (1). Because raltegravir undergoes nonhepatic glucuronidation, the potential for drug-drug interactions is decreased (2). Raltegravir lipophilicity is reduced as pH increases from 5 to 9 (3). Cellular permeativity of raltegravir is lower in the presence of magnesium and calcium. Cellular retention of raltegravir is increased by inhibiting the ABCB1 transporter and by lowering extracellular pH from pH 8 to 5 (3). Raltegravir binds to the IN active site, causing it to disengage from the deoxyadenosine at the 3'-end of viral DNA (4). This inhibitor occupy contacts the β_4-α_2 loop of the catalytic core domain (4). **1**. Hare, Vos, Clayton, *et al.* (2010) *Proc. Natl. Acad. Sci. U.S.A.* **107**, 20057. **2**. Hicks & Gulick (2009) *Clin. Infect. Dis.* **48**, 931. **3**. Moss, Siccardi, Murphy, *et al.* (2012) *Antimicrob. Agents Chemother.* **56**, 3020. **4**. Hare, Smith, Métifiot, *et al.* (2011) *Mol. Pharmacol.* 2011, 80(4), 565.

Raltitrexed

This antifolate and antineoplastic agent (FW = 456.48 g/mol; CAS 112887-68-0), known also as Tomudex™ as well as ZD-1694, and systematically named as N-(5-(N-(3,4-dihydro-2-methyl-4-oxoquinazolin-6-ylmethyl)-N-methylamino)-2-thenoyl)-L-glutamic acid, inhibits thymidylate synthase (1-4). Once within cells, raltitrexed is rapidly polyglutamated, and the pentaglutamate analogue is also inhibitory. **1**. Tong, Liu-Chen, Ercikan-Abali, *et al.* (1998) *J. Biol. Chem.* **273**, 11611. **2**. Marsham, Hughes, Jackman, *et al.* (1991) *J. Med. Chem.* **34**, 1594. **3**. Bijnsdorp, Comijn, Padron, Gmeiner & Peters (2007) *Oncol. Rep.* **18**, 287. **4**. Hekmat-Nejad & Rathod (1996) *Antimicrob. Agents Chemother.* **40**, 1628.

Ramipril

This long-lasting antihypertensive prodrug (FW = 416.52 g/mol; CAS 87333-19-5), variously known as H,OE-498, Acovil™, Altace™, Carasel™, Delix™, Ramace™, Triatec™, Tritace™, Vesdil™, Zabien™, and named systematically as 2-[N-[(S)-1-ethoxycarbonyl-3-phenylpropyl]-L-alanyl]-(1S,3S,5S)-2-azabicyclo[3.3.0]octane-3-carboxylic acid, undergoes hepatic conversion to ramiprilat, a potent inhibitor of peptidyl-dipeptidase A, *or* angiotensin I-converting enzyme, IC_{50} = 4 nM (1-3). (**See active drug** *Ramiprilat*) **Key Pharmacokinetic Parameters:** *See* Appendix II in Goodman & Gilman's THE PHARMACOLOGICAL BASIS OF THERAPEUTICS,

12th Edition (Brunton, Chabner & Knollmann, eds.) McGraw-Hill Medical, New York (2011). **1**. van Griensven, Schoemaker, Cohen, *et al.* (1995) *Eur. J. Clin. Pharmacol.* **47**, 513. **2**. Bender, Rangoonwala, Rosenthal & Vasmant (1990) *Clin. Physiol. Biochem.* **8** Suppl. 1, 44. **3**. Levitt & Schoemaker (2006) *BMC Clin. Pharmacol.* **6**, 1.

Ramiprilat

This long-lasting antihypertensive (FW$_{free-acid}$ = 388.46 g/mol; CAS 87269-97-4), also known as 2-[N-[(S)-1-carboxy-3-phenylpropyl]-L-alanyl]-(1S,3S,5S)-2-azabicyclo[3.3.0]octane-3-carboxylate, is not assimilated directly and must be formed by hepatic metabolism of its prodrug form ramipril. Ramiprilat exists as slowly interconverting *cis* and *trans* geometric isomers with respect to the amide bond. The *trans* isomer is the most potent enzyme inhibitor of angiotensin-converting enzyme (K_i = 7 pM), first rapidly binding to the enzyme to form an initially weak enzyme-inhibitor complex, which then undergoes a slow isomerization. The *cis* isomer exhibits a very low inhibitory effect. (**See prodrug** *Ramipril*) **Target(s):** peptidyl-dipeptidase A, *or* angiotensin I-converting enzyme (1-6); and Xaa-Pro aminopeptidase, *or* aminopeptidase P (7-9). **1**. Baudin & Beneteau-Burnat (1999) *J. Enzyme Inhib.* **14**, 447. **2**. Bunning (1987) *J. Cardiovasc. Pharmacol.* **10** Suppl. 7, S31. **3**. Skoglof, Gothe & Deinum (1990) *Biochem. J.* **272**, 415. **4**. Skoglof, Nilsson, Gustafsson, Deinum & Gothe (1990) *Biochim. Biophys. Acta* **1041**, 22. **5**. Deddish, Wang, Jackman, *et al.* (1996) *J. Pharmacol. Exp. Ther.* **279**, 1582. **6**. Levitt & Schoemaker (2006) *BMC Clin. Pharmacol.* **6**, 1. **7**. Hooper, Hryszko, Oppong & Turner (1992) *Hypertension* **19**, 281. **8**. Orawski & Simmons (1995) *Biochemistry* **34**, 11227. **9**. Lloyd, Hryszko, Hooper & Turner (1996) *Biochem. Pharmacol.* **52**, 229.

Ramoplanins

These glycolipodepsipeptide antibiotics (FW = 2254.06; CAS 76168-82-6 for Ramoplanin A$_1$), including Ramoplanin A$_1$: R$_1$ = octa-2(Z),4(E)-dienoyl and R$_2$ = OH; Ramoplanin A$_2$: R$_1$ = 7 methylocta-2(Z),4(E)-dienoyl and R$_2$ = OH; Ramoplanin A3: R$_1$ = 9-methyldeca-2(Z),4(E)-dienoyl and R$_2$ = OH; and Ramoplanose: R$_1$ = 7-methylocta-2(Z),4(E)-dienoyl; R$_2$ = D-mannose, are isolated from the cell-culture broths of actinomycetes (genus *Actinoplanes*, strain ATCC 33076) and their enduracidin congenitors. Ramoplanins are powerful inhibitors of cell wall biosynthesis (1-4), exhibiting promising properties for treating vancomycin- and methicillin-resistant enterococci as well as *Clostridia difficile*. These inhibitors sequester peptidoglycan biosynthetic lipid intermediates by blocking late-stage cell wall synthesis catalyzed by MurG (*or* undecaprenyldiphospho-muramoylpentapeptide β-N-acetyl-glucosaminyltransferase) and related transglycosylases. Ramoplanins are functionally related to mersacidin and vancomycin. **Target(s):** transglycosylases, bacterial (1,5); undecaprenyl-diphosphomuramoylpentapeptide β-N-acetyl-glucosaminyltransferase, *or* MurG transferase (1,2,6). 1. McCafferty, Cudic, Frankel, *et al.* (2002) *Biopolymers* **66**, 261. 2. Crouvoisier, Mengin-Lecreulx & van Heijenoort (1999) *FEBS Lett.* **449**, 289. 3. Kurz & Guba (1996) *Biochemistry* **35**, 12570. 4. Somner & Reynolds (1990) *Antimicrob. Agents Chemother.* **34**, 413. 5. Hu, Helm, Chen, Ye & Walker (2003) *J. Amer. Chem. Soc.* **125**, 8736. 6. Branstrom, Midha, Longley, *et al.* (2000) *Anal. Biochem.* **280**, 315.

Ramosetron

This 5-HT₃ receptor antagonist and antiemetic (FW = 279.33 g/mol; CAS 132907-72-3; IUPAC: (1-methyl-1H-indol-3-yl)[(5R)-4,5,6,7-tetrahydro-1H-benzimidazol-5-yl]methanone) targets the 5-hydroxytryptamine type-3 (serotonin) receptor (K_d = 0.15 nM, B_{max} = 653 ± 30 fmol/mg protein), with higher affinity than [³H]granisetron (K_d = 1.17 nM, B_{max} = 427 ± 43 fmol/mg protein). Kinetic studies revealed that dissociation rate constant (k_{off}) for [³H]ramosetron was also lower than that for [³H]granisetron (1). These results suggest that ramosetron is a highly potent 5-HT₃-receptor antagonist. Ramosetron is used to treat irritable bowel syndrome (IBS), a highly prevalent functional bowel disorder. Serotonin (5-HT) regulates gastrointestinal function, and, in experimental studies, 5-HT₃ receptor antagonists slow colon transit, blunt gastrocolonic reflex, and reduce rectal sensitivity (2). 1. Akuzawa, Ito & Yamaguchi (1998) *Japanese J. Pharmacol.* **78**, 381. 2. Min & Rhee (2015) *Therap. Adv. Gastroenterol.* **8**, 136.

Ramucirumab

This VEGF receptor 2 inhibitor (MW = 143.6 kDa; CAS 947687-13-0), also known by its code name IMC-1121B, is a fully human IgG₁ monoclonal antibody intended for the treatment of solid tumors, particularly as a second-line option for patients with advanced gastric or gastroesophageal junction cancer.

Ranakinin

This undecapeptide amide KPNPERFYGLM-NH₂ (MW = 1349 g/mol; CAS 139446-71-2), first isolated from the brain of the frog *Rana ridibunda*, binds to the NK₁ tachykinin receptor, inhibiting the binding of mammalian substance P (1). Ranakinin also stimulates phospholipase C activity in the frog adrenal gland. The stimulatory action of ranakinin on inositol phosphate formation (with cocommitant decrease in membrane polyphosphoinositides) and corticosteroid secretion is mediated through activation of a phospholipase C positively coupled to a pertussis toxin-sensitive G protein (2). 1. O'Harte, Burcher, Lovas, *et al.* (1991) *J. Neurochem.* **57**, 2086. 2. Kodjo, Desrues, Lavagno, *et al.* (1998) *Endocrinology* **139**, 505.

Ranibizumab

This first-in-class angiostatic agent (MW = 48 kDa; CAS 347396-82-1), known by the trade name Lucentis®, is a humanized, affinity-matured anti-VEGF Fab antibody fragment with anti-angiogenic properties. The splice variant VEGF₁₆₅ increases the permeability of retinal endothelial cells, often resulting in diabetic macular edema; however, capillary barrier function can be restored by treating affected patients with the ranibizumab. Derived from the same parent mouse antibody used to prepare bevacizumab (Avastin®), ranibizumab is the first FDA-approved anti-angiogenic agent for the treatment of "wet" type, age-related macular degeneration. Ranibizumab is administered as an intravireal injection on a monthly schedule.

Ranitidine

This furanyl derivative (FW$_{free-base}$ = 314.41 g/mol; CAS 66357-35-5; M.P. = 69-70°C; Soluble to 50 mM in water), also known as Zantac™ and N-[2-[[[5-[(dimethylamino)methyl]-2-furanyl]methyl]thio]ethyl]-N'-methyl-2-nitro-1,1-ethanediamine hydrochloride, is a potent, competitive and selective antagonist (IC₅₀ = 3.3 µM) of histamine H₂ receptors, thereby blocking gastric acid secretion. By 1982, only three years after receiving FDA approval, Zantac sales reached $1 billion, making it the largest selling prescription drug in the world. Zantac has since largely been superseded by the even more effective proton pump inhibitors, with omeprazole becoming the biggest-selling drug. [Ranitidine was widely used to treat ulcers prior to the discovery of *Helicobacter pylori* (formerly *Campilobacter pylori*).] **Key Pharmacokinetic Parameters:** *See* Appendix II in Goodman & Gilman's *THE PHARMACOLOGICAL BASIS OF THERAPEUTICS*, 12ᵗʰ Edition (Brunton, Chabner & Knollmann, eds.) McGraw-Hill Medical, New York (2011). **Target(s):** acetylcholine esterase, K_i = 2.1 µM (1); butyrylcholinesterase, K_i = 61 µM (1); histamine H₂ receptor (2-4); and monoamine oxidase B (5). 1. Laine-Cessac, Turcant, Premel-Cabic, Boyer & Allain (1993) *Res. Commun. Chem. Pathol. Pharmacol.* **79**, 185. 2. van der Goot & Timmerman (2000) *Eur. J. Med. Chem.* **35**, 5. 3. Smit, Leurs, Alewijnse, *et al.* (1996) *Proc. Natl. Acad. Sci. U.S.A.* **93**, 6802. 4. Hill, Ganellin, Timmerman, *et al.* (1997) *Pharmacol. Rev.* **49**, 253. 5. Lu & Silverman (1993) *J. Enzyme Inhib.* **7**, 43.

Ranolazine

This orally active antianginal (FW$_{free-base}$ = 427.54 g/mol; CAS 95635-55-5), known by its trade name Ranexa™, its code name RS-43285 and its systematic name (RS)-N-(2,6-dimethylphenyl)-2-[4-[2-hydroxy-3-(2-methoxyphenoxy)propyl]piperazin-1-yl]acetamide, also reduces myocardial infarct size and cardiac troponin T release in the rat. **Action on Inward Sodium Current:** The depolarizing inward sodium current (I_{Na}) initiates the cardiac action potential, and the main ion channel that conducts I_{Na} in cardiac cells is Na$_V$1.5, the tetrodotoxin-resistant, pore-forming α-subunit of the voltage-gated sodium channel (VGSC). In cardiac myocytes, ranolazine not only blocks peak Na$_V$ current, but also inhibits the response of Na$_V$1.5 channels to mechanical stimulation, very likely by altering membrane partitioning (1). Its effects on Na$_V$1.7 and Na$_V$1.8 sodium channels suggest its potential utility in treating neuropathic pain. Ranolazine is also a partial fatty acid β-oxidation inhibitor, inhibiting 3-ketoacyl coenzyme A thiolase and thereby shifting the energy substrate preference away from fatty acid metabolism and toward glucose metabolism (2). It also improves left ventricular function by blocking uptake of free fatty acids by the heart while shifting metabolism to anaerobic glycolysis during myocardial ischemia. Note that ranolazine activates pyruvate dehydrogenase. Persistent current blockers of voltage-gated sodium channels, like ranolazine, are also receiving greater attention in the face of mounting evidence that metastatic progression in carcinomas is accompanied, perhaps even preceded, by upregulation of voltage-gated sodium channels, which enhance invasiveness (3). **Target(s):** acetyl-CoA C-acyltransferase (2); β-adrenoceptor, weak antagonist (4); fatty acid β-oxidation (2,5); HERG and IsK currents (6); late sodium current (7); NADH dehydrogenase, *or* complex I, uncompetitive inhibition with respect to ubiquinone-1, IC₅₀ > 350 µM (8). (Note that rotenone displaces ranolazine from its binding site, a property that may explain ranolazine's anti-ischemic activity.) 1. Beyder, Strege, Reyes, *et al.* (2012) *Circulation* **125**, 2698. 2. Fragasso, Spoladore, Cuko & Palloshi (2007) *Curr. Clin. Pharmacol.* **2**, 190. 3. Djamgoz & Onkal (2013) *Recent*

Pat. Anticancer Drug Discov. **8**, 66. **4**. Letienne, Vie, Puech, *et al.* (2001) *Naunyn Schmiedebergs Arch. Pharmacol.* **363**, 464. **5**. Zacharowski, Blackburn & Thiemermann (2001) *Eur. J. Pharmacol.* **418**, 105. **6**. Schram, Zhang, Derakhchan, *et al.* (2004) *Brit. J. Pharmacol.* **142**, 1300. **7**. Undrovinas, Belardinelli, Undrovinas & Sabbah (2006) *J. Cardiovasc. Electrophysiol.* **17** Suppl. 1, S169. **8**. Wyatt, Skene, Veitch, Hue & McCormack (1995) *Biochem. Pharmacol.* **50**, 1599.

Ranunculeate, *See* Columbinate

Rapamycin

This macrocyclic antifungal agent and immunosuppressant (FW = 914.19 g/mol; CAS 53123-88-9; Soluble to 50 mM in DMSO), also known as Sirolimus and Rapamune, from *Streptomyces hygroscopicus* potently inhibits interleukin-2- (*or* IL-2-) dependent T-cell proliferation (1-9). Its chief advantage is that, unlike calcineurin inhibitors, transplant patients can be maintained long-term on the latter without developing impaired kidney function or chronic renal failure. **Primary Mode of Inhibitory Action:** Mammalian target of rapamycin (*or* mTOR) is a Ser/Thr protein kinase that is conserved in all eukaryotes that plays a key role in cell growth and is likewise a central effector of several pathways regulating essential cell functions. Hyperactivation of the mTOR-dependent signalling pathway occurs in many human diseases, making it a druggable target. However, the dual nature of mTOR, existing in two multiprotein complexes mTORC1 and mTORC2, which is driven by different feedback loops, modulating rapamycin's therapeutic effects. Rapamycin's antiproliferative effect is mediated through the formation of an active complex with the cytosolic receptor protein FKBP12, inhibiting the translation of mRNAs encoding ribosomal proteins and elongation factors, thereby decreasing protein biosynthesis. Rapamycin also inhibits the cyclin-dependent kinase cdk2-cyclin E complex, which functions as a crucial regulator of G_1/S cell cycle transition. Therefore, although it exerts a suppressive effect on the immune system that is similar to that of tacrolimus (FK506), sirolimus's (rapamycin's) action directly binds the mTOR Complex1 (mTORC1) without involving calcineurin. Recent work suggests that the amino acid sequence comprising a phosphorylation site is a likely factor in determining why certain mTORC1 sites are rapamycin-insensitive (10). Rapamycin also strongly augments lentiviral (LV) transduction of human stem cells (HSCs) *in vitro* and *in vivo*, dramatically enhancing marking frequency within long-term engrafting cells in mice (11). Mechanistically, rapamycin enhances post-binding endocytic events, leading to increased levels of LV cytoplasmic entry, reverse transcription, and genomic integration. **Use in Combination Chemotherapy:** Inhibition of the mammalian target of rapamycin (mTOR) signaling pathway promotes the initiation of autophagy, and the Mitogen-Activated Protein Kinase/Extracellular signal-Regulated protein Kinase (*or* MAPK/ERK) is well known to induce autophagy, a self-defense mechanism of cancer cells subjected to antitumor agents. Blocking autophagy, rapamycin triggers apoptosis, and application of the MEK inhibitor U0126 offers an effective treatment for Malignant Fibrous Histiocytoma, *or* MFH (12). Significantly, Rapamycin reverses insulin resistance (IR) in high-glucose medium without causing IR in normoglycemic medium (13). **Key Pharmacokinetic Parameters:** *See* Appendix II in Goodman & Gilman's THE PHARMACOLOGICAL BASIS OF

THERAPEUTICS, 12th Edition (Brunton, Chabner & Knollmann, eds.) McGraw-Hill Medical, New York (2011). **Target(s):** cyclin-dependent kinase, by binding FK506-rapamycin complex (1-3); p33[edk2] kinase (2); p34[edk2] kinase (2,3); p70 S6 kinase (4-6); and peptidylprolyl isomerase (7-9). **1**. MacKintosh & MacKintosh (1994) *Trends Biochem. Sci.* **19**, 444. **2**. Morice, Wiederrecht, Brunn, Siekierka & Abraham (1993) *J. Biol. Chem.* **268**, 22737. **3**. Morice, Brunn, Wiederrecht, Siekierka & Abraham (1993) *J. Biol. Chem.* **268**, 3734. **4**. Kuo, Chung, Fiorentino, *et al.* (1992) *Nature* **358**, 70. **5**. Eriksson, Toivola, Sahlgren, Mikhailov & Härmälä-Braskén (1998) *Meth. Enzymol.* **298**, 542. **6**. Price, Grove, Calvo, Avruch & Bierer (1992) *Science* **257**, 973. **7**. Breiman, Fawcett, Ghirardi & Mattoo (1992) *J. Biol. Chem.* **267**, 21293. **8**. Monaghan & Bell (2005) *Mol. Biochem. Parasitol.* **139**, 185. **9**. Leuzzi, Serino, Scarselli, *et al.* (2005) *Mol. Microbiol.* **58**, 669. **10**. Wang, Sather, Wang, *et al.* (2014) *Blood* **124**, 913. **11**. Yoon & Roux (2013) *Curr Biol.* **23**, R880. **12**. Nakamura, Hitora, Yamagami, *et al.* (2014) *Int. J. Mol. Med.* **33**, 1491. **13**. Leontieva, Demidenko & Blagosklonny (2014) *Cell Death Dis.* **5**, e1214.

Ras Inhibitory Peptide, *See* L-Valyl-L-prolyl-L-prolyl-L-prolyl-L-valyl-L-prolyl-L-prolyl-L-arginyl-L-arginyl-L-arginine

Rasagiline

This irreversible monoamine oxidase inhibitor and early Parkinson monotherapy (FW = 171.24 g/mol; CAS 1875-50-9), also known as AGN 1135, TVP 1012, Azilect®, and (*R*)-*N*-(prop-2-ynyl)-2,3-dihydro-1*H*-inden-1-amine, forms stoichiometric N^5-flavocyanine adducts with the FAD moieties within monoamine oxidase, Types A and B (1). The telling observation that no H_2O_2 is produced during monamine oxidase A and B inactivation indicates that covalent addition occurs within a single catalytic round. Rasagiline has a specificity index (k_{inact}/K_i) that is ninety times greater for MAO B than MAO A (2). Rasagiline rescues degenerating dopamine neurons by inhibiting death signal transduction that is initiated by increased mitochondria permeability and induction of antiapoptotic Bcl-2 (signaling through PKC) and G̲lial cell line-d̲erived n̲eurotrophic f̲actor (GDNF) (3). When given prior to the dopaminergic neurotoxin 1-methyl-4-phenyl-1,2,3,6-tetrahydropyridine (MPTP), rasagiline protected mice against neurotoxicity, supporting the oxidation of MPTP to 1-methyl-4-phenyl-pyridinium (MPP⁺) as an important feature of the neurotoxic process (4). (*See also* Ladostigil) **1**. Youdim & Finberg (1987) *Adv. Neurol.* **45**, 127. **2**. Hubálek, Binda, Li, *et al.* (2004) *J. Med. Chem.* **47**, 1760. **3**. Maruyama, Akao, Carrillo, *et al.* (2002) *Neurotoxicol. Teratol.* **24**, 675. **4**. Heikkila, Duvoisin, Finberg & Youdim (1983) *Eur. J. Pharmacol.* **116**, 313.

Raubasine

This alkaloid (FW_{free-base} = 352.43 g/mol; CAS 483-04-5), also known as δ-yohimbine, ajmalicine, and (19α)-16,17-didehydro-19-methyloxayohimban-16-carboxylic acid methyl ester, isolated from *Corynanthe johimbe* and *Rauwolfia serpentina* is an α_1-adrenergic blocker (1-3) that is structurally similar to α-yohimbine (*or* rauwolscine). Like corynanthine, it acts as a α_1-adrenergic receptor antagonist with preferential actions over α_2-adrenergic receptors, underlying its hypotensive rather than hypertensive effects (4). **1**. Rimele, Rooke, Aarhus & Vanhoutte (1983) *J. Pharmacol. Exp. Ther.* **226**, 668. **2**. Demichel, Gomond & Roquebert (1981) *Brit. J. Pharmacol.* **74**, 739. **3**. Roquebert (1986) *Arch. Int. Pharmacodyn. Ther.* **282**, 252. **4**. Roquebert & Demichel (1984) *Eur. J. Pharmacol.* **106**, 203.

Rauwolfia Serpentina Root

This tranquillizing and antihypertensive powder, prepared from the root of the Indian snakeroot (*sarpagandha* in Hindi and *shégēn mù* in Chinese), also known as Insanity Herb, contains variable amounts of psychoactive substances, including reserpine, yohimbine, ajmaline, deserpidine, rescinnamine, and serpentinine. As the major alkaloid of the root, reserpine was one of the earliest recognized tranquilizers and also found broad application in the treatment of paranoia and schizophrenia. Its antihypertensive properties are well described (1). The root enjoys a long history in India, where medicine men still use it to treat both mental illness and snakebite. *Note*: Although having received FDA action on January 28, 1954 (NDA 009276), there are no Indications of Approved Use or designated Therapeutic Equivalents. **1**. Genest, Adamkiewics, Robillard & Tremblay (1955) *London Med. Asoc. J.* **72**, 483.

Rauwolscine

This alkaloid (FW$_{free-base}$ = 356.42 g/mol; CAS 6211-32-1), also known as α-yohimbine, from the tree bark of *Rauwolfia serpentina*, *R. canescens*, and *Corynanthe yohimbe* is an α$_2$-adrenergic antagonist as well as a 5-HT$_{1A}$ serotonergic receptor agonist. Rauwolscine diastereoisomers include yohimbine, corynanthine, and *allo*-yohimbine. **See also Yohimbines** **Target(s)**: α$_2$-adrenergic receptor. **1**. Perry & U'Prichard (1981) *Eur. J. Pharmacol.* **76**, 461.

Raventoxin

These venom toxins from the spider *Macrothele raveni* target voltage-gated sodium channels, shifting to a hyperpolarized state and slowing recovery. They include: Raventoxin-I (MW = 4840.11 Da); Raventoxin-II (MW = 3021.56 Da); Raventoxin-III (MW = 3286.58 Da); Raventoxin-V (MW = 3133.48 Da); and Raventoxin-VI (MW = 5371.60 Da). The crude venom was found to be neurotoxic in mice (LD$_{50}$ = 2.8 mg/kg) (1). Raventoxin-I appears to exert an effect of first exciting and then inhibiting the contraction of mouse diaphragm muscle caused by electrically stimulating the phrenic nerve (1). Venom potently and selectively suppresses the growth of K562 chronic myelogenous leukemia cells, inducing apoptosis via Caspase 3- and Caspase 8- mediated signaling pathways (2). **1**. Zeng, Xiao & Liang (2003) *Toxicon.* **41**, 651. **2**. Liu, Zhao, Li, *et al.* (2012) *Leuk. Res.* **36**, 1063.

Ravuconazole

This antifungal agent (FW = 437.47 g/mol; CAS 170864-29-6; *Abbreviation*: RAV), also known as ER-30306 and systematically named as 3-(4-(4-cyanophenyl)thiazol-2-yl)-2-(2,4-difluorophenyl)-1-(1*H*-1,2,4-triazol-1-yl)-2-butanol, inhibits sterol 14-demethylase, *or* CYP51 (1). Ravuconazole is a broad-spectrum triazole with *in vitro* activity against most *Candida* and *Aspergillus* species, some non-*Aspergillus* species of filamentous fungi, *Cryptococcus*, dermatophytes, and fungi that cause the endemic mycoses. The sterol biosynthesis pathway has received attention as a target for the development of new drugs for Chagas disease, and inhibitors

of sterol 14-demethylase are shown to be extremely active on *Trypanosoma cruzi*, both in *in vitro* and in animal experiments. **Prodrug**: In the orally administered, water soluble ravuconazole prodrug known as E1224, the hydrogen of the –OH group is replaced by –CH$_2$OPO$_3$H$^-$. **1**. Mellado, Garcia-Effron, Buitrago, *et al.* (2005) *Antimicrob. Agents Chemother.* **49**, 2536.

Razoxane

This racemic iron-chelating antimetastatic agent (FW$_{free-base}$ = 268.27 g/mol; CAS 21416-87-5), also known as ICRF-159, is a cyclized methyl-EDTA derivative that inhibits topoisomerase II by blocking the enzyme reaction cycle at the stage where the enzyme forms a closed clamp conformation around DNA, thereby interfering with transcription (or other DNA metabolic processes) and resulting in cell death. *Note*: Razoxane has a central chiral carbon, and its (+)- and (–)-enantiomers, often called dexrazoxane (Zinecard®) and levorazoxane, have the pharmaceutical code names ICRF-187 and ICRF-186, respectively. **Target(s)**: collagen-peptidase (1); and DNA topoisomerase II (2-10). **1**. Boggust & McGauley (1978) *Brit. J. Cancer* **38**, 329. **2**. Hammonds, Maxwell & Jenkins (1998) *Antimicrob. Agents Chemother.* **42**, 889. **3**. Tanabe, Ikegami, Ishida & Andoh (1991) *Cancer Res.* **51**, 4903. **4**. Ishida, Hamatake, Wasserman, *et al.* (1995) *Cancer Res.* **55**, 2299. **5**. Hasinoff, Kuschak, Yalowich & Creighton (1995) *Biochem. Pharmacol.* **50**, 953. **6**. Jensen, Nitiss, Rose, *et al.* (2000) *J. Biol. Chem.* **275**, 2137. **7**. Vaughn, Huang, Wessel, *et al.* (2005) *J. Biol. Chem.* **280**, 11920. **8**. Skladanowski, Come, Sabisz, Escargueil & Larsen (2005) *Mol. Pharmacol.* **68**, 625. **9**. Sorensen, Grauslund, Jensen, Sehested & Jensen (2005) *Biochem. Biophys. Res. Commun.* **334**, 853. **10**. Renodon-Corniere, Jensen, Nitiss, Jensen & Sehested (2003) *Biochemistry* **42**, 9749.

RB 38A

This peptidase inhibitor (FW$_{free-acid}$ = 370.41 g/mol; CAS 105831-46-7; (*R*)-[3-(*N*-hydroxycarboxamido)-2-benzylpropanoyl]-L-phenylalanine) targets neprilysin (1,2) and membrane alanyl aminopeptidase, *or* aminopeptidase N, with respective K$_i$ values of 0.9 and 120 nM. RB 38A is relatively poor at crossing the blood-brain barrier. Membrane alanyl aminopeptidase degrades a number of small secreted peptides, among them, amyloid β-peptide which has been implicated as a causative factor of Alzheimer's disease. Aminopeptidase N is responsible for the processing of peptide hormones, such as angiotensin III and IV, neuropeptides, and chemokines. It also cleaves antigen peptides bound to major histocompatibility complex class II molecules of presenting cells and degrades certain neurotransmitters at synaptic junctions. **1**. Roques, Noble, Crine & Fournié-Zaluski (1995) *Meth. Enzymol.* **248**, 263. **2**. Bauvois & Dauzonne (2006) *Med. Res. Rev.* **26**, 88. **3**. Noble, Luciani, Da Nacimento, *et al.* (2000) *FEBS Lett.* **467**, 81. **4**. Luciani, Marie-Claire, Ruffet, *et al.* (1998) *Biochemistry* **37**, 686.

RB 101

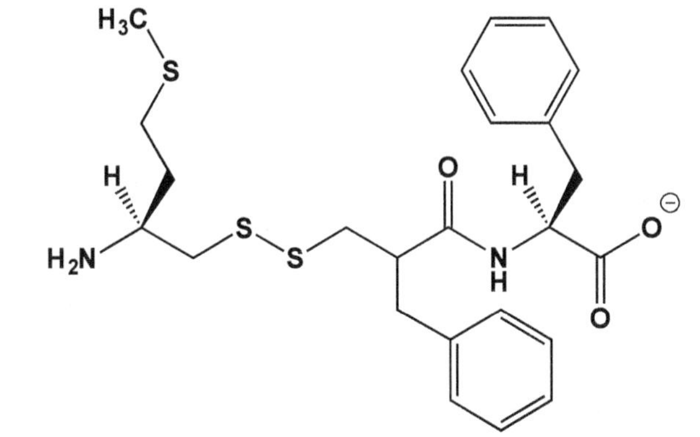

This double-prodrug (FW = 492.73 g/mol; CAS 135949-60-9), also called N-(R,S)-2-benzyl-3[(S)-(2-amino-4-methylthiobutyldithio)-1-oxopropyl]-L-phenylalanine, generates the following two specific inhibitors upon cleavage of its disulfide bond: N-[(R,S)-2-mercaptomethyl-1-oxo-3-phenylpropyl]-L-phenylalanine, a neprilysin inhibitor (K_i = 2 nM), and (S)-2-amino-1-mercapto-4-methylthiobutane, an aminopeptidase N (*or* membrane alanyl aminopeptidase) inhibitor (K_i = 11 nM). It was the first systemically-active mixed inhibitor of enkephalin-degrading enzymes. *Note*: The benzyl ester of the prodrug has also been referred to as RB 101. **Target(s):** membrane alanyl aminopeptidase, *or* aminopeptidase N (1-3); and neprilysin (1-3). **1**. Roques, Noble, Crine & Fournié-Zaluski (1995) *Meth. Enzymol.* **248**, 263. **2**. Ruiz-Gayo, Baamonde, Turcaud, Fournie-Zaluski & Roques (1992) *Brain Res.* **571**, 306. **3**. Bauvois & Dauzonne (2006) *Med. Res. Rev.* **26**, 88.

RB 104

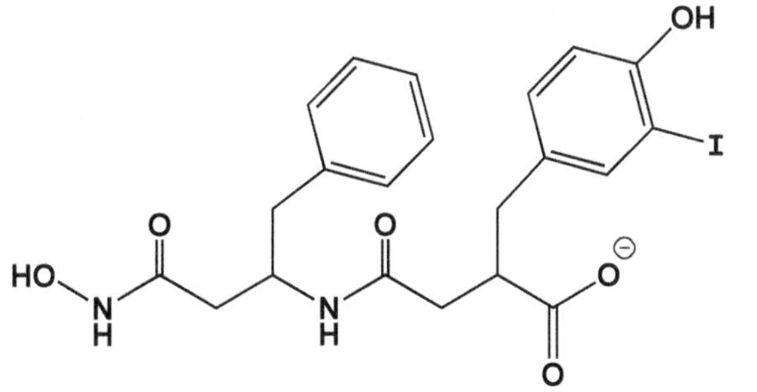

This peptide analogue (FW$_{free-acid}$ = 526.33 g/mol; CAS 145253-48-1), also called 2-((3-iodo-4-hydroxy)phenylmethyl)-4-N-[3-(hydroxyamino-3-oxo-1-phenylmethyl)propyl]amino-4-oxobutanoic acid, inhibits neprilysin (K_i = 27 pM). **Target(s):** *Aplysia* neutral endopeptidase (1); neprilysin (2-4); peptidyl-dipeptidase A, *or* angiotensin I-converting enzyme; weakly inhibited, K_i = 15 µM (3). **1**. Bawab, Aloyz, Crine, Roques & DesGroseillers (1993) *Biochem. J.* **296**, 459. **2**. Roques, Noble, Crine & Fournié-Zaluski (1995) *Meth. Enzymol.* **248**, 263. **3**. Fournié-Zaluski, Soleilhac, Turcaud, *et al.* (1992) *Proc. Natl. Acad. Sci. U.S.A.* **89**, 6388. **4**. Landry, Santagata, Bawab, *et al.* (1993) *Biochem. J.* **291**, 773.

Reactive Blue 4

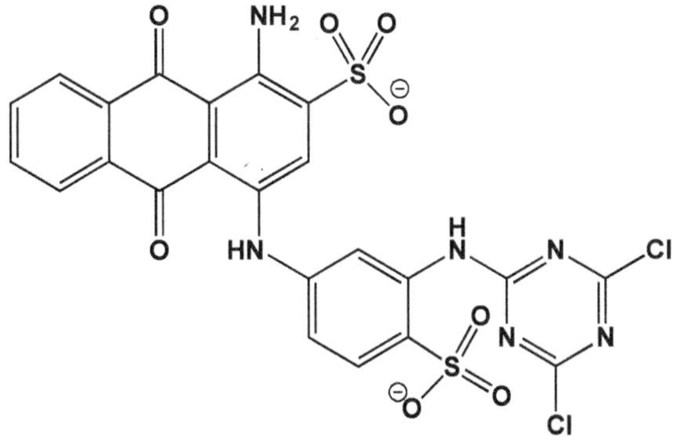

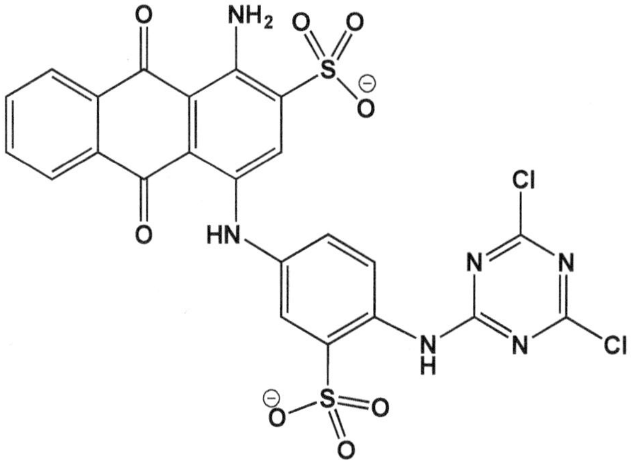

This blue dye (FW$_{free-acid}$ = 637.44 g/mol; CAS 13324-20-4; λ_{max} = 595 nm), also known as Procion Blue MX-R, Procion Royal Blue MX-R, Procion Blue MX-3G, Procion Blue M-3GS, and Procion Brilliant Blue RS, is a triazine stain. *Note*: There is some confusion concerning the nomenclature and structure. Some sources, including commercial vendors, such as Aldrich, refer to the upper structure, whereas many of the literature sources refer to the lower structure. For example, the anthraquinone dye (bottom) is the electrophilic dichlorotriazine precursor of Cibacron Blue. It is an irreversible inhibitor of NAD(P)H:(quinone-acceptor) oxidoreductase, *or* quinone reductase; K_i = 16 nM, k_3 = 0.03 min^{-1} (1). This inactivation is blocked by Cibacron Blue. **Target(s):** alcohol dehydrogenase (2); cholinephosphate cytidylyltransferase (3); α1,6-fucosyltransferase (4); glucokinase (5); glucose-6-phosphate dehydrogenase (6); glycerol kinase (5); NAD(P)H:(quinone-acceptor) oxidoreductase, inhibited by Procion Blue M-3GS (1); polyphosphate:glucose phosphotransferase (7); pyruvate kinase (8,9); and ricin A chain (10). **1**. Prestera, Prochaska & Talalay (1992) *Biochemistry* **31**, 824. **2**. Small, Lowe, Atkinson & Bruton (1982) *Eur. J. Biochem.* **128**, 119. **3**. Hunt & Postle (1986) *Biochem. Soc. Trans.* **14**, 1279. **4**. Kaminska, Dzieciol & Koscielak (1999) *Glycoconj. J.* **16**, 719. **5**. Goward, Scawen & Atkinson (1987) *Biochem. J.* **246**, 83. **6**. Reuter, Metz, Lorenz & Kopperschlager (1990) *Biomed. Biochim. Acta* **49**, 151. **7**. Hsieh, Shenoy, Jentoft & Phillips (1993) *Protein Expr. Purif.* **4**, 76. **8**. Byford & Bloxham (1984) *Biochem. J.* **223**, 359. **9**. Guy & Giles (1986) *Biochem. Soc. Trans.* **14**, 152. **10**. Alderton, Thatcher & Lowe (1995) *Eur. J. Biochem.* **233**, 880.

Reboxetine

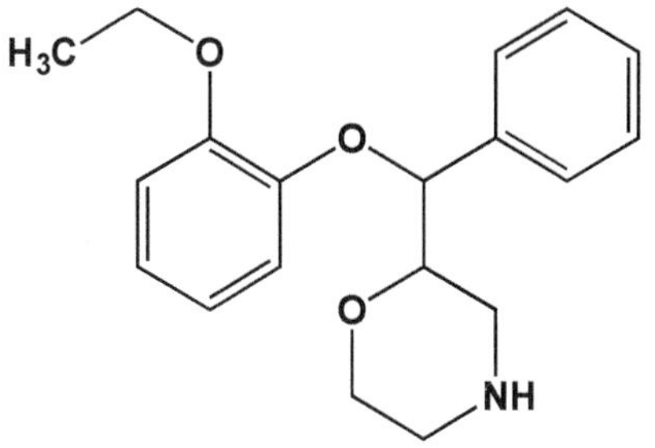

This oral antidepressant (FW$_{free-base}$ = 313.40 g/mol; CAS 98769-81-4, for mesylate salt; Solubility: 50 mM in H$_2$O; 50 mM in DMSO), known also as Vestra™ and named systematically as 2-((2-ethoxyphenoxy)benzyl)morpholine methanesulfonate, is the first selective noradrenaline (norepinephrine) reuptake inhibitor. Reboxetine is a potent and selective inhibitor of noradrenalin transporter, *or* NET (K_i = 1.1 nM), with weaker action against SERT (K_i = 129 nM) and DAT (K_i > 10000 nM), and displaying > 1000x selectivity over α-adrenoceptors, 5-HT, dopamine and muscarinic ACh receptors. It also prevents clonidine-induced hypothermia in mice. **1**. Kasper, el Giamal & Hilger (2000) *Expert Opin. Pharmacother.* **1**, 771. **2**. Sacchetti, *et al.* (1999) *Brit. J. Pharmacol.* **128**, 1332. **3**. Wong, *et al.* (2000) *Biol. Psychiatry* **47**, 818. **4**. Millan, *et al.* (2001) *J. Pharmacol. Exp. Ther.* **298**, 565. **5**. Owen & Whitton (2003) *Neurosci. Lett.* **348**, 171.

Recoverin

This retinal/pineal-specific calcium binding protein (MW = 23 kDa; CAS 135844-11-0) participates in visual signal transduction and is tethered to membranes by means of its *N*-terminal myristoyl group. A surge in intracellular calcium ion forms Ca^{2+}-bound recoverin, which then combines with rhodopsin kinase to form a Ca^{2+}-recoverin-rhodopsin kinase ternary complex that is stabilized by its membrane association, resulting in effective suppression of rhodopsin kinase activity (1-12). Half-maximal inhibition occurs at 1.5-3 µM free Ca^{2+}, with a Hill coefficient of ~2. A drop in free calcium ion following illumination is thus a key regulatory step during photoresponse recovery and light adaptation. **See also** S-Modulin 1. Chen & Hurley (2000) *Meth. Enzymol.* **315**, 404. **2**. Kawamura, Hisatomi, Kayada, Tokunaga & Kuo (1993) *J. Biol. Chem.* **268**, 14579. **3**. Gorodovikova, Gimelbrant, Senin & Philippov (1994) *FEBS Lett.* **349**, 187. **4**. Satpaev, Chen, Scotti, *et al.* (1998) *Biochemistry* **37**, 10256. **5**. Senin, Höppner-Heitmann, Polkovnikova, *et al.* (2004) *J. Biol. Chem.* **279**, 48647. **6**. Komolov, Zinchenko, Churumova, *et al.* (2005) *Biol. Chem.* **386**, 285. **7**. Higgins, Oprian & Schertler (2006) *J. Biol. Chem.* **281**, 19426. **8**. Chen, Inglese, Lefkowitz & Hurley (1995) *J. Biol. Chem.* **270**, 18060. **9**. Ames, Levay, Wingard, Lusin & Slepak (2006) *J. Biol. Chem.* **281**, 37237. **10**. Chen (2002) *Adv. Exp. Med. Biol.* **514**, 101.

Reduced 1-(2-Carboxyphenylamino)-1-deoxy-D-ribulose 5'-Phosphate, *See* N-(5'-Phosphoribityl)-anthranilate

Reduced Lapachol, *See* 3-(Methyl-2-butenyl)-1,2,4-trihydroxynaphthalene

Reductone, *See* specific reductone; e.g., 2,3-Dihydroxy-2-propenal

Regitine, *See* Phentolamine

Regorafenib

This orally active, multi-kinase inhibitor (FW = 482.82 g/mol; CAS 755037-03-7), also known as BAY 73-4506, Stivarga®, and 4-[4-({[4-chloro-3-(trifluoromethyl)phenyl]carbamoyl}amino)-3-fluorophenoxy]-*N*-methylpyridine-2-carboxamide hydrate, targets angiogenic, stromal, and oncogenic receptor tyrosine kinase-mediated processes, mainly through its inhibitory action on the anti-angiogenic activity of VEGFR2-TIE2 tyrosine kinases. Treatment of advanced-stage cancer patients with regorafenib has adverse hepatic, cardiovascular, cutaneous, gastrointestinal, thyroid, neurological and haematological effects, as well as infections and bleeding (2). These adverse effects are severe in about 40% of patients. A Phase I dose escalation study suggests that egorafenib has an acceptable safety profile as well as antitumor activity in patients with solid tumors (3). **Target(s):** VEGFR1 (IC_{50} = 13 nM), VEGFR2 (IC_{50} = 4.2 nM), VEGFR3 (IC_{50} = 46 nM), PDGFRβ (IC_{50} = 22 nM), Kit (IC_{50} = .7 nM), RET (IC_{50} = 1.5 nM) and Raf-1 (IC_{50} = 2.5 nM). **1**. Wilhelm, Dumas, Adnane, *et al.* (2011) *Int. J. Cancer* **129**, 245. **2**. No Authors (2014) *Prescrire Int.* **23**, 8. **3**. Mross, Frost, Steinbild, *et al.* (2012) *Clin. Cancer Res.* **18**, 2658.

REGN727, *See* Alirocumab

Regucalcin
This Ca^{2+} and GTP binding protein (FW = 33250 g/mol; Accession: BAA84082.1; Symbol: RGN) participates in maintaining intracellular Ca^{2+} homeostasis due to activating Ca^{2+} pump enzymes in the plasma membrane (basolateral membrane), microsomes (endoplasmic reticulum), and the mitochondria of many cell types. Because its expression is down-regulated with aging, RGN is also known as senescence marker protein-30. Beyond its well-known role in Ca^{2+} homeostasis, RGN has been linked to the control of as oxidative stress, cell proliferation, apoptosis, and signal transduction (4). RGN has antioxidant properties, reducing ROS production while increasing antioxidant defenses. RGN also suppresses cell proliferation trough regulation of oncogene and tumor suppressor gene expression (4). **Target(s):** phosphoprotein phosphatase (1); protein kinase C (2); and protein-tyrosine-phosphatase (2,3). **1**. Ichikawa, Tsurusaki & Yamaguchi (2004) *Int. J. Mol. Med.* **13**, 289. **2**. Yamaguchi (2005) *Int. J. Mol. Med.* **15**, 371. **3**. Fukaya & Yamaguchi (2004) *Int. J. Mol. Med.* **14**, 427. **4**. Vaz, Correia, Cardoso, *et al.* (2016) *Curr. Mol. Med.* **16**, 607.

Remazol Brilliant Blue R

This reactive anthraquinone dye ($FW_{disodium-salt}$ = 626.55 g/mol; CAS 2580-78-1; λ_{max} = 592 nm), also known as reactive blue 19 and named systematically as, 1-amino-9,10-dihydro-9,10-dioxo-4-((3-((2-(sulfooxy)ethyl)sulfonyl)phenyl)amino)-2-anthracenesulfonic acid, is frequently linked to polysaccharides or polymeric proteinase substrates to form chromagenic substrates. **Target(s):** lactate dehydrogenase (1,2); and Na^+/K^+-exchanging ATPase (2). **1**. Bohacova, Docolomansky, Breier, Gemeiner & Ziegelhoffer (1998) *J. Chromatogr. B Biomed. Sci. Appl.* **715**, 273. **2**. Durisova, Vrbanova, Ziegelhoffer & Breier (1990) *Gen. Physiol. Biophys.* **9**, 519.

Remikiren

This peptide ($FW_{free-base}$ = 630.85 g/mol; CAS 135669-48-6), also known as Ro 42-5892 and named systematically as (*S*)-α-((*S*)-α-((*t*-butylsulfonyl)-methyl)hydrocinnamamido)-*N*-((1*S*,2*R*,3*S*)-1-(cyclohexylmethyl)-3-cyclo-propyl-2,3-dihydroxypropyl)imidazole-4-propionamide, is a potent inhibitor of human renin, IC_{50} = 0.7 nM (1,2). It was the first orally active renin inhibitor, albeit with low oral absorption. **1**. Fischli, Clozel, el Amrani, *et al.* (1991) *Hypertension* **18**, 22. **2**. Mathews, Doebli, Pruschy, *et al.* (1996) *Protein Expr. Purif.* **7**, 81.

Remoxipride

This now discontinued atypical antipsychotic (FW = 371.27 g/mol; CAS 117591-79-4), also named Roxiam® and systematically as 3-bromo-*N*-[[(2*S*)-1-ethylpyrrolidin-2-yl]methyl]-2,6-dimethoxybenzamide, had been used in Europe to treat schizophrenia and acute mania. It also has marked affinity for central sigma receptors. Remoxipride was withdrawn in 1993 due to high incidence of aplastic anemia in treated patients. Remoxipride acts as a selective D_2 and D_3 receptor antagonist and also has high affinity for the sigma receptor, possibly playing a role in its atypical neuroleptic action. **1**. Wadworth & Heel (1990) *Drugs* **40**, 863. (Erratum: *Drugs* (1991) **41**, 532.)

Renilla Luciferin, *See* Coelenterazine

Renin Inhibitor H142

This pseudopeptide-containing inhibitor (MW = 1208.5 g/mol; *Sequence* = PHPFHL[CH₂-NH]VIHK, where the linkage between Leu-6 and Val-7 is a "reduced" peptide bond (as a mimic of the presumptive tetrahedral intermediate)) targets endothiapepsin, K_i = 160 nM (1-4) and renin, K_i = 10 nM (3-6). **1**. Hallett, Jones, Atrash, *et al.* (1985) in *Aspartic Proteinases and Their Inhibitors* (Kostka, ed.), *Proc. FEBS Advanced Course* 84/07, pp. 467-478, Walter de Gruyter & Co., Berlin. **2**. Bailey & Cooper (1994) *Protein Sci.* **3**, 2129. **3**. Foundling, Cooper, Watson, *et al.* (1987) *Nature* **327**, 349. **4**. Hemmings, Foundling, Sibanda, *et al.* (1985) *Biochem. Soc. Trans.* **13**, 1036. **5**. Szelke, Leckie, Hallett, *et al.* (1982) *Nature* **299**, 555. **6**. Kay, Jupp, Norey, *et al.* (1988) *Adv. Exp. Med. Biol.* **240**, 1.

Rentiapril

This sulfhydryl-containing agent (FW$_{free-acid}$ = 313.40 g/mol; CAS 72679-47-1), also known as (2R,4R)-2-(o-hydroxyphenyl)-3-(3-mercaptopropionyl)-4-thiazolidinecarboxylate and SA 446, inhibits peptidyl-dipeptidase A, *or* angiotensin-converting enzyme (1-4) and Xaa-Trp aminopeptidase, *or* aminopeptidase W (5-7). **1**. Takase, Ikuse, Aono & Okahara (1995) *Arzneimittelforschung* **45**, 15. **2**. Yamauchi, Nishimura, Nakata, *et al.* (1987) *Arzneimittelforschung* **37**, 157. **3**. Iso, Yamauchi, Suda, *et al.* (1981) *Jpn. J. Pharmacol.* **31**, 875. **4**. Mendelsohn, Csicsmann & Hutchinson (1981) *Clin. Sci. (London)* **61** Suppl. 7, 277s. **5**. Hooper (1998) in *Handb. Proteolytic Enzymes* (Barrett, Rawlings & Woessner, eds.), p. 1513, Academic Press, San Diego. **6**. Tieku & Hooper (1992) *Biochem. Pharmacol.* **44**, 1725. **7**. Tieku & Hooper (1993) *Biochem. Soc. Trans.* **21**, 250S.

RepSox

This potent and selective ALK inhibitor (FW = 287.33 g/mol; CAS 446859-33-2; Solubility: 57 mg/mL DMSO; <1 mg/mL H₂O), also known as E-616452, SJN-2511, and 2-[3-(6-methyl-2-pyridinyl)-1H-pyrazol-4-yl]-1,5-naphthyridine, targets TGFβR-1/ALK5 with IC₅₀ values of 23 nM and 4 nM for ATP binding to ALK5 and ALK5 autophosphorylation (1,2). RepSox replaces Sox2 by inhibiting Tgf-β signaling in cultures containing stable intermediate cells trapped in a partially reprogrammed state. The resulting sustained transcription of Nanog demonstrates the feasibility of replacing the central reprogramming transgenes with small molecules that modulate discrete cellular pathways or processes rather than by globally altering chromatin structure (2). **1**. Gellibert, Woolven, Fouchet, *et al.* (2004) *J. Med. Chem.* **47**, 4494. **2**. Ichida, Blanchard, Lam, *et al.* (2009) *Cell Stem Cell* **5**, 491.

RES 701-1

This endothelin type B receptor antagonist (MW = 2045.20 g/mol; CAS 151308-34-8), a cyclopeptide from *Streptomyces* sp RE-701, has the sequence GNWHGTAPDWFFNYYW, wherein an isopeptide bond joins Glycine-1 and Aspartate-9. RES-701-1 blocks endothelin binding (IC₅₀ = 10 nM) (1-3). **1**. Morishita, Chiba, Tsukuda, *et al.* (1994) *J. Antibiot.* **47**, 269. **2**. Yamasaki, Yano, Yoshida, Matsuda & Yamaguchi (1994) *J. Antibiot.* **47**, 276. **3**. Tanaka, Tsukuda, Nozawa, *et al.* (1994) *Mol. Pharmacol.* **45**, 724.

Reserpine

This catecholamine-depleting alkaloid (FW$_{free-base}$ = 608.69 g/mol CAS 50-55-5; Source: *Rauwolfia*), also kown as (3β,16β,17α,18β,20α)-11,17-dimethoxy-18-[(3,4,5-trimethoxy-benzoyl)oxy]yohimban-16-carboxylic acid methyl ester, is both an antihypertensive agent and a tranquilizer. Reserpine irreversibly blocks the Vesicular Monoamine Transporter (VMAT) that transports free intracellular norepinephrine, serotonin, and dopamine within the presynaptic nerve terminal into presynaptic vesicles for their subsequent exocytotic release into the synaptic cleft. Reserpine thereby reduces catecholamine and serotonin stores, altering catecholaminergic- and serotoninergic-dependent neuronal processes. **Target(s):** aldehyde dehydrogenase (1); Ca²⁺ transport (5,6); dopamine uptake (4;) hyaluronoglucosaminidase, *or* hyaluronidase (9); multidrug efflux pumps (5,8); NAD⁺ ADP-ribosyltransferase, *or* poly(ADP-ribose) polymerase (10); Na⁺/K⁺-exchanging ATPase (6); oxidative phosphorylation (3); and xenobiotic-transporting ATPase, *or* multidrug-resistance protein (8). **1**. Youdim & Sandler (1968) *Eur. J. Pharmacol.* **4**, 105. **2**. Login, Judd, Cronin, Yasumoto & MacLeod (1985) *Amer. J. Physiol.* **248**, E15. **3**. Balzer, Makinose & Hasselbach (1968) *Naunyn Schmiedebergs Arch. Exp. Pathol. Pharmakol.* **260**, 444. **4**. Metzger, Brown, Sandoval, *et al.* (2002) *Eur. J. Pharmacol.* **456**, 39. **5**. Schmitz, Fluit, Luckefahr, *et al.* (1998) *J. Antimicrob. Chemother.* **42**, 807. **6**. Kanoh (1994) *Laryngoscope* **104**, 197. **7**. Weinbach, Costa, Claggett, Fay & Hundal (1983) *Biochem. Pharmacol.* **32**, 1371. **8**. Van Veen, Venema, Bolhuis, *et al.* (1996) *Proc. Natl. Acad. Sci. U.S.A.* **93**, 10668. **9**. Girish & Kemparaju (2005) *Biochemistry (Moscow)* **70**, 948. **10**. Banasik, Komura, Shimoyama & Ueda (1992) *J. Biol. Chem.* **267**, 1569.

Resiquimod

This immune response modifier (FW = 314.39 g/mol; CAS 144875-48-9; Solubility: 200 mM in DMSO), also known as R-848 and 4-amino-2-(ethoxymethyl)-α,α-dimethyl-1H-imidazo[4,5-c]quinoline-1-ethanol, is a potent toll-like receptor-7 (TLR 7) and -8 (TLR8) agonist. Resiquimod mimics effects of the T-dependent CD40 signal in both mouse and human B cell lines. Toll-like receptors (TLRs) play an important role in the innate immune response to pathogens. TLRs detect PAMPs (pathogen-associated molecular patterns) and stimulate immune cells *via* the MyD88-dependent interleukin-1 receptor (IL-1R)-TLR signaling pathway, which leads to activation of the transcription factor NF-κB. Like CD40, resiquimod stimulates antibody secretion, cytokine production, protects cells from apoptosis, and upregulates CD80. In addition, it shows synergy with signals delivered by the B-cell antigen receptor and heightens CD40-mediated B cell activation, demonstrating that resiquimod can enhance antigen-specific responses in B lymphocytes. **1**. Bishop, Ramirez, Baccam, *et al.* () *Cellular Immunol.* **208**, 9.

Resorcinol

This common redox-active reagent (FW = 110.11 g/mol; CAS 108-46-3; white solid; M.P. = 109-111°C) is less frequently known as 1,3-quinol, 1,3-hydroquinone, resorcin, and 1,3-dihydroxybenzene. *Note*: Resorcinol decomposes (turning pink) upon exposure to light, air, and iron. Solutions should always be freshly prepared and protected from light. Resorcinol is also a key component in the Seliwanoff test for the qualitative determinaion of ketoses. **Target(s):** γ-butyrobetaine hydroxylase, or γ-butyrobetaine:2-oxoglutarate dioxygenase (1); cAMP phosphodiesterase (2); carboxylase (3,4); catalase (5); catechol oxidase (6,7); complement component C'3, guinea pig (8); 2,6-dihydroxypyridine 3-monooxygenase (9); esterase (10); glyoxalase I, or lactoylglutathione lyase (11); lipase, weakly inhibited (12); lipoxygenase (13); papain (14); pyrogallol hydroxytransferase (15); starch phosphorylase (16); tannase (17); tyrosinase, alternative substrate (6,18); and tyrosine phenol-lyase (19). **1.** Ng, Hanauske-Abel & Englard (1991) *J. Biol. Chem.* **266**, 1526. **2.** Shinohara, Fujiki, Hidaka, *et al.* (1985) *Acta Vitaminol. Enzymol.* **7**, 99. **3.** Karrer & Visconti (1947) *Helv. Chim. Acta* **30**, 268. **4.** Massart (1950) *The Enzymes*, 1st ed. (Sumner & Myrbäck, eds.), **1** (part 1), 307. **5.** Blaschko (1935) *Biochem. J.* **29**, 2303. **6.** Webb (1966) *Enzyme and Metabolic Inhibitors*, vol. **2**, Academic Press, New York. **7.** Richter (1934) *Biochem. J.* **28**, 901. **8.** Shin & Mayer (1968) *Biochemistry* **7**, 3003. **9.** Baitsch, Sandu, Brandsch & Igloi (2001) *J. Bacteriol.* **183**, 5262. **10.** Glick & King (1932) *J. Biol. Chem.* **97**, 675. **11.** Iio, Okabe & Omura (1976) *J. Nutr. Sci. Vitaminol.* **22**, 53. **12.** Weinstein & Wynne (1936) *J. Biol. Chem.* **112**, 649. **13.** Holman & Bergström (1951) *The Enzymes*, 1st ed. (Sumner & Myrbäck, eds.), **2** (part 1), 559. **14.** Bahadur & Atreya (1960) *Enzymologia* **21**, 238. **15.** Brune & Schink (1990) *J. Bacteriol.* **172**, 1070. **16.** Singh & Sanwal (1976) *Phytochemistry* **15**, 1447. **17.** Iibuchi, Minoda & Yamada (1972) *Agric. Biol. Chem.* **36**, 1553. **18.** Gortner (1911) *J. Biol. Chem.* **10**, 113. **19.** Nagasawa, Utagawa, Goto, *et al.* (1981) *Eur. J. Biochem.* **117**, 33.

β-Resorcylaldehyde, See *2,4-dihydroxy-benzaldehyde*

α-Resorcylate (α-Resorcylic Acid)

This aromatic acid (FW_{free-acid} = 154.12 g/mol; CAS 99-10-5; 64887-61-2, Cu²⁺ salt)), also known as 3,5-dihydroxybenzoic acid, inhibits γ-butyrobetaine hydroxylase, or γ-butyrobetaine:2-oxoglutarate dioxygenase, K_i = 0.05 mM (1), dihydroorotate dehydrogenase (2), glutamate decarboxylase (3), and tannase (4). **1.** Ng, Hanauske-Abel & Englard (1991) *J. Biol. Chem.* **266**, 1526. **2.** Palfey, Bjornberg & Jensen (2001) *J. Med. Chem.* **44**, 2861. **3.** Youngs & Tunnicliff (1991) *Biochem. Int.* **23**, 915. **4.** Iibuchi, Minoda & Yamada (1972) *Agric. Biol. Chem.* **36**, 1553.

β-Resorcylate (β-Resorcylic Acid)

This aromatic acid (FW_{free-acid} = 154.12 g/mol; CAS 89-86-1; M.P. = 213°C; pK_a = 3.30 at 25°C), also known as β-resorcylic acid and 2,4-dihydroxybenzoate, inhibits γ-butyrobetaine hydroxylase, or γ-butyrobetaine:2-oxoglutarate dioxygenase, K_i = 1.08 mM (1); 3-galactosyl-N-acetylglucosaminide 4-α-L-fucosyl-transferase (2), glycoprotein 3-α-L-fucosyltransferase, weakly inhibited (2), 4-hydroxybenzoate:CoA ligase, or 4-hydroxybenzoyl-CoA synthetase (3), 4-hydroxybenzoate decarboxylase (4,5), 4-hydroxybenzoate 3-monooxygenase (6), orsellinate decarboxylase (7,8); prephenate dehydratase (9), protocatechuate 3,4-dioxygenase (10), and o-pyrocatechuate decarboxylase (11-14). **1.** Ng, Hanauske-Abel & Englard (1991) *J. Biol. Chem.* **266**, 1526. **2.** Niu, Fan, Sun, *et al.* (2004) *Arch. Biochem. Biophys.* **425**, 51. **3.** Biegert, Altenschmidt, Eckerskorn & Fuchs (1993) *Eur. J. Biochem.* **213**, 555. **4.** Gallert & Winter (1992) *Appl. Microbiol. Biotechnol.* **37**, 119. **5.** He & Wiegel (1995) *Eur. J. Biochem.* **229**, 77. **6.** Fujii & Kaneda (1985) *Eur. J. Biochem.* **147**, 97. **7.** Petterson

(1965) *Acta Chem. Scand.* **19**, 2013. **8.** Mosbach & Schultz (1971) *Eur. J. Biochem.* **22**, 485. **9.** Schmit, Artz & Zalkin (1970) *J. Biol. Chem.* **245**, 4019. **10.** Stanier & Ingraham (1954) *J. Biol. Chem.* **210**, 799. **11.** Rao, Moore & Towers (1970) *Meth. Enzymol.* **17A**, 514. **12.** Rao, Moore & Towers (1967) *Arch. Biochem. Biophys.* **122**, 466. **13.** Santha, Rao & Vaidyanathan (1996) *Biochim. Biophys. Acta* **1293**, 191. **14.** Kamath, Dasgupta & Vaidyanathan (1987) *Biochem. Biophys. Res. Commun.* **145**, 586.

γ-Resorcylate

This aromatic acid (FW_{free-acid} = 154.12 g/mol; CAS 303-07-1), also known as γ-resorcylic acid and 2,6-dihydroxybenzoic acid, inhibits γ-butyrobetaine hydroxylase, or γ-butyrobetaine:2-oxoglutarate dioxygenase, K_i = 0.26 mM (1), deacetylcephalosporin-C acetyltransferase (2), o-pyrocatechuate decarboxylase (3,4), and tannase (5). **1.** Ng, Hanauske-Abel & Englard (1991) *J. Biol. Chem.* **266**, 1526. **2.** Scheidegger, Gutzwiller, Kuenzi, Fiechter & Nuesch (1985) *J. Biotechnol.* **3**, 109. **3.** Rao, Moore & Towers (1970) *Meth. Enzymol.* **17A**, 514. **4.** Santha, Rao & Vaidyanathan (1996) *Biochim. Biophys. Acta* **1293**, 191. **5.** Iibuchi, Minoda & Yamada (1972) *Agric. Biol. Chem.* **36**, 1553.

Restrictocin. See *Ribotoxins*

Resveratrol

This *trans*-stilbene derivative (FW = 228.25 g/mol; CAS 501-36-0, for *trans*-isomer), also known as 3,4',5-trihydroxy-*trans*-stilbene, is the phenolic antioxidant found in red wine, grape skins, peanuts, and cranberries. Resveratrol modulates lipid metabolism and inhibits platelet aggregation. It is also an antagonist of the aryl hydrocarbon (Ah) receptor. Other studies indicate resveratrol exhibits vasodilative, and anti-inflammatory properties. **Target(s):** aldehyde dehydrogenase (1); Ca²⁺ influx (2); casein kinase II (31); creatine kinase (3); cyclooxygenase, or peroxidase activity, IC₅₀ = 15 µM for COX-1 and 200 µM for COX-2 (4-9); CYP1A1 (10,11); CYP1A2 (10); CYP3A5 (12); CYP1B1 (10,13-15); CYP2B1 (16); CYP2C19 (17); CYP2D (18); CYP2E1 (19,20); CYP3A4 (12,17,20); DNA-directed DNA polymerase (21); DNA topoisomerase II (37); electron transport (22); estrone sulfotransferase (41,42); fatty-acid synthase (47); F₀F₁ ATPase, or ATP synthase, or H⁺-transporting two-sector ATPase (22-25); glutathione S-transferase (46); IκB kinase (43); lipoxygenase (6,26); mitogen-activated protein kinase (27); monoamine oxidase A (28); peptidyl-dipeptidase A, or angiotensin I-converting enzyme; weakly inhibited (39); phenylpyruvate tautomerase (38); 1-phosphatidylinositol 4-kinase (45); prostaglandin H synthase (I7); protein kinase C (29,30); protein kinase CKII, or casein kinase II (31); protein kinase D (30,32); non-specific protein-tyrosine kinase (33,34,44); steroid 21-monooxygenase (49); sterol esterase, or cholesterol esterase (40); succinate dehydrogenase, weakly inhibited (22); tyrosinase, or monophenol monooxygenase (35,36,50,51); and xanthine oxidase (28,48). **1.** Kitson, Kitson & Moore (2001) *Chem. Biol. Interact.* **130-132**, 57. **2.** Dobrydneva, Williams & Blackmore (1999) *Brit. J. Pharmacol.* **128**, 149. **3.** Miura, Muraoka & Fujimoto (2002) *Pharmacol. Toxicol.* **90**, 66. **4.** Sylvia, Del Toro, Dean, *et al.* (2001) *J. Cell Biochem.* **81**, 32. **5.** Subbaramaiah, Michaluart, Chung, *et al.* (1999) *Ann. N. Y. Acad. Sci.* **889**, 214. **6.** MacCarrone, Lorenzon, Guerrieri & Agro (1999) *Eur. J. Biochem.* **265**, 27. **7.** Johnson & Maddipati (1998) *Prostaglandins Other Lipid Mediat.* **56**,

131. **8.** Subbaramaiah, Chung, Michaluart, *et al.* (1998) *J. Biol. Chem.* **273**, 21875. **9.** Lee, Shin, Kang, *et al.* (1998) *Planta Med.* **64**, 204. **10.** Chang, Chen & Lee (2001) *J. Pharmacol. Exp. Ther.* **299**, 874. **11.** Chun, Kim & Guengerich (1999) *Biochem. Biophys. Res. Commun.* **262**, 20. **12.** Chang & Yeung (2001) *Can. J. Physiol. Pharmacol.* **79**, 220. **13.** Guengerich, Chun, Kim, Gillam & Shimada (2003) *Mutat. Res.* **523-524**, 173. **14.** Chang, Lee & Ko (2000) *Can. J. Physiol. Pharmacol.* **78**, 874. **15.** Ciolino & Yeh (1999) *Mol. Pharmacol.* **56**, 760. **16.** Huynh & Teel (2002) *Anticancer Res.* **22**, 1699. **17.** Yu, Shin, Kosmeder, Pezzuto & van Breemen (2003) *Rapid Commun. Mass Spectrom.* **17**, 307. **18.** Skaanild & Friis (2002) *Pharmacol. Toxicol.* **91**, 198. **19.** Mikstacka, Gnojkowski & Baer-Dubowska (2002) *Acta Biochim. Pol.* **49**, 917. **20.** Piver, Berthou, Dreano & Lucas (2001) *Toxicol. Lett.* **125**, 83. **21.** Sun, Woo, Cassady & Snapka (1998) *J. Nat. Prod.* **61**, 362. **22.** Zini, Morin, Bertelli, Bertelli & Tillement (1999) *Drugs Exp. Clin. Res.* **25**, 87. **23.** Kipp & Ramirez (2001) *Endocrine* **15**, 165. **24.** Zheng & Ramirez (2000) *Brit. J. Pharmacol.* **130**, 1115. **25.** Zheng & Ramirez (1999) *Biochem. Biophys. Res. Commun.* **261**, 499. **26.** Pinto, Garcia-Barrado & Macias (1999) *J. Agric. Food Chem.* **47**, 4842. **27.** El-Mowafy & White (1999) *FEBS Lett.* **451**, 63. **28.** Zhou, Kong, Ye, Cheng & Tan (2001) *Planta Med.* **67**, 158. **29.** Slater, Seiz, Cook, Stagliano & Buzas (2003) *Biochim. Biophys. Acta* **1637**, 59. **30.** Haworth & Avkiran (2001) *Biochem. Pharmacol.* **62**, 1647. **31.** Yoon, Kim, Ghim, Song & Bae (2002) *Life Sci.* **71**, 2145. **32.** Stewart, Christman & O'Brian (2000) *Biochem. Pharmacol.* **60**, 1355. **33.** Palmieri, Mameli & Ronca (1999) *Drugs Exp. Clin. Res.* **25**, 79. **34.** Jayatilake, Jayasuriya, Lee, *et al.* (1993) *J. Nat. Prod.* **56**, 1805. **35.** Kim, Yun, Lee, *et al.* (2002) *J. Biol. Chem.* **277**, 16340. **36.** Gilly, Mara, Oded & Zohar (2001) *J. Agric. Food Chem.* **49**, 1479. **37.** Jo, Gonzalez de Mejia & Lila (2005) *J. Agric. Food Chem.* **53**, 2489. **38.** Molnar & Garai (2005) *Int. Immunopharmacol.* **5**, 849. **39.** Actis-Goretta, Ottaviani & Fraga (2006) *J. Agric. Food Chem.* **54**, 229. **40.** Sbarra, Ristorcelli, Petit-Thévenin, *et al.* (2005) *Biochim. Biophys. Acta* **1736**, 67. **41.** Furimsky, Green, Sharp, *et al.* (2008) *Drug Metab. Dispos.* **36**, 129. **42.** Otake, Nolan, Walle & Walle (2000) *J. Steroid Biochem. Mol. Biol.* **73**, 265. **43.** Luo, Kamata & Karin (2005) *J. Clin. Invest.* **115**, 2625. **44.** Hollósy & Kéri (2004) *Curr. Med. Chem. Anticancer Agents* **4**, 173. **45.** Srivastava, Ratheesh, Gude, *et al.* (2005) *Biochem. Pharmacol.* **70**, 1048. **46.** Hayeshi, Mutingwende, Mavengere, Masiyanise & Mukanganyama (2007) *Food Chem. Toxicol.* **45**, 286. **47.** Li, Ma, Wang & Tian (2005) *J. Biochem.* **138**, 679. **48.** Huang, Li, Shi, *et al.* (2008) *Chem. Biodivers.* **5**, 636. **49.** Supomsilchai, Svechnikov, Seidlova-Wuttke, Wuttke & Söder (2005) *Horm. Res.* **64**, 280. **50.** Shin, Ryu, Choi, *et al.* (1998) *Biochem. Biophys. Res. Commun.* **243**, 801. **51.** Likhitwitayawuld, Sritularak & De-Eknamkul (2000) *Planta Med.* **66**, 275.

Retapamulin

This pleuromutilin-class topical antibiotic (FW = 517.77 g/mol; CAS 224452-66-8), also named SB-275833, Altabax®, Altargo®, and mutilin 14-(*exo*-8-methyl-8-azabicyclo[3.2.1]oct-3-yl)sulfanylacetate, is a translational inhibitor that binds to domain V of 23S rRNA on the bacterial ribosome's 50S subunit, thereby directly blocking peptide formation by interfering with substrate binding (1). Retapamulin has demonstrated efficacy against certain Gram-positive bacteria, including methicillin-resistant *Staphylococcus aureus* (MRSA). The MIC$_{90}$ for retapamulin is 0.12 µg/mL for *Staphylococcus aureus* and ≤ 0.03 µg/mL for *Streptococcus pyogenes*. There is no cross-resistance for microorganism subsets that are resistant to oxacillin, erythromycin, or mupirocin. Compared to tiamulin, retapamulin appears to be uniformly eightfold more potent. **1.** Jones, Fritsche, Sader & Ross (2006) *Antimicrob. Agents Chemother.* **50**, 2583.

all-trans-Retinal

This aldehyde (FW = 284.44 g/mol; CAS 116-31-4; orange solid; M.P = 61-64°C; light exposure generates various *cis*-isomers), commonly known as vitamin A aldehyde and systematically named as (2*E*,4*E*,6*E*,8*E*)-3,7-dimethyl-9-(2,6,6-trimethylcyclohex-1-en-1-yl)nona-2,4,6,8-tetraenal, is an essential component of the visual cycle. Retinal is a substrate of the reaction catalyzed by retinal dehydrogenase, retinal oxidase, and retinal isomerase. Retinal formation is catalyzed by retinol dehydrogenase and β-carotene 15,15'-dioxygenase. **Target(s):** CYP1A1 (1); NAD$^+$ ADP-ribosyltransferase, *or* poly(ADP-ribose) polymerase; IC$_{50}$ = 0.45 mM (2) **1.** Inouye, Mae, Kondo & Ohkawa (1999) *Biochem. Biophys. Res. Commun.* **262**, 565. **2.** Banasik, Komura, Shimoyama & Ueda (1992) *J. Biol. Chem.* **267**, 1569.

Retinal2, See *3,4-Didehydroretinal*

all-trans-Retinaldehyde, See *all-trans-Retinal*

Retinene, See *all-trans-Retinal*

all-trans-Retinoate

This bioactive acid (FW$_{\text{free-acid}}$ = 300.44 g/mol; CAS 302-79-4; yellow solid; M.P. = 180-182°C; soluble in methanol, ethanol, and dimethyl sulfoxide; λ_{max} = 350 nm in ethanol (ε = 45400 M^{-1}cm^{-1}), also known as *all-trans*-retinoic acid, (2*E*,4*E*,6*E*,8*E*)-3,7-dimethyl-9-(2,6,6-trimethylcyclohex-1-en-1-yl)nona-2,4,6,8-tetraenoic acid and vitamin A$_1$ acid, regulates gene expression by binding to the retinoic acid receptor. Retenoate plays a major role in development, cell proliferation, an in morphogenic changes. Retinoate is produced from retinal by either retinal dehydrogenase or retinal oxidase. It is also a substrate for a number of cytochrome P450 monooxygenases (*e.g.*, CYP2C8). **Target(s):** CYP1A1 (1); diferric-transferrin reductase (2); estrone sulfotransferase (3); β-glucuronidase (4); glutamate dehydrogenase (4); guanidinobenzoatase (5); lipoxygenase (6); prostaglandin-D synthase (7); retinol acyltransferase (8); retinyl-palmitate esterase (9); and ribonuclease P (10). **1.** Inouye, Mae, Kondo & Ohkawa (1999) *Biochem. Biophys. Res. Commun.* **262**, 565. **2.** Sun, Toole-Simms, Crane, *et al.* (1987) *Biochem. Biophys. Res. Commun.* **146**, 976. **3.** Adams & Ellyard (1972) *Biochim. Biophys. Acta* **260**, 724. **4.** Tappel & Dillard (1967) *J. Biol. Chem.* **242**, 2463. **5.** Anees & Steven (1994) *J. Enzyme Inhib.* **8**, 51. **6.** Goldreich, Grossman, Sofer, Breitbart & Sklan (1997) *Int. J. Vitam. Nutr. Res.* **67**, 4. **7.** Tanaka, Urade, Kimura, *et al.* (1997) *J. Biol. Chem.* **272**, 15789. **8.** Muller & Norum (1986) *Brit. J. Nutr.* **55**, 37. **9.** Ritter & Smith (1996) *Biochim. Biophys. Acta* **1291**, 228. **10.** Papadimou, Georgiou, Tsambaos & Drainas (1998) *J. Biol. Chem.* **273**, 24375.

13-cis-Retinoate

This naturally occurring vitamin A metabolite (FW$_{\text{free-acid}}$ = 300.44 g/mol; CAS 4759-48-2; M.P. = 174-175°C; λ_{max} = 354 nm; ε = 39800 M^{-1}cm^{-1}), also known as 13-*cis*-retinoic acid, isotretinoin and (2*Z*,4*E*,6*E*,8*E*)-3,7-dimethyl-9-(2,6,6-trimethylcyclohex-1-en-1-yl)nona-2,4,6,8-tetraenoic acid, inhibits sebum production and is used as an antiacne agent. **Target(s):** lipoxygenase (1); retinol dehydrogenase, *or* 3α-hydroxysteroid dehydrogenase (2,3); retinyl-palmitate esterase (4); ribonuclease P (5); and thioredoxin reductase (6,7). **1**. Goldreich, Grossman, Sofer, Breitbart & Sklan (1997) *Int. J. Vitam. Nutr. Res.* **67**, 4. **2**. Karlsson, Vahlquist, Kedishvili & Torma (2003) *Biochem. Biophys. Res. Commun.* **303**, 273. **3**. Gamble, Mata, Tsin, Mertz & Blaner (2000) *Biochim. Biophys. Acta* **1476**, 3. **4**. Ritter & Smith (1996) *Biochim. Biophys. Acta* **1291**, 228. **5**. Papadimou, Georgiou, Tsambaos & Drainas (1998) *J. Biol. Chem.* **273**, 24375. **6**. Arnér, Zhong & Holmgren (1999) *Meth. Enzymol.* **300**, 226. **7**. Rigobello, Callegaro, Barzon, Benetti & Bindoli (1998) *Free Radic. Biol. Med.* **24**, 370.

all-trans-Retinol

This essential nutrient and potential signaling molecule (FW = 286.46 g/mol; CAS 11103-57-4; λ_{max} = 325 nm in ethanol (ε = 52480 $M^{-1}cm^{-1}$; photolabile; readily oxidized in air), commonly called vitamin A, can be obtained from carotenoids. It is also unstable in acidic conditions. Retinol is stabilized by the presence of antioxidants such as hydroquinone or α-tocopherol. Retinol is also stabilized by esterification (*e.g.*, retinyl palmitate). **Target(s):** arylsulfatase A, *or* cerebroside-sulfatase (1); cathepsin D (2); CYP1A1 (3); β-glucuronidase (4); glutamate dehydrogenase (4); GTPase (5); hemoglobin S polymerization (6); and lipoxygenase (7). **1**. Bleszynski (1967) *Enzymologia* **32**, 169. **2**. Woessner & Shamberger (1971) *J. Biol. Chem.* **246**, 1951. **3**. Inouye, Mae, Kondo & Ohkawa (1999) *Biochem. Biophys. Res. Commun.* **262**, 565. **4**. Tappel & Dillard (1967) *J. Biol. Chem.* **242**, 2463. **5**. Zhao, Morre, Paulik, Yim & Morre (1990) *Biochim. Biophys. Acta* **1055**, 230. **6**. Freedman, Weissmann, Gorman & Cunningham-Rundles (1973) *Biochem. Pharmaacol.* **22**, 667. **7**. Goldreich, Grossman, Sofer, Breitbart & Sklan (1997) *Int. J. Vitam. Nutr. Res.* **67**, 4.

all-trans-Retinyl Bromoacetate

This retinol ester (FW = 407.39 g/mol) is a potent irreversible inhibitor of phosphatidylcholine:retinol *O*-acyltransferase: K_i = 12.1 μM and the pseudo-first-order inactivation rate constant (k_{inh} = 8.2 x 10^{-4} s^{-1}). **Target(s):** phosphatidylcholine:retinol *O*-acyltransferase (1-6); retinol *O*-fatty-acyltransferase (5). **1**. Ruiz & Bok (2000) *Meth. Enzymol.* **316**, 400. **2**. Shi, Furuyoshi, Hubacek & Rando (1993) *Biochemistry* **32**, 3077. **3**. Trehan, Canada & Rando (1990) *Biochemistry* **29**, 309. **4**. Ruiz, Winston, Lim, *et al.* (1999) *J. Biol. Chem.* **274**, 3834. **5**. Kaschula, Jin, Desmond-Smith & Travis (2005) *Exp. Eye Res.* **82**, 111. **6**. Trevino, Schuschereba, Bowman & Tsin (2005) *Exp. Eye Res.* **80**, 897.

Retrothiorphan, *See* 3-[(R)-1-(Mercaptomethyl)-2-phenylethyl]amino-3-oxopropanoate

Retrovir, *See* 3'-Azido-3'-deoxythymidine

REV 5367

This aryl methyl phenyl ether (FW = 328.41 g/mol) is a structural analogue of 15-hydroxy-(5Z,8Z,11Z,13E)-eicosatetraenoic acid (15-HETE), which is produced from arachidonate in the 15-lipoxygenase pathway. REV-5367 inhibits 5-lipoxygenase with I_{50} = 3 μM. *Note:* 15-HETE also inhibits 5-lipoxygenase. **1**. Coutts, Khandwala, Van Inwegen, *et al.* (1985) in *Prostaglandins, Leukotrienes, and Lipoxins* (Bailey, ed.), p. 627, Plenum Press, New York.

Revaprazan

This novel proton pump antagonist ($FW_{free-base}$ = 362.44 g/mol; CAS 178307-42-1; FW_{HCl} = 398.90 g/mol), also known by the code name YH1885 and systematically as *N*-(4-fluorophenyl)-4,5-dimethyl-6-(1-methyl-3,4-dihydro-1*H*-isoquinolin-2-yl)pyrimidin-2-amine, targets H^+/K^+-ATPase, thereby controlling gastric acid secretion. It also exerts anti-inflammatory action against *Helicobacter pylori*-induced COX-2 expression by inhibiting IκB-α degradation as well as Akt inactivation, resulting in attenuation of COX-2 expression (1). Plasma concentrations of YH1885 reached peak levels 1.3 to 2.5 hours after single-dose administration and then declined monoexponentially with a terminal half-life of 2.3 hours in dosage groups up to 200 mg in the single-dose study. YH1885 showed little accumulation after multiple administrations (2). The parent drug was not detected in urine, indicating its likely first-pass absorption and metabolism (2). Revaprazan also alleviates ER stress (*i.e.*, accumulation of unfolded proteins in the endoplasmic reticulum), often accompanied by drastic apoptosis (3). **1**. Lee, Cho, Song, Kim & Hahm (2012) *J. Clin. Biochem. Nutr.* **51**, 77. **2**. Kim, Park, Cheung, *et al.* (2010) *J. Gastroenterol. Hepatol.* **25**, 1618. **3**. Kim, Kim, Ock, *et al.* (2012) *J. Gastroenterol. Hepatol.* **27**, 120.

Reveromycin A

This polyketide antibiotic ($FW_{free-acid}$ = 660.80 g/mol; CAS 134615-37-5), isolated from a species of *Streptomyces*, inhibits cell-cycle G_1 phase and also induces apoptosis specifically in osteoclasts (1,2). Reveromycin A also inhibits isoleucyl-tRNA synthetase (3). **1**. Takahashi, Osada, Koshino, *et al.* (1992) *J. Antibiot.* **45**, 1414. **2**. Takahashi, Osada, Koshino, *et al.* (1992) *J. Antibiot.* **45**, 1409. **3**. Miyamoto, Machida, Mizunuma, *et al.* (2002) *J. Biol. Chem.* **277**, 28810.

RG108

This potent, specific and cell-permeant DNA methylation inhibitor and epigenetic reprogramming agent (FW = 334.33 g/mol; CAS 48208-26-0; Solubility = 67 mg/mL DMSO, <1 mg/mL H_2O), also known as *N*-phthalyl-L-tryptophan, targets human DNA methyltransferase, *or* DNMT, IC_{50} = 115 nM (1-3). RG108 inhibits the free DNA methyltransferase, whereas 5-azacytidine needs to be incorporated into DNA. Other experiments showed that increasing concentrations of RG108 resulted in increasing amounts of unprotected BstUI restriction fragments, representing the first example for pharmacologic inhibition of a purified DNA methyltransferase *in vitro* (1). Treatment of various cancer cell lines with RG108 results in net demethylation of DNA, attended by reactivation of tumor suppressor genes and inhibition of human tumor cell proliferation (3). **1**. Brueckner, Boy, Siedlecki, *et al.* (2005) *Cancer Res.* **65**, 6305. **2**. Schirrmacher, *et al.* (2006) *Bioconjug. Chem.* **17**, 261. **3**. Pasha, *et al.* (2011) *PLoS One* **6**, e23667.

RG760, *See Venetoclax*

RG2833

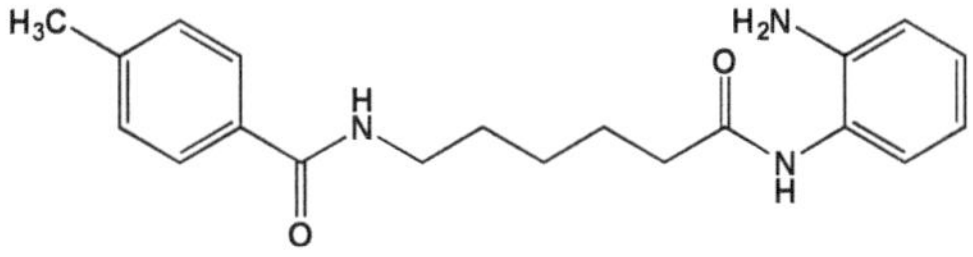

This brain-penetrant HDAC inhibitor (FW = 339.43 g/mol; CAS 1215493-56-3; Solubility: 68 mg/mL DMSO; <1 mg/mL H_2O), also known as *N*-(6-(2-aminophenylamino)-6-oxohexyl)-4-methylbenzamide and RGFP109, targets HDAC1 (IC_{50} = 60 nM; K_i = 32 nM) and HDAC3 (IC_{50} = 50 nM; K_i = 5 nM). Friedreich's ataxia (FRDA), the most common recessive ataxia in Caucasians, is linked to severely reduced levels of frataxin resulting from a large GAA triplet repeat expansion within the first intron of the *FXN* gene. In FRDA patients' peripheral blood mononuclear cells, RG2833 is highly potent in upregulating frataxin, supporting the concept that HDAC3 selectivity correlates with this property (1). RG2833 also shows a good dose-response correlation, with a 4-9x up-regulation in frataxin mRNA at 10 μM (1). These findings suggest that RG2833 decondenses the chromatin structure at the *FXN* gene, restoring frataxin levels in cells from FRDA patients. In a GAA-repeat expansion mutation-containing YG8R FRDA mouse model, RG2833 treatment improves motor coordination and increases locomotor activity (2). RGFP109 also attenuates L-DOPA-induced dyskinesia in the MPTP-lesioned marmoset (3). **1**. Rai, Soragni, Chou, *et al.* (2010) *PLoS One* **5**, e8825. **2**. Sandi, Pinto, Al-Mahdawi, *et al.* (2011) *Neurobiol. Dis.* **42**, 496. **3**. Johnston, Huot, Damude, *et al.* (2013) *Parkinsonism Relat. Disord.* **19**, 260.

RG7204, *See Vemurafenib*

RG7227, *See Danoprevir*

RG7603, *See GDC-0349*

RG7604, *See Taselisib*

RG13022, *See Tyrphostin RG13022*

RG 14355, *See Lavendustin A*

RG 14620, *See Tyrphostin RG14620*

RG 80267, *See 1,6-Bis(cyclohexyloximino-carbonylamino)hexane*

RGFP109, *See RG2833*

RGH-188, *See Cariprazine*

RH-34

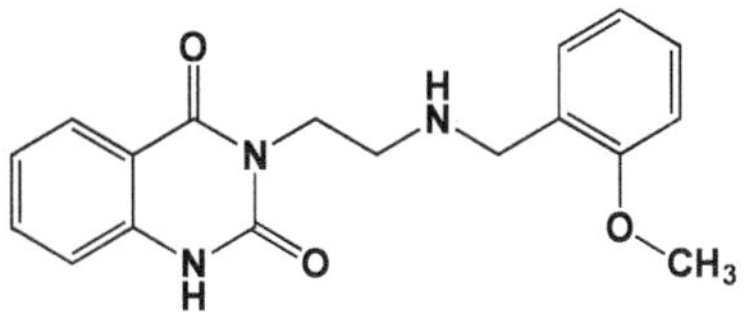

This potent and selective 5-HT$_{2A}$ partial agonist (FW = 325.36 g/mol; CAS 1028307-48-3) targets the serotonin subtype-2A receptor. RH-34 is a ketanserin derivative, with the latter's 4-(*p*-fluorobenzoyl)-piperidine moiety replaced by the *N*-(2-methoxybenzyl) pharmacophore already described for the potent 5-HT$_{2A}$ agonists known as NBOMe-2C-B and NBOMe-2C-I. *See also Ketanserin; Pimavanserin; Risperidone; Ritanserin; Setoperone; Volinanserin* **1**. Silva, Heim, Strasser, Elz && Dove (2011) *J. Comput. Aided Mol. Des.* **25**, 51.

RH-7281, *See Zoxamide*

L-Rhamnopyranose, *See L-Rhamnose*

N-(α-Rhamnopyranosyloxyhydroxyphosphinyl)-L-leucyl-L-tryptophan, *See Phosphoramidon*

L-Rhamnose

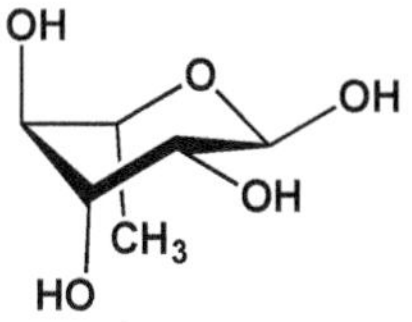

This deoxyhexose (FW = 164.16 g/mol; CAS 3615-41-6), also known as 6-deoxy-L-mannose, is a bacterial polysaccharide that is a common constituent in plant glycosides as well as gums and mucilages. Rhamnose is also found in free form in a number of plants (*e.g.*, poison sumac, *Rhus toxicodendron*). The α-pyranose form, shown above, is the prevalent species in aqueous solutions (60% at 44°C in D_2O; 87% in dimethyl sulfoxide). The α-anomer has a very sweet taste. The β-crystal is hygroscopic and undergoes mutarotation upon exposure to moist air. L-Rhamnose is very soluble in water (57 g/100 mL at 18°C; 109 g/100 mL at 40°C). The epimers of L-rhamnose are 6-deoxy-L-glucose (L-quinovose), 6-deoxy-L-altrose, 6-deoxy-L-talose, and 6-deoxy-D-gulose. **Target(s)**: β-fructofuranosidase, *or* invertase, inhibited by β-anomer (1); α-L-fucosidase (2); β-glucosidase, K_i = 110 mM for the sweet almond enzyme (3); α-L-rhamnosidase, product inhibition (4-16). **1**. Neuberg & Mandl (1950) *The Enzymes*, 1st ed. (Sumner & Myrbäck, eds.), **1** (part 1), 527. **2**. Grove & Serif (1981) *Biochim. Biophys. Acta* **662**, 246. **3**. Dale, Ensley, Kern, Sastry & Byers (1985) *Biochemistry* **24**, 3530. **4**. Park, Kim & Kim (2005) *J. Microbiol. Biotechnol.* **15**, 519. **5**. Jang & Kim (1996) *Biol. Pharm. Bull.* **19**, 1546. **6**. Bourbouze, Percheron & Courtois (1976) *Eur. J. Biochem.* **63**, 331. **7**. Manzanares, de Graaff & Visser (1997) *FEMS Microbiol. Lett.* **157**, 279. **8**. Zverlov, Hertel, Bronnenmeier, *et al.* (2000) *Mol. Microbiol.* **35**, 173. **9**. Romero, Manjon, Bastida & Iborra (1985) *Anal. Biochem.* **149**, 566. **10**. Gallego, Piñaga, Ramón & Vallés (2001) *J. Food Sci.* **66**, 204. **11**. Hashimoto, Miyake, Nankai & Murata (2003) *Arch. Biochem. Biophys.* **415**, 235. **12**. Yanai & Sato (2000) *Biosci. Biotechnol. Biochem.* **64**, 2179. **13**. Birgisson, Hreggvidsson, Fridjonsson, *et al.* (2004) *Enzyme Microb. Technol.* **34**, 561. **14**. Soria & Ellenrieder (2002) *Biosci. Biotechnol. Biochem.* **66**, 1442. **15**. Manzanares, Orejas, Ibañez, Vallés & Ramón (2000) *Lett. Appl. Microbiol.* **31**, 198. **16**. Manzanares, van den Broeck, de Graaff & Visser (2001) *Appl. Environ. Microbiol.* **67**, 2230.

D-*epi*-Rhamnose, *See Quinovose*

D-*iso*-Rhamnose, *See Quinovose*

Rh-ATP, *See Rhodium and Rhodium Ions*

RHC-80267, *See 1,6-Bis(cyclohexyloximinocarbonyl-amino)hexane*

Rhein

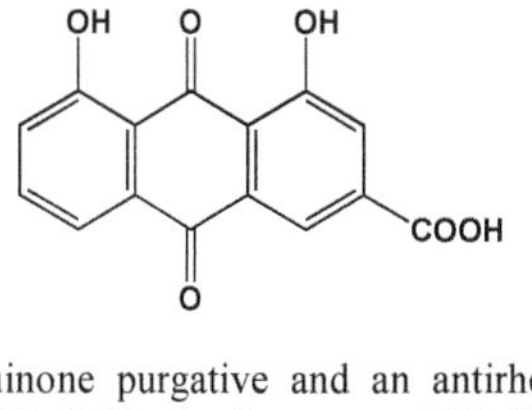

This yellow anthraquinone purgative and an antirheumatic (FW$_{free-acid}$ = 284.22 g/mol; CAS 478-43-3), also known as 9,10-dihydro-4,5-dihydroxy-9,10-dioxo-2-anthracenecarboxylic acid and 1,8-dihydroxyanthraquinone-2-carboxylic acid, is found in rhubarb, Senna leaves, and other higher plants. **Target(s)**: activator protein-1 activity and cell transformation through the inhibition of a c-Jun NH$_2$-terminal kinase-dependent mechanism (1); arylamine *N*-acetyltransferase (2); CYP1A1 (3); electron transport (4,5); glucose uptake (6); inositol-trisphosphate 3-kinase (7); NADH dehydrogenase (4,8); and Na$^+$/K$^+$-exchanging ATPase (9). **1**. Lin, Li, Fujii & Hou (2003) *Int. J. Oncol.* **22**, 829. **2**. Chung, Tsou, Wang, *et al.* (1998) *J. Appl. Toxicol.* **18**, 117. **3**. Sun, Sakakibara, Ashida, Danno & Kanazawa (2000) *Biosci. Biotechnol. Biochem.* **64**, 1373. **4**. Singer (1979) *Meth. Enzymol.* **55**, 454. **5**. Floridi, Castiglione & Bianchi (1989) *Biochem. Pharmacol.* **38**, 743. **6**. Floridi, Castiglione, Bianchi & Mancini (1990) *Biochem. Pharmacol.* **40**, 217. **7**. Mayr, Windhorst & Hillemeier (2005) *J.*

Biol. Chem. **280**, 13229. **8**. Zakharova (2002) *Biochemistry (Moscow)* **67**, 651. **9**. Wanitschke & Karbach (1988) *Pharmacology* **36** (*Suppl.* 1), 98.

Rhazinilam

This microtubule poison (FW = 294.39 g/mol; CAS 36193-36-9; IUPAC Name: (3*aR*)-3*a*-ethyl-2,3,3*a*,4,5,7-hexahydroindolizino[8,1-*ef*][1]benzazon in-6(1*H*)-one) from the *Melodinus australis* and *Rhazya stricta* protects assembled microtubules from cold-induced disassembly, but not from Ca^{2+}-induced disassembly (1). Incubation of tubulin with rhazinilam at both 373 K and 410 K induced formation of anomalous tubulin polymeric spirals. The latter is prevented by maytansine and vinblastine, but not colchicine. Preferential saturable and stoichiometric incorporation of radiolabeled rhazinilam into these spirals, K_d = 5 μM, but binding is blocked by vinblastine and maytansine (1). Specific binding of radioactive rhazinilam to microtubules is not observed. Omission of GTP, causes a major change in the morphology in (-)-rhazinilam-induced tubulin polymers, with half of the observed polymer microtubule-like and the other half spirals (2). **See** *Colchicine; Epothilone; Maytansine; Paclitaxel (Taxol); Podophyllotoxin; Taxotere; Vinblastine* **1**. David, Sévenet, Morgat, *et al.* (1994) *Cell Motil. Cytoskeleton.* **28**, 317. **2**. Edler, Yang, Jung, *et al.* (2009) *Arch. Biochem. Biophys.* **487**, 98.

Rhizobitoxine

This toxic substance ($FW_{dihydrochloride}$ = 263.12 g/mol), also known as L-2-amino-4-(2'-amino-3'-hydroxypropoxy)-*trans*-3-butenoic acid, is produced by *Rhizobium japonicum* that inhibits cystathionine β-lyase, apparent K_i = 220 nM. Rhizobitoxine inhibits a number of pyridoxal-5'-phosphate-dependent enzymes. This phytotoxin also inhibits ethylene production. **Target(s):** 1-aminocyclopropane-1-carboxylate synthase (1-3); and cystathionine β-lyase (4-10). **1**. Adams & Yang (1981) *Trends Biochem. Sci.* **6**, 161. **2**. Yasuta, Satoh & Minamisawa (1999) *Appl. Environ. Microbiol.* **65**, 849. **3**. Jakubowicz (2002) *Acta Biochim. Pol.* **49**, 757. **4**. Guggenheim (1971) *Meth. Enzymol.* **17B**, 439. **5**. Giovanelli (1987) *Meth. Enzymol.* **143**, 443. **6**. Davis & Metzler (1972) *The Enzymes*, 3rd ed. (Boyer, ed.), **7**, 33. **7**. Giovanelli, Owens & Mudd (1971) *Biochim. Biophys. Acta* **227**, 671. **8**. Owens, Guggenheim & Hilton (1968) *Biochim. Biophys. Acta* **158**, 219. **9**. Giovanelli, Owens & Mudd (1972) *Plant Physiol.* 51, 492. **10**. Ravanel, Job & Douce (1996) *Biochem. J.* **320**, 383.

Rhizoxin

This macrocyclic lactone (FW = 625.76 g/mol; CAS 90996-54-6) inhibits mitosis by interfering with microtubule self-assembly (1-4) and appears to be effective against vincristine-resistant human and murine tumor cells (5). In evidence of its substantial effect on tubulin conformation, rhizoxin blocks intrachain covalent cross-linking in the β-tubulin subunit by *N*,*N*-methylene-*bis*(iodoacetamide). β-tubulin may also be irreversibly modified with the photoaffinity agent azidodansyl-rhizoxin. Plant and fungal tubulins also bind rhizoxin in a manner that inhibits *de novo* microtubule assembly and promotes depolymerization of preexisting microtubules. Rhizoxin is stable for up to several years, when stored at 0°C. Solutions may be prepared in dimethyl sulfoxide or ethanol and are stable for several weeks in the cold. (**See also** *Maytansine, Vinblastine, Vincristine*) **1**. Sawada, Kobayashi, Hashimoto & Iwasaki (1993) *Biochem. Pharmacol.* **45**, 1387. **2**. Takahashi, Iwasaki, Kobayashi, *et al.* (1987) *Biochim. Biophys. Acta* **926**, 215. **3**. Takahashi, Iwasaki, Kobayashi, *et al.* (1987) *J. Antibiot.* **40**, 66. **4**. Sullivan, Prasad, Roach, *et al.* (1990) *Cancer Res.* **50**, 4277. **5**. Tsuruo, Oh-hara, Iida, *et al.* (1986) *Cancer Res.* **46**, 381.

Rhodionin

This flavone glycoside (FW = 432.38 g/mol; CAS 85571-15-9), isolated from species of *Rhodiola*, inhibits (IC_{50} = 22 μM) prolyl oligopeptidase, *or* PEP (EC. 3.4.21.26), an enzyme that plays a role in the metabolism of proline-containing neuropeptidase that is involved in learning and memory. 1. Fan, Tezuka, Ni & Kadota (2001) *Chem. Pharm. Bull. (Tokyo)* **49**, 396.

Rhodiosin, *similar in action to Rhodionin*

Rhodium(II) Acetate Dimer, See *Tetrakis(m-acetato)dirhodium(II,II)*

Rhodomycins

These DNA-intercalating antibiotics ($FW_{Rhodomycin-A}$ = 701.78 g/mol; CAS 1401-16-7) from *Streptomyces purpurascens* and a red-pigmented mutant of *Streptomyces griseus*, selectively inhibit DNA and RNA biosynthesis by forming complexes with the polynucleic acids (1,2). It is likely that protonation of the tertiary amine facilitates DNA binding, with the flattened anthracycline ring intercalating between base-pairs, and the positively charged glycone binding to the phosphodiester backbone. Rhodomycins inhibit cancer cell proliferation, clonogenicity, motility, and invasiveness, underscoring that they are likely to be nonspecific in their actions. Indeed, rhodomycin A renders gefitinib-resistant lung adenocarcinoma cells more sensitive to gefitinib treatment, implying a synergistic effect of the combination therapy. Our data also reveal that the inhibitory effect of rhodomycin A on lung cancer progression may act through suppressing the Src-related multiple signalling pathways, including PI3K, JNK, Paxillin, and p130cas (3). Related anthracycline antitumor drugs include daunorubicin (DNR) and doxorubicin (DXR). **1**. Shockman & Waksman (1951) *Antibiot. Chemother.* (Northfield) **1**, 68. **2**. Oki (1977) *Jpn. J. Antibiot.* **30** Suppl., 70. **3**. Lai, Chen, Lin, *et al.* (2015) *Oncotarget* **6**, 26252.

Rhodoviolascin, See *Spirilloxanthin*

Rho GDP Dissociation Inhibitor (RhoGDI)

These guanine dissociation inhibitors ($MW_{Human\text{-}Isoform\text{-}A}$ = 23.2 kDa; $MW_{Human\text{-}Isoform\text{-}B}$ = 18.2 kDa; $MW_{Human\text{-}Isoform\text{-}C}$ = 25.8 kDa; $MW_{Human\text{-}Isoform\text{-}D}$ = 36.6 kDa) block GDP dissociation from Rho GTPases, locking these small GTPases in an inactive GDP-bound state (1-4). RhoGDI interacts with lipid-modified GDP-RhoA and GDP-RhoB. Other targets include Rac1, Rac2, and Cdc42 (**See also** *Rab GDP Dissociation Inhibitor*). RhoGDI is made up of two domains: a flexible N-terminal domain of about 70 residues and a folded 134-residue C-terminal domain (4). Two regions (residues 9-120 and 36-58) tend to form helices, and *in vitro* and *in vivo* functional assays with truncated proteins indicate that first 30 residues are not needed for inhibition of GDP dissociation; instead, they appear to be important for GTP hydrolysis (4) Removal of the first 41 residues completely abolishes RhoGDI inhibition of GDP dissociation from Rho. Structural and functional studies also explain why RhoGDI and D4GDI can interact in similar ways with GDP-bound GTPase, yet differ in their regulation of GTP-bound forms. Two transient helices appear to exert different effects, providing an example of structure-activity relationships within a flexible protein domain (4). **Role in Phagocyte NADPH Oxidase System:** 'Professional' phagocytic cells employ the the NADPH Oxidase System (*or* Phage Ox System) to form superoxide in the lumen of phagocytes, thereby promoting microbial killing by myeloperoxidase. GDI-1 indirectly activates the phagocyte superoxide-generating NADPH oxidase, a process requiring the membrane-associated cytochrome b_{559} with the cytosolic components, p47-phox, p67-phox and GDI-1, the latter forming a heterodimeric p21rac1·GDI complex, *or* σ_1 (5). **1**. Ohga, Kikuchi, Ueda, Yamamoto & Takai (1989) *Biochem. Biophys. Res. Commun.* **163**, 1523. **2**. Fukumoto, Kaibuchi, Hori, *et al.* (1990) *Oncogene* **5**, 1321. **3**. Lucero, Stack, Bresnick & Shuster (2006) *Mol. Biol. Cell* **17**, 4093. **4**. Golovanov, Chuang, DerMardirossian, *et al.* (2001) *J. Mol. Biol.* **305**, 121. **5**. Pick, Gorzalczany & Engel (1993) *Eur. J. Biochem.* **217**, 441.

Ribavirin 5'-Monophosphate

This antiviral nucleotide ($FW_{free\text{-}acid}$ = 324.19 g/mol), also known as virazole 5'-monophosphate and formed by enzymatic phosphorylation of the nucleoside ribavirin (CAS 36791-04-5), inhibits adenylosuccinate lyase (1) and IMP dehydrogenase (2). **Key Pharmacokinetic Parameters:** *See* Appendix II in Goodman & Gilman's *THE PHARMACOLOGICAL BASIS OF THERAPEUTICS*, 12th Edition (Brunton, Chabner & Knollmann, eds.) McGraw-Hill Medical, New York (2011). **1**. Shaw, Thomas, Patey & Thomas (1979) *J. Chem. Soc. Perkin Trans.* **1**, 1415. **2**. Prosise, Wu & Luecke (2002) *J. Biol. Chem.* **277**, 50654.

Ribavirin 5'-Triphosphate

This antiviral nucleotide ($FW_{free\text{-}acid}$ = 484.15 g/mol), formed by metabolic phosphorylation of the nucleoside ribavirin, inhibits mRNA guanylyltransferase (1), mRNA (nucleoside-2'-*O*-)-methyltransferase (2), nucleoside-triphosphatase/ DNA helicases (3-6), RNA-directed DNA polymerase (7), and visna virus reverse transcriptase (7). **1**. Bougie & Bisaillon (2004) *J. Biol. Chem.* **279**, 22124. **2**. Benarroch, Egloff, Mulard, *et al.* (2004) *J. Biol. Chem.* **279**, 35638. **3**. Borowski, Lang, Niebuhr, *et al.* (2001) *Acta Biochim. Pol.* **48**, 739. **4**. Borowski, Niebuhr, Mueller, *et al.* (2001) *J. Virol.* **75**, 3220. **5**. Borowski, Mueller, Niebuhr, *et al.* (2000) *Acta Biochim. Pol.* **47**, 173. **6**. Borowski, Niebuhr, Schmitz, *et al.* (2002) *Acta*

Biochim. Pol. **49**, 597. **7**. Frank, McKernan, Smith & Smee (1987) *Antimicrob. Agents Chemother.* **31**, 1369.

RHPS 4

This acridine-based DNA intercalator and quadruplex-stabilizing telomerase inhibitor (FW = 458.11 g/mol; CAS 390362-78-4; Solubility: 10 mM in H_2O with gentle warming; 20 mM in DMSO), also named 3,11-difluoro-6,8,13-trimethylquino[4,3,2-*kl*]acridinium methyl sulfate, induces telomere injury and promotes apoptosis (1-3). RHPS-4 inhibits growth of UXF1138L cells (uterus carcinoma cells with short telomeres), medulloblastoma and glioblastoma cells *in vitro* (with IC_{50} of 2-3 μM, depending on cell type) and blocks growth of CG5 breast cancer xenografts in mice. **Mechanism of Antiproliferative Action:** The pentacyclic acridinium methosulfate salt RHPS4 induces the 3' single-stranded guanine-rich telomeric overhang to fold into a G-quadruplex structure. Stabilization of the latter is incompatible with an attachment of telomerase to the telomere and thus G-quadruplex ligands can effectively inhibit both the catalytic and capping functions of telomerase. Repetitive TTAGGG sequences at the ends of chromosomes allow cells to distinguish between natural chromosome ends and double-strand DNA breaks, and the perpetual maintenance of this telomeric DNA allows cancer cells to have unlimited replicative potential. Activated telomerase maintains telomere length homeostasis in ~85% of human cancers, motivating anti-cancer strategies that target components of the telomerase holoenzyme. One strategy is to use a G-quadruplex (G4) ligand to critically erode chromosome end(s), resulting in telomere shortening and/or telomere uncapping. Because telomerase requires the 3'-telomeric end to be in a single-stranded configuration, sequestering of the telomere in a four-stranded structure by small molecules that can compete with telomere-associated proteins, should inhibit telomerase binding to telomere ends. The resulting loss of telomere maintenance should activate DNA damage response and arrest growth. **1**. Phatak, *et al.* (2007) *Br. J. Cancer* **96**, 1223. **2**. Salvati, *et al.* (2007) *J. Clin. Invest.* **117**, 3236. **3**. Lagah, Tan, Radhakrishnan, *et al.* (2014) *PLoS One* **9**, e86187.

9-(β-D-Ribofuranuronic)adenine Monophosphate Mixed Anhydride

This carboxylic-phosphoric mixed anhydride ($FW_{free\text{-}acid}$ = 361.21 g/mol) is an AMP isostere that inactivates adenylosuccinate lyase (1). A carboxylic-phosphoric mixed anhydride was synthesized, in which the 5'-methylene group of ATP is replaced by a carbonyl group (2). This compound does not inhibit rabbit adenylate kinase, but rapidly inactivates rabbit pyruvate kinase. The latter effect was prevented by ATP, ADP and phosphoenolpyruvate, indicating that it appears to be ATP-site-directed. The analogous anhydride isosteric with adenosine 5'-phosphate (AMP) rapidly inactivated rabbit AMP aminohydrolase, with inactivation prevented by an equimolar level of AMP (2). Such findings suggest these anhydrides are powerful and potentially useful reagents for adenine nucleotide binding sites of enzymes. **1.** Hampton & Harper (1971) *Arch. Biochem. Biophys.* **143**, 340. **2.** Hampton, Harper, Sasaki, Howgate & Preston (1975) *Biochem. Biophys. Res. Commun.* **65**, 945.

9-(β-D-Ribofuranuronic)adenine Triphosphate Mixed Anhydride

This carboxylic-phosphoric mixed anhydride ($FW_{free-acid}$ = 521.17 g/mol) is an ATP isostere that inhibits pyruvate kinase. **1.** Hampton, Harper, Sasaki, Howgate & Preston (1975) *Biochem. Biophys. Res. Commun.* **65**, 945.

Ribonuclease Inhibitor RPI

This acidic cytoplasmic protein (MW = 49,973; pI = 4.7) forms a high-affinity (K_i = 35-44 fM) with pancreatic RNase. The porcine inhibitor has a value of 4 fM for porcine ribonuclease-4. Ribonuclease protein inhibitor (RPI) is inactivated by *p*-hydroxymercuribenzoate. *Note*: Each animal species studied to date contains only a single ribonuclease inhibitor. **Target(s):** angiogenin (1,2); Onconase™ (1); pancreatic ribonuclease, *or* ribonuclease A, also known as seminal ribonuclease (1-13); ribonuclease-2, eosinophil-derived neurotoxin (1); ribonuclease-3 (1); ribonuclease-4, porcine (1); ribonuclease-6, *or* poly-(U)-specific ribonuclease (14); and RNA lariat debranching enzyme (15). **1.** Shapiro (2001) *Meth. Enzymol.* **341**, 611. **2.** Lee & Vallee (1993) *Prog. Nucl. Acid Res. Mol. Biol.* **44**, 1. **3.** Blackburn & Moore (1982) *The Enzymes*, 3rd ed. (Boyer, ed.), **15**, 317. **4.** Futami, Tsushima, Murato, *et al.* (1997) *DNA Cell Biol.* **16**, 413. **5.** Kobe & Deisenhofer (1996) *J. Mol. Biol.* **264**, 1028. **6.** Sorrentino, Barone, Bucci, *et al.* (2000) *FEBS Lett.* **466**, 35. **7.** Sica, Di Fiore, Merlino & Mazzarella (2004) *J. Biol. Chem.* **279**, 36753. **8.** Landré, Hewett, Olivot, *et al.* (2002) *J. Cell. Biochem.* **86**, 540. **9.** Klink & Raines (2000) *J. Biol. Chem.* **275**, 17463. **10.** Hofsteenge, Servis & Stone (1991) *J. Biol. Chem.* **266**, 24198. **11.** Neumann & Hofsteenge (1994) *Protein Sci.* **3**, 248. **12.** Bartholeyns & Baudhuin (1977) *Biochem. J.* **163**, 675. **13.** Bosch, Benito, Ribo, *et al.* (2004) *Biochemistry* **43**, 2167. **14.** Sana, Ghosh, Saha & Mukherjee (2008) *Microbiol. Res.* **163**, 31. **15.** Ooi, Dann, Nam, *et al.* (2001) *Meth. Enzymol.* **342**, 233.

Ribonuclease Inhibitors

Given the pervasive presence of numerous ribonucleases in nearly all cell types and their ability to rapidly hydrolyze messenger RNA, ribonuclease inhibitors are of paramount significance in experiments requiring intact mRNA (1). In fact, ribonucleases are surprisingly difficult to remove completely from samples or to inactivate during RNA isolation. RNases are so ubiquitous that they are often introduced into samples, even in deionized laboratory water. **Reversible Low-Molecular-Weight RNase Inhibitors:** These substances bind within RNase active sites, including: (a) Vanadyl ribonucleoside complexes, which resemble transition-state analogue and consist of a 1:1 complex of oxovanadium ion and a ribonucleosides. These complexes may be added before or during cell lysis solutions to inhibit the endogenous RNase activity of the cells. Such complexes bind to a broad range of RNases, typically exhibiting a K_i of 10 μM for pancreatic RNase, *or* RNase A. However, these vanadyl complexes also inhibit reverse transcriptase (thereby affecting RT-PCR assays) as well as *in vitro* translation. (b) 5'-Diphosphoadenosine 3'-phosphate exhibits a 1-μM K_i, among the strongest reversible low-molecular-weight inhibitor. (c) Simple nucleotides (including 2'-CMP, 2'-UMP, 4-thiouridine 3'-phosphate, thymidine 3',5'-diphosphate, and pAp) that have K_i values of 1-10 μM, depending on assay conditions. (d) Other RNase A inhibitors are di-ribonucleoside 2'-5' monophosphates (2'-5'-GpG, 2'-5'-CpG, and 2'-5'-UpG, as well as such di-ribonucleoside 3'-5' monophosphates as ApU, ApC, and GpU. (e) Guanylyl-2',5'-guanosine (2',5'-GpG) and 2'-guanylic acid (2'-GMP), arguably the most potent RNase inhibitors, especially for RNase T1. **Irreversible Low-Molecular-Weight RNase Inhibitors:** These agents covalently modify active-site residues required for RNase catalysis. The most widely applied reagent is diethyl pyrocarbonate, which inactivates RNases. DEPC is typically added to buffers and water used in mRNA preparations, usually

with 0.1% v/v DEPC for >1 hour at 37°C, followed by autoclaving at least 15 min to inactivate any remaining traces of DEPC. **High-Molecular-Weight RNase Inhibitors:** The angiogenic RNase inhibitor PRI binds very tightly to RNase A and to angiogenin, exhibiting K_i values in the 10^{-14} to 10^{-16} M range. In the presence of sodium dodecyl sulfate, proteinase K will digest many RNases, with consequential loss of enzymatic activity. **1.** Pasloske (2000) *Meths. Molec. Biol.* (Schein, ed.) **160**, 105.

D-Ribose

β-D-ribopyranose   open chain   β-D-ribofuranose

This aldopentose (FW = 150.13 g/mol; CAS 50-69-1), a component of RNA, nucleotides, certain coenzymes and glycosides, and intermediates in carbohydrate metabolism. Note that the majority of free D-ribose is in the pyranose forms rather than the furanose structure typically observed in nucleotides. In D_2O at 31°C, D-ribose is 21.5% α-pyranose, 58.5% β-pyranose, 6.5% α-furanose, and 13.5% β-furanose (the corresponding percentages in dimethyl sulfoxide at 25°C are 18%, 57%, 6% and 9%). The normal crystal form is β-D-ribopyranose (M.P. = 86-87°C). **Target(s):** aldose 1-epimerase (1); aspartate carbamoyltransferase (2); cytidine deaminase, K_i = 12 mM (3); formaldehyde transketolase (4); β-galactosidase (5); α-glucosidase (6,7); ketohexokinase, *or* hepatic fructokinase (8); lactose synthase (9); orotidine-5'-phosphate decarboxylase, weakly inhibited (10); prodigiosin formation (11); purine nucleosidase, product inhibition (12-14); ribose-5-phosphate isomerase (15); superoxide dismutase, Cu/Zn (16); uridine nucleosidase (17,18); and xylose isomerase (19,20). **1.** Bailey, Fishman, Kusiak, Mulhern & Pentchev (1975) *Meth. Enzymol.* **41**, 471. **2.** Burns, Mendz & Hazell (1997) *Arch. Biochem. Biophys.* **347**, 119. **3.** Carlow & Wolfenden (1998) *Biochemistry* **37**, 11873. **4.** Waites & Quayle (1983) *J. Gen. Microbiol.* **129**, 935. **5.** Huber, Roth & Bahl (1990) *J. Protein Chem.* **15**, 621. **6.** Giblin, Kelly & Fogarty (1987) *Can. J. Microbiol.* **33**, 614. **7.** Peruffo, Renosto & Pallavicini (1978) *Planta* **142**, 195. **8.** Raushel & Cleland (1977) *Biochemistry* **16**, 2169. **9.** Ebner (1973) *The Enzymes*, 3rd ed. (Boyer, ed.), **9**, 363. **10.** Miller, Butterfoss, Short & Wolfenden (2001) *Biochemistry* **40**, 6227. **11.** Lawanson & Sholeye (1976) *Experientia* **32**, 439. **12.** Parkin (1996) *J. Biol. Chem.* **271**, 21713. **13.** Parkin, Horenstein, Abdulah, Estupinan & Schramm (1991) *J. Biol. Chem.* **266**, 20658. **14.** Hansen & Dandanell (2005) *Biochim. Biophys. Acta* **1723**, 55. **15.** Agosin & Aravena (1960) *Enzymologia* **22**, 281. **16.** Ukeda, Hasegawa, Ishi & Sawamura (1997) *Biosci. Biotechnol. Biochem.* **61**, 2039. **17.** Magni (1978) *Meth. Enzymol.* **51**, 290. **18.** Magni, Fioretti, Ipata & Natalini (1975) *J. Biol. Chem.* **250**, 9. **19.** Vartak, Srinivasan, Powar, Rele & Khire (1984) *Biotechnol. Lett.* **6**, 493. **20.** Danno (1970) *Agric. Biol. Chem.* **34**, 1805.

D-Ribose 1,5-Bisphosphate

This bisphosphorylated aldopentose ($FW_{free-acid}$ = 310.09 g/mol) is a regulator of carbohydrate metabolism that relieves phosphofructokinase from ATP inhibition and increased the affinity for fructose 6-phosphate. This sugar phosphate is also a cofactor of phosphoribomutase (phosphopentomutase). **Target(s):** fructose-1,6-bisphosphatase (1); and pyruvate kinase (2). **1.** Ozaki, Mitsui, Sugiya & Furuyama (2000) *Comp. Biochem. Physiol. B Biochem. Mol. Biol.* **125**, 97. **2.** Lin, Turpin & Plaxton (1989) *Arch. Biochem. Biophys.* **269**, 228.

D-Ribose 5-Diphosphate

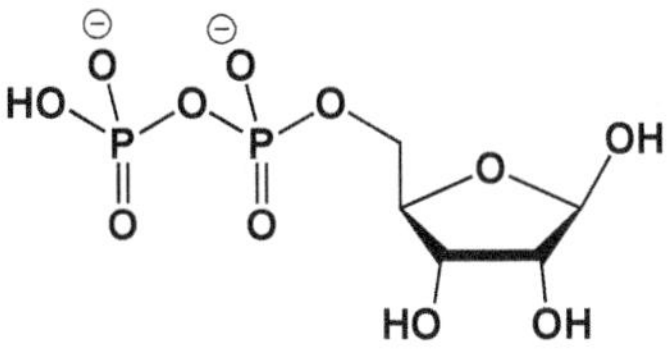

This phosphorylated aldopentose (FW$_{\text{free-acid}}$ = 326.09 g/mol; β-anomer shown above) inhibits RNA polymerase. Ribose and deoxyribose 5' pyro- and tri-phosphates are potent substrate-competitive inhibitors for an *in vitro* transcription system containing either calf thymus or T$_7$ DNA as template, and the E. coli RNA polymerase [E.C.2.7.7.6]. Each analogue gave K_i ~ 25 μM, essentially the same as the K_m values (~15 μM) for the substrates. In contrast, the ribose and deoxyribose 5'-monophosphates, ribonucleosides, deoxynucleoside mono and triphosphates gave no significant inhibition. The data are consistent with the interpretation that the enzyme binds the substrate primarily through the 3'-endo ribose polyphosphate moiety. **1.** Sylvester & Dennis (1977) *Biochem. Biophys. Res. Commun.* **75**, 667.

D-Ribose 1-Phosphate

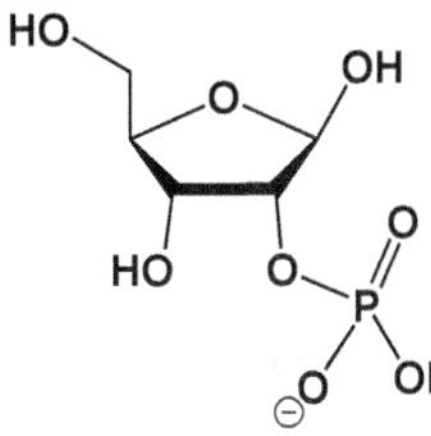

This phosphorylated aldopentose (FW$_{\text{free-acid}}$ = 230.11 g/mol) is produced by a number of enzymes (*e.g.*, uridine phosphorylase and purine-nucleoside phosphorylase) and is a product inhibitor in many of these catalyzed reactions. The α-anomer shown above and the product of the phosphorylases is not as stable as the β-anomer and is rapidly hydrolyzed in strong acid (50% in 3.5 hours at 20°C in 10 mM HCl or 2.5 minutes at 25°C in 500 mM HCl). The β-epimer is 50% hydrolyzed in 4 hours at 26°C in 100 mM HCl. **Target(s):** *S*-methyl-5'-thioadenosine phosphorylase (1); ribose-phosphate diphosphokinase, *or* phosphoribosyl-pyrophosphate synthetase, weakly inhibited (2); thymidine phosphorylase (3,4); and triose-phosphate isomerase (5). **1.** Garbers (1978) *Biochim. Biophys. Acta* **523**, 82. **2.** Fox & Kelley (1972) *J. Biol. Chem.* **247**, 2126. **3.** Schwartz (1978) *Meth. Enzymol.* **51**, 442. **4.** Blank & Hoffee (1975) *Arch. Biochem. Biophys.* **168**, 259. **5.** Noltmann (1972) *The Enzymes*, 3rd ed., **6**, 271.

D-Ribose 5-Phosphate

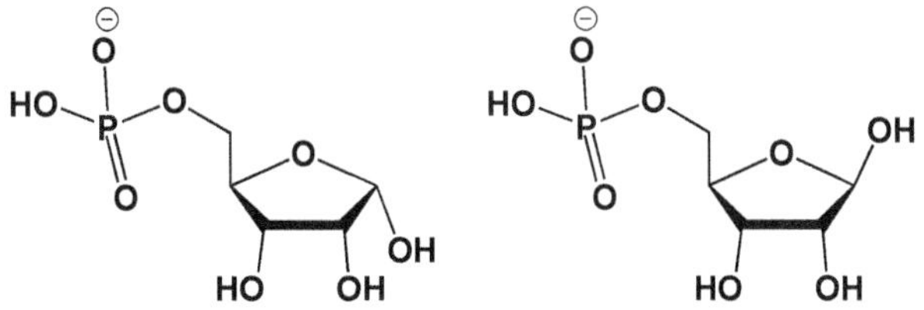

α-anomer   β-anomer

This phosphorylated aldopentose (FW$_{\text{free-acid}}$ = 230.11 g/mol), a component of many nucleotides, is also a intermediate in the pentose phosphate pathway and in photosynthesis. This reducing sugar is slightly soluble in water and more soluble in dilute mineral acids. It is slowly hydrolyzed by acids: 59% is hydrolyzed in two hours at 100°C by 1.0 M HCl. The half-life in 10 mM acid at 100°C is about seventeen hours. The sodium salt is hygroscopic and should be stored in a desiccator. D-Ribose 5-P is a product of the reactions catalyzed by transketolase, ribokinase, AMP nucleosidase, pyrimidine-5'-nucleotide nucleosidase, inosinate nucleosidase, NMN nucleosidase, ADP-ribose pyrophosphatase, and phosphopentomutase. Ribose 5-P is a substrate for acetoin:ribose-5-phosphate transaldolase, phosphoribokinase, ribose-phosphate pyrophosphokinase, ribose-5-phosphate adenylyltransferase, pseudouridylate synthase, ribose 5-phosphate epimerase (ribose-5-phopsphate isomerase), and ribose-5-phosphate:ammonia ligase. Ribose 5-P is an activator of *Bacillus*

stearothermophilus pyruvate kinase but an inhibitor of the chloroplastic *Selenastrum minutum* enzyme. **Target(s):** D-amino-acid oxidase, weakly inhibited (1); arabinose-5-phosphate isomerase (2); aspartate carbamoyl-transferase (3,4); 2-carboxy-D-arabinitol-1-phosphatase, weakly inhibited (5); 2-dehydro-3-deoxyphosphooctonate aldolase (6); 3-deoxy-8-phosphooctulonate synthase, *or* 3-deoxy-D-*manno*-2-octulosonate-8-phosphate synthase (7,8); fructose-1,6-bisphosphate aldolase (9,10); glucose dehydrogenase (11-13); glucose-6-phosphate dehydrogenase, weakly inhibited (13); glucose-6-phosphate isomerase (14,15); glycerate kinase (16); IMP dehydrogenase (17); *myo*-inositol-1-phosphate synthase (18); L-*myo*-inositol-1-phosphatase (19); mannose-6-phosphate isomerase (14); NMN Nucleosidase, product inhibition (19); orotate phosphoribosyl-transferase (20); orotidine-5'-phosphate decarboxylase, *or* orotidylate decarboxylase (21-23); 6-phosphogluconate dehydrogenase (24); phosphoglycolate phosphatase (25,26); phospho-protein phosphatase (27); pyruvate kinase, chloroplastic *Selenastrum minutum*, an activator for other pyruvate kinases (28); ribulose-bisphosphate carboxylase, oxygenase activity (29); and uridine nucleosidase (30). **1.** Walaas & Walaas (1956) *Acta Chem. Scand.* **10**, 122. **2.** Bigham, Gragg, Hall, *et al.* (1984) *J. Med. Chem.* **27**, 717. **3.** Adair & Jones (1978) *Meth. Enzymol.* **51**, 51. **4.** Burns, Mendz & Hazell (1997) *Arch. Biochem. Biophys.* **347**, 119. **5.** Salvucci & Holbrook (1989) *Plant Physiol.* **90**, 679. **6.** Ray (1982) *Meth. Enzymol.* **83**, 525. **7.** Kohen, Jakob & Baasov (1992) *Eur. J. Biochem.* **208**, 443. **8.** Ray (1980) *J. Bacteriol.* **141**, 635. **9.** Moorhead & Plaxton (1990) *Biochem. J.* **269,**·133. **10.** Bais, James, Rofe & Conyers (1985) *Biochem. J.* **230**, 53. **11.** Strecker (1955) *Meth. Enzymol.* **1**, 335. **12.** Webb (1966) *Enzyme and Metabolic Inhibitors*, vol. 2, p. 410, Academic Press, New York. **13.** Horne, Anderson & Nordlie (1970) *Biochemistry* **9**, 610. **14.** Noltmann (1972) *The Enzymes*, 3rd ed. (Boyer, ed.), **6**, 271. **15.** Sangwan & Singh (1989) *J. Biosci.* **14**, 47. **16.** Saharan & Singh (1993) *Plant Physiol. Biochem.* **31**, 559. **17.** Nichol, Nomura & Hampton (1967) *Biochemistry* **6**, 1008. **18.** Charalampous & Chen (1966) *Meth. Enzymol.* **9**, 698. **19.** Foster (1981) *J. Bacteriol.* **145**, 1002. **20.** Victor, Greenberg & Sloan (1979) *J. Biol. Chem.* **254**, 2647. **21.** Jones, Kavipurapu & Traut (1978) *Meth. Enzymol.* **51**, 155. **22.** Porter & Short (2000) *Biochemistry* **39**, 11788. **23.** Miller & Wolfenden (2002) *Ann. Rev. Biochem.* **71**, 847. **24.** Sokolov, Luchin & Trotsenko (1980) *Biokhimiia* **45**, 1371. **25.** Husic & Tolbert (1984) *Arch. Biochem. Biophys.* **229**, 64. **26.** Christeller & Tolbert (1978) *J. Biol. Chem.* **253**, 1780. **27.** Polya & Haritou (1988) *Biochem. J.* **251**, 357. **28.** Lin, Turpin & Plaxton (1989) *Arch. Biochem. Biophys.* **269**, 228. **29.** Ryan & Tolbert (1975) *J. Biol. Chem.* **250**, 4234. **30.** Magni (1978) *Meth. Enzymol.* **51**, 290.

Ribostamycin

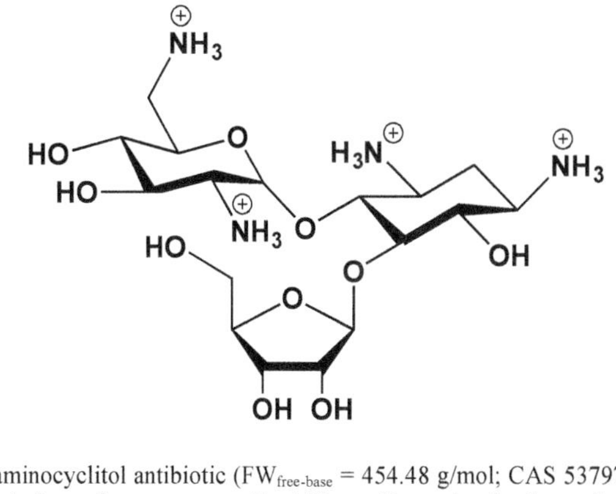

This aminocyclitol antibiotic (FW$_{\text{free-base}}$ = 454.48 g/mol; CAS 53797-35-6), isolated from *Streptomyces ribosidificus* (formerly *S. thermoflavus*), is soluble in water and slightly unstable in acidic conditions. **Target(s):** exoribonuclease H (1); gentamicin 3'-*N*-acetyltransferase, K_i = 70 μM (2); nonenzymatic reduction of cytochrome *c* by FeSO$_4$ (3); protein-disulfide isomerase, chaperone activity only, K_d = 0.32 mM (isomerase activity not inhibited) (4); protein-synthesizing GTPase, elongation factor (5); and ribonuclease H (1). **1.** Li, Barbieri, Lin, *et al.* (2004) *Biochemistry* **43**, 9732. **2.** Williams & Northrop (1978) *J. Biol. Chem.* **253**, 5902 and 5908. **3.** Yamabe (1980) *Chemotherapy* **26**, 28. **4.** Horibe, Nagai, Sakakibara, Hagiwara & Kikuchi (2001) *Biochem. Biophys. Res. Commun.* **289**, 967. **5.** Campuzano & Modolell (1981) *Eur. J. Biochem.* **117**, 27.

7-Ribosyloxipurinol 5'-Monophosphate, *See* *7-Oxipurinol*

Ribonucleoside 5'- Monophosphate

D-Ribosyl Phosphate, *See* *D-Ribose 1-Phosphate*

9-(β-D-Ribosyl)purine, *See* *Nebularine*

Ribosylthymine 5'-Monophosphate, *See* *Ribothymidine 5'-Monophosphate*

5-β-D-Ribosyluracil, *See* *Pseudouridine*

3-*N*-Ribosylxanthine 5'-Monophosphate

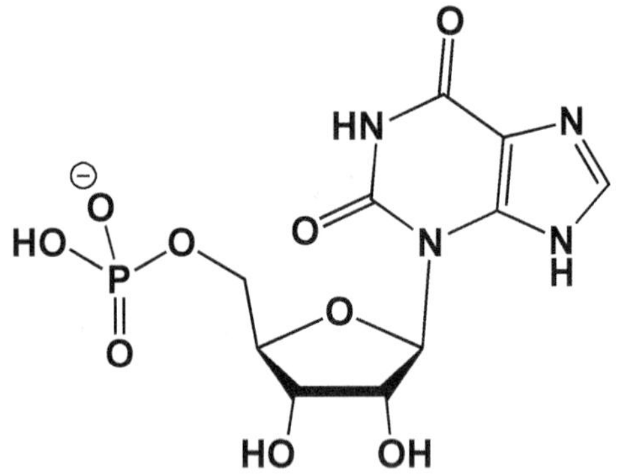

This unusual nucleotide (FW$_{\text{free-acid}}$ = 364.21 g/mol) is analogous to xanthosine 5'-monophosphate, but linked to 6-membered ring of the purine base). **Target(s):** dioxotetrahydropyrimidine phosphoribosyltransferase (1); orotate phosphoribosyltransferase (2); orotidylate decarboxylase (2); pyrimidine nucleoside phosphorylase (1); and uridine phosphorylase (1). **1**. Hatfield & Wyngaarden (1964) *J. Biol. Chem.* **239**, 2580. **2**. Silva & Hatfield (1978) *Meth. Enzymol.* **51**, 143.

9-*N*-Ribosylxanthine 5'-Monophosphate, *See* *Xanthosine 5'-Monophosphate*

Ribosylzeatin, *See* *Zeatin Riboside*

Ribothymidine, *See* *Thymine Riboside*

Ribothymidine 2',3'-Cyclic-Monophosphate, *See* *Ribothymidine 2',3'-Phosphate*

Ribothymidine 3',5'-Cyclic-Monophosphate, *See* *Ribothymidine 3',5'-Phosphate*

Ribozymes, *See* *RNA*

D-Ribulose 1,5-Bisphosphate

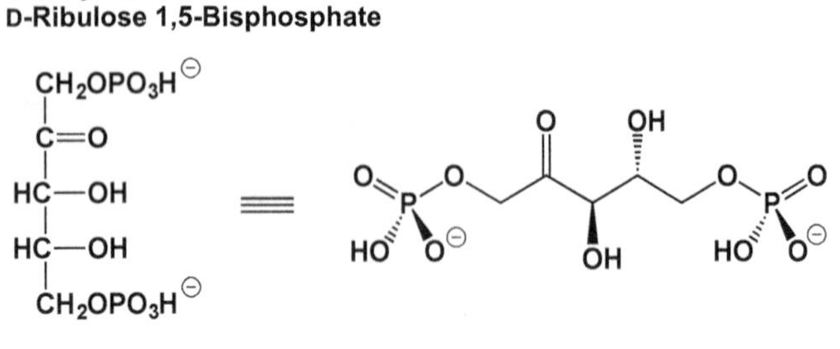

This bis-phosphorylated ketopentose (FW$_{\text{free-acid}}$ = 310.09 g/mol), known formerly as D-ribulose 1,5-diphosphate, is a product of the reaction catalyzed by phosphoribulokinase and is a substrate for ribulose-bisphosphate carboxylase/ oxygenase in the carbon dioxide fixation segment of the Calvin cycle. The more acid-labile C1-phosphate monoester exhibits a 12-min half-life for hydrolysis in 0.1 M acid at 100°C. **Target(s):** fructose-1,6-bisphosphatase (1); fructose-1,6-bisphosphate aldolase (2); glycerate kinase (3); pyruvate kinase, chloroplastic (4); ribose-5-phosphate isomerase (5); and sedoheptulose-bisphosphatase (6). **1**. Amachi & Bowien (1979) *J. Gen. Microbiol.* **113**, 347. **2**. Nakahara, Yamamoto, Miyake & Yokota (2003) *Plant Cell Physiol.* **44**, 326. **3**. Kleczkowski, Randall & Zahler (1985) *Arch. Biochem. Biophys.* **236**, 185. **4**. Lin, Turpin & Plaxton (1989) *Arch. Biochem. Biophys.* **269**, 228. **5**. MacElroy & Middaugh (1982) *Meth. Enzymol.* **89**, 571. **6**. Schimkat, Heineke & Heldt (1990) *Planta* **181**, 97.

D-Ribulose 1,5-Diphosphate, *See* *D-Ribulose 1,5-Bisphosphate*

D-Ribulose 5-Phosphate

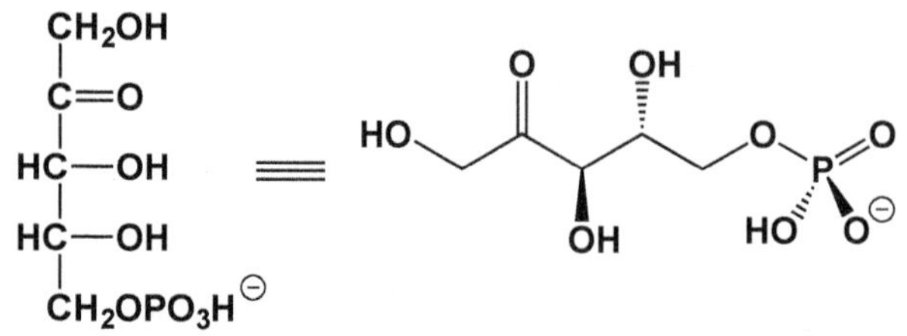

This phosphorylated ketose (FW$_{\text{free-acid}}$ = 230.11 g/mol), a component of the pentose phosphate pathway and the CO_2 assimilation pathway in photosynthesis, is a product of the reactions catalyzed by 6-phosphogluconate dehydrogenase (decarboxylating), ribitol-5-phosphate 2-dehydrogenase, D-ribulokinase, ribose 5-phosphate epimerase, and arabinose-5-phosphate isomerase. It is a substrate for phosphoribulokinase, ribulose-phosphate 3-epimerase, and transketolase. The half-life for the hydrolysis of the phosphate group in 0.5 M H_2SO_4 at 100°C is 40 minutes. Note that the L-enantiomer also occurs naturally. **Target(s):** glucose-6-phosphate isomerase (1-4); 6-phosphogluconate dehydrogenase, product inhibition (5-6); and uridine nucleosidase (8). **1**. Noltmann (1972) *The Enzymes*, 3rd ed. (Boyer, ed.), **6**, 271. **2**. Sangwan & Singh (1990) *Indian J. Biochem. Biophys.* **27**, 23. **3**. Sangwan & Singh (1989) *J. Biosci.* **14**, 47. **4**. Thomas (1981) *J. Gen. Microbiol.* **124**, 403. **5**. Price & Cook (1996) *Arch. Biochem. Biophys.* **336**, 215. **6**. Moritz, Striegel, De Graaf & Sahm (2000) *Eur. J. Biochem.* **267**, 3442. **7**. Menezes, Kelkar & Kaklij (1989) *Indian J. Biochem. Biophys.* **26**, 329. **8**. Weisz, Schofield & Edwards (1985) *J. Neurochem.* **44**, 510. **9**. Magni (1978) *Meth. Enzymol.* **51**, 290.

9-β-D-Riburonosyladenine, *See* *9-(β-D-Ribofuranuronic)adenine*

Riccardin C

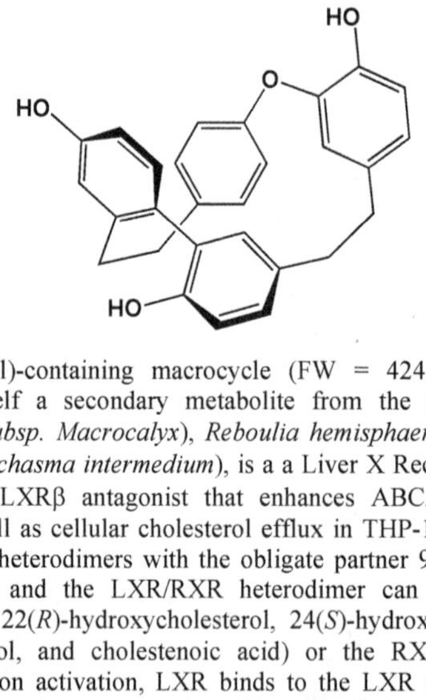

This *bis*(bibenzyl)-containing macrocycle (FW = 424.49 g/mol; CAS 84575-08-6), itself a secondary metabolite from the Siberian cowslip (*Primula veris subsp. Macrocalyx*), *Reboulia hemisphaerica*, and Chinese liverwort (*Plagiochasma intermedium*), is a a Liver X Receptor (or LXRα) agonist and an LXRβ antagonist that enhances ABCA1 and ABCG1 expression as well as cellular cholesterol efflux in THP-1 cells (1). LXRα and LXRβ form heterodimers with the obligate partner 9-cis retinoic acid receptor (RXR), and the LXR/RXR heterodimer can be activated by oxysterols (*e.g.*, 22(*R*)-hydroxycholesterol, 24(*S*)-hydroxycholesterol, 27-hydroxycholesterol, and cholestenoic acid) or the RXR agonist 9-*cis*-retinoic acid. Upon activation, LXR binds to the LXR response element (LXRE) within target gene promoters. Riccardin C thus provides a novel tool for identifying LXR subtype and probing receptor function. **1**. Tamehiro, Sato, Suzuki, *et al.* (2005) *FEBS Lett.* **579**, 5299.

Rice α-Amylase/Subtilisin Inhibitor

This plant protein (MW = 21 kDa; often abbreviated RASI, isoelectric point = 9.05) inhibits subtilisin strongly but inhibites germinating rice seed α-amylase from weakly. RASI inhibits rice α-amylase more than barley α-amylase, and the inhibition of rice α-amylase was greater at higher pHs. RASI did not inhibit trypsin, chymotrypsin, cucumisin, or mammalian α-amylase (1-3). **1**. Nielsen, Bonsager, Fukuda & Svensson (2004) *Biochim. Biophys. Acta* **1696**, 157. **2**. Ohtsubo & Richardson (1992) *FEBS Lett.* **309**, 68. **3**. Yamagata, Kunimatsu, Kamasaka, Kuramoto & Iwasaki (1998) *Biosci. Biotechnol. Biochem.* **62**, 978.

Ricin

This heterodimeric toxin and type-II ribosome-inactivating protein (CAS 9009-86-3) from the seeds of *Ricinus communis* consists of an enzymatic A chain (267 residues; MW = 30 kDa) and a Gal/GalNAc-binding B chain (262 residues; MW = 32 kDa), joined by a single disulfidc bond. **Mode of**

Cell Entry: Ricin's A-chain is te active *N*-glycosidase, and Ricin B-chain is a lectin that facilitates tight binding to the cell surface glycoconjugates of target cells and facilitates internalization/translocation of the toxin to the cytosol. The toxin binds to the cell surface via its B-chain, followed by retrograde trafficking through intracellular compartments to the ER, where the A-chain is transported across the ER membrane using cellular proteins involved in the disposal of aberrant ER proteins, a process known as Retrograde Translocation. **Mode of Ribosomal Inactivation:** The ricin A chain possesses rRNA *N*-glycosidase activity that catalyzes depurination of A4256 in the 28S ribosomal RNA, interfering with the interaction of eEF-2 with the ribosome and inhibiting protein biosynthesis (1). When separated from the B-chain and administered parenterally to animals, the toxicity of the ricin A-chain is reduced >1,000-times compared to ricin holotoxin. Upon depurination, changes occur in domains I, II and V of 28 S rRNA as well as in 5 S rRNA. Changes were found in functional regions of the rRNAs, such as the regions involved in peptidyltransferase activity, subunit interaction and in the binding of elongation factors. Most of these structural changes make the rRNAs less accessible for chemical modification, suggesting that the ribosomal particles becomes less flexible after ricin treatment (2). Depurination in the ricin loop (5'-AGUAC**GA**GAGGA-3', where the bold-type shows A4256) also affects the structure of the 3' major domain in 18 S rRNA. Ricin is about 60-100x less potent than the related plant toxin (**See** *Abrin*) having a maximal ribosome-inactivation rate constant of around 1500 s^{-1}. **Discovery of Ricin's *N*-Glycosidase Actvity:** Ricin was first recognized as a protein toxin in 1888, when Richard Stillmark isolated a toxin-enriched extract. Three years later, Paul Ehrlich showed that mice could be immunized against ricin. Even so, the work of Endo & Tsurugi (3) was definitive. To determine whether ricin A-chain release any bases from 28 S rRNA, they incubated rat liver ribosomes with a catalytic amount of the toxin, and a fraction containing free bases and nucleosides was prepared by means of ion-exchange column chromatography. Thin-layer chromatographic analysis of this fraction revealed a release of one mol of adenine per mol of ribosome. When the ribosomes or naked total RNAs were treated with ricin A-chain in the presence of [^{32}P]-phosphate, little incorporation of the radioactivity into 28 S rRNA was observed, indicating that the release is not mediated by phosphorolysis. This evidence and the results of an earlier paper (4) indicated that the ricin A-chain inactivates the ribosomes by cleaving the *N*-glycosidic bond of A4324 of 28 S rRNA in a hydrolytic fashion. This catalytic activity of ricin A-chain was also observed when naked 28 S rRNA is used as a substrate, showing that the toxin directly acts on the RNA. Similar activity on 28 S rRNA is also exhibited by ricin-related toxins (abrin and modeccin), suggesting a general mechanistic pathway for ribosome inactivation by these lectin toxins. **Catalytic Mechanism:** The sarcin/ricin loop containing A4256 of the ribosome docks into a cleft comprising ricin's active site, such that its adenine is sandwiched between Tyrosine-80 and Tyrosine-123 in a base-stacking interaction. The leaving group (*i.e.*, adenine) is protonated by Arginine-180, thereby promoting C-1'–N-9 bond scission, forming an oxacarbenium ion on the ribose of A4256. The resulting transition-state is stabilized by an ion-pairing interaction with Glutamate-177. A water molecule (presumably polarized by Arginine-180) is then delivered to the reaction center, attacking the sugar carbonium intermediate and completing the catalytic round. **Transition-State Inhibition:** These features agree with kinetic and structural studies the potent inhibitor cyclic-G(9-DA)GA-2'-OMe, a transition state mimic that contains the tetranucleotide sequence of the GAGA sarcin–ricin loop, but with its ricin-susceptible adenosine replaced by DADMe-Immucillin-A (5). A methylene bridge between aza-sugar and adenine group (circled below) in the transition state analogue mimics the geometry of the dissociative transition state.

cyclic-G(9-DA)GA-2-OMe

The design of this inhibitor is based on studies of the catalytic properties of a related plant toxin (**See** *Saporin-L1 Toxin*), which exhibits robust depurinating activity against defined nucleic acid substrates and mammalian ribosomes (6). Low-molecular-weight transition-state analogues, with adenosine replaced with the transition-state mimic 9-deazaadenine-9-methylene-*N*-hydroxy-pyrrolidine (DADMeA), were found to be powerful, slow-onset inhibitors. Linear, cyclic, and stem-loop oligonucleotide inhibitors containing DADMeA and based on the GAGA sarcin-ricin tetranucleotide loop gave slow-onset tight-binding, with inhibition constants K_i^* values of 2.3–8.7 nM under physiological conditions, or about 40000-times tighter than corresponding RNA substrates (6). **Caution:** Ricin is extremely toxic, and care must be exercised when working with this agent. Within 12 hours of ingestion, symptoms of ricin toxicity are nonspecific (*e.g.*, nausea, vomiting, diarrhea, and abdominal pain), often progressing to hypotension, liver failure, renal dysfunction, and death due to multi-organ failure and/or cardiovascular collapse. Within 8 hours after ricin inhalation, symptoms include cough, dyspnea, arthralgia, and fever, eventually progressing to respiratory distress and death, often without involving other organs. The U.S. Center for Disease Control advises that ricin may cause severe allergic reactions and that exposure to milligram quantities of ricin can prove fatal. Ricin toxin is classified as a select agent, requiring registration with CDC for ricin possession, use, storage or transfer. **1.** Jiménez (1976) *Trends Biochem. Sci.* **1**, 28. **2.** Lord, Roberts & Robertus (1994) *FASEB J.* **8**, 201. **2.** Holmberg & Nygård (1996) *J. Mol. Biol.* **259**, 81. **3.** Endo & Tsurugi (1987) *J. Biol. Chem.* **262**, 8128. **4.** Endo, Mitsui, Motizuki & Tsurugi (1987) *J. Biol. Chem.* **262**, 5908. **5.** Ho, Sturm, Almo & Schramm (2009) *Proc. Natl. Acad. Sci.* **106**, 20276. **6.** Sturm, Tyler, Evans & Schramm (2009) *Biochemistry* **48**, 9941.

Ricinine

This toxic alkaloid (FW = 164.16 g/mol; CAS 524-40-3; M.P. = 201°C), also known as ricinin and 1,2-dihydro-4-methoxy-1-methyl-2-oxo-3-pyrimidinecarbonitrile, from the leaves and seeds of the castor plant (*Ricinus communis*) induces seizures, when administered to mice at doses higher than 20 mg/kg. Ricinine inhibits toyocamycin nitrile hydratase. **1.** Uematsu & Suhadolnik (1975) *Meth. Enzymol.* **43**, 759.

Ricinolate, *See Ricinoleate*

Ricinoleate

This hydroxylated derivative (FW$_{free-acid}$ = 298.47 g/mol; CAS 5323-95-5) of oleic acid, also known as ricinoleic acid and (9*Z*,12*R*)-12-hydroxy-9-octadecenoic acid and ricinolic acid, is found in large quantities (as the triacylglycerol triricinolein) in castor oil (indeed, ~90% of the total triacylglycerols in castor oil is triricinolein). The free acid is a liquid at room temperature (melting point of the α-form is 5.5°C while the β-crystal form melts at 16°C). The *trans*-isomer of ricinoleic acid, also known as ricinelaidic acid, melts at 53°C. Note: Castor oil, which is obtained by expressing castor seeds (*Ricinus communis*), contains some triacylglycerols with oleoyl, linoleoyl, palmitoyl, stearoyl, and dihydroxystearoyl residues. **Target(s):** alanine absorption (1); cholesterol biosynthesis (2); chondroitin AC lyase (3); chondroitin B lyase (3); and hyaluronate lyase (3). **1.** Hajjar,

Murphy & Scheig (1979) *Amer. J. Physiol.* **236**, E534. **2**. Kuroda & Endo (1976) *Biochim. Biophys. Acta* **486**, 70. **3**. Suzuki, Terasaki & Uyeda (2002) *J. Enzyme Inhib. Med. Chem.* **17**, 183.

Ridaforolimus

This mTOR inhibitor (FW = 958.22 g/mol; CAS 572924-54-0), formerly as deforolimus and also known by AP23573 and MK-8669, inhibits mammalian Target of Rapamycin (or mTOR), a protein kinase within the phosphatidylinositol 3-kinase (PI3K)/Akt signalling cascade. The latter plays a central role in controlling cell proliferation, survival, mobility and angiogenesis. Dysregulation of mTOR pathway has been observed in many human tumors, making everolimus a promising new anticancer drug whose action leads to cell cycle arrest in the G_1 phase, with inhibition of tumor angiogenesis by reducing VEGF formation. **1**. Rubio-Viqueira & Hidalgo (2006) *Curr, Opin. Investig. Drugs* **7**, 501. **2**. Wan, Shen, Mendoza, Khanna & Helman (2006) *Neoplasia* **6**, 394.

Ridogrel

This oral antithrombotic agent (FW$_{free-acid}$ = 366.34 g/mol; CAS 110140-89-1), also known as R 68070, inhibits human thromboxane-A synthase, IC$_{50}$ = 4-14.7 nM (1-5) and is also a thromboxane receptor antagonist, K_d = 1.7-8 µM (1-3,5). In humans, ridogrel is absorbed within 30-60 min, with a half-life of 6-9 hours. At the oral dose of 300 mg b.i.d., steady state is reached within 3 days. **1**. Takeuchi, Kohn, True, *et al.* (1998) *J. Med. Chem.* **41**, 5362. **2**. De Clerck, Beetens, de Chaffoy de Courcelles, *et al.* (1989) *Prog. Clin. Biol. Res.* **301**, 567. **3**. De Clerck, Beetens, de Chaffoy de Courcelles, Freyne & Janssen (1989) *Thromb. Haemost.* **61**, 35. **4**. Vanden Bossche, Willemsens, Bellens & Janssen (1992) *Biochem. Pharmacol.* **43**, 739. **5**. Soyka, Guth, Weisenberger, Luger & Müller (1999) *J. Med. Chem.* **42**, 1235.

Rifalazil

This antimicrobial agent (FW = 941.08 g/mol; CAS 129791-92-0), a semisynthetic derivative of rifamycin S, inhibits bacterial DNA-directed RNA polymerase (1,2). **See** *Rifamycin* **1**. Chopra (2007) *Curr. Opin. Invest. Drugs* **8**, 600. **2**. Fujii, Saito, Tomioka, Mae & Hosoe (1995) *Antimicrob. Agents Chemother.* **39**, 1489.

Rifamide, See *Rifamycin B N,N-Diethylamide*

Rifampicin

This semisynthetic antibiotic (FW = 822.95 g/mol; CAS 13292-46-1), also called rifampin and rifamycin AMP, is prepared chemically from one of the rifamycins isolated from *Streptomyces mediterranei* (formerly called *Nocardia mediterranei*). Rifampicin inhibits bacterial transcription at very low concentrations (*e.g.*, 0.01 µg/mL). It was first marketed as a treatment for tuberculosis in 1957. RNA biosynthesis is inhibited by the binding of rifampicin to the β subunit of RNA polymerase. At higher concentrations (*e.g.*, 1 µg/mL), mitochondrial and chloroplast transcription can be affected as well. No effect is observed with eukaryotic RNA polymerase at 100 µg/mL. Rifampicin will also inhibit the assembly of DNA and protein into mature virus particles and is used in the removal of plasmids in bacteria (*i.e.*, plasmid curing). It also activates the human glucocorticoid receptor and acts as an immunodepressant. **Drug Interactions:** Rifampicin is a powerful known inducer of hepatic cytochrome P450 enzyme system, increasing the cellular content of CYP2B6, CYP2C8, CYP2C9, CYP2C19, CYP3A4, CYP3A5, and CYP3A7. Aa consequence, rifampicin speeds up the metabolism of many drugs (including cyclosporine (CYP3A4), simvastatin (CYP3A4), rosiglitazone (CYP2C9), warfarin (CYP2C9), and bupropion (CYP2B6)) making them less effective or ineffective. **Properties:** Rifampicin is a stable, red-orange solid (decomposes at 183–188°C) that is soluble in dimethyl sulfoxide, ethyl acetate, and chloroform. It is slightly soluble in water (at pH < 6). It should be protected from light and stored in the dark. It is zwitterionic in nature with a 4-hydroxy pK_a of 1.7 and 3-piperazine pK_a of 7.9. The λ_{max} values at pH 7.38 are 237, 255, 334, and 475 nm (the corresponding molar extinction coefficients are 33200, 32100, 27000, and 15400 M^{-1}cm^{-1}, respectively). **Key Pharmacokinetic Parameters:** *See* Appendix II in Goodman & Gilman's THE PHARMACOLOGICAL BASIS OF THERAPEUTICS, 12th Edition (Brunton, Chabner & Knollmann, eds.) McGraw-Hill Medical, New York (2011). **Target(s):** DNA-directed DNA polymerase (1); DNA-directed RNA polymerase (2-10); 16α-hydroxy-progesterone dehydratase (11); polynucleotide adenylyltransferase, *or* poly(A) polymerase (12-14); RNA polymerases, certain mitochondrial (15); RNA polymerases, certain plant (15); tubulin polymerization (16). **1.** Müller, Yamazaki & Zahn (1972) *Enzymologia* **43**, 1. **2.** Burgess (1971) *Ann. Rev. Biochem.* **40**, 711. **3.** Wehrli, Knuesel, Schmid & Staehelin (1968) *Proc. Natl. Acad. Sci. U.S.A.* **61**, 667. **4.** Wehrli (1983) *Rev. Infect. Dis.* **5**, S407. **5.** Chamberlin (1974) *The Enzymes*, 3rd ed. (Boyer, ed.), **10**, 333. **6.** Straat & Ts'o (1970) *Biochemistry* **9**, 926. **7.** Sethi (1971) *Prog. Biophys. Mol. Biol.* **23**, 67. **8.** Allan & Kropinski (1987) *Biochem. Cell Biol.* **65**, 776. **9.** Pich & Baghl (1991) *J. Bacteriol.* **173**, 2120. **10.** Fujita & Amemura (1992) *Biosci. Biotechnol. Biochem.* **56**, 1797. **11.** Glass & Burley (1984) *J. Steroid Biochem.* **21**, 65. **12.** Mans (1973) *FEBS Lett.* **33**, 245. **13.** Rose & Jacob (1976) *Eur. J. Biochem.* **67**, 11. **14.** Cheung & Newton (1978) *J. Biol. Chem.* **253**, 2254. **15.** Chambon (1974) *The Enzymes*, 3rd ed. (Boyer, ed.), **10**, 261. **16.** Rajagopalan & Gurnani (1985) *Biochem. Pharmacol.* **34**, 3415.

Rifampin, *See Rifampicin*

Rifamycins

These antibiotics (CAS 14487-05-9 for rifamycin B) from *Streptomyces mediterranei* are highly effective against Gram-positive organisms and *Mycobacterium tuberculosis*, with significantly weaker action on Gram-negative bacteria. Rifamycins bind to the β-subunit of prokaryotic RNA polymerase and inhibit chain initiation (but not elongation after elongation has commenced). **Target(s):** DNA-directed DNA polymerase (1); DNA-directed RNA polymerase (2,3); ribonuclease H (4,5); and RNA-directed DNA polymerase (6,7). **1.** Müller, Yamazaki & Zahn (1972) *Enzymologia* **43**, 1. **2.** Lal & Lal (1994) *Bioessays* **16**, 211. **3.** Riva & Silvestri (1972) *Ann. Rev. Microbiol.* **26**, 199. **4.** Haberkern & Cantoni (1973) *Biochemistry* **12**, 2389. **5.** Sekeris & Roewekamp (1972) *FEBS Lett.* **23**, 34. **6.** Ting, Yang & Gallo (1972) *Nature New Biol.* **236**, 163. **7.** Gurgo (1980) in *Inhibitors of DNA and RNA Polymerases* (Sarin & Gallo, eds.), p. 159, Pergamon Press, New York.

Rifamycin AF/05

This semi-synthetic antibiotic (FW = 907.03 g/mol), also known as rifamycin SV 3'-formyldiphenylmethyloxime, inhibits certain nucleic acid polymerases. Solutions of rifamycin AF/05 should contain reducing reagents (such as ascorbic acid) that suppress formation of the quinone. **Target(s):** DNA polymerase (1); polynucleotide adenylyltransferase, *or* poly(A) polymerase (1); ribonuclease H (2); and RNA polymerase (1,3). **1.** Sethi & Okano (1976) *Biochim. Biophys. Acta* **454**, 230. **2.** Sekeris & Roewekamp (1972) *FEBS Lett.* **23**, 34. **3.** Meilhac, Tysper & Chambon (1972) *Eur. J. Biochem.* **28**, 291.

Rifamycin AF/013

This semi-synthetic antibiotic (FW = 853.02 g/mol), derived from rifamycin SV and known as rifamycin SV 3'-formyl-*n*-octyloxime, inhibits certain nucleic acid polymerases. Solutions of rifamycin AF/013 should contain

reducing reagents (such as ascorbic acid) to prevent formation of the quinone. Rifamycin AF/013 inhibits initiation and destabilizes the nucleosome. **Target(s):** cyclic-nucleotide independent, heparin-sensitive nuclear protein kinase (1); DNA polymerase (2-5); polynucleotide adenylyltransferase, *or* poly(A) polymerase (2,6,7,15-18); ribonuclease H (8,9); RNA polymerase (2,7,10-14); RNA polymerase I (11); RNA polymerase II (7,10,11); and RNA polymerase III (10). **1.** Rose, Bell, Siefken & Jacob (1981) *J. Biol. Chem.* **256**, 7468. **2.** Sethi & Okano (1976) *Biochim. Biophys. Acta* **454**, 230. **3.** Schmidt, Bollum & Litwack (1982) *Proc. Natl. Acad. Sci. U.S.A.* **79**, 4555. **4.** Yagura, Kozu & Seno (1981) *J. Biochem.* **90**, 1397. **5.** Byrnes, Downey, Black & So (1976) *Biochemistry* **15**, 2817. **6.** Blakesley & Boezi (1975) *Biochim. Biophys. Acta* **414**, 133. **7.** Rose, Ruch & Jacob (1975) *Biochemistry* **14**, 3598. **8.** Roewekamp & Sekeris (1974) *Eur. J. Biochem.* **43**, 405. **9.** Sekeris & Roewekamp (1972) *FEBS Lett.* **23**, 34. **10.** Logan, Zhang, Davis & Ackerman (1989) *DNA* **8**, 595. **11.** Logan & Ackerman (1988) *DNA* **7**, 483. **12.** Voets, Lagrou, Hilderson, Van Dessel & Dierick (1982) *Int. J. Biochem.* **14**, 405. **13.** Towle, Jolly & Boezi (1975) *J. Biol. Chem.* **250**, 1723. **14.** Cooper & Keir (1975) *Biochem. J.* **145**, 509. **15.** Kurl, Holmes, Verney & Sidransky (1988) *Biochemistry* **27**, 8974. **16.** Jacob & Rose (1974) *Nucl. Acids Res.* **1**, 1549. **17.** Rose & Jacob (1976) *Eur. J. Biochem.* **67**, 11. **18.** Rose, Morris & Jacob (1975) *Biochemistry* **14**, 1025.

Rifamycin SV

This semi-synthetic antibiotic (FW = 697.78 g/mol) inhibits certain nucleic acid polymerases and is derived from rifamycin S. Solutions of rifamycin SV should contain reducing reagents (such as ascorbic acid) to prevent reformation of rifamycin S (*i.e.*, the quinone form). **Target(s):** DNA-directed RNA polymerase (1-3); polynucleotide adenylyltransferase, *or* poly(A) polymerase (4,5); polynucleotide phosphorylase, *or* polyribonucleotide nucleotidyltransferase (6,7); and RNA-directed DNA polymerase (8). **1.** Stender & Scheit (1977) *Eur. J. Biochem.* **76**, 591. **2.** Deshko, Kudelina & Surin (1984) *Bioorg. Khim.* **10**, 641. **3.** Arora (1983) *Mol. Pharmacol.* **23**, 133. **4.** Jacob & Rose (1974) *Nucl. Acids Res.* **1**, 1549. **5.** Rose & Jacob (1976) *Eur. J. Biochem.* **67**, 11. **6.** Littauer & Soreq (1982) *The Enzymes*, 3rd ed. (Boyer, ed.), **15**, 517. **7.** Erickson & Grosch (1977) *J. Bacteriol.* **130**, 869. **8.** Ting, Yang & Gallo (1972) *Nature New Biol.* **236**, 163.

Rifaximin

This semisynthetic rifamycin-class antibiotic (FW = 885.89 g/mol; CAS 80621-81-4), also known as Xifaxan®, binds to the β-subunit of bacterial RNA polymerase, thereby blocking transcription by interfering with the translocation step that must follow the formation of the first phosphodiester bond (*See Rafamycin*). The poor absorption properties of rifaximin commend its use in gut-specific infections, including those associated with irritable bowel disorder (with diarrhea). Rifaximin has a low level of resistance selection, but may select stable highly resistant mutants in a single step; periodic surveillance of rifaximin resistance should detect the appearance of rifaximin-resistant clinical isolates (1). **Antibacterial Spectrum:** *Campylobacter jejun*, MIC_{90} = 32 µg/mL *Clostridium dfficile*, MIC_{90} = 128 µg/mL; *Escherichia coli*, MIC_{90} = 32 µg/mL; *Helicobacter pylori*, MIC_{90} = 4 µg/mL; *Salmonella jejun*, MIC_{90} = 64 µg/mL; *Sigella*, MIC_{90} = 64 µg/mL; *Vibrio cholerae*, MIC_{90} = 4 µg/mL; *Yersinia enterocolitica*, MIC_{90} = 32 µg/mL; *Campylobacter jejun*, MIC_{90} = 16 µg/mL. **1.** Ruiz, Mensa, Pons, Vila & Gascon (2008) *J. Antimicrob. Chemother.* **61**, 1016.

Rigoserib

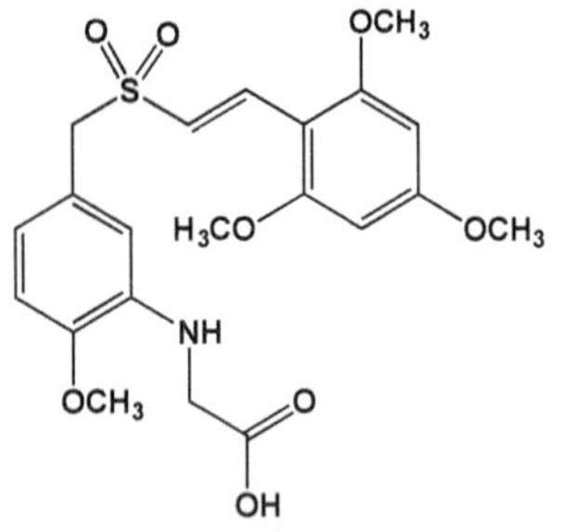

Rigoserib (ON 01910) 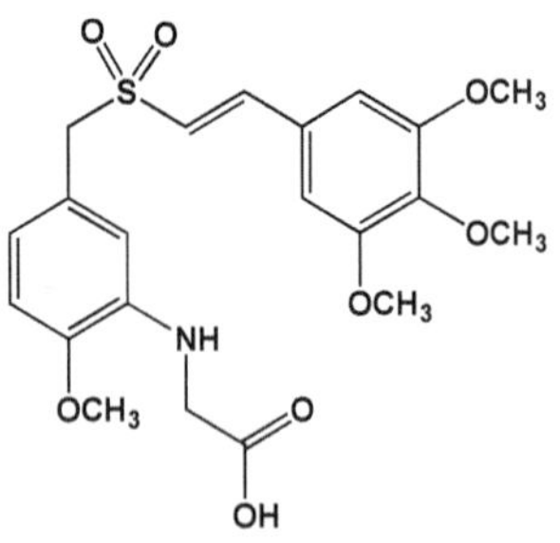   **Inactive Control (ON 01911)**

This multi-mode chemotherapeutic agent ($FW_{free-acid}$ = 451.50 g/mol; $FW_{Sodium-Salt}$ = 473.47 g/mol; CAS 1225497-78-8), also known as ON-01910, is likely to inhibit PI3K/Akt and PLK signaling (1,2) as well as to disrupt mitosis, the latter by reducing microtubule (MT) dynamics (3). PI3K signaling promotes the growth/survival under stressful conditions, such as the low oxygen levels found in solid tumors. If the PI3K pathway is over-active, apoptosis is diminished, leading to excessive proliferation. By inhibiting the PI3K pathway, rigosertib promotes tumor cell apoptosis. Rigosertib also influences signals along the PI3K pathway, such as those leading to the production of cyclin D_1. The PLK pathway plays a critical role in maintaining proper organization and sorting of chromosomes during cell division. Excessive PLK activity in cancer cells can result in uncontrolled proliferation. By inhibiting polo-like kinase 1 (IC_{50} = 9 nM) in cancer cells, rigosertib also inhibits cellular division, leading to chromosome disorganization and death. Recent work on Rigoserib and Cent-1 (*See Centimitor-1*) provides new clues about the likely mode of action of these agents during M phase (3). Both cause centrosome fragmentation and reduced the amount of centrosome-associated γ-tubulin. They also significantly retard MT dynamics in interphase cells, shorten spindle length in mitosis, and decrease tension across sister kinetochores in mitotic chromosomes. They also cause the delocalization of NuMA and EB1 in mitotic cells. Taken together, these data suggest Cent-1 and rigosertib impair MT-mediated processes during M-phase. Acentrosomal spindle poles, reduced interkinetochore tension, mislocalization of MT end-tracking proteins, and chromosome misalignment are consequences of treating cells with low doses of well-studied MT drugs, again supporting the inference that these compounds alter microtubule dynamics. It is noteworthy that neither Rigoserib nor Cent-1 perturb MT self-assembly *in vitro*. Current findings do not distinguish between direct and indirect drug-induced mechanisms for altering microtubule processes and thereby hindering mitosis. Adding to the dispute over the mode of action is the finding that ON 01910 does not induce consistent activation or only delayed activation of proteins traditionally associated with this DNA damage-response, whereas both doxorubicin and camptothecin provided prompt activation of Chk1, Chk2, and histone H2A.X (4). Instead, ON-01910 induced hyperphosphorylation of RanGAP1·SUMO1, and the latter correlated with accumulation of mitotic cells, prolonged M-phase arrest, and apoptotic cell death. A pharmacologically inactive analogue (ON-01911, *See structure above*) fails to induce RanGAP1·SUMO1 phosphorylation (4). Given ON 01910's inhibitory action on cancer cells *in vitro*, on tumor xenografts, and on some clinical cancers, determination of its molecular target(s) is of paramount importance. **1.** Gumireddy, Reddy, Cosenza, *et al.* (2005) *Cancer Cell* **7**, 275. **2.** Zimmerman & Erikson (2007) *Cell Cycle* **6**, 1314. **3.** Mäki-Jouppila, Laine, Rehnberg, *et al.* (2014) *Mol. Cancer Ther.* **13**, 1054. **4.** Oussenko, Holland, Reddy & Ohnuma (2011) *Cancer Res.* **71**, 4968.

Rilonacept

This soluble receptor decoy for IL-1 and long-acting anti-inflammatory (MW ≈ 251 kDa; CAS 501081-76-1), also known as Arcalyst® and IL-1 Trap®, binds IL-1, thereby neutralizing the latter's pro-inflammatory actions. Rilonacept is a dimeric fusion protein consisting of the ligand-binding domains in the extracellular domains of the human interleukin-1 receptor component (IL-$1R_1$) and IL-1 receptor accessory protein (IL-1RAcP), both linked in tandem to the Fc region of a human IgG1 that binds IL-1. Rilonacept binds IL-1β, IL-1α, and IL-1 receptor antagonist (IL-1ra) with high affinity, thereby preventing the activation of IL-1 receptors, attenuating inflammatory responses and other effects secondary to IL-1 excess. Rilonacept (final concentration = 80 mg/mL; initial dose = 4 mL; maximum weekly dose = 2 mL) is administered subcutaneously. It is indicated for the treatment of cryopyrin-associated periodic syndromes (*or* CAPS). The latter are rare, inherited, inflammatory disorders resulting from a mutation in the *NLRP3* gene, which codes for cryopyrin, a regulator in the acute immune reaction. Rilonacept's terminal half-life is 8.6 days in adults and 6.3 days in pediatric patients. Because an IL-1 blockade may interfere with immune response to infection, some added risk of infection may occur during rilonacept treatment.

Rilpivirine

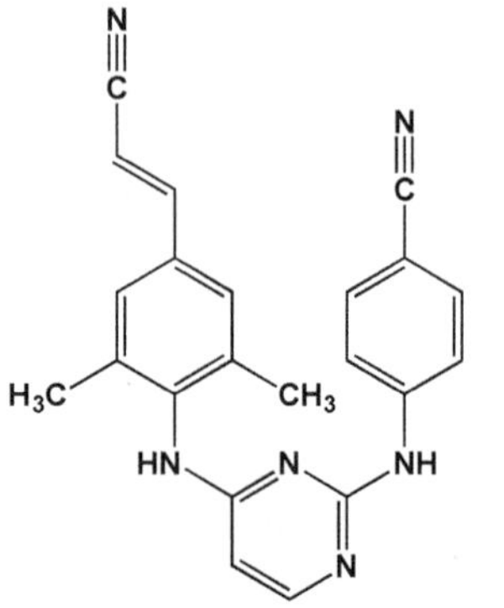

This substituted pyrimidine (FW = 366.43 g/mol; CAS 500287-72-9), also known as TMC278, is a non-nucleoside inhibitor that targets HIV-1 reverse transcriptase and RNA-directed DNA polymerase (1-3). Rilpivirine undergoes oxygenation reactions catalyzed primarily by CYP3A4 and CYP3A5 (4). **1.** El Safadi, Vivet-Boudou & Marquet (2007) *Appl. Microbiol. Biotechnol.* **75**, 723. **2.** Janssen, Lewi, Arnold, *et al.* (2005) *J. Med. Chem.* **48**, 1901. **3.** Fang, Bauman, Das, *et al.* (2008) *Proc. Natl. Acad. Sci. U.S.A.* **105**, 1472. **4.** Lade, Avery & Bumpus (2013) *Antimicrob. Agents Chemother.* **57**, 5067.

Riluzole

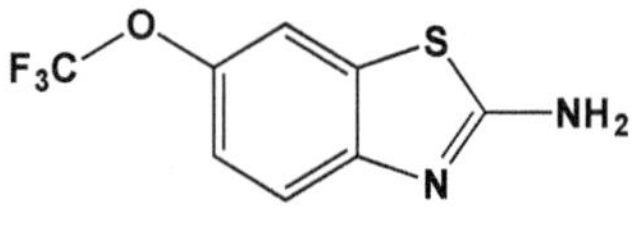

This neuroprotecting glutamate antagonist (FW = 234.20 g/mol; CAS 1744-22-5), also known as 2-amino-6-(trifluoromethoxy)benzothiazole, exhibits anti-excitotoxic action and protects against nonexcitotoxic oxidative neuronal injury. Riluzole affects neurons by inhibiting excitatory amino acid release, inhibiting events following stimulation of excitatory amino acid receptors, and stabilizing the inactivated state of voltage-dependent, tetrodotoxin-sensitive sodium channels. Riluzole (as Rilutek® tablets) was the first drug to be approved by the FDA for treating amyotrophic lateral sclerosis. **Key Pharmacokinetic Parameters:** *See* Appendix II in Goodman & Gilman's *THE PHARMACOLOGICAL BASIS OF THERAPEUTICS*, 12th Edition (Brunton, Chabner & Knollmann, eds.) McGraw-Hill Medical, New York (2011). **Target(s):** γ-aminobutyrate (GABA) uptake (1); glutamate and aspartate release (2); Na^+ channels (3,4); and protein kinase C (5). **1.** Mantz, Laudenbach, Lecharny, Henzel & Desmonts (1994) *Eur. J. Pharmacol.* **257**, R7. **2.** Martin, Thompson & Nadler (1993) *Eur. J. Pharmacol.* **250**, 473. **3.** Yokoo, Shiraishi, Kobayashi, *et al.* (1998) *Naunyn Schmiedebergs Arch. Pharmacol.* **357**, 526. **4.** Benoit & Escande (1991) *Pflugers Arch.* **419**, 603. **5.** Noh, Hwang, Shin & Koh (2000) *Neurobiol. Dis.* **7**, 375.

Rimantadine

This antiviral agent ($FW_{hydrochloride}$ = 215.77 g/mol; CAS 13392-28-4), also known as remantadine and α-methyl-1-adamantanemethylamine, inhibits the influenza virus M2 ion channel, a transmembrane protein that regulates pH during infection (1-3). Reversal potentials were close to equilibrium potentials for transmembrane pH gradients and not to those for Na^+, K^+ or Cl^- concentration gradients. M2 permeability to Na^+ relative to H^+ was estimated to be less than 6 x 10^{-7}. The M2 conductance increased as external pH decreased below 8.5 and approached saturation at an external pH of 4, effects attributable to increased permeability due to increased driving potential and to activation by low external pH. *See also Amantadine* 1. Chizhmakov, Geraghty, Ogden, *et al.* (1996) *J. Physiol.* **494**, 329. **2.** Karginov, Blinov, Safronov, Mamaev & Golovin (1987) *Bioorg. Khim.* **13**, 1638. **3.** Lin, Heider & Schroeder (1997) *J. Gen. Virol.* **78**, 767.

Rimonabant

This once-popular anorectic anti-obesity drug (FW = 462.08 g/mol; CAS 158681-13-1), also known as SR141716, Acomplia®, Zimulti®, and 5-(4-chlorophenyl)-1-(2,4-dichlorophenyl)-4-methyl-*N*-(piperidin-1-yl)-1*H*-pyrazole-3-carboxamide, is an inverse agonist for the CB_1 cannabinoid receptor (K_i = 2 nM), with no discernible action at CB_2 receptors. Rimonabant is also an inverse agonist of adenylyl cyclase, reversing inhibition by WIN 55,212-2. Its undesirable clinical side-effects (*e.g.*, severe depression and suicidal thoughts) that led to its withdrawal from the EU market. Rimonabant is the first drug to target the endocannabinoid (CB) pathway by inhibiting the actions of anandamide and 2-archidonyl-glycerol on CB1 receptors. Rimonabant is also.a μ-opioid receptor antagonist. CB_1 receptor bockade decreases in appetite and has direct actions in adipose tissue and liver to improve glucose, fat and cholesterol metabolism, thereby improving insulin resistance, triglycerides and high-density lipoprotein cholesterol (HDL-C) and in some patients, blood pressure. Despite its removal from the market, rimonabant remains a valuable tool for the pharmacologic investigation of cannabinoid receptots. Rimonabant is also an inverse agonist of adenylyl cyclase, reversing inhibition by WIN 55,212-2. **1.** Rinaldi-Carmona, *et al.* (1994) *FEBS Lett.* 350 240.. **2.** Rinaldi-Carmona, *et al.* (1995) *Life Sci.* **56**, 1941. **3.** Colombo, *et al.* (1998) *Life Sci.* 63, 113. **4.** Janiak, *et al.* (2007) *Kidney Int.* **72**, 1345.

Riociguat

This first-in-class, oral vasodilator (FW = 422.22 g/mol; CAS 625115-55-1; Solubility: 84 mg/mL DMSO; <1 mg/mL H_2O), also known by its code name BAY 63-2521, its trade name Adempas® and its systematic name, methyl   *N*-[4,6-diamino-2-[1-[(2-fluorophenyl)methyl]-1*H*-pyrazolo[3,4-*b*]pyridin-3-yl]-5-pyrimidinyl]-*N*-methylcarbaminate, stimulates soluble guanylate cyclase, both independently of endogenously generated vasodilator nitric oxide and in synergy with NO to produce its anti-aggregatory, anti-proliferative, and vasodilatory effects. Riociguat stimulates soluble guanylate cyclase (sGC) in a dose-dependent manner, increasing its catalysis by as much as 73-fold (1). In October 2013, riociguat received FDA approval to reduce mean pulmonary arterial pressure in the treatment of persistent/recurrent Chronic Thromboembolic Pulmonary Hypertension, *or* CTEPH (WHO Group 4) after surgical treatment or inoperable CTEPH to improve exercise capacity. Riociguat is rapidly absorbed, reaching a maximum plasma concentration within 0.5–1.5 h, after which it is eliminated with a 5-10 hour half-life. **Targets:** P-gp (f2-value = 12 μM), BCRP (IC_{50} = 46 μM), OATP1B1 (IC_{50} = 34 μM), OATP1B3 (IC_{50} = 50 μM), CYP2D6 (IC_{50} = 12 μM), and CYP2C19 (IC_{50} = 46 μM). Riociguat also induces mRNA expression of BCRP/ABCG2 (3-times at 20 μM) and to a lesser extent of CYP3A4 (2.3-times at 20 μM), UGT1A4, and ABCB11 (2). **1.** Mittendorf, Weigand, Alonso-Alija, *et al.* (2009) *ChemMedChem.* **4**, 853. **2.** Rickert, Haefeli & Weiss (2014) *Pulm. Pharmacol. Ther.* **28**, 130.

Ripasudil

This potent ROCK inhibitor (FW = 395.88 g/mol; CAS 887375-67-9; 4-fluoro-5-[[(2S)-hexahydro-2-methyl-1*H*-1,4-diazepin-1-yl]sulfonyl]isoquinoline and K-115, which targets the Rho-associated, coiled-coil containing protein kinases ROCK1 (IC_{50} = 51 nM) and ROCK2 (IC_{50} = 19 nM), respectively, is used in the treatment of glaucoma and ocular hypertension. ROCK is a key regulator of actin organization and thus a regulator of cell migration. **1.** Nakagawa, Koizumi, Okumura, Suganami & Kinoshita (2015) *PLoS One* **10**, e0136802.

Risedronate

This bisphosphonate ($FW_{free-acid}$ = 283.11 g/mol; CAS 115436-72-1), also known by the trade names Actonel®, Atelvia®, and Benet®, as well as resedronate and (1-hydroxy-1-phosphono-2-pyridin-3-ylethyl)phosphonate, targets farnesyl pyrophosphate synthase. Like analogous bisphosphonates (ibandronate, pamidronate, alendronate, and zoledronate) risedronate mimics isoprenoid diphosphate lipids, thereby inhibiting FPP synthase, an enzyme in the mevalonate pathway. Bisphosphonates show remarkable affinity for hydroxyapatite, accounting for their bone-homing propensity. Inhibition of FPP synthase in osteoclasts reduces the levels of FPP and GGPP, which are needed for the post-translational farnesylation and geranylgeranylation of small GTPase signalling proteins and essential for osteoclast podosome assembly, V-ATPase acidification of bone surfaces, and consequential resorption/turnover of bone. In postmenopausal women, risedronate reduces the elevated rate of bone turnover, often resulting in a regain of bone mass. Risedronate is thus is an osteoclast-mediated bone-resorption inhibitor and is used to treat Paget's disease. **Target(s):** dimethylallyl-*trans*transferase, *or* geranyl-diphosphate synthase (1); farnesyl-*trans*transferase, *or* geranylgeranyl-diphosphate synthase (2,3); geranyl-*trans*transferase, *or* farnesyl-diphosphate synthase (3-13); hypoxanthine:guanine phosphoribosyltransferase (14); protein geranylgeranyltransferase (15); protein geranylgeranyltransferase type II (16); pyrophosphatase, *or* inorganic diphosphatase (17); pyrophosphate-dependent phosphofructokinase, *or* diphosphate:fructose-6-phosphate 1-phosphotransferase (18); and vacuolar H^+-transporting pyrophosphatase (17). **1**. Burke, Klettke & Croteau (2004) *Arch. Biochem. Biophys.* **422**, 52. **2**. Szabo, Matsumura, Fukura, *et al.* (2002) *J. Med. Chem.* **45**, 2185. **3**. Ling, Li, Miranda, Oldfield & Moreno (2007) *J. Biol. Chem.* **282**, 30804. **4**. van Beek, Pieterman, Cohen, Lowik & Papapoulos (1999) *Biochem. Biophys. Res. Commun.* **264**, 108. **5**. Grove, Brown & Watts (2000) *J. Bone Miner. Res.* **15**, 971. **6**. Dunford, Thompson, Coxon, *et al.* (2001) *J. Pharmacol. Exp. Ther.* **296**, 235. **7**. Montalvetti, Bailey, Martin, *et al.* (2001) *J. Biol. Chem.* **276**, 33930. **8**. Montalvetti, Fernandez, Sanders, *et al.* (2003) *J. Biol. Chem.* **278**, 17075. **9**. Sanders, Song, Chan, *et al.* (2005) *J. Med. Chem.* **48**, 2957. **10**. Bergstrom, Bostedor, Masarachia, Reszka & Rodan (2000) *Arch. Biochem. Biophys.* **373**, 231. **11**. Glickman & Schmid (2007) *Assay Drug Dev. Technol.* **5**, 205. **12**. Dunford, Kwaasi, Rogers, *et al.* (2008) *J. Med. Chem.* **51**, 2187. **13**. Sigman, Sánchez & Turjanski (2006) *J. Mol. Graph. Model.* **25**, 345. **14**. Fernández, Wenck, Craig III & Delfino (2004) *Bioorg. Med. Chem. Lett.* **14**, 4501. **15**. Coxon, Helfrich, Van't Hof, *et al.* (2000) *J. Bone Miner. Res.* **15**, 1467. **16**. Coxon, Ebetino, Mules, *et al.* (2005) *Bone* **37**, 349. **17**. McIntosh & Vaidya (2002) *Int. J. Parasitol.* **32**, 1. **18**. Bruchhaus, Jacobs, Denart & Tannich (1996) *Biochem. J.* **316**, 57.

Risperidone

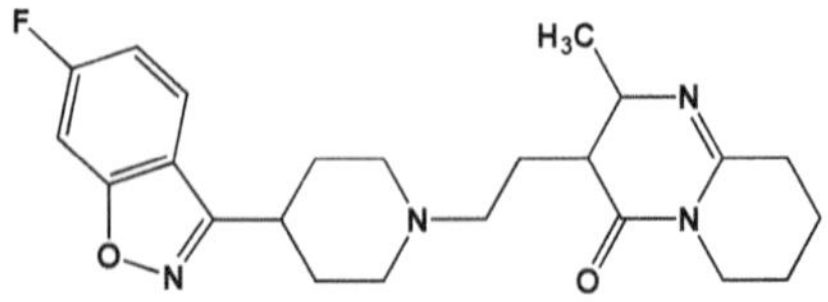

This atypical antipsychotic/antischizophrenic drug (FW = 410.49 g/mol; CAS 106266-06-2; Solubility: 3 mg/mL DMSO, 1 mg/mL H_2O), also known by its code name R-64766, its trade name Risperdal® and its systematic name, 4-[2-[4-(6-fluorobenzo[*d*]isoxazol-3-yl)-1-piperidyl]-ethyl]-3-methyl-2,6-diazabicyclo[4.4.0]deca-1,3-dien-5-one, is a dopamine antagonist that also exhibits antiserotonergic, antiadrenergic, and antihistaminergic properties (1,2). Risperidone shows very high binding affinity for Serotonin-2, *or* 5-HT$_2$, receptors, K_i = 0.16 nM, with a slow dissociation, $t_{0.5}$ = 31 min (1). Risperidone likewise potently blocks serotonin-induced phosphatidic acid formation in human blood platelets (IC$_{50}$ = 0.5 nM). Risperidone also shows high-affinity binding for dopamine-D$_2$ receptors, K_i = 3.1 nM, and rapid dissociation, $t_{0.5}$ = 2.7 min (1). It also displays high-affinity binding to α$_1$-adrenergic (K_i = 0.8 nM), histamine-H$_1$ (K_i = 2.2 nM) and α$_1$-adrenergic (K_i = 7.5 nM) receptors (1). In rats (2), risperidone's potent central 5-HT$_2$ and catecholamine actions account for its potent and complete antagonism of an LSD (dose = 0.16 mg/kg). The CYP2D6 poor metabolizer phenotype appears to be associated with risperidone adverse drug response and discontinuation (4,5). (**See** *Paliperidone*) **Note:** Atypical neuroleptics are defined as dopamine (DA) receptor blockers that differ from typical neuroleptics in that they show a markedly lower propensity to induce parkinsonian side-effects or tardive dyskinesias. Some, but not all, are also more effective in treating schizophrenic patients with negative symptoms or who resist classical treatments. **Key Pharmacokinetic Parameters:** *See* Appendix II in Goodman & Gilman's THE PHARMACOLOGICAL BASIS OF THERAPEUTICS, 12th Edition (Brunton, Chabner & Knollmann, eds.) McGraw-Hill Medical, New York (2011). **See also** *Ketanserin; Pimavanserin; RH-34; Ritanserin; Setoperone; Volinanserin*. **Targets:** Serotonin 5-HT$_{1A}$ Receptor, K_i = 423 nM (as antagonist); Serotonin 5-HT$_{1B}$ Receptor, K_i = 14.9 nM (as antagonist); Serotonin 5-HT$_{1D}$ Receptor, K_i = 84.6 nM (as antagonist); Serotonin 5-HT$_{2A}$ Receptor, K_i = 0.17 nM (as inverse agonist); Serotonin 5-HT$_{2B}$ Receptor, K_i = 61.9 nM (as inverse agonist); Serotonin 5-HT$_{2C}$ Receptor, K_i = 12.0 nM (as inverse agonist); Serotonin 5-HT$_{5A}$ Receptor, K_i = 206 nM (as antagonist); Serotonin 5-HT$_6$ Receptor, K_i = 2,060 nM (as antagonist); Serotonin 5-HT$_7$ Receptor, K_i = 6.60 nM (as irreversible antagonist); Dopamine D$_1$ Receptor, K_i = 244 nM (as antagonist); Dopamine D$_2$ Receptor, K_i = 3.57 nM (as antagonist); Dopamine D$_{2S}$ Receptor, K_i = 4.73 nM (as antagonist); Dopamine D$_{2L}$ Receptor, K_i = 4.16 nM (as antagonist); Dopamine D$_3$ Receptor, K_i = 3.6 nM (as inverse agonist); Dopamine D$_4$ Receptor, K_i = 4.66 nM (as antagonist); Dopamine D$_5$ Receptor, K_i = 290 nM (as antagonist); Adenergic α$_{1A}$ Receptor, K_i = 5.0 nM (as antagonist); Adenergic α$_{1B}$ Receptor, K_i = 9.0 nM (as antagonist); α$_{2A}$ Receptor, K_i = 16.5 nM (as antagonist); α$_{2B}$ Receptor, K_i = 108 nM (as antagonist); Adenergic α$_{2C}$ Receptor, K_i = 1.3 nM (as antagonist); muscarinic Acetylcholine Receptors (mAChRs) >10,000 nM); Histamine H$_1$ Receptor, K_i = 20.1 nM (as inverse agonist); Histamine H$_2$ Receptor, K_i = 120 nM (as inverse agonist). **1**. Leysen, Gommeren, Eens, *et al.* (1988) *J. Pharmacol. Exp. Ther.* **247**, 661. **2**. Meert, de Haes & Janssen (1989) *Psychopharmacology (Berlin)* **97**, 206. **3**. Leysen, Janssen, Megens & Schotte (1994) *J. Clin. Psychiatry* **55** Suppl., 5. **4**. Shin, Soukhova & Flockhart (1999) *Drug Metab. Dispos.* **27**, 1078. **5**. de Leon, Susce, Pan *et al.* (2005) *J. Clin. Psychiat.* **66**, 15.

Ristocetin

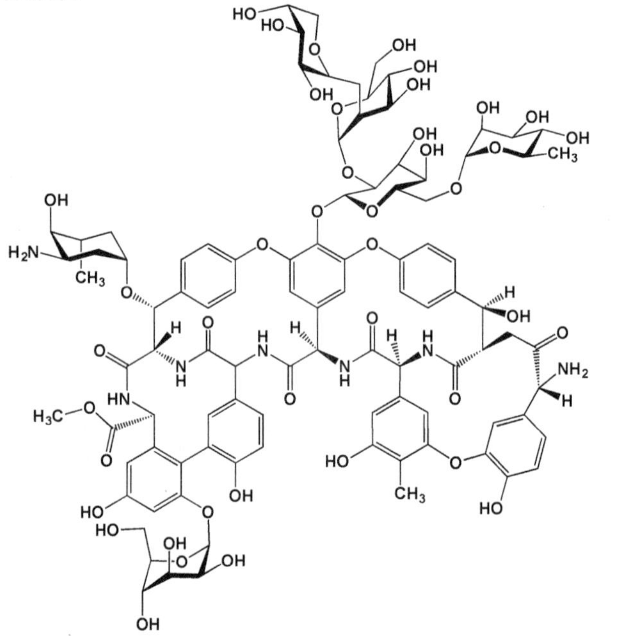

These *Nocardia lurida* glycopeptide antibiotics (Ristocetin A (*shown above*): FW$_{free-base}$ = 2067.95 g/mol; CAS 1404-55-3; unstable above pH 7), also known as ristomycin, containing a variable number of carbohydrate substituents, inhibits cell wall biosynthesis in Gram-positive bacteria (1). Ristocetin is also a platelet aggregation agent. **Target(s):** CDP-glycerol glycerol-phosphotransferase (1); CDP-ribitol ribitolphospho-transferase (2); muramoylpentapeptide carboxypeptidase (3); penicillin-binding protein 1b (4); peptidoglycan biosynthesis (5); peptidoglycan glycosyltransferase (4); phospho-*N*-acetylmuramoyl-pentapeptide-transferase (6); and undeca-prenyldiphospho-muramoyl-pentapeptide β-*N*-acetylglucosaminyltransferase (7). **1**. Burger & Glaser (1964) *J. Biol. Chem.* **239**, 3168. **2**. Ishimoto & Strominger (1966) *J. Biol. Chem.* **241**, 639. **3**. Leyh-Bouille, Ghuysen, Nieto, *et al.* (1970) *Biochemistry* **9**, 2971. **4**. Chandrakala, Shandil, Mehra, *et al.* (2004) *Antimicrob. Agents Chemother.* **48**, 30. **5**. Izaki, Matsuhashi & Strominger (1966) *Meth. Enzymol.* **8**, 487. **6**. Struve, Sinha & Neuhaus (1966) *Biochemistry* **5**, 82. **7**. Meadow, Anderson & Strominger (1964) *Biochem. Biophys. Res. Commun.* **14**, 382.

RITA

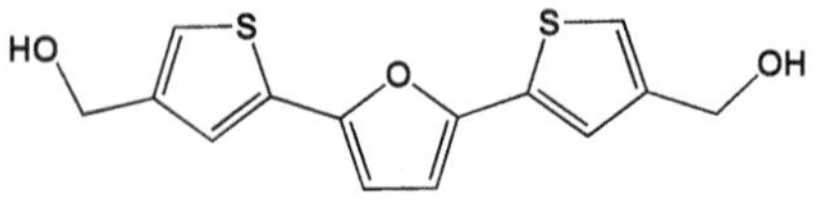

This symmetrical DNA-damaging thiophene (FW = 292.37 g/mol; CAS 213261-59-7; Miscible with Water; Protect from light and heat) also known as NSC 652287, SOS Bismethanol, and systematically as 2,5-bis(5-hydroxymethyl-2-thienyl)furan, inhibits hypoxia-inducible factor-1α and vascular endothelial growth factor expression *in vivo* and induces tumor cell apoptosis in normoxia and hypoxia, the latter by activation of a novel p53-dependent S-phase checkpoint involving CHK-1. In studies with a renal carcinoma cell line, RITA induced both DNA-protein and DNA-DNA cross-links with no detectable DNA single-strand breaks. Because NSC 652287 does not cross-link purified DNA or mammalian topoisomerase I, it is likely that metabolic activation is required for cross-linking (1). **Toxic Agent**: Moderate to severe irritant to the skin and eyes. This pharmaceutically active agent should only be handled by trained personnel. 1. Nieves-Neira, Rivera, Kohlhagen, *et al.* (1999) *Mol. Pharmacol.* **56**, 478.

Ritalin, *See* Methylphenidate

Ritanserin

This antidepressant (FW$_{free-base}$ = 477.58 g/mol; CAS 87051-43-2) is a potent 5-HT$_{2A}$ serotonin receptor antagonist that can cross the blood-brain barrier (1,2). *See also Ketanserin; Pimavanserin; RH-34; Risperidone; Setoperone; Volinanserin.* 1. Leysen, Gommeren, Van Gompel, *et al.* (1985) *Mol. Pharmacol.* **27**, 600. 2. Shi, Nathaniel & Bunney (1995) *J. Pharmacol. Exp. Ther.* **274**, 735.

Ritodrine

This tocolytic drug (FW$_{free-base}$ = 287.36 g/mol; CAS 26652-09-5) is a β$_2$-adrenergic agonist that also inhibits uterine muscle contraction and the Ca^{2+}-ATPase, inhibiting labor, prolonging pregnancy, and allowing time for the fetus (mainly lungs) to mature. Maternal side-effects include metabolic hyperglycemia, hyperinsulinemia, hypokalemia, antidiuresis, altered thyroid function, physiologic tremor, palpitations, nervousness, nausea, vomiting, fever, and hallucinations. Fetal and neonatal side-effects include tachycardia, hypotension, hypoglycemia, hypocalcemia, hyperbilirubinemia, and intraventricular hemorrhage. 1. Plenge-Tellechea, Soler & Fernandez-Belda (1998) *Arch. Biochem. Biophys.* **357**, 179.

Ritonavir

This pseudopeptide bond-containing peptide analogue (FW = 720.96 g/mol; CAS 155213-67-5), also known as Norvir™, Abbott 84538 and ABT-538, is a antiviral agent that inhibits HIV-1 (K_i = 0.1 nM *in vitro*, EC$_{50}$ = 22-130 nM) and HIV-2 (EC$_{50}$ = 160 nM) proteases, blocking polyprotein cleavage in human immunodeficiency virus (1-12). Ritonavir shows high and sustained plasma concentrations after oral administration. Ritonavir produces CYP3A inhibition equivalent to or greater than ketoconazole, and is the best index CYP3A inhibitor alternative to ketoconazole (13). **Target(s):** calpain-1, *or* μ-calpain (1); calpain-2, *or* m-calpain (1); candidapepsin (2); CYP3A, *or* cytochrome P450 (3); glucuronosyltransferases, *or* UDP-glucuronosyltransferases known as UGT1A1, UGT1A3, and UGT1A4 (4); and HIV-1 retropepsin (2,5-12). **Key Pharmacokinetic Parameters:** See Appendix II in Goodman & Gilman's *THE PHARMACOLOGICAL BASIS OF THERAPEUTICS*, 12th Edition (Brunton, Chabner & Knollmann, eds.) McGraw-Hill Medical, New York (2011). 1. Wan & DePetrillo (2002) *Biochem. Pharmacol.* **63**, 1481. 2. Tossi, Benedetti, Norbedo, *et al.* (2003) *Bioorg. Med. Chem.* **11**, 4719. 3. Kempf, Marsh, Kumar, *et al.* (1997) *Antimicrob. Agents Chemother.* **41**, 654. 4. Zhang, Chando, Everett, *et al.* (2005) *Drug Metab. Dispos.* **33**, 1729. 5. Kempf, Marsh, Denissen, *et al.* (1995) *Proc. Natl. Acad. Sci. U.S.A.* **92**, 2484. 6. Clemente, Moose, Hemrajani, *et al.* (2004) *Biochemistry* **43**, 12141. 7. Wilson, Phylip, Mills, *et al.* (1997) *Biochim. Biophys. Acta* **1339**, 113. 8.

Clemente, Hemrajani, Blum, Goodenow & Dunn (2003) *Biochemistry* **42**, 15029. 9. Pettit, Everitt, Choudhury, Dunn & Kaplan (2004) *J. Virol.* **78**, S477. 10. Clemente, Coman, Thiaville, *et al.* (2006) *Biochemistry* **45**, 5468. 11. Surleraux, de Kock, Verschueren, *et al.* (2005) *J. Med. Chem.* **48**, 1965. 12. Shuman, Haemaelaeinen & Danielson (2004) *J. Mol. Recognit.* **17**, 106. 13. Greenblatt & Harmatz (2015) *Br. J. Clin. Pharmacol.* **80**, 342.

Rivaroxaban

This synthetic, first-in-class, orally available anticoagulant (F.W. = 435.88 g/mol; CAS 366789-02-8; Solubility = 12.5 mg/mL DMSO), also known by tradenames BAY 59-7939 and Xarelto®, and its IUPAC name, (*S*)-5-chloro-*N*-{[2-oxo-3-[4-(3-oxomorpholin-4-yl)-phenyl]oxazolidin-5-yl]methyl}-thiophene-2-carboxamide, competitively inhibits blood-clotting Factor Xa (K_i = 0.4 nM), showing >10,000 times greater selectivity for it over other serine proteases (1-3). Rivaroxiban also inhibits prothrombinase activity (IC$_{50}$ = 2.1 nM). **Mode of Action:** The prothrombinase complex (consisting of Factor Xa, Factor Va, Ca^{2+}, and phospholipids) rapidly converts prothrombin to thrombin. The membrane-dependent protease activation process requires multiple binding steps including Factor Xa binding to tissue factor-bearing cells and platelets through specific binding sites and receptors, limiting inhibitor access to the two dimensional membrane surface (4). Factor Xa catalyzes peptide-bond cleavage after an arginine residue, with a preferred cleavage site: Ile-(Glu or Asp)-Gly-Arg.) Rivaroxaban's anticoagulant action is exerted by inhibiting thrombin formation, thereby preventing thrombi formation. Rivaroxaban also appears to provide more consistent anticoagulation *versus* vitamin antagonists (VKAs) in atrial fibrillation as well as for the treatment and prevention of venous thromboembolism (3). Without the need for mandatory monitoring and with only few drug interactions, rivaroxaban and other direct Factor Xa inhibitors are appealing alternatives to VKAs in patients for these indications (4). There is a close correlation between the plasma rivaroxaban concentration and prolongation of prothrombinase-induced clotting time and reduction in endogenous thrombin potential (4). Rivaroxaban strongly inhibits platelet-induced thrombin generation, after activation of either platelets or the coagulation pathway, even in the presence of minimal Factor Xa inhibition in plasma (5). **Reversal of Inhibition:** Although there is currently no reliable way to reverse rivaroxaban's anticoagulant action acutely, polyphosphate is a promising agent. This potent and naturally occurring hemostatic regulator accelerates blood coagulation by activating the contact pathway and by promoting Factor V activation, in turn attenuating the anticoagulant function of tissue factor pathway inhibitor (6). Indeed, 100 μM polyphosphate significantly shortens the clotting time of normal plasma containing 1 μg/mL rivaroxaban (6). *See other direct Factor Xa inhibitors: Apixaban, or Eliquis®; Betrixaban; Darexaban (discontinued); Edoxaban, or Lixiana®; and Otamixaban injectable (discontinued).* 1. Roehrig, Straub, Pohlmann *et al.* (2005) *J. Med. Chem.* **48**, 5900. 2. Kubitza, Becka, Wensing, Voith & Zuehlsdorf (2005) *Eur. J. Clin. Pharmacol.* **61**, 873. 3. Steffel & Luscher (2012) *Hamostaseologie* **32**, 249. 4. Mahaffey & Becker (2006) *Circulation* **114**, 2313. 5. Graff, von Hentig, Misselwitz, *et al.* (2007) *J. Clin. Pharmacol.* **47**, 1398. 6. Smith & Morrissey (2008) *J. Thromb. Haemost.* **6**, 1750.

Rivastigmine

This unusual ester (FW$_{free-base}$ = 250.34 g/mol; CAS 123441-03-2), also known as Exelon™, is a brain cholinesterase inhibitor used to treat mild to moderate Alzheimer's disease. *Torpedo californica* acetylcholinesterase is carbamoylated very slowly, whereas the bimolecular rate constant for reaction of rivastigmine with human acetylcholinesterase is much higher. For human butyrylcholinesterase and for *Drosophila melanogaster* acetylcholinesterase, carbamoylation is even more rapid. For all four enzymes, spontaneous reactivation is very slow. The Exelon Patch™ consists of four layers: an outer shell, an inner layer containing rivastigmine (4.6 mg initial dose; 9.5 mg maintenance dose), an adhesive layer, and a peel-off layer protecting the adhesive. **Key Pharmacokinetic Parameters:** *See* Appendix II in Goodman & Gilman's *THE PHARMACOLOGICAL BASIS OF THERAPEUTICS*, 12ᵗʰ Edition (Brunton, Chabner & Knollmann, eds.) McGraw-Hill Medical, New York (2011). (*See also Ladostigil*) **Target(s):** acetylcholine esterase (1-5); cholinesterase, *or* butyrylcholinesterase (2-5). **1.** Amstutz, Enz, Marzi, Boelsterli & Walkinshaw (1990) *Helv. Chim. Acta* **73**, 739. **2.** Darvesh, Walsh, Kumar, *et al.* (2003) *Alzheimer Dis. Assoc. Disord.* **17**, 117. **3.** Luo, Yu, Zhan, *et al.* (2005) *J. Med. Chem.* **48**, 986. **4.** Giacobini (2003) *Neurochem. Res.* **28**, 515. **5.** Bar-On, Millard, Harel, *et al.* (2002) *Biochemistry* **41**, 3555.

RK-682

This cell-impermeable agent (FW = 368.51 g/mol; CAS 154639-24-4) from species of *Streptomyces*, inhibits protein-tyrosine phosphatases (*e.g.*, CD45, IC_{50} = 54 µM, and vaccinia VHR, IC_{50} = 2 µM, *in vitro* (1-3). RK-682 also exerts mammalian cell cycle arrest in the G_1 phase. **1.** Fujii, Kato, Furuse, *et al.* (1995) *Neurosci. Lett.* **187**, 130. **2.** Fujii, Ito, Osada, *et al.* (1995) *Neurosci. Lett.* **187**, 133. **3.** Hamaguchi, Sudo & Osada (1995) *FEBS Lett.* **372**, 54.

RKI-1447

This potent, cell-permeable protein kinase inhibitor (FW = 326.37 g/mol; CAS 1342278-01-6; Solubility: 65 mg/mL DMSO; < 1 mg/mL H_2O), also named *N*-[(3-hydroxyphenyl)methyl]-*N'*-[4-(4-pyridinyl)-2-thiazolyl]urea, targets the Rho-associated kinases ROCK1 (IC_{50} = 14.5 nM) and ROCK2 (IC_{50} = 6.2 nM) by binding the ATP binding site through interactions with its hinge region and DFG motif (1,2). Importantly, RKI-1447 suppresses phosphorylation of ROCK substrates (MLC-2 and MYPT-1) in human cancer cells, but is without effect on AKT, MEK, and S6 kinase at concentrations as high as 10 µM (1). Consistent with Rho's involvement in actin-based motility, RKI-1447 inhibits migration, invasion and anchorage-independent tumor growth of breast cancer cells. RKI-1447 is also a highly effective inhibitor of mammary tumor outgrowth in a transgenic mouse model (1). **1.** Patel, Forinash, Pireddu, *et al.* (2012) *Cancer Res.* **72**, 5025. **2.** Piredd, Forinash, Sun, *et al.* (2012) *Med. Chem. Comm.* **3**, 699.

RMI 14514, *See TOFA*

RMI 12330A, *See MDL 12330A*

RMI 12936, *See 17β-Hydroxy-7α-methylandrost-5-en-3-one*

RNA

Aside from the occasional report that of mRNA or tRNA inhibiting enzyme catalysis by blocking the binding a negatively charged substrate to an oppositely charged active site, there are several far more important and extremely generalizable forms of inhibitory RNA. **Antisense RNA:** These small RNA molecules hybridize with specific messenger RNA molecules to form high-affinity complexes that block translation (1-3). Such a mode of action is possible because mRNA is a single-stranded polynucleotide, often referred to as the "sense" strand, because its translation results in a gene product (protein). The unpaired nucleotides of mRNA must hybidize by forming hydrogen-bonded hybrids with the anticodons of aminoacylated transfer RNA molecules as the ribosome proceeds in translating the mRNA sequence into a defined polypeptide sequence. By increasing their length, one can increase the binding energy associated with antisense RNA interaction with its cognate mRNA, thereby increasing affinity and lowering its K_i value. For example, because tomatoes must be shipped to distant markets before they ripen, and because ripening gas ethylene is required for ripening, transgenic tomatoes have been engineered to carry an artificial

gene within their genome, and transcription of that gene into an antisense RNA that complementary to the mRNA for an enzyme involved in ethylene production. Such tomatoes make <10% of the normal amount of the enzyme, thereby greatly prolonging ripening and potential spoilage. One limitation of antisense RNA is that some are degraded by intracellular nucleases. A second limitation is that sufficient antisense RNA must be present to sequester the mRNA as an inactive hybridization complex with its specific antisense RNA. **Catalytic RNA:** These low-molecular-weight RNA molecules, also known as ribozymes, catalyze hydrolase-type reactions that can be engineered to optimize their action within living cells. RNA-cleaving ribozymes achieve their target specificity from Watson-Crick base-pairing between the ribozyme's specificity-conferring binding-arm sequences as well as other sequences flanking the cleavage site within the target RNA (4). Once bound, the cleavage mechanism involves attack of the 2'-OH lying 5' to the scissile bond in the target, thus destabilizing the target RNA's phosphate backbone. Upon cleavage, the hydrolysis products dissociate from the ribozyme complex and the ribozyme is released and may bind and cleave other targets again. The cleavage event renders the mRNA untranslatable and leads to further degradation of the target by cellular ribonucleases. Liu & Altman (5), for example, converted catalytic RNA subunit (known as M1 RNA) of *Escherichia coli* RNase P to an endoribonuclease that specifically cleaves the mRNA encoding the thymidine kinase (TK) of herpes simplex virus 1 (HSV-1). Covalent attachment to the 3' end of M1 RNA of a sequence complementary to TK mRNA resulted in very efficient cleavage of the target RNA *in vitro*. When cultured mouse cells expressing the novel M1 RNA construct were infected with HSV-1, the levels of both TK mRNA and TK protein were reduced to ~20% that found in mock-treated HSV-1-infected cells. Introduction of a library of active ribozymes into specific cells and subsequent screening for phenotypic changes, allows the rapid identification of gene function (6). By destroying a targetted mRNA or directly disrupting their genes, catalytic RNA and DNA are particularly attractive tools for rational design. For the determination of gene function, ribozyme technology complements another RNA-based tool that is based on libraries of small interfering RNAs. **Small Interference RNA:** This class of inhibitory RNA, known as small interfering RNA (siRNA) or silencing RNA, is double-stranded RNA molecules involved in the RNA interference pathway, wherein expression of a specific gene is greatly, but most often incompletely, suppressed (7-10). This technique for inactivating genes within cell relies on the introduction of double-stranded RNA (dsRNA), because animal cells selectively destroy dsRNA. Therefore, if a segment of dsRNA has the same sequence as part of a gene, its introduction into a cell will trigger mRNA destruction, thereby silencing gene expression. This inhibitory mechanism may have evolved as a defense against retroviruses, including HIV, in which genetic information is stored as double-stranded RNA. Small interfering RNAs typically consist of a short (most often 21-nucleotide) double-strand of RNA, with two-nucleotide 3'-overhangs located on either end. Each strand also has a 5'-phosphomonoester group and a 3'-hydroxyl group, resulting from processing by the enzyme, known as dicer, which either converts long double-stranded RNAs or small hairpin RNA's into siRNAs. Because they are so small, siRNA enjoy the advantages that they can be rapidly transcribed from their genes, and there is no requirement for translation into a protein product for inhibition to take place. When introduced into cells by transfection, siRNAs specifically block the expression of a target gene, with targeting based on sequence complementarity with appropriately tailored siRNA. By using two or three different siRNAs, each designed for the very same target, their combined effect is often >80-90% reduction in target gene expression at 48 hours post-transfection, when the latter is carried out at final concentrations of ~5 nM of both siRNAs. **Targets:** Because antisense RNA, catalytic RNA, and small interference RNA are most effective when optimized for a particular enzyme target in a given organism and cell type, and because these techniques do not target conserved binding sites on enzyme targets *per se*, the reader is advised to search PubMed for details on specific applications of RNA inhibitors for a desired enzyme target. Commercial sources (*e.g.*, Ambion.com, Dharmacon.com, GenScript.com, Origene.com, Scbt.com, siRNA.com, *etc.*) also rely on their own well-tested proprietary approaches, including statistical and clustering approaches, for designing highly effective inhibitory RNAi and ribozymes. Finally, one should also consider the possibility of employing RNA therapeutics to increase the therapeutic efficacy of active-site-directed enzyme inhibitors (*i.e.*, reversible and irreversible inhibitors, transition-state mimics, *etc.*). **1.** Green, Pines & Inouye (1986) *Annu. Rev. Biochem.* **55**, 569. **2.** Toulmé & Hélène (1988) *Gene.* **72**, 51. **3.** Phillips (1999) *Meth. Enzymol. vols.* **313** (pp. 580) and 314 (pp. 647). (two entire monographs focusing on general methods of design, preparation and delivery of antisense RNA) **4.** Ussman

& Blatt (2000) *J. Clin. Invest.* **106**, 1197. **5.** Liu & Altman (1995) *Genes Dev.* **9**, 471. **6.** Akashi, Matsumota & Taira (2005) *Nature Rev. Molec. Cell Biol.* **6**, 413. **7.** Hutvágner & Zamore (2002) *Curr Opin Genet Dev.* **12**, 225. **8.** Micura (2002) *Angew. Chem. Int. Ed. Engl.* **41**, 2265. **9.** Arenz & Schepers (2003) *Naturwissenschaften* **90**, 345. **10.** Engelke & Rossi **(2005)** *Meth. Enzymol. vol.* **392**, pp. 496 (entire monograph focusing on the design, preparation and delivery interference RNA).

Ro, *See Rhizobitoxine*

Ro 4-4602, *See Benserazide*

Ro 8-4304

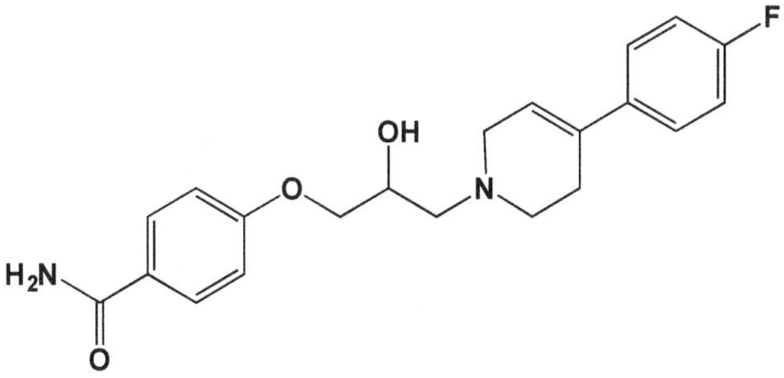

This voltage-independent, non-competitive NMDA receptor antagonist (FW$_{\text{free-base}}$ = 370.42 g/mol; CAS 195988-65-9) exhibits a state-dependent mode of action similar to that described for ifenprodil, when tested in cultured rat cortical neurons. Its apparent affinity for the NMDA receptor increases in a manner that depends on NMDA concentration, inhibiting 10 and 100 μM NMDA responses with respective IC$_{50}$ values of 2.3 and 0.36 μM. Currents elicited by 1 μM NMDA were slightly potentiated in the presence of 10 μM Ro 8-4304, and Ro 8-4304 binding slowed the rate of glutamate dissociation from NMDA receptors. Such behavior were suggests that Ro 8-4304 exhibits a 14 and 23 times higher affinity for the activated and desensitized states of the NMDA receptor, respectively, relative to the agonist-unbound resting state. Ro 8-4304 binding also results in a 3–4 fold increase in receptor affinity for glutamate site agonists. **1.** Kew, Trube & Kemp (1998) *Brit. J. Pharmacol.* **123**, 463.

Ro-8-6837, *See Methiothepin*

RO9021

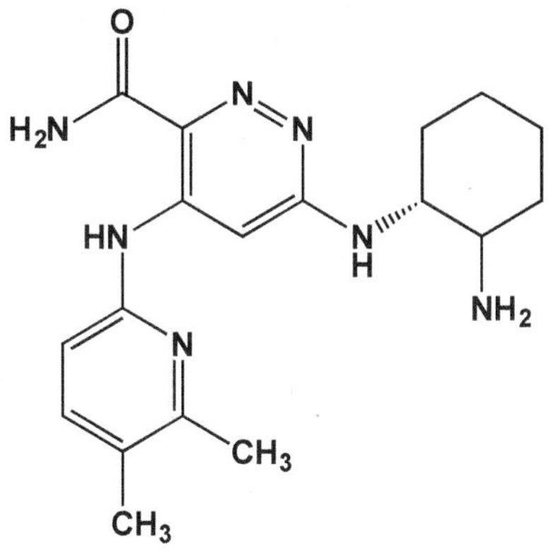

This novel ATP-competitive protein tyrosine kinase inhibitor (FW = 355.45 g/mol), also named systematically as 6-((1*R*,2*S*)-2-aminocyclohexylamino)-4-(5,6-dimethylpyridin-2-ylamino)pyridazine-3-carboxylic acid amide, targets <u>S</u>pleen <u>ty</u>rosine <u>k</u>inase, or SYK (*K*$_i$ = 5.6 nM), a key integrator of intracellular signals triggered by activated immunoreceptors, including B-cell and Fc receptors that are important for lymphoid cell development and function. Significantly, RO9021 blocks osteoclastogenesis from mouse bone marrow macrophages *in vitro*. Toll-like Receptor-9 (TLR9) signaling in human B-cells is inhibited by RO9021, decreasing levels of plasmablasts, immunoglobulin (Ig) M and IgG upon B-cell differentiation. RO9021 also potently inhibits type I interferon production by human plasmacytoid dendritic cells (pDC) upon TLR9-specific activation. RO9021 does not inhibit TLR4- or JAK-STAT-mediated signaling. **1.** Liao, Hsu, Kim, *et al.* (2013) *Arthritis Res. Ther.* **15**, R146.

Ro 13-9904, *See Ceftriaxone*

Ro 15-4513

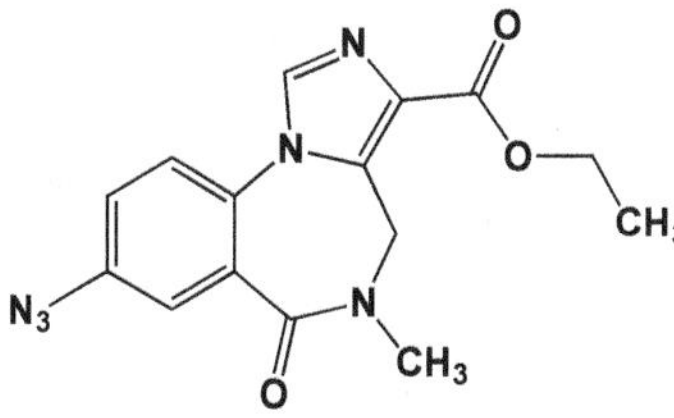

This benzodiazepine-based ethanol antagonist (FW = 326.31 g/mol; CAS 91917-65-6), also known as ethyl 8-azido-6-dihydro-5-methyl-6-oxo-4*H*-imidazo[1,5-*a*][1,4]benzodiazepine-3-carboxylate, is a partial, inverse agonist of benzodiazepine receptors and a potent antagonist of ethanol-stimulated chloride uptake into brain vesicles (1,2). However, Ro 15-4513 fails to antagonize either pentobarbital- or muscimol-stimulated ^{36}Cl$^-$ uptake. In addition, horse liver alcohol dehydrogenase is inactivated after forming a binary complex with LADH (*K*$_d$ = 8.6 mM at pH 7.0). **1.** Langeland & McKinley-McKee (1994) *Arch. Biochem. Biophys.* **308**, 367. **2.** Suzdak, Glowa, Crawley, *et al.* (1986) *Science* **234**, 1243.

Ro 18-0647, *See Orlistat*

Ro 19-6327, *See Lazabemide*

Ro 20-1724, *See 4-(3-Butoxy-4-methoxybenzyl)-2-imidazolidinone*

Ro 24-5913, *See Cinalukast*

Ro 25-6981

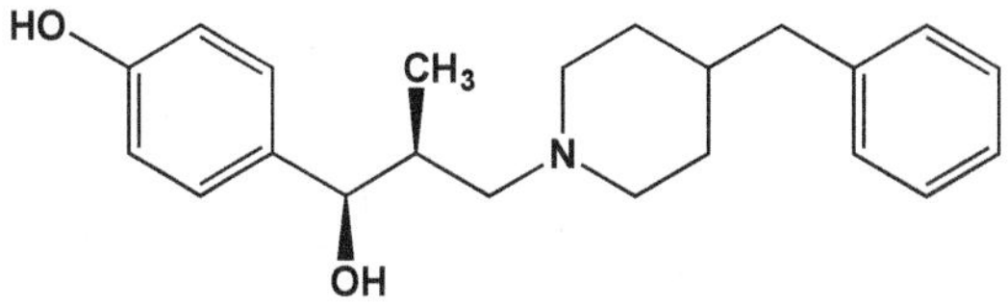

This NR2B subunit-selective NMDA receptor antagonist (FW$_{\text{free-base}}$ = 339.48 g/mol; FW$_{\text{maleate-salt}}$ = 455.55 g/mol; CAS 1312991-76-6; Soluble to 10 mM in water with gentle warming and to 100 mM in DMSO), systematically named (α*R*,β*S*)-α-(4-hydroxyphenyl)-β-methyl-4-(phenyl-methyl)-1-piperidinepropanol maleate, is a potent and selective activity-dependent antagonist for the NR1C/NR2B-type NMDA receptor (IC$_{50}$ = 9 nM), with much weaker action against the NR1C/NR2A-type NMDA receptor (IC$_{50}$ = 52 μM), when both were cloned and expressed in *Xenopus* oocytes. Ro 25-6981 exerts neuroprotectant effects both *in vitro* and *in vivo*, albeit little or no protection against kainate toxicity (exposure to 500 μM for 20 h) and only weak activity in blocking Na$^+$ and Ca^{2+} channels, activated by exposure of cortical neurons to veratridine (10 μM) and potassium (50 mM), respectively (1). Ro-25-6981 shares structural determinants with ifenprodil and other modulators in the NR2B subunit (2). Ro 25-6981 (10 mg/kg i.p. in rats) significantly potentiated the nicotine-induced dopamine (DA) release, although it has no effect on DA release when given alone, suggesting that, compared with other subunits of the NMDA receptor, the NR2B subunit might play a different role in the reinforcing effects of nicotine (3). **1.** Fischer, Mutel, Trube, *et al.* (1997) *J. Pharmacol. Exp. Ther.* **283**, 1285. **2.** Lynch, Shim, Seifert, *et al.* (2000) *Eur. J. Pharmacol.* **416**, 185. **3.** Kosowski & Liljequist (2004) *J. Pharmacol. Exp. Ther.* **311**, 560.

Ro 31-8220

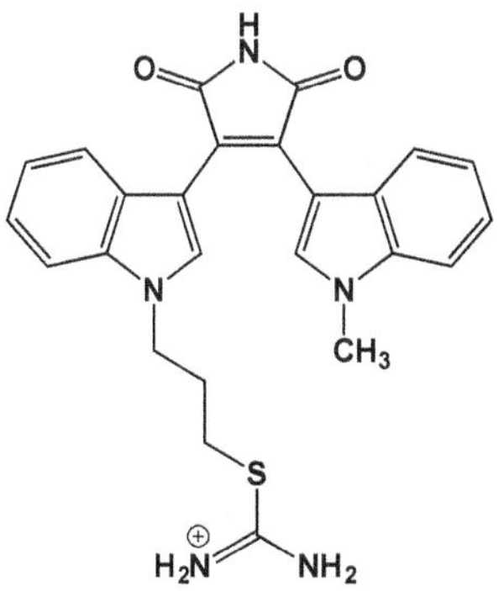

This maleimide derivative ($FW_{free-base}$ = 457.56 g/mol; CAS 125314-64-9), also known as bisindolylmaleimidine IX, Ro 318220, and 2-(1-[3-(amidinothio)propyl]-1*H*-indol-3-yl)-3-(1-methylindo-3-yl)maleimide, is a strong inhibitor of protein kinase C, exhibiting some preference for certain isozymes (*e.g.*, the IC_{50} value for PKCα is 5 nM and the value for PKCγ is 27 nM). In addition, Ro 31-8220inhibits the growth factor-stimulated expression of MAP kinase phosphatase-1 and c-Fos, but strongly stimulates c-Jun expression and also activates Jun N-terminal kinase. Ro 31-8220 is often supplied as the methanesulfonate salt. **Target(s):** Akt-3 serine/threonine protein kinase (1); glycogen-synthase kinase III, *or* [tau protein] kinase (2,3); [mitogen-activated protein kinase]-activated protein kinase-1 (3,4); mitogen- and stress-activated protein kinase (3); Na^+-channels, voltage-dependent (5); non-specific serine/threonine protein kinase (1,3,4); protein kinase C (3,6-8); and ribosomal protein S6 kinase, p70 (3,4). **1**. Masure, Haefner, Wesselink, *et al.* (1999) *Eur. J. Biochem.* **265**, 353. **2**. Hers, Tavare & Denton (1999) *FEBS Lett.* **460**, 433. **3**. Davies, Reddy, Caivano & Cohen (2000) *Biochem. J.* **351**, 95. **4**. Alessi (1997) *FEBS Lett.* **402**, 121. **5**. Lingameneni, Vysotskaya, Duch & Hemmings, Jr. (2000) *FEBS Lett.* **473**, 265. **6**. Beltman, McCormick & Cook (1996) *J. Biol. Chem.* **271**, 27018. **7**. Wilkinson, Parker & Nixon (1993) *Biochem. J.* **294**, 335. **8**. Chen & Exton (2004) *J. Biol. Chem.* **279**, 22076.

Ro 31-8472

This cilazapril derivative ($FW_{free-acid}$ = 405.45 g/mol; CAS 21802-37-9), also known as caerulomycin A and systematically as 9-[1-carboxy-3-(4-hydroxy-phenyl)propylamino]octahydro-10-oxo-6*H*-pyridazo[1,2-*a*][1,2]diazepine-1-carboxylic acid, inhibits peptidyl-dipeptidase A, *or* angiotensin-I converting enzyme, IC_{50} = 0.6 nM (1-3). **1**. Corvol, Williams & Soubrier (1995) *Meth. Enzymol.* **248**, 283. **2**. Bevilacqua, Vago, Rogolino, *et al.* (1996) *J. Cardiovasc. Pharmacol.* **28**, 494. **3**. Perich, Jackson & Johnston (1994) *Eur. J. Pharmacol.* **266**, 201.

Ro 31-8624

This pseudopeptide bond-containing peptide analogue ($FW_{free-base}$ = 620.77 g/mol) potently inhibits HIV-1 protease, K_i = 4.5 nM. **1**. Wilson, Phylip, Mills, *et al.* (1997) *Biochim. Biophys. Acta* **1339**, 113.

Ro 31-8875

This pseudopeptide bond-containing peptide analogue ($FW_{free-base}$ = 649.83 g/mol) inhibits HIV-1 protease, K_i = 8.5 nM. **1**. Wilson, Phylip, Mills, *et al.* (1997) *Biochim. Biophys. Acta* **1339**, 113.

Ro 31-8959, *See* Saquinavir

Ro 31-9790

This peptidomimetic hydroxamate (FW = 315.41 g/mol) inhibits a number of matrix metalloproteinases, including gelatinase A, *or* matrix metalloproteinase 2 (1,2); gelatinase B, *or* matrix metalloproteinase 9 (2); interstitial collagenases, *or* matrix metalloproteinase 1, K_i = 11 nM (2,3); membrane-type matrix metalloproteinase 1, *or* matrix metalloproteinase 14 (2); stromelysin 1, *or* matrix metalloproteinase 3, K_i = 616 nM (3). **1**. Steinmann-Niggli, Lukes & Marti (1997) *J. Amer. Soc. Nephrol.* **8**, 395. **2**. Yamamoto, Tsujishita, Hori, *et al.* (1998) *J. Med. Chem.* **41**, 1209. **3**. Knauper, Patterson, Gomis-Rüth, *et al.* (2001) *Eur. J. Biochem.* **268**, 1888.

Ro 32-0432

This photosensitive bisindolylmaleimide ($FW_{free-base}$ = 452.56 g/mol; CAS 151342-35-7), also known as bisindolylmaleimide XI, is a potent protein kinase C inhibitor, exhibiting the folowing a selectivity: PKCα > PKCβI > PKCβII > PKCγ > PKCε, with respective IC_{50} values of 9, 28, 31, 37, and 108 nM (1-4). Other kinases (*e.g.*, G-protein-coupled receptor kinases, GRK2, GRK5, and GRK63) are also inhibited, albeit with significantly higher IC_{50} values (1). **1**. Aiyar, Disa, Dang, *et al.* (2000) *Eur. J. Pharmacol.* **403**, 1. **2**. Wilkinson, Parker & Nixon (1993) *Biochem. J.* **294**, 335. **3**. Saraiva, Fresco, Pinto & Goncalves (2003) *J. Enzyme Inhib. Med. Chem.* **18**, 475. **4**. Ranganathan, Liu, Migliorini *et al.* (2004) *J. Biol. Chem.* **279**, 40536.

Ro 32-3555

This matrix metalloproteinase inhibitor (FW = 436.55 g/mol; CAS 190648-49-8), known systematically as 3(*R*)-(cyclopentylmethyl)-2(*R*)-[(3,4,4-trimethyl-2,5-dioxo-1-imidazolidinyl)methyl]-4-oxo-4-piperidinobutyro-hydroxamate, is an orally bioavailable inhibitor of interstitial collagenases (K_i = 3 nM), neutrophil collagenases (K_i = 4 nM), collagenase-3 (K_i = 3 nM), gelatinase A (K_i = 154 nM), gelatinase B (K_i = 59 nM), and stromelysin 1 (K_i = 527 nM) (1,2). **1**. Leung, Abbenante & Fairlie (2000) *J. Med. Chem.* **43**, 305. **2**. Lewis, Bishop, Bottomley, *et al.* (1997) *Brit. J. Pharmacol.* **121**, 540.

Ro 40-4388

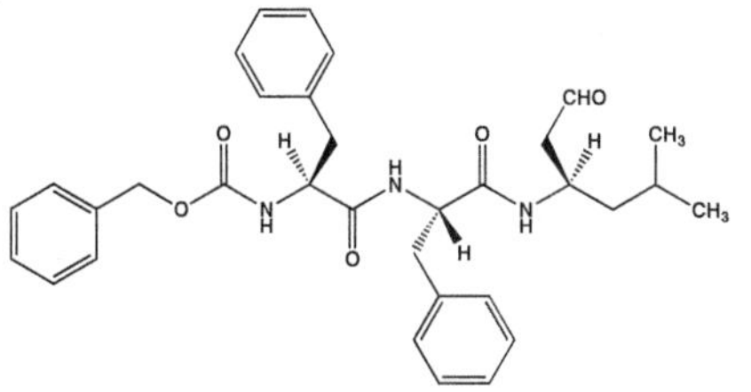

This peptide analogue (FW = 557.69 g/mol; CAS 90042-57-2) inhibits cathepsin D (1); cathepsin E (1); gastricsin, weakly (1); pepsin, weakly (1); plasmepsin I, K_i = 9 nM; and plasmepsin II, K_i = 700 nM (1-4). **1.** Tyas, Gluzman, Moon, *et al.* (1999) *FEBS Lett.* **454**, 210. **2.** Moon, Tyas, Certa, *et al.* (1997) *Eur. J. Biochem.* **244**, 552. **3.** Moon, Bur, Loetscher, *et al.* (1998) *Adv. Exp. Med. Biol.* **436**, 397. **4.** Tyas, Moon, Loetscher, *et al.* (1998) *Adv. Exp. Med. Biol.* **436**, 407.

Ro 40-5576

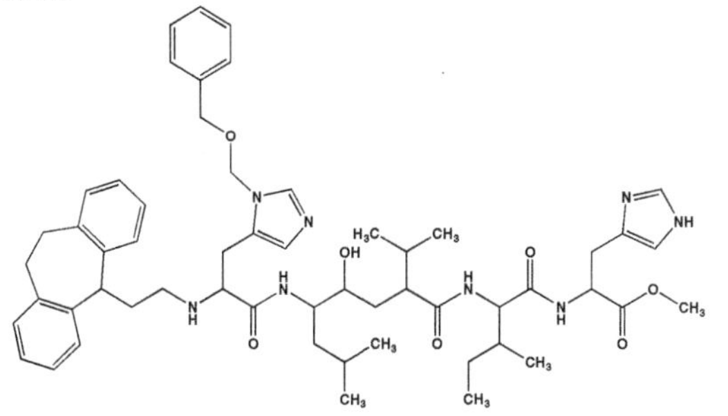

This peptide analogue (FW = 973.27 g/mol) inhibits both plasmepsin I and II, with respective K_i values of 8 and 250 nM, with substantially weaker inhibition of cathepsin, cathepsin E, gastricsin, and pepsin. **1.** Tyas, Gluzman, Moon, *et al.* (1999) *FEBS Lett.* **454**, 210. **2.** Moon, Tyas, Certa, *et al.* (1997) *Eur. J. Biochem.* **244**, 552. **3.** Berry (2000) *Curr. Opin. Drug Discov. Devel.* **3**, 624. **4.** Moon, Bur, Loetscher, *et al.* (1998) *Adv. Exp. Med. Biol.* **436**, 397. **5.** Tyas, Moon, Loetscher, *et al.* (1998) *Adv. Exp. Med. Biol.* **436**, 407.

Ro 40-5967, *See Mibefradil*

Ro 40-7592, *See Tolcapone*

Ro 41-0960

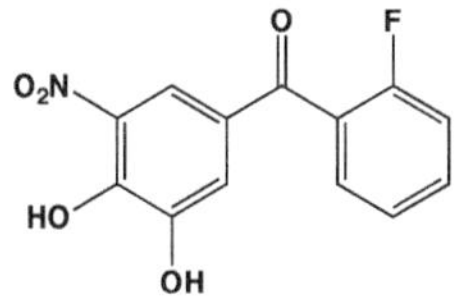

This photosensitive benzophenone (FW = 277.21 g/mol; CAS 125628-97-9), also known as 2'-fluoro-3,4-dihydroxy-5-nitrobenzophenone, is an orally active inhibitor of catechol *O*-methyltransferase (1-4). This inhibitor is easily photodegraded. **1.** Borgulya, Bruderer, Bernauer, Zurcher & Daprada (1989) *Helv. Chim. Acta* **72**, 952. **2.** Miller, Shukitt-Hale, Villalobos-Molina, *et al.* (1997) *Clin. Neuropharmacol.* **20**, 55. **3.** van Duursen, Sanderson, de Jong, Kraaij & van den Berg (2004) *Toxicol. Sci.* **81**, 316. **4.** Percy, Kaye, Lambert, *et al.* (1999) *Brit. J. Pharmacol.* **128**, 774.

Ro 41-1049

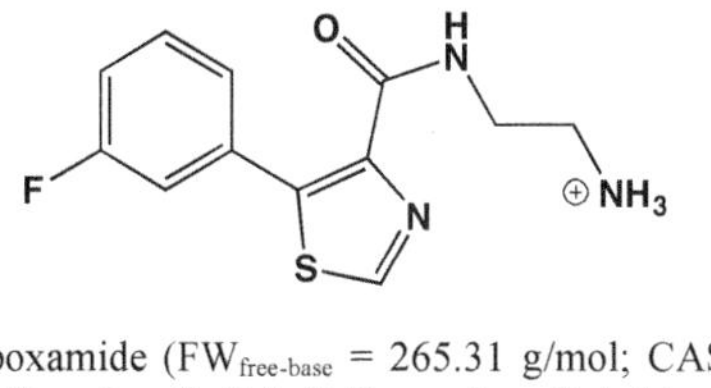

This thiazolecarboxamide (FW$_{free-base}$ = 265.31 g/mol; CAS 127500-84-9), also known as *N*-(2-aminoethyl)-5-(3-fluorophenyl)-4-thiazolecarboxamide, is a potent and selective reversible inhibitor of monoamine oxidase A (1,2). **1.** Da Prada, Kettler, Keller, *et al.* (1990) *J. Neural Transm.* Suppl. **29**, 279. **2.** Fernandes & Soares-da-Silva (1990) *J. Pharmacol. Exp. Ther.* **255**, 1309.

Ro 42-5892, *See Remikiren*

Ro 48-8071

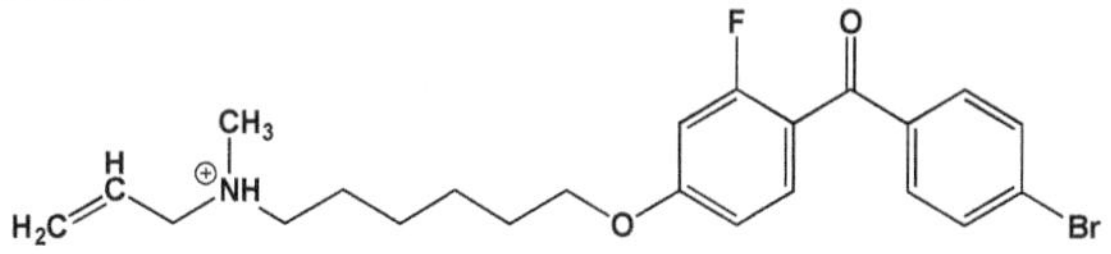

This orally active benzophenone (FW$_{free-base}$ = 448.38 g/mol; CAS 161582-11-2; typically supplied as the 1:1 fumarate salt) inhibits human lanosterol synthase, *or* 2,3-oxidosqualene:lanosterol cyclase, IC$_{50}$ ≈ 6.5 nM (1-5) and squalene:hopene cyclase (6-9). Oxidosqualene cyclase (OSC) catalyzes the cyclization of monooxidosqualene to form lanosterol in the cholesterol synthetic pathway. Ro 48-8071 is an inhibitor of oxidosqualene cyclase (OSC) that has LDL cholesterol-lowering activity similar to the HMG-CoA inhibitor simvastatin. It inhibits human liver microsomal OSC (IC$_{50}$ = 6.5 nM) and HepG2 cells (IC$_{50}$ = 1.5 nM). Ro 48-8071 is readily photodegraded. **1.** Morand, Aebi, Dehmlow, *et al.* (1997) *J. Lipid Res.* **38**, 373. **2.** Dehmlow, Aebi, Jolidon, *et al.* (2003) *J. Med. Chem.* **46**, 3354. **3.** Wu, Huang, Ko, *et al.* (2004) *Arch. Biochem. Biophys.* **421**, 42. **4.** Thoma, Schulz-Gasch, D'Arcy, *et al.* (2004) *Nature* **432**, 118. **5.** Ruf, Muller, D'Arcy, *et al.* (2004) *Biochem. Biophys. Res. Commun.* **315**, 247. **6.** Dang, Abe, Zheng & Prestwich (1999) *Chem. Biol.* **6**, 333. **7.** Lenhart, Weihofen, Pleschke & Schulz (2002) *Chem. Biol.* **9**, 639. **8.** Cravotto, Balliano, Tagliapietra, Palmisano & Penoni (2004) *Eur. J. Med. Chem.* **39**, 917. **9.** Zheng, Abe & Prestwich (1998) *Biochemistry* **37**, 5981.

Ro 61-8048, *See 3,4-Dimethoxy-[N-4-(nitrophenyl)thiazol-2-yl]benzenesulfonamide*

Ro 67-0565, *See Avosentan*

Ro-1130830, *See RS-130830*

Ro 4929097

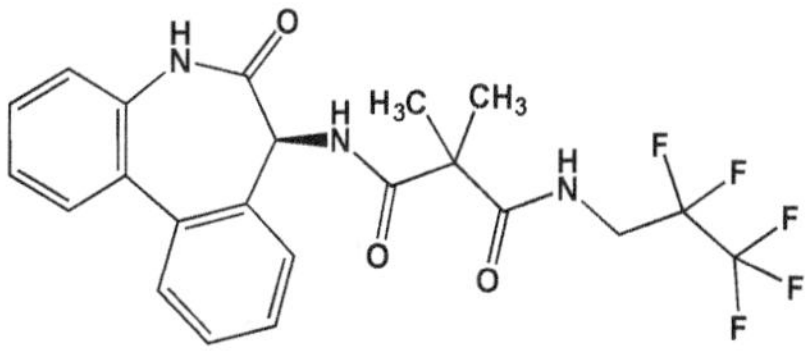

This selective protease inhibitor (FW = 469.4 g/mol; CAS 847925-91-1), also known as N^1-[(7S)-6,7-dihydro-6-oxo-5H-dibenz[b,d]azepin-7-yl]-2,2-dimethyl-N^3-(2,2,3,3,3-pentafluoropropyl)propanediamide, targets γ-secretase (GS), with IC$_{50}$ = 4 nM and decreases the amount of Aβ peptides secreted into the culture medium in HEK293 cells, with EC$_{50}$ = 14 nM. **1.** Li, *et al.* (2000) *Proc. Natl. Acad Sci. USA* **97**, 6138.

RO5190591, *See Danoprevir*

RO5424802, *See Alectinib*

Rociletinib

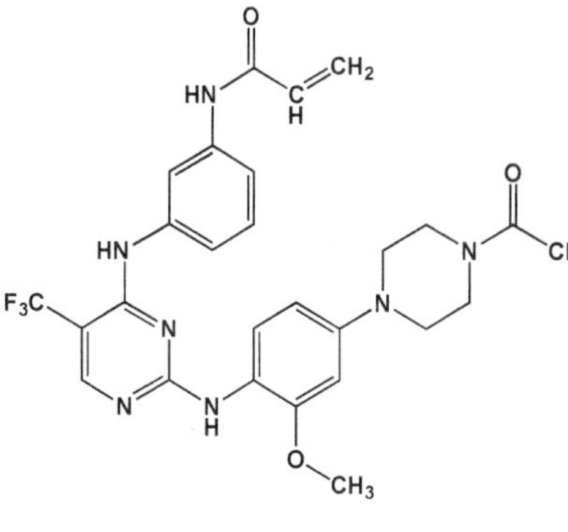

This orally available mutant-selective EGFR kinase inhibitor (FW = 555.56 g/mol; CAS 1374640-70-6; Solubility: 100 mg/mL DMSO; <1 mg/mL H$_2$O), named CO-1686, CNX-419, AVL-301, and *N*-[3-[[2-[[4-(4-acetyl-1-piperazinyl)-2-methoxyphenyl]amino]-5-(trifluoromethyl)-4-pyrimidinyl]amino]phenyl]-2-propenamide, irreversibly targets EGFR$^{L858R/T790M}$ (K_i = 21.5 nM) and EGFR^{WT} (K_i = 21.5 nM) respectively. After inhibitor binding, the enzyme is covalently modified at the conserved Cys797 within the ATP-binding pocket of EGFR's kinase domain. Potency is expressed by the ratio (k_{inact}/K_i), where k_{inact} is the inactivation rate constant and K_i is the binding constant. For EGFR$^{L858R/T790M}$ kinase, k_{inact}/K_i equals 2.41×10^5 M^{-1}s^{-1} and is approximately 22x more selective than with WT EGFR, for which k_{inact}/K_i equals 1.12×10^4 M^{-1}s^{-1}. For comparison, erlotinib, a first-generation tyrosine kinase inhibitor, potently inhibits EGFR^{WT} ($K_i = 0.40$ nM), but with little activity against EGFR$^{L858R/T790M}$ ($K_i = 98.0$ μM). CO-1686 also demonstrates a favorable selectivity profile when profiled against 434 kinases. Twenty-three targets (representing 14 different kinases) were identified to be inhibited more than 50% at 0.1 μM CO-1686. EGFR Δ19-, T790M-, L858R/T790M-, and L858R-mutant kinases demonstrate the highest degree of inhibition, indicating the specificity of CO-1686. Other kinase targets are inhibited at lower potency, including focal adhesion kinase (FAK), CHK2, ERBB4, and Janus-activated kinase 3 (JAK3). **1.** Walter, Sjin, Haringsma, *et al.* (2013) *Cancer Discov.* 3, 1304.

Rofecoxib

This once widely used anti-inflammatory agent (FW = 314.36 g/mol; CAS 162011-90-7), also known as Vioxx™ and MK-0966 and systematically as 4-(4'-methylsulfonylphenyl)-3-phenyl-2-(5H)-furanone, is a potent, orally active cyclooxygenase-2 inhibitor (1,2). No significant inhibition of cyclooxygenase-1 is observed, even at oral doses of 1000 mg (1,2). On September 30, 2004, the FDA issued a Public Health Advisory on Vioxx, announcing that Merck was withdrawing the drug after the safety monitoring board overseeing a long-term study of this drug recommended identified the increased risk of serious cardiovascular events, including heart attacks and strokes, in patients taking Vioxx compared to those receiving placebo. This unfortunate episode in modern drug discovery teaches the risks of assuming a particular enzyme only has a single major role in human metabolism and that extremely high affinity drugs can be safely developed absent a detailed understanding of the diverse and often vital roles of the target enzyme in different tissues. **Likely Effects on Aterial Elasticity:** The actions of orally administered [^{14}C]-rofecoxib and two other COX-2 inhibitors, [^{14}C]-celecoxib and [^{14}C]-apricoxib, were assessed in rats by whole-body autoradioluminography and quantitative determination of the tissue concentrations (3). These experiments showed that considerable radioactivity is retained by and accumulated in the thoracic aorta of rats after oral administration of [^{14}C]-rofecoxib, but not [^{14}C]-celecoxib or [^{14}C]-apricoxib. Acid, organic solvent, and proteolytic enzyme treatments of aorta retaining high levels of radioactivity from [^{14}C]-rofecoxib demonstrated that most of the radioactivity is covalently bound to elastin (3). In agreement with this result, the radioactivity was found to be highly localized on the elastic fibers in the aorta by microautoradiography. The retention of radioactivity on the elastic fibers was also observed in the aortic arch and the coronary artery. These findings indicate that rofecoxib and/or its metabolite(s) are covalently bound to elastin in the arteries. Subsequent work demonstrated that rofecoxib, but not other COX-2 inhibitors, is capable of covalently binding to the aldehyde group of allysine in human elastin (4). These data are consistent with the suggestion of modified arterial elasticity leading to an increased risk of cardiovascular impairment after long-term treatment with rofecoxib (3). **1.** Doret, Mellier, Benchaib, *et al.* (2002) *BJOG* 109, 983. **2.** Prasit, Wang, Brideau, *et al.* (1999) *Bioorgan. Med. Chem. Lett.* 9, 1773. **3.** Oitate, Hirota, Koyama, *et al.* (2006) *Drug Metab Dispos.* 34, 1417. **4.** Oitate, Hirota, Murai, Miura & Ikeda (2007) *Drug Metab. Dispos.* 35, 1846.

Rogletimide

This aminoglutethimide analogue (FW = 218.26 g/mol; CAS 121840-95-7), also known as pyridoglutethimide and 3-ethyl-3-(4-pyridyl)piperidine-2,6-dione, is a potent aromatase inhibitor. Both enantiomers inhibit, with IC$_{50}$

values of 24.58 and 24.43 μM for the (+)-R- and (−)-S-stereoisomers, respectively (1). **Target(s):** aromatase, *or* CYP19 (1,2); cholesterol monooxygenase, side-chain-cleaving, *or* CYP11A1, weakly inhibited (3). **1.** Ogbunude & Aboul-Enein (1994) *Chirality* 6, 623. **2.** Kitawaki, Yamamoto, Urabe, *et al.* (1990) *Acta Endocrinol. (Copenhagen)* 122, 592. **3.** Ahmed (2000) *Biochem. Biophys. Res. Commun.* 274, 821.

Rolapitant

This selective CNS-penetrant NK$_1$R antagonist (FW$_{free-base}$ = 500.48 g/mol; FW$_{HCl-salt}$ = 536.94 g/mol; CAS 552292-08-7 (free base); 914462-92-3 (HCl monohydrate); Soluble in DMSO, not in water), also known as Varubi$^®$, SCH619734, SCH-619734, SCH 619734, and (5S,8S)-8-(((R)-1-(3,5-bis(trifluoromethyl)phenyl)ethoxy)methyl)-8-phenyl-1,7-diazaspiro[4.5] decan-2-one, targets neurokinin (tachykinin) NK$_1$ receptor and shows behavioral effects in animals models of emesis. *In vitro* studies suggest that rolapitant has a high affinity for the human NK1 receptor ($K_i = 0.66$ nM), showing a high selectivity of >1000-fold over the human NK2 and NK3 subtypes. Rolapitant is a functionally competitive antagonist, as measured by calcium efflux, with a calculated K_b of 0.17 nM. Varubi is an FDA-approved medication that is specifically indicated for use in combination with other antiemetic agents in adults to prevent the delayed nausea and vomiting associated with initial and repeat courses of emetogenic cancer chemotherapy. **See also** Aprepitant **1.** Melton, Nielsen, Tucker, Klein & Gan (2014) *Anesthesiol. Clin.* 32, 505. **2.** Morrow, Navari & Rugo (2014) *Clin. Adv. Hematol. Oncol.* 12, 1. **3.** Navari (2013) *Drugs* 73, 249. **4.** Duffy, Morgan, Naylor, *et al.* (2012) *Pharmacol. Biochem. Behav.* 102, 95. **5.** Gan, Gu, Singla, *et al.* (2011) *Anesth Analg.* 112, 804.

Rolipram

This potent cell-permeable 3',5'-cGMP phosphodiesterase inhibitor and antidepressant (FW = 275.35g/mol; CASs = 61413-54-5, 85416-74-6; Solubility: 55 mg/mL DMSO; <1 mg/mL H$_2$O), also known as ZK 62711, SB 95952 and 4-[3-(cyclopentyloxy)-4-methoxyphenyl]-2-pyrrolidinone, targets the Type-4 phosphodiesterase isoform (PDE4), IC$_{50}$ = 1 μM (1-5). Rolipram also promotes neuronal survival (6) and exerts antipsychotic effects in mice (7). **Target(s):** reduces inflammation by suppressing leukocyte function, blocking C5a-stimulated leukotrienc C$_4$ synthesis in eosinophils, IC$_{50}$ = 200 nM (8); inhibits lipopolysaccharide-induced TNF synthesis in monocytes, IC$_{50}$ = 360 nM (9); prevents bone loss in ovariectomized rats (10). **1.** Thomas, Francis & Corbin (1990) *J. Biol. Chem.* 265, 14964. **2.** Manganiello, Degerman & Elks (1988) *Meth. Enzymol.* 159, 504. **3.** Wachtel (1983) *Neuropharmacology* 22, 267. **4.** Schneider, Schmiechen, Brezinski & Seidler (1986) *Eur. J. Pharmacol.* 127, 105. **5.** Sudo, Tachibana, Toga, *et al.* (2000) *Biochem. Pharmacol.* 59, 347. **6.** Sasaki, Kitagawa, Omura-Matsuoka *et al.* (2007) *Stroke* 38, 1597. **7.** Kanes, Tokarczyk, Siegel, *et al.* (2007) *Neuroscience* 144, 239. **8.** Tenor, Hatzelmann, Church, Schudt & Shute (1996) *Brit. J. Pharmacol.* 118, 1727. **9.** Souness, Griffin, Maslen, *et al.* (1996) *Brit. J. Pharmacol.* 118, 646. **10.** Yao, Tian, Chen, *et al.* (2007) *J. Musculoskelet. Neuronal Interact.* 7, 119.

Romidepsin

This depsipeptide pro-drug (FW = 540.70 g/mol; CAS 128517-07-7); Solubility: 10 mg/mL DMSO; <1 mg/mL H_2O), also known as FK228, FR901228, NSC 630176, and systematically as cyclo[(2Z)-2-amino-2-butenoyl-L-valyl-(3S,4E)-3-hydroxy-7-mercapto-4-heptenoyl-D-valyl-D-cysteinyl], cyclic (3→5)-disulfide, forms a histone deacetylase inhibitor that targets HDAC1 and HDAC2 with IC_{50} values of 36 nM and 47 nM, respectively. **Mechanism of Action**: The disulfide-containing pro-drug is reduced within the cytoplasm to release a zinc-binding thiol that combines with a catalytically essential zinc within histone deacetylase. **1.** Furumai, et al. (2002) *Cancer Res.* **62**, 4916. **2.** Sandor, et al. (2000) *Brit. J. Cancer* **83**, 817. **3.** Blagosklonny, et al. (2002) *Mol. Cancer Ther.* **1**, 937. **4.** Kwon, et al. (2002) *Int. J. Cancer* **97**, 290. **5.** Sasakawa, et al. (2002) *Biochem. Pharmacol.* **64**, 1079. **6.** Aron, et al. (2003) *Blood* **102**, 652.

Rondomycin, See Methacycline

Roscovitine

This diaminopurine derivative (FW = 354.45 g/mol; CAS 186692-46-6; Solubility: 30 mg/mL DMSO; <1 mg/mL H_2O; Formulation: Dissolved in methanol or DMSO, then dilute in 10% Tween 80, 20% *N,N*-dimethylacetamide, and 70% PEG-400), also known as Seliciclib, CYC202, R-roscovitine, and systematically as (R)-2-(6-(benzylamino)-9-isopropyl-9H-purin-2-ylamino)butan-1-ol, targets Cdc2/cyclin B, CDK2/cyclin A, CDK2/cyclin E and CDK5/p53 with IC_{50} values of 0.65 μM, 0.7 μM, 0.7 μM and 0.16 μM, respectively. **Cyclin Target Selectivity:** Cdk1 (weak, if any), Cdk2 (++++), Cdk3 (weak, if any), Cdk4 (weak, if any), Cdk5 (weak, if any), Cdk6 (weak, if any), Cdk7 (++++), Cdk8 (weak, if any), Cdk9 (++++), Cdk10 (weak, if any). **Target(s):** cAMP-dependent protein kinase, *or* protein kinase A, weakly inhibited (1); cyclin-dependent kinases: cdk2/cyclin A, IC_{50} = 0.25-0.7 μM; cdk1, IC_{50} = 0.65 μM; cdk5, IC_{50} = 0.2 μM; also cdk7 and cdk9 (1-17); casein kinase 1δ, IC_{50} = 17 μM (1); dual-specificity kinase, *or* dual-specificity tyrosine-phosphorylated and regulated kinase 1A, *or* DYRK1A, IC_{50} = 3.1 μM (1); mitogen-activated protein kinase (1); non-specific serine/threonine protein kinase (1); pyridoxal kinase (17,18); [RNA-polymerase]-subunit kinase, *or* CTD kinase (9); and [tau protein] kinase, *or* [glycogen synthase] kinase-3 (7,8,10). **1.** Bain, McLauchlan, Elliott & Cohen (2003) *Biochem. J.* **371**, 199. **2.** Meijer & Kim (1997) *Meth. Enzymol.* **283**, 113. **3.** Nikolic & Tsai (2000) *Meth. Enzymol.* **325**, 200. **4.** De Azevedo, Leclerc, Meijer, et al. (1997) *Eur. J. Biochem.* **243**, 518. **5.** Meijer, Borgne, Mulner, et al. (1997) *Eur. J. Biochem.* **243**, 527. **6.** Meijer (1996) *Trends Cell Biol.* **6**, 393. **7.** Hamdane, Sambo, Delobel, et al. (2003) *J. Biol. Chem.* **278**, 34026. **8.** Darios, Muriel, Khondiker, Brice & Ruberg (2005) *J. Neurosci.* **25**, 4159. **9.** Tamrakar, Kapasi & Spector (2005) *J. Virol.* **79**, 15477. **10.** Gompel, Soulié, Ceballos-Picot & Meijer (2004) *Neurosignals* **13**, 134. **11.** Habran, Bontems, Di Valentin, Sadzot-Delvaux & Piette (2005) *J. Biol. Chem.* **280**,

29135. **12.** Sanchez, McElroy, Yen, et al. (2004) *J. Virol.* **78**, 11219. **13.** Zhu, Saito, Asada, Maekawa & Hisanaga (2005) *J. Neurochem.* **94**, 1535. **14.** Hisanaga & Saito (2003) *Neurosignals* **12**, 221. **15.** Kesavapany, Li, Amin, et al. (2004) *Biochim. Biophys. Acta* **1697**, 143. **16.** Bukczynska, Klingler-Hoffmann, Mitchelhill, et al. (2004) *Biochem. J.* **380**, 939. **17.** Bach, Knockaert, Reinhardt, et al. (2005) *J. Biol. Chem.* **280**, 31208. **18.** Tang, Li, Cao, et al. (2005) *J. Biol. Chem.* **280**, 31220.

Rosenthal's Inhibitor, See Dimethyl-DL-2,3-distearoyloxypropyl(2-hydroxyethyl)-ammonium Acetate

Rosiglitazone

This orally active antidiabetic agent (FW = 357.43 g/mol; CAS 122320-73-4; Solubility: 100 mM in DMSO and to 100 mM in 1eq. HCl), also known as BRL-49653 and 5-[[4-[2-(methyl-2-pyridinylamino)ethoxy]phenyl]-methyl]-2,4-thiazolidinedione, is a PPAR-γ inhibitor (EC_{50} = 60 nM) that targets Peroxisome Proliferator Activated Receptor (Type γ) in fat cells, rendering them more responsive to insulin (1,2). (Rosiglitazone exhibits no activity at PPARα and PPARβ.) Rosiglitazone is marketed as a stand-alone drug Avandia™, in combination with metformin, as Avandamet™, and in combination with glimeperide (as Avandaryl™). In 2007, an Advisory Committee of the Food and Drug Administration concluded that rosiglitazone use appeared to carry a greater risk of myocardial ischemic events and heart attacks than a placebo; however, several long-term, prospective clinical trials showed that, when compared to metformin or sulfonylurea, there was added risk of heart attack. The FDA issued a finding that the association between rosiglitazone and myocardial ischemia was inconclusive. Thus, despite its sustained effects on glycemic control, rising concerns about an elevated risk of myocardial infarction and other cardiopathy have reduced the clinical use of rosiglitazone. Rosiglitazone also appears to exert anti-inflammatory effects. **Target(s):** CYPC17, weakly inhibited (3); 3β-hydroxysteroid dehydrogenase type II, weakly inhibited (3); lipoprotein lipase (4); long-chain-fatty-acyl-CoA synthetase, *or* long-chain-fatty-acid:CoA ligase (5). **1.** Lehmann, et al. (1995) *J. Biol. Chem.* **270**, 12953. **2.** Willson, et al. (1996) *J. Med. Chem.* **39**, 665. **3.** Arlt, Auchus & Miller (2001) *J. Biol. Chem.* **276**, 16767. **4.** Ranganathan & Kern (1998) *J. Biol. Chem.* **273**, 26117. **5.** Kim, Lewin & Coleman (2001) *J. Biol. Chem.* **276**, 24667.

Rosilate, See 10-Hydroxystearate

Rosinduline, See Azocarmine G

Rosolic Acid, See Aurin

Rosuvastatin

This 3-hydroxy-3-methylglutaryl-CoA reductase inhibitor ($FW_{free-acid}$ = 481.55 g/mol; CAS 287714-41-4), also known as Crestor® and (3R,5S,6E)-7-[4-(4-fluorophenyl)-2-(N-methylmethanesulfonamido)-6-(propan-2-yl)pyrimidin-5-yl]-3,5-dihydroxyhept-6-enoate, is employed in the chronic treatment of hyperlipidemia, inhibits (1,2), Rosuvastatin reduces low-density lipoprotein-cholesterol (LDL-C) and very low-density lipoprotein-cholesterol (VLDL-C) levels, thereby retarding progression of atherosclerosis and improving the outcomes of coronary heart disease. **Mode of Inhibitor Action:** HMG-CoA reductase is the rate-controlling step in the biosynthesis of cholesterol and other isoprenoids. Reversible binding of rsuvastatin to HMG-CoA reductase involves formation of an initial complex ($K_i \approx 1$ nM), which then undergoes a slow transition over 0.5-6 min to yield a tighter complex ($K_i^* \approx 0.1$ nM). At steady state, rosuvastatin is at least as potent as atorvastatin, cerivastatin, and simvastatin, while being more potent than fluvastatin and pravastatin (2). **Effects on Vascular Cell Dynamics:** Rosuvastatin significantly suppresses PDGF-BB-induced migration of vascular smooth muscle cells (VSMCs), with likely inhibitory effects on expression of matrix metalloproteinases MMP2 and MMP9 (3). It also inhibits mitogen-activated protein kinase (MAPK) signaling by downregulating extracellular signal-regulated kinase (ERK) and p38 MAPK, although the phosphorylation level of c-Jun N-terminal kinase (c-JNK) is not altered (3). **Key Pharmacokinetic Parameters:** The approximate elimination half-life is 19 hours, and the time to peak plasma concentration is reached in 3–5 hours following oral administration. *See also* Appendix II in Goodman & Gilman's THE PHARMACOLOGICAL BASIS OF THERAPEUTICS, 12th Edition (Brunton, Chabner & Knollmann, eds.) McGraw-Hill Medical, New York (2011). **1.** Istvan & Deisenhofer (2001) *Science* **292**, 1160. **2.** Holdgate, Ward & McTaggart (2003) *Biochem. Soc. Trans.* **31**, 528. **3.** Gan, Li, Wang, *et al.* (2013) *Exp. Ther. Med.* **6**, 899.

Rotenone

This insecticide and fish poison (FW = 394.42 g/mol; CAS 83-79-4), which is found in a number of East Asian and South American plants and used by indigenous tribes as an arrow tip poison, interferes with ubiquinone reduction by binding to (saturable, K_d = 15-55 nM; Hill coefficient = 1) and inhibiting on the "oxygen side" of NADH dehydrogenase (*or* Complex I) by blocking electron transfer from the Fe-S centers in Complex I to ubiquinone. By impeding mitochondrial electron transport rotenone poisons terminal respiration, and ATP depletion induces oxidative stress in cells. The latter accounts for rotenone's ability to induce apoptosis in a variety of cell types through activation of the Jun N-terminal kinase pathway, caspase-activated DNAase, p53 relocalization, and activation of Bad. *Note:* Rotenone readily decomposes upon exposure to light and air. **Plant Sources:** Rotenone producers include: Hoary pea, *or* goat's rue (*Tephrosia virginiana*); Jícama (*Pachyrhizus erosus*); Cubé plant, *or* lancepod (*Lonchocarpus utilis*); Barbasco (*Lonchocarpus urucu*); Tuba plant (*Derris elliptica*); Jewel vine (*Derris involuta*); Cork-bush (*Mundulea sericea*); Florida fishpoison tree (*Piscidia piscipula*). Little is known about how plants escape rotenone poisoning. **Target(s):** alliin lyase (1); corticosterone 18-monooxygenase (2); electron transport (3-8); glutamate dehydrogenase (9); 25-hydroxyvitamin D 1α-hydroxylase, *or* calcidiol 1-monooxygenase (10); 2-ketoisovalerate dehydrogenase (11); 4-methoxybenzoate monooxygenase, *O*-demethylating (12); NADH dehydrogenase, *or* complex I, *or* NADH:ubiquinone reductase (3-8,13); NADH dehydrogenase, plant inner surface (14); and squalene monooxygenase, *or* squalene epoxidase (15); tubulin polymerization, IC$_{50}$ = 3 μM (16). **1.** Jansen, Muller & Knobloch (1989) *Planta Med.* **55**, 440. **2.** Rosenthal & Narasimhulu (1969) *Meth. Enzymol.* **15**, 596. **3.** Slater (1967) *Meth. Enzymol.* **10**, 48. **4.** Hatefi & Rieske (1967) *Meth. Enzymol.* **10**, 235. **5.** Izawa & Good (1972) *Meth. Enzymol.* **24**, 355. **6.** Hatefi (1978) *Meth. Enzymol.* **53**, 11. **7.** Singer (1979) *Meth. Enzymol.* **55**, 454. **8.** Rapoport & Schewe (1977) *Trends Biochem. Sci.* **2**, 186. **9.** Butow (1967) *Biochemistry* **6**, 1088. **10.** Lobaugh, Almond & Drezner (1986) *Meth. Enzymol.* **123**, 159. **11.** Connelly, Danner & Bowden (1970) *Meth. Enzymol.* **17A**, 818. **12.** Bernhardt, Pachowsky & Staudinger (1975) *Eur. J. Biochem.* **57**, 241. **13.** Hatefi & Stiggall (1976) *The Enzymes*, 3rd ed. (Boyer, ed.), **13**, 175. **14.** Palmer & Møller (1982) *Trends Biochem. Sci.* **7**, 258. **15.** Ryder & Dupont (1984) *Biochim. Biophys. Acta* **794**, 466. **16.** Srivastava & Panda (2007) *FEBS Lett.* **274**, 4688.

Rottlerin

This natural dye (FW = 518.52 g/mol; CAS 82-08-6) from *Mallotus philippinensis* (or *Rottlera tinctoria*), is a reddish-brown solid also named mallotoxin and (*E*)-1-[6-[(3-acetyl-2,4,6-trihydroxy-5-methylphenyl)-methyl]-5,7-dihydroxy-2,2-dimethyl-2*H*-1-benzopyran-8-yl]-3-phenyl-2-propen-1-one. Rottlerin inhibits different isozymes of protein kinase C, with an IC$_{50}$ value of 3-6 μM for the δ-isozyme, whereas the IC$_{50}$ value is 30-42 μM for the α-, β-, and γ-proteins, and 80-100 μM for protein kinase C-ε, -η, and -ζ. The IC$_{50}$ value for calmodulin-dependent kinase III (also called elongation factor-2 kinase) is 5.3 μM. Rottlerin potently stabilizes the channel-open conformation of large conductance potassium channel (BKCa^{++}), which is found in the inner mitochondrial membrane of cardiomyocytes. Opening of these channels is beneficial for post-ischemic changes in vasodilation. **Target(s):** Abl1 kinase (1); CaM kinase-III (2-4); chymotrypsin (1); [glycogen-synthase] kinase III, *or* [tau protein] kinase (4); inositol-trisphosphate 3-kinase (5); β-lactamase (1); malate dehydrogenase (1); [mitogen-activated protein kinase]-activated protein kinase (2-4); mitogen- and stress-activated protein kinase (1-4); p38-regulated/activated protein kinase (4); and protein kinase C (2,4,6-9). **1.** McGovern & Shoichet (2003) *J. Med. Chem.* **46**, 1478. **2.** Gschwendt, Muller, Kielbassa, *et al.* (1994) *Biochem. Biophys. Res. Commun.* **199**, 93. **3.** schwendt, Kittstein & Marks (1994) *FEBS Lett.* **338**, 85. **4.** Davies, Reddy, Caivano & Cohen (2000) *Biochem. J.* **351**, 95. **5.** Mayr, Windhorst & Hillemeier (2005) *J. Biol. Chem.* **280**, 13229. **6.** Saraiva, Fresco, Pinto & Goncalves (2003) *J. Enzyme Inhib. Med. Chem.* **18**, 475. **7.** Spitaler & Cantrell (2004) *Nat. Immunol.* **5**, 785. **8.** Harper & Poole (2007) *Biochem. Soc. Trans.* **35**, 1005. **9.** Johnson, Guptaroy, Lund, Shamban & Gnegy (2005) *J. Biol. Chem.* **280**, 10914.

Round-up™, *See Glyphosate*

Roussin's Black Salt

This black polynuclear NO-containing, iron-sulfur compound (FW$_{Potassium-Salt}$ = 568.73 g/mol; Chemical Formula: K[Fe$_4$S$_3$(NO)$_7$]; CAS 12518-87-5) is photolabile, water soluble, and can release nitric oxide when exposed to light (1). **Target(s):** alcohol dehydrogenase, yeast (2); carbonic anhydrase, *or* carbonate dehydratase (3); cellular respiration (4). **1.** Bourassa, Lee, Bernard, Schoonover & Ford (1999) *Inorg. Chem.* **38**, 2947. **2.** Sund & Theorell (1963) *The Enzymes*, 2nd ed. (Boyer, Lardy & Myrbäck, eds.), **7**, 25. **3.** Lindskog, Henderson, Kannan, *et al.* (1971) *The Enzymes*, 3rd. ed. (Boyer, ed.), **5**, 587. **4.** Hurst, Chowdhury & Clark (1996) *J. Neurochem.* **67**, 1200.

Roxarsone

This feed-efficiency enhancer and coccidiostatic agent (FW = 263.04 g/mol; CAS 121-19-7), also known as 4-hydroxy-3-nitrobenzenearsonic acid and (4-hydroxy-3-nitrophenyl)arsonic acid, targets bile salt hydrolase (BSH), an intestinal bacterial enzyme (taurodeoxycholate hydrolase and taurocholate hydrolase activities) that reduces host fat digestion and utilization (1). Synthesized from cholesterol and conjugated to either glycine or taurine in the liver, bile acids pass into the intestine, where it is hydrolyzed by BSH (produced by *Bacteroides*, *Clostridium*, *Enterococcus*, *Bifidobacterium*, and *Lactobacillus*), freeing taurine to function as an electron acceptor. BSHs may also decrease the toxicity of conjugated bile acids for bacteria. Deconjugated bile acids also have decreased solubility and diminished detergent activity, and maybe less toxic to certain GI bacteria. Although one of four FDA-approved arsenicals (*others*: nitarsone, arsanilic acid, and carbarsone) for use in poultry and/or swine, roxarsone is metabolized to arsenate in chickens, leading to arsenate accumulation in the liver (2). In 2013, the USFDA withdrew approval of roxarsone, carbarsone, and arsanilic but is reviewing nitarsone. **1**. Moser & Savage (2001) *Appl. Environ. Microbiol.* **67**, 3476. **2**. U.S. Food and Drug Administration (September 20, 2011) *FDA Response to Citizen Petition on Arsenic-based Animal Drugs.*

Roxithromycin

This semi-synthetic antibiotic (FW$_{\text{free-base}}$ = 837.06 g/mol; CAS 80214-83-1), which is prepared from erythromycin, is similar in action to that parent antibiotic and is predominantly bacteriostatic and is ineffective against Gram-negative bacilli. **Target(s):** CYP3A4 (1); and protein biosynthesis (2). **1**. Yamazaki & Shimada (1998) *Drug Metab. Dispos.* **26**, 1053. **2**. Champney, Tober & Burdine (1998) *Curr. Microbiol.* **37**, 412.

RP 67580

This substance P receptor antagonist (FW$_{\text{free-base}}$ = 438.57 g/mol; CAS 135911-02-3), also known as (3a*R*,7a*R*)-7,7-diphenyl-2-[1-imino-2-(2-methoxyphenyl)ethyl]perhydroisoindol-4-one, targets the neurokinin-1 receptor (K_i = 2.9 nM), exerting antidepressant, anxiolytic, and antiemetic properties (1). Its enantiomer (RP 67581) is without effect. *Note:* RP-67580 has high affinity for the NK$_1$ receptor in rats and mice, but not in humans. RP 67580 also inhibits dactylysin, K_i = 190 μM (2). **1**. Garret, Carruette, Fardin, *et al.* (1991) *Proc. Natl. Acad. Sci. U.S.A.* **88**, 10208. **2**. Joudiou, Carvalho, Camarao, Boussetta & Cohen (1993) *Biochemistry* **32**, 5959.

RP-3735, *See Lucanthone*

RP-60475, *See Intoplicine*

RP 62203, *See Fananserin*

RP 64206, *See Sparfloxacin*

RP 73401, *See Piclamilast*

RPC1063, *See Ozanimod*

RPL-554

This experimental bronchodilator and antiinflammatory drug (FW = 477.55 g/mol), also known as LS-193,855 and named *N*-{2-[(2*E*)-2-(mesitylimino)-9,10-dimethoxy-4-oxo-6,7-dihydro-2*H*-pyrimido[6,1-*a*]isoquinolin-3(4*H*)-yl]ethyl}urea is a long-acting inhibitor of PDE-3 and PDE-4, dual cAMP and cGMP phosphodiesterases, with clinically significant roles in regulating cardiac and vascular smooth muscle as well as platelet aggregation. **1**. Boswell-Smith, Spina, Oxford, *et al.* (2006) *J. Pharmacol. Exp. Ther.* **318**, 840.

RPR 107393

This quinuclidine derivative (FW = 340.43 g/mol; CAS 89567-03-38) inhibits squalene synthase, IC$_{50}$ = 68 nM for the rat enzyme (1). It significantly reduced both plasma total cholesterol and plasma triglyceride levels following oral dosing to rats with a reduced tendency to elevate plasma transaminase levels. **1**. Ishihara, Kakuta, Moritani, Ugawa & Yanagisawa (2004) *Bioorg. Med. Chem.* **12**, 5899.

RPR 120844

This substituted pyrrolidinone (FW = 460.58 g/mol) is a potent inhibitor of coagulation factor Xa (K_i = 7 nM), but a weaker inhibitor of thrombin, trypsin, protein C (activated), plasmin, and t-plasminogen activator, with respective K_i values of 1, 0.53, 2.4, 4.4, and 8.6 μM. **1**. Leadley (2001) *Curr. Top. Med. Chem.* **1**, 151. **2**. Ewing, Becker, Manetta, *et al.* (1999) *J. Med. Chem.* **42**, 3557.

RPR 130737

This substituted benzamidine (FW = 471.56 g/mol), also known as 4-hydroxy-3-[2-oxo-3(*S*)-(5-pyridin-3-ylthiophene-2-sulfonylamino)pyrrol-idin-1-ylmethyl]benzamidine, is a potent inhibitor of human coagulation factor Xa (K_i = 2.4 nM) and a weak inhibitor of thrombin, activated protein C, plasmin, t-plasminogen activator, and trypsin, with respective K_i values of 2.8 μM, > 18.5 μM, 2.9 μM, 6.7 μM, and 2.9 μM. **1**. Chu, Brown, Colussi, *et al.* (2000) *Thromb. Res.* **99**, 71.

RS-21607, See *Azalanstat*

RS 25259, See *Palonosetron*

RS-43285, See *Ranolazine*

RS 61443, See *Mycophenolate mofetil*

RS-130830

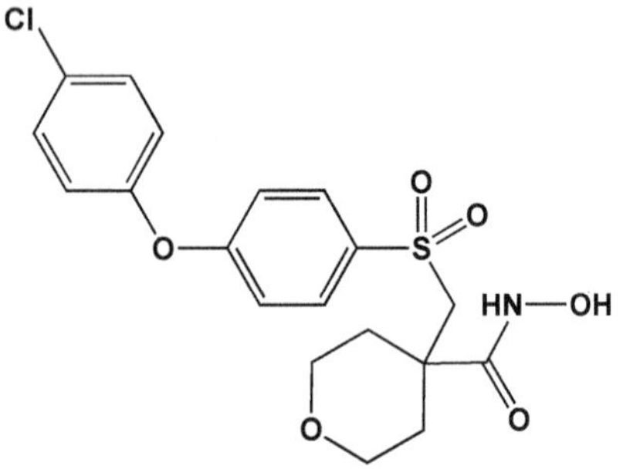

This second-generation, hydroxamate-containing MMP inhibitor (FW = 425.88 g/mol; CAS 193022-04-7), also known as 4-[[4-(4-chlorophenoxy)phenyl]sulfonylmethyl]-*N*-hydroxyoxane-4-carboxamide, Ro-1130830, and CTS-1027, targets MMP-2 (*Substrates*: basement membrane and nonfibrillar collagens (types IV, V, VII, X), Fibronectin, Elastin), IC_{50} = 0.4 nM; MMP-3 (*Substrates*: Proteoglycan, Laminin, Fibronectin, Collagen (Types III, IV, V, IX), Gelatins, pro-MMP-1), MMP-8 (*Substrates*: Fibrillar Collagens (Types I, II, III), MMP-9 (*Substrates*: Basement Membrane Collagens (Types IV, V), Gelatins) Basement Membrane Collagens (Types IV and V), Gelatins), MMP-12 (*Substrates*: Elastin), MMP-13 (*Substrates*: Fibrillar Collagens (Types I, II, III), Gelatins) IC_{50} = 0.8 nM; and MMP-14 (*Substrates*: pro-72 kDa Gelatinase). RS-130830 shows >1,000 times more potency, as compared to matrix metalloproteinases MMP-1, IC_{50} = 800 nM. **1.** Leung, Abbenante & Fairlie (2000) *J. Med. Chem.* **43**, 305. **2.** Barta, *et al.* (2001) *Bioorg. Med. Chem. Lett.* **11**, 2481. **3.** Toth, Sohail & Fridman (2012) *Methods Mol. Biol.* **878**, 121. **4.** Roomi, *et al.* (2009) *Oncol. Rep.* **21**, 1323. **5.** Kahraman, *et al.* (2009) *Hepatol. Res*, **39**, 805.

RS 25259-197, See *Palonosetron*

RS-25560-197, See *Nepicastat*

RSVA 405

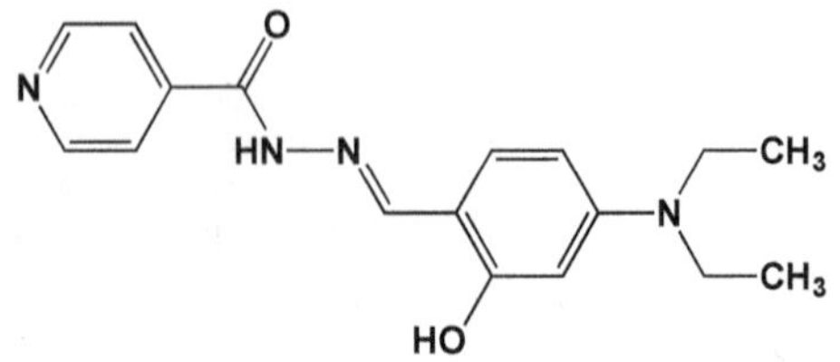

This small-molecule resveratrol analogue and AMPK activator (FW = 312.37 g/mol; CAS 140405-36-3; Solubility: 50 mM in DMSO), also named 2-[[4-(diethylamino)-2-hydroxyphenyl]methylene]hydrazide-4-pyridinecarboxylic acid, facilitates CaMKKβ-dependent activation of AMP-activated protein Kinase (EC_{50} = 1 μM), inhibits mTOR (mammalian Target of Rapamycin), and promotes autophagy to increase Alzheimer Aβ-peptide degradation by lysosomes, with an apparent EC_{50} ~ 1 μM (1). RSVA405 thus inhibits acetyl-CoA carboxylase (ACC), a target of AMPK and a key regulator of fatty acid biogenesis in nondifferentiated and proliferating adipocytes (2). RSVA 405 also potently inhibits Signal Transducer and Activator of Transcription 3 (or STAT3), a key mediator of the inflammatory responses of macrophages and other immune cell types. It also inhibits constitutive STAT3 activity in HEK293 cells and stimulated STAT3 activity lipopolysaccharide-stimulated RAW 264.7 macrophages (3). These effects on STAT3 phosphorylation are prevented by inhibitors of protein tyrosine phosphatases, indicating that RSVA405 promotes STAT3 dephosphorylation by protein tyrosine phosphatases. **1.** Vingtdeux, Chandakkar, Zhao, *et al.* (2011) *FASEB J.* **25**, 219. **2.** Vingtdeux,

Chandakkar, Zhao, Davies & Marambaud (2011) *Mol. Med.* **17**, 1022. **3.** Capiralla, Vingtdeux, Venkatesh (2012) *FEBS J.* **279**, 3791.

RTA-408, See *Omaveloxolone*

RTI-55

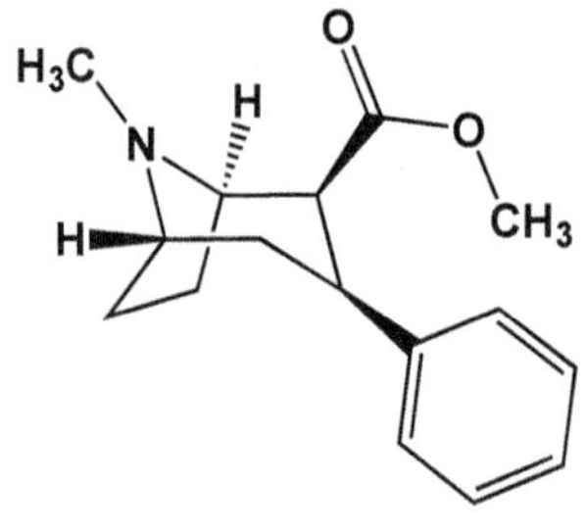

This methylecgonidine derivative (FW = 385.24 g/mol; CAS 136794-86-0), also known systematically as methyl (1*R*,2*S*,3*S*)-3-(4-iodophenyl)-8-methyl-8-azabicyclo[3.2.1]octane-2-carboxylate, is a non-selective dopamine reuptake inhibitor. **1.** Boja, Patel, Carroll, *et al.* (1991) *Eur. J. Pharmacol.* **194**, 133.

RTI-229

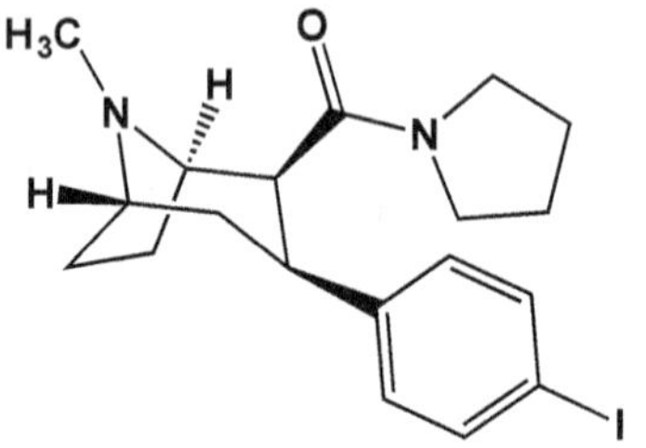

This neurotransmitter reuptake inhibitor (FW = 424.32 g/mol; CAS 160948-17-4), also known systematically as [(1*R*,2*S*,3*S*,5*S*)-3-(4-iodophenyl)-8-methyl-8-azabicyclo[3.2.1]octan-2-yl]-pyrrolidin-1-ylmethanone, exhibits a K_i value of 0.35 nM for dopamine and a K_i value of 20 nM for norepinephrine. **1.** Rothman, Baumann, Dersch, *et al.* (2001) *Synapse* **39**, 32.

RTX Toxin

These cytotoxic Gram-negative bacterial proteins encompass >1000 members exhibit characteristic glycine and aspartate-rich repeats and extracellular secretion by the type I secretion system (T1SS), which prevents RTX toxin accumulation in the periplasmic space. RTX toxins may be divided into pore-forming leukotoxins and multifunctional autoprocessing RTX toxins (MARTX). The prototypical RTX toxin is α-hemolysin (HlyA), a virulence factor produced by uropathogenic *Escherichia coli* (UPEC). Enterohemorrhagic *E. coli* (EHEC Serotype O157:H7) produces the RTX toxin known as EHEC hemolysin (EHEC-Hly). Likewise, *Vibrio* RtxA inhibits the actin cytoskeleton by targeting Rho GTPases. **1.** Lally, Hill, Kieba & Korostoff (1999) *Trends Microbiol.* **7**, 356. **2.** Benz (2015) *Biochim. Biophys. Acta* **1858**, 526.

RU 486, RU-486, or RU486, See *Mifepristone*

RU2390, See *Nilutamide*

RU 23345

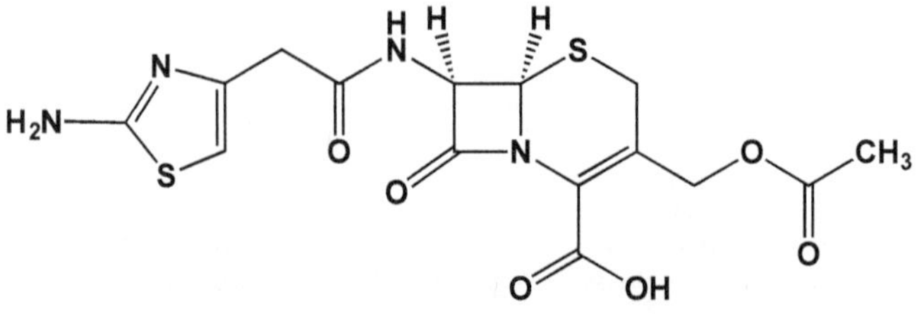

This desoxyimino-3-cephem antibiotic (FW = 412.44 g/mol) inhibits serine-type D-Ala-D-Ala carboxypeptidase and is a β-lactamase substrate. Interestingly, the *Streptomyces albus* G Zn^{2+}-containing D-alanyl-D-alanine-cleaving peptidase (a highly penicillin-resistant enzyme) remains highly resistant to all compounds tested. **1.** Laurent, Durant, Frère, Klein & Ghuysen (1984) *Biochem. J.* **218**, 933.

RU-28318

This potent and selective mineralocorticoid receptor (MR) antagonist ($FW_{free-acid}$ = 402.58 g/mol; $FW_{potassium-salt}$ = 440.67 g/mol; CAS 76676-34-1), also known as potassium oxprenoate and ($7\alpha,17\alpha$)-17-hydroxy-3-oxo-7-propylpregn-4-ene-21-carboxylate, inhibits aldosterone production and secretion. A dose-related inhibition of aldosterone production is observed for doses ranging from 10^{-5} to 10^{-3} M of RU 28318, and an intermediate dose (10^{-4} M) causes 70% inhibition (1). Long-term infusion (8 hours) results in to significant, stable and reversible inhibition of aldosterone production. Acute treatment of adrenalectomized rats with RU28318 (10-50 mg/kg) selectively decreased ex-vivo available MR binding in the hippocampus (2). Intracerebroventricular injection of RU28318 causes significant decrease in systolic blood pressure within 8 hours, demonstrating that brain mineralocorticoid receptors participate in blood pressure and renal function control. 1. Perroteau, Netchitailo, Delarue, *et al.* (1984) *J. Steroid Biochem.* **20**, 853. 2. Kim, Cole, Kalman & Spencer (1998) *J. Steroid Biochem. Mol. Biol.* **67**, 213. 3. Rahmouni, Barthelmebs, Grima, Imbs & De Jong (1999) *Eur. J. Pharmacol.* **385**, 199.

RU 38486, *See* Mifepristone

RU 44004

This peptide analogue (FW = 280.39 g/mol) inhibits neprilysin (K_i = 200 nM), an endothelial membrane-bound Zn^{2+} metallopeptidase found int brain, kidneys, lungs, gastrointestinal tract, heart, and peripheral vasculature. NEP degrades and inactivates a number of endogenous peptides, including enkephalins, circulating bradykinin, angiotensin peptides, and natriuretic peptides. 1. Roques, Noble, Crine & Fournié-Zaluski (1995) *Meth. Enzymol.* **248**, 263.

Rubeanic Acid, *See* Dithiooxamide
Rubidazone, *See* Zorubicin
Rubomycin, *See* Daunorubicin

Ruboxistaurin

This orally available *N-N'*-bridged *bis*-indolylmaleimide-based macrocycle (FW = 468.56 g/mol; CAS 169939-94-0; *Symbol*: RBX), also known by its trade name Arxxant® and systematic name (9*S*)-9-[(dimethylamino)methyl]-6,7,10,11-tetrahydro-9*H*,18*H*-5,21:12,17-di(metheno)-dibenzo[*e,k*]pyrrole-[3,4-*h*][1,4,13]oxadiazacyclohexadecine-18,20-dione, is an investigational drug for the treatment of diabetic peripheral retinopathy. Ruboxistaurin inhibits the PKC βI (IC_{50} = 4.7 nM) and PKC βII (IC_{50} = 5.9 nM) isozymes and is 76x and 61x more selective than compared to PKC α (2). PKC β is a druggable target for treating diabetic peripheral neuropathy (DPN). This drug is also likely to exert beneficial effects in preventing the development of diabetic nerve dysfunction by its action on the endoneurial microvasculature (3). The benefit of RBX for peripheral neuropathy has not been successfully demonstrated in Phase III trials. 1. 2. Jirousek, Gillig, Gonzalez, *et al.* (1996) *J. Med. Chem.* **39**, 2664. 3. Nakamura, Kato, Hamada, *et al.* (1999) *Diabetes* **48**, 2090.

Rubratoxin B

This lactone-containing bisanhydride metabolite (FW = 518.52 g/mol; CAS 21794-01-4) of certain toxigenic molds (*e.g.*, *Penicillium rubrum* and *P. purpurogenum*) is a hepatotoxin that inhibits gap junction intracellular communication, induces nuclear and internucleosomal fragmentation, and induces apoptosis. **Target(s):** cytochrome P450 (1); electron transport (2); and Na^+/K^+-exchanging ATPase (3). 1. Siraj & Hayes (1979) *Toxicol. Appl. Pharmacol.* **48**, 351. 2. Hayes (1976) *Toxicology* **6**, 253. 3. Phillips, Hayes, Ho & Desaiah (1978) *J. Biol. Chem.* **253**, 3487.

Rubrofusarin B

This phytotoxin, isolated from the fungus *Guanomyces polythrix*, inhibits calmodulin-dependent cyclic-nucleotide phosphodiesterase and NAD^+ kinase, with respective IC_{50} values of 4.7 and 13.3 μM, as well as NAD^+ kinase (1) and xanthine oxidase (2). 1. Mata, Gamboa, Macias, *et al.* (2003) *J. Agric. Food Chem.* **51**, 4559. 2. Song, Li, Ye, *et al.* (2004) *FEMS Microbiol Lett.* **241**, 67.

Rucaparib

This PARP inhibitor (FW = 323.37 g/mol; $FW_{Phosphate-Salt}$ = 421.36 g/mol; CAS 459868-92-9; Solubility = 200 mM in DMSO), also known as AG-014699, PF-01367338, and 8-fluoro-5-(4-((methylamino)methyl)phenyl)-3,4-dihydro-2*H*-azepino[5,4,3-*cd*]indol-1(6*H*)-one, targets the poly(ADP-ribose) polymerase PARP-1 (K_i = 1.4 nM), a nuclear enzyme that promotes the base excision repair within DNA breaks. Rucaparib inhibition of PARP-1 greatly enhances the efficacy of DNA alkylating agents, topoisomerase I poisons, and ionizing radiation. Rucaparib binding is also detected with eight other PARP domains (*e.g.*, PARP-2, PARP-3, PARP-4, PARP-10,

PARP-15, PARP-16, TNKS1 and TNKS2). The phosphate salt has improved aqueous solubility and is used in clinical trials. Rucaparib inhibits myosin light chain kinase 10x more potently than ML-9, producing an additive relaxation above that achievable with ML-9 (2). This finding suggested MLCK inhibition is not solely responsible for dilation. That rucaparib possesses a nicotinamide-like pharmacophore also raises the possibility it may inhibit other NAD^+-dependent processes (2). **1**. Thomas, Calabrese, Batey, *et al.* (2007) *Mol. Cancer Ther.* **6**, 945. **2**. McCrudden, O'Rourke, Cherry, *et al.* (2015) *PLoS One* **10**, e0118187.

Rufinamide

This anticonvulsant drug (FW = 238.19 g/mol; CAS 106308-44-5), also known as CGP-33101, RUF-331, Inovelon™, Banzel™ and 1-(2,6-difluorobenzyl)-1*H*-1,2,3-triazole-4-carboxamide, is indicated as an adjunctive, second/third-line treatment for seizures associated with Lennox-Gastaut syndrome in adults and children who are four years and older. Rufinamide appears to limit the firing of sodium-dependent action potentials in neurons, consistent with the stabilization of the sodium channel inactive state, thereby maintaining these ion channels in a closed state. **1**. Perucca, Cloyd, Critchley & Fuseau (2008) *Epilepsia* **49**, 1123. **2**. Glauser, Kluger, Sachdeo, *et al.* (2008) *Neurology* **70**, 1950. **3**. Cheng-Hakimian, Anderson & Miller (2006) *Internat. J. Clin. Pract.* **60**, 1497.

Rufloxacin

This fluoroquinolone-class antibiotic (FW = 363.41 g/mol; CAS 101363-10-4 and 106017-08-7; IUPAC Name: 9-fluoro-10-(4-methylpiperazin-1-yl)-7-oxo-2,3-dihydro-7*H*-[1,4]thiazino[2,3,4-*ij*]quinoline-6-carboxylate), marketed under the trade names Ruflox®, Monos®, Qari®, Tebraxin®, Uroflox®, and Uroclar®, is a broad-spectrum systemic antibacterial agent that targets Type II DNA topoisomerases (gyrases), which are required for bacterial replication and transcription. Rufloxacin has a broad spectrum of activity in vitro against clinically important gram-positive and gram-negative aerobes (37, 54). The pharmacokinetics are characterized by a mean level of plasma protein binding of 60%, a long elimination half-life of about 30-35 hours, low renal clearance, and good tissue penetration. In rats, ~60% of the dose was absorbed. After an oral dose of 200 mg, humans reach maximum plasma levels of ~4 µg/mL in 3-5 hours, and steady-state serum concentrations within 4-5 days. Urine excretion of rufloxacin is 25-50% of the dose, and renal clearance is about 20 mL/min. For the prototypical member of this antibiotic class. **See** *Ciprofloxacin* **1**. Schito, Acar, Bauernfeind, *et al.* () *J. Antimicrob. Chemother.* **38**, 627. **2**. Wise, Andrews, Matthews & Wolstenholme (1992) *J. Antimicrob. Chemother.* **29**, 649.

Rupatadine

This orally active, dual-receptor antagonist and second-generation antihistamine (FW = 532.03 g/mol (fumarate salt); CAS 182349-12-8;

Solubility: 10 mg/mL DMSO; <1 mg/mL H_2O), named systematically as 8-chloro-6,11-dihydro-11-[1-[(5-methyl-3-pyridinyl)methyl]-4-piperidinylidene]-5*H*-benzo[5,6]cyclohepta[1,2-*b*]pyridine), targets the platelet-activating factor receptor PAFR (a G-protein coupled receptor that binds platelet-activating factor) and histamine (H_1) receptor (a Rhodopsin-like G-protein coupled receptor that binds histamine) with K_i values of 550 and 102 nM, respectively. Rupatadine competitively inhibited histamine-induced guinea pig ileum contraction, *without* affecting contraction induced by ACh, serotonin or leukotriene D_4. It also competitively inhibited PAF-induced platelet aggregation in washed rabbit platelets and in human platelet-rich plasma (IC_{50} = 0.68 µM), while not affecting ADP- or arachidonic acid-induced platelet aggregation. **1**. Merlos, *et al.* (1997) *J. Pharmacol. Exp. Ther.* **280**, 114. **2**. Queralt, et al. (2000) *Inflamm. Res.* **49**, 355. **3**. Barbanoj, *et al.* (2004) *Neuropsychobiology* **50**, 311. **4**. Sudhakara Rao, *et al.* (2009) *Indian J. Otolaryngol. Head Neck Surg.* **61**, 320.

Rupintrivir

This peptidomimetic (FW = 598.67 g/mol; CAS 223537-30-2), also known as AG 7088, is a potent irreversible inhibitor of picornain 3C, with a $k_{obs}/[I]$ value of 1.47 x 10^6 $M^{-1}s^{-1}$ for HRV 14 (1-3). Rupintrivir exhibits potent *in vitro* activity against all HRV serotypes and HEV strains. **1**. Dragovich, Prins, Zhou, *et al.* (1999) *J. Med. Chem.* **42**, 1213. **2**. Patick, Binford, Brothers, *et al.* (1999) *Antimicrob. Agents Chemother.* **43**, 2444. **3**. Matthews, Dragovich, Webber, *et al.* (1999) *Proc. Natl. Acad. Sci. U.S.A.* **96**, 11000.

RU-SKI 43

This Sonic Hedgehog (Shh) signaling pathway inhibitor ($FW_{free-base}$ = 386.55 g/mol; $FW_{hydrochloride}$ = 423.01 g/mol; CAS 1043797-53-0; Solubility: 93 mM in DMSO), also named 2-(2-methylbutylamino)-1-(4-(*m*-tolyloxymethyl)-6,7-dihydrothieno[3,2-*c*]pyridin-5(4*H*)-yl)ethanone-HCl, targets hedgehog acyltransferase (HhAT), blocking Sonic Hedgehog signal transduction (**See also** *Wnt-C59*). **Mode of Inhibitor Action:** Several lines of evidence suggest RU-SKI 43 is a specific inhibitor of Shh palmitoylation. First, neither palmitoylation of H-Ras and Fyn nor myristoylation of c-Src is affected by when cells are treated RU-SKI 43. Second, RU-SKI 43 treatment is without effect on fatty acylation of Wnt3a12 by Porcupine, another member of the MBOAT family, whereas Wnt C59 (itself a Porcupine inhibitor) blocked radiolabel incorporation. Third, overexpression of Hhat reduced the ability of RU-SKI 43 to inhibit Shh palmitoylation in transfected COS-1 cells, whereas overexpression of Porcupine had no effect. Fourth, RU-SKI 43 inhibited palmitoylation of Shh by endogenous Hhat in COS-1 cells. Lastly, RU-SKI 43 did not alter Shh autoprocessing, steady-state levels of Shh and Hhat, or subcellular localization of Shh and Hhat. **1**. Petrova, Rios-Esteves, Ouerfelli, Glickman, & Resh (2013) *Nature Chem Biol.* **9**, 24.

Russell's Viper Venom Proteinase Inhibitors

These Kunitz-type proteinase inhibitors and exert antihemorrhagic agents (MW ≈ 7.2 kDa), isolated from the venom of Russell's viper (*Daboia russelli*), show K_i values of 0.8 nM for bovine trypsin, 0.1 nM for bovine chymotrypsin, 0.3 nM for bovine plasma kallikrein, and 1 nM for bovine plasmin. **See also Trypsin Inhibitors** 1. Iwanaga, Takahashi & Suzuki (1976) *Meth. Enzymol.* **45**, 874. 2. Kurachi & Davie (1981) *Meth. Enzymol.* **80**, 211.

Rustmicin

This lactone (FW = 380.48 g/mol; CAS 100227-57-4) from a *Micromonospora* exhibits potent antifungal activity against phytopathogenic fungi but has no effect on lipid biosynthesis in mammalian cells. Rustmicin has a relatively short half-life in aqueous solutions (<1 hour below pH 4 or above pH 7; longer at pH 5.5). **Target(s):** inositolphosphoryl-ceramide synthase, *or* ceramide inositol phosphoryltransferase (1,2). 1. Fischl, Liu, Browdy & Cremesti (2000) *Meth. Enzymol.* **311**, 123. 2. Mandala & Harris (2000) *Meth. Enzymol.* **311**, 335.

Rutaecarpine

This quinazolinocarboline alkaloid (FW = 287.32 g/mol; CAS 20575-76-2), also called rutecarpine, from the fruit of *Evodia rutaecarpa* and *Hortia arborea* is a vasodilator and a delayed rectifier K⁺ channel blocker. **Target(s):** COX-2, *or* cyclooxygenase (1); CYP1A1 (2); CYP1A2 (2); CYP3A4 (3); and delayed rectifier K⁺ channel (4). 1. Moon, Murakami, Kudo, *et al.* (1999) *Inflamm. Res.* **48**, 621. 2. Ueng, Jan, Lin, *et al.* (2002) *Drug Metab. Dispos.* **30**, 349. 3. Iwata, Tezuka, Kadota, Hiratsuka & Watabe (2005) *Drug Metab. Pharmacokinet.* **20**, 34. 4. Wu, Lo, Chen, Li & Chiang (2001) *Neuropharmacology* **41**, 834.

Rutamycin A

This macrolide antibiotic (FW = 777.05 g/mol; CAS 1404-59-7), also known as oligomycin D and 26-demethyloligomycin A, strongly inhibits fungal mitochondrial F_oF_1 ATP synthase, *or* H⁺-transporting two-sector ATPase (1-5). Rutamycin B, isolated from *Streptomyces aureofaciens*, likewise inhibits mitochondrial ATP synthase. 1. Stiggall, Galante & Hatefi (1979) *Meth. Enzymol.* **55**, 308. 2. Racker (1979) *Meth. Enzymol.* **55**, 315. 3. Linnett & Beechey (1979) *Meth. Enzymol.* **55**, 472. 4. Lardy, Witonsky & Johnson (1965) *Biochemistry* **4**, 552. 5. Kageyama, Tamura, Nantz, *et al.* (1990) *J. Amer. Chem. Soc.* **112**, 7407.

Rutecarpine, See Rutaecarpine

Rutin

This remarkably abundant plant flavonoid (FW = 610.53 g/mol; CAS 153-18-4; hygroscopic solid that darkens upon light exposure) is the 3'-rutinoside of quercetin found in many plants (*e.g.*, buckwheat, hydrangea, eucalyptus, pansies, tobacco, and forsythia). This natural free-radical scavenger, occasionally referred to as vitamin P, has a protective antioxidant effect in cells and protects cells with carcinogen-induced DNA damage. **Target(s):** aldose reductase, *or* aldehyde reductase (1,2); ATPase, weakly inhibited (3); Ca²⁺-exporting ATPase (4); carbonyl reductase (5,6); catechol *O*-methyltransferase (24); chitin synthase (7); DNA-directed DNA polymerase (23); dopamine β-hydroxylase (8); hyaluronoglucosaminidase, *or* hyaluronidase (9-11); iodothyronine deiodinase, type I (12); leucoanthocyanidin reductase (26); [myosin light-chain] kinase (22); NADH oxidase, weakly inhibit (3); NAD(P)H:quinone reductase (13); P-glycoprotein (14); phosphoenolpyruvate carboxylase (15,16); phospholipase A₂ (17,18); protein kinase C (19); pyruvate kinase (20); succinoxidase, weakly inhibited (3); tyrosinase, *or* monophenol monooxygenase (10,27,28); xanthine oxidase (21,25); and xenobiotic-transporting ATPase, *or* multidrug-resistance protein (14). 1. Wermuth & von Wartburg (1982) *Meth. Enzymol.* **89**, 181. 2. Branlant (1982) *Eur. J. Biochem.* **129**, 99. 3. Bohmont, Aaronson, Mann & Pardini (1987) *J. Nat. Prod.* **50**, 427. 4. Barzilai & Rahamimoff (1983) *Biochim. Biophys. Acta* **730**, 245. 5. Atalla, Breyer-Pfaff & Maser (2000) *Xenobiotica* **30**, 755. 6. Wermuth (1981) *J. Biol. Chem.* **256**, 1206. 7. Hwang, Ahn, Lee, *et al.* (2001) *Planta Med.* **67**, 501. 8. Denke, Schempp, Weiser & Elstner (2000) *Arzneimittelforschung* **50**, 415. 9. Beiler & Martin (1947) *J. Biol. Chem.* **171**, 507. 10. Rodney, Swanson, Wheeler, Smith & Worrel (1950) *J. Biol. Chem.* **183**, 739. 11. Meyer & Rapport (1952) *Adv. Enzymol.* **13**, 199. 12. Ferreira, Lisboa, Oliveira, *et al.* (2002) *Food Chem. Toxicol.* **40**, 913. 13. Merk, Jugert, Bonnekoh & Mahrle (1991) *Skin Pharmacol.* **4**, 183. 14. Sharom, Liu, Romsicki & Lu (1999) *Biochim. Biophys. Acta* **1461**, 327. 15. Moraes & Plaxton (2000) *Eur. J. Biochem.* **267**, 4465. 16. Pairoba, Colombo & Andreo (1996) *Biosci. Biotechnol. Biochem.* **60**, 779. 17. Lindahl & Tagesson (1997) *Inflammation* **21**, 347. 18. Fawzy, Vishwanath & Franson (1988) *Agents Actions* **25**, 394. 19. End, Look, Shaffer, Balles & Persico (1987) *Res. Commun. Chem. Pathol. Pharmacol.* **56**, 75. 20. Smith, Knowles & Plaxton (2000) *Eur. J. Biochem.* **267**, 4477. 21. Beiler & Martin (1951) *J. Biol. Chem.* **192**, 831. 22. Jinsart, Ternai & Polya (1991) *Biol. Chem. Hoppe-Seyler* **372**, 819. 23. Spampinato, Pairoba, Colombo, Benediktsson & Andreo (1994) *Biosci. Biotechnol. Biochem.* **58**, 822. 24.

Bonifácio, Palma, Almeida & Soares-da-Silva (2007) *CNS Drug Reviews* **13**, 352. **25**. Dew, Day & Morgan (2005) *J. Agric. Food Chem.* **53**, 6510. **26**. Tanner, Francki, Abrahams, *et al.* (2003) *J. Biol. Chem.* **278**, 31647. **27**. Lin, Hsu, Chen, Chern & Lee (2007) *Phytochemistry* **68**, 1189. **28**. Karioti, Protopappa, Magoulas & Skaltsa (2007) *Bioorg. Med. Chem.* **15**, 2708.

Ruxolitinib

This Janus-associated kinase inhibitor (F.Wt. = 306.43 g/mol; CAS 941678-49-5; Soluble in DMSO; Water < 1 mg/mL), also known as (*R*)-3-(4-(7*H*-pyrrolo[2,3-*d*]pyrimidin-4-yl)-1*H*-pyrazol-1-yl)-3-cyclopentylpropanenitrile and NCB018424, acts selectively, with IC$_{50}$ values of 2.7 nM, 4.5 nM, and 322 nM for JAK1, JAK2, and JAK3 (1). The rationale for its therapeutic action is the observation that most polycythemia vera patients carry the JAK2-Val617Phe mutation (2). Ruxolitinib is an oral Janus kinase (JAK) 1/JAK2 inhibitor approved in the US for the treatment of intermediate-or high-risk myelofibrosis (MF); however, because thrombopoietin and erythropoietin signal through JAK2, dose-dependent cytopenias are expected with treatment (3). Ruxolitinib treatment dose-dependently alleviates CCl$_4$-induced hepatic injury and necroinflammation, as indicated by biochemical markers of injury and histopathology (4). **1**. Williams, *et al.* (2008) *Ann. Rheum. Dis.* **67** (Suppl II), 62. **2**. Quintás-Cardama, Vaddi, Liu, Manshouri, *et al.* (2010) *Blood* **115**, 3109. **3**. Verstovsek, Gotlib, Gupta, *et al.* (2013) *Onco Targets Ther.* **7**, 13. **4**. Hazem, Shaker, Ashamallah & Ibrahim (2014) *Chem. Biol Interact.* **220**, 116.

RWJ-17021, *See* Topiramate

RWJ-37947

This topiramate analogue (FW = 361.35 g/mol), also known as RWJ-37497, inhibits carbonic anhydrase, *or* carbonate dehydratase, with IC$_{50}$ = 36 nM for the human II isozyme (1,2). **Key Pharmacokinetic Parameters:** *See* Appendix II in Goodman & Gilman's THE PHARMACOLOGICAL BASIS OF THERAPEUTICS, 12th Edition (Brunton, Chabner & Knollmann, eds.) McGraw-Hill Medical, New York (2011). **1**. Recacha, Constanzo, Maryanhoff & Chatto-padhyay (2002) *Biochem. J.* **361**, 437. **2**. Maryanoff, Costanzo, Nortey, *et al.* (1998) *J. Med. Chem.* **41**, 1315.

RWJ-50353

This arginine-containing derivative (FW = 551.72 g/mol), known as *N*-methyl-D-phenylalanyl-*N*-[5-[(aminoiminomethyl)amino]-1-[[(2-benzothi-azolyl)carbonyl]butyl]-L-prolinamide, inhibits thrombin, K_i = 0.19 nM (1-3), trypsin, K_i = 3.1 nM (2,4), and tryptase K_i = 6.5 nM (4). **1**. Matthews, Krishnan, Costanzo, Maryanoff & Tulinsky (1996) *Biophys. J.* **71**, 2830. **2**. Costanzo, Maryanoff, Hecker, *et al.* (1996) *J. Med. Chem.* **39**, 3039. **3**. Recacha, Costanzo, Maryanoff, *et al.* (2000) *Acta Crystallogr. Sect. D* **56**, 1395. **4**. Costanzo, Yabut, Almond, Jr., *et al.* (2003) *J. Med. Chem.* **46**, 3865.

RWJ-51084

This arginine-containing derivative (FW = 387.51 g/mol), a dipeptide-based transition-state analogue containing a benzothiazole ketone, inhibits trypsin (K_i = 30 nM) and tryptase (K_i = 88 nM) (1,2). Aerosol administration of RWJ-58643 to allergic sheep effectively antagonized antigen-induced asthmatic responses, with substantial blockade of the early response and complete ablation of the late response and airway hyperresponsiveness. **1**. Costanzo, Yabut, Almond, Jr., *et al.* (2003) *J. Med. Chem.* **46**, 3865. **2**. Recacha, Carson, Costanzo, *et al.* (1999) *Acta Crystallogr. D Biol. Crystallogr.* **55**, 1785.

RWJ-51438

This phenylalanylprolylarginine-containing benzothiazole (FW = 593.69 g/mol) is thrombin inhibitor (K_i = 1.1 nM) that becomes covalently linked to the hydroxyl group of Ser195, forming a tetrahedral intermediate, hemiketal structure. **1**. Recacha, Costanzo, Maryanoff, *et al.* (2000) *Acta Crystallogr. Sect. D* **56**, 1395.

RWJ-355871, *See* 2-{3-[N-Methyl-N-(N-(b-naphthoyl)piperidin-4-yl)carbamoyl]-naphthalene-2-yl}-1-(naphthalene-1-yl)-2-oxoethylphosphonate

RXP 407

This phosphinic peptide (FW = 440.44 g/mol), also known as *N*-Acetyl-α-aspartyl-3-[(1-amino-2-phenylethyl)hydroxyphosphinyl]-2-methylpropan-oylalaninamide or *N*-acetyl-L-aspartyl-L-phenylalanylΨ[PO$_2$-CH$_2$]-L-alanyl-L-alaninamide (where ψ indicates a pseudopeptide linkage), is a potent inhibitor of peptidyl-dipeptidase A, *or* angiotensin I-converting enzyme. RXP 407 is the first inhibitor reported to differentiate between the two active sites of the two homologous domains of this enzyme, with a K_i value of 12 nM for the active site in the N-terminal domain and a K_i value of 25 μM for the C-terminal domain (1) **1**. Cotton, Hayashi, Cuniasse, *et al.* (2002) *Biochemistry* **41**, 6065. **2**. Dive, Cotton, Yiotakis, *et al.* (1999) *Proc. Natl. Acad. Sci. U.S.A.* **96**, 4330.

Ryanodine

This poisonous alkaloid and potent insecticide (FW = 493.55 g/mol; CAS 15662-33-6) from *Ryania speciosa* binds to large, high conductance Ca^{2+} channels (ryanodine receptors) controlling the release of Ca^{2+} from an intracellular compartment (*i.e.*, the sarco/endoplasmic reticulum). At nM concentrations, ryanodine locks the receptor in a half-open state; however, at μM concentrations, the receptor closes fully. Ryanodine binds with such high affinity that it was used as a radiolabel the receptor, thereby permitting purification of this class of ion channels. Mammalian tissues express three closely related ryanodine receptors. **1**. Feher & Lipford (1985) *Biochim. Biophys. Acta* **813**, 77. **2**. Marban & Wier (1985) *Circ. Res.* **56**, 133.

– S –

S3QEL 2

This cell-permeable substituted allopurinol (FW = 323.44 g/mol; CAS 890888-12-7; Solubility: 20 mM in DMSO, with gentle warming), also named 1-(3,4-dimethylphenyl)-*N*,*N*-dipropyl-1*H*-pyrazolo[3,4-*d*]pyrimidin-4-amine, suppresses superoxide production by mitochondrial Complex III (IC$_{50}$ = 1.7 µM), without affecting normal electron flux or cellular oxidative phosphorylation. S3QEL-2 also suppresses HIF-1α accumulation and enhances survival and function of primary pancreatic islets *in vitro*. Superoxide production by respiratory complex III is implicated in diverse signaling events and pathologies, but its role remains controversial. **1**. Orr, Vargas, Turk, *et al.* (2015) *Nature Chem. Biol.* **11**, 834.

S-13, *See* *2',5-Dichloro-3-tert-butyl-4'-nitrosalicylanilide*

S-2720, *See* *6-Chloro-3,3-dimethyl-4-(isopropenyl-oxycarbonyl)-3,4-dihydroquinoxalin-2(1H)-thione*

S 2101

This LSD1 inhibitor (FW$_{HCl-Salt}$ = 311.75 g/mol; CAS 1239262-36-2; Soluble to 50 mM in H$_2$O; 100 mM in DMSO), also named (1*R*,2*S*)-*rel*-2-[3,5-difluoro-2-(phenylmethoxy)phenyl]cyclopropanamine HCl, targets Lysine-Specific Demethylase 1A, a flavin-dependent monoamine oxidase that catalyzes the demethylation of mono- and di-methylated lysines, specifically acting on histone 3, lysines 4 and 9 (IC$_{50}$ = 990 nM; K_i = 610 nM). S-2101 exhibits selectivity for LSD1, as compared to monoamine oxidases MAO-A (K_i = 110 µM) and MAO-B (K_i = 17 µM). LSD1 plays critical roles in embryogenesis and tissue-specific differentiation, as well as oocyte growth. **1**. Mimasu, Umezawa, Sato, *et al.* (2010) *Biochemistry* **49**, 6494.

S-3483

This chlorogenic acid derivative (FW$_{free-acid}$ = 502.14 g/mol) is a reversible, linear competitive inhibitor of the glucose-6-phosphatase (Glc-6-Pase) system within rat renal microsomes and rat and human liver microsomes (1). The K_i value of 129 nM for S-3483 in rat liver microsomes is three orders of magnitude lower than that for chlorogenic acid. At concentrations up to 100 µM, S-3483 had no direct effect on the Glc-6-Pase enzyme activity or on its inorganic pyrophosphatase activity (*i.e.*, on T$_2$, the P$_i$/inorganic pyrophosphate transporter). Thus, like chlorogenic acid, S-3483 appears to be a site-specific inhibitor of T$_1$, the Glc-6-P transporter of renal and liver microsomes (2). **1**. Simon, Herling, Preibisch & Burger (2000) *Arch. Biochem. Biophys.* **373**, 418. **2**. Arion, Canfield, Ramos, *et al.* (1998) *Arch. Biochem. Biophys.* **351**, 279.

S 07662

This CAR inverse agonist (FW = 300.38 g/mol; CAS 883226-64-0; Solubility = 100 mM in DMSO), also named *N*-[(2-methyl-3-benzofuranyl)methyl]-*N*-(2-thienylmethyl)urea, S-07662, and S07662, targets the orphan nuclear Constitutive Androstane Receptor (*or* CAR) via recruitment of nuclear receptor co-repressor (*or* NCoR). In addition to being one of the key regulators of xenobiotic and endobiotic metabolism, CAR also influences a variety of physiological functions, such as gluconeogenesis, metabolism of xenobiotics, fatty acids, bilirubin, bile acids, hormonal regulation, *etc.* **1**. Küblbeck, Jyrkkärinne, Molnár, *et al.* (2011) *Mol. Pharm.* **8**, 2424.

S-16257, *See* *Ivabradine*

S 17625

This quinolinone-phosphonic acid (FW$_{free-acid}$ = 294.03 g/mol), also known as 6,7-dichloro-2(1*H*)-oxoquinoline-3-phosphonic acid, is an AMPA (*or* α-amino-3-hydroxy-5-methylisoxazole-4-propionic acid) receptor antagonist. Replacement of the chlorine in position-6 by a sulfonylamine led to very potent AMPA antagonists endowed with good in vivo activity and lacking nephrotoxicity potential. **1**. Cordi, Desos, Ruano, *et al.* (2002) *Farmaco* **57**, 787.

S 23515

This imidazoline-like agent (FW = 271.11 g/mol) has a high affinity for non-adrenergic imidazoline receptor binding sites and will induce hypotension (1). S23515 also specifically inhibits cholesterol synthesis in cultured rodent and primate hepatocytes in a dose-dependent manner, and this hypocholesterolemic effect was likely due to the inhibition of the oxido:lanosterol cyclase (OSC), a rate-limiting enzyme in the cholesterol biosynthetic pathway (2). Partial OSC inhibition induced by S23515 led to the generation of 24(*S*),25-epoxycholesterol, a potent ligand for the liver X receptor (LXR). Such results suggest that S-23515, and potentially other imidazoline-like drugs, exert hypolipidemic effects in addition to their hypotensive activities. **1**. Venteclef, Guillard & Issandou (2005) *Biochem.*

Pharmacol. **69**, 1041. **2**. Ventecief, Guillard & Issandou (2005) *Biochem. Pharmacol.* **69**, 1041.

S 25585

This potent NPY receptor antagonist (FW = 528.50 g/mol; CAS 263849-50-9; Soluble to 100 mM in DMSO), also systematically named 1-benzoyl-2-[[*trans*-4-[[[[2-nitro-4-(trifluoro-methyl)phenyl]sulfonyl]amino]methyl]-cyclohexyl]carbonyl]hydrazine, targets neuropeptide Y_5 receptor (IC$_{50}$ = 5.4 nM), with weaker action against Y_1 (IC$_{50}$ > 1000 nM), Y_2 (IC$_{50}$ > 10000 nM), and Y_4 (IC$_{50}$ > 10000 nM) receptors. S-25585 displays no affinity for over 40 other receptors, channels or uptake systems (1). S 25585 does not induce a conditioned taste aversion, does not significantly alter need-induced sodium appetite, and does not induce pica, suggesting that at this dose, it did not induce illness or malaise. Significantly, S25585 inhibits NPY-induced feeding. Although S 25585 appears to influence a physiological system controlling appetite, its action doe not involve the Y_5 receptor, because it markedly reduces food intake in the NPY Y_5 knockout mouse (1). Such results show that the activity of any new NPY Y_5 antagonist must be assessed in a Y_5 knockout mouse, before assuming an effect is due to blockade of this receptor. **1**. Della-Zuana, Revereault, Beck-Sickinger, *et al* (2004) *Int. J. Obes.* **28**, 628.

S 35047

This cytotoxic indenoindole derivative (FW$_{free-base}$ = 495.58 g/mol) inhibits human DNA topoisomerase II. The growth of human leukemia cells is strongly reduced in the presence of the S 35047, with an IC$_{50}$ value in the 50 nM range. Cytoxicity is attributed to cell cycle arrest at the G2/M junction as well as apoptosis, the latter evidenced by extensive internucleosomal DNA cleavage. **1**. Bal, Baldeyrou, Moz, *et al.* (2004) *Biochem. Pharmacol.* **68**, 1911.

S49076

This orally active anticancer agent (FW = 424.53 g/mol), named 3-[(3-{[4-(4-morpholinylmethyl)-1*H*-pyrrol-2-yl]methylene}-2-oxo-2,3-dihydro-1*H*-indol-5-yl)methyl]-1,3-thiazolidine-2,4-dione, inhibits MET, AXL and FGFRs, receptor tyrosine kinase that are associated with tumor progression in a wide variety of human malignancies. With MET, the on-rate constant was 1.27 x 10^6 M^{-1}s^{-1}, the off-rate constant was 1.62 s^{-1}, and the calculated K_d was 1.3 nM. With the other tyrosine kinase, the IC$_{50}$ was about 200 nM. Attesting to the selectivity of S49076 was the observation that VEGFR2 was not inhibited. S49076 also lowers viability, motility, and three-dimensional colony formation in tumor cell lines bearing MET, AXL and FGFR. **1**. Burbridge, Bossard, Saunier *et al.* (2013) *Molec. Cancer Therap.* [Published On-line]

S-265744, See *GSK1265744*

SA 57

This dual FAAH/MAGL inhibitor (FW = 338.83 g/mol; CAS 1346169-63-8; Soluble to 25 mM in DMSO), systematically named 4-[2-(4-chlorophenyl)ethyl]-1-piperidinecarboxylic acid 2-(methylamino)-2-oxoethyl ester, targets human and mouse fatty acid amide hydrolase, *or* FAAH (IC$_{50}$ < 10 nM) as well as mouse and human monoacylglycerol lipase, *or* MAGL (IC$_{50}$ = 410 nM and 1.4 μM, respectively) (1). FAAH and MAGL catalyze the hydrolysis of anandamide (AEA) and 2-arachidonoylglycerol (2-AG), two major endocannabinoids. FAAH and MAGL inhibitors elevate brain AEA and 2-AG levels, respectively, thereby reducing pain, anxiety, and depression in rodents, but without the psychotropic behavioral effects observed with direct cannabinoid receptor-1 (CB$_1$) agonists. Significantly, SA-57 reduces abrupt withdrawal in morphine-dependent mice and is likewise effective in an *in vitro* model of opioid withdrawal (2). **1**. Niphakis, Johnson, Ballard, Stiff & Cravatt (2012) *ACS Chem. Neurosci.* **3**, 418. **2**. Ramesh, Gamage, Vanuytsel, *et al.* (2013) *Neuropsychopharmacol.* **38**, 1039.

SA751

This peptidomimetic (FW$_{free-acid}$ = 465.55 g/mol), also known as *N*-(1(*R*)-carboxyethyl)-α-(*S*)-(4-phenyl-3-butynyl)glycyl-L-*O*-methyltyrosine *N*-methylamide, is a selective inhibitor of neutrophil collagenases, *or* matrix metalloproteinase 8, K_i = 2 nM, with K_i of >10,000 nm against MMP-3 and MMP-1 (1). The latter are part of the 'aggricanase' activity that degrades the aggregating cartilage proteoglycan, or aggrecan, which along with type II collagen gives rise to the mechanical properties of articular cartilage. Aggrecan core protein cleavage at the Glu^{373}-Ala^{374} bond apparently plays a critical role in cartilage matrix degradation. **1**. Arner, Decicco, Cherney & Tortorella (1997) *J. Biol. Chem.* **272**, 9294.

SA 862

This antithrombotic benzamidine derivative, also known as 3-(3-carbamimidoyl-phenyl)-isoxazole-4-carboxylic acid (2'-sulfamoylbiphenyl-

4-yl)amide, noncovalently inhibits coagulation factor Xa (*i.e.,* Stuart-Prower factor or prothrombinase (EC 3.4.21.6), K_i = 0.15 nM, but is relatively non-inhibitory for the closely related proteases thrombin (K_i = 2800 nM) and trypsin (K_i = 21 nM). Further optimization in the pyrazole series resulted in the discovery of fXa inhibitors such as SN429 with picomolar factor Xa affinity. **1.** Pinto, Orwat, Wang, *et al.* (2001) *J. Med. Chem.* **44**, 566.

Saccharate, *See* D-Glucarate

Saccharin

This nonnutritive sweetener (FW = 183.19 g/mol; CAS: 81-07-2 and 128-44-9; Solubility (sodium salt) = 1 g/mL), also known as 1,2-benzisothiazol-3(2*H*)-one 1,1-dioxide and 2*H*-1λ^6,2-benzothiazol-1,1,3-trione, is ~500 times sweeter than sucrose. The hydrogen atom on the nitrogen atom is quite acidic (pK_a ~2), readily forming sodium and calcium salts. **ACME:** Because saccharin is nearly fully ionized at physiological pH in body fluids, it is more completely absorbed from the stomachs of animals having low pH (guinea pig, pH 1.4; rabbit, pH 1.9) than from those with higher pH (rat, pH 4.2). When present at the higher pH of the intestine, it is slowly absorbed. Saccharin is rapidly eliminated in the urine. Following administration of a single oral dose of saccharin in humans, peak plasma levels of saccharin are rapidly achieved; however, saccharin clearance from the plasma is prolonged. Saccharin binds reversibly to plasma proteins. The presence of food in the gut is associated with a reduced initial peak plasma concentration. **Carcinogenesis:** Modest promotion of bladder carcinogens by high dietary concentrations of sodium saccharin is only observed in rats. Malignant transformation of cultured human foreskin fibroblasts occurred in cells exposed to a non-toxic concentration of sodium saccharin (50 µg/mL) after release from the G_1 cell-cycle phase, followed by exposure to either *N*-ethyl- or *N*-methyl-nitrosourea. Despite suggestions that saccharin induces and/or promotes carcinogenesis in experimental animals, subsequent work suggests it would be inappropriate to consider the bladder tumors induced in male rats by sodium saccharin to be a relevant to hazard to humans. **Target(s):** adenylyl cyclase (1); carbonic anhydrase (2); glucose-6-phosphatase (competitively) (3,4); tubulin polymerization, *or* microtubule assembly (5); urease (6); guanylate cyclase (7). **1.** Dib, Wrisez, el Jamali, Lambert & Correze (1997) *Cell Signal* **9**, 431. **2.** Wilson, Tanko, Wendland, *et al.* (1998) *Physiol. Chem. Phys. Med. NMR* **30**, 149. **3.** Lygre (1976) *Can. J. Biochem.* **54**, 587. **4.** Lygre (1974) *Biochim. Biophys. Acta* **341**, 291. **5.** Albertini, Friederich, Holderegger & Wurgler (1988) *Mutat. Res.* **201**, 283. **6.** Lok, Iverson & Clayson (1982) *Cancer Lett.* **16**, 163. **7.** Vesely DL, Levey (1978) *Biochem. Biophys. Res. Commun.* **81**, 1384.

Saccharo-1,4-Lactone, *See* D-Glucaro-1,4-Lactone

Saccharomyces cerevisiae Killer Toxins
These toxic proteins, which are encoded by a double-stranded RNA virus residing within the cytoplasm of *S. cerevisiae*, first appear as translated preproteins that proteolytically processed and released to the medium, where they then may then bind to and kill susceptible yeast. In *S. cerevisiae*, the two most studied toxins are: K1 toxin (*or* M1-1 Toxin), which binds to the β-1,6-D-glucan receptor on the target cell wall, entering the host where it then binds to the plasma membrane receptor Kre1p to form a cation-selective ion channel that is lethal to the cell; and K28 toxin, which uses the α-1,6-mannoprotein receptor to gain entry to target cells. K28 then utilizes the secretory pathway in reverse by displaying the endoplasmic reticulum HDEL signal. From the ER, K28 moves into the cytoplasm and shuts down DNA synthesis in the nucleus, triggering apoptosis. K1, K2, and K28 killer toxins are genetically encoded by medium-sized, double-stranded RNA (dsRNA) viruses grouped into three types: M1 (1.6 kb), M2 (1.5 kb), and M28 (1.8 kb). Only the positive strand has coding capacity, and the 5′-end region contains an open reading frame that encodes the preprotoxin (*or* pptox), which also provides immunity. The three toxin-coding M dsRNAs show no sequence homology to each other. The phenomenon of killer yeast attacking susceptible brewing yeast cultures was first noted by Pasteur in 1877, the system has provided great insights into the nature of yeast viruses

as well as their remarkable pathogenic mechanisms. **1.** Schmitt & Breinig (2006) *Nature Rev. Microbiol.* **4**, 212.

(*R*)-Saclofen

This competitive GABA antagonist (FW = 249.72 g/mol; CAS 125464-42-8; Soluble to 10 mM in H_2O), also named (*R*)-3-amino-2-(4-chlorophenyl)propylsulfonic acid, targets $GABA_B$, seven-helix receptor that couples to the $G_{i/o}$ class of heterotrimeric G-proteins. Saclofen is an analogue of the $GABA_B$ agonist baclofen. The *S*-enantiomer is inactive. The *S*-enantiomer is inactive. Saclofen's antiepileptic action is likey to be the consequence of its $GABA_B$ effect, which is coupled to excitation in the thalamo-cortical circuits, lowering the threshold for T-type Ca^{2+} channel opening. **1.** Kerr, Ong, Vaccher (1996) *European Journal of Pharmacology* **308**, R1.

Sacubitril

This oral antihypertensive pro-drug (FW = 411.49 g/mol; CAS 149709-62-6; Solubility: 100 mM in DMSO; compounded as the hemi-calcium salt), also known as AHU.377 and 4-{[(2*S*,4*R*)-1-(4-biphenylyl)-5-ethoxy-4-methyl-5-oxo-2-pentanyl]amino}-4-oxobutanoate, undergoes metabolic desterification to form the active neprilysin inhibitor, IC_{50} = 5 nM (1). Neprilysin is a neutral endopeptidase and its inhibition increases bioavailability of natriuretic peptides, bradykinin, and substance P, resulting in natriuretic, vasodilatatory, and anti-proliferative effects. Entresto® (Code name: LCZ696) is a drug combining the actions of valsartan and sacubitril, as a first-in-class angiotensin II-receptor and neprilysin inhibitor for the treatment of heart failure and as a drug that may exert a preferential effect on systolic pressure (2). (*See* Valsartan). **1.** Ksander, Ghai, de Jesus, *et al.* (1995) *J. Med. Chem.* **38**, 1689. **2.** Bavishi, Messerli, Kadosh, Ruilope & Kario (2015) *Eur. Heart J.* **36**, 1967.

SADH, *See* Daminozide

Safingol, *See* threo-Dihydrosphingosine

SAHA, *See* Vorinostat

Sal 003

This ER-stress/UPR modulator (FW = 463.21 g/mol; CAS 1164470-53-4; Soluble to 100 mM in DMSO), also named 3-phenyl-*N*-(2,2,2-trichloro-1-((((4-chlorophenyl)amino)carbonothioyl)amino)ethyl)acrylamide, targets enzymes (EC$_{50}$ ~ 20 μM) catalyzing the dephosphorylation of eukaryotic translation Initiation Factor 2 subunit α (eIF2α), thereby protecting cells from Endoplasmic Reticulum stress-induced (also called Unfolded Protein Response-$_{induced}$) apoptosis (1). Sal-003 is a salubrinal analogue possessing improved aqueous solubility (*See Salubrinol*). Sal-003 prevents long-term potentiation (LTP) and long-term memory formation (LTM) in mice. **1.** Baltzis, Pluquet, Papadakis, *et al.* (2007) *J. Biol. Chem.* **282**, 31675. **2.** Costa-Mattioli, Gobert, Stern, *et al.* (2007) *Cell* **129**, 195.

Salacinol

This sulfonium ion-containing monosaccharide analogue (FW = 320.34 g/mol; CAS 200399-47-9) from the stems and roots of kothalahimbutu (*Salacia reticulata*) inhibits α-glucosidase activities. Water left overnight in a cup made of kothalahimbutu wood is used as a traditional medicine in Sri Lanka and India for treating diabetes. **Target(s):** α-amylase (1); glucoamylase (2); α-glucosidase (3-5); isomaltase (3); maltase (3); α-mannosidase (5); mannosyl-oligosaccharide 1,3-1,6-α-mannosidase, *or* mannosidase II (6); and sucrase (3). **1.** Ghavami, Johnston, Jensen, Svensson & Pinto (2001) *J. Amer. Chem. Soc.* **123**, 6268. **2.** Johnson, Jensen, Svensson & Pinto (2003) *J. Amer. Chem. Soc.* **125**, 5663. **3.** Yoshikawa, Murakami, Shimada, *et al.* (1997) *Tetrahedron Lett.* **38**, 8367. **4.** Matsuda, Murakami, Yashiro, Yamahara & Yoshikawa (1999) *Chem. Pharm. Bull.* **47**, 1725. **5.** Yuasa, Takada & Hashimoto (2001) *Bioorg. Med. Chem. Lett.* **11**, 1137. **6.** Kuntz, Ghavami, Johnston, Pinto & Rose (2005) *Tetrahedron Asymmetry* **16**, 25.

Salermide

This SIRT inhibitor (FW = 394.47 g/mol; CAS 1105698-15-4; Soluble to 100 mM in DMSO), also named *N*-[3-[[(2-hydroxy-1-naphthalenyl)-methylene]amino]phenyl]-α-methylbenzeneacetamide, targets class III histone/protein deacetylases (sirtuins) SIRT1 and SIRT2, with a stronger inhibitory effect on SIRT2 than on SIRT1 *in vitro*. Salermide induces the reactivation of proapoptotic genes repressed by SIRT1, resulting in massive apoptosis in cancer cells within 24 hours (1,2). Salermide likewise prompts tumor-specific cell death in a wide range of human cancer cell lines that is independent of global tubulin and K16H4 acetylation, ruling out a putative Sirt2-mediated apoptotic pathway and suggesting an *in vivo* mechanism involving Sirt1. Indeed, RNA interference-mediated knockdown of Sirt1, but not Sirt2, induces apoptosis in cancer cells. **1.** Mai, Massa, Lavu, *et al.* (2005) *J. Med. Chem.* **48**, 7789. **2.** Lara, Mai, Calvanese, *et al.* (2009) *Oncogene* **28**, 781.

Salicin

This naturally occurring, non-reducing glucoside (FW = 286.28 g/mol; CAS 138-52-3; M.P. = 199-202°C; water soluble 40 mg/mL at 25°C), also known as saligenin β-D-glucopyranoside, from the bark of poplar and willow, is frequently used as a alternative substrate for β-glucosidases and often acts as a competitive inhibitor of glucosidases. **Target(s):** cellobiase, *Neocallimastix frontalis* (1); cyclomaltodextrin glucanotransferase (2); β-glucosidase, *Thermomyces lanuginosus* (3); phosphoenolpyruvate-dependent phosphotransferase system, *Staphylococcus carnosus* (4); sucrose synthase (5); thioglucosidase, *or* myrosinase; *or* sinigrinase (6-9); and α,α-trehalase (10). **1.** Li & Calza (1991) *Biochim. Biophys. Acta* **1080**, 148. **2.** Bovetto, Villette, Fontaine, Sicard & Bouquelet (1992) *Biotechnol. Appl. Biochem.* **15**, 59. **3.** Lin, Pillay & Singh (1999) *Biotechnol. Appl. Biochem.* **30**, 81. **4.** Christiansen & Hengstenberg (1999) *Microbiology* **145**, 2881. **5.** Wolosiuk & Pontis (1974) *Arch. Biochem. Biophys.* **165**, 140. **6.** Ohtsuru & Hata (1973) *Agric. Biol. Chem.* **37**, 2543. **7.** Ohtsuru, Tsuruo & Hata (1969) *Agric. Biol. Chem.* **33**, 1315. **8.** Tani, Ohtsuru & Hata (1974) *Agric. Biol. Chem.* **38**, 1623. **9.** Tsuruo & Hata (1968) *Agric. Biol. Chem.* **32**, 1420. **10.** Silva, Terra & Ferreira (2004) *Insect Biochem. Mol. Biol.* **34**, 1089.

Salicylaldoxime

This metal ion chelator (FW = 137.14 g/mol; CAS 94-67-7), also known as 2-hydroxybenzaldehyde oxime and 2-hydroxybenzaldoxime, forms high-affinity complexes with Ni^{2+}, Cu^{2+}, and Co^{2+} ($K_{formation}$= 6300 M^{-1}, 16000 M^{-1}, and 1.3 x 10^8 M^{-1}, respectively). The solid (melting point = 57°C) is slightly soluble in cold water, more soluble in hot water, and has pK_a values of 8.9 and 11.1. Salicylaldoxime decomposes on heating to form salicylaldehyde and hydroxylamine. **Target(s):** alcohol dehydrogenase (1); anthranilate 3-monooxygenase (deaminating), *or* anthranilate 2,3-dioxygenase (deaminating) (2-4); arylformamidase (5); ascorbate oxidase (6,7); catechol oxidase (8-10); complement component C'3, guinea pig (11); cytochrome *c* oxidase, complex IV (12); 2,3-dihydroxybenzoate 2,3-dioxygenase (13); DNA topoisomerase II, inhibited by Co(II)salicylaldoxime complex (14); electron transport, chloroplast (15-20); Hill reaction (15); hydroxylamine reductase (21); K$^+$ and Ca^{2+} currents (22); nitroaryl nitroreductase (23); photosynthesis (15,24); tyrosinase, *or* monophenol monooxygenase (10,25,26); and xanthine oxidase (27). **1.** Sund & Theorell (1963) *The Enzymes*, 2nd ed. (Boyer, Lardy & Myrbäck, eds.), **7**, 25. **2.** Subba Rao, Sreeleela, Prem Kumar & Vaidyanathan (1970) *Meth. Enzymol.* **17A**, 510. **3.** Kumar, Streeleela, Rao & Vaidyanathan (1973) *J. Bacteriol.* **113**, 1213. **4.** Streeleela, SubbaRao, Premkumar & Vaidyanathan (1969) *J. Biol. Chem.* **244**, 2293. **5.** Seifert & Casida (1979) *Pestic. Biochem. Physiol.* **12**, 273. **6.** Stark & Dawson (1963) *The Enzymes*, 2nd ed. (Boyer, Lardy & Myrbäck, eds.), **8**, 297. **7.** McCarthy, Green & King (1939) *J. Biol. Chem.* **128**, 455. **8.** Dawson & Magee (1955) *Meth. Enzymol.* **2**, 817. **9.** Asghar & Siddiqi (1970) *Enzymologia* **39**, 289. **10.** Ben-Shalom, Kahn, Harel & Mayer (1977) *Phytochemistry* **16**, 1153. **11.** Shin & Mayer (1968) *Biochemistry* **7**, 3003. **12.** Yonetani (1963) *The Enzymes*, 2nd ed. (Boyer, Lardy & Myrbäck, eds.), **8**, 41. **13.** Sharma & Vaidyanathan (1975) *Eur. J. Biochem.* **56**, 163. **14.** Jayaraju, Gopal & Kondapi (1999) *Arch. Biochem. Biophys.* **369**, 68. **15.** Izawa & Good (1972) *Meth. Enzymol.* **24**, 355. **16.** Milton & Rienits (1972) *FEBS Lett.* **22**, 335. **17.** Golbeck (1980) *Arch. Biochem. Biophys.* **202**, 458. **18.** Berg & Izawa (1976) *Biochim. Biophys. Acta* **440**, 483. **19.** Renger, Vater & Witt (1967) *Biochem. Biophys. Res. Commun.* **26**, 477. **20.** Katoh & San Pietro (1966) *Biochem. Biophys. Res. Commun.* **24**, 903. **21.** Zucker & Nason (1955) *Meth. Enzymol.* **2**, 416. **22.** Karhu, Perttula, Weckstrom, Kivisto & Sellin (1995) *Eur. J. Pharmacol.* **279**, 7. **23.** Higgins (1961) *Enzymologia*

23, 176. **24**. Green, McCarthy & King (1939) *J. Biol. Chem.* **128**, 447. **25**. Dawson & Tarpley (1951) *The Enzymes*, 1st ed. (Sumner & Myrbäck, eds.), **2** (part 1), 454. **26**. Ley & Bertram (2001) *Bioorg. Med. Chem.* **9**, 1879. **27**. Roussos (1967) *Meth. Enzymol.* **12A**, 5.

Salicylamide

This analgesic (FW = 137.14 g/mol; CAS 65-45-2; M.P. = 140°C; water soluble 2 g/L at 30°C), also known as 2-hydroxybenzamide, inhibits acetyltransferase (1), lactate dehydrogenase (2,3), medium-chain fatty-acyl-CoA synthetase, *or* butyryl-CoA synthetase (4), monoamine oxidase, *or* amine oxidase (5) NAD$^+$ ADP-ribosyltransferase, *or* poly(ADP-ribose) polymerase (6), nicotinate glucosyltransferase (7), salicylate monooxygenase (8), sulfanilamide acetylase (9), and UDP-glucuronosyltransferase (10). **1**. Voice, Manis & Weber (1993) *Drug Metab. Dispos.* **21**, 181. **2**. von Euler (1942) *Ber.* **75**, 1876. **3**. Webb (1966) *Enzyme and Metabolic Inhibitors*, vol. 2, p. 501, Academic Press, New York. **4**. Londesborough & Webster, Jr. (1974) *The Enzymes*, 3rd ed. (Boyer, ed.), **10**, 469. **5**. Byczkowski & Korolkiewicz (1976) *Pharmacol. Res. Commun.* **8**, 477. **6**. Banasik, Komura, Shimoyama & Ueda (1992) *J. Biol. Chem.* **267**, 1569. **7**. Taguchi, Sasatani, Nishitani & Okumura (1997) *Biosci. Biotechnol. Biochem.* **61**, 720. **8**. Massey & Hemmerich (1975) *The Enzymes*, 3rd ed. (Boyer, ed.), **12**, 191. **9**. Johnson (1955) *Can. J. Biochem. Physiol.* **33**, 107. **10**. Boutin, Thomassin, Siest & Batt (1984) *Pharmacol. Res. Commun.* **16**, 227.

Salicylate (Salicylic Acid)

This sweet-tasting phytohormone and aspirin breakdown product (FW$_\text{free-acid}$ = 138.12 g/mol; CAS 69-72-7; White Solid; M.P. = 159 °C; pK_a = 2.9 and 13.1; sodium salt Solubility: 1.25 g/mL H$_2$O) is often found as an ester in many plants, especially sweet birch bark and wintergreen leaves. In plants, salicylate plays diverse roles, affecting growth and development, photosynthesis, transpiration, ion transport, and morphogenetically by regulating leaf anatomy and chloroplast structure. Although derived chiefly from phenylalanine, some plants (*e.g. Arabidopsis thaliana*) synthesize salicylate by a second pathway. Salicylate also activates adenosine monophosphate-activated protein kinase (AMPK), a signal-transducing enzyme that controls the activity of many energy-consuming biosynthetic reactions. The net effect of AMPK activation is enhancement of hepatic fatty acid oxidation, ketogenesis, skeletal muscle fatty acid oxidation as well as inhibition of cholesterol synthesis, lipogenesis, and triglyceride formation, adipocyte lipolysis and lipogenesis, as well as modulation of insulin secretion by pancreatic β-cells. **Metal Ion Interactions:** Salicylate is also a metal ion chelator, forming reasonably high-affinity chelate complexes with Mn^{2+} (K_assoc = 7.9 x 10^5 M^{-1}), Be^{2+} (K_assoc = 20000 M^{-1}), iron (K_assoc = 4.0 x 10^6 M^{-1}), Co^{2+} (K_assoc = 5.0 x 10^6 M^{-1}), Ni^{2+} (K_assoc = 1.0 x 10^7 M^{-1}), and Cu^{2+} (K_assoc = 4.0 x 10^{10} M^{-1}). As a consequence, one must always entertain the possibility that its inhibitory action results from its metal ion-sequestering ability and depletion of catalytically essential active-site metal ion. (*For Pharmacokinetics & Metabolism, See Acetylsalicylate*) **Target(s):** alanine aminotransferase (1); alcohol dehydrogenase (2,3); aldose reductase (4); D-amino-acid oxidase (5,6); L-amino-acid oxidase (7-9); arginase (10,11); arylamine *N*-acetyltransferase (12); ascorbate peroxidase (13); aspartate aminotransferase (1,14); ATPase, mitochondrial (15); benzoyl-CoA synthetase, *or* benzoate:CoA ligase, also weak alternative substrate (69); γ-butyrobetaine hydroxylase, *or* γ-butyrobetaine:2-oxoglutarate dioxygenase; K_i = 1.3 mM (16); cathepsin L, weakly inhibited (80);

cholesterol sulfotransferase (17); creatine kinase (18,19); L-3-cyanoalanine synthase (72); 4,5-dihydroxyphthalate decarboxylase (75); esterase (20); estrone sulfotransferase (89,90); gentisate 1,2-dioxygenase (21); glucose transport (22); glutamate decarboxylase (23); glutamate dehydrogenase (24,25); L-glutamine:D-fructose-6-phosphate aminotransferase (26,27); glutamyl-tRNA synthetase (28); glycerol dehydratase (29); 3-hydroxyanthranilate 3,4-dioxygenase (30); *N*-hydroxyarylamine *O*-acetyltransferase (31); 4-hydroxybenzoate 4-*O*-β-glucosyltransferase (93); 4-hydroxybenzoate 3-monooxygenase (NAD(P)H) (32); γ-hydroxybutyrate dehydrogenase (33); IκB kinase, weakly inhibited (91); imidazoleacetate phosphoribosyltransferase (34); lactate dehydrogenase (5,24,25,35,36); lipase, *or* triacylglycerol lipase (37); long-chain 3-hydroxyacyl-CoA dehydrogenase (38); mandelate racemase (70,71); medium-chain fatty-acyl-CoA synthetase, *or* butyryl-CoA synthetase (39-43); 6-methylsalicylate decarboxylase (44); monoamine oxidase (45); myeloperoxidase (46); NAD$^+$ nucleosidase, *or* NADase (82,83); NAD(P)H dehydrogenase (quinone), *or* DT diaphorase (47); NADPH oxidase, neutrophil (48); nicotinate *N*-methyltransferase (95); nicotinate phospho-ribosyltransferase (92); orsellinate decarboxylase (49); pectate lyase (73,74); peptidyl-dipeptidase A, *or* angiotensin I-converting enzyme (81); peroxidase (50); phenol sulfotransferase, *or* aryl sulfotransferase (12,51,52,90); phenylalanine ammonia-lyase (53); phosphatase, acid (85); 3-phosphoglycerate kinase (54,55); phospholipase A$_2$ (87); phylloquinone epoxide reductase (56); polyphenol oxidase (96); propionyl-CoA synthetase (57); prostaglandin H$_2$ synthase (58); protein-tyrosine-phosphatase 1B, weakly inhibited (84); *o*-pyrocatechuate decarboxylase (59,76-78); RNA polymerase (60); short-chain 3-hydroxyacyl-CoA dehydrogenase (38); steryl sulfatase (61); sulfate transport (62); tannase (86); thyroxine 5'-deiodinase (63,64,79); tyrosinase, *or* monophenol monooxygenase (5,96); tyrosine-ester sulfotransferase (88); urease (65); vitamin K epoxide reductase (66); and xanthine oxidase (67,68). Salicylic acid also inhibits enzymatic browning of fresh-cut Chinese chestnut (*Castanea mollissima*) by competitively inhibiting polyphenol oxidase (97). **1**. Hanninen & Hartiala (1965) *Biochem. Pharmacol.* **14**, 1073. **2**. Brändén, Jörnvall, Eklund & Furugren (1975) *The Enzymes*, 3rd ed. (Boyer, ed.), **11**, 103. **3**. Winberg, Thatcher & McKinley-McKee (1982) *Biochim. Biophys. Acta* **704**, 17. **4**. Sharma & Cotlier (1982) *Exp. Eye Res.* **35**, 21. **5**. Webb (1966) *Enzyme and Metabolic Inhibitors*, vol. 2, Academic Press, New York. **6**. Yagi, Osawa & Okada (1959) *Biochim. Biophys. Acta* **35**, 102. **7**. Ratner (1955) *Meth. Enzymol.* **2**, 204. **8**. Zeller & Maritz (1945) *Helv. Chim. Acta* **28**, 365. **9**. Krebs (1951) *The Enzymes*, 1st ed. (Sumner & Myrbäck, eds.), **2** (part 1), 499. **10**. Greenberg (1951) *The Enzymes*, 1st ed. (Sumner & Myrbäck, eds.), **1** (part 2), 893. **11**. Hunter & Downs (1948) *J. Biol. Chem.* **173**, 31. **12**. Rao & Duffel (1991) *Drug Metab. Dispos.* **19**, 543. **13**. Durner & Klessig (1995) *Proc. Natl. Acad. Sci. U.S.A.* **92**, 11312. **14**. Braunstein (1973) *The Enzymes*, 3rd ed. (Boyer, ed.), **9**, 379. **15**. Chatterjee & Stefanovich (1976) *Arzneimittelforschung* **26**, 499. **16**. Ng, Hanauske-Abel & Englard (1991) *J. Biol. Chem.* **266**, 1526. **17**. Epstein, Jr., Bonifas, Barber & Haynes (1984) *J. Invest. Dermatol.* **83**, 332. **18**. Watts (1973) *The Enzymes*, 3rd ed. (Boyer, ed.), **8**, 383. **19**. Chegwidden & Watts (1980) *Biochem. Pharmacol.* **29**, 2113. **20**. Weber & King (1935) *J. Biol. Chem.* **108**, 131. **21**. Harpel & Lipscomb (1990) *Meth. Enzymol.* **188**, 101. **22**. Scharff, Badr & Doyle (1982) *Fundam. Appl. Toxicol.* **2**, 168. **23**. McArthur & Smith (1969) *J. Pharm. Pharmacol.* **21**, 21. **24**. Baker (1967) *Design of Active-Site-Directed Irreversible Enzyme Inhibitors*, Wiley, New York. **25**. Baker, Lee, Tong, Ross & Martinez (1962) *J. Theoret. Biol.* **3**, 446. **26**. Chan & Lee (1975) *J. Pharm. Sci.* **64**, 1182. **27**. Chatterjee & Stefanovich (1976) *Arzneimittelforschung* **26**, 502. **28**. Vadeboncoeur & Lapointe (1980) *Brain Res.* **188**, 129. **29**. Johnson, Stroinski & Schneider (1975) *Meth. Enzymol.* **42**, 315. **30**. Nishizuka, Ichiyama & Hayaishi (1970) *Meth. Enzymol.* **17A**, 463. **31**. Yamamura, Sayama, Kakikawa, *et al.* (2000) *Biochim. Biophys. Acta* **1475**, 10. **32**. Fujii & Kaneda (1985) *Eur. J. Biochem.* **147**, 97. **33**. Kaufman & Nelson (1987) *J. Neurochem.* **48**, 1935. **34**. Moss, De Mello, Vaughan & Beaven (1976) *J. Clin. Invest.* **58**, 137. **35**. von Euler (1942) *Ber.* **75**, 1876. **36**. Cheshire & Park (1971) *Biochem. J.* **125**, 45P. **37**. Karamac & Amarowicz (1996) *Z. Naturforsch. [C]* **51**, 903. **38**. Glasgow, Middleton, Moore, Gray & Hill (1999) *Biochim. Biophys. Acta* **1454**, 115. **39**. Londesborough & Webster, Jr. (1974) *The Enzymes*, 3rd ed. (Boyer, ed.), **10**, 469. **40**. Kasuya, Hiasa, Kawai, Igarashi & Fukui (2001) *Biochem. Pharmacol.* **62**, 363. **41**. Kasuya, Igarashi & Fukui (1999) *Chem. Biol. Interact.* **118**, 233. **42**. Kasuya, Igarashi, Fukui & Nokihara (1996) *Drug Metab. Dispos.* **24**, 879. **43**. Kasuya, Igarashi & Fukui (1996) *Biochem. Pharmacol.* **52**, 1643. **44**. Light & Vogel (1975) *Meth. Enzymol.* **43**, 530. **45**. Faraj, Caplan, Lolies & Buchanan (1987) *J. Pharm. Sci.* **76**, 423. **46**.

Kettle & Winterbourn (1991) *Biochem. Pharmacol.* **41**, 1485. **47.** Hildebrandt & Suttie (1983) *J. Pharm. Pharmacol.* **35**, 421. **48.** Umeki & Soejima (1990) *Kansenshogaku Zasshi* **64**, 1184. **49.** Petterson (1965) *Acta Chem. Scand.* **19**, 2013. **50.** Lück (1958) *Enzymologia* **19**, 227. **51.** Vietri, Pietrabissa, Mosca, Rane & Pacific (2001) *Xenobiotica* **31**, 153. **52.** Vietri, De Santi, Pietrabissa, Mosca & Pacifici (2000) *Eur. J. Clin. Pharmacol.* **56**, 81. **53.** Subba Rao & Towers (1970) *Meth. Enzymol.* **17A**, 581. **54.** Khamis & Larsson-Raznikiewicz (1987) *Acta Chem. Scand. B* **41**, 348. **55.** Larsson-Raznikiewicz & Wiksell (1978) *Biochim. Biophys. Acta* **523**, 94. **56.** Takahashi (1988) *Biochem. Pharmacol.* **37**, 2857. **57.** Krahenbuhl & Brass (1991) *Biochem. Pharmacol.* **41**, 1015. **58.** Aronoff, Boutaud, Marnett & Oates (2003) *J. Pharmacol. Exp. Ther.* **304**, 589. **59.** Rao, Moore & Towers (1970) *Meth. Enzymol.* **17A**, 514. **60.** Janakidevi & Smith (1970) *J. Pharm. Pharmacol.* **22**, 58. **61.** Dibbelt & Kuss (1991) *Biol. Chem. Hoppe Seyler* **372**, 173. **62.** Shennan & Boyd (1986) *Brit. J. Obstet. Gynaecol.* **93**, 522. **63.** Shields & Eales (1986) *Gen. Comp. Endocrinol.* **63**, 334. **64.** Huang, Chopra, Beredo, Solomon & Chua Teco (1985) *Endocrinology* **117**, 2106. **65.** Onodera (1915) *Biochem. J.* **9**, 544. **66.** Thijssen, Janssen & Vervoort (1994) *Biochem. J.* **297**, 277. **67.** Bray (1975) *The Enzymes*, 3rd ed. (Boyer, ed.), **12**, 299. **68.** Booth (1935) *Biochem. J.* **29**, 1732. **69.** Wischgoll, Heintz, Peters, *et al.* (2005) *Mol. Microbiol.* **58**, 1238. **70.** Kenyon & Hegeman (1979) *Adv. Enzymol. Relat. Areas Mol. Biol.* **50**, 325. **71.** Maggio, Kenyon, Mildvan & Hegeman (1975) *Biochemistry* **14**, 1131. **72.** Wen, Huang, Liang & Liang (1997) *Plant Sci.* **125**, 147. **73.** Tardy, Nasser, Robert-Baudouy & Hugouvieux-Cotte-Pattat (1997) *J. Bacteriol.* **179**, 2503. **74.** Pissavin, Robert-Baudouy & Hugouvieux-Cotte-Pattat (1998) *Biochim. Biophys. Acta* **1383**, 188. **75.** Nakazawa & Hayashi (1978) *Appl. Environ. Microbiol.* **36**, 264. **76.** Subba Rao, Moore & Towers (1967) *Arch. Biochem. Biophys.* **122**, 466. **77.** Santha, Rao & Vaidyanathan (1996) *Biochim. Biophys. Acta* **1293**, 191. **78.** Kamath, Dasgupta & Vaidyanathan (1987) *Biochem. Biophys. Res. Commun.* **145**, 586. **79.** Goswami, Leonard & Rosenberg (1982) *Biochem. Biophys. Res. Commun.* **104**, 1231. **80.** Raghav & Singh (1993) *Indian J. Med. Res.* **98**, 188. **81.** Oshima & Nagasawa (1977) *J. Biochem.* **81**, 57. **82.** Schuber, Pascal & Travo (1978) *Eur. J. Biochem.* **83**, 205. **83.** Travo, Muller & Schuber (1979) *Eur. J. Biochem.* **96**, 141. **84.** Sarmiento, Wu, Keng, *et al.* (2000) *J. Med. Chem.* **43**, 146. **85.** Belfield, Ellis & Goldberg (1972) *Enzymologia* **42**, 91. **86.** Iibuchi, Minoda & Yamada (1972) *Agric. Biol. Chem.* **36**, 1553. **87.** Hendrickson, Trygstad, Loftness & Sailer (1981) *Arch. Biochem. Biophys.* **212**, 508. **88.** Duffel (1994) *Chem. Biol. Interact.* **92**, 3. **89.** Prusakiewicz, Harville, Zhang, Ackermann & Voorman (2007) *Toxicology* **232**, 248. **90.** King, Ghosh & Wu (2006) *Curr. Drug Metab.* **7**, 745. **91.** Burke (2003) *Curr. Opin. Drug Discov. Devel.* **6**, 720. **92.** Gaut & Solomon (1971) *Biochem. Pharmacol.* **20**, 2903. **93.** Katsumata, Shige & Ejiri (1989) *Phytochemistry* **28**, 359. **94.** Paul & Ratledge (1973) *Biochim. Biophys. Acta* **320**, 9. **95.** Taguchi, Nishitani, Okumura, Shimabayashi & Iwai (1989) *Agri. Biol. Chem.* **53**, 2867. **96.** Güllcin, Küfrevioglu & Oktay (2005) *J. Enzyme Inhib. Med. Chem.* **20**, 297. **97.** Zhou, Li, Wu, Fan & Ouyang (2015) *Food Chem.* **171**, 19.

Salicylazosulfapyridine, *See* Sulfasalazine

Salicylhydroxamic Acid

This hydroxamic acid (FW = 153.14 g/mol; CAS 89-73-6; pK_a value of 4.19 at 25°C), also known as SHAM and 2-hydroxybenzhydroxamic acid, inhibits a number of oxidoreductases. **Target(s):** alternative oxidase in plant and algal mitochondria (cyanide-insensitive respiration) (1-4); cytochrome bc_1 complex (5); cytochrome *c* oxidase (6); Δ^6-desaturase (7); electron transport (8); glycerol-3-phosphate dehydrogenase (9); lathosterol oxidase, *or* δ^7-sterol C$^{5(6)}$-desaturase (10,11); lipoxygenase (12-16); mucopolysaccharide sulfotransferase (17); myeloperoxidase (18-22); NAD(P)H dehydrogenase (23); NAD(P)H:quinone oxidoreductase (24); peroxidase (18-22,25,26); phenol sulfotransferase (17); polyphenol oxidase (27); ribonucleoside-diphosphate reductase (28,29); tyrosinase, *or* monophenol monooxygenase (26,30,31); and xanthine dehydrogenase (32,33). **1.** Shepherd, Chin & Sullivan (1978) *Arch. Microbiol.* **116**, 61. **2.** Drabikowska (1978) *Acta Biochim. Pol.* **25**, 71. **3.** Schonbaum, Bonner, Jr.,

Storey & Bahr (1971) *Plant Physiol.* **47**, 124. **4.** Atkin, Villar & Lambers (1995) *Plant Physiol.* **108**, 1179. **5.** Turrens, Bickar & Lehninger (1986) *Mol. Biochem. Parasitol.* **19**, 259. **6.** Goyal & Srivastava (1995) *J. Helminthol.* **69**, 13. **7.** Khozin-Goldberg, Bigogno & Cohen (1999) *Biochim. Biophys. Acta* **1439**, 384. **8.** Stoppani, Docampo, de Boiso & Frasch (1980) *Mol. Biochem. Parasitol.* **2**, 3. **9.** Wells, Xu, Washburn, Cirrito & Olson (2001) *J. Biol. Chem.* **276**, 2404. **10.** Rahier, Benveniste, Husselstein & Taton (2000) *Biochem. Soc. Trans.* **28**, 799. **11.** Taton & Rahier (1996) *Arch. Biochem. Biophys.* **325**, 279. **12.** Schewe, Wiesner & Rapoport (1981) *Meth. Enzymol.* **71**, 430. **13.** Park & Pariza (2001) *Biochim. Biophys. Acta* **1534**, 27. **14.** Vianello, Braidot, Bassi & Macri (1995) *Biochim. Biophys. Acta* **1255**, 57. **15.** Macri, Braidot, Petrussa & Vianello (1994) *Biochim. Biophys. Acta* **1215**, 109. **16.** Ishiura, Yoshimoto & Villee (1986) *FEBS Lett.* **201**, 87. **17.** Foye & Kulapaditharom (1985) *J. Pharm. Sci.* **74**, 355. **18.** Jerlich, Fritz, Kharrazi, *et al.* (2000) *Biochim. Biophys. Acta* **1481**, 109. **19.** Ikeda-Saito, Shelley, Lu, *et al.* (1991) *J. Biol. Chem.* **266**, 3611. **20.** Taylor, Guzman, Pohl & Kinkade (1990) *J. Biol. Chem.* **265**, 15938. **21.** Humphreys, Davies, Hart & Edwards (1989) *J. Gen. Microbiol.* **13**, 1187. **22.** Davies & Edwards (1989) *Biochem. J.* **258**, 801. **23.** Matsuo, Endo & Asada (1998) *Plant Cell Physiol.* **39**, 263. **24.** Matsuo, Endo & Asada (1998) *Plant Cell Physiol.* **39**, 751. **25.** Wisniewski, Rathbun, Knox & Brewin (2000) *Mol. Plant Microbe Interact.* **13**, 413. **26.** Rich, Wiegand, Blum, Moore & Bonner, Jr. (1978) *Biochim. Biophys. Acta* **525**, 325. **27.** Mazzafera & Robinson (2000) *Phytochemistry* **55**, 285. **28.** Larsen, Sjöberg & Thelander (1982) *Eur. J. Biochem.* **125**, 75. **29.** Elford, Van't Riet, Wampler, Lin & Elford (1981) *Adv. Enzyme Regul.* **19**, 151. **30.** Zhang, van Leeuwen, Wichers & Flurkey (1999) *J. Agric. Food Chem.* **47**, 374. **31.** Flurkey, Cooksey, Reddy, *et al.* (2008) *J. Agric. Food Chem.* **56**, 4760. **32.** Montalbini (1998) *Plant Sci.* **134**, 89. **33.** Nguyen & Feierbend (1978) *Plant Sci. Lett.* **13**, 125.

Salicylyl Phosphate, *See* 2-Carboxyphenyl Phosphate

Saligenin β-D-Glucopyranoside, *See* Salicin

Salinomycin

This polyether ionophore antibiotic and oxidative phosphorylation inhibitor (FW$_{free-acid}$ = 791.84 g/mol; CAS 53003-10-4), from *Streptomyces albus*, binds monovalent cations such as potassium ion. Salinomycin is an effective agent against coccidiosis and is used as a feed additive in animal husbandry. The unusual tricyclic spiroketal moiety represents a daunting challenge for chemical synthesis of salinomycin. *Note:* The fact that salinomycin is some 100 times more effective than taxol in killing human breast cancer cells suggests that this ionophore may have toxic side effects in mammals. **Target(s):** oxidative phosphorylation (1-3). **1.** Reed (1979) *Meth. Enzymol.* **55**, 435. **2.** Mitani, Yamanishi, Miyazaki & Otake (1976) *Antimicrob. Agents Chemother.* **9**, 655. **3.** Mitani, Yamanishi & Miyazaki (1975) *Biochem. Biophys. Res. Commun.* **66**, 1231.

Salinosporamide A

This cytotoxic marine natural product (FW = 313.78 g/mol), also named Marizomib, NPI-0052, and (4*R*,5*S*)-4-(2-chloroethyl)-1-((1*S*)-cyclohex-2-enyl(hydroxy)methyl)-5-methyl-6-oxa-2-azabicyclo[3.2.0]heptane-3,7-dione, from *Salinospora* strains present in ocean sediments, shows high affinity (IC$_{50}$ = 1.3 nM) for 20S proteasomes, covalently modifying an active-site threonine residue (1). The first stereoselective synthesis was reported in 2004 (2). **1**. Feling, Buchanan, Mincer, *et al.* (2003) *Angewan. Chem., Inter. Ed.* **42**, 355. **2**. Leleti & Corey (2004) *J. Am. Chem. Soc.* **126**, 6230.

Salirasib

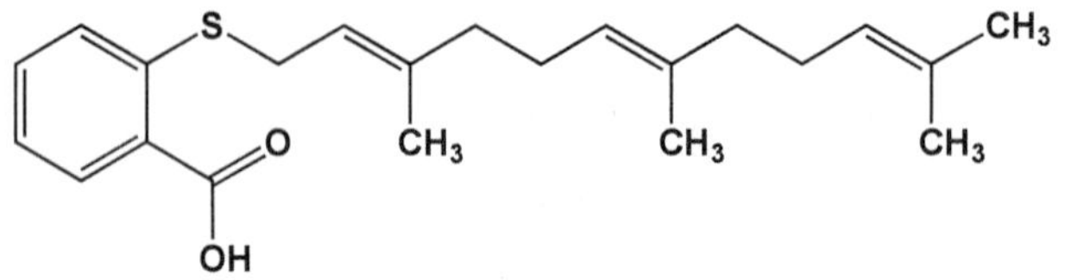

This Ras protein inhibitor (FW = 358.54 g/mol; CAS 162520-00-5; Soluble to 100 mM in DMSO and to 50 mM in 1eq. NaOH), also named 2-[[(2E,6E)-3,7,11-trimethyl-2,6,10-dodecatrien-1-yl]thio]benzoic acid, displaces active Ras (EC$_{50}$ = 10-15 µM) from the plasma membrane, thereby impairing downstream signaling and inhibiting endometrial carcinoma cell proliferation. Salirasib also facilitates Ras degradation and likewise induces autophagy in several human cancer cell lines. **1**. Haklai, Weisz, Elad, *et al.* (1998) *Biochemistry* **37**, 1306. **2**. Faigenbaum, Haklai, Ben-Baruch & Kloog (2013) *Oncotarget* **4**, 316. **3**. Schmukler, Grinboim, Schokoroy, *et al.* (2013) *Oncotarget* **4**, 142.

Salmeterol

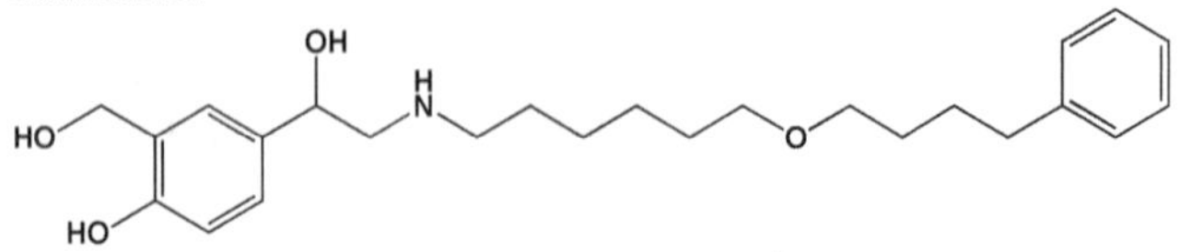

This bronchodilator (FW = 415.57 g/mol; CAS 89365-50-4), also named (*RS*)-2-(hydroxymethyl)-4-{1-hydroxy-2-[6-(4-phenylbutoxy)hexylamino]-ethyl}phenol, is a structural analogue of albuterol and is a β$_2$-adrenergic agonist. Its lipophilic aryl alkyl group makes it β$_2$-selective. Like other long-acting β$_2$-adrenoceptor agonists, which include formoterol and bambuterol, salmeterol is used in the management of asthma and chronic obstructive pulmonary disease (COPD). **Target(s):** aldehyde oxidase (1); and CYP2C8 (2). **1**. Obach, Huynh, Allen & Beedham (2004) *J. Clin. Pharmacol.* **44**, 7. **2**. Walsky, Gaman & Obach (2005) *J. Clin. Pharmacol.* **45**, 68.

Salsolinol

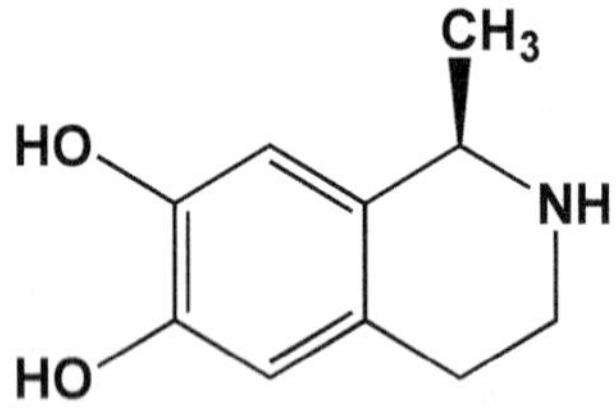

This dopamine-derived alkaloid (FW$_{free-base}$ = 179.22 g/mol; CAS 57256-34-5; Solubility: 25 mM in H$_2$O), also known as 6,7-dihydroxy-1-methyl-1,2,3,4-tetrahydroisoquinoline, induces DNA damage and chromosomal aberrations and also inhibits calcium binding. Both the (+)-(*R*)-salsolinol (shown above) and (−)-(*S*)-salsolinol of this tetrahydroisoquinoline are alternative substrates for soluble and membrane-bound catechol *O*-methyltransferases, thereby accounting for their action as competitive inhibitors of dopamine methylation. **Target(s):** catechol *O*-methyltransferase (1-3); complex II, mitochondrial, *or* succinate:ubiquinone reductase (4); dihydropteridine reductase (5); monoamine oxidase (2,6-8); tryptophan 5-monooxygenase, inhibited by both enantiomers (9); and tyrosine 3-monooxygenase, inhibited by both enantiomers (8,10). **1**. Hotzl & Thomas (1997) *Chirality* **9**, 367. **2**. Giovine, Renis & Bertolino (1976) *Pharmacology* **14**, 86. **3**. Tunnicliff & Ngo (1983) *Int. J. Biochem.* **15**, 733. **4**. Storch, Kaftan, Burkhardt & Schwarz (2000) *Brain Res.* **855**, 67. **5**. Shen, Smith, Davis, Brubaker & Abell (1982) *J. Biol. Chem.* **257**, 7294. **6**. Minami, Maruyama, Dostert, Nagatsu & Naoi (1993) *J. Neural. Transm. Gen. Sect.* **92**, 125. **7**. Bembenek, Abell, Chrisey, *et al.* (1990) *J. Med. Chem.* **33**, 147. **8**. Patsenka & Antkiewicz-Michaluk (2004) *Pol. J. Pharmacol.* **56**, 727. **9**. Ota, Dostert, Hamanaka, Nagatsu & Naoi (1992) *Neuropharmacology* **31**, 337. **10**. Minami, Takahashi, Maruyama, *et al.* (1992) *J. Neurochem.* **58**, 2097.

Salubrinol

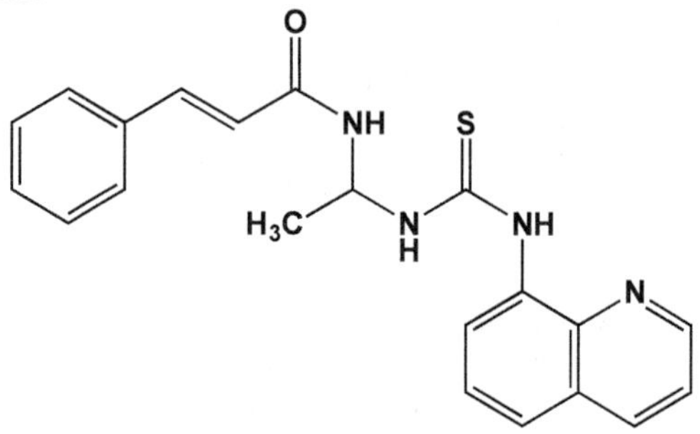

This cell-permeable, ER-stress/UPR modulator (FW = 479.81 g/mol; CAS 405060-95-9; Soluble to 100 mM in DMSO), also known as 3-phenyl-*N*-[2,2,2-trichloro-1-[[(8-quinolinylamino)thioxomethyl]amino]ethyl]-2-propenamide, targets enzymes (EC$_{50}$ ~ 15 µM) catalyzing the dephosphorylation of eukaryotic translation Initiation Factor 2 subunit α (eIF2α), thereby protecting cells from Endoplasmic Reticulum stress-induced (also called Unfolded Protein Response-induced) apoptosis (1). Salubrinal blocks eIF2α dephosphorylation mediated by a Herpes Simplex virus protein, thereby inhibiting viral replication (1). Salubrinal also interacts with the anti-apoptotic protein Bcl-2, thereby inhibiting binding of the non-peptidic antagonist HA14-1 and of a porphycene that catalyzes Bcl-2 photodamage. As a result, salubrinal offers protection from the apoptotic and autophagic effects that can result from loss of Bcl-2 function (2). It also delays the formation of insoluble aggregates of the mutant superoxide dismutase-1 (SOD1) and suppresses the mutant-induced cell death (3). The hypothesis that sleep is facilitated by signals associated with the ER stress response is supported, in part, by the finding that intracerebroventricular (ICV) infusion of salubrinal increases phospho-eIF2α by inhibiting its dephosphorylation and likewise increases deep slow-wave sleep by 255%, reduces active waking by 49%, increases non-rapid eye movement (NREM) sleep, and reduces power in the σ, β, and γ bands during NREM sleep (4). Cyclosporine-triggered ER stress induces endothelial cells to undergo endothelial phenotypic changes suggestive of a partial endothelial-to-mesenchymal transition, and salubrinal stabilizes the endothelial phenotype (5). Taken together, saubrinol should be cytoprotective against cytostatic/cytotoxic effects of drugs inducing ER stress. **1**. Boyce, Bryant, Jousse, *et al.* (2005) *Science* **307**, 935. **2**. Kessel (2006) *Biochem. Biophys. Res. Commun.* **346**, 1320. **3**. Oh, Shin, Yuan & Kang (2008) *J. Neurochem.* **104**, 993. **4**. Methippara, Bashir, Kumar, *et al.* (2009) *Am. J. Physiol Regul. Integr. Comp. Physiol.* **296**, R178. **5**. Bouvier, Flinois, Gilleron, *et al.* (2009) *Am. J. Physiol. Renal Physiol.* **296**, F160.

(+)-Salutaridine

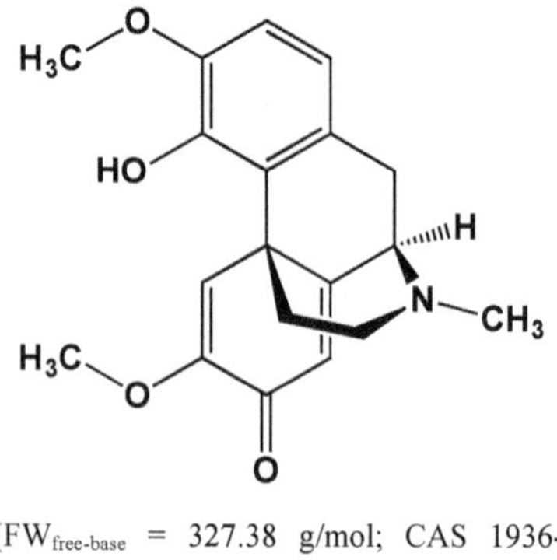

This alkaloid (FW$_{free-base}$ = 327.38 g/mol; CAS 1936-18-1), itself an intermediate in the biosynthesis of morphine, inhibits salutaridinol 7-*O*-acetyltransferase. The (−)-isomer is also called sinoaculine. Salutaridinol-7-*O*-acetyltransferase catalyzes the stoichiometric transfer of the acetyl group from acetyl-CoA to the 7-OH group of salutaridinol yielding salutaridinol-7-*O*-acetate, which is a new intermediate in morphine biosynthesis. Salutaridinol-7-*O*-acetate undergoes a subsequent spontaneous allylic elimination at pH 8-9, leading to the formation of thebaine, the first morphinan alkaloid with the complete pentacyclic ring system, or at pH 7 leading to dibenz[d,f]azonine alkaloids that contain a nine-membered ring. Acetylation and subsequent allylic elimination is a new enzymic mechanism in alkaloid biosynthesis, which in the poppy plant can transform one precursor into alkaloids possessing markedly different ring systems, depending on the reaction pH. **1**. Lenz & Zenk (1995) *J. Biol. Chem.* **270**, 31091.

Salubrinal

This phosphoprotein phosphatase inhibitor (FW = 479.81 g/mol; CAS 405060-95-9; Solubility: 96 mg/mL DMSO; <1 mg/mL H_2O), also named 3-phenyl-N-[2,2,2-trichloro-1-[[(8-quinolinylamino)thioxomethyl]amino]-ethyl]-(2E)-propenamide, protects cells (EC$_{50}$ = ~15 μM) from ER stress-mediated apoptosis (1). Salubrinal likewise inhibits viral replication by blocking herpes simplex virus protein-mediated eIF2α dephosphorylation (1). **Mode of Inhibitory Action:** ER-assisted folding (ERAF) within the lumen of the endoplasmic reticulum (ER) is tightly coupled to protein-synthesis pathways operating on the outer surface of the ER, a process that is regulated in part by the translation initiation factor eIF2α. Imbalance of these events is associated with numerous misfolding diseases that directly or indirectly trigger the Unfolded Protein Response (UPR). By selectively blocking dephosphorylation of the ribosomal initiation protein eIF2α, salubrinal reprograms the kinetics of protein synthesis and chaperone-assisted folding, thereby providing a novel approach for rebalancing ER-protein load with cellular-folding capacity, in some cases correcting an underlying mechanism of pathogenesis (2). Arrest of protein synthesis is thus a first-line defense against the accumulation of unfolded or misfolded proteins in the ER. **Other Actions of Salubrinal:** Selective inhibition of eIF2α dephosphorylation also potentiates fatty acid-induced ER stress, causing pancreatic β-cell dysfunction and apoptosis (3). By interacting with the anti-apoptotic protein Bcl-2, salubrinal inhibits the actions of HA14-1, a non-peptidic Bcl-2 antagonist (**See HA14-1**) and porphycene, which catalyzes Bcl-2 photodamage (**See Porphycene**), thereby protecting cells from the apoptotic and autophagic effects resulting from loss of Bcl-2 function (4). Salubrinal also increases deep slow-wave sleep by 255%, reduces active waking by 49%, and increases the amount of p-eIF2α in both cholinergic and noncholinergic neurons (5). The latter is more widespread among the noncholinergic neurons. Salubrinal also induces c-Fos activation of GABAergic neurons within the sleep-promoting rostral median preoptic nucleus, while simultaneously reducing c-Fos activation of wake-promoting lateral hypothalamic orexin-expressing neurons and magnocellular BF cholinergic neurons (6). Such findings suggest ER stress plays a role in NREM sleep homeostasis in response to sleep deprivation and provides a mechanistic explanation for the sleep modulation by molecules signaling the need for brain protein synthesis (6). Accumulation α-synuclein within ER leads to chronic ER stress conditions that contribute to neurodegeneration in α-synucleinopathies, and salubrinal significantly attenuates disease manifestations in both the A53TαS Tg mouse model and the adeno-associated virus-transduced rat model of A53TαS-dependent dopaminergic neurodegeneration (7). **1.** Boyce, Bryant, Jousse, *et al.* (2005) *Science* **307**, 935. **2.** Wiseman & Balch (2005) *Trends Mol. Med.* **11**, 347. **3.** Cnop, Ladriere, Hekerman, *et al.* (2007) *J. Biol. Chem.* **282**, 3989. **4.** Kessel (2006) *Biochem. Biophys. Res. Commun.* 346, 1320. **5.** Methippara, Bashir, Kumar, *et al.* (2009) *Am. J. Physiol. Regul. Integr. Comp. Physiol.* **296**, R178. **6.** Methippara, Mitrani, Schrader, Szymusiak & McGinty (2012) *Neuroscience* **209**, 108. **7.** Colla, Coune, Liu, *et al.* (2012) *J. Neurosci.* **32**, 3306.

Salvicine

This cytotoxic agent (FW = 330.42 g/mol; CAS 133276-80-9), which can induce apoptosis, is a structurally modified *o*-quinone derived from *Salvia prionitis*. It promotes the formation of the [DNA topoisomerase II]-DNA complex and inhibits pre- and post-strand DNA religation, IC$_{50}$ = 3 μM (1-3). **1.** Meng, Zhang & Ding (2001) *Biochem. Pharmacol.* **62**, 733. **2.** Meng, He, Zhang & Ding (2001) *Acta Pharmacol. Sin.* **22**, 741. **3.** Hu, Zuo, Xiong, *et al.* (2006) *Mol. Pharmacol.* **70**, 1593.

Salyrgan, See *Mersalyl*

SAM, See *S-Adenosyl-L-methionine*

Samarium-ATP, See *Sm^{3+}-ATP*

Sampatrilat

This substituted tyrosine (FW = 584.68 g/mol; CAS 129981-36-8), also known as UK 81252, is a potent inhibitor of both angiotensin-converting enzyme (*or* peptidyl-dipeptidase A) and neprilysin (1,2). **1.** Wallis, Ramsay & Hettiarachchi (1998) *Clin. Pharmacol. Ther.* **64**, 439. **2.** Kirk & Wilkins (1996) *Brit. J. Pharmacol.* **119**, 943.

Sangivamycin

This nucleoside antibiotic (FW = 309.28 g/mol; CAS 18417-89-5), also known as 4-amino-5-carboxamide-7-(D-ribofuranosyl)pyrrolo[2,3-d] pyrimidine and 7-deaza-7-carbamoyl-adenosine, is similar in structure to toyocamycin, but the latter is not a strong protein kinase C inhibitor. **Target(s):** adenosine kinase (1); [β-adrenergic-receptor] kinase (2); histone H1 kinase, *or* nuclear protein kinase (3); protein biosynthesis (4); protein kinase C, K_i = 11 μM (5-8); and rhodopsin kinase, K_i = 180 nM (7,9-11). Sangivamycin's amidine derivative (FW$_{free-base}$ = 308.30 g/mol) inhibits rhodopsin kinase: K_i = 2.2 μM (10-11). Its corresponding amidoxime derivative (FW = 324.30 g/mol) has a K_i = 1.0 μM (10-11). **1.** Long & Parker (2006) *Biochem. Pharmacol.* **71**, 1671. **2.** Benovic (1991) *Meth. Enzymol.* **200**, 351. **3.** Saffer & Glazer (1981) *Mol. Pharmacol.* **20**, 211. **4.** Cohen & Glazer (1985) *Mol. Pharmacol.* **27**, 349. **5.** Loomis & Bell (1988) *J. Biol. Chem.* **263**, 1682. **6.** Greene, Williams & Newton (1995) *J. Biol. Chem.* **270**, 6710. **7.** Lebioda, Hargrave & Palczewski (1990) *FEBS Lett.* **266**, 102. **8.** Osada, Sonoda, Tsunoda & Isono (1989) *J. Antibiot. (Tokyo)* **42**, 102. **9.** Palczewski, Rispoli & Detwiler (1992) *Neuron* **8**, 117. **10.** Palczewski, Carruth, Adamus, McDowell & Hargrave (1990) *Vision Res.* **30**, 1129. **11.** Palczewski, Kahn & Hargrave (1990) *Biochemistry* **29**, 6276.

Sanguinarine

This alkaloid (FW$_{HCl-Salt}$ = 367.78 g/mol; CAS 2447-54-3), also known as pseudochelerythrine (or, ψ-chelerythrine) and 13-methyl[1,3]benzodioxolo [5,6-c]-1,3-dioxolo[4,5-i]phenanthridinium, is widely distributed in plants, particularly in poppy-fumaria species. The chloride dihydrate exhibits antiplaque activity and is frequently incorporated into mouthrinses and toothpaste. Sanguinarine intercalates DNA, thereby inhibiting DNA biosynthesis and reverse transcriptase. **Target(s):** acetylcholinesterase (1,2); alanine aminotransferase (3); amine oxidase (copper-containing), *or* diamine oxidase (4,5); aminopeptidase N, *or* membrane alanyl aminopeptidase (6); α-amylase (7); aromatic-amino-acid decarboxylase (8); butyryl-cholinesterase (2); Ca^{2+}-dependent ATPase (9); cation-transporting ATPases (10); choline acetyltransferase (2); deoxyribonuclease I (11); dipeptidyl-peptidase IV (6,12); DNA ligase (ATP), human, IC_{50} = 127 µM (13); elastase (14); glutamate decarboxylase (15); lipase, *Candida rugosa* (16); 5-lipoxygenase (17); 12-lipoxygenase (17); monoamine oxidase (18); myosin-light-chain kinase (19); Na^+/K^+-exchanging ATPase (20-24); NF-κB activation (25); protein kinase A (19); protein kinase C (19,26); and reverse transcriptase, *or* RNA-directed DNA polymerase (2). **1.** Kuznetsova, Nikol'skaia, Sochilina & Faddeeva (2001) *Tsitologiia* **43**, 1046. **2.** Schmeller, Latz-Bruning & Wink (1997) *Phytochemistry* **44**, 257. **3.** Walterova, Ulrichova, Preininger, *et al.* (1981) *J. Med. Chem.* **24**, 1100. **4.** Luhova, Frebort, Ulrichova, *et al.* (1995) *J. Enzyme Inhib.* **9**, 295. **5.** Vaidya, Rajagopalan, Kale & Levine (1980) *J. Postgrad. Med.* **26**, 28. **6.** Sedo, Vlasicova, Bartak, *et al.* (2002) *Phytother. Res.* **16**, 84. **7.** Zajoncova, Kosina, Vicar, Ulrichova & Pec (2005) *J. Enzyme Inhib. Med. Chem.* **20**, 261. **8.** Drsata, Ulrichova & Walterova (1996) *J. Enzyme Inhib.* **10**, 231. **9.** Faddeeva & Beliaeva (1988) *Tsitologia* **30**, 685. **10.** Faddeeva & Beliaeva (1997) *Tsitologiia* **39**, 181. **11.** Beliaeva, Sedova & Faddeeva (1984) *Biokhimiia* **49**, 222. **12.** Sedo, Malik, Vicar, Simanek & Ulrichova (2003) *Physiol. Res.* **52**, 367. **13.** Tan, Lee, Lee, *et al.* (1996) *Biochem. J.* **314**, 993. **14.** Tanaka, Metori, Mineo, *et al.* (1993) *Planta Med.* **59**, 200. **15.** Netopilova, Drsata & Ulrichova (1996) *Pharmazie* **51**, 589. **16.** Grippa, Valla, Battinelli, *et al.* (1999) *Biosci. Biotechnol. Biochem.* **63**, 1557. **17.** Vavreckova, Gawlik & Muller (1996) *Planta Med.* **62**, 397. **18.** Lee, Kai & Lee (2001) *Phytother. Res.* **15**, 167. **19.** Wang, Lu & Polya (1997) *Planta Med.* **63**, 494. **20.** Tandon, Das & Khanna (1993) *Nat. Toxins* **1**, 235. **21.** Faddeeva, Beliaeva & Sokolovskaia (1987) *Tsitologiia* **29**, 576. **22.** Seifen, Adams & Riemer (1979) *Eur. J. Pharmacol.* **60**, 373. **23.** Meyerson, McMurtrey & Davis (1978) *Neurochem. Res.* **3**, 239. **24.** Cohen, Seifen, Straub, Tiefenbach & Stermitz (1978) *Biochem. Pharmacol.* **27**, 2555. **25.** Chaturvedi, Kumar, Darnay, *et al.* (1997) *J. Biol. Chem.* **272**, 30129. **26.** Gopalakrishna, Chen & Gundimeda (1995) *Meth. Enzymol.* **252**, 132.

SANT-1

This Smo antagonist (FW = 373.49 g/mol; CAS 304909-07-7), also named *N*-[(3,5-dimethyl-1-phenyl-1*H*-pyrazol-4-yl)methylene]-4-(phenylmethyl)-1-piperazinamine, potently and selectively inhibits cyclopamine- and jervine-mediated ciliary translocation of Smoothened (IC_{50} = 20 nM; K_i = 1.2 nM), a distant relative of G protein-coupled receptors, which mediates Hedgehog (Hh) signaling during embryonic development and initiates/transmits ligand-independent pathway activation in tumorigenesis (1,2). With neoplastic pancreatic cells, SANT-1 enhances the induction of apoptosis, cell-cycle arrest in the G_0/G_1 junction, and ductal epithelial differentiation (3). (*Other Hedgehog inhibitors include: GANT61; BMS-833923; Purmorphamine; PF-5274857; LY2940680; Cyclopamine, LDE225, or NVP-LDE225, or Erismodegib; Vismodegib, or GDC-0449*) **1.** Chen, Taipale, Young, Maiti & Beachy (2002) *Proc. Natl. Acad. Sci. USA* **99**, 14071. **2.** Wilson, Chen & Chuang (2009) *PLoS One* **4**, e5182. **3.** Chun, Zhou & Yee (2009) *Cancer Biol. Ther.* **8**, 1328.

SANT-2

This allosteric Smo antagonist and Sonic hedgehog signal disruptor (FW = 479.96 g/mol; CAS 329196-48-7; Soluble to 100 mM in DMSO), also known as *N*-[3-(1*H*-benzimidazol-2-yl)-4-chlorophenyl]-3,4,5-triethoxy-benzamide, potently and selectively inhibits Smoothened receptor (K_i = 12 nM), a distant relative of G protein-coupled receptors. SANT-2 displaces smo-[^{3}H]SAG-1.3 and smo-[^{3}H]cyclopamine binding, with respective K_i values of 7.8 nM and 8.4 nM (1,2). The finding that SANT-1 and SANT-2 only partially inhibit SAG-1.3 binding to Smo was rationalized with an allosteric model accounting for their action (2). Subtle differences in binding of allosteric ligands such as SANT-1 and SANT-2 may result in unique phenotypes because they may change the way smoothened receptors interact with accessory proteins (2). **1.** Chen, *et al.* (2002) *Proc. Natl. Acad. Sci.* **99**, 14071. **2.** Rominger, Bee, Copeland, *et al.* (2009) *J. Pharmacol. Exp. Ther.* **329**, 995.

Sapanisertib

This ATP-competitive mTOR inhibitor (FW = 308.14 g/mol; CAS 1224844-38-5; Soluble in DMSO), also known as TAK-228, MLN0128, INK 128, and 1-isopropyl-3-(2-methylbenzo[*d*]oxazol-5-yl)-1*H*-pyrazolo[3,4-*d*]pyrimidin-4-amine, potently and selectively targets mammalian Target of Rapamycin (IC_{50} = 1 nM), with some 200-times weaker action against Class I PI3K isoforms. As TORC1/2 inhibitor, sapanisertib inhibits the phosphorylation of both the translational regulator S6 Kinase 1 (S6K1) and 4EBP1, the latter a translational suppressor eIF4E binding protein that functions as a key regulator in cellular growth, differentiation, apoptosis and survival. Both S6 and 4EBP1 are downstream substrates of TORC1, and selectively inhibits AKT phosphorylation at Ser473, the downstream substrate of TORC2. Sapanisertib also shows potent inhibition in cell lines resistant to rapamycin and pan-PI3K inhibitors. In a ZR-75-1 breast cancer xenograft model, it shows tumor growth inhibition efficacy at a dose of 0.3 mg/kg/day. Daily oral administration also inhibits angiogenesis and tumor growth in multiplexenograft models (2). Functional evaluation of rapamycin and sapanisertib on HRG-dependent signaling activities, uncovered a necessary role for mTORC2 in the regulation of the AKT/TSC2/mTORC1 axis by affecting the phosphorylation of AKT at the PDK1(PDPK1)-dependent site (Thr-308) as well as at the mTORC2-dependent site (Ser-473) (3). These findings suggest the potential benefits of targeting mTORC2 in HRG/ErbB2-induced breast cancer (3). mTOR inhibitors (*i.e.*, both rapalogs and kinase inhibitors) has been investigated as a co-therapy (with either trastuzumab or lapatinib) for cancers that are refractory to ErbB2-directed monotherapies, inasmuch as aberrant PI3K/AKT/mTOR activity is a hallmark of resistance to ErbB2 therapy resistance. **1.** García-García, Ibrahim, Serra, *et al.* (2012) *Clin. Cancer Res.* **18**, 2603. **2.** Liu, Thoreen, Wang, *et al.* (2009) *Drug Disc. Today: Therapeut. Strat.* 6(2), 47-55. **3.** Lin, Rojas, Cerione & Wilson (2014) *Mol. Cancer Res.* **12**, 940.

Saposins

This group of lipid-transfer and sphingolipid activator glycoproteins (which include Saposin A, 84 residues; Saposin B, 84 residues; Saposin C, 80 residues; and Saposin D, 83 residues) are required for the lysosomal degradation of sphingolipids and for the loading of lipid antigens onto antigen-presenting molecules of the CD1 type. Sap-A assists in the degradation of galactosylceramide by galactosylceramide-beta-galactosidase *in vivo*, which takes place at the surface of intra-endosomal/intra-lysosomal vesicles. Sap-A is believed to mediate the interaction between the enzyme and its membrane-bound substrate. The stimulatory effect of saposins A, C, and D on acid sphingomyelinase activity is greatly influenced by the physical environment of the enzyme. Saposin D is an inhibitor of alkaline ceramidase, *or N*-acylsphingosine deacetylase and an activator of acid ceramidase (1). In the absence of a detergent, saposin D inhibited the purified acid sphingomyelinase (2). Prosaposin is synthesized as a 53-kDa precxursor protein that is post-translationally modified to form a 65 kDa form that is then glycosylated to yield a 70 kDa secretory product. This latter undergoes partial proteolysis to produce saposin A, B, C, and D (MW = 12-14 kDa). Each member acts in conjunction with hydrolase enzymes to facilitate the breakdown of glycosphingolipids within the lysosome. The saposins modify the environment of target lipids to make them accessible to the active sites of specific enzymes. Saposin B also binds and transfers coenzyme Q_{10} (CoQ_{10}), an essential,but highly hydrophobic, antioxidant and redox component that is required for ATP production. **1.** Nikolova-Karakashian & Merrill, Jr. (2000) *Meth. Enzymol.* **311**, 194. **2.** Tayama, Soeda, Kishimoto, *et al.* (1993) *Biochem. J.* **290**, 401.

Sapropterin, *See (6R)-Tetrahydrobiopterin*

Saquinavir

This statine-containing HIV protease inhibitor (FW = 670.85 g/mol; CAS 127779-20-8), also known as Ro 31-8959, inhibits many HIV-1 and HIV-2 proteases with a K_i of 0.3 nM. The methanesulfonate salt is known as Invirase®. The K_i values showed that its potency towards the V82F, L90M, I84V and G48V mutant proteinases respectively was 2-, 3-, 17- and 27-fold less than against the wild-type proteinase. Designed on the basis of the protease's three-dimensional structure, saquinavir became the first pharmaceutical approved for human use in 1996. Saquinavir inhibition prevents cleavage of the viral polyproteins resulting in the formation of immature non-infectious viral particles. Protease inhibitors are almost always used in combination with at least two other anti-HIV drugs. Saquinavir is poorly absorbed and is rapidly degraded by CYP3A4. **Target(s):** glucuronosyl-transferase, *or* UDP glucuronosyltransferases: UGT1A1, UGT1A3, and UGT1A4 (1); HIV-I retropepsin, *or* human immunodeficiency virus I protease (2-12); HIV-2 retropepsin, *or* human immunodeficiency virus 2 protease (7); and proteasome (13). **1.** Zhang, Chando, Everett, *et al.* (2005) *Drug Metab. Dispos.* **33**, 1729. **2.** Vacca (1994) *Meth. Enzymol.* **241**, 311. **3.** Knight (1995) *Meth. Enzymol.* **248**, 85. **4.** Roberts, Martin, Kinchington, *et al.* (1990) *Science* **248**, 358. **5.** Craig, Duncan, Hockley, *et al.* (1991) *Antiviral Res.* **16**, 295. **6.** Phylip, Mills, Parten, Dunn & Kay (1992) *FEBS Lett.* **314**, 449. **7.** Griffiths, Tomchak, Mills, *et al.* (1994) *J. Biol. Chem.* **269**, 4787. **8.** Wilson, Phylip, JMills, *et al.* (1997) *Biochim. Biophys. Acta* **1339**, 113. **9.** Clemente, Coman, Thiaville, *et al.* (2006) *Biochemistry* **45**, 5468. **10.** Surleraux, de Kock, Verschueren, *et al.* (2005) *J. Med. Chem.* **48**, 1965. **11.** Ermolieff, Lin & Tang (1997) *Biochemistry* **36**, 12364. **12.** Hong, Zhang, Hartsuck & Tang (2000) *Protein Sci.* **9**, 1898. **13.** Pajonk, Himmelsbach, Riess, Sommer & McBride (2002) *Cancer Res.* **62**, 5230.

SAR156497

This highly potent tricyclic aurora kinase inhibitor (FW = 467.52 g/mol) targets Aurora A, Aurora B, and Aurora C, shows exquisite selectivity *versus* other protein kinases (1). When tested in female SCID mice bearing human colon adenocarcinoma HCT116 xenografts, continuous subcutaneous infusion with ALZET® mini-pump (24 or 48 h at 25 or 50 mg/kg/day) was highly effective in suppressing tumor growth. The X-ray

structure of the bound inhibitor indicates that the DFG motif of Aurora A displays an "in" conformation and that the enzyme adopts an α_C-Helix "Glu out" (or inactivated) conformation. Gln-185, an amino acid residue within the αC Helix, is highly selective for Aurora kinases, making it a major determinant for target selectivity. **1.** Carry, Clerc, Minoux, *et al.* (2014) *J. Med. Chem.* **58**, 362.

SAR 216471

This reversible antithrombotic agent (FW = 599.51 g/mol; CAS 1279829-64-9; Solubility: 20 mM in DMSO, with gentle warming), also named 5-chloro-N-[6-[5-methyl-4-(1-oxobutyl)-1H-pyrazol-1-yl]-3-pyridazinyl]-1-[2-(4-methyl-1-piperazinyl)-2-oxoethyl]-1H-indole-3-carboxamide·HCl, targets the P2Y12 receptor, acting as an antagonist and blocking ADP-induced platelet aggregation. Adenosine 5'-diphosphate (ADP) liberated from damaged red blood cells or endothelial cells is a key mediator of activation and aggregation of platelets. ADP could bind two purinergic receptors P2Y1 and P2Y12. P2Y1 is implicated in platelet shape changes, whereas P2Y12 is involved in platelet aggregation, thrombin generation, microparticles release, and thrombus stabilization. **1.** Boldron, Besse, Bordes, *et al.* (2014) *J. Med. Chem.* **57**, 729.

SAR236553, *See Alirocumab*

SAR245409, *See Voxtalisib*

SAR302503, *See Fedratinib*

Saracatinib

This orally active tyrosine protein kinase inhibitor (FW = 542.03 g/mol; CAS 379231-04-6; Solubility: 184 mg/mL DMSO, <1 mg/mL H_2O), also known as AZD0530 and N-(5-chlorobenzo[d][1,3]dioxol-4-yl)-7-(2-(4-methylpiperazin-1-yl)ethoxy)-5-(tetrahydro-2H-pyran-4-yloxy)quinazolin-4-amine, is an investigational drug that targets Src (IC_{50} = 2.7 nM) and also potently inhibits Fyn, c-Yes, Lyn, Blk, Fgr and Lck, with weaker action against Abl and the L858R and L861Q mutants of EGFR. Because the tyrosine kinase Fyn is activated *via* cell surface binding of Alzheimer-related Aβ-oligomers to cellular prion protein, and because Fyn also interacts with both Aβ and the microtubule-associated axonal protein Tau, AZD0530 is likewise under investigation for its potentially beneficial effects in Alzheimer's Disorder (8). **1.** Chang, *et al.* (2008) *Oncogene* **27**, 6365. **2.** Green, *et al.* (2009) *Mol. Oncol*, **3**, 248. **3.** Purnell, et al. (2009) *J. Thorac. Oncol.* **4**, 448. **4.** de Vries, *et al.* (2009) *Mol. Cancer Res.* **7**, 476. **5.** Gwanmesia, et al. (2009) *BMC Cancer* **9**, 53. **6.** Hiscox, et al. (2009) *Breast Cancer Res. Treat.* **115**, 57. **7.** Chen, *et al.* (2009) *Clin. Cancer Res.* **15**, 3396. **8.** Nygaard, van Dyck & Strittmatter (2014) *Alzheimers Res. Ther.* **6**, 8.

Saralasin

This peptidic angiotensin antagonist (FW = 909.11 g/mol; CAS 34273-10-4; Sequence: XRVYVHPA (Modification: X = Sar); Soluble to 1 mg/mL in

H_2O) competitively and non-selectively targets the angiotensin II receptor, enlarging microvascular diameters and increasing glomerular blood flow (1). Saralasin also exerts antiarrhythmic effects in myocardial reperfusion by a mechanism independent of circulatory and central actions (2). **1.** Steinhausen, Kücherer, Parekh, et al. (1986) Kidney Int. **30**, 56. **2.** Thomas, Howlett & Ferrier (1995) J. Pharmacol. Exp. Ther. **274**, 1379.

Sarcine, See Hypoxanthine

Sarcodictyin A & B

These microtubule-stabilizing agents (Sarcodictyin A [R = –CH$_3$], MW$_{SD-A}$ = 496.60 g/mol; Sarcodictyin B [R = –CH$_2$CH$_3$], MW$_{SD-B}$ = 510.63 g/mol) from the Japanese soft coral Bellonella albiflora exhibit IC$_{50}$ values in the 200-500 nM range, considerably more weakly than paclitaxel (IC$_{50}$ <10 nM, except in resistant cell lines) and eleutherobin and epothilone A, (IC$_{50}$ = 10-40 nM). **1.** Hamel, Sackett, Vourloumis & Nicolaou (1999) Biochemistry **38**, 5490.

Sarcolysine, See Melphalan

Sarcosine

This methylated amino acid and endogenous glycine transporter GlyT1 inhibitor (FW = 89.09 g/mol; CAS 127779-20-8; pK_a's = 2.12 and 10.2; Soluble to 100 mM in water; Abbreviation: Sar), also known as N-methylglycine, is an intermediate in choline metabolism and a regulator of the S-adenosyl-L-methionine and S-adenosyl-L-homocysteine concentrations via the actions of glycine N-methyltransferase, dimethylglycine dehydrogenase, sarcosine oxidase, and sarcosine dehydrogenase. A sarcosyl moiety is also present in the structure of actinomycin D. **Target(s):** L-alanine dehydrogenase (1,2); dimethylglycine N-methyltransferase (3); glycine transport (4); 4-hydroxyproline epimerase (5,6); proline racemase (7); and prolyl-tRNA synthetase (8). **1.** Yoshida & Freese (1970) Meth. Enzymol. **17A**, 176. **2.** Yoshida & Freese (1965) Biochim. Biophys. Acta **96**, 248. **3.** Waditee, Tanaka, Aoki, et al. (2003) J. Biol. Chem. **278**, 4932. **4.** Hammer, Kolajova, Leveille, Claman & Baltz (2000) Hum. Reprod. **15**, 419. **5.** Adams (1971) Meth. Enzymol. **17B**, 266. **6.** Finlay & Adams (1970) J. Biol. Chem. **245**, 5248. **7.** Cardinale & Abeles (1968) Biochemistry **7**, 3970. **8.** Norris & Fowden (1972) Phytochemistry **11**, 2921.

Sarcosyllaurate, See N-Lauroylsarcosine

Sarkosyl, See N-Lauroylsarcosine

Sarilesin

This angiotensin II receptor analogue (FW = 968.15 g/mol; CAS 37827-06-8; Sequence: Sar-Arg-Val-Phe-Ile-His-Pro-Ile, where Sar represents sarcosine (i.e., N-methylglycine), also inhibits cAMP phosphodiesterase (1-

4). **1.** Koziarz & Moore (1989) Proc. West Pharmacol. Soc. **32**, 79. **2.** Matsoukas, Agelis, Wahhab, et al. (1995) J. Med. Chem. **38**, 4660. **3.** Sharma (2003) Ind. J. Biochem. Biophys. **40**, 77. **4.** Sharma, Smith & Moore (1991) Biochem. Biophys. Res. Commun. **179**, 85.

Sarin

This lethal chemical warfare (CW) agent (FW = 140.09 g/mol; CAS 107-44-8; clear and odorless liquid, M.P. = –57°C; B.P. = 147°C), also known as methylisopropoxyfluorophosphine oxide and O-isopropyl methylphos-phonofluoridate, frequently called Agent GB, was an insecticide first synthesized in 1937 by Schrader, Ambrose, Rüdiger, and van der Linde in Germany, hence the name. Sarin was soon found to be an extremely potent nerve agent and was produced in large quantities during World War II. Germany, however, refrained from using sarin as a chemical warfare weapon, most likely out of fear of worldwide condemnation and certain reprisal. (On March 20, 1995, members of the cult movement Aum Shinrikyo carried out the Tokyo Subway Sarin Incident (地下鉄サリン事件), an act of domestic terrorism that killed 12, severely injured 50, and temporarily impaired the vision of 5,000 others.) **Caution:** Sarin is extremely toxic and vapors can be readily absorbed through the skin and eyes. The transdermal LD$_{50}$ in human is 100-300 mg, death almost always occurring within fifteen minutes. The i.p. LD$_{50}$ in mice is 0.42 mg/kg. The LC$_{50}$ (i.e., the concentration in air that is lethal in 50% of exposed individuals) is 0.6-1.2 ppm. Sarin is a substrate for a number of carboxylesterases and is a potent inhibitor of acetylcholinesterase and cholinesterase. At low concentrations, sarin inhibits the evoked release of γ-aminobutyrate in rat hippocampal slices (1). Interestingly, sarin has been used in the treatment of leprosy and of AIDS. **Cyclosarin:** This structurally related toxin (FW = 180.16 g/mol; B.P. = 239 °C), also named cyclohexyl methyl-phoshonofluoridate and (fluoro-methyl-phosphoryl)-oxycyclo-hexane, has comparable properties to sarin as a nerve gas, albeit far less volatile. **Target(s):** acetylcholinesterase (2-12); carboxylesterase (9); cholinesterase, or butyrylcholinesterase (2,6,9,12-15); chymotrypsin (9); α-lytic endopeptidase (16); thrombin (17,18); and trypsin (9). **1.** Chebabo, Santos & Albuquerque (1999) Neurotoxicology **20**, 871. **2.** Holmstedt (1959) Pharmacol. Rev. **11**, 567. **3.** Khan, Dechkovskaia, Herrick, Jones & Abou-Donia (2000) Toxicol. Sci. **57**, 112. **4.** Tripathi & Dewey (1989) J. Toxicol. Environ. Health **26**, 437. **5.** Gray & Dawson (1987) Toxicol. Appl. Pharmacol. **91**, 140. **6.** Boter & van Dijk (1969) Biochem. Pharmacol. **18**, 2403. **7.** Millard, Kryger, Ordentlich, et al. (1999) Biochemistry **38**, 7032. **8.** Froede & Wilson (1971) The Enzymes, 3rd ed. (Boyer, ed.), **5**, 87. **9.** Cohen, Oosterbaan & Berends (1967) Meth. Enzymol. **11**, 686. **10.** Smith & Usdin (1966) Biochemistry **5**, 2914. **11.** Worek, Thiermann, Szinicz & Eyer (2004) Biochem. Pharmacol. **68**, 2237. **12.** Schopfer, Voelker, Bartels, Thompson & Lockridge (2005) Chem. Res. Toxicol. **18**, 747. **13.** Brown, Kalow, Pilz, Whittaker & Woronick (1981) Adv. Clin. Chem. **22**, 1. **14.** Ashani, Segev & Balan (2004) Toxicol. Appl. Pharmacol. **194**, 90. **15.** Cerasoli, Griffiths, Doctor, et al. (2005) Chem. Biol. Interact. **157-158**, 363. **16.** Whitaker (1970) Meth. Enzymol. **19**, 599. **17.** Thompson (1970) Biochim. Biophys. Acta **198**, 392. **18.** Lundblad, Kingdon & Mann (1976) Meth. Enzymol. **45**, 156.

Sarkosyl, See N-Lauroylsarcosine

Sarpogrelate

This serotonin 5HT$_{2A}$ and 5-HT$_{2B}$ receptor antagonist (FW = 429.51 g/mol; CAS 125926-17-2), also known by code names MCI-9042 and LS-187118, trade name Anplag®, and systematic name, 4-[2-(dimethylamino)-1-({2-[2-

(3-methoxyphenyl)ethyl]phenoxy}methyl)ethoxy]-4-oxobutanoic acid, blocks serotonin-induced platelet aggregation and has been used to treat diabetic nephropathy and neuropathy, Buerger's disease (or thromboangiitis obliterans), Raynaud's disease, coronary artery disease, angina, and atherosclerosis. Sarpogrelate has a K_i value of 8.4 nM with 5HT$_2$ receptors (1). (The pK_i values are 8.52, 7.43 and 6.57 for antagonist action on 5-HT$_{2A}$, 5-HT$_{2C}$ and 5-HT$_{2B}$ receptors, respectively.) (R,S)-1-[2-[2-(3-Methoxyphenyl)ethyl]phenoxy]-3-(dimethylamino)-2-propanol (or M-1), its major metabolite, is more active (K_i = 1.7 nM). Both sarpogrelate and M-1 show no affinity for 5-HT$_{1A}$ receptors, but these substances, at a concentration of 10 μM, displace the specific binding of [^{125}I]iodocyanopindolol to the 5-HT$_{1B}$ receptors, with respective K_i values of 0.88 and 0.86 μM. Sarpogrelate, (R,S)-M-1, (R)-M-1 and (S)-M-1 are competitive antagonists of rat tail artery 5-HT$_{2A}$ receptors, showing pA$_2$ values of 8.53, 9.04, 9.00 and 8.81, respectively (2). **1.** Nishio, Inoue & Nakata (1996) *Arch. Int. Pharmacodyn. Ther.* **331**, 189. **2.** Pertz & Elz (1995) *J. Pharm. Pharmacol.* **47**, 310.

Saucerneol B

This lignan (FW = 534.60 g/mol) inhibits cholesterol *O*-acyltransferase, IC$_{50}$ = 43 μM for human ACAT-1 (1). **Target(s):** HIV-1 protease (1); sterol *O*-acyltransferase, or cholesterol *O*-acyltransferase, also known as ACAT (2,3). **1.** Lee, Huh, Kim, Hattori & Otake (2010) *Antiviral Res.* **85**, 425. **2.** Jeong, Kim, Yu, et al. (2005) *Bioorg. Med. Chem. Lett.* **15**, 385. **3.** Lee, Lee, Baek, et al. (2004) *Bioorg. Med. Chem. Lett.* **14**, 3109.

Sauvagine

This corticotropin-releasing factor (CRF) receptor agonist (FW = 4599.35 g/mol; CAS 74434-59-6; Sequence: XGPPISIDLSLELLRKMIEIEKQEK EKQQAANNRLLLDTI (Modifications: X-1 = pGlu, Ile-40 = C-terminal amide); Store desiccated at –20°C) inhibits of ^{125}I-[D-Tyr^1]astressin binding to hCRF-R1 (K_i = 9.4 nM), rCRF-R2a (K_i = 9.9 nM), and mCRF-R2b (K_i = 3.8 nM). Sauvagine was first isolated from from the skin of *Phyllomedusa sauvagei*, a frog of Central and South America (1). **1.** Montecucchi & Henschen (1981) *Int. J. Pept. Protein Res.* **18**, 113. **2.** Eckart, et al. (1999) *Curr. Med. Chem.* **6**, 1035. **3.** Perrin & Vale (1999) *Ann. N.Y. Acad. Sci.* **885**, 312.

Savignin

This coagulation-modulating protein (MW = 12.4 kDa) from the tick *Ornithodoros savingnyi* is a potent thrombin inhibitor, K_i = 4.9 pM for α-thrombin and 22 nM for γ-thrombin (1,2). Savignin is a competitive, slow, tight-binding inhibitor, requiring thrombin's fibrinogen-binding exo-site for optimal inhibition. **1.** Nienaber, Gaspar & Neitz (1999) *Exp. Parasitol.* **93**, 82. **2.** Mans, Louw & Neitz (2002) *Insect Biochem. Mol. Biol.* **32**, 821.

Saxagliptin

This substituted pyrrolidone (FW = 315.41 g/mol; CAS 361442-04-8), also known as BMS-477118 and marketed under the tradename Onglyza™, is a potent, selective, and orally active inhibitor (K_i = 0.6 nM) of dipeptidyl-peptidase IV, or DPP-4 (1,2). By limiting proteolysis of the gastrointestinal

hormones, glucagon-like peptides 1 and 2 (GLP-1 and GLP-2), saxagliptin is a so-called incretin enhancer that indirectly maintains insulin secretion and suppresses glucagon release. Saxagliptin has good oral bioavailability and enjoys the added advantage of not being significantly influenced by food intake. Moreover, its prolonged *in vivo* half-life and its efficient inhibition of DPP-4 permit a single oral dose for daily management of Type-2 diabetes (3). Saxagliptin is a P450 substrate that is metabolized by CYP3A4/A5. Finally, because dipeptidyl-peptidase IV has diverse physiologic roles (*e.g.*, angiogenesis, vascular remodeling, inflammation, regulation of Na$^+$/H$^+$ exchanger isoform activity in proximal tubule cells, attenuation of blood pressure, *etc.*), one may anticipate that high-potency DPP-4 inhibitors will elicit a range of side-effects. **See** *Sitagliptin, Vildagliptin* **1.** Augeri, Robl, Betebenner, et al. (2005) *J. Med. Chem.* **48**, 5025. **2.** Metzler, Yanchunas, Weigelt, et al. (2008) *Protein Sci.* **17**, 240. **3.** Scheen (2010) *Diabetes Obes. Metab.* **12**, 648.

Saxitoxin

This selective Na$^+$ channel blocker (FW$_{dihydrochloride}$ = 370.2 g/mol; CAS 35523-89-8; hygroscopic), produced by red tide dinoflagellates of *Gonyaulax* species, is a paralytic toxin that accumulates in shellfish, often reaching levels that can result in severe poisoning of humans consuming affected shellfish. Saxitoxin inhibits sodium ion transport by binding to a site at/near the extracellular face of the sodium channel's pore (1-4). In each of the six species studied, a single class of high-affinity receptor sites was detected with B_{max} values of 1.7–12 fmol/mg (wet weight) and K_d values of 0.3–7.5 nM for STX, compared to a value of 3.5–10.4 nM for tetrodotoxin. For example, saxitoxin inhibits the action potential Na$^+$ ionophore of electrically excitable neuroblastoma cells with a K_i of 3.7 nM. Binding experiments detect a single class of saturable binding sites with K_d of 3.9 nM and a B_{max} of 156 fmol/mg of cell protein (~78 sites per μmeter^2 of cell surface) (5). Saturable binding is completely inhibited by tetrodotoxin but is unaffected by scorpion toxin or batrachotoxin. No saturable binding is observed in cultures of clone N103, a variant neuroblastoma clone lacking the action potential Na+ response (5). Saxitoxin is thus a valuable tool in neurochemical investigations on voltage-gated channels. The oral LD$_{50}$ for humans is 5.7 μg/kg. **1.** Catterall (1986) *Ann. Rev. Biochem.* **55**, 953. **2.** Cheymol, Bourillet, Long & Roch-Arveiller (1968) *Arch. Int. Pharmacodyn. Ther.* **174**, 393. **3.** Evans (1969) *Brit. Med. Bull.* **25**, 263. **4.** Tanaka, Doyle & Barr (1984) *Biochim. Biophys. Acta* **775**, 203. **5.** Catterall & Morrow (1978) *Proc.Natl. Acad. Sci. U.S.A.* **75**, 218.

Sazetidine A

This subtype-selective nicotinic acetylcholine receptor ligand (FW = 333.25 g/mol; CAS 1197329-42-2; Soluble to 50 mM in H$_2$O), also named 6-[5-[(2S)-2-azetidinylmethoxy]-3-pyridinyl]-5-hexyn-1-ol di-HCl, targets α$_4$β$_2$

receptor (K_i = 0.26), with 200x greater affinity than for the $\alpha_3\beta_4$ receptor (K_i = 54 nM). Depending on subunit composition, Sazetidine A acts either as a silent desensitizer or as an agonist, with an EC_{50} of 1.1 nM for nAChR-stimulated dopamine release. It also exhibits analgesic activity *in vivo* and significantly reduces nicotine self-administration in a rat model. **1.** Xiao, *et al.* (2006) *Mol. Pharmacol.* **70**, 1454. **2.** Cucchiaro, *et al.* (2008) *Anesthesiology* **109**, 512. **3.** Zwart, *et al.* (2008) *Mol. Pharmacol.* **73**, 1843. **4.** Levin, *et al.* (2010) *J. Pharmacol. Exp. Ther.* **332**, 933.

SB1518, *See* *Pacritinib*

SB2343, *See* *VS-5584*

SB 95952, *See* *Rolipram*

SB 200646

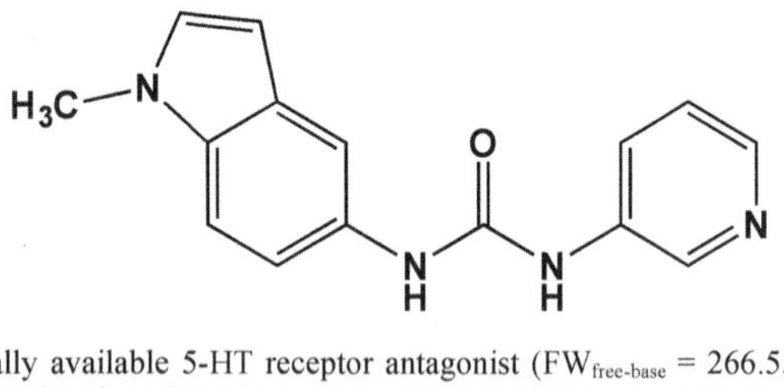

This orally available 5-HT receptor antagonist (FW$_{free-base}$ = 266.55 g/mol; CAS 143797-63-1; Soluble to 100 mM in DMSO), also known as *N*-(1-methyl-5-indolyl)-*N'*-(3-pyridyl)urea, is a 5-HT$_{2C/2B}$ serotonin receptor antagonist (1,2). Affinities are 7.4 (pA$_2$), 6.9 (pK$_i$) and 5.2 (pK$_i$) for 5-HT$_{2B}$, 5-HT$_{2C}$ and 5-HT$_{2A}$ respectively. **1.** Kennett, Wood, Glen, *et al.* (1994) *Brit. J. Pharmacol.* **111**, 797. **2.** Forbes, Kennett, Gadre, *et al.* (1993) *J. Med. Chem.* **36**, 1104. **3.** Kennett, *et al.* (1994) *Brit. J. Pharmacol.* **111**, 797. **4.** Kennett, *et al.* (1995) *Psychopharmacol.* (Berl.) **118**, 178.

SB 201076, *See* *(2S)-2-[(2R)-8-(2,4-Dichlorophenyl)-2-hydroxyoctyl]-2 hydroxysuccinate and (2S)-2-[(2R)-8-(2,4-Dichlorophenyl)-2-hydroxyoctyl]-2-hydroxysuccinic Acid*

SB 202190

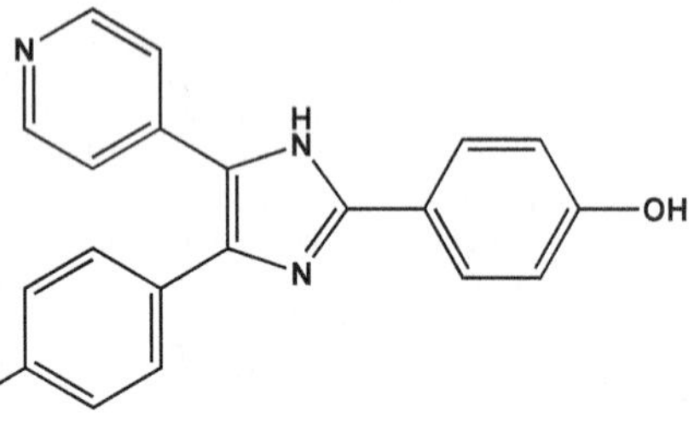

This substituted imidazole (FW = 331.35 g/mol; CAS 152121-30-7), also known as (4-(4-fluorophenyl)-2-(4-hydroxyphenyl)-5-(4-pyridyl)-1*H*-imidazole, is structurally similar to SB 203580. SB 202190 inhibits p38 stress activated protein kinase 2a and p38 stress-activated protein kinase 2b, IC_{50} values of 50 and 100 nM, respectively (1-5). *Note*: SB 202474 (4-ethyl-2(*p*-methoxyphenyl)-5-(4'-pyridyl)-1*H*-imidazole), in which the 4-fluorophenyl group of SB 202190 has been replaced with an ethyl moiety and the hydroxyl group is methylated, is often used as a negative control in investigations of mitogen-activated protein kinase. **1.** Davies, Reddy, Caivano & Cohen (2000) *Biochem. J.* **351**, 95. **2.** Ratner, Bryan, Weber, *et al.* (2001) *J. Biol. Chem.* **276**, 19267. **3.** Lee, Laydon, McDonnell, *et al.* (1994) *Nature* **372**, 739. **4.** Kong, Klassen & Rabkin (2005) *Mol. Cell. Biochem.* **278**, 39. **5.** Hofmann, Zaper, Bernd, *et al.* (2004) *Biochem. Biophys. Res. Commun.* **316**, 673.

SB 203186

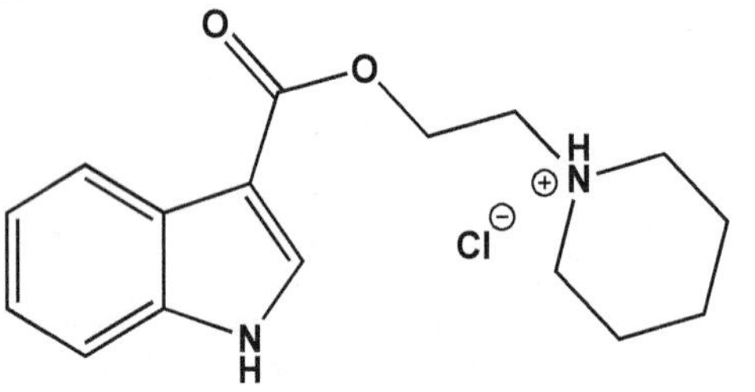

This potent serotonin antagonist (FW = 308.81 g/mol; CAS 207572-69-8; Soluble to 100 mM in water), also named 1-piperidinylethyl-1*H*-indole-3-carboxylate hydrochloride, targets 5-HT$_4$ receptors, with high affinity for human atrial receptors. **1.** Midhurst & Kaumann (1993) *Brit. J. Pharmacol.* **110**, 1023. **2.** McLean & Coupar (1995) *Naunyn. Schmiedebergs Arch. Pharmacol.* **352**, 132. **3.** Parker, *et al.* (1995) *Naunyn. Schmiedebergs Arch. Pharmacol.* **353**, 28.

SB 203207

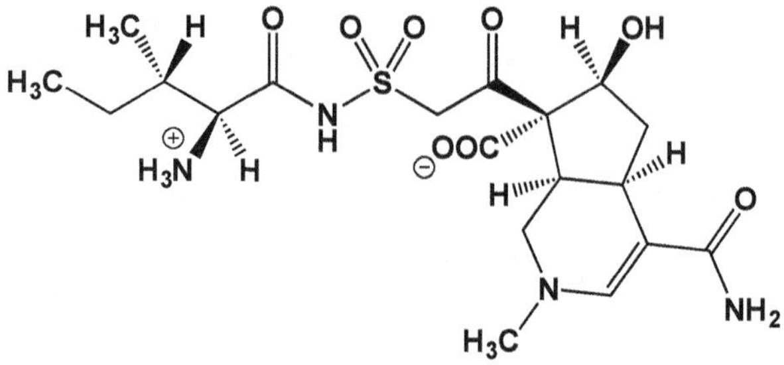

This isoleucyl derivative (FW = 474.54 g/mol), from a strain of *Streptomyces*, inhibits *Staphyococcus aureus* and rat liver isoleucyl-tRNA synthetases, IC_{50} = 1.7 and < 2 nM, respectively (1-3). Note that the semi-synthetic leucyl and valyl analogues inhibit their cognate aminoacyl-tRNA synthetases. **Target(s):** isoleucyl-tRNA synthetase (1-3); leucyl-tRNA synthetase, moderately inhibited (1); and valyl-tRNA synthetase, moderately inhibited (1,2). **1.** Crasto, Forrest, Karoli, *et al.* (2003) *Bioorg. Med. Chem.* **11**, 2687. **2.** Stefanska, Cassels, Ready & Warr (2000) *J. Antibiot.* **53**, 357. **3.** Houge-Frydrych, Gilpin, Skett & Tyler (2000) *J. Antibiot.* **53**, 364.

SB 203347

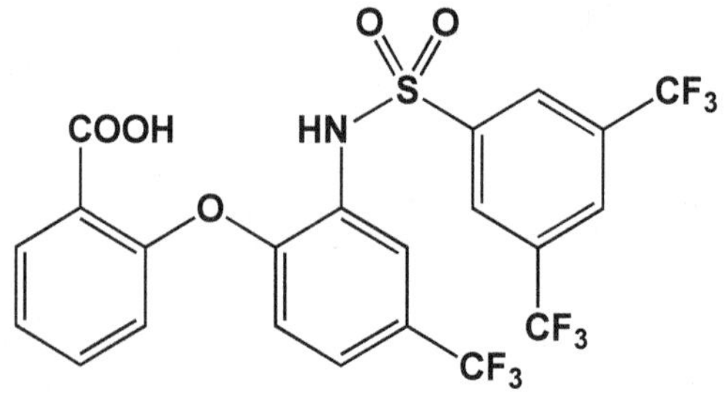

This sulfonamine (FW = 573.39 g/mol; CAS 169527-42-8) inhibits human type II phospholipase A$_2$ (IC_{50} = 0.5 μM) inhibits human glycerophospholipid arachidonyltransferase (CoA-independent), IC_{50} = 23 μM (1) and phospholipase A$_2$ (2,3). **1.** Winkler, Eris, Sung, *et al.* (1996) *J. Pharmacol. Exp. Ther.* **279**, 956. **2.** Anderson, Roshak, Winkler, McCord & Marshall (1997) *J. Biol. Chem.* **272**, 30504. **3.** Marshall, Hall, Winkler, *et al.* (1995) *J. Pharmacol. Exp. Ther.* **274**, 1254.

SB 203386

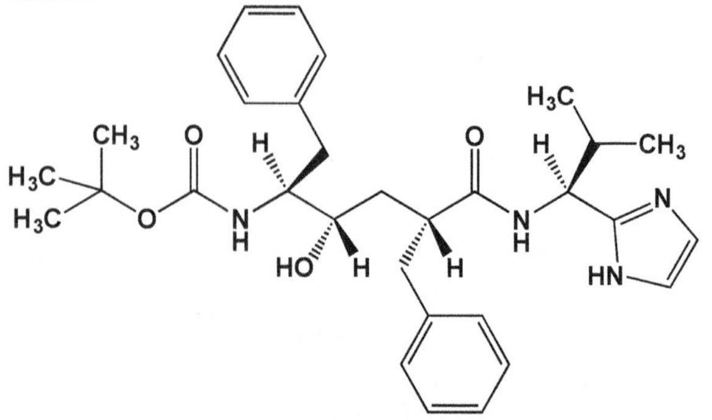

This tripeptide analogue (FW = 507.68 g/mol) inhibits HIV-1 and HIV-2 retropepsin, K_i = 18 and 1280 nM, respectively (1). SB 203386 inhibition prevents cleavage of the viral polyproteins resulting in the formation of immature non-infectious viral particles. Protease inhibitors are almost always used in combination with at least two other anti-HIV drugs. **1.** Swairjo, Towler, Debouck & Abdel-Meguid (1998) *Biochemistry* **37**, 10928.

SB 203580

This anti-inflammatory (FW$_{free-base}$ = 377.44 g/mol; CAS 152121-47-6; Solubility: 25 mM in DMSO as free base; 100 mM as mono-HCl salt), also known as 4-(4-fluorophenyl)-2-(4-methylsulfinylphenyl)-5-(4-pyridyl)-1H-imidazole, is a selective inhibitor of p38 mitogen-activated protein kinase SAPK2a/p38 (IC$_{50}$ = 50 nM) and SAPK2b/p38β2 (IC$_{50}$ = 500 nM), displaying 100-500x selectivity over LCK, GSK3β and PKBα (1-15). SB 203580 inhibits interleukin-2-induced T-cell proliferation, cyclooxygenase-1 and -2, and thromboxane synthase. SB 203580 suppresses the activation of MAPKAP kinase-2 and inhibits the phosphorylation of heat shock protein 27. As a SAPK2 pathway inhibitor, SB 203580 treatment results in the phosphorylation of such AP1 transcription factors c-Jun. Its solubility in dimethyl sulfoxide is 50 mg/mL. **Target(s):** cRaf protein kinase (16); cyclooxygenases, COX-I and II (17); [glycogen-synthase] kinase III (4); Jun N-terminal kinase, especially JNK2 and JNK3 (18); lymphocyte kinase (4); mitogen-activated protein kinase (*or* p38 MAP kinase/reactivating kinase), p38 stress-activated protein kinase 2a, and p38 stress-activated protein kinase 2b, with IC$_{50}$ values of 50 and 500 nM, respectively (1-15); protein kinase B (4); and thromboxane-A synthase (17). 1. Cuenda, Rouse, Doza, *et al.* (1995) *FEBS Lett.* **364**, 229. 2. Hazzalin, Cano, Cuenda, *et al.* (1996) *Curr. Biol.* **6**, 1028. 3. Eriksson, Toivola, Sahlgren, Mikhailov & Härmälä-Braskén (1998) *Meth. Enzymol.* **298**, 542. 4. Davies, Reddy, Caivano & Cohen (2000) *Biochem. J.* **351**, 95. 5. Whitmarsh & Davis (2001) *Meth. Enzymol.* **332**, 319. 6. Ward, Parry, Matthews & O'Neill (1997) *Biochem. Soc. Trans.* **25**, 304S. 7. Fitzgerald, Patel, Becker, *et al.* (2003) *Nature Struct. Biol.* **10**, 764. 8. Szafranska & Dalby (2005) *FEBS J.* **272**, 4631. 9. Tudan, Jackson, Higo, *et al.* (2004) *Cell. Signal.* **16**, 211. 10. Samuvel, Jayanthi, Bhat & Ramamoorthy (2005) *J. Neurosci.* **25**, 29. 11. Shimo, Koyama, Sugito, *et al.* (2005) *J. Bone Miner. Res.* **20**, 867. 12. Lee, Yu & Chung (2005) *Free Radic. Res.* **39**, 399. 13. Hofmann, Zaper, Bernd, *et al.* (2004) *Biochem. Biophys. Res. Commun.* **316**, 673. 14. Uddin, Ah-Kang, Ulaszek, Mahmud & Wickrema (2004) *Proc. Natl. Acad. Sci. U.S.A.* **101**, 147. 15. Bukhtiyarova, Northrop, Chai, *et al.* (2004) *Protein Expr. Purif.* **37**, 154. 16. Hall-Jackson, Goedert, Hedge & Cohen (1999) *Oncogene* **18**, 2047. 17. Borsch-Haubold, Pasquet & Watson (1998) *J. Biol. Chem.* **273**, 28766. 18. Murray, Bennett & Sasaki (2001) *Meth. Enzymol.* **332**, 432.

SB 203580 Sulfone

This SB 203580 derivative (FW = 393.44 g/mol), known systematically as 4-(4-fluorophenyl)-2-(4-methylsulfonylphenyl)-5-(4-pyridyl)-1H-imidazole, is a strong inhibitor of mitogen-activated protein kinase, *or* p38 stress-activated protein kinase (IC$_{50}$ = 30 nM). 1. Gallagher, Seibel, Kassis, *et al.* (1997) *Bioorg. Med. Chem.* **5**, 49.

SB 204070

This potent and selective serotonin receptor antagonist (FW = 382.88 g/mol; CAS 148688-01-1; Soluble to 100 mM in DMSO), also named (1-Butyl-4-piperidinyl)methyl 8-amino-7-chloro-1,4-benzodioxane-5-carboxylate hydrochloride, 5-HT$_4$ receptor antagonist (IC$_{50}$ = 90 nM). SB 204070 displays >5000x selectivity for 5-HT$_4$ *versus* 5-HT$_{1A}$, 5-HT$_{1D}$, 5-HT$_{1E}$, 5-HT$_{2A}$, 5-HT$_{2C}$, and 5-HT$_3$ receptors. SB 204070 exhibits anxiolytic activity upon systemic administration *in vivo*. 1. Gaster et al (1993) *J. Med. Chem.* **36**, 4121. 2. Bingham, *et al.* (1995) *J. Pharm. Pharmacol.* **47**, 219. 3. Prins, *et al.* (2000) *Brit. J. Pharmacol.* **129**, 1601.

SB 204144

This phosphinate-containing peptidomimetic (FW = 768.85 g/mol), also named (1R)-bis[2-phenyl-1-[[(carbobenzoxy)valyl]amino]ethyl]phosphinic acid (FW$_{free-acid}$ = 770.86 g/mol; melting point = 236-238°C), is a potent tetrahedral-adduct transition-state mimic that inhibits HIV-I protease, K_i = 2.8 nM, at pH 6.0. 1. Ringe (1994) *Meth. Enzymol.* **241**, 157. 2. Kempf (1994) *Meth. Enzymol.* **241**, 334. 3. Abdel-Meguid, Zhao, Murthy, *et al.* (1993) *Biochemistry* **32**, 7972.

SB 204741

This serotonin antagonist (FW = 286.35 g/mol; CAS 152239-46-8; Soluble to 100 mM in DMSO), also named N-(1-methyl-1H-indolyl-5-yl)-N''-(3-methyl-5-isothiazolyl)urea, targets 5-HT$_{2B}$ receptor antagonist (pA$_2$ = 7.95), displaying greater than 100x selectivity versus 5-HT$_{2C}$ (pK$_i$ = 5.82), 5-HT$_{2A}$ (pK$_i$ < 5.2), 5-HT$_{1A}$, 5-HT$_{1D}$, 5-HT$_{1E}$, 5-HT$_3$ and 5-HT$_4$ receptors. 1. Bonhaus, *et al.* (1995) *Brit. J. Pharmacol.* **115**, 622. 2. Forbes, *et al.* (1995) *J. Med. Chem.* **38**, 855. 3. Glusa & Pertz (2000) *Brit. J. Pharmacol.* **130**, 692. 4. Ebrahimkhani, *et al.* (2011) *Nature Med.* **17**, 1668.

SB 205952

This pseudomonic/monic acid analogue (FW = 478.50 g/mol) is a strong inhibitor of isoleucyl-tRNA synthetase as well as a substrate for bacterial NAD(P)H-dependent reductases (1-4). 1. Pope, Moore, McVey, *et al.* (1998) *J. Biol. Chem.* **273**, 31691. 2. Pope, McVey, Fantom & Moore (1998) *J. Biol. Chem.* **273**, 31702. 3. Broom, Cassels, Cheng, *et al.* (1996) *J. Med. Chem.* **39**, 3596. 4. Wilson, Oliva, Cassels, O'Hanlon & Chopra (1995) *Antimicrob. Agents Chemother.* **39**, 1925.

SB 212047

This β-lactam (FW = 497.37 g/mol), also named 4-methoxybenzyl (3S,4R)-6-bromo-6-[(1-methyl-1,2,3-triazol-4-yl)hydroxymethyl]penicillanate, irreversibly inhibits human CoA-independent glycerophospholipid arachidonoyltransferase (*or* CoA-IT), a key mediator of arachidonate remodeling that moves arachidonate into 1-ether-containing phospholipids. Blockade of CoA-IT blocks release of arachidonate in stimulated neutrophils and inhibits production of eicosanoids and platelet-activating factor. Evidence of irreversibility (namely that extensive washing does not restore activity) is unconvincing. **1**. Winkler, Sung, Chabot-Flecher, *et al.* (1998) *Mol. Pharm.* **53**, 322.

SB 214357

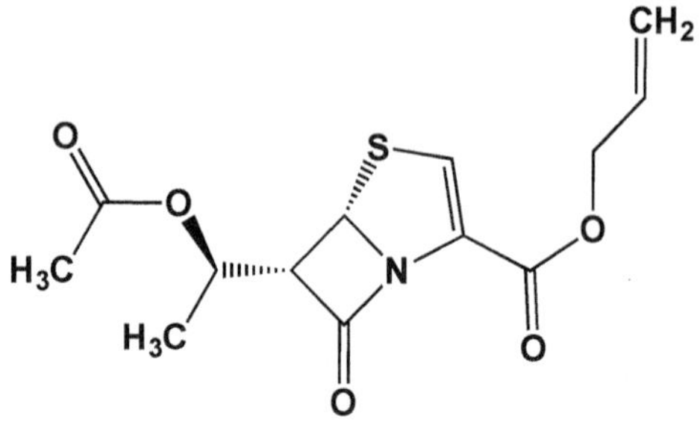

This penem antibiotic derivative (FW = 297.33 g/mol) inhibits certain signal peptidases. *Note*: Some literature sources have mislabeled this inhibitor as SB 216357. **Target(s):** chloroplast and cyanobacterial signal peptidase 1 (1,2); C-terminal processing peptidase, *or* tsp peptidase (3); and *Escherichia coli* leader peptidase (4). **1**. Howe & Floyd (2002) *The Enzymes*, 3rd ed. (Dalbey & Sigman, eds.), **22**, 101. **2**. Barbrook, Packer & Howe (1996) *FEBS Lett.* **398**, 198. **3**. Chaal, Ishida & Green (2003) *Plant Mol. Biol.* **52**, 463. **4**. Perry, Ashby & Elsmere (1995) *Biochem. Soc. Trans.* **23**, 548S.

SB 216357, See SB 214357

SB 216477, See 4-(Phenylthio)-N-(4-phenyl-2-oxobutyl)azetidin-2-one

SB 216763

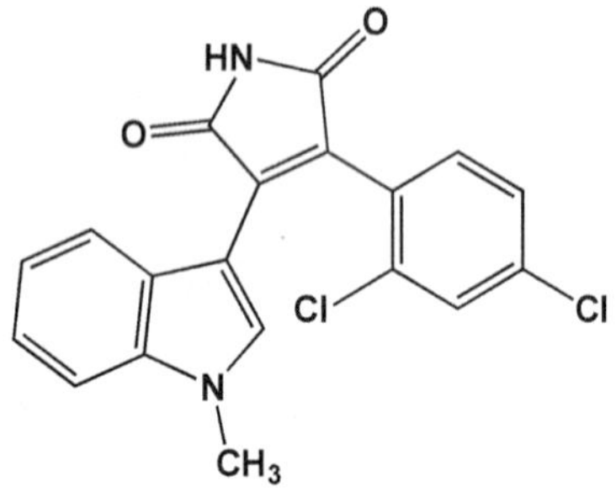

This substituted maleimide (FW = 371.22 g/mol; CAS 280744-09-4) inhibits [glycogen synthase] kinase-3 (GSK3), IC_{50} = 34 nM, protecting both central and peripheral nervous system neurons in culture from death induced by reduced PI3-kinase pathway activity. The inhibition of neuronal death mediated by SB-216763 correlates with inhibition of GSK-3 activity and modulation of GSK-3 substrates Tau protein and β-catenin. **1**. Cross, Culbert, Chalmers, *et al.* (2001) *J. Neurochem.* **77**, 94. **2**. Coghlan, Culbert, Cross, *et al.* (2000) *Chem. Biol.* **7**, 793.

SB 220025

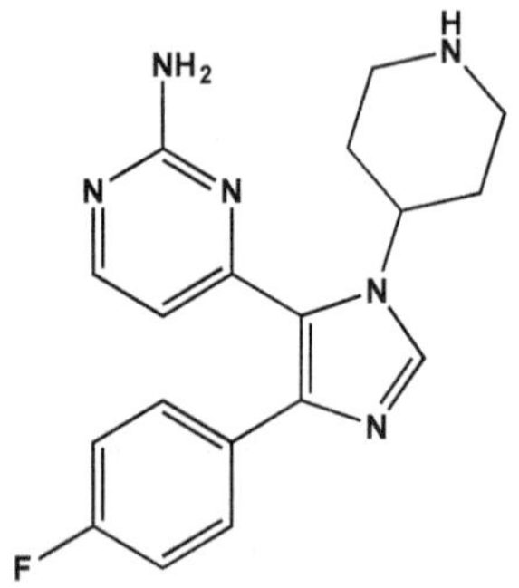

This photosensitive p38 inhibitor (FW = 338.39 g/mol; IUPAC: 5-(2-amino-4-pyrimidinyl)-4-(4-fluorophenyl)-1-(4-piperidinyl)imidazole, targets p38 MAP kinase, IC_{50} = 60 nM (human), with and 50x to 1000x

selectivity versus other kinases tested (1,2). SB 220025 binds in an extended pocket within the active site and is complementary to the open domain structure of the low-activity form of p38. **1**. Jackson, Bolognese, Hillegass, *et al.* (1998) *J. Pharmacol. Exp. Ther.* **284**, 687. **2**. Wang, Canagarajah, Boehm, *et al.* (1998) *Structure* **6**, 1117.

SB 222657 & SB 223777

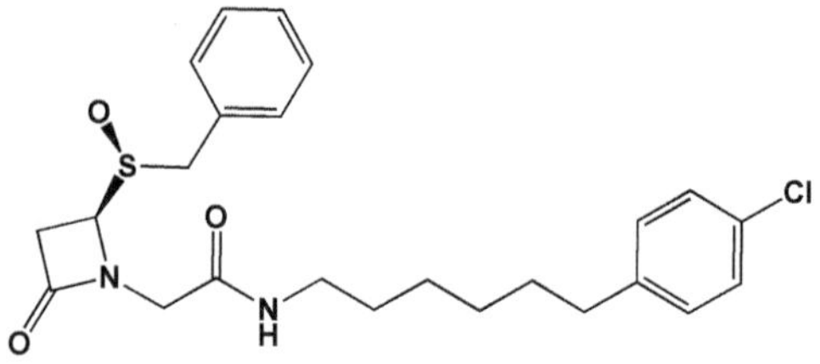

SB-222657

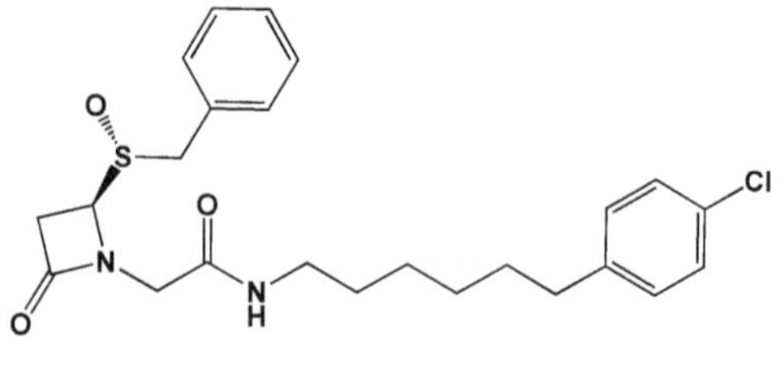

SB-223777

These mechanism-based azetidinones (FW = 460.02 g/mol) inhibit lipoprotein-associated phospholipase A_2, platelet-activating factor acetylhydrolase, with K_i values of 40 nM and 6.3 μM for SB-222657 and SB-223777, respectively, based on the reversible phase prior to inactivation. Stereoselectivity is also observed in the second-order rate constants for inactivation: $k_{obs}/[I]$ = 6.6 × 10^5 $M^{-1} \cdot s^{-1}$ for SB-222657, and $k_{obs}/[I]$ = 1.6 × 10^4 $M^{-1} \cdot s^{-1}$ for SB-223777. SB-222657 is completely inactive against the human synovial fluid Type IIa PLA_2, which, unlike Lp-PLA_2, is a Ca^{2+}-dependent PLA_2 and lacks an active-site serine residue and is therefore not expected to be inhibited. These inhibitors are unique tools for investigating the role of Lp-PLA_2 in the altered phospholipid metabolism during oxidative modification of lipoproteins. These findings also demonstrate unequivocally that Lp-PLA_2 is solely responsible for the elevation of lyso-PtdCho found within oxidized LDL. **1**. MacPhee, Moores, Boyd, *et al.* (1999) *Biochem. J.* **338**, 479.

SB 225002

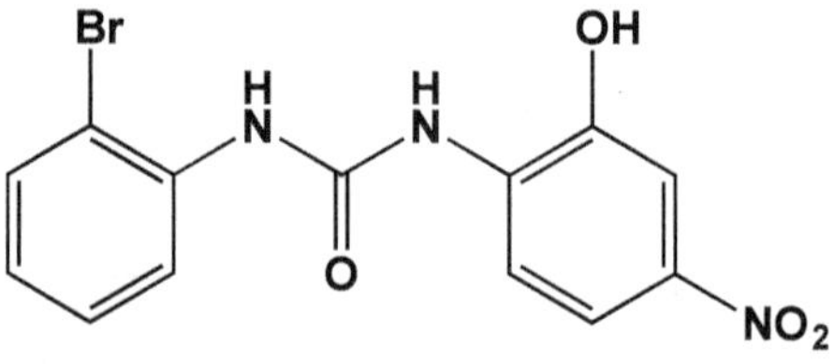

This potent and selective CXCR2 antagonist (FW = 352. 14 g/mol; CAS 182498-32-4; Solubility: 100 mM in DMSO), systematically named *N*-(2-bromophenyl)-*N'*-(2-hydroxy-4-nitrophenyl)urea, blocks Interleukin-8 (IL-8) binding to the CXCR2 chemokine receptor (IC_{50} = 22 nM). SB 225002 displays more than 150x selectivityfor CXCR2 over CXCR1 receptors. CXC chemokine receptors are G-coupled, seven-helix integral membrane proteins that bind cytokines in the CXC chemokine family. *In vitro*, SB 225002 potently inhibits human and rabbit neutrophil chemotaxis induced by both IL-8 and GROα. *In vivo*, SB 225002 selectively blocks IL-8-induced neutrophil margination in rabbits (1). SB 225002 also inhibits HIV-1 replication in both T lymphocytes and macrophages, indicating potential therapeutic uses for CXCR2 antagonists in limiting HIV-1 infection and AIDS (2). This inhibitory effect on Aβ40 and Aβ42 formation is mediated by γ-secretase through reduction in expression of presenilin, a γ-secretase substrate. SB225002 also inhibits angiogenic activity of IL-8 by blocking its binding with CXCR2 in the early stage of experimentally induced choroidal neovascularization (CNV), which may provide a potential new therapeutic strategy for CNV (3). Cell cycle analysis, immunocytometry, immunoblotting, and RNA interference reveal that SB225002 induces mitotic checkpoint serine/threonine-protein kinase BubR1-dependent mitotic arrest (4). SB225002 also possesses a microtubule-destabilizing activity, as evidenced by Bcl_2 and Bcl_{xL} hyperphosphorylation, suppression of microtubule polymerization, and induction of a prometaphase arrest. Molecular docking studies suggest that SB225002 should a good affinity toward vinblastine binding site on β-tubulin subunit. By contrast, SB265610, exhibits potent IL-8RB antagonistic activity but does not exhibit a similar antimitotic activity. The antitumor activity of SB225002 is unaffected by P-glycoprotein-mediated multidrug resistance. **1.** White, Lee, Young, et al. (1998) *J. Biol. Chem.* **273**, 10095. **2.** Lane, Lore, Bock, et al. (2001) *J. Virol.* **75**, 8195. **3.** Qu, Zhou & Xu. (2009) *Zhonghua Yan Ke Za Zhi.* **45**, 742. **4.** Goda, Koyama, Sowa, et al. (2013) *Biochem. Pharmacol.* **85**, 1741.

SB 239063

This highly substituted imidazole (FW = 368.41 g/mol), also known as *trans*-1-(4-hydroxycyclohexyl)-4-(4-fluorophenyl)-5-(2-methoxypyrimidin-4-yl)imidazole, is a strong inhibitor of the α- and β-isoforms of p38 mitogen-activated protein kinase (IC_{50} = 44 nM) and exhibits anti-inflammatory/antiallergic activity, commending it for the treatment of asthma and other inflammatory disorders. **1.** Underwood, Osborn, Kotzer, et al. (2000) *J. Pharmacol. Exp. Ther.* **293**, 281.

SB 243545

This L-tyrosine-containing metabolite ($FW_{free-base}$ = 487.51 g/mol) is a potent inhibitor of bacterial tyrosyl-tRNA synthetase, IC_{50} = 0.2 nM for the *Staphylococcus aureus* enzyme. The inhibitor occupies the known substrate binding sites as well as an affinity-enhancing butyl binding pocket. **1.** Qiu, Janson, Smith, et al. (2001) *Protein Sci.* **10**, 2008.

SB 252218, See *Tranilast*

SB 269970

This brain-penetrant serotonin antagonist (FW = 388.95 g/mol; CAS 261901-57-9; Soluble to 20 mM in H_2O and to 100 mM in DMSO), also named (2R)-1-[(3-hydroxyphenyl)sulfonyl]-2-[2-(4-methyl-1-piperidinyl)-ethyl]pyrrolidine hydrochloride, potently and selectively targets 5-HT_{7A} receptors (pK_i = 8.9), with weaker action against 5-HT_{5A} (pK_i = 7.2), and 5-HT_{5B} (pK_i = 6.0) receptors and showing pK_i values < 6.0 for 5-HT_{1A}, 5-HT_{1D}, 5-HT_{1E}, 5-HT_{1F}, 5-HT_{2A}, 5-HT_{2B}, 5-HT_{2C}, 5-HT_4 and 5-HT_6 receptors. **1.** Hagan, et al. (2000) *Brit. J. Pharmacol.* **130**, 539. **2.** Lovell, et al. (2000) *J. Med. Chem.* **43**, 342. **3.** Kogan, et al. (2002) *Eur. J. Pharmacol.* **449**, 105.

SB-275833, See *Retapamulin*

SB 284485

This L-tyrosine-containing metabolite (FW = 400.39 g/mol) is a potent inhibitor of bacterial tyrosyl-tRNA synthetase, with an IC_{50} of 4 nM for the *Staphylococcus aureus* enzyme. The inhibitor occupies the known substrate binding sites in unique ways that reveal a so-called butyl binding pocket. **1.** Qiu, Janson, Smith, et al. (2001) *Protein Sci.* **10**, 2008.

SB 332235

This CXCR2 antagonist (FW = 410.66 g/mol; CAS 276702-15-9; Solubility: 100 mM in DMSO), also named 6-chloro-3-[[[(2,3-dichlorophenyl)amino]carbonyl]amino]-2-hydroxybenzenesulfonamide, targets <u>C</u>-<u>X</u>-<u>C</u> motif Chemokine Receptor 2 (1). The CXCR2 binds interleukin 8 (IL8) with high affinity, transducing the signal through a G-protein-activated second messenger system. This receptor also binds to chemokine (C-X-C motif) ligand 1 (*or* CXCL1/MGSA), a protein with melanoma growth-stimulating activity required for serum-dependent melanoma cell growth. CXCR2 mediates neutrophil migration to sites of inflammation, and receptor knockout studies in mice suggest it controls oligodendrocyte precursor positioning in developing spinal cord by arresting their migration. SB-332235 inhibits binding of ^{125}I-labeled human IL-8 (IC_{50} = 40.5 nM) as well as human IL-8-induced calcium mobilization by rabbit CXCR2 (IC_{50} = 7.7 nM), but not the corresponding binding of human IL-8 to rabbit CXCR1 (IC_{50} > 1000) or human IL-8-induced calcium mobilization by rabbit CXCR1 (IC_{50} = 2200 nM). Such findings suggest the rabbit is an appropriate species for examining anti-inflammatory effects of human CXCR2-selective antagonists (1). When tested *in vitro*, SB332235 potently inhibited human IL-8-induced chemotaxis of rabbit neutrophils (IC_{50} = 0.75 nM), suggesting that inhibition of leukocyte migration into the knee joint is a likely mechanism by which the CXCR2 antagonist modulates disease (1). CXCR2 is also significantly increased in MDS CD34$^+$ cells from a large clinical cohort and was predictive of increased transfusion dependence (2). High CXCR2 expression is also an adverse prognostic factor in the Cancer Genome Atlas AML cohort, again pointing to a critical role of the IL8-CXCR2 axis in AML/MDS. CXCR2 knockdown and pharmacologic inhibition significantly reduce proliferation in several leukemic cell lines and in primary MDS/AML samples by inducing G_0/G_1 cell-cycle arrest. **1.** Podolin, Bolognese, Foley, et al. (2002) *J. Immunol.* **169**, 6435. **2.** Schinke, Giricz, Li, et al. (2015) *Blood* **125**, 3144.

SB 334867, Similar in action to *SB 408124*

SB 408124

This non-peptide, orexin receptor antagonist (FW = 356.37 g/mol; CAS 288150-92-5), also known as Tocris-1963 and named systematically as N-(6,8-difluoro-2-methyl-4-quinolinyl)-N'-[4-(dimethylamino)phenyl]urea, selectively targets the human OX_1 receptor (K_i = 22 nM), with much low affinity for the OX_2 receptor (K_i = 1.4 nM). Orexin, a peptide neurotransmitter (known also as hypocretin) produced by 10,000-20,000 cells within the hypothalamus, regulates arousal, wakefulness, and appetite. SB 408124 blocks the orexin-A induced grooming in rats following oral administration *in vivo*. 1. Langmead, Jerman, Brough, *et al.* (2004) *Brit. J. Pharmacol.* **141**, 340.

SB 415286

This substituted maleimide (FW = 359.73 g/mol; CAS 264218-23-7; Soluble in DMSO), also known as 3-(3-chloro-4-hydroxyphenylamino)-4-(2-nitrophenyl)-1H-pyrrole-2,5-dione, inhibits both GSKα and GSKβ, *or* [glycogen synthase] kinase, with IC_{50} and K_i values of 71 nM and 31 nM, respectively (1). **Mode of Inhibitory Action:** GSK is a serine/threonine protein kinase, that is inhibited by a variety of extracellular stimuli including insulin, growth factors, cell specification factors and cell adhesion. As a result, GSK-3 inhibition is a druggable target for altering numerous signalling pathways that control a wide range of cellular responses. Although lithium ion has been most commonly used to investigate the effects of GSK-3β inhibition, it is nonspecific and may produce results due to other kinase inhibition (*See Lithium Ion*); in contrast, SB415286 is potent and highly specific GSK-3β inhibitor, acting in an ATP-competitive manner (1). **Cellular Effects:** SB-415286 stimulates glycogen synthesis in human liver cells and induces expression of a β-catenin-LEF/TCF regulated reporter gene in HEK293 cells (1). It inhibits cellular GSK-3 activity, as assessed by activation of glycogen synthase (GS), which is a direct target of this kinase (1). SB415286 also protects both central and peripheral nervous system neurons in culture from cell death induced by reducing PI 3-kinase pathway activity, and its inhibitory action correlates with inhibition of GSK-3 phosphorylation of the mictorubule-associated Tau protein and β-catenin (2). In L6 myotubes, SB-415286 induces a much greater activation of GS (6.8x) compared to that elicited by insulin (4.2x) or lithium ion (4x) (3). In adipocytes, insulin, lithium ion and SB-415286 all cause comparable activation of GS, despite a substantial differentiation-linked reduction in GSK3 expression, indicating that GSK3 is an important determinant of GS activation in fat cells (3). SB-415286 was also used to show that GSK3β plays an important role in rapamycin-mediated cell cycle regulation and chemosensitivity, thereby potentiating rapamycin's antitumor effects (4). Likewise, this inhibitor aided in showing that that Coxsackievirus B3 (CVB3) infection stimulates GSK3β activity by a tyrosine kinase-dependent mechanism that contributes to CVB3-induced cytopathic effects and apoptosis through dysregulation of β-catenin (5). SB-415286 exerts neuroprotective effects on hydrogen peroxide-induced cell death in B65 rat neuroblastoma cells and neurons (6). GSK-3 inhibitors like SB-415286 also provide new ways to enhance the anti-tumor

activity of Tumor necrosis factor-Related Apoptosis-Inducing Ligand (TRAIL) in hepatocellular carcinoma (HCC) (7). SB-415286 effects also suggest that GSK-3 regulates multiple myeloma cell growth and bortezomib-induced cell death (8) as well as cell death, apoptosis, and in vivo tumor growth delay in neuroblastoma Neuro-2A cell line (9). 1. Coghlan, Culbert, Cross, *et al.* (2000) *Chem. Biol.* **7**, 793. 2. Cross, Culbert, Chalmers *et al.* (2001) *J. Neurochem.* **77**, 94. 3. MacAulay, Hajduch, Blair, *et al.* (2003) *Eur. J. Biochem.* **270**, 3829. 4. Dong, Pweng, Zhang, *et al.* (2005) *Cancer Res.* **65**, 1961. 5. Yuan, Zang, Wang, *et al.* (2005) *Cell Death Differ.* **12**, 1097. 6. Pizarro, Yeste-Velasco, Rimbau, *et al.* (2008) *Int. J. Dev. Neurosci.* **26**, 269. 7. Beurel, Blivet-Van Eggelpoël, Kornprobst, *et al.* (2009) *Biochem. Pharmacol.* **77**, 54. 8. Piazza, Manni, Tubi, *et al.* (2010) *BMC Cancer* **10**, 526. 9. Dickey, Schleicher, Leahy, *et al.* (2011) *J. Neurooncol.* **104**, 145.

SB 418011

This substituted indole carboxylate (FW = 443.33 g/mol) inhibits *Escherichia coli*, *Streptococcus pneumoniae*, and *Hemophilus influenzae* β-ketoacyl-[acyl-carrier-protein] synthase, with IC_{50} values of 1.2, 0.016, and 0.59 μM, respectively. In the bacterial type II fatty acid synthase system, β-ketoacyl-acyl carrier protein (ACP) synthase III (FabH) catalyzes the condensation of acetyl-CoA with malonyl-ACP. 1. Khandekar, Gentry, Van Aller, *et al.* (2001) *J. Biol. Chem.* **276**, 30024.

SB 431542

This potent antitumor agent (FW = 384.39 g/mol; CAS 301836-41-9; Solubility: 77 mg/mL DMSO; <1 mg/mL H$_2$O), also named 4-(4-(benzo[*d*][1,3]dioxol-5-yl)-5-(pyridin-2-yl)-1*H*-imidazol-2-yl)benzamide, inhibits transforming growth factor-β superfamily type 1 activin receptor-like kinase (ALK) receptors ALK-4, -5, and -7, selectively targetting ALK5 with an IC_{50} of 94 nM in a cell-free assay. SB431542 shows 100-fold more selective for ALK5 over p38 MAPK and other kinases (1-3). SB-431542 is a selective inhibitor of endogenous activin and TGF-β signaling, but has no effect on bone morphogenetic proteins (BMPs) signaling. B-431542 is without effect on ERK, JNK, or p38 MAP kinase pathways or on signaling pathways activated in response to serum. 1. Laping, Grygielko, Mathur, *et al.* (2002) *Mol. Pharmacol.* **62**, 58. 2. Inman, Nicolas, Callahan, *et al.* (2002) *Mol. Pharmacol.* **62**, 65. 3. Halder, Beauchamp & Datta (2005) *Neoplasia* **7**, 509.

SB505124

This selective TGFβR inhibitor (FW = 335.40 g/mol; CAS 694433-59-5; Solubility: 67 mg/mL DMSO; <1 mg/mL H₂O; 2-(4-(benzo[*d*][1,3]dioxol-5-yl)-2-*tert*-butyl-1*H*-imidazol-5-yl)-6-methylpyridine) potently targets the Activin receptor-Like Kinase activities of Transforming Growth Factor-β (TGF- β) receptors: ALK4 (IC$_{50}$ = 129 nM) and ALK5 (IC$_{50}$ = 47 nM), also inhibiting ALK7, but not ALK1, 2, 3, or 6 (1). Docking studies point to specificity-conferring hydrogen bond interactions between SB-505124 and amino acids His-283 and Ser-280 of ALK-5 (2). SB505124 brings about concentration-dependently inhibitions ALK4-, ALK5-, and ALK 7-dependent activation of downstream cytoplasmic signal transducers, Smad2 and Smad3, and of TGF- β-induced mitogen-activated protein kinase pathway components but does not alter ALK1, ALK2, ALK3 or ALK6-induced Smad signaling. SB-505124 also inhibits more complex endpoints of TGF- β action, as evidenced by its ability to abrogate cell death caused by TGF-β1 treatment (1). SB-505124 blocks TGF- β-induced cytoskeletal alterations (*i.e.*, filipodia formation and F-actin assembly) in endothelial cells linked to nitric oxide generation and cell-cell interactions (3). Likewise, migration and invasion by MCF-7-M5 cells are suppressed by SB-505124, whereas anti-estrogens do not suppress the motile properties of these cells (4). **1**. DaCosta, Byfield, Major, Laping & Roberts (2004) *Mol. Pharmacol.* **65**, 744. **2**. Sapitro, Dunmire, Scott, *et al.* (2010) *Mol. Vis.* **16**, 1880. **2**. Hu, Ramachandrarao, Siva, *et al.* (2005) *Am. J. Physiol. Renal Physiol.* **289**, F816. **3**. Goto, Hiyoshi, Ito, *et al.* (2011) *Cancer Sci.* **102**, 1501.

SB 706504

This p38 inhibitor (FW = 492.46 g/mol; CAS 911110-38-8; Solubility: 100 mM in DMSO), also named *N*-cyano-*N'*-[2-[[8-(2,6-difluorophenyl)-4-(4-fluoro-2-methylphenyl)-7,8-dihydro-7-oxopyrido[2,3-*d*]pyrimidin-2-yl]amino]ethyl]guanidine, targets p38 mitogen-activated protein kinase (MAPK), a signaling enzyme known to be increased in macrophages of individuals with chronic obstructive pulmonary disease (COPD). Extracellular stimuli, including the TOLL-like receptor (TLR) 4 ligand lipopolysaccharide and cytokines activate p38 MAPK signaling, up-regulating production of proinflammatory cytokines and chemokines by activating such transcription factors as NFκB and Activating Transcription Factor 2 or by altering chromatin structure to allow NFκB binding to the promoter regions of inflammatory genes.. When LPS-stimulated monocyte-derived macrophages (MDMs) and alveolar macrophages (AMs) are cultured with dexamethasone and/or SB706504, the latter causes transcriptional inhibition of a range of cytokines and chemokines in COPD MDMs. Combined use of SB706504 and dexamethasone caused greater suppression of gene expression (–8.90), when compared to SB706504 alone (–2.04) or dexamethasone (–3.39). SB706504 significantly inhibits LPS-stimulated TNFα production from COPD and smoker AMs, with near-maximal suppression caused by targeting a subset of inflammatory macrophage genes and when used with dexamethasone causes effective suppression of these genes. **1**. Kent, Smyth, Plumb, *et al.* (2009) *J. Pharmacol. Exp. Ther.* **328**, 458.

SB-715992, See Ispinesib

SB 743921

This selective cell-cycle inhibitor (FW = 553.52 g/mL; CAS 940929-33-9 (hydrochloride), 618430-39-0 (free base); Solubility: 110 mg/mL DMSO; 20 mg/mL H₂O), also known as (*R*)-*N*-(3-aminopropyl)-*N*-(1-(3-benzyl-7-chloro-4-oxo-4*H*-chromen-2-yl)-2-methylpropyl)-4-methylbenzamide hydrochloride, inhibits Kinesin Spindle Protein (KSP), a member of the kinesin superfamily of microtubule-based motors that plays a critical role in mitosis by mediating centrosome separation and bipolar spindle assembly and maintenance. Inhibition of KSP function leads to cell cycle arrest at mitosis with the formation of monoastral microtubule arrays, and ultimately, to cell death. The K_i values for SB 743921 inhibition of human and mouse KSP are 0.1 nM and 0.12 nM, respectively, while the K_i values for inhibition of other kinesins (including MKLP1, and Kin2) exceed 70 µM. **Note**: The target KSP must not be confused with the neurofilament-directed protein kinases that phosphorylate KSPXXXX (or Lys-Ser-Pro-Xaa-Xaa-Xaa-Xaa) sequences in light, medium- and high-molecular weight neurofilament proteins. **1**. Sakowicz, Finer, Beraud et al. (2004) *Cancer Res.* **64**, 3276.

SB-3CT, See *(4-Phenoxyphenylsulfonyl)methylthiirane*

SB-RA-2001

This taxane (FW = 319.32 g/mol) inhibits the proliferation of *Bacillus subtilis* 168 and *Mycobacterium smegmatis*, with minimum inhibitory concentrations of 38 and 60 µM, respectively. The lengths of these microorganisms increase remarkably in the presence of SB-RA-2001, indicating inhibition of cytokinesis. SB-RA-2001 also perturbs formation of FtsZ ring in *B. subtilis* 168 cells and also altered the mid-cell localization of the late cell division protein, DivIVA. Flow cytometric analysis of the SB-RA-2001-treated *B. subtilis* cells indicated that the compound did not affect DNA replication. SB-RA-2001 does not affect the localization of the chromosome-partitioning protein, Spo0J, along the two ends of the nucleoids and is without discernable effect on the nucleoid segregation. *In vitro*, SB-RA-2001 binds to FtsZ with modest affinity, promotes assembly and bundling of FtsZ protofilaments, and reduces FtsZ GTPase activity. GTP did not alter SB-RA-2001 binding to FtsZ. Instead of binding to the nucleotide site, SB-RA-2001 binds in a cleft region between the C-terminal domain and H^7 helix, resembling FtsZ onteractions with PC190723, a well-characterized inhibitor of FtsZ. The findings suggest SB-RA-2001 inhibits bacterial proliferation by targeting FtsZ assembly dynamics of and this can be exploited further to develop potent FtsZ targeted antimicrobials. (*See* *PC190723*) **1**. Singh, Bhattacharya, Rai, *et al.* (2014) *Biochemistry* **53**, 2979.

SC-236

This sulfonamide (FW = 401.80 g/mol), known systematically as 4-[5-(4-chlorophenyl)-3-(trifluoromethyl)-1*H*-pyrazol-1-yl]benzenesulfonamide, is a strong inhibitor of cyclooxygenase-2 (IC$_{50}$ = 0.01 μM), exhibiting a preference over cyclooxygenase-1, IC$_{50}$ = 17.8 μM (1-4). SC-236 also inhibits angiogenesis in a dose-dependent manner. **1**. Castano, Bartrons & Gil (2000) *J. Pharmacol. Exp. Ther.* **293**, 509. **2**. Masferrer, Koki & Seibert (1999) *Ann. N. Y. Acad. Sci.* **889**, 84. **3**. Penning, Talley, Bertenshaw, *et al.* (1997) *J. Med. Chem.* **40**, 1347. **4**. Gierse, McDonald, Hauser, *et al.* (1996) *J. Biol. Chem.* **271**, 15810.

SC-514

This orally active, ATP-competitive IKK-2 inhibitor (FW = 224.30 g/mol; CAS 354812-17-2; Solubility: 44 mg/mL DMSO, <1 mg/mL H$_2$O), also named 4-amino[2,3'-bithiophene]-5-carboxamide, blocks NF-κB-dependent gene expression (IC$_{50}$ = 3-12 μM), without affecting other IKK isoforms or other serine-threonine and tyrosine kinases. **Mode of Inhibitor Action:** NFκB-induced gene expression contributes significantly to the pathogenesis of inflammatory diseases, such as arthritis. The IκB kinase is an upstream component of the NF-κB signal transduction cascade, IκBα (*or Inhibitor of κB*) is a protein that inactivates NF-κB transcription factor by masking its nuclear localization signals. SC-514 does not inhibit the phosphorylation and activation of the IKK complex. Its action is characterized by a delay, but not a complete blockade, in IκBα phosphorylation and degradation. There is also slightly slowed p65 nuclear import as well as faster export from the nucleus. IKK-2I has comparable K_m and k_{cat} values for κBα and p65 as substrates, and SC-514 likewise inhibits their phosphorylation to a simiar degree. **1**. Kishore, Sommers, Mathialagan, *et al.* (2003) *J. Biol. Chem.* **278**, 32861.

SC-558

This substituted pyrazole (FW = 446.25 g/mol), also known as 4-[5-(4-bromophenyl)-3-(trifluoromethyl)-1*H*-pyrazol-1-yl]benzenesulfonamide, is a strong inhibitor of cycloxygenase-2, IC$_{50}$ = 9.3 nM, exhibiting a preference over cyclooxygenase-1, IC$_{50}$ = 17.7 μM (1,2). **1**. Chavatte, Yous, Marot, Baurin & Lesieur (2001) *Bioorg. Med. Chem.* **44**, 3223. **2**. Kurumbail, Stevens, Gierse, *et al.* (1996) *Nature* **384**, 644.

SC-560

This orally active agent (FW = 352.74 g/mol; CAS 188817-13-2), also known as 5-(4-chlorophenyl)-1-(4-methoxyphenyl)-3-trifluoromethyl pyrazole, is a selective inhibitor of cyclooxygenase-1, IC$_{50}$ = 9 nM, showing markedly weaker action against cyclooxygenase-2, IC$_{50}$ = 6.3 μM. SC-560 does not inhibit COX-2-derived PGs in the lipopolysaccharide-induced rat air pouch. Therapeutic or prophylactic administration of SC-560 in the rat carrageenan footpad model does not affect acute inflammation or hyperalgesia at doses that markedly inhibits *in vivo* COX-1 activity. **1**. Smith, Zhang, Koboldt, *et al.* (1998) *Proc. Natl. Acad. Sci. U.S.A.* **95**, 13313.

SC 19220

This selective EP$_1$ receptor antagonist (FW = 331.76 g/mol; CAS 19395-87-0; Soluble to 100 mM in DMSO), also named 8-chlorodibenz[*b,f*][1,4]oxazepine-10(11*H*)-carboxylic 2-acetylhydrazide, targets human prostaglandin EP$_1$ receptor, with an IC$_{50}$ value of 6.7 μM for inhibiting binding of [^{3}H]-PGE$_2$ to EP$_1$ transfected COS cells. **1**. Sanner (1972) *Intra-Science Chem. Report* **6**, 1. **2**. Funk et al (1993) *J. Biol. Chem.* **268**, 26767. **3**. Botella, *et al.* (1995) *J. Pharmacol. Exp. Ther.* **273**, 1008.

SC12267, *See 4SC-101*

SC 26196

This desaturase inhibitor (FW = 423.55 g/mol; CAS 218136-59-5; Soluble to 25 mM in DMSO), also named α,α-diphenyl-4-[(3-pyridinylmethylene)amino]-1-piperazinepentanenitrile, targets Δ^6-desaturase (IC$_{50}$ = 0.2 μM, *in vitro*), the rate-limiting step in the polyunsaturated fatty acid (PUFA) biosynthetic pathway. It displays 1000x selectivity over Δ^5 and Δ^9 desaturases (IC$_{50}$ · values are >200 μM). That the antiinflammatory properties of SC-26196 are consistent with its action *in vivo* as a Δ^6-desaturase inhibitor is evidenced by: (a) the strong correlation between liver Δ^6-desaturase inhibition and decreased edema; (b) the time-dependent decreased in edema upon SC-26196 administration; (c) SC-26196's dose-dependent action in reducing arachidonate in liver, plasma and peritoneal cells; and (d) the reversal of changes resulting from SC-26196 inhibition upon controlled refeeding of arachidonic acid, but not oleic acid (1). SC-26196 also impedes intestinal tumorigenesis (2). **1.** Obukowicz, Welsch, Salsgiver, *et al.* (1998) *J. Pharmacol. Exp. Ther.* **287**, 157. **2.** Hansen-Petrik, McEntee, Johnson, *et al.* (2002) *Cancer Lett.* 175, 157.

SC-29333, *See* Misoprostol

SC 39026, *See* 2-Chloro-4-(1-hydroxyoctadecyl)benzoate

SC 40827

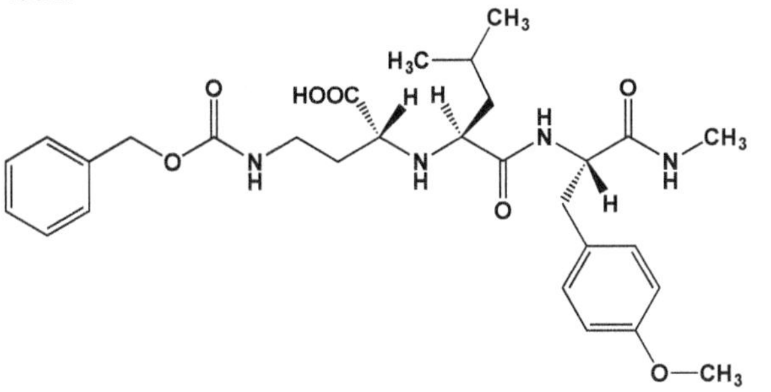

This peptidomimetic (FW$_{free-acid}$ = 555.65 g/mol), also known systematically as *N*-[(3-*N*-benzyloxycarbonyl)amino-1-(*R*)-carboxypropyl]-L-leucyl-(*O*-methyl)-L-tyrosine-*N*-methyl-amide, inhibits matrilysin, *or* matrix metalloproteinase 7 (1,2) and tissue collagenases (3,4). **1.** Woessner (1995) *Meth. Enzymol.* **248**, 485. **2.** Abramson, Conner, Nagase, Neuhaus & Woessner, Jr. (1995) *J. Biol. Chem.* **270**, 16016. **3.** Brannstrom, Woessner, Jr., Koos, Sear & LeMaire (1988) *Endocrinology* **122**, 1715. **4.** Delaisse, Eeckhout, Sear, *et al.* (1985) *Biochem. Biophys. Res. Commun.* **133**, 483.

SC 44463

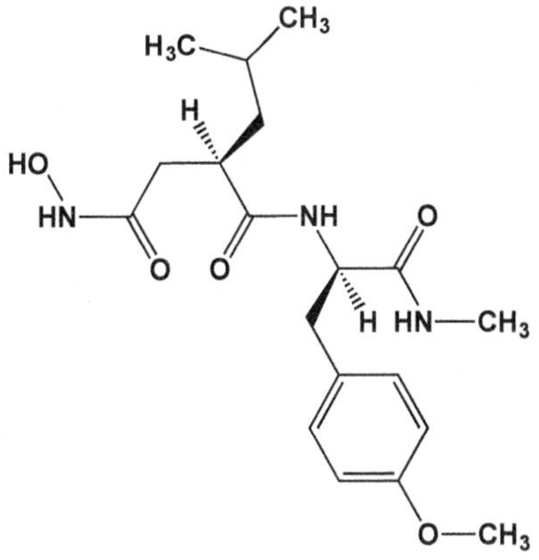

This peptide analogue (FW = 379.46 g/mol; CAS 104408-38-0), also known as *N*⁴-hydroxyl-*N*¹-{1*S*-[(4-methoxphenyl)methyl]-2-(methylamino)-2-oxoethyl}-2*R*-(2-methylpropyl)butanediamide, inhibits a number of metalloproteinases, with an IC$_{50}$ value of 10 nM for matrilysin. **Target(s):** atrolysin C (1,2); matrilysin, *or* matrix metalloproteinase 7 (3,4); and tissue collagenases (5). **1.** Zhang, Botos, Gomis-Ruth, *et al.* (1994) *Proc. Natl. Acad. Sci. U.S.A.* **91**, 8447. **2.** Shannon, Baramova, Bjarnason & Fox (1989) *J. Biol. Chem.* **264**, 11575. **3.** Woessner, Jr. (1995) *Meth. Enzymol.* **248**, 485. **4.** Abramson, Conner, Nagase, Neuhaus & Woessner, Jr. (1995) *J. Biol. Chem.* **270**, 16016. **5.** Butler, Zhu, Mueller, *et al.* (1991) *Biol. Reprod.* **44**, 1183.

SC-50083, *See* N-[N-(t-Butoxycarbonyl)-L-phenyl-alanyl-L-leucyl]-2(S)-amino-3(S),6(S)-dihydroxy-5(R)-methyl-1-phenyldecane

SC-52551

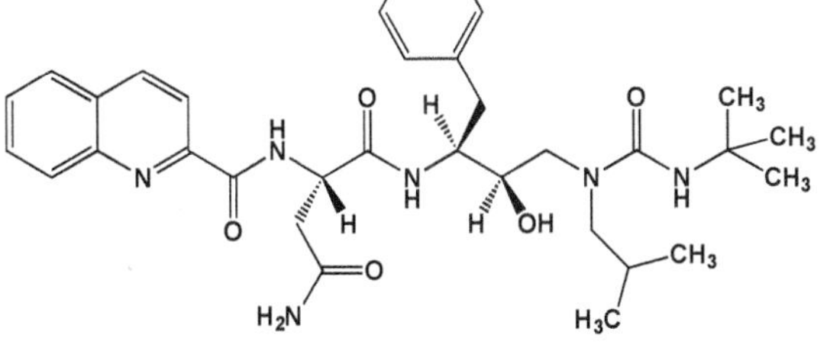

This pseudopeptide-containing peptidomimic (FW = 604.75 g/mol) is a potent inhibitor of human immunodeficiency virus I retropepsin, *or* HIV-I protease, with a K_i value for the *R*-stereoisomer is 6.1 nM (1,2). SC-52551 inhibition prevents cleavage of the viral polyproteins resulting in the formation of immature non-infectious viral particles. Protease inhibitors are almost always used in combination with at least two other anti-HIV drugs. **1.** Vacca (1994) *Meth. Enzymol.* **241**, 311. **2.** Getman, DeCrescenzo, Heintz, *et al.* (1993) *J. Med. Chem.* **36**, 288.

SC-58125

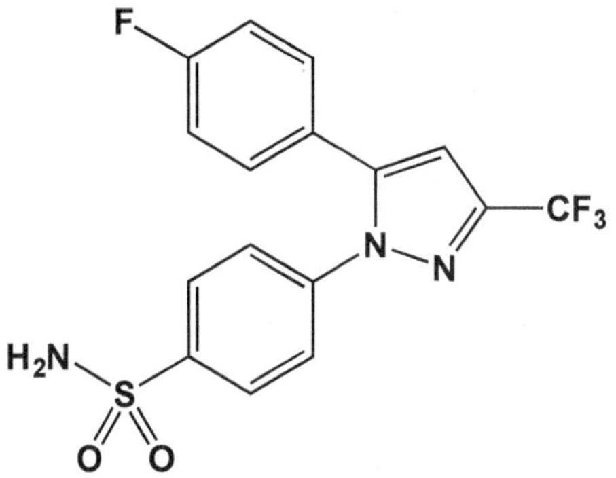

This substituted pyrazole (FW = 384.35 g/mol), also known as 1-[(4-methylsulfonyl)phenyl]-3-trifluoromethyl-5-(4-fluorophenyl)pyrazole, irreversibly inhibits cyclooxygenase-2, IC$_{50}$ = 0.05 μM, showing a preference over cyclooxygenase-1, IC$_{50}$ > 10 μM (1-3). **1.** Gierse, McDonald, Hauser, *et al.* (1996) *J. Biol. Chem.* **271**, 15810. **2.** Seibert, Zhang, Leahy, *et al.* (1994) *Proc. Natl. Acad. Sci. U.S.A.* **91**, 12013. **3.** Patrignani, Panara, Sciulli, *et al.* (1997) *J. Physiol. Pharmacol.* **48**, 623.

SC-58272

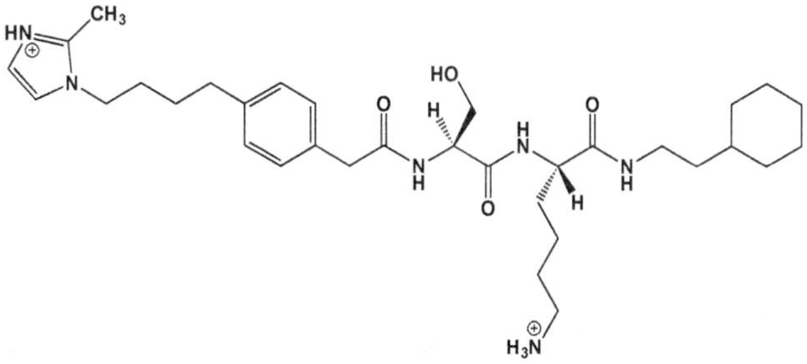

This substituted dipeptide (FW$_{free-base}$ = 596.81 g/mol) is a potent inhibitor of glycylpeptide *N*-tetradecanoyltransferase, *or* protein *N*-myristoyltransferase, IC$_{50}$ = 56 nM (1-3). **1.** Bhatnagar, Ashrafi, Fütterer, Waksman & Gordon (2001) *The Enzymes*, 3rd ed. (Tamanoi & Sigman, ed.), **21**, 241. **2.** Zhang, Jackson-Machelski & Gordon (1996) *J. Biol. Chem.* **271**, 33131. **3.** Sogabe, Masubuchi, Sakata, *et al.* (2002) *Chem. Biol.* **9**, 1119.

SC-58635, *See* Celecoxib

SC-59383

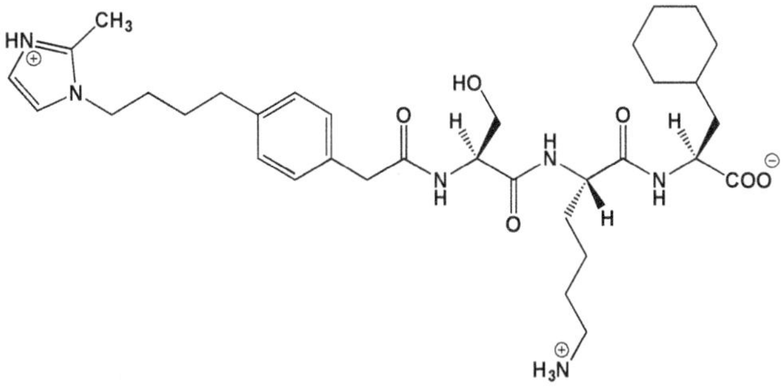

This substituted tripeptide (FW$_{inner salt}$ = 640.82 g/mol) strongly inhibits protein *N*-myristoyltransferase, IC$_{50}$ = 1.45 μM for the *Candida albicans* enzyme (1,2). **1.** Bhatnagar, Ashrafi, Fütterer, Waksman & Gordon (2001) *The Enzymes*, 3rd ed. (Tamanoi & Sigman, ed.), **21**, 241. **2.** Lodge, Jackson-Machelski, Devadas, *et al.* (1997) *Microbiology* **143**, 357.

SC-ααδ9

This highly substituted glutamine derivative (FW$_{\text{free-acid}}$ = 680.84 g/mol), also known as 4-(benzyl-(2-[(2,5-diphenyloxazole-4-carbonyl)amino]-ethyl)carbamoyl)-2-decanoylaminobutyric acid, inhibits Cdc25A protein phosphatase, which dephosphorylates and activates cyclin-dependent kinases (CDKs), thereby effecting the progression from one phase of the cell cycle to the next. SC-ααδ9 als induces IGF-1-resistant apoptosis (1,2). **1**. Vogt, Wang, Johnson, *et al.* (2000) *J. Pharmacol. Exp. Ther.* **292**, 530. **2**. Vogt, Rice, Settineri, *et al.* (1998) *J. Pharmacol. Exp. Ther.* **287**, 806.

4SC-101

This uridine biosynthesis inhibitor (FW = 355.37 g/mol), also named 2-(3-fluoro-3'-methoxybiphenyl-4-carbamoyl)cyclopent-1-enecarboxylic acid and formerly as SC12267, targets dihydroorotate dehydrogenase, *or* DHODH, which catalyzes the fourth step in the *de novo* pyrimidine biosynthesis, converting dihydroorotate to orotate. *De novo* pyrimidine biosynthesis is essential for supplying DNA precursors in rapidly proliferating cells, and DHODH inhibition suppresses the rapid expansion of autoreactive lymphocytes in autoimmune diseases. 4SC-101 also inhibits the proliferation of phytohemagglutinin-stimulated lymphocytes (IC$_{50}$ ≈ 13 μM) and phytohemagglutinin-induced interleukin-17 secretion from human peripheral blood mononuclear cells, *or* PBMCs (IC$_{50}$ ≈ 6 μM), but independently of lymphocyte proliferation (1,2). 4SC-101 acts as a novel immunosuppressive drug, inhibits IL-17, attenuating colitis in two murine models of inflammatory bowel disease (3). **Mode of Inhibitory Action:** Structural studies indicate that the α- and the β-barrel domains of DHODH form a tunnel through which substrates gain access to the active site, and compounds interacting with the α- and β-barrel domains are known to block DHODH activity. (**See also** *Leflunomide*) 4SC-101 likewise binds in this region, IC$_{50}$ = 134 nM (1,2). **1**. Leban, Kralik, Mies, *et al.* (2005) *Bioorg. Med. Chem. Lett.* **15**, 4854. **2**. Kulkarni, Sayyed, Kantner, *et al.* (2010) *Am. J. Pathol.* **176**, 2840. **3**. Fitzpatrick, Deml, Hofmann, *et al.* (2010) *Inflamm. Bowel Dis.* **16**, 1763.

4SC-207

This microtubule-directed antineoplastic agent (FW = 382.44 g/mol; Solubility: 25 mM in DMSO), also known as (*E*)-ethyl 3-cyano-2-(3-(pyridin-3-yl)acrylamido)-4,5-dihydrothieno[2,3-*c*]pyridine-6(7*H*)-ca rboxylate, inhibits tubulin polymerization *in vivo*, strongly reducing both the rate and extent of microtubule self-assembly. Cells treated with 4SC-207 are delayed in mitosis, unable to complete chromosome alignment, or to maintain a metaphase plate. In many cells, spindles are disorganized, unstable, and marked by the appearance of extra poles. 4SC-207 also shows strong anti-proliferative activity in a large panel of tumor cell lines (average GI$_{50}$ ≈ 11nM). Importantly, 4SC-207 is active in multi-drug-resistant cell lines (*e.g.*, HCT-15 and ACHN), suggesting it is a poor substrate for drug

efflux pumps. 4SC-207 inhibits microtubule growth *in vivo* and promotes mitotic delay/arrest in a dose-dependent manner, followed by apoptosis or aberrant divisions due to chromosome misalignment and multi-polar spindle formation. 4SC-207 exhibits a low propensity towards bone marrow toxicity at concentrations that inhibit tumor growth in paclitaxel-resistant xenograft models. **1**. Bausch, Kohlhof, Hamm, *et al.* (2013) *PLoS One* **8**, e79594.

12-*epi*-Scalaradial

This dialdehyde-containing steroid (FW = 428.61 g/mol) is the position-12 epimer of a marine natural product isolated from the sponge *Cacospongia* sp. 12-*epi*-Scalaradial possesses anti-inflammatory properties *in vivo* and *in vitro* and is a potent irreversible inactivator of bee venom phospholipase A$_2$, IC$_{50}$ = 70 nM (1,2). **1**. de Carvalho & Jacobs (1991) *Biochem. Pharmacol.* **42**, 1621. **2**. Potts, Faulkner & Jacobs (1992) *J. Nat. Prod.* **55**, 1701.

Scandium & Scandium Ions

This Group 3 (or Group IIIA) element (Symbol = Sc; Atomic Number = 21; Atomic Weight = 44.955912) has a ground-state, neutral atom electronic configuration of $1s^2 2s^2 2p^6 3s^2 3p^6 4s^2 3d^1$. The ionic radius of Sc^{3+} is 0.732 Å. There is only one stable isotope: ^{45}Sc, and the radionuclide ^{46}Sc has the longest half-life (83.81 days). **Target(s):** acetylcholinesterase (1,2); calcium channels (3); Ferroxidase (4,5); myosin ATPase (traps MgADP or ScADP within myosin subfragments), by fluoroscandium anions ScF$_x$ (6,7); and phosphoprotein phosphatase (8). **1**. Marquis & Black (1985) *Biochem. Pharmacol.* **34**, 533. **2**. Marquis & Lerrick (1982) *Biochem. Pharmacol.* **31**, 1437. **3**. Beedle, Hamid & Zamponi (2002) *J. Membr. Biol.* **187**, 225. **4**. Huber & Frieden (1970) *J. Biol. Chem.* **245**, 3979. **5**. Frieden & Hsieh (1976) *Adv. Enzymol. Rel. Areas Mol. Biol.* **44**, 187. **6**. Gopal & Burke (1995) *J. Biol. Chem* **270**, 19282. **7**. Park, Ajtai & Burghardt (1999) *Biochim. Biophys. Acta* **1430**, 127. **8**. Zhou, Clemens, Hakes, Barford & Dixon (1993) *J. Biol. Chem.* **268**, 17754.

Scandium Fluoride, See *Scandium and Scandium Ions*

Scarlet Red, See *Biebrich Scarlet*

Scepter, See *Imazaquin*

Sceptrin

This antimicrobial and antifungal agent (FW$_{\text{free-base}}$ = 620.31 g/mol) from the South Pacific sponge *Agelas mauritiana* and related *A. nakamurai* species (1,2) is believed to disrupt cell membranes of both prokaryotic and eukaryotic cells (1). Sceptrin is also competitive antagonist of muscarinic acetylcholine receptors (2). **1**. Bernan, Roll, Ireland, *et al.* (1993) *J. Antimicrob. Chemother.* **32**, 539. **2**. Rosa, Silva, Escalona de Motta, *et al.* (1992) *Experientia* **48**, 885.

SCH 23388, See *SCH 23390*

SCH 23390

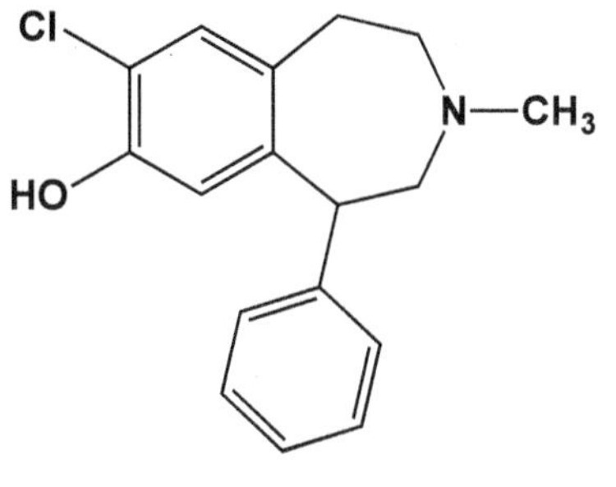

This photosensitive benzazepine (FW$_{\text{free-base}}$ = 287.79 g/mol), known systematically as [R(+)]-7-chloro-8-hydroxy-3-methyl-1-phenyl-2,3,4,5-tetrahydro-1H-3-benzazepine, was the first selective dopamine D$_1$-like receptor antagonist (1-3). The S(–)-enantiomer SCH 23388 is less active. **Target(s):** D$_1$ dopamine receptor, K_i = 0.2 nM (1-3); and D$_5$ dopamine receptor, K_i = 0.3 nM (2). **1**. O'Boyle & Waddington (1987) *J. Neurochem.* **48**, 1039. **2**. Bourne (2001) *CNS Drug Rev.* **7**, 399. **3**. Hyttel (1983) *Eur. J. Pharmacol.* **91**, 153.

Sch 25298, See *Florfenicol*

SCH 28080

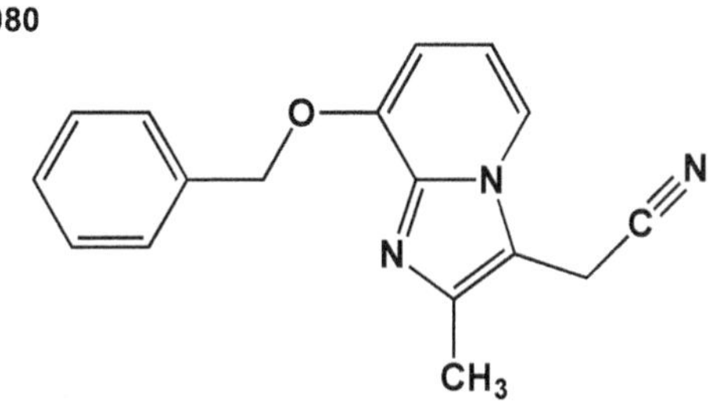

This hydrophobic imidazo[1,2-a]pyridine (FW = 277.33 g/mol), known systematically as 2-methyl-8-(phenylmethoxy)imidazo[1,2-a]pyridine-3-acetonitrile, binds near the M5-6 luminal loop of gastric H$^+$/K$^+$ ATPase, thereby preventing K$^+$ access to the ion binding domain (1-9). **1**. Wallmark, Briving, Fryklund, *et al.* (1987) *J. Biol. Chem.* **262**, 2077. **2**. Scott & Sundell (1985) *Eur. J. Pharmacol.* **112**, 268. **3**. Vagin, Denevich, Munson & Sachs (2002) *Biochemistry* **41**, 12755. **4**. Keeling, Taylor & Schudt (1989) *J. Biol. Chem.* **264**, 5545. **5**. Chatterjee & Das (1995) *Mol. Cell. Biochem.* **148**, 95. **6**. Planelles, Anagnostopoulos, Cheval & Doucet (1991) *Amer. J. Physiol.* **260**, F806. **7**. Adams, Tillekeratne, Yu, Pestov & Modyanov (2001) *Biochemistry* **40**, 5765. **8**. Silver, Frindt, Mennitt & Satlin (1997) *J. Exp. Zool.* **279**, 443. **9**. Vagin, Munson, Lambrecht, Karlish & Sachs (2001) *Biochemistry* **40**, 7480.

SCH 29851, See *Loratadine*

SCH 32615

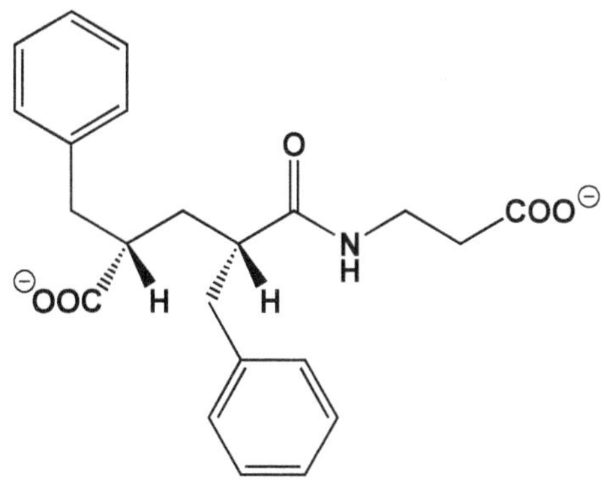

This small peptide (FW$_{\text{free-acid}}$ = 384.43 g/mol), also known as [N-(L-(1-carboxy-2-phenyl)ethyl]-L-phenylalanyl-β-alanine, inhibits neprilysin, K_i = 20 nM (1-4). **1**. Roques, Noble, Crine & Fournié-Zaluski (1995) *Meth. Enzymol.* **248**, 263. **2**. Chipkin, Berger, Billard, *et al.* (1988) *J. Pharmacol.*

Exp. Ther. **245**, 829. **3**. Oshita, Yaksh & Chipkin (1990) *Brain Res.* **515**, 143. **4**. Salles, Rodewald, Chin, Reinherz & Shipp (193) *Proc. Natl. Acad. Sci. U.S.A.* **90**, 7618.

SCH 34826

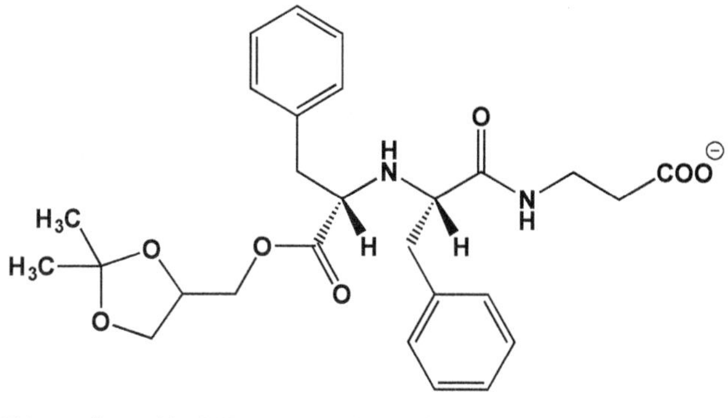

This small peptide (FW$_{\text{free-acid}}$ = 498.58 g/mol), also named (S)-N-(N-(2-[(2,2-dimethyl-1,3-dioxolan-4-yl)methoxy]-2-oxo-1-(phenylmethyl)ethyl)-phenylalanyl)-β-alanine, is a prodrug ester that upon hydrolysis generates the active constituent SCH 32615, a potent and selective neprilysin inhibitor (1,2). **1**. Monopoli, Ongini, Cigola & Olivetti (1992) *J. Cardiovasc. Pharmacol.* **20**, 496. **2**. Chipkin, Berger, Billard, *et al.* (1988) *J. Pharmacol. Exp. Ther.* **245**, 829.

SCH 39370

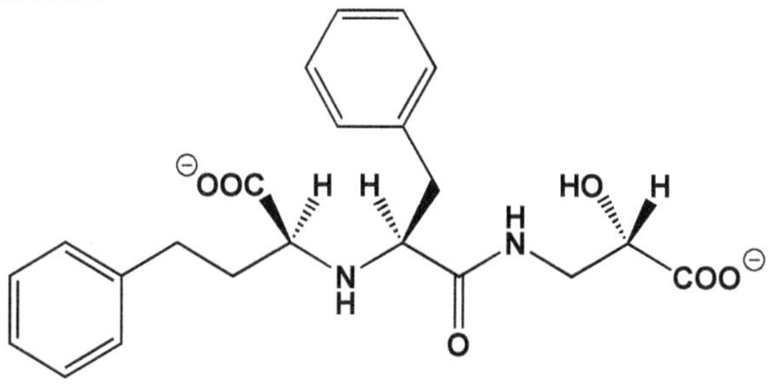

This small peptide (FW$_{\text{free-acid}}$ = 414.46 g/mol), also named N-[N-[1-(S)-carboxyl-3-phenylpropyl]-(S)-phenylalanyl]-(S)-isoserine, inhibits neutral endopeptidase (neprilysin), K_i = 11 nM (1-4). **1**. Roques, Noble, Crine & Fournié-Zaluski (1995) *Meth. Enzymol.* **248**, 263. **2**. Sybertz, Chiu, Vemulapalli, *et al.* (1989) *J. Pharmacol. Exp. Ther.* **250**, 624. **3**. Johnson, Arik & Foster (1989) *J. Biol. Chem.* **264**, 11637. **4**. Kanazawa, Casley, Sybertz, Haslanger & Johnston (1992) *J. Pharmacol. Exp. Ther.* **261**, 1231.

SCH 44342

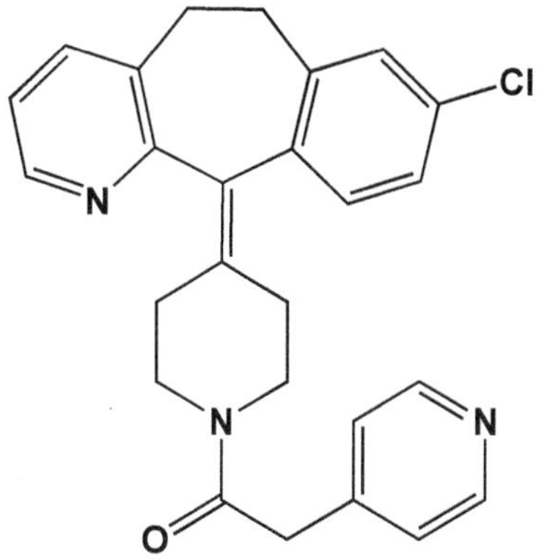

This substituted piperidine (FW = 429.95 g/mol), known systematically as (1-(4-pyridylacetyl)-4-(8-chloro-5,6-dihydro-11H-benzo[5,6]cyclohepta[1,2-b]pyridin-11-ylidene)piperidine, was one of the earliest nonpeptidic, nonthiol inhibitors of both rat brain and recombinant human protein farnesyltransferases with an IC$_{50}$ value of ~250 nM. SCH 44342 also weakly inhibits rat brain protein geranylgeranyltransferase type I, IC$_{50}$ > 115 μM. SCH 44342 behaves as a competitive inhibitor with respect to the protein substrate (1-3). **1**. Gibbs (2001) *The Enzymes*, 3rd ed., **21**, 81. **2**.

Bishop, Bond, Petrin, *et al.* (1995) *J. Biol. Chem.* **270**, 30611. **3.**
Yokoyama, Trobridge, Buckner, *et al.* (1998) *J. Biol. Chem.* **273**, 26497.

SCH 47896

This small peptide (FW$_{free-acid}$ = 430.46 g/mol), known systematically as *N*-[*N*-[1-(*S*)-carboxyl-3-(4-hydroxyphenyl)propyl]-(*S*)-phenylalanyl]-(*S*)-isoserine, inhibits neutral endopeptidase, *or* neprilysin (K_i = 85.1 nM), the neutral metalloendopeptidase (EC 3.4.24.11) that rapidly degrades atrial natriuretic peptide (ANP), with the kidney being a major site of ANP clearance. **1.** Kanazawa, Casley, Sybertz, Haslanger & Johnston (1992) *J. Pharmacol. Exp. Ther.* **261**, 1231.

SCH48446

This peptidomimetic (FW$_{free-acid}$ = 682.25 g/mol), known systematically as *N*-[*N*-[1-(*S*)-carboxyl-3-(4-hydroxy-3,5-diiodophenyl)propyl]-(*S*)-phenyl-alanyl]-(*S*)-isoserine, inhibits neutral endopeptidase, *or* neprilysin (K_i = 43.3 nM), the enzyme responsible for degrading Atrial natriuretic peptide (ANP) in the kidney (1). Neprilysin activity in rat kidney was inhibited by specific NEP inhibitors (phosphoramidon, thiorphan, SCH39370, SCH47896 and SCH48446) at µM concentrations. SCH48446 is the di-iodo analogue of SCH47896, which itself is a phenolic derivative of SCH39370. Autoradiography with [^{125}I]SCH47896 demonstrated that maximal binding occurs in the outer stripe of the outer medulla and to the inner cortex, a finding that is was consistent with binding to the deep proximal tubules. Moreover, these results indicate that degradation of ANP by neprilysin occurs mainly in the deep proximal tubules and that the proximal convoluted tubule within the outer cortex is unlikely to be a major site of neprilysin. **1.** Kanazawa, Casley, Sybertz, Haslanger & Johnston (1992) *J. Pharmacol. Exp. Ther.* **261**, 1231. **2.** Kanazawa, Yasujima, Kohzuki & Abe (1994) *Nihon. Jinzo. Gakkai. Shi.* **36**, 791.

SCH 51866

This purine derivative (FW = 389.15 g/mol) inhibits 3',5'-cyclic-GMP phosphodiesterases, *or* cGMP phospho-diesterase (1-8); cGMP phosphodiesterase-12 (4,8), cGMP phosphodiesterase-5 (3,4); cGMP phoisphodiesterase-7 (5,6), and cGMP phosphodiesterase-9 (1,2,7). **1.** Wang, Wu, Egan & Billah (2003) *Gene* **314**, 15. **2.** Soderling, Bayuga & Beavo (1998) *J. Biol. Chem.* **273**, 15553. **3.** Vemulapalli, Watkins, Chintala, *et al.* (1996) *J. Cardiovasc. Pharmacol.* **28**, 862. **4.** Yan, Zhao, Bentley & Beavo (1996) *J. Biol. Chem.* **271**, 25699. **5.** Hetman, Soderling, Glavas & Beavo (2000) *Proc. Natl. Acad. Sci. U.S.A.* **97**, 472. **6.** Sasaki, Kotera, Yuasa & Omori (2000) *Biochem. Biophys. Res. Commun.* **271**, 575. **7.** Diederen, La Heij, Markerink-van Ittersum, *et al.* (2007) *Brit. J. Ophthalmol.* **91**, 379. **8.** Dunkern & Hatzelmann (2007) *FEBS J.* **274**, 4812.

SCH 53239

This nucleotide exchange inhibitor (FW = 520.55 g/mol; CAS 188480-51-5) targets Ras-GDP, binding to the critical switch II region in the Ras protein, based on intra- and inter-molecular NOE distance constraints. Ras protein exists in two conformations that are in slow exchange on the NMR timescale. Reversible conformational changes occur mainly in the GDP-binding pocket, within the switch I and the switch II regions. **1.** Ganguly, Wang, Pramanik, *et al.* (1998) *Biochemistry* **37**, 15631.

SCH58235, *See Ezetimibe*

SCH58261

This adenosine receptor antagonist (FW = 345.36 g/mol; CAS 160098-96-4; Solubility: 69 mg/mL; < 1 mg/mL H$_2$O), also named [7-(2-phenylethyl)-5-amino-2-(2-furyl)pyrazolo[4,3-*e*]-1,2,4-triazolo[1,5-*c*]pyrimidine], potently and selectively targets A$_{2A}$ receptors (K_i = 2.3 nM, rat; K_i = 2 nM, bovine) (1). SCH 58261 does not show affinity for either the A$_3$ receptors or other receptors at concentrations up to 1 µM (1). Saturation experiments on rat A$_1$ and A$_{2a}$ adenosine receptors indicated the competitive nature of the antagonism. SCH 58261 antagonizes competitively the effects induced by the A$_{2a}$ adenosine-selective agonist CGS 21680 as a rabbit platelet aggregation inhibitor and porcine coronary artery relaxation (1). SCH58261 protects against spinal cord injury through peripheral and central effects (2). **1.** Zocchi, Ongini, Conti, *et al.* (1996) *J. Pharmacol. Exp. Ther.* **276**, 398. **2.** Paterniti, Melani, Cipriani, *et al.* (2011) *J. Neuroinflam.* **8**, 31.

SCH66336, *See Lonafarnib*

SCH79797

This nonpeptide receptor antagonist (FW$_{free-base}$ = 371.21 g/mol; FW$_{dihydrochloride}$ = 444.41 g/mol; CAS 245520-69-8; Solubility: 50 mM in DMSO), also known as *N*^{3}-cyclopropyl-7-[[4-(1-methylethyl)phenyl]methyl]-7*H*-pyrrolo[3,2-*f*]quinazoline-1,3-diamine dihydrochloride, potently and selectively targets the Protease-Activated Receptor-1, *or* PAR$_1$ (IC$_{50}$ = 70 nM), inhibiting haTRAP-induced platelet aggregation, but having no effect on γ-thrombin-, ADP- or collagen-induced aggregation. SCH79797 also selectively blocks PAR$_1$ agonist- or thrombin-induced increases in cytosolic Ca^{2+} in vascular smooth muscle cells (1-4). **Antiamnestic Properties:** Minimal Traumatic Brain injury (mTBI) is associated with retrograde amnesia and microscopic bleeds, the latter containing activated coagulation factors. In an mTBI model, intracerebroventricular injection of thrombin or a PAR-1 agonist (1 hour after memory acquisition) induces amnesia through its receptor PAR_1. Long-term potentiation, measured in hippocampal slices 24 h after mTBI, ICV thrombin or the PAR-1 agonist, was significantly impaired and this effect was completely reversed by the PAR-1 antagonist. Such findings reveal a novel therapeutic target for the cognitive effects of brain trauma. Significantly, SCH79797 completely blocks the amnestic effects of thrombin and the PAR-1 agonist in a Minimal Traumatic Brain injury (mTBI). **1**. Ahn, Arik, Boykow, *et al.* (1999) *Bioorg. Med. Chem. Lett.* **9**, 2073. **2**. Ahn, Foster, Boykow, *et al.* (2000) *Biochem. Pharmacol.* **60**, 1425. **3**. Lidington, Steinberg, Kinderlerer, *et al.* (2005) *Am. J. Physiol. Cell Physiol.* **289**, C1437. **4**. Itzekson, Maggio, Milman, *et al.* (2013) *J. Mol. Neurosci.* **53**, 87.

SCH 351125

This oral first-in-class retroviral entry inhibitor (FW = 557.50) is a HIV-1 co-receptor CCR antagonist. While SCH-C exhibits potent activity against clinical HIV-1 R5 isolates, both *in vitro* and in mouse models, it shows no effectiveness against HIV-1 X4 isolates. In view its dose-dependent prolongation of the corrected cardiac QT interval, further development of SCH 351125 was abandoned in favor of developing more effective agents. Other similarly acting viral entry inhibitors include: aplaviroc, vivriviroc, TAK-779, and Maraviroc. By contrast, the entry inhibitor PRO 140 is a human monoclonal antibody to the CCR5 receptor. **1**. Strizki, Xu, Wagner, *et al.* (2001) *Proc. Natl. Acad. Sc. U.S.A.* **98**, 12718.

SCH 530348, See *Vorapaxar*

SCH619734, See *Rolapitant*

SCH727965, See *Dinaciclib*

SCH 772984

This novel ATP-competitive protein kinase inhibitor (FW = 587.67 g/mol; CAS 942183-80-4; Solubility: <1 mg/mL DMSO or H_2O), also named (*R*)-1-(2-oxo-2-(4-(4-(pyrimidin-2-yl)phenyl)piperazin-1-yl)ethyl)-*N*-(3-(pyridin-4-yl)-1*H*-indazol-5-yl)pyrrolidine-3-carboxamide, specifically targets both Extracellular-signal-Regulated Kinase ERK1 (IC_{50} = 4 nM) and ERK2 (IC_{50} = 1 nM). SCH772984 has nanomolar cellular potency in tumor cells with mutations in BRAF, NRAS, or KRAS and induces tumor regressions in xenograft models at tolerated doses. Importantly, SCH772984 effectively inhibited MAPK signaling and cell proliferation in BRAF or MEK inhibitor-resistant models as well as in tumor cells resistant to concurrent

treatment with BRAF and MEK inhibitors. **1**. Morris, Jha, Restaino, *et al.* (2013) *Cancer Discov.* **3**, 742. **2**. Nissan, Rosen & Solit (2013) *Cancer Discov.* **3**, 719.

SCH 900776

This cyclin kinase-directed inhibitor (FW = 376.25 g/mol; CAS 891494-63-6; Solubility: 75 mg/mL DMSO; <1 mg/mL H_2O; Formulation: Dissolve in 20% hydroxypropyl β-cyclodextrin), systematically named 6-bromo-3-(1-methyl-1*H*-pyrazol-4-yl)-5-((*R*)-piperidin-3-yl)pyrazolo[1,5-*a*]pyrimidin-7-amine, selectively targets the serine-threonine, cell-cycle checkpoint kinase Chk1 inhibitor with IC_{50} of 3 nM and also inhibits CDK2 with IC_{50} of 160 nM. Chk1 is essential for maintenance of replication fork viability during exposure to DNA antimetabolites. **Cyclin Target Selectivity:** Cdk1 (weak, if any), Cdk2 (weak, if any), Cdk3 (weak, if any), Cdk4 (weak, if any), Cdk5 (weak, if any), Cdk6 (weak, if any), Cdk7 (weak, if any), Cdk8 (weak, if any), Cdk9 (weak, if any), Cdk10 (weak, if any), CHK1 (+++). **1**. Guzi, *et al.* (2011) *Mol. Cancer Ther.* **10**, 591. **2**. Montano, *et al.* (2012) *Mol. Cancer Ther* **11**, 427. **3**. Schenk, *et al.* (2012) *Clin. Cancer Res.* **18**, 5364.

Scopolamine

(–)-Scopolamine

This nonselective antimuscarinic agent ($FW_{\text{free-base}}$ = 303.36 g/mol; CAS 51-34-3), also known as L-hyoscine and scopine tropate, is the competitive antagonist (IC_{50} = 55 nM) of acetylcholine and other muscarinic agonists. Scopolamine induces amnesia and has also been used as a truth serum for extracting information. The free base decomposes upon standing at room temperature, whereas the hydrochloride is much more stable at room temperature. **Target(s):** cholinesterase, *or* butyrylcholinesterase (1); muscarinic cholinergic receptors (2). **1**. Nagasawa, Sugisaki, Tani & Ogata (1976) *Biochim. Biophys. Acta* **429**, 817. **2**. Frey & Howland (1992) *J. Pharmacol. Exp. Ther.* **263**, 1391.

Scopolamine *N*-Oxide

This scopolamine derivative (FW = 319.36 g/mol; CAS 97-75-6; HBr salt is soluble to 10 g/100 mL H_2O), also known as hyoscine *N*-oxide, and systematically as [(1α,2β,4β,5α)-9-methyl-3-oxa-9-azatricyclo[3.3.1.0^{2,4}]-nonane 9-oxide]-7β-yl (*S*)-α-(hydroxy-methyl)-benzeneacetate, is as a muscarinic receptor antagonist. Under the same conditions, the pK_a of hyoscine *N*-oxide is 5.78, with the unprotonated form inactive on muscarinic receptors in the ileum. By contrast, the protonated form is highly active with $-\log K$ estimated to be 9.9, at least as active as hyoscine methobromide ($-\log K$ = 9.85). **1**. Barlow & Winter (1981) *Brit. J. Pharmacol.* **72**, 657.

Scopoletin

This hydroxycoumarin (FW = 192.17 g/mol; CAS 92-61-5; Slightly soluble in H_2O; λ_{max} at 230 nm (ε_M = 12900 $M^{-1}cm^{-1}$), 254 nm (ε_M = 4790 $M^{-1}cm^{-1}$), 260 nn (ε_M = 4270 $M^{-1}cm^{-1}$), 298 nm (ε_M = 4790 $M^{-1}cm^{-1}$), and 346 nm (ε_M = 11700 $M^{-1}cm^{-1}$)), also known as 7-hydroxy-6-methoxy-2*H*-1-benzopyran-2-one, inhibits cell proliferation and induces apoptosis. **Target(s):** arylamine *N*-acetyltransferase (1,2); monoamine oxidase (3); nitric-oxide synthase (4); phenylalanine ammonia-lyase (5); phenylpyruvate tautomerase (6); and xanthine oxidase (7,8). **1**. Kukongviriyapan, Phromsopha, Tassaneeyakul, *et al.* (2006) *Xenobiotica* **36**, 15. **2**. Makarova (2008) *Curr. Drug Metab.* **9**, 538. **3**. Yun, Lee, Ryoo & Yoo (2001) *J. Nat. Prod.* **64**, 1238. **4**. Kim, Pae, Ko, *et al.* (1999) *Planta Med.* **65**, 656. **5**. Dubery (1990) *Phytochemistry* **29**, 2107. **6**. Molnar & Garai (2005) *Int. Immunopharmacol.* **5**, 849. **7**. Chang & Chiang (1995) *Anticancer Res.* **15**, 1969. **8**. Chang, Chang, Lu & Chiang (1994) *Anticancer Res.* **14**, 501.

SCR7

This DNA ligase IV and NHEJ inhibitor (FW = 332.38 g/mol; CAS 14892-97-8; Solubility: 100 mM in DMSO), also named 2,3-dihydro-6,7-diphenyl-2-thioxo-4(1*H*)-pteridinone, when present at 1 μM, targets <u>N</u>on<u>h</u>omologous <u>E</u>nd-<u>J</u>oining, thereby significantly enhancing the efficiency of <u>C</u>lustered <u>R</u>egularly <u>I</u>nterspaced <u>S</u>hort <u>P</u>alindromic <u>R</u>epeat-Cas9 (*or* CRISPR-Cas9) -mediated homology-directed repair (HDR) *in vitro* (1-3). CRISPR-Cas9 is a prokaryotic immune system that confers resistance to foreign genetic elements such as plasmids and phages, and provides a form of acquired immunity. Insertion of precise genetic modifications by genome editing tools such as CRISPR-Cas9 is limited by the relatively low efficiency of homology-directed repair (HDR) compared with the higher efficiency of the NHEJ pathway (1). Indeed, NHEJ is an alternative DNA repair pathway that directly competes with homology-directed repair (HDR) (2). **Selectivity:** Scr7 targets the DNA binding domain of DNA Ligase IV, reducing its affinity for DSBs and inhibiting its function (4). It also inhibits DNA Ligase III (but not DNA Ligase I), albeit less efficiently (1) Treatment of mice with Scr7 affects lymphocyte development, as DNA Ligase IV plays a key role in the joining of coding ends during V(D)J recombination via C-NHEJ (1). **1**. Srivastava, *et al.* (2012) *Cell.* 151:1474. **2**. Maruyama, Dougan, Truttmann, *et al.* (2015) *Nature Biotechnol.* 33, 538. **3**. Chu, Weber, Wefers, *et al.* (2015) *Nature Biotechnol.* **33**, 543.

Scriptaid

This active-site-directed, high-affinity hydroxamate (FW = 326.35 g/mol; CAS 287383-59-9), also known as Scriptide®, GCK 1026, and *N*-hydroxy-1,3-dioxo-1*H*-benz[*de*]isoquinoline-2(3*H*)-hexanamide, inhibits histone

deacetylase, showing a greater effect on acetylated H4 than H3 (1). Use of Scriptaid (optimally at 6–8 μM) results in a one hundred times increase in histone acetylation in cultured cells, confirming HDAC inhibition. The hydroxamic acid group coordinates to the zinc atom in HDAC's polar pocket within the HDAC-Zn^{2+}-TSA complex (2). The hydroxamic acid group on a Scriptaid analogue extends from a 5-carbon aliphatic chain spanning a narrow tube-like pit formed by the surface of HDAC (2). Nanomolar concentrations of suberoylanilide hydroxamic acid (SAHA), suberic bishydroxamic acid (SBHA), Scriptaid (IC_{50} = 39 nM), and trichostatin A (TSA) inhibited *Toxoplasma gondii* tachyzoite proliferation. Scriptaid was the most potent hydroxamic acid inhibitor. For comparison, other carboxylate histone deacetylase inhibitors (such as sodium valproate, sodium butyrate, and 4-phenylbutyrate) were far less potent, IC_{50} range = 1-5 mM (3). Quantitative Real-Time PCR showed 2000-20,000x increase of estrogen receptor α (ER) mRNA transcript in three cell lines after a 48-hour treatment with Scriptaid (4). Moreover, Scriptaid and AZA co-treatment is more effective in inducing ER than Scriptaid or AZA alone. Scriptaid induces the growth inhibition of endometrial and ovarian cancer cells, decreasing the proportion of cells in the S phase and increased the proportion in the G_0/G_1 and/or G_2/M phases (5). Induction of apoptosis is confirmed by annexin V staining of externalized phosphatidylserine and loss of the transmembrane potential of mitochondria. Scriptaid elicited a dose-dependent decrease in lesion size at 1.5 to 5.5 mg/kg and a concomitant attenuation in motor and cognitive deficits when delivered 30 minutes postinjury in a model of moderate traumatic brain injury, *or* TBI (6). **1**. Su, Sohn, Ryu & Kern (2000) *Cancer Res.* **60**, 3137. **2**. Finnin, Donigian, Cohen, *et al.* (1999) *Science* **401**, 188. **3**. Strobl, Cassell, Mitchell, *et al.* (2007) *J. Parasitol.* **93**, 694. **4**. Keen, Yan, Mack, *et al.* (2003) *Breast Cancer Res. Treat.* **81**, 177. **5**. Takai, Ueda, Nishida, Nasu & Narahar (2006) *Int. J. Mol. Med.* **17**, 323. **6**. Wang, Jiang, Pu, *et al.* (2013) *Neurotherapeutics* **10**, 124.

SCS

This GABA receptor antagonist (FW = 256.26 g/mol; CAS 3232-36-8; Soluble to 50 mM in DMSO) shows sub-type selectivity in targeting the β_1 subunit-containing $GABA_A$ receptors, with respective IC_{50} values of 4.5, 5.3, and 7.9 nM for $\alpha_2\beta_1\gamma_1\theta$, $\alpha_2\beta_1\gamma_1$ and $\alpha_2\beta_1\gamma_{2s}$ $GABA_A$ receptor sub-types. **1**. Thompson, Wheat, Brown, *et al.* (2004) *Brit. J. Pharmacol.* **142**, 97.

Scutellarein

This yellow flavonol (FW = 286.24 g/mol; CAS 529-53-3), also known as 4',5,6,7-tetrahydroxyflavone, is the aglycon of scutellarin. Care should be exercised when searching the literature to avoid confusing scutellarein and scutellarin with scutelarin, an endopeptidase from *Oxyuranus scutellatus*. **Target(s):** avian myeloblastosis reverse transcriptase (1); cAMP phosphodiesterase (2); Maloney murine leukemia virus reverse transcriptase (1); RNA-directed DNA polymerase (1); and Rous-associated virus-2 reverse transcriptase (1). **1**. Spedding, Ratty & Middleton, Jr. (1989) *Antiviral Res.* **12**, 99. **2**. Kuppusamy & Das (1992) *Biochem. Pharmacol.* **44**, 1307.

Scutellarioside II, *See* 10-(4-Hydroxycinnamoyl)-catalpol

Scyllitol, *See* scyllo-Inositol

Scyllo-Inosose, See myo-Inosose-2

Scyphostatin

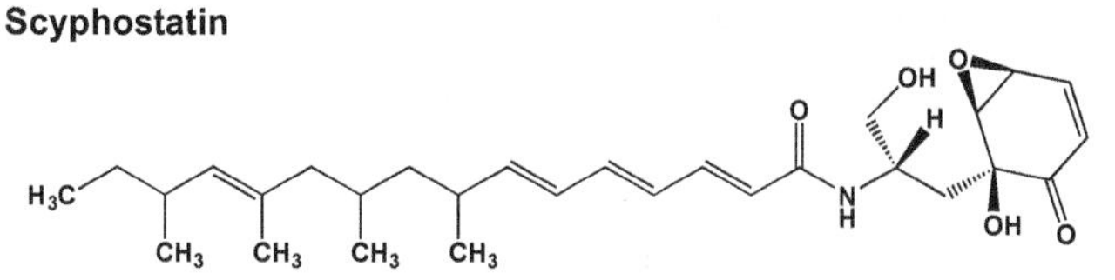

This natural product (FW = 485.66 g/mol; CAS 87413-60-3) from extracts of *Dasyscyphus mollissimus* is a potent inhibitor of neutral sphingomyelinase, with 95% inhibition observed at 20 μM. Scyphostatin is a much weaker inhibitor of acid sphingomyelinase. **Target(s):** phospholipase C (1); and sphingomyelin phosphodiesterase, *or* sphingomyelinase (2-8). **1.** Hanada, Palacpac, Magistrado, *et al.* (2002) *J. Exp. Med.* **195**, 23. **2.** Bernardo, Krut, Wiegmann, *et al.* (2000) *J. Biol. Chem.* **275**, 7641. **3.** Nara, Tanaka, Hosoya, Suzuki-Konagai & Ogita (1999) *J. Antibiot.* **52**, 525. **4.** Tanaka, Nara, Suzuki-Konagai, Hosoya & Ogita (1997) *J. Amer. Chem. Soc.* **119**, 7871. **5.** Goñi & Alonso (2002) *FEBS Lett.* **531**, 38. **6.** Krut, Wiegmann, Kashkar, Yazdanpanah & Krönke (2006) *J. Biol. Chem.* **281**, 13784. **7.** Hanada, Mitamura, Fukasawa, *et al.* (2000) *Biochem. J.* **346**, 671. **8.** Czarny & Schnitzer (2004) *Amer. J. Physiol.* **287**, H1344.

Scyptolin A and B

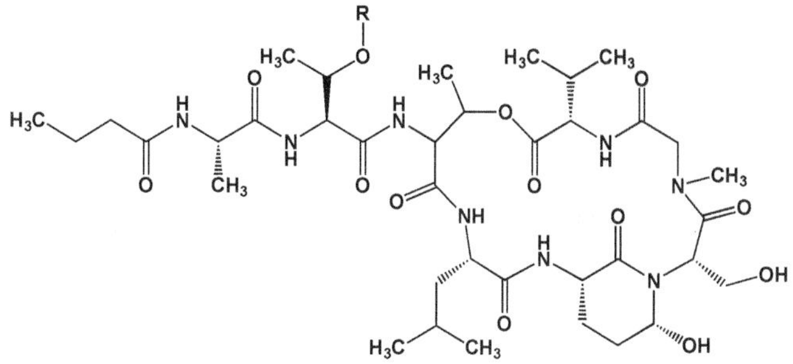

These depsipeptides (for Scyptolin A, FW = 981.54 g/mol), where R = H; for Scyptolin B (FW = 1139.72 g/mol), where R = *N*-butyroyl-L-alanine) from the terrestrial cyanobacterium *Scytonema hofmanni* PCC 7110, inhibit pancreatic elastase (1,2). Note the presence of a *cis*-peptide bond between the L-threonyl and *N*-methyl-3'-chloro-L-tyrosyl residues. **1.** Matern, Schleberger, Jelakovic, Weckesser & Schulz (2003) *Chem. Biol.* **10**, 997. **2.** Matern, Oberer, Falchetto, *et al.* (2001) *Phytochemistry* **58**, 1087.

Scytonemin

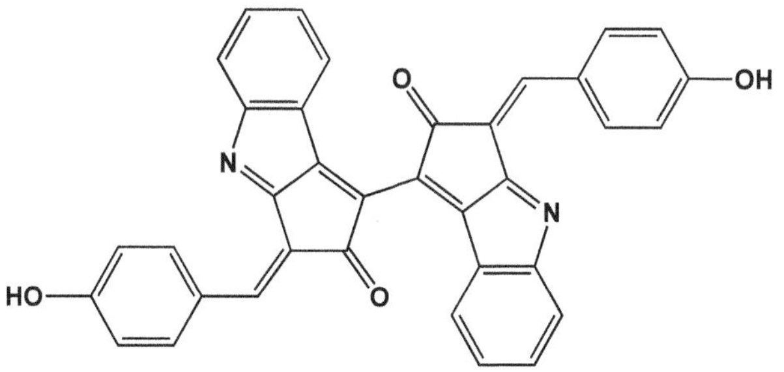

This cyanobacter-derived pigment (FW = 544.57 g/mol; CAS 112793-66-5) inhibits human polo-like kinase-1 (IC$_{50}$ = 2 μM). Scytonemin has an extracellular photoprotective role in certain cyanobacteria, absorbing strongly and broadly. **1.** Zhang, Su, Feng, *et al.* (2007) *Mol. Reprod. Dev.* **74**, 1247. **2.** Stevenson, Capper, Roshak, *et al.* (2002) *Inflamm. Res.* **51**, 112. **3.** McInnes, Mezna & Fischer (2005) *Curr. Top. Med. Chem.* **5**, 181.

SD 906, See 3-[(3,4-Dimethoxyphenyl)sulfonyl]-1-(3,4-dimethylphenyl)imidazolidine-2,4-dione

SD 8339, See *N*^{6}-Benzyl-*N*^{9}-(2-tetrahydropyranyl)-adenine

SD 8447, See Tetrachlorvinphos

SDS, See Sodium Dodecyl Sulfate

SDX-105, See Bendamustine

SDZ 242-484, See *N*^{4}-(2,2-Dimethyl-1(S)-methylcarbamoylpropyl)-*N*^{1}-hydroxy-2(R)-hydroxymethyl-3(S)-(4-methoxy-phenyl)-succinamide

S-2E, See S-2E-CoA

S-2E-CoA

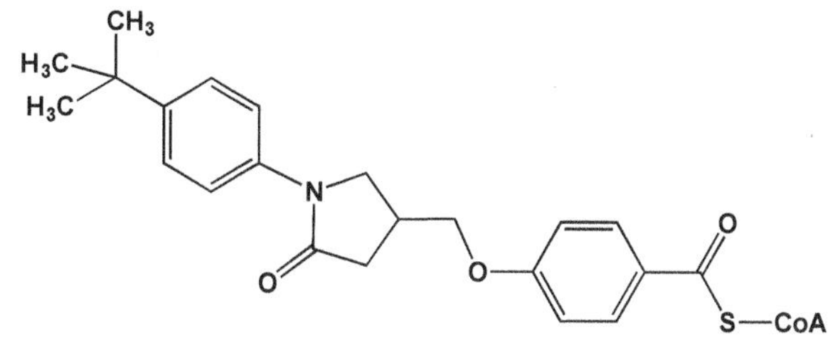

This acyl-coenzyme A derivative (FW = 1117.07 g/mol), which is formed by metabolic conversion of commercially available (*S*)-(+)-4-[1-(4-*tert*-butylphenyl)-2-oxopyrrolidin-4-yl]methoxybenzoic acid within the liver, noncompetitively inhibits acetyl-CoA carboxylase (K_i = 69 μM) and 3-hydroxy-3-methylglutaryl-CoA reductase (K_i = 18 μM). **1.** Ohmori, Yamada, Yasuda, *et al.* (2003) *Eur. J. Pharmacol.* **471**, 69.

Secobarbital

amido tautomer   imido tautomer   sodium salt

This barbiturate (FW$_{free-acid}$ = 238.29 g/mol; CAS 76-73-3; sodium salt, very soluble in H$_2$O), also known as Seconal and 5-(1-methylbutyl)-5-(2-propenyl)-2,4,6(1*H*,3*H*,5*H*)-pyrimidinetrione, is a short-acting hypnotic that is often used as a sedative. It also possesses anesthetic, anticonvulsant, and anxiolytic properties. By depressing sensory cortex and motor activity, secobarbital produces drowsiness, sedation, and hypnosis. Barbiturates of this type also inhibit AP-1 complex formation as well as AP-1-dependent gene expression at clinically relevant doses. They act by suppressing the activity of mitogen-activated protein (MAP) kinase, which in turn regulates AP-1 complex formation. (In the United Kingdom, Seconal is known as Quinalbarbitone.) **Caution:** This once widely abused and habit-forming drug is now classified as a Schedule-2 controlled substance. It is also prescribed under physician-assisted euthanasia laws in Oregon, Washington, Vermont, and New Mexico. **Target(s):** acetylcholine receptor (1,2); CYP2B (3); enkephalinase A, *or* neprilysin (4,5); NADH dehydrogenase, *or* complex I, *or* NADH:ubiquinone reductase (6-8); phosphatidylinositol kinase (9); phosphatidylinositol 4-phosphate kinase (9); and protein kinase C (9). **1.** Yost & Dodson (1993) *Cell Mol. Neurobiol.* **13**, 159. **2.** Dodson, Braswell & Miller (1987) *Mol. Pharmacol.* **32**, 119. **3.** Hanioka, Hamamura, Kakino, *et al.* (1995) *Xenobiotica* **25**, 1207. **4.** van Amsterdam, van Buuren, de Jong & Soudijn (1983) *Life Sci.* **33** Suppl. 1, 109. **5.** Altstein, Blumberg & Vogel (1982) *Adv. Biochem. Psychopharmacol.* **33**, 261. **6.** Hatefi & Stiggall (1978) *Meth. Enzymol.* **53**, 5. **7.** Hatefi (1978) *Meth. Enzymol.* **53**, 11. **8.** Hatefi & Stiggall (1976) *The Enzymes*, 3rd ed., **13**, 175. **9.** Deshmukh, Kuizon, Chauhan & Brockerhoff (1989) *Neuropharmacology* **28**, 1317.

5,10-Secoestr-5-yne-3,10-dione-17β-ol, See 17β-Hydroxy-5,10-secoestr-5-yne-3,10-dione

5,10-Secoestr-5-yne-3,10,17-trione

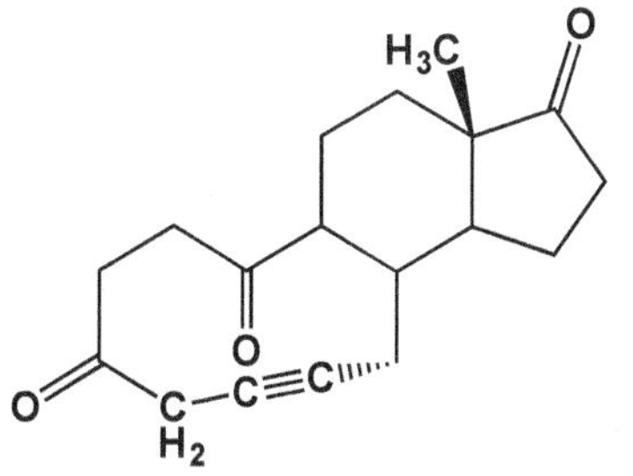

This steroid analogue (FW = 286.37 g/mol; M.P. = 142-145°C) is a suicide inhibitor of both the C19- and the C21-Δ^{5-3}-ketosteroid isomerase (*or* steroid Δ-isomerase) activities of beef adrenal cortex microsomes (1-5). *Note*: This derivative has been reported to convert to (4*R*)-5,10-secoestr-4,5-diene-3,10,17-trione under standard buffered conditions. **1.** Thomas, Strickler, Myers & Covey (1992) *Biochemistry* **31**, 5522. **2.** Thomas, Evans & Strickler (1997) *Biochemistry* **36**, 9029. **3.** Batzold, Benson, Covey, Robinson & Talalay (1977) *Meth. Enzymol.* **46**, 461. **4.** Batzold & Robinson (1975) *J. Amer. Chem. Soc.* **97**, 2576, 6607, and (1976) **98**, 641. **5.** Penning, Westbrook & Talalay (1980) *Eur. J. Biochem.* **105**, 461.

Secologanin

This plant metabolite (FW = 388.37 g/mol; CAS 19351-63-4), an intermediate in the biosynthesis of antihypertensive drug ajmalicine (itself an alkaloid from *Rauwolfia* spp., *Catharanthus roseus*, and *Mitragyna speciosa*), inhibits deacetylvindoline *O*-acetyltransferase, which catalyzes the final step in vindoline biosynthesis in *Catharanthus roseus*. **1.** Power, Kurz & de Luca (1990) *Arch. Biochem. Biophys.* **279**, 370.

Seconal, *See Secobarbital*

5,10-Seco-19-norpregn-5-yne-3,10,20-trione

This steroid analogue (FW = 314.42 g/mol) inhibits steroid Δ-isomerase. It is a suicide inhibitor of both the C19- and the C21-Δ^{5-3}-ketosteroid isomerase activities of beef adrenal cortex microsomes (1-4). Note that this derivative has been reported to convert to (4*R*)-5,10-seco-19-norpregn-4,5-diene-3,10,20-trione under standard buffered conditions. **1.** Batzold, Benson, Covey, Robinson & Talalay (1977) *Meth. Enzymol.* **46**, 461. **2.** Covey & Robinson (1976) *J. Amer. Chem. Soc.* **98**, 5038. **3.** Penning (1982) *Steroids* **39**, 301. **4.** Penning, Covey & Talalay (1981) *J. Biol. Chem.* **256**, 6842.

β-Secretase Inhibitor

This term refers to a class of moderately high-affinity β-secretase inhibitors, of which many incorporate a statine (Sta) moiety into a substrate analogue sequence, such as: L-Lys-L-Thr-L-Glu-L-Glu-L-Ile-L-Ser-L-Glu-L-Val-L-Asn-[Sta]-L-Val-L-Ala-L-Glu-L-Phe, and often exhibiting IC$_{50}$ values of 10–30 nM (1,2) **See specific inhibitor** **1.** Sinha, Anderson, Barbour, *et al.* (1999) *Nature* **402**, 537. **2.** Tung, Davis, Anderson, *et al.* (2002) *J. Med. Chem.* **45**, 259.

β-Secretase Inhibitor II, See N-(Benzyloxycarbonyl)-L-valyl-L-leucyl-L-leucinal

β-Secretase Inhibitor III, See L-Glutamyl-L-valyl-L-asparaginyl-statinyl-L-valyl-L-alanyl-L-glutamyl-L-phenylalaninamide

γ-Secretase Inhibitor I, See N$^\alpha$-(Benzyloxy-carbonyl)-L-leucyl-L-leucyl-L-norleucinal

γ$_{40}$-Secretase Inhibitor I, See N-trans-3,5-Dimethoxy-cinnamoyl)-L-isoleucyl-L-leucinal

γ-Secretase Inhibitor II, See N-[N$^\alpha$-(t-Butoxy-carbonyl)-L-valyl-L-isoleucyl]-[(4S)-4-amino-2,2-difluoro-3-oxopentanoyl]-L-valyl-L-isoleucine Methyl Ester

γ$_{40}$-Secretase Inhibitor II, See N-(tert-Butyloxy-carbonyl)glycyl-L-valyl-L-valinal

γ-Secretase Inhibitor III, See N$^\alpha$-(Benzyloxy-carbonyl)-L-leucyl-L-leucinal

γ-Secretase Inhibitor IV, See N$^\alpha$-(2-Naphthoyl)-L-valyl-L-phenylalaninal

γ-Secretase Inhibitor V, See N$^\alpha$-(Benzyloxycarbonyl)-L-leucyl-L-phenylalaninal

γ-Secretase Inhibitor VI, See 1-(S)-endo-N-(1,3,3)-Trimethylbicyclo[2.2.1]hept-2-yl)-4-fluorophenyl Sulfonamide

γ-Secretase Inhibitor VII, See N^a-(Menthyloxy-carbonyl)-L-leucyl-L-leucinamide

γ-Secretase Inhibitor IX, See N-[N-(3,5-Difluorophen-acetyl-L-alanyl)]-(S)-phenylglycine t-Butyl Ester

γ-Secretase Inhibitor X, See {1S-Benzyl-4R-[1-(1S-carbamoyl-2-phenethylcarbamoyl)-1S-3-methylbutyl-carbamoyl]-2R-hydroxy-5-phenylpentyl}carbamic Acid tert-Butyl Ester

γ-Secretase Inhibitor XI, See 7-Amino-4-chloro-3-methoxyisocoumarin

γ-Secretase Inhibitor XII, See N$^\alpha$-(Benzyloxy-carbonyl)-L-isoleucyl-L-leucinal

γ-Secretase Inhibitor XIII, See N$^\alpha$-(Benzyloxy-carbonyl)-L-tyrosyl-L-isoleucyl-L-leucinal

γ-Secretase Inhibitor XIV, See N$^\alpha$-(Benzyloxy-carbonyl)-L-(S-t-butyl)cysteinyl-L-isoleucyl-L-leucinal

γ-Secretase Inhibitor XVI, See N-[N-3,5-Difluoro-phenacetyl]-L-alanyl-(S)-phenylglycine Methyl Ester

γ-Secretase Inhibitor XVII, See WPE-III-31C

Secretory Leukocyte Protease Inhibitor

This 132-residue serine proteinase inhibitor (MW = 14,326 g; GenBank = CAA28187.1; Isoelectric Point = 9.11; Symbol = SLPI) is a non-glycosylated serpin that prevents neutrophils from attacking the underlying fibronectin or elastin components in the extracellular matrix (1,2). SLPI is an antimicrobial peptide and an anti-inflammatory molecule that targets elastase and cathepsin G from neutrophils, chymase and tryptase from mast cells, as well as trypsin and chymotrypsin from pancreatic acinar cells. **Primary Mechanism of Action:** Human neutrophils undergo actin-based chemotaxis and degranulation of proteolytic enzymes in response to changes in cytosolic calcium Ca^{2+}, with the latter mediated inositol 1,4,5-triphosphate (IP$_3$) production in response to G-protein-coupled receptor stimuli. Once released, neutrophil proteases rapidly degrade extracellular matrix proteins, especially fibronectin or elastin, by directing the protease's hydrolytic action to sites of cell-substrate contact (3). SLPI (~0.5 μM) significantly inhibits fMet-Leu-Phe- (fMLP-) and interleukin-8-induced neutrophil chemotaxis and decreased degranulation of matrix metalloprotease-9 (MMP-9), hCAP-18, and myeloperoxidase (3). SLPI is released by cells possessing mucosal surfaces (*e.g.*, bronchia, cervix, nasal mucosa) as well as saliva and seminal fluid. Its inhibitory effect contributes to the immune response by protecting epithelial surfaces from attack by endogenous proteases. As such, Secretory Leukocyte Protease Inhibitor is a major barrier to tissue destruction by polymorphonuclear leukocytes (PMNs). SLPI is a significant component of the anti-neutrophil elastase (anti-NE) shield in the lung, with a different reactivity from, and is therefore complementary to, the anti-NE action of α_1-proteinase inhibitor (4). 17β-Estradiol inhibits IL-8 in cystic fibrosis by upregulating secretory leucoprotease inhibitor (5). SLPI also modulates monocyte/macrophage function through inhibition of NF-κB signaling. **Structural Features:** SLPI has a four-disulfide core that is thought to stabilize that compensates for the absence of a typical hydrophobic core and helix or sheet secondary structure. X-ray crystallography (6) and site-directed mutagenesis (7) identify Leu72-Met73 in human SLPI as the reactive site for inhibition of elastase, trypsin, and chymotrypsin. **Other Actions:** SLPI modulates the inflammatory and immune responses after bacterial infection, and after infection by the intracellular parasite *Leishmania major*. SLPI down-regulates host cell response to bacterial lipopolysaccharide (LPS); Plays a role in regulating the activation of NFκB and inflammatory response. SLPI exhibits antimicrobial activity against *Mycobacteria*, but not *Salmonella*. Finally, SLPI is required for wound healing, most likely by preventing tissue damage by limiting protease activity. **Targets(s):** human leukocyte elastase (EC 3.4.21.37); cathepsin G (EC 3.4.21.20); human trypsin (EC 3.4.21.4). **1.** Thompson & Ohlsson (1986) *Proc. Natl. Acad. Sci. U.S.A.* **83**, 6692. **2.** Doumas, Kolokotronis & Stefanopoulos (2005) *Infect. Immun.* **73**, 1271. **3.** Reeves, Banville, Ryan, *et al.* (2013) *Biomed. Res. Int.* **2013**, 560141. **4.** Bingle & Tetley (1996) *Thorax* **51**, 1273. **5.** Chotirmall, Greene, Oglesby, *et al.* (2010) *Am. J. Respir. Crit. Care Med.* **182**, 62. **6.** Grutter, Fendrich, Huber & Bode (1988) *EMBO J,* **7**, 345. **7.** Eisenberg, Hale, Heimdal & Thompson (1998) *J. Biol. Chem,* **265**, 7976.

Secukinumab

This whole human monoclonal G1κ immunoglobulin (FW = 147.9 kDa; CASs = 875356-43-7 (heavy chain), 875356-44-8 (light chain)), also known by the trade name Cosentyx®, binds to interleukin IL-17A, blocking its receptor signaling and demonstrating efficacy in the systemic treatment of moderate-to-severe plaque psoriasis. The latter is a chronic and disabling inflammatory skin disease, the pathogenesis of which is linked to several cytokines and chemokines, particularly IL-17A. Increased helper T cells (T$_H$17) and elevated levels of IL-17A are found in psoriatic plaques, and increased levels of T$_H$17 are likewise present in the plasma of psoriasis patients. Elevated IL-17A also induces neutrophilia, inflammation, and angiogenesis. IL-17A also plays a key role in host defense, particularly in mucocutaneous immunity against *Candida albicans*, as well as in hematopoiesis, the latter through stimulation of granulopoiesis and neutrophil trafficking. While IL-17A is a key product of T$_H$17 cells, it is also produced by neutrophils, mast cells and Tc17 cells. Each of these cell types is found in psoriatic lesions. IL-17A acts on keratinocytes to increase expression of chemokines (*e.g.* CCL20, CXCL1, CXCL3, CXCL5, CXCL6 and CXCL8) involved in recruiting myeloid dendritic cells, Th17 cells and neutrophils to the lesion site. IL-17A induces production of antimicrobial peptides and proinflammatory cytokines that, in turn, may help sustain immune responses in the skin. Secukinumab is also under investigation for treating uveitis, rheumatoid arthritis, ankylosing spondylitis, and psoriatic arthritis. **1.** Yamauchi & Bagel (2015) *J. Drugs Dermatol.* **14**, 244.

Securines A and B, *See Securamines A and B*

Sedanolide

This mosquitocidal, nematicidal, and antifungal agent (FW = 194.27 g/mol; CAS 6415-59-4) from celery seed oil, known systematically as 3-butyl-3a,4,5,6-tetrahydro-1(3*H*)-isobenzofuranone, topoisomerases I and II, and prostaglandin H endoperoxide synthase-I. **1.** Momin & Nair (2002) *Phytomedicine* **9**, 312.

D-Sedoheptulose 1,7-Bisphosphate

α-D-**sedoheptulopyranose 1,7-bisphosphate**

This phosphorylated sedoheptulose derivative (FW = 366.11 g/mol), an alternative product of fructose-bisphosphate aldolase and a substrate of sedoheptulose-bisphosphatase, inhibits 3-deoxy-7-phosphoheptulonate synthase, *or* 2-keto-3-deoxy-D-*arabino*-heptonate-7-phosphate synthase (1,2). fructose-1,6-bisphosphatase (3,4), ribose-5-phosphate isomerase (5), and ribulose-bisphosphate carboxylase (6). **1.** Webb (1966) *Enzyme and Metabolic Inhibitors*, vol. **2**, p. 413, Academic Press, New York. **2.** Srinivasan & Sprinson (1954) *J. Biol. Chem.* **234**, 716. **3.** Bonsignore, Mangiarotti, Pontremoli, Deflora & Mangiarotti (1963) *J. Biol. Chem.* **238**, 3151. **4.** Springgate & Stachow (1972) *Arch. Biochem. Biophys.* **152**, 1. **5.** Skrukrud, Gordon, Dorwin, *et al.* (1991) *Plant Physiol.* **97**, 730. **6.** Pettersson & Ryde-Pettersson (1988) *Eur. J. Biochem.* **177**, 351.

D-Sedoheptulose 7-Phosphate

α-D-**sedoheptulopyranose 7-phosphate**

This phosphorylated seven-carbon ketose (FW$_{ion}$ = 288.15 g/mol), an intermediate in the pentose phosphate pathway, inhibits 3-deoxy-7-phosphoheptulonate synthase, *or* 2-keto-3-deoxy-D-*arabino*-heptonate-7-phosphate synthase (1-3), glucose-6-phosphate isomerase (4-6), and sedoheptulose-bisphosphatase (7). **1.** Webb (1966) *Enzyme and Metabolic Inhibitors*, vol. **2**, p. 413, Academic Press, New York. **2.** Xia & Chiao (1989) *Biochim. Biophys. Acta* **991**, 1. **3.** Srinivasan & Sprinson (1954) *J. Biol. Chem.* **234**, 716. **4.** Noltmann (1972) *The Enzymes*, 3rd ed. (Boyer, ed.), **6**, 271. **5.** Zera (1987) *Biochem. Genet.* **25**, 205. **6.** Thomas (1981) *J. Gen. Microbiol.* **124**, 403. **7.** Schimkat, Heineke & Heldt (1990) *Planta* **181**, 97.

Sefotel, *See CGS 19755*

Selachyl Diacetate, *See 1,2-Diacetyl-3-oleylglycerol*

Selamectin

This topical veterinary endectocide, parasiticide and antihelminthic (FW = 769.96 g/mol; CAS 220119-17-5), also known by the tradenames Revolution® and Stronghold® and systematically as 25-cyclohexyl-25-de(1-methylpropyl)-5-deoxy-22,23-dihydro-5-(hydroxyimino)avermectin B$_1$ monosaccharide, activates the chloride current without desensitization, thereby permitting chloride ions to enter the nerve cells and causing neuromuscular paralysis, impaired muscular contraction, and eventual death. It. prevents infections of heartworms (*Dirofilaria immitis*), fleas (*Ctenocephalides felis felis*), ear mites, scabies, and certain types of ticks (*Ixodes ricinus*) in dogs as well as adult ascarids (*Toxocara cati*) and adult hookworms (*Ancylostoma tubaeforme*) in cats. Consult reference (3) for the pharmacokinetics of selamectin following intravenous, oral and topical administration in cats and dogs. **1.** Jacobs (2000) *Vet. Parasitol.* **91**, 161. **2.** Pacey, Dutton, Monday, Ruddock & Smith (2000) *J. Antibiot.* (Tokyo) 53, 301. **3.** Sarasola, Jernigan, Walker, *et al.* (2010) *J. Vet. Pharmacol. Ther.* 25, 265.

Selective Serotonin Re-uptake Inhibitors

This class of antidepressants (*or* SSRIs), also called serotonin-specific reuptake inhibitors, are most often used to treat depression, anxiety disorders, and some personality disorders. Drugs in this class include (trade names in parentheses): *Citalopram* (Celexa®, Cipramil®, Cipram®, Cital®, Citox®, Dalsan®, Emocal®, Recital®, Sepram®, Seropram®); *Dapoxetine* (Priligy®); *Escitalopram* (Lexapro®, Cipralex®, Seroplex®, Esertia®); *Fluoxetine* (Depex®, Prozac®, Fontex®, Fluctin®, Fluox® (New Zealand), Flutop®, Ladose®, Lovan® (Australia), Motivest®, Prodep® (India), Sarafem®, Seromex®, Seronil®); *Fluvoxamine* (Dumyrox®, Faverin®, Favoxil®, Fevarin®, Floxyfral®, Luvox®, Movox®); *Indalpine* (Upstene®, discontinued); *Paroxetine* (Paxil®, Seroxat®, Sereupin®, Aropax®, Deroxat®, Divarius®, Rexetin®, Xetanor®, Paroxat®, Loxamine®, Deparoc®); *Sertraline* (Zoloft®, Lustral®, Serlain®, Asentra®, Tresleen®); and *Zimelidine* (Zelmid®, Normud®).

Selegiline, *See* Deprenyl

Selenalysine, *See* Se-(β-Aminoethyl)-L-seleno-cysteine

Selenate

The free acid H_2SeO_4 (FW = 144.97 g/mol; deliquescent solid; M.P. = 58°C; B.P. = 260°C) is very soluble in water. Selenate may be reduced by many common reagents, including HBr, HI, H_2S, phenylhydrazine, formic acid, oxalic acid, malonic acid, pyruvic acid, and several metals. Selenate salts (SeO_4^{2-}) are isomorphous with the corresponding sulfates, but selenate is toxic substance that has been used as an insecticide. **Target(s):** arsenite methyltransferase (1) 6-phosphofructo-2-kinase/fructose-2,6-bisphosphatase (2); 3-phosphoglycerate kinase (3); polynucleotide 5'-hydroxyl-kinase (4); protein kinase C (5); ribose-5-phosphate adenylyltransferase (6); sulfate adenylyltransferase (7); sulfate adenylyltransferase (ADP-forming) (8); and sulfate transport (9). **1.** Zakharyan, Wu, Bogdan & Aposhian (1995) *Chem. Res. Toxicol.* **8**, 1029. **2.** Kountz, McCain, el-Maghrabi & Pilkis (1986) *Arch. Biochem. Biophys.* **251**, 104. **3.** Joao & Williams (1993) *Eur. J. Biochem.* **216**, 1. **4.** Zimmerman & Pheiffer (1981) *The Enzymes*, 3rd ed. (Boyer, ed.), **14**, 315. **5.** Su, M. Shoji, Mazzei, Vogler & Kuo (1986) *Cancer Res.* **46**, 3684. **6.** Evans & Pietro (1966) *Arch. Biochem. Biophys.* **113**, 236. **7.** Shaw & Anderson (1974) *Biochem. J.* **139**, 37. **8.** Grunberg-Manago, Del Campillo-Campbell, Dondon & Michelson (1966) *Biochim. Biophys. Acta* **123**, 1. **9.** Shennan & Boyd (1986) *Biochim. Biophys. Acta* **859**, 122.

Selenazofurin

Selenazofurin   **Selenazofurin 5'-Triphosphate**

Selenazofurin Adenine Dinucleotide

This selenazole analogue of tiazofurin (FW = 307.16 g/mol; CAS 83705-13-9), also known as 2-β-D-ribofuranosylselenazole-4-carboxamide, is transported into animal cells, where it is metabolically converted to its nucleoside mono-, di, and tri-phosphate analogue s as well as the corresponding NAD^+ analogue. (**See also** Benzamide Riboside; Tiazofurin) Selenazofurin 5'-monophosphate is an IMP analogue that specifically inhibits IMP dehydrogenase activity in cultured cells. The dose-dependence for this inhibition correlates with the relative cytotoxicity of the drug, indicating that the inhibition of IMP dehydrogenase by the selenazole analog is primarily responsible for its cytotoxicity (1). Although

selenazofurin is a highly potent antiviral agent *in vitro*, its antiviral activity is readily reversed by the addition of exogenous guanosine (2). Selenazofurin adenine dinucleotide competes with NAD^+ and inhibits many oxidoreductase reactions as well as the NAD-dependent enzyme, poly(ADP-ribose) polymerase (3). **1.** Streeter & Robins (1983) *Biochem. Biophys. Res. Commun.* **115**, 544. **2.** Robins, Revankar & McKernan (1985) *Adv. Enzyme Regul.* **24**, 29. **3.** Berger, Berger, Catino, Petzold & Robins (1985) *J. Clin. Invest.* **75**, 702.

Selenite

The free acid H_2SeO_3 (FW = 128.97 g/mol; pK_1 = 2.46 and pK_2 = 8.31 at 25°C), also known as selenous acid, is a deliquescent solid that decomposes at 70°C with the release of water and selenium dioxide. note that it has a significant vapor pressure at room temperature. **Target(s):** alcohol dehydrogenase, yeast (1); arsenite methyltransferase (3,34); aryl hydrocarbon hydroxylase (2); Ca^{2+}-dependent ATPase (4); caspase-3 (5); c-Jun N-terminal kinase/stress-activated protein kinase (6); deoxyribonuclease I (7); DNA methylases, *or* DNA-cytosine methyltransferase (8-10); DNA polymerase, inhibited in the presence of sulfhydryl compounds (11,12); glycerol-3-phosphate dehydrogenase (13); IκB kinase (14); Na^+/K^+-exchanging ATPase (15); nuclear factor-κB (14); pancreatic ribonuclease, *or* ribonuclease A (23,32); phosphatase, acid (33); porphobilinogen synthase, *or* δ-aminolevulinate dehydratase (16,17); prostaglandin-D synthase, *or* prostaglandin H_2:D-isomerase (18-20); protein disulfide reduction by thioredoxin systems (21); protein kinase C (22); ribonucleotide reductase (21); RNA polymerase, inhibited in the presence of sulfhydryl compounds (11,12,24); squalene monooxygenase (25,26); succinate dehydrogenase, *or* succinate oxidase (27-29); tubulin polymerization, *or* microtubule assembly (30); and UDP-*N*-acetylglucosamine 2-epimerase (31). **1.** Sund & Theorell (1963) *The Enzymes*, 2nd ed. (Boyer, Lardy & Myrbäck, eds.), **7**, 25. **2.** Asokan, Das, Dixit & Mukhtar (1985) *Acta Pharmacol. Toxicol. (Copenhagen)* **57**, 72. **3.** Walton, Waters, Jolley, *et al.* (2003) *Chem. Res. Toxicol.* **16**, 261. **4.** Hightower & McCready (1991) *J. Curr. Eye Res.* **10**, 299. **5.** Park, Huh, Kim, *et al.* (2000) *J. Biol. Chem.* **275**, 8487. **6.** Park, Park, Kim, *et al.* (2000) *J. Biol. Chem.* **275**, 2527. **7.** McDonald (1955) *Meth. Enzymol.* **2**, 437. **8.** Cox (1985) *Biochem. Int.* **10**, 63. **9.** Fiala, Staretz, Pandya, El-Bayoumy & Hamilton (1998) *Carcinogenesis* **19**, 597. **10.** Cox (1986) *Toxicol. Pathol.* **14**, 477. **11.** Frenkel, Walcott & Middleton (1987) *Mol. Pharmacol.* **31**, 112. **12.** Frenkel (1985) *Toxicol. Lett.* **25**, 219. **13.** Kim, Park, Kang, Choi & Kim (2002) *Biochim. Biophys. Acta* **1569**, 67. **14.** Gasparian, Yao, Lu, *et al.* (2002) *Mol. Cancer Ther.* **1**, 1079. **15.** Bergad & Rathbun (1986) *Curr. Eye Res.* **5**, 919. **16.** Barbosa, Rocha, Zeni, *et al.* (1998) *Toxicol. Appl. Pharmacol.* **149**, 243. **17.** Farina, Brandao, Lara, *et al.* (2003) *Toxicol. Lett.* **139**, 55. **18.** Takahata, Matsumura, Kantha, *et al.* (1993) *Brain Res.* **623**, 65. **19.** Matsumura, Takahata & Hayaishi (1991) *Proc. Natl. Acad. Sci. U.S.A.* **88**, 9046. **20.** Islam, Watanabe, Morii & Hayaishi (1991) *Arch. Biochem. Biophys.* **289**, 161. **21.** Björnstedt, Kumar & Holmgren (1995) *Meth. Enzymol.* **252**, 209. **22.** Su, Shoji, Mazzei, Vogler & Kuo (1986) *Cancer Res.* **46**, 3684. **23.** Ganther & Corcoran (1969) *Biochemistry* **8**, 2557. **24.** Frenkel & Falvey (1989) *Biochem. Pharmacol.* **38**, 2849. **25.** Gupta & Porter (2002) *J. Biochem. Mol. Toxicol.* **16**, 18. **26.** Laden, Tang & Porter (2000) *Arch. Biochem. Biophys.* **374**, 381. **27.** Potter & DuBois (1943) *J. Gen. Physiol.* **26**, 391. **28.** Potter & Elvehjem (1937) *J. Biol. Chem.* **117**, 341. **29.** Kun & Abood (1949) *J. Biol. Chem.* **180**, 813. **30.** Leynadier, Peyrot, Codaccioni & Briand (1991) *Chem. Biol. Interact.* **79**, 91. **31.** Zeitler, Banzer, Bauer & Reutter (1992) *Biometals* **5**, 103. **32.** Anfinsen & White (1961) *The Enzymes*, 2nd ed. (Boyer, Lardy & Myrbäck, eds.) **5**, 95. **33.** Andrews & Pallavicini (1973) *Biochim. Biophys. Acta* **321**, 197. **34.** Zakharyan, Wu, Bogdan & Aposhian (1995) *Chem. Res. Toxicol.* **8**, 1029.

Seleno-coenzyme M

This selenium-containing coenzyme ($FW_{free-acid}$ = 189.09 g/mol) irreversibly inhibits *Methanobacterium marburgensis* coenzyme-B sulfoethylthio-transferase, *or* methyl-CoM reductase. The natural substance, Ccoenzyme M, is the C1 donor in bacterial methanogenesis, where it is converted to propyl coenzyme M-thioester. Nickel-requiring methyl-coenzyme M reductase (MCR) catalyses the reduction of methyl-coenzyme M (CH$_3$-S-CoM) with coenzyme B (HS-CoB) to methane and CoM-S-S-CoB. Coenzyme M reacts with Coenzyme B, 7-thioheptanoylthreonine-phosphate, to give a homodisulfide, releasing methane (*Reaction:* CH$_3$–S–CoM + HS–CoB → CH$_4$ + CoB–S–S–CoM). This enzyme contains the nickel porphyrinoid F$_{430}$ prosthetic group that, for catalysis, must be Ni(I). **1**. Goenrich, Mahlert, Duin, *et al.* (2004) *J. Biol. Inorg. Chem.* **9**, 691.

Selenocystamine

This selenium-containing cystamine analogue (FW$_{free-base}$ = 246.06 g/mol) catalyzes the oxygen-mediated oxidation of excess glutathione to glutathione disulfide at neutral pH and ambient pO$_2$. Selenocystamine can also catalyze the decomposition of such *S*-nitrosothiols as *S*-nitroso-glutathione and *S*-nitroso-*N*-acetyl-DL-penicillamine in the presence of different thiols (such as glutathione) to generate nitric oxide. **Target(s):** γ-glutamylcysteine synthetase (1); and hydroxyindole *O*-methyltransferase (2). **1**. Seelig & Meister (1984) *J. Biol. Chem.* **259**, 3534. **2**. Sugden & Klein (1987) *J. Biol. Chem.* **262**, 6489.

L-Selenocysteine

This selenium-containing cysteine analogue (FW = 168.05 g/mol; CAS 3614-08-2) is found at the active site of a number of redox-active selenoproteins. Selenocysteine is readily oxidized to selenocystine. R–SeH has a pK_a = 5.2, the much lower than 8.3 value for cysteine's thiol. At neutral pH, selenocysteine is largely present as R–Se$^-$, whereas cysteine is largely R–SH. **Metabolic Properties:** Selenocysteine is widely recognized as the 21st amino acid, participating in ribosome-mediated protein synthesis. Its specific incorporation into polypeptides is directed by the UGA codon. Unique tRNAs that have complementary UCA anticodons are aminoacylated with serine, with seryl-tRNA converted to selenocysteyl-tRNA (1). The latter binds specifically to a special elongation factor and is delivered to the ribosome. Recognition elements within the mRNAs are essential for translation of UGA as selenocysteine. A reactive oxygen-labile compound, selenophosphate, is the selenium donor required for synthesis of selenocysteyl-tRNA. Selenophosphate synthetase, which forms selenophosphate from selenide and ATP, is found in various prokaryotes, eukaryotes, and archaebacterial (1). **Roles in Catalysis:** Selenocysteine-containing enzymes (*e.g.*, glutathione peroxidase, thioredoxin reductase, and iodothyronine deiodinase) play prominent roles in metabolism. Artificial selenoenzymes, such as selenosubtilisin, have been produced by chemical modification. Genetic engineering techniques also have been used to replace cysteine residues with selenocysteine. Selenocysteine plays mechanistic roles in glutathione peroxidase, the 5'-deiodinases, formate dehydrogenases, glycine reductase, and a few hydrogenases. In a few cases, a marked decrease in catalytic activity of an enzyme is observed when a selenocysteine residue is replaced with cysteine. This substitution causes complete loss of glycine reductase selenoprotein A activity. **Target(s):** cysteine desulfurase, also as an alternative substrate (2); cysteinyl-tRNA synthetase, also as an alternative substrate (3); glycerol-3-phosphate dehydrogenase (4); and L-serine dehydratase, by DL-selenocysteine (5). **1**. Stadtman (1996) *Ann. Rev. Biochem.* **65**, 83. **2**. Lacourciere & Stadtman (1998) *J. Biol. Chem.* **273**, 30921. **3**. Burnell & Whatley (1977) *Biochim. Biophys. Acta* **481**, 266. **4**. Kim, Park, Kang, Choi & Kim (2002) *Biochim. Biophys. Acta* **1569**, 67. **5**. Suda & Nakagawa (1971) *Meth. Enzymol.* **17B**, 346.

L-Selenocystine

This cystine analogue (FW = 334.09 g/mol; CAS 1464-43-3) is readily produced by oxidation of selenocysteine and inhibits cystine transport (1) and squalene monooxygenase, IC$_{50}$ = 65 μM (2). **1**. Greene, Marcusson, Morell & Schneider (1990) *J. Biol. Chem.* **265**, 9888. **2**. Gupta & Porter (2001) *J. Nutr.* **131**, 1662.

Selenoethanolamine, *See* Selenocysteamine

Selenolysine, *See* Se-(β-Aminoethyl)-L-seleno-cysteine

L-Selenomethionine

This selenium-containing L-methionine analogue (FW = 196.11 g/mol) can replace L-methionyl residues in proteins. The presence of selenomethionyl residues in proteins is an important tool in X-ray crystallography, where its anomalous dispersion solves the phase problem. L-Selenomethionine has an antioxidant activity, can activate glutathione peroxidase, and is soluble in water. **Target(s):** cystine transport (1); glycerol-3-phosphate dehydrogenase (2); homoserine *O*-acetyltransferase (3); methionine *S*-adenosyltransferase (4); methionine *S*-methyltransferase (5); methionyl-tRNA synthetase, also acts as an alternative substrate (6); squalene monooxygenase, weakly inhibited (7); and tRNA (guanine-N^2-)-methyltransferase (8). **1**. Greene, Marcusson, Morell & Schneider (1990) *J. Biol. Chem.* **265**, 9888. **2**. Kim, Park, Kang, Choi & Kim (2002) *Biochim. Biophys. Acta* **1569**, 67. **3**. Shiio & Ozaki (1981) *J. Biochem.* **89**, 1493. **4**. Yarlett, Garofalo, Goldberg, *et al.* (1993) *Biochim. Biophys. Acta* **1181**, 68. **5**. James, Nolte & Hanson (1995) *J. Biol. Chem.* **270**, 22344. **6**. Burnell (1981) *Plant Physiol.* **67**, 325. **7**. Gupta & Porter (2001) *J. Nutr.* **131**, 1662. **8**. Taylor & Gantt (1979) *Biochemistry* **18**, 5253.

Selenopurine

This hypoxanthine analogue (FW = 199.08 g/mol; CAS 226-088-2) is five times more active than mercaptopurine in inhibiting the growth of mouse leukemia L5178 cells in tissue culture (1). Selenopurine is considerably more active than mercaptopurine as an inhibitor of chicken kidney xanthine dehydrogenase (2). When injected into mice bearing L5178-Y mouse lymphoma cells or S-180 tumor cells, 6-selenopurine had lower antitumor activity and greater host toxicity than equimolar amounts of 6-mercaptopurine (3). However, when injected into L1210-bearing mice, 6-selenopurine and 6-mercaptopurine exhibit similar antitumor activities at all dose levels (3). Like 6-mercaptopurine, methylation decreases 6-selenopurine's antitumor activity. **1**. Clarke, Philips, Sternberg, *et al.* (1953) *Cancer Res.* **13**, 593. **2**. Galton (1957) in *The Chemistry and Biology of Purines*. Boston: Little, Brown & Co. **2**. Mautner & Jaffe (1958) *Cancer Res.* **18**, 294.

Seleno-Salacinol

This salacinol analogue (FW = 367.23 g/mol) inhibits mannosyl-oligosaccharide 1,3-1,6-α-mannosidase, *or* mannosidase II. **1**. Kuntz, Ghavami, Johnston, Pinto & Rose (2005) *Tetrahedron Asymmetry* **16**, 25.

Selenotaurine

This selenium-containing taurine analogue (FW = 172.04 g/mol; CAS 16698-41-2) inhibits hypotaurine aminotransferase. At alkaline pH, selenocystamine undergoes autoxidation, entirely transformed into selenohypotaurine in the presence of cupric ions. **1.** Fellman (1987) *Meth. Enzymol.* **143**, 183.

Selenous Acid, *See Selenite*

Selinexor

This orally bioavailable and selective CRM1 inhibitor (FW = 443.31 g/mol; CAS 1393477-72-9; Solubility = <1 mg/mL DMSO or H$_2$O), also known as KPT-330 and (Z)-3-(3-(3,5-bis(trifluoromethyl)phenyl-1H-1,2,4-triazol-1-yl)-N'-(pyrazin-2-yl)acrylohydrazide, targets the nuclear export Chromosome Region Maintenance-1, or CRM1 (also called Exportin-1, or XPO1) and displays selective anticancer activity in preclinical models of T-cell acute lymphoblastic leukemia and acute myeloid leukemia (ED$_{50}$ < 200 nM), reducing growth in MOLT-4, Jurkat, HBP-ALL, KOPTK-1, SKW-3, and DND-41 cell lines, with IC$_{50}$ values ranging from 34 to 203 nM (1,2). KPT-330 suppresses the growth of T-acute lymphoblastic leukemia (T-ALL) cells (*i.e.,* MOLT-4) and acute myelogenous leukemia (AML) cells (*i.e.,* MV4–11) *in vivo*, with little toxicity to normal haematopoietic cells (1). In SCID mice with diffuse human multiple myeloma (MM) bone lesions, KPT-330 inhibits tumor-induced bone lysis, prolonging survival. KPT-330 also impairs osteoclastogenesis and bone resorption by blocking RANKL-induced NF-κB and NFATc1, with minimal impact on osteoblasts and bone marrow stem cells (BMSCs). Tumor-suppressor proteins are inactivated by nuclear exclusion by CRM-1, and increased tumor levels of CRM-1 correlate with poor prognosis for pancreatic cancer patients. Selective Inhibitors of Nuclear Export (SINEs) bind to CRM-1 to irreversibly inhibit (most probably by alkylation of the essential thiol at Cysteine-528) its ability to export proteins. Oral administration of KPT-330 to mice reduced growth of subcutaneous and orthotopic xenograft tumors without major toxicity (3). Analysis of tumor remnants showed that KPT-330 disrupted the interaction between CRM-1 and PAR-4, activated PAR-4 signaling, and reduced proliferation of tumor cells (3). **1.** Etchin, Sanda, Mansour, *et al.* (2013) *Brit. J. Hematol.* **161**, 117. **2.** Tai, Landesman, Acharya, *et al.* (2013) *Leukemia* **28**, 155. **3.** Azmi, Aboukameel, Bao, *et al.* (2013) *Gastroenterology* **144**, 447.

Selumetinib

This orally bioavailable, potent and highly selective protein kinase inhibitor (FW = 457.68; CAS 606143-52-6; Solubility (25°C): 92 mg/mL DMSO, <1 mg/mL Water), also known as AZD6244, ARRY-142886 and 6-(4-bromo-2-chloro-phenylamino)-7-fluoro-N-(2-hydroxyethoxy)-3-methyl-3H-benzo[d]imidazole-5-carboxamide, targets MEK1 (IC$_{50}$ = 14 nM) and ERK1/2 phosphorylation (tight-binding, uncompetitive inhibitor; IC$_{50}$ = 10 nM), with little or no inhibition of p38α, MKK6, EGFR, ErbB2, ERK2, or B-Raf (1-3). Treatment of primary hepatocellular carcinoma (HCC) cells with AZD6244 led to growth inhibition, elevation of the cleavage of caspase-3 and caspase-7, and cleaved poly(ADP)ribose polymerase, but inhibition of ERK1/2 and p90RSK phosphorylation (1). AZD6244 also exhibited sub-mM IC$_{50}$ values in 5 of 31 human breast cancer cell lines and

15 of 43 human non-small cell lung cancer (NSCLC) cell lines, with a correlation between sensitivity and raf mutations in breast cancer cell lines (P = 0.022) and ras mutations in NSCLC cell lines (P = 0.045) (2). Treatment with AZD6244 results in the growth inhibition of several cell lines containing B-Raf and Ras mutations but has no effect on a normal fibroblast cell line (3). When dosed orally, ARRY-142886 was capable of inhibiting both ERK1/2 phosphorylation and growth of HT-29 xenograft tumors in nude mice. Treatment of primary HCC cells with the mitogen-activated protein/extracellular signal-regulated kinase (ERK) kinase 1/2 inhibitor AZD6244 (ARRY-142886) plus doxorubicin led to synergistic growth inhibition and apoptosis (4). AZD6244 has potential to inhibit proliferation and induce apoptosis and differentiation, but the response varies between different xenografts (5). Enhanced antitumor efficacy can be obtained by combining AZD6244 with the cytotoxic drugs irinotecan or docetaxel (5). *In vitro*, treatment with selumetinib and vorinostat synergistically inhibited proliferation and spheroid formation in both colorectal cancer (CRC) cell lines (6). This inhibition was associated with an increase in apoptosis, cell-cycle arrest in G$_1$, reduced cellular migration, and VEGF-A secretion (6). *In vivo*, the combination resulted in additive tumor growth inhibition. Treatment of renal cell carcinoma RCC 786-0 cells with combined sorafenib/AZD6244 resulted in G$_1$ cell cycle arrest and blockade of serum-induced cell migration (7). Sorafenib/AZD6244 also induced apoptosis in primary RCC 08-0910 cells at low concentrations. *In vivo* addition of AZD6244 to sorafenib significantly augmented the antitumor activity of sorafenib and allowed dose reduction of sorafenib without compromising its antitumor activity. Thyroid cancers with BRAF mutation are preferentially sensitive to AZD6244 and other MEK inhibitors (8). In melanoma cells, most mutations conferring resistance to the MEK inhibitor AZD6244 *in vitro* populated the allosteric drug binding pocket or α-helix C and showed robust (~100x) resistance (9). BRAF Val-600-Glu disrupts AZD6244-induced abrogation of negative feedback pathways between extracellular signal-regulated kinase and Raf proteins (10). AZD6244 enhances the anti-tumor activity of sorafenib in ectopic and orthotopic models of human hepatocellular carcinoma (HCC) cells (11). **1.** Huynh, Soo, Chow & Tran (2007) *Mol. Cancer Therapy* **6**, 138. **2.** Garon, Finn, Hosmer, *et al.* (2010) *Mol. Cancer Ther.* **9**, 1985. **3.** Yeh, Marsh, Bernat, *et al.* (2007) *Clin. Cancer Res.* **13**, 1576. **4.** Huynh, Chow & Soo (2007) *Mol. Cancer Ther.* **6**, 2468. **5.** Davies, Logie, McKay, *et al.* (2007) *Mol. Cancer Ther.* **6**, 2209. **6.** Morelli, Tentler, Kulikowski, *et al.*, (2012) *Clin. Cancer Res.* **18**, 1051. **7.** Yuen, Sim, Sim, *et al.* (2012) *Int. J. Oncol.* **41**, 712. **8.** Leboeuf, Baumgartner, Benezra, *et al.* (2008) *J. Clin. Endocrinol. Metab.* **93**, 2194. **9.** Emery, Vijayendran, Zipser, *et al.* (2009) *Proc. Natl. Acad. Sci. U.S.A.* **106**, 20411. **10.** Friday, Yu, Dy, *et al.* (2008) *Cancer Res.* **68**, 6145. **11.** Huynh, Ngo, Koong, *et al.* (2010) *J. Hepatol.* **52**, 79.

Semagacestat

This γ-secretase inhibitor (FW = 361.43 g/mol; CAS 425386-60-3), known also as LY-450139 and named systematically as 2-hydroxy-3-methyl-N-[1-methyl-2-oxo-2-[(2,3,4,5-tetrahydro-3-methyl-2-oxo-1H-3-benzazepin-1-yl)amino]ethyl]butanamide, was once a leading AD drug candidate because a single oral dose (0.1-10 mg/kg) reduced CSF levels of Aβ40 and Aβ43 peptides, the chief constituents of amyloid plaques (1,2). Its inhibitory target, γ-secretase, mediates intramembrane proteolysis of Notch and other essential proteins. Semagacestat is rapidly absorbed (t_{max} ≈ 0.5 hour), and eliminated from systemic circulation, with a $t_{1/2}$ of 2.4 hour (3). Semagacestat failed to slow AD progression in large clinical trials comparing semagacestat with placebo in more than 2600 patients with mild-to-moderate AD. Semagacestat actually worsened both cognition and the ability to perform activities of daily living, leading its developer Lilly to withdraw this experimental agent from further clinical examination. While casting doubt on the amyloid plaque hypothesis, γ-secretase may have other poorly understood roles within the brain. **1.** Best, Jay, Otu, *et al.* (2005) *J.*

Pharmacol. Exp. Ther. **313**, 902. **2**. Lanz, Karmilowicz, Wood, *et al.* (2006) *J. Pharmacol. Exp. Ther.* **319**, 924. **3**. Yi, Hadden, Kulanthaivel, et al. (2010) *Drug Metab Dispos.* **38**, 554.

Semapimod

This macrophage arginine transporter inhibitor (FW$_{tetra-HCl}$ = 890.72 g/mol; FW$_{free-base}$ = 744.91 g/mol; CAS 164301-51-3 and 352513-83-8), also named CNI-1493 and *N,N'*-bis[3,5-bis[*N*-(diaminomethylideneamino)-*C*-methylcarbonimidoyl]phenyl]decanediamide, also exhibits anti-inflammatory, immunomodulatory, anti-cytokine, antiviral, and antimalarial properties. **Mode of Action:** Upon activation by cytokines, macrophages increase L-arginine transport to support the production of NO by NOS. Because endothelial cells do not require additional arginine transport to produce NO, semapimoda is a competitive inhibitor of cytokine-inducible L-arginine transport that does not inhibit Endothelial-Derived Relaxing Factor (EDRF) activity in blood vessels. CNI-1493 to be a selective inhibitor of cytokine-inducible arginine transport and NO production, but does not inhibit EDRF activity (1). **Other Actions:** CNI-1493 prevents lipopolysaccharide (LPS)-induced NO production, even when added in concentrations 10-fold less than required to competitively inhibit L-arginine uptake (2). CNI-1493's action is independent of extracellular L-arginine and NO production, and it is not restricted to induction by LPS. As a selective inhibitor of macrophage activation that prevents TNF production, this tetravalent guanylhydrazone shows promise as a cytokine-suppressive agent for treating diseases mediated by overproduction of cytokines (2). Semapimoda inhibits the biosynthesis of the 26-kDa membrane form of TNF, indicating altered translational regulation (3). Both 5'- and 3'-untranslated regions of the TNF gene are required to elicit maximal translational suppression by CNI-1493 (3). **1**. Bianchi, Ulrich, Bloom, *et al.* (1995) *Mol. Med.* **1**, 254. **2**. Bianchi, Bloom, Raabe, *et al.* (1996) *J. Exp. Med.* **183**, 927. **3**. Cohen, Nakshatri, Dennis, *et al.* (1996) *Proc. Natl. Acad. Sci. USA* **93**, 3967.

Semaxinib

This experimental-stage cancer therapeutic (FW = 238.29 g/mol; CAS 194413-58-6), also known as SU5416 and named systematically as (3*Z*)-3-[(3,5-dimethyl-1*H*-pyrrol-2-yl)methylidene]-1,3-dihydro-2*H*-indol-2-one, is a potent and selective synthetic inhibitor of the Flk-1/KDR vascular endothelial growth factor (VEGF) receptor tyrosine kinase. SU5416 was shown to inhibit vascular endothelial growth factor-dependent mitogenesis of human endothelial cells without inhibiting the growth of a variety of tumor cells in vitro. In contrast, systemic administration of SU5416 at nontoxic doses in mice resulted in inhibition of subcutaneous tumor growth of cells derived from various tissue origins. It targets the VEGF pathway and has demonstrated antiangiogenic potential both *in vivo* and *in vitro*. The antitumor action of SU5416 was accompanied by the appearance of pale white tumors that were resected from drug-treated animals, supporting the antiangiogenic property of this agent. **Status:** Based on consequence of discouraging results, Pharmacia,/Sugen terminated Phase-III clinical trials of Semaxinib in the treatment of advanced colorectal cancer in early 2002. 1. Fong, Shawver, Sun, *et al.* (1999) *Cancer Res.* **59**, 99

Semicarbazide

This common laboratory reagent and urea analogue (FW$_{hydrochloride}$ = 111.53 g/mol; CAS 537-47-3) reacts with aldehydes and ketones, the product of which is a semicarbazone. By reacting with free and enzyme-bound pyridoxal 5-P, semicarbazide characteristically inhibits nearly all pyridoxal-phosphate-requiring enzymes (1). Semicarbazide breaks the Schiff base of PLP by transamination and is is nearly independent of pH over the physiologic range. The highly stable semicarbazone of PLP, which emits strongly at 460 nm, when excited at 380 nm. The Schiff base is inherently 3o-times more reactive than pyridoxal phosphate toward semicarbazide, but is much less reactive than protonated Schiff bases, as judged from studies on morpholine-catalyzed pyridoxal phosphate semicarbazone formation. The latter reaction proceeds via a cationic imine intermediate (1). Semicarbazide also reacts with pyrroloquinoline quinone (PQQ) found in enzymes like methylamine dehydrogenase (primary-amine:(acceptor) oxidoreductase (deaminating), EC 1.4.99.3 (2). **1**. Cordes and Jencks (1962 *Biochemistry* **1**, 773. **2**. Van der Meer, Jongejan, and Duine (1987) *FEBS Lett.* **2**, 299.

Senexin A

This ATP-competitive cdk inhibitor (FW = 310.78 g/mol; Solubility: 100 mM in DMSO), also named 4-[(2-phenylethyl)amino]-6-quinazolinecarbonitrile, targets cyclin-dependent kinase 8, *or* CDK8 (IC$_{50}$ = 280 nM) and also binds to cdk19 isoform. This CDK8 inhibitor suppresses damage-induced tumor-promoting paracrine activities of tumor cells and normal fibroblasts and reverses the increase in tumor engraftment and serum mitogenic activity in mice pretreated with a chemotherapeutic drug. It also increases the efficacy of chemotherapy against xenografts formed by tumor cell/fibroblast mixtures. **1**. Porter, Farmaki, Altilia, *et al.* (2012) *Proc. Natl. Acad. Sci. U.S.A.* **109**, 13799.

Sepantronium bromide

This small-molecule, pro-apoptotic antineoplastic agent (FW$_{HBr salt}$ = 443.29 g/mol; CAS 781661-94-7 and 753440-91-4 (free base); Solubility: 60 mg/mL DMSO; 90 mg/mL H$_2$O), also known as YM155 and named systematically as 4,9-dihydro-1-(2-methoxyethyl)-2-methyl-4,9-dioxo-3-(2-pyrazinylmethyl)-1*H*-naphth[2,3-*d*]imidazolium bromide, selectively inhibits (*IC$_{50}$* = 0.54 nM) survivin gene expression in tumor cells, resulting in inhibition of survivin's anti-apoptotic activity (via extrinsic and intrinsic apoptotic pathways) as well as tumor cell apoptosis. A member of the inhibitor-of-apoptosis (IAP) gene family, survivin is expressed during embryonal development, but is absent in most normal, terminally differentiated tissues. Survivin is also up-regulated in a number of human cancers, with its expression associated with aggressive-phenotype tumors as well as shorter survival times and lower responsiveness to chemotherapy. YM155 potently inhibits human cancer cell lines (mutated or truncated p53) including PC-3, PPC-1, DU145, TSU-Pr1, 22Rv1, SK-MEL-5 and A375 with IC_{50} from 2.3 to 11 nM, respectively. **1**. Nakahara, *et al.* (2007) *Cancer Res.* **67**, 8014. **2**. Iwasa, *et al.* (2008) *Clin. Cancer Res.* **14**, 6496.

Sephin 1

This protein phosphatase localization inhibitor (FW$_{carbonate-salt}$ = 258.66 g/mol; Soluble to 100 mM in DMSO), also named (2*E*)-2-[(2-chlorophenyl)methylene]hydrazinecarboximidamide, selectively targets the stress-induced PPP1R15A, which recruits the serine/threonine-protein phosphatase PP1 to dephosphorylate the translation initiation factor eIF-2A/EIF2S1, thereby reversing the shut-off of protein synthesis initiated by stress-inducible kinases and facilitating recovery from stress. Sephin-1 is without effect on the related and constitutive PPP1R15B. Upon oral administration at 1 or 10 mg/kg, Sephin 1 rapidly disappeared from plasma, and was concentrated in the nervous system, reaching concentrations 7–44 times higher in the brain and sciatic nerve (up to ~ 1 μM) than in plasma. In mice, chronic treatment (1 mg/kg for 1 month) was tolerable, with no measurable adverse effects on body weight. When tested *in vivo*, Sephin1 safely prevents the motor, morphological, and molecular defects of two otherwise unrelated protein-misfolding diseases in mice, Charcot-Marie-Tooth 1B, and amyotrophic lateral sclerosis. **1**. Das, Krzyzosiak, Schneider, *et al.* (2015) *Science* **348**, 239.

Sepiapterin

This pterin (FW = 237.22 g/mol; CAS 17094-01-8) is a precursor in the biosynthesis of tetrahydrobiopterin via the pterin salvage pathway. **Target(s):** dihydropteridine reductase (1); GTP cyclohydrolase I (1-4); nitric-oxide synthase (5); and tRNA-guanine transglycosylase (6). **1**. Shen, Zhang & Perez-Polo (1989) *J. Enzyme Inhib.* **3**, 119. **2**. Jacobson & Manos (1989) *Biochem. J.* **260**, 135. **3**. Shen, Alam & Zhang (1988) *Biochim. Biophys. Acta* **965**, 9. **4**. Blau & Niederwieser (1986) *Biochim. Biophys. Acta* **880**, 26. **5**. Jorens, van Overveld, Bult, Vermeire & Herman (1992) *Brit. J. Pharmacol.* **107**, 1088. **6**. Farkas, Jacobson & Katze (1984) *Biochim. Biophys. Acta* **781**, 64.

Sepimostat

This aminobenzoate derivative (FW$_{free-base}$ = 373.41 g/mol), known in brief as FUT-187 and systematically as 6-amidino-2-naphthyl [4-(4,5-dihydro-1*H*-imidazol-2-yl)amino]benzoate, is an orally active serine proteinase inhibitor, acting competitively toward trypsin (K_i = 0.097 μM), pancreatic kallikrein (K_i = 0.029 μM), plasma kallikrein (K_i = 0.61 μM), plasmin (K_i = 0.57 μM), thrombin (K_i = 2.5 μM), coagulation factor Xa (K_i = 20.4 μM), and complement subcomponent C1r (K_i = 6.4 μM). Sepimostat is a noncompetitive inhibitor for coagulation factor XIIa (K_i = 0.021 μM) and an uncompetitive inhibitor for complement subcomponent C1s (K_i = 0.18 μM). The complement-mediated hemolyses in the classical and alternative

pathways were also inhibited by sepimostat with IC_{50} values of 0.17 and 3.5 μM, respectively. **Target(s):** alternative-complement-pathway C3/C5 convertase (1); classical-complement-pathway C3/C5 convertase (1); coagulation factor Xa (2); coagulation factor XIIa (1); complement subcomponent C1r (1,2); complement subcomponent C1s (1,2); kallikrein (1,2); plasmin (1,2); thrombin (1,2); and trypsin (1,2). **1**. Nakamura, Johmura, Oda, *et al.* (1995) *Yakugaku Zasshi* **115**, 201. **2**. Oda, Ino, Nakamura, *et al.* (1990) *Jpn. J. Pharmacol.* **52**, 23.

Septamycin

This polyether ionophore and antibiotic (FW$_{free-acid}$ = 913.20 g/mol; CAS 55924-40-8), from *Streptomyces hygroscopicus* NRRL 5678, forms complexes with metal ions and inhibits oxidative phosphorylation (1,2), especially active against Gram-positive bacteria and *Eimeria tenella* (chicken coccidiosis). **1**. Reed (1979) *Meth. Enzymol.* **55**, 435. **2**. Keller-Juslen, King, Kis & von Wartburg (1975) *J. Antibiot.* **28**, 854.

[Ser4,Glu8]-Oxytocin, *See Glumitocin*

D-Serine

This neuroactive amino acid (FW = 105.09 g/mol; CAS 312-84-5), systematically known as (*R*)-2-amino-3-hydroxypropanoic acid, is found in silkworms and earthworms as well as in such antibiotics as tolaasin and the polymyxins. D-Serine is most likely the endogenous ligand for the glycine modulatory binding site on the NR1 subunit of *N*-methyl-D-aspartate receptors, serving as a neuromodulator by coactivating NMDA receptors and facilitating their opening when they bind glutamate. The D-serine concentrations within neurons are regulated by its synthesis and degradation by serine racemase (a fully reversible enzymatic reaction) as well as by its oxidation by D-amino acid oxidase. (*See* L-*Serine*) **Target(s):** D-alanine aminotransferase (1); D-alanyl-poly(phosphoribitol) synthetase, *or* D-alanine:poly(phosphoribitol) ligase (2); 1-aminocyclopropane-1-carboxylate deaminase (3-5); aspartate 1-decarboxylase (6,7); aspartate 4-decarboxylase, *or* aminomalonate decarboxylase (8); diphosphate:serine phosphotransferase (9); kynureninase (10); phosphoserine phosphatase (11-13); pyruvate kinase (14); L-serine ammonia-lyase, *or* L-serine dehydratase (15-20); serine:glyoxylate aminotransferase (21); and threonine ammonia-lyase, *or* threonine dehydratase (22,23). **1**. Martinez-Carrion & Jenkins (1965) *J. Biol. Chem.* **240**, 3547. **2**. Reusch, Jr., & Neuhaus (1971) *J. Biol. Chem.* **246**, 6136. **3**. Honma (1986) *Agric. Biol. Chem.* **50**, 3189. **4**. Minami, Uchiyama, Murakami, *et al.* (1998) *J. Biochem.* **123**, 1112. **5**. Honma, Shimomura, Shiraishi, Ichihara & Sakamura (1979) *Agric. Biol. Chem.* **43**, 1677. **6**. Williamson (1985) *Meth. Enzymol.* **113**, 589. **7**. Williamson & Brown (1979) *J. Biol. Chem.* **254**, 8074. **8**. Thanassi & Fruton (1962) *Biochemistry* **1**, 975. **9**. Cagen & Friedmann (1972) *J. Biol. Chem.* **247**, 3382. **10**. Jakoby & Bonner (1953) *J. Biol. Chem.* **205**, 709. **11**. Schmidt & Laskowski, Sr. (1961) *The Enzymes*, 2nd ed. (Boyer, Lardy & Myrbäck, eds.), **5**, 3. **12**. Byrne (1961) *The Enzymes*, 2nd ed. (Boyer, Lardy & Myrbäck, eds.), **5**, 73. **13**. Bridgers (1967) *J. Biol. Chem.* **242**, 2080. **14**. Schering, Eigenbrodt, Linder & Schoner (1982) *Biochim. Biophys. Acta* **717**, 337. **15**. Gannon, Bridgeland & Jones (1977) *Biochem. J.* **161**, 345. **16**. Zinecker, Andreesen & Pich (1998) *J. Basic Microbiol.* **38**,

147.   **17**. Newman & Kapoor (1980) *Can. J. Biochem.* **58**, 1292.   **18**. Hofmeister, Grabowski, Linder & Buckel (1993) *Eur. J. Biochem.* **215**, 341. **19**. Morikawa, Nakamura & Kimura (1974) *Agric. Biol. Chem.* **38**, 531.   **20**. Velayudhan, Jones, Barrow & Kelly (2004) *Infect. Immun.* **72**, 260.   **21**. Paszkowski (1991) *Acta Biochim. Pol.* **38**, 437.   **22**. Hirata, Tokushige, Inigaki & Hayaishi (1965) *J. Biol. Chem.* **240**, 1711.   **23**. Laakmann-Ditges & Klemme (1988) *Arch. Microbiol.* **149**, 249.

L-Serine

This hydrophilic proteagenic amino acid (FW = 105.09 g/mol; CAS 56-45-1; pK_1 = 2.19, and pK_2 = 9.21 at 25°C; $pK_3 \approx 15$), systematically known as (*S*)-2-amino-3-hydroxypropanoic acid is en coded by six codons (UCU, UCC, UCA, UCG, AGU, ad AGC) is usually found on the surfaces of proteins. When buried in the interior of a protein, the hydroxyl side-chain is usually hydrogen-bonded to an oxygen atom of a peptide bond. Seryl residues tend to favor α-helical structures and are often located at the ends of helices or reverse turns. L-Serine has a specific rotation of +15.1° at 25°C of the D line for 2.0 g/100 mL in 5 M HCl (–7.5° in distilled water) and is soluble in water (25 g/100 mL at 20°C). Many enzymes use D- or L-serine as alternative substrates (and hence competitive inhibitors), including L-amino-acid dehydrogenase, the amino acid oxidases, D-amino-acid dehydrogenase, alanopine dehydrogenase, alanine:oxo-acid aminotransferase, ornithine:oxo-acid aminotransferase, asparagine:oxo-acid aminotransferase, tryptophanase, tyrosine phenol-lyase, threonine dehydratase, diaminopropionate ammonia-lyase, and amino-acid racemase. ***See also*** *D-Serine; Serine-Borate Complex* **Target(s):** D-alanine 2-hydroxymethyl-transferase (1,63); D-amino-acid aminotransferase (53); 1-aminocyclopropane-1-carboxylate deaminase (38-40); aminomalonate decarboxylase, *or* aspartate β-decarboxylase (2;) 8-amino-7-oxononanoate synthase (3,60,61); asparagine:oxo-acid aminotransferase (54,55); aspartate 1-decarboxylase (4,37); aspartate 4-decarboxylase (2,36);   choline kinase (5,51); cystathionine γ-lyase, *or* cysteine desulfhydrase (6-8); cysteine lyase (9); cysteine synthase (58); glutamine synthetase (10-18); γ-glutamyl transpeptidase, *or* γ-glutamyltransferase (19,20,59); D-glycerate dehydrogenase (21); glycine aminotransferase (57); glycine dehydrogenase (cytochrome) (22); glycine:oxaloacetate aminotransferase (52); glycyl-tRNA synthetase (23); homoserine dehydrogenase (24); homoserine kinase (49,50); kynureninase (25); ornithine aminotransferase (56); ornithine carbamoyltransferase (62); 3-phosphoglycerate dehydrogenase (21,26,27); phosphoglycerate phosphatase (41); phosphoglycolate phosphatase (42); phosphoserine phosphatase (43-48); propionyl-CoA carboxylase, *Myxococcus Xanthus* (28); D-serine ammonia-lyase, *or* D-serine dehydratase (29-31); threonine ammonia-lyase, *or* threonine dehydratase (32-34); and UDP-*N*-acetylmuramate:L-alanine ligase (35). **1**. Miles (1971) *Meth. Enzymol.* **17B**, 341. **2**. Thanassi & Fruton (1962) *Biochemistry* **1**, 975.   **3**. Izumi, Tani & Ogata (1979) *Meth. Enzymol.* **62**, 326.   **4**. Williamson (1985) *Meth. Enzymol.* **113**, 589.   **5**. Ishidate & Nakazawa (1992) *Meth. Enzymol.* **209**, 121.   **6**. Flavin & Segal (1964) *J. Biol. Chem.* **239**, 2220.   **7**. Fromageot & Grand (1942) *Enzymologia* **11**, 81.   **8**. Chu, Ebersole, Kurzban & Holt (1999) *Clin. Infect. Dis.* **28**, 442.   **9**. Braunstein & Goryachenkova (1984) *Adv. Enzymol. Relat. Areas Mol. Biol.* **56**, 1.   **10**. Meister (1985) *Meth. Enzymol.* **113**, 185.   **11**. Rhee, Chock & Stadtman (1985) *Meth. Enzymol.* **113**, 213.   **12**. Meister (1974) *The Enzymes*, 3rd ed. (Boyer, ed.), **10**, 699.   **13**. Stadtman & Ginsburg (1974) *The Enzymes*, 3rd ed. (Boyer, ed.), **10**, 755.   **14**. Orr & Haselkorn (1981) *J. Biol. Chem.* **256**, 13099.   **15**. Blanco, Alana, Llama & Serra (1989) *J. Bacteriol.* **171**, 1158. **16**. Krishnan, Singhal & Dua (1986) *Biochemistry* **25**, 1589.   **17**. Ertan (1992) *Arch. Microbiol.* **158**, 35.   **18**. Liaw, Pan & Eisenberg (1993) *Proc. Natl. Acad. Sci. U.S.A.* **90**, 4996.   **19**. Allison (1985) *Meth. Enzymol.* **113**, 419.   **20**. Thompson & Meister (1977) *J. Biol. Chem.* **252**, 6792.   **21**. Uhr & Sneddon (1971) *FEBS Lett.* **17**, 137.   **22**. Goldman & Wagner (1962) *Biochim. Biophys. Acta* **65**, 297.   **23**. Boyko & Fraser (1964) *Can. J. Biochem.* **42**, 1677.   **24**. Mankovitz & Segal (1969) *Biochemistry* **8**, 3765. **25**. Jakoby & Bonner (1953) *J. Biol. Chem.* **205**, 709.   **26**. Pizer & Sugimoto (1971) *Meth. Enzymol.* **17B**, 325.   **27**. Slaughter & Davis (1975) *Meth. Enzymol.* **41**, 278.   **28**. Kimura, Kojyo, Kimura & Sato (1998) *Arch. Microbiol.* **170**, 179.   **29**. Robinson & Labow (1971) *Meth. Enzymol.* **17B**, 356.   **30**. Federiuk & Shafer (1981) *J. Biol. Chem.* **256**, 7416.   **31**.

Schnackerz, Tai, Potsch & Cook (1999) *J. Biol. Chem.* **274**, 36935.   **32**. Greenberg (1962) *Meth. Enzymol.* **5**, 942.   **33**. Greenberg (1961) *The Enzymes*, 2nd ed. (Boyer, Lardy & Myrbäck, eds.), **5**, 563.   **34**. Davis & Metzler (1972) *The Enzymes*, 3rd ed. (Boyer, ed.), **7**, 33.   **35**. Liger, Blanot & van Heijenoort (1991) *FEMS Microbiol. Lett.* **80**, 111.   **36**. Wong (1985) *Neurochem. Int.* **7**, 351.   **37**. Williamson & Brown (1979) *J. Biol. Chem.* **254**, 8074.   **38**. Honma (1985) *Agric. Biol. Chem.* **49**, 567.   **39**. Minami, Uchiyama, Murakami, *et al.* (1998) *J. Biochem.* **123**, 1112.   **40**. Honma, Shimomura, Shiraishi, Ichihara & Sakamura (1979) *Agric. Biol. Chem.* **43**, 1677.   **41**. Pestka & Delwiche (1981) *Can. J. Microbiol.* **27**, 808.   **42**. Verin-Vergeau, Baldy & Cavalie (1980) *Phytochemistry* **19**, 763.   **43**. Hawkinson, Acosta-Burruel, Ta & Wood (1997) *Eur. J. Pharmacol.* **337**, 315.   **44**. Bridgers (1967) *J. Biol. Chem.* **242**, 2080.   **45**. Ho, Noji & Saito (1999) *J. Biol. Chem.* **274**, 11007.   **46**. Bridgers (1969) *Arch. Biochem. Biophys.* **133**, 201.   **47**. Kearney & Holloway (1987) *FEMS Microbiol. Lett.* **42**, 109.   **48**. Knox, Herzfeld & Hudson (1969) *Arch. Biochem. Biophys.* **132**, 397.   **49**. Burr, Walker, Truffa-Bachi & Cohen (1976) *Eur. J. Biochem.* **62**, 519.   **50**. Thoen, Rognes & Aarnes (1978) *Plant Sci. Lett.* **13**, 103.   **51**. Setty & Krishnan (1972) *Biochem. J.* **126**, 313.   **52**. Gibbs & Morris (1966) *Biochem. J.* **99**, 27P.   **53**. Martinez-Carrion & Jenkins (1965) *J. Biol. Chem.* **240**, 3547.   **54**. Maul & Schuster (1986) *Arch. Biochem. Biophys.* **251**, 585. **55**. Cooper (1977) *J. Biol. Chem.* **252**, 2032.   **56**. Goldberg, Flescher & Lengy (1979) *Exp. Parasitol.* **47**, 333.   **57**. Yamaguchi, Ohtani, Amachi, Shinoyama & Fujii (2003) *Biosci. Biotechnol. Biochem.* **67**, 783.   **58**. Burnell & Whatley (1977) *Biochim. Biophys. Acta* **481**, 246.   **59**. Moallic, Dabonne, Colas & Sine (2006) *Protein J.* **25**, 391.   **60**. Izumi, Morita, Tani & Ogata (1973) *Agric. Biol. Chem.* **37**, 1327.   **61**. Izumi, Sato, Tani & Ogata (1973) *Agric. Biol. Chem.* **37**, 1335.   **62**. Legrain & Stalon (1976) *Eur. J. Biochem.* **63**, 289.   **63**. Wilson & Snell (1962) *J. Biol. Chem.* **237**, 3171.

Serine-Borate Complex

This molecular entity, which forms upon binding of serine to γ-glutamyl transpeptidase in the presence of borate anion (10 mM), represents a transition state-like bridge complex between the serine hydroxyl group and a putative active-site hydroxyl group. L-Serine binds roughly 8.5-fold more tightly than with D-serine (K_i values of 20 and 170 μM). **Target(s):** D-alanine γ-glutamyltransferase (1); and γ-glutamyl transpeptidase, *or* γ-glutamyltransferase (2-4) **1**. Kawasaki, Ogawa & Sasaoka (1982) *Biochim. Biophys. Acta* **716**, 194.   **2**. Tate & Meister (1978) *Proc. Natl. Acad. Sci. U.S.A.* **75**, 4806.   **3**. Allison (1985) *Meth. Enzymol.* **113**, 419.   **4**. Revel & Ball (1959) *J. Biol. Chem.* **234**, 577.

L-Serine Hydroxamate

This *N*-acylhydroxamate analogue of serine (FW$_{ion}$ = 121.12 g/mol; CAS 31697-35-5) reacts with ATP on seryl-tRNA synthetase to form a tightly bound seryl-hydroxamate-adenylate (1-5). This inhibitor design should suffice for other amino acyl tRNA synthases, provided the Because acyl hydroxamates is available. Note: Acylhydroxamates readily chelate Fe^{3+}, such experiments should be conducted in the presence of a chelating agent. Alternatively, the hydroxamate should be freed of metal ions by passage over a cation exchange column or Chelex resin. **1**. Söll & Schimmel (1974) *The Enzymes*, 3rd ed. (Boyer, ed.), **10**, 489.   **2**. Landeka, Filipic-Rocak, Zinic & Weygand-Durasevic (2000) *Biochim. Biophys. Acta* **1480**, 160. **3**. Weygand-Durasevic, Ban, Jahn & Söll (1993) *Eur. J. Biochem.* **214**, 869. **4**. Tosa & Pizer (1971) *J. Bacteriol.* **106**, 972.   **5**. Ahel, Slade, Mocibob, Söll & Weygand-Durasevic (2005) *FEBS Lett.* **579**, 4344.

L-Serine *O*-Phosphate, See *O-Phospho-L-serine*

D-Serine O-Sulfate

This *O*-serine ester (FW$_{ion}$ = 184.15 g/mol), also known as *O*-sulfo-D-serine, inhibits serine-sulfate ammonia-lyase and D-glutamate:D-amino acid aminotransferase. **Target(s):** D-glutamate:D-amino acid aminotransferase (1); D-lactate-2-sulfatase, as alternative substrate inhibitor (2); and serine-sulfate ammonia-lyase (3-5). **1**. Jones, Soper, Ueno & Manning (1985) *Meth. Enzymol.* **113**, 108. **2**. Crescenzi, Dodgson & White (1984) *Biochem. J.* **223**, 487. **3**. Tudball & Thomas (1971) *Meth. Enzymol.* **17B**, 361. **4**. Murakoshi, Sanda & Haginiwa (1977) *Chem. Pharm. Bull.* **25**, 1829. **5**. Thomas & Tudball (1967) *Biochem. J.* **105**, 467.

L-Serine O-Sulfate

This *O*-serine ester (FW = 184.15 g/mol), also known as *O*-sulfo-L-serine, is a substrate for serine-sulfate ammonia-lyase, a reversible inhibitor of many enzymes, and a mechanism-based inactivator glutamate racemase. **Target(s):** *O*-acetylhomoserine sulfhydrylase (1); 2-aminoadipate aminotransferase (2); aspartate aminotransferase (3-8); glutamate carboxypeptidase II, *or* *N*-acetylated-γ-linked-acidic dipeptidase (9); glutamate decarboxylase (7,10,11); glutamate dehydrogenase (12); glutamate racemase (13); kynurenine aminotransferase (2); and serine racemase (14). **1**. Yamagata (1987) *Meth. Enzymol.* **143**, 465. **2**. Tobes & Mason (1977) *J. Biol. Chem.* **252**, 4591. **3**. Rando (1977) *Meth. Enzymol.* **46**, 28. **4**. Sigman & Mooser (1975) *Ann. Rev. Biochem.* **44**, 889. **5**. Braunstein (1973) *The Enzymes*, 3rd ed. (Boyer, ed.), **9**, 379. **6**. John & Fasella (1969) *Biochemistry* **8**, 4477. **7**. Ator & Ortiz de Montellano (1990) *The Enzymes*, 3rd ed. (Sigman & Boyer, eds.), **19**, 213. **8**. Recasens, Benezra & Mandel (1980) *Biochemistry* **19**, 4583. **9**. Robinson, Blakely, Couto & Coyle (1987) *J. Biol. Chem.* **262**, 14498. **10**. Likos, Ueno, Feldhaus & Metzler (1982) *Biochemistry* **21**, 4377. **11**. Fonda (1985) *Meth. Enzymol.* **113**, 11. **12**. Tudball & Thomas (1971) *Biochem. J.* **123**, 421. **13**. Ashiuchi, Yoshimura, Esaki, Ueno & Soda (1993) *Biotechnol. Biochem.* **57**, 1978. **14**. Cook, Galve-Roperh, Martinez del Pozo & Rodriguez-Crespo (2002) *J. Biol. Chem.* **277**, 27782.

[Ser4,Ile8]-Oxytocin, *See* Isotocin

Serotonin

This photosensitive neurotransmitter (FW$_{free-base}$ = 176.22 g/mol; CAS 153-98-0 (hydrochloride); pK_1 = 9.8 (R–NH$_2$); pK_2 = 11.1 (R-OH); λ_{max} = 275 nm at pH 7, ε = 5800 M^{-1}cm^{-1}), also known as 5-hydroxytryptamine (5-HT), is also a vasoconstricting agent that increases vascular permeability. *Note*: Serotonin slowly oxidizes after only a few hours at neutral pH, and oxidation is even more rapid at more alkaline pH values. **Target(s):** acetylcholinesterase (1,2); aromatic-L-amino-acid decarboxylase, *or* dopa decarboxylase (3-6); aryl-acylamidase (2,7-12); chymase (13); CYP2C9 (14); diacylglycerol cholinephosphotransferase (15); dihydropteridine reductase (16); [elongation factor 2] kinase (17); ethanolaminephosphotransferase (15); glutamate dehydrogenase (18); γ-glutamylhistamine synthetase (19); histamine *N*-methyltransferase (20,21); Na$^+$/K$^+$-exchanging ATPase (22); pyridoxal kinase (23-25); thyroxine deiodinase, *or* iodothyronine deiodinase (26); tryptophan 2,3-dioxygenase

(27); tryptophan 5-monooxygenase (28); and tyrosine aminotransferase (29). **1**. Krupka (1965) *Biochemistry* **4**, 429. **2**. Costagli & Galli (1998) *Biochem. Pharmacol.* **55**, 1733. **3**. Barboni, Voltattorni, D'Erme, *et al.* (1981) *Biochem. Biophys. Res. Commun.* **99**, 576. **4**. Fellman (1959) *Enzymologia* **20**, 366. **5**. Jung (1986) *Bioorg. Chem.* **14**, 429. **6**. Bender & Coulson (1977) *Biochem. Soc. Trans.* **5**, 1353. **7**. Paul & Halaris (1976) *Biochem. Biophys. Res. Commun.* **70**, 207. **8**. Hsu, Paul, Halaris & Freedman (1977) *Life Sci.* **20**, 857. **9**. Oommen & Balasubramanian (1979) *Eur. J. Biochem.* **94**, 135. **10**. George & Balasubramanian (1980) *Eur. J. Biochem.* **111**, 511. **11**. Fujimoto (1976) *FEBS Lett.* **71**, 121. **12**. Fujimoto (1974) *Biochem. Biophys. Res. Commun.* **61**, 72. **13**. Yurt & Austen (1977) *J. Exp. Med.* **146**, 1405. **14**. Gervasini, Martinez, Agundez, Garcia-Gamito & Benitez (2001) *Pharmacogenetics* **11**, 29. **15**. Strosznajder, Radominska-Pyrek & Horrocks (1979) *Biochim. Biophys. Acta* **574**, 48. **16**. Shen (1985) *J. Enzyme Inhib.* **1**, 61. **17**. Carroll, Warren, Fan & Sossin (2004) *J. Neurochem.* **90**, 1464. **18**. Smith, Austen, Blumenthal & Nyc (1975) *The Enzymes*, 3rd ed. (Boyer, ed.), **11**, 293. **19**. Stein & Weinreich (1982) *J. Neurochem.* **38**, 204. **20**. Axelrod (1971) *Meth. Enzymol.* **17B**, 766. **21**. Brown, Tomchick & Axelrod (1959) *J. Biol. Chem.* **234**, 2948. **22**. Stepp & Novakoski (1997) *Arch. Biochem. Biophys.* **337**, 43. **23**. Takeuchi, Tsubouchi & Shibata (1985) *Biochem. J.* **227**, 537. **24**. Neary, Meneely, Grever & Diven (1972) *Arch. Biochem. Biophys.* **151**, 42. **25**. Gäng & von Collins (1973) *Int. J. Vitam. Nutr. Res.* **43**, 318. **26**. Nakagawa & Ruegamer (1967) *Biochemistry* **6**, 1249. **27**. Webb (1966) *Enzyme and Metabolic Inhibitors*, vol. 2, p. 325, Academic Press, New York. **28**. Park, Stone, Kim & Joh (1994) *Mol. Cell. Neurosci.* **5**, 87. **29**. Jacoby & La Du (1964) *J. Biol. Chem.* **239**, 419.

SERP-1, *See* Myxoma Virus Serine Proteinase Inhibitor

Serpasil, *See* Reserpine

Serpentine

This alkaloid (FW = 348.40 g/mol; CAS 18786-24-8) from the roots of *Rauwolfia serpentina* is a DNA-intercalating alkaloid that stabilizes the formation of topoisomerase II-DNA covalent complex, thereby stimulating DNA scission without any preference for cutting at a specific base (1). Serpentine also inhibits CYP2D6 (2) and 3-α-(*S*)-strictosidine β-glucosidase (3,4). **1**. Dassonneville, Bonjean, De Pauw-Gillet, *et al.* (1999) *Biochemistry* **38**, 7719. **2**. Usia, Watabe, Kadota & Tezuka (2005) *Biol. Pharm. Bull.* **28**, 1021. **3**. Luijendijk, Stevens & Verpoorte (1998) *Plant Physiol. Biochem.* **36**, 419. **4**. Gerasimenko, Sheludko, Ma & Stockigt (2002) *Eur. J. Biochem.* **269**, 2204.

SERPIN A1, *See* α$_1$-Antitrypsin

SERPIN B5, *See* Maspin

Serpin b6b

This 377-residue serpin (*Official Name*: Serine (or cysteine) Proteinase Inhibitor, Clade B, Member 6b) is specific intracellular inhibitor of mouse Granzyme A (mGzmA), but not human Granzyme A (hGzmA). Serpinb6b utilizes an unusual exosite interaction that depends on the dimeric nature of mGzmA, and its mechanism of action indicates evolutionary divergence in cytotoxic potential and physiological function between mouse and human enzymes. (*For a discussion of serpin mechanisms,* **See also** *Serpins; α$_1$-Antichymotrypsin*) **1**. Kaiserman, Stewart, Plasman, *et al.* (2014) *J. Biol. Chem.* **289**, 9408.

Serpin b12

This 405-residue serine protease inhibitor (MW$_{predicted}$ = 46,276), expressed by numerous tissues (*e.g.,* brain, bone marrow, lymph node, heart, lung, liver, pancreas, testis, ovary, and intestine) inhibits plasmin and trypsin. (*For a discussion of the likely mechanism of inhibitory action,* **See** *Serpins; also* α$_1$-

Antichymotrypsin) **1**. Askew, Pak, Luke, *et al.* (2001) *J. Biol. Chem.* **276**, 49320

SERPINB13, See *Headpin*

Serpin Proteinase Inhibitor 9, See *PI9*

Serpins

These <u>ser</u>ine proteinase <u>in</u>hibitors form catalytically inactive complexes with their target protease, and, upon binding to the target enzyme's active site, undergo P^1–$P^{1'}$ cleavage to form high-affinity complexes that dissociate extremely slowly. (*See specific serpin:* α_1-Antitrypsin, Antithrombin, Bomapin, Complement C1 inhibitor, crmA, Plasminogen Activator Inhibitors, *etc.*) **RCL Recognition:** A key structural element in the action of this class of inhibitors is thus a conformationally mobile reactive-center loop that enables the serpin to form a virtually irreversible, high-affinity complex with their target proteinase. The inhibitory mechanism of serpin-class inhibitors relies on its exposed region, called the Reactive Center Loop (*or* RCL) to mimic the specific substrate of the target protease. RCL cleavage by the target protease initiates a conformational change in the serpin fold that captures the protease in a 1:1 covalent complex (1). Canonical sequences serpin RCLs are thus critical determinants of protease inhibition, and they closely resemble the substrate's sequence an cleavage site. **Detailed Serpin Mechanism for Antichymotrypsin Interactions with Chymotrypsin:** O'Malley *et al.* (2) demonstrated that antichymotrypsin (I) binds to chymotrypsin (E) to form an E*I* complex via a three-step mechanism: $E + I \rightleftharpoons E \cdot I \rightleftharpoons E \cdot I' \rightleftharpoons E^* - I^*$. In this scheme, EI′ retains the P^1–$P^{1'}$ linkage and is formed in a partly, or largely, rate-determining step. Using rapid quench-flow assay for post-complex fragment formation, Nair & Cooperman (3) determined that the E·I encounter complex of serpin and enzyme forms both E*–I* and the post-complex fragment with the same rate constant, indicating that both species arise from EI′ conversion to E*–I*. These results indicate (a) that the peptide bond remains intact within the EI′ complex, and (b) that E*–I* is likely to be either the acyl-enzyme or the tetrahedral intermediate formed after water attack on acyl-enzyme (4). (*For further discussion of the likely mechanism of inhibitor action, See* α_1-*Antichymotrypsin*) **1**. Gettins (2002) *Chem. Rev.* **102**, 4751. **2**. O'Malley, Nair, Rubin & Cooperman (1997) *J. Biol. Chem.* **272**, 5354. **3**. Nair & Cooperman (1998) *J. Biol. Chem.* **273**, 17459. **4**. Purich (2010) *Enzyme Kinetics: Catalysis & Control*, Academic Press, San Diego.

***Serratia marcescens* Proteinase Inhibitor**

This *Serratia marcescens* polypeptide (MW = 10 kDa) from is localized in the periplasmic space, where it inhibits the proteinase serralysin (1-3). **1**. Kim, Kim, Kim, Byun & Shin (1995) *Appl. Environ. Microbiol.* **61**, 3035. **2**. Suh & Benedik (1992) *J. Bacteriol.* **174**, 2361. **3**. Bae, Kim, Kim, Shin & Byun (1998) *Arch. Biochem. Biophys.* **352**, 37.

Sertraline

This widely used antidepressant (FW$_{\text{free-base}}$ = 306.23 g/mol; CAS 79617-96-2; pK_a = 9.48), known also as Zoloft® and its IUPAC name (1*S*,4*S*)-*N*-methyl-4-(3,4-dichlorophenyl)-1,2,3,4-tetrahydro-1-naphthylamine, is a selective (and competitive) serotonin reuptake inhibitor, *or* SSRI, K_i = 3.3 nM. Also available in generic forms (*i.e.*, the U.S. patent expired in 2006), sertraline is prescribed throughout the world for the treatment of depression, obsessive compulsive disorder (OCD), panic disorder, posttraumatic stress disorder (PTSD), social anxiety disorder, and premenstrual dysphoric disorder (PMDD). **Likely Mechanism of Action:** Sertraline's antidepressant action is presumed to be rooted in its inhibition of CNS neuronal uptake of serotonin (5HT) (1-7). Studies at clinically relevant doses show that Sertraline blocks serotonin uptake into human platelets. *In vitro* studies in animals also suggest that Sertraline is a potent and selective inhibitor of neuronal serotonin reuptake and has only very weak effects on

norepinephrine and dopamine neuronal reuptake. *In vitro* studies confirm that Sertraline has no significant affinity for α_1-adrenergic receptors, α_2-adrenergic receptors, β-adrenergic receptors, cholinergic receptors, GABA receptors, dopaminergic receptors, histaminergic receptors, serotonergic (5-HT$_{1A}$, 5-HT$_{1B}$, 5-HT$_2$) receptors, or benzodiazepine receptors. On the other hand, Sertraline preferentially does block the inwardly rectifying astroglial K$^+$ channel Kir4.1, but not Kir1.1 or Kir2.1 channels, an action that may be implicated in its therapeutic and/or adverse actions (8). **Key Pharmacokinetic Parameters:** *See* Appendix II in Goodman & Gilman's *THE PHARMACOLOGICAL BASIS OF THERAPEUTICS*, 12th Edition (Brunton, Chabner & Knollmann, eds.) McGraw-Hill Medical, New York (2011). **Target(s):** Sertraline is a mixed-type inhibitor of Dynamin-1 with respect to both GTP and L-α-phosphatidyl-L-serine (PS) *in vitro*, also inhibiting synaptic vesicle endocytosis via inhibition of Dynamin GTPase in human neuroblastoma SH-Sy5Y cells and HeLa cells (9). In mitochondria isolated from rat liver, sertraline uncoupled mitochondrial oxidative phosphorylation and inhibited the activities of oxidative phosphorylation complexes I and V (10). Additionally, sertraline induced Ca^{2+}-mediated mitochondrial permeability transition (MPT), and the induction was prevented by bongkrekic acid (BA), a specific MPT inhibitor targeting adenine nucleotide translocator (ANT), implying that the MPT induction is mediated by ANT (10). Sertraline (1.5–25 µM) progressively inhibits the rise in Na$^+$ and the release of pre-loaded [^{3}H]Glu as well as the release of endogenous 5-HT (11). Glu and GABA (as detected by HPLC) induced by veratridine depolarization either under external Ca^{2+}-free conditions or in the presence of external Ca^{2+} (10). Sertraline also inhibits insulin-induced Tyr phosphorylation of insulin receptor substrate (IRS)-2 protein and the activation of its downstream targets Akt and the ribosomal protein S6 kinase-1 (S6K1) (12). Inhibition was dose-dependent with half-maximal effects at ~15-20 µM. It correlated with a rapid dephosphorylation and activation of the IRS kinase GSK3β. **Drug Interactions:** When assayed in the presence of NADPH, incubation of human liver microsomes with sertraline markedly decreases testosterone 6β-hydroxylation activities, indicating that sertraline metabolism leads to CYP3A4 inactivation (13). This inactivation required NADPH and was not protected by glutathione. No significant inactivation was observed for other P450 enzymes. Spectroscopic evaluation revealed that microsomes with and without sertraline in the presence of NADPH gave a Soret peak at 455 nm, suggesting formation of complexes of sertraline metabolite(s) with reduced form of P450 (13). No significant inactivation was observed with other P450 enzymes. **1**. Koe, Weissman, Welch & Browne (1983) *J. Pharmacol. Exp. Ther.* **226**, 686. **2**. Bogetto, Albert & Maina (2002) *Eur. Neuropsychopharmacol.* **12**, 181. **3**. Schwartz & Rothbaum (2002) *Expert Opin. Pharmacother.* **3**, 1489. **4**. O'Reardon, Peshek & Allison (2005) *CNS Drugs* **19**, 997. **5**. Stunkard, Allison & O'Reardon (2005) *Appetite* **45**, 182. **6**. O'Reardon, Stunkard & Allison (2004) *Int. J. Eat. Disord.* **35**, 16. **7**. O'Reardon, Allison, Martino, *et al.* (2006) *Amer. J. Psychiatry* **163**, 893. **8**. Ohno, Hibino, Lossin, Inanobe & Kurachi (2007) *Brain Res.* **1178**, 44. **9**. Takahashi, Miyoshi, Otomo, *et al.* (2010) *Biochem. Biophys. Res. Commun.* **391**, 382. **10.** Li, Couch, Higuchi, Fang & Guo (2012) *Toxicol Sci.* **127**, 582. **11**. Aldana & Sitges ()*J Neurochem.* **12**, 197. **12**. Isaac R, Boura-Halfon S, Gurevitch, *et al.* (2013) *J. Biol. Chem.* **288**, 5682. **13**. Masubuchi & Kawaguchi (2013) *Biopharm. Drug Dispos.* **34**, 423.

Sertindole

This phenylindole-class antipsychotic (FW = 440.94 g/mol; CAS 106516-24-9), also known by thetrade names Serdolect® and Serlect® as well as 1-[2-[4-[5-chloro-1-(4-fluorophenyl)indol-3-yl]-1-piperidyl]ethyl]imidazole-idin-2-one, is active at both dopamine and serotonin receptors and is used to treat schizophrenia. (*See also Quetiapine; Remoxipride; Risperidone; Zotepine*) **Targets:** Dopamine D$_2$ Receptor, 2.4 nM; Dopamine D$_3$ Receptor, 2.3 nM; Dopamine D$_2$ Receptor, 4.9 nM; Serotonin 5-HT$_{1A}$ Receptor, K_i = 280 nM; Serotonin 5-HT$_{1B}$ Receptor, K_i = 60 nM; Serotonin 5-HT$_{1D}$ Receptor, K_i = 96 nM; Serotonin 5-HT$_{1E}$ Receptor, K_i = 430 nM; Serotonin 5-HT$_{1F}$

Receptor, K_i = 360 nM; Serotonin 5-H_{2A} Receptor, K_i = 0.4 nM; Serotonin 5-HT_{2C} Receptor, K_i = 0.9 nM; Serotonin 5-HT_6 Receptor, K_i = 5.4 nM; Serotonin 5-HT_7 Receptor, K_i = 28 nM. **1**. Muscatello, Bruno, Micali-Bellinghieri, Pandolfo & Zoccali (2014) *Expert Opin. Pharmacother.* **15**, 1943.

L-Seryl-L-alanyl-L-alanyl-(5(S)-amino-4(S)-hydroxy-6-(4'-hydroxyphenyl)hexanoyl)-L-valyl-L-valine Methyl Ester

This hydroxyethylene isostere-containing peptidomimetic (FW$_{free-base}$ = 680.80 g/mol) inhibits HIV-1 protease, *or* HIV-1 retropepsin, apparent K_i = 79 nM. This peptide prevents cleavage of the viral polyproteins resulting in the formation of immature non-infectious viral particles. Protease inhibitors are almost always used in combination with at least two other anti-HIV drugs. **1**. Dreyer, Metcalf, Tomaszek, Jr., *et al.* (1989) *Proc. Natl. Acad. Sci. U.S.A.* **86**, 9752.

L-Seryl-L-alanyl-L-alanyl-(5(S)-amino-4(S)-hydroxy-6-phenylhexanoyl)-L-valyl-L-valine Methyl Ester

This hydroxyethylene isostere-containing peptidomimetic (FW$_{free-base}$ = 664.80 g/mol) inhibits HIV-1 protease, *or* HIV-1 retropepsin, apparent K_i = 62 nM. This peptide prevents cleavage of the viral polyproteins resulting in the formation of immature non-infectious viral particles. Protease inhibitors are almost always used in combination with at least two other anti-HIV drugs. **1**. Dreyer, Metcalf, Tomaszek, Jr., *et al.* (1989) *Proc. Natl. Acad. Sci. U.S.A.* **86**, 9752.

L-Seryl-L-alanyl-L-alanyl-(4(S)-amino-3(S)-hydroxy-5-phenylpentanoyl)-L-valyl-L-valine Methyl Ester

This pseudopeptide bond-containing peptide analogue (FW$_{free-base}$ = 650.77 g/mol) inhibits HIV-1 protease, *or* HIV-1 retropepsin, apparent K_i = 0.81 µM. This peptide prevents cleavage of the viral polyproteins resulting in the formation of immature non-infectious viral particles. Protease inhibitors are almost always used in combination with at least two other anti-HIV drugs. **1**. Dreyer, Metcalf, Tomaszek, *et al.* (1989) *Proc. Natl. Acad. Sci. U.S.A.* **86**, 9752.

Sesamin

This water-soluble natural product (FW = 354.36 g/mol; CAS 607-80-7), known systematically as 5,5'-(tetrahydro-1*H*,3*H*-furo[3,4-*c*]furan-1,4-

diyl)bis-1,3-benzodioxole, is isolated from sesame oil, the barks of species of *Fagara*, and from fruit of *Piper lowong*. Sesamin is a noncompetitive inhibitor of fatty acid Δ^5-desaturase (1-3), lysophosphatidylcholine acyltransferase (4), and tocopherol ω-hydroxylase, *or* CYP4F2 (5). **1**. Shimizu, Akimoto, Shinmen, *et al.* (1991) *Lipids* **26**, 512. **2**. Umeda-Sawada, Takahashi & Igarashi (1995) *Biosci. Biotechnol. Biochem.* **59**, 2268. **3**. Fujiyama-Fujiwara, Umeda & Igarashi (1992) *J. Nutr. Sci. Vitaminol. (Tokyo)* **38**, 353. **4**. Chatrattanakunchai, Fraser & Stobart (2000) *Biochem. Soc. Trans.* **28**, 718. **5**. Sontag & Parker (2002) *J. Biol. Chem.* **277**, 25290.

Sespendole

This fungal indole (FW = 517.75 g/mol), isolated from *Pseudobotrytis terrestris*, inhibits sterol *O*-acyltransferase, *or* cholesterol *O*-acyltransferase, *or* ACAT (1). Sespendole also inhibits inhibitor of lipid droplet formation in macrophages by inhibiting the synthesis of cholesteryl ester and triacylglycerol with IC$_{50}$ values of 4.0 and 3.2 µM, respectively. **1**. Ohshiro, Rudel, Omura & Tomoda (2007) *J. Antibiot.* **60**, 43.

Sethoxydim

This selective post-emergence herbicide (FW = 326.48 g/mol; CAS 74051-80-2), also known as 2-[1-(ethoxyimino)butyl]-5-[2-(ethylthio)propyl]-3-hydroxy-2-cyclohexen-1-one and Poast (FW = 327.49 g/mol), is a potent inhibitor of acetyl-CoA carboxylase (1-8), waith an IC$_{50}$ value of 0.96 µM for the wheat enzyme. Note that commercial preparations often are the (*RS*),(*EZ*)-mixture. Sethoxydim is often used to control annual and perennial grass weeds. **Target(s):** acetyl-CoA carboxylase (1-8); and 4-hydroxyphenylpyruvate dioxygenase (9). **1**. Lin & Yang (1999) *Bioorg. Med. Chem. Lett.* **9**, 551. **2**. Burton, Gronwald, Somers, *et al.* (1987) *Biochem. Biophys. Res. Commun.* **148**, 1039. **3**. Harwood (1988) *Trends Biochem. Sci.* **13**, 330. **4**. Dotray, DiTomaso, Gronwald, Wyse & Kochian (1993) *Plant Physiol.* **103**, 919. **5**. Egli, Gengenbach, Gronwald, Somers & Wyse (1993) *Plant Physiol.* **101**, 499. **6**. Rendina, Craig-Kennard, Beaudoin & Breen (1990) *J. Agric. Food Chem.* **38**, 1282. **7**. Rendina & Felts (1988) *Plant Physiol.* **86**, 983. **8**. Zagnitko, Jelenska, Tevadze, Haselkorn & Gornicki (2001) *Proc. Natl. Acad. Sci. U.S.A.* **98**, 6617. **9**. Lin & Yang (1999) *Bioorg. Med. Chem. Lett.* **9**, 551.

Setipiprant

This potent CRTH2 antagonist (FW = 402.43 g/mol; CAS 866460-33-5), also named ACT-129968 and 2-(2-(1-naphthoyl)-8-fluoro-3,4-dihydro-1*H*-pyrido[4,3-*b*]indol-5(2*H*)-yl)acetic acid, targets Chemoattractant Receptor on Th-2 cells (CRTH2), a G-coupled, prostaglandin D_2 (PGD$_2$) receptor with important roles in allergic inflammation. **Mode of Inhibitory Action:** T helper type 2 cells (*or* Th2 cells) are a distinct lineage of CD4$^+$ effector T cells that secrete interleukins (*e.g.*, IL-4, IL-5, IL-9, IL-13, and IL-17E/IL-

25) and coordinate immune responses to large extracellular pathogens. Produced by the mast cells, PGD$_2$ is a key mediator in various inflammatory diseases, including allergy and asthma. PGD$_2$ binds CRTH2 on the surface of blood-borne cells, inducing chemotaxis in Th2 cells, basophils, and eosinophils, and stimulating cytokine release from these cells. Pre-clinical data showed that setipiprant potently inhibits migration of eosinophils towards PGD$_2$ in vitro as well as in an in vivo rat model of lung eosinophilia. **1**. Baldoni, Mackie, Gutierrez, Theodor & Dingemanse (2013) *Clin. Ther.* **35**, 1842. **2**. Satoh, Moroi, Aritake, Urade, *et al.* (2006) *J. Immunol.* 177, 2621. **3**. Diamant, Singh, O'Connor, et al. (2012) *Am. J. Respir. Crit. Care Med.* **185**, A3957.

Setoperone

This serotonin antagonist (FW = 401.50 g/mol; CAS 86487-64-1), also known as 6-[2-[4-(4-fluorobenzoyl)piperidin-1-yl]ethyl]-7-methyl-2,3-dihydro[1,3]thiazolo[3,2-*a*]pyrimidin-5-one, 5-HT$_{2A}$ receptor is pharmacologically characterized by its potent serotonin and moderate dopamine receptor blocking properties. Blockade of serotonin receptors may be related to improvement of autistic behaviour, dysphoria, and parkinson-like symptoms. **See also** *Ketanserin; Pimavanserin; RH-34; Risperidone; Ritanserin; Volinanserin*. **1**. Ceulemans, Gelders, Hoppenbrouwers, Reyntjens & Janssen (1985) *Psychopharmacology* (Berlin) **85**, 329.

Sevin, *See Carbaryl*

Sevoflurane

This sweet-smelling inhalation anesthetic (FW = 200.06 g/mol; CAS 28523-86-6), also known as 1,1,1,3,3,3-hexafluoro-2-(fluoromethoxy)-propane, is a nonflammable, highly fluoroalkyl ether that induces and maintains general anesthesia. Along with desflurane, sevoflurane is replacing isoflurane and halothane in modern anesthesiology, and is frequently co-administered with nitrous oxide and oxygen. With an oil/gas partition coefficient of 47.2, the Minimum Alveolar Concentration (MAC) of sevoflurane is 2.05%. Its potency is considerably lower than that of halothane and isoflurane, but is about three times more potent than desflurane (1). About 2-5% of the drug taken up by the liver is metabolized. Demonstrating less irritation to mucous membranes, sevoflurane is now preferred over desflurane. **1**. Behne, Wilke & Harder (1999) *Clin. Pharmacokinet.* **36**, 13.

SF2809 Compounds

These indole derivatives: SF2809-I (FW = 387.46 g/mol), R$_1$ = –H and R$_2$ = –NHCOCH$_3$), SF2809-II (FW = 362.55 g/mol), R$_1$ = –C$_6$H$_4$(*p*-OH) and R$_2$ = –NHCOCH$_3$); SF2809-III (FW = 346.41 g/mol), R$_1$ = –H and R$_2$ = –OH); SF2809-IV (FW = 438.50 g/mol, R$_1$ = –C$_6$H$_4$(*p*-OH) and R$_2$ = –OH); SF2809-V (FW = 463.56 g/mol, R$_1$ = –C$_6$H$_5$ and R$_2$ = –NHCOCH$_3$); and SF2809-VI (FW = 422.50 g/mol, where R$_1$ = –C$_6$H$_5$ and R$_2$ = –OH)) from the fermentation broth of *Dactylosporangium* sp. strain SF2809, inhibit chymase, with IC$_{50}$ values of 7.3, 0.041, 2.1, 0.081, 0.043, and 0.014 mM

for SF2809-I, -II, -III, -IV, -V, and -VI, respectively. **1**. Tani, Gyobu, Sasaki, *et al.* (2004) *J. Antibiot.* **57**, 83.

SF6847, *See* *3,5-Di-tert-butyl-4-hydroxybenzylidenemalononitrile*

SGC707

This potent, selective, and cell-active PRMT3 inhibitor (FW = 298.34 g/mol; CAS 1687736-54-4; Solubility: 59 mg/mL in DMSO), also named *N*-6-isoquinolinyl-*N*'-[2-oxo-2-(1-pyrrolidinyl)ethyl]urea, targets protein arginine methyltransferase 3 (IC$_{50}$ = 31 nM; K_d = 53 nM). SGC707 exhibits outstanding selectivity *versus* 31 other methyltransferases and >250 non-epigenetic targets. Mechanism of action studies and crystal structure of the PRMT3-SGC707 complex confirm its allosteric inhibition mode. SGC707 engages PRMT3 and potently inhibits its methyltransferase activity in cells. It is also bioavailable and suitable for animal studies, making it an excellent tool to investigate the role of PRMT3 in health and disease. **1**. Kaniskan, Szewczyk, Yu, *et al.* (2015) *Angew. Chem. Int. Ed. Engl.* **54**, 5166.

SGI-110

This 5-aza-2'-dC-containing demethylating pro-drug (FW = 579.39 g/mol; CAS 929901-49-5), also known as S-110 and 2'-deoxy-5-azacytidylyl-(3'→5')-2'-deoxyguanosine, is a novel DNA hypomethylating dinucleotide antimetabolite, consisting of a decitabine (*See Decitabine*) joined via a phosphodiester bond to 2'-deoxyguanosine (1-5). SGI-110 is also a "chemosensitizer", making cancer cells more susceptible to the action of other anticancer agents like Cisplatin, especially those cells that are Cisplatin-resistant. For example, in a mouse xenoplant model for human ovarian cancer, DNA damage induced by Cisplatin was increased by SGI-110, as measured by ICP-mass spectrometry analysis of DNA adduct formation and repair of Cisplatin-induced damage (6). **Mechanism of Inhibitory Action:** DNA methyltransferases (DNMTs) are a family of nuclear enzymes catalyzing methylation of CpG dinucleotides, resulting in epigenetic methylomes distinguishing normal cells from those observed in diseases, such as cancer. SGI-110 is a potential anticancer agent that, following metabolic phosphorylation and incorporation into DNA, inhibits DNA methyltransferase, thereby causing genome-wide and non-specific hypomethylation, inducing cell cycle arrest at S-phase. Significantly, SGI-110 is resistant to cytidine deaminase, resulting in gradual decitabine release, both extracellularly and intracellularly, to provide prolonged decitabine exposure to susceptible cells. SGI-110 treatment results in the up-regulation of Melanoma Antigen family A members (*e.g.*, MAGE-A1, MAGE-A2, MAGE-A3, MAGE-A4, MAGE-A10) and G-antigen family members (*e.g.*, GAGE 1-2, GAGE 1-6), as well as NY-ESO-1 and SSX 1-5 in all cancer cell lines studied, both at mRNA and at protein levels. (*Note:* The chemical nature of the pharmacologically active form(s) remain uncertain, and SGI-110 may first undergo hydrolysis to generate the well known DNMT inhibitor 5-aza-2'-deoxycytidine, *or* Decitabine). **1**. Foulks, Parnell, Ni, *et al.* (2012) J. Biomol. Screen. **17**, 2. **2**. Foulks & Parnell, Nix, *et al.* (2012) *J. Biomol. Screen.* **17**, 2. **3**. Coral, Parisi, Nicolay, *et al.* (2013) *Cancer Immunol. Immunother.* 62, 605. **4**. Singh, Sharma, Capalash (2013) *Curr. Cancer Drug Targets* 13, 379. **5**. Tellez, Grimes, Picchi, *et al.* (2014) *Int. J. Cancer.* **135**, 2223. **6**. Fang, Munck, Tang, *et al.* (2014) *Clin Cancer Res.* **20**, 6504.

SGI-1027

This DNMT inhibitor (FW = 461.52 g/mol; CAS 1020149-73-8; Solubility: 92 mg/mL DMSO; <1 mg/mL H_2O), also known as N-[4-[(2-amino-6-methyl-4-pyrimidinyl)amino]phenyl]-4-(4-quinolinylamino)benzamide, targets DNA (cytosine-5)-methyltransferases DNMT1 (IC_{50} = 6 µM), DNMT3A (IC_{50} = 8 µM), and DNMT3B (IC_{50} = 7.5 µM), enzymes that help establish and regulate tissue-specific DNA methylation patterns, thereby maintaining tissue integrity and preventing both tumorigenesis and developmental abnormalities. In cancer, tumor suppressor genes are silenced by oncomethylation, and reactivation of them occurs when 5-azacytidine and 5-aza-2'-deoxycytidine selectively and rapidly induce proteasome-mediated degradation of maintenance DNA methyltransferase-1. Treatment of various cancer cell lines with SGI-1027 results in selective degradation of DNMT1, with minimal or no effects on DNMT3A and DNMT3B (1). At a concentration of 2.5-5 µM (similar range as that of decitabine), complete degradation of DNMT1 protein is achieved within 24 hours, without significantly affecting its mRNA level. MG132 blocked SGI-1027-induced depletion of DNMT1, indicating the involvement of proteasomal pathway (1). SGI-1027 4 can now be synthesized using as palladium-catalyzed Ar–N bond formation reactions carried out in a sequential or convergent manner (2). **1**. Datta, Ghoshal, Denny, *et al.* (2009) *Cancer Res.* **69**, 4277. **2**. García-Domínguez, Dell'aversana, Alvarez, Altucci & de Lera (2013) *Bioorg. Med. Chem. Lett.* **23**, 1631.

SGI-1776

This ATP-competitive protein kinase inhibitor (FW = 405.42 g/mol; CAS 1025065-69-3; Solubility = 80 mg/mL DMSO, <1 mg/mL H_2O), also known as N-((1-methylpiperidin-4-yl)methyl)-3-(3-(trifluoromethoxy)-phenyl)imidazo[1,2-*b*]pyridazin-6-amine, targets the proto-oncogene serine/threonine-protein kinase Pim1 (IC_{50} = 7 nM), Pim2 (IC_{50} = 363 nM), Pim3 (IC_{50} = 69 nM), and the Fms-like tyrosine kinase 3, *or* Flt3 (IC_{50} = 44 nM). Expressed mainly in spleen, thymus, bone marrow, prostate, oral epithelial, hippocampus and fetal liver cells, Pim-1 is present at high levels in human tumor cell cultures. SGI-1776 induces apoptosis in chronic lymphocytic leukemia (CLL) cells (1). Unlike in replicating cells, phosphorylation of traditional Pim-1 kinase targets, phospho-Bad (Ser112) and histone H3 (Ser10), and cell-cycle proteins were unaffected by SGI-1776, suggesting an alternative mechanism in CLL. Levels of antiapoptotic proteins Bcl-2, Bcl-X(L), XIAP, and proapoptotic Bak and Bax were unchanged; however, a significant reduction in Mcl-1 was observed that was not caused by caspase-mediated cleavage of Mcl-1 protein. Later work showed that PIM-3 kinase is a positive regulator of STAT3 signaling, suggest that PIM-3 inhibitors like SGI-1776 cause growth inhibition of cancer cells by down-regulating pSTAT3(Tyr705) expression. Treatment of acute myeloid leukemia (AML) cells with SGI-1776 results in a concentration-dependent induction of apoptosis, again significantly reducing Mcl-1 expression (3). SGI-1776 decreases cell surface expression of P-glycoprotein (ABCB1) and breast cancer resistance protein (ABCG2) and drug transport by Pim-1-dependent and -independent mechanisms (4). By contrast, SGI-1776 treatment in myeloma cell lines and CD138+ myeloma cells elicits its deleterious effects through inhibition of translation and induction of autophagy (5). **1**. Chen, Redkar, Bearss, Wierda & Gandhi (2009) *Blood* **114**, 4150. **2**. Chang, Kanwar, Feng, *et al.* (2010) *Mol. Cancer Ther.* **9**, 2478. **3**. Chen, Redkar, Taverna, Cortes & Gandhi (2011) *Blood* **118**, 693. **4**. Natarajan, Bhullar, Shukla *et al.* (2013) *Biochem Pharmacol.* **85**, 514. **5**. Cervantes-Gomez, Chen, Orlowski & Gandhi (2013) *Clin. Lymphoma Myeloma Leuk.* Suppl 2, S317.

S/GSK1349572, *See* Dolutegravir

Shermilamine C

This pyridoacridine alkaloid (FW = 430.53 g/mol) from two marine species of *Cystodytes* and *Trididemnum* inhibits DNA topoisomerase II. These results suggest that disruption of the function of TOPO II, subsequent to intercalation, is a probable mechanism by which pyridoacridines inhibit the proliferation of HCT cells by disrupting DNA and RNA synthesis with little effect on protein biosynthesis. DNA is the primary cellular target of the pyridoacridine alkaloids, a behavior that is consistent with the action of DNA intercalators. **1**. McDonald, Eldredge, Barrows & Ireland (1994) *J. Med. Chem.* **37**, 3819.

Shiga Toxins

This family of bacterial cytotoxins Stx1 and Stx2 are produced by the bacteria *Shigella dysenteriae* and the Shigatoxigenic group of *Escherichia coli* (STEC), respectively. The structurally related toxin has five β-subunits (MW = 7.7 kDa) and one α-subunit (MW = 32 kDa), linked together by a disulfide bond. The β-subunits combine to form a pentameric ring that binds binding with the globotriaosylceramide receptor (Galα(1→4)Galβ(1→4)-glucosyl ceramide (*or* Gb3) in the peripheral membranes of target cells. (A variant binds to the globotetraosylceramide receptor (*or* Gb4), accounting for its ability to target other cell types.) The complex enters by way of clarthrin-coated cavities and is taken up by the endoplasmic reticula (ER) via retrograde transport. Once within the ER, the disulfide bond releasing the α-subunit, an active N-glycosidase that catalyzes ricin-like depurination of 28S rRNA of the 60S subunit of host cell ribosomes, and, in the process, releasing adenine and inactivating the modified ribosome. *Escherichia coli* strains that produce Stx2 are isolated from hemolytic-uremic syndrome (HUS) cases more frequently than are strains that produce both Stx1 and Stx2, whereas strains that produce only Stx1 are rarely isolated from HUS cases. Studies have implicated Stx2 as the sole contributor to acute kidney failure and other systemic complications in humans. The most common Shiga toxin-producing organism causing disease in North America is *E. coli* O157:H7, accounting for nearly 90% of all diarrhea-associated HUS cases.

Shikimate (Shikimic Acid)

This metabolite (FW = 174.15 g/mol; CAS 138-59-0; λ_{max} = 213 nm, ε = 8900 $M^{-1}cm^{-1}$ in ethanol) is a key intermediate in the biosynthesis of tyrosine, phenylalanine, tryptophan, folic acid, the tocopherols, ubiquinone, and certain antibiotics. Known systematically as (3R,4S,5R)-3,4,5-trihydroxy-1-cyclohexene-1-carboxylic acid, shikimate inhibits chitin synthase (1) and 3-deoxy-7-phosphoheptulonate synthase (2). **1**. Hwang, Ahn, Lee, *et al.* (2001) *Planta Med.* **67**, 501. **2**. Wu, Sheflyan & Woodard (2005) *Biochem. J.* **390**, 583.

Shikimate-5-enolpyruvate 3-Phosphate, *See* 5-O-(1-Carboxyvinyl)-3-phosphoshikimate

Shikonin

This naphthoquinone (FW = 288.30 g/mol; CAS 517-89-5), also known as (+)-5,8-dihydroxy-2-[(1R)-1-hydroxy-4-methyl-3-pentenyl]-1,4-naphthalenedione, from the Chinese herb Shiunko exhibits significant antibacterial, anti-inflammatory, and antitumor activities (1). Shikonin inhibits TNF-α-induced and B-16 melanoma-induced angiogenesis in mice (2). It also blocks the expression of integrin avb3 and inhibits endothelial cell proliferation and migration *in vitro*. Shikonin will induce apoptosis in HL60 human promyelocytic leukemia cell line (3). **Target(s):** alkaline phosphatase (4); chemokine binding to CC chemokine receptor-1 (5); DNA topoisomerase I (6); leukotriene B4 biosynthesis (7); photosynthetic electron transfer (8); ribonuclease H (9); sterol O-acyltransferase, *or* ACAT (10); and telomerase (11). **1**. Li, Luo & Zhou (1999) *Phytother. Res.* **13**, 236. **2**. Hisa, Kimura, Takada, Suzuki & Takigawa (1998) *Anticancer Res.* **18**, 783. **3**. Yoon, Kim, Lim, Jeon & Sung (1999) *Planta Med.* **65**, 532. **4**. Gorgees, Naqvi, Rashan & Zakaria (1978) *Folia Histochem. Cytochem. (Krakow)* **16**, 51. **5**. Chen, Oppenheim & Howard (2001) *Int. Immunopharmacol.* **1**, 229. **6**. Plyta, Li, Papageorgiou, *et al.* (1998) *Bioorg. Med. Chem. Lett.* **8**, 3385. **7**. Wang, Bai, Liu, Xue & Zhu (1994) *Yao Xue Xue Bao* **29**, 161. **8**. Frigaard, Tokita & Matsuura (1999) *Biochim. Biophys. Acta* **1413**, 108. **9**. Min, Miyashiro & Hattori (2002) *Phytother. Res.* **16** Suppl. 1, S57. **10**. An, Park, Paik, Jeong & Lee (2007) *Bioorg. Med. Chem. Lett.* **17**, 1112. **11**. Lu, Liu, Ding, Cai & Duan (2002) *Bioorg. Med. Chem. Lett.* **12**, 1375.

SHK-186, *See* Dalazatide

ShK-Dap22

This semi-synthetic potassium channel toxin and immunosuppressive polypeptide (MW = 4.1 kDa), also named Potassium Channel Toxin κ-Stichotoxin-She1a, targets $K_v1.3$, the voltage-gated potassium channel in T lymphocytes, an important molecular target for immunosuppressive agents. It is prepared from the parent polypeptide ShK (from the sea anemone *Stichodactyla helianthus*) by site-directed substitution of the non-natural amino acid diaminopropionic acid in place of lysine at the critical residue-22. ShK-Dap22 potently inhibits $K_v1.3$ ($K_d = 11$ pM) and also blocks $K_v1.1$, $K_v1.4$, and $K_v1.6$ at sub-nM concentrations. At sub-nM concentrations, ShK-Dap22 also suppresses anti-CD3-induced human T-lymphocyte [³H]thymidine incorporation *in vitro*. Its toxicity is low in a rodent model (median paralytic i.v. dose ≈ 200 mg/kg body weight). **1**. Kalman, Pennington, Lanigan, *et al.* (1998) *J. Biol. Chem.* **273**, 32697.

Showdomycin

This water-soluble nucleoside-like antibiotic (FW = 229.19 g/mol; CAS 16755-07-0), also known as 2-(β-D-ribofuranosyl)maleimide, from *Streptomyces showdoensis* is an structurally related to uridine and pseudouridine. The double-bond in the maleimide moiety of showdomycin reacts with catalytically essential sulfhydryl groups, therby inactivating target enzymes. **Target(s):** N-acetylglucosaminyldiphosphodolichol N-acetylglucos-aminyltransferase (1-3); adenosine transport, Na⁺-linked (4); amino acid transport (5); DNA-directed RNA polymerase (6); dolichyl-phosphate β-glucosyltransferase (7-10); dolichyl-phosphate β-D-

mannosyltransferase (11); 1,3-β-glucan synthase (12); GTP cyclohydrolase II (13); Na⁺/K⁺-exchanging ATPase (14,15); poly(ADPribosyl) polymerase, *or* NAD⁺ ADP-ribosyltransferase (16); protein glycosylation (7-9,17); sugar transport (5); thymidylate synthase (18,19); UDP-N-acetylglucosamine:dolichyl-phosphate N-acetylglucosaminephosphotransferase (20); UDP-glucose dehydrogenase (8,9); and UDP-glucose 4-epimerase (21). **1**. Kean & Wei (1998) *Glycoconjugate J.* **15**, 405. **2**. Kean & Niu (1998) *Glycoconjugate J.* **15**, 11. **3**. Kean, Wei, Anderson, Zhang & Sayre (1999) *J. Biol. Chem.* **274**, 34072. **4**. Dagnino, Bennett, Jr., & Paterson (1991) *J. Biol. Chem.* **266**, 6312. **5**. Roy-Burman, Huang & Visser (1971) *Biochem. Biophys. Res. Commun.* **42**, 445. **6**. Maryanka & Johnston (1970) *FEBS Lett.* **7**, 125. **7**. Schwarz & Datema (1982) *Meth. Enzymol.* **83**, 432. **8**. Elbein (1983) *Meth. Enzymol.* **98**, 135. **9**. Elbein (1987) *Meth. Enzymol.* **138**, 661. **10**. Kang, Spencer & Elbein (1979) *J. Biol. Chem.* **254**, 10037. **11**. Kean (1983) *Biochim. Biophys. Acta* **752**, 488. **12**. Cabib & Kang (1987) *Meth. Enzymol.* **138**, 637. **13**. Elstner & Suhadolnik (1975) *Meth. Enzymol.* **43**, 515. **14**. Tobin & Akera (1975) *Biochim. Biophys. Acta* **389**, 126. **15**. Hara, Hara, Nakao & Nakao (1981) *Biochim. Biophys. Acta* **644**, 53. **16**. Muller & Zahn (1975) *Experientia* **31**, 1014. **17**. Elbein (1984) *CRC Crit. Rev. Biochem.* **16**, 21. **18**. Kalman (1972) *Biochem. Biophys. Res. Commun.* **49**, 1007. **19**. Danenberg (1977) *Biochim. Biophys. Acta* **473**, 73. **20**. Kaushal & Elbein (1986) *Plant Physiol.* **82**, 748. **21**. Geren, Geren & Ebner (1977) *J. Biol. Chem.* **252**, 2089.

SHP099

This selective and orally bioavailable small-molecule SHP2 inhibitor ($FW_{free-base}$ = 352.26 g/mol; $FW_{HCl-Salt}$ = 388.72 g/mol; CAS 1801747-11-4 (HCl salt) and 801747-42-1 (Free Base); Soluble in DMSO; IUPAC: 6-(4-amino-4-methylpiperidin-1-yl)-3-(2,3-dichlorophenyl)pyrazin-2-amine) targets the non-receptor protein tyrosine phosphatase SHP2 (IC_{50} = 71 nM), also known as Protein Tyrosine-Phosphatase Non-receptor type 11 (PTPN11), Protein-Tyrosine Phosphatase 1D (PTP-1D) and Protein-Tyrosine Phosphatase 2C (PTP-2C). Indeed, SHP2 is ubiquitously expressed and regulates cell survival and proliferation primarily through activation of the RAS–ERK signalling pathway. This enzyme plays important roles in signal transduction occurring downstream of growth factor receptor signaling. SHP2 was the first reported oncogenic tyrosine phosphatase, and it is also a key mediator in the Programmed cell Death-1 (PD-1) and B- and T-Lymphocyte Attenuator (BTLA) immune-checkpoint pathways. **Mode of Inhibitory Action:** SHP099 stabilizes SHP2 in an auto-inhibited conformation by binding concurrently to the interface of the N-terminal SH2, C-terminal SH2, and protein tyrosine phosphatase domains, thereby allosterically inhibiting SHP2's phosphatase activity. It also suppresses RAS–ERK signaling, thereby inhibiting proliferation of receptor-tyrosine-kinase-driven human cancer cells *in vitro* and in mouse tumor xenograft models, the latter achieved after a single oral dose (100 mg per kg) of SHP099. **1**. Chen, LaMarche, Chane, *et al.* (2016) *Nature* **535**, 148.

Siastatin B

This *Streptomyces*-derived piperidine (FW = 218.21 g/mol; CAS 54795-58-3), known systematically as (3S,4S,5R,6R)-6-(acetylamino)-4,5-dihydroxy-

3-piperidinecarboxylic acid, is a broad-spectrum inhibitor of bacterial exo-α-sialidase, *or* neuraminidase (1-3). The charge distribution in the siastatin B zwitterion is thought to resemble that in the *N*-acetylneuraminate oxocarbenium ion, a proposed intermediate in the sialidase-catalyzed reaction. **1**. Miyagi, Hata, Hasegawa & Aoyagi (1993) *Glycoconj. J.* **10**, 45. **2**. Umezawa, Aoyagi, Komiyama, Morishima & Hamada (1974) *J. Antibiot.* **27**, 963. **3**. Knapp & Zhao (2000) *Org. Lett.* **2**, 4037.

Sibutramine

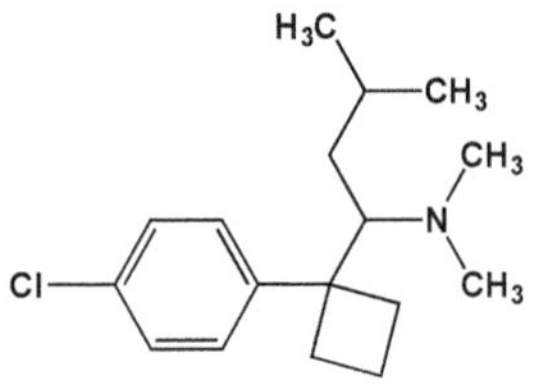

This centrally-acting neurotransmitter reuptake inhibitor and anorectic agent (FW = 279.85 g/mol; CAS 106650-56-0), also known by the trade names Reductil®, Meridia®, and Sibutrex®, as well as the systematic name (±)-dimethyl-1-[1-(4-chlorophenyl)cyclobutyl]-*N*,*N*,3-trimethylbutan-1-amine, is a β-phenethylamine that targets the serotonin transporter (K_i = 0.3 μM), norepinephrine transporter (K_i = 5.4 μM), and dopamine transporter (K_i = 0.94 μM), thereby increasing their levels within synaptic clefts and enhancing satiety. Sibutramine is metabolized by CYP3A4 into two pharmacologically-active primary and secondary amines (called active metabolites 1 and 2) with half-lives of 14 and 16 hours, respectively. Although sibutramine had been widely marketed, an associated high risk of cardiovascular events, strokes, and death led to its withdrawal in many countries. **1**. Araújo, Silva, Cruz & Falcão-Reis (2014) *J. Glaucoma* **23**, 415.

Sideromycins

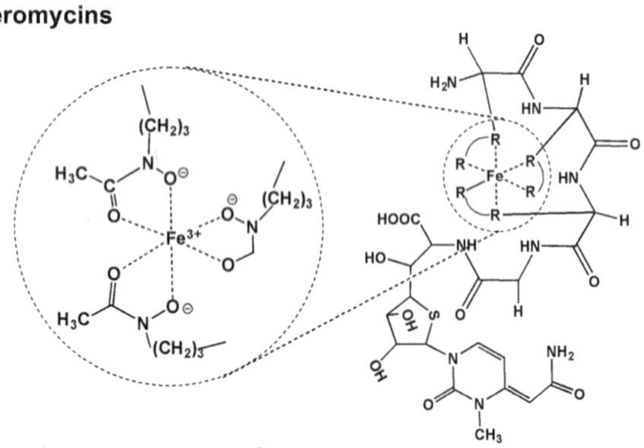

These naturally occurring Fe^{3+}-siderophores and antibacterial agents are covalently linked to an antibiotic moiety, the former accounting for their efficient cellular uptake and the latter explaining their substantial cytotoxic/cytostatic antibacterial action (1). Shown above is Albomycin δ₂. Siderophores and sideromycins compete for common uptake systems, which transport of Fe^{3+}-siderophores across the outer membrane, driven by the cytoplasmic membrane's proton-motive force and transduced by the Ton protein complex, consisting of TonB, ExbB, and ExbD proteins. As a consequence, their minimal inhibitory concentration is at least 100-fold lower than that of antibiotics entering cells by diffusion. One such sideromycin is albomycin (the hexacoordinate, Fe^{3+}-hydroxamate-containing complex shown above), which consists of a siderophore that is structurally similar to ferrichrome and a thioribosyl pyrimidine antibiotic. 1. Braun, Pramanik, Gwinner, Köberle & Bohn (2009) *Biometals* **22**, 3.

Siguazodan, *See SKF 94836*

Silencing RNA, *See Small Interfering RNA*

Sildenafil

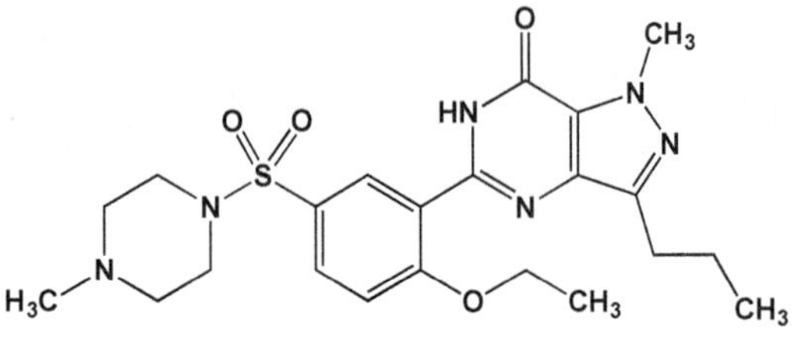

This prototypical erectile dysfunction (ED) drug ($FW_{free-base}$ = 474.58 g/mol; CAS 139755-83-2; M.P. = 187-189°C), also named 1-[4-ethoxy-3-(6,7-dihydro-1-methyl-7-oxo-3-propyl-1*H*-pyrazolo[4,3-*d*]pyrimidin-5-yl)phenyl sulfonyl]-4-methylpiperazine, targets the type-5 isozyme of 3',5'-cyclicGMP phosphodiesterase, resulting in the sustained phosphorylation of protein targets of 3',5'-cyclicGMP-stimulated protein kinase, attended by prolonged vasodilation (1-8). **Mode of Inhibitory Action:** Both sildenafil and zaprinast are competitive inhibitors, and double-inhibition analysis shows that they bind to PDE5 in a mutually exclusive manner (2). Site-directed mutagenesis of each of 23 conserved amino acid residues in the catalytic domain of PDE5 show that the pattern of change in the IC_{50} values for sildenafil or UK-122764 were found to be similar to that for zaprinast. However, among the three inhibitors, sildenafil exhibits the most similar pattern of changes in the IC_{50} to that found for the affinity of cGMP, implying similar interactions with the catalytic domain (2). This may explain the stronger inhibitory potency of sildenafil for wild-type PDE5 (*i.e.*, sildenafil (K_i = 1 nM) > UK-122764 (K_i = 5 nM) > zaprinast (K_i = 130 nM) (2). The affinity of each inhibitor is much higher than for cGMP (K_m = 2 μM). **Key Pharmacokinetic Parameters:** *See* Appendix II in Goodman & Gilman's *THE PHARMACOLOGICAL BASIS OF THERAPEUTICS*, 12th Edition (Brunton, Chabner & Knollmann, eds.) McGraw-Hill Medical, New York (2011). Sildenafil citrate is the formulation known by the proprietary name Viagra™. It is also a potent stimulator of angiogenesis, achieved through upregulation of pro-angiogenic factors and by regulation of cGMP concentration. **1**. Gopal, Francis & Corbin (2001) *Eur. J. Biochem.* **268**, 3304. **2**. Turko, Ballard, Francis & Corbin (1999) *Mol. Pharmacol.* **56**, 124. **3**. Sung, Hwang, Jeon, *et al.* (2003) *Nature* **425**, 98. **4**. Mochida, Noto, Inoue, Yano & Kikkawa (2004) *Eur. J. Pharmacol.* **485**, 283. **5**. Zhang, Kuvelkar, Wu, *et al.* (2004) *Biochem. Pharmacol.* **68**, 867. **6**. Wang, Wu, Myers, *et al.* (2001) *Life Sci.* **68**, 1977. **7**. Day, Dow, Houslay & Davies (2005) *Biochem. J.* **388**, 333. **8**. Sopory, Kaur & Visweswariah (2004) *Cell. Signal.* **16**, 681.

Silibinin

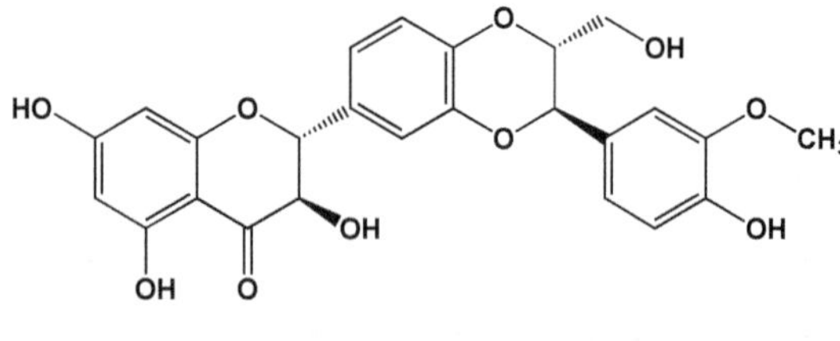

This benzodioxin (FW = 482.44 g/mol; CAS 22888-70-6), also known as silybin and named systematically as 2,3-dihydro-3-(4-hydroxy-3-methoxyphenyl)-2-(hydroxymethyl)-6-(3,5,7-trihydroxy-4-oxobenzopyran-2-yl)benzodioxin, is the principal component of silymarin isolated from the seeds of the milk thistle (*Silybum marianum*). This flavonoid possesses antioxidant, anti-inflammatory and cancer-preventive activities. **Target(s):** CYP3A4 (1,2); CYP2C9 (2); CYP2D6 (1,2); CYP2E1 (1); β-glucuronidase (3); and glucuronosyltransferase (4). Silibinin also exhibits inhibitory effects on basal and cytokine-induced expression of Secreted Phospholipase A_2 (sPLA₂) in cancer cells (5). **1**. Zuber, Modriansky, Dvorak, *et al.* (2002) *Phytother. Res.* **16**, 632. **2**. Beckmann-Knopp, Rietbrock, Weyhenmeyer, *et al.* (2000) *Pharmacol. Toxicol.* **86**, 250. **3**. Kim, Jin, Jung, Han & Kobashi (1995) *Biol. Pharm. Bull.* **18**, 1184. **4**. D'Andrea, Peréz & Sánchez Pozzi (2005) *Life Sci.* **77**, 683. **5**. Hagelgans, Nacke, Zamaraeva, Siegert & Menschikowski (2014) *Anticancer Res.* **34**, 1723.

Silodosin

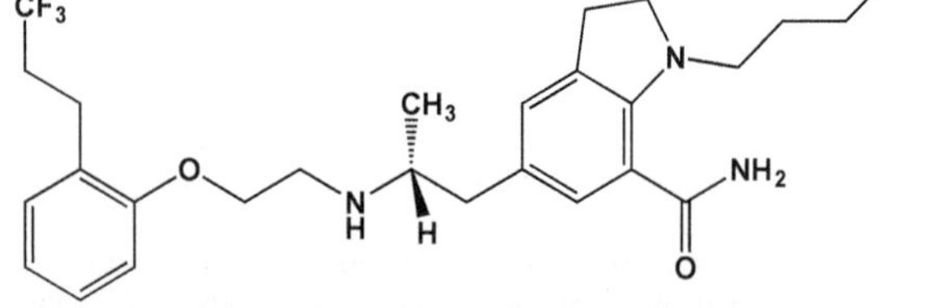

This α₁ₐ-adrenoceptor antagonist (FW = 495.53 g/mol; CAS 160970-54-7), also known by its code name KMD-3213, its trade names Rapaflo™ (in USA) and Silodyx™ (in Europe), Rapilif™ (in India), and Urief™ (in Japan) as well as its systematic name, 1-(3-hydroxypropyl)-5-[(2*R*)-({2-[2-[2-(2,2,2-trifluoroethoxy)phenoxy]ethyl}amino)propyl]indoline-7-

carboxamide) is a uroselective muscle relaxant used to treat reduced urinary flow associated with benign prostatic hyperplasia, or BPH (1,2). Silodosin binds to cloned human α_{1a}-AR (K_i = 0.036 nM), but exhibits 583x and 56x lower potency α_{1b}-AR and α_{1a}-AR, respectively. KMD-3213 also inhibits norepinephrine-induced increases in intracellular $Ca2+$ concentrations in α_{1a}-AR-expressing Chinese hamster ovary (CHO) cells, with an IC_{50} of 0.32 nM, again with much weaker effects on α_{1b}- and α_{1a}-ARs (1). **1**. Shibata, Foglar, Horie, et al (1995) Mol. Pharmacol. **48**, 250. **2**. Moriyama, Akiyama, Murata et al. (1997) Eur. J. Pharmacol. **331**, 39.

Siltuximab

This mouse/human chimeric anti-IL-6 monoclonal antibody (MW = 145.0 kDa; CAS 541502-14-1), also known as CNTO 328 and Sylvant®, binds to human interleukin-6, a multifunctional, proinflammatory cytokine produced by various cells such as T-cells, B-cells, monocytes, fibroblasts, and endothelial cells. IL-6 overproduction in activated B cells within affected lymph nodes is linked to multicentric Castleman's disease. Siltuximab is indicated for treating Castleman's patients, who are HIV-/HHV-8-negative. Treatment of therapy-resistant prostate cancer cells with Sylvant induces apoptosis and down-regulation of Myeloid cell leukemia-1 (Mcl-1) protein, an antiapoptotic member in the Bcl-2 family (1). IL-6 also stimulates inflammatory cytokine production, tumor angiogenesis, and the tumor macrophage infiltrate in ovarian cancer, actions that can be inhibited by Sylvant (2). Siltuximab also completely inhibits STAT3 tyrosine phosphorylation in H1650 cells, resulting in inhibition of lung cancer cell growth in vivo (3). **1**. Cavarretta, Neuwirt, Untergasser, et al. (2007) Oncogene **26**, 2822. **2**. Coward, Kulbe, Chakravarty, et al. (2011) Clin. Cancer Res. **17**, 6083. **3**. Song, Rawal, Nemeth & Haura (2011) Mol. Cancer Ther. **10**, 481.

Silver Ion

This well-known bactericidal agent Ag^+ (most often supplied as the soluble (2.16 g/mL at 20 °C), but nonhygroscopic mononitrate salt, $AgNO_3$; FW = 169.87 g/mol) is frequently used to stain proteins and RNA in polyacrylamide gels. Silver ions react with sulfhydryl groups, accounting for its inhibitory action on a large number of enzymes requiring free active-site thiols for catalysis. Silver ion will often replace a Cu(I) in certain copper-dependent proteins, and a Ag,Zn-superoxide dismutase has been reported. The antimicrobial properties of silver were first recognized thousands of years ago, when silver containers were found to be highly effective for preserving potable water. (**See also** Argyrol, for use in prevention of gonorrheal blindness) Another product, SILVADUR™ ET, is Dow Chemical's silver-based antimicrobial that uses a patented delivery system for transporting and binding silver to articles of clothing, thereby inhibiting the growth of microbes. This proprietary technology controls the release of the silver, thereby avoiding discoloration and reducing unnecessary loss of the bound silver. SILVADUR ™ ET may be applied to both organic and inorganic surfaces using standard coating methods such as saturation, spray, foam, printing or exhaust. Approved applications are the preservation of non-food contact coatings and films, as well as woven and nonwoven fibers commonly used in industry and homes.

Silver Sulfadiazine, See Sulfadiazine

Simazine

This pre-emergence herbicide and algacide (FW = 201.66 g/mol; CAS 122-34-9; water solubility ≈ 30 μM)), widely known as Princep™ and named systematically either as 2-chloro-4,6-bis(ethylamino)-s-triazine, 2,4-bis(ethylamino)-6-chloro-s-triazine or 6-chloro-N,N'-diethyl-1,3,5-triazine-2,4-diamine, inhibits photosynthetic electron transport. Simazine is absorbed through the roots and accumulates in the leaves and the meristem of plants. Its toxicity rquires exposure to light, and plants kept in the dark are virtually unaffected. Simazine was introduced as a herbicide in the early 1950s, but its use is now restricted to reduce ground-water contamination **Target(s)**: alcohol dehydrogenase (1); and photosynthetic electron transport (2,3). **1**. Leblova, Galociova & Cerovska (1983) Environ. Res. **30**,

389. **2**. Izawa & Good (1972) Meth. Enzymol. **24**, 355. **3**. Dodge (1977) Spec. Publ., Chem. Soc. **29**, 7.

Simeprevir

This hepatitis C proteinase inhibitor (FW = 748.95 g/mol; CAS 923604-59-5), also known as TMC435, TMC435350, Olysio® and (2R,3aR,10Z,11aS,12aR,14aR)-N-(cyclopropylsulfonyl)-2-{[2-(4-isopropyl-1,3-thiazol-2-yl)-7-methoxy-8-methyl-4-quinolinyl]oxy}-5-methyl-4,14-dioxo-2,3,3a,4,5,6,7,8,9,11a,12,13,14,14a-tetradecahydrocyclopenta[c]cyclopropa[g][1,6]diaza-cyclotetradecine-12a(1H)-carboxamide, is an FDA-approved treatment for chronic hepatitis C (Genotype 1) infection, when used in combination with pegylated-interferon α and ribavirin (two other FDA-approved treatments for chronic hepatitis C) in adult patients with compensated liver disease (including cirrhosis) who are treatment-naïve or who have failed previous interferon therapy (pegylated or non-pegylated) with or without ribavirin. Simeprevir is a potent inhibitor of HCV NS3/4A protease (K_i = 0.36 nM) and viral replication (replicon EC_{50} = 7.8 nM). **1**. Raboisson, de Kock, Rosenquist, et al. (2008) Bioorg. Med. Chem. Lett. **18**, 4853

Simvastatin

This cholesterol-reducing statin (FW = 418.57 g/mol; CAS 79902-63-9; tradename: Zocor™), a synthetic derivative of a fermentation product of Aspergillus terreus, targets 3-hydroxy-3-methylglutaryl-CoA reductase (IC_{50} = 11 nM), the rate-controlling reaction in the biosynthesis of cholesterol and other isoprenoids (1-3). When taken along with a low-fat diet, simvastatin reduces the risk of heart attack, stroke, certain kinds of heart surgeries, as well as chest pain in patients with heart disease or several common heart disease risk factors (i.e., family history of heart disease, high blood pressure, age, low HDL cholesterol, and smoking). Simvastatin has also been reported to modify the cholesterol exchange between plasma and the erythrocyte membrane. Note: Because Simvastatin is metabolized by CYP3A4, co-administration of a CYP3A4 inhibitor can lead to adverse drug interactions. Strong CYP3A4 inhibitors include itraconazole, ketoconazole, posaconazole, erythromycin, clarithromycin, telithromycin, HIV protease inhibitors, boceprevir, telaprevir, nefazodone), gemfibrozil, cyclosporine, and danazol. **Antineoplastic Action:** Commonly used statins (including lovastatin, fluvastatin, and simvastatin) are known to inhibit proliferation of cancer cells. Simvastatin inhibits cell proliferation in a dose-dependent manner in both endometrial cancer cell lines and 5/8 primary cultures of endometrial cancer cells. Simvastatin treatment arrests cell cycle at G_1, reduces HMG-CoA reductase activity, induces apoptosis as well as DNA damage and cellular stress (4). Simvastatin treatment also inhibits MAPK pathway signalling and shows differential effects on the AKT/mTOR pathway in the ECC-1 and Ishikawa cells. Minimal change in AKT phosphorylation was seen in both cell lines. Treatment also reduces cell adhesion and invasion in these cell lines (4). Simvastatin enhances phosphorylation of the endogenous Akt substrate endothelial nitric oxide synthase (eNOS), inhibited apoptosis and accelerated vascular structure formation in vitro in an Akt-dependent manner (5). Similar to vascular endothelial growth factor (VEGF) treatment, simvastatin administration and enhanced Akt signaling in the endothelium promoted angiogenesis in ischemic limbs of normocholesterolemic rabbits. **Key Pharmacokinetic Parameters:** *See* Appendix II in Goodman & Gilman's THE PHARMACOLOGICAL BASIS OF THERAPEUTICS, 12th Edition (Brunton, Chabner & Knollmann, eds.) McGraw-Hill Medical, New York (2011). **Target(s):** 3-hydroxy-3-methylglutaryl-CoA reductase (1-3); cyclin-dependent kinase, *or* Cdk2 (6); acetylcholinesterase (7); cholinesterase, *or* butyrylcholinesterase (7); CYP3A4 (8); CYP2C9 (8); CYP2D6 (8). **1.** Istvan & Deisenhofer (2001) *Science* **292**, 1160. **2.** Hoeg & Brewer, Jr. (1987) *JAMA* **258**, 3532. **3.** Rabini, Polenta, Staffolani, *et al.* (1993) *Exp. Mol. Pathol.* **59**, 51. **4.** Schointuch, Gilliam, Stine, *et al.* (2014) *Gynecol. Oncol.* **134**, 346. **5.** Kureishi, Luo, Shiojima, et al. (2000) *Nature Med.* **6**, 1004. (*Erratum in: Nature Med.* (2001) **7**, 129.) **6.** Andrés (2004) *Cardiovasc. Res.* **63**, 11. **7.** Chiou, Lai, Tsai, *et al.* (2005) *J. Chin. Chem. Soc.* **52**, 843. **8.** Transon, Leemann & Dayer (1996) *Eur. J. Clin. Pharmacol.* **50**, 209.

SIN-1, *See 3-Morpholinosydnonimine*

Sinapate

This coumarate derivative (FW$_{free-acid}$ = 224.21 g/mol; CAS 530-59-6), also known as sinapinic acid and 4-hydroxy-3,5-dimethoxycinnamic acid, antioxidant-exhibiting precursor in the biosynthesis of sinapyl alcohol. Sinapic acid is often used as a matrix in MALDI-TOF mass spectroscopy. **Target(s):** activator-protein 1, *or* AP-1 (1); α-amylase, *weakly* (2); lipase, *or* triacylglycerol lipase, weakly (3); peroxidase, horseradish (4); phenol sulfotransferase, *or* aryl sulfotransferase (5); and phenylalanine ammonia-lyase, weakly (6). **1.** Maggi-Capeyron, Ceballos, Cristol, *et al.* (2001) *J. Agric. Food Chem.* **49**, 5646. **2.** Funke & Melzig (2005) *Pharmazie* **60**, 796. **3.** Karamac & Amarowicz (1996) *Z. Naturforsch. [C]* **51**, 903. **4.** Gamborg, Wetter & Neish (1961) *Can. J. Biochem. Physiol.* **39**, 1113. **5.** Yeh & Yen (2003) *J. Agric. Food Chem.* **51**, 1474. **6.** Jorrin & Dixon (1990) *Plant Physiol.* **92**, 447.

Sinefungin

This S-adenosyl-L-homocysteine analogue and antibiotic (FW = 381.39 g/mol; CAS 58944-73-3; pK_1 = 2.9, pK_2 =3.9, pK_3 =8.9, and pK_4 = 10.2 in *N,N*-dimethylformamide, 66% vol/vol; λ$_{max}$ = 259 nm; ε = 15,400 M^{-1}cm^{-1}), also known as A-9145, adenosylornithine and 6,9-diamino-1-(6-amino-9*H*-purin-9-yl)-1,5,6,7,8,9-hexadeoxy-D-glycero-α-L-*talo*-decofuranuronic acid, is produced by *Streptomyces griseoleus* NRRL 3739. Sinefungin inhibits a many methyltransferases, including those catalyzing methylation of

residues in mRNA and DNA. **Target(s):** acyl-homoserine-lactone synthase (41); adenosyl-fluoride synthase, weakly (40); adenosyl-homocysteinase, *or* S-adenosylhomocysteine hydrolase (1,38); 1-aminocyclopropane-1-carboxylate synthase (2,3,36,37); calmodulin-lysine *N*-methyltransferase (62); CheR methyltransferase (4); catechol *O*-methyltransferase (5,66,68); cycloartenol 24-*C*-methyltransferase (32); cyclopropane-fatty-acyl-phospholipid synthase (6,54-56); [cytochrome *c*]-arginine *N*-methyltransferase (46); [cytochrome *c*]-methionine *S*-methyltransferase (46); demethylmacrocin *O*-methyltransferase, weakly (47); 3'-demethystaurosporine *O*-methyltransferase (42); DNA (adenine)-methyltransferase M.*Dam* (7); DNA (adenine)-methyltransferase M.*Eca* (18); DNA (adenine)-methyltransferase M.*Eco*RI (9,57); DNA (adenine)-methyltransferase *Mme*I (10); DNA (adenine)-methyltransferase M.*Taq*I (11_; DNA (adenine)-methyltransferase, *Streptomyces* (12); DNA (cytosine-5)-methyltransferase M.*Msp*I (13); DNA (cytosine-)-methyltransferase, *Streptomyces* (12); DNA methyltransferase M.*Kpn*I (14); guanidinoacetate *N*-methyltransferase (15,69,70); histamine methyltransferase (16,66); histone-arginine *N*-methyltransferase (43-45); histone-lysine *N*-methyltransferase (17,18); homarine synthase (19); macrocin *O*-methyltransferase (48); magnesium protoporphyrin IX methyltransferase (67); meromycolic acid-[acyl-carrier protein] methyltransferase (20); 5'-methylthioadenosine nucleosidase (21,39); mRNA (guanine-N^7-)-methyltransferase (18,22,63, 64); mRNA (nucleoside-2'-*O*-)-methyltransferase (18,22,63); [myelin basic protein]-arginine *N*-methyltransferase (43); norepinephrine *N*-methyltransferase (18,66); phenylethanolamine *N*-methyltransferase (16, 66); protein-arginine methyltransferase (18,23,24); protein-glutamate *O*-methyltransferase (18,25,53); protein-*S*-isoprenylcysteine *O*-methyltransferase (18,26,27,49-52); RNA (adenine-1)-methyltransferase, *Streptomyces*28; RNA, *or* adenine-N^6)-methyltransferase, *Streptomyces* (28); RNA (guanine-7)-methyltransferase, *Streptomyces* (28); rRNA (adenine-N^6)-methyltransferase *Erm*C'29; site-specific DNA-methyltransferase (adenine-specific) (7-12,57-61); spermidine synthase (30); splice-leader RNA methyltransferase (31); sterol 24-*C*-methyltransferase (32,65); thioether *S*-methyltransferase (33); thiopurine methyltransferase (34); and type IV pilin *N*-methyltransferase (35). **1.** Fabianowska-Majewska, Duley & Simmonds (1994) *Biochem. Pharmacol.* **48**, 897. **2.** Adams & Yang (1987) *Meth. Enzymol.* **143**, 426. **3.** Ickeson & Apelbaum (1983) *Biochem. Biophys. Res. Commun.* **113**, 586. **4.** Subbaramaiah & Simms (1992) *J. Biol. Chem.* **267**, 8636. **5.** Dhar & Rosazza (2000) *Appl. Environ. Microbiol.* **66**, 4877. **6.** Smith, Jr., & Norton (1980) *Biochem. Biophys. Res. Commun.* **94**, 1458. **7.** Wenzel, Moulard, Lobner-Olesen & Guschlbauer (1991) *FEBS Lett.* **280**, 147. **8.** Szilak, Venetianer & Kiss (1992) *Eur. J. Biochem.* **209**, 391. **9.** Reich & Mashhoon (1990) *J. Biol. Chem.* **265**, 8966. **10.** Tucholski, Zmijewski & Podhajska (1998) *Gene* **223**, 293. **11.** Schluckebier, Kozak, Bleimling, Weinhold & Saenger (1997) *J. Mol. Biol.* **265**, 56. **12.** Barbes, Sanchez, Yebra, Robert-Gero & Hardisson (1990) *FEMS Microbiol. Lett.* **57**, 239. **13.** Zingg, Shen & Jones (1998) *Biochem. J.* **332**, 223. **14.** Finta, Sulima, Venetianer & Kiss (1995) *Gene* **164**, 65. **15.** Im, Chiang & Cantoni (1979) *J. Biol. Chem.* **254**, 11047. **16.** Vedel, Robert-Gero, Legraverend, Lawrence & Lederer (1978) *Nucl. Acids Res.* **5**, 2979. **17.** Tuck, Farooqui & Paik (1985) *J. Biol. Chem.* **260**, 7114. **18.** Robert-Gero, Pierre, Vedel, *et al.* (1980) in *Enzyme Inhibitors* (Brodbeck, ed.), p. 61, Verlag Chemie, Weinheim. **19.** Nishitani, Kikuchi, Okumura & Taguchi (1995) *Arch. Biochem. Biophys.* **322**, 327. **20.** Yuan, Mead, Schroeder, Zhu & Barry III (1998) *J. Biol. Chem.* **273**, 21282. **21.** Guranowski, Chiang & Cantoni (1983) *Meth. Enzymol.* **94**, 365. **22.** Pugh, Borchardt & Stone (1978) *J. Biol. Chem.* **253**, 4075. **23.** Tuck & Paik (1984) *Meth. Enzymol.* **106**, 268. **24.** Rawal, Rajpurohit, Paik & Kim (1994) *Biochem. J.* **300**, 483. **25.** Rollins & Dahlquist (1980) *Biochemistry* **19**, 4627. **26.** De Busser, Van Dessel & Lagrou (2000) *Int. J. Biochem. Cell Biol.* **32**, 1007. **27.** Perez-Sala, Tan, Canada & Rando (1991) *Proc. Natl. Acad. Sci. U.S.A.* **88**, 3043. **28.** Yebra, Sanchez, Martin, Hardisson & Barbes (1991) *J. Antibiot. (Tokyo)* **44**, 1141. **29.** Schluckebier, Zhong, Stewart, Kavanaugh & Abad-Zapatero (1999) *J. Mol. Biol.* **289**, 277. **30.** Cacciapuoti, Porcelli, Carteni-Farina, Gambacorta & Zappia (1986) *Eur. J. Biochem.* **161**, 263. **31.** McNally & Agabian (1992) *Mol. Cell Biol.* **12**, 4844. **32.** McCammon & Parks (1981) *J. Bacteriol.* **145**, 106. **33.** Mozier, McConnell & Hoffman (1988) *J. Biol. Chem.* **263**, 4527. **34.** Van Loon, Szumlanski & Weinshilboum (1992) *Biochem. Pharmacol.* **44**, 775. **35.** Strom & Lory (2002) *The Enzymes*, 3rd ed. (Dalbey & Sigman, eds.), **22**, 127. **36.** Jakubowicz (2002) *Acta Biochim. Pol.* **49**, 757. **37.** Miyazaki & Yang (1987) *Phytochemistry* **26**, 2655. **38.** Impagnatiello, Franceschini, Oratore & Bozzi (1996) *Biochimie* **78**, 267. **39.** Guranowski, Chiang & Cantoni (1981) *Eur. J. Biochem.* **114**, 293. **40.** Schaffrath, Deng & O'Hagan (2003) *FEBS Lett.* **547**, 111. **41.**

Parsek, Val, Hanzelka, Cronan, Jr., & Greenberg (1999) *Proc. Natl. Acad. Sci. U.S.A.* **96**, 4360. **42**. Yang, Lin, Cordell, Wang & Corley (1999) *J. Nat. Prod.* **62**, 1551. **43**. Ghosh, Paik & Kim (1988) *J. Biol. Chem.* **263**, 19024. **44**. Gupta, Jensen, Kim & Paik (1982) *J. Biol. Chem.* **257**, 9677. **45**. Han, Hong, Han, *et al.* (1999) *Biochem. Arch.* **15**, 45. **46**. Farooqui, Tuck & Paik (1985) *J. Biol. Chem.* **260**, 537. **47**. Kreuzman, Turner & Yeh (1988) *J. Biol. Chem.* **263**, 15626. **48**. Bauer, Kreuzman, Dotzlaf & Yeh (1988) *J. Biol. Chem.* **263**, 15619. **49**. Pillinger, Volker, Stock, Weissmann & Philips (1994) *J. Biol. Chem.* **269**, 1486. **50**. Hasne & Lawrence (1999) *Biochem. J.* **342**, 513. **51**. Giner & Rando (1994) *Biochemistry* **33**, 15116. **52**. Li, Kowluru & Metz (1996) *Biochem. J.* **316**, 345. **53**. Kim (1984) *Meth. Enzymol.* **106**, 295. **54**. Gulanvarc'h, Guangqi, Drujon, *et al.* (2008) *Biochim. Biophys. Acta* **1784**, 1652. **55**. Gulanvarc'h, Drujon, Leang, Courtois & Ploux (2006) *Biochim. Biophys. Acta* **1764**, 1381. **56**. Wang, Grogan & Cronan, Jr. (1992) *Biochemistry* **31**, 11020. **57**. Mashhoon, Carroll, Pruss, *et al.* (2004) *J. Biol. Chem.* **279**, 52075. **58**. Evdokimov, Zinoviev, Malygin, Schlagman & Hattman (2002) *J. Biol. Chem.* **277**, 279. **59**. Horton, Liebert, Hattman, Jeltsch & Cheng (2005) *Cell* **121**, 349. **60**. Thomas, Scavetta, Gumport & Churchill (2003) *J. Biol. Chem.* **278**, 26094. **61**. Bheemanaik, Reddy & Rao (2006) *Biochem. J.* **399**, 177. **62**. Wright, Bertics & Siegel (1996) *J. Biol. Chem.* **271**, 12737. **63**. Pugh & Borchardt (1982) *Biochemistry* **21**, 1535. **64**. Zheng, Hausmann, Liu, *et al.* (2006) *J. Biol. Chem.* **281**, 35904. **65**. Jayasimha & Nes (2008) *Lipids* **43**, 681. **66**. Fuller & Nagarajan (1978) *Biochem. Pharmacol.* **27**, 1981. **67**. Vothknecht, Willows & Kannangara (1995) *Plant Physiol. Biochem.* **33**, 759. **68**. Dhar & Rosazza (2000) *Appl. Environ. Microbiol.* **66**, 4877. **69**. Konishi & Fujioka (1991) *Arch. Biochem. Biophys.* **289**, 90. **70**. Takata & Fujioka (1992) *Biochemistry* **31**, 4369.

Sirtinol

This cell-permeable hydroxynapthaldehyde-containing sirtuin (FW = 394.47 g/mol; CAS 410536-97-9; Soluble to 100 mM in DMSO), also named 2-[[(2-hydroxy-1-naphthalenyl)-methylene]amino]-*N*-(1-phenyl-ethyl)benzamide, inhibits members of the sirtuin family of NAD$^+$-dependent deacetylases (SIRT1, IC$_{50}$ = 131 µM, and SIRT2, IC$_{50}$ = 38 µM), including the yeast Sir2p transcriptional silencing activity (1). Sirtinol inhibits Hepatitis A Virus (HAV) replication by inhibiting HAV internal ribosomal entry site activity (2). Spectroscopic and structural data reveal that this compound is also an iron chelator forming high-spin ferric species *in vitro* and in cultured leukemia cells (3). Sirtinol-induced senescence-like growth arrest was accompanied by impaired activation of mitogen-activated protein kinase (MAPK) pathways, namely, extracellular-regulated protein kinase, c-jun N-terminal kinase and p38 MAPK, in response to epidermal growth factor (EGF) and insulin-like growth factor-I (IGF-I) (4). Active Ras was reduced in sirtinol-treated senescent cells compared with untreated cells. However, tyrosine phosphorylation of the receptors for EGF and IGF-I and Akt/PKB activation were unaltered by sirtinol treatment (4). **Other Target(s):** inhibits neutrophil elastase activity, attenuating lipopolysaccharide-mediated acute lung injury in mice (5); attenuates trauma hemorrhage-induced hepatic injury through Akt-dependent pathway in rats (6); modulates adhesion molecule expression and monocyte adhesion on activated primary human dermal microvascular endothelial cells (HDMEC) (7). **1**. Grozinger, Chao, Blackwell, Moazed & Schreiber (2001) *J. Biol. Chem.* **276**, 38837. **2**. Kanda, Sasaki, Nakamoto, *et al.* (2015) *Biochem. Biophys. Res. Commun.* **466**, 567. **3**. Gautam, Akam, Astashkin, Loughrey & Tomat (2015) *Chem. Commun.* (Camb) **51**, 5104. **4**. Ota, Tokunaga, Chang, *et al.* (2006) *Oncogene* **25**, 176. **5**. Tsai, Yu, Chang, *et al.* (2015) *Sci. Rep.* **5**, 8347. **6**. Liu, Tsai & Yu (2013) *J. Trauma Acute Care Surg.* **74**, 1027. **7**. Orecchia, Scarponi, Di Felice, *et al.* (2011) *PLoS One* **6**, e24307.

Sisomicin

This aminoglycoside antibiotic (FW$_{free-base}$ = 447.53 g/mol; CAS 53179-09-2) from *Micromonospora inyoesis* consists of 2-deoxystreptamine linked covalently to sisosamine and garosamine. Sisomicin exhibits bactericidal activity against a wide range of Gram-negative bacteria (*e.g.*, *Escherichia coli*, *Pseudomonas aeruginosa*, and speicies of *Enterobacter*, *Klebsiella*, and *Proteus*). Although active against staphylococci, most streptococci are usually resistant (except when β-lactam antibiotics are present). Anaerobic organisms are not sensitive. Sisomicin is structurally closely related to gentamicin CIa, which contains a purpurosamine component in place of sisosamine. *In vitro* studies have shown it to have superior activity to gentamicin against *P. aeruginosa*, while still possessing the high activity of gentamicin against *Serratia* and other Gram-negative rods. Sisomicin is inactivated by virtually all bacterial enzymes which inactivate gentamicin and tobramycin. **Target(s):** gentamicin 3'-acetyltransferase (1); phospholipases (2); and protein biosynthesis (3). **1**. Haas & Dowding (1975) *Meth. Enzymol.* **43**, 611. **2**. Carlier, Laurent, Claes, Vanderhaeghe & Tulkens (1983) *Antimicrob. Agents Chemother.* **23**, 440. **3**. Cavallo & Martinetto (1981) *G. Batteriol. Virol. Immunol.* **74**, 335.

Sitafloxacin

This fluoroquinolone-class antibiotic (FW = 409.82 g/mol; CAS 127254-12-0), also known as DU-6859a, inhibits bacterial DNA topoisomerase II. (For the prototypical member of this antibiotic class, *See Ciprofloxacin*) **Target(s):** DNA topoisomerase II, *or* DNA gyrase (1-3); and DNA topoisomerase IV (2,3). **1**. Akasaka, Kurosaka, Uchida, *et al.* (1998) *Antimicrob. Agents Chemother.* **42**, 1284. **2**. Onodera, Uchida, Tanaka & Sato (1999) *J. Antimicrob. Chemother.* **44**, 533. **3**. Onodera, Okuda, Tanaka & Sato (2002) *Antimicrob Agents Chemother.* **46**, 1800.

Sitagliptin

This orally bioavailable antihyperglycemic agent (FW = 407.31 g/mol; CAS 654671-77-9, 486460-32-6 (free base), 654671-78-0 (phosphate); Solubility: 100 mg/mL DMSO; 100 mg/mL Water; Formulation: 0.5% aqueous hydroxyethylcellulose), also known as MK-0431 or Januvia® and named systematically as (2*R*)-4-oxo-4-[3-(trifluoromethyl)-5,6-dihydro[1,2,4]triazolo[4,3-*a*]pyrazin-7(8*H*)-yl]-1-(2,4,5-trifluorophenyl)butan-2-amine, is a potent competitive inhibitor of dipeptidyl-peptidase IV, *or* DDP-

4, IC_{50} = 18 nM. DDP-4 cleaves the incretins, a group of gastrointestinal hormones that increase insulin release, even before blood glucose levels become elevated. By limiting proteolysis of the gastrointestinal hormones glucagon-like protein-1 and -2 (GLP-1 and GLP-2), sitagliptin indirectly maintains insulin secretion and suppresses glucagon release (1). **Key Pharmacokinetic Parameters:** *See* Appendix II in Goodman & Gilman's THE *PHARMACOLOGICAL BASIS OF THERAPEUTICS*, 12th Edition (Brunton, Chabner & Knollmann, eds.) McGraw-Hill Medical, New York (2011). *See Incretin Hormone Enhancers; Saxagliptin* 1. Kim, Wang, Beconi, *et al.* (2005) *J. Med. Chem.* **48**, 141.

Sitaxentan

This orally active, highly selective ET_A antagonist (FW$_{free-acid}$ = 454.90 g/mol; FW$_{Na-salt}$ = 476.89 g/mol; CAS 210421-74-2; Solubility: 40 mg/mL DMSO; <1 mg/mL H_2O), also known as TBC11251 and *N*-(4-chloro-3-methyl-1,2-oxazol-5-yl)-2-[2-(6-methyl-2*H*-1,3-benzodioxol-5-yl)acetyl] thiophene-3-sulfonamide, targets the endothelin A receptor (IC_{50} = 1.4 nM; K_i = 0.43 nM), exhibiting 7000x selectivity over ET_B receptors (IC_{50} = 9800 nM) and inhibiting ET-1-induced stimulation of phosphoinositide turnover with a K_i of 0.686 nM (1). Three weeks after rats were first treated with a single subcutaneous injection of monocrotaline (MCT), sitaxsentan dose dependently (10 and 50 mg/kg per day in the drinking water) attenuated right ventricular systolic pressure, right heart hypertrophy, and pulmonary vascular remodeling (1-3). [*Note:* Administration of small doses of MCT or its active metabolite, monocrotaline pyrrole (MCTP), causes delayed and progressive lung injury characterized by pulmonary vascular remodeling, pulmonary hypertension, and compensatory right heart hypertrophy (4).] Systemic administration of this ET_A receptor antagonist significantly attenuates cerebral vasospasm after subarachnoid hemorrhage (SAH), thus providing additional support for the role of ET-1 in vasospasm (5). 1. Wu, Chan, Stavros, *et al.* (1997) *J. Med. Chem.* **40**, 1690. 2. Tilton, Munsch, Sherwood, *et al.* (2000) *Pulm. Pharmacol. Ther.* **13**, 87. 3. Holm (1997) *Scand. Cardiovasc. J.* (Suppl.) **46**, 1. 4. Schultze & Roth (1998) *J. Toxicol. Environ. Health B Crit. Rev.* **1**, 271. 5. Wanebo, Arthur, Louis, *et al.* (1998) *Neurosurgery* **43**, 1409.

β-Sitosterol

This phytosterol (FW = 414.72 g/mol), also known as (3β)-stigmast-5-en-3-ol, is commonly found in many plants (*e.g.*, soy beans, rice embryos, wheat germ, corn oil, rye germ oil, cottonseed oil, pine bark, potatoes, sugar cane wax, tobacco, *etc.*). **Target(s):** cholesterol 7α-hydroxylase, weak alternative substrate (1); cholesterol uptake (2-4); complement component C1 (5); cycloartenol 24-*C*-methyltransferase (6,7); protein-tyrosine kinase activation (8,9); steroid 5α-reductase (10); sterol acyltransferase, *or* cholesterol acyltransferase (11); sterol 27-monooxygenase (12); and steryl-β-glucosidase (13). 1. Shefer, Salen, Nguyen, *et al.* (1988) *J. Clin. Invest.* **82**, 1833. 2. Wong (2001) *Can. J. Cardiol.* **17**, 715. 3. Ikeda, Tanaka, Sugano, Vahouny & Gallo (1988) *J. Lipid Res.* **29**, 1573. 4. Ikeda & Sugano (1983) *Biochim. Biophys. Acta* **732**, 651. 5. Cerqueira, Watanadilok, Sonchaeng, *et al.* (2003) *Planta Med.* **69**, 174. 6. Zhao & Nes (2003) *Arch. Biochem. Biophys.* **420**, 18. 7. Zhao, Song, Kanagasabai, *et al.*

(2004) *Molecules* **9**, 185. 8. Zhang, Morita, Zhang, *et al.* (2000) *Planta Med.* **66**, 119. 9. Zhang, Morita, Shao, Yao & Murota (2000) *Planta Med.* **66**, 114. 10. Cabeza, Bratoeff, Heuze, *et al.* (2003) *Proc. West Pharmacol. Soc.* **46**, 153. 11. Billheimer (1985) *Meth. Enzymol.* **111**, 286. 12. Nguyen, Shefer, Salen, Tint & Batta (1998) *Proc. Assoc. Amer. Physicians* **110**, 32. 13. Kalinowska & Wojciechowski (1978) *Phytochemistry* **17**, 1533.

Sivelestat

This peptide analogue (FW$_{free-acid}$ = 434.46 g/mol; CAS 201677-61-4), also known by the code name ONO-5046, the trade name Elaspol®, and *N*-[2-[4-(2,2-dimethylpropionyloxy)phenyl-sulfonylamino]aminoacetic acid, inhibits neutrophil elastase, K_i = 200 nM, IC_{50} = 44 nM (1) and is used to treat pulmonary fibrosis and idiopathic interstitial pneumonia. Human neutrophil elastase (HNE) is a serine protease, which plays a major role in the COPD inflammatory process. Sivelestat also inhibits the growth of human lung cancer cell lines in SCID mice. 1. Kawabata, Suzuki, Sugitani, *et al.* (1991) *Biochem. Biophys. Res. Commun.* **177**, 814.

Sizukacillin, *See Suzukacillin*

SJ 172550

This reversible MDMX inhibitor (FW = 428.87 g/mol; CAS 431979-47-4; Solubility: 100 mM DMSO), also named 2-[2-chloro-4-[(1,5-dihydro-3-methyl-5-oxo-1-phenyl-4*H*-pyrazol-4-ylidene)methyl]-6-ethoxyphenoxy]-acetic acid methyl ester, binds MdmX (EC_{50} = 2.3 μM), which inhibits p53 through Mdm2-dependent and independent mechanisms. MDM2 and MDMX (also known as HDMX and MDM4) proteins are dysregulated in many human cancers, exerting oncogenic activity by inhibiting the p53 tumor suppressor. MdmX is regulated by DNA damage signaling events, indicating its involvement in the p53 response to DNA damage SJ-172550 inhibits the MDMX-p53 interaction in cultured retinoblastoma cells, freeing p53 to induce apoptosis. The effect of SJ-172550 is additive when combined with an MDM2 inhibitor. Results from a series of biochemical and structural modeling studies suggest SJ-172550 binds to the p53-binding pocket of MDMX, thereby displacing p53. 1. Reed, Shen, Shelat, *et al.* (2010) *J. Biol. Chem.* **285**, 10786.

Skepinone-L

This protein kinase inhibitor (FW = 425.42; CAS 1221485-83-1), also named 2-[(2,4-difluorophenyl)amino]-7-[(2R)-2,3-dihydroxypropoxy]-10,11-dihydro-5H-dibenzo[a,d]-cyclohepten-5-one and CEP-32496, targets p38α-MAPK (IC$_{50}$ = 5 nM), wild-type BRAF (K_d = 36 nM), BRAF^{V600E} (K_d = 14 nM) and c-Raf (K_d = 39 nM). Skepinone-L inhibits HSP27 phosphorylation at Ser82 by the p38 MAPK (IC$_{50}$ = 25 nM) (1,2). Skepinone-L (at 1 μM) almost completely blocked phosphorylation of platelet p38 MAPK substrate Hsp27 after stimulation with C-Reactive Protein (CRP) at 1 μg/ml, thrombin at 5 mU/mL, or the thromboxane A$_2$ analogue U-46619 at 1 μM (2). Skepinone-L significantly blunted activation-dependent platelet secretion and aggregation following threshold concentrations of CRP, thrombin and thromboxane A2 analogue U-46619. Skepinone-L did not impair platelet Ca^{2+} signaling but prevented agonist-induced thromboxane A$_2$ synthesis through abrogation of p38 MAPK-dependent phosphorylation of platelet cytosolic phospholipase A$_2$, or cPLA$_2$ (2). Skepinone-L further markedly blunted thrombus formation under low and high arterial shear rates. **1**. Koeberle, Romir, Fischer, *et al.* (2011) *Nature Chem. Biol.* **8**, 141. **2**. Borst, Walker, Münzer, *et al.* (2013) *Cell Physiol. Biochem.* **31**, 914.

SK549

This antithrombotic agent (FW$_{ion}$ = 546.59 g/mol) inhibits coagulation factor Xa, K_i = 0.52 nM, but is a weaker inhibitor of thrombin, trypsin, and plasmin, with respective K_i values of 400 nM, 45 nM, and 890 nM. **1**. Leadley, Jr. (2001) *Curr. Top. Med. Chem.* **1**, 151. **2**. Quan, Ellis, Liauw, *et al.* (1999) *J. Med. Chem.* **42**, 2760. **3**. Wong, Quan, Crain, *et al.* (2000) *J. Pharmacol. Exp. Ther.* **292**, 351. **Alternate Identifiers:** SK549, SK-549

Skatole

This foul-smelling indole (FW = 131.18 g/mol; CAS 83-34-1; M.P. = 95°C), also known as 3-methyl-1H-indole, is a degradation product of indoleacetate and tryptophan that is soluble in hot water, ethanol, and benzene. Skatole has a fecal odor that readily permeates clothing. While many bacteria metabolize tryptophan to indole and indole acetic acid (IAA), only a few gut bacteria, mainly from the *Clostridium* and *Bacteroides* genera, catalyze the conversion of IAA to skatole. **Target(s):** chymotrypsin (1,2); lipid peroxidation (3); and tryptophanase (4). **1**. Webb (1966) *Enzyme and Metabolic Inhibitors*, vol. **2**, p. 374, Academic Press, New York. **2**. Foster (1961) *J. Biol. Chem.* **236**, 2461. **3**. Adams, Jr., Heins & Yost (1987) *Biochem. Biophys. Res. Commun.* **149**, 73. **4**. Kazarinoff & Snell (1980) *J. Biol. Chem.* **255**, 6228.

SKF *trans*-385, See *Tranylcypromine*

SKF 525A, See *Proadifen*

SKF 3301A

This proadifen analogue (FW$_{free-base}$ = 311.47 g/mol), named systematically as 2,2-diphenyl-1-(β-dimethylaminoethoxy)pentane or N,N-dimethylamino-ethyl 2,2-diphenylpentyl ether, inhibits *ent*-copalyl-diphosphate synthase (1), isopentenyl-diphosphate Δ-isomerase (2,3), and *ent*-kaurene synthase (1,4). **1**. Frost & West (1977) *Plant Physiol.* **59**, 22. **2**. Zabkiewicz, Keates & Brooks (1969) *Phytochemistry* **8**, 2087. **3**. Holmes & DiTullio (1963) *Amer. J. Clin. Nutr.* **10**, 310. **4**. Fall & West (1971) *J. Biol. Chem.* **246**, 6913. **Alternate Identifiers:** SKF3301A, SKF-3301A

SKF 64139, See *7,8-Dichloro-1,2,3,4-tetrahydroisoquinoline*

SKF 86466

This benzazepine (FW$_{free-base}$ = 195.69 g/mol), also known as 6-chloro-2,3,4,5-tetrahydro-3-methyl-1H-3-benzazepine, is a potent and selective α$_2$ adrenoceptor antagonist (1-3). **1**. DeMarinis, Hieble & Matthews (1983) *J. Med. Chem.* **26**, 1213. **2**. Hieble, DeMarinis, Fowler & Matthews (1986) *J. Pharmacol. Exp. Ther.* **236**, 90. **3**. Roesler, McCafferty, DeMarinis, Matthews & Hieble (1986) *J. Pharmacol. Exp. Ther.* **236**, 1. **Alternate Identifiers:** SKF86466, SKF-86466

SKF 89976A

This agent (FW = 335.45 g/mol), which can cross the blood-brain barrier, inhibits the γ-aminobutyrate (GABA) transporter type 1 (GAT-1), also known as sodium- and chloride-dependent GABA transporter 1 (1-3). **1**. Zuiderwijk, Veenstra, Lopes da Silva & Ghijsen (1996) *Eur. J. Pharmacol.* **307**, 275. **2**. Zorn & Enna (1985) *Life Sci.* **37**, 1901. **3**. Larsson, Falch, Krogsgaard-Larsen & Schousboe (1988) *J. Neurochem.* **50**, 818.

SKF 91488

This reagent (FW$_{free-base}$ = 175.30 g/mol), also known as 4-(N,N-dimethylamino)butylisothiourea, is a potent inhibitor of histamine N-methyltransferase (1-5). **Target(s):** diamine oxidase (1); and histidine decarboxylase, weakly (1). **1**. Beaven & Roderick (1980) *Biochem. Pharmacol.* **29**, 2897. **2**. Beaven & Shaff (1979) *Agents Actions* **9**, 455. **3**. Klein & Gertner (1981) *J. Pharmacol. Exp. Ther.* **216**, 315. **4**. Yamauchi, Sekizawa, Suzuki, *et al.* (1994) *Amer. J. Physiol.* **267**, L342. **5**. Dent, Nilam & Smith (1982) *Biochem. Pharmacol.* **31**, 2297.

SKF 93944, *See* Temelastine

SKF 94836

This substituted guanidine (FW = 284.32 g/mol), also known as Siguazodan, inhibits cAMP phosphodiesterase III, K_i = 1-3 μM (1,2). Time course studies indicated that SK&F 94836-induced relaxation of trachealis strips contracted with 0.1 μM methacholine was accompanied by an activation of cAMP-dependent protein kinase (cAMP-PK). In subsequent experiments, trachealis strips were contracted with three concentrations of methacholine (0.1, 1.0 or 3.0 μM) or two concentrations of histamine (10 or 300 μM) before being relaxed by the cumulative addition of SKF-94836. The relaxant response to SKF-94836 (EC_{50} = 1-10 μM) decreases progressively as tissues are contracted with higher concentrations of methacholine. In parallel with its inhibitory effect on SKF-94836-induced relaxation, methacholine suppresses the ability of SKF-94836 to activate cAMP-PK. Inhibition of cAMP-PK activity is not accompanied by a significant inhibition of SK&F 94836-stimulated cAMP accumulation. The concentration of histamine used to contract tissues has no effect on SKF-94836-induced relaxation or cAMP-PK activation. To determine the effect of SKF-94836 on the mechanical and biochemical responses to the β-adrenoceptor agonist isoproterenol, tissues are first contracted with 3.0 μM methacholine and then incubated with 0, 0.3, 3.0 or 30 μM SKF-94836 before being relaxed by the cumulative addition of isoproterenol. SKF-94836 potentiates isoproterenol-induced relaxation, cAMP accumulation and cAMP-PK activation in a concentration-dependent manner. **1**. Torphy, Burman, Huang & Tucker (1988) *J. Pharmacol. Exp. Ther.* **246**, 843. **2**. Tomkinson, Karlsson & Raeburn (1993) *Brit. J. Pharmacol.* **108**, 57.

SKF 95282, *See* Zolantidine

SKF 96022, *See* Pantoprazole

SKF 96365

This calcium ion antagonist and Store-operated Ca^{2+} entry (*or* SOCE) inhibitor ($FW_{hydrochloride}$ = 402.92 g/mol; CAS 130495-35-1; Solubility: 100 mM in H_2O, as HCl salt), also named 1-(β-[3-(4-methoxyphenyl)propoxy]-4-methoxyphenethyl)-1*H*-imidazole, is structurally distinct from known Ca^{2+} antagonists and selectively blocks <u>R</u>eceptor-<u>M</u>ediated <u>C</u>alcium ion <u>E</u>ntry, *or* RMCE, compared with receptor-mediated internal Ca^{2+} release (1). SK&F 96365 gave IC_{50} values of 8.5 μM or 11.7 μM for ADP- and thrombin-activated platelets, but does not affect internal Ca^{2+} release (1). Similar effects of SK&F 96365 are observed in suspensions of neutrophils and endothelial cells. SK&F 96365 also inhibits agonist-stimulated Mn^{2+} entry in platelets and neutrophils. Voltage-gated Ca^{2+} entry in fura-2-loaded

GH3 pituitary cells and rabbit ear-artery smooth-muscle cells held under voltage-clamp is also inhibited by SK&F 96365, but the ATP-gated Ca^{2+}-permeable channel of arterial smooth-muscle cells is unaffected by SK&F 96365. Unlike 'organic Ca^{2+} antagonists', SK&F 96365 shows no selectivity between voltage-gated Ca^{2+} entry and RMCE, although the lack of effect on ATP-gated channels indicates that it discriminates between different types of RMCE. Because SK&F 96365 is not as potent (IC_{50} ~10 μM) and nonselectively inhibits voltage-gated Ca^{2+} entry), caution must be exercised for unambiguous use of this agent. SKF-96365 exhibits potent anti-neoplastic activity by inducing cell-cycle arrest and apoptosis in colorectal cancer cells (2). It also induces cytoprotective autophagy to delay apoptosis by preventing the release of cytochrome *c* (cyt *c*) from the mitochondria into the cytoplasm (2). Mechanistically, SKF-96365 treatment inhibited the calcium/calmodulin-dependent protein kinase IIγ (CaMKIIγ)/AKT signaling cascade *in vitro* and *in vivo*. Overexpression of CaMKIIγ or AKT abolished the effects of SKF-96365 on cancer cells, suggesting a critical role of the CaMKIIγ/AKT signaling pathway in SFK-96365-induced biological effects (2). (*See also* Maitotoxin; Palytoxin) SKF 96365 also inhibits tubulin polymerization (microtubule self-assembly) *in vitro* (3). **1**. Merritt, Armstrong, Benham, *et al.* (1990) *Biochem. J.* **271**, 515. **2**. Jing, Sui, Yao, *et al.* (2016) *Cancer Lett.* **372**, 226. **3**. Mitsui-Saito, Nakahata & Ohizumi (2000) *Jpn. J. Pharmacol.* **82**, 269.

SKF 98625

This phosphonate diester (FW_{ion} = 531.66 g/mol) is a strong inhibitor of human glycerophospholipid arachidonyl-transferase (CoA-independent), IC_{50} = 9 μM (1-4). **1**. Winkler, Eris, Sung, *et al.* (1996) *J. Pharmacol. Exp. Ther.* **279**, 956. **2**. Winkler, Sung, Marshall & Chilton (1996) *Adv. Exp. Med. Biol.* **416**, 11. **3**. Winkler, Fonteh, Sung, *et al.* (1995) *J. Pharmacol. Exp. Ther.* **274**, 1338. **4**. Chilton, Fonteh, Sung, *et al.* (1995) *Biochemistry* **34**, 5403.

SK&F D-39162, *See* Auranofin

SKI-1, *See* 6,7-Dimethoxy-4-(4'-phenoxyanilino)-quinazoline

SKI II

This highly selective ATP-noncompetitive inhibitor (FW = 302.74 g/mol; CAS 312636-16-1), also known as 2-(4-hydroxyanilino)-4-(4-chlorophenyl)thiazole, targets the sphingosine-1-phosphate receptor kinase (IC_{50} = 0.5 μM), with little or no inhibitory action on other kinases, including PI3K, PKCα and ERK2. SK inhibitors were anti-proliferative toward a panel of tumor cell lines, including lines with the multidrug resistance phenotype because of overexpression of either P-glycoprotein or multidrug resistance phenotype 1, and were shown to inhibit endogenous human SK activity in intact cells (1). SKI II-inhibited human cancer cell lines include: T-24 (IC_{50} = 4.6 μM), MCF-7 (IC_{50} = 1.2 μM), MCF-7/VP (IC_{50} = 0.9 μM), and NCI/ADR (IC_{50} = 1.3 μM). In mice, SKI-II is orally bioavailable and can be detected in the blood for >8 hours after administration (2). SKI II ameliorates antigen-induced bronchial smooth muscle hyper-responsiveness in mice, but not airway inflammation (3). SKI-II appears to be a novel activator of Nuclear factor erythroid 2-related factor 2 (Nrf2) signalling via the inactivation of Keap1. The latter is a negative regulator of Nrf2 expression, and the effect of SKI II is independent of sphingosine kinase inhibition (4). **1**. French, Schrecengost, Lee, *et al.* (2003) *Cancer Res.* **63**, 5962. **2**. French, Upson, Keller, *et al.* (2006) *J. Pharmacol. Exp. Ther.* **318**, 596. **3**. Chiba, Takeuchi, Sakai &

Misawa (2010) *J. Pharmacol. Sci.* **114**, 304. **4**. Mercado, Kizawa, Ueda, *et al.* (2014) *PLoS One* **9**, e88168.

SKI-606, See *Bosutinib*

SL 327

These photosensitive isomeric phenylacetonitriles (FW = 335.35 g/mol) inhibit MEK1 and MEK2, with IC$_{50}$ values of 180 and 220 nM for the *E*- and *Z*-isomers, respectively, and block fear conditioning and learning in rats. *Note*: Commercial sources often supply a mixture of the *E*- and *Z*-isomers. **1**. Salzmann, Marie-Claire, Le Guen, Roques & Noble (2003) *Brit. J. Pharmacol.* **140**, 831. **2**. Selcher, Atkins, Trzaskos, Paylor & Sweatt (1999) *Learn Mem.* **6**, 478.

Slotoxin

This 37-residue peptide (MW = 4092 g/mol; CAS 144026-79-9; *Sequence*: TFIDVDCTVSKECWAPCKAAFGVDRGKCMGKKCKCYV) from the venom of the scorpion *Centruroides noxius* blocks the MaxiK (hSlo) pore-forming α subunit reversibly, K_d = 1.5 nM (1). **1**. Garcia-Valdes, Zamudio, Toro, Possani & Possan (2001) *FEBS Lett.* **505**, 369.

µ-SLPTX-Ssm6a

This unique 46-residue peptide and neurotoxin from the venom of the predatory centipede *Scolopendra suvspinipes mutilans* potently inhibits the voltage-gated sodium channel, Na$_v$1.7. **Mode of Inhibitory Action:** The Na$_v$1.7 channel amplifies membrane depolarizations, and when the membrane potential difference reaches a specific threshold, the neuron fires. In sensory neurons, multiple voltage-dependent sodium currents are differentiated by their voltage dependence and by their sensitivity to tetrodotoxin, the classical voltage-gated sodium-channel blocker. µ-SLPTX-Ssm6a has high affinity for the Na$_v$1.7 channel (IC$_{50}$ ≈ 23 nM), exhibiting >150x selectivity over all other human Na$_v$ subtypes, with the exception of Na$_v$1.2, for which its selectivity is 32x. µ-SLPTX-Ssm6a is a more potent analgesic than morphine in a rodent chemical-induced pain model and is equipotent with morphine in rodent thermal and acid-induced pain models. **Structural Features:** µ-SLPTX-Ssm6a contains three disulfide bonds, exhibiting a unique disulfide connectivity pattern. µ-SLPTX-Ssm6a has no significant sequence homology with any known peptide or protein. **1**. Yang, Xiao, Kang, *et al.* (2013) *Proc. Natl. Acad. Sci. U.S.A.* **110**, 17534.

SM-7338, See *Meropenem*

SM-13496, See *Lurasidone*

SM-19712

This sulfonamide (FW = 437.84 g/mol), named systematically as 4-chloro-*N*-[[(4-cyano-3-methyl-1-phenyl-1*H*-pyrazol-5-yl)amino]carbonyl]benzenesulfonamide, sodium salt, inhibits rat lung endothelin-converting enzyme, IC$_{50}$ = 42 nM. At concentrations ranging from 10 to 100 µM, no effect was reported for other metalloproteases. **1**. Umekawa, Hasegawa, Tsutsumi, *et al.* (2000) *Jpn. J. Pharmacol.* **84**, 7.

SM-324405

This potent TLR7 agonist (FW = 385.42 g/mol; CAS 677773-91-0; Solubility: 100 mM in DMSO), also named methyl 3-[(6-amino-2-butoxy-7,8-dihydro-8-oxo-9*H*-purin-9-yl)methyl]benzeneacetate, targets Toll-like Receptor-7 (EC$_{50}$ = 50 nM), which recognises single-stranded RNA in endosomes, the latter a common feature of viral genomes that are internalized by macrophages and dendritic cells. Exhibits selectivity for TLR7 over TLR8. SM-324405 treatment induces both IFN-α and IFN-γ expression in human peripheral blood mononuclear cells (PBMCs) as well as mouse splenocytes, and italso inhibits IL-5 production in human PBMCs *in vitro*. SM-324405 is rapidly metabolized to acid metabolite ($t_{1/2}$ = 2.6 min in human plasma). **1**. Kurimoto, Hashimoto, Nakamura, *et al.* (2010) *J. Med. Chem.* **53**, 2964.

Small Interfering RNA, See *RNA*

SMER3

This <u>S</u>mall-<u>M</u>olecule <u>E</u>nhancer of <u>R</u>apamycin (FW = 224.14 g/mol; CAS 67200-34-4) is an inhibitor of the Skp1-Cullin-F-box, *or* SCF (Met30) ubiquitin ligase (IC$_{50}$ = 51 nM) *in vivo* and *in vitro*, but not the closely related SCF(Cdc4). SMER3 also diminishes binding of the F-box subunit Met30 to the SCF core complex *in vivo* by binding directly to Met30. **1**. Aghajan, Jonai, Flick, *et al.* (2010) *Nature Biotechnol.* **28**, 738.

SMI-4a

This novel small-molecule protein kinase inhibitor (FW = 273.23 g/mol; CAS 438190-29-5; Solubility = 55 mg/mL DMSO) targets Pim1 kinase (IC$_{50}$ = 17 nM), a serine/threonine protein kinase that is up-regulated in specific hematologic neoplasms. A screen of 60 other serine, threonine, and tyrosine protein kinases indicates that SMI-4a and related benzylidene-thiazolidine-2,4-diones are highly specific inhibitors of the Pim protein kinase. SMI-4a is cytotoxic against both myeloid and lymphoid cell lines, particularly precursor T-cell lymphoblastic leukemia/lymphoma cells (*or* pre-T-LBL cells). Incubation of such cells with SMI-4a induced G$_1$ phase cell-cycle arrest and dose-dependent induction of p27^{Kip1}, mitochondrial pathway-mediated apoptosis, and inhibition of the mammalian target of rapamycin C1 (mTORC1) pathway. The phosphatidylinositol/Akt/mTOR pathway is often activated in leukemia and lymphoma downstream of a variety of oncogenes including receptor tyrosine kinases. Treatment of these cells with SMI-4a also induces phosphorylation of extracellular signal-related kinase1/2 (ERK1/2), and the combination of SMI-4a and a mitogen-activated protein kinase kinase 1/2 (MEK1/2) inhibitor was highly synergistic in killing pre-T-LBL cells. Addition of Pim inhibitors to several prostate and leukemic cell lines induces a G$_1$ cell cycle block, although, in other cell lines, especially in low-serum conditions, these inhibitors are capable of inducing apoptosis. **1**. Beharry, Zemskova, Mahajan, *et al.*

(2009) *Mol. Cancer Ther.* **8**, 1473. **2**. Lin, Beharry, Hill, *et al.* (2010) *Blood* **115**, 824.

SMIFH2

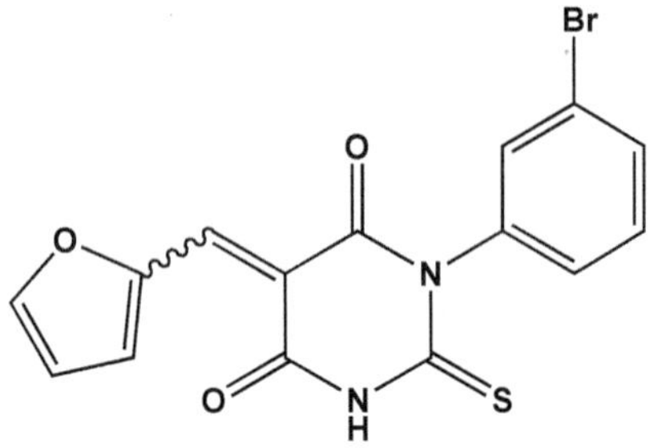

This actin cytoskeletal motor inhibitor (FW = 377.21 g/mol; CAS 340316-62-3; Soluble to 100 mM in DMSO), also named 1-(3-bromophenyl)-5-(2-furanylmethylene)dihydro-2-thioxo-4,6(1*H*,5*H*)-pyrimidinedione, targets Formin Homology-2 (FH2) domains, preventing both formin-mediated actin nucleation (IC$_{50}$ = 5-10 µM) and processive barbed end elongation as well as decreasing formin affinity for filament barbed-ends (1,2). Formins are actoclampin-type (+)-end-tracking molecular motors that exploit processive actin monomer insertion to generate propulsive forces (3,4). At low-µM concentrations, SMIFH2 disrupts formin-dependent, but not Arp2/3 complex-dependent, actin cytoskeletal structures in fission yeast and mammalian NIH 3T3 fibroblasts. **1**. Rizvi, Neidt, Cui, *et al.* (2009) *Chem. Biol.* **16**, 1158. **2**. Poincloux, Collin, Lizárraga, *et al.* (2011) *Proc. Natl. Acad. Sci. U.S.A.* **108**, 1943. **3**. Dickinson & Purich (2002) *Biophys. J.* **82**, 605. **4**. Dickinson, Caro & Purich (2004) *Biophys. J.* **87**, 2838.

SMP-797, See *N-(4-Amino-2,6-diisopropylphenyl)-N'-[1-butyl-4-[3-(3hydroxypropoxy)-phenyl]-2-oxo-1,2-dihydro-1,8-naphthyridin-3-yl]urea*

SMPB, See *4-(4-Maleimidophenyl)butyric Acid N-Hydroxysuccinimide Ester*

SN-38

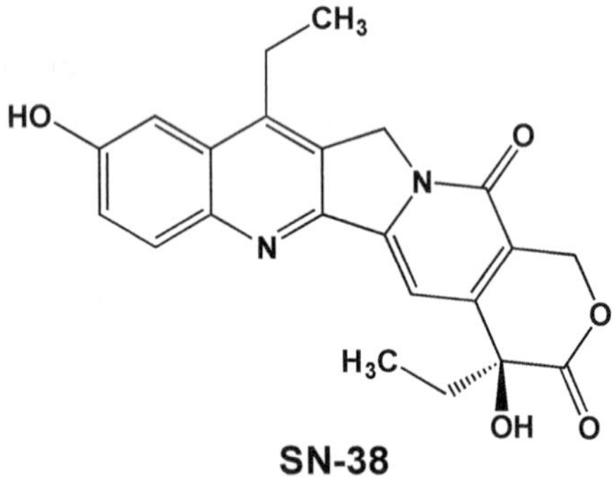

SN-38

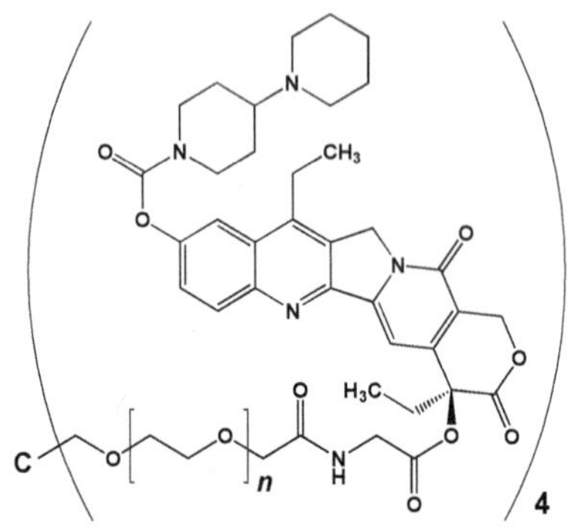

Etirinotecan Pegol

This topoisomerase I inhibitor (FW = 392.40 g/mol; CAS 86639-52-3), also named 7-ethyl-10-hydroxycamptothecin, is the active drug formed via hydrolysis of its prodrug Etirinotecan Pegol (MW = 23 kDa; CAS 1193151-09-5), a polyethylene glycol-encapsulated irinotecan that is eventually hydrolyzed by carboxylesterases. The active drug, SN-38, inhibits topoisomerase I activity by stabilizing the cleavable Topoisomerase·DNA complex, resulting in DNA breaks that inhibit DNA replication and trigger apoptosis. Eventually glucuronidated by UGT1A1, the resulting SN38-glucuronide is incorporated into the bile and feces. Pegylation improves drug penetration into tumors and also retards drug clearance, thereby increasing therapeutic exposure and duration, while reducing its toxicity. Complications include diarrhea and myelosuppression in a fourth of those receiving the prodrug. **1**. Atyabi, Farkhondehfai, Esmaeili & Dinarvand (2009) *Acta Pharm.* **59**, 133. **2**. Liu, Robinson, Sun & Dai (2008) *J. Am. Chem. Soc.* **130**, 10876. **3**. Scott, Yao, Benson, *et al.* (2009) *Cancer Chemother. Pharmacol.* **63**, 363.

SN390, See *Quinacrine*

SN 429

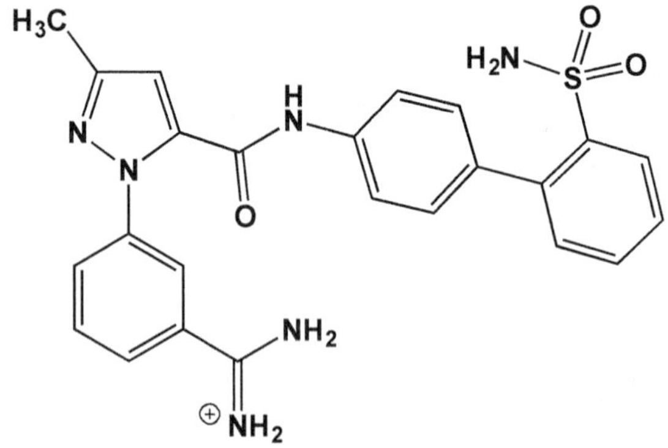

This benzamidine derivative (FW$_{ion}$ = 475.55 g/mol), also known as 1-(3-[amino(imino)methyl]phenyl)-N-[2'-(aminosulfonyl)-[1,1'-biphenyl]-4-yl]-3-methyl-1*H*-pyrazole-5-carboxamide, noncovalently inhibits coagulation factor Xa, thrombin, and trypsin, with K_i values of 0.013, 300, and 16 nM, respectively (1). **See also SA 862** **1**. Pinto, Orwat, Wang, *et al.* (2001) *J. Med. Chem.* **44**, 566.

SN1796, See *9-(2-Diamylamino-1-hydroxyethyl)-1,2,3,4-tetrahydrophenanthrene*

SN5949, See *2-Hydroxy-3-(2'-methyloctyl)-1,4-naphthoquinone*

SN6911, See *7-Chloro-4-(4-diethylamino-1-methylbutylamino)-3-methylquinoline*

SN7135, See *7-Chloro-4-(4-diethylamino-1-methyl-butylamino)-2-methylquinoline*

SN7618, See *Chloroquine*

SN8285-4, See *3-Diethylamino-7-(di-n-butylamino)-1-methylphenazthionium Chloride*

SN12710-6029, See *Novalauramine*

(*S*)-SNAP-5114

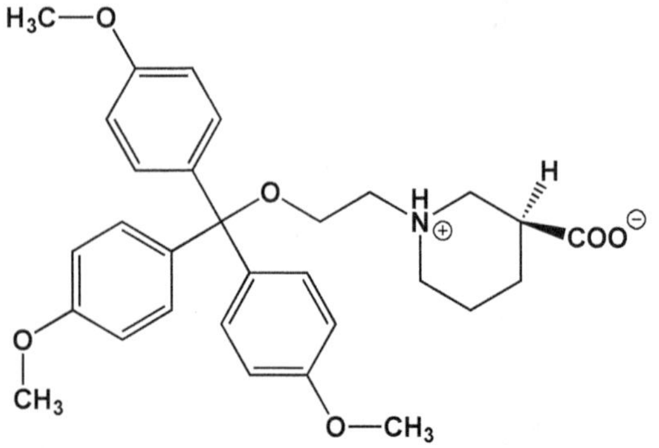

This synthetic amino acid (FW = 505.61 g/mol; CAS 157604-55-2; Soluble to 100 mM in DMSO), also named 1-[2-[*tris*(4-methoxyphenyl)methoxy]ethyl]-(*S*)-3-piperidinecarboxylic acid, inhibits GABA (γ-aminobutyrate) transport, with an IC$_{50}$ value of 5 µM at GAT-3 and 21 µM at GAT-2 (1,2).The latter are specific, high-affinity, sodium- and chloride-dependent transporters thought to be located on presynaptic terminals and surrounding glial cells. **1**. Borden (1996) *Neurochem. Int.* **29**, 335. **2**. Borden, Dhar, Smith, *et al.* (1994) *Receptors Channels* **2**, 207.

SNDX-275, See *Entinostat*

SNK-860, See *Fidarestat*

SNS-032

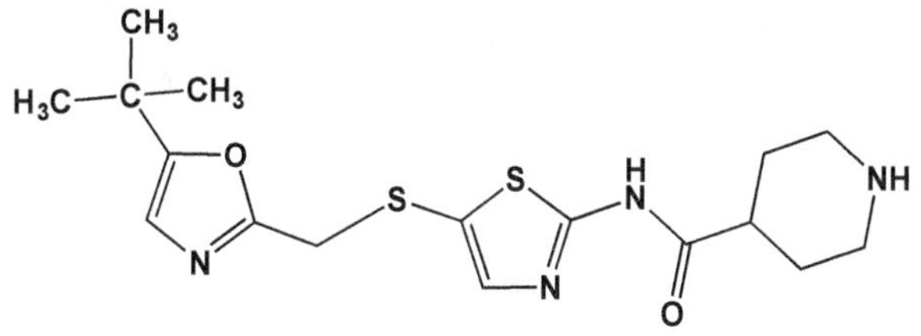

This cyclin kinase-directed inhibitor (FW = 380.53 g/mol; CAS 345627-80-7, 345627-90-9 (HCl); Solubility: 75 mg/mL DMSO; <1 mg/mL H_2O), also known as BMS-387032 and systematically named *N*-(5-((5-*tert*-butyloxazol-2-yl)methylthio)thiazol-2-yl)piperidine-4-carboxamide, targets CDK2, CDK7 and CDK9 with IC_{50} values of 38 nM, 62 nM and 4 nM, respectively. **Cyclin Target Selectivity:** Cdk1 (weak, if any), Cdk2 (++), Cdk3 (weak, if any), Cdk4 (weak, if any), Cdk5 (weak, if any), Cdk6 (weak, if any), Cdk7 (+), Cdk8 (weak, if any), Cdk9 (+++), Cdk10 (weak, if any). **1**. Chen, *et al.* (2009) *Blood* **113**, 4637. **2**. Ali, *et al.* (2007) *Neoplasia* **9**, 370. **3**. Conroy, *et al.* (2009) *Cancer Chemother. Pharmacol.* **64**, 723. **4**. Walsby, *et al.* (2011) *Leukemia* **25**, 411.

SNX-5422, See *PF-4929113*

Sodium Dodecyl Sulfate

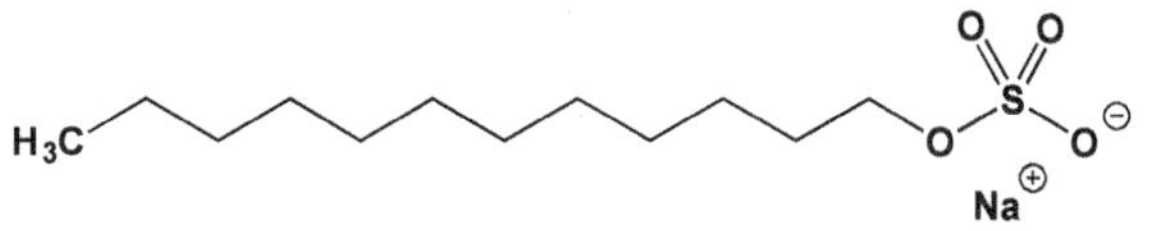

This common laboratory detergent (FW = 288.38 g/mol; CAS 151-21-3; Abbreviation: SDS), also known as sodium lauryl sulfate, is a powerful denaturing agent. Originally used to solubilize proteins from cell membranes, the denaturing power of SDS often precludes such use. This anionic detergent is a white powder with a cmc (critical micelle concentration) value of about 8.2 mM in water (0.52 mM in 0.5 M NaOH). SDS also exhibits a bactericidal activity, particularly at low pH. Note also that enzymes inhibited by free fatty acids are often inhibited by non-denaturing concentrations of SDS. Sodium dodecyl sulfate is also used in polyacrylamide gel electrophoresis to obtain a rough estimate of protein molecular mass. This is possible because the alkyl portion of SDS binds tightly and extensively to most proteins (1.4 g SDS per gram of protein), such that the detergent-bound polypeptides minimize electrostatic reulsion by adopting an elongated rod-like shape that is hydrodynamically well-behaved during electrophoresis.

Sodium Borate, See *Borax*

Sodium Dodecyl Sulfonate

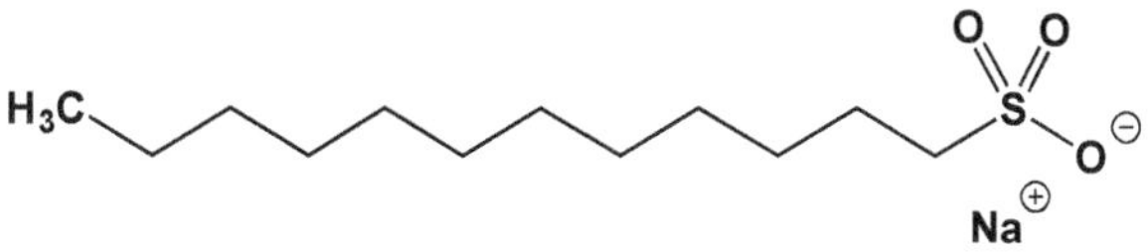

This ionic surfactant (FW = 272.38 g/mol), also called sodium lauryl sulfonate, inhibits various lipases by directly binding in place of lipd substrates and/or by altering lipid and phospholipid micellization in a manner that changes enzyme binding at the lipid-water interface. **See also** *Sodium Dodecyl Sulfate* **Target(s):** arylsulfatase (1); lipid-phosphate phosphatase (2); lysozyme (3); soluble epoxide hydrolase, *or* phosphatase activity (2). **1**. Beil, Kehrli, James, *et al.* (1995) *Eur. J. Biochem.* **229**, 385. **2**. Tran, Aronov, Tanaka, *et al.* (2005) *Biochemistry* **44**, 12179. **3**. Jollès (1960) *The Enzymes*, 2nd ed. (Boyer, Lardy & Myrbäck, eds.), **4**, 431.

Sodium Tetraborate, See *Borax*

Sofosbuvir

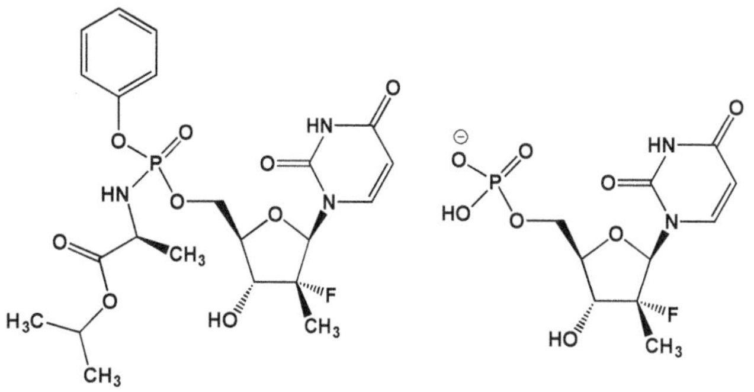

This cell-penetrating HCV pro-drug (FW = 529.46 g/mol; CAS 1190307-88-0), also known as isopropyl (2*S*)-2-[[[(2*R*,3*R*,4*R*,5*R*)-5-(2,4-dioxopyrimidin-1-yl)-4-fluoro-3-hydroxy-4-methyl-tetrahydrofuran-2-yl]methoxyphenoxyphosphoryl]amino]propanoate, PSI-7977, GS-7977, and Sovaldi®, is taken up by cells and efficiently hydrolyzed through the action of human cathepsin A and/or carboxylesterase 1 (1,2). The resulting 2'-deoxy-2'-α-fluoro-β-*C*-methyluridine-5'-monophosphate is itself a pro-drug that must undergo metabolic phosphorylation to its triphosphate form, which then inhibits Hepatitis C viruse NS5B RNA polymerase, causing chain termination during replication of the HCV genome. Sofosbuvir is a pan-genotype, direct-acting antiviral for hepatitis C virus infection. Oral sofosbuvir has been approved by the FDA and the European Medicines Agency's Committee for Medicinal Products for Human Use for the treatment of chronic hepatitis C. Despite its high barrier to developing drug resistance, sofosbuvir is almost always used in combination with another HCV inhibitor, most often daclatasvir. (Harvoni®, for instance, contains ledipasvir and sofosbuvir.) High rates of sustained virological response (SVR) are observed with interferon-free drug combinations, particularly with genotypes 1 and 2. In patients with hepatocellular carcinoma who received sofosbuvir for extended treatment duration, while awaiting liver transplantation, the Leu-159-Phe mutation emerged in those infected with Genotype 1a or 2b. Sofosbuvir's safety profile is favorable in both cirrhotic and noncirrhotic patients (3). While sofosbuvir's game-changing 90% success rate commends it as the treatment-of-choice for HCV, its controversial high cost (*i.e.*, $84,000 for a 12-week treatment course in 2014) has placed it well beyond the affordability of many patients and healthcare systems. That the high price is unrelated to production costs is strongly suggested by a recent agreement between its manufacturer and India, where the same 12-week course will be 50-60 times less expensive. **1**. Sofia, Bao, Chang, *et al.* (2010) *J. Med. Chem.* **53**, 7202. **2**. Murakami, Tolstykh, Bao, *et al.* (2010) *J. Biol. Chem.* **285**, 34337. **3**. Stedman (2014) *Therap. Adv. Gastroenterol.* **7**, 131.

Solanidine

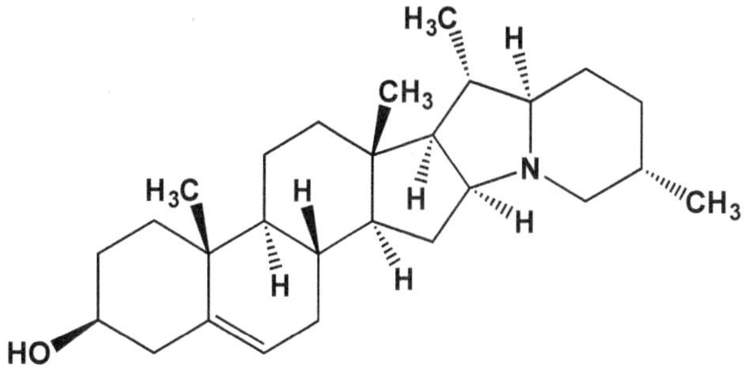

This α-solanine aglycone (FW$_{free-base}$ = 397.64 g/mol; CAS 80-78-4; M.P. = 218-219°C; soluble in benzene and chloroform, but insoluble in water), named for its abundance in sprouts of the common potato *Solanum tuberosum*, inhibits butyrylcholinesterase (1) and sterol 24-*C*-methyltransferase, *or* cycloartenol 24-*C*-methyl-transferase; K_i = 2 μM (2-4). **1**. Nigg, Ramos, Graham, *et al.* (1996) *Fundam. Appl. Toxicol.* **33**, 272. **2**. Nes, Guo & Zhou (1997) *Arch. Biochem. Biophys.* **342**, 68. **3**. Mangla & Nes (2000) *Bioorg. Med. Chem.* **8**, 925. **4**. Nes (2000) *Biochim. Biophys. Acta* **1529**, 63.

α-Solanine

This glycoalkaloid (FW_free-base = 867.09 g/mol; CAS 20562-02-1; soluble in hot ethanol, but poorly soluble in water (25 mg/L)), found in several potato species (genus *Solanum*), contains the branched trisaccharide solatriose comprised of D-glucose, D-galactose, and L-rhamnose). Because α-solanine is toxic and is present in elevated concentrations potato shoots, the latter cannot be used as a food source. **Target(s):** acetylcholinesterase (1); and butyrylcholinesterase (2-4), **1.** Nigg, Ramos, Graham, *et al.* (1996) *Fundam. Appl. Toxicol.* **33**, 272. **2.** Nes, Guo & Zhou (1997) *Arch. Biochem. Biophys.* **342**, 68. **3.** Mangla & Nes (2000) *Bioorg. Med. Chem.* **8**, 925. **4.** Nes (2000) *Biochim. Biophys. Acta* **1529**, 63.

Solasodine

This steroidal alkaloid (FW = 413.64 g/mol; CAS 126-17-0), also known as (22R,25R)-spirosol-5-ene-3β-ol is an aglycone of α-solasonine present in several species of *Solanum* that inhibits sterol 24-*C*-methyltransferase, *or* cycloartenol 24-*C*-methyltransferase (1-3). **1.** Nes, Guo & Zhou (1997) *Arch. Biochem. Biophys.* **342**, 68. **2.** Mangla & Nes (2000) *Bioorg. Med. Chem.* **8**, 925. **3.** Nes (2000) *Biochim. Biophys. Acta* **1529**, 63.

Solithromycin

This oral fluoroketolide antibiotic (FW = 845.01 g/mol; CAS 760981-83-7), also known by the code names CEM-101 and OP-1068, is under development for treating community-acquired pneumonia as well as infections by other Gram-positive pathogens. Solithromycin binds tightly to the large 50S subunit of the ribosome, thereby inhibiting both protein biosynthesis and further synthesis of the 50S subunit (1). Among 1363 strains of *Streptococcus pneumoniae*, 99.9% display a minimal inhibitory concentration (MIC) ≤ 0.5 μg/mL, and 100% were inhibited at 1 μg/mL (2). Solithromycin demonstrates activity and potency against *Haemophilus influenzae* (MIC$_{50\%}$ = 1 mg/L and MIC$_{90\%}$ = 2 μg/mL), comparable to that of azithromycin (2). **1.** Rodgers, Frazier & Champney (2013) *Antimicrob. Agents Chemother.* **57**, 1632. **2.** Farrell, Castanheira, Sader & Jones (2010) *J. Infect.* **61**, 476.

Soman

This chemical warfare agent and acetylcholinesterase inhibitor (FW = 182.18 g/mol; CAS 96-64-0), often designated GD and also known as methylpinacolyloxyfluorophosphine oxide and pinacolyl methylphospho-fluoridate, has four stereoisomers, of which the two *P*(–)-stereoisomers are more effective than the *P*(+)-isomers (1). Nobelist Richard Kuhn's group discovered soman at the Kaiser Wilhelm Institute for Medical Research (Heidelberg). Soman inhibits AChE, cholinesterase, and certain carboxylesterases. A soman-hydrolyzing enzyme has been reported (2). **Caution:** Soman is extremely toxic (even exceeding tabun and sarin). As a vapor, it is readily absorbed through the skin and eyes. Inhalation or absorption of a lethal dose will result in death in one to ten minutes (LD$_{50}$ = 0.35 g). The LC$_{50}$ (*i.e.*, the concentration in air that proves lethal to 50% of exposed individuals) is 0.5-0.9 ppm. To improve survival (even if the smallest drop contacts the skin), one must immediately flood the affected area with bleach and dilute alkali. In mice, the i.p. LD$_{50}$ is 0.6 mg/kg. **Target(s):** acetylcholinesterase (1,3-11); carboxylesterase (3,12); choline acetyltransferase (13); cholinesterase, *or* butyrylcholinesterase (3,9,10,14-17); chymotrypsin (3,9); and trypsin (3). **1.** Benschop, Konings, van Genderen & de Jong (1984) *Fundam. Appl. Toxicol.* **4**, S84. **2.** Wang, Sun, Zhang & Huang (1998) *J. Biochem. Mol. Toxicol.* **12**, 213. **3.** Cohen, Oosterbaan & Berends (1967) *Meth. Enzymol.* **11**, 686. **4.** Millard, Kryger, Ordentlich, *et al.* (1999) *Biochemistry* **38**, 7032. **5.** Qian & Kovach (1993) *FEBS Lett.* **336**, 263. **6.** Forsberg & Puu (1984) *Eur. J. Biochem.* **140**, 153. **7.** Lenz & Maxwell (1981) *Biochem. Pharmacol.* **30**, 1369. **8.** Froede & Wilson (1971) *The Enzymes*, 3rd ed. (Boyer, ed.), **5**, 87. **9.** Segall, Waysbort, Barak, *et al.* (1993) *Biochemistry* **32**, 13441. **10.** Schopfer, Voelker, CBartels, Thompson & Lockridge (2005) *Chem. Res. Toxicol.* **18**, 747. **11.** Kaplan, Barak, Ordentlich, *et al.* (2004) *Biochemistry* **43**, 3129. **12.** Cohen (1981) *Arch. Toxicol.* **49**, 105. **13.** Thompson & Thomas (1985) *Experientia* **41**, 1437. **14.** Baille, Dorandeu, Carpentier, *et al.* (2001) *Pharmacol. Biochem. Behav.* **69**, 561. **15.** Clery, Bec, Balny, Mozhaev & Masson (1995) *Biochim. Biophys. Acta* **1253**, 85. **16.** Ashani, Segev & Balan (2004) *Toxicol. Appl. Pharmacol.* **194**, 90. **17.** Cerasoli, Griffiths, Doctor, *et al.* (2005) *Chem. Biol. Interact.* **157-158**, 363.

Somatoliberin, *See Growth Hormone Releasing Factor*

Somatostatin-14 & Somatostatin-28

AGCKNFFWKTFTSC

SANSNPALAPRERKAGAGCKNFFWKTFTSC

These inhibitory hormones Somatostatin-14 (MW = 1640; CAS 38916-34-6; pI = 8.9) and Somatostatin-28 (MW = 3261; CAS 75037-27-3; pI = 9.85) inhibit the release of somatotropin as well as other secretory proteins. Somatostatin is also known as Growth Hormone-Inhibiting Hormone (GHIH), Somatotropin Release-Inhibiting Factor (SRIF), and Somatotropin Release-Inhibiting Hormone. The SRIF system, including SRIF ligand and receptors, regulates anterior pituitary gland function. Its main role is to inhibit hormone secretion. SRIF-14 binds to G-protein-coupled receptor subtypes 1-5, activating many targets, among them adenylate cyclase, MAPK, ion channel-dependent pathways. **Target(s):** Somatostatin's actions within the anterior pituitary gland include: inhibition of Growth Hormone release (thereby opposing the effects of Growth Hormone-

Releasing Hormone), inhibition of thyroid-stimulating hormone (TSH), inhibition of adenylyl cyclase in parietal cells, and inhibition of prolactin release. Its actions within the gastrointestinal tract include inhibition of gastrin release, inhibition of cholecystokinin release, inhibition of secretin release, inhibition of motilin release, inhibition of vasoactive intestinal peptide release, inhibition of gastric inhibitory polypeptide release, inhibition of gastrointestinal hormone(s) release, and inhibition of enteroglucagon. cystinyl aminopeptidase, *or* oxytocinase (1); protein-disulfide isomerase (2); and serotonin *N*-acetyltransferase (3). **1**. Burbach, De Bree, Terwel, *et al.* (1993) *Peptides* **14**, 807. **2**. Morjana & Gilbert (1991) *Biochemistry* **30**, 4985. **3**. Klein & Namboodiri (1982) *Trends Biochem. Sci.* **7**, 98.

Sonidegib (Erismodegib)

This orally active Hedgehog signaling pathway inhibitor (FW = 485.50 g/mol; CAS 1218778-77-8), also known as LDE225, Odomzo®, and *N*-[6-[(2*S*,6*R*)-2,6-dimethylmorpholin-4-yl]pyridin-3-yl]-2-methyl-3-[4-(trifluoromethoxy)phenyl]benzamide, is a Smoothened antagonist and anticancer agent, first identified by a cell-based phenotypic high-throughput screen (1). Blockade of aberrant hedgehog (Hh) signaling shows great promise in cancer chemotherapy, and LDE225 represses tumor growth and prolongs survival in a transgenic mouse model of islet cell neoplasias (2). **1**. Pan, Wu, Jiang, *et al.* (2010) *ACS Med. Chem. Lett.* **1**, 130. **2**. Fendrich, Wiese, Waldmann, *et al.* (2011) *Ann. Surg.* **254**, 818.

Sophoraflavanone G

This biodefensive flavanone (FW = 432.38 g/mol; IUPAC: (2S)-2-(2,4-dihydroxyphenyl)-5,7-dihydroxy-8-[(2*R*)-5-methyl-2-(prop-1-en-2-yl)hex-4-en-1-yl]-2,3-dihydro-4*H*-chromen-4-one) noncompetitively inhibits invasive protozoa, fungi, and bacteria affecting its organism of origin (*Sophora flavescens*). Sophoraflavanone G inhibits tyrosinase, *or* monophenol monooxygenase, K_i = 7.7 µM. **1**. Ryu, Westwood, Kang, *et al.* (2008) *Phytomedicine* **15**, 612.

Sophorose

This rare disaccharide (FW = 342.30 g/mol; CAS 534-46-3; M.P. = 196-198°C), also known as 2-*O*-β-D-glucopyranosyl-D-glucopyranose, from *Sophora japonica*, is a component of the sweetener stevioside. Sophorose is a substrate for many β-glucosidases. **Target(s)**: membrane-oligosaccharide glycerophospho-transferase (1). **1**. Goldberg, Rumley & Kennedy (1981) *Proc. Natl. Acad. Sci. U.S.A.* **78**, 5513.

Sorafenib

This oral apoptosis-promoting agent (FW = 464.83 g/mol; CAS 284461-73-0), also known as BAY 43-9006, inhibits Raf kinase and has been used in the treatment of colorectal and breast cancers, hepatocellular carcinomas, and non-small-cell lung cancer. **Key Pharmacokinetic Parameters:** *See* Appendix II in Goodman & Gilman's THE PHARMACOLOGICAL BASIS OF THERAPEUTICS, 12th Edition (Brunton, Chabner & Knollmann, eds.) McGraw-Hill Medical, New York (2011). **Target(s)**: c-kit (1); p38 (1); platelet-derived growth factor receptor B kinase (1); polo kinase (2); Raf kinase (1,3-5); vascular endothelial growth factor receptor 2 kinase (1); and vascular endothelial growth factor receptor 3 kinase (1). **1**. Ahmad & Eisen (2004) *Clin. Cancer Res.* **10**, 6388S. **2**. Johnson, Stewart, Woods, Giranda & Luo (2007) *Biochemistry* **46**, 9551. **3**. Hilger, Kredke, Hedley, *et al.* (2002) *Int. J. Clin. Pharmacol. Ther.* **40**, 567. **4**. Lyons, Wilhelm, Hibner & Bollag (2001) *Endocr. Relat. Cancer* **8**, 219. **5**. Lee & McCubrey (2003) *Curr. Opin. Investig. Drugs* **4**, 757.

Sorangicins

Sorangin A

These antimicrobial agents (FW = 773.98 g/mol for Soragicin A; CAS 100415-25-6), isolated from the myxobacterium *Sorangium cellulosum*, inhibit bacterial DNA-directed RNA polymerase (1,2). This antibiotic acts mainly against Gram-positive bacteria, including myocobacteria, with MIC values between 0.01-0.1 µg/mL. Higher concentrations (MIC 3-30 µg/ml) Gram-negatives are also inhibited. Yeasts and molds are completely resistant. Sorangicin is a specific inhibitor of eubacterial RNA polymerase which it blocks, but, only if added before RNA polymerization has commenced. **1**. Chopra (2007) *Curr. Opin. Invest. Drugs* **8**, 600. **2**. Irschik, Jansen, Gerth, Höfle & Reichenbach (1987) *J. Antibiot.* **40**, 7.

Sorbic Acid

This unsaturated carboxylic acid (FW$_{free-acid}$ = 112.13 g/mol; CAS 16577-94-9; pK_a = 4.76 at 25°C; M. P. = 134.5°C; water solubility = 2.5 g/L at 30°C), also known as (2*E*,4*E*)-hexadienoic acid, inhibits the growth of molds and yeast, and it is frequently used as a fungastatic agent in food containers. Sorbate should be stored at temperatures below 40°C. **Target(s)**: catalase (1). **1**. Troller (1965) *Can. J. Microbiol.* **11**, 611.

Sorbinil

This spirohydantoin (FW = 236.20 g/mol; CAS 68367-52-2), also known as (4S)-6-fluorospiro(chroman-4,4'-imidazolidine)-2',5'-dione, is a strong inhibitor of aldose reductase, the enzyme responsible for formation and accumulation of sorbitol in the human lens and retina. Sorbinil acts primarily by binding to aldose reductase complexed with oxidized nicotinamide dinucleotide phosphate (E•NADP$^+$) to form a ternary dead-end complex, thereby preventing its turnover in the steady state. **Target(s):** alcohol dehydrogenase, NADP$^+$-requiring (1); aldose reductase, aldehyde reductase (2-5); 4-aminobutyrate aminotransferase (6); hexonate dehydrogenase (7); and succinate-semialdehyde dehydrogenase (6). **1.** Whittle & Turner (1981) *Biochem. Pharmacol.* **30**, 119. **2.** Wermuth & von Wartburg (1982) *Meth. Enzymol.* **89**, 181. **3.** Ramana, Chandra, Srivastava, Bhatnagar & Srivastava (2003) *Chem. Biol. Interact.* **143-144**, 587. **4.** Bohren & Grimshaw (2000) *Biochemistry* **39**, 9967. **5.** De Jongh, Schofield & Edwards (1987) *Biochem. J.* **242**, 143. **6.** Whittle & Turner (1978) *J. Neurochem.* **31**, 1453. **7.** Poulsom (1986) *Biochem. Pharmacol.* **35**, 2955.

L-Sorbinose, *See* L-Sorbose

Sorbitan Laurate

This nonionic liquid surfactant (FW = 346.46 g/mol; CAS 1338-39-2), also known as Span 20 and 1,4-anhydro-D-glucitol 6-dodecanoate, is often used to solubilize membrane-interacting proteins. **Target(s):** bis(2-ethylhexyl)phthalate esterase (1); and steryl-sulfatase, *or* arylsulfatase C (2). **1** Gniot-Szulzycka & Komoszynski (1972) *Enzymologia* **42**, 11. **2.** Krell & Sandermann, Jr. (1984) *Eur. J. Biochem.* **143**, 57.

D-Sorbitol

This common sugar alcohol (FW = 182.17 g/mol; CAS 50-70-4), also known as D-glucitol (note that D-glucitol is identical to L-gulitol), is produced in the aldose reductase by the reduction of the C1 aldehyde of D-glucose. The latter reaction is problematic in diabetic hyperglycemia, where excessive accumulation of sorbitol results in osmotic damage in diabetic retinopathy. When heated in the presence of acids, D-sorbitol converts to several cyclic ethers (*e.g.*, 1,4-sorbitan). **Target(s):** acetyl-CoA synthetase, *or* acetate:CoA ligase, weakly (1); aldose 1-epimerase, *or* mutarotase (2-4); α-amylase (5); D-arabinose isomerase (6); dihydrodiol dehydrogenase (7); ferredoxin:NADP$^+$ reductase, inhibited by elevated concentrations (8); L-fucose isomerase (6); glucan 1,4-α-glucosidase, *or* glucoamylase; weakly inhibited (9); mannonate dehydratase (10,11); naringinase, weakly (12); sorbitol-6-phosphatase (13); thioglucosidase, *or* myrosinase (14); and xylose isomerase, *or* glucose isomerase (15-25). **1.** Roughan & Ohlrogge

(1994) *Anal. Biochem.* **216**, 77. **2.** Bentley (1962) *Meth. Enzymol.* **5**, 219. **3.** Bailey, Fishman, Kusiak, Mulhern & Pentchev (1975) *Meth. Enzymol.* **41**, 471. **4.** Webb (1966) *Enzyme and Metabolic Inhibitors*, vol. **2**, p. 413, Academic Press, New York. **5.** Ali & Abdel-Moneim (1989) *Zentralbl. Mikrobiol.* **144**, 615. **6.** Boulter & Gielow (1973) *J. Bacteriol.* **113**, 687 **7.** Matsuura, Hara, Nakayama, Nakagawa & Sawada (1987) *Biochim. Biophys. Acta* **912**, 270. **8.** Satoh (1981) *Biochim. Biophys. Acta* **638**, 327. **9.** Buettner, Bode & Birnbaum (1987) *J. Basic Microbiol.* **27**, 299. **10.** Robert-Baudouy, Jimeno-Abendano & Stoeber (1982) *Meth. Enzymol.* **90**, 288. **11.** Robert-Baudouy & Stoeber (1973) *Biochim. Biophys. Acta* **309**, 473. **12.** Nomura (1965) *Enzymologia* **29**, 272. **13.** Zhou, Cheng & Wayne (2003) *Plant Sci.* **165**, 227. **14.** Tani, Ohtsuru & Hata (1974) *Agric. Biol. Chem.* **38**, 1623. **15.** Yamanaka (1966) *Meth. Enzymol.* **9**, 588. **16.** Sanchez & Smiley (1975) *Appl. Microbiol.* **29**, 745. **17.** Noltmann (1972) *The Enzymes*, 3rd ed. (Boyer, ed.), **6**, 271. **18.** Smith, Rangarajan & Hartley (1991) *Biochem. J.* **277**, 255. **19.** Ananichev, Ulezlo, Egorov, Bezborodov & Berezin (1980) *Biokhimiia* **45**, 992. **20.** Rangarajan & Hartley (1992) *Biochem. J.* **283**, 223. **21.** Khire, Lachke, Srinivasan & Vartak (1990) *Appl. Biochem. Biotechnol.* **23**, 41. **22.** Inyang, Gebhart, Obi & Bisswanger (1995) *Appl. Microbiol. Biotechnol.* **43**, 632. **23.** Danno (1970) *Agric. Biol. Chem.* **34**, 1805. **24.** Henrick, Collyer & Blow (1989) *J. Mol. Biol.* **208**, 129. **25.** Callens, Kersters-Hilderson, van Opstal & de Bruyne (1986) *Enzyme Microb. Technol.* **8**, 696.

D-Sorbitol 1,6-Bisphosphate

This phosphorylated sugar alcohol (FW$_{free-acid}$ = 342.13 g/mol), also known as D-glucitol 1,6-bisphosphate and D-glucitol 1,6-diphosphate, inhibits fructose-1,6-bisphosphate aldolase, K_i = 12 μM (1,2). **1.** Hartman & Barker (1965) *Biochemistry* **4**, 1068. **2.** Ginsburg & Mehler (1966) *Biochemistry* **5**, 2623.

D-Sorbitol 6-Phosphate

This phosphorylated sugar alcohol (FW$_{free-acid}$ = 262.15 g/mol), also known as D-glucitol 6-phosphate, is a substrate for sorbitol-6-phosphate 2-dehydrogenase, aldose-6-phosphate reductase (NADPH), and sorbitol-6-phosphatase. **Target(s):** glucokinase, in the presence of the regulatory protein (1-3); glucose-6-phosphate isomerase (4-7); glutamine:fructose-6-phosphate aminotransferase, isomerizing (7-9); inositol-3-phosphate synthase, *or* inositol-1-phosphate synthase (10-12); mannose-6-phosphate isomerase (4); phosphofructokinase-2 (13); pyrophosphate-dependent phosphofructokinase, *or* diphosphate:fructose-6-phosphate 1-phosphotransferase (14). **1.** Vandercammen, Detheux & Van Schaftingen (1992) *Biochem. J.* **286**, 253. **2.** Van Schaftingen, Vandercammen, Detheux & Davies (1992) *Adv. Enzyme Regul.* **32**, 133. **3.** Detheux, Vandercammen & Van Schaftingen (1991) *Eur. J. Biochem.* **200**, 553. **4.** Noltmann (1972) *The Enzymes*, 3rd ed. (Boyer, ed.), **6**, 271. **5.** Parr (1957) *Biochem. J.* **65**, 34P. **6.** Hansen, Schlichting, Felgendreher & Schonheit (2005) *J. Bacteriol.* **187**, 1621. **7.** Milewski, Janiak & Wojciechowski (2006) *Arch. Biochem. Biophys.* **450**, 39. **8.** Leriche, Badet-Denisot & Badet (1997) *Eur. J. Biochem.* **245**, 418. **9.** Floquet, Richez, Durand, *et al.* (2007) *Bioorg. Med.*

Chem. Lett. **17**, 1966. **10.** Barnett, Rasheed & Corina (1973) *Biochem. J.* **131**, 21. **11.** RayChaudhuri, Hait, das Gupta, *et al.* (1997) *Plant Physiol.* **115**, 727. **12.** Barnett, Rasheed & Corina (1973) *Biochem. Soc. Trans.* **1**, 1267. **13.** Campos, Guixe & Babul (1984) *J. Biol. Chem.* **259**, 6147. **14.** Bertagnolli, Younathan, Voll, Pittman & Cook (1986) *Biochemistry* **25**, 4674.

Sorivudine

This antiviral agent (FW = 349.14 g/mol; CAS 77181-69-2), also known systematically as 1-β-D-arabinofuranosyl-5-[(1*E*)-2-bromoethenyl]-2,4(1*H*, 3*H*)-pyrimidinedione, is a synthetic analogue of thymidine. Sorivudine exhibits antiviral activity against varicella zoster virus, herpes simplex type 1 virus, and Epstein-Barr virus. While shown to be a potent agent for treating varicella zoster infections, sorivudine's use has been discontinued in a number of nations in view of toxic responses and deaths, particularly when administered along with fluoropyrimidine drugs. **Target(s):** dihydropyrimidine dehydrogenase (1,2); DNA-directed DNA polymerase, as the 5'-triphosphate (3); DNA polymerase I, as the 5'-triphosphate (3); thymidine kinase (4,5); varicella zoster virus replication (6,7); viral DNA polymerase, as the 5'-triphosphate (5). **1.** Diasio (1998) *Brit. J. Clin. Pharmacol.* **46**, 1. **2.** Ogura, Nishiyama, Takubo, *et al.* (1998) *Cancer Lett.* **122**, 107. **3.** Suzutani, Machida & Honess (1993) *Microbiol. Immunol.* **37**, 511. **4.** Balzarinia, Degreve, Zhu, *et al.* (2001) *Biochem. Pharmacol.* **61**, 727. **5.** Gustafson, Chillemi, Sage & Fingeroth (1998) *Antimicrob. Agents Chemother.* **42**, 2923. **6.** Prisbe & Chen (1996) *Meth. Enzymol.* **275**, 425. **7.** Yokota, Konno, Mori, *et al.* (1989) *Mol. Pharmacol.* **36**, 312.

Sotalol

This water-soluble sulfonamide (FW$_{free-base}$ = 272.37 g/mol; CAS 3930-20-9), also known as Sotacor and *N*-[4-[1-hydroxy-2-[(1-methylethyl)amino]-ethyl]phenyl]methanesulfonamide, is a potent β-adrenergic receptor antagonist (1-5), predominantly as the (–)-enantiomer (*i.e.*, *l*-sotalol). The dextrorotatory isomer of sotalol (*i.e.*, (+)- or *d*-sotalol) is a class III anti-arrhythmic that prolongs cardiac repolarization by inhibiting the fast component of the delayed outward rectifying potassium channel. The *l*-(–)-enantiomer has both β-blocking (class II) activity and potassium-channel-blocking (class III) properties. The *d*-(+)-enantiomer has class III properties similar to those of *l*-sotalol; however, the affinity of *d*-sotalol for β-adrenergic receptors is 30-60 times lower than the affinity of *l*-sotalol. **Key Pharmacokinetic Parameters:** *See* Appendix II in Goodman & Gilman's *THE PHARMACOLOGICAL BASIS OF THERAPEUTICS*, 12th Edition (Brunton, Chabner & Knollmann, eds.) McGraw-Hill Medical, New York (2011). **1.** Salimi (1975) *Pharmacology* **13**, 441. **2.** Lish, Shelanski, LaBudde & Williams (1967) *Curr. Ther. Res. Clin. Exp.* **9**, 311. **3.** Cavusoglu & Frishman (1995) *Prog. Cardiovasc. Dis.* **37**, 423. **4.** Funck-Brentano (1983) *Eur. Heart J.* **14** Suppl. H, 30. **5.** Doggrell (1993) *Chirality* **5**, 8.

Sotrastaurin

This potent pan-Protein Kinase C (pan-PKC) inhibitor (F.Wt. = 438.48 g/mol; CAS 425637-18-9 and 1058706-32-3 (HCl); Solubility: DMSO = 87 mg/mL; Water < 1 mg/mL), also known as AEB071 and systematically as 3-(1*H*-indol-3-yl)-4-(2-(4-methylpiperazin-1-yl)quinazolin-4-yl)-1*H*-pyrrole-2,5-dione, inhibits the PKCθ isozyme slightly more potently (1). **Targets:** PKCα (IC$_{50}$ = 0.95 nM); PKCβ (IC$_{50}$ = 0.64 nM); PKCδ (IC$_{50}$ = 2.1 nM); PKCε (IC$_{50}$ = 3.2 nM); PKCη (IC$_{50}$ = 1.8 nM); and PKCθ (IC$_{50}$ = 0.22 nM). Sotrastaurin (<10μM) blocks early T-cell activation (indicated by interleukin-2 secretion and CD25 expression) in primary human and mouse T cells. AEB071 (200 nM) inhibits the CD3/CD28 antibody- and alloantigen-induced T-cell proliferation responses in the absence of nonspecific antiproliferative effects. AEB071 (< 3 μM) markedly inhibits lymphocyte function-associated antigen-1-mediated T-cell adhesion (1). AEB071 (<20 μM) selectively impairs CD79 mutant ABC-DLBCL cell proliferation, with decreased NF-κB signaling (1). AEB071 (5 μM) induces G$_1$ cell-cycle arrest and/or cell death in CD79 mutant cells (2). **1.** Evenou, *et al.* (2009) *J. Pharmacol. Exp. Ther.* **330**, 792. **2.** Naylor, *et al.* (2011) *Cancer Res.* **71**, 2643. **3.** Weckbecker, *et al.* (2010) *Transpl. Int.* **23**, 543.

Soybean Trypsin Inhibitor (Kunitz)

This 181-aminoacyl residue polypeptide (Mr = 28,964; CAS 9035-81-8; pI = 5.35) is a nutrient protein in soybeans that strongly inhibits trypsin (K_i = ~ 10 pM). They will also inhibit chymotrypsin, albeit with a larger K_i value (approximately 10 μM). Note that soybeans and other plants contain other trypsin and chymotrypsin inhibitors that are not members of this Kunitz family; hence, the designation Kunitz should be present. ***See*** *Trypsin Inhibitor; Bowman-Birk Inhibitor* **1.** Birk (1996) *Arch. Latinoam. Nutr.* **44**, Suppl. 1, 26S.

SP 600125

This cell-permeable, c-Jun *N*-terminal kinase inhibitor (FW = 220.23 g/mol (HCl Salt); CAS 129-56-6, 67072-00-8 (potassium salt); Solubility: 44 mg/mL DMSO; <1 mg/mL Water), also known as JNK inhibitor II, anthra[1,9-*cd*]pyrazol-6(2*H*)-one, and 1,9-pyrazoloanthrone, targets JNK1, JNK2, JNK3, Aurora A, Flt3, and TRKA kinases, with IC$_{50}$ of 40 nM, 40 nM, 90 nM, 60 nM, 90 nM, and 70 nM, respectively. SP600125 inhibits the phosphorylation of c-Jun with IC$_{50}$ of 10 μM (1-7). It also inhibits Mps1 (*or* Monopolar spindle-1), also known as TKK, a dual-specificity protein kinase that phosphorylates tyrosine, serine, or threonine residues. Mps1 is activated during mitosis and is essential for centrosome duplication, mitotic checkpoint signaling, and the maintenance of chromosomal instability, *or* CIN (8). **Target(s):** AMP-activated protein kinase (1); casein kinase 1δ (1); c-Jun N-terminal kinase α1, IC$_{50}$ = 5.8 μM (1,2); c-Jun N-terminal kinase α2, *IC$_{50}$* = 6.1 μM; cyclin-dependent protein kinase; dual-specificity, tyrosine-phosphorylated and regulated kinase; p70 ribosomal protein S6 kinase; and serum- and glucocortico-induced kinase (1); ERK2, $IC_{50} > 10$ µM; p38-2, $IC_{50} > 10$ µM; p56Lck, $IC_{50} = 4.3$ µM; Chk1, $IC_{50} > 10$ µM; EGF-TK, $IC_{50} > 10$ µM; IKK1, $IC_{50} > 10$ µM; IKK2, $IC_{50} > 10$ µM; MEKK1, $IC_{50} > 10$ µM; MKK3, $IC_{50} = 1.5$ µM; MKK4, $IC_{50} = 0.4$ µM; ERK2, $IC_{50} > 10$ µM; ERK2, $IC_{50} > 10$ µM; ERK2, $IC_{50} > 10$ µM; ERK2, $IC_{50} > 10$ µM; MKK6, $IC_{50} = 1$ µM; MKK7, $IC_{50} = 5$ µM; ERK2, $IC_{50} > 10$ µM; PKA, $IC_{50} > 10$ µM; PKB/AKT, $IC_{50} = 1$ µM; PKC, $IC_{50} > 10$ µM; PKCα, $IC_{50} > 10$ µM; PKCθ, $IC_{50} > 10$ µM. **1.** Bain, McLauchlan, Elliott & Cohen (2003) *Biochem. J.* **371**, 199. **2.** Bennett, *et al.* (2001) *Proc. Natl. Acad. Sci. USA* **98**, 13681. **3.** Joiakim, *et al.* (2003) *Drug Metab. Dispos.* **31**, 1279. **4.** Schmidt, *et al.* (2005) *EMBO Rep.* **6**, 866. **5.** Colombo, *et al.* (2010) *Cancer Res.* **70**, 10255. **6.** Vaishnav, *et al.* (2003) *Biochem. Biophys. Res. Communs.* **307**, 855. **7.** Kim, *et al.* (2010) *Oncogene* **29**, 1702. **8.** Kusakabe, Ide, Naigo, *et al.* (2013) *J. Med. Chem.* **56**, 4343.

Spantide I

This D-aminoacid-containing undecapeptide amide (M_r = 1499; CAS 91224-37-2; *Sequence*: D-Arg-L-Pro-L-Lys-L-Pro-L-Gln-L-Gln-D-Trp-L-Phe-D-Trp-L-Leu-L-Leu-NH$_2$), also known as [D-Arg^1,D-Trp^{7,9},Leu^{11}]-substance P, is a potent NK-1 tachykinin receptor antagonist. **1.** Folkers, Hakanson, Horig, Xu & Leander (1984) *Brit. J. Pharmacol.* **83**, 449.

Sparangiomycin, *See Sporangiomycin*

Sparfloxacin

This quinolone-based antimicrobial agent (FW = 343.40 g/mol; CAS 111542-93-9), also known as AT-4140 and RP 64206, inhibits bacterial DNA topoisomerase, *or* DNA gyrase (1-3). **1.** Tanaka, Sato, Kimura, *et al.* (1991) *Antimicrob. Agents Chemother.* **35**, 1489. **2.** Pan & Fisher (1999) *Antimicrob. Agents Chemother.* **43**, 1129.

Sparsomycin

This antiviral (FW = 361.44 g/mol; CAS 1404-64-4) from *Streptomyces sparsogenes* and *S. cuspidosporus*, inhibits protein biosynthesis by causing the formation of an inert complex between a peptidyl donor and the 70S ribosomal subunit. Sparsomycin binds to the ribosome strongly in the presence of an N-blocked tRNA substrate, which it stabilizes on the ribosome. Sparsomycin is the first example of a slow-binding inhibitor of the eukaryotic peptidyltransferase (1-13). Structures of the large ribosomal subunit of *Haloarcula marismortui* at 3.0Å resolution indicate the poses taken by bound anisomycin, chloramphenicol, sparsomycin, blasticidin S, and virginiamycin M (14). Two hydrophobic crevices, one at the peptidyl transferase center and the other at the entrance to the peptide exit tunnel play roles in antibiotic binding. Sparsomycin contacts primarily a P-site bound substrate, but also extends into the active-site hydrophobic crevice. Most ribotoxic antibiotics bind to sites that overlap those of either peptidyl-tRNA or aminoacyl-tRNA, consistent with their functioning as competitive inhibitors of peptide bond formation (14). (For the prototypical member of this antibiotic class, *See Ciprofloxacin*) **1.** Ioannou, Coutsogeorgopoulos & Synetos (1998) *Mol. Pharmacol.* **53**, 1089. **2.** Gottesman (1971) *Meth.*

Enzymol. **20**, 490. **3.** Pestka (1974) *Meth. Enzymol.* **30**, 261. **4.** Carrasco, Battaner & Vazquez (1974) *Meth. Enzymol.* **30**, 282. **5.** Scott & Tomkins (1975) *Meth. Enzymol.* **40**, 273. **6.** Jiménez (1976) *Trends Biochem. Sci.* **1**, 28. **7.** Lucas-Lenard & Beres (1974) *The Enzymes*, 3rd ed. (Boyer, ed.), **10**, 53. **8.** Fernandez-Muñoz, Monro & Vazquez (1971) *Meth. Enzymol.* **20**, 481. **9.** Goldberg & Mitsugi (1967) *Biochemistry* **6**, 372 and 383. **10.** Ioannou, Coutsogeorgopoulos & Drainas (1997) *Anal. Biochem.* **247**, 115. **11.** Polacek & Mankin (2005) *Crit. Rev. Biochem. Mol. Biol.* **40**, 285. **12.** Tate & Caskey (1974) *The Enzymes*, 3rd ed. (Boyer, ed.), **10**, 87. **13.** Campuzano & Modolell (1981) *Eur. J. Biochem.* **117**, 27. **14.** Hansen, Moore & Steitz (2003) *J. Mol. Biol.* **330**, 1061.

(–)-Sparteine

This alkaloid (FW$_{free-base}$ = 234.38 g/mol; CAS 90-39-1; liquid; B.P. = 137-138°C at 1 mm Hg; water solubility = 1 g/325 mL water), also known as lupinidine, found in lupin beans (*Lupinus luteus* and *L. niger*) as well as *Sarothamnus scoparius*, is a substrate for CYP2D6. **Target(s):** arginyl-tRNA synthetase (1); cholinesterase (2); and K$^+$-currents, ATP-regulated (3). **1.** Zwierzynski, Joachimiak, Barciszewska, Kulinska & Barciszewski (1982) *Chem. Biol. Interact.* **42**, 107. **2.** Augustinsson (1950) *The Enzymes*, 1st ed. (Sumner & Myrbäck, eds.), **1** (part 1), 443. **3.** Ashcroft, Kerr, Gibson & Williams (1991) *Brit. J. Pharmacol.* **104**, 579.

SPD754, *See Apricitabine*

Spectinomycin

This water-soluble, broad-spectrum antibiotic (FW$_{free-base}$ = 332.35 g/mol; CAS 1695-77-8), from *Streptomyces spectabilis* is similar in action to streptomycin. Spectinomycin interacts specifically with the residues G1064 and C1192 in *Escherichia coli* 16S rRNA, thereby locking the ribosomal translation machinery in an inactive conformation. Spectinomycin is highly effective in the treatment of uncomplicated gonorrhea since it is especially active against *Neisseria gonorrhoeae*. **Target(s):** protein biosynthesis, *or* peptidyl-tRNA translocation (1-6); and self-splicing intron RNA (7,8). **1.** Pestka (1974) *Meth. Enzymol.* **30**, 261. **2.** Reusser (1976) *J. Antibiot. (Tokyo)* **29**, 1328. **3.** Kostiashkina, Asatrian, Gavrilova & Spirin (1975) *Mol. Biol. (Moscow)* **9**, 775. **4.** Wallace, Tai & Davis (1974) *Proc. Natl. Acad. Sci. U.S.A.* **71**, 1634. **5.** Davies, Anderson & Davis (1965) *Science* **149**, 1096. **6.** Brink, Brink, Verbeet & de Boer (1994) *Nucl. Acids Res.* **22**, 325. **7.** Park & Sung (2000) *Biochim. Biophys. Acta* **1492**, 94. **8.** Park, Kim, Lim & Shin (2000) *Biochem. Biophys. Res. Commun.* **269**, 574.

Spermidine

This cationic polyamine (FW$_{free-base}$ = 145.25 g/mol; CAS 124-20-9; MP$_{trihydrochloride}$ = 256-258°C), also known as *N*-(3-aminopropyl)-1,4-diaminobutane, found typically in millimolar concentrations in numerous cell types, binds preferentially to duplex DNA or RNA and interacts at K$^+$ and Na$^+$ binding sites on numerous proteins. Spermidine also stimulates T$_4$ polynucleotide kinase. The fact that spermidine has so many inhibitory effects suggests that within cells the free (unbound) spermidine level must be quite low to prevent such promiscuous inhibition. Measurements of the free spermidine concentration in tissues have not been reported. **Target(s):** adenosine deaminase, or RNA-adenosine deaminase (74); adenosylmethionine decarboxylase (54-56); adenylate cyclase (1-3); [β-adrenergic-receptor] kinase (4); agmatine deaminase (75-78); *Agropyron elongatum* nuclease (82); amino-acid *N*-acetyltransferase (100); 1-aminocyclopropane-1-carboxylate synthase (53); arginine decarboxylase (5,57-66); arginine *N*-succinyltransferase (98); arginyltransferase (6); *Bam*HI restriction endonuclease (7); Ca^{2+}-dependent ATPase (8); carbamoyl-phosphate synthetase, glutamine hydrolyzing (9,10); carnosine synthetase (50); deoxyhypusine monooxygenase (11); CDP-glycerol glycerophosphotransferase (89); DNA (cytosine-5-)-methyltransferase (105,106); DNA ligase, ATP-dependent (12-14); DNA-3-methyladenine glycosylase I (81); DNA polymerase (15,16); DNA polymerase III, however, stimulation of the complex or holoenzyme (17); DNA topoisomerase I (51,52;) dsRNA-dependent protein kinase (18); *Eco*RI restriction endonuclease (7,19); epidermal-growth-factor-receptor protein-tyrosine kinase (20); exopolyphosphatase (73); F$_1$ ATPase, or H$^+$-transporting two-sector ATPase (21); glutamate dehydrogenase (22,23); *Hind*III restriction endonuclease (7); histone acetyltransferase (99); homospermidine synthase (93); *Hpa*I restriction endonuclease (7); inositol 1,4,5-trisphosphate receptor (24); K$_{ATP}$ channels (25); lipoxygenase-1 (26); methionine *S*-adenosyltransferase (27,94); 3-methyladenine-DNA glycosylase (28); Na$^+$/K$^+$-exchanging ATPase (29); nitric-oxide synthase(30); OmpF and OmpC bacterial porins (31); ornithine aminotransferase (32); ornithine carbamoyl-transferase (102-104); ornithine decarboxylase (33,67-72); papain (34); pepsin (35,36); phosphatidate phosphatase, weakly inhibited (86); phospholipase C (37); phosphoprotein phosphatase (38,39); photosystem II (40); polynucleotide adenylyltransferase, or poly(A) polymerae (91); protein-arginine *N*-methyltransferase (41); protein-glutamine γ-glutamyltransferase, or transglutaminase, also alternative substrate (95-97); protein kinase C (42,43); protein-tyrosine-phosphatase (85); *Pst*I restriction endonuclease (7); putrescine carbamoyltransferase (101); pyridoxal kinase, weakly inhibited (92); rhodopsin kinase (88); ribonuclease H (83); ribonuclease U$_2$, or *Ustilago sphaerogena* ribonuclease (48); [RNA-polymerase]-subunit kinase (87); semenogelase (44,79); T$_4$ polynucleotide ligase (45); tRNA (adenine-N^1-)-methyltransferase (107); tRNA adenylyltransferase (46,90); tRNA cytidylyltransferase (90); tRNA (cytosine-5-)-methyltransferase (109); tRNA nucleotidyltransferase (90); tRNA (uracil-5-)-methyltransferase (108); *Trypanosoma cruzi* protein kinases (47); tubulinyl-tyrosine carboxypeptidase, inhibited above 1 mM (80); vacuolar ion channels (49); and yeast ribonuclease, weakly inhibited (84). 1. Khan, Quemener & Moulinoux (1990) *Life Sci.* **46**, 43. 2. Wright, Buehler, Schott & Rennert (1978) *Pediatr. Res.* **12**, 830. 3. Rennert & Shukla (1978) *Adv. Polyamine Res.* **2**, 195. 4. Benovic, Stone, Caron & Lefkowitz (1989) *J. Biol. Chem.* **264**, 6707. 5. Boeker & Snell (1971) *Meth. Enzymol.* **17B**, 657. 6. Kato (1983) *J. Biochem.* **94**, 2015. 7. Kuosmanen & Poso (1985) *FEBS Lett.* **179**, 17. 8. Hughes, Starling, East & Lee (1994) *Biochemistry* **33**, 4745. 9. Mori & Tatibana (1978) *Meth. Enzymol.* **51**, 111. 10. Szondy, Matyasi & Elodi (1989) *Acta Biochim. Biophys. Hung.* **24**, 107. 11. Abbruzzese, Park, Beninati & Folk (1989) *Biochim. Biophys. Acta* **997**, 248. 12. Engler & Richardson (1982) *The Enzymes*, 3rd ed. (Boyer, ed.), **15**, 3. 13. Teraoka, Sawai & Tsukada (1983) *Biochim. Biophys. Acta* **747**, 117. 14. Ferretti & Sgaramella (1981) *Nucl. Acids. Res.* **9**, 3695. 15. Weissbach (1981) *The Enzymes*, 3rd ed. (Boyer, ed.), **14**, 67. 16. Osland & Kleppe (1978) *Biochim. Biophys. Acta* **520**, 317. 17. McHenry & Kornberg (1981) *The Enzymes*, 3rd ed. (Boyer, ed.), **14**, 39. 18. Samuel, Knutson, Berry, Atwater & Lasky (1986) *Meth. Enzymol.* **119**, 499. 19. Pingoud, Urbanke, Alves, *et al.* (1984) *Biochemistry* **23**, 5697. 20. Faaland, Laskin & Thomas (1995) *Cell Growth Differ.* **6**, 115. 21. Igarashi, Kashiwagi, Kobayashi, *et al.* (1989) *J. Biochem.* **106**, 294. 22. Kuo, Michalik & Erecinska (1994) *J. Neurochem.* **63**, 751. 23. Jarzyna, Lietz & Bryla (1994) *Biochem. Pharmacol.* **47**, 1387. 24. Sayers & Michelangeli (1993) *Biochem. Biophys. Res. Commun.* **197**, 1203. 25. Niu & Meech (1994) *J. Physiol.* **508**, 401. 26. Maccarrone, Baroni & Finazzi-Agro (1998) *Arch. Biochem. Biophys.* **356**, 35. 27. Geller, Legros, Wherry & Kotb (1997) *Arch. Biochem. Biophys.* **345**, 97. 28. Quarfoth, Ahmed & Foster (1978) *Biochim. Biophys. Acta* **526**, 580. 29. Gallagher & Brent (1984) *Biochim. Biophys. Acta* **782**, 394. 30. Hu, Mahmoud & el-Fakahany (1994) *Neurosci. Lett.* **175**, 41. 31. Iyer & Delcour (1997) *J. Biol. Chem.* **272**,

18595. **32.** Deshmukh & Srivastava (1984) *Experientia* **40**, 357. **33.** Boeker & Snell (1972) *The Enzymes*, 3rd ed. (Boyer, ed.), **6**, 217. **34.** Mekras, Lawton & Washington (1989) *J. Pharm. Pharmacol.* **41**, 22. **35.** Mekras (1989) *Int. J. Biol. Macromol.* **11**, 207. **36.** Lawton & Mekras (1985) *J. Pharm. Pharmacol.* **37**, 396. **37.** Takahashi, Sugahara & Ohsaka (1981) *Meth. Enzymol.* **71**, 710. **38.** Ballou & Fischer (1986) *The Enzymes*, 3rd ed. (Boyer & Krebs, eds.), **17**, 311. **39.** Khandelwal & Enno (1985) *J. Biol. Chem.* **260**, 14335. **40.** Bograh, Gingras, Tajmir-Riahi & Carpentier (1997) *FEBS Lett.* **402**, 41. **41.** Yoo, Park, Okuda, *et al.* (1999) *Amino Acids* **17**, 391. **42.** Kikkawa & Nishizuka (1986) *The Enzymes*, 3rd ed. (Boyer & Krebs, eds.), **17**, 167. **43.** Thams, Capito & Hedeskov (1986) *Biochem. J.* **237**, 131. **44.** Chao (1998) in *Handb. Proteolytic Enzymes* (Barrett, Rawlings & Woessner, eds.), p. 102, Academic Press, San Diego. **45.** Raae, Kleppe & Kleppe (1975) *Eur. J. Biochem.* **60**, 437. **46.** Deutscher (1974) *Meth. Enzymol.* **29**, 706. **47.** Walter & Ebert (1979) *Tropenmed. Parasitol.* **30**, 9. **48.** Glitz & Dekker (1964) *Biochemistry* **3**, 1391. **49.** Dobrovinskaya, Muniz & Pottosin (1999) *J. Membr. Biol.* **167**, 127. **50.** Seely & Marshall (1982) in *Peptide Antibiotics: Biosynthesis and Functions: Enzymatic Formation of Bioactive Peptides and Related Compounds* (Kleinkauf & von Döhren, eds.) W. de Gruyter, Berlin, pp. 347. **51.** Alkorta, Park, Kong, *et al.* (1999) *Arch. Biochem. Biophys.* **362**, 123. **52.** Yang, Lu & Rubin (1996) *Gene* **178**, 63. **53.** Miyazaki & Yang (1987) *Phytochemistry* **26**, 2655. **54.** Yamanoha & Cohen (1985) *Plant Physiol.* **78**, 784. **55.** Yang & Cho (1991) *Biochem. Biophys. Res. Commun.* **181**, 1181. **56.** Sakai, Hori, Kano & Oka (1979) *Biochemistry* **18**, 5541. **57.** Blethen, Boeker & Snell (1968) *J. Biol. Chem.* **243**, 1671. **58.** Smith (1979) *Phytochemistry* **18**, 1447. **59.** Ramakrishna & Adiga (1975) *Eur. J. Biochem.* **59**, 377. **60.** Das, Bhaduri, Bose & Ghosh (1996) *J. Plant Biochem. Biotechnol.* **5**, 123. **61.** Balasundaram & Tyagi (1989) *Eur. J. Biochem.* **183**, 339. **62.** Choudhuri & Ghosh (1982) *Agric. Biol. Chem.* **46**, 739. **63.** Rosenfeld & Roberts (1976) *J. Bacteriol.* **125**, 601. **64.** Nam, Lee & Lee (1997) *Plant Cell Physiol.* **38**, 1150. **65.** Winer, Vinkler & Apelbaum (1984) *Plant Physiol.* **76**, 233. **66.** Li, Regunathan & Reis (1995) *Ann. N.Y. Acad. Sci.* **763**, 325. **67.** Koromilas & Kyriakidis (1988) *Phytochemistry* **27**, 989. **68.** Arteaga-Nieto, Villagomez-Castro, Calvo-Mendez & Lopez-Romero (1996) *Int. J. Parasitol.* **26**, 253. **69.** Guirard & Snell (1980) *J. Biol. Chem.* **255**, 5960. **70.** Schaeffer & Donatelli (1990) *Biochem. J.* **270**, 599. **71.** Pandit & Ghosh (1988) *Phytochemistry* **27**, 1609. **72.** Ono, Inoue, Suzuki & Takeda (1972) *Biochim. Biophys. Acta* **284**, 285. **73.** Wurst & Kornberg (1994) *J. Biol. Chem.* **269**, 10996. **74.** Hough & Bass (1994) *J. Biol. Chem.* **269**, 9933. **75.** Chaudhuri & Ghosh (1985) *Phytochemistry* **24**, 2433. **76.** Yanagisawa & Suzuki (1981) *Plant Physiol.* **67**, 697. **77.** Sindhu & Desai (1979) *Phytochemistry* **18**, 1937. **78.** Park & Cho (1991) *Biochem. Biophys. Res. Commun.* **174**, 32. **79.** Watt, Lee, Timkulu, Chan & Loor (1986) *Proc. Natl. Acad. Sci. U.S.A.* **83**, 3166. **80.** Barra & Argarana (1982) *Biochem. Biophys. Res. Commun.* **108**, 654. **81.** Thomas, Yang & Goldthwait (1982) *Biochemistry* **21**, 1162. **82.** Yupsanis, Symeonidis, Kalemi, Moustaka & Yupsani (2004) *Plant Physiol. Biochem.* **42**, 795. **83.** Kitahara, Sawai & Tsukada (1982) *J. Biochem.* **92**, 855. **84.** Nakao, Lee, Halvorson & Bock (1968) *Biochim. Biophys. Acta* **151**, 114. **85.** Boivin & Galand (1986) *Biochem. Biophys. Res. Commun.* **134**, 557. **86.** Nanjundan & Possmayer (2000) *Exp. Lung Res.* **26**, 361. **87.** Guilfoyle (1989) *Plant Cell* **1**, 827. **88.** Kikkawa, Yoshida, Nakagawa, Iwasa & Tsuda (1998) *J. Biol. Chem.* **273**, 7441. **89.** Burger & Glaser (1964) *J. Biol. Chem.* **239**, 3168. **90.** Evans & Deutscher (1976) *J. Biol. Chem.* **251**, 6646. **91.** Rose & Jacob (1976) *Arch. Biochem. Biophys.* **175**, 748. **92.** Gäng & von Collins (1973) *Int. J. Vitam. Nutr. Res.* **43**, 318. **93.** Srivenugopal & Adiga (1980) *Biochem. J.* **190**, 461. **94.** Schröder, Eichel, Breinig & Schröder (1997) *Plant Mol. Biol.* **33**, 211. **95.** Siegel & Khosla (2007) *Pharmacol. Ther.* **115**, 232. **96.** Wu, Lai & Tsai (2005) *Int. J. Biochem. Cell Biol.* **37**, 386. **97.** Jeon, Lee, Jang, *et al.* (2004) *Exp. Mol. Med.* **36**, 576. **98.** Tricot, Vander Wauven, Wattiez, Falmagne & Stalon (1994) *Eur. J. Biochem.* **224**, 853. **99.** Libby (1980) *Arch. Biochem. Biophys.* **203**, 384. **100.** Haas & Leisinger (1974) *Biochem. Biophys. Res. Commun.* **60**, 42. **101.** Wargnies, Lauwers & Stalon (1979) *Eur. J. Biochem.* **101**, 143. **102.** Tricot, Schmid, Baur, *et al.* (1994) *Eur. J. Biochem.* **221**, 555. **103.** Sainz, Tricot, Foray, *et al.* (1998) *Eur. J. Biochem.* **251**, 528. **104.** Tricot, Villeret, Sainz, Dideberg & Stalon (1998) *J. Mol. Biol.* **283**, 695. **105.** Theiss, Schleicher, Schimpff-Weiland & Follmann (1987) *Eur. J. Biochem.* **167**, 89. **106.** Cox (1979) *Biochem. Biophys. Res. Commun.* **86**, 594. **107.** Mutzel, Malchow, Meyer & Kersten (1986) *Eur. J. Biochem.* **160**, 101. **108.** Ny, Lindström, Hagervall & Björk (1988) *Eur. J. Biochem.* **177**, 467. **109.** Nau, Pham-Coeur-Joly & Dubert (1983) *Eur. J. Biochem.* **130**, 261.

Spermidine Diacridine, *See Diacridines*

Spermine

This polyamine (FW$_{free-base}$ = 202.34 g/mol; CAS 71-44-3; M.P. = 55-60°C; deliquescent solid), known systematically as N,N'-bis(3-aminopropyl)-1,4-butanediamine, is found in all tissues and is required for growth. Spermine is a mixed NMDA (N-methyl-D-aspartate) receptor agonist/antagonist, participates in cellular proliferation and differentiation, and also exhibits neuroprotective effects. (The fact that spermine has so many inhibitory effects suggests that within cells the free (unbound) spermine level must be quite low to prevent such promiscuous inhibition. NMR measurements of the free spermine concentration in tissues have not been reported.) Note: Solutions of the water-soluble free base will absorb atmospheric carbon dioxide; thus, containers of spermine must be kept tightly closed, and spermine solutions should be freshly prepared. **Target(s):** acetylspermidine deacetylase, alternative product (1,99,100); adenosyl-methionine decarboxylase (2,79-81); adenylate cyclase (3-6); [β-adrenergic-receptor] kinase, weakly inhibited (7,8); agmatine deaminase (95-98); *Agropyron elongatum* nuclease (103); alfalfa leaf protease (9); amino-acid N-acetyltransferase (128); 1-aminocyclopropane-1-carboxylate synthase (78); arginine decarboxylase (10,82-87); arginyltransferase (11); *Bam*HI restriction endonuclease (12); bis(5'-adenosyl)-triphosphatase (94); Ca^{2+}-dependent ATPase (13,14); carbamoyl-phosphate synthetase, glutamine-hydrolyzing (15,16); cyclic-nucleotide phosphodiesterase (17-19); deoxyhypusine monooxygenase (20); deoxyhypusine synthase, weakly inhibited (119); DNA (cytosine-5-)-methyltransferase (129,130); DNA ligase (ATP) (21,22); DNA polymerase (23,24); dsRNA-dependent protein kinase (25); *Eco*RI restriction endonuclease (12,26); epidermal-growth-factor-receptor protein-tyrosine kinase (27); exoribonuclease II, *or* ribonuclease II, weakly inhibited (105); F$_1$ ATPase, *or* H$^+$-transporting two-sector ATPase (28); glutamate dehydrogenase (29,30); [glycogen synthase] kinase 3 (31); glycyl-tRNA synthetase (77); GTPase activity of G$_i$ proteins (32); H$^+$ ATPase (33); *Hind*III restriction endonuclease (12); homospermidine synthase, spermidine-specific (120); *Hpa*I restriction endonuclease (12); inositol 1,3,4-trisphosphate 5/6-kinase (34); inositol 1,4,5-trisphosphate receptor (35); K$_{ATP}$ channels (36); K$^+$-dependent phosphatase (37); lipoxygenase-1 (38); methionine S-adenosyltransferase (39,122); 3-methyladenine-DNA glycosylase (40); Na$^+$/K$^+$ exchanging ATPase (41-43); nardilysin (44,101); nicotinamide mononucleotide glycohydrolase (45); nicotinic acetylcholine receptor (46); nitric-oxide synthase (47,48); nuclear protein kinase NI (49); OmpF and OmpC bacterial porins (50); ornithine aminotransferase (51); ornithine decarboxylase (52,88-93); papain (53,54); pepsin (55); peptidyltransferase (56,126,127); phosphatidate phosphatase (109,110); phosphatidylinositol 3-kinase (57); 1-phosphatidylinositol-4-phosphate 5-kinase (117); phosphoinositide phospholipase C (106); phospholipase A$_2$ (58); phospholipase C (58,59); phosphoprotein phosphatase, including [phosphorylase] phosphatase (60-63,108); photosystem II (64); polynucleotide adenylyltransferase, *or* poly(A) polymerase (65,114,115); polynucleotide 5'-hydroxyl-kinase, activated at 1 mM spermine, but inhibited at elevated concentrations (116); protein-arginine N-methyltransferase (66); protein-glutamine γ-glutamyltransferase, *or* transglutaminase, also alternative substrate (123-125); protein kinase C (67,68); protein-tyrosine-phosphatase (107); *Pst*I restriction endonuclease (12); pyridoxal kinase, weakly inhibited (118); rhodopsin kinase, also activated at low concentrations (111); ribonuclease H (104); ribonuclease U$_2$, *Ustilago sphaerogena* ribonuclease (75); RNA-directed RNA polymerase (112,113); RNA ligase ribozyme (69); spermidine synthase (121); T$_4$ polynucleotide ligase (70); transglutaminase (71); tRNA adenylyltransferase (72); tRNA (cytosine-5-)-methyltransferase (131,132); *Trypanosoma cruzi* protein kinases (73); tubulinyl-tyrosine carboxypeptidase, inhibited above 0.06 mM (102); tyrosine monooxygenase (74); and vacuolar ion channels (76). **1.** Libby (1983) *Meth. Enzymol.* **94**, 329. **2.** Sakai, Hori, Kano & Oka (1979) *Biochemistry* **18**, 5541. **3.** Khan, Quemener & Moulinoux (1990) *Life Sci.* **46**, 43. **4.** Clo, Tantini, Sacchi & Caldarera (1988) *Adv. Exp. Med. Biol.* **250**, 535. **5.** Wright, Buehler, Schott & Rennert (1978) *Pediatr. Res.* **12**, 830. **6.** Rennert & Shukla (1978) *Adv. Polyamine Res.* **2**, 195. **7.** Benovic (1991) *Meth. Enzymol.* **200**, 351. **8.** Benovic, Stone, Caron & Lefkowitz (1989) *J. Biol. Chem.* **264**, 6707. **9.** Balestreri, Cioni, Romagnoli, *et al.* (1987) *Arch. Biochem. Biophys.* **255**, 460. **10.** Ramakrishna & Adiga (1975) *Eur. J. Biochem.* **59**, 377. **11.** Kato (1983) *J. Biochem.* **94**, 2015. **12.** Kuosmanen & Poso (1985) *FEBS Lett.*

179, 17. **13.** Palacios, Sepulveda & Mata (2003) *Biochim. Biophys. Acta* **1611**, 197. **14.** Hughes, Starling, East & Lee (1994) *Biochemistry* **33**, 4745. **15.** Mori & Tatibana (1978) *Meth. Enzymol.* **51**, 111. **16.** Szondy, Matyasi & Elodi (1989) *Acta Biochim. Biophys. Hung.* **24**, 107. **17.** Kincaid & Vaughan (1988) *Meth. Enzymol.* **159**, 557. **18.** Walters & Johnson (1988) *Biochim. Biophys. Acta* **957**, 138. **19.** Shah & Sheth (1978) *Experientia* **34**, 980. **20.** Abbruzzese, Park, Beninati & Folk (1989) *Biochim. Biophys. Acta* **997**, 248. **21.** Engler & Richardson (1982) *The Enzymes*, 3rd ed. (Boyer, ed.), **15**, 3. **22.** Teraoka, Sawai & Tsukada (1983) *Biochim. Biophys. Acta* **747**, 117. **23.** Marcus, Kopelman, Koll & Bacchi (1982) *Mol. Biochem. Parasitol.* **5**, 231. **24.** Osland & Kleppe (1978) *Biochim. Biophys. Acta* **520**, 317. **25.** Samuel, Knutson, Berry, Atwater & Lasky (1986) *Meth. Enzymol.* **119**, 499. **26.** Pingoud, Urbanke, Alves, *et al.* (1984) *Biochemistry* **23**, 5697. **27.** Faaland, Laskin & Thomas (1995) *Cell Growth Differ.* **6**, 115. **28.** Igarashi, Kashiwagi, Kobayashi, *et al.* (1989) *J. Biochem.* **106**, 294. **29.** Kuo, Michalik & Erecinska (1994) *J. Neurochem.* **63**, 751. **30.** Jarzyna, Lietz & Bryla (1994) *Biochem. Pharmacol.* **47**, 1387. **31.** Hegazy, Schlender, Wilson & Reimann (1989) *Biochim. Biophys. Acta* **1011**, 198. **32.** Daeffler, Chahdi, Gies & Landry (1999) *Brit. J. Pharmacol.* **127**, 1021. **33.** Schwarcz de Tarlovsky, Rilo, Hernandez, *et al.* (1995) *Cell. Mol. Biol. (Noisy-le-grand)* **41**, 861. **34.** Hughes, Kirk & Michell (1994) *Biochim. Biophys. Acta* **1223**, 57. **35.** Sayers & Michelangeli (1993) *Biochem. Biophys. Res. Commun.* **197**, 1203. **36.** Niu & Meech (1998) *J. Physiol.* **508**, 401. **37.** Tashima, Hasegawa, Mizunuma & Sakagishi (1977) *Biochim. Biophys. Acta* **482**, 1. **38.** Maccarrone, Baroni & Finazzi-Agro (1998) *Arch. Biochem. Biophys.* **356**, 35. **39.** Geller, Legros, Wherry & Kotb (1997) *Arch. Biochem. Biophys.* **345**, 97. **40.** Gallagher & Brent (1984) *Biochim. Biophys. Acta* **782**, 394. **41.** Robinson, Leach & Robinson (1986) *Biochim. Biophys. Acta* **856**, 536. **42.** Quarfoth, Ahmed & Foster (1978) *Biochim. Biophys. Acta* **526**, 580. **43.** Heinrich-Hirsch, Ahlers & Peter (1977) *Enzyme* **22**, 235. **44.** Csuhai, Juliano, Juliano & Hersh (1999) *Arch. Biochem. Biophys.* **362**, 291. **45.** Imai (1987) *J. Biochem.* **101**, 153. **46.** Shao, Mellor, Brierley, Harris & Usherwood (1998) *J. Pharmacol. Exp. Ther.* **286**, 1269. **47.** Hu, Mahmoud & el-Fakahany (1994) *Neurosci. Lett.* **175**, 41. **48.** Blachier, Mignon & Soubrane (1997) *Nitric Oxide* **1**, 268. **49.** Verma & Chen (1986) *J. Biol. Chem.* **261**, 2890. **50.** Iyer & Delcour (1997) *J. Biol. Chem.* **272**, 18595. **51.** Deshmukh & Srivastava (1984) *Experientia* **40**, 357. **52.** Jänne & Williams-Asham (1971) *J. Biol. Chem.* **246**, 1725. **53.** Mekras, Lawton & Washington (1989) *J. Pharm. Pharmacol.* **41**, 22. **54.** Lawton & Mekas (1985) *J. Pharm. Pharmacol.* **37**, 396. **55.** Mekras (1989) *Int. J. Biol. Macromol.* **11**, 207. **56.** Kalpaxis & Drainas (1993) *Arch. Biochem. Biophys.* **300**, 629. **57.** Singh, Chauhan, Brockerhoff & Chauhan (1995) *Life Sci.* **57**, 685. **58.** Sechi, Cabrini, Landi, Pasquali & Lenaz (1978) *Arch. Biochem. Biophys.* **186**, 248. **59.** Takahashi, Sugahara & Ohsaka (1981) *Meth. Enzymol.* **71**, 710. **60.** Ballou & Fischer (1986) *The Enzymes*, 3rd ed. (Boyer & Krebs, eds.), **17**, 311. **61.** Khandelwal & Enno (1985) *J. Biol. Chem.* **260**, 14335. **62.** Killilea, Mellgren, Aylward & Lee (1978) *Biochem. Biophys. Res. Commun.* **81**, 1040. **63.** Nakai & Glinsmann (1977) *Mol. Cell. Biochem.* **15**, 141. **64.** Bograh, Gingras, Tajmir-Riahi & Carpentier (1997) *FEBS Lett.* **402**, 41. **65.** Edmonds (1982) *The Enzymes*, 3rd ed. (Boyer, ed.), **15**, 217. **66.** Yoo, Park, Okuda, *et al.* (1999) *Amino Acids* **17**, 391. **67.** Kikkawa & Nishizuka (1986) *The Enzymes*, 3rd ed. (Boyer & Krebs, eds.), **17**, 167. **68.** Thams, Capito & Hedeskov (1986) *Biochem. J.* **237**, 131. **69.** Glasner, Bergman & Bartel (2002) *Biochemistry* **41**, 8103. **70.** Raae, Kleppe & Kleppe (1975) *Eur. J. Biochem.* **60**, 437. **71.** Korner, Schneider, Purdon & Bjornsson (1989) *Biochem. J.* **262**, 633. **72.** Deutscher (1974) *Meth. Enzymol.* **29**, 706. **73.** Walter & Ebert (1979) *Tropenmed. Parasitol.* **30**, 9. **74.** Kiuchi, Kiuchi, Togari & Nagatsu (1987) *Biochem. Biophys. Res. Commun.* **148**, 1460. **75.** Glitz & Dekker (1964) *Biochemistry* **3**, 1391. **76.** Dobrovinskaya, Muniz & Pottosin (1999) *J. Membr. Biol.* **167**, 127. **77.** Dignam & Dignam (1984) *J. Biol. Chem.* **259**, 4043. **78.** Miyazaki & Yang (1987) *Phytochemistry* **26**, 2655. **79.** Ferioli, Candiani, Rocca & Scalabrino (1989) *Biogenic Amines* **6**, 513. **80.** Yang & Cho (1991) *Biochem. Biophys. Res. Commun.* **181**, 1181. **81.** Persson, Aslund, Grahn, Hanke & Heby (1998) *Biochem. J.* **333**, 527. **82.** Smith (1979) *Phytochemistry* **18**, 1447. **83.** Das, Bhaduri, Bose & Ghosh (1996) *J. Plant Biochem. Biotechnol.* **5**, 123. **84.** Choudhuri & Ghosh (1982) *Agric. Biol. Chem.* **46**, 739. **85.** Nam, Lee & Lee (1997) *Plant Cell Physiol.* **38**, 1150. **86.** Winer, Vinkler & Apelbaum (1984) *Plant Physiol.* **76**, 233. **87.** Li, Regunathan & Reis (1995) *Ann. N.Y. Acad. Sci.* **763**, 325. **88.** Koromilas & Kyriakidis (1988) *Phytochemistry* **27**, 989. **89.** Arteaga-Nieto, Villagomez-Castro, Calvo-Mendez & Lopez-Romero (1996) *Int. J. Parasitol.* **26**, 253. **90.** Guirard & Snell (1980) *J. Biol. Chem.* **255**, 5960. **91.** Schaeffer & Donatelli (1990) *Biochem. J.* **270**, 599. **92.** Pandit & Ghosh (1988) *Phytochemistry* **27**, 1609. **93.** Ono, Inoue, Suzuki & Takeda

(1972) *Biochim. Biophys. Acta* **284**, 285. **94**. Jakubowski & Guranowski (1983) *J. Biol. Chem.* **258**, 9982. **95**. Chaudhuri & Ghosh (1985) *Phytochemistry* **24**, 2433. **96**. Yanagisawa & Suzuki (1981) *Plant Physiol.* **67**, 697. **97**. Sindhu & Desai (1979) *Phytochemistry* **18**, 1937. **98**. Park & Cho (1991) *Biochem. Biophys. Res. Commun.* **174**, 32. **99**. Libby (1978) *Arch. Biochem. Biophys.* **188**, 360. **100**. Santacroce & Blankenship (1982) *Proc. West. Pharmacol. Soc.* **25**, 113. **101**. Csuhai, Juliano, Juliano & Hersh (1999) *Arch. Biochem. Biophys.* **362**, 291. **102**. Barra & Argarana (1982) *Biochem. Biophys. Res. Commun.* **108**, 654. **103**. Yupsanis, Symeonidis, Kalemi, Moustaka & Yupsani (2004) *Plant Physiol. Biochem.* **42**, 795. **104**. Kitahara, Sawai & Tsukada (1982) *J. Biochem.* **92**, 855. **105**. Kumagai, Igarshi, Tanaka, Nakao & Hirose (1979) *Biochim. Biophys. Acta* **566**, 192. **106**. Ochocka & Pawelczyk (2003) *Acta Biochim. Pol.* **50**, 1097. **107**. Harder, Owen, Wong, *et al.* (1994) *Biochem. J.* **298**, 395. **108**. Polya & Haritou (1988) *Biochem. J.* **251**, 357. **109**. Nanjundan & Possmayer (2000) *Exp. Lung Res.* **26**, 361. **110**. Jamdar & Osborne (1983) *Biochim. Biophys. Acta* **752**, 79. **111**. Kikkawa, Yoshida, Nakagawa, Iwasa & Tsuda (1998) *J. Biol. Chem.* **273**, 7441. **112**. Lazarus & Itin (1973) *Arch. Biochem. Biophys.* **156**, 154. **113**. Schiebel, Haas, Marinkovic, Klanner & Sänger (1993) *J. Biol. Chem.* **268**, 11851. **114**. Kurl, Holmes, Verney & Sidransky (1988) *Biochemistry* **27**, 8974. **115**. Rose & Jacob (1976) *Arch. Biochem. Biophys.* **175**, 748. **116**. Bosdal & Lillehaug (1985) *Biochim. Biophys. Acta* **840**, 280. **117**. Vancurova, Choi, Lin, Kuret & Vancura (1999) *J. Biol. Chem.* **274**, 1147. **118**. Gäng & von Collins (1973) *Int. J. Vitam. Nutr. Res.* **43**, 318. **119**. Jakus, Wolff, Park & Folk (1993) *J. Biol. Chem.* **268**, 13151. **120**. Böttcher, Adolph & Hartmann (1993) *Phytochemistry* **32**, 679. **121**. Graser & Hartmann (2000) *Planta* **211**, 239. **122**. Schröder, Eichel, Breinig & Schröder (1997) *Plant Mol. Biol.* **33**, 211. **123**. Wu, Lai & Tsai (2005) *Int. J. Biochem. Cell Biol.* **37**, 386. **124**. Jeon, Lee, Jang, *et al.* (2004) *Exp. Mol. Med.* **36**, 576. **125**. Korner, Schneider, Purdon & Bjornsson (1989) *Biochem. J.* **262**, 633. **126**. Karahalios, Mamos, Karigiannis & Kalpaxis (1998) *Eur. J. Biochem.* **258**, 437. **127**. Michelinaki, Spanos, Coutsogeorgopoulos & Kalpaxis (1997) *Biochim. Biophys. Acta* **1342**, 182. **128**. Haas & Leisinger (1974) *Biochem. Biophys. Res. Commun.* **60**, 42. **129**. Theiss, Schleicher, Schimpff-Weiland & Follmann (1987) *Eur. J. Biochem.* **167**, 89. **130**. Cox (1979) *Biochem. Biophys. Res. Commun.* **86**, 594. **131**. Nau, Pham-Coeur-Joly & Dubert (1983) *Eur. J. Biochem.* **130**, 261. **132**. Hurwitz, Gold & Anders (1964) *J. Biol. Chem.* **239**, 3474.

Spheroidine, *See* Tetrodotoxin

Sphinganine

This long-chain alkylamino diol precursor of sphingosine (FW$_{free-base}$ = 301.51 g/mol; CAS 764-22-7), also known as D-*erythro*-2-amino-1,3-octadecanediol and D-*erythro*-dihydrosphingosine, has surfactant properties, binds to membranes, and inhibits various lipid-peocessing enzymes. *Note:* Because this substance readily forms micelles, one must take care to discriminate whether the monomeric and/or micelle form(s) is(are) inhibitory, when present at concentrations above the critical micelle concentration (*cmc*). ***See also*** threo-Dihydro-sphingosine **Target(s):** 2-acylglycerol *O*-acyltransferase, *or* monoacylglycerol *O*-acyltransferase (1); CDP-diacylglycerol:serine *O*-phosphatidyltransferase, *or* phosphatidylserine synthase (2); ceramidase (3-6); ceramide kinase (7); 1,3-β-glucan synthase (8); [myosin light-chain] kinase (9); NADPH oxidase (10); phosphatidate phosphatase (11-14); phospholipase A$_2$ (15); phospholipase D (15); protein kinase C, IC$_{50}$ = 2.8 μM (16,17); and sphinganine kinase, *or* sphingosine kinase, inhibited by D-(+)-*threo*-, L-(–)-*erythro*-, and L-*threo*-(–)-sphinganine (18,19). **1**. Bhat, Wang & Coleman (1995) *Biochemistry* **34**, 11237. **2**. Yamashita & Nikawa (1997) *Biochim. Biophys. Acta* **1348**, 228. **3**. Usta, El Bawab, Roddy, *et al.* (2001) *Biochemistry* **40**, 9657. **4**. Tani, Okino, Mitsutake, *et al.* (2000) *J. Biol. Chem.* **275**, 3462. **5**. Okino, Tani, Imayama & Ito (1998) *J. Biol. Chem.* **273**, 14368. **6**. Nieuwenhuizen, van Leeuwen, Jack, Egmond & Gotz (2003) *Protein Expr. Purif.* **30**, 94. **7**. Van Overloop, Gijsbers & Van Veldhoven (2006) *J. Lipid Res.* **47**, 268. **8**. Abe, Nishida, Minemura, *et al.* (2001) *J. Biol. Chem.* **276**, 26923. **9**. Jinsart, Ternai & Polya (1991) *Plant Sci.* **78**, 165. **10**. Sasaki, Yamaguchi, Saeki, *et al.* (1996) *J. Biochem.* **120**, 705. **11**. Wu & Carman (2000) *Meth. Enzymol.* **312**, 373. **12**. Wu & Carman (1996) *Biochemistry* **35**, 3790. **13**. Wu, Lin, Wang, Merrill, Jr., &

Carman (1993) *J. Biol. Chem.* **268**, 13830. **14**. English, Martin, Harvey, *et al.* (1997) *Biochem. J.* **324**, 941. **15**. Franson, Harris, Ghosh & Rosenthal (1992) *Biochim. Biophys. Acta* **1136**, 169. **16**. Smith, Merrill, Jr., Obeid & Hannun (2000) *Meth. Enzymol.* **312**, 361. **17**. Merrill, Jr., Nimkar, Menaldino, *et al.* (1989) *Biochemistry* **28**, 3138. **18**. Stoffel, Hellenbroich & Heimann (1973) *Hoppe-Seyler's Z. Physiol. Chem.* **354**, 1311. **19**. Buehrer & Bell (1992) *J. Biol. Chem.* **267**, 3154.

Sphingomyelins

This entire class of sphingolipids (general structure: *N*-acyl-4-sphingenyl-1-*O*-phosphorylcholine) first identified in animal cell membranes, especially those comprising the myelin sheath surrounding the axons of neurons. Sphingomyelins contain a saturated or unsaturated fatty acid linked as an amide to the nitrogen of sphingosine, with the terminal hydroxyl group esterified to phosphocholine. Often the fatty acyl group is twenty carbons long or longer. In view of the spectrum of acyl groups, the inhibitory properties are likely to depend on the source and homogeneity of a particular sphingomyelin preparation. *Note:* Because this substance readily forms micelles, one must take care to determine whether the monomeric and/or micelle form(s) is(are) inhibitory, when present at concentrations above the critical micelle concentration (*cmc*). ***See*** *specific sphingomyelin* **Target(s):** β-*N*-acetylhexosaminidase (1); acyl-CoA synthetase (2); acylglycerol kinase, *or* monoacylglycerol kinase (10); ceramidase (3,4); cerebroside-sulfatase, *or* arylsulfatase A (5); choline-phosphate cytidylyltransferase (6); glycosyl-phosphatidylinositol phospholipase D (7); 3-hydroxybutyrate dehydrogenase (8); lysophospholipase (9); monoacylglycerol kinase (10); phosphatidate phosphatase (11); phosphatidylcholine:sterol *O*-acyltransferase, *or* LCAT (12,13); phosphatidylinositol diacylglycerol-lyase (14); phosphoinositide phospholipase C (14-17); phospholipase C, inhibited by increasing interlipid hydrogen bonding and by decreasing membrane hydration (15-18); and protein-tyrosine sulfotransferase, weakly inhibited (19). **1**. Frohwein & Gatt (1967) *Biochemistry* **6**, 2783. **2**. Koshlukova, Momchilova-Pankova, Markovska & Koumanov (1992) *J. Membr. Biol.* **127**, 113. **3**. El Bawab, Birbes, Roddy, *et al.* (2001) *J. Biol. Chem.* **276**, 16758. **4**. Yada, Higuchi & Imokawa (1995) *J. Biol. Chem.* **270**, 12677. **5**. Stinshoff & Jatzkewitz (1975) *Biochim. Biophys. Acta* **377**, 126. **6**. Drobnies, Van der Ende, Thewalt & Cornell (1999) *Biochemistry* **38**, 15606. **7**. Rhode, Schulze, Cumme, *et al.* (2000) *Biol. Chem.* **381**, 471. **8**. Gotterer (1967) *Biochemistry* **6**, 2147. **9**. Wright, Payne, Santangelo, *et al.* (2004) *Biochem. J.* **384**, 377. **10**. Shim, Lin & Strickland (1989) *Biochem. Cell Biol.* **67**, 233. **11**. Nanjundan & Possmayer (2000) *Exp. Lung Res.* **26**, 361. **12**. Subbaiah, Horvath & Achar (2006) *Biochemistry* **45**, 5029. **13**. Jonas (2000) *Biochim. Biophys. Acta* **1529**, 245. **14**. Dawson, Hemington & Irvine (1985) *Biochem. J.* **230**, 61. **15**. Scarlata, Gupta, Garcia, *et al.* (1996) *Biochemistry* **35**, 14882. **16**. Pawelczyk & Lowenstein (1997) *Biochimie* **79**, 741. **17**. Ochocka & Pawelczyk (2003) *Acta Biochim. Pol.* **50**, 1097. **18**. Hanada, Palacpac, Magistrado, *et al.* (2002) *J. Exp. Med.* **195**, 23. **19**. Kasinathan, Sundaram, Slomiany & Slomiany (1993) *Biochemistry* **32**, 1194.

Sphingosine

This long-chain alkylamino diol (FW$_{free-base}$ = 299.50 g/mol; CAS 1670-26-4; M.P. = 82.5-83°C; essentially insoluble in water and soluble in acetone, chloroform, and ethanol), also known as D-*erythro*-sphingosine, D-*erythro*-4-*trans*-sphingenine and (2S,3R,4E)-2-amino-4-octadecene-1,3-diol, is an important component of biomembranes. Sphingosine is typically not found in free form in high concentrations but as a component of sphingomyelins, cerebrosides, gangliolipids, *etc*. *Note*: Because this substance readily forms micelles, one must take care to discriminate whether the monomeric and/or micelle form(s) is(are) inhibitory, when present at concentrations above the critical micelle concentration (*cmc*). **See also *Phytosphingosine* Target(s):** β-*N*-acetylhexosaminidase (1); 2-acylglycerol *O*-acyltransferase, *or* monoacylglycerol *O*-acyltransferase (19); [β-adrenergic-receptor] kinase (2); Ca^{2+}/calmodulin-dependent phosphodiesterase (3); Ca^{2+}/calmodulin-dependent protein kinase (3); ceramidase, inhibited by all four stereoisomers (4,41-48); ceramide glycanase (5); ceramide kinase (63,64); cholesterol monooxygenase, *or* cholesterol side-chain-cleavage enzyme (6); choline-phosphate cytidylyltransferase (7,8,61,62); diacylglycerol *O*-acyltransferase (66); diacylglycerol kinase (9); DNA primase (10,11); 7-ethoxycoumarin *O*-deethylase, *or* cytochrome P450 (12); β-galactosidase (13); galactosylceramidase (13-15); glucosylceramidase, the *threo*-analogue also inhibits (5,16,49,50); 2-hydroxyacylsphingosine 1β-galactosyltransferase, *or* cerebroside synthase (17,65); insulin receptor protein-tyrosine kinase (18); [myosin light-chain] kinase (3,59); Na$^+$/K$^+$-exchanging ATPase (7,20); nitric-oxide synthase (21); phosphatidate phosphatase (7,22,23,54-58); phospholipase A$_2$ (24); phosphoinositide phospholipase C (53); phospholipase D (24); protein kinase C, IC$_{50}$ = 2.8 μM (3,7,25-36); protein-tyrosine kinase, insulin receptor (18); protein-tyrosine sulfotransferase (37); sodium-calcium exchange (39); sphingomyelin phosphodiesterase, *or* sphingomyelinase (39,51,52); sphingomyelin synthase (60); telomerase, weakly inhibited (40). **1.** Frohwein & Gatt (1969) *Meth. Enzymol.* **14**, 161. **2.** Benovic (1991) *Meth. Enzymol.* **200**, 351. **3.** Jefferson & Schulman (1988) *J. Biol. Chem.* **263**, 15241. **4.** Usta, El Bawab, Roddy, *et al.* (2001) *Biochemistry* **40**, 9657. **5.** Basu, Kelly, Girzadas, Li & Basu (2000) *Meth. Enzymol.* **311**, 287. **6.** Rabe, Weidenhammer & Runnebaum (1983) *J. Steroid Biochem.* **18**, 333. **7.** Hannun & Bell (1989) *Science* **243**, 500. **8.** Sohal & Cornell (1990) *J. Biol. Chem.* **265**, 11746. **9.** Kanoh, Sakane & Yamada (1992) *Meth. Enzymol.* **209**, 162. **10.** Tamiya-Koizumi, Murate, Suzuki, *et al.* (1997) *Biochem. Mol. Biol. Int.* **41**, 1179. **11.** Simbulan, Tamiya-Koizumi, Suzuki, *et al.* (1994) *Biochemistry* **33**, 9007. **12.** Muller-Enoch, Fintelmann, Nicolaev & Gruler (2001) *Z. Naturforsch. [C]* **56**, 1082. **13.** Gatt (1969) *Meth. Enzymol.* **14**, 156. **14.** Radin (1972) *Meth. Enzymol.* **28**, 834. **15.** Radin (1972) *Meth. Enzymol.* **28**, 844. **16.** Gatt (1969) *Meth. Enzymol.* **14**, 152. **17.** Neskovic, Mandel & Gatt (1981) *Meth. Enzymol.* **71**, 521. **18.** Arnold & Newton (1991) *Biochemistry* **30**, 7747. **19.** Bhat, Wang & Coleman (1995) *Biochemistry* **34**, 11237. **20.** Oishi, Zheng & Kuo (1990) *J. Biol. Chem.* **265**, 70. **21.** Viani, Giussani, Riboni, Bassi & Tettamanti (1999) *FEBS Lett.* **454**, 321. **22.** Wu & Carman (2000) *Meth. Enzymol.* **312**, 373. **23.** Wu, Lin, Wang, Merrill, Jr., & Carman (1993) *J. Biol. Chem.* **268**, 13830. **24.** Franson, Harris, Ghosh & Rosenthal (1992) *Biochim. Biophys. Acta* **1136**, 169. **25.** Hannun, Merrill, Jr., & Bell (1991) *Meth. Enzymol.* **201**, 316. **26.** Smith, Merrill, Jr., Obeid & Hannun (2000) *Meth. Enzymol.* **312**, 361. **27.** Cardenas, Fabila, Yum, *et al.* (2000) *Cell Signal* **12**, 649. **28.** Keenan, Goode & Pears (1997) *FEBS Lett.* **415**, 101. **29.** Senisterra & Epand (1992) *Biochem. Biophys. Res. Commun.* **187**, 635. **30.** Khan, Mascarella, Lewin, *et al.* (1991) *Biochem. J.* **278**, 387. **31.** Bottega, Epand & Ball (1989) *Biochem. Biophys. Res. Commun.* **164**, 102. **32.** Bazzi & Nelsestuen (1987) *Biochem. Biophys. Res. Commun.* **146**, 203. **33.** Hannun, Loomis, Merrill, Jr., & Bell (1986) *J. Biol. Chem.* **261**, 12604. **34.** Merrill, Jr., Nimkar, Menaldino, *et al.* (1989) *Biochemistry* **28**, 3138. **35.** Saraiva, Fresco, Pinto & Goncalves (2003) *J. Enzyme Inhib. Med. Chem.* **18**, 475. **36.** Khan, Dobrowsky, el Touny & Hannun (1990) *Biochem. Biophys. Res. Commun.* **172**, 683. **37.** Kasinathan, Sundaram, Slomiany & Slomiany (1993) *Biochemistry* **32**, 1194. **38.** Condrescu & Reeves (2001) *J. Biol. Chem.* **276**, 4046. **39.** Gatt & Barenholz (1969) *Meth. Enzymol.* **14**, 144. **40.** Ku, Cheng & Wang (1997) *Biochem. Biophys. Res. Commun.* **241**, 730. **41.** El Bawab, Birbes, Roddy, *et al.* (2001) *J. Biol. Chem.* **276**, 16758.

42. Mao, Xu, Szulc, *et al.* (2003) *J. Biol. Chem.* **278**, 31184. **43.** Okino, Tani, Imayama & Ito (1998) *J. Biol. Chem.* **273**, 14368. **44.** Nieuwenhuizen, van Leeuwen, Jack, Egmond & Gotz (2003) *Protein Expr. Purif.* **30**, 94. **45.** Gatt (1966) *J. Biol. Chem.* **241**, 3724. **46.** Yada, Higuchi & Imokawa (1995) *J. Biol. Chem.* **270**, 12677. **47.** Bawab, Bielawska & Hannun (1999) *J. Biol. Chem.* **274**, 27948. **48.** Nieuwenhuizen, van Leeuwen, Gotz & Egmond (2002) *Chem. Phys. Lipids* **114**, 181. **49.** Osiecki-Newman, Fabbro, Legler, Desnick & Grabowski (1987) *Biochim. Biophys. Acta* **915**, 87. **50.** Greenberg, Merrill, Liotta & Grabowski (1990) *Biochim. Biophys. Acta* **1039**, 12. **51.** Barnholz, Roitman & Gatt (1966) *J. Biol. Chem.* **241**, 3731. **52.** Ella, Qi, Dolan, Thompson & Meier (1997) *Arch. Biochem. Biophys.* **340**, 101. **53.** Ochocka & Pawelczyk (2003) *Acta Biochim. Pol.* **50**, 1097. **54.** Fleming & Yeaman (1995) *Biochem. J.* **308**, 983. **55.** Kanoh, Imai, Yamada & Sakane (1992) *J. Biol. Chem.* **267**, 25309. **56.** Nanjundan & Possmayer (2000) *Exp. Lung Res.* **26**, 361. **57.** English, Martin, Harvey, *et al.* (1997) *Biochem. J.* **324**, 941. **58.** Roberts, Sciorra & Morris (1998) *J. Biol. Chem.* **273**, 22059. **59.** Jinsart, Ternai & Polya (1991) *Plant Sci.* **78**, 165. **60.** Vivekananda, Smith & King (2001) *Amer. J. Physiol.* **281**, L98. **61.** Jackowski & Fagone (2005) *J. Biol. Chem.* **280**, 853. **62.** Yeo, Larvor, Ancelin & Vial (1997) *Biochem. J.* **324**, 903. **63.** Van Overloop, Gijsbers & Van Veldhoven (2006) *J. Lipid Res.* **47**, 268. **64.** Sugiura, Kono, Liu, *et al.* (2002) *J. Biol. Chem.* **277**, 23294. **65.** Basu, Schultz, Basu & Roseman (1971) *J. Biol. Chem.* **246**, 4272. **66.** Kamisaka, Mishra & Nakahara (1997) *J. Biochem.* **121**, 1107.

Sphingosine 1-Phosphate

This powerful sphingolipid agonist (FW = 378.47 g/mol) of receptor-mediated signal transduction culminates in endothelial cell proliferation and chemotaxis. *Note*: Because this substance readily forms micelles, one must take care to discriminate whether the monomeric and/or micelle form(s) is(are) inhibitory, when present at concentrations above the critical micelle concentration (*cmc*). **Target(s):** acid sphingomyelinase (1); alkylglycerophosphoethanolamine phosphodiesterase, *or* lysophospholipase D (2); and phosphatidate phosphatase (3). **1.** Gomez-Munoz, Kong, Salh & Steinbrecher (2003) *FEBS Lett.* **539**, 56. **2.** van Meeteren, Ruurs, Christodoulou, *et al.* (2005) *J. Biol. Chem.* **280**, 21155. **3.** Nanjundan & Possmayer (2000) *Exp. Lung Res.* **26**, 361.

Sphingosylphosphocholine

This lipid mediator (FW = 465.64 g/mol), also known as lysosphingomyelin, potently attenuates calcium entry via voltage-operated calcium channels. Sphingosyl-phosphocholine also stimulates mitogen-activated protein kinase and casein kinase II and inhibits platelet aggregation. *Note*: Because this substance readily forms micelles, one must take care to discriminate whether the monomeric and/or micelle form(s) is(are) inhibitory, when present at concentrations above the critical micelle concentration (*cmc*). **Target(s):** choline-phosphate cytidylyltransferase (1); phosphatidylinositol diacylglycerol-lyase, *or* phosphatidylinositol-specific phospholipase C (2); sphingomyelin phosphodiesterase *or* sphingomyelinase (3,4); and voltage-operated calcium channels (5). **1.** Sohal & Cornell (1990) *J. Biol. Chem.* **265**, 11746. **2.** Dawson, Hemington & Irvine (1985) *Biochem. J.* **230**, 61. **3.** Sloan (1972) *Meth. Enzymol.* **28**, 874. **4.** Callahan, Gerrie, Jones & Shankaran (1981) *Biochem. J.* **193**, 275. **5.** Tornquist, Pasternack & Kaila (1995) *Endocrinology* **136**, 4894.

4-Sphingynine

This sphingosine analogue (FW = 376.45 g/mol) inhibits glucosylceramidase (1). *Note*: Because this substance readily forms micelles, one must take care to discriminate whether the monomeric and/or micelle form(s) is(are) inhibitory, when present at concentrations above the critical micelle concentration (*cmc*). **1.** Greenberg, Merrill, Liotta & Grabowski (1990) *Biochim. Biophys. Acta* **1039**, 12.

SPI-3

This serpin produced by orthopoxviruses modulates cell-cell fusion during virus infection. It is a glycoprotein with a molecular mass of about 50 kDa (the amino acid sequence predicts a mass of 42 kDa). **Target(s):** plasmin (1); tissue-type plasminogen activator (1); and urokinase-type plasminogen activator. (*For further discussion of the likely mechanism of inhibitor action, See* α_1-*Antichymotrypsin*) **1.** Turner, Baquero, Yuan, Thoennes & Moyer (2000) *Virology* **272**, 267.

Spilanthol

This bioactive *N*-isobutylamide (FW = 221.34 g/mol; CAS 25394-57-4) from the traditional medicinal plant *Acmella oleracea*, also named (2*E*,6*Z*,8*E*)-*N*-isobutyl-2,6,8-decatrienamide, is a fatty acid amide with antimalarial proerties. In bioassays against the yellow fever mosquito *Aedes aegypti*, spilanthol has a 24-hour LD_{100} of 12 µg/mL and a LD_{50} of 6 µg/mL. Spilanthol is likely to account for the local anesthetic properties of plant extracts, often imparting a tingling or numbing sensation when imbibed as a flower bud extract. **1.** Ramsewak, Erickson & Nair (1999) *Phytochem.* **51**, 729. **2.** Spelman, Depoix, McCray, *et al.* (2011) *Phytother. Res.* **25**, 1098.

Spinorphin

This endogenous nonclassical opioid (*Heptapeptide Sequence*: LVVYPWT; MW = 877.05 g/mol; CAS 137201-62-8; Storage: Desiccate at -20°C), also called LVV-hemorphin-4, from bovine spinal cord is a potent P2X3 receptor antagonist, IC_{50} = 8.3 pM (1). Spinorphin also inhibits enkephalin-degrading enzymes and blocks fMet-Leu-Phe-induced neutrophil chemotaxis by acting as a specific antagonist at the *N*-formylpeptide receptor subtype FPR. It also displays antinociceptive effects in mice. Rather remarkably, the sequence of spinorphin exactly matches a highly conserved sequence in β-hemoglobin (32-LVVYPWT-39). **Target(s):** dipeptidyl-peptidase III (2-4); peptidyl-dipeptidase A (angiotensin I-converting enzyme (2); and neprilysin (2). **1.** Jung, Moon, Lee, *et al.* (2007) *J. Med. Chem.* **50**, 4543. **2.** Nishimura & Hazato (1993) *Biochem. Biophys. Res. Commun.* **194**, 713. **3.** Yamamoto, Ono, Ueda, *et al.* (2002) *Curr. Protein Pept. Sci.* **3**, 587. **3.** Chiba, Li, Yamane, *et al.* (2003) *Peptides* **24**, 773.

Spiperone

This antipsychotic (FW$_{free-base}$ = 395.48 g/mol; CAS 749-02-0; Store at RT), also called spiroperidol, is a selective D_2 receptor antagonist (respective K_i values are 0.06, 0.6, 0.08, ~ 350, ~ 3500 nM for D_2, D_3, D_4, D_1 and D_5 receptors) as well as an α_{1B}-adrenergic receptor antagonist and a mixed 5-HT$_{2A}$/5-HT$_1$ serotonin antagonist. **Target(s):** α_{1B}-adrenoceptor (1); dopamine D_2-like receptors (2); dopamine-stimulated adenylate cyclase (3,4); and 5-HT$_{2A}$ and 5-HT$_1$ serotonin receptor (5). **1.** Testa, Guarneri, Poggesi, *et al.* (1995) *Brit. J. Pharmacol.* **114**, 745. **2.** Amenta, Bronzetti,

Felici, Ricci & Tayebati (1999) *J. Auton. Pharmacol.* **19**, 151. **3.** Iversen (1975) *Science* **188**, 1084. **4.** Iversen (1976) *Trends Biochem. Sci.* **1**, 121. **5.** Metwally, Dukat, Egan, *et al.* (1998) *J. Med. Chem.* **41**, 5084.

Spiramycin

These structurally related antibiotics: (CAS 8025-81-8; Spiramycin I: FW$_{free-base}$ = 843.07 g/mol, R = H; Spiramycin II: FW$_{free-base}$ = 885.10 g/mol, R = COCH$_3$; and Spiramycin III: FW$_{free-base}$ = 899.13 g/mol, R = COCH$_2$CH$_3$), also called foronacidins, are produced by *Streptomyces ambofaciens*. These antibiotics inhibit the peptidyltransferase activity in prokaryotic protein biosynthesis (1-6) by inhibiting the binding of the donor and the acceptor substrates. They also induce rapid turnover of polyribosomes as well as stimulate the peptidyl-tRNA dissociation from ribosomes during translocation (7). **1.** Poulsen, Kofoed & Vester (2000) *J. Mol. Biol.* **304**, 471. **2.** Dinos, Synetos & Coutsogeorgopoulos (1993) *Biochemistry* **32**, 10638. **3.** Pestka (1974) *Meth. Enzymol.* **30**, 261. **4.** Jiménez (1976) *Trends Biochem. Sci.* **1**, 28. **5.** Brisson-Noel, Trieu-Cuot & Courvalin (1988) *J. Antimicrob. Chemother.* **22** Suppl. B, 13. **6.** Ahmed (1968) *Biochim. Biophys. Acta* **166**, 205. **7.** Cundliffe (1969) *Biochemistry* **8**, 2063.

Spironolactone

This orally bioavailable, K$^+$-sparing, synthetic steroid and aldosterone receptor antagonist (FW = 416.58 g/mol; CAS 50-01-7), also called Aldactone and known systematically as (7α,17α)-7-(acetylthio)-17-hydroxy-3-oxopregn-4-ene-21-carboxylic acid γ-lactone and, exhibits anti-mineralocorticoid activity (1), and is used as a diuretic and antagonist of aldosterone receptors. Spironolactone is used to treat high blood pressure. Administration often prevents bone loss. Spironolactone is metabolized to canrenone (also an anti-mineralocorticoid agent) via loss of the acetylthio moiety and insertion of a double bond at position-6,7. **Key Pharmacokinetic Parameters:** *See* Appendix II in Goodman & Gilman's THE PHARMACOLOGICAL BASIS OF THERAPEUTICS, 12th Edition (Brunton, Chabner & Knollmann, eds.) McGraw-Hill Medical, New York (2011). **Target(s):** aldosterone receptor (2); carbonic anhydrase (3); CYP3A (4); dehydroepiandrosterone sulfotransferase (5); 3α/β,20β-hydroxysteroid dehydrogenase (6); steroid 11β-monooxygenase (7-10); steroid 17α-monooxygenase (1); and steroid 18-monooxygenase (9). **1.** Scott & Tomkins (1975) *Meth. Enzymol.* **40**, 273. **2.** Little (1966) *Ann. N. Y. Acad. Sci.* **139**, 466. **3.** Puscas, Coltau, Baican, Domuta & Hecht (1999) *Drugs Exp. Clin. Res.* **25**, 271. **4.** Cook, Hauswald, Oppermann & Schoenhard

(1993) *J. Pharmacol. Exp. Ther.* **266**, 1. **5.** Bamforth, Dalgliesh & Coughtrie (1992) *Eur. J. Pharmacol.* **228**, 15. **6.** Itoda, Takase & Nakajin (2002) *Biol. Pharm. Bull.* **25**, 1220. **7.** Katagiri, Takemori, Itagaki & Suhara (1978) *Meth. Enzymol.* **52**, 124. **8.** Wada, Ohnishi, Nonaka & Okamoto (1988) *J. Steroid Biochem.* **31**, 803. **9.** Cheng, Suzuki, Sadee & Harding (1976) *Endocrinology* **99**, 1097. **10.** Denner, Vogel, Schmalix, Doehmer & Bernhardt (1995) *Pharmacogenetics* **5**, 89. **11.** Ruiz de Galarreta, Fanjul, Meidan & Hsueh (1983) *J. Biol. Chem.* **258**, 10988.

Spiro-17β-oxiranyl-4-androsten-3-one

This steroid (FW = 300.44 g/mol) inactivates steroid Δ-isomerase, a key enzyme in the biosynthesis and metabolic transformation of steroids. **1.** Kayser, Bounds, Bevins & Pollack (1983) *J. Biol. Chem.* **258**, 909.

Spiro-17β-oxiranylestra-1,3,5(10),6,8-pentaen-3-ol, *See* (17S)-Spiro[estra-1,3,5(10),6,8-pentaene-17,2'-oxiran]-3-ol

3β-Spirooxiranyl-5α-pregnan-20β-ol

This synthetic steroid (FW = 332.53 g/mol) inactivates steroid Δ-isomerase in a time-dependent manner. The inactivated enzyme may be dialyzed without regain of activity, indicating a stable covalent bond forms between the inhibitor and the enzyme. Inactivation follows pseudo-first-order kinetics, and at higher inhibitor concentrations, saturation was observed. The competitive inhibitor 17 β-oestradiol protects against the inactivation. **1.** Penning (1985) *Biochem. J.* **226**, 469.

Spiroperidol, *See* Spiperone

(3β,25R)-Spirost-5-en-3-ol, *See* Diosgenin

Spiruchostatin A

This novel HDAC prodrug (FW = 473.61 g/mol; CAS 328548-11-4; Soluble in DMSO, not in H$_2$O), also known as OBP-801 and YM753 as well as systematically as (1S,5S,6R,9S,15E,20R)-5-hydroxy-20-methyl-6-(propan-2-yl)-2-oxa-11,12-dithia-7,19,22-triazabicyclo[7.7.6]docos-15-ene-3,8,18,21-tetrone, targets histone deacetylase *in vitro* (IC$_{50}$ = 2.0 nM, in the presence of dithiothreitol). Upon entry into tumor cells, the disulfide

linkage in Spiruchostatin A is reductively cleaved, whereupon the product inhibits HDAC, inducing the accumulation of acetylated histones. The latter is attended by p21WAF1/Cip1 gene expression, tumor cell growth inhibition, and tumor-selective cell death. In an *in vitro* washout experiment, YM753 prolongs the accumulation of acetylated histones in WiDr human colon carcinoma cells. YM753 administration to mice harboring WiDr xenografts likewise reduces tumor growth. While YM753 rapidly disappears from the plasma, its reduced form remains within the tumor tissue. **1.** Shindoh, Mori, Terada, *et al.* (2014) *Int. J. Oncol.* **32**, 545.

SPL 334

This *S*-nitrosoglutathione modulator (FW = 433.50 g/mol; CAS 688347-51-5; Solubility: 100 mM in DMSO; 4-[[2-[[(2-cyanophenyl) methyl]thio]-4-oxothieno[3,2-*d*]pyrimidin-3(4*H*)-yl]methyl]benzoate, inhibits *S*-nitroso-glutathione reductase, an enzyme that was originally identified in plants and other organisms as formaldehyde dehydrogenase (FALDH), a type III alcohol dehydrogenase. Nitric oxide (NO) reacts with glutathione (GSH) to form *S*-nitrosoglutathione (GSNO), which in turn transfers its NO group to other cellular thiols to form *S*-nitrosothiols (SNOs). This NO adduct has been proposed to contribute significantly to NO regulatory mechanisms, particularly in *S*-nitrosylation of proteins. Interstitial lung disease (ILD), a disorder characterized by pulmonary fibrosis and inflammation. Daily administration of SPL-334 (at 0.3, 1.0, or 3.0 mg/kg i.p.) in a mouse model of bleomycin-induced ILD is without effect on animal body weight, appearance, behavior, total and differential bronchoalveolar lavage (BAL) cell counts, or collagen accumulation in the lungs, demonstrating little or no toxicity of this agent. Similar administration of SPL-334 for 7 days before, followedby an additional 14 days after bleomycin exposure, results in a preventive protective effect on the bleomycin challenge-induced decline in total body weight and changes in total and differential BAL cellularity. Administration of SPL-334 at days 7-21 after bleomycin challenge attenuates levels of profibrotic cytokines interleukin-6, monocyte chemoattractant protein-1, and transforming growth factor-β (TGF-β). Related experiments in cell cultures of primary normal human lung fibroblast also demonstrate attenuation of TGF-β-induced up-regulation in collagen, suggesting SPL-334 is a potential therapeutic agent for ILD. **1.** Luzina, Lockatell, Todd, *et al.* (2015) *J. Pharmacol. Exp. Ther.* **355**, 13.

sPLA2 Inhibitor, *See* specific secretory phospho-lipase A$_2$ inhibitor; e.g., 5-(4-Benzyloxyphenyl)-4S-(7-phenylheptanoylamino) pentanoate

Spleen Inhibitor II, *See* Deoxyribonuclease I Calf Spleen Inhibitory Protein II

Spliceostatin A

This FR901464 derivative (FW = 505.65 g/mol; CAS 391611-36-2) targets the spliceosome, a multicomponent complex that assembles on the newly synthesized pre-mRNA and catalyzes the removal of intervening sequences from transcripts (*See also E7107; FR901464; Meayamycin; Pladienolide B*). Spliceostatin A inhibits *in vitro* splicing and promotes pre-mRNA accumulation b;y binding to SF3b, a subcomplex of the U2 small nuclear ribonucleoprotein in the spliceosome. Treatment of cells with this agent results in leakage of (unspliced) pre-mRNA to the cytoplasm, where it is translated. Knockdown of SF3b by small interfering RNA induced phenotypes similar to those seen with spliceostatin A treatment. SF3b appears to have two functions, first in the splicing pre-mRNA and in retaining it. **1**. Kaida, Motoyoshi, Tashiro, *et al.* (2007) *Nature Chem. Biol.* **3**, 576.

Splitomicin

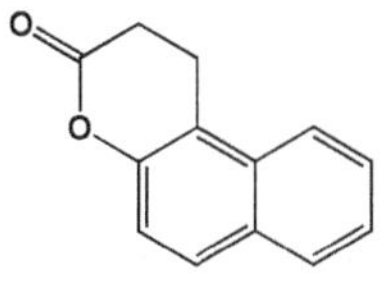

This cell-permeable lactone (FW = 198.22 g/mol; CAS 5690-03-9; Soluble to 100 mM in DMSO), also named 1,2-dihydro-3*H*-naphtho[2,1-*b*]pyran-3-one, inhibits the histone deacetylase activity of Sir2p, *or* silent information regulator (IC$_{50}$ = 60 μM), required for chromatin-dependent silencing in yeast. **1**. Bedalov, Gatbonton, Irvine, Gottschling & Simon (2001) *Proc. Natl. Acad. Sci. U.S.A.* **98**, 15113.

Spongoadenosine, See *Adenine 9- β-D-Arabino-furanoside*

Spongothymidine, See *Thymine 1- β-D-Arabino-furanoside*

Spongouridine, See *Uracil 1- β-D-Arabinofuranoside*

SPP100, See *Aliskiren*

SPP301, See *Avosentan*

SQ-14225, See *Captopril*

SQ 14603, See *(RS)-2-Benzyl-3-mercaptopropionate*

SQ-11725, See *Nadolol*

SQ 20009, See *Etazolate*

SQ 22536, See *9-(Tetrahydro-2-furanyl)-9H-purin-6-amine*

SQ 24798, See *(RS)-2-Mercaptomethyl-5-guanidinopentanoate*

SQ 28133

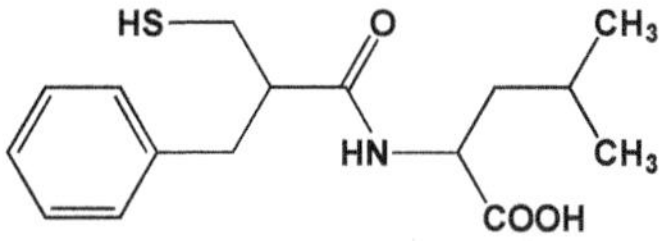

This high-affinity proteinase inhibitor (FW$_{free-acid}$ = 309.43 g/mol), also named *N*-[2-(mercaptomethyl)-1-oxo-3-phenylpropyl]-L-leucine, inhibits neprilysin, K_i = 4.5 nM, and peptidyl-dipeptidase A, K_i = 55 nM. SQ-28133 elicits significant depressor activities in both conscious spontaneously hypertensive rats (SHRs) and in DOCA/salt hypertensive rats during inhibition of angiotensin-converting enzyme (ACE) by captopril or SQ 27,519 (*i.e.*, the free acid form of the prodrug fosinopril) (2). **1**. Roques, Noble, Crine & Fournié-Zaluski (1995) *Meth. Enzymol.* **248**, 263. **2**. Seymour, Swerdel & Abboa-Offei (1991) *J. Cardiovasc. Pharmacol.* **17**, 456.

SQ 29548

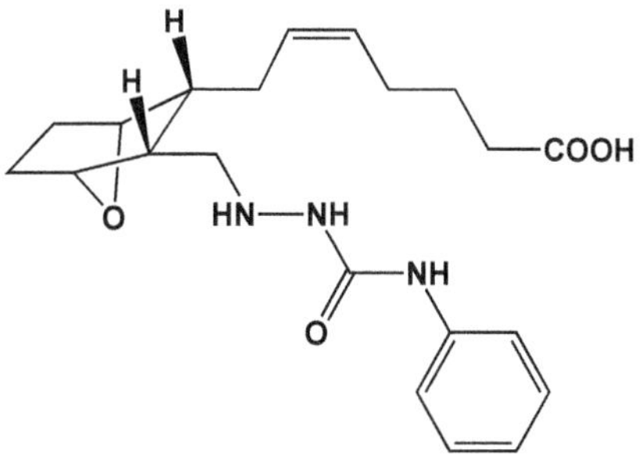

This hygroscopic acid (FW$_{free-acid}$ = 387.48 g/mol), also known as 1*S*-(1α,2β(5*Z*),3β,4α)]-7-[3-[[2-[(phenylamino)carbonyl]hydrazino]methyl]-7-oxabicyclo[2.2.1]hept-2-yl]-5-heptenoic acid, is a potent and specific thromboxane A$_2$ receptor antagonist (1-3). **1**. Ogletree, Harris, Greenberg, Haslanger & Nakane (1985) *J. Pharmacol. Exp. Ther.* **234**, 435. **2**.

Monshizadegan, Hedberg & Webb (1992) *Life Sci.* **51**, 431. **3**. Naka, Mais, Morinelli, *et al.* (1992) *J. Pharmacol. Exp. Ther.* **262**, 632.

SQ-34676, See *Entecavir*

Squalamine

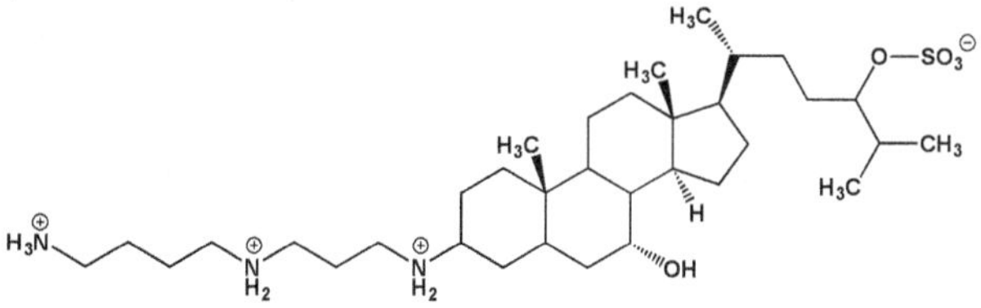

This aminosterol (FW = 601.94 g/mol; CAS 148717-90-2) is an antibiotic isolated from shark tissue (*e.g.*, from the stomach of the spiney dogfish shark [*Squalus acanthias*]). It is water soluble and is effective against protozoa and Gram-positive and Gram-negative bacteria. It also causes changes in endothelial cell shape and is a natural antiangiogenic sterol (1). **Target(s):** Na$^+$/H$^+$ exchanger isoform NHE32. **1**. Marwick (2001) *J. Natl. Cancer Inst.* **93**, 1685. **2**. Akhter, Nath, Tse, *et al.* (1999) *Amer. J. Physiol.* **276**, C136.

Squalene-maleimide

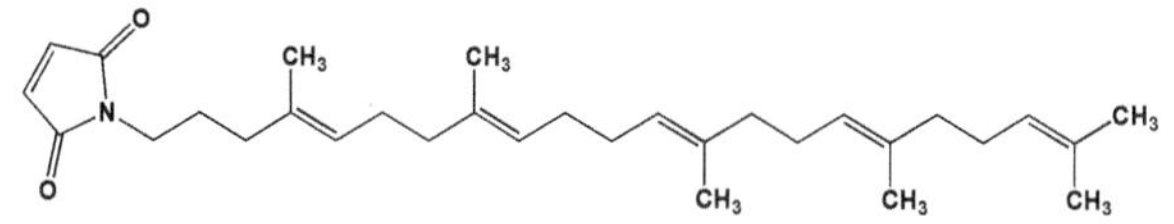

This thiol-reactive squalene analogue (FW = 465.72 g/mol) inhibits lanosterol synthase (1) and squalene:hopene cyclase (2). **1**. Cattel, Ceruti, Balliano, *et al.* (1995) *Lipids* **30**, 235. **2**. Milla, Lenhart, Grosa, *et al.* (2002) *Eur. J. Biochem.* **269**, 2108.

Squalestatin 1, See *Zaragozates*

Squamous Cell Carcinoma Antigen 2

This serpin (often abbreviated SCCA2) from human cervical squamous carcinoma cells inhibits a number of chymotrypsin-like serine proteinases. (*For further discussion of the likely mechanism of inhibitor action, See α$_1$-Antichymotrypsin*) **Target(s):** cathepsin G (1,2); and mast cell chymase (1,2). **1**. Silverman, Bartuski, Cataltepe, *et al.* (1998) *Tumour Biol.* **19**, 480. **2**. Schick, Kamachi, Bartuski, *et al.* (1997) *J. Biol. Chem.* **272**, 1849.

Squarate (Squaric Acid)

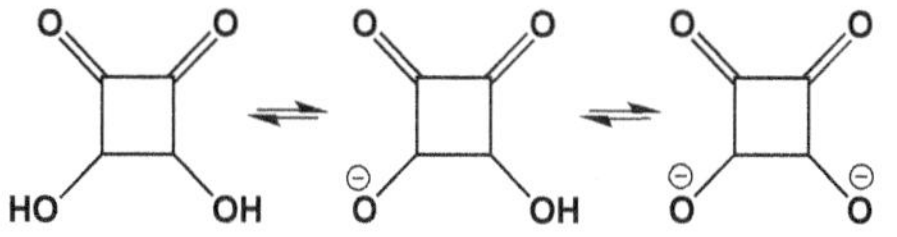

This diacid (FW = 114.06 g/mol; CAS 2892-51-5), also known as 3,4-dihydroxy-3-cyclobutene-1,2-dione, inhibits glyoxalase I, *or* lactoylglutathione lyase (1-3). **1**. Kraus & Castaing (1989) *Res. Commun. Chem. Pathol. Pharmacol.* **65**, 105. **2**. Kraus, Pernice & Ponce (1988) *Res. Commun. Chem. Pathol. Pharmacol.* **59**, 419. **3**. Douglas & Nadvi (1979) *FEBS Lett.* **106**, 393.

SR 202, See *Mifobate*

SR-4233, See *Tirapazamine*

SR 8278

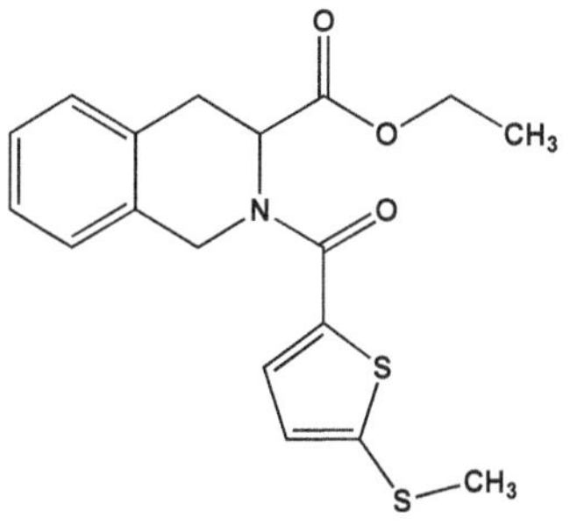

This Rev-Erbα antagonist (FW = 361.48 g/mol; CAS 1254944-66-5; Soluble to 100 mM in DMSO), also named 1,2,3,4-tetrahydro-2-[[5-(methylthio)-2-thienyl]carbonyl]-3-isoquinoline carboxylic acid ethyl ester, inhibits Rev-Erbα transcriptional repression, EC$_{50}$ = 0.47 μM. REV-ERBα is a member of the nuclear receptor superfamily that functions as a receptor for the porphoryin heme. REV-ERBα suppresses transcription of its target genes in a heme-dependent manner. In HEK293 cells, SR 8278 blocks activity of Rev-Erbα agonist GSK 4112, the first nonporphyrin synthetic ligand for REV-ERBα. It also increases expression of the glucose-regulating enzymes, G6Pase and PEPCK, in HepG2 cells. **1.** Kojetin, Wang, Kamenecka & Burris (2011) *ACS Chem. Biol.* **6**, 131.

SR 9238

This LXR inverse agonist (FW = 595.73 g/mol; CAS 1416153-62-2; Soluble to 100 mM in DMSO), also named ethyl 5-[[[[3'-(methyl-sulfonyl)[1,1'-biphenyl]-4-yl]methyl]-[(2,4,6-trimethylphenyl)sulfonyl]-amino]methyl]-2-furancarboxylate, potently and selectively targets Liver X Receptors LXRα (IC$_{50}$ = 43 nM) and LXRβ (IC$_{50}$ = 214 nM), a subfamily of the nuclear receptors that form obligate heterodimers with retinoid X receptors (RXRs) and regulate expression of target genes containing LXR response elements (1). Teatment of diet-induced obese mice with SR9238 suppressed plasma cholesterol levels (1). SR9238 is also effective in reducing hepatic steatosis, inflammation and fibrosis in an animal model of NASH. These results have important implications for the development of therapeutics for treatment of non-alcoholic steatohepatitis (NASH) in humans (2). **1.** Griffett, Solt, El-Gendy, Kamenecka & Burris (2013) *ACS Chem. Biol.* **8**, 559. **2.** Griffett, Welch, Flaveny, *et al.* (2015) *Mol. Metab.* **4**, 353.

SR 12813

This bisphosphonate ester (FW = 504.34 g/mol; CAS 126411-39-0; Store at +4°C; Soluble to 100 mM in DMSO), also known as tetraethyl 2-(3,5-di-*tert*-butyl-4-hydroxyphenyl)ethenyl-1,1-bisphosphonate, indirectly inhibits cholesterol biosynthesis by enhancing the degradation of 3-hydroxy-3-methylglutaryl-CoA reductase. SR 12813 *does not* inhibit the reductase directly. SR 12813 is also an agonist for the human nuclear xenobiotic PXR (NR 112) receptor (EC$_{50}$ values are 200 and 700 nM for human and rabbit PXR respectively). SR 12813 also activates the farnesoid X receptor (FXR) at μM concentrations. **1.** Berkhout, Simon, Patel, *et al.* (1996) *J. Biol. Chem.* **271**, 14376.

SR 13800

This MCT1 inhibitor (FW = 435.58 g/mol; CAS 227321-12-2; Solubility: 100 mM DMSO), named 2,6-dihydro-7-[(3-hydroxypropyl)thio]-2-methyl-4-(2-methylpropyl)-6-(1-naphthalenylmethyl)-1*H*-pyrrolo[3,4-*d*]pyridazin-1one, targets H$^+$-coupled Monocarboxylate transporter 1 (Gene Name: *SLC16A1*), blocking proliferation of Raji lymphoma cells *in vitro* (IC$_{50}$ = 0.5 nM). MCT1 catalyzes rapid transport across the plasma membrane of many monocarboxylates (*e.g.*, lactate, pyruvate, branched-chain oxo-acids derived from leucine, valine and isoleucine, as well as acetoacetate, β-hydroxybutyrate and acetate. Depending on the tissue and on cicumstances, MCT1 mainly mediates the import or export of lactic acid and ketone bodies. SR-13800 also inhibits lactate uptake in breast cancer cells *in vitro*. Inhibition of MCT1 during T lymphocyte activation results in selective and profound inhibition of the extremely rapid phase of T cell division essential for an effective immune response (1). **1.** Murray, Hutchinson, Bantick *et al.* (2005) *Nature Chem. Biol.* **1**, 371.

SR 33557, *See Fantofarone*

SR-46349, *See Eplivanserin*

SR 59230A

This aryloxypropanolaminotetralin (FW = 415.49 g/mol), also known as 3-(2-ethylphenoxy)-1-[(1S)-1,2,3,4-tetrahydronaphth-1-ylaminol]-(2S)-2-propanol , was the first β$_3$-adrenergic receptor antagonist (1-3). The *RR*-enantiomer (SR 59483) is inactive. **1.** De Ponti, Gibelli, Croci, *et al.* (1996) *Brit. J. Pharmacol.* **117**, 1374. **2.** Manara, Badone, Baroni, *et al.* (1996) *Brit. J. Pharmacol.* **117**, 435. **3.** Nisoli, Tonello, Landi & Carruba (1996) *Mol. Pharmacol.* **49**, 7.

SR 90107A, *See Fondaparinux*

SR141617A

This selective CB$_1$ receptor antagonist (FW = 463.93 g/mol), also named as (*N*-(piperidiny-1-yl)-5-(4-chlorophenyl)-1-(2,4-dichlorophenyl)-4-methyl-1*H*-pyrazole-3-carboxamide), inhibits the binding of 2-arachidonylglycerol to the Cannabinoid receptor Type-1 receptor (1,2) and permanently prevents milk ingestion in a dose-dependent manner, when administered to mouse pups, within one day of birth (3). Cannabinoids affected performance in a dose-x-delay-dependent manner, with the synthetic cannabinoid WIN-2 showing a potency more than four times that of Δ^9-THC, but these effects were eliminated if SR141617A was preadministered, but doses of the antagonist alone had no effect on performance (1). (**Sea also** *Arachidonyl-2'-chloroethylamide*) **1**. Hampson & Deadwyler (2000) *J. Neurosci.* **20**, 8932. **2**. Esteban & García-Sevilla (2012) *Prog. Neuro-psychopharmacol. Biol. Psych.* **38**, 78. **3**. Fride, Foox, Rosenberg, *et al.* (2003) *Eur. J. Pharmacol.* **461**, 27.

SR141716, *See* Rimonaban

SR 144528

This orally bioavailable CB$_2$ receptor-selective inverse agonist (FW = 476.05 g/mol; CAS 192703-06-3; Soluble to 100 mM in DMSO), also named 5-(4-chloro-3-methylphenyl)-1-[(4-methylphenyl)methyl]-*N*-[(1S,2S, 4R)-1,3,3-trimethylbicyclo[2.2.1]hept-2-yl]-1H-pyrazole-3-carboxamide, targets rat spleen and human CB$_2$ inverse agonist (K$_i$ = 0.6 nM). SR 144528 exhibits >700-fold selectivity for CB2 (K$_i$ = 400 nM) over CB$_1$ receptors (1). Based on both its binding and functional properties, SR 144528 as the first-in-class, highly potent, selective and orally active CB2 receptor antagonist (1). It also blocks the effects of CP 55,940, a cannabinoid agonist that is considerably more potent than Δ^9-THC, on forskolin-sensitive adenylyl cyclase activity and MAPK in CHO cells expressing CB$_2$ receptors. Also blocks CP 55,940-induced B-cell activation. By means of its binding to the CB$_2$ receptors, SR 144528 blocks the direct activation of the G$_i$ protein by mastoparan analog in Chinese hamster ovary CB2 cell membranes (2). Sustained treatment with SR 144528 induced an up-regulation of the cellular G$_i$ protein level, as demonstrated in Western blotting as well as in confocal microscopic experiments (2). This up-regulation occurred with a concomitant loss of SR 144528 ability to inhibit the insulin or lysophosphatidic acid-induced MAPK activation (2). Subsequent studies demonstrated the importance of hydrogen bonding and aromatic stacking in SR 144528's affinity and efficacy as a cannabinoid receptor ligand (3). **1**. Rinaldi-Carmona, Barth, Millan, *et al.* (1998) *J. Pharmacol. Exp. Ther.* **284**, 644. **2**. Bouaboula, Desnoyer, Carayon, Combes & Casellas (1999) *Mol. Pharmacol.* **55**, 473. **3**. Kotsikorou, Navas, 3rd, Roche, *et al.* (2003) *J. Med. Chem.* **56**, 6593.

Src I1

This competitive dual-site protein tyrosine kinase inhibitor (FW = 373.40 g/mol; CAS 179248-59-0; Soluble to 100 mM in DMSO), also named 6,7-dimethoxy-*N*-(4-phenoxyphenyl)-4-quinazolinamine, targets human pp60c-src kinase, with IC$_{50}$ values of 44 and 88 nM for Src and Lck, respectively, by binding to ATP and peptide binding sites. Src-I1 inhibits VEGFR2 and *c*-fms at much higher concentrations, exhibiting IC$_{50}$ values are 0.32 and 30 μM, respectively. Structural modeling of Src tyrosine kinase (Src TK) with inhibitor and peptide substrate bound indicated a direct atomic-level clash between the bulky 4-position group and the hydroxy of the peptide tyrosyl to which the γ-phosphoryl of ATP is transferred during the kinase reaction (1). This atomic conflict would likely prevent simultaneous binding of both inhibitor and peptide, an action that is consistent with the observed kinetic competitiveness of the inhibitor with peptide. Such dual-site inhibitors appear to have enhanced potency and selectivity for Src TK (1,2). **1**. Tian, Cory, Smith & Knight (2001) *Biochemistry* **40**, 7084. **2**. Bain, Plater, Elliott, *et al.* (2007) *Biochem. J.* **408**, 297.

SRIF (Somatotropin Release-Inhibiting Factor), *See* Somatostatin

SRPIN340

This SRPK arginine (FW = 349.35 g/mol; CAS 218156-96-8; Soluble to 100 mM in DMSO and to 50 mM in 1 eq. HCl), also named *N*-[2-(1-piperidinyl)-5-(trifluoromethyl)-phenyl]-4-pyridinecarboxamide, selectively targets Serine arginine protein kinase-1, *or* SRPK1 (K$_i$ = 0.89 μM), with action against SRPK2 only at much higher concentrations. SRPIN340 does not significantly inhibit other SRPKs, such as CLK1 and CLK4, or other classes of SR kinases. SRPIN340 down-regulates expression of VEGF$_{165}$ but is without effect on VEGF$_{165b}$ expression. **1**. Fukuhara, Hosoya, Shimizu, *et al.* (2006) *Proc. Natl. Acad. Sci. U.S.A.* **103**, 11329. **2**. Oltean, Gammons, Hulse, *et al.* (2012) *Biochem. Soc. Trans.* **40**, 831. **3**. Gammons, Dick, Harper & Bates (2013) *Invest. Ophthalmol. Vis. Sci.* **54**, 5797.

SRS, *See* Leukotriene C$_4$; Leukotriene D$_4$; Leukotriene E$_4$

SRT1720

This putative SIRT1 activator, or STAC (FW = 469.56 g/mol; CAS 925434-55-5), also named *N*-[2-[3-(piperazin-1-ylmethyl)imidazo[2,1-*b*][1,3]thiazol-6-yl]phenyl]quinoxaline-2-carboxamide, was thought to target the NAD$^+$-dependent deacetylase Sirtuin-1 (1). However, SRT1720 and structurally related compounds (SRT2183 and SRT1460), as well as resveratrol, do not activate SIRT1 with native peptide or full-length protein substrates, but instead activate SIRT1 with peptide substrate containing a covalently attached fluorophore. NMR, surface plasmon resonance (SPR), and isothermal calorimetry evidence indicates that these compounds directly interact with fluorophore-containing peptide substrates. Furthermore, SRT1720 neither lowers plasma glucose nor improves mitochondrial capacity in mice fed a high fat diet. SRT1720, SRT2183, SRT1460, and resveratrol exhibit multiple off-target activities against receptors, enzymes, transporters, and ion channels. Taken together, SRT1720, SRT2183, SRT1460, and resveratrol are not direct activators of SIRT1 (2). **1**. Milne, Lambert, Schenk, *et al.* (2007) *Nature* **450**, 712. **2**. Pacholec, Bleasdale, Chrunyk, *et al.* (2010) *J. Biol. Chem.* **285**, 8340.

SSR69071

This substituted pyrido[1,2-a]pyrimidine (FW = 580.66 g/mol; CAS 344930-95-6), known systematically as 2-(9-(2-piperidinoethoxy)-4-oxo-4H-pyrido[1,2-a]pyrimidin-2-yloxymethyl)-4-(1-methylethyl)-6-methoxy-1,2-benzisothiazol-3(2H)-one-1,1-dioxide, is a potent inhibitor of human leukocyte elastase, or HLE (K_i = 17 pM). SSR-69071 is a potent inhibitor of HLE, with the inhibition constant of 168 pM, an association rate constant of $0.183 \ 10^6 \ M^{-1}s^{-1}$, and a dissociation rate constant of $3.1 \times 10^{-6} \ s^{-1}$. Bronchoalveolar lavage fluid from mice orally treated with SSR69071 inhibits HLE (ex vivo), and in this model, SSR69071 has a dose-dependent efficacy with an ED_{50} = 10.5 mg/kg p.o. **1**. Kapui, Varga, Urban-Szabó, et al. (2003) J. Pharmacol. Exp. Ther. **305**, 451. **Alternate Identifiers:** SSR69071, SSR-69071

SSR128129E

This orally-active allosteric FGFR1 inhibitor ($FW_{free-acid}$ = 324.24 g/mol; FW_{NaSalt} = 346.31 g/mol; CAS 848318-25-2; Solubility: 69 mg/mL DMSO, 1 mg/mL H_2O), also named 2-amino-5-[(1-methoxy-2-methyl-3-indolizinyl)carbonyl]benzoate, targets Fibroblast Growth Factor Receptor 1 (FGFR1), also known as Basic Fibroblast Growth Factor Receptor 1 (IC_{50} = 1.9 μM), with little effect on other related RTKs. **1**. Bono, De Smet, Herbert, et al. (2013) Cancer Cell **23**, 477. **2**. Dol-Gleizes F1, Delesque-Touchard N, Marès A, et al. (2013) PLoS One **8**, e80027.

SSR 180711

This α_7 nAChR partial agonist and antidepressant ($FW_{free-base}$ = 325.21 g/mol; CAS 446031-79-4; Soluble to 100 mM in H_2O (HCl Salt) or DMSO (free base)), also named 4-bromophenyl 1,4-diazabicyclo[3.2.2]nonane-4-carboxylate, targets α_7 nicotinic acetylcholine receptors, enhancing episodic memory and reversing MK-801-induced deficits in retention of episodic memory (1). Selectivity of SSR180711 was confirmed as these effects were abolished by 3 mg/kg (i.p.) or 1 mg/kg (i.v.) methyllycaconitine, a selective α_7 n-AChR antagonist (1). SR180711 displays high affinity for rat α7 n-AChRs (K_i = 22 nM) and human α_7 n-AChRs (K_i = 14 nM) (2). Ex vivo [^{3}H]-α-bungarotoxin binding experiments demonstrate that SSR180711 rapidly penetrates into the brain (ID_{50} = 8 mg/kg p.o.). In functional studies performed with human α_7 n-AChRs expressed in Xenopus oocytes or GH4C1 cells, SSR180711 shows partial agonism (Intrinsic activities of 51 and 36%, and EC_{50} values of 4.4 and 0.9 μM, respectively) (2). **1**. Pichat, Bergis, Terranova, et al. (2007) Neuropsychopharmacol. **32**, 17. **2**. Biton, Bergis, Galli, et al. (2007) Neuropsychopharmacol. **32**, 1.

SSR-591813, See Dianicline

SSRI, See Selective Serotonin Re-uptake Inhibitors

ST 638

This tyrosine kinase inhibitor (FW = 354.43 g/mol; photosensitive), also known as α-cyano-(3-ethoxy-4-hydroxy-5-phenylthiomethyl)cinnamide, exhibits an IC_{50} value of 0.37 μM. ST 638 also inhibits (hepatocyte growth factor)-induced mitogen-activated protein (MAP) kinase activation in hepatocytes and phospholipase D activity in human neutrophils. **Target(s):** (hepatocyte growth factor)-induced mitogen-activated protein (MAP) kinase activation (1); inositol-trisphosphate 3-kinase (2); and protein-tyrosine kinase (1,3). **1**. Adachi, Nakashima, Saji, Nakamura & Nozawa (1996) Hepatology **23**, 1244. **2**. Mayr, Windhorst & Hillemeier (2005) J. Biol. Chem. **280**, 13229. **3**. Shiraishi, Kameyama, Imai, et al. (1988) Chem. Pharm. Bull. **36**, 974.

Stachybotrylactone

This spirocyclic drimane (FW = 348.44 g/mol; CAS 149691-31-6), originally isolated from the culture broth of the fungus Stachybotrys cylindrospora, is a strong inhibitor of rat β-galactoside α-2,3-sialyltransferase (K_i = 10 μM). **1**. Lin, Chang, Chen & Tsai (2005) Biochem. Biophys. Res. Commun. **331**, 953.

Stallimycin, See Distamycin A

Stampidine

This nucleotide analogue (FW = 550.28 g/mol) is a prodrug form of stavudine 5'-triphosphate, which inhibits HIV-1 reverse transcriptase (1,2). **1**. El Safadi, Vivet-Boudou & Marquet (2007) Appl. Microbiol. Biotechnol. **75**, 723. **2**. Uckun, Pendergrass, Venkatachalam, Qazi & Richman (2002) Antimicrob. Agents Chemother. **46**, 3613.

Stanolone, See 5a-Androstan-17β-ol-3-one

Stanozolol

This anabolic steroid (FW = 328.50 g/mol; CAS 10418-03-8), also known as androstanazole and 17β-hydroxy-17α-methylandrostano[3,2-c]pyrazole, is a controlled substance exhibiting one of the largest anabolic/androgenic ratios of the common steroids. Stanozolol also possesses substantial fibrinolytic properties. **Target(s):** steroid 17α-hydroxylase/C17,20-lyase, or CYP17A1. **1.** Nakajin, Takahashi & Shinoda (1989) *Chem. Pharm. Bull.* **37**, 1855.

Staphcillin, See Methicillin

Staphylococcus aureus Exfoliative Toxins

These "epidermolytic" toxins (MW = 30 kDa) are extremely specific serine proteases (serotypes ETA and ETB) that recognize and cleave desmosomal cadherins within the superficial layers of the skin and are directly responsible for the clinical manifestation of staphylococcal scalded skin syndrome (SSSS). The target protein, desmoglein 1, is recognized through an interaction at the classical P1 site and by additional features in the tertiary structure, located away from the site of hydrolysis. The cleavage of Dsg-1 results in the destruction of desmosomal cell-cell attachments in a superficial layer of the skin. **1.** Bukowski, Wladyka & Dubin (2010) *Toxins* (Basel) **2**, 1148.

Statin: A Confusing Nomenclature

Use of 'statin' as a suffix is both confusing and arbitrary. Even so, there is little likelihood for a self-consistent remedy. In the broadest sense, a statin is an agent that maintains or holds steady a metabolic or physiologic process through some activity-modulating effect. In layman usage, statin most often refers to an LDL-cholesterol-reducing drug acting as an inhibitor of the hydroxymethyl-CoA reductase reaction (*See Statin Drugs: LDL Cholesterol-Reducing*). Many other peptidomimetics (*e.g.*, pepstatin) are protease inhibitors, particularly those containing the tetrahedral transition-state mimic known as statine (*See Statine*). Still other inhibitors contain "statin" in their suffix (*e.g.*, pancreastatin is derived by proteolytic processing of the glycoprotein chromogranin A). Readers must therefore exercise care to avoid confusing the modes of action of these agents.

Statins: LDL Cholesterol-Reducing Drugs

These high-affinity inhibitors target hydroxymethylglutaryl-coenzyme A (*or* HMGCoA) reductase (*Reaction*: (*S*)-3-Hydroxy-3-methylglutaryl-CoA + 2 NADPH $\rightleftharpoons$ (*R*)-Mevalonate + 2 NADP$^+$), which catalyzes the rate-determining step in the mevalonate pathway. Most manufacturers recommend that statins be taken at night, on the basis of physiological studies show most cholesterol is synthesised when dietary cholesterol intake is at its nadir. **METABOLIC ASPECTS:** Formation of mevalonate leads to the synthesis of cholesterol and other isoprenoids, including various steroids, coenzyme Q$_{10}$, isopentenyl pyrophosphate (IPP), dolichol pyrophosphate, farnesyl pyrophosphate (FPP) and geranylgeranyl pyrophosphate (GGPP). The last two are especially critical in the membrane-anchoring posttranslational modification reactions involving proteins essential for cell proliferation and differentiation. The latter include the small GTP-binding proteins known as Ras, Rac, and Rho. The mevalonate pathway also contributes to multidrug resistance (MDR), a major challenge to durable tumor eradication by chemotherapy. **COMMONLY USED STATINS:** Advicor®, a combination of lovastatin and extended-release niacin; Altoprev®, *or* extended-release lovastatin; Caduet®, *or* amlodipine and atorvastatin; Crestor®, *or* rosuvastatin; Juvisync®, *or* sitagliptin and simvastatin; Lescol®, *or* fluvastatin; Lescol XL®, extended-release fluvastatin; Lipitor®, *or* atorvastatin; Livalo®, *or* pitavastatin; Mevacor®, *or* lovastatin; Pravachol®, *or* pravastatin; Simcor®, a combination of simvastatin and extended-release niacin; Vytorin®, *or* ezetimibe/simvastatin; and Zocor®, *or* simvastatin. **SIDE-EFFECTS OF STATIN THERAPY:** Often exhibiting significant biochemical and physiologic side effects, HMGCoA reductase inhibitors are far from being entirely benign. Mild effects include: constipation, stomach pain, nausea, headache, memory loss, forgetfulness, and confusion. More serious side effects include: muscle pain, tenderness, or weakness; decreased urination; lack of energy, tiredness, or weakness; loss of appetite; stomach pain; fever, chills, and/or flushing; skin blisters, rash, hives, itching; swelling of the face, throat, tongue, lips, eyes, hands, feet, ankles, or lower legs; difficulty breathing or swallowing; hoarseness; joint pain; as well as sensitivity to light. **Impact of Mevalonate Depletion:** On average, humans produce about 20 mmol mevalonate each day, with about 80-90% ultimately converted to cholesterol, and the remainder forming other isoprenoids. By greatly reducing mevalonate biosynthesis, statins deplete cellular stores of steroids, geranyl-PP, geranylgeranyl-PP, farnesyl-PP, Coenzyme Q$_{10}$, and other less abundant isoprenoids. Cellular processes relying on these metabolites are therefore negatively impacted during statin therapy. For example, all statins cause an accumulation of unprenylated signal transduction GTP-regulatory proteins, thereby blocking their membrane tethering and limiting their effectiveness in signal transduction. **Physiologic Effects:** Among the most common physiologic effects are myalgia/rhabdomyolysis, gastrointestinal intolerance, and an increase in hepatic aminotransferases. Symptoms of statin-induced myopathy include fatigue, muscle pain, muscle tenderness, muscle weakness, nocturnal cramping, and tendon pain. There is increased muscle cell apoptosis, often taxing the kidney as a consequence of muscle protein turnover. **Lipophilic versus Hydrophilic Statins:** Statin pharmacodynamics and pharmacokinetics are also influenced by the inhibitors lipophilic or hydrophilic character. Lipophilic statins appear to evoke the greatest complaint of muscle pain. Lovastatin and simvastatin are the most lipophilic, followed by atorvastatin, fluvastatin, rosuvastatin and pravastatin. *Note*: The mevalonate pathway is also inhibited by aminobisphosphonates (NBPs), which target FPP synthase (FPPS) selectively. NBPs are commonly used to inhibit the activity of osteoclasts in patients affected by bone metastases or other causes of bone fragility (e.g., osteoporosis). Zoledronic acid (ZA) is the most potent NBP currently available for clinical use.

Statine

This unusual γ-amino acid (FW = 175.23 g/mol; CAS 49642-07-1), when incorporated in peptides/polypeptides, becomes a potent transition-state inhibitor for many aspartic proteinases (1,2). The inhibitory action of these peptidomimetics often becomes selective, when the sequence matches the specificity-conferring subsites in the target proteinase (3). (*See Pepstatin, Pancrestatin*) **1.** Umezawa, Aoyagi, Morishima, Matsuzaki & Hamada (1970). *J. Antibiot.* **23**, 259. **2.** Marciniszyn, Hartsuck & Tang (1976) *J. Biol. Chem.* **251**, 7088. **3.** Jupp, Dunn, Jacobs, *et al.* (1990) *Biochem. J.* **265**, 871.

Stattic

This small-molecule protein kinase inhibitor (FW = 211.19 g/mol; CAS 19983-44-9; Solubility = 45 mg/mL DMSO, <1 mg/mL H$_2$O), also named 6-nitrobenzo[*b*]thiophene 1,1-dioxide, inhibits (IC$_{50}$ = 5.1 μM) STAT3 (*or* Signal Transducer and Activator of Transcription factor 3) activation, dimerization, and nuclear translocation, thereby increasing apoptosis in STAT3-dependent breast cancer cells (1). Later work provided preclinical proof-of-principle that STAT3 is a good druggable target (2). Direct inhibition of STAT3 with short hairpin RNA suppressed cancer cell growth in a manner associated with apoptosis, repression of STAT3 target genes, and inhibition of a tumor-promoting microenvironment. Such results indicate STAT3 inhibition may be a promising approach in its therapy. Binding of Stattic to the unphosphorylated STAT3βtc (U-STAT3) protein is attended by alkylation of at least four cysteine residues (3). This finding not only bears on its mechanism of inhibitory action but also raises concerns about specificity and clinical suitability. Indeed, a preview to the original paper mentions this inhibitor's likely susceptibility to Michael addition (4). A study of signal transduction pathways within dendritic cells recently demonstrated that cell type and/or cytokine-specific effects can modulate the selectivity of Stattic toward STAT3 (5). These investigators emphasize the need for interrogating an inhibitor's action against anticancer targets within a microenvironment likely to surround the tumor *in situ* (5; *See also* reply, 6). Given its modest size, Stattic is likely to lack sufficient specificity-conferring determinants seen in inhibitors with formula weights 2-3x higher. 1. Schust, Sperl, Hollis, Mayer & Berg (2006) *Chem. Biol.* 13, 1235. 2. Scuto, Kujawski, Kowolik, *et al.* (2011) *Cancer Res.* 71, 3182. 3. Heidelberger, Zinzalla, Antonow, *et al.* (2013) *Bioorg. Med. Chem. Lett.* 23, 4719. 4. McMurray (2006) *Chem. Biol.* 13, 1123. 5. Sanseverino, Purificato, Gauzzi & Gessani (2012) *Chem. Biol.* 19, 1213. 6. Berg (2012) *Chem. Biol.* 19, 1215.

Staurosporine

This antibiotic alkaloid (FW = 466.54 g/mol; CAS 62996-74-1; water-insoluble; soluble to 50 mM in DMSO; photosensitive), produced by a number of *Streptomyces* species, is a broadly acting and potent protein kinase inhibitor, with strongest effects (IC_{50} = 3 nM) on the PKC βII isoform. Other inhibited include protein kinase A (IC_{50} = 7 nM), $p^{60v-src}$ tyrosine protein kinase (IC_{50} = 6 nM) and CaM kinase II (IC_{50} = 20 nM). Staurosporine is effective against fungi and yeasts, but in effective against bacteria. In higher organisms, it wreaks havoc by reducing the phosphorylation of multiple proteins. For example, even when used in low concentrations (1-10 nM), staurosporine causes the complete collapse and loss of hair bundles in postnatal mouse cochlear cultures, without provoking hair-cell death, as judged by lack of TUNEL labelling or reactivity to anti-activated caspase-3 (1). Staurosporine exposure results in the fusion of the hair bundle's stereocilia, resorption of the parallel actin filament bundles of the stereocilia into the cytoplasm of the hair cell, detachment of the apical, non-stereociliary membrane of the hair cell from the underlying cuticular plate, and severing of hair-bundle's rootlets from the actin cores with stereocilia. Staurosporine does not block membrane retrieval at the apical pole of the hair cells, nor does it elicit the externalisation of phosphatidylserine. Notably, it reduces the levels of the phosphorylated forms of ezrin, radixin and moesin in cochlear cultures during the period of hair-bundle loss, indicating the integrity of the hair bundle may be maintained by the phosphorylation status of these proteins. The take-home lesson is that staurosporine is evoking many changes in the phosphorylation state of numerous proteins (1). **IUPAC Name:** [9S-(9α,10β,11β,13α)]-2,3,10,11,12,13-hexahydro-10-methoxy-9-methyl-11-(methylamino)-9,13-epoxy-1*H*,9*H*-diindolo[1,2,3-*gh*:3',2',1'-*lm*]pyrrolo[3,4-*j*][1,7]-benzo-diazonin-1-one. **Target(s):** AfsK serine/ threonine protein kinase (28); Akt-3 serine/threonine protein kinase (29); Bruton's protein-tyrosine kinase (32); Ca^{2+}/calmodulin-dependent protein kinase (2,19); cAMP-dependent protein kinase, *or* protein kinase A (11,12,24,25); cGMP-dependent protein kinase, *or* protein kinase G (23); cyclin-dependent kinases Cdk1 and Cdk2, IC_{50} = 9 and 7 nM, respectively (3-5); dual-specificity kinase (6,7); IκB kinases (8); inositol-trisphosphate 3-kinase (35); Lyn protein-tyrosine kinase (31); mitogen-activated protein kinase (9); non-specific serine/threonine protein kinase (26-30); phosphatidylinositol 3-kinase (34); phosphorylase kinase (10); PknH serine/threonine protein kinase (27); polo kinase (33); protein kinase C (7,11-18,20-22); protein-tyrosine kinase (7,11,12,31,32); PstA serine/threonine protein kinase (26); receptor protein-tyrosine kinase (7); Rho-associated kinase (17); and StkP serine/threonine protein kinase (30). 1. Goodyear, Ratnayaka, Warchol & Richardson (2014) *J. Comp. Neurol.* 522, 3281. 2. Yanagihara, Tachikawa, Izumi, *et al.* (1991) *J. Neurochem.* 56, 294. 3. Meijer & Kim (1997) *Meth. Enzymol.* 283, 113. 4. Woodard, Li, Kathcart, *et al.* (2003) *J. Med. Chem.* 46, 3877. 5. Lawrie, Noble, Tunnah, *et al.* (1997) *Natre Struct. Biol.* 4, 796. 6.

Rudrabhatla & Rajasekharan (2004) *Biochemistry* 43, 12123. 7. MacKintosh & MacKintosh (1994) *Trends Biochem. Sci.* 19, 444. 8. Peet & Li (1999) *J. Biol. Chem.* 274, 32655. 9. Toda, Shimanuki & Yanagida (1991) *Genes Dev.* 5, 60. 10. Elliott, Wilkinson, Sedgwick, *et al.* (1990) *Biochem. Biophys. Res. Commun.* 171, 148. 11. Rosenshine, Ruschkowski & Finlay (1994) *Meth. Enzymol.* 236, 467. 12. Tamaoki (1991) *Meth. Enzymol.* 201, 340. 13. Tamaoki, Nomoto, Takahashi, *et al.* (1986) *Biochem. Biophys. Res. Commun.* 135, 397. 14. Couldwell, Hinton, He, *et al.* (1994) *FEBS Lett.* 345, 43. 15. Wilkinson, Parker & Nixon (1993) *Biochem. J.* 294, 335. 16. Schachtele, Seifert & Osswald (1988) *Biochem. Biophys. Res. Commun.* 151, 542. 17. Amano, Fukata, Shimokawa & Kaibuchi (2000) *Meth. Enzymol.* 325, 149. 18. Spitaler & Cantrell (2004) *Nat. Immunol.* 5, 785. 19. Boulton, Gregory & Cobb (1991) *Biochemistry* 30, 278. 20. Beltowski, Marciniak, Jamroz-Wisniewska, Borkowska & Wojcicka (2004) *Acta Biochim. Pol.* 51, 757. 21. Ranganathan, Liu, Migliorini, *et al.* (2004) *J. Biol. Chem.* 279, 40536. 22. Toda, Shimanuki & Yanagida (1993) *EMBO J.* 12, 1987. 23. Baker & Deng (2005) *Front. Biosci.* 10, 1229. 24. Patel, Soulages, Wells & Arrese (2004) *Insect Biochem. Mol. Biol.* 34, 1269. 25. Prade, Engh, Girod, *et al.* (1997) *Structure* 5, 1627. 26. Peirs, De Wit, Braibant, Huygen & Content (1997) *Eur. J. Biochem.* 244, 604. 27. Sharma, Chandra, Gupta, *et al.* (2004) *FEMS Microbiol. Lett.* 233, 107. 28. Horinouchi (2003) *J. Ind. Microbiol. Biotechnol.* 30, 462. 29. Masure, Haefner, Wesselink, *et al.* (1999) *Eur. J. Biochem.* 265, 353. 30. Novakova, Saskova, Pallova, *et al.* (2005) *FEBS J.* 272, 1243. 31. Merciris, Claussen, Joiner & Giraud (2003) *Pflugers Arch.* 446, 232. 32. Dinh, Grunberger, Ho, *et al.* (2007) *J. Biol. Chem.* 282, 8768. 33. Johnson, Stewart, Woods, Giranda & Luo (2007) *Biochemistry* 46, 9551. 34. Gray, Olsson, Batty, Priganica & Downes (2003) *Anal. Biochem.* 313, 234. 35. Mayr, Windhorst & Hillemeier (2005) *J. Biol. Chem.* 280, 13229.

Stavudine

This thymidine analogue (FW = 224.22 g/mol; CAS 3056-17-5; M.P. = 165-166°C; λ_{max} = 266 nm, with ε = 10150 $M^{-1}cm^{-1}$), known systematically as 2',3'-didehydro-2',3'-dideoxythymidine and 2',3'-didehydro-3'-deoxythymidine, is an antiviral agent that inhibits HIV-1 replication. **Target(s):** deoxynucleoside kinase (1); viral DNA polymerases, as the 5'-triphosphate (2); HIV-1 reverse transcriptase, as the 5'-triphosphate (2-6); RNA-directed DNA polymerase, as the 5'-triphosphate (2-6); thymidine kinase (1,7,8). 1. Johansson, Van Rompay, Degrève, Balzarini & Karlsson (1999) *J. Biol. Chem.* 274, 23814. 2. Wilson, Porter & Reardon (1996) *Meth. Enzymol.* 275, 398. 3. Balzarini & De Clercq (1996) *Meth. Enzymol.* 275, 472. 4. Cherrington, Fuller, Mulato, *et al.* (1996) *Antimicrob. Agents Chemother.* 40, 1270. 5. Duan, Poticha, Stoeckli, *et al.* (2001) *J. Infect. Dis.* 184, 1336. 6. El Safadi, Vivet-Boudou & Marquet (2007) *Appl. Microbiol. Biotechnol.* 75, 723. 7. Lock, Thorley, Teo & Emery (2002) *J. Antimicrob. Chemother.* 49, 359. 8. Gustafson, Chillemi, Sage & Fingeroth (1998) *Antimicrob. Agents Chemother.* 42, 2923.

Stavudine 5'-Triphosphate

This nucleotide derivative (FW$_{free-acid}$ = 464.16 g/mol), also known as 2',3'-didehydro-2',3'-dideoxythymidine 5'-triphosphate and often abbreviated d4TTP, inhibits HIV-1 reverse transcriptase *in vitro*, K_i = 5 nM. The nucleotide will not enter cells, and the action reported below is for *in vitro* assays. *See Staudine for in vivo agent* **Target(s):** avian myeloblastosis virus reverse transcriptase (1); DNA helicases (2); DNA-directed DNA polymerase (1,3); DNA polymerase I (1); DNA polymerase β, $IC_{50} = 1.0$ μM (1,3); DNA polymerase γ (3); HIV-1 reverse transcriptase (3-5); nucleoside-triphosphatase, as alternative substrate (2); Rauscher murine leukemia virus reverse transcriptase (6); Raus sarcoma virus reverse transcriptase (1), *or* RNA-directed DNA polymerase (1,3-6); and terminal deoxynucleotidyltransferase (1). **1.** Dyatkina, Minassian, Kukhanova, *et al.* (1987) *FEBS Lett.* **219**, 151. **2.** Locatelli, Gosselin, Spadari & Maga (2001) *J. Mol. Biol.* **313**, 683. **3.** De Clercq (1992) *AIDS Res. Human Retrovir.* **8**, 119. **4.** El Safadi, Vivet-Boudou & Marquet (2007) *Appl. Microbiol. Biotechnol.* **75**, 723. **5.** Nissley, Radzio, Ambrose, *et al.* (2007) *Biochem. J.* **404**, 151. **6.** Ono, Nakane, Herdewijn, Balzarini & De Clerq (1988) *Nucl. Acids Res.* **20**, 5.

Stearamide

This amide (FW = 283.49 g/mol; CAS 124-26-5) inhibits microsomal epoxide hydrolase, $IC_{50} = 20$ μM. *Note*: Because this substance readily forms micelles, one must take care to discriminate whether the monomeric and/or micelle form(s) is(are) inhibitory, when present at concentrations above the critical micelle concentration (*cmc*). **1.** Morisseau, Newman, Dowdy, Goodrow & Hammock (2001) *Chem. Res. Toxicol.* **14**, 409.

Stearate (Stearic Acid)

This carboxylic acid (FW = 284.48 g/mol; CAS 57-11-4; $pK_a \approx 4.8$; M.P. = 69-70°C; B.P. = 383°C; water insoluble; slightly soluble in benzene (2.5 g per 100 mL), also called *n*-octadecanoic acid, is a long-chain fatty acid widely distributed in nature. Stearic acid is a solid at room temperature (benzene (2.5 g per 100 mL), ethyl acetate, and acetone. It is very soluble in diethyl ether. Chemist Paul Sabatier (Nobel Laureate, 1912), used catalytic hydrogenation to chemically convert oleic acid to stearic acid, forming the basis of the margarine and oil hydrogenation in the food industry. *Note*: Because this substance readily forms micelles, one must take care to discriminate whether the monomeric and/or micelle form(s) is(are) inhibitory, when present at concentrations above the critical micelle concentration (*cmc*). **Target(s):** alcohol dehydrogenase (1); D-amino-acid oxidase (2,3); choline oxidase (3); chondroitin AC lyase (18); chondroitin B lyase, weakly inhibited (18); chymase (4,20); dihydroorotate dehydrogenase (5); dimethylaniline monooxygenase (6); DNA-directed DNA polymerase (7,8); DNA polymerases α, γ, δ, and ε (7,8); 7-ethoxycoumarin *O*-deethylase, *or* cytochrome P450 (9); firefly luciferase, *or* *Photinus*-luciferin 4-monooxygenase, $IC_{50} = 0.63$ μM (10,27); β-galactoside α-2,6-sialyltransferase (25); 1,3-β-glucan synthase (26); glucose-6-phosphate dehydrogenase (11); glutathione *S*-transferase (12); glycerol-3-phosphate dehydrogenase (13); hexokinase (24); hyaluronate lyase (18); leukocyte elastase, *or* neutrophil elastase (19); Lysophospholipase (23); microbial collagenases (19); palmitoyl-CoA synthetase, *or* palmitate:CoA ligase (14); phospholipase D (22); *Photinus*-luciferin 4-monooxygenase (10); prostaglandin dehydrogenase (15,16); prostaglandin D_2 dehydrogenase (16); prostaglandin D synthetase (16); sphingomyelin phosphodiesterase, weakly inhibited (21); succinate oxidase, weakly inhibited (3); and L-threonine dehydrogenase (17). **1.** Winer & Theorell (1960) *Acta Chem. Scand.* **14**, 1729. **2.** Brachet, Carreira & Puigserver (1990) *Biochem. Int.* **22**, 837. **3.** Bernheim (1940) *J. Biol. Chem.* **133**, 291. **4.** Kido, Fukusen & Katunuma (1984) *Arch. Biochem. Biophys.* **230**, 610. **5.** Miller (1978) *Meth. Enzymol.* **51**, 63. **6.** Ziegler & Poulsen (1978) *Meth. Enzymol.* **52**, 142. **7.** Mizushina, Tanaka, Yagi, *et al.* (1996) *Biochim. Biophys. Acta* **1308**, 256. **8.** Mizushina, Tsuzuki, Eitsuka, *et al.* (2004) *Lipids* **39**, 977. **9.** Muller-Enoch, Fintelmann, Nicolaev & Gruler (2001) *Z. Naturforsch. [C]* **56**, 1082. **10.** Ueda & Suzuki (1998) *Biophys. J.* **75**, 1052. **11.** Criss & McKerns (1969) *Biochim. Biophys. Acta* **184**, 486. **12.** Mitra, Govindwar, Joseph & Kulkarni (1992) *Toxicol. Lett.* **60**, 281. **13.** McLoughlin, Shahied & MacQuarrie (1978) *Biochim. Biophys. Acta* **527**, 193. **14.** Bhushan & Singh (1987) *Prep. Biochem.* **17**, 173. **15.** Mibe, Nagai, Oshige & Mori (1992) *Prostaglandins Leukot. Essent. Fatty Acids* **46**, 241. **16.** Osama, Narumiya, Hayaishi, *et al.* (1983) *Biochim. Biophys. Acta* **752**, 251. **17.** Guerranti, Pagani, Neri, *et al.* (2001) *Biochim. Biophys. Acta* **1568**, 45. **18.** Suzuki,

Terasaki & Uyeda (2002) *J. Enzyme Inhib. Med. Chem.* **17**, 183. **19.** Rennert & Melzig (2002) *Planta Med.* **68**, 767. **20.** Kido, Fukusen & Katunuma (1985) *Arch. Biochem. Biophys.* **239**, 436. **21.** Liu, Nilsson & Duan (2002) *Lipids* **37**, 469. **22.** Okamura & Yamashita (1994) *J. Biol. Chem.* **269**, 31207. **23.** Zhang & Dennis (1988) *J. Biol. Chem.* **263**, 9965. **24.** Morris, DeBruin, Yang, *et al.* (2006) *Eukaryot. Cell* **5**, 2014. **25.** Hickman, Ashwell, Morell, van den Hamer & Scheinberg (1970) *J. Biol. Chem.* **245**, 759. **26.** Ko, Frost, Ho, Ludescher & Wasserman (1994) *Biochim. Biophys. Acta* **1193**, 31. **27.** Matsuki, Suzuki, Kamaya & Ueda (1999) *Biochim. Biophys Acta* **1426**, 143.

Stearidonate

This long-chain fatty acid ($FW_{\text{free-acid}}$ = 276.42 g/mol; CAS 20290-75-9), also called stearidonic acid and named systematically as (Z,Z,Z,Z)-6,9,12,15-octadecatetraenoic acid, is an ω–3 fatty acid that is biosynthesized from α-linolenic acid. Stearidonic acid is also a precursor to a number of long-chain unsaturated acids (*e.g.*, eicosapentaenoic acid). *Note*: Because this substance readily forms micelles, one must take care to discriminate whether the monomeric and/or micelle form(s) is(are) inhibitory, when present at concentrations above the critical micelle concentration (*cmc*). **Target(s):** DNA topoisomerase II, $IC_{50} = 60$ μM (1); and 5-lipoxygenase (2). **1.** Mizushina, Tsuzuki, Eitsuka, *et al.* (2004) *Lipids* **39**, 977. **2.** Guichardant, Traitler, Spielmann, Sprecher & Finot (1993) *Lipids* **28**, 321.

Stearoyl-CoA

This thioester (FW = 1034.01 g/mol; CAS 362-66-3), also known as *n*-octadecanoyl-CoA, is a substrate of stearoyl-CoA 9-desaturase, icosanoyl-CoA synthase (or, eicosanoyl-CoA synthase), and many acyltransferases. *Note*: Because this substance readily forms micelles, one must take care to discriminate whether the monomeric and/or micelle form(s) is(are) inhibitory, when present at concentrations above the critical micelle concentration (*cmc*). **Target(s):** acetyl-CoA carboxylase, $K_i = 0.71$ μM for the rat liver enzyme (1,2); acetyl-CoA synthetase, *or* acetate:CoA ligase (3); AMP deaminase (4); aralkylamine *N*-acetyltransferase (23); ATP:citrate lyase, *or* ATP:citrate synthase (5,6); choline kinase (7-9); cyclooxygenase (10); ethanolamine kinase (8); fatty-acid synthase (11); glucose-6-phosphatase (12,13); glucose-6-phosphate dehydrogenase (14,15); glycerol-3-phosphate dehydrogenase (16); isocitrate dehydrogenase ($NADP^+$) (17); lipoxygenase (10); phosphate-activated glutaminase (18,19); protein-palmitoyl acyltransferase (20); [pyruvate dehydrogenase (acetyl-transferring)] kinase (21); and RNA polymerase (22). **1.** Tanabe, Nakanishi, Hashimoto, *et al.* (1981) *Meth. Enzymol.* **71**, 5. **2.** Bortz & Lynen (1963) *Biochem. Z.* **337**, 505. **3.** Satyanarayana & Klein (1973) *J. Bacteriol.* **115**, 600. **4.** Skladanowski, Kaletha & Zydowo (1978) *Int. J. Biochem.* **9**, 43. **5.** Shashi, Bachhawat & Joseph (1990) *Biochim. Biophys. Acta* **1033**, 23. **6.** Boulton & Ratledge (1983) *J. Gen. Microbiol.* **129**, 2863. **7.** Ishidate & Nakazawa (1992) *Meth. Enzymol.* **209**, 121. **8.** Brophy, Choy, Toone & Vance (1977) *Eur. J. Biochem.* **78**, 491. **9.** Rodríguez-González, Ramírez de Molina, Benítez-Rajal & Lacal (2003) *Prog. Cell Cycle Res.* **5**, 191. **10.** Fujimoto, Tsunomori, Sumiya, *et al.* (1995) *Prostaglandins Leukot. Essent. Fatty Acids* **52**, 255. **11.** Roncari (1975) *Can. J. Biochem.* **53**, 135. **12.** Fulceri, Gamberucci, Scott, *et al.* (1995) *Biochem. J.* **307**, 391. **13.** Methieux & Zitoun (1996) *Eur. J. Biochem.* **235**, 799. **14.** Cacciapuoti & Morse (1980) *Can. J. Microbiol.* **26**, 863. **15.** Eger-Neufeldt, Teinzer, Weiss & Wieland (1965) *Biochem. Biophys. Res. Commun.* **19**, 43. **16.** Edgar & Bell (1979) *J. Biol. Chem.* **254**, 1016. **17.** Farrell, Jr., Wickham & Reeves (1995) *Arch. Biochem. Biophys.* **321**, 199. **18.** Kvamme, Torgner & Svenneby (1985) *Meth. Enzymol.* **113**, 241. **19.** Kvamme & Torgner (1975) *Biochem. J.* **149**, 83. **20.** Das, Dasgupta, Bhattacharya & Basu (1997) *J. Biol. Chem.* **272**, 11021. **21.** Rahmatullah & Roche (1985) *J. Biol. Chem.*

260, 10146. **22**. Yokokawa, Fujiwara, Shimada & Yasumasu (1983) *J. Biochem.* **94**, 415. **23**. Ferry, Loynel, Kucharczyk, *et al.* (2000) *J. Biol. Chem.* **275**, 8794.

1-Stearoylglycerol

This monoacylglycerol (FW = 358.56 g/mol; CAS 22610-63-5; M.P. = 74.4°C), also known as 1-monostearoylglycerol and 1-monostearin, is often supplied commercially as the racemic mixture (*e.g.,* 1-stearoyl-*rac*-glycerol). *Note*: Because this substance readily forms micelles, one must take care to discriminate whether the monomeric and/or micelle form(s) is(are) inhibitory, when present at concentrations above the critical micelle concentration (*cmc*). *See also 2-Stearoylglycerol* **Target(s):** alcohol *O*-acetyltransferase (1); 7-ethoxycoumarin *O*-deethylase, *or* cytochrome P450 (2); sphingomyelin phosphodiesterase, inhibited by the racemic mixture (3). **1**. Yoshioka & Hashimoto (1981) *Agric. Biol. Chem.* **45**, 2183. **2**. Muller-Enoch, Fintelmann, Nicolaev & Gruler (2001) *Z. Naturforsch. [C]* **56**, 1082. **3**. Liu, Nilsson & Duan (2002) *Lipids* **37**, 469.

2-Stearoylglycerol

This monoacylglycerol (FW = 358.56 g/mol; M.P. = 74.4°C), also known as 2-monostearoylglycerol and 2-monostearin, inhibits diacylglycerol lipase. Note that 2-acylglycerols are achiral and that acyl migration can readily occur. *Note*: Because this substance readily forms micelles, one must take care to discriminate whether the monomeric and/or micelle form(s) is(are) inhibitory, when present at concentrations above the critical micelle concentration (*cmc*). **Target(s):** diacylglycerol lipase (1). **1**. Moriyama, Urade & Kito (1999) *J. Biochem.* **125**, 1077.

Steviol

This diterpene, (FW$_{\text{free-acid}}$ = 318.46 g/mol; CAS 471-80-7; M.P. = 215°C.) also known as (4α)-13-hydroxykaur-16-en-18-oate, is the aglycon of stevioside *See Stevioside* **Target(s):** *p*-aminohippurate transport (1); ATPase (2); glucose absorption (3); glutamate dehydrogenase (2); oxidative phosphorylation (2); and succinate dehydrogenase (2). **1**. Chatsudthipong & Jutabha (2001) *J. Pharmacol. Exp. Ther.* **298**, 1120. **2**. Kelmer Bracht, Alvarez & Bracht (1985) *Biochem. Pharmacol.* **34**, 873. **3**. Toskulkao, Sutheerawattananon & Piyachaturawat (1995) *Toxicol. Lett.* **80**, 153.

Stevioside

This non-nutritive glycosylated diterpene sweetener (FW = 804.88 g/mol; CAS 57817-89-7; hygroscopic solid; M.P. = 198°C) from the leaves of the South America bush *Stevia rebaudiana* is about 300-times as sweet as cane

sugar. **Target(s):** *p*-aminohippurate transport (1,2); ATPase (3); Ca^{2+} influx (4); glutamate dehydrogenase (3); long-chain fatty acid transport (5); oxidative phosphorylation (3); and succinate dehydrogenase (3). **1**. Chatsudthipong & Jutabha (2001) *J. Pharmacol. Exp. Ther.* **298**, 1120. **2**. Jutabha, Toskulkao & Chatsudthipong (2000) *Can. J. Physiol. Pharmacol.* **78**, 737. **3**. Kelmer Bracht, Alvarez & Bracht (1985) *Biochem. Pharmacol.* **34**, 873. **4**. Lee, Wong, Liu, *et al.* (2001) *Planta Med.* **67**, 796. **5**. Constantin, Ishii-Iwamoto, Ferraresi-Filho, Kelmer-Bracht & Bracht (1991) *Braz. J. Med. Biol. Res.* **24**, 767.

STF-118804

This highly specific NAMPT inhibitor (FW = 461.53 g/mol; CAS 894187-61-2; Solubility = <1 mg/mL DMSO or H$_2$O), also known as 4-[5-methyl-4-[[(4-methylphenyl)sulfonyl]methyl]-2-oxazolyl]-*N*-(3-pyridinylmethyl) benzamide, targets nicotinamide phospho-ribosyl transferase, *or* NAMPT (*Reaction*: Nicotinamide + PRPP ⇌ NMN + PP$_i$), a rate-limiting enzyme in the biosynthesis of NAD$^+$, a coenzyme for many biochemical processes, including cytoplasmic and mitochondrial redox reactions, telomere poly(adenosine diphosphate-ribose) polymerase (*or* PARP), and the chromatin-silencing NAD-dependent histone deacetylases (*or* Sirtuins). STF-118804 is a highly specific NAMPT inhibitor, improves survival in an orthotopic xenotransplant model of high-risk acute lymphoblastic leukemia, and targets leukemia stem cells (IC$_{50}$ range: 3-60 nM). **1**. Matheny, Wei, Bassik, *et al.* (2013) *Chem. Biol.* **20**, 1352. **2**. Mei & Brenner (2013) *Chem. Biol.* **20**, 1307.

Stibogluconate, Sodium

Pentostam, with Sb(V)   **Triostam, with Sb(III)**

This phlebotoxic antileishmanial drug (FW = 336.87 g/mol, but 907.88 g/mol, also known as Pentostam® and 2,4:2',4'-*O*-(oxydistibylidyne)bis[D-gluconic acid] *Sb,Sb*'-dioxide trisodium salt nonahydrate, irreversibly inhibits protein-tyrosine phosphatase. When formulated as the water-soluble trisodium Sb(III) monogluconate salt, it is known as Triostam™. Pentavalent antimonials have been used to treat leishmaniasis for well over a half-century. While Sb(V) is thought to be a prodrug that is converted to the toxic Sb(III), direct involvement by pentavalent antimony has not been excluded. **Target(s):** DNA topoisomerase I (1,2); protein-tyrosine phosphatase (3); and trypanothione reductase (4). **1**. Chakraborty & Majumder (1988) *Biochem. Biophys. Res. Commun.* **152**, 605. **2**. Walker & Saravia (2004) *J. Parasitol.* **90**, 1155. **3**. Pathak & Yi (2001) *J. Immunol.* **167**, 3391. **4**. Cunningham & Fairlamb (1995) *Eur. J. Biochem.* **230**, 460.

Stichodactyla Toxin

RSCIDTIPKSRCTAFQCKHSMKYRLSFCRKTCGTC

This naturally occurring 35-residue peptide (FW = 4054.84 g/mol; CAS 172450-46-3; *Sequence*: RSCIDTIPKSRCTAFQCKHSMKYRLSFCRKTC GTC, with disulfide linkages indicated above), also known as ShK peptide, from the Caribbean Sea anemone, *Stichodactyla heliantus*, is among the most potent blockers of Kv1.3 voltage-gated potassium ion channel, IC$_{50}$ ≈ 10 pM (1). Its therapeutic application in autoimmune diseases is compromised by its lack of selectivity against other channels, particularly Kv1.1 channels, IC$_{50}$ ≈ 10 pM. Several analogues with altered selectivity (*e.g.,* ShK-Dap22, ShK-F6CA, and ShK-192), have been developed by using non-natural amino acids and/or adducts. Another analogue, ShK-186, contains a phosphorylated Tyr residue. **See also** *Dalazitide* **1**. Pennington, Mahnir, Krafte, *et al.* (1996) *Biochemistry* **35**, 16407. **2**. Tudor, Pallaghy, Pennington & Norton (1996) *Nature Struct. Biol.* **3**, 317.

Stigmast-5-ene-3β-ol, *similar in action to* Stigmast-5-ene-3β,26-diol

Stigmasterol

This phytosterol (FW = 412.70 g/mol; CAS 83-48-7; M.P. = 170°C; very low water solubility), also known as stigmasta-5,22-dien-3β-ol, found in calabar beans, soybeans, carrots, and almond hulls, inhibits cholesterol absorption (1); Na^+/K^+-exchanging ATPase (2); sterol acyltransferase, *or* cholesterol acyltransferase (3); and sterol Δ^{24}-reductase (4). **1**. Vahouny, Connor, Subramaniam, Lin & Gallo (1983) *Amer. J. Clin. Nutr.* **37**, 805. **2**. Hirano, Homma & Oka (1994) *Planta Med.* **60**, 30. **3**. Billheimer (1985) *Meth. Enzymol.* **111**, 286. **4**. Fernandez, Suarez, Ferruelo, Gomez-Coronado & Lasuncion (2002) *Biochem. J.* **366**, 109.

Stigmatellin A

This antibiotic (FW = 514.66 g/mol; CAS 91682-96-1; decomposes below pH 5) from *Stigmatella auraniaca* inhibits microbial electron transport at the bc_1 complex by binding to the heme b_1 domain of cytochrome *b* as well as to the iron-sulfur protein. In plants, stigmatellin inhibits electron transport by binding to the quinol oxidation (Q$_o$) site of the cytochrome bc_1 complex in mitochondria and the cytochrome b_6f complex of thylakoid membranes. (**See also** *Myxothiazol*) **Target(s):** inositol-trisphosphate 3-kinase (1); nitrate reductase A (2); plastoquinol:plastocyanin reductase, *or* cytochrome b_6/f complex (3,4); and ubiquinol:cytochrome *c* reductase, *or* complex III (5-10). **1**. Mayr, Windhorst & Hillemeier (2005) *J. Biol. Chem.* **280**, 13229. **2**. Magalon, Rothery, Lemesle-Meunier, *et al.* (1998) *J. Biol. Chem.* **273**, 10851. **3**. Malkin (1988) *Meth. Enzymol.* **167**, 341. **4**. Roberts, Bowman & Kramer (2002) *Biochemistry* **41**, 4070. **5**. von Jagow & Link (1986) *Meth. Enzymol.* **126**, 253. **6**. Schägger, Brandt, Gencic & von Jagow (1995) *Meth. Enzymol.* **260**, 82. **7**. Gutierrez-Cirlos & Trumpower (2002) *J. Biol. Chem.* **277**, 1195. **8**. Matsuno-Yagi & Hatefi (2001) *J. Biol. Chem.* **276**, 19006. **9**. Matsuno-Yagi & Hatefi (1999) *J. Biol. Chem.* **274**, 9283. **10**. Yang & Trumpower (1986) *J. Biol. Chem.* **261**, 12282.

Stilbamidine

This antiprotozoal agent (FW$_{free-base}$ = 264.33 g/mol; CAS 122-06-5), known also as 4,4'-(1,2-ethenediyl)bisbenzenecarboximidamide (*E*-isomer (above); *Z*-isomer (below)), inhibits acetylcholinesterase (1); *S*-adenosylmethionine decarboxylase (2); diamine oxidase (3); putrescine uptake (4); and thymidylate synthetase (5). **1**. Bergman, Wilson & Nachmansohn (1950) *Biochim. Biophys. Acta* **6**, 217. **2**. Klein, Favreau, Alexander-Bowman, *et al.* (1997) *Exp. Parasitol.* **87**, 171. **3**. Zeller (1963) *The Enzymes*, 2nd ed. (Boyer, Lardy & Myrbäck, eds.), **8**, 313. **4**. Reguera, Balana Fouce, Cubria, Alvarez Bujidos & Ordonez (1994) *Biochem. Pharmacol.* **47**, 1859. **5**. Kaplan & Myers (1977) *J. Pharmacol. Exp. Ther.* **201**, 554.

cis- and *trans-*Stilbene Oxide

This oxidized stilbene (FW = 196.25 g/mol; M.P. = 38-40°C (*cis*-isomer); M.P. = 65-67°C (*trans*-isomer), also known as *trans*-1,2-diphenyloxirane, inhibits cholesterol-5,6-oxide hydrolase, by *trans* isomer (1); microsomal epoxide hydrolase, by *cis* isomer (2); soluble epoxide hydrolase, also alternative substrate (3,4). **1**. Levin, Michaud, Thomas & Jerina (1983) *Arch. Biochem. Biophys.* **220**, 485. **2**. Oesch (1974) *Biochem. J.* **139**, 77. **3**. Morisseau, Archelas, Guitton, *et al.* (1999) *Eur. J. Biochem.* **263**, 386. **4**. Blée & Schuber (1992) *Biochem. J.* **282**, 711.

Stilbestrol

This notorious endocrine-disrupting chemical (FW = 212.25 g/mol; CAS 13425-53-1), also known as stilboestrol, 4,4'-dihydroxystilbene, and 4,4-[(1*E*)-1,2-ethenediyl]bisphenol, is a nonsteroidal estrogenic agent that was once used as a growth promoter in farm animals, but is now banned in most nations. *Note*: The term stilbestrol has also been used as a synonym for diethylstilbestrol. **See** *Diethylstilbestrol* **Target(s):** glucose-6-phosphate dehydrogenase (1,2); 3α/β-hydroxysteroid dehydrogenase (3,4); and steroid 11β-monooxygenase (5). **1**. McKerns & Kaleita (1960) *Biochem. Biophys. Res. Commun.* **2**, 344. **2**. Criss & McKerns (1969) *Biochim. Biophys. Acta* **184**, 486. **3**. Nakajin, Ishii & Shinoda (1994) *Biol. Pharm. Bull.* **17**, 1155. **4**. Nakajin, Fujita, Ohno, *et al.* (1994) *J. Steroid Biochem. Mol. Biol.* **48**, 249. **5**. Suzuki, Sanga, Chikaoka & Itagaki (1993) *Biochim. Biophys. Acta* **1203**, 215.

Stilboestrol, *See* Stilbestrol; Diethylstilbestrol

STO 609

This 1,8-naphthoylene benzimidazole-3-carboxylic acid (FW = 315.01 g/mol), also known as 7-oxo-7*H*-benzimidazo[2,1-*a*]benz[*de*]isoquinoline-3-carboxylic acid, inhibits Ca^{2+}/calmodulin-dependent protein kinase kinase (1,2). STO-609 inhibits the activities of recombinant CaM-KKα and CaM-KKβ isoforms, with K_i values of 0.25 and 0.05 μM, respectively, and also inhibits their autophosphorylation activities. Comparison of the inhibitory potency of the compound against various protein kinases revealed that STO-609 is highly selective for CaM-KK without any significant effect on the downstream CaM kinases (CaM-KI and -IV), and the IC$_{50}$ against CaM-KII is ~0.03 μM. STO-609 inhibits constitutively active CaM-KKα (glutathione S-transferase (GST)-CaM-KK-(84-434)) as well as the wild-type enzyme. Kinetic analysis indicates STO-609 is a competitive inhibitor of ATP. In transfected HeLa cells, it suppresses the Ca^{2+}-induced activation of CaM-KIV in a dose-dependent manner. In agreement with this observation, the inhibitor significantly reduces the endogenous activity of CaM-KK in SH-SY5Y neuroblastoma cells at a concentration of 1 microg/ml (~80% inhibitory rate). Such results indicate that STO-609 is a selective and cell-permeable inhibitor of CaM-KK and that it may be a useful tool for evaluating the physiological significance of the CaM-KK-mediated pathway *in vivo* as well as *in vitro* (1). The distinct sensitivity of CaM-KK isoforms to STO-609 arises from a single Val/Leu substitution in the ATP-binding pocket, as confirmed by the generatiom of a STO-609-resistant CaM-KK mutant that might be useful for validating the pharmacological effects and specificity of STO-609 *in vivo* (2). **1.** Tokumitsu, Inuzuka, Ishikawa, *et al.* (2002) *J. Biol. Chem.* **277**, 15813. **2.** Ishikawa, Tokumitsu, Inuzuka, *et al.* (2003) *FEBS Lett.* **550**, 57.

Streptavidin

This tetrameric biotin-binding protein (M_r = 18,834; CAS 9013-20-1; pI = 5.5) from *Streptomyces avidinii* binds biotin (K_d = 40 fM, at pH 7 and 25°C) (1-4). Streptavidin is frequently used in biotin-binding immunosorbant assays, protein blotting, radioimmuno-assays, and as a DNA probe, when the macromolecule of interest has been biotinylated *See also* Avidin **Target(s):** glycoprotein 6-α-L-fucosyltransferase (4); 3-methylcrotonoyl-CoA carboxylase (5); and urease, ATP-hydrolyzing (1). **1.** Chaiet & Wolf (1964) *Arch. Biochem. Biophys.* **106**, 1. **2.** Green (1990) *Meth. Enzymol.* **184**, 51. **3.** Bayer, Ben-Hur & Wilchek (1990) *Meth. Enzymol.* **184**, 80. **4.** Shao, Sokolik & Wold (1994) *Carbohydr. Res.* **251**, 163. **5.** Baldet, Alban, Axiotis & Douce (1992) *Plant Physiol.* **99**, 450.

Streptidine

This component of the antibiotic streptomycin ($FW_{free-base}$ = 262.27 g/mol; CAS 85-17-6), also known as 1,3-diguanido-2,4,5,6-cyclohexanetetrol, inhibits diamine oxidase (EC 1.4.3.6), the first enzymatic reaction in putrescine degradation. **1.** Cubria, Ordonez, Alvarez-Bujidos, Negro & Ortiz (1991) *Comp. Biochem. Physiol. B* **100**, 543.

Streptococcal NAD⁺-Glycohydrolase Inhibitor

This thermostable 161-residue protein (MW = 18.8 kDa), also known as IFS, competitively inhibits streptococcal NAD^+ glycohydrolase, *or* NADase (1). *Streptococcus pyogenes* NAD^+-glycohydrolase (SPN) is itself a toxic enzyme, introduced into *Streptococcus*-infected host cells by the cytolysin-mediated translocation pathway. *S. pyogenes* protects itself from SPN self-toxicity by means of NAD^+-glycohydrolase inhibitor, the small inhibitory protein localized in the bacterial cytoplasmic compartment. IFS combines with one-to-one stoichiometry relative to SPN, thereby inhibiting SPN's otherwise toxic NAD^+-glycohydrolase activity. IFS is the first molecularly characterized endogenous inhibitor of a bacterial β-NAD^+-consuming toxin, the protective mechanism underlying SPN-mediated streptococcal pathogenesis (2). **1.** Kimoto, Fujii, Hirano, Yokota & Taketo (2006) *J. Biol. Chem.* **281**, 9181. **2.** Meehl, Pinkner, Anderson, Hutgren & Caparon (2005) *PLoS* **1**, E35.

Streptolydigin

This antibiotic (FW = 600.71 g/mol; CAS 7229-50-7), also known as portamycin, from *Streptomyces lydicus* inhibits bacterial RNA polymerase without affecting the fidelity of transcription. The product of the polymerase reaction does not dissociate from the inhibitor-bound translocated ternary complex. **Target(s):** DNA-directed RNA polymerase (1-9); DNA nucleotidylexotransferase, *or* terminal deoxyribonucleotidyl-transferase (10); and oxidative phosphorylation (11). **1.** Chamberlin (1974) *The Enzymes*, 3rd ed., **10**, 333. **2.** Jin & Zhou (1996) *Meth. Enzymol.* **273**, 300. **3.** von der Helm & Krakow (1972) *Nature New Biol.* **235**, 82. **4.** Cassani, Burgess, Goodman & Gold (1971) *Nature New Biol.* **230**, 197. **5.** Logan & Ackerman (1988) *DNA* 7, 483. **6.** McClure (1980) *J. Biol. Chem.* **255**, 1610. **7.** Sethi (1971) *Prog. Biophys. Mol. Biol.* **23**, 67. **8.** Chopra (2007) *Curr. Opin. Invest. Drugs* **8**, 600. **9.** Allan & Kropinski (1987) *Biochem. Cell Biol.* **65**, 776. **10.** DiCioccio & Srivastava (1976) *Biochem. Biophys. Res. Commun.* **72**, 1343. **11.** Reusser (1969) *J. Bacteriol.* **100**, 1335.

Streptomyces Subtilisin Inhibitor

This serine proteinase inhibitor ($MW_{calculated}$ = 11,483 g; Abbreviation: SSI) is a homodimeric protein, with each subunit consisting of 113 residues and two disulfide bridges. many protein protease inhibitors are susceptible to hydrolysis inflicted by the protease. Ganz *et al.* (1) engineered SSI to resist proteolysis by adding an interchain -S–S- bond and removing a subtilisin cleavage site at Leu-63. When combined with changes optimizing subtilisin affinity SSI affinity, the resulting inhibitor provided complete protease stability for at least 5 months at 31°C in a subtilisin-containing liquid laundry detergent, while allowing full recovery of subtilisin activity upon dilution in the was cycle of North American washing machine. **Target(s):** *Aspergillus* serine proteinase (2); aqualysin I (3,4); mycolysin (5); oryzin (6); streptogrisin A (7,8); streptogrisin B (8,9); and subtilisin (10-13). **1.** Ganz, Bauer, Sun, *et al.* (2004) *Prot. Eng. Des. Select.* **17**, 333. **2.** Markaryan, Beall & Kolattukudy (1996) *Biochem. Biophys. Res. Commun.* **220**, 372. **3.** Suefuji, Lin & Matsuzawa (1998) in *Handb. Proteolytic Enzymes* (Barrett, Rawlings & Woessner, eds.), p. 315, Academic Press, San Diego. **4.** Matsuzawa, Tokugawa, Hamaoki, *et al.* (1988) *Eur. J. Biochem.* **171**, 441. **5.** Ishii & Kumazaki (1998) in *Handb. Proteolytic Enzymes* (Barrett, Rawlings & Woessner, eds.), p. 1078, Academic Press, San Diego. **6.** Kolattukudy & Sirakova (1998) in *Handb. Proteolytic Enzymes* (Barrett, Rawlings & Woessner, eds.), p. 320, Academic Press, San Diego. **7.** Qasim (1998) in *Handb. Proteolytic Enzymes* (Barrett, Rawlings & Woessner, eds.), p. 241, Academic Press, San Diego. **8.** Christensen, Ishida, Ishii, *et al.* (1985) *J. Biochem.* **98**, 1263. **9.** Qasim (1998) in *Handb. Proteolytic Enzymes* (Barrett, Rawlings & Woessner, eds.), p. 236, Academic Press, San Diego. **10.** Ballinger & Wells (1998) in *Handb. Proteolytic Enzymes* (Barrett, Rawlings & Woessner, eds.), p. 289, Academic Press, San Diego. **11.** Takeuchi, Satow, Nakamura & Mitsui (1991) *J. Mol. Biol.* **221**, 309. **12.** Mitsui, Satow, Watanabe, Hirono & Iitaka (1979) *Nature* **277**, 447. **13.** Kourteva & Boteva (1989) *FEBS Lett.* **247**, 468.

Streptomycin

This aminoglycoside antibiotic ($FW_{free-base}$ = 581.58 g/mol; CAS 57-92-1) inhibits protein biosynthesis, binding to the 16S ribosomal RNA of the 30S ribosomal subunit, and inhibits initiation and causes misreading. This antibiotic is composed of three components: streptidine (*i.e.*, 1,3-diguanido-2,4,5,6-cyclohexanetetrol), streptose (5-deoxy-3-*C*-formyl-L-lyxose), and *N*-methyl-L-glucosamine, with the last two components linked glycosidically are occasionally referred to as streptobiosamine. Streptomycin was the first antibiotic identified by Waksman (Nobel

Laureate, 1952) in the soil bacterium *Streptomyces griseus*, but has been isolated from other *Streptomyces* species. *S. biliniensis, S. olivaceus, S. mashuensis, S. galbus, S. rameus,* and *S. glovisporus streptomicine).* Streptomycin first associates with the cell surface of the microorganism and diffuses through the outer membrane into the periplasmic space. It then enters the cytosol using the electron-transport system in an energy-dependent process. **Antibiotic Resistance:** Once inside the microorganism, streptomycin binds to the 30S ribosomal subunit (specifically nucleotides 911-915 of the 530 loop within 16S ribosomal RNA of *Escherichia coli*). An abnormal initiation complex is subsequently formed, fixing the ribosomal complex at the start codon of the mRNA. Streptomycin resistance arises from alteration of the 530 loop of the 16S ribosomal RNA, one of the most highly conserved part of the aminoacyl-tRNA binding site (A-site), and is manifested by greatly attenuated affinity for streptomycin. **Chemical Properties:** Streptomycin is usually supplied as the salt (*e.g.,* the trihydrochloride, pantothenate, or sesquisulfate). These salts are hygroscopic and deliquesce when exposed to air. As expected, these formulations are very soluble in water (for example, more than 20 mg of the trihydrochloride or sesquisulfate can dissolve in 1 mL of water at 28° C). **Target(s):** alcohol dehydrogenase, yeast (1); D-amino acid oxidase (2); aspartyltransferase (3); CDP-glycerol glycerophospho-tramsferase (4); diamine oxidase (5,6); 2,5-diaminovalerate aminotransferase, weakly inhibited (7); group I ribozymes (8,9); peptidyl-tRNA hydrolase, *or* aminoacyl-tRNA hydrolase (10); phosphatidylinositol-specific phospholipase C (11,12); phospholipase D (13); protein biosynthesis (10,14-17); ribonuclease (18); and urate oxidase (19). **1.** Sund & Theorell (1963) *The Enzymes,* 2nd ed. (Boyer, Lardy & Myrbäck, eds.), **7,** 25. **2.** Yagi (1971) *Meth. Enzymol.* **17B,** 608. **3.** Jayaram, Ramakrishnan & Vaidyanathan (1969) *Indian J. Biochem.* **6,** 106. **4.** Burger & Glaser (1964) *J. Biol. Chem.* **239,** 3168. **5.** Zeller (1951) *The Enzymes,* 1st ed. (Sumner & Myrbäck, eds.), **2** (part 1), 536. **6.** Zeller, Owen, Jr., & Karlson (1951) *J. Biol. Chem.* **188,** 623. **7.** Roberts (1954) *Arch. Biochem. Biophys.* **48,** 395. **8.** von Ahsen & Schroeder (1991) *Nucl. Acids Res.* **19,** 2261. **9.** Hertweck, Hiller & Mueller (2002) *Eur. J. Biochem.* **269,** 175. **10.** Tate & Caskey (1974) *The Enzymes,* 3rd ed. (Boyer, ed.), **10,** 87. **11.** Lipsky & Lietman (1982) *J. Pharmacol. Exp. Ther.* **220,** 287. **12.** Schwertz, Kreisberg & Venkatachalam (1984) *J. Pharmacol. Exp. Ther.* **231,** 48. **13.** Liscovitch, Chalifa, Danin & Eli (1991) *Biochem. J.* **279,** 319. **14.** Pestka (1974) *Meth. Enzymol.* **30,** 261. **15.** Jiménez (1976) *Trends Biochem. Sci.* **1,** 28. **16.** Coutsogeorgopoulos (1967) *Biochemistry* **6,** 1704. **17.** Flaks, Cox & White (1962) *Biochem. Biophys. Res. Commun.* **7,** 385. **18.** Massart, Peeters & Lagrain (1948) *Arch. intern. pharmacodynamie* **76,** 72. **19.** Rainforth & Truscoe (1969) *Enzymologia* **37,** 185.

Streptonigrin

This antibiotic and antineoplastic agent (FW$_{free-acid}$ = 508.48 g/mol; CAS 3930-19-6; dark brown solid; M.P. = 262-263°C; slightly soluble in water) from *Streptomyces flocculus* requires iron activation for its antibiotic action. In the presence of certain metal ions, streptonigrin causes DNA cleavage and chromosome damage via the production of free radicals. Streptonigrin also induces apoptosis in the presence of NF-κB. **Target(s):** avian myeloblastosis virus reverse transcriptase (1,2); DNA topoisomerase II (3); phosphoenolpyruvate carboxykinase (4); and RNA-directed DNA polymerase (1,2,5). **1.** Hafuri, Takemori, Oogose, *et al.* (1988) *J. Antibiot.* **41,** 1471. **2.** Okada, Inouye & Nakamura (1987) *J. Antibiot.* **40,** 230. **3.** Anderberg & Harding (2001) *Anticancer Drug Des.* **16,** 143. **4.** Merryfield & Lardy (1982) *Biochem. Pharmacol.* **31,** 1123. **5.** Inouye, Oogose, Take, Kubo & Nakamura (1987) *J. Antibiot.* **40,** 702.

Streptothricin

Streptothricin F

These antibiotics from *Actinomyces lavendulae* (FW = 502.526 g/mol; CAS 11076-96-3), which differ with respect to the number (*i.e.,* one to seven) of β-lysyl residues, were originally thought to be a single substance, when isolated by Nobel Laureate Selman Waksman. For streptothricin B, *n* = 5, and for streptothricin F, *n* = 1. *Note:* The sugar component is a derivative of D-gulosamine. **Target(s):** EF-G-dependent translocation, inhibited by streptothricin F (1); and EF-Tu-dependent binding of aminoacyl-tRNA to the ribosome, inhibited by streptothricin F (1); and phosphatidylinositol phospholipase C, inhibited by streptothricin B (2). **1.** Haupt, Jonak, Rychlik & Thrum (1980) *J. Antibiot.* **33,** 636. **2.** Hill, Bonjouklian, Powis, *et al.* (1994) *Anticancer Drug Des.* **9,** 353.

Streptovaricins

These ansamysin-type macrolide antibiotics (CASs = 1404-74-6 for general class; for Streptovaricin C: FW = 769.84 g/mol; CAS 23344-17-4) from *Streptomyces spectabilis* inhibit RNA and protein synthesis in microorganisms, particularly in Gram-positive bacteria and also inhibits in vitro RNA synthesis catalyzed by DNA-dependent RNA polymerase. Unlike other RNA polymerase-directed antibiotics, the streptovaricins target the polymerase and not its DNA template. This antibiotic group includes: Streptovaricin A (FW = 770.81 g/mol): R = COOCH$_3$, R' = OCOCH$_3$, W = OH, X = Ac, Y = OH, Z = OH); Streptovaricin B (FW = 754.81 g/mol): R = COOCH$_3$, R' = OCOCH$_3$, W = OH, X = Ac, Y = OH, Z = H; Streptovaricin C (the antibiotic component; (FW = 729.78 g/mol): R = COOCH$_3$, R' = OCOCH$_3$, W = OH, X = H, Y = OH, Z = H; Streptovaricin D (FW = 713.78 g/mol): R = COOCH$_3$, R' = OCOCH$_3$, W = H, X = H, Y = OH, Z = H; Streptovaricin E (FW = 628.77 g/mol): R = COOCH$_3$, R' = OCOCH$_3$, W = OH, X = H, Y = O, Z = H; Streptovaricin G (FW = 7766.77 g/mol): R = COOCH$_3$, R' = OCOCH$_3$, W = OH, X = H, Y = OH, Z = OH; Streptovaricin J (FW = 755.82 g/mol): R = COOCH$_3$, R' = OCOCH$_3$, W = OH, X = H, Y = OAc, Z = H; and Streptovaricin K (FW = 773.67 g/mol): (R = COOCH$_3$, R' = OCOCH$_3$, W = OH; X = H, Y = OAc, Z = OH). **Target(s):** DNA-directed RNA polymerase (1-8); and RNA-directed DNA polymerase, RNA tumor viruses (2,9,10). **1.** Chamberlin (1974) *The Enzymes,* 3rd ed. (Boyer, ed.), **10,** 333. **2.** Ooka, Tanaka, Ishikawa & Kato (1999) *Biol. Pharm. Bull.* **22,** 107. **3.** Logan, Zhang, Davis & Ackerman (1989) *DNA* **8,** 595. **4.** Mizuno, Yamazaki, Nitta & Umezawa (1968)

Biochim. Biophys. Acta **157**, 322. **5**. Mizuno, Yamazaki, Nitta & Umezawa (1968) *Biochem. Biophys. Res. Commun.* **30**, 379. **6**. Mizuno, Yamazaki, Nitta & Umezawa (1968) *J. Antibiot. (Tokyo)* **21**, 66. **7**. Sethi (1971) *Prog. Biophys. Mol. Biol.* **23**, 67. **8**. Allan & Kropinski (1987) *Biochem. Cell Biol.* **65**, 776. **9**. Milavetz, Horoszewicz, Rinehart, Jr., & Carter (1978) *Antimicrob. Agents Chemother.* **13**, 435. **10**. Brockman, Carter, Li, Reusser & Nichol (1971) *Nature* **230**, 249.

Streptovirudins

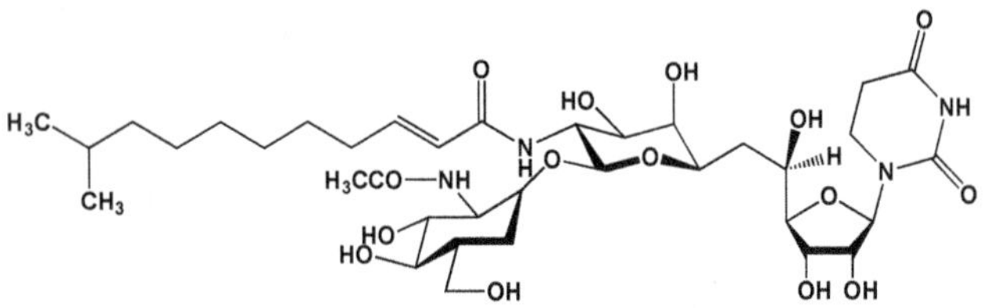

These nucleoside-containing antibiotics (Streptovirudin A₁ (shown above); FW$_{\text{Streptovirudin-A1}}$ = 788.89 g/mol; CAS 51330-28-0), consisting of at least ten metabolites from *Streptomyces griseoflavus*, are structurally related to tunicamycin. There are two distinct subcategories of streptovirudins: series 1 contain a dihydrouracil moiety (Streptovirudin A₁, B₁, C₁, and D₁) whereas series 2 contains a uracil (Streptovirudins A₂, B₂, C₂, and D₂). *Note*: Streptovirudin C₂ is identical to tunicamycin A. All are relatively stable antibiotics that are slightly soluble in water and have an inhibitory effect on Gram-positive microorganisms, mycobacteria, and a number of viruses. **Target(s):** protein glycosylation (1-4); and UDP-*N*-acetylglucosamine: dolichyl-phosphate *N*-acetylglucosamine phosphotrans-ferase (1,2,4) **1**. Elbein (1983) *Meth. Enzymol.* **98**, 135. **2**. Elbein (1987) *Meth. Enzymol.* **138**, 661. **3**. Elbein (1984) *CRC Crit. Rev. Biochem.* **16**, 21. **4**. Keenan, Hamill, Occolowitz & Elbein (1981) *Biochemistry* **20**, 2968.

Streptozotocin

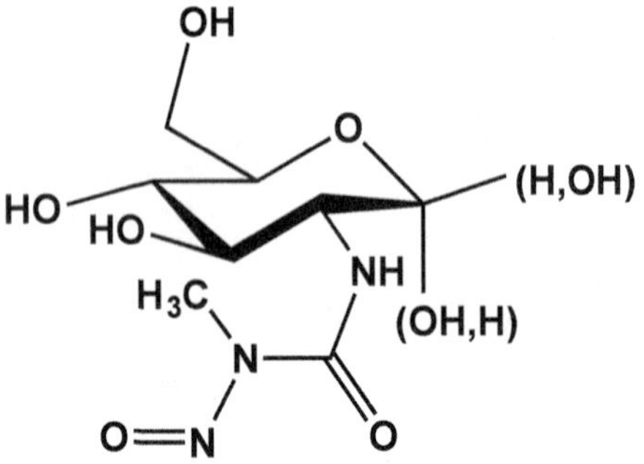

This nitrosourea-containing antibiotic (FW = 265.22 g/mol; CAS 18883-66-4; abbreviated STZ), also known as streptozocin and 2-deoxy-2-[[(methylnitrosoamino)carbonyl]amino]-D-glucopyranose, is also an antineoplastic agent and a diabetes-causing agent originally isolated from *Streptomyces achromogenes*. Its action in inducing experimental diabetes in laboratory animals has been attributed to its selective uptake by the GLUT2 transporter, the main glucose transporter of pancreatic β-cells. Although the inactivation of superoxide dismutase should be toxic, the exact lethal action of streptozotocin on β-cells has not been unambiguously established. STZ is also a nonspecific DNA-alkylating agent *in vitro* and is an FDA-approved treatment for metastatic islet cell cancer. **Induction of Experimental Diabetes:** Males of various mouse strains are differentially susceptible to STZ-induced diabetes. The following multiple low-dose STZ protocol induces diabetes in susceptible males. Mice receive daily IP injections (50 mg STZ/kg body weight) for five consecutive days, with age-matched controls receiving only buffer injections. Because mice metabolize STZ for at least 24 hours post-injection, experimental and control mice are housed separately in disposable cages, using appropriate absorbent bedding and supplying food and water *ad libitum*. At post-injection day-10, pancreatic islet cells are totally absent or severely depleted. Mice are weighed and their blood glucose levels determined. They are deemed to be diabetic, if their non-fasted blood glucose levels exceed 300-400 mg/dL. **Target(s):** *N*-acetyl-β-D-glucosaminidase, *O*-GlcNAc-selective (1); β-glucosidase (2); stearoyl-CoA desaturase (3); and superoxide dismutase (4). **1**. Konrad, Mikolaenko, Tolar, Liu & Kudlow (2001) *Biochem. J.* **356**, 31. **2**. Dale,

Ensley, Kern, Sastry & Byers (1985) *Biochemistry* **24**, 3530. **3**. Nishida, Sasaki, Terada & Kawada (1988) *Experientia* **44**, 756. **4**. Crouch, Gandy, Kimsey, *et al.* (1981) *Diabetes* **30**, 235.

Strobilurin A

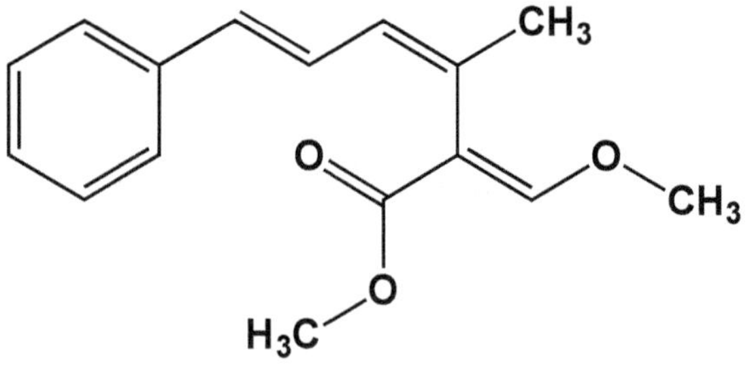

This antifungal antibiotic (FW = 258.32 g/mol; CAS 65105-52-4), commonly known as mucidin and systematically as (*E,Z,E*)-2-(methoxymethylene)-3-methyl-6-phenyl-3,5-hexadienoic acid methyl ester, from wood-rotting mushrooms (*e.g.*, *Strobilurus tenacellus*) inhibits mitochondrial respiration at the cytochrome *bc*1 step. It has a low solubility in water and is soluble in many organic solvents. The molar extinction coefficients at 230, 237, and 294 nm in ethanol are 16900, 15330, and 21850 M⁻¹cm⁻¹, respectively. Strobilurin A is the parent compound of a large class of naturally occurring and synthetic compounds frequently used as agricultural antifungal agents. **Target(s):** chitin synthase (1); and ubiquinol:cytochrome *c* reductase, *or* complex III (2-4). **1**. Pfefferle, Anke, Bross & Steglich (1990) *Agric. Biol. Chem.* **54**, 1381. **2**. Von Jagow & Link (1986) *Meth. Enzymol.* **126**, 253. **3**. Wood & Hollomon (2003) *Pest Manag. Sci.* **59**, 499. **4**. Von Jagow, Gribble & Trumpower (1986) *Biochemistry* **25**, 775.

Strobilurin B

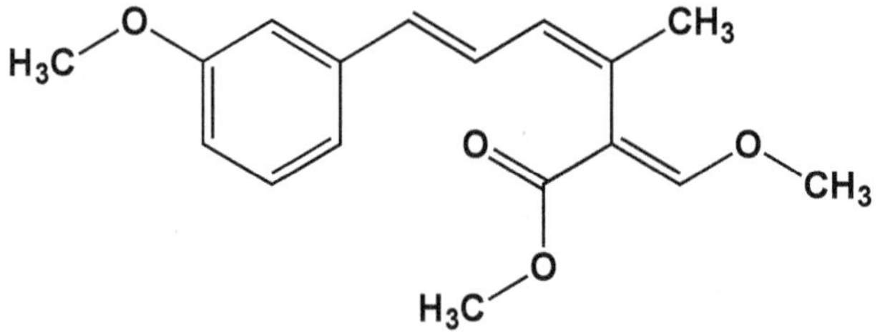

This antifungal antibiotic (FW = 321.78 g/mol) inhibits mitochondrial respiration at the cytochrome *bc*1 step. **Target(s):** chitin synthase (1); and ubiquinol:cytochrome *c* reductase (2). **1**. Pfefferle, Anke, Bross & Steglich (1990) *Agric. Biol. Chem.* **54**, 1381. **2**. Von Jagow & Link (1986) *Meth. Enzymol.* **126**, 253.

Stromelysin I Inhibitor, *See* specific inhibitor; e.g., *N*ᵃ-Acetyl-L-arginyl-L-cysteinylglycyl-L-valyl-L-prolyl-L-aspartate a-Amide

Strontium (II) (*or* Sr²⁺) Ion

This divalent cation from the Group II alkaline earth element acts either directly as an inhibitory metal ion and/or Sr²⁺-ligand/metabolite complex. Strontium commonly occurs in nature and is the 15ᵗʰ most abundant element (~360 parts per million) in the Earth's crust. Strontium(II) acts as an alternative metal ion cofactor for many (*but not all*) calcium ion-dependent biological processes. For example, Sr²⁺ is reported to have similar kinetics with respect to the sarcoplasmic reticulum calcium ATPase (1) and can induce Ca²⁺ release via cyclic ADP-ribose (2). **Target(s):** *N*-acylneuraminate cytidylyltransferase (63); adenylate cyclase (3); alkaline phosphatase (57,58); 4-aminobutyrate aminotransferase (67); α-amylase (51-53); aryldialkyl-phosphatase, *or* paraoxonase (61); arylesterase (62); asparagine synthetase (4); calcidiol 1-monooxygenase (5); calpain-1, *or* μ-calpain (16); chitinase (47); creatine kinase (7,8); dextranase (48); dextransucrase (69); dipeptidyl-peptidase IV (9); diphosphate:serine phosphotransferase, weakly inhibited (64); DNA-directed DNA polymerase (10); enolase (11,12); exo-α-sialidase, *or* neuraminidase, *or* sialidase (45); glucan 1,4-α-glucosidase, *or* glucoamylase (49,50); glucan 1,4-α-maltohexaosidase (13,14); β-glucosidase (43,44); glucose-6-phosphate isomerase (15); glycine *C*-acetyltransferase (70); GTP cyclohydrolase I (16); H⁺/K⁺-exchanging ATPase (17); isocitrate lyase (18,19); 3-

ketovalidoxylamine C-N-lyase (20); α-mannosidase (42); methylaspartate ammonia-lyase, *or* β-methylaspartase (21,22); micrococcal nuclease (23); Na$^+$/K$^+$-exchanging ATPase (24); α-neoagaro-oligosaccharide hydrolase (25); nicotinate-nucleotide diphosphorylase, carboxylating (68); oryzin (26); oxidative phosphorylation (27); pantothenate synthetase, *or* pantoate:β-alanine ligase, slightly inhibited (28); pectate lyase, weakly inhibited (29); phosphatidylinositol diacylglycerol-lyase (30); phospholipase A$_2$ (31-33,59,60); phospholipase C (34,35); phospholipase D (54); 3- or 4-phytase (55); polygalacturonidase, *or* pectinase (46); pullulanase (36); pyruvate decarboxylase, slightly inhibited (37); pyruvate kinase (65); riboflavin kinase (66); ribonuclease, plant (38); RNA ligase ribozyme (39); thioglucosidase, *or* myrosinase (40); thymidine kinase (41); trehalose-phosphatase (56); and trehalose-phosphate synthase (56). **1.** Fujimori & Jencks (1992) *J. Biol. Chem.* **267**, 18466. **2.** Lee (1993) *J. Biol. Chem.* **268**, 293. **3.** Johnson & Sutherland (1974) *Meth. Enzymol.* **38**, 135. **4.** Lea & Fowden (1975) *Proc. R. Soc. Lond. B Biol. Sci.* **192**, 13. **5.** Henry (1980) *Meth. Enzymol.* **67**, 445. **6.** Inomata, Nomoto, Hayashi, *et al.* (1984) *J. Biochem.* **95**, 1661. **7.** O'Sullivan & Morrison (1963) *Biochim. Biophys. Acta* **77**, 142. **8.** Kuby & Noltmann (1962) *The Enzymes*, 2nd ed. (Boyer, Lardy & Myrbäck, eds.), **6**, 515. **9.** Shibuya-Saruta, Kasahara & Hashimoto (1996) *J. Clin. Lab. Anal.* **10**, 435. **10.** Fansler & Loeb (1974) *Meth. Enzymol.* **29**, 53. **11.** Bücher (1955) *Meth. Enzymol.* **1**, 427. **12.** Malmström (1961) *The Enzymes*, 2nd ed. (Boyer, Lardy & Myrbäck, eds.), **5**, 471. **13.** Takasaki (1982) *Agric. Biol. Chem.* **46**, 1539. **14.** Nakakuki, Hayashi, Monma, Kawashima & Kainuma (1983) *Biotechnol. Bioeng.* **25**, 1095. **15.** Sangwan & Singh (1989) *J. Biosci.* **14**, 47. **16.** Suzuki, Yasui & Abe (1979) *J. Biochem.* **86**, 1679. **17.** Hango, Nojima & Setaka (1990) *Jpn. J. Pharmacol.* **52**, 295. **18.** Tanaka, Yoshida, Watanabe, Izumi & Mitsunaga (1997) *Eur. J. Biochem.* **249**, 820. **19.** Hoyt, Johnson & Reeves (1991) *J. Bacteriol.* **173**, 6844. **20.** Takeuchi, Asano, Kameda & Matsui (1985) *J. Biochem.* **98**, 1631. **21.** Hanson & Havir (1972) *The Enzymes*, 3rd ed. (Boyer, ed.), **7**, 75. **22.** Bright (1967) *Biochemistry* **6**, 1191. **23.** Cuatrecasas, Fuchs & Anfensen (1967) *J. Biol. Chem.* **242**, 1541. **24.** Knudsen, Berthelsen & Johansen (1990) *Brit. J. Pharmacol.* **100**, 453. **25.** Suzuki, Sawai, Suzuki & Kawai (2002) *J. Biosci. Bioeng.* **93**, 456. **26.** Ohara & Nasuno (1972) *Agric. Biol. Chem.* **36**, 1797. **27.** Fagian, da Silva & Vercesi (1986) *Biochim. Biophys. Acta* **852**, 262. **28.** Miyatake, Nakano & Kitaoka (1978) *J. Nutr. Sci. Vitaminol.* **24**, 243. **29.** Tardy, Nasser, Robert-Baudouy & Hugouvieux-Cotte-Pattat (1997) *J. Bacteriol.* **179**, 2503. **30.** Abdel-Latif, Luke & Smith (1980) *Biochim. Biophys. Acta* **614**, 425. **31.** Deems & Dennis (1981) *Meth. Enzymol.* **71**, 703. **32.** Reynolds & Dennis (1991) *Meth. Enzymol.* **197**, 359. **33.** de Haas, Bonson, Pieterson & van Deenen (1971) *Biochim. Biophys. Acta* **239**, 252. **34.** Hayaishi (1955) *Meth. Enzymol.* **1**, 660. **35.** Zeller (1951) *The Enzymes*, 1st ed. (Sumner & Myrbäck, eds.), **1** (part 2), 986. **36.** Odibo & Obi (1988) *J. Ind. Microbiol.* **3**, 343. **37.** Leblova, Malik & Fojta (1989) *Biologia (Bratisl.)* **44**, 329. **38.** Anfinsen & White, Jr. (1961) *The Enzymes*, 2nd ed. (Boyer, Lardy & Myrbäck, eds.), **5**, 95. **39.** Glasner, Bergman & Bartel (2002) *Biochemistry* **41**, 8103. **40.** Tani, Ohtsuru & Hata (1974) *Agric. Biol. Chem.* **38**, 1623. **41.** Bresnick (1978) *Meth. Enzymol.* **51**, 360. **42.** Matta & Bahl (1972) *J. Biol. Chem.* **247**, 1780. **43.** Plant, Oliver, Patchett, Daniel & Morgan (1988) *Arch. Biochem. Biophys.* **262**, 181. **44.** Yoshioka & Hayashida (1980) *Agric. Biol. Chem.* **44**, 1729. **45.** De Martínez & Olavarria (1973) *Biochim. Biophys. Acta* **320**, 301. **46.** Miyairi, Okuno & Sawai (1985) *Agric. Biol. Chem.* **49**, 111. **47.** Bhushan & Hoondal (1998) *Biotechnol. Lett.* **20**, 157. **48.** Madhu & Prabhu (1984) *Enzyme Microb. Technol.* **6**, 217. **49.** Tsao, Hsu, Chao & Jiang (2004) *Fish. Sci.* **70**, 174. **50.** Marlida, Saari, Hassan, Radu & Bakar (2000) *Food Chem.* **71**, 221. **51.** Takasaki (1983) *Agric. Biol. Chem.* **47**, 2193. **52.** Takasaki (1985) *Agric. Biol. Chem.* **49**, 1091. **53.** Tsao, Hsu, Chao & Jiang (2004) *Fish. Sci.* **70**, 174. **54.** Abousalham, Riviere, Teissere & Verger (1993) *Biochim. Biophys. Acta* **1158**, 1. **55.** Powar & Jagannathan (1982) *J. Bacteriol.* **151**, 1102. **56.** Silva, Alarico & da Costa (2005) *Extremophiles* **9**, 29. **57.** Sakurai, Toda & Shiota (1981) *Agric. Biol. Chem.* **45**, 1959. **58.** Romero-Saravia & Hamdorf (1983) *Biochim. Biophys. Acta* **729**, 90. **59.** Slotboom, Jansen, Vlijm, *et al.* (1978) *Biochemistry* **17**, 4593. **60.** Pieterson, Volwerk & de Haas (1974) *Biochemistry* **13**, 1439. **61.** Mende & Moreno (1975) *Biochemistry* **14**, 3913. **62.** Toshimitsu, Hamada & Kojima (1986) *J. Ferment. Technol.* **64**, 459. **63.** Bravo, Barrallo, Ferrero, *et al.* (2001) *Biochem. J.* **358**, 585. **64.** Cagen & Friedmann (1972) *J. Biol. Chem.* **247**, 3382. **65.** Sakai, Suzuki & Imahori (1986) *J. Biochem.* **99**, 1157. **66.** Merrill, Jr., & McCormick (1980) *J. Biol. Chem.* **255**, 1335. **67.** Schousboe, Wu & Roberts (1974) *J. Neurochem.* **23**, 1189. **68.** Taguchi & Iwai (1975) *Agric. Biol. Chem.* **39**, 1599. **69.** Miller & Robyt (1986) *Arch.*

Biochem. Biophys. **248**, 579. **70.** Mukherjee & Dekker (1987) *J. Biol. Chem.* **262**, 14441.

Strophanthidin

This toxic sterol (FW = 404.50 g/mol; CAS 66-28-4), also known as (3β,5β)-3,5,14-trihydroxy-19-oxocard-20(22)-enolide, is the aglycon on *k*-strophanthin (Note: γ-Strophanthin is ouabain), from *Strophanthus kombe* inhibits Ca^{2+}-transporting ATPase (1), H$^+$/K$^+$-transporting ATPase (1), H$^+$-transporting ATPase (1), and Na$^+$/K$^+$-exchanging ATPase (2-4). **1.** Xu (1992) *Biochim. Biophys. Acta* **1159**, 109. **2.** Ruoho & Kyte (1977) *Meth. Enzymol.* **46**, 523. **3.** Glynn (1964) *Pharmacol. Rev.* **16**, 381. **4.** Hokin & Yoda (1964) *Proc. Natl. Acad. Sci. U.S.A.* **52**, 454.

Strophanthidin-3-bromoacetate

This steryl haloacetate (FW = 525.44 g/mol) is an irreversible inhibitor of Na$^+$/K$^+$-exchanging ATPase, binding at the same site where cardiotonic steroids bind. **1.** Hokin, Mokotoff & Kupchan (1966) *Proc. Natl. Acad. Sci. U.S.A.* **55**, 797.

Strychnine

This toxic, bitter-tasting alkaloid and seizurogenic agent (FW$_{free-base}$ = 334.42 g/mol; CAS 57-24-9) from the seeds of *Strychnos nux-vomica* (the dog button plant) and *Strychnos ignatti* (Saint Ignatius bean) exhibits convulsion-inducing properties. Rather than directly affecting excitatory neurotransmission, strychnine (2 mg/kg in mice) blocks the inhibitory regulation of neurotransmission by antagonism of inhibitory glycine synapses (primarily) and GABA synapses (secondarily). Strychnine-sensitive glycine receptors are primarily localized in the brainstem and spinal cord where they are the major mediators of postsynaptic inhibition (1). Such receptors are ligand-gated ion channels arranged as five 4-helix subunits surrounding a central pore. LD$_{50}$ values are 0.16 mg/kg in rats and 1–2 mg/kg in humans (orally). In most cases of lethal strychnine poisoning, the patient dies before ever reaching the hospital. Little or no activity against seizures is produced by other chemical convulsants (bicuculline;

quinolinic acid; mercaptopropionic acid); by electrical stimuli (maximal electroshock); or by sensory stimuli (audiogenic seizure susceptible mice) (1) The glycine receptor agonist MDL 27,531 (or 4-methyl-3-methylsulphonyl-5-phenyl-4H-1,2,4-triazole), however, successfully blocks strychnine-induced tonic extensor seizures in mice following either intraperitoneal (ED_{50} = 12.8 mg/kg; 30 min) or oral (ED_{50} = 7.3 mg/kg; 30 min) administration (1). **Target(s):** cholinesterase, or acetylcholinesterase (2,3); esterase, nonspecific (3); glycine receptor (4-8); and lactate dehydrogenase (9). **1.** Kehne, Kane, Miller, et al. (1992) Brit. J. Pharmacol. **106**, 910. **2.** Augustinsson (1950) The Enzymes, 1st ed. (Sumner & Myrbäck, eds.), **1** (part 1), 443. **3.** Nachmansohn & Schneemann (1945) J. Biol. Chem. **159**, 239. **4.** Taepavarapruk, McErlane & Soja (2002) J. Neurosci. **22**, 5777. **5.** Nguyen, Malgrange, Belachew, et al. (2002) Eur. J. Neurosci. **15**, 1299. **6.** Rasmussen, Rasmussen, Triller & Vannier (2002) Mol. Cell. Neurosci. **19**, 201. **7.** Zhou (2001) J. Neurosci. **21**, 5158. **8.** Lopez-Corcuera, Geerlings & Aragon (2001) Mol. Membr. Biol. **18**, 13. **9.** Wade & Whiteley (1987) Biochem. Int. **14**, 189.

Stylomycin, See Puromycin

Styrene

This hazardous and versatile organochemical starting reagent (FW = 104.15 g/mol; CAS 100-42-5; slightly soluble in water, but soluble in most methanol, diethyl ether, ethanol, and acetone) for organic chemical synthesis is a colorless liquid (M.P. = –30.6°C; B.P. = 145-146°C). On exposure to light, styrene slowly undergoes free radical-mediated polymerization, and many commercial sources of styrene contaon the styrene polymerization inhibitor 4-(t-butyl)catechol) as an additive. **Target(s):** acetylcholinesterase (1); ATPase (2,3); δ-aminolevulinate dehydratase, or porphobilinogen synthase, observed inhibition may occur by contaminating styrene oxide (4,5); dopamine transport (6); monoamine oxidase A (7); and Na^+/K^+-exchanging ATPase (8). **1.** Korpela & Tahti (1986) Arch. Toxicol. Suppl. **9**, 320. **2.** Vaalavirta & Tahti (1995) Clin. Exp. Pharmacol. Physiol. **22**, 293. **3.** Vaalavirta & Tahti (1995) Life Sci. **57**, 2223. **4.** Fujita, Ishida & Akagi (1995) Nippon Rinsho **53**, 1408. **5.** Mori, Fujishiro & Inoue (1988) J. UOEH **10**, 269. **6.** Chakrabarti (1999) Biochem. Biophys. Res. Commun. **255**, 70. **7.** Egashira, Takayama, Sakai & Yamanaka (2000) Toxicol. Lett. **117**, 115. **8.** Singh & Srivastava (1986) J. Environ. Pathol. Toxicol. Oncol. **6**, 29.

Styrene Oxide

This oxidized styrene metabolite FW = 120.15 g/mol; CAS 96-09-3), also known as oxidostyrene, phenyloxirane, and styrene 7,8-oxide, inhibits a variety of enzymes, including alcohol dehydrogenase (1). Styrene oxide also stimulates formation of alkali-sensitive, DNA single-strand breaks and is a cancer-suspect agent. **Target(s):** δ-aminolevulinate dehydratase, or porphobilinogen synthase (2,3); dopamine transport (4); glutathione S-transferase (5); mandelate racemase (6); microsomal epoxide hydrolase, also as a substrate (7); Na^+/K^+-exchanging ATPase (8); soluble epoxide hydrolase, weak alternative substrate (9-14). **1.** Klinman, Welsh & Hogue-Angeletti (1977) Biochemistry **16**, 5521. **2.** Mori, Fujishiro & Inoue (1988) J. UOEH **10**, 269. **3.** Fujita, Koizumi, Hayashi & Ikeda (1986) Biochim. Biophys. Acta **867**, 89. **4.** Chakrabarti (1999) Biochem. Biophys. Res. Commun. **255**, 70. **5.** Ansari, Singh, Gan & Awasthi (1987) Toxicol. Lett. **37**, 57. **6.** Kenyon & Hegeman (1977) Meth. Enzymol. **46**, 541. **7.** Oesch (1974) Biochem. J. **139**, 77. **8.** Singh & Srivastava (1986) J. Environ. Pathol. Toxicol. Oncol. **6**, 29. **9.** Wixtrom & Hammock (1985) Biochem. Pharmacol. Toxicol. **1**, 1. **10.** Hammock, Prestwich, Loury, et al. (1986) Arch. Biochem. Biophys. **244**, 292. **11.** Chang & Gill (1991) Arch. Biochem. Biophys. **285**, 276. **12.** Morisseau, Archelas, Guitton, et al. (1999) Eur. J. Biochem. **263**, 386. **13.** Schladt, Thomas, Hartmann & Oesch (1988) Eur. J. Biochem. **176**, 715. **14.** Blée & Schuber (1992) Biochem. J. **282**, 711.

STZ, See Streptozotocin

SU 101, See Leflunomide

SU 1498

This acrylonitrile (FW = 390.53 g/mol), also known as (E)-3-(3,5-diisopropyl-4-hydroxyphenyl)-2-[(3-phenyl-n-propyl)aminocarbonyl]acrylonitrile, SU1498, and SU-1498, is a potent and selective inhibitor of Flk-1 kinase, IC_{50} = 0.7 μM (1-3), a vascular-endothelial-growth-factor-receptor (VEGFR) protein-tyrosine kinase. SU 1498 also reduces the expression of ets-1, a transcription factor stimulated by the vascular endothelial growth factor receptor (1). It has a weak inhibitory effect on PDGF-receptor (IC_{50} > 50 mM), EGF-receptor (IC_{50} > 100 mM), and HER2 (IC_{50} > 100 mM) kinases. It has also been shown to act as an angiogenesis inhibitor. **1.** Arbiser, Larsson, Claesson-Welsh, et al. (2000) Amer. J. Pathol. **156**, 1469. **2.** Strawn, McMahon, App, et al. (1996) Cancer Res. **56**, 3540. **3.** Jin, Zhu, Sun, et al. (2002) Proc. Natl. Acad. Sci. U.S.A. **99**, 11946.

SU 4312

This substituted indolinone (FW = 264.33 g/mol), also named SU4312 and SU-4312, inhibits EGF receptor protein-tyrosine kinase. Note that commercial sources often supply a mixture of the E and Z isomers. **Target(s):** endothelial growth factor receptor protein-tyrosine kinase (1); platelet derived growth factor receptor protein-tyrosine kinase (1); protein-tyrosine kinase (1). **1.** Kendall, Rutledge, Mao, et al. (1999) J. Biol. Chem. **274**, 6453.

SU 4885, See Metyrapone

SU 5271, See 4-[(3-Bromophenyl)amino]-6,7-dimethoxy-quinazoline

SU 5402

This potent multi-target receptor tyrosine kinase inhibitor (FW = 296.32 g/mol; CAS 215543-92-3; Solubility: 59 mg/mL DMSO; <1 mg/mL H_2O), also named 3-(4-methyl-2-((2-oxoindolin-3-ylidene)methyl)-1H-pyrrol-3-yl)propanoic acid, SU5402 and SU-5402, acts on VEGFR2 (IC_{50} = 20 nM), FGFR1 (IC_{50} = 30 nM), and PDGF-Rβ (IC_{50} = 510 nM) (1). SU5416 inhibits vascular endothelial growth factor-dependent mitogenesis in human endothelial cells without inhibiting the growth of a variety of tumor cells in vitro (2). Systemic administration of SU5416 at nontoxic doses in mice results in inhibition of subcutaneous tumor growth of cells derived from various tissue origins (2). The antitumor effect of SU5416 was accompanied by the appearance of pale white tumors that were resected from drug-treated animals, supporting the antiangiogenic property of this agent (2). **1.** Sun,

Tran, Liang, *et al.*. (1999) *J. Med. Chem.* **42**, 5120. **2**. Fong, Shawver, Sun, *et al.* (1999) *Cancer Res.* **59**, 99.

SU5416, *See* Semaxinib

SU 5614

This indolinone (FW = 272.73 g/mol), also known as 5-chloro-3-[(3,5-dimethylpyrrol-2-yl)methylene]-2-indolinone, SU5614, and SU-5614, is a selective inhibitor of vascular-endothelial-growth-factor receptor protein-tyrosine kinase Flk-1, IC_{50} = 1.2 μM (1,2) and PDGF receptor protein-tyrosine kinase, IC_{50} = 2.9 μM (1). SU 5614 is without effect on the EGF and IGF receptor protein tyrosine kinases. SU 5614 also inhibits the VEGF-driven mitogenesis of human umbilical vein endothelial cells, $IC_{50} \approx 0.7$ μM). **1**. Sun, Tran, Tang, *et al.* (1998) *J. Med. Chem.* **41**, 2588. **2**. Yee, O'Farrell, Smolich, *et al.* (2002) *Blood* **100**, 2941. **3**. Spiekermann, Dirschinger, Schwab, *et al.* (2003) *Blood* **101**, 1494.

SU 6656

This photosensitive sulfonamide (FW = 371.46 g/mol), also known as 2-oxo-3-(4,5,6,7-tetrahydro-1*H*-indol-2-ylmethylene)-2,3-dihydro-1*H*-indole-5-sulfonic acid dimethylamide, is a selective src family kinase inhibitor used to probe growth factor signaling. **Target(s):** AMP-activated protein kinase (1); dual-specificity, tyrosine phosphorylated and regulated kinase 1A (1); Fyn protein-tyrosine kinase, IC_{50} = 0.17 μM (2); lymphocyte protein kinase, IC_{50} = 0.04 μM (1,3); phosphorylase kinase (1); platelet-derived growth factor receptor protein-tyrosine kinase, weakly inhibited, IC_{50} > 10 μM (3); protein-tyrosine kinase (1-6); ribosomal S6 kinase 1 (1); src protein-tyrosine kinase, IC_{50} = 0.28 μM (3-6); Yes protein tyrosine kinase, IC_{50} = 0.02 μM (2,5). **1**. Bain, McLauchlan, Elliott & Cohen (2003) *Biochem. J.* **371**, 199. **2**. Sanguinetti, Cao & Mastick (2003) *Biochem. J.* **376**, 159. **3**. Blake, Broome, Liu, *et al.* (2000) *Mol. Cell Biol.* **20**, 9018. **4**. Okutani, Lodyga, Han & Liu (2006) *Amer. J. Physiol. Lung Cell Mol. Physiol.* **291**, L129. **5**. Meyn III, Schreiner, Dumitrescu, Nau & Smithgall (2005) *Mol. Pharmacol.* **68**, 1320. **6**. Sirvent, Boureux, Simon, Leroy & Roche (2007) *Oncogene* **26**, 7313.

SU6668

This tyrosine kinase inhibitor, *or* TKI (FW = 323.14 g/mol; CAS 210644-62-5), also known as (Z)-3-[2,4-dimethyl-5-(2-oxo-1,2-dihydroindol-3-ylidenemethyl)-1*H*-pyrrol-3-yl]propionate, is aa ATP-competitive inhibitor of Flk-1 transphosphorylation (K_i = 2.1 μM), FGFR1 trans-phosphorylation (K_i = 1.2 μM), and PDGFR autophosphorylation (K_i = 0.008 μM). SU6668 has greatest potency against PDGFR autophosphorylation, but also strongly inhibits inhibits Flk-1 and FGFR1 trans-phosphorylation (1). In contrast, SU6668 does not inhibit EGFR kinase activity at concentrations up to 100 μM. The biochemical IC_{50} values of SU6668 against EGFR, insulin-like

growth factor I receptor, Met, Src, Lck, Zap70, Abl, and cyclin-dependent kinase 2 are at least 10 μM, indicating that it shows a high level of selectivity against other tyrosine and serine/threonine kinases (1). Although HUVECs stimulated by VEGF exhibit an increase in tyrosine phosphorylation of KDR, treatment with SU6668 inhibits this increase in a dose-dependent manner. SU6668 also inhibits PDGF-stimulated PDGFRβ tyrosine phosphorylation in NIH-3T3 cells overexpressing PDGFRβ at a minimum concentration of 0.03–0.1 μM. SU6668 inhibits acidic FGF-induced phosphorylation of the FGFR1 substrate 2 (FRS-2) at concentrations of 10 μM. However, SU6668 has no detectable effect on epidermal growth factor-stimulated EGFR tyrosine phosphorylation in NIH-3T3 cells overexpressing EGFR at concentrations of up to 100 μM. These cellular data demonstrate that SU6668 inhibits Flk-1/KDR, PDGFR, and FGFR but has no activity against EGFR at the concentrations tested (1). As would be expected of an inhibitor of Flk-1, FGFR1, and PDGFR kinase activity, SU6668 demonstrated significant antitumor activity against a wide range of xenografts. SU6668 may influence tumor growth by multiple mechanisms including inhibition of endothelial cell proliferation and/or survival as well as tumor cell and stromal cell proliferation. **1**. Laird, Vajkoczy, Shawver, *et al.* (2000) *Cancer Res.* **60**, 4152.

SU 9516

This substituted indoline (FW = 241.25 g/mol), also known as 3-[1-(3*H*-imidazol-4-yl)-meth-(Z)-ylidene]-5-methoxy-1,3-dihydroindol-2-one, SU9516, and SU-9516, inhibits cyclin-dependent kinase-2 (Cdc2), IC_{50} = 22 nM (1-3). SU 9516 also induces apoptosis and promotes the accumulation of high molecular weight E2F complexes in human colon carcinoma cells. It will also kill human leukemia cells via down-regulation of Mcl-1 through a transcriptional mechanism. **1**. Lane, Yu, Rice, *et al.* (2001) *Cancer Res.* **61**, 6170. **2**. Moshinsky, Bellamacina, Boisvert, *et al.* (2003) *Biochem. Biophys. Res. Commun.* **310**, 1026. **3**. Bukczynska, Klingler-Hoffmann, Mitchelhill, *et al.* (2004) *Biochem. J.* **380**, 939.

SU 10603, *See* 3-(1,2,3,4-Tetrahydro-1-oxo-7-chloro-2-naphthyl)pyridine; 3-(1,2,3,4-Tetrahydro-4-oxo-7-chloro-2-naphthyl)pyridine

SU11652

This potent cell-permeable, multi-targeted receptor tyrosine kinase inhibitor, *or* RTKI (FW = 416.95 g/mol; CAS 326914-10-7), also known by its systematic name *N*-(2-diethylaminoethyl)-5-[(Z)-(5-chloro-2-oxo-1*H*-indol-3-ylidene)methyl]-2,4-dimethyl-1*H*-pyrrole-3-carboxamide, targets PDGF and VEGF receptors. (*See also* Sunitinib) SU11652 acts as an ATP-competitive tyrosine kinase receptor and angiogenic inhibitor that exhibits greater selectivity for PDGFRβ (IC_{50} = 3 nM), VEGFR2 (IC_{50} = 27 nM), FGFR1 (IC_{50}= 170 nM), and Kit family members (IC_{50} ~ 10 to 500 nM) over EGFR (IC_{50} >20 μM) (1). FLT3-ITD and FLT3-TKD mutations are frequently found in acute myeloid leukemia (AML), and inhibition of these mutants by makes SU11652 a highly attractive inhibitor for AML drug development (7). **Modes of Inhibitory Action:** SU11652 inhibits all receptors for platelet-derived growth factor (PDGF) and vascular endothelial growth factor (VEGF), both of which play essential roles in tumor angiogenesis and tumor cell proliferation. Simultaneous inhibition of these targets reduces tumor vascularization and promotes cancer cell death. SU11652 and its fluoro-substituted congener, Sunitinib, kill HeLa cervix carcinoma, U-2-OS osteosarcoma, Du145 prostate carcinoma and WEHI-S fibrosarcoma cells at low micromolar concentrations by accumulating rapidly in lysosomes, disturbing their pH regulation and ultrastructure, and ultimately leading to lysosomal protease leakage into the cytosol (8). Lysosomal destabilization was preceded by an early inhibition of acid sphingomyelinase (ASM), a lysosomal lipase that promotes lysosomal membrane stability (8). **1.** Johnson, Stewart, Woods, Giranda & Luo (2007) *Biochemistry* **46**, 9551. **2.** Schmidt-Arras, Böhmer, Markova, *et al.* (2005) *Mol. Cell. Biol.* **25**, 3690. **3.** O'Farrell, Abrams, Yuen, *et al.* (2003) *Blood* **101**, 3597. **4.** Mendel, Laird, Xin, *et al.* (2003) *Clin. Cancer Res.* **9**, 327. **5.** Medinger & Drevs (2005) *Curr. Pharm. Des.* **11**, 1139. **6.** Liao, Chien, Shenoy, *et al.* (2002) *Blood* **100**, 585. **2.** Guo, Chen, Xu, Fu & Zhao (2012) *J. Hematol. Oncol.* **5**, 72. **3.** Ellegaard, Groth-Pedersen, Oorschot, *et al.* (2013) *Mol. Cancer Ther.* **12**, 2018.

SU11248, See *Sunitinib*

SU 11274

This sulfonamide (FW = 568.10 g/mol), also known as (3*Z*)-*N*-(3-chlorophenyl)-3-({3,5-dimethyl-4-[(4-methylpiperazin-1-yl)carbonyl]-1*H*-pyrrol-2-yl}methylene)-*N*-methyl-2-oxo-2,3-dihydro-1*H*-indole-5-sulfonamide, inhibits Met protein-tyrosine kinase, K_i = 10 nM. **Target(s):** FGFR-1 protein-tyrosine kinase, mildly inhibited (1); Flk protein-tyrosine kinase (1); and Met protein-tyrosine kinase (1,2). **1.** Sattler, Pride, Ma, *et al.* (2003) *Cancer Res.* **63**, 5462. **2.** Berthou, Aebersold, Schmidt, *et al.* (2004) *Oncogene* **23**, 5387.

SUAM-1221, See *(N-[N-(4-Phenylbutanoyl)-L-prolyl]pyrrolidone)*

Suberic Acid Bis(3-sulfo-*N*-hydroxysuccinimide Ester), See *Bis(sulfosuccinimidyl) Suberate*

Suberoylanilide Hydroxamic Acid

This hydroxamic acid (FW = 264.32 g/mol; *Symbol*: SAHA) inhibits histone deacetylase, thereby inducing tumor cell differentiation, apoptosis, and/or growth arrest in several *in vitro* and *in vivo* experimental models. SAHA also enhances radiosensitivity and suppresses lung metastasis of triple-negative breast cancer (TNBC), the latter defined by the absence of an estrogen receptor, progesterone receptor, and human epidermal growth factor receptor 2 expression (4). Other studies have demonstrated that the expression of matrix metalloproteinase-9 (MMP-9) has been associated with a high potential of metastasis in several human carcinomas including breast cancer, and some HDAC inhibitors retard lung cancer cell migration via reduced activities of MMPs, RhoA, and focal adhesion complex. **Target(s):** amidase, *or* histone-deacetylase-like amidohydrolase (1); and histone deacetylase (2,3). **1.** Hildmann, Ninkovic, Dietrich, *et al.* (2004) *J. Bacteriol.* **186**, 2328. **2.** Butler, Agus, Scher, *et al.* (2000) *Cancer Res.* **60**, 5165. **3.** Richon, Emiliani, Verdin, *et al.* (1998) *Proc. Natl. Acad. Sci. U.S.A.* **95**, 3003. **4.** Chiu, Yeh, Wang, *et al.* (2013) *PLoS One* **8**, e76340.

Substance K, See *α-Neurokinin*

Substance P

This undecapeptide amide (MW = 1348 g/mol; *Sequence*: RPKPQQFFGLM-NH2; CAS 33507-63-0; Symbol: SP), first discovered by von Euler and Gaddum in 1931, has a major role in hypertension and vasodilation in many vertebrates. The endogenous substance P receptor is the Neurokinin 1 Receptor (NK1 Receptor, NK1R), belonging to the tachykinin receptor sub-family of G-Protein-Coupled Receptors. Three second-messenger pathways are activated upon agonist binding to NK-1 receptors: (a) phospholipase C-mediated stimulation of phosphatidyl inositol turnover, thereby mobilizing calcium ion release from both intra- and extracellular stores; (b) phospholipase A2-mediated mobilization of arachidonate, and adenylate cyclase-mediated stimulation of cAMP synthesis. Substance P also stimulates intestinal contractions as well as secretion of saliva. It is also a neuromodulator participating in the transmission of pain (1) and is a tachykinin NK-1 receptor agonist (2). Because SP promotes the proliferation of tumor cells, angiogenesis and the migration of tumor cells, SP receptor antagonists show promise in comatting cancer. **Target(s):** carboxypeptidase E (3); glutamyl aminopeptidase, *or* aminopeptidase A (4); membrane alanyl amino-peptidase, *or* aminopeptidase N, K_i = 0.44 μM (5,6); neurolysin, alternative substrate (7,8); and peptidyl-dipeptidase A (9). **1.** Snijdelaar, Dirksen, Slappendel & Crul (2000) *Eur. J. Pain* **4**, 121. **2.** Saria (1999) *Eur. J. Pharmacol.* **375**, 51. **3.** Hook & LaGamma (1987) *J. Biol. Chem.* **262**, 12583. **4.** Goto, Hattori, Ishii, Mizutani & Tsujimoto (2006) *J. Biol. Chem.* **281**, 23503. **5.** Xu, Wellner & Scheinberg (1995) *Biochem. Biophys. Res. Commun.* **208**, 664. **6.** Bauvois & Dauzonne (2006) *Med. Res. Rev.* **26**, 88. **7.** Vincent, Dauch, Vincent & Checler (1997) *J. Neurochem.* **68**, 837. **8.** Barelli, Vincent & Checler (1993) *Eur. J. Biochem.* **211**, 79. **9.** Kase, Hazato, Shimamura, Kiuchi & Katayama (1985) *Arch. Biochem. Biophys.* **240**, 330.

[D-Arg¹,D-Phe⁵,D-Trp⁷,⁹,Leu¹¹]-Substance P, See *D-Arginyl-L-prolyl-L-lysyl-L-prolyl-D-phenylalanyl-L-glutaminyl-L-tryptophanyl-L-phenylalanyl-D-tryptophanyl-L-leucyl-L-leucinamide*

[D-Arg¹,D-Pro²,D-Phe⁷,D-His⁹]-Substance P, See *D-Arginyl-D-prolyl-L-lysyl-L-prolyl-L-glutaminyl-L-glutaminyl-D-phenylalanyl-L-phenylalanyl-D-histidyl-L-leucyl-L-methioninamide*

[D-Arg¹,D-Pro²,D-Trp⁷,⁹,Leu¹¹]-Substance P, See *D-Arginyl-D-prolyl-L-lysyl-L-prolyl-L-glutaminyl-L-glutaminyl-D-tryptophanyl-L-phenylalanyl-D-tryptophanyl-L-leucyl-L-leucinamide*

[D-Arg¹,D-Trp⁷,⁹,Leu¹¹]-Substance P, See *Spantide I*

[D-Pro²,D-Phe⁷,D-Trp⁹]-Substance P, See *L-Arginyl-D-prolyl-L-lysyl-L-prolyl-L-glutaminyl-L-glutaminyl-D-phenylalanyl-L-phenylalanyl-D-tryptophanyl-L-leucyl-L-methioninamide*

[D-Pro²,D-Trp⁷,⁹]-Substance P, See *L-Arginyl-D-prolyl-L-lysyl-L-prolyl-L-glutaminyl-L-glutaminyl-D-tryptophanyl-L-phenylalanyl-D-tryptophanyl-L-leucyl-L-methioninamide*

[Arg⁶,D-Trp⁷,⁹,N-Me-Phe⁹]-Substance P Fragment 6-11, See *L-Arginyl-D-tryptophanyl-N-methyl-L-phenylalanyl-D-tryptophanyl-L-leucyl-L-methioninamide*

[D-Pro⁴,D-Trp⁷,⁹,¹⁰]-Substance P Fragment 4-11, See *D-Prolyl-L-glutaminyl-L-glutaminyl-D-tryptophanyl-L-phenylalanyl-D-tryptophanyl-D-tryptophanyl-L-methioninamide*

[pGlu⁵,D-Trp⁷,⁹,¹⁰]-Substance P Fragment 5-11, See *L-Pyroglutamyl-L-glutaminyl-D-tryptophanyl-L-phenylalanyl-D-tryptophanyl-D-tryptophanyl-L-methioninamide*

[D-Pro⁴,D-Trp⁷,⁹,¹⁰,Phe¹¹]-Substance P Fragment 4-11, See *D-Prolyl-L-glutaminyl-L-glutaminyl-D-tryptophanyl-L-phenylalanyl-D-tryptophanyl-D-tryptophanyl-L-phenylalaninamide*

Substance P Fragment 7-11 Amide, See *L-Phenylalanyl-L-phenylalanylglycyl-L-leucyl-L-methioninamide*

Substance P Fragment 8-11 Amide, See *L-Phenylalanylglycyl-L-leucyl-L-methioninamide*

Subtilisin Inhibitor, See *specific inhibitor; e.g., Nᵃ-(t-Butoxycarbonyl)-L-alanyl-L-prolyl-L-phenylalanine O-(Benzoyl)hydroxylamine*

Succimer, See *meso-2,3-Dimercaptosuccinate and meso-2,3-Dimercaptosuccinic Acid*

Succinate (Succinic Acid)

This dicarboxylic acid (FW = 118.09 g/mol; CAS 110-15-6), also known as butanedioic acid and ethanedicarboxylic acid, is a key intermediate in the tricarboxylic acid cycle and the glyoxylate cycle. The free acid is soluble in water (6.8 g/100 mL at 20°C; the sodium and potassium salts are considerably more soluble), has a melting point of 185-187°C, and decomposes at 235°C with a partial conversion to succinic anhydride. The pK_a values are 4.21 and 5.72 at 25°C (the dpK_a/dT value for pK_{a1} is −0.0018). **Target(s):** 2-(acetamidomethylene)-succinate hydrolase (92); N^4-(β-N-acetylyglucosaminyl)-L-asparaginase (33); aconitase, or aconitate hydratase (74); adenine phosphoribosyltransferase (1); adenosine-phosphate deaminase (2,88); adenylosuccinate synthetase (3,57-59); aerobactin synthase (61); alanine aminotransferase (110); alanopine dehydrogenase (5); β-alanyl-CoA ammonia-lyase (6); amidase, or half-amidase (94); amino-acid N-acetyltransferase (116); D-amino-acid aminotransferase, or D-alanine aminotransferase (4,107); 4-aminobutyrate aminotransferase (7,108); aminolevulinate aminotransferase (106); AMP deaminase (2); AMP nucleosidase (98); argininosuccinate lyase (65,66); aspartate aminotransferase (8-10,110-113); aspartate ammonia-lyase, weakly inhibited (68-70); aspartate carbamoyltransferase (11-14,117-123); aspartate 1-decarboxylase (15,84); aspartate 4-decarboxylase (16,82,83); aspartate kinase (103); D-aspartate oxidase (17-19); γ-butyrobetaine hydroxylase, or γ-butyrobetaine:2-oxoglutarate dioxygenase, K_i = 0.07 mM (20); carbamoyl-phosphate synthetase (21,56); carboxymethylene-butenolidase (102); carboxy-cis,cis-muconate cyclase (63); carboxypeptidase A, K_i = 4 mM (22); carboxypeptidase B, K_i = 28 mM (22,96); choline acetyltransferase (23); creatinase (89); cutinase (24); cyanate hydratase, or cyanase (71); dehydro-L-gulonate decarboxylase (25); dehydrogluconate dehydrogenase (26); 3-dehydroquinate dehydratase (27,73); deoxyribonuclease II, or spleen acid deoxyribonuclease (52,99); dichloromuconate cycloisomerase (62); 5-formyltetrahydrofolate cyclo-ligase, or 5,10-methenyltetrahydrofolate synthetase (60); fumarase, or fumarate hydratase (28-30,75); glutamate decarboxylase, K_i = 0.29 mM (31,32); glutamate formimidoyltransferase (124); glutaminase (95); (S)-2-hydroxy-acid oxidase (34); α-ketoglutarate dehydrogenase (35); kynurenine:2-oxoglutarate aminotransferase (36); L-lactate dehydrogenase, particularly LDH-X (37); leucine aminotransferase (109); maleimide hydrolase, alternative product inhibition (90); malic enzyme, or malate dehydrogenase (decarboxylating) (38-40); 2-methyleneglutarate mutase (41); N-methylglutamate dehydrogenase (42); (S)-2-methylmalate dehydratase (43,72); myosin ATPase (87); nicotinate-nucleotide diphosphorylase (carboxylating) (114,115); oxaloacetate decarboxylase (85,86); pantothenase (93); peptidyl-dipeptidase A, or angiotensin I-converting enzyme (97); phosphoenolpyruvate carboxykinase (GTP), weakly inhibited (77); phosphoenolpyruvate carboxylase (44,78-81); phosphoenolpyruvate phosphatase (100); 6-phosphofructokinase (45-47,105); 3-phosphoglycerate kinase (48,104); [phosphorylase] phosphatase (101); pyrophosphatase, or inorganic diphosphatase (49); pyruvate kinase (50); serine-sulfate ammonia-lyase, weakly inhibited (67); tartrate decarboxylase (76); tauropine dehydrogenase (53); theanine hydrolase (91); thiosulfate sulfurtransferase, or rhodanese (51); triose-phosphate isomerase (54,64); and urocanate hydratase (55).

1. Arnold & Kelley (1978) Meth. Enzymol. 51, 568. 2. Zielke & Suelter (1971) The Enzymes, 3rd ed. (Boyer, ed.), 4, 47. 3. Markham & Reed (1977) Arch. Biochem. Biophys. 184, 24. 4. Martinez-Carrion & Jenkins (1970) Meth. Enzymol. 17A, 167. 5. Fields & Hochachka (1981) Eur. J. Biochem. 114, 615. 6. Vagelos (1962) Meth. Enzymol. 5, 587. 7. Sytinsky & Vasilijev (1970) Enzymologia 39, 1. 8. Jenkins, Yphantis & Sizer (1959) J. Biol. Chem. 234, 51. 9. Velick & Vavra (1962) The Enzymes, 2nd ed. (Boyer, Lardy & Myrbäck, eds.), 6, 219. 10. Braunstein (1973) The Enzymes, 3rd ed. (Boyer, ed.), 9, 379. 11. Jacobson & Stark (1973) The Enzymes, 3rd ed. (Boyer, ed.), 9, 225. 12. Burns, Mendz & Hazell (1997) Arch. Biochem. Biophys. 347, 119. 13. Gerhart & Pardee (1964) Fed. Proc. 23, 727. 14. Changeux, Gerhart & Schachman (1968) Biochemistry 7, 531. 15. Williamson (1985) Meth. Enzymol. 113, 589. 16. Miles & Sparrow (1970) Meth. Enzymol. 17A, 689. 17. Dixon (1970) Meth. Enzymol. 17A, 713. 18. Dixon & Kenworthy (1967) Biochim. Biophys. Acta 146, 54. 19. Rinaldi (1971) Enzymologia 40, 314. 20. Ng, Hanauske-Abel & Englard (1991) J. Biol. Chem. 266, 1526. 21. Jones (1962) Meth. Enzymol. 5, 903. 22. Asante-Appiah, Seetharaman, Sicheri, Yang & Chan (1997) Biochemistry 36, 8710. 23. Korey, de Braganza & Nachmansohn (1951) J. Biol. Chem. 189, 705. 24. Maeda, Yamagata, Abe, et al. (2005) Appl. Microbiol. Biotechnol. 67, 778. 25. Kagawa & Shimazono (1970) Meth. Enzymol. 18A, 46. 26. Shinagawa & Ameyama (1982) Meth. Enzymol. 89, 194. 27. Chaudhuri, Duncan & Coggins (1987) Meth. Enzymol. 142, 320. 28. Hill & Bradshaw (1969) Meth. Enzymol. 13, 91. 29. Massey (1953) Biochem. J. 55, 172. 30. Hill & Teipel (1971) The Enzymes, 3rd. ed. (Boyer, ed.), 5, 539. 31. Fonda (1972) Biochemistry 11, 1304. 32. Gerig & Kwock (1979) FEBS Lett. 105, 155. 33. Risley, Huang, Kaylor, Malik & Xia (2001) J. Enzyme Inhib. 16, 269. 34. Jorns (1975) Meth. Enzymol. 41, 337. 35. Bunik & Pavlova (1997) Biochemistry (Moscow) 62, 1012. 36. Tobes (1987) Meth. Enzymol. 142, 217. 37. Holbrook, Liljas, Steindel & Rossmann (1975) The Enzymes, 3rd ed. (Boyer, ed.), 11, 191. 38. Kun (1963) The Enzymes, 2nd ed. (Boyer, Lardy & Myrbäck, eds.), 7, 149. 39. Stickland (1959) Biochem. J. 73, 654. 40. Veiga Salles & Ochoa (1950) J. Biol. Chem. 187, 849. 41. Kung & Stadtman (1971) J. Biol. Chem. 246, 3378. 42. Hersh, Stark, Worthen & Fiero (1972) Arch. Biochem. Biophys. 150, 219. 43. Wang & Barker (1969) Meth. Enzymol. 13, 331. 44. Utter & Kolenbrander (1972) The Enzymes, 3rd ed. (Boyer, ed.), 6, 117. 45. Bloxham & Lardy (1973) The Enzymes, 3rd ed. (Boyer, ed.), 8, 239. 46. Webb (1966) Enzyme and Metabolic Inhibitors, vol. 2, p. 385, Academic Press, New York. 47. Passonneau & Lowry (1963) Biochem. Biophys. Res. Commun. 13, 372. 48. Khamis & Larsson-Raznikiewicz (1987) Acta Chem. Scand. B 41, 348. 49. Naganna (1950) J. Biol. Chem. 183, 693. 50. Ambasht, Malhotra & Kayastha (1997) Indian J. Biochem. Biophys. 34, 365. 51. Oi (1975) J. Biochem. 78, 825. 52. Bernardi & Griffe (1964) Biochemistry 3, 1419. 53. Gäde (1986) Eur. J. Biochem. 160, 311. 54. Noltmann (1972) The Enzymes, 3rd ed. (Boyer, ed.), 6, 271. 55. Phillips & George (1971) Meth. Enzymol. 17B, 73 56. Garcia-Espana, Alonso & Rubio (1991) Arch. Biochem. Biophys. 288, 414. 57. Stayton, Rudolph & Fromm (1983) Curr. Top. Cell. Regul. 22, 103. 58. Ogawa, Shiraki, Matsuda, Kakiuchi & Nakagawa (1977) J. Biochem. 81, 859. 59. Gorrell, Wang, Underbakke, Hou, Honzatko & Fromm (2002) J. Biol. Chem. 277, 8817. 60. Grimshaw, Henderson, Soppe, et al. (1984) J. Biol. Chem. 259, 2728. 61. Appanna & Viswanatha (1986) FEBS Lett. 202, 107. 62. Kuhm, Schlömann, Knackmuss & Pieper (1990) Biochem. J. 266, 877. 63. Thatcher & Cain (1975) Eur. J. Biochem. 56, 193. 64. Tomlinson & Turner (1979) Phytochemistry 18, 1959. 65. Raushel & Nygaard (1983) Arch. Biochem. Biophys. 221, 143. 66. Yu & Howell (2000) Cell. Mol. Life Sci. 57, 1637. 67. Tudball & Thomas (1973) Eur. J. Biochem. 40, 25. 68. Falzone, Karsten, Conley & Viola (1988) Biochemistry 27, 9089. 69. Karsten & Viola (1991) Arch. Biochem. Biophys. 287, 60. 70. Mizuta & Tokushige (1975) Biochim. Biophys. Acta 403, 221. 71. Anderson, Johnson, Endrizzi, Little & Korte (1987) Biochemistry 26, 3938. 72. Wang & Barker (1969) J. Biol. Chem. 244, 2516. 73. Chaudhuri, Lambert, McColl & Coggins (1986) Biochem. J. 239, 699. 74. Eprintsev, Semenova & Popov (2002) Biochemistry (Moscow) 67, 795. 75. Behal & Oliver (1997) Arch. Biochem. Biophys. 348, 65. 76. Furuyoshi, Nawa, Kawabata, Tanaka & Soda (1991) J. Biochem. 110, 520. 77. Hebda & Nowak (1982) J. Biol. Chem. 257, 5503. 78. Schwitzguebel & Ettlinger (1979) Arch. Microbiol. 122, 109. 79. Gold & Smith (1974) Arch. Biochem. Biophys. 164, 447. 80. Yoshida, Tanaka, Mitsunaga & Izumi (1995) Biosci. Biotechnol. Biochem. 59, 140. 81. Peak & Peak (1981) Biochim. Biophys. Acta 677, 390. 82. Wilson & Kornberg (1963) Biochem. J. 88, 578. 83. Shibatani, Kakimoto, Kato, Nishimura & Chibata (1974) J. Ferment. Technol. 52, 886. 84. Williamson & Brown (1979) J. Biol. Chem. 254, 8074. 85. Sender, Martin, Peiru & Magni (2004) FEBS Lett. 570, 217. 86. Jetten & Sinskey (1995) Antonie Leeuwenhoek 67, 221. 87. Bolognani, Buttafoco, Ferrari, Venturelli & Volpi (1992) Biochem. Int. 26, 231. 88. Yates (1969) Biochim. Biophys. Acta 171, 299. 89. Coll, Knof, Ohga, et al. (1990) J. Mol. Biol. 214, 597. 90. Ogawa, Soong, Honda & Shimizu (1997) Eur. J. Biochem. 243, 322. 91. Tsushida & Takeo (1985) Agric. Biol. Chem. 49, 2913. 92. Huynh & Snell (1985) J. Biol. Chem. 260, 2379. 93. Airas (1976) Biochim. Biophys. Acta 452, 201. 94. Soong, Ogawa & Shimizu (2000) Appl. Environ. Microbiol. 66, 1947. 95. Ardawi & Newsholme (1984) Biochem. J. 217, 289. 96. Asante-Appiah & Chan (1996) Biochem. J. 320, 17. 97. Oshima & Nagasawa (1977) J. Biochem. 81, 57. 98. Yoshino, Ogasawara, Suzuki & Kotake (1967) Biochim. Biophys. Acta 146, 620. 99. Bernardi (1971) The Enzymes, 3rd ed. (Boyer, ed.), 4, 271. 100. Malhotra & Kayastha (1990) Plant Physiol. 93, 194. 101. Wingender-Drissen & Becker (1983) Biochim. Biophys. Acta 743, 343. 102. Schlömann, Ngai, Ornston & Knackmuss (1993) J. Bacteriol. 175, 2994. 103. Keng & Viola (1996) Arch. Biochem. Biophys. 335, 73. 104. Joao & Williams (1993) Eur. J. Biochem. 216, 1. 105. Van Praag, Zehavi & Goren (1999) Biochem. Mol. Biol. Int. 47, 749. 106. Neuberger & Turner (1963) Biochim. Biophys. Acta 67, 342. 107. Martinez-Carrion & Jenkins (1965) J. Biol. Chem. 240, 3547. 108. Liu, Peterson, Langston, et al. (2005) Biochemistry 44, 2982. 109. Pathre, Singh, Viswanathan & Sane (1987) Phytochemistry 26, 2913. 110. Eze & Echetebu (1980) J. Gen. Microbiol. 120, 523. 111. Owen & Hochachka (1974) Biochem. J. 143, 541. 112. Martins, Mourato & de Varennes (2001) J. Enzyme Inhib. 16, 251. 113.

Rakhmanova & Popova (2006) *Biochemistry (Moscow)* 71, 211. **114.** Shibata & Iwai (1980) *Biochim. Biophys. Acta* **611**, 280. **115.** Iwai, Shibata & Taguchi (1979) *Agric. Biol. Chem.* **43**, 351. **116.** Shigesada & Tatibana (1978) *Eur. J. Biochem.* **84**, 285. **117.** Achar, Savithri, Vaidyanathan & Rao (1974) *Eur. J. Biochem.* **47**, 15. **118.** Jacobson & Stark (1975) *J. Biol. Chem.* **250**, 6852. **119.** Burns, Mendz & Hazell (1997) *Arch. Biochem. Biophys.* **347**, 119. **120.** Savithri, Vaidyanathan & Rao (1978) *Proc. Indian Acad. Sci. Sect. B* **87B**, 81. **121.** Masood & Venkitasubramanian (1988) *Biochim. Biophys. Acta* **953**, 106. **122.** Ong & Jackson (1972) *Biochem. J.* **129**, 571. **123.** Xu, Zhang, Liang, *et al.* (1998) *Microbiology* **144**, 1435. **124.** Miller & Waelsch (1957) *J. Biol. Chem.* **228**, 397.

Succinate Semialdehyde

This monoaldehyde of succinic acid (FW$_{free-acid}$ = 102.09 g/mol), also known as 4-oxobutanoic acid, is an intermediate in the metabolism of γ-aminobutyric acid. Succinate semialdehyde is a product of the reversible reactions catalyzed by 4-hydroxybutyrate dehydrogenase and 4-aminobutyrate aminotransferase as well as a product of 2-(acetamidomethylene)succinate hydrolase and α-ketoglutarate decarboxylase. It is a substrate of the succinate-semialdehyde dehydrogenases. This water-soluble and viscous liquid (B.P. = 134-136°C at 14 mm Hg) tends to polymerize; nevertheless, aqueous solutions are stable below pH 6 at 2°C for at least three months. **Target(s):** aminolevulinate aminotransferase (1,2); *N*-carbamoylsarcosine amidase (3); dihydrodipicolinate synthase (4); and malonate-semialdehyde dehydrogenase (5). **1.** Varticovski, Kushner & Burnham (1980) *J. Biol. Chem.* **255**, 3742. **2.** Bajkowski & Friedmann (1982) *J. Biol. Chem.* **257**, 2207. **3.** Zajc, Romano, Turk & Huber (1996) *J. Mol. Biol.* **263**, 269. **4.** Karsten (1997) *Biochemistry* **36**, 1730. **5.** Nakamura & Bernheim (1961) *Biochim. Biophys. Acta* **50**, 147.

Succinic Acid 2,2-Dimethylhydrazide, *See* Daminozide

Succinic Acid Monomethyl Ester, *See* Monomethyl Succinate

Succinic Acid Semialdehyde, *See* Succinate Semialdehyde

Succinic Monohydroxamic Acid

This hydroxamic acid (FW$_{free-acid}$ = 133.10 g/mol) inhibits ADAM 17 endopeptidase, *or* tumor necrosis factor-α converting enzyme; TACE (1) and 3-ketoacid CoA-transferase (2). **1.** Pickart & Jencks (1979) *J. Biol. Chem.* **254**, 9120. **2.** Becherer, Lambert & Andrews (2000) *Handbook of Experimental Pharmacology* (von der Helm, Korant & Cheronis, eds.) **140**, 235.

Succinic Semialdehyde, *See* Succinate Semialdehyde

Succinimide-(2-Naphthoxy)acetyl Ester, *See* (2-Naphthoxy)acetic Acid N-Hydroxysuccinimide Ester

Succinimide-Phenoxyacetyl Ester, *See* Phenoxyacetic Acid N-Hydroxysuccinimide Ester

***N*-Succinimidyl 3-(4-Hydroxyphenyl)propionate, *See* 3-(4-Hydroxyphenyl)propionic Acid N-Hydroxysuccinimide Ester**

***N*-Succinimidyl 4-(4-Maleimidophenyl)butyrate, *See* 4-(4-Maleimidophenyl)butyric Acid N-Hydroxysuccinimide Ester**

Succinimidyl (2-Naphthoxy)acetate, *See* (2-Naphthoxy)acetic Acid N-Hydroxysuccinimide Ester

Succinimidyl Phenoxyacetate, *See* Phenoxyacetic Acid N-Hydroxysuccinimide Ester

***N*⁶-Succinoadenosine 5'-Monophosphate, *See* Adenylosuccinate**

4-(*N*-Succinocarboxamide)-5-aminoimidazole Ribotide, *See* Phosphoribosylaminoimidazole-succinocarboxamide

Succinylacetone, *See* 4,6-Diketoheptanoate

N^α-Succinyl-L-alanyl-L-alanyl-L-prolyl-L-phenylalanine Chloromethyl Ketone

This tetrapeptide halomethyl ketone (FW$_{free-acid}$ = 537.01 g/mol) inhibits cuticle degrading endopeptidase, *Metarhizium* (1). Note that *N*^{α}-succinyl-L-alanyl-L-alanyl-L-prolyl-L-phenylalanine *p*-nitroanilide is a substrate for chymotrypsin, cathepsin G, and peptidyl-prolyl isomerase. **1.** St. Leger (1998) in *Handb. Proteolytic Enzymes* (Barrett, Rawlings & Woessner, eds.), p. 325, Academic Press, San Diego.

Succinylcholine

This potent nicotinic cholinergic receptor antagonist (FW$_{dication}$ = 234.38 g/mol; CAS 6101-15-1, for the dichloride dehydrate; Halide salts are water-soluble (~1 g/mL), also known by its generic name Suxamethonium chloride (*or* "SUX"), its trade names Anectine® and Quelicin®, and 2,2'-[(1,4-dioxobutane-1,4-diyl)bis(oxy)]bis(*N,N,N*-trimethylethanaminium ion), induces persistent receptor activation, blocking synaptic transmission at myoneural junctions (1-3). Succinylcholine is often used to as a muscle relaxant to induce short-term paralysis, especially to facilitate tracheal intubation. Neuromuscular transmission is inhibited as long as an adequate succinylcholine concentration remains at the receptor site. Onset of flaccid paralysis is rapid (<1 minute after intravenous administration), and with single administration lasts approximately 4–6 minutes. Its utility as a muscle relaxant was first investigated by Nobel Laureate Daniel Bovet (Physiology and Medicine, 1957). Aqueous solutions are unstable (particularly above pH 9 and at elevated temperature) and should be prepared freshly. Succinylmonocholine (SMC), which is the product of the inefficient hydrolysis of succinylcholine by butyrylcholinesterase, is a forensic marker for succinylcholine poisoning. The post-mortem half-life for plasma SMC is 1-3 hours, putting the reliable detection interval of 8-24 hours (4). The following drugs enhance the neuromuscular blocking by succinylcholine: promazine, oxytocin, aprotinin, certain non-penicillin antibiotics, quinidine, β-adrenergic blockers, procainamide, lidocaine, trimethaphan, lithium carbonate, quinine, chloroquine, isoflurane, and desflurane. Its neuromuscular effects may be enhanced by drugs reducing plasma cholinesterase activity (*e.g.*, oral contraceptives, glucocorticoids, and monoamine oxidase inhibitors) or by agents that inhibit plasma cholinesterase irreversibly. **1.** Lanks & Sklar (1976) *Res. Commun. Chem. Pathol. Pharmacol.* **14**, 269. **2.** Desire, Blanchet, Definod & Arnaud (1975) *Biochimie* **57**, 1359. **3.** Sung & Johnstone (1967) *Biochem. J.* **105**, 497. **4.** Kuepper, Musshoff, Hilger, Herbstreit & Madea (2011) *J. Anal. Toxicol.* **35**, 302.

Succinyl-CoA

This citric acid cycle intermediate (FW$_{ion}$ = 866.61 g/mol) is a potent product inhibitor of the α-ketoglutarate dehydrogenase complex. As with most acyl derivatives of coenzyme A, succinyl-CoA is unstable at neutral pH at room temperature ($t_{1/2}$ = 1-2 hours in bicarbonate buffer at pH 7.5).

Complete hydrolysis occurs within two minutes at 100°C. On the other hand, succinyl-CoA solutions are relatively stable at pH 1. **Target(s):** acyl-CoA carboxylase (1,12); arylamine *N*-acetyltransferase (30); butyrate:acetoacetate CoA-transferase (21); carnitine *O*-palmitoyltransferase (2,3); citrate (*si*)-synthase (4,5); 3-hydroxy-3-methylglutaryl-CoA lyase (6); 3-hydroxy-3-methylglutaryl-CoA reductase, weakly inhibited (7); 3-hydroxy-3-methylglutaryl-CoA synthase (27,28); α-ketoglutarate dehydrogenase complex (8,9); malonyl-CoA decarboxylase (10-13); [acyl-carrier-protein] *S*-malonyltransferase (29); malyl-CoA synthetase, *or* malate:CoA ligase, altenative product inhibition (14); methylmalonyl-CoA decarboxylase (15); methylmalonyl-CoA epimerase (16); (*S*)-methylmalonyl-CoA hydrolase (24); [3-methyl-2-oxobutanoate dehydrogenase, *or* 2-methylpropanoyl-transferring)] phosphatase ([branched-chain-α-keto-acid dehydrogenase] phosphatase (22,23); oxaloacetate decarboxylase, weakly inhibited (17); 3-oxoacid CoA-transferase (25); pantothenate kinase (26); phosphoprotein phosphatase (23); propionyl-CoA carboxylase (18); and pyruvate kinase (19,20). **1.** Chuakrut, Arai, Ishii & Igarashi (2003) *J. Bacteriol.* **185**, 938. **2.** Kashfi, Mynatt & Cook (1994) *Biochim. Biophys. Acta* **1212**, 245. **3.** Zierz, Neumann-Schmidt & Jerusalem (1993) *Clin. Investig.* **71**, 763. **4.** Smith & Williamson (1971) *FEBS Lett.* **18**, 35. **5.** Lee, Park & Yim (1997) *Mol. Cells* **7**, 599. **6.** Deana, Rigoni & Galzigna (1979) *Clin. Sci. (London)* **56**, 251. **7.** Kirtley & Rudney (1967) *Biochemistry* **6**, 230. **8.** Kiselevsky, Ostrovtsova & Strumilo (1990) *Acta Biochim. Pol.* **37**, 135. **9.** Hamada, Koike, Nakaula, Hiraoka & Koike (1975) *J. Biochem.* **77**, 1047. **10.** Kolattukudy, Poulose & Kim (1981) *Meth. Enzymol.* **71**, 150. **11.** Kim, Kolattukudy & Boos (1979) *Arch. Biochem. Biophys.* **196**, 543. **12.** Koeppen, Mitzen & Ammoumi (1974) *Biochemistry* **13**, 3589. **13.** Hunaiti & Kolattukudy (1984) *Arch. Biochem. Biophys.* **229**, 426. **14.** Hersh (1974) *J. Biol. Chem.* **249**, 6264. **15.** Galivan & Allen (1968) *Arch. Biochem. Biophys.* **126**, 838. **16.** Stabler, Marcell & Allen (1985) *Arch. Biochem. Biophys.* **241**, 252. **17.** Horton & Kornberg (1964) *Biochim. Biophys. Acta* **89**, 381. **18.** Hügler, Krieger, Jahn & Fuchs (2003) *Eur. J. Biochem.* **270**, 736. **19.** Kayne (1973) *The Enzymes*, 3rd ed. (Boyer, ed.), **8**, 353. **20.** Waygood, Mort & Sanwal (1976) *Biochemistry* **15**, 277. **21.** Sramek & Frerman (1975) *Arch. Biochem. Biophys.* **171**, 14. **22.** Damuni, Merryfield, Humphreys & Reed (1984) *Proc. Natl. Acad. Sci. U.S.A.* **81**, 4335. **23.** Reed & Damuni (1987) *Adv. Protein Phosphatases* **4**, 59. **24.** Kovachy, Copley & Allen (1983) *J. Biol. Chem.* **258**, 11415. **25.** Hersh & Jencks (1967) *J. Biol. Chem.* **242**, 3468. **26.** Vallari, Jackowski & Rock (1987) *J. Biol. Chem.* **262**, 2468. **27.** Lowe & Tubbs (1985) *Biochem. J.* **232**, 37. **28.** Reed, Clinkenbeard & Lane (1975) *J. Biol. Chem.* **250**, 3117. **29.** Guerra & Ohlrogge (1986) *Arch. Biochem. Biophys.* **246**, 274. **30.** Kawamura, Graham, Mushtaq, *et al.* (2005) *Biochem. Pharmacol.* **69**, 347.

O-Succinylhomoserine

This metabolite (FW$_{hydrochloride}$ = 219.19 g/mol; CAS 1492-23-5), an intermediate in the biosynthesis of L-methionine, inhibits *O*-acetylhomoserine aminocarboxypropyltransferase, inhibited by the racemic mixture (1,2), *O*-acetylserine/*O*-acetylhomoserine sulfhydrylase, inhibited by the racemic mixture (3), and homoserine *O*-acetyltransferase, weakly inhibited (4). **1.** Murooka, Kakihara, Miwa, Seto & Harada (1977) *J. Bacteriol.* **130**, 62. **2.** Yamagata (1971) *J. Biochem.* **70**, 1035. **3.** Yamagata (1987) *Meth. Enzymol.* **143**, 478. **4.** Yamagata (1987) *J. Bacteriol.* **169**, 3458.

N^α-Succinyl-L-prolyl-L-leucyl-L-phenylalanine Chloromethyl Ketone

This halomethyl ketone (FW$_{free-acid}$ = 508.01 g/mol) inhibits cathepsin G (1,2); chymase, mast cell protease I (1-3), chymotrypsin (1,2), and tryptase (1,2). **1.** Powers, Tanaka, Harper, *et al.* (1985) *Biochemistry* **24**, 2048. **2.** Yoshida, Everitt, Neurath, Woodbury & Powers (1980) *Biochemistry* **19**, 5799. **3.** Woodbury, Everitt & Neurath (1981) *Meth. Enzymol.* **80**, 588.

Sucralfate

This cytoprotective agent and oral gastrointestinal medication (FW = 2111 g/mol; CAS 54182-58-0) is a basic aluminum sucrose sulfate complex (R = –SO$_3$[Al$_2$(OH)$_5$]) is an antiulcer drug that inhibits peptic hydrolysis. Sucralfate binds to the mucosa, creating a physical barrier that impairs diffusion of hydrochloric acid within the gastrointestinal tract and preventing degradation of mucus by acid hydrolysis. Sucralfate also inhibits Ca^{2+} channels (1,2), glycoprotein *N*-palmitoyltransferase, K_i = 0.91 μM (3), glycosulfatase (4), and urease (5). **1.** Liu, Slomiany & Slomiany (1992) *Gen. Pharmacol.* **23**, 1129. **2.** Slomiany, Liu & Slomiany (1992) *Biochem. Int.* **28**, 1125. **3.** Slomiany, Liau, Mizuta & Slomiany (1987) *Biochem. Pharmacol.* **36**, 3273. **4.** Slomiany, Murty, Piotrowski, Grabska & Slomiany (1992) *Amer. J. Gastroenterol.* **87**, 1132. **5.** Slomiany, Piotrowski & Slomiany (1997) *Biochem. Mol. Biol. Int.* **42**, 155.

Sucralose

This non-nutritive sweetener (FW = 397.64 g/mol; CAS 56038-13-2), also known as Splenda®, Zerocal®, Sukrana®, SucraPlus®, Candys®, Cukren®, Nevella®, and 1,6-dichloro-1,6-dideoxy-β-D-fructofuranosyl-4-chloro-4-deoxy-α-D-galactopyranoside, is 320 to 1,000 sweeter than sucrose. **Inhibitory Target(s):** fructosyltransferase and glucosyltransferase. **1.** Young & Bowen (1990) *J. Dent. Res.* **69**, 1480.

Sucrose

This nonreducing disaccharide (FW = 342.30 g/mol; CAS 57-50-1) is the common commercial sugar used as a sweetener and is obtained primarily from sugar cane (*Saccharum officinarum*) or sugar beet (*Beta vulgaris*). **Target(s):** acetyl-CoA:α-glucosaminide *N*-acetyltransferase, *or* heparan-α-glucosaminide *N*-acetyltransferase (1); acetyl-CoA synthetase, *or* acetate:CoA ligase (2); β-N-acetylhexosaminidase (27); aldose 1-epimerase (3); α-amylase (4); calcidiol 1-monooxygenase (64); concanavalin A (5); cyclomaltodextrin glucanotransferase (61); cytochrome c oxidase, weakly inhibited (6); exoribonuclease II, ribonuclease II, weakly inhibited (46); 2,1-fructan:2,1-fructan 1-fructosyltransferase (58,59); fructan β-fructosidase (7,8); fructan β-(2,1)-fructosidase (8-11); fructan β-(2,6)-fructosidase (11-13); β-fructofuranositase, *or* invertase; inhibited by elevated concentrations of sucrose (14,15); β-D-fucosidase (28); fucosterol-epoxide lyase (16); α-galactosidase (38); β-galactosidase, weakly inhibited (36,37); glucan 1,4-α-glucosidase, *or* glucoamylase (41,45); α-glucosidase (40-43); β-glucosidase (39); β-glucuronidase (29); glycogen phosphorylase b (17); guanylate cyclase (18); hyaluronan synthase (57); inulinase (19); levanase, weakly inhibited (20); naringinase (21); pectinesterase (54); α,α-phosphotrehalase, *or* trehalose-6-phosphate hydrolase (22); polynucleotide phosphorylase, *or* polyribonucleotide nucleotidyltransferase (56); procollagen galactosyltransferase (23); procollagen glucosyltransferase (23,60); sucrose-phosphatase (47-53); sucrose-6-phosphate hydrolase (24); sucrose-phosphate synthase (62,63); thioglucosidase, *or* myrosinase (25); α,α-trehalase (26,30-35); and UTP:glucose-1-phosphate uridylyltransferase (55). **1**. Bame & Rome (1987) *Meth. Enzymol.* **138**, 607. **2**. Roughan & Ohlrogge (1994) *Anal. Biochem.* **216**, 77. **3**. Fishman, Pentchev & Bailey (1975) *Meth. Enzymol.* **41**, 484. 4. Webb (1966) *Enzyme and Metabolic Inhibitors*, vol. **2**, p. 420, Academic Press, New York. **5**. Goldstein, Hollerman & Smith (1965) *Biochemistry* **4**, 876. **6**. Hasinoff & Davey (1989) *Biochem. J.* **258**, 101. **7**. Van den Ende, De Coninck & Van Laere (2004) *Trends Plant Sci.* **9**, 523. **8**. De Roover, Van Laere, De Winter, Timmermanns & Van den Ende (1999) *Physiol. Plant* **106**, 28. **9**. Marx, Nosberger & Frehner (1997) *New Phytol.* **135**, 267. **10**. Henson & Livingston III (1998) *Plant Physiol. Biochem.* **36**, 715. **11**. Bonnett & Simpson (1995) *New Phytol.* **131**, 199. **12**. Van den Ende, De Coninck, Clerens, Vergauwen & Van Laere (2003) *Plant J.* **36**, 697. **13**. Marx, Nosberger & Frehner (1997) *New Phytol.* **135**, 279. **14**. Ruchti & McLaren (1964) *Enzymologia* **27**, 185. **15**. Nelson & Schubert (1928) *J. Amer. Chem. Soc.* **50**, 2188. **16**. Prestwich, Angelastro, de Palma & Perino (1985) *Anal. Biochem.* **151**, 315. **17**. Tsitsanou, Oikonomakos, Zographos, *et al.* (1999) *Protein Sci.* **8**, 741. **18**. Janssens & de Jong (1988) *Biochem. Biophys. Res. Commun.* **150**, 405. **19**. Avigad & Bauer (1966) *Meth. Enzymol.* **8**, 621. **20**. Miasnikov (1997) *FEMS Microbiol. Lett.* **154**, 23. **21**. Nomura (1965) *Enzymologia* **29**, 272. **22**. Helfert, Gotsche & Dahl (1995) *Mol. Microbiol.* **16**, 111. **23**. Spiro & Spiro (1972) *Meth. Enzymol.* **28**, 625. **24**. Chassy & Porter (1982) *Meth. Enzymol.* **90**, 556. **25**. Tani, Ohtsuru & Hata (1974) *Agric. Biol. Chem.* **38**, 1623. **26**. Sasajima, Kawachi, Sato & Sugimura (1975) *Biochim. Biophys. Acta* **403**, 139. **27**. Sakai, Narihara, Kasama, Wakayama & Moriguchi (1994) *Appl. Environ. Microbiol.* **60**, 2911. **28**. Nunoura, Ohdan, Yano, Yamamoto & Kumagai (1996) *Biosci. Biotechnol. Biochem.* **60**, 188. **29**. Dean (1974) *Biochem. J.* **138**, 395. **30**. Nakano, Moriwaki, Washino, *et al.* (1994) *Biosci. Biotechnol. Biochem.* **58**, 1430. **31**. Lopez & Torrey (1985) *Arch. Microbiol.* **143**, 209. **32**. Talbot, Muir & Huber (1975) *Can. J. Biochem.* **53**, 1106. **33**. Wisser, Guttenberger, Hampp & Nehls (2000) *New Phytol.* **146**, 169. **34**. Terra, Ferreira & de Bianchi (1978) *Biochim. Biophys. Acta* **524**, 131. **35**. Galand (1984) *Biochim. Biophys. Acta* **789**, 10. **36**. Itoh, Suzuki & Adachi (1982) *Agric. Biol. Chem.* **46**, 899. **37**. Etcheberrigaray, Vattuone & Sampietro (1986) *Phytochemistry* **20**, 49. **38**. Schuler, Mudgett & Mahoney (1985) *Enzyme Microb. Technol.* **7**, 207. **39**. Ferreira & Terra (1983) *Biochem. J.* **213**, 43. **40**. Peruffo, Renosto & Pallavicini (1978) *Planta* **142**, 195. **41**. Pereira & Sivakami (1989) *Biochem. J.* **261**, 43. **42**. Yamasaki, Suzuki & Ozawa (1977) *Agric. Biol. Chem.* **41**, 1451. **43**. Chadalavada & Sivakami (1997) *Biochem. Mol. Biol. Int.* **42**, 1051. **44**. Khan & Eaton (1967) *Biochim. Biophys. Acta* **146**, 173. **45**. Yamasaki, Suzuki & Ozawa (1977) *Agric. Biol. Chem.* **41**, 1443. **46**. Spahr (1964) *J. Biol. Chem.* **239**, 3716. **47**. Hawker & Hatch (1966) *Biochem. J.* **99**, 102. **48**. Hawker & Smith (1984) *Phytochemistry* **23**, 245. **49**. Echeverria & Salerno (1993) *Physiol. Plant.* **88**, 434. **50**. Lunn, Ashton, Hatch & Heldt (2000) *Proc. Natl. Acad. Sci. U.S.A.* **97**, 12914. **51**. Whitaker (1984) *Phytochemistry* **23**, 2429. **52**. Hawker (1971) *Phytochemistry* **10**, 2313. **53**. Cumino, Ekeroth & Salerno (2001) *Planta* **214**, 250. **54**. Lim & Chung (1993) *Arch. Biochem. Biophys.* **307**, 15. **55**. Aksamit & Ebner (1972) *Biochim. Biophys. Acta* **268**, 102. **56**. Nolden & Richter (1982) *Z. Naturforsch.* **37c**, 600. **57**. Tlapak-Simmons, Baron & Weigel (2004) *Biochemistry* **43**, 9234. **58**. Van den Ende, Van Wonterghem, Verhaert, Dewil & Van Laere (1996) *Planta* **199**, 493. **59**. Van den Ende, De Roover & Van Laere (1996) *Physiol. Plant.* **97**, 346. **60**. Smith, Wu & Jamieson (1977) *Biochim. Biophys. Acta* **483**, 263. **61**. Bovetto, Villette, Fontaine, Sicard & Bouquelet (1992) *Biotechnol.*

Appl. Biochem. **15**, 59. **62**. Siegl & Stitt (1990) *Plant Sci.* **66**, 205. **63**. Salerno & Pontis (1978) *FEBS Lett.* **86**, 263. **64**. Gray, Omdahl, Ghazarian & DeLuca (1972) *J. Biol. Chem.* **247**, 7528.

Sulbactam

This water-soluble, semi-synthetic β-lactam (FW$_{\text{free-acid}}$ = 233.25 g/mol; CAS 68373-14-8; M.P. = 148-151°C), also known as penicillanic acid sulfone and penicillanic acid 1,1-dioxide, is an irreversible β-lactamase inhibitor that is often used to maintain the therapeutic action of other β-lactam antibiotics, most often penicillin. (In the U.S., sulbactam is combined with cefoperazone to form cefoperazone/sulbactam (MAGNEX®) and ampicillin to form ampicillin/sulbactam (Unasyn®).) While effective against most forms of β-lactamase, sulbactam does not inhibit ampC cephalosporinase, conferring little protection against *Pseudomonas aeruginosa*, *Citrobacter*, *Enterobacter*, and *Serratia*. **1**. Walsh (1983) *Trends Biochem. Sci.* **8**, 254. **2**. Mantoku & Ogawara (1981) *J. Antibiot.* **34**, 1347. **3**. Kemal & Knowles (1981) *Biochemistry* **20**, 3688. **4**. Neu & Fu (1980) *Antimicrob. Agents Chemother.* **18**, 582. **5**. Retsema, English & Girard (1980) *Antimicrob. Agents Chemother.* **17**, 615. **6**. Labia, Lelievre & Peduzzi (1980) *Biochim. Biophys. Acta* **611**, 351. **7**. Imtiaz, Billings, Knox & Mobashery (1994) *Biochemistry* **33**, 5728. **8**. Ator & Ortiz de Montellano (1990) *The Enzymes*, 3rd ed. (Sigman & Boyer, eds.), **19**, 213. **9**. Eliasson, Kamme, Vang & Waley (1992) *Eur. J. Clin. Microbiol. Infect. Dis.* **11**, 313. **10**. Brown, Young & Amyes (2005) *Clin. Microbiol. Infect.* **11**, 15. **11**. Hedberg, Lindqvist, Tuner & Nord (1992) *Eur. J. Clin. Microbiol. Infect. Dis.* **11**, 1100. **12**. Poirel, Brinas, Verlinde, Ide & Nordmann (2005) *Antimicrob. Agents Chemother.* **49**, 3743.

Sulconazole

This antifungal agent (FW = 397.75 g/mol; CAS 61318-90-9), also known as 1-[2-[[(4-chlorophenyl)methyl]thio]-2-(2,4-dichlorophenyl)ethyl]-1H-imidazole, inhibits CYP1A2 (1), CYP2B6 (1), CYP2C9 (1), CYP2C19 (1), CYP2D6 (1), and CYP19, *or* aromatase (2). Its mechanisn of action is unknown. Sulconazole is the indicated topical treatment for *tinea corporis* (Ringworm-of-the-Body), tinea *cruris* (ringworm of the groin, *or* Jock Itch), and tinea pedis (ringworm of the foot, *or* Athlete's Foot) caused by *Trichophyton rubrum*, *T. mentagrophytes*, *Epidermophyton floccosum*, and *Microsporum canis*. **1**. Zhang, Ramamoorthy, Kilicarslan, *et al.* (2002) *Drug Metab. Dispos.* **30**, 314. **2**. Kragie, Turner, Patten, Crespi & Stresser (2002) *Endocr. Res.* **28**, 129.

Sulfacetamide

This sulfonamide antibiotic (FW = 214.25 g/mol; CAS 144-80-9; pK_1 = 1.8 for amino group, and pK_1 = 5.4 for sulfonamide), also known as N-[(4-aminophenyl)sulfonyl]acetamide and N^1-acetylsulfanilamide, has a low lipid solubility. The sodium salt is very soluble in water (one part in 1.5 parts water). When formulated as complexes with Ag(I) and Zn(II), sulfacetamide is an effective against *Aspergillus* and *Candida spp.*, with minimum inhibitory concentrations in the 0.3 – 0.5 μg/mL range. The mechanism of antifungal action of these complexes does not appear to be connected with inhibition of lanosterol-14-α-demethylase, since sterol levels in the fungi cultures are unaffected by Ag(I) sulfacetamide complex or Zn(II) sulfacetamide complex. *Note*: Sulfacetamide should not be confused with N^4-acetylsulfanilamide. **Target(s)**: carbonic anhydrase, *or* carbonate dehydratase (1); and dihydropteroate synthase (2-5). **1**. Lanir & Navon (1972) *Biochemistry* **11**, 3536. **2**. Prabhu, Lui & King (1997) *Phytochemistry* **45**, 23. **3**. Kasekarn, Sirawaraporn, Chahomchuen, Cowman & Sirawaraporn (2004) *Mol. Biochem. Parasitol.* **137**, 43. **4**. Berglez, Iliades, Sirawaraporn, Coloe & Macreadie (2004) *Int. J. Parasitol.* **34**, 95. **5**. Iliades, Meshnick & Macreadie (2005) *Antimicrob. Agents Chemother.* **49**, 741.

Sulfachloropyridazine

This antibacterial sulfonamide (FW = 284.73 g/mol; CAS 80-32-0), also known as sulfachlorpyridazine and 4-amino-*N*-(6-chloro-3-pyridazinyl)-benzenesulfonamide, is frequently used clinically to combat enteric infections. **Target(s)**: dihydropteroate synthase (1-5). **1**. Hong, Hossler, Calhoun & Meshnick (1995) *Antimicrob. Agents Chemother.* **39**, 1756. **2**. Kasekarn, Sirawaraporn, Chahomchuen, Cowman & Sirawaraporn (2004) *Mol. Biochem. Parasitol.* **137**, 43. **3**. Berglez, Iliades, Sirawaraporn, Coloe & Macreadie (2004) *Int. J. Parasitol.* **34**, 95. **4**. Iliades, Meshnick & Macreadie (2005) *Antimicrob. Agents Chemother.* **49**, 741. **5**. Fernley, Iliades & Macreadie (2007) *Anal. Biochem.* **360**, 227.

Sulfadiazine

This antibacterial agent (FW = 250.28 g/mol; CAS 68-35-9), also known as 4-amino-*N*-2-pyrimidinylbenzenesulfonamide, was developed by Roblin et al. (1) in 1940 and is most often used along with other antibiotics, such as trimethoprim. Sulfadiazine is a white solid (M.P. = 252-256°C) that is slightly soluble in water (200 mg/100 mL at 37°C and pH 7.5). **Target(s)**: D-amino-acid oxidase (2); ATPase, inhibited by silver sulfadiazine (3); bontoxilysin, inhibited by silver sulfadiazine (4); dihydroorotase, $K_i = 0.19$ mM (5); dihydropteroate synthase (6-12); glucose-6-phosphate dehydrogenase, weakly inhibited (13); mannose-6-phosphate isomerase, inhibited by silver sulfadiazine (14) **1**. Roblin, Williams, Winnek & English (1940) *J. Amer. Chem. Soc.* **62**, 2002. **2**. Hellerman, Lindsay & Bovarnick (1946) *J. Biol. Chem.* **163**, 553. **3**. Nechay & Saunders (1984) *J. Environ. Pathol. Toxicol. Oncol.* **5**, 119. **4**. Burnett, Schmidt, Stafford, et al. (2003) *Biochem. Biophys. Res. Commun.* **310**, 84. **5**. Pradhan & Sander (1973) *Life Sci.* **13**, 1747. **6**. Ho (1980) *Meth. Enzymol.* **66**, 553. **7**. Prabhu, Lui & King (1997) *Phytochemistry* **45**, 23. **8**. Chio, Bolyard, Nasr & Queener (1996) *Antimicrob. Agents Chemother.* **40**, 727. **9**. McCullough & Maren (1974) *Mol. Pharmacol.* **10**, 140. **10**. Kasekarn, Sirawaraporn, Chahomchuen, Cowman & Sirawaraporn (2004) *Mol. Biochem. Parasitol.* **137**, 43. **11**. Iliades, Meshnick & Macreadie (2005) *Antimicrob. Agents Chemother.* **49**, 741. **12**. Eagon & McManus (1989) *Antimicrob. Agents Chemother.* **33**, 1936. **13**. Altman (1946) *J. Biol. Chem.* **166**, 149. **14**. Wells, Scully, Paravicini, Proudfoot & Payton (1995) *Biochemistry* **34**, 7896.

Sulfadimethoxine

This water-soluble sulfonamide (FW = 310.33 g/mol; CAS 122-11-2), also known as 4-amino-*N*-(2,6-dimethoxy-4-pyrimidinyl)benzenesulfonamide, is frequently used as an antibacterial agent, in conjunction with other pharmaceuticals such as trimethoprim. **Target(s)**: CYP2C9 (1); and dihydropteroate synthase (2-4). **1**. Zweers-Zeilmaker, Horbach & Witkamp (1997) *Xenobiotica* **27**, 769. **2**. Kasekarn, Sirawaraporn, Chahomchuen, Cowman & Sirawaraporn (2004) *Mol. Biochem. Parasitol.* **137**, 43. **3**. Berglez, Iliades, Sirawaraporn, Coloe & Macreadie (2004) *Int. J. Parasitol.* **34**, 95. **4**. Iliades, Meshnick & Macreadie (2005) *Antimicrob. Agents Chemother.* **49**, 741.

Sulfadoxine

This water-soluble sulfonamide (FW = 310.33 g/mol; CAS 2447-57-6) is a structural analogue of sulfadimethoxine and is frequently used as an antibacterial agent in conjunction with other antibiotics. Sulfadoxine/pyrimethamine is widely used in Africa for treating chloroquine-resistant *Plasmodium falciparum* malaria. (*See also Dapsone; Sulfadiazine*) **Target(s)**: Dihydropteroate synthase (DHPS) is the primary target of sulfone and sulfonamide drugs, which are used extensively in the control of many infections including *P. falciparum* (1-5). Sequencing of the DHPS gene in 10 isolates from Thailand identified a new allele of DHPS that has a previously unidentified amino acid difference (5). Eight alleles of *P. falciparum* PPPK-DHPS have been expressed in *Escherichia coli* and purified to homogeneity (5). Strikingly, the K_i for sulfadoxine varies by almost three orders of magnitude from 0.14 μM for the DHPS allele from sensitive isolates to 112 μM for an enzyme expressed in a highly resistant isolate. Comparison of the K_i values of different sulfonamides and the sulfone dapsone suggests that amino acid differences in DHPS are likely to confer cross-resistance to these compounds (5). Mutations in the DHPS gene of sulfadoxine-resistant isolates of *P. falciparum* are thus central to the mechanism of resistance to sulfones and sulfonamides (5). Other targets include 2-amino-4-hydroxy-6-hydroxymethyldihydropteridine diphospho-kinase (1), CYP2C9 (6), and dihydrofolate reductase (7). **1**. Kasekarn, Sirawaraporn, Chahomchuen, Cowman & Sirawaraporn (2004) *Mol. Biochem. Parasitol.* **137**, 43. **2**. Triglia, Menting, Wilson & Cowman (1997) *Proc. Natl. Acad. Sci. U.S.A.* **94**, 13944. **3**. Berglez, Iliades, Sirawaraporn, Coloe & Macreadie (2004) *Int. J. Parasitol.* **34**, 95. **4**. Iliades, Meshnick & Macreadie (2005) *Antimicrob. Agents Chemother.* **49**, 741. **5**. Triglia, Menting, Wilson & Cowman (1997) *Proc. Natl. Acad. Sci. U.S.A.* **94**, 13944. **6**. Zweers-Zeilmaker, Horbach & Witkamp (1997) *Xenobiotica* **27**, 769. **7**. Sirawaraporn & Yuthavong (1986) *Antimicrob. Agents Chemother.* **29**, 899.

Sulfaethidole

This thiadiazole antibacterial (FW = 284.36 g/mol; CAS 94-19-9), known systematically as sulfaethylthiadiazole and 4-amino-*N*-(aminoimino-methyl)benzenesulfonamide, also inhibits glucose-6-phosphatase. **1**. Jasmin & Johnson (1959) *J. Amer. Pharm. Assoc.* **48**, 113.

Sulfaguanidine

This sulfonamide derivative (FW$_{free-base}$ = 214.25 g/mol; M.P. = 190-193°C; CAS 57-67-0; water solubility = 1 g/L at 25°C), also known as N^1-amidinosulfanilamide, N^1-guanylsulfanilamide, and 4-amino-*N*-(aminoiminomethyl)benzenesulfonamide, is an antibacterial agent.

Target(s): L-amino-acid oxidase (1); 2-amino-4-hydroxy-6-hydroxy-methyldihydropteridine pyrophosphokinase (2); carbonic anhydrase, *or* carbonate dehydratase, $K_i = 1.9$ mM (3); and dihydropteroate synthase (2,4). **1.** Zeller & Maritz (1945) *Helv. Chim. Acta* **28**, 365. **2.** Walter & Königk (1980) *Meth. Enzymol.* **66**, 564. **3.** Pocker & Stone (1968) *Biochemistry* **7**, 2936. **4.** Walter & Königk (1974) *Hoppe-Seyler's Z. Physiol. Chem.* **355**, 431.

Sulfamethoxazole

This membrane-permeant sulfonamide (FW = 253.28 g/mol; CAS 723-46-6), also known as 4-amino-*N*-(5-methyl-3-isoxazolyl)benzenesulfonamide, targets dihydropteroate synthase (*Reaction*: (2-amino-4-hydroxy-7,8-dihydropteridin-6-yl)methyldiphosphate + 4-aminobenzoate $\rightleftharpoons$ dihydropteroate + PP_i) and is highly effective antibiotic for treating intracellular pathogens (*e.g.*, *Listeria* and *Shigella*), especially when used in combination with trimethoprim (1). **Pharmacokinetics:** Sulfamethoxazole is rapidly absorbed when administered orally, distributing into most tissues and accumulating in sputum, vaginal fluid, and middle-ear fluid. Seventy percent of circulating drug is bound to plasma proteins. The time to reach maximum plasma concentration is 1–4 hours after, and the mean serum half-life is 10 hours. About 50-70% of oral sulfamethoxazole undergoes hepatic metabolism, forming at N^4-acetyl-, N^4-hydroxy-, 5-methylhydroxy-, N^4-acetyl-5-methylhydroxy-derivatives as well as its *N*-glucuronide conjugate (15-20%). CYP2C9 catalyzes the formation of the N^4-hydroxy derivative. Sulfamethoxazole is not a substrate of the P-glycoprotein transporter. It is primarily excreted as glomerular filtrate and tubular secretion. *See also* Appendix II in Goodman & Gilman's THE *PHARMACOLOGICAL BASIS OF THERAPEUTICS*, 12th Edition (Brunton, Chabner & Knollmann, eds.) McGraw-Hill Medical, New York (2011). **Target(s):** CYP2C9 (2); and dihydropteroate synthase (3-12). **1.** Southwick & Purich (1996) *New Engl. J. Med.* **334**, 770. **2.** Wen, Wang, Backman, Laitila & Neuvonen (2002) *Drug Metab. Dispos.* **30**, 631. **3.** Hong, Hossler, Calhoun & Meshnick (1995) *Antimicrob. Agents Chemother.* **39**, 1756. **4.** Zhang & Meshnick (1991) *Antimicrob. Agents Chemother.* **35**, 267. **5.** Nopponpunth, Sirawaraporn, Greene & Santi (1999) *J. Bacteriol.* **181**, 6814. **6.** Triglia, Menting, Wilson & Cowman (1997) *Proc. Natl. Acad. Sci. U.S.A.* **94**, 13944. **7.** Kasekarn, Sirawaraporn, Chahomchuen, Cowman & Sirawaraporn (2004) *Mol. Biochem. Parasitol.* **137**, 43. **8.** Berglez, Iliades, Sirawaraporn, Coloe & Macreadie (2004) *Int. J. Parasitol.* **34**, 95. **9.** Iliades, Meshnick & Macreadie (2005) *Antimicrob. Agents Chemother.* **49**, 741. **10.** Fernley, Iliades & Macreadie (2007) *Anal. Biochem.* **360**, 227. **11.** Vinnicombe & Derrick (1999) *Biochem. Biophys. Res. Commun.* **258**, 752. **12.** Roland, Ferone, Harvey, Styles & Morrison (1979) *J. Biol. Chem.* **254**, 10337.

Sulfamethoxypyridazine

This sulfonamide (FW = 280.31 g/mol; CAS 80-35-3), also known as 4-amino-*N*-(6-methoxy-3-pyridazinyl)benzenesulfonamide, inhibits dihydropteroate synthase (1-4). **1.** Hong, Hossler, Calhoun & Meshnick (1995) *Antimicrob. Agents Chemother.* **39**, 1756. **2.** Nopponpunth, Sirawaraporn, Greene & Santi (1999) *J. Bacteriol.* **181**, 6814. **3.** Kasekarn, Sirawaraporn, Chahomchuen, Cowman & Sirawaraporn (2004) *Mol. Biochem. Parasitol.* **137**, 43. **4.** Iliades, Meshnick & Macreadie (2005) *Antimicrob. Agents Chemother.* **49**, 741.

Sulfamidochrysoidine

This red dye and antibiotic prodrug (FW$_{free-base}$ = 291.33 g/mol; CAS 103-12-8; IUPAC Name: 2,4-diaminophenylazobenzene-4'-sulfonamide) was first synthesized by Bayer chemists Klarer and Mietzsch and found by Domagk (Nobel Prize, 1939) to be a lethal antibiotic for many Gram-positive bacteria. In 1935, Pasteur Institute researchers Fourneau, Tréfouël, Bovet (Nobel, 1955; for discovery of neurotransmitters and antihistamines) and Nitti found that it is metabolized to sulfanilamide (*p*-aminobenzenesulfonamide), a potent competitive inhibitor of enzymes acting on *p*-aminobenzoic acid. The hydrochloride (FW = 327.79 g/mol), known as Prontosil, is an orange-red solid that is slightly soluble in water (1 g/400 mL). This dye binds tightly to carbonic anhydrase and is often used as a stain for that enzyme in polyacrylamide gels. Prontosil-dextrans are also inhibitors of carbonic anhydrase. **Target(s):** carbonic anhydrase, *or* carbonate dehydratase (1,2). **1.** Siffert, Teske, Gartner & Gros (1983) *J. Biochem. Biophys. Methods* **8**, 331. **2.** van Goor (1948) *Enzymologia* **13**, 73.

Sulfamisterin

This fungal metabolite (FW$_{ion}$ = 438.56 g/mol), first obtained from a strain of *Pycinidiella*, suppresses sphingolipid biosynthesis by inhibiting serine *C*-palmitoyltransferase, IC$_{50}$ = 3 nM. Sulfamisterin markedly inhibits sphingolipid biosynthesis in CHO cells and *Saccharomyces cerevisiae*, as monitored by incorporation of radioactive precursors into sphingolipids. Unlike cell-free experiments, 10 µM sulfamisterin was needed for complete inhibition of sphingolipid biosynthesis in intact cells. **1.** Yamaji-Hasegawa, Takahashi, Tetsuka, Senoh & Kobayashi (2005) *Biochemistry* **44**, 268.

Sulfamoxole

This antibacterial (FW = 267.31 g/mol; CAS 729-99-7; Solubility: 160 mg/100 mL in 10 mM HCl; 200 mg/100 mL in 10 mM NaOH), also known as 4-amino-*N*-(4,5-dimethyl-2-oxazolyl)benzenesulfonamide, inhibits CYP2C (1) and dihydropteroate synthase (2-4). **1.** Zweers-Zeilmaker, Horbach & Witkamp (1997) *Xenobiotica* **27**, 769. **2.** Kasekarn, Sirawaraporn, Chahomchuen, Cowman & Sirawaraporn (2004) *Mol. Biochem. Parasitol.* **137**, 43. **3.** Berglez, Iliades, Sirawaraporn, Coloe & Macreadie (2004) *Int. J. Parasitol.* **34**, 95. **4.** Iliades, Meshnick & Macreadie (2005) *Antimicrob. Agents Chemother.* **49**, 741.

(R)-N-Sulfamoyl-α-benzyl-β-alanine, See (R)-2-Benzyl-3-(N-sulfamoyl)-aminopropanoate

Sulfanilamide

This *p*-aminobenzoate mimic and antibacterial (FW = 172.21 g/mol; CAS 63-74-1; Soluble in boiling water = 0.5 g/mL), also known as 4-aminobenzenesulfonamide, exhibits wide-spectrum of action against most Gram-positive and many Gram-negative organisms. Exogenous folinic acid restores optimal growth in most strains. Sulfanilamide was first synthesized by Austrian chemist Paul Gelmo in 1908, but its antibacterial activity was first noted by Domagk (Nobel Laureate, 1939). Many bacteria were later found to develop sulfanilamide resistance. By examining the effects of sulfanilamide on dihydropteroate synthetase activity in genetically transformed mutants of *Diplococcus pneumoniae*, Hotchkiss and Evans (1) concluded that drug resistance was achieved by changes in dihydropteroate synthetase structure, resulting in reduced binding affinity for sulfanilamide, but not for the natural substrate, *p*-aminobenzoic acid (*or* PABA). This conclusion was later confirmed by kinetic measurements on dihydropteroate synthetase fom naturally occurring clinical isolates of *Neisseria meningitidis* and *Neisseria gonorrhoeae* (2). Additional resistance mechanisms have been now been identified, incuding increased expression of the pABA synthase and dihydropteroate synthetase genes. **Target(s):** alkaline phosphatase (3,4); D-amino-acid oxidase (5); 2-amino-4-hydroxy-6-hydroxymethyl-dihydropteridine pyrophosphokinase (6); aminopeptidase B (7); carbonic anhydrase, *or* carbonate dehydratase (8-26,40-42); catalase (27); catechol oxidase (28); dihydrofolate synthase (29,39); dihydroorotase, K_i = 3.4 mM (30); dihydropteroate synthase, *clinical target* (6,29,31,32,39,43-48); glucose-6-phosphate dehydrogenase (33,34); glutamate decarboxylase, mildly inhibited (35); histidine decarboxylase (36); lactate dehydrogenase, weakly inhibited (34); luciferase, bacterial, *or* alkanal monooxygenase (37); and peroxidase, horseradish (38). **1.** Hotchkiss & Evans (1958) *Cold Spring Harbor Symp. Quant. Biol.* **23**, 85. **2.** Ho, Corman, Morse & Artenstein (1974) *Antimicrob. Agents and Chemother.* **5**, 388. **3.** Fauré-Fremiet, Stolkowski & Ducornet (1948) *Biochim. Biophys. Acta* **2**, 668. **4.** Benesch, Chance & Glynn (1945) *Nature* **155**, 203. **5.** Hellerman, Lindsay & Bovarnick (1946) *J. Biol. Chem.* **163**, 553. **6.** Walter & Königk (1980) *Meth. Enzymol.* **66**, 564. **7.** Mäkinen & Hopsu-Havu (1967) *Enzymologia* **32**, 333. **8.** Waygood (1955) *Meth. Enzymol.* **2**, 836. **9.** Mann & Keilin (1940) *Nature* **146**, 164. **10.** Krebs (1948) *Biochem. J.* **43**, 525. **11.** Roughton & Clark (1951) *The Enzymes*, 1st ed. (Sumner & Myrbäck, eds.), **1** (part 2), 1250. **12.** Davis (1961) *The Enzymes*, 2nd ed. (Boyer, Lardy & Myrbäck, eds.), **5**, 545. **13.** Whitney, Folsch, Nyman & Malmstrom (1967) *J. Biol. Chem.* **242**, 4206. **14.** Lindskog, Henderson, Kannan, *et al.* (1971) *The Enzymes*, 3rd. ed. (Boyer, ed.), **5**, 587. **15.** Leibman & Greene (1967) *Proc. Soc. Exp. Biol. Med.* **125**, 106. **16.** Feldstein & Silverman (1984) *J. Biol. Chem.* **259**, 5447. **17.** Olander & Kaiser (1970) *J. Amer. Chem. Soc.* **92**, 5758. **18.** Maren, A. Parcell & Malik (1960) *J. Pharmacol. Exp. Ther.* **130**, 389. **19.** Maren (1963) *J. Pharmacol. Exp. Ther.* **139**, 129 and 140. **20.** Feldstein & Silverman (1984) *J. Biol. Chem.* **259**, 5447. **21.** Davenport (1945) *J. Biol. Chem.* **158**, 567. **22.** Main & Locke (1941) *J. Biol. Chem.* **140**, 909. **23.** Main & Locke (1942) *J. Biol. Chem.* **143**, 729. **24.** van Goor (1944) *Enzymologia* **11**, 174. **25.** van Goor (1948) *Enzymologia* **13**, 73. **26.** Pocker & Stone (1968) *Biochemistry* **7**, 2936. **27.** Shinn, Main & Mellon (1940) *Proc. Soc. Exp. Biol. Med.* **44**, 596. **28.** Shacter (1950) *J. Biol. Chem.* **184**, 697. **29.** Ortiz & Hotchkiss (1966) *Biochemistry* **5**, 67. **30.** Pradhan & Sander (1973) *Life Sci.* **13**, 1747. **31.** Ho (1980) *Meth. Enzymol.* **66**, 553. **32.** Prabhu, Lui & King (1997) *Phytochemistry* **45**, 23. **33.** Noltmann & Kuby (1963) *The Enzymes*, 2nd ed. (Boyer, Lardy & Myrbäck, eds.), **7**, 223. **34.** Altman (1946) *J. Biol. Chem.* **166**, 149. **35.** Ambe & Sohonie (1963) *Enzymologia* **26**, 98. **36.** Schales (1951) *The Enzymes*, 1st ed. (Sumner & Myrbäck, eds.), **2** (part 1), 216. **37.** Harvey (1951) *The Enzymes*, 1st ed. (Sumner & Myrbäck, eds.), **2** (part 1), 581. **38.** Lipmann (1941) *J. Biol. Chem.* **139**, 977. **39.** Ortiz (1970) *Biochemistry* **9**, 355. **40.** Sanyal, Pessah & Maren (1981) *Biochim. Biophys. Acta* **657**, 128. **41.** Gill, Fedorka-Cray, Tweten & Sleeper (1984) *Arch. Microbiol.* **138**, 113. **42.** Baird, Waheed, Okuyama, Sly & Fierke (1997) *Biochemistry* **36**, 2669. **43.** Walter & Königk (1974) *Hoppe-Seyler's Z. Physiol. Chem.* **355**, 431. **44.** Kasekarn, Sirawaraporn, Chahomchuen, Cowman & Sirawaraporn (2004) *Mol. Biochem. Parasitol.* **137**, 43. **45.** Berglez, Iliades, Sirawaraporn, Coloe & Macreadie (2004) *Int. J. Parasitol.* **34**, 95. **46.** Iliades, Meshnick & Macreadie (2005) *Antimicrob. Agents Chemother.* **49**, 741. **47.** Vinnicombe & Derrick (1999) *Biochem. Biophys. Res. Commun.* **258**, 752. **48.** Roland, Ferone, Harvey, Styles & Morrison (1979) *J. Biol. Chem.* **254**, 10337.

Sulfanilate

This arylsulfonic acid (FW$_{free-acid}$ = 173.19 g/mol; CAS 121-57-3; acid has low water solubility (0.01 g/mL at 20°C), but sodium salt is freely soluble), also known as sulfanilic acid and 4-aminobenzenesulfonic acid, inhibits α-amylase, *Bacillus subtilis* (1,2) dihydropteroate synthase (3); β-fructofuranosidase, *or* invertase (4); and glutathione *S*-transferase (5). **1.** Mäkinen & Hopsu-Havu (1967) *Enzymologia* **32**, 333. **2.** Takagi, Toda & Isemura (1971) *The Enzymes*, 3rd ed. (Boyer, ed.), **5**, 235. **3.** Walter & Königk (1974) *Hoppe-Seyler's Z. Physiol. Chem.* **355**, 431. **4.** Neuberg &

Mandl (1950) *The Enzymes*, 1st ed. (Sumner & Myrbäck, eds.), **1** (part 1), 527. **5.** Dierickx & Yde (1982) *Res. Commun. Chem. Pathol. Pharmacol.* **37**, 385.

N-(Sulfanylacetyl)-L-phenylalanyl-L-alanine, See *N*-(Mercaptoacetyl)-L-phenylalanyl-L-alanine

Sulfanilyl-dicyandiamide

This substituted sulfonamide (FW = 239.26 g/mol), produced by reaction of dicyandiamide with a sulfonyl halide, has K_i values of 3 μM for thrombin and 15 μM for trypsin. **1.** Clare, Scozzafava & Supuran (2001) *J. Enzyme Inhib.* **16**, 1.

N-(2(*R,S*)-Sulfanylheptanoyl)-L-phenylalanyl-L-alanine, See *N*-(2(*R,S*)-Mercapto-heptanoyl)-L-phenylalanyl-L-alanine

N-(2(*S*)-Sulfanyl-3-phenylpropanoyl)-L-alanine, See *N*-(2(*S*)-Mercapto-3-phenylpropionyl)-L-alanine

N-(2(*S*)-Sulfanyl-3-phenylpropanoyl)-L-alanyl-L-proline, See *N*-(2(*S*)-Mercapto-3-phenylpropionyl)-L-alanyl-L-proline

N-(2(*S*)-Sulfanyl-3-phenylpropanoyl)glycyl-5-phenyl-L-proline, See *N*-(2(*S*)-Mercapto-3-phenylpropionyl)glycyl-5-phenyl-L-proline

N-(2(*S*)-Sulfanyl-3-phenylpropanoyl)-L-phenylalanyl-L-alanine, See *N*-(2(*S*)-Mercapto-3-phenylpropionyl)-L-phenylalanyl-L-alanine

N-(2(*S*)-Sulfanyl-3-phenylpropanoyl)-L-phenylalanyl-L-tyrosine, See *N*-(2(*S*)-Mercapto-3-phenylpropionyl)-L-phenylalanyl-L-tyrosine

Sulfaphenazole

This antibacterial sulfonamide (FW = 314.37 g/mol; CAS 526-08-9; low water solubility (0.15 g/100 mL at 25°C and pH 7.0); sodium salt is water soluble), also known as 4-amino-*N*-(1-phenyl-1*H*-pyrazol-5-yl)benzenesulfonamide, inhibits CYP2C9 (1-5), CYP2C19, weakly inhibited (4,6), 4-nitrophenol 2-monooxygenase, *or* CYP2E1, weakly inhibited (7), and quinine 3-monooxygenase (8). **1.** Rodrigues (1996) *Meth. Enzymol.* **272**, 186. **2.** Hyland, Jones & Smith (2003) *Drug Metab. Dispos.* **31**, 540. **3.** Michaelis, Fisslthaler, Medhora, *et al.* (2003) *FASEB J.* **17**, 770. **4.** Miyazawa, Shindo & Shimada (2002) *Drug Metab. Dispos.* **30**, 602. **5.** Baldwin, Bloomer, Smith, *et al.* (1995) *Xenobiotica* **25**, 261. **6.** Hartter, Tybring, Friedberg, Weigmann & Hiemke (2002) *Pharm. Res.* **19**, 1034. **7.** Tassaneeyakui, Veronese, Birkett, Gonzalez & Miners (1993) *Biochem. Pharmacol.* **46**, 1975. **8.** Zhao & Ishizaki (1997) *J. Pharmacol. Exp. Ther.* **283**, 1168.

Sulfapyridine

This sulfanilamide derivative (FW = 249.29 g/mol; CAS 144-83-2; Sparingly soluble, 280 mg/L H$_2$O), also named 4-amino-*N*-2-pyridinylbenzenesulfonamide and N^1-(2-pyridyl)sulfonilamide, is an antibacterial agent. **Target(s):** D-amino-acid oxidase (1); carbonic anhydrase, K_i ≈ 21 mM (2); dihydropteroate synthase (3-5); glucose-6-phosphate dehydrogenase (6); lyso-(platelet-activating factor):acetyl-CoA acetyltransferase (7); myeloperoxidase (8,9); and peroxidase, horseradish (10). **1.** Hellerman, Lindsay & Bovarnick (1946) *J. Biol. Chem.* **163**, 553. **2.** Pocker & Stone (1968) *Biochemistry* **7**, 2936. **3.** Kasekarn, Sirawaraporn, Chahomchuen, Cowman & Sirawaraporn (2004) *Mol.*

Biochem. Parasitol. **137**, 43.   **4**. Berglez, Iliades, Sirawaraporn, Coloe & Macreadie (2004) *Int. J. Parasitol.* **34**, 95.   **5**. Iliades, Meshnick & Macreadie (2005) *Antimicrob. Agents Chemother.* **49**, 741.   **6**. Altman (1946) *J. Biol. Chem.* **166**, 149.   **7**. Faison & White (1992) *Prostaglandins* **44**, 245.   **8**. Kettle & Winterbourn (1991) *Biochem. Pharmacol.* **41**, 1485. **9**. Kazmierowski, Ross, Peizner & Wuepper (1984) *J. Clin. Immunol.* **4**, 55. **10**. Lipmann (1941) *J. Biol. Chem.* **139**, 977.

Sulfaquinoxaline

This sulfonamide (FW = 300.34 g/mol; CAS 967-80-6), also known as 4-amino-*N*-2-quinoxalinylbenzenesulfonamide   and N^1-(2-quinoxalinyl)-sulfanilamide, is a potent (K_i = 1 μM) inhibitor of the dithiothreitol-dependent reduction of both vitamin K epoxide and vitamin K quinone by rat liver microsomes *in vitro*. The inhibition by sulfaquinoxaline resembles inhibition by coumarin anticoagulants and by hydroxynaphthoquinones. Sulfaquinoxaline has a low solubility in water (0.75 mg/100 mL at pH 7.0; however, the deliquescent sodium salt is much more soluble. Note that aqueous solutions absorb atmospheric carbon dioxide resulting in decreased solubility of the sulfonamide. **Target(s):** dihydropteroate synthase (1-4); vitamin K epoxide reductase (5); and vitamin K quinone reductase (5). **1**. Kasekarn, Sirawaraporn, Chahomchuen, Cowman & Sirawaraporn (2004) *Mol. Biochem. Parasitol.* **137**, 43.   **2**. Berglez, Iliades, Sirawaraporn, Coloe & Macreadie (2004) *Int. J. Parasitol.* **34**, 95.   **3**. Iliades, Meshnick & Macreadie (2005) *Antimicrob. Agents Chemother.* **49**, 741.   **4**. Vinnicombe & Derrick (1999) *Biochem. Biophys. Res. Commun.* **258**, 752.   **5**. Preusch, Hazelett & Lemasters (1989) *Arch. Biochem. Biophys.* **269**, 18.

Sulfasalazine

This brownish-yellow sulfonamide and efficient free radical scavenger (FW$_{free-acid}$ = 398.40 g/mol; CAS 599-79-1; Low H_2O solubility), also known as 2-hydroxy-5-[[4-[(2-pyridinylamino)sulfonyl]phenyl]azo]benzoic acid and salicylazosulfapyridine, is widely used for the treatment of ileitis and colitis. Its pharmacologic action has been attributed to 5-aminosalicylic acid, a primary breakdown product. **Target(s):** cystine transport (1); dihydrofolate reductase (2); γ-glutamyl hydrolase (3); glutathione *S*-transferase and leukotriene C$_4$ synthase (4-6); HIV-1 integrase (7); 9-hydroxyprostaglandin   dehydrogenase   (8);   15-hydroxyprostaglandin dehydrogenase (9,10); IκB kinases α/β (11,26,27); lipoxygenase (12); lipoxygenase, soybean (13, however, reference 1 reports no inhibition); lyso-(platelet-activating   factor):acetyl-CoA   acetyltransferase   (15); myeloperoxidase (16); NF-κB (nuclear factor κB) activation (17-20); 3-phosphoglycerate   kinase   (21,28);   AICAR   transformylase,   *or* phosphoribosylaminoimidazole carboxamide formyltransferase (2); serine hydroxymethyl-transferase, *or* glycine hydroxymethyl-transferase (22); thiopurine *S*-methyltransferase (23,24,29,30); thromboxane synthase (25); and folate transport (30,31). **1**. Gout, Buckley, Simms & Bruchovsky (2001) *Leukemia* **15**, 1633.   **2**. Baggott, Morgan, Ha, Vaughn & Hine (1992) *Biochem. J.* **282**, 197.   **3**. Reisenauer & Halsted (1981) *Biochim. Biophys. Acta* **659**, 62.   **4**. He, Awasthi, Singhal, Trent & Boor (1998) *Toxicol. Appl. Pharmacol.* **152**, 83.   **5**. Tsuchida, Izumi, Shimizu, *et al.* (1987) *Eur. J. Biochem.* **170**, 159.   **6**. Bach, Brashler & Johnson (1985) *Biochem. Pharmacol.* **34**, 2695.   **7**. Neamati, Mazumder, Zhao, *et al.* (1997) *Antimicrob. Agents Chemother.* **41**, 385.   **8**. Moore, Griffiths & Lofts (1983) *Biochem. Pharmacol.* **32**, 2813.   **9**. Berry, Hoult, Peers & Agback (1983) *Biochem. Pharmacol.* **32**, 2863.   **10**. Hoult & Moore (1978) *Brit. J. Pharmacol.* **64**, 6.   **11**. Weber, Liptay, Wirth, Adler & Schmid (2000) *Gastroenterology* **119**, 1209.   **12**. Stenson & Lobos (1982) *J. Clin. Invest.* **69**, 494.   **13**. Sircar, Schwender & Carethers (1983) *Biochem. Pharmacol.* **32**, 170.   **14**. Marcinkiewicz, Duniec & Robak (1985) *Biochem. Pharmacol.*

34, 148.   **15**. Faison & White (1992) *Prostaglandins* **44**, 245.   **16**. Kettle & Winterbourn (1991) *Biochem. Pharmacol.* **41**, 1485.   **17**. Egan & Sandborn (1998) *Gastroenterology* **115**, 1295.   **18**. Wahl, Liptay, Adler & Schmid (1998) *J. Clin. Invest.* **101**, 1163.   **19**. Liptay, Bachem, Hacker, *et al.* (1999) *Brit. J. Pharmacol.* **128**, 1361.   **20**. Merlo, Freudenthal & Romano (2002) *Neuroscience* **112**, 161.   **21**. Joao, Williams, Littlechild, Nagasuma & Watson (1992) *Eur. J. Biochem.* **205**, 1077.   **22**. Baum, Selhub & Rosenberg (1981) *J. Lab. Clin. Med.* **97**, 779.   **23**. Lennard (1998) *Ther. Drug Monit.* **20**, 527.   **24**. Szumlanski & Weinshilboum (1995) *Brit. J. Clin. Pharmacol.* **39**, 456.   **25**. Stenson & Lobos (1983) *Biochem. Pharmacol.* **32**, 2205.   **26**. Burke (2003) *Curr. Opin. Drug Discov. Devel.* **6**, 720.   **27**. Luo, Kamata & Karin (1993) *J. Clin. Invest.* **115**, 2625.   **28**. Joao & Williams (1993) *Eur. J. Biochem.* **216**, 1.   **29**. Lysaa, Giverhaug, Wold & Aarbakke (1996) *Eur. J. Clin. Pharmacol.* **49**, 393.   **30**. Zhou & Chowbay (2007) *Curr. Pharmacogenomics* **5**, 103.   **31**. Franklin & Rosenberg (1973) *Gastroenterology* **64**, 517.   **32**. Rosenberg (1976) *Clin. Haematol.* **5**, 589.

Sulfate

This inorganic anion SO_4^{2-} (FW$_{ion}$ = 96.06 g/mol) is an essential and ubiquitous metabolite in all living organisms, participates in a wide variety of processes, and is a product and/or inhibitor of many sulfatases. In renal failure, sulfoesters accumulate and hypersulfatemia contributes directly to the unmeasured anion gap characteristic of the condition. Sulfate weakly inhibits over 150 enzymes, with inhibition constants ranging in the 0.1–5 mM range.

Sulfathiazole

This sulfanilamide derivative (FW = 255.32 g/mol; CAS 72-14-0), also called N^1-2-thiazolylsulfanilamide   and   4-amino-*N*-2-thiazolylbenzene sulfonamide, inhibits folic acid biosynthesis and exerts antibacterial activity. **Target(s):** D-amino-acid oxidase (1); 2-amino-4-hydroxy-6-hydroxymethyldihydropteridine diphosphokinase (2,3); carbonic anhydrase, *or* carbonate dehydratase, weakly inhibited (4); catechol oxidase (5); dihydropteroate synthase (2,3,6-14); glucose-6-phosphate dehydrogenase (15); γ-glutamyl transpeptidase (16); peroxidase, horseradish (17); pyruvate carboxylase (18); and tyrosinase (19). **1**. Hellerman, Lindsay & Bovarnick (1946) *J. Biol. Chem.* **163**, 553.   **2**. Walter & Königk (1980) *Meth. Enzymol.* **66**, 564.   **3**. Kasekarn, Sirawaraporn, Chahomchuen, Cowman & Sirawaraporn (2004) *Mol. Biochem. Parasitol.* **137**, 43.   **4**. Pocker & Stone (1968) *Biochemistry* **7**, 2936.   **5**. Dawson & Magee (1955) *Meth. Enzymol.* **2**, 817.   **6**. Richey & Brown (1971) *Meth. Enzymol.* **18B**, 765.   **7**. Thijssen (1980) *Meth. Enzymol.* **66**, 570.   **8**. Hong, Hossler, Calhoun & Meshnick (1995) *Antimicrob. Agents Chemother.* **39**, 1756.   **9**. Allegra, Boarman, Kovacs, *et al.* (1990) *J. Clin. Invest.* **85**, 371.   **10**. Walter & Königk (1974) *Hoppe-Seyler's Z. Physiol. Chem.* **355**, 431.   **11**. Triglia, Menting, Wilson & Cowman (1997) *Proc. Natl. Acad. Sci. U.S.A.* **94**, 13944.   **12**. Berglez, Iliades, Sirawaraporn, Coloe & Macreadie (2004) *Int. J. Parasitol.* **34**, 95. **13**. Iliades, Meshnick & Macreadie (2005) *Antimicrob. Agents Chemother.* **49**, 741.   **14**. Roland, Ferone, Harvey, Styles & Morrison (1979) *J. Biol. Chem.* **254**, 10337.   **15**. Altman (1946) *J. Biol. Chem.* **166**, 149.   **16**. Binkley (1961) *J. Biol. Chem.* **236**, 1075.   **17**. Lipmann (1941) *J. Biol. Chem.* **139**, 977.   **18**. Vennesland (1951) *The Enzymes*, 1st ed. (Sumner & Myrbäck, eds.), **2** (part 1), 183.   **19**. Dawson & Tarpley (1951) *The Enzymes*, 1st ed. (Sumner & Myrbäck, eds.), **2** (part 1), 454.

Sulfide Ions

This dianion S^{2-} (Na$_2$S; FW = 78.05 g/mol; very hygroscopic; solubility = 18.6 g/100 g H_2O at 20°C; K$_2$S; FW = 110.26 g/mol; unstable and may explode on percussion or upon rapid heating), decomposes in air to produce hydrogen sulfide. The sulfide ion can be thought of as a conjugate base of hydrogen sulfide. H_2S has pK_a values of 7.04 and 14.9, and at typical alkaline pH values, the main species in aqueous solutions is HS$^-$. **See also** *Hydrogen Sulfide* **Target(s):** aconitase (1,2); alcohol dehydrogenase (3,4); alkaline phosphatase (5); amine oxidase (6); ascorbate oxidase (7); bacterial leucyl aminopeptidase (60); carbonic anhydrase, *or* carbonate dehydratase (8-15,54); carboxypeptidase A (16-18); catalase (19,20); cysteinyl-tRNA synthetase (21,22); cytochrome *c* oxidase, *or* complex IV (23-26); cytosol alanyl aminopeptidase (59); formate dehydrogenase (27);

L-galactonolactone oxidase (28); glutamate dehydrogenase (29); γ-glutamyl cyclotransferase (30); glycyl-tRNA synthetase (31); lactate dehydrogenase (32); L-lactate dehydrogenase (33); leucine dehydrogenase (34); leucyl aminopeptidase (35); lipase (36); lipoxygenase (37); membrane dipeptidase (38); peroxidase (39); peroxidase, horseradish (40); phosphonoacetaldehyde hydrolase (57); phosphonoacetate hydrolase (56); phosphonopyruvate hydrolase (55); photosystem II (41); ribonuclease T_1 (42-44); ribonuclease T_2 (44,45); sulfate adenylyltransferase, *or* ATP sulfurylase (46,47); sulfite reductase (48); thiosulfate:thiol sulfurtransferase (49); urocanate hydratase, *or* urocanase (50); Xaa-His Dipeptidase, *or* carnosinase (35,51,52,58); and xanthine oxidase (53). **1.** Ochoa (1951) *The Enzymes*, 1st ed. (Sumner & Myrbäck, eds.), **1** (part 2), 1217. **2.** Dickman (1961) *The Enzymes*, 2nd ed. (Boyer, Lardy & Myrbäck, eds.), **5**, 495. **3.** Sund & Theorell (1963) *The Enzymes*, 2nd ed. (Boyer, Lardy & Myrbäck, eds.), **7**, 25. **4.** von Wartburg, Bethune & Vallee (1964) *Biochemistry* **3**, 1775. **5.** Plocke, Levinthal & Vallee (1962) *Biochemistry* **1**, 373. **6.** Dooley & Cote (1984) *J. Biol. Chem.* **259**, 2923. **7.** Dawson & Magee (1955) *Meth. Enzymol.* **2**, 831. **8.** Waygood (1955) *Meth. Enzymol.* **2**, 836. **9.** Keilin & Mann (1940) *Biochem. J.* **34**, 1163. **10.** Davis (1961) *The Enzymes*, 2nd ed. (Boyer, Lardy & Myrbäck, eds.), **5**, 545. **11.** Lindskog, Henderson, Kannan, *et al.* (1971) *The Enzymes*, 3rd. ed. (Boyer, ed.), **5**, 587. **12.** Schwimmer (1969) *Enzymologia* **37**, 163. **13.** Kiese & Hastings (1940) *J. Biol. Chem.* **132**, 281. **14.** Main & Locke (1942) *J. Biol. Chem.* **143**, 729. **15.** van Goor (1948) *Enzymologia* **13**, 73. **16.** Pétra (1970) *Meth. Enzymol.* **19**, 460. **17.** Axelrod, Purvis & Hofmann (1948) *J. Biol. Chem.* **176**, 695. **18.** Smith, Hanson & Wendelboe (1949) *J. Biol. Chem.* **179**, 803. **19.** Nicholls & Schonbaum (1963) *The Enzymes*, 2nd ed. (Boyer, Lardy & Myrbäck, eds.), **8**, 147. **20.** Carlsson, Berglin, Claesson, Edlund & Persson (1988) *Mutat. Res.* **202**, 59. **21.** Burnell & Whatley (1977) *Biochim. Biophys. Acta* **481**, 266. **22.** Burnell & Shrift (1977) *Plant Physiol.* **60**, 670. **23.** Slater (1967) *Meth. Enzymol.* **10**, 48. **24.** Wharton & Tzagoloff (1967) *Meth. Enzymol.* **10**, 245. **25.** Errede, Kamen & Hatefi (1978) *Meth. Enzymol.* **53**, 40. **26.** Wilson & Erecinska (1978) *Meth. Enzymol.* **53**, 191. **27.** Kanamori & Suzuki (1968) *Enzymologia* **35**, 185. **28.** Bleeg & Christensen (1982) *Eur. J. Biochem.* **127**, 391. **29.** Frieden (1963) *The Enzymes*, 2nd ed. (Boyer, Lardy & Myrbäck, eds.), **7**, 3. **30.** Orlowski & Meister (1971) *The Enzymes*, 3rd. ed. (Boyer, ed.), **4**, 123. **31.** Freist, Logan & Gauss (1996) *Biol. Chem. Hoppe-Seyler* **377**, 343. **32.** Boyer (1959) *The Enzymes*, 2nd ed. (Boyer, Lardy & Myrbäck, eds.), **1**, 511. **33.** Schwert & Winer (1963) *The Enzymes*, 2nd ed. (Boyer, Lardy & Myrbäck, eds.), **7**, 142. **34.** Zink & Sanwal (1970) *Meth. Enzymol.* **17A**, 799. **35.** Smith (1951) *The Enzymes*, 1st ed. (Sumner & Myrbäck, eds.), **1** (part 2), 793. **36.** Weinstein & Wynne (1936) *J. Biol. Chem.* **112**, 649. **37.** Holman & Bergström (1951) *The Enzymes*, 1st ed. (Sumner & Myrbäck, eds.), **2** (part 1), 559. **38.** Yudkin & Fruton (1947) *J. Biol. Chem.* **169**, 521. **39.** Guibault, Brignac, Jr., & Zimmer (1968) *Anal. Chem.* **40**, 190. **40.** Getchell & Walton (1931) *J. Biol. Chem.* **91**, 419. **41.** Oren, Padan & Malkin (1979) *Biochim. Biophys. Acta* **546**, 270. **42.** Uchida & Egami (1967) *Meth. Enzymol.* **12A**, 228. **43.** Egami, Takahashi & Uchida (1964) *Progress in Nucleic Acid Research and Molecular Biology* **3**, 59. **44.** Uchida & Egami (1971) *The Enzymes*, 3rd ed. (Boyer, ed.), **4**, 205. **45.** Uchida & Egami (1967) *Meth. Enzymol.* **12A**, 239. **46.** Stadtman (1973) *The Enzymes*, 3rd ed. (Boyer, ed.), **8**, 1. **47.** Peck, Jr. (1974) *The Enzymes*, 3rd ed. (Boyer, ed.), **10**, 651. **48.** Hatefi & Stiggall (1976) *The Enzymes*, 3rd ed. (Boyer, ed.), **13**, 175. **49.** Uhteg & Westley (1979) *Arch. Biochem. Biophys.* **195**, 211. **50.** Hug, O'Donnell & Hunter (1979) *J. Biol. Chem.* **253**, 7622. **51.** Bauer (1998) in *Handb. Proteolytic Enzymes* (Barrett, Rawlings & Woessner, eds.), p. 1525, Academic Press, San Diego. **52.** Hanson & Smith (1949) *J. Biol. Chem.* **179**, 789. **53.** Roussos (1967) *Meth. Enzymol.* **12A**, 5. **54.** Engberg, Millqvist, Pohl & Lindskog (1985) *Arch. Biochem. Biophys.* **241**, 628. **55.** Ternan, Hamilton & Quinn (2000) *Arch. Microbiol.* **173**, 35. **56.** McMullan & Quinn (1994) *J. Bacteriol.* **176**, 320. **57.** La Nauze, Rosenberg & Shaw (1970) *Biochim. Biophys. Acta* **212**, 332. **58.** Kunze, Kleinkauf & Bauer (1986) *Eur. J. Biochem.* **160**, 605. **59.** Garner & Behal (1974) *Biochemistry* **13**, 3227. **60.** Prescott & Wilkes (1966) *Arch. Biochem. Biophys.* **117**, 328.

Sulfinpyrazone

This potent uricosuric drug (FW = 404.48 g/mol; CAS 57-96-5), also known as 1,2-diphenyl-4-[2-(phenylsulfinyl)ethyl]-3,5-pyrazolidinedione, scavenges free radicals and inhibits platelet aggregation by blocking degranulation. Sulfinpyrazone may be used to treat gout by competitively inhibiting uric acid reabsorption by the kidney proximal tubule. **Target(s):** cadmium-transporting ATPase (1); cathepsin G (2,3); cyclooxygenase (4-10); CYP2C9 (11,12); CYP2C6 (13); elastase (2,3); glutathione-transporting ATPase (1); urate transport (14); and xenobiotic-transporting ATPase, *or* multidrug-resistance protein (15). **1.** Rebbeor, Connolly, Dumont & Ballatori (1998) *J. Biol. Chem.* **273**, 33449. **2.** Steinmeyer & Kalbhen (1996) *Inflamm. Res.* **45**, 324. **3.** Steinmeyer & Kalbhen (1989) *Arzneimittelforschung* **39**, 1208. **4.** Janero, Burghardt, Lopez & Cardell (1989) *Biochem. Pharmacol.* **38**, 4381. **5.** Patrono, Ciabattoni, Patrignani, *et al.* (1985) *Circulation* **72**, 1177. **6.** Cerletti (1985) *Int. J. Tissue React.* **7**, 309. **7.** Pedersen & FitzGerald (1985) *Clin. Pharmacol. Ther.* **37**, 36. **8.** Del Maschio, Livio, Cerletti & De Gaetano (1984) *Eur. J. Pharmacol.* **101**, 209. **9.** Cerletti, Livio & De Gaetano (1982) *Biochim. Biophys. Acta* **714**, 122. **10.** Ali & McDonald (1978) *Thromb. Res.* **13**, 1057. **11.** Miners & Birkett (1998) *Brit. J. Clin. Pharmacol.* **45**, 525. **12.** He, Kunze & Trager (1995) *Drug Metab. Dispos.* **23**, 659. **13.** Lakshmi, Zenser & Davis (1997) *Drug Metab. Dispos.* **25**, 481. **14.** Dan & Koga (1990) *Eur. J. Pharmacol.* **187**, 303. **15.** Wijnholds, Mol, van Deemter, *et al.* (2000) *Proc. Natl. Acad. Sci. U.S.A.* **97**, 7476.

Sulfisoxazole

This antibacterial agent (FW = 267.31 g/mol; CAS 127-69-5), also known as 4-amino-*N*-(3,4-dimethyl-5-isoxazolyl)benzenesulfonamide, inhibits folate biosynthesis and is a selective endothelin receptor ETA antagonist (also an antagonist at ETB at higher concentrations). It has a water solubility of 0.13 mg/mL at 25°C; however, the salt is far more soluble. **Target(s):** dihydropteroate synthase (1-3); and endothelin receptor, *or* ETA (4). **1.** Kasekarn, Sirawaraporn, Chahomchuen, Cowman & Sirawaraporn (2004) *Mol. Biochem. Parasitol.* **137**, 43. **2.** Berglez, Iliades, Sirawaraporn, Coloe & Macreadie (2004) *Int. J. Parasitol.* **34**, 95. **3.** Iliades, Meshnick & Macreadie (2005) *Antimicrob. Agents Chemother.* **49**, 741. **4.** Chan, Okun, Stavros, *et al.* (1994) *Biochem. Biophys. Res. Commun.* **201**, 228.

Sulfite Ion

This redox-active dianion (FW$_{ion}$ = 80.06 g/mol; FW$_{disodium-salt}$ = 126.06 g/mol; CAS 7757-83-7) inhibits enzymes in a multiple ways, often by acting as a sulfate analogue or as a reducing agent. Sodium sulfite reacts reversibly with FAD and FMN (1,2), and the ability to form covalent N^5-flavin adducts with oxidases distinguishes these enzymes from other classes of flavoproteins. At high concentration, sulfite reacts with disulfide bonds, especially at elevated pH, to produce an *S*-substituted sulfite (R–S–SO$_3^-$) and a free thiol. Metal ions (particularly Cu^{2+}) facilitate sulfitolysis by forming a complex with the thiol anion product. Sulfite is also a substrate in reactions catalyzed by the sulfite reductases, sulfite dehydrogenase, sulfite oxidase, UDP-sulfoquinovose synthase, and cysteine lyase. It is also a product of the reactions catalyzed by the phosphoadenylyl-sulfate reductases, the adenylyl-sulfate reductases, sulfur dioxygenase, taurine dioxygenase, 4-sulfobenzoate 3,4-dioxygenase, thiosulfate sulfurtransferase (rhodanese), thiosulfate:thiol sulfurtransferase, thiosulfate:dithiol sulfurtransferase, 2-(2-hydroxyphenyl)-benzenesulfinate hydrolase, and sulfoacetaldehyde lyase. **Target(s):** 2-(acetamido-methylene)succinate hydrolase (3); *N*-acetylgalactosamine-4-sulfatase, *or* arylsulfatase B (4,5); *N*-acetylgalactosamine-6-sulfatase (67); *N*-acetylglucosamine-6-sulfatase (6); acid phosphatase (77); adenylylsulfatase (54); adenylylsulfate reductase (7); alcohol dehydrogenase (8); AMP nucleosidase (63); Arylsulfatase (5,9-14,70-75); aspartate 4-decarboxylase (15); catalase (16,17); cerebroside-sulfatase, *or* arylsulfatase A (4,5,12,65); chaperonin ATPase (53); choline-sulfatase (13,18,19,66); creatine kinase (20); cyanate hydratase, *or* cyanase

(43); cyclamate sulfohydrolase (52); diiodophenylpyruvate reductase (21); disulfoglucosamine-6-sulfatase (64); dopamine β-monooxygenase (88); F$_1$F$_o$ ATP synthase, *or* H$^+$-transporting two-sector ATPase (22,23); formate dehydrogenase (24); β-fructofuranosidase, *or* invertase (25); fructose-1,6-bisphosphate aldolase (26); L-galactono-1,4-lactone oxidase (27); glucose-6-phosphate dehydrogenase (17); glucosinolate sulfatase (9); glutathione peroxidase (17); glyceraldehyde-3-phosphate dehydrogenase, *or* acetylphosphatase activity (28); D-β-hydroxybutyrate dehydrogenase (29); 2-(hydroxymethyl)-3-(acetamidomethylene)succinate hydrolase (61); 4-hydroxyphenyllactate dehydrogenase, *or* aromatic α-keto acid reductase (30); iodothyronine-5'-deiodinase (31); malate dehydrogenase (32-34); methanol:5-hydroxybenzimidazolylcobamide *Co*-methyltransferase (87); oxalate decarboxylase (50); peptidylglycine monooxygenase (88); peroxidase (35); phosphoenolpyruvate mutase (36); phosphonate dehydrogenase, *or* phosphite dehydrogenase (57-59); phosphono-acetaldehyde hydrolase (51); propanediol-phosphate dehydrogenase (37); pyrophosphate-dependent phosphofructokinase, *or* diphosphate:fructose-6-phosphate 1-phosphotransferase (86); rubredoxin:NAD$^+$ reductase (60); steryl-sulfatase (9,68,69); streptopain (62); sucrose carrier (39); sulfate adenylyltransferase (85); sulfoacetaldehyde lyase, product inhibition (40); sulfur reductase (56); superoxide dismutase (17); tetrachloroethene reductive dehalogenase (55); thiosulfate:dithiol sulfurtransferase, *or* thiosulfate reductase (41); thiosulfate sulfurtransferase, *or* rhodanese (38,80-84); thiosulfate:thiol sulfurtransferase, *or* thiosulfate reductase (42,78,79); and urocanate hydratase, *or* urocanase (44-49). **1.** Ghisla & Massey (1991) in *Chemistry and Biochemistry of Flavoenzymes* (Müller, ed.) Vol. **2**, p. 243, CRC Press, Boca Raton, FL. **2.** Lederer (1991) in *Chemistry and Biochemistry of Flavoenzymes* (Müller, ed.) Vol. **2**, p. 153, CRC Press, Boca Raton, FL. **3.** Huynh & Snell (1985) *J. Biol. Chem.* **260**, 2379. **4.** Fluharty & Edmond (1978) *Meth. Enzymol.* **50**, 537. **5.** Bleszynski & Leznicki (1967) *Enzymologia* **33**, 373. **6.** Basner, Kresse & Figura (1979) *J. Biol. Chem.* **254**, 1151. **7.** Yagi & Ogata (1996) *Biochimie* **78**, 838. **8.** Sund & Theorell (1963) *The Enzymes*, 2nd ed. (Boyer, Lardy & Myrbäck, eds.), **7**, 25. **9.** Roy (1987) *Meth. Enzymol.* **143**, 361. **10.** Roy (1953) *Biochem. J.* **55**, 653. **11.** Roy (1954) *Biochem. J.* **57**, 465. **12.** Nicholls & Roy (1971) *The Enzymes*, 3rd ed. (Boyer, ed.), **5**, 21. **13.** Webb (1966) *Enzyme and Metabolic Inhibitors*, vol. **2**, pp. 443-444, Academic Press, New York. **14.** Dodgson (1959) *Enzymologia* **20**, 301. **15.** Soda, Novogrodsky & Meister (1964) *Biochemistry* **3**, 1450. **16.** Veljovic-Jovanovic, Milovanovic, Oniki & Takahama (1999) *Free Radic. Res.* **31** Suppl, S51. **17.** Khan, Schuler & Coppock (1987) *J. Toxicol. Environ. Health* **22**, 481. **18.** Roy (1971) *The Enzymes*, 3rd ed. (Boyer, ed.), **5**, 1. **19.** Takebe (1961) *J. Biochem.* **50**, 245. **20.** Morin (1977) *Clin. Chem.* **23**, 646. **21.** Zannoni (1970) *Meth. Enzymol.* **17A**, 665. **22.** Pacheco-Moises, Minauro-Sanmiguel, Bravo & Garcia (2002) *J. Bioenerg. Biomembr.* **34**, 269. **23.** Das & Ljungdahl (2003) *J. Bacteriol.* **185**, 5527. **24.** Kanamori & Suzuki (1968) *Enzymologia* **35**, 185. **25.** Neuberg & Mandl (1950) *The Enzymes*, 1st ed. (J. B. Sumner & K. Myrbäck, eds.), **1** (part 1), 527. **26.** Chan (1968) *Biochemistry* **7**, 4247. **27.** Bleeg & Christensen (1982) *Eur. J. Biochem.* **127**, 391. **28.** Allison & Conners (1970) *Arch. Biochem. Biophys.* **136**, 383. **29.** Phelps & Hatefi (1981) *Biochemistry* **20**, 453. **30.** Zannoni (1970) *Meth. Enzymol.* **17A**, 665. **31.** Visser (1980) *Biochim. Biophys. Acta* **611**, 371. **32.** Ziegler (1974) *Biochim. Biophys. Acta* **364**, 28. **33.** Pfleiderer, Deckel & Wieland (1956) *Biochem. Z.* **328**, 187. **34.** Kun (1963) *The Enzymes*, 2nd ed. (Boyer, Lardy & Myrbäck, eds.), **7**, 149. **35.** Guibault, Brignac, Jr., & Zimmer (1968) *Anal. Chem.* **40**, 190. **36.** Seidel & Knowles (1994) *Biochemistry* **33**, 5641. **37.** Miller (1966) *Meth. Enzymol.* **9**, 336. **38.** Sörbo (1955) *Meth. Enzymol.* **2**, 334. **39.** Maurousset, Lemoine, Gallet, Delrot & Bonnemain (1992) *Biochim. Biophys. Acta* **1105**, 230. **40.** Kondo & Ishimoto (1975) *J. Biochem.* **78**, 317. **41.** Akagi, Drake, Kim & Gevertz (1994) *Meth. Enzymol.* **243**, 260. **42.** Uhteg & Westley (1979) *Arch. Biochem. Biophys.* **195**, 211. **43.** Anderson, Johnson, Endrizzi, Little & Korte (1987) *Biochemistry* **26**, 3938. **44.** Hug, O'Donnell & Hunter (1978) *Biochem. Biophys. Res. Commun.* **81**, 1435. **45.** Gerlinger & Retey (1987) *Z. Naturforsch. C* **42**, 349. **46.** O'Donnell, Hug & Hunter (1980) *Arch. Biochem. Biophys.* **202**, 242. **47.** Hacking, Bell & Hassall (1978) *Biochem. J.* **171**, 41. **48.** Hug, O'Donnell & Hunter (1979) *J. Biol. Chem.* **253**, 7622. **49.** O'Donnell & Hug (1989) *J. Photochem. Photobiol. B, Biol.* **3**, 429. **50.** Emiliani & Riera (1968) *Biochim. Biophys. Acta* **167**, 414. **51.** La Nauze, Rosenberg & Shaw (1970) *Biochim. Biophys. Acta* **212**, 332. **52.** Nimura, Tokiedo & Yamaha (1974) *J. Biochem.* **75**, 407. **53.** Taguchi, Konishi, Ishi & Yoshida (1991) *J. Biol. Chem.* **266**, 22411. **54.** Li & Schiff (1991) *Biochem. J.* **274**, 355. **55.** Magnuson, Stern, Gossett, Zinder & Burris (1998) *Appl. Environ. Microbiol.* **64**, 1270. **56.** Sugio, Oda,

Matsumoto, *et al.* (1998) *Biosci. Biotechnol. Biochem.* **62**, 705. **57.** Costas, White & Metcalf (2001) *J. Biol. Chem.* **276**, 17429. **58.** Relyea & van der Donk (2005) *Bioorg. Chem.* **33**, 171. **59.** Relyea, Vrtis, Woodyer, Rimkus & van der Donk (2005) *Biochemistry* **44**, 6640. **60.** Chen, Liu, Legall, *et al.* (1993) *Eur. J. Biochem.* **216**, 443. **61.** Huynh & Snell (1985) *J. Biol. Chem.* **260**, 2379. **62.** Kortt & Liu (1973) *Biochemistry* **12**, 320. **63.** Yoshino, Ogasawara, Suzuki & Kotake (1967) *Biochim. Biophys. Acta* **146**, 620. **64.** Weissmann, Chao & Chow (1980) *Biochem. Biophys. Res. Commun.* **97**, 827. **65.** Farooqui & Bachhawat (1972) *Biochem. J.* **126**, 1025. **66.** Scott & Spencer (1968) *Biochem. J.* **106**, 471. **67.** Glössl, Truppe & Kresse (1979) *Biochem. J.* **181**, 37. **68.** Iwamori, Moser & Kishimoto (1976) *Arch. Biochem. Biophys.* **174**, 199. **69.** Norkowska & Gniot-Szulzycka (2002) *J. Steroid Biochem. Mol. Biol.* **81**, 263. **70.** Ueki, Sawada, Fukagawa & Oki (1995) *Biosci. Biotechnol. Biochem.* **59**, 1062 and 1069. **71.** Knoess & Glombitza (1993) *Phytochemistry* **32**, 1119. **72.** Matusiewicz, Krzystek-Korpacka & Dabrowski (2005) *J. Exp. Mar. Biol. Ecol.* **317**, 175. **73.** Okamura, Yamada, Murooka & Harada (1976) *Agric. Biol. Chem.* **40**, 2071. **74.** Murooka, Yim & Harada (1980) *Appl. Environ. Microbiol.* **39**, 812. **75.** Amoscato, Brumfield, Sansoni, Herberman & Chambers (1991) *J. Immunol.* **147**, 950. **76.** Kato & Bishop (1972) *J. Biol. Chem.* **247**, 7420. **77.** Andrews & Pallavicini (1973) *Biochim. Biophys. Acta* **321**, 197. **78.** Chauncey & Westley (1983) *J. Biol. Chem.* **258**, 15037. **79.** Aird, Heinrikson & Westley (1987) *J. Biol. Chem.* **262**, 17327. **80.** Oi (1975) *J. Biochem.* **78**, 825. **81.** Westley (1973) *Adv. Enzymol. Relat. Areas Mol. Biol.* **39**, 327. **82.** Turkowsky, Blotevogel & Fischer (1991) *FEMS Microbiol. Lett.* **81**, 251. **83.** Vandenbergh & Berk (1980) *Can. J. Microbiol.* **26**, 281. **84.** Vazquez, Gazzaniga, Polo & Batlle (1997) *Cancer Biochem. Biophys.* **15**, 285. **85.** Renosto, Martin, Wailes, Daley & Segel (1990) *J. Biol. Chem.* **265**, 10300. **86.** Mahajan & Singh (1989) *Plant Physiol.* **91**, 421. **87.** van der Meijden, te Brömmelstroet, Poirot, van der Drift & Vogels (1984) *J. Bacteriol.* **160**, 629. **88.** Merkler, Kulathila, Francisco, Ash & Bell (1995) *FEBS Lett.* **366**, 165.

3-Sulfo-L-alanine, See *L-Cysteate*

S-Sulfocysteine

Both enantiomers of *S*-sulfocysteine (FW = 200.22 g/mol), also known as cysteine *S*-sulfate, rapidly inactivate rat kidney γ-glutamylcysteine synthetase, but are reversible inhibitors of sheep brain glutamine synthetase (1). The binding of the inhibitor to the enzyme induces formation of a very stable enzyme-inactivator complex, stabilized from interactions involving the sulfenyl sulfur atom of the *S*-sulfo amino acid and an active-site thiol group of the enzyme. **Target(s):** cysteine synthase (1-3); glutamine synthetase (4); γ-glutamylcysteine synthetase (4-7); and serine-sulfate ammonia-lyase, weakly inhibited (8). **1.** Nakamura, Iwahashi & Eguchi (1984) *J. Bacteriol.* **158**, 1122. **2.** Hensel & Trüper (1981) *Arch. Microbiol.* **130**, 228. **3.** Hensel & Trüper (1983) *Arch. Microbiol.* **134**, 227. **4.** Moore, Weiner & Meister (1987) *J. Biol. Chem.* **262**, 16771. **5.** Meister (1995) *Meth. Enzymol.* **252**, 26. **6.** Huang, Moore & Meister (1988) *Proc. Natl. Acad. Sci. U.S.A.* **85**, 2464. **7.** Griffith & Mulcahy (1999) *Adv. Enzymol. Relat. Areas Mol. Biol.* **73**, 209. **8.** Tudball & Thomas (1973) *Eur. J. Biochem.* **40**, 25.

6-Sulfo-6-deoxy-D-glucose, See *6-Sulfoquinovose*

N^d-(N'-Sulfodiaminophosphinyl)-L-ornithine

This potent toxin (FW = 290.24 g/mol), also known as octicidine, is produced from phaseolotoxin and is a strong inhibitor of ornithine carbamoyltransferase (1-3). **See also** *Phaseolotoxin* **1.** Langley, Templeton, Fields, Mitchell & Collyer (2000) *J. Biol. Chem.* **275**, 20012. **2.** Templeton, Mitchell, Sullivan & Shepherd (1985) *Biochem. J.* **228**, 347. **3.** Jahn, Sauerstein & Reuter (1987) *Arch. Microbiol.* **147**, 174.

Sulfo DSPD, *See Disulfodisalicylidene propanediamine*

1-*O*-(3-*O*-Sulfo-β-D-galactopyranosyl)ceramide, *See Galactosylceramide 3-O-Sulfate*

Sulfoglycolate

This bisulfite addition compound ($FW_{disodium-salt}$ = 200.08 g/mol), formed by the combination of of sodium glyoxylate and sodium bisulfite, inhibits glycolate oxidase. *See also Hydroxymethanesulfonate; Glycolate 2-Sulfate* **1.** Zelitch (1957) *J. Biol. Chem.* **224**, 251.

S-Sulfohomocysteine

This thiol-reactive amino acid analogue (FW = 215.25 g/mol; CAS 28715-19-7) rapidly inactivates rat kidney γ-glutamylcysteine synthetase, and both enantiomers reversibly inhibit sheep brain glutamine synthetase (1). The binding of the inhibitor to γ-glutamylcysteine synthetase induces formation of a very stable enzyme-inactivator complex, stabilized from interactions involving the sulfenyl sulfur atom of the *S*-sulfo amino acid and an active-site thiol group of the enzyme. **Target(s):** glutamine synthetase (1); and γ-glutamylcysteine synthetase (1-4). **1.** Moore, Weiner & Meister (1987) *J. Biol. Chem.* **262**, 16771. **2.** Meister (1995) *Meth. Enzymol.* **252**, 26. **3.** Huang, Moore & Meister (1988) *Proc. Natl. Acad. Sci. U.S.A.* **85**, 2464. **4.** Griffith & Mulcahy (1999) *Adv. Enzymol. Relat. Areas Mol. Biol.* **73**, 209.

Sulfometuron-methyl

This nonselective pre- and post-emergence herbicide (FW = 364.38 g/mol; pK_a = 5.2), also known as DPX-5648 and Oust, inhibits yeast acetolactate synthase, I_{50} = 120 nM, and tobacco acetolactate synthase, I_{50} = 8 nM (1-8). The herbicidal action appears to be due to an accumulation of 2-ketobutyrate. **1.** LaRossa & Van Dyk (1988) *Meth. Enzymol.* **166**, 97. **2.** Schloss & Van Dyk (1988) *Meth. Enzymol.* **166**, 445. **3.** Barak, Calvo & Schloss (1988) *Meth. Enzymol.* **166**, 455. **4.** Epelbaum, Chapman & Barak (2000) *Meth. Enzymol.* **324**, 10. **5.** Ray (1986) *Trends Biochem. Sci.* **11**, 180. **6.** Zohar, Einav, Chipman & Barak (2003) *Biochim. Biophys. Acta* **1649**, 97. **7.** McCourt, Pang, Guddat & Duggleby (2005) *Biochemistry* **44**, 2330. **8.** LaRossa & Schloss (1984) *J. Biol. Chem.* **259**, 8753.

***p*-Sulfonamidobenzoate,** *See Carzenide*

Sulfonated Hesperidin, *See Hesperidin*

12-Sulfonoxy-*trans*-9-octadecenoate, *similar in action to 10-Sulfonooxy-octadecanoate*

3-Sulfopropionate

This succinate analogue ($FW_{free-acid}$ = 154.14 g/mol), also known as 3-sulfopropanoate, inhibits aspartate carbamoyltransferase, weakly inhibited, K_i = ~ 100 mM (1), 3-oxoacid CoA-transferase (2); succinate dehydrogenase, succinate oxidase (3,4). **1.** Foote, Lauritzen & Lipscomb (1985) *J. Biol. Chem.* **260**, 9624. **2.** Fenselau & Wallis (1974) *Biochemistry* **13**, 3884. **3.** Klotz & Tietze (1947) *J. Biol. Chem.* **168**, 399. **4.** Webb

(1966) *Enzyme and Metabolic Inhibitors*, vol. **2**, p. 242, Academic Press, New York.

Sulfopyruvate

This metabolite ($FW_{free-acid}$ = 168.13 g/mol), an intermediate in the biosynthesis of plant sulfolipids and coenzyme M, is produced by the action of (*R*)-2-hydroxyacid dehydrogenase and is a substrate of sulfopyruvate decarboxylase. *Note*: Sulfopyruvate is an isoelectronic analogue of oxaloacetate and phosphoenolpyruvate. It is an alternative substrate of malate dehydrogenase. **Target(s):** phosphoenolpyruvate carboxykinase (GTP) (1,2); phosphoenolpyruvate mutase, K_i = 22 μM (3); phosphonopyruvate decarboxylase, sulfopyruvate is also a slow alternative substrate (4); and phosphonopyruvate hydrolase (5). **1.** Jabalquinto & Cardemil (1993) *Biochim. Biophys. Acta* **1161**, 85. **2.** Ash, Emig, Chowdhury, Satoh & Schramm (1990) *J. Biol. Chem.* **265**, 7377. **3.** Liu, Lu, Jia, Dunaway-Mariano & Herzberg (2002) *Biochemistry* **41**, 10270. **4.** Zhang, Dai, Lu & Dunaway-Mariano (2003) *J. Biol. Chem.* **278**, 41302. **5.** Chen, Han, Niu, *et al.* (2006) *Biochemistry* **45**, 11491.

Sulforhodamine 101 Acid Chloride, *See Texas Red*

***O*-Sulfo-D-serine,** *See D-Serine O-Sulfate*

***O*-Sulfo-L-serine,** *See L-Serine O-Sulfate*

***O*-Sulfo-L-tyrosine,** *See L-Tyrosine O-Sulfate*

Sulfoxine, *See 8-Hydroxyquinoline-5-sulfonate*

Sulfur Dioxide

This colorless gas SO_2 (FW = 64.065 g/mol; CAS 7446-09-5; M.P. = –72°C; B.P. = –10°C) reacts with water to form sulfurous acid H_2SO_3 and bisulfite HSO_3^-. SO_2 reversibly inhibits noncyclic photophosphorylation in isolated, envelope-free chloroplasts (1). Inhibition is competitive relative to phosphate, K_i = 0.8 mM. The same inhibition characteristics are observed when phosphoglycerate (PGA)- or ribulose-1,5-bisphosphate (RuBP)-dependent oxygen evolution is examined in a reconstituted chloroplast system in the presence of SO_3^{2-}. *See also Bisulfite* **Target(s):** ascorbate oxidase (2); ATPase (3); DNA polymerase I (4); and sucrose carrier (5). **1.** Cerović, Kalezić, Plesničar, *et al.* (1982) *Planta* **156**, 249. **2.** Stark & Dawson (1963) *The Enzymes*, 2nd ed. (Boyer, Lardy & Myrbäck, eds.), **8**, 297. **3.** Carrete, Vidal, Bordons & Constanti (2002) *FEMS Microbiol. Lett.* **211**, 155. **4.** Mallon & Rossman (1983) *Chem. Biol. Interact.* **46**, 101. **5.** Maurousset, Lemoine, Gallet, Delrot & Bonnemain (1992) *Biochim. Biophys. Acta* **1105**, 230.

Sulfur Mustard, *See Mustard Gas*

Sulindac

This anti-inflammatory agent (FW$_{free-acid}$ = 356.42 g/mol; CAS 38194-50-2; pK_a = 4.7 at 25°C), also known as Clinoril® and systematically as (1Z)-5-fluoro-2-methyl-1-[[4-(methylsulfinyl)phenyl]methylene]-1H-indene-3-acetic acid, is slightly soluble in water (3 mg/mL at pH 7). Sulindac is actually a prodrug that is converted in the liver to an active nonsteroidal anti-inflammatory drugs (NSAID) that is excreted in the bile and subsequently reabsorbed from the intestine. This route maintains a relatively constant blood level of the resulting COX-1 and/or COX-2 inhibitor. *See also Sulindac Sulfide; Sulindac Sulfone* **Target(s):** aldose reductase (1-5); arylamine N-acetyltransferase (6); collagenases (7); cyclic-nucleotide phosphodiesterase (8); cyclooxygenase (9-12); dihydrofolate reductase (13); elastase (14); estrone sulfotransferase (15); 15-hydroxyprostaglandin dehydrogenase, NAD$^+$-dependent (16); phenol sulfo-transferase, *or* aryl sulfotransferase (15); and phospho-ribosylaminoimidazolecarboxamide formyltransferase, *or* AICAR transformylase (13). **1**. Kwee & Nakada (1989) *NMR Biomed.* **2**, 44. **2**. Williams & Odom (1987) *Exp. Eye Res.* **44**, 717. **3**. Crabbe, Freeman, Halder & Bron (1985) *Ophthalmic Res.* **17**, 85. **4**. Chaudhry, Cabrera, Juliani & Varma (1983) *Biochem. Pharmacol.* **32**, 1995. **5**. Sharma & Cotlier (1982) *Exp. Eye Res.* **35**, 21. **6**. Lo, Hsieh & Chung (1998) *Microbios.* **93**, 159. **7**. Barracchini, Franceschini, Amicosante, *et al.* (1998) *J. Pharm. Pharmacol.* **50**, 1417. **8**. Silvola, Kangasaho, Tokola & Vapaatalo (1982) *Agents Actions* **12**, 516. **9**. Pouplana, Lozano, Perez & Ruiz (2002) *J. Comput. Aided Mol. Des.* **16**, 683. **10**. Cheng, Finckenberg, Louhelainen, *et al.* (2003) *Eur. J. Pharmacol.* **461**, 159. **11**. Fosslien (2000) *Crit. Rev. Clin. Lab. Sci.* **37**, 431. **12**. Dunn (1984) *Ann. Rev. Med.* **35**, 411. **13**. Baggott, Morgan, Ha, Vaughn & Hine (1992) *Biochem. J.* **282**, 197. **14**. Lentini, Ternai & Ghosh (1987) *Biochem. Int.* **15**, 1069. **15**. King, Ghosh & Wu (2006) *Curr. Drug Metab.* **7**, 745. **16**. Jarabak (1988) *Prostaglandins* **35**, 403.

Sulindac Sulfide

This reduced form of sulindac (FW$_{free-acid}$ = 340.42 g/mol; CAS 32004-67-4), known systematically as (Z)-5-fluoro-2-methyl-1-[p-(methylthio)-benzylidene]indene-3-acetic acid, inhibits cell growth and induces apoptosis, perhaps as a consequence of its ability to block Ras activation of Raf-1. Suldinac sulfide also inhibits cyclooxygenase-1, IC$_{50}$ = 0.5 µM, exhibiting a selectivity over cyclooxygenase-2, IC$_{50}$ = 14 µM. Sulindac sulfide also prevents cell transformation and slow cancer cell growth by triggering the endoplasmic reticulum stress response (ERSR), as revealed by upregulation of molecular chaperones such as GRP78 and C/EBP homologous protein, *or* CHOP (1). Because sulindac sulfide has good brain bioavailability, lower COX-2 inhibition, and no mitochondrial effects, it represents an appealing candidate for treating malignant gliomas through gliotoxicity arising from activation of the ERSR (1). *See also Sulindac; Sulindac Sulfone* **Target(s):** aldose reductase (2); cyclooxygenase, *or* prostaglandin-endoperoxide synthase (3-6); cyclic-nucleotide phosphodiesterase (7); elastase (8); epidermal growth factor-induced phosphorylation of extracellular-regulated kinase and Bad (9); 15-hydroxyprostaglandin dehydrogenase, NAD$^+$-dependent (10); IκB kinase (11); and phospholipase A$_2$, weakly inhibited (12); Ras signaling (13); and γ-secretase (14). **1**. **2**. Williams & Odom (1987) *Exp. Eye Res.* **44**, 717. **3**. Riendeau, Charleson, Cromlish, *et al.* (1997) *Can. J. Physiol. Pharmacol.* **75**, 1088. **4**. Smith, Meade & DeWitt (1994) *Ann. N. Y. Acad. Sci.* **744**, 50. **5**. Meade, Smith & DeWitt (1993) *J. Biol. Chem.* **268**, 6610. **6**. Egan, Gale, VandenHeuvel, Baptista & Kuehl (1980) *J. Biol. Chem.* **255**, 323. **7**. Silvola, Kangasaho, Tokola & Vapaatalo (1982) *Agents Actions* **12**, 516. **8**. Lentini, Ternai & Ghosh (1987) *Biochem. Int.* **15**, 1069. **9**. Rice, Washington, Schleman, *et al.* (2003) *Cancer Res.* **63**, 616. **10**. Jarabak (1988) *Prostaglandins* **35**, 403. **11**. Luo, Kamata & Karin (2005) *J. Clin. Invest.* **115**, 2625. **12**. Marshall, Bauer, Sung & Chang (1991) *J.*

Rheumatol. **18**, 59. **13**. Herrmann, Block, Geisen, *et al.* (1998) *Oncogene* **17**, 1769. **14**. Takahashi, Hayashi, Tominari, *et al.* (2003) *J. Biol. Chem.* **278**, 18664.

Sulmazole

This inotropic agent (FW = 287.34 g/mol; M.P. = 203-205°C), also known as 2-[2-methoxy-4-(methylsulfinyl)phenyl]-1H-imidazo[4,5-b]pyridine, is often used as a cardiotonic agent. **Target(s):** A$_1$ adenosine receptor (1); and cyclic-nucleotide phosphodiesterase (2-4). **1**. Parsons, Ramkumar & Stiles (1988) *Mol. Pharmacol.* **33**, 441. **2**. Schuhmacher, Greeff & Noack (1982) *Arzneimittel-forschung* **32**, 80. **3**. el Allaf, D'Orio & Carlier (1984) *Arch. Int. Physiol. Biochim.* **92**, S69. **4**. Scholz & Meyer (1986) *Circulation* **73**, III 99.

(S)-(–)-Sulpiride

This dopamine antagonist (FW = 341.43 g/mol; CAS 15676-16-1; Solubility: 100 mM in DMSO), also known as S)-(–)-5-aminosulfonyl-N-[(1-ethyl-2-pyrrolidinyl)methyl]-2-methoxybenzamide, selectively targets D$_2$, D$_3$, D$_4$, D$_1$ and D$_5$ receptors, with respective K_i values of ~15 nM, ~13 nM, 1 µM, ~45 µM, and ~ 77 µM. Dopamine receptors are the primary targets in the treatment of schizophrenia, Parkinson's disease, and Huntington's chorea. **Primary Mode of Inhibitory Action:** Although sulpiride has a high affinity for dopamine receptors involved in emesis and prolactin secretion, it lacks part of the behavioral and biochemical profiles of the classical dopamine receptor antagonist neuroleptics (1). In the cardiovascular system, sulpiride is a potent prejunctional dopamine receptor antagonist, but has variable effectiveness in post-junctional dopamine receptor models (1). Sulpiride, but not haloperidol, up-regulates γ-hydroxybutyrate receptors *in vivo* and in cultured cells (2). **Pharmacokinetics:** Oral bioavailability is low (25–35% with marked inter-individual differences), absorbed slowly from the GI tract and reaching peak plasma concentrations in 4–5 hours. Sulpiride's *in vivo* half-life is 6–8 hours, with 92% excreted unchanged in the urine. The serum half-life of the terminal slope following intravenous administration averaged 5.3 hours (range 3.7–7.1 hours) according to the two-compartment model. In two subjects the half-lives were 11.0 and 13.9 hours, when the three-compartment model was applied (3). **Target(s):** mesolimbic and mesocortical D$_2$ dopamine receptors (1-10); D$_3$ dopamine receptors (9,10); carbonic anhydrase, 1-20 nM, depending on enzyme isotype (11,12). **1**. O'Connor & Brow (1982) *Gen. Pharmacol.* **13**, 185. **2**. Ratomponirina, Gobaille, Hodé, *et al.* (1998) *Eur. J. Pharmacol.* **346**, 331. **3**. Wiesel, Alfredsson, Ehrnebo & Sedvall (1980) *Eur. J. Clin. Pharmacol.* **17**, 385. **4**. Berretta, Sachs & Graybiel (1999) *Eur. J. Neurosci.* **11**, 4309. **5**. Hauber & Lutz (1999) *Behav. Brain Res.* **106**, 143. **6**. Jenner, Clow, Reavill, Theodorou & Marsden (1980) *J. Pharm. Pharmacol.* **32**, 39. **7**. O'Connor & Brown (1982) *Gen. Pharmacol.* **13**, 185. **8**. Duarte, Biala, Le Bihan, Hamon & Thiebot (2003) *Psychopharmacology* **166**, 19. **9**. Gerlach (1991) *Schizophr. Bull.* **17**, 289. **10**. Spano, Stefanini, Trabucchi & Fresia (1979) in *Sulpiride and Other Benzamides* (Spano, Trabucchi, Corsini & Gessa, eds.), pp. 11-31, Italian Brain Research Foundation Press, Milan. **11**. Inui, Konishi, Fukuda & Kaneko (1979) *Arzneimittelforschung.* **29**, 668. **12**.

Abbatea, Coetzeeb, Casini, *et al.* (2004) *Bioorg. & Med. Chem. Lett.* **14**, 337.

Sumithion, *See Fenitrothion*

Sunitinib

This orally bioavailable multi-targeted receptor tyrosine kinase inhibitor, *or* RTKI (FW = 400.23 g/mol; CAS 341031-54-7), also known by its code name SU11248 and its systematic name *N*-(2-diethylaminoethyl)-5-[(*Z*)-(5-fluoro-2-oxo-1*H*-indol-3-ylidene)methyl]-2,4-dimethyl-1*H*-pyrrole-3-carboxamide, targets all receptors for platelet-derived growth factor (PDGF) and vascular endothelial growth factor (VEGF), both of which play essential roles in tumor angiogenesis and tumor cell proliferation (1). (**See also** *SU11652*) **Primary Mode of Inhibitory Action:** Simultaneous inhibition of these targets reduces tumor vascularization and promotes cancer cell death. SU11248 treatment of mice bearing subcutaneously implanted NCI-H526 tumors resulted in significant tumor growth inhibition, with the reduced levels of phospho-KIT levels correlating with efficacy (3). This RTK inhibitor does not induce regression of primary tumors but slows the progression of tumor growth, suggesting the greatest role for VEGF antagonists may be to prevent the formation of new blood vessels, during and after conventional therapy is given to existing neoplastic disease (4). Sunitinib and its chloro-substituted congener, SU11652, kill HeLa cervix carcinoma, U-2-OS osteosarcoma, Du145 prostate carcinoma and WEHI-S fibrosarcoma cells at low micromolar concentrations by accumulating rapidly in lysosomes and disturbing their pH regulation and ultrastructure, and eventually leading to the leakage of lysosomal proteases into the cytosol (5). Lysosomal destabilization was preceded by an early inhibition of acid sphingomyelinase (ASM), a lipase that promotes lysosomal membrane stability (5). **Drug Interactions:** Sunitinib interacts with CYP3A4 inducers, inhibitors and substrates as well as with P-glycoprotein and ABCG2 substrates (6). **Key Pharmacokinetic Parameters:** *See* Appendix II in Goodman & Gilman's THE PHARMACOLOGICAL BASIS OF THERAPEUTICS, 12th Edition (Brunton, Chabner & Knollmann, eds.) McGraw-Hill Medical, New York (2011). **1.** Mendel, Laird, Xin, *et al.* (2003) *Clin. Cancer Res.* **9**, 327. **2.** Sun, Liang, Shirazian, *et al.* (2003) *J. Med. Chem.* **46**, 1116. **3.** Abrams, Lee, Murray, Pryer & Cherrington (2003) *Mol. Cancer Ther.* **2**, 471. **4.** Osusky, Hallahan, Fu, *et al.* (2004) *Angiogenesis* **7**, 225. **5.** Ellegaard, Groth-Pedersen, Oorschot, *et al.* (2013) *Mol. Cancer Ther.* **12**, 2018. **6.** Bilbao-Meseguer, Jose, Lopez-Gimenez, *et al.* (2014) *J. Oncol. Pharm. Pract.* **21**, 52.

Superinone, *See Tyloxapol*

Superoxide Anion

This paramagnetic radical anion $O_2^{\cdot-}$, which is generated in a number of biochemical processes, is formally equivalent to the addition of an unpaired electron to dioxygen. Because of the spin restriction, univalent O_2 reduction to $O_2^{\cdot-}$ is facile, and $O_2^{\cdot-}$ production by spontaneous and enzyme-catalyzed reactions is well-documented. $O_2^{\cdot-}$ is the product of the reaction catalyzed by xanthine oxidase. Superoxide is a strong oxidizing agent and can have significant effects on a cell (1-3). Although $O_2^{\cdot-}$ can initiate and propagate free radical oxidations of leukoflavins, tetrahydropterins, catecholamines, and related compounds and can inactivate [4Fe-4S]-containing dehydratases, it does not significantly attack polyunsaturated lipids or DNA. In what Fridovich (4) referred to as the "*in vivo* Haber-Weiss reaction," $O_2^{\cdot-}$ increases "free" iron by oxidizing the [4Fe-4S] center of dehydrases such as dihydroxy acid dehydrase, 6-phosphogluconate dehydrase, fumarases A and B, and aconitase. The released iron is maintained in its reduced state by cellular reductants, and the Fe(II) reacts with H_2O_2, as in the Fenton

reaction, to yield Fe(III) + HO· or its formal equivalent, Fe(II)O. This was proposed to provide an explanation for the enhanced O_2-dependent mutagenesis exhibited by *sodA sodB E. coli*, and it has been experimentally verified. The importance of release of iron by $O_2^{\cdot-}$ from [4Fe-4S] clusters of dehydrases was underscored by the recent observation of the complementation of *sodA sodB E. coli* by insertion and overexpression of rubredoxin reductase. Several superoxide dismutases (EC 1.15.1.1) catalyze the dismutation of superoxide: $2\ O_2^{\cdot-} + 2\ H^+ \rightarrow O_2 + H_2O_2$. In aqueous solutions, $O_2^{\cdot-}$ is highly hydrated. The oxygen–oxygen bond length is slightly longer (0.133 nm) in superoxide than in dioxygen. Stable ionic superoxides are formed by certain alkali metals. Transient levels of superoxide in aqueous solution can be generated a number of ways. **Target(s):** guanylate cyclase (1); and 6-phosphogluconate dehydratase (2,3). **1.** Bruene, Schmidt & Ullrich (1990) *Eur. J. Biochem.* **192**, 683. **2.** Gardner & Fridovich (1991) *J. Biol. Chem.* **266**, 1478. **3.** Rodriguez, Wedd & Scopes (1996) *Biochem. Mol. Biol. Int.* **39**, 783. **4.** Fridovich (1997) *J. Biol. Chem.* **272**, 18515.

Suplatast

This orally active T-helper cell-2 (*or* Th2) cytokine inhibitor and anti-allergy drug (FW = 209.20 g/mol; CAS 94055-76-2 (for tosylate salt); Solubiity: 100 mM in H_2O; 100 mM in DMSO), also named [3-[[4-(3-ethoxy-2-hydroxypropoxy)phenyl]amino]-3-oxopropyl]dimethylsulfonium ion, attenuates IL-2, IL-5 and IL-13 production, but is without effect on IFN-γ production. Suplatast is an immunoregulator that suppresses IgE production, eosinophil infiltration and histamine release. It also exhibits anti-asthmatic, anti-inflammatory and anti-fibrotic activity *in vivo*. Suplatast tosylate inhibits histamine signaling by direct and indirect down-regulation of histamine H_1 receptor gene expression through suppression of histidine decarboxylase and IL-4 gene transcription (4). **1.** Kurokawa, Kawazu, Asano, *et al.* (2001) *Mediators Inflamm.* **10**, 333. **2.** Hsu, Yang, Goto, *et al.* (2007) *Transpl. Immunol.* **18**, 108. **3.** Furonaka, Hattori, Tanimoto, *et al.* (2009) *J. Pharmacol. Exp. Ther.* **328**, 55. **4.** Shahriar, Mizuguchi, Maeyama, *et al.* (2009) J Immunol. **183**, 2133.

Suprafenacine

This colchicine site-directed ligand (FW = 282.35 g/mol; IUPAC: 4,5,6,7-tetrahydro-1*H*-indazole-3-carboxylic acid [1*p*-tolyl-meth-(*E*)-ylidene]-hydrazide; Abbreviation: SRF), first identified by *in silico* screening of annotated chemical libraries, binds to tubulin and inhibits its polymerization (*i.e.*, microtubule self-assembly). Suprafenacine treatment of human cancer cell lines (*e.g.*, HeLa (IC_{50} = 0.20 μM), MDA-MB-231 (IC_{50} = 0.25 μM), A549 (IC_{50} = 0.23 μM), HCT15 (IC_{50} = 0.24 μM), Jurkat (IC_{50} = 0.21 μM), and SH-SY5Y(IC_{50} = 0.08 μM)) as well as CCL-116 (IC_{50} = 0.21 μM) and WI-38 (IC_{50} = 0.22 μM) human fibroblast cell lines leads to G_2/M cell cycle arrest and subsequent cell death by a mitochondria-mediated apoptotic pathway. Significantrly, SRF bypasses the p-glycoprotein multi-drug resistance system, explaining why it is more potent than taxol in inhibiting SH-SY5Y neuroblastoma and HCT-15 colorectal adenocarcinoma cells, as both these cancer cell types express elevated levels of P-gp protein. Despite the structural dissimilarity, SRF and colchicine share tubulin binding site, suggesting a similar role in disrupting microtubule self-assembly (*See* *Colchicine*). **1**. Choi, Chattopadhaya, Thanh, *et al.* (2014) *PLoS One* **9**, e110955.

Suprofen

This anti-inflammatory agent ($FW_{free-acid}$ = 260.31 g/mol; pK_a = 3.91; soluble in methanol and ethanol), also known as α-methyl-4-(2-thienylcarbonyl)benzeneacetic acid, is an orally effective non-narcotic analgesic having a potent inhibitory action on prostaglandin biosynthesis. **Target(s):** anandamide deamidase (1); carbonyl reductase (2); cyclooxygenase, *or* prostaglandin-endoperoxide synthase (3-6); and 3α-hydroxysteroid dehydrogenase (7). **1**. Fowler, Tiger & Stenstrom (1997) *J. Pharmacol. Exp. Ther.* **283**, 729. **2**. Imamura, Koga, Higuchi, *et al.* (1997) *J. Enzyme Inhib.* **11**, 285. **3**. Nguyen & Lee (1993) *Exp. Eye Res.* **57**, 97. **4**. Dubinsky & Schupsky (1984) *Prostaglandins* **28**, 241. **5**. Hahn, Carraher & McGuire (1982) *Prostaglandins* **23**, 1. **6**. Barnett, Chow, Ives, *et al.* (1994) *Biochim. Biophys. Acta* **1209**, 130. **7**. Yamamoto, Matsuura, Shintani, *et al.* (1998) *Biol. Pharm. Bull.* **21**, 1148.

Suramin

This hygroscopic hexasulfonated naphthylurea ($FW_{hexasodium-salt}$ = 1429.19 g/mol; CAS 129-46-4), also known as Bayer 205, is a strong inhibitor of a wide variety of enzymes. Suramin is also known to inhibit the binding of growth factors (*e.g.*, epidermal growth factor, platelet-derived growth factor, and tumor growth factor-β) to their receptors and thus antagonize the ability of these factors to stimulate growth of tumor cells *in vitro*. Suramin readily binds to plasma proteins, suggesting that it may enter *T. brucei* by some form of receptor-mediated endocytosis, possibly bound to LDL. That said, suramin's actual mode of inhibitory action remains uncertain, as do its therapeutic targets. Suramin is also a non-selective P_2 purinergic receptor antagonist. It likewise blocks calmodulin binding to recognition sites and G-protein coupling to GCPRs. Increases open probability of ryanodine receptor (RyR) channels. **Clinical Applications:** Suramin is a trypanocide used to combat the early stages of African trypanosomiasis. It is also an effective anthelmintic used to treat onchoceriasis. Suramin is ineffective against early stage of *Trypanosoma brucei* gambiense and late-stages of both Human African Trypanosomiases, *or* HATs. As a result of campaigns of coordinated surveillance and treatment programs using suramin and other drugs (*See also Pentamidine, Melarsoprol, and Eflornithine*), the number of people succumbing to HAT has fallen considerably, now to around 70,000 annually. (*See also NF062, NF127, etc.*) **Target(s):** acid amyloglucosidase (1); acrosin (92); O^6-alkylguanine-DNA alkyltransferase (2); apyrase (3,85-89); arylsulfatase A, weakly inhibited (4); ascorbate oxidase (5); ATPase, Mg^{2+} (6,7); bis(5'-adenosyl)-triphosphatase (82); bis(5'-nucleosyl)-tetraphosphatase, asymmetrical (82); cathepsin G (8); cellulose 1,4-β-cellobiosidase, weakly inhibited (93); choline acetyltransferase (9); choline dehydrogenase (5); chymase10; chymosin, *or* renin (5); citrate synthase (11); complement factor I (91); cyclin-dependent kinases (12); cytochrome *c* oxidase (5); diadenosine polyphosphate hydrolase (13,14); DNA-dependent protein kinase (15); DNA polymerase (16-20); DNA primase (17,20); DNA topoisomerase II (21-23,75); duck hepatitis B virus reverse

transcriptase (18); ecto-ATPase (24-31); ectonucleotide diphosphohydrolase, *or* nucleotide diphosphatase (32,83); folylpolyglutamate synthetase (33); β-fructofuranosidase, *or* invertase (34,35); fructose-1,6-bisphosphate aldolase (77-79); fumarase (36); β-galactocerebrosidase, weakly inhibited (4); α-galactosidase, weakly inhibited (1,4); glucocerebrosidase, weakly inhibited (4); glucose-6-phosphate dehydrogenase (37); glucose-6-phosphate isomerase (38); glyceraldehyde-3-phosphate dehydrogenase (39); glycerol-3-phosphate dehydrogenase (40); GM_3-sialidase (4,41); heparanase (42,43); heparin-sulfate lyase (76); hexokinase (5,44); β-hexosaminidase (4,45); HIV reverse transcriptase (46-49); hyaluronidase (50); iduronate-2-sulfatase (94); lactate dehydrogenase (51,52); lysozyme (53); malate dehydrogenase (51); malic enzyme (54); methenyltetrahydrofolate cyclohydrolase (90); monoamine oxidase (55); NADPH oxidase (56); Na^+/K^+-exchanging ATPase (80); neutrophil elastase (8); nucleotide diphosphatase (32,83,84); papain (5); pepsin (5); 6-phosphogluconate dehydrogenase (57); 3-phosphoglycerate kinase (58,97-99); phospholipase D (59); phospholipid-translocating ATPase, *or* flippase (81); prostate-specific membrane antigen, *or* glutamate exocarboxypeptidase (60); proteinase (38); protein kinase I (61); protein kinase C (62,63); protein-tyrosine-phosphatase (64,65,95); ribonuclease H (48); RNA-directed DNA polymerase, *or* reverse transcriptase (17,18,46-49,66,96); RNA polymerase (17); simian immunodeficiency virus reverse transcriptase (96); sphingomyelinase (4); steroid 5α-reductase (67); succinate dehydrogenase (5,68); telomerase (69); thymidine kinase (100); thymidylate synthase (70); triose-phosphate isomerase (71); trypsin (5,72); vacuolar H^+-ATPase (73); and urease (5,74). **1**. Panagiotidis, Salehi & Lundquist (1991) *Pharmacology* **43**, 163. **2**. Link & Tempel (1991) *J. Cancer Res. Clin. Oncol.* **117**, 549. **3**. Bonan, Roesler, Quevedo, *et al.* (1999) *Pharmacol. Biochem. Behav.* **63**, 153. **4**. Constantopoulos, Rees, Cragg, Barranger & Brady (1981) *Res. Commun. Chem. Pathol. Pharmacol.* **32**, 87. **5**. Wills & Wormall (1950) *Biochem. J.* **47**, 158. **6**. Smolen & Weissmann (1978) *Biochim. Biophys. Acta* **512**, 525. **7**. Higa & Cazzulo (1981) *Mol. Biochem. Parasitol.* **3**, 357. **8**. Cadene, Duranton, North, *et al.* (1997) *J. Biol. Chem.* **272**, 9950. **9**. Swamy (1989) *Mol. Biochem. Parasitol.* **35**, 259. **10**. Takao, Takai, Ishihara, Mita & Miyazaki (1999) *Jpn. J. Pharmacol.* **81**, 404. **11**. Salvarrey & Cazzulo (1982) *Comp. Biochem. Physiol. B* **72**, 165. **12**. Meijer & Kim (1997) *Meth. Enzymol.* **283**, 113. **13**. Asensio, Rodriguez-Ferrer, Oaknin & Rotllan (2000) *Acta Biochim. Pol.* **47**, 435. **14**. Mateo, Rotllan & Miras-Portugal (1996) *Brit. J. Pharmacol.* **119**, 1. **15**. Hosoi, Matsumoto, Tomita, *et al.* (2002) *Brit. J. Cancer* **86**, 1143. **16**. Jindal, Anderson, Davis & Vishwanatha (1990) *Cancer Res.* **50**, 7754. **17**. Ono, Nakane & Fukushima (1988) *Eur. J. Biochem.* **172**, 349. **18**. Offensperger, Walter, Offensperger, *et al.* (1988) *Virology* **164**, 48. **19**. Basu & Modak (1985) *Biochem. Biophys. Res. Commun.* **128**, 1395. **20**. Ono, Nakane & Fukushima (1985) *Nucleic Acids Symp. Ser.* **1985** (16), 249. **21**. Negri, Bernardi, Donzelli & Scovassi (1995) *Biochimie* **77**, 893. **22**. Funayama, Nishio, Takeda, *et al.* (1993) *Anticancer Res.* **13**, 1981. **23**. Bojanowski, Lelievre, Markovits, *et al.* (1992) *Proc. Natl. Acad. Sci. U.S.A.* **89**, 3025. **24**. Jesus, Lopes & Meyer-Fernandes (2002) *Vet. Parasitol.* **103**, 29. **25**. Caldwell, Hornyak, Pendleton, Campbell & Knowles (2001) *Arch. Biochem. Biophys.* **387**, 107. **26**. Meyer-Fernandes, Lanz-Mendoza, Gondim, Willott & Wells (2000) *Arch. Biochem. Biophys.* **382**, 152. **27**. Barros, De Menezes, Pinheiro, *et al.* (2000) *Arch. Biochem. Biophys.* **375**, 304. **28**. Dowd, Li & Zeng (1999) *Arch. Oral Biol.* **44**, 1055. **29**. Schwarzbaum, Frischmann, Krumschnabel, Rossi & Wieser (1998) *Amer. J. Physiol.* **274**, R1031. **30**. Chen, Lee & Lin (1996) *Brit. J. Pharmacol.* **119**, 1628. **31**. Ziganshin, Ziganshina, King & Burnstock (1995) *Pflugers Arch.* **429**, 412. **32**. dos Passos Lemos, de Sa Pinheiro, de Berredo-Pinho, *et al.* (2002) *Parasitol. Res.* **88**, 905. **33**. McGuire & Haile (1996) *Arch. Biochem. Biophys.* **335**, 139. **34**. Myrbäck (1960) *The Enzymes*, 2nd ed. (Boyer, Lardy & Myrbäck, eds.), **4**, 379. **35**. Quastel & Yates (1936) *Enzymologia* **1**, 60. **36**. Quastel (1931) *Biochem. J.* **25**, 1121. **37**. Titanji, Muluh & Tchoupe (1988) *Parasitol. Res.* **74**, 380. **38**. Marchand, Kooystra, Wierenga, *et al.* (1989) *Eur. J. Biochem.* **184**, 455. **39**. Lambeir, Loiseau, Kuntz, *et al.* (1991) *Eur. J. Biochem.* **198**, 429. **40**. Marche, Michels & Opperdoes (2000) *Mol. Biochem. Parasitol.* **106**, 83. **41**. Schneider-Jakob & Cantz (1991) *Biol. Chem. Hoppe Seyler* **372**, 443. **42**. Marchetti, Reiland, Erwin & Roy (2003) *Int. J. Cancer* **104**, 167. **43**. Nakajima, DeChavigny, Johnson, *et al.* (1991) *J. Biol. Chem.* **266**, 9661. **44**. McDonald (1955) *Meth. Enzymol.* **1**, 269. **45**. Constantopoulos, Rees, Cragg, Barranger & Brady (1981) *Biochem. Biophys. Res. Commun.* **101**, 1345. **46**. De Clercq (1987) *Antiviral Res.* **7**, 1. **47**. Jeffries (1989) *J. Infect.* **18** Suppl. 1, 5. **48**. Andréola, Tharaud, Litvak & Tarrago-Litvak (1993) *Biochimie* **75**, 127. **49**. Jentsch, Hunsmann, Hartmann & Nickel (1987) *J.*

Gen. Virol. **68**, 2183. **50.** Beiler & Martin (1948) *J. Biol. Chem.* **174**, 31. **51.** Walter & Schulz-Key (1980) *Tropenmed. Parasitol.* **31**, 55. **52.** Walter (1979) *Tropenmed. Parasitol.* **30**, 463. **53.** Imoto, Johnson, North, Phillips & Rupley (1972) *The Enzymes*, 3rd ed. (Boyer, ed.), **7**, 665. **54.** Walter & Albiez (1981) *Mol. Biochem. Parasitol.* **4**, 53. **55.** Agarwal, Shukla, Ghatak & Tekwani (1990) *Int. J. Parasitol.* **20**, 873. **56.** Heyneman (1987) *Vet. Res. Commun.* **11**, 149. **57.** Hanau, Rippa, Bertelli, Dallocchio & Barrett (1996) *Eur. J. Biochem.* **240**, 592. **58.** Misset & Opperdoes (1987) *Eur. J. Biochem.* **162**, 493. **59.** Gratas & Powis (1993) *Anticancer Res.* **13**, 1239. **60.** Slusher, Tiffany, Merion, Lapidus & Jackson (2000) *Prostate* **44**, 55. **61.** Walter (1980) *Mol. Biochem. Parasitol.* **1**, 139. **62.** Mahoney, Azzi & Huang (1990) *J. Biol. Chem.* **265**, 5424. **63.** Lopez-Lopez, Langeveld, Pizao, *et al.* (1994) *Anticancer Drug Des.* **9**, 279. **64.** Ghosh & Miller (1993) *Biochem. Biophys. Res. Commun.* **194**, 36. **65.** Zhang, Keng, Zhao, Wu & Zhang (1998) *J. Biol. Chem.* **273**, 12281. **66.** De Clercq (1979) *Cancer Lett.* **8**, 9. **67.** Taylor, Bhattacharyya & Collins (1995) *Steroids* **60**, 452. **68.** Singer & Kearney (1963) *The Enzymes*, 2nd ed. (Boyer, Lardy & Myrbäck, eds.), **7**, 383. **69.** Lonnroth, Andersson & Lundholm (2001) *Int. J. Oncol.* **18**, 929. **70.** Chalabi & Gutteridge (1977) *Biochim. Biophys. Acta* **481**, 71. **71.** Lambeir, Opperdoes & Wierenga (1987) *Eur. J. Biochem.* **168**, 69. **72.** Coltrini, Rusnati, Zoppetti, *et al.* (1993) *Eur. J. Biochem.* **214**, 51. **73.** Moriyama & Nelson (1988) *FEBS Lett.* **234**, 383. **74.** Varner (1960) *The Enzymes*, 2nd ed. (Boyer, Lardy & Myrbäck, eds.), **4**, 247. **75.** Andersen, Bendixen & Westergaard (1996) *DNA Replication in Eucaryotic Cells, Cold Spring Harbor Laboratory Press*, p. 587. **76.** Graham, Mitchell & Underwood (1995) *Biochem. Mol. Biol. Int.* **37**, 239. **77.** de Walque, Opperdoes & Michels (1999) *Mol. Biochem. Parasitol.* **103**, 279. **78.** Willson, Callens, Kuntz, Perié & Opperdoes (1993) *Mol. Biochem. Parasitol.* **59**, 201. **79.** Döbeli, Trzeciak, Gillessen, *et al.* (1990) *Mol. Biochem. Parasitol.* **41**, 259. **80.** Fortes, Ellory & Lew (1973) *Biochim. Biophys. Acta* **318**, 262. **81.** Beleznay, Zachowski, Devaux & Ott (1997) *Eur. J. Biochem.* **243**, 58. **82.** Rotllan, Rodriguez-Ferrer, Asensio & Oaknin (1998) *FEBS Lett.* **429**, 143. **83.** Grobben, Claes, Roymans, *et al.* (2000) *Brit. J. Pharmacol.* **130**, 139. **84.** Vollmayer, Clair, Goding, *et al.* (2003) *Eur. J. Biochem.* **270**, 2971. **85.** Kukulski & Komoszynski (2003) *Eur. J. Biochem.* **270**, 3447. **86.** Alves-Ferreira, Dutra, Lopes, *et al.* (2003) *Curr. Microbiol.* **47**, 265. **87.** Ziganshin, Zigashina, Bodin, Bailey & Burnstock (1995) *Biochem. Mol. Biol. Int.* **36**, 863. **88.** Oses, Cardoso, Germano, *et al.* (2004) *Life Sci.* **74**, 3275. **89.** Demenis, Furriel & Leone (2003) *Biochim. Biophys. Acta* **1646**, 216. **90.** Jaffe, Chrin & Smith (1980) *Comp. Biochem. Physiol. B* **66**, 597. **91.** Tsiftsoglou & Sim (2004) *J. Immunol.* **173**, 367. **92.** Hermans, Haines, James & Jones (2003) *FEBS Lett.* **544**, 119. **93.** Singh, Agrawal, Abidi & Darmwal (1990) *J. Gen. Appl. Microbiol.* **36**, 245. **94.** Constantopoulos, Rees, Cragg, Barranger & Brady (1980) *Proc. Natl. Acad. Sci. U.S.A.* **77**, 3700. **95.** McCain, Wu, Nickel, *et al.* (2004) *J. Biol. Chem.* **279**, 14713. **96.** Lüke, Hoefer, Moosmayer, *et al.* (1990) *Biochemistry* **29**, 1764. **97.** Boyle, Fairbrother & Williams (1989) *Eur. J. Biochem.* **184**, 535. **98.** Zomer, Allert, Chevalier, *et al.* (1998) *Biochim. Biophys. Acta* **1386**, 179. **99.** Pal, Pybus, Muccio & Chattopadhyay (2004) *Biochim. Biophys. Acta* **1699**, 277. **100.** Chello & Jaffe (1972) *Comp. Biochem. Physiol. B* **43**, 543.

Surfactin

This lipodepsipeptide antibiotic (FW$_{\text{free-acid}}$ = 1036.34 g/mol; CAS 24730-31-2) from *Bacillus subtilis* is a potent surfactant and can cause lysis of erythrocytes and bacteria (1). Surfactin is produced and contains β-hydroxy-fatty acid of 13-to-15 carbon atoms in length, depending upon the composition of the fermentation medium. **Target(s):** phospholipase A_2, platelet (2). **1.** Arima, Kakinuma & Tamura (1968) *Biochem. Biophys. Res. Commun.* **31**, 488. **2.** Kim, Jung, Lee, *et al.* (1998) *Biochem. Pharmacol.* **55**, 975.

Survivin

This caspase-inhibiting homodimer and anti-apoptotic factor (MW = 34 kDa) is a structurally unique member of the Inhibitors-of-Apoptosis Protein (IAP) family and plays a role in controlling cell division and inhibiting apoptosis. The respective K_i values for surviving survivin inhibition of caspase-3 and caspase-7 are 36 and 16 nM (1,2). The shared structural feature of all IAP family members is a 70-amino acid motif, termed BIR,

that is present in one to three copies. Survivin expression is a predictor of both poor prognosis and decreased survival time in cancer therapy. In view of its likely role in conferring chemotherapeutic and radiotherapeutic resistance, survivin expression should prove to be a druggable target. **1.** Deveraux, Welsh & Reed (2000) *Meth. Enzymol.* **322**, 154. **2.** Shin, Sung, Cho, *et al.* (2001) *Biochemistry* **40**, 1117.

Suvorexan

This orally bioavailable, dual orexin (hypocretin) receptor antagonist (FW = 450.92 g/mol; CAS 1030377-33-3; Solubility: 10 mg/mL DMSO; <1 mg/mL H_2O), also known by the proprietary name Belsomra®, its codename MK-4305 and IUPAC name 5-chloro-2-[(5R)-5-methyl-4-[5-methyl-2-(2H-1,2,3-triazol-2-yl)benzoyl]-1,4-diazepan-1-yl]-1,3-benzoxazole, targets OX_1 and OX_2 receptors, with K_i values of 0.55 nM and 0.35 nM, respectively (1,2). **Pharmacokinetics:** Bioavailability is 82%, with circulating suvorexant highly protein-bound. Food delays the time to maximum plasma concentration. Suvorexant is metabolized by the liver It is excreted renally (23% unchanged) and through feces (66% unchanged), with an elimination half-life of 12 hours. EMPA, suvorexant, almorexant and TCS-OX-29 all bind to the OX_2 receptor with moderate to high affinity, with pK_i values ≥ 7.5, whereas the primarily OX1 selective antagonists SB-334867 and SB-408124 displayed low affinity, with pK_i values ~6 (2). Competition kinetic analysis showed that the compounds displayed a range of dissociation rates from very fast (TCS-OX2-29, k_{off} = 0.22 min^{-1}) to very slow (almorexant, k_{off} = 0.005 min^{-1}). There is a clear correlation between association rate and affinity (2). Given CYP3A's role in suvorexant metabolism, dosage should be reduced when other drugs (*e.g.*, verapamil, erythromycin, diltiazem, or dronedarone) are in use. **Structural Basis of Suvorexant-Receptor Interactions:** A 2.5-Å resolution structure reveals that suvorexant adopts a π-stacked horseshoe-like conformation when bound deep in the orthosteric pocket of the receptor (3). Those interactions are stabilized by a network of extracellular salt bridges and blocking transmembrane helix motions necessary for activation. Computational docking suggests that other classes of synthetic antagonists may interact with the receptor at a similar position in an analogous π-stacked fashion (3). The hOX_1R structures also reveal a conserved amphipathic α-helix, in the extracellular N-terminal region, that interacts with orexin-A and is essential for high-potency neuropeptide activation at both receptors (4). Human receptors OX_1R and OX_2R are 64% identical in sequence and have overlapping, but distinct, physiological functions and therapeutic profiles (4). (*Other Orexin Receptor Antagonists include: Almorexant, Filorexant, Lemborexant, and SB-649868*) **1.** Cox, Breslin, Whitman, *et al.* (2010) *J. Med. Chem.* **53**, 5320. **2.** Mould, Brown, Marshall & Langmead (2014) *Brit. J. Pharmacol.* **171**, 351. **3.** Yin, Mobarec, Kolb & Rosenbaum (2015) *Nature* **519**, 247. **4.** Yin, Babaoglu, Brautigam, *et al.* (2016) *Nature Struct. Mol. Biol.* **23**, 293.

Suxamethonium Ion, *See* Succinylcholine

SW 033291

This high-affinity 15-PGDH inhibitor (FW = 412.58 g/mol; CAS 459147-39-8; Soluble to 100 mM in DMSO), also named 2-(butylsulfinyl)-4-phenyl-6-(2-thienyl)thieno[2,3-b]pyridin-3-amine, targets 15-hydroxy-prostaglandin dehydrogenase (K_i = 0.1 nM), increasing PGE_2 levels in A549 cells *in vitro* and in bone marrow, colon, lung and liver in mice. Prostaglandin PGE2, a lipid signaling molecule that supports expansion of several types of tissue stem cells, is a candidate therapeutic target for promoting tissue regeneration *in vivo*. SW-033291 also promotes tissue regeneration in mouse models of colon and liver injury. Moreover, tissues from 15-PGDH-null mice demonstrate similar increased regenerative capacity. **1**. Zhang, Desai, Yang, *et al.* (2015) *Science* **348**, 2340.

Swainsonine

This indolizidine alkaloid ($FW_{free-base}$ = 173.21 g/mol; CAS 72741-87-8), isolated from *Swainsona canescens* and spotted locoweed (*Astragalus lentiginosus*), inhibits α-mannosidases, leading to the accumulation of mannose-rich oligosaccharides in the lysosomal system. Its epimers are milder inhibitors. Swainsonine prevents complex glycoprotein formation in cultured cells, apparently by inhibiting the mannosyl trimming process. It blocks the synthesis of complex *N*-linked glycoproteins by inhibiting Golgi mannosidase, an enzyme that normally acts on glycoproteins containing $GlcNAcMan_5GlcNAc_2$ groups. Ingestion of *Swainsona* sp. induces mannosidosis, a condition resembling hereditary lysosomal storage disease. Swainsonine and swainsonine *N*-oxide are likewise believed to be causative agents of locoism, a chronic neurological disorder. **Target(s):** β-D-arabinosidase, moderately inhibited (1); arylmannosidase (2,3); β-galactosidase, moderately inhibited (1); α-glucosidase, moderately inhibited (1); β-glucuronidase, moderately inhibited (1); glycoprotein processing (4-10); α-mannosidase (1,2,4-6,11-18,36-45); α-1,2-mannosidase, *Trypanosoma cruzi* and *Vigna umbellate* (34,35); mannosyl-oligosaccharide 1,3-1,6-α-mannosidase, mannosidase II (1,2,4-6,19-33,35); and β-D-xylosidase, moderately inhibited (1). **1**. Siriwardena, Strachan, El-Daher, *et al.* (2005) *Chem. Bio. Chem.* **6**, 845. **2**. Kaushal & Elbein (1994) *Meth. Enzymol.* **230**, 316. **3**. Pastuszak, Kaushal, Wall, *et al.* (1990) *Glycobiology* **1**, 71. **4**. Elbein (1983) *Meth. Enzymol.* **98**, 135. **5**. Elbein (1987) *Meth. Enzymol.* **138**, 661. **6**. Schwarz & Datema (1984) *Trends Biochem. Sci.* **9**, 32. **7**. Elbein, Solf, Dorling & Vosbeck (1981) *Proc. Natl. Acad. Sci. U.S.A.* **78**, 7393. **8**. Tulsiani, Harris & Touster (1982) *J. Biol. Chem.* **257**, 7936. **9**. Moremen (2002) *Biochim. Biophys. Acta* **1573**, 225. **10**. Elbein (1991) *FASEB J.* **5**, 3055. **11**. Kishimoto, Hori, Takano, *et al.* (2001) *Physiol. Plant* **112**, 15. **12**. Nok, Shuaibu, Kanbara & Yanagi (2000) *Parasitol. Res.* **86**, 923. **13**. Bonay & Fresno (1999) *Glycobiology* **9**, 423. **14**. Wongvithoonyaporn, Bucke & Svasti (1998) *Biosci. Biotechnol. Biochem.* **62**, 613. **15**. Bonay & Hughes (1991) *Eur. J. Biochem.* **197**, 229. **16**. Cenci di Bello, Fleet, Namgoong, Tadano & Winchester (1989) *Biochem. J.* **259**, 855. **17**. Dorling, Huxtable & Colegate (1980) *Biochem. J.* **191**, 649. **18**. Tulsiani, Broquist & Touster (1985) *Arch. Biochem. Biophys.* **236**, 427. **19**. Kaushal & Elbein (1989) *Meth. Enzymol.* **179**, 452. **20**. Presper & Heath (1983) *The Enzymes*, 3rd ed., **16**, 449. **21**. van den Elsen, Kuntz & Rose (2001) *EMBO J.* **20**, 3008. **22**. Rabouille, Kuntz, Lockyer, *et al.* (1999) *J. Cell Sci.* **112**, 3319. **23**. Kaushal, Szumilo, Pastuszak & Elbein (1990) *Biochemistry* **29**, 2168. **24**. Shah, Kuntz & Rose (2003) *Biochemistry* **42**, 13812. **25**. Ren, Castellino & Bretthauer (1997) *Biochem. J.* **324**, 951. **26**. Numao, Kuntz, Withers & Rose (2003) *J. Biol. Chem.* **278**, 48074. **27**. Strasser, Schoberer, Jin, *et al.* (2006) *Plant J.* **45**, 789. **28**. Altmann & März (1995) *Glycoconjugate J.* **12**, 150. **29**. Kawar, Karaveg, Moremen & Jarvis (2001) *J. Biol. Chem.* **276**, 16335. **30**. Tropea, Kaushal, Pastuszak, *et al.* (1990) *Biochemistry* **29**, 10062. **31**. Moremen & Robbins (1991) *J. Cell Biol.* **115**, 1521. **32**. Oh-Eda, Nakagawa, Akama, *et al.* (2001) *Eur. J. Biochem.* **268**, 1280. **33**. Moremen, Touster & Robbins (1991) *J. Biol. Chem.* **266**, 16876. **34**. Bonay & Fresno (1999) *Glycobiology* **9**, 423. **35**. Wongvithoonyaporn, Bucke & Svasti (1998) *Biosci. Biotechnol. Biochem.* **62**, 613. **36**. Ikeda, Kato, Adachi, Haraguchi & Asano (2003) *J. Agric. Food Chem.* **51**, 7642. **37**. Fiaux, Popowycz, Favre, *et al.* (2005) *J. Med. Chem.* **48**, 4237. **38**. Vázquez-Reyna, Balcázar-Orozco & Flores-Carreón (1993) *FEMS Microbiol. Lett.* **106**, 321. **39**. Vázquez-Reyna, Ponce-Noyola, Calvo-Méndez, López-Romero & Flores-Carreón (1999) *Glycobiology* **9**, 533. **40**. Yamashiro, Itoh, Yamagishi, *et al.* (1997) *J. Biochem.* **122**, 1174. **41**. Waln & Poulton (1987) *Plant Sci.* **53**, 1. **42**. Conti, Carratu & Giannattasio (1987) *Phytochemistry* **26**, 2909. **43**. Kishimoto, Hori, Takano, *et al.* (2001) *Physiol. Plant.* **112**, 15. **44**. Nakajima, Imamura, Shoun & Wakagi (2003) *Arch. Biochem. Biophys.* **415**, 87. **45**. Liao, Lal & Moremen (1996) *J. Biol. Chem.* **271**, 28348.

Swertifrancheside

This flavonoxanthone *C*-glucoside (FW = 720.60 g/mol; CAS 155740-03-7), known systematically as 1,5,8-trihydroxy-3-methoxy-7-(5',7',3'',4''-tetrahydroxy-6'-*C*-β-D-glucopyranosyl-4'-oxy-8'-flavyl)xanthone, from *Swertia franchetiana* inhibits DNA ligase (ATP), human, IC_{50} = 11 μM (1), HIV-1 reverse transcriptase, IC_{50} = 43 μM (2,3), and RNA-directed DNA polymerase (2,3). **1**. Tan, Lee, Lee, *et al.* (1996) *Biochem. J.* **314**, 993. **2**. Pengsuparp, Cai, Constant, *et al.* (1995) *J. Nat. Prod.* **58**, 1024. **3**. Wang, Hou, Liu, *et al.* (1994) *J. Nat. Prod.* **57**, 211.

Swinholide A

This marine toxin (FW = 1387.92 g/mol; CAS 95927-67-6) from the sponge *Theonella swinhoei* is a 44-carbon, dimeric dilactone macrolide with a C_2 axis of symmetry. At concentrations as low as 80 nm, swinholide A causes rounding of cultured mouse embryo fibroblast cells within one hour, attended by massive loss of the actin cytoskeletal network, as evidenced by treatment with fluorescently labeled phalloidin, the latter binding only to filamtous actin (1). At 10–50 nm swinholide, partial cell retraction and diminution of filament bundles (stress fibers) commences after 2–4 hours, with complete loss by 5–7 hours. Swinholide action is the consequence of efficient actin filament severing (1). At much higher concentrations, swinholide A binds very tightly to actin dimers with unit stoichiometry, a behavior that may underlie its filament-severing action. **1**. Bubb & Spector (1998) *Meth. Enzymol.* **298**, 26. **2**. Bubb, Spector, Bershadsky & Korn (1995) *J. Biol. Chem.* **270**, 3463. **3**. Saito, Watabe, Ozaki, *et al.* (1998) *J. Biochem.* **123**, 571.

SX 001

This protein kinase inhibitor (FW = 483.96 g/mol; CAS 309913-42-6; Soluble to 100 mM in DMSO), also named 6-chloro-5-[[4-[(4-fluorophenyl)methyl]-1-piperidinyl]carbonyl-*N*,*N*,1-trimethyl-α-oxo-1*H*-indole-3-acetamide, selectively targets p38α (IC$_{50}$ = 9 nM), with weaker inhibition of p38β (IC$_{50}$ = 90 nM) and JNK-2 (IC$_{50}$ = 100 nM) and insignificant inhibition of p38γ, p38δ, ERK-2, and JNK-1. **1**. Lee & Dominguez (2005) *Curr. Med. Chem.* **12**, 2979.

SXN101959

This recombinant Targeted Secretion Inhibitor (MW = 105 kDa) is a proprietary recombinant protein that inhibits growth hormone (GH) hyper-secretion from the benign pituitary tumor responsible for acromegaly. **Mode of Inhibitor Action:** SXN10195 consists of covalently linked botulinum toxin light designed to target GHRH receptor-rich somatotrophs, thereby allowing internalization of the endopeptidase (light chain) and leading to potent inhibition of vesicular secretion of growth hormone. The expressed single-chain polypeptide is activated by proteolytic cleavage into a 100-kDa heavy chain (HC) and a 50-kDa light chain (LC), linked together by a single disulfide bond. In TSIs, three distinct protein domains play key roles in inhibiting vesicle fusion in target neurons: a binding domain and a membrane-translocation domain (both located in the HC) and an endopeptidase domain (located in the LC) (1,2). The binding domain confers high-affinity selective binding to the target neurons, linking to the membrane translocation domain via a peptide bond. Noncovalent interactions and a single disulfide linking translocation and endopeptidase domains, facilitating entry assist into a target cell's cytosol, where it is released (presumably through the action of reduced glutathione). The endopeptidase then cleaves SNARE proteins, which play an essential role in vesicle docking and enable secretion of its contents. By cleaving them, the endopeptidase prevents formation of a productive SNARE complex and inhibits neurotransmitter secretion (3-5). The duration of that inhibition can range from less than a month to over six months depending on the LC and the specific SNARE protein that is cleaved. In acromegaly, GH hypersecretion is presently treated with somatostatins to reduce secretion, but over one-half of the patients are refractory to this treatment (6). In principle, Syntaxin's TSI strategy, which provides a unique and potentially longer-lasting treatment for inhibiting GH release, is generalizable to other vesicle secretion disorders. **1**. Swaminathan (2011) *FEBS J.* **278**, 4467. **2**. Lacy & Stevens (1999) *J. Mol. Biol.* **291**, 1091. **3**. Simpson (1980) *J. Pharmacol. Exp. Ther.* **212**, 16. **4**. Schiavo, et al. (1992) *Nature* **359**, 832. **5**. Montecucco & Schiavo (1994) *Mol. Microbiol.* **13**, 1. **6**. Melmed (2006) *New Engl. J. Med.* **355**, 2558.

SYM 2206

This AMP$_A$ receptor antagonist and anticonvulsant (FW = 366.42 g/mol; CAS 173952-44-8; Soluble to 100 mM in DMSO), also named (±)-4-(4-aminophenyl)-1,2-dihydro-1-methyl-2-propylcarbamoyl-6,7-methylene-dioxyphthalazine, targets the α-amino-3-hydroxy-5-methyl-4-isoxazolepropionic acid receptor, *or* AMP$_A$ receptor (IC$_{50}$ = 2.8 μM), acting allosterically at the same regulatory site as GYKI 52466, GYKI 53655, and other benzodiazepines, but does not bind directly to the diazepine binding site. SYM 2206 is selective for AMP$_A$ relative to kainate receptor sub-types. **1**. Pelletier, *et al.* (1996) *J. Med. Chem.* **39**, 343. **2**. Li, *et al.* (1999) *Nature* **397**, 161. **3**. Bleakman, *et al.* (2002) *Curr. Pharm. Des.* **8**, 873.

Symadex

This DNA topoisomerase II inhibitor, also known as C-1311 (FW$_{free-base}$ = 350.42 g/mol; CAS 138154-39-9), induces cell death by mitotic catastrophe in human colon carcinoma cells (1,2). Although most topoisomerase II inhibitors are recognized by the P-glycoprotein, a few are either poor substrates for the P-glycoprotein (amsacrine, mitoxantrone) or, in the case of Symadex, not recognized at all (1). Vincristine-resistant cells were 4- to 20-times cross-resistant to tested topoisomerase II inhibitors than are parental cells. Vincristine-resistant cells were also cross-resistant to ICRF-187, a topoisomerase II inhibitor that stabilizes noncovalent, rather than covalent complexes between DNA and topoisomerase II (2). **1**. Skladanowski, Plisov, Konopa & Larsen (1996) *Mol. Pharmacol.* **49**, 772. **2**. Skladanowski, Come, Sabisz, Escargueil & Larsen (2005) *Mol. Pharmacol.* **68**, 625.

Syn-A$_2$

This synthetic, amphiphicic peptide amide (FW = 2774 g/mol; Sequence: MLSRLSLRLLSRLSLRLLSRYLL-NH$_2$ inhibits bovine heart mitochondrial F$_1$ ATPase, *or* H$^+$-transporting two-sector ATPase, IC$_{50}$ = 40 nM (***See also*** *Syn-C*). The order of effectiveness of the peptide inhibitors (expressed as IC$_{50}$ values is: Syn-A$_2$ (40 nM), Syn-C (54 nM), and melittin (5 μM). Rhodamine B, rhodamine 123, dequalinium, melittin, and Syn-A$_2$ all showed noncompetitive inhibition, whereas other inhibitors (rhodamine 6G, rosaniline, malachite green, coriphosphine, acridine orange, and Syn-C) showed mixed inhibition. With the exception of Syn-C, for which the slope replot was hyperbolic and the intercept replot was parabolic, steady-state kinetic analyses indicated that inhibition by the other inhibitors was complete. The inhibition constants obtained by steady-state kinetic analyses were in agreement with the IC$_{50}$ values for each inhibitor. Rhodamine 6G, rosaniline, dequalinium, melittin, Syn-A$_2$, and Syn-C all protected F$_1$ against inactivation by the aziridinium form of quinacrine mustard in agreement with experimentally determined IC$_{50}$ values. **1**. Bullough, Ceccarelli, Roise & Allison (1989) *Biochim. Biophys. Acta* **975**, 377.

Syncurine, *See Decamethonium Bromide*

p-Synephrine

This adrenergic receptor agonist (FW$_{free-base}$ = 167.09 g/mol; CAS 94-07-5), also known as sympatol, oxedrine, and 4-[1-hydroxy-2-(methylamino)-ethyl]phenol, is a simple alkaloid that is biosynthesized in various *Citrus* species (*Likely Pathway*: Tyrosine → Tyramine → *N*-Methyltyramine → Synephrine) through the successive action of tyrosine decarboxylase, tyramine *N*-methyltransferase, and *N*-methyl-tyramine-β-hydroxylase. Synephrine exhibits distinct preference for the α_1 over the α_2 subtype, but its potency at these receptors is relatively low. It also has weak activity at 5-HT receptors and the Trace Adrenergic Amine Receptors TAAR1. Synephrine is a weak substrate for rat brain mitochondrial monoamine oxidase (K_m = 0.25 mM). Synephrine significantly inhibits IL-4-induced eotaxin-1 expression, an effect that is mediated through the inhibition of STAT6 phosphorylation in the JAK/STAT signaling pathway (1). *p*-Synephrine also significantly stimulates glycogenolysis, glycolysis, gluconeogenesis, and oxygen uptake, and these effects are mediated by both α- and β-adrenergic signaling, requiring simultaneous participation of both Ca^{2+} and cAMP (2). **1**. Roh, Kim, Kim, *et al.* (2014) *Molecules* **19**, 11883. **2**. de Oliveira, Comar, de Sá-Nakanishi, Peralta & Bracht (2014) *Mol. Cell Biochem.* **388**, 135.

Synergistin A

This antibiotic (CAS 11126-45-7), also known as PA114A, is a polyunsaturated cyclic macrolactone from *Streptomyces olivaceus* that binds to the 50S ribosomal subunit and inhibits protein biosynthesis (chain elongation; peptidyltransferase). Synergistin A is reportedly identical to virginiamycin M₁. **See Virginiamycin M₁** **1**. Pestka (1974) *Meth. Enzymol.* **30**, 261.

Synergistin B

This antibiotic (CAS 3131-03-1), also known as PA114B, is a cyclic hexadepsipeptide from *Streptomyces olivaceus* that binds to the 50S ribosomal subunit and inhibits protein biosynthesis (chain elongation; peptidyltransferase). Synergistin B is reportedly identical to pristinamycin IA. **See Pristinamycin IA** **1**. Pestka (1974) *Meth. Enzymol.* **30**, 261.

Synip

This 553-residue cytoplasmic regulatory protein (MW = 61,662 Da), also known as Syntaxin 4-interacting protein and STX4-interacting protein, is an insulin-regulated syntaxin 4-binding protein that mediates the translocation of GLUT4-laden vesicles within adipocytes. Synip arrests SNARE-dependent membrane fusion, thereby preventing initiation of ternary SNARE complex assembly (1). It binds to GLUT4 exocytic SNAREs and inhibits the docking, lipid mixing, and content mixing phases of the vesicle fusion reaction. Synip is also phosphorylated on Ser-99 by PKB/Akt2 after treatment with insulin. Ser-99-Ala-Synip, a phosphorylation-deficient mutant, inhibits insulin-stimulated Glut4 translocation and 2-deoxyglucose uptake in adipocytes, demonstrating that Ser-99 phosphorylation is required for Glut4 translocation as well as glucose uptake in both adipocytes and podocytes (3) This study further suggests that defects in Synip phosphorylation may underlie insulin resistance and associated diabetic nephropathy (3). **1**. Min, Okada, Kanzaki, *et al.* (1999) *Mol. Cell* **3**, 751. **2**. Yu, Rathore & Shen (2013) *J. Biol. Chem.* **288**, 18885. **3**. Yamada, Saito, Okada, *et al.* (2014 *Endocr J.* **61**, 523.

Synthalin

This cationic reagent ($FW_{dihydrochloride}$ = 329.32 g/mol; CAS 301-15-5; Soluble to 5 mM in H_2O; Symbol: bisG10), also known as decamethylenediguanidine, *N*,*N*-1,10-decanediyl-bis(guanidine), and 1,10-*bis*(guanidino)-*n*-decane, will act as a potassium ion channel blocker and inhibit diamine oxidase. **Target(s):** Ca^{2+} release (1); cytochrome *c* oxidase (2); diamine oxidase (3-5); K^+ channels (6,7); *N*-methyl-D-aspartate receptor (8,9); and photophosphorylation (10). **1**. Allard, Moutin & Ronjat (1992) *FEBS Lett.* **314**, 81. **2**. Slater (1967) *Meth. Enzymol.* **10**, 48. **3**. Zeller (1938) *Helv. Chim. Acta* **21**, 1645. **4**. Massart (1950) *The Enzymes*, 1st ed., **1** (part 1), 307. **5**. Zeller (1951) *The Enzymes*, 1st ed., **2** (part 1), 536. **6**. Allard, Fournet, Rougier, Descans & Vivaudou (1995) *FEBS Lett.* **375**, 215. **7**. Wang & Best (1994) *J. Physiol.* **477**, 279. **8**. Reynolds (1992) *J. Pharmacol. Exp. Ther.* **263**, 632. **9**. Reynolds, Baron & Edwards (1991) *J. Pharmacol. Exp. Ther.* **259**, 626. **10**. Izawa & Good (1972) *Meth. Enzymol.* **24**, 355.

Syringate

This benzoic acid derivative ($FW_{free-acid}$ = 198.18 g/mol; M.P. = 206-209°C), also known as 4-hydroxy-3,5-dimethoxybenzoic acid and 3,5-dimethoxy-4-hydroxybenzoic acid, is found in most angiosperms. **Target(s):** 3-galactosyl-*N*-acetylglucosaminide 4-α-L-fucosyltransferase (1); glycoprotein 3-α-L-fucosyltransferase, weakly inhibited (1); hemoglobin S polymerization (2); lipase, *or* triacylglycerol lipase (3); and phenol sulfotransferase, *or* aryl sulfotransferase (4). **1**. Niu, Fan, Sun, *et al.* (2004) *Arch. Biochem. Biophys.* **425**, 51. **2**. Gamaniel, Samuel, Kapu, *et al.* (2000) *Phytomedicine* **7**, 105. **3**. Karamac & Amarowicz (1996) *Z. Naturforsch.* [C] **51**, 903. **4**. Yeh & Yen (2003) *J. Agric. Food Chem.* **51**, 1474.

Syringolin A

This plant pathogen virulence factor and irreversible proteasome inhibitor (FW = 493.60 g/mol; CAS 212115-96-3; Symbol: SylA), also systematically named *N*-[(1-{[(3*E*,9*Z*)-5-isopropyl-2,7-dioxo-1,6-diazacyclododeca-3,9-dien-8-yl]carbamoyl}-2-methylpropyl)carbamoyl]-valine, isolated from the bacterial pathogen *Pseudomonas syringae*, is a nonribosomal cyclic peptide that targets the β1 caspase-like activity that cleaves after acidic residues, the β2 trypsin-like activity that cleaves after basic residues, and the β5 chymotrypsin-like activity that cleaves after hydrophobic residues (1). Syringolin A also preferentially inhibits β2 and β5 of the proteasome from *Arabidopsis thaliana*, both *in vitro* and *in vivo* (2). Syringolin A-negative mutants, which induce stomatal closure, can be complemented by exogenous addition of not only syringolin A but also MG132, a well-characterized and structurally unrelated proteasome inhibitor (**See MG132**), demonstrating that proteasome activity is crucial for guard cell function. The dipeptide tail of SylA contributes to β2 specificity, and the subunit selectivity can be explained on the basis of crystallographic findings (2). **1**. Groll, Schellenberg, Bachmann, *et al.* (2008) *Nature* **452**, 755. **2**. Schellenberg, Ramel & Dudler (2010) *Mol. Plant Microbe Interact.* **23**, 1287.

SZL P1-41

This Skp2 inhibitor (FW = 420.52 g/mol; CAS 222716-34-9; Soluble to 20 mM in 1eq. HCl and to 5 mM in DMSO with gentle warming), also named 3-(2-benzothiazolyl)-6-ethyl-7-hydroxy-8-(1-piperidinylmethyl)-4*H*-1-benz opyran-4-one, interferes with the assembly of Skp2-Skp1 complexes, selectively suppressing Skp2 SCF E_3 ligase activity, but without effect on the activity of other SCF complexes. SZL P1-41 also inhibits Skp2-mediated p27 and Akt ubiquitination *in vivo* and *in vitro*. It phenocopies the effects observed upon genetic Skp2 deficiency, such as suppressing survival and Akt-mediated glycolysis and triggering p53-independent cellular senescence. SZL P1-41 also suppresses survival of cancer cell and cancer stem cells by triggering cell senescence and inhibiting glycolysis. SZL P1-41 exhibits antitumor effects in multiple animal models and cancer cell lines. 1. Chan, Morrow, Li, *et al.* (2013) *Cell* **154**, 556.

– T –

T, See *L-Threonine; Thymine; Tocopherol*

2,4,5-T, See *(2,4,5-Trichlorophenoxy)acetate*

T2AA

This nonhormonal T_3 thyronine analogue (FW = 511.10 g/mol; CAS 1380782-27-3), also named (βS)-β-amino-4-(4-hydroxyphenoxy)-3,5-diiodobenzenepropanol, inhibits (IC$_{50}$ ~1 µM) the binding of a PIP-box sequence peptide (*i.e.*, *N*-terminal 5-carboxyfluorescein-labeled SAVLQKKITDYFHPKK) to Proliferating Cell Nuclear Antigen (*or* PCNA) protein by competing for the same binding site, as evidenced by the co-crystal structure of the PCNA-T_3 complex at 2.1-Å resolution (1). PCNA is an essential component of the DNA replication apparatus and the DNA damage response pathway, supporting translesion DNA synthesis (TLS) by interacting with TLS polymerases through PIP-box interaction. The impetus for synthesizing this analogue was the discovery that 3,3′,5-triiodothyronine (T_3) inhibits binding of a PIP-box sequence peptide to PCNA. T2AA abolishes PCNA binding to DNA polymerase δ within cellular chromatin. *De novo* DNA synthesis is also inhibited by T2AA, arresting cells in S-phase. T2AA inhibited growth of cancer cells with induction of early apoptosis. In cells lacking nucleotide-excision repair activity, T2AA inhibits reactivation of a reporter plasmid that is globally damaged by cisplatin, suggesting that the inhibitors blocked the translesion DNA synthesis (TLS) that allows replication of the plasmid. **1.** Punchihewa, Inoue, Hishiki, *et al.* (2012) *J. Biol. Chem.* **287**, 14289. **2.** Actis, Inoue, Evison, *et al.* (2013) *Bioorg. Med. Chem.* **21**, 1972.

T-156, See *2-(2-Methylpyridin-4-yl)methyl-4-(3,4,5-trimethoxyphenyl)-8-(pyrimidin-2-yl)methoxy-1,2-dihydro-1-oxo-2,7-naphthyridine-3-carboxylic Acid Methyl Ester*

T-705, See *Favipiravir-β-ribosyl-triphosphate*

T70907

This potent PPAR inhibitor (FW = 277.66 g/mol; CAS 313516-66-4) targets nuclear Peroxisome Proliferator-Activated Receptor γ, PPARγ (IC$_{50}$ = 1 nM) by reacting irreversibly with an active-site thiol at Cys-313 (1). T0070907 displays >800x selectivity over PPARα and PPARδ. In experiments on MDA-MB-231 and MCF-7 breast cancer cells, T0070907 inhibits proliferation, invasion and migration, but does not significantly affect apoptosis (2). T0070907 also mediates a dose-dependent decrease in PPARγ phosphorylation, DNA binding, and mitogen-activated protein kinase signaling. **1.** Lee, Elwood, McNally, *et al.* (2002) *J. Biol. Chem.* **277**, 19649. **2.** Zaytseva, Wallis, Southard & Kilgore (2011) *Anticancer Res.* **31**, 813.

T0901317

This cholesterol absorption inhibitor (FW = 481.30 g/mol; CAS 293754-55-9; λ$_{max}$ at 227, 265, 272 nm; Solubility: 96 mg/mL DMSO), also named *N*-(2,2,2-trifluoroethyl)-*N*-[4-[2,2,2-trifluoro-1-hydroxy-1-(trifluoromethyl) ethyl]phenyl]benzenesulfonamide, is a potent and selective *agonist* for the Liver X Receptor (LXR), EC$_{50}$ = ~50 nM, the Farnesoid X receptor (FXR), EC$_{50}$ = 5 µM, and the receptor pregnane X receptor (PXR). **Role as Cholesterol Absorption Inhibitor:** LXRα and LXRβ are nuclear hormone receptors that bind 22(*R*)-hydroxy-cholesterol and regulate the oxysterol-induced expression of cholesterol 7α-hydroxylase (CYP7A1), which catalyzes the rate-limiting step in bile acid synthesis (1). Acting through LXR, T0901317 increases expression of the ABCA1 reverse cholesterol transporter, thus increasing cholesterol efflux from enterocytes and having the overall effect of inhibiting cholesterol absorption. T0901317 activates (EC$_{50}$ = ~5 µM) bile acid FXRs, with ~10x more potency than natural ligand chenodeoxycholate (2). T0901317 induces PXR target gene expression in cells and animals, including the scavenger receptor CD36 (3). It also decreases amyloid-β production in cultured primary neurons (4). By binding to RORα (K_i = 132 nM) and RORγ (K_i = 51 nM), T0901317 modulates their interactions with transcriptional cofactor proteins (5). **1.** Repa, Turley, Lobaccaro, *et al.* (2000) *Science* **289**, 1524. **2.** Houck, Borchert, Hepler, *et al.* (2004) *Mol. Genet. Metab.* **83**, 184. **3.** Mitro, Vargas, Romeo, Koder & Saez (2007) *FEBS Lett.* **581**, 1721. **4.** Koldamova, Lefterov, Staufenbiel, *et al.* (2005) *J Biol Chem.* **280**, 4079. **5.** Kumar, Solt, Conkright, *et al.* (2010) *Mol. Pharmacol.* **77**, 228.

T-5601640

This selective LIM kinase inhibitor (FW = 389.33 g/mol; CAS 924473-59-6; Solubility: 50 mM in DMSO), systematically named 3-methyl-*N*-[3-[[[3-(trifluoromethyl)phenyl]amino]carbonyl]phenyl]-5-isoxazolecarboxamide, targets LIM Kinase 2, *or* LIMK2. The latter directly phosphorylates and thereby inactivates cofilin family members, resulting in stabilization of filamentous (F)-actin. Some 40 eukaryotic LIM proteins are so named, because they share a highly conserved cysteine-rich structures containing 2 zinc fingers. LIMK1 and LIMK2 belong to a small subfamily with a unique combination of 2 N-terminal LIM motifs and a C-terminal protein kinase domain. Lim kinases are activated by signaling through small GTPases of the Rho family. T5601640 reduces cofilin phosphorylation in cells overexpressing LIMK2, but not LIMK1. It also attenuates the growth of several cancer cell lines (IC$_{50}$ = 10-30 µM), reducing phospho-cofilin levels and Panc-1 tumor size in a mouse xenograft model. **1.** Rak, Haklai, Elad-Tzfadia, *et al.* (2014) *Oncoscience* **1**, 39.

Tabtoxinine β-Lactam

This lactam-containing glutamine analogue (FW = 188.18 g/mol), also named 2-amino-4-[3-hydroxy-2-oxo-azacyclobutan-3-yl]butanoate, from produced from tabtoxin in several *Pseudomonas syringae* pathovars, is an active site-directed, irreversible inhibitor of glutamine synthetase, wherein ATP is required for inactivation (1-3). The lactam ring is essential for inhibition. Tabtoxinine-β-lactam is generated by combining a 4-carbon fragment, a 2-carbon fragment, and a single carbon atom. The 4-carbon

1168

fragment arises from aspartic acid, and the 2-carbon unit is donated from carbons 2 and 3 of pyruvate. The 6-carbon backbone of tabtoxinine-β-lactam arises from the condensation of fragments from aspartate and pyruvate, most likely by using reactions analogous to the initial steps in the pathway of lysine biosynthesis. **1.** Langston-Unkefer, Robinson, Knight & Durbin (1987) *J. Biol. Chem.* **262**, 1608. **2.** Knight, Durbin & Langston-Unkefer (1986) *J. Bacteriol.* **166**, 224. **3.** Thomas & Durbin (1985) *J. Gen. Microbiol.* **131**, 1061.

Tabun

This nerve poison (FW = 162.13 g/mol; CAS 77-81-6; M.P. = –50°C; B.P. = 240-246°C), also known as ethyl-*N*-dimethyl phosphoramidocyanidate, was first developed as an insecticide in Germany in 1936. The first nerve agent to be synthesized, tabun's primary targets are acetylcholinesterase (1-7) and cholinesterase, *or* butyrylcholinesterase (1,2,7-9). Also known by its chemical warfare designation GA, tabun is a clear liquid with a slightly fruity odor (similar to that of bitter almonds, perhaps originating from the presence of contaminating cyanide). Tabun is extremely toxic, and vapors are readily absorbed through the skin, eyes, and nasal mucosa. The skin absorption LD_{50} in man is 1–1.5 g. The i.p. LD_{50} in mice is considerably lower (0.6 mg/kg). The LC_{50} (*i.e.*, the concentration in air that is lethal to 50% of exposed individuals) is 1-2 ppm. This chemical warfare agent was reportedly used by Iraq in its war with Iran in March of 1984. A United Nations investigation team verified tabun use, and Iran had reported prior use of nerve agents, dating back to 1980. **1.** Holmstedt (1959) *Pharmacol. Rev.* **11**, 567. **2.** Hoskins, Fernando, Dulaney, *et al.* (1986) *Toxicol Lett.* **30**, 121. **3.** Gray & Dawson (1987) *Toxicol. Appl. Pharmacol.* **91**, 140. **4.** Tripathi & Dewey (1989) *J. Toxicol. Environ. Health* **26**, 437. **5.** Froede & Wilson (1971) *The Enzymes*, 3rd ed. (Boyer, ed.), **5**, 87. **6.** Worek, Thiermann, Szinicz & Eyer (2004) *Biochem. Pharmacol.* **68**, 2237. **7.** Schopfer, Voelker, Bartels, Thompson & Lockridge (2005) *Chem. Res. Toxicol.* **18**, 747. **8.** Brown, Kalow, Pilz, Whittaker & Woronick (1981) *Adv. Clin. Chem.* **22**, 1. **9.** Cerasoli, Griffiths, Doctor, *et al.* (2005) *Chem. Biol. Interact.* **157-158**, 363.

Tacedinaline

An orally bioavailable, antineoplastic agent (FW = 269.32 g/mol; CAS 112522-64-2, 1299346-14-7 (HCl), 1353653-79-8 (TFA)); Solubility: 50 mg/mL DMSO; <1 mg/mL H$_2$O), also known as CI994, PD-123654 and systematically named 4-(acetylamino)-*N*-(2-aminophenyl)benzamide, inhibits histone deacetylation (IC_{50} = 0.57 μM), resulting in histone hyperacetylation, induction of cell differentiation, inhibition of cell proliferation, and apoptosis in susceptible tumor cell populations. **1.** Methot, *et al.* (2008) *Bioorg. Med. Chem. Lett.* **18**, 973. **2.** Loprevite, *et al.* (2005) *A. Oncol. Res.* **15**, 39. **3.** Gediya, *et al.* (2008) *Bioorg. Med. Chem.* **16**, 3352. **4.** LoRusso, *et al.* (1996) *Invest. New Drugs* **14**, 349. **5.** Hubeek, *et al.* (2008) *Oncol. Rep.* **19**, 1517.

Tacrine, *See 9-Amino-1,2,3,4-tetrahydroacridine*

(*RS*)-(±)-Tacrine(10)-hupyridone

This dicationic alkylene-linked alkaloid (FW = 502.74 g/mol), consisting of tacrine (*See 9-Amino-1,2,3,4-tetrahydroacridine*) and hupyridone, inhibits acetylcholinesterase (AChE) with far greater affinity than with either tacrine or hupyridone alone. One unit of the pseudodimeric inhibitor is bound at the "anionic" subsite of the active site, situated near the bottom of the active-site gorge, as observed for the *Torpedo californica* AChE complex. The second unit interacts with the peripheral "anionic" site (PAS) located at the top of the gorge, where it is hydrogen-bonded to the side-chains of residues belonging to an adjacent, symmetry-related AChE molecule covering the gorge entrance. **1.** Haviv, Wong, Greenblatt, *et al.* (2005) *J. Amer. Chem. Soc.* **127**, 11029.

Tacrolimus, *See FK-506*

Tadalafil

This orally active phosphodiesterase inhibitor (FW = 377.40 g/mol; CAS 171596-29-5), systematically named (6*R*-*trans*)-6-(1,3-benzodioxol-5-yl)-2,3,6,7,12,12*a*-hexahydro-2-methylpyrazino[1',2':1,6]pyrido[3,4-*b*]indole-1,4-dione) and marketed under the tradename Cialis™, is a potent, long-lasting vasodilator that is widely used to treat erectile dysfunction and pulmonary arterial hypertension. Tadalafil exhibits high selectivity for 3',5'-cyclic-GMP phosphodiesterase Type 5 over the other phosphodiesterases. PDE5 inhibition within the corpus cavernosum increases intracellular cGMP levels, facilitating relaxation of smooth muscle and resulting in penile erection. Tadalafil is also cardiomyoprotective in cases of ischemia/reperfusion injury, doxorubicin cardiotoxicity, ischemic/diabetic cardiomyopathy, and cardiac hypertrophy, acting to increase the expression of nitric oxide synthases, activating protein kinase G (PKG), to stimulate PKG-dependent hydrogen sulfide generation, and to enhance phosphorylation of glycogen synthase kinase-3β, the latter a master switch immediately proximal to mitochondrial permeability transition pore as well as the end-effector of cardioprotection. An increase in Bcl2/Bax during early-phase and transcriptional upregulation of PKG-I by STAT3 during late-phase were responsible for stem cell protection by tadalafil against ischemic injury (11). Tadalafil also appears to beneficially modulate the tumor micro- and macro-environment in patients with head and neck squamous cell carcinoma (HNSCC) by lowering myeloid-derived suppressor cells (MDSC) and regulatory T (Treg) cells and by increasing tumor-specific CD8(+) T cells in a dose-dependent fashion (12). **Pharmacokinetic Parameters:** In healthy subjects, tadalafil is absorbed rapidly, with a mean C_{max} of 378 μg·Liter^{-1} for a 20-mg dose at 2 hours, declining exponentially ($t_{1/2}$ = 17.5 hours) (12). The mean oral clearance (CL/F) is 2.48 Liters·hour^{-1} and apparent volume of distribution is 62.6 Liters. There is no clinically meaningful effect of BMI, age, gender or smoking. Exposure is also substantially unaffected by time of dosing. Food has negligble effects on bioavailability, as assessed by C_{max} and AUC (12). Parameters were proportional to dose, indicating that doubling the dose doubled exposure. Steady-state is attained by day-5 after once-daily administration, with an accumulation that is consistent with its $t_{1/2}$ (13). **1.** Rotella (2002) *Nat. Rev. Drug Discov.* **1**, 674. **2.** Corbin & Francis (2002) *Int. J. Clin. Pract.* **56**, 453. **3.** Daugan, Grondin, Ruault, *et al.* (2003) *J. Med. Chem.* **46**, 4525 and 4533. **4.** Sung, Hwang, Jeon, *et al.* (2003) *Nature* **425**, 98. **5.** Blount, Beasley, Zoraghi, *et al.* (2004) *Mol. Pharmacol.* **66**, 144. **6.** Bischoff (2004) *Int. J. Imp. Res.* **16**, S11. **7.** Weeks, Zoraghi, Beasley, *et al.* (2005) *Int. J. Imp. Res.* **17**, 5. **8.** Mancina, Filippi, Marini, *et al.* (2005) *Mol. Hum. Reprod.* **11**, 107. **9.** Loughney, Taylor & Florio (2005) *Int. J. Imp. Res.* **17**, 320. **10.** Pomara & Morelli (2005) *Int. J. Imp. Res.* **17**, 385. **11.** Kumar & Ashraf (2015) *Stem Cells Dev.* **24**, 1332. **12.** Weed, Vella, Reis, *et al.* (2015) *Clin. Cancer Res.* **21**, 39. 13. Forgue, Patterson, Bedding, *et al.* (2006) *Brit. J. Clin. Pharmacol.* **61**, 280.

TAE684

This ALK protein kinase inhibitor (FW = 614.21 g/mol; CAS 761439-42-3; Solubility: 3 mg/mL DMSO; < 1 mg/mL H_2O), also known as NVP-TAE684 and systematically named 5-chloro-N^4-(2-(isopropylsulfonyl)phenyl)-N^2-(2-methoxy-4-(4-(4-methylpiperazin-1-yl)piperidin-1-yl)phenyl)pyrimidine-2,4-diamine, targets Nucleophosmin-Anaplastic Lymphoma Kinase (NPM-ALK), an oncogenic fusion protein that ius formed by (2;5)(p23;q35) translocation and is constitutively overexpressed and/or activated in anaplastic large-cell lymphomas (ALCLs) and thereby drives tumor cell survival and proliferation. NVP-TAE684 blocks the growth of ALCL-derived and ALK-dependent cell lines with IC_{50} values between 2-10 nM (1). NVP-TAE684 treatment results in rapid and sustained inhibition of NPM-ALK phosphorylation and its downstream effectors, with subsequent induction of apoptosis and cell-cycle arrest. *In vivo*, TAE684 suppresses lymphomagenesis in two independent models of ALK-positive ALCL and induces regression of established Karpas-299 lymphomas (1). TAE-684 is also active against crizotinib-resistant (even at 1 μM) cells harboring the L1196M gatekeeper mutation within the kinase domain. Other constitutively active ALK mutations (*e.g.*, G1128A, I1171N, F1174L, R1192P, F1245C and R1275Q) can be blocked by crizotinib and/or TAE684, although some mutants require high levels of inhibitor (3). Unlike crizotinib, the sulfonated aniline moiety of TAE684 makes hydrophobic interactions with the Gly1123-His1124 segment, a property likely to account for the latter's actions (4). NPM-ALK) is also a Hsp90-client tyrosine kinase, whose expression and tyrosine phosphorylation is down-regulated in ALK$^+$ CD30$^+$ lymphoma cells treated with the Hsp90 antagonist 17-allylamino,17-demethoxygeldanamycin (5). **1.** Galkin, Melnick, Kim, *et al.* (2007) *Proc. Natl. Acad. Sci. U.S.A.* **104**, 270. *Erratum: Proc. Natl. Acad. Sci. U.S.A.* **104**, 2025. **2.** Katayama, Khan, Benes, *et al.* (2011) *Proc. Natl. Acad. Sci. U.S.A.* **108**, 7535. **3.** Schönherr, Ruuth, Yamazaki, *et al.* (2011) *Biochem. J.* **440**, 405. **4.** Bossi, Saccardo, Ardini, *et al.* (2010) *Biochemistry* **49**, 6813. **5.** Bonvini, Gastaldi, Falini & Rosolen (2002) *Cancer Res.* **62**, 1559.

Tafamidis

This TTR amyloidosis drug (FW = 308.12 g/mol; CAS 594839-88-0), also named Fx-1006A, Vyndaqel®, and 2-(3,5-dichlorophenyl)-1,3-benzoxazole-6-carboxylate, binds to stabilizes the natively folded tetrameric form of the transthyretin (TTR). Tafamidis ameliorates transthyretin-related hereditary amyloidosis, also called familial amyloid polyneuropathy (FAP). In this rare but deadly neurodegenerative disorder, TTR dissociates in the rate-limiting step for amyloid fibril formation, causing failure of the autonomic nervous system and/or the peripheral nervous system (neurodegeneration) initially and later failure of the heart. (**See also** *Tolcapone*) **1.** Bulawa, Connelly, DeVit, *et al.* (2012) *Proc. Natl. Acad. Sci. U.S.A.* **109**, 9629.

Tafluposide

This antineoplastic agent (FW$_{free-acid}$ = 1116.72 g/mol; CAS 179067-42-6), also known as the 2",3"-bispentafluorophenoxyacetyl-4",6"-ethylidene-β-D-glucoside of 4'-phosphate-4'-demethylepipodophyllotoxin and F11782, is a dual inhibitor of DNA topoisomerases I and II (1,2). While most topo inhibitors typically favor either topo I or topo II, a smaller number that can act against both enzymes fall three classes: first, those drugs that bind to DNA via intercalation (*e.g.*, benzopyridoindole intoplicine, TAS-103, XR11576, and NSC 366140; second, hybrid molecules, such as camptothecin-epipodophyllotoxin and ellipticine-distamycin hybrids, that consist of physically linked topo I and topo II pharmacophores; third, those that recognize structural motifs present in both topo I and II (*e.g.*, NK 109, BN 80927 and F-11782). While DNA intercalation also factors into the action of some topo inhibitors, F11782 does not intercalate into DNA. **1.** Kruczynski, Barret, van Hille, *et al.* (2004) *Clin. Cancer Res.* **10**, 3156. **2.** Perrin, van Hille, Barret, *et al.* (2000) *Biochem. Pharmacol.* **59**, 807.

D-Tagatose 1,6-Bisphosphate

This substrate analogue (FW$_{free-acid}$ = 340.11 g/mol; CAS 55529-38-9) weakly inhibits pyrophosphate-dependent phosphofructokinases from the facultative anaerobic bacterium *Propionibacterium freudenreichii* (K_i = 0.82 mM) and the mung bean *Phaseolus aureus*. The dissociation constants for D-psicose 6-phosphate, D-tagatose 6-phosphate, and L-sorbose 6-phosphate are >10x that for D-fructose 6-phosphate, suggesting the stringent steric requirement for the D-*threo* (trans) configuration at the two nonanomeric furan ring hydroxyl groups. **1.** Bertagnolli, Younathan, Voll, Pittman & Cook (1986) *Biochemistry* **25**, 4674.

Tagetitoxin

This bacterial phytotoxin (FW = 416.30 g/mol; CAS 87913-21-1; *Symbol*: TGT) from *Pseudomonas syringae pv. tagetis* (1) inhibits multi-subunit RNA polymerases chloroplasts and bacteria (2) as well as eukaryotic pol III (3), all with low μM inhibition constants. This agent disrupts the ability of these enzymes to catalyze nucleotide addition reaction cycles consisting of phosphodiester bond formation, translocation, and binding of the subsequent nucleotide (4-6). A unique property of TGT is that it retards nucleotide addition in a sequence-dependent manner, but only at certain positions of the template (7). It neither alters the chemistry of RNA synthesis nor induces backward translocation. What's more, TGT does not compete with the nucleoside triphosphate (NTP) in the active center. Instead, it appears to increase the stability of the pre-translocated state of elongation complex, thus slowing down addition of the next nucleotide (7). The extent of inhibition directly depends on the intrinsic stability of the pre-translocated state on the target enzyme. What's more, the dependence of translocation equilibrium on the transcribed sequence results in a wide distribution ($1-10^3$ times) the inhibitory effect of TGT at different positions of the template, thus explaining sequence-specificity of TGT action (7). *Note*: Pralidoxime (*or* 2-[(hydroxyimino)methyl]-1-methylpyridin-1-ium) is often used, along with the muscarinic antagonist atropine, to reduce the parasympathetic effects of organophosphate poisoning. Pralidoxime is only effective in organophosphate toxicity and has not effect if the acetylcholinesterase is carbamylated, as occurs with neostigmine or physostigmine. **1.** Mitchell & Durbin (1981) *Physiol. Plant Pathol.* **18**, 157. **2.** Mathews & Durbin (1990) *J. Biol. Chem.* **265**, 493. **3.** Steinberg, Mathews, Durbin & Burgess (1990) *J. Biol. Chem.* **265**, 499. **4.** Vassylyev, Svetlov, Vassylyeva, *et al.* (2005) *Nature Struct. Mol. Biol.* **12**, 1086. **5.** Artsimovitch, Svetlov, Nemetski, *et al.* (2011) *J. Biol. Chem.* **286**, 40395. **6.** Malinen, Turtola & Parthiban (2012) *Nucleic Acids Res.* **40**, 7442. **7.** Yuzenkova, Roghanian, Bochkareva & Zenkin (2013) *Nucleic Acids Res.* 41, 9257.

Tagit, See *3-(4-Hydroxyphenyl)propionic Acid N-Hydroxysuccinimide Ester*

TAK-063

This orally active PDE inhibitor and potential schizophrenia drug (FW = 428.43 g/mol), also known as 1-[2-fluoro-4-(1*H*-pyrazol-1-yl)phenyl]-5-methoxy-3-(1-phenyl-1*H*-pyrazol-5-yl)pyridazin-4(1*H*)-one, selectively targets (IC$_{50}$ = 0.3 nM) phosphodiesterase 10a (PDE10A), a dual substrate (cAMP/cGMP) enzyme in the medium spiny neurons of the striatum, a region strongly associated with motor and cognitive functions. TAK-063 shows favorable pharmacokinetics, including high brain penetration, in mice. Oral administration elevated striatal 3',5'-cyclic adenosine monophosphate (cAMP) and 3',5'-cyclic guanosine monophosphate (cGMP) levels at 0.3 mg/kg (in mice), showing potent suppression of phencyclidine (PCP)-induced hyperlocomotion at a minimum effective dose (MED) of 0.3 mg/kg. Other PDE10A-selective inhibitors include PF-2545920 (*or* MP-10), OMS824, RG7203, and FRM-6308 (*formerly* EVP-6308). **Target(s)**: PDE1A, IC$_{50}$ = 12000 nM (range: 9800–14000 nM); PDE2A$_3$, IC$_{50}$ > 30000 nM; PDE3A, IC$_{50}$ = >30000 nM; PDE4D2, IC$_{50}$ = 5500 nM (range: 5000–6100 nM); PDE5A1, IC$_{50}$ = 5700 nM (range: 4700–6900 nM); PDE6AB, IC$_{50}$ = 6800 nM (range: 5300–8900 nM); PDE7B, IC$_{50}$ = 19000 nM (range: 18000–21000 nM); PDE8A$_1$, IC$_{50}$ > 30000 nM; PDE9A$_2$, IC$_{50}$ > 30000 nM; PDE10A$_2$, IC$_{50}$ = 0.30 nM (range: 0.22–0.40 nM); PDE11A$_4$, IC$_{50}$ > 30000 nM. **1.** Kunitomo, Yoshikawa, Fushimi, *et al.* (2014) *J. Med.Chem.* **57**, 9627.

TAK 165

This (FW = 470.50 g/mol; CAS 366017-09-6; Soluble to 75 mM in DMSO and to 5 mM in Ethanol), also named 1-[4-[4-[[2-[(1*E*)-2-[4-(trifluoromethyl)phenyl]ethenyl]-4-oxazolyl]methoxy]phenyl]butyl]-1*H*-1,2,3-triazole, is a potent, irreversible human epithelial growth factor receptor 2 (ErbB2) inhibitor (IC$_{50}$ = 6 nM), displaying > 4000x selectivity over EGFR, FGFR, PDGFR, JAK1 and Src. TAK-165 also exhibits potent antiproliferative effects in ErbB2-overexpressing cancer in BT474 breast cancer cells (IC$_{50}$ = 5 nM) and significantly inhibits growth of bladder, breast and prostate cancer xenografts. **1.** Nagasawa, *et al.* (2006) *Int. J. Urol.* **13**, 585. **2.** Spector, *et al.* (2007) *Breast Cancer Res.* **9**, 205.

TAK-228, See *Sapanisertib*

TAK-438, See *Vonoprazan*

TAK-475, See *Lapaquistat Acetate*

TAK-491, See *Azilsartan & Azilsartan Medoxomil*

TAK-536, See *Azilsartan & Azilsartan Medoxomil*

TAK-599, See *Ceftaroline Fosamil*

TAK-632

This novel 1,3-benzothiazole and potent pan-RAF inhibitor (FW = 554.51 g/mol; CAS 1228591-30-7), also known as *N*-(7-cyano-6-(4-fluoro-3-(2-(3-(trifluoromethyl)phenyl)acetamido)-[phenoxy)benzo[*d*]thiazol-2-yl)cyclopropanecarboxamide, targets BRAF$^{wild-type}$ (IC$_{50}$ = 8.3 nM(, with slow off-rate (assayed *in vitro*), BRAF^{V600E} (IC$_{50}$ = 2.4 nM, *in vitro*), CRAF (IC$_{50}$ = 1.4 nM, *in vitro*), pMEK, *or* A375, (IC$_{50}$ = 12 nM *in vitro*), and pMEK, *or* HMVII (IC$_{50}$ = 49 nM *in vitro*). TAK-632 demonstrates antitumor efficacy against both A375 (BRAF^{V600E}) and HMVII (NRAS^{Q61K}) tumor xenografts in rats. **1.** Okaniwa, Hirose, Arita, *et al.* (2013) *J. Med. Chem.* **56**, 6478.

TAK-715

This orally available MAPK inhibitor (FW = 399.51 g/mol; CAS 303162-79-0), also named *N*-[4-[2-ethyl-4-(3-methylphenyl)-5-thiazolyl]-2-pyridinyl]benzamide, targets p38 mitogen-activated protein kinase, an enzyme in the proinflammatory cytokine signal pathway associated with chronic inflammatory disorders, including rheumatoid arthritis and inflammatory bowel disease. TAK-715 potently inhibits p38α (IC$_{50}$ = 7.1 nM), LPS-stimulated release of TNF-α from THP-1 (IC$_{50}$ = 48 nM), LPS-induced TNF-α production in mice, 87.6% inhibition at 10 mg/kg, oral dose, with no inhibitory action against major CYPs, including CYP3A4 (1). Although TAK-715 had been thought to affect Wnt-3a-stimulated β-catenin signaling, two highly selective and equally potent p38 inhibitors, VX-745 and Scio-469, do not. Moreover, profiling of TAK-715 against 200 kinases revealed inhibition of casein kinases Iδ and ε, enzymes that are known activators of Wnt/β-catenin signaling. This lack of TAK-715 specificity weighs against a role of p38 in Wnt/β-catenin signaling. **1.** Miwatashi, Arikawa, Kotani, *et al.* (2005) *J. Med. Chem.* **48**, 5966. **2.** Verkaar, van der Doelen, Smits, Blankesteijn & Zaman (2011) *Chem. Biol.* **18**, 485.

TAK-773

This potent allosteric-site inhibitor (FW = 504.23 g/mol; CAS 1035555-63-5; Solubility: 100 mg/mL DMSO), also named (R)-3-(2,3-dihydroxypropyl)-6-fluoro-5-(2-fluoro-4-iodophenylamino)-8-methylpyrido[2,3-d]pyrimidine-4,7(3H,8H)-dione, selectively targets MEK1 (IC$_{50}$ = 3.2 nM), with no discernible effect on Abl1, AKT3, c-RAF, CamK1, CDK2, or c-Met. TAK-733. It also exhibits broad antitumor activity in mouse xenograft models for human cancer including models of melanoma, colorectal, NSCLC, pancreatic and breast cancer. TAK-733 is also well tolerated, with pharmacokinetics and pharmacodynamics that support once-daily oral dosing in humans. **1**. Dong, Dougan, Gong et al. (2011) Bioorg. Med. Chem. Lett. **21**, 1315.

TAK-779

This nonpeptide CCR5 receptor antagonist (FW = 531.13 g/mol) prevents HIV-1 from entering and infecting peripheral blood CD4$^+$ immune cells, when present at 2-4 nM (1). CCR5 is a member of G-protein-coupled, seven transmembrane segment receptors, also known as C-C motif chemokine receptor-5. TAK-779 also antagonizes the binding of RANTES (or Regulated on Activation, Normal T cell Expressed and Secreted) cytokine to CCR5-expressing Chinese hamster ovary cells and blocked CCR5-mediated Ca^{2+} signaling, again at nanomolar concentrations (2). **Primary Mechanism of Inhibitory Action:** TAK-779 inhibits HIV-1 replication at the membrane fusion stage by blocking the interaction of the viral surface glycoprotein gp120 with CCR5, and alanine scanning mutagenesis of the transmembrane domains revealed that the binding site for TAK-779 on CCR5 is located near the extracellular surface of the receptor, within a cavity formed between transmembrane helices 1, 2, 3, and 7 (2). The efficiency of entry of cell-attached infectious HIV-1 appears to be controlled by three different kinetic processes (3). The first shows lag-phase kinetics, caused in part by the concentration-dependent reversible association of virus with CD4 and CCR5, forming an equilibrium manifold of complexes. In the second, this assembly step lowers, but does not eliminate, a large activation energy barrier for a rate-limiting, CCR5-dependent conformational change in gp41. The rate of infection therefore depends on the fraction of infectious virions that have enough CCR5 to undergo this conformational change and surmount the energy barrier. While only a small fraction of fully assembled viral complexes overcome this barrier per hour, the later steps of entry are rapidly completed within 5-10 min. Thus, this barrier limits the overall rate for virion attachment and entry, but it is without effect on the already mentioned lag time. Third, there follows a relatively rapid and kinetically dominant process of viral inactivation, which partly involves endocytosis and competes with infectious viral entry. Other similarly acting viral entry inhibitors include aplaviroc, vivriviroc, maraviroc, and SCH-C (or SCH 351125). By contrast, the entry inhibitor PRO-140 is a human monoclonal antibody to the CCR5 receptor. This chemokine receptor antagonist is protective against ischemic brain injury (4). **1**. Baba, Nishimura, Kanzaki, et al. (1999) Proc. Natl. Acad. Sci. U.S.A. **96**, 5698. **2**. Dragic, Trkola, Thompson, et al. (2000) Proc. Natl. Acad. Sci. U.S.A. **97**, 5639. **3**. Platt, Durnin & Kabat (2005) J. Virol. **79**, 4347. **4**. Takami, Minami, Katayama, et al. (2002) J. Cereb. Blood Flow Metab. **22**, 780.

TAK-802

This novel acetylcholine esterase inhibitor (FW$_{hydrochloride}$ = 456.99 g/mol), also known as 8-[3-[1-[(3-fluoro-phenyl)methyl]-4-piperidinyl]-1-oxo-propyl]-1,2,5,6-tetrahydro-4H-pyrrolo[3,2,1-ij]-quinolin-4-one, shows high selectivity for acetylcholinesterase over butyryl-cholinesterase. TAK-802 inhibits human-erythrocyte-derived acetylcholinesterase activity (IC$_{50}$ = 1.5 nM), with 30x and 250x higher affinity than neostigimine and distigmine, respectively (1). TAK-802 increases isovolumetric bladder contractions, with a minimum effective dose (MED) of 10 µg/kg i.v. in rats and guinea pigs, and this effects is completely abolished by atropine. Such results suggest TAK-802 and can effectively increase reflex bladder contractions by increasing the efficacy of acetylcholine released by nerve impulses. **1**. Nagabukuro, Okanishi, Imai, et al. (2004) Eur. J. Pharmacol. **485**, 299.

Talampanel

This oral benzodiazepine (FW = 337.38 g/mol; CAS 161832-65-1; Soluble to 100 mM in DMSO), also known as GYKI 53773, LY 300164, and (8R)-7-acetyl-5-(4-aminophenyl)-8,9-dihydro-8-methyl-7H-1,3-dioxolo[4,5-h][2,3]benzodiazepine, is a noncompetitive antagonist of the α-amino-3-hydroxy-5-methylisoxazolepropionic acid (AMPA) receptor (1). LY 300164 potentiates the protective action of diazepam in some animal models of seizures, suggesting a new approach for treating drug-resistant epilepsy or status epilepticus (2) **1**. Solyom & Tarnawa (2002) Curr. Pharm. Des. **8**, 913. **2**. Borowicz, Kleinrok & Czuczwar (2000) Naunyn. Schmiedebergs Arch. Pharmacol. **361**, 629.

Talniflumate

This CaCC blocker and anti-inflammatory agent (FW = 414.34 g/mol; CAS 66898-62-2; Soluble to 25 mM in DMSO), also named 1,3-dihydro-3-oxo-1-isobenzofuranyl 2-[[3-(trifluoromethyl)phenyl]amino]-3-pyridine carboxylate, targets calcium-activated chloride channels (CaCC), including hCLCA1 and mCLCA3, reducing mucin synthesis and release in cell culture and animal models (1,2). Talniflumate's anti-inflammatory actions relate to its inhibitory effects on cyclooxygenases. It also inhibits the Cl$^-$/HCO$_3^-$ exchanger activity, increasing survival in a cystic fibrosis mouse model of distal intestinal obstructuve syndrome (3). **1**. Sanchez, et al. (1982) J. Appl. Toxicol. **2**, 42. **2**. Knight (2004) Curr. Opin. Invest. Drugs **5**, 557. **3**. Walker, et al. (2006) J. Pharmacol. Exp. Ther. **317**, 275.

Talopram

This potent and selective transport inhibitor and antidepressant (FW$_{HCl\text{-}Salt}$ = 331.88 g/mol; CAS 7013-41-4; Solubility: 50 mM in H$_2$O, 100 mM in DMSO), known also as Lu 3-010 and 1,3-dihydro-N,3,3-trimethyl-1-phenyl-1-isobenzofuranpropanamine, targets the noradrenalin transporter, *or* norepinephrine transporter NET (IC$_{50}$ = 2.9 nM), showing selectivity for NET *versus* SERT (5-HT transporters) and DAT (dopamine transporters) (1-3). The basis for such selectivity is the observation that only small amines are accommodated at the 5-HT-uptake site, whereas larger amines such as piperazine are accommodated both at the DA-and NE-uptake sites. The synthesis and evaluation of [^{11}C]talopram is relevant to its use in positron emission tomography (2). (*See Talsupram*) 1. Bøgesø, Christensen, Hyttel & Liljefors (1985) *J. Med. Chem.* 28, 1817. 2. Schou, Sóvágó, Pike, *et al.* (2006) *Mol. Imaging Biol.* 8, 1. 3. Eildal, Andersen, Kristensen, *et al.* (2008). *J. Med. Chem.* 51, 3045.

Talsupram

This potent and selective transport inhibitor and antidepressant (FW$_{HCl\text{-}Salt}$ = 347.95 g/mol; CAS 25487-28-9; Solubility: 20 mM in H$_2$O, 100 mM in DMSO; 1,3-dihydro-N,3,3-trimethyl-1-phenylbenzo[c]thiophene-1-propanamine) targets the noradrenalin transporter, *or* norepinephrine transporter, NET (IC$_{50}$ = 0.8 nM), showing selectivity for NET *versus* SERT, *or* 5-HT transporters (IC$_{50}$ = 850 nM) and DAT, *or* dopamine transporters (IC$_{50}$ = 9300 nM). (*See Talopram*). 1. Schou, Sóvágó, Pike, *et al.* (2006) *Mol. Imaging Biol.* 8, 1.

TAME

This amino-substituted L-arginine ester and well-known synthetic protease inhibitor (FW$_{hydrochloride}$ = 378.88 g/mol; CAS 1784-03-8; hydrolyzes nonenzymatically above pH 9.), also known as N^{α}-(p-Tosyl)-L-arginine methyl ester, is an alternative substrate that is useful for quantifying the esterase activity of many neutral proteinases, including trypsin, plasmin, and papain (1-10). Depending on the target enzyme, TAME actions as a slow alternative substrate (*i.e.*, both low K_m and low k_{cat}) results in competitive inhibition the peptide-hydrolyzing activities of proteases. **Pharmacologic Inhibition of Anaphase-Promoting Complex in Mitosis:** Microtubule inhibitors that induce mitotic arrest by activating the Spindle Assembly Checkpoint (SAC), which in turn inhibits the ubiquitin ligase activity of the Anaphase-Promoting Complex (APC). TAME binds to the APC, preventing proteolytic activation by Cdc20 and Cdh1. A cell-permeable prodrug (*See pro-TAME*) arrests cells in metaphase without perturbing the spindle, properties that commend its development as a anticancer drug (11). **Target(s):** barrierpepsin, above pH 6 (1); N-benzoyl-D-arginine-4-nitroanilide amidase (2); dactylysin, weakly inhibited (3); mitochondrial intermediate peptidase (4); scutelarin (5); signal peptidase II (6,7); thrombin (8,9); u-plasminogen activator, *or* urokinase (10). 1. Nath (1993) *Biochimie* 75, 467. 2. Gofshtein-Gandman, Keynan & Milner (1988) *J. Bacteriol.* 170, 5895. 3. Delporte, Carvalho, Leseney, *et al.* (1992) *Biochem. Biophys. Res. Commun.* 182, 158. 4. Kalousek, Isaya & Rosenberg (1992) *EMBO J.* 11, 2803. 5. Walker, Owen & Esmon (1980) *Biochemistry* 19, 1020. 6. Sankaran (1998) in *Handb. Proteolytic Enzymes* (Barrett, Rawlings & Woessner, eds.), p. 982, Academic Press, San Diego. 7. Sankaran & Wu (1995) *Meth. Enzymol.* 248, 169. 8. Okamoto & Hijikata-Okunomiya (1993) *Meth. Enzymol.* 222, 328. 9. Lorand & Yudkin (1957) *Biochim. Biophys. Acta* 25, 437. 10. Lorand & Condit (1965) *Biochemistry* 4, 265. 11. Zeng, Sigoillot, Gaur, *et al.* (2010) *Cancer Cell* 18, 382.

Tamoxifen

R = H   Tamoxifen
R = OH   Afimoxifene

This mixed antagonist/partial-agonist and breast cancer drug (FW = 371.52 g/mol; CAS 10540-29-1; Hygroscopic; UV-sensitive; Soluble to 100 mM in DMSO with gentle warming; IUPAC Name: (Z)-2-[4-(1,2-diphenyl-1-butenyl)phenoxy]-N,N-dimethylethanamine), also known as Nolvadex™, Istubal™, and Valodex™, is an estrogen receptor (ER) antagonist, with effects that vary, depending on an organism's species and tissue type. In addition to its effects on estrogen homeostasis, tamoxifen inhibits many enzymes and induces apoptosis in susceptible cell lines. Tamoxifen is also neuroprotective in female rats. **Metabolic Activation:** A prodrug with little affinity for the estrogen receptor, tamoxifen is metabolized by liver cytochromes CYP2D6 and CYP3A4 to **Afimoxifene** (FW = 387.51 g/mol; CAS 68392-35-8; Alternative Name: 4-hydroxytamoxifen) and **Endoxifen** (FW = 373.49 g/mol; CAS 112093-28-4; Alternative Name: N-demethyl-4-hydroxytamoxifen). With 30x-100x greater estrogen receptor affinity than tamoxifen, these metabolites compete with estrogen for the estrogen receptor. In breast tissue, 4-hydroxytamoxifen acts as an estrogen receptor antagonist that inhibits transcription of estrogen-responsive genes. Tamoxifen is currently used to treat both early and advanced ER$^+$ (estrogen receptor-positive) breast cancer in pre- and post-menopausal women and as the most common hormone treatment for male breast cancer. Tamoxifen is a high-affinity agonist at the membrane estrogen receptor GPR30, an orphan receptor unrelated to nuclear estrogen receptors, has all the binding and signaling characteristics of a membrane ER. GPR30 is structurally unrelated to the recently discovered family of GPCR-like membrane progestin receptors. **Key Pharmacokinetic Parameters:** See Appendix II in Goodman & Gilman's *THE PHARMACOLOGICAL BASIS OF THERAPEUTICS*, 12th Edition (Brunton, Chabner & Knollmann, eds.) McGraw-Hill Medical, New York (2011). **Target(s):** [β-adrenergic-receptor] kinase (1); aldehyde oxidase (2); aromatase, *or* CYP19 (3,4); ceramide glycanase (5,6); cholestenol Δ-isomerase, *or* sterol Δ^{7,8}-isomerase (7-11a) at the so-called emopamil-insensitive site (11b); CYP3A4 (12); CYP3B1 (12); glucuronosyltransferase (13); Na$^+$/K$^+$-exchanging ATPase (14); nitric-oxide synthase (15); protein kinase C (16-18); sterol Δ^{14}-reductase (8); and sterol Δ^{24}-reductase (8). 1. Benovic (1991) *Meth. Enzymol.* 200, 351. 2. Obach, Huynh, Allen & Beedham (2004) *J. Clin. Pharmacol.* 44, 7. 3. Toi, Bando & Saji (2001) *Breast Cancer* 8, 329. 4. Manni (1993) *J. Cell. Biochem. Suppl.* 17G, 242. 5. Basu, Kelly, Girzadas, Li & Basu (2000) *Meth. Enzymol.* 311, 287. 6.

Basu, Kelly, O'Donnell, *et al.* (1999) *Biosci. Rep.* **19**, 449.   **7.** Moebius, Reiter, Bermoser, *et al.* (1998) *Mol. Pharmacol.* **54**, 591.   **8.** Cho, Kim & Paik (1998) *Mol. Cells* **8**, 233.   **9.** Paul, Silve, De Nys, *et al.* (1998) *J. Pharmacol. Exp. Ther.* **285**, 1296.   **10.** Nes, Zhou, Dennis, *et al.* (2002) *Biochem. J.* **367**, 587.   **11a.** Bae, Seong & Paik (2001) *Biochem. J.* **353**, 689. **11b.** Paul, Silve, De Nys, *et al.* (1998) *J. Pharmacol. Exp. Ther.* **285**, 1296. **12.** Zhao, Jones, Wang, Grimm & Hall (2002) *Xenobiotica* **32**, 863.   **13.** Hara, Nakajima, Miyamoto & Yokoi (2007) *Drug Metab. Pharmacokinet.* **22**, 103. **14.** Repke & Matthes (1994) *J. Enzyme Inhib.* **8**, 207.   **15.** Renodon, Boucher, Sari, *et al.* (1997) *Biochem. Pharmacol.* **54**, 1109.   **16.** O'Brian, Liskamp, Solomon & Weinstein (1985) *Cancer Res.* **45**, 2462.   **17.** Schwartz, Sylvia, Guinee, Dean & Boyan (2002) *J. Steroid Biochem. Mol. Biol.* **80**, 401.   **18.** Saraiva, Fresco, Pinto & Goncalves (2003) *J. Enzyme Inhib. Med. Chem.* **18**, 475.

Tamsulosin

This selective α_{1a}/α_{1d}-receptor antagonist and muscle relaxant (FW$_{\text{free-base}}$ = 408.51 g/mol; FW$_{\text{hydrochloride}}$ = 433.96; CAS 106133-20-4), widely known as Flomax® and (*R*)-5-(2-{[2-(2-ethoxyphenoxy)ethyl]amino}propyl)-2-methoxybenzene-1-sulfonamide, is used to treat benign prostatic hyperplasia (BPH) symptomatically and to promote passage of moderately sized kidney stones, by increasing blood flow in small blood vessels as well as urine flow through the ureter (1). Tamsulosin decreases resting maximal urethral pressure, with negligable effect on mean arterial blood pressure. The specific binding of [^{3}H]tamsulosin in human prostatic membranes is saturable, showing high affinity (K_d = 0.04 nM), and the density of binding sites (measured as B$_{\text{max}}$ in Scatchard plots) was 409 fmol/mg protein (2). This antagonist has pK_i values are 9.97, 9.64 and 8.86 for α_{1A}, α_{1B} and α_{1D} subtypes, respectively. Tamsulosin is metabolized by the liver cytochrome system and should not be used in combination with warfarin or strong CYP3A4 inhibitors (*e.g.*, ketoconazole, erythromycin, fluconazole, grapefruit juice, and omeprazole). **Key Pharmacokinetic Parameters:** *See* Appendix II in Goodman & Gilman's THE PHARMACOLOGICAL BASIS OF THERAPEUTICS, 12th Edition (Brunton, Chabner & Knollmann, eds.) McGraw-Hill Medical, New York (2011).   **1.** Yazawa, Takanashi, Sudoh, Inagaki & Honda (1992) *J. Pharmacol. Exp. Ther.* **263**, 201.   **2.** Yamada, Tanaka, Ohkura, *et al.* (1994) Urol. Res. **22**, 273.

Tandospirone

This serotonin receptor partial agonist andanxiolytic agent (FW = 419.95 g/mol; CAS 99095-10-0), also named (3a*R*,4*S*,7*R*,7a*S*)-*rel*-hexahydro-2-[4-[4-(2-pyrimidinyl)-1-piperazinyl]butyl]-4,7-methano-1*H*-isoindole-1,3(2*H*)-dione, targets 5-HT$_{1A}$ (K_i = 27 nM), displaying great selectivity over 5-HT$_2$, 5-HT$_{1C}$, α_1, α_2, D$_1$ and D$_2$ receptors, for which K_i values range from 1.3 to 41 μM (1). Saturation and competition studies using 3H-tandospirone also suggest that the drug interacts with 5-HT1A receptor binding sites in rat cortical membranes (K_D = 4.5 nM; B$_{\text{max}}$ = 2.2 pmol/g tissue). Based on adenylate cyclase studies, which measure 5-HT$_{1A}$ receptor-mediated effects, tandospirone displays ~60% of the agonist effect of 8-OH-DPAT, a selective 5-HT$_{1A}$ agonist. Tandospirone is qualitatively similar to the other azapirone anxiolytics buspirone, gepirone and ipsapirone, but is different from sedative-hypnotic anxiolytics (2). Serotonin-1A receptors are distributed throughout the brain with their highest concentrations in the frontal cortex, subthalamic nucleus and entopeduncular nucleus as well as the dorsal and median raphe nucleus. There is growing evidence that 5-HT1A receptor agonists have an antidepressant effect in individuals with major depressive disorders (3). **1.** Hamik, Oksenberg, Fischette & Peroutka (1990) *Biol. Psychiatry* **28**, 99. **2.** Pollard, Nanry & Howard (1992) *Eur. J. Pharmacol.* **221**, 297. **3.** Matsubara, Shimizu, Suno, *et al.* (2006) *Brain Res.* **1112**, 126.

Tanespimycin

This potent heat shock protein inhibitor and antineoplastic agent (FW = 585.69 g/mol; CAS 75747-14-7; Solubility = 100 mg/mL DMSO, and <1 mg/mL H$_2$O; Symbol: 17-AAG), also known as 17-demethoxy-17-(2-propenylamino)geldanamycin and [(3*S*,5*S*,6*R*,7*S*,8*E*,10*R*,11*S*,12*E*,14*E*)-21-(allylamino)-6-hydroxy-5,11-dimethoxy-3,7,9,15-tetramethyl-16,20,22-tri-oxo-17-azabicyclo[16.3.1]docosa-8,12,14,18,21-pentaen-10-yl] carbamate, targets HSP90, displaying an IC$_{50}$ value of 5 nM. The natural product HSP90 inhibitors, like geldanamycin and radicicol, exert their antitumour effects by inhibiting HSP90's intrinsic ATPase activity, resulting in degradation of HSP90 client proteins by the ubiquitin/proteasome pathway. Indeed, tanespimycin inhibition of this chaperone protein results in the depletion of such oncogenic proteins as Raf-1, mutant p53, and Bcr-Abl (1). Tanespimycin also depletes N-ras, Ki-ras, and c-Akt and also inhibits c-AKT phosphorylation in human colon adenocarcinoma cells, thereby inducing cytostasis and apoptosis (2). Tanespimycin also enhances tumor cell radiosensitivity, often enhancing the killing action by factors ranging from 1.3 to 1.7. Enzyme-catalyzed reduction of 17-AAG by NAD(P)H:quinone oxidoreductase 1 (NQO1) generates 17-allylaminodemethoxygeldanamycin hydroquinone (17-AAGH$_2$), a relatively stable hydroquinone that exhibits superior Hsp90 inhibition (4). Significantly, tanespimycin also associates with the mitochondrial membrane voltage-dependent anion channel (VDAC) via a hydrophobic interaction that is independent of HSP90 (5). *In vitro*, tanespimycin functions as a Ca^{2+} mitochondrial regulator akin to benzoquinone-ubiquinones like Ub0. In contrast, the HSP90 inhibitor radicicol, which lacks a bezoquinone moiety, has no measurable effect on cationic current and is less effective in influencing intercellular Ca^{2+} concentration (5). Recent work suggests that tanespimycin mainly causes a cytostatic antiproliferative effect, rather than cell death, and further suggests that BAX status may not alter the overall clinical response to HSP90 inhibitors (6). (**See also** *Ganetespib*) **1.** Bagatell, Paine-Murrieta, Taylor, *et al.* (2000) *Clin. Cancer Res.* **6**, 3312.   **2.** Hostein, Robertson, DiStefano, Workman & Clarke (2001) *Cancer Res.* **61**, 4003.   **3.** Russell, Burgan, Oswald, Camphausen & Tofilon (2003) *Clin. Cancer Res.* **9**, 3749.   **4.** Guo, Reigan, Siegel, *et al.* (2005) *Cancer Res.* **65**, 10006.   **5.** Xie, Wondergem, Shen, *et al.* (2011) *Proc. Natl. Acad. Sci. U.S.A.* **108**, 4105.   **6.** Powers, Valenti, Miranda, *et al.* (2013) *Oncotarget* **4**, 1963.

Tanomastat

This synthetic inhibitor (FW$_{\text{free-acid}}$ = 410.92 g/mol; CAS 179545-77-8), also known as BAY 12-9566, targets a matrix metalloproteinases and is relatively selective for gelatinase A, *or* matrix metalloproteinase-2 (K_i = 11 nM). Higher BAY 12-9566 concentrations also inhibit stromelysin 1 (K_i = 140 nM), gelatinase B (K_i = 300 nM), collagenase-3 (K_i = 1.5 μM), and interstitial collagenase (K_i = 5 μM). Tanomastat induces extracellular matrix degradation, and inhibits angiogenesis, tumor growth and invasion, and metastasis. **1.** Leung, Abbenante & Fairlie (2000) *J. Med. Chem.* **43**, 305.

Tanshinone

This antioxidant and anti-inflammatory agent (FW = 294.34 g/mol; CAS 568-72-9; Soluble to 5 mM in DMSO with gentle warming), also named 6,7,8,9-tetrahydro-1,6,6-trimethylphenanthro[1,2-*b*]furan-10,11-dione, from *Salvia miltiorrhiza*, inhibits NF-κB and AP-1 binding to DNA and cytotoxic activity for a number of different cancer cells (1,2). Tanshinone also inhibits human aortic smooth muscle cell migration and MMP-9 activity through AKT signaling pathway (2,3). Tanshinone inhibits β-amyloid aggregation and protects PC12 cells from β-amyloid$_{25-35}$-induced apoptosis via PI3K/Akt signaling pathway (4). **1.** Wang, Wang, Jiang, *et al.* (2007) *J. Neurooncol.* **82**, 11. **2.** Jin, Suh, Chang, *et al.* (2008) *J. Cell Biochem.* **104**, 15. **3.** Lee, Sher, Chen, *et al.* (2008) *Mol. Cancer Ther.* **7**, 3527. **4.** Dong, Mao, Wei, *et al.* (2012) *Mol. Biol. Rep.* **39**, 6495.

Tapentadol

This powerful analgesic (FW$_{free-base}$ = 221.34 g/mol; CAS 175591-09-0), also known by the trade names Nucynta®, Palexia®, and Tapal® as well as the systematic name 3-[(1*R*,2*R*)-3-(dimethylamino)-1-ethyl-2-methylpropyl]-phenol, is a CNS analgesic with dual action as a μ-opioid receptor agonist and a norepinephrine reuptake inhibitor (1,2). Tapentadol is also an agonist of the σ$_2$ receptor, but the role of this receptor is unclear. It shows high efficacy in acute nociceptive, acute and chronic inflammatory, as well as in chronic neuropathic pain. **Pharmacokinetics:** In four healthy male subjects receiving a single 100-mg oral dose of 3-[^{14}C]-labeled tapentadol, the serum pharmacokinetics and excretion kinetics of tapentadol and its conjugates were assessed, as was tolerability (3). Absorption was rapid, with a mean serum [C$_{max}$] of 2.45 μg-eq/mL and a time-to-C$_{max}$ value of 75-90 minutes. The drug was present mainly (24:1) as conjugated metabolites. Excretion of radiocarbon was rapid and complete (>95% within 24 hours; 99.9% within 5 days) and almost exclusively renal (99%: 69% as conjugates; 27% as other metabolites; 3% in unchanged form). No severe adverse events or clinically relevant changes in vital signs, laboratory measurements, electrocardiogram recording, or physical examination findings were reported (2). Tapentadol is metabolized to *N*-demethyltapentadol. With no active metabolites and minimal protein binding, tapentadol provides improved tolerability with a lower potential for pharmacokinetic drug-drug interactions or accumulation with impaired renal or hepatic function when compared with oxycodone (4). (*See Oxycodone; Tramadol*) **Other Actions:** Tapentadol also inhibits K⁺-stimulated Calcitonin Gene-Related Peptide (CGRP) release from the rat brainstem in vitro through a mechanism involving an increase in 5-HT levels in the system and the subsequent activation of 5-HT3 receptors (5). Notably, the α$_2$-antagonist yohimbine does not counteract the effects of tapentadol. Moreover, neither NA nor the α$_2$-agonist clonidine per se inhibits K⁺-stimulated CGRP release, thereby indicating that the effects of tapentadol are nor mediated through the block of NA reuptake. 5-HT and tramadol (the latter inhibiting both NA and 5-HT reuptake) significantly reduces K⁺-stimulated CGRP release. **See also Calcitonin 1.** Tzschentke, Christoph, Kögel, *et al.* (2007) *J. Pharmacol. Exp. Ther.* **323**, 265. **2.** Greco, Lisi, Currò, Navarra & Tringali (2014) *Peptides* **56**, 8. **3.** Terlinden, Ossig, Fliegert, Lange & Göhler

(2007) *Eur. J. Drug Metab. Pharmacokinet.* **32**, 163. **4.** Hartrick (2010) *Expert Rev. Neurother.* **10**, 861. **5.** Greco, Navarra & Tringali (2015) *Life Sci.* **145**, 161.

Taraxast-20-ene-3β,16β-diol, *See Faradiol*

Taraxast-20(30)-ene-3β,16β-diol. *See Amidiol*

Tarichatoxin, *See Tetrodotoxin*

Tartradiamide

This diamide of tartaric acid (FW = 148.12 g/mol), whicn can exist as one of three possible stereoisomers (D, L, and *meso*) inhibits acid phosphatase (IC$_{50}$ ≈ 17 mM). **Target(s):** acid phosphatase (1,2). **1.** Hollander (1971) *The Enzymes*, 3rd ed. (Boyer, ed.), **4**, 449. **2.** Kilsheimer & Axelrod (1957) *J. Biol. Chem.* **227**, 879.

Tartronate

This α-hydroxy acid (FW$_{free-acid}$ = 120.06 g/mol), also known as hydroxymalonic acid and hydroxypropanedioic acid, is a water-soluble dicarboxylic acid with pK_a values of 2.42 and 4.54 at 25°C. **Target(s):** acid phosphatase, weakly inhibited (1,2); D-aspartate oxidase (3,4); cyanate hydratase, *or* cyanase (5); D-2-hydroxyacid dehydrogenase (6); isocitrate lyase (7-9); D-lactate dehydrogenase (cytochrome) (10); L-lactate dehydrogenase (11-13); L-malate dehydrogenase (14-18); malic enzyme (19-23); phosphoenolpyruvate carboxykinase (GTP), weakly inhibited (24,25); 3-phosphoglycerate kinase (26); serine 3-dehydrogenase (27); D(−)-tartrate dehydratase (28); L(+)-tartrate dehydratase (29,30); *meso*-tartrate dehydratase (28); tartronate semialdehyde reductase, *or* 2-hydroxy-3-oxopropionate reductase (31); D-threonine dehydrogenase (32); L-threonine dehydrogenase (33). **1.** Hollander (1971) *The Enzymes*, 3rd ed. (Boyer, ed.), **4**, 449. **2.** Kilsheimer & Axelrod (1957) *J. Biol. Chem.* **227**, 879. **3.** de Marco & Crifò (1967) *Enzymologia* **33**, 325. **4.** Rinaldi (1971) *Enzymologia* **40**, 314. **5.** Anderson, Johnson, Endrizzi, Little & Korte (1987) *Biochemistry* **26**, 3938. **6.** Cremona (1964) *J. Biol. Chem.* **239**, 1457. **7.** Hoyt, Robertson, Berlyn & Reeves (1988) *Biochim. Biophys. Acta* **966**, 30. **8.** Hoyt, Johnson & Reeves (1991) *J. Bacteriol.* **173**, 6844. **9.** Reiss & Rothstein (1974) *Biochemistry* **13**, 1796. **10.** Nygaard (1963) *The Enzymes*, 2nd ed. (Boyer, Lardy & Myrbäck, eds.), **7**, 557. **11.** Chang, Huang & Chiou (1991) *Arch. Biochem. Biophys.* **284**, 285. **12.** Schwert & Winer (1963) *The Enzymes*, 2nd ed. (Boyer, Lardy & Myrbäck, eds.), **7**, 142. **13.** Green & Brosteaux (1936) *Biochem. J.* **30**, 1489. **14.** Yoshida (1969) *Meth. Enzymol.* **13**, 141. **15.** Green (1936) *Biochem. J.* **30**, 2095. **16.** Scholefield (1955) *Biochem. J.* **59**, 177. **17.** Harada & Wolfe (1968) *J. Biol. Chem.* **243**, 4123 and 4131. **18.** Banaszak & Bradshaw (1975) *The Enzymes*, 3rd ed. (Boyer, ed.), **11**, 369. **19.** Chang, Huang, Wang, *et al.* (1992) *Arch. Biochem. Biophys.* **296**, 468. **20.** Park, Kiick, Harris & Cook (1984) *Biochemistry* **23**, 5446. **21.** Iwakura, Hattori, Arita, Tokushige & Katsuki (1979) *J. Biochem.* **85**, 1355. **22.** Stickland (1959) *Biochem. J.* **73**, 646 and 654. **23.** Schimerlik & Cleland (1977) *Biochemistry* **16**, 565. **24.** Guidinger & Nowak (1990) *Arch. Biochem. Biophys.* **278**, 131. **25.** Harlocker, Kapper, Greenwalt & Bishop (1991) *J. Exp. Zool.* **257**, 285. **26.** Tompa, Hong & Vas (1986) *Eur. J. Biochem.* **154**, 643. **27.** Chowdhury, Higuchi, Nagata & Misono (1997) *Biosci. Biotechnol. Biochem.* **61**, 152. **28.** Shilo (1957) *J. Gen. Microbiol.* **16**, 472. **29.** Hurlbert & Jakoby (1965) *J. Biol. Chem.* **240**, 2772. **30.** Kelly & Scopes (1986) *FEBS Lett.* **202**, 274. **31.** Gotto & Kornberg (1961) *Biochem. J.* **81**, 273. **32.** Misono, Kato, Packdibamrung, Nagata & Nagasaki (1993) *Appl. Environ. Microbiol.* **59**, 2963. **33.** Guerranti, Pagani, Neri, *et al.* (2001) *Biochim. Biophys. Acta* **1568**, 45.

D-Tartronate Semialdehyde 2-Phosphate

This metabolite (FW$_{free-acid}$ = 184.04 g/mol), named systematically as 3-oxo-2-phosphonooxy-propanoic acid, is produced by the action of pyruvate kinase on β-hydroxypyruvate, first as the enol form, which then tautomerizes to the aldehyde and becomes hydrated. Both the enolate and the aldehyde forms are potent inhibitors of yeast enolase (1-4). Brewer *et al.* (5) determined the kinetics of the reaction of human neuronal enolase and yeast enolase 1 with the slowly-reacting chromophoric substrate D-tartronate semialdehyde phosphate (TSP). All data were biphasic, fitting either two successive first-order reactions or two independent first-order reactions, with higher Mg^{2+} concentrations reducing the relative magnitude of the slower reaction. The results suggest a catalytically significant interaction between the enzyme's two subunits (5). **1**. Spring (1970) Ph.D. Thesis, Univ. of Minnesota, Minneapolis. **2**. Wold (1975) *Meth. Enzymol.* **41**, 120. **3**. Weiss, Boerner & Cleland (1989) *Biochemistry* **28**, 1634. **4**. Wold (1971) *The Enzymes*, 3rd. ed. (Boyer, ed.), **5**, 499. **5**. Brewer, McKinnon & Phillips (2010) *FEBS Lett.* **584**, 979.

Tartronic Semialdehyde, See *Tartronate Semialdehyde*

TAS-102, See *Tipiracil*

TAS-106, See *3'-Ethynyl-Cytosine*

TAS 301

This synthetic indole derivative (FW = 357.40 g/mol; CAS 193620-69-8; Solubility: 40 mg/mL DMSO) inhibits smooth muscle cell migration and proliferation, inhibiting intimal thickening after balloon injury to rat carotid arteries (1). TAS-301 also blocks receptor-operated calcium influx and inhibits rat vascular smooth muscle cell proliferation induced by basic fibroblast growth factor and platelet-derived growth factor. **1**. Muranaka, Yamasaki, Nozawa, *et al.* (1998) *J. Pharmacol. Exp. Ther.* **285**, 1280. **2**. Sasaki, Nozawa, Miyoshi, *et,al.* (2000) *Jpn. J. Pharmacol.* **84**, 252.

Taselisib

This orally bioavailable phosphatidylinositol 3-kinase (*or* PI3K) inhibitor (FW = 459.55 g/mol; CAS 1282512-48-4), also called GDC-0032, RG7604, and 2-(4-(2-(1-isopropyl-3-methyl-1*H*-1,2,4-triazol-5-yl)-5,6-dihydrobenzo-[*f*]imidazo[1,2-*d*][1,4]oxazepin-9-yl)-1*H*-pyrazol-1-yl)-2-methylpropan-amide, selectively targets PI3Kα (K_i = 0.2 nM), with reduced inhibitory activity against PI3Kβ (1,2). Mean single-dose pharmacokinetic parameters, determined by an LC-MS/MS method, for a Phase I/II clinical trial were: C_{max} = 35.2 ng/mL, AUC$_{0-inf}$ = 1570 ng·hour/mL, and $t_{1/2}$ = 39.3 hours (3). **1**. Lopez, Schwab, Cocco, *et al.* (2014) *Gynecol. Oncol.* **135**, 312. **2**. Ndubaku, Heffron, Staben, *et al.* (2013) *J. Med. Chem.* **56**, 4597. **3**. Ding, Faber, Shi, *et al.* (2016) *J. Pharm. Biomed. Anal.* **126**, 117.

TAT-*cyclo*-(CLLFVY)

TAT penetratin sequence

This disulfide-linked membrane-penetrating derivative of *cyclo*-(Cys-Leu-Leu-Phe-Val-Tyr) (FW = g/mol; CAS 1446322-66-2) is a selective HIF-1 dimerization inhibitor that blocks protein-protein interaction of recombinant HIF-1α (but not HIF-2α) with HIF-1β. TAT-*cyclo*-(CLLFVY) exploits the presence of the TAT protein sequence, which facilitates direct cellular uptake. Once within cells, TAT-*cyclo*-(CLLFVY) is cleaved by reduced glutathione to produce *cyclo*-(CLLFVY), which then inhibits hypoxia-induced HIF-1 activity, and decreases VEGF and CAIX expression in osteosarcoma and breast cancer cells *in vitro*. TAT-*cyclo*-(CLLFVY) also inhibits capillary tubularization of hypoxic HUVECs. Hypoxia inducible factor-1 (HIF-1) is a heterodimeric transcription factor that acts as the master regulator of cellular response to reduced oxygen levels, thus playing a key role in the adaptation, survival, and progression of tumors. Identified from a library of 3.2 million cyclic hexapeptides, this inhibitor blocks (IC$_{50}$ = 1.3 µM) the HIF-1α/HIF-1β protein-protein interaction *in vitro* and in cells. **1**. Packham, Eccles & Tavassoli (2013) *J. Am. Chem. Soc.* **135**, 10418.

Taurine

This water-soluble cysteine metabolite (FW = 125.15 g/mol; CAS 107-35-7; pK_a = 1.5 & 8.74), also known as 2-aminoethanesulfonate, is a β-aminosulfonic acid that is abundant in brain. The thiol group of cysteine is oxidized to cysteine sulfinic acid (by cysteine dioxygenase), decarboxylated to hypotaurine (by sulfinoalanine decarboxylase), and then oxidized to taurine spontaneously or enzymatically. Taurine functions in many metabolic reactions and processes, the latter including osmoregulation, bile salt formation, and as an apoptosis modulator in some cells. Taurine is conjugated via its amino terminal group with chenodeoxycholic acid and cholic acid to form the bile salts known as sodium taurochenodeoxycholate and sodium taurocholate. Taurine also exhibits diuretic properties, faciliating the movement of potassium, sodium, and calcium in and out of the cell. **Target(s):** 4-aminobutyrate aminotransferase (1); 2-aminohexano-6-lactam racemase (2); bile acid acyl glucuronosyltransferase (3); γ-glutamyl transpeptidase, *or* weakly inhibited (4); K$^+$ channel, ATP-sensitive (5); Na$^+$/K$^+$-exchanging ATPase (6); pantothenate synthetase, *or* pantoate:β-alanine ligase (7,8); partial agonist at the inhibitory glycine receptor, with low concentrations competitively inhibiting glycine responses, whereas higher concentrations elicit a significant membrane current (9). **1**. Sulaiman, Suliman & Barghouthi (2003) *J. Enzyme Inhib. Med. Chem.* **18**, 297. **2**. Ahmed, Esaki, Tanaka & Soda (1985) *Agric. Biol. Chem.* **49**, 2991. **3**. Mano, Nishimura, Narui, Ikegawa & Goto (2002) *Steroids* **67**, 257. **4**. Lisy, Dutton & Currie (1980) *Life Sci.* **27**, 2615. **5**. Satoh (1996) *Gen. Pharmacol.* **27**, 625. **6**. Mrsny & Meizel (1985) *Life Sci.* **36**, 271. **7**. Miyatake, Nakano & Kitaoka (1979) *Meth. Enzymol.* **62**, 215. **8**. Miyatake, Nakano & Kitaoka (1978) *J. Nutr. Sci. Vitaminol.* **24**, 243. **9**. Schmieden and Betz (1995) *Mol. Pharmacol.* **48**, 919.

Taurine Chloramine, See *N-Chlorotaurine*

Taurochenodeoxycholate

This bile salt (FW$_{sodium-salt}$ = 521.69 g/mol; CAS 6009-98-9), an amide conjugate of chenodeoxycholic acid with taurine, is a major bile constituent and detergent in most mammals. **Target(s):** alkaline phosphatase (1); γ-glutamyl transpeptidase, weakly inhibited (2); glutathione *S*-transferase (3); kinesin (4); lipase, pancreatic (5,6); and pepsin (7). **1.** Martins, Negrao, Hipolito-Reis & Azevedo (2000) *Clin. Biochem.* **33**, 611. **2.** Abbott & Meister (1983) *J. Biol. Chem.* **258**, 6193. **3.** Boyer, Vessey, Holcomb & Saley (1984) *Biochem. J.* **217**, 179. **4.** Marks, LaRusso & McNiven (1995) *Gastroenterology* **108**, 824. **5.** Patton & Carey (1981) *Amer. J. Physiol.* **241**, G328. **6.** Bosc-Bierne, Rathelot, Perrot & Sarda (1984) *Biochim. Biophys. Acta* **794**, 65. **7.** Eto & Tompkins (1985) *Amer. J. Surg.* **150**, 564.

Taurocholate

This amphipathic bile salt and bio-detergent (FW$_{sodium-salt}$ = 537.69 g/mol; CAS 145-42-6; pK_a = 1.9), an amide conjugate of cholic acid with taurine, has a critical micelle concentration of 3-11 mM and an aggregation number of four. For many of the following enzymes, the inhibitor is the micellar form, not monomeric taurocholate. **Target(s):** β-*N*-acetylhexosaminidase (1); 2-acylglycerol *O*-acyltransferase, *or* monoacylglycerol *O*-acyltransferase (59,60); alkaline phosphatase (2); bile-acid-CoA:amino acid *N*-acyltransferase (57,58); bis(2-ethylhexyl)phthalate esterase (46); carbonic anhydrase (3-5); carboxylesterase, *or* esterase (13,14,52); cholesterol acyltransferase (6); cholesterol esterase (7); cholesterol 7α-monooxygenase, *or* CYP7A1 (8-10); choloyl-CoA synthetase, *or* cholate:CoA ligase (36); CYP7B1, *or* oxysterol 7α-monooxygenase (11); diacylglycerol *O*-acyltransferase (61); dolichyl esterase (12); α-L-fucosidase (15); α-galactosidase (39); galactosylceramidase (37); galactosylgalactosylglucosyl-ceramidase (16); glucose-6-phosphatase (17); β-glucosidase (40-42); γ-glutamyl transpeptidase, weakly inhibited (18); *sn*-glycerol-3-phosphate dehydrogenase (19); glycine *N*-choloyltransferase, alternative product (20); glycolipid 2-α-mannosyltransferase (55); guanidoacetate methyltransferase (21); 2-hydroxyacylsphingosine 1β-galactosyltransferase, *or* cerebroside synthase (22); 3-hydroxy-3-methylglutaryl-CoA reductase (9,23); lipase, *or* triacylglycerol lipase (24-26,51); lipoprotein lipase (27); lipoxygenase (28); lysophospholipase (29); lysozyme (30); β-mannosidase (31,38); phosphoglyceride: lysophosphatidylglycerol *O*-acyltransferase (32); phospholipase A$_2$ (50); phospholipase D (45); pyridoxine-5'-β-D-glucoside β-glucosidase (33); retinol *O*-fatty-acyltransferase (34,56); sphingomyelin phosphodiesterase, *or* sphingomyelinase (44); sterol esterase, *or* cholesterol esterase (47-49); sterol 3β-glucosyltransferase (53,54); and steryl-sulfatase (35,43). **1.** Johnson, Mook & Brady (1972) *Meth. Enzymol.* **28**, 857. **2.** Martins, Negrao, Hipolito-Reis & Azevedo (2000) *Clin. Biochem.* **33**, 611. **3.** Milov, Jou, Shireman & Chun (1992) *Hepatology* **15**, 288. **4.** Salomoni, Zuccato, Granelli, *et al.* (1989) *Scand. J. Gastroenterol.* **24**, 28. **5.** Kivilaakso (1982) *Amer. J. Surg.* **144**, 554. **6.** Swell & Treadwell (1950) *J. Biol. Chem.* **185**, 349. **7.** Poon & Simon (1975) *Biochim. Biophys. Acta* **384**, 138. **8.** Twisk, Lehmann & Princen (1993) *Biochem. J.* **290**, 685. **9.** Pandak, Vlahcevic, Chiang, Heuman & Hylemon (1992) *J. Lipid Res.* **33**, 659. **10.** Shefer, Nguyen, Salen, *et al.* (1990) *J. Clin. Invest.* **85**, 1191. **11.** Pandak,

Hylemon, Ren, *et al.* (2002) *Hepatology* **35**, 1400. **12.** Sumbilla & Waechter (1985) *Meth. Enzymol.* **111**, 471. **13.** Glick & King (1932) *J. Biol. Chem.* **97**, 675. **14.** Nachlas & Seligman (1949) *J. Biol. Chem.* **181**, 343. **15.** Dawson & Tsay (1977) *Arch. Biochem. Biophys.* **184**, 12. **16.** Johnson & Brady (1972) *Meth. Enzymol.* **28**, 849. **17.** Wallin & Arion (1972) *Biochem. Biophys. Res. Commun.* **48**, 694. **18.** Abbott & Meister (1983) *J. Biol. Chem.* **258**, 6193. **19.** Edgar & Bell (1979) *J. Biol. Chem.* **254**, 1016. **20.** Czuba & Vessey (1980) *J. Biol. Chem.* **255**, 5296. **21.** Sourkes (1951) *The Enzymes*, 1st ed. (J. B. Sumner & Myrbäck, eds.), **1** (part 2), 1068. **22.** Neskovic, Mandel & Gatt (1981) *Meth. Enzymol.* **71**, 521. **23.** Nguyen, S. Shefer, G. Salen, *et al.* (1994) *Metabolism* **43**, 1446. **24.** Desnuelle (1972) *The Enzymes*, 3rd ed. (Boyer, ed.), **7**, 575. **25.** Fodor (1948) *Enzymologia* **12**, 333. **26.** Fodor & Chari (1948) *Enzymologia* **13**, 258. **27.** Korn (1962) *Meth. Enzymol.* **5**, 542. **28.** Holman & Bergström (1951) *The Enzymes*, 1st ed. (J. B. Sumner & Myrbäck, eds.), **2** (part 1), 559. **29.** Sun, Tang, Huang & MacQuarrie (1987) *Neurochem. Res.* **12**, 451. **30.** Vantrappen, Ghoos & Peeters (1976) *Amer. J. Dig. Dis.* **21**, 547. **31.** LaBadie & Aronson, Jr. (1973) *Biochim. Biophys. Acta* **321**, 603. **32.** Hostetler, Huterer & Wherrett (1992) *Meth. Enzymol.* **209**, 104. **33.** Trumbo, Banks & Gregory III (1990) *Proc. Soc. Exp. Biol. Med.* **195**, 240. **34.** Helgerud, Petersen & Norum (1982) *J. Lipid Res.* **23**, 609. **35.** Dibbelt & Kuss (1991) *Biol. Chem. Hoppe Seyler* **372**, 173. **36.** Simion, Fleischer & Fleischer (1983) *Biochemistry* **22**, 5029. **37.** Rushton & Dawson (1975) *Biochim. Biophys. Acta* **388**, 92. **38.** Dawson (1982) *J. Biol. Chem.* **257**, 3369. **39.** Dean & Sweeley (1979) *J. Biol. Chem.* **254**, 9994. **40.** Daniels, Coyle, Chiao, Glew & Labow (1981) *J. Biol. Chem.* **256**, 13004. **41.** Maret, Salvayre, Negre & Douste-Blazy (1983) *Eur. J. Biochem.* **133**, 283. **42.** Glew, Peters & Christopher (1976) *Biochim. Biophys. Acta* **422**, 179. **43.** Iwamori, Moser & Kishimoto (1976) *Arch. Biochem. Biophys.* **174**, 199. **44.** Barnholz, Roitman & Gatt (1966) *J. Biol. Chem.* **241**, 3731. **45.** Taki & Kanfer (1979) *J. Biol. Chem.* **254**, 9761. **46.** Krell & Sandermann (1984) *Eur. J. Biochem.* **143**, 57. **47.** Stoddard Hatch, Brown, Deck, *et al.* (2002) *Biochim. Biophys. Acta* **1596**, 381. **48.** Natarajan, Ghosh & Grogan (1996) *Biochem. Biophys. Res. Commun.* **225**, 413. **49.** Hajjar, Minick & Fowler (1983) *J. Biol. Chem.* **258**, 192. **50.** Yang, Farooqui & Horrocks (1996) *Adv. Exp. Med. Biol.* **416**, 309. **51.** Finkelstein, Strawich & Sonnino (1970) *Biochim. Biophys. Acta* **206**, 380. **52.** McGhee (1987) *Biochemistry* **26**, 4101. **53.** Wojciechowski, Zimowski, Zimowski & Lyznik (1979) *Biochim. Biophys. Acta* **570**, 363. **54.** Kalinowska & Wojciechowski (1986) *Phytochemistry* **25**, 45. **55.** Schutzbach, Springfield & Jensen (1980) *J. Biol. Chem.* **255**, 4170. **56.** Chaudhary & Nelson (1987) *Biochim. Biophys. Acta* **917**, 24. **57.** Kimura, Okuno, Inada, Ohyama & Kido (1983) *Hoppe-Seyler's Z. Physiol. Chem.* **364**, 637. **58.** Czuba & Vessey (1980) *J. Biol. Chem.* **255**, 5296. **59.** Coleman & Haynes (1986) *J. Biol. Chem.* **261**, 224. **60.** Bhat, Bardes & Coleman (1993) *Arch. Biochem. Biophys.* **300**, 663. **61.** Parthasarathy, Murari, Crilly & Baumann (1981) *Biochim. Biophys. Acta* **664**, 249.

Taurodehydrocholate

This bile acid analogue (FW = 509.66 g/mol; CAS 517-37-3), also known as 5β-cholanic acid-3,7,12-trione *N*-(2-sulfoethyl)amide, inhibits bile-salt sulfotransferase. **1.** Chen (1982) *Biochim. Biophys. Acta* **717**, 316.

Taurodeoxycholate

This bile salt and bio-detergent (FW$_{sodium-salt}$ = 521.69 g/mol; CAS 81-24-3), a product of the conjugation of deoxycholic acid with taurine, is a major constituent of bile in most mammals. Taurodeoxycholate is frequently employed in the isolation of membrane proteins. Taurodeoxycholate has a critical micelle concentration of 1-4 mM and an aggregation number of six. For many of the following enzymes, the inhibitor is the micellar form, not monomeric taurocholate. **Target(s):** β-N-acetylhexosaminidase (1); alkaline phosphatase (2); bile-acid-CoA:amino acid N-acyltransferase (3); ceramidase (4); CYP7B1, *or* oxysterol 7α-monooxygenase (5); endoglycosylceramidase (6,7); γ-glutamyl transpeptidase, weakly inhibited (8); 2-hydroxyacylsphingosine 1β-galactosyltransferase, *or* cerebroside synthase (9); lipase, *or* triacylglycerol lipase (10-14); N-(long-chain-acyl)ethanolamine deacylase (15); *trans*-octaprenyl*trans*transferase (16); phospholipase A$_1$ (17); phospholipase D (18); proteasome endopeptidase complex (19); and squalene:hopene cyclase (20-22). **1**. Uda & Itoh (1983) *J. Biochem.* **93**, 847. **2**. Martins, Negrao, Hipolito-Reis & Azevedo (2000) *Clin. Biochem.* **33**, 611. **3**. Kimura, Okuno, Inada, Ohyama & Kido (1983) *Hoppe-Seyler's Z. Physiol. Chem.* **364**, 637. **4**. Kita, Okino & Ito (2000) *Biochim. Biophys. Acta* **1485**, 111. **5**. Pandak, Hylemon, Ren, *et al.* (2002) *Hepatology* **35**, 1400. **6**. Horibata, Okino, Ichinose, Omori & Ito (2000) *J. Biol. Chem.* **275**, 31297. **7**. Horibata, Sakaguchi, Okino, *et al.* (2004) *J. Biol. Chem.* **279**, 33379. **8**. Abbott & Meister (1983) *J. Biol. Chem.* **258**, 6193. **9**. Neskovic, Mandel & Gatt (1981) *Meth. Enzymol.* **71**, 521. **10**. Patton & Carey (1981) *Amer. J. Physiol.* **241**, G328. **11**. Momsen & Brockman (1976) *J. Biol. Chem.* **251**, 384. **12**. Desnuelle (1972) *The Enzymes*, 3rd ed. (Boyer, ed.), **7**, 575. **13**. Yang & Lowe (1998) *Protein Expr. Purif.* **13**, 36. **14**. Van Bennekum, Fisher, Blaner & Harrison (2000) *Biochemistry* **39**, 4900. **15**. Schmid, Zuzarte-Augustin & Schmid (1985) *J. Biol. Chem.* **260**, 14145. **16**. Teclebrhan, Olsson, Swiezewska & Dallner (1993) *J. Biol. Chem.* **268**, 23081. **17**. Nieder & Law (1983) *J. Biol. Chem.* **258**, 4304. **18**. Kobayashi & Kanfer (1991) *Meth. Enzymol.* **197**, 575. **19**. Klinkradt, Naude, Muramoto & Oelofson (1997) *Int. J. Biochem. Cell Biol.* **29**, 611. **20**. Ochs, Tappe, Gaertner, Kellner & Poralla (1990) *Eur. J. Biochem.* **194**, 75. **21**. Kleemann, Kellner & Poralla (1994) *Biochim. Biophys. Acta* **1210**, 317. **22**. Tippelt, Jahnke & Poralla (1998) *Biochim. Biophys. Acta* **1391**, 223.

Taurolithocholate 3-Sulfate

This sulfated bile salt (FW$_{ion}$ = 549.75 g/mol) inhibits the phosphatase activity of soluble epoxide hydrolase (IC$_{50}$ = 5 μM) and lipid-phosphate phosphatase. Soluble epoxide hydrolase is involved in the metabolism of arachidonic acid epoxides, endogenous chemical mediators that play important roles in blood pressure regulation, cell growth, and inflammation. This sulfated bile salt inhibits through a noncompetitive inhibition mechanism involving a new binding site on the C-terminal domain. **1**. Tran, Aronov, Tanaka, *et al.* (2005) *Biochemistry* **44**, 12179.

Tautomycin

This spiroketal-containing polyketide (FW = 809.05 g/mol), a dialkyl maleic anhydride derivative from *Streptomyces verticillatus*, exhibits IC$_{50}$ values of 0.21 nM and 0.94 nM, respectively, for phosphoprotein phosphatases 1 and 2A (1-11). Tautomycin also induces apoptosis (12) and inhibit smooth muscle endogenous phosphatase (13). **Target(s):** [myosin-light-chain] phosphatase (14); phosphoprotein phosphatases (1-11); protein phosphatase (11); protein phosphatase 2A (1-3); protein phosphatase (48); smooth muscle endogenous phosphatase (13). **1**. Eriksson, Toivola, Sahlgren, Mikhailov & Härmälä-Braskén (1998) *Meth. Enzymol.* **298**, 542. **2**. McCluskey, Sim & Sakoff (2002) *J. Med. Chem.* **45**, 1151. **3**. MacKintosh & Klumpp (1990) *FEBS Lett.* **277**, 137. **4**. Stubbs, Tran, Atwell, *et al.* (2001) *Biochim. Biophys.*

Acta **1550**, 52. **5**. Andrioli, Zaini, Viviani & da Silva (2003) *Biochem. J.* **373**, 703. **6**. Solow, Young & Kennelly (1997) *J. Bacteriol.* **179**, 5072. **7**. Sayed, Whitehouse & Jones (1997) *J. Endocrinol.* **154**, 449. **8**. Cohen, Philp & Vázquez-Martin (2005) *FEBS Lett.* **579**, 3278. **9**. Pilecki, Grzyb, Zien, Sekula & Szyszka (2000) *J. Basic Microbiol.* **40**, 251. **10**. Cohen (2004) in *Topics in Current Genetics* (J. Arino & D. R. Alexander, eds.) Springer **5**, 1. **11**. Plummer, Perreault, Holmes & Posse de Chaves (2005) *Biochem. J.* **385**, 685. **12**. Kikuchi, Shima, Mitsuhashi, Suzuki & Oikawa (1999) *Int. J. Mol. Med.* **4**, 395. **13**. Gong, Cohen, Kitazawa, *et al.* (1992) *J. Biol. Chem.* **267**, 14662. **14**. Essler, Amano, Kruse, *et al.* (1998) *J. Biol. Chem.* **273**, 21867.

Taxifolin

This phytoflavonoid and superoxide scavenger (FW = 304.26 g/mol; CAS 480-18-2), also called dihydroquercetin and 3,3′,4′,5,7-pentahydroxyflavanone, also inhibits numerous enzymes, including aldose reductase (1); creatine kinase (2); fatty-acid synthase (3); inositol-trisphosphate 3-kinase (4); lipoxygenase (15); Moloney murine leukemia virus reverse transcriptase (6); protein kinase C (7); and RNA-directed DNA polymerase (8). **1**. Haraguchi, Ohmi, Fukuda, *et al.* (1997) *Biosci. Biotechnol. Biochem.* **61**, 651. **2**. Miura, Muraoka & Fujimoto (2003) *Food Chem. Toxicol.* **41**, 759. **3**. Li & Tian (2004) *J. Biochem.* **135**, 85. **4**. Mayr, Windhorst & Hillemeier (2005) *J. Biol. Chem.* **280**, 13229. **5**. Ratty, Sunamoto & Das (1988) *Biochem. Pharmacol.* **37**, 989. **6**. Chu, Hsieh & Lin (1992) *J. Nat. Prod.* **55**, 179. **7**. Ferriola, Cody & Middleton, Jr. (1989) *Biochem. Pharmacol.* **38**, 1617.

Taxol, *See Paclitaxel*

Taxotere, *See Docetaxel*

Tay-Sachs' Ganglioside, *See Ganglioside* G$_{M2}$

Tazobactam

This triazolyl-substituted penicillanic acid sulfone (FW$_{free-acid}$ = 300.30 g/mol; CAS 89786-04-9), often used clinically with a broad-spectrum penicillin (*e.g.*, piperacillin), is an effective inhibitor against class A β-lactamases and some Class-C β-lactamases (1-12). Tazobactam first acylates the reactive seryl residue of its target, and the acylated enzymes then partitions between a hydrolysis and inactivation pathway involving interaction with other enzyme nucleophiles. Because the hydrolytic rate is high, thousands of hydrolytic turnovers may actually occur until inhibition is achieved. ***See also*** *LN-1-255* **1**. Perilli, Franceschini, Bonfiglio, *et al.* (2000) *J. Enzyme Inhib.* **15**, 1. **2**. Bonomo, Liu, Chen, *et al.* (2001) *Biochim. Biophys. Acta* **1547**, 196. **3**. Kuzin, Nukaga, Nukaga, *et al.* (2001) *Biochemistry* **40**, 1861. **4**. Yang, Rasmussen & Shlaes (1999) *Pharmacol. Ther.* **83**, 141. **5**. Eliasson, Kamme, Vang & Waley (1992) *Eur. J. Clin. Microbiol. Infect. Dis.* **11**, 313. **6**. Power, Galleni, Ayala & Gutkind (2006) *Antimicrob. Agents Chemother.* **50**, 962. **7**. Caporale, Franceschini, Perilli, *et al.* (2004) *Antimicrob. Agents Chemother.* **48**, 3579. **8**. Mariotte-Boyer, Nicolas-Chanoine & Labia (1996) *FEMS Microbiol. Lett.* **143**, 29. **9**. Voha, Docquier, Rossolini & Fosse (2006) *Antimicrob. Agents Chemother.* **50**, 2673. **10**. Hedberg, Lindqvist, Tuner & Nord (1992) *Eur. J. Clin. Microbiol. Infect. Dis.* **11**, 1100. **11**. Poirel, Brinas, Verlinde, Ide & Nordmann (2005) *Antimicrob. Agents Chemother.* **49**, 3743. **12**. Pagan-Rodriguez, Zhou, Simmons, *et al.* (2004) *J. Biol. Chem.* **279**, 19494.

TB21007

This brain-penetrant $GABA_A$ receptor inverse agonist (FW = 339.50 g/mol; CAS 207306-50-1; Soluble to 100 mM in DMSO), also named 6,7-dihydro-3-[(2-hydroxyethyl)thio]-6,6-dimethyl-1-(2-thiazolyl)benzo[c]thiophen-4(5H)-one, selectively targets selective for the α_5-subtype (K_i = 1.6 nM), with weaker affinity for subtypes α_2 (K_i = 16 nM), α_1 (K_i = 20 nM) and α_3 (K_i = 20 nM). TB-21007 enhances cognitive performance in rats in the delayed 'matching-to-place' Morris water maze test following i.p. administration. **1.** Chambers, Atack, Broughton, *et al.* (2003) *J Med Chem.* **46**, 2227.

TBB, *See* *4,5,6,7-Tetrabromobenzotriazole*

TBC11251, *See* *Sitaxentan*

DL-TBOA

This amino acid transport inhibitor (FW = 239.23 g/mol; CAS 205309-81-5; Soluble to 100 mM in DMSO and to 5 mM in H_2O with gentle warming), also known as DL-*threo*-β-benzyloxy-L-aspartic acid (with * marking the second asymmetric carbon), is a competitive, non-transportable blocker of excitatory amino acid transporters $EAAT_1$ (IC_{50} = 70 μM), $EAAT_2$ (IC_{50} = 70 μM), $EAAT_3$ (IC_{50} = 6 μM), $EAAT_4$ (K_i = 4.4 μM) and $EAAT_5$ (K_i = 3.2 μM), with low selectivity for ionotropic and metabotropic glutamate receptors.

TBPS, *See* *t-Butylbicyclo[2.2.2]phosphorothionate*

TC 1

This high-affinity σ_1 receptor ligand (FW = 297.37 g/mol; CAS 362512-81-0; Soluble to 50 mM in DMSO and to 50 mM in ethanol, both with gentle warming), also named (N-(3'-fluorophenyl)ethyl-4-azahexacyclo[5.4.1.$0^{2,6}.0^{3,10}.0^{5,9}.0^{8,11}$]dodecan-3-ol, displays selectivity, with K_i values of 10 and 370 nM, respectively, for σ_1 and σ_2 receptors (1). TC-1 exhibits low affinity for: dopamine transporters, *or* DAT (K_i > 10 μM); serotonin transporters, *or* SERT (K_i > 10 μM); noradrenalin transporters, *or* NET (K_i > 10 μM); and dopamine D_2 receptors (K_i = 1226 nM). TC-1 attenuates cocaine-induced locomotor activity in mice. **1.** Liu, Banister, Christie, *et al.* (2007) *Eur. J. Pharmacol.* **555**, 37.

TC 1698

This neuroprotective nicotinic receptor agonist (FW = 275.22 g/mol; CAS 787587-06-8; Soluble to 100 mM in water), also named 2-(3-pyridinyl)-1-azabicyclo[3.2.2]nonane dihydrochloride, selectively targets α_7 receptor agonist (EC_{50} = 440 nM) and displays weak partial agonist/antagonist activity at β-subunit-containing receptors (1,2). Its neuroprotective effect is prevented

through angiotensin receptor activation of the protein tyrosine phosphatase SHP-1. (1). **1.** Marrero, Papke, Bhatti, Shaw & Bencheri (2004) *J. Pharmacol. Exp. Ther.* **309**, 16. **2.** Papke, McCormack, Jack, *et al.* (2005) *Eur. J. Pharmacol.* **524**, 11.

TC 2559

This partial nicotinic acetylcholine receptor agonist ($FW_{free-base}$ = 206.14 g/mol; $FW_{di-Fumarate-Salt}$ = 438.43 g/mol; CAS 212332-35-9), also named 4-(5-ethoxy-3-pyridinyl)-N-methyl-(3E)-3-buten-1-amine, is subtype-selective for $\alpha_4\beta_2$ receptor (EC_{50} = 0.18 μM), showing weaker interactions with $\alpha_4\beta_4$ (EC_{50} = 12.5 μM), $\alpha_2\beta_4$ (EC_{50} = 14.0 μM), $\alpha_3\beta_4$ (EC_{50} > 30 μM), $\alpha_3\beta_2$ (EC_{50} > 100 μM), and α_7 (EC_{50} > 100 μM) subtypes. TC 2559 displays selectivity for $(\alpha_4)_2(\beta_2)_3$ receptor stoichiometry and enhanced central versus peripheral nervous system (CNS-PNS) selectivity ratio. TC-2559 competes effectively with nicotine binding (K_i = 5 nM), but not with bungarotoxin (K_i > 50 μM) (1-3). Dopamine release from striatal synaptosomes and ion flux from thalamic synaptosomes indicate that TC-2559 is potent and efficacious in the activation of CNS receptors and significantly reduced glutamate-induced neurotoxicity *in vitro*. TC-2559 has no detectable effects on muscle and ganglion-type nicotinic acetylcholine receptors at concentrations up to 1 mM. TC-2559 significantly attenuates scopolamine-induced cognitive deficits in a step-through passive avoidance task (1). TC 2559 attenuates scopolamine-induced cognitive deficits in a step-through passive avoidance task. Two glutamate receptor antagonists (6-cyano-7-nitroquinoxaline-2,3-dione and D(-)-2-amino-5-phosphonopentanoic acid) fail to reduce TC-2559-induced responses, suggesting that the TC-2559-induced increase in DA cell firing is caused by direct postsynaptic depolarization *via* activation of $\alpha_4\beta_2$ receptors, and not by enhancement of glutamate release (2). The EC_{50} for TC-2559 is unchanged in *Xenopus* oocytes that manipulated to express exclusively $(\alpha4)_2(\beta2)_3$ nicotinic acetylcholine receptors, however, the maximum effect of TC-2559 was dramatically enhanced. These results suggest that TC-2559 is a selective agonist of $(\alpha_4)_2(\beta2)_3$ nicotinic acetylcholine receptor. **1.** Bencherif, Bane, Miller, Dull & Gatto (2000) *Eur. J. Pharmacol.* **409**, 45. **2.** Chen, Sharples, Phillips, *et al.* (2003) *Neuropharmacology* **45**, 334. **3.** Zwart, Broad, Xi, *et al.* (2006) *Eur. J. Pharmacol.* **539**, 10.

TC-2153

This potent STEP inhibitor ($FW_{free-base}$ = 354.86 g/mol), also named 8-(trifluoromethyl)-1,2,3,4,5-benzopentathiepin-6-amine, targets STriatal-Enriched protein tyrosine Phosphatase, *or* STEP (IC_{50} = 25 nM), a neuron-specific enzyme regulating N-methyl-D-aspartate receptor (NMDAR) and α-amino-3-hydroxy-5-methyl-4-isoxazolepropionic acid receptor (AMPAR) trafficking, as well as ERK1/2, p38, Fyn, and Pyk2 activity. STEP is overactive in several neuropsychiatric and neurodegenerative disorders, including Alzheimer's disease (AD). Although the elemental sulfur allotrope S_8 was found to be a STEP inhibitor, it suffers from low water-solubility and an inability to be modified with the goal of improving its physicochemical properties, redox activity, binding affinity, and selectivity. By contrast, TC-2153 is a derivative with more desirable properties that may even serve as a lead for inhibitor development. This cyclic polysulfide forms a reversible covalent bond with the catalytic cysteine, thereby blocking STEP catalysis. TC-2153 increased tyrosine phosphorylation of STEP substrates ERK1/2, Pyk2, and GluN2B, and exhibited no toxicity in cortical cultures. In cell-based assays, TC-2153 increased tyrosine phosphorylation of STEP substrates ERK1/2, Pyk2, and GluN2B, and exhibited no toxicity in cortical cultures. TC-2153 improves cognitive function in several cognitive tasks in 6 and 12 month-old triple transgenic AD (3xTg-AD) mice, with no change in β-amyloid and phospho-Tau protein levels. **1.** Xu, Chatterjee, Baguley, *et al.* (2014) *PLoS Biol.* **12**, e1001923.

TC 14012

Arg-Arg-Nal-Cys-Tyr-Cit-Lys-D-Cit-Pro-Tyr-Arg-Cit-Cys-Arg-NH$_2$

This type-specific, peptidomimetic CXCR antagonist/ agonist (FW = 2066.43 g/mol; CAS 368874-34-4; Soluble to 1 mg/mL H$_2$O), containing 3-(2-naphthyl)alanine (Nal) and L-citrulline (Cit) replacements to reduce non-specific binding and cytotoxicity, is a polyphemusin derivative that acts an inverse agonist of CXCR4 as well as an agonist on CXCR7. A seven-transmembrane domain chemokine receptor, CXCR4 is involved in stem cell homing to bone marrow niches, cancer biology and metastasis, as well as HIV infection. CXCR7 is an atypical chemokine receptor that signals through β-arrestin in response to agonists without detectable activation of heterotrimeric G-proteins. The potency of β-arrestin recruitment to CXCR7 by TC14012 is much higher than that of the previously reported nonpeptidic antagonist Plerixafor (or AMD3100) and differs only by one log from that of the natural ligand CXCL12 (EC$_{50}$ 350 nM for TC14012, as compared with 30 nM for CXCL12 and 140 μM for Plerixafor). (**See** *Plerixafor*) **1**. Gravel, Malouf, Boulais, *et al.* (2010) *J. Biol. Chem.* **285**, 37939.

TC-A 2317

This potent, cell-permeable protein kinase inhibitor (FW$_{HCl-Salt}$ = 392.93 g/mol; CAS 1245907-03-2; Soluble to 100 mM in DMSO), also named 2-[(5-hydroxy-1,5-dimethylhexyl)amino]-4-methyl-6-[(5-methyl-1*H*-pyrazol-3-yl)amino]-3-pyridinecarbonitrile, targets Aurora kinase A (K_i = 1.2 nM) and Aurora kinase B (K_i = 101 nM), with IC$_{50}$ values > 1000 nM for >60 other kinases. **1**. Ando, Ikegami, Sakiyama, *et al.* (2010) *Bioorg. Med. Chem. Lett.* **20**, 4709.

TC AQP1 1

This water channel blocker (FW = 218.21 g/mol; CAS 37710-81-9), also named 3,3'-(1,3-phenylene)*bis*(2-propenoic acid), targets human Aquaporin 1, *or* hAQP1, (IC$_{50}$ = 8 μM), inhibiting water flux in *Xenopus laevis* oocyte swelling assays. Site-directed mutagenesis showed that Lys36, which is not conserved among the hAQP family, is crucially for hAQP1 binding. **1**. Seeliger, Zapater, Krenc, *et al.* (2013) *ACS Chem. Biol.* **8**, 249.

TC ASK 10

This orally available and potent protein kinase inhibitor (FW = 359.43 g/mol; CAS 1005775-56-3; Soluble to 100 mM in H$_2$O and to 100 mM in DMSO), also named 4-(1,1-dimethylethyl)-*N*-[6-(1*H*-imidazol-1-yl)imidazo[1,2-*a*]pyridin-2-yl]benzamide, selectively targets Apoptosis signal-regulating kinase 1, *or* ASK1 (IC$_{50}$ = 14 nM), displaying 36x weaker binding to ASK2 (IC$_{50}$ = 0.51 μM) and much weaker (IC$_{50}$ > 10 μM) binding to MEKK1, TAK1, IKKβ, ERK1, JNK1, p38α, GSK-3β, PKCθ and B-raf. TC ASK 10 blocks downstream JNK1/p38 phosphorylation in cells. **1**. Terao, Suzuki, Yoshikawa, *et al.* (2012) *Bioorg. Med. Chem. Lett.* **22**, 7326.

TcdA-10463, *See Large Clostridial Cytotoxins*

TC-DAPK 6

This potent ATP-competitive protein kinase inhibitor (FW = 276.29 g/mol; CAS 315694-89-4; Soluble to 50 mM in DMSO), also named (4*Z*)-2-[(*E*)-2-phenylethenyl]-4-(3-pyridinylmethylene)-5(4*H*)-oxazolone, selectively targets the death-associated protein kinase 1 (DAPK1), with IC$_{50}$ values of 69 and 225 nM for DAPK1 and DAPK3 and sharply lower affinity (IC$_{50}$ > 10 μM) for 48 other kinases, including Abl, AMPK, Chk1, Met and Src. **1**. Okamoto, Takayama, Shimizu, *et al.* (2009) *J. Med. Chem.* **52**, 7323.

TCDD

This carcinogenic agent (FW = 321.97 g/mol; CAS 1746-01-6), known also as dioxin and 2,3,7,8-tetrachlorodibenzo[*b,e*][1,4]dioxin, is an minor by-product in the chemical synthesis of chlorinated phenol herbicides, such as the defoliant Agent Orange, consisting of a 50:50 mixture of the butyl esters of (2,4,5-trichlorophenoxy)acetic acid (*or* 2,4,5-T) and (2,4-dichlorophenoxy)acetic acid (*or* 2,4-D). Great care must be exercised in handling dioxins, which are extremely toxic, causing dermatitis, liver damage, reproductive defects, porphyria, ulcers, and lesions. TCDD is a potent inducer of CYP1A (1). Most of the effects caused by TCDD are mediated by the aryl hydrocarbon receptor (AhR) which binds TCDD tightly. The liganded receptor translocates from the cytoplasm to the nuclei where it switches its partner molecule from Hsp90 to Arnt. Thus formed, the AhR/Arnt heterodimer binds to a specific DNA sequence in the promoter region of the target genes which include the genes for CYP1A1, UDP-glucuronosyltransferase, **Target(s):** retinol *O*-fatty-acyltransferase (2). **1**. Mimura & Fujii-Kuriyama (2003) *Biochim. Biophys. Acta* **1619**, 263. **2**. Nilsson, Hanber, Trossvik & Haekansson (1996) *Environ. Toxicol. Pharmacol.* **2**, 17.

TC-E 5001

This dual tankyrase inhibitor and Wnt pathway inhibitor (FW = 409.46 g/mol; CAS 865565-29-3; Soluble to 100 mM in DMSO), also named 3-(4-methoxyphenyl)-5-[[[4-(4-methoxyphenyl)-5-methyl-4*H*-1,2,4-triazol-3-yl]thio]methyl]-1,2,4-oxadiazole, targets TNKS1 (K_d = 79 nM), TNKS2 (K_d = 28 nM; IC$_{50}$ = 33 nM) but is without effect on PARP1 and PARP2 (IC$_{50}$ >19 μM). TC-E 5001 inhibits Wnt signaling and stabilizes Axin2 levels. **1**. Shultz, *et al.* (2012) *J. Med. Chem.* **55**, 1127.

TC-E 5003

This PRMT inhibitor (FW = 401.26 g/mol; CAS 17328-16-4; Soluble to 50 mM in DMSO), also named N,N'-(sulfonyldi-4,1-phenylene)bis(2-chloroacetamide), selectively targets protein arginine methyl-transferase-1, or PRMT1 (IC$_{50}$ = 1.5 µM), with no inhibitory activity against CARM1 and the lysine methyltransferase Set7/9. PRMT catalyzes methylation of arginine residues, transferring methyl groups from S-adenosyl-L-methionine to its terminal guanidino nitrogen atoms. TC-E 5003 inhibits growth of MCF7a breast cancer cells and LNCaP prostate cancer cells. It also attenuates androgen-induced gene expression in LNCaP cells. Given the susceptibility of α-chloroacetamido groups to nucleophilic displacement by active-site thiol and imidazole groups, TC-E 5003 most likely irreversibly inactivates PRMT1. **1.** Bissinger, Heinke, Spannhoff (2011) *Bioorg. Med. Chem.* **19**, 3717.

TC-E 5005

This potent and selective PDE inhibitor (FW = 270.33 g/mol; CAS 959705-64-7; Soluble in H$_2$O to 100 mM as HCl salt), also named 2-methoxy-6,7-dimethyl-9-propyl-imidazo[1,5-a]pyrido[3,2-e]pyrazine, selectively targets phosphodiesterase PDE10A (IC$_{50}$ = 7.28 nM), with far lower affinity for PDE2A (IC$_{50}$ = 239 nM), PDE11A (IC$_{50}$ = 779 nM), PDE5A (IC$_{50}$ = 919 nM), PDE7B (IC$_{50}$ = 3100 nM) and PDE3A (IC$_{50}$ = 3700 nM), with IC$_{50}$ values >5000 nM for PDE1B, PDE4A, PDE6, PDE8A and PDE9A). **1.** Höfgen, Stange, Schindler, *et al.* (2010) *J. Med. Chem.* **53**, 4399.

TC-F 2

This potent, reversible FAAH inhibitor (FW = 439.51 g/mol; CAS 1304778-15-1; Soluble to 100 mM in DMSO), also named 1-[(3S)-1-[4-(2-benzofuranyl)-2-pyrimidinyl]-3-piperidinyl]-3-ethyl-1,3-dihydro-2H-benzimidazol-2-one, targets fatty acid amide hydrolase, or FAAH, with respective IC$_{50}$ values of 28 and 100 nM for human and rat enzymes. Also known as oleamide hydrolase and anandamide amidohydrolase, FAAH plays a central role in inactivating endocannabinoids, such as N-arachidonyl-ethanolamide (anandamide). TC-F 2 shows far lower affinity (IC$_{50}$ > 20 µM) for cannabinoid-related targets like the CB$_1$ and CB$_2$ cannabinoid receptors as well as the <u>T</u>ransient <u>R</u>eceptor <u>P</u>otential cation channel subfamily <u>V</u> member <u>1</u> (TrpV1). **1.** Gustin, Ma, Min, *et al.* (2011) *Bioorg. Med. Chem. Lett.* **21**, 2492. **2.** Min, Thibault, Porter, *et al.* (2011) *Proc. Natl. Acad. Sci. U.S.A.* **108**, 7379.

TC-G 1004

This potent receptor antagonist (FW = 421.50 g/mol; CAS 1061747-72-5; Soluble to 100 mM (as HCl salt) and to 100 mM in DMSO), also named N-[2-(3,5-dimethyl-1H-pyrazol-1-yl)-6-[6-(4-methoxy-1-piperidinyl)-2-pyridinyl]-4-pyrimidinyl]acetamide, targets adenosine A$_{2A}$ receptors (K_i = 0.44 nM), with >100-fold selectivity compared to A$_1$ receptors (K_i = 85 nM)

and potentiates L-DOPA-induced rotational behavior in 6-hydroxydopamine-(or 6-OHDA-) lesioned rats. **1.** Zhang, Tellew, Luo, *et al.* (2008) *J. Med. Chem.* **51** 7099.

TC-G 24

This potent, brain-penetrating protein kinase inhibitor (FW = 330.73 g/mol; CAS 1257256-44-2; Soluble to 10 mM in DMSO, with gentle warming), also named N-(3-chloro-4-methylphenyl)-5-(4-nitrophenyl)-1,3,4-oxadiazol-2-amine, targets GSK-3β (IC$_{50}$ = 17 nM), displaying selectivity *versus* CDK2 (~22% inhibition at 10 µM). TC-G 24 increases liver glycogen reserves in rodents. **1.** Khanfar, Hill, Kaddoumi & El Sayed (2001) *J. Med. Chem.* **53**, 8534.

TC-H 106

This brain-penetrant, Class I HDAC inhibitor (FW = 339.43 g/mol; CAS 937039-45-7; Soluble to 100 mM in DMSO and to 50 mM in Ethanol) targets HDAC1 (IC$_{50}$ = 150 nM), HDAC3 (IC$_{50}$ = 370 nM), HDAC2 (IC$_{50}$ = 760 nM) and HDAC8 (IC$_{50}$ = 5000 nM), with no activity against class II HDACs. TC-H 106 is a slow, tight-binding inhibitor for HDAC1, with k_1 = 4.9 × 10^4 M^{-1} min^{-1}, k_{-1} = 0.0072 min^{-1}; K_i = 148 nM (1). **1.** Chou, *et al.* (2008) *J. Biol. Chem.* **283**, 35402. **2.** Rai, *et al.* (2008) *PLoS ONE* **3**, e1958. **3.** Xu, *et al.* (2009) *Chem. Biol.* **16**, 980.

TC HSD 21

This potent 17β-hydroxysteroid dehydrogenase inhibitor (FW = 422.32 g/mol; CAS 330203-01-5; Soluble: 100 mM in DMSO), also named 5-[(3-bromo-4-hydroxyphenyl)methylene]-3-(4-methoxyphenyl)-2-thioxo-4-thiazolidinone, targets human 17β-HSD isozyme-3 (IC$_{50}$ = 6 nM) and mouse 17β-HSD3 (IC$_{50}$ = 40 nM), with no activity toward 17β-HSD1, 17β-HSD2, or the estrogen receptor α (ERα), androgen receptors, or glucocorticoid receptors. **1.** Harada, Kubo, Tanaka & Nishioka (2012) *Bioorg. Med. Chem. Lett.* **22**, 504.

TC-I 15

This potent α$_2$β$_1$ integrin inhibitor (FW = 520.62 g/mol; CAS 916734-43-5; Solubility: 100 mM in DMSO; 25 mM as the sodium salt), also named N-[[(4R)-5,5-dimethyl-3-(phenylsulfonyl)-4-thiazolidinyl]carbonyl]-3-[[[(phenylmethyl)amino]carbonyl]amino]-L-alanine, blocks human platelet adhesion to type I collagen under static (IC$_{50}$ = 12 nM) and under flow (IC$_{50}$ = 715 nM) conditions). TC-I 15 displays selectivity for α$_2$β$_1$, with an IC$_{50}$ value >1000 nM for α$_v$β$_3$, α$_5$β$_1$, α$_6$β$_1$ and α$_{IIb}$β$_3$. It also reduces collagen IV production in mesangial cells. **1.** Miller, Basra, Kulp, *et al.* (2009) *Proc. Natl. Acad. Sci. U.S.A.* **106**, 719. **2.** Borza, Su, Chen, *et al.* (2012) *J. Am. Soc. Nephrol.* **23**, 1027.

TCID

This membrane-permeable UCH inhibitor (FW = 283.92 g/mol; CAS 30675-13-9; Solubility: 50 mM in DMSO), also named 4,5,6,7-tetrachloro-1H-indene-1,3(2H)-dione, selectively targets ubiquitin C-terminal hydrolase-L3 (UCH-L3) inhibitor (IC$_{50}$ = 0.6 μM), exhibiting >100x selectivity for UCH-L3 over UCH-L1 (1). **Pharmacologic Action:** Inhibitory glycinergic neurotransmission is terminated by Na$^+$/Cl$^-$-dependent glycine transporters (*or* GlyT's), with the glial GlyT1 mainly responsible for completion of inhibitory neurotransmission, and the neuronal GlyT2 mediating reuptake of the neurotransmitter. Protein Kinase C accelerates GlyT2 endocytosis by increasing its ubiquitination, and this process is in turn regulated by the deubiquitinating enzyme (*or* DUB) known as ubiquitin C-terminal hydrolase-L1 (UCHL1). TCID diminishes glycine transporter GlyT2 ubiquitination in brain stem and spinal cord primary neurons (2). **1**. Liu, Lashuel, Choi, *et al.* (2003) *Chem Biol.* **10**, 837. **2**. de Juan-Sanz, Núñez, López-Corcuera & Aragó (2013) *PLoS One* **8**, e58863.

TC LPA5 4

This LPA receptor antagonist (FW = 410.89 g/mol; CAS 1393814-38-4; Soluble to 100 mM in DMSO), also named 5-(3-chloro-4-cyclohexylphenyl)-1-(3-methoxyphenyl)-1H-pyrazole-3-carboxylic acid, targets the G protein-coupled, lysophosphatidic acid receptor-5 (LPA$_5$) receptor antagonist (IC$_{50}$ = 0.8 μM), with great selectivity compared to eighty other screened targets. TC-LPA5-4 also inhibits LPA-induced aggregation of isolated human platelets. **1**. Kozian, Evers, Florian, *et al.* (2012) *Bioorg. Med. Chem. Lett.* **22**, 5239.

TC-MCH 7c

This orally active, brain-penetrant MCH$_1$R antagonist (FW = 408.47 g/mol; CAS 864756-35-4; Solubility: 100 mM in water (as HCl salt); 50 mM in DMSO), also named 4-[(4-fluorophenyl)methoxy]-1-[4-[2-(1-pyrrolidinyl)ethoxy]phenyl]-2(1H)-pyridinone, potently and selectively targets the melanin-concentrating hormone receptor 1 (*or* MCH$_1$R) antagonist (IC$_{50}$= 5.6 nM, in hMCH$_1$R-expressing CHO cells). Melanin-concentrating hormone (MCH), which is a neuropeptide expressed in the hypothalamus of the brain, is involved in regulating feeding behavior and energy homeostasis. TC-MCH 7c displays selectivity for MCH$_1$R over MCH$_2$R receptors (IC$_{50}$ => 10 μM). It also decreases body weight in a mouse model of diet-induced obesity. **1**. Ito, Ishihara, Gomori, *et al.* (2009) *Eur. J. Pharmacol.* **624**, 77. **2**. Haga, Mizutani, Naya, *et al.* (2011) *Bioorg. Med. Chem.* **19**, 883.

TC Mps1 12

This orally active Mps1 inhibitor (FW = 324.38 g/mol; CAS 1206170-62-8; Solubility: 100 mM in DMSO), also named 4-[[4-amino-5-cyano-6-[(1,1-dimethylethyl)amino]-2-pyridinyl]amino]benzamide, potently and selectively targets monopolar spindle-1, *or* Mps1 (IC$_{50}$ = 6.4 nM), a kinetochore-associated protein kinase involved in both chromosome attachment and spindle checkpoint control. MPS1 kinases also function at centrosomes, with wider impact on development, cytokinesis, and several different signaling pathways. By virtue of its unusual flipped-peptide conformation, TC Mps1 12 exhibits high selectivity for Mps1when tested with a panel of 95 kinases, including JNK. It also inhibits A549 lung carcinoma cell proliferation and attenuates A549 cell xenograft growth in mice. **1**. Kusakabe, Ide, Daigo, *et al.* (2012) *ACS Med. Chem. Lett.* **3**, 560.

TCN-201

This NMDA receptor antagonist (FW = 461.89 g/mol; CAS 852918-02-6; Soluble to 100 mM in DMSO) display sub-μM and μM potency at NR1 and NR2A receptors, respectively, although they did not show activity at NR2B-containing receptor up to 50 μM concentration (1-4). NR1 and NR2A represent a subtype of *N*-methyl-D-aspartate receptors (NMDARs) that are glutamate- and glycine-gated Ca^{2+}-permeable channels and are highly expressed in the central nervous system. Antagonists that are sufficiently selective to preferentially block GluN2A-containing NMDA receptors (NMDARs) over GluN2B-containing NMDARs are few in number. Electrophysiological recordings from chimeric and mutant rat NMDA receptors suggest that TCN-201 binds to a novel allosteric site located at the dimer interface between the GluN1 and GluN2 agonist binding domains. Furthermore, we demonstrate that occupancy of this site by TCN-201 inhibits NMDA receptor function by reducing glycine potency. Addition of 1 mM glycine, but not 1 mM L-glutamate, surmounts the inhibitory effects of TCN-201 in the FLIPR NR1/NR2A assay. However, TCN-201 displaces the glutamate-site antagonist CGP 39653 to a greater extent than the glycine site antagonist MDL 105,519 in rat brain cortex binding assay. **1**. Bettini, Sava, Griffante, *et al.* (2010) *J. Pharm. Exp. Ther.* **335**, 644. **2**. Shin, Kim & Thayer (2012) *Brit. J. Pharmacol.* **166**, 1002. **3**. Edman, McKay, Macdonald, *et al.* (2012) *Neuropharmacology* **63**, 441. **4**. Hansen, Ogden & Traynelis (2012) *J. Neurosci.* **32**, 6197.

TC-N 1752

This channel blocker and experimental analgesic (FW = 516.52 g/mol; CAS 1211866-85-1), also named *N*-[2-methyl-3-[[4-[4-[[4-(trifluoromethoxy)phenyl]methoxy]-1-piperidinyl]-1,3,5-triazin-2-yl]amino]phenyl]acetamide, targets voltage-gated sodium channels: NaV1.7 channels (IC$_{50}$ = 0.17 μM) values are 0.17, 0.3, 0.4 and 1.1 μM at hNa$_V$1.7, hNa$_V$1.3 (IC$_{50}$ = 0.3 μM), hNa$_V$1.4 (IC$_{50}$ = 0.4 μM) and hNa$_V$1.5 (IC$_{50}$ = 1.1 μM). TC-N 1752 also inhibits tetrodotoxin-sensitive sodium channels. **1**. Bregman H1, Berry L, Buchanan

Tcnα-19402, *See Large Clostridial Cytotoxins*

TC OT 39

This nonpeptide OT/V2 partial agonist (FW = 600.78 g/mol; CAS 479232-57-0; Solubility: 100 mM in 1eq. HCl; 100 mM in DMSO), also named (2*S*)-*N*-[[4-[(4,10-dihydro-1-methylpyrazolo[3,4-*b*][1,5]benzodiazepin-5(1*H*)-yl)carbonyl]-2-methylphenyl]methyl]-2-[(hexahydro-4-methyl-1*H*-1,4-diazepin-1-yl)thioxomethyl]-1-pyrrolidinecarboxamide, targets oxytocin (*or* OT) (K_i = 147 nM) and vasopressin V$_2$ (K_i >1000 nM) receptors, while acting as an antagonist of V$_{1a}$ receptors, (K_i = 300 nM). TC OT 39 stimulates uterine contraction in rats. **1.** Pitt, Batt, Haigh, *et al.* (2004) *Bioorg. Med. Chem. Lett.* **14**, 4585. **2.** Frantz, Rodrigo, Boudier, *et al.* (2010) *J. Med. Chem.* 2010 **53**, 1546.

TC-P 262

This diaminopyrimidine (FW = 258.32 g/mol; CAS 873398-67-5; Soluble to 100 mM in DMSO), also named 5-[5-methyl-2-(1-methylethyl)phenoxy]-2,4-pyrimidinediamine, is a selective purinergic receptor antagonist that targets P2X$_3$ (pIC$_{50}$ = 7.39) and P2X$_{2/3}$ (pIC$_{50}$ = 6.68) receptors. The latter are cation-permeable ligand gated ion channels that open in response to the binding of extracellular ATP. TC-P-262 displays no detectable activity at P2X$_1$, P2X$_2$, P2X$_4$ and P2X$_7$ receptors, for which pIC$_{50}$ values are less than 4.7. Whole-cell voltage clamp electrophysiology confirmed these results. **1.** Ballini, Virginio, Medhurst, *et al.* (2011) *Brit. J. Pharmacol.* **163**, 1315.

TCS 46b

This NMDA receptor antagonist (FW = 345.44 g/mol; CAS 302799-86-6; Soluble to 100 mM in DMSO), also named 1,3-dihydro-5-[3-[4-(phenylmethyl)-1-2*H*-benzimidazol-2-one, targets the NR1A/NR2B (IC$_{50}$ = 5.3 nM), NR1A/2A (IC$_{50}$ = 35000 nM), and NR1A/2C (IC$_{50}$ = 100000 nM) subtypes of the *N*-methyl-D-aspartate receptor. Oral administration of at 10 and 30 mg/kg TCS 46b also potentiates the effect of L-DOPA in unilaterally 6-hydroxydopamine-lesioned (6-OHDA) rats. **1.** Wright, Gregory, Kesten, *et al.* (2000) *J. Med. Chem.* **43**, 3408. **2.** Roger, Lagnel, Besret, *et al.* (2003) *Bioorg. Med. Chem.* **11**, 5401.

TCS 183

This short peptide (FW = 1425.58 g/mol; Sequence: MSGRPRTTSFAES-NH$_2$; CAS unassigned; Soluble to 1 mg/mL H$_2$O), corresponding to residues 1-13 of human and mouse GSK-3β, is a competitive inhibitor (possibly even an alternative phosphoryl-accepting substrate) for enzymes catalyzing phosphorylation of Ser-9 in GSK-3β. This GSK isoform is unusual, because its kinase activity is primarily determined by the phosphorylation status of Ser-9. Dephosphorylation of this residue by Ser/Thr protein phosphatases activates GSK3β, whereas phosphorylation of Ser-9 inhibits its activity. In glycogen metabolism, for example, insulin stimulates PI3K, which leads to activation of Akt (*or* Protein Kinase B), which catalyzes GSK3β phosphorylation of GSK3β, favoring dephosphorylation of glycogen synthase and stimulation of glycogen synthesis. The sequence-scrambled control peptide, TCS 184 (FW = 1425.58 g/mol; Sequence: TAESTFMRPSGSR-NH$_2$) is a commercially available sequence-scrambled version of TCS 183 that serves as a control for nonspecific effects. **1.** Peineau, Taghibiglou, Bradley, *et al.* (2007) *Neuron* **53**, 703.

TCS 359

This potent receptor-type tyrosine-protein kinase inhibitor (FW = 360.43 g/mol; CAS 301305-73-7; Solubility: 15 mg/mL DMSO; <1 mg/mL H$_2$O), also named 2-[(3,4-dimethoxybenzoyl)amino]-4,5,6,7-tetrahydro-benzo[b]thiophene-3-carboxamide, targets Fetal Liver Kinase FLT3, with an IC$_{50}$ value of 27 nM for the isolated enzyme and 42 nM for the proliferation of MV4-11 cells. **1.** Patch, Baumann, Liu, *et al.* (2006) *Bioorg. Med. Chem Lett.* **16**, 3282.

TCS 401

This orally available and selective protein-tyrosine phosphatase inhibitor (FW$_{HCl-Salt}$ = 306.72 g/mol; CAS 243966-09-8), also named 2-[(carboxycarbonyl)amino]-4,5,6,7-tetrahydrothieno[2,3-*c*]pyridine-3-carboxylic acid hydrochloride, targets PTP1B (K_i = 0.29 nM), with far less affinity for CD45 D1D2 (K_i = 59 nM), PTPβ (K_i = 560 nM), PTPε D1 (K_i = 1100 nM), SHP-1 (K_i > 2000 nM), PTPα D1 (K_i > 2000 nM), and LAR D1D2 (K_i > 2000 nM). **1.** Iversen, *et al.* (2000) *J. Biol. Chem.* **275**, 10300. **2.** Andersen et al (2002) *J. Med. Chem.* **45**, 4443.

TCS 1102

This potent, dual orexin receptor antagonist, *or* DORA (FW = 470.59 g/mol; CAS 916141-36-1; Soluble to 100 mM in DMSO and to 100 mM in Ethanol), also named *N*-[1,1'-biphenyl]-2-yl-1-[2-[(1-methyl-1*H*-benzimidazol-2-yl)thio]acetyl-2-pyrrolidinedicarboxamide, targets OX$_2$ (K_i = 0.2 nM) and

OX$_1$ (K_i = 3 nM) receptors. **1**. Bergman, Roecker, Mercer, *et al.* (2008) *Bioorg. Med. Chem. Letts.* **18**, 1425. **2**. Winrow, Tanis, Reiss, *et al.* (2010) *Neuropharmacology* 58, 185.

TCS 2002

This potent and orally available GSK-3β inhibitor (FW = 338.38 g/mol; CAS 1005201-24-0; Soluble to 100 mM in DMSO), systematically named 2-methyl-5-[3-[4-(methylsulfinyl)phenyl]-5-benzofuranyl]-1,3,4-oxadiazole, targets Glycogen Synthase Kinase-3β (IC$_{50}$ = 35 nM). TCS-2202 shows a good pharmacokinetic profile, including favorable blood-brain barrier (BBB) penetration. TCS-2202 inhibits stress-induced hyperphosphorylation of the microtubule-associated protein Tau in mouse brain. **1**. Saitoh, Kunitomo, Kimura, *et al.* (2009) *J. Med. Chem.* **52**, 6270.

TCS 2312

This protein kinase inhibitor (FW = 412.49 g/mol; CAS 838823-31-7; Solubility: 20 mM in DMSO), also named 4'-[5-[[3-[(cyclopropylamino)-methyl]phenyl]amino]-1*H*-pyrazol-3-yl]-[1,1'-biphenyl]-2,4-diol, potently and selectively targets Checkpoint Kinase 1, *or* chk1 (K_i = 0.38 nM, EC$_{50}$ = 60 nM). TCS 2312 enhances the cell killing activity of gemcitabine in breast and prostate cancer cell lines and displays antiproliferative effects *in vitro*. **1**. Teng, Zhu, Johnson, Chen, *et al.* (2007) *J. Med. Chem.* **50**, 5253.

TCS 2314

This integrin very late antigen-4 (VLA-4; α$_4$β$_1$) antagonist (FW = 522.59 g/mol; CAS 317353-73-4; Soluble to 100 mM in DMSO and to 50 mM in ethanol), also systematically named 1-[[(3*S*)-4-[2-[4-[[[(2-methylphenyl) amino]carbonyl]amino]phenyl]acetyl]-3-morpholinyl]carbonyl]-4-piperi-dinediacetic acid, blocks (IC$_{50}$ = 4.4 nM) the activation of inflammatory cells. The integrin very late antigen-4 (VLA-4,R4β1,CD49d/CD29) is intrinsically involved in the pathogenesis of inflammatory and autoimmune diseases such as asthma, rheumatoid arthritis, type 1 diabetes, psoriasis, multiple sclerosis (MS), inflammatory bowel disease, and hepatitis C. A heterodimeric glycoprotein receptor that is constitutively expressed on the surface of nearly all leukocytes, VLA-4 binds vascular cell adhesion molecule-1 (VCAM-1, CD106) expressed on cytokine-stimulated endothelial cells and the alternatively spliced portion of the type-III connecting segment of fibronectin (FN), and mediates the process of adhesion, migration, and activation of inflammatory cells at the site of inflammation. Therefore, the blockade of the

interaction of VLA-4 with the ligands should alleviate the inappropriate cellular signaling processes and would thus be expected to be useful in the treatment of inflammatory and autoimmune diseases. TCS 2312 demonstrates excellent efficacy in models of asthma and PK profiles in rats, dogs, and monkeys. **1**. Muro, Iimura, Sugimoto, *et al.* (2009) *J. Med. Chem.* **52**, 7974.

TC-S 7003

This orally active, potent and selective protein kinase inhibitor (FW = 530.62 g/mol; CAS 847950-09-8; Soluble to 100 mM in DMSO) targets <u>L</u>ymphocyte specific <u>k</u>inase, *or* Lck (IC$_{50}$ = 7 nM), Lyn (IC$_{50}$ = 21 nM), Src (IC$_{50}$ = 42 nM), and Syk (IC$_{50}$ = 200 nM). TC-S 7003 displays >1000x selectivity for Lck *versus* MAPK, CDK and RSK family representatives. TC-S 7003 inhibits T-cell proliferation *in vitro* and inhibits arthritis in two *in vivo* models. **1**. Martin, Newcomb, Nunes, *et al.* (2008) *J. Med. Chem.* **51**, 1637.

TC-S 7005

This potent and selective serine/threonine protein kinase inhibitor (FW = 359.38 g/mol; CAS 1082739-92-1; Soluble to 100 mM in DMSO), also named 3-(1,3-benzodioxol-5-yl)-*N*-[(1*S*)-1-phenylethyl]isoxazolo[5,4-*c*]pyridin-5-amine, targets Polo-Like Kinase 2, *or* PLK2 (IC$_{50}$ = 4 nM), with weaker action against PLK3 (IC$_{50}$ = 24 nM), and PLK1 (IC$_{50}$ = 214 nM). Polo-like kinases play fundamental roles in regulating cell division, and treatment of cells with PLK inhibitors results in multiple mitotic defects, often attended by cell death. These enzymes are thus promising targets for cancer therapy, especially in view of their frequent overexpression in human tumors. TC-S 7005 induces mitotic arrest and cell death in HCT 116 colorectal cells. **1**. Hanan, Fucini, Romanowski, *et al.* (2008) *Bioorg. Med. Chem. Lett.* 18, 5186.

TCS 21311

This potent JAK inhibitor (FW = 526.51 g/mol; CAS 1260181-14-3; Soluble to 100 mM in DMSO), also named 3-[5-[4-(2-hydroxy-2-methyl-1-oxopropyl)-1-piperazinyl]-2-(trifluoromethyl)phenyl]-4-(1*H*-indol-3-yl)-1*H*-pyrrole-2,5-dione, targets Janus Kinase-3, *or* JAK3 (IC$_{50}$ = 8 nM), with much weaker action against JAK1 (IC$_{50}$ = 1017 nM), JAK2 (IC$_{50}$ = 2550 nM), and TYK2 (IC$_{50}$ = 8055 nM). TCS 21311 is only moderately selective inasmuch as it also inhibits GSK-3β (IC$_{50}$ = 3 nM), PKCα (IC$_{50}$ = 13 nM) and PKCθ (IC$_{50}$ = 68 nM). **1**. Thoma, Nuninger, Falchetto, *et al.* (2010) *J. Med.Chem.* **54**, 284.

TCS 5861528

This TRP$_{A1}$ channel blocker (FW = 369.42 g/mol; CAS 332117-28-9; Soluble to 100 mM in DMSO and to 10 mM in ethanol), also named 2-(1,3-dimethyl-2,6-dioxo-1,2,3,6-tetrahydro-7H-purin-7-yl)-N-[4-(1-methylpropyl)phenyl]-acetamide, antagonizes AITC- and 4-HNE-evoked calcium influx (IC$_{50}$ values of 14.3 and 18.7 μM respectively) and attenuates diabetic hypersensitivity in *in vivo* rat model. The TRP$_{A1}$ ion channel modulates excitability of nociceptors, and it may be activated by compounds resulting from oxidative insults. Diabetes mellitus produces oxidative stress and sensory neuropathy. **1**. Wei, Hämäläinen, Saarnilehto, Koivisto & Pertovaara (2009) *Anesthesiology* **111**, 147. **2**. Wei, Hämäläinen, Saarnilehto, Koivisto & Pertovaara (2010) *Neuropharmacology* **58**, 578. **3**. Wei, Koivisto & Pertovaara (2010) *Neurosci. Letts.* **479** 253.

TCS ERK 11e

This orally bioavailable and potent ERK inhibitor (FW = 500.35 g/mol; CAS 896720-20-0; Soluble to 100 mM in DMSO) selectively targets Extracellular signal-Related Kinase 2, *or* ERK2 (K_i < 2 nM), with much weaker action against GSK-3 (K_i = 395 nM), Aurora Kinase A (K_i - 540 nM) and Cdk2 (K_i = 852 nM). Specificity for ERK was optimized following the discovery of a conformational change for this drug, when bound to ERK2, relative to anti-target GSK3. A selective subnanomolar ERK2 inhibitor with potent cellular activity was obtained by stabilizing the ERK-bound conformation through the introduction of additional functionality. TCS ERK 11e also potently blocks proliferation of HT29 cells (IC$_{50}$ = 48 nM). **1**. Aronov, Tang, Martinez-Botella, *et al.* (2009) *J. Med. Chem.* **52**, 6362.

TcsH-82, See *Large Clostridial Cytotoxins*

TCS JNK 5a

This selective protein inhibitor (FW = 332.42 g/mol; CAS 312917-14-9; Soluble to 100 mM in DMSO), also named N-(3-cyano-4,5,6,7-tetrahydrobenzo[b]thienyl-2-yl)-1-naphthalenecarboxamide, targets c-Jun N-terminal kinases JNK3 (pIC$_{50}$ = 6.7), JNK2 (pIC$_{50}$ = 6.5), JNK1 (pIC$_{50}$ < 5.0), and p38α (pIC$_{50}$ < 4.8), without significant activity at a range of other protein kinases, including EGFR, ErbB2, cdk2, PLK-1 and Src, for which pIC$_{50}$ is <5.0. **1**. Angell, Atkinson, Brown, *et al.* (2007) *Bioorg. Med. Chem. Letts.* **17**, 1296.

TCS JNK 6o

This selective ATP-competitive protein inhibitor (FW = 356.38 g/mol; CAS 894804-07-0; Solubility: 100 mM in DMSO), also named N-(4-amino-5-cyano-6-ethoxy-2-pyridinyl)-2,5-dimethoxybenzeneacetamide, targets c-Jun N-terminal kinase (JNK) inhibitor (IC$_{50}$ = 2 nM) , 4 and 52 nM for JNK1 (IC$_{50}$ = 2 nM), JNK2 (IC$_{50}$ = 4 nM), and JNK3 (IC$_{50}$ = 52 nM), with >1000x selectivity over other kinases, including ERK2 and p38. TCS JNK 6o also inhibits c-Jun phosphorylation (EC$_{50}$ = 920 nM) and prevents collagen-induced platelet aggregation *in vitro*. **1**. Szczepankiewicz, *et al.* (2006) *J. Med. Chem.* **49**, 3563. **2**. Kauskot, *et al.* (2007) *J. Biol. Chem.* **282**, 31990.

TcsL-82, See *Large Clostridial Cytotoxins*

TCS OX2 29

This potent OX$_2$ antagonist (FW$_{HCl-Salt}$ = 433.97 g/mol; CAS 372523-75-6; Soluble to 100 mM in H$_2$O and to 25 mM in DMSO), also named (2S)-1-(3,4-dihydro-6,7-dimethoxy-2(1H)-isoquinolinyl)-3,3-dimethyl-2-[(4-pyridinylmethyl)amino]-1-butanone, targets orexin 2 receptor, *or* OX$_2$R (IC$_{50}$ = 40 nM), displaying >250c selectivity over OX$_1$ and fifty other receptors, ion channels and transporters. 1. Hirose, Egashira, Goto, *et al.* (2003) *Bioorg. Med. Chem. Lett.* **13**, 4497.

TCS PIM-1 1

This selective, ATP-competitive Pim kinase inhibitor (FW = 367.20 g/mol; CAS 491871-58-0; Soluble to 100 mM in DMSO), also named 3-cyano-4-phenyl-6-(3-bromo-6-hydroxy)phenyl-2(1H)-pyridone, targets Pim-1 (IC$_{50}$ = 50 nM), displaying 400x selectivity over Pim-2 (IC$_{50}$ = 20000 nM) and MEK1/2 (IC$_{50}$ > 20000 nM). 1. Cheney, Yan, Appleby, *et al.* (2007) *Bioorg. Med. Chem. Lett.* **17**, 1679.

TCS PIM-1 4a

This selective, ATP-competitive Pim kinase inhibitor (FW = 273.23 g/mol; CAS 438190-29-5; Soluble to 100 mM in DMSO), also named SMI-4a and (5Z)-5-[[3-(trifluoromethyl)phenyl]-methylene]-2,4-thiazolidinedione, targets Pim-1 (IC_{50} = 24 nM) and Pim-2 (IC_{50} = 100 nM), displaying >150x selectivity over a panel of fifty other kinases tested. TCS PIM-1 4a inhibits (IC_{50} = 17 μM) cell growth and is cytotoxic to PC3 prostate carcinoma cells *in vitro*. **1**. Xia, Knaak, Ma, *et al.* (2009) *J. Med. Chem.* **52**, 74.

TCS PrP Inhibitor 13

This antiprion agent (FW = 281.27 g/mol; CAS 34320-83-7; Soluble to 100 mM in DMSO), also named 2,4-dihydro-5-(4-nitrophenyl)-2-phenyl-3*H*-pyrazol-3-one, potently inhibits (IC_{50} = 3 nM) accumulation of protease-resistant prion protein (PrP-res) in two models for prion-infected mouse neuroblastoma cell lines. Although its mechanism of action remains uncertain, the antiprion action of this compound does not correlate with any antioxidant activities, any hydroxyl radical-scavenging activities, or any SOD-like activities. **1**. Kimata, Nakagawa, Ohyama, *et al.* (2007) *J. Med. Chem.* **50**, 5053.

TC-T 6000

This potent ENT inhibitor (FW = 504.71 g/mol; CAS 949467-71-4; Soluble to 20 mM in DMSO), targets the Equilibrative Nucleoside Transporter 4, *or* ENT4 passive transporter (IC_{50} = 74.4 nM), with 20- and 80-fold selectivity over ENT2 and ENT1, respectively. Structure-activity relationship suggests that nitrogen-containing monocyclic rings and noncyclic substituents at the 4- and 8-positions of the pyrimido[5,4-d]pyrimidine are important for the inhibitory activity against hENT4. The most potent and selective hENT4 inhibitors tend to have a 2,6-di(*N*-monohydroxyethyl) substitution on the pyrimidopyrimidine ring system. ENT4 exhibited a novel, pH-dependent adenosine transport activity optimal at acidic pH, with apparent K_m values 0.78 and 0.13 mM, respectively, at pH 5.5, which was absent at pH 7.4. Its cardiac abundance suggests contribution to regulation of extracellular adenosine concentrations under ischemia conditions. **1**. Wang, Lin, Playa, *et al.* (2013) *Biochem. Pharmacol.* **86**, 1531.

dTDP-D-Glucose

This nucleoside diphosphosugar (FW = 564.33 g/mol; CAS 2196-62-5), also known as thymidine 5'-diphospho-α-D-glucopyranose, is formed from dTTP and α-D-glucose 1-P by glucose-1-phosphate thymidylyl-transferase. **Target(s):** CMP-*N*-acylneuraminate phospho-diesterase (1); 1,3-β-glucan

synthase (2); glucose-1-phosphate thymidylyltransferase, by product inhibition (3); glucose-1-phosphate uridylyltransferase, *or* UDPglucose pyrophosphorylase (4); glycogenin glucosyltransferase (5); glycogen phosphorylase (6,7); β-1,4-mannosyl-glycoprotein 4-β-*N*-acetylglucosaminyltransferase (8); UDP-glucuronate 4-epimerase (9); xyloglucan 4-glucosyltransferase, weakly inhibited (11). **1**. Kean & Bighouse (1974) *J. Biol. Chem.* **249**, 7813. **2**. Orlean (1982) *Eur. J. Biochem.* **127**, 397. **3**. Zuccotti, Zanardi, C. Rosano, Sturla, Tonetti & Bolognesi (2001) *J. Mol. Biol.* **313**, 831. **4**. Turnquist & Hansen (1973) *The Enzymes*, 3rd ed., **8**, 51. **5**. Meezan, Manzella & Roden (1995) *Trends Glycosci. Glycotechnol.* **7**, 303. **6**. Chen & Segel (1968) *Arch. Biochem. Biophys.* **127**, 175. **7**. Robson & Morris (1974) *Biochem. J.* **144**, 513. **8**. Ikeda, Koyota, Ihara, *et al.* (2000) *J. Biochem.* **128**, 609. **9**. Gaunt, Ankel & Schutzbach (1972) *Meth. Enzymol.* **28**, 426. **10**. Gaunt, Maitra & Ankel (1974) *J. Biol. Chem.* **249**, 2366. **11**. Hayashi & Matsuda (1981) *J. Biol. Chem.* **256**, 11117.

TDZD-8, *See 4-Benzyl-2-methyl-1,2,4-thiadiazolidine-3,5-dione*

TEA, *See Triethanolamine; Tetraethylammonium Chloride*

Tebipenem & Tebi-pivoxil

Tebipenem (R = H)
Tebi-pivoxil (R = CH₂–O–CO–*t*-Butyl)

This first-in-class, orally available carbapenem antibiotic (1) (FW = 383.48 g/mol; CAS 161715-21-5) is a slow substrate that inhibits *Mycobacterium tuberculosis* β-lactamase, K_m < 2 μM and k_{cat} ≈ 0.04 min⁻¹ (2). FT-ICR mass spectrometry showed that the tebipenem acyl-enzyme complex remains stable for >90 min, existing as mixture of covalently bound drug and bound retro-aldol cleavage product. This low rate of deacylation (via hydrolysis of the lactamase–tebipenem covalent adduct) permits structural analysis of the covalent intermediate by soaking apoenzyme crystals with tebipenem. Its prodrug form tebi-pivoxil (FW = 497.62 g/mol; CAS 211558-19-9) is membrane-permeant and is hydrolyzed to tebipenem. **1**. Sato, Kijima, Koresawa, *et al.* (2008) *Drug Metab. Pharmacokin.* **23**, 434. **2**. Hazra, Xu & Blanchard (2014) *Biochemistry* **53**, 3671.

Tecoviramat

This oral synthetic antiviral (FW = 376.33 g/mol; CAS 816458-31-8), also named *N*-{3,5-dioxo-4-azatetracyclo[5.3.2.0²,⁶.0⁸,¹⁰]dodec-11-en-4-yl}-4-(trifluoromethyl)benzamide and ST-246, is active against orthopoxviruses (including smallpox, vaccinia, monkeypox, camelpox, cowpox, mousepox (ectromelia), and variola viruses) by targeting virus p37, which is needed for vaccinia virus envelopment and secretion of extracellular (infective) forms (1-3). Proteins associated with the late endosome (LE) appear to play a central role in the envelopment of a number of taxonomically diverse viruses. ST-246 prevents formation of extracellular virus by inhibiting the formation of a putative wrapping complex derived from virus-modified LE membranes. The LE is enriched for Rab9 protein that mediates recycling of Cation-Dependent Mannose-6-Phosphate Receptor (CD-MPR) and Cation-Independent Mannose-6-Phosphate Receptor (CI-MPR) from the LE to the Trans-Golgi Network (TGN). The wrapping complex catalyzes the envelopment of intracellular mature virus particles to produce egress-competent forms of the virus. Rab9-dependent recycling is mediated through interactions with the tail interacting protein of 47 kD (TIP47), a Rab9-specific effector. ST-246 was discovered by high throughput screening to identify inhibitors of vaccinia virus-induced cytopathic effects. The EC_{50} values for inhibition of viral replication ranged from 0.01 μM for vaccinia virus to 0.07 μM for ectromelia virus. Unrelated viruses have EC_{50} values >40 μM. Cowpox appears to be 5 to 50-times less susceptible to ST-246 (1). The basis for reduced susceptibility may reflect a different mode of virus spread. ST-246 was active against a CDV-resistant (CDVr) cowpox virus (EC_{50} = 0.05 μM), suggesting that the mechanism by which ST-246 inhibits virus replication is distinct from that of CDV. Furthermore, ST-246 inhibited clinical isolates from both of the major clades of monkeypox and variola viruses in cell culture (3). Tecovirimat has been mass produced and maintained in the US Strategic National Stockpile as a defense against a smallpox outbreak. **1.** Yang, Pevear, Davies, *et al.* (2005) *J. Virol.* **79**, 13139. **2.** Quenelle, Buller, Parker, *et al.* (2007) *Antimicrob. Agents Chemother.* **51**, 689. **3.** Jordan, Leeds, Tyavanagimatt & Hruby (2010) *Viruses* **2**, 2409.

Tectochrysin, *See* *5-Hydroxy-7-methoxyflavone*
Tectoquinone, *See* *2-Methylanthraquinone*

Tectorigenin

This naturally occurring isoflavone (FW = 300.27 g/mol; CAS 13111-57-4; M.P. = 225-226°C), known systematically as 5,7-dihydroxy-3-(4-hydroxy-phenyl)-6-methoxy-4*H*-1-benzopyran-4-one, from *Pueraria thunbergiana* (kudzu) and the rhizomes of *Iris tectorum*, induces cell differentiation and apoptosis in human promyelocytic leukemia HL-60 cells (1). **Target(s):** aldose reductase (2); and cyclooxygenase (3). **1.** Lee, Sohn, Kim, *et al.* (2001) *Biol. Pharm. Bull.* **24**, 1117. **2.** Jung, Lee, Lee, *et al.* (2002) *Arch. Pharm. Res.* **25**, 306. **3.** You, Jong & Kim (1999) *Arch. Pharm. Res.* **22**, 18.

ψ-Tectorigenin

This *O*-methylated isoflavone (FW = 300.27 g/mol; CAS 548-77-6), systematically named as 5,7-dihydroxy-3-(4-hydroxyphenyl)-8-methoxy-4*H*-1-benzopyran-4-one, is a structural isomer of tectorigenin and inhibits the epidermal growth factor-induced activation of phospholipase C (1) and inhibits phosphatidylinositol turnover (2) **Target(s):** catechol *O*-methyltransferase (3); histidine decarboxylase (3); epidermal growth factor receptor (EGFR) protein-tyrosine kinase, K_i = 0.33 μM (4). **1.** Imoto, Shimura & Umezawa (1991) *J. Antibiot.* **44**, 915. **2.** Imoto, Yamashita, Sawa, *et al.* (1988) *FEBS Lett.* **230**, 43. **3.** Umezawa, Tobe, Shibamoto, Nakamura & Nakamura (1975) *J. Antibiot.* **28**, 947. **4.** Filipeanu, Brailoiu, Huhurez, *et al.* (1995) *Eur. J. Pharmacol.* **281**, 29.

2-TEDC

This novel caffeic acid derivative (FW = 315.34 g/mol; CAS 132465-10-2; Solubiity: 100 mM in DMSO), also named 2-(1-thienyl)ethyl 3,4-dihydroxybenzylidenecyanoacetate, is a potent inhibitor of 5-lipoxygenase (IC_{50} = 0.09 μM), 12-lipoxygenase (IC_{50} = 0.013 μM), and 15-lipoxygenase

(IC_{50} = 0.5 μM). **1.** Cho, Ueda, Tamaoka, *et al.* (1991) *J. Med. Chem.* **34**, 1503.

Tedizolid

This orally available, 1,3-oxazolidinone-containing antibiotic (FW = 370.34 g/mol; CAS 856866-72-3), also code-named TR-700 and systematically named (5*R*)-3-{3-fluoro-4-[6-(2-methyl-2*H*-tetrazol-5-yl)pyridin-3-yl]phenyl}-5-(hydroxymethyl)-1,3-oxazolidin-2-one, and formerly known as torezolid, targets translation at an early stage, most likely by preventing the formation of the initiation complex (1). (For details on similar mode of action, *See* *Linezolid*) Tedizolid is active against most disease-causing Gram-positive bacteria, including vancomycin-resistant enterococci (VRE), and methicillin-resistant *Staphylococcus aureus* (MRSA). With a favorable pharmacokinetic profile that is suited for once-daily dosing and facile i.v.-to-oral conversion, tedizolid offers the advantage that it does not interact with serotonergi agents. Impotantly, tedizolid's potency (based on the MIC_{90} values) against coagulase-negative staphylococci was >16x greater than that of linezolid (2). Tedizolid also retained activity against most of the linezolid-resistant staphylococci tested, including multidrug-resistant isolates with elevated linezolid MICs of 32 to >128 mg/L (2). **1.** Bae, Yang, Shin, *et al.* (2007) *J. Pharm. Pharmacol.* **59**, 955. **2.** Rodríguez-Avial, Culebras, Betriu, *et al.* (2012) *J. Antimicrob. Chemother.* **67**, 167.

α-TEG, *See* *Ethyl Thio-α-D-glucoside*

β-TEG, *See* *Ethyl Thio-β-D-glucoside*

Tegafur, *See* *5-Fluoro-1-(tetrahydro-2-furfuryl)uracil*

TEI-5624, *See* *7-[N-(4-Chlorophenylsulfonyl)-L-glutamyl]amino-5-methyl-2-isopropylamino-4H-3,1-benzoxazin-4-one*

TEI-6344, *See* *7-[N-(4-Chlorophenylsulfonyl)-L-lysyl]amino-5-methyl-2-isopropylamino-4H-3,1-benzoxazin-4-one*

TEI-6720, *See* *Febuxostat*

Teicoplanins

These orally bioavailable semisynthetic glycopeptide antibiotics (FW = 1709.39 g/mol; CAS 61036-64-4) were first isolated from fermentation broths of *Actinoplanes teichomyceticus*. Known individually by the their unique fatty acyl side-chains (listed below) and collectively by the trade name Targocid™, teicoplanins exhibit vancomycin-like bacteriocidal action against Gram-positive bacteria, including methicillin-resistant *Staphylococcus aureus* (MRSA) and *Enterococcus faecalis* (1). A_2-1, R = –C(=O)-CH_2-CH_2-CH=CH-CH_2-CH_2-CH_2-CH_2-CH_3, with *trans* double-bond; A_2-2, R = –C(=O)-CH_2-CH_2-CH=CH-CH_2-CH_2-CH_2-$(CH_3)_2$, with *trans* double-bond; A_2-3, R = –C(=O)-CH_2-CH_2-CH_2-CH_2-CH_2-CH_2-CH_2-CH_2-CH_3; A_2-4,

R = –C(=O)-CH$_2$-CH$_2$-CH$_2$-CH$_2$-CH$_2$(CH$_3$)-CH$_3$-CH$_3$; A$_2$-5, R = –C(=O)-CH$_2$-CH$_2$-CH$_2$-CH$_2$-CH$_2$-CH$_2$-CH$_2$-(CH$_3$)$_2$. Teicoplanin variants possess a common aglycone core, consisting of seven amino acids that are bound by peptide and ether bonds, thereby forming a four-ring system. Teicoplanin has a longer half-life than vancomycin and can be administered intravenously (as a bolus dose) or by intramuscular injection, with rare nephrotoxicity and ototoxicity. **Primary Mode of Inhibitory Action:** Teicoplanins inhibit peptidoglycan polymerization, resulting in inhibition of Gram-positive bacteria cell walls synthesis and consequent cell death. When given as an intravenous bolus dose, serum and urine teicoplanin concentrations exceeded the minimal inhibitory concentration (typical MIC ≈ 0.02–2 µg/mL) on many pathogenic organisms for at least one day after administration (2). With side effects that do not require close monitoring, its longer serum half-life commends teicoplanin as a valuable vancomycin alternative, making it the glycopeptide antibiotic of choice in many hospitals. **1**. Cynamon & Granato (1982) *Antimicrob. Agents Chemother.* **21**, 504. **2**. Traina & Bonati (1984) *J. Pharmacokinet. Biopharm.* **12**, 119.

Teijin Compound 1

This chemokine receptor antagonist (FW = 476.32 g/mol; CAS 1313730-14-1; Soluble to 10 mM in water and to 100 mM in DMSO), also named *N*-[2-[[(3*R*)-1-[(4-chlorophenyl)methyl]-3-pyrrolidinyl]amino]-2-oxoethyl]-3-(trifluoromethyl)benzamide hydrochloride, targets the CCR2b receptor (IC$_{50}$ = 180 nM) and potently inhibits cell chemotaxis induced by MCP-1 (EC$_{50}$ = 24 nM). **1**. Moree, *et al.* (2004) *Bioorg. Med. Chem. Letts.* **14**, 5413. **2**. Moree, Kataoka, Ramirez-Weinhouse, *et al.* (2008) *Bioorg. Med. Chem. Letts.* **18**, 1869. **3**. Hall, Mao, Nicolaidou, *et al.* (2009) *Mol. Pharmacol.* **75**, 1325.

Teixobactin

This first-in-class, low/no-resistance antibiotic (FW = 1242.47 g/mol; CAS 1613225-53-8), isolated from the previously unculturable bacteria *Eleftheria terrae*, exhibits broad action against Gram-positive bacteria by binding to a highly conserved motif of Lipid II and Lipid III, essential peptidoglycan precursors involved in cell wall biosynthesis. In wide-ranging studies, no mutants of *Staphylococcus aureus* or *Mycobacterium tuberculosis* were found to be resistant to teixobactin (1). That resistance has not developed suggests that the target is not a protein, inasmuch as a teixobactin binding site on a protein would be subject to affinity-reducing mutations (1). Given the presence of a metal ion-chelating macrolide ring, another likely mechanism is sequestratiuon of essential metal ions, either in their free or bound forms. **Minimal Inhibitory Concentrations:** (*from* Reference 1) *Staphylococcus aureus* (MSSA), 0.25 µg/mL; *S. aureus* + 10% serum, 0.25 µg/mL; *S. aureus* (MRSA), 0.25 µg/mL; *Enterococcus faecalis* (VRE), 0.5 µg/mL; *Enterococcus faecium* (VRE), 0.5 µg/mL; *Streptococcus pneumoniae* (penicillin-resistant), ≤ 0.03 µg/mL; *Streptococcus pyogenes*, 0.06 µg/mL; *Streptococcus agalactiae*, 0.12 µg/mL; *Viridans group streptococci*, 0.12 µg/mL; *Bacillus anthracis*, ≤ 0.06 µg/mL; *Clostridium difficile*, 0.005 µg/mL; *Propionibacterium acnes*, 0.08 µg/mL; *Mycobacterium tuberculosis*, 0.125 µg/mL; *Haemophilus influenza*, 4 µg/mL; *Moraxella catarrhalis*, 2 µg/mL; *Escherichia coli*, 25 µg/mL; *Escherichia coli* (asmB1), 2.5 µg/mL; *Pseudomonas aeruginosa*, > 32

µg/mL; and *Klebsiella pneumoniae*, > 32 µg/mL. **1**. Ling, Schneider, Peoples, *et al.* (2015) *Nature* **517**, 455.

Telaprevir

This orally available serine protease-directed antiviral (FW = 697.84 g/mol; CAS 402957-28-2 and 569364-34-7; Solubility: 136 mg/mL DMSO; <1 mg/mL Water), also known as VX-950 and systematically as (1*S*,3a*R*,6a*S*)-2-[(2*S*)-2-[[(2*S*)-2-cyclohexyl-2-(pyrazine-2-carbonylamino)acetyl]amino]-3,3-dimethylbutanoyl]-*N*-[(3*S*)-1-(cyclopropylamino)-1,2-dioxohexan-3-yl]-3,3a,4,5,6,6a-hexahydro-1*H*-cyclopenta[*c*]pyrrole-1-carboxamide, inhibits hepatitis C virus NS3.4A serine protease (1). In studies with protease domain mutants A156T, D168V and R155Q of full-length NS3 protease, VX-950 was the only known NS3-directed antiviral to inhibit the D168V enzyme more efficiently than the wild type NS3 (2). Telaprevir induces a median 4-log decline of HCV RNA after two weeks of therapy (3) and, when combined with cyclosporine, is highly effective against the recurrence of hepatitis C in liver transplant patients (4). One must avoid co-administration of telaprevir and simvastatin, inasmuch as there is an undesirable drug interaction due to telaprevir inhibition of CYP3A4-mediated simvastatin clearance, increasing plasma concentration by as much as 30x. **1**. Lin, Lin, Luong, *et al.* (2004) *J. Biol. Chem.* **279**, 17508. **2**. Dahl, Sandström, Akerblom & Danielson (2007) *Antivir. Ther.* **12**, 733. **3**. Gentile, Viola, Borgia *et al.* (2009) *Curr. Med. Chem.* **16**, 1115. **4**. Kikuchi, Okuda, Ueda, *et al.* (2013) *Biol. Pharm. Bull.* **37**, 417.

Telavancin

This semi-synthetic vancomycin derivative (FW = 1755.63 g/mol; CAS 372151-71-8), also known by its trade name Vibativ®, is a bactericidal lipoglycopeptide used to treat methicillin-resistant *Staphylococcus aureus* (MRSA) and other Gram-positive infections, including hospital-acquired and ventilator-associated pneumonia caused by *S. aureus*. Telavancinexhibits concentration-dependent bactericidal effects *in vitro* via a dual mechanism of action of inhibition of bacterial cell wall synthesis and disruption of bacterial cell membrane barrier functions (1). Like its parent compound, telavancin inhibits bacterial cell wall synthesis by binding to the D-Ala-D-Ala terminus of the peptidoglycan in the growing cell wall (**See** *Vancomycin*). Telavancin is approved in the United States and Canada for the treatment of adult patients with complicated skin and skin structure infections (cSSSI) due to susceptible gram-positive pathogens, especially hospital-acquired bacterial pneumonia (HABP) such as ventilator-associated bacterial pneumonia (VABP) due to susceptible isolates of *S. aureus* (MRSA). Telavancin shows a higher rate of kidney failure than vancomycin. **1**. Lunde, Hartouni, Janc, *et al.* (2009) *Antimicrob. Agents Chemother.* **53**, 3375.

Telbivudine

This synthetic L-nucleoside analogue of thymidine (FW = 242.23 g/mol; CAS 3424-98-4), also named Tyzeka® and 1-(2-deoxy-β-L-*erythro*-pentofuranosyl)-5-methylpyrimidine-2,4(1*H*,3*H*)-dione, is converted metabolically to telbivudine-triphosphate, a potent inhibitor of Hepatitis B reverse transcriptase. Removal or substitution of the 3'OH group results in loss of inhibition. Moreover, Telbivudine is highly selective for HBV DNA and inhibits viral DNA synthesis with no effect on human DNA or other viruses. Compared to Lamivudine, Telbivudine showed a statistically significantly greater reduction in HBV DNA, a greater proportion of alanine aminotransferase normalization, a greater histological response, fewer cases of treatment failure, and less virological resistance. (**See also** *Adefovir & Adefovir Dipivoxil; Entecavir; Lamivudine; Tenofovir; Emtricitabine*) **1**. Nash (2009) *Adv. Ther.* **26**, 155.

Telenzepine

This selective M_1 muscarinic receptor antagonist (FW = 370.48 g/mol; CAS 80880-90-6) inhibits gastric acid secretion. Telenzepine specifically blocks the so-called M_1 receptors located in the ganglia of the myenteric plexus. It represents an attractive alternative to the H_2 blockers, because M_1 receptor blockers are virtually devoid of the untoward reactions typical of atropine-like compounds, *e.g.*, dry mouth, mydriasis, tachycardia, *etc.* (1). Telenzepine binds with high affinity ($K_{assoc} = 3 \times 10^9$ M^{-1}) to a subpopulation of muscarinic binding sites in rat cerebral cortex, which has high affinity for pirenzepine (2). When compared head to head with pirenzepine (*i.e.*, on a molar basis), telenzepine proved to be 25x and 50x more potent as an inhibitor of gastric and salivary secretion, respectively (3). **1**. Bertaccini & Coruzzi (1989) *Pharmacol. Res.* **21**, 339. **2**. Eveleigh, Hulme, Schudt & Birdsall (1989) *Mol. Pharmacol.* **35**, 477. **3**. Londong, Londong, Meierl, & Voderholzer (1987) *Gut* **28**, 888.

Telitromycin

This first-in-class ketolide macrolide antibiotic and nAChR ligand (FW = 790.10 g/mol; CAS 191114-48-4), also known as RU-66647 and Ketek®, inhibits translation by binding to the bacterial ribosome 50S subunit, blocks progression of the growing polypeptide chain. Telithromycin exhibits >10x higher affinity than erythromycin. It also binds strongly to two domains in the 23S RNA core, whereas older macrolides bind strongly to one domain,

and only weakly to the second domain. Telithromycin also inhibits the assembly of 50S and 30S subunits. Unlike erythromycin, telithromycin is acid-stable resists gastric acids (1). RU-66647 is more active than erythromycin and azithromycin against oxacillin-resistant *Staphylococcus* spp., vancomycin-resistant enterococci, and *Klebsiella pneumoniae*. Adverse effects have limited the clinical use of telithromycin. Telithromycin preferentially inhibits nicotinic acetylcholine receptors located at the neuromuscular junction ($\alpha_3\beta_2$ AChR), ciliary ganglion of the eye ($\alpha_3\beta_4$ and α_7 AChRs), and vagus nerve innervating the liver (α_7 AChR), accounting in part for its capacity to exacerbate myasthenia gravis, induce visual disturbances, and give rise to liver failure (2). Telithromycin also reduces *N*-demethylation of oxycodone by inhibiting CYP3A4, suggesting that telithromycin use in patients receiving multiple doses of oxycodone for pain relief may increase the risk of opioid adverse effects (3). Some 121 drugs (represented by 341 brand and generic products) are known to have major interactions with telithromycin. **Key Pharmacokinetic Parameters:** *See* Appendix II in Goodman & Gilman's THE PHARMACOLOGICAL BASIS OF THERAPEUTICS, 12th Edition (Brunton, Chabner & Knollmann, eds.) McGraw-Hill Medical, New York (2011). **1**. Jones & Biedenbach (1997) *Diagn. Microbiol. Infect Dis.* **27**, 7. **2**. Bertrand, Bertrand, Neveu & Fernandes (2010) *Antimicrob. Agents Chemother.* **54**, 5399. **3**. Grönlund, Saari, Hagelberg, *et al.* (2010) *J. Clin. Pharmacol.* **50**, 101.

Tellurium(IV) Bis(citrate)

This organotellurium(IV) compound (FW = 507.81 g/mol) inhibits papain, presumably by means of a nucleophilic displacement reaction, leading to formation of a Te–S bond to its active-site thiol. **1**. Albeck, Weitman, Sredni & Albeck (1998) *Inorg. Chem.* **37**, 1704.

Tellurium Dioxide, *See Tellurite*

Telmisartan

This substituted benzimidazole (FW$_{free-acid}$ = 514.63 g/mol; CAS 144701-48-4; Soluble to 20 mM in DMSO), also named 4'-[(1,4'-dimethyl-2'-propyl[2,6'-bi-1*H*-benzimidazol]-1'-yl)methyl]-[1,1'-biphenyl]-2-carboxylic acid, is an angiotensin II type 1 (AT$_1$) receptor antagonist, K_i = 3.7 nM (1-3). Telmisartan also has an insulin-sensitizing action, involving adipose tissue-independent down-regulation of peroxisome proliferator-activated receptor-γ- (PPAR-γ-) regulated genes, thereby ameliorating fatty liver (4). Intriguingly, telmisartan diminishes deleterious effects of chronic restraint stress on memory in a statistically significant manner (p<0.01), as evaluated in a passive avoidance (PA) situation and in an object recognition test (ORT) (5). Surprisingly, telmisartan (at 5mg/kg) also displays antiinflammatory activity by modulating TNF-α, IL-10, and RANK/RANKL in a rat model of ulcerative colitis (6). (**See also** *Candesartan; Eprosartan; Irbesartan; Losartan; Olmesartan; Valsartan*) **Pharmacological Parameters:** Bioavailability = 43 %; Food Effect? NO; Drug $t_{1/2}$ = 24 hours; Metabolite $t_{1/2}$ = not determined; Drug's Protein Binding = >99 %; Route of Elimination = 0.5 % Renal and >97 % Hepatic. **1**. Ries, Mihm, Narr, *et al.* (1993) *J. Med. Chem.* **36**, 4040. **2**. Wienen, Hauel, Van Meel, *et al.* (1993) *Brit. J. Pharmacol.* **110**, 245. **3**. Maillard, Perregaux, Centeno, *et al.* (2002) *J. Pharmacol. Exp. Ther.* **302**, 1089. **4**. Rong, Li, Ebihara, *et al.* (2009) *J. Pharmacol. Exp. Ther.* **331**, 1096. **5**. Wincewicz & Braszko (2014) *Pharmacol. Rep.* **66**, 436. **6**. Guerra, Araújo, Lira, *et al.* (2015) *Pharmacol. Rep.* **67**, 520.

Temelastine

This nonsedating agent (FW = 444.38 g/mol; CAS 86181-42-2), also known as SK&F 93944, is a potent and selective histamine H_1 receptor antagonist, $K_i = 0.32$ nM (1-3). Temelastine was found to be a weak noncompetitive antagonist of carbachol on the guinea-pig ileum, but was devoid of measurable anticholinergic activity *in vivo*. 1. Hill, Ganellin, Timmerman, *et al.* (1997) *Pharmacol. Rev.* **49**, 253. 2. Alexander, Stote, Allison, *et al.* (1986) *J. Int. Med. Res.* **14**, 200. 3. Ter Laak, Donne-Op den Kelder, Bast & Timmerman (1993) *Eur. J. Pharmacol.* **232**, 199.

Temozolomide

TMZ   **MTIC**

This orally bioavailable radio-sensitizing/chemotherapeutic agent (FW = 194.15 g/mol; CAS 85622-93-1; Symbol: TMZ), also known by the trade names Temodar™, Temodal™ and Temcad™ and by its systematic names 3,4-dihydro-3-methyl-4-oxoimidazo[5,1-*d*][1,2,3,5]tetrazine-8-carboxamide and 4-methyl-5-oxo-2,3,4,6,8-pentazabicyclo[4.3.0]nona-2,7,9-triene-9-carboxamide, is taken up by cells and metabolized to 3-methyl(triazen-1-yl)imidazole-4-carboxamide (MTIC), the latter a highly effective DNA-alkylating agent that preferentially methylates guanine at its O^6 and N^7 atoms, thereby triggering cell death, especially in the presence of a alkylation-sensitizing PARP inhibitor (**See also** *Veliparib*). Because of its small size and lipophilic properties, TMZ crosses the blood-brain barrier, entering the cerebrospinal fluid. *In vitro*, temozolomide has demonstrated schedule-dependent antitumor activity against highly resistant malignancies, including high-grade glioma (1). In clinical studies, TMZ demonstrates noncumulative minimal myelosuppression that is rapidly reversible, with activity against a variety of solid tumors in both children and adults. The standard therapy for glioblastoma multiforme (GBM), the most common primary brain tumour in adults, is maximal surgical resection followed by radiotherapy with concurrent and adjuvant TMZ. This agent has been associated with anemia, lymphopenia, neutropenia, and severe thrombocytopenia. **Primary Mode of Inhibitory Action:** Temozolomide is [an analogue of 5-(3,3-dimethyl-1-triazeno)imidazole-4-carboxamide that does not require metabolic activation. Instead, it rearranges nonenzymatically to produce methyldiazonium ion, a highly reactive alkylating agent is formed by temozolomide breakdown, mainly forming O^6- and N^7-methylguanine (2,3).

Of these, O^6-methylguanine formation is more cytotoxic, because DNA mismatch repair enzymes try, but fail, to excise O^6-methylguanine, in the process generating single- and double-strand breaks that lead to the activation of apoptotic pathways. Among the damage to DNA is methylation at the N^7 position of guanine, followed by methylation at the O^3 position of adenine and the O^6 position of guanine. Although the N^7-methylguanine and O^3-methyladenine adducts both contribute to TMZ's antitumor activity in some, if not all, sensitive cells, their role is controversial. Because TMZ does not cross-link DNA, it is less toxic to bone marrow hematopoietic progenitor cells than DNA cross-linkers like nitrosoureas (*i.e.*, carmustine and lomustine), platinum compounds (*i.e.*, cisplatin and carboplatin), and procarbazine. **Resistance:** Inherent and acquired glioma cell resistance to TMZ is caused by enhanced levels of O^6-methylguanine DNA methyltransferase (MGMT) as well as overexpression of epidermal growth factor receptor (EGFR), galectin-1, murine double minute 2 (Mdm2), p53 and phosphatase and tensin homolog (PTEN) mutations (4). To reduce acquired resistance, different therapeutic molecules have been developed, for example, O^6-benzyl-guanine which inhibits MGMT, tyrosine kinase inhibitors that act on EGFR, nutlin-3 which inhibits Mdm2, which contributes to the restoration of p53 activity (4). **Pharmacokinetics:** Peak plasma MTIC concentrations of 0.07–0.61 μg/mL were observed approximately 1 hour after an oral TMZ dose (5). Appearance and disappearance ($t_{1/2} = 88$ min) of the reactive metabolite paralleled the appearance and disappearance of TMZ in plasma. The mean values of the metabolite peak plasma concentration and AUC were 2.6% (*range: 1.6–4.6%*) and 2.2% (*range: 0.8–3.6%*), respectively, of the TMZ values. MTIC does not accumulate in plasma, even after five consecutive daily doses of TMZ. **2**. O'Reilly, Newlands, Glaser, *et al.* (1993) *Eur. J. Cancer* **29**, 940. **1**. Friedman, Kerby & Calvert (2000) *Clin. Cancer Res.* **6**, 2585. **3**. Newlands, Stevens, Wedge, *et al.* (1997) *Cancer Treat. Rev.* **23**, 35. **4**. Messaoudi, Clavreul & Lagarce (2015) *Drug Discov. Today.* **20**, 899. **5**. Reid, Stevens, Rubin & Ames (1997) *Clin. Cancer Res.* **3**, 2393.

Temsirolimus

This intravenously administered antineoplastic agent (FW = 1030.28 g/mol; CAS 162635-04-3), also known as CCI-779 and Torisel®, inhibits mammalian Target of Rapamycin (or mTOR), a protein kinase operating within the phosphatidylinositol 3-kinase (PI3K)/Akt signalling cascade. The latter plays a central role in controlling cell proliferation, survival, mobility and angiogenesis. Dysregulation of mTOR pathway has been observed in many human tumors, making temsirolimus a promising new anticancer drug, the action of which leads to cell cycle arrest in the G_1 phase as well as inhibition of tumor angiogenesis by reducing VEGF formation. Temsirolimus binds to FKBP, a cytosolic protein that subsequently inhibits mTOR. Inhibition of mTOR blocks a number of signal transduction pathways that suppress translation of several key proteins regulating the cell cycle. **Pharmacokinetic Parameters:** In the treatment of advanced renal cell carcinoma, temsirolimus is administered as a 30- to 60-minute IV infusion once weekly at a flat dose of 25 mg. This dosage results in high peak temsirolimus concentrations and limited immuno-suppressive activity. Temsirolimus shows activity on its own, but it is also known to be converted to rapamycin *in vivo*. Its activity may thus be more attributed to its metabolite rather than the prodrug itself. [*See also* Appendix II in Goodman & Gilman's THE PHARMACOLOGICAL BASIS OF THERAPEUTICS, 12th Edition (Brunton, Chabner & Knollmann, eds.) McGraw-Hill Medical, New York (2011).] **1**. Wan, Shen, Mendoza, Khanna & Helman (2006) *Neoplasia* **6**, 394. **2**. Rubio-Viqueira & Hidalgo (2006) *Curr, Opin. Investig. Drugs* **7**, 501.

Tenascin

This 2199-residue major chondroitin sulfate proteoglycan (MW = 240,718 g/mol; EMBL Accession: X56160.1), produced abundantly in the extracellular matrix of developing vertebrate embryos and reappearing around healing wounds and in the stroma of some tumors (1), inhibits axon regeneration within the injured central nervous system, presumably by binding to growth cone receptors (Rc), via inhibition of Rho family GTP-regulatory proteins and disruption of the actin cytoskeleton, resulting in growth cone collapse/repulsion (2). Tenascin C promiscuously binds growth factors via its fifth fibronectin type III-like domain (3). **Structural Features:** The monomer consists of four major domains: the N-terminal domain, which mediates hexamerization by forming a coiled structure with interchain disulfide bonds; a series of 14.5 epidermal growth factor-like (EGFL) repeats, each 30-50 amino acids long and containing six cysteines; a series of up to 15 fibronectin type III-like (TNCIII) repeats, each ~90 amino acids long that form two sheets of antiparallel β-strands, serving as cell-adhesive, integrin-binding sites that promote cell adhesion; and a fibrinogen-like globular domain at the C-terminus. **1**. Dou & Levine (1994) *J. Neurosci.* **14**, 7616. **2**. Sandvig, Berry, Barrett, *et al.* (2004) *Glia* **46**, 225. **3**. De Laporte, Rice, Tortelli & Hubbell (2013) *PLoS One* **8**, e62076.

Tendamistat

This 74-residue polypeptide (MW = 7961.77 g) from *Streptomyces tendae* forms a 1:1 complex with mammalian α-amylase, inhibiting the latter with K_i = 9 pM (1-4). It is, however, without effect on plant and microbial α-amylases. Tendamistat contains six β-sheets and two disulfide bridges between Cys11-Cys27 and Cys45-Cys73. **1**. Wiegand, Epp & Huber (1995) *J. Mol. Biol.* **247**, 99. **2**. Vertesy, Oeding, Bender, Zepf & Nesemann (1984) *Eur. J. Biochem.* **141**, 505. **3**. Ono, Umezaki, Tojo, *et al.* (2001) *J. Biochem.* **129**, 783. **4**. Virolle, Long, Chang & Bibb (1988) *Gene* **74**, 321.

Tenidap

This novel cytokine-modulating anti-inflammatory and antiarthritic agent (FW = 320.75 g/mol; CAS 120210-48-2; Soluble to 100 mM in DMSO), also known by its code name CP-66,248 and its systematic name 5-chloro-2-hydroxy-3-(2-thienylcarbonyl)-1*H*-indole-1-carboxamide, displays symptomatic efficacy superior to nonsteroidal anti-inflammatory drugs (NSAIDs) and is equivalent to combinations of NSAIDs and second-line agents (1-3). Tenidap preferentially inhibits COX-1 (IC_{50} < 0.03 μM), COX-2 (IC_{50} = 1.2 μM), and 5-lipoxygenase (IC_{50} > 30 μM). Tenidap inhibits formation of pro-inflammatory arachidonic acid metabolites in isolated human peripheral polymorphonuclear leukocytes. It potently inhibits leukotriene B$_4$ and prostanoid synthesis in human polymorphonuclear leucocytes *in vitro* (1). Tenidap, but not NSAIDs, also causes rapid and sustained acidification of the cytoplasmic compartment, but does not act as a proton ionophore or dissipate the low lysosomal pH (2). Tenidap does not alter pH$_{internal}$ via inhibition of the sodium-proton antiporter, but instead inhibits chloride-bicarbonate exchangers, as does UK5099, a known anion-transport inhibitor that likewise lowers pH$_{internal}$, suggesting that the pH$_{internal}$ change is coupled to anion transport inhibition. Tenidap also inhibits mannose 6-phosphate receptor-mediated endocytosis and protein synthesis, stimulating leucine accumulation (2). Tenidap is a potent channel opener, acting on the inwardly rectifying K$^+$ channel hKir2.3 (4). Down-regulation of the Kir2.3 channel expression is believed to contribute to the pathogenesis of pilocarpine-induced temporal lobe epilepsy (TLE), a condition that may be ameliorated by the administration of tenidap (5). **1**. Moilanen, Alanko, Asmawi & Vapaatalo (1988) *Eicosanoids* **1**, 35. **2**. McNiff, Svensson, Pazoles & Gabel (1994) *J. Immunol.* **153**, 2180. **3**. Wylie, Appelboom, Bolten, *et al.* (1995) *Brit. J. Rheumatol.* **34**, 554. **4**. Liu, Liu, Printzenhoff, *et al.* (2002) *Eur. J. Pharmacol.* **435**, 153. **5**. Xu, Hao, Wu, *et al.* (2013) *Neurol. Res.* **35**, 561.

Teniposide

This semi-synthetic podophyllotoxin derivative (FW = 656.66 g/mol; CAS 29767-20-2), also known as VM-26, is a potent cytotoxic agent that is primarily directed at DNA winding/unwinding but also reduces microtubule self-assembly. **Target(s):** CYP3A4 (1); DNA topoisomerase II (2-8); tubulin polymerization (9). **1**. Baumhakel, Kasel, Rao-Schymanski, *et al.* (2001) *Int. J. Clin. Pharmacol. Ther.* **39**, 517. **2**. Kellner, Rudolph & Parwaresch (2000) *Onkologie* **23**, 424. **3**. Galande & Muniyappa (1997) *Biochem. Pharmacol.* **53**, 1229. **4**. Minocha & Long (1984) *Biochem. Biophys. Res. Commun.* **122**, 165. **5**. Melendy & Ray (1989) *J. Biol. Chem.* **264**, 1870. **6**. Vosberg (1985) *Curr. Top. Microbiol. Immunol.* **114**, 19. **7**. Hammonds, Maxwell & Jenkins (1998) *Antimicrob. Agents Chemother.* **42**, 889. **8**. Gruger, Nitiss, Maxwell, *et al.* (2004) *Antimicrob. Agents Chemother.* **48**, 4495. **9**. Loike, Brewer, Sternlicht, Gensler & Horwitz (1978) *Cancer Res.* **38**, 2688.

Tenofovir

This reverse transcriptase inhibitor and antiretroviral drug (FW = 287.21 g/mol; CAS 147127-20-6), also known as Viread® and (*R*)-9-[2-(phosphonomethoxy)propyl]adenine, and its membrane-permeant prodrug (FW = 635.51 g/mol; CAS 202138-50-9), also known as tenofovir disoproxil and Reviro®, target the reverse transcriptases of Human Immunodeficiency Virus-1 (HIV-1) and hepatitis B virus. Tenofovir alone, or in combination with emtricitabine, has enjoyed growing demand as a pre-exposure prophylactic, one that significantly decreases the risk of contracting HIV. In adult humans, tenofovir has a volume of distribution of 0.813 L/kg, is minimally bound to plasma protein (7.2%), has a plasma elimination half-life of 12-14 hours, and is mainly excreted unchanged in urine (70%-80%) (2). **Key Pharmacokinetic Parameters:** *See* Appendix II in Goodman & Gilman's *THE PHARMACOLOGICAL BASIS OF THERAPEUTICS*, 12th Edition (Brunton, Chabner & Knollmann, eds.) McGraw-Hill Medical, New York (2011). **1**. Kearney, Yale, Shah, Zhong & Flaherty (2006) *Clin. Pharmacokinet.* **45**, 1115. **2**. Fung, Stone & Piacenti (2002) *Clin. Ther.* **24**, 1515.

Tenovin-1

This small-molecule p53 activator (FW = 369.38 g/mol; CAS 380315-80-0; Solubility: 74 mg/mL DMSO) binds to and protects (when present at 10μM) p53 against MDM2-mediated degradation, a process involving ubiquitination and acting through inhibition of protein-deacetylating activities of NAD$^+$-

dependent histone deacetylases, Sirtuins-1 and -2 (*or* SirT1 and SirT2). **1.** Lain, Hollick, Campbell, *et al.* (2008) *Cancer Cell* **13**, 454.

Tenoxicam

This anti-inflammatory drug (FW = 337.38 g/mol; CAS 59804-37-4; M.P. = 209-213°C; Solubility = 45 μg/mL; IUPAC Name: 4-hydroxy-2-methyl-*N*-2-pyridinyl-2*H*-thieno[2,3-*e*]-1,2-thiazine-3-carboxamide) inhibits cyclooxygenase 1 and 2 (1,2). **1.** Lora, Morisset, Menard, Leduc & de Brum-Fernandes (1997) *Prostaglandins Leukot. Essent. Fatty Acids* **56**, 361. **2.** Yamada, Niki, Yamashita, Mue & Ohuchi (1997) *J. Pharmacol. Exp. Ther.* **281**, 1005.

Tenox PG, *See n-Propyl Gallate*

Tentoxin

This cyclic tetrapeptide antibiotic (FW = 414.50 g/mol; CAS 28540-82-1; *Sequence*: cyclo-(L-MeAla^1-L-Leu^2-MeΔ^{Z}Phe^3-Gly^4) from the phytopathogenic fungus *Alternaria alternate* induces chlorosis (*i.e.,* insufficient chlorophyll production, with lightening of leaf color) in several germinating seedlings (1-3). Tentoxin inactivates photophosphorylation in a sensitive species (lettuce) by binding (K_d = 50-750 nM) to Coupling Factor-1 (CF$_1$) ATPase at a single site (1). Tentoxin increases the binding of ATP even under conditions where the ATPase activity is activated by light plus dithiothreitol (2). The effect of tentoxin on CF$_1$, however, is more complex than simply decreasing the efficiency of the conversion of State I to State II. Tentoxin markedly decreases the magnitude of the rapid phase ($t_{1/2}$ = 20 msec) of binding in the dark associated with the back reaction of State II to State I and decreases the total amount of adenine nucleotides that bind in the dark. Moreover, at concentrations greater than 5 x 10^{-7} M selectively increases potassium ion conductivity across lipid bilayers, whereas dihydrotentoxin, another naturally occurring derivative, has no such effect. Current-voltage curves, zero-current potential and charge-pulse measurements suggest that tentoxin dimerize in order to bind and transport K$^+$ ion (4). This model contradicts the earlier conclusion that tentoxin is a pore-forming peptide antibiotic (5). Tentoxin is without effect on the mammalian mitochondrial complex (6-9). **Target(s):** chloroplast ATP synthase, *or* chloroplast H$^+$-transporting two-sector ATPase (6-9); and 4-phytase, weakly inhibited (10). **1.** Steele, Uchytil, Durbin, Bhatnagar & Rich (1976) *Proc. Natl. Acad. Sci. U.S.A.* **73**, 2245. **2.** Shoshan & Selman (1979) *J. Biol. Chem.* **254**, 8808. **3.** Lax, Shepard & Edwards (1988) *Weed Technol.* **2**, 540. **4.** Klotz, Müller & Liebermann (1987) *Biophys. Chem.* **27**, 183. **5.** Heitz, Jacquier, Kaddari & Verducci (1986) *Biophys Chem.* **23**, 245. **6.** Linnett & Beechey (1979) *Meth. Enzymol.* **55**, 472. **7.** Minoletti, Santolini, Haraux, Pothier & Andre (2002) *Proteins* **49**, 302. **8.** Santolini, Minoletti, Gomis, *et al.* (2002) *Biochemistry* **41**, 6008. **9.** Arntzen (1972) *Biochim. Biophys. Acta* **283**, 539. **10.** Gibson & Ullah (1988) *Arch. Biochem. Biophys.* **260**, 503.

Tenuazonic Acid

This *Alternaria tenuis* and *A. alternata* mycotoxin (FW = 196.23 g/mol; CAS 610-88-8; low water solubility; freely soluble in chloroform, acetone, ethanol), named systematically as 3-acetyl-1,5-dihydro-4-hydroxy-5-(1-methylpropyl)-2*H*-pyrrol-2-one, inhibits mammalian protein biosynthesis by suppressing the release of newly formed polypeptide from the ribosomes. It is without effect on yeast. **Target(s):** DNA synthesis (1); protein biosynthesis (elongation) (2,3). **1.** Friedman, Aggarwal & Lester (1975) *Res. Commun. Chem. Pathol. Pharmacol.* **11**, 311. **2.** Carrasco, Battaner & Vazquez (1974) *Meth. Enzymol.* **30**, 282. **3.** Jiménez (1976) *Trends Biochem. Sci.* **1**, 28. *See also (–)-Altenuene*

Temafloxacin

This fluoroquinolone-class antibiotic (FW = 417.38 g/mol; CAS 108319-06-8), also named 1-(2,4-difluorophenyl)-6-fluoro-7-(3-methylpiperazin-1-yl)-4-oxoquinoline-3-carboxylate, is a broad-spectrum systemic antibacterial agent that targets Type II DNA topoisomerases (gyrases), which are required for bacterial replication and transcription. FDA-approved and marketed as Omniflox®, temafloxacin was withdrawn within the same year, after it was shown to give rise to serious allergic reactions and hemolytic anemia. For the prototypical member of this antibiotic class, *See Ciprofloxacin*

Tenuiorin

This tridepside secondary 3-orcinol metabolite (FW = 467.41 g/mol; CAS 570-07-0), derived from, from species of *Pseudocyphellaria*, *Lobaria* and *Peltiger* inhibits 5-lipoxygenase and 15-lipoxygenase. The term has been used to refer to either the trimer of *o*-orsellinic acid (or the methyl ester of this trimer) or to the polymer. **1.** Ingolfsdottir, Gudmundsdottir, Ogmundsdottir, *et al.* (2002) *Phytomedicine* **9**, 654.

Tepoxalin

This nonsteroidal anti-inflammatory (FW = 385.85 g/mol), also known as 5-(4-chlorophenyl)-*N*-hydroxy-(4-methoxyphenyl)-*N*-methyl-1*H*-pyrazole-3-ropanamide and marketed under the brand name Zubrin™, inhibits both COX-1 and COX-2 cyclooxygenases as well as 5-lipoxygenase (1,2). Tepoxalin also indirectly inhibits nuclear factor-κB activation (3). **1.** Kirchner, Aparicio, Argentieri, Lau & Ritchie (1997) *Prostaglandins Leukot. Essent. Fatty Acids* **56**, 417. **2.** Tam, Lee, Wang, Munroe & Lau (1995) *J. Biol. Chem.* **270**, 13948. **3.** J. I. Lee & G. J. Burckart (1998) *J. Clin. Pharmacol.* **38**, 981.

TEPP, See *Tetraethyl Pyrophosphate*

Teprotide, See *Bradykinin Potentiating Peptide 9a*

Terazosin

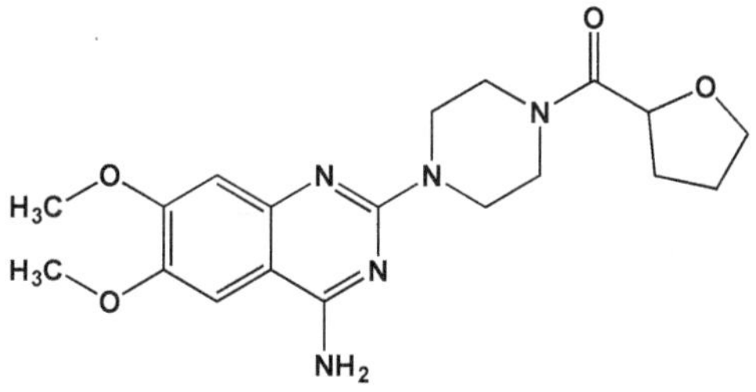

This orally available α_1-adrenergic receptor antagonist (FW$_{free-base}$ = 387.44 g/mol; CAS 63590-64-7; water solubility = 24 mg/mL for the hydrochloride dihydrate), known systematically as [4-(4-amino-6,7-dimethoxyquinazolin-2-yl)piperazin-1-yl]-(oxolan-2-yl)methanone, targets α_{1A}, α_{1B}, α_{1D}, α_{2B}, α_{2A} and α_{2C} receptors, with K_i values are 3.3, 0.7, 1.1, 7.7, 1510 and 78.2 nM, respectively (1-3). Terazosin induces apoptosis and reduces prostate tumor vascularity in benign and malignant prostate cells. Terzosin also suppresses prostate growth without affecting cell proliferation (3-6). **Key Pharmacokinetic Parameters:** *See* Appendix II in Goodman & Gilman's *THE PHARMACOLOGICAL BASIS OF THERAPEUTICS*, 12th Edition (Brunton, Chabner & Knollmann, eds.) McGraw-Hill Medical, New York (2011). **Target(s):** α_1-adrenergic receptor (2). **1**. Anglin, Glassman & Kyprianou (2002) *Prostate Cancer Prostatic Dis.* **5**, 88. **2**. Kyncl (1993) *J. Clin. Pharmacol.* **33**, 878. **3**. Maruyama, *et al.* (1994) *Biol. Pharm. Bull.* **17**, 1126. **4**. Kyncl (1986) *Am. J. Med.* **80**, 12. **5**. Maruyama, *et al.* (1994) *Biol. Pharm. Bull.* **17**, 1126. **6**. Hancock, *et al.* (1995) *J. Recept. Signal Transduct. Res.* **15**, 863.

Terbacil

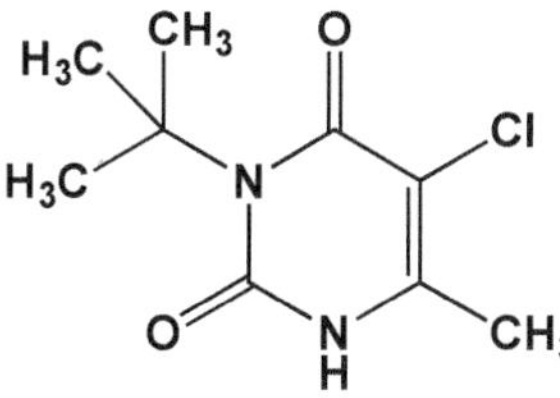

This uracil derivative (FW = 216.67 g/mol; CAS 5902-51-2; Properties: white, crystalline, odorless solid, which is wettable, noncorrosive and nonflammable), also known as Sinbar® and 5-chloro-3-(1,1-dimethylethyl)-6-methyl-2,4(1*H*,3*H*)-pyrimidinedione, is a selective herbicide for controlling the growth of annual grasses, broad-leaved weeds, and some perennial weeds. Its principal mode of action the inhibition of photosynthetic electron transport (1). **1**. Dodge (1977) *Spec. Publ., Chem. Soc.* **29**, 7.

Terbogrel

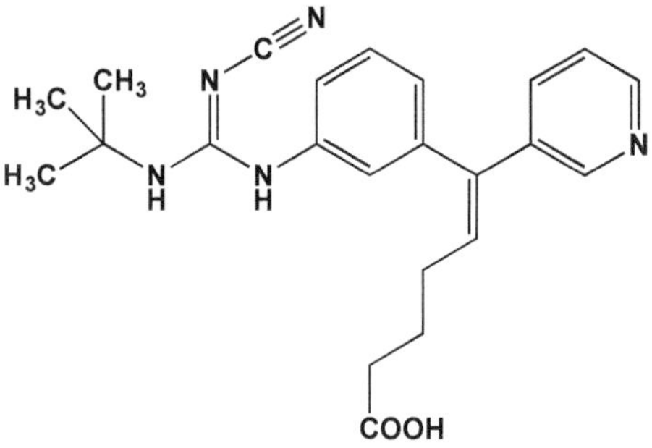

This guanidine derivative (FW$_{free-acid}$ = 405.50 g/mol; CAS 149979-74-8), also known as (5*E*)-6-(3-(2-cyano-3-*tert*-butylguanidino)phenyl)-6-(3-pyridyl)hex-5-enoic acid, inhibits human thromboxane-A synthase (IC$_{50}$ = 4 nM) and is a thromboxane receptor antagonist (IC$_{50}$ = 11 nM). Note that the 5*Z*-isomer is a weaker inhibitor (IC$_{50}$ = 1.6 and 2.6 µM for the synthase and receptor, respectively). **Target(s):** thromboxane-A synthase (1,2); thromboxane receptor (1,2). **1**. Soyka, Guth, Weisenberger, Luger & Müller (1999) *J. Med. Chem.* **42**, 1235. **2**. Michaux, Norberg, Dogne, *et al.* (2000) *Acta Crystallogr. C* **56**, 1265.

Terbutryn

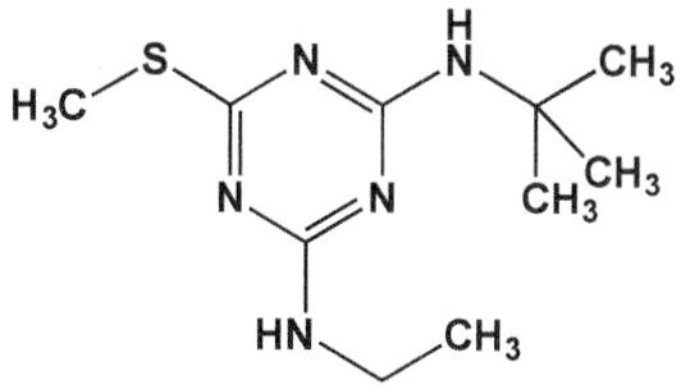

This triazine-based plant control agent (FW = 241.36 g/mol; CAS 886-50-0), also known as terbutryne and *N*-(1,1-dimethylethyl)-*N'*-ethyl-6-(methylthio)-1,3,5-triazine-2,4-diamine, is a selective pre- and post emergent herbicide for most grasses and many annual broadleaf weeds. It is used in crops of wheat, barley, sorghum, sugarcane, sunflowers, peas, and potatoes. Terbutryn is a competitive inhibitor of quinone binding at bacterial photosynthetic reaction centers. **Target(s):** photosynthetic electron transport (1); photosynthetic reaction center (2,3). **1**. Dodge (1977) *Spec. Publ., Chem. Soc.* **29**, 7. **2**. Kirmaier, Laible, Hanson & Holten (2003) *Biochemistry* **42**, 2016. **3**. Stein, Castellvi, Bogacz & Wraight (1984) *J. Cell Biochem.* **24**, 243.

Terephthalate

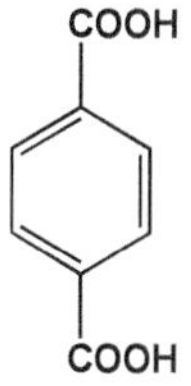

This aromatic dicarboxylic acid (FW$_{free-acid}$ = 166.13 g/mol; CAS 3198-30-9), also called *p*-carboxybenzoic acid and *p*-phthalic acid is only soluble in slightly alkaline solutions. **Target(s):** D-amino-acid oxidase (1-3); γ-butyrobetaine hydroxylase, *or* γ-butyrobetaine:2-oxoglutarate dioxygenase, weakly inhibited (4); 4,5-dihydroxyphthalate decarboxylase (5); glutamate decarboxylase, weakly inhibited (6); kynurenine aminotransferase, weakly inhibited (7); monophenol monooxygenase (1,8); procollagen-proline:2-oxoglutarate 4-dioxygenase (9); serine-sulfate ammonia-lyase, *or* weakly inhibited (10); succinate dehydrogenase (1,11). **1**. Webb (1966) *Enzyme and Metabolic Inhibitors*, vol. 2, Academic Press, New York. **2**. Bartlett (1948) *J. Amer. Chem. Soc.* **70**, 1010. **3**. Frisell, Lowe & Hellerman (1956) *J. Biol. Chem.* **223**, 75. **4**. Ng, Hanauske-Abel & England (1991) *J. Biol. Chem.* **266**, 1526. **5**. Nakazawa & Hayashi (1978) *Appl. Environ. Microbiol.* **36**, 264. **6**. Fonda (1972) *Biochemistry* **11**, 1304. **7**. Mason (1959) *J. Biol. Chem.* **234**, 2770. **8**. Krueger (1955) *Arch. Biochem. Biophys.* **57**, 52. **9**. Tschank, Hanauske-Abel & Peterkofsky (1988) *Arch. Biochem. Biophys.* **261**, 312. **10**. Tudball & Thomas (1973) *Eur. J. Biochem.* **40**, 25. **11**. Dietrich, Monson, Williams & Elvehjem (1952) *J. Biol. Chem.* **197**, 37.

Terfenadine

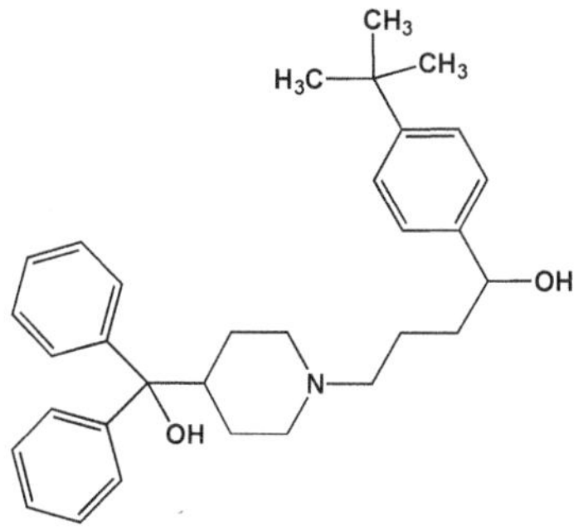

This first-in-class, nonsedating antihistamine (FW = 471.68 g/mol; CAS 50679-08-8; Soluble to 100 mM in DMSO; solid has three polymorphic forms, with melting points at 142-144°C, 146-148°C, and 149-152°C; sparingly soluble in H$_2$O (1 mg/100 mL at 30°C, but has higher solubility (0.11 g/100 mL) in 0.1 M citric acid), also known as α-[4-(1,1-dimethylethyl)phenyl]-4-(hydroxy-diphenylmethyl)-1-piperidinebutanol and Seldane™, is a histamine-H$_1$ receptor antagonist. In addition to its antihistamine effects, terfenadine also acts as a K$_V$11.1potassium channel blocker (encoded by the gene *hERG*). Terfenadine is also an alternative substrate for CYP3A4 and CYP2D6. **See also** *Loratadine; Fexofenadine* **Target(s):** calcium channel (1); histamine H_1 receptor, K_i = 80 nM (2-5); 5-lipoxygenase (6); potassium currents (7,8); and sodium currents (9). **1**. Liu, Melchert & Kennedy (1997) *Circ. Res.* **81**, 202. **2**. Masheter (1993) *Clin. Rev. Allergy* **11**, 5. **3**. Zhang & Timmerman (1993) *Pharm. World Sci.* **15**, 186. **4**. Hill, Ganellin, Timmerman, *et al.* (1997) *Pharmacol. Rev.* **49**, 253. **5**. Ter Laak, Donne-Op den Kelder, Bast & Timmerman (1993) *Eur. J. Pharmacol.* **232**, 199. **6**. Hamasaki, Kita, Hayasaki, *et al.* (1998) *Prostaglandins Leukot. Essent. Fatty Acids* **58**, 265. **7**. Ohtani, Hanada, Hirota, *et al.* (1999) *J. Pharm. Pharmacol.* **51**, 1059. **8**. Nishio, Habuchi, Tanaka, *et al.* (1998) *J. Pharmacol. Exp. Ther.* **287**, 293. **9**. Lu & Wang (1999) *J. Cardiovasc. Pharmacol.* **33**, 507.

Tergitol

This class of nonionic detergents (MW = 264.41 for n = 7, but depending on value of *n*), known also as *n*-nonylphenoxypolyethoxyethanols, was originally produced by Union Carbide. Tergitol NP-10, a light yellow liquid (M. P. = 6°C; B. P. = 250°C) is a polyoxyethylene, with a *p*-nonylphenyl moiety located on the α-position and a hydroxyl group in the ω-position. *Note*: Tergitol is a skin irritant; avoid contact with skin. **Target(s):** 5-aminolevulinate synthase (1); diacylglycerol *O*-acyltransferase, inhibited by Tergitol NP-40 (2); and triacylglycerol lipase (3). **1**. Murthy & Woods (1974) *Biochim. Biophys. Acta* **350**, 240. **2**. Grigor & Bell (1982) *Biochim. Biophys. Acta* **712**, 464. **3**. Khoo & Steinberg (1975) *Meth. Enzymol.* **35**, 181. **See also** *Triton N-101; Niaproof 4*

Teriflunomide

This orally active MS drug (FW = 270.21 g/mol; CAS 163451-81-8), also known by the code name A771726, A771726 HMR-1726, trade name Aubagio®, and its systematic name (2*Z*)-2-cyano-3-hydroxy-*N*-[4-(trifluoromethyl)phenyl]but-2-enamide, is *de novo* pyrimidine nucleotide synthesis inhibitor, targeting dihydroorotate dehydrogenase (*DHODH Reaction*: Dihydroorotate + NAD$^+$ ⇌ Orotate + NADH). Inhibition of DHODH has a cytostatic effect on actively dividing B- and T-cells. Salvage pathways allow the proliferation of more slowly dividing cells (*e.g.*, memory T-cells and haematopoietic cells) that are not as susceptible to DHODH inhibition and are able to sustain pyrimidine synthesis, thus maintaining the patient's immune response. Teriflunomide reduces lymphocyte proliferation and exerts immunomodulatory affecting the progression of multiple sclerosis. While cytostatic, teriflunomide is not cytotoxic, and does not affect resting or slowly dividing lymphocytes. Teriflunomide inhibits T cell receptor (TCR)/CD3-mediated calcium mobilization, with little effect on other critical T cell signaling events, such as MAPK activation and NF-κB activation (1,2). Teriflunomide interfered strikingly with TCR/CD3-triggered integrin-$\beta_{I,II}$ avidity and integrin-mediated costimulation (*i.e.*, outside-in signaling). Teriflunomide is the active metabolite of leflunomide, a potent disease-modifying antirheumatic drug that exhibits antiinflammatory, antiproliferative, and immunosuppressive effects (**See** *Leflunomide*). **1**. Zeyda, Poglitsch, Geyeregger, *et al.* (2005) *Arthritis Rheum.* **52**, 2730. **2**. Garnock-Jones (2013) *CNS Drugs* **27**, 1103.

Terpendole C and E

Terpendole C

Terpendole E

These fungal metabolites, particularly terpendole C (FW = 519.67 g/mol; CAS 164323-42-6) and terpendole E (FW = 437.60 g/mol; CAS 167427-23-8; Alternate Name: 1'α-hydroxy-paspaline), abbreviated TerC and TerE, from *Albophoma yamanashiensis* (Strain FO-2546) arrest at cell-cycle M-phase by inhibiting kinesin Eg5. TerE does not affect microtubule integrity in interphase, but induces formation of a monoastral spindle. TerE also inhibits both kinesin motor Eg5, but does not affect conventional kinesins from either *Drosophila* or bovine brain. Although terpendoles have been reported as inhibitors of acyl-CoA:cholesterol *O*-acyltransferase (ACAT), the Eg5 inhibitory activity of TerE was independent of ACAT inhibition. **Target(s):** microtubule-stimulated ATPase, inhibited by terpendole E (1); sterol *O*-acyltransferase, *or* cholesterol *O*-acyltransferase (ACAT) (2,3). **1**. Nakazawa, Yajima, Usui, *et al.* (2003) *Chem. Biol.* **10**, 131. **2**. Ohshiro, Rudel, Omura & Tomoda (2007) *J. Antibiot.* **60**, 43. **3**. Huang, Tomoda, Nishida, Masuma & Omura (1995) *J. Antibiot.* **48**, 1.

Terpestacin

This fungal metabolite (FW = 402.58 g/mol; CAS 146436-22-8), isolated from *Arthrinium sp.* FA1744, binds to the 13.4-kDa subunit (UQCRB) of mitochondrial Complex III, resulting in inhibition of hypoxia-induced reactive oxygen species generation, blocking hypoxia-inducible factor activation and tumor angiogenesis *in vivo*, without inhibiting mitochondrial respiration. **1**. Jung, Shim, Lee, *et al.* (1999) *Bioorg. Med. Chem. Lett.* **9**, 2279.

Tertiapin

This basic honey bee venom toxin (MW = 2459.1 g/mol; CAS 252198-49-5; *Sequence*: ALCNCNRIIIPHMCWKKCGKK-NH$_2$) is a potent inhibitor of the inward-rectifier K$^+$ channels, binds specifically to different subunits of the inward rectifier potassium channel (Kir), namely GIRK1 (Kir 3.1), GIRK4 (Kir 3.4) and ROMK1 (Kir 1.1), inducing a dose-dependent block of the potassium current. It is thought that tertiapin binds to the Kir channel with its α-helix situated at the C-terminal of the peptide. This α-helix is plugged into the external end of the conduction pore, thereby blocking the channel. The N-terminal of the peptide sticks out of the extracellular side. Tertiapin has a high affinity for Kir channels with approximately K_d = 8 nM for GIRK1/4 channels and K_d = 2 nM for ROMK1 channels. **Target(s):** calmodulin activating activity (3); inward-rectifier K$^+$ channels (1,2). **1**. Jin & Lu (1998) *Biochemistry* **37**, 13291. **2**. Schultz, Czachurski, Volk, Ehmke & Seller (2003) *Brain Res.* **963**, 113. **3**. Miroshnikov, Boikov, Snezhkova, Severin & Shvets (1983) *Bioorg. Khim.* **9**, 26.

Tertiapin-Q

This air oxidation-resistant analogue (FW = 2295.82 g/mol; CAS 252198-49-5) of honey bee venom toxin tertiapin (*Sequence*: ALCNCNRIIIPHQ CWKKCGKK-NH$_2$, with Cys^3-S-S-Cys^{14} and Cys^5-S-S-Cys^{18} disulfide bridges and with Q replacing M at residue-13) blocks the ROMK1 (K$_{ir}$1.1) channel and the G-protein-gated K$^+$ channel, *or* GIRK1/4 (K$_{ir}$3.1/3.4) with K_i values are 1.3 and 13.3 nM respectively (1,2). Tertiapin-Q shows selective over K$_{ir}$2.1 channels. Note that the methionine residue in Tertiapin is sensitive to oxidation, reducing the ability to block the ionic channels; this problem is alleviated in Tertiapin-W, wherein the methionine is replaced by glutamine. **1**. Sackin, Vasilyev, Palmer & Krambis (2003) *Biophys. J.* **84**, 910. **2**. Jin, Klem, Lewis & Lu (1999) *Biochemistry* **38**, 14294.

Tessulin

This 81-residue cysteine-rich polypeptide (MW = 9 kDa) from the duck leech *Theromyzon tessulatum* inhibits trypsin, K_i = 1 pM, and chymotrypsin, K_i =

150 pM (1). Tessulin also decreases the lipopolysaccharide-induced activation of human monocytes and granulocytes, but does not inhibit elastase or cathepsin G. When acting in conjunction with other *Theromyzon* serine-protease inhibitors, namely therin and theromin, tessulin significantly diminishes the level of human granulocyte and monocyte activation induced by lipopolysaccharides (10 μg), and the combined level of inhibition is higher than that of aprotinin. Such properties commend its use as an anti-inflammatory agent. **1.** Chopin, Stefano & Salzet (1998) *Eur. J. Biochem.* **258**, 662.

Testane, *See* 5α-Androstane

Testosterone

This androgenic steroid (FW = 288.43 g/mol; CAS 58-22-0; water solubility = 0.13 mM; λ_{max} = 239 nm; ε = 17000 $M^{-1}cm^{-1}$), also known as Δ^4-androsten-17β-ol-3-one, is synthesized in Leydig cells of the testes. Testosterone is converted to 5α-dihydrotestosterone by steroid 5α-reductase, a metabolite that is required for normal male sexual differentiation. Testosterone also has an antiglucocorticoid activity (1). **Target(s):** aldehyde dehydrogenase (2); D-amino-acid oxidase (3); cortisol sulfotransferase (4,5); F_1F_o ATPase, *or* H^+-transporting two-sector ATPase (6); glucose-6-phosphate dehydrogenase (7); glucuronosyltransferase (8-10); glutamate dehydrogenase (2); hydroxyindole *O*-methyltransferase (11); 21-hydroxysteroid dehydrogenase (NAD^+) (12-14); isocitrate dehydrogenase, weakly inhibited (15); 6-phosphogluconate dehydrogenase, weakly inhibited (15); steroid Δ-isomerase (16-18); steroid 21-monooxygenase (19); steroid sulfotransferase (20); sterol *O*-acyltransferase, *or* cholesterol *O*-acyltransferase; ACAT (21,22); sterol demethylase (23); sterol esterase, *or* cholesterol esterase (24); and steryl sulfatase (25). **1.** Scott & Tomkins (1975) *Meth. Enzymol.* **40**, 273. **2.** Douville & Warren (1968) *Biochemistry* **7**, 4052. **3.** Hayano, Dorfman & Yamada (1950) *J. Biol. Chem.* **186**, 603. **4.** Singer (1979) *Arch. Biochem. Biophys.* **196**, 340. **5.** Singer & Bruns (1980) *Can. J. Biochem.* **58**, 660. **6.** Zheng & Ramirez (1999) *Eur. J. Pharmacol.* **368**, 95. **7.** Criss & McKerns (1969) *Biochim. Biophys. Acta* **184**, 486. Matern, Matern, Schelzig & Gerok (1980) *FEBS Lett.* **118**, 251. **9.** Yoshigae, Konno, Takasaki & Ikeda (2000) *J. Toxicol. Sci.* **25**, 433. **10.** Rao, Rao & Breuer (1976) *Biochim. Biophys. Acta* **452**, 89. **11.** Morton & Forbes (1989) *J. Pineal Res.* **6**, 259. **12.** Monder & Furfine (1969) *Meth. Enzymol.* **15**, 667. **13.** Talalay (1963) *The Enzymes*, 2nd ed. (Boyer, Lardy & Myrbäck, eds.), **7**, 177. **14.** Monder & White (1962) *Biochem. Biophys. Res. Commun.* **8**, 383. **15.** McKerns (1963) *Biochim. Biophys. Acta* **73**, 507. **16.** Benisek (1977) *Meth. Enzymol.* **46**, 469. **17.** Jones & S. Ship (1972) *Biochim. Biophys. Acta* **258**, 800. **18.** Weintraub, Vincent, Baulieu & Alfsen (1977) *Biochemistry* **16**, 5045. **19.** Sharma (1964) *Biochemistry* **3**, 1093. **20.** Singer, Gebhart & Hess (1978) *Can. J. Biochem.* **56**, 1028. **21.** Billheimer (1985) *Meth. Enzymol.* **111**, 286. **22.** Simpson & Burkhart (1980) *Arch. Biochem. Biophys.* **200**, 79. **23.** Gaylor, Chang, Nightingale, Recio & Ying (1965) *Biochemistry* **4**, 1144. **24.** Tomita, Sawamura, Uetsuka, *et al.* (1996) *Biochim. Biophys. Acta* **1300**, 210. **25.** Notation & Ungar (1969) *Biochemistry* **8**, 501.

***cis*-Testosterone, *See* Epitestosterone**

Testosterone Acetate

This steroid ester (FW = 330.47 g/mol; CAS = 1045-69-8; M.P. = 140-141°C), also known as 4-androsten-17β-ol-3-one 17-acetate and inhibits *aromatase or* CYP19 (1) and steroid Δ-isomerase (2,3). **1.** Danisova,

Sebokova & Kolena (1987) *Exp. Clin. Endocrinol.* **89**, 165. **2.** Jones & Ship (1972) *Biochim. Biophys. Acta* **258**, 800. **3.** Weintraub, Vincent, Baulieu & Alfsen (1977) *Biochemistry* **16**, 5045.

Tetanus Toxin

This powerful bacterial neurotoxin ($MW_{pre-protein}$ = 150,317 Da; GenBank: AAK72964.2; *Symbol:* TeTx), also known as tetanospasmin, from the Gram-positive spore-forming bacillus *Clostridium tetani* has an LD_{50} of 1 ng/mL, second only to Botulinum Toxin in its deadliness. The toxin induces violent recurrent spasms/paralysis, *or* tetany, marked by inhibition of glycine-mediated and GABA-ergic neurotransmission (1) as well as involuntary contraction of peripheral skeletal muscles, with death often arising from respiratory failure. **Mechanism of Inhibitory Action:** For activation, a single precursor is proteolyzed to form a disulfide-linked heterodimer consisting of 50 kDa A-Chain and a 100 kDa B-Chain (2,3). The A-chain is a Zn^{2+}-metalloprotease (4), whereas B-chain is a membrane-docking component that binds to neuron and facilitates endocytosis. The main target of TeTx is the central nervous system, where it blocks synaptic function by several mechanisms. The first involves selective proteolysis of the integral membrane proteins synaptobrevin II (also known as the Vesicle-Associated Membrane Protein (VAMP) or its homologue cellubrevin (5). A second entails TeTx activation Ca^{2+}-dependent, GTP-modulated transglutaminases that cross-link synapsin, a phosphoprotein involved in neurotransmission (6,7). Two forms of TeTx-activated transglutaminase are found in nerve endings (*i.e.,* one in the cytosol and the other on synaptic vesicles. A third mechanism exploits TeTx's ability to block synapsin phosphorylation after depolarizing rat brain synaptosomes (8). TeTx also markedly decreased the translocation of synapsin I from the small synaptic vesicles and the cytoskeleton into the cytosol, on depolarization of synaptosomes. The effect of TeTx on synapsin I phosphorylation was both time and TeTx concentration dependent and required active toxin. A fourth mechanism relates to the discovery that TeTx exerts a time- and dose-dependent inhibition of Na^+-dependent serotonin (5-HT) uptake in CNS synaptosomes, an effect found in all CNS tryptaminergic areas, but highest in the hippocampus and occipital cortex (9). Notably, fenfluramine (IC_{50} = 11 μM), paroxetine (IC_{50} = 33 μM), and imipramine (IC_{50} = 90 μM) are 1,000 to 10,000 times less potent serotonin transport inhibitors than TeTx (IC_{50} = 9 nM). The TeTx effect on serotonin accumulation may be responsible for some tetanic symptoms (9). The Hc fragment of TeTx inhibits both basal and stimulated serotonin uptake in primary neuronal cultures, effects that parallel the activation of phosphoinositide-phospholipase C activity and PKC activation (10). **1.** Pearce, Gard & Dutton (1983) *J. Neurochem.* **40**, 887. **2.** Niemann (1992) *Toxicon* **30**, 223. **3.** Montecucco & Schiavo (1995) *Quart. Rev. Biophys.* **28**, 423. **4.** Jongeneel, Bouvier & Bairoch (1989) *FEBS Lett.* **242**, 211. **5.** Schiavo, Benfenati, Poulain, *et al.* (1992) *Nature* **359**, 832. **6.** Folk (1980) *Annu. Rev. Biochem.* **49**, 517. **7.** Facchiano, Benfenati, Valtorta & Luini (1993) *J. Biol. Chem.* **268**, 4588. **8.** Presek, Kessen, Dreyer, *et al.* (1992) *J. Neurochem.* **59**, 1336. **9.** Inserte, Najib, Pelliccioni, Gil & Aguilera (1999) *Biochem. Pharmacol.* **57**, 111. **10.** Pelliccioni, Gil, Najib, *et al.* (2001) *J. Mol. Neurosci.* **17**, 303.

1,2,3,4-Tetra(adenosine-5'-*O*-phospho)erythritol

This unusual oligonucleotide cojugate (FW_{ion} = 1385.95 g/mol) inhibits both symmetrical and asymmetrical bis(5'-nucleosyl)tetraphosphatases. Dinucleoside 5',5'''-P^1,P''-polyphosphates (particularly the diadenosine compounds) have been implicated in extracellular purinergic signalling and in various intracellular processes, including DNA metabolism, tumour suppression and stress responses. If permitted to accumulate, they may also be toxic. Adenosine-5'-*O*-phosphorothioylated polyols were generally stronger inhibitors than their adenosine-5'-*O*-phosphorylated counterparts. **1.** Guranowski, Starzynska, McLennan, Baraniak & Stec (2003) *Biochem. J.* **373**, 635.

Tetra(*p*-amidinophenoxy)-*neo*-pentane

This aromatic tetrabenzamidine derivative (FW$_{free-base}$ = 608.70 g/mol; Symbol: TAPP) inhibits coagulation factor Xa (1); plasmin (1); thrombin (1,2); tissue kallikrein (1,2); and trypsin (1,2). Over the pH range from 2 to 8, values of K_i for TAPP binding to α-thrombin, Factor Xa, Lys77-Plasmin and β-Kallikrein-B may be described as depending on the pK-shift, upon inhibitor association, of two equivalent proton-binding amino acid residues. **1.** Menegatti, Ferroni, Scalia, *et al.* (1987) *J. Enzyme Inhib.* **2**, 23. **2.** Menegatti, Ferroni, Nastruzzi, *et al.* (1991) *Farmaco* **46**, 1297.

Tetrabenazine

This orally available dopamine-depleting drug (FW = 317.43 g/mol; CAS 58-46-8; *Symbol* = TBZ; Soluble to 30 mM in ethanol and to 100 mM in DMSO), also known by the trade names Nitoman® and Xenazine® and as (*SS,RR*)-3-isobutyl-9,10-dimethoxy-1,3,4,6,7,11*b*-hexahydro-pyrido[2,1-*a*]isoquinolin-2-one, is a Vesicular Monoamine Transporter-2 (*or* VMAT2) inhibitor that is especially effective in promoting dopamine depletion within the striatum. Tetrabenazine is an FDA-approved drug used to treat the symptomatic hyperkinesia associated with Huntington's chorea. **Mode of Inhibitory Action:** α-DTBZ inhibits monoamine uptake, as driven by transmembrane proton electrochemical gradients generated by an ATP-hydrolyzing proton pump. TBZ is a reversible high-affinity inhibitor of monoamine uptake into granular vesicles of presynaptic neurons (1). It is also a weak D$_2$ postsynaptic receptor blocker in high doses (2), decreasing voluntary movements. TBZ depletes serotonin, norepinephrine, and especially dopamine (3). In rats and mice, TBZ also induces tremulous jaw movements (TJMs), which have many characteristics of Parkinsonian tremors in humans (4). Treatment of TJM, with deprenyl, a selective monoamine oxidase-B inhibitor, shows a dose-related suppression of TBZ-induced TJMs. (*See Deprenyl*) **1.** Pettibone, Totaro & Pflueger (1984) *Eur. J. Pharmacol.* **102**, 425. **2.** Reches, Burke, Kuhn, *et al.* (1983) *J. Pharmacol. Exp. Ther.* **225**, 515. **3.** Pletscher, Brossi & Gey (1962) *Internat. Rev. Neurobiol.* **4**, 275. **4.** Podurgiel, Yohn, Dortche, Correa & Salamone (2015) *Behav. Brain Res.* **298**, 188.

Tetrabromoaurate(III), Potassium

This KAuBr$_4$ dihydrate (FW = 591.71 g/mol; CAS 14323-32-1; reddish-brown solid) is a planar d^8 tetracoordinate Au(III) complex formed by slow hydrolysis/aquation of HAu(III)Cl$_4$·(H$_2$O)$_3$ in the presence of excess HBr

(typically requiring >24 hours incubation prior to use) to form the inhibitor (1,2). **Target(s):** acid phosphatase (2); β-glucuronidase (2); and L-malate dehydrogenase (3). **1.** Catalini, Chessa, Michelon, *et al.* (1985) *Inorg. Chem.* **24**, 3409. **2.** Lee, Ahmed & Friedman (1989) *J. Enzyme Inhib.* **3**, 23. **3.** Friedman, Chasteen, Shaw, Allen & McAuliffe (1982) *Chem. Biol. Interact.* **42**, 321.

4,5,6,7-Tetrabromobenzotriazole

This aromatic triazole (FW = 434.71 g/mol; CAS 17374-26-4; *Symbol*: TBB; Soluble to 100 mM in DMSO and to 15 mM in Ethanol), also known as 4,5,6,7-tetrabromo-2-azabenzimidazole, is a potent inhibitor of casein kinase-2: IC$_{50}$ = 0.9 μM for the rat liver enzyme. TBB acts in an ATP/GTP-competitive manner, with 10x to 100x selectivity over a panel of 33 protein kinases. **Target(s):** casein kinase-2 (1-6); DNA helicases (7); and non-specific serine/threonine protein kinase (1-6). **1.** Sarno, Reddy, Meggio, *et al.* (2001) *FEBS Lett.* **496**, 44. **2.** Szyszka, Grankowski, Felczak & Shugar (1995) *Biochem. Biophys. Res. Commun.* **208**, 418. **3.** Battistutta, De Moliner, Sarno, Zanotti & Pinna (2001) *Protein Sci.* **10**, 2200. **4.** Pagano, Cesaro, Meggio & Pinna (2006) *Biochem. Soc. Trans.* **34**, 1303. **5.** Duncan & Litchfield (2008) *Biochim. Biophys. Acta* **1784**, 33. **6.** Mottet, Ruys, Demazy, Raes & Michiels (2005) *Int. J. Cancer* **117**, 764. **7.** Borowski, Deinert, Schalinski, *et al.* (2003) *Eur. J. Biochem.* **270**, 1645.

Tretrabromocinnamic Acid

This CK2 inhibitor (FW = 463.74 g/mol; CAS 934358-00-6), also known as TBCA and (2*E*)-3-(2,3,4,5-tetrabromophenyl)-2-propenoate, targets Casein Kinase-2 (IC$_{50}$ = 0.11 μM) and exhibits selectivity over CK1, DYRK1A, *or* Dual-specificity tyrosine phosphorylation-regulated kinase-1A (IC$_{50}$ = 24.5 μM)), and a panel of 27 other kinases. The most widely employed CK2 inhibitor is 4,5,6,7-tetrabromobenzotriazole, *or* TBB (IC$_{50}$ = 0.54 μM), which exhibits a comparable efficacy toward another kinase, DYRK1a. Abnormally high constitutive activity of protein kinase CK2 are suspected to underlie its pathogenesis of vertrain cancers. TBCA reduces the viability of Jurkat cells more efficiently than TBB through enhancement of apoptosis. **1.** Pagano, Poletto, Di Maira, *et al.* (2007) *Chembiochem.* **8**, 129.

Tetracaine

This local anesthetic (FW$_{hydrochloride}$ = 300.83 g/mol; CAS 136-47-0; solubility = 1 part in 7.5 parts water at 20°C), also known as 4-(butylamino)benzoic acid 2-(dimethylamino)ethyl ester, blocks the voltage-sensitive release of calcium ions from the sarcoplasmic reticulum. **Target(s):** acetylcholinesterase (1-8); 1-acylglycero-phosphocholine *O*-acyltransferase, *or* lysophosphatidyl-choline *O*-acyltransferase (49); β$_2$-adrenergic receptor (9); amino acid transport (10); Ca^{2+}-dependent cyclic nucleotide phosphodiesterase (11); Ca^{2+}/Mg^{2+}-dependent ATPase (12); Ca^{2+} release channel (13-15); choline-phosphate cytidylyltransferase, *or* CTP:phosphocholine cytidylyltransferase (42,48); cholinesterase (5,19); cytochrome *c* oxidase (20), electron transport (21,22); F$_1$ ATP synthase, *or* mitochondrial ATPase; H⁺-transporting two-sector ATPase (7,23-28); GABA$_A$ receptors (29); D-3-hydroxybutyrate dehydrogenase (30); K⁺/Cl⁻ cotransport (31); kinesin (32); lecithin:cholesterol acyltransferase, *or* LCAT (33); luciferase (20,34); *N*-methyl-D-aspartate receptor (35); monoamine oxidase (36); NADPH:ferrihemoprotein reductase (37); Na⁺/K⁺-exchanging ATPase (7,38); nicotinic acetylcholine receptors (39,40); μ-, κ-, and δ-opioid receptors (41); phospholipase A$_2$ (43); phospholipid-dependent protein kinase (44); protein kinase C (45,46); quaternary-amine-transporting ATPase (47); stearoyl-CoA Desaturase (37); and sterol *O*-acyltransferase, *or* cholesterol *O*-acyltransferase, ACAT (16-18). **1.** al-Jafari, Kamal, Duhaiman & Alhomida (1996) *J. Enzyme Inhib.* **11**, 123. **2.** Duhaiman, Alhomida, Rabbani, Kamal & al-Jafari (1996) *Biochimie* **78**, 46. **3.** al-Jafari (1993) *Comp. Biochem. Physiol. C* **105**, 323. **4.** Spinedi, Pacini & Luly (1989) *Biochem. J.* **261**, 569. **5.** Perez-Guillermo, Delgado & Vidal (1987) *Biochem. Pharmacol.* **36**, 3593. **6.** Haque & Poddar (1985) *Biochem. Pharmacol.* **34**, 2599. **7.** Sidek, Nyquist-Battie & Vanderkooi (1984) *Biochim. Biophys. Acta* **801**, 26. **8.** Agbaji, Gerassimidis & Hider (1984) *Comp. Biochem. Physiol. C* **78**, 211. **9.** Butterworth, James & Grimes (1997) *Anesth. Analg.* **85**, 336. **10.** Bieger, Peter, Volkl & Kern (1977) *Cell Tissue Res.* **180**, 45. **11.** Hidaka, Inagaki, Nishikawa & Tanaka (1988) *Meth. Enzymol.* **159**, 652. **12.** Garcia-Martin & Gutierrez-Merino (1986) *J. Neurochem.* **47**, 668. **13.** Komai & Lokuta (1999) *Anesthesiology* **90**, 835. **14.** Overend, Eisner & O'Neill (1997) *J. Physiol.* **502**, 471. **15.** Antoniu, Kim, Morii & Ikemoto (1985) *Biochim. Biophys. Acta* **816**, 9. **16.** Chang & Doolittle (1983) *The Enzymes*, 3rd ed. (Boyer, ed.), **16**, 523. **17.** Bell (1981) *Biochim. Biophys. Acta* **666**, 58. **18.** Bell & Hubert (1980) *Biochim. Biophys. Acta* **619**, 302. **19.** Traynor & Kunze (1975) *Biochim. Biophys. Acta* **409**, 68. **20.** Vanderkooi & Chazotte (1982) *Proc. Natl. Acad. Sci. U.S.A.* **79**, 3749. **21.** Semin, Ivanov & Rubin (1990) *Gen. Physiol. Biophys.* **9**, 65. **22.** Chazotte & Vanderkooi (1981) *Biochim. Biophys. Acta* **636**, 153. **23.** Vanderkooi & Adade (1986) *Biochemistry* **25**, 7118. **24.** Kresheck, Adade & Vanderkooi (1985) *Biochemistry* **24**, 1715. **25.** Adade, Chignell & Vanderkooi (1984) *J. Bioenerg. Biomembr.* **16**, 353. **26.** Laikind, Goldenberg & Allison (1982) *Biochim. Biophys. Res. Commun.* **109**, 423. **27.** Chazotte, Vanderkooi & Chignell (1982) *Biochim. Biophys. Acta* **680**, 310. **28.** Vanderkooi, Shaw, Storms, Vennerstrom & Chignell (1981) *Biochim. Biophys. Acta* **635**, 200. **29.** Sugimoto, Uchida, Fukami, *et al.* (2000) *Eur. J. Pharmacol.* **401**, 329. **30.** Gotterer (1969) *Biochemistry* **8**, 641. **31.** Sachs (1994) *Amer. J. Physiol.* **266**, C997. **32.** Miyamoto, Muto, Mashimo, *et al.* (2000) *Biophys. J.* **78**, 940. **33.** Bell & Hubert (1980) *Lipids* **15**, 811. **34.** Ueda, Kamaya & Eyring (1976) *Proc. Natl. Acad. Sci. U.S.A.* **73**, 481. **35.** Nishizawa, Shirasaki, Nakao, Matsuda & Shingu (2002) *Anesth. Analg.* **94**, 325. **36.** Yasuhara, Wada, Sakamoto & Kamijo (1982) *Jpn. J. Pharmacol.* **32**, 213. **37.** S. Umeki & Y. Nozawa (1986) *Biol. Chem. Hoppe Seyler* **367**, 61. **38.** Hudgins & Bond (1984) *Biochem. Pharmacol.* **33**, 1789. **39.** Papke, Horenstein & Placzek (2001) *Mol. Pharmacol.* **60**, 1365. **40.** Gentry & Lukas (2001) *J. Pharmacol. Exp. Ther.* **299**, 1038. **41.** Hirota, Okawa, Appadu, Grandy & Lambert (2000) *Brit. J. Anaesth.* **85**, 740. **42.** Mansbach & Arnold (1986) *Biochim. Biophys. Acta* **875**, 516. **43.** Jackson, Jones & Harris (1984) *Biosci. Rep.* **4**, 581. **44.** Mori, Takai, Minakuchi, Yu & Nishizuka (1980) *J. Biol. Chem.* **255**, 8378. **45.** Kikkawa, Minakuchi, Takai & Nishizuka (1983) *Meth. Enzymol.* **99**, 288. **46.** Kikkawa & Nishizuka (1986) *The Enzymes*, 3rd ed. (Boyer & Krebs, eds.), **17**, 167. **47.** van der Heide, Stuart & Poolman (2001) *EMBO J.* **20**, 7022. **48.** Pelech, Jetha & Vance (1983) *FEBS Lett.* **158**, 89. **49.** Sanjanwala, Sun & MacQuarrie (1989) *Arch. Biochem. Biophys.* **271**, 407.

1,2,3,4-Tetracarboxycyclopentane, *See* 1,2,3,4-Cyclopentanetetracarboxylate *and* 1,2,3,4-Cyclopentanetetracarboxylic Acid

2,3,5,6-Tetrachloro-1,4-benzenedicarboxylic Acid Dimethyl Ester, *See* DCPA

2,3,5,6-Tetrachloro-1,4-benzoquinone, *See* Chloranil

2,2',3,3'-Tetrachlorobiphenyl

This polychlorinated biphenyl *or* PCB congener (FW = 291.99 g/mol; CAS 52663-59-9; Abbreviation: 2,2',3,3'-TCB and 2,2',3,3'-TCBP) is a membrane-permeant environmental pollutant that inhibits uroporphyrinogen decarboxylase (1). In all, there are 169 cogeners that contain four or more chlorine atoms. Virtually all are toxic. PCBs were widely used for many applications, especially as dielectric fluids in transformers and capacitors, but their toxicity and biological persistence to led to current bans on PCB production in the United States Congress since 1979. **1.** Kawanishi, Seki & Sano (1983) *J. Biol. Chem.* **258**, 4285.

2,3,7,8-Tetrachlorodibenzo-*p*-dioxin, *See* TCDD

Tetrachloro-*p*-hydroquinone

This quinol and porphyrin synthesis inhibitor (FW = 247.89 g/mol; CAS 87-87-6), itself a hydroxylation metabolite of pentachlorophenol, targets both porphyrinogen carboxy-lyase (1,2) and uroporphyrinogen decarboxylase (3). **1.** Billi, Koss & San Martin de Viale (1986) *IARC Sci. Publ.* (77), 471. **2.** Billi, Koss & San Martin de Viale (1986) *Res. Commun. Chem. Pathol. Pharmacol.* **51**, 325. **3.** Koss, Losekam, Seidel, Steinbach & Koransky (1987) *Ann. N. Y. Acad. Sci.* **514**, 148.

4,5,6,7-Tetrachloroindan-1,3-dione

This substituted indane (FW = 283.93 g/mol; CAS 606-23-5 and 30675-13-9; orange solid) is a novel inhibitor of ubiquitin C-terminal hydrolase L3, *or* UCH-L3 (IC$_{50}$ = 0.6 μM), displaying 125x greater selectivity for UCH-L3 than for UCH-L1 (IC$_{50}$ = 75 μM). **1.** Liu, Lashuel, Choi, *et al.* (2003) *Chem. Biol.* **10**, 837. **2.** Love, *et al.* (2007) *Nature Chem. Biol.* **3**, 697.

Tetrachloropalladate dianion

These inorganic reagents [Na$_2$PdCl$_4$ (FW = 294.21 g/mol; CAS 14349-67-8) and K$_2$PdCl$_4$ (FW = 326.43 g/mol)] are planar d^8-tetracoordinate Pd(II) dianionic complexes that inhibit the sarcoplasmic reticulum Ca²⁺/Mg²⁺-dependent ATPase (1) and and electron transport (2). **1.** Tat'ianenko, Sokolova & Moshkovskii (1982) *Vopr. Med. Khim.* **28**, 126. **2.** Biagini, Moorman & Winston (1982) *Toxicol Lett.* **12**, 165.

Tetrachloroplatinate, Potassium

This water-soluble, ruby-red, planar d^8-tetracoordinate Pt(II) dianionic complexes complex K$_2$[PtCl$_4$] (FW = 415.09 g/mol) is an irreversible inhibitor of dihydropteridine reductase (1), fructose-1,6-bisphosphate aldolase (2), glucose-6-phosphate dehydrogenase (2), glyceraldehyde-3-phosphate dehydrogenase (2), and malate dehydrogenase (3,4). **1.** Armarego & Ohnishi (1987) *Eur. J. Biochem.* **164**, 403. **2.** Aull, Allen, Bapat, *et al.* (1979) *Biochim. Biophys. Acta* **571**, 352. **3.** Friedman, Otwell & Teggins (1975) *Biochim. Biophys. Acta* **391**, 1. **4.** Friedman & Teggins (1974) *Biochim. Biophys. Acta* **341**, 277.

4,5,6,7-Tetrachloro-2'-trifluoromethylbenzimidazole

This weakly acidic, membrane-permeant halogenated benzimidazole (FW = 323.92 g/mol; pK_a = 5.6), often abbreviated TTFB or TCFB, is a potent proton gradient-dissipating agent that uncouples mitochondrial oxidative phosphorylation and chloroplast electron transport (1-3). **See Uncouplers** 1. Heytler (1979) *Meth. Enzymol.* **55**, 462. **2.** Hanstein (1976) *Trends Biochem. Sci.* **1**, 65. **3.** Beechey (1966) *Biochem. J.* **98**, 284.

Tetracycline

The antibiotic (FW = 444.44 g/mol; CAS 60-54-8; solubility = 1.7 mg/mL at 28°C), systematically named [4S-(4α,4aα,5aα,6β,12aα)]-4-(dimethylamino)-1,4,4a,5,5a,6,11,12a-octahydro-3,6,10,12,12a-pentahydroxy-6-methyl-1,11-dioxo-2-naphthacenecarboxamide, from *Streptomyces* spp. inhibits binding of aminoacyl-tRNA to ribosomal acceptor site (A-site). Tetracycline-sensitive organisms are those that are most readily affected by this class of antibiotics. Tetracycline-resistant organisms exhibit diminished cellular accumulation of the antibiotic. Two transposons in enteric bacteria that carry genes associated with tetracycline resistance. In Gram-positive bacteria, resistance reportedly involves a translation attenuation mechanism. **Target(s):** catalase (1); fructose-1,6-bisphosphate aldolase (2); gingipain R (3); glutamate racemase (4); GTP diphosphokinase (5,6); guanosine-3',5'-bis(diphosphate) 3'-diphosphatase (7); guanosine 3',5'-polyphosphate synthetase (6); lipase, *or* triacylglycerol lipase (8-11); nucleotide diphosphokinase (12); peptidyl-tRNA hydrolase, *or* aminoacyl-tRNA hydrolase (13); protein biosynthesis, at elongation and termination (14-19); protein-synthesizing GTPase (elongation factor) (19); streptomycin 3"-adenylyltransferase (20). **See also Chlortetracycline; Oxytetracycline** 1. Ghatak, Saxena & Agarwala (1958) *Enzymologia* **19**, 261. **2.** Ghatak & Shrivastava (1958) *Enzymologia* **19**, 237. **3.** Houle, Grenier, Plamondon & Nakayama (2003) *FEMS Microbiol. Lett.* **221**, 181. **4.** Tanaka, Kato & Kinoshita (1961) *Biochem. Biophys. Res. Commun.* **4**, 114. **5.** Knutsson Jenvert & Holmberg Schiavone (2005) *FEBS J.* **272**, 685. **6.** Fehr & Richter (1981) *J. Bacteriol.* **145**, 68. **7.** Richter (1980) *Arch. Microbiol.* **124**, 229. **8.** Weaber, Freedman & Eudy (1971) *Appl. Microbiol.* **21**, 639. **9.** Hassing (1971) *J. Invest. Dermatol.* **56**, 189. **10.** Shalita & Wheatley (1970) *J. Invest. Dermatol.* **54**, 413. **11.** Mates (1973) *J. Invest. Dermatol.* **60**, 150. **12.** Nishino & Murao (1975) *Agric. Biol. Chem.* **39**, 1827. **13.** Tate & Caskey (1974) *The Enzymes*, 3rd ed. (Boyer, ed.), **10**, 87. **14.** Fresno & Vázquez (1979) *Meth. Enzymol.* **60**, 566. **15.** Jiménez (1976) *Trends Biochem. Sci.* **1**, 28. **16.** Bread, Armentrout & Weisberger (1969) *Pharmacol. Rev.* **21**, 213. **17.** Coutsogeorgopoulos (1967) *Biochemistry* **6**, 1704. **18.** Laskin & Chan (1964) *Biochem. Biophys. Res. Commun.* **14**, 137. **19.** Mesters, Potapov, de Graaf & Kraal (1994) *J. Mol. Biol.* **242**, 644. **20.** Haas & Dowding (1975) *Meth. Enzymol.* **43**, 611.

n-Tetradecanoate, See *Myristate*

1-Tetradecanol, See *Myristyl Alcohol*

2-Tetradecanoylaminohexanol-1-phosphoethanolamine, *action similar to: 2-Tetradecanoylaminohexanol-1-phosphocholine*

2-Tetradecanoylaminohexanol-1-phosphoglycol, *action similar to: 2-Tetradecanoylaminohexanol-1-phosphocholine*

Tetradecanoyl-CoA, See *Myristoyl-CoA*

12-O-Tetradecanoylphorbol 13-Acetate, See *Phorbol 12-Myristate 13-Acetate*

2-Tetradecanylglutarate, See *2-Tetradecylglutarate*

5Z-Tetradecenoate and 5Z-Tetradecenoic Acid, See *Physeteroate*

9Z-Tetradecenoate and 9Z-Tetradecenoic Acid, See *Myristoleate*

O-(n-Tetradecyl) Methylphosphonofluoridate

This *n*-alkyl methylphosphonofluoridate (FW = 278.39 g/mol) irreversibly inhibits acetylcholinesterase and 1-alkyl-2-acetylglycerophosphocholine esterase, *or* platelet-activating-factor acetylhydrolase. Platelet-activating factor (PAF) is a potent endogenous phospholipid modulator of diverse biological activities, including inflammation and shock. PAF levels are primarily regulated by PAF acetylhydrolases (PAF-AHs). These enzymes are candidate secondary targets of organophosphorus (OP) pesticides and related toxicants. Ultra-high potency and selectivity are achieved with *N*-alkyl methylphosphonofluoridates (long-chain sarin analogues): mouse brain and testes IC_{50} < or = 5 nM for C_8-C_{18} analogues and 0.1-0.6 nM for C_{13} and C_{14} compounds; human plasma IC_{50} ≤ 2 nM for C_{13}-C_{18} analogues. Acetylcholine esterase (AChE) inhibitory potency decreases as chain length increased with maximum brain PAF-AH/AChE selectivity (>3000x) for C_{13}-C_{18} compounds. 1. Quistad, Fisher, Owen, Klintenberg & Casida (2005) *Toxicol. Appl. Pharmacol.* **205**, 149.

5-(Tetradecyloxy)-2-furanoyl-CoA

This acylated coenzyme A (FW = 1091.00 g/mol), produced by the action of fatty acyl-CoA synthetases on 5-(tetradecyloxy)-2-furoic acid, inhibits acetyl-CoA carboxylase, thereby indirectly stimulating fatty acid oxidation and inhibiting glycolysis. **Target(s):** acetyl-CoA carboxylase (1,2); and adenine nucleotide translocase (1). 1. Harris & McCune (1981) *Meth. Enzymol.* **72**, 552. **2.** Halvorson & McCune (1984) *Lipids* **19**, 851.

Tetradecyl Sulfate, See *Myristyl Sulfate*

Tetradecyltriphenylphosphonium Bromide

This phosphonium salt (FW_{ion} = 539.58 g/mol) inhibits glycylpeptide *N*-tetradecanoyltransferase, *or* protein *N*-myristoyltransferase (1). **1.** Selvakumar, Lakshmi-kuttyamma, Shrivastav, *et al.* (2007) *Prog. Lipid Res.* **46**, 1.

Tetraeicosanoate, See *Lignocerate*

cis-15-Tetraeicosenoate, See *Nervonate*

Tetraeicosanoyl-CoA, See *Lignoceroyl-CoA*

Tetraethylammonium Chloride

This quaternary amine salt (FW = 165.71 g/mol; CAS 56-34-8; soluble in water, ethanol, ether, and acetone; *Symbol* = TEA chloride) is the prototypical ganglionic blocking agent that inhibits K^+ channels and nicotinic acetylcholine neurotransmission by blocking the receptor-mediated K^+

currents (**See** *Ganglionic Blocking Agents*). Tetraethylammonium bromide (FW = 210.16 g/mol; CAS 71-91-0; *Symbol* = TEA bromide) is likewise a ganglionic blocking agent and has been shown to inhibit acetylcholinesterase. **Target(s):** acetylcholinesterase (1-6); acid phosphatase (7); choline kinase (8;) cholinesterase, *or* butyrylcholinesterase (9-12); K⁺ channels (13) **1**. Seto & Shinohara (1988) *Arch. Toxicol.* **62**, 37. **2**. Bergmann & Shimoni (1953) *Biochim. Biophys. Acta* **10**, 49. **3**. Krupka (1965) *Biochemistry* **4**, 429. **4**. Krupka (1966) *Biochemistry* **5**, 1988. **5**. Wilson (1952) *J. Biol. Chem.* **197**, 215. **6**. Wilson (1954) *J. Biol. Chem.* **208**, 123. **7**. Garrido, Lisa & Domenech (1988) *Mol. Cell. Biochem.* **84**, 41. **8**. Ishidate & Nakazawa (1992) *Meth. Enzymol.* **209**, 121. **9**. Stojan, Golicnik, Froment, Estour & Masson (2002) *Eur. J. Biochem.* **269**, 1154. **10**. Moritoki, Shinohara, Yamauchi & Ishida (1978) *Eur. J. Pharmacol.* **47**, 95. **11**. Fitch (1963) *Biochemistry* **2**, 1221. **12**. Masson, Xie, Froment & Lockridge (2001) *Biochim. Biophys. Acta* **1544**, 166. **13**. Tan & Llano (1999) *J. Physiol.* **520**, 65.

Tetraethyldithiocarbamyl Disulfide, See *Tetraethylthiuram Disulfide*

Tetraethylene Glycol Monododecyl Ether, See *Brij 30*

Tetraethylenepentamine

This toxic oligoamine (FW_{free-base} = 189.30 g/mol; CAS 112-57-2), often abbreviated TEPA, usually supplied as the pentahydrochloride) is a metal chelator that tightly binds zinc and copper ions, often depriving metalloenzymes of a required metal ion. Containing both primary and secondary amines, TEPA also reacts with aldehydes, acids and chlorinated hydrocarbons. **Target(s):** adenosylhomocysteinase (1); *Aeromonas caviae* metalloproteinase (2); *Agkistrodon acutus* (hundred-pace snake) Ac3-proteinase (3); *Agkistrodon halys* hemorrhagic toxin I (4); *Calloselasma rhodostoma* hemorrhagic protease (5); fibrolase (6,7); leucyl aminopeptidase (8); β-lytic metalloendopeptidase (9); pitrilysin, *or* protease Pi (10); pseudolysin (12); *Pseudomonas aeruginosa* LasA, *or* extracellular proteinase (12); serralysin (13); *Serratia marcescens* 56K protease (14); staphylolysin (15); and Cu/Zn superoxide dismutase (16,17). **1**. Birkaya & Aletta (2005) *J. Neurobiol.* **63**, 49. **2**. Kawakami, Toma & Honma (2000) *Microbiol. Immunol.* **44**, 34. **3**. Yagihashi, Niaki, Mori, Kishida & Sugihara (1986) *Int. J. Biochem.* **18**, 885. **4**. Nikai (1998) in *Handb. Proteolytic Enzymes* (Barrett, Rawlings & Woessner, eds.), p. 1274, Academic Press, San Diego. **5**. Bando, Nikai & Sugihara (191) *Int. J. Biochem.* **23**, 1193. **6**. Swenson & Markland, Jr. (1998) in *Handb. Proteolytic Enzymes* (Barrett, Rawlings & Woessner, eds.), p. 1254, Academic Press, San Diego. **7**. Guan, Retzios, Henderson & Markland (1991) *Arch. Biochem. Biophys.* **289**, 197. **8**. Cahan, Axelrad, Safrin, Ohmann & Kessler (2001) *J. Biol. Chem.* **276**, 43645. **9**. Kessler (1995) *Meth. Enzymol.* **248**, 740. **10**. Anastasi, Knight & Barrett (1993) *Biochem. J.* **290**, 601. **11**. Kessler & Ohman (1998) in *Handb. Proteolytic Enzymes* (Barrett, Rawlings & Woessner, eds.), p. 1058, Academic Press, San Diego. **12**. Kessler, Safrin, Abrams, Rosenbloom & Ohman (1997) *J. Biol. Chem.* **272**, 9884. **13**. Kim & Kim (1993) *Biosci. Biotechnol. Biochem.* **57**, 29. **14**. Matsumoto, Maeda & Okamura (1984) *J. Biochem.* **96**, 1165. **15**. Kessler & Ohman (1998) in *Handb. Proteolytic Enzymes* (Barrett, Rawlings & Woessner, eds.), p. 1476, Academic Press, San Diego. **16**. Ishiyama, Ogino, Hobara, *et al.* (1991) *Pharmacol. Toxicol.* **69**, 215. **17**. Kelner, Bagnell, Hale & Alexander (1989) *Free Radic. Biol. Med.* **6**, 355.

Tetra(ethylene sulfonate), See *Polyethylene Sulfonate*

Tetraethyllead

This toxic organolead compound (FW = 323.4 g/mol; CAS 78-00-2; B.P. ≈ 200°C) was previously employed as a gasoline additive that increases the effective octane number and reduces engine knocking. When absorbed by an organism, tetraethyllead is converted to triethyllead cation *via* cytochrome P450 systems, and this metabolite is the cause of the symptoms of tetraethyllead poisoning. By the mid-1970s, however, when tetraethyllead

usage reached its peak, enviromentalists succeeded in banning this additive in the United States and Europe. **Target(s):** γ-aminobutyrate (GABA) uptake, Na⁺-dependent, inhibited by triethyllead cation (1,2); Ca²⁺-channel, voltage-dependent, inhibited by triethyllead cation (3); cerebroside biosynthesis (4); cytokinesis (5,6); DNA biosynthesis (4); oxidative phosphorylation (7); porphobilinogen synthase, *or* δ-aminolevulinate dehydratase, inhibited by triethyllead cation (8); sulfatide biosynthesis (4); tubulin polymerization (microtubule assembly), inhibited by triethyllead cation (9,10) **See also** *Triethyl lead Chloride* **1**. Seidman, Olsen & Verity (1987) *J. Neurochem.* **49**, 415. **2**. Seidman & Verity (1987) *J. Neurochem.* **48**, 1142. **3**. Busselberg, Evans, Rahmann & Carpenter (1991) *Neurotoxicology* **12**, 733. **4**. Ammitzboll, Kobayasi, Grundt & Clausen (1978) *Arch. Toxicol. Suppl.* (1), 319. **5**. Roderer & Schnepf (1977) *Naturwissenschaften* **64**, 588. **6**. Giri & Devi (1992) *Indian J. Exp. Biol.* **30**, 201. **7**. Kauppinen, Komulainen & Taipale (1988) *J. Neurochem.* **51**, 1617 and (1989) **52**, 661. **8**. Bondy (1986) *J. Toxicol. Environ. Health* **18**, 639. **9**. Zimmermann, Doenges & Roderer (1985) *Exp. Cell Res.* **156**, 140. **10**. Zimmermann, Faulstich, Hansch, Doenges & Stournaras (1988) *Mutat. Res.* **201**, 293.

Tetraethyl Pyrophosphate

This toxic insecticide (FW = 290.19 g/mol; CAS 107-49-3), abbreviated TEPP and systematically named tetraethyl diphosphate, is a hygroscopic liquid that inhibits cholinesterases. TEPP also reacts very slowly with chymotrypsin (and even slower with trypsin). Once inactivated by TEPP, acetylcholine esterase will slowly reactivate, regaining 75% of its activity in 3 days, when incubated with saturating choline. Reactivation is faster, when carried out in the presence of 1-2 M freshly prepared neutral hydroxylamine. In the latter case, the likely nucleophile is the hydroxyl group of hydroxylamine. *Note*: TEPP is miscible with water, but quickly hydrolyzes to ethanol and orthophosphate ($t_{1/2}$ = 7 hours at 25°C of a 1:1 solution). **Target(s):** acetylcholinesterase (1-9); acetylesterase (10,11); arylformamidase (12); cholinesterase (3,13,14); chymotrypsin (14-16); esterase (11); glucose-6-phosphate dehydrogenase, weakly inhibited (17); lipase (14); lipoprotein lipase (18); 6-phosphogluconate dehydrogenase (17); and trypsin (15). **1**. Wilson (1960) *The Enzymes*, 2nd ed. (Boyer, Lardy & Myrbäck, eds.) **4**, 501. **2**. Cohen, Oosterbaan & Berends (1967) *Meth. Enzymol.* **11**, 686. **3**. Augustinsson (1950) *The Enzymes*, 1st ed. (Sumner & Myrbäck, eds.), **1** (part 1), 443. **4**. Froede & Wilson (1971) *The Enzymes*, 3rd ed. (Boyer, ed.), **5**, 87. **5**. Berry (1971) *Biochem. Pharmacol.* **20**, 1333. **6**. Augustinsson & Nachmansohn (1949) *J. Biol. Chem.* **179**, 543. **7**. Wilson & Bergmann (1950) *J. Biol. Chem.* **185**, 479. **8**. Wilson (1951) *J. Biol. Chem.* **190**, 111. **9**. Wilson (1952) *J. Biol. Chem.* **199**, 113. **10**. Jansen, Fellows Nutting & Balls (1948) *J. Biol. Chem.* **175**, 975. **11**. Jansen, Fellows Nutting, Jang & Balls (1949) *J. Biol. Chem.* **179**, 189. **12**. Seifert & Casida (1979) *Pestic. Biochem. Physiol.* **12**, 273. **13**. Moritoki, Shinohara, Yamauchi & Ishida (1978) *Eur. J. Pharmacol.* **47**, 95. **14**. Mounter, Tuck, Alexander & Dien (1957) *J. Biol. Chem.* **226**, 873. **15**. Jansen, Fellows Nutting, Jang & Balls (1950) *J. Biol. Chem.* **185**, 209. **16**. Jansen, Curl & Balls (1951) *J. Biol. Chem.* **190**, 557. **17**. Chefurka (1957) *Enzymologia* **18**, 209. **18**. Korn (1962) *Meth. Enzymol.* **5**, 542.

Tetraethylthiuram Disulfide

This disulfide (FW = 296.55 g/mol; Solubility: 2.82 g/100 mL ethanol and 7.14 g/100 mL diethyl ether), also known as Antabuse, disulfiram, and tetraethyldithiocarbamyl disulfide, is is a natural product of the mushroom *Coprinus atramentarius* (inky cap) that is used in aversion therapy for controlling alcoholism. Tetraethylthiuram disulfide inhibits monomeric mitochondrial aldehyde dehydrogenase by inducing the formation of an intramolecular disulfide bond (1). When alcohol is consumed, acetaldehyde is formed in the alcohol dehydrogenase, accumulating to toxic levels and manifested as a set of unpleasant sensations (referred to as the Disulfiram-Ethanol Reaction, or DER) that include flushing, tachycardia, nausea, and vomiting. The perceived threat of DER is thought to be dominant psychological factor in dissuading alcohol use. **Target(s):** alcohol dehydrogenase (2,3); aldehyde dehydrogenase (1,4-12); aldehyde reductase (13); D-amino-acid oxidase (14); ascorbate oxidase (15); caspase 1 (16); caspase 3 (16); carboxylesterase (17); cholesterol 7α-monooxygenase (18,19); CYP2E1 (20-24); dopamine β-monooxygenase (25,26); glucose-6-phosphate dehydrogenase (27); glutathione S-transferase (28); guanylate cyclase (29); L-gulonolactone oxidase (30); inositol-phosphate phosphatase, or myo-inositol monophosphatase (31); inositol-polyphosphate 5-phosphatase, or myo-inositol-1,4,5-trisphosphate 5-phosphatase (31,32); lactate dehydrogenase (33); 5-lipoxygenase (34); microtubule assembly (tubulin polymerization) (35); 6-phosphogluconate dehydrogenase (27); 1-pyrroline-5-carboxylate dehydrogenase (36); salicylhydroxamic acid reductase (37); succinate dehydrogenase (38-41); Cu/Zn superoxide dismutase (42); 4-N'-trimethylamino-butyraldehyde dehydrogenase (43); and xanthine oxidase (41,44). **1.** Lipsky, Shen & Naylor (2001) *Chem. Biol. Interact.* **130-132**, 81 and 93. **2.** Sund & Theorell (1963) *The Enzymes*, 2nd ed. (Boyer, Lardy & Myrbäck, eds.), **7**, 25. **3.** Langeland & McKinley-McKee (1997) *Comp. Biochem. Physiol. C Pharmacol. Toxicol. Endocrinol.* **117**, 55. **4.** Racker (1955) *Meth. Enzymol.* **1**, 514. **5.** Eckfeldt & Yonetani (1982) *Meth. Enzymol.* **89**, 474. **6.** Vallari & Pietruszko (1983) *Pharmacol. Biochem. Behav.* **18** Suppl. 1, 97. **7.** Taguchi, Murase & Miwa (2002) *Cell Biochem. Funct.* **20**, 223. **8.** Shen, Johnson, Mays, Lipsky & Naylor (2001) *Biochem. Pharmacol.* **61**, 537. **9.** Shen, Lipsky & Naylor (2000) *Biochem. Pharmacol.* **60**, 947. **10.** Veverka, Johnson, Mays, Lipsky & Naylor (1997) *Biochem. Pharmacol.* **53**, 511. **11.** Keung & Vallee (1993) *Proc. Natl. Acad. Sci. U.S.A.* **90**, 1247. **12.** Deitrich & Hellerman (1963) *J. Biol. Chem.* **238**, 1683. **13.** Pozniakova, Konoplitskaia, Zakashun & Khustochka (1987) *Ukr. Biokhim. Zh.* **59**, 19. **14.** Hellerman, Coffey & Neims (1965) *J. Biol. Chem.* **240**, 290. **15.** Stark & Dawson (1963) *The Enzymes*, 2nd ed. (Boyer, Lardy & Myrbäck, eds.), **8**, 297. **16.** Nobel, Kimland, Nicholson, Orrenius & Slater (1997) *Chem. Res. Toxicol.* **10**, 1319. **17.** Nousiainen & Torronen (1984) *Gen. Pharmacol.* **15**, 223. **18.** Norlin & Wikvall (1998) *Biochim. Biophys. Acta* **1390**, 269. **19.** Waxman (1986) *Arch. Biochem. Biophys.* **247**, 335. **20.** Frye & Branch (2002) *Brit. J. Clin. Pharmacol.* **53**, 155. **21.** Kharasch, Hankins, Fenstamaker & K. Cox (2000) *Eur. J. Clin. Pharmacol.* **55**, 853. **22.** Frye, Tammara, Cowart & Bramer (1999) *J. Clin. Pharmacol.* **39**, 1177. **23.** Emery, Jubert, Thummel & Kharasch (1999) *J. Pharmacol. Exp. Ther.* **291**, 213. **24.** Kharasch, Hankins, Jubert, Thummel & Taraday (1999) *Drug Metab. Dispos.* **27**, 717. **25.** Weppelman (1987) *Meth. Enzymol.* **142**, 608. **26.** Hashimoto, Ohi & Imaizumi (1965) *Jpn. J. Pharmacol.* **15**, 445. **27.** Chefurka (1957) *Enzymologia* **18**, 209. **28.** Ploemen, van Iersel, Wormhoudt, *et al.* (1996) *Biochem. Pharmacol.* **52**, 197. **29.** Akiyama, Kuo, Hayashi & Miki (1980) *Gann* **71**, 356. **30.** Chatterjee (1970) *Meth. Enzymol.* **18A**, 28. **31.** Wang, Akhtar & Abdel-Latif (1994) *Biochim. Biophys. Acta* **1222**, 27. **32.** Fowler, Brannstrom, Ahlgren, Florvall & Akerman (1993) *Biochem. J.* **289**, 853. **33.** Lukrec, Phillips & Howard (1989) *Biochem. Pharmacol.* **38**, 3132. **34.** Choo & Riendeau (1987) *Can. J. Physiol. Pharmacol.* **65**, 2503. **35.** Potchoo, Braguer, Peyrot, *et al.* (1986) *Int. J. Clin. Pharmacol. Ther. Toxicol.* **24**, 499. **36.** Farres, Julia & Pares (1988) *Biochem. J.* **256**, 461. **37.** Katsura, Kitamura & Tatsumi (1993) *Arch. Biochem. Biophys.* **302**, 356. **38.** Jay (1991) *J. Bioenerg. Biomembr.* **23**, 335. **39.** Boyer (1959) *The Enzymes*, 2nd ed. (Boyer, Lardy & Myrbäck, eds.), **1**, 511. **40.** Keilin & Hartree (1940) *Proc. Roy. Soc. B* **129**, 277. **41.** Richert, Vanderlinde & Westerfeld (1950) *J. Biol. Chem.* **186**, 261. **42.** Marikovsky, Nevo, Vadai & Harris-Cerruti (2002) *Int. J. Cancer* **97**, 34. **43.** Hulse & Henderson (1980) *J. Biol. Chem.* **255**, 1146. **44.** Bray (1975) *The Enzymes*, 3rd ed. (Boyer, ed.), **12**, 299.

S-(1,1,2,2-Tetrafluoroethyl)-L-cysteine

This toxic cysteine conjugate (FW = 221.18 g/mol), which is formed metabolically in mammalians exposed to the industrial gas tetrafluoroethylene, is further metabolized to difluorothioacetyl fluoride. The latter, which is thought to inhibit mitochondrial aconitase and the α-

ketoglutarate dehydrogenase complex, also rapidly inactivates branched-chain-amino-acid aminotransferase. **Target(s):** aconitase, or aconitate hydratase (1); branched-chain-amino-acid aminotransferase (2); dihydrolipoamide dehydrogenase (3); α-ketoglutarate dehydrogenase complex (3); and lipoamide succinyltransferase (3). **1.** James, Gygi, Adams, *et al.* (2002) *Biochemistry* **41**, 6789. **2.** Cooper, Bruschi, Conway & Hutson (2003) *Biochem. Pharmacol.* **65**, 181. **3.** Bruschi, Lindsay & Crabb (1998) *Proc. Natl. Acad. Sci. U.S.A.* **95**, 13413.

DL-4,5,6,7-Tetrafluorotryptophan

This halogenated tryptophan (FW = 276.19 g/mol) competitively inhibits tryptophanyl-tRNA synthetase (1-3). **1.** Kisselev, Favorova & Kolaleva (1979) *Meth. Enzymol.* **59**, 174. **2.** Favorova & Lavrik (1975) *Biokhimiia* **40**, 368. **3.** Knorre, Lavrik, Petrova, Savchenko & Yakobson (1971) *FEBS Lett.* **12**, 204.

1,2,3,6-Tetra-*O*-galloyl-β-D-glucose

This substituted glucoside (FW = 740.54 g/mol) inhibits aldose reductase (1), electron transport, complex II (2), poly(ADP-ribose) glycohydrolase (3), prolyl oligopeptidase, IC$_{50}$ = 0.025 μM (4), and succinate dehydrogenase (2). **1.** Aida, Tawata, Shindo, *et al.* (1989) *Planta Med.* **55**, 22. **2.** Konishi, Adachi, Kita, *et al.* (1989) *Chem. Pharm. Bull. (Tokyo)* **37**, 2533. **3.** Tsai, Aoki, Maruta, *et al.* (1992) *J. Biol. Chem.* **267**, 14436. **4.** Fan, Tezuka, Ni & Kadota (2001) *Chem. Pharm. Bull. (Tokyo)* **49**, 396.

Tetrahydrobenzo[*h*][1,6]naphthyridine-6-chlorotacrine Hybrids

These novel AChE inhibitors (Hybrid 5a: FW$_{n=3}$ = 596.56 g/mol; Hybrid 5b: FW$_{n=4}$ = 610.58 g/mol; Hybrid 5c: FW$_{n=5}$ = 624.60 g/mol; Hybrid 5d: FW$_{n=8}$ = 680.72 g/mol), which bind simultaneously at the catalytic anionic site (CAS) and the peripheral anionic site (PAS) of cholinesterases, were designed by molecular hybridization of a novel 3,4-dihydro-2*H*-pyrano[3,2-

c]quinoline scaffold and the potent AChE CAS-directed inhibitor, 6-chlorotacrine (1). Hybrid 5a, the trimethylene-bridged analogue, is a remarkably tight inhibitor for human acetylcholine esterase, *or* hAChE (IC_{50} = 0.006 nM), with much weaker effects on human butyrylcholine esterase, *or* hBChE (IC_{50} = 120 nM), Aβ42 aggregation in *Escherichia coli* (52% inhibition at 10 μM), and microtubule-associated protein Tau aggregation in *E. coli* (41% inhibition at 10 μM). Hybrid 5b, the tetramethylene-bridged analogue, inhibits hAChE (IC_{50} = 14.2 nM), hBChE (IC_{50} = 205 nM), Aβ42 aggregation (56%), and Tau aggregation (59%). Hybrid 5c, the pentamethylene-bridged analogue, inhibits hAChE (IC_{50} = 14.2 nM), hBChE (IC_{50} = 337 nM), Aβ42 aggregation (54%), and Tau aggregation (58%). Hybrid 5d, the octamethylene-bridged analogue, inhibits hAChE (IC_{50} = 2 nM), hBChE (IC_{50} = 286 nM), Aβ42 aggregation (77%), and Tau aggregation (69%). *In vitro* studies using artificial membranes predict good brain permeability, indicating a potential for these hybrid to reach their intended CNS targets. **1.** Di Pietro, Pérez-Areales, Juárez-Jiménez, *et al.* (2014) *Eur. J. Med. Chem.* **84**, 107.

Tetrahydroberberine, *See Canadine*

(6*R*)-Tetrahydrobiopterin

This biopterin derivative (FW = 241.25 g/mol; λ_{max} = 265 nm in 0.1 M HCl (ε = 14000 $M^{-1}cm^{-1}$) and 300 nm at pH 6.8), also known as BH_4 and (6*R*)-L-*erythro*-5,6,7,8-tetrahydrobiopterin, is a cofactor for phenylalanine 4-monooxygenase, tyrosine 3-monooxygenase, anthranilate 3-monooxygenase, tryptophan 5-monooxygenase, glyceryl-ether monooxygenase, mandelate 4-monooxygenase, and nitric oxide synthase. In addition, BH_4 stimulates dopamine release from the rat striatum and PC12 pheochromocytoma cells as well as Ca^{2+} channels in rat brain. PC12 cells are activated by tetrahydrobiopterin via a cAMP-dependent protein kinase pathway. Note that excess tetrahydrobiopterin will inhibit tyrosine 3-monooxygenase (1) and phenylalanine 4-monoxygenase. **Solution Preparation:** Tetrahydropterin stock solutions should be prepared daily, usually by dissolving the dihydrochloride salt in deionized water. BH_4 concentration is determined from the absorbance at 265 nm in 2 M perchloric acid (ε = 18,000 $M^{-1}cm^{-1}$). BH_4 is rapidly oxidized in aqueous solutions, with a 6-min half-life at pH 6.8 and 24°C. **Other Target(s):** dihydropteridine reductase (2); GTP cyclohydrolase I (2-11); [phenylalanine monooxygenase] kinase (12); tRNA-guanine transglycosylase (13); and xanthine oxidase (14). **1.** Alterio, Ravassard, Haavik, *et al.* (1998) *J. Biol. Chem.* **273**, 10196. **2.** Shen, Zhang & Perez-Polo (1989) *J. Enzyme Inhib.* **3**, 119. **3.** Werner, Bahrami, Heller & Werner-Felmayer (2002) *J. Biol. Chem.* **277**, 10129. **4.** Yoneyama, Wilson & Hatakeyama (2001) *Arch. Biochem. Biophys.* **388**, 67. **5.** Harada, Kagamiyama & Hatakeyama (1993) *Science* **260**, 1507. **6.** Shen, Alam & Zhang (1988) *Biochim. Biophys. Acta* **965**, 9. **7.** Blau & Niederwieser (1986) *Biochim. Biophys. Acta* **880**, 26. **8.** Maita, Okada, Hatakeyama & Hakoshima (2002) *Proc. Natl. Acad. Sci. U.S.A.* **99**, 1212. **9.** Yoneyama & Hatakeyama (2001) *Protein Sci.* **10**, 871. **10.** Maita, Hatakeyama, Okada & Hakoshima (2004) *J. Biol. Chem.* **279**, 51534. **11.** Schoedon, Redweik, Frank, Cotton & Blau (1982) *Eur. J. Biochem.* **210**, 561. **12.** Kaufman (1987) *The Enzymes*, 3rd ed. (Boyer & Krebs, eds.), **18**, 217. **13.** Farkas, Jacobson & Katze (1984) *Biochim. Biophys. Acta* **781**, 64. **14.** Wede, Altindag, Widner, Wachter & Fuchs (1998) *Free Radic. Res.* **29**, 331.

Δ⁹-Tetrahydrocannabinol

This much abused psychotropic substance and CB_1/CB_2 partial agonist (FW = 314.47 g/mol; CAS 1972-08-3; λ_{max} = 283 and 276 nm (ε = 1620 and 1580

$M^{-1}cm^{-1}$, respectively, in Ethanol; Abbreviations: Δ^9-THC and THC), also called (−)-Δ^1-3,4-*trans*-tetrahydrocannabinol, is the psychoactive principle in *Cannabis* (e.g., marijuana, hashish, or "pot"). Δ^9-THC has a K_i of 10 nM (partial agonist) for the CB_1 receptor and a K_i of 24 nM (partial agonist) for the CB_2 receptor. These G-protein-coupled receptors are activated by three major ligand types: endocannabinoids (*e.g.*, oleoylethanolamine), plant cannabinoids (*e.g.*, THC), and synthetic cannabinoids (*e.g.*, HU-210). The two known cannabinoid receptors subtypes, termed CB_1 and CB_2, with the former mainly expressed in brain (but also in lung, liver, and kidney), and the latter in the immune and hematopoietic cells. Other effects include relaxation, alteration of visual, auditory, and olfactory senses, fatigue, and appetite stimulation. THC has marked antiemetic properties, the latter commending its use to suppress vomiting in HIV-AIDS patients. **Psychotomimetic Effects:** Δ^9-THC is classified as a psychotomimetic (*or* psychotogenic) agent, one whose actions mimic the symptoms of psychosis, including delusions and/or delirium, as opposed to only hallucinations. In the case of Δ^9-THC, the effects are numerous, a finding that suggests the drug's widespread influence over the Central Nervous System. Indeed, a recent study (1) found that Δ^9-THC induces transient effects in healthy individuals, including positive symptoms, negative symptoms, perceptual alterations, euphoria, anxiety, as well as deficits in working memory, recall, and the executive control of attention without altering general orientation. Other symptoms include suspiciousness, paranoid and grandiose delusions, conceptual disorganization, and illusions. It also produces depersonalization, derealization, distorted sensory perceptions, altered body perception, feelings of unreality and extreme slowing of time. Δ^9-THC produces blunted affect, reduced rapport, lack of spontaneity, psychomotor retardation, and emotional withdrawal (1). In the case of pot smoking, the main route for THC entry is via the lungs, eventually reaching the central nervous system, where it can bind to and stimulate CB_1 receptors. That said, the lipophilic nature leads to almost immediate sequestration of THC within lipid-rich biological membranes, followed by slow release, the latter accounting for its long-lasting psychotropic effects. THC binding gives rise to anxiolytic- and antidepressant-like effects. **Target(s):** adenylate cyclase (2,3); anandamide amidohydrolase (4); ATPase (5); cholesterol esterase (6,7); lysolecithin acyltransferase (8); NADH oxidase (9); and Na^+/K^+-exchanging ATPase (10). **1.** D'Souza, Perry, MacDougall, *et al.* (2004) *Neuropsychopharmacology* **29**, 1558. **2.** Howlett, Scott & Wilken (1989) *Biochem. Pharmacol.* **38**, 3297. **3.** Howlett (1987) *Neuropharmacol.* **26**, 507. **4.** Watanabe, Ogi, Nakamura, *et al.* (1998) *Life Sci.* **62**, 1223. **5.** Gilbert, Pertwee & Wyllie (1977) *Brit. J. Pharmacol.* **59**, 599. **6.** Shoupe, Hunter, Burstein & Hubbard (1980) *Enzyme* **25**, 87. **7.** Burstein, Hunter & Shoupe (1978) *Life Sci.* **23**, 979. **7.** Greenberg, Saunders & Mellors (1977) *Science* **197**, 475. **9.** Bartova & Birmingham (1976) *J. Biol. Chem.* **251**, 5002. **10.** Laurent, Roy & Gailis (1974) *Can. J. Physiol. Pharmacol.* **52**, 1110.

(+)-*S*-4,5,6,7-Tetrahydro-9-chloro-5-methyl-6-(3-methyl-2-butenyl)imidazo[4,5,1-*jk*][1,4]benzodiazepine-2(1*H*)-thione

This benzodiazepine (FW = 321.87 g/mol), also known as R82913 and 9-chloro-TIBO, inhibits a number of viral reverse transcriptases by binding within a hydrophobic pocket in the enzyme-DNA complex near the active-site catalytic residues (1). **Target(s):** HIV-1 reverse transcriptase (1-6); RNA-directed DNA polymerase (1-6); and viral reverse transcriptase (2). **1.** Spence, Kati, Anderson & Johnson (1995) *Science* **267**, 988. **2.** Tucker, Lumma & Culberson (1996) *Meth. Enzymol.* **275**, 440. **3.** Esnouf, Stuart, De Clercq, Schwartz & Balzarini (1997) *Biochem. Biophys. Res. Commun.* **234**, 458. **4.** Balzarini & De Clercq (1996) *Meth. Enzymol.* **275**, 472. **5.** Shao, Rytting, Ekstrand, *et al.* (1998) *Antivir. Chem. Chemother.* **9**, 167. **6.** De Clercq (1992) *AIDS Res. Human Retrovir.* **8**, 119.

Tetrahydrofolate

This coenzyme ($FW_{free-acid}$ = 445.44 g/mol; CAS 29347-89-5), named systematically as (−)-5,6,7,8-tetrahydropteroyl-L-glutamate or *N*-[4-[[(2-amino-3,4,5,6,7,8-hexahydro-4-oxopteridin-6-yl)methyl]amino]benzoyl]-L-glutamate, is a crucial one-carbon donor in the metabolism of L-methionine, glycine, L-serine, L-histidine, and the purines and pyrimidines. Tetrahydrofolate it is slowly oxidized in solid form and must be stored in the dark and cold under vacuum conditions or an inert gas. The presence of heavy metal ions or extremes of pH accelerates decomposition, as will light. The half-life of a solution of tetrahydrofolate in phosphate buffer at 23°C is about 40 min; if a stabilizer, such as ascorbate or 2-mercaptoethanol, is added, the half-life is increased by 2-3 times. Stability also depends on the buffer: relatively unstable in phosphate buffers and more stable in Tris or triethanolamine. The most common biological form of tetrahydrofolate contains between three and seven glutamyl residues, linked via their γ-carboxyl groups. **Target(s):** dihydropteroate synthase (1); formiminotetrahydrofolate cyclodeaminase, *or* formimidoyl-tetrahydrofolate cyclodeaminase (2-4); glutamate dehydrogenase (5); GTP cyclohydrolase I, weakly inhibited (6); guanine deaminase (7-9); indoleacetaldoxime dehydratase (10); and methionyl-tRNA formyltransferase (11). **See also** *Tetrahydropteroyl-triglutamate* 1. Mouillon, Ravanel, Douce & Rebeille (2002) *Biochem. J.* **363**, 313. 2. Uyeda & Rabinowitz (1963) *Meth. Enzymol.* **6**, 380. 3. Rader & Huennekens (1973) *The Enzymes*, 3rd ed. (Boyer, ed.), **9**, 197. 4. Paquin, Baugh & MacKenzie (1985) *J. Biol. Chem.* **260**, 14925. 5. White, Yielding & Krumdieck (1976) *Biochim. Biophys. Acta* **429**, 689. 6. Shen, Alam & Zhang (1988) *Biochim. Biophys. Acta* **965**, 9. 7. Glantz & Lewis (1978) *Meth. Enzymol.* **51**, 512. 8. Gupta & Glantz (1985) *Arch. Biochem. Biophys.* **236**, 266. 9. Lewis & Glantz (1974) *J. Biol. Chem.* **249**, 3862. 10. Shulka & Mahadevan (1970) *Arch. Biochem. Biophys.* **137**, 166. 11. Dickerman & Smith (1970) *Biochemistry* **9**, 1247.

Tetrahydrohomofolate

The homologues of tetrahydrofolate and tetrahydrofolate polyglutamates can serve as inhibitors of a number of folate-dependent systems. **Target(s):** methionyl-tRNA formyltransferase (1,2); phosphoribosylglycinamide formyltransferase (3,4); thymidylate synthase (4-7). 1. Rader & Huennekens (1973) *The Enzymes*, 3rd ed. (Boyer, ed.), **9**, 197. 2. Dickerman & Smith (1970) *Biochemistry* **9**, 1247. 3. Thorndike, Kisliuk, Gaumont, Piper & Nair (1990) *Arch. Biochem. Biophys.* **277**, 334. 4. Thorndike, Gaumont, Kisliuk, *et al.* (1989) *Cancer Res.* **49**, 158. 5. Baker (1967) *Design of Active-Site-Directed Irreversible Enzyme Inhibitors*, Wiley, New York. 6. Livingston, Crawford & Friedkin (1968) *Biochemistry* **7**, 2814. 7. Plante, Crawford & Friedkin (1967) *J. Biol. Chem.* **242**, 1466.

Tetrahydrolipstatin

This L-leucine derivative (FW = 495.74 g/mol; CAS 96829-58-2; M.P. = 43°C), also known by its proprietary name Orlistat®, has been used clinically to inhibit digestive lipases in individuals requiring weight reduction.

Tetrahydrolipstatin is a potent irreversible inhibitor of lipoprotein lipase, but it slowly turned over. **Target(s):** cholesterol esterase (1); diacylglycerol lipase (2); fatty-acid synthase (3); lipase, *or* triacylglycerol lipase (1,4-7); lipase-related protein (2); pancreatic (5); lipoprotein lipase (8-10); microbial lipases from *Chromobacterium viscosum* and *Rhizopus oryzae* (1). 1. Haalck & Spener (1997) *Meth. Enzymol.* **286**, 252. 2. Lee, Kraemer & Severson (1995) *Biochim. Biophys. Acta* **1254**, 311. 3. Lupu & Menendez (2006) *Curr. Pharm. Biotechnol.* 7, 483. 4. Ransac, Gargouri, Marguet, *et al.* (1997) *Meth. Enzymol.* **286**, 190. 5. Roussel, Yang, Ferrato, *et al.* (1998) *J. Biol. Chem.* **273**, 32121. 6. Hadváry, Lengsfeld & Wolfer (1988) *Biochem. J.* **256**, 357. 7. Haiker, Lengsfeld, Hadváry & Carrière (2004) *Biochim. Biophys. Acta* **1682**, 72. 8. Olivecrona & Lookene (1997) *Meth. Enzymol.* **286**, 102. 9. Lookene, Skottova & Olivecrona (1994) *Eur. J. Biochem.* **222**, 395. 10. Rinninger, Brundert, Brosch, *et al.* (2001) *J. Lipid Res.* **42**, 1740.

Tetrahydromesobilane-IXa. See *Stercobilinogen*

Tetrahydromesobilene-(*b*)-IXa, See *Stercobilin*

1,2,3,4-Tetrahydro-2-methyl-4-phenyl-8-isoquinolinamine, See *Nomifensine*

β-Tetrahydronaphthylamine, See *2-Aminotetralin*

2-(5,6,7,8-Tetrahydronaphthyl)-*N*-hexyl Carbamate

This carbamate (FW = 275.39 g/mol) is a pseudo-substrate inhibitor of cholesterol esterase, in which decarbamoylation of the enzyme is slow, yielding an approximate K_d value of 16 nM in the presence of taurocholate (1,2). 1. Feaster & Quinn (1997) *Meth. Enzymol.* **286**, 231. 2. Feaster, Lee, Baker, Hui & Quinn (1996) *Biochemistry* **35**, 16723.

1,4,5,6-Tetrahydronicotinamide Adenine Dinucleotide

This NADH analogue ($FW_{free-acid}$ = 448.24 g/mol), often abbreviated H₂NADH, binds to the coenzyme-binding site of a number of oxidoreductases. **Target(s):** alcohol dehydrogenase (1,2); estradiol 17β-dehydrogenase (3); glutamate dehydrogenase (4); and malate dehydrogenase (5). 1. Biellmann & Jung (1971) *Eur. J. Biochem.* **19**, 130. 2. Dunn, Biellmann & Branlant (1975) *Biochemistry* **14**, 3176. 3. Biellmann & Hirth (1975) *Eur. J. Biochem.* **56**, 557. 4. Thevenot, Godinot, Gautheron, Branlant & Biellman (1975) *FEBS Lett.* **54**, 206. 5. Chapman, Cortes, Dafforn, Clarke & Brady (1999) *J. Mol. Biol.* **285**, 703.

6,7,8,9-Tetrahydro-5-nitro-1*H*-benz[*g*]indole-2,3-dione-3-oxime, See NS 102
1,2,3,4-Tetrahydro-6-nitro-2,3-dioxo-benzo[*f*]quinoxaline-7-sulfonamide, See NBQX

3-(1,2,3,4-Tetrahydro-4-oxo-7-chloro-2-naphthyl)pyridine

This pyridine derivative (FW = 257.72 g/mol; CAS 786-97-0), also known as SU 10603 and systematically named as 3-(1,2,3,4-tetrahydro-4-keto-7-chloro-2-naphthyl)pyridine, inhibits steroid 17α-monooxygenase (1). This oxidoreductase [EC 1.14.99.9] is a microsomal cytochrome P-450 that catalyzes steroid hydroxylation at the 17-position by accepting electrons from

NADPH via a flavoprotein. It participates in the synthesis of steroid hormones, including cortisol, androgens, and estrogens. The enzyme also possesses 17α-hydroxyprogesterone aldolase activity. 17α-Hydroxylase deficiency is an autosomal recessive trait that causes congenital adrenal hyperplasia (type V). SU-10603 produces a concentration-dependent (0.01-1.0 mM) inhibition of ethylmorphine demethylation, aniline hydroxylation, and benzo[a]pyrene hydroxylation. A concentration of 0.1-0.2 mM decreases the metabolism of all three substrates by approximately 50%. SU-10603 is a more potent inhibitor of ethylmorphine metabolism than metyrapone, and its relative potency is even greater with respect to aniline and benzo[a]pyrene metabolism. Similar results are obtained with guinea pig liver microsomes. *Note*: The very same code name (SU 10603) has been used to refer to the structural isomer 3-(1,2,3,4-tetrahydro-4-oxo-2-naphthyl)pyridine (**See next entry**). **1**. Shikita, Ogiso & Tamacki (1965) *Biochim. Biophys. Acta* **105**, 516.

5,6,7,8-Tetrahydropteroylglutamate, *See* Tetrahydrofolate

β-(2-Tetrahydropyryloxy)ethylcobalamin

This vitamin B$_{12}$ derivative (FW = 1242.34 g/mol), often abbreviated AB$_{12}$, inhibits ethanolamine ammonia-lyase (1-3). **1**. Abeles (1971) *The Enzymes*, 3rd ed. (Boyer, ed.), **5**, 481. **2**. Stadtman (1972) *The Enzymes*, 3rd ed. (Boyer, ed.), **6**, 539. **3**. Babior (1969) *J. Biol. Chem.* **244**, 2917.

6,7,8,9-Tetrahydro-5*H*-tetrazolo[1,5-*a*]azepine, *See* Pentylenetetrazole

Tetrahydrothiamin Pyrophosphate

This thiazolidine ring derivative of thiamin pyrophosphate (FW = 426.32 g/mol), also called tetrahydrothiamin diphosphate, is produced by borohydride reduction of the thiazolium ring, yielding four stereoisomers, due to the introduction of two chiral centers on the thiazolidine ring. The resulting racemic mixture inhibits the pyruvate decarboxylase component of the pyruvate dehydrogenase complex of *Escherichia coli* (1). **Target(s):** acetolactate synthase (2); α-ketoglutarate dehydrogenase complex (3); pyruvate decarboxylase (1,4,5); and pyruvate dehydrogenase complex (1,4,5). **1**. Lowe, Leeper & Perham (1983) *Biochemistry* **22**, 150. **2**. Schloss & Van Dyk (1988) *Meth. Enzymol.* **166**, 445. **3**. Marley, Meganathan & Bentley (1986) *Biochemistry* **25**, 1304. **4**. Strumilo, Czygier & Markiewicz (1996) *J. Enzyme Inhib.* **10**, 65. **5**. Ostrovskii, Nemeria, Zabrodskaia & Gerashchenko (1992) *Ukr. Biokhim. Zh.* **64**, 46.

3,4,5,6-Tetrahydrouridine

This tetrahedral transitition-state analogue (FW = 248.24 g/mol; CAS 18771-50-1) rapidly inhibits bacterial cytidine deaminase, whereas the human liver enzyme is inactivated slowly (1-8). With the latter, the rate constants for binding and release (k_{on} = 2.4 x 10^4 M^{-1}sec^{-1} and k_{off} = 5.6 x 10^{-4} sec^{-1}) are in reasonable agreement with the K_i value of 29 nM measured separately under steady-state conditions. The slow onset of inhibition has been suggested to be due to structural reorganization preceding the formation of a stable enzyme-inhibitor complex (1). **Target(s):** cytidine deaminase, slow-binding inhibition (1-8); deoxycytidine deaminase (9); deoxycytidine kinase (10); and uracil phosphoribosyl-transferase (11). **1**. Wentworth & Wolfenden (1975) *Biochemistry* **14**, 5099. **2**. Cohen & Wolfenden (1971) *J. Biol. Chem.* **246**, 7561. **3**. Wentworth & Wolfenden (1978) *Meth. Enzymol.* **51**, 401. **4**. Morrison (1982) *Trends Biochem. Sci.* **7**, 102. **5**. Carlow, Carter, Mejlhede, Neuhard & Wolfenden (1999) *Biochemistry* **38**, 12258. **6**. Barchi, Haces, Marquez & McCormack (1992) *Nucleosides Nucleotides* **11**, 1781. **7**. Vincenzetti, Cambi, Neuhard, *et al.* (1999) *Protein Expr. Purif.* **15**, 8. **8**. Vincenzetti, Cambi, Neuhard, Garattini & Vita (1996) *Protein Expr. Purif.* **8**, 247. **9**. Heinemann & Plunkett (1989) *Biochem. Pharmacol.* **38**, 4115. **10**. Wachsman & Morgan (1973) *Appl. Microbiol.* **25**, 506. **11**. Dai, Lee & O'Sullivan (1995) *Int. J. Parasitol.* **25**, 207.

2,2',4,4'-Tetrahydroxybenzophenone

This endocrine disruptor (FW = 246.22 g/mol; CAS 131-55-5; Symbol: BP2) interferes with the thyroid hormone (TH) axis. As a symmetric ketone, it also inhibits aldose reductase (2) and xanthine oxidase, IC$_{50}$ = 48 μM (3,4). **1**. Zhang, Li, Gupta, Nam & Andersson (2016) *Chem. Res. Toxicol.* **29**, 1345. **2**. Ono & Hayano (1982) *Nippon Ganka Gakkai Zasshi* **86**, 353. **3**. Sheu, Tsai & Chiang (1999) *Anticancer Res.* **19**, 1131. **4**. Chang, Yan & Chiang (1995) *Anticancer Res.* **15**, 2097.

3,4,2',4'-Tetrahydroxychalcone

This photosensitive plant pigment (FW = 272.26 g/mol; CAS 21849-70-7), also called butein, exhibits a potent inhibitory effect on H$_2$O$_2$-induced hemolysis by acting as an antioxidant. **Target(s):** aldose reductase (1); cAMP-specific phosphodiesterase, type IV (2); Ca^{2+}-transporting ATPase (3); cyclic-nucleotide phosphodiesterase types I, III, and V, weakly inhibited (2); glutathione-disulfide reductase (4); glutathione *S*-transferase (5,6); 12-lipoxygenase (7-10); and protein-tyrosine kinase, *or* epidermal growth factor receptor kinase and p60c-*src* kinase (11,12). **1**. Lim, Jung, J. Shin & Keum (2001) *J. Pharm. Pharmacol.* **53**, 653. **2**. Yu, Cheng & Kuo (1995) *Eur. J. Pharmacol.* **280**, 69. **3**. Thiyagarajah, Kuttan, Lim, Teo & Das (1991) *Biochem. Pharmacol.* **41**, 669. **4**. Zhang, Yang, Tang, Wong & Mack (1997) *Biochem. Pharmacol.* **54**, 1047. **5**. Zhang & Wong (1997) *Biochem. J.* **325**, 417. **6**. Zhang & Das (1994) *Biochem. Pharmacol.* **47**, 2063. **7**. Nakadate, Yamamoto, Aizu & Kato (1989) *Carcinogenesis* **10**, 2053. **8**. Yamamoto, Sasakawa, Kiyoto, Nakadate & Kato (1987) *Eur. J. Pharmacol.* **144**, 101. **9**. Aizu, Nakadate, Yamamoto & Kato (1986) *Carcinogenesis* **7**, 1809. **10**.

Nakadate, Aizu, Yamamoto & Kato (1985) *Prostaglandins* **30**, 357.   **11.** Yang, Zhang, Cheng & Mack (1998) *Biochem. Biophys. Res. Commun.* **245**, 435.   **12.** Yang, Guo, Zhang, Chen & Mack (2001) *Biochim. Biophys. Acta* **1550**, 144.

Tetraisopropyl Pyrophosphoramide

This hazardous, skin-penetrating chemical warfare agent (FW = 342.36 g/mol; CAS 513-00-8), also known as iso-OMPA and tetra(monoisopropyl) pyrophosphortetramide, is a selective butyrylcholinesterase inhibitor. **Target(s):** acetylcholine esterase, weakly inhibited (1-3); aryl-acylamidase (4-6); carboxylesterase, weakly inhibited (7); cholinesterase, *or* butyrylcholinesterase (1,4,8-17). **1.** Giacobini (2003) *Neurochem. Res.* **28**, 515.   **2.** Sanders, Mathews, Sutherland, *et al.* (1996) *Comp. Biochem. Physiol. B* **115**, 97.   **3.** Earle & Barclay (1986) *FEMS Microbiol. Lett.* **35**, 83.   **4.** Costagli & Galli (1998) *Biochem. Pharmacol.* **55**, 1733.   **5.** Oommen & Balasubramanian (1979) *Eur. J. Biochem.* **94**, 135.   **6.** George & Balasubramanian (1980) *Eur. J. Biochem.* **111**, 511.   **7.** Chambers, Hartgraves, Murphy, *et al.* (1991) *Neurosci. Biobehav. Rev.* **15**, 85.   **8.** Augustinsson (1960) *The Enzymes*, 2nd ed. (Boyer, Lardy & Myrbäck, eds.), **4**, 521.   **9.** Austin & Berry (1953) *Biochem. J.* **54**, 695.   **10.** Aldridge (1953) *Biochem. J.* **53**, 62.   **11.** Davison (1953) *Brit. J. Pharmacol.* **8**, 208.   **12.** deVos & Dick (1992) *Comp. Biochem. Physiol. C* **103**, 129.   **13.** Mehrani (2004) *Proc. Biochem.* **39**, 877.   **14.** Rodriguez-Fuentes & Gold-Bouchot (2004) *Mar. Environ. Res.* **58**, 505.   **15.** Treskatis, Ebert & Layer (1992) *J. Neurochem.* **58**, 2236.   **16.** García-Ayllón, Sáez-Valero, Muñoz-Delgado & Vidal (2001) *Neuroscience* **107**, 199.   **17.** Brown, Davies, Moffat, Redshaw & Craft (2003) *Mar. Environ. Res.* **57**, 155.

N,N,N',N'-Tetrakis(2-pyridylmethyl)ethylenediamine

This cell-penetrant chelator (FW = 424.55 g/mol; CAS 16858-02-9; IUPAC: *N,N,N',N'*-Tetrakis(2-pyridylmethyl)ethylenediamine) binds divalent cations, with selectively toward Zn^{2+}, Cu^{2+}, and Fe^{2+}, but not greatly perturbing Ca^{2+} and Mg^{2+} concentrations. K_d values are 0.26 x 10^{-15} M for Zn^{2+}, 2.4 x 10^{-15} M for Fe^{2+}, 54 x 10^{-12} M for Mn^{2+}, 4 x 10^{-5} M Ca^{2+}, and 20 mM for Mg^{2+}. TPEN was useful in demonstrating that mitotic progression and nuclear assembly are Ca^{2+}-mediated processes that are not attributable to other ions. Related nonpenetrating sulfonates include (4-{[2-(*bis*-pyridin-2-ylmethylamino)ethylamino]methyl}phenyl)methanesulfonic acid (DPESA) and [4-({[2-(*bis*-pyridin-2-ylmethylamino)ethyl]pyridin-2-ylmethylamino}-methyl)phenyl]methanesulfonic acid (TPESA). **Target(s):** Cu^{2+}-dependent ATPase (1); picornain 2A (2); protein kinase C translocation (3); quorum-quenching *N*-acyl-homoserine lactonase (4). TPEN also induces axon and dendrite degeneration (5), strip metalloproteins of zinc cofactors (6), and promote apoptosis (7-9). **1.** Burlando, Evangelisti, Dondero, *et al.* (2002) *Biochem. Biophys. Res. Commun.* **291**, 476.   **2.** Glaser, Triendl & Skern (2003) *J. Virol.* **77**, 5021.   **3.** Baba, Etoh & Iwata (1991) *Brain Res.* **557**, 103.   **4.** Kim, Choi, Kang *et al.* (2005) *Proc. Natl. Acad. Sci. U.S.A.* **102**, 17606.   **5.** Yang, Kawataki, Fukui, & Koike (2007) *J. Neurosci. Res.* **85**, 2844.   **6.** Meeusen, Nowakowski & Petering (2012) *Inorg. Chem.* **51**, 3625.   **7.** Lee, Kim, Ra, *et al.* (2008) *FEBS Lett.* **582**, 1871.   **8.** Makhov, Golovine, Uzzo, *et al.* (2008) *Cell Death Differ.* **15**, 1745.   **9.** Carraway & Dobner (2012) *Biochim. Biophys. Acta* **1823**, 544.

Tetralin, See *1,2,3,4-Tetrahydronaphthalene*

Tetram, See *Amiton*

Tetramethylammonium Salts

This hygroscopic organic cation (FW = 109.60 g/mol; CAS 75-57-0); bromide salt (FW = 154.05 g/mol; CAS 15625-56-6); iodide (FW = 201.05 g/mol; CAS 75-58-1), as well as fluoride, acetate, hydroxide, perchlorate, and peroxynitrate salts) is a toxic agent often used as an extracellular space marker. **Target(s):** acetylcholinesterase (1-15); acid phosphatase (16); choline kinase (17); cholinesterase, *or* butyrylcholinesterase (6,18,19); choline-sulfatase (20); choline sulfotransferase (21); H^+/K^+-exchanging ATPase (22); K^+/HCO_3^- cotransporter (23); and trimethylamine dehydrogenase (24-26). **1.** Wilson (1960) *The Enzymes*, 2nd ed. (Boyer, Lardy & Myrbäck, eds.), **4**, 501.   **2.** Froede & Wilson (1971) *The Enzymes*, 3rd ed. (Boyer, ed.), **5**, 87.   **3.** Krupka (1963) *Biochemistry* **2**, 76.   **4.** Stojan, Marcel & Fournier (1999) *Chem. Biol. Interact.* **119-120**, 137.   **5.** Cohen, Chishti, Bell, *et al.* (1991) *Biochim. Biophys. Acta* **1076**, 112.   **6.** Seto & Shinohara (1988) *Arch. Toxicol.* **62**, 37.   **7.** Metzger & Wilson (1964) *Biochemistry* **3**, 926.   **8.** Wilson & Alexander (1962) *J. Biol. Chem.* **237**, 1323.   **9.** Krupka (1965) *Biochemistry* **4**, 429.   **10.** Krupka (1966) *Biochemistry* **5**, 1988.   **11.** Wilson (1952) *J. Biol. Chem.* **197**, 215.   **12.** Wilson (1952) *J. Biol. Chem.* **199**, 113.   **13.** Wilson (1954) *J. Biol. Chem.* **208**, 123.   **14.** Bergman, Wilson & Nachmansohn (1950) *Biochim. Biophys. Acta* **6**, 217.   **15.** Ciliv & Oezand (1972) *Biochim. Biophys. Acta* **284**, 136.   **16.** Garrido, Lisa & Domenech (1988) *Mol. Cell. Biochem.* **84**, 41.   **17.** Ishidate & Nakazawa (1992) *Meth. Enzymol.* **209**, 121.   **18.** Fitch (1963) *Biochemistry* **2**, 1221.   **19.** Masson, Xie, Froment & Lockridge (2001) *Biochim. Biophys. Acta* **1544**, 166.   **20.** Lucas, Burchiel & Segel (1972) *Arch. Biochem. Biophys.* **153**, 664.   **21.** Renosto & Segel (1977) *Arch. Biochem. Biophys.* **180**, 416.   **22.** Hango, Nojima & Setaka (1990) *Jpn. J. Pharmacol.* **52**, 295.   **23.** Davis, Hogan, Cooper, *et al.* (2001) *J. Membr. Biol.* **183**, 25.   **24.** McIntire (1990) *Meth. Enzymol.* **188**, 250.   **25.** Pace & Stankovich (1991) *Arch. Biochem. Biophys.* **287**, 97.   **26.** Steenkamp & Beinert (1982) *Biochem. J.* **207**, 233.

N,N,N',N'-Tetramethylazodicarboxamide

This unselective thiol oxidant (FW = 172.19 g/mol), commonly referred to as diamide and known systematically as 1,1'-azobis(*N,N*-dimethylformamide), converts thiol-containing substances to their disulfide form, such as the oxidation of glutathione to glutathione disulfide (1). **Target(s):** *N*-acetylglucosamine kinase (2); *N*-acylmannosamine kinase (2); β-amino acid transport (3); Ca^{2+}-dependent ATPase (4,5); diphosphomevalonate decarboxylase (6); ferrochelatase (7); glyceraldehyde-3-phosphate dehydrogenase (8-10); D-β-hydroxybutyrate dehydrogenase (11); 3-hydroxy-3-methylglutaryl-CoA reductase (12); inositol-1,2,3,5,6-pentakisphosphate 5-phosphatase (13); insulysin (14); iodothyronine 5'-deiodinase (15,16); leukotriene-A_4 hydrolase, weakly inhibited (17); Na^+/Ca^{2+} exchange (18); phospholipid-translocating ATPase, *or* flippase (19); phosphoprotein phosphatase (20); protein kinase (21-23); protein phosphatase 7 (20); protein-tyrosine phosphatase (24-27); and UDP-glucose 4-epimerase (28). **1.** Kosower, Kosower, Wertheim & Correa (1969) *Biochem. Biophys. Res. Commun.* **37**, 593.   **2.** Hinderlich, Nöhring, Weise, *et al.* (1998) *Eur. J. Biochem.* **252**, 133.   **3.** Chesney, Gusowski & Albright (1985) *Pediatr. Pharmacol.* **5**, 63.   **4.** Hightower & McCready (1991) *Curr. Eye Res.* **10**, 299.   **5.** Scutari, Ballestrin & Covaz (1979) *Boll. Soc. Ital. Biol. Sper.* **55**, 1283.   **6.** Alvear, Jabalquinto & Cardemil (1989) *Biochim. Biophys. Acta* **994**, 7.   **7.** Blom, Klasen & Van Steveninck (1990) *Biochim. Biophys. Acta* **1039**, 339.   **8.** Iglesias & Losada (1988) *Arch. Biochem. Biophys.* **260**, 830.   **9.** Schuppe-Koistinen, Moldeus, Bergman & Cotgreave (1994) *Eur. J. Biochem.* **221**, 1033.   **10.** Iglesias, Serrano Guerrero & Losada (1987) *Biochim. Biophys. Acta* **925**, 1.   **11.** Phelps & Hatefi (1981) *Biochemistry* **20**, 453 and 459.   **12.** Lippe, Deana, Cavallini & Galzigna (1985) *Biochem.*

Pharmacol. **34**, 3293. **13**. Bandyopadhyay, Kaiser, Rudolf, *et al.* (1997) *Biochem. Biophys. Res. Commun.* **240**, 146. **14**. Chowdhary, Smith & Peters (1985) *Biochim. Biophys. Acta* **840**, 180. **15**. Bhat, Iwase, Hummel & Walfish (1989) *Biochem. J.* **258**, 785. **16**. St. Germain (1988) *Endocrinology* **122**, 1860. **17**. Mueller, Samuelsson & Haeggström (1995) *Biochemistry* **34**, 3536. **18**. Antolini, Debetto, Trevisi & Luciani (1991) *Pharmacol. Res.* **23**, 163. **19**. Connor & Schroit (1991) *Biochim. Biophys. Acta* **1066**, 37. **20**. Andreeva, Solov'eva, Kakuev & Kutuzov (2001) *Arch. Biochem. Biophys.* **396**, 65. **21**. Harry, Pines & Applebaum (1978) *Arch. Biochem. Biophys.* **191**, 325. **22**. McClung & Miller (1977) *Biochem. Biophys. Res. Commun.* **76**, 910. **23**. Pillion, Leibach, von Tersch & Mendicino (1976) *Biochim. Biophys. Acta* **419**, 104. **24**. Chen & Geng (2001) *Biochem. Biophys. Res. Commun.* **286**, 831. **25**. Chen & Geng (2001) *Biochem. Biophys. Res. Commun.* **286**, 609. **26**. Zipser, Piade & Kosower (1997) *FEBS Lett.* **406**, 126. **27**. Monteiro, Ivaschenko, Fischer & Stern (1991) *FEBS Lett.* **295**, 146. **28**. Ray & Bhaduri (1980) *J. Biol. Chem.* **255**, 10777.

Tetramethyl-*p*-benzoquinone, *See* Duroquinone

2,3,5,6-Tetramethyl-2,5-cyclohexadiene-1,4-dione, *See* Duroquinone

Tetramethylenediamine, *See* Putrescine

3,3-Tetramethyleneglutarate

This reagent (FW$_{free-acid}$ = 186.21 g/mol), also known as 1,1-cyclopentanediacetate, inhibits acetoin dehydrogenase (1) alcohol dehydrogenase, NADP$^+$-dependent (1); aldehyde reductase (2), aldose reductase (3-5), carbonyl reductase (6,7), and glutamate decarboxylase, K_i = 3.0 mM (8). **1**. Hara, Seiriki, Nakayama & Sawada (1985) *Prog. Clin. Biol. Res.* **174**, 291. **2**. Gabbay & Kinoshita (1975) *Meth. Enzymol.* **41**, 159. **3**. Wermuth & von Wartburg (1982) *Meth. Enzymol.* **89**, 181. **4**. Williams & Odom (1987) *Exp. Eye Res.* **44**, 717. **5**. Branlant (1982) *Eur. J. Biochem.* **129**, 99. **6**. Usui, Hara, Nakayama & Sawada (1984) *Biochem. J.* **223**, 697. **7**. Wermuth (1981) *J. Biol. Chem.* **256**, 1206. **8**. Fonda (1972) *Biochemistry* **11**, 1304.

3,7,11,15-Tetramethylhexadecanoate, *See* Phytanate

(*E,E,E*)-*O*-(3,7,11,15-Tetramethyl-2,6,10,14-hexadecatetraenyl)-*N*-(aminosulfonyl)urethane

This geranylgeranyl diphosphate analogue (FW = 412.59 g/mol) inhibits Rap 1 geranylgeranylation (IC$_{50}$ = 28.2 μM) by targeting protein geranylgeranyltransferase I (1,2). **1**. Macchia, Jannitti, Gervasi & Danesi (1996) *J. Med. Chem.* **39**, 1352. **2**. Yokoyama & Gelb (2001) *The Enzymes*, 3rd. ed. (Tamanoi & Sigman, eds.), **21**, 105.

N-(2,2,5,5-Tetramethylpyrrolin-1-oxyl-3-carbonyl)-L-alanyl-L-alanyl-L-phenylalanine Methyl Ketone, *similar in action to:* N-(2,2,5,5-tetramethylpyrrolin-1-oxyl-3-carbonyl)-L-alanyl-L-alanyl-L-phenylalanine Methyl Ketone

3,3,14,14-Tetramethylthapsate, *See* 3,3,14,14-Tetramethylhexadecanedioate

Tetramethylthiuram Disulfide (*or* Thiuram)

This disulfide (FW = 240.44 g/mol; low water-solubility; slightly soluble in acetone), also called thiuram and bis(dimethyl)thiocarbamoyl disulfide, is frequently used as an antiseptic, fungicide, and bacteriostat. **Target(s):** aldehyde dehydrogenase (1,2); dopamine β-monooxygenase (3); glucose-6-phosphate dehydrogenase (4); and 6-phosphogluconate dehydrogenase (4).

1. Sanny & Weiner (1987) *Biochem. J.* **242**, 499. **2**. Deitrich & Hellerman (1963) *J. Biol. Chem.* **238**, 1683. **3**. Weppelman (1987) *Meth. Enzymol.* **142**, 608. **4**. Chefurka (1957) *Enzymologia* **18**, 209.

Tetranactin

This antibiotic macrotetralide (FW = 793.05 g/mol; CAS 33956-61-5) is a neutral ionophore and homologue of nonactin that binds and transports monovalent cations (1). **Target(s):** interleukin 1β- and cAMP-induced nitric-oxide synthase (2) and phospholipase A$_2$ expression (3); and oxidative phosphorylation (1). **1**. Pressman (1976) *Ann. Rev. Biochem.* **45**, 501. **2**. Kunz, Walker, Wiesenberg & Pfeilschifter (1996) *Brit. J. Pharmacol.* **118**, 1621. **3**. Walker, Kunz, Pignat, *et al.* (1996) *Eur. J. Pharmacol.* **306**, 265.

Tetrandrine

This bis-benzylisoquinoline alkaloid (FW = 622.76 g/mol; CAS 518-34-3) from the Chinese medicinal *Radix stephania tetrandrae* inhibits vascular contraction induced by membrane depolarization with KCl or adrenoceptor activation with phenylephrine. It also inhibits the release of endothelium-derived nitric oxide as well as NO production by inducible nitric oxide synthase. Tetrandrine also inhibits multiple Ca^{2+} entry pathways. **Target(s):** Ca^{2+}-dependent ATPase (1,2); calcium channel (2-5); nicotinic acetylcholine receptors (6); tyrosine monooxygenase (7). **1**. Chen, Chen, Tian & Yu (2000) *Brit. J. Pharmacol.* **131**, 530. **2**. Liu, Li, Gang, Karpinski & Pang (1995) *J. Pharmacol. Exp. Ther.* **273**, 32. **3**. Low, Berdik, Sormaz, *et al.* (1996) *Life Sci.* **58**, 2327. **4**. Weinsberg, Bickmeyer & Wiegand (1994) *Neuropharmacology* **33**, 885. **5**. King, Garcia, Himmel, *et al.* (1988) *J. Biol. Chem.* **263**, 2238. **6**. Slater, Houlihan, Cassels, Lukas & Bermudez (2002) *Eur. J. Pharmacol.* **450**, 213. **7**. Zhang & Fang (2001) *Planta Med.* **67**, 77.

1,3,5,7-Tetrazacyclopent-[*f*]-azulene, *See* Parazoanthoxanthin A; Zoanthoxanthin

α-Tetrazole, *See* DL-4-Amino-4-(5'-tetrazolyl)butanoic Acid

γ-Tetrazole, *See* DL-2-Amino-4-(5'-tetrazolyl)butanoic Acid

Tetrazolium Nitroblue Dichloride, *See* Nitroblue Tetrazolium Dichloride

Tetrazolium Red, *See* Triphenyltetrazolium Chloride

Tetrodotoxin

This natural product (FW$_{free-base}$ = 319.27 g/mol; CAS 4368-28-9; Soluble to 3 mM in acidic buffer, pH 4.8), which is destroyed by strong acid or alkaline solutions, also called fugu toxin, spheroidine, and tarichatoxin, is a toxin that exists in two tautomeric forms in the ovaries, skin, and liver of the fugu or puffer fish. The structure was determined by the Nobelist R. B. Woodward (1). Tetrodotoxin essentially prevents affected nerve cells from firing by binding to the pores of the voltage-gated, fast sodium ion channels blocking the channels used in the process. Tetrodotoxin is a specific blocker of voltage-gated sodium channels. This extremely toxic agent is also found in the newt, octopus, and several starfish. Tetrodotoxin is slightly soluble in water and has a pK_a value of 8.76. **Target(s):** Na^+ channels (2-6); and rubredoxin:NAD^+ reductase (7). **1**. Woodward (1964) *Pure Appl. Chem.* **9**, 49. **2**. Conn (1983) *Meth. Enzymol.* **103**, 401. **3**. Prince (1988) *Trends Biochem. Sci.* **13**, 76. **4**. Suehiro (1996) *Rev. Hist. Pharm. (Paris)* **44**(312 suppl), 379. **5**. Hille (1975) *Biophys. J.* **15**, 615. **6**. Catterall (1980) *Ann. Rev. Pharmacol. Toxicol.* **20**, 15. **7**. Ueda & Coon (1972) *J. Biol. Chem.* **247**, 5010.

TFA, See *Trifluoroacetate*

TFPI, See *S-Ethyl-N-[4-(trifluoromethyl)phenyl] isothiourea*

TG101348, See *Fedratinib*

TGX-221

This potent ATP site-competitive inhibitor (F.W. = 364.44 g/mol; CAS 663619-89-4; Solubility: 12 mg/mL DMSO; <1 mg/mL H_2O), also known as 7-methyl-2-morpholino-9-(1-(phenylamino)ethyl)-4H-pyrido[1,2-a]pyrimidin-4-one, targets the phosphoinositide-3-kinase-β (PI3Kβ), also known as p110β, (IC$_{50}$ = 7 nM), with IC$_{50}$ values of 0.1, 3.5 and 5 μM, respectively, for the δ, γ and α isoforms. TGX-221 also shows >1,000x selectively for PIKβ over a wide range of other kinases. It also inhibits thrombus formation in animal models. **1**. Chaussade, *et al.* (2007) *Biochem. J.* **404**, 449.

TH-302

This anticancer prodrug (FW = 449.04 g/mol; CAS 918633-87-1; Solubility: 90 mg/mL DMSO or 10 mg/mL H_2O, the latter with warming) is reductively activated in hypoxic cells to generate the bromo-isophosphoramide mustard.

Nitrogen mustards form cyclic aminium ions (aziridinium rings) by intramolecular displacement of the chloride by the amine nitrogen. This aziridinium group then alkylates DNA once it is attacked by the N-7 nucleophilic center on the guanine base. Upon displacement of the second chloride, a second alkylation results in the formation of interstrand cross-links (so-called 1,3-ICLs) in 5'-d(GNC) sequences. The The strong cytotoxic effect of ICL formation elicits cell cycle arrest as well as formation of histone

γH2AX, a marker for DNA double-strand breaks. Under hypoxic conditions, TH-302 is cytotoxic in CHO cells deficient in homology-dependent DNA repair, but not in lines deficient in base excision, nucleotide excision, or nonhomologous end-joining repair. TH-302 is also effective against lines deficient in BRCA1, BRCA2, and FANCA. In studies with three-dimensional spheroid and multicellular layer models, TH-302 shows enhanced potency in H460 spheroids, when tested under normoxia. **1**. Duan, Jiao, Kaizerman, *et al.* (2008) *J. Med. Chem.* **51**, 2412. **2**. Meng, Evans, Bhupathi, *et al.* (2012) *Mol. Cancer Ther.* **11**, 740.

Thalidomide

S(–)-Thalidomide

R(+)-Thalidomide

This notorious sedative and immunomodulator (FW = 258.23 g/mol; CAS 50-35-1), named systematically as (±)-2-(2,6-Dioxo-3-piperidinyl)-1H-isoindole-1,3(2H)-dione, also blocks basic fibroblast growth factor (bFGF)-induced angiogenesis in cornea (1). Thalidomide also inhibits replication of the HIV-I virus (2) and has been used to treat AIDS-related oral ulcers. Both enantiomers inhibit tumor necrosis factor α (TNF-α) release (3,4). Racemic thalidomide causes severe fetal abnormalities (thalidomide embryopathy), particularly stunted limb growth, when administered during pregnancy. Of the >10,000 thalidomide babies world-wide, two-thirds were in Germany, where thalidomide was approved for over-the-counter sale. The ability of the *S*-stereoisomer to intercalate into GC-rich regions of DNA appears to give rise to its teratogenic properties, whereas the *R*-stereoisomer exhibits the sedative properties. Even so, thalidomide racemizes rapidly *in vivo*, making it impossible for the human body to favor only one enantiomer. **Key Pharmacokinetic Parameters:** *See* Appendix II in Goodman & Gilman's *THE PHARMACOLOGICAL BASIS OF THERAPEUTICS*, 12th Edition (Brunton, Chabner & Knollmann, eds.) McGraw-Hill Medical, New York (2011). **Target(s):** HIV-I virus replication (2,5); IκB kinase (6). **1**. D'Amato, Loughnan, Flynn & Folkman (1994) *Proc. Natl. Acad. Sci. U.S.A.* **91**, 4082. **2**. S. Makonkawkeyoon, Limson-Pobre, Moreira, Schauf & Kaplan (1993) *Proc. Natl. Acad. Sci. U.S.A.* **90**, 5974. **3**. Tramontana, Utaipat, Molloy *et al.* (1995) *Mol. Med.* **1**, 384. **4**. Wnendt, Finkam, Winter, *et al.* (1996) *Chirality* **8**, 390. **5**. Moreira, Corral, Ye, *et al.* (1997) *AIDS Res. Hum. Retroviruses* **13**, 857. **6**. Keifer, Guttridge, Ashburner & Baldwin (2001) *J. Biol. Chem.* **276**, 22382.

Thallium

This Period 6, Group-13 "post-transition" element (Atomic Symbol: Tl; Atomic Number: 81; Atomic Weight: 204.38) is a nonessential metal in humans, with two principal oxidation states: thallous ion (Tl(I) *or* Tl$^+$) and thallic ion (Tl(III) *or* Tl^{3+}). **Thallium Toxicity:** Both Tl$^+$ and Tl^{3+} are highly toxic. Given the ability of Tl$^+$ ions (radius = 1.47 Å) to mimic K$^+$ ions (radius = 1.33 Å), mammals absorb thallium, as do plants. Once within cells, Th$^+$ interferes with potassium ion-dependent signaling and metabolism. Agonist-induced Tl$^+$ influx into acetylcholine receptor-rich membrane vesicles occurs on a millisecond time scale. Tl^{3+} compounds resemble corresponding Al^{3+} compounds, suggesting the likelihood of forming exchange-inert metal ion-ligand complexes with ATP and other di- and tri-phosphate-containing metabolites. Many thallium compounds are colorless, odorless and tasteless, properties commending their use in rodenticides. The toxic mechanism is incompletely understood. Thallium inhibits glycolysis, TCA cycle, and oxidative phosphorylation. Toxicity is attended by neuropathy and alopecia. Acute-exposure symptoms in humans include pain, nausea, vomiting, and angina. Chronic-exposure issues are hair loss and birth defects. Thallium poisoning is readily diagnosed and detected (the latter by atomic and mass spectroscopy). Forensic awarenes of thallium toxicity is high, after a series of notorious murders in the U.S. and People's Republic of China. **Thallium Stress Test:** This nuclear imaging method evaluates how well blood flows into the heart muscle, both at rest and during supervised activity, usually on an exercycle or treadmill. Radioactive thallium-201 is employed to image blood flow characteristics within patients suffering coronary artery disease. Thallium scans are also made to evaluate the success of bypass surgery or angioplasty. With a half-life of 73 hours, ^{201}Tl decays by electron capture, emitting Hg X-rays at ~70–80 keV and photons at 135 and 167 keV. This radionuclide thus provides good imaging characteristics without excessive radiation exposure.

THAM, *See* *Tris(hydroxymethyl)aminomethane*

Thapsigargin

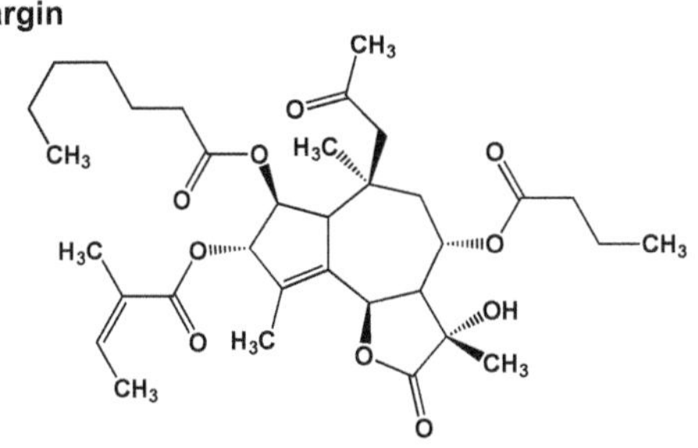

This cell-penetrating and IP$_3$-independent sesquiterpene lactone (FW = 650.76 g/mol; CAS 67526-96-8) from the roots of the umbelliferous plant *Thapsia garganica* is a potent inhibitor of the sarcoplasmic reticulum Ca^{2+}-ATPases in mammalian cells, IC$_{50}$ = 4-13 nM (1-4), but is without effect on the activities of plasma membrane Ca^{2+}-dependent ATPases, inositol-1,4,5-trisphosphate production, or protein kinase C. Calcium ion signaling relies on an intricate network of cellular channels and transporters to maintain a low resting calcium concentrations in the cytosol, with release/uptake producing amplitude- and/or frequency-encoded spikes in ion concentration in response to receptor-bound hormones and growth factors. Thapsigargin is often used to manipulate intracellular calcium ion levels and to increase Ca^{2+}-dependent Na$^+$ influx. It also depletes the endoplasmic reticulum calcium ion stores, inducing the ER stress response. **Target(s):** ATP-diphosphohydrolases, wherein only the ATRPase activity is inhibited (5); Ca^{2+}-transporting ATPase (1,4,6-8); Mg^{2+}-importing ATPase (8); and photophosphorylation (9); catalytic domain ErbB subfamily of protein tyrosine kinases 1. Treiman, Caspersen & Christensen (1998) *Trends Pharmacol. Sci.* **19**, 131. 2. Korge & Weiss (1999) *Eur. J. Biochem.* **265**, 273. 3. Thomas, Sanderson & Duncan (1999) *J. Physiol.* **516**, 191. 4. Inesi & Sagara (1992) *Arch. Biochem. Biophys.* **298**, 313. 5. Martins, Torres & Ferreira (2000) *Biosci. Rep.* **20**, 369. 6. Fierro & Parekh (2000) *J. Physiol.* **522**, 247. 7. Almeida, Benchimol, De Souza & Okorokov (2003) *Biochim. Biophys. Acta* **1615**, 60. 8. Taffet & Tate (1992) *Arch. Biochem. Biophys.* **299**, 287. 9. Santarius, Falsone & Haddad (1987) *Toxicon* **25**, 389.

Theaflavin

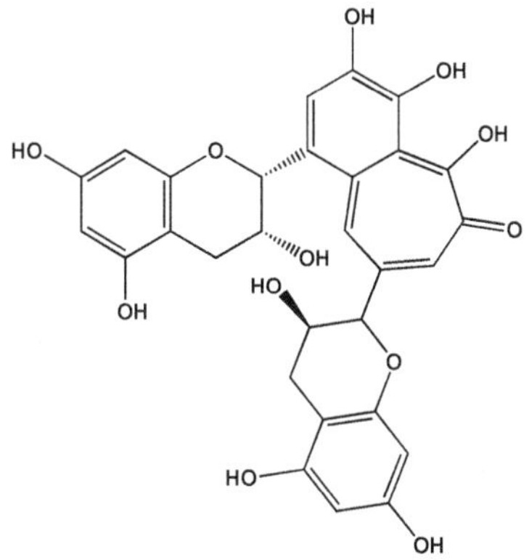

This natural product (FW = 564.50 g/mol; CAS 4670-05-7), an antioxidant isolated from black tea, targets 1-alkylglycero-phosphocholine *O*-acetyltransferase (1), collagenases, type IV (2), α-glucosidase, *or* maltase (3), glucosyltransferase (4), and phenol sulfotransferase, *or* aryl sulfotransferase (5). 1. Sugatani, Fukazawa, Ujihara, *et al.* (2004) *Int. Arch. Allergy Immunol.* **134**, 17. 2. Sazuka, Imazawa, Shoji, *et al.* (1997) *Biosci. Biotechnol. Biochem.* **61**, 1504. 3. Matsui, Tanaka, Tamura, *et al.* (2007) *J. Agric. Food Chem.* **55**, 99. 4. Hattori, Kusumoto, Namba, Ishigami & Hara (1990) *Chem. Pharm. Bull. (Tokyo)* **38**, 717. 5. Nishimuta, Ohtani, Tsujimoto, *et al.* (2007) *Biopharm. Drug Dispos.* **28**, 491.

Theaflavin digallate

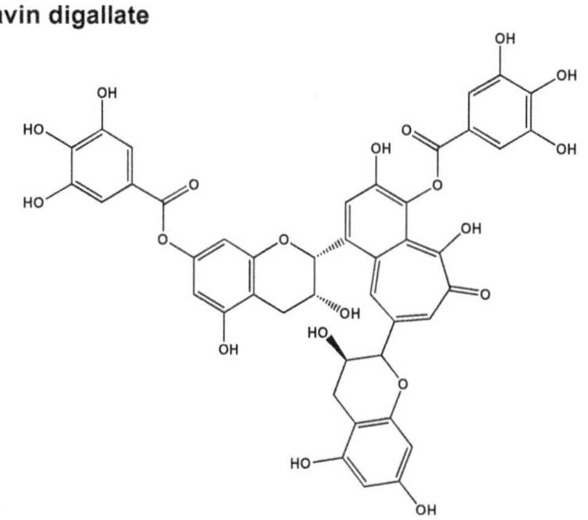

This polyphenol antioxidant (FW = 868.71 g/mol), a constituent in black tea, often inhibits many viral and cellular processes *in vitro*. (**See** *Theaflavin*) Theaflavin digallate (1-10 μM) inhibits the infectivity of both influenza A virus and influenza B virus in Madin-Darby canine kidney (MDCK) cells *in vitro* (1). Electron microscopy also revealed that EGCg and TF3 agglutinated influenza viruses as well as did antibody, and that they prevented the viruses from adsorbing to MDCK cells. Theaflavin digallate xanthine oxidase and suppression of intracellular Reactive Oxygen Species (ROS) in HL-60 cells (2). It also inhibits 3CL^{Pro} (IC$_{50}$ = 7 μM), the virally encoded 3C-like protease that is critical for the viral replication of SARS-CoV, the causative agent of severe acute respiratory syndrome, or SARS (3). Theaflavin digallate inhibits SAM-mediated DNA methylation by DNA methyltransferase 3a (Dnmt3a), an enzyme that regulates osteoclastogenesis via epigenetic repression of anti-osteoclastogenic genes (4). In this context, theaflavin digallate abrogates bone loss in osteoporosis models. 1. Nakayama, Suzuki, Toda, *et al.* (1993) *Antiviral Res.* **21**, 289. 2. Lin, Chen, Ho & Lin-Shiau (2000) *J. Agricult. & Food Chem.* **48**, 2736. 3. Chen, Coney, Lin, *et al.* (2005) *Evidence-based Compl. and Alt. Med.* **2**, 209. 4. Nishikawa, Iwamoto, Kobayashi, *et al.* (2015) *Nature Med.* **21**, 281.

L-Theanine

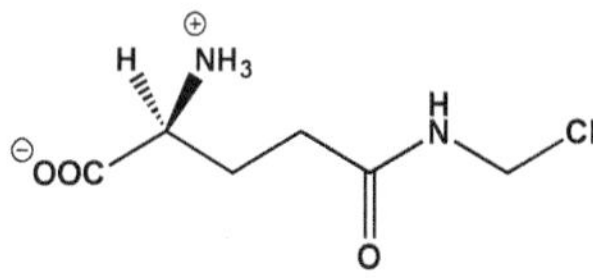

This L-glutamine derivative (FW = 174.20 g/mol; CAS 34271-54-0), also called N^5-ethylglutamine and L-glutamic γ-ethylamide, is enriched in green tea. **Target(s):** glutamate transporter via the glutathione-*S*-conjugate export pump (1,2); glutaminase (3,4); glutaminyl-tRNA synthetase (5); 4-methylene-glutaminase, weakly inhibited (6). 1. Sadzuka, Sugiyama, Suzuki & Sonobe (2001) *Toxicol. Lett.* **123**, 159. 2. Sugiyama, Sadzuka, Tanaka & Sonobe (2001) *Toxicol. Lett.* **121**, 89. 3. Hartman (1970) *Meth. Enzymol.* **17A**, 941. 4. Hartman (1971) *The Enzymes*, 3rd ed. (Boyer, ed.), **4**, 79. 5. Lea & Fowden (1973) *Phytochemistry* **12**, 1903. 6. Ibrahim, Lea & Fowden (1984) *Phytochemistry* **23**, 1545.

Thebaine

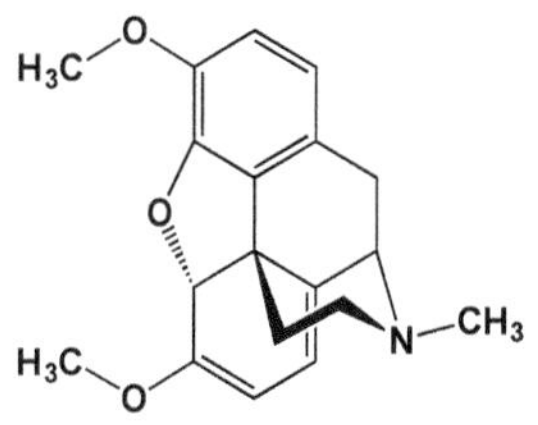

This (–)-benzylisoquinoline alkaloid (FW = 311.38 g/mol; CAS 115-37-7), which is a precursor to codeine and morphine and is a constituent of opium, binds to μ and δ opioid receptors. Thebaine is also the starting reagent for the synthesis of such controlled substances as oxycodone, oxymorphone, nalbuphine, naloxone, naltrexone, buprenorphine and etorphine. *Note*: Thebaine is itself a controlled substance, and care must be exercised in its use. **Target(s):** salutaridinol 7-*O*-acetyltransferase (1). 1. Lenz & Zenk (1995) *J. Biol. Chem.* **270**, 31091. **Theelin,** *See* *Estrone*

Theelol, *See* Estriol

1-(2'-Thenoyl)-3,3,3-trifluoroacetone

This high-affinity heavy metal ion chelator (FW = 222.19 g/mol; CAS 326-91-0; Melting Point = 40-44°C), also known as 4,4,4-trifluoro-1-(2-thienyl)-1,3-butanedione, is a strong inhibitor of complex II of the mitochondrial electron transport chain, but does not inhibit solubilized succinate dehydrogenase. **Target(s):** dihydroorotate dehydrogenase (1); electron transport, mitochondrial (2,3); fumarate reductase (4); glycerol-3-phosphate dehydrogenase (5); hexacyanoferrate III reductase (6); 4-hydroxyphenylpyruvate dioxygenase (7); malate dehydrogenase (8,9); NAD(P)H dehydrogenase (10); oleoyl-CoA Desaturase (11); photosystem II (12); semidehydroascorbate reductase, NADH-dependent (13); steroid 21-monooxygenase (14); succinate dehydrogenase, *or* complex II (2-4,15-20). 1. Miller (1978) *Meth. Enzymol.* **51**, 63. 2. Rapoport & Schewe (1977) *Trends Biochem. Sci.* **2**, 186. 3. Singer (1979) *Meth. Enzymol.* **55**, 454. 4. Armson, Grubb & Mendis (1995) *Int. J. Parasitol.* **25**, 261. 5. Dawson & Thorne (1975) *Meth. Enzymol.* **41**, 254. 6. Doring & Luthje (2001) *Protoplasma* **217**, 3. 7. Fellman (1987) *Meth. Enzymol.* **142**, 148. 8. Gutman & Hartstein (1977) *Biochim. Biophys. Acta* **481**, 33. 9. Gutman & Hartstein (1974) *FEBS Lett.* **49**, 170. 10. Serrano, Cordoba, Gonzalez-Reyes, Navas & Villalba (1994) *Plant Physiol.* **106**, 87. 11. Wilson, Adams & Miller (1980) *Can. J. Biochem.* **58**, 97. 12. Ikezawa, Ifuku, Endo & Sato (2002) *Biosci. Biotechnol. Biochem.* **66**, 1925. 13. Nishino & Ito (1986) *J. Biochem.* **100**, 1523. 14. Rosenthal & Narasimhulu (1969) *Meth. Enzymol.* **15**, 596. 15. Maklashina & Cecchini (1999) *Arch. Biochem. Biophys.* **369**, 223. 16. Ramsay, Ackrell, Coles, *et al.* (1981) *Proc. Natl. Acad. Sci. U.S.A.* **78**, 825. 17. Ingledew & Ohnishi (1977) *Biochem. J.* **164**, 617. 18. Hatefi & Stiggall (1976) *The Enzymes*, 3rd ed. (Boyer, ed.), **13**, 175. 19. Ziegler & Rieske (1967) *Meth. Enzymol.* **10**, 231. 20. Hatefi & Stiggall (1978) *Meth. Enzymol.* **53**, 21.

Theobromine

This dimethyl xanthine (FW = 180.17 g/mol; CAS 83-67-0; sparingly soluble in water), also called 3,7-dimethylxanthine and 3,7-dihydro-3,7-dimethyl-1*H*-purine-2,6-dione, is enriched in cacao beans, cola nuts, and tea. As a 3',5'-cyclicAMP phosphodiesterase inhibitor, theobromine prevents the conversion of cyclicAMP to 5'-AMP. Theobromine is a cardiac stimulator and vasodilator, a smooth muscle relaxant, and a diuretic. It is also a weak adenosine receptor antagonist. **Target(s):** adenosine deaminase (1); 3',5'-cyclic-nucleotide phosphodiesterase (2,3); hydroxyacylglutathione hydrolase, *or* glyoxalase II (4); inosine Nucleosidase (5); lactoylglutathione lyase, *or* glyoxalase I (4); NAD+-dependent ADP-ribosyltransferase, poly(ADP-ribose) polymerase (6,7); NAD+ Nucleosidase, *or* NADase (8,9); and purine-nucleoside phosphorylase (5,10). 1. Jun, Kim & Yeeh (1994) *Biotechnol. Appl. Biochem.* **20**, 265. 2. Kemp & Huang (1974) *Meth. Enzymol.* **38**, 240. 3. Butcher & Sutherland (1962) *J. Biol. Chem.* **237**, 1244. 4. Oray & Norton (1980) *Biochem. Biophys. Res. Commun.* **95**, 624. 5. Koch (1956) *J. Biol. Chem.* **223**, 535. 6. Rankin, Jacobson, Benjamin, Moss & Jacobson (1989) *J. Biol. Chem.* **264**, 4312. 7. Burtscher, Klocker, Schneider, *et al.* (1987) *Biochem. J.* **248**, 859. 8. Alivisatos, Kashket & Denstedt (1956) *Can. J. Biochem. Physiol.* **34**, 46. 9. Webb (1966) *Enzyme and Metabolic Inhibitors*, vol. 2, p. 492, Academic Press, New York. 10. Friedkin & Kalckar (1961) *The Enzymes*, 2nd ed. (Boyer, Lardy & Myrbäck, eds.), **5**, 237.

Theophylline

This naturally occurring dimethyl xanthine and smooth muscle relaxant (FW = 180.17 g/mol; CAS 58-55-9 ; M.P. = 274°C; Soluble to 25 mM in water and to 100 mM in DMSO), also known as 1,3-dimethylxanthine and 3,7-dihydro-1,3-dimethyl-1*H*-purine-2,6-dione, is found in tea and other plants. Theophylline produces vasodilation, cardiac stimulation, and diuresis. **Target(s):** adenine deaminase (1); adenosine deaminase (2-4); alkaline phosphatase (5,6); 3',5'-cyclic-GMP phosphodiesterase (7,8,31,32,35); 2',3'-cyclic nucleotide 3'-phosphodiesterase (9,10); 3',5'-cyclic-nucleotide phosphodiesterase, K_i = 0.1 mM (11-18,31-35); CYP1A2 (19); cytokinin 7β-glucosyltranferase (47); glycerol-1,2-cyclic-phosphate phosphodiesterase (20); glycogen synthase (48); hydroxyacylglutathione hydrolase, *or* glyoxalase II (22); 15-hydroxyprostaglandin dehydrogenase (21); hypoxanthine(guanine) phosphoribosyltransferase (1,46); lactoylglutathione lyase, *or* glyoxalase I (22); myeloperoxidase (23); NAD+-dependemnt ADP-ribosyltransferaase, *or* poly(ADP-ribose) polymerase (42-45); NAD+ Nucleosidase, *or* NADase (24,25); NAD+:protein-arginine ADP-ribosyltransferase (41); nucleoside-diphosphate kinase (37); 1-phosphatidylinositol 4-kinase (38); poly(ADP-ribose) polymerase (26); purine nucleosidase (1); purine-nucleoside phosphorylase (27,28); pyridoxal kinase (29,39); rhodopsin kinase (36); thymidine kinase (40); urate oxidase, *or* uricase (30). **See also Aminophylline** 1. Nolan & Kidder (1979) *Biochem. Biophys. Res. Commun.* **91**, 253. 2. Challa, Johnson, Robertson & Gunasekaran (1999) *J. Basic Microbiol.* **39**, 97. 3. Singh & Sharma (2000) *Mol. Cell. Biochem.* **204**, 127. 4. Jun, Kim & Yeeh (1994) *Biotechnol. Appl. Biochem.* **20**, 265. 5. Fawaz & Tejirian (1972) *Hoppe-Seyler's Z. Physiol. Chem.* **353**, 1779. 6. Glogowski, Danforth & Ciereszko (2002) *J. Androl.* **23**, 783. 7. Liebman & Evanczuk (1982) *Meth. Enzymol.* **81**, 532. 8. Morishima (1975) *Biochim. Biophys. Acta* **410**, 310. 9. Sprinkle (1989) *CRC Crit. Rev. Clin. Neurobiol.* **4**, 235. 10. Sims & Carnegie (1978) *Adv. Neurochem.* **3**, 1. 11. Kemp & Huang (1974) *Meth. Enzymol.* **38**, 240. 12. Lin, Liu & Cheung (1974) *Meth. Enzymol.* **38**, 262. 13. Hosono (1988) *Meth. Enzymol.* **159**, 497. 14. Drummond & Yamamoto (1971) *The Enzymes*, 3rd ed. (Boyer, ed.), **4**, 355. 15. Stefanovich, von Polnitz & Reiser (1974) *Arzneimittelforschung* **24**, 1747. 16. Ferretti, Coppi, Blengio & Genazzani (1992) *Int. J. Tissue React.* **14**, 31. 17. Cheung (1967) *Biochemistry* **6**, 1079. 18. Butcher & Sutherland (1962) *J. Biol. Chem.* **237**, 1244. 19. Murray, Odupitan, Murray, Boobis & Edwards (2001) *Xenobiotica* **31**, 135. 20. Clarke & Dawson (1978) *Biochem. J.* **173**, 579. 21. Pace-Asciak & Smith (1983) *The Enzymes*, 3rd ed. (Boyer, ed.), **16**, 543. 22. Oray & Norton (1980) *Biochem. Biophys. Res. Commun.* **95**, 624. 23. van Zyl, Kriegler & van der Walt (1992) *Int. J. Biochem.* **24**, 929. 24. Alivisatos, Kashket & Denstedt (1956) *Can. J. Biochem. Physiol.* **34**, 46. 25. Webb (1966) *Enzyme and Metabolic Inhibitors*, vol. 2, p. 492, Academic Press, New York. 26. Villamil, Podesta, Molina Portela & Stoppani (2001) *Mol. Biochem. Parasitol.* **115**, 249. 27. Friedkin & Kalckar (1961) *The Enzymes*, 2nd ed. (Boyer, Lardy & Myrbäck, eds.), **5**, 237. 28. Koch (1956) *J. Biol. Chem.* **223**, 535. 29. Ubbink, Bissbort, Vermaak & Delport (1990) *Enzyme* **43**, 72. 30. Hurst, Griffiths & Vayianos (1985) *Clin. Biochem.* **18**, 247. 31. Methven, Francis & Bhoola (1980) *Biochem. J.* **186**, 491. 32. Thomas, Francis & Corbin (1990) *J. Biol. Chem.* **265**, 14964. 33. Lim, Palanisamy & Ong (1986) *Arch. Microbiol.* **146**, 142. 34. Lim, Woon, Tan & Ong (1989) *Int. J. Biochem.* **21**, 909. 35. Pannbacker, Fleischman & Reed (1972) *Science* **175**, 757. 36. Weller, Virmaux & Mandel (1975) *Proc. Natl. Acad. Sci. U.S.A.* **72**, 381. 37. Lam & Packham (1986) *Biochem. Pharmacol.* **35**, 4449. 38. Buckley (1977) *Biochim. Biophys. Acta* **498**, 1. 39. Lainé-Cessac, Cailleux & Allain (1997) *Biochem. Pharmacol.* **54**, 863. 40. Sandlie & Kleppe (1980) *FEBS Lett.* **110**, 223. 41. Moss, Stanley & Watkins (1980) *J. Biol. Chem.* **255**, 5838. 42. Rankin, Jacobson, Benjamin, Moss & Jacobson (1989) *J. Biol. Chem.* **264**, 4312. 43. Kofler, Wallraff, Herzog, *et al.* (1993) *Biochem. J.* **293**, 275. 44. Burtscher, Klocker, Schneider, Auer, Hirsch-Kauffmann & Schweiger (1987) *Biochem. J.* **248**, 859. 45. Werner, Sohst, Gropp, Simon, Wagner & Kröger (1984) *Eur. J. Biochem.* **139**, 81. 46. Miller, Ramsey, Krenitsky & Elion (1972)

Biochemistry **11**, 4723. **47**. Parker, Entsch & Letham (1986) *Phytochemistry* **25**, 303. **48**. Moses, Bashan & Gutman (1972) *Eur. J. Biochem.* **30**, 205.

13-Thiaarachidonate

This long-chain, thioether-containing acid (FW$_{\text{free-acid}}$ = 322.51 g/mol) is a irreversible mechanism-based inhibitor of arachidonate 15-lipoxygenase. O$_2$ is required for inactivation. The corresponding sulfoxide is a competitive inhibitor but not an inactivator. **1**. Ator & Ortiz de Montellano (1990) *The Enzymes*, 3rd ed. (Sigman & Boyer, eds.), **19**, 213.

Thiabendazole

This common fungicide and anthelmintic (FW = 201.25 g/mol; CAS 148-79-8; M.P. = 304-305°C; λ_{max} = 298 nm (ε = 23330 M^{-1}cm^{-1} in methanol); fluorescence emission maximum in acid is 370 nm, when excited at 310 nm) inhibits growth of *Penicillium atrovenetum*, when present at 8–10 µg/mL (1). TBZ completely inhibits the following systems of isolated heart or fungus mitochondria: reduced nicotinamide adenine dinucleotide oxidase, succinic oxidase, reduced nicotinamide adenine dinucleotide-cytochrome *c* reductase, and succinic-cytochrome *c* reductase at concentrations of 10, 167, 10, and 0.5 µg/mL, respectively (1). Cytochrome *c* oxidase was not inhibited (1). **Target(s):** CYP1A2 (2); fumarate reductase (3); glucose-6-phosphate dehydrogenase (4); malate dehydrogenase (4-6); methionyl aminopeptidase, K_i = 0.4 µM (7); Cu/Zn superoxide dismutase (8,9); triacylglycerol lipase (4); and tubulin polymerization (10,11). **1**. Allen & Gottlieb (1970) *Applied Environ. Microbiol.* **20**, 919. **2**. Bapiro, Egnell, Hasler & Masimirembwa (2001) *Drug Metab. Dispos.* **29**, 30. **3**. Armson, Grubb & Mendis (1995) *Int. J. Parasitol.* **25**, 261. **4**. Sarwal, Sanyal & Khera (1989) *J. Parasitol.* **75**, 808. **5**. Tejada, Sanchez-Moreno, Monteoliva & Gomez-Banqueri (1987) *Vet. Parasitol.* **24**, 269. **6**. Sharma, Singh, Saxena & Saxena (1986) *Angew. Parasitol.* **27**, 175. **7**. Schiffmann, Neugebauer & Klein (2006) *J. Med. Chem.* **49**, 511. **8**. Castellanos-Gonzalez, Jimenez & Landa (2002) *Int. J. Parasitol.* **32**, 1175. **9**. Sanchez-Moreno, Entrala, Janssen, *et al.* (1996) *Pharmacology* **52**, 61. **10**. Brunner, Albertini & Wurgler (1991) *Mutagenesis* **6**, 65. **11**. Lubega & Prichard (1990) *Mol. Biochem. Parasitol.* **38**, 221.

Thiacetazone

This well-known tuberculostatic (FW = 236.30 g/mol; CAS 104-06-3), also called 4'-formyl-acetanilide thiosemicarbazone, is a potent noncompetitive inhibitor of influenza A/WSN/33 neuraminidase, *or sialidase* (K_i = 4 µM). Note that thiacetazone is highly specific in its inhibition; neuraminidases from other influenza strains are not inhibited (1). Thiacetazone should be stored in light-tight containiner, as it darkens upon exposure to light. **1**. Wu, Peet, Coutts, *et al.* (1995) *Biochemistry* **34**, 7154.

6-Thia-7-dehydro-(*RS*)-2,3-oxidosqualene

This oxidosqualene analogue (FW = 432.75 g/mol), also called S-6, inhibits rat liver lanosterol cyclase and *Alicyclobacillus acidocaldarius* squalene:hopene cyclase (K_i = 520 and 127 nM, respectively). (*See also 6,10-*

Dimethyl-9,10-epoxy-(5E)-undecen-2-yl   4',8',12'-Trimethyl-(3E,7E)-tridecatrienyl Sulfide) **1**. Zheng, Abe & Prestwich (1998) *Biochemistry* **37**, 5981.

10-Thia-11-dehydro-(*RS*)-2,3-oxidosqualene

This oxidosqualene analogue also called S-10 (FW = 432.75 g/mol), inhibits rat liver lanosterol cyclase and *Alicyclobacillus acidocaldarius* squalene:hopene cyclase (K_i = 2.1 and 0.97 µM, respectively). **Target(s):** lanosterol synthase (1); and squalene:hopene cyclase (1). *See also 6,10-Dimethyl-9,10-epoxy-(5E)-undecen-2-yl   4',8',12'-Trimethyl-(3E,7E)-tridecatrienyl Sulfide* **1**. Zheng, Abe & Prestwich (1998) *Biochemistry* **37**, 5981.

(7*E*,11*Z*,14*Z*,17*Z*)-5-Thiaeicosa-7,11,14,17-tetraenoate

This long-chain, thioether-containing acid (FW$_{\text{free-acid}}$ = 322.51 g/mol) inhibits cyclooxygenase and prostaglandin biosynthesis. The most potent COX-2-catalyzed prostaglandin biosynthesis inhibitor was all-(*Z*)-5-thia-8,11,14,17-eicosatetraenoic acid, followed by eicosapentaenoic acid (EPA), docosahexaenoic acid (DHA), α-linolenic acid (α-LNA), linoleic acid (LA), (7*E*,11*Z*,14*Z*,17*Z*)-5-thiaeicosa-7,11,14,17-tetraenoic acid, all-(*Z*)-3-thia-6,9,12,15-octadecatetraenoic acid, and (5*E*,9*Z*,12*Z*,15*Z*,18*Z*)-3-oxaheneicosa-5,9,12,15,18-pentaenoic acid, with IC$_{50}$ values ranging from 3.9 to180 µM. **1**. Ringbom, Huss, Stenholm, *et al.* (2001) *J. Nat. Prod.* **64**, 745.

Thiamet G

This oral and blood brain barrier-permeable hexosamine analogue (FW = 248.30 g/mol; CAS 1009816-48-1; Soluble to 50 mM in DMSO; (3a*R*,5*R*, 6*S*,7*R*,7a*R*)-2-(ethylamino)-3a,6,7,7a-tetrahydro-5-(hydroxymethyl)-5*H*-pyrano[3,2-*d*]thiazole-6,7-diol, is a sterically constrained mechanism-inspired inhibitor that potently and selectively targets human Protein *O*-GlcNAcase (*Reaction*:   [Protein]-3-*O*-(*N*-acetyl-β-D-glucosaminyl)-L-(serine/threonine) + H$_2$O $\rightleftharpoons$ [Protein]-L-(serine/threonine) + *N*-acetyl-D-glucosamine), with a K_i value of 21 nM. Given the observed reciprocal relationship between the states phosphorylation and *O*-GlcNAc modification of the microtubule-associated protein Tau as well as the reduced extent of *O*-GlcNAc in Alzheimer Tau, there is justifiably great motivation for generating potent and selective *O*-GlcNAcase inhibitors. Thiamet-G decreases Tau phosphorylation in PC-12 pheochromocytoma cells, an NGF-activatable cell line of neural crest origin. Moreover, attenuated phosphorylation occurs at the pathologically relevant sites, Thr-231 and Ser-396 (1). Thiamet-G also efficiently reduces Tau phosphorylation at Thr-231, Ser-396 and Ser-422 in both rat cortex and hippocampus, revealing the rapid and dynamic relationship between *O*-GlcNAc-Tau and Tau phosphorylation *in vivo*. Indeed, Thiamet G-induced increases in Tau *O*-GlcNAcylation reduces pathological Tau without affecting its normal phosphorylation in a mouse tauopathy model (2). **1**. Yuzwa, Macauley, Heinonen, *et al.* (2008) *Nature Chem. Biol.* **4**, 483. **2**. Graham, Gray, Joyce, *et al.* (2014) *Neuropharmacology* **79**, 307.

Thiamethoxam

This broad-spectrum systemic insecticide (FW = 291.71 g/mol; CAS 153719-23-4), also named *N*-3-(2-chloro-5-thiazolylmethyl)tetrahydro-5-methyl-*N*-nitro-4*H*-1,3,5-oxadiazin-4-imine, is a neonicotinoid that blocks nicotinic acetylcholine receptors, paralyzing the muscles and resulting in death. As a systemic insecticide, it is absorbed quickly by plants and is transported throughout, including pollen, thereby deterring insects from feeding on susceptible plant parts. Neonicotinoid-class insecticides generally have favorable safety profiles, because they preferentially target nicotinic receptor (nAChR) subtypes in insects and poorly penetrate the mammalian blood-brain barrier. Low application rates also limit risk of toxicity in humans. **1.** Sheets, Li, Minnema, *et al.* (2016) *Crit. Rev. Toxicol.* **46**, 153.

Thiamin

This essential nutrient (FW$_{HCl-Salt}$ = 300.81 g/mol; CAS 59-43-8), also known as vitamin B$_1$, is a water-soluble vitamin containing UV-absorbing pyrimidine and thiazole rings, the latter essential for its biochemical function (1-3). At pH 7 and above, thiamin has two bands (λ_{max} at 235 and 267 nm, with ε = 11300 and 8300 M^{-1}cm^{-1}, respectively). At and below pH 5.5, there is one UV band (λ_{max} = 247 nm; ε = 14200 M^{-1}cm^{-1}). **Metabolic Forms:** Thiamin is the transport form of the vitamin, mainly associated with albumin. Within cells, thiamin is converted to: thiamine monophosphate (ThMP); thiamine diphosphate (ThDP), also called thiamine pyrophosphate (TPP); thiamine triphosphate (ThTP), adenosine thiamine triphosphate (AThTP), and adenosine thiamine diphosphate (AThDP).**Chemical Properties:** Thiamin chloride is readily hydrated, and aqueous solutions are stable at pH 3.5; however, above pH 5, solutions are less stable to heat and, above pH 7, solutions are unstable, even at room temperature. Accordingly, thiamin solutions should always be freshly prepared and kept cold. **Target(s):** acetylcholinesterase (4,5); acid phosphatase (6); ascorbate oxidase (7); [branched-chain α-keto-acid dehydrogenase] kinase, *or* [3-methyl-2-oxobutanoate dehydrogenase (acetyl-transferring)] kinase (8); cholinesterase (4); diamine oxidase (9,10); β-fructofuranosidase, *or* invertase (11); glycogen phosphorylase *b* (12); intron (group I) self-splicing (13); sulfoacetaldehyde acetyltransferase (14); thiamin-phosphate kinase (15); thiamin pyridinylase, *or* thiaminase I, reversibly inactivated by the natural substrate (16-19); and thiamin pyrophosphatase (20,21). **1.** Clarke & Gurin (1935) *J. Amer. Chem. Soc.* **57**, 1876. **2.** Williams (1936) *J. Amer. Chem. Soc.* **58**, 1063. **3.** Williams & Cline (1936) *J. Amer. Chem. Soc.* **58**, 1504. **4.** Augustinsson (1950) *The Enzymes*, 1st ed. (J. B. Sumner & Myrbäck, eds.), 1 (part 1), 443. **5.** Alspach & Ingraham (1977) *J. Med. Chem.* **20**, 161. **6.** Westenbrink & Van Dorp (1941) *Enzymologia* **10**, 212. **7.** Stark & Dawson (1963) *The Enzymes*, 2nd ed. (Boyer, Lardy & Myrbäck, eds.), **8**, 297. **8.** Lau, Fatania & Randle (1982) *FEBS Lett.* **144**, 57. **9.** Zeller (1951) *The Enzymes*, 1st ed. (J. B. Sumner & Myrbäck, eds.), **2** (part 1), 536. **10.** Zeller (1963) *The Enzymes*, 2nd ed. (Boyer, Lardy & Myrbäck, eds.), **8**, 313. **11.** Neuberg & Mandl (1950) *The Enzymes*, 1st ed. (Sumner & Myrbäck, eds.), 1 (part 1), 527. **12.** Klinova, Klinov, Kurganov, Mikhno & Baliakina (1988) *Bioorg. Khim.* **14**, 1520. **13.** Ahn & Park (2003) *Int. J. Biochem. Cell Biol.* **35**, 157. **14.** Kondo & Ishimoto (1974) *J. Biochem.* **76**, 229. **15.** Nishino (1972) *J. Biochem.* **72**, 1093. **16.** Agee & Airth (1973) *J. Bacteriol.* **115**, 957. **17.** Suzuki & Ooba (1973) *Biochim. Biophys. Acta* **293**, 111. **18.** Suzuki & Ooba (1972) *J. Biochem.* **72**, 1053. **19.** Wittliff & Airth (1968) *Biochemistry* **7**, 736. **20.** Westenbrink, Van Dorp, Gruber & Veldman (1940) *Enzymologia* **9**, 73. **21.** Westenbrink, Steyn Parvé & Goudsmit (1943) *Enzymologia* **11**, 26.

Thiamin Diphosphate, *See* Thiamin Pyrophosphate

Thiamin Monophosphate

Thiamin monophosphate (FW$_{free-acid}$ = 380.79 g/mol; CAS 22457-89-2), abbreviated ThMP, is an intermediate in thiamin pyrophosphate metabolism that can also be prepared by sulfuric acid hydrolysis of thiamin pyrophosphate. **Target(s):** intron (group I) self-splicing (1); thiamin diphosphokinase (2-4); thiamin pyrophosphatase (5); thiamin transport (6,7); and thiamin-triphosphatase (8). **1.** Ahn & Park (2003) *Int. J. Biochem. Cell Biol.* **35**, 157. **2.** Mitsuda, Takii, Iwami, Yasumoto & Nakajima (1979) *Meth. Enzymol.* **62**, 107. **3.** Mitsuda, Takii, Iwami & Yasumoto (1975) *J. Nutr. Sci. Vitaminol.* **21**, 189. **4.** Mitsuda, Takii, KimikazuK. Yasumoto (1975) *J. Nutr. Sci. Vitaminol.* **21**, 103. **5.** Westenbrink, Van Dorp, Gruber & Veldman (1940) *Enzymologia* **9**, 73. **6.** Casirola, Ferrari, Gastaldi, Patrini & Rindi (1988) *J. Physiol.* **398**, 329. **7.** Shigeoka, Onishi, Maeda, Nakano & Kitaoka (1987) *Biochim. Biophys. Acta* **929**, 247. **8.** Hashitani & Cooper (1972) *J. Biol. Chem.* **247**, 2117.

Thiamin Pyrophosphate

This key metabolite (FW$_{ion}$ = 425.31 g/mol; CAS 532-40-1), also known as thiamin diphosphate and abbreviated TPP, serves as a coenzyme for many enzymes. TPP absorbs light significantly in the ultraviolet: in phosphate buffer at pH 8, it has λ_{max} values of 233 and 267 nm (ε = 10800 and 7800 M^{-1}cm^{-1}, respectively). Thiamin pyrophosphate is soluble in water and is slightly less stable than thiamin; therefore, solutions should always be freshly prepared and stored in the cold until needed. **Target(s):** [branched-chain α-ketoacid dehydrogenase] kinase, *or* 3-methyl-2-oxobutanoate dehydrogenase (acetyl-transferring)] kinase (1-6); 3',5'-cyclic-nucleotide phosphodiesterase (7); diamine oxidase (8,9); FAD diphosphatase (10); glycogen phosphorylase *b* (11); intron (group I) self-splicing (12); nucleotide diphosphatase (13); [pyruvate dehydrogenase (acetyl-transferring)] kinase (14-21); thiamin diphosphokinase, product inhibition (22-24); thiamin pyridinylase, *or* thiaminase I (25). **1.** Randle, Patston & Espinal (1987) *The Enzymes*, 3rd ed. (Boyer & Krebs, eds.), **18**, 97. **2.** Espinal, Beggs & Randle (1988) *Meth. Enzymol.* **166**, 166. **3.** Lau, Fatania & Randle (1982) *FEBS Lett.* **144**, 57. **4.** Reed, Damuni & Merryfield (1985) *Curr. Top. Cell. Regul.* **27**, 41. **5.** Li, Wynn, Machius, *et al.* (2004) *J. Biol. Chem.* **279**, 32968. **6.** Hawes, Schnepf, Jenkins, *et al.* (1995) *J. Biol. Chem.* **270**, 31071. **7.** Chassy & Porter (1974) *Meth. Enzymol.* **38**, 244. **8.** Zeller (1942) *Adv. Enzymol.* **2**, 93. **9.** Massart (1950) *The Enzymes*, 1st ed. (Sumner & Myrbäck, eds.), 1 (part 1), 307. **10.** Ragab, Brightwell & Tappel (1968) *Arch. Biochem. Biophys.* **123**, 179. **11.** Klinova, Klinov, Kurganov, Mikhno & Baliakina (1988) *Bioorg. Khim.* **14**, 1520. **12.** Ahn & Park (2003) *Int. J. Biochem. Cell Biol.* **35**, 157. **13.** Byrd, Fearney & Kim (1985) *J. Biol. Chem.* **260**, 7474. **14.** Pettit, Yeaman & Reed (1983) *Meth. Enzymol.* **99**, 331. **15.** Randle (1978) *Trends Biochem. Sci.* **3**, 217. **16.** Sheu, Lai & Blass (1984) *J. Neurochem.* **42**, 230. **17.** Reed & Yeaman (1987) *The Enzymes*, 3rd ed. (Boyer & Krebs, eds.), **18**, 77. **18.** Baker, Yan, Peng, Kasten & Roche (2000) *J. Biol. Chem.* **275**, 15773. **19.** Schuller & Randall (1990) *Arch. Biochem. Biophys.* **278**, 211. **20.** Robertson, Barron & Olson (1986) *J. Biol. Chem.* **261**, 76. **21.** Robertson, Barron & Olson (1990) *J. Biol. Chem.* **265**, 16814. **22.** Molin & Fites (1980) *Plant Physiol.* **66**, 308. **23.** Sanemori & Kawasaki (1980) *J. Biochem.* **88**, 223. **24.**

Howle & Fites (1991) *Physiol. Plant* **81**, 24. **25**. Suzuki & Ooba (1973) *Biochim. Biophys. Acta* **293**, 111.

2-Thiazole-DL-alanine, *See* β-(2-Thiazolyl)-DL-alanine

Thiazolidine-4-carboxylate

This heterocyclic proline analogue (FW = 133.17 g/mol; soluble in hot water and in acids and bases), also known as thiaproline, thioproline, and timonacic, is prepared upon reaction of formaldehyde and cysteine. Thiaproline exhibits the ability to reverse the transformation of certain tumor cells to a normal cellular pheotype. **Target(s):** cysteinyl-tRNA synthetase (1); γ-glutamyl kinase, *or* glutamate 5-kinase, weakly inhibited (2); proline transport (3); prolyl-tRNA synthetase, as alternative substrate (4); and pyrroline-5-carboxylate reductase (5,6). **1**. Bunjun, Stathopoulos, Graham, *et al.* (2000) *Proc. Natl. Acad. Sci. U.S.A.* **97**, 12997. **2**. Krishna & Leisinger (1979) *Biochem. J.* **181**, 215. **3**. Law & Mukkada (1979) *J. Protozool.* **26**, 295. **4**. Norris & Fowden (1972) *Phytochemistry* **11**, 2921. **5**. Strecker (1971) *Meth. Enzymol.* **17B**, 258. **6**. Krishna, Beilstein & Leisinger (1979) *Biochem. J.* **181**, 233.

1-(2-Thiazolinyl)-4-(1-(4-bromophenylsulfonyl)piperazin-4-ylcarbonyl)piperidine

This piperidine derivative (FW = 501.47 g/mol potently and selectively inhibits rat 2,3-oxidosqualene cyclase–lanosterol synthase (OSC) (IC$_{50}$ = 0.4 μM for rat and 0.11 μM for human enzyme), resulting in selective oral inhibition of rat cholesterol biosynthesis (ED$_{80}$ = 1.2 ± 0.3 mg/kg. Piperidinopyrimidine OSC inhibitors have a significantly lower pK_a than the corresponding pyridine or the previously reported quinuclidine OSC inhibitor series. **1**. Brown, Hollinshead, Stokes, *et al.* (2000) *J. Med. Chem.* **43**, 4964.

N^1-2-Thiazolylsulfanilamide, *See* Sulfathiazole

Thiazovivin

This potent protein kinase inhibitor and vasodilator (FW = 432.37 g/mol for the monohydrochloride; CAS 864082-47-3; Solubility: 15 mg/mL DMSO, <1 mg/mL H$_2$O), also known systematically as *N*-benzyl-2-(pyrimidin-4-ylamino)thiazole-4-carboxamide, targets RhoA/Rho kinases (IC$_{50}$ = 0.5 μM). Abbreviated ROCK, these kinases play important roles in mediating vasoconstriction and vascular remodeling in the pathogenesis of pulmonary hypertension. **1**. Xu Y, *et al.* (2010) *Proc. Natl. Acad. Sci. USA* **107**, 8129. **2**. Lin, *et al.* (2009) *Nat. Methods* **6**, 805. **3**. Hu, *et al.* (2011) *Blood* **117**, e109.

Thienamycin

This Δ^2-penem-class antibiotic (FW = 273.33 g/mol; CAS 59995-64-1) from *Streptomyces cattleya*, is the first member of the penem family of antibiotics containing a thioethylamine moiety on an enamine five-membered ring. The β-lactam ring is unusually sensitive to hydrolysis above pH 8 and to reaction with nucleophiles, such as hydroxylamine and cysteine. The primary amine

of the antibiotic itself may also act as a nucleophile (1). **Target(s):** β-lactamase (2-4); muramoyltetrapeptide carboxypeptidase (*or* murein tetrapeptide LD-carboxypeptidases) from *Escherichia coli* and *Gaffkya homari* (5,6); and peptidoglycan transpeptidase (7,8). *See also* Imipenem **1**. Kahan, Kahan, Goegelman, *et al.* (1979) *J. Antibiot. (Tokyo)* **32**, 1. **2**. Walsh (1983) *Trends Biochem. Sci.* **8**, 254. **3**. Hashizume, Yamaguchi & Sawai (1988) *Chem. Pharm. Bull. (Tokyo)* **36**, 676. **4**. Cullmann (1985) *Chemotherapy* **31**, 272. **5**. Templin & Höltje (1998) in *Handb. Proteolytic Enzymes* (Barrett, Rawlings & Woessner, eds.), p. 1574, Academic Press, San Diego. **6**. Hammes & Seidel (1978) *Eur. J. Biochem.* **91**, 509. **7**. Oka, Hashizume & Fujita (1980) *J. Antibiot. (Tokyo)* **33**, 1357. **8**. Moore, Jevons & Brammer (1979) *Antimicrob. Agents Chemother.* **15**, 831.

Thienodiazaborine, *See* Diazaborines

β-(2-Thienyl)alanine

The L-enantiomer of this derivative of histidine isostere and L-phenylalanine analogue (FW = 171.22 g/mol) is an alternative substrate for phenylalanyl-tRNA synthetase and phenylalanine oxidase. **Target(s):** alkaline phosphatase (1); arogenate dehydratase (2,3); chorismate mutase (4,5); 3-deoxy-7-phiosphoheptulonate synthase (6,7); histidine ammonia-lyase (8); isoleucyl-tRNA synthetase (9); and phenylalanine monooxygenase (10-13). **1**. Fishman & Sie (1971) *Enzymologia* **41**, 141. **2**. Fischer & Jensen (1987) *Meth. Enzymol.* **142**, 495. **3**. Zamir, Tiberio, Fiske, Berry & Jensen (1985) *Biochemistry* **24**, 1607. **4**. Brown & Dawes (1990) *Mol. Gen. Genet.* **220**, 283. **5**. Sugimoto & Shiio (1980) *J. Biochem.* **88**, 167. **6**. Simpson & Davidson (1976) *Eur. J. Biochem.* **70**, 509. **7**. McCandliss, Poling & Herrmann (1978) *J. Biol. Chem.* **253**, 4259. **8**. Peterkofsky & Mehler (1963) *Biochim. Biophys. Acta* **73**, 159. **9**. Stulberg & Novelli (1962) *The Enzymes*, 2nd ed. (Boyer, Lardy & Myrbäck, eds.), **6**, 401. **10**. Wapnir & Moak (1979) *Biochem. J.* **177**, 347. **11**. Beerstecher & Shive (1947) *J. Biol. Chem.* **167**, 49. **12**. Udenfriend & Cooper (1952) *J. Biol. Chem.* **194**, 503.

4S-[4-(3-Thienyl)benzyl]-D-glutamic Acid, *See* (2R,4S)-2-Amino-4-[4-(3-thienyl)benzyl]pentanedioate *and* (2R,4S)-2-Amino-4-[4-(3-thienyl)benzyl]pentanedioic Acid

4-(3-Thienyl)-2-ketobutenoate, *similar in action to: 4-(2-Thienyl)-2-ketobutenoate*

4S-(2-Thienyl)methyl-D-glutamic Acid, *See* (2R,4S)-2-Amino-4-(2-thienyl)methylpentanedioate

Thioacetamide

Thioacetamide (TA)   **TA S-Oxide**   **TA S,S-Dioxide**

This hepatotoxic/hepatocarcinogenic agent (FW = 75.13 g/mol; CAS 62-55-5; M.P. = 113-114°C; Solubility = 0.17 g/mL H$_2$O at 25°C; Symbol: TA; LD$_{50}$ of 300 mg/kg in rats and mice) was once widely used in qualitative inorganic analysis as a source for sulfide ions. Treatment of aqueous solutions of many metal cations (M = Ni, Pb, Cd, Hg) with aqueous thioacetamide yields the corresponding metal sulfide (*Reaction:* M^{2+} + CH$_3$C(S)NH$_2$ + H$_2$O → MS + CH$_3$C(O)NH$_2$ + 2 H$^+$). **Toxicity & Experimental Cirrhosis:** Cirrhosis is a slowly progressing disease, wherein healthy tissue is replaced with scar tissue, eventually preventing the liver from functioning properly. In humans, it is most often associated with excessive and chronic consumption to alcohol, but is chronic and not easily induced. Thioacetamide may be employed to induce experimental cirrhosis in rats. Even a single dose can often cause hepatic centrolobular necrosis, attended by dramatic increases in plasma transaminases and bilirubin. TA is oxidatively metabolized (bioactivated) to its *S*-oxide (FW = 91.13 g/mol; CAS 2669-09-2; IUPAC: thioacetamide-*S*-oxide; Symbol: TASO) and then to its highly reactive *S,S*-dioxide (FW = 106.12 g/mol; IUPAC: 1-iminoethanesulfinate; Symbol: TASO$_2$), with the latter capable of modifying amine-lipids and proteins (1). TA itself is nontoxic, even at concentrations up to 50 mM for 40 hours (1). TASO, however, is highly toxic to isolated hepatocytes, as indicated by LDH release, changes in cellular morphology, as well as loss of vital staining with Hoechst 33342/propidium iodide. TASO toxicity was partially blocked by the CYP2E1 inhibitors diallyl sulfide and 4-methylpyrazole and was strongly inhibited by TA (1). **Other Targets:** adenylylsulfatase (2); aldehyde dehydrogenase (3); amidase (4); δ-aminolevulinate synthase (5,6); methane monooxygenase (7); nitrile hydratase (8); trimethylamine oxidase (9). **1.** Hajovsky, Hu, Koen, *et al.* (2012) *Chem. Res. Toxicol.* **25**, 1955. **2.** Li & Schiff (1991) *Biochem. J.* **274**, 355. **3.** RPatel, Hou, Derelanko & Felix (1980) *Arch. Biochem. Biophys.* **203**, 654. **4.** Maestracci, Thiery, Bui, Arnaud & Galzy (1984) *Arch. Microbiol.* **138**, 315. **5.** Matsuura, Takizawa, Fukuda, Yoshida & Kuroiwa (1983) *J. Pharmacobiodyn.* **6**, 340. **6.** Yoshida & Neal (1978) *Biochem. Pharmacol.* **27**, 2095. **7.** Tonge, Harrison & Higgins (1977) *Biochem. J.* **161**, 333. **8.** Moreau, Azza, Arnaud & Galzy (1993) *J. Basic Microbiol.* **33**, 323. **9.** Pearson, Greenwood, Butler & Fenwick (1982) *Comp. Biochem. Physiol. C* **73**, 389.

2-Thioacetate, *See* Thioglycolate

Thiobarbituric Acids

2-Thiobarbituric acid   6-Thiobarbituric acid

Two thiobarbituric acids (FW = 144.15 g/mol) contain a single sulfur atom in replace of the oxygen atom in barbituric acid. 2-Thiobarbituric acid, also known as 4,6-dihydroxy-2-mercaptopyrimidine (decomposes at 245°C), is frequently used to detect lipid hydroperoxides and lipid oxidation as well as to quantitate lipopolysaccharides, sialic acids, and carrageenans. **Target(s):** dihydroorotate dehydrogenase, inhibited by 6-thiobarbiturate (1); γ-glutamyl transpeptidase (γ-glutamyltransferase (2,3); orotate phosphoribosyl-transferase, inhibited by 2-thiobarbiturate, K_i = 41 μM (4). **1.** Gero, O'Sullivan & Brown (1985) *Biochem. Med.* **34**, 60. **2.** Sachdev, D. S. Leahy & K. V. Chace (1983) *Biochim. Biophys. Acta* **749**, 125. **3.** Allison (1985) *Meth. Enzymol.* **113**, 419. **4.** Javaid, el Kouni & Iltzsch (1999) *Biochem. Pharmacol.* **58**, 1457.

Thiocapillarisin, *See* Capillarisin

Thiocarbamide, *See* Thiourea

Thiocholine

This sulfur-containing choline analogue (FW$_{chloride}$ = 155.69 g/mol; CAS 625-00-3) is an alternative product inhibitor for reactions catalyzed by acetylcholinesterase (1), choline kinase (2,3), choline sulfotransferase (4,5), and glycerophosphocholine cholinephosphodiesterase (6-8). **1.** Turdean, Popescu, Oniciu & Thevenot (2002) *J. Enzyme Inhib. Med. Chem.* **17**, 107. **2.** Ishidate & Nakazawa (1992) *Meth. Enzymol.* **209**, 121. **3.** Reinhardt, Wecker & Cook (1984) *J. Biol. Chem.* **259**, 7446. **4.** Orsi & Spencer (1964) *J. Biochem.* **56**, 81. **5.** Renosto & Segel (1977) *Arch. Biochem. Biophys.* **180**, 416. **6.** Sok & Kim (1995) *Neurochem. Res.* **20**, 151. **7.** Sok (1998) *Neurochem. Res.* **23**, 1061. **8.** Lee, Kim, Kim, Myung & Sok (1997) *Neurochem. Res.* **22**, 1471.

L-Thiocitrulline

This citrulline analogue (FW = 191.25 g/mol; water soluble; acid stable) is a strong inhibitor (not substrate) of nitric-oxide synthase, K_i = 60 nM for the rat neuronal enzyme and 3.6 μM for the inducible enzyme. **Target(s):** Arginase (1); nitric-oxide synthase (2-7); protein-arginine deaminase (8) **1.** Colleluori & Ash (2001) *Biochemistry* **40**, 9356. **2.** Narayanan & Griffith (1994) *J. Med. Chem.* **37**, 885. **3.** Griffith & Kilbourn (1996) *Meth. Enzymol.*

268, 375. **4.** Griffith & Stuehr (1995) *Ann. Rev. Physiol.* **57**, 707. **5.** Salerno, Frey, McMillan, *et al.* (1995) *J. Biol. Chem.* **270**, 27423. **6.** Joly, Narayanan, Griffith & Kilbourn (1995) *Brit. J. Pharmacol.* **115**, 491. **7.** Frey, Narayanan, McMillan, *et al.* (1994) *J. Biol. Chem.* **269**, 26083. **8.** McGraw, Potempa, Farley & Travis (1999) *Infect. Immun.* **67**, 3248.

Thioctate, *See* Lipoate; Dihydrolipoate

Thioctic Acid Amide, *See* Lipoamide; Dihydrolipoamide

Thiocyanate

resonance

This cyanate isostere (FW$_{Na-salt}$ = 80.07 g/mol; CAS 540-72-7; FW$_{K-salt}$ = 97.18 g/mol; CAS 333-20-0) is a potent competitive inhibitor of the thyroid sodium-iodide symporter, decreasing iodide transport into the thyroid follicle and reducing thyroxine production. Also called rhodanide, thiocyanate is formed from cyanide in the rhodanese reaction. *Note*: Inasmuch as thiocyanate shares its negative charge almost equally between its sulfur and nitrogen atoms, it acts as a nucleophile at either sulfur or nitrogen, making it an ambidentate ligand **1.** Braverman, He, Pino, *et al.* (2005) *J. Clin. Endocrinol. Metab.* **90**, 700.

Thiocyanate Radical Anion, *See* Di(thiocyanate) Radical Anion
6-Thio-7-deaza-2'-deoxyguanosine 5'-Triphosphate, *See* 7-Deaza-2'-deoxy-6-thioguanosine 5'-Triphosphate

6-Thio-2'-deoxyguanosine

This thiopurine deoxynucleoside (FW = 283.31 g/mol; CAS 789-61-7; Solubility: 59 mg/mL DMSO; <1mg/mL H$_2$O) is taken up by many cells and metabolically phosphorylated to its 5'-triphosphate, which is then bound by telomerase and incorporated into *de novo*-synthesized telomeres, leading to telomere dysfunction in cells expressing telomerase. 6-Thio-dG, but not 6-thioguanine, induces telomere dysfunction in telomerase-positive human cancer cells and hTERT-expressing human fibroblasts, but not in telomerase-negative cells. Treatment with 6-thio-dG results in rapid cell death for the vast majority of the cancer cell lines tested, whereas normal human fibroblasts and human colonic epithelial cells are largely unaffected. In A549 lung cancer cell-based mouse xenograft studies, 6-thio-dG caused a decrease in the tumor growth rate in a manner that is superior to that observed with 6-thioguanine. In addition, 6-thio-dG increased telomere dysfunction in tumor cells in vivo. Such findings suggest that 6-thio-dG may provide a new telomere-targeting anticancer strategy. *Note*: 6-thioguanosine monophosphate is further metabolized to 6-thio-2'-deoxyguanosine 5'-triphosphate by kinases and RNA reductases, which eventually may be incorporated into DNA strands during DNA replication. DNA-incorporated 6-thioguanine may also generate reactive oxygen species, which may cause additional damage to DNA, proteins and other cellular macromolecules, and thus block cellular replication. **1.** Mender, Gryaznov, Dikmen, Wright & Shay (2015) *Cancer Discov.* 5, 82.

4-Thio-2'-deoxythymidylate, *See* 4-Thiothymidylate

5-Thio-2'-deoxyuridine 5'-Triphosphate, *See* 5-Mercapto-2'-deoxyuridine 5'-Triphosphate

2,2'-Thiodiacetate and 2,2'-Thiodiacetic Acid, *See* Thiodiglycolate

2,2'-Thiodiethanol, *See* Bis(hydroxyethyl)sulfide

Thiodiglycolate (Thiodiglycolic Acid)

This dicarboxylic acid (FW$_{free-acid}$ = 150.16 g/mol; soluble in water and ethanol), also known as 2,2'-thiodiacetic acid and 2,2'-thiobis(acetic acid), is often used in the detection of copper, lead, mercury, and silver, is produced in such detoxification of breakdown products of vinyl chlorides. **Target(s):** glutamate dehydrogenase (1,2); pyrimidine-deoxynucleoside 2'-dioxygenase, *or* thymidine:2-oxoglutarate dioxygenase (3). **1**. Smith, Austen, Blumenthal & Nyc (1975) *The Enzymes*, 3rd ed. (Boyer, ed.), **11**, 293. **2**. Rogers (1971) *J. Biol. Chem.* **246**, 2004. **3**. Bankel, Lindstedt & Lindstedt (1972) *J. Biol. Chem.* **247**, 6128.

N-(1-Thiododecyl)-4α,10-dimethyl-8-aza-*trans*-decal-3β-ol

This thioamide (FW = 381.67 g/mol) inhibits cholesterol biosynthesis in HepG2 cells via inhibition of 2,3-oxidosqualene cyclase. While the initial strategy for making a high-affinity inhibitor focused on amine-containing analogues of the putative transition-state intermediate, later work determined that the amine is not required. **1**. Wannamaker, Waid, Van Sickle, *et al.* (1992) *J. Med. Chem.* **35**, 3581.

2-Thioethanesulfonate, See Coenzyme M

2-Thioflavin Mononucleotide

This FMN analogue (FW = 472.42 g/mol) inhibits rat FAD synthetase (ATP:FMN adenylyltransferase, EC 2.7.7.2), K_i = 106 μM. The C=O group at position-2, the NH group at position-3, and a five-carbon side chain at the N^{10} position seem to be most crucial for flavin substrate binding to enzyme. **1**. Bowers-Komro, Yamada & McCormick (1989) *Biochemistry* **28**, 8439.

Thioflavin T

This well-known histologic staining agent (FW = 318.86 g/mol; CAS 2390-54-7), also named 4-(3,6-dimethyl-1,3-benzothiazol-3-ium-2-yl)-*N*,*N*-dimethylaniline chloride, binds to so-called cross-β structures consisting of small and densely packed β-sheet structures, including those in amyloid- β and Tau aggregates. Upon binding to such aggregates, thioflavin T displays enhanced fluorescence and a characteristic red shift of its emission spectrum. The related compound Thioflavin S, a homogeneous mixture of compounds, resulting from methylation of dehydrothiotoluidine, is also used to stain amyloid plaques. Like Thioflavin T, it binds to amyloid fibrils – not monomers, increasing in fluorescence. Unlike Thioflavin T, however, it does not produce a characteristic shift in the excitation or emission spectra, resulting in high background fluorescence and making ioflavin S unsuitable for quantifying of fibril formation. Note that thioflavin T also dimerizes, a property allowing it to intercalate into DNA as well as bind to the mior and major grooves of DNA (2). **1**. LeVine (1999) *Methods in Enzymology.* **309**, 274. **2**. Biancardi, Biver, Burgalassi *et al.* (2014) *Phys. Chem. Chem. Phys.* **16**, 20061.

5-Thio-L-fucose

This sugar analogue (FW = 182.22 g/mol) inhibits α-L-fucosidase, K_i = 0.05 mM. Peracetylated 5-thio-L-fucose (5T-Fuc) is taken up by cancer cells and converted to GDP-5T-Fuc, which blocks α1,3-fucosyltrasferase (FUT) activity and limits sLe^X presentation on HepG2 cells, with a low-μM EC$_{50}$ (2). 5T-Fuc also impairs adhesion to immobilized adhesion molecules and human endothelial cells (2). 5T-Fuc is thus a useful probe for modulating sLe^X levels in cells, allowing one to evaluate the consequences of inhibiting FUT-mediated sLe^X formation (2). (*Note*: sLe^X is a tetrasaccharide that serves as a ligand for cell adhesion proteins known as selectins, enabling adhesion of leukocytes and cancer cells to endothelial cells within capillaries, resulting in their extravasation into tissues. The last step in sLe^X biosynthesis is the FUT-catalyzed transfer of an L-fucose residue to carbohydrate acceptors. GDP-5T-Fuc is not transferred by either FUT3 or FUT7.) **1**. Dumas, Kajimoto, Liu, *et al.* (1992) *Bioorg. Med. Chem. Lett.* **2**, 33. **2**. Zandberg, Kumarasamy, Pinto & Vocadlo (2012) *J. Biol. Chem.* **287**, 40021.

5-Thio-D-glucose

5-thio-α-D-glucopyranose

This sulfur-containing D-glucose analogue (FW = 196.22 g/mol; melting point = 135-136°C), also known as 5-deoxy-5-thio-D-glucose, interferes with glucose transport (1) and, other than a hormone or alkylating reagent, was the first reagent to alter spermatogenesis. **Target(s):** amylo-α-1,6-glucosidase/4-α-glucanotransferase, *or* glycogen debranching enzyme (2); catechol oxidase (2); glucose transport (3,4); β-glucosidase (5); hexokinase (3,6-8); protein-$N^π$-phosphohistidine:sugar phosphotransferase (9); and xylose isomerase (10). **1**. Gillard & Nelson (1977) *Biochemistry* **16**, 3978. **2**. Prabhakaran (1976) *Experientia* **32**, 152. **3**. Tielens, Houweling & Van den Bergh (1985) *Biochem. Pharmacol.* **34**, 3369. **4**. Pitts, Chemielewski, Chen, el-Rahman & Whistler (1975) *Arch. Biochem. Biophys.* **169**, 384. **5**. Dale, Ensley, Kern, Sastry & Byers (1985) *Biochemistry* **24**, 3530. **6**. Wilson & Chung (1989) *Arch. Biochem. Biophys.* **269**, 517. **7**. Machado de Domenech & Sols (1980) *FEBS Lett.* **119**, 174. **8**. Racagni & Machado de Domenech (1983) *Mol. Biochem. Parasitol.* **9**, 181. **9**. Hüdig & Hengstenberg (1980) *FEBS Lett.* **114**, 103. **10**. Rangarajan & Hartley (1992) *Biochem. J.* **283**, 223.

Thioglucosoaurate, See Aurothioglucose

6-Thioguanine

This protypical anticancer and immunosuppressive agent (FW = 167.19 g/mol; CAS 154-42-7; Water-insoluble), also known as 2-amino-6-mercaptopurine, is incorporated into DNA, where it inhibits DNA replication. Thioguanine is a prodrug that requires uptake and conversion to its 6-thio-GMP through the purine salvage enzyme, hypoxanthine:guanine phosphoribosyltransferase. 6-TG is often used to treat both immune disorders and leukemia during their rapid proliferative phases. 6-TG displays cytotoxic and antineoplastic properties and disrupts cytosine methylation by DNA methyltransferases, after incorporation into DNA. 6-Thioguanine allso selectively kills BRCA2-defective tumors in a xenograft model. It also facilitates proteasome-mediated degradation of DNA (cytosine-5)-methyltransferase 1 (DNMT1). The immunosuppressive action of this thiopurine arises because, like liver, immune cells are major sites of de novo nucleotide biosynthesis, the latter needed for rapid clonal expansion of these cell types. Thioguanine was first synthesized by Elion and Hitchings (1955), who shared the 1988 Nobel Prize in Medicine and Physiology for their germinal work in cancer chemotherapy. **Target(s):** cytidine deaminase (2); guanosine phosphorylase, *or* purine-nucleoside phosphorylase, by alternative product inhibition (3); hypoxanthine(guanine) phosphoribosyltransferase, as weak alternative substrate (4-6); tRNA-guanine transglycosylase (7); xanthine oxidase (8,9); and xanthine phosphoribosyltransferase (4). **1.** Elion & Hitchings (1955) *J. Amer. Chem. Soc.* **77**, 1676. **2.** Kara, Bartova, Ryba, *et al.* (1982) *Collect. Czech. Chem. Commun.* **47**, 2824. **3.** Baker & Schaeffer (1971) *J. Med. Chem.* **14**, 809. **4.** Naguib, Iltzsch, el Kouni, Panzica & el Kouni (1995) *Biochem. Pharmacol.* **50**, 1685. **5.** Walter & König (1974) *Tropenmed. Parasitol.* **25**, 227. **6.** Nussbaum & Caskey (1981) *Biochemistry* **20**, 4584. **7.** Farkas, Jacobson & Katze (1984) *Biochim. Biophys. Acta* **781**, 64. **8.** Baker & Hendrickson (1967) *J. Pharmaceu. Sci.* **56**, 955. **9.** Kela & Vijayvargiya (1981) *Experientia* **37**, 175.

6-Thioguanosine

This sulfur-containing guanosine analogue (FW 299.31 g/mol; CAS 85-31-4; λ_{max} at 257 ($\varepsilon = 8820$ $M^{-1}cm^{-1}$) and 342 nm ($\varepsilon = 24800$ $M^{-1}cm^{-1}$)), also known as 6-mercaptoguanosine, inhibits adenosine deaminase (1) and sorbitol dehydrogenase (2). **1.** Agarwal & Parks, Jr. (1978) *Meth. Enzymol.* **51**, 502. **2.** Lindstad & McKinley-McKee (1996) *Eur. J. Biochem.* **241**, 142.

6-Thioguanosine 5'-Monophosphate

This sulfur-containing guanine nucleotide analogue ($FW_{free-acid}$ = 395.4 g/mol), also known as 6-thioguanylate and 2-amino-6-mercaptopurine ribonucleoside 5'-monophosphate, inhibits adenylosuccinate synthetase (1,2), amidophospho-ribosyltransferase (3) GMP reductase, the latter *via* formation of a disulfide-linked enzyme adduct (4,5), GMP synthetase (6), and guanylate kinase (7-10). **1.** Spector, Jones & Elion (1979) *J. Biol. Chem.* **254**, 8422. **2.** Spector & Miller (1976) *Biochim. Biophys. Acta* **445**, 509. **3.** Hill & Bennett, Jr. (1969) *Biochemistry* **8**, 122. **4.** Shaw (1970) *The Enzymes*, 3rd ed. (Boyer, ed.), **1**, 91. **5.** Brox & Hampton (1968) *Biochemistry* **7**, 398. **6.** Spector, Miller, Fyfe & Krenitsky (1974) *Biochim. Biophys. Acta* **370**, 585. **7.** Agarwal & Parks, Jr. (1975) *Biochem. Pharmacol.* **24**, 791. **8.** Anderson (1973) *The Enzymes*, 3rd ed. (Boyer, ed.), **9**, 49. **9.** Miech, R. York & Parks, Jr. (1969) *Mol. Pharmacol.* **5**, 30. **10.** Buccino, Jr., & Roth (1969) *Arch. Biochem. Biophys.* **132**, 49.

6-Thioguanylate, See *6-Thioguanosine 5'-Monophosphate*

6-Thiohypoxanthine, See *6-Mercaptopurine*

6-Thioinosine

This inosine analogue (FW = 284.31 g/mol; CAS 574-25-4), also known as 6-mercaptopurine ribonucleoside, is formed upon conversion of 6-mercaptopurine by guanine(hypoxanthine) phosphoribosyltransferase. **Target(s):** adenosine deaminase, K_i = 370 μM for calf muscle enzyme (1-3); adenosine kinase (4,5); GMP synthetase, weakly (6); hypoxanthine(guanine) phosphoribosyltransferase (7); inosine nucleosidase

(8); methionine *S*-adenosyltransferase (9); 5'-nucleotidase (10,11); and sorbitol dehydrogenase, weakly (12). **1.** Agarwal & Parks, Jr. (1978) *Meth. Enzymol.* **51**, 502. **2.** Cory & Suhadolnik (1965) *Biochemistry* **4**, 1729 and 1733. **3.** Baker (1967) *Design of Active-Site-Directed Irreversible Enzyme Inhibitors*, Wiley, New York. **4.** Long & Parker (2006) *Biochem. Pharmacol.* **71**, 1671. **5.** Kidder (1982) *Biochem. Biophys. Res. Commun.* **107**, 381. **6.** Spector & Beecham III (1975) *J. Biol. Chem.* **250**, 3101. **7.** Schimandle, Mole & Sherman (1987) *Mol. Biochem. Parasitol.* **23**, 39. **8.** Le Floc'h & Lafleuriel (1981) *Phytochemistry* **20**, 2127. **9.** Berger & Knodel (2003) *BMC Microbiol.* **3**, 12. **10.** Bolling, Olszanski, Bove & Childs (1992) *J. Thorac. Cardiovasc. Surg.* **103**, 73. **11.** Carter & Tipton (1986) *Phytochemistry* **25**, 33. **12.** Lindstad & McKinley-McKee (1996) *Eur. J. Biochem.* **241**, 142.

6-Thioinosine 5'-Monophosphate

This sulfur-containing nucleotide analogue ($FW_{free-acid}$ = 364.28 g/mol), also known 6-mercaptopurine ribonucleoside 5'-monophosphate and 6-thio-IMP, strongly inhibits IMP dehydrogenase. **Target(s):** adenylosuccinate lyase, especially in the presence of Cu^{2+} or Hg^{2+} (1-3); adenylosuccinate synthetase (2-9); amidophospho-ribosyltransferase (10,11); GMP reductase (12,13); IMP cyclohydrolase (14,15); and IMP dehydrogenase (3,4,12,16,17). **1.** Ratner (1972) *The Enzymes*, 3rd ed. (Boyer, ed.), **7**, 167. **2.** Atkinson, Morton & Murray (1964) *Biochem. J.* **92**, 398. **3.** Baker (1967) *Design of Active-Site-Directed Irreversible Enzyme Inhibitors*, Wiley, New York. **4.** Webb (1966) *Enzyme and Metabolic Inhibitors*, vol. **2**, Academic Press, New York. **5.** Hampton (1962) *Fed. Proc.* **21**, 370. **6.** Spector, Jones & Elion (1979) *J. Biol. Chem.* **254**, 8422. **7.** Van der Weyden & Kelly (1974) *J. Biol. Chem.* **249**, 7282. **8.** Spector & Miller (1976) *Biochim. Biophys. Acta* **445**, 509. **9.** Stayton, Rudolph & Fromm (1983) *Curr. Top. Cell. Regul.* **22**, 103. **10.** Hill & Bennett, Jr. (1969) *Biochemistry* **8**, 122. **11.** Holmes, McDonald, McCord, Wyngaarden & Kelley (1973) *J. Biol. Chem.* **248**, 144. **12.** Shaw (1970) *The Enzymes*, 3rd ed. (Boyer, ed.), **1**, 91. **13.** Brox & Hampton (1968) *Biochemistry* **7**, 398. **14.** Szabados, Hindmarsh, Phillips, Duggleby & Christopherson (1994) *Biochemistry* **33**, 14237. **15.** Christopherson, Williams, Schoettle, *et al.* (1995) *Biochem. Soc. Trans.* **23**, 888. **16.** Atkinson, Morton & Murray (1963) *Biochem. J.* **89**, 167. **17.** Hampton & Nomura (1967) *Biochemistry* **6**, 679.

6-Thioinosine 5'-Triphosphate

This sulfur-containing ITP analogue ($FW_{free-acid}$ = 524.24 g/mol) inhibits DNA-directed RNA polymerase (1), leucyl-tRNA synthetase (2), and RNA ligase (3). **1.** Kawahata, Chuang, Holmberg, Osburn & Chuang (1983) *Cancer Res.* **43**, 3655. **2.** Marutzky, Flossdorf & Kula (1976) *Nucl. Acids Res.* **3**, 2067. **3.** Iuodka, Labeikite & Sasnauskene (1993) *Biokhimiia* **58**, 857.

Thiolacetate, See *Thioglycolate*

Thiolactomycin

This broad-spectrum, naturally occurring antibiotic (FW = 210.30 g/mol), first isolated from *Nocardia*, is active against both Gram-positive and Gram-negative bacteria as well as *Mycobacterium tuberculosis*. Thiolactomycin (TLM) inhibits β-ketoacyl-[acyl-carrier-protein] synthase in fatty-acid synthase II and the elongation step involved in the synthesis of α-mycolates and oxygenated mycolates (1-3). **Target(s):** [acyl-carrier-protein] *S*-acetyltransferase (4,5); fatty-acid synthase, acyl carrier-protein dependent (1,6-9); fatty-acyl-CoA synthase (9); *N*-hydroxyarylamine *O*-acetyltransferase (10); β-ketoacyl-[acyl-carrier-protein] synthase I (2,3,11,12); β-ketoacyl-[acyl-carrier protein] synthase II (1,2,12-15); and β-ketoacyl-[acyl-carrier-protein] synthase III (4,11,12,16-20). **1**. Slayden, Lee, Armour, *et al.* (1996) *Antimicrob. Agents Chemother.* **40**, 2813. **2**. Kremer, Douglas, Baulard, *et al.* (2000) *J. Biol. Chem.* **275**, 16857. **3**. Schaeffer, Agnihotri, Volker, *et al.* (2001) *J. Biol. Chem.* **276**, 47029. **4**. Gulliver & Slabas (1994) *Plant Mol. Biol.* **25**, 179. **5**. Lobo, Florova & Reynolds (2001) *Biochemistry* **40**, 11955. **6**. Hayashi, Yamamoto, Sasaki, Kawaguchi & Okazaki (1983) *Biochem. Biophys. Res. Commun.* **115**, 1108. **7**. Hayashi, Yamamoto, Sasaki, Okazaki & Kawaguchi (1984) *J. Antibiot.* **37**, 1456. **8**. Jones, Dancer & Harwood (1994) *Biochem. Soc. Trans.* **22**, 258S. **9**. Jones, Herbert, Rutter, Dancer & Harwood (2000) *Biochem. J.* **347**, 205. **10**. Wang, Soisson, Young, *et al.* (2006) *Nature* **441**, 358. **11**. Khandekar, Gentry, Van Aller, *et al.* (2001) *J. Biol. Chem.* **276**, 30024. **12**. Price, Choi, Heath, *et al.* (2001) *J. Biol. Chem.* **276**, 6551. **13**. Saito, Shinohara, Kamataki & Kato (1985) *Arch. Biochem. Biophys.* **239**, 286. **14**. Sridharan, Wang, Brown, *et al.* (2007) *J. Mol. Biol.* **366**, 469. **15**. Lack, Homberger-Zizzari, Folkers, Scapozza & Perozzo (2006) *J. Biol. Chem.* **281**, 9538. **16**. Castillo & Perez (2008) *Mini Rev. Med. Chem.* **8**, 36. **17**. He & Reynolds (2002) *Antimicrob. Agents Chemother.* **46**, 1310. **18**. Choi, Kremer, Besra & Rock (2000) *J. Biol. Chem.* **275**, 28201. **19**. Jones, Gane, Herbert, *et al.* (2003) *Planta* **216**, 752. **20**. Han, Lobo & Reynolds (1998) *J. Bacteriol.* **180**, 4481.

Thiomersal, *See* Thimerosal

Thiomethyl β-D-Galactopyranoside, *See* Methyl Thio-β-D-galactopyranoside

2-Thiomethyl-6-phenyl-4-(4'-hydroxybutyl)-1,2,4,-triazole-(5,1-c)(1,2,4)triazine-7-one Triphosphate

This novel acyclic nucleotide analogue (FW = 571.33 g/mol) inhibits DNA polymerase β (K_i = 68 μM, fully competitive (FC) versus dTTP), and DNA polymerase λ (K_i = 2.8 μM, FC versus dTTP), polymerase λ^{Y505A} (K_i = 1.8 μM, FC versus dTTP). **1**. Crespan, Alexandrova, Khandazhinskaya, *et al.* (2007) *Nucl. Acids Res.* **35**, 45.

Thiomorpholide Amido Ferrocene, *See* Ferrocenylthioacetylmorpholide

Thio-NAD⁺, *See* Thionicotinamide Adenine Dinucleotide

Thio-NADP⁺, *See* Thionicotinamide Adenine Dinucleotide Phosphate

Thioneine, *See* Ergothioneine

Thionicotinamide

This nicotinamide analogue (FW = 138.19 g/mol; decomposes at 190-191°C; water-soluble; λ_{max} = 400 nm), also known as 3-pyridinethiocarboxamide, is an alternative substrate and competitive inhibitor for nicotinamide phosphoribosyltransferase. **Target(s):** NAD⁺ ADP-ribosyltransferase (poly(ADP-ribose) polymerase, IC₅₀ = 1.8 mM (1); nicotinamidase (2); nicotinamide phosphoribosyltransferase, as alternative substrate (3,4). **1**. Banasik, Komura, Shimoyama & Ueda (1992) *J. Biol. Chem.* **267**, 1569. **2**.

Johnson & Gadd (1974) *Int. J. Biochem.* **5**, 633. **3**. Dietrich (1971) *Meth. Enzymol.* **18B**, 144. **4**. Dietrich & Muniz (1972) *Biochemistry* **11**, 1691.

Thionicotinamide Adenine Dinucleotide

This sulfur analogue of NAD⁺ (FW$_{free-acid}$ = 680.50 g/mol; often abbreviated S-NAD⁺) is an alternative substrate for many oxidoreductases (*e.g.*, glutamate dehydrogenase). In most of the cases, the reaction velocity is slower with this modified coenzyme. Note that the reduced form, *i.e.* S-NADH, has an absorbance maxima at 400 nm, offering a convenient assay. **Target(s):** ADP-ribose diphosphatase (1); alcohol dehydrogenase (2,3); aldehyde dehydrogenase (4,5); glucose-6-phosphate dehydrogenase (6); L-lactate dehydrogenase (7-9); NAD⁺ nucleosidase, *or* NADase (10-13); NMN nucleosidase (14); nicotinamide phosphoribosyltransferase (15,16); 1-pyrroline-5-carboxylate reductase (17); rubredoxin:NAD⁺ reductase (18); and UDP-glucuronate decarboxylase (19). **1**. Wu, Lennon & Suhadolnik (1978) *Biochim. Biophys. Acta* **520**, 588. **2**. Rudolph & Fromm (1970) *Biochemistry* **9**, 4660. **3**. Purich, Fromm & Rudolph (1973) *Adv. Enzymol.* **39**, 249. **4**. Monder, Purkaystha & Pietruszko (1982) *J. Steroid Biochem.* **17**, 41. **5**. Sidhu & Blair (1975) *J. Biol. Chem.* **250**, 7899. **6**. Anderson, Wise & Anderson (1997) *Biochim. Biophys. Acta* **1340**, 268. **7**. Schwert & Winer (1963) *The Enzymes*, 2nd ed. (Boyer, Lardy & Myrbäck, eds.), **7**, 142. **8**. Chang, Huang & Chiou (1991) *Arch. Biochem. Biophys.* **284**, 285. **9**. Okabe, Hayakawa, Hamada & Koike (1968) *Biochemistry* **7**, 79. **10**. Everse, Everse & Simeral (1980) *Meth. Enzymol.* **66**, 137. **11**. Okayama, Ueda & Hayaishi (1980) *Meth. Enzymol.* **66**, 151. **12**. Nakazawa, Ueda, Honjo, *et al.* (1968) *Biochem. Biophys. Res. Commun.* **32**, 143. **13**. Stathakos, Isaakidou & Thomou (1973) *Biochim. Biophys. Acta* **302**, 80. **14**. Foster (1981) *J. Bacteriol.* **145**, 1002. **15**. Dietrich (1971) *Meth. Enzymol.* **18B**, 144. **16**. Dietrich & Muniz (1972) *Biochemistry* **11**, 1691. **17**. Krishna, Beilstein & Leisinger (1979) *Biochem. J.* **181**, 223. **18**. Ueda & Coon (1972) *J. Biol. Chem.* **247**, 5010. **19**. Ankel & Feingold (1966) *Meth. Enzymol.* **8**, 287.

Thionicotinamide Adenine Dinucleotide Phosphate

This sulfur-containing NADP⁺ analogue (FW$_{free-acid}$ = 759.47 g/mol) is a weak alternative substrate for many oxidoreductases (*e.g.*, glutamate dehydrogenase and transhydrogenase). Note that the reduced form, *i.e.* S-NADPH, has an absorbance maxima at 400 nm. S-NADP⁺ also blocks nicotinate adenine dinucleotide phosphate-induced calcium release (1). **Target(s):** glucose-6-phosphate dehydrogenase (2,3); nicotinate adenine dinucleotide phosphate-induced calcium release (1); and 1-pyrroline-5-carboxylate reductase (4). **1**. Gupta, Quirk, Venning, *et al.* (1998) *Biochim. Biophys. Acta* **1409**, 25. **2**. Noltmann & Kuby (1963) *The Enzymes*, 2nd ed. (Boyer, Lardy & Myrbäck, eds.), **7**, 223. **3**. Anderson, Wise & Anderson (1997) *Biochim. Biophys. Acta* **1340**, 268. **4**. Krishna, Beilstein & Leisinger (1979) *Biochem. J.* **181**, 223.

Thionine

This chromatin- and mucin-staining thiazine dye (FW$_{chloride}$ = 263.75 g/mol; CAS 78338-22-4; λ_{max} = 602.5 nm), also known as thionin, Lauth's violet, and 3,7-diaminophenothiazine (also formerly referred to as 3,6-

diaminophenothiazine), is structurally similar to proflavine. Thionine is a blackish-green solid that is soluble in hot water, initially producing a blue solution that eventually turns violet. Solutions become bluer upon addition of HCl. *Note*: (a) Care must be exercised when encountering this term in the literature; (b) thionin also refers to a class of small and toxic plant proteins; and (c) thionein is the apoprotein of metallothionein and would be expected to inhibit a number of metalloenzymes. **Target(s):** glutathione-disulfide reductase (1); and trypsin (2-4). **1.** Luond, McKie, Douglas, Dascombe & Vale (1998) *J. Enzyme Inhib.* **13**, 327. **2.** Walsh (1970) *Meth. Enzymol.* **19**, 41. **3.** Glazer (1967) *J. Biol. Chem.* **242**, 3326. **4.** Melo, Rigden, Franco, *et al.* (2002) *Proteins Struct. Funct. Genet.* **48**, 311.

Thio-o-nitrophenyl β-D-Galactoside, *See o-Nitrophenyl β-D-Thiogalactoside*

1-(2'-Thionyl)-3,3,3-trifluoroacetone, *See 1-(2'-Thenoyl)-3,3,3-trifluoroacetone*

Thiooxamilate, *See Thiooxanilate*

2-Thio-6-oxo-1,6-dihydropyrimidine

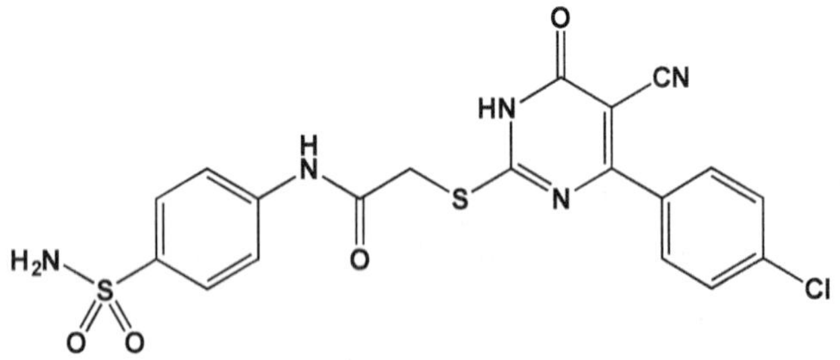

This dihydropyrimidine (FW = 475.94 g/mol) inhibits lactate dehydrogenase isozyme A, or LDH_A (IC_{50} = 8.8 μM), an enzyme believed to support glycolysis in tumor cells. Consistent with its known ordered ternary complex kinetic mechanism (with the coenzyme binding first), the inhibitor required simultaneous binding of the NADH cofactor. Although bound within the enzyme's active site, this and structurally related inhibitors made no direct contacts with the majority of enzyme residues involved in LDH_A catalysis. **1.** Dragovich, Fauber, Corson, *et al.* (2013) *Bioorg. Med. Chem. Lett.* **23**, 3186.

2-Thio-6-oxypurine, *See 2-Thioxanthine*

Thiopental

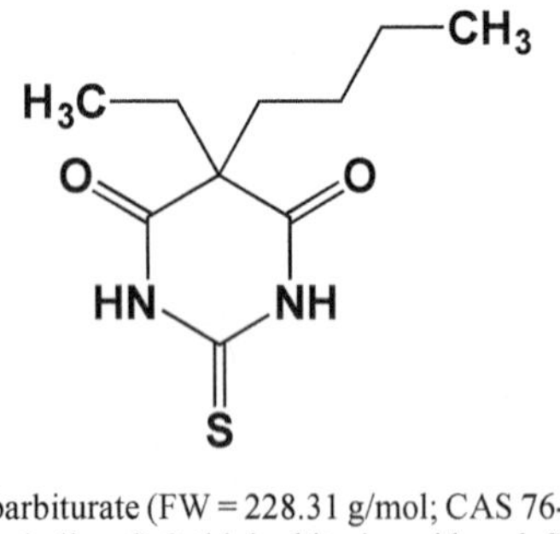

This short-acting barbiturate (FW = 228.31 g/mol; CAS 76-75-5), also known as 5-ethyl-5-(1-methylbutyl)-2-thiobarbituric acid and Sodium Pentothal® (Abbott), is a used to rapidly induce anesthesia. Thiopental crosses the blood brain barrier rapidly, and its short duration of action arises from its redistribution to muscle and fat tissue, mainly the latter. Hepatic metabolism forms pentobarbital, 5-ethyl-5-(1'-methyl-3'-hydroxybutyl)-2-thiobarbiturate, and 5-ethyl-5-(1'-methyl-3'-carboxypropyl)-2-thiobarbiturate. **Target(s):** acetylcholinesterase (1); γ-aminobutyrate aminotransferase (2); ATP/K^+ channel (3); Na^+/K^+-exchanging ATPase (4,5); $α_7$ nicotinic acetylcholine receptor (6); and phosphofructokinase (7). **1.** Pastuszko (1980) *Neurochem. Res.* **5**, 769. **2.** Cheng & Brunner (1979) *Biochem. Pharmacol.* **28**, 105. **3.** Kozlowski & Ashford (1991) *Brit. J. Pharmacol.* **103**, 2021. **4.** Kutchai, Geddis & Farley (1999) *Pharmacol. Res.* **40**, 469. **5.** Mazzanti, Rabini, Staffolani, *et al.* (1990) *Biochem. Biophys. Res. Commun.* **173**, 1248. **6.** Coates, Mather, Johnson & Flood (2001) *Anesth. Analg.* **92**, 930. **7.** Bielicki, Krieglstein & Wever (1980) *Arzneimittelforschung* **30**, 594.

Thiopentone Sodium, *See Thiopental*

Thiopeptin

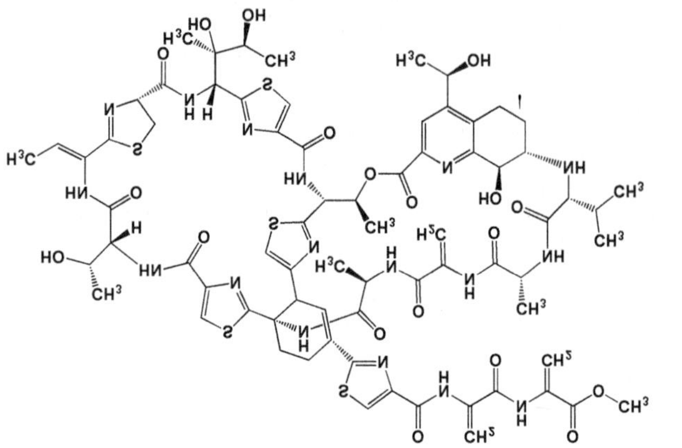

This sulfur-containing peptide antibiotic complex ($FW_{Thipeptin-B}$ = 1670.05 g/mol; CAS 37339-66-5) potently inhibits protein biosynthesis. The complex consists of several components, .the major component being thiopeptin B, which contains dehydroalanyl residues. inhibits elongation factor (EF)-Tu-dependent GTP hydrolysis and binding of aminoacyl-tRNA to the ribosome. The peptidyl transferase-catalyzed puromycin reaction is not significantly affected by the antibiotic. Thiopeptin inhibits EF-G-associated GTPase reaction, and translocation of peptidyl-tRNA and mRNA from the acceptor site to the donor site. Thiopeptin Bb is a faint yellow solid that is insoluble in water and hexane and soluble in methanol, acetone, and ethyl acetate. *See also Thiostrepton* **Target(s):** protein biosynthesis, elongation and GTPase activity (1-4). **1.** Jiménez (1976) *Trends Biochem. Sci.* **1**, 28. **2.** Lucas-Lenard & Beres (1974) *The Enzymes*, 3rd ed. (Boyer, ed.), **10**, 53. **3.** Liou, Kinoshita & Tanaka (1976) *Jpn. J. Microbiol.* **20**, 233. **4.** Kinoshita, Liou & Tanaka (1971) *Biochem. Biophys. Res. Commun.* **44**, 859.

Thiophosphate, Sodium

This corrosive inorganic reagent (Na_3SPO_3; FW = 180.01 g/mol; CAS 12609-84-6), also called sodium phosphorothioate, cleaves disulfide bonds in proteins and also reacts with cystine and glutathione disulfide (1). Thiophosphate is also an orthophosphate analogue and competitively inhibits such enzymes such protein-tyrosine phosphatase (2). **Target(s):** alkaline phosphatase (3,4); cytochrome *c* oxidase (5); inositol-phosphatae phosphatase (6); photophosphorylation (7); and protein-tyrosine phosphatase (2). **1.** Neumann & Smith (1967) *Arch. Biochem. Biophys.* **122**, 354. **2.** Zhao (1996) *Biochem. Biophys. Res. Commun.* **218**, 480. **3.** Denier, Vergnes, Brisson-Lougarre, *et al.* (1996) *Ann. Clin. Biochem.* **33**, 215. **4.** Reid & Wilson (1971) *The Enzymes*, 3rd ed. (Boyer, ed.), **4**, 373. **5.** Manon, Camougrand & Guerin (1989) *J. Bioenerg. Biomembr.* **21**, 387. **6.** Fauroux, MLee, Cullis, *et al.* (2002) *J. Med. Chem.* **45**, 1363. **7.** Izawa & Good (1972) *Meth. Enzymol.* **24**, 355.

Thioproline, *See Thiazolidine-4-carboxylate*

2-Thiopurine, *See 2-Mercaptopurine*

6-Thiopurine, *See 6-Mercaptopurine*

Thioridazine

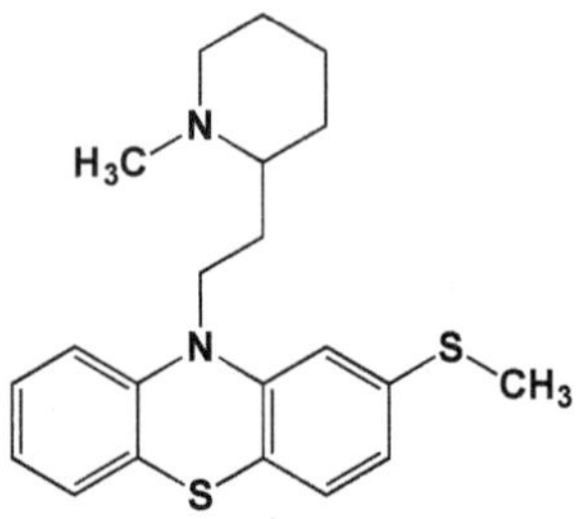

This phenothiazine antipsychotic and peroxisomal β-oxidation inhibitor ($FW_{free-base}$ = 370.58 g/mol; CAS 50-52-2) is a dopamine D_2 antagonist and Ca^{2+} channel blocker (1). The hydrochloride salt is soluble in water (one part in nine) and has an ultraviolet spectra (in 0.1 N HCl, $λ_{max}$ = 264 (ε = 42371 $M^{-1}cm^{-1}$) and 305 nm (ε = 5495 $M^{-1}cm^{-1}$). Phenothiazine antipsychotics such as thioridazine also prevent Tousled-like kinase- (TLK-) mediated phosphorylation of Rad9(S328), impairing checkpoint recovery and double-strand break (*or* DSB) repair as well as potentiating radiomimetic tumor killing, especially when used in combination with other chemotherapeutics (1). The Tousled-like kinases (TLKs) are involved in chromatin assembly, DNA repair, and transcription, thereby maintaining genomic stability. Two human TLKs exist, often with dysregulated gene expression in cancer, phosphorylate Asf1 and Rad9 and regulate DSB repair and the DNA damage response (*or* DDR). *In vivo* administration of thioridazine, inhibits hepatic peroxisomal β-oxidation in mice. In starving animals, 3-hydroxybutyrate concentration is not decreased by thioridazine treatment, suggesting its use as a tool for simulating pathological conditions in which peroxisomal β-oxidation is impaired. **Receptor Binding:** Thioridazine also shows substantial affinity for dopamine receptors (D_1, IC_{50} = 94.5 nM; D_2, IC_{50} = 0.4 nM (responsible for antipsychotic action); D_3, IC_{50} = 1.5 nM; D_4, IC_{50} = 1.5 nM; and D_1, IC_{50} = 258 nM) (3). **Target(s):** acetylcholinesterase (4); adenosyl-methionine decarboxylase (5); adenylate cyclase, dopamine-stimulated (6); aryl hydrocarbon hydroxylase (17); ATPase (8,9); calcineurin (10); cAMP phosphodiesterase (11); cholesterol 7α-hydroxylase (12); cholinesterase, *or* butyrylcholinesterase (13); CYP1A2 (14); CYP3A2 (14); CYP2D6 (15); cytochrome *c* oxidase (16); cytochrome P450-dependent activities, such as 7-pentoxyresorufin *O*-depentylase and 7-ethoxyresorufin *O*-deethylase (17); Na^+/K^+-exchanging ATPase (8,18); pyruvate dehydrogenase (19); [pyruvate dehydrogenase (acetyl-transferring)]-phosphatase (20); retinyl ester hydrolase (21); and peroxisomal β-oxidation (22). **1.** Miller (2009) *Curr. Neuropharmacol.* **7**, 315. **2.** Ronald, Awate, Rath, *et al.* (2013) *Genes Cancer* **4**, 39. **3.** Brunton, Chabner & Knollman (2010). *Goodman and Gilman's The Pharmacological Basis of Therapeutics*, 12th ed., New York: McGraw-Hill Professional. **4.** Desire & Blanchet (1975) *C. R. Acad. Sci. Hebd. Seances Acad. Sci. D* **281**, 1135. **5.** Hietala, Lapinjoki & Pajunen (1984) *Biochem. Int.* **8**, 245. **6.** Iversen (1975) *Science* **188**, 1084. **7.** Stohs & Wu (1982) *Pharmacology* **25**, 237. **8.** Shacoori, Leray, Guenet, *et al.* (1988) *Res. Commun. Chem. Pathol. Pharmacol.* **59**, 161. **9.** Corbett, Christian, Monti & McClain (1974) *Res. Commun. Chem. Pathol. Pharmacol.* **8**, 607. **10.** Mukai, Ito, Kishima, Kuno & Tanaka (1991) *J. Biochem.* **110**, 402. **11.** Janiec, Pytlik & Piekarska (1980) *Pol. J. Pharmacol. Pharm.* **32**, 297. **12.** Holsztynska & Waxman (1987) *Arch. Biochem. Biophys.* **256**, 543. **13.** Bailey & Briggs (2005) *Amer. J. Clin. Pathol.* **124**, 226. **14.** Daniel, Syrek, Rylko & Kot (2001) *Pol. J. Pharmacol.* **53**, 615. **15.** Shin, Soukhova & Flockhart (1999) *Drug Metab. Dispos.* **27**, 1078. **16.** Moubarak & Muhoberac (1991) *Biochem. Biophys. Res. Commun.* **179**, 1063. **17.** Murray & Reidy (1989) *Biochem. Pharmacol.* **38**, 4359. **18.** Davis & Brody (1966) *Biochem. Pharmacol.* **15**, 703. **19.** Sacks, Esser & Sacks (1991) *Biol. Psychiatry* **29**, 176. **20.** Bak, Huh, Hong & Song (1999) *Biochem. Mol. Biol. Int.* **47**, 1029. **21.** Schindler (2001) *Lipids* **36**, 543. **22.** Van den Branden & Roels (1985) *FEBS Lett.* **187**, 331.

Thiosemicarbazide

This thioamide (FW = 91.14 g/mol; CAS 79-19-6; MP = 182-184°C; soluble in water and ethanol), also called hydrazine carbothioamide, is used in the detection and extraction of metal ions. Semicarbazide derivatives (*e.g.*, semicarbazones and thiosemicarbazones) display antiviral, antiinfective and antineoplastic actions, presumably through binding to copper or iron within cells. **Target(s):** alanine aminotransferase (1); alcohol dehydrogenase, yeast (2); aromatic-L-amino-acid decarboxylase, *or* dopa decarboxylase (3); aspartate aminotransferase (4-6); branched-chain amino-acid aminotransferase (7); catalase (8,9); glutamate decarboxylase (10-12); leucine aminotransferase (13); lysyl oxidase, *or* protein-lysine 6-oxidase, collagen cross-linking (14,15); methane monooxygenase (16); phosphatidylserine decarboxylase (17); and RNA-directed DNA polymerase (18). **1.** Turano, Bossa, Fasella & Rossi Fanelli (1966) *Enzymologia* **30**, 185. **2.** Sund & Theorell (1963) *The Enzymes*, 2nd ed. (Boyer, Lardy & Myrbäck, eds.), **7**, 25. **3** Borri Voltattorni, Giartosio & Turano (1987) *Meth. Enzymol.* **142**, 179. **4.** Sizer & Jenkins (1962) *Meth. Enzymol.* **5**, 677. **5.** Jenkins & D'Ari (1966) *Biochemistry* **5**, 2900. **6.** Lain-Guelbenzu, Muñoz-Blanco & Cárdenas (1990) *Eur. J. Biochem.* **188**, 529. **7.** Taylor & Jenkins (1966) *J. Biol. Chem.* **241**, 4396. **8.** Nicholls & Schonbaum (1963) *The Enzymes*, 2nd. ed. (Boyer, Lardy & Myrbäck, eds.) **8**, 147. **9.** Feinstein, Seaholm & Ballonoff (1964) *Enzymologia* **27**, 30. **10.** Tapia & Salazar (1991) *Neurochem. Res.* **16**, 263. **11.** Diaz-Munoz & Tapia (1988) *J. Neurosci. Res.* **20**, 376. **12.** Tunnicliff (1990) *Int. J. Biochem.* **27**, 1235. **13.** Pathre, Singh, Viswanathan & Sane (1987) *Phytochemistry* **26**, 2913. **14.** Levene, Sharman

& Callingham (1992) *Int. J. Exp. Pathol.* **73**, 613. **15.** Tanzer, Monroe & Gross (1966) *Biochemistry* **5**, 1919. **16.** Tonge, Harrison & Higgins (1977) *Biochem. J.* **161**, 333. **17.** Suda & Matsuda (1974) *Biochim. Biophys. Acta* **369**, 331. **18.** Levinson, Faras, Woodson, Jackson & Bishop (1973) *Proc. Natl. Acad. Sci. U.S.A.* **70**, 164.

Thiostrepton

This sulfur- and dehydroalanyl-containing peptide antibiotic (FW = 1664.9 g/mol; CAS 1393-48-2), also known as bryamycin and thiactin, from *Streptomyces azureus* inhibits protein biosynthesis and peptide translocation by preventing GTP binding to the 50S ribosomal subunit, thus inhibiting the mechanochemical events linked to GTP hydrolysis. Thiostrepton also prevents the binding of elongation factor G to the ribosome. ***See also Thiopeptin*** **Target(s):** GTP diphosphokinase (4,5); guanosine-3',5'-bis(diphosphate 3'-diphosphatase (6); guanosine 3',5'-polyphosphate synthetase I (5); IF-2-dependent ribosomal GTPase (7); initiation factor (18); and protein biosynthesis (1-3,9). **1.** Pestka (1974) *Meth. Enzymol.* **30**, 261. **2.** Jiménez (1976) *Trends Biochem. Sci.* **1**, 28. **3.** Lucas-Lenard & Beres (1974) *The Enzymes*, 3rd ed. (Boyer, ed.), **10**, 53. **4.** Knutsson Jenvert & Holmberg Schiavone (2005) *FEBS J.* **272**, 685. **5.** Fehr & Richter (1981) *J. Bacteriol.* **145**, 68. **6.** Richter (1980) *Arch. Microbiol.* **124**, 229. **7.** Grunberg-Manago, Dondon & Graffe (1972) *FEBS Lett.* **22**, 217. **8.** Sarkar, Stringer & Maitra (1974) *Proc. Natl. Acad. Sci. U.S.A.* **71**, 4986. **9.** McConkey, Rogers & McCutchan (1997) *J. Biol. Chem.* **272**, 2046.

Thiosulfate

This dianionic reducing agent and sulfate analogue (FW_{NaSalt} = 158.11 g/mol; CAS 7772-98-7; S–O bond-length = 147 pm; S–S bond-length = 201 pm) is a redox-active substrate for thiosulfate dehydrogenase, thiosulfate sulfurtransferase, *or* rhodanese, thiosulfate:thiol sulfurtransferase, and thiosulfate:dithiol sulfurtransferase as well as a product of the reaction catalyzed by trithionate hydrolase. Thiosulfate is an intermediate in the sulfur cycle in anoxic marine and freshwater sediments, where partakes in reduction, oxidation, and disproportionation pathways. It is also a potential respiratory electron acceptor for bacteria living in anoxic environments or at the anoxic/oxic interface. The ability to respire thiosulfate is conferred by the enzyme thiosulfate reductase (*Reaction*: $S_2O_3^{2-} + 2H^+ + 2e^- \rightarrow HS^- + HSO_3^-$) (1,2). In sulfate-reducing bacteria, the sulfite produced in the thiosulfate reductase reaction is further reduced to sulfide by sulfite reductase. For *Salmonella enterica* (serovar: *Typhimurium*), thiosulfate reduction is catalyzed by the membrane-bound enzyme thiosulfate reductase. Experiments with quinone biosynthesis mutants show that menaquinol is the sole electron donor to thiosulfate reductase (2). The reduction of thiosulfate by menaquinol is highly endergonic under standard conditions ($\Delta E^{o'} = -328$ mV), and thiosulfate reductase activity was found to depend on the proton motive force (PMF) across the cytoplasmic membrane (2). Sulfite reduction is an energy-yielding reaction that is also the final step in sulfate respiration. (***See also Dithionite***) **Target(s):** *N*-acetylgalactosamine-4-sulfatase, *or* arylsulfatase B (3); *N*-acetylgalactosamine-6-sulfatase (4); *N*-acetylglucosamine-6-sulfatase (5); adenylylsulfatase (6); alcohol dehydrogenase (7); aminopeptidase B (8); cyanate hydratase, *or* cyanase (9);

glyceraldehyde-3-phosphate dehydrogenase, acetylphosphatase activity (10); iodothyronine-5'-deiodinase (11); lipase, *or* triacylglycerol lipase (12); nitrile hydratase (13); pyrophosphate-dependent phosphofructokinase, *or* diphosphate:fructose-6-phosphate 1-phosphotransferase (14); sulfate adenylyltransferase, *or* ATP sulfurylase (15-19); sulfur reductase (20); thrombin (21); and trithionate hydrolase, product inhibition (22). **1.** Barrett & Clark (1987) *Microbiol. Rev.* **51**, 192. **2.** Stoffels, Krehenbrink, Berks & Unden (2012) *J. Bacteriol.* **194**, 475. **3.** Bleszynski & Leznicki (1967) *Enzymologia* **33**, 373. **4.** Glössl, Truppe & Kresse (1979) *Biochem. J.* **181**, 37. **5.** Basner, Kresse & Figura (1979) *J. Biol. Chem.* **254**, 1151. **6.** Li & Schiff (1991) *Biochem. J.* **274**, 355. **7.** Sund & Theorell (1963) *The Enzymes*, 2nd ed. (Boyer, Lardy & Myrbäck, eds.), **7**, 25. **8.** Hopsu, Mäkinen & Glenner (1966) *Arch. Biochem. Biophys.* **114**, 567. **9.** Anderson, Johnson, Endrizzi, Little & Korte (1987) *Biochemistry* **26**, 3938. **10.** Allison & Conners (1970) *Arch. Biochem. Biophys.* **136**, 383. **11.** Visser (1980) *Biochim. Biophys. Acta* **611**, 371. **12.** Yang & Hsu (1944) *J. Biol. Chem.* **155**, 137. **13.** Nagasawa, Takeuchi & Yamada (1991) *Eur. J. Biochem.* **196**, 581. **14.** Mahajan & Singh (1989) *Plant Physiol.* **91**, 421. **15.** Hanna, Ng, MacRae, *et al.* (2004) *J. Biol. Chem.* **279**, 4415. **16.** Renosto, Martin, Wailes, Daley & Segel (1990) *J. Biol. Chem.* **265**, 10300. **17.** Seubert, Hoang, Renosto & Segel (1983) *Arch. Biochem. Biophys.* **225**, 679. **18.** Yu, Martin, Jain, Chen & Segel (1989) *Arch. Biochem. Biophys.* **269**, 156. **19.** Renosto, Schultz, Re, *et al.* (1985) *J. Bacteriol.* **164**, 674. **20.** Sugio, Oda, Matsumoto, *et al.* (1998) *Biosci. Biotechnol. Biochem.* **62**, 705. **21.** Seegers (1951) *The Enzymes*, 1st ed. (Sumner & Myrbäck, eds.), **1** (part 2), 1106. **22.** Meulenberg, Pronk, Frank, *et al.* (1992) *Eur. J. Biochem.* **209**, 367.

Thiosulfatoaurate (I), Sodium, *See* Aurothiosulfate, Sodium

Thiotaurine, *See* 2-Aminoethane Thiosulfate

2-Thiouracil

This minor tRNA constituent (FW = 128.15 g/mol; slightly water soluble; decomposes at 340°C; λ_{max} = 268 nm (ε = 11800 $M^{-1}cm^{-1}$) at pH 7.4), also known as 2-mercapto-4-hydroxypyrimidine, inhibits the iodination of thyroxine precursors and is employed in the treatment of hyperthyroidism. 2-Thiouracil is also incorporated into RNA, but not DNA, resulting in translation errors in protein biosynthesis. **Target(s):** 4-aminobutyrate amino-transferase, weakly inhibited (1); catechol oxidase (2); glycine amidinotransferase (3); nitric-oxide synthase (4); orotate phosphoribosyltransferase (5); peroxidase, horseradish (6); ribonuclease, pancreatic (7); thyroid peroxidase (8); thyroxine 5'-deiodinase, *or* iodothyronine 5'-deiodinase (9-12); tyrosinase (13); and urate-ribonucleotide phosphorylase (14). **1.** Tamaki, Kubo, Aoyama & Funatsuka (1983) *J. Biochem.* **93**, 955. **2.** Dawson & Magee (1955) *Meth. Enzymol.* **2**, 817. **3.** Ratner (1962) *The Enzymes*, 2nd ed. (Boyer, Lardy & Myrbäck, eds.), **6**, 267. **4.** Palumbo, d'Ischia & Cioffi (2000) *FEBS Lett.* **485**, 109. **5.** Javaid, el Kouni & Iltzsch (1999) *Biochem. Pharmacol.* **58**, 1457. **6.** Zaton & Ochoa de Aspuru (1995) *FEBS Lett.* **374**, 192. **7.** Richards & Wyckoff (1971) *The Enzymes*, 3rd ed. (Boyer, ed.), **4**, 647. **8.** Davidson, Soodak, Neary, *et al.* (1978) *Endocrinology* **103**, 871. **9.** Harbottle & Richardson (1984) *Biochem. J.* **217**, 485. **10.** Visser (1980) *Trends Biochem. Sci.* **5**, 222. **11.** Chopra, Chua Teco, Eisenberg, Wiersinga & Solomon (1982) *Endocrinology* **110**, 163. **12.** Ericksen, Cavalieri & Rosenberg (1981) *Endocrinology* **108**, 1257. **13.** Lerner, Fitzpatrick, Calkins & Summerson (1950) *J. Biol. Chem.* **187**, 793. **14.** Laster & Blair (1963) *J. Biol. Chem.* **238**, 3348.

2-Thioxanthine

This urate analogue (FW = 168.18 g/mol; CAS 2487-40-3; pK_a = 5.9), also called 2-thio-6-oxypurine, inhibits glycine amidino-transferase (1,2); hypoxanthine(guanine) phosphoribosyl-transferase (3,4); urate oxidase, *or* uricase (5); urate-ribonucleotide phosphorylase (6); and xanthine phosphoribosyltransferase (3,7). **Mechanism-based Inactivation of Myeloperoxidase (MPO):** 2-Thioxanthines are mechanism-based inactivators of MPO that modify the heme prosthetic groups of the enzyme, exhibiting good specificity and undergoing limited oxidation before the enzyme is fully inactivated (8). They exert only a moderate effect on bacterial killing by neutrophils even at concentrations that completely blocked extracellular production of hypochlorous acid. 2-Thioxanthines inhibit hypochlorous acid production by both the purified enzyme and isolated neutrophils. The IC_{50} values with neutrophils were better than those for other inhibitors, including numerous nonsteroidal anti-inflammatory drugs. Indeed, TX2 and TX4 are the best known inhibitors of the chlorination activity of MPO (IC_{50} = 0.2 μm) that do not act by converting the enzyme to compound II (8). 2-Thioxanthines are mechanism-based inactivators of MPO, because they promote inactivation of the enzyme to a far greater extent in the presence of added hydrogen peroxide than in its absence. Inactivation results from modification of the heme through covalent attachment of the 2-thioxanthine, as evident from the increase in mass of the modified heme and the X-ray crystal structure of the inactivated enzyme (8). **1.** Ratner (1962) *The Enzymes*, 2nd ed. (Boyer, Lardy & Myrbäck, eds.), **6**, 267. **2.** Walker (1957) *J. Biol. Chem.* **224**, 57. **3.** Naguib, Iltzsch, el Kouni, Panzica & el Kouni (1995) *Biochem. Pharmacol.* **50**, 1685. **4.** Miller, Ramsey, Krenitsky & Elion (1972) *Biochemistry* **11**, 4723. **5.** Bergmann, Kwietny-Govrin, Ungar-Waron, Kalmus & Tamari (1963) *Biochem. J.* **86**, 567. **6.** Laster & Blair (1963) *J. Biol. Chem.* **238**, 3348. **7.** Miller, Adamczyk, Fyfe & Elion (1974) *Arch. Biochem. Biophys.* **165**, 349. **8.** Tidén, Sjögren, Svensson, *et al.* (2011) *J. Biol. Chem.* **286**, 37578.

6-Thioxanthine

This urate analogue (FW = 168.18 g/mol; CAS 2002-59-7; pK_a = 6.5), also known as 6-thio-2-oxopurine, inhibits hypoxanthine (guanine) phosphoribosyltransferase (1,2); thiopurine *S*-methyltransferase (3,4); urate oxidase, *or* uricase (5); and xanthine phosphoribosyltransferase (1,6). **1.** Naguib, Iltzsch, el Kouni, Panzica & el Kouni (1995) *Biochem. Pharmacol.* **50**, 1685. **2.** Miller, Ramsey, Krenitsky & Elion (1972) *Biochemistry* **11**, 4723. **3.** Kroplin, Fischer & Iven (1999) *Eur. J. Clin. Pharmacol.* **55**, 285. **4.** Jackson, Hughes & Stupans (1996) *Biochem. Mol. Biol. Int.* **38**, 357. **5.** Bergmann, Kwietny-Govrin, Ungar-Waron, Kalmus & Tamari (1963) *Biochem. J.* **86**, 567. **6.** Miller, Adamczyk, Fyfe & Elion (1974) *Arch. Biochem. Biophys.* **165**, 349.

6-Thioxanthosine 5'-Monophosphate

This XMP analogue and guanine nucleotide-depleting agent ($FW_{free-acid}$ = 380.28 g/mol; CAS 3237-49-8), also named 6-thioxo-6,9-dihydro-1*H*-purin-2(3*H*)-one-β-D-ribofuranose 5'-phosphate, inhibits *Escherichia coli* GMP synthetase, acting as a weak alternative substrate (1,2), as well as IMP dehydrogenase (3). **1.** Spector, Miller, Fyfe & Krenitsky (1974) *Biochim. Biophys. Acta* **370**, 585. **2.** Spector (1975) *J. Biol. Chem.* **250**, 7372. **3.** Pfefferkorn, Bzik & Honsinger (2001) *Exp. Parasitol.* **99**, 235.

L(+)-Threonate (*or* L(+)-Threonic Acid)

This aldonic acid ($FW_{free\text{-}acid}$ = 136.10 g/mol; CAS 7306-96-9), also known as (2R,3S)-Trihydroxybutyric acid, inhibits acid phosphatase (1,2) and L(+)-tartrate dehydratase (3). **1**. Hollander (1971) *The Enzymes*, 3rd ed. (Boyer, ed.), **4**, 449. **2**. Kilsheimer & Axelrod (1957) *J. Biol. Chem.* **227**, 879. **3**. Hurlbert & Jakoby (1965) *J. Biol. Chem.* **240**, 2772.

L-Threoninamide

This L-threonine amide ($FW_{free\text{-}base}$ = 118.14 g/mol; CAS 2280-40-2) inhibits aspartokinases (1) and bacterial leucyl aminopeptidase (2). L-Threonine amide also substitutes for L-threonine as an allosteric feedback inhibitor of aspartokinase. **1**. McCarron & Chang (1978) *J. Bacteriol.* **134**, 483. **2**. Baker, Wilkes, Bayliss & Prescott (1983) *Biochemistry* **22**, 2098.

allo-Threonine, *See* D-Allothreonine; L-Allothreonine

L-Threonyl-L-seryl-L-lysyl-L-tyrosyl-L-arginine, *See* Neo-Kyotorphin

L-Threonyl-L-threonyl-L-tyrosyl-L-alanyl-L-alanyl-L-phenylalanyl-L-isoleucyl-L-alanyl-L-serylglycyl-L-arginyl-L-threonylglycyl-L-arginyl-L-arginyl-L-asparaginyl-L-alanyl-L-isoleucinamide

This octadecapeptide amide (MW = 1927.17 g/mol; Sequence = TTYAAFIASGRTGRRNAI-NH₂) inhibits cAMP-dependent protein kinase: K_i = 8 nM (1). **1**. Walsh & Glass (1991) *Meth. Enzymol.* **201**, 304.

L-Threonyl-L-threonyl-L-tyrosyl-L-alanyl-L-aspartyl-L-phenylalanyl-L-isoleucyl-L-alanyl-L-serylglycyl-L-arginyl-L-threonylglycyl-L-arginyl-L-arginyl-L-asparaginyl-L-alanyl-L-isoleucinamide

This octadecapeptide (MW = 1968.0 g/mol; TTYADFIASGRTGRRNAI-NH₂), known commercially as WIPTIDE, is a potent cAMP-dependent protein kinase inhibitor, K_i = 3.1 nM (1). **1**. Walsh & Glass (1991) *Meth. Enzymol.* **201**, 304.

L-Threonyl-L-threonyl-L-tyrosyl-L-alanyl-L-aspartyl-L-phenylalanyl-L-isoleucyl-L-alanyl-L-serylglycyl-L-arginyl-L-threonylglycyl-L-arginyl-L-arginyl-L-asparaginyl-L-alanyl-L-isoleucyl-L-histidyl-L-aspartate, *See* Protein Kinase A Inhibitor Fragment 5-24

D-Threose 2,4-Bisphosphate

This doubly phosphorylated form of D-threose ($FW_{free\text{-}acid}$ = 280.06 g/mol), formerly called D-threose 2,4-diphosphate is a noncompetitive inhibitor of glyceraldehyde-3 phosphate dehydrogenase (1-4). **1**. Fluharty & Ballou (1958) *J. Biol. Chem.* **234**, 2517. **2**. Wolfenden (1977) *Meth. Enzymol.* **46**, 15. **3**. Velick & Furfine (1963) *The Enzymes*, 2nd ed. (Boyer, Lardy & Myrbäck, eds.), **7**, 243. **4**. Webb (1966) *Enzyme and Metabolic Inhibitors*, vol. **2**, p. 408, Academic Press, New York.

Thrombin-Activatable Fibrinolysis Inhibitor

This plasma pro-protease (MW = 48.4 kDa; M_r = 60,000; NCBI Reference Sequence: NP 001863.3; Symbol: TAFI), also known as plasma carboxypeptidase B, *or* pCPB (so designated for its ability to cleave after basic amino residues), carboxypeptidase B₂, *or* CPB2, and carboxypeptidase U (*or* CPU), is synthesized in the liver but circulates as a plasminogen-bound zymogen. When activated by proteolysis at Arg92, as catalyzed by the thrombin/thrombomodulin complex, TAFI exhibits carboxypeptidase activity that reduces fibrinolysis by removing the fibrin C-terminal residues needed for binding and activating plasminogen. The pre-protein consists of a 22-residue signal peptide, a 92-residue activation peptide, and a 309-residue catalytic domain. 6-Amino-*n*-hexanoic acid, *or* εACA (*See* 6-*Aminohexanoate*), a well-known competitive inhibitor of basic carboxypeptidases, selectively limits trypsin cleavage of pro-pCPB (2). In the presence of εACA, trypsin cleavage at Arg-330 is significantly limited while the cleavage at Arg-92 is unaffected (2). Elevated TAFI levels form a mild risk factor for venous thrombosis (3). Such levels are found in 9% of healthy controls and in 14% of patients with a first deep vein thrombosis. Elevated TAFI levels do not enhance the thrombotic risk associated with Factor V (Leiden) but may interact with high Factor VIII levels (3). **1**. Eaton, Malloy, Tsai, Henzel & Drayna (1991) *J. Biol. Chem.* **266**, 21833. **2**. Tan & Eaton (1995) *Biochemistry* **34**, 5811. **3**. Van Tiiburg, Rosendaal & Bertina (2000) *Blood* **95**, 2855.

Thrombospondin-1

This 1170-residue, homodimeric platelet glycoprotein (MW = 129.4 kDa; Isoelectric Point = 4.71) is released from platelet α-granules in response to thrombin stimulation and inhibits cathepsin G, K_i = 1.2 nM and leukocyte elastase, K_i = 60 nM. Upon release from platelets, thrombospondin-1 binds to platelet membranes, and the resulting complex acts as an adhesion protein participating in cell-cell and cell-matrix interactions. Thrombospondin-1 also binds to fibrinogen, type V collagen, fibronectin, and laminin (1). **Target(s):** cathepsin G (2); leukocyte elastase (3,4). **1**. Sage & Bornstein (1991) *J. Biol. Chem.* **266**, 14831. **2**. Hogg, Owensby & Chesterman (1993) *J. Biol. Chem.* **268**, 21811. **3**. Bieth (1998) in *Handb. Proteolytic Enzymes* (Barrett, Rawlings & Woessner, eds.), p. 5. **4**, Academic Press, San Diego. **4**. Hogg, Owensby, Mosher, Misenheimer & Chesterman (1993) *J. Biol. Chem.* **268**, 7139.

α-Thujaplicin

This hydroxylated tropolone (FW = 164.20 g/mol; CAS 1946-74-3), also known as 2-hydroxy-3-(1-methylethyl)-2,4,6-cycloheptatrien-1-one, inhibits carboxypeptidase A (1), exoribonuclease H (2), HIV-1 reverse transcriptase (2), ribonuclease H (2), and RNA-directed DNA polymerase (2). **1**. Morita, Matsumura, Tsujibo, *et al.* (2001) *Biol. Pharm. Bull.* **24**, 607. **2**. Budihas, Gorshkova, Gaidamakov, *et al.* (2005) *Nucl. Acids Res.* **33**, 1249.

β-Thujaplicin

This hydroxytropoloneone (FW = 164.20 g/mol), also known as hinokitiol, 4-isopropyltropolone, and 2-hydroxy-4-(1-methylethyl)-2,4,6-cyclohepta-trien-1-one, is a major component of *Thujopsis dolabrata*. β-Thujaplicin is also an iron chelator and induces cell differentiation and apoptosis in teratocarcinoma F9 cells. **Target(s):** arachidonate 12-lipoxygenase, *or* simply 12-lipoxygenase (1); exoribonuclease H, weakly (2); *S*-lactoylglutathione lyase (3); ribonuclease H, weakly (2); and tyrosine monooxygenase (4). **1**. Suzuki, Ueda, Juranek, *et al.* (2000) *Biochem. Biophys. Res. Commun.* **275**, 885. **2**. Budihas, Gorshkova, Gaidamakov, *et al.* (2005) *Nucl. Acids Res.* **33**, 1249. **3**. Barnard & Honek (1989) *Biochem. Biophys. Res. Commun.* **165**, 118. **4**. Goldstein, Gang & Anagnoste (1967) *Life Sci.* **6**, 1457.

γ-Thujaplicin

This hydroxytropolone (FW = 164.20 g/mol), also known as 2-hydroxy-5-(1-methylethyl)-2,4,6-cycloheptatrien-1-one, inhibits exoribonuclease H and ribonuclease H (1). **1**. Budihas, Gorshkova, Gaidamakov, Wamiru, Bona, Parniak, Crouch, McMahon, Beutler & Le Grice (2005) *Nucl. Acids Res.* **33**, 1249.

Thymidine

This DNA constituent (FW = 242.23 g/mol) absorbs light in the ultraviolet (λ_{max} = 267 nm at pH 7 (ε = 9700 $M^{-1}cm^{-1}$), abbreviated dT, inhibits acid phosphatase (1), ADP-ribose diphosphatase (2), aspartate carba)moyltransferase (3,4), cytidine deaminase (5-8), cytosine deaminase (9,10), dCMP deaminase (11); deoxycytidine kinase, weakly (12), dihydroorotase (13), dTMP kinase, or thymidylate kinase (14-20), NAD^+ ADP-ribosyltransferase, or poly(ADP-ribose) synthetase (21-28); NAD^+ nucleosidase (29); NAD^+:protein-arginine ADP-ribosyltransferase (30), 5'-nucleotidase (3,31-33), pancreatic ribonuclease, or ribonuclease A (34); ribonuclease T_2 (35), ribonucleoside diphosphate reductase (36), and urate-ribonucleotide phosphorylase (37). **Double-Thymidine Block Method for G_1/S Cell Synchronization:** The following seven-step protocol is optimized to achieve nearly complete synchronization of HeLa cells at the G_1/S border, by keeping all cells in G_1. *Procedure: Step-1.* Grow HeLa cells to about 40% confluence in standard Dulbecco's Minimal Essential Medium (DMEM) supplemented with 10% Fetal Bovine Serum (FBS). *Step-2.* Add sufficient thymidine (20 mM in Phosphate-Buffered Saline, or PBS) to a final concentration of 2mM in the media. *Step-3.* Incubate culture at 37° C for 19 hours. (Note: Strict time keeping is required in this step.) *Step-4.* Remove media, and wash cells three times with PBS. *Step-5.* Add fresh thymidine-free media, and incubate for 9 hours at 37° C. *Step-6.* Add thymidine (again at 2 mM final concentration) and incubate for additional 16 hours. *Step-7.* Wash cells with PBS, and add fresh thymidine-free media. The cells are now in G_1 and will be "released" to progress through the cell cycle over the ensuing 15 hours. The cells typically remain synchronized for about 1-2 cell divisions, after which their growth becomes asynchronous. *Note:* This protocol will not work on cell lines with intact p53 apoptotic responses (*See Methotrexate for another synchronization protocol*). **1.** Uerkvitz (1988) *J. Biol. Chem.* **263**, 15823. **2.** Wu, Lennon & Suhadolnik (1978) *Biochim. Biophys. Acta* **520**, 588. **3.** Webb (1966) *Enzyme and Metabolic Inhibitors*, vol. **2**, Academic Press, New York. **4.** Bresnick (1963) *Biochim. Biophys. Acta* **67**, 425. **5.** Vita, Amici, Cacciamani, Lanciotti & Magni (1985) *Biochemistry* **24**, 6020. **6.** Hosono & Kuno (1973) *J. Biochem.* **74**, 797. **7.** Cacciamani, Vita, Cristalli, Vincenzetti, Natalini, Ruggieri, Amici & Magni (1991) *Arch. Biochem. Biophys.* **290**, 285. **8.** Vita, Amici, Lanciotti, Cacciamani & Magni (1986) *Ital. J. Biochem.* **35**, 145A. **9.** Ipata & Cercignani (1978) *Meth. Enzymol.* **51**, 394. **10.** Ipata, Marmocchi, Magni, Felicioli & Polidoro (1971) *Biochemistry* **10**, 4270. **11.** Xu & Plunkett (1992) *Biochem. Pharmacol.* **44**, 1819. **12.** Wang, Kucera & Capizzi (1993) *Biochim. Biophys. Acta* **1202**, 309. **13.** Bresnick & Blatchford (1964) *Biochim. Biophys. Acta* **81**, 150. **14.** Jong & Campbell (1984) *J. Biol. Chem.* **259**, 14394. **15.** Vanheusden, Munier-Lehmann, Froeyen, *et al.* (2004) *J. Med. Chem.* **47**, 6187. **16.** Munier-Lehmann, Pochet, Dugué, *et al.* (2003) *Nucleosides Nucleotides Nucleic Acids* **22**, 801. **17.** Haouz, Vanheusden, Munier-Lehmann, *et al.* (2003) *J. Biol. Chem.* **278**, 4963. **18.** Cheng & Prusoff (1973) *Biochemistry* **12**, 2612. **19.** Pochet, Dugué, Labesse, Delepierre & Munier-Lehmann (2003) *ChemBioChem* **4**, 742. **20.** Lee & Cheng (1977) *J. Biol. Chem.* **252**, 5686. **21.** Shizuta, Ito, Nakata & Hayaishi (1980) *Meth. Enzymol.* **66**, 159. **22.** August, Cooper & Prusoff (1991) *Cancer Res.* **51**, 1586. **23.** Banasik, Komura, Shimoyama & Ueda (1992) *J. Biol. Chem.* **267**, 1569. **24.** Rankin, Jacobson, Benjamin, Moss & Jacobson (1989) *J. Biol. Chem.* **264**, 4312. **25.** Kofler, Wallraff, Herzog, *et al.* (1993) *Biochem. J.* **293**, 275. **26.** Burtscher, Klocker, Schneider, *et al.* (1987) *Biochem. J.* **248**, 859. **27.** Werner, Sohst, Gropp, *et al.* (1984) *Eur. J. Biochem.* **139**, 81. **28.** Furneaux & Pearson (1980) *Biochem. J.* **187**, 91. **29.** Ueda, Fukushima, Okayama & Hayaishi (1975) *J. Biol. Chem.* **250**, 7541. **30.** Moss, Stanley & Watkins (1980) *J. Biol. Chem.* **255**, 5838. **31.** Klein (1957) *Z. Physiol. Chem.* **307**, 254. **32.** Garvey, Lowen & Almond (1998) *Biochemistry* **37**, 9043. **33.** Olson & Fraser (1974) *Biochim. Biophys. Acta* **334**, 156. **34.** Richards & Wyckoff (1971) *The Enzymes*, 3rd ed. (Boyer, ed.), **4**, 647. **35.** Irie & Ohgi (1976) *J. Biochem.* **80**, 39. **36.** Scott & Tomkins (1975) *Meth. Enzymol.* **40**, 273. **37.** Laster & Blair (1963) *J. Biol. Chem.* **238**, 334.

Thymidine 3',5'-Bisphosphate

This synthetic nucleotide analogue ($FW_{free-acid}$ = 402.19 g/mol) inhibits exoribonuclease II (1), micrococcal nuclease, or staphyococcal nuclease, K_i = 0.2 μM (2-8), ribonuclease A, K_i = 1 μM (9), and *Serratia marcescens* nuclease (10). **1.** Sporn, Lazarus, Smith & Henderson (1969) *Biochemistry* **8**, 1698. **2.** Cuatrecasas, Fuchs & Anfinsen (1967) *J. Biol. Chem.* **242**, 1541, 3063, and 4759. **3.** Cotton & Hazen, Jr. (1971) *The Enzymes*, 3rd ed. (Boyer, ed.), **4**, 153. **4.** Anfinsen, Cuatrecasas & Taniuchi (1971) *The Enzymes*, 3rd ed. (Boyer, ed.), **4**, 177. **5.** Cuatrecasas, Wilchek & Anfinsen (1969) *Biochemistry* **8**, 2277. **6.** Cuatrecasas, Wilchek & Anfinsen (1969) *Science* **162**, 1491. **7.** Omenn, Ontjes & Anfinsen (1970) *Biochemistry* **9**, 304. **8.** Cotton, Hazen, Jr., & Legg (1979) *Proc. Natl. Acad. Sci. U.S.A.* **76**, 2551. **9.** Russo, Acharya & Shapiro (2001) *Meth. Enzymol.* **341**, 629. **10.** Filimonova, Krause & Benedik (1994) *Biochem. Mol. Biol. Int.* **33**, 1229.

Thymidine 5'-Diphosphate

This 2'-deoxy-pyrimidine nucleotide ($FW_{free-acid}$ = 402.19 g/mol; λ_{max} = 267 nm at pH 7, ε = 9600 $M^{-1}cm^{-1}$), usually abbreviated dTDP, is a component of certain nucleoside diphosphosugars as well as an intermediate in the formation of dTTP (*i.e.*, a substrate for nucleoside diphosphate kinase). The ribonucleotide is far less common than the 2'-deoxy derivative. **Target(s):** α-*N*-acetylneuraminate α-2,8-sialyltransferase (1); cellulose synthase, UDP-forming (41); cytidylate kinase (2); dCMP deaminase, weakly inhibited (3); dCTP deaminase (4); deoxycytidine kinase (5-7); deoxyguanosine kinase (8,9); DNA-directed DNA polymerase (10); dTDP-glucose 4,6-dehydratase (11-13); dTMP kinase, or thymidylate kinase, by product inhibition (14-20); glucose-1-phosphate cytidylyltransferase, weakly (21); glycoprotein 3-α-L-fucosyltransferase, weakly (22); glycoprotein 6-α-L-fucosyltransferase (40); α-1,3-mannosyl-glycoprotein 2-β-*N*-acetylglucosaminyl-transferase (39); β-1,4-mannosyl-glycoprotein 4-β-*N*-acetylglucosaminyl-transferase (23); nicotinate-nucleotide diphosphorylase (carboxylating), weakly (24); 5'-nucleotidase (25); [protein-PII] uridylyltransferase (26); ribose-phosphate diphosphokinase, or phosphoribosyl-pyrophosphate synthetase (27,28); sinapate 1-glucosyltransferase (29); sucrose synthase (30); thymidine kinase (31-34); UDP-glucuronate decarboxylase (35); UDP-glucuronate 4-epimerase (36,37); xanthine phospho-ribosyltransferase, weakly (38). **1.** Eppler, Morré & Keenan (1980) *Biochim. Biophys. Acta* **619**, 332. **2.** Ruffner & Anderson (1969) *J. Biol. Chem.* **244**, 5994. **3.** Ellims, Kao & Chabner (1981) *J. Biol. Chem.* **256**, 6335. **4.** Tomita & Takahashi (1969) *Biochim. Biophys. Acta* **179**, 18. **5.** Ives & Wang (1978) *Meth. Enzymol.* **51**, 337. **6.** Anderson (1973) *The Enzymes*, 3rd ed. (Boyer, ed.), **9**, 49. **7.** Kozai, Sonoda, Kobayashi & Sugino (1972) *J. Biochem.* **71**, 485. **8.** Yamada, Goto & Ogasawara (1982) *Biochim. Biophys. Acta* **709**, 265. **9.** Barker & Lewis (1981) *Biochim. Biophys. Acta* **658**, 111. **10.** Bollum (1968) *Meth. Enzymol.* **12B**, 591. **11.** Yoo, Han & Sohng (1999) *J. Microbiol. Biotechnol.* **9**, 206. **12.** Glaser & Zarkowsky (1971) *The Enzymes*, 3rd ed. (Boyer, ed.), **5**, 465. **13.** Vara & Hutchinson (1988) *J. Biol. Chem.* **263**, 14992. **14.** Jong & Campbell (1984) *J. Biol. Chem.* **259**, 14394. **15.** Cheng & Prusoff (1973) *Biochemistry* **12**, 2612. **16.** Chen, Walker & Prusoff (1979) *J. Biol. Chem.* **254**, 10747. **17.** Smith & Eakin (1975) *Arch. Biochem. Biophys.* **167**, 61. **18.** Lee & Cheng (1977) *J. Biol. Chem.* **252**, 5686. **19.** Tamiya, Yusa,

Yamaguchi, *et al.* (1989) *Biochim. Biophys. Acta* **995**, 28. **20**. Petit & Koretke (2002) *Biochem. J.* **363**, 825. **21**. Kimata & Suzuki (1966) *J. Biol. Chem.* **241**, 1099. **22**. Murray, Takayama, Schultz & Wong (1996) *Biochemistry* **35**, 11183. **23**. Ikeda, Koyota, Ihara, *et al.* (2000) *J. Biochem.* **128**, 609. **24**. Shibata & Iwai (1980) *Agric. Biol. Chem.* **44**, 119. **25**. Burger & Lowenstein (1970) *J. Biol. Chem.* **245**, 6274. **26**. Engleman & Francis (1978) *Arch. Biochem. Biophys.* **191**, 602. **27**. Fox & Kelley (1972) *J. Biol. Chem.* **247**, 2126. **28**. Green & Martin (1973) *Proc. Natl. Acad. Sci. U.S.A.* **70**, 3698. **29**. Wang & Ellis (1998) *Phytochemistry* **49**, 307. **30**. Avigad & Milner (1962) *Meth. Enzymol.* **8**, 341. **31**. Swinton & Chiang (1979) *Mol. Gen. Genet.* **175**, 399. **32**. Chraibi & Wright (1983) *J. Biochem.* **93**, 323. **33**. Gröbner (1979) *J. Biochem.* **86**, 1607. **34**. Her & Momparler (1971) *J. Biol. Chem.* **246**, 6152. **35**. Suzuki, Watanabe, Masumura & Kitamura (2004) *Arch. Biochem. Biophys.* **431**, 169. **36**. Gaunt, Ankel & Schutzbach (1972) *Meth. Enzymol.* **28**, 426. **37**. Gaunt, Maitra & Ankel (1974) *J. Biol. Chem.* **249**, 2366. **38**. Miller, Adamczyk, Fyfe & Elion (1974) *Arch. Biochem. Biophys.* **165**, 349. **39**. Nishikawa, Pegg, Paulsen & Schachter (1988) *J. Biol. Chem.* **263**, 8270. **40**. Ihara, Ikeda & Taniguchi (2006) *Glycobiology* **16**, 333. **41**. Kuribayashi, Kimura, Morita & Igaue (1992) *Biosci. Biotechnol. Biochem.* **56**, 388.

Thymidine 3'-Monophosphate

This nucleotide, also called 3'-thymidylic acid and abbreviated as 3'-dTMP (FW$_{free-acid}$ = 322.21 g/mol; λ_{max} = 267 nm at pH 7, with ε = 9500 M^{-1}cm^{-1}) inhibits ribonuclease A, *or* pancreatic ribonuclease (1,2) and spleen exonuclease (3). **1**. Russo, Acharya & Shapiro (2001) *Meth. Enzymol.* **341**, 629. **2**. Richards & Wyckoff (1971) *The Enzymes*, 3rd ed. (Boyer, ed.), **4**, 647. **3**. Bernardi & Bernardi (1971) *The Enzymes*, 3rd ed. (Boyer, ed.), **4**, 329.

Thymidine 5'-Monophosphate

This nucleotide (FW$_{free-acid}$ = 322.21 g/mol; λ_{max} at pH 7 is 267 nm with ε = 9600 M^{-1}cm^{-1}), also called 5'-thymidylic acid and usually abbreviated dTMP, is a relatively weak (K_i > 0.1 mM). **Target(s):** α-*N*-acetylneuraminate α-2,8-sialyltransferase (1); Amidophosphoribosyltransferase (35); aspartate carbamoyltransferase, weakly (2,3); cellulose synthase, UDP-forming (4); dCMP deaminase, *or* deoxycytidylate deaminase (2,5-10); dCTP deaminase (11); cytidine deaminase (12); deoxyguanosine kinase, weakly (13); dihydroorotase (14); φ$_{29}$ DNA polymerase (15); DNA-directed DNA polymerase (16); fructose-1,6-bisphosphatase (17); micrococcal nuclease, *or* staphylococcal nuclease, K_i = 0.19 mM (18-21); NAD$^+$ ADP-ribosyltransferase (22); NAD$^+$ nucleosidase (23); orotidylate decarboxylase (24); phenylethanolamine *N*-methyltransferase (25); phosphodiesterase I, *or* 5'-exonuclease (26-28); [protein-PII] uridylyltransferase (29,30); ribonuclease A (31); ribonuclease T$_2$ (32); UDP-glucuronate decarboxylase (33); undecaprenyl-phosphate galactose phosphotransferase, weakly (34), and uridine phosphorylase (36). **1**. Eppler, Morré & Keenan (1980) *Biochim. Biophys. Acta* **619**, 332. **2**. Webb (1966) *Enzyme and Metabolic Inhibitors*, vol. 2, Academic Press, New York. **3**. Bresnick (1963) *Biochim. Biophys. Acta* **67**, 425. **4**. Kuribayashi, Kimura, Morita & Igaue (1992) *Biosci. Biotechnol. Biochem.* **56**, 388. **5**. Maley (1967) *Meth. Enzymol.* **12A**, 170. **6**. Scarano, Bonaduce & de Petrocellis (1960) *J. Biol. Chem.* **235**, 3556. **7**. Maley & Maley (1959) *J. Biol. Chem.* **234**, 2975. **8**. Geraci, Rossi & Scarano (1967) *Biochemistry* **6**, 183. **9**. Rossi, Geraci & Scarano (1967) *Biochemistry* **6**, 3640. **10**. Ellims, Kao & Chabner (1981) *J. Biol. Chem.* **256**, 6335. **11**.

Tomita & Takahashi (1969) *Biochim. Biophys. Acta* **179**, 18. **12**. Ipata & Cercignani (1978) *Meth. Enzymol.* **51**, 394. **13**. Green & Lewis (1979) *Biochem. J.* **183**, 547. **14**. Bresnick & Blatchford (1964) *Biochim. Biophys. Acta* **81**, 150. **15**. Lázaro, Blanco & Salas (1995) *Meth. Enzymol.* **262**, 42. **16**. Bollum (1968) *Meth. Enzymol.* **12B**, 591. **17**. Fujita & Freese (1979) *J. Biol. Chem.* **254**, 5340. **18**. Cuatrecasas, Wilchek & Anfinsen (1969) *Biochemistry* **8**, 2277. **19**. Cuatrecasas, Wilchek & Anfinsen (1969) *Science* **162**, 1491. **20**. Cuatrecasas, Fuchs & Anfensen (1967) *J. Biol. Chem.* **242**, 1541, 3063, and 4759. **21**. Anfinsen, Cuatrecasas & Taniuchi (1971) *The Enzymes*, 3rd ed. (Boyer, ed.), **4**, 177. **22**. Ueda, Migakawa & Hayaishi (1972) *Z. Physiol. Chem.* **353**, 844. **23**. Ueda, Fukushima, Okayama & Hayaishi (1975) *J. Biol. Chem.* **250**, 7541. **24**. Jones, Kavipurapu & Traut (1978) *Meth. Enzymol.* **51**, 155. **25**. Pohorecky & Baliga (1973) *Arch. Biochem. Biophys.* **156**, 703. **26**. Razzell (1963) *Meth. Enzymol.* **6**, 236. **27**. Sabatini & Hotchkiss (1969) *Biochemistry* **8**, 4831. **28**. Razzell (1961) *J. Biol. Chem.* **236**, 3031. **29**. Engleman & Francis (1978) *Arch. Biochem. Biophys.* **191**, 602. **30**. Francis & Engleman (1978) *Arch. Biochem. Biophys.* **191**, 590. **31**. Russo, Acharya & Shapiro (2001) *Meth. Enzymol.* **341**, 629. **32**. Irie & Ohgi (1976) *J. Biochem.* **80**, 39. **33**. Suzuki, Watanabe, Masumura & Kitamura (2004) *Arch. Biochem. Biophys.* **431**, 169. **34**. Osborn & Tze-Yuen (1968) *J. Biol. Chem.* **243**, 5145. **35**. Holmes, McDonald, McCord, Wyngaarden & Kelley (1973) *J. Biol. Chem.* **248**, 144. **36**. Avraham, Yashphe & Grossowicz (1988) *FEMS Microbiol. Lett.* **56**, 29.

Thymidine 3'-Monophosphate *p*-Nitrophenyl Ester, *See* Thymidine 3'-*p*-Nitrophenyl Phosphate

Thymidine 5'-Monophosphate *p*-Nitrophenyl Ester, *See* Thymidine 5'-*p*-Nitrophenyl Phosphate

Thymidine 3'-*p*-Nitrophenyl Phosphate

This pyrimidine 3'-nucleotide derivative (FW$_{free-acid}$ = 443.31 g/mol; λ_{max} = 270.5 nm, ε = 15400 M^{-1}cm^{-1}), also known as thymidine 3'-monophosphate *p*-nitrophenyl ester and *p*-nitrophenyl thymidine 3'-phosphate, is an artificial chromogenic substrate for 3'-phosphodiesterases and micrococcal nuclease. Spleen phosphodiesterase readily hydrolyzes this alternative substrate to dT-3'-MP and *p*-nitrophenol (1), whereas it is not a substrate for venom phosphodiesterases. The sodium and potassium salts are readily soluble in water. The *p*-nitrophenol product, released via the action of spleen phosphodiesterase, has a yellow color in alkaline solutions (λ_{max} = 400 nm at pH 9-10; ε = 18300 M^{-1}cm^{-1}). **See also** *Thymidine 3'-p-Nitrophenyl Phosphate 5'-Phosphate* **Target(s):** nucleotide diphosphatase (2); and staphylococcal nuclease, *or* micrococcal nuclease (3). **1**. Razzell & Khorana (1961) *J. Biol. Chem.* **236**, 1144. **2**. Byrd, Fearney & Kim (1985) *J. Biol. Chem.* **260**, 7474. **3**. Anfinsen, Cuatrecasas & Taniuchi (1971) *The Enzymes*, 3rd ed. (Boyer, ed.), **4**, 177.

Thymidine 5'-*p*-Nitrophenyl Phosphate

This pyrimidine nucleotide derivative (FW$_{free-acid}$ = 443.31 g/mol; water-soluble; λ_{max} = 270 nm (pH 2); ε = 16250 M^{-1}cm^{-1} at), also known as thymidine 5'-monophosphate *p*-nitrophenyl ester, is an artificial chromogenic substrate for 5'-phosphodiesterases. Venom phosphodiesterases readily hydrolyze this substrate to dTMP and *p*-

nitrophenol (1). The. The *p*-nitrophenol product, released via the action of phosphodiesterase, has a yellow color in alkaline solutions ($\lambda_{max} = 400$ nm at pH 9-10; $\varepsilon_M = 18300$ M^{-1}cm^{-1}) **Target(s):** alkylglycerophosphoethanolamine phosphodiesterase, *or* lysophospholipase D (2); and nucleotide diphosphatase (3). **1.** Razzell & Khorana (1959) *J. Biol. Chem.* **234**, 2105 & 2114. **2.** Xie & Meier (2004) *Cell. Signal.* **16**, 975. **3.** Byrd, Fearney & Kim (1985) *J. Biol. Chem.* **260**, 7474.

Thymidine 5'-Triphosphate

This nucleotide (FW$_{free-acid}$ = 482.17 g/mol; $\lambda_{max} = 267$ nm; $\varepsilon_M = 9600$ M^{-1}cm^{-1} at pH 7), abbreviated dTTP, is a required DNA precursor. Although dTTP inhibits many enzymes, the most trivial mode of nucleoside 5'-triphosphate inhibition is the binding of a divalent ions, such as Mg^{2+}, Mn^{2+}, or Ca^{2+}, that is required for enzymatic activity. **Target(s):** α-*N*-acetylneuraminate α-2,8-sialyltransferase (113); Amidophosphoribosyltransferase, weakly inhibited (1,116); ATP:citrate lyase, *or* ATP:citrate synthase (120); CDP-diacylglycerol:inositol 3-phosphatidyltransferase (65,66); CDP reductase (2); cellulose synthase, UDP-forming (119); citrate (*si*)-synthase, weakly inhibited (121); 3',5'-cyclic-nucleotide phosphodiesterase (3); cytidine deaminase (4,5,61,62); cytidylate kinase (70); cytosine deaminase (4,6); dCMP deaminase, *or* deoxycytidylate deaminase (7-13,48-60); dCTP deaminase (14,15,45-47); dCTP deaminase, dUMP-forming (43); deoxycytidine kinase (86-89); deoxyguanosine kinase (83); deoxynucleoside kinase (77-82); deoxyribonuclease X (64); DNA-directed DNA polymerase (16); DNA helicases (40); DNA polymerase I, in the presence of ultraviolet light (16); dTDP-dihydrostreptose:streptidine-6-phosphate dihydrostreptosyltransferase (17); dTDP-glucose 4,6-dehydratase (18-20); dTMP kinase, *or* thymidylate kinase (71-76); dUTP diphosphatase (21-23); exopolyphosphatase (41); flavanone 7-*O*-β-glucosyltransferase (118); fucokinase (90); glucokinase, also weak alternative substrate (110,111); glucose-1-phosphate cytidylyltransferase (24); glucuronokinase (91); GTP cyclohydrolase I (44); lactose synthase (25); methionine *S*-adenosyltransferase (112); NADH peroxidase (26); nicotinate-nucleotide diphosphorylase (carboxylating), weakly inhibited (115); nucleoside-triphosphatase, alternative substrate (40); nucleotidases (41); 3'-nucleotidase (41); 5'-nucleotidase (27,41); nucleotide diphosphatase (42); polynucleotide 5'-hydroxyl-kinase (84,85); [protein-PII] uridylyltransferase (67); pyruvate carboxylase (29); *recA* enzyme of *Escherichia coli* (30); ribonucleoside-diphosphate reductase (31); ribose-phosphate diphosphokinase, *or* phosphoribosyl-pyrophosphate synthetase (28,69); RNA-directed RNA polymerase (68); *Serratia marcescens* nuclease (63); thymidine kinase (32-38,92-109); UDP-glucuronate decarboxylase (39); uracil phosphoribosyltransferase (117); and xanthine phosphoribosyltransferase, weakly inhibited (114). **1.** Hill & Bennett (1969) *Biochemistry* **8**, 122. **2.** Langelier, Dechamps & Buttin (1978) *J. Virol.* **26**, 547. **3.** Cheung (1967) *Biochemistry* **6**, 1079. **4.** Ipata & Cercignani (1978) *Meth. Enzymol.* **51**, 394. **5.** Vita, Cacciamani, Natalini, Ruggieri, Raffaelli & Magni (1989) *Comp. Biochem. Physiol. B* **93**, 591. **6.** Ipata, Marmocchi, Magni, Felicioli & Polidoro (1971) *Biochemistry* **10**, 4270. **7.** Maley (1967) *Meth. Enzymol.* **12A**, 170. **8.** Maley (1978) *Meth. Enzymol.* **51**, 412. **9.** Geraci, Rossi & Scarano (1967) *Biochemistry* **6**, 183. **10.** Scarano, Geraci & Rossi (1967) *Biochemistry* **6**, 192. **11.** Scarano, Geraci & Rossi (1964) *Biochem. Biophys. Res. Commun.* **16**, 239. **12.** Rossi, Geraci & Scarano (1967) *Biochemistry* **6**, 3640. **13.** Scarano, Geraci & Rossi (1967) *Biochemistry* **6**, 3645. **14.** Neuhard (1978) *Meth. Enzymol.* **51**, 418. **15.** Tomita & Takahashi (1969) *Biochim. Biophys. Acta* **179**, 18. **16.** Abraham & Modak (1984) *Biochemistry* **23**, 1176. **17.** Kniep & Grisebach (1980) *Eur. J. Biochem.* **105**, 139. **18.** Yoo, Han & Sohng (1999) *J. Microbiol. Biotechnol.* **9**, 206. **19.** Glaser & Zarkowsky (1971) *The Enzymes*, 3rd ed. (Boyer, ed.), **5**, 465. **20.** Vara & Hutchinson (1988) *J. Biol. Chem.* **263**, 14992. **21.** Nord, Larsson, Kvassman, Rosengren & Nyman (1997) *FEBS Lett.* **414**, 271. **22.** Bergman, Nyman & Larsson (1998) *FEBS Lett.* **441**, 327. **23.** Huffman, Li, White & Tainer (2003) *J. Mol. Biol.* **331**, 885. **24.** Kimata & Suzuki (1966) *J. Biol. Chem.* **241**, 1099. **25.** Babad & Hassid (1966) *Meth. Enzymol.* **8**, 346. **26.** Badwey

& Karnovsky (1979) *J. Biol. Chem.* **254**, 11530. **27.** Drummond & Yamamoto (1971) *The Enzymes*, 3rd ed. (Boyer, ed.), **4**, 337. **28.** Wong & Murray (1969) *Biochemistry* **8**, 1608. **29.** Scrutton, Olmsted & Utter (1969) *Meth. Enzymol.* **13**, 235. **30.** McEntee & Weinstock (1981) *The Enzymes*, 3rd ed. (Boyer, ed.), **14**, 445. **31.** Moore (1967) *Meth. Enzymol.* **12A**, 155. **32.** Chen & Prusoff (1978) *Meth. Enzymol.* **51**, 354. **33.** Bresnick (1978) *Meth. Enzymol.* **51**, 360. **34.** Cheng (1978) *Meth. Enzymol.* **51**, 365. **35.** Anderson (1973) *The Enzymes*, 3rd ed. (Boyer, ed.), **9**, 49. **36.** Okazaki & Kornberg (1964) *J. Biol. Chem.* **239**, 275. **37.** Webb (1966) *Enzyme and Metabolic Inhibitors*, vol. 2, p. 470, Academic Press, New York. **38.** Bresnick, Thompson, Morris & Liebelt (1964) *Biochem. Biophys. Res. Commun.* **16**, 278. **39.** Suzuki, Watanabe, Masumura & Kitamura (2004) *Arch. Biochem. Biophys.* **431**, 169. **40.** Locatelli, Gosselin, Spadari & Maga (2001) *J. Mol. Biol.* **313**, 683. **41.** Proudfoot, Kuznetsova, Brown, *et al.* (2004) *J. Biol. Chem.* **279**, 54687. **42.** Krishnan & Rao (1972) *Arch. Biochem. Biophys.* **149**, 336. **43.** Li, Xu, Graham & White (2003) *J. Biol. Chem.* **278**, 11100. **44.** Blau & Niederwieser (1984) *Biochem. Clin. Aspects Pteridines* **3**, 77. **45.** Beck, Eisenhardt & J. Neuhard (1975) *J. Biol. Chem.* **250**, 609. **46.** Price (1974) *J. Virol.* **14**, 1314. **47.** Bjornberg, Neuhard & Nyman (2003) *J. Biol. Chem.* **278**, 20667. **48.** Xu & Plunkett (1992) *Biochem. Pharmacol.* **44**, 1819. **49.** Sergott, Debeer & Bessman (1971) *J. Biol. Chem.* **246**, 7755. **50.** Maley, Lobo & Maley (1993) *Biochim. Biophys. Acta* **1162**, 161. **51.** Raia, Nucci, Vaccaro, Sepe & Rella (1982) *J. Mol. Biol.* **157**, 557. **52.** Ellims, Kao & Chabner (1981) *J. Biol. Chem.* **256**, 6335. **53.** Nucci, Raia, Vaccaro, *et al.* (1978) *J. Mol. Biol.* **124**, 133. **54.** Mastrantonio, Nucci, Vaccaro, Rossi & Whitehead (1983) *Eur. J. Biochem.* **137**, 421. **55.** Nucci, Raia, Vaccaro, Rossi & Whitehead (1991) *Arch. Biochem. Biophys.* **289**, 19. **56.** Whitehead, Nucci, Vaccaro & Rossi (1991) *Arch. Biochem. Biophys.* **289**, 12. **57.** Maley, MacColl & Maley (1972) *J. Biol. Chem.* **247**, 940. **58.** Rolton & Keir (1974) *Biochem. J.* **141**, 211. **59.** Moore, Ciesla, Changchien, Maley & Maley (1994) *Biochemistry* **33**, 2104. **60.** Maley & Maley (1990) *Prog. Nucl. Acid Res. Mol. Biol.* **39**, 49. **61.** Hosono & Kuno (1973) *J. Biochem.* **74**, 797. **62.** Cacciamani, Vita, Cristalli, Vincenzetti, Natalini, Ruggieri, Amici & Magni (1991) *Arch. Biochem. Biophys.* **290**, 285. **63.** Filimonova, Krause & Benedik (1994) *Biochem. Mol. Biol. Int.* **33**, 1229. **64.** Ghosh & DasGupta (1984) *Curr. Trends Life Sci.* **12**, 79. **65.** Antonsson & Klig (1996) *Yeast* **12**, 449. **66.** Antonsson (1994) *Biochem. J.* **297**, 517. **67.** Engleman & Francis (1978) *Arch. Biochem. Biophys.* **191**, 602. **68.** Hung, Gibbs & Tsiang (2002) *Antiviral Res.* **56**, 99. **69.** Fox & Kelley (1972) *J. Biol. Chem.* **247**, 2126. **70.** Liou, Dutschman, Lam, Jiang & Cheng (2002) *Cancer Res.* **62**, 1624. **71.** Jong & Campbell (1984) *J. Biol. Chem.* **259**, 14394. **72.** Cheng & Prusoff (1973) *Biochemistry* **12**, 2612. **73.** Chen, Walker & Prusoff (1979) *J. Biol. Chem.* **254**, 10747. **74.** Smith & Eakin (1975) *Arch. Biochem. Biophys.* **167**, 61. **75.** Lee & Cheng (1977) *J. Biol. Chem.* **252**, 5686. **76.** de Groot & Schweiger (1983) *J. Cell Sci.* **64**, 13. **77.** Balzarini, Degrève, Hatse, *et al.* (2000) *Mol. Pharmacol.* **57**, 811. **78.** Munch-Petersen, Piskur & Søndergaard (1998) *Adv. Exp. Med. Biol.* **431**, 465. **79.** Knecht, Petersen, Munch-Petersen & Piskur (2002) *J. Mol. Biol.* **315**, 529. **80.** Eriksson, Munch-Petersen, Johansson & Eklund (2002) *Cell. Mol. Life Sci.* **59**, 1327. **81.** Mikkelsen, Johansson, Karlsson, *et al.* (2003) *Biochemistry* **42**, 5706. **82.** Welin, Skovgaard, Knecht, *et al.* (2005) *FEBS J.* **272**, 3733. **83.** Green & Lewis (1979) *Biochem. J.* **183**, 547. **84.** Austin, Sirakoff, Roop & Moyer (1978) *Biochim. Biophys. Acta* **522**, 412. **85.** Levin & Zimmerman (1976) *J. Biol. Chem.* **251**, 1767. **86.** Kozai, Sonoda, Kobayashi & Sugino (1972) *J. Biochem.* **71**, 485. **87.** Durham & Ives (1970) *J. Biol. Chem.* **245**, 2276. **88.** Wang, Kucera & Capizzi (1993) *Biochim. Biophys. Acta* **1202**, 309. **89.** Ives & Durham (1970) *J. Biol. Chem.* **245**, 2285. **90.** Kilker, Shuey & Serif (1979) *Biochim. Biophys. Acta* **570**, 271. **91.** Gillard & Dickinson (1978) *Plant Physiol.* **62**, 706. **92.** Cheng & Prusoff (1974) *Biochemistry* **13**, 1179. **93.** Swinton & Chiang (1979) *Mol. Gen. Genet.* **175**, 399. **94.** Chraibi & Wright (1983) *J. Biochem.* **93**, 323. **95.** Ellims & Van der Weyden (1981) *Biochim. Biophys. Acta* **660**, 238. **96.** Iwasuki (1977) *J. Biochem.* **82**, 1347. **97.** Barroso, Carvalho & Flatmark (2005) *Biochemistry* **44**, 4886. **98.** Wang, Saada & Eriksson (2003) *J. Biol. Chem.* **278**, 6963. **99.** Tzeng, Chang, Peng, Wang, J.-Y. Lin, G.-H. Kou & C.-F. Lo (2002) *Virology* **299**, 248. **100.** Chello & Jaffe (1972) *Comp. Biochem. Physiol. B* **43**, 543. **101.** Kit, Leung & Trkula (1973) *Arch. Biochem. Biophys.* **158**, 503. **102.** Taylor, Stafford & Jones (1972) *J. Biol. Chem.* **247**, 1930. **103.** Gröbner (1979) *J. Biochem.* **86**, 1607. **104.** Her & Momparler (1971) *J. Biol. Chem.* **246**, 6152. **105.** Birringer, Claus, Folkers, *et al.* (2005) *FEBS Lett.* **579**, 1376. **106.** Welin, Kosinska, Mikkelsen, *et al.* (2004) *Proc. Natl. Acad. Sci. U.S.A.* **101**, 17970. **107.** Li, Lu, Ke & Chang (2004) *Biochem. Biophys. Res. Commun.* **313**, 587. **108.** Kizer & Hollman (1974) *Biochim. Biophys. Acta* **350**, 193. **109.** Roberts, Fyfe, McKee, *et al.* (1993) *Biochem. Pharmacol.* **46**, 2209. **110.** Porter, Chassy & Holmlund

(1982) *Biochim. Biophys. Acta* **709**, 178. **111**. Doelle (1982) *Eur. J. Appl. Microbiol. Biotechnol.* **14**, 241. **112**. Chou & Talalay (1973) *Biochim. Biophys. Acta* **321**, 467. **113**. Eppler, Morré & Keenan (1980) *Biochim. Biophys. Acta* **619**, 332. **114**. Miller, Adamczyk, Fyfe & Elion (1974) *Arch. Biochem. Biophys.* **165**, 349. **115**. Shibata & Iwai (1980) *Agric. Biol. Chem.* **44**, 119. **116**. Holmes, McDonald, McCord, Wyngaarden & Kelley (1973) *J. Biol. Chem.* **248**, 144. **117**. Natalini, Ruggieri, Santarelli, Vita & Magni (1979) *J. Biol. Chem.* **254**, 1558. **118**. McIntosh & Mansell (1990) *Phytochemistry* **29**, 1533. **119**. Kuribayashi, S. Kimura, T. Morita & I. Igaue (1992) *Biosci. Biotechnol. Biochem.* **56**, 388. **120**. Antranikian, Herzberg & Gottschalk (1982) *J. Bacteriol.* **152**, 1284. **121**. Okabayashi & Nakano (1979) *J. Biochem.* **85**, 1061.

L-Thymidine 5'-Triphosphate

This rare nucleotide ($FW_{\text{free-acid}}$ = 482.17 g/mol) contains 2'-deoxy-L-ribose as the sugar component. It inhibits a number of DNA polymerases. Although L-dTTP inhibits many enzymes, the most trivial mode of nucleoside 5'-triphosphate inhibition is the binding of a divalent ions, such as Mg^{2+}, Mn^{2+}, or Ca^{2+}, that is required for enzymatic activity. **Target(s):** DNA-directed DNA polymerase (2); DNA helicases (1); DNA nucleotidylexotransferase, *or* terminal deoxyribonucleotidyltransferase (2); DNA polymerase α (2); DNA polymerase γ (2); DNA polymerase δ (2); DNA polymerase ε (2); DNA polymerase I (2); HIV-1 reverse transcriptase (2); HSV-1 DNA polymerase (2); nucleoside-triphosphatase (1); RNA-directed DNA polymerase (2); telomerase (3). **1**. Locatelli, Gosselin, Spadari & Maga (2001) *J. Mol. Biol.* **313**, 683. **2**. Focher, Maga, Bendiscioli, *et al.* (1995) *Nucl. Acids Res.* **23**, 2840. **3**. Yamaguchi, Yamada, Tomikawa, *et al.* (2001) *Nucleosides, Nucleotides, Nucleic Acids* **20**, 1243.

Thymine

This major DNA constituent (FW = 126.11 g/mol; λ_{max} = 264.5 nm (at pH 7) ε = 7900 $M^{-1}cm^{-1}$); solubility= 4 g/L water at 25°C), known systematically as 2,6-dihydroxy-5-methylpyrimidine, is also found at low levels in certain tRNA. Thymine is an alternative substrate and competitive inhibitor of uracil dehydrogenase. **Target(s):** dihydrofolate reductase, weakly (1); hydroxyacylglutathione hydrolase, *or* glyoxalase II (2); lactoylglutathione lyase, *or* glyoxalase I (2); NADase (3-5); pancreatic ribonuclease, weakly inhibited (6); uracil dehydrogenase, as poor alternative substrate (7); urate-ribonucleotide phosphorylase (8); and uridine phosphorylase (9). **1**. Bertino, Perkins & Johns (1965) *Biochemistry* **4**, 839. **2**. Oray & Norton (1980) *Biochem. Biophys. Res. Commun.* **95**, 624. **3**. Alivisatos, Kashket & Denstedt (1956) *Can. J. Biochem. Physiol.* **34**, 46. **4**. Webb (1966) *Enzyme and Metabolic Inhibitors*, vol. 2, p. 492, 493, Academic Press, New York. **5**. Hofmann & Rapaport (1957) *Biochem. Z.* **329**, 437. **6**. Richards & Wyckoff (1971) *The Enzymes*, 3rd ed. (Boyer, ed.), **4**, 647. **7**. Hayaishi & Kornberg (1952) *J. Biol. Chem.* **197**, 717 **8**. Laster & Blair (1963) *J. Biol. Chem.* **238**, 3348. **9**. Vita, Amici, Cacciamani, Lanciotti & Magni (1986) *Int. J. Biochem.* **18**, 431.

Thymine Deoxyriboside, *See Thymidine*

Thymine Deoxyriboside 3'-Monophosphate, *See Thymidine 3'-Monophosphate*

Thymine Deoxyriboside 5'-Monophosphate, *See Thymidine 5'-Monophosphate*

Thyminose, *See 2-Deoxy-D-ribose*

Thymoquinone

This superoxide scavenger (FW = 164.20 g/mol; CAS 490-91-5; Symbol: TQ), also known as 2-methyl-5-isopropyl-*p*-benzoquinone and 5-isopropyltoluquinone, is a major constituent of black cumin (*Nigella sativa*) seeds. TQ inhibits leukotriene C_4 synthase (1), 5-lipoxygenase (1), and pyruvate decarboxylase (2). In triple-negative breast cancer (TNBC) cells lacking functional p53 tumor suppressor, thymoquinone induces G_1 phase cell cycle arrest and apoptosis, the latter characterized by the loss of mitochondrial membrane integrity (3). TQ also inhibits tamoxifen-induced hepatic GSH depletion and renal lipid peroxidation (LPO) (4). It likewise normalizes the activity of SOD, inhibits the rise in TNF-α, and ameliorates histopathological changes. **1**. Mansour & Tornhamre (2004) *J. Enzyme Inhib. Med. Chem.* **19**, 431. **2**. Vovk & Murav'eva (1993) *Ukr. Biokhim. Zh.* **65**, 42. **3**. Sutton, Greenshields, Hoskin, *et al.* (2014) *Cancer Lett.* **357**, 129. **4**. Suddek (2014) *Can. J. Physiol. Pharmacol.* **92**, 640.

Thymosin β4

This highly abundant naturally occurring actin monomer-sequestering peptide (FW = 4963.49 g/mol; CAS 77591-33-4; Symbol: Tβ4; Sequence: Acetyl-SDKPDMAEIEKFDKSKLKKTETQEKNPLPSKETIEQKQAGES) binds G-Actin (*i.e.,* Actin monomer) as a 1:1 binary complex, K_d = 0.7–1.0 μM (1-3). Thymosin β4 serves to buffer the intracellular actin monomer levels, preferring to bind Actin·ATP over Actin·ADP. The Thymosin-β4 concentration in human platelets is in the range of 200–500 μM, which when combined with the high intracellular profilin concentrations, is more than enough to account for the pool of unpolymerized Actin within the cytosol. There was an extraordinary claim that Profilin promotes *nonequilibrium* filament assembly by accelerating the hydrolysis of filament-bound ATP, a property that was said to be more pronounced in the presence of Thymosin-β4 (4). Using right-angle light scattering to directly measure polymer weight concentration, it proved possible to use Spectrin-4.1·Actin seeds to investigate (+)-end elongation solely, allowing profilin-mediated filament assembly to be examined in the absence and presence of Thymosin-β4 (4). KINSIM-based modeling and Global Statistical Analysis allowed the evaluation of all pertinent rate and equilibrium constants (5). In the absence of Tβ4, both Actin·ATP and Profilin·Actin·ATP are equally competent in (+)-end polymerization, with k_{on} ~10 x 10^6 $M^{-1}s^{-1}$. When measured in the absence of Profilin, Actin assembly curves (over a 0.7–4 μM Tβ4 concentration range) fit a simple monomer-sequestering model (K_d = 1 μM) for Tβ4·Actin (5). The corresponding constant for Tβ4·pyrenyl-Actin, however, was significantly higher (K_d = ~9–10 μM), suggesting that actin pyrenylation markedly weakens Tβ4 binding, and selective enrichment of pyrenyl-Actin in the presence of Tβ4 most likely caused erroneous fluorescence readings in the earlier study (5). Notably, actin filament assembly in presence of Tβ4 and Profilin fits a simple thermodynamic energy cycle, without any hint of the nonequilibrium effects (5). These findings and earlier work carried out in the absence of Tβ4 (6) indicate Profilin·Actin is directly incorporated at the (+)-end of actively polymerizing actin filaments, without selective facilitation of filament formation by Profilin. **1**. Safer, Elzinga & Nachmias (1991) *J. Biol. Chem.* **266**, 4029. **2**. Weber, Nachmias, Pennise, Pring & Safer (1992) *Biochemistry* **31**, 6179. **3**. Nachmias (1993) *Curr. Opin. Cell Biol.* **5**, 56. **4**. Pantaloni & Carlier (1993) *Cell* **75**, 1007. **5**. Kang, Purich & Southwick (1999) *J. Biol. Chem.* **274**, 36963. **6**. Korenbaum, Nordberg, Björkegren-Sjögren, *et al.* (1998) *Biochemistry* **37**, 9274.

L-Thyroxine

This tetra-iodinated amino acid (FW = 776.87 g/mol; CAS 51-48-9; λ_{max} = 295 nm (ε = 4160 $cm^{-1}M^{-1}$; Symbol = Thx = T_4) in 40 mM HCl; pK_a values of 2.2, 6.45 (phenol OH), and 10.1; photolabile), also known as L-3,3',5,5'-tetraiodothyronine and O^4-(4-hydroxy-3,5-diiodophenyl)-3,5-diiodotyrosine, is obtained by the proteolysis of thyroglobulin. T_4 exhibits about one-fifth the biological activity of 3,5,3'-triiodothyronine (or T_3). In mature organisms, the thyroid hormones increase oxygen consumption, elevate metabolic rates, increase thermogenesis, and induce RNA and protein synthesis. **See also** *3,5,3'-Triiodothyronine* **Target(s):** acid phosphatase, no inhibition reported when *p*-nitrophenyl phosphate is the substrate (45); acylphosphatase (1,43); alanine aminotransferase (2); alcohol dehydrogenase (3-7); aromatase, or CYP19 (8); aspartate aminotransferase (9,10); bisphosphoglycerate mutase (11); cAMP-dependent protein kinase (12); cAMP phosphodiesterase (13,14); creatine kinase (15,16); cystathionine γ-lyase (17,18); dihydropteridine reductase (19); GABA$_A$ receptor (20); glucose-6 phosphate dehydrogenase (21,22); glutamate dehydrogenase (23,24); IMP dehydrogenase (25); inositol trisphosphate 3-kinase (46); iodothyronine deiodinase, type II (26); lipoprotein lipase (27); lysozyme (28); malate dehydrogenase (29-31); monoamine oxidase (32); NADPH:cytochrome c_2 reductase (33); NAD(P)H dehydrogenase, or DT diaphorase (34-37); NAD(P)$^+$ transhydrogenase (38); nicotinamidase (39,44); phenylalanyl-tRNA synthetase (40); prostaglandin-D synthase (41); and tyrosine aminotransferase (42). **1**. Grisolia, Caravaca & Joyce (1958) *Biochim. Biophys. Acta* **29**, 432. **2**. Horvath (1958) *Enzymologia* **19**, 297. **3**. McCarthy, Lovenberg & Sjoerdsma (1968) *J. Biol. Chem.* **243**, 2754. **4**. Mardh, Auld & Vallee (1987) *Biochemistry* **26**, 7585. **5**. Sund & Theorell (1963) *The Enzymes*, 2nd ed. (Boyer, Lardy & Myrbäck, eds.), **7**, 25. **6**. Brändén, Jörnvall, Eklund & Furugren (1975) *The Enzymes*, 3rd ed. (Boyer, ed.), **11**, 103. **7**. von Wartburg, Bethune & Vallee (1964) *Biochemistry* **3**, 1775. **8**. Bullion & Osawa (1988) *Endocr. Res.* **14**, 21. **9**. Smejkalová & Smejkal (1966) *Enzymologia* **30**, 254. **10**. Smejkal & Smejkalová (1967) *Enzymologia* **33**, 320. **11**. Joyce & Grisolia (1959) *J. Biol. Chem.* **234**, 1330. **12**. Friedman, Lang & Burke (1978) *Endocr. Res. Commun.* **5**, 109. **13**. Law & Henkin (1984) *Res. Commun. Chem. Pathol. Pharmacol.* **43**, 449. **14**. Nagasaka & Hidaka (1976) *Biochim. Biophys. Acta* **438**, 449. **15**. Kuby, Noda & Lardy (1954) *J. Biol. Chem.* **210**, 65. **16**. Kuby & Noltmann (1962) *The Enzymes*, 2nd ed. (Boyer, Lardy & Myrbäck, eds.), **6**, 515. **17**. Fernández & Horvath (1963) *Enzymologia* **26**, 113. **18**. González & Horvath (1964) *Enzymologia* **26**, 364. **19**. Hasegawa & Nakanishi (1987) *Meth. Enzymol.* **142**, 111. **20**. Martin, Padron, Newman, *et al.* (2004) *Brain Res.* **1004**, 98. **21**. McKerns (1962) *Biochim. Biophys. Acta* **62**, 402. **22**. Criss & McKerns (1969) *Biochim. Biophys. Acta* **184**, 486. **23**. Frieden (1963) *The Enzymes*, 2nd ed. (Boyer, Lardy & Myrbäck, eds.), **7**, 3. **24**. Smith, Austen, Blumenthal & Nyc (1975) *The Enzymes*, 3rd ed. (Boyer, ed.), **11**, 293. **25**. al-Mudhaffar & Ackerman (1968) *Endocrinology* **82**, 912. **26**. Köhrle (2002) *Meth. Enzymol.* **347**, 125. **27**. Hofstee (1960) *The Enzymes*, 2nd ed. (Boyer, Lardy & Myrbäck, eds.), **4**, 485. **28**. Imoto, Johnson, North, Phillips & Rupley (1972) *The Enzymes*, 3rd ed. (Boyer, ed.), **7**, 665. **29**. Maggio & Ullman (1978) *Biochim. Biophys. Acta* **522**, 284. **30**. Thorne & Kaplan (1963) *J. Biol. Chem.* **238**, 1861. **31**. Stern, Ochoa & Lynen (1952) *J. Biol. Chem.* **198**, 313. **32**. Knopp (1982) *Endokrinologie* **80**, 367. **33**. Sabo & Orlando (1968) *J. Biol. Chem.* **243**, 3742. **34**. Ernster (1967) *Meth. Enzymol.* **10**, 309. **35**. Lind, Cadenas, Hochstein & Ernster (1990) *Meth. Enzymol.* **186**, 287. **36**. Martius (1963) *The Enzymes*, 2nd ed. (Boyer, Lardy & Myrbäck, eds.), **7**, 517. **37**. Ernster, Ljunggren & Danielson (1960) *Biochem. Biophys. Res. Commun.* **2**, 48 and 88. **38**. Keister & San Pietro (1963) *Meth. Enzymol.* **6**, 434. **39**. Su & Chaykin (1971) *Meth. Enzymol.* **18B**, 185. **40**. Nielsen & Haschemeyer (1976) *Biochemistry* **15**, 348. **41**. Beuckmann, Aoyagi, Okazaki, *et al.* (1999) *Biochemistry* **38**, 8006. **42**. Diamondstone & Litwack (1963) *J. Biol. Chem.* **238**, 3859. **43**. Raijman, Grisolia & Edelhoch (1960) *J. Biol. Chem.* **235**, 2340. **44**. Gillam, Watson & Chaykin (1973) *Arch. Biochem. Biophys.* **157**, 268. **45**. Kuo & Blumenthal (1961) *Biochim. Biophys. Acta* **52**, 13. **46**. Mayr, Windhorst & Hillemeier (2005) *J. Biol. Chem.* **280**, 13229.

Tiagabine

This anti-convulsant (FW = 375.55 g/mol; CAS 115103-54-3), also known by its code name NO-328, its trade name Gabitril®, and its systematic name (*R*)-1-[4,4-*bis*(3-methylthiophen-2-yl)but-3-enyl]piperidine-3-carboxylic acid, targets the GABA (γ-amino-butyrate) transporter GAT-1, a high-affinity, sodium- and chloride-dependent transporter located on presynaptic terminals and surrounding glial cells (1-4). GABA is the major inhibitory neurotransmitter in brain, and inhibition of its re-uptake from the synaptic cleft is an important way to enhance its neurotransmitter activity. Because the brain fails to take up radioiodine-labeled tiagabine to a sufficient degree, it cannot be used to localize cerebral GABA transporters (5). (**See also** *N-(1-Benzyl-4-piperidinyl)-2,4-dichlorobenzamide*) Gabitril is the first and currently only FDA-approved GAT inhibitor and is intended to treat partial seizures in adults and children age 12 and older. **1**. Nielsen, Suzdak, Andersen, *et al.* (1991) *Eur. J. Pharmacol.* **196**, 257. **2**. Andersen, Braestrup, Grønwald, *et al.* (1993) *J. Med. Chem.* **36**, 1716. **3**. Borden, Murali-Dhar, Smith, *et al.* (1994) *Eur. J. Pharmacol.* **269**, 219. **4**. Borden (1996) *Neurochem Int.* **29**, 335. **5**. Schijns, van Kroonenburgh, Beekman, *et al.* (2013) *Nucl. Med. Commun.* **34**, 175.

Tiazofurin

Tiazofurin

Tiazole-4-carboxamide Adenine Dinucleotide

This pyridine nucleoside analogue (FW = 260.27 g/mol; CAS 60084-10-8) is an antineoplastic agent that inhibits IMP dehydrogenase (1-3), but only after undergoing phosphorylation by adenosine kinase, followed by ATP-dependent adenylylation by NMN adenylyltransferase (NMNAT) to form the NAD$^+$-like analogue known as tiazole-4-carboxamide adenine dinucleotide (TAD). (**See** *Selenazofurin; Benzamide Riboside*) IMPDH is a rate-limiting step in the de novo synthesis of guanylates, including GTP and dGTP. TAD$^+$ inhibition of IMPDH depletes cells of GTP, thereby down-regulating G-protein function, inhibiting cell growth, and/or inducing apoptosis. Resistance to tiazofurin and related IMPDH inhibitors relates mainly to a decrease in NMNAT activity. **1**. Witters, Mendel & Colliton (1987) *Arch. Biochem. Biophys.* **252**, 130. **2**. Jayaram, Cooney & Grusch (1999) *Curr. Med. Chem.* **6**, 561. **3**. Tricot, Jayaram, E. Lapis, *et al.* (1989) *Cancer Res.* **49**, 3696.

Tibolone

This synthetic steroid (FW = 312.45 g/mol; CAS 5630-53-5), also named 17-hydroxy-7α-methyl-19-nor-17α-pregn-5(10)-en-20-yn-3-one, is used in the treatment of menopause to prevent bone-loss and relieve symptoms (often as effectively as estrogen). **Target(s):** aromatase (CYP19), however, earlier reports suggest stimulation (1); estrone sulfatase (2-7); estrone sulfotransferase, activated at low concentrations (8,9); and 17β-hydroxysteroid dehydrogenase (10). **1**. Raobaikady, Parsons, Reed & Purohit (2007) *J. Steroid Biochem. Mol. Biol.* **104**, 15. **2**. Chetrite,

Kloosterboer & Pasqualini (1997) *Anticancer Res.* **17**, 135. **3**. Pasqualini & Chetrite (1999) *J. Steroid Biochem. Mol. Biol.* **69**, 28. **4**. de Gooyer, Kleyn, Smits, *et al.* (2001) *Mol. Cell. Endocrinol.* **183**, 55. **5**. Purohit, Malini, Hooymans & Newman (2002) *Horm. Metab. Res.* **34**, 1. **6**. van de Ven, Donker, Sprong, Blankenstein & Thijssen (2002) *J. Steroid Biochem. Mol. Biol.* **81**, 237. **7**. Raobaikady, Day, Purohit, Potter & Reed (2005) *J. Steroid Biochem. Mol. Biol.* **94**, 229. **8**. Pasqualini & Chetrite (2007) *Anticancer Res.* **27**, 3219. **9**. Chetrite, Kloosterboer, Philippe & Pasqualini (1999) *Anticancer Res.* **19**, 269. **10**. Chetrite, Kloosterboer, Philippe & Pasqualini (1999) *Anticancer Res.* **19**, 261.

TIC10

This orally active dual Akt/ERK inhibitor (FW = 386.49 g/mol; CAS 41276-02-2; IUPAC: 2,6,7,8,9,10-hexahydro-10-[(2-methylphenyl)methyl]-7-(phenylmethyl)imidazo[1,2-*a*]pyrido[4,3-*d*]-pyrimidin-5(3*H*)-one) targets Foxo3a nuclear translocation, TRAIL gene induction, and displays potent antitumor effects. TRAIL is <u>T</u>umor necrosis factor-<u>R</u>elated <u>A</u>poptosis-<u>I</u>nducing <u>L</u>igand, an antitumor protein with considerable potential for therapeutic targeting in cancer. TIC10 inhibits kinases Akt and extracellular signal-regulated kinase (ERK), leading to the translocation of Foxo3a into the nucleus, where it binds to the TRAIL promoter to up-regulate gene transcription. A member of the forkhead transcription factor famiy, Foxo3a protects cells from oxidative stress by upregulating catalase and MnSOD. Like other members of its family, Foxo3a also is a pr-apoptotic factor. Significantly, TIC10 exhibits the capacity to cross the blood/brain barrier, offering the advantage of accessing CNS targets. **1**. Allen, Krigsfeld, Mayes, *et al.* (2013) *Sci. Transl. Med.* **5**, 171.

Ticagrelor

This ADP receptor (subtype P2Y$_{12}$) antagonist (FW = 522.56 g/mol; CAS 274693-27-5), known as AZD6140 and Brilinta™, is a potent platelet aggregation inhibitor with clinical outcome properties thought to be superior to clopidogrel (1). Unlike clopidogrel and prasugrel, ticagrelor is not a thienopyridine, does not require *in vivo* biotransformation, and has a reversible mode of action ($t_{1/2}$ = 12 h). The first drug of a new chemical class called cyclopentyltriazolopyrimidines, ticagrelor is administered orally and is characterized by a rapid, greater and consistent antiplatelet effect with a favorable safety profile. By inhibiting adenosine uptake by red blood cells, ticagrelor increases adenosine plasma concentration (APC) in acute coronary syndrome (ACS) patients, compared with clopidogrel (2). Ticagrelor's ability to promote very rapid and reversible platelet inhibition makes it a promising treatment option. Side-effects include dyspnea, bradyarrhythmia, and increased serum levels of uric acid and creatinine (3). **1.** Wallentin, Becker, Budaj, Cannon, *et al.* (2009) *New Engl. J. Med.* **361**, 1045. **2**. Bonello, Laine, Kipson, *et al.* (2014) *J. Am. Coll. Cardiol.* **63**, 872. **3**. Schömig (2009) *New Engl. J. Med.* **361**, 1108.

Ticarcillin

This carboxypenicillin (FW = 384.43 g/mol; CAS 34787-01-4), also named (2*S*,5*R*,6*R*)-6-{[(2*R*)-2-carboxy-2-(3-thienyl)acetyl]amino}-3,3-dimethyl-7-oxo-4-thia-1-azabicyclo[3.2.0]heptane-2-carboxylate, is an IV antibiotic for the treatment of Gram-negative bacteria, particularly *Pseudomonas aeruginosa*. Given its susceptibility to β-lactamases, ticarcillin is also often paired with a β-lactamase inhibitor, such as clavulanic acid (*e.g.*, Timentin®). It is also one of the few antibiotics that can be used to treat *Stenotrophomonas maltophilia* infections. Tolerability is good, with hypokalaemia the most frequently reported side-effect. **See also** *Carbenicillin; Carindacillin; Temocillin*

Tick Anticoagulant Peptide

This peptide (TAP) from the South African soft tick (*Ornithodoros moubata*), a vector for swine fever, strongly inhibits coagulation factor Xa and trypsin (1-5). The peptide consists of a single chain of sixty residues. Two isoforms of the inhibitor have been reported, differing in only four residues. Both isoforms are slow, tight-binding inhibitors with K_i values of 0.2 nM. A 15-kDa peptide isolated from the camel tick *Hyalomma dromedarii* also inhibits coagulation factor Xa (K_i = 134 nM). **1**. Dinwiddie, L. Waxman, Vlasuk & Friedman (1993) *Meth. Enzymol.* **223**, 291. **2**. Leadley, Jr. (2001) *Curr. Top. Med. Chem.* **1**, 151. **3**. Ibrahim, Ghazy, Maharem & M. I. Khalil (2001) *Comp. Biochem. Physiol. B* **130B**, 501. **4**. Krishnaswamy, Vlasuk & Bergum (1994) *Biochemistry* **33**, 7897. **5**. Waxman, Smith, Arcuri & Vlasuk (1990) *Science* **248**, 593.

Ticlopidine

This thienotetrahydropyridine (FW = 263.79 g/mol; CAS 55142-85-3), also known as 5-(2-chlorobenzyl)-4,5,6,7-tetrahydrothieno[3,2-*c*]pyridine and Ticlid®, is a prodrug that upon metabolic activation reduces ADP-mediated platelet activation (1).

By interfering with platelet function, ticlopidine prevents clots from forming on the inside of blood vessels. Anti-platelet effects begin within 2 days, reaching a maximum after 6 days of therapy. Ticlopidine's effects generally persist for 3 days upon discontinuing ticlopidine, but may take 1–2 weeks for normal platelet function to return. Because this medication affects platelets irreversibly, platelets must be formed before platelet function normalizes.UR-4509, the bioactive metabolite formed from ticlopidine, also inhibits CYP2Y12 (1); CYP2C19 (2-4); CYP2D6 (3). The main biliary metabolite of ticlopidine was characterized as a glutathione (GSH) conjugate of ticlopidine S-oxide, in which conjugation had occurred at carbon 7a in the thienopyridine moiety. The main biliary metabolite of ticlopidine was characterized as a glutathione (GSH) conjugate of ticlopidine S-oxide, in which conjugation had occurred at carbon 7a in the thienopyridine moiety (5).. **1**. Kam & Nethery (2003) *Anaesthesia* **58**, 28. **2**. Ha-Duong, Dijols, Macherey, Dansette & D. Mansuy (2001) *Adv. Exp. Med. Biol.* **500**, 145. **3**. Ko, Desta, Soukhova, Tracy & Flockhart (2000) *Brit. J. Clin. Pharmacol.* **49**, 343. **4**. Donahue, Flockhart, Abernethy & Ko (1997) *Clin. Pharmacol. Ther.* **62**, 572. **5**. Shimizu, Atsumi, Nakazawa, *et al.* (2009) *Drug Metab. Dispos.* **37**, 1904.

Tideglusib

This potent ATP-noncompetitive GSK3 inhibitor (FW = 334.39; CAS 865854-05-3; Solubility (25°C): 10 mg/mL DMSO, <1 mg/mL H_2O), also known as NP031112 and NP-12, as well as systematically as 4-benzyl-2-(naphthalen-1-yl)[1,2,4]thiadiazolidine-3,5-dione, targets Glycogen Synthase Kinase-3 (or "Tau Kinase"), with an IC_{50} value of 60 nM. Sustained oral administration of tideglusib in a variety of animal models has the effect of decreasing Tau hyper-phosphorylation, lowering brain amyloid plaque, improving memory and learning, and preventing neuronal loss (1). Once bound, the inhibitor is held tightly by the enzyme – so strongly that there is no recovery of enzyme activity, even after the unbound drug has been removed from the reaction medium (2). Such behavior suggests tideglusib is either an irreversible enzyme inhibitor or a slow, tight-binding inhibitor with extreme affinity. Tideglusib protects neural stem cells against NMDA receptor overactivation and NMDA receptor-induced cell death (3). Tideglusib significantly decreased ROS production and membrane degradation, but did not change intracellular Ca^{2+} levels following NMDA receptor activation. Moreover, the PPARγ antagonist GW9662 significantly inhibits the protective effect of tideglusib in NMDA/d-serine-treated cells. **1**. Luna-Medina, Cortes-Canteli, Sanchez-Galiano, et al. (2007) *J. Neurosci.* **27**, 5766. **2**. Domínguez, Fuertes, Orozco, *et al.* (2012) *J. Biol. Chem.* **287**, 893. **3**. Armagan, Keser, Atalayın & Dagcı (2015) *Pharmacol. Rep.* **67**, 823.

Tiglate

This necic acid (FW_{free-acid} = 100.12 g/mol; CAS 80-59-1; M.P. = 63.5-64°C, B.P. = 198.5°C; Soluble in hot water), also known as (2*E*)-2-methyl-2-butenoic acid and *cis*-dimethylacrylic acid, is a volatile, branched-chain fatty acid that is also found as an ester in many alkaloids. Angelic acid is its *trans* isomer. (*Note*: Tiglate is also a powerful vesicant.) **See also Angelate (or Angelic Acid) Target(s)**: D-amino-acid oxidase (1); esterase (2); and malic enzyme, NAD⁺-dependent (3). **1**. Webb (1966) *Enzyme and Metabolic Inhibitors*, vol. **2**, p. 343, Academic Press, New York. **2**. Weber & King (1935) *J. Biol. Chem.* **108**, 131. **3**. Landsperger & Harris (1976) *J. Biol. Chem.* **251**, 3599.

Tiglyl-CoA

This coenzyme A-based thiolester (FW = 849.74 g/mol; CAS 6247-62-7), also known as (*E*)-2-methyl-2-butenoyl-CoA and tigloyl-CoA, is an intermediate in isoleucine metabolism, where it is formed from methylbutyryl-CoA and subsequently converted to succinyl-CoA and acetyl-CoA. Tiglyl-CoA inhibits amino-acid N-acetyltransferase, *or* N-acetylglutamate synthase (1), glycine cleavage complex, *or* glycine synthase (2), and 3-methylcrotonoyl-CoA carboxylase, the latter weakly (3). **1**. Coude, Sweetman & Nyhan (1979) *J. Clin. Invest.* **64**, 1544. **2**. Hayasaka &

Tada (1983) *Biochem. Int.* **6**, 225. **3**. Diez, Wurtele & Nikolau (1994) *Arch. Biochem. Biophys.* **310**, 64.

Tilfrinib

This potent ATP-competitive Brk inhibitor (FW = 275.30 g/mol; CAS 1600515-49-8; Solubility: 100 mM in DMSO), systematically named 3-(9*H*-pyrido[2,3-*b*]indol-4-ylamino)phenol, selectively targets breast tumor kinase (IC_{50} = 3.15 nM), *or* PTK6, a non-receptor tyrosine-protein kinase that operates in signaling pathways controlling differentiation and maintenance of normal epithelia, as well as tumor growth. PTK6 substrates include certain RNA-binding proteins (*e.g.*, KHDRBS1/SAM68, KHDRBS2/SLM1, KHDRBS3/SLM2 and SFPQ/PSF), transcription factors (*e.g.*, STAT3 and STAT5A/B), as well as signaling proteins, such as STAP2/BKS, PXN/paxillin, BTK/ATK, and ARHGAP35/ p190RhoGAP. **1**. Mahmoud, Krug, Wersig, *et al.* (2014) *Bioorg. Med. Chem. Lett.* **24**, 1948.

Tilorone

This fluorenone (FW_{free-base} = 410.56 g/mol; CAS 27591-97-5), also known as 2,7-bis-[2-(diethylamino)ethoxy]-9*H*-fluoren-9-one and bis-DEAE-fluorenone, which was the first synthetic interferon-inducer (1,2), inhibits DNA polymerase (3,4). **1**. Mayer & Krueger (1970) *Science* **169**, 1214. **2**. Krueger & Mayer (1970) *Science* **169**, 1213. **3**. Chandra, Will, Gericke & Gotz (1974) *Biochem. Pharmacol.* **23**, 3259. **4**. Schafer, Chirigos & Papas (1974) *Cancer Chemother. Rep.* **58**, 821.

Tilorone R10,556 DA

This symmetrical diester inducer of interferon (FW_{free-base} = 518.65 g/mol), also called 2,7-bis(piperidinobutyryl)-9*H*-fluoren-9-one, inhibits poly(ADP-ribose) glycohydrolase (1). **1**. Tavassoli, Tavassoli & Shall (1985) *Biochim. Biophys. Acta* **827**, 228.

Tiludronate

This bisphosphonate-based bone resorption inhibitor (FW_{free-acid} = 318.61 g/mol; CAS 155453-10-4) inhibits protein-tyrosine phosphatase (1); pyrophosphatase, *or* inorganic diphosphatase (2); vacuolar H⁺-transporting ATPase (3); and vacuolar H⁺-transporting pyrophosphatase (2). **1**. Murakami, Takahashi, Tanaka, *et al.* (1997) *Bone* **20**, 399. **2**. McIntosh & Vaidya (2002) *Int. J. Parasitol.* **32**, 1. **3**. David, Nguyen, Barbier & Baron (1996) *J. Bone Miner. Res.* **11**, 1498.

Timolol

This common vasodilator and antiglaucoma agent (FW$_{free-base}$ = 316.43 g/mol; CAS 26839-75-8), also known as Timoptic® and Blocadren®, is a β$_1$/β$_2$-adrenergic antagonist, with K_i values of 1.97 nM and 2.0 nM, respectively. Timolol is typically supplied as the water-/ethanol-soluble maleate salt. It has been proposed for use as an antihypertensive, antiarrhythmic, antiangina, and antiglaucoma agent. **Target(s):** β-adrenergic receptor (1); ATPase (2); carbonic anhydrase (3); cholinesterase, weakly inhibited (4); Na$^+$/K$^+$-exchanging ATPase (2,5); and phospholipase A$_1$, weakly inhibited (6). **1.** Karhuvaara, Kaila & Huupponen (1989) *J. Pharm. Pharmacol.* **41**, 649. **2.** Whikehart, Montgomery & Sorna (1992) *J. Ocul. Pharmacol.* **8**, 107. **3.** Puscas, Reznicek, Moldovan, Puscas & Sturzu (1985) *Med. Interne.* **23**, 185. **4.** Alkondon, Ray & Sen (1986) *J. Pharm. Pharmacol.* **38**, 848. **5.** Whikehart, Montgomery, Sorna & Wells (1991) *J. Ocul. Pharmacol.* **7**, 195. **6.** Pappu, Yazaki & Hostetler (1985) *Biochem. Pharmacol.* **34**, 521.

Timonacic, See Thiazolidine-4-carboxylate

Tingitanine, See Lathyrine

Tinidazole

This 5-nitroimidazole antibiotic pro-drug (FW = 247.27 g/mol; CAS 19387-91-8; IUPAC: 1-(2-ethylsulfonylethyl)-2-methyl-5-nitroimidazole), also known by the trade names Tindamax®, Fasigyn®, Simplotan®, and Sporinex®, is active *in vitro* against a wide variety of anaerobic bacteria and protozoa. It is indicated for the treatment of infections arising from amoebae, giardia and trichomonas. Tinidazole is one of the most active antibacterial agents against *Bacteroides fragilis*, one of the most resistant species of anaerobic bacteria. Tinidazole is an FDA- approved treatment of infections caused by *Trichomonas vaginalis*, *Entamoeba histolytica* and *Giardia lamblia*. Since tinidazole has no activity against aerobic bacteria, it must be combined with other antibacterial agents in the treatment of mixed infections involving aerobic and anaerobic bacteria. *See Metronidazole for details on pro-drug activation and mechanisms of action.* **1.** Nord & Kager (1983) *Infection* **11**, 54.

Tin-mesoporphyrin

This porphyrin derivative (FW$_{free-acid}$ = 683.39 g/mol; CAS 106344-20-1) competitively inhibits heme oxygenase (1-3), the rate-limiting enzyme in the heme catabolic pathway. It may be administered as a single intramuscular dose to very-low-birth-weight infants experiencing severe hemolytic hyperbilirubinemia. Tin-heme complexes (particularly Sn(II)-Protoporhyrin) accumulate in humans during severe iron deficiency anemia, wherein ferrochelatase inserts Sn^{2+} in the face of extremely low Fe^{2+} concentrations. **1.** Wagner, Hua, de Courten-Myers, *et al.* (2000) *Cell Mol. Biol. (Noisy-le-grand)* **46**, 597. **2.** Boni, Huch Boni, Galbraith, Drummond & Kappas (1993) *Pharmacology* **47**, 318. **3.** Delaney, Mauzerall, Drummond & Kappas (1988) *Pediatrics* **81**, 498.

Tin-Protoporphyrin IX

This tin-containing porphyrin derivative (FW$_{free-acid}$ = 679.36 g/mol) inhibits ferrochelatase (1) and heme oxygenase (2-5). The extent to which tin is incorporated into hemoglobin-bound protoporphryn IX is an indicator of iron scarcity during bouts of iron deficiency anemia. **1.** Dailey, Jones & Karr (1989) *Biochim. Biophys. Acta* **999**, 7. **2.** Anderson, Simionatto, Drummond & Kappas (1984) *J. Pharmacol. Exp. Ther.* **228**, 327. **3.** Sardana & Kappas (1987) *Proc. Natl. Acad. Sci. U.S.A.* **84**, 2464. **4.** Delaney, Mauzerall, Drummond & Kappas (1988) *Pediatrics* **81**, 498. **5.** Jozkowicz, Huk, Nigisch, *et al.* (2003) *Antioxid. Redox Signal* **5**, 155.

Tinzaparin

This low-molecular-weight heparin, *or* LMWH (MW$_{average}$ ≈ 6600 g/mol; CAS 9005-49-6 (free acid) and 9041-08-1 (sodium salt)), marketed under the trade name Innohep™, is an FDA-approved, once-daily antithrombotic drug for the treatment and prophylaxis of deep-vein thrombosis and pulmonary embolism. Tinzaparin is a LMWH produced by heparinase digestion of heparin. Tinzaparin is believed to be safe during pregnancy and for treating critically ill patients experiencing renal failure. (*For a listing of likely off-target inhibitory effects, see Heparin*) **1.** Friedel & Balfour (1994) *Drugs* **48**, 638.

Tiopronin

This FDA-registered orphan drug (FW = 163.19 g/mol; CAS 217-778-4), also known as Tiopronin (TSH), chemically reduces cystine to cysteine (*Reactions:* Cys-S–S-Cys + T-SH$_{excess}$ ⇌ Cys-S-S-T + Cys-SH; Cys-S–S-T + T-SH$_{excess}$ ⇌ T-S-S-T + Cys-SH), thereby forming cysteine (Cys-SH), the more soluble amino acid, and lowering otherwise supersaturating levels of cystine (Cys-S–S-Cys) that drive the formation of cystine stones in the kidneys of patients with cystinuria. Owing to the chronic nature of this disorder, tiopronin treatment requires many months, and often years, of administration. It is also a ROS scavenger and is an investigational drug for treating rheumatoid arthritis. Tiopronin also is an inhibitor of glutathione peroxidase (3), a finding that explains the otherwise confounding finding that, despite its action an antioxidant, tiopronin can actually elevate oxidative stress. Tiopronin also elicits collateral sensitivity (*or* CS), a phenomenon of drug hypersensitivity that is defined by the ability of certain compounds to selectively target MDR cells, but not the drug-sensitive parent cells from which they were derived (4). Unlike other CS-promoting agents, however, tiopronin's action does not depend on the expression of P-gp in MDR cells (4). Instead, its CS activity is mediated through the generation of reactive oxygen species (ROS) and can be reversed by a variety of ROS-scavenging compounds. Moreover, tiopronin's selective toxicity of towards MDR cells is achieved by inhibition of glutathione peroxidase (GPx). The pharmacokinetics of tiopronin and its principal metabolite, 2-mercaptopropionic acid (2-MPA) are complex, with delayed maximal absorption of tiopronin and high protein and tissue binding of the drug, making its bioavailability variable (5,6). **1.** Joly, Rieu, Méjean, *et al.* (1999) *Pediatr. Nephrol.* **13**, 945. **2.** Lindell, Denneberg & Jeppsson (1995) *Nephron.* **71**, 328. **3.** Chaudiere, Wilhelmsen & Tappel (1984) *J. Biol. Chem.* **259**, 1043. **4.** Hall, Marshall, Kwit, *et al.* (2014) *J. Biol. Chem.* **289**, 21473. **5.** Hercelin, Leroy, Nicolas, *et al.* (1992) *Eur. J. Clin. Pharmacol.* **43**, 93. **6.** Carlsson, Denneberg, Emanuelsson, Kågedal & Lindgren (1993) *Eur. J. Clin. Pharmacol.* **45**, 79.

Tiotidine

This cyanoguanidinium derivative (FW$_{free-base}$ = 312.42 g/mol; CAS 69014-14-8; Soluble to 100 mM DMSO), also known as ICI 125,211 and is a histamine H$_2$ receptor inverse agonist, showing negligible activity against H$_1$- and H$_3$- receptors (1-3). Tiotidine binds with high affinity to a form of the receptor coupled to G-protein that then remains inactive. It binds with high affinity to a site that no longer exists in the presence of GTPγS or in desensitized cells (4). Tiotidine binding leads to a dose-dependent decrease in cAMP basal levels in U-937 cells, an effect that is evident in both homologous and heterologous overexpression systems, as well as in forskolin pretreated cells (4). **Target(s):** aldehyde dehydrogenase (5); cytochrome P450 (6); and histamine H$_2$ receptor (7,8). **1.** Hill (1990) *Pharmacol. Rev.* **42**, 45. **2.** Trzeciakowski & Levi (1980) *J. Pharmacol. Exp. Ther.* **214**, 629. **3.** Yellin, *et al.* (1979) *Life Sci.* **25**, 2001. **4.** Monczor, Fernández, Legnazzi, *et al.* (2003) *Molec. Pharmacol.* **64**, 512. **5.** Kikonyogo & Pietruszko (1997) *Mol. Pharmacol.* **52**, 267. **6.** Bast, Savenije-Chapel & Kroes (1984) *Xenobiotica* **14**, 399. **7.** Domschke & Domschke (1980) *Hepatogastroenterology* **27**, 163. **8.** Richardson, Feldman, Brater & Welborn (1981) *Gastroenterology* **80**, 301.

Tiotropium Bromide

This long-acting muscarinic receptor antagonist (FW = 472.43 g/mol; CAS 136310-93-5; Soluble to 20 mM in water and to 100 mM in DMSO), also named (1α,2β,4β,7β)-7-[(hydroxydi-2-thienylacetyl)oxy]-9,9-dimethyl-3-oxa-9-azoniatricyclo[3.3.1.0^{2,4}]nonane bromide and Spiriva®, binds to human muscarinic receptors: M$_1$ (found mediating slow EPSP at the ganglion in the postganglionic nerve, commonly in exocrine glands and in the CNS, M$_2$, (found on cholinergic nerve terminals) and M$_3$ (found on airway smooth muscle), with K_d values in the sub-nanomolar concentration range. Tiotropium bromide is widely used as a bronchodilator for treating chronic obstructive pulmonary disease. The dissociation rate constants k_{off} are extremely low: 1.3×10^{-5} s^{-1}, 5.3×10^{-5} s^{-1}, and 5.6×10^{-6} s^{-1} for human receptors M$_1$, M$_2$, and M$_3$ (1). In this respect, tiotropium bromide exhibits "kinetic subtype selectivity", manifested by a more rapid dissociation from human M$_2$ than from the M$_1$ and M$_3$ receptors. When examined *in vitro*, tiotropium exerts potent inhibitory effects against cholinergic nerve-induced contraction of guinea-pig and human airways, with a slower onset than atropine or ipratropium bromide. Upon washout, tiotropium bromide dissociates extremely slowly compared with atropine and ipratropium bromide. Spiriva demonstrates a prolonged protective effect against cholinergic agonists and cholinergic nerve stimulation in animal and human airways (2). **1.** Disse, Reichl, Speck, *et al.* (1993) *Life Sci.* **52**, 537. **2.** Barnes (2001) *Expert Opin. Investig. Drugs* **10**, 733.

Tipiracil

This novel thymidine phosphorylase inhibitor (FW = 242.67 g/mol; CAS 183204-74-2), also named 5-chloro-6-[(2-imino-1-pyrrolidinyl)methyl]-2,4(1H,3H)-pyrimidinedione, is a uracil derivative that is used in use in combination with trifluridine (*i.e.*, as the binary drug TAS-102 or Lonsurf®). TAS-102 is a novel oral nucleoside antitumor agent containing both trifluridine and tipiracil to improve overall survival of patients with unresectable advanced or recurrent colorectal cancer that are insensitive to standard chemotherapies. Trifluridine incorporation into DNA induces DNA dysfunction, including DNA strand breaks (**See** *5-(Trifluoromethyl)-2'-deoxyuridine*). Thymidine phosphorylase inhibition prevents metabolic inactivation of trifluridine. **1.** Tanaka, Sakamoto, Okabe, *et al.* (2014) *Oncol. Rep.* **32**, 2319.

Tipifarnib

This orally bioavailable farnesyltransferase inhibitor (FW = 489.40 g/mol; CAS 192185-72-1; Solubility = 14 mg/mL DMSO), also known by its code name R115777, the trade name Zarnestra®, and its systematic name (*R*)-6-(amino(4-chlorophenyl)(1-methyl-1H-imidazol-5-yl)methyl)-4-(3-chlorophenyl)-1-methyl-quinolin-2(1H)-one, is intended as a treatment for acute myeloid leukemia (AML) in elderly patients who are not candidates for standard chemotherapy. **Mode of Action:** Ras is the most common human oncogene, and mutations that activate Ras are found in 20-25% of all human tumors, with up to 90% in certain cancers. Tipifarnib targets prenylation (IC$_{50}$ = 0.6 nM) of the CAAX motif (where C is cysteine residue, A and A are two aliphatic residues, and X is any C-terminal amino acid, depending on a particular prenyltransferase's substrate specificity) required for membrane localization of small GTPase Ras to bind to the membrane, where it subsequently switches on other proteins that ultimately turn on genes involved in cell growth, differentiation and survival. **Inhibition of Tumor Cell Growth:** When tested *in vitro*, R115777 competitively inhibits the farnesylation of lamin B and K-RasB peptide substrates, with IC$_{50}$ values of 0.86 nM and 7.9 nM, respectively (1). In a panel of 53 human tumor cell lines, the growth of nearly three-fourths were found to be sensitive to Tipifarnib. The majority of sensitive cell lines had a wild-type ras gene. Tumor cell lines bearing H-ras or N-ras mutations were among the most sensitive of the cell lines tested, with responses observed at nanomolar concentrations of R115777 (1). Tumor cell lines bearing mutant K-ras genes required higher concentrations for inhibition of cell growth, with 50% of the cell lines resistant to R115777 up to concentrations of 500 nM (1). Inhibition of H-Ras, N-Ras, and lamin B protein processing was observed at concentrations of R115777 that inhibited cell proliferation. R115777 also induces dose-dependent growth inhibition of the three myeloma cell lines, attended by significant and time-dependent apoptosis. Phosphorylation of both STAT3 and ERK1/2 induced by IL-6 was totally blocked at 15 μM of R115777 and partially blocked when R115777 was used at 10 and 5 μM. **1.** End, Smets, Todd, *et al.* (2001) *Cancer Res.* **61**, 131. **2.** Le Gouill, Pellat-Deceunynck & Harousseau (2002) *Leukemia* **16**, 1664.

TIPP

This conformationally constrained tetrapeptide (FW = 634.73 g/mol), also known as Tyr-Tic-Phe-Phe, is a δ-opioid receptor antagonist (1). This report also demonstrated that the tetrapeptide amide Tyr-D-Phe-Phe-Phe-NH$_2$ is a potent μ-selective opioid agonist, consisting entirely of aromatic amino acid residues that can be conformationally restricted in a number of ways. Substitution of the D- and L- enantiomers of the conformationally restricted phenylalanine analogues N^α-methylphenylalanine and tetrahydro-3-isoquinoline carboxylic acid (Tic) for D-Phe produces astonishing changes in receptor affinity and intrinsic activity. These structure-activity relationship studies defined a class of potent and selective antagonists, characterized by the N-terminal sequence H-Tyr-Tic-Phe-. **1.** Schiller, Nguyen, Weltrowska, *et al.* (1992) *Proc. Natl. Acad. Sci. U.S.A.* **89**, 11871.

Tipranavir

This nonpeptidic HIV-1 protease inhibitor (FW = 602.66 g/mol; CAS 174484-41-4), also known as PNU-140690E, Aptivus™, and systematically as N-{3-[(1R)-1-[(2R)-6-hydroxy-4-oxo-2-(2-phenylethyl)-2-propyl-3,4-dihydro-2H-pyran-5-yl]propyl]phenyl}-5-(trifluoromethyl)pyridine-2-sulfonamide, targets the virus-specific processing of the viral Gag and Gag-Pol polyproteins in HIV-1 infected cells, thereby preventing the formation of mature virions (1-6). Tipranavir is orally available, especially potent (K_i = 8 pM), and inhibits protease inhibitor-resistant clinical HIV-1 isolates. While the presence of five or fewer protease gene mutations or 1-2 protease inhibitor resistance-associated mutations (PRAMs) is enough to reduce most currently available protease inhibitors, 16-20 mutations (including three or more PRAMs) are needed to achieve resistance to tipranavir (7). Aptivus is used in combination with ritonavir to treat adult and pediatric HIV-1 infected patients infected with HIV-1 strains that are resistant to more than one protease inhibitor. **1.** Poppe, Slade, Chong, *et al.* (1997) *Antimicrob. Agents Chemother.* **41**, 1058. **2.** Turner, Strohbach, Tommasi, *et al.* (1998) **41**, 3467. **3.** Thaisrivongs & Strohbach (1999) *Biopolymers* **51**, 51. **4.** Rusconi, La Seta Catamancio, Citterio, *et al.* (2000) *Antimicrob. Agents Chemother.* **44**, 1328. **5.** Specker, Böttcher, Brass, *et al.* (2006) *Chem. Med. Chem.* **1**, 106. **6.** Surleraux, de Kock, Verschueren, *et al.* (2005) *J. Med. Chem.* **48**, 1965. **7.** Plosker & Figgitt (2003) *Drugs* **63**, 1611.

TIQ-A, See *Thieno[2,3-c]isoquinolin-5-one*

Tirapazamine

This hypoxia-induced cytotoxin and antineoplastic agent (FW = 178.15 g/mol; CAS 27314-97-2), also known as 3-amino-1,2,4-benzotriazine 1,4-dioxide and SR-4233, is a bioreductive drug that exploits tumor hypoxia to interrupt cell cycle progression and induce apoptosis. Tirapazamine is metabolized to form a transient oxidizing radical. In the absence of oxygen, this radical reacts with DNA to form DNA radicals, largely at C4' on the ribose ring, culminating in DNA strand breaks. **Target(s):** DNA replication (1); DNA topoisomerase II (2); and oxidative phosphorylation (3). **1.** Peters, Wang, Brown & Iliakis (2001) *Cancer Res.* **61**, 5425. **2.** Peters & Brown (2002) *Cancer Res.* **62**, 5248. **3.** Ara, Coleman & Teicher (1994) *Cancer Lett.* **85**, 195.

Tiratricol, See *3,3',5-Triiodothyroacetate*

Tirofiban

This platelet aggregation inhibitor (FW = 440.60 g/mol; CAS 144494-65-5), known initially as MK-383 and L-700,462, marketed under the trade names Aggrastat™, and systematically named as N-(butylsulfonyl)-O-(4-(4-piperidyl)butyl)-L-tyrosine, is a synthetic, nonpeptide antagonist of platelet glycoprotein (GP) IIb/IIIa receptors, thereby accounting for its anticoagulant action by blocking platelet aggregation (1,2). Based on a peptide in an anticoagulant found in the venom of the saw-scaled viper *Echis carinatus*, tirofiban was the first is the first nonpeptide GPIIb/IIIa antagonist of this new class of antiplatelet agents. The plasma IC_{50} for MK-383 in healthy volunteers is approximately 13 ng/mL, with a Hill coefficient n_H > 5 (3). **1.** Hartman, Egbertson, Halczenko, *et al.* (1992) *J. Med. Chem.* **35**, 4640. **2.** Peerlinck, de Lepeleire, Goldberg *et al.* (1993) *Circulation* **88**, 1512. **3.** Barrett, Murphy, Peerlinck *et al.* (1994) *J. Clin. Pharmacol. Ther.* **56**, 377.

Tiron

This metal ion chelator ($FW_{disodium\ salt}$ = 314.20 g/mol; CAS 149-45-1), also known as 4,5-dihydroxybenzene-1,3-disulfonate and catechol-3,5-disulfonate, forms high-affinity complexes with many heavy metal ions, particularly with Fe^{3+}, and is thus a likely inhibitor of metalloenzymes. Tiron is used in colorimetric assays for iron, manganese, titanium, and molybdenum, but also binds cobalt, nickel, copper, and zinc ions. Complexes with heavy metals are water soluble and are typically colored: ferric chloride produces a deep blue color below pH 5, titanium salt complexes are orange, copper salts will generate a greenish yellow color, and Mo(VI) produces a complex with a canary yellow color. Tiron is a cell-permeable superoxide scavenger (1), and is oxidized to an EPR-visible semiquinone by the superoxide radical (2). The latter property provides a probe for determining whether the superoxide anion or molecular oxygen is the true substrate for a particular mono- or dioxygenase. Tiron likewise inhibits those processes that require the participation of the superoxide anion. **Target(s):** aliphatic aldoxime dehydratase (3); arylacetonitrilase, weakly inhibited (4); catechol 1,2-dioxygenase (5); catechol 2,3-dioxygenase (6); CMP-N-acetylneuraminate hydroxylase (7); glycerol-3-phosphate dehydrogenase (8); 3-hexulose-6-phosphate synthase (9); indoleamine 2,3-dioxygenase (10); lactate racemase (11); NADH peroxidase (12); nitrile hydratase (13); nitrogenase (14,15); tryptophan 5-monooxygenase (16); and vanillate hydroxylase (17). **1.** Greenstock & Miller (1975) *Biochim. Biophys. Acta* **396**, 11. **2.** Ledenev, Konstantinov, Popova & Ruuge (1986) *Biochem. Int.* **13**, 391. **3.** Xie, Kato, Komeda, Yoshida & Asano (2003) *Biochemistry* **42**, 12056. **4.** Nagasawa, Mauger & Yamada (1990) *Eur. J. Biochem.* **194**, 765. **5.** Nakazawa & Nakazawa (1970) *Meth. Enzymol.* **17A**, 518. **6.** Klecka &Gibson (1981) *Appl. Environ. Microbiol.* **41**, 1159. **7.** Shaw, Schneckenburger, Carlsen, Christiansen & Schauer (1992) *Eur. J. Biochem.* **206**, 269. **8.** Dawson & Thorne (1975) *Meth. Enzymol.* **41**, 254. **9.** Kato (1990) *Meth. Enzymol.* **188**, 397. **10.** Hirata & Hayaishi (1975) *J. Biol. Chem.* **250**, 5960. **11.** Hiyama, Fukui & Kitahara (1968) *J. Biochem.* **64**, 99. **12.** Badwey & Karnovsky (1979) *J. Biol. Chem.* **254**, 11530. **13.** Nagasawa, Nanba, Ryuno, Takeuchi & Yamada (1987) *Eur. J. Biochem.* **162**, 691. **14.** Bulen & LeComte (1972) *Meth. Enzymol.* **24**, 456. **15.** Burns & Hardy (1972) *Meth. Enzymol.* **24**, 480. **16.** Massey & Hemmerich (1975) *The Enzymes*, 3rd ed. (Boyer, ed.), **12**, 191. **17.** Buswell & Eriksson (1988) *Meth. Enzymol.* **161**, 274.

Δ^7-Tirucallol

This triterpene (FW = 425.72 g/mol; CAS 514-46-5), also known as tirucalla-7,24-dien-3β-ol and a diastereoisomer of lanost-7,24-dien-3β-ol, isolated from the flowers of the genus *Compositae*, inhibits both chymotrypsin, K_i = 72 μM, and trypsin, K_i = 152 μM (1). **1.** Rajic, Akihisa, Ukiya, *et al.* (2001) *Planta Med.* **67**, 599.

Tissue Factor Pathway Inhibitor

This 276-residue glycoprotein (MW = 34-40 kDa, depending on site/extent of proteolysis; *Symbol*: TFPI), also known as Lipoprotein-Associated Coagulation Inhibitor, is a multivalent protease inhibitor that arrests coagulation factor Xa with slow, tight-binding kinetics. TFPI is itself a complex protein, consisting of a highly anionic N-terminal region, three tandem Kunitz domains, and a highly cationic C-terminus. It blocks the initiation of coagulation by inhibiting TF-activated factor VII, activated factor X, and early prothrombinase. Coagulation factor VIIa is also inhibited by the formation of a complex consisting of VIIa, Xa, tissue factor pathway inhibitor, and tissue factor (1-8). While TFPI inhibits the activity of factor Xa and the complex of tissue factor and factor VIIa, TFPI-2 does not inhibit factor Xa; it is instead a strong inhibitor of factor XIa, plasma kallikrein, plasmin, and trypsin (1,4,8). Prothrombinase inhibition by TFPIα is mediated through a high-affinity exosite interaction between the basic region of TFPIα and the FV acidic region, which is retained in FXa-activated FVa and platelet FVa (9). This inhibitory action is not mediated by TFPIβ and is lost upon removal of the acidic region of FVa by thrombin. These properties define an isoform-specific anticoagulant function for TFPIα, and, when combined with earlier evidence of differential expression patterns of TFPIα and TFPIβ in platelets and endothelial cells, suggest that TFPI isoforms have distinct mechanisms for inhibiting the initial stages of intravascular coagulation: TFPIβ dampens TF present on the surface of vascular cells, and TFPIα dampens the initial prothrombinase on the surface of activated platelets. Humans produce the 3' splice variants TFPIα and TFPIβ that are differentially expressed in endothelial cells and platelets and that possess distinct structural features affecting their inhibitory function. TFPI also undergoes alternative splicing of Exon-2 within its 5'-untranslated region (10). Exon-2 behaves as a molecular switch preventing translation of TFPIβ, and a 5'-untranslated region alternative splicing event alters translation of isoforms by independent 3' splicing events within the same gene. **Effect of LDL:** Addition of full-length free TFPI (TFPIα) to plasma prolongs the clotting time, whereas low-density lipoprotein particles (LDL-TFPI) exclusively decreased the peak height of thrombin without effecting the lag phase (11). Steady-state and transient kinetics showed that LDL-TFPI was a more potent inhibitor of FXa than TFPIα and TFPI$_{1-161}$, indicating that FXa inhibition was not rate-determining for the lag phase, whereas it affects thrombin generation during the propagation phase (11). Note: Activated coagulation factor XI (FXIa) neutralizes both endothelium- and platelet-derived Tissue Factor Pathway Inhibitor, *or* TFPI (itself an essential reversible inhibitor of activated Factor X (FXa) and the FVIIa-TF complex), cleaving the protein between the Kunitz K1 and K2 domains (at Lys86/Thr87) and at the active sites of the K2 domain (at Arg107/Gly108) and K3 domain (at Arg199/Ala200) (12). **Target(s):** cathepsin G (1); chymotrypsin (1); coagulation factor VIIa, when in the presence of factor Xa (2-4); coagulation factor Xa (5-7); coagulation factor Xia (8); kallikrein (1,8); plasmin (8); and trypsin (1,4,8). **1.** Konduri, Rao, Chandrasekar, *et al.* (2001) *Oncogene* **20**, 6938. **2.** Morrissey (1998) in *Handb. Proteolytic Enzymes* (Barrett, Rawlings & Woessner, eds.), p. 161, Academic Press, San Diego. **3.** Mann & Lorand (1993) *Meth. Enzymol.* **222**, 1. **4.** Girard & Broze (1993) *Meth. Enzymol.* **222**, 195. **5.** Stenflo (1998) in *Handb. Proteolytic Enzymes* (Barrett, Rawlings & Woessner, eds.), p. 163, Academic Press, San Diego. **6.** Manithody, Yang & Rezaie (2002) *Biochemistry* **41**, 6780. **7.** Franssen, Salemink, Willems, *et al.* (1997) *Biochem. J.* **323**, 33. **8.** Sierko, Zawadzki & Wojtukiewicz (2002) *Pol. Merkuriusz. Lek.* **13**, 66. **9.** Wood, Bunce, Maroney, *et al.* (2013) *Proc. Natl. Acad. Sci. U.S.A.* **110**, 17838. **10.** Ellery, Maroney, Martinez, Wickens & Mast (2013) *Arterioscler. Thromb. Vasc. Biol.* **34**, 187. **11.** Augustsson, Hilden & Petersen (2014) *Thromb. Res.* **134**, 132. **12.** Puy, Tucker, Matafonov, *et al.* (2015) *Blood* **125**, 1488.

Tissue Inhibitor of Metalloproteinase

This family of naturally occurring glycoproteins (Symbol: TIMP) target metalloproteinases (MMPs), the Zn^{2+}-activated peptidases that degrade extracellular matrix (ECM) components, thereby promoting tumor growth, angiogenesis and metastatic disease in a wide range of cell types. **Human TIMP1** (MW = 23171 g/mol; Accession: CAG46779) forms binary complexes with target metalloproteinases, such as collagenases, irreversibly inactivating the latter by interacting with their catalytic zinc cofactor. TIMP1 inhibits MMP1, MMP2, MMP3, MMP7, MMP8, MMP9, MMP10, MMP11, MMP12, MMP13 and MMP16; however, it fails to inhibit MMP14. In some cases, TIMP1 acts as a growth factor, regulating cell differentiation, migration and cell death and activating cellular signaling cascades via CD63 and ITGB1. TIMP1 lays a role in integrin signaling. TIMP1 also promotes erythropoiesis *in vitro*, stimulating the growth and differentiation of human erythroid progenitor cells. **Human TIMP2** (MW = 16082 g/mol; Accession:

EAW89541) uniquely activate MMP-2, using its N-terminus to bind and inhibit MT1-MMP and employing its C-terminus to bind to secreted zymogen, pro-MMP-2. Upon binding of TIMP-2 to MT1-MMP, a neighboring uninhibited MMP-14 subunit can cleave and activate a bound pro-MMP-2, when bound to $α_vβ_3$ integrin receptor on the cell surface. **Human TIMP3** (MW = 24145 g/mol; Accession: CAG30479) inhibiting Tumor-Necrosis Factor α (TNF-α)-Converting Enzyme (*or* TACE, also known as ADAM17) controlling levels of TNF in the liver. **Human TIMP4** (MW = 25503 g/mol; Accession: AAC34422) is involved in regulation of platelet aggregation and recruitment and may play role in hormonal regulation and endometrial tissue remodeling. **1.** Gomez, Alonso, Yoshiji & Thorgeirsson (1997) *Eur. J. Cell Biol.* **74**, 111.

Tissue Inhibitor of Metalloproteinases 1, See *TIMP-1*

Tissue Inhibitor of Metalloproteinases 2, See *TIMP-2*

Tityustoxin-Kα

This oligopeptide toxin (MW = 3942 Da) from the Brazilian scorpion *Tityus serrulatus* increases Na^+ permeability (1) and potently and selectively inhibits the voltage-gated non-inactivating K^+ channel, K_d = 0.21 nM (2). TsTX-Kα does not interact with the gating of $K_V1.3$ since it did not affect the voltage-dependence of channel activation (3). Given its photosensitivity, tityustoxin-Kα should be stored in a light-tight plastic tube. **1.** Conn (1983) *Meth. Enzymol.* **103**, 401. **2.** Werkman, Gustafson, Rogowski, Blaustein & Rogawski (1993) *Mol. Pharmacol.* **44**, 430. **3.** Rodrigues, *et al.* (2003) *Brit. J. Pharmacol.* **139**, 1180.

Tivantinib

This synthetic non-ATP-competitive protein kinase inhibitor (FW = 369.32 g/mol; CAS 905854-02-6, 1000873-98-2, and 1228508-24-4; Solubility: 74 mg/mL DMSO, <1mg/mL H_2O), also known as ARQ 197 and (3*R*,4*R*)-3-(2,3-dihydro-1*H*-pyrrolo[3,2,1-*ij*]quinolin-6-yl)-4-(1*H*-indol-3-yl)pyrrolidine-2,5-dione, is an orally avalible selective human c-Met inhibitor (IC_{50} = 0.1 μM). Exposure of c-Met-expressing cancer cell lines to ARQ 197 inhibited proliferation and induced caspase-dependent apoptosis (1). These cellular responses to ARQ 197 were phenocopied by RNAi-mediated c-Met depletion (2). Combined use of of tivantinib and its bisphosphate, zoledronate, also prevents tumor bone engraftment and inhibits progression of established bone metastases in a breast xenograft model (3). Tivantinib also suppresses oral squamous cell carcinoma (OSCC) cell proliferation and colony formation, but its anti-tumor activity is independent of the inhibition of MET signaling pathway (4). Tivantinib causes G_2/M cell cycle arrest and caspase-dependent apoptosis in OSCC cell lines, dose-dependently suppressing FAK activation and expression (4). **1.** Munshi, Jeay, Li, *et al.* (2010) *Mol. Cancer Ther.* 9, 1544. **2.** Comoglio, Giordano & Trusolino (2008) *Nat. Rev. Drug Discov.* **7**, 504. **3.** Previdi, Scolari, Chilà, *et al.* (2013) *PLoS One* **8**, e79101. **4.** Xi, Yang, Cao & Qian (2015) *Biochem. Biophys. Res. Commun.* **457**, 723.

Tivirapine

This non-nucleoside reverse transcriptase inhibitor (FW = 321.87 g/mol; CAS 137232-54-8), also known as R86183, 8-chloro-TIBO, and (+)-*S*-

4,5,6,7-tetrahydro-8-chloro-5-methyl-6-(3-methyl-2-butenyl)imidazo[4,5,1-*jk*][1,4]benzodiazepine-2(1*H*)-thione, inhibits HIV-1, HIV-2, and SIV retroviruses by binding within a hydrophobic pocket near critical active-site catalytic residues in the enzyme-DNA complex. **Target(s):** HIV-1 reverse transcriptase (1-3); HIV-2 and SIV reverse transcriptase (4); RNA-directed DNA polymerase (1-3); viral entry (3); and viral reverse transcriptase (1-3). **1.** Tucker, Lumma & Culberson (1996) *Meth. Enzymol.* **275**, 440. **2.** Balzarini & De Clercq (1996) *Meth. Enzymol.* **275**, 472. **3.** De Vreese, Reymen, Griffin, *et al.* (1996) *Antiviral Res.* **29**, 209. **4.** Witvrouw, Pannecouque, Van Laethem, *et al.* (1999) *AIDS* **13**, 1477.

Tivozanib

This oral VEGF receptor tyrosine kinase inhibitor (FW = 454.86 g/mol; CAS 475108-18-0), also known as AV-951, KRN-951, and 1-{2-chloro-4-[(6,7-dimethoxy-quinolin-4-yl)oxy]phenyl}-3-(5-methylisoxazol-3-yl)urea, targets ligand-induced phosphorylation of VEGFR-1, VEGFR-2 and VEGFR-3, all at picomolar concentrations, with an *in vivo* half-life of 4 days (1). VEGF(165), but not IGF-1 or basic Fibroblast Growth Factor (bFGF), is mainly responsible for increased cellular permeability in retinal endothelial cells (REC), a property that is partially inhibited by tivozanib (2). **1.** Eskens, de Jonge, Bhargava, *et al.* (2011) *Clin. Cancer Res.* **17**, 7156. **2.** Deissler, Deissler & Lang (2011) *Brit. J. Ophthalmol.* **95**, 1151.

TKI258, See *CHIR-258*

T-Kininogen, See *Kininogens*

TLCK, See *N^α-Tosyl-L-lysine Chloromethyl Ketone*

TLK199, See *Ezatiostat*

TM6008

This orally active prolyl hydroxylase inhibitor (FW = 387.40 g/mol), also named 6-amino-1,3-dimethyl-5-(2-pyridin-2-yl-quinoline-4-carbonyl)-1*H*-pyrimidine-2,4-dione, chelates transition metals *in vitro* and likewise inhibits copper-catalyzed, nonenzymatic oxidation of ascorbic acid (1). TM6008 binds to the PHD2 active site, where it chelates the catalytically essential Fe(II) ion. Pharmacokinetics studies in rats (oral dose = 50 mg/kg) gave plasma T_{max}, C_{max}, and $t_{1/2}$ values of 3.5 hour, 0.9 μg/mL and 1.5 hour (1). The protective effect of TM6008 against ischemia-induced cerebral lesions also suggests it may have potential for treating ischemic disorders of heart and/or kidney. **Mode of Inhibitory Action:** Under normoxic conditions, hypoxia-inducible factor (HIF) is constitutively transcribed and translated, but its stability is drastically reduced by the oxygen-dependent enzymatic hydroxylation of proline residues by prolyl hydroxylases (PHD). Hydroxylated HIF is ubiquitinylated and subsequently degraded. Under hypoxic conditions, HIF is not hydroxylated and binds to its heterodimeric partner HIF-1β to transactivate genes involved in hypoxic stress adaptation. By inhibiting oxygen-dependent enzymatic hydroxylation of HIF, TM6008 stimulates HIF activity in various organs of transgenic rats expressing a hypoxia-responsive reporter vector. Notably, reatment of cultured SHSY-5Y cells with TM6008 increased expression levels of heme oxygenase 1, erythropoietin, and glucose transporter-3, all of which were genes downstream of HIF-1α (2). **1.** Nangaku, Izuhara, Takizawa, *et al.* (2007)

Arterioscler. Thromb. Vasc. Biol. **27**, 2548. **2.** Kontani, Nagata, Uesugi, *et al.* (2013) *Neurochem. Res.* **38**, 2588.

TM 5275

This selective, orally active PAI-1 inhibitor and anti-inflammatory agent (FW = 543.97 g/mol; CAS 1103926-82-4; Soluble in DMSO) targets Plasminogen Activator Inhibitor-1 (IC_{50} = 6.95 μM) in an assay of tissue plasminogen activator-dependent peptide hydrolysis (1). Plasminogen activator inhibitor (PAI)-1, a serine protease inhibitor, is involved in numerous processes including thrombosis and fibrosis, TM-5275 provides antithrombotic benefits devoid of bleeding effect in nonhuman primates (1). TM 5275 also prolongs tissue plasminogen activator (tPA) retention and enhances plasmin generation on the vascular endothelial cell surface (2). TM-5275 represents a novel class of anti-inflammatory agents targeting macrophage migration by the inhibition of the interaction of PAI-1 with low-density lipoprotein receptor-related protein, with an IC_{50} = 3 μM for LRP1 protein Cl II or Cl IV (3). **Pharmacokinetics:** An oral dose of 50 mg/kg of TM5275, administered in rats, yields calculated plasma T_{max}, C_{max}, and $t_{1/2}$ of 2 hours, 34 μmol/L, and 2.5 hours, respectively, *versus* 18 hours, 8.8 μmol/L, and 124 hours, respectively, in rats administered the same dose of TM5007 (1). TM5275 thus increases C_{max} fourfold, and markedly shortens both T_{max} and $t_{1/2}$. In mice, an oral dose of 50 mg/kg of TM5275 yields the following values for these parameters: 1 hour, 6.9 μmol/L, and 6.5 hours, respectively (1). In monkeys, an oral dose of 1 mg/kg of TM5275 yields T_{max}, C_{max}, and $t_{1/2}$ values of 6 hours, 10.5 μmol/L, and 114.7 hours, respectively, with a bioavailability of 96% in monkeys (1). **1.** Izuhara, Yamaoka, Kodama, *et al.* (2010) *J. Cereb. Blood Flow Metab.* **30**, 904. **2.** Yasui, *et al.* (2013) *Thromb. Res.* **132**, 100. **3.** Ichimura, *et al.* (2013) *Arterioscler. Thromb. Vasc. Biol.* **33**, 935.

TMB-8, See *8-(Diethylamino)octyl 3,4,5-Trimethoxybenzoate*

TMC114, See *Darunavir*

TMC125, See *Etravirine*

TMC207, See *Bedaquiline*

TMC278, See *Rilpivirine*

TNP-470

This semisynthetic fumagillin analogue (FW = 401.89 g/mol; CAS 129298-91-5; Solubility: >15 mg/mL DMSO), also known as AGM-1470, NSC 642492, and *O*-chloroacetylcarbamoyl fumagillol, is an angiogenesis inhibitor that targets methionyl aminopeptidase (IC_{50} = 2 μM), a cytosolic metalloenzyme that catalyzes the hydrolytic removal of N-terminal methionine residues from nascent proteins (1-3). TNP-470 also blocks posttranslational modification and translocation of nitric-oxide synthase to the peripheral membrane (4). (**See also** *Fumagillin; Beloranib*) **1.** Zhang, Huang, Cali, *et al.* (2005) *Folia Parasitol.* **52**, 182. **2.** Liu, Widom, Kemp, C. M. Crews & Clardy (1998) *Science* **282**, 1324. **3.** Brdlik & Crews (2004) *J. Biol. Chem.* **279**, 9475. **4.** Yoshida, Kaneko, Tsukamoto, *et al.* (1998) *Cancer Res.* **58**, 3751.

TNP-ATP, See *2',3'-O-(2,4,6-Trinitrophenyl)adenosine 5'-Triphosphate*

Tobramycin

This aminoglycoside antibiotic ($FW_{free-base}$ = 467.52 g/mol; CAS 32986-56-4; Soluble in 1.5 parts water; pK_a values indicated for each amino group), a component of the nebramycin complex from *Streptomyces tenebrarius*, inhibits bacterial protein synthesis (IC_{50} = 9.7 μM) as well as enzymes of various classes. Tobramycin binds tightly to 30S and 50S ribosome subunits, thereby preventing formation of the 70S complex and inhibiting synthesis of bacterial proteins. (***For additional mechanistic details***) See Neomycins. **Target(s):** *N*-acetyl-β-D-glucosaminidase, weakly inhibited (1); DNA-directed DNA polymerase, Klenow fragment (2); hepatitis delta virus genomic ribozyme self-cleavage (3); kanamycin kinase (4); leukocyte elastase, slightly inhibited (5); lysozyme (6;) phospholipase A_1, inhibition may be due to substrate depletion (7-11); phospholipase A_2 (8-10,12,13); phospholipase C (9,10,14,15); poly(A)-specific ribonuclease (2); bacterial protein biosynthesis, primarily by binding to the A site in the ribosome (16); protein kinase C (17); ribonuclease P (18); streptomycin 3''-adenylyltransferase (19). **1.** Mozer, Gibey, Dupond & Henry (1988) *Pathol. Biol. (Paris)* **36**, 230. **2.** Ren, Martínez, Kirsebom & Virtanen (2002) *RNA* **8**, 1393. **3.** Chia, Wu, Wang, Chen & Chen (1997) *J. Biomed. Sci.* **4**, 208. **4.** Haas & Dowding (1975) *Meth. Enzymol.* **43**, 611. **5.** Jones, Elphick, Pettitt, Everard & Evans (2002) *Eur. Respir. J.* **19**, 1136. **6.** Shiono & Hayasaka (1985) *Arch. Ophthalmol.* **103**, 1747. **7.** Hostetler & Jellison (1990) *J. Pharmacol. Exp. Ther.* **254**, 188. **8.** Carlier, Laurent & Tulkens (1984) *Arch. Toxicol. Suppl.* **7**, 282. **9.** Hostetler & Hall (1982) *Biochim. Biophys. Acta* **710**, 506. **10.** Hostetler & Hall (1982) *Proc. Natl. Acad. Sci. U.S.A.* **79**, 1663. **11.** Uchiyama, Miyazaki, Amakasu, *et al.* (1999) *J. Biochem.* **125**, 1001. **12.** Makela, Kuusi & Schroder (1997) *Scand. J. Clin. Lab. Invest.* **57**, 401. **13.** Carrier, Bou Khalil & Kealey (1998) *Biochemistry* **37**, 7589. **14.** Schwertz, Kreisberg & Venkatachalam (1984) *J. Pharmacol. Exp. Ther.* **231**, 48. **15.** Lipsky & Lietman (1982) *J. Pharmacol. Exp. Ther.* **220**, 287. **16.** Jiménez (1976) *Trends Biochem. Sci.* **1**, 28. **17.** Hagiwara, Inagaki, Kanamura, Ohta & Hidaka (1988) *J. Pharmacol. Exp. Ther.* **244**, 355. **18.** Tekos, Tsagla, Stathopoulos & Drainas (2000) *FEBS Lett.* **485**, 71. **19.** Jana & Deb (2005) *Biotechnol. Lett.* **27**, 519.

Tocilizumab

This humanized anti-IL-6 receptor monoclonal antibody (MW = 145 kDa; CAS 375823-41-9), marketed by Genentech under the trade name Actemra®, binds specifically to interleukin-6 (IL-6) receptors, which bind the versatile pro-inflammatory cytokine IL-6, itself produced by a variety of cell types, including T- and B-cells, lymphocytes, monocytes and fibroblasts. IL-6 is implicated in the pathogenesis of autoimmune disorders, multiple myeloma, and prostate cancer. Unlike other cytokines, IL-6 activates target cells through membrane-bound (IL-6R) and soluble receptors (sIL-6R), widening the number of responsive cell types. Tocilizumab binds to both sIL-6R and mIL-6R, inhibiting IL-6 binding and blocking both sIL-6R and mIL-6R signaling but not that of other IL-6 family cytokines. Administered intravenously on a monthly schedule, tocilizumab has an *in vivo* half-life of 8-14 days, commending its use in the treatment of moderate to severe rheumatoid arthritis, especially in combination with methotrexate. **1.** Mihara, Kasutani, Okazaki, *et al.* (2005) *Int. Immunopharmacol.* **5**, 1731. **2.** Alten & Maleitzke (2013) *Ann. Med.* **45**, 357.

α-Tocopherol

This methylated tocol (FW = 430.71 g/mol; CAS 1406-18-4; MP = 2.5-3.5°C; $λ_{max}$ = 292 nm (ε = 3260 $M^{-1}cm^{-1}$), strongly fluorescent with Emission$_{max}$ at 340 nm) is also called 5,7,8-trimethyltocol and (2*R*)-3,4-dihydro-2,5,7,8-tetramethyl-2-[(4*R*,8*R*)-4,8,12-trimethyl-tridecyl]-2*H*-1-benzopyran-6-ol. **Target(s):** ascorbate oxidase (1); ATPase (2); cyclooxygenase (3); glutathione *S*-transferase (4); lipoxygenase (5-9); 5-lipoxygenase (10,11); phospholipase A_2 (12); protein kinase C (12,13-17); retinyl-palmitate esterase, *or* retinyl esterase (18-21); sphingomyelinase (22); and urokinase, *or* u-plasminogen activator (23). **1.** Stark & Dawson (1963) *The Enzymes*, 2nd ed. (Boyer, Lardy & Myrbäck, eds.), **8**, 297. **2.** Kawai, Nakao, Nakao & Katsui (1974) *Amer. J. Clin. Nutr.* **27**, 987. **3.** Ali, Gudbranson & McDonald (1980) *Prostaglandins Med.* **4**, 79. **4.** van Haaften, Evelo, Haenen & Bast (2001) *Biochem. Biophys. Res. Commun.* **280**, 631. **5.** Tappel (1962) *Meth. Enzymol.* **5**, 539. **6.** Schewe, Wiesner & Rapoport (1981) *Meth. Enzymol.* **71**, 430. **7.** Gwebu, Trewyn, Cornwell & Panganamala (1980) *Res. Commun. Chem. Pathol. Pharmacol.* **28**, 361. **8.** Holman & Bergström (1951) *The Enzymes*, 1st ed. (Sumner & Myrbäck, eds.), **2** (part 1), 559. **9.** Tappel (1963) *The Enzymes*, 2nd ed. (Boyer, Lardy & Myrbäck, eds.), **8**, 275. **10.** Ricciarelli, Zingg & Azzi (2001) *IUBMB Life* **52**, 71. **11.** Reddanna, Rao & Reddy (1985) *FEBS Lett.* **193**, 39. **12.** Douglas, Chan & Choy (1986) *Biochim. Biophys. Acta* **876**, 639. **13.** Chan, Monteiro, Schindler, Stern & Junqueira (2001) *Free Radic. Res.* **35**, 843. **14.** Ricciarelli, Tasinato, Clement, *et al.* (1998) *Biochem. J.* **334**, 243. **15.** Maltseva, Palmina & Burlakova (1998) *Membr. Cell Biol.* **12**, 251. **16.** Boscoboinik, Chatelain, Bartoli, Stauble & Azzi (1994) *Biochim. Biophys. Acta* **1224**, 418. **17.** Kikkawa & Nishizuka (1986) *The Enzymes*, 3rd ed. (Boyer & Krebs, eds.), **17**, 167. **18.** Napoli & Beck (1984) *Biochem. J.* **223**, 267. **19.** Prystowsky, Smith & Goodman (1981) *J. Biol. Chem.* **256**, 4498. **20.** Napoli, McCormick, O'Meara & Dratz (1984) *Arch. Biochem. Biophys.* **230**, 194. **21.** Schindler, Mentlein & Feldheim (1998) *Eur. J. Biochem.* **251**, 863. **22.** Martin, Navarro, Forthoffer, Navas & Villalba (2001) *J. Bioenerg. Biomembr.* **33**, 143. **23.** Ogston (1982) *Acta Haematol.* **67**, 114.

α-Tocopherol Phosphate

This phosphoester ($FW_{free-acid}$ = 510.69 g/mol), also known as α-tocopheryl phosphate, inhibits cytochrome *c* oxidase (1); NAD^+ nucleosidase, *or* NADase (2); phenylalanine 4-monooxygenase (3); succinate dehydrogenase (1,4,5); and succinate oxidase (1). **1.** Rabinovitz & Boyer (1950) *J. Biol. Chem.* **183**, 111. **2.** Spaulding & Graham (1947) *J. Biol. Chem.* **170**, 711. **3.** Woo, Gillam & Woolf (1974) *Biochem. J.* **139**, 741. **4.** Basinski & Hummel (1947) *J. Biol. Chem.* **167**, 339. **5.** Ames (1947) *J. Biol. Chem.* **169**, 50.

α-Tocopherol Succinate

This α-tocopherol succinate ester ($FW_{free-acid}$ = 530.79 g/mol), also known as α-tocopherol acid succinate, is practically insoluble in water and is soluble in organic solvents such as diethyl ether, acetone, and chloroform. It is unstable in alkaline conditions. ***See also*** α-Tocopherol **Target(s):** adenylate cyclase (1); NF-κB transcriptional activity (2); and succinate dehydrogenase (3). **1.** Sahu, Edwards-Prasad & Prasad (1987) *J. Cell Physiol.* **133**, 585. **2.** Nakamura, Goto, Matsumoto & Tanaka (1998) *Biofactors* **7**, 21. **3.** Basinski & Hummel (1947) *J. Biol. Chem.* **167**, 339.

Tocris-1963, *See SB408124*

TOFA

This irreversible fatty acid synthesis inhibitor (FW = 324.59 g/mol; CAS 54857-86-2; $λ_{max}$ = 278 nm), also named RMI 14514 and 5-(tetradecyloxy)-2-furancarboxylic acid, targets the synthesis of malonyl-CoA by acetyl-CoA carboxylase, *or* ACC, reduced food intake and body weight in mice treated with fatty acid synthase inhibitors (1) and resulting in cytotoxicity and apoptosis in human cancer cell lines (2). (**See also** *Cerulenin*) Both cerulenin (~10 µg/ml) and TOFA (~1 µg/ml) are effective in blocking the incorporation of radiolabeled acetate into palmitate; however, TOFA reduces malonyl-CoA levels, rather than elevating them. TOFA is also relatively non-toxic to various cancer cell lines. TOFA attenuates inhibition of feeding, when FAS inhibitors like cerulenin are administered to obese *ob/ob* mice. **1**. Loftus, Jaworsky, Frehywot, *et al.* (2000) *Science* **288**, 2379.

Tofacitinib

This Janus kinase 3 inhibitor and immunosuppressant (FW = 311.39 g/mol; CAS 477600-75-2; Solubility: 100 mg/mL DMSO, <1 mg/mL H_2O), known formerly as tasocitinib and also by its code name CP-690550, its trade name Xeljanz®, and its IUPAC name 3-[(3*R*,4*R*)-4-methyl-3-[methyl(7*H*-pyrrolo[2,3-*d*]pyrimidin-4-yl)amino]piperidin-1-yl]-3-oxopropanenitrile, is an FDA-approved drug for the treatment of rheumatoid arthritis and is under investigation for treatment of psoriasis, inflammatory bowel disease, and prevention of transplant organ rejection. **Mechanism of Inhibitory Action:** Janus kinase 3 (*or* JAK3) is expressed during allograft rejection, and, with a 1-nM K_i for JAK3, tofacitinib delays allograft rejection, significantly prolonging kidney survival. Tofacitinib citrate inhibits interleukin-6-induced phosphorylation of STAT1 and STAT3 with IC_{50} values of 23 nM and 77 nM, respectively. Tofacitinib inhibits IL-2-mediated human T-cell blast proliferation and IL-15-induced CD69 expression with IC_{50} of 11 nM and 48 nM, respectively. When used in combination with the IMP dehydrogenase inhibitor mycophenolate mofetil to block *de novo* guanine nucleotide biosynthesis in leukocytes, Tofacitinib is even more efficient in preventing organ rejection (2). Co-administration of CP-690550 and methotrexate is also well tolerated in patients with rheumatoid arthritis without need for dose adjustment (3) (**See also** *Baricitinib*). **Pharmacokinetics in Humans:** Tofacitinib is rapidly absorbed, with plasma concentrations and total radioactivity peaking ~1 hour after oral administration (4). The mean terminal-phase $t_{1/2}$ was ~3.2 h for both parent drug and total radioactivity. Two-thirds of circulating radioactivity in plasma is the parent drug, with each tofacitinib metabolite <10% of total circulating radioactivity. Hepatic clearance is ~70% of total clearance, with renal clearance accounting for the remainder. Tofacitinib undergoes oxidation of its pyrrolopyrimidine and piperidine rings, oxidation of the piperidine ring side chain, *N*-demethylation, and glucuronidation (4). Consistent with clinical reports of likely drug interactions, tofacitinib is mainly metabolized by CYP3A4, albeit with some contribution from CYP2C19. **1**. Changelian, Flanagan, Ball, *et al.* (2003) *Science* **302**, 875. **2**. Borie, Larson, Flores, *et al.* (2005) *Transplantation* **80**, 1756. **3**. Cohen, Zwillich, Chow, Labadie & Wilkinson (2010) *Brit. J. Clin. Pharmacol.* **69**, 143. **4**. Dowty, Lin, Ryder, *et al.* (2014) *Drug Metab. Dispos.* **42**, 759.

Tofogliflozin

This Type-2 diabetes drug (FW = 404.45 g/mol; CAS 1201913-82-7; Code Name: CSG452; IUPAC Name: (1*S*,3'*R*,4'*S*,5'*S*,6'*R*)-6-(4-ethylbenzyl)-6'-(hydroxymethyl)-3',4',5',6'-tetrahydro-3*H*-spiro[2-benzofuran-1,2'-pyran]-3',4',5'-triol monohydrate) inhibits the sodium-glucose transporter (SGLT2), Subtype 2, a carrier responsible for >90% of kidney glucose reabsorption (1,2). Importantly, such reduction in renal glucose reabsorption only occurs during hyperglycemia, but not under hypo- or euglycemia. The other SGLT2 inhibitors include canagliflozin, dapagliflozin, empagliflozin, and luseogliflozin (3). **1**. Suzuki, Honda, Fukazawa, *et al.* (2012) *J. Pharmacol. Exp. Ther.* **341**, 692. **2**. Ohtake, Sato, Kobayashi, *et al.* (2012) *J. Med. Chem.*

55, 7828. **3**. Nagata, Fukazawa, Honda, *et al.* (2013) *Am. J. Physiol. Endocrinol. Metab.* **304**, E414.

Tolazoline

This vasodilator (FW = 160.22 g/mol; CAS 59-98-3) is an α_2-adrenergic blocker, typically formulated as the water-soluble hydrochloride Priscol, inhibits acetylcholinesterase (1,2) and α_2-adrenergic receptor (3,4). **1**. Augustinsson (1950) *The Enzymes*, 1st. ed. (Sumner & Myrbäck, eds.), **1** (part 1), 443. **2**. Schär-Wüthrich (1943) *Helv. Chim. Acta* **26**, 1836. **3**. Schwartz & Clark (1998) *J. Vet. Pharmacol. Ther.* **21**, 342. **4**. Angel, Niddam & Langer (1990) *J. Pharmacol. Exp. Ther.* **254**, 877.

Tolbutamide

This oral hypoglycemic sulfonylurea (FW = 270.35 g/mol; CAS 64-77-7; M.P.= 128.5-129.5°C), also known as Orinase®, *N*-butyl-*N'*-*p*-toluenesulfonylurea, 1-(*p*-tolylsulfonyl)-3-butylurea, and D-860, stimulates insulin release and is also a substrate for a number of cytochrome P450 systems. Tolbutamide activates 6-phosphofructo-2-kinase while inhibiting fructose-2,6-bisphosphate 2-phosphatase. **Target(s):** acetohexamide reductase, heart (1); ATP-sensitive K^+-channel (2); cAMP-dependent protein kinase, *or* protein kinase A (3,4); cAMP phosphodiesterase (5,6); carnitine acyltransferase (7,8); carnitine palmitoyl-transferase (8,9); cholesterol biosynthesis (10,11); fructose-2,6-bisphosphate 2-phosphatase (12); glucose-6-phosphatase (13-16); glutamine:fructose-6-phosphate aminotransferase, isomerizing (17); hexose-phosphate aminotransferase (17); insulysin (18); lathosterol oxidase (19); lipase (20); lipoprotein lipase, via inhibition of the activation of the enzyme (21); phosphoenolpyruvate carboxykinase (22); protein-glutamine γ-glutamyl-transferase, *or* transglutaminase (23); protein phosphatase (24); and urease (25). **1**. Imamura, Koga, Migita, *et al.* (1997) *J. Biochem.* **121**, 705. **2**. Bryan & Aguilar-Bryan (1999) *Biochim. Biophys. Acta* **1461**, 285. **3**. Kanamori, Hayakawa & Nagatsu (1976) *Biochim. Biophys. Acta* **429**, 147. **4**. Wray & Harris (1973) *Biochem. Biophys. Res. Commun.* **53**, 291. **5**. Goldfine, Perlman & Roth (1971) *Nature* **234**, 295. **6**. Brooker & Fichman (1971) *Biochem. Biophys. Res. Commun.* **42**, 824. **7**. Broadway & Saggerson (1995) *FEBS Lett.* **371**, 137. **8**. Cook (1987) *J. Biol. Chem.* **262**, 4968. **9**. Patel (1986) *Amer. J. Physiol.* **251**, E241. **10**. McDonald & Dalidowicz (1962) *Biochemistry* **1**, 1187. **11**. Dalidowicz & McDonald (1965) *Biochemistry* **4**, 1138. **12**. Kaku, Matsuda, Matsutani & Kaneko (1986) *Biochem. Biophys. Res. Commun.* **139**, 687. **13**. Nordlie (1971) *The Enzymes*, 3rd ed., **4**, 543. **14**. Jasmin & Johnson (1959) *J. Amer. Pharm. Asso.* **48**, 113. **15**. Weber & Cantero (1958) *Metab. Clin. Exptl.* **7**, 333. **16**. Mohnike & Knitsch (1956) *Naturwissenschaften* **43**, 449. **17**. Malathy & Kurup (1972) *Indian J. Biochem. Biophys.* **9**, 310. **18**. Burghen, Kitabchi & Brush (1972) *Endocrinology* **91**, 633. **19**. Dempsey (1969) *Meth. Enzymol.* **15**, 501. **20**. Shepherd & Fain (1977) *Fed. Proc.* **36**, 2732. **21**. Agardh, Bjorgell & Nilsson-Ehle (1999) *Diabetes Res. Clin. Pract.* **46**, 99. **22**. Emoto, Inoue, Kaku & Kaneko (1993) *Biochem. Biophys. Res. Commun.* **191**, 465. **23**. Gomis, Mathias, Lebrun, *et al.* (1984) *Res. Commun. Chem. Pathol. Pharmacol.* **46**, 331. **24**. Gagliardino, Rossi & Garcia (1997) *Acta Diabetol.* **34**, 6. **25**. Herrera, Sabater & Molina (1977) *Rev. Esp. Fisiol.* **33**, 37.

Tolcapone

This orally bioavailable nitrocatechol (FW = 273.25 g/mol; CAS 134308-13-7; Soluble to 100 mM in ethanol and to 100 mM in DMSO), also known as Ro 40-7592, inhibits catechol *O*-methyltransferase (1,2) and is and FDA-approved drug in the treatment of Parkinson Disease. Tolcapone also binds specifically to transthyretin (TTR) in human plasma (K_{d1} = 21 nM; K_{d2} = 58 nM), stabilizing the native tetramer in vivo in mice and humans and inhibiting TTR cytotoxicity. Crystal structures of tolcapone bound to wild-type TTR and to the V122I cardiomyopathy-associated variant show that it docks better into the TTR T4 pocket than tafamidis, the only drug on the market for treating TTR amyloidosis (3). (**See** *Tafamidis*) Already in clinical trials for familial amyloid polyneuropathy, tolcapone is a strong candidate for therapeutic intervention in such diseases, including those affecting the central nervous system (3). **1**. Bonifácio, Palma, Almeida & Soares-da-Silva (2007) *CNS Drug Reviews* **13**, 352. **2**. Zürcher, Keller, Kettler, *et al.* (1990) *Adv. Neurol.* **53**, 497. **3**. Sant'Anna, Gallego, Robinson, *et al.* (2016) *Nature Commun.* **7**, 10787.

Tolfenamate

This oral nonsteroidal anti-inflammatory drug (FW$_{free-acid}$ = 261.71 g/mol; CAS 13710-19-5), also named 2-[(3-chloro-2-methylphenyl)amino] benzoate, is a derivative of anthranilic acid that inhibits COX, thereby blocking prostaglandin biosynthesis. **Target(s):** aromatic-amino-acid decarboxylase (1); Ca^{2+} influx (2); cyclooxygenase (3-7); phenol sulfotransferase (8); and thiopurine *S*-methyltransferase (9). **1**. Pribova, Gregorova & Drsata (1992) *Pharmacol. Res.* **25**, 271. **2**. Kankaanranta, Wuorela, Siltaloppi, *et al.* (1995) *Eur. J. Pharmacol.* **291**, 17. **3**. Ricketts, Lund & Seibel (1998) *Amer. J. Vet. Res.* **59**, 1441. **4**. Kay-Mugford & Conlon (1999) *Amer. J. Vet. Res.* **60**, 275. **5**. Kay-Mugford, Benn, LaMarre & Conlon (2000) *Amer. J. Vet. Res.* **61**, 802. **6**. Proudman & McMillan (1991) *Agents Actions* **34**, 121. **7**. Kauppila, Puolakka & Ylikorkala (1979) *Prostaglandins* **18**, 655. **8**. Vietri, De Santi, Pietrabissa, Mosca & Pacifici (2000) *Xenobiotica* **30**, 111. **9**. Oselin & Anier (2007) *Drug Metab. Dispos.* **35**, 1452.

Tolnaftate

This nonprescription thiocarbamate-based antifungal (FW = 307.42 g/mol; CAS 2398-96-1), also known as *O*-2-naphthyl methyl(3-methylphenyl)thiocarbamate and Tinactin™, inhibits the biosynthesis of sterols, an essential component of fungal membranes (1). Tolnaftate is essentially insoluble in water and soluble in organic solvents (*e.g.*, chloroform and acetone). **Target(s):** aflatoxin biosynthesis (2); and squalene monooxygenase, *or* squalene epoxidase (3-5). **1**. Robinson & Raskin (1965) *Arch. Dermatol.* **91**, 372. **2**. Khan, Maggon & Venkitasubramanian (1978) *Appl. Environ. Microbiol.* **36**, 270. **3**. Favre & Ryder (1996) *Antimicrob. Agents Chemother.* **40**, 443. **4**. Barrett-Bee & Dixon (1995) *Acta Biochim. Pol.* **42**, 465. **5**. Georgopapadakou & Bertasso (1992) *Antimicrob. Agents Chemother.* **36**, 1779.

Tolonium Chloride, *See* *Toluidine Blue O*

Toloxatone

This orally bioavailable antidepressant (FW = 207.23 g/mol; CAS 29218-27-7), also known as 5-hydroxymethyl-3-*m*-tolyloxazolidin-2-one, reversibly inhibits monoamine oxidase A, *or* MAO (1). After administration of toloxatone, cerebral concentrations of MAO substrates noradrenaline, dopamine and 5-hydroxy-tryptamine increased, whereas their metabolite concentrations were reduced (2). Synaptosomal uptake processes of these amines were not changed by toloxatone. **1**. Coston, Gouret & Raynaud (1975) *Therapie* **30**, 725. **2**. Keane, Kan, Sontag & Benedetti (1979) *J. Pharm. Pharmacol.* **31**, 752.

Tolserine

This cymserine analogue (FW$_{free-base}$ = 351.45 g/mol) crosses the blood-brain barrier and inhibits acetylcholine esterase and butyrylcholinesterase (IC$_{50}$ = 10.3 and 1950 nM, respectively) and is used in the treatment of individuals with Alzheimer's diesease. **Target(s):** acetylcholinesterase (1,2); and cholinesterase, *or* butyrylcholinesterase (1,3). **1**. Yu, Holloway, Utsuki, Brossi & Greig (1999) *J. Med. Chem.* **42**, 1855. **2**. Kamal, Greig, Alhomida & Al-Jafari (2000) *Biochem. Pharmacol.* **60**, 561. **3**. Cerasoli, Griffiths, Doctor, *et al.* (2005) *Chem. Biol. Interact.* **157-158**, 363.

Tolterodine

This muscarinic receptor antagonist (FW$_{L-tartrate-Salt}$ = 475.57 g/mol; CAS 124937-52-6), also named 2-[(1*R*)-3-[bis(1-methylethyl)amino]-1-phenylpropyl]-4-methylphenol, is active at all muscarinic receptors (*i.e.*, subtypes M_1 through M_5), showing K_i values of ~3 nM. Tolterodine exhibits a greater effect on the bladder than salivary glands *in vivo*. Radioligand binding experiments showed that tolterodine binds with high affinity to muscarinic receptors in urinary bladder (K_i = 2.7 nM), heart (K_i = 1.6 nM), cerebral cortex (K_i = 0.75 nM) and parotid gland (K_i = 4.8 nM) from guinea pigs and in urinary bladder from humans (K_i = 3.3 nM). This muscarinic receptor antagonist is intended for the treatment of urinary urge incontinence and other symptoms related to an overactive bladder. **1**. Nilvebrant, Andersson, Gillberg, Stahl & Sparf (1997) *Eur. J. Pharmacol.* **327**, 195.

Toluene-3,4-dithiol

This reagent (FW = 156.27 g/mol; MP = 31°C; soluble in benzene and modestly alkaline solutions), also called 1,2-dimercapto-4-methylbenzene, 3,4-dimercaptotoluene, 3,4-dithiotoluene, binds heavy metal ions to form colored complexes that are slightly soluble in water. This reagent is also used in the detection of metals such as bismuth, molybdenum, rhenium, tin, and tungsten, and one should anticipate that toluene-3,4-dithiol would inhibit a number of metalloproteins. **Target(s):** alcohol dehydrogenase, yeast (1); carbonic anhydrase (2); cyclooxygenase (3); endothelin-converting enzyme (4,5); and lipoxygenase (3,6,7). **1**. Sund & Theorell (1963) *The Enzymes*, 2nd ed. (Boyer, Lardy & Myrbäck, eds.), **7**, 25. **2**. Davis (1961) *The Enzymes*, 2nd ed. (Boyer, Lardy & Myrbäck, eds.), **5**, 545. **3**. Miyazawa, Iimori, Makino, Mikami & Miyasaka (1985) *Jpn. J. Pharmacol.* **38**, 199. **4**. Ashizawa, Okumura, Kobayashi, *et al.* (1994) *Biol. Pharm. Bull.* **17**, 212. **5**. Ashizawa, Okumura, Kobayashi, *et al.* (1994) *Biol. Pharm. Bull.* **17**, 207. **6**. Peterson & Gerrard (1983) *Prostaglandins Leukot. Med.* **10**, 107. **7**. Aharony, Smith & Silver (1981) *Prostaglandins Med.* **6**, 237.

N-(4-Toluenesulfonylaminocarbonyl)-*N*-(5*H*-dibenzo[*a,d*] cyclohepten-5-yl)glycine, *See* *N*-(5*H*-Dibenzo[*a,d*] cyclohepten-5-yl)-*N*-(4-toluenesulfonylaminocarbonyl)glycine

α-*N*-Toluene-*p*-sulfonyl-L-arginine Methyl Ester, *See* *N*$^\varepsilon$-(*p*-Tosyl)-L-arginine Methyl Ester

α-Toluenesulfonyl Fluoride, *See* *Phenylmethane-sulfonyl Fluoride*

N-4-Toluenesulfonyl-N-4-nitrobenzyl-β-alanine Hydroxamate,
See N-4-Nitrobenzyl-N-4-toluenesulfonyl-β-alanine Hydroxamate

2-(p-Toluidino)-6-naphthalenesulfonate

This substituted naphthalene (FW$_{\text{potassium salt}}$ = 351.47 g/mol), abbreviated as TNS and also known as 6-(p-toluidino)-2-naphthalenesulfonate, is a frequently used extrinsic fluorescent probe, especially in conformational studies, *e.g.,* fluoresces strongly when bound to hydrophobic regions (1). TNS has also been used to assess binding of inhibitors to proteins, *e.g.,* binding of pesticides to glutathione *S*-transferase (2). **Target(s):** alcohol dehydrogenase (3); chymotrypsin (4); firefly luciferase, *or* luciferin 4-monooxygenase (5); and pyruvate decarboxylase (6). **1**. McClure & Edelman (1966) *Biochemistry* **5**, 1908. **2**. Di Ilio, Sacchetta, Iannarelli & Aceto (1995) *Toxicol. Lett.* **76**, 173. **3**. Brändén, Jörnvall, Eklund & Furugren (1975) *The Enzymes*, 3rd ed. (Boyer, ed.), **11**, 103. **4**. McClure & Edelman (1967) *Biochemistry* **6**, 559. **5**. DeLuca (1969) *Biochemistry* **8**, 160. **6**. Zehender, Trescher & Ullrich (1987) *Eur. J. Biochem.* **167**, 149.

2-p-Toluidylnaphthalene-6-sulfonate, *See 2-(p-Toluidino)-6-naphthalenesulfonate*

Toluquinone

This quinone (FW = 122.12 g/mol; MP = 67-70°C), also known as methyl-1,4-benzoquinone, is an inhibits alkaline phosphatase (1); α-amylase (2); β-amylase (2); catalase (3); α-glycerophosphatase (4); homogentisate oxidase, *or* homogentisate 1,2-dioxygenase (5); papain (6); pyruvate decarboxylase (7); succinate dehydrogenase (8); and succinate oxidase (8,9). **1**. Anderson (1961) *Biochim. Biophys. Acta* **54**, 110. **2**. Owens (1953) *Contrib. Boyce Thompson Inst.* **17**, 273. **3**. Hoffmann-Ostenhof & Biach (1947) *Monatsh. Chem.* **76**, 319. **4**. Hoffmann-Ostenhof & Putz (1948) *Monatsh. Chem.* **79**, 421. **5**. Schepartz (1953) *J. Biol. Chem.* **205**, 185. **6**. Hoffmann-Ostenhof & Biach (1946) *Experentia* **2**, 405. **7**. Kuhn & Beinert (1947) *Ber.* **80**, 101. **8**. Herz (1954) *Biochem. Z.* **325**, 83. **9**. Redfern & Whittaker (1962) *Biochim. Biophys. Acta* **56**, 440.

Tolvaptan

This potent and orally active, vasopressin receptor antagonist (FW = 448.95 g/mol; CAS 150683-30-0; Solubility: 90 mg/mL DMSO, <1 mg/mL H$_2$O), also known by the code name OPC-41061 and systematically named N-(4-{[(5R)-7-chloro-5-hydroxy-2,3,4,5-tetrahydro-1H-1-benzazepin-1-yl]carbonyl}-3-methylphenyl)-2-methylbenzamide, selectively and competitively targets Arginine Vasopressin Receptor 2 (*or* AVP2), with an IC$_{50}$ value of 1.28 μM for inhibition of AVP-induced platelet aggregation. OPC-41061 antagonizes [^{3}H]-AVP binding to human V2-receptors (K_i = 0.43 nM) more potently than AVP (K_i = 0.78 nM) or OPC-31260 (K_i = 9.42 nM). OPC-41061 also inhibited [^{3}H]-AVP binding to human V1a-receptors (K_i = 12.3 nM) but not to human V1b-receptors, indicating that OPC-41061 was 29 times more selective for V2-receptors than for V1a-receptors (2). Tolvaptan induces free water excretion without increasing sodium excretion, safely correcting mild or moderate hyponatremia in patients with syndrome of inappropriate secretion of antidiuretic hormone (SIADH). **1**. Yamamura,

Nakamura, Itoh, *et al.* (1998) *J. Pharmacol. Exp. Ther.* **287**, 860. **2**. Kondo, Ogawa H, Yamashita, *et al.* (1999) *Bioorg. Med. Chem.* **7**, 1743.

Tomoxetine, *See Atomoxetine*

Tomudex, *See Raltitrexed*

Topa, *See 2,4,5-Trihydroxyphenylalanine*

Topiramate

This orally active and structurally novel antiepileptic drug (FW = 339.36 g/mol; CAS 97240-79-4; Soluble to 100 mM in DMSO), also known as McN-4853 and RWJ-17021 and named systematically as 2,3:4,5-bis-O-(1-methylethylidene)-β-D-fructopyranose sulfamate, has at least five distinct modes of action: (a) modulation of the blocking of voltage-activated sodium ion channels; (b) potentiation of GABA-mediated inhibition of neurotransmission; (c) antagonism of APMA glutamate receptors; (d) inhibition of L-type calcium ion channels, and (e) inhibition of carbonic anhydrase isozymes II (K_i = 0.2 μM) and IV (K_i = 0.2 μM), thereby lowering intraneuronal pH. In mice pretreated with a cytochrome P450 enzyme inhibitor (SKF-525A), TPM's anticonvulsant potency was either increased or unaffected 0.5-2 h post-i.p. administration, suggesting topiramate itself (and not a derived metabolite) is likely to be the active agent. **1**. Maryanoff, Nortey, Gardocki, Shank & Dodgson (1987) *J. Med. Chem.* **30**, 880. **2**. Shank, Gardocki, Vaught *et al.* (1994) *Epilepsia* **35**, 450. **3**. Taverna, Sancini, Mantegazza, Franceschetti & Avanzini (1999) *J. Pharmacol. Exp. Ther.* **288**, 960.

Torcetrapib

This CETP inhibitor (FW = 600.47 g/mol; CAS 262352-17-0) targets cholesterylester transfer protein (CETP), EC$_{50}$ = 43 nM, which transfers cholesterol from HDL cholesterol to very low density lipoproteins (VLDL) or low density lipoproteins (LDL), thereby raising HDL levels (1). Significantly, CETP inhibition by torcetrapib increased with escalating dose, leading to elevations of HDL-C of 16% to 91%. It binds specifically to CETP with 1:1 stoichiometry and blocks both neutral lipid and phospholipid (PL) transfer activities. Torcetrapib increases CETP affinity for HDL by ~5x, most likely representing a shift to a lipid transfer-impermissive binding state (2). When used alone, torcetrapib reduces VLDL, IDL, and LDL apoB100 levels primarily by increasing the rate of apoB100 clearance (3). In contrast, when added to atorvastatin treatment, torcetrapib reduces apoB100 levels mainly by enhancing VLDL apoB100 clearance and reducing production of IDL and LDL apoB100 (3). The ambitious Phase-III ILLUMINATE study, which enrolled 15000 subjects, was designed to determine the clinical outcome of raising HDL by torcetrapib (60 mg) plus atorvastatin *versus* atorvastatin alone (10 to 80 mg). The study was abruptly terminated in view of an excess of deaths in the torcetrapib/atorvastatin *versus* atorvastatin groups (82 versus 51, respectively) and upon noting increased heart failure and angina for those on the drug combination (4). A novel, non-tetrahydroquinoline-containing torcetrapib analogue shows improved aqueous solubility (5). **1**. Clark, Sutfin, Ruggeri, *et al.* (2004) *Arterioscler. Thromb. Vasc. Biol.* **24**, 490. **2**. Clark, Ruggeri, Cunningham & Bamberger (2006) *J. Lipid Res.* **47**, 537. **3**. Millar, Brousseau, Diffenderfer, *et al.* (2006) *Arterioscler. Thromb. Vasc. Biol.* **26**,

1350. **4.** Tall, Yvan-Charvet & Wang (2007) *Arterioscler. Thromb. Vasc. Biol.* **27**, 257. **5.** Kalgutkar, Frederick, Hatch, *et al.* (2013) *Xenobiotica* **44**, 591.

Topotecan

This water-soluble camptothecin derivative and antineoplastic agent ($FW_{free-base}$ = 421.45 g/mol; CAS 119413-54-6; Solubility: 90 mg/mL DMSO, 90 mg/mL Water, <1 mg/mL), also known as Hycamtin, NSC 609699, and systematically as (*S*)-10-[(dimethylamino)methyl]-4-ethyl-4,9-dihydroxy-1*H*-pyrano[3',4':6,7]indolizino[1,2-*b*]quinolone-3,14(4*H*,12*H*)-dione monohydrochloride, inhibits DNA topoisomerase I (1,2). This highly effective topoisomerase I inhibitor for MCF-7 Luc cells and DU-145 Luc cells with IC_{50} values of 13 nM and 2 nM, respectively. Topotecan inhibition of poly(ADP-ribose) polymerase-1 converts PARP1 into a dominant-negative molecule that poisons the ability of DNA repair machinery to participate in either PARP1-dependent or PARP1-independent repair of topotecan-induced DNA damage (3). **1.** Iyer & Ratain (1998) *Cancer Chemother. Pharmacol.* **42** Suppl., S31. **2.** Kingsbury, Boehm, Jakas, *et al.* (1991) *J. Med. Chem.* **34**, 98. **3.** Patel, Flatten, Schneider, *et al.* (2012) *J. Biol. Chem.* **287**, 4198.

Torbafylline

This xanthine derivative (FW = 338.40 g/mol; CAS 105102-21-4), also named and 7-(ethoxymethyl)-1-(5-hydroxy-5-methylhexyl)-3-methyl-3,7-dihydro-1*H*-purine-2,6-dione and HWA-448, inhibits proteasome-mediated proteolysis and prevents accumulation of polyubiquitinated proteins in the myofibrillar fraction of muscles from cachectic rats, supporting the idea that the Ub pathway degrades actin and myosin. HWA 448 did not alter peripheral blood mononuclear cell viability, and has a longer serum half-life and presumably lower toxicity than pentoxifylline, which is commonly used in humans (**See** *Pentoxifylline*). **1.** Combaret, Tilignac, Claustre, *et al.* (2002) *Biochem. J.* (2002) **361**, 185.

Toremifene

This orally active selective estrogen receptor modulator, *or* SERM (FW = 405.97 g/mol; CAS 89778-26-7), also named 2-{4-[(1*Z*)-4-chloro-1,2-diphenyl-but-1-en-1-yl]phenoxy}-*N,N*-dimethylethanamine, is a tamoxifen derivative that competitively inhibits [³H]estradiol binding to to the Estrogen Receptor (IC_{50} = 0.3 μM), reducing MCF-7 cell growth in a concentration-dependent manner and exhibiting cell-killing effects above 3 μM (1). (Marketed in the United States under the trade name Fareston®, toremifene is FDA-approved for use in advanced (metastatic) breast cancer.) The overall pharmacokinetic profile is remarkably similar to tamoxifen. Toremifene is

highly metabolized in the liver and is eliminated primarily in the feces following enterohepatic circulation. Some metabolites retain biological activity. Toremifene inhibits the growth of ER-negative, glucocorticoid sensitive, mouse uterine sarcoma in a dose-dependent manner. Its pharmacokinetics and metabolism resemble those of tamoxifen, but this chlorinated drug forms different metabolites. **See also** *Tamoxifen; Lasofoxifene* **1.** Kangas (1990) *J. Steroid Biochem.* **36**, 191.

Torin-1

This potent, cell-permeable, and ATP-competitive mTOR inhibitor (FW = 607.64 g/mol; CAS 1222998-36-8; Solubility = 2 mg/mL DMSO, <1 mg/mL H_2O), also named 1-[4-[4-(1-oxopropyl)-1-piperazinyl]-3-(trifluoromethyl)-phenyl]-9-(3-quinolinyl)benzo[*h*]-1,6-naphthyridin-2(1*H*)-one, targets the mammalian target of rapamycin complexes, designated mTORC1 and mTORC2, with IC_{50} of 2 nM and 10 nM, respectively, while exhibiting 1000x lower action against PI3K (1). **Mode of Inhibitory Action:** mTORC1 integrates mitogen and nutrient signals to control cell proliferation, size, cell cycle progression, and viability. Rapamycin is a potent allosteric inhibitor of mTORC1 with promising clinical applications as an immunosuppressant and anti-cancer agent. Torin1 directly inhibits both complexes, impairing cell growth and proliferation to a far greater degree than achievable with rapamycin (2). These effects are independent of mTORC2 inhibition and are instead, because of suppression of rapamycin-resistant functions of mTORC1 required for cap-dependent translation and suppression of autophagy (2). Notably, mTORC1inhibition by rapamycin (a mTORC1 inhibitor), torin-1 (targeting both mTORC1 and mTORC2) or short hairpin RNA-mediated knockdown of mTOR, regulatory associated protein of mTOR (RAPTOR), and p70 S6 kinase (p70S6K), all increased basal NT release via upregulating NT gene expression in the human endocrine cell line BON cells. (3). mTORC1-specific inhibition helped to demonstrate that control of cell size and cell cycle progression appear to be independent in mammalian cells, whereas in lower eukaryotes, translation initiation factor 4E-binding proteins (4E-BPs)influence both cell growth and proliferation (4). **1.** Liu, Chang, Wang, *et al.* (2010) *J. Med. Chem.* **53**, 7146. **2.** Thoreen, Kang, Chang, *et al.* (2009) *J. Biol. Chem.* **284**, 8023. **3.** Li, Liu, Song, *et al.* (2011) *Am. J. Physiol. Cell Physiol.* **301**, C213. **4.** Dowling, *et al.* (2010) *Science* **328**, 1172.

Tosufloxacin

This fluoroquinolone-class antibiotic (FW = 404.34 g/mol; CAS 100490-36-6), also named Ozex® (marketed in Japan) and 7-(3-aminopyrrolidin-1-yl)-1-(2,4-difluorophenyl)-6-fluoro-4-oxo-1,4-dihydro-1,8-naphthyridine-3-carboxylate, is a broad-spectrum systemic antibacterial agent that targets Type II DNA topoisomerases (gyrases), which are required for bacterial replication and transcription (1). Depending on dosage, tosufloxacin can cause severe thrombocytopenia and nephritis (2). For the prototypical member of this antibiotic class, **See** *Ciprofloxacin* **1.** Niki (2002) *J. Infect. Chemother.* **8**, 1. **2.** Rubinstein (2001) *Chemotherapy* **47** (Supplement 3), 3.

*N*ᵅ-(*p*-Tosyl)-L-arginine Methyl Ester, *See* TAME

Tosyl Chloride, *See* p-Toluenesulfonyl Chloride

Tosyl Fluoride, *See* p-Toluenesulfonyl Fluoride

N^α-Tosylglycyl-3-amidino-DL-phenylalanine Methyl Ester

This dipeptide analogue (FW = 433.15 g/mol), also known as Pefabloc Xa, inhibits several serine proteinases, particularly coagulation factor Xa, K_i = 0.84 µM (1-4), complement factor I (5), thrombin (2-4), and trypsin (4). **1.** Vieweg & Wagner (1987) *Pharmazie* **42**, 268. **2.** Hauptmann, Kaiser, Nowak, Stürzebecher & Markwardt (1990) *Thromb. Haemost.* **63**, 220. **3.** Stürzebecher, Stürzebecher, Vieweg, *et al.* (1989) *Thromb. Res.* **54**, 245. **4.** Sperl, Bergner, Stürzebecher, *et al.* (2000) *Biol. Chem.* **381**, 321. **5.** Tsiftsoglou & Sim (2004) *J. Immunol.* **173**, 367.

N^α-(p-Tosyl)-L-lysine Chloromethyl Ketone

This widely used haloketone-based protease inhibitor (FW$_{hydrochloride}$ = 369.31 g/mol), abbreviated TLCK also known as as 1-chloro-3-tosylamido-7-amino-L-2-heptanone, was a product of the pioneering work of biochemist Eliot Shaw, who in the mid-1960s developed specific active-site-directed irreversible inhibitors of of trypsin and trypsin-like enzymes (1-3). TLCK alkylates the histidyl residue in trypsin's catalytic triad. *Note:* This reagent also inhibits some enzymes reversibly, *e.g.*, butyrylcholinesterase (4). TLCK also alkylates the sulfhydryl group in picornain 3C. Because this reagent is unstable above pH 7.5, stock solutions (10 mM) are prepared by dissolving TLCK in 1.0 mM HCl or methanol. **Target(s):** acrosin (5-8,130,164-168); adenain (94,96,97, however, see 95); adenosine deaminase, *or* RNA-adenosine deaminase (82); adenylate cyclase (9,10); AMP deaminase, *Candida albicans* (81); α-amylase (182); arylesterase (186); arylsulfate sulfotransferase (80); *Aspergillus oryzae* aminoacylase (11); assemblin, herpesvirus (12); *N*-benzoyl-D-arginine-4-nitroanilide amidase (84); bleomycin hydrolase (93); brachyurins (3,159-162); bromelain, stem (14-17); butyrylcholinesterase (4); Ca^{2+}-independent endoprotease (18); calpain-1, *or* µ-calpain (89); calpain-2, *or* m-calpain (88,89); cAMP-dependent protein kinase (19); carboxylesterase (187); carboxypeptidase C, weakly inhibited (175,176); carboxypeptidase D (174); cathepsin B (118-128); cathepsin H (106-110); cathepsin L (111-115); cathepsin T (20,21,105); cephalosporin-C deacetylase (185); cGMP-dependent protein kinase (22); clostripain (6,23,24,116,117); cocoonase (6,25); complement subcomponent C1r, *or* moderately inhibited (157); crotalase (26); cruzipain (27); cystinyl aminopeptidase (181); *Dictyostelium discoideum* cysteine proteinase-7 (28); enteropeptidase (29,169); envelysin, weakly inhibited (87); *Fasciola* cysteine endopeptidases (30); ficain, *or* ficin (16,31,32); furin (139); gabonase, *or* venombin AB (33,151); gametolysin, weakly inhibited (86); *Giardia lamblia* proteinase (34); gingipain K (90-92); gingipain R (91,92,98); glucose-6-phosphatase (35); glucose-6-phosphate translocase (183); γ-glutamyl transpeptidase, *or* γ-glutamyltransferase (190,191); granzyme K (36); hepacivirin, *or* hepatitis C virus NS3 serine proteinase (131-133); histolysain, *or* histolysin (99); HtrA2 peptidase (129); kallikrein, urinary (37); lactosylceramide α-2,3-sialyltransferase (189); leukocyte elastase, weakly inhibited (158); Lysophospholipase (38); *Locusta migratoria* serine proteinases (39,40); lysyl endopeptidase (41,152-156); 3-mercaptopyruvate sulfurtransferase (188); *Metarhizium anisopliae* cuticle-degrading protease (42); 5-methyltetrahydropteroyltriglutamate:homocysteine *S*-methyltransferase (193); mitochondrial intermediate peptidase (85); mouse submandibular nerve growth factor protease, possibly γ-renin (43); non-acrosin *p*-aminobenzamidine-sensitive acrosomal protease (44); okinaxobin II (150); oligopeptidase B (45-47,135-137); pancreatic elastase II (140); papain (3,16,17,48-51); peptidyl-glycinamidase, *or* carboxamidopeptidase

(173); picornain 2A (100-102); picornain 3C (52,103,104); plasmin (53-56); polyneuridine-aldehyde esterase, weakly inhibited (184); prolyl aminopeptidase, weakly inhibited (179); prolyl oligopeptidase (163); proprotein convertase 1 (138); protease IV (134); proteasome (57); protein-arginine deaminase (83); protein-tyrosine kinase (58); salivary gland proteinase K (59); sea urchin cortical granule protease (60); signal peptidase I1 (34); thrombin (3,6,61-65,170); tricorn protease (66); tripeptide aminopeptidase (180); tripeptidyl-peptidase I, weakly inhibited (178); trypsin (1-3,6,16,67-73,171,172); trypsin, crayfish (74); trypsin, *Streptomyces erythraeus* (75); tryptase (76,141-149); tulip mosaic virus NIa protease (77); tymovirus endopeptidase (78); u-plasminogen activator (79); venombin AB (33,150,151); vinorine synthase (192); Xaa-Xaa-Pro tripeptidyl-peptidase, *or* prolyltripeptidyl aminopeptidase (177). **1.** Shaw (1967) *Meth. Enzymol.* **11**, 677. **2.** Shaw (1972) *Meth. Enzymol.* **25**, 655. **3.** Shaw, Mares-Guia & Cohen (1965) *Biochemistry* **4**, 2219. **4.** Cengiz, Cokugras, Kilinc & Tezcan (1997) *Biochem. Mol. Med.* **61**, 52. **5.** Schleuning & Fritz (1976) *Meth. Enzymol.* **45**, 330. **6.** Powers (1977) *Meth. Enzymol.* **46**, 197. **7.** Anderson, Jr., Beyler, Mack & Zaneveld (1981) *Biochem. J.* **199**, 307. **8.** Connors, Greenslade & Davanzo (1973) *Biol. Reprod.* **9**, 57. **9.** Mork & Geisler (1991) *Arch. Int. Physiol. Biochim. Biophys.* **99**, 161. **10.** Abramowitz & Birnbaumer (1979) *Biol. Reprod.* **21**, 213. **11.** Gentzen, Loffler & Schneider (1980) *Z. Naturforsch. [C]* **35**, 544. **12.** Darke (1998) in *Handb. Proteolytic Enzymes* (Barrett, Rawlings & Woessner, eds.), p. 470, Academic Press, San Diego. **13.** Tsu & Craik (1998) in *Handb. Proteolytic Enzymes* (Barrett, Rawlings & Woessner, eds.), p. 25. Academic Press, San Diego. **14.** Murachi (1970) *Meth. Enzymol.* **19**, 273. **15.** Murachi & Kato (1967) *J. Biochem.* **62**, 627. **16.** Shaw (1970) *The Enzymes*, 3rd ed. (Boyer, ed.), **1**, 91. **17.** Glazer & Smith (1971) *The Enzymes*, 3rd ed. (Boyer, ed.), **3**, 501. **18.** Bendjennat, Bahbouhi & Bahraoui (2001) *Biochemistry* **40**, 4800. **19.** Roach (1984) *Meth. Enzymol.* **107**, 81. **20.** Gohda & Pitot (1998) in *Handb. Proteolytic Enzymes* (Barrett, Rawlings & Woessner, eds.), p. 774, Academic Press, San Diego. **21.** Pitot & Gohda (1987) *Meth. Enzymol.* **142**, 279. **22.** Mackenzie III, Tse & Donnelly, Jr. (1981) *Life Sci.* **29**, 1235. **23.** Ullmann & Bordusa (1998) in *Handb. Proteolytic Enzymes* (Barrett, Rawlings & Woessner, eds.), p. 759, Academic Press, San Diego. **24.** Mitchell & Harrington (1971) *The Enzymes*, 3rd. ed. (Boyer, ed.), **3**, 699. **25.** Hruska & Law (1970) *Meth. Enzymol.* **19**, 221. **26.** Markland, Jr., (1976) *Meth. Enzymol.* **45**, 223. **27.** Cazzulo (1998) in *Handb. Proteolytic Enzymes* (Barrett, Rawlings & Woessner, eds.), p. 591, Academic Press, San Diego. **28.** Mehta & Freeze (1998) in *Handb. Proteolytic Enzymes* (Barrett, Rawlings & Woessner, eds.), p. 602, Academic Press, San Diego. **29.** Lu & Sadler (1998) in *Handb. Proteolytic Enzymes* (Barrett, Rawlings & Woessner, eds.), p. 50, Academic Press, San Diego. **30.** Wijffels (1998) in *Handb. Proteolytic Enzymes* (Barrett, Rawlings & Woessner, eds.), p. 606, Academic Press, San Diego. **31.** Liener & Friedenson (1970) *Meth. Enzymol.* **19**, 261. **32.** Stein & Liener (1967) *Biochem. Biophys. Res. Commun.* **26**, 376. **33.** Pirkle & Markland, Jr. (1998) in *Handb. Proteolytic Enzymes* (Barrett, Rawlings & Woessner, eds.), p. 223, Academic Press, San Diego. **34.** Hare, Jarroll & Lindmark (1989) *Exp. Parasitol.* **68**, 168. **35.** Speth & Schulze (1992) *Biochem. Biophys. Res. Commun.* **183**, 590. **36.** Babé & Schmidt (1998) in *Handb. Proteolytic Enzymes* (Barrett, Rawlings & Woessner, eds.), p. 83, Academic Press, San Diego. **37.** Chao (1981) *Hoppe Seylers Z. Physiol. Chem.* **362**, 1113. **38.** Weller (1988) *Meth. Enzymol.* **163**, 31. **39.** Hanzon, Smirnoff, Applebaum, Mattoo & Birk (2003) *Arch. Biochem. Biophys.* **410**, 83. **40.** Lam, Coast & Rayne (2000) *Insect Biochem. Mol. Biol.* **30**, 85. **41.** Sakiyama & Masaki (1994) *Meth. Enzymol.* **244**, 126. **42.** Pei, Ji, Yang, Lu & Xia (2000) *Wei Sheng Wu Xue Bao* **40**, 306. **43.** Young & Koroly (1980) *Biochemistry* **19**, 5316. **44.** Yamagata, Murayama, Kohno, Kashiwabara & Baba (1998) *Zygote* **6**, 311. **45.** Tsuru (1998) in *Handb. Proteolytic Enzymes* (Barrett, Rawlings & Woessner, eds.), p. 375, Academic Press, San Diego. **46.** Burleigh & Andrews (1998) in *Handb. Proteolytic Enzymes* (Barrett, Rawlings & Woessner, eds.), p. 376, Academic Press, San Diego. **47.** Tsuru & Yoshimoto (1994) *Meth. Enzymol.* **244**, 201. **48.** Arnon (1970) *Meth. Enzymol.* **19**, 226. **49.** Cohen (1970) *The Enzymes*, 3rd ed. (Boyer, ed.), **1**, 147. **50.** Wolthers (1969) *FEBS Lett.* **2**, 143. **51.** Whitaker & Perez-Villaseñor (1968) *Arch. Biochem. Biophys.* **124**, 70. **52.** Orr, Long, Kay, Dunn & Cameron (1989) *J. Gen. Virol.* **70**, 2931. **53.** Castellino (1998) in *Handb. Proteolytic Enzymes* (Barrett, Rawlings & Woessner, eds.), p. 190, Academic Press, San Diego. **54.** Robbins & Summaria (1970) *Meth. Enzymol.* **19**, 184. **55.** Robbins & Summaria (1976) *Meth. Enzymol.* **45**, 257. **56.** Castellino & Sodetz (1976) *Meth. Enzymol.* **45**, 273. **57.** Gonzalez-Flores, Guerra-Araiza, Cerbon, Camacho-Arroyo & Etgen (2004) *Endocrinology* **145**, 2328. **58.** Richert, Davies, Jay & Pastan (1979) *Cell* **18**, 369. **59.** Chao (1998) in *Handb. Proteolytic Enzymes* (Barrett, Rawlings & Woessner, eds.), p. 113, Academic Press, San Diego. **60.** Carroll, Jr. (1976)

Meth. Enzymol. **45**, 343. **61**. Magnusson (1970) Meth. Enzymol. **19**, 157. **62**. Lundblad, Kingdon & Mann (1976) Meth. Enzymol. **45**, 156. **63**. Magnusson (1971) The Enzymes, 3rd ed. (Boyer, ed.), **3**, 277. **64**. Gorman (1975) Biochim. Biophys. Acta **412**, 273. **65**. Baird & Elmore (1968) FEBS Lett. **1**, 343. **66**. Tamura & Baumeister (1998) in Handb. Proteolytic Enzymes (Barrett, Rawlings & Woessner, eds.), p. 465, Academic Press, San Diego. **67**. Halfon & Craik (1998) in Handb. Proteolytic Enzymes (Barrett, Rawlings & Woessner, eds.), p. 12, Academic Press, San Diego. **68**. Walsh (1970) Meth. Enzymol. **19**, 41. **69**. Keil (1971) The Enzymes, 3rd ed. (Boyer, ed.), **3**, 249. **70**. Baker (1967) Design of Active-Site-Directed Irreversible Enzyme Inhibitors, Wiley, New York. **71**. Valaitis, Augustin & Clancy (1999) Insect Biochem. Mol. Biol. **29**, 405. **72**. Tudela, Garcia-Canovas, Garcia-Carmona, Iborra & Lozano (1986) Int. J. Biochem. **18**, 285. **73**. Travis & Roberts (1969) Biochemistry **8**, 2884. **74**. Zwilling & Neurath (1981) Meth. Enzymol. **80**, 633. **75**. Norioka & Sakiyama (1998) in Handb. Proteolytic Enzymes (Barrett, Rawlings & Woessner, eds.), p. 22, Academic Press, San Diego. **76**. Johnson (1998) in Handb. Proteolytic Enzymes (Barrett, Rawlings & Woessner, eds.), p. 70, Academic Press, San Diego. **77**. Kim & Choi (1998) in Handb. Proteolytic Enzymes (Barrett, Rawlings & Woessner, eds.), p. 721, Academic Press, San Diego. **78**. Rozanov (1998) in Handb. Proteolytic Enzymes (Barrett, Rawlings & Woessner, eds.), p. 687, Academic Press, San Diego. **79**. Radek, Davidson & Castellino (1993) Meth. Enzymol. **223**, 145. **80**. Kim, Konishi & Kobashi (1986) Biochim. Biophys. Acta **872**, 33. **81**. Thompson, Hall & Gunasekaran (1998) Microbios **96**, 133. **82**. Hough & Bass (1994) J. Biol. Chem. **269**, 9933. **83**. McGraw, Potempa, Farley & Travis (1999) Infect. Immun. **67**, 3248. **84**. Gofshtein-Gandman, Keynan & Milner (1988) J. Bacteriol. **170**, 5895. **85**. Kalousek, Isaya & Rosenberg (1992) EMBO J. **11**, 2803. **86**. Jaenicke, Kuhne, Spessert, Wahle & Waffenschmidt (1987) Eur. J. Biochem. **170**, 485. **87**. Fan & Katagiri (2001) Eur. J. Biochem. **268**, 4892. **88**. Ladrat, Verrez-Bagnis, Noel & Fleurence (2002) Mar. Biotechnol. **4**, 51. **89**. Inomata, Nomoto, Hayashi, et al. (1984) J. Biochem. **95**, 1661. **90**. Abe, Kadowaki, Okamoto, et al. (1998) J. Biochem. **123**, 305. **91**. Pike, McGraw, Potempa & Travis (1994) J. Biol. Chem. **269**, 406. **92**. Fujimura, Hirai, Shibata, Nakayama & Nakamura (1998) FEMS Microbiol. Lett. **163**, 173. **93**. Nishimura, Tanaka, Suzuki & Tanaka (1987) Biochemistry **26**, 1574. **94**. Webster & Kemp (1993) J. Gen. Virol. **74**, 1415. **95**. Tihanyi, Bourbonniere, Houde, Rancourt & Weber (1993) J. Biol. Chem. **268**, 1780. **96**. McGrath, Abola, Toledo, Brown & Mangel (1996) Virology **217**, 131. **97**. Bhatti & Weber (1979) Virology **96**, 478. **98**. Chen, Potempa, Polanowski, Wikstrom & Travis (1992) J. Biol. Chem. **267**, 18896. **99**. Moncada, Keller & Chadee (2003) Infect. Immun. **71**, 838. **100**. Sommergruber, Ahorn, Zöphel, et al. (1992) J. Biol. Chem. **267**, 22639. **101**. Wang, Johnson, Sommergruber & Shepherd (1998) Arch. Biochem. Biophys. **356**, 12. **102**. König & Rosenwirth (1988) J. Virol. **62**, 1243. **103**. Hata, Sato, Sorimachi, Ishiura & Suzuki (2000) J. Virol. Methods **84**, 117. **104**. Davis, Wang, Cox, et al. (1997) Arch. Biochem. Biophys. **346**, 125. **105**. Gohda & Pitot (1981) J. Biol. Chem. **256**, 2567. **106**. Aranishi, Hara & Ishinara (1992) Comp. Biochem. Physiol. **102B**, 499. **107**. Raghav, Kamboj, Parnami & Singh (1995) Indian J. Biochem. Biophys. **32**, 279. **108**. Schwartz & Barrett (1980) Biochem. J. **191**, 487. **109**. Matsuishi, Saito, Okitani & Kato (2003) Int. J. Biochem. Cell Biol. **35**, 474. **110**. Yamamoto, Kamata & Kato (1984) J. Biochem. **95**, 477. **111**. Kirschke, Langner, Wiederanders, Ansorge & Bohley (1977) Eur. J. Biochem. **74**, 293. **112**. Okitani, Matsukura, Kato & Fujimaki (1980) J. Biochem. **87**, 1133. **113**. Mason, Taylor & Etherington (1984) Biochem. J. **217**, 209. **114**. McDonald & Kadkhodayan (1988) Biochem. Biophys. Res. Commun. **151**, 827. **115**. Völkel, Kurz, Linder, et al. (1996) Eur. J. Biochem. **238**, 198. **116**. Gilles, Imhoff & Keil (1979) J. Biol. Chem. **254**, 1462. **117**. Kembhavi, Buttle, Rauber & Barrett (1991) FEBS Lett. **283**, 277. **118**. Towatari, Kawabata & Katunuma (1979) Eur. J. Biochem. **102**, 279. **119**. Takahashi, Murakami & Miyake (1981) J. Biochem. **90**, 1677. **120**. Banno, Yano & Nozawa (1983) Eur. J. Biochem. **132**, 563. **121**. Etherington (1976) Biochem. J. **153**, 199. **122**. Kamboj, Pal & Singh (1990) J. Biosci. **15**, 397. **123**. Aranishi, Hara, Osatomi & Ishihara (1997) Comp. Biochem. Physiol. B **117B**, 579. **124**. Hirao, Hara & Takahashi (1984) J. Biochem. **95**, 871. **125**. Okitani, Matsuishi, Matsumoto, et al. (1988) Eur. J. Biochem. **171**, 377. **126**. Jiang, Lee & Chen (1999) J. Agric. Food Chem. **42**, 1073. **127**. Kawada, Hara, Morimoto, Hiruma & Ishibashi (1995) Int. J. Biochem. Cell Biol. **27**, 175. **128**. Swanson, Martin & Spicer (1974) Biochem. J. **137**, 223. **129**. Gray, Ward, Karran, et al. (2000) Eur. J. Biochem. **267**, 5699. **130**. Sawada, Yokosawa & Ishii (1984) J. Biol. Chem. **259**, 2900. **131**. Mori, Yamada, Kimura, et al. (1996) FEBS Lett. **378**, 37. **132**. D'Souza, Grace, Sangar, Rowlands & Clarke (1995) J. Gen. Virol. **76**, 1729. **133**. Markland, Petrillo, Fitzgibbon, et al. (1997) J. Gen. Virol. **78**, 39. **134**. Engel, Hill, Caballero, Green & O'Callaghan (1998) J. Biol. Chem. **273**, 16792. **135**. Kanatani,

Masuda, Shimoda, et al. (1991) J. Biochem. **110**, 315. **136**. Pacaud & Richaud (1975) J. Biol. Chem. **250**, 7771. **137**. Pacaud (1978) Eur. J. Biochem. **82**, 439. **138**. Jean, Basak, Rondeau, et al. (1993) Biochem. J. **292**, 891. **139**. Molloy, Bresnahan, Leppla, Klimpel & Thomas (1992) J. Biol. Chem. **267**, 16396. **140**. Szilagyi, Sarfati, Pradayrol & Morisset (1995) Biochim. Biophys. Acta **1251**, 55. **141**. Muramatu, Itoh, Takei & Endo (1988) Biol. Chem. Hoppe-Seyler **369**, 617. **142**. Braganza & Simmons (1991) Biochemistry **30**, 4997. **143**. Cromlish, Seidah, Marcinkiewicz, et al. (1987) J. Biol. Chem. **262**, 1363. **144**. Walls, Bennett, Sueiras-Diaz & Olsson (1992) Biochem. Soc. Trans. **20**, 260S. **145**. Fiorucci, Erba & Ascoli (1992) Biol. Chem. Hoppe-Seyler **373**, 483. **146**. Butterfield, Weiler, Hunt, Wynn & Roche (1990) J. Leukoc. Biol. **47**, 409. **147**. Harvima, Schechter, Harvima & Fräki (1988) Biochim. Biophys. Acta **957**, 71. **148**. Schechter, Slavin, Fetter, Lazarus & Fräki (1988) Arch. Biochem. Biophys. **262**, 232. **149**. Caughey, Viro, Ramachandran, et al. (1987) Arch. Biochem. Biophys. **258**, 555. **150**. Nose, Shimohigashi, Hattori, Kihara & Ohno (1994) Toxicon **32**, 1509. **151**. Pirkle, Theodor, Miyada & Simmons (1986) J. Biol. Chem. **261**, 8830. **152**. Fujimura, Shibata & Nakamura (1993) FEMS Microbiol. Lett. **113**, 133. **153**. Masaki, Tanabe, Nakamura & Soejima (1981) Biochim. Biophys. Acta **660**, 44. **154**. Chohnan, Nonaka, Teramoto, et al. (2002) FEMS Microbiol. Lett. **213**, 13. **155**. Elliott & Cohen (1986) J. Biol. Chem. **261**, 11259. **156**. Masaki, Nakamura, Isono & Soejima (1978) Agric. Biol. Chem. **42**, 1443. **157**. Andrews & Baillie (1979) J. Immunol. **123**, 1403. **158**. Starkey & Barrett (1976) Biochem. J. **155**, 265. **159**. Sakharov & Litvin (1994) Comp. Biochem. Physiol. B **108**, 561. **160**. Kim, Park, Kim & Shahidi (2002) J. Biochem. Mol. Biol. **35**, 165. **161**. Grant & Eisen (1980) Biochemistry **19**, 6089. **162**. Eisen, Henderson, Jeffrey & Bradshaw (1973) Biochemistry **12**, 1814. **163**. Besedin & Rudenskaya (2003) Russ. J. Bioorg. Chem. **29**, 1. **164**. Adekunle, Storey & Teuscher (1989) Biol. Reprod. **40**, 127. **165**. Polakoski & McRorie (1973) J. Biol. Chem. **248**, 8183. **166**. Kobayashi, Matsuda, Oshio, et al. (1991) Arch. Androl. **27**, 9. **167**. Takano, Yanagimachi & Urch (1993) Zygote **1**, 79. **168**. Gilboa, Elkana & Rigbi (1973) Eur. J. Biochem. **39**, 85. **169**. Maroux, Baratti & Desnuelle (1971) J. Biol. Chem. **246**, 5031. **170**. Bezaud & Guillin (1988) J. Biol. Chem. **263**, 3576. **171**. Olafson & Smilie (1975) Biochemistry **14**, 1161. **172**. Asgeirsson, Fox & Bjarnason (1989) Eur. J. Biochem. **180**, 85. **173**. Simmons & Walter (1980) Biochemistry **19**, 39. **174**. Latchinian-Sadek & Thomas (1993) J. Biol. Chem. **268**, 534. **175**. Liu, Tachibana, Taira, et al. (2004) J. Ind. Microbiol. Biotechnol. **31**, 572. **176**. Liu, Tachibana, Taira, Ishihara & Yasuda (2004) J. Ind. Microbiol. Biotechnol. **31**, 23. **177**. Fujimura, Ueda, Shibata & Hirai (2003) FEMS Microbiol. Lett. **219**, 305. **178**. Page, Fuller, Chambers & Warburton (1993) Arch. Biochem. Biophys. **306**, 354. **179**. Basten, Moers, van Ooyen, & Schaap (2005) Mol. Gen. Genet. **272**, 673. **180**. Hayashi & Oshima (1980) J. Biochem. **87**, 1403. **181**. Chapot-Chartier, Rul, Nardi & Gripon (1994) J. Biochem. **224**, 497. **182**. Mäntsälä & Zalkin (1979) J. Biol. Chem. **254**, 8540. **183**. Van Schaftingen & Gerin (2002) Biochem. J. **362**, 513. **184**. Mattern-Dogru, Ma, Hartmann, Decker & Stöckigt (2002) Eur. J. Biochem. **269**, 2889. **185**. Takimoto, Mitsushima, Yagi & Sonoyama (1994) J. Ferment. Bioeng. **77**, 17. **186**. Ryan, Keegan, McMartin & Dickerman (1984) Biochim. Biophys. Acta **800**, 87. **187**. Valkova, Lepine, Labrie, Dupont & Beaudet (2003) J. Biol. Chem. **278**, 12779. **188**. Alphey, Williams, Mottram, Coombs & Hunter (2003) J. Biol. Chem. **278**, 48219. **189**. Melkerson-Watson & Sweeley (1991) Biochem. Biophys. Res. Commun. **175**, 325. **190**. Suzuki, Kumagai & Tochikura (1986) J. Bacteriol. **168**, 1325. **191**. Chu, Xu, Dong, Cappelli & Ebersole (2003) Infect. Immun. **71**, 335. **192**. Bayer, Ma & Stöckigt (2004) Bioorg. Med. Chem. **12**, 2787. **193**. González, Banerjee, Huang, Sumner & Matthews (1992) Biochemistry **31**, 6045.

N^{α}-(p-Tosyl)-DL-norleucine Diazomethyl Ketone

This active site-directed irreversible inhibitor (FW = 310.39 g/mol), also known as DL-1-diazo-3-tosylamido-2-heptanone, inhibits aspergillopepsin I and II, with concomitant covalent incorporation of approximately two molecules per molecule of protein. **1**. Chang, Horiuchi, Takahashi, Yamasaki & Yamada (1976) *J. Biochem.* **80**, 975.

N^α-(p-Tosyl)-L-phenylalanine Chloromethyl Ketone

This widely used haloketone-based protease inhibitor (FW = 351.85 g/mol; CAS 329-30-6; Symbol: TPCK; also known as (*S*)-1-chloro-3-tosylamido-4-phenyl-2-butanone and L-(1-tosylamido-2-phenyl)ethylchloro-methyl ketone, was a product of the pioneering work of biochemist Eliot Shaw, who in the mid-1960s developed specific active-site-directed irreversible inhibitors of chymotrypsin and chymotrypsin-like enzymes, *via* alkylation of the catalytic histidyl residue (1-4). TPCK also reportedly alkylates the SH group in picornain 3C protease. TPCK also blocks apoptosis (typically by inactivating the caspase "death" cascade); however, TPCK-modified caspase-11 still activates Aip1:cofilin-mediated (–)-end actin filament depolymerization and cell motility (5). *Note:* This chloromethyl ketone inhibits some enzymes reversibly, *e.g.*, butyrylcholinesterase (6). **Target(s):** acid carboxypeptidases (7); adenain (71,73-75), however, see (72); aminopeptidase B (8); aminopeptidase I, weakly inhibited (124); aminotripeptidase (9); assemblin, herpesvirus (10); brachyurin, *or Paralithodes camtschatica* protease C (110); bromelain, stem (4,11-13); butyrylcholinesterase (6); calpain-2, *or* m-calpain (69); carboxylesterase (129,130); carboxypeptidase C (7,119,120); cathepsin B (7,88-96); cathepsin G (14,113,114); cathepsin H (81-84); cathepsin L (85,86); chymase, inhibited slowly (15,16,107,108); chymotrypsin (1-5,7,17-21,116,117); chymotrypsin C, *or* caldecrin (22); clostripain, weakly inhibited (87); *Coccidiodes* endopeptidase (23); cystinyl aminopeptidase (127); *Dictyostelium discoideum* cysteine proteinase-7 (24); dipeptidyl-peptidase III (121,122); elongation factor, thus, inhibiting protein synthesis (25-32); endopeptidase La (106); endopeptidase So (33,34,102); envelysin (4,46-66); ficin, *or* ficin (4,35); gametolysin, weakly inhibited (62); *Giardia lamblia* proteinase (36); gingipain K (70); gingipain R (76); glucose-6-phosphatase (37); glucose-6-phosphate translocase (128); glutamyl endopeptidase II (100); γ-glutamyl transpeptidase, *or* γ-glutamyltransferase (132); glycerophospholipid arachidonyltransferase, CoA-independent (134); *hepacivirin, orhepatitis* C virus NS3 serine proteinase (97-99); histolysain, *or* histolysin (77); 2-hydroxymuconate-semialdehyde hydrolase (38); isoprenylated protein endopeptidase (39,40); kinin inactivating serine endopeptidase H2, *or* kininase (41,42); lactosylceramide α-2,3-sialyltransferase (131); legumain (78); firefly luciferase, *or Photinus*-luciferin 4-monooxygenase (4,30,43); membrane alanyl aminopeptidase (121); 5-methyltetrahydropteoyl-triglutamate:homocysteine *S*-methyltransferase (135); metridin (18,44,115); mitochondrial intermediate peptidase (61); nardilysin (45); omptin (23,67); pancreatic elastase II (109); papain (4,7,13,46-48); peptidyl-glycinamidase, *or* carboxamidopeptidase (118); picornain 2A (79,80); picornain 3C (49-51); polyporopepsin (68); prolyl aminopeptidase, *or* weakly inhibited (125); prolyl oligopeptidase (111,112); proteasome endopeptidase complex (52,53); proteinase I of *Sulfolobus solfataricus* (54); proteinase II of *Sulfolobus solfataricus* (54); semenogelase (55,101); subtilisin Sendai (103); thimet oligopeptidase, weakly inactivated (63); thromboxane synthase (56); trepolisin (57); tricorn protease (58); tripeptide aminopeptidase (126); tryptase, *or* weakly inhibited (104,105); tsp protease, *or* protease Re (34); tulip mosaic virus NIa protease (59); tymovirus endopeptidase (60); vignain (78); vinorine synthase (133); Xaa-His Dipeptidase, *or* carnosinase, weakly inhibited (123). **1**. Schoellmann & Shaw (1963) *Biochemistry* **2**, 252. **2**. Shaw (1967) *Meth. Enzymol.* **11**, 677 and (1972) **25**, 655. **3**. Shaw (1970) *The Enzymes*, 3rd ed. (Boyer, ed.), **1**, 91. **4**. Hess (1971) *The Enzymes*, 3rd ed. (Boyer, ed.), **3**, 213. **5**. Li, Brieher, Scimone, *et al.* (2007) *Nat. Cell Biol.* **9**, 276. **6**. Cengiz, Cokugras, Kilinc & Tezcan (1997) *Biochem. Mol. Med.* **61**, 52. **7**. Powers (1977) *Meth. Enzymol.* **46**, 197. **8**. DeLange & Smith (1971) *The Enzymes*, 3rd ed. (Boyer, ed.), **3**, 81. **9**. Sachs & Marks (1982) *Biochim. Biophys. Acta* **706**, 229. **10**. Darke (1998) in *Handb. Proteolytic Enzymes* (Barrett, Rawlings & Woessner, eds.), p. 470, Academic Press, San Diego. **11**. Murachi (1970) *Meth. Enzymol.* **19**, 273. **12**. Murachi & Kato (1967) *J. Biochem.* **62**, 627. **13**. Glazer & Smith (1971) *The Enzymes*, 3rd

ed. (Boyer, ed.), **3**, 501. **14**. Rindler-Ludwig & Braunsteiner (1975) *Biochim. Biophys. Acta* **379**, 606. **15**. Woodbury, Everitt & Neurath (1981) *Meth. Enzymol.* **80**, 588. **16**. Hultsch, Ennis & Heidtmann (1988) *Adv. Exp. Med. Biol.* **240**, 133. **17**. Carpenter (1967) *Meth. Enzymol.* **11**, 237. **18**. Wilcox (1970) *Meth. Enzymol.* **19**, 64. **19**. Baker (1967) *Design of Active-Site-Directed Irreversible Enzyme Inhibitors*, Wiley, New York. **20**. Sakal, Applebaum & Birk (1988) *Int. J. Pept. Protein Res.* **32**, 590. **21**. Glick (1968) *Biochemistry* **7**, 3391. **22**. Szilágyi (1998) in *Handb. Proteolytic Enzymes* (Barrett, Rawlings & Woessner, eds.), p. 38, Academic Press, San Diego. **23**. Rawlings & Barrett (1994) *Meth. Enzymol.* **244**, 19. **24**. Mehta & Freeze (1998) in *Handb. Proteolytic Enzymes* (Barrett, Rawlings & Woessner, eds.), p. 602, Academic Press, San Diego. **25**. Pestka (1974) *Meth. Enzymol.* **30**, 261. **26**. Jonak, Pokorna, Meloun & Karas (1986) *Eur. J. Biochem.* **154**, 355. **27**. Lucas-Lenard & Beres (1974) *The Enzymes*, 3rd ed. (Boyer, ed.), **10**, 53. **28**. Young & Neidhardt (1978) *J. Bacteriol.* **135**, 675. **29**. Rychlik, Jonak & Sedlacek (1974) *Acta Biol. Med. Ger.* **33**, 86. **30**. Hartman (1977) *Meth. Enzymol.* **46**, 130. **31**. Richman & Bodley (1973) *J. Biol. Chem.* **248**, 381. **32**. Sedlacek, Jonak & Rychlik (1971) *Biochim. Biophys. Acta* **254**, 478. **33**. Chung & Goldberg (1998) in *Handb. Proteolytic Enzymes* (Barrett, Rawlings & Woessner, eds.), p. 541, Academic Press, San Diego. **34**. Goldberg, Sreedhara Swamy, Chung & Larimore (1981) *Meth. Enzymol.* **80**, 680. **35**. ILiener & Friedenson (1970) *Meth. Enzymol.* **19**, 261. **36**. Hare, Jarroll & Lindmark (1989) *Exp. Parasitol.* **68**, 168. **37**. Speth & Schulze (1992) *Biochem. Biophys. Res. Commun.* **183**, 590. **38**. Díaz & Timmis (1995) *J. Biol. Chem.* **270**, 6403. **39**. Rando (1998) in *Handb. Proteolytic Enzymes* (Barrett, Rawlings & Woessner, eds.), p. 770, Academic Press, San Diego. **40**. Young, Ambroziak, Kim & Clarke (2001) *The Enzymes*, 3rd. ed. (Tamanoi & Sigman, eds.), **21**, 155. **41**. Quinto, Juliano, Juliano, *et al.* (1999) *Immunopharmacology* **45**, 223. **42**. Casarini, Stella, Araujo & Sampaio (1993) *Braz. J. Med. Biol. Res.* **26**, 15. **43**. Lee & McElroy (1969) *Biochemistry* **8**, 130. **44**. Barrett (1998) in *Handb. Proteolytic Enzymes* (Barrett, Rawlings & Woessner, eds.), p. 24, Academic Press, San Diego. **45**. Chesneau, Pierotti, Barre, *et al.* (1994) *J. Biol. Chem.* **269**, 2056. **46**. Arnon (1970) *Meth. Enzymol.* **19**, 226. **47**. Shaw, Mares-Guia & Cohen (1965) *Biochemistry* **4**, 2219. **48**. Wolthers (1969) *FEBS Lett.* **2**, 143. **49**. Bergmann (1998) in *Handb. Proteolytic Enzymes* (Barrett, Rawlings & Woessner, eds.), p. 709, Academic Press, San Diego. **50**. Orr, Long, Kay, Dunn & Cameron (1989) *J. Gen. Virol.* **70**, 2931. **51**. Jewell, Swietnicki, Dunn & Malcolm (1992) *Biochemistry* **31**, 7862. **52**. Drexler (1997) *Proc. Natl. Acad. Sci. U.S.A.* **94**, 855. **53**. Klinkradt, Naude, Muramoto & Oelofson (1997) *Int. J. Biochem. Cell Biol.* **29**, 611. **54**. Vanoni & Tortora (1998) in *Handb. Proteolytic Enzymes* (Barrett, Rawlings & Woessner, eds.), p. 334, Academic Press, San Diego. **55**. Chao (1998) in *Handb. Proteolytic Enzymes* (Barrett, Rawlings & Woessner, eds.), p. 102, Academic Press, San Diego. **56**. Yahn & Feinstein (1981) *Prostaglandins* **21**, 243. **57**. Uitto (1998) in *Handb. Proteolytic Enzymes* (Barrett, Rawlings & Woessner, eds.), p. 313, Academic Press, San Diego. **58**. Tamura & Baumeister (1998) in *Handb. Proteolytic Enzymes* (Barrett, Rawlings & Woessner, eds.), p. 465, Academic Press, San Diego. **59**. Kim & Choi (1998) in *Handb. Proteolytic Enzymes* (Barrett, Rawlings & Woessner, eds.), p. 721, Academic Press, San Diego. **60**. Rozanov (1998) in *Handb. Proteolytic Enzymes* (Barrett, Rawlings & Woessner, eds.), p. 687, Academic Press, San Diego. **61**. Kalousek, Isaya & Rosenberg (1992) *EMBO J.* **11**, 2803. **62**. Jaenicke, Kuhne, Spessert, Wahle & Waffenschmidt (1987) *Eur. J. Biochem.* **170**, 485. **63**. Barrett & Brown (1990) *Biochem. J.* **271**, 701. **64**. Fan & Katagiri (2001) *Eur. J. Biochem.* **268**, 4892. **65**. D'Aniello, Denuce, de Vincentiis, di Fiore & Scippa (1997) *Biochim. Biophys. Acta* **1339**, 101. **66**. Post, Schuel & Schuel (1988) *Biochem. Cell. Biol.* **66**, 1200. **67**. Sugimura & Nishihara (1988) *J. Bacteriol.* **170**, 5625. **68**. Kobayashi, Kusakabe & Murakami (1985) *Agric. Biol. Chem.* **49**, 2393. **69**. Ladrat, Verrez-Bagnis, Noel & Fleurence (2002) *Mar. Biotechnol.* **4**, 51. **70**. Abe, Kadowaki, Okamoto, *et al.* (1998) *J. Biochem.* **123**, 305. **71**. Webster & Kemp (1993) *J. Gen. Virol.* **74**, 1415. **72**. Tihanyi, Bourbonniere, Houde, Rancourt & Weber (1993) *J. Biol. Chem.* **268**, 1780. **73**. McGrath, Abola, Toledo, Brown & Mangel (1996) *Virology* **217**, 131. **74**. Bhatti & Weber (1979) *Virology* **96**, 478. **75**. Bhatti & Weber (1979) *J. Biol. Chem.* **254**, 12265. **76**. Chen, Potempa, Polanowski, Wikstrom & Travis (1992) *J. Biol. Chem.* **267**, 18896. **77**. Moncada, Keller & Chadee (2003) *Infect. Immun.* **71**, 838. **78**. Kembhavi, Buttle, Knight & Barrett (1993) *Arch. Biochem. Biophys.* **303**, 2078. **79**. Sommergruber, Ahorn, Zöphel, *et al.* (1992) *J. Biol. Chem.* **267**, 22639. **80**. König & Rosenwirth (1988) *J. Virol.* **62**, 1243. **81**. Aranishi, Hara & Ishinara (1992) *Comp. Biochem. Physiol.* **102B**, 499. **82**. Raghav, Kamboj, Parnami & Singh (1995) *Indian J. Biochem. Biophys.* **32**, 279. **83**. Schwartz & Barrett (1980) *Biochem. J.* **191**, 487. **84**. Matsuishi, Saito, Okitani & H. Kato (2003) *Int. J. Biochem. Cell Biol.* **35**, 474. **85**. H. Kirschke, J. Langner, Wiederanders,

Ansorge & Bohley (1977) *Eur. J. Biochem.* **74**, 293. **86.** Okitani, Matsukura, Kato & Fujimaki (1980) *J. Biochem.* **87**, 1133. **87.** Kembhavi, Buttle, Rauber & Barrett (1991) *FEBS Lett.* **283**, 277. **88.** Agarwal, Choudhury, Lamsal & Khan (1997) *Biochem. Mol. Biol. Int.* **42**, 1215. **89.** Towatari, Kawabata & Katunuma (1979) *Eur. J. Biochem.* **102**, 279. **90.** Takahashi, Murakami & Miyake (1981) *J. Biochem.* **90**, 1677. **91.** Takahashi, Isemura & Ikenaka (1979) *J. Biochem.* **85**, 1053. **92.** Banno, Yano & Nozawa (1983) *Eur. J. Biochem.* **132**, 563. **93.** Kamboj, Pal & Singh (1990) *J. Biosci.* **15**, 397. **94.** Aranishi, Hara, Osatomi & Ishihara (1997) *Comp. Biochem. Physiol. B* **117B**, 579. **95.** Hirao, Hara & Takahashi (1984) *J. Biochem.* **95**, 871. **96.** Okitani, Matsuishi, Matsumoto, *et al.* (1988) *Eur. J. Biochem.* **171**, 377. **97.** Mori, Yamada, Kimura, *et al.* (1996) *FEBS Lett.* **378**, 37. **98.** D'Souza, Grace, Sangar, Rowlands & Clarke (1995) *J. Gen. Virol.* **76**, 1729. **99.** Markland, Petrillo, Fitzgibbon, *et al.* (1997) *J. Gen. Virol.* **78**, 39. **100.** Yoshida, Tsuruyama, Nagata, *et al.* (1988) *J. Biochem.* **104**, 451. **101.** Watt, P. Lee, Timkulu, Chan & Loor (1986) *Proc. Natl. Acad. Sci. U.S.A.* **83**, 3166. **102.** Chung & Goldberg (1983) *J. Bacteriol.* **154**, 231. **103.** Yamagata, Isshiki & Ichishima (1995) *Enzyme Microb. Technol.* **17**, 653. **104.** Fiorucci, Erba & Ascoli (1992) *Biol. Chem. Hoppe-Seyler* **373**, 483. **105.** Harvima, Schechter, Harvima & Fräki (1988) *Biochim. Biophys. Acta* **957**, 71. **106.** Waxman & Goldberg (1985) *J. Biol. Chem.* **260**, 12022. **107.** Woodward, Fraser, Winkler, *et al.* (1998) *J. Immunol.* **160**, 4988. **108.** Yurt & Austen (1977) *J. Exp. Med.* **146**, 1405. **109.** Ardelt (1974) *Biochim. Biophys. Acta* **341**, 318. **110.** Sakharov & Litvin (1994) *Comp. Biochem. Physiol. B* **108**, 561. **111.** Besedin & Rudenskaya (2003) *Russ. J. Bioorg. Chem.* **29**, 1. **112.** Walter (1976) *Biochim. Biophys. Acta* **422**, 138. **113.** Biggs, Yang, Gullberg, *et al.* (2001) *Proc. Natl. Acad. Sci. U.S.A.* **98**, 3814. **114.** Starkey & Barrett (1976) *J. Biochem.* **155**, 273. **115.** Gibson & Dixon (1969) *Nature* **222**, 753. **116.** Asgeirsson & Bjarnason (1991) *Comp. Biochem. Physiol. B* **99B**, 327. **117.** Peterson, Fernando & Wells (1995) *Insect Biochem. Mol. Biol.* **25**, 765. **118.** Simmons & Walter (1980) *Biochemistry* **19**, 39. **119.** Liu, Tachibana, Taira, *et al.* (2004) *J. Ind. Microbiol. Biotechnol.* **31**, 572. **120.** Liu, Tachibana, Taira, Ishihara & Yasuda (2004) *J. Ind. Microbiol. Biotechnol.* **31**, 23. **121.** Vanha-Perttula (1988) *Clin. Chim. Acta* **177**, 179. **122.** Abramic, Schleuder, Dolovcak, *et al.* (2000) *Biol. Chem.* **381**, 1233. **123.** Margolis, Grillo, Grannot-Reisfeld & Farbman (1983) *Biochim. Biophys. Acta* **744**, 237. **124.** Moriyasu, Sakano & Tazawa (1987) *Plant Physiol.* **84**, 720. **125.** Basten, Moers, van Ooyen, & Schaap (2005) *Mol. Gen. Genet.* **272**, 673. **126.** Sachs & Marks (1982) *Biochim. Biophys. Acta* **706**, 229. **127.** Chapot-Chartier, Rul, Nardi & Gripon (1994) *J. Biochem.* **224**, 497. **128.** Van Schaftingen & Gerin (2002) *Biochem. J.* **362**, 513. **129.** Valkova, Lepine, Labrie, Dupont & Beaudet (2003) *J. Biol. Chem.* **278**, 12779. **130.** Wood, Fernandez-Lafuente & Cowan (1995) *Enzyme Microb. Technol.* **17**, 816. **131.** Melkerson-Watson & Sweeley (1991) *Biochim. Biophys. Res. Commun.* **175**, 325. **132.** Suzuki, Kumagai & Tochikura (1986) *J. Bacteriol.* **168**, 1325. **133.** Bayer, Ma & Stöckigt (2004) *Bioorg. Med. Chem.* **12**, 2787. **134.** Winkler, Sung, Bennett & Chilton (1991) *Biochim. Biophys. Acta* **1081**, 339. **135.** González, Banerjee, Huang, Sumner & Matthews (1992) *Biochemistry* **31**, 6045.

N-Tosyl-L-phenylalanine Diazomethyl Ketone, See *N-(p-Tosyl)-L-phenylalanyldiazomethane*

Toxin A, *Pseudomonas aeruginosa*, See *Pseudomonas Exotoxin A*

α-Toxin, *Clostridium perfringens*

This toxic zinc metalloenzyme (MW = 45,545 g/mol; GenBank: CAA35186.1), *or* Alpha Toxin, is a phospholipase C [*Reaction*: Phosphatidylinositol 4,5-bisphosphate (PIP$_2$) + H$_2$O → Diacylglycerol (DAG) + Inositol 1,4,5-trisphosphate (IP$_3$)] from *Clostridium perfringens*, the anaerobic pathogen responsible for gas gangrene and myonecrosis in infected tissues. The cytokine storm induced by α-toxin (mainly via the release of TNF-α), plays an important role in the death and massive hemolysis. The toxin-induced release of TNF-α from neutrophils and macrophages is dependent on the activation of ERK1/2 signal transduction via the TrkA receptor (1). Significantly, there is complete suppression of α-toxin activity by tetracycline, when *C. perfringens* is treated with metronidazole, rifampin, clindamycin, and chloramphenicol at concentrations equal to their respective MIC values, whereas α-toxin activity persists at concentrations of penicillin equal to and above the MIC (2). **1.** Oda, Shiihara, Ohmae, *et al.* (2012) *Biochim. Biophys. Acta* **1822**, 1581. **2.** Stevens, Maier & Mitten (1987) Antimicrob. Agents Chemother. **31**, 213.

α-Toxin, *Staphylococcus aureus*, See α-Hemolysin, *Staphylococcus aureus*

Toxoflavin

This flavin-like yellow pigment (FW = 193.16 g/mol; CAS 84-82-2; $E_0' = -$0.049Volt), named systematically as 1,6-dimethylpyrimido[5,4-*e*][1,2,4]triazine-5,7(1*H*,6*H*)-dione, was first isolated from *Bacterium bongkrek* in 1934. Through the use of high-throughput screening (HTS), toxoflavins and deazaflavins were identified as the first reported submicromolar, selective inhibitors of the recently discovered enzyme tyrosyl-DNA phosphodiesterase 2 (TDP2), which has been implicated in the topoisomerase-mediated repair of DNA damage. **1.** Raoof, Depledge, Hamilton, *et al.* (2013) *J. Med. Chem.* **56**, 6352.

Toyocamycin

This antifungal antibiotic (FW = 291.27 g/mol; CAS: 606-58-6; Soluble to 100 mM in DMSO) also known as Antibiotic 1037 and 4-amino-7-β-D-ribofuranosyl-7*H*-pyrrolo[2,3-*d*]pyrimidine-5-carbonitrile), isolated from *Streptomyces toyocaensis* (1), inhibits specific RNA self-cleavage, EC$_{50}$ = 0.18 μM (Compare to 8-azaadenosine, EC$_{50}$ = 3.6 μM; Sangivamycin, EC$_{50}$ = 1.1 μM; Tubericidin, EC$_{50}$ = 2.5 μM; EC$_{50}$ = 2.5 μM; Tubericidin monophosphate, EC$_{50}$ = 2.2 μM; tubericidin triphosphate, EC$_{50}$ = 1.9 μM; Nebularine, EC$_{50}$ = 10.2 μM; Tricyclic nucleotide, EC$_{50}$ = 39.5 μM) (2). It also suppresses thapsigargin-, tunicamycin- and 2-deoxyglucose-induced X-Box-binding Protein-1(XBP1) mRNA splicing in HeLa cells without affecting activating transcription factor 6 (ATF6) and PKR-like ER kinase (PERK) activation (3). (*Note*: XBP1 acts as a transcription factor during endoplasmic reticulum (ER) stress by regulating the unfolded protein response, *or* UPR. Indeed, the IRE1α-XBP1 pathway is a key component of ER stress response and is considered to be a critical regulator for survival of multiple myeloma (MM) cells.) Furthermore, although toyocamycin cannot inhibit IRE1α phosphorylation, it prevents IRE1α-induced XBP1 mRNA cleavage *in vitro*. Thus, toyocamycin is an inhibitor of IRE1α-induced XBP1 mRNA cleavage (3). Toyocamycin inhibits not only ER stress-induced but also constitutive activation of XBP1 expression in MM lines as well as primary samples from patients. It shows synergistic effects with bortezomib, and induced apoptosis of MM cells including bortezomib-resistant cells at nanomolar levels in a dose-dependent manner (3). It also inhibits growth of xenografts in an *in vivo* model of human MM (3). **Target(s):** adenosine kinase, also weak alternative substrate (4); 1-phosphatidylinositol 4-kinase (5); rhodopsin kinase (6); and rRNA maturation (7-9). **1.** Aszalos, Lemanski, Robison, Davis & Berk (1966) *J. Antibiot.* (Tokyo) **19**, 285. **2.** Yen, Magnier, Weissleder, Stockwell & Mulligan (2006) *RNA* **12**, 797. **3.** Ri, Tashiro, Oikawa, *et al.* (2012) *Blood Cancer J.* **2**, e79. **4.** Long & Parker (2006) *Biochem. Pharmacol.* **71**, 1671. **5.** Nishioka, Sawa, Hamada, *et al.* (1990) *J. Antibiot.* **43**, 1586. **6.** Palczewski, Kahn & Hargrave (1990) *Biochemistry* **29**, 6276. **7.** Hadjiolova, Naydenova & Hadjiolov (1981) *Biochem. Pharmacol.* **30**, 1861. **9.** Weiss & Pitot (1974) *Cancer Res.* **34**, 581. **9.** Cohen & Glazer (1985) *Mol. Pharmacol.* **27**, 349.

Toyomycin, See *Chromomycin A$_3$*

Tozasertib, See *MK-0457*

TP-434, See *Eravacycline*

TPCA-1

This signal transduction inhibitor and antineoplastic agent (FW = 279.29 g/mol; CAS 507475-17-4; Solubility: 50 mg/mL DMSO; <1 mg/mL Water), systematically named 2-[(aminocarbonyl)amino]-5-(4-fluoro-phenyl)-3-thiophenecarboxamide, targets IKK-2 (inhibitor of nuclear factor κ-B kinase) with IC_{50} of 17.9 nM. TPCA-1 exhibits IC_{50} values of 400 and 3600 nM, respectively, against IKK-1 and JNK3. It also inhibits the production of TNF-α, IL-6, and IL-8 in a concentration-dependent manner, with respective IC_{50} values of 170, 290, and 320 nM (1). TPCA-1 inhibits glioma cell proliferation, as well as TNF-induced RelA (p65) nuclear translocation and NFκB-dependent IL8 gene expression. TPCA-1 also inhibits IFN-induced gene expression, completely suppressing MX1 and GBP1 gene expression, with only a minor effect on ISG15 expression (2). **1**. Podolin, *et al.* (2005) *J. Pharmacol. Exp. Ther.* **312**, 373. **2**. Du, *et al.* (2012) *J. Interferon Cytokine Res.* **32**, 368.

TRAM-34

This clotrimazole-like ion channel blocker (FW = 344.84 g/mol; CAS 289905-88-0; Soluble to 2 mg/mL in DMSO; Water-insoluble), also named 1-[(2-chlorophenyl)diphenylmethyl]-1*H*-pyrazole, is a potent inhibitor of the intermediate-conductance Ca^{2+}-activated K^+ channel, IC_{50} = 20 nM (1). Based on the use of the ratiometric calcium indicator Fura-2 AM, TRAM-34 was found to increase the frequency of rhythmical depolarizations and $[Ca^{2+}]_{intracellular}$ in smooth muscle cells (2). Structure-activity studies (3) suggest that the side chain methyl groups of Thr(250) and Val(275) may lock the triarylmethanes in place via hydrophobic interactions with the π-electron clouds of the phenyl rings. Importantly, TRAM-34's reported advantage over clotrimazole (*i.e.*, lack of cytochrome P450 inhibition) has been convincingly refuted by findings that TRAM-34 inhibits recombinant rat CYP2B1, CYP2C6 and CYP2C11 and human CYP2B6, CYP2C19 and CYP3A4 with IC_{50} values ranging from 0.9 μM to 12.6 μM, but is without effect on recombinant rat CYP1A2, human CYP1A2, or human CYP19A1 at TRAM-34 levels up to 80 μM (4). TRAM-34 treatment inhibits proliferation of human endometrial cancer (EC) cells, and blocks EC cell cycle at G_0/G_1 phase (5). **1**. Wulff, Miller, Hansel, *et al.* (2000) *Proc. Natl. Acad. Sci. U.S.A.* **97**, 8151. **2**. Haddock & Hill (2002) *J. Physiol.* **545** (Part 2), 615. **3**. Wulff, Gutman, Cahalan & Chandy (2001) *J. Biol. Chem.* **276**, 32040. **4**. Agarwal, Zhu, Zhang, Mongin & Hough (2013) *PLoS One* **8**, e63028. **5**. Wang ZH, Shen B, Yao, *et al.* (2007) *Oncogene* **26**, 5107.

Tramadol

(1R,2R)-tramadol   **(1S,2S)-tramadol**

This centrally acting analgesic (FW = 263.4 g/mol; CAS 27203-92-5), also known as Ultram®, Conzip®, Ryzolt®, and Ultracet® (when combined with paracetamol), as well as 2-[(dimethylamino)methyl]-1-(3-methoxyphenyl)-cyclohexanol, is a orally, rectally, intravenously, epidurally or intramuscularly administered benzenoid-class opioid agonist that binds to the μ-opioid receptor but exerts many other effects (1-3). Intravenous tramadol (50-150 mg) is equivalent in analgesic efficacy to morphine (5-15 mg) in patients with moderate post-operative pain (1). When administered epidurally, tramadol was one-thirtieth as potent as morphine. Orally administered tramadol is an effective analgesic at Step 2 of WHO guidelines for treating patients with cancer pain (1). Tramadol is also a norepinephrine reuptake inhibitor (3), a serotonin reuptake inhibitor and releasing agent (4-7), a NMDA receptor antagonist (IC_{50} = 16.5 μM) (8), 5-HT_{2C} receptor antagonist (EC_{50} = 26 nM) (9), α7-nicotinic acetylcholine receptor antagonist (10), TRPV1 receptor agonist (11), and muscarinic M_1 and M_3 acetylcholine receptor antagonist (12-13). Tramadol is rapidly distributed in the body; plasma protein binding is about 20%. It undergoes hepatic metabolism by cytochromes CYP2B6, CYP2D6 and CYP3A4, and one metabolite, *O*-demethyltramadol, has 200x the μ-opioid receptor affinity of (+)-tramadol, with an elimination half-life of nine hours, compared with six hours for tramadol. Clinical experience suggests that tramadol has a low potential for abuse or addiction. **1**. Lee, McTavish & Sorkin (1993) *Drugs* **46**, 313. **2**. Hennies, Friderichs & Schneider (1988) *Arzneimittel-Forschung* **38**, 877. **3**. Frink, Hennies, Englberger, Haurand & Wilffert (1996). Arzneimittel-Forschung **46**, 1029. **4**. Gobbi, Moia, Pirona, *et al.* (2002) *J. Neurochem.* **82**, 1435. **5**. Driessen & Reimann (1992) *Brit. J. Pharmacol.* **105**, 147. **6**. Bamigbade, Davidson, Langford & Stamford (1997) *Brit. J. Anaesth.* **79**, 352. **7**. Reimann & Schneider (1998) *Eur. J. Pharmacol.* **349**, 199. **8**. Hara, Minami & Sata (2005) *Anesth. Analges.* **100**, 1400. **9**. Ogata, Minami, Uezono, *et al.* (2004) *Anesth. & Analges* **98**, 1401. **10**. Shiraishi, Minami, Uezono, *et al.* (2002) Brit. J. Pharmacol. **136**, 207. **11**. Marincsák, Tóth, Czifra, *et al.* (2008) *Anesth. Analges.* **106**, 1890. **12**. Shiraishi, Minami, Uezono, Yanagihara & Shigematsu (2001) *J. Pharmacol. Exper. Therapeut.* **299**, 255. **13**. Shiga, Minami, Shiraishi, *et al.* (2002) *Anesth. Analges* **95**, 1269.

Trametinib

This orally bioavailable and selective protein kinase inhibitor (FW = 615.39 g/mol; CAS 871700-17-3; Solubility: 5 mg/mL DMSO, <1 mg/mL Water; Drug Formulation: Dissolve in 10% Cremophor EL - 10% PEG400), also known as JTP-74057, GSK1120212 and Mekinist®, targets mitogen-activated protein kinase kinase (*or* MEK, *or* MAP2K) and is indicated for treatment of unresectable or metastatic melanoma with BRAF^{V600E} or BRAF^{V600K} mutation. Trametinib inhibits the phosphorylation of myelin basic protein, regardless of the Raf and MEK isotype, with IC_{50} ranging from 0.92 nM to 3.4 nM, but does not inhibit c-Raf, B-Raf, ERK1 and ERK2 kinases. Attesting to its selectivity, GSK1120212 show weak or nor inhibitory potential toward 98 other protein kinases. In enzymatic and cellular studies, GSK1120212 inhibits MEK1/2 kinase activity and prevents Raf-dependent MEK phosphorylation (at Ser-217 for MEK1), producing prolonged p-ERK1/2 inhibition (1). Potent cell growth inhibition was evident in most tumor lines with mutant BRAF or Ras. In xenografted tumor models, GSK1120212 orally dosed once daily had a long circulating half-life and sustained suppression of p-ERK1/2 for more than 24 hours; GSK1120212 also reduced tumor Ki67, increased p27(Kip1/CDKN1B), and caused tumor growth inhibition in multiple tumor models (1). The greatest antitumor effect was among tumors harboring mutant BRAF or Ras (1). In patients with advanced BRAF-mutated melanoma, treatment with a single daily oral dose of trametinib significantly decreased the rate of tumor progression by about 3 months, as compared with patients receiving cytotoxic chemotherapy (2). Significantly, trametinib exhibits antitumour activity in patients with BRAF^{K601E} and BRAF^{L597Q} mutation-positive metastatic melanoma (3). **1**. Gilmartin, Bleam, Groy, *et al.* (2011) *Clin. Cancer Res.* **17**, 989. (*Erratum*: Clin *Clin. Cancer Res.* **18**, 2413.) **2**. Flaherty, Robert, Hersey, *et al.* (2012) *New Engl. J. Med.* **367**, 107. **3**. Bowyer, Rao, Lyle, *et al.* (2014) *Melanoma Res.* **24**, 504.

Trandolapril

This ACE inhibitor and antihypertensive prodrug ($FW_{\text{free-acid}}$ = 430.54 g/mol; CAS 87679-37-6; MP = 125°C) is hydrolyzed physiologically to the active diacid trandolaprilat, a potent inhibitor of peptidyl-dipeptidase A, *or* angiotensin-converting enzyme, IC_{50} = 0.93 nM, and exerting its antihypertensive effect through the renin-angiotensin-aldosterone system (1-6). **See** *Trandolaprilat* **1**. Peters, Noble & Plosker (1998) *Drugs* **56**, 871. **2**. Miyazaki, Kawamoto & Okunishi (1995) *Amer. J. Hypertens.* **8**, 63S. **3**. Wiseman & McTavish (1994) *Drugs* **48**, 71. **4**. Conen & Brunner (1993) *Amer. Heart J.* **125**, 1525. **5**. Duc & Brunner (1992) *Amer. J. Cardiol.* **70**, 27D. **6**. Brown, Badel, Benzoni, *et al.* (1988) *Eur. J. Pharmacol.* **148**, 79.

Trandolaprilat

This dicarboxylic acid ($FW_{\text{free-acid}}$ = 402.49 g/mol; CAS 87679-71-8), formed by hydrolysis of trandolapril, is the primary agent in the inhibition of angiotensin I-converting enzyme (IC_{50} = 0.93 nM (1-5) and protein biosynthesis (6). **See** *Trandolapril* **1**. Corvol, Williams & Soubrier (1995) *Meth. Enzymol.* **248**, 283. **2**. Wiseman & McTavish (1994) *Drugs* **48**, 71. **3**. Conen & Brunner (1993) *Amer. Heart J.* **125**, 1525. **4**. Brown, Badel, Benzoni, *et al.* (1988) *Eur. J. Pharmacol.* **148**, 79. **5**. Wei, Clauser, Alhenc-Gelas & Corvol (1992) *J. Biol. Chem.* **267**, 13398. **6**. Uehara, Numabe, Kawabata, *et al.* (1993) *J. Hypertens.* **11**, 1073.

Tranexamate (Tranexamic Acid)

This antifibrinolytic agent (FW = 157.21 g/mol; CAS 1197-18-8; Solubility: 170 mg/mL H_2O); Decomposes at 386-392°C), also called *trans*-4-(aminomethyl)cyclohexanecarboxylate, is a lysine analogue that blocks lysyl binding sites of plasminogen, thus inhibiting plasminogen activation (1). Tranexamic acid prevents plasmin binding to and preventing fibrinolysis, thereby stabilizing clots. Tranexamic acid possesses ~8x the antifibrinolytic activity of ε-aminocaproic acid. **Target(s):** lysine carboxypeptidase, *or* lysine(arginine) carboxypeptidase, *or* carboxypeptidase N (2); plasmin (3-5); u-plasminogen activator, *or* urokinase (5). **1**. Petersen, Brender & Suenson (1985) *Biochem. J.* **225**, 149. **2**. Juillerat-Jeanneret, Roth & Bargetzi (1982) *Hoppe-Seyler's Z. Physiol. Chem.* **363**, 51. **3**. Robbins & Summaria (1970) *Meth. Enzymol.* **19**, 184. **4**. Longstaff (1994) *Blood Coagul. Fibrinolysis* **5**, 537. **5**. Johnson, Skoza & Tse (1969) *Thromb. Diath. Haemorrh. Suppl.* **32**, 105.

Tranilast

This antiallergic, anti-inflammatory, and antiproliferative anthranilate derivative ($FW_{\text{free-acid}}$ = 327.34 g/mol; CAS 53902-12-8; Soluble in dimethyl sulfoxide; Photosensitive) potent inhibits vascular endothelial growth factor- (VEGF-) and vascular permeability factor-induced angiogenesis as well as collagen synthesis. Tranilast inhibits VEGF- and PMA-stimulated (phorbol myristate acetate-stimulated) protein kinase C activity in retinal capillary endothelial cells without affecting the VEGF binding or VEGF receptor phosphorylation. It induces Ca^{2+} mobilization in vascular smooth muscle. *Note*: Whether the active agent is the *E* or *Z* geometric isomer remains unclear. **Target(s):** cyclin-dependent kinases Cdk2 and Cdk4 (1); glucuronosyltransferase (2); prostaglandin D synthase (3); thioltransferase (4); VEGF- and PMA-stimulated protein kinase C activity (5); and VEGF- and vascular permeability factor-induced angiogenesis and collagen synthesis (5-7). **1**. Andrés (2004) *Cardiovasc. Res.* **63**, 11. **2**. Katoh, Matsui & Yokoi (2007) *Drug Metab. Dispos.* **35**, 583. **3**. Ikai, Ujihara, Fujii & Urade (1989) *Biochem. Pharmacol.* **38**, 2673. **4**. Mizoguchi, Nishnaka, Uchida, *et al.* (1993) *Biol. Pharm. Bull.* **16**, 840. **5**. Koyama, Takagi, Otani, *et al.* (1999) *Brit. J. Pharmacol.* **127**, 537. **6**. Isaji, Miyata, Ajisawa & Yoshimura (1998) *Life Sci.* **63**, PL71. **7**. Isaji, Miyata, Ajisawa, Takehana & Yoshimura (1997) *Brit. J. Pharmacol.* **122**, 1061.

Trans-24

This cell-permeable, mitotic inhibitor (FW = 423.47 g/mol; CAS 869304-55-2; Solubility: 10 mg/mL DMSO; Photosensitive), also known as (5*R*,11a*S*)-2-benzyl-5-(3-hydroxyphenyl)-6*H*-1,2,3,5,11,11a-hexahydroimidazo[1,5-*b*]-β-carboline-1,3-dione, targets Eg5 (IC_{50}= 650 nM), a mitotic kinesin that plays an essential role in centrosome separation and bipolar spindle formation. Unlike classical mitotic spindle poisons (*e.g.*, colchicine; vinblastine; podophyllotoxin, etc.), Trans-24 does not show any tendency to bind to tubulin. (**See also** *Monastrol; HR22C16; Trityl-L-cysteine*) **1**. Sunder-Plassmann, Sarli, Gartner, *et al.* (2005) *Bioorg. Med. Chem.* **13**, 6094.

Transplatin

This light yellow solid (FW = 300.05 g/mol; CAS 14913-33-8), first synthesized by the Nobelist Alfred Werner, and also known as *trans*-diamminedichloroplatinum (II) and *trans*-dichlorodiammine platinum (II), forms slowly from cisplatin, a widely known anticancer drug (**See also** *Cisplatin*), in aqueous solutions. (*Note*: The *trans*-isomer is less soluble than the *cis* enantiomer.) While a weaker mutagen than cisplatin, transplatin interacts with DNA, but produces different adducts (*i.e.*, cisplatin forms 1,2-intrastrand crosslink's, whereas transplatin displays a greater variation in nucleobase donor sites). The 1,2-intrastrand adducts produced by cisplatin causes DNA kinking and leads to thermal destabilization. The effects of transplatin adducts are of greater variability, causing either thermal stabilization or destabilization. **Target(s):** adenosylhomocysteinase, *or* S-adenosyl-homocysteine hydrolase (1); chymotrypsin (2); dihydropteridine reductase (3); glyceraldehyde-3-phosphate dehydrogenase (4); lactate dehydrogenase (5); peroxidase (2); ribonucleoside-diphosphate reductase, weakly inhibited (6); thioredoxin reductase (7); thymidylate synthase (8); and trypsin (2). **1**. Impagnatiello, Franceschini, Oratore & Bozzi (1996) *Biochimie* **78**, 267. **2**. Muginova, Vil'ms, Shekhovtsova & Ivanov (1999) *Biochemistry (Moscow)* **64**, 399. **3**. Armarego & Ohnishi (1987) *Eur. J. Biochem.* **164**, 403. **4**. Aull, Allen, Bapat, *et al.* (1979) *Biochim. Biophys. Acta* **571**, 352. **5**. Hannemann & Baumann (1988) *Res. Commun. Chem. Pathol. Pharmacol.* **60**, 371. **6**. Smith & Douglas (1989) *Biochem. Biophys. Res. Commun.* **162**, 715. **7**. Arner, Nakamura, Sasada, *et al.* (2001) *Free Radic. Biol. Med.* **31**, 1170. **8**. Aull, Rice & Tebbetts (1977) *Biochemistry* **16**, 672.

Transvaalin, *See Scillaren A*

Tranylcypromine

This membrane-penetrating antidepressant/anxiolytic agent ($FW_{\text{free-base}}$ = 133.19 g/mol; CAS 95-62-5; Soluble to 100 mM in water and to 100 mM in

DMSO; Abbreviation: 2-PCPA), also known by its systematic name, (1*R*,2*S*)-*trans*-2-phenylcyclopropylamine, code name SKF *trans*-385, and trade names Parnate® and Jatrosom® (in Germany), is a mechanism-based, irreversible inhibitor of monoamine oxidase (1-9) and LSD1, a histone demethylase catalyzing demethylation of Lys-4 on histone H3 (10,11). Parnate also inhibits phenylethanolamine *N*-methyltransferase, which catalyzes the conversion of norepinephrine to epinephrine (8). The most likely basis for its antidepressant action is inhibition of monoamine oxidase, thereby preventing the breakdown of epinephrine, norepinephrine, and serotonin neurotransmitters and thereby increasing their availability and action. It is indicated for the treatment of major depressive episode without melancholia. This drug is best suited for patients who failed to respond to commonly administered drugs for depression. (Tranylcypromine was temporarily withdrawn from the market in 1964 due to deaths involving hypertensive crises and intracranial bleeding, but was reintroduced that same year with more limited indications and specific warnings.) Because MAOs have *in vivo* turnover times of 10-14 days, recovery from the action of irreversible MAOIs typically exhibits a 3–5 day delay. **MAO Inhibition:** The apparent second order rate constant of inactivation of monoamine oxidase by D-PCPA at 30°C and pH 7.2, derived from pseudo-first order kinetics was 5 x 10^5 $M^{-1}min^{-1}$, with an apparent K_i of 4 µM (6). Because the nonprotonated form of the inhibitor is the active species (pK_a = 8.2), the actual K_i is 0.36 µM. For L-PCPA, the apparent second order inactivation rate constant was 6.5 x 10^3 $M^{-1}min^{-1}$, with an apparent K_i of 300 µM, and a pH-corrected value of 30 µM. It was suggested that the imine (or ketone) formed on oxidation of the inhibitor by the flavin combines with an -SH group at the substrate binding site to form a thioaminoketal or thiohemiketal, and this structure is then stabilized by noncovalent interactions in the native enzyme but becomes labile and dissociable when the enzyme is unfolded (6). The fact that the flavin moiety of the enzyme remains reduced, even in air, may represent steric hindrance to the access of oxygen to the reduced flavin by the bulky phenylcyclopropyl group. Silverman (9,10) suggested an alternative mechanism that exploits the one-electron chemistry of MAOs as well as the propensity of cyclopropylaminyl radicals to undergo homolytic cleavage. **LSD1 Inhibition:** The catalytic domain of the flavin-dependent human histone Lysine-Specific Demethylase 1 (LSD1) belongs to the same enzyme family as monoamine oxidase. A survey of MAOIs as potential LSD1 inhibitors demonstrated that *trans*-2-phenylcyclopropylamine was highly potent, IC$_{50}$ < 2 µM (10). 2-PCPA is a time-dependent, mechanism-based irreversible inhibitor of LSD1 with a K_i of 242 µM and a k_{inact} of 0.0106 s^{-1} (12). 2-PCPA shows limited selectivity for human MAOs *versus* LSD1, with k_{inact}/K_i values only 16x and 2.4x higher for MAO B and MAO A, respectively (12). Profiles of LSD1 activity and inactivation by 2-PCPA as a function of pH are consistent with a mechanism of inactivation dependent upon enzyme catalysis. Mass spectrometric data indicates that FAD is covalent modified by 2-PCPA. In *Arabidopsis thaliana*, 2-PCPA mimics the loss-of-function phenotype for FLOWERING LOCUS D (RSI1; *alias* FLD) a homologue of human histone demethylase required for Systemic Acquired Resistance, *or* SAR (13). (*See also cis-2-Phenylcyclopropylamine*) **Target(s):** cholesterol 24-hydroxylase (14); CYP2A6 (15,16); CYP2C19 (17,18); histamine and 5-hydroxytryptamine uptake (19); lysyl oxidase, *or* protein-lysine monooxygenase (20); prostaglandin-1 synthase, *or* prostacyclin synthase, moderately inhibited (21-23).

1. Blaschko (1963) *The Enzymes*, 2nd ed. (Boyer, Lardy & Myrbäck, eds.), **8**, 337. **2.** Rando (1977) *Meth. Enzymol.* **46**, 28. **3.** Ator & Ortiz de Montellano (1990) *The Enzymes*, 3rd ed. (Sigman & Boyer, eds.), **19**, 213. **4.** Brush & Kozarich (1992) *The Enzymes*, 3rd ed. (Sigman, ed.), **20**, 317. **5.** Guha (1966) *Biochem. Pharmacol.* **15**, 161. **6.** Paech, Salach & Singer (1980) *J. Biol. Chem.* **255**, 2700. **7.** McEwan, Sasaki & Jones (1969) *Biochemistry* **8**, 3963. **8.** Axelrod (1971) *Meth. Enzymol.* **17B**, 761. **9.** Silverman (1983) *J. Biol. Chem.* **258**, 14766. **10.** Silverman (1997) *The Organic Chemistry of Enzyme-Catalyzed Reactions*, pp. 132-136, Academic Press, New York. **11.** Lee, Wynder, Schmidt, McCafferty & Shiekhattar (2006) *Chem. Biol.* **13**, 563. **12.** Schmidt & McCafferty (2007) *Biochemistry* **46**, 4408. **13.** Singh, Banday & Nandi (2014) *Plant Signal Behav.* **9**, e29658. **14.** Mast, White, Bjorkhem, *et al.* (2008) *Proc. Natl. Acad. Sci. U.S.A.* **105**, 9546. **15.** Zhang, Kilicarslan, Tyndale & Sellers (2001) *Drug Metab. Dispos.* **29**, 897. **16.** Taavitsainen, Juvonen & Pelkonen (2001) *Drug Metab. Dispos.* **29**, 217. **17.** Bu, Knuth, Magis & Teitelbaum (2001) *J. Pharm. Biomed. Anal.* **25**, 437. **18.** Yin, Racha, Li, *et al.* (2000) *Xenobiotica* **30**, 141. **19.** Tuomisto & Walaszek (1974) *Med. Biol.* **52**, 255. **20.** Shah, Trackman, Gallop & Kagan (1993) *J. Biol. Chem.* **268**, 11580. **21.** Terashita, Nishikawa, Terao, Nakagawa & Hino (1979) *Biochem. Biophys. Res. Commun.* **91**, 72. **22.** Haurand & Ullrich (1992) *Front. Biotransform.* **6**, 183.

23. McNamara, Hussey, Kerstein, *et al.* (1984) *Biochem. Biophys. Res. Commun.* **118**, 33.

TRAP-Inhibitory Protein, *See Anti-TRAP Protein*

Trapoxin

This cyclotetrapeptide HDAC inhibitor (FW = 602.72 g/mol; CAS 133155-89-2), also named cyclo[(*S*)-phenylalanyl-(*S*)-phenylalanyl-(*R*)-pipecolinyl-(2*S*,9*S*)-2-amino-8-oxo-9,10-epoxydecanoyl-], first isolated from culture broth of *Helicoma ambiens* RF-1023, exhibits detransformation activities against v-sis oncogene-transformed NIH3T3 cells (*or* sis/NIH3T3) as antitumor agents (1,2). Actin stress fibers are evident after trapoxin treatment. Almost complete reversion into the flat phenotype was observed at 6 hours after the administration of trapoxin (2). This effect is reversible, when cultured cells are incubated for 24 hours after drug removal. The intracellular level of sis-mRNA did not decrease with trapoxin treatment at a concentration (50 ng/ml), sufficient to reverse the transformed morphology (2). Trapoxin was later found to cause accumulation of highly acetylated core histones in a variety of mammalian cell lines (3). *In vitro* experiments showed that a low concentration of trapoxin irreversibly inhibited deacetylation of acetylated histones. Chemical reduction of an epoxide group in trapoxin completely abolished the inhibitory activity, suggesting that trapoxin employs the epoxide to bind covalently to the histone deacetylase. By contrast, trichostatin A inhibits histone deacetylase reversibly. (*See Trichostatin A; Valproate*) **1.** Itazaki, Nagashima, Sugita, *et al.* (1990) *J. Antibiot.* (Tokyo) **43**, 1524. **2.** Yoshida & Sugita (1992) *Japan J. Cancer Res.* **83**, 324. **3.** Kijima, Yoshida, Sugita, Horinouchi & Beppu (1993) *J. Biol. Chem.* **268**, 22429.

Trappin-2, *See Elafin*

Trazodone

This 5-HT$_{2A}$ receptor antagonist and antidepressant (FW = 371.87 g/mol; CAS 19794-93-5; IUPAC Name: 2-{3-[4-(3-chlorophenyl)piperazin-1-yl]propyl}[1,2,4]triazolo[4,3-*a*]pyridin-3(2*H*)-one), marketed under the tradenames Depyrel®, Desyrel®, Mesyrel®, Molipaxin®, Oleptro®, Trazodil®, Trazorel®, Trialodine®, and Trittico®, is a Serotonin Antagonist/Reuptake Inhibitor, *or* SARI, with anxiolytic and hypnotic effects. Lacking anticholinergic side effects, trazodone is especially useful in situations where antimuscarinic effects are problematic, such as in patients with prostatic hypertrophy, closed-angle glaucoma, or severe constipation. Trazodone's side effects and toxicity different considerably from monoamine oxidase inhibitors and tricyclic antidepressants. **ADME Properties:** After oral administration, trazodone is nearly completely absorbed; although food delays absorption and reduces peak serum concentration, total area under the plasma concentration-time curve is not altered. Trazodone has biphasic elimination, with a redistribution half-life of about one hour and an elimination half-life of 10-12 hours. Trazodone is nearly completely metabolized hepatically by hydroxylation and oxidation to metabolites that are probably inactive. **Targets:** Serotonin transporter (SERT), 370 nM; norepinephrine transporters (NET), >10000 nM; dopamine transporters DAT, >7000 nM; 5-serotonin HT$_{1A}$ receptor, 120 nM; serotonin 5-HT$_{1B}$ receptor, >10000 nM; serotonin 5-HT$_{1D}$ receptor, 110 nM; serotonin 5-HT$_{1E}$ receptor, >10000 nM; 5-serotonin 5-HT$_{2A}$ receptor, 40 nM; serotonin 5-HT$_{2B}$ receptor, 80 nM; serotonin 5-HT$_{2C}$ receptor, 220 nM; serotonin 5-HT$_3$ receptor, >10000 nM; serotonin 5-HT$_{5A}$ receptor, >10000 nM; serotonin 5-HT$_6$ receptor, >10000 nM; serotonin 5-HT$_7$ receptor, 1800 nM; α$_{1A}$ receptor, 150 nM; α$_{1B}$

adrenergic receptor, ND; α_{2A} adrenergic receptor, 730 nM; α_{2B} adrenergic receptor, ND; α_{2C} adrenergic receptor, 155 nM; β_1 adrenergic receptor, >10000 nM; β_2 adrenergic receptor, >10000 nM; dopamine D_1 receptor, 3700 nM; dopamine D_2 receptor, 4100 nM; dopamine D_3 receptor, ND, dopamine D_4 receptor, 700 nM, dopamine D_5 receptor, >10000 nM; benzodiazepine (BDZ) receptors, >10000 nM; Ca^{2+} channel, >10000 nM; cannabinoid receptors, >10000 nM; σ_1 receptor, >10000 nM; σ_2 receptor, 540 nM; H_1 receptor, 220 nM; H_2 receptor, 3300 nM; H_3 receptor, >10000 nM; H_4 receptor, >10000 nM; μ-opioid receptor, >10000 nM; δ-opioid receptor, >10000 nM; κ-opioid receptor, >10000 nM; E_3 receptor, >10000 nM; E_4 receptor, >10000 nM. **1**. Bryant & Ereshefsky (1982) *Clin. Pharm.* **1**, 406. **2**. Garattini (1985) *Clin. Pharmacokinet.* **10**, 216.

Trastuzumab

This humanized recombinant IgG1 monoclonal antibody (MW = 145531.5 g/mol; CAS 180288-69-1), also known as Herceptin™ (Genentech/ Hoffman-LaRoche), is a human epidermal growth factor receptor-2 (HER2) dimerization inhibitor (HER dimerization inhibitor) that lengthens remission time for breast cancers (1-3). Recent also suggests that the HER2-targeting antibodies Trastuzumab and Pertuzumab act synergistically to inhibit the survival of breast cancer cells growing *in vitro* (4). **Mechanism of Action:** Trastuzumab disrupts ligand-independent HER2/HER3 interactions in HER2-amplified cells, and the observed kinetics of dissociation parallel HER3 dephosphorylation and uncoupling from PI3K activity, leading to down-regulation of proximal and distal AKT signaling in a manner that correlates with trastuzumab's antiproliferative effects. The selective and potent PI3K inhibitor, GDC-0941, is highly efficacious both in combination with trastuzumab and in the treatment of trastuzumab-resistant cells and tumors (5). *For more details on mechanism of action, see Pertuzumab* **1**. Adams, Allison, Flagella, *et al.* (2006) *Cancer Immunol. Immunother.* **55**, 717. **2**. Agus, Gordon, Taylor, *et al.* (2005) *J. Clin. Oncol.* **23**, 2534. **3**. Ng, Lum, Gimenez, Kelsey & Allison (2006) *Pharm. Res.* **23**, 1275. **4**. Nahta, Hung & Esteva (2004) *Cancer. Res.* **64**, 2343. **5**. Junttila, Akita, Parsons, *et al.* (2009) *Cancer Cell.* **15**, 429.

Trastuzumab emtansine

This antineoplastic agent (Molecular Mass = 148.5 kDa; CAS 1018448-65-1) is an antibody-drug conjugate consisting of the monoclonal antibody trastuzumab (itself directed against recombinant anti-epidermal growth factor receptor 2, or HER2; *See Trastuzumab; Maytansine*) and the maytansinoid DM1 (a potent macrolide inhibitor of tubulin polymerization; *See DM1*) via a nonreducible thioether linkage (1). The trastuzumab component directs the conjugate to tumor cells possessing HER2, and upon internalization, undergoes cleavage to release the DM1 component, thereby disrupting microtubule assembly/disassembly dynamics and inhibiting proliferation of cancer cells overexpressing HER2. This binary antibody-drug therefore selectively inhibits cells rich in HER2, such as seen in HER2$^+$ metastatic breast cancer (2). Pharmacokinetics (3). Physicochemical stability (4). **1**. Lewis-Phillips, Li, Dugger, *et al.* (2008) *Cancer Res.* **68**, 9280. **2**. LoRusso, Weiss, Guardino, Girish, and Sliwkowski (2011) Clin. Cancer Res. 17, 6437. **3**. Shen, Bumbaca, Saad, *et al.* (2012) *Curr Drug Metab.* **13**, 901. **4**. Wakankar, Feeney, Rivera, *et al.* (2010) *Bioconjug Chem.* **21**, 1588.

Trasylol, *See Aprotinin*

Trehazolin

This antibiotic pseudosaccharide and trehalose analogue (FW = 366.33 g/mol; CAS 132729-37-4), produced by *Micromonospora coriacea*, acts as an insecticide by inhibiting insect flight muscle trehalose, with an apparent K_i value of 10–100 nM. Pig kidney trehalase is also inhibited, IC_{50} = 19 nM. **Target(s):** glucosidase I, plant (*Note*: Trehazolin does not inhibit plant glucosidase II) (1); trehalase, slow tight-binding inhibition (2-5). **1**. Zeng & Elbein (1998) *Arch. Biochem. Biophys.* **355**, 26. **2**. Ando, Nakajima, Kifune, Fang & Tanzawa (1995) *Biochim. Biophys. Acta* **1244**, 295. **3**. Kyosseva, Kyossev & Elbein (1995) *Arch. Biochem. Biophys.* **316**, 821. **4**. Wegener, Tschiedel, Schloder & Ando (2003) *J. Exp. Biol.* **206**, 1233. **5**. Ando, Satake, Itoi, *et al.* (1991) *J. Antibiot.* **44**, 1165.

Treosulfan

This antineoplastic drug (FW = 278.30 g/mol; CAS 299-75-2), also known as (2*S*,3*S*)-2,3-dihydroxybutane-1,4-diyl dimethanesulfonate, is a cell cycle non-specific DNA alkylating agent that penetrates cancer cells and induces apoptosis. (*For mechanism of action, See Busulfan*). Treosulfan is better tolerated by elderly patients. **1**. Thigpen, Vance, Balducci & Khansur (1984) *Semin. Oncol.* **11**, 314. **2**. Galaup & Paci (2013) *Expert Opin. Drug Metab. Toxicol.* **9**, 333.

Trequinsin

This orally active cAMP phosphodiesterase inhibitor and antihypertensive agent ($FW_{free-base}$ = 405.50 g/mol; $FW_{HCl-Salt}$ = 441.95 g/mol; CAS 78416-81-6; Soluble to 100 mM in DMSO), also known by the code name HL-725 and systematically as 2,3,6,7-tetrahydro-9,10-dimethoxy-3-methyl-2-[(2,4,6-trimethylphenyl)imino]-4*H*-pyrimido[6,1-*a*]isoquinolin-4-one hydrochloride, is an extremely potent inhibitor of human platelet aggregation, as induced *in vitro* by ADP, collagen, thrombin, and epinephrine. The aggregation induced by 0.5 mM arachidonic acid is inhibited about 50% with 50 pM HL-725, a potency that exceed that of prostacyclin, the most active natural aggregation inhibitor. Trequinsin targets cAMP phosphodiesterase-3, *or* PDE3 (IC_{50} = 250 pM), reducing systemic blood pressure in both normotensive and hypertensive animal models (1,2). If whole blood is pretreated with adenosine deaminase, there is no inhibitory effect of dipyridamole plus HL 725 on platelet aggregation, demonstrating that plasma adenosine also plays a crucial role in the anti-aggregatory actions of Trequinsin and several other inhibitors of cAMP PDE, both in human and rat blood (3). **1**. Ruppert & Weithmann (1982) *Life Sci.* **31**, 2037. **2**. Lal, Dohadwalla, Dadkar, D'Sa & de Souza (1984) *J. Med. Chem.* **27**, 1470. **3**. Agarwal, Buckley & Parks (1987) *Thromb. Res.* **47**, 191. PMID: 2821650.

5-(3α,7α,12α-Triacetoxy-5β-cholanamido)-1,3,4-thiadiazole-2-sulfonamide

This sulfonamide (FW = 694.60 g/mol) inhibits carbonic anhydrase, *or* carbonate dehydratase, IC_{50} = 66 nM for the human isozyme II (1,2). **1**. Bülbül, Hisar, Beydemir, Çiftçi & Küfrevioglu (2003) *J. Enzyme Inhib. Med. Chem.* **18**, 371. **2**. Bülbül, Saraçoglu, Küfrevioglu & Çiftçi (2002) *Bioorg. Med. Chem.* **10**, 2561.

N,N',N''-Triacetylchitotriose

This trisaccharide (FW = 627.60 g/mol) of N-acetyl-D-glucosamine has β1→4 linkages and can form a stable complex with lysozyme. **Target(s):** N-acetyl-D-glucosamine-specific lectin from the ascidian *Didemnum ternatanum* (1); β-N-acetylhexosaminidase (2,3); *Bandeiraea simplicifolia* lectin II (4); cephalochordate *Branchiostoma lanceolatum* lectin (5); chitinase (6,7); chitin synthase (8); gorse (*Ulex europeus*) phytohemaagglutinin II (9); lysozyme (10-15); and stinging nettle (*Urtica dioica*) rhizome lectin (16). **1.** Belogortseva, Molchanova, Glazunov, Evtushenko & Luk'yanov (1998) *Biochim. Biophys. Acta* **1380**, 249. **2.** Ohtakara (1988) *Meth. Enzymol.* **161**, 462. **3.** Ohtakara, Yoshida, Murakami & Izumi (1981) *Agric. Biol. Chem.* **45**, 239. **4.** Ebisu & Goldstein (1978) *Meth. Enzymol.* **50**, 350. **5.** Mock & Renwrantz (1991) *Comp. Biochem. Physiol. B* **99**, 699. **6.** Bhushan (2000) *J. Appl. Microbiol.* **88**, 800. **7.** Molano, Polacheck, Duran & Cabib (1979) *J. Biol. Chem.* **254**, 4901. **8.** Mayer, Chen & DeLoach (1980) *Insect Biochem.* **10**, 549. **9.** Osawa & Matsumoto (1972) *Meth. Enzymol.* **28**, 323. **10.** Carlström (1962) *Biochim. Biophys. Acta* **59**, 361. **11.** Imoto, Johnson, North, Phillips & Rupley (1972) *The Enzymes*, 3rd ed. (Boyer, ed.), **7**, 665. **12.** Tompkins, O'Neill, Cafarella & Germaine (1991) *Infect. Immun.* **59**, 655. **13.** Bernard, Canioni, Cozzone, Berthou & Jolles (1990) *Int. J. Protein Res.* **36**, 46. **14.** Ito, Yamada, Nakamura, *et al.* (1993) *Eur. J. Biochem.* **213**, 649. **15.** Ito, Yamada, Nakamura & Imoto (1990) *J. Biochem.* **107**, 236. **16.** Shibuya, Goldstein, Shafer, Peumans & Broekaert (1986) *Arch. Biochem. Biophys.* **249**, 215.

Triacetylglycerol, *See* Triacetin

Triacontanoate, *See* Melissate

Triacsin A

This naturally occurring long-chain fatty acid analogue (FW = 209.29 g/mol; CAS 76896-80-5; Soluble to 25 mM in DMSO), also named (2E,4E,7E)-2,4,7-undecatrienal nitrosohydrazone, is one of four structurally related compounds from *Streptomyces* sp. SK-1894. Triacsin A specifically inhibits long-chain fatty acyl-CoA synthetases from widely different sources. Inhibition is competitive with respect to long-chain fatty acids. The N-hydroxytriazene moiety, common to all triacsins, is essential for inhibition. Triacsins differ with respect to the number and sites of unsaturation (1-3). These inhibitors are especially useful in demonstrating an essential role of long chain acyl-CoA synthetase in animal cell proliferation. **See also** *Triacsins B, C, and D* **1.** Omura, Tomoda, Xu, Takahashi & Iwai (1986) *J. Antibiot. (Tokyo)* **39**, 1211. **2.** Tomoda, Igarashi & Omura (1987) *Biochim. Biophys. Acta* **921**, 595. **3.** Tomoda, Igarashi, Cyong & Omura (1991) *J. Biol. Chem.* **266**, 4214.

Triacsin B

This naturally occurring long-chain fatty acid analogue (FW = 205.26 g/mol; CAS 105201-47-6; MP = 144-147°C), one of four structurally related compounds from *Streptomyces* sp. SK-1894, inhibits specifically long-chain fatty acyl-CoA synthetases from widely different sources. Triacsin A inhibition is competitive with respect to long-chain fatty acids, and weaker than othere triacsins. The N-hydroxytriazene moiety, common to all triacsins, is essential for inhibition. Triacsins differ with respect to the number and sites of unsaturation (1,2).. **See also** *Triacsins A, C, and D* **1.** Tomoda, Igarashi & Omura (1987) *Biochim. Biophys. Acta* **921**, 595. **2.** Tomoda, Igarashi, Cyong & Omura (1991) *J. Biol. Chem.* **266**, 4214.

Triacsin C

This naturally occurring long-chain fatty acid analogue (FW = 207.28 g/mol; CAS 76896-80-5), one of four structurally related compounds from *Streptomyces* sp. SK-1894, potently inhibits specifically long-chain fatty acyl-CoA synthetases from widely different sources. Inhibition is competitive with respect to long-chain fatty acids. The N-hydroxytriazene moiety, common to all triacsins, is essential for inhibition. Triacsins differ with respect to the number and sites of unsaturation. Note that one of the sites of unsaturation is not conjugated. In addition to blocking the biosynthesis of triacylglycerols and cholesterol esters, triacsin C will also block β-cell

apoptosis induced by fatty acids. **See also** *Triacsins A, B, and D* **Target(s):** arachidonyl-CoA synthetase, *or* arachidonate:CoA ligase (1-3); long-chain-fatty-acyl-CoA synthetase, *or* long-chain-fatty-acid:CoA ligase (1-9); medium-chain-fatty-acyl-CoA synthetase, *or* butyrate:CoA ligase, weakly inhibited (9). **1.** Lewin, Kim, Granger, Vance & Coleman (2001) *J. Biol. Chem.* **276**, 24674. **2.** Kim, Lewin & Coleman (2001) *J. Biol. Chem.* **276**, 24667. **3.** Hartman, Omura & Laposata (1989) *Prostaglandins* **37**, 655. **4.** Igal, Wang & Coleman (1997) *Biochem. J.* **324**, 529. **5.** Omura, Tomoda, Xu, Takahashi & Iwai (1986) *J. Antibiot. (Tokyo)* **39**, 1211. **6.** Tomoda, Igarashi & Omura (1987) *Biochim. Biophys. Acta* **921**, 595. **7.** Van Horn, Caviglia, Li, *et al.* (2005) *Biochemistry* **44**, 1635. **8.** Tomoda, Igarashi, Cyong & Omura (1991) *J. Biol. Chem.* **266**, 4214. **9.** Vessey, Kelley & Warren (2004) *J. Biochem. Mol. Toxicol.* **18**, 100.

Triacsin D

This naturally occurring long-chain fatty acid analogue (FW = 207.28 g/mol), one of four structurally related compounds from *Streptomyces* sp. SK-1894, weakly inhibits specifically long-chain fatty acyl-CoA synthetases from widely different sources. Inhibition is competitive with respect to long-chain fatty acids (1,2). The N-hydroxytriazene moiety, common to all triacsins, is essential for inhibition. Triacsins differ with respect to the number and sites of unsaturation. **1.** Tomoda, Igarashi & Omura (1987) *Biochim. Biophys. Acta* **921**, 595. **2.** Tomoda, Igarashi, Cyong & Omura (1991) *J. Biol. Chem.* **266**, 4214.

1,2,3-Tri(adenosine-5'-O-phospho)glycerol

This unusual oligonucleotide (FW = 1078.71 g/mol) competitively inhibits bis(5'-nucleosyl)tetraphosphatase (K_i = 80 nM), a hydrolase that degrades ApppppA (*or* P^1,P^4-di(adenosine-5')tetraphosphate). The latter has been implicated in extracellular purinergic signalling and in various intracellular processes, including DNA metabolism, tumor suppression, and cellular stress responses. If permitted to accumulate, ApppppA and related metabolites may also be cytotoxic. **1.** Guranowski, Starzynska, McLennan, Baraniak & Stec (2003) *Biochem. J.* **373**, 635.

Triadimefon

This common fungicide (FW = 293.75 g/mol; CAS 43121-43-3; MP = 82°C; slightly soluble in water) inhibits aromatase, *or* CYP19 (1), (*S*)-canadine synthase (2), and sterol 14-demethylase, *or* CYP51 (3,4). **1.** Vinggaard, Hnida, Breinholt & Larsen (2000) *Toxicol. In Vitro* **14**, 227. **2.** Rueffer & Zenk (1994) *Phytochemistry* **36**, 1219. **3.** Joseph-Horne, Hollomon, Loeffler & Kelly (1995) *FEBS Lett.* **374**, 174. **4.** Trösken, Adamska, Arand, *et al.* (2006) *Toxicology* **228**, 24.

Triadimenol

This common fungicide (FW = 295.77 g/mol; CAS 55219-65-3; MP = 112-117°C) is the reduced derivative of triadimefon. The (1S,2R)-stereoisomer is the strongest inhibitor of sterol 14-demethylase. **Target(s):** aromatase, *or* CYP19 (1) and sterol 14-demethylase, CYP51 (2-5). **1.** Vinggaard, Hnida, Breinholt & Larsen (2000) *Toxicol. In Vitro* **14**, 227. **2.** Trösken, Adamska, Arand, *et al.* (2006) *Toxicology* **228**, 24. **3.** Yoshida & Aoyama (1990) *Chirality* **2**, 10. **4.** Lamb, Kelly, Manning, Hollomon & Kelly (1998) *FEMS Microbiol. Lett.* **169**, 369. **5.** Lamb, Cannieux, Warrilow, *et al.* (2001) *Biochem. Biophys. Res. Commun.* **284**, 845.

Triallate

This thiocarbamate (FW = 304.67 g/mol; CAS 2303-17-5; MP = 20-30°C), known commercially as Avadex BW and systematically as bis(1-methylethyl)carbamothioic acid *S* (2,3,3-trichloro-2-propenyl) ester, is a volatile and selective pre-emergence herbicide used to control grass weeds in field and pulse crops, particularly in the control of wild oats, black grass, and annual meadow grass. **Target(s):** aldehyde dehydrogenase (1); and fatty acid biosynthesis (2). **1.** Quistad, Sparks & Casida (1994) *Life Sci.* **55**, 1537. **2.** Dodge (1977) *Spec. Publ., Chem. Soc.* **29**, 7.

Triamcinolone

This synthetic glucocorticosteroid (FW = 394.44 g/mol; CAS 124-94-7), known systematically as 9α-fluoro-11β,16α,17α,21-tetrahydroxypregna-1,4-diene-3,20-dione, binds to glucocorticoid receptors, exerting powerful anti-inflammatory activity and affecting protein biosynthesis. Triamcinolone will also induce apoptosis and impair the tumor necrosis factor-induced degradation of κB-α. **Target(s):** chymotrypsin (1); corticosteroid acetyltransferase (2); pyruvate kinase (3); and trypsin (1). **Key Pharmacokinetic Parameters:** *See* Appendix II in Goodman & Gilman's *THE PHARMACOLOGICAL BASIS OF THERAPEUTICS*, 12th Edition (Brunton, Chabner & Knollmann, eds.) McGraw-Hill Medical, New York (2011). **1.** Mayer, Neufeld & Finci (1982) *Biochem. Pharmacol.* **31**, 2989. **2.** Purdy & Rao (1970) *Steroids* **16**, 649. **3.** Johnson & Veneziale (1980) *Biochemistry* **19**, 2195.

Triamterene

This pteridine derivative (FW = 253.27 g/mol; CAS 2303-17-5), also called 2,4,7-triamino-6-phenylpteridine, is a mild 'potassium-sparing' diuretic usually employed in combination with other more potent diuretics in the treatment of hypertension. Triamterene blocks sodium ion reuptake in the kidney. **Target(s):** carbonic anhydrase, *or* carbonate dehydratase (1,2); dihydrofolate reductase, K_i = 13 nM (3-7); folate transport (8); histamine *N*-methyltransferase (9); kallikrein, weakly inhibited (10); Na⁺ channels (11-14); Na⁺/K⁺-exchanging ATPase (15,16); phosphodiesterase (17); tyrosine monooxygenase (18). **1.** Puscas, Coltau, Baican, Pasca & Domuta (1999) *Res. Commun. Mol. Pathol. Pharmacol.* **105**, 213. **2.** Puscas, Coltau, Baican, Domuta & Hecht (1999) *Drugs Exp. Clin. Res.* **25**, 271. **3.** Sidhom & Velez (1989) *J. Pharm. Biomed. Anal.* **7**, 1551. **4.** Schalhorn & Wilmanns (1979) *Arzneimittelforschung* **29**, 1409. **5.** Roberts & Hall (1968) *J. Clin. Pharmacol. J. New Drugs* **8**, 217. **6.** Bertino, Perkins & Johns (1965) *Biochemistry* **4**, 839. **7.** Perkins & Bertino (1966) *Biochemistry* **5**, 1005. **8.** Zimmerman, Selhub & Rosenberg (1986) *J. Lab. Clin. Med.* **108**, 272. **9.** Thithapandha & Cohn (1978) *Biochem. Pharmacol.* **27**, 263. **10.** Margolius & Chao (1980) *J. Clin. Invest.* **65**, 1343. **11.** Busch, Suessbrich, Kunzelmann, *et al.* (1996) *Pflugers Arch.* **432**, 760. **12.** Li & Lindemann (1983) *J. Membr. Biol.* **76**, 235. **13.** Kramer, Rorig & Volger (1981) *Pharmacology* **23**, 149. **14.** Knauf, Wais, Albiez & Lubcke (1976) *Arzneimittelforschung* **26**, 484. **15.** Reznik & Miazina (1985) *Vopr. Med. Khim.* **31**, 122. **16.** Ozegovic & Milkovic (1982) *Arzneimittelforschung* **32**, 1279. **17.** Schuhmacher, Greeff & Noack (1983) *Eur. J. Pharmacol.* **95**, 71. **18.** Steinberg & Rubio (1984) *Naunyn Schmiedebergs Arch. Pharmacol.* **327**, 119.

Triapine

This substituted pyridine (FW = 195.25 g/mol; CAS 236392-56-6; *Symbol*: 3-AP), also known as 3-aminopyridine-2-carboxaldehyde thiosemicarbazone, is a anticancer agent that induces apoptosis in cancer cells (1-5). Treatment of colon cancer cells with 3-AP activates the PERK, IRE1a and ATF6 endoplasmic reticulum stress pathways by phosphorylation of eIF2α and upregulation of ATF4 and ATF6 gene expression (6). After induction of the Unfolded Protein Response (UPR), significant upregulation of pro-apoptotic proteins is detected, including higher levels of the transcription factor CHOP as well as the BH3-only member protein Bim, itself an essential factor for ER stress-related apoptosis. Despite a recent report (7) suggesting that the structurally related thiosemicarbazone Dp44mT may, in part, exert cytotoxicity by targeting DNA topoisomerase IIα, the results of subsequent application of a variety of assays proved otherwise (8). Triapine was without effect on inhibited topoisomerase IIα decatenation, induced cleavage of pBR322 DNA in the presence of enzyme. In cells, Dp44mT did not stabilize topoisomerase IIα covalent binding to DNA using an immunoblot band depletion assay, an ICE assay, and a protein-DNA covalent complex forming assay. Dp44mT also failed to display cross-resistance to etoposide-resistant K562 cells containing reduced topoisomerase IIα levels. **Target(s):** ribonucleoside-diphosphate reductase (1-5); and ribonucleoside-triphosphate reductase (1,2). **1.** Karp, Giles, Gojo, *et al.* (2008) *Leuk. Res.* **32**, 71. **2.** Odenike, Larson, Gajiria, *et al.* (2008) *Invest. New Drugs* **26**, 233. **3.** Cory, Cory, Rappa, *et al.* (1994) *Biochem. Pharmacol.* **48**, 335. **4.** Finch, Liu, Grill, *et al.* (2000) *Biochem. Pharmacol.* **59**, 983. **5.** Finch, Liu, Cory, Cory & Sartorelli (1999) *Adv. Enzyme Regul.* **39**, 3. **6.** Trondl, Flocke, Kowol, *et al.* (2013) *Mol. Pharmacol.* **85**, 415. **7.** Rao, Klein, Agama, *et al.*, (2009) *Cancer Res.* **69**, 948. **8.** Yalowich, Wu, Zhang, *et al.* (2012) *Biochem. Pharmacol.* **84**, 52.

3-(1,2,4-Triazol-3-yl)-DL-alanine

This histidine isostere (FW = 156.14 g/mol) is a synthetic allosteric feedback inhibitor of ATP phosphoribosyl-transferase (1), the enzyme catalyzing the first step in bacterial histidine biosynthesis. 3-(1,2,4-Triazol-3-yl)-alanine also competitively inhibits histidine ammonia-lyase (2,3) and histidyl-tRNA synthetase (4,5). **1.** Ohta, Fujimori, Mizutani, *et al.* (2000) *Plant Physiol.* **122**, 907. **2.** Shibatani, Kakimoto & Chibata (1975) *Eur. J. Biochem.* **55**, 263. **3.** Brand & Harper (1976) *Biochemistry* **15**, 1814. **4.** Kalousek & Konigsberg (1974) *Biochemistry* **13**, 999. **5.** Lepore, di Natale, Guarini & de Lorenzo (1975) *Eur. J. Biochem.* **56**, 369.

6-[(4H-1,2,4-Triazol-4-ylamino)methyl]uracil

This uracil derivative (FW = 208.18 g/mol) inhibits *Escherichia coli* thymidine phosphorylase, K_i = 0.5 µM, as well as thymidine phosphorylase/platelet derived endothelial cell growth factor (TP/PD-ECGF). **1.** Kalman & Lai (2005) *Nucleosides Nucleotides Nucleic Acids* **24**, 367.

17-(1H-1,2,3-Triazol-1-yl)androsta-4,16-dien-3-one

This synthetic steroid (FW = 337.47 g/mol) is a potent inhibitor of steroid 17α-monooxygenase, *or* CYP17, K_i = 8 nM for human enzyme and steroid 5α-reductase, IC_{50} = 198 nM for human enzyme (1). The 1H-1,2,4-triazol-1-yl analogue also inhibits steroid 17α-monooxygenase, K_i = 41 Nm, for human enzyme and steroid 5α-reductase, IC_{50} = 152 nM for human enzyme. **1.** Njar, Kato, Nnane, *et al.* (1998) *J. Med. Chem.* **41**, 902.

4-(1H-1,2,3-Triazol-5-yl)aniline, *See 5-(4-Amino-phenyl)-1H-1,2,3-triazole*

Tribufos

This phosphorothioate-based plant defoliant (FW = 272.44 g/mol; CAS 78-48-8), also known as S,S,S-tributyl phosphorotrithioate, inhibits acylglycerol lipase, *or* monoacylglycerol lipase (1), 1-alkyl-2-acetylglycerophosphocholine esterase, *or* platelet-activating-factor acetylhydrolase (2,3), acylpeptide hydrolase (4), butyrylcholinesterase (4), carboxylesterase (1,5), fatty acid amide hydrolase (1,6), lysophospholipase (1,7), neuropathy target esterase, weakly inhibited (1), and thrombin (3,8). **1.** Quistad, Klintenberg, Caboni, Liang & Casida (2006) *Toxicol. Appl. Pharmacol.* **211**, 78. **2.** Quistad, Fisher, Owen, Klintenberg & Casida (2005) *Toxicol. Appl. Pharmacol.* **205**, 149. **3.** Casida & Quistad (2005) *Chem. Biol. Interact.* **157-158**, 277. **4.** Quistad, Klintenberg & Casida (2005) *Toxicol. Sci.* **86**, 291. **5.** Valles, Oi & Strong (2001) *Insect Biochem. Mol. Biol.* **31**, 715. **6.** Quistad, Sparks & Casida (2001) *Toxicol. Appl. Pharmacol.* **173**, 48. **7.** Quistad & Casida (2004) *Toxicol. Appl. Pharmacol.* **196**, 319. **8.** Quistad & Casida (2000) *J. Biochem. Mol. Toxicol.* **14**, 51.

(Tributyl)tin Chloride & (Tributyl)tin Acetate

These membrane-penetrating organometallic salts (FW$_{chloride}$ = 325.51 g/mol; CAS 688-73-3; colorless liquid; BP = 172°C at 5 mm Hg; moisture sensitive; FW$_{acetate}$ = 349.10 g/mol; Structure: CH$_3$CH$_2$CH$_2$CH$_2$)$_3$Sn$^+$ CH$_3$COO$^-$; white powder; MP = 86-87°C) are uncouplers of oxidative phosphorylation uncoupler. (Tributyl)tin acetate that is moisture sensitive. **See also** *Bis(tributyltin)oxide; Uncouplers* **Target(s):** aromatase, *or* CYP19 (1,2); ATP synthase, *or* F$_1$F$_o$ ATPase, *or* H$^+$-transporting two-sector ATPase (3-7); calmodulin (8,9); carboxylesterase (10); coproporphyrinogen oxidase (11); cytochrome *c* is released by the presence of these agents (12); electron transport, chloroplast (13); estrone sulfotransferase (33); ferrochelatase (14); glutathione S-transferase, by (tributyl)tin acetate (15,16,34-37); H$^+$-exporting ATPase (17,18); H$^+$/K$^+$-ATPase (19); K$^+$-ATPase (19,20); lipoxygenase (21); mitochondrial inner membrane anion channel (22); Na$^+$-dependent ATPase (20,23); NAD(P)$^+$ transhydrogenase (24); Na$^+$/K$^+$-exchanging ATPase (20,25,26); Na$^+$-transporting two-sector ATPase (27); oxidative phosphorylation (4-6,12); phenol sulfotransferase, *or* aryl sulfotransferase (33); prostaglandin-D synthase (28); steroid 5α-reductase (29); tubulin polymerization, *or* microtubule assembly (30,31); and tyrosine monooxygenase (32). **1.** Cooke (2002) *Toxicol. Lett.* **126**, 121. **2.** Heidrich, Steckelbroeck & Klingmuller (2001) *Steroids* **66**, 763. **3.** Linnett & Beechey (1979) *Meth. Enzymol.* **55**, 472. **4.** Matsuno-Yagi & Hatefi (1993) *J. Biol. Chem.* **268**, 6168. **5.** Matsuno-Yagi & Hatefi (1993) *J. Biol. Chem.* **268**, 1539. **6.** Snoeij, Punt, Penninks & Seinen (1986) *Biochim. Biophys. Acta* **852**, 234. **7.** Fillingame (1981) *Curr. Top. Bioenerg.* **11**, 35. **6.** Yallapragada, Vig, Kodavanti & Desaiah (1991) *J. Toxicol. Environ. Health* **34**, 229. **7.** Yallapragada, Vig & Desaiah (1990) *J. Toxicol. Environ. Health* **29**, 317. **10.** Al-Ghais, Ahmad & Ali (2000) *Ecotoxicol. Environ. Saf.* **46**, 258. **11.** Rossi, Attwood & Garcia-Webb (1992) *Biochim. Biophys. Acta* **1135**, 262. **12.** Gogvadze, Stridh, Orrenius & Cotgreave (2002) *Biochem. Biophys. Res. Commun.* **292**, 904. **13.** Izawa & Good (1972) *Meth. Enzymol.* **24**, 355. **14.** Rossi, Attwood, Garcia-Webb & Costin (1990) *Clin. Chim. Acta* **188**, 1. **15.** Asakura, Ohkawa, Takahashi, *et al.* (1997) *Brit. J. Cancer* **76**, 1333. **16.** Mosialou & Morgenstern (1990) *Chem. Biol. Interact.* **74**, 275. **17.** Uchida, Ohsumi & Anraku (1988) *Meth. Enzymol.* **157**, 544. **18.** Apps & Webster (1996) *Biochem. Biophys. Res. Commun.* **227**, 839. **19.** Matsuya, Okamoto, Ochi, *et al.* (2000) *Biochem. Pharmacol.* **60**, 1855. **20.** Cameron, Kodavanti, Pentyala & Desaiah (1991) *J. Appl. Toxicol.* **11**, 403. **21.** Josephson, Lindsay & Stuiber (1989) *J. Environ. Sci. Health B* **24**, 539. **22.** Powers & Beavis (1991) *J. Biol. Chem.* **266**, 17250. **23.** Kaim & Dimroth (1994) *Eur. J. Biochem.* **222**, 615. **24.** Singh & Bragg (1979) *Can. J. Biochem.* **57**, 1384. **25.** Maier & Costa (1990) *Toxicol. Lett.* **51**, 175. **26.** Zucker, Elstein, Easterling, *et al.* (1998) *Toxicol. Appl. Pharmacol.* **96**, 393. **27.** Müller, Aufurth & Rahlfs (2001) *Biochim. Biophys. Acta* **1505**, 108. **28.** Thomson, Meyer & Hayes (1998) *Biochem. J.* **333**, 317. **29.** Doering, Steckelbroeck, Doering & Klingmuller (2002) *Steroids* **67**, 859. **30.** Cima, Ballarin, Bressa & Burighel (1998) *Ecotoxicol. Environ. Saf.* **40**, 160. **31.** Jensen, Onfelt, Wallin, Lidums & Andersen (1991) *Mutagenesis* **6**, 409. **32.** Kim, Lee, Yin, *et al.* (2002) *Neurosci. Lett.* **332**, 13. **33.** Martin-Skilton, Coughtrie & Porte (2006) *Aquat. Toxicol.* **79**, 24. **34.** Guthenberg, Warholm, Rane & Mannervik (1986) *Biochem. J.* **235**, 741. **35.** Stockman, McLellan & Hayes (1987) *Biochem. J.* **244**, 55. **36.** Warholm, Jensson, Tahir & Mannervik (1986) *Biochemistry* **25**, 4119. **37.** Alin, Jensson, Guthenberg, *et al.* (1985) *Anal. Biochem.* **146**, 313.

Tricarballylate

This tricarboxylic acid analogue of isocitric and citric acid (FW$_{free-acid}$ = 176.13 g/mol; CAS 99-14-9; pK_a values of 3.49, 4.58, and 5.83 at 30°C), also known as 1,2,3-propanetricarboxylic acid, is a powerful chelator of calcium and magnesium ions. **Target(s):** aconitase, *or* aconitate hydratase (1-3); ATP:citrate lyase, *or* ATP:citrate synthase (4); glucose-1,6-bisphosphate synthase (5); NADP$^+$-dependent isocitrate dehydrogenase (6,7); kynurenine aminotransferase, weakly inhibited (8); 2-methylcitrate dehydratase (9); phosphofructokinase (10); and succinate dehydrogenase (11-13). **1.** Glusker (1971) *The Enzymes*, 3rd ed. (Boyer, ed.), **5**, 413. **2.** Uhrigshardt, Walden, John & Anemuller (2001) *Eur. J. Biochem.* **268**, 1760. **3.** Gawron, Kennedy & Rauner (1974) *Biochem. J.* **143**, 717. **4.** Antranikian, Herzberg & Gottschalk (1982) *J. Bacteriol.* **152**, 1284. **5.** Rose, Warms & Wong (1977) *J. Biol. Chem.* **252**, 4262. **6.** Head (1980) *Eur. J. Biochem.* **111**, 581. **7.** Seelig & Colman (1978) *Arch. Biochem. Biophys.* **188**, 394. **8.** Mason (1959) *J. Biol. Chem.* **234**, 2770. **9.** Aoki & Tabuchi (1981) *Agric. Biol. Chem.* **45**,

2831. **10.** Passonneau & Lowry (1963) *Biochem. Biophys. Res. Commun.* **13**, 372. **11.** Quastel & Woolbridge (1928) *Biochem. J.* **22**, 689. **12.** Massart (1950) *The Enzymes*, 1st ed. (J. B. Sumner & Myrbäck, eds.), **1** (part 1), 307. **13.** Webb (1966) *Enzyme and Metabolic Inhibitors*, vol. 2, p. 240, Academic Press, New York.

Triciribine

This tricyclic adenine nucleoside analogue (F.Wt. = 320.3; CAS 1191951-57-1 and 35943-35-2; Solubility (25°C): 60 mg/mL DMSO; <1 mg/mL Water), also known as NSC-154020, TCN and API-2, and named systematically as 6-amino-4-methyl-8-(β-D-ribofuranosyl)-4*H*,8*H*-pyrrolo [4,3,2-*de*]pyrimido[4,5-*c*]pyridazine, is a late-phase inhibitor (IC_{50} = 20 nM) of human immunodeficiency virus type 1 (HIV-1) replication by a unique mechanism not involving inhibition of viral replication enzymes (1). Inhibition requires TCN phosphorylation to its 5'-phosphomonoester by intracellular adenosine kinase. Triciribine also inhibits the Akt (IC_{50} = 130 nM). TCN is highly selective for Akt and does not inhibit the activation of phosphatidylinositol 3-kinase, phosphoinositide-dependent kinase-1, protein kinase C, serum and glucocorticoid-inducible kinase, protein kinase A, signal transducer and activators of transcription 3, extracellular signal-regulated kinase-1/2, or c-Jun N-terminal kinase (3). TCN-5'-P also weakly inhibits (GTP more than ATP) purine nucleotide biosynthesis at reactions catalyzed by amidophosphoribosyltransferase, IC_{50} = 20 mM, and IMP dehydrogenase, IC_{50} = 5-8 μM (4,5). **1.** Kucera, *et al.* (1993) *AIDS Res Hum Retroviruses* **9**, 307. **2.** Gursel DB, *et al.* (2011) *Neuro. Oncol.* **13**, 610. **3.** Yang, *et al.* (2004) *Cancer Res.* **64**, 4394. **4.** Moore, Hurlbert & Massia (1989) *Biochem. Pharmac.* **38**, 4037. **5.** Moore, Hurlbert, Boss & Massia (1989) *Biochem. Pharmacol.* **38**, 4045.

Trichloracetaldehyde, *See Chloral Hydrate*

Trichlorfon

This membrane-permeable organophosphate (FW = 257.44 g/mol; CAS 52-68-6; white solid; MP = 83-84°C; soluble in water (15.4 g/100 mL at 25°C), benzene, and ethanol), also known as *O,O*-dimethyl-2,2,2-trichloro-1-hydroxyethyl phosphonate, Neguvon, and metrifonate, is an anthelmintic agent. It has also been found to be effective in the treatment of the cognitive symptoms of Alzheimer's disease. The active metabolite of trichlorfon is actually *O,O*-dimethyl-2,2-dichlorovinyl phosphonate which irreversibly inhibits acetylcholinesterase. **Target(s):** acetylcholinesterase (1-4); arylesterase (5); carboxylesterase (6); cholinesterase, *or* butyrylcholinesterase, weakly inhibited (1,3,4,7); electron transport (8); glucose-6-phosphate dehydrogenase (9). **1.** Reiner, Krauthacker, Simeon & Skrinjaric-Spoljar (1975) *Biochem. Pharmacol.* **24**, 717. **2.** Ringman & Cummings (1999) *J. Clin. Psychiatry* **60**, 776. **3.** Darvesh, Walsh, Kumar, *et al.* (2003) *Alzheimer Dis. Assoc. Disord.* **17**, 117. **4.** Giacobini (2003) *Neurochem. Res.* **28**, 515. **5.** Moon & Smith (2005) *Soil Biol. Biochem.* **37**, 1211. **6.** McGhee (1987) *Biochemistry* **26**, 4101. **7.** Bueding, Liu & Rogers (1972) *Brit. J. Pharmacol.* **46**, 480. **8.** Roshchina, Solomatkin & Mutuskin (1982) *Biokhimiia* **47**, 937. **9.** Chefurka (1957) *Enzymologia* **18**, 209.

Trichlormethine

This membrane-penetrating nitrogen mustard (FW_{free-base} = 204.53 g/mol; CAS 555-77-1; MP = –4°C), abbreviated as HN-3, and tris(2-chloroethyl)amine, is a DNA and RNA alkylating agent and inhibits both transcription and translation. The hydrochloride is an antineoplastic drug used the treatment of Hodgkin's disease and leukemias), and the free base is a chemical warfare agent. **Caution:** The free base should only be used in a chemical hood or with a gas mask. Protective clothing should be worn to prevent contact with skin. Trichlormethine **See also** *Mechlorethamine; Bis(2-chloroethyl)ethylamine* **Target(s):** cholinesterase (1); and hexokinase (2). **1.** Peters (1947) *Nature* **159**, 149. **2.** McDonald (1955) *Meth. Enzymol.* **1**, 269.

Trichloroacetate

This trihaloacid (FW = 163.39 g/mol; CAS 76-03-9; pK_a = 0.66 at 25°C; MP = 57-58°C; BP = 196-197°C) is frequently used precipitate proteins. *Note*: This extremely deliquescent solid should be kept in a tightly closed container and stored in cool, well-ventillated location. Dilute aqueous solutions should not be stored, because the acid slowly decomposes. **Target(s):** acid phosphatase (1); adenylosuccinate synthetase (2); creatine kinase (3); fumarylacetoacetase (4); glucose-6-phosphate dehydrogenase (5); pancreatic ribonuclease, *or* ribonuclease A (6); pantothenate synthetase (7); tyrosine decarboxylase (8); urease (9); and wax biosynthesis (10). **1.** Belfanti, Contardi & Ercoli (1935) *Biochem. J.* **29**, 517. **2.** Markham & Reed (1977) *Arch. Biochem. Biophys.* **184**, 24. **3.** Watts (1973) *The Enzymes*, 3rd ed. (Boyer, ed.), **8**, 383. **4.** Braun & Schmidt (1973) *Biochemistry* **12**, 4878. **5.** Slein (1950) *J. Biol. Chem.* **186**, 753. **6.** Sagar & Pandit (1983) *Biochim. Biophys. Acta* **743**, 303. **7.** Van Oorschot & Hilton (1963) *Arch. Biochem. Biophys.* **100**, 289. **8.** Sundaresan & Coursin (1970) *Meth. Enzymol.* **18A**, 509. **9.** Onodera (1915) *Biochem. J.* **9**, 544. **10.** Kolattukudy (1965) *Biochemistry* **4**, 1844.

2-([3-[(2,3,5-Trichlorobenzyl)amino]propyl]amino)quinolin-4(1*H*)-one

This substituted quinolinone (FW = 410.73 g/mol), first identified through analysis of Structure Activity Relationships, enantioselectively inhibits methionyl-tRNA synthetase, IC_{50} < 3 nM for the *Staphylococcus aureus* enzyme. **1.** Jarvest, Armstrong, Berge, *et al.* (2004) *Bioorg. Med. Chem. Lett.* **14**, 3937.

1,1,1-Trichloro-2,2-bis(*p*-chlorophenyl)ethane

This highly toxic and notorious polychlorinated pesticide (FW = 354.49 g/mol; MP = 108.5-109°C), widely known as DDT (for Dichloro Diphenyl Trichloroethane) was first synthesized in 1873. DDT was found to be a highly effective insecticide, and during World War II, DDT became extremely valuable in combating typhus and malaria. When DDT came into wider use in agriculture, its toxicity (LD_{50} = ~115 mg/kg in rats) quickly became evident, and Paul Müller was awarded the 1948 Nobel Prize in Physiology or Medicine for his discovery of its high efficiency as a contact poison. DDT accumulates in lipid-rich foodstuffs (*e.g.*, solubility ≈ 11 g/100 mL peanut oil) as a consequence of its nonpolar nature. Resistant to oxidation and ultraviolet/visible light, DDT persists within the food-chain, where its steroidomimmetic properties reduce egg shell biomineralization to such an extent that the fragile eggs are prone to cracking and microbial infection. **Caution:** DDT is readily absorbed orally, transdermally and by inhalation.

Extreme caution must be exercised when handling DDT. **Target(s):** ATPase (1-5); Ca^{2+}-dependent ATPase (6-11); calmodulin-stimulated cyclic nucleotide phosphodiesterase (12); carbonic anhydrase, *or* carbonate dehydratase (13); chitin synthase (14); F_oF_1 ATP synthase, *or* H^+-transporting two-sector ATPase (15-20); Hill reaction (21); K^+ uptake (22); lipase (23); Na^+/K^+-exchanging ATPase (24-27); photosynthetic electron transport and photophosphorylation (21,28-31); xanthine oxidase (32). **1.** Hennighausen, Hehl & Rychly (1985) *Arch. Toxicol. Suppl.* **8**, 500. **2.** Esher, Wolfe & Koch (1980) *Comp. Biochem. Physiol. C* **65C**, 43. **3.** Koch (1969) *Chem. Biol. Interact.* **1**, 199. **4.** Koch, Cutkomp & Do (1969) *Life Sci.* **8**, 289. **5.** Koch (1969) *J. Neurochem.* **16**, 269. **6.** Treinen & Kulkarni (1986) *Toxicol. Lett.* **30**, 223. **7.** Ghiasuddin & Matsumura (1981) *Biochem. Biophys. Res. Commun.* **103**, 31. **8.** Doherty, Salem, Jr., Lauter & Trams (1981) *Comp. Biochem. Physiol. C* **69**, 185. **9.** Elder, Morre & Yunghans (1979) *Eur. J. Cell Biol.* **19**, 231. **10.** Kolaja & Hinton (1977) *J. Toxicol. Environ. Health* **3**, 699. **11.** Kolaja & Hinton (1977) *Bull. Environ. Contam. Toxicol.* **17**, 591. **12.** Hagmann (1982) *FEBS Lett.* **143**, 52. **13.** Bitman, Cecil & Fries (1970) *Science* **168**, 594. **14.** Leighton, Marks & Leighton (1981) *Science* **213**, 905. **15.** McEnery & Pedersen (1986) *Meth. Enzymol.* **126**, 470. **16.** Younis, Abo-El-Saad, Abdel-Razik & Abo-Seda (2002) *Biotechnol. Appl. Biochem.* **35**, 9. **17.** Chefurka (1983) *Comp. Biochem. Physiol. C* **74**, 259. **18.** Ohyama, Takahashi & Ogawa (1982) *Biochem. Pharmacol.* **31**, 397. **19.** Cutkomp, Yap, Vea & Koch (1971) *Life Sci. II* **10**, 1201. **20.** Koch, Cutkomp & Yap (1971) *Biochem. Pharmacol.* **20**, 3243. **21.** Owen, Rogers & Hayes (1971) *Biochem. J.* **121**, 6P. **22.** Chefurka, Zahradka & Bajura (1980) *Biochim. Biophys. Acta* **601**, 349. **23.** Christensen & Riedel (1981) *Arch. Environ. Contam. Toxicol.* **10**, 357. **24.** Iturri & Wolff (1982) *Comp. Biochem. Physiol. C* **71C**, 131. **25.** Schneider (1975) *Biochem. Pharmacol.* **24**, 939. **26.** Davis & Wedemeyer (1971) *Comp. Biochem. Physiol. B* **40**, 823. **27.** Akera, Brody & Leeling (1971) *Biochem. Pharmacol.* **20**, 471. **28.** Owen & Rogers (1971) *Biochem. J.* **125**, 43P. **29.** Delaney, Owen & Rogers (1971) *Biochem. J.* **124**, 24P. **30.** Bowes & Gee (1971) *J. Bioenerg.* **2**, 47. **31.** Lawler & Rogers (1968) *Biochem. J.* **110**, 381. **32.** Sumbayev (2000) *Biochemistry (Moscow)* **65**, 972.

2,6,3'-Trichloroindophenol, See *2,6,3'-Trichlorophenolindophenol*

2,4,5-Trichlorophenol

This fungicide and bacteriocide (FW = 197.45 g/mol; CAS 95-95-4; MP = 67°C, has a low water solubility (< 0.2 g per 100 g H_2O; pK_a = 7.37 at 25°C); highly soluble in acetone (615 g per 100 g solvent) and methanol (also 615 g per 100 g solvent). **Target(s):** acetylcholinesterase (1); D-amino-acid oxidase (2); lipase (3); malate dehydrogenase (4); and NADPH dehydrogenase (quinone) activity of ζ-crystallin (5) **1.** Matsumura, Matsuoka, Igisu & Ikeda (1997) *Arch. Toxicol.* **71**, 151. **2.** Krahl, Keltch & Clowes (1940) *J. Biol. Chem.* **136**, 563. **3.** Christensen & Riedel (1981) *Arch. Environ. Contam. Toxicol.* **10**, 357. **4.** Wedding, Hansch & Fukuto (1967) *Arch. Biochem. Biophys.* **121**, 9. **5.** Shehu, al-Hamidi, Rabbani & Duhaiman (1998) *J. Enzyme Inhib.* **13**, 229.

2,6,3'-Trichlorophenolindophenol

This widely used synthetic electron acceptor (FW$_{sodium\ salt}$ = 324.53 g/mol; E_o' = +0.26 V), also called 2,6,3'-trichloroindophenol and abbreviated TCIP, functions as such for a wide variety of oxidation/reduction enzymes. At pH 7, aqueous solutions are blue when the reagent is oxidized, but colorless when reduced. The oxidized form inhibits photophosphorylation, perhaps acting both as a proton gradient dissipator as well as a redox mediator. *Note*: The solid will slowly decompose in light; hence, it should be stored in the dark. Aqueous solutions will decompose upon standing, particularly under acidic conditions. **1.** Izawa & Good (1972) *Meth. Enzymol.* **24**, 355.

(2,4,5-Trichlorophenoxy)acetate

This herbicide (FW$_{free-acid}$ = 255.48 g/mol; MP = of 153°C; free acid is water-insoluble; salts are water soluble) is a synthetic auxin and component of the notorious jungle defoliant Agent Orange. 2,4,5-T kills broad-leaf plants, but not grasses. *Note*: Caution should be exercised in its handling to avoid contact dermatitis and liver damage. The use of 2,4,5-T as a herbicide became controversial as many regarded it unsafe to animals, humans, and the environment. The U. S. Environmental Protection Agency restricted its use to rangelands and rice fields, but, in 1985, the same agency completely banned its use as a herbicide in the U.S. **Target(s):** benzoyl-CoA synthetase, *or* benzoate:CoA ligase (1); Ca^{2+}-dependent ATPase (2); glutathione S-transferase (3,4); lactate dehydrogenase (5,6); ATP-dependent palmitoyl-CoA synthetase (7). **1.** Gregus, Halaszi & Klaassen (1999) *Xenobiotica* **29**, 547. **2.** Thebault & Decaris (1983) *Comp. Biochem. Physiol. C* **75**, 369. **3.** Singh & Awasthi (1985) *Toxicol. Appl. Pharmacol.* **81**, 328. **4.** Vessey & Boyer (1984) *Toxicol. Appl. Pharmacol.* **73**, 492. **5.** Baker (1967) *Design of Active-Site-Directed Irreversible Enzyme Inhibitors*, Wiley, New York. **6.** Ottolenghi & Denstedt (1958) *Can. J. Biochem. Physiol.* **36**, 1075. **7.** Roberts & Knights (1992) *Biochem. Pharmacol.* **44**, 261.

2,6,8-Trichloropurine

This urate analogue (FW = 223.45 g/mol; MP = 159-161°C; slightly soluble in water) inhibits urate oxidase, *or* uricase, K_i = 0.8 μM (1-3). **1.** Baum, Hübscher & Mahler (1956) *Biochim. Biophys. Acta* **22**, 514. **2.** Mahler (1963) *The Enzymes*, 2nd ed. (Boyer, Lardy & Myrbäck, eds.), **8**, 285. **3.** Webb (1966) *Enzyme and Metabolic Inhibitors*, vol. **2**, p. 284, Academic Press, New York.

O-(2,3,5-Trichloropyrid-6-yl)-(O-methyl-O-n-butyl)phosphate, See *O-(n-Butyl) O-Methyl O-(2,3,5-Trichloropyrid-6-yl)phosphate*

Trichokirin

This ricin-like toxic 27-kDa enzyme from the seeds of *Trichosanthes kirilowii*, exhibits rRNA N-glycosylase activity and is classified as a type I ribosome-inactivating protein (1). Trichokirin catalyzes endohydrolysis of the N-glycosidic bond at one specific adenosine on the 28S rRNA. Trichokirin-S1 is the smaller 11-kDa protein (2). **1.** Casellas, Dussossoy, Falasca, *et al.* (1988) *Eur. J. Biochem.* **176**, 581. **2.** Li, Yang, Hu, *et al.* (2003) *Sheng Wu Hua Xue Yu Sheng Wu Wu Li Xue Bao (Shanghai)* **35**, 841.

Trichostatin A

This iron-binding *Streptomyces hygroscopicus*-derived antifungal (FW = 302.37 g/mol; CAS 58880-19-6; Symbol: TSA), known systematically as [*R-(E,E)*]-7-[4-(dimethylamino)phenyl]-N-hydroxy-4,6-dimethyl-7-oxo-2,4-heptadienamide, is active against trychophytons and some fungi (1) and noncompetitively inhibits histone deacetylase, *or* HDAC (K_i = 3.4 nM for the mouse mammary tumor enzyme) (2-5), disrupting chromatin function (6,7). HDAC8 is the only known HDAC that is unaffected by TSA. (*For further details, see Vorinostat*) Trichostatin A induces a variety of biological responses of cells (*e.g.*, induced differentiation, reduced migration, and cell cycle arrest). TSA prevents new actin filament formation in hepatic stellate cells by down-regulating express of Arp2/3 nucleating proteins and by up-regulating expression of the filament-capping proteins adducin-like protein 70 (ADDL70) and gelsolin (8). RhoA also decreases following TSA exposure. Trichostatin A has proved to be a useful probe of histone deacetylase structure as well as the roles of HDAC in chromatin structure and function. Trichostatin B (FW = 959.93 g/mol) is the ferric chelate formed by three trichostatin A molecules. Trichostatin C (FW = 464.51 g/mol) is the glucosyl hydroxamate of trichostatin A. **Target(s):** histone-deacetylase-like amidohydrolase (1); and histone deacetylase (2-6). **1.** Tsuji, Kobayashi, Nagashima, Wakisaka & Koizumi (1976) *J. Antibiot.* (Tokyo) **29**, 1. **2.** Hildmann, Ninkovic, Dietrich, *et al.* (2004) *J. Bacteriol.* **186**, 2328. **3.** Takahashi, Miyaji, Yoshida, Sato & Mizukami (1996) *J. Antibiot. (Tokyo)* **49**, 453. **4.** Yoshida, Horinouchi & Beppu (1995) *Bioessays* **17**, 423. **5.** Yoshida, Kijima, Akita & Beppu (1990) *J. Biol. Chem.* **265**, 17174. **6.** Finnin, Donigian, Cohen, *et al.* (1999) *Nature* **401**, 188. **7.** Bonamy, Guiochon-Mantel & Allison (2005) *Mol. Endocrinol.* **19**, 1213. **8.** Rombouts, Knittel, Machesky, *et al.* (2002) *J. Hepatol.* **37**, 788.

Trichothecene, See *specific mycotoxin; e.g., Nivalenol; T-2 Toxin; Trichothecin*

Trichothecin

This sesquiterpene antibiotic and mycotoxin (FW = 332.40 g/mol; CAS 6379-69-7; soluble in most organic solvents) from *Trichothecium roseum* inhibits peptide bond formation in protein biosynthesis, chain elongation (1,2). Aqueous solutions are stable for at least two days at pH 1-10. **1.** Barbacid, Fresno & Vazquez (1975) *J. Antibiot. (Tokyo)* **28**, 453. **2.** Jimenez & Vazquez (1975) *Eur. J. Biochem.* **54**, 483.

Tricine

This zwitterionic Good's buffer (FW = 179.17 g/mol; CAS 5704-04-1; pK_{a1} = 2.1 and pK_{a2} = 8.15 at 20°C; dpK_{a2}/dT = –0.021 ph units/°C), known systematically as *N*-[tris(hydroxymethyl)methyl]glycine, also weakly chelates Zn^{2+} and Ca^{2+}, the latter with a formation constant of 16 M^{-1} (1). *Note:* Chloroethane sulfonate is an essential ingedient in the chemical synthesis of ethane sulfonic acid-containing Good's Buffers, giving rise to variable contaminating levels oligovinylsulfonate (OVS) that inhibit ribonuclease (2) and are likely to act similarly with other enzymes that bind polyanions. **Target(s):** carbonic anhydrase, *or* carbonate dehydratase (3); H⁺-transporting inorganic pyrophosphatase (4); mannosyl-oligosaccharide glucosidase, *or* glucosidase I (5); ornithine decarboxylase (6); quinate *O*-hydroxycinnamoyltransferase (7); and shikimate *O*-hydroxycinnamoyltransferase (7). **1.** Good, Winget, Winter, *et al.* (1966) *Biochemistry* **5**, 467. **2.** Smith, Soellner & Raines (203) *J. Biol. Chem.* **278**, 20934. **3.** Hatch (1991) *Anal. Biochem.* **192**, 85. **4.** Gordon-Weeks, Koren'kov, Steele & Leigh (1997) *Plant Physiol.* **114**, 901. **5.** Szumilo, Kaushal & Elbein (1986) *Arch. Biochem. Biophys.* **247**, 261. **6.** Morley & Ho (1976) *Biochim. Biophys. Acta* **438**, 551. **7.** Lotfy, Fleuriet & Macheix (1992) *Phytochemistry* **31**, 767.

Triclosan

This cell-permeant, antiseptic agent (FW = 289.54 g/mol; CAS 3380-34-5; IUPAC name: 2-hydroxy-2',4,4'-trichlorodiphenyl ether), which exhibits both bacteriostatic and fungicidal activities, is a common additive in many cosmetic and personal care products, including toothpaste. Triclosan inhibits (K_i = 7 pM) fatty acid synthase at the enoyl-[acyl-carrier protein] reductase step by forming an enzyme–NAD⁺–triclosan ternary complex that is stabilized by well-placed hydrogen bonds and hydrophobic interactions

between triclosan and the enzyme and coenzyme (1-3). Triclosan is also both an alternative substrate and inhibitor of human liver 3'-phosphoadenosine 5'-phosphosulfate-sulfotransferases and UDP-glucuronosyltransferases (4). Apparent K_m and V_{max} values for triclosan sulfonation were 9 μM and 0.1 nmol/min/mg protein, whereas K_m and V_{max} values for glucuronidation were 100 μM and 0.7 nmol/min/mg protein. Triclosan inhibits hepatic cytosolic sulfonation of 3-hydroxybenzo(*a*)pyrene, bisphenol A, *p*-nitrophenol, and acetaminophen with IC_{50} concentrations of 2.9, 3.0, 6.5, and 18 μM, respectively (4). Triclosan is also a potent inhibitor of both estradiol and estrone sulfonation, exhibiting an IC_{50} of 0.6 nM for the latter (5). Given its structure and pK_a value of 7.9, triclosan is also a likely to be an effective protonophoric uncoupler of oxidative phosphorylation. Isolated liver mitochondria (at 60 nmol triclosan per mg mitochondrial protein) exhibited decreased oxygen consumption typical of State-3 respiration, as well as a decreased Δψ (6). Indeed, mitochondrial swelling, as driven by K⁺ diffusion potential in the presence of valinomycin, was also inhibited at triclosan concentrations greater than 10 nmol/mg protein (6). Triclosan also alters thyroid hormone receptor α transcript levels in premetamorphic tadpole brain and induces transient weight loss (7). Triclosan is chemically and functionally related to the antiseptic hexachlorophene (*or* 2,2'-methylenebis[3,4,6-trichlorophenol]). Note also that triclosan resists biological turnover and is likely to accumulate in lipid-rich tissues. Long-term human health risks remain uncertain. **Target(s):** CYP1A or CYP2B (1); enoyl-{acyl-carrier protein] reductase (2,3,8-10); estrogen sulfotransferase (5); NADPH-dependent enoyl-[acyl-carrier protein] reductase (11); and fatty-acid synthase (12,13). **1.** Hanioka, Omae, Nishimura, *et al.* (1996) *Chemosphere* **33**, 265. **2.** Heath, Rubin, Holland, *et al.* (1999) *J. Biol. Chem.* **274**, 11110. **3.** Stewart, Parikh, Xiao, Tonge & Kisker (1999) *J. Mol. Biol.* **290**, 859. **4.** Wang, Falany & James (2004) *Drug Metab. & Dispos.* **32**, 1162. **5.** James, Li, Summerlot, *et al.*(2010) *Environ. Internat.* **36**, 942. **6.** Newton, Cadena, Rocha, Carnieri & de Oliviera (2005) *Toxicol. Lett.* **160**, 49. **7.** Veldhoen, Skirrow & Osachoff (2006) *Aquatic Toxicology* **80**, 217. **8.** Ward, Holdgate, Rowsell, *et al.* (1999) *Biochemistry* **38**, 12514. **9.** Roujeinikova, Levy, Rowsell, *et al.* (1999) *J. Mol. Biol.* **294**, 527. **10.** Muralidharan, Suguna, Surolia & Surolia (2003) *J. Biomol. Struct. Dyn.* **20**, 589. **11.** Heath, Li, Roland & Rock (2000) *J. Biol. Chem.* **275**, 4654. **12.** Schmid, Rippmann, Tadayyon & Hamilton (2005) *Biochem. Biophys. Res. Commun.* **328**, 1073. **13.** Lupu & Menendez (2006) *Curr. Pharm. Biotechnol.* **7**, 483.

Tri-o-cresyl Phosphate, See *Tri-o-tolyl Phosphate*

1,1,3-Tricyano-2-amino-1-propene, See *2-Amino-1,1,3-tricyano-1-propene*

Tricyclazole

This fungicide (FW = 189.24 g/mol; CAS 41814-78-2), also known as 5-methyl-1,2,4-triazolo[3,4-*b*]benzothiazole, inhibits melanin biosynthesis and has been used to treat rice blast disease. **Target(s):** scytalone dehydratase (1); tetrahydroxynaphthalene reductase (2,3); and trihydroxynaphthalene reductase (2,4,5). **1.** Wheeler & Greenblatt (1988) *Exp. Mycol.* **12**, 151. **2.** Liao, Thompson, Fahnestock, Valent & Jordan (2001) *Biochemistry* **40**, 8696. **3.** Thompson, Fahnestock, Farrall, *et al.* (2000) *J. Biol. Chem.* **275**, 34867. **4.** Andersson, Jordan, Schneider & Lindqvist (1996) *Structure* **4**, 1161. **5.** Andersson, Jordan, Schneider, Valent & Lindqvist (1996) *Proteins* **24**, 525.

O-Tricyclo[5.2.1.0²,⁶]dec-9-yl Dithiocarbonate

This hygroscopic dithiocarbonate (FW = 228.38 g/mol), frequently referred to as D609 or tricyclodecan-9-yl xanthogenate, is a selective inhibitor of phosphatidyl-choline-specific phospholipase C, but does not inhibit phosphatidylinositol-specific phospholipase C or phospholipase A. D609 inhibits angiogenesis by preventing the synthesis of basement membrane. *Note*: Avoid using this reagent with HEPES buffer, because HEPES renders the reagent toxic. **Target(s):** basement membrane biosynthesis (1); herpes simplex virus type 1 replication (2); HIV-1 replication (3); Lysophospholipase (4); phospholipase C (5-7); phospholipase D, weakly inhibited (8); protein kinase C (2); and sphingomyelin synthase (9-13). **1**. Maragoudakis, Missirlis, Karakiulakis, *et al.* (1993) *Kidney Int.* **43**, 147. **2**. Walro & Rosenthal (1997) *Antiviral Res.* **36**, 63. **3**. Mellert, Amtmann, Erfle & Sauer (1988) *AIDS Res. Hum. Retroviruses* **4**, 71. **4**. Prabhakaran, Harris & Randhawa (1996) *J. Basic Microbiol.* **36**, 341. **5**. Muller-Decker (1989) *Biochem. Biophys. Res. Commun.* **162**, 198. **6**. Monick, Carter, Gudmundsson, *et al.* (1999) *J. Immunol.* **162**, 3005. **7**. Ramoni, Spadaro, Menegon & Podo (2001) *J. Immunol.* **167**, 2642. **8**. Gratas & Powis (1993) *Anticancer Res.* **13**, 1239. **9**. Nikolova-Karakashian (2000) *Meth. Enzymol.* **311**, 31. **10**. Luberto, Stonehouse, Collins, *et al.* (2003) *J. Biol. Chem.* **278**, 32733. **11**. Luberto & Hannun (1998) *J. Biol. Chem.* **273**, 14550. **12**. Huitema, van den Dikkenberg, Brouwers & Holthuis (2004) *EMBO J.* **23**, 33. **13**. Meng, Luberto, Meier, *et al.* (2004) *Exp. Cell Res.* **292**, 385.

Tricyclohexyltin Hydroxide

This UV-sensitive organotin acaricide (FW = 385.18 g/mol; white solid; MP = 195-198°C), also called Plictran, tricyclohexylstannol, and cyhexatin, inhibits the adrenergic stimulated calcium pump, *or* Ca^{2+}-dependent ATPase (1,2), ecto-ATPase, by tricyclohexyltin chloride (3); F$_o$F$_1$ ATP synthase (4,5), and glutathione *S*-transferase (6). **1**. Sahib & Desaiah (1987) *Cell. Biochem. Funct.* **5**, 149. **2**. Sahib & Desaiah (1986) *J. Biochem. Toxicol.* **1**, 55. **3**. Hennighausen, Hehl & Rychly (1985) *Arch. Toxicol. Suppl.* **8**, 500. **4**. McEnery & Pedersen (1986) *Meth. Enzymol.* **126**, 470. **5**. McEnery & Pedersen (1986) *J. Biol. Chem.* **261**, 1745. **6**. Hennighausen & Merkord (1985) *Arch. Toxicol.* **57**, 67.

Tridemorph

This fungicide (FW$_{free-base}$ = 297.52 g/mol; CAS 24602-86-6), also known as 2,6-dimethyl-*N*-tridecylmorpholine, inhibits chitin synthase (1); cholestenol Δ-isomerase, *or* sterol Δ^{7,8}-isomerase, K_i = 90 pM (2-5), cholesterol oxidase (6), cycloeucalenol cycloisomerase, *or* cycloeucalenol:obtusifoliol isomerase (7,8), and electron transport (9,10). **1**. Lyr & Seyd (1978) *Z. Allg. Mikrobiol.* **18**, 721. **2**. Nes, Zhou, Dennis, *et al.* (2002) *Biochem. J.* **367**, 587. **3**. Moebius, Reiter, Bermoser, *et al.* (1998) *Mol. Pharmacol.* **54**, 591. **4**. Paul, Silve, De Nys, *et al.* (1998) *J. Pharmacol. Exp. Ther.* **285**, 1296. **5**. Moebius, Bermoser, Reiter, Hanner & Glossmann (1996) *Biochemistry* **35**, 16871. **6**. Hesselink, Kerkenaar & Witholt (1990) *J. Steroid Biochem.* **35**, 107. **7**. Rahier, Bouvier, Cattel, Narula & Benveniste (1983) *Biochem. Soc. Trans.* **11**, 537. **8**. Bladocha & Benveniste (1983) *Plant Physiol.* **71**, 756. **9**. Rapoport & Schewe (1977) *Trends Biochem. Sci.* **2**, 186. **10**. Müller & Schewe (1976) *Acta Biol. Med. Germ.* **35**, 693 and 1737.

Tridiphane

This herbicide (FW = 320.43 g/mol; CAS 58138-08-2) is a substrate of epoxide hydrolases and glutathione *S*-transferases. The glutathione conjugate of tridiphane is a strong inhibitor of glutathione *S*-transferase (1,2) and

tabersonine 16-hydroxylase (3). **1**. Deng & Hatzios (2002) *Pestic. Biochem. Physiol.* **72**, 10. **2**. Lamoureux & Rusness (1986) *Pestic. Biochem. Physiol.* **26**, 323. **3**. St-Pierre & De Luca (1995) *Plant Physiol.* **109**, 131.

Triethanolamine

This frequently pH buffer component (FW$_{free-base}$ = 149.19 g/mol; CAS 102-71-6) is a hygroscopic and viscous liquid (melting point = 21.57°C; B.P. = 335.4°C) and has a pK_a value of 7.76 at 25°C (dpK_a/dT = –0.020 pH units/°C). **Target(s):** choline sulfotransferase (1); dipeptidase (2); formimino-tetrahydrofolate cyclodeaminase (3); nardilysin (4); oligo-1,6-glucosidase, *or* isomaltase (5); sucrose α-glucosidase, *or* sucrase-isomaltase (5). **1**. Orsi & Spencer (1964) *J. Biochem.* **56**, 81. **2**. Patterson (1976) *Meth. Enzymol.* **45**, 386. **3**. Uyeda & Rabinowitz (1963) *Meth. Enzymol.* **6**, 380. **4**. Csuhai, Juliano, Juliano & Hersh (1999) *Arch. Biochem. Biophys.* **362**, 291. **5**. Kano, Usami, Adachi, Tatematsu & Hirano (1996) *Biol. Pharm. Bull.* **19**, 341.

2-(*N,N,N*-Triethylamino)-2-methyl-1,3-propanediol, *See* 2-Amino-2-methyl-*N,N*-triethyl-1,3-propanediol

Triethylenethiophosphoramide, *See* ThioTEPA

(Triethyl)lead Chloride

This cytotoxic alkylated lead (FW = 329.8 g/mol; CAS 5224-23-7), which is the major metabolite of the once widely used gasoline anti-knocking additive (tetraethyl)lead, is known to disrupt neurofilaments within neurons. *Caution*: Exposure to this agent is known to cause permanent neurologic deficiencies in experimental animals. *See also* Tetraethyllead   **Target(s):** γ-aminobutyrate (GABA) uptake (1,2); δ-aminolevulinate dehydratase, *or* porphobilinogen synthase (3-5); Ca^{2+}-channel, voltage-dependent (6); F$_o$F$_1$ ATPase, *or* H$^+$-transporting two-sector ATPase (7); glutathione *S*-transferase, *or* ligandin (8); Na$^+$/K$^+$-exchanging ATPase (7); oxidative phosphorylation (7,9); tubulin polymerization, *or* microtubule assembly (10,11). **1**. Seidman & Verity (1987) *J. Neurochem.* **48**, 1142. **2**. Seidman, Olsen & Verity (1987) *J. Neurochem.* **49**, 415. **3**. Yagminas & Villeneuve (1987) *J. Appl. Toxicol.* **2**, 115. **4**. Burns & Godwin (1991) *J. Appl. Toxicol.* **11**, 103. **5**. SBondy (1986) *J. Toxicol. Environ. Health* **18**, 639. **6**. Busselberg, Evans, Rahmann & Carpenter (1991) *Neurotoxicology* **12**, 733. **7**. Munter, Athanasiou & Stournaras (1989) *Biochem. Pharmacol.* **38**, 3941. **8**. Byington & Hansbrough (1979) *J. Pharmacol. Exp. Ther.* **208**, 248. **9**. Kauppinen, Komulainen & Taipale (1988) *J. Neurochem.* **51**, 1617 and (1989) **52**, 661. **10**. Zimmermann, Doenges & Roderer (1985) *Exp. Cell Res.* **156**, 140. **11**. Zimmermann, Faulstich, Hansch, Doenges & Stournaras (1988) *Mutat. Res.* **201**, 293.

Triethyloxonium Tetrafluoroborate

This moisture-sensitive reagent (FW = 189.99 g/mol; CAS 368-39-8) reacts, under acidic conditions, with carboxyl groups within proteins and enzymes to form ethyl esters. To control which carboxyl groups are esterified, one may vary pH, ionic strength, or degree of active-site occupancy by a substrate, coenzyme, metal ion or regulatory effector. If the unmodified COOH group is essential for catalytic activity, then triethyloxonium tetrafluoroborate will behave as an active site-directed irreversible enzyme inhibitor. **Target(s):** β-lactamase I (1); lysozyme (2,3); trypsin (4); and β-xylosidase (5). **1.** Waley (1975) *Biochem. J.* **149**, 547. **2.** Parsons, Jao, Dahlquist, *et al.* (1969) *Biochemistry* **8**, 700. **3.** Parsons & Raftery (1969) *Biochemistry* **8**, 4199. **4.** Ben Avraham & Shalitin (1985) *FEBS Lett.* **180**, 239. **5.** Gomez, Isorna, Rojo & Estrada (2001) *Biochimie* **83**, 961.

Trifluoperazine

This phenothiazine antipsychotic (FW$_{free-base}$ = 407.50 g/mol; CAS 117-89-5) is a dopamine D$_2$ receptor and calmodulin antagonist, an inhibitor of cAMP-gated cation channels, and an inhibitor of the calmodulin stimulation of cAMP phosphodiesterase. The dihydrochloride is hygroscopic and freely soluble in water. The ultraviolet spectra (in ethanol) has λ$_{max}$ values of 258 nm (ε = 31600 M^{-1}cm^{-1}) and 307.5 nm (ε = 3200 M^{-1}cm^{-1}). **Target(s):** adenylate cyclase (1,2); [β-adrenergic-receptor] kinase (3); 1-alkylglycerophosphocholine *O*-acetyltransferase (42); 1-aminocyclopropane-1-carboxylate synthase (4); apyrase (5); ATPase and electron transport, plant (6); Ca^{2+}/calmodulin-dependent protein kinase II (34); Ca^{2+} channel, type 1 inositol 1,4,5-trisphosphate-sensitive (7); calmodulin (8-10); calmodulin-dependent glycogen synthase kinase (11); Ca^{2+}-transporting ATPase (12); cholestenol Δ-isomerase, *or* sterol Δ^{7,8}-isomerase (13); choline-phosphate cytidylyltransferase (35); 3',5'-cyclic-nucleotide phosphodiesterase (8,9,14); diacylglycerol *O*-acyltransferase (44); fatty-acid$_0$-methyltransferase (45); G-protein coupled receptor kinase 2 (15); helicase/ATPase (16); H$^+$-exporting ATPase (17); [3-hydroxy-3-methylglutaryl-CoA reductase (NADPH)] kinase (30); [acyl-carrier-protein] *S*-malonyltransferase (43); [myosin light-chain] kinase (11,18,33); NAD$^+$ kinase (38-41); nucleoside-triphosphatase (16); 1-phosphatidylinositol 4-kinase (37); 1-phosphatidylinositol-4-phosphate 5-kinase (36); phospholipase A$_2$ (19); phosphoprotein phosphatase (21,25-27); phosphorylase kinase (31,32); phosphoserine phosphatase (28,29); protein kinase C (9-11,20); protein phosphatase 2B (21,26); protein-tyrosine-phosphatase, weakly inhibited (22); stearoyl-CoA 9-desaturase (46); ubiquityl-calmodulin synthetase, *or* ubiquitin:calmodulin ligase (23); Xaa-methyl-His Dipeptidase, *or* anserinase, *or* N-acetylhistidine deacetylase (24). **1.** Iversen (1975) *Science* **188**, 1084. **2.** Asbury, Cook & Wolff (1978) *J. Biol. Chem.* **253**, 5286. **3.** Benovic (1991) *Meth. Enzymol.* **200**, 351. **4.** Mattoo, Adams, Patterson & Lieberman (1983) *Plant Sci. Lett.* **28**, 173. **5.** Schetinger, Vieira, Morsch & Balz (2001) *Comp. Biochem. Physiol. B Biochem. Mol. Biol.* **128**, 731. **6.** Dunn, Slabas, Cottingham & Moore (1984) *Arch. Biochem. Biophys.* **229**, 287. **7.** Khan, Dyer & Michelangeli (2001) *Cell Signal* **13**, 57. **8.** Weiss (1983) *Meth. Enzymol.* **102**, 171. **9.** Hidaka, Inagaki, Nishikawa & Tanaka (1988) *Meth. Enzymol.* **159**, 652. **10.** Kikkawa, Minakuchi, Takai & Nishizuka (1983) *Meth. Enzymol.* **99**, 288. **11.** Roach (1984) *Meth. Enzymol.* **107**, 81. **12.** Wright & van Houten (1990) *Biochim. Biophys. Acta* **1029**, 241. **13.** Moebius, Reiter, Bermoser, *et al.* (1998) *Mol. Pharmacol.* **54**, 591. **14.** Chaudry & Casillas (1988) *Arch. Biochem. Biophys.* **262**, 439. **15.** Kassack, Hogger, Gschwend, *et al.* (2000) *AAPS PharmSci.* **2**, E2. **16.** Borowski, Niebuhr, Schmitz, *et al.* (2002) *Acta Biochim. Pol.* **49**, 597. **17.** Schneider & Chin (1988) *Meth. Enzymol.* **157**, 591. **18.** Sobieszek (1999) *Biochim. Biophys. Acta* **1450**, 77. **19.** Bartolf & Franson (1984) *Biochim. Biophys. Acta* **793**, 379. **20.** Kikkawa & Nishizuka (1986) *The Enzymes*, 3rd ed. (Boyer & E. G. Krebs, eds.), **17**, 167. **21.** Cohen (1991) *Meth. Enzymol.* **201**, 389. **22.** Aguirre-Garcia, Escalona-Montano, Bakalara, *et al.* (2006) *Parasitology* **132**, 641. **23.** Parag, Dimitrovsky, Raboy & Kulka (1993) *FEBS Lett.* **325**, 242. **24.** Lenney, Baslow & Sugiyama (1978) *Comp. Biochem. Physiol. B* **61**, 253. **25.** Cheng, Wang, Gong, *et al.* (2001) *Neurochem. Res.* **26**, 425. **26.** Dobson, May, Berriman, *et al.* (1999) *Mol. Biochem. Parasitol.* **99**, 167. **27.** Ichikawa, Tsurusaki & Yamaguchi (2004) *Int. J. Mol. Med.* **13**, 289. **28.** Shetty & Shetty (1991) *Neurochem. Res.* **16**, 1203. **29.** Hawkinson, Acosta-Burruel, Ta & Wood (1997) *Eur. J.*

Pharmacol. **337**, 315. **30.** Beg, Stonik & Brewer (1987) *J. Biol. Chem.* **262**, 13228. **31.** Picton, Shenolikar, Grand & Cohen (1983) *Meth. Enzymol.* **102**, 219. **32.** Chan & Graves (1982) *J. Biol. Chem.* **257**, 5956. **33.** Bailin (1984) *Experientia* **40**, 1185. **34.** Rodriguez-Mora, LaHair, Howe, McCubrey & Franklin (2005) *Exp. Opin. Ther. Targets* **9**, 791. **35.** Pelech, Jetha & Vance (1983) *FEBS Lett.* **158**, 89. **36.** Husebye & Flatmark (1989) *Biochim. Biophys. Acta* **1010**, 250. **37.** Husebye, Letcher, Lander & Flatmark (1990) *Biochim. Biophys. Acta* **1042**, 330. **38.** Gallais, de Crescenzo & Laval-Martin (2001) *Aust. J. Plant Physiol.* **28**, 363. **39.** Turner, Waller, Vanderbeld & Snedden (2004) *Plant Physiol.* **135**, 1243. **40.** Muto (1983) *Z. Pflanzenphysiol.* **109**, 385. **41.** Williams & Jones (1985) *Arch. Biochem. Biophys.* **237**, 80. **42.** Gomez-Cambronero, Nieto & Mato & Sanchez-Crespo (1985) *Biochim. Biophys. Acta* **845**, 511. **43.** Sinha & Dick (2004) *J. Antimicrob. Chemother.* **53**, 1072. **44.** Kamisaka, Mishra & Nakahara (1997) *J. Biochem.* **121**, 1107. **45.** Safayhi, Anazodo & Ammon (1991) *Int. J. Biochem.* **23**, 769. **46.** Griffiths, Griffiths & Stobart (1998) *Phytochemistry* **48**, 261.

Trifluopromazine, See *Triflupromazine*

***N*-(Trifluoroacetyl)-L-valyl-L-alanine *p*-(trifluoromethyl)anilide,** *similar in action to N-(Trifluoroacetyl)-L-lysyl-L-leucine p-(Isopropyl)anilide*

3,3,3-Trifluoroalanine

This modified alanine (FW = 143.07 g/mol; CAS 10065-69-7) is a mechanism-based inactivator. For example, inactivation of the alanine racemases from the Gram-negative organism *Salmonella typhimurium* and Gram-positive organism *Bacillus stearothermophilus* with 2,2,2-trifluoroalanine occurs with the same rate constant as that for formation of a broad 460-490-nm enzyme chromophore (1). Loss of two fluoride ions per mole of inactivated enzyme and retention of [1-^{14}C]trifluoroalanine label accompany inhibition, suggesting a monofluoro enzyme adduct. Partial denaturation (1 M guanidine) leads to rapid return of the initial 420-nm chromophore, followed by a slower ($t_{1/2}$ ≈30 min^{-1}) loss of the fluoride ion and ^{14}CO$_2$ release. At this point, reduction by NaBH$_4$ and tryptic digestion yielded a single radiolabeled peptide, with lysine-38 covalently attached to the PLP cofactor. The likely mechanism for enzyme inactivation by trifluoroalanine entails nucleophilic attack of released aminoacrylate on the PLP aldimine leads to enzyme inactivation (1). For trifluoroalanine inactivation, nucleophilic attack of lysine-38 on the electrophilic 2,2-difluoro-α,β-unsaturated imine provides an alternative mode of inhibition for these enzymes. **Target(s):** alanine racemase (1-3); 8-amino-7-oxononanoate synthase (4); cystathionine β-lyase (5-8); cystathionine γ-lyase (8-12); cystathionine synthase (8); threonine dehydrase (8); tryptophan indole-lyase, *or* tryptophanase (8,13); tryptophan synthase (8,13); UDP-*N*-acetylmuramate:L-alanine ligase, inhibited by the DL-mixture (14,15). **1.** Faraci & Walsh (1989) *Biochemistry* **28**, 431. **2.** Ator & Ortiz de Montellano (1990) *The Enzymes*, 3rd ed. (Sigman & Boyer, eds.), **19**, 213. **3.** Wang & Walsh (1981) *Biochemistry* **20**, 7539. **4.** Alexeev, Baxter, Campopiano, *et al.* (2006) *Org. Biomol. Chem.* **4**, 1209. **5.** Uren (1987) *Meth. Enzymol.* **143**, 483. **6.** Clausen, Huber, Laber, Pohlenz & Messerschmidt (1996) *J. Mol. Biol.* **262**, 202. **7.** Dwivedi, Ragin & Uren (1982) *Biochemistry* **21**, 3064. **8.** Silverman & Abeles (1976) *Biochemistry* **15**, 4718. **9.** Steegborn, Clausen, Sondermann, *et al.* (1999) *J. Biol. Chem.* **274**, 12675. **10.** Fearon, Rodkey & Abeles (1982) *Biochemistry* **21**, 3790. **11.** Silverman & Abeles (1977) *Biochemistry* **16**, 5515. **12.** Alston, Maramatsu, Ueda & Bright (1981) *FEBS Lett.* **128**, 293. **13.** Phillips & Dua (1992) *Arch. Biochem. Biophys.* **296**, 489. **14.** Liger, Blanot & van Heijenoort (1991) *FEMS Microbiol. Lett.* **80**, 111. **15.** Liger, Masson, Blanot, van Heijenoort & Parquet (1995) *Eur. J. Biochem.* **230**, 80.

1,1,1-Trifluoro-3-(decylsulfanyl)propan-2-one, See *3-(Decylthio)-1,1,1-trifluoropropan-2-one*

1,1,1-Trifluoro-3-(dodecylsulfanyl)propan-2-one, See *3-(Dodecylthio)-1,1,1-trifluoropropan-2-one*

2,2,2-Trifluoroethyl Glutamate, See *Glutamic Acid γ-(2,2,2-Trifluoroethyl) Ester*

α-(3'-Trifluoromethylbenzylamino)benzylphosphonate

This substituted phosphonic acid (FW$_{free-acid}$ = 345.26 g/mol) inhibits human prostatic acid phosphatase, IC$_{50}$ = 4.9 nM, amounting to a 3500-times improvement in potency over the carbon analogue, α-phenylethyl. This enhanced potency is likely to result from a combination of favorable interactions with the phosphate binding region, the presence the hydrophobic moieties of the benzylamino and phenylphosphonic acid, as well as a rigid conformer produced by an internal salt bridge between the phosphonate and the α-amino group. Replacement of the phosphonic acid moiety with a phosphinic or carboxylic acid or the deletion of the benzyl substitution of the α-amino group led to great reductions in potency. 1. Beers, Schwender, Loughney, et al. (1996) Bioorg. Med. Chem. 4, 1693.

4-Trifluoromethyl-Celecoxib

This non-steroidal anti-inflammatory agent and celecoxib analogue (FW = 435.34 g/mol; Symbol: TFM-C), also named 4-[5-(4-trifluoromethylphenyl)-3-(trifluoromethyl)-1H-pyrazol-1-yl]benzenesulfonamide, displays 205-fold lower COX2-inhibition than celecoxib, yet is equally effective in inhibiting the secretion, but not transcription, of Interleukin-12 (IL-12 p35/p40 heterodimer) and p80 (p40/p40 homodimer) by a mechanism involving altered cytokine-chaperone interaction in the endoplasmic reticulum (ER) (1). Such behavior is consistent with findings that celecoxib inhibits experimental autoimmune encephalomyelitis (EAE) in COX-2-deficient mice, indicating action by a COX2-independent pathway. Noting that IL-12, p80, and IL-23 are structurally related cytokines sharing common p40 subunits, later investigation demonstrated that celecoxib and TFM-C also block secretion of IL-23 p40/p19 heterodimers (2). Significantly, TFM-C and celecoxib ameliorate EAE mouse model for multiple sclerosis with equal potency, coinciding with reduced IL-17 and IFN-γ secretion by MOG-reactive T-cells and secretion of IL-23 and associated inflammatory cytokines by bone marrow-derived dendritic cells. These findings suggest that TFM-C is a promising drug candidate, one that retains celecoxib's beneficial effects in the EAE, but with greatly decreased COX2 inhibitory activity and less likelihood of exerting adverse cardiovascular effects. 1. Alloza, Baxter, Chen, Matthiesen & Vandenbroeck (2006) Mol. Pharmacol. 69, 1579. 2. Di Penta, Chiba, Alloza, et al. (2013) PLoS One 8, e83119.

5-(Trifluoromethyl)-2'-deoxyuridine

This thymidine nucleoside analogue (FW = 296.20 g/mol; CAS 70-00-8), also known by its trade name Viroptic® and by its alternative names trifluorothymidine and trifluridine, and the IUPAC name 1-[4-hydroxy-5-(hydroxymethyl)oxolan-2-yl]-5(trifluoromethyl)pyrimidine-2,4-dione, is an antiherpetic agent that inhibits viral DNA polymerase-catalyzed elongation. The active drug, trifluridine 5'-triphosphate, is formed after cellular uptake and enzyme-catalyzed phosphorylation. A likely contributing factor to its cytotoxicity is the failure to make correct base-pairing interactions. FTD incorporation into DNA induces DNA dysfunction, including DNA strand breaks. When combined with tipiracil to inhibit thymidine phosphorylase (which metabolically deactivates trifluridine), the resultant binary drug (known as Lonsurf®) is a promising treatment option for patients resistant to or intolerant of 5-FU-based fluoropyrimidines (See Tipiracil). **Target(s):** deoxycytidine kinase (1); viral DNA polymerase, via the 5'-triphosphate (2-4); thymidylate kinase, or dTMP kinase (5,6); and thymidylate synthase (7). 1. Krenitsky, Tuttle, Koszalka, et al. (1976) J. Biol. Chem. 251, 4055. 2. Prisbe & Chen (1996) Meth. Enzymol. 275, 425. 3. Satake, Takeda, Matsumura, et al. (1992) Nucleic Acids Symp. Ser. 1992, 189. 4. Satake, Takeda & Wataya (1991) Nucleic Acids Symp. Ser. 1991, 37. 5. Munier-Lehmann, Pochet, Dugué, et al. (2003) Nucleosides Nucleotides Nucleic Acids 22, 801. 6. Pochet, Dugué, Labesse, Delepierre & Munier-Lehmann (2003) Chem. Bio. Chem. 4, 742. 7. Sigman & Mooser (1975) Ann. Rev. Biochem. 44, 889.

5-Trifluoromethyl-2'-deoxyuridine 5'-Monophosphate

This modified deoxynucleotide (FW$_{free-acid}$ = 376.18 g/mol), also known as trifluorothymidine 5'-monophosphate, is a mechanism-based inhibitor of Lactobacillus casei thymidylate synthase. The inhibitor binds to the active site of the enzyme in the absence of the cofactor N^5,N^{10}-methylenetetrahydrofolate. Cys198 attacks C6 of the pyrimidine, thereby activating release of fluoride ion from the trifluoromethyl group. The activated heterocycle adduct then reacts with a nucleophile of the enzyme (Tyr146) to form a moderately stable covalent complex (1-5). 1. Eckstein, Foster, Finer-Moore, Wataya & Santi (1994) Biochemistry 33, 15086. 2. Danenberg (1977) Biochim. Biophys. Acta 473, 73. 3. Fridland, Langenbach & Heidelberger (1971) J. Biol. Chem. 246, 7110. 4. Carpenter (1974) J. Insect Physiol. 20, 1389. 5. Capco, Krupp & Mathews (1973) Arch. Biochem. Biophys. 158, 726.

5-Trifluoromethyl-2'-deoxyuridine 5'-Triphosphate

This modified deoxynucleotide (FW$_{free-acid}$ = 536.14 g/mol), also known as trifluorothymidine 5'-triphosphate, inhibits DNA polymerases (1-4). 1. Prisbe & Chen (1996) Meth. Enzymol. 275, 425. 2. Satake, Takeda, Matsumura, et al. (1992) Nucleic Acids Symp. Ser. 1992 (27), 189. 3. Satake, Takeda & Wataya (1991) Nucleic Acids Symp. Ser. 1991 (25), 37. 4. Tone & Heidelberger (1973) Mol. Pharmacol. 9, 783.

Trifluorperazine, See Trifluoperazine

Trifluperidol

This antipsychotic drug (FW = 445.88 g/mol; typically supplied as the hydrochloride) inhibits dopamine receptor DA-2 (1), cholestenol Δ-isomerase, *or* sterol $\Delta^{7,8}$-isomerase (2), and *N*-methyl-D-aspartate (NMDA) receptor (3-5). **1**. Felder, McKelvey, Gitler, Eisner & Jose (1989) *Kidney Int.* **36**, 183. **2**. Moebius, Reiter, Bermoser, *et al.* (1998) *Mol. Pharmacol.* **54**, 591. **3**. Shim, Grant, Singh, Gallagher & Lynch (1999) *Neurochem. Int.* **34**, 167. **4**. Whittemore, Ilyin & Woodward (1997) *J. Pharmacol. Exp. Ther.* **282**, 326. **5**. Coughenour & Cordon (1997) *J. Pharmacol. Exp. Ther.* **280**, 584.

Triflupromazine

This tranquilizing, antipsychotic phenothiazine (FW = 352.42 g/mol; CAS 146-54-3), also known as (*N*,*N*-dimethyl-2-(trifluoromethyl)-10*H*-phenothiazine-10-propanamine, is similar, both in structure and action, to the more widely used trifluoperazine. **Target(s):** adenosylmethionine decarboxylase (1); apyrase (2); calmodulin (3); calmodulin-dependent 3'5'-cAMP phosphodiesterase (3); K^+ channel, ATP-sensitive (4); Na^+/K^+-exchanging ATPase (5); [pyruvate dehydrogenase (acetyl-transferring)]-phosphatase, weakly inhibited (6). **1**. Hietala, Lapinjoki & Pajunen (1984) *Biochem. Int.* **8**, 245. **2**. Komoszynski (1996) *Comp. Biochem. Physiol. B* **113**, 581. **3**. Weiss (1983) *Meth. Enzymol.* **102**, 171. **4**. Muller, De Weille & Lazdunski (1991) *Eur. J. Pharmacol.* **198**, 101. **5**. Davis & Brody (1966) *Biochem. Pharmacol.* **15**, 703. **6**. Bak, Huh, Hong & Song (1999) *Biochem. Mol. Biol. Int.* **47**, 1029.

Trifluralin

This pre-emergence herbicide (FW = 335.28 g/mol; CAS 1582-09-8), also known as Treflan and α,α,α-trifluoro-2,6-dinitro-*N*,*N*-dipropyl-*p*-toluidine, is a potent inhibitor of plant microtubule assembly and plant cell division. Dinitroaniline herbicides (such as trifluralin and oryzalin) have been developed for the selective control of weeds in arable crops. No effect of trifluralin on animal microtubule systems has been reported, suggesting that trifluralin's action reveals a pharmacological difference between plant and animal tubulins (1). Likewise, this agent is without effect on yeasts such as *Schizosaccharomyces pombe* (2). Mutation of a single residue (Thr239-to-Ile), situated at the dimer interface in the microtubule protofilaments, confers resistance to dinitroaniline herbicides (3). Trifluralin also inhibits photosynthetic electron transport (4). **1**. Hess & Bayer (1977) *J. Cell Sci.* **24**, 351. **2**. Walker (1982) *J. Gen. Microbiol.* **128**, 61. **3**. Anthony, Waldin, Ray, Bright & Hussey (1998) *Nature* **393**, 260. **4**. Trebst (1980) *Meth. Enzymol.* **69**, 675.

Trifluridine, *See 5-(Trifluoromethyl)-2'-deoxyuridine*

Trifop

This herbicide (FW = 326.27 g/mol; CAS 58594-74-4), also named (*RS*)-2-[4-(α,α,α-trifluoro-*p*-tolyloxy)phenoxy]propionic acid, is a strong reversible inhibitor of acetyl-CoA carboxylase, with an IC$_{50}$ value of 0.26 μM for the wheat enzyme (1,2). **1**. Rendina, Craig-Kennard, Beaudoin & Breen (1990) *J. Agric. Food Chem.* **38**, 1282. **2**. Rendina, Felts, Boudoin, *et al.* (1988) *Arch. Biochem. Biophys.* **265**, 219.

Trigocherrierin A

This daphnane diterpenoid orthoester (FW = 598.69 g/mol) from leaves of the critically endangered New Caledonian plant *Trigonostemon cherrieri* is a potent and selective replication inhibitor of Chikungunya Virus, *or* CHIKV (EC$_{50}$ = 0.6 μM). Its complex structure and stereochemical features were determined by mass spectroscopy, 1-D and 2-D NMR, as well as structural data already in the literature. **1**. Bourjot, Leyssen, Neyts, Dumontet & Litaudon (2014) *Molecules* **19**, 3617.

Trigonelline

This zwitterion nicotinate catabolite (FW = 137.14 g/mol; FW$_{hydrochloride}$ = 173.60 g/mol; CAS 535-83-1), also known as 3-carboxy-1-methylpyridinium inner salt and nicotinic acid *N*-methylbetaine, accumulates in plants, sea urchin, jellyfish, as well as the urine of humans after ingestion of nicotinic acid. Trigonelline is a cell cycle regulator in a number of organisms and causes G$_2$ cell-cycle arrest (1). It also inhibits D-amino-acid oxidase (2), acetylcholinesterase (3), DNA replication (1) and is a product of the reaction catalyzed by nicotinate *N*-methyltransferase. **1**. Mazzuca, Bitonti, Innocenti & Francis (2000) *Planta* **211**, 127. **2**. Nishina, Sato, Miura & Shiga (2000) *J. Biochem.* **128**, 213. **3**. Bergmann, Wilson & Nachmansohn (1950) *J. Biol. Chem.* **186**, 693.

Trihexyphenidyl Hydrochloride

This anti-parkinson disease agent (FW = 337.93 g/mol; hydrochloride is slightly soluble (1.0 g/100 mL at 25°C)), also known as 3-(1-piperidyl)-1-cyclohexyl-1-phenyl-1-propanol hydrochloride, is an anticholinergic agent (1-3). Both stereoisomers are biologically active and different isotypes of the muscarinic receptor display different degrees of preference for the two enantiomers. **1**. Onali, Aasen & Olianas (1994) *Brit. J. Pharmacol.* **113**, 775. **2**. Dorje, Wess, Lambrecht, *et al.* (1991) *J. Pharmacol. Exp. Ther.* **256**, 727. **3**. Hudkins & DeHaven-Hudkins (1991) *Life Sci.* **49**, 1229.

2,10,11-Trihydroxyaporphine

R(–)-stereomer   *S*(+)-stereomer

This hygroscopic and photosensitive compound, also known as 2-hydroxyapomorphine (FW$_{free-base}$ = 283.33 g/mol), is a dopamine receptor agonist. **Target(s):** amyloid-beta fibril formation (1); and dihydropteridine reductase, inhibited by both the *R*(–)- and *S*(+)-stereoisomers (2). **1**. Lashuel,

Hartley, Balakhaneh, *et al.* (2002) *J. Biol. Chem.* **277**, 42881. **2.** Shen, Smith, Davis & Abell (1984) *J. Biol. Chem.* **259**, 8994.

2,3,4-Trihydroxybenzoate

This aromatic acid (FW$_{free-acid}$ = 170.12 g/mol; CAS 56128-66-6) inhibits 4-hydroxybenzoate decarboxylase (1,2), protocatechuate decarboxylase, *or* 3,4-dihydroxybenzoate decarboxylase (3), *o*-pyrocatechuate decarboxylase (4,5), and xanthine oxidase (6). **1.** Gallert & Winter (1992) *Appl. Microbiol. Biotechnol.* **37**, 119. **2.** He & Wiegel (1995) *Eur. J. Biochem.* **229**, 77. **3.** He & Wiegel (1996) *J. Bacteriol.* **178**, 3539. **4.** Subba Rao, Moore & Towers (1967) *Arch. Biochem. Biophys.* **122**, 466. **5.** Santha, Rao & Vaidyanathan (1996) *Biochim. Biophys. Acta* **1293**, 191. **6.** Chang, Yan & Chiang (1995) *Anticancer Res.* **15**, 2097.

3,4,5-Trihydroxybenzoate (*or* 3,4,5-Trihydroxybenzoic Acid), *See Gallate*

5-(3α,7α,12α-Trihydroxy-5β-cholanamido)-1,3,4-thiadiazole-2-sulfonamide

This sulfonamide (FW = 570.77 g/mol) inhibits carbonic anhydrase: IC$_{50}$ = 49 nM for the rainbow trout (*Oncorhynchus mykiss*) enzyme and 66 nM for human carbonic anhydrase II (1,2). **1.** Bülbül, Hisar, Beydemir, Çiftçi & Küfrevioglu (2003) *J. Enzyme Inhib. Med. Chem.* **18**, 371. **2.** Bülbül, Saraçoglu, Küfrevioglu & Çiftçi (2002) *Bioorg. Med. Chem.* **10**, 2561.

3α,7α,12α-Trihydroxycholanate, *See Cholate*

5,6,7-Trihydroxyflavone, *See Baicalein*

5,7,4'-Trihydroxyflavone, *See Apigenin*

2,4,6-Triiodophenol

This iodinated phenol (FW = 471.80 g/mol; melting point = 157-159°C) inhibits estrone sulfotransferase (1), glutamate dehydrogenase (2), thyroxine deiodinase (3), as well as transthyretin amyloid fibril formation (4). **1.** Kester, Bulduk, Tibboel, *et al.* (2000) *Endocrinology* **141**, 1897. **2.** Frieden (1963) *The Enzymes*, 2nd ed. (Boyer, Lardy & Myrbäck, eds.), 7, 3. **3.** Fekkes, Hennemann & Visser (1982) *Biochem. Pharmacol.* **31**, 1705. **4.** Miroy, Lai, Lashuel, *et al.* (1996) *Proc. Natl. Acad. Sci. U.S.A.* **93**, 15051.

3,3',5-Triiodothyroacetate

This thyroid hormone analogue (FW$_{free-acid}$ = 621.94 g/mol), also called triac and tiratricol, is a metabolite of thyroxine and 3,3',5-triiodothyronine. **Target(s):** alcohol dehydrogenase (1,2); aldehyde dehydrogenase (3); GABA$_A$ receptor (4); glucose-6-phosphate dehydrogenase (5); glutamate dehydrogenase (6); 15-hydroxyprostaglandin dehydrogenase (7); oxidative phosphorylation (8); prostaglandin E$_2$ 9-reductase (9); and protein-disulfide isomerase (10). **1.** Mardh, Auld & Vallee (1987) *Biochemistry* **26**, 7585. **2.**

Brändén, Jörnvall, Eklund & Furugren (1975) *The Enzymes*, 3rd ed. (Boyer, ed.), **11**, 103. **3.** Zhou & Weiner (1997) *Eur. J. Biochem.* **245**, 123. **4.** Martin, Padron, Newman, *et al.* (2004) *Brain Res.* **1004**, 98. **5.** McKerns (1962) *Biochim. Biophys. Acta* **62**, 402. **6.** Frieden (1963) *The Enzymes*, 2nd ed. (Boyer, Lardy & Myrbäck, eds.), **7**, 3. **7.** Pace-Asciak & Smith (1983) *The Enzymes*, 3rd ed. (Boyer, ed.), **16**, 543. **8.** Brodie (1963) *Meth. Enzymol.* **6**, 284. **9.** Tai & Yuan (1982) *Meth. Enzymol.* **86**, 113. **10.** Guthapfel, Gueguen & Quemeneur (1996) *Eur. J. Biochem.* **242**, 315.

3,3',5-Triiodo-L-thyronine

This amino acid (FW = 650.98 g/mol; pK_a values of 2.2, 8.40 (phenol OH), and 10.1), symbolized by T$_3$, is the principal thyroid hormone derived by proteolysis of thyroglobulin. T$_3$ is about five-fold more potent than T$_4$. Much of its physiological role is associated with its binding to the thyroid hormone receptor, with resulting regulation of gene expression. In mature organisms, the thyroid hormones increase oxygen consumption, elevate metabolic rates, increase thermogenesis, and induce the biosynthesis of RNA and proteins (*e.g.*, mitochondrial glycerol-1-phosphate dehydrogenase). T$_3$ also inhibits oxidative phosphorylation (1,2) They also participate in embryogenesis, growth and development, and sexual maturation. Enzymes that can utilize 3,3',5-triodo-L-thyronine as a substrate or alternative substrate include thyroid-hormone aminotransferase, diiodotyrosine aminotransferase, and certain phenol sulfotransferases. This hormone absorbs UV light: λ$_{max}$ = 295 nm (ε = 4090 cm^{-1}M^{-1}) in 40 mM HCl; also, λ$_{max}$ = 224 nm (ε = 49200 cm^{-1}M^{-1}) and 320 nm (ε = 4660 cm^{-1}M^{-1}) in dilute NaOH. T$_3$ is also photosensitive and decomposes with attendant deiodination. **Target(s):** alcohol dehydrogenase (3-5); aldehyde dehydrogenase (6); aromatase, *or* CYP19 (7); aspartate aminotransferase (8); cAMP-dependent protein kinase (9); 3',5'-cyclic-nucleotide phosphodiesterase (10-12); cystathionine γ-lyase (13,14); 15-hydroxyprostaglandin dehydrogenase (15); IMP dehydrogenase (16); glucose-6-phosphate dehydrogenase (17); glutamate dehydrogenase (18); malate dehydrogenase (19-21); monoamine oxidase (22); NAD(P)H dehydrogenase, *or* DT diaphorase (23-25); NAD(P)$^+$ transhydrogenase (26); Na$^+$/K$^+$-exchanging ATPase (27); oxidative phosphorylation (1,2); phenylalanyl-tRNA synthetase (28); prostaglandin-D synthase (29); protein-disulfide isomerase, inhibited by both D- and L-isomers (30,31); pyruvate kinase (32); thyroxine 5-deiodinase (33,34); and tyrosine aminotransferase (35). **1.** Pullman & Penefsky (1963) *Meth. Enzymol.* **6**, 277. **2.** Maley & Lardy (1953) *J. Biol. Chem.* **204**, 435. **3.** Gilleland & Shore (1969) *J. Biol. Chem.* **244**, 535. **4.** Mardh, Auld & Vallee (1987) *Biochemistry* **26**, 7585. **5.** Brändén, Jörnvall, Eklund & Furugren (1975) *The Enzymes*, 3rd ed. (Boyer, ed.), **11**, 103. **6.** Zhou & Weiner (1997) *Eur. J. Biochem.* **245**, 123. **7.** Bullion & Osawa (1988) *Endocr. Res.* **14**, 21. **8.** Smejkal & Smejkalová (1967) *Enzymologia* **33**, 320. **9.** Friedman, Lang & Burke (1978) *Endocr. Res. Commun.* **5**, 109. **10.** Law & Henkin (1984) *Res. Commun. Chem. Pathol. Pharmacol.* **43**, 449. **11.** Nagasaka & Hidaka (1976) *Biochim. Biophys. Acta* **438**, 449. **12.** Drummond & Yamamoto (1971) *The Enzymes*, 3rd ed. (Boyer, ed.), **4**, 355. **13.** Fernández & Horvath (1963) *Enzymologia* **26**, 113. **14.** González & Horvath (1964) *Enzymologia* **26**, 364. **15.** Pace-Asciak & Smith (1983) *The Enzymes*, 3rd ed. (Boyer, ed.), **16**, 543. **16.** al-Mudhaffar & Ackerman (1968) *Endocrinology* **82**, 912. **17.** McKerns (1962) *Biochim. Biophys. Acta* **62**, 402. **18.** Frieden (1963) *The Enzymes*, 2nd ed. (Boyer, Lardy & Myrbäck, eds.), **7**, 3. **19.** Maggio & Ullman (1978) *Biochim. Biophys. Acta* **522**, 284. **20.** Thorne & Kaplan (1963) *J. Biol. Chem.* **238**, 1861. **21.** Stern, Ochoa & Lynen (1952) *J. Biol. Chem.* **198**, 313. **22.** Knopp (1982) *Endokrinologie* **80**, 367. **23.** Ernster (1967) *Meth. Enzymol.* **10**, 309. **24.** Martius (1963) *The Enzymes*, 2nd ed. (Boyer, Lardy & Myrbäck, eds.), **7**, 517. **25.** Ernster, Ljunggren & Danielson (1960) *Biochem. Biophys. Res. Commun.* **2**, 48 and 88. **26.** Rydström, Hoek & Ernster (1976) *The Enzymes*, 3rd ed. (Boyer, ed.), **13**, 51. **27.** Sarkar & Ray (1998) *Neuroreport* **9**, 1149. **28.** Nielsen & Haschemeyer (1976) *Biochemistry* **15**, 348. **29.** Beuckmann, Aoyagi, Okazaki, *et al.* (1999) *Biochemistry* **38**, 8006. **30.** Guthapfel, Gueguen & Quemeneur (1996) *Eur. J. Biochem.* **242**, 315. **31.** Barbouche, Miquelis, Jones & Fenouillet (2003) *J. Biol. Chem.* **278**, 3131. **32.** Ashizawa, McPhie, Lin & Cheng (1991) *Biochemistry* **30**, 7105. **33.** Wynn & Gibbs (1963) *J. Biol. Chem.* **238**, 3490. **34.** Shintani, Nohira, Hikosaka & Kawahara (2002) *Dev. Growth Differ.* **44**, 327. **35.** Diamondstone & Litwack (1963) *J. Biol. Chem.* **238**, 3859.

Trilostane

This synthetic epoxy sterol (FW = 329.44 g/mol; CAS 13647-35-3, 27107-98-8, and 28414-46-2; ε_{252nm} = 8300 $M^{-1}cm^{-1}$ in ethanol; Solubility: 70 mg/mL DMSO, <1 mg/mL Water), also known as Vetoryl and systematically as 4α,5-epoxy-17β-hydroxy-3-oxo-5α-androstane-2α-carbonitrile, is an inhibitor of steroid biosynthesis and is used clinically as an adrenocortical suppressant and in the treatment of breast cancer. As a potent inhibitor of 3 β-hydroxysteroid dehydrogenase (*Reaction:* 3β-hydroxy-Δ^5-steroid (*or* pregnenolone) + NAD^+ ⇌ 3-oxo-Δ^5-steroid (*or* progesterone) + NADH + H^+), trilostane is used in the treatment of Cushing's syndrome, *or* hypercortisolism. **Target(s):** estrogen receptor (1); estrone sulfate sulfatase, weakly inhibited (2); 15-hydroxysteroid dehydrogenase (3); 3β-hydroxysteroid dehydrogenase/steroid Δ-isomerase (4-12); and 3-ketosteroid reductase (13). **1.** Puddefoot, Barker, Glover, Malouitre & Vinson (2002) *Int. J. Cancer* **101**, 17. **2.** Carlstrom, Doberl, Pousette, Rannevik & Wilking (1984) *Acta Obstet. Gynecol. Scand. Suppl.* **123**, 107. **3.** Patel & Challis (2002) *J. Clin. Endocrinol. Metab.* **87**, 700. **4.** Furster (1999) *Biochim. Biophys. Acta* **1436**, 343. **5.** Port, Bowen, Keyes & Townson (2000) *J. Reprod. Fertil.* **119**, 93. **6.** Cooke (1996) *J. Steroid. Biochem. Mol. Biol.* **58**, 95. **7.** Young, Corpechot, Perche, *et al.* (1996) *Steroids* **61**, 144. **8.** Luu-The, Takahashi, de Launoit, *et al.* (1991) *Biochemistry* **30**, 8861. **9.** Luu-The, Takahashi & Labrie (1991) *J. Steroid Biochem. Mol. Biol.* **40**, 545. **10.** Hiwatashi, Hamamoto & Ichikawa (1985) *J. Biochem.* **98**, 1519. **11.** Potts, Creange, Hardomg & Schane (1978) *Steroids* **32**, 257. **12.** Thomas, Boswell, Scaccia, Pletnev & Umland (2005) *J. Biol. Chem.* **280**, 21321. **13.** De Launoit, Zhao, Belanger, Labrie & Simard (1992) *J. Biol. Chem.* **267**, 4513.

Trimetazidine

This antianginal agent (FW$_{free-base}$ = 266.34 g/mol) inhibits acetyl-CoA *C*-acyltransferase (1,2), carnitine palmitoyltransferase (3,4), fatty acid oxidation (1-4), and Na^+/K^+-exchanging ATPase (5). By assessing insulin resistance (measured by glucose/insulin tolerance test) and lipid metabolite content in the gastrocnemius muscle of trimetazidine-treated obese mice (15 mg/kg/day), a recent study (6) found a mild shift in substrate preference towards carbohydrates as an oxidative fuel source, as evidenced by an increase in the respiratory exchange ratio. This shift was accompanied by an accumulation of long-chain acyl CoA and an increase in triacylglycerol content in gastrocnemius muscle, without exacerbating high-fat diet (HFD) - induced insulin resistance. **1.** Fragasso, Spoladore, Cuko & Palloshi (2007) *Curr. Clin. Pharmacol.* **2**, 190. **2.** Kantor, Lucien, Kozak & Lopaschuk (2000) *Circ. Res.* **86**, 580. **3.** Willoughby, Chirkov, Kennedy, *et al.* (1998) *Eur. J. Pharmacol.* **356**, 207. **4.** Kennedy & Horowitz (1998) *Cardiovasc. Drugs Ther.* **12**, 359. **5.** Hisatome, Ishiko, Tanaka, *et al.* (1991) *Eur. J. Pharmacol.* **195**, 381. **6.** Ussher, Keung, Fillmore, *et al.* (2014) *J. Pharmacol. Exp. Ther.* **349**, 487.

Trimethoprim

This broad-spectrum, cell-penetrating antibiotic (FW = 290.32 g/mol; CAS 738-70-5), also known systematically as 2,4-diamino-5-(3',4',5'-trimethoxybenzyl)pyrimidine, inhibits enzymatic formylation and is a strong inhibitor of dihydrofolate reductase (*Reaction*: Dihydrofolate + NADPH ⇌

Tetrahydrofolate + $NADP^+$), exhibiting 50,000–100,000-times greater affinity for the bacterial reductase than its mammalian counterpart (1-12). By denying bacteria of sufficient tetrahydrofolate to support the synthesis of required DNA precursors (especially dTMP), replication cannot proceed and death ensues. Trimethoprim is readily absorbed by the oral route and is widely distributed in body fluids and tissues, where it is active against a wide range of Gram-positive and Gram-negative aerobes. When combined with sulfamethoxazole, a cell-penetrating structural analogue of *p*-aminobenzoic acid that inhibits dihydropteroate synthetase, trimethoprim is highly effective against *Listeria monocytogenes* and *Shigella flexneri*, intracellular pathogens that rely on induced phagocytosis by nonprofessional phagocytes and subsequent lysis of the resulting pathogen-laden phagolysosome to enter the cytoplasm of host cells, where they multiply and undergo actin-based motility and cell-to-cell spread that are cardinal features in both listeriosis and shigellosis (13). Bacterial resistance to trimethoprim is due to a variety of mechanisms, including mobile genetic elements (plasmids, transposons and integrons) that encode trimethoprim resistance factors, some generating additional mutants in DHFR that are less sensitive to trimethoprim inhibition. Trimethoprim-resistance genes (*dfrA17* and *dfrA12*) are often found in clinical isolates of trimethoprim-resistant *Escherichia coli* and are also used as markers in cloning vectors. **Target(s):** CYP2C8 and CYP2C9 (1); glutathione synthetase (14); thymidylate synthase (15,16). **Key Pharmacokinetic Parameters:** *See* Appendix II in Goodman & Gilman's *THE PHARMACOLOGICAL BASIS OF THERAPEUTICS*, 12th Edition (Brunton, Chabner & Knollmann, eds.) McGraw-Hill Medical, New York (2011). **1.** Wen, Wang, Backman, Laitila & Neuvonen (2002) *Drug Metab. Dispos.* **30**, 631. **2.** Roth & Burchall (1971) *Meth. Enzymol.* **18B**, 779. **3.** Dams & Jaenicke (2001) *Meth. Enzymol.* **331**, 305. **4.** Matthews, Bolin, Burridge, *et al.* (1985) *J. Biol. Chem.* **260**, 392. **5.** Baker (1967) *Design of Active-Site-Directed Irreversible Enzyme Inhibitors*, Wiley, New York. **6.** Baker (1967) *J. Med. Chem.* **10**, 912. **7.** Kompis, Then, Wick & Montavon (1980) in *Enzyme Inhibitors* (Brodbeck, ed.), pp. 177 Verlag Chemie, Weinheim. **8.** Burchall & Hitchings (1965) *Mol. Pharmacol.* **1**, 126. **9.** Gangjee, Li, Yang & Kisliuk (2008) *J. Med. Chem.* **51**, 68. **10.** Gangjee, Qiu, Li & Kisliuk (2008) *J. Med. Chem.* **51**, 5789. **11.** Pattanakitsakul & Ruenwongsa (1984) *Int. J. Parasitol.* **14**, 513. **12.** Aboge, Jia, Terkawi, *et al.* (2008) *Antimicrob. Agents Chemother.* **52**, 4072. **13.** Southwick & Purich (1996) *New Engl. J. Med.* **334**, 770. **14.** Kato, Chihara, Nishioka, *et al.* (1987) *J. Biochem.* **101**, 207. **15.** So, Wong & Ko (1994) *Exp. Parasitol.* **79**, 526. **16 .** Chalabi & Gutteridge (1977) *Biochim. Biophys. Acta* **481**, 71.

3,4,5-Trimethoxybenzamidoxime

This hydroxyl radical scavenger and Fe^{3+} chelator (FW = 184.2 g/mol; CAS 95933-74-7; λ_{max} = 214 and 261 nm; stability > 2 years, when stored desiccated at –20° C), also known as trimidox and VF 233, potently inhibits ribonucleotide reductase (K_i = 5 μM) or ~100x more effective than hydroxyurea (1) and induces apoptosis in cultured cells (2). Trimidox is also a candidate antineoplastic agent (3). **1.** Szekeres, Fritizer, Strobl, *et al.* (1994) *Blood* **84**, 4316. **2.** Kanno, Uwai, Tomizawa, *et al.* (2006) *Basic Clin. Pharmacol. Toxicol.* **98**, 44. **3.** Szekeres, Gharehbaghi, Fritzer, *et al.* (1994) *Cancer Chemother. & Pharmacol.* **34**, 63.

3,4,5-Trimethoxybenzoic Acid 8-(Dimethylamino)octyl Ester, *See* 8-(N,N-Diethylamino)octyl 3,4,5-Trimethoxybenzoate

4S-(3,4,5-Trimethoxy)benzyl-D-glutamic Acid, *See* (2R,4S)-2-Amino-4-(3,4,5-trimethoxy)benzylpentanedioate

Trimethylacetyl Phosphate, *See* 2,2-Dimethylpropionyl Phosphate

Trimethylamine

This naturally occurring tertiary amine (FW = 59.11 g/mol; pK_a = 9.9; Colorless gas; M.P. = –117.08 °C; B.P. = 2.87°C) is formed by gut bacteria during the metabolic degradation of carnitine, choline, and other trimethylamine-containing metabolites (**See** *Trimethylamine N-oxide*). The free base has the odor of rotting fish. Trimethylammonium hydrochloride is a deliquescent solid that decomposes at 277-278°C and is very soluble in water and ethanol. **Target(s):** acetylcholinesterase, inhibited by the cation (1-8); acid phosphatase (9); anthranilate synthase (10); choline dehydrogenase (11); choline-sulfatase (12); choline sulfotransferase (13); flavin monooxygenase, *or* N,N-dimethylaniline N-oxygenase, *or* thiobenzamide S-monooxygenase (14-16); β-glucosidase (17); quinoprotein methylamine dehydrogenase (18); trimethylamine-oxide aldolase (19-21), however, reference 21 reports no inhibition. **1.** Wilson (1960) *The Enzymes*, 2nd ed. (Boyer, Lardy & Myrbäck, eds.), **4**, 501. **2.** Froede & Wilson (1971) *The Enzymes*, 3rd ed. (Boyer, ed.), **5**, 87. **3.** Krupka (1964) *Biochemistry* **3**, 1749. **4.** Krupka (1965) *Biochemistry* **4**, 429. **5.** Krupka (1966) *Biochemistry* **5**, 1988. **6.** Wilson (1952) *J. Biol. Chem.* **197**, 215. **7.** Wilson (1952) *J. Biol. Chem.* **199**, 113. **8.** Wilson (1954) *J. Biol. Chem.* **208**, 123. **9.** Garrido, Lisa & Domenech (1988) *Mol. Cell. Biochem.* **84**, 41. **10.** Zalkin & Kling (1968) *Biochemistry* **7**, 3566. **11.** Mann, Woodward & Quastel (1938) *Biochem. J.* **32**, 1024. **12.** Lucas, Burchiel & Segel (1972) *Arch. Biochem. Biophys.* **153**, 664. **13.** Renosto & Segel (1977) *Arch. Biochem. Biophys.* **180**, 416. **14.** Peters, Livingstone, Shenin-Johnson, Hines & Schlenk (1995) *Xenobiotica* **25**, 121. **15.** Schlenk & Li-Schlenk (1994) *Comp. Biochem. Physiol. B Biochem. Mol. Biol.* **109**, 655. **16.** Duffel & Gillespie (1984) *J. Neurochem.* **42**, 1350. **17.** Dale, Ensley, Kern, Sastry & Byers (1985) *Biochemistry* **24**, 3530. **18.** Davidson & Kumar (1990) *Biochim. Biophys. Acta* **1016**, 339. **19.** Parkin & Hultin (1986) *J. Biochem.* **100**, 77. **20.** Myers & Zatman (1971) *Biochem. J.* **121**, 10P. **21.** Large (1971) *FEBS Lett.* **18**, 297.

Trimethylamine N-Oxide

This water- and ethanol-soluble alkyl-nitrogen oxide (FW = 75.111 g/mol; pK_a = 4.65 at 20°C), a natural osmolyte formed in the degradation of nitrogenous substances in many organisms, is a deliquescent solid that decomposes on heating to dimethylamine and formaldehyde. **Target(s):** α_1-antitrypsin, polymerization of M and Z variants (1); electron transport (2); and phosphorylase kinase (3). **1.** Devlin, Parfrey, Tew, Lomas & Bottomley (2001) *Amer. J. Respir. Cell Mol. Biol.* **24**, 727. **2.** Suzuki, Kubo, Shinano & Takama (1992) *Microbios* **71**, 145. **3.** Chebotareva, Andreeva, Makeeva, Livanova & Kurganov (2004) *J. Mol. Recognit.* **17**, 426.

N,N,N,-Trimethylaminomethaneboronic Acid

This boronate analogue of betaine (FW = 118.97 g/mol) competitively inhibits rat liver betaine:homocysteine S-methyltransferase (*or* BHMT), K_i = 45 μM. Since the K_m for betaine measured with the purified enzyme is near 0.1 mM, the boronic acid analogue of betaine appears to function more effectively as a substrate analogue inhibitor of BHMT. **1.** Lee, Cava, Amiri, Ottoboni & Lindquist (1992) *Arch. Biochem. Biophys.* **292**, 77.

p-(Trimethylammonio)benzenediazonium Fluoroborate

This quaternary amine and diazonium salt is ligand for acetylcholine receptor (1) and an irreversible inhibitor of acetylcholinesterase (2-5). **1.** Changeux, Podleski & Wofsy (1967) *Proc. Natl. Acad. Sci. U.S.A.* **58**, 2063. **2.** Shaw (1970) *The Enzymes*, 3rd ed. (Boyer, ed.), **1**, 91. **3.** Froede & Wilson (1971) *The Enzymes*, 3rd ed. (Boyer, ed.), **5**, 87. **4.** Meunier & Changeux (1969)

FEBS Lett. **2**, 224. **5.** Wofsy & Michaeli (1967) *Proc. Natl. Acad. Sci. U.S.A.* **58**, 2296.

4-(Trimethylammonio)butyrate, *See γ-Butyrobetaine*

5-Trimethylammonio-2-pentanone, *See 4-Oxopentyltrimethylammonium Chloride*

3-(Trimethylammonio)-1-propanol, *See Homocholine*

m-(Trimethylammonio)trifluoroacetophenone

This quaternary amine (FW$_{chloride}$ = 267.68 g/mol), also known as m-(N,N,N-trimethylammonio)-2,2,2-trifluoroacetophenone and TMTFA, is a highly potent, reversible, time-dependent, transition-state inhibitor of *Electrophorus electricus* acetylcholinesterase (K_i = 1.3 fM) and *Torpedo californica* acetylcholinesterase (K_i = 15 fM) (1-4). TMTFA therefore exhibits about 10^{10}-times tighter binding to acetylcholinesterase than acetylcholine. **1.** Brodbeck, Schweikert, Gentinetta & Rottenberg (1979) *Biochim. Biophys. Acta* **567**, 357. **2.** Massiah, Viragh, Reddy, *et al.* (2001) *Biochemistry* **40**, 5682. **3.** Ordentlich, Barak, Kronman, *et al.* (1998) *J. Biol. Chem.* **273**, 19509. **4.** Szegletes, Mallender & Rosenberry (1998) *Biochemistry* **37**, 4206.

N-(1,5,9-Trimethyldecyl)-4α,10-dimethyl-8-aza-trans-decal-3β-ol

This isoquinoline derivative (FW$_{free-base}$ = 365.64 g/mol), also known as (4aα,5a,6a,8aβ)-decahydro-5,8a-dimethyl-2-(1,5,9-trimethyldecyl)-6-isoquinolinol, 4a,10-dimethyl-8-aza-trans-decal-3β-ol, and MDL 28815, inhibits phytosterol and cholesterol biosynthesis. It mimics the carbocation intermediates postulated for the enzymic cyclization of 2,3-oxidosqualene. Note MDL 28815 does not inhibit β-amyrin synthase. **Target(s):** cholestenol Δ-isomerase, *or* sterol $\Delta^{7,8}$-isomerase (1); cycloartenol synthase (2); cycloeucalenol cycloisomerase (3); and lanosterol synthase (2,4). **1.** Moebius, Reiter, Bermoser, *et al.* (1998) *Mol. Pharmacol.* **54**, 591. **2.** Taton, Benveniste & Rahier (1986) *Biochem. Biophys. Res. Commun.* **138**, 764. **3.** Taton, Benveniste & Rahier (1987) *Phytochemistry* **26**, 385. **4.** Barth, Binet, Thomas, *et al.* (1996) *J. Med. Chem.* **39**, 2302.

3,5,5-Trimethyloxazolidine-2,4-dione, *See Trimethadione*

4,5',6-Trimethylpsoralen, *See Trioxsalen*

Trimethylquinone, *See Cumoquinone*

N,N,N-Trimethyl-1-(4-trans-stilbenoxy)-2-propylammonium Iodide

This oxystilbene derivative (FW = 422.33 g/mol), often referred to simply as F3, is a relatively potent and apparently competitive antagonist of nicotinic acetylcholine receptors. The R-stereoisomer is more potent (IC$_{50}$ = 0.35 μM) than the S-enantiomer (IC$_{50}$ = 1.5 µM). **1**. Di Angelantonio, Nistri, Moretti, Clementi & Gotti (2000) *Brit. J. Pharmacol.* **129**, 1771.

(Trimethyl)tin Chloride

This synthetic organotin compound (FW = 199.27 g/mol; CAS 1066-45-1; B.P. = 148 °C; Soluble in chloroform and organic solvents, Miscible in H$_2$O; *Symbol*: TMT; colorless-to-white granular solid; unpleasant odor), also known as chlorotrimethylstannane; chlorotrimethyltin, and trimethyl chlorostannane, is an insect, bacteria and fungus control agent that is used primarily to preserve wood, textiles, leather, and paints. **Caution: TMT is neurotoxic and causes irrevsible brain damage in test animals.** A factor contributing to organotin toxicity is its ability to pass through membranes. TMT-induced cognitive dysfunction and exhibited significant decreases in JAK2/STAT3 signaling and M$_1$ muscarinic acetylcholine receptor (mAChR) expression, as well as other cholinergic parameters (1). TMT decreased intracellular pH and opened K$^+$ channels in renal intercalated cells, suggesting that it can directly inhibit the activity of H$^+$/K$^+$-ATPases in renal intercalated cells, reducing urine K$^+$ reabsorption and inducing hypokalemia (2). Trimethyltin inhibits the uptake of γ-aminobutyric acid (GABA), norepinephrine and serotonin by mouse forebrain synaptosomes with IC$_{50}$ values of 75, 43 and 24 µM, respectively (3). **Target(s):** γ-aminobutyrate uptake (1); ATPases (5-9), ATP synthase (6,8,10), Ca^{2+}-dependent ATPase (7,9), calmodulin (11,12), K$^+$-dependent ATPase (5), mitochondrial inner-membrane anion channel (6); Na$^+$/K$^+$-exchanging ATPase (5,8); protein biosynthesis (13); *and* tubulin polymerization, *or* microtubule assembly (14). **1**. Kim, Tran, Shin, *et al.* (2013) *Cell Signal.* **25**, 1348. **2**. Tang, Yang, Lai, *et al.* (2010) *Toxicology* **271**, 45. **3**. Doctor, Costa, Kendall & Murphy (1982) *Toxicology* **25**, 213. **4**. Costa (1985) *Toxicol. Appl. Pharmacol.* **79**, 471. **5**. Cameron, Kodavanti, Pentyala & Desaiah (1991) *J. Appl. Toxicol.* **11**, 403. **6**. Powers & Beavis (1991) *J. Biol. Chem.* **266**, 17250. **7**. Kodavanti, Cameron, Yallapragada, Vig & Desaiah (1991) *Arch. Toxicol.* **65**, 311. **8**. Stine, Reiter & Lemasters (1988) *Toxicol. Appl. Pharmacol.* **94**, 394. **9**. Chambers, Rizopoulos, Armstrong, Wayner & Valdes (1987) *Brain Res. Bull.* **18**, 569. **10**. Linnett & Beechey (1979) *Meth. Enzymol.* **55**, 472. **11**. Yallapragada, Vig, Kodavanti & Desaiah (1991) *J. Toxicol. Environ. Health* **34**, 229. **12**. Yallapragada, Vig & Desaiah (1990) *J. Toxicol. Environ. Health* **29**, 317. **13**. Costa & Sulaiman (1986) *Toxicol. Appl. Pharmacol.* **86**, 189. **14**. Jensen, Onfelt, Wallin, Lidums & Andersen (1991) *Mutagenesis* **6**, 409.

Trimetrexate

This lipophilic methotrexate analogue (FW = 369.42 g/mol; CAS 52128-35-5), also known as 2,4-diamino-5-methyl-6-[(3,4,5-trimethoxyanilino)methyl] quinazoline, is a tight-binding inhibitor (K_i = 40 pM) of dihydrofolate reductase (1-4). It differs from methotrexate in its transport and intracellular retention, and may be useful against tumors resistant to methotrexate because of impaired transport or deficient polyglutamylation. **1**. Elslager, Johnson & Werbel (1983) *J. Med. Chem.* **26**, 1753. **2**. Bertino, Sawicki, Moroson, Cashmore & Elslager (1979) *Biochem. Pharmacol.* **28**, 1983. **3**. Jackson, Fry, Boritzki, *et al.* (1984) *Adv. Enzyme Regul.* **22**, 187. **4**. Sirawaraporn, Edman & Santi (1991) *Protein Expr. Purif.* **2**, 313.

Trimipramine

This tricyclic antidepressant (FW$_{free-base}$ = 294.44 g/mol; CAS 739-71-9) increases rapid eye movement sleep and nocturnal prolactin secretion, while suppressing nocturnal cortisol secretion. Although trimipramine shares equivalent efficacy with doxepin, imipramine, maprotiline and amitriptyline, it possesses a different side-effect profile, including lower cardiotoxicity, lower epileptogenic potential, and differing neurotransmitter function/reuptake (1). Moreover, trimipramine only weakly inhibits noradrenaline and serotonin reuptake and does not down-regulate β$_1$-adrenoceptors and does not down-regulate β$_1$-adrenoceptors (1). **Target(s):** γ-aminobutyrate transport (2); Ca^{2+}-dependent ATPase (3,4); lysosomal acid sphingomyelinase (5); and monoamine oxidase (6,7); Human serotonin transporter (SERT), K_i = 150 nM; Human norepinephrine transporter (NET), K_i = 2.4 µM; Human dopamine transporter (DAT), K_i = 3.8 µM; Human histamine H$_1$ receptor, K_i = 1.4 nM; Rat adrenergic α$_1$ receptor, K_i = 24 nM; Rat adrenergic α$_{2A}$ receptor, K_i = 1.4 nM; Rat adrenergic α$_{2B}$ receptor, K_i = 0.4 µM; Human muscarinic acetyl-choline receptors (mAChRs), K_i = 59 nM; Rat serotonin 5-HT$_{2A}$ receptor, K_i = 20 nM; Rat serotonin 5-HT$_{2C}$ receptor, K_i = 540 nM; Rat serotonin 5-HT$_3$ receptor, K_i = 9.1 µM; Rat dopamine D$_1$ receptor, K_i = 350 nM; Rat dopamine D$_2$ receptor, K_i = 58 nM. **1**. Gastpar (1989) *Drugs* **38**, Supplement 1, 43. **2**. Nakashita, Sasaki, Sakai & Saito (1997) *Neurosci. Res.* **29**, 87. **3**. Soler, Plenge-Tellechea, Fortea & Fernandez-Belda (2000) *J. Bioenerg. Biomembr.* **32**, 133. **4**. Plenge-Tellechea, Soler & Fernandez-Belda (1999) *Arch. Biochem. Biophys.* **370**, 119. **5**. Jaffrezou, Chen, Duran, *et al.* (1995) *Biochim. Biophys. Acta* **1266**, 1. **6**. Gnerre, Kosel, Baumann, Carrupt & Testa (2001) *J. Pharm. Pharmacol.* **53**, 1125. **7**. Yu & Boulton (1990) *Prog. Neuropsychopharmacol. Biol. Psychiatry* **14**, 409.

2',3'-O-(2,4,6-Trinitrophenyl)uridine 5'-Monophosphate

This modified UMP (FW = 534.26 g/mol), also known as 2',3'-O-(2,4,6-trinitrocyclohexadienyl)uridine 5'-monophosphate, is a strong inhibitor of UDP-glucose 4-epimerase. The extreme sensitivity of the fluorescence emission spectrum of this analog to solvent polarity makes it an excellent probe for the study of the environment at the active site. **1**. Ray, Ali & Bhaduri (1992) *Indiana J. Biochem. Biophys.* **29**, 209.

Triolein

This water-insoluble fatty acid triester (FW = 885.45 g/mol; MP = –4° C), also called 1,2,3-tri(cis-9-octadecenoyl)-glycerol and glyceryl trioleate, is a triacylglycerol, in which all three fatty acyl components are oleoyl residues. Triolein inhibits retinyl-palmitate esterase (1), all-trans-retinyl-palmitate hydrolase (1); 11-cis-retinyl-palmitate hydrolase (1); sphingomyelin phosphodiesterase (2). **1**. Mata, Mata & Tsin (1996) *J. Lipid Res.* **37**, 1947. **2**. Liu, Nilsson & Duan (2002) *Lipids* **37**, 469.

Tripelennamine

This pyridine/ethylenediamine-class antihistamine and psychoactive drug (FW = 255.36 g/mol; CAS 91-81-6 (free base) and 154-69-8 (monohydrochloride)), also known by the trade name Pyribenzamine® and systematic name N-benzyl-N',N'-dimethyl-N-2-pyridylethylenediamine, is a histamine H₁ receptor antagonist used to treat hay fever and other severe allergies. Tripelennamine was the first antihistamine ever synthesized, a feat accomplished by Djerassi, Huttrer & Scholz in 1943 (U.S. Patent No. 2,406,594). A recent analysis of cationic amphiphilic drugs that are potent inhibitors of yeast sporulation revealed that tripelennamine induces sporulation-specific cytotoxicity and a strong inhibition of meiotic M phase (1). Moreover, chemical-genomic screening identified genes involved in autophagy as hypersensitive to tripelennamine (1). 1. Schlecht, St. Onge, Walther, François & Davis (2012) *PLoS One* 7, e42853.

Tris(hydroxymethyl)aminomethane

This widely used primary amine-containing buffer (FW$_{\text{free-base}}$ = 121.14 g/mol; CAS 77-86-1), known commonly as Tris, is a weak base with a pK_a of 8.3 at 20°C and 8.06 at 25°C, with dpK_a/dT value of –0.028. Many enzymes that are inhibited by Tris. Moreover, certain pH electrodes are adversely affected by prolonged exposure to elevated Tris concentrations. Substrate and products can also interact with this buffer. The lack of enzymatic activity in Tris buffer for 2-dehydro-3-deoxyglucarate aldolase, for example, is likely due to the reaction of the buffer with the substrate (1).
Target(s): acetyl-CoA *C*-acyltransferase (236); acetyl-CoA synthetase, *or* acetate:CoA ligase (2); acetylenecarboxylate hydratase (101); *N*-acetylmuramoyl-L-alanine amidase (109); acid phosphatase (207); adenosylhomocysteine nucleosidase (124); adenylosuccinate synthetase (3); alcohol dehydrogenase (4); alternansucrase (220); D-amino-acid aminotransferase (213); aminolevulinate aminotransferase (210,211); 5-aminolevulinate synthase (234); aminopeptidase PS (5); α-amylase (6); amylo-α-1,6-glucosidase/4-α-glucanotransferase,*or* glycogen debranching enzyme; K_i = 12 mM (87,147-153); amylosucrase (229); D-arabinose isomerase (7,92); arylamine glucosyltransferase (223); asparaginyl-tRNA synthetase (88); aspartate-semialdehyde dehydrogenase (8); ATP Sulfurylase, *or* sulfate adenylyltransferase (9); bacterial leucyl aminopeptidase (122); biuret amidohydrolase (107); branched-chain-amino-acid aminotransferase (212); branched-chain-α-keto-acid dehydrogenase, yeast (10); carbamoyl-phosphate synthetase I, ammonia-utilizing (11); carnitine dehydratase (100); cellulose 1,4-β-cellobiosidase (136); ceramide glucosyltransferase (222); cholesterol-5,6-oxide hydrolase (123); clostripain (12,13); cystathionine γ-lyase (14); cytochrome b_5 reductase (15); 2-dehydro-3-deoxyglucarate aldolase, *or* 2-keto-3-deoxyglucarate aldolase (1); 2-dehydro-3-deoxy-D-gluconate 6-dehydrogenase, *or* 2-keto-3-deoxy-D-gluconate dehydrogenase (16); 2-dehydropantoate aldolase (103); 3-dehydroquinate synthase (17); 2-deoxyglucosidase (131); 3-deoxy-*manno*-octulosonate-8-phosphatase (18); deoxyribonuclease I (203,204); deoxyribonuclease X (202); dextransucrase (19); 4,5-dihydroxyphthalate decarboxylase (104); dipeptidase (20,21); dipeptidyl-peptidase II (22,113-121); diphosphate:serine phosphotransferase, weakly (209); DNA β-glucosyltransferase (23); DNA nucleotidylexotransferase, *or* terminal deoxyribonucleotidyltransferase (24,208); dodecenoyl-CoA isomerase, *or* Δ^{3,2}-enoyl-CoA isomerase (95); ecdysone 20-monooxygenase (242); electron transport, chloroplast (25); enoyl-CoA hydratase (95); flavone 7-*O*-β-glucosyltransferase (721); fructan β-fructosidase (139); β-fructofuranosidase, *or* invertase (26,163-165); L-fucose isomerase (7,92); β-D-fucosidase (144-146); α-L-fucosidase (141); galactinol:raffinose galactosyltransferase, *or* stachyose synthase (224); galactose oxidase (27); α-

galactosidase (28,168); β-galactosidase (167); glucan 1,3-α-glucosidase, *or* glucosidase II; (32) mannosyl-oligosaccharide glucosidase II, (32,137,138); glucan 1,4-α-glucosidase, *or* glucoamylase (29,200,201); 4-α-glucanotransferase (225); α-glucosidase (30,31,176-196); β-glucosidase (169-175); glucosylceramidase (33); β-glucuronidase (154); glutamate dehydrogenase (34); glutamate formimidoyltransferase (35,241); glutamate mutase, *or* methylaspartate mutase (36); γ-glutamyl kinase, *or* glutamate 5-kinase (37); glyoxylate dehydrogenase, acylating (38); glyceraldehyde-3-phosphate dehydrogenase (39); glycine:oxaloacetate aminotransferase (40); glycosylceramidase (33); GMP synthetase, glutamine hydrolyzing, *or* XMP amidotransferase (41); guanosine phosphorylase (42); hemoglobin S polymerization (43); histidine *N*-acetyltransferase, weakly inhibited (235); H⁺-transporting inorganic pyrophosphatase (44); (*S*)-2-hydroxy-acid oxidase (45); 3-hydroxyacyl-CoA dehydrogenase (95); 3-hydroxybutyryl-CoA epimerase, *or* 3-hydroxyacyl-CoA epimerase (95); 16α-hydroxyprogesterone dehydratase (99); isomaltase (46); isomaltulose synthase (90); 3-ketoacyl-CoA thiolase (95); α-ketoisocaproate dioxygenase (47); kynureninase (48); kynurenine:oxoglutarate aminotransferase (214); lactase/phlorizin hydrolase, *or* glycosylceramidase (132,133,140); lactosylceramide α-2,3-sialyltransferase (216); levanase (49); levansucrase (50); β-lysine 5,6-aminomutase (51,52); maltase (53); maltose α-D-glucosyltransferase, *or* trehalose synthase (89); malyl-CoA lyase (54,102); β-mannanase (55); α-mannosidase (166); mannosyl-oligosaccharide glucosidase, *or* glucosidase I (134,135); mannosyl-oligosaccharide 1,2-α-mannosidase, *or* mannosidase IA and IB (32,56,127-130); mannotetraose 2-α-*N*-acetylglucosaminyltransferase (57); membrane alanine aminotransferase (58); methylamine:glutamate methyltransferase (59); *N*-methyl-2-oxoglutaramate hydrolase (108); 5'-methylthioadenosine/*S*-adenosylhomocysteine Nucleosidase (124); NAD⁺ nucleosidase (60); naringenin-chalcone synthase (233); nicotinate-nucleotide diphosphorylase, carboxylating (218); nucleoside deoxyribosyltransferase (61,219); oligo-1,6-glucosidase, *or* isomaltase (46,62,198,199); ornithine carbamoyltransferase (63,240); ornithine decarboxylase (64); 3-oxoacid CoA-transferase, *or* 3-ketoacid CoA-transferase (106); phenylalanine 4-monooxygenase (65,243); phenylalanyl-tRNA synthetase (66); phosphate acetyltransferase (237,238); phosphoenolpyruvate carboxykinase, diphosphate (67,105); phosphofructo-kinase (68); phospholipase C (69,205,206); polygalacturonidase, *or* pectinase (197); porphobilinogen synthase, *or* δ-aminolevulinate dehydratase (70); procollagen C-endopeptidase, inhibited at high concentrations (110); procollagen N-endopeptidase (111); pyroglutamyl-peptidase II (112); pyruvate carboxylase (71); pyruvate synthase (72); quinate *O*-hydroxycinnamoyltransferase (230-232); ribonuclease, *Ustilago sphaerogena* (73); sarcosine dehydrogenase (74); D-serine ammonia-lyase, *or* D-serine dehydratase (97,98); L-serine dehydratase (75); strictosidine synthase (96); styrene-oxide isomerase (91); sucrose α-glucosidase, *or* sucrase-isomaltase (46,76,142,143); sucrose-phosphate synthase, weakly inhibited (226); sucrose synthase (227,228); sulfite:cytochrome *c* oxidoreductase (77); transaldolase (239); α,α-trehalase (78,79,155-162); trehalose synthase (80); trimethylamine dehydrogenase (81); tryptophanase (82); uracilylalanine synthase (215); uridine Nucleosidase (125,126); vanillate hydroxylase (83); xylan 1,4-β-xylosidase (84); xyloglucan 6-xylosyltransferase (217); and xylose isomerase (85,86,93,94). 1. Wood (1972) *The Enzymes*, 3rd ed. (Boyer, ed.), 7, 281. 2. Londesborough & Webster, Jr. (1974) *The Enzymes*, 3rd ed. (Boyer, ed.), 10, 469. 3. Lieberman (1963) *Meth. Enzymol.* 6, 100. 4. Li, Ulmer & Vallee (1963) *Biochemistry* 2, 482. 5. Dando & Barrett (1998) in *Handb. Proteolytic Enzymes* (Barrett, Rawlings & Woessner, eds.), p. 1013, Academic Press, San Diego. 6. Franzini, Bonini & Sola (1969) *Enzymologia* 36, 117. 7. Yamanaka & Izumori (1975) *Meth. Enzymol.* 41, 462. 8. Hegeman, Cohen & Morgan (1970) *Meth. Enzymol.* 17A, 708. 9. Peck, Jr. (1974) *The Enzymes*, 3rd ed. (Boyer, ed.), 10, 651. 10. Dickinson (2000) *Meth. Enzymol.* 324, 389. 11. Lund & Wiggins (1987) *Biochem. J.* 243, 273. 12. Ullmann & Bordusa (1998) in *Handb. Proteolytic Enzymes* (Barrett, Rawlings & Woessner, eds.), p. 759, Academic Press, San Diego. 13. Mitchell & Harrington (1970) *Meth. Enzymol.* 19, 635. 14. Metaxas & Delwiche (1955) *J. Bacteriol.* 69, 735. 15. Hultquist (1978) *Meth. Enzymol.* 52, 463. 16. Preiss & Ashwell (1962) *J. Biol. Chem.* 237, 317. 17. Sprinson, Srinivasan & Rothschild (1962) *Meth. Enzymol.* 5, 398. 18. Ray & Benedict (1980) *J. Bacteriol.* 142, 60. 19. Miller & Robyt (1986) *Arch. Biochem. Biophys.* 248, 579. 20. Patterson (1976) *Meth. Enzymol.* 45, 377. 21. Patterson (1976) *Meth. Enzymol.* 45, 386. 22. McDonald (1998) in *Handb. Proteolytic Enzymes* (Barrett, Rawlings & Woessner, eds.), p. 408, Academic Press, San Diego. 23. Josse (1968) *Meth. Enzymol.* 12B, 496. 24. Ratliff (1981) *The Enzymes*, 3rd ed. (Boyer, ed.), 14, 105. 25. Izawa & Good (1972) *Meth. Enzymol.* 24, 355. 26. Goldstein &

Lampen (1975) *Meth. Enzymol.* **42**, 504. **27**. Malmström, Andréasson & Reinhammar (1975) *The Enzymes*, 3rd ed. (Boyer, ed.), **12**, 507. **28**. Pederson & Goodman (1980) *Can. J. Microbiol.* **26**, 978. **29**. De Mot & Verachtert (1987) *Eur. J. Biochem.* **164**, 643. **30**. Halvorson (1966) *Meth. Enzymol.* **8**, 559. **31**. Halvorson & Ellias (1958) *Biochem. Biophys. Acta* **30**, 28. **32**. Presper & Heath (1983) *The Enzymes*, 3rd ed. (Boyer, ed.), **16**, 449. **33**. Brady & Kanfer (1966) *Meth. Enzymol.* **8**, 591. **34**. Smith, Austen, Blumenthal & Nyc (1975) *The Enzymes*, 3rd ed. (Boyer, ed.), **11**, 293. **35**. MacKenzie (1980) *Meth. Enzymol.* **66**, 626. **36**. Barker (1972) *The Enzymes*, 3rd ed. (Boyer, ed.), **6**, 509. **37**. Vasáková & Stefl (1982) *Coll. Czech. Chem. Commun.*, No. 42, 349. **38**. Quayle (1966) *Meth. Enzymol.* **9**, 342. **39**. Velick (1955) *Meth. Enzymol.* **1**, 401. **40**. Gibbs & Morris (1970) *Meth. Enzymol.* **17A**, 981. **41**. Patel, Moyed & Kane (1977) *Arch. Biochem. Biophys.* **178**, 652. **42**. Yamada (1961) *J. Biol. Chem.* **236**, 3043. **43**. Freedman, Weissmann, Gorman & Cunningham-Rundles (1973) *Biochem. Pharmacol.* **22**, 667. **44**. Gordon-Weeks, Koren'kov, Steele & Leigh (1997) *Plant Physiol.* **114**, 901. **45**. Baker & Tolbert (1966) *Meth. Enzymol.* **9**, 338. **46**. Kano, Usami, Adachi, Tatematsu & Hirano (1996) *Biol. Pharm. Bull.* **19**, 341. **47**. Sabourin & Bieber (1988) *Meth. Enzymol.* **166**, 288. **48**. Jakoby & Bonner (1953) *J. Biol. Chem.* **205**, 699. **49**. Avigad & Bauer (1966) *Meth. Enzymol.* **8**, 621. **50**. Deponder (1966) *Meth. Enzymol.* **8**, 500. **51**. Stadtman & Grant (1971) *Meth. Enzymol.* **17B**, 206. **52**. Stadtman (1972) *The Enzymes*, 3rd ed. (Boyer, ed.), **6**, 539. **53**. Wang & Hartman (1976) *Appl. Environ. Microbiol.* **31**, 108. **54**. Hacking & Quayle (1990) *Meth. Enzymol.* **188**, 379. **55**. Burke & Khan (2000) *Biomacromolecules* **1**, 688. **56**. Tulsiani & Touster (1989) *Meth. Enzymol.* **179**, 446. **57**. Douglas & Ballou (1982) *Biochemistry* **21**, 1561. **58**. Pfleiderer (1970) *Meth. Enzymol.* **19**, 514. **59**. Shaw & Stadtman (1970) *Meth. Enzymol.* **17A**, 868. **60**. Kuwahara (1980) *Meth. Enzymol.* **66**, 123. **61**. Cardinaud (1978) *Meth. Enzymol.* **51**, 446. **62**. Larner (1960) *The Enzymes*, 2nd ed. (Boyer, Lardy & Myrbäck, eds.), **4**, 369. **63**. Carunchio, Girelli & Messina (1999) *Biomed. Chromatogr.* **13**, 65. **64**. Morley & Ho (1976) *Biochim. Biophys. Acta* **438**, 551. **65**. Shiman (1987) *Meth. Enzymol.* **142**, 17. **66**. Kull & Jacobson (1971) *Meth. Enzymol.* **20**, 220. **67**. Wood, Davies & Willard (1969) *Meth. Enzymol.* **13**, 297. **68**. Gottschalk & Hugo (1982) *Meth. Enzymol.* **90**, 82. **69**. Ottolenghi (1969) *Meth. Enzymol.* **14**, 188. **70**. Shemin (1972) *The Enzymes*, 3rd ed. (Boyer, ed.), **7**, 323. **71**. Charles & Willer (1984) *Can. J. Microbiol.* **30**, 532. **72**. Rabinowitz (1975) *Meth. Enzymol.* **41**, 334. **73**. Glitz & Dekker (1964) *Biochemistry* **3**, 1391. **74**. Mackenzie & Hoskins (1962) *Meth. Enzymol.* **5**, 738. **75**. Sagers & Carter (1971) *Meth. Enzymol.* **17B**, 351. **76**. Vasseur, Frangne, Cauzac, Mahmood & Alvarado (1990) *J. Enzyme Inhib.* **4**, 15. **77**. Suzuki (1994) *Meth. Enzymol.* **243**, 447. **78**. Terra, Ferreira & de Bianchi (1978) *Biochim. Biophys. Acta* **524**, 131. **79**. Sasajima, Kawachi, Sato & Sugimura (1975) *Biochim. Biophys. Acta* **403**, 139. **80**. Pan, Edavana, Jourdian, *et al.* (2004) *Eur. J. Biochem.* **271**, 4259. **81**. McIntire (1990) *Meth. Enzymol.* **188**, 250. **82**. Gopinathan & DeMoss (1968) *Biochemistry* **7**, 1685. **83**. Buswell & Eriksson (1988) *Meth. Enzymol.* **161**, 274. **84**. Kersters-Hilderson, Claeyssens, Van Doorslaer, Saman & De Bruyne (1982) *Meth. Enzymol.* **83**, 631. **85**. Slein (1962) *Meth. Enzymol.* **5**, 347. **86**. Topper (1961) *The Enzymes*, 2nd ed. (Boyer, Lardy & Myrbäck, eds.), **5**, 429. **87**. Nelson, Kolb & Larner (1969) *Biochemistry* **8**, 1419. **88**. Lea & Fowden (1973) *Phytochemistry* **12**, 1903. **89**. Nishimoto, Nakano, Nakada, *et al.* (1996) *Biosci. Biotechnol. Biochem.* **60**, 640. **90**. Nagai, Sugitani & Tsuyuki (1994) *Biosci. Biotechnol. Biochem.* **58**, 1789. **91**. Hartmans, Smits, van der Werf, Volkering & de Bont (1989) *Appl. Environ. Microbiol.* **55**, 2850. **92**. Yamanaka & Izumori (1976) *Agric. Biol. Chem.* **40**, 439. **93**. Danno (1970) *Agric. Biol. Chem.* **34**, 1805. **94**. Callens, Kersters-Hilderson, van Opstal & de Bruyne (1986) *Enzyme Microb. Technol.* **8**, 696. **95**. Pramanik, Pawar, Antonian & Schulz (1979) *J. Bacteriol.* **137**, 469. **96**. Stevens, Giroud, Pennings & Verpoorte (1993) *Phytochemistry* **33**, 99. **97**. Federiuk & Shafer (1981) *J. Biol. Chem.* **256**, 7416. **98**. Dupourque, Newton & Snell (1966) *J. Biol. Chem.* **241**, 1233. **99**. Watkins & Glass (1991) *J. Steroid Biochem. Mol. Biol.* **38**, 257. **100**. Jung, Jung & Kleber (1989) *Biochim. Biophys. Acta* **1003**, 270. **101**. Yamada & Jakoby (1959) *J. Biol. Chem.* **234**, 941. **102**. Hacking & Quayle (1974) *Biochem. J.* **139**, 399. **103**. McIntosh, Purko & Wood (1957) *J. Biol. Chem.* **228**, 499. **104**. Nakazawa & Hayashi (1978) *Appl. Environ. Microbiol.* **36**, 264. **105**. Lochmueller, Wood & Davis (1966) *J. Biol. Chem.* **241**, 5678. **106**. Hersh & Jencks (1967) *J. Biol. Chem.* **242**, 3468. **107**. Cook, Beilstein, Grossenbacher & Hütter (1985) *Biochem. J.* **231**, 25. **108**. Hersh, Tsai & Stadtman (1969) *J. Biol. Chem.* **244**, 4677. **109**. Li, Norioka & Sakiyama (2000) *J. Biochem.* **127**, 1033. **110**. Hojima, van der Rest & Prockop (1985) *J. Biol. Chem.* **260**, 15996. **111**. Hojima, McKenzie, van der Rest & Prockop (1980) *J. Biol. Chem.* **264**, 11336. **112**. Gallagher & O'Connor (1998) *Int. J. Biochem. Cell Biol.* **30**, 115. **113**. McDonald, Leibach, Grindeland & Ellis (1968) *J. Biol. Chem.* **243**, 4143. **114**. McDonald & Schwabe (1980) *Biochim. Biophys. Acta* **616**, 68. **115**. Eisenhauer & McDonald (1986) *J. Biol. Chem.* **261**, 8859. **116**. Sakai & Kojima (1987) *J. Chromatogr.* **416**, 131. **117**. Mentlein & Struckhoff (1989) *J. Neurochem.* **52**, 1284. **118**. DiCarlantonio, Talbot & Dudenhausen (1986) *Gamete Res.* **15**, 161. **119**. Fukasawa, Fukasawa, Hiraoka & Harada (1983) *Biochim. Biophys. Acta* **745**, 6. **120**. Gossrau & Lojda (1980) *Histochemistry* **70**, 53. **121**. McDonald, Reilly, Zeitman & Ellis (1968) *J. Biol. Chem.* **243**, 2028. **122**. Desmarais, Bienvenue, Bzymek, *et al.* (2002) *Structure* **10**, 1063. **123**. Levin, Michaud, Thomas & Jerina (1983) *Arch. Biochem. Biophys.* **220**, 485. **124**. Cornell, Swarts, Barry & Riscoe (1996) *Biochem. Biophys. Res. Commun.* **228**, 724. **125**. Magni, Fioretti, Ipata & Natalini (1975) *J. Biol. Chem.* **250**, 9. **126**. Vita, Natalini, Ipata & Magni (1974) *Boll. Soc. Ital. Biol. Sper.* **50**, 1077. **127**. Schutzbach & Forsee (1990) *J. Biol. Chem.* **265**, 2546. **128**. Kimura, Yamaguchi, Suehisa & Tagaki (1991) *Biochim. Biophys. Acta* **1075**, 6. **129**. Tabas & Kornfeld (1979) *J. Biol. Chem.* **254**, 11655. **130**. Tulsiani, Hubbard, Robbins & Touster (1982) *J. Biol. Chem.* **257**, 3660. **131**. Canellakis, Bondy, May, Jr., Myers-Robfogel & Sartorelli (1984) *Eur. J. Biochem.* **143**, 159. **132**. Skovbjerg, Sjöström & Noren (1981) *Eur. J. Biochem.* **114**, 653. **133**. Skovbjerg, Noren, Sjöström, Danielsen & Enevoldsen (1982) *Biochim. Biophys. Acta* **707**, 89. **134**. Herscovics (1999) *Biochim. Biophys. Acta* **1426**, 275. **135**. Szumilo, Kaushal & Elbein (1986) *Arch. Biochem. Biophys.* **247**, 261. **136**. Ubhayasekera, Muñoz, Vasella, Ståhlberg & Mowbray (2005) *FEBS J.* **272**, 1952. **137**. Burns & Touster (1982) *J. Biol. Chem.* **257**, 9991. **138**. Brada & Dubach (1984) *Eur. J. Biochem.* **141**, 149. **139**. Takahashi, Mizuno & Takamori (1985) *Infect. Immun.* **47**, 271. **140**. Kraml, Kolínská, Ellederová & Hirsová (1972) *Biochim. Biophys. Acta* **258**, 520. **141**. Miletti, Almeida-de-Faria, Colli & Alves (2003) *Braz. J. Med. Biol. Res.* **36**, 595. **142**. Matsushita (1983) *Comp. Biochem. Physiol. B* **76**, 465. **143**. Sigrist, Ronner & Semenza (1975) *Biochim. Biophys. Acta* **406**, 433. **144**. Chinchetru, Cabezas & Calvo (1983) *Comp. Biochem. Physiol.* **75**, 719. **145**. Colas (1980) *Biochim. Biophys. Acta* **613**, 448 **146**. Surarit, Matsui, Chiba, Svasti & Srisomsap (1996) *Biosci. Biotechnol. Biochem.* **60**, 1265. **147**. Gillard & Nelson (1977) *Biochemistry* **16**, 3978. **148**. Fitzgerald & Madsen (1986) *J. Crystal Growth* **76**, 600. **149**. Becker, Long & Fischer (1977) *Biochemistry* **16**, 291. **150**. Ryman & Whelan (1971) *Adv. Enzymol.* **34**, 285. **151**. Gordon, Brown & Brown (1972) *Biochim. Biophys. Acta* **289**, 97. **152**. Nelson & Larner (1970) *Biochim. Biophys. Acta* **198**, 538. **153**. Nelson, Kolb & Larner (1968) *Biochim. Biophys. Acta* **167**, 212. **154**. Diez & Cabezas (1979) *Eur. J. Biochem.* **93**, 301. **155**. Silva, Terra & Ferreira (2004) *Insect Biochem. Mol. Biol.* **34**, 1089. **156**. Killick (1983) *Arch. Biochem. Biophys.* **222**, 561. **157**. Talbot, Muir & Huber (1975) *Can. J. Biochem.* **53**, 1106. **158**. Nakano & Sacktor (1985) *J. Biochem.* **97**, 1329. **159**. Nakano, Sumi & Miyakawa (1977) *J. Biochem.* **81**, 1041. **160**. Terra, Ferreira & de Bianchi (1978) *Biochim. Biophys. Acta* **524**, 131. **161**. Galand (1984) *Biochim. Biophys. Acta* **789**, 10. **162**. Yoneyama (1987) *Arch. Biochem. Biophys.* **255**, 168. **163**. Ishimoto & Nakamura (1997) *Biosci. Biotechnol. Biochem.* **61**, 599. **164**. Rojo, Quiroga, Vattuone & Sampietro (1998) *Phytochemistry* **49**, 965. **165**. Lopez, Vattuone & Sampietro (1988) *Phytochemistry* **27**, 3077. **166**. Jelinek-Kelly & Herscovics (1988) *J. Biol. Chem.* **263**, 14757. **167**. Miyazaki (1988) *Agric. Biol. Chem.* **52**, 625. **168**. Grossmann & Terra (2001) *Comp. Biochem. Physiol. B* **128**, 109. **169**. Park, Bae, Sung, Lee & Kim (2001) *Biosci. Biotechnol. Biochem.* **65**, 1163. **170**. Ferreira & Terra (1983) *Biochem. J.* **213**, 43. **171**. Sano, Amemura & Harada (1975) *Biochim. Biophys. Acta* **377**, 410. **172**. Patchett, Daniel & Morgan (1987) *Biochem. J.* **243**, 779. **173**. Santos & Terra (1985) *Biochim. Biophys. Acta* **831**, 179. **174**. Dion, Fourage, Hallet & Colas (1999) *Glycoconjugate J.* **16**, 27. **175**. Ait, Creuzet & Cattaneo (1982) *J. Gen. Microbiol.* **128**, 569. **176**. Bravo-Torres, Villagómez-Castro, Calvo-Méndez, Flores-Carreón & López-Romero (2004) *Int. J. Parasitol.* **34**, 455. **177**. Berthelot & Delmotte (1999) *Appl. Environ. Microbiol.* **65**, 2907. **178**. Suzuki & Uchida (1984) *Agric. Biol. Chem.* **48**, 1343. **179**. Yamasaki & Suzuki (1978) *Agric. Biol. Chem.* **42**, 971. **180**. Giblin, Kelly & Fogarty (1987) *Can. J. Microbiol.* **33**, 614. **181**. Suzuki, Yuki, Kishigami & Abe (1976) *Biochim. Biophys. Acta* **445**, 386. **182**. Yamasaki & Suzuki (1979) *Agric. Biol. Chem.* **43**, 481. **183**. Peruffo, Renosto & Pallavicini (1978) *Planta* **142**, 195. **184**. Yamasaki & Suzuki (1974) *Agric. Biol. Chem.* **38**, 443. **185**. Yamasaki, Miyake & Suzuki (1973) *Agric. Biol. Chem.* **37**, 251. **186**. Yamasaki, Suzuki & Ozawa (1977) *Agric. Biol. Chem.* **41**, 1451 and 1559. **187**. Yamasaki, Suzuki & Ozawa (1976) *Agric. Biol. Chem.* **40**, 669. **188**. Kelly, Giblin & Fogarty (1986) *Can. J. Microbiol.* **32**, 342. **189**. Thirunavukkarasu & Priest (1984) *J. Gen. Microbiol.* **130**, 3135. **190**. Silva & Terra (1995) *Insect Biochem. Mol. Biol.* **25**, 487. **191**. Chadalavada & Sivakami (1997) *Biochem. Mol. Biol. Int.* **42**, 1051. **192**. Kanaya, Chiba, Shimomura & Nishi (1976) *Agric. Biol. Chem.* **40**, 1929. **193**. Takahashi, Shimomura & Chiba (1971) *Agric. Biol. Chem.* **35**, 2015. **194**. De Cort,

Shantha Kumara & Verachtert (1994) *Appl. Environ. Microbiol.* **60**, 3074. **195**. Reiss & Sacktor (1981) *Arch. Biochem. Biophys.* **209**, 342. **196**. Killilea & Clancy (1978) *Phytochemistry* **17**, 1429. **197**. Lim, Yamasaki, Suzuki & Ozawa (1980) *Agric. Biol. Chem.* **44**, 473. **198**. Suzuki & Tomura (1986) *Eur. J. Biochem.* **158**, 77. **199**. Suzuki, Aoki & Hayashi (1982) *Biochim. Biophys. Acta* **704**, 476. **200**. Yamasaki, Suzuki & Ozawa (1977) *Agric. Biol. Chem.* **41**, 1443 and 2149. **201**. Yamasaki, Tsuboi & Suzuki (1977) *Agric. Biol. Chem.* **41**, 2139. **202**. Ghosh & DasGupta (1984) *Curr. Trends Life Sci.* **12**, 79. **203**. Junowicz & Spencer (1973) *Biochim. Biophys. Acta* **312**, 72. **204**. Parisi & De Petrocellin (1972) *Biochem. Biophys. Res. Commun.* **49**, 706. **205**. Sonoki & Ikezawa (1975) *Biochim. Biophys. Acta* **403**, 412. **206**. Shiloach, Bauer, Vlodavsky & Selinger (1973) *Biotechnol. Bioeng.* **15**, 551. **207**. Garrido, Lisa & Domenech (1988) *Mol. Cell. Biochem.* **84**, 41. **208**. Coleman (1977) *Arch. Biochem. Biophys.* **182**, 525. **209**. Cagen & Friedmann (1972) *J. Biol. Chem.* **247**, 3382. **210**. Varticovski, Kushner & Burnham (1980) *J. Biol. Chem.* **255**, 3742. **211**. Hoare & Datta (1990) *Arch. Biochem. Biophys.* **277**, 122. **212**. Yennawar, Dunbar, Conway, Hutson & Farber (2001) *Acta Crystallogr. Sect. D* **57**, 506. **213**. Ro (2002) *J. Biochem. Mol. Biol.* **35**, 306. **214**. Han & Li (2004) *Eur. J. Biochem.* **271**, 4804. **215**. Ahmmad, Maskall & Brown (1984) *Phytochemistry* **23**, 265. **216**. Fishman, Bradley & Henneberry (1976) *Arch. Biochem. Biophys.* **172**, 618. **217**. Hayashi & Matsuda (1981) *J. Biol. Chem.* **256**, 11117. **218**. Gholson, Ueda, Ogasawara & Henderson (1964) *J. Biol. Chem.* **239**, 1208. **219**. Roush & Betz (1958) *J. Biol. Chem.* **233**, 261. **220**. Côté & Robyt (1982) *Carbohydr. Res.* **101**, 57. **221**. Sutter, Ortmann & Grisebach (1972) *Biochim. Biophys. Acta* **258**, 71. **222**. Shukla & Radin (1990) *Arch. Biochem. Biophys.* **283**, 372. **223**. Frear (1968) *Phytochemistry* **7**, 381. **224**. Gaudreault & Webb (1981) *Phytochemistry* **20**, 2629. **225**. Schmidt & John (1979) *Biochim. Biophys. Acta* **566**, 100. **226**. Amir & Preiss (1982) *Plant Physiol.* **69**, 1027. **227**. Morell & Copeland (1985) *Plant Physiol.* **78**, 149. **228**. Ross & Davies (1992) *Plant Physiol.* **100**, 1008. **229**. MacKenzie, Johnson & McDonald (1977) *Can. J. Microbiol.* **23**, 1303. **230**. Rhodes, Wooltorton & Lourencq (1979) *Phytochemistry* **18**, 1125. **231**. Rhodes & Wooltorton (1976) *Phytochemistry* **15**, 947. **232**. Lotfy, Fleuriet & Macheix (1992) *Phytochemistry* **31**, 767. **233**. Ozeki, Sakano, Komamine, *et al.* (1985) *J. Biochem.* **98**, 9. **234**. Tait (1973) *Biochem. J.* **131**, 389. **235**. Yamada, Tanaka & Furuichi (1995) *Biochim. Biophys. Acta* **1245**, 239. **236**. Pramanik, Pawar, Antonian & Schulz (1979) *J. Bacteriol.* **137**, 469. **237**. Bergmeyer, Holz, Klotsch & Lang (1963) *Biochem. Z.* **338**, 114. **238**. Robinson & Sagers (1972) *J. Bacteriol.* **112**, 465. **239**. Sprenger, Schörken, Sprenger & Sahm (1995) *J. Bacteriol.* **177**, 5930. **240**. Lee, Jun, Kim & Kwon (1998) *Planta* **205**, 375. **241**. Beaudet & Mackenzie (1975) *Biochim. Biophys. Acta* **410**, 252. **242**. Greenwood & Rees (1984) *Biochem. J.* **223**, 837. **243**. Martínez, Andersson, Haavik & Flatmark (1991) *Eur. J. Biochem.* **198**, 675.

Trisnorsqualene Cyclopropylamine

This squalene analogue (FW = 423.73 g/mol) inhibits squalene monooxygenase, *or* squalene epoxidase, IC$_{50}$ = 2 μM. The shorter homologue tetranorsqualene cyclopropyl-amine also inhibits, IC$_{50}$ = 4 μM. The *N*-methylated analogue trisnorsqualene *N*-methylcyclopropylamine is a significantly weaker inhibitor (IC$_{50}$ = 100 μM). Note that the corresponding *N*-oxide is also a weak inhibitor of oxidosqualene cyclase (IC$_{50}$ = 40 μM). **1**. Sen & Prestwich (1989) *J. Amer. Chem. Soc.* **111**, 8761.

O,O,O-Tri-(o-tolyl)phosphate

This phosphoric triester (FW = 368.37 g/mol; M.P. = 25.6°C; B.P. = 410°C), also called tri(*o*-cresyl) phosphate and tris(*o*-cresyl) phosphate, poisoned an estimated 50,000 in the U.S. in 1930, when an alcoholic extract of Jamaican

ginger was adulterated with this triester, with many crippled by irreversible paralysis. **Target(s):** ATPase, synapto-somal (1); carboxylesterase (2-4); Ca^{2+}-transporting ATPase (5); cholinesterase (6-8); and fatty-acyl ethyl-ester synthase (9,10). **1**. Brown & Sharma (1976) *Experientia* **32**, 1540. **2**. Chemnitius, Haselmeyer & Zech (1983) *Arch. Toxicol.* **53**, 235. **3**. Clement & Erhardt (1990) *Arch. Toxicol.* **64**, 414. **4**. Krisch (1971) *The Enzymes*, 3rd ed. (Boyer, ed.), **5**, 43. **5**. Barber, Hunt & Ehrich (2001) *J. Toxicol. Environ. Health A* **63**, 101. **6**. Augustinsson (1950) *The Enzymes*, 1st. ed. (Sumner & Myrbäck, eds.), **1** (part 1), 443. **7**. Mendel & Rudney (1944) *Science* **100**, 499. **8**. Hollingsworth (1988) *Eur. J. Pharmacol.* **153**, 167. **9**. Mericle, Kaphalia & Ansari (2002) *Toxicol. Appl. Pharmacol.* **179**, 119. **10**. Kaphalia, Green & Ansari (1999) *Toxicol. Appl. Pharmacol.* **159**, 134.

Triton CF-54

This synthetic polyethylene glycol detergent (FW = indefinite polymer; CAS 51938-58-0) inhibits *N*-acetyllactosaminide α-2,3-sialyltransferase (1), globoside α-*N*-acetylgalactosaminyltransferase (2), lactosylceramide α-2,3-sialyltransferase, activated at low concentrations (3). **1**. Kono, Ohyama, Lee, *et al.* (1997) *Glycobiology* **7**, 469. **2**. Kijimoto, Ishibashi & Makita (1974) *Biochem. Biophys. Res. Commun.* **56**, 177. **3**. Iber, van Echten & Sandhoff (1991) *Eur. J. Biochem.* **195**, 115.

Triton N-101

This proprietary mixture of *n*-nonylphenoxypoly-ethoxyethanols (FW = 616.83 and 660.89 g/mol) has a critical micelle concentration of about 0.085 mM. Note that this detergent absorbs UV light and should be stored in glass containers at 0°C to minimize peroxidation. **Target(s):** diacylglycerol kinase (1-3);and reverse transcriptase, RNA-directed DNA polymerase (4) **1**. Daleo, Piras & Piras (1976) *Eur. J. Biochem.* **68**, 339. **2**. Yada, Ozeki, Kanoh & Nozawa (1990) *J. Biol. Chem.* **265**, 19237. **3**. Lundberg & Sommarin (1992) *Biochim. Biophys. Acta* **1123**, 177. **4**. Baltimore (1970) *Nature* **226**, 1209.

Triton X-100

This proprietary surfactant (FW = indefinite composition CAS 9002-93-1; Critical Micelle Concentration = 0.22–0.24 mM; Aggregation Number = 100–155), consisting of two *t*-octylphenoxy-polyethoxy-ethanols, was originally formulated for secondary recovery of petroleum, but is widely used in solubilizing membrane proteins without solubilizing much membrane lipid. Triton X-100 has the virtue of exerting mild nondenaturing effects on non-membrane proteins. The *cloud point* (*i.e.*, the temperature at which micellar aggregates begin to form and come out of solution) is 64°C. Triton X-100 has proved to be of great value in the surface phospholipid dilution strategy for defining lipase action at water-phospholipid interfaces (1,2). In this experimental approach, the surface concentration of phospholipid in mixed micelles is reduced by the addition Triton as a neutral diluent, thereby increasing the average distance between phospholipids. This allows one to draw mechanistic inferences about the binding interactions of lipases and phospholipases with their lipid substrates (1). This detergent absorbs light in the ultraviolet region of the spectrum (reduced Triton X-100 is the cyclohexane analogue and does not absorb significantly in the ultraviolet). Triton should be stored in glass containers at 0°C to minimize peroxidation. *See* Cutscum **Target(s):** 2-acetyl-1-alkyl-glycerophosphocholine esterase, *or* platelet-activating factor acetylhydrolase (3); acetyl-CoA hydrolase (114); acetyl-CoA:long-chain-base acetyl-transferase (4); α-*N*-acetylgalactosaminide α-2,6-sialyltransferase (5); acetyl-galactosaminyl-*O*-glycosyl-glycoprotein β-1,3-*N*-acetyl-glucosaminyl-transferase (222,223); *N*-acetylglucosaminyl-diphosphoundecaprenol *N*-acetyl-β-D-mannosamino-transferase (216); *N*-acetyl-neuraminate 4-*O*-acetyl-transferase (249); *N*-acetyl-neuraminate 7-*O*(or, 9-*O*)-acetyltransferase (6,247,248); *N*-acetyl-neuraminylgalactosylglucosylceramide β-1,4-*N*-acetylgalactos-aminyl-transferase (220,221); acid phosphatase, *or* tadpole isozyme 1 (104); acyl-[acyl-carrier-protein]:UDP-*N*-acetyl-glucosamine *O*-acyltransferase (240); 2-acylglycerol *O*-acyltransferase, *or* monoacyl-glycerol *O*-acyltransferase (262,263); acylglycerol lipase (monoacylglycerol lipase, weakly inhibited (135,164); 1-acylglycerol-3-phosphate *O*-acyltransferase (244,245); 1-acylglycerophosphocholine *O*-acyltransferase, *or* lysophos-phatidylcholine *O*-acyltransferase (241,259-261); [β-adrenergic-receptor] kinase (7); aldehyde oxidase (8,9); 1-alkenylglycerophosphocholine *O*-acyltransferase (241); alkenylglycerophosphocholine hydrolase (71); alkenyl-glycerophosphoethanolamine hydrolase (70-72); 1-alkyl-2-acetylglycerophosphatase (94); 1-alkyl-2-acetylglycero-phosphocholine esterase, *or* platelet-activating-factor acetylhydrolase (128); 1-alkylglycerophosphocholine *O*-acetyltransferase (243); alkylglycerophos-phoethanolamine phosphodiesterase, *or* lysophospholipase D (77); arylamine *N*-acetyltransferase (279); arylesterase (166); asparaginase (63); bis(2-ethylhexyl)phthalate esterase (125); carboxylesterase (167); carnitine *O*-palmitoyltransferase (264-267); carotene 7,8-desaturase (286); catechol *O*-methyltransferase (285); CDP-diacylglycerol:glycerol-3-phosphate 3-phosphatidyltransferase (177,178); CDP-diacylglycerol:inositol-3-phosphatidyltransferase (10,174-176); ceramidase, activated at low concentrations (60); ceramide glucosyltransferase (226-228); ceramide kinase (11); chloramphenicol acetyltransferase (12); chlorophyllase (137,138); chlorophyllide *a* prenyltransferase (13); choline-phosphate cytidylyltransferase (186); cholinesterase, *or* butyrylcholinesterase (145,146); choloyl-CoA synthetase, *or* cholate:CoA ligase (44); coagulation factor Xa (65); 3',5'-cyclic-GMP phosphodiesterase, *or* cGMP phospho-diesterase (78); CYP1A (114); cytochrome *c* oxidase (15,16); dehydrodolichyl diphosphate synthase (17); diacylglycerol *O*-acyltransferase (268-272); diacylglycerol cholinephosphotransferase (179-183); diacylglycerol kinase (18,190,191); diacylglycerol lipase (164); dimethylallyl*trans*transferase, *or* geranyl-diphosphate synthase (22,209); dipeptidyl-peptidase IV (66,67); DNA-(apurinic or apyrimidinic site) lyase (52); dolichol kinase, activated at low concentrations (189); dolichyldiphosphatase, activated at low concentrations (54,55,96); dolichyl esterase (19); dolichyl-phosphatase, inhibited at elevated concentrations (96); ecdysone 20-monooxygenase (287); endo-β-*N*-acetylglucosaminidase L (20); endopeptidase La (64); fatty acyl-CoA reductase, hexadecanal dehydrogenase (21); ferredoxin:NADP⁺ reductase (58); flavonoid 3',5'-hydroxylase (295); β-D-fucosidase (73); fucosterol-epoxide lyase (53); galactolipase (132-134); galactoside 2-α-L-fucosyltransferase (230); ganglioside galactosyltransferase (231); geranyl-*trans*transferase, *or* farnesyl-diphosphate synthase (207,208); globotriaosylceramide α-galactosidase (23); glucan 1,4-α-glucosidase, *or* glucoamylase (76); 1,3-β-glucan synthase (233,234); glucomannan 4-β-mannosyltransferase (235); β-glucosidase (74); glucuronosyltransferase (236-238); glutaminase (62); glycerol-3-phosphate *O*-acyltransferase (273-278); glycerone-phosphate *O*-acyltransferase (dihydroxyacetone-phosphate *O*-acyltransferase (251-257); glyceryl-ether monooxygenase (293); glycoprotein 3-α-L-fucosyl-transferase (211); glycosaminoglycan galactosyltransferase (229); glycosyl-phosphatidylinositol diacylglycerol-lyase (50); histone acetyltransferase (246); (hydroxyamino)-benzene mutase (49); 4-hydroxybenzoate nonaprenyltransferase (202-204); inositol-polyphosphate 5-phosphatase (95); isopenicillin-N synthase (57); juvenile hormone epoxide hydrolase (68); juvenile-hormone esterase (126,127); lipase (triacylglycerol lipase (40,158-165); long-chain-acyl-CoA hydrolase (24); *N*-(long-chain-acyl)ethanolamine deacylase (59); long-chain-fatty-acid:[acyl-carrier-protein] ligase (41-43); long-chain-fatty-acyl-CoA synthetase, *or* long-chain-fatty-acid:CoA ligase (45-47); lysophospholipase (25,147-154); microsomal epoxide hydrolase (68,69); mRNA (guanine-*N*⁷-)-methyltransferase (283); NADH dehydrogenase, ubiquinone (26,27); NAD⁺:protein-arginine ADP-ribosyltransferase (210); nicotinamidase (61); nuatigenin 3β-glucosyl-transferase (213-215); *trans*-octaprenyl*trans*transferase (206); palmitoyl-CoA hydrolase, *or* acyl-CoA hydrolase (105-113); penicillin-binding protein 1b (224); peptidoglycan glycosyltransferase (224,225); phenylacetone monooxygenase (294); phosphatidate cytidylyltransferase (28,184,185); phosphatidate phosphatase, weakly inhibited (103); phosphatidyl-choline:dolichol *O*-acyltransferase (242); phosphatidyl-choline:sterol *O*-acyltransferase (LCAT) (250); phosphatidylcholine synthase (171);

phosphatidylethanolamine *N*-methyltransferase (10,284); phosphatidyl-glycerophosphatase, inhibited at elevated concentrations (99-102); phosphatidylinositol diacyl-glycerol-lyase (29,51); 1-phosphatidylinositol 3-kinase (188); 1-phosphatidylinositol 4-kinase (197-201); phosphatidyl-inositol-4-phosphate 3-kinase (187); 1-phosphatidylinositol-4-phosphate 5-kinase (196,197); phosphatidyl-*N*-methylethanolamine *N*-methyltransferase (281,282); phospho-*N*-acetylmuramoyl-pentapeptide-transferase (173); phosphoglyceride: lysophosphatidyl-glycerol *O*-acyltransferase (30); phosphoinositide 5-phosphatase (97,98); phosphoinositide phospholipase C, *or* phosphatidylinositol-specific phospholipase C (83); phospholipase A, exhibits both phospholipase A₁ and A₂ activities (131); phospholipase A₁ (129,130); phospholipase A₂ (155-157); phospholipase C (91-93); phospholipase D (31,32,84-90); phospholipid:lysophospholipid *O*-acyltransferase (33); phytoene synthase (205); polygalacturonidase, *or* pectinase (75); poly(3-hydroxybutyrate) depolymerase (120-124); poly(3-hydroxyoctanoate) depolymerase (115-119); protein-*S*-isoprenylcysteine *O*-methyltransferase (34,280); protein-tyrosine sulfotransferase (168); pyrophosphatase, *or* inorganic diphosphatase (56); retinol dehydrogenase, microsomal (35); Sac1 phosphatase (36); sarsapogenin 3β-glucosyl-transferase (212); sphinganine kinase, *or* sphingosine kinase (192-195); sphingomyelin phospho-diesterase, *or* sphingomyelinase, Triton X-100 is an activator for many sphingomyelinases (79-82); squalene:hopene cyclase (48); squalene monooxygenase (288-290); steroid 5α-reductase (37); sterol *O*-acyltransferase (cholesterol *O*-acyltransferase; ACAT (38,258); sterol esterase, *or* cholesterol esterase (139-144); sterol 3β-glucosyltransferase (217-219); tannase136; thiosulfate sulfurtransferase, *or* rhodanese; weakly inhibited (170); γ-tocopherol methyltransferase39; α,α-trehalose-phosphate synthase (UDP-forming) (239); tyrosinase, *or* monophenol monooxygenase; polyphenol oxidase (291,292); tyrosine-ester sulfotransferase (169); UDP-*N*-acetylglucosamine:dolichyl-phosphate *N*-acetylglucosaminephosphotrans-ferase (172); and undecaprenyl-phosphate mannosyltransferase (232). **1.** Warner & Dennis (1975) *J. Biol. Chem.* **250**, 8044. **2.** Deems, Eaton & Dennis (1975) *J. Biol. Chem.* **250**, 9013. **3.** Stafforini, Prescott & McIntyre (1991) *Meth. Enzymol.* **197**, 411. **4.** Barenholz & Gatt (1975) *Meth. Enzymol.* **35**, 242. **5.** Sadler, Beyer, Oppenheimer, *et al.* (1982) *Meth. Enzymol.* **83**, 458. **6.** Diaz, Higa & Varki (1989) *Meth. Enzymol.* **179**, 416. **7.** Benovic (1991) *Meth. Enzymol.* **200**, 351. **8.** Rajagopalan & Handler (1966) *Meth. Enzymol.* **9**, 364. **9.** Rajagopalan, Fridovich & Handler (1962) *J. Biol. Chem.* **237**, 922. **10.** Moore, Jr. (1981) *Meth. Enzymol.* **71**, 596. **11.** Bajjalieh & Batchelor (2000) *Meth. Enzymol.* **311**, 207. **12.** Lu & Jiang (1993) *Biochem. Biophys. Res. Commun.* **196**, 12. **13.** Camara (1985) *Meth. Enzymol.* **110**, 274. **14.** Inouye, Kondo, Yamamura, Nakanishi & Sakaki (2001) *Biochem. Biophys. Res. Commun.* **280**, 1346. **15.** Maeshima, Hattori & Asahi (1987) *Meth. Enzymol.* **148**, 491. **16.** Sinjorgo, Durak, Dekker, *et al.* (1987) *Biochim. Biophys. Acta* **893**, 241. **17.** Bukhtiyarov, Shabalin & Kulaev (1993) *J. Biochem.* **113**, 721. **18.** Kanoh, Sakane & Yamada (1992) *Meth. Enzymol.* **209**, 162. **19.** Sumbilla & Waechter (1985) *Meth. Enzymol.* **111**, 471. **20.** Trimble, Tarentino, Aumick & Maley (1982) *Meth. Enzymol.* **83**, 603. **21.** Kolattukudy, Rogers & Larson (1981) *Meth. Enzymol.* **71**, 263. **22.** Sagami & Ogura (1985) *Meth. Enzymol.* **110**, 188. **23.** Kusiak, Quirk & Brady (1978) *Meth. Enzymol.* **50**, 533. **24.** Miyazawa, Furuta & Hashimoto (1981) *Eur. J. Biochem.* **117**, 425. **25.** Karasawa & Nojima (1991) *Meth. Enzymol.* **197**, 437. **26.** Chappell (1964) *Biochem. J.* **90**, 225. **27.** Ushakova, Grivennikova, Ohnishi & Vinogradov (1999) *Biochim. Biophys. Acta* **1409**, 143. **28.** Moore, Jr. (1987) *Meth. Enzymol.* **148**, 585. **29.** Low (1981) *Meth. Enzymol.* **71**, 741. **30.** Hostetler, Huterer & Wherrett (1992) *Meth. Enzymol.* **209**, 104. **31.** Taki & Kanfer (1981) *Meth. Enzymol.* **71**, 746. **32.** Kobayashi & Kanfer (1991) *Meth. Enzymol.* **197**, 575. **33.** Sugiura & Waku (1992) *Meth. Enzymol.* **209**, 72. **34.** Philips & Pillinger (1995) *Meth. Enzymol.* **256**, 49. **35.** Leo & Lieber (1990) *Meth. Enzymol.* **189**, 520. **36.** Woscholski (2002) *Meth. Enzymol.* **345**, 335. **37.** Houston, Chisholm & Habib (1985) *J. Steroid Biochem.* **22**, 461. **38.** Billheimer (1985) *Meth. Enzymol.* **111**, 286. **39.** Camara (1985) *Meth. Enzymol.* **111**, 544. **40.** Khoo & Steinberg (1975) *Meth. Enzymol.* **35**, 181. **41.** Shen, Fice & Byers (1992) *Anal. Biochem.* **204**, 34. **42.** Byers & Holmes (1990) *Biochem. Cell. Biol.* **68**, 1045. **43.** Fice, Shen & Byers (1993) *J. Bacteriol.* **175**, 1865. **44.** Falany, Xie, Wheeler, *et al.* (2002) *J. Lipid Res.* **43**, 2062. **45.** Philipp & Parsons (1979) *J. Biol. Chem.* **254**, 10785. **46.** Reddy, Sprecher & Bazan (1984) *Eur. J. Biochem.* **145**, 21. **47.** Van Horn, Caviglia, Li, *et al.* (2005) *Biochemistry* **44**, 1635. **48.** Kleemann, Kellner & Poralla (1994) *Biochim. Biophys. Acta* **1210**, 317. **49.** He, Nadeau & Spain (2000) *Eur. J. Biochem.* **267**, 1110. **50.** Bütikofer & Brodbeck (1993) *J. Biol. Chem.* **268**, 17794. **51.** Schwertz, JKreisberg & Venkatachalam (1983) *Arch. Biochem. Biophys.* **224**, 555. **52.** Thibodeau & Verly (1980) *Eur. J. Biochem.* **107**, 555. **53.** Prestwich, Angelastro, de Palma & Perino (1985) *Anal.*

Biochem. **151**, 315. **54**. Applekvist, Chojnacki & Dallner (1981) *Biosci. Rep.* **1**, 619. **55**. Belocopitow & Boscoboinik (1982) *Eur. J. Biochem.* **125**, 167. **56**. Morita & Yasui (1985) *Agric. Biol. Chem.* **49**, 1397. **57**. Palissa, von Dohren, Kleinkauf, Ting & Baldwin (1989) *J. Bacteriol.* **171**, 5720. **58**. Bojko, Kruk & Wieckowski (2003) *Phytochemistry* **64**, 1055. **59**. Schmid, Zuzarte-Augustin & Schmid (1985) *J. Biol. Chem.* **260**, 14145. **60**. Mitsutake, Tani, Okino, *et al.* (2001) *J. Biol. Chem.* **276**, 26249. **61**. Wintzerith, Dierich & Mandel (1980) *Biochim. Biophys. Acta* **613**, 191. **62**. Nelson, Rumsey & Erecinska (1992) *Biochem. J.* **282**, 559. **63**. Raha, Roy, Dey & Chakrabarty (1990) *Biochem. Int.* **21**, 987. **64**. Desautels & Goldberg (1982) *J. Biol. Chem.* **257**, 11673. **65**. Pejler, Lunderius & Tomasini-Johansson (2000) *Thromb. Haemost.* **84**, 429. **66**. Jobin, Martinez, Motard, Gottschalk & Grenier (2005) *J. Bacteriol.* **187**, 795. **67**. Fujimura, Shibata, Hirai & Ueda (2005) *Eur. J. Med. Res.* **10**, 278. **68**. Touhara & Prestwich (1993) *J. Biol. Chem.* **268**, 19604. **69**. Vogel-Bindel, Bentley & Oesch (1982) *Eur. J. Biochem.* **126**, 425. **70**. J. Gunawan & H. Debuch (1982) *J. Neurochem.* **39**, 693. **71**. Jurkowitz, Horrocks & Litsky (1999) *Biochim. Biophys. Acta* **1437**, 142. **72**. Jurkowitz-Alexander, Ebata, Mills, Murphy & Horrocks (1989) *Biochim. Biophys. Acta* **1002**, 203. **73**. Chinchetru, Cabezas & Calvo (1983) *Comp. Biochem. Physiol.* **75**, 719. **74**. Patchett, Daniel & Morgan (1987) *Biochem. J.* **243**, 779. **75**. Kaur, Kumar & Satyanarayana (2004) *Biores. Technol.* **94**, 239. **76**. Kumar & Satyanarayana (2003) *Biotechnol. Prog.* **19**, 936. **77**. Wykle, Kraemer & Schremmer (1977) *Arch. Biochem. Biophys.* **184**, 149. **78**. Utarabhand (1993) *Comp. Biochem. Physiol. B* **104**, 577. **79**. Wu, Nilsson, Jonsson, *et al.* (2006) *Biochem. J.* **394**, 299. **80**. Ella, Qi, Dolan, Thompson & Meier (1997) *Arch. Biochem. Biophys.* **340**, 101. **81**. Samet & Barenholz (1999) *Chem. Phys. Lipids* **102**, 65. **82**. Duan, Nyberg & Nilsson (1995) *Biochim. Biophys. Acta* **1259**, 49. **83**. Litosch (2000) *Biochemistry* **39**, 7736. **84**. Kokusho, Kato, Machida & Iwasaki (1987) *Agric. Biol. Chem.* **51**, 2515. **85**. Virto, Svensson & Adlercreutz (2000) *Chem. Phys. Lipids* **106**, 41. **86**. Ueda, Liu & Yamanaka (2001) *Biochim. Biophys. Acta* **1532**, 121. **87**. Taki & Kanfer (1979) *J. Biol. Chem.* **254**, 9761. **88**. Okamura & Yamashita (1994) *J. Biol. Chem.* **269**, 31207. **89**. Salvador & Giusto (1998) *Lipids* **33**, 853. **90**. Vinggaard & Hansen (1995) *Biochim. Biophys. Acta* **1258**, 169. **91**. Okawa & Yamaguchi (1975) *J. Biochem.* **78**, 537. **92**. Lucchesi & Domenech (1994) *Int. J. Biochem.* **26**, 155. **93**. Alam, Banno & Nozawa (1993) *J. Eukaryot. Microbiol.* **40**, 775. **94**. Tellis & Lekka (2000) *J. Eukaryot. Microbiol.* **47**, 122. **95**. Lemos, Dumont & Erneux (1989) *FEBS Lett.* **249**, 321. **96**. Belocopitow & Boscoboinik (1982) *Eur. J. Biochem.* **125**, 167. **97**. Palmer (1981) *Can. J. Biochem.* **59**, 469. **98**. Cooper & Hawthorne (1975) *Biochem. J.* **150**, 537. **99**. Icho & Raetz (1983) *J. Bacteriol.* **153**, 722. **100**. Chang & Kennedy (1967) *J. Lipid Res.* **8**, 456. **101**. Funk, Zimniak & Dowhan (1992) *J. Bacteriol.* **174**, 205. **102**. Larson, Hirabayashi & Dowhan (1976) *Biochemistry* **15**, 974. **103**. Casola & Possmayer (1981) *Biochim. Biophys. Acta* **664**, 298. **104**. Filburn (1973) *Arch. Biochem. Biophys.* **159**, 683. **105**. Broustas & Hajra (1995) *J. Neurochem.* **64**, 2345. **106**. Sanjanwala, Sun & MacQuarrie (1987) *Arch. Biochem. Biophys.* **258**, 299. **107**. Berge & Dossland (1979) *Biochem. J.* **181**, 119. **108**. Ohkawa, Shiga & Kageyama (1979) *J. Biochem.* **86**, 643. **109**. Miyazawa, Furuta & Hashimoto (1981) *Eur. J. Biochem.* **117**, 425. **110**. Berge, Slinde & Farstad (1981) *Biochim. Biophys. Acta* **666**, 25. **111**. Akao, Kusaka & Kobashi (1981) *J. Biochem.* **90**, 1661. **112**. Diczfalusy & Alexson (1996) *Arch. Biochem. Biophys.* **334**, 104. **113**. Svensson, Alexson & Hiltunen (1995) *J. Biol. Chem.* **270**, 12177. **114**. Prass, Isohashi & Utter (1980) *J. Biol. Chem.* **255**, 5215. **115**. Elbanna, Lütke-Eversloh, Jendrossek, Luftmann & Steinbüchel (2004) *Arch. Microbiol.* **182**, 212. **116**. Kim, Kim, Kim & Rhee (2005) *J. Microbiol.* **43**, 285. **117**. Kim, Kim, Nam, Bae & Rhee (2003) *Antonie Leeuwenhoek* **83**, 183. **118**. Gao, Maehara, Yamane & Ueda (2001) *FEMS Microbiol. Lett.* **196**, 159. **119**. Kim, Nam & Rhee (2002) *Biomacromolecules* **3**, 291. **120**. Brucato & Wong (1991) *Arch. Biochem. Biophys.* **290**, 497. **121**. Handrick, Reinhardt, Kimmig & Jendrossek (2004) *J. Bacteriol.* **186**, 7243. **122**. Abe, Kobayashi & Saito (2005) *J. Bacteriol.* **187**, 6982. **123**. Kobayashi, Sugiyama, Kawase, *et al.* (1999) *J. Environ. Polym. Degrad.* **7**, 9. **124**. Kobayashi, Nishikori & Saito (2004) *Curr. Microbiol.* **49**, 199. **125**. Krell & Sandermann, Jr. (1984) *Eur. J. Biochem.* **143**, 57. **126**. Kamita, Hinton, Wheelock, *et al.* (2003) *Insect Biochem. Mol. Biol.* **33**, 1261. **127**. De Kort & Granger (1981) *Ann. Rev. Entomol.* **26**, 1. **128**. Stafforini, Prescott & McIntyre (1987) *J. Biol. Chem.* **262**, 4223. **129**. Nishijima, Akamatsu & Nojima (1974) *J. Biol. Chem.* **249**, 5658. **130**. Nieder & Law (1983) *J. Biol. Chem.* **258**, 4304. **131**. Cao, Tam, Arthur, Chen & Choy (1987) *J. Biol. Chem.* **262**, 16927. **132**. Matsuda & Hirayama (1979) *Agric. Biol. Chem.* **43**, 697. **133**. O'Sullivan, Warwick & Dalling (1987) *J. Plant Physiol.* **131**, 393. **134**. Matsuda & Hirayama (1979) *Agric. Biol. Chem.* **43**, 463. **135**. Imamura & Kitaura (2000) *J. Biochem.* **127**, 419. **136**. Kar, Banerjee & Bhattacharyya (2003) *Proc. Biochem.* **38**, 1285. **137**. Moll & Stegwee (1978) *Planta* **140**, 75. **138**. Benedetti & Arruda (2002) *Plant Physiol.* **128**, 1255. **139**. van Berkel, Vaandrager, Kruijt & Koster (1980) *Biochim. Biophys. Acta* **617**, 446. **140**. Durham III & Grogan (1984) *J. Biol. Chem.* **259**, 7433. **141**. Taketani, Nishino & Katsuki (1981) *J. Biochem.* **89**, 1667. **142**. Tuháckovà, Kríz & Hradec (1980) *Biochim. Biophys. Acta* **617**, 439. **143**. Calero-Rueda, Plou, Ballesteros, Martínez & Martínez (2002) *Biochim. Biophys. Acta* **1599**, 28. **144**. Sugihara, Shimada, Nomura, *et al.* (2002) *Biosci. Biotechnol. Biochem.* **66**, 2347. **145**. Yildiz, Bodur, Cokugras & Ozer (2004) *Protein J.* **23**, 143. **146**. Boeck, Schopfer & Lockridge (2002) *Biochem. Pharmacol.* **63**, 2101. **147**. Doi & Nojima (1975) *J. Biol. Chem.* **250**, 5208. **148**. Ichimasa, Morooka & Niimura (1984) *J. Biochem.* **95**, 137. **149**. Tamura, Ajayi, Allmond, *et al.* (2004) *Biochem. Biophys. Res. Commun.* **316**, 323. **150**. de Jong, van den Bosch, Rijken & van Deenen (1974) *Biochim. Biophys. Acta* **369**, 50. **151**. Wright, Payne, Santangelo, *et al.* (2004) *Biochem. J.* **384**, 377. **152**. Victoria & Korn (1975) *Arch. Biochem. Biophys.* **171**, 255. **153**. Ichimasa & Shiobara (1985) *Agric. Biol. Chem.* **49**, 1083. **154**. Chen, Wright, Golding & Sorrell (2000) *Biochem. J.* **347**, 431. **155**. Yang, Farooqui & Horrocks (1996) *Adv. Exp. Med. Biol.* **416**, 309. **156**. Kawauchi, Takasaki, Matsuura & Masuho (1994) *J. Biochem.* **116**, 82. **157**. Rahman, Cerny & Peraino (1973) *Biochim. Biophys. Acta* **321**, 526. **158**. Quyen, Le, Nguyen, Oh & Lee (2005) *Protein Expr. Purif.* **39**, 97. **159**. Sinchaikul, Sookkheo, Phutrakul, Pan & Chen (2001) *Protein Expr. Purif.* **22**, 388. **160**. Quyen, Schmidt-Dannert & Schmid (2003) *Protein Expr. Purif.* **28**, 102. **161**. Huang, Locy & Weete (2004) *Lipids* **39**, 251. **162**. Lima, Krieger, Mitchell & Fontana (2004) *Biochem. Eng. J.* **18**, 65. **163**. Deb, Daniel, Sirakova, *et al.* (2006) *J. Biol. Chem.* **281**, 3866. **164**. Stam, Broekhoven-Schokker & Hülsmann (1986) *Biochim. Biophys. Acta* **875**, 76. **165**. Ibrik, Chahinian, Rugani, Sarda & Comeau (1998) *Lipids* **33**, 377. **166**. Toshimitsu, Hamada & Kojima (1986) *J. Ferment. Technol.* **64**, 459. **167**. McGhee (1987) *Biochemistry* **26**, 4101. **168**. William, Ramaprasad & Kasinathan (1997) *Arch. Biochem. Biophys.* **338**, 90. **169**. Vargas, Frerot, Tuong & Schwartz (1985) *Biochemistry* **24**, 5938. **170**. Vandenbergh & Berk (1980) *Can. J. Microbiol.* **26**, 281. **171**. de Rudder, Sohlenkamp & Geiger (1999) *J. Biol. Chem.* **274**, 20011. **172**. Chandra, Doody & Bretthauer (1991) *Arch. Biochem. Biophys.* **290**, 345. **173**. Umbreit & Strominger (1972) *Proc. Natl. Acad. Sci. U.S.A.* **69**, 1972. **174**. Antonsson & Klig (1996) *Yeast* **12**, 449. **175**. Bleasdale, Wallis, MacDonald & Johnston (1979) *Biochim. Biophys. Acta* **575**, 135. **176**. Sexton & Moore (1978) *Plant Physiol.* **62**, 978. **177**. Bleasdale & Johnston (1982) *Biochim. Biophys. Acta* **710**, 377. **178**. Jiang, Kelly, Hagopian & Greenberg (1998) *J. Biol. Chem.* **273**, 4681. **179**. Coleman & Bell (1977) *J. Biol. Chem.* **252**, 3050. **180**. Kanoh & Ohno (1981) *Meth. Enzymol.* **71**, 536. **181**. Kanoh & Ohno (1976) *Eur. J. Biochem.* **66**, 201. **182**. Ishidate, Matsuo & Nakazawa (1993) *Lipids* **28**, 89. **183**. Cornell & MacLennan (1985) *Biochim. Biophys. Acta* **821**, 97. **184**. Sribney & Hegadorn (1982) *Can. J. Biochem.* **60**, 668. **185**. Monaco & Feldman (1997) *Biochem. Biophys. Res. Commun.* **239**, 166. **186**. Choy & Vance (1978) *J. Biol. Chem.* **253**, 5163. **187**. Ono, Nakagawa, Saito, *et al.* (1998) *J. Biol. Chem.* **273**, 7731. **188**. Shibasaki, Homma & Takenawa (1991) *J. Biol. Chem.* **266**, 8108. **189**. Rip & Carroll (1980) *Can. J. Biochem.* **58**, 1051. **190**. Wissing, Heim & Wagner (1989) *Plant Physiol.* **90**, 1546. **191**. Gómez-Merino, Brearley, Ornatowska, *et al.* (2004) *J. Biol. Chem.* **279**, 8230. **192**. Taha, Hannun & Obeid (2006) *J. Biochem. Mol. Biol.* **39**, 113. **193**. Coursol, Le Stunff, Lynch, *et al.* (2005) *Plant Physiol.* **137**, 724. **194**. Liu, Sugiura, Nava, *et al.* (2000) *J. Biol. Chem.* **275**, 19513. **195**. Baumruker, Bornancin & Billich (2005) *Immunol. Lett.* **96**, 175. **196**. Kai, Salway & Hawthorne (1968) *Biochem. J.* **106**, 791. **197**. Cooper & Hawthorne (1976) *Biochem. J.* **160**, 97. **198**. Li, F. Porter, Hoffman & Deuel (1989) *Biochem. Biophys. Res. Commun.* **160**, 202. **199**. Hou, Zhang & Tai (1988) *Biochim. Biophys. Acta* **959**, 67. **200**. Kanoh, Banno, Hirata & Nozawa (1990) *Biochim. Biophys. Acta* **1046**, 120. **201**. Collins & Wells (1983) *J. Biol. Chem.* **258**, 2130. **202**. Kalén, Applekvist, Chojnacki & Dallner (1990) *J. Biol. Chem.* **265**, 1158. **203**. Melzer & Heide (1994) *Biochim. Biophys. Acta* **1212**, 93. **204**. Buron, Herman, Alcain & Villalba (2006) *Anal. Biochem.* **353**, 15. **205**. Schofield & Paliyath (2005) *Plant Physiol. Biochem.* **43**, 1052. **206**. Teclebrhan,Olsson, Swiezewska & Dallner (1993) *J. Biol. Chem.* **268**, 23081. **207**. Sen, Trobaugh, Béliveau, Richard & Cusson (2007) *Insect Biochem. Mol. Biol.* **37**, 1198. **208**. Dhiman, Schulbach, Mahapatra, *et al.* (2004) *J. Lipid Res.* **45**, 1140. **209**. Sagami & Ogura (1981) *J. Biochem.* **89**, 1573. **210**. Zheng, Morrison, Chung, Moss & Bortell (2006) *J. Cell. Biochem.* **98**, 851. **211**. Staudacher, Altmann, Glössl, *et al.* (1991) *Eur. J. Biochem.* **199**, 745. **212**. Paczkowski & Wojciechowski (1988) *Phytochemistry* **27**, 2743. **213**. Kalinowska & Wojciechowski (1987) *Phytochemistry* **26**, 353. **214**. Kalinowska & Wojciechowski (1988) *Plant Sci.* **55**, 239. **215**. Kalinowska & Wojciechowski (1986) *Phytochemistry* **25**, 2525. **216**. Murazumi, Kumita, Araki & Ito (1988) *J. Biochem.* **104**, 980.

217. Staver, Glick & Baisted (1978) *Biochem. J.* **169**, 297. **218**. Madina, Sharma, Chaturvedi, Sangwan & Tuli (2007) *Biochim. Biophys. Acta* **1774**, 392. **219**. Wojciechowski, Zimowski & Tyski (1977) *Phytochemistry* **16**, 911. **220**. Takeya, Hosomi & Kogure (1987) *J. Biochem.* **101**, 251. **221**. Malagolini, Dall'Olio, Guerrini & Serafini-Cessi (1994) *Glycoconjugate J.* **11**, 89. **222**. Brockhausen, Matta, Orr & Schachter (1985) *Biochemistry* **24**, 1866. **223**. Vavasseur, Yang, Dole, Paulsen & Brockhausen (1995) *Glycobiology* **5**, 351. **224**. Nakagawa, Tamaki, Tomioka & Matsuhashi (1984) *J. Biol. Chem.* **259**, 13937. **225**. Park, Seto, Hakenbeck & Matsuhashi (1985) *FEMS Microbiol. Lett.* **27**, 45. **226**. Shah (1973) *Arch. Biochem. Biophys.* **159**, 143. **227**. Coste, Martel, Azzar & Got (1985) *Biochim. Biophys. Acta* **814**, 1. **228**. Durieux, Martel & Got (1990) *Biochim. Biophys. Acta* **1024**, 263. **229**. Andersson & Eriksson (1979) *Biochim. Biophys. Acta* **570**, 239. **230**. Trinchera & Bozzaro (1996) *FEBS Lett.* **395**, 68. **231**. Neskovic, Mandel & Gatt (1978) *Adv. Exp. Med. Biol.* **101**, 613. **232**. Forsee & Elbein (1973) *J. Biol. Chem.* **248**, 2858. **233**. Kamat, Garg & Sharma (1992) *Arch. Biochem. Biophys.* **298**, 731. **234**. Beaulieu, Tang, Yan, *et al.* (1994) *Antimicrob. Agents Chemother.* **38**, 937. **235**. Piro, Zuppa, Dalessandro & Northcote (1993) *Planta* **190**, 206. **236**. Matern, Matern, Schelzig & Gerok (1980) *FEBS Lett.* **118**, 251. **237**. Kurkela, Mörsky, Hirvonen, Kostiainen & Finel (2004) *Mol. Pharmacol.* **65**, 826 **238**. Kurkela, García-Horsman, Luukkanen, *et al.* (2003) *J. Biol. Chem.* **278**, 3536. **239**. Killick (1979) *Arch. Biochem. Biophys.* **196**, 121. **240**. Anderson & Raetz (1987) *J. Biol. Chem.* **262**, 5159. **241**. Arthur & Choy (1986) *Biochem. J.* **236**, 481. **242**. Keenan & Kruczek (1976) *Biochemistry* **15**, 1586. **243**. Wykle, Malone & Snyder (1980) *J. Biol. Chem.* **255**, 10256. **244**. Mizuno, Sugiura & Okuyama (1984) *J. Lipid Res.* **25**, 843. **245**. Yamashita, Nakaya, Miki & Numa (1975) *Proc. Natl. Acad. Sci. U.S.A.* **72**, 600. **246**. Wiegand & Brutlag (1981) *J. Biol. Chem.* **256**, 4578. **247**. Vandamme-Feldhaus & Schauer (1998) *J. Biochem.* **124**, 111. **248**. Schauer (1970) *Hoppe-Seyler's Z. Physiol. Chem.* **351**, 595. **249**. Iwersen, Vandamme-Feldhaus & Schauer (1998) *Glycoconjugate J.* **15**, 895. **250**. Smith & Kuksis (1980) *Can. J. Biochem.* **58**, 1286. **251**. Hajra (1968) *J. Biol. Chem.* **243**, 3458. **252**. Hardeman & Van den Bosch (1988) *Biochim. Biophys. Acta* **963**, 1. **253**. Schlossman & Bell (1977) *Arch. Biochem. Biophys.* **182**, 737. **254**. Declercq, Haagsman, Van Veldhoven, *et al.* (1984) *J. Biol. Chem.* **259**, 9064. **255**. Datta & Hajra (1984) *FEBS Lett.* **176**, 264. **256**. Schlossman & Bell (1978) *J. Bacteriol.* **133**, 1368. **257**. Schutgens, Romeyn, Ofman, *et al.* (1986) *Biochim. Biophys. Acta* **879**, 286. **258**. Erickson, Shrewsbury, Brooks & Meyer (1980) *J. Lipid Res.* **21**, 930. **259**. Weltzien, Richter & Ferber (1979) *J. Biol. Chem.* **254**, 3652. **260**. Hasegawa-Sasaki & Ohno (1980) *Biochim. Biophys. Acta* **617**, 205. **261**. Yamashita, Nakaya, Miki & Numa (1975) *Proc. Natl. Acad. Sci. U.S.A.* **72**, 600. **262**. Coleman & Haynes (1986) *J. Biol. Chem.* **261**, 224. **263**. Tumaney, Shekar & Rajasekharan (2001) *J. Biol. Chem.* **276**, 10847. **264**. Zierz & Engel (1987) *Biochem. J.* **245**, 205. **265**. Murthy & Pande (1987) *Biochem. J.* **248**, 727. **266**. Woeltje, Esser, Weis, *et al.* (1990) *J. Biol. Chem.* **265**, 10714. **267**. Zhu, Shi, Cregg & Woldegiorgis (1997) *Biochem. Biophys. Res. Commun.* **239**, 498. **268**. Grigor & Bell (1982) *Biochim. Biophys. Acta* **712**, 464. **269**. Akao & Kusaka (1976) *J. Biochem.* **80**, 723. **270**. Weselake, Taylor, Pomeroy, Lawson & Underhill (1991) *Phytochemistry* **30**, 3533. **271**. Triki, Ben Hamida & Mazliak (2000) *Biochem. Soc. Trans.* **28**, 689. **272**. Polokoff & Bell (1980) *Biochim. Biophys. Acta* **618**, 129. **273**. Kume, Shimizu & Seyama (1987) *J. Biochem.* **101**, 653. **274**. Carroll, Morris, Grosjean, Anzalone & Haldar (1982) *Arch. Biochem. Biophys.* **214**, 17. **275**. Snider & Kennedy (1977) *J. Bacteriol.* **130**, 1072. **276**. Schlossman & Bell (1978) *J. Bacteriol.* **133**, 1368. **277**. Kessels & Van Den Bosch (1982) *Biochim. Biophys. Acta* **713**, 570. **278**. Datta & Hajra (1984) *FEBS Lett.* **176**, 264. **279**. Barenholz, Edelman & Gatt (1974) *Biochim. Biophys. Acta* **358**, 262. **280**. De Busser, Van Dessel & Lagrou (2000) *Int. J. Biochem. Cell Biol.* **32**, 1007. **281**. Gaynor & Carman (1990) *Biochim. Biophys. Acta* **1045**, 156. **282**. Schneider & Vance (1979) *J. Biol. Chem.* **254**, 3886. **283**. Ramadevi, Burroughs, Mertens, Jones & Roy (1998) *Proc. Natl. Acad. Sci. U.S.A.* **95**, 13537. **284**. Guan, Wang, Xiao, Hu & Liu (1999) *Neurochem. Int.* **34**, 41. **285**. Jeffery & Roth (1984) *J. Neurochem.* **42**, 826. **286**. Albrecht, Linden & Sandmann (1996) *Eur. J. Biochem.* **236**, 115. **287**. Greenwood & Rees (1984) *Biochem. J.* **223**, 837. **288**. Favre & Ryder (1996) *Antimicrob. Agents Chemother.* **40**, 443. **289**. Ryder & Dupont (1984) *Biochim. Biophys. Acta* **794**, 466. **290**. Satoh, Horie, Watanabe, Tsuchiya & Kamei (1993) *Biol. Pharm. Bull.* **16**, 349. **291**. Liu, Zhang, Wang, *et al.* (2004) *Lett. Appl. Microbiol.* **39**, 407. **292**. Saeidian, Keyhani & Keyhani (2007) *J. Agric. Food Chem.* **55**, 3713. **293**. Kosar-Hashemi & Armarego (1993) *Biol. Chem. Hoppe-Seyler* **374**, 9. **294**. Fraaije, Kamerbeek, Heidekamp, Fortin & Janssen (2004) *J. Biol. Chem.* **279**, 3354. **295**. Menting, Scopes & Stevenson (1994) *Plant Physiol.* **106**, 633.

Triton X-114

This proprietary mixture of two *t*-octylphenoxy-polyethoxyethanols (FW = 514.70 and 558.75 g/mol), also called oxynols has a low cmc value of 0.17-0.20 mM and is widely used to solubilize membrane proteins without extensively solubilizing much membrane lipid. Triton X-114 has a greater solubility in water at lower temperatures compared to Triton X-100 and has a low cloud point (*i.e.*, the temperature at which micellar aggregates begin to form and come out of solution) of 22-28°C. For this reason, Triton X-114 separates from water at higher temperatures to form a separate detergent phase, and this property has been exploited in the purification of detergent-solubilized proteins (1-3). Note that this detergent absorbs UV light and should be stored in glass containers at 0°C to minimize peroxidation. **Target(s):** ceramide kinase (4); flavonoid 3',5'-hydroxylase (5); and glutaconyl-CoA decarboxylase (6). **1**. Bordier (1981) *J. Biol. Chem.* **256**, 1604. **2**. Brusca & Radolf (1994) *Meth. Enzymol.* **228**, 182. **3**. Pryde (1986) *Trends Biochem. Sci.* **11**, 160. **4**. Bajjalieh & Batchelor (2000) *Meth. Enzymol.* **311**, 207. **5**. Menting, Scopes & Stevenson (1994) *Plant Physiol.* **106**, 633. **6**. Buckel (1986) *Meth. Enzymol.* **125**, 547.

S-Trityl-L-Cysteine

This potent tumor growth inhibitor (FW$_{free-base}$ = 363.47g/mol; CAS 2799-07-7; Abbreviation STLC), commonly used in protective-group chemistry required in solid-phase peptide/protein synthesis, targets the mitosis-associated kinesin molecular motor Eg5. STLC does not prevent cell-cycle progression at the S or G$_2$ phases but instead inhibits both separation of the duplicated centrosomes as well as bipolar spindle formation, thereby stalling cells in the M phase with single-aster spindles. Upon removal of *S*-trityl-L-cysteine, cells exit mitosis normally. *In vitro*, *S*-trityl-L-cysteine targets Eg5's catalytic domain, inhibiting its basal and microtubule- (MT-) activated ATPase activity as well as the release of ADP. *S*-Trityl-L-cysteine qualifies as a tight binding inhibitor ($K_{i,app} < 150$ nM) and binds more tightly than monastrol. STLC has an ~8-fold faster on-rate (*i.e.*, 6.1 µM^{-1} s^{-1} *versus* 0.78 µM^{-1} s^{-1}) and ~4-fold slower off-rate (*i.e.*, 3.6 s^{-1} *versus* 15 s^{-1}). (**See** *Monastrol*) *S*-Trityl-L-cysteine also reversibly inhibits Eg5-driven MT-sliding rate (IC$_{50}$ of 500 nM). HeLa cells cultured for several weeks at increasing concentrations of STLC generate clones that can grow in the presence of 20 µM STLC, a concentration sufficient to fully inhibit Eg5 (2). Further analysis indicated that the majority of cells in these clones were able to assemble a bipolar spindle even in the presence in the presence 20 µM STLC, owing to compensating increase in nuclear envelope dynein. **Target(s):** Among ten human kinesins (*e.g.*, Eg5, CENP-E, MKLP1, RabK6, KIFC1, Kid, KIF2A and KIF2C) that were tested, *S*-trityl-L-cysteine was specific for Eg5 (1). **1**. Skoufias, DeBonis, Saoudi, *et al.* (2006) *J. Biol. Chem.* **281**, 17559. **2**. Raaijmakers, van Heesbeen, Meaders, *et al.* () *EMBO* . **31**, 4179.

Troglitazone

This once widely studied peroxisome PARP agonist (FW = 441.55 g/mol) showed promising properties (*e.g.*, improving insulin sensitivity and decreasing hepatic glucose production) for treating type 2 diabetes, but was withdrawn when its hepatotoxicity became evident. **Target(s):** aromatase, or CYP19 (1,2); CYP3A4 (3); CYP2C8 (3); CYP2C9 (3); CYP2C19 (3); β-hydroxysteroid dehydrogenase (4); K_{ATP} channel (5); glucuronosyltransferase (6,7); and long-chain-fatty-acyl-CoA synthetase, or long-chain-fatty-acid CoA ligase (8-10). **1.** Rubin, Duong, Clyne, et al. (2002) Endocrinology **143**, 2863. **2.** Mu, Yanase, Nishi, et al. (2000) Biochem. Biophys. Res. Commun. **271**, 710. **3.** Yamazaki, Suzuki, Tane, et al. (2000) Xenobiotica **30**, 61. **4.** Gasic, Nagamani, Green & Urban (2001) Amer. J. Obstet. Gynecol. **184**, 575. **5.** Lee, Ibbotson, Richardson & Boden (1996) Eur. J. Pharmacol. **313**, 163. **6.** Yoshigae, Konno, Takasaki & Ikeda (2000) J. Toxicol. Sci. **25**, 433. **7.** Ito, Yamamoto, Sato, Fujiyama & Bamba (2001) Eur. J. Clin. Pharmacol. **56**, 893. **8.** Kim, Lewin & Coleman (2001) J. Biol. Chem. **276**, 24667. **9.** Lewin, Kim, Granger, Vance & Coleman (2001) J. Biol. Chem. **276**, 24674. **10.** Fulgencio, Kohl, Girard & Pegorier (1996) Diabetes **45**, 1556.

Troleandomycin

This semisynthetic antibiotic (FW = 813.98 g/mol; CAS 2751-09-9), also called oleandomycin triacetate ester, inhibits CYP3A4/5 (1-4) and quinine 3-monooxygenase (4-6). **1.** Koyama, Sasabe & Miyamoto (2002) Xenobiotica **32**, 573. **2.** Fowler, Taylor, Friedberg, Wolf & Riley (2002) Drug Metab. Dispos. **30**, 452. **3.** von Rosensteil & Adam (1995) Drug Saf. **13**, 105. **4.** Zhao & Ishizaki (1997) J. Pharmacol. Exp. Ther. **283**, 1168. **5.** Zhao, Yokoyama, Chiba, Wanwimolnuk & Ishizaki (1996) J. Pharmacol. Exp. Ther. **279**, 1327. **6.** Zhao, Kawashiro & Ishizaki (1998) Drug Metab. Dispos. **26**, 188.

Trolox

This water-soluble vitamin E analogue and antioxidant (FW = 250.29 g/mol; CAS 53188-07-1), also known as 3,4-dihydro-6-hydroxy-2,5,7,8-tetramethyl-2H-1-benzopyran-2-carboxylic acid, potently inhibited interleukin-1-induced osteoclast formation in bone marrow cell-osteoblast coculture by abrogating induction by Receptor Activator of NF-κB Ligand (or RANKL) of c-Fos protein by suppressing its translation (1). Consonant with this action is the finding that ectopic overexpression of c-Fos overcomes trolox inhibition of osteoclastogenesis in bone marrow macrophages (1). Trolox also suppresses interleukin-1-induced osteoclast formation and bone loss in mouse calvarial bone (1). Taken together, such findings show that trolox prevents osteoclast formation and bone loss by inhibiting both RANKL induction in osteoblasts and c-Fos expression in osteoclast precursors. Trolox also prevents the oxidative reactions initiated upon oxidation of the hemoprotein by H_2O_2 and involving tyrosyl radicals, inhibiting covalent binding of protein to the heme group, while also inhibited dimerization of myoglobin, the latter process entailing the intermolecular covalent binding of tyrosines (2). Inhibition of both processes requires 20-50x higher trolox levels than needed to reduce the hypervalent heme iron to its ferric form. **Trolox Equivalent Antioxidant Capacity:** This analytical biochemical method employs 2,2'-azino-di-[3-ethylbenzthiazoline sulphonate], or ABTS, as a redox-active chromophore in a plate-based colorimetric assay at 405 or 750 nm to quantify the total antioxidant capacity of substances within foodstuffs, plasma, serum, urine, saliva, cell lysates, etc. The assay relies on the ability

of antioxidants within a sample to inhibit the oxidation of ABTS to the radical-cation $ABTS^{\bullet+}$ upon reaction with met-myoglobin (3-6). The reliability of this method becomes quantitative by including measured quantities of trolox as an internal reference standard. This procedure has also been applied to the determination of physiological antioxidant compounds, including radical-scavenging drugs. **1.** Lee, Kim, Yang, et al. (2009) J. Biol. Chem. **284**, 13725. **2.** Giulivi & Cadenas (1993) Arch. Biochem. Biophys. **303**, 152. **3.** Miller, Rice-Evans, Davies, et al. (1993) Clin. Sci. **84**, 407. **4.** Miller & Rice-Evans (1997) Free Radical Res. **26**, 195. **5.** Miller, Rice-Evans & Davies (1993) Biochem. Soc. Trans. **21**, 95S. **6.** Rice-Evans & Miller (1994) Meth. Enzymol. **234**, 279.

3-Tropanyl 3,5-Dichlorobenzoate

This 5-HT₃ receptor antagonist (FW$_{free-base}$ = 314.21 g/mol; CAS 40796-97-2), also known as MDL 72222, targets peripheral neuronal 5-hydroxytryptamine receptors of mammals and suppresses voluntary ethanol consumption that is mediated by the 5-HT₃ receptor (1,2). **1.** Higgins, Kilpatrick, Bunce, Jones & Tyers (1989) Brit. J. Pharmacol. **97**, 247. **2.** Fozard (1984) Naunyn Schmiedebergs Arch. Pharmacol. **326**, 36.

Tropicamide

This muscarinic receptor antagonist (FW = 284.36 g/mol; CAS 1508-75-4; Soluble to 100 mM in DMSO), systematically named N-ethyl-3-hydroxy-2-phenyl-N-(pyridinylmethyl)propanamide, is M₄-selective, with a pIC50 of 7.62. Its action is distinct from that expected for M₁-selective compounds (e.g., pirenzepine and telenzepine), M₂-selective compounds (e.g., himbacine and methoctramine), and M₃-selective compounds (e.g., hexahydro-siladifenidol and 4-diphenylacetoxy-N-methylpiperidine methiodide). **1.** Lazareno, Buckley & Roberts (1990) Mol. Pharmacol. **38**, 805. **2.** Hernández, Símonsen, Prieto, et al. (1993) Brit. J. Pharmacol. **110**, 1413.

Tropisetron

This serotonin 5-HT₃ receptor antagonist (FW = 284.35 g/mol; CAS 89565-68-4 (free base) and 105826-92-4 (hydrochloride)), also known as Navoban®, 3-tropanylindole-3-carboxylate, and (1R,5S)-8-methyl-8-azabicyclo[3.2.1] octan-3-yl-1-methylindole-3-carboxylate, is an anti-emetic that is often used to reduce vomiting and nausea following a round of chemotherapy and to treat pain associated with fibromyalgia (1,2). Tropisetron is highly potent and selective competitive antagonist of 5-HT₃ receptors, the latter comprising a subclass of serotonin receptors located on peripheral neurons and within the central nervous system. Navoban has a 24-hour duration of action, allowing convenient once-daily administration. Its inhibitory effects on the glycine receptor chloride channel occur in the μM range, whereas its potentiating effects occur at femptomolar concentrations at the homomeric α₁ receptor. Tropisetron is also a partial agonist of α₇ nicotinic acetyl-choline receptors, or nAChR (3). **1.** Shearman & Tolcsvai (1987) Psychopharmacology (Berlin) **92**, 520. **2.** Yang, Ney, Cromer, et al. (2007) J. Neurochem. **100**, 758. **3.** Papke, Porter-Papke & Rose (2004) Bioorg. Med. Chem. Lett. **14**, 1849.

Tropolone

This α-hydroxy cyclic ketone (FW = 122.12 g/mol; CAS 533-75-5; M.P. = 51-54°C), also known as 2-hydroxy-2,4,6-cycloheptatrien-1-one, is a structural element of many drugs and metabolites, including colchicine and is also found in some organisms. **Target(s):** (S)-canadine synthase (1); catechol O-methyltransferase (2-10); 8-dimethyl-alltlnaringenin 2'-hydroxylase (11); L-dopachrome isomerase (12); dopamine β-monooxygenase (13); exoribonuclease H (14); peroxidase (15); polyphenol oxidase (16); ribonuclease H (14); thiol S-methyltransferase (5); thiopurine S-methyltransferase, K_i = 0.85 mM (5,17,18); tyrosinase, or monophenol monooxygenase (16,19-25); and tyrosine 3-monooxygenase (26). **1.** Rueffer & Zenk (1994) *Phytochemistry* **36**, 1219. **2.** Borchardt (1981) *Meth. Enzymol.* **77**, 267. **3.** Vieira-Coelho & Soares-da-Silva (1999) *Brain Res.* **821**, 69. **4.** Borchardt (1973) *J. Med. Chem.* **16**, 377. **5.** Otterness, Keith, Kerremans & Weinshilboum (1986) *Drug Metab. Dispos.* **14**, 680. **6.** Bonifácio, Palma, Almeida & Soares-da-Silva (2007) *CNS Drug Reviews* **13**, 352. **7.** Veser (1987) *J. Bacteriol.* **169**, 3696. **8.** Tong & D'Iorio (1977) *Can. J. Biochem.* **55**, 1108. **9.** Jeffery & Roth (1984) *J. Neurochem.* **42**, 826. **10.** Axelrod & Tomchick (1958) *J. Biol. Chem.* **233**, 702. **11.** Yamamoto, Yatou & Inoue (2001) *Phytochemistry* **58**, 651. **12.** Palumbo, d'Ischia, Misuraca, De Martino & Prota (1994) *Biochem. J.* **299**, 839. **13.** Goldstein, Lauber & McKereghan (1965) *J. Biol. Chem.* **240**, 2066. **14.** Budihas, Gorshkova, Gaidamakov, *et al.* (2005) *Nucl. Acids Res.* **33**, 1249. **15.** Lee, Hegarty & Christie (1979) *Chem. Biol. Interact.* **27**, 17. **16.** Pérez-Gilabert & García Carmona (2000) *J. Agric. Food Chem.* **48**, 695. **17.** Woodson & Weinshilboum (1983) *Biochem. Pharmacol.* **32**, 819. **18.** Szumlanski, Honchel, Scott & Weinshilboum (1992) *Pharmacogenetics* **2**, 148. **19.** Espin & Wichers (1999) *J. Agric. Food Chem.* **47**, 2638. **20.** Jackson, Smith & Peddie (1993) *Dev. Comp. Immunol.* **17**, 97. **21.** Kahn & Andrawis (1985) *Phytochemistry* **24**, 905. **22.** Kim & Uyama (2005) *Cell. Mol. Life Sci.* **62**, 1707. **23.** Shiino, Watanabe & Umezawa (2008) *J. Enzyme Inhib. Med. Chem.* **23**, 16. **24.** Flurkey, Cooksey, Reddy, *et al.* (2008) *J. Agric. Food Chem.* **56**, 4760. **25.** Orenes-Pinero, Garcia-Carmona & Sanchez-Ferrer (2007) *J. Mol. Catal. B* **47**, 143. **26.** Yamamoto, Kobayashi, Yoshitama, Teramoto & Komamine (2001) *Plant Cell Physiol.* **42**, 969.

Trovafloxacin

This fluoroquinolone antibiotic (FW = 512.46 g/mol; CAS 147059-75-4; Solubility: 50 mM in H_2O; 100 mM in DMSO), also known by CP-99,219 and 7-[(1α,5α,6α)-6-amino-3-azabicyclo[3.1.0]hex-3-yl]-1-(2,4-difluoro-phenyl)-6-fluoro-1,4-dihydro-4-oxo-1,8-naphthyridine-3-carboxylic acid, inhibits bacterial DNA topoisomerase IV and DNA gyrase by forming a stable quinolone-DNA adduct with these enzymes, thereby reversibly inhibiting DNA synthesis (1,2). (For the prototypical member of this antibiotic class, *See Ciprofloxacin*) Trovafloxacin displays potent activity against Gram-positive and Gram-negative bacteria. Ninety percent of the *Proteus vulgaris*, *Providencia rettgeri*, *Providencia stuartii*, and *Serratia marcescens* isolates are inhibited by 0.5 to 2 µg/mL. CP-99,219 inhibited 90% of the *Pseudomonas aeruginosa* and *Haemophilus influenzae* isolates at 1 and 0.015 µg/mL, respectively. The drug inhibited methicillin-susceptible *Staphylococcus aureus* at 0.06 µg/mL, whereas a ciprofloxacin concentration of 1 µg/mL was required to inhibit these organisms. [See Table I of (1) for complete listing and comparison with Sparfloxacin, Tosufloxacin, Ciprofloxacin & Temafloxacin.] Trovafloxacin also increases the production

of mitochondrial NO in immortalized hepatocytes and increases mitochondrial Ca^{2+} (2). Somewhat surprisingly, the plasma membrane pannexin-1 channel, known as PANX1, is a direct target of trovafloxacin at drug concentrations ($IC_{50} \sim 4µM$) observed in human plasma (3). Inhibition of PANX1 led to dysregulated fragmentation of apoptotic cells. Genetic loss of PANX1 phenocopied trovafloxacin effects, revealing a non-redundant role for pannexin channels in regulating cellular disassembly during apoptosis. (3). **1.** Neu & Chin (1994) *Antimicrob. Agents Chemother.* **38**, 2615. **2.** Gootz, Zaniewski, Haskell, *et al.* (1996) *Antimicrob. Agents Chemother.* **40**, 2691. **3.** Poon, Chiu, Armstrong, *et al.* (2014) *Nature* **507**, 329.

Trp Peptide Amide

This acetylated oligopeptide amide Ac-RRKWQKTGHAVRAIGRL-NH$_2$ inhibits [myosin light-chain] kinase, K_d = 8.6 pM (1). Substitution of a tyrosyl residue for Trp4 weakens the inhibition (K_d = 7.3 nM). The deacetylated peptide ARRKWQKTGHAVRAIGRLSS has a K_d value of 0.11 nM while RRKWQKTGHAVRAIGRLSSS has a value of 1.6 nM. **1.** Török, Cowley, Brandmeier, *et al.* (1998) *Biochemistry* **37**, 6188.

TRU-016, *See Otlertuzumab*

Trusopt, *See Dorzolamide*

Trypan Blue

This so-called vital stain (FW$_{tetrasodium\ salt}$ = 960.82 g/mol) is a water-soluble blue dye that is excluded up by living cells, but is taken up by dead cells. Note, however, that this dye stains viable macrophages. When dissolved in water, this acid dye forms a deep blue solution with a violet tinge (λ_{max} = 607 nm). **Target(s):** acid phosphatase (1-3); β-fructofuranosidase, or invertase (4); fumarase (5,6); guanidinobenzoatase (7); hemolytic activity of oligonucleotide-streptolysin S complex (8); thrombin (9); and triose-phosphate isomerase (10). **1.** Hollander (1971) *The Enzymes*, 3rd ed. (Boyer, ed.), **4**, 449. **2.** Lloyd (1968) *Enzymologia* **35**, 75. **3.** Igarashi & Hollander (1968) *J. Biol. Chem.* **243**, 6084. **4.** Quastel & Yates (1936) *Enzymologia* **1**, 60. **5.** Quastel (1931) *Biochem. J.* **25**, 898. **6.** Quastel (1931) *Biochem. J.* **25**, 1121. **7.** Anees (1996) *J. Enzyme Inhib.* **10**, 203. **8.** Taketo & Taketo (1982) *Z. Naturforsch. [C]* **37**, 385. **9.** Seegers (1951) *The Enzymes*, 1st ed. (Sumner & Myrbäck, eds.), **1** (part 2), 1106. **10.** Joubert, Neitz & Louw (2001) *Proteins* **45**, 136.

Trypanothione

This N^1,N^{10}-bis(glutathionyl)spermidine disulfide (FW = 721.86 g/mol; CASA Registry Number = 96304-42-6), found in trypanosomatids, is reduced to the dithiol by trypanothione reductase. This product is one of the major thiols of trypanosomatids and functions in maintaining the redox balance of the cell. Trypanothione participates in the detoxification of hydrogen peroxide, utilizing the enzyme trypanothione reductase, tryparedoxin, and tryparedoxin peroxidase. Tryparedoxin is a thiol protein similar in action to thioredoxin and glutaredoxin. **Target(s):** glutathionylspermidine synthetase, inhibited by the reduced form of trypanothione (1); and tryparedoxin activity (2). **1.** Smith, Nadeau, Bradley, Walsh & Fairlamb (1992) *Protein Sci.* **1**, 874. **2.** Krauth-Siegel & Schmidt (2002) *Meth. Enzymol.* **347**, 259.

Trypan Red

This water-soluble red dye (FW$_{\text{pentasodium salt}}$ = 1002.81 g/mol) is frequently used as a biological stain and a trypanocide. Trypan red was produced by Paul Ehrlich in the early twentieth century and shown by his assistant Shiga to be effective against trypanosomes. **Target(s):** fumarase (1); hemolytic activity of oligonucleotide-streptolysin S complex (2); neuraminidase, *or* sialidase (3). **1.** Quastel (1931) *Biochem. J.* **25**, 1121. **2.** Taketo & Taketo (1982) *Z. Naturforsch. [C]* **37**, 385. **3.** Gottschalk & Bhargava (1971) *The Enzymes*, 3rd ed. (Boyer, ed.), **5**, 321.

Tryparsamide

This light-sensitive antiprotozoal arsenical (FW$_{\text{sodium salt}}$ = 296.09 g/mol; solubility = 0.5 g/mL water), with particular action against the sleeping sickness-causing pathogen *Trypanosoma gambiense*, inhibits glyceraldehyde-3-phosphate dehydrogenase (1), lipase, *or* triacylglycerol lipase, weakly inhibited (2); urease, weakly inhibited (1). **1.** Chen (1948) *J. Infect. Diseases* **82**, 226. **2.** Gordon & Quastel (1948) *Biochem. J.* **42**, 337.

Trypsin-Chymotrypsin Inhibitor, *See* Trypsin Inhibitors

Trypsin Inhibitors

These proteins form highly affine, albeit reversible, complex with trypsin and trypsin-like proteases to form complexes devoid of any proteolytic activity. Soybean trypsin inhibitors (Kunitz-type has K_i = 10 pM, with 1,000,000-times weaker binding to chymotrypsin) are single polypeptide chains with 181-aminoacyl residues and two disulfide bonds. The 71-residue soybean Bowman-Birk inhibitor, itself containing seven disulfide bonds, has two active sites, one for trypsin and one for chymotrypsin. **See also** a$_1$-Antitrypsin; Aprotinin; a$_1$-Antichymotrypsin; etc. **Target(s):** acrosin (1-6,91,126-135); activated protein C (7); adenain (86); bacillopeptidase F (8); brachyurins (9,10,115-120); cathepsin G (12-14,123-125); cerevisin, weakly inhibited (103,104); chymase (15-17); chymotrypsin (2,3,6,18-29,151-154); chynotrypsin C, *or* caldecrin (11); clostripain (30,87); coagulation factor VIIa (31,122); coagulation factor Xa (32,142,143); coagulation factor Xia (33,34); coagulation factor XIIa (35,36,106); cocoonase (19,37,82); complement factor I (105); diacylglycerol *O*-acyltransferase (159); duodenase (38); enteropeptidase, (43,137-141), however, see also ref. 136; envelysin (83-85); follipsin (44); granzyme A (45,46); granzyme B (47,93); kexin (99); leukocyte elastase (39-42,107,124); leucyl endopeptidase (53); limulus clotting enzyme (54); lysosomal Pro-Xaa carboxypeptidase (156,157); oviductin (88); pancreatic elastase (25,107,109); pancreatic elastase II (97,108); peptidyl-dipeptidase Dcp (158); peptidyl-glycinamidase, *or* carboxamidopeptidase (155); plasma kallikrein (2,18,19,28,29,48-50,112-114); plasmin (2-5,18,19,28,29,55-58,145); plasminogen activator (59); prolyl oligopeptidase (121); proteinase A, mouse (60); proteinase K, salivary gland (61); scutelarin (62,100); sea urchin cortical granule protease (63); semenogelase (64,94); sheep mast cell proteinase (138); snake venom factor V activator, *or* thrombocytin in the presence of heparin (65); spermosin (91); stratum corneum chymotryptic enzyme (66,89,90); streptogrisin A (92); streptogrisin B (92); subtilisin (21,29,98); thermomycolin (67); thrombin, weakly inhibited (144); tissue kallikrein (2,18,50-52,110,111); tonin (68); trepolisin (69); trypsin (2-6,18-29,42,58,70-82,145-150,154); tryptase (16,101,102,145); u-plasminogen activator, *or* urokinase (96); and venombin A (95). **1.** Schleuning & Fritz (1976) *Meth. Enzymol.* **45**, 330. **2.** Tschesche & Dietl (1976) *Meth. Enzymol.* **45**, 772. **3.** Cechová (1976) *Meth. Enzymol.* **45**, 806. **4.** Fink & Fritz (1976) *Meth. Enzymol.* **45**, 825. **5.** Fritz, Tschesche & Fink (1976) *Meth. Enzymol.* **45**, 834. **6.** Schiessler, Fink & Fritz (1976)

Meth. Enzymol. **45**, 847. **7.** Kisiel & Davie (1981) *Meth. Enzymol.* **80**, 320. **8.** Hageman (1998) in *Handb. Proteolytic Enzymes* (Barrett, Rawlings & Woessner, eds.), p. 301, Academic Press, San Diego. **9.** Tsu & Craik (1998) in *Handb. Proteolytic Enzymes* (Barrett, Rawlings & Woessner, eds.), p. 25, Academic Press, San Diego. **10.** Grant, Eisen & Bradshaw (1981) *Meth. Enzymol.* **80**, 722. **11.** Tomomura (1998) in *Handb. Proteolytic Enzymes* (Barrett, Rawlings & Woessner, eds.), p. 40, Academic Press, San Diego. **12.** Salvesen (1998) in *Handb. Proteolytic Enzymes* (Barrett, Rawlings & Woessner, eds.), p. 60, Academic Press, San Diego. **13.** Fioretti, M. Angeletti, M. Coletta, *et al.* (1993) *J. Enzyme Inhib.* **7**, 57. **14.** Barrett (1981) *Meth. Enzymol.* **80**, 561. **15.** Caughey (1998) in *Handb. Proteolytic Enzymes* (Barrett, Rawlings & Woessner, eds.), p. 66, Academic Press, San Diego. **16.** Woodbury, Everitt & Neurath (1981) *Meth. Enzymol.* **80**, 588. **17.** Knight (1995) *Meth. Enzymol.* **248**, 85. **18.** Kassell (1970) *Meth. Enzymol.* **19**, 844. **19.** Kassell (1970) *Meth. Enzymol.* **19**, 853. **20.** Kassell (1970) *Meth. Enzymol.* **19**, 862. **21.** Kassell (1970) *Meth. Enzymol.* **19**, 872. **22.** Birk (1976) *Meth. Enzymol.* **45**, 697. **23.** Birk (1976) *Meth. Enzymol.* **45**, 700. **24.** Birk (1976) *Meth. Enzymol.* **45**, 707. **25.** Birk (1976) *Meth. Enzymol.* **45**, 710. **26.** Birk (1976) *Meth. Enzymol.* **45**, 716. **27.** Birk (1976) *Meth. Enzymol.* **45**, 723. **28.** Tschesche (1976) *Meth. Enzymol.* **45**, 792. **29.** Laskowski, Jr., & Sealock (1971) *The Enzymes*, 3rd ed. (Boyer, ed.), **3**, 375. **30.** Ullmann & Bordusa (1998) in *Handb. Proteolytic Enzymes* (Barrett, Rawlings & Woessner, eds.), p. 759, Academic Press, San Diego. **31.** Nemerson (1966) *Biochemistry* **5**, 601. **32.** Church, Messier, Tucker & Mann (1988) *Blood* **72**, 1911. **33.** Walsh (1998) in *Handb. Proteolytic Enzymes* (Barrett, Rawlings & Woessner, eds.), p. 153, Academic Press, San Diego. **34.** Kurachi & Davie (1981) *Meth. Enzymol.* **80**, 211. **35.** Ratnoff (1998) in *Handb. Proteolytic Enzymes* (Barrett, Rawlings & Woessner, eds.), p. 144, Academic Press, San Diego. **36.** Griffin & Cochrane (1976) *Meth. Enzymol.* **45**, 56. **37.** Hruska & Law (1970) *Meth. Enzymol.* **19**, 221. **38.** Pemberton (1998) in *Handb. Proteolytic Enzymes* (Barrett, Rawlings & Woessner, eds.), p. 76, Academic Press, San Diego. **39.** Bieth (1998) in *Handb. Proteolytic Enzymes* (Barrett, Rawlings & Woessner, eds.), p. 54, Academic Press, San Diego. **40.** Ascenzi, Amiconi, Bolognesi, *et al.* (1991) *J. Enzyme Inhib.* **5**, 207. **41.** Barrett (1981) *Meth. Enzymol.* **80**, 581. **42.** Carrell & Travis (1985) *Trends Biochem. Sci.* **10**, 20. **43.** Lu & Sadler (1998) in *Handb. Proteolytic Enzymes* (Barrett, Rawlings & Woessner, eds.), p. 50, Academic Press, San Diego. **44.** Takahashi & Ohnishi (1998) in *Handb. Proteolytic Enzymes* (Barrett, Rawlings & Woessner, eds.), p. 202, Academic Press, San Diego. **45.** Froelich & Salvesen (1998) in *Handb. Proteolytic Enzymes* (Barrett, Rawlings & Woessner, eds.), p. 77, Academic Press, San Diego. **46.** Simon & Kramer (1994) *Meth. Enzymol.* **244**, 68. **47.** Peitsch & Tschopp (1994) *Meth. Enzymol.* **244**, 80. **48.** Colman (1998) in *Handb. Proteolytic Enzymes* (Barrett, Rawlings & Woessner, eds.), p. 147, Academic Press, San Diego. **49.** Colman & Bagdasarian (1976) *Meth. Enzymol.* **45**, 303. **50.** Greenbaum (1971) *The Enzymes*, 3rd ed. (Boyer, ed.), **3**, 475. **51.** Webster & Prado (1970) *Meth. Enzymol.* **19**, 681. **52.** Chao & Chao (1988) *Meth. Enzymol.* **163**, 128. **53.** Ascenzi, Amiconi, Ballio, *et al.* (1991) *J. Enzyme Inhib.* **4**, 283. **54.** Kawabata, Muta & Iwanaga (1998) in *Handb. Proteolytic Enzymes* (Barrett, Rawlings & Woessner, eds.), p. 213, Academic Press, San Diego. **55.** Castellino (1998) in *Handb. Proteolytic Enzymes* (Barrett, Rawlings & Woessner, eds.), p. 90, Academic Press, San Diego. **56.** Robbins & Summaria (1970) *Meth. Enzymol.* **19**, 184. **57.** Castellino & Sodetz (1976) *Meth. Enzymol.* **45**, 273. **58.** Greene, Pubols & Bartelt (1976) *Meth. Enzymol.* **45**, 813. **59.** Ascenzi, Amiconi, Bolognesi, Menegatti & Guarneri (1990) *J. Enzyme Inhib.* **4**, 51. **60.** Chao (1998) in *Handb. Proteolytic Enzymes* (Barrett, Rawlings & Woessner, eds.), p. 104, Academic Press, San Diego. **61.** Chao (1998) in *Handb. Proteolytic Enzymes* (Barrett, Rawlings & Woessner, eds.), p. 113, Academic Press, San Diego. **62.** Rosing (1998) in *Handb. Proteolytic Enzymes* (Barrett, Rawlings & Woessner, eds.), p. 227, Academic Press, San Diego. **63.** Carroll, Jr. (1976) *Meth. Enzymol.* **45**, 343. **64.** Chao (1998) in *Handb. Proteolytic Enzymes* (Barrett, Rawlings & Woessner, eds.), p. 102, Academic Press, San Diego. **65.** Kisiel & Canfield (1981) *Meth. Enzymol.* **80**, 275. **66.** Egelrud (1998) in *Handb. Proteolytic Enzymes* (Barrett, Rawlings & Woessner, eds.), p. 87, Academic Press, San Diego. **67.** Gaucher & Stevenson (1976) *Meth. Enzymol.* **45**, 415. **68.** Chao (1998) in *Handb. Proteolytic Enzymes* (Barrett, Rawlings & Woessner, eds.), p. 115, Academic Press, San Diego. **69.** Uitto (1998) in *Handb. Proteolytic Enzymes* (Barrett, Rawlings & Woessner, eds.), p. 313, Academic Press, San Diego. **70.** Halfon & Craik (1998) in *Handb. Proteolytic Enzymes* (Barrett, Rawlings & Woessner, eds.), p. 12, Academic Press, San Diego. **71.** Kassell (1970) *Meth. Enzymol.* **19**, 840. **72.** Burck (1970) *Meth. Enzymol.* **19**, 906. **73.** Menegatti, Boggian, Ascenzi & Luisi (1987) *J. Enzyme Inhib.* **2**, 67. **74.** Steinbuch (1976) *Meth. Enzymol.* **45**, 760. **75.** Desnuelle (1960) *The Enzymes*, 2nd ed. (Boyer, Lardy & Myrbäck, eds.), **4**, 93. **76.** Keil (1971)

The Enzymes, 3rd ed. (Boyer, ed.), **3**, 249. **77**. Wu & Scheraga (1962) *Biochemistry* **1**, 698. **78**. Kassell & Chow (1966) *Biochemistry* **5**, 3449. **79**. Horwitt (1944) *J. Biol. Chem.* **156**, 427. **80**. Laskowski, Jr., & Laskowski (1951) *J. Biol. Chem.* **190**, 563. **81**. Banerji & Sohonie (1969) *Enzymologia* **36**, 137. **82**. Hixson, Jr., & Laskowski, Jr. (1970) *Biochemistry* **9**, 166. **83**. Fan & Katagiri (2001) *Eur. J. Biochem.* **268**, 4892. **84**. D'Aniello, Denuce, de Vincentiis, di Fiore & Scippa (1997) *Biochim. Biophys. Acta* **1339**, 101. **85**. Fodor, Ako & Walsh (1975) *Biochemistry* **14**, 4923. **86**. McGrath, Abola, Toledo, Brown & Mangel (1996) *Virology* **217**, 131. **87**. Siffert, Emöd & Keil (1976) *FEBS Lett.* **66**, 114. **88**. Hardy & Hedrick (1992) *Biochemistry* **31**, 4466. **89**. Franzke, Baici, Bartels, Christophers & Wiedow (1996) *J. Biol. Chem.* **271**, 21886. **90**. El Moujahed, Gutman, Brillard & Gauthier (1990) *FEBS Lett.* **265**, 137. **91**. Sawada, Yokosawa & Ishii (1984) *J. Biol. Chem.* **259**, 2900. **92**. Christensen, Ishida, Ishii, *et al.* (1985) *J. Biochem.* **98**, 1263. **93**. Poe, Blake, Boulton, *et al.* (1991) *J. Biol. Chem.* **266**, 98. **94**. Watt, Lee, Timkulu, Chan & Loor (1986) *Proc. Natl. Acad. Sci. U.S.A.* **83**, 3166. **95**. Hung & Chiou (1994) *Biochem. Biophys. Res. Commun.* **201**, 1414. **96**. Chao (1983) *J. Biol. Chem.* **258**, 4434. **97**. Szilagyi, Sarfati, Pradayrol & Morisset (1995) *Biochim. Biophys. Acta* **1251**, 55. **98**. Durham (1993) *Biochem. Biophys. Res. Commun.* **194**, 1365. **99**. Bessmertnaya, Loiko, Goncharova, *et al.* (1997) *Biochemistry (Moscow)* **62**, 850. **100**. Walker, Owen & Esmon (1980) *Biochemistry* **19**, 1020. **101**. Fiorucci, Erba, Falasca, Dini & Ascoli (1995) *Biochim. Biophys. Acta* **1243**, 407. **102**. Fiorucci, Erba, Coletta & Ascoli (1995) *FEBS Lett.* **363**, 81. **103**. Kominami, Hoffschulte & Holzer (1981) *Biochim. Biophys. Acta* **661**, 124. **104**. Fujishiro, Sanada, Tanaka & Katunuma (1980) *J. Biochem.* **87**, 1321. **105**. Tsiftsoglou & Sim (2004) *J. Immunol.* **173**, 367. **106**. Davie, Fujikawa, Kurachi & Kisiel (1979) *Adv. Enzymol.* **48**, 277. **107**. Starkey & Barrett (1976) *Biochem. J.* **155**, 265. **108**. Ardelt (1974) *Biochim. Biophys. Acta* **341**, 318. **109**. Gildberg & Overbo (1990) *Comp. Biochem. Physiol. B* **97**, 775. **110**. Xiong, Chen, Woodley-Miller, Simson & Chao (1990) *Biochem. J.* **267**, 639. **111**. Ohnishi, Ikekita, Atomi, *et al.* (1992) *Protein Seq. Data Anal.* **5**, 1. **112**. Paquin, Benjannet, Sawyer, *et al.* (1989) *Biochim. Biophys. Acta* **999**, 103. **113**. Nagase & Barrett (1981) *Biochem. J.* **193**, 187. **114**. Sampaio, Wong & Shaw (1974) *Arch. Biochem. Biophys.* **165**, 133. **115**. Roy, Colas & Durand (1996) *Comp. Biochem. Physiol. B* **115**, 87. **116**. Sakharov & Litvin (1994) *Comp. Biochem. Physiol. B* **108**, 561. **117**. Kim, Park, Kim & Shahidi (2002) *J. Biochem. Mol. Biol.* **35**, 165. **118**. Kristjansson, Gudmundsdottir, Fox & Bjarnason (1995) *Comp. Biochem. Physiol. B* **110**, 707. **119**. Grant & Eisen (1980) *Biochemistry* **19**, 6089. **120**. Eisen, Henderson, Jeffrey & Bradshaw (1973) *Biochemistry* **12**, 1814. **121**. Yoshimoto, Simmons, Kita & Tsuru (1981) *J. Biochem.* **90**, 325. **122**. Neuenschwander & Morrissey (1995) *Biochemistry* **34**, 8701. **123**. Maison, Villiers & Colomb (1991) *J. Immunol.* **147**, 921. **124**. Virca, Metz & Schnebli (1984) *Eur. J. Biochem.* **144**, 1. **125**. Berlov, Lodygin, Andreeva & Kokryakov (2001) *Biochemistry (Moscow)* **66**, 1008. **126**. Hardy, Schoots & Hedrick (1989) *Biochem. J.* **257**, 447. **127**. Polakoski & McRorie (1973) *J. Biol. Chem.* **248**, 8183. **128**. Anderson, Jr., Beyler, Mack & Zaneveld (1981) *Biochem. J.* **199**, 307. **129**. Kobayashi, Matsuda, Oshio, *et al.* (1991) *Arch. Androl.* **27**, 9. **130**. Richardson, Korn, Bodine & Thurston (1992) *Poult. Sci.* **71**, 1789. **131**. Brown, Andani & Hartree (1975) *Biochem. J.* **149**, 133. **132**. Takano, Yanagimachi & Urch (1993) *Zygote* **1**, 79. **133**. Gilboa, Elkana & Rigbi (1973) *Eur. J. Biochem.* **39**, 85. **134**. Polakoski, McRorie & Williams (1973) *J. Biol. Chem.* **248**, 8178. **135**. Schiessler, Fritz, Arnold, Fink & Tschesche (1972) *Hoppe-Seyler's Z. Physiol. Chem.* **353**, 1638. **136**. Maroux, Baratti & Desnuelle (1971) *J. Biol. Chem.* **246**, 5031. **137**. Light & Fonseca (1984) *J. Biol. Chem.* **259**, 13195. **138**. Light & Janska (1989) *Trends Biochem. Sci.* **14**, 110. **139**. Liepnieks & Light (1979) *J. Biol. Chem.* **254**, 1677. **140**. Fonseca & Light (1983) *J. Biol. Chem.* **258**, 14516. **141**. Gasparian, Ostapchenko, Schulga, Dolgikh & Kirpichnikov (2003) *Protein Expr. Purif.* **31**, 133. **142**. Chan, Chan, Liu, *et al.* (2002) *J. Immunol.* **168**, 5170. **143**. Lorentsen, Moller, Etzerodt, Thogersen & Holtet (2003) *Org. Biomol. Chem.* **1**, 1657. **144**. De Cristofaro, Akhavan, Altomare, *et al.* (2004) *J. Biol. Chem.* **279**, 13035. **145**. Hijikata-Okunomiya, Tamao, Kikumoto & Okamoto (2000) *J. Biol. Chem.* **275**, 18995. **146**. Jeohn, Serizawa, Iwamatsu & Takahashi (1995) *J. Biol. Chem.* **270**, 14748. **147**. Ahsan & Watabe (2001) *J. Protein Chem.* **20**, 49. **148**. Volpicella, Ceci, Cordewener, *et al.* (2003) *Eur. J. Biochem.* **270**, 10. **149**. Asgeirsson, Fox & Bjarnason (1989) *Eur. J. Biochem.* **180**, 85. **150**. Bode (1979) *Naturwissenschaften* **66**, 251. **151**. Scheidig, Hynes, Pelletier, Wells & Kossiakoff (1997) *Protein Sci.* **6**, 1806. **152**. Al-Ajlan & Bailey (1997) *Arch. Biochem. Biophys.* **348**, 363. **153**. Al-Ajlan & Bailey (2000) *Mol. Cell. Biochem.* **203**, 73. **154**. Ryan, Clary & Tomimatsu (1965) *Arch. Biochem. Biophys.* **110**, 175. **155**. Simmons & Walter (1980) *Biochemistry* **19**, 39. **156**. Shariat-Madar, Mahdi & Schmaier (2002) *J. Biol. Chem.* **277**, 17962. **157**. Moreira, Schmaier, Mahdi, *et al.* (2002) *FEBS Lett.* **523**, 167.

158. Henrich, Becker, Schroeder & Plapp (1993) *J. Bacteriol.* **175**, 7290. **159**. Lung & Weselake (2006) *Lipids* **41**, 1073.

Trypstatin

This mast cell protease inhibitor (CAS 97502-03-9), first purified from rat peritoneal mast cells, is a proteolytic fragment of the inter-α-trypsin inhibitor light chain, *or* bikunin (1). **Target(s):** chymase, partial inhibition (2); coagulation factor Xa (3); and tryptase (2-4). **1**. Itoh, Ide, Ishikawa & Nawa (1994) *J. Biol. Chem.* **269**, 3818. **2**. Kido, Fukusen & Katunuma (1985) *Arch. Biochem. Biophys.* **239**, 436. **3**. Kido, Yokogoshi & Katunuma (1988) *J. Biol. Chem.* **263**, 18104. **4**. Johnson (1998) in *Handb. Proteolytic Enzymes* (Barrett, Rawlings & Woessner, eds.), p. 70, Academic Press, San Diego.

Tryptamine

This biogenic amine ($FW_{hydrochloride}$ = 196.68 g/mol; CAS 61-54-1; pK_a = 10.2 at 25°C; (λ_{max} at 282 and 290 nm; ε = 6000 and 5100 $M^{-1}cm^{-1}$ in 95% ethanol, respectively), known systematically as 2-(3-indolyl)ethylamine, is formed by the pyridoxal-phosphate-dependent decarboxylation of tryptophan. While there is, as yet, no compelling evidence to support persistent speculation that tryptamine is itself a neurotransmitter, it is nonetheless the direct precursor of serotonin. Moreover, tryptamine is a weak alternative substrate for the serotonin transporter. Weak transport substrates can, in principle, have higher K_m value (suggesting a lower affinity than the preferred transport substrate) and/or lower k_{cat} values (indicating a longer than normal transport cycle). If the k_{cat} is lower for tryptamine transport, it may play a neuromodulatory role by reducing the efficiency of serotonin transport. *Note*: The term "tryptamine" also refers to a broad class of classical or serotonergic hallucinogens (*e.g.*, psilocybin, LSD, and recent designer hallucinogens) that, in humans, produce profound changes in sensory perception, mood and thought, primarily by acting as 5-HT$_{2A}$ receptor agonists. This term is based on the presence of a tryptamine nucleus in their structures, although some synthetic tryptamines do not. **Target(s):** acetylcholinesterase (1); agmatine deaminase (2-4); anthranilate synthase (5); aryl-acylamidase, weakly inhibited (6-8); chymotrypsin (9-11); CYP1A2 (12); CYP2A6 (12); CYP2E1 (13); deacetylvindoline *O*-acetyltransferase (14); glutamate dehydrogenase (15); γ-glutamylhistamine synthetase (16); histamine *N*-methyltransferase (17); kynureninase, weakly inhibited (18); phenylalanine decarboxylase (19); pyridoxal kinase, perhaps combining with pyridoxal (20,21); 3-α-(*S*)-strictosidine β-glucosidase, weakly inhibited (22); tryptophanase, weakly inhibited (9,23); tryptophan 2,3-dioxygenase (9); tryptophanyl-tRNA synthetase (24-29); and tyramine *N*-feruloyltransferase, also alternative substrate (30). **1**. Krupka (1965) *Biochemistry* **4**, 429. **2**. Chaudhuri & Ghosh (1985) *Phytochemistry* **24**, 2433. **3**. Yanagisawa & Suzuki (1981) *Plant Physiol.* **67**, 697. **4**. Sindhu & Desai (1979) *Phytochemistry* **18**, 1937. **5**. Poulsen, Bongaerts & Verpoorte (1993) *Eur. J. Biochem.* **212**, 431. **6**. Hsu, Paul, Halaris & Freedman (1977) *Life Sci.* **20**, 857. **7**. Oommen & Balasubramanian (1979) *Eur. J. Biochem.* **94**, 135. **8**. Fujimoto (1974) *Biochem. Biophys. Res. Commun.* **61**, 72. **9**. Webb (1966) *Enzyme and Metabolic Inhibitors*, vol. **2**, Academic Press, New York. **10**. Foster (1961) *J. Biol. Chem.* **236**, 2461. **11**. Huang & Niemann (1952) *J. Amer. Chem. Soc.* **74**, 101. **12**. Zhang, Kilicarslan, Tyndale & Sellers (2001) *Drug Metab. Dispos.* **29**, 897. **13**. Albano, Tomasi, Persson, *et al.* (1991) *Biochem. Pharmacol.* **41**, 1895. **14**. Power, Kurz & de Luca (1990) *Arch. Biochem. Biophys.* **279**, 370. **15**. Smith, Austen, Blumenthal & Nyc (1975) *The Enzymes*, 3rd ed. (Boyer, ed.), **11**, 293. **16**. Stein & Weinreich (1982) *J. Neurochem.* **38**, 204. **17**. Fuhr & Kownatzki (1986) *Pharmacology* **32**, 114. **18**. Jakoby & Bonner (1953) *J. Biol. Chem.* **205**, 709. **19**. Nakazawa, Sano, Kumagai & Yamada (1977) *Agric. Biol. Chem.* **41**, 2241. **20**. Takeuchi, Tsubouchi & Shibata (1985) *Biochem. J.* **227**, 537. **21**. Gäng & von Collins (1973) *Int. J. Vitam. Nutr. Res.* **43**, 318. **22**. Hemscheid & Zenk (1992) *FEBS Lett.* **110**, 187. **23**. Gooder & Happold (1954) *Biochem. J.* **57**, 369. **24**. Davie (1962) *Meth. Enzymol.* **5**, 718. **25**. Kisselev, Favorova & Kolaleva (1979) *Meth. Enzymol.* **59**, 174. **26**. Lowe & Tansley (1984) *Eur. J. Biochem.* **138**, 597. **27**. Stulberg & Novelli (1962) *The Enzymes*, 2nd ed. (Boyer, Lardy & Myrbäck, eds.), **6**, 401. **28**. Sharon & Lipmann (1957) *Arch. Biochem. Biophys.* **69**, 219. **29**. Paley (1999) *Cancer Lett.* **137**, 1. **30**. Ishihara, Kawata, Matsukawa & Iwamura (2000) *Biosci. Biotechnol. Biochem.* **64**, 1025.

Tryptamine-4,5-dione

This neurotoxin (FW$_{free-base}$ = 191.21 g/mol), formed via the oxidation of 5-hydroxytryptamine by reactive oxygen and reactive nitrogen species, is a strong electrophile that can covalently modify cysteinyl residues. Tryptamine-4,5-dione is formed by oxidation of 5-hydroxytryptamine by reactive oxygen and reactive nitrogen species. It is also a powerful electrophile that can covalently modify cysteinyl residues of proteins, perhaps deactivating key enzymes that play roles in the degeneration of serotonergic neurons in brain disorders, such as Alzheimer's disease or evoked by amphetamine drugs. Tryptamine-4,5-dione reacts with free cycteine, glutathione, N-acetylcysteine, and cysteamine. **Target(s):** α-ketoglutarate dehydrogenase (1); pyruvate dehydrogenase (1); and tryptophan hydroxylase (2). 1. Jiang & Dryhurst (2002) *Chem. Res. Toxicol.* **15**, 1242. 2. Wrona & Dryhurst (2001) *Chem. Res. Toxicol.* **14**, 1184.

D-Tryptophan

This nonproteogenic amino acid (FW = 204.23 g/mol), known systematically as (R)-2-Amino-3-(1H-indol-3-yl)propanoic acid, is found in the antibiotics tyrocidin C and tyrocidin D. (**See** L-Tryptophan) **Target(s):** anthranilate synthase (1); aromatic-L-amino-acid decarboxylase, *or* dopa decarboxylase (2,3); carboxypeptidase A (4,5); prephenate dehydratase (6); tryptophanase (7); tryptophan 2,3-dioxygenase (8); tryptophan synthase (9); tryptophanyl-tRNA synthetase (10-12); tyrosine phenol-lyase (13). 1. Poulsen, Bongaerts & Verpoorte (1993) *Eur. J. Biochem.* **212**, 431. 2. Jung (1986) *Bioorg. Chem.* **14**, 429. 3. Facchini, Huber-Allanach & Tari (2000) *Phytochemistry* **54**, 121. 4. Coleman & Vallee (1964) *Biochemistry* **3**, 1874. 5. Bradley, Naudé, Muramoto, Yamauchi & Oelofsen (1994) *Int. J. Biochem.* **26**, 555. 6. Bode, Melo & Birnbaum (1985) *J. Basic Microbiol.* **25**, 291. 7. Gunsalus, Galeener & Stamer (1955) *Meth. Enzymol.* **2**, 238. 8. Webb (1966) *Enzyme and Metabolic Inhibitors*, vol. 2, p. 325, Academic Press, New York. 9. Miles, Bauerle & Ahmed (1987) *Meth. Enzymol.* **142**, 398. 10. Kisselev, Favorova & Kolaleva (1979) *Meth. Enzymol.* **59**, 174. 11. Stulberg & Novelli (1962) *The Enzymes*, 2nd ed. (Boyer, Lardy & Myrbäck, eds.), **6**, 401. 12. Sharon & Lipmann (1957) *Arch. Biochem. Biophys.* **69**, 219. 13. Faleev, Ruvinov, Demidkina, *et al.* (1988) *Eur. J. Biochem.* **177**, 395.

L-Tryptophan

This naturally occurring aromatic amino acid (Symbol = Trp = W; FW = 204.23 g/mol; pK_a values of 2.46 and 9.41 at 25°C; UV spectrum: λ_{max} = 218 nm (ε = 33500 M^{-1}cm^{-1}), 278 nm (ε = 5550 M^{-1}cm^{-1}), and 287.5 nm (ε = 4550 M^{-1}cm^{-1}) in 0.1 M HCl; Specific Rotation = +2.8° at 25°C for the D line for 1.0-2.0 g/100 mL tryptophan in 1 M HCl and –33.7°; solubility = 1.14 g/100 mL water at 25°C), known systematically as (S)-2-amino-3-(1H-indol-3-yl)propanoic acid, is the least abundant of the twenty proteogenic amino acids and is a nutritionally essential amino acid in mammals. Most proteins have very few tryptophanyl residues, making tryptophan fluorescence a powerful probe of protein conformational changes. Many enzymes also use D- and L-tryptophan as weak alternative substrates (and thus competitive inhibitors for the natural substrate). These include L-amino-acid dehydrogenase, amino acid oxidases, peptide-tryptophan 2,3-dioxygenase, aspartate aminotransferase, ornithine:oxo-acid aminotransferase, γ-glutamyl transpeptidase (L-tryptophan is a weak γ-glutamyl acceptor substrate), aromatic amino acid transferase, and amino-acid racemase. **Target(s):** acetylcholinesterase (1); alkaline phosphatase (2,61); 2-aminohexano-6-

lactam racemase, weakly inhibited (36); aminopeptidase B, weakly inhibited (58); anthranilate phosphoribosyltransferase (3,14,86-89); anthranilate synthase (3-10,44-53); anthranilate synthase/anthranilate phosphoribosyl-transferase complex (4,87,88); arginase (11,55); argininosuccinate synthetase (12); arylformamidase (56); β-cyclopiazonate dehydrogenase (54); cystathionine γ-lyase (13); cytosol alanyl aminopeptidase (57); 3'-demethylstaurosporine O-methyltransferase, activated at low concentrations (92); 3-deoxy-7-phosphoheptulonate synthase, *or* phospho-2-dehydro-3-deoxyheptonate aldolase (25,72-85); diphosphomevalonate decarboxylase (63); L-dopachrome isomerase (14); glutamate decarboxylase (15); glutamine synthetase, weakly inhibited (16-20,35); [glutamine-synthetase] adenylyltransferase (62); γ-glutamyl transpeptidase, weakly inhibited (21,22); hemoglobin S polymerization (23); histidine decarboxylase (9); hydroxyacylglutathione hydrolase, *or* glyoxalase II (37); kynureninase, weakly inhibited (24); kynurenine:oxoglutarate aminotransferase (67-71); lactoylglutathione lyase, *or* glyoxalase I (37); membrane alanyl aminopeptidase, *or* aminopeptidase N (59,60); phenylalanine 4-monooxygenase (93); prephenate dehydratase (26,40-42); prephenate dehydrogenase (27); propionyl-CoA carboxylase, *Myxococcus xanthus* (28); pyridoxal 5'-phosphate hydrolase, *or* alkaline phosphatase (29); pyrophosphatase, *or* inorganic diphosphatase (30); pyruvate kinase (64-66); L-serine ammonia-lyase, *or* L-serine dehydratase (38,39); shikimate kinase (31); starch phosphorylase, inhibited by DL-tryptophan (90,91); tyrosine 3-monooxygenase (32,33); and tyrosine phenol-lyase (34,43). 1. Bergmann, Wilson & Nachmansohn (1950) *J. Biol. Chem.* **186**, 693. 2. Fishman & Sie (1971) *Enzymologia* **41**, 141. 3. Egan & Gibson (1970) *Meth. Enzymol.* **17A**, 380. 4. Bauerle, Hess & French (1987) *Meth. Enzymol.* **142**, 366. 5. Zalkin (1985) *Meth. Enzymol.* **113**, 287. 6. Tamir & Srinivasan (1970) *Meth. Enzymol.* **17A**, 401. 7. Tang, Ezaki, Atomi & Imanaka (2001) *Biochem. Biophys. Res. Commun.* **281**, 858. 8. Malik (1980) *Trends Biochem. Sci.* **5**, 68. 9. Webb (1966) *Enzyme and Metabolic Inhibitors*, vol. 2, Academic Press, New York. 10. Zalkin & Kling (1968) *Biochemistry* **7**, 3566. 11. Hunter & Downs (1945) *J. Biol. Chem.* **157**, 427. 12. Takada, Saheki, Igarashi & Katsunuma (1979) *J. Biochem.* **85**, 1309. 13. Metaxas & Delwiche (1955) *J. Bacteriol.* **69**, 735. 14. Aroca, Garcia-Borron, Solano & Lozano (1990) *Biochim. Biophys. Acta* **1035**, 266. 15. Roberts & Frankel (1951) *J. Biol. Chem.* **190**, 505. 16. Rhee, Chock & Stadtman (1985) *Meth. Enzymol.* **113**, 213. 17. Shapiro & Stadtman (1970) *Meth. Enzymol.* **17A**, 910. 18. Stadtman & Ginsburg (1974) *The Enzymes*, 3rd ed. (Boyer, ed.), **10**, 755. 19. Dahlquist & Purich (1975) *Biochemistry* **14**, 1980. 20. Kumar & Nicholas (1984) *J. Gen. Microbiol.* **130**, 959. 21. Allison (1985) *Meth. Enzymol.* **113**, 419. 22. Thompson & Meister (1977) *J. Biol. Chem.* **252**, 6792. 23. Noguchi & Schechter (1978) *Biochemistry* **17**, 5455. 24. Jakoby & Bonner (1953) *J. Biol. Chem.* **205**, 709. 25. Gaertner & DeMoss (1970) *Meth. Enzymol.* **17A**, 387. 26. Fischer & Jensen (1987) *Meth. Enzymol.* **142**, 507. 27. Fischer & Jensen (1987) *Meth. Enzymol.* **142**, 503. 28. Kimura, Kojyo, Kimura & Sato (1998) *Arch. Microbiol.* **170**, 179. 29. Lumeng & Li (1979) *Meth. Enzymol.* **62**, 574. 30. Naganna (1950) *J. Biol. Chem.* **183**, 693. 31. De Feyter (1987) *Meth. Enzymol.* **142**, 355. 32. Nagatsu, Levitt & Udenfriend (1964) *J. Biol. Chem.* **239**, 2910. 33. McGeer, McGeer & Peters (1967) *Life Sci.* **6**, 2221. 34. Faleev, Ruvinov, Demidkina, *et al.* (1988) *Eur. J. Biochem.* **177**, 395. 35. Ross & Ginsburg (1969) *Biochemistry* **8**, 4690. 36. Ahmed, Esaki, Tanaka & Soda (1985) *Agric. Biol. Chem.* **49**, 2991. 37. Oray & Norton (1980) *Biochem. Biophys. Res. Commun.* **95**, 624. 38. Morikawa, Nakamura & Kimura (1974) *Agric. Biol. Chem.* **38**, 531. 39. Kubota, Yokozeki & Ozaki (1989) *J. Ferment. Bioeng.* **67**, 391. 40. Hagino & Nakayama (1974) *Agric. Biol. Chem.* **38**, 2367. 41. Friedrich, Friedrich & Schlegel (1976) *J. Bacteriol.* **126**, 723. 42. Jensen, D'Amato & Hochstein (1988) *Arch. Microbiol.* **148**, 365. 43. Demidkina, Barbolina, Faleev, *et al.* (2002) *Biochem. J.* **363**, 745. 44. Matsukawa, Ishihara & Iwamura (2002) *Z. Naturforsch. [C]* **57**, 121. 45. Zalkin & Hwang (1971) *J. Biol. Chem.* **246**, 6899. 46. Francis, Vining & Westlake (1971) *J. Bacteriol.* **134**, 10. 47. Schmauder & Gröger (1976) *Biochem. Physiol. Pflanz.* **169**, 471. 48. Hertel, Hieke & Gröger (1991) *Biochem. Physiol. Pflanz.* **187**, 121. 49. Widholm (1972) *Biochim. Biophys. Acta* **279**, 48. 50. Poulsen, Bongaerts & Verpoorte (1993) *Eur. J. Biochem.* **212**, 431. 51. Baker & Crawford (1966) *J. Biol. Chem.* **241**, 5577. 52. Caligiuri & Bauerle (1991) *Science* **252**, 1845. 53. Robb, Hutchinson & Belser (1971) *J. Biol. Chem.* **246**, 6908. 54. Steenkamp, Schabort & Ferreira (1973) *Biochim. Biophys. Acta* **309**, 440. 55. Fujimoto, Kameji, Kanaya & Hagihira (1976) *J. Biochem.* **79**, 441. 56. Serrano & Nagayama (1991) *Comp. Biochem. Physiol. B Comp. Biochem.* **99**, 281. 57. Garner & Behal (1977) *Arch. Biochem. Biophys.* **182**, 667. 58. Kawata, Takayama, Ninomiya & Makisumi (1980) *J. Biochem.* **88**, 1601. 59. McCaman & M. Villarejo (1982) *Arch. Biochem. Biophys.* **213**, 384. 60.

Tokioka-Terao, Hiwada & Kokubu (1984) *Enzyme* **32**, 65. **61**. Le Du, Stigbrand, Taussig, Menez & Stura (2001) *J. Biol. Chem.* **276**, 9158. **62**. Ebner, Wolf, Gancedo, Elsässer & Holzer (1970) *Eur. J. Biochem.* **14**, 535. **63**. Shama Bhat & Ramasarma (1979) *Biochem. J.* **181**, 143. **64**. Guderley & Hochachka (1980) *J. Exp. Zool.* **212**, 461. **65**. Baysdorfer & Bassham (1984) *Plant Physiol.* **74**, 374. **66**. Feksa, Cornelio, Dutra-Filho, *et al.* (2005) *Int. J. Dev. Neurosci.* **23**, 509. **67**. Guidetti, Okuno & Schwarcz (1997) *J. Neurosci. Res.* **50**, 457. **68**. Wejksza, Rzeski, Okuno, *et al.* (2005) *Neurochem. Res.* **30**, 963. **69**. Milart, Urbanska, Turski, Paszkowski & Sikorski (2001) *Placenta* **22**, 259. **70**. Zarnowski, Rejdak, Zagorski, *et al.* (2004) *Ophthalmic Res.* **36**, 124. **71**. Baran, Amann, Lubec & Lubec (1997) *Pediatr. Res.* **41**, 404. **72**. Görisch & Lingens (1971) *Biochim. Biophys. Acta* **242**, 630. **73**. Rubin & Jensen (1985) *Plant Physiol.* **79**, 711. **74**. Nimmo & Coggins (1981) *Biochem. J.* **199**, 657. **75**. Wu, Sheflyan & Woodard (2005) *Biochem. J.* **390**, 583. **76**. Whitaker, Fiske & Jensen (1982) *J. Biol. Chem.* **257**, 12789. **77**. Gosset, Bonner & Jensen (2001) *J. Bacteriol.* **183**, 4061. **78**. Ip & Doy (1979) *Eur. J. Biochem.* **98**, 431. **79**. Tianhui & Chiao (1989) *Biochim. Biophys. Acta* **991**, 1. **80**. Jain & Bhalla-Sarin (2000) *Indian J. Biochem. Biophys.* **37**, 235. **81**. Walker, Dunbar, Hunter, Nimmo & Coggins (1996) *Microbiology* **142**, 1973. **82**. Akowski & Bauerle (1997) *Biochemistry* **36**, 15817. **83**. Euverink, Hessels, Franke & Dijkhuizen (1995) *Appl. Environ. Microbiol.* **61**, 3796. **84**. Nimmo & Coggins (1981) *Biochem. J.* **197**, 427. **85**. Stuart & Hunter (1993) *Biochim. Biophys. Acta* **1161**, 209. **86**. O'Gara & Dunican (1995) *Appl. Environ. Microbiol.* **61**, 4477. **87**. Egan & Gibson (1972) *Biochem. J.* **130**, 847. **88**. Ito & Yanofsky (1969) *J. Bacteriol.* **97**, 734. **89**. Grieshaber (1978) *Z. Naturforsch. C* **33c**, 235. **90**. Kumar & Sanwal (1982) *Biochemistry* **21**, 4152. **91**. Singh & Sanwal (1976) *Phytochemistry* **15**, 1447. **92**. Weidner, Kittelmann, Goeke, Ghisalba & Zähner (1998) *J. Antibiot.* **51**, 697. **93**. Ledley, Grenett & Woo (1987) *J. Biol. Chem.* **262**, 2228.

Tryptophol

This naturally occurring plant auxin (FW = 161.20 g/mol), also known as indole-3-ethanol and 2-(3-indolyl)ethanol, is found in a wide variety of microorganisms, sponges (*Ircinia spinulosa*), plants and mammals. It is an effector of indoleamine 2,3-dioxygenase, which catalyzes the degradation of L-tryptophan to form *N*-formylkynurenine, with superoxide anion serving as an oxygen donor. Indoleamine 2,3-dioxygenase is the first and rate-limiting enzyme of tryptophan catabolism. In *Saccharomyces cerevisiae*, tryptophol is a quorum-sensing molecule that allows the yeast cells to synchronously sense changes in cell density and other growth cues and to adapt collectively to new conditions through pre-programmed physiological and morphological responses. *Note*: Tryptophol must not be confused with tryptophanol. **Target(s):** aralkylamine *N*-acetyltransferase, *or* serotonin *N*-acetyltransferase (2); chymotrypsin (3); and Na⁺/K⁺-exchanging ATPase (4). **1**. Sono (1989) *Biochemistry* **28**, 5400. **2**. De Angelis, Gastel, Klein & Cole (1998) *J. Biol. Chem.* **273**, 3045. **3**. Valenzuela & Bender (1970) *Biochemistry* **9**, 2440. **4**. Stepp & Novakoski (1997) *Arch. Biochem. Biophys.* **337**, 43.

Trytophol Phosphate, *See Indole-3-ethanol Phosphate*

TS-071, *See Luseogliflozin*

TS-2 and TS-4

These 1,3-selenazines TS-2 (FW = 266.25 g/mol), known systematically as 4-ethyl-4-hydroxy-2-*p*-tolyl-5,6-dihydro-4*H*-1,3-selenazine, and TS-4 (FW

= 293.29 g/mol), known systematically as 4-hydroxy-6-isopropyl-4-methyl-2-*p*-tolyl-5,6-dihydro-4*H*-1,3-selenazine, inhibit [elongation factor 2] kinase, with IC₅₀ values of 0.36 and 0.31 μM (1,2). Both are much weaker inhibitors of protein kinase A, protein kinase C, protein-tyrosine kinase, calmodulin-dependent protein kinase, and v-*src* kinase. **1**. Gutzkow, Låhne, Naderi, *et al.* (2003) *Cell. Signal.* **15**, 871. **2**. Cho, Koketsu, Ishihara, *et al.* (2000) *Biochim. Biophys. Acta* **1475**, 20.

Tsushimycin

This acidic acylpeptide antibiotic (FW = 1302.49 g/mol; CAS 11054-63-0; Structural Components: Itd = Δ³-isotetradecenoic acid; Dab = 2,3-diaminobutyric acid; Pip = pipecolinic acid; Map, β-methylaspartate.) is structurally related to amphomycin. **Target(s):** dolichyl-phosphate β-D-mannosyltransferase (1-4); phospho-*N*-acetylmuramoyl-pentapeptide-transferase (5); and protein glycosylation (1-4). **1**. Elbein (1983) *Meth. Enzymol.* **98**, 135. **2**. Elbein (1987) *Meth. Enzymol.* **138**, 661. **3**. Elbein (1984) *CRC Crit. Rev. Biochem.* **16**, 21. **4**. Elbein (1981) *Biochem. J.* **193**, 477. **5**. Tanaka, Oiwa, Matsukura, Inokoshi & Omura (1982) *J. Antibiot.* **35**, 1216.

TTP 22

This high-affinity, ATP-competitive casein kinase inhibitor (FW = 330.42 g/mol; CAS 329907-28-0; Solubility: 100 mM in DMSO; 20 mM in Ethanol), also named 3-[[5-(4-methylphenyl)thieno[2,3-*d*]pyrimidin-4-yl]thio]-propanoic acid, targets casein kinase 2 (CK2) (IC₅₀ = 0.1 μM, Kᵢ = 40 nM), displays no inhibitory effects, even at 10 μM, JNK3, ROCK1 and MET. **1**. Golub, Bdzhola, Briukhovetska, *et al.* (2011) *Eur. J. Med. Chem.* **46**, 870.

TTT-3002

This novel protein tyrosine kinase inhibitor (FW_free-base = 465.51 g/mol) targets the FMS-like tyrosine kinase-3, *or* FLT3 (IC₅₀ = 100–250 pM), a receptor tyrosine kinase that is expressed in hematopoietic stem/progenitor cells. Upon binding the FLT3 ligand (*or* FL), the receptor dimerizes, thereby activating its kinase domain and undergoing autophosphorylation as well as catalyzing the phosphorylation of target proteins in the STAT5, PI3K/AKT, and RAS/MAPK pathways. Notably, >35% of acute myeloid leukemia (AML) patients harbor a constitutively activating mutation in FLT3. TTT-3002 shows potent cytostatic activity (IC$_{50}$ = 490–920 pM) against these FLT3-associated leukemias, both *in vitro* and *in vivo*. TTT-3002 also potently inhibits *in vitro* kinase activity of Leucine-Rich Repeat Kinase-2, *or* LRRK2, wild-type and mutant proteins, attenuated phosphorylation of cellular LRRK2 and rescued neurotoxicity of mutant LRRK2 in transfected cells. Mutations in LRRK2 are most frequently associated with late-onset Parkinson disease (PD). TTT-3002 is ineffective against the neurodegenerative phenotype in transgenic *Caenorhabditis elegans* carrying the inhibitor-resistant A2016T mutation of LRRK2, suggesting it elicits its neuroprotective effects *in vivo* by targeting LRRK2 specifically. **1**. Ma, Nguyen, Li, *et al.* (2014) *Blood* **123**, 1525. **2**. Yao, Johnson, Gao, *et al.* (2013) *Hum. Mol. Genet.* **22**, 328.

Tubacin

This selective tubulin deacetylase inhibitor (FW = 721.86 g/mol; CAS 1350555-93-9; Soluble to 10 mM in DMSO), also named *N*-[4-[(2*R*,4*R*,6*S*)-4-[[(4,5-diphenyl-2-oxazolyl)thio]methyl]-6-[4-(hydroxymethyl)phenyl]-1,3-dioxan-2-yl]phenyl]-*N'*-hydroxyoctanediamide, reversibly inhibits Histone Deacetylase-6, *or* HDAC6 (EC$_{50}$ = 2.5 μM), which catalyzes the deacetylation of α-tubulin, thereby increasing α-tubulin acetylation levels in cells (1). Tubacin does not alter histone acetylation, gene-expression patterns, or cell-cycle progression. In cultured cells, tubacin (10 μM) induces up to a 3x increase in the relative α-tubulin-acetylation level. Niltubacin, the carboxylate analogue of tubacin, is without effect on α-tubulin or histone acetylation and therefore may be used as a negative control. Requirement of the hydroxamic acid for inhibitory activity suggests that tubacin targets a metal-dependent hydrolase. Levels of total α-tubulin and α-tubulin containing a C-terminal tyrosine (which is decreased in stabilized microtubules) are unaffected by tubacin (1). Unlike taxol, which increases α-tubulin acetylation by directly stabilizing assembled microtubules, tubacin has no overall effect on cell morphology. Tubacin treatment does not affect the stability of microtubules but does decrease cell motility (1). Endothelial barrier dysfunction (EBD) involves both microtubule disassembly and enhanced cell contractility, and, when HDAC6 deacetylates α-tubulin, microtubules are destabilized. HDAC6 inhibition attenuates thrombin-induced microtubule disassembly, attended by an increase in Phosphorylated Myosin Light Chain 2 (P-MLC2) and contractility, reducing endothelial barrier dysfunction induced by thrombin (2). Tubacin completely blocks inflammatory tolerance, revealing the opposing effects between HDAC6 and Glycogen Synthase Kinase-3 (GSK3) in regulating tolerance and indicating how it may be possible to rebalance these opposing forces to obliterate or enhance tolerance to lipopolysaccharide (LPS) in astrocytes (3). Tubacin blocks deacetylation of α-tubulin (and not other HDAC6 substrates) reduces cell motility to an extent that is equivalent to the effects of agents that block all NAD-independent HDACs (4). This selective action of tubacin helped to show that microtubule acetylation influences the subcellular distribution of vesicles associated with the kinesin KIF1C, as well as their directionality, velocity and run length (5). Tubulin acetylation alters the targeting frequency of microtubule plus ends on podosomes and influences the number of podosomes per cell and thus the matrix-degrading capacity of macrophages (5). **1**. Haggarty, Koeller, Wong, Grozinger & Schreiber (2003) *Proc. Natl. Acad. Sci. U.S.A.* **100**, 4389. **2**. Saito, Lasky, Guo, *et al.* (2011) *Biochem. Biophys. Res. Commun.* **408**, 630. **3**. Beurel (2011) PLoS One **6**, e25804. **4**. Tran, Marmo, Salam, *et al.* (2007) *J. Cell Sci.* **120**, 1469. **5**. Bhuwania, Castro-Castro & Linder (2014) *Eur. J. Cell Biol.* **93**, 424.

Tubastatin A

This potent HDAC inhibitor (FW = 335.31 g/mol; CAS 1239262-52-2), named systematically *N*-hydroxy-4-[(1,2,3,4-tetrahydro-2-methyl-5*H*-pyrido[4,3-*b*]indol-5-yl)methyl]benzamide, targets histone deacetylase-6, *or* HDAC6 (IC$_{50}$ = 15 nM), demonstrating >1,000x selectivity versus all other HDAC isoforms (IC$_{50}$s >16 μM), whereas HDAC8 (IC$_{50}$ = 0.9 μM) (1). Tubastatin A also conferred dose-dependent protection in primary cortical neuron cultures against glutathione depletion-induced oxidative stress (1). When present at 2.5 μM, tubastatin A also induces hyperacetylation of α-tubulin in primary cortical neuron cultures, consistent with its HDAC6 selectivity (1). In a model of oxidative stress induced by glutathione depletion, 5-10 μM tubastatin A displays dose-dependent neuronal protection of primary cortical neuron cultures. **1**. Butler, Kalin, Brochier, *et al.* (2010) *J. Am. Chem. Soc.* **132**, 10842.

Tuberactinomycins

This group of cyclic peptide antibiotics from *Streptomyces griseoverticillatus* are potent inhibitors of prokaryotic protein biosynthesis. In tuberactinomycin A, R$_1$ = OH, and R$_2$ = OH; in tuberactinomycin B, R$_1$ = H, and R$_2$ = OH; in tuberactinomycin C, R$_1$ = OH, and R$_2$ = H; and in tuberactinomycin D, R$_1$ = H, and R$_2$ = H. Tuberactinomycin A (FW = 701.70 g/mol) is soluble in water and practically insoluble in most organic solvents. Tuberactinomycin B is also known as viomycin and the N antibiotic is enviomycin. **Target(s):** group I intron RNA splicing (1); hammerhead ribozyme, inhibited by tuberactinomycin A with positive cooperativity (2). **1**. Wank, Rogers, Davies & Schroeder (1994) *J. Mol. Biol.* **236**, 1001. **2**. Jenne, Hartig, Piganeau, *et al.* (2001) *Nat. Biotechnol.* **19**, 56.

Tubercidin

This cytotoxic adenosine analogue (FW = 266.26 g/mol; CAS 69-33-0; λ$_{max}$ of 270 nm, ε = 12100 M^{-1}cm^{-1}; readily soluble in acidic or alkaline solutions), also known as 7-deazaadenosine, from *Streptomyces tubercidicus* is an antifungal and antibacterial agent that interferes with purine metabolism and glycolysis. Tubercidin also inhibits leukocyte adhesion to TNF-treated endothelial cells. Tubercidin is often a weak alternative substrate for adenosine kinase and other adenosine-dependent enzymes. It inhibits protein biosynthesis and the growth of many microorganisms, including *Mycobacterium tuberculosis*. Nucleic acids containing tubercidin are biologically inactive. *See also* **5-Iodotubercidin** **Target(s):** adenosine kinase, also alternative substrate (1-6); adenosylhomocysteinase, *or* S-adenosylhomocysteine hydrolase (7-9); 2',3'-cyclic-nucleotide 3'-

phosphodiesterase (10); estrone sulfotransferase (11); GMP synthetase (12); purine Nucleosidase, *or* purine-specific nucleoside *N*-ribohydrolase (13-15); toyocamycin nitrile hydratase (16); and tRNA methyltransferases (17,18). **1.** Long & Parker (2006) *Biochem. Pharmacol.* **71**, 1671. **2.** Galazka, Striepen & Ullman (2006) *Mol. Biochem. Parasitol.* **149**, 223. **3.** Kidder (1982) *Biochem. Biophys. Res. Commun.* **107**, 381. **4.** Datta, Bhaumik & Chatterjee (1987) *J. Biol. Chem.* **262**, 5515. **5.** Henderson, Mikoshiba, Chu & Caldwell (1972) *J. Biol. Chem.* **247**, 1972. **6.** Drabikowska, Halec & Shugar (1985) *Z. Naturforsch. C* **40**, 34. **7.** Fabianowska-Majewska, Duley & Simmonds (1994) *Biochem. Pharmacol.* **48**, 897. **8.** Shimizu, Shiozaki, Ohshiro & Yamada (1984) *Eur. J. Biochem.* **141**, 385. **9.** Bozzi, Parisi & Martini (1993) *J. Enzyme Inhib.* **7**, 159. **10.** Sims & Carnegie (1978) *Adv. Neurochem.* **3**, 1. **11.** Horwitz, Misra, Rozhin, *et al.* (1978) *Biochim. Biophys. Acta* **525**, 364. **12.** Spector & Beecham III (1975) *J. Biol. Chem.* **250**, 3101. **13.** Parkin (1996) *J. Biol. Chem.* **271**, 21713. **14.** Versées, Decanniere, Pellé, *et al.* (2001) *J. Mol. Biol.* **307**, 1363. **15.** Parkin, Horenstein, Abdulah, Estupinan & Schramm (1991) *J. Biol. Chem.* **266**, 20658. **16.** Uematsu & Suhadolnik (1975) *Meth. Enzymol.* **43**, 759. **17.** Kerr (1974) *Meth. Enzymol.* **29**, 716. **18.** Kerr & Borek (1973) *The Enzymes*, 3rd ed. (Boyer, ed.), **9**, 167.

Tubercidin 5'-Monophosphate

This AMP analogue (FW$_{\text{free-acid}}$ = 346.24 g/mol), also known as 7-deazaadenosine 5'-monophosphate and 4-amino-7-(β-D-ribofuranosyl) pyrrolo[2,3-*d*]pyrimidine 5'-monophosphate, inhibits adenylate kinase (1), amidophosphoribosyltransferase (2), AMP Nucleosidase (3,4), and rhodopsin kinase (5), and toyocamycin nitrile hydratase (6). **1.** Munier-Lehmann, Burlacu-Miron, Craescu, Mantsch & Schultz (1999) *Proteins Struct. Funct. Genet.* **36**, 238. **2.** Hill & Bennett, Jr. (1969) *Biochemistry* **8**, 122. **3.** DeWolf, Jr., Fullin & Schramm (1979) *J. Biol. Chem.* **254**, 10868. **4.** Schramm (1976) *J. Biol. Chem.* **251**, 3417. **5.** Palczewski, Kahn & Hargrave (1990) *Biochemistry* **29**, 6276. **6.** Uematsu & Suhadolnik (1975) *Meth. Enzymol.* **43**, 759.

Tubercidin 5'-Triphosphate

This ATP analogue (FW$_{\text{free-acid}}$ = 506.20 g/mol), also known as 7-deazaadenosine 5'-triphosphate, inhibits lysyl- and threonyl-tRNA synthetases (*See Tubercidin*). 1. Freist, Sternbach, von der Haar & Cramer (1978) *Eur. J. Biochem.* **84**, 499.

S-Tubercidinyl-L-homocysteine

This *S*-adenosyl-L-homocysteine analogue (FW = 383.43 g/mol; CAS 57344-98-6), also known as *S*-(7-deazaadenosyl)-L-homocysteine, inhibits thiol methyltransferase (1) and is an inhibitor for many methyltransferases (2). It also inhibits 5'-methylthioadenosine/*S*-adenosylhomocysteine nucleosidase, K_i = 1.9 µM (3), catechol *O*-methyltransferase (2), histamine *N*-methyltransferase (2), hydroxyindole *O*-methyltransferase (2), indole-ethylamine *N*-methyl-transferase (2), 5'-methylthioadenosine/*S*-adenosyl-homocysteine nucleosidase (3), mRNA (guanine-N^7-)-methyltransferase

(4,5), mRNA (nucleoside-2'-*O*-)-methyltransferase (4), phenylethanolamine *N*-methyltransferase (2), and thiol *S*-methyltransferase (1). **1.** Borchardt & Cheng (1978) *Biochim. Biophys. Acta* **522**, 340. **2.** Borchardt, Huber & Wu (1976) *J. Med. Chem.* **19**, 1094. **3.** Della Ragione, Porcelli, Carteni-Farina, Zappia & Pegg (1985) *Biochem. J.* **232**, 335. **4.** Pugh & Borchardt (1982) *Biochemistry* **21**, 1535. **5.** Pugh, Borchardt & Stone (1977) *Biochemistry* **16**, 3928.

S-Tubercidinyl-L-methionine

This *S*-adenosyl-L-methionine analogue (FW$_{\text{chloride}}$ = 433.92 g/mol), also known as *S*-(7-deazaadenosyl)-L-methionine, inhibits spermidine synthase (1), spermine synthase (1), and thiol *S*-methyltransferase (2). **1.** Hibasami, Borchardt, Chen, Coward & Pegg (1980) *Biochem. J.* **187**, 419. **2.** Borchardt & Cheng (1978) *Biochim. Biophys. Acta* **522**, 340.

Tubocurarine

This *bis*-benzyl-isoquinoline curare alkaloid (FW = 681.66 g/mol; CAS 57-94-3; Symbol: dTC), also known as dextrotubocurarine chloride (*d*-tubocurarine chloride, *not* D) from *Chondodendron tomentosum*, is the principal paralyzing agent in arrow-tip poisons used by certain indigenous South Americans, as first reported by European explorers in the 16th century. That curare produces a neuromuscular block was later determined by the French physiologist Claude Bernard in the mid-19th century. Since then and extending into modern times, tubocurarine has helped to establish the role of acetylcholine in neuromuscular transmission. Its use as a muscle relaxant during surgical anaesthesia in World War II. **Classification:** Tubucurare is considered to be the prototypical neuromuscular-blocking drug, the potency of which is related to the separation distance between the two quaternary ammonium head-groups; LD$_{50}$ (dog) = 0.5 mg/kg. **Mode of Inhibitory Action:** Tubocurarine interacts with acetylcholine receptors and as a 5-HT$_3$ and GABA$_A$ receptor antagonist. Tubocurarine binds non-equivalently to the two agonist sites (K_d values of 30 nM and 8 µM) within the *Torpedo* nicotinic acetylcholine receptor (1). When nAChR-rich membranes equilibrated with [^{3}H]dTC are irradiated with 254-nm UV light, [^{3}H]dTC is covalently incorporated into the α-, γ-, and δ-subunits in a concentration-dependent and agonist-inhibitable manner, consistent with the localization of the high and low affinity dTC binding sites at the α-γ- and α-δ-subunit interfaces (1). amino acids contributing to high- and low-affinity d-tubocurarine sites have been identified (2). **Target(s):** acetylcholinesterase (3,4); acid phosphatase (5); choline acetyltransferase (6); cholinesterase (7); diamine oxidase (8); histamine *N*-methyltransferase, inhibited by tubocurare of which the primary active component is tubocurarine (9). **1.** Pedersen & Cohen (1990) *Proc. Natl. Acad. Sci. U.S.A.* **87**, 2785. **2.** Chiara & Cohen (1997) *J. Biol. Chem.* **272**, 32940. **3.** Zorko & Pavlic (1986) *Biochem. Pharmacol.* **35**, 2287. **4.** Bergman, Wilson & Nachmansohn (1950) *Biochim. Biophys. Acta* **6**, 217. **5.** Lisa, Garrido & Domenech (1984) *Mol. Cell. Biochem.* **63**, 113. **6.** Kambam, Janson, Day & Sastry (1990) *Can. J. Anaesth.* **37**, 690. **7.** Fitch (1963) *Biochemistry* **2**, 1221. **8.** Sattler, Hesterberg, Lorenz, *et al.* (1985) *Agents Actions* **16**, 91. **9.** Brown, Tomchick & Axelrod (1959) *J. Biol. Chem.* **234**, 2948. **IUPAC Name:** 2,3,13*a*,14,15,16,25,25*a*-Octahydro-9,19-dihydroxy-18,29-dimethoxy-1,14,14-trimethyl-13*H*-4,6:21,24-dietheno-8,12-metheno-1*H*-pyrido[3',2':14,15][1,11]dioxacycloeicosino[2,3,4-*ij*]-isoquinolinium ion.

Tubulysin A

This highly cytotoxic antimitotic peptide (FW = 842.11 g/mol; CAS 205304-86-5) induces microtubule (MT) depletion, triggering apoptosis, and stalls cells in the G_2/M phase (1,2). Tubulysin A inhibits tubulin polymerization more efficiently than vinblastine and induces depolymerization of MTs in cell-free systems or in cells, even after assembled MTs or cells are pre-incubated with the MT-stabilizing agents epothilone B and paclitaxel (2). In competition experiments, Tubulysin A strongly inhibits (K_i = 3 μM) the binding of vinblastine to tubulin. Electron microscopy revealed that Tubulysin A induces the formation of rings, double rings, and pinwheel structures (2). The mode of action of Tubulysin A resembles that of peptide antimitotics Dolastatin 10, Phomopsin A, Hemiasterlin, and Symplostatin. Folate-conjugated tubulysins are highly cytostatic in cancer cells containing large numbers of folate receptors (4). **1.** Sasse, Steinmetz, Heil, Höfle & Reichenbach (2000) *J. Antibiot.* (Tokyo) **53**, 879. **2.** Kaur, Hollingshead, Holbeck, *et al.* (2006) *Biochem J.* **396**, 235. **3.** Khalil, Sasse, Lünsdorf, *et al.* (2006) *ChemBioChem* **7**, 678. **4.** Reddy, Dorton, Dawson, *et al.* (2009) *Mol. Pharm.* **6**, 1518.

Tumstatin

This 28-kDa fragment of type IV collagen is an endothelial cell-specific inhibitor of protein synthesis. Through a requisite interaction with αVβ3 integrin, tumstatin inhibits activation of focal adhesion kinase, phosphatidylinositol 3-kinase, protein kinase B, and the mammalian target of rapamycin. In addition, it prevents the dissociation of eukaryotic initiation factor 4E protein (eIF4E) from 4E-binding protein 1 (1). **1.** Maeshima, Sudhakar, Lively, *et al.* (2002) *Science* **295**, 140.

Tungstate (Tungstic Acid)

This conjugate base of tungstic acid ($FW_{sodium\text{-}salt}$ = 271.84 g/mol; CAS 11120-01-7; $FW_{methylamine\text{-}salt}$ = 311.97; CAS 55979-60-7), which is only slightly soluble in water, is often used to precipitate proteins. Indeed, through its strong association with proteins, its methyl amine salt serves as an contrast agent for visualizing protein structures by means of electron microscopy. **Target(s):** acid phosphatase (1,2,26-29); alkaline phosphatase (2,3,30,31); aryl sulfatase (2); carbonic anhydrase, *or* carbonate dehydratase (4); dimethylsulfoxide reductase (5); fatty-acyl-CoA synthase, inhibited by the methylamine salt (33); glucan endo-1,3-β-D-glucosidase (6); glucose-6-phosphatase (7,25); nitrogenase (8); phloretin hydrolase (9); 6-phosphofructo-2-kinase (10); phosphonoacetaldehyde hydrolase (11); phosphoprotein phosphatase (16,23,24); 3-phytase (12); 4-phytase (12-14); polynucleotide 5'-hydroxyl-kinase (15); protein-phosphatase 2B (24); protein-tyrosine phosphatase (17,18); pyrophosphate-dependent phosphofructokinase, *or* diphosphate:fructose-6-phosphate 1-phosphotransferase (32); sabinene-hydrate synthase (19); and selenate reductase (20-22). **1.** Van Etten, Waymack & Rehkop (1974) *J. Amer. Chem. Soc.* **96**, 6782. **2.** Stankiewicz & Gresser (1988) *Biochemistry* **27**, 206. **3.** Chen & Zhou (1999) *J. Enzyme Inhib.* **14**, 251. **4.** Innocenti, Vullo, Scozzafava & Supuran (2005) *Bioorg. Med. Chem. Lett.* **15**, 567. **5.** Bilous & Weiner (1985) *J. Bacteriol.* **162**, 1151. **6.** Yamamoto & Nagasaki (1975) *Agric. Biol. Chem.* **39**, 2163. **7.** Foster, Young, Brandt & Nordlie (1998) *Arch. Biochem. Biophys.* **354**, 125. **8.** Siemann, Schneider, Oley & Müller (2003) *Biochemistry* **42**, 3846. **9.** Minamikawa, Jayasankar, Bohm, Taylor & Towers (1970) *Biochem. J.*

116, 889. **10.** Kountz, McCain, el-Maghrabi & Pilkis (1986) *Arch. Biochem. Biophys.* **251**, 104. **11.** Zhang, Mazurkie, Dunaway-Mariano & Allen (2002) *Biochemistry* **41**, 13370. **12.** Greiner (2002) *J. Agric. Food Chem.* **50**, 6858. **13.** Greiner, Konietzny & Jany (1998) *J. Food Biochem.* **22**, 143. **14.** Greiner & Alminger (1999) *J. Sci. Food Agric.* **79**, 1453. **15.** Zimmerman & Pheiffer (1981) *The Enzymes*, 3rd ed. (Boyer, ed.), **14**, 315. **16.** Reiter, White & Rusnak (2002) *Biochemistry* **41**, 1051. **17.** Aguirre-Garcia, Escalona-Montano, Bakalara, *et al.* (2006) *Parasitology* **132**, 641. **18.** Waheed, Laidler, Wo & Van Etten (1988) *Biochemistry* **27**, 4265. **19.** Hallahan & Croteau (1988) *Arch. Biochem. Biophys.* **264**, 618. **20.** Watts, Ridley, Condie, *et al.* (2003) *FEMS Microbiol. Lett.* **228**, 273. **21.** Watts, Ridley, Dridge, *et al.* (2005) *Biochem. Soc. Trans.* **33**, 173. **22.** Stolz & Oremland (1999) *FEMS Microbiol. Rev.* **23**, 615. **23.** Zhou, Clemens, Hakes, Barford & Dixon (1993) *J. Biol. Chem.* **268**, 17754. **24.** Haddy & Rusnak (1994) *Biochem. Biophys. Res. Commun.* **200**, 1221. **25.** Van Schaftingen & Gerin (2002) *Biochem. J.* **362**, 513. **26.** Lawrence & van Etten (1981) *Arch. Biochem. Biophys.* **206**, 122. **27.** Andrews & Pallavicini (1973) *Biochim. Biophys. Acta* **321**, 197. **28.** Jing, Li, Li, Xie & Zhang (2006) *Comp. Biochem. Physiol. B* **143**, 229. **29.** Reilly, Felts, Henzl, Calcutt & Tanner (2006) *Protein Expr. Purif.* **45**, 132. **30.** Sakurai, Toda & Shiota (1981) *Agric. Biol. Chem.* **45**, 1959. **31.** Xiao, Xie, Lin, *et al.* (2002) *J. Mol. Catal. B* **17**, 65. **32.** Mahajan & Singh (1989) *Plant Physiol.* **91**, 421. **33.** Stoops & Wakil (1978) *J. Biol. Chem.* **253**, 4464.

21-Tungsto-9-antimoniate Ammonium Salt

This transition metal complex (Formula: $(NH_4)_{17}Na[NaW_{21}Sb_9O_{86}]\cdot14H_2O$ (FW = 6937.5 g/mol), also known as HPA23, is an antiviral agent that has been shown to potently inhibit both human and murine DNA polymerase α and murine DNA polymerase γ. 21-Tungsto-9-antimoniate also noncompetitively inhibits murine DNA polymerase α (K_i = 24 nM) and competitively inhibits the γ enzyme, K_i = 20 nM (1-3). **1.** Ono, Nakane, Barre-Sinoussi & Chermann (1988) *Eur. J. Biochem.* **176**, 305. **2.** Ono, Nakane, Matsumoto, Barre-Sinoussi & Chermann (1984) *Nucl. Acids Symp. Ser.* **1984** (15), 169. **3.** Chermann, Sinoussi & Jasmin (1975) *Biochem. Biophys. Res. Commun.* **65**, 1229.

Tunicamycin

These lipophilic analogues of UDP-*N*-acetylglucosamine and antiviral antibiotics (Tunicamycin A_1, *n* = 8, FW = 816.90 g/mol; CAS 11089-65-9; Tunicamycin B_2, *n* = 9, FW = 830.93 g/mol; Tunicamycin C_1, *n* = 10, FW = 844.95 g/mol; CAS 73942-07-1; and Tunicamycin D_2, *n* = 11, FW = 858.99 g/mol) from *Streptomyces lysosuperificus* inhibit protein glycosylation and protein palmitoylation, making them powerful tools in investigations on glycoprotein processing. Tunicamycin also inhibits capillary endothelial cell proliferation by inducing apoptosis. *Note*: The exact composition of commercial tunicamycin will vary from batch to batch, requiring an investigator to save the certificate of analysis for each preparation purchased. **Target(s):** cellulose synthase, UDP-forming (1); dolichyl-phosphate β-glucosyltransferase (2,3); dolichyl-phosphate β-D-mannosyltransferase (2); lactose synthase (4); peptidoglycan glycosyltransferase (5); phosphatidyl-choline:sterol *O*-acyltransferase, LCAT (6); phospho-*N*-acetylmuramoyl-pentapeptide-transferase (7-10); protein *N*-acetylglucosaminyltransferase (11); protein glycosylation (12-17); protein palmitoylation (16); UDP-*N*-acetylglucosamine:dolichyl-phosphate *N*-acetylglucosamine phosphotrans-ferase (12-15,17-23); undecaprenyl-diphosphomuramoylpentapeptide β-*N*-acetylglucosaminyl-transferase, *or* MurG-transferase (24,25). **1.** Haass, Hackspacher & Franz (1985) *Plant Sci.* **41**, 1. **2.** Riedell & Miernyk (1988) *Plant Physiol.* **87**, 420. **3.** Miernyk & Riedell (1991) *Phytochemistry* **30**, 2865. **4.** West (1985) *Biochem. Soc. Trans.* **13**, 694. **5.** Ramachandran, Chandrakala, Kumar, *et al.* (2006) *Antimicrob. Agents Chemother.* **50**, 1425. **6.** Collet & Fielding (1991) *Biochemistry* **30**, 3228. **7.** Bouhss, Crouvoisier, Blanot & Mengin-Lecreulx (2004) *J. Biol. Chem.* **279**, 29974. **8.** Brandish, Kimura, Inukai, *et al.* (1996) *Antimicrob. Agents Chemother.* **40**, 1640. **9.** Stachyra, Dini, Ferrari, *et al.* (2004) *Antimicrob. Agents Chemother.* **48**, 897.

10. Ikeda, Wachi, Jung, Ishino & Matsuhashi (1991) *J. Bacteriol.* **173**, 1021. 11. Khalkhali, Marshall, Reuvers, Habets-Willems & Boer (1976) *Biochem. J.* **160**, 37. 12. Schwarz & Datema (1982) *Meth. Enzymol.* **83**, 432. 13. Elbein (1983) *Meth. Enzymol.* **98**, 135. 14. Elbein (1981) *Trends Biochem. Sci.* **6**, 219. 15. Elbein (1987) *Meth. Enzymol.* **138**, 661. 16. Patterson & Skene (1995) *Meth. Enzymol.* **250**, 284. 17. Schwarz & Datema (1980) *Trends Biochem. Sci.* **5**, 65. 18. Presper & Heath (1983) *The Enzymes*, 3rd ed. (Boyer, ed.), **16**, 449. 19. Kaushal & Elbein (1986) *Plant Physiol.* **82**, 748. 20. Shailubhai, Dong-Yu, Saxena & Vijay (1988) *J. Biol. Chem.* **263**, 15964. 21. Chandra, Doody & Bretthauer (1991) *Arch. Biochem. Biophys.* **290**, 345. 22. Heifetz, Keenan & Elbein (1979) *Biochemistry* **18**, 2186. 23. Sharma, Lehle & Tanner (1982) *Eur. J. Biochem.* **126**, 319. 24. Ravishankar, Kumar, Chandrakala, *et al.* (2005) *Antimicrob. Agents Chemother.* **49**, 1410. 25. Branstrom, Midha, Longley, *et al.* (2000) *Anal. Biochem.* **280**, 315.

Turanose

This sweet-tasting reducing disaccharide (FW = 342.30 g/mol; CAS 547-25-1), known systematically as 3-*O*-α-D-glucopyranosyl-D-fructose, is an unusual sucrose isomer that is found in the plant trisaccharide melezitose found. **Target(s):** concanavalin A (1); β-fructofuranosidase, *or* invertase (2); glucan 1,3-α-glucosidase, *or* glucosidase II, *or* mannosyl-oligosaccharide glucosidase II (3,4); α-glucosidase, K_i = 2.8 mM for the rat liver enzyme (5-18); and sucrose-phosphatase (19). 1. Goldstein, Hollerman & Smith (1965) *Biochemistry* **4**, 876. 2. Isla, Vattuone, Gutierrez & Sampietro (1988) *Phytochemistry* **27**, 1993. 3. Saxena, Shailubhai, Dong-Yu & Vijay (1987) *Biochem. J.* **247**, 563. 4. Brada & Dubach (1984) *Eur. J. Biochem.* **141**, 149. 5. Brown, Brown & Jeffrey (1972) *Meth. Enzymol.* **28**, 805. 6. Webb (1966) *Enzyme and Metabolic Inhibitors*, vol. 2, p. 416, Academic Press, New York. 7. Jeffrey, Brown & Brown (1970) *Biochemistry* **9**, 1416. 8. Yamasaki & Suzuki (1978) *Agric. Biol. Chem.* **42**, 971. 9. Yamasaki & Suzuki (1980) *Agric. Biol. Chem.* **44**, 707. 10. Yamasaki & Suzuki (1979) *Agric. Biol. Chem.* **43**, 481. 11. Peruffo, Renosto & Pallavicini (1978) *Planta* **142**, 195. 12. Yamasaki & Suzuki (1974) *Agric. Biol. Chem.* **38**, 443. 13. Yamasaki, Miyake & Suzuki (1973) *Agric. Biol. Chem.* **37**, 251. 14. Yamasaki, Suzuki & Ozawa (1977) *Agric. Biol. Chem.* **41**, 1451 and 1559. 15. Yamasaki, Suzuki & Ozawa (1976) *Agric. Biol. Chem.* **40**, 669. 16. Chadalavada & Sivakami (1997) *Biochem. Mol. Biol. Int.* **42**, 1051. 17. Martiniuk & Hirschhorn (1981) *Biochim. Biophys. Acta* **658**, 248. 18. Banno & Nozawa (1985) *J. Biochem.* **97**, 409. 19. Hawker (1971) *Phytochemistry* **10**, 2313.

ar-Turmerone

This natural sesquiterpenoid oil (FW = 216.32 g/mol; CAS 38142-58-4; B.P. = 159-160°C), isolated from the oil of turmeric (*Curcuma longa*), inhibits the prostaglandin cyclooxygenase COX-2 and inducible nitric-oxide synthase. 1. Lee, Hong, Huh, *et al.* (2002) *J. Environ. Pathol. Toxicol. Oncol.* **21**, 141.

TW-37

This novel nonpeptide (FW = 573.7; CAS 877877-35-5; Solubility (25°C): 110 mg/mL DMSO), also known as 5-(2-isopropylbenzyl)-*N*-(4-(2-*tert*-butylphenylsulfonyl)phenyl)-2,3,4-trihydroxybenzamide, inhibits the pro-survival Bcl (B-cell lymphoma) proteins Bcl-2 (K_i = 0.29 μM), Bcl-xL (K_i = 1.1 μM), and Mcl-1 (K_i = 0.26 μM), targets the Bcl-2 Homology domain- (*or* BH-) binding groove in Bcl-2, where proapoptotic Bcl-2 proteins are known to bind. 1. Mohammad, *et al.* (2007) *Clin. Cancer Res.* **13**, 2226. 2. Zeitlin, *et al.* (2006) *Cancer Res.* **66**, 8698.

TX-1918

This potent eEF2K inhibitor (FW = 228.25 g/mol; CAS 503473-32-3; Solubility: 10 mg/mL in DMSO), also named 2-((3,5-dimethyl-4-hydroxyphenyl)methylene)-4-cyclopentene-1,3-dione, is a cell-permeable arylidenecyclopentenedione-derived tyrphostin that potent inhibits eukaryotic elongation factor-2 kinase (eEF2K, IC_{50} = 440 nM). At higher concentrations, TX-1918 inhibits other kinases, including proto-oncogene tyrosine-protein kinase Src (c-SRC), IC_{50} = 4.4 μM, protein kinases A and C (IC_{50} = 44 μM in both cases), and Epidermal Growth Factor Receptor Kinase (EGFR-K, IC_{50} = 440 μM). 1. Hori H, Nagasawa H, Ishibashi *et al.* (2002) *Bioorg. Med. Chem.* **10**, 3257.

Tylenol™, *See Acetaminophen*

Tylocrebrine

This alkaloid and antileukemic agent ($FW_{free-base}$ = 393.48 g/mol; CAS 6879-02-3) inhibits protein biosynthesis at the translocation step (1). *Note:* The (−)-stereoisomer is isolated from *Tylophora crebriflora* whereas the (+)-isomer is reported in *Ficus septica*. 1. Jiménez (1976) *Trends Biochem. Sci.* **1**, 28. 2. Donaldson, Atkinson & Murray (1968) *Biochem. Biophys. Res. Commun.* **31**, 104. 3. Huang & Grollman (1972) *Mol. Pharmacol.* **8**, 538. 4. Gupta, Krepinsky & Siminovitch (1980) *Mol. Pharmacol.* **18**, 136.

Tylosin

This macrolide antibiotic ($FW_{free-base}$ = 916.11 g/mol; CAS 1401-69-0) from a strain of *Streptomyces fradiae* inhibits protein biosynthesis by binding to the 50S ribosomal subunit (1). The disaccharide present at position 5 in the lactone 16-member ring of the macrolide is essential for inhibition of peptidyltransferase (2). In acidic conditions (pH < 4), tylosin is converted to desmycosin, which does not inhibit the peptidyltransferase of ribosomes (2). **Target(s):** peptidyltransferase (2); protein biosynthesis (1-3); and rRNA (guanine-N^1-)-methyltransferase (4). 1. Pestka (1974) *Meth. Enzymol.* **30**, 261. 2. Poulsen, Kofoed & Vester (2000) *J. Mol. Biol.* **304**, 471. 3. Dinos & Kalpaxis (2000) *Biochemistry* **39**, 11621. 4. Douthwaite, Crain, Liu & Poehlsgaard (2004) *J. Mol. Biol.* **337**, 1073.

Tyloxapol

This nonionic surfactant (FW = indefinite polymer; CAS 1401-69-0; R = $(CH_2CH_2O)_xH$, $x = 8–10$, $m < 6$), also known as Tyloxapol, Triton WR-1339, and Triton A-20, is slowly miscible with water. **Target(s):** 1-acylglycerophosphocholine O-acyltransferase (1); 1-alkenylglycerophosphocholine O-acyltransferase (1); cholesterol-5,6-oxide hydrolase (2); diacylglycerol O-acyltransferase (3); diacylglycerol cholinephospho-transferase (4); ethanolaminephosphotransferase (4); lipoprotein lipase (5-8); phosphoglyceride: lysophosphatidyl-glycerol O-acyltransferase (9); sterol O-acyltransferase, or cholesterol O-acyltransferase, or ACAT (10,11). **1.** Arthur & Choy (1986) *Biochem. J.* **236**, 481. **2.** Watabe, Ozawa, Ishii, Chiba & Hiratsuka (1986) *Biochem. Biophys. Res. Commun.* **140**, 632. **3.** Akao & Kusaka (1976) *J. Biochem.* **80**, 723. **4.** Coleman & Bell (1977) *J. Biol. Chem.* **252**, 3050. **5.** Korn (1962) *Meth. Enzymol.* **5**, 542. **6.** Sato, Akiba & Horiguchi (1997) *Comp. Biochem. Physiol. A Physiol.* **118**, 855. **7.** Sato, Akiba, Kimura & Horiguchi (1995) *Comp. Biochem. Physiol. C Pharmacol. Toxicol. Endocrinol.* **112**, 315. **8.** Sheorain, Rao & Subrahmanyam (1980) *Enzyme* **25**, 81. **9.** Hostetler, Huterer & Wherrett (1992) *Meth. Enzymol.* **209**, 104. **10.** Davis, Showalter & Kern, Jr. (1978) *Biochem. J.* **174**, 45. **11.** Vajda, Ferguson, Shand, Noble & Speake (1999) *Comp. Biochem. Physiol. B* **122**, 301.

Tynorphin

This pentapeptide VVYPW (FW$_{zwitterion}$ = 647.77 g/mol) inhibits dipeptidyl-peptidase III, K_i = 75 nM (1-4). **1.** Mazzocco, Fukasawa, Auguste & Puiroux (2003) *Eur. J. Biochem.* **270**, 3074. **2.** Yamamoto, Hashimoto, Shimamura, Yamaguchi & Hazato (2000) *Peptides* **21**, 503. **3.** Mazzocco, Gillibert-Duplantier, Neaud, *et al.* (2006) *FEBS J.* **273**, 1056. **4.** Chiba, Li, Yamane, *et al.* (2003) *Peptides* **24**, 773.

Tyramine

This tyrosine decarboxylation product (FW$_{free-base}$ = 137.18 g/mol; CAS 51-67-2; M.P. = 164-165°C; pK_a values of 9.74 and 10.52 at 25°C) also known as β-(p-hydroxyphenyl)ethylamine, does not cross the blood-brain-barrier, limiting its catecholamine-releasing effects to peripheral neurons, where it displaces stored monoamines (*e.g.*, dopamine, norepinephrine, and epinephrine) from pre-synaptic vesicles. Tyramine is also synthesisized by the decarboxylation of tyrosine within the CNS, where it may serve as a neurotransmitter. **Target(s):** aromatic-L-amino-acid decarboxylase, or dopa decarboxylase (1); arylsulfatase (2-4); dihydropteridine reductase (5,6); dopamine β-monooxygenase (7-9); γ-glutamylhistamine synthetase (10); histamine N-methyltransferase (11); pyridoxal kinase (12); tubulin:tyrosine ligase, also, weak alternative substrate (13); tyrosine aminotransferase (14); tyrosine decarboxylase (15); tyrosyl-tRNA synthetase, K_i = 5.6 μM for the *Escherichia coli* enzyme (16-19). **1.** Nakazawa, Kumagai & Yamada (1987) *Agric. Biol. Chem.* **51**, 2531. **2.** Okamura, Yamada, Murooka & Harada (1976) *Agric. Biol. Chem.* **40**, 2071. **3.** Beil, Kehrli, James, *et al.* (1995) *Eur. J. Biochem.* **229**, 385. **4.** Murooka, Yim & Harada (1980) *Appl. Environ. Microbiol.* **39**, 812. **5.** Shen (1984) *Biochim. Biophys. Acta* **785**, 181. **6.** Shen (1983) *Biochim. Biophys. Acta* **743**, 129. **7.** Webb (1966) *Enzyme and Metabolic Inhibitors*, vol. **2**, p. 320, Academic Press, New York. **8.** Goldstein & Contrera (1961) *Experentia* **17**, 267. **9.** Goldstein & Contrera (1962) *J. Biol. Chem.* **327**, 1898. **10.** Stein & Weinreich (1982) *J. Neurochem.* **38**, 204. **11.** Fuhr & Kownatzki (1986) *Pharmacology* **32**, 114. **12.** Gäng & von Collins (1973) *Int. J. Vitam. Nutr. Res.* **43**, 318. **13.** Raybin & Flavin (1977) *Biochemistry* **16**, 2189. **14.** Jacoby & La Du (1964) *J. Biol. Chem.* **239**, 419. **15.** Moreno-Arribas & Lonvaud-Funel (1999) *FEMS Microbiol. Lett.* **180**, 55. **16.** Schweet (1962) *Meth. Enzymol.* **5**, 722. **17.** Stulberg & Novelli (1962) *The Enzymes*, 2nd ed. (Boyer, Lardy & Myrbäck, eds.), **6**, 401. **18.** Calendar & Berg (1966) *Biochemistry* **5**, 1690. **19.** Santi & Peña (1973) *J. Med. Chem.* **16**, 273.

Tyrocidines

This mixture consists of at least four cyclic decapeptide antibiotics by *Bacillus brevis* and related organisms. The predominent form tyrocidine A (MW = 1373.70 g/mol; CAS 8011-61-8) is cyclo(L-Leu-D-Phe-L-Pro-L-Phe-D-Phe-L-Asn-L-Gln-L-Tyr-L-Val-L-Orn-), where L-Orn refers to L-ornithine. In tyrocidine B (MW = 1312.03 g/mol), the L-phe at position-four is replaced with an L-trp. Tyrocidine C (MW = 1351.57 g/mol) is identical to tyrocidine B, except for D-trp in place of D-phe at position-five. Tyrocidine D (MW = 1374.61 g/mol) is identical to tyrocidine C, except for L-trp in place of L-tyr at position-eight. The relative yields of the four tyrocidines appears to depend on the concentrations of the aromatic amino acids in the growth media. Dubos first isolated tyrocidines in 1939 from *Bacillus brevis*, which itself is immune to their effects. Tyrocidines are cationic detergents that increase the permeability of bacterial cell membranes, rendering them particularly bactericidal with respect to Gram-positive microorganisms. The tyrocidines also uncouple oxidative phosphorylation (1) and inhibit RNA polymerase (2). **See Uncouplers** **1.** Hunter & Schwartz (1967) in *Antibiotics I: Mode of Action* (Gottlieb & Shaw, eds.), p. 142, Springer-Verlag, New York. **2.** Schazschneider, Ristow & Kleinkauf (1974) *Nature* **249**, 757.

Tyromycin A

This natural product (FW = 446.58 g/mol; CAS 141364-77-4), also known as 1,16-bis-(4-methyl-2,5-dioxo-3-furyl)hexadecane, from *Tyromyces lacteus* and inhibits cystinyl and leucyl aminopeptidases. **1.** Weber, Semar, Anke, Bross & Steglich (1992) *Planta Med.* **58**, 56.

Tyropeptin A

This peptide aldehyde (FW = 511.62 g/mol), also known as N-(n-butyryl)-L-tyrosyl-L-leucyl-DL-tyrosinal, and isolated from the culture broth of a strain of *Kitasatospora*, is a cell-permeable peptide aldehyde that competitively inhibits the chymotrypsin-like and trypsin-like activities of the 20S proteasome with IC$_{50}$ values of 0.1 μg/mL and 1.5 μg/mL, respectively (1,2). Tyropeptin A does not inhibit the peptidylglutamyl-peptide hydrolyzing activity. **Target(s):** calpain (1); cathepsin L (1); α-chymotrypsin (1); and proteasome (1-3). **1.** Momose, Sekizawa, Hashizume, *et al.* (2001) *J. Antibiot.* **54**, 997. **2.** Momose, Sekizawa, Hirosawa, *et al.* (2001) *J. Antibiot.* **54**, 1004. **3.** Momose, Sekizawa, Iinuma & Takeuchi (2002) *Biosci. Biotechnol. Biochem.* **66**, 2256.

Tyropeptin B

This peptide aldehyde (FW = 511.62 g/mol), also known as N-(n-butyryl)-L-tyrosyl-L-leucyl-DL-tyrosinal, and isolated from the culture broth of a strain of *Kitasatospora*, is a cell-permeable peptide aldehyde that competitively inhibits the chymotrypsin-like and trypsin-like activities of the 20S proteasome with an activity about two-fold less than that of tyropeptin A

(1,2). **1**. Momose, Sekizawa, Hashizume, *et al.* (2001) *J. Antibiot.* **54**, 997 **2**. Momose, Sekizawa, Hirosawa, *et al.* (2001) *J. Antibiot.* **54**, 1004.

L-Tyrosinamide

This amino acid amide (FW$_{free-base}$ = 180.21 g/mol; CAS 4985-46-0; pK_a values of 7.33 (amino) and 10.4 (phenolic OH) at 25°C), which is frequently used in the chemical synthesis of natural products, inhibits bacterial leucyl aminopeptidase (1), tyramine *N*-feruloyltransferase (2), and tyrosyl-tRNA synthetase, K_i = 0.31 mM for the *Escherichia coli* enzyme (3,4). **1**. Baker, Wilkes, Bayliss & Prescott (1983) *Biochemistry* **22**, 2098. **2**. Negrel & Javelle (1997) *Eur. J. Biochem.* **247**, 1127. **3**. Calendar & Berg (1966) *Biochemistry* **5**, 1690. **4**. Santi & Peña (1973) *J. Med. Chem.* **16**, 273.

D-Tyrosine

This enantiomer of L-tyrosine (FW = 181.19 g/mol; CAS 556-02-5), known systematically as (*R*)-2-amino-3-(4-hydroxyphenyl)propanoic acid, is present as its *O*-methyl ether in the antibiotic cycloheptamycin. *See L-Tyrosine* **Target(s):** arogenate dehydrogenase (1); carboxypeptidase A (2,3); phenylalanine ammonia-lyase (4,5); phenylalanine dehydrogenase (6); prephenate dehydrogenase (7,8); tyrosine aminotransferase, weakly inhibited (9). **1**. Bonner & Jensen (1987) *Meth. Enzymol.* **142**, 488. **2**. Giusti, Carrara, Cima & Borin (1985) *Eur. J. Pharmacol.* **116**, 287. **3**. Elkins-Kaufman & Neurath (1948) *J. Biol. Chem.* **175**, 893. **4**. Hanson & Havir (1972) *The Enzymes*, 3rd ed. (Boyer, ed.), **7**, 75. **5**. Camm & Towers (1973) *Phytochemistry* **12**, 961. **6**. Misono, Yonezawa, Nagata & Nagasaki (1989) *J. Bacteriol.* **171**, 30. **7**. Cotton & Gibson (1970) *Meth. Enzymol.* **17A**, 564. **8**. Fischer & Jensen (1987) *Meth. Enzymol.* **142**, 503. **9**. Jacoby & La Du (1964) *J. Biol. Chem.* **239**, 419.

L-Tyrosine

This proteogenic aromatic amino acid (FW = 181.19 g/mol; CAS 60-18-4; pK_a values of 2.20, 9.21, and 10.46 (phenolic OH) at 25°C; Symbol = Tyr = Y), known systematically as (*S*)-2-amino-3-(4-hydroxyphenyl)propanoic acid, absorbs UV light (λ_{max} = 223 nm (ε = 8200 M^{-1}cm^{-1}) and 274.5 nm (ε = 1340 M^{-1}cm^{-1}) in 0.1 M HCl; λ_{max} = 240 nm (ε = 11050 M^{-1}cm^{-1}) and 293.5 nm (ε = 2330 M^{-1}cm^{-1}) in 0.1 M NaOH). Enzymes utilizing L-tyrosine as a substrate or product include: cyclohexadienyl dehydrogenase, tyrosine *N*-monooxygenase, phenylalanine 4-monooxygenase, tyrosine 3-monooxygenase, monophenol monooxygenase, tyrosine aminotransferase, tubulin:tyrosine ligase, tubulinyl-Tyr carboxypeptidase, tyrosine decarboxylase, tyrosine phenol-lyase, tyrosine 2,3-aminomutase,tyrosyl-tRNA synthetase, and tyrosylarginine synthetase. Many enzymes use D- or L-tyrosine as alternative substrates, including L-amino-acid dehydrogenase, phenylalanine dehydrogenase, the amino acid oxidases, aspartate aminotransferase, aromatic amino acid transferase, aromatic-amino-acid:glyoxylate aminotransferase, phenylalanine decarboxylase, phenylalanine ammonia-lyase (albeit, not from all sources), and amino-acid racemase. **Target(s):** alkaline phosphatase (1,2); chorismate mutase (3-7); cyclohexadienyl dehydrogenase *or* arogenate dehydrogenase (8,9); cystinyl aminopeptidase, *or* oxytocinase (29); cytosol alanyl aminopeptidase (27,28); 3-deoxy-7-phosphoheptulonate synthase, *or* 3-deoxy-D-*arabino*heptulosonate-7-phosphate synthetase, *or* phospho-2-dehydro-3-deoxyheptonate aldolase (10,11,21,37-50); dihydropteridine reductase (12,13); diphosphomevalonate decarboxylase (34); glutamine synthetase (14); histidine ammonia-lyase (15); histidine decarboxylase (16); leucyl aminopeptidase (32); membrane alanyl aminopeptidase, *or* aminopeptidase N (30,31); mimosinase (17); phenylalanine ammonia-lyase (16,18-20); prephenate dehydratase (3,22); prephenate dehydrogenase (3,4); propionyl-CoA carboxylase, *Myxococcus Xanthus* (23); protein-tyrosine-phosphatase (33); pyrophosphatase, *or* inorganic diphosphatase (24); pyruvate kinase

(35,36); shikimate kinase (25); starch phosphorylase (51-53); and tryptophan 5-monooxygenase (26,54). **1**. Fernley (1971) *The Enzymes*, 3rd ed. (Boyer, ed.), **4**, 417. **2**. Fishman & Sie (1971) *Enzymologia* **41**, 141. **3**. Cotton & Gibson (1970) *Meth. Enzymol.* **17A**, 564 and (1965) *Biochim. Biophys. Acta* **100**, 76. **4**. Davidson & Hudson (1987) *Meth. Enzymol.* **142**, 440. **5**. Gilchrist & Connelly (1987) *Meth. Enzymol.* **142**, 450. **6**. Sugimoto & Shiio (1980) *J. Biochem.* **88**, 167. **7**. Baker (1966) *Biochemistry* **5**, 2655. **8**. Gaines, Byng & Whitaker (1982) *Planta* **156**, 233. **9**. Xie, Bonner & Jensen (2000) *Comp. Biochem. Physiol. C Toxicol. Pharmacol.* **125**, 65. **10**. Simpson & Davidson (1976) *Eur. J. Biochem.* **70**, 509. **11**. Previc & Binkley (1964) *Biochem. Biophys. Res. Commun.* **16**, 162. **12**. Shen (1984) *Biochim. Biophys. Acta* **785**, 181. **13**. Shen (1983) *Biochim. Biophys. Acta* **743**, 129. **14**. Blanco, Alana, Llama & Serra (1989) *J. Bacteriol.* **171**, 1158. **15**. Rechler & Tabor (1971) *Meth. Enzymol.* **17B**, 63. **16**. Webb (1966) *Enzyme and Metabolic Inhibitors*, vol. 2, Academic Press, New York. **17**. Tangendjaja, Lowry & Wills (1986) *J. Sci. Food Agric.* **37**, 523. **18**. Koukol & Conn (1961) *J. Biol. Chem.* **236**, 2692 **19**. Camm & Towers (1973) *Phytochemistry* **12**, 961. **20**. Jorrin, Lopez-Valbuena & Tena (1988) *Biochim. Biophys. Acta* **964**, 73. **21**. Gaertner & DeMoss (1970) *Meth. Enzymol.* **17A**, 387. **22**. Bode, Melo & Birnbaum (1985) *J. Basic Microbiol.* **25**, 291. **23**. Kimura, Kojyo, Kimura & Sato (1998) *Arch. Microbiol.* **170**, 179. **24**. Naganna (1950) *J. Biol. Chem.* **183**, 693. **25**. De Feyter (1987) *Meth. Enzymol.* **142**, 355. **26**. Freedland, Wadzinski & Waisman (1961) *Biochem. Biophys. Res. Commun.* **6**, 227. **27**. Garner & Behal (1977) *Arch. Biochem. Biophys.* **182**, 667. **28**. Garner & Behal (1975) *Biochemistry* **14**, 3208. **29**. Krishna & Kanagasabapathy (1989) *J. Endocrinol.* **121**, 537. **30**. McCaman & Villarejo (1982) *Arch. Biochem. Biophys.* **213**, 384. **31**. Tokioka-Terao, Hiwada & Kokubu (1984) *Enzyme* **32**, 65. **32**. Machuga & Ives (1984) *Biochim. Biophys. Acta* **789**, 26. **33**. Hörlein, Gallis, Brautigan & Bornstein (1982) *Biochemistry* **21**, 5577. **34**. Shama Bhat & Ramasarma (1979) *Biochem. J.* **181**, 143. **35**. Lin, Turpin & Plaxton (1989) *Arch. Biochem. Biophys.* **269**, 228. **36**. Schering, Eigenbrodt, Linder & Schoner (1982) *Biochim. Biophys. Acta* **717**, 337. **37**. Hu & Sprinson (1977) *J. Bacteriol.* **129**, 177. **38**. McCandliss, Poling & Herrmann (1978) *J. Biol. Chem.* **253**, 4259. **39**. Wu, Sheflyan & Woodard (2005) *Biochem. J.* **390**, 583. **40**. Whitaker, Fiske & Jensen (1982) *J. Biol. Chem.* **257**, 12789. **41**. Ramilo & Evans (1997) *Protein Expr. Purif.* **9**, 253. **42**. Friedrich & Schlegel (1975) *Arch. Microbiol.* **103**, 133. **43**. Dusha & Dénes (1976) *Biochim. Biophys. Acta* **438**, 563. **44**. Schnappauf, Hartmann, Künzler & Braus (1998) *Arch. Microbiol.* **169**, 517. **45**. Wu, Howe & Woodard (2003) *J. Biol. Chem.* **278**, 27525. **46**. Euverink, Hessels, Franke & Dijkhuizen (1995) *Appl. Environ. Microbiol.* **61**, 3796. **47**. Liao, Lin, Chien & Hsu (2001) *FEMS Microbiol. Lett.* **194**, 59. **48**. Schoner & Herrmann (1976) *J. Biol. Chem.* **251**, 5440. **49**. Reinink & Borstlap (1982) *Plant Sci. Lett.* **26**, 167. **50**. Bracher & Schweingruber (1977) *Biochim. Biophys. Acta* **485**, 446. **51**. Kumar & Sanwal (1982) *Biochemistry* **21**, 4152. **52**. Singh & Sanwal (1976) *Phytochemistry* **15**, 1447. **53**. Kumar (1984) *Indian J. Plant Physiol.* **27**, 209. **54**. Naoi, Maruyama, Takahashi, Ota & Parvez (1994) *Biochem. Pharmacol.* **48**, 207.

m-Tyrosine

This naturally occurring tyrosine isomer (FW = 181.19 g/mol; CAS 775-06-4), also called (3-hydroxyphenyl)alanine, is an alternative intermediate in the biosynthesis of certain catecholamines as well as many plant quinones. *m*-Tyrosine is also a marker of oxidative damage in motor neuron disease (1) as well as a marker of hydroxyl radical generation (2). **Target(s):** aromatic-L-amino-acid decarboxylase, *also* alternative substrate (3,4); 3-deoxy-7-phosphoheptulonate synthase, *or* 3-deoxy-D-*arabino*heptulosonate-7-phosphate synthetase (5); phenylalanine ammonia-lyase (6); phenylalanine 4-monooxygenase (7); prephenate dehydratase (8); and tyrosine phenol-lyase, inhibited by the L-enantiomer (9,10). **1**. Poljak, Pamphlett, Gurney & Duncan (2000) *Redox Rep.* **5**, 137. **2**. Nair, Nair, Friesen, Bartsch & Ohshima (1995) *Carcinogenesis* **16**, 1195. **3**. O'Leary & Baughn (1977) *J. Biol. Chem.* **252**, 7168. **4**. Nakazawa, Kumagai & Yamada (1987) *Agric. Biol. Chem.* **51**, 2531. **5**. Simpson & Davidson (1976) *Eur. J. Biochem.* **70**, 509. **6**. Camm & Towers (1973) *Phytochemistry* **12**, 961. **7**. Letendre, Dickens & Guroff (1975) *J. Biol. Chem.* **250**, 6672. **8**. Dopheide, Crewther & Davidson (1972) *J. Biol. Chem.* **247**, 4447. **9**. Muro, Nakatani, Hiromi, Kumagai & Yamada (1978) *J. Biochem.* **84**, 633. **10**. Demidkina, Myagkikh & Azhayev (1987) *Eur. J. Biochem.* **170**, 311.

o-Tyrosine

This naturally occurring tyrosine isomer (FW = 181.19 g/mol), also called (2-hydroxyphenyl)alanine, is a marker of oxidative damage in motor neuron disease (1), a marker of hydroxyl radical generation (2), and is most abudant in the human lens (3). **Target(s):** 3-deoxy-7-phosphoheptulonate synthase, or 3-deoxy-D-*arabino*heptulosonate-7-phosphate synthetase (4); glucose-6-phosphatase (5); phenylalanine ammonia-lyase (6); and prephenate dehydratase (7). **1**. Poljak, Pamphlett, Gurney & Duncan (2000) *Redox Rep.* **5**, 137. **2**. Nair, Nair, Friesen, Bartsch & Ohshima (1995) *Carcinogenesis* **16**, 1195. **3**. Wells-Knecht, Huggins, Dyer, Thorpe & Baynes (1993) *J. Biol. Chem.* **268**, 12348. **4**. Simpson & BDavidson (1976) *Eur. J. Biochem.* **70**, 509. **5**. Nordlie (1971) *The Enzymes*, 3rd ed. (Boyer, ed.), **4**, 543. **6**. Hodgins (1972) *Arch. Biochem. Biophys.* **149**, 91. **7**. Dopheide, Crewther & Davidson (1972) *J. Biol. Chem.* **247**, 4447.

L-Tyrosine Benzyl Ester

This esterified amino acid (FW$_{free-base}$ = 271.32 g/mol) inhibits tyramine *N*-feruloyltransferase (1,2) and *Escherichia coli* tyrosyl-tRNA synthetase, K_i = 0.03 mM (3). **1**. Negrel & Javelle (1997) *Eur. J. Biochem.* **247**, 1127. **2**. Negrel & Javelle (2001) *Phytochemistry* **56**, 523. **3**. Santi & Peña (1973) *J. Med. Chem.* **16**, 273.

L-Tyrosine Ethyl Ester

This amino acid ester (FW$_{hydrochloride}$ = 245.71 g/mol; CAS 949-67-7; M.P. = 166-168°C) is an alternative substrate for chymotrypsin. Hydrolysis of the ester is accompanied by a decrease in absorbance at 233.5 nm at pH 6.5. The ester undergoes slow nonenzymatic hydrolysis (4% in 24 hours at pH 7 and 4°C) in aqueous solution. **Target(s):** L-amino-acid oxidase (1); pepsin, K_i = 18 mM (2); tyramine *N*-feruloyltransferase (3); tyrosyl-tRNA synthetase, K_i = 0.1 mM for the *Escherichia coli* enzyme (4,5). **1**. Zeller & Maritz (1945) *Helv. Chim. Acta* **28**, 365. **2**. Inouye & Fruton (1968) *Biochemistry* **7**, 1611. **3**. Negrel & Javelle (1997) *Eur. J. Biochem.* **247**, 1127. **4**. Calendar & Berg (1966) *Biochemistry* **5**, 1690. **5**. Santi & Peña (1973) *J. Med. Chem.* **16**, 273.

L-Tyrosine Methyl Ester

This esterified amino acid (FW$_{free-base}$ = 195.22 g/mol) inhibits tyramine *N*-feruloyltransferase (1) tyrosyl-tRNA synthetase, K_i = 0.09 mM for the *Escherichia coli* enzyme (2,3). **1**. Negrel & Javelle (1997) *Eur. J. Biochem.* **247**, 1127. **2**. Calendar & Berg (1966) *Biochemistry* **5**, 1690. **3**. Santi & Peña (1973) *J. Med. Chem.* **16**, 273.

L-Tyrosine β-Naphthylamide

This derivative of L-tyrosine (FW$_{free-base}$ = 306.36 g/mol), used frequently in the assay of tyrosine aminopeptidases, is a potent inhibitor of tobacco and potato tyramine *N*-feruloyltransferase, K_i values of 0.66 and 0.3 μM, respectively, and *Escherichia coli* tyrosyl-tRNA synthetase, K_i = 96 μM (2). **1**. Negrel & Javelle (1997) *Eur. J. Biochem.* **247**, 1127. **2**. Santi & Peña (1973) *J. Med. Chem.* **16**, 273.

Tyrosine-Specific Protein Kinase Inhibitor

This oligopeptide (FW = 2482.74 g/mol; Sequence: VAPSDSIQAEEWYFG KITRRE), corresponding to a sequence in noncatalytic domain of p60^{v-src}, inhibits the autophosphorylation of epidermal growth factor receptor/kinase and the tyrosine-specific protein phosphorylation in the acetylcholine receptor-rich membranes isolated from electroplax of *Narke japonica*. **Target(s):** p60^{v-src} protein-tyrosine kinase (1). **1**. Sato, Miki, Tachibana, *et al.* (1990) *Biochem. Biophys. Res. Commun.* **171**, 1152.

L-Tyrosinol

This reduced derivative of L-tyrosine (FW$_{free-base}$ = 167.21 g/mol; CAS 87745-27-5) binds within the active-site of tyrosyl-tRNA synthetase (K_i = 16 μM for the *Escherichia coli* enzyme), but cannot undergo adenylylation. **Target(s):** tyramine *N*-feruloyltransferase (1); and tyrosyl-tRNA synthetase (2-4). **1**. Negrel & Javelle (1997) *Eur. J. Biochem.* **247**, 1127. **2**. Monteilhet, Blow & Brick (1984) *J. Mol. Biol.* **173**, 477. **3**. Calendar & Berg (1966) *Biochemistry* **5**, 1690. **4**. Santi & Peña (1973) *J. Med. Chem.* **16**, 273.

D-Tyrosinol Adenylate

This synthetic aminoacyl adenylate mimic (FW = 496.42 g/mol), also known as D-tyrosinyl adenylate, is a potent isosteric inhibitor of tyrosyl-tRNA synthetase. **1**. Santi & Peña (1973) *J. Med. Chem.* **16**, 273.

L-Tyrosinol 2'-Deoxyadenylate

This aminoacyl adenylate mimic (FW = 480.42 g/mol), also known as L-tyrosinyl 2'-deoxyadenylate, inhibits tyrosyl-tRNA synthetase (IC$_{50}$ = 100 μM for the *Staphylococcus aureus* enzyme and 18 μM for the *Bacillus stearothrmophilus* protein) (1). **See** *L-Tyrosinol Adenylate* **1**. Brown, Richardson, Mensah, *et al.* (1999) *Bioorg. Med. Chem.* **7**, 2473.

L-Tyrosinol 3'-Deoxyadenylate

This aminoacyl adenylate mimic, also known as L-tyrosinyl 2'-deoxyadenylate (FW = 480.42 g/mol), inhibits tyrosyl-tRNA synthetase (IC$_{50}$ = 140 μM for the *Staphylococcus aureus* enzyme and 36 μM for the *Bacillus stearothrmophilus* protein) (1). **See** *L-Tyrosinol Adenylate* **1**. Brown, Richardson, Mensah, *et al.* (1999) *Bioorg. Med. Chem.* **7**, 2473.

L-Tyrosinol 2',3'-O-Isopropylideneadenylate

This aminoacyl adenylate mimic, also known as L-tyrosinyl 2',3'-O-isopropylidene adenylate (FW = 536.48 g/mol), inhibits *Staphylococcus aureus* tyrosyl-tRNA synthetase (IC$_{50}$ = 26 nM). **See** *L-Tyrosinol Adenylate* 1. Brown, Richardson, Mensah, *et al.* (1999) *Bioorg. Med. Chem.* 7, 2473.

L-Tyrosinol 1β-Naphthyl-1,4-Anhydro-D-ribitol-5-O-phosphate

This aminoacyl adenylate mimic (FW = 489.46 g/mol), also known as L-tyrosinyl 1,4-anhydro-D-ribitol-5-O-phosphate, inhibits *Staphylococcus aureus* tyrosyl-tRNA synthetase (IC$_{50}$ = 150 nM). **See** *L-Tyrosinol Adenylate* 1. Brown, Richardson, Mensah, *et al.* (1999) *Bioorg. Med. Chem.* 7, 2473.

L-Tyrosinol Uridylate

This aminoacyl adenylate mimic (FW = 473.38 g/mol), also known as L-tyrosinyl uridine-5'-O-phosphate, inhibits *Staphylococcus aureus* tyrosyl-tRNA synthetase (IC$_{50}$ = 210 nM). **See** *L-Tyrosinol Adenylate* 1. Brown, Richardson, Mensah, *et al.* (1999) *Bioorg. Med. Chem.* 7, 2473.

Tyrostatin

This peptide aldehyde (FW = 525.65 g/mol) from *Kitasatospora*, also known as *N*-isovaleroyl-L-tyrosyl-L-leucyl-L-tyrosinal (FW = 525.65 g/mol), was found to be a competitive inhibitor for kumamolysin, K_i = 2.8 μM (1), pseudomonalisin, K_i = 2.6 nM (2-8), and xanthomonalisin, K_i = 2.1 nM (5,8-10). Tyrostatin forms a covalent bond to an active-site seryl residue *via* its hydrated aldehyde moiety, with the side-chains occupying subsites S1-S4 of the enzymes. **See also** *Pseudo-tyrostatin; Pseudo-iodotyrostatin; Iodotyrostatin* 1. Oyama, Hamada, Ogasawara, *et al.* (2002) *J. Biochem.* 131, 757. 2. Murao (1998) in *Handb. Proteolytic Enzymes* (Barrett, Rawlings & Woessner, eds.), p. 975, Academic Press, San Diego. 3. Oda, Nakatani & Dunn (1992) *Biochim. Biophys. Acta* 1120, 208. 4. Oda, Takahashi, Tokuda, Shibano & Takahashi (1994) *J. Biol. Chem.* 269, 26518. 5. Ito, Dunn & Oda (1996) *J.*

Biochem. 120, 845. 6. Wlodawer, Li, Gustchina, *et al.* (2001) *Biochemistry* 40, 15602. 7. Ito, Narutaki, Uchida & Oda (1999) *J. Biochem.* 125, 210. 8. Oda, Fukuda, Murao, Uchida & Kainosho (1989) *Agric. Biol. Chem.* 53, 405. 9. Oda (1998) in *Handb. Proteolytic Enzymes* (Barrett, Rawlings & Woessner, eds.), p. 976, Academic Press, San Diego. 10. Oda, Ito, Uchida, *et al.* (1996) *J. Biochem.* 120, 564.

L-Tyrosyl-L-alanine

This dipeptide (FW = 252.27 g/mol) inhibits peptidyl-dipeptidase A (angiotensin I-converting enzyme; K_i = 0.06 mM). It also inhibits bovine tubulin:tyrosine ligase (K_i = 0.19 mM) and *Escherichia coli* tyrosyl-tRNA synthetase (K_i = 5 μM). **Target(s):** alanine carboxypeptidase, alternative substrate (1); dipeptidyl-peptidase III, weakly inhibited (2); peptidyl-dipeptidase A, *or* angiotensin I-converting enzyme (3,4); tubulin:tyrosine ligase (5); and tyrosyl-tRNA synthetase (6). 1. Levy & Goldman (1969) *J. Biol. Chem.* 244, 4467. 2. Lee & Snyder (1982) *J. Biol. Chem.* 257, 12043. 3. Das & Soffer (1975) *J. Biol. Chem.* 250, 6762. 4. Soffer (1976) *Ann. Rev. Biochem.* 45, 73. 5. Raybin & Flavin (1977) *Biochemistry* 16, 2189. 6. Santi & Peña (1973) *J. Med. Chem.* 16, 273.

L-Tyrosyl-L-alanyl-L-aspartyl-L-phenylalanyl-L-isoleucyl-L-alanyl-L-serylglycyl-L-arginyl-L-threonylglycyl-L-arginyl-L-arginyl-L-asparaginyl-L-alanyl-L-isoleucyl-L-histidine-L-glutamate

This 18-residue peptide (FW = 2034.22 g/mol; YADFIASGRTGRRNAIHE) inhibits cAMP-dependent protein kinase, K_i = 2 nM. 1. Walsh & Glass (1991) *Meth. Enzymol.* 201, 304.

L-Tyrosyl-L-aspartyl-L-prolyl-L-prolyl-L-alanyl-L-isoleucyl-statinyl-L-isoleucyl-L-isoleucine

This short peptide (FW = 1069.31 g/mol; Sequence: RDPPAI(Sta)II) inhibits bovine leukemia virus retropepsin (1-3) at nanomolar concentrations. If the statin residue is replaced with L-leucyl-L-prolyl, the resulting peptide is hydrolyzed by the same aspartic proteinase. 1. Ménard & Guillemain (1998) in *Handb. Proteolytic Enzymes* (Barrett, Rawlings & Woessner, eds.), p. 940, Academic Press, San Diego. 2. Ménard, Mamoun, Geoffre, *et al.* (1993) *Virology* 193, 680. 3. Précigoux, Geoffre, Léonard, *et al.* (1993) *FEBS Lett.* 326, 237.

N-[3-(D-Tyrosyl-L-glutamyl-L-lysyl-L-glutamyl-L-arginyl-L-seryl-L-lysyl-L-arginyl)propionyl]-L-alanyl-L-leucyl-L-arginyl-L-aspartate

This peptide analogue contains an ArgΨ(COCH$_2$)Gly replacing ArgSer in a peptide substrate for furin and proprotein convertase 1, with K_i values of 2.4 and 7.2 μM, respectively. 1. Jean, Basak, DiMaio, Seidah & Lazure (1995) *Biochem. J.* 307, 689.

D-Tyrosyl-L-glutamyl-L-phenylalanyl-L-lysyl-L-arginine Chloromethyl Ketone

This halomethyl ketone inhibits plasma kallikrein (1) and tryptase (2). 1. Paquin, Benjannet, Sawyer, *et al.* (1989) *Biochim. Biophys. Acta* 999, 103. 2. Cromlish, Seidah, Marcinkiewicz, *et al.* (1987) *J. Biol. Chem.* 262, 1363.

L-Tyrosylglycine

This dipeptide (FW = 222.24 g/mol) is a γ-glutamyl acceptor substrate and inhibitor of γ-glutamyl transpeptidase (1,2), tubulin:tyrosine ligase (3), and *Escherichia coli* tyrosyl-tRNA synthetase, K_i = 12 μM (4). 1. Allison (1985) *Meth. Enzymol.* 113, 419. 2. Thompson & Meister (1977) *J. Biol. Chem.* 252, 6792. 3. Raybin & Flavin (1977) *Biochemistry* 16, 2189. 4. Santi & Peña (1973) *J. Med. Chem.* 16, 273.

L-Tyrosylglycyl-L-arginine Chloromethyl Ketone

This halomethyl ketone ($FW_{free-base}$ = 426.90 g/mol; Symbol: YGR-*CMK*) inhibits certain mammalian proteasome activities. **1**. Rivett, Savory & Djaballah (1994) *Meth. Enzymol.* **244**, 331.

L-Tyrosyl-L-leucyl-L-tyrosyl-L-glutamyl-L-isoleucyl-L-alanyl-L-arginine

This heptapeptide (FW = 927.07 g/mol; Sequence: YLYEIAR), with a sequence corresponding to residues to 138–144 of human serum albumin (also known as acein-1), inhibits peptidyl-dipeptidase A *or* angiotensin I-converting enzyme, IC_{50} = 16 μM (1-2). The The hexapeptide lacking the C-terminal arginine is a weaker inhibitor (IC_{50} = 500 μM), as is the octapeptide with an additional C-terminal arginine (IC_{50} = 86 μM). **1**. Nakagomi, Yamada, Ebisu, *et al.* (2000) *FEBS Lett.* **467**, 235. **2**. Nakagomi, Fujimura, Ebisu, *et al.* (1998) *FEBS Lett.* **438**, 255.

L-Tyrosyl-L-tyrosine

This dipeptide (FW = 344.37 g/mol), often abbreviated YY, inhibits dipeptidyl-peptidase III1-4; tubulin:tyrosine ligase, IC_{50} = 23 μM (5); and staphylococcal tyrosyl-tRNA synthetase, IC_{50} = 3.8 μM (6,7). **1**. Chan, Jones, Sweeney & Toursarkissian (1987) *Exp. Mycol.* **11**, 27. **2**. Sentandreu & Toldrá (1998) *J. Agric. Food Chem.* **46**, 3977. **3**. Lee & Snyder (1982) *J. Biol. Chem.* **257**, 12043. **4**. Smyth & O'Cuinn (1994) *J. Neurochem.* **63**, 1439. **5**. Raybin & Flavin (1977) *Biochemistry* **16**, 2189. **6**. Jarvest, Berge, Houge-Frydrych, *et al.* (1999) *Bioorg. Med. Chem. Lett.* **9**, 2859. **7**. Santi & Peña (1973) *J. Med. Chem.* **16**, 273.

Tyrphostins

This class of protein-tyrosine kinase inhibitors, of which many are benzylidenemalononitrile derivatives, inhibit other processes, including lipoxygenases and oxidative phosphorylation. Most penetrate cell membranes slowly. A few examples are provided in the entries below. *See Uncouplers* **Target(s):** protein-tyrosine kinases (1,2) **1**. Levitzki, Gazit, Osherov, Posner & Gilon (1991) *Meth. Enzymol.* **201**, 347. **2**. Gazit, Yaish, Gilon & Levitzki (1989) *J. Med. Chem.* **32**, 2344.

Tyrphostin 1

This light-sensitive tyrosine phosphorylation inhibitor (FW = 184.20 g/mol; melting point = 115°C), also known as (4-methoxybenzylidene) malononitrile, tyrphostin A1, and tyrphostin AG9, is such a poor inhibitor of protein-tyrosine kinase, IC_{50} > 1.25 mM (1), is frequently used as a negative control. **1**. Gazit, Yaish, Gilon & Levitzki (1989) *J. Med. Chem.* **32**, 2344.

Tyrphostin 9, *See Tyrphostin A9*

Tyrphostin 23

This photosensitive tyrosine phosphorylation inhibitor (FW = 186.17 g/mol; melting point = 225°C), also known as (3,4-dihydroxybenzylidene) malononitrile, tyrphostin A23, tyrphostin AG18, and RG-50810, has an IC_{50} for the EGF receptor protein-tyrosine kinase of 35 μM and K_i value of 11 μM). Other protein-tyrosine kinases with relatively low IC_{50} values include platelet-derived growth factor receptor kinase (IC_{50} = 25 μM) and p210$^{bcr-abl}$

(IC_{50} = 75 μM). **Target(s):** calcineurin, ior phosphoprotein phosphatase (1); DNA topoisomerase I, IC_{50} = 750 μM (2,3); EGF receptor protein-tyrosine kinase (4-6); guanylyl cyclase (7); PDGF receptor protein-tyrosine kinase (4,6); p210$^{bcr-abl}$ protein-tyrosine kinase (4); protein-tyrosine kinase (4,8,9); and receptor protein-tyrosine kinase (4-6,9). **1**. Martin (1998) *Biochem. Pharmacol.* **56**, 483. **2**. Markovits, Larsen, Segal-Bendirdjian, *et al.* (1994) *Biochem. Pharmacol.* **48**, 549. **3**. Bendetz-Nezer, Gazit & Priel (2004) *Mol. Pharmacol.* **66**, 627. **4**. Levitzki, Gazit, Osherov, Posner & Gilon (1991) *Meth. Enzymol.* **201**, 347. **5**. Rosenshine, Ruschkowski & Finlay (1994) *Meth. Enzymol.* **236**, 467. **6**. Gazit, Yaish, Gilon & Levitzki (1989) *J. Med. Chem.* **32**, 2344. **7**. Jaleel, Shenoy & Visweswariah (2004) *Biochemistry* **43**, 8247. **8**. Gao, Lau, Wong & Li (2004) *Cell. Signal.* **16**, 333. **9**. Du, Gao, Lau, *et al.* (2004) *J. Gen. Physiol.* **123**, 427.

Tyrphostin 25

This photosensitive tyrosine phosphorylation inhibitor (FW = 202.17 g/mol; MP = 245°C), also known as (3,4,5-trihydroxybenzylidene)malononitrile, tyrphostin A25, and tyrphostin AG82, has an IC_{50} for the EGF receptor protein-tyrosine kinase of 3 μM and it is one of the most frequently used inhibitors for that kinase. Tyrphostin also inhibits the GTPase activity of transducin (IC_{50} = 7 μM) and the kinase activity of Abl p210$^{bcr-abl}$ (IC_{50} = 3.6 μM). **Target(s):** Abl p210$^{bcr-abl}$ protein-tyrosine kinase (1); aldehyde dehydrogenase (2); Bruton's protein-tyrosine kinase (10); epidermal growth factor receptor protein-tyrosine kinase (3-6); guanylyl cyclase (7); neuromedin B-induced phosphorylation of p125(FAK) (focal adhesion kinase) (8); protein-tyrosine kinase (1,9,10); receptor protein-tyrosine kinase (3-6,9); and transducin GTPase (11). **1**. Anafi, Gazit, Gilon, Ben-Neriah & Levitzki (1992) *J. Biol. Chem.* **267**, 4518. **2**. Poole, Bowden & Halestrap (1993) *Biochem. Pharmacol.* **45**, 1621. **3**. Rosenshine, Ruschkowski & Finlay (1994) *Meth. Enzymol.* **236**, 467. **4**. Gazit, Yaish, Gilon & Levitzki (1989) *J. Med. Chem.* **32**, 2344. **5**. Piontek, Hengels, Porschen & Strohmeyer (1993) *Anticancer Res.* **13**, 2119. **6**. Yaish, Gazit, Gilon & Levitzki (1988) *Science* **242**, 933. **7**. Jaleel, Shenoy & Visweswariah (2004) *Biochemistry* **43**, 8247. **8**. Tsuda, Kusui & Jensen (1997) *Biochemistry* **36**, 16328. **9**. Du, Gao, Lau, *et al.* (2004) *J. Gen. Physiol.* **123**, 427. **10**. Dinh, Grunberger, Ho, *et al.* (2007) *J. Biol. Chem.* **282**, 8768. **11**. Wolbring, Hollenberg & Schnetkamp (1994) *J. Biol. Chem.* **269**, 22470.

Tyrphostin 46

This photosensitive tyrosine phosphorylation inhibitor (FW = 204.19 g/mol; MP = 247°C), also known as 3,4-dihydroxy-α-cyanocinnamamide, tyrphostin A46, tyrphostin B40, and tyrphostin AG99, has an IC_{50} value for the EGF receptor protein-tyrosine kinase of 10 μM (1-3) and also inhibits ErbB2/neu protein-tyrosine kinase (3). **1**. Rosenshine, Ruschkowski & Finlay (1994) *Meth. Enzymol.* **236**, 467. **2**. Gazit, Yaish, Gilon & Levitzki (1989) *J. Med. Chem.* **32**, 2344. **3**. Gazit, Osherov, Posner, *et al.* (1991) *J. Med. Chem.* **34**, 1896.

Tyrphostin 47

This photosensitive tyrosine phosphorylation inhibitor (FW = 220.25 g/mol; MP = 213°C), also known as α-cyano-3,4-dihydroxythiocinnamamide, tyrphostin A47, and tyrphostin AG213, has an IC_{50} for the EGF (epidermal growth factor) receptor protein-tyrosine kinase of 2.4 μM, but is a relatively weak inhibitor of protein kinase C (IC_{50} = 60 μM). **Target(s):** aldehyde dehydrogenase (1); Ca^{2+} entry channels (2); cell proliferation (3); dual-specificity kinase (4); EGF receptor protein-tyrosine kinase (5-9); guanylyl cyclase (10); PDGF receptor protein-tyrosine kinase (5,7); $p210^{bcr-abl}$ protein-tyrosine kinase, IC_{50} = 5.8 μM (5); protein-glutamine γ-glutamyltransferase, *or* transglutaminase (11); and protein-tyrosine kinase (5,12). **1.** Poole, Bowden & Halestrap (1993) *Biochem. Pharmacol.* **45**, 1621. **2.** Ohta, Yasuda, Hasegawa, Ito & Nakazato (2000) *Eur. J. Pharmacol.* **387**, 211. **3.** Twaddle, Turbov, Liu & Murthy (1999) *J. Surg. Oncol.* **70**, 83. **4.** Rudrabhatla & Rajasekharan (2004) *Biochemistry* **43**, 12123. **5.** Levitzki, Gazit, Osherov, Posner & Gilon (1991) *Meth. Enzymol.* **201**, 347. **6.** Rosenshine, Ruschkowski & Finlay (1994) *Meth. Enzymol.* **236**, 467. **7.** Gazit, Yaish, Gilon & Levitzki (1989) *J. Med. Chem.* **32**, 2344. **8.** Lyall, Zilberstein, Gazit, *et al.* (1989) *J. Biol. Chem.* **264**, 14503. **9.** Yaish, Gazit, Gilon & Levitzki (1988) *Science* **242**, 933. **10.** Jaleel, Shenoy & Visweswariah (2004) *Biochemistry* **43**, 8247. **11.** Lai, Liu, Tucker, *et al.* (2008) *Chem. Biol.* **15**, 969. **12.** Merciris, Claussen, Joiner & Giraud (2003) *Pflugers Arch.* **446**, 232.

Tyrphostin 48

This tyrosine phosphorylation inhibitor (FW = 236.23 g/mol; melting point = 225°C), also known as 3-amino-2,4-dicyano-5-(4'-hydroxyphenyl)penta-2,4-dienonitrile, tyrphostin A48, and tyrphostin AG112, is a broad-range inhibitor of EGF (epidermal growth factor) receptor protein-tyrosine kinase (IC_{50} = 0.125 μM). **Target(s):** calcineurin, *or* phosphoprotein phosphatase (1); and EGF receptor protein-tyrosine kinase (2). **1.** Martin (1998) *Biochem. Pharmacol.* **56**, 483. **2.** Gazit, Yaish, Gilon & Levitzki (1989) *J. Med. Chem.* **32**, 2344.

Tyrphostin 51

This photosensitive tyrosine phosphorylation inhibitor (FW = 268.23 g/mol; MP = 275°C), also known as 2-amino-1,1,3-tricyano-4-(3',4',5'-trihydroxyphenyl)buta-1,3-diene, tyrphostin A51, and tyrphostin AG183, has an IC_{50} for the EGF (epidermal growth factor) receptor protein tyrosine kinase of 0.8 μM. **Target(s):** EGF receptor protein-tyrosine kinase (1-3); inositol-trisphosphate 3-kinase (4); and receptor protein-tyrosine kinase (1-3,5). **1.** Rosenshine, Ruschkowski & Finlay (1994) *Meth. Enzymol.* **236**, 467. **2.** Gazit, Yaish, Gilon & Levitzki (1989) *J. Med. Chem.* **32**, 2344. **3.** Levitzki (1990) *Biochem. Pharmacol.* **40**, 913. **4.** Mayr, Windhorst & Hillemeier (2005) *J. Biol. Chem.* **280**, 13229. **5.** Mergler, Dannowski, Bednarz, *et al.* (2003) *Exp. Eye Res.* **77**, 485.

Tyrphostin 63

This photosensitive tyrosine phosphorylation inhibitor (FW = 172.19 g/mol), also known as (4-hydroxybenzyl)malononitrile, tyrphostin A63, and tyrphostin AG43, has such a high IC_{50} value for EGF receptor protein-tyrosine kinase of 6.5 mM that it is frequently used as a negative control (1). **1.** Gazit, Yaish, Gilon & Levitzki (1989) *J. Med. Chem.* **32**, 2344.

Tyrphostin A1, See *Tyrphostin 1*

Tyrphostin A8, See *Tyrphostin AG10*

Tyrphostin A9

This tyrosine phosphorylation inhibitor (FW = 282.39 g/mol; melting point = 135°C), also known as tyrphostin 9, tyrphostin AG17, RG 50852, and [[3,5-bis(1,1-dimethylethyl)-4-hydroxyphenyl]methylene]propanedinitrile, is a selective inhibitor of platelet-derived growth factor (PDGF) receptor protein-tyrosine kinase (IC_{50} = 0.5 μM). Tyrphostin also reportedly uncouples oxidative phosphorylation. **Target(s):** PDGF receptor protein-tyrosine kinase (1,2); and receptor protein-tyrosine kinase (1,2). **1.** Bilder, Krawiec, McVety, *et al.* (1991) *Amer. J. Physiol.* **260**, C721. **2.** Wang, Buck, RYang, Macey & Neve (2005) *J. Neurochem.* **93**, 899.

Tyrphostin A23, See *Tyrphostin 23*

Tyrphostin A24, See *Tyrphostin AG34*

Tyrphostin A25, See *Tyrphostin 25*

Tyrphostin A46, See *Tyrphostin 46*

Tyrphostin A47, See *Tyrphostin 47*

Tyrphostin A48, See *Tyrphostin 48*

Tyrphostin A51, See *Tyrphostin 51*

Tyrphostin A63, See *Tyrphostin 63*

Tyrphostin AG9, See *Tyrphostin 1*

Tyrphostin AG10

This tyrosine phosphorylation inhibitor (FW = 170.17 g/mol), also known as tyrphostin A8, α-cyano-(4-hydroxy)cinnamonitrile, and 4-hydroxybenzylidene-malononitrile, inhibits the GTPase activity of transducin, IC_{50} = 45 μM (1) and also inhibits calcineurin activity, IC_{50} = 21 μM (2). **1.** Wolbring, Hollenberg & Schnetkamp (1994) *J. Biol. Chem.* **269**, 22470. **2.** Martin (1998) *Biochem. Pharmacol.* **56**, 483.

Tyrphostin AG17, See *Tyrphostin A9*

Tyrphostin AG18, See *Tyrphostin 23*

Tyrphostin AG30

This tyrosine phosphorylation inhibitor ($FW_{free-acid}$ = 205.17 g/mol), also known as α-cyano-(3,4-dihydroxy)cinnamic acid, inhibits c-ErbB protein-tyrosine kinase (1). **1.** Wessely, Mellitzer, von Lindern, *et al.* (1997) *Cell Growth Differ.* **8**, 481.

Tyrphostin AG34

This tyrosine phosphorylation inhibitor (FW = 216.20 g/mol; MP = 235°C), also called tyrphostin A24, inhibits casein kinase II (1); EGF receptor protein-tyrosine kinase (2); and protein-tyrosine kinase (2,3). **1.** Kang, Huang & Liang (1997) *Zhongguo Yao Li Xue Bao* **18**, 56. **2.** Gazit, Yaish, Gilon &

Levitzki (1989) *J. Med. Chem.* **32**, 2344. **3**. Schwartz, Lamprecht, Polak-Charcon, Niv & Kim (1995) *Oncol. Res.* **7**, 277.

Tyrphostin AG43, See *Tyrphostin 63*

Tyrphostin AG82, See *Tyrphostin 25*

Tyrphostin AG99, See *Tyrphostin 46*

Tyrphostin AG112, See *Tyrphostin 48*

Tyrphostin AG126

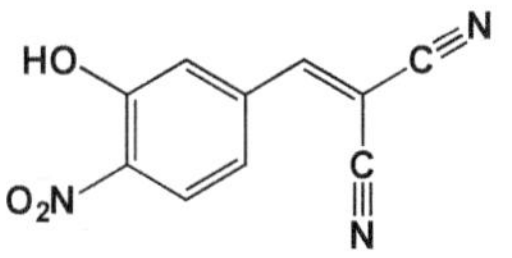

This tyrosine phosphorylation inhibitor (FW = 215.17 g/mol), also known as (3-hydroxy-4-nitrobenzylidene)-malononitrile, blocks the production of tumor necrosis factor-α and nitric oxide in macrophages. This inhibitory effect correlates with the ability of this tyrphostin to block the lipopolysaccharide-induced protein-tyrosine phosphorylation of a macrophage p42MAPK protein substrate (1). **Target(s):** EGF (epidermal growth factor) receptor protein-tyrosine kinase, IC_{50} = 450 μM (2); protein-tyrosine kinase (1,2); and TNF-α production (1). **1**. Novogrodsky, Vanichkin, Patya, *et al.* (1994) *Science* **264**, 1319. **2**. Gazit, Yaish, Gilon & Levitzki (1989) *J. Med. Chem.* **32**, 2344.

Tyrphostin AG183, See *Tyrphostin 51*
Tyrphostin AG213, See *Tyrphostin 47*

Tyrphostin AG370

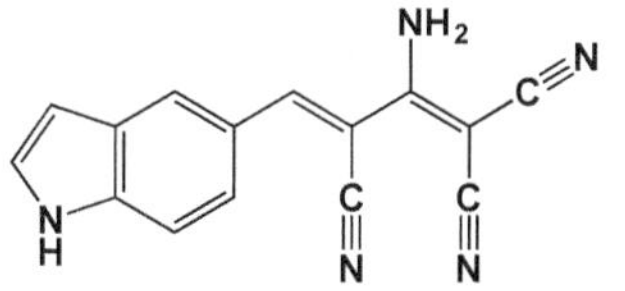

This tyrosine phosphorylation inhibitor (FW = 259.27 g/mol; melting point = 242°C), also known as 2-amino-4-(1*H*-indo-5'-yl)-1,1,3-tricyanobuta-1,3-diene and tyrphostin B7, inhibits certain protein-tyrosine kinases. AG370 was found to be the a potent blocker of PDGF-induced (platelet-derived growth factor-induced) mitogenesis, IC_{50} = 20-25 μM (1). **Target(s):** PDGF protein-tyrosine kinase (1,2); and receptor protein-tyrosine kinase (1-3). **1**. Bryckaert, Eldor, Fontenay, *et al.* (1992) *Exp. Cell Res.* **199**, 255. **2**. Wang, Buck, Yang, Macey & Neve (2005) *J. Neurochem.* **93**, 899. **3**. Levitzki, Gazit, Osherov, Posner & Gilon (1991) *Meth. Enzymol.* **201**, 347.

Tyrphostin AG473

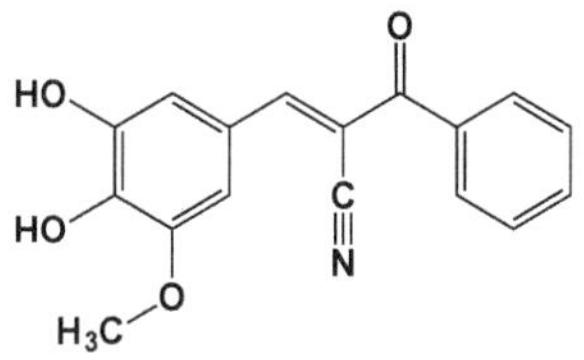

This tyrosine phosphorylation inhibitor (FW = 265.27 g/mol), also known as 3,4-dihydroxy(α-benzoyl)*cis*-cinnamonitrile, inhibits certain protein-tyrosine kinases (1). **1**. Levitzki, Gazit, Osherov, Posner & Gilon (1991) *Meth. Enzymol.* **201**, 347.

Tyrphostin AG490

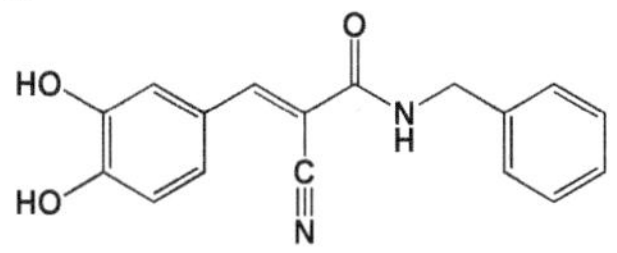

This photosensitive tyrosine phosphorylation inhibitor (FW = 294.31 g/mol; MP = 215°C), also known as 2-cyano-3-(3,4-dihydroxyphenyl)-*N*-(benzyl)-2-propenamide and tyrphostin B42, is a selective inhibitor of Jak-2 protein-tyrosine kinase, IC_{50} = 100 nM. Pretreating cells with tyrophostin AG490 abolishes protein-tyrosine phosphorylation of the p85 subunit of PI3 kinase induced by granulocyte-macrophage colony-stimulating factor, or GM-CSF (1). Tyrphostin AG490 also causes cells to arrest at late G1 and during S phase (2). **Target(s):** Cdk2 activation3; EGF (epidermal growth factor) receptor protein-tyrosine kinase, IC_{50} = 0.1 μM (4); guanylyl cyclase (5); inositol-trisphosphate 3-kinase (6); and Jak-2 protein-tyrosine kinase (1,7,8). **1**. Al-Shami & Naccache (1999) *J. Biol. Chem.* **274**, 5333. **2**. Kleinberger-Doron, Shelah, Capone, Gazit & Levitzki (1998) *Exp. Cell Res.* **241**, 340. **3**. Osherov & Levitzki (1997) *FEBS Lett.* **410**, 187. **4**. Gazit, Osherov, Posner, *et al.* (1991) *J. Med. Chem.* **34**, 1896. **5**. Jaleel, Shenoy & Visweswariah (2004) *Biochemistry* **43**, 8247. **6**. Mayr, Windhorst & Hillemeier (2005) *J. Biol. Chem.* **280**, 13229. **7**. Meydan, Grunberger, Dadi, *et al.* (1996) *Nature* **379**, 645. **8**. Anastasiadou & Schwaller (2003) *Curr. Opin. Hematol.* **10**, 40.

Tyrphostin AG494

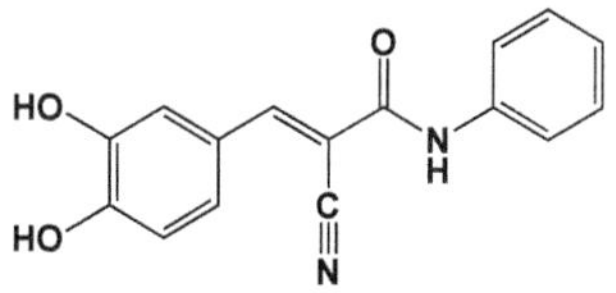

This tyrosine phosphorylation inhibitor (FW = 280.28 g/mol; MP = 258°C), also known as tyrphostin B48, *N*-phenyl-3,4-dihydroxybenzylidene-cyanoacetamide, and 3,4-dihydroxy(α-benzamido)-*cis*-cinnamonitrile, inhibits EGF-receptor phosphorylation (albeit, not in intact cells): the IC_{50} is 1.24 μM. Tyrphostin AG494 also causes cells to arrest at late G1 and during S phas (1). **Target(s):** Cdk2 activation (2); cell proliferation (3); epidermal growth factor receptor protein-tyrosine kinase (1,4); and guanylyl cyclase (5). **1**. Kleinberger-Doron, Shelah, Capone, Gazit & Levitzki (1998) *Exp. Cell Res.* **241**, 340. **2**. Osherov & Levitzki (1997) *FEBS Lett.* **410**, 187. **3**. Twaddle, Turbov, Liu & Murthy (1999) *J. Surg. Oncol.* **70**, 83. **4**. Levitzki, Gazit, Osherov, Posner & Gilon (1991) *Meth. Enzymol.* **201**, 347. **5**. Jaleel, Shenoy & Visweswariah (2004) *Biochemistry* **43**, 8247.

Tyrphostin AG527

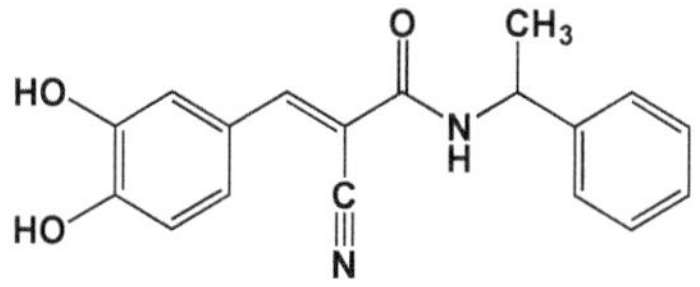

This tyrosine phosphorylation inhibitor (FW = 308.34 g/mol; MP = 135°C), also known as tyrphostin B44 and (−)-(*R*)-*N*-(α-methylbenzyl)-3,4-dihydroxybenzylidenecyano-acetamide, inhibits EGF-receptor phosphorylation: the IC_{50} is 2.5 μM (1). *Note*: The (+)-(*S*)-enantiomer of tyrphostin AG527 is tyrphostin AG835. **1**. Gazit, Osherov, Posner, *et al.* (1991) *J. Med. Chem.* **34**, 1896.

Tyrphostin AG528

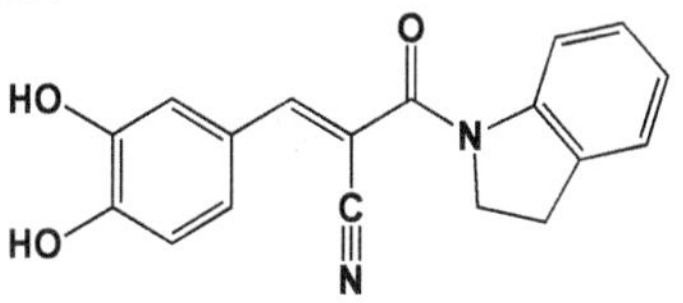

This photosensitive tyrosine phosphorylation inhibitor (FW = 306.32 g/mol; MP = 215°C), also known as tyrphostin B66, inhibits EGF-receptor phosphorylation, IC_{50} is 12 μM (1). **1**. Gazit, Osherov, Posner, *et al.* (1991) *J. Med. Chem.* **34**, 1896.

Tyrphostin AG537

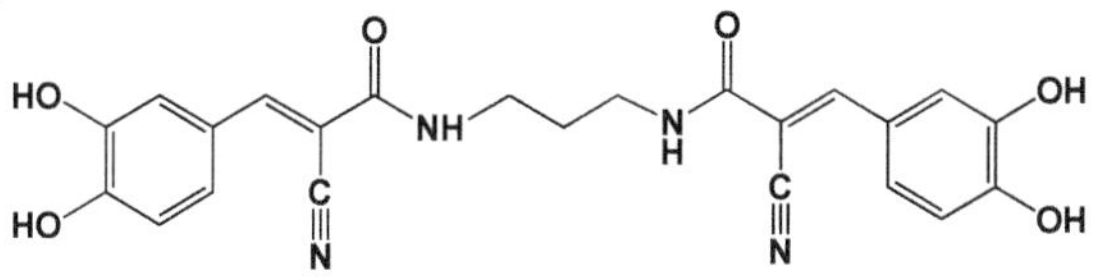

This photosensitive tyrosine phosphorylation inhibitor (FW = 448.44 g/mol), also known as bis-tyrphostin, inhibits EGF (epidermal growth factor) receptor protein-tyrosine kinase, IC_{50} = 0.4 μM and K_i = 60 nM. **Target(s):** EGF receptor protein-tyrosine kinase (1); inositol-trisphosphate 3-kinase (2); and protein-tyrosine kinase (3). **1**. Levitzki & Gilon (1991) *Trends*

Pharmaceut. Sci. **12**, 171. **2**. Mayr, Windhorst & Hillemeier (2005) *J. Biol. Chem.* **280**, 13229. **3**. Tardif, Dubé & Bailey (2003) *Biol. Reprod.* **68**, 207.

Tyrphostin AG538

This tyrosine phosphorylation inhibitor (FW = 297.27 g/mol), also known as α-cyano-(3,4-dihydroxy)cinnamoyl-(3',4'-dihydroxyphenyl)ketone, inhibits insulin-like growth factor receptor protein-tyrosine kinase, IC$_{50}$ = 0.4 µM (1). **1**. Blum, Gazit & Levitzki (2000) *Biochemistry* **39**, 15705.

Tyrphostin AG538 I-Ome, *See 5-Iodo-3-Methoxy-tyrphostin AG538*

Tyrphostin AG555

This photosensitive tyrosine phosphorylation inhibitor (FW = 322.36 g/mol; melting point = 165°C), also known as tyrphostin B46 and *N*-(3'-phenylpropyl)-3,4-dihydroxy-benzylidenecyanoacetamide, inhibits epidermal growth factor receptor protein-tyrosine kinase: IC$_{50}$ = 0.7 µM. **Target(s):** Cdk2 activation (1); cell proliferation (2); DNA topoisomerase I, IC$_{50}$ = 100 µM (3,4); EGF receptor protein-tyrosine kinase (5); and inositol-trisphosphate 3-kinase (6). **1**. Osherov & Levitzki (1997) *FEBS Lett.* **410**, 187. **2**. Sion-Vardy, Vardy, Rodeck, *et al.* (1995) *J. Surg. Res.* **59**, 675. **3**. Markovits, Larsen, Segal-Bendirdjian, *et al.* (1994) *Biochem. Pharmacol.* **48**, 549. **4**. Bendetz-Nezer, Gazit & Priel (2004) *Mol. Pharmacol.* **66**, 627. **5**. Gazit, Osherov, Posner, *et al.* (1991) *J. Med. Chem.* **34**, 1896. **6**. Mayr, Windhorst & Hillemeier (2005) *J. Biol. Chem.* **280**, 13229.

Tyrphostin AG556

This photosensitive tyrosine phosphorylation inhibitor (FW = 336.39 g/mol; melting point = 180°C), also known as tyrphostin B56 and *N*-(4-phenylbutyl)-3,4-dihydroxy-benzylidenecyanoacetamide, inhibits EGF (epidermal growth factor) receptor protein-tyrosine kinase: IC$_{50}$ = 5 µM. Tyrphostin AG556 also blocks lipopolysaccharide-induced TNF-α production and ErbB-1 kinase. **Target(s):** EGF receptor protein-tyrosine kinase (1,2); guanylyl cyclase (3); and receptor protein-tyrosine kinase (1,2). **1**. Gazit, Osherov, Posner, *et al.* (1991) *J. Med. Chem.* **34**, 1896. **2**. Du, Gao, Lau, *et al.* (2004) *J. Gen. Physiol.* **123**, 427. **3**. Jaleel, Shenoy & Visweswariah (2004) *Biochemistry* **43**, 8247.

Tyrphostin AG658

This photosensitive tyrosine phosphorylation inhibitor (FW = 340.40 g/mol; MP = 184°C), also known as tyrphostin C6 and 4-hydroxy-3-methoxy-5-(phenylthiomethyl)benz-ylidenecyanoacetamide, inhibits EGF (epidermal

growth factor) receptor protein-tyrosine kinase, IC$_{50}$ = 0.4 µM (1). **1**. Gazit, Osherov, Posner, *et al.* (1993) *J. Med. Chem.* **36**, 3556.

Tyrphostin AG698

This photosensitive tyrosine phosphorylation inhibitor (FW = 308.34 g/mol; melting point = 185°C), also known as tyrphostin B52 and *N*-(2-phenylethyl)-3,4-dihydroxy-benzylidenecyanoacetamide, inhibits EGF receptor protein-tyrosine kinase, IC$_{50}$ = 0.93 µM. **1**. Gazit, Osherov, Posner, *et al.* (1991) *J. Med. Chem.* **34**, 1896.

Tyrphostin AG808

This tyrosine phosphorylation inhibitor (FW = 304.30 g/mol), also known as 2-cyano-3-(3',4'-dihydroxyphenyl)-1-(3"-indolyl)-3-oxo-1-propene, inhibits protein-tyrosine kinase. **1**. Romer, McLean, Turner & Burridge (1994) *Mol. Biol. Cell* **5**, 349.

Tyrphostin AG825

This tyrosine phosphorylation inhibitor (FW = 397.48 g/mol; MP = 268°C), also known as tyrphostin C15 and 5-[(benzothiazol-2-yl)thiomethyl]-4-hydroxy-3-methoxy-benzylidenecyanoacetamide, inhibits EGF (epidermal growth factor) receptor protein-tyrosine kinase (IC$_{50}$ = 19 µM). Note that this tyrphostin does not inhibit the IGF-1 receptor protein-tyrosine kinase activity. **Target(s):** EGF receptor protein-tyrosine kinase (1,2); and HER2 kinase, p185^{neu} (2,3). **1**. Gazit, Osherov, Posner, *et al.* (1993) *J. Med. Chem.* **36**, 3556. **2**. Osherov, Gazit, Gilon & Levitzki (1993) *J. Biol. Chem.* **268**, 11134. **3**. Tsai, Levitzki, Wu, *et al.* (1996) *Cancer Res.* **56**, 1068.

Tyrphostin AG835

This photosensitive tyrosine phosphorylation inhibitor (FW = 308.34 g/mol; MP = 135°C), also known as tyrphostin B50 and (+)-(*S*)-*N*-(α-methylbenzyl)-3,4-dihydroxy-benzylidenecyanoacetamide, inhibits EGF-receptor (epidermal growth factor receptor) phosphorylation, IC$_{50}$ = 0.86 µM. *Note*: The (−)-(*R*)-enantiomer of tyrphostin AG835 is tyrphostin AG527. **Target(s):** cell proliferation (1); and EGF receptor protein-tyrosine kinase (2). **1**. Twaddle, Turbov, Liu & Murthy (1999) *J. Surg. Oncol.* **70**, 83. **2**. Gazit, Osherov, Posner, *et al.* (1991) *J. Med. Chem.* **34**, 1896.

Tyrphostin AG879

This tyrosine phosphorylation inhibitor (FW = 316.47 g/mol; MP = 210°C), also known as α-cyano-(3,5-di-*t*-butyl-4-hydroxy)thiocinnamide, inhibits protein-tyrosine kinase activity of the nerve factor receptor. **Target(s):** HER2 kinase (1,2); and nerve growth factor receptor p140c-*trk* protein-tyrosine kinase (1,2). **1**. Nie, Mei, Malek, *et al.* (1999) *Mol. Pharmacol.* **56**, 947. **2**. Ohmichi, Pang, Ribon, *et al.* (1993) *Biochemistry* **32**, 4650.

Tyrphostin AG957

This tyrosine phosphorylation inhibitor (FW = 273.29 g/mol), also known as methyl 4-[*N*-(2',5'-dihydroxybenzyl)amino]benzoate, inhibits human p210 protein-tyrosine kinase, *or* p210$^{bcr-abl}$, IC$_{50}$ = 0.75 μM (1). Tyrphostin AG957 also inhibits epidermal growth factor receptor protein-tyrosine kinase, IC$_{50}$ = 0.25 μM. **See also** *Lavendustin C; Lavendustin C Methyl Ester* **Target(s):** EGF receptor protein-tyrosine kinase (1); and p210 protein-tyrosine kinase, human (1,2). **1**. Anafi, Gazit, Gilon, Ben-Neriah & Levitzki (1992) *J. Biol. Chem.* **267**, 4518. **2**. Sun, Layton, Elefanty & Lieschke (2001) *Blood* **97**, 2008.

Tyrphostin AG974

This tyrosine phosphorylation inhibitor (FW = 312.07 g/mol; melting point = 105°C) inhibits protein-tyrosine kinase (1) and DNA topoisomerase I, IC$_{50}$ = 750 μM (2); and protein-tyrosine kinase (1). **1**. Ohmichi, Pang, Ribon, *et al.* (1993) *Biochemistry* **32**, 4650. **2**. Bendetz-Nezer, Gazit & Priel (2004) *Mol. Pharmacol.* **66**, 627.

Tyrphostin AG1024

This tyrosine phosphorylation inhibitor (FW = 305.17 g/mol) inhibits protein-tyrosine kinases, particularly the insulin-like growth factor-1 (IGF-1) protein-tyrosine kinase, IC$_{50}$ = 57 μM (1-2) and receptor protein-tyrosine kinase (1-3). **1**. Ohmichi, Pang, Ribon, *et al.* (1993) *Biochemistry* **32**, 4650. **2**. Parrizas, Gazit, Levitzki, Wertheimer & LeRoith (1997) *Endocrinology* **138**, 1427. **3**. Mehdi, Azar & Srivastava (2007) *Cell Biochem. Biophys.* **47**, 1.

Tyrphostin AG1222

This potent tyrosine phosphorylation inhibitor (FW = 274.28 g/mol) targets p210$^{bcr-abl}$ protein-tyrosine kinase (1) and casein kinase II (2). **1**. Anafi, Gazit, Zehavi, Ben-Neriah & Levitzki (1993) *Blood* **82**, 3524. **2**. Kang, Huang & Liang (1997) *Zhongguo Yao Li Xue Bao* **18**, 56.

Tyrphostin AG1288

This tyrosine phosphorylation inhibitor (FW = 231.17 g/mol), also known as (3,4-dihydroxy-5-nitrobenzylidene)malononitrile, blocks the cytotoxicity associated with tumor necrosis factor-α (1) and inhibits guanylyl cyclase (1); protein-tyrosine kinase (2). **1**. Jaleel, Shenoy & Visweswariah (2004) *Biochemistry* **43**, 8247. **2**. Novogrodsky, Vanichkin, Patya, *et al.* (1994) *Science* **264**, 1319.

Tyrphostin AG1295

This tyrosine phosphorylation inhibitor (FW = 234.30 g/mol), also known as 6,7-dimethyl-2-phenylquinoxaline, is a selective inhibitor of PDGF receptor protein-tyrosine kinase, IC$_{50}$ = 0.5 μM (1). Tyrphostin AG1295 also inhibits PDGF-dependent DNA synthesis in 3T3 cells and in porcine aorta endothelial cells. **Target(s):** platelet-derived growth factor (PDGF) receptor protein-tyrosine kinase (1); and receptor protein-tyrosine kinase (1,2). **1**. Kovalenko, Gazit, Bohmer, *et al.* (1994) *Cancer Res.* **54**, 6106. **2**. Mehdi, Azar & Srivastava (2007) *Cell Biochem. Biophys.* **47**, 1.

Tyrphostin AG1296

This tyrosine phosphorylation inhibitor (FW = 266.30 g/mol), also known as 6,7-dimethoxy-2-phenylquinoxaline, is a selective inhibitor of PDGF receptor protein-tyrosine kinase, IC$_{50}$ = 1 μM). Tyrphostin AG1296 also inhibits PDGF-dependent DNA synthesis in 3T3 cells and in porcine aorta endothelial cells (1-3). In addition, AG1296 inhibits signaling of human PDGF α- and β-receptors as well as of the related stem cell factor receptor (c-Kit). **1**. Kovalenko, Gazit, Bohmer, *et al.* (1994) *Cancer Res.* **54**, 6106. **2**. Schmidt-Arras, Bohmer, Markova, *et al.* (2005) *Mol. Cell. Biol.* **25**, 3690. **3**. Ingram & Bonner (2006) *Curr. Mol. Med.* **6**, 409.

Tyrphostin AG1433

This tyrosine phosphorylation inhibitor (FW$_{free-base}$ = 266.30 g/mol), also known as SU1433 and 2-(3',4'-dihydroxyphenyl)-6,7-dimethylquinoxaline, inhibits PDGFβ (platelet-derived growth factor-β) receptor protein-tyrosine kinase, IC$_{50}$ = 5 μM) and Flk-1/KDR (receptor for vascular endothelial growth factor) protein-tyrosine kinase, IC$_{50}$ = 9.3 μM). Tyrphostin AG1433 also inhibits angiogenesis. **Target(s):** Flk-1/KDR (a receptor for vascular endothelial growth factor) protein-tyrosine kinase, IC$_{50}$ = 9.3 μM (1,2); and PDGFβ receptor protein-tyrosine kinase (1,2). **1.** Strawn, McMahon, App, *et al.* (1996) *Cancer Res.* **56**, 3540. **2.** Kroll & Waltenberger (1997) *J. Biol. Chem.* **272**, 32521.

Tyrphostin AG1478

This tyrosine phosphorylation inhibitor (FW = 315.76 g/mol), also known as *N*-(3-chlorophenyl)-6,7-dimethoxy-4-quinazolinamine and 4-(3-chloroanilino)-6,7-dimethoxyquinazoline, is a selective inhibitor of epidermal growth factor receptor protein-tyrosine kinase, IC$_{50}$ = 3 nM. **Target(s):** EGF receptor protein-tyrosine kinase (1-7); fructose-1,6-bisphosphatase, IC$_{50}$ = 1.3 μM (8); and receptor protein-tyrosine kinase (1-7,9,10). **1.** Osherov & Levitzki (1994) *Eur. J. Biochem.* **225**, 1047. **2.** Rosenshine, Ruschkowski & Finlay (1994) *Meth. Enzymol.* **236**, 467. **3.** El-Obeid, Hesselager, Westermark & Nister (2002) *Biochem. Biophys. Res. Commun.* **290**, 349. **4.** Yoshinaga, Murayama & Nomura (2000) *Biochem. Pharmacol.* **60**, 111. **5.** Ingram & Bonner (2006) *Curr. Mol. Med.* **6**, 409. **6.** Benter, Juggi, Khan, *et al.* (2005) *Mol. Cell. Biochem.* **268**, 175. **7.** Wang, Buck, Yang, Macey & Neve (2005) *J. Neurochem.* **93**, 899. **8.** Wright, Hageman, McClure, *et al.* (2001) *Bioorg. Med. Chem. Lett.* **11**, 17. **9.** Mehdi, Azar & Srivastava (2007) *Cell Biochem. Biophys.* **47**, 1. **10.** Werry, Gregory, Sexton & Christopoulos (2005) *J. Neurochem.* **93**, 1603.

Tyrphostin AG1517, *See* 4-[(3-Bromophenyl)amino]-6,7-dimethoxyquinazoline

Tyrphostin AG1879, *See* PP2

Tyrphostin B7, *See* Tyrphostin AG370

Tyrphostin B40, *See* Tyrphostin 46

Tyrphostin B42, *See* Tyrphostin AG490

Tyrphostin B44, *See* Tyrphostin AG527

Tyrphostin B46, *See* Tyrphostin AG555

Tyrphostin B48, *See* Tyrphostin AG494

Tyrphostin B50, *See* Tyrphostin AG835

Tyrphostin B52, *See* Tyrphostin AG698

Tyrphostin B56, *See* Tyrphostin AG556

Tyrphostin B66, *See* Tyrphostin AG528

Tyrphostin C6, *See* Tyrphostin AG658

Tyrphostin C15, *See* Tyrphostin AG825

Tyrphostin I-OMe-AG538, *See* 5-Iodo-3-Methoxy-tyrphostin AG538

Tyrphostin RG13022

This tyrosine phosphorylation inhibitor (FW = 266.30 g/mol), also known as 2-(3',4'-dimethoxyphenyl)-1-(3''-pyridinyl)acrylonitrile, inhibits EGF receptor protein-tyrosine kinase (IC$_{50}$ = 1-3 μM), cell proliferation (2); and DNA synthesis (3). **1.** Yoneda, Lyall, Alsina, *et al.* (1991) *Cancer Res.* **51**, 4430. **2.** Twaddle, Turbov, Liu & Murthy (1999) *J. Surg. Oncol.* **70**, 83. **3.** McLeod, Brunton, Eckardt, *et al.* (1996) *Brit. J. Cancer* **74**, 1714.

Tyrphostin RG14620

This antiproliferative tyrosine phosphorylation inhibitor (FW = 275.14 g/mol), also known as 2-(3',5'-dichlorophenyl)-1-(3''-pyridinyl)acrylonitrile, inhibits EGF receptor protein-tyrosine kinase, IC$_{50}$ = 3 μM (1,2). **1.** Yoneda, Lyall, Alsina, *et al.* (1991) *Cancer Res.* **51**, 4430. **2.** Sion-Vardy, Vardy, Rodeck, *et al.* (1995) *J. Surg. Res.* **59**, 675.

Tyrphostin SU1498

This selective receptor-linked tyrosine phosphorylation inhibitor (FW = 390.53 g/mol), also known as (*E*)-*N*-(3''-phenylpropyl)-α-cyano-3',5'-diisopropyl-4'-hydroxy-cinnamamide, targets VEGF receptor protein-tyrosine kinase Flk-1, IC$_{50}$ = 0.7 μM (1). **1.** Strawn, McMahon, App, *et al.* (1996) *Cancer Res.* **56**, 3540.

– U –

U, *See Uracil; Uridine*

U-104

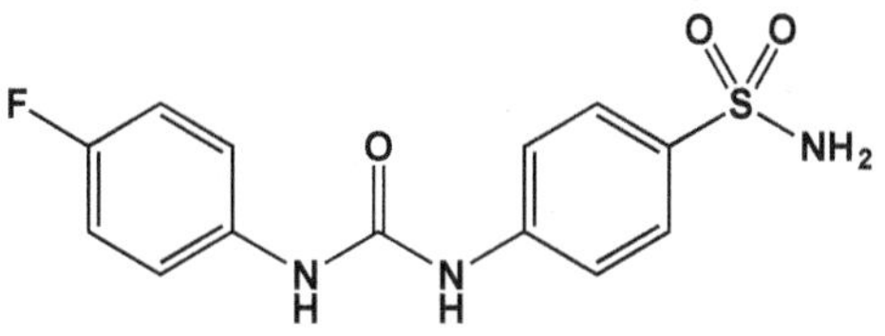

This potent inhibitory sulfonamide (FW = 309.32 g/mol; CAS 178606-66-1; Soluble to 100 mM in DMSO) targets carbonic anhydrases CA-XII (K_i = 4.5 nM), CA-IX (K_i = 45 nM), CA-I (K_i = 5.1 μM) and CA-II (K_i = 9.6 μM). Carbonic anhydrase IX (CA-IX) is a hypoxia and HIF-1-inducible protein that regulates intra- and extracellular pH under hypoxic conditions and promotes tumor cell survival and invasion within hypoxic microenvironments. Stable depletion of CAIX activity in MDA-MB-231 human breast cancer xenografts attenuates primary tumor growth. CAIX depletion in the 4T1 cells leads to caspase-independent cell death and reversal of extracellular acidosis under hypoxic conditions *in vitro*. 1. Lou, McDonald, Oloumi, *et al.* (2011) *Cancer Res.* **71**, 3364.

U0126

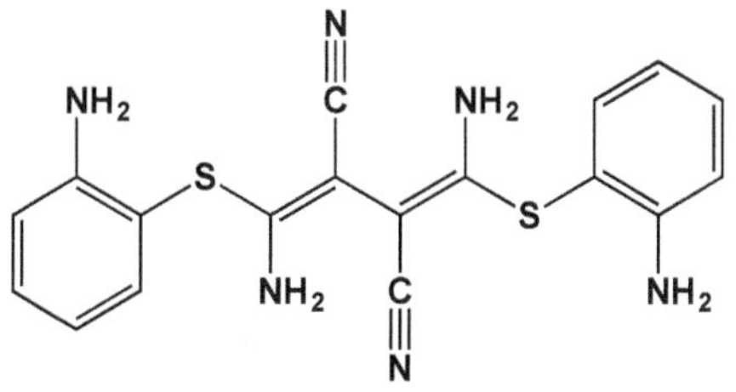

This synthetic rotationally symmetric agent (FW$_{\text{free-base}}$ = 380.50 g/mol; CAS 1173097-76-1; Solubility: 85 mg/mL DMSO; <1 mg/mL H$_2$O), also named 1,4-diamino-2,3-dicyano-1,4-bis(*o*-aminophenylmercapto)butadiene, targets the intracellular Raf/MEK/ERK signaling pathway, inhibiting MEK1 (IC$_{50}$ = 0.07 μM), MEK2 (IC$_{50}$ = 0.06 μM), with 100-times higher affinity for MEK$^{\Delta N3\text{-}S218E/S222D}$ than observed with the MAPKK inhibitor PD098059. U0126 induces apoptosis in cancer cells. (*See Rapamycin*)
Target(s): mitogen-activated protein kinase (1); [mitogen-activated protein kinase] kinase (2-10); [mitogen-activated protein kinase kinase] kinase (1); and stress-activated protein kinase 2a, mildly inhibited (4). U0126 also exhibits antiviral activity against pandemic H1N1v swine influenza A virus and highly pathogenic avian influenza virus (AIV), both *in vitro* and *in vivo* (13). Indeed, treatment of mice with U0126 via the aerosol route led to inhibition of MEK activation in the lung, reduction of progeny IAV titers compared to untreated controls, protection of IAV-infected mice against a 100x lethal viral challenge. 1. Otsuka, Goto, Tsuchiya & Aramaki (2005) *Biol. Pharm. Bull.* **28**, 1707. 2. Favata, Horiuchi, Manos, *et al.* (1998) *J. Biol. Chem.* **273**, 18623. 3. Ahn, Nahreini, Tolwinski & Resing (2001) *Meth. Enzymol.* **332**, 417. 4. Davies, Reddy, Caivano & Cohen (2000) *Biochem. J.* **351**, 95. 5. Liang, Ting, Yin, *et al.* (2006) *Biochem. Pharmacol.* **71**, 806. 6. Tan, Tamori, Egami, *et al.* (2004) *Oncol. Rep.* **11**, 993. 7. Karim, Hu, Adwanikar, Kaplan & Gereau (2005) *Mol. Pain* **2**, 1. 8. Lee, Chang, Kuo, *et al.* (2006) *Eur. J. Oral Sci.* **114**, 154. 9. Ross, Corey, Dunn & Kelley (2007) *Cell. Signal.* **19**, 923. 10. Feinstein & Linstedt (2007) *Mol. Biol. Cell* **18**, 594. 11. Xu, Stippec, Lenertz, *et al.* (2004) *J. Biol. Chem.* **279**, 7826. 12. Droebner, Pleschka, Ludwig & Planz (2011) *Antiviral Res.* **92**, 195.

U-7984, *See Decoyinine*

U-9189, *See 2-Amino-1,1,3-tricyano-1-propene*

U-11100A, *See Nafoxidine*

U-18666A, *See 3b-(2-Diethylaminoethoxy)androst-5-en-17-one*

U-19718, *See Kalafungin*

U-24522

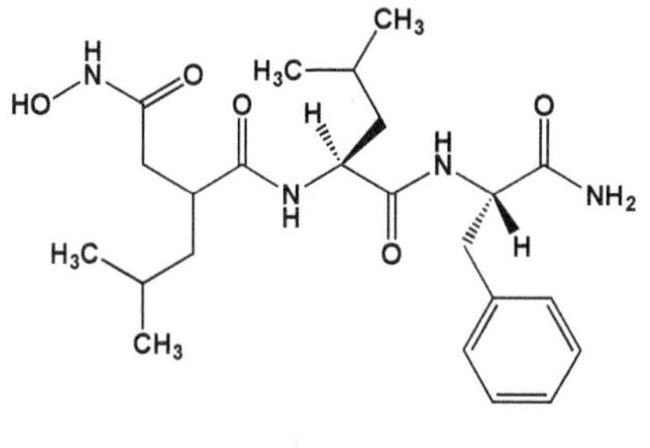

This synthetic hydroxamate-containing derivative (FW = 448.56 g/mol), also known as (*R,S*)-*N*-[2-[2-(hydroxyamino)-2-oxoethyl]-4-methyl-1-oxopentyl]-L-leucyl-L-phenylalaninamide, inhibits stromelysin 1, *or* matrix metalloproteinase 3 (1,2). 1. Doughty, Goldberg, Ganu, *et al.* (1993) *Agents Actions* **39** Spec No, C151. 2. Johnson, Pavlovsky, Johnson, *et al.* (2000) *J. Biol. Chem.* **275**, 11026.

U27810, *See Berninamycin A*

U-46619, *See 9,11-Dideoxy-9α,11α-epoxymethanoprostaglandin F$_{2a}$*

U60257B, *See Piriprost*

U-62168E, *See L-Prolyl-L-histidyl-L-prolyl-L-phenylalanyl-L-histidyl-L-phenylalanyl-L-phenylalanyl-L-isoleucyl-L-histidyl-L-lysine*

U63557a, *See 5-(3'-Pyridinylmethyl)benzofuran-2-carboxylate*

U-70531E, *See D-Histidyl-L-prolyl-L-phenylalanyl-L-histidyl-(2S-amino-3-phenylpropyl)-L-phenylalanyl-L-valyl-L-tyrosine*

U-71017

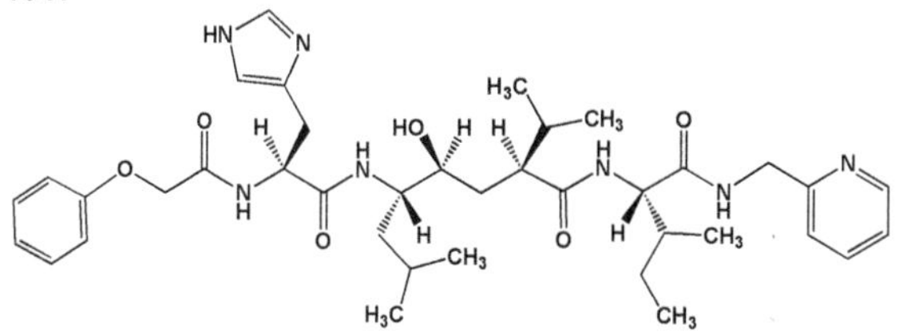

This synthetic peptide analogue (FW$_{\text{free-base}}$ = 705.90 g/mol), also known as Phenyloxyacetyl-His-Leu[CH(OH)CH$_2$]Val-Ile-aminomethylpyridine, inhibits renin (1) and HIV-1 protease, with IC$_{50}$ values of 10 nM (1,2). 1. Sawyer, Staples, Liu, *et al.* (1992) *Int. J. Pept. Protein Res.* **40**, 274. 2. Mildner, Rothrock, Leone, *et al.* (1994) *Biochemistry* **33**, 9405.

U-71038, *See Ditekiren*

U-71908E, *See N-Acetyl-L-(N^{indole}-formyl)tryptophanyl-L-prolyl-L-phenylalanyl-L-histidyl-(2S-amino-3-phenylpropyl)-L-phenylalaninamide*

U-71909E, *See N-Acetyl-L-prolyl-L-phenylalanyl-L-histidyl-(2S-amino-3-phenylpropyl)-L-phenylalaninamide*

U-72409E

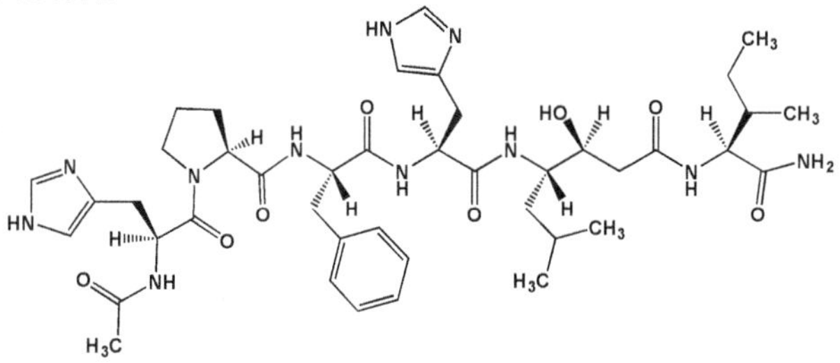

This statine-containing peptidomimetic (FW$_{\text{free-base}}$ = 848.02 g/mol) inhibits renin (K_d = 23 nM). Importantly, although the general mode of binding of these renin inhibitors is primarily driven by hydrophobic interactions, significant deviations from thermodynamic additivity and independent subsite model constraints are observed. Another determinant for binding is the conformation assumed by the peptide inhibitor in solution, suggesting that caution be exercised in using affinity constants to assess the interactions of peptide inhibitors with human renin and possibly with other enzymes having extended binding sites. The thermodynamic parameters of a class of compounds provide more information as to the mode of binding of ligands to their respective receptors than do dissociation constants. **1.** Epps, Cheney, Schostarez, *et al.* (1990) *J. Med. Chem.* **33**, 2080.

U-73122

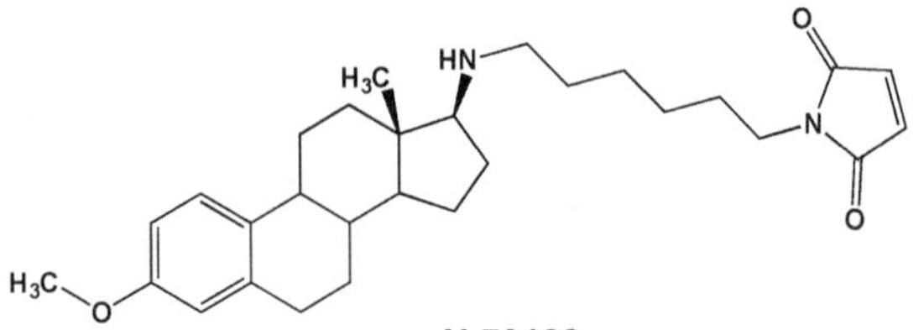

U-73122

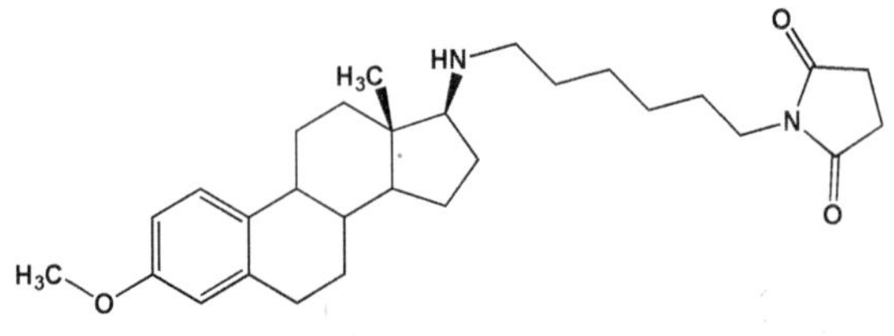

U-73343

This thiol-reactive steroid analogue (FW = 464.65 g/mol; CAS 142878-12-4; IUPAC: 1-[6-((17β-3-methoxyestra-1,3,5(10)-trien-17-yl)amino) hexyl]-1*H*-pyrrole-2,5-dione; *photosensitive*, protect from light), blocks G-protein coupling to human platelet and neutrophil phospholipase C, thereby inhibiting the latter. The thiol-unreactive analogue U-73343 (FW = 466.66 g/mol; CAS 142878-12-4; IUPAC: 1-[6-[((17β)-3-methoxyestra-1,3,5[10]-trien-17-yl)amino]hexyl]-2,5-pyrrolidinedione) may be used as a negative control. **1.** Hou, Kirchner, Singer, *et al.* (2004) *J. Pharmacol. Exp. Ther.* **309**, 697. **2.** Piacentini, Piatti, Fraternale, *et al.* (2004) *Biochimie* **86**, 343. **3.** Bleasdale, Thakur, Gremban, *et al.* (1990) *J. Pharmacol. Exp. Ther.* **255**, 756.

U-73343, *See U-73122*

U-76088

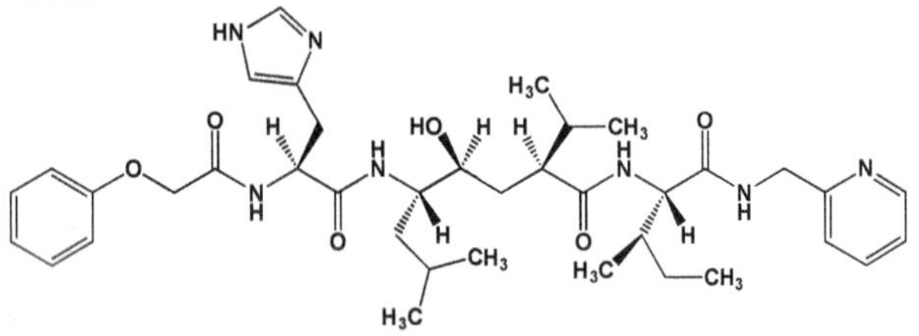

This synthetic peptide analogue (FW$_{free-base}$ = 705.90 g/mol) inhibits HIV-1 protease, *or* HIV-1 retropepsin (K_i = 0.3 nM at pH 5). **1.** Lin, Lin, Hong, *et al.* (1995) *Biochemistry* **34**, 1143.

U-77455E, *See* N^a-Isovaleryl-L-histidyl-L-prolyl-L-phenylalanyl-L-histidyl-statinyl-L-isoleucyl-L-phenylalaninamide

U-77646E

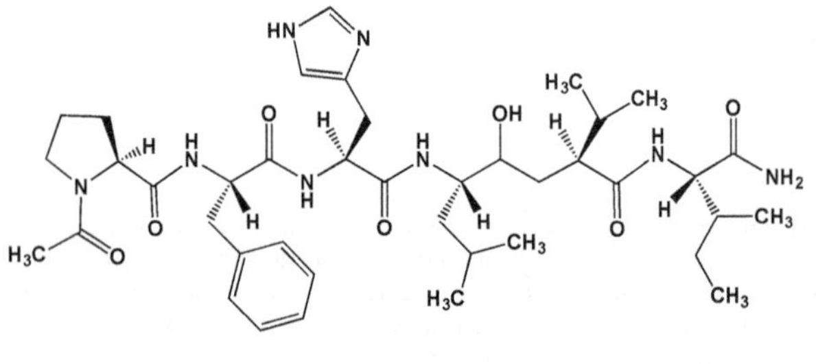

This substrate analogue (FW$_{free-base}$ = 766.98 g/mol) inhibits renin (K_d = 5.4 nM). The corresponding analogue containing a D-prolyl residue (U-77647E)

is a slightly better inhibitor (K_d = 1.3 nM). **1.** Epps, Cheney, Schostarez, *et al.* (1990) *J. Med. Chem.* **33**, 2080.

U-77647E, *See U-77646E*

U-81749, *See N-(tert-Butylacetyl)-(5S-amino-6-cyclohexyl-4S-hydroxy-2S-isopropylhexanoyl)-L-isoleucine 2-Pyridylmethylamide*

U-85548

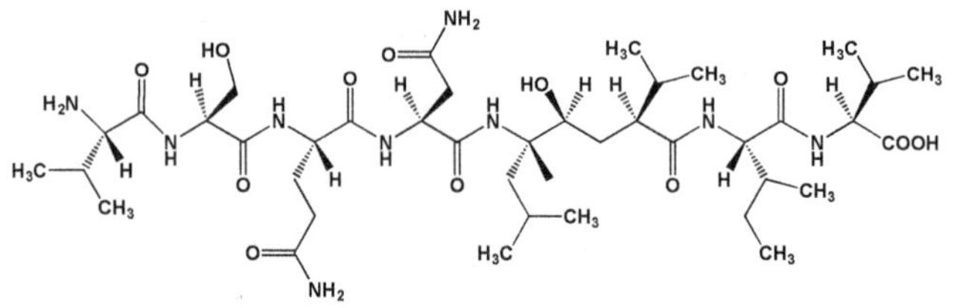

This substrate-based HIV inhibitor (FW = 872.07 g/mol), also known as Val-Ser-Gln-Asn-Leuψ[CH(OH)CH$_2$]Val-Ile-Val and *N*-[*N*-(Val-Ser-Gln-Asn)-5-amino-4-hydroxy-2-isopropyl-7-methyloctanoyl]-Ile-Val, contains a hydroxyethylene isostere in place of the scissile bond and is believed to mimic the tetrahedral transition state of the proteolytic reaction catalyzed by HIV-1 protease (K_i = 3 nM at pH 5). U-85548 inhibits HIV-2 protease (K_i = 0.042 nM), human pepsin (K_i = 9 nM), and rhizopuspepsin (K_i = 0.34 nM). **Target(s):** HIV-1 protease, *or* human immunodeficiency virus 1 retropepsin (1-4); HIV-2 protease (3); pepsin (3); and rhizopuspepsin (3,5). **1.** Ringe (1994) *Meth. Enzymol.* **241**, 157. **2.** Harte, Jr., & Beveridge (1994) *Meth. Enzymol.* **241**, 178. **3.** Sawyer, Staples, Liu, *et al.* (1992) *Int. J. Pept. Protein Res.* **40**, 274. **4.** Lin, Lin, Hong, *et al.* (1995) *Biochemistry* **34**, 1143. **5.** Lowther, Sawyer, Staples, *et al.* (1992) in *Peptides: Chemistry and Biology: Proc. of the 12th Amer. Peptide Symp.* (June 16-21, 1991) (Smith & Rivier, eds.), pp. 413-414.

U-85777E

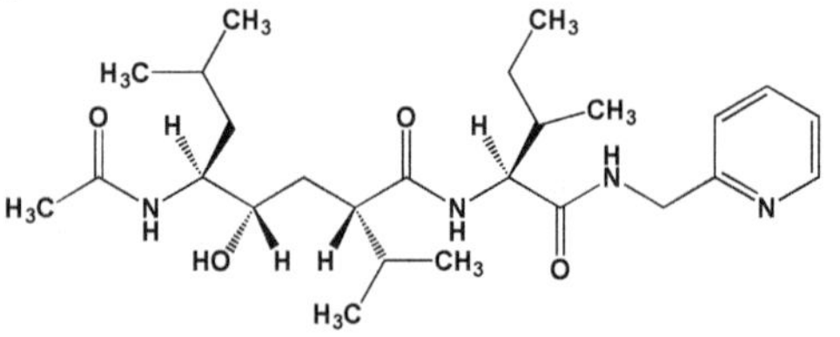

This synthetic peptide analogue (FW = 476.66 g/mol), also known as Ac-Leuψ[CH(OH)CH$_2$]Val-Ile-Amp and *N*-(5-acetylamino-4-hydroxy-2-isopropyl-7-methyloctanoyl)-L-isoleucine 2-pyridylmethylamide, contains a hydroxyethylene isostere in place of the scissile bond and potently inhibits HIV-1 protease (K_i = 70 nM). Sawyer, Staples, Liu, *et al.* (1992) *Int. J. Pept. Protein Res.* **40**, 274.

U-89360

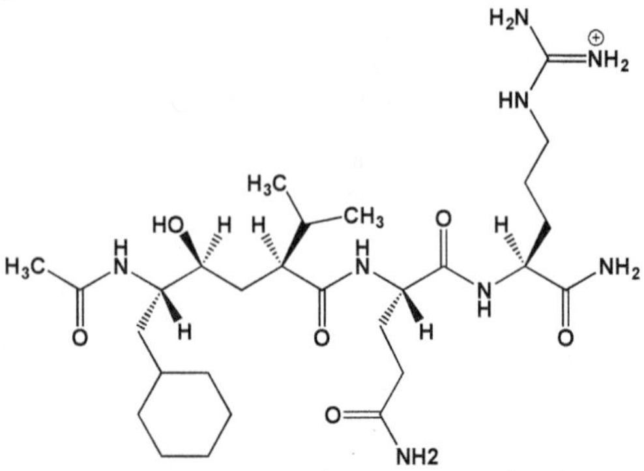

This peptide analogue (FW$_{free-base}$ = 596.77 g/mol) possesses a hydroxyethylene isostere in place of the scissile bond, allowing it to potently inhibit HIV-1 protease with a K_i of 20 nM (1-4). **1.** Lin, Lin, Hong, *et al.* (1995) *Biochemistry* **34**, 1143. **2.** Bardi, Lugue & Freire (1997) *Biochemistry* **36**, 6588. **3.** Hong, Zhang, Hartsuck, Foundling & Tang (1998) *Adv. Exp. Med. Biol.* **436**, 59. **4.** Hong, Hartsuck, Foundling, Ermolieff & Tang (1998) *Protein Sci.* **7**, 300.

U-89920E

This hydroxyethylene isostere-containing peptide analogue (FW = 867.02 g/mol), also known as Ac-Gly-Ser-His-Leuψ[CH(OH)CH$_2$]Val-Glu-Ala-Leu-NH$_2$ and N-[N-(N-acetylglycyl-L-seryl-L-histidyl)-5-amino-4-hydroxy-2-isopropyl-7-methyloctanoyl]-L-glutamyl-L-alanyl-L-leucinamide (FW$_{HCl}$ = 903.47 g/mol), mimics a tetrahedral reaction intermediate in HIV-1 protease catalysis, thereby potently inhibiting this enzyme (K_i = 10 nM). **1.** Sawyer, Staples, Liu, *et al.* (1992) *Int. J. Pept. Protein Res.* **40**, 274.

U 93631

This GABA$_A$ receptor antagonist (FW = 299.33 g/mol; CAS 152273-12-6), also named 4,5-dihydro-4,4-dimethylimidazo[1,5-a]quinoxaline-3-carboxylic acid 1,1-dimethylethyl ester, binds at the picrotoxin site (K_d = 2 μM), stabilizing the inactive form of the channel, and allosterically decreasing the probability of single-channel opening. U-93631 accelerates the decay of γ-aminobutyric acid-induced chloride ion currents, with little effect on peak amplitude (1,2). It also inhibits 5-HT$_{3A}$ receptors by an apparently similar mechanism (3). **1.** Dillon, Im, Hamilton, *et al.* (1993) *Mol. Pharmacol.* **44**, 860. **2.** Dillon, Im, Pregenzer, Carter & Hamilton (1995) *J. Pharmacol. Exp. Ther.* **272**, 597. **3.** Das, Bell-Horner, Machu & Dillon (2003) *Neuropharmacol.* **44**, 431.

U-93965

This synthetic statine-containing peptide analogue (FW = 802.05 g/mol) inhibits human immunodeficiency virus type 1 polyprotein protease (K_i = 1.6 nM at pH 5). Key site-directed mutants (*e.g.*, V82N, V82E, V82A, V82S, V82D, V82Q, D30F, D30W, G48H, G48D, G48Y, and K45E) were analyzed to investigate mutational effects on enzyme kinetics and inhibition. The k_{cat} and K_m values of many Val82 mutants and Lys45 mutant are comparable to the native enzyme. Surprisingly, Gly48 mutations produce enzymes with catalytic efficiency superior to that of the wild-type enzyme by as much as 10-fold. Modeling of the structure of the mutants suggests that the high catalytic efficiency of some substrates is related to increase rigidity in the flap region. **1.** Lin, Lin, Hong, *et al.* (1995) *Biochemistry* **34**, 1143.

U 99194

This selective D$_3$ antagonist (FW$_{free-base}$ = 277.41 g/mol; FW$_{maleate-salt}$ = 393.48 g/mol; CAS 234757-41-6; Soluble to 25 mM in water), also known as PNU 99194 and 2,3-dihydro-5,6-dimethoxy-N, N-dipropyl-1H-inden-2-amine, targets human dopamine D$_3$ (K_i = 160 nM), D$_2$ (K_i = 2280 nM), and D$_4$ (K_i = > 10 μM) receptors. **1.** Audinot et al (1998) *J. Pharmacol. Exp.*

Ther. **287**, 187. **2.** Clifford and Waddington (1998) *Psychopharmacol.* **136**, 284. **3.** LaHoste et al (2000) *J. Neurosci.* **20**, 6666.

UA 62784

This microtubule assembly inhibitor (FW = 353.38 g/mol; CAS 313367-92-9; Soluble to 25 mM in DMSO), also named 4-[5-(4-methoxyphenyl)-2-oxazolyl]-9H-fluoren-9-one, interacts with tubulin dimers 10x more potently than colchicine, vinblastine, or nocodazole. Competition experiments revealed that UA62784 binds at or near the colchicine-binding site. Nanomolar doses of UA62784 promote the accumulation of mammalian cells in mitosis, forming aberrant mitotic spindles. Treatment of cancer cell lines with UA62784 activates apoptosis. When used at low concentrations, UA62784 and vinblastine potentiate each other's ability to inhibit cell proliferation. Contrary to an earlier report (2), UA62784 is NOT an inhibitor of centromere protein E kinesin-like protein (1). **1.** Tcherniuk, Deshayes, Sarli, Divita & Abrieu (2011) *Chem. Biol.* **18**, 631. **2.** Henderson, Shaw, Wang, *et al.* (2009) *Mol. Cancer Ther.* **8**, 36.

UAMC 00039

This potent, orally available dipeptidyl peptidase inhibitor (FW = 382.76 g/mol; CAS 697797-51-6), also known as (2S)-2-amino-4-[[(4-chlorophenyl)methyl]amino]-1-(1-piperidinyl)-1-butanone dihydrochloride, selectively targets proline-specific dipeptidyl peptidase DPP-II (IC$_{50}$ = 0.48 nM), with much weaker action against DPP-9 (IC$_{50}$ = 79 nM), DPP-8 (IC$_{50}$ = 142 nM) and DPP-IV (IC$_{50}$ = 165 μM). **1.** Maes, Dubois, Brandt, *et al.* (2007) *J. Leukoc. Biol.* **81**, 1252. **2.** Van Goethem, Matheeussen, Joossens, *et al.* (2011) *J. Med. Chem.* **54**, 5737.

Ubiquinone

This isoprenoid-derived, lipid-soluble electron carrier (FW$_{Ubiquinone-50}$ = 863.36 g/mol; CAS 303-98-0), also known as Coenzyme Q, is a polyprenylated benzoquinone that, along with its reduced form Ubiquinol (*or* Coenzyme QH$_2$), is a required component in the the Electron tTransport Chain. These ubiquinones are 2,3-dimethoxy-5-methylbenzoquinones that contain a terpenoid chain at position-6, consisting of 1–12 *trans*-isoprenoid units. Ubiquinone-0 is 2,3-dimethoxy-5-methyl-p-benzoquinone. The most common ubiquinone contains a fifty-carbon terpenoid component and is designated either Ubiquinone 50 (also called coenzyme Q_{50}) or Ubiquinone 10 (coenzyme Q_{10}), since there are ten isoprene units present in Ubiquinone 50. **Chemical & Physical Properties:** All ubiquinones are sensitive to dioxygen, sunlight, and ultraviolet light. They are insoluble in water but highly soluble in most organic solvents, including ethanol, petroleum ether, diethyl ether, and cyclohexane. They absorb light in the ultraviolet portion of the spectrum: ubiquinone 50 has a λ_{max} of 275 nm in ethanol (ε = 14600 $M^{-1}cm^{-1}$). The melting points of ubiquinones increase with the size of the isoprenoid side-chain: Ubiquinone-30 (19–20°C), Ubiquinone-35 (31–32°C), Ubiquinone-40 (37–38°C), Ubiquinone-45 (44–45°C), and Ubiquinone-50 (49°C). Note that Ubiquinone-0 has a melting point of 59–60°C. **Role of Ubiquinone in Cellular Redox Reactions:** In the NADH:ubiquinone oxidoreductase (*or* NADH dihydro-genase) of Mitochondrial Complex I, two electrons are removed from NADH and transferred to ubiquinone (Q). The resulting reduced product, *or* ubiquinol (QH_2) freely diffuses within the membrane, allowing Complex I to translocate four protons across the membrane, producing a proton gradient. Complex I is the main site for premature electron leakage to oxygen occurs and is thus being one of the main sites of superoxide production. Quinones are thus mobile, lipid-soluble carriers that shuttle electrons (and protons) between large, relatively immobile macromolecular complexes embedded in the membrane. Bacteria use ubiquinone and related molecules, including menaquinone. **Ubiquinone Inhibitors:** Structural analogues of ubiquinone and ubiquinol are often inhibitors of electron transport. Compounds that inhibit the NADH-ubiquinone reductase activity of Complex I are classified as follows: Type A are antagonists of the ubiquinone substrate; Type B displace the ubisemiquinone intermediate; and Type C are antagonists of the ubiquinol product (1). **Target(s):** Complex II, by Ubiquinone-0 (2); horseradish peroxidase, by Ubiquinone-0 (3); sphingomyelinase (4); vitamin K-dependent γ-glutamylcarboxylase, by Ubiquinone-9 and Ubiquinone-10 (5); vitamin K-epoxide reductase, by Ubiquinone-9 and -10 (5). **1.** Degli Esposti (1998) *Biochim. Biophys Acta* 1364, 222. **2.** Jacobs & Crane (1960) *Biochem. Biophys. Res. Commun.* 3, 333. **3.** Klapper & Hackett (1963) *J. Biol. Chem.* 238, 3736. **4.** Martin, Navarro, Forthoffer, Navas & Villalba (2001) *J. Bioenerg. Biomembr.* 33, 143. **5.** Ronden, Soute, Thijssen, Saupe & Vermeer (1996) *Biochim. Biophys. Acta* 1298, 87.

Ubiquitin Aldehyde

This form of recombinant ubiquitin (MW = 8.5 kDa; CAS 79586-22-4), containing an aldehyde group at the C-terminus, inhibits (as the respective aldehyde hydrate) the hydrolysis of ubiquitin conjugates of small molecules *via* ubiquitin C-terminal hydrolase and inhibits several enzymes that regenerate free ubiquitin from adducts with proteins and intermediates in protein degradation. **Handling:** Stable for at least two months when stored at 4°C; Soluble and stable in aqueous solution at pH < 7.0; Avoid presence of amino-containing compounds; Do not lyophilize; Do not neutralise until immediately prior to use. **Target(s):** ubiquitin-dependent proteolysis (1-3); ubiquitin thiolesterase, *or* ubiquitinyl hydrolase (1); ubiquitin C-terminal hydrolase, K_i = 2.5 nM (1,2,4-11); and U1p1 peptidase (12). **1.** Hershko & Rose (1987) *Proc. Natl. Acad. Sci. U.S.A.* 84, 1829. **2.** Pickart & Rose (1986) *J. Biol. Chem.* 261, 10210. **3.** Wilkinson (1990) *Meth. Enzymol.* 185, 387. **4.** Woo, Baek, Lee, et al. (1997) *J. Biochem.* 121, 684. **5.** Moskovitz (1994) *Biochem. Biophys. Res. Commun.* 205, 354. **6.** Larsen, Krantz & Wilkinson (1998) *Biochemistry* 37, 3358. **7.** Baek, Woo, Lee, et al. (1997) *Biochem. J.* 325, 325. **8.** Rose (2005) *Proc. Natl. Acad. Sci. U.S.A.* 102, 11575. **9.** Mayer & Wilkinson (1989) *Biochemistry* 28, 166. **10.** Case & Stein (2006) *Biochemistry* 45, 2443. **11.** Melandri, Grenier, Plamondon, Huskey & Stein (1996) *Biochemistry* 35, 12893. **12.** Lima (2004) in *Handb. Proteolytic Enzymes* (Barrett, Rawlings & Woessner, eds.) 2, p. 1340, Academic Press, San Diego.

Ubiquitin Hydroxamate

This ubiquitin inhibitor (MW = 8579.86) targets both ubiquitin C-terminal hydrolase and the isopeptidase (1,2). Indeed, the ubiquitin C-terminal hydrolase (hereafter Enz) is also inactivated by millimolar concentrations of hydroxylamine, but only if Ub is present (1). Such a result suggests that the hydrolase mechanism is one of nucleophilic catalysis with an acyl-Ub-Enz intermediate. Inactivation in the presence of hydroxylamine of hydrolase occurs once during hydrolysis of 1200 molecules of Ub-hydroxamate by the enzyme. The hydrolysis/inactivation ratio is constant over the range of 10-50 mM hydroxylamine showing that forms of Enz-Ub with which hydroxylamine and water react are different and not in rapid equilibrium (1). The inactive enzyme may be an acylhydroxamate formed from an Enz-Ub mixed anhydride generated from the Enz-Ub (thiol) ester, the existence

of which was inferred from parallel borohydride trapping experiments (1,2). **1.** Pickart & Rose (1986) *J. Biol. Chem.* 261, 10210. **2.** Purich (2002) *Meth. Enzymol.* 354, 174.

UBP 282

This N^3-substituted willardiine analogue and AMPA receptor antagonist (FW = 333.30 g/mol; CAS 544697-47-4; Soluble to 100 mM in 1 equivalent NaOH and to 25 mM in 1 equivalent HCl), also known as 3-CBW and (α*S*)-α-amino-3-[(4-carboxyphenyl)methyl]-3,4-dihydro-2,4-dioxo-1(2*H*)-pyrimidinepropanoic acid, targets α-amino-3-hydroxy-5-methyl-4-isoxazolepropionic acid (*or* AMPA) receptor, but not kainate receptor-mediated currents on spinal neonatal motoneurons, yet antagonizes kainate-induced responses on dorsal root C-fibers. **1.** More, Troop & Jane (2002) *Brit. J. Pharmacol.* 137, 1125. **2.** More, Troop, Dolman & Jane (2003) *Brit. J. Pharmacol.* 138, 1093.

UBP 310

This receptor antagonist (FW = 353.35 g/mol; CAS 902464-46-4; Soluble to 100 mM in DMSO), also named (*S*)-1-(2-amino-2-carboxyethyl)-3-(2-carboxythiophene-3-ylmethyl)-5-methylpyrimidine-2,4-dione, targets the ionotropic kainate glutamate GLU_{K5} receptor (IC_{50} = 130 nM) and also blocks recombinant homomeric GLU_{K7} receptors (1-3). Crytsal structures indicate that this antagonist binds by a novel mechanism that does not directly contact the E723 side-chain, as suggested in all previously solved AMPA and kainate receptor agonist and antagonist complexes (1). There is instead a hyperextension of the ligand binding core, and, in dimer assemblies, there is a 22-Å extension of the ion channel linkers in the transition from antagonist- to glutamate-bound forms. This large conformational change suggests that glutamate receptors are capable of much larger movements than previously thought (1). It has an apparent K_D value is 18 nM for depression of kainate responses on the dorsal root. UBP 310 displays 12,700x selectivity for GLU_{K5} over GLU_{K6}. UBP 310 exhibits no activity at mGlu group I or NMDA receptors at concentrations of up to 10 μM. **1.** Mayer, Ghosal, Dolman & Jane (2006) *J. Neurosci.* 26, 2852. **2.** Dolman, More, Alt, et al. (2007) *J. Med. Chem.* 50, 1558. **3.** Perrais, Pinheiro, Jane & Mulle (2009) *Neuropharmacology* 56, 131.

UCB 35625

UCB 35625   J 113863

This potent chemokine CCR1/CCR3 receptor antagonist (FW = 655.44 g/mol; CAS 301648-08-8; Soluble to 100 mM in DMSO and to 50 mM in ethanol), also known as 1,4-*trans*-1-(1-cycloocten-1-ylmethyl)-4-[[(2,7-

dichloro-9H-xanthen-9-yl)carbonyl]amino]-1-ethyl-piperidinium iodide, inhibits MIP-1α-induced chemotaxis in CCR1 transfectants, IC_{50} = 9.6 nM, and eotaxin-induced chemotaxis in CCR3 transfectants, IC_{50} = 94 nM (1,2). UCB 35625 antagonizes CCR3-mediated entry of HIV-1 isolate 89.6 into NP-2 cells (IC_{50} = 57 nM). Its enantiomer, **J 113863** (FW = 655.44 g/mol; CAS 353791-85-2; IUPAC: 1,4-cis-1-(1-cycloocten-1-ylmethyl)-4-[[(2,7-dichloro-9H-xanthen-9-yl)carbonyl]amino]-1-ethylpiperidinium iodide), is also a CCR1 antagonist, with IC_{50} values are 0.9 and 5.8 nM for human and mouse CCR1 receptors respectively. It displays high selectivity for human, IC_{50} = 0.58 nM, but not mouse CCR3 receptors, IC_{50} = 460 nM (3,4). J-113863 improves paw inflammation, joint damage and dramatically reduces cell infiltration into joints in collagen-induced arthritis in mice (4). **1.** Sabroe, Peck, Van Keulen, et al. (2000) J. Biol. Chem. **275**, 25985. **2.** de Mendonça, da Fonseca, Phillips, et al. (2005) J. Biol. Chem. **280**, 4808. **3.** Naya, Sagara, Ohwaki, et al. (2001) J. Med. Chem. **44**, 1429. **4.** Amat, Benjamim, Williams, et al. (2006) Brit. J. Pharmacol. **149**, 666.

UCB L059, See Levetiracetam

UCF1-C, See Manumycin A

UCF-101

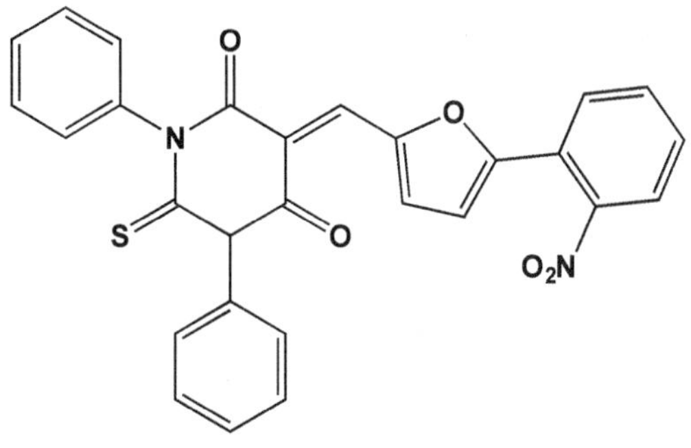

This 2-thiobarbituric acid derivative (FW = 494.52 g/mol) inhibits the mammalian high-temperature requirement protein A_2 serine proteinase (IC_{50} = 9.5 μM). It is a weaker inhibitor of coagulation factor Xa, t-plasminogen activator, u-plasminogen activator, and protein C (activated). UCF-102, -103, and -104 are weaker inhibitors (the 2-nitrophenyl component being replaced with 3-chloro-4-methoxyphenyl, 3-carboxyphenyl, and 2-methyl-5-nitrophenyl, respectively). **Target(s):** coagulation factor Xa; HtrA2 peptidase; protein C (activated); t-plasminogen activator; u-plasminogen activator. **1.** Cilenti, Lee, Hess, et al. (2003) J. Biol. Chem. **278**, 11489.

UCL 1684

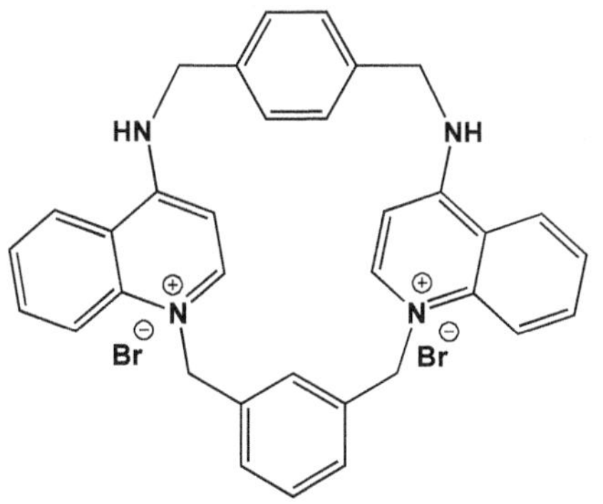

This nonpeptidic, bis-quinolinium cyclophane blocker (FW = 654.44 g/mol; CAS 199934-16-2; Soluble to 10 mM in DMSO), also named 6,12,19,20,25,26-hexahydro-5,27:13,18:21,24-trietheno-11,7-metheno-7H-dibenzo[b,n][1,5,12,16]tetraazacyclotricosine-5,13-diium dibromide, targets apamin-sensitive Ca^{2+}-activated K^+ channel ($K_{Ca}2.1$), with an IC_{50} value of 3 nM in rat sympathetic neurons (1). UCL 1684 also blocks $hK_{Ca}2.1$ and $rK_{Ca}2.2$ channels expressed in HEK 293 cells with IC_{50} values of 762 and 364 pM respectively (2). UCL 1407's inhibitory action is characterized by (a) a Hill slope greater than unity, (b) sensitivity to an increase in external potassium ion concentration, and (c) a time-course of onset suggesting use-dependence. UCL 1684 also blocks small-conductance SK channels hSK1 with an IC_{50} value of 762 pM and rSK2 with an IC_{50} value of 364 pM. **1.**

Campos, Galanakis, Piergentili, et al. (2000) J. Med. Chem. **43**, 420. **2.** Malik-Hall, Ganellin, Galanakis & Jenkinson (2000) Br. J. Pharmacol. **129**, 1431. **3.** Strøbaek, Jørgensen, Christophersen, Ahring & Olesen (2000) Br. J. Pharmacol. **129**, 991.

UCL 2077

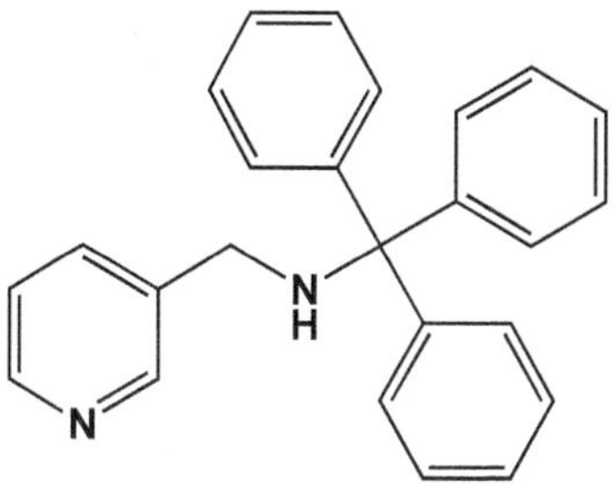

This sAHP channel blocker (FW = 350.46 g/mol; CAS 918311-87-2), also named N-trityl-3-pyridinemethanamine, reduces calcium-activated, slow after-hyperpolarization, a potassium conductance implicated in memory, aging, and epilepsy. The significance of sAHP stems from its exceedingly long time-course, a property that integrates action potential-induced calcium signals, controls neuronal excitability, and prevents runaway channel firing (2). UCL 2077 suppresses the sAHP present in hippocampal neurons in culture (IC_{50} = 0.5 μM) and in the slice preparation (IC_{50} ≈ 10 μM). UCL2077 was selective, having minimal effects on Ca^{2+} channels, action potentials, input resistance and the medium afterhyperpolarization. UCL2077 also had little effect on heterologously expressed small conductance Ca^{2+}-activated K^+ (or SK) channels. **1.** Shah, Javadzadeh-Tabatabaie, Benton, et al. (2006) Mol. Pharmacol. **70**, 1494. **2.** Kim, Kobayashi, Takamatsu & Tzingounis (2012) Biophys. J. **103**, 2446. **3.** Zhang, Ouyang, Ganellin & Thomas (2013) J. Neurosci. **33**, 5006.

UCM 707

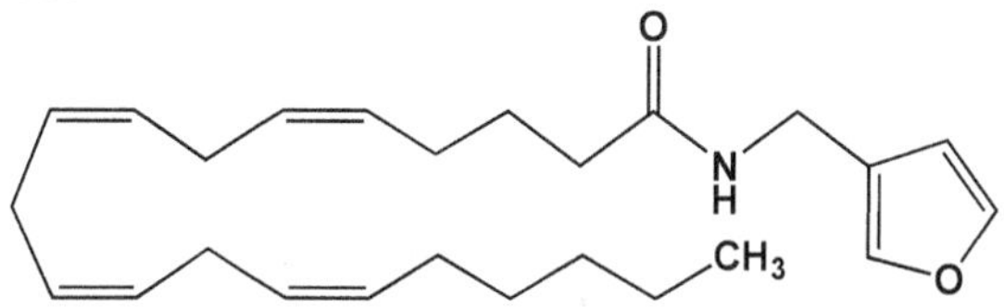

This potent endocannabinoid transport inhibitor (FW = 383.57 g/mol; CAS 390824-20-1; Solubility: 100 mM in DMSO), also named (5Z,8Z,11Z,14Z)-N-(3-Furanylmethyl)-5,8,11,14-eicosatetraenamide, blocks anandamide transport (IC_{50} = 0.8 μM), but more weakly inhibits (IC_{50} = 0.8 μM) Fatty Acid Amide Hydrolase (FAAH, EC 3.5.1.99), a member of the serine hydrolase family, also known as oleamide hydrolase and anandamide amidohydrolase. UCM-707 is an effective inhibitor of the cannabinoid receptor CB_2 (K_i = 0.07 μM), and displays K_i values of 4.7 and >5 μM for CB_1 and VR1 receptors respectively. UCM-707 also potentiates hypokinetic and the antinociceptive effects of anandamide in vivo. **1.** López-Rodríguez, Viso, Ortega-Gutiérrez, et al. (2003) Eur. J. Med. Chem. **38**, 403. **2.** López-Rodríguez, Viso, Ortega-Gutiérrez, et al. (2003) J. Med. Chem. **46**, 1512.

UCM 17197

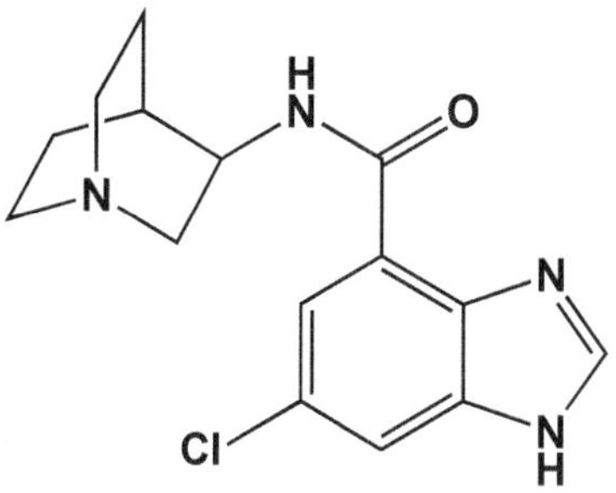

This substituted benzimidazole ($FW_{free-base}$ = 304.78 g/mol), also known as (S)-(–)-N-(1-azabicyclo[2.2.2]oct-3-yl)-6-chlorobenzimidazole-4-carbox-amide, is a serotonin 5-HT$_3$ receptor antagonist, K_i = 0.13 nM. **1.** Lopez-Rodriguez, Benhamu, Morcillo, et al. (1999) J. Med. Chem. **42**, 5020.

UCN-01

This staurosporine analogue and checkpoint inhibitor (FW = 482.54 g/mol; CAS 112953-11-4; Alternate Name: 7-hydroxystaurosporine) from *Streptomyces* (1) is more selective than staurosporine, with IC_{50} values of 29, 34, 30, 590, and 530 nM for α, βI, γ, δ, and ε protein kinase C forms, respectively, and K_i values of 42 and 45 nM for protein kinase A and p60v-src protein-tyrosine kinase are 42 and 45 nM (2,3). UCN-01 has low water-solubility but dissolves in dimethyl sulfoxide and dimethylformamide. UCN-01 also targets cyclin-dependent kinases (4-6); protein kinase C (2,6); protein-tyrosine kinase (2). UCN-01 also inhibits nucleotide excision repair in response to the DNA damage initiated by cisplatin (7), and UCN-01 may interfere with required protein phosphoryl-ation of DNA repair enzymes (8). **1.** Takahashi, Kobayashi, Asano, Yoshida & Nakano (1987) *J. Antibiot. (Tokyo)* **40**, 1782. **2.** Tamaoki (1991) *Meth. Enzymol.* **201**, 340. **3.** Seynaeve, Kazanietz, Blumberg, Sausville & Worland (1994) *Mol. Pharmacol.* **45**, 1207. **4.** Meijer & Kim (1997) *Meth. Enzymol.* **283**, 113. **5.** Kawakami, Futami, Takahara & Yamaguchi (1996) *Biochem. Biophys. Res. Commun.* **219**, 778. **6.** VanderWel, Harvey, McNamara, *et al.* (2005) *J. Med. Chem.* **48**, 2371. **7.** Jiang & Yang (1999) *Cancer Res.* **59**, 4529. **8.** Yamauchi, Keating & Plunkett (2002) *Mol. Cancer Ther.* **1**, 287.

UCPH 101

This selective non-substrate EAAT inhibitor (FW = 422.48 g/mol; CAS 1118460-77-7; Soluble to 25 mM in DMSO), also named 2-amino-5,6,7,8-tetrahydro-4-(4-methoxyphenyl)-7-(naphthalen-1-yl)-5-oxo-4H-chromene-3-carbonitrile, targets Excitatory Amino Acid Transporter 1, or EAAT1 (IC_{50} = 660 nM), with far lower affinity for EAAT2 (IC_{50} > 300000 nM) and EAAT3 (IC_{50} > 300000 nM). UCPH 101 also demonstrates no significant inhibition at EAAT4 or EAAT5 in a patch-clamp electrophysiology assay at 10 µM (2). The allosteric mode of UCPH-101 inhibition underlines the functional importance of the trimerization domain of the EAAT and demonstrates the feasibility of modulating transporter function through ligand binding to regions distant from its "transport domain" (3). **1.** Jensen, Erichsen, Nielsen, *et al.* (2009) *J. Med. Chem.* **52**, 912. **2.** Erichsen, Huynh, Abrahamsen, *et al.* (2010) *J. Med. Chem.* **53**, 7180. **3.** Abrahamsen, Schneider, Erichsen, *et al.* (2013) *J. Neurosci.* **33**, 1068.

UCS15A

This nonpeptide, protein–protein interaction disruptor (FW = 480.55 g/mol) inhibits src-specific phosphorylation of tyrosine residues within numerous proteins in v-src-transformed cells. Earlier disruptors of SH3-mediated protein–protein interactions were synthetic 10-residue, or longer, peptides based on the native target sequence or peptides selected from screening with degenerate peptide and phage display libraries. T consisted of 10 or more amino acid residues. UCS15A inhibits the following SH3-mediated protein-protein interactions *in vivo*: Src–Sam68, Sam68–Grb2, Sam68–PLCγ, cortactin–ZO1 and Grb2–Sos, as well as atypical SH3-mediated interactions, such as Grb2–Gab1. UCS15A fails to inhibit non-SH3-mediated protein–protein interactions found in the E-cadherin. UCS15A exerts its src-inhibitory effects by disrupting protein-protein interactions mediated by src. One of the biological consequences of src-inhibition by UCS15A was its ability to inhibit osteoclast-mediated bone resorption *in vitro* (1) UCS15A's effects are not restricted to Src-SH3 mediated protein-protein interactions, since it can disrupt the *in vivo* interactions of Sam68 with other SH3 domain containing proteins, such as Grb2 and PLCγ. UCS15A can also disrupt other typical SH3-mediated protein-protein interactions, such as Grb2–Sos1, cortactin–ZO1, as well as atypical SH3-mediated protein-protein interactions such as Grb2–Gab1 (2). UCS15A cannot, however, disrupt the non-SH3-mediated protein-protein interactions of β-catenin, with E-cadherin and α-catenin. In addition, UCS15A is without effect on the SH2-mediated interactions between Grb2 and activated Epidermal Growth Factor receptor (2). Thus, the ability of UCS15A, to disrupt protein-protein interactions appeared to be restricted to SH3-mediated protein-protein interactions. (1). (**See also** *AP22161*) **1.** Sharma, Oneyama, Yamashita, *et al.* (2001) *Oncogene* **20**, 2068. **2.** Oneyama, Nakano & Sharma (2002) *Oncogene* **21**, 2037.

Udenafil

This PDE-5 inhibitor and erectile dysfunction drug (FW = 516.66 g/mol; CAS 268203-93-6; Solubility: >20 mg/mL DMSO), also known by its code name DA-8159, its trade name Zydena®, and its systematic name 3-(1-methyl-7-oxo-3-propyl-4,7-dihydro-1H-pyrazolo[4,3-d]pyrimidin-5-yl)-N-[2-(1-methylpyrrolidin-2-yl)ethyl]-4-propoxybenzenesulfonamide, targets the Type-5 isozyme of human 3',5'-cGMP phosphodiesterase. Reaching maximal plasma concentration within 1.0 to 1.5 hours and possessing a $t_{1/2}$ of 11 to 13 hours, udenafil enjoys properties commending its use in treating erectile dysfunction. Udenafil is metabolized to its major metabolite DA-8164 by CYP3A4, and ketoconazole, a known CYP3A4 inhibitor, alters udenafil pharmacokinetics, increasing its plasma concentration (2). Daily dosing with this PDE5 inhibitor also improves cognitive function, depression and somatization (3). **1.** Oh, Kang, Ahn, Yoo & Kim (2000) *Arch. Pharm. Res.* **23**, 471. **2.** Shin, Chung, Kim, *et al.* (2010) *Brit. J. Clin. Pharmacol.* **69**, 307. **3.** Shim, Pae, Cho, *et al.* (2013) *Int. J. Impot Res.* **26**, 76.

UDP-N-acetyl-D-galactosamine

This sugar nucleotide metabolite ($FW_{free-acid}$ = 607.36 g/mol; λ_{max} = 261 nm; ε = 10100 $M^{-1}cm^{-1}$ at pH 1 and 7 and ε = 7500 $M^{-1}cm^{-1}$ at pH 12), also known as uridine diphosphate N-acetyl-D-galactosamine, is a key intermediate in the biosynthesis of sialic acid, certain glycoproteins, and lipopolysaccharides. **Target(s):** fucosylgalactoside 3-α-galactosyl-transferase (1); hyaluronan synthase (2); lactosylceramide α-2,3-sialyltransferase, weakly inhibited (3); α-1,3-mannosyl-glycoprotein 2-β-N-acetylglucosaminyltransferase, weakly inhibited (4); α-1,3-mannosyl-glycoprotein 4-β-*N*-acetyl-glucosaminyltransferase (5); and protein *N*-acetylglucosaminyltransferase (6). **1**. Carne & Watkins (1977) *Biochem. Biophys. Res. Commun.* **77**, 700. **2**. Tlapak-Simmons, Baron & Weigel (2004) *Biochemistry* **43**, 9234. **3**. Cambron & Leskawa (1993) *Biochem. Biophys. Res. Commun.* **193**, 585. **4**. Nishikawa, Pegg, Paulsen & Schachter (1988) *J. Biol. Chem.* **263**, 8270. **5**. Oguri, Minowa, Ihara, *et al.* (1997) *J. Biol. Chem.* **272**, 22721. **6**. Haltiwanger, Blomberg & Hart (1992) *J. Biol. Chem.* **267**, 9005.

UDP-*N*-acetyl-D-glucosamine

This sugar nucleotide (FW$_{free-acid}$ = 607.36 g/mol; λ_{max} = 261 nm; ε = 10100 M^{-1}cm^{-1} at pH 1 and 7; ε = 7500 M^{-1}cm^{-1} at pH 12), is a key intermediate in the biosynthesis of sialic acid, chitin, glycoprotein, and lipopolysaccharides. Although UDP-*N*-acetyl-D-glucosamine weakly inhibits many enzymes, its physiologic concentration is unlikely to be sufficient for substantial inhibition. **Target(s):** α-*N*-acetylgalactosaminide α-2,6-sialyltransferase (1); *N*-acetylglucosamine kinase (2,3); *N*-acetylglucosamine-6-phosphate deacetylase (4); *N*-acetylglucosamine-1-phosphodiester α-*N*-acetyl-glucosaminidase (5-8); *N*-acylneuraminate cytidylyltransferase, weakly inhibited (9); CMP-*N*-acylneuraminate phospho-diesterase (10,11); β-galactoside α-2,3-sialyltransferase (1); globoside α-*N*-acetylgalactos-aminyltransferase (27); glucosamine-6-phosphate deaminase, weakly inhibited (12); glucuronosyltransferase (28-30); glutamine:fructose-6-phosphate aminotransferase (isomerizing), *or* fructose-6-phosphate amidotransferase (13-22); glycogenin glucosyltransferase (self-glucosylation) (23); glycolipid 2-α-mannosyltransferase, weakly inhibited (24); glycolipid 3-α-mannosyltransferase, weakly inhibited (25); and UDP-*N*-acetylglucosamine diphosphorylase, *or* *N*-acetylglucos-amine-1-phosphate uridylyltransferase, weakly inhibited (26). **1**. Sadler, Beyer, Oppenheimer, *et al.* (1982) *Meth. Enzymol.* **83**, 458. **2**. Datta (1975) *Meth. Enzymol.* **42**, 58. **3**. Datta (1970) *Biochim. Biophys. Acta* **220**, 51. **4**. Gopal, Sullivan & Shepherd (1982) *J. Gen. Microbiol.* **128**, 2319. **5**. Varki & Kornfeld (1980) *J. Biol. Chem.* **255**, 8398. **6**. Mullis, Huynh & Kornfeld (1994) *J. Biol. Chem.* **269**, 1718. **7**. Varki & Kornfeld (1981) *J. Biol. Chem.* **256**, 9937. **8**. Lee & Pierce (1995) *Arch. Biochem. Biophys.* **319**, 413. **9**. Bravo, Barrallo, Ferrero, *et al.* (2001) *Biochem. J.* **358**, 585. **10**. van Dijk, Maier & van den Eijnden (1976) *Biochim. Biophys. Acta* **444**, 816. **11**. Kean & Bighouse (1974) *J. Biol. Chem.* **249**, 7813. **12**. Weidanz, Campbell, DeLucas, *et al.* (1995) *Brit. J. Haematol.* **91**, 72. **13**. Zalkin (1985) *Meth. Enzymol.* **113**, 278. **14**. Kornfeld (1967) *J. Biol. Chem.* **242**, 3135. **15**. Mendicino & Rao (1975) *Eur. J. Biochem.* **51**, 547. **16**. Huynh, Gulve & Dian (2000) *Arch. Biochem. Biophys.* **379**, 307. **17**. Ellis & Sommar (1972) *Biochim. Biophys. Acta* **267**, 105. **18**. Trujillo & Gan (1974) *Int. J. Biochem.* **5**, 515. **19**. Moriguchi, Yamamoto, Kawai & Tochikura (1976) *Agric. Biol. Chem.* **40**, 1655. **20**. Hosoi, Kobayashi & Ueha (1978) *Biochem. Biophys. Res. Commun.* **85**, 558. **21**. Kikuchi & Tsuiki (1976) *Biochim. Biophys. Acta* **422**, 241. **22**. Vessal & Hassid (1972) *Plant Physiol.* **49**, 977. **23**. Manzella, Ananth, Oegema, *et al.* (1995) *Arch. Biochem. Biophys.* **320**, 361. **24**. Schutzbach, Springfield & Jensen (1980) *J. Biol. Chem.* **255**, 4170. **25**. Jensen & Schutzbach (1981) *J. Biol. Chem.* **256**, 12899. **26**. Mengin-Lecreulx & van Heijenoort (1994) *J. Bacteriol.* **176**, 5788. **27**. Ishibashi, Atsuta & Makita (1976) *Biochim. Biophys. Acta* **429**, 759. **28**. Koster & Noordhoek (1983) *Biochim. Biophys. Acta* **761**, 76. **29**. Gregory & Strickland (1973) *Biochim. Biophys. Acta* **327**, 36. **30**. Matern, Matern & Gerok (1982) *J. Biol. Chem.* **257**, 7422.

UDP-*N*-acetyl-D-glucosamine 6-Phosphate

This sugar nucleotide (FW$_{free-acid}$ = 687.34 g/mol; ; λ_{max} = 261 nm; ε = 10100 M^{-1}cm^{-1} at pH 1 and 7 and ε = 7500 M^{-1}cm^{-1} at pH 12), also known as uridine diphosphate *N*-acetyl-D-glucosamine 6-phosphate, is a UDP derivative of the natural product of the reaction catalyzed by *N*-acetylglucosamine kinase. **1**. Datta (1975) *Meth. Enzymol.* **42**, 58.

[(UDP-*N*-acetyl-D-glucosamine-3-yl)methyl](2-carboxypropyl) Phosphinate

This transition-state mimic (FW$_{free-acid}$ = 774.03 g/mol; λ_{max} = 261 nm; ε = 10100 M^{-1}cm^{-1} at pH 1 and 7 and ε = 7500 M^{-1}cm^{-1} at pH 12, based on uridine's UV spectrum) inhibits *Escherichia coli* UDP-*N*-acetylmuramate:L-alanine ligase (IC$_{50}$ = 49 nM), the L-alanine adding enzyme (also known as MurC) of bacterial peptidoglycan biosynthesis. Note that the phosphinate moiety resembles the likely tetrahedral reaction intermediate. **1**. Reck, Marmor, Fisher & Wuonola (2001) *Bioorg. Med. Chem. Lett.* **11**, 1451.

UDP-*N*-acetyl-D-muramate (UDP-*N*-acetyl-D-muramic Acid)

This sugar nucleotide (FW$_{free-acid}$ = 679.4 g/mol; λ_{max} = 261 nm; ε = 10100 M^{-1}cm^{-1} at pH 1 and 7 and ε = 7500 M^{-1}cm^{-1} at pH 12, based on uridine's UV spectrum), also known as uridine diphosphate *N*-acetyl-D-muramic acid, is is a product of the reaction catalyzed by UDP-*N*-acetylmuramate dehydrogenase and is the substrate for UDP-*N*-acetylmuramate:L-alanine ligase and is a key intermediate in the biosynthesis of a number of compounds, including peptidoglycans and lipopolysaccharides. **Target(s):** glucosamine-1-phosphate *N*-acetyltransferase (1); UDP-*N*-acetylglucos-amine 1-carboxyvinyltransferase (2-4). **1**. Mengin-Lecreulx & van Heijenoort (1994) *J. Bacteriol.* **176**, 5788. **2**. Wickus & Strominger (1973) *J. Bacteriol.* **113**, 287. **3**. Venkateswaran, Lugtenberg & Wu (1973) *Biochim. Biophys. Acta* **293**, 570. **4**. Mizyed, Oddone, Byczynski, Hughes & Berti (2005) *Biochemistry* **44**, 4011.

UDP-*N*-acetylmuramoyl-L-alanyl-γ-(D-glutamyl)-(*meso*-2,6-diaminopimelate)

This metabolite ($FW_{hydrochloride}$ = 1088.3 g/mol; λ_{max} = 261 nm; ε = 10100 M^{-1}cm^{-1} at pH 1 and 7 and ε = 7500 M^{-1}cm^{-1} at pH 12, based on uridine's UV spectrum) is a feedback inhibitor of bacterial cell wall peptidoglycan biosynthesis. **Target(s):** UDP-*N*-acetylglucosamine 1-carboxyvinyl-transferase (1,2). **1**. Zemell & Anwar (1975) *J. Biol. Chem.* **250**, 3185. **2**. Venkateswaran, Lugtenberg & Wu (1973) *Biochim. Biophys. Acta* **293**, 570.

UDP-*N*-acetylmuramoyl-L-alanyl-γ-(D-glutamyl)-(L)-(*meso*-2,6-diaminopimeloyl)-D-alanyl-D-alanine

This bacterial metabolite ($FW_{hydrochloride}$ = 1230.4 g/mol; λ_{max} = 261 nm; ε = 10100 M^{-1}cm^{-1} at pH 1 and 7 and ε = 7500 M^{-1}cm^{-1} at pH 12, based on uridine's UV spectrum), also called Park nucleotide (pimelate form), is a feedback inhibitor in the biosynthesis of cell wall peptidoglycan. Note that there are two forms of Park nucleotide, containing either an L-lysyl or *meso*-diaminopimelate residue. Gram-negative bacteria typically contain the pimelate form. **Target(s):** UDP-*N*-acetylglucosamine 1-carboxyvinyl-transferase (1,2). **1**. Zemell & Anwar (1975) *J. Biol. Chem.* **250**, 3185. **2**. Venkateswaran, Lugtenberg & Wu (1973) *Biochim. Biophys. Acta* **293**, 570.

[1-(UDP-*N*-acetylmuramido)ethyl](2,4-dicarboxybutyl) Phosphinate

This substrate analogue ($FW_{free-acid}$ = 914.60 g/mol; λ_{max} = 261 nm; ε = 10100 M^{-1}cm^{-1} at pH 1 and 7 and ε = 7500 M^{-1}cm^{-1} at pH 12, based on uridine's UV spectrum) inhibits UDP-*N*-acetylmuramoylalanine:D-glutamate ligase, IC$_{50}$ < 1 nM. **1**. Gegnas, Waddella, Chabinb, Reddyb & Wong (1998) *Bioorg. Med. Chem. Lett.* **8**, 1643.

UDP-L-arabinose

This sugar nucleotide ($FW_{free-acid}$ = 536.28 g/mol; λ_{max} = 261 nm; ε = 10100 M^{-1}cm^{-1} at pH 1 and 7 and ε = 7500 M^{-1}cm^{-1} at pH 12, based on uridine's UV spectrum), also known as uridine diphosphate L-arabinose, is a key intermediate in the biosynthesis of arabinoxylans. It is formed from UDP-D-xylose and is a structural analogue of UDP-D-galactose. **Target(s):** UDP-glucuronate decarboxylase, weakly inhibited (1); UDP-glucuronate 4-

epimerase, but enantiomer was unspecified (2). **1**. Ankel & Feingold (1965) *Biochemistry* **4**, 2468. **2**. Gu & Bar-Peled (2004) *Plant Physiol.* **136**, 4256.

UDP Chloroacetol, *See* Uridine 5′-Diphosphate Chloroacetol

UDP-2-deoxy-D-glucose

This structural analogue of UDPglucose ($FW_{free-acid}$ = 550.31 g/mol; λ_{max} = 261 nm; ε = 10100 M^{-1}cm^{-1} at pH 1 and 7 and ε = 7500 M^{-1}cm^{-1} at pH 12, based on uridine's UV spectrum), also known as uridine diphosphate 2-deoxy-D-glucose, inhibits the formation of dolichyl-phospho-D-glucose but not the formation of dolichyl-phospho-D-mannose. **Target(s):** dolichyl-phosphate glucosyltransferase (1); and protein glycosylation (1,2). **1**. Schwarz & Datema (1980) *Trends Biochem. Sci.* **5**, 65. **2**. Schwarz, Schmidt & Lehle (1978) *Eur. J. Biochem.* **85**, 163.

UDP-D-galactosamine

This naturally occurring sugar nucleotide ($FW_{hydrochloride}$ = 361.92 g/mol; λ_{max} = 261 nm; ε = 10100 M^{-1}cm^{-1} at pH 1 and 7 and ε = 7500 M^{-1}cm^{-1} at pH 12, based on uridine's UV spectrum), also known as uridine diphosphate D-galactosamine, inhibits UDPglucose dehydrogenase (1). **1**. Bauer & Reutter (1973) *Biochim. Biophys. Acta* **293**, 11.

UDP-D-galactose

This naturally occurring alkali-labile sugar nucleotide ($FW_{free-acid}$ = 566.31 g/mol; λ_{max} = 261 nm; ε = 10100 M^{-1}cm^{-1} at pH 1 and 7 and ε = 7500 M^{-1}cm^{-1} at pH 12, based on uridine's UV spectrum; Symbol: UDPGal), also known as uridine diphosphate D-galactose, is a key intermediate in D-galactose metabolism as well as in the formation of a number of other cellular components (*e.g.*, cell wall polysaccharides, rhamnogalacturonans, *etc.*). Although UDP-*N*-galactose weakly inhibits many enzymes *in vitro*, it is unlikely that its physiologic concentration is sufficient for substantial inhibition within cells. **Target(s):** α-*N*-acetylgalactosaminide α-2,6-sialyltransferase (1); *N*-acetyllactosaminide β-1,6-*N*-acetylglucosaminyl-transferase (2); (*N*-acetylneuraminyl)-galactosylglucosylceramide *N*-acetylgalactosaminyltransferase (20); CMP-*N*-acyl-neuraminate phospho-diesterase (3,4); β-galactoside α-2,3-sialyltransferase (1); globoside α-*N*-acetylgalactos-aminyltransferase (21); glucuronosyltransferase (23); glycogenin glucosyltransferase (self-glucosylation) (5); glycogen synthase (24); glycolipid 2-α-mannosyl-transferase, weakly inhibited (19); glycoprotein-fucosyl-galactoside α-*N*-acetylgalactosaminyltransferase (22); hyaluronan synthase (6); lactosylceramide α-2,3-sialyltransferase, weakly inhibited (7); α-1,3-mannosyl-glycoprotein 4-β-*N*-acetylglucosaminyl-transferase (18); NDP-glucose:starch glucosyltransferase (8); nucleotide diphosphatase, also alternative substrate (9); UDP-glucose dehydrogenase (10); UDP-glucose:hexose-1-phosphate uridylyltransferase, product inhibition (11); UDP-glucuronate 4-epimerase (12); and UTP:glucose-1-phosphate uridylyltransferase (13-17). **1**. Sadler, Beyer, Oppenheimer, *et* *al.* (1982) *Meth. Enzymol.* **83**, 458. **2**. Sakamoto, Taguchi, Tano *et al.* (1998) *J. Biol. Chem.* **273**, 27625. **3**. van Dijk, Maier & van den Eijnden (1976) *Biochim. Biophys. Acta* **444**, 816. **4**. Kean & Bighouse (1974) *J. Biol. Chem.* **249**, 7813. **5**. Manzella, Ananth, Oegema, *et al.* (1995) *Arch. Biochem. Biophys.* **320**, 361. **6**. Tlapak-Simmons, Baron & Weigel (2004) *Biochemistry* **43**, 9234. **7**. Cambron & Leskawa (1993) *Biochem. Biophys. Res. Commun.* **193**, 585. **8**. Zea & Pohl (2005) *Biopolymers* **79**, 106. **9**. Evans, Hood & Gurd (1973) *Biochem. J.* **135**, 819. **10**. Zalitis, Uram, Bowser & Feingold (1972) *Meth. Enzymol.* **28**, 430. **11**. Ruzicka, Geeganage & Frey (1998) *Biochemistry* **37**, 11385. **12**. Munoz, Lopez, de Frutos & Garcia (1999) *Mol. Microbiol.* **31**, 703. **13**. Lee, Kimura & Tochikura (1979) *J. Biochem.* **86**, 923. **14**. Turnquist, Gillett & Hansen (1974) *J. Biol. Chem.* **249**, 7695. **15**. Hopper & Dickinson (1973) *Biochim. Biophys. Acta* **309**, 307. **16**. Hopper & Dickinson (1972) *Arch. Biochem. Biophys.* **148**, 523. **17**. Turnquist, Turnquist, Bachmann & Hansen (1974) *Biochim. Biophys. Acta* **364**, 59. **18**. Oguri, Minowa, Ihara, *et al.* (1997) *J. Biol. Chem.* **272**, 22721. **19**. Schutzbach, Springfield & Jensen (1980) *J. Biol. Chem.* **255**, 4170. **20**. Senn, Cooper, Warnke, Wagner & Decker (1981) *Eur. J. Biochem.* **120**, 59. **21**. Ishibashi, Atsuta & Makita (1976) *Biochim. Biophys. Acta* **429**, 759. **22**. Takeya, Hosomi & Ishiura (1990) *J. Biochem.* **107**, 360. **23**. Matern, Matern & Gerok (1982) *J. Biol. Chem.* **257**, 7422. **24**. Zea & Pohl (2005) *Biopolymers* **79**, 106.

UDP-D-galacturonate

This sugar nucleotide (FW$_{free-acid}$ = 580.29 g/mol; λ_{max} = 261 nm; ε = 10100 M^{-1}cm^{-1} at pH 1 and 7 and ε = 7500 M^{-1}cm^{-1} at pH 12, based on uridine spectrum), also known as uridine diphosphate D-galacturonate, is formed via a reversible epimerization of UDP-D-glucuronate. It is required for capsular biosynthesis in a number of microorganisms and in the formation of rhamnogalacturonans. **Target(s):** chitin synthase (1); CMP-*N*-acylneuraminate phosphodiesterase (2); glucuronosyltransferase (3,4); procollagen galactosyl-transferase (5); UDP-D-apiose/UDP-D-xylose synthase (6); UTP:glucose-1-phosphate uridylyltransferase, *or* glucose-1-phosphate uridylyltransferase (7-9). **1**. Jan (1974) *J. Biol. Chem.* **249**, 1973. **2**. Kean & Bighouse (1974) *J. Biol. Chem.* **249**, 7813. **3**. Bock, Josting, Lilienblum & Pfeil (1979) *Eur. J. Biochem.* **98**, 19. **4**. Matern, Matern & Gerok (1982) *J. Biol. Chem.* **257**, 7422. **5**. Risteli (1978) *Biochem. J.* **169**, 189. **6**. Molhoj, Verma & Reiter (2003) *Plant J.* **35**, 693. **7**. Roach, Warren & Atkinson (1975) *Biochemistry* **14**, 5445. **8**. Hopper & Dickinson (1973) *Biochim. Biophys. Acta* **309**, 307. **9**. Hopper & Dickinson (1972) *Arch. Biochem. Biophys.* **148**, 523.

UDP-D-glucose

This sugar nucleotide (FW$_{free-acid}$ = 566.31 g/mol; λ_{max} = 261 nm; ε = 10100 M^{-1}cm^{-1} at pH 1 and 7 and ε = 7500 M^{-1}cm^{-1} at pH 12; Symbol: UDPGlc), also known as uridine diphosphate D-glucose, is a key intermediate in the formation of polysaccharides, including glycogen, from D-glucose. Although UDP-glucose weakly inhibits many enzymes *in vitro*, it is unlikely that its physiologic concentration is sufficient for substantial inhibition within cells. **Target(s):** α-*N*-acetylgalactosaminide α-2,6-sialyltransferase (1); *N*-acetyl-glucosamine-1-phosphodiester α-*N*-acetylglucosaminidase (2); *N*-acetylglucosaminyldiphosphodolichol *N*-acetyl-glucosaminyltransferase (39,40); *N*-acetyllactosaminide β-1,6-*N*-acetylglucosaminyltransferase (38); ATP diphosphatase, weakly inhibited (3); carbamoyl-phosphate synthetase II (glutamine hydrolyzing) (4); chitin synthase (49); CMP-*N*-acylneuraminate phosphodiesterase (5); β-galactoside α-2,3-sialyltransferase (1); galactosyltransferase (6); globoside

α-*N*-acetylgalactosaminyltransferase (45,46); glucose-1-phosphate adenyl-yltransferase (30); glucose-1-phosphate thymidylyltransferase (7); glucuronosyl-transferase (48); glutamine:fructose-6-phosphate amino-transferase (isomerizing), *or* fructose-6-phosphate amidotransferase (8,35); glycogen phosphorylase (9,10,50-60); glycolipid 2-α-mannosyltransferase, weakly inhibited (43); glycolipid 3-α-mannosyltransferase, weakly inhibited (42); glycoprotein 2-β-D-xylosyltransferase (36); hexokinase (34); hyaluronan synthase (37); lactose synthase (6); maltose synthase (41); α-1,3-mannosyl-glycoprotein 2-β-*N*-acetylglucosaminyltransferase, weakly inhibited (44); α-1,3-mannosyl-glycoprotein 4-β-*N*-acetylglucosaminyl-transferase (38); nucleotide diphosphatase, alternative substrate (11,12); [phosphorylase] phosphatase (13,14); procollagen galactosyltransferase (47); [protein-PII] uridylyltransferase (29); starch phosphorylase (52,61-67); UDP-*N*-acetylglucosamine:dolichyl-phosphate *N*-acetyl-glucosamine-phosphotransferase (15,16); UDP-*N*-acetylglucosamine 2-epimerase (17); UDP-*N*-acetyl-glucosamine:lysosomal-enzyme *N*-acetylglucosamine-phosphotransferase (18-20); UDP-glucuronate decarboxylase (21-24); UDP-glucuronate 4-epimerase (25-27); UTP:glucose-1-phosphate uridylyltransferase, product inhibition (31-34); and zeatin *O*-β-D-xylosyltransferase (28). **1**. Sadler, Beyer, Oppenheimer, *et al.* (1982) *Meth. Enzymol.* **83**, 458. **2**. Varki & Kornfeld (1980) *J. Biol. Chem.* **255**, 8398. **3**. Torp-Pedersen, Flodgaard & Saermark (1979) *Biochim. Biophys. Acta* **571**, 94. **4**. Mori & Tatibana (1978) *Meth. Enzymol.* **51**, 111. **5**. Kean & Bighouse (1974) *J. Biol. Chem.* **249**, 7813. **6**. Ebner (1973) *The Enzymes*, 3rd ed. (Boyer, ed.), **9**, 363. **7**. Robbins & Bernstein (1966) *Meth. Enzymol.* **8**, 253. **8**. Mendicino & Rao (1975) *Eur. J. Biochem.* **51**, 547. **9**. Thomas & Wright (1976) *J. Biol. Chem.* **251**, 1253. **10**. Madsen (1961) *Biochem. Biophys. Res. Commun.* **6**, 310. **11**. Byrd, Fearney & Kim (1985) *J. Biol. Chem.* **260**, 7474. **12**. Decker & Bischoff (1972) *FEBS Lett.* **21**, 95. **13**. Madsen (1986) *The Enzymes*, 3rd ed. (Boyer & Krebs, eds.), **17**, 365. **14**. Detwiler, D. Gratecos & Fischer (1977) *Biochemistry* **16**, 4818. **15**. Kaushal & Elbein (1986) *Plant Physiol.* **82**, 748. **16**. Shailubhai, Dong-Yu, Saxena & Vijay (1988) *J. Biol. Chem.* **263**, 15964. **17**. Kikuchi & Tsuiki (1973) *Biochim. Biophys. Acta* **327**, 193. **18**. Waheed, Hasilik & von Figura (1982) *J. Biol. Chem.* **257**, 12322. **19**. Bao, Elmendorf, Booth, Drake & Canfield (1996) *J. Biol. Chem.* **271**, 31446. **20**. Zhao, Yeh & Miller (1992) *Glycobiology* **2**, 119. **21**. Ankel & Feingold (1966) *Meth. Enzymol.* **8**, 287. **22**. Grisebach, Baron, Sandermann & Wellmann (1972) *Meth. Enzymol.* **28**, 439. **23**. Ankel & Feingold (1965) *Biochemistry* **4**, 2468. **24**. Suzuki, Watanabe, Masumura & Kitamura (2004) *Arch. Biochem. Biophys.* **431**, 169. **25**. Gaunt, Ankel & Schutzbach (1972) *Meth. Enzymol.* **28**, 426. **26**. Gaunt, Maitra & Ankel (1974) *J. Biol. Chem.* **249**, 2366. **27**. Munoz, Lopez, de Frutos & Garcia (1999) *Mol. Microbiol.* **31**, 703. **28**. Martin, Cloud, Mok & Mok (2000) *Plant Growth Regul.* **32**, 289. **29**. Engleman & Francis (1978) *Arch. Biochem. Biophys.* **191**, 602. **30**. Amir & Cherry (1972) *Plant Physiol.* **49**, 893. **31**. Lee, A. Kimura & Tochikura (1979) *J. Biochem.* **86**, 923. **32**. Hopper & Dickinson (1973) *Biochim. Biophys. Acta* **309**, 307. **33**. Hopper & Dickinson (1972) *Arch. Biochem. Biophys.* **148**, 523. **34**. Gao & Leary (2003) *J. Amer. Soc. Mass Spectrom.* **14**, 173. **35**. Hosoi, Kobayashi & Ueha (1978) *Biochem. Biophys. Res. Commun.* **85**, 558. **36**. Bencúr, Steinkellner, Svoboda, *et al.* (2005) *Biochem. J.* **388**, 515. **37**. Tlapak-Simmons, Baron & Weigel (2004) *Biochemistry* **43**, 9234. **38**. Oguri, Minowa, Ihara, *et al.* (1997) *J. Biol. Chem.* **272**, 22721. **39**. Kean & Niu (1998) *Glycoconjugate J.* **15**, 11. **40**. Kaushal & Elbein (1986) *Plant Physiol.* **81**, 1086. **41**. Schilling (1982) *Planta* **154**, 87. **42**. Jensen & Schutzbach (1981) *J. Biol. Chem.* **256**, 12899. **43**. Schutzbach, Springfield & Jensen (1980) *J. Biol. Chem.* **255**, 4170. **44**. Nishikawa, Pegg, Paulsen & Schachter (1988) *J. Biol. Chem.* **263**, 8270. **45**. Kijimoto, Ishibashi & Makita (1974) *Biochem. Biophys. Res. Commun.* **56**, 177. **46**. Ishibashi, Atsuta & Makita (1976) *Biochim. Biophys. Acta* **429**, 759. **47**. Risteli (1978) *Biochem. J.* **169**, 189. **48**. Bock, Josting, Lilienblum & Pfeil (1979) *Eur. J. Biochem.* **98**, 19. **49**. Jan (1974) *J. Biol. Chem.* **249**, 1973. **50**. Ariki & Fukui (1975) *J. Biochem.* **78**, 1191. **51**. Ercan-Fang, Taylor, Treadway, *et al.* (2005) *Amer. J. Physiol. Endocrinol. Metab.* **289**, E366. **52**. Kumar & Sanwal (1982) *Biochemistry* **21**, 4152. **53**. Schultz & Ankel (1970) *Biochim. Biophys. Acta* **215**, 39. **54**. Tanabe, Kobayashi & Matsuda (1987) *Agric. Biol. Chem.* **51**, 2465. **55**. Chen & Segel (1968) *Arch. Biochem. Biophys.* **127**, 175. **56**. Tanabe, Kobayashi & Matsuda (1988) *Agric. Biol. Chem.* **52**, 757. **57**. van Marrewijk, Van den Broek & Beenakkers (1988) *Insect Biochem.* **18**, 37. **58**. Hata, Yokoyama, Suda, Hata & Matsuda (1988) *Comp. Biochem. Physiol. B Comp. Biochem.* **87**, 747. **59**. Robson & Morris (1974) *Biochem. J.* **144**, 513. **60**. Takata, Takaha, Okada, Takagi & Imanaka (1998) *J. Ferment. Bioeng.* **85**, 156. **61**. Nakamura & Imamura (1983) *Phytochemistry* **22**, 835. **62**. Kokesh,

Stephenson & Kakuda (1977) *Biochim. Biophys. Acta* **483**, 258. **63**. Singh & Sanwal (1976) *Phytochemistry* **15**, 1447. **64**. Matheson & Richardson (1973) *Phytochemistry* **17**, 195. **65**. Hsu, Yang, Su & Lee (2004) *Bot. Bull. Acad. Sin.* **45**, 187. **66**. Chang & Su (1986) *Plant Physiol.* **80**, 534. **67**. Yu & Pedersen (1991) *Physiol. Plant.* **81**, 149.

UDP-D-glucos-6-yl, *See* *P¹-5'-Uridine-P²-glucose-6-yl Diphosphate*

UDP-D-glucuronate

This sugar nucleotide (FW$_{\text{free-acid}}$ = 580.29 g/mol; λ_{max} = 261 nm; ε = 10100 M^{-1}cm^{-1} at pH 1 and 7 and ε = 7500 M^{-1}cm^{-1} at pH 12, based on uridine spectrum), also known as uridine diphosphate D-glucuronate, is a key intermediate in the formation of glucuronosides and is a substrate of several UDPglucuronosyltransferases. It is slightly unstable and slowly decomposes at room temperature. UDP-D-glucuronate is a product inhibitor for UDP-glucose dehydrogenase. **Target(s):** *N*-acetyllactosaminide β-1,6-*N*-acetylglucosaminyltransferase (1); chitin synthase (2); CMP-*N*-acyl-neuraminate phosphodiesterase (3); dolichylphosphate-glucose phosph-odiesterase (4); dolichyl-phosphate β-glucosyltransferase (5); glucuronokinase (6); glycogenin glucosyltransferase (self-glucosylation) (7); α-1,3-mannosyl-glycoprotein 4-β-*N*-acetylglucosaminyl-transferase (8); procollagen galactosyltransferase (9); α,α-trehalose-phosphate synthase (UDP-forming) (10); UDP-galactose 4-epimerase, in the presence of D-fucose or L-arabinose (11); UTP:glucose-1-phosphate uridylyl-transferase, *or* glucose-1-phosphate uridylyltransferase (12-14); and 1,4-β-D-xylan synthase (15). **1**. Sakamoto, Taguchi, Tano, *et al.* (1998) *J. Biol. Chem.* **273**, 27625. **2**. Jan (1974) *J. Biol. Chem.* **249**, 1973. **3**. Kean & Bighouse (1974) *J. Biol. Chem.* **249**, 7813. **4**. Crean (1984) *Biochim. Biophys. Acta* **792**, 149. **5**. Miernyk & Riedell (1991) *Phytochemistry* **30**, 2865. **6**. Leibowitz, Dickinson, Loewus & Loewus (1977) *Arch. Biochem. Biophys.* **179**, 559. **7**. Manzella, Ananth, Oegema, *et al.* (1995) *Arch. Biochem. Biophys.* **320**, 361. **8**. Oguri, Minowa, Ihara, *et al.* (1997) *J. Biol. Chem.* **272**, 22721. **9**. Risteli (1978) *Biochem. J.* **169**, 189. **10**. Londesborough & Vuorio (1991) *J. Gen. Microbiol.* **137**, 323. **11**. Blackburn & Ferdinand (1976) *Biochem. J.* **155**, 225. **12**. Roach, Warren & Atkinson (1975) *Biochemistry* **14**, 5445. **13**. Hopper & Dickinson (1973) *Biochim. Biophys. Acta* **309**, 307. **14**. Hopper & Dickinson (1972) *Arch. Biochem. Biophys.* **148**, 523. **15**. Baydoun, Waldron & Brett (1989) *Biochem. J.* **257**, 853.

UDP-hexanolamine, *See* *UDP-hexylamine*

UDP-hexylamine

This UDP derivative (FW$_{\text{hydrochloride}}$ = 539.80 g/mol; λ_{max} = 261 nm; ε = 10100 M^{-1}cm^{-1} at pH 1 and 7 and ε = 7500 M^{-1}cm^{-1} at pH 12, based on uridine spectrum), also known as uridine diphosphate *O*-hexyl-4-amine and UDP-hexanolamine, weakly inhibits *Knyveromyces lactis* mannotetraose 2-α-*N*-acetylglucosaminyltransferase (IC$_{50}$ = 0.6 mM). **Target(s):** *N*-acetyllactosaminide β-1,6-*N*-acetylglucos-aminyltransferase (1); α-1,3-mannosyl-glycoprotein 4-β-*N*-acetylglucosaminyltransferase (2); α-1,6-mannosyl-glycoprotein 4-β-*N*-acetylglucosaminyltransferase (3); manno-tetraose 2-α-*N*-acetylglucosaminyltransferase (4); UDP-*N*-acetylglucos-amine:dolichyl-phosphate *N*-acetyl-glucosaminephosphotrans-ferase (5). **1**. Sakamoto, Taguchi, Tano, *et al.* (1998) *J. Biol. Chem.* **273**, 27625. **2**. Oguri, Minowa, Ihara, *et al.* (1997) *J. Biol. Chem.* **272**, 22721. **3**. Brockhausen, Hull, Hindsgaul, *et al.* (1989) *J. Biol. Chem.* **264**, 11211. **4**.

Douglas & Ballou (1982) *Biochemistry* **21**, 1561. **5**. Shailubhai, Dong-Yu, Saxena & Vijay (1988) *J. Biol. Chem.* **263**, 15964.

UDP-D-mannose

This sugar nucleotide (FW$_{\text{free-acid}}$ = 566.31 g/mol; λ_{max} = 261 nm; ε = 10100 M^{-1}cm^{-1} at pH 1 and 7 and ε = 7500 M^{-1}cm^{-1} at pH 12, based on uridine spectrum), also known as uridine diphosphate D-mannose (abbreviation: UDPMan), inhibits UDP-glucuronate 4-epimerase. **Target(s):** chitin synthase (1); CMP-*N*-acylneuraminate phosphodiesterase (2); sinapate 1-glucosyltransferase, weakly inhibited (3); sterol 3β-glucosyltransferase (4); UDP-glucuronate 4-epimerase (5); and UTP:glucose-1-phosphate uridylyltransferase (6,7). **1**. Jan (1974) *J. Biol. Chem.* **249**, 1973. **2**. Kean & Bighouse (1974) *J. Biol. Chem.* **249**, 7813. **3**. Wang & B. E. Ellis (1998) *Phytochemistry* **49**, 307. **4**. Warnecke & Heinz (1994) *Plant Physiol.* **105**, 1067. **5**. Munoz, Lopez, de Frutos & Garcia (1999) *Mol. Microbiol.* **31**, 703. **6**. Hopper & Dickinson (1973) *Biochim. Biophys. Acta* **309**, 307. **7**. Hopper & Dickinson (1972) *Arch. Biochem. Biophys.* **148**, 523.

UDP-pyridoxal

This pyridinyl nucleotide (FW$_{\text{hydrochloride}}$ = 589.77 g/mol), also known as uridine diphosphopyridoxal and pyridoxal 5'-diphospho-5'-uridine, is an active-site probe of UDPsugar-utilizing enzymes that reacts with a specific lysyl residue in rabbit muscle glycogen synthase. **Target(s):** glycogen synthase (1-4); UDP-glucose pyrophosphorylase (1). **1**. Fukui & Tanizawa (1997) *Meth. Enzymol.* **280**, 41. **2**. Colman (1990) *The Enzymes*, 3rd ed. (Sigman & Boyer, eds.), **19**, 283. **3**. Pitcher, Smythe & Cohen (1988) *Eur. J. Biochem.* **176**, 391. **4**. Jiao, Shashkina, Shashkin, Hansson & Katz (1999) *Biochim. Biophys. Acta* **1427**, 1.

UDP-D-xylose

This sugar nucleotide (FW$_{\text{free-acid}}$ = 536.28 g/mol; *Symbol* = UDPX; λ_{max} = 261 nm; ε = 10100 M^{-1}cm^{-1} at pH 1 and 7 and ε = 7500 M^{-1}cm^{-1} at pH 12, based on uridine spectrum), also known as uridine diphosphate D-xylose, is a key intermediate in the biosynthesis of xylans, xyloglucans, arabinoxylans, and xylosylzeatin. UDPX is also the donor substrate in the xylosylation of proteins, flavonols, and dolichols. UDPX is a product of the reaction catalyzed by UTP:xylose-1-phosphate uridylyltransferase and is a substrate for 1,4-β-D-xylan synthase, protein xylosyltransferase, dolichyl-phosphate D-xylosyltransferase, flavonol-3-*O*-glycoside xylosyltransferase, glycoprotein 2-β-D-xylosyltransferase, xyloglucan 6-xylosyltransferase, and zeatin *O*-β-D-xylosyltransferase. **Target(s):** chitin synthase (1); glucuronosyltransferase (2); glycogenin glucosyltransferase, in which UDP-xylose is also an alternative substrate (3,4); sinapate 1-glucosyltransferase, weakly inhibited (5); UDP-*N*-acetylglucosamine:dolichyl-phosphate *N*-

acetylglucosaminephosphotransferase (6); UDP-glucose dehydrogenase (7-10); UDP-glucuronate decarboxylase (11,12); UDP-glucuronate 4-epimerase (13-16); UTP:glucose-1-phosphate uridylyltransferase (17-20); UTP:hexose-1-phosphate uridylyltransferase (18); and xyloglucan 4-glucosyltransferase (21). **1.** Jan (1974) *J. Biol. Chem.* **249**, 1973. **2.** Matern, Matern & Gerok (1982) *J. Biol. Chem.* **257**, 7422. **3.** Manzella, Ananth, Oegema, *et al.* (1995) *Arch. Biochem. Biophys.* **320**, 361. **4.** Meezan, Manzella & Roden (1995) *Trends Glycosci. Glycotechnol.* **7**, 303. **5.** Wang & Ellis (1998) *Phytochemistry* **49**, 307. **6.** Shailubhai, Dong-Yu, Saxena & Vijay (1988) *J. Biol. Chem.* **263**, 15964. **7.** Zalitis, Uram, Bowser & Feingold (1972) *Meth. Enzymol.* **28**, 430. **8.** Gainey & Phelps (1975) *Biochem. J.* **145**, 129. **9.** Neufeld & Hall (1965) *Biochem. Biophys. Res. Commun.* **19**, 456. **10.** Ankel, Ankel & Feingold (1966) *Biochemistry* **5**, 1864. **11.** Ankel & Feingold (1966) *Meth. Enzymol.* **8**, 287. **12.** Grisebach, Baron, Sandermann & Wellmann (1972) *Meth. Enzymol.* **28**, 439. **13.** Gaunt, Ankel & Schutzbach (1972) *Meth. Enzymol.* **28**, 426. **14.** Gu & Bar-Peled (2004) *Plant Physiol.* **136**, 4256. **15.** Munoz, Lopez, de Frutos & Garcia (1999) *Mol. Microbiol.* **31**, 703. **16.** Molhoj, Verma & Reiter (2004) *Plant Physiol.* **135**, 1221. **17.** Roach, Warren & Atkinson (1975) *Biochemistry* **14**, 5445. **18.** Lee, Kimura & Tochikura (1979) *J. Biochem.* **86**, 923. **19.** Hopper & Dickinson (1973) *Biochim. Biophys. Acta* **309**, 307. **20.** Hopper & Dickinson (1972) *Arch. Biochem. Biophys.* **148**, 523. **21.** Hayashi, Koyama & Matsuda (1988) *Plant Physiol.* **87**, 341.

UFP-101

This potent and selective antinociceptive peptide (FW = 1908.19 g/mol; Sequence: *N*-(Bn)GGGFTGARKSARKRKNQ-NH$_2$; CAS 849024-68-6) is a competitive silent antagonist for the Nociceptin Receptor, *or* NOP Opioid Receptor (**See** *Nociceptin*). Nociceptin/orphanin FQ (N/OFQ) modulates several biological functions by activating a specific G-protein coupled receptor (NOP). UFP-101 binds to NOP with high affinity (pK$_i$ = 10.24), displaying > 3000x selectivity over δ, μ and κ opioid receptors. In isolated peripheral tissues of mice, rats and guinea-pigs, and in rat cerebral cortex synaptosomes pre-loaded with [^{3}H]-5-HT, UFP-101 competitively antagonized the effects of N/OFQ with pA$_2$ values in the range of 7.3 - 7.7 (1). In the same preparations, the peptide was inactive alone and did not modify the effects of classical opioid receptor agonists. UFP-101 is also active in mice where it prevented the depressant action on locomotor activity and the pronociceptive effect induced by N/OFQ (1 nmol, as an intracerebroventricular injection (i.c.v.) (1). In male Swiss mice, i.c.v. injection of UFP-101 (1-10 nmol) dose-dependently reduced the immobility time, *or* 192 sec for control *versus* 91 sec for UFP-101 (2). The effect of 3 or 10 nmol UFP-101 was fully or partially reversed, respectively, by the coadministration of 1 nmol N/OFQ, which was inactive *per se* (2). UFP-101 does not stimulate GTPγS binding on its own, but does produce a concentration dependent and parallel rightward shift in the concentration response curves to all agonists (3). **1.** Calo, Rizzi, Rizzi, *et al.* (2002) *Brit. J. Pharmacol.* **136**, 303. **2.** Gavioli, Marzola, Guerrini, *et al.* (2003) *Eur. J. Neurosci.* **17**, 1987. **3.** McDonald, Calo, Guerrini & Lambert (2003) *Naunyn Schmiedebergs Arch. Pharmacol.* **367**, 183.

UK5099

This transport inhibitor (FW = 288.31 g/mol; CAS 56396-35-1; Solubility = >20 mg/mL DMSO), also known as UK-5099, UK 5099, and 2-cyano-3-(1-phenyl-1*H*-indol-3-yl)-2-propenoic acid, is a potent, cell-permeable inhibitor (IC$_{50}$ = 1-5 nM) of the mitochondrial pyruvate carrier (MPC), inhibiting pyruvate-dependent O$_2$ consumption rat heart mitochondria (IC$_{50}$ = 50 nM) and displaying excellent selectivity over monocarboxylates and anions. In castor bean mitochondria, MPC exhibits Michaelis-Menten kinetics, with a K$_m$ of 0.10 mM for pyruvate and a V$_m$ of 0.95 nmol/min per mg of mitochondrial protein (1). (**See also** *Zaprinast*) Two proteins, Mpc1 and Mpc2, appear to be essential for mitochondrial pyruvate transport in

yeast, Drosophila, and humans (2). Mpc1 and Mpc2 associate to form an ~150-kDa complex in the inner mitochondrial membrane. Yeast and *Drosophila* mutants lacking the *MPC1* gene display impaired pyruvate metabolism, with an accumulation of upstream metabolites and a depletion of tricarboxylic acid cycle intermediates. Mpc1 appears to be a key target for UK-5099. By screening for *MPC1* mutants that could grow in the presence of UK-5099, an Asp-118-Gly resistance-conferring substitution was identified in Mpc1 (2). (**See also** *α-Cyano-4-hydroxycinnamate, another mitochondrial pyruvate transport inhibitor*) **1.** Brailsford, Thompson, Kaderbhai & Beechey (1986) *Biochem J.* **239**, 355. **2.** Bricker, Taylor, Schell, *et al.* (2012) *Science* **337**, 96.

UK33274, *See* Doxazosin

UK 69578

This dicarboxylic acid (FW$_{free-acid}$ = 399.48 g/mol) is a selective inhibitor of neprilysin (also called enkephalinase): K$_i$ = 28 nM. **1.** Roques, Noble, Crine & Fournié-Zaluski (1995) *Meth. Enzymol.* **248**, 263. **2.** Danilewicz, Barclay, Barnish, *et al.* (1989) *Biochem. Biophys. Res. Commun.* **164**, 58.

UK 76654, *See* Zamifenacin

UK 81252, *See* Sampatrilat

UK 156406

This peptidomimetic (FW$_{inner-salt}$ = 539.66 g/mol), which has a plasma half-life of 48 minutes, inhibits thrombin (K$_i$ = 0.46 nM) as well as thrombin-induced fibroblast proliferation, procollagen production, and connective tissue growth factor mRNA levels, when used at equimolar concentration to thrombin. **Target(s):** thrombin (1,2); and trypsin (1). **1.** Allen, Abel, Barber, *et al.* (1998) *215th ACS Natl. Meeting, Dallas, TX,* Abst. MEDI 200. **2.** Howell, Goldsack, Marshall *et al.* (2001) *Amer. J. Pathol.* **159**, 1383.

UK-427,857, *See* Maraviroc

Ulinastatin

This 352-residue serine protease inhibitor (MW$_{unglycosylated}$ = 38,999 g; Isoelectric Point = 5.95) also known as α-1-microglobulin, AMBP, urinary trypsin inhibitor (UTI), urinary trypstatin, Bikunin® and Urinastatin®, is a urine-derived glycoprotein used to treat acute pancreatitis, chronic pancreatitis, toxic shock, Stevens–Johnson syndrome, burn patients, severe sepsis and toxic epidermal necrolysis. Circulating α-1-microglobulin is a monomer that forms complexes with IgA and albumin. In mast cells, α-1-microglobulin occurs as a complex with tryptase. **Backgound:** Inter-α-trypsin inhibitor, *or* ITI, is an abundant 220-kDa human plasma serine protease inhibitor (0.4–0.5 g/L) that is a complex of two heavy chains and a light one, the latter corresponding to the urinary trypsin inhibitor. Naturally occurring UTI is a proteoglycan containing a chondroitin sulfate side chain mainly synthetized in liver. UTI inhibits trypsin, elastase, and cathepsin G, the latter released by stimulated neutrophils. ITI is degraded into lower molecular weight derivatives present in plasma, among which UTI excreted in urine. During infection, cancer, tissue injury during surgery, kidney disease, vascular disease, coagulation, and diabetes, the concentrations of

Bikunin in plasma and urine are increased. **Other Targets:** Broadly active against serine proteases, including trypsin, thrombin, chymotrypsin, kallikrein, plasmin, elastase, cathepsin, Factors IXa, Xa, XIa, and XIIa. UTI's physiologic spectrum of inhibitory action is most likely determined by location and extent of biosynthesis. **1.** Pugia, Valdes & Jortani (2007) *Adv. Clin. Chem.* **44**, 223.

Ulipristal acetate

This selective progesterone receptor modulator, *or* SPRM (FW = 475.62 g/mol; CAS 126784-99-4; *Symbol:* UPA; IUPAC Name: (8*S*,11*S*,13*S*,14*R*,17*R*)-17-acetoxy-11-[4-(dimethylamino)phenyl]-19-norpregna-4,9-diene-3,20-dione), also known as the contraceptive EllaOne® (in EU) and Ella® (in U.S) as well as Esmya® for uterine fibroid therapy, is effective and well-tolerated oral hormonal emergency contraceptive, when administered within 48-120 hours after unprotected intercourse (1,2). This agent delays or inhibits ovulation, but is without effect once luteinizing hormone has begun to increase. Animal studies show ulipristal acetate to be embryotoxic, making it necessary to be certain the recipient is not already pregnant. Ulipristal acetate is metabolized by CYP3A4 *in vitro* to two other pharmacologically active, but less potent forms. Because it is also likely to interact with other CYP3A4 substrates (*e.g.,* rifampicin, phenytoin, erythromycin, St John's wort, carbamazepine, and ritonavir), concomitant use alters UPA pharmacokinetics. *See also Levonorgestrel and Mifepristone* **1.** Fine, Mathé, Ginde, *et al.* (2010) *Obstet Gynecol.* **115**, 257. **2.** Bouchard, Chabbert-Buffet & Fauser (2011) *Fertil. Steril.* **96**, 1175.

Umbelliferone

This coumarin derivative (FW = 162.14 g/mol), also known as 7-hydroxycoumarin, is frequently found in plants (*e.g.,* the aglucon of skimmin) and is often used as a fluorescent label (solutions will exhibit a blue fluorescence) and a pH indicator. **Target(s):** chymotrypsin (1); cyclooxygenase (2); 17β-hydroxysteroid dehydrogenase (3); lipoxygenase (4); phenylalanine ammonia-lyase (5); phenylpyruvate tautomerase (6); and xanthine oxidase (7). **1.** Ghani, Ng, Atta-ur-Rahman, *et al.* (2001) *J. Mol. Biol.* **314**, 519. **2.** Lee, Bykadi & Ritschel (1981) *Arzneimittelforschung* **31**, 640. **3.** Le Lain, Barrell, Saeed, *et al.* (2002) *J. Enzyme Inhib. Med. Chem.* **17**, 93. **4.** Sekiya, Okuda & Arichi (1982) *Biochim. Biophys. Acta* **713**, 68. **5.** Dubery (1990) *Phytochemistry* **29**, 2107. **6.** Molnar & Garai (2005) *Int. Immunopharmacol.* **5**, 849. **7.** Chang & Chiang (1995) *Anticancer Res.* **15**, 1969.

UNC549

This orally bioavailable pyrazolopyrimidine (FW = 396.50 g/mol; Solubility = 100 mg/mL DMSO), also known as 1-((*trans*-4-aminocyclohexyl)-methyl)-*N*-butyl-3-(4-fluorophenyl)-1*H*-pyrazolo[3,4-*d*]pyrimidin-6-amine, MERTK Inhibitor I, RP38 Inhibitor I, TAM Family RTK Inhibitor I, and

UNC-569, is a potent and reversible ATP-competitive Mer receptor tyrosine kinase inhibitor (IC$_{50}$ = 2.9 nM; K_i = 4.3 nM), exhibiting ~10x greater selectivity over the TAM family kinases Axl (IC$_{50}$ = 37 nM) and Tyro3 (IC$_{50}$ = 48 nM). UNC549 also blocks Mer auto-phosphorylation in human Pre-B leukemia 697 cells, IC$_{50}$ = 141 nM. When present at 30 nM, UNC549 inhibits Flt3, MAPKAPK2, RET and Ret-Y791F by 82%, 92%, 59% and 56%, respectively. Treatment of acute lymphoblastic leukemia (ALL) cells with UNC569 reduces proliferation and survival in liquid culture, decreases colony formation in methylcellulose/soft agar, and increases sensitivity to cytotoxic chemotherapies (1). Upon activation by the ligands Gas6 and Protein S, Tyro3/Axl/Mer (TAM) receptor tyrosine kinases promote phagocytic clearance of apoptotic cells and downregulate immune responses initiated by Toll-like receptors and type I interferons, *or* IFNs (2). Many enveloped viruses display the phospholipid phosphatidylserine on their membranes, through which they bind Gas6 and Protein S and engage TAM receptors. Such findings suggest that agents like UNC-549 will be effective in modulating these processes (2). **1.** Christoph, Deryckere, Schlegel, *et al.* (2013) *Mol. Cancer Ther.* **12**, 2367. **2.** Bhattacharyya, Zagórska, Lew, *et al.* (2013) *Cell Host Microbe* **14**, 136.

UNC669

This malignant brain tumor inhibitor (FW = 338.24 g/mol; CAS 1314241-44-5; Solubility: 11 mg/mL DMSO), also named 2-(phenylamino)-1,4-phenylene)bis((4-(pyrrolidin-1-yl)piperidin-1-yl)methanone, interfere with the binding of methyl-lysine binding proteins to *N*-CH$_3$-lysine-containing regulatory proteins, such as p53 and Rb, thereby altering chromatin compaction and repressing transcription. **1.** Herold, Wigle, Norris, *et al.* (2011) *J. Med. Chem.* **54**, 2504.

UNC1215

This MBT antagonist (FW$_{free-base}$ = 529.73 g/mol; CAS 1415800-43-9; Solubility: 100 mg/mL DMSO or Ethanol; <1 mg/mL H$_2$O), also known as 1,1'-[2-(phenylamino)-1,4-phenylene]bis[1-[4-(1-pyrrolidinyl)-1-piperidinyl]methanone, potently and selectively targets the <u>M</u>alignant <u>B</u>rain <u>T</u>umor L3MBTL3 with IC$_{50}$ of 40 nM and K$_d$ of 120 nM, exhibiting 50x tighter binding *versus* other members of the human MBT family. In cells, UNC1215 is nontoxic and binds directly to L3MBTL3 via the Kme-binding pocket of its MBT domains. UNC1215 increases the cellular mobility of GFP-L3MBTL3 fusion proteins. Moreover, point mutations disrupting the Kme-binding function of GFP-L3MBTL3 phenocopy the effects of UNC1215 on localization. UNC1215 was used to reveal a new Kme-dependent interaction of L3MBTL3 with BCLAF1, a protein implicated in DNA damage repair and apoptosis. **1.** James, Barsyte-Lovejoy, Zhong, *et al.* (2013) *Nature Chem. Biol.* **9**, 184.

Uncouplers (Gradient-Dissipating Proton Carriers)

These membrane-permeant proton carriers (protonophores) and mobile, membrane-permeant ion carriers (ionophores) shuttle protons or ions across excitable biomembranes (often with bilayer crossing frequenies of 10^7 sec^{-1}, or higher), thereby dissipating their respective transmembrane proton or ion gradients and frequently defeating life-sustaining metabolic processes (1,2). An important property of uncouplers is that they do not interacting with respiratory chain components or the ATP synthase. Instead, protonophoric uncouplers inhibit oxidative phosphorylation by defeating the productive mechanochemical coupling of the proton gradient (formed by the NADH oxidation-dependent Electron Transport System) to ATP synthase. Mitochondrial uncoupling refers to a condition in which protons cross the inner membrane back into the matrix, while bypassing ("short-circuiting") ATP synthase, with consequential reduction in the transmembrane potential, $\delta\Psi$. The unprotonated uncoupler (U$^-$) combines with a proton H$^+$ at the proton-rich face of the membrane, forming the protonated uncoupler (HU), which then moves across to the other face of the membrane, whereupon protonated uncoupler readily dissociates, thereby increasing the proton concentration on the proton-poor side of the membrane and dissipating the proton gradient.

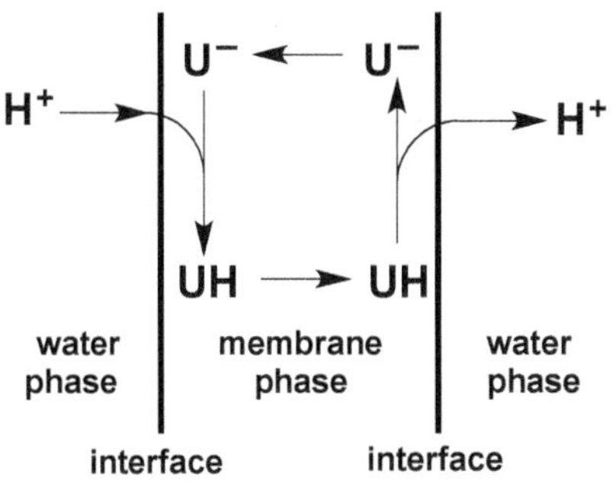

A key feature of effective uncouplers is their ability to delocalize the negative charge of U$^-$ to an extent that it can recross the membrane to reprotonate (2). Consider the case of salicylanilide:

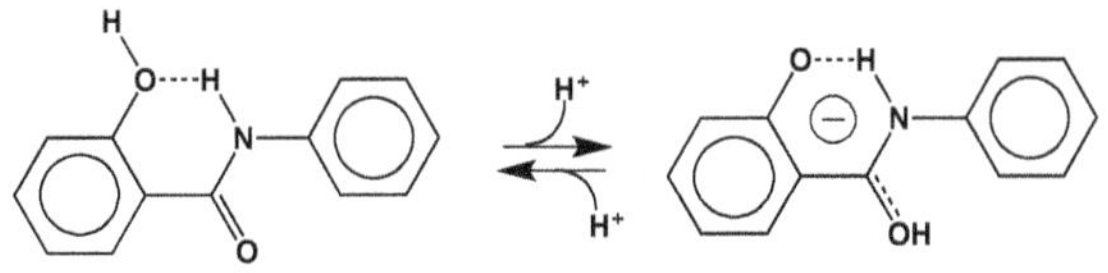

Notice how the negative charge is spread throughout the central, hydrogen bond-containing ring. Uncouplers like SF6847 and S13 are highly effective in delocalizing their negative charge, and they shuttle protons with respective cycling rates of 800 and 400 s^{-1}, at or near the theoretical limit of 1000 s^{-1} (2). Note that interfacial proton transfer is extremely rapid, occurring on μsec timescales, with an estimated bimolecular rate constant of 2×10^{11} M^{-1} s^{-1} (3). This property indicates that there is little or no kinetic barrier for proton transfer from the aqueous phase to the membrane-water interface. Absent the proton gradient acting as a driving force, ATP synthesis is inhibited. In a futile effort to maintain the transmembrane potential in the presence of a protonophoric uncoupler, ATP synthase operates in reverse, catalyzing ATP hydrolysis to liberate ADP, P$_i$, and a proton, a process attended by release of considerable heat. **Frequently Used Uncouplers:** *Carbonyl cyanide m-chlorophenylhydrazone* (Abbreviation: CCCP; Functional Group: Aminoguanidine; Effective Concentration: 110 nM, pK_a = 6.0); *3,5-di-tert-Butyl-4-hydroxybenzylidene-malononitrile* (Abbreviation: SF-6847; Functional Group: Phenol; Effective Concentration: 10 nM, pK_a = 6.8); *2,4-Dinitrophenol* (Abbreviation: 2,4-DNP; Functional Group: Phenol; Effective Concentration: 24 μM, pK_a = 4.1); *Pentachlorophenol* (Abbreviation: PCP; Functional Group: Phenol; Effective Concentration: 1 μM, pK_a = 4.6); *Salicylanilide* (Functional Group: 2-Hydroxybenzanilide; Effective Concentration: 200 nM, pK_a = 7.0); *4,5,6,7-Tetrafluoro-2-trifluoromethylbenzimidazole* (Abbreviation: TTFB; Functional Group: Imidazole; Effective Concentration: 30 nM, pK_a = 5.5); *Trifluorocarbonylcyanide phenylhydrazone* (Abbreviation: FCCP; Functional Group: Aminoguanidine; Effective Concentration: 70 nM, pK_a = 6.2). These agents are moderately hydrophobic, membrane-permeant weak acids, with pK_a values ranging from 3.5 to 7.5. Optimal uncoupling is observed with compounds having pK_a values near 4.0. For example, S-133 and SF6847 are highly effective uncouplers, even at 10 nM. Free fatty acids (p$K_a \approx 4.8$) may also operate as mild uncouplers. Brown adipose tissue of in newborn and hibernating mammals as well as irisin-responsive white-fat adipose tissue utilizes its uncoupling protein *thermogenin* (*See Uncoupling Protein 1*), a protein that stimulates ATP hydrolysis, thereby warming the animal. Thermogenin is activated by fatty acids and inhibited by nucleotides. The anti-inflammatory activity of several drugs is also believed to depend on their ability to uncouple oxidative phosphorylation. Note also that some uncouplers dissipate transmembrane proton gradients and stimulate

vacuolar proton-translocating ATPases (*i.e.,* the so-called V-ATPases). Ionophoric uncouplers defeat productive mechanochemical coupling of the ion gradient formed by ATP-dependent membrane pumps. An ionophoric uncoupler combines with a ion at the ion-rich side of the membrane, moves across to the other face of the membrane, where the ion readily dissociates, thereby increasing the ion concentration on the ion-poor side of the membrane. An example is valinomycin, a membrane-permeant natural product that binds K$^+$ roughly 10^4 times more tightly than Na$^+$ and each valinomycin carries up to 10^4 K$^+$ ions per sec, thereby dissipating potassium ion gradients in many cells. (The antibiotic gramicidin, for comparison, forms a transmembrane channel that transport up to 10^7 K$^+$ ions per sec.) Valinomycin uncouples both oxidative- and photo-phosphorylation, especially when cells contain H$^+$/K$^+$-exchanging channels or ion pumps (4). By shuttling ions forth an back across their target membrane, uncouplers dissipate proton gradients, thus inhibiting ATP synthesis without directly interacting with respiratory chain components or the ATP synthase. In a futile effort to maintain the transmembrane potential, ATP synthase operates in reverse, catalyzing ATP hydrolysis. An example of a mixed uncoupler is nigericin, an antibiotic derived from *Streptomyces hygroscopicus*, that acts both as a protonophore and a K$^+$-ionophore. An important factor, when considering the action of a particular uncoupler, is its effect on ΔpH and $\Delta\psi$ components of the chemisosmotic driving force (3). Consider, for example, the action of nigericin, which when added to a suspension of vescicles containing the electron transport system, will exchange H$^+$ for K$^+$, thus abolishing the pH gradient, while leaving the transmembrane potential intact. For similarly treated suspensions of mitochondria or chloroplast thylakoids, ATP synthesis is effectively uncoupled by nigericin; however, in bacterial preparations, such is not the case, because the system still retains sufficient membrane potential to drive ATP synthesis. Now consider the action of valinomycin: by moving potassium ions across a membrane, valinomycin destroys the membrane potential, while leaving the ΔpH intact. Therefore, in the presence of K$^+$, valinomycin does not uncouple ATP synthesis unless a ΔpH-dissipating agent, such as nigericin is also included. (**See also** *Dinitrophenol (includes additional information on low-molecular-weight protonophores); Carbonyl cyanide m-chlorophenylhydrazone (CCCP)*) Finally, as discussed by Terada (2), not all uncouplers are proton carriers. 1. Skidmore & Whitehouse (1965) Biochem. Pharmacol. **14**, 547. 2. Terada (1990) *Environ. Health. Perspect.* **87**, 213. 3. Maity & Krishnamoorthy (2007) *J. Biosci.* **20**, 573. 4. Sybesma (1989) *Biophysics: An Introduction*, 320 pp., Kluwer, Inc., Dordrecht, Holland.

Uncoupling Protein-1

This mammalian mitochondrial protein (MW$_{human}$ = 33004.50 g/mol; Official Symbol: UCP1), also known as thermogenin, is found in Brown Adipose Tissue (BAT), where it uncouples the transmembrane proton gradient, dissipating the considerable Gibbs free energy of NADH oxidation as thermal energy, a process often referred called a proton leak. UCPs facilitate the transfer of anions from the inner to the outer mitochondrial compartment as well as the return transfer of protons from the outer to the inner mitochondrial compartment, thereby reducing the transmembrane potential. See also *Uncouplers (Gradient-Dissipating Proton Carriers)*

Undecanoyl-CoA

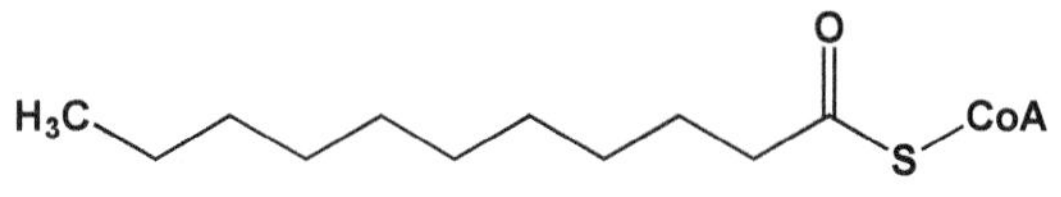

This fatty-acyl-CoA (FW$_{free-acid}$ = 935.82 g/mol) inhibits glucose-6-phosphatase. Fatty-acyl-CoA esters with chain lengths ≤ 9 carbons do not inhibit Glc6Pase. Medium-chain fatty-acyl-CoA esters (10-14 carbons) inhibit Glc6Pase of untreated microsomes in a dose-dependent manner, with K_i values in the range 1-20 microM. The inhibitory effect also depends on the acyl-chain length. The higher the chain-length, the stronger the inhibitory effect. 1. Methieux & Zitoun (1996) *Eur. J. Biochem.* **235**, 799.

2-Undecyl-4H-1,3,2-benzodioxaphosphorin 2-Oxide

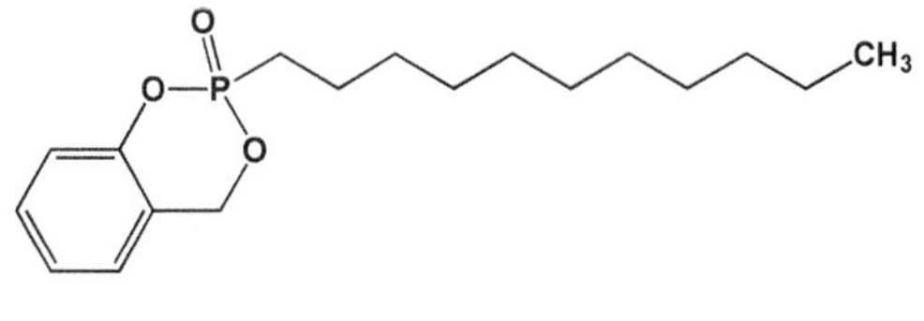

This 2-substituted benzodioxaphosphorin oxide (FW = 324.40 g/mol) inhibits 1-alkyl-2-acetyl-glycerophospho-choline esterase, *or* platelet-activating-factor acetyl-hydrolase. Platelet-activating factor (PAF) is a potent endogenous phospholipid modulator of diverse biological activities, including inflammation and shock. PAF levels are mainly regulated by PAF acetylhydrolases, which are secondary targets of sarin-type organophosphorus (OP) pesticides (*See Sarin*). Inhibitors exhibiting ultrahigh potency and selectivity were achieved with *n*-alkyl methylphosphonofluoridates (*i.e.,* long-chain sarin analogues): mouse brain and testes $IC_{50} \leq 5$ nM for C_8-C_{18} analogues, and 0.1-0.6 nM for C(13) and C(14) compounds; human plasma IC50 < or = 2 nM for C_{13}-C_{18} analogues. AChE inhibitory potency decreased as chain length increased, with maximum brain PAF-AH/AChE selectivity (>3000-fold) for C_{13}-C_{18} compounds. The toxicity of i.p.-administered PAF ($LD_{50} \approx 0.5$ mg/kg) was increased less than 2-fold by pretreatment with tribufos or the C_{13} *n*-alkyl methylphosphonofluoridate. Such studies indicate that PAF-AH is not a major secondary target of OP pesticide poisoning. **1**. Quistad, Fisher, Owen, Klintenberg & Casida (2005) *Toxicol. Appl. Pharmacol.* **205**, 149.

O-(*n*-Undecyl) Ethylphosphonofluoridate

This *n*-alkyl methylphosphonofluoridate (FW = 266.34 g/mol) inhibits acetylcholinesterase and 1-alkyl-2-acetylglycerophosphocholine esterase, *or* platelet-activating-factor acetylhydrolase. *See also* *2-Undecyl-4H-1,3,2-benzodioxaphosphorin 2-Oxide* 1. Quistad, Fisher, Owen, Klintenberg & Casida (2005) *Toxicol. Appl. Pharmacol.* **205**, 149.

5-(*n*-Undecyl)-6-hydroxy-4,7-dioxobenzothiazole

This ubiquinone analogue (FW = 335.47 g/mol; $pK_a \approx 6.5$; Symbol: UHDBT) inhibits complex III of electron transport (1) and *Escherichia coli* cytochrome *o*-type oxidase, $K_i = 0.3$ µM (2). **Target(s):** cytochrome bo_3 ubiquinol oxidase complex (3); cytochrome *o*-type oxidase (2,3); plastoquinol:plastocyanin reductase (cytochrome b_6/f complex) (4); sulfide:quinone reductase (5); and ubiquinol:cytochrome *c* reductase (1). **1**. von Jagow & Link (1986) *Meth. Enzymol.* **126**, 253. **2**. Matsushita, Patel & Kaback (1986) *Meth. Enzymol.* **126**, 113. **3**. Musser, Stowell, Lee, Rumbley & Chan (1997) *Biochemistry* **36**, 894. **4**. Malkin (1988) *Meth. Enzymol.* **167**, 341. **5**. Arieli, Padan & Shahak (1991) *J. Biol. Chem.* **266**, 104.

Undecylhydroxynaphthoquinone

This ubiquinone analogue (FW = 328.45 g/mol) inhibits the reoxidation of the iron-sulfur center and the reduction of cytochrome b_1 in complex III of electron transport. **1**. von Jagow & Link (1986) *Meth. Enzymol.* **126**, 253.

Undecylsuccinate (Undecylsuccinic Acid)

This dicarboxylic acid ($FW_{free-acid}$ = 272.38 g/mol), also called 3-carboxytetradecanoic acid and 3-carboxy-physeteroic acid, inhibits succinate dehydrogenase (1). *Note:* Because this substance readily forms micelles, one must take care to discriminate whether the monomeric and/or

micelle form(s) is(are) inhibitory, when present at concentrations above the critical micelle concentration (*cmc*). **1**. Franke (1944) *Z. Physiol. Chem.* **280**, 76.

S-(3-Undecynoyl)-*N*-acetylcysteamine

This substituted cysteamine, also known as *N*-acetylcysteamine 3-undecynoate thioester (FW = 283.44 g/mol), is an analogue of an acylated acyl-carrier protein and inhibits 3-hydroxydecanoyl-[acyl-carrier-protein] dehydratase of the fatty acid synthase complex (1,2). **1**. Bloch (1971) *The Enzymes*, 3rd ed. (Boyer, ed.), **5**, 441. **2**. Helmkamp, Rando, Brock & Bloch (1968) *J. Biol. Chem.* **243**, 3229.

Unguinol

This herbicide (FW = 326.35 g/mol; CAS 36587-59-4), systematically named 6-[(2*E*)-2-Buten-2-yl]-3,8-dihydroxy-1,9-dimethyl-11*H*-dibenzo[*b,e*][1,4]dioxepin-11-one, originally isolated from *Aspergillus unguis* and *A. nidulans*, inhibits pyruvate, orthophosphate dikinase. **1**. Motti, Bourne, Burnell, *et al.* (2007) *Appl. Environ. Microbiol.* **73**, 1921.

Uniconazole

This plant growth retardant (FW = 291.78 g/mol; CAS 83657-22-1; IUPAC Name: (*E*)-(*RS*)-1-(4-chlorophenyl)-4,4-dimethyl-2-(1*H*-1,2,4-triazol-1-yl)pent-1-en-3-ol) is strong competitive inhibitor ($K_i = 8$ nM) of *Arabidopsis* abscisic acid (ABA) 8'-hydroxylase (1). Uniconazole-treated plants showed enhanced drought tolerance. Uniconazole also inhibits trans-zeatin biosynthesis and targets CYP735As in *Arabidopsis* (2). **1**. Saito, Okamoto, Shinoda, *et al.* (2006) *Biosci. Biotechnol. Biochem.* **70**, 1731. **2**. Sasaki, Ogura, Takei, *et al.* (2013) *Phytochemistry* **87**, 30.

Untenmone A

This marine natural product (FW = 379.56 g/mol; methyl (1*R*,2*R*)-2-hexadecyl-2-hydroxy-5-oxo-3-cyclopentene-1-carboxylate) inhibits DNA polymerases α and β as well as DNA nucleotidylexoytansferase. **Target(s):** DNA-directed DNA polymerase; DNA polymerase α; DNA polymerase β; DNA nucleotidylexotransferase. *Note:* Because this substance readily forms micelles, one must take care to discriminate whether the monomeric and/or micelle form(s) is(are) inhibitory, when present at concentrations above the critical micelle concentration (*cmc*). **1.** Doncaster, Etchells, Kershaw, *et al.* (2006) *Bioorg. Med. Chem. Lett.* **16**, 2877.

UPF596, See *(S)-2-(3'-Carboxybicyclo[1.1.1]-pentyl)glycine*

UPF 1069

This PARP-2 inhibitor (FW = 279.30 g/mol; CAS 1048371-03-4; Soluble to 100 mM in DMSO), also named 5-(2-oxo-2-phenylethoxy)-3,4-dihydroisoquinolin-1(2*H*)-one, selectively targets poly(ADP-ribose) polymerase-2 (IC$_{50}$ = 0.3 μM), with 27x weaker action against PARP-1 (IC$_{50}$ = 8.0 μM). Selective PARP-2 inhibitors increase post-Oxygen-Glucose Deprivation (post-OGD) cell death in a model characterized by loss of neurons through a caspase-dependent, apoptosis-like process (hippocampal slice cultures), but they reduced post-OGD damage and increased cell survival in a model characterized by a necrosis-like process (cortical neurons). UPF-1069 may be a valuable tool for exploring the function of PARP-2 in biological systems and examining the roles of PARP isoenzymes in cell death and survival. **1.** Pellicciari, Camaioni, Costantino, *et al.* (2008) *Chem. Med. Chem.* **3**, 914. **2.** Moroni, Formentini, Gerace, *et al.* (2009) *Brit. J. Pharmacol.* **157**, 854.

Uprosertib

This potent and orally bioavailable pan-AKT inhibitor (FW = 429.20 g/mol; CAS 1047634-65-0 (free base) 1047635-80-2 (HCl salt); Solubility: 93 mg/mL DMSO), also known as GSK2141795 and GSK795, is ATP-competitive and targets AKT1 (*K*$_i$ = 0.066 nM), AKT1^{E17K} (*K*$_i$ = 0.2 nM), AKT2 (*K*$_i$ = 1.4 nM), and AKT3 (*K*$_i$ = 1.5 nM), together comprising a family of PI3K-AKT pathway serine-threonine kinases that are often dysregulated in many human malignancies, leading to increased cell survival, growth and proliferation (1). PIP3 lipids tether these kinases to the membrane via their plextrin homology domain, enabling activation by Thr308 phosphorylation by PDK1 and Ser473 phosphorylation by the mTORC2 complex. Activated AKT in turn phosphorylates FOXO, TSC1/2, PRAS40, and GSK3β. Combination of Lapatinib with Uprosertib (or Afuresertib) is synergistic in HER2$^+$/PIK3CA^{mut} cell lines, but not in HER2$^+$/PIK3CA$^{wild-type}$ cell lines (2). **See also** Afuresertib Changes in phosphoprotein levels in 15 cell types revealed that p-S6RP levels were less well attenuated by lapatinib in HER2$^+$/PIK3CA^{mut} cells, as compared to HER2$^+$/PIK3CA$^{wild-type}$ cells, and that Lapatinib plus Uprosertib (or Afuresertib) reduce p-S6RP levels to those achieved in HER2$^+$/PIK3CA$^{wild-type}$ cells with lapatinib alone. There is also compensatory up-regulation of p-HER3, and p-HER2 is blunted in PIK3CA(mut) cells, following Lapatinib plus Uprosertib (or Afuresertib) treatment (2). Identification of AKT pathway dysregulation in ovarian cancer as a platinum resistance-specific event led to a comprehensive analysis of GSK2141795 *in vitro*, *in vivo* and in the clinic (3). Proteomic analysis of GSK2141795 *in vitro* and *in vivo* identified changes in AKT and p38 phosphorylation and total Bim, IGF1R, AR and YB1 levels. In patient biopsies, prior to treatment with GSK2141795 in a Phase 1 clinical trial, this signature was predictive of post-treatment changes in the response marker CA125, demonstrating the clinical importance of AKT inhibition for re-sensitisation of platinum resistant ovarian cancer to platinum (3). **Other Targets:** P70SK6 (IC$_{50}$ = 50 nM), PKA (IC$_{50}$ = 2.0 nM), PKCα (IC$_{50}$ = >1000 nM), PKCβ1 (IC$_{50}$ = 56 nM), PKCβ2 (IC$_{50}$ = 86 nM), PKCδ (IC$_{50}$ = 69 nM), PKCγ (IC$_{50}$ = 200 nM), PKCε (IC$_{50}$ = >1000 nM), PKCθ (IC$_{50}$ = 64 nM), PKCη (IC$_{50}$ = 49

nM), PKCξ (IC$_{50}$ = >1000 nM), PKG1α (IC$_{50}$ <1 nM), PKG1β (IC$_{50}$ <1 nM), ROCK (IC$_{50}$ = 126 nM), RSK1 (IC$_{50}$ = 200 nM) (1). **1.** Dumble, Crouthamel, Zhang, *et al.* (2014) *PLoS One* **9**, e100880. **2.** Korkola, Collisson, Heiser, *et al.* (2015) *PLoS One* **10**, e0133219. **3.** Cheraghchi-Bashi, Parker, Curry, *et al.* (2015) *Oncotarget* **6**, 41736.

Uracil

This pyrimidine base (FW = 112.09 g/mol), known systematically as 2,4(1*H*,3*H*)-pyrimidinedione, absorbs UV light (λ$_{max}$ = 259.5 nm (ε = 8200 M^{-1}cm^{-1}) at pH 4 and 7; 284 nm (ε = 6200 M^{-1}cm^{-1}) at pH 12), consistent with its predominant diketo tautomeric straucture. **Target(s):** D-amino-acid oxidase, weakly inhibited (1); aspartate carbamoyltransferase (2); carbamoyl-phosphate synthetase (glutamine-hydrolyzing) (3); DNA-uracil glycosylase (4); dUTP diphosphatase (5,6); hydroxyacylglutatahione hydrolase (glyoxalase II) (7); hypoxanthine(guanine) phosphoribosyltransferase (8); lactoylglutathione lyase (glyoxalase I) (7); orotate phosphoribosyltransferase (9-11); ribonuclease, pancreatic (12); thymidine phosphorylase (13-16); thymine,α-ketoglutarate dioxygenase, *or* thymine 7-hydroxylase (17); tRNA-guanine transglycosylase, weakly inhibited (18); urate-ribonucleotide phosphorylase (19); xanthine phosphoribosyltransferase (8,20). **1.** Walaas & Walaas (1956) *Acta Chem. Scand.* **10**, 122. **2.** Rave, Hunter & Shive (1959) *Biochem. Biophys. Res. Commun.* **1**, 115. **3.** Llamas, Suarez, Quesada, Bejar & Del Moral (2003) *Extremephiles* **7**, 205. **4.** Duncan (1981) *The Enzymes*, 3rd ed. (Boyer, ed.), **14**, 565. **5.** Pardo & Gutierrez (1990) *Exp. Cell Res.* **186**, 90. **6.** Bergman, Nyman & Larsson (1998) *FEBS Lett.* **441**, 327. **7.** Oray & Norton (1980) *Biochem. Biophys. Res. Commun.* **95**, 624. **8.** Naguib, Iltzsch, el Kouni, Panzica & el Kouni (1995) *Biochem. Pharmacol.* **50**, 1685. **9.** Jones, Kavipurapu & Traut (1978) *Meth. Enzymol.* **51**, 155. **10.** Javaid, el Kouni & Iltzsch (1999) *Biochem. Pharmacol.* **58**, 1457. **11.** Krungkrai, Aoki, Palacpac, Sato, Mitamura, Krungkrai & Horii (2004) *Mol. Biochem. Parasitol.* **134**, 245. **12.** Richards & Wyckoff (1971) *The Enzymes*, 3rd ed. (Boyer, ed.), **4**, 647. **13.** Baker & Kelley (1971) *J. Med. Chem.* **14**, 812. **14.** Baker & Kawazu (1967) *J. Med. Chem.* **10**, 311. **15.** Blank & Hoffee (1975) *Arch. Biochem. Biophys.* **168**, 259. **16.** Miszczak-Zaborska & Wozniak (1997) *Z. Naturforsch. C* **52**, 670. **17.** Hayaishi, Nozaki & Abbott (1975) *The Enzymes*, 3rd ed. (Boyer, ed.), **12**, 119. **18.** Farkas, Jacobson & Katze (1984) *Biochim. Biophys. Acta* **781**, 64. **19.** Laster & Blair (1963) *J. Biol. Chem.* **238**, 3348. **20.** Miller, Adamczyk, Fyfe & Elion (1974) *Arch. Biochem. Biophys.* **165**, 349.

Uracil-4-acetate, See *6-Carboxymethyluracil*

Uracil 1-β-D-Arabinofuranoside

This uridine analogue (FW = 244.20 g/mol; CAS 3083-77-0; λ$_{max}$ = 261 nm; ε = 10100 M^{-1}cm^{-1} at pH 1 and 7 and ε = 7500 M^{-1}cm^{-1} at pH 12, based on uridine's UV spectrum), also known as 1-β-D-arabinofuranosyluracil (ara-U) and spongouridine, was first isolated from the Caribbean sponge *Cryptotethya crypta*. **Target(s):** cytidine deaminase (1,2); deoxycytidine kinase, weakly inhibited (3,4); and deoxynucleoside kinase (5). **1.** Hosono & Kuno (1973) *J. Biochem.* **74**, 797. **2.** Cacciamani, Vita, Cristalli, Vincenzetti, Natalini, Ruggieri, Amici & Magni (1991) *Arch. Biochem. Biophys.* **290**, 285. **3.** Krenitsky, Tuttle, Koszalka, *et al.* (1976) *J. Biol. Chem.* **251**, 4055. **4.** Wang, Kucera & Capizzi (1993) *Biochim. Biophys. Acta* **1202**, 309. **5.** Johansson, Van Rompay, Degrève, Balzarini & Karlsson (1999) *J. Biol. Chem.* **274**, 23814.

Uracil-6-carboxylate, See *Orotate*

6-Uracil Methyl Sulfone

This nonionizable orotate analogue (FW = 190.18 g/mol) only weakly inhibits orotate phosphoribosyltransferase (*Reaction*: Orotate + p-Rib-pp $\rightleftharpoons$ OMP + PP$_i$), K_i = 0.71 mM (1-3). **1.** Flaks (1963) *Meth. Enzymol.* **6**, 136. **2.** Holmes (1956) *J. Biol. Chem.* **223**, 677. **3.** Webb (1966) *Enzyme and Metabolic Inhibitors*, vol. **2**, p. 473, Academic Press, New York.

6-Uracilsulfonamide

This nonionizable orotate analogue (FW = 191.17 g/mol) inhibits orotate phosphoribosyltransferase, K_i = 7 μM (1-3). 6-Uracilsulfonamide and 6-uracil methylsulfone noncompetitively inhibit the growth of two strains of *Lactobacillus bulgaricus* in media supplemented with orotate or with certain precursors of this compound, including carbamyl-DL-aspartate or L-dihydroorotate. In contrast to the marked effect of this compound on the growth of these organisms, the same analogues had little effect on the growth of *Leuconostoc citrovorum* (*Pediococcus cerevisiae*) 8081 and *Streptococcus jaecalis* 8043, organisms that do not need added orotate for growth. **1.** Flaks (1963) *Meth. Enzymol.* **6**, 136. **2.** Holmes (1956) *J. Biol. Chem.* **223**, 677. **3.** Webb (1966) *Enzyme and Metabolic Inhibitors*, vol. **2**, p. 473, Academic Press, New York.

Uramil

This barbituric acid derivative (FW = 143.10 g/mol; CAS 118-78-5), also known as 5-mino-2,4,6-trihydroxypyrimidine and 5-aminobarbituric acid, is obtained from derivatives of uric acid. Uramil inhibits glucokinase, a finding that suggests yet another off-target effect of barbiturates (1). **1.** Lenzen, Brand & Panten (1988) *Brit. J. Pharmacol.* **95**, 851.

Uranium (and Uranium Ions)

This unstable element (atomic symbol U and atomic number 92) is the heaviest of all naturally occurring elements All isotopes are radioactive: ^{234}U ($t_{1/2}$ = 2.45 x 10^5 years) has a natural abundance of 0.0055 %; ^{235}U ($t_{1/2}$ = 7.04 x 10^8 years) has a natural abundance of 0.720 %; and ^{238}U ($t_{1/2}$ = 4.46 x 10^9 years) has a natural abundance of 99.2745 %. Uranium exhibits oxidation states of 3, 4, 5, or 6 (the ionic radii for U^{4+} and U^{6+} are 0.97 and 0.80 Å, respectively). Uranyl cation (UO$_2^{2+}$) inhibits numerous enzymes (1). **Target(s):** acid phosphatase (2,3); *S*-adenosylmethionine synthase, by uranyl cations (4); ATPases, by uranyl nitrate (5); creatine kinase, by UO$_2^{2+}$ (6); dextranase, by UO$_2^{2+}$ (7); exopolyphosphatase, by UO$_2^{2+}$ (8); β-fructofuranosidase, *or* invertase, by UO$_2^{2+}$ (9); α-glucosidase, by UO$_2^{2+}$ (10); hydroxylamine reductase, by uranyl acetate (11); hydroxymethylbilane synthase, by UO$_2^{2+}$ (12); lysozyme, by UO$_2^{2+}$ (13); NADH oxidase (14); peroxidase, horseradish (15); photosystem II, by uranyl cations (16); pyrophosphatase, by uranyl cations (17); tryptophan transport, by uranyl nitrate (18); and urease, by UO$_2^{2+}$ (19). **1.** Takusagawa, Kamitori, Misaki & Markham (1996) *J. Biol. Chem.* **271**, 136. **2.** Joyce & Grisolia (1960) *J. Biol. Chem.* **235**, 2278. **3.** Andrews & Pallavicini (1973) *Biochim. Biophys. Acta* **321**, 197. **4.** McQueney & Markham (1995) *J. Biol. Chem.* **270**, 18277. **5.** Nechay, Thompson & Saunders (1980) *Toxicol. Appl. Pharmacol.* **53**, 410. **6.** O'Sullivan & Morrison (1963) *Biochim. Biophys. Acta* **77**, 142. **7.** Arnold, Nguyen & Mann (1998) *Arch. Microbiol.* **170**, 91. **8.** Afansieva & Kulaev (1973)

Biochim. Biophys. Acta **321**, 336. **9.** Myrbäck (1960) *The Enzymes*, 2nd ed. (Boyer, Lardy & Myrbäck, eds.), **4**, 379. **10.** Olusanya & Olutiola (1986) *FEMS Microbiol. Lett.* **36**, 239. **11.** Bernheim (1972) *Enzymologia* **43**, 168. **12.** Farmer & Hollebone (1984) *Can. J. Biochem. Cell Biol.* **62**, 49. **13.** Imoto, Johnson, North, Phillips & Rupley (1972) *The Enzymes*, 3rd ed. (Boyer, ed.), **7**, 665. **14.** Regueiro, Amelunxen & Grisolia (1962) *Biochemistry* **1**, 553. **15.** Getchell & Walton (1931) *J. Biol. Chem.* **91**, 419. **16.** Ananyev, Murphy, Abe & Dismukes (1999) *Biochemistry* **38**, 7200. **17.** Bienwald & Hohne (1978) *Acta Biol. Med. Ger.* **37**, 1129. **18.** Kotyk & Dvorakova (1990) *Folia Microbiol. (Praha)* **35**, 209. **19.** Schmidt (1928) *J. Biol. Chem.* **78**, 53.

Urantide

This UT antagonist (FW = 1075.26 g/mol; CAS 669089-53-6; Sequence: DXFWXYCV, with the following modifications: X-2 = Pen, X-5 = Orn, Trp-4 = D-Trp, disulfide bridge between positions 2 and 7; Solubility: 2 mg/mL in H$_2$O) selectively and competitively targets urotensin-II (K_i = 5 x 10^{-8} M). Urantide blocks hU-II induced contractions in thoracic aorta *ex vivo*. It exhibits no effect on noradrenaline or endothelin 1-induced contraction or on acetylcholine-induced relaxation. Urantide as a partial agonist in a calcium mobilization assay in CHO cells expressing hUT receptors. **1.** Patacchini, Santicioli, Giuliani, *et al.* (2003) *Brit. J. Pharmacol.* **140**, 1155. **2.** Carotenuto, Auriemma, Merlino, *et al.* (2014) *J. Med. Chem.* **57**, 5965.

URB597

This carbamate (FW = 338.41 g/mol; CAS 546141-08-6; Soluble to 50 mM in DMSO), also named cyclohexylcarbamic acid 3'-(aminocarbonyl)-[1,1'-biphenyl]-3-yl ester, selectively inhibits fatty acid amide hydrolase, with IC$_{50}$ values of 3 and 5 nM for human liver and rat brain enzymes, respectively (1,2). URB597 exhibits no significant inhibitory activity against a variety of receptors, ion channels and enzymes, including human cannabinoid receptors and rat monoacylglycerol lipase. It also displays antiallodynic and antihyperalgesic activity in an inflammatory pain model. When treated with URB597, adolescent and adult Sprague-Dawley rats displayed higher levels of social behavior and emitted more 50-kHz USVs than Wistar rats, enhancing social play behavior in adolescent Wistar rats under all experimental conditions (3). **1.** Fegley, Gaetani, Duranti, *et al.* (2005) *J. Pharmacol. Exp. Ther.* **313**, 352. **2.** Mor, Rivara, Lodola *et al.* (2004) *J. Med. Chem.* **47**, 4998. **2.** Manduca, Servadio, Campolongo, *et al.* (2014) *Eur. Neuropsychopharmacol.* **24**, 1337.

Urea-1-carboxylate, p-[----See *Allophanate and Allophanic Acid*

Ureidoacetate and Ureidoacetic Acid, See *N-Carbamoylglycine*

5-Ureidohydantoin, See *Allantoin*

β-Ureidoisobutyrate

This metabolite (FW$_{free-acid}$ = 146.15 g/mol), also known as 2-(ureidomethyl)propionic acid, 2-methyl-*N*-carbamoyl-β-alanine, and 2-methyl-3-ureidopropionic acid, is an intermediate in the degradation of deoxythymidine and a precursor to β-aminoisobutyrate. β-Ureido-isobutyrate is also an alternative substrate and competitive inhibitor of maize β-ureidopropionase (1,2). **1**. Walsh, Green, Larrinua & Schmitzer (2001) *Plant Physiol.* **125**, 1001. **2**. Wasternack, Lippmann & Reinbotte (1979) *Biochim. Biophys. Acta* **570**, 341.

Ureidosuccinate and Ureidosuccinic Acid, *See* N-Carbamoyl-L-aspartate and N-Carbamoyl-L-aspartic Acid

5-Ureidouracil

This pyrimidine and H(G)PRT inhibitor (FW = 170.13 g/mol), often misnamed 2,4-dihydroxy-5-carboxamidopyrimidine, targets hypoxanthine (guanine) phosphoribosyl-transferase. **1**. Miller, Ramsey, Krenitsky & Elion (1972) *Biochemistry* **11**, 4723.

Urethane

This simple carbamate (FW = 89.09 g/mol; M.P. = 48–50°C; B.P. 182–184°C; highly water-soluble), also known as urethan, ethyl carbamate, ethylurethan, and carbamic acid ethyl ester, is cytotoxic and inhibits mitosis. Urethan also forms etheno and oxoethyl adducts when combined with DNA. (**Caution:** Carbamate is toxic, and one should avoid inhalation of the dry powder or contact of the powder or urethane solutions with skin.) **Target(s):** arginine decarboxylase (1,2); bacterial luciferase, *or* alkanal monooxygenase (3); carbonic anhydrase (4); esterase (5); firefly luciferase, *or Photinus*-luciferin 4-monooxygenase (6); formate hydrogen lyase (7); glycerol-3-phosphate dehydrogenase (8); hydrogenase (9,10); D-3-hydroxybutyrate dehydrogenase (11); L-lactate dehydrogenase (12); lipoxygenase (13); pyrophosphatase (14); and succinate dehydrogenase, *or* succinate oxidase (15-18). **1**. Das, Bhaduri, Bose & Ghosh (1996) *J. Plant Biochem. Biotechnol.* **5**, 123. **2**. Choudhuri & Ghosh (1982) *Agric. Biol. Chem.* **46**, 739. **3**. Harvey (1951) *The Enzymes*, 1st ed. (Sumner & Myrbäck, eds.), **2** (part 1), 581. **4**. Whitney, Nyman & Malmström (1967) *J. Biol. Chem.* **242**, 4212. **5**. Stedman & Stedman (1932) *Biochem. J.* **26**, 1214. **6**. M. Nehls & Bittar (1989) *Life Sci.* **45**, 2225. **7**. Stephenson & Stickland (1932) *Biochem. J.* **26**, 712. **8**. Green (1936) *Biochem J.* **30**, 629. **9**. Umbreit (1951) *The Enzymes*, 1st ed. (Sumner & Myrbäck, eds.), **2** (part 1), 329. **10**. Stephenson & Strickland (1931) *Biochem. J.* **25**, 205. **11**. Gotterer (1969) *Biochemistry* **8**, 641. **12**. Green & JBrosteaux (1936) *Biochem. J.* **30**, 1489. **13**. Holman & Bergström (1951) *The Enzymes*, 1st ed. (Sumner & Myrbäck, eds.), **2** (part 1), 559. **14**. Naganna (1950) *J. Biol. Chem.* **183**, 693. **15**. Bonner (1955) *Meth. Enzymol.* **1**, 722. **16**. Singer & Kearney (1963) *The Enzymes*, 2nd ed. (Boyer, Lardy & Myrbäck, eds.), **7**, 383. **17**. Stoppani (1948) *Enzymologia* **13**, 165. **18**. Sen (1931) *Biochem. J.* **25**, 849.

Uric Acid

This purine catabolic end-product (FW = 168.11 g/mol; CAS 69-93-2; Solubility: 2 mg/100 mL H_2O at 25°C; pK_a values of 5.4 and 11.3) is formed in reactions catalyzed by xanthine oxidase (*Reaction:* Xanthine + H_2O + $O_2 \rightleftharpoons$ Uric acid + H_2O_2) or xanthine dehydrogenase (*Reaction:* Xanthine + NAD^+ + $H_2O \rightleftharpoons$ Urate + NADH + H^+). While uric acid plays only a minor role in human nitrogen excretion, birds and reptiles are uricotelic, meaning that urate is the primary end-product. Although an unobvious acid, uric acid exists as tautomers (*see above*), and the imido species has a pK_a of 5.4, only slightly less acidic than acetic acid (pK_a = 4.8).

While sodium urate is quite soluble in physiologic fluids, uric acid is not and forms paracrystalline masses that accumulate in joints, especially in the big toe.. Indeed, in humans, excess uric acid formation leads to Gout, a disorder that is characterized by soreness in toe joints as well as the formation of tofi, sack-like malformations beneath the skin that are filled with crystalline, chaulk-like deposits of uric acid. In gout, uric acid accumulation in joints initiates an inflammasome response, resulting in excruciating pain. Excessive uric acid formation and gouty deposits are also observed in Lesch-Nyhans Disorder, a condition associated with defects in the gene encoding hypoxanthine:guanine phosphoribosyltransferase (HGPRT). In humans, uric acid also affects endothelial and adipose cell function and has been linked to hypertension, metabolic syndrome, and cardiovascular disease. Hyperuricemia can, in some instances, stimulate onset of hypertension, most likely by generating an inflammatory cascade, where endothelial dysfunction, smooth muscle proliferation and development of renal afferent arteriolosclerosis appear. **Target(s):** In the presence of peroxynitrite, uric acid is converted nonenzymatically to triuret (*or* 1,1'-carbonyl-bis-urea), which inhibits dipeptidyl peptidase IV *in vitro* (1). Uric acid inhibits arginase (2); methionine *S*-adenosyltransferase (3); phosphorylase a_3; protein-tyrosine phosphatase, *or* low-molecular-mass phosphotyrosine protein phosphatase (4,5); sepiapterin deaminase, weakly inhibited (6); and xanthine phosphoribosyltransferase (7). **1**. Mohandas, Sautina, Beem, *et al.* (2014) *Exp. Cell Res.* **326**, 136. **2**. Rosenfeld, Dutta, Chheda & Tritsch (1975) *Biochim. Biophys. Acta* **410**, 164. **3**. Berger & Knodel (2003) *BMC Microbiol.* **3**, 12. **4**. Ercan-Fang, Nuttall & Gannon (2001) *Amer. J. Physiol. Endocrinol. Metab.* **280**, E248. **5**. Granjeiro, Ferreira & Granjeiro, *et al.* (2002) *J. Enzyme Inhib. Med. Chem.* **17**, 345. **6**. Tsusue (1971) *J. Biochem.* **69**, 781. **7**. Miller, Adamczyk, Fyfe & Elion (1974) *Arch. Biochem. Biophys.* **165**, 349.

Uridine

This pyrimidine nucleoside and RNA precursor (FW = 244.20 g/mol; (λ_{max} is 261 nm; ε = 10100 $M^{-1}cm^{-1}$ at pH 1 and 7 and ε = 7500 $M^{-1}cm^{-1}$ at pH 12; pK_a values = 9.2 and 12.5 at 25°C) is a component of ribonucleotides, RNA, and sugar nucleotides. Uridine is prepared from cytidine by treatment with nitrous acid to form the diazonium salt, followed by hydrolysis. *Note*: Xuriden® (FW_uridine-triacetate = 370.3 g/mol; CAS 4105-38-8; IUPAC: (2R,3R,4R,5R)-2-(acetoxymethyl)-5-(2,4-dioxo-3,4-dihydropyrimidin-1(2H)-yl)tetrahydrofuran-3,4-diyl diacetate), is a membrane-permeant pro-drug that, upon oral administration, is deacetylated by nonspecific esterases throughout the body. Xuriden provides uridine to patients with hereditary orotic aciduria and cannot synthesize adequate quantities of uridine due to a genetic defect in OMP decarboxylase. **Target(s):** Acid phosphatase (1,2); adenosine kinase (3,4); adenosylhomocysteinase (5); AMP:thymidine kinase (6); CMP-*N*-acylneuraminate phosphodiesterase, weakly inhibited (7); cytidine deaminase (8,9); deoxycytidine deaminase (10); deoxycytidine kinase, weakly inhibited (11); dihydroorotase (12); globoside α-*N*-acetylgalactosaminyltransferase (27,28); glycogen synthase (30); glycoprotein-fucosylgalactoside α-*N*-acetylgalactos-aminyltransferase (29); hydroxyacylglutathione hydrolase, *or* glyoxalase II (13); lactoylglutathione lyase, *or* glyoxalase II (13); NDP-glucose:starch glucosyltransferase14; nucleoside phosphotransferase (15); 5'-nucleotidase, weakly inhibited (16-18); pancreatic ribonuclease, *or* ribonuclease A (19); purine nucleosidase (20,21); ribonuclease T₂ (22); thymidine phosphorylase (23,24); UDP-*N*-acetyl-glucosamine diphosphorylase, *or N*-acetylglucosamine-1-phosphate uridylyltransferase (25); and urate-ribonucleotide phosphorylase (26). **1**. Belfield, Ellis & Goldberg (1972) *Enzymologia* **42**, 91. **2**. Uerkvitz (1988)

J. Biol. Chem. **263**, 15823. **3.** Long & Parker (2006) *Biochem. Pharmacol.* **71**, 1671. **4.** Palella, Andres & Fox (1980) *J. Biol. Chem.* **255**, 5264. **5.** Knudsen & Yall (1972) *J. Bacteriol.* **112**, 569. **6.** Grivell & Jackson (1976) *Biochem. J.* **155**, 571. **7.** Kean & Bighouse (1974) *J. Biol. Chem.* **249**, 7813. **8.** Frick, Yang, Marquez & Wolfenden (1989) *Biochemistry* **28**, 9423. **9.** Wentworth & Wolfenden (1978) *Meth. Enzymol.* **51**, 401. **10.** Le Floc'h & Guillot (1974) *Phytochemistry* **13**, 2503. **11.** Krenitsky, Tuttle, Koszalka, *et al.* (1976) *J. Biol. Chem.* **251**, 4055. **12.** Bresnick & Blatchford (1964) *Biochem. Biophys. Acta* **81**, 150. **13.** Oray & Norton (1980) *Biochem. Biophys. Res. Commun.* **95**, 624. **14.** Zea & Pohl (2005) *Biopolymers* **79**, 106. **15.** Brunngraber (1978) *Meth. Enzymol.* **51**, 387. **16.** Webb (1966) *Enzyme and Metabolic Inhibitors*, vol. 2, p. 472, Academic Press, New York. **17.** Klein (1957) *Z. Physiol. Chem.* **307**, 254. **18.** Segal & Brenner (1960) *J. Biol. Chem.* **235**, 471. **19.** Richards & Wyckoff (1971) *The Enzymes*, 3rd ed. (Boyer, ed.), **4**, 647. **20.** Ogawa, Takeda, Xie, *et al.* (2001) *Appl. Environ. Microbiol.* **67**, 1783. **21.** Atkins, Shelp & Storer (1989) *J. Plant Physiol.* **134**, 447. **22.** Irie & Ohgi (1976) *J. Biochem.* **80**, 39. **23.** Panova, Alexeev, Kuzmichov, *et al.* (2007) *Biochemistry (Moscow)* **72**, 21. **24.** Blank & Hoffee (1975) *Arch. Biochem. Biophys.* **168**, 259. **25.** Yamamoto, Moriguchi, Kawai & Tochikura (1980) *Biochim. Biophys. Acta* **614**, 367. **26.** Laster & Blair (1963) *J. Biol. Chem.* **238**, 3348. **27.** Ishibashi, Ohkubo & Makita (1977) *Biochim. Biophys. Acta* **484**, 24. **28.** Ishibashi, Atsuta & Makita (1976) *Biochim. Biophys. Acta* **429**, 759. **29.** Navaratnam, Findlay, Keen & Watkins (1990) *Biochem. J.* **271**, 93. **30.** Zea & Pohl (2005) *Biopolymers* **79**, 106.

Uridine 5'-(3-Acetamido-2,6-anhydro-1,3-dideoxy-D-*arabino*-hept-2-enitol-1-yl phosphono)-Phosphate

This phosphinate-containing bisubstrate analogue (FW$_{\text{free-acid}}$ = 603.6 g/mol) inhibits UDP-acetylglucosamine 2-epimerase by mimicking the transition state for the second step of the reaction. To mimic the assumed first transition state of this reaction, novel UDP-exo-glycal derivatives were designed and synthesized. (*See Uridine 5'-[(Z)-3-Acetamido-2,6-anhydro-1,3-dideoxy-D-gluco-hept-1-enitol-1-yl phosphono] Phosphate*) **1.** Stolz, Reiner, Blume, Reutter & Schmidt (2004) *J. Org. Chem.* **69**, 665.

Uridine 6-Carboxylate, *See* Orotidine

Uridine 2',3'-Dialdehyde 5'-Diphosphate, *See* Uridine 5'-Diphosphate 2',3'-Dialdehyde

Uridine 5'-Diphosphate

This pyrimidine nucleotide (FW$_{\text{free-acid}}$ = 404.16 g/mol), abbreviated UDP, is a biosynthetic precursor of UTP and serves as the activating group in many nucleoside diphospho-sugars. UDP absorbs ultraviolet light strongly (λ_{max} is 262 nm with ε = 10000 M^{-1}cm^{-1} at pH 2 and 7; and λ_{max} = 261 nm, ε = 7900 M^{-1}cm^{-1} at pH 11). Note: UDP is a product of many glycosyltransferase reactions and acts as a product inhibitor. UDP is also an alternative substrate of guanosine-diphosphatase. **Target(s):** α-*N*-acetylgalactosaminide α-2,6-sialyltransferase (1); *N*-acetylglucosaminyl-diphospho-dolichol *N*-acetylglucosaminyltransferase (2); α-*N*-acetylneuraminate α-2,8-sialyltransferase, weakly inhibited (76); adenine phosphoribosyltransferase, weakly inhibited (83); adenylosuccinate synthetase (3); alcohol *O*-cinnamoyltransferase (96); amidophosphoribosyltransferase, weakly inhibited (80); *Bacillus subtilis* ribonuclease, weakly inhibited (41); bis(5'-nucleosyl)-tetraphosphatase (39); [branched-chain α-keto-acid dihydrogenase] kinase, *or* [3-methyl-2-oxobutanoate dehydrogenase (acetyl-transferring)] kinase (50); carbamoyl-phosphate synthetase (glutamine-hydrolyzing) (5-7); CDP-diacylglycerol:inositol 3-phosphatidyltransferase (56,57); chitobiosyldiphospho-dolichol β-mannosyltransferase (2); CMP-*N*-acylneuraminate phosphodiesterase (43); cytidylate kinase (63); deoxycytidine kinase (8,9,65-69); deoxyguanosine kinase (10); diacylglycerol kinase (64); dolichyl-phosphate β-D-mannosyltransferase (91); dUTP diphosphatase (37); fructose-1,6-bisphosphatase (47-49); galactolipid galactosyltransferase (87); β-galactoside α-2,3-sialyltransferase (1); 1,3-β-D-glucan synthase (11,12); glucomannan 4-β-mannosyltransferase, weakly inhibited (93); glucose-6-phosphate dehydrogenase, weakly inhibited (35); glucuronokinase (70); glycogen phosphorylase (94,95); glycogen synthase (13,14); glycolipid 2-α-mannosyltransferase, weakly inhibited (89); glycolipid 3-α-mannosyltransferase, weakly inhibited (88); glycoprotein 3-α-L-fucosyltransferase, weakly inhibited (84-86); glycoprotein 6-α-L-fucosyltransferase (92); hexokinase, weakly inhibited (71-74); hypoxanthine(guanine) phosphoribosyltransferase, weakly inhibited (83); lactosyl-ceramide α-2,3-sialyltransferase (75); [3-methyl-2-oxobutanoate dehydrogenase, *or* [branched-chain α-keto-acid dehydrogenase] phosphatase (4,45); NAD$^+$ diphosphatase (38); nicotinate-nucleotide diphosphorylase (carboxylating), weakly inhibited (79); 5'-nucleotidase (15); nucleotide diphosphatase (40); oligonucleotidase (42); orotate phosphoribosyltransferase (16,81); orotidylate decarboxylase (17); phosphodiesterase I, *or* 5'-exonuclease (44); phosphoglucomutase (17); phosphoprotein phosphatase (46); procollagen galactosyltransferase (19); [protein-PII] uridylyltransferase (59); protein xylosyl-transferase (77); pyruvate,orthophosphate dikinase (51); ribose-phosphate diphosphokinase, *or* phosphoribosyl-pyrophosphate synthetase (18,60-62); sucrose:sucrose fructosyltransferase (90); UDP-*N*-acetylglucosamine: dolichyl-phosphate *N*-acetylglucosaminephosphotransferase (54,55); UDP-*N*-acetylglucosamine 2-epimerase (20,28,29); UDP-*N*-acetylglucosamine:lysosomal-enzyme *N*-acetyl-glucosaminephosphotransferase (21,52,53); UDP-galactopyranose mutase (22); UDP-glucose 4-epimerase, inhibition often enhanced in the presence of a sugar (23,33,34); UDP-glucuronate decarboxylase (24-26,36); UDP-glucuronate 4-epimerase (27,30-32); undecaprenyl-phosphate galactose phosphotransferase, weakly inhibited (58); uracil phosphoribosyltransferase (82); and xanthine phosphoribosyltransferase (78); competitive antagonist at P2Y$_{14}$ receptors, K_i = 2 nM (no antagonist activity was observed with ADP, CDP, or GDP, and other uracil analogs) (97).

1. Sadler, Beyer, Oppenheimer, *et al.* (1982) *Meth. Enzymol.* **83**, 458. **2.** Sharma, Lehle & Tanner (1982) *Eur. J. Biochem.* **126**, 319. **3.** Van der Weyden & Kelly (1974) *J. Biol. Chem.* **249**, 7282. **4.** Damuni & Reed (1988) *Meth. Enzymol.* **166**, 321. **5.** Anderson, Wellner, Rosenthal & Meister (1970) *Meth. Enzymol.* **17A**, 235. **6.** Mori & Tatibana (1978) *Meth. Enzymol.* **51**, 111. **7.** Anderson & Meister (1966) *Biochemistry* **5**, 3164. **8.** Ives & Wang (1978) *Meth. Enzymol.* **51**, 337. **9.** Anderson (1973) *The Enzymes*, 3rd ed. (Boyer, ed.), **9**, 49. **10.** Barker Lewis (1981) *Biochim. Biophys. Acta* **658**, 111. **11.** Taft, Zugel & Selitrennikoff (1991) *J. Enzyme Inhib.* **5**, 41. **12.** Marechal, Goldemberg & Marver (1964) *J. Biol. Chem.* **239**, 3163. **13.** Roach & Larner (1976) *Trends Biochem. Sci.* **1**, 110. **14.** Steiner, Younger & King (1965) *Biochemistry* **4**, 740. **15.** Madrid-Marina & Fox (1986) *J. Biol. Chem.* **261**, 444. **16.** Jones, Kavipurapu & Traut (1978) *Meth. Enzymol.* **51**, 155. **17.** Ray & Peck (1972) *The Enzymes*, 3rd ed. (Boyer, ed.), **6**, 407. **18.** Wong & Murray (1969) *Biochemistry* **8**, 1608. **19.** Spiro & Spiro (1972) *Meth. Enzymol.* **28**, 625. **20.** Kikuchi & Tsuiki (1973) *Biochim. Biophys. Acta* **327**, 193. **21.** Reitman, Lang & Kornfeld (1984) *Meth. Enzymol.* **107**, 163. **22.** Caravano, Sinay & Vincent (2006) *Bioorg. Med. Chem. Lett.* **16**, 1123. **23.** Glaser (1972) *The Enzymes*, 3rd ed. (Boyer, ed.), **6**, 355. **24.** Ankel & Feingold (1966) *Meth. Enzymol.* **8**, 287. **25.** Grisebach, Baron, Sandermann & Wellmann (1972) *Meth. Enzymol.* **28**, 439. **26.** Ankel & Feingold (1965) *Biochemistry* **4**, 2468. **27.** Gaunt, Ankel & Schutzbach (1972) *Meth. Enzymol.* **28**, 426. **28.** Kawamura, Kimura, Yamamori & Ito (1978) *J. Biol. Chem.* **253**, 3595. **29.** Sommar & Ellis (1972) *Biochim. Biophys. Acta* **268**, 590. **30.** Gu & Bar-Peled (2004) *Plant Physiol.* **136**, 4256. **31.** Gaunt, Maitra & Ankel (1974) *J. Biol. Chem.* **249**, 2366. **32.** Munoz, Lopez, de Frutos & Garcia (1999) *Mol. Microbiol.* **31**, 703. **33.** Geren, Geren & Ebner (1977) *J. Biol. Chem.* **252**, 2089. **34.** Lee, Kimura & Tochikura (1978) *Agric. Biol. Chem.* **42**, 731. **35.** Horne, Anderson & Nordlie (1970) *Biochemistry* **9**, 610. **36.** Bar-Peled, Griffith & Doering (2001) *Proc. Natl. Acad. Sci. U.S.A.* **98**, 12003. **37.** Bergman, Nyman & Larsson (1998) *FEBS Lett.* **441**, 327. **38.** Nakajima, Fukunaga,

Sasaki & Usami (1973) *Biochim. Biophys. Acta* **293**, 242. **39**. Moreno, Lobaton, Sillero & Sillero (1982) *Int. J. Biochem.* **14**, 629. **40**. Wise, Anderson & Anderson (1997) *Vet. Microbiol.* **58**, 261. **41**. Yamasaki & Arima (1970) *Biochim. Biophys. Acta* **209**, 47542. Futai & Mizuno (1967) *J. Biol. Chem.* **242**, 5301. **43**. Kean & Bighouse (1974) *J. Biol. Chem.* **249**, 7813. **44**. Futai & Mizuno (1967) *J. Biol. Chem.* **242**, 5301. **45**. Damuni, Merryfield, Humphreys & Reed (1984) *Proc. Natl. Acad. Sci. U.S.A.* **81**, 4335. **46**. Nakai & Thomas (1974) *J. Biol. Chem.* **249**, 6459. **47**. Fujita & Freese (1979) *J. Biol. Chem.* **254**, 5340. **48**. Kruger & Beevers (1984) *Plant Physiol.* **76**, 49. **49**. Rittmann, Schaffer, Wendisch & Sahm (2003) *Arch. Microbiol.* **180**, 285. **50**. Reed, Damuni & Merryfield (1985) *Curr. Top. Cell. Regul.* **27**, 41. **51**. Tjaden, Plagens, Dörr, Siebers & Hensel (2006) *Mol. Microbiol.* **60**, 287. **52**. Bao, Elmendorf, Booth, Drake & Canfield (1996) *J. Biol. Chem.* **271**, 31446. **53**. Zhao, Yeh & Miller (1992) *Glycobiology* **2**, 119. **54**. Shailubhai, Dong-Yu, Saxena & Vijay (1988) *J. Biol. Chem.* **263**, 15964. **55**. Sharma, Lehle & Tanner (1982) *Eur. J. Biochem.* **126**, 319. **56**. Antonsson & Klig (1996) *Yeast* **12**, 449. **57**. Antonsson (1994) *Biochem. J.* **297**, 517. **58**. Osborn & Tze-Yuen (1968) *J. Biol. Chem.* **243**, 5145. **59**. Engleman & Francis (1978) *Arch. Biochem. Biophys.* **191**, 602. **60**. Fox & Kelley (1972) *J. Biol. Chem.* **247**, 2126. **61**. Tatibana, Kita, Taira, *et al.* (1995) *Adv. Enzyme Regul.* **35**, 229. **62**. Switzer & Sogin (1973) *J. Biol. Chem.* **248**, 1063. **63**. Seagrave & Reyes (1987) *Arch. Biochem. Biophys.* **254**, 518. **64**. Wissing & Wagner (1992) *Plant Physiol.* **98**, 1148. **65**. Durham & Ives (1970) *J. Biol. Chem.* **245**, 2276. **66**. Datta, Shewach, Mitchell & Fox (1989) *J. Biol. Chem.* **264**, 9359. **67**. Wang, Kucera & Capizzi (1993) *Biochim. Biophys. Acta* **1202**, 309. **68**. Ives & Durham (1970) *J. Biol. Chem.* **245**, 2285. **69**. Hughes, Hahn, Reynolds & Shewach (1997) *Biochemistry* **36**, 7540. **70**. Gillard & Dickinson (1978) *Plant Physiol.* **62**, 706. **71**. Renz & Stitt (1993) *Planta* **190**, 166. **72**. Yamashita & Ashihara (1988) *Z. Naturforsch. C* **43**, 827. **73**. Gao & Leary (2003) *J. Amer. Soc. Mass Spectrom.* **14**, 173. **74**. Dörr, Zaparty, Tjaden, Brinkmann & Siebers (2003) *J. Biol. Chem.* **278**, 18744. **75**. Cambron & Leskawa (1993) *Biochem. Biophys. Res. Commun.* **193**, 585. **76**. Eppler, Morré & Keenan (1980) *Biochim. Biophys. Acta* **619**, 332. **77**. Casanova, Kuhn, Kleesiek & Götting (2008) *Biochem. Biophys. Res. Commun.* **365**, 678. **78**. Miller, Adamczyk, Fyfe & Elion (1974) *Arch. Biochem. Biophys.* **165**, 349. **79**. Taguchi & Iwai (1976) *Agric. Biol. Chem.* **40**, 385. **80**. Holmes, McDonald, McCord, Wyngaarden & Kelley (1973) *J. Biol. Chem.* **248**, 144. **81**. Ashihara (1978) *Z. Pflanzenphysiol.* **87**, 225. **82**. Natalini, Ruggieri, Santarelli, Vita & Magni (1979) *J. Biol. Chem.* **254**, 1558. **83**. Nagy & Ribet (1977) *Eur. J. Biochem.* **77**, 77. **84**. de Vries, Storm, Rotteveel, *et al.* (2001) *Glycobiology* **11**, 711. **85**. Murray, Takayama, Schultz & Wong (1996) *Biochemistry* **35**, 11183. **86**. Shinoda, Morishita, Sasaki, *et al.* (1997) *J. Biol. Chem.* **272**, 31992. **87**. Heemskerk, Jacobs, Scheijen, Helsper & Wintermans (1987) *Biochim. Biophys. Acta* **918**, 189. **88**. Jensen & Schutzbach (1981) *J. Biol. Chem.* **256**, 12899. **89**. Schutzbach, Springfield & Jensen (1980) *J. Biol. Chem.* **255**, 4170. **90**. Satyanarayana (1976) *Indian J. Biochem. Biophys.* **13**, 261. **91**. Villagómez-Castro, Calvo-Méndez, Vargas-Rodríguez, Flores-Carreón & López-Romero (1998) *Exp. Parasitol.* **88**, 111. **92**. Ihara, Ikeda & Taniguchi (2006) *Glycobiology* **16**, 333. **93**. Piro, Zuppa, Dalessandro & Northcote (1993) *Planta* **190**, 206. **94**. Tanabe, Kobayashi & Matsuda (1988) *Agric. Biol. Chem.* **52**, 757. **95**. Madsen (1961) *Biochem. Biophys. Res. Commun.* **6**, 310. **96**. Mock & Strack (1993) *Phytochemistry* **32**, 575. **97**. Fricks, Maddileti, Carter, *et al.* (2008) *J. Pharmacol. Exp. Ther.* **325**, 588.

Uridine 5'-Diphosphate Chloroacetol

This modified nucleotide (FW$_\text{free-acid}$ = 494.67 g/mol) is an irreversible inhibitor of *Escherichia coli* UDPgalactose 4-epimerase: K_D = 0.110 mM and k_inact = 0.84 min^{-1} at pH 8.5. A nucleotide-NAD$^+$ adduct is proposed to form *via* alkylation of the nicotinamide ring by the chloroacetol enolate. **Target(s):** UDP-glucose dehydrogenase (1); UDP-glucose 4-epimerase, *or* UDP-galactose 4-epimerase (2). **1**. Campbell, Sala, I. van de Rijn & Tanner (1997) *J. Biol. Chem.* **272**, 3416. **2**. Flentke & Frey (1990) *Biochemistry* **29**, 2430.

Uridine 5'-Diphosphate 2',3'-Dialdehyde

This UDP derivative (FW$_\text{free-acid}$ = 402.15 g/mol), which is readily generated by treating UDP with stoichiometric amounts of periodate, followed by incubation with 0.1 M ethylene glycol to inactivate any unreacted periodate, binds to and irreversibly modifies several enzymes. **Target(s):** galactosyltransferase (1); lactosylceramide α-2,3-sialyltransferase, *or* CMP-*N*-acetyl-neuraminate:lactosylceramide (α2→3) sialyltransferase (2); sterol 3β-glucosyltransferase (3); UDP-*N*-acetylglucos-amine 2-epimerase (4). **1**. Powell & Brew (1976) *Biochemistry* **15**, 3499. **2**. Cambron & Leskawa (1993) *Biochem. Biophys. Res. Commun.* **193**, 585. **3**. Ullmann, Ury, Rimmele, Benveniste & Bouvier-Navé (1993) *Biochimie* **75**, 713. **4**. Blume, Chen, Reutter, Schmidt & Hinderlich (2002) *FEBS Lett.* **521**, 127.

Uridine Diphosphate D-Glucose, See *UDP-D-glucose*

Uridine 5'-Diphosphate, Periodate-Oxidized, See *Uridine 5'-Diphosphate 2',3'-Dialdehyde*

Uridine Diphospho-*N*-acetylgalactosamine, See *UDP-D-N-acetylgalactosamine*

Uridine Diphospho-*N*-acetylglucosamine, See *UDP-D-N-acetylglucosamine*

Uridine Diphosphogalactose, See *UDP-D-galactose*

Uridine Diphosphoglucose, See *UDP-D-glucose*

Uridine Diphosphoglucuronate, See *UDP-D-glucuronate*

[1-[6-(Uridinediphospho)hexanamido]ethyl](2,4-dicarboxybutyl) Phosphinate, See *[1-[6-(UDP)hexanamido]ethyl](2,4-dicarboxybutyl) Phosphinate*

Uridine Diphosphopyridoxal, See *UDP-pyridoxal*

Uridine Diphosphoxylose, See *UDP-D-xylose*

Uridine 2'-Monophosphate

This nucleotide (FW$_\text{free-acid}$ = 324.18 g/mol), which is obtained upon hydrolysis of RNA, is slightly more labile than 5'-UMP under alkaline conditions. 2'-UMP absorbs light strongly in the ultraviolet region of the spectrum (λ_max is 260 nm with ε = 9900 and 10000 M^{-1}cm^{-1} at pH 2 and 7, respectively; and λ_max = 261 nm, ε = 7300 M^{-1}cm^{-1} at pH 12). **Target(s):** 2',3'-cyclic-nucleotide 3'-phospho-diesterase (1); pancreatic ribonuclease, *or* ribonuclease A (2-4); ribonuclease M (5); ribonuclease Rh (5); ribonuclease T$_1$ (6); ribonuclease T$_2$ (5,7). **1**. Díaz & Heredia (1996) *Biochim. Biophys. Acta* **1290**, 135. **2**. Russo, Acharya & Shapiro (2001) *Meth. Enzymol.* **341**, 629. **3**. Richards & Wyckoff (1971) *The Enzymes*, 3rd ed., **4**, 647. **4**. Blackburn & Moore (1982) *The Enzymes*, 3rd ed., **15**, 317. **5**. Irie & Ohgi (2001) *Meth. Enzymol.* **341**, 42. **6**. Takahashi & Moore (1982) *The Enzymes*, 3rd ed. (Boyer, ed.), **15**, 435. **7**. Irie & Ohgi (1976) *J. Biochem.* **80**, 39.

Uridine 3'-Monophosphate

This nucleotide (FW$_{free-acid}$ = 324.18 g/mol) is obtained as the limit-hydrolysate of RNase action and upon hydrolysis of RNA under alkaline conditions. Uridine 3'-Monophosphate is stable under alkaline conditions (*e.g.*, 0.3 M KOH at 37°C for 16 hours). 3'-UMP absorbs light strongly in the ultraviolet region of the spectrum (λ_{max} is 262 nm with ε = 10000 M^{-1}cm^{-1} at pH 1 and 7; and λ_{max} = 261 nm, ε = 7800 M^{-1}cm^{-1} at pH 13). **Target(s):** pancreatic ribonuclease, *or* ribonuclease A (1-3); phosphodiesterase I, *or* 5'-exonuclease, weakly inhibited (4); ribonuclease M (5); ribonuclease Rh (5); ribonuclease T$_1$ (6); ribonuclease T$_2$ (5,7); and ribose-5-phosphate isomerase (8). **1.** Park, Kelemem, Klink, *et al.* (2001) *Meth. Enzymol.* **341**, 81. **2.** Russo, Acharya & Shapiro (2001) *Meth. Enzymol.* **341**, 629. **3.** Richards & Wyckoff (1971) *The Enzymes*, 3rd ed. (Boyer, ed.), **4**, 647. **4.** Futai & Mizuno (1967) *J. Biol. Chem.* **242**, 5301. **5.** Irie & K. Ohgi (2001) *Meth. Enzymol.* **341**, 42. **6.** Takahashi & Moore (1982) *The Enzymes*, 3rd ed. (Boyer, ed.), **15**, 435. **7.** Irie & Ohgi (1976) *J. Biochem.* **80**, 39. **8.** Horitsu, Sasaki, Kikuchi, *et al.* (1976) *Agric. Biol. Chem.* **40**, 257.

Uridine 3',5'-Monophosphate, *See* Uridine 3',5'-Cyclic Monophosphate

Uridine 5'-Monophosphate

This nucleotide (FW$_{free-acid}$ = 324.18 g/mol), a key intermediate in the biosynthesis of UTP, is produced by the action of orotidylate decarboxylase (on which UMP acts as a product inhibitor) and in a salvage pathway from uridine. UMP absorbs light strongly in the ultraviolet region of the spectrum (λ_{max} is 262 nm with ε = 10000 M^{-1}cm^{-1} at pH 2 and 7; and λ_{max} = 261 nm, ε = 7800 M^{-1}cm^{-1} at pH 11). **Target(s):** α-N-acetylgalactosaminide α-2,6-sialyl-transferase (1); N-acetylglucosaminyldiphospho-dolichol N-acetylglucosaminyltransferase (2,96,97); N-acetylglucosaminyldiphospho-undecaprenol N-acetyl-β-D-mannos-aminyltransferase, weakly inhibited (90); β-N-acetyl-glucosaminylglycopeptide β-1,4-galactosyltransferase (112); N-acetyllactosamine synthase (102); N-acetyl-lactosaminide β-1,6-N-acetylglucosaminyltransferase (95); α-N-acetylneuraminate α-2,8-sialyltransferase, weakly inhibited (78); adenine phosphoribosyltransferase (84); adenylosuccinate synthetase (3); adenylyl-sulfate kinase (74); amidophosphoribosyltransferase, weakly inhibited (82); D-amino-acid oxidase (4); AMP Nucleosidase (51); asparagine synthetase, weakly inhibited (5); aspartate carbamoyltransferase (6-8); carbamate kinase (9); carbamoyl-phosphate synthetase II (glutamine-hydrolyzing) (10-15,37); CDP-diacylglycerol diphosphatase, weakly inhibited (46); cellulose synthase (UDP-forming) (119,120); chitin synthase (117); CMP-N-acylneuraminate phosphodiesterase (55,56); cytidylate kinase, alternative substrate inhibition (71); deoxycytidine kinase (73); deoxyguanosine kinase (72); diacylglycerol choline-phosphotransferase, weakly inhibited (65); dihydroorotase (16); dolichyl phosphate β-glucosyltransferase (17,98,99); dolichyl-phosphate β-D-mannosyltransferase (104); ethanolaminephospho-transferase, weakly inhibited (65); FAD diphosphatase, weakly inhibited

(47); fructokinase (75); fructose-1,6-bisphosphatase (61); galactolipid galactosyltransferase (91); β-galactoside α-2,3-sialyl-transferase (1); globoside α-N-acetylgalactosaminyl-transferase (103); 1,3-β-glucan synthase (113,114); glucuronosyltransferase1(16); glutamine synthetase (38); glycogen synthase (18-22,121,122); glycoprotein-fucosyl-galactoside α-N-acetylgalactosaminyltransferase (1,110, 111); hexokinase, weakly inhibited (76,77); hyaluronan synthase (87); hydroxyacylglutathione hydrolase, *or* glyoxalase II (45); lactose synthase (23); lactoylglutathione lyase, *or* glyoxalase I (45); α-1,3-mannosyl-glycoprotein 2-β-N-acetylglucosaminyltransferase (1,24,100); NDP-glucose:starch glucosyl-transferase (85); nuatigenin 3β-glucosyltransferase (89); nucleotide diphosphatase, weakly inhibited (48-50); 1,3-β-oligoglucan phosphorylase (115); oligonucleotidase (54); orotate phosphoribosyltransferase (25,83); pancreatic ribonuclease, *or* ribonuclease A(26); phosphodiesterase I, *or* 5'-exonuclease (57-59); phosphoprotein phosphatase (60); 5-phosphoribosylamine synthetase, *or* ribose-5-phosphate:ammonia ligase, weakly inhibited (27); polygalacturonate 4-α-galacturonosyl-transferase (106); polypeptide N-acetylgalactosaminyl-transferase (107-109); procollagen galactosyltransferase (105); protein N-acetyl-glucosaminyltransferase (101); [protein-PII] uridylyltransferase (66,67); protein xylosyl-transferase (79); pyrimidine-nucleoside phosphorylase (28); retinyl-phosphate glycosyltransferase (29); ribonuclease T$_2$ (53); ribose-phosphate diphosphokinase, *or* phosphoribosyl-pyrophosphate synthetase (69,70); ribose-5-phosphate isomerase (39); sarsasapogenin 3β-glucosyltransferase (88); [Skp1-protein]-hydroxyproline N-acetylglucosaminyl-transferase (86); sphingomyelin phosphodiesterase, *or* sphingomyelinase (30); sterol 3β-glucosyltransferase (92,93); sucrose 6^F-α-galactosyltransferase (94); sucrose synthase (31); α,α-trehalose-phosphate synthase (118); UDP-N-acetylglucosamine diphosphorylase, *or* N-acetylglucosamine-1-phosphate uridylyltransferase (68); UDP-N-acetylglucosamine: dolichyl-phosphate N-acetylglucosaminephospho-transferase (62,63); UDP-N-acetylglucosamine 2-epimerase (32); UDP-glucose 4-epimerase, inhibition often enhanced in the presence of a sugar (33,42-44); UDP-glucuronate decarboxylase (34,35); UDP-glucuronate 4-epimerase (36,40,41); undecaprenyl-phosphate galactose phosphotransferase (64); uridine Nucleosidase (52); and 1,4-β-D-xylan synthase (80,81). **1.** Sadler, Beyer, Oppenheimer, *et al.* (1982) *Meth. Enzymol.* **83**, 458. **2.** Sharma, Lehle & Tanner (1982) *Eur. J. Biochem.* **126**, 319. **3.** Van der Weyden & Kelly (1974) *J. Biol. Chem.* **249**, 7282. **4.** McCormick, Chassy & Tsibris (1964) *Biochim. Biophys. Acta* **89**, 447. **5.** Hongo, Matsumoto & Sato (1978) *Biochim. Biophys. Acta* **522**, 258. **6.** Jacobson & Stark (1973) *The Enzymes*, 3rd ed. (Boyer, ed.), **9**, 225. **7.** Webb (1966) *Enzyme and Metabolic Inhibitors*, vol. **2**, p. 468, Academic Press, New York. **8.** Bresnick (1963) *Biochim. Biophys. Acta* **67**, 425. **9.** Raijman & Jones (1973) *The Enzymes*, 3rd ed. (Boyer, ed.), **9**, 97. **10.** Kaseman & Meister (1985) *Meth. Enzymol.* **113**, 305. **11.** Anderson, Wellner, Rosenthal & Meister (1970) *Meth. Enzymol.* **17A**, 235. **12.** Trotta, Burt, Pinkus, *et al.* (1978) *Meth. Enzymol.* **51**, 21. **13.** Ingraham & Abdelal (1978) *Meth. Enzymol.* **51**, 29. **14.** Williams & Davis (1978) *Meth. Enzymol.* **51**, 105. **15.** Anderson & Meister (1966) *Biochemistry* **5**, 3164. **16.** Bresnick & Blatchford (1964) *Biochem. Biophys. Acta* **81**, 150. **17.** Villemez & Carlo (1979) *J. Biol. Chem.* **254**, 4814. **18.** Leloir & Goldemberg (1962) *Meth. Enzymol.* **5**, 145. **19.** Leloir & Cardini (1962) *The Enzymes*, 2nd ed. (Boyer, Lardy & Myrbäck, eds.), **6**, 317. **20.** Leloir & Goldemberg (1960) *J. Biol. Chem.* **235**, 919. **21.** Mied & Bueding (1979) *J. Parasitol.* **65**, 14. **22.** Rothman & Cabib (1967) *Biochemistry* **6**, 2107. **23.** Babad & Hassid (1966) *Meth. Enzymol.* **8**, 346. **24.** Presper & Heath (1983) *The Enzymes*, 3rd ed. (Boyer, ed.), **16**, 449. **25.** Silva & Hatfield (1978) *Meth. Enzymol.* **51**, 143. **26.** Richards & Wyckoff (1971) *The Enzymes*, 3rd ed. (Boyer, ed.), **4**, 647. **27.** Westby & Tsai (1974) *J. Bacteriol.* **117**, 1099. **28.** Hatfield & Wyngaarden (1964) *J. Biol. Chem.* **239**, 2580. **29.** Rask & Peterson (1980) *Meth. Enzymol.* **67**, 270. **30.** Callahan, Jones, Davidson & Shankaran (1983) *J. Neurosci. Res.* **10**, 151. **31.** Avigad & Milner (1966) *Meth. Enzymol.* **8**, 341. **32.** Kikuchi & Tsuiki (1973) *Biochim. Biophys. Acta* **327**, 193. **33.** Glaser (1972) *The Enzymes*, 3rd ed. (Boyer, ed.), **6**, 355. **34.** Ankel & Feingold (1966) *Meth. Enzymol.* **8**, 287. **35.** Ankel & Feingold (1965) *Biochemistry* **4**, 2468. **36.** Gaunt, Ankel & Schutzbach (1972) *Meth. Enzymol.* **28**, 426. **37.** Yang, Park, Nolan, Lu & Abdelal (1997) *Eur. J. Biochem.* **249**, 443. **38.** Singh & Singh (1990) *Arch. Int. Physiol. Biochim.* **98**, 95. **39.** Horitsu, Sasaki, Kikuchi, *et al.* (1976) *Agric. Biol. Chem.* **40**, 257. **40.** Gaunt, Maitra & Ankel (1974) *J. Biol. Chem.* **249**, 2366. **41.** Munoz, Lopez, de Frutos & Garcia (1999) *Mol. Microbiol.* **31**, 703. **42.** Ray, Ali & Bhaduri (1992) *Indian J. Biochem. Biophys.* **29**, 209. **43.** Geren, Geren & Ebner (1977) *J. Biol. Chem.* **252**,

2089. **44.** Ray & Bhaduri (1973) *Biochim. Biophys. Acta* **302**, 129. **45.** Oray & Norton (1980) *Biochem. Biophys. Res. Commun.* **95**, 624. **46.** Raetz, Hirschberg, Dowhan, Wickner & Kennedy (1972) *J. Biol. Chem.* **247**, 2245. **47.** Mistuda, Tsuge, Tomozawa & Kawai (1970) *J. Vitaminol.* **16**, 31. **48.** Bachorik & Dietrich (1972) *J. Biol. Chem.* **247**, 5071. **49.** Decker & Bischoff (1972) *FEBS Lett.* **21**, 95. **50.** Bischoff, Tran-Thi & Decker (1975) *Eur. J. Biochem.* **51**, 353. **51.** DeWolf, Fullin & Schramm (1979) *J. Biol. Chem.* **254**, 10868. **52.** Raggi-Ranieri & Ipata (1971) *Ital. J. Biochem.* **20**, 27. **53.** Irie & Ohgi (1976) *J. Biochem.* **80**, 39. **54.** Futai & Mizuno (1967) *J. Biol. Chem.* **242**, 5301. **55.** Kean & Bighouse (1974) *J. Biol. Chem.* **249**, 7813. **56.** Spik, Six, Bouquelet, Sawicka & Montreuil (1979) *Glycoconjugate Res.* **2**, 933. **57.** Futai & Mizuno (1967) *J. Biol. Chem.* **242**, 5301. **58.** Sabatini & Hotchkiss (1969) *Biochemistry* **8**, 4831. **59.** Luthje & Ogilvie (1985) *Eur. J. Biochem.* **149**, 119. **60.** Nakai & Thomas (1974) *J. Biol. Chem.* **249**, 6459. **61.** Fujita & Freese (1979) *J. Biol. Chem.* **254**, 5340. **62.** Shailubhai, Dong-Yu, Saxena & Vijay (1988) *J. Biol. Chem.* **263**, 15964. **63.** Sharma, Lehle & Tanner (1982) *Eur. J. Biochem.* **126**, 319. **64.** Osborn & Tze-Yuen (1968) *J. Biol. Chem.* **243**, 5145. **65.** Percy, Carson, Moore & Waechter (1984) *Arch. Biochem. Biophys.* **230**, 69. **66.** Engleman & Francis (1978) *Arch. Biochem. Biophys.* **191**, 602. **67.** Francis & Engleman (1978) *Arch. Biochem. Biophys.* **191**, 590. **68.** Bulik, Lindmark & Jarroll (1998) *Mol. Biochem. Parasitol.* **95**, 135. **69.** Fox & Kelley (1972) *J. Biol. Chem.* **247**, 2126. **70.** Tatibana, Kita, Taira, *et al.* (1995) *Adv. Enzyme Regul.* **35**, 229. **71.** Pasti, Gallois-Montbrun, Munier-Lehmann, *et al.* (2003) *Eur. J. Biochem.* **270**, 1784. **72.** Green & Lewis (1979) *Biochem. J.* **183**, 547. **73.** Datta, Shewach, Mitchell & Fox (1989) *J. Biol. Chem.* **264**, 9359. **74.** J. D. Schwenn & H. G. Jender (1984) *Arch. Microbiol.* **138**, 9. **75.** Yamashita & Ashihara (1988) *Z. Naturforsch. C* **43**, 827. **76.** Renz & Stitt (1993) *Planta* **190**, 166. **77.** Yamashita & Ashihara (1988) *Z. Naturforsch. C* **43**, 827. **78.** Eppler, Morré & Keenan (1980) *Biochim. Biophys. Acta* **619**, 332. **79.** Casanova, Kuhn, Kleesiek & Götting (2008) *Biochem. Biophys. Res. Commun.* **365**, 678. **80.** Dalessandro & Northcote (1981) *Planta* **151**, 61. **81.** Bailey & Hassid (1966) *Proc. Natl. Acad. Sci. U.S.A.* **56**, 1586. **82.** Holmes, McDonald, McCord, Wyngaarden & Kelley (1973) *J. Biol. Chem.* **248**, 144. **83.** Ashihara (1978) *Z. Pflanzenphysiol.* **87**, 225. **84.** Hirose & Ashihara (1983) *Z. Pflanzenphysiol.* **110**, 135. **85.** Zea & Pohl (2005) *Biopolymers* **79**, 106. **86.** Teng-umnuay, van der Wel & West (1999) *J. Biol. Chem.* **274**, 36392. **87.** Tlapak-Simmons, Baron & Weigel (2004) *Biochemistry* **43**, 9234. **88.** Paczkowski & Wojciechowski (1988) *Phytochemistry* **27**, 2743. **89.** Kalinowska & Wojciechowski (1988) *Plant Sci.* **55**, 239. **90.** Murazumi, Kumita, Araki & Ito (1988) *J. Biochem.* **104**, 980. **91.** Heemskerk, Jacobs, MScheijen, Helsper & Wintermans (1987) *Biochim. Biophys. Acta* **918**, 189. **92.** Staver, Glick & Baisted (1978) *Biochem. J.* **169**, 297. **93.** Paczkowski, Zimowski, Krawczyk & Wojciechowski (1990) *Phytochemistry* **29**, 63. **94.** Hopf, Spanfelner & Kandler (1984) *Z. Pflanzenphysiol.* **114**, 485. **95.** Sakamoto, Taguchi, Tano, *et al.* (1998) *J. Biol. Chem.* **273**, 27625. **96.** Kean & Niu (1998) *Glycoconjugate J.* **15**, 11. **97.** Kaushal & Elbein (1986) *Plant Physiol.* **81**, 1086. **98.** Arroyo-Flores, Rodríguez-Bonilla, Villagómez-Castro, *et al.* (2000) *Fungal Genet. Biol.* **30**, 127. **99.** Rodríguez-Bonilla, Vargas-Rodríguez, Calvo-Méndez, Flores-Carreón & López-Romero (1998) *Antonie Leeuwenhoek* **73**, 373. **100.** Nishikawa, Pegg, Paulsen & Schachter (1988) *J. Biol. Chem.* **263**, 8270. **101.** Haltiwanger, Blomberg & Hart (1992) *J. Biol. Chem.* **267**, 9005. **102.** Rao, Garver & Mendicino (1976) *Biochemistry* **15**, 5001. **103.** Ishibashi, Atsuta & Makita (1976) *Biochim. Biophys. Acta* **429**, 759. **104.** Villagómez-Castro, Calvo-Méndez, Vargas-Rodríguez, Flores-Carreón & López-Romero (1998) *Exp. Parasitol.* **88**, 111. **105.** LRisteli (1978) *Biochem. J.* **169**, 189. **106.** Villemez, Swanson & Hassid (1966) *Arch. Biochem. Biophys.* **116**, 446. **107.** Takeuchi, Yoshikawa, Sasaki & Chiba (1985) *Agric. Biol. Chem.* **49**, 1059. **108.** Sugiura, Kawasaki & Yamashina (1982) *J. Biol. Chem.* **257**, 9501. **109.** Mendicino & Sangadala (1998) *Mol. Cell. Biochem.* **185**, 135. **110.** Navaratnam, Findlay, Keen & Watkins (1990) *Biochem. J.* **271**, 93. **111.** Schwyzer & Hill (1977) *J. Biol. Chem.* **252**, 2338. **112.** Rao, Garver & Mendicino (1976) *Biochemistry* **15**, 5001. **113.** Kamat, Garg & Sharma (1992) *Arch. Biochem. Biophys.* **298**, 731. **114.** Beauvais, Drake, Ng, Diaquin & Latgé (1993) *J. Gen. Microbiol.* **139**, 3071. **115.** Marechal (1967) *Biochim. Biophys. Acta* **146**, 417. **116.** Winsnes (1972) *Biochim. Biophys. Acta* **289**, 88. **117.** Ruiz-Herrera, Lopez-Romero & Bartnicki-Garcia (1977) *J. Biol. Chem.* **252**, 3338. **118.** Londesborough & Vuorio (1991) *J. Gen. Microbiol.* **137**, 323. **119.** Haass, Hackspacher & Franz (1985) *Plant Sci.* **41**, 1. **120.** Kuribayashi, Kimura, Morita & Igaue (1992) *Biosci. Biotechnol. Biochem.* **56**, 388. **121.** Zea & Pohl (2005) *Biopolymers* **79**, 106. **122.** Sølling (1979) *Eur. J. Biochem.* **94**, 231.

Uridine 5'-Monophosphoric (1-Hexadecane-sulfonic) Anhydride

This UMP derivative ($FW_{free\text{-}acid}$ = 612.68 g/mol), also known as UMP-hexadecanesulfonic anhydride, inhibits pig polypeptide *N*-acetylgalactosaminyltransferase, IC_{50} = 160 μM (1). *Note:* Because this substance readily forms micelles, one must take care to discriminate whether the monomeric and/or micelle form(s) is(are) inhibitory, when present at concentrations above the critical micelle concentration (*cmc*). **1.** Hatanaka, Slama & Elbein (1991) *Biochem. Biophys. Res. Commun.* **175**, 668.

Uridine 5'-Triphosphate

This pyrimidine nucleotide ($FW_{free\text{-}acid}$ = 484.14 g/mol) is a precursor in the biosynthesis of CTP, nucleic acids, and nucleotide sugars. UTP activates chloride channels in epithelial cells. Solid sodium and potassium salts of UTP are very soluble in water. The phosphoanhydride bonds are labile in acids (*e.g.*, UTP is hydrolyzed quantitatively in fifteen minutes at 100°C in 0.1 N H_2SO_4, forming UMP and orthophosphate); the phosphate ester bond is not as labile. UTP absorbs light strongly in the ultraviolet region of the spectrum (λ_{max} is 262 nm with ε = 10000 $M^{-1}cm^{-1}$ at pH 2 and 7; and λ_{max} = 261 nm, ε = 8100 $M^{-1}cm^{-1}$ at pH 11). UTP forms tight complexes with many metal ions. The stability constants are reported to be similar to those for ATP (1), for which stability constants for are 73000, 35000, and 100000 M^{-1} for $MgATP^{2-}$, $CaATP^{2-}$, and $MnATP^{2-}$, respectively. **Target(s):** α-*N*-acetylgalactosaminide α-2,6-sialyltransferase (2); *N*-acetylglucosamine kinase (3,138); *N*-acetylglucosamine-6-phosphate deacetylase (94); *N*-acetylglucosaminyldiphosphodolichol *N*-acetylglucos-aminyltransferase (178,179); *N*-acetyllactosaminide β-1,6-*N*-acetylglucosaminyltransferase (177); *N*-acylneuraminate cytidylyltransferase (111); α-*N*-acetylneuram-inate α-2,8-sialyltransferase, weakly inhibited (155); adenylate cyclase (73); adenylylsulfatase (78); amidophosphoribosyl-transferase, weakly inhibited (4,165); AMP deaminase (88); arylsulfate sulfotransferase (5); asparagine synthetase (64); aspartate carbamoyltransferase (6-10); ATP:citrate lyase (210); carbamate kinase (12); carbamoyl-phosphate synthetase (glutamine-hydrolyzing) (13-18); CDP-diacylglycerol:inositol 3-phosphatidyl-transferase (108,109); cellulose synthase (UDP-forming) (19,204,205); chitin synthase (193-196); citrate (*si*)-synthase, weakly inhibited (211); CMP-*N*-acylneuraminate phosphodiesterase (97); 3',5'-cyclic-nucleotide phosphodiesterase (20,21); cytidine deaminase (22,89-91); cytidylate kinase (23,129); cytosine deaminase (92,93); deoxycytidine kinase (135); deoxyguanosine kinase (130,131); diacylglycerol kinase (24,132); dCMP deaminase, weakly inhibited (87); dolichyl-phosphate β-D-mannosyltransferase (183); exopolyphosphatase (80); 5'→3' exoribonuclease (25); flavanone 7-*O*-glucoside 2"-*O*-β-L-rhamnosyltransferase (169); flavanone 7-*O*-β-glucosyltransferase (173); β-fructofuranosidase, *or* invertase (95); fructose-1,6-bisphosphatase (26,105); fumarase (27); β-galactoside α-2,3-sialyltransferase (2,156); β-galactoside α-2,6-sialyltransferase (157); 1,3-β-glucan synthase (189,190); glucokinase, also poor alternative substrate (150,151); glucomannan 4-β-mannosyltransferase, weakly inhibited (191); glucose-1-phosphate cytidylyltransferase (115); glucose-6-phosphate dehydrogenase (28,74); glucuronosyl-transferase (192); [glutamine-synthetase] adenylyl-transferase (29,112,113); glycogen synthase (30-32,206-208); glycolipid 3-α-mannosyltransferase, weakly inhibited (180); glycoprotein-fucosyl-galactoside α-*N*-acetylgalactos-aminyltransferase (2); GTP cyclohydrolase (I83-86); guanylate cyclase (33,71,72); hexokinase (34); hyaluronan synthase (35,171); hydroxyacylglutathione hydrolase, or glyoxalase II (75); 5-hydroxypentanoate CoA-transferase, weakly inhibited (82); myo-inositol oxidase (36); isocitrate dehydrogenase (NAD⁺) (37,38); lactose synthase (39); lactoylglutathione lyase (glyoxalase I) (75); α-1,3-mannosylglycoprotein 2-β-N-acetylglucosaminyl-transferase (2,40,181); membrane dipeptidase, or renal dipeptidase (41); methionine S-adenosyltransferase (152,153); [3-methyl-2-oxobutanoate dehydrogenase (2-methylpropanoyl-transferring)] phosphatase, or [branched-chain α-keto-acid dehydrogenase] phosphatase (11,101); multiple inositol-polyphosphate phosphatase, or inositol-1,3,4,5-tetrakisphosphate 3-phosphatase (100); NAD⁺ diphosphatase (79); NADH peroxidase (42); NAD(P)H dehydrogenase (43); nicotinate glucosyltransferase (172); nicotinate-nucleotide diphosphorylase (carboxylating) (163,164); nicotinate phosphoribosyltransferase, weakly inhibited (166); nucleotidases (80); 3'-nucleotidase (80); 5'-nucleotidase (44-46,80); nucleotide diphosphatase (81); oligonucleotidase, weakly inhibited (96); orotate phosphoribosyl-ransferase (47); orotidylate decarboxylase (47); phosphatidate cytidylyltransferase (114); 1-phosphatidylinositol 4-kinase (136,137); phosphodiesterase I, or 5'-exonuclease (98,99); phosphoenolpyruvate carboxylase (77); 1-phosphofructokinase (139); 6-phosphofructokinase (148,149); phosphoglucomutase (48,65,66); phosphopantothenoylcysteine decarboxylase (49,76); phosphoprotein phosphatase (103,104); phosphorylase kinase, weakly inhibited (106); [phosphorylase] phosphatase (102); polygalacturonate 4-α-galacturonosyltransferase (185,186); polynucleotide adenylyltransferase, or poly(A) polymerase (51,52,116-118); polynucleotide 5'-hydroxyl-kinase (133,134); polypeptide N-acetylgalactosaminyltransferase (187,188); porphobilinogen synthase (δ-aminolevulinate dehydratase) (53); procollagen glucosyltransferase (54,184); protein N-acetylglucosaminyltransferase (182); protein-glutamine γ-glutamyltransferase, or transglutaminase (212); protein xylosyltransferase (159); pyruvate carboxylase (55); pyruvate kinase (146.147); ribose-5-phosphate diphosphokinase, or phosphoribosyl-pyrophosphate synthetase (50,119,120); [Skp1-protein]-hydroxyproline N-acetylglucosaminyl-transferase (170); starch phosphorylase (209); sterol 3β-glucosyltransferase (174-176); sucrose-phosphate synthase (198-200); sucrose synthase (56,201-203); thiamin-phosphate diphosphorylase (57,154); α,α-trehalose-phosphate synthase (UDP-forming) (197); UDP-N-acetylglucosamine 2-epimerase (67); UDP-N-acetylglucosamine:lysosomal enzyme N-acetylglucosamine-phosphotransferase (107); UDP-glucose 4-epimerase (58,69,70); UDP-glucuronate decarboxylase (59); UDP-glucuronate 4-epimerase (60,68); UMP kinase (121-128); undecaprenyl-phosphate galactose phospho-transferase, weakly inhibited (110); uracil phosphoribosyl-transferase (167,168); uridine kinase (23,61-63,140-145); xanthine phosphoribosyltransferase (162); 1,4-β-D-xylan synthase (160,161); xyloglucan 4-glucosyltransferase (158); and xyloglucan 6-xylosyltransferase (158).

1. O'Sullivan & Smithers (1979) Meth. Enzymol. **63**, 294. **2.** Sadler, Beyer, Oppenheimer, et al. (1982) Meth. Enzymol. **83**, 458. **3.** Datta (1975) Meth. Enzymol. **42**, 58. **4.** Hill & L. L. Bennett, Jr. (1969) Biochemistry **8**, 122. **5.** Konishi-Imamura, Kim, Koizumi & Kobashi (1995) J. Enzyme Inhib. **8**, 233. **6.** Adair & Jones (1978) Meth. Enzymol. **51**, 51. **7.** Jacobson & Stark (1973) The Enzymes, 3rd ed. (Boyer, ed.), **9**, 225. **8.** Purcarea (2001) Meth. Enzymol. **331**, 248. **9.** Webb (1966) Enzyme and Metabolic Inhibitors, vol. **2**, p. 468, Academic Press, New York. **10.** Bresnick (1963) Biochim. Biophys. Acta **67**, 425. **11.** Damuni & Reed (1988) Meth. Enzymol. **166**, 321. **12.** Raijman & Jones (1973) The Enzymes, 3rd ed. (Boyer, ed.), **9**, 97. **13.** Anderson, Wellner, Rosenthal & Meister (1970) Meth. Enzymol. **17A**, 235. **14.** Williams & Davis (1978) Meth. Enzymol. **51**, 105. **15.** Mori & Tatibana (1978) Meth. Enzymol. **51**, 111. **16.** Lyons & Christopherson (1985) Eur. J. Biochem. **147**, 587. **17.** Mori, Ishida & Tatibana (1975) Biochemistry **14**, 2622. **18.** Anderson & Meister (1966) Biochemistry **5**, 3164. **19.** Potter & Weisman (1972) Meth. Enzymol. **28**, 581. **20.** Drummond & Yamamoto (1971) The Enzymes, 3rd ed. (Boyer, ed.), **4**, 355. **21.** Cheung (1967) Biochemistry **6**, 1079. **22.** Ipata & Cercignani (1978) Meth. Enzymol. **51**, 394. **23.** Anderson (1973) The Enzymes, 3rd ed. (Boyer, ed.), **9**, 49. **24.** Daleo, Piras & Piras (1976) Eur. J. Biochem. **68**, 339. **25.** Slobin (2001) Meth. Enzymol. **342**, 282. **26.** Taketa & Pogell (1963) Biochem. Biophys. Res. Commun. **12**, 229. **27.** Hill & Teipel (1971) The Enzymes, 3rd. ed. (Boyer, ed.), **5**, 539. **28.** Domagk, Chilla, Domschke, Engel & Sörensen (1969) Hoppe Seyler's Z. physiol. Chem. **350**, 626. **29.** Shapiro & Stadtman (1970) Meth. Enzymol. **17A**, 910. **30.** Stalmans & Hers (1973) The Enzymes, 3rd ed. (Boyer, ed.), **9**, 309. **31.** Rosell-Perez & Larner (1964) Biochemistry **3**, 773. **32.** Rothman & Cabib (1967) Biochemistry **6**, 2107. **33.** Hardman & Sutherland (1969) J. Biol. Chem. **244**, 6363. **34.** Purich & Fromm (1971) J. Biol. Chem. **246**, 3456.

35. Markovitz & Dorfman (1962) Meth. Enzymol. **5**, 155. **36.** Charalampous (1962) Meth. Enzymol. **5**, 329. **37.** Plaut (1969) Meth. Enzymol. **13**, 34. **38.** Chen & Plaut (1963) Biochemistry **2**, 1023. **39.** Babad & Hassid (1966) Meth. Enzymol. **8**, 346. **40.** Presper & Heath (1983) The Enzymes, 3rd ed. (Boyer, ed.), **16**, 449. **41.** Harper, Rene & Campbell (1971) Biochim. Biophys. Acta **242**, 446. **42.** Badwey & Karnovsky (1979) J. Biol. Chem. **254**, 11530. **43.** Koli, Yearby, Scott &Donaldson (1969) J. Biol. Chem. **244**, 621. **44.** Drummond & Yamamoto (1971) The Enzymes, 3rd ed. (Boyer, ed.), **4**, 337. **45.** Madrid-Marina & Fox (1986) J. Biol. Chem. **261**, 444. **46.** Ipata (1968) Biochemistry **7**, 507. **47.** Jones, Kavipurapu & Traut (1978) Meth. Enzymol. **51**, 155. **48.** Ray & Peck (1972) The Enzymes, 3rd ed. (Boyer, ed.), **6**, 407. **49.** Abiko (1970) Meth. Enzymol. **18A**, 354. **50.** Wong & Murray (1969) Biochemistry **8**, 1608. **51.** Edmonds (1990) Meth. Enzymol. **181**, 161. **52.** Edmonds (1982) The Enzymes, 3rd ed. (Boyer, ed.), **15**, 217. **53.** Tigier, Batlle & Locascio (1970) Enzymologia **38**, 43. **54.** Spiro & Spiro (1972) Meth. Enzymol. **28**, 625. **55.** Scrutton, Olmsted & Utter (1969) Meth. Enzymol. **13**, 235. **56.** Avigad & Milner (1966) Meth. Enzymol. **8**, 341. **57.** Penttinen (1979) Meth. Enzymol. **62**, 68. **58.** Glaser (1972) The Enzymes, 3rd ed. (Boyer, ed.), **6**, 355. **59.** Grisebach, Baron, Sandermann & Wellmann (1972) Meth. Enzymol. **28**, 439. **60.** Gaunt, Ankel & J. Schutzbach (1972) Meth. Enzymol. **28**, 426. **61.** Orengo & Kobayashi (1978) Meth. Enzymol. **51**, 299. **62.** Valentin-Hansen (1978) Meth. Enzymol. **51**, 308. **63.** Anderson (1978) Meth. Enzymol. **51**, 314. **64.** Hongo, Matsumoto & Sato (1978) Biochim. Biophys. Acta **522**, 258. **65.** Popova, Matasova & Lapot'ko (1998) Biochemistry (Moscow) **63**, 697. **66.** Maino & Young (1974) J. Biol. Chem. **249**, 5176. **67.** Kikuchi & Tsuiki (1973) Biochim. Biophys. Acta **327**, 193. **68.** Gaunt, Maitra & Ankel (1974) J. Biol. Chem. **249**, 2366. **69.** Geren, Geren & Ebner (1977) J. Biol. Chem. **252**, 2089. **70.** Lee, Kimura & Tochikura (1978) Agric. Biol. Chem. **42**, 731. **71.** Krishnan, Fletcher, Chader & Krishna (1978) Biochim. Biophys. Acta **523**, 506. **72.** Zwiller, Basset & Mandel (1981) Biochim. Biophys. Acta **658**, 64. **73.** Ide (1971) Arch. Biochem. Biophys. **144**, 262. **74.** Horne, Anderson & Nordlie (1970) Biochemistry **9**, 610. **75.** Oray & Norton (1980) Biochem. Biophys. Res. Commun. **95**, 624. **76.** Abiko (1967) J. Biochem. **61**, 300. **77.** Singal & Singh (1986) Plant Physiol. **80**, 369. **78.** Stokes, Denner & Dodgson (1973) Biochim. Biophys. Acta **315**, 402. **79.** Nakajima, Fukunaga, Sasaki & Usami (1973) Biochim. Biophys. Acta **293**, 242. **80.** Proudfoot, Kuznetsova, Brown, et al. (2004) J. Biol. Chem. **279**, 54687. **81.** Krishnan & Rao (1972) Arch. Biochem. Biophys. **149**, 336. **82.** Eikmanns & Buckel (1990) Biol. Chem. Hoppe-Seyler **371**, 1077. **83.** Blau & Niederwieser (1984) Biochem. Clin. Aspects Pteridines **3**, 77. **84.** Ferre, JYim & Jacobson (1986) J. Chromatogr. **357**, 283. **85.** Yoo, Han, Ko & Bang (1998) Arch. Pharm. Res. **21**, 692. **86.** de Saizieu, Vankan & van Loon (1995) Biochem. J. **306**, 371. **87.** Ellims, Kao & Chabner (1981) J. Biol. Chem. **256**, 6335. **88.** Yabuki & Ashihara (1992) Phytochemistry **31**, 1905. **89.** Cacciamani, Vita, Cristalli, Vincenzetti, Natalini, Ruggieri, Amici & Magni (1991) Arch. Biochem. Biophys. **290**, 285. **90.** Vincenzetti, Cambi, Neuhard, et al. (1999) Protein Expr. Purif. **15**, 8. **91.** Vincenzetti, Cambi, Neuhard, Garattini & Vita (1996) Protein Expr. Purif. **8**, 247. **92.** Yu, Kim, Kasuragi, Sakai & Tonomura (1991) J. Ferment. Bioeng. **72**, 266. **93.** Yu, Kim & Kim (1998) J. Microbiol. **36**, 39. **94.** Gopal, Sullivan & Shepherd (1982) J. Gen. Microbiol. **128**, 2319. **95.** Lee & Sturm (1996) Plant Physiol. **112**, 1513. **96.** Datta & Niyogi (1975) J. Biol. Chem. **250**, 7313. **97.** Kean & Bighouse (1974) J. Biol. Chem. **249**, 7813. **98.** Futai & Mizuno (1967) J. Biol. Chem. **242**, 5301. **99.** Picher & Boucher (2000) Amer. J. Respir. Cell Mol. Biol. **23**, 255. **100.** Höer, Höer & Oberdisse (1990) Biochem. J. **270**, 715. **101.** Damuni, Merryfield, Humphreys & Reed (1984) Proc. Natl. Acad. Sci. U.S.A. **81**, 4335. **102.** Detwiler, Gratecos & Fischer (1977) Biochemistry **16**, 4818. **103.** Nakai & Thomas (1974) J. Biol. Chem. **249**, 6459. **104.** Polya & Haritou (1988) Biochem. J. **251**, 357. **105.** Fujita & Freese (1979) J. Biol. Chem. **254**, 5340. **106.** Chan & Graves (1982) J. Biol. Chem. **257**, 5948. **107.** Bao, Elmendorf, Booth, Drake & Canfield (1996) J. Biol. Chem. **271**, 31446. **108.** Antonsson & Klig (1996) Yeast **12**, 449. **109.** Antonsson (1994) Biochem. J. **297**, 517. **110.** Osborn & Tze-Yuen (1968) J. Biol. Chem. **243**, 5145. **111.** Rodríguez-Aparicio, Luengo, Gonzalez-Clemente & Reglero (1992) J. Biol. Chem. **267**, 9257. **112.** Wolf & Ebner (1972) J. Biol. Chem. **247**, 4208. **113.** Ebner, Wolf, Gancedo, Elsässer & Holzer (1970) Eur. J. Biochem. **14**, 535. **114.** Morii, Nishihara & Koga (2000) J. Biol. Chem. **275**, 36568. **115.** Kimata & Suzuki (1966) J. Biol. Chem. **241**, 1099. **116.** Sarkar, Cao & Sarkar (1997) Biochem. Mol. Biol. Int. **41**, 1045. **117.** Pellicer, Salas & Salas (1978) Biochim. Biophys. Acta **519**, 149. **118.** Roggen & Slegers (1985) Eur. J. Biochem. **147**, 225. **119.** Hove-Jensen & McGuire (2004) Eur. J. Biochem. **271**, 4526. **120.** Tatibana, Kita, Taira, et al. (1995) Adv. Enzyme Regul. **35**,

229. **121.** Evrin, Straut, Slavova-Azmanova, *et al.* (2007) *J. Biol. Chem.* **282**, 7242. **122.** Jensen, Johansson & Jensen (2007) *Biochemistry* **46**, 2745. **123.** Serina, Blondin, Krin, *et al.* (1995) *Biochemistry* **34**, 5066. **124.** Bucurenci, Serina, Zaharia, *et al.* (1998) *J. Bacteriol.* **180**, 473. **125.** Fassy, Krebs, Lowinski, *et al.* (2004) *Biochem. J.* **384**, 619. **126.** Serina, Bucurenci, Gilles, *et al.* (1996) *Biochemistry* **35**, 7003. **127.** Briozzo, Evrin, Meyer, *et al.* (2005) *J. Biol. Chem.* **280**, 25533. **128.** Sakamoto, Landais, Evrin, *et al.* (2004) *Microbiology* **150**, 2153. **129.** Liou, Dutschman, Lam, Jiang & Cheng (2002) *Cancer Res.* **62**, 1624. **130.** Yamada, Goto & Ogasawara (1982) *Biochim. Biophys. Acta* **709**, 265. **131.** Green & Lewis (1979) *Biochem. J.* **183**, 547. **132.** Wissing & Wagner (1992) *Plant Physiol.* **98**, 1148. **133.** Austin, Sirakoff, Roop & Moyer (1978) *Biochim. Biophys. Acta* **522**, 412. **134.** Levin & Zimmerman (1976) *J. Biol. Chem.* **251**, 1767. **135.** Datta, Shewach, Mitchell & Fox (1989) *J. Biol. Chem.* **264**, 9359. **136.** Steinert, Wissing & Wagner (1994) *Plant Sci.* **101**, 105. **137.** Yamakawa & Takenawa (1988) *J. Biol. Chem.* **263**, 17555. **138.** Datta (1970) *Biochim. Biophys. Acta* **220**, 51. **139.** Van Hugo & Gottschalk (1974) *Eur. J. Biochem.* **48**, 455. **140.** Vidair & Rubin (2005) *Proc. Natl. Acad. Sci. U.S.A.* **102**, 662. **141.** Cihak (1975) *FEBS Lett.* **51**, 133. **142.** Appleby, Larson, Cheney, *et al.* (2005) *Acta Crystallogr. Sect. D* **61**, 278. **143.** Suzuki, Koizumi, Fukushima, Matsuda & Inagaki (2004) *Structure* **12**, 751. **144.** Payne, Cheng & Traut (1985) *J. Biol. Chem.* **260**, 10242. **145.** Orengo (1969) *J. Biol. Chem.* **244**, 2204. **146.** Gupta & Singh (1989) *Plant Physiol. Biochem.* **27**, 703. **147.** Singh, Malhotra & Singh (2000) *Indian J. Biochem. Biophys.* **37**, 51. **148.** Isaac & Rhodes (1986) *Phyto-chemistry* **25**, 339. **149.** Sapico & Anderson (1969) *J. Biol. Chem.* **244**, 6280. **150.** Porter, Chassy & Holmlund (1982) *Biochim. Biophys. Acta* **709**, 178. **151.** Doelle (1982) *Eur. J. Appl. Microbiol. Biotechnol.* **14**, 241. **152.** Chou & Talalay (1973) *Biochim. Biophys. Acta* **321**, 467. **153.** Porcelli, Cacciapuoti, Cartení-Farina & Gambacorta (1988) *Eur. J. Biochem.* **177**, 273. **154.** Kayama & Kawasaki (1973) *Arch. Biochem. Biophys.* **158**, 242. **155.** Eppler, Morré & Keenan (1980) *Biochim. Biophys. Acta* **619**, 332. **156.** Kurosawa, Hamamoto, Inoue & Tsuji (1995) *Biochim. Biophys. Acta* **1244**, 216. **157.** Scudder & Chantler (1981) *Biochim. Biophys. Acta* **660**, 136. **158.** Hayashi & Matsuda (1981) *J. Biol. Chem.* **256**, 11117. **159.** Casanova, Kuhn, Kleesiek & Götting (2008) *Biochem. Biophys. Res. Commun.* **365**, 678. **160.** Dalessandro & Northcote (1981) *Planta* **151**, 61. **161.** Bailey & Hassid (1966) *Proc. Natl. Acad. Sci. U.S.A.* **56**, 1586. **162.** Miller, Adamczyk, Fyfe & Elion (1974) *Arch. Biochem. Biophys.* **165**, 349. **163.** Shibata & Iwai (1980) *Agric. Biol. Chem.* **44**, 119. **164.** Taguchi & Iwai (1976) *Agric. Biol. Chem.* **40**, 385. **165.** Holmes, McDonald, McCord, Wyngaarden & Kelley (1973) *J. Biol. Chem.* **248**, 144. **166.** Imsande & Handler (1961) *J. Biol. Chem.* **236**, 525. **167.** Linde & Jensen (1996) *Biochim. Biophys. Acta* **1296**, 16. **168.** Plunkett & Moner (1978) *Arch. Biochem. Biophys.* **187**, 264. **169.** Bar-Peled, Lewinsohn, Fluhr & Gressel (1991) *J. Biol. Chem.* **266**, 20953. **170.** Teng-umnuay, van der Wel & West (1999) *J. Biol. Chem.* **274**, 36392. **171.** Tlapak-Simmons, Baron & Weigel (2004) *Biochemistry* **43**, 9234. **172.** Taguchi, Sasatani, Nishitani & Okumura (1997) *Biosci. Biotechnol. Biochem.* **61**, 720. **173.** McIntosh & Mansell (1990) *Phytochemistry* **29**, 1533. **174.** Madina, Sharma, Chaturvedi, Sangwan & Tuli (2007) *Biochim. Biophys. Acta* **1774**, 392. **175.** Paczkowski, Zimowski, Krawczyk & Wojciechowski (1990) *Phytochemistry* **29**, 63. **176.** Wojciechowski, Zimowski & Tyski (1977) *Phytochemistry* **16**, 911. **177.** Sakamoto, Taguchi, Tano, *et al.* (1998) *J. Biol. Chem.* **273**, 27625. **178.** Kean & Niu (1998) *Glycoconjugate J.* **15**, 11. **179.** Kaushal & Elbein (1986) *Plant Physiol.* **81**, 1086. **180.** Jensen & Schutzbach (1981) *J. Biol. Chem.* **256**, 12899. **181.** Nishikawa, Pegg, Paulsen & Schachter (1988) *J. Biol. Chem.* **263**, 8270. **182.** Haltiwanger, Blomberg & Hart (1992) *J. Biol. Chem.* **267**, 9005. **183.** Villagómez-Castro, Calvo-Méndez, Vargas-Rodríguez, Flores-Carreón & López-Romero (1998) *Exp. Parasitol.* **89**, 111. **184.** Smith, Wu & Jamieson (1977) *Biochim. Biophys. Acta* **483**, 263. **185.** Takeuchi & Tsumuraya (2001) *Biosci. Biotechnol. Biochem.* **65**, 1519. **186.** Villemez, Swanson & Hassid (1966) *Arch. Biochem. Biophys.* **116**, 446. **187.** Takeuchi, Yoshikawa, Sasaki & Chiba (1985) *Agric. Biol. Chem.* **49**, 1059. **188.** Sugiura, Kawasaki & Yamashina (1982) *J. Biol. Chem.* **257**, 9501. **189.** Kamat, Garg & Sharma (1992) *Arch. Biochem. Biophys.* **298**, 731. **190.** Beauvais, Drake, Ng, Diaquin & Latgé (1993) *J. Gen. Microbiol.* **139**, 3071. **191.** Piro, Zuppa, Dalessandro & Northcote (1993) *Planta* **190**, 206. **192.** Winsnes (1972) *Biochim. Biophys. Acta* **289**, 88. **193.** De Rousett-Hall & Gooday (1975) *J. Gen. Microbiol.* **89**, 146. **194.** Ruiz-Herrera, Lopez-Romero & Bartnicki-Garcia (1977) *J. Biol. Chem.* **252**, 3338. **195.** Cohen & Casida (1982) *Pestic. Biochem. Physiol.* **17**, 301. **196.** Mayer, Chen & DeLoach (1980) *Insect Biochem.* **10**, 549. **197.** Londesborough & Vuorio (1991) *J. Gen. Microbiol.* **137**, 323. **198.** Chen, Huang, Liu, *et al.* (2001) *Bot. Bull. Acad. Sin.* **42**, 123. **199.** Salerno & Pontis (1978) *Planta* **142**, 41. **200.** Harbron, Foyer & Walker (1981) *Arch. Biochem. Biophys.* **212**, 237. **201.** Tsai (1974) *Phytochemistry* **13**, 885. **202.** Morell & Copeland (1985) *Plant Physiol.* **78**, 149. **203.** Hisajima & Ito (1981) *Biol. Plant.* **23**, 356. **204.** Haass, Hackspacher & Franz (1985) *Plant Sci.* **41**, 1. **205.** Kuribayashi, Kimura, Morita & Igaue (1992) *Biosci. Biotechnol. Biochem.* **56**, 388. **206.** Nakai & Thomas (1975) *J. Biol. Chem.* **250**, 4081. **207.** Sølling (1979) *Eur. J. Biochem.* **94**, 231. **208.** Schlender & Larner (1973) *Biochim. Biophys. Acta* **293**, 73. **209.** Nakamura & Imamura (1983) *Phytochemistry* **22**, 835. **210.** Antranikian, Herzberg & Gottschalk (1982) *J. Bacteriol.* **152**, 1284. **211.** Okabayashi & Nakano (1979) *J. Biochem.* **85**, 1061. **212.** Kawashima (1991) *Experientia* **47**, 709.

Uridine-Vanadate Complex

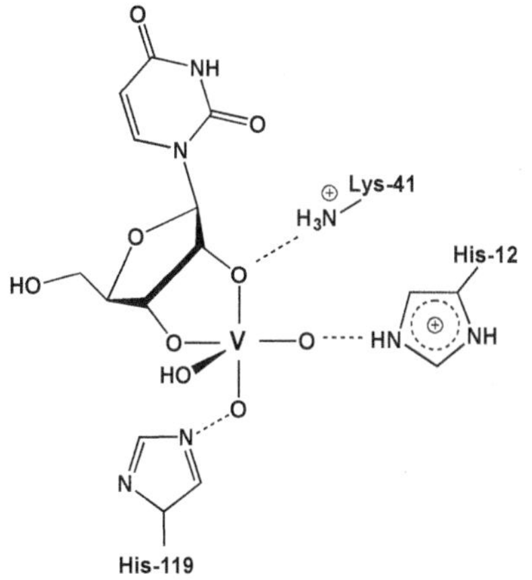

This somewhat unstable one-to-one complex of uridine and vanadium(V) forms and remains bound within the active site of pancreatic ribonuclease, showing moderate affinity, with a K_i value of 0.45 µM (1-4). When ribonuclease A is exposed to mixture of 10 mM uridine and 0.1 mM vanadium(V) at pH 7 complex formation results in occupancy of about 80% of the enzyme. The structure shown above is based on experiments employing vanadium NMR, X-ray crystallography, and neutron diffraction. Note that some minor distortion of the trigonal bipyramid geometry has been reported. **Target(s):** pancreatic ribonuclease, *or* ribonuclease A (1-3); and yeast Pac1 ribonuclease (4). **1.** Lindquist, Lynn & Lienhard (1973) *J. Amer. Chem. Soc.* **95**, 8762. **2.** Russo, Acharya & Shapiro (2001) *Meth. Enzymol.* **341**, 629. **3.** Blackburn & Moore (1982) *The Enzymes*, 3rd ed. (Boyer, ed.), **15**, 317. **4.** Rotondo & Frendewey (2001) *Meth. Enzymol.* **342**, 168. *See also Vanadyl Ribonucleoside Complex*

5'-Uridylylcobalamin, *See 5'-Deoxyuridylylcobalamin*

URMC-099

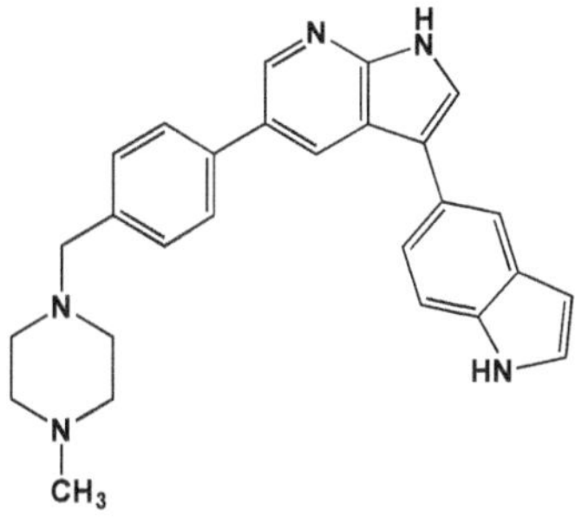

This orally bioavailable, brain-penetrant MLK inhibitor (FW = 421.54 g/mol; CAS 1229582-33-5), also named 3-(1*H*-indol-5-yl)-5-[4-[(4-methyl-1-piperazinyl)methyl]phenyl]-1*H*-pyrrolo[2,3-*b*]pyridine, targets the Mixed Lineage Kinases (with IC$_{50}$ values of 19 nM, 42 nM, and 14 nM for MLK1, MLK2, and MLK3) as well as DLK (IC$_{50}$ = 150 nM) (1). It also inhibits Leucine-Rich Repeat Kinase 2 (LRRK2), a novel regulator of microglial phagocytosis (2). MLK3 activation is associated with many of the pathologic hallmarks of HIV-associated neurocognitive disorders, making this enzyme a prime target for adjunctive therapy using small-molecule kinase inhibitors. **1.** Marker, Tremblay, Puccini, *et al.* (2013) *J. Neurosci.* **33**, 9998. **2.** Goodfellow, Loweth, Ravula, *et al.* (2013) *J. Med. Chem.* **56**, 8032.

Urocanate

trans-urocanate   **cis-urocanate**

This metabolite (FW$_{zwitterion}$ = 138.13 g/mol), also known as 4-imidazoleacrylic acid, is an intermediate in the degradation of L-histidine. Urocanate has *cis* and *trans* isomers, and the *trans*-isomer is the product of histidine ammonia-lyase. *trans*-Urocanic acid has pK_a values at 25°C of 3.5 and 5.8 whereas the *cis*-isomer has values of 3.0 and 6.7. Irradiation of *trans*-urocanate with ultraviolet light will generate *cis*-urocanate. **Target(s):** histidine decarboxylase (1-4). **1**. Huynh & Snell (1986) *J. Biol. Chem.* **261**, 4389. **2**. Snell (1986) *Meth. Enzymol.* **122**, 128. **3**. Snell & Guirard (1986) *Meth. Enzymol.* **122**, 139. **4**. Yamagata & Snell (1979) *Biochemistry* **18**, 2964.

[Orn5]-URP

This ornithine-containing urotensin analogue (FW = 1003.20 g/mol; CAS 782485-03-4; Solubility: 1 mg/mL H$_2$O) retains much of receptor binding affinity (EC$_{50}$ = 75 nM) of Urotensin II (UII), the most potent endogenous vasoconstricting peptide ligand of the orphan G-protein-coupled urotensinergic receptor GPR14 (1). [Orn5]-URP behaves as a pure antagonist (*i.e.*, exerts zero agonist action) in both rat aortic ring contraction and astrocyte cytosolic calcium ion concentration mobilization assays (2). **1**. Chatenet, Dubessy, Leprince, *et al.* (2004) *Peptides* **25**, 1819. **2**. Diallo, Jarry, Desrues, *et al.* (2008) *Peptides* **29**, 813.

Urs-12-ene-2,28-diol, See *Uvaol*

Urs-12-ene-3β,16β-diol, See *Brein*

Ursin, See *Arbutin*

Ursocholanate, See *5β-Cholanate*

Ursodeoxycholate

This bile acid (FW$_{free-acid}$ = 392.58 g/mol; CAS 128-13-2; slightly water-soluble), also known as ursodiol and 3α,7β-dihydroxy-5β-cholanic-24-acid, found in relatively high concentrations in bear bile (as is its taurine conjugate) and humans, inhibits bilirubin UDP-glucuronosyltransferase (1); glycochenodeoxycholate sulfotransferase (2-4); and pepsin (5). **1**. Sanchez Pozzi, Luquita, Catania, Rodriguez Garay & Mottino (1994) *Life Sci.* **55**, 111. **2**. Barnes, Buchina, King, McBurnett & Taylor (1989) *J. Lipid Res.* **30**, 529. **3**. Barnes, Burhol, Zander, *et al.* (1979) *J. Lipid Res.* **20**, 952. **4**. Suckling, Murphy & Higgins (1993) *Biochem. Soc. Trans.* **21**, 447S. **5**. Eto & Tompkins (1985) *Amer. J. Surg.* **150**, 564.

Ursodiol, See *Ursodeoxycholate (UrsodeoxycholicAcid)*

Ursolate (Ursolic Acid)

This naturally occurring pentacyclic triterpenoid carboxylate (FW$_{free-acid}$ = 456.71 g/mol; CAS 77-52-1; IUPAC: 3β-hydroxyurs-12-en-28-oic acid) is widely distributed in leaves and in the skins of fruits (*e.g.*, in the waxy skin layer of apples, pears, cherries, and cranberries), appearing as the free acid, as esters, or as a saponin component. Ursolic acid is slightly soluble in methanol, moderately soluble in acetone, and has a very low solubility in water. Ursolic acid inhibits intracellular trafficking of proteins and reduces IL-1α-induced cell-surface ICAM-1 expression in human cancer cell lines and human umbilical vein endothelial cells (1). By contrast, ursolic acid exerts weak inhibitory effects on the IL-1α-induced ICAM-1 expression at the protein level. Remarkably, ursolic acid decreases the apparent molecular weight of ICAM-1 and altered the structures of ICAM-1-bound *N*-linked oligosaccharides. Upon treatment of cells with ursolic acid, high-mannose-type glycan-containing ICAM-1 accumulates in the ER, and the Golgi apparatus is fragmented into pieces and distributed throughout the cells (1). *Note*: Corosolic acid (*or* 2α-hydroxyursolic acid), a pentacyclic triterpene acid from *Lagerstroemia speciosa*, is likely to have similar inhibitory targets. **Target(s):** α-amylase (2); cAMP-dependent protein kinase (3); chitin synthase II (4); DNA-directed DNA polymerase (6-9); human DNA ligase (ATP), IC$_{50}$ = 216 μM (5); DNA polymerase α (6,7); DNA polymerase β (6-8); DNA topoisomerase I (6,9); DNA topoisomerase II (6,7,9); elastase (10-12); glycogen phosphorylase (13); HIV-I retropepsin (14-16); leukocyte elastase (11,12); 5-lipoxygenase (17); 15-lipoxygenase (17); plant DNA polymerase II (6); and protein-tyrosine-phosphatase 1B (18,19). **1**. Mitsuda, Yokomichi, Yokoigawa & Kataoka (2014) *FEBS Open Bio.* **4**, 229. **2**. Ali, Houghton & Soumyanath (2006) *J. Ethnopharmacol.* **107**, 449. **3**. Wang & Polya (1996) *Phytochemistry* **41**, 55. **4**. Jeong, Hwang, Lee, *et al.* (1999) *Planta Med.* **65**, 261. **5**. Tan, Lee, Lee, *et al.* (1996) *Biochem. J.* **314**, 993. **6**. Mizushina, Iida, Ohta, Sugawara & Sakaguchi (2000) *Biochem. J.* **350**, 757. **7**. Mizushina, Ikuta, Endoh, *et al.* (2003) *Biochem. Biophys. Res. Commun.* **305**, 365. **8**. Deng, Starck & Hecht (1999) *J. Nat. Prod.* **62**, 1624. **9**. Syrovets, Buchele, Gedig, J. R. Slupsky & T. Simmet (2000) *Mol. Pharmacol.* **58**, 71. **10**. Safayhi, Rall, Sailer & Ammon (1997) *J. Pharmacol. Exp. Ther.* **281**, 460. **11**. Ying, Rinehart, Simon & Cheronis (1991) *Biochem. J.* **277**, 521. **12**. Becker, Franz & Alban (2003) *Thromb. Haemost.* **89**, 915. **13**. Wen, Sun, Liu, *et al.* (2008) *J. Med. Chem.* **51**, 3540. **14**. Min, Jung, Lee, *et al.* (1999) *Planta Med.* **65**, 374. **15**. Quere, Wenger & Schramm (1996) *Biochem. Biophys. Res. Commun.* **227**, 484. **16**. Xu, Zeng, Wan & Sim (1996) *J. Nat. Prod.* **59**, 643. **17**. Simon, Najid, Chulia, Delage & Rigaud (1992) *Biochim. Biophys. Acta* **1125**, 68. **18**. Zhang, Hong, Zhou, *et al.* (2006) *Biochim. Biophys. Acta* **1760**, 1505. **19**. Na, Yang, He, *et al.* (2006) *Planta Med.* **72**, 261.

Ursolate β-D-Glucopyranosyl Ester

This triterpene glycoside (FW = 618.85 g/mol) inhibits rabbit muscle glycogen phosphorylase (IC$_{50}$ = 97 μM). X-ray analysis revealed that this inhibitor binds at AMP allosteric activator site, making it and related compounds promising antidiabetic agents by exerting hypoglycemic effects, through inhibition of glycogenolysis. **1**. Wen, Sun, Liu, *et al.* (2008) *J. Med. Chem.* **51**, 3540.

Ursonate (Ursonic Acid)

This triterpene ($FW_{\text{free-acid}} = 454.69$ g/mol) inhibits rabbit muscle glycogen phosphorylase ($IC_{50} = 57$ μM) by binding to the allosteric activator site for AMP. **1**. Wen, Sun, Liu, *et al.* (2008) *J. Med. Chem.* **51**, 3540.

Urushiols

R = –$(CH_2)_{14}CH_3$
R = –$(CH_2)_7CH=CH(CH_2)_5CH_3$
R = –$(CH_2)_7CH=CHCH_2CH=CH(CH_2)_2CH_3$
R = –$(CH_2)_7CH=CHCH_2CH=CHCH=CHCH_3CH_3$
R = –$(CH_2)_7CH=CHCH_2CH=CHCH_2CH=CH_2$

These skin-adherent oleo-resins (CAS 53237-59-5) and closely related substances are oily plant catechols that give rise to plant-associated allergic contact dermatitis resulting from contact of skin and clothing with the leaves, stems and fruit of poison ivy, poison oak, and poison sumac (1,2). Based on a thorough examination of the effects of various leaf extract constituents (3), the following compounds were found to be strong allergens: 3-pentadecyl catechol, 4-pentadecyl catechol, "urushiol" dimethyl ether, 3-pentadecenyl-1'-veratrole, 3-methyl catechol, and hydrourushiol dimethyl ether. **Mechanism of Action:** Poison oak urushiols consist of catechol rings with a hydrocarbon side chain at the 3-position. Most poison oak urushiols have 17-carbon chains containing 1–3 double bonds. Their long side-chains facilitate adsorption to the lipid-rich epidermis, allowing subsequent, deeper penetration into the dermis itself. Polyunsaturated side-arms are associated with the most severe allergic responses. An oxidized form of the catechol ring is thought to be responsible for adduct formation with target proteins, which then undergo proteolysis to form the true immunogen. Shortening of the aliphatic chain by β-oxidation (within peroxisomes and/or mitochondria) may also be required for antigen presentation. In this respect, the active agent responsible for severe dermatitis is not urushiol *per se*. Instead, the immunogen is a combination of an urushiol (acting as a hapten) and one or more self proteins (acting as an immunoadjuvant). This mechanism accounts for delayed eruption, beginning at sites of greatest exposure and only later at less afeected sites. **1**. Symes & Dawson (1953) *Nature* **171**, 841. **2**. Byck & Dawson (1967) *J. Org. Chem.* 32, 1084. **3**. Keil, Wasserman & Dawson (1944) *J. Exp. Med.* **80**, 275.

Usnate (Usnic Acid)

This dibenzofurandione (FW = 344.32 g/mol; CAS 125-46-2) is found both as (+)- and (–)-stereoisomers in a number of lichens: occur in nature. (+)- and (–)-Usnic acid exhibits low solubility in water, but recrystallize as yellow orthorhombic prisms from acetone (solubility of 0.77 g/100 mL at 25°C). **Target(s):** deoxyribonuclease I in the presence of Co^{2+} (1); plant 4-hydroxyphenylpyruvate dioxygenase (2); oxidative phosphorylation, uncoupler (3); protophorphyrinogen oxidase (2); and urease (4,5). *See Uncouplers* **1**. McDonald (1955) *Meth. Enzymol.* **2**, 437. **2**. Romagni, Meazza, Nanayakkara & Dayan (2000) *FEBS Lett.* **480**, 301. **3**. Abo-Khatwa, al-Robai & al-Jawhari (1996) *Nat. Toxins* **4**, 96. **4**. Cifuentes, Garcia & Vicente (1983) *Z. Naturforsch. [C]* **38**, 273. **5**. Garcia, Cifuentes & Vicente (1980) *Z. Naturforsch. [C]* **35**, 1098.

Ussuristatins

These 71-residue, cysteine-rich disintegrins from the Chinese viper *Agkistrodon ussuriensis* contain a KGD loop that facilitates tight binding to integrins. Ussuristatin 1 inhibits platelet aggregation as well as platelet adhesion to fibronectin, $IC_{50} = 17$-33 nM. **1**. Oshikawa & Terada (1999) *J. Biochem.* **125**, 31.

Ustiloxins

These structurally related cyclic peptides ($FW_{\text{Ustilotoxin-A}} = 673.73$ g/mol; CAS 143557-93-1), designated Ustilotoxins A through F, from the rice pathogen *Ustilaginoidea virens* inhibit tubulin polymerization by binding to a sub-site of the Vincblastine site that can also be occupied by phomopsin A and rhizoxin (1). Ustilotoxin A and B share the same cyclic peptide nucleus and R_1 side-chain (*e.g.*, –$S(O)CH_2CH(OH)CH_2CH(NH_3^+)COO^-$), but differ with respect to R_2 (*i.e.*, Ustiloxin A possessing –$CH(CH_3)_2$ and Ustiloxin B possessing –CH_3 side-chains). Ustiloxin C has the same cyclic peptide nucleus with R_1 = –$S(O)CH_2CH_2OH$ and R_2 = –CH_3 side-chains, respectively. Ustiloxin D likewise has the same cyclic peptide nucleus, but R_1 = –H and R_1 = –$CH(CH_3)_2$. **1**. Li, Koiso, Kobayashi, Hashimoto & Iwasaki (1995) *Biochem. Pharmacol.* **49**, 1367. **2**. Ranaivoson, Gigant, Berritt, Joullié & Knossow (2012) *Acta Crystallogr. D Biol. Crystallogr.* **68**, 927.

Ustekinumab

This human monoclonal antibody (MW = 145.64 kDa; CAS 815610-63-0), also known as Stelara®, binds to the shared p40 subunit of interleukin (IL)-12 and IL-23, blocking signaling of their cognate receptors. It is approved for the treatment of moderate to severe plaque psoriasis and also shows promise for treating sarcoidosis. Indeed, extensive immunologic and genomic research identified IL-12 and IL-23 of the Th1 and Th17 inflammatory pathways, respectively, as key mediators in psoriasis, a complex, multigenic immune/inflammatory-mediated disorder that variably affects the skin, nails, and joints.

Uteroglobin

This low-MW protein, also known as blastokinin, is a secretoglobin family 1A member 1 (SCGB1A1) that strongly inhibits phospholipase A_2 (1-4). UG is a secreted, disulfide-bridged, progesterone-induced homodimer with structural similarity with phospholipase A_2 and other PLA_2 inhibitory proteins like the lipocortins. These properties suggest that its physiological function as a PLA_2 inhibitor may be mainly immunomodulatory (5). **1**. Chowdhury, Mantile-Selvaggi, Kundu, *et al.* (2000) *Ann. N. Y. Acad. Sci.* **923**, 307. **2**. Miele, Cordella-Miele, Mantile, Peri & Mukherjee (1994) *J. Endocrinol. Invest.* **17**, 679. **3**. Miele, Cordella-Miele, Facchiano & Mukherjee (1990) *Adv. Exp. Med. Biol.* **279**, 137. **4**. Levin, Butler, Schumacher, Wightman & Mukherjee (1986) *Life Sci.* **38**, 1813. **5**. Miele, Cordella-Miele & Mukherjee (1987) *Endocr. Rev.* **8**, 474.

UVI 3003

This RXR antagonist (FW = 436.58 g/mol; CAS 847239-17-2; Solubility: 100 mM in DMSO; IUPAC Name: 3-[4-hydroxy-3-[5,6,7,8-tetrahydro-5,5,8,8-tetramethyl-3-(pentyloxy)-2-naphthalenyl]phenyl]-2-propenoate) targets the Retinoid X Receptor (*or* RXR), a nuclear receptor that is activated by 9-*cis* retinoic acid. Binding of an agonist (referred to as a rexinoid) to the RX Receptor triggers the dissociation of a corepressor and recruitment of coactivator protein, promoting transcription of the downstream target gene into mRNA and eventually protein biosynthesis. The antagonist UVI-3003 does not affect the corepressor interaction capacity of the RARα subunit within the context of the RAR-RXR heterodimer. **1**. Nahoum, Pérez, Germain, *et al.* (2007) *Proc. Natl. Acad. Sci. U.S.A.* **104**, 17323. **2**. Santin, *et al.* (2009) *J. Med. Chem.* **52**, 3150.

Uzarin

This cardenolide (FW = 698.81 g/mol; CAS 20231-81-6) from *Gomphocarpus* sp. and the milkweed *Asclepias asperula* inhibits porcine kidney Na^+/K^+-exchanging ATPase, K_d = 4 μM (1). **1**. Abbott, Holoubek & Martin (1998) *Biochem. Biophys. Res. Commun.* **251**, 256.

– V –

V, *See* L-Valine; Vanadium and Vanadium Ions

Vaccenate

cis-Vaccenate

trans-Vaccenate

These fatty acids (FW$_{free-acid}$ = 282.47 g/mol; CAS 1937-63-9), corresponding to *cis* or *trans* isomers of octadec-11-enoic acid, or a mixture thereof, inhibit numerous enzymes. *cis* or (*Z*)-octadec-11-enoic acid (18:1Δ^{11},*cis*), has a melting point of 13.5°C and, while found in many tissues) is the major unsaturated fatty acid of *Escherichia coli*. *trans* or (*E*)-octadec-11-enoic acid (18:1Δ^{11},*trans*), also reported in many tissues, has a melting point of 44°C. **Target(s):** chondroitin AC lyase (1); chondroitin B lyase (1); Δ^9-desaturase, by *cis* vaccenate (2); DNA-directed DNA polymerase (3); DNA polymerases α and β, both inhibited by *cis*-vaccenate (3); DNA α-polymerase, inhibited by *trans*-isomer (3); DNA topoisomerase I (4); DNA topoisomerase II (4); hyaluronate lyase (1); 4-hydroxybenzoate nonaprenyltransferase, by *cis*-vaccenate (5); linoleate isomerase, by both *cis*- and *trans*-isomers (6). **1.** Suzuki, Terasaki & Uyeda (2002) *J. Enzyme Inhib. Med. Chem.* **17**, 183. **2.** Rosenthal & M. C. Whitehurst (1983) *Biochim. Biophys. Acta* **753**, 450. **3.** Mizushina, Tanaka, H. Yagi, *et al.* (1996) *Biochim. Biophys. Acta* **1308**, 256. **4.** Suzuki, Shono, Kai, Uno & Uyeda (2000) *J. Enzyme Inhib.* **15**, 357. **5.** Kawahara, Koizumi, Kawaji, *et al.* (1991) *Agric. Biol. Chem.* **55**, 2307. **6.** Kepler, Tucker & Tove (1970) *J. Biol. Chem.* **245**, 3612.

Vacor, *See* Pyriminil

Vadimezan

This vascular-disrupting agent (VDA) and chemotherapeutic agent (FW = 282.30 g/mol; CAS 117570-53-3; λ_{max} at 242, 270, and 338 nm), also known by its code names DMXAA and ASA404 as well as its systematic name, 2-(5,6-dimethyl-9-oxo-9*H*-xanthen-4-yl)acetic acid, is a multi-kinase inhibitor, with anti-VEGFR activity that exerts non-immune-mediated effects on the vasculature (capillary formation and endothelial barrier function), denying the availability of essential nutrients (including O$_2$) to the core of solid tumors. Indeed, the tumor vascular endothelium is characterized by increased permeability, abnormal morphology, disorganized vascular networks, and variable density. VDAs induce rapid shutdown of tumor blood supply, causing subsequent tumor death from hypoxia and nutrient deprivation. Vadimezan potently activates the TANK-binding kinase 1-interferon regulatory factor 3 signaling pathway in leukocytes, inducing type-I-interferon production (1-3). In addition, DMXAA significantly inhibits several other kinases in endothelial cells, including the vascular endothelial growth factor receptors VEGFR1 (IC$_{50}$ = 119 μM) and VEGFR2 (IC$_{50}$ = 11 μM) (4). ASA404 also competitively inhibits DT-diaphorase (*Reaction*: NAD(P)H + H$^+$ + Quinone $\rightleftharpoons$ Substrate NAD(P)$^+$ + Hydroquinone Product), with a K_i value of 20 μM and an IC$_{50}$ value of 63 μM, respectively, without significantly affecting the activity of cytochrome b_5 reductase and cytochrome P450 reductase. When evaluated for their *in vivo* antitumor activity and cytokine production by stromal or cancer cells in xenografts, mono-methyl XAA analogues with substitutions at the seventh and eighth positions were the most active in stimulating IL-6

and IL-8 production in human cell lines, whereas 5- and 6-substituted analogues were the most active in murine systems, highlighting the need to use appropriate *in vivo* animal models in selecting promising clinical candidates (5). **1.** Roberts, Goutagny, Perera, *et al.* (2007) *J. Exp. Med.* **204**, 1559. **2.** Tang, Aoshi, Jounai, *et al.* (2013) *PLoS One* **8**, e60038. **3.** Wallace, LaRosa, Kapoor, *et al.* (2007) *Cancer Res.* **67**, 7011. **4.** Buchanan, Shih, Aston, *et al.* (2012) *Clin. Sci.* (London) **122**, 449. **5.** Tijono, Guo, Henare, *et al.* (2013) *Brit. J. Cancer* **108**, 1306.

Valdecoxib

This non-steroidal anti-inflammatory agent (FW = 314.36 g/mol; CAS 181695-72-7), also known as Bextra® and 4-[5-methyl-3-phenylisoxazol-4-yl]benzenesulfonamide, inhibits cyclooxygenase-2, IC$_{50}$ = 5 nM. Concerns about its unacceptable risk of heart attack and stroke led to its voluntary withdrawal from the market in 2005. Indeed, the manufacturer (Pfizer) acknowledged its cardiotoxicity, responsibly suggesting that Bextra cannot be ethically tested in patients at high risk for heart disease. Its water-soluble and injectable prodrug Parecoxib (FW = 370.42 g/mol; CAS 202409-33-4; IUPAC: *N*-{[4-(5-methyl-3-phenylisoxazol-4-yl)phenyl]-sulfonyl}pro-panamide) is sold as Dynastat® in the European Union. **1.** Talley, Brown, Carter, *et al.* (2000) *J. Med. Chem.* **43**, 775.

Valepotriatum, *See* Valtratum

Valerate (Valeric Acid)

This short-chain fatty acid (FW$_{free-acid}$ = 102.13 g/mol; CAS 109-52-4; F.P. = –34.5°C; B.P.= 186-187°C, pK_a = 4.81 at 25°C; unpleasant odor; Solubility: 0.03 mg/mL H$_2$O), also known as *n*-pentanoic acid, is not a natural human metabolite, but is found naturally in the valerian, a perennial flowering plant (*Valeriana officinalis*), from which it derived its name. Valerate is likewise formed and degraded by fermentative microorganisms in ruminant animals. In 1904, the German chemist Franz Knoop elucidated the steps in β-oxidation by feeding dogs odd- and even-chain ω-phenyl fatty acids, including ω-phenylvaleric acid and and ω-phenylbutyric acid, cleverly using the phenyl group as a metabolic marker at a time when stable isotope and radioisotope tracers were entirely unavailable. Indeed, valeryl-CoA is an intermediate in the β-oxidation of medium-chain fatty acids found in butter and coconut oil. **Target(s):** alcohol dehydrogenase, weakly inhibited (1); amidase (2); D-amino-acid oxidase (3-6); γ-aminobutyrate aminotransferase, weakly inhibited (7); *p*-aminohippurate transport (8); cutinase (9); cytosol alanyl aminopeptidase (10); glutamate decarboxylase (11); guanidinobutyrase (12); (*S*)-2-hydroxy-acid oxidase (13); L-lactate dehydrogenase (cytochrome), weakly inhibited (14); mandelonitrile lyase (15); *N*-methylglutamate dehydrogenase (16); nitrile hydratase (17); nitroalkane oxidase, *or* 2-nitropropane dioxygenase (18); ornithine aminotransferase (19); valine decarboxylase, *or* leucine decarboxylase (20). **1.** Winer & Theorell (1960) *Acta Chem. Scand.* **14**, 1729. **2.** Maestracci, Thiery, Bui, Arnaud & Galzy (1984) *Arch. Microbiol.* **138**, 315. **3.** Dixon & Kleppe (1965) *Biochim. Biophys. Acta* **96**, 368 and 383. **4.** Brachet, Carreira & Puigserver (1990) *Biochem. Int.* **22**, 837. **5.** Brown & Scholefield (1953) *Proc. Soc. Exp. Biol. Med.* **82**, 34. **6.** Klein & Austin (1953) *J. Biol. Chem.* **205**, 725. **7.** Sytinsky & Vasilijev (1970) *Enzymologia* **39**, 1. **8.** Ullrich, Rumrich & Kloss (1987) *Pflugers Arch.* **409**, 547. **9.** Maeda, Yamagata, Abe, *et al.* (2005) *Appl. Microbiol. Biotechnol.* **67**, 778. **10.** Garner & Behal (1977) *Arch. Biochem. Biophys.* **182**, 667. **11.** Fonda (1972) *Arch. Biochem. Biophys.* **153**, 763. **12.** Yorifuji, Shimizu, Hirata, *et al.* (1992) *Biosci. Biotechnol. Biochem.* **56**,

773.   **13.** Jorns (1975) *Meth. Enzymol.* **41**, 337.   **14.** Dikstein (1959) *Biochim. Biophys. Acta* **36**, 397.   **15.** Jorns (1980) *Biochim. Biophys. Acta* **613**, 203.   **16.** Hersh, Stark, Worthen & Fiero (1972) *Arch. Biochem. Biophys.* **150**, 219.   **17.** Kopf, Bonnet, Artaud, Pétré & Mansuy (1996) *Eur. J. Biochem.* **240**, 239.   **18.** Gadda, Banerjee, Fleming & Fitzpatrick (2001) *J. Enzyme Inhib.* **16**, 157.   **19.** Kalita, Kerman & Strecker (1976) *Biochim. Biophys. Acta* **429**, 780.   **20.** Webb (1966) *Enzyme and Metabolic Inhibitors*, vol. **2**, p. 352, Academic Press, New York.

Valeryl-Coenzyme A

This short-chain fatty acyl thiolester (FW = 852.67 g/mol; CAS 4752-33-4), also known as pentanoyl-CoA (CoA = Coenzyme A) and adenosine 3'-phosphoric acid, 5'-[diphosphoric acid P^2-[2,2-dimethyl-3-hydroxy-3-[[2-[[2-(valerylthio)-ethyl]-aminocarbonyl]-ethyl]-aminocarbonyl]-propyl]] ester, is a β-oxidative intermediate that also inhibits acetyl-CoA hydrolase (1,2), 3-hydroxyisobutyryl-CoA hydrolase (3,4), 3-methylcrotonoyl-CoA carboxylase, weakly (5) and (S)-methylmalonyl-CoA hydrolase (6).   **1.** Nakanishi, Isohashi, Ebisuno & Sakamoto (1988) *Biochemistry* **27**, 4822.   **2.** Nakanishi, Isohashi, Matsunaga & Sakamoto (1985) *Eur. J. Biochem.* **152**, 337.   **3.** Shimomura, Murakami, Nakai, *et al.* (2000) *Meth. Enzymol.* **324**, 229.   **4.** Shimomura, Murakami, Fujitsuka, *et al.* (1994) *J. Biol. Chem.* **269**, 14248.   **5.** Diez, Wurtele & Nikolau (1994) *Arch. Biochem. Biophys.* **310**, 64.   **6.** Kovachy, Copley & Allen (1983) *J. Biol. Chem.* **258**, 11415.

Validamine

This amino-pyranose analogue (FW$_{free-base}$ = 177.20 g/mol; CAS 32780-32-8), also known as 6-amino-4-(hydroxymethyl)-4-cyclohexane-1,2,3-triol, inhibits glucan 1,3-α-glucosidase, *or* glucosidase II, *or* mannosyl-oligosaccharaide glucosidase II (1), glucoamylase (2), isomaltase (2), maltase (2), mannosyl-oligosaccharide glucosidase, *or* glucosidase I (1), sucrase (2), and trehalose (2).   **1.** Takeuchi, Kamata, Yoshida, Kameda & Matsui (1990) *J. Biochem.* **108**, 42.   **2.** Takeuchi, Takai, Asano, Kameda & Matsui (1990) *Chem. Pharm. Bull.* **38**, 1970.

Validamycin

Validamycin A

These structurally related antifungal agents from *Streptomyces hygroscopicus*, represented by validamycin A (FW = 497.50 g/mol; CAS 37248-47-8; IUPAC (1R,2R,3S,4S,6R)-2,3-dihydroxy-6-(hydroxymethyl)-4-{[(1S,4S,5S,6S)-4,5,6-trihydroxy-3-(hydroxymethyl)cyclohex-2-en-1-yl]amino}cyclohexyl β-D-glucopyranoside) consist of a trideoxyglucosyl residue, a *chiro*-aminoinositol residue, and a hydroxylated cyclohexane. Validamycin A reduces the maximum rate of hyphal extension and increases hyphal branching without affecting the growth rate. **Target(s):** α,α-trehalase, by validoxylamine A (1-7); α,α trehalose phosphorylase (8); and α,α-trehalose phosphorylase, configuration-retaining (9).   **1.** Asano, Yamaguchi, Kameda & Matsui (1987) *J. Antibiot.* **40**, 526.   **2.** Kyosseva, Kyossev & Elbein (1995) *Arch. Biochem. Biophys.* **316**, 821.   **3.** Temesvari & Cotter (1997) *Biochimie* **79**, 229.   **4.** Garcia, Iribarne, Lopez, Herrera-Cervera & Lluch (2005) *Plant Physiol. Biochem.* **43**, 355.   **5.** Salleh & Honek (1990) *FEBS Lett.* **262**, 359.   **6.** Müller, Aeschbacher, Wingler, Boller & Wiemken (2001) *Plant Physiol.* **125**, 1086.   **7.** Wisser, Guttenberger, Hampp & Nehls (2000) *New Phytol.* **146**, 169.   **8.** Kizawa,

Miyagawa & Sugiyama (1995) *Biosci. Biotechnol. Biochem.* **59**, 1908.   **9.** Wannet, Op den Camp, Wisselink, *et al.* (1998) *Biochim. Biophys. Acta* **1425**, 177. **See also** *Validoxylamine A*

Validoxylamine A

This pseudodisaccharide (FW = 335.35 g/mol; CAS 38665-10-0) component of validamycin A strongly inhibits the trehalases of *Rhizoctonia solani* and other fungi (K_i = 1.9 nM). The pig kidney trehalase has a K_i value of 0.52 nM. **Target(s):** α,α-trehalase (1-5); α,α-trehalose phosphorylase (6); α,α-trehalose phosphorylase, configuration-retaining (7).   **1.** Asano, Yamaguchi, Kameda & Matsui (1987) *J. Antibiot.* **40**, 526.   **2.** Kameda, Asano, Yamaguchi & Matsui (1987) *J. Antibiot.* **40**, 563.   **3.** Asano, Kato & Matsui (1996) *Eur. J. Biochem.* **240**, 692.   **4.** Kyosseva, Kyossev & Elbein (1995) *Arch. Biochem. Biophys.* **316**, 821.   **5.** Salleh & Honek (1990) *FEBS Lett.* **262**, 359.   **6.** Kizawa, Miyagawa & Sugiyama (1995) *Biosci. Biotechnol. Biochem.* **59**, 1908.   **7.** Goedl, Griessler, Schwarz & Nidetzky (2006) *Biochem. J.* **397**, 491.

Valienamine

This amino-hexose analogue (FW = 175.18 g/mol; CAS 38231-86-6), also known as (1S,2S,3R,6S)-6-amino-4-(hydroxymethyl)-6-cyclohexene-1,2,3-triol, is derived from *Streptomyces hygroscopicus* and inhibits glucosidases I and II. The presence of a double-bond in the six-membered ring doubtlessly imposes a flattened, half-chair conformation that mimics *oxa*-carbenium ion intermediates typically observed in glycosidase catalysis. **Target(s):** glucan 1,3-α-glucosidase *or* glucosidase II *or* mannosyl-oligosaccharide glucosidase II (1); glucoamylase (4); α-glucosidase (2,3); isomaltase (4); maltase (4); mannosyl oligosaccharide glucosidase *or* glucosidase I (1); sucrose α-glucosidase *or* sucrase (4-6); trehalose (4).   **1.** Takeuchi, Kamata, Yoshida, Kameda & Matsui (1990) *J. Biochem.* **108**, 42.   **2.** Kameda, Asano, Yoshikawa, *et al.* (1984) *J. Antibiot.* **37**, 1301.   **3.** Kameda, Asano, Yoshikawa & Matsui (1980) *J. Antibiot.* **33**, 1575.   **4.** Takeuchi, Takai, Asano, Kameda & Matsui (1990) *Chem. Pharm. Bull.* **38**, 1970.   **5.** Zheng, Shentu & Shen (2005) *J. Enzyme Inhib. Med. Chem.* **20**, 49.   **6.** Zheng, Shentu & Shen (2005) *Chin. J. Chem. Engin.* **13**, 429.

D-Valine

This non-protein-forming amino acid (FW = 117.15 g/mol; CAS 640-68-6), also known as (R)-valine, is present in a number of peptide antibiotics (*e.g.*, actinomycin, valinomycin, tolaasin, and gramicidins A, B, C, and D). It is known to inhibit acetolactate synthase (1), actinomycin biosynthesis (2,3), and 3-methyl-2 ketobutanoate hydroxymethyltransferase *or* 3-methyl-2-oxobutanoate hydroxymethyltransferase (4). (**See** *L-Valine*) **1.** Huppatz & Casida (1985) *Z. Naturforsch. [C]* **40**, 652.   **2.** Beaven, Barchas, Katz & Weissbach (1967) *J. Biol. Chem.* **242**, 657.   **3.** Katz (1960) *J. Biol. Chem.* **235**, 1090.   **4.** Powers & Snell (1976) *J. Biol. Chem.* **251**, 3786.

L-Valine

This hydrophobic proteinogenic amino acid (FW = 117.15 g/mol; CAS 72-18-4; pK_1 = 2.29; pK_2 = 9.74; Codons: GUU, GUC, GUA, and GUG), known systematically as (*S*)-2-amino-3-methylbutanoic acid, inhibits numerous enzymes. **Target(s):** acetolactate synthase (1-5); 2-aminohexanoate transaminase (39); arginase (6,10); argininosuccinate synthetase (11); arylformamidase, weakly inhibited (12); aspartate 4-decarboxylase, weakly inhibited (13); bacterial leucyl aminopeptidase (33); creatine kinase (35); cystathionine γ-lyase *or* cysteine desulfhydrase (14); cytosol nonspecific dipeptidase (15); glutamine synthetase (16); γ-glutamyl transpeptidase (17,18); homoserine kinase (19,37,38); isoleucyl-tRNA synthetase (20); β-isopropylmalate dehydrogenase (21); leucyl aminopeptidase (34); 3-methyl-2-ketobutanoate hydroxymethyltransferase, *or* 3-methyl-2-oxobutanoate hydroxymethyltransferase (22); ornithine aminotransferase (40-43); pantothenate synthetase, *or* pantoate:β-alanine ligase (23); propionyl-CoA carboxylase, *Myxococcus Xanthus* (24); pyruvate kinase (36); saccharopine dehydrogenase, weakly inhibited (25); L-serine ammonia-lyase, *or* L-serine dehydratase (26); threonine ammonia-lyase, *or* threonine dehydratase (27-31); Xaa-Pro dipeptidase, *or* prolidase isozyme I (32). **1.** Huppatz & Casida (1985) *Z. Naturforsch. [C]* **40**, 652. **2.** Pang & Duggleby (2001) *Biochem. J.* **357**, 749. **3.** De Felice, Levinthal, Iaccarino & Guardiola (1979) *Microbiol. Rev.* **43**, 42. **4.** Barak, Calvo & Schloss (1988) *Meth. Enzymol.* **166**, 455. **5.** Malik (1980) *Trends Biochem. Sci.* **5**, 68. **6.** Hunter & Downs (1945) *J. Biol. Chem.* **157**, 427. **7.** Webb (1966) *Enzyme and Metabolic Inhibitors*, vol. **2**, p. 335, Academic Press, New York. **8.** Patchett, Daniel & Morgan (1991) *Biochim. Biophys. Acta* **1077**, 291. **9.** Kaysen & Strecker (1973) *Biochem. J.* **133**, 779. **10.** Singh & Singh (1990) *Arch. Int. Physiol. Biochim.* **98**, 411. **11.** Takada, Saheki, Igarashi & Katsunuma (1979) *J. Biochem.* **85**, 1309. **12.** Serrano & Nagayama (1991) *Comp. Biochem. Physiol. B Comp. Biochem.* **99**, 281. **13.** Wong (1985) *Neurochem. Int.* **7**, 351. **14.** Fromageot & Grand (1942) *Enzymologia* **11**, 81. **15.** Wang, Liu, Yamashita, Manabe & Kodama (2004) *Clin. Chem. Lab. Med.* **42**, 1102. **16.** Caballero, Cejudo, Florencio, Cárdenas & Castillo (1985) *J. Bacteriol.* **162**, 804. **17.** Allison (1985) *Meth. Enzymol.* **113**, 419. **18.** Thompson & Meister (1977) *J. Biol. Chem.* **252**, 6792. **19.** Wormser & Pardee (1958) *Arch. Biochem. Biophys.* **78**, 416. **20.** Bergmann (1962) *Meth. Enzymol.* **5**, 708. **21.** Bode (1991) *Antonie Van Leeuwenhoek* **60**, 125. **22.** Powers, Snell (1976) *J. Biol. Chem.* **251**, 3786. **23.** Miyatake, Nakano & Kitaoka (1978) *J. Nutr. Sci. Vitaminol.* **24**, 243. **24.** Kimura, Kojyo, Kimura & Sato (1998) *Arch. Microbiol.* **170**, 179. **25.** Fujioka & Nakatani (1972) *Eur. J. Biochem.* **25**, 301. **26.** Kubota, Yokozeki & Ozaki (1989) *J. Ferment. Bioeng.* **67**, 391. **27.** Hatfield & Umbarger (1971) *Meth. Enzymol.* **17B**, 561. **28.** Datta (1971) *Meth. Enzymol.* **17B**, 566. **29.** Nath & Sanwal (1972) *Arch. Biochem. Biophys.* **151**, 420. **30.** Muramatsu & Nosoh (1996) *J. Microbiol.* **80**, 485. **31.** Proteau & Silver (1980) *Can. J. Microbiol.* **26**, 385. **32.** Liu, Nakayama, Sagara, *et al.* (2005) *Clin. Biochem.* **38**, 625. **33.** Baker, Wilkes, Bayliss & Prescott (1983) *Biochemistry* **22**, 2098. **34.** Machuga & Ives (1984) *Biochim. Biophys. Acta* **789**, 26. **35.** Pilla, Cardozo, Dornelles, *et al.* (2003) *Int. J. Dev. Neurosci.* **21**, 145. **36.** Schering, Eigenbrodt, Linder & Schoner (1982) *Biochim. Biophys. Acta* **717**, 337. **37.** Burr, Walker, Truffa-Bachi & Cohen (1976) *Eur. J. Biochem.* **62**, 519. **38.** Thoen, Rognes & Aarnes (1978) *Plant Sci. Lett.* **13**, 103. **39.** Der Garabedian & Vermeersch (1987) *Eur. J. Biochem.* **167**, 141. **40.** Strecker (1965) *J. Biol. Chem.* **240**, 1225. **41.** Yasuda, Tanizawa, Misono, Toyama & Soda (1981) *J. Bacteriol.* **148**, 43. **42.** Kalita, Kerman & Strecker (1976) *Biochim. Biophys. Acta* **429**, 780. **43.** Matsuzawa (1974) *J. Biochem.* **75**, 601.

L-Valine Hydroxamate

This aminoacylhydroxamate ($FW_{free-base}$ = 132.16 g/mol) inhibits a number of aminopeptidases, among them atrolysin A (4), bacterial leucyl aminopeptidase (1-3), leucyl aminopeptidase (1), and membrane alanyl aminopeptidase (1-3). In metal ion-coordinating enzymes, *N*-substituted hydroxamic acid functional groups often exist as a mixture of *syn* and *anti* rotamers, with relative abundances depending on their pK_a, that might influence their effectiveness as enzyme ligands and inhibitors. Sulfonylated derivatives of L-valine hydroxamate also inhibit *Clostridium histolyticum* collagenases (5). **1.** Wilkes & Prescott (1983) *J. Biol. Chem.* **258**, 13517. **2.** Chevier & D'Orchymont (1998) in *Handb. Proteolytic Enzymes* (Barrett, Rawlings & Woessner, eds.), p. 1433, Academic Press, San Diego. **3.** Baker, Wilkes, Bayliss & Prescott (1983) *Biochemistry* **22**, 2098. **4.** Fox, Campbell, Beggerly & Bjarnason (1986) *Eur. J. Biochem.* **156**, 65. **5.** Supuran & Scozzafava (2000) *Eur. J. Pharm. Sci.* **10**, 67.

Valinol

This optically active alcohol analogue of valine ($FW_{free-base}$ = 103.16 g/mol; CAS 2026-48-4) also known as 2-amino-3-methyl-1-butanol, reacts with aldehydes to form imines as well as nitriles to form oxazolines. **Target(s):** isoleucyl-tRNA synthetase, by the L-enantiomer (1); nitric oxide synthase, by the L-enantiomer (2); and valyl-tRNA synthetase, by the L-enantiomer (3). **1.** Freist & Cramer (1983) *Eur. J. Biochem.* **131**, 65. **2.** Hrabak, Bajor & Temesi (1996) *Comp. Biochem. Physiol. B Biochem. Mol. Biol.* **113**, 375. **3.** Kakitani, Tonomura & Hiromi (1987) *Biochem. Int.* **14**, 597.

Valinomycin

This cyclic, twelve-residue depsipeptide (FW = 1111.3 g/mol; CAS 2001-95-8) from *Streptomyces fulvissimus* is an ionophorous uncoupler of transmembrane potentials involving potassium ion (1-8). Valinomycin may be regarded as a cyclic trimer, with each unit consisting of a D-valine, L-valine, D-hydroxyvalerate, and L-lactate residue. Valinomycin has the effect of uncoupling both mitochondrial oxidative phosphorylation (3-10 µg/mL) and chloroplast photophosphorylation (~2 µg/mL). When treated with small doses (~30 ng/mL) of valinomycin, human cells quickly show evidence of mitochondrial swelling and reduced viability. **Antibiotic Action Against *Mycobacterium tuberculosis*:** Valinomycin is a particularly effective antibiotic against the human pathogen *M. tuberculosis*, an organism that exhibits reduced ability to maintain internal pH towards neutrality at very acidic pH conditions (pH 3–5). This property and its valinomycin sensitivity may result from an increased proton permeability of the *M. tuberculosis* membrane or a decreased proton extrusion by the membrane-embedded ATPase (9). **Primary Mechanism of Action:** Valinomycin catalyzes the transport of K^+ and other monovalent cations of similar radius (*e.g.*, Rb^+ or Cs^+, but not Na^+) as a charged species across biomembranes. Because rubidium, cesium, silver, and thallium ions are not abundant in living organisms, valinomycin is mainly a potassium ionophore (*Ion Selectivity*: $Rb^+ > K^+ > Cs^+ > Ag^+ > Tl^+ >> NH_4^+ > Na^+ \approx Li^+$. Potassium ion binds roughly 10^4 times tighter than sodium ion. Each valinomycin crosses the bilayer at a rate of roughly 10^5 per second. Transmembrane transfer of neutral, metal ion-free valinomycin between the membrane's surfaces is slower than the transfer of charged potassium ion-valinomycin complexes (10). Transfer of the complex is hastened either by the deformation of the membrane or transfer of a conformational change in the free carrier (10).

Valinomycin Effects on Mitochondria: in In the presence of valinomycin, mitochondria will take up K^+ at the expense of the proton gradient driven by coupled electron transfer or ATP hydrolysis. In the presence of excess K^+, the electrical component $\Delta\psi$ of the transmembrane proton gradient collapse, at least during the time required to accumulate a substantial K^+ gradient. Accumulation of potassium inside the mitochondria promotes an increase in the volume of matrix (as evidenced by swelling) and contacts between inner and outer mitochondrial membranes decrease. In valinomycin-induced stimulation of mitochondrial energy-dependent reversible swelling (supported by succinate oxidation, cytochrome c and sulfite oxidase (normally present in the mitochondrial intermembrane space) are released by mitochondria (11). This effect can be observed at valinomycin concentrations as low as 1 nM, and the rate of cytosolic NADH/cytochrome-c electron transport pathway is also greatly stimulated. Magnesium ions prevent at least in part the valinomycin effects. Rather than to the dissipation of membrane potential, the pro-apoptotic property of valinomycin can be ascribed to both the release of cytochrome-c from mitochondria to cytosol and the increased rate of cytosolic NADH coupled with an increased availability of energy in the form of glycolytic ATP, useful for the correct execution of apoptotic program (11). **Effect on Red Blood Cells:** Exposure of red blood cells to valinomycin, causes loss of KCl and water, resulting in cell dehydration and increasing cell density. While almost all normal valinomycin-treated RBC dehydrate, in sickle cell anemia a fraction of RBCs fail to dehydrate, indicating the existence of valinomycin-resistant RBC (12). *See Uncouplers* **Target(s):** ATP synthase (5); nitric-oxide synthase (13); photosynthetic electron flow (14,15). **1.** Reed (1979) *Meth. Enzymol.* **55**, 435. **2.** Slater (1967) *Meth. Enzymol.* **10**, 48. **3.** Heytler (1979) *Meth. Enzymol.* **55**, 462. **4.** Izawa & Good (1972) *Meth. Enzymol.* **24**, 355. **5.** Stiggall, Galante & Hatefi (1979) *Meth. Enzymol.* **55**, 308. **6.** Pressman (1976) *Ann. Rev. Biochem.* **45**, 501. **7.** Karlish & Avron (1968) *FEBS Lett.* **1**, 21. **8.** Moore & Pressman (1964) *Biochem. Biophys. Res. Commun.* **15**, 562. **9.** Zhang, Zhang & Sun (2003) *J. Antimicrob. Chemother.* **52**, 56. **10.** Hladky, Leung & Fitzgerald (1995) *Biophys J.* **69**, 1758. **11.** Lofrumento, La Piana, Abbrescia, *et al.* (2011) *Apoptosis* **16**, 1004. **12.** Amer, Etzion, Bookchin & Fibach (2006) *Biochim. Biophys. Acta* **1760**, 793. **13.** Griffith, Edwards, Newby, Lewis & Henderson (1986) *Cardiovasc. Res.* **20**, 7. **14.** Trebst (1980) *Meth. Enzymol.* **69**, 675. **15.** Voegell, O'Keeffe, Whitmarsh & Dilley (1977) *Arch. Biochem. Biophys.* **183**, 333.

Valiolamine

This glucose analogue (FW$_{free-base}$ = 193.20 g/mol; CAS 83465-22-9), named systematically as (1(OH),2,4,5/1,3)-5-amino-1-C-(hydroxymethyl) 1,2,3,4-cyclohexanetetrol, from *Streptomyces hygroscopicus* is a strong inhibitor of rat intestinal sucrase, maltase, glucoamylase, isomaltase, and trehalase (K_i = 0.32 μM, 2.9 μM, 1.2 μM, 0.91 μM, and 49 μM, respectively). *See also Valienamine* **Target(s):** glucan 1,3-α-glucosidase, *or* glucosidase II (1); glucoamylase (2); α-glucosidase (1,3); isomaltase (2,3); lysosomal α-glucosidase, *or* glucan 1,4-α-glucosidase (1); maltase (1-3); mannosyl oligosaccharide glucosidase, *or* glucosidase I (1); sucrase (2,3); and trehalase (2). **1.** Takeuchi, Kamata, Yoshida, Kameda & K. Matsui (1990) *J. Biochem.* **108**, 42. **2.** Takeuchi, Takai, Asano, Kameda & Matsui (1990) *Chem. Pharm. Bull.* **38**, 1970. **3.** Kameda, Asano, Yoshikawa, *et al.* (1984) *J. Antibiot.* **37**, 1301.

Valium *See Diazepam*

Valproate (*or* Valproic Acid)

This broad-spetrum antiseizure drug and HDAC inhibitor (FW = 144.21 g/mol; CAS 1069-66-5 (sodium salt); colorless liquid; slightly soluble in water; pK_a ≈ 4.7 at 25°C; *Abbreviation:* VPA), also known as 2-propylpentanoic acid and di(*n*-propyl)acetic acid, is a γ-aminobutyrate antagonist at plasma valproate concentrations of 50–100 μg/mL and is also effective against pentylenetetrazol-induced seizures. Glutamatergic transmission is another major target for VPA, which decreases brain levels of glutamate and aspartate, while increasing its taurine content (1). **Mechanism of Action:** Among its many actions, valproate inhibits sustained repetitive firing via depolarization of cortical and spinal neurons. While without effect on GABA-regulated responses, valproate also increases GABA levels by stimulating glutamate decarboxylase. Valproate is FDA-approved for treating certain types of epilepsy, manic episodes in bipolar disorder, and migraine headaches. Its pharmacokinetics include a 1-4 hour delay before reaching peak plasma concentration, with 90% bound to plasma proteins. In humans, its volume of distribution is roughly 0.2 L/kg. (Chronic use can be toxic to liver and pancreas, attended by excessive tiredness, lack of energy, weakness, stomach pain, appetite loss, nausea/vomiting, or swelling of the face.) **Role as a Histone Deacetylase Inhibitor:** The finding that VPA inhibits HDAC was demonstrated by analyzing the degree of histone acetylation *in vitro* and *in vivo* using an antibody against hyperacetylated histones H3 or H4 (15). While low-level histone acetylation is detected in untreated teratocarcinoma or HeLa cells, 0.25 mM VPA increases the amount of acetylated histone H4. Massive acetylation is found with 2 mM VPA. Such a degree of acetylation is observed at 5 mM butyrate or 100 nM trichostatin A. Valproic acid relieves HDAC-dependent transcriptional repression and causes hyperacetylation of histones in cultured cells and *in vivo*. VPA inhibition of HDAC activity *in vitro* occurs most likely by binding to the active sites of HDACs (15). Corepressor-associated HDACs mediate repression via transcription factors, and inappropriate repression of target genes is responsible for transformation of leukemic cells which harbor disease-associated translocations, leading to expression of the fusion proteins. VPA is also a potent class-selective HDAC inhibitor at nontoxic therapeutic concentrations, relieving HDAC-dependent transcriptional repression and causing histone hyperacetylation in cultured cells and *in vivo*. Valproic acid also induces differentiation of hematopoietic progenitor cells and leukemic blasts obtained from AML patients. **Key Pharmacokinetic Parameters:** *See* Appendix II in Goodman & Gilman's THE PHARMACOLOGICAL BASIS OF THERAPEUTICS, 12th Edition (Brunton, Chabner & Knollmann, eds.) McGraw-Hill Medical, New York (2011). **Target(s):** acyl-CoA dehydrogenases (2); NADP$^+$-alcohol dehydrogenase (3); aldehyde reductase (3-6); 4-aminobutyrate aminotransferase (7-9,24,25); carbonyl reductase (10); CYP2C9 (11); glycine synthase (12,13); [glycogen synthase] kinase-3 (14); histone deacetylase (15); inositol-3 phosphate synthase, *or* inositol-1-phosphate synthase (16); α-ketoglutarate dehydrogenase (17); 3-methylbutanal reductase, weakly inhibited (18); β-oxidation (19); palmitoyl-CoA hydrolase, *or* acyl CoA hydrolase (20); succinate-semialdehyde dehydrogenase (9,21,22); and vitamin D$_3$ 25-monooxygenase (cytochrome P450) (23); HDAC6 (IC$_{50}$ = 0.3 mM), leading to tubulin hyper-acetylation and providing a rationale for VPA enhancement of paclitaxel's anticancer action (26). **1.** Slevin & Ferrara (1985) *Neurology* **35**, 728. **2.** Anderson, Acheampong & Levy (1994) *Neurology* **44**, 742. **3.** Whittle & Turner (1981) *Biochem. Pharmacol.* **30**, 1191. **4.** Daly & Mantle (1982) *Biochem. J.* **205**, 381. **5.** Morjana & Flynn (1989) *J. Biol. Chem.* **264**, 2906. **6.** De Jongh, Schofield & Edwards (1987) *Biochem. J.* **242**, 143. **7.** Fushiya, Kanazawa & Nozoe (1997) *Bioorg. Med. Chem.* **5**, 2089. **8.** Fowler, Beckford & John (1975) *Biochem. Pharmacol.* **24**, 1267. **9.** Whittle & Turner (1978) *J. Neurochem.* **31**, 1453. **10.** Wermuth (1981) *J. Biol. Chem.* **256**, 1206. **11.** Wen, Wang, Kivisto, Neuvonen & Backman (2001) *Brit. J. Clin. Pharmacol.* **52**, 547. **12.** Martin, Benavides & Ugarte (1982) *Rev. Esp. Fisiol.* **38** Suppl, 59. **13.** Mortensen, Kolvraa & Christensen (1980) *Epilepsia* **21**, 563. **14.** Chen, Huang, Jiang & Manji (1999) *J. Neurochem.* **72**, 1327. **15.** Göttlicher, Minucci, Zhu, *et al.* (2001) *EMBO J.* **20**, 6969. **276**, 36734. **16.** Ju & Greenberg (2004) *Clin. Neurosci. Res.* **4**, 181. **17.** Luder, Parks, Frerman & Parker (1990) *J. Clin. Invest.* **86**, 1574. **18.** Van Nedervelde, Verlingen, Philipp & Debourg (1997) *Proc. Congr. Eur. Brew. Conv.* **26**, 447. **19.** Silva, Ruiter, Ijlst, *et al.* (2001) *Chem. Biol. Interact.* **137**, 203. **20.** Dixon, Osterloh & Becker (1990) *J. Pharm. Sci.* **79**, 103. **21.** van der Laan, de Boer & Bruinvels (1979) *J. Neurochem.* **32**, 1769. **22.** Cash, Maitre, Rumigny, Weissman-Nanopoulos & Mandel (1982) *Prog. Clin. Biol. Res.* **114**, 379. **23.** Tomita, Ohnishi, Nakano & Ichikawa (1991) *J. Steroid Biochem. Mol. Biol.* **39**, 479. **24.** Maître, Ciesielski, Cash & Mandel (1978) *Biochim. Biophys. Acta* **522**, 385. **25.** Choi & Lee (2006) *Chem. Pharm. Bull.* **54**, 1720. **26.** Catalano, Poli, Pugliese, Fortunati & Boccuzzi (2007) *Endocr. Relat. Cancer* **14**, 839.

Valrubicin

This antineoplastic drug (FW = 723.65 g/mol; CAS 56124-62-0), also named *N*-trifluoroacetyl-adriamycin 14-valerate, *N*-trifluoroacetyl-doxorubicin 14-valerate, and 2-oxo-2-[(2*S*,4*S*)-2,5,12-trihydroxy-7-methoxy-6,11-dioxo-4-({2,3,6-trideoxy-3-[(trifluoroacetyl)amino]hexopyranosyl}oxy)-1,2,3,4,6,11-hexahydrotetracen-2-yl]ethyl pentanoate, is a lipophilic analogue of adriamycin. It has been used in the treatment of urinary bladder carcinoma. **Target(s):** DNA topoisomerase II (1,2); protein kinase C (3); and RNA polymerase II (4). **1.** Insaf, Danks & Witiak (1996) *Curr. Med. Chem.* **3**, 437. **2.** Christmann-Franck, Bertrand, Goupil-Lamy, *et al.* (2004) *J. Med. Chem.* **47**, 6840. **3.** Chuang, Kung, Israel & Chuang (1992) *Biochem. Pharmacol.* **43**, 865. **4.** Chuang, Kawahata & Chuang (1980) *FEBS Lett.* **117**, 247.

Valsartan

This angiotensin receptor blocker, *or* ARB (FW = 435.52 g/mol; CAS 137862-53-4), also named (*S*)-3-methyl-2-(*N*-{[2'-(2*H*-1,2,3,4-tetrazol-5-yl)biphenyl-4-yl]methyl}pentanamido)butanoic acid, is is a potent, specific, highly selective antagonist of angiotensin II (AII) at the Type I (AT1) angiotensin receptor, but does not possess agonistic activity. It is an efficacious, orally active, blood pressure-lowering agent in conscious renal hypertensive rats and in conscious normotensive, sodium-depleted primates. The hypotensive effects of valsartan last longer than those of losartan. (*See also* Candesartan; Eprosartan; Irbesartan; Losartan; Olmesartan; Telmisartan) **Pharmacological Parameters:** Bioavailability = 25 %; Food Effect? YES; Drug $t_{1/2}$ = 9 hours; Drug's Protein Binding = 95 %; Route of Elimination = 13 % Renal, 85 % Hepatic. **Key Pharmacokinetic Parameters:** See Appendix II in Goodman & Gilman's THE PHARMACOLOGICAL BASIS OF THERAPEUTICS, 12th Edition (Brunton, Chabner & Knollmann, eds.) McGraw-Hill Medical, New York (2011). **1.** Criscione, de Gasparo, Bühlmayer, *et al.* (1993) *Brit. J. Pharmacol.* **110**, 761.

Valspodar

This P-gp-inhibiting chemosensitizer (FW = 1214.64 g/mol; CAS 121584-18-7), also known as PSC-833, which lacks the immunosuppressive and nephrotoxic properties of cyclosporin A, binds tightly to P-glycoprotein, inhibits the latter's multidrug resistance-reversing activity, and is superior to cyclosporin A and verapamil. Valspodar is also a substrate of CYP3A (cytochrome P450 3A). **Target(s):** P-glycoprotein (1-3); and xenobiotic-transporting ATPase multidrug-resistance protein (2,3). **1.** Achira, Suzuki, Ito & Sugiyama (1999) *AAPS PharmSci.* **1**, E18. **2.** Paul, Breuninger & Kruh (1996) *Biochemistry* **35**, 14003. **3.** Garrigues, Escargueil & Orlowski (2002) *Proc. Natl. Acad. Sci. U.S.A.* **99**, 10347.

L-Valyl-L-asparaginyl-[(2R,4S,5S)-5-amino-4-hydroxy-2,7-dimethyloctanoyl]-L-alanyl-L-glutamyl-L-phenylalanine

This heptapeptide analogue (FW = 763.41 g/mol; CAS 121584-18-7), also known as OM99-1, L-valyl-L-asparaginyl-L-leucyl-Ψ[CH(OH)CH2]-L-alanyl-L-alanyl-L-glutamyl-L-phenylalanine (*or* VNLΨAAEF), has a nonhydrolyzable hydroxyethylene peptide-bond isostere (–(*S*)-CH(OH)CH2–), denoted by Ψ, between Leu-3 and Ala-4 residues. OM99-1 inhibits memapsin-2, *or* β-secretase, with a K_i of 0.3 nM. The inihitor sequence is based on the β-secretase cleavage site of Swedish β-amyloid precursor protein, with an Asp residue replaced by Ala. **1.** Ghosh, Hong & Tang (2002) *Curr. Med. Chem.* **9**, 1135.

L-Valyl-3-cyclohexyl-L-alanine *N*-butylamide

This dipeptide amide (FW = 325.50 g/mol) inhibits tripeptidyl-peptidase II (K_i = 57 μM), an intermediate exopeptidase (EC 3.4.14.10) required for efficient protein turnover, showing broad sequence specificity and releasing tripeptides from the N-terminus of oligopeptides. **1.** Ganellin, Bishop, Bambal, *et al.* (2000) *J. Med. Chem.* **43**, 664.

L-Valyl-L-glutamyl-L-glutamate

This tripeptides (FW = 375.38 g/mol; *Sequence*: VEE) inhibits glutamate carboxypeptidase II, also known as *N*-acetylated-γ-linked-acidic dipeptidase. **1.** Robinson, Blakely, Couto & Coyle (1987) *J. Biol. Chem.* **262**, 14498.

L-Valyl-L-leucyl-L-isoleucyl-L-valyl-L-proline

This pentapeptide (FW = 539.72 g/mol; *Sequence*: VLIVP), obtained from hydrolysis of glycinin, inhibits peptidyl-dipeptidase A, also known as angiotensin I-converting enzyme (K_i = 4.5 μM). Mallikarjun Gouda, Gowda, Appu Rao & Prakash (2006) *J. Agric. Food Chem.* **54**, 4568.

L-Valyl-L-prolyl-L-leucine

This naturally occurring hydrophobic tripeptides (FW = g/mol), also called diprotin B and VPL, inhibits dipeptidyl aminopeptidase IV (1). Other investigators have reported that this tripeptide is actually an alternative substrate and is slowly hydrolyzed by the peptidase. **Target(s):** dipeptidyl-peptidase IV, *or* dipeptidyl-aminopeptidase IV (1-6); and Xaa-Pro dipeptidyl-peptidase (7).   1. Umezawa, Aoyagi, Ogawa, Naganawa, Hamada & Takeuchi (1984) *J. Antibiot.* **37**, 422.   2. Rahfeld, Schierhorn, Hartrodt, Neubert & Heins (1991) *Biochim. Biophys. Acta* **1076**, 314.   3. Malík, Busek, Mares, Sevcík, Kleibl & Sedo (2003) *Adv. Exp. Med. Biol.* **524**, 95.   4. Davy, Thomsen, Juliano, Alves, Svendsen & Simpson (2000) *Plant Physiol.* **122**, 425.   5. Hanski, Hihle & Reutter (1985) *Biol. Chem. Hoppe-Seyler* **366**, 1169.   6. Piazza, Callanan, Mowery & Hixson (1989) *Biochem. J.* **262**, 327.   7. Rigolet, Xi, Rety & Chich (2005) *FEBS J.* **272**, 2050.

1-(L-Valyl)pyrrolidone

This dipeptide analogue, also known as L-valyl-pyrrolidide (FW$_{free-base}$ = 170.25 g/mol), inhibits dipeptidyl-peptidases II (1) and IV (1-7) and Xaa-Pro dipeptidyl-peptidase (2).   1. Stöckel-Maschek, Mrestani-Klaus, Stiebitz, Demuth & Neubert (2000) *Biochim. Biophys. Acta* **1479**, 15.   2. Rigolet, Xi, Rety & Chich (2005) *FEBS J.* **272**, 2050.   3. Lambeir, Rea, Fulop, *et al.* (2003) *Adv. Exp. Med. Biol.* **524**, 29.   4. Stöckel-Maschek, Stiebitz, Born, *et al.* (2000) *Adv. Exp. Med. Biol.* **477**, 117.   5. Leiting, Pryor, Wu, *et al.* (2003) *Biochem. J.* **371**, 525.   6. Rosenblum & Kozarich (2003) *Curr. Opin. Chem. Biol.* **7**, 496.   7. Rasmussen, Branner, Wiberg & Wagtmann (2003) *Nature Struct. Biol.* **10**, 19.

L-Valyl-L-seryl-L-glutaminyl-L-asparaginyl-(5-amino-4-hydroxy-2-isopropyl-7-methyloctanoyl)-L-isoleucyl-L-valine, See U-85548E

L-Valyl-L-tryptophan

This dipeptide (FW = 302.35 g/mol) inhibits peptidyl-dipeptidase A (angiotensin I-converting enzyme), K_i = 0.3 μM (1-3).   1. Cheung, Wang, Ondetti, Sabo & Cushman (1980) *J. Biol. Chem.* **255**, 401.   2. Kawamura, Kikuno, Oda & Muramatsu (2000) *Biosci. Biotechnol. Biochem.* **64**, 2193.   3. Ono, Hosokawa, Miyashita & Takahashi (2006) *Int. J. Food Sci. Technol.* **41**, 383.

N-(L-Valyl-L-valyl)-(5-amino-4-hydroxy-2-isopropyl-7-methyloctanoyl)-L-isoleucine 2-pyridyl-methylamide

This peptide analogue (FW = 632.49 g/mol), also known as Val-Val-Leuψ[CH(OH)CH₂]Val-Ile-Amp, inhibits HIV 1 and HIV-2 protease, with K_i values of 7 and 3.5 nM, respectively.   Sawyer, Staples, Liu, Tomasselli, *et al.* (1992) *Int. J. Pept. Protein Res.* **40**, 274.

L-Valyl-L-valyl-L-seryl-L-valyl-L-leucyl-L-threonine

This hexapeptide (MW = 616.76 g/mol; Sequence: VVSVLT), also known as Cabin-3, corresponds to residues 185-190 of human immunoglobulin G γ-chain and inhibits human cathepsin B, IC$_{50}$ = 20 μM (1).   1. Nakagomi, Fujimura, Maeda, Sadakane, Fujii, Akizawa, Tanimura & Hatanaka (2002) *Biol. Pharm. Bull.* **25**, 564.

L-Valyl-L-valyl-L-tyrosyl-L-prolyl-L-tryptophan, See Tynorphin

L-Valyl-L-valyl-L-tyrosyl-L-prolyl-L-tryptophanyl-L-threonyl-L-glutaminyl-L-arginyl-L-phenylalanine

This nonapeptide (MW = 1195.29 g/mol; *Sequence*: VVYPWTQRF), also known as VV-hemorphin-7 (hemorphins are small β-globin-derived peptides), inhibits peptidyl-dipeptidase A, *or* angiotensin I-converting enzyme, with a K_i of 14 μM (1). This nonapeptide and the corresponding truncated hepta- and octapeptides (*i.e.*, VVYPWTQ and VVYPWTQR) also inhibit dipeptidyl-peptidase IV (2).   1. Fruitier-Arnaudin, Cohen, Bordenave, Sannier & Piot (2002) *Peptides* **23**, 1465.   2. Cohen, Fruitier-Arnaudin & Piot (2004) *Biochimie* **86**, 31.

Vanadate, See Metavanadate; Orthovanadate; Pervanadate

Vanadate-Uridine, See Uridine-Vanadate; Vanadyl Ribonucleoside Complex

Vanadyl-Ribonucleoside Complex

This RNase inhibitor cocktail, consisting of VO₂⁺:nucleotide binary complexes formed upon reaction of vanadyl sulfate (VOSO₄) with an equimolar mixture of 2',3'-cyclicAMP, 2',3'-cyclicGMP, 2',3'-cyclicCMP, and 2',3'-cyclicUMP), inhibits a Pac1 ribonuclease (1); pancreatic ribonuclease, *or* ribonuclease A (2); and the splicing enzyme RNA 3' phosphate cyclase (3,4). (*See Uridine-Vanadate Complex; Metavanadate; Orthovanadate; Pervanadate*)   1. Rotondo & Frendewey (2001) *Meth. Enzymol.* **342**, 168.   2. Blackburn & Moore (1982) *The Enzymes*, 3rd ed. (Boyer, ed.), **15**, 317.   3. Filipowicz & Vicente (1990) *Meth. Enzymol.* **181**, 499.   4. Reinberg, Arenas & Hurwitz (1985) *J. Biol. Chem.* **260**, 6088.

Vanadyl Sulfate

This water-soluble, blue solid VOSO₄ (FW = 163.00 g/mol) forms VO₂⁺:nucleotide binary complexes that are pentacoordinate mimics of ribonuclease reaction transition-states. (*Note*: In this form, vanadium has an unpaired electron and is paramagnetic). (*See Uridine-Vanadate Complex; Metavanadate; Orthovanadate; Pervanadate*) **Target(s):** alkaline phosphatase (1,2); creatine kinase (3); dolichyl-phosphatase (4,5); phospholipid-translocating *ATPase or* flippase (6); and ribonuclease (1,7).   1. Macara (1980) *Trends Biochem. Sci.* **5**, 92.   2. Lopez, Stevens & Lindquist (1976) *Arch. Biochem. Biophys.* **175**, 31.   3. O'Sullivan & Morrison (1963) *Biochim. Biophys. Acta* **77**, 142.   4. Adrian & Keenan (1981) *Biochem. J.* **197**, 233.   5. Adrian & Keenan (1979) *Biochim. Biophys. Acta* **575**, 431.   6. Daleke & Lyles (2000) *Biochim. Biophys. Acta* **1486**, 108.   7. Lienhard, Secemski, Koehler & Lindquist (1971) *Cold Spring Harbor Symp. Quant. Biol.* **36**, 45. See *Vanadyl-Ribonucleoside Complex*

Vancomycin

This glycopeptide antibiotic (FW = 1449.27 g/mol; CAS 1404-93-9), once known as vancocin, inhibits cell wall peptidoglycan biosynthesis in Gram-positive bacteria (1,2). With D-Alanyl-D-alanine essential for bacterial cell wall synthesis (*i.e*, in assembling one of the subunits used for peptidoglycan cross-linking), vancomycin binds (largely by means of specific hydrogen bonds) to the aminoacyl-D-Ala-D-Ala strand, thereby interfering with cross-linking required in the biosynthesis of bacterial cell walls. **Key Pharmacokinetic Parameters:** *See* Appendix II in Goodman & Gilman's *THE PHARMACOLOGICAL BASIS OF THERAPEUTICS*, 12th Edition (Brunton, Chabner & Knollmann, eds.) McGraw-Hill Medical, New York (2011). **Guidelines for Clinical Use:** Given the emergence of vancomycin-resistant enterococci, the U.S. Centers for Disease Control Hospital Infection Control Practices Advisory Committee issued guidelines restricting the vancomycin use to the following indications: (*a*) Treatment of serious infections caused by methicillin-resistant *Staphylococcus aureus* [MRSA], Multiresistant *S. epidermidis* (MRSE), or in individuals with life-threatening penicillin allergies; (*b*) Treatment of pseudomembranous colitis caused by *Clostridium difficile*, especially relapse infections that are unresponsive to metronidazole; (*c*) Treatment of infections caused by Gram-positive microorganisms in patients with serious allergies to β-lactam antimicrobials; (*d*) Antibacterial prophylaxis for endocarditis in penicillin-hypersensitive individuals; (*e*) Surgical prophylaxis for major procedures involving implantation of prostheses in institutions with high rates of MRSA or MRSE; and (*f*) Preemptive treatment for likely MRSA infection, while awaiting identification of the organism. **Target(s):** CDP-glycerol glycerophospho-transferase (3,4); CDP-ribitol ribitolphosphotransferase, weakly inhibited (5); muramoylpentapeptide carboxy-peptidase (6); penicillin-binding protein 1b (7,8); peptidoglycan glycosyltransferase (7-11); phospho-*N* acetylmuramoyl-pentapeptide-transferase (12); MurG transferase, *or* undecaprenyl-diphosphomuramoyl-pentapeptide β-*N*-acetylglucosaminyltransferase, IC$_{50}$ = 15.7 μM (13-15). **1**. Izaki, M. Matsuhashi & Strominger (1966) *Meth. Enzymol.* **8**, 487. **2**. Hammes & Neuhaus (1974) *Antimicrob. Agents Chemother.* **6**, 722. **3**. Burger & Glaser (1966) *Meth. Enzymol.* **8**, 430. **4**. Burger & Glaser (1964) *J. Biol. Chem.* **239**, 3168. **5**. Ishimoto & Strominger (1966) *J. Biol. Chem.* **241**, 639. **6**. Leyh-Bouille, Ghuysen, Nieto, *et al.* (1970) *Biochemistry* **9**, 2971. **7**. Nakagawa, Tamaki, Tomioka & Matsuhashi (1984) *J. Biol. Chem.* **259**, 13937. **8**. Chandrakala, Shandil, Mehra, *et al.* (2004) *Antimicrob. Agents Chemother.* **48**, 30. **9**. Park, Seto, Hakenbeck & Matsuhashi (1985) *FEMS Microbiol. Lett.* **27**, 45. **10**. Zawadzka-Skomial, Markiewicz, Nguyen-Distèche, *et al.* (2006) *J. Bacteriol.* **188**, 1875. **11**. Taku, Stuckey & Fan (1982) *J. Biol. Chem.* **257**, 5018. **12**. Struve, Sinha & Neuhaus (1966) *Biochemistry* **5**, 82. **13**. Meadow, Anderson & Strominger (1964) *Biochem. Biophys. Res. Commun.* **14**, 382. **14**. Ravishankar, Kumar, Chandrakala, *et al.* (2005) *Antimicrob. Agents Chemother.* **49**, 1410. **15**. Liu, Ritter, Sadamoto, *et al.* (2003) *ChemBioChem* **4**, 603.

Vandetanib

This PKI and anti-cancer agent (FW = 475.36 g/mol; CAS 443913-73-3; Code Name: ZD6474; IUPAC Name: *N*-(4-bromo-2-fluorophenyl)-6-methoxy-7-[(1-methylpiperidin-4-yl)methoxy]quinazolin-4-amine; Trade Name: Caprelsa®) targets the intrinsic protein tyrosine kinase activity of Vascular Endothelial Growth Factor Receptor-2 (IC$_{50}$ = 40 nM for VEGFR2 *versus* IC$_{50}$ = 110 nM for VEGFR3); Epidermal Growth Factor Receptor (EGFR3, IC$_{50}$ = 500 nM), and RET-tyrosine kinase. Vandetanib is not an effective inhibitor for PDGFRβ, Flt1, Tie-2 and FGFR1, with IC$_{50}$ values ranging from 1.1 to 3.6 μM. It shows almost no inhibition of MEK, CDK2, c-Kit, erbB2, FAK, PDK1, Akt and IGF-1R, all with IC$_{50}$ values >10 μM. Once-daily oral administration of ZD6474 to growing rats for 14 days produced a dose-dependent increase in the femoro-tibial epiphyseal growth plate zone of hypertrophy, which is consistent with inhibition of VEGF signaling and angiogenesis *in vivo* (1). Administration of 50 mg/kg/day Vandetanib (once-daily, p.o.) to athymic mice with intradermally implanted A549 tumor cells also inhibited tumor-induced neovascularization significantly. Oral administration of Vandetanib to athymic mice bearing established, histologically distinct (*e.g.,* human lung, prostate, breast, ovarian, colon, or vulval) tumor xenografts or after implantation of aggressive syngeneic rodent tumors (*e.g.,* lung, melanoma) in immunocompetent mice, produced a dose-dependent inhibition of tumor growth in *all* cases (1). Significantly, while the human ABCG2 plasma membrane glycoprotein is a xenobiotic transporter that confers multidrug resistance against gefitinib and pelitinib, ABCG2 is without effect on the intracellular action of vandetanib and neratinib (2). Such finding suggest that development of ABCG2 inhibitors might well improve the efficacy of other protein tyrosine kinase inhibitors. Combination of high-dose vandetanib and high-dose docetaxel resulted in antiproliferative effects that were lower than expected from the sum of individual drug effects. This finding further suggests vandetanib should not be used along with docetaxel in treatment-naive or docetaxel-resistant prostate cancer (3). Vandetanib suppresses VEGFR-2 phosphorylation in HUVECs and EGFR phosphorylation in hepatoma cells and inhibited cell proliferation (4). In tumor-bearing mice, vandetanib suppresseds phosphorylation of VEGFR-2 and EGFR in tumor tissues, significantly reducing tumor vessel density, enhancing tumor cell apoptosis, suppressed tumor growth, improved survival, reduced number of intrahepatic metastases, and upregulated VEGF, TGF-α, and EGF in tumor tissues (4). Beyond inhibiting endothelial cell proliferation by blocking VEGF-induced signaling, vandetanib may also be able to inhibit cancer cell growth by blocking EGFR autocrine signaling. Such results provide also a rationale commending the clinical evaluation of vandetanib, especially when combined with taxanes in cancer patients (5). **1**. Wedge, Ogilvie, Dukes, *et al.* (2002) *Cancer Res.* **62**, 4645. **2**. Hegedüs, Truta-Feles, Antalffy, *et al.* (2012) *Biochem. Pharmacol.* **84**, 260. **3**. Guérin, Etienne-Grimaldi, Monteverde, *et al.* (2013) *Urol. Oncol.* **31**, 1567. Inoue, Torimura, Nakamura, *et al.* (2012) *Clin. Cancer. Res.* **18**, 3924. **5**. Ciardiello F, *et al.* (2003) *Clin. Cancer Res.* **9**, 1546.

Vanillate (Vanillic Acid)

This natural product (FW$_{free-acid}$ = 168.15 g/mol; CAS 121-34-6; M.P. = 210°C; highly water-soluble), also known as 4-hydroxy-3-methoxybenzoic acid, inhibits a variety of mechanistically unrelated enzymes. **Target(s):** chorismate lyase (1); α-glucosidase (2); glycogen phosphorylase (2); 4-hydroxybenzoate 4-*O*-β-glucosyltransferase (3); phenol sulfotransferase, *or* aryl sulfotransferase (4); phenylalanine ammonia-lyase (5); protocatechuate 3,4-dioxygenase (6); shikimate dehydrogenase (7). **1**. Holden, Mayhew, Gallagher & Vilker (2002) *Biochim. Biophys. Acta* **1594**, 160. **2**. Li, Lu, Su, *et al.* (2008) *Planta Med.* **74**, 287. **3**. Katsumata, Shige & Ejiri (1989) *Phytochemistry* **28**, 359. **4**. Yeh & Yen (2003) *J. Agric. Food Chem.* **51**, 1474. **5**. Hanson & Havir (1972) *The Enzymes*, 3rd ed. (Boyer, ed.), 7, 75. **6**. Que, Jr., Lipscomb, Münck & Wood (1977) *Biochim. Biophys. Acta* **485**, 60. 7. Balinsky & Dennis (1970) *Meth. Enzymol.* **17A**, 354.

Vanillin

This aromatic aldehyde (FW = 152.15 g/mol; CAS 121-33-5; M.P. = 80-81°C), also known as 4-hydroxy-3-methoxybenzaldehyde, has a solubility of about one gram per 100 mL of water at room temperature. Vanillin is photosensitive and should be stored in light-resistant containers. It is frequently used in analytical biochemistry; for example, in the detection of catecholamines, ornithine, sugar alcohols, and catechins. **Target(s):** 4-aminobutyrate aminotransferase (1); DNA-dependent protein kinase (2); hemoglobin S polymerization, mildly inhibited (3); malonyl-CoA decarboxylase (4); shikimate dehydrogenase (5,6); succinate dehydrogenase (7); and succinate-semialdehyde dehydrogenase (1). **1.** Tao, Yuan, Tang, Xu & Yang (2006) *Bioorg. Med. Chem. Lett.* **16**, 592. **2.** Durant & Karran (2003) *Nucl. Acids Res.* **31**, 5501. **3.** Zaugg, Walder & Klotz (1977) *J. Biol. Chem.* **252**, 8542. **4.** Scholte (1973) *Biochim. Biophys. Acta* **309**, 457. **5.** Balinsky & Dennis (1970) *Meth. Enzymol.* **17A**, 354. **6.** Balinsky & Davies (1961) *Biochem. J.* **80**, 296. **7.** Sen (1931) *Biochem. J.* **25**, 849.

Vanillinoxime, See *4-Hydroxy-3-methoxybenzaldoxime*

3-O-Vanilloylveracevine, See *Veracevine*

Vanillylmandelate and Vanillylmandelic Acid, See *4-Hydroxy-3-methoxymandelate and 4-Hydroxy-3-methoxymandelic Acid*

Vanilmandelate and Vanilmandelic Acid, See *4-Hydroxy-3-methoxymandelate and 4-Hydroxy-3-methoxymandelic Acid*

Vanoxerine

This piperazine-class dopamine reuptake inhibitor (FW$_{free-base}$ = 450.56 g/mol; FW$_{di-HCl}$ = 523.49 g/mol; CAS 67469-78-7), also known as GBR-12909 and systematically as 1-[2-[bis(4-fluorophenyl)methoxy]ethyl]-4-(3-phenylpropyl)piperazine, binds to transporter some 400-500 times higher affinity than cocaine, but also inhibits dopamine release. The net effect is that vanoxerine does not elevate synaptic dopamine levels significantly. Vanoxerine is a potent and selective inhibitor of synaptosomal dopamine uptake (K_i = 1 nM), with a 20-fold lower affinity for the histamine H$_1$-receptor and a more than 100-fold affinity for the noradrenaline and 5-HT uptake carriers, the dopamine D$_1$ receptor, D$_2$ receptor (1). 5-HT$_2$ receptor, 5-HT$_{1A}$ receptor, and α$_1$-receptor as well as voltage-dependent sodium channels. GBR 12909 (3 μM) was without effect on muscarinic, α$_2$, β$_1$ and β$_2$, γ-aminobutyric acid (GABA) and benzodiazepine receptors, and on choline and GABA uptake carriers. The selective dopamine uptake inhibitory profile of GBR 12909 was confirmed by *ex vivo* uptake experiments (1). GBR 12909 inhibited uptake *in vitro* in a competitive manner. GBR 12935 binding was competitively inhibited by GBR 12909 as well as by dopamine, cocaine and methylphenidate. GBR 12909 inhibits dopamine uptake with equal potency in nucleus accumbens and striatum. GBR 12909 is the only compound with this neurochemical profile, making it a valuable experimental tool and antidepressant. *See also 1-(2-[Bis(4-fluorophenyl)methoxy]ethyl)-4-(3-phenylpropyl)piperazine* **1**. Andersen (1989) *Eur. J. Pharmacol.* **166**, 493.

Vardenafil

This purine derivative (FW = 474.59 g/mol; CAS: 224785-90-4), known commercially as Levitra™ and systematically as 2-[2-ethoxy-5-(4-ethylpiperazine-1-sulfonyl)phenyl]-5-methyl-7-propyl-3*H*-imidazo[5,1-*f*][1,2,4]triazin-4-one, strongly inhibits 3',5'-cyclic-GMP phospho-diesterase, *or* cGMP phosphodiesterase, with greatest affinity for the Type-5 isozyme (1-7); cGMP phosphodiesterase Type-6 (6); cGMP phosphodiesterase Type-9 (6); cGMP phosphodiesterase Type-11, slightly inhibited (6,7). **Key Pharmacokinetic Parameters:** *See* Appendix II in Goodman & Gilman's THE PHARMACOLOGICAL BASIS OF THERAPEUTICS, 12th Edition (Brunton, Chabner & Knollmann, eds.) McGraw-Hill Medical, New York (2011). **1.** Sung, Hwang, Jeon, *et al.* (2003) *Nature* **425**, 98. **2.** Porst, Rosen, Padma-Nathan, *et al.* (2001) *Int. J. Impot. Res.* **13**, 192. **3.** Kim, Huang, Goldstein, Bischoff & Traish (2001) *Life Sci.* **69**, 2249. **4.** Saenz de Tejada, Angulo, Cuevas, *et al.* (2001) *Int. J. Impot. Res.* **13**, 282. **5.** Blount, Beasley, Zoraghi, *et al.* (2004) *Mol. Pharmacol.* **66**, 144. **6.** Bischoff (2004) *Int. J. Imp. Res.* **16**, S11. **7.** Weeks, Zoraghi, Beasley, *et al.* (2005) *Int. J. Imp. Res.* **17**, 5.

Varenicline

This smoking-cessation drug (FW = 211.27 g/mol; CAS 249296-44-4; 375815-87-5), known by the trade name Chantix® and systematic name 6,7,8,9-tetrahydro-6,10-methano-6*H* pyrazino[2,3-*h*][3]benzazepine, is a α$_4$β$_2$ type-specific partial agonist of nicotinic receptors (EC$_{50}$ = 2.3 μM), with an efficacy of 13.4 %, relative to acetylcholine (1-3). Whereas varenicline is a partial agonist at other heteromeric neuronal nicotinic receptors (*e.g.*, EC$_{50}$ = 55 μM; Efficacy = 75 %, relative to acetylcholine for α$_3$β$_4$ nicotinic receptors), it is a full agonist at the homomeric α$_7$ receptor (EC$_{50}$ = 18 μM; Efficacy = 93 %, relative to acetylcholine). Varenicline also seems to be a weak partial agonist of α$_3$β$_2$ and α$_6$-containing receptors, with an efficacy <10%. Varenicline can be considered as a structural analogue of cytisine (**See** *Cytisine*). **Rationale:** Repeated exposure to nicotine develops neuroadaptation, resulting in tolerance to many of the effects of nicotine. A confounding behavior of nicotine's reinforcing effects is its stimulation of brain nicotinic receptors, especially α$_4$β$_2$ receptors, with release of dopamine in the meso-limbic area. Nicotine abstinence reduces dopamine release, accompanied by withdrawal symptoms and nicotine craving. By blocking nicotine receptors and simultaneously stimulating dopamine, varenicline is the first drug to modulate both nicotinic receptor action and dopamine release. Involvement of presynaptic α$_4$- and α$_6$-containing receptors in regulating dopamine release in the striatum makes the partial agonism of varenicline at these receptors an attractive potential mechanism (2). **Key Pharmacokinetic Parameters:** *See* Appendix II in Goodman & Gilman's THE PHARMACOLOGICAL BASIS OF THERAPEUTICS, 12th Edition (Brunton, Chabner & Knollmann, eds.) McGraw-Hill Medical, New York (2011). **1.** Coe, Brooks, Vetelino, *et al.* (2005) *J. Med. Chem.* **48**, 3474. **2.** Coe, Vetelino, Bashore, *et al.* (2005) *Bioorg. Med. Chem. Lett.* **15**, 2974. **3.** Mihalak, Carroll & Luetje (2006) *Mol. Pharmacol.* **70**, 801.

Varespladib

Varespladib   Varespladib Methyl

This intravenously administered anti-inflammatory agent and experimental drug (FW = 380.40 g/mol; CAS 172732-68-2; Solubility: 75 mg/mL DMSO, <1 mg/mL H$_2$O), also known as A-001, previously LY315920, S-5920, and 2-(3-(2-amino-2-oxoacetyl)-1-benzyl-2-ethyl-1*H*-indol-4-yloxy)acetic acid, targets IIa, V, and X isoforms of secretory phospholipase A$_2$ (sPLA$_2$) and human non-secretory phospholipase A$_2$, *or* hnsPLA (IC$_{50}$ = 7 nM), thereby inhibiting a key first step of the arachidonic acid pathway needed to trigger inflammation in atherosclerotic disease. The orally bioavailable, prodrug varespladib methyl (FW = 394.43 g/mol; CAS 172733-08-3) is rapidly metabolized to varespladib; both compounds potently inhibit human secretory phospholipase groups IIa, V and X (2). In patients with recent Acute Coronary Syndrome, *or* ACS (*i.e.,* sudden blockage of blood flow to heart muscles), varespladib did not reduce the risk of recurrent cardiovascular events and actuallu increased significantly the risk of myocardial infarction (3). When administered at a once-daily dose of 500 mg (in addition to atorvastatin and other established therapies), varespladib does reduce adverse cardiovascular outcomes after ACS and may even prove to be harmful. **1.** Snyder, Bach, Dillard, *et al.* (1999) *J. Pharmacol. Exp. Ther.* **288**, 1117. **2.** Rosenson, Fraser, Goulder & Hislop (2011) *Cardiovasc. Drugs Ther.* **25**, 539. **3.** Nicholls, Kastelein, Schwartz, *et al.* (2014) *JAMA* **311**, 252.

Vasopressin

Cys-Tyr-Phe-Gln-Asn-Cys-Pro-Arg-Gly-NH₂

[Arg⁸]-Vasopressin

This peptide amide hormone (FW = 1083.22 g/mol; CAS 9034-50-8), also known as antidiuretic hormone (ADH), which contains nine aminoacyl residues with an internal disulfide bridge and an amide at the C-terminal amide, regulates water balance by exerting antidiuretic action, and by inducing aeriole contraction (1,2). **Target(s):** carboxypeptidase E (2); hydroxyindole *O*-methyltransferase (3); serotonin *N*-acetyltransferase (4). **1.** Alescio-Lautier & Soumireu-Mourat (1998) *Prog. Brain Res.*, 501. **2.** Hook & LaGamma (1987) *J. Biol. Chem.* **262**, 12583. **3.** Sugden & Klein (1987) *J. Biol. Chem.* **262**, 6489. **4.** Klein & Namboodiri (1982) *Trends Biochem. Sci.* **7**, 98.

[deamino-Cys1,D-Arg8]-Vasopressin, *See Desmopressin*

Vasotocin

Cys-Tyr-Ile-Gln-Asn-Cys-Pro-Arg-Gly-NH₂

[Arg⁸]-Vasotocin

This vasoactive peptide amide hormone (MW = 1108.25 g/mol; CAS 113-80-4), which is closely related to vasopressin and oxytocin, also contains nine aminoacyl residues with an internal disulfide bridge as well as a C-terminal amide, also inhibits adenylate cyclase (1,2); androgen biosynthesis (3); arylalkylamine *N*-acetyltransferase, serotonin *N*-acetyltransferase (4,5); and hydroxyindole *O*-methyltransferase (6). **1.** Ammar, Roseau & Butlen (1995) *Gen. Comp. Endocrinol.* **98**, 102. **2.** Guibbolini & Lahlou (1992) *Peptides* **13**, 865. **3.** Adashi & Hsueh (1982) *J. Biol. Chem.* **257**, 1301. **4.** Klein & Namboodiri (1982) *Trends Biochem. Sci.* **7**, 98. **5.** Namboodiri, Weller & Klein (1980) *J. Biol. Chem.* **255**, 6032. **6.** Sugden & Klein (1987) *J. Biol. Chem.* **262**, 6489.

Vatalanib

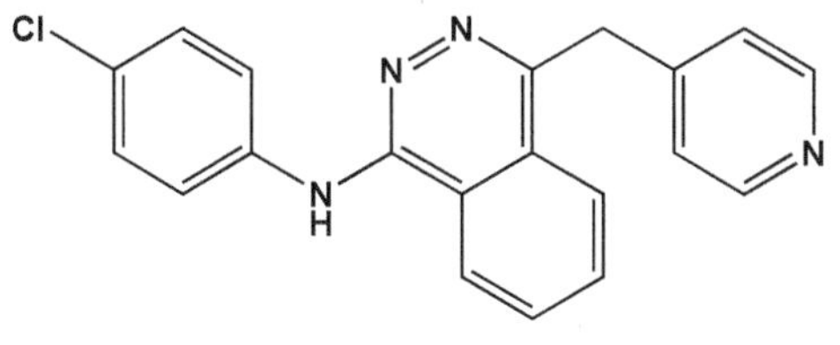

This orally available TKI (FW = 346.82 g/mol; CAS 212142-18-2; Soluble to 100 mM in DMSO), also known as CGP 79787D, ZK 222584, PTK787, and 1-[4 chloroanilino]-4-[4-pyridylmethyl]-phthalazine succinate, blocks phosphorylation by vascular endothelial growth factor receptors (IC₅₀ = 77 nM for VEGFR-1; IC₅₀ = 37 nM for VEGFR-2) and is also active against platelet-derived growth factor receptor β protein-tyrosine kinase, c-Kit, and c-Fms at higher concentrations. Vatalanib also inhibits proliferation, migration and survival of HUVECs *in vitro* as well as the growth, vascularization and metastasis of VEGFR-expressing tumors in mouse xenograft models. It is also a potent aromatase inhibitor (IC₅₀ = 50 nM). **Target(s):** platelet-derived growth factor receptor protein-tyrosine kinase (1,2); receptor protein-tyrosine kinase (1-3); and vascular endothelial cell growth factor (VEGF) receptor protein-tyrosine kinase (1-3). **1.** Ozaki, Seo, Ozaki, Yamada, Yamada, Okamoto, Hofmann, Wood & Campochiaro (2000) *Amer. J. Pathol.* **156**, 697. **2.** Wood, Bold, Buchdunger, Cozens, *et al.* (2000) *Cancer Res.* **60**, 2178. **3**. Medinger & Drevs (2005) *Curr. Pharm. Des.* **11**, 1139.

VDM-11

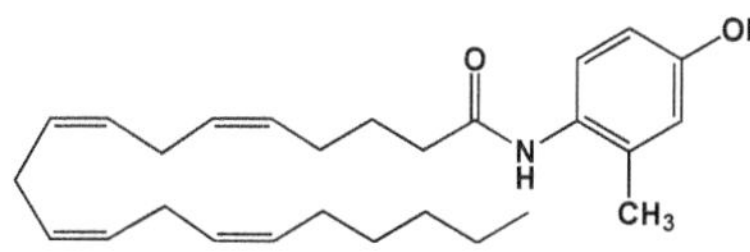

This potent AMT inhibitor (FW = 409.61 g/mol; CAS 313998-81-1; most often dissolved in Ethanol or as an emulsion in Tocrisolve™ 100 (Tocris Biosciences Cat. No. 1686)), also named (5Z,8Z,11Z,14Z)-N-(4-hydroxy-2-methylphenyl)-5,8,11,14-eicosatetraenamide, selectively targets the <u>A</u>nandamide (*N*-arachidonoylethanolamine) <u>M</u>embrane <u>T</u>ransporter (IC₅₀ = 4-11 μM), displaying negligible activity at the human vanilloid (capsaicin), *or* hVR₁, receptor. VDM-11 also shows very weak agonist action at CB₁ and CB₂ receptors, with K_i values are greater than 5-10 μM. This anandamide membrane transporter inhibitor can be used to distinguish between CB₁ or VR₁ receptor-mediated actions of anandamide (2). **1.** De Petrocellis, Bisogno, Davis, Pertwee & Di Marzo (2000) *FEBS Lett.* **483**, 52. **2.** De Petrocellis, Bisogno, Maccarrone, *et al.* (2001) *J. Biol. Chem.* **276**, 12856

VE

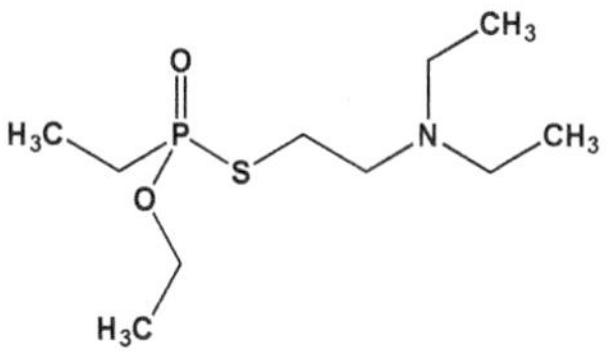

This synthetic acetylcholinesterase inhibitor (FW₍free-base₎ = 253.35 g/mol), also named *O*-ethyl *S*-(2-diethylamino)ethyl ethylphosphonothioate, is a second-generation chemical warfare (CW) agent. **Caution:** Extreme care must be exercised to avoid contact or inhalation. A droplet contacting the skin is sufficient to kill most humans. Ellison (2000) *Handbook of Chemical and Biological Warfare Agents*, CRC Press, Boca Raton.

VE-821

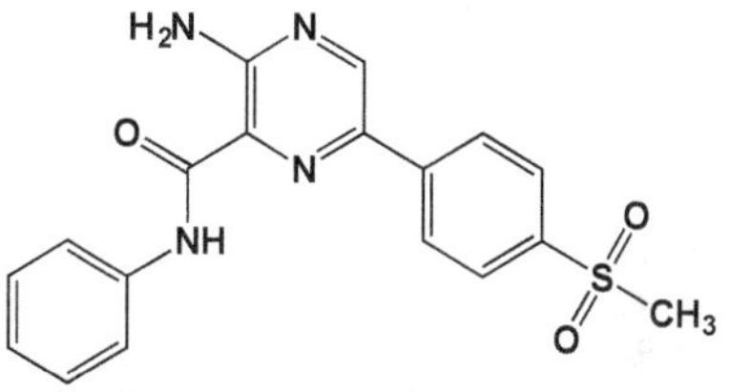

This potent and selective ATP-competitive inhibitor (FW = 368.41 g/mol; CAS 1232410-49-9), also named 3-amino-6-[4-(methylsulfonyl)phenyl]-*N*-phenyl- 2-pyrazinecarboxamide, targets the <u>A</u>taxia <u>T</u>elangiectasia and <u>Rad3</u>-related protein kinase, *or* ATR (K_i = 13 nM; IC₅₀ = 26 nM), inhibiting the phosphorylation of histon H2AX (*or* H2A histone family, member X), with inhibitory action against PIKKs ATM, DNA-PK, mTOR and PI3Kγ. The differential responses of cancer and normal cells highlights the great potential for ATR as a novel target for dramatically increasing the efficacy of many established drugs and ionizing radiation. **1.** Reaper, Griffiths, Long, *et al.* (2011) *Nature Chem. Biol.* **7**, 428. **2.** Charrier, Durrant & Golec, *et al.* (2011) *J. Med. Chem.* **54**, 2320.

VE-822

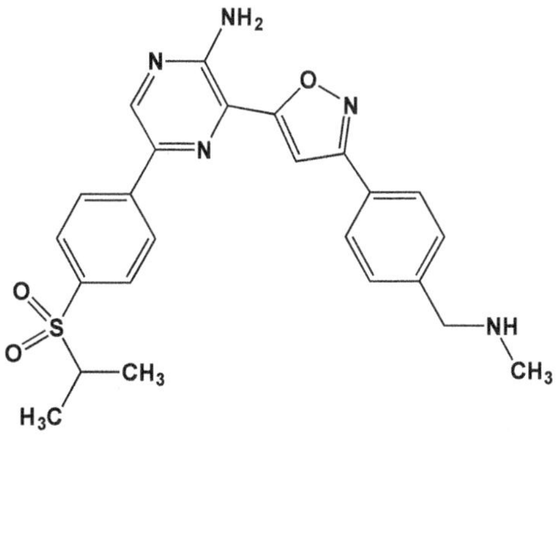

This novel potent inhibitor (FW = 463.56 g/mol; CAS 1232416-25-9; Solubility: 36 mg/mL DMSO; <1 mg/mL H_2O or Ethanol), also named 3-[3-[4-[(methylamino)methyl]phenyl]-5-isoxazolyl]-5-[4-[(1-methylethyl)-sulfonyl]phenyl]-2-pyrazinamine, selectively targets ATR kinase, a serine/threonine-protein kinase also known as ataxia telangiectasia and Rad3-related protein (or ATR) kinase with IC_{50} of 19 nM. VE-822 decreases cell-cycle checkpoint maintenance, increases persistent DNA damage, and decreases homologous recombination in irradiated cancer cells (1). VE-822 decreases survival of pancreatic cancer cells but not normal cells in response to XRT or gemcitabine. VE-822 markedly prolongs growth delay of pancreatic cancer xenografts after XRT and gemcitabine-based chemoradiation without augmenting normal cell or tissue toxicity (1). 1. Fokas, Prevo, Pollard, et al. (2012) Cell Death Dis. **3**, e441. **2**. Fokas, Prevo, Hammond, et al. (2013) Cancer Treat. Rev. S0305-7372.

VE-13045, *See N-(Benzyloxycarbonyl)-L-valyl-L-alanyl-L-aspartic(O-ethyl ester) 2,6-Dichlorobenzoyl-oxymethyl Ketone*

Vecuronium Bromide

This well known paralyzing agent (FW = 637.74 g/mol; CAS 50700-72-6), also known as Norcuron® and 1-[(2β,3α,5α,16β,17β)-3,17-bis(acetyloxy)-2-(1-piperidinyl) androstan-16-yl]-1-methylpiperidinium bromide, is a muscle blocker that competitively inhibits cholinesterase (1,2) as well as histamine *N*-methyltransferase (3,4). By competing for the choline receptors on the motor end-plate, vercuronium bromide's muscle-relaxing properties can be used to aid general anesthetics. **See also** *Tubocurarine* 1. Whittaker & Britten (1980) Clin. Chim. Acta **108**, 89. **2**. Popovic & Kunec-Vajic (1990) Lijec Vjesn. **112**, 142. **3**. Futo, Kupferberg & Moss (1990) Biochem. Pharmacol. **39**, 415. **4**. Futo, Kupferberg, Moss, et al. (1988) Anesthesiology **69**, 92.

Vegolysen, *See Hexamethonium Bromide*

Veliparib

This orally active and brain-permeable PARP inhibitor and anticancer drug (FW = 244.39 g/mol; CAS 912444-00-9), also known as ABT 888 and (*R*)-2-(2-methylpyrrolidin-2-yl)-1*H*-benzo[*d*]imidazole-4-carboxamide, targets poly(ADP-ribose) polymerase 1 and 2 (PARP1, K_i = 5.2 nM; PARP2, K_i = 2.9 nM), thereby potentiating the DNA-damaging effects of anticancer treatments using temozolomide, platinums, cyclophosphamide, as well as radiation in syngeneic and xenograft tumor models (**See** *Temozolomide*). Veliparib is inactive in SIRT2. Under acute hypoxia, it radiosensitizes malignant cells to a level similar to oxic radiosensitivity, demonstrating that inhibition of PARP activity can sensitize hypoxic cancer cells and the combination of IR-PARPi has the potential to improve the therapeutic ratio of radiotherapy (2). 1. Donawho, Luo, Luo, et al. (2007) Clin. Cancer Res. **13**, 2728. **2**. Liu, Coackley, Krause, et al. (2008) Radiother. Oncol. **88**, 258.

Velnacrine

This acetylcholinesterase inhibitor (FW = 214.3 g/mol; CAS 321-64-2), also known as 1-hydroxytacrine, has been used to treat Alzheimer's disease by increasing the strength and duration of acetylcholine that accumulates in the synaptic groove. There is compelling evidence for the selective degeneration in AD of acetylcholine-releasing neurons, the cell bodies of which lie in the basal forebrain and provide widespread innervation of the cerebral cortex and related structures affecting cognitive functions, especially memory. **Target(s):** acetylcholinesterase and cholinesterase (1-3); cytochrome P450 (4); and histamine methyltransferase (5). 1. Ebmeier, Hunter, Curran, et al. (1992) Psychopharmacol. **108**, 103. **2**. Wesnes, Simpson, White, et al. (1991) Ann. N. Y. Acad. Sci. **640**, 268. **3**. Shutske, Pierrat, Kapples, et al. (1989) J. Med. Chem. **32**, 1805. **4**. Eccles, Danbury, Ford & Roberts (1997) Eur. J. Drug Metab. Pharmacokinet. **22**, 121. **5**. Prell, Morrishow, Duoyon & Lee (1997) J. Neurochem. **68**, 142.

Velsicol 1068, *See Chlordan*

Vemurafenib

This first-in-class B-Raf protein kinase inhibitor (FW = 489.92; CAS 1029872-54-5), known as PLX-4032, RG7204, and Zelboraf®, as well as its systematic name *N*-(3-{[5-(4-chlorophenyl)-1*H*-pyrrolo[2,3-*b*]pyridin-3-yl]carbonyl}-2,4-difluorophenyl)propane-1-sulfonamide, is the FDA- and European Commission-approved monotherapy for unresectable metastatic melanoma in adult BRAF^{V600E}-positive patients (1-2). The BRAF^{V600E} mutation leads to constitutive activation of the MAPK signaling pathway. Vemurafenib actually stimulates cells possessing the wild-type B-Raf protein kinase, thereby activating the ERK pathway and enhancing cell migration and proliferation of BRAF^{WT} melanoma cells (3). **Vemurafenib Resistance:** Most BRAF^{V600E}-positive melanoma patients treated with vemurafenib initially show good clinical responses, but eventually relapse from acquired resistance (4). *In vitro* induction of vemurafenib resistance in melanoma cells is associated with an increased malignancy phenotype, and such cells show a higher level of expression of genes coding for cancer stem cell markers (including JARID1B, CD271 and Fibronectin), genes involved in drug resistance (ABCG2), cell invasion, and genes associated with the promotion of metastasis (MMP-1 and MMP-2) (4). Drug-resistant melanoma cells also adhere better to and transmigrate more efficiently through lung endothelial cells than drug-sensitive cells. They also alter their microenvironment in a different manner from that of drug-sensitive cells. Among the established mechanisms of vemurafenib resistance are: (a) overexpression of β-type platelet-derived growth factor receptor, thereby affording an alternative survival pathway; (b) mutations in the NRAS oncogene that reactivate the normal BRAF survival pathway; (c) enhanced secretion or up-regulation of RTK ligands (*e.g.*, Hepatocyte Growth Factor (HGF), Epidermal Growth Factor (EGF), Fibroblast Growth Factor (FGF), Platelet-Derived Growth Factor (PDGF), Neuregulin-1 (NRG1) and Insulin-like Growth Factor (IGF)) that are widely expressed in tumors. In a papillary thyroid carcinoma model, long-term vemurafenib treatment drives inhibitor resistance through a spontaneous KRAS^{G12D} mutation (5) in subpopulation of KTC1 cells after five months of treatmen. Increases in activated AKT, ERK1/2, and EGFR are also observed in these cells. Resistant cells are less sensitive to combinations. *Note*: Vemurafenib is the first FDA-approved drug designed by fragment-based lead approaches. 1. Halaban, Zhang, Bacchiocchi, et. al. (2010) Pigment Cell Melanoma Res. **23**, 190 (*Erratum in*: **25**, 402). **2**. Smalley (2010) Curr. Opin. Investig. Drugs **11**, 699. **3**. Hatzivassiliou, Song, Yen, et. al. (2010) Nature **464**, 431. **4**. Zubrilov, Sagi-Assif, Izraely, et al. (2015) Cancer Lett. **361**, 86. **5**. Danysh, Rieger, Sinha, et al. (2016) Oncotarget **7**, 30907.

Venetoclax

This highly potent, orally bioavailable small molecule (FW = 868.44 g/mol; CAS 1257044-40-8; Solubility: 100 mg/mL DMSO; <1 mg/mL H_2O), also known by the trade name Venclexta® and codenames ABT-199, GDC-0199, and RG760, mimics BH3-only proteins, thereby inhibiting antiapoptotic members of the Bcl-2 family. Venetoclax received FDA Breakthrough Drug Status to treat Chronic Lymphocytic Leukemia (CLL) patients who have either relapsed or have been refractory to previous treatment (most often fludarabine) and have the 17p deletion genetic mutation (1). (*Note*: Drug resistance in CLL patients with the 17p deletion is most likely caused by the inactivation of p53 by mutation of the remaining *TP53* allele.) (*Note*: BCL-2 family proteins (sharing at least one of the four domains of BCL-2 homology: BH1–BH4) regulate the balance between cell life and death by facilitating mitochondria-to-cytoplasm transfer of apoptogenic factors such as Cytochrome *c*. Prosurvival members of the family (*e.g.*, BCL-2, BCL-X$_L$, BCL-W, MCL-1, and A1) sequester BAX and BAK, the pro-apoptotic members that act as "executioner" molecules. All prosurvival proteins sequester BAX, whereas only BCL-X$_L$ and MCL-1 bind to BAK. This anti-apoptotic activity is antagonized by the BH3-only members (e.g., NOXA, BAD and BIM), so named because they only possess the BH3 domain). These proteins insert their α-helical BH3 domain into the hydrophobic groove of prosurvival proteins, displacement BAX and BAK.) Although navitoclax is a selective inhibitor of BCL-2 and BCL-2-like-1, thrombocytopenia limits its efficacy. Venetoclax was therefore developed by re-engineering navitoclax into a BCL-2-selective inhibitor that restores apoptosis by binding directly to the BCL-2 protein, displacing pro-apoptotic proteins like BIM, triggering mitochondrial outer membrane permeabilization, and activating caspases, thereby inhibiting the growth of BCL-2-dependent tumors *in vivo*, while sparing platelets (1). Venetoclax targets Bcl-2 (K_i < 0.01 nM), with a >4800x selectivity over Bcl-xL and Bcl-w, but showing no activity toward Mcl-1. A single dose of ABT-199 in three patients with refractory chronic lymphocytic leukemia resulted in tumor lysis within 24 hours. Although treatment of CLL patients with navitoclax (ABT-263) is complicated by thrombocytopenia as a consequence of BCL-XL inhibition, venetoclax's enhanced specificity avoids this problem and also overcomes stroma-mediated resistance to apoptosis (2). **1**. Souers, Leverson, Boghaert, *et al.* (2013) *Nature Med.* **19**, 202. **2**. Davids, Letai & Brown (2013) *Leuk. Lymphoma* **54**, 1823. **IUPAC Name:** 4-(4-{[2-(4-Chlorophenyl)-4,4-dimethyl-1-cyclohexen-1-yl]-methyl}-1-piperazinyl)-*N*-({3-nitro-4-[(tetrahydro-2*H*-pyran-4-ylmethyl)-amino]-phenyl}-sulfonyl)-2-(1*H*-pyrrolo[2,3-*b*]-pyridin-5-yloxy)-benzamide

Venlafaxine

This SNRI-type antidepressant (FW$_{\text{free-base}}$ = 277.41 g/mol; CAS 99300-78-4, hydrochloride), also known by trade names Effexor® or Efexor® and its IUPAC name (*RS*)-1-[2-dimethylamino-1-(4-methoxyphenyl)ethyl]-cyclo-hexanol, selectively inhibits reuptake of serotonin, norepinephrine, and, to a lesser extent, dopamine. In animal models, venlafaxine does not significantly inhibit muscarinic, histaminic, or adrenergic receptor activity and does not inhibit monoamine oxidase. Venlafaxine is rapidly absorbed and metabolized in the liver to its active metabolite, *O*-

desmethylvenlafaxine (ODV). Time-to-peak concentration is 1-2 h for Venlafaxine and 4-5 hours for ODV. Venlafaxine and its major active metabolite, *O*-desmethylvenlafaxine, exhibit linear kinetics with respective elimination half-lives of 5 and 11 hours. **Target(s):** CYP2D6, *or* cytochrome P450 2D6 (1,2); monoamine uptake (3); nicotinic acetylcholine receptor (4); and the sodium channel (5). **1**. Alfaro, Lam, Simpson & Ereshefsky (2000) *J. Clin. Pharmacol.* **40**, 58. **2**. Ball, Ahern, Scatina & Kao (1997) *Brit. J. Clin. Pharmacol.* **43**, 619. **3**. Muth, Haskins, Moyer, *et al.* (1986) *Biochem. Pharmacol.* **35**, 4493. **4**. Fryer & Lukas (1999) *J. Neurochem.* **72**, 1117. **5**. Khalifa, Daleau & Turgeon (1999) *J. Pharmacol. Exp. Ther.* **291**, 280.

Venturicidins

Venturicidin A, R = NH$_2$CO; Venturicidin B, R = H

These antifungal antibiotics (Venturicidin A: FW = 749.98 g/mol, CAS 33538-71-5; and Venturicidin B: FW = 706.96 g/mol, CAS 33538-72-6), from *Streptomyces aureofaciens*, inhibit ATP synthase (K_i = 2-5 nM), binding to the F$_O$ subunit and therefore acting as a partial inhibitor. Venturicidin appears to alter k_{cat}, allowing ATP hydrolysis and synthesis to proceed, albeit at a much lower rate (1). It specifically blocks proton translocation through F$_O$, binding at the interface of subunit *a* and *c*-ring oligomer in a manner like oligomycin. Its action is not limited to mitochondrial ATP synthase, and it is also effective against bacterial and chloroplast ATP synthases as well as Na$^+$-translocating ATP synthases. When F$_O$-F$_1$ coupling is good, venturicidin blocks the F$_1$ activity, making its action a test of F$_O$F$_1$ coupling efficiency. Venturicidin may be stored as a concentrated stock solution for a long time without loss of its inhibitory power. (**For comparison of Modes of Action, See** Oligomycin, Tentoxin, Azide, DCCD, Efrapeptin, AlF$_4$-, *or* Fluoro-aluminate) **Target(s):** ATP synthase, *or* F$_1$F$_O$ ATPase, *or* H$^+$-transporting two-sector ATPase (1-6); H$^+$-transporting ATPase (7); Na$^+$-transporting two-sector ATPase (8). **1**. Matsuno-Yagi & Hatefi (1993) *J. Biol. Chem.* **268**, 6168. **2**. Stiggall, Galante & Hatefi (1979) *Meth. Enzymol.* **55**, 308. **3**. Soper & Pedersen (1979) *Meth. Enzymol.* **55**, 328. **4**. Linnett & Beechey (1979) *Meth. Enzymol.* **55**, 472. **5**. McEnery & Pedersen (1986) *Meth. Enzymol.* **126**, 470. **6**. Criddle, Johnston & Stack (1979) *Curr. Top. Bioenerg.* **9**, 89. **7**. Perlin, Latchney & Senior (1985) *Biochim. Biophys. Acta* **807**, 238. **8**. Müller, Aufurth & Rahlfs (2001) *Biochim. Biophys. Acta* **1505**, 108.

VER 155008

This ATP-competitive chaperonin inhibitor (FW = 556.40 g/mol; CAS 1134156-31-2), also known as 5'-*O*-[(4-cyanophenyl)methyl]-8-[[(3,4-dichlorophenyl)methyl]amino]adenosine, targets Heat Shock Protein 70, *or* Hsp70 (IC$_{50}$ = 0.5 μM), impairing the folding of client proteins and inhibiting the *in vitro* cell proliferation of multiple human tumor cell lines (1-3). Grp78, the 78-kDa glucose-regulated HSP binds VER-155008, K_d = 60 nM (3) Phenotypically, Hsc70/Hsp70-interaction inhibitors mimic the cellular mode of action of a small molecule Hsp90 inhibitors, potentiating the apoptotic potential of Hsp90 inhibitors in certain cell lines (2). VER-155008 also binds to Hsc70 and Grp78, displaying low affinity against Hsp90β, IC_{50} >200 μM (1). VER-155008 inhibited the proliferation of human breast and colon cancer cell lines with GI_{50} values in the range 5.3-14.4 μM, and induced Hsp90 client protein degradation in both HCT116 and BT474 cells (2). As a single agent, VER-155008 induced caspase-3/7 dependent apoptosis in BT474 cells and non-caspase dependent cell death in HCT116 cells. VER-155008 potentiated the apoptotic potential of a small molecule Hsp90 inhibitor in HCT116, but not HT29 or MDA-MB-468 cells. *In vivo*, VER-155008 demonstrated rapid metabolism and clearance, along with tumor levels below the predicted pharmacologically active level. **1**. Williamson, Borgognoni, Clay, *et al.* (2009) *J. Med. Chem.* **52**, 1510. **2**. Massey, Williamson, Browne, *et al.* (2009) *Cancer Chemother. Pharmacol.* **66**, 535. **3**. Massey (2010) *J. Med. Chem.* **53**, 7280. **4**. Macias, Williamson, Allen, *et al.* (2011) *J. Med. Chem.* **54**, 4034.

Verapamil

This phenylalkylamine derivative and vasodilator ($FW_{free-base}$ = 454.61 g/mol; $FW_{HCl-salt}$ = 489.16 g/mol; CAS 52-53-9): IUPAC: α-[3-[[2-(3,4-dimethyloxyphenyl)ethyl]methylamino]propyl]-3,4-dimethoxy-α-(1-methylethyl)benzeneacetonitrile, exhibits calcium ion blocking activity by inhibiting the movement of Ca^{2+} across biological membranes. Verapamil enjoys broad use in the treatment of supraventricular tachyarrhythmias as well as for hypertension and control of symptoms in angina pectoris, and cluster headaches. Its action resembles propranolol inasmuch as it reduces the O_2 requirement of the heart and abolishes certain arrhythmias. Unlike propranolol, however, it does not antagonize cardiac β-adrenoreceptors. Verapimil is a P-glycoprotein inhibitor and, as such, is likely to interfere with other drugs that are transported by this multidrug resistance factor. **Key Pharmacokinetic Parameters:** After either intravenous or oral drug administration, plasma concentrations correlate with both electrophysiological and haemodynamic activity, with considerable intra- and intersubject variation in the intensity of its pharmacological effects (1). Its actions depend on the route of administration, and, while verapamil is widely distributed throughout body tissues, drug distribution to target organs and tissues is different for parenteral versus oral administration. The drug is eliminated by hepatic metabolism, with excretion of inactive products in the urine and/or feces. Norverapamil, its *N*-demethylated metabolite, exerts weaker vasodilator effects *in vitro*. Upon intravenous administration, systemic clearance appears to approach liver blood flow. The high hepatic extraction results in low systemic bioavailability (20%) after oral drug administration. Because of the complex pharmacokinetics associated with multiple-dose administration and the variation in individual patient responsiveness to the drug, verapamil must be titrated to achieve a clinical end-point (1). Single-day sustained release verapamil formulations for oral administration are as effective in lowering blood pressure over 24 hours as equivalent doses of conventional verapamil formulations given three times daily (2). *See* Appendix II in Goodman & Gilman's THE PHARMACOLOGICAL BASIS OF THERAPEUTICS, 12th Edition (Brunton, Chabner & Knollmann, eds.) McGraw-Hill Medical, New York (2011). **Target(s):** aldehyde oxidase (3); cadmium-transporting ATPase (4); calcium channel (5); cholestenol Δ-isomerase, *or* sterol $Δ^{7,8}$-isomerase (6); CYP3A (7); nucleoside triphosphate diphosphohydrolase (8); phospholipid-translocating ATPase, *or* flippase (13); protein kinase C (14,15); xenobiotic-transporting ATPase, *or* multidrug-resistance protein (10-13,16,17). **1**. Hamann, Blouin & McAllister (1984) *Clin. Pharmacokinet.* **9**, 26. **2**. McTavish & Sorkin (1989) *Drugs* **38**, 19. **3**. Obach, Huynh, Allen & Beedham (2004) *J. Clin. Pharmacol.* **44**, 7. **4**. Li, Szczypka, Lu, Thiele & Rea (1996) *J. Biol. Chem.* **271**, 6509. **5**. Atlas & Adler (1981) *Proc. Natl. Acad. Sci. U.S.A.* **78**, 1237. **6**. Moebius, Reiter, Bermoser, *et al.* (1998) *Mol. Pharmacol.* **54**, 591. **7**. Kovarik, Beyer, Bizot, Jiang, Allison & Schmouder (2005) *Brit. J. Clin. Pharmacol.* **60**, 434. **8**. Gendron, Latour, Gravel, Wang & Beaudoin (2000) *Biochem. Pharmacol.* **60**, 1959. **9**. Achira, Suzuki, Ito & Sugiyama (1999) *AAPS PharmSci.* **1**, E18. **10**. Sharom, Liu, Romsicki & Lu (1999) *Biochim. Biophys. Acta* **1461**, 327. **9**. Lu, Liu & Sharom (2001) *Eur. J. Biochem.* **268**, 1687. **11**. Wang, Casciano, Clement & Johnson (2001) *Drug Metab. Dispos.* **29**, 1080. **12**. Romsicki & Sharom (2001) *Biochemistry* **40**, 6937.

13. Kikkawa, Minakuchi, Takai & Nishizuka (1983) *Meth. Enzymol.* **99**, 288. **14**. Kikkawa & Nishizuka (1986) *The Enzymes*, 3rd ed. (Boyer & Krebs, eds.), **17**, 167. **14**. Paul, Breuninger & Kruh (1996) *Biochemistry* **35**, 14003. **15**. Van Veen, Venema, Bolhuis, *et al.* (1996) *Proc. Natl. Acad. Sci. U.S.A.* **93**, 10668.

Veratridine

This toxic alkaloid (FW = 673.80 g/mol; CAS 71-62-5), named systematically as: 4,9-epoxycevane-3,4,12,14,16,17,20-heptol 3-(3,4-dimethoxybenzoate), from the seeds of the lily family member *Schoenocaulon officinalis* and the rhizome of *Veratrum album*, causes Na^+ channels to remain open during sustained membrane depolarization by abolishing inactivation. The consequential Na^+ influx, either by itself or by causing a maintained depolarization, leads to many secondary effects such as increasing pump activity, Ca^{2+} influx, and in turn exocytosis. If the membrane is voltage-clamped in the presence of the alkaloid, a lasting depolarizing impulse induces, following the "normal" transient current, another much more slowly developing Na^+ current that reaches a constant level after a few seconds. Repolarization then is followed by an inward tail current that slowly subside. **1**. Ulbricht (1998) *Rev. Physiol. Biochem. Pharmacol.* **133**, 1. **2**. McKinney, Chakraverty & De Weer (1986) *Anal. Biochem.* **153**, 33. **3**. Ulbricht (1998) *Rev. Physiol. Biochem. Pharmacol.* **133**, 1. **4**. Conn (1983) *Meth. Enzymol.* **103**, 401.

Veratrine, *See Cevadine*

Verbenalol, *See Verbenalin*

Verdinexor

This orally bioavailable nuclear export inhibitor (FW = 442.32 g/mol; CAS 1392136-43-4), also known as KPT-335 and (2Z)-3-(3-(3,5-bis(trifluoromethyl)phenyl)-1*H*-1,2,4-triazol-1-yl)-*N*'-(pyridin-2-yl)prop-2-enehydrazide, blocks the activity of XPO1/CRM1, itself one of seven nuclear export proteins, forcing the nuclear retention of key tumor suppressor proteins (TSP) and leading to selective apoptosis of tumor cells. Canine tumor cell lines derived from non-Hodgkin lymphoma, mast cell tumor, melanoma and osteosarcoma exhibited growth inhibition and apoptosis in response to nanomolar concentrations of KPT-335, with IC_{50} values ranging from 2 to 42 nM (1). Nuclear export of influenza virus ribonucleoprotein (vRNP) from infected cells is mediated by Exportin 1 interaction with viral nuclear export protein tethered to vRNP. Verdinexor limits the replication of various strains of influenza A and B viruses, including a pandemic H1N1 influenza virus strain, a highly pathogenic H5N1 avian influenza virus strain, and an emerging H7N9 influenza virus strain (2). Orally administered verdinexor is efficacious in limiting lung virus burdens in influenza virus-infected mice, as well as limiting lung proinflammatory cytokine expression, pathology, and death. **1**. London, Bernabe, Barnard, *et al.* (2014) *PLoS One* **9**, e87585. **2**. Perwitasari, Johnson, Yan, *et al.* (2014) *J. Virol.* **88**, 10228.

Vernamycin A, *See Virginiamycin M1*

Vernolate

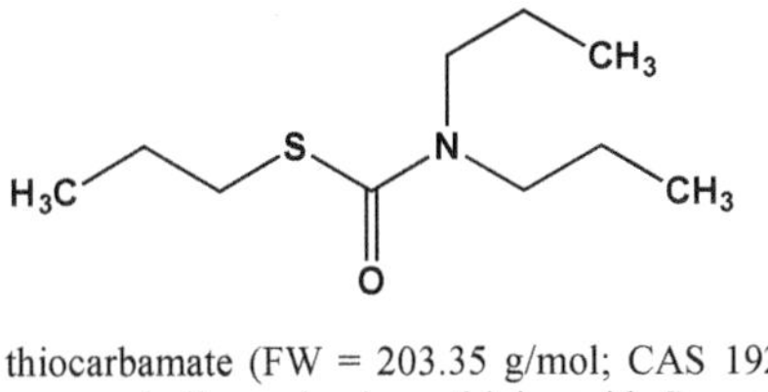

This alkylated thiocarbamate (FW = 203.35 g/mol; CAS 1929-77-7), also known as Vernam and dipropylcarbamothioic acid S-propyl ester, is a selective herbicide developed by the Stauffer Chemical Co. in the late 1950s (1). A liquid at room temperature, vernolate is slightly soluble in water (107 mg/L at 25°C), and readily soluble in organic solvents. Vernolate is toxic to germinating broadleaf and grassy weeds and is frequently used to control weeds in fields of soybeans, peanuts, and sweet potatoes. Care should be exercised when encountering this name in the literature since the term vernolate also refers to the conjugate base of vernolic acid. **Target(s):** aldehyde dehydrogenase (2,3). **1**. Tilles (1959) *J. Amer. Chem. Soc.* **81**, 714. **2**. Hart & Faiman (1995) *Biochem. Pharmacol.* **49**, 157. **3**. Quistad, Sparks & Casida (1994) *Life Sci.* **55**, 1537.

Verrucarin A

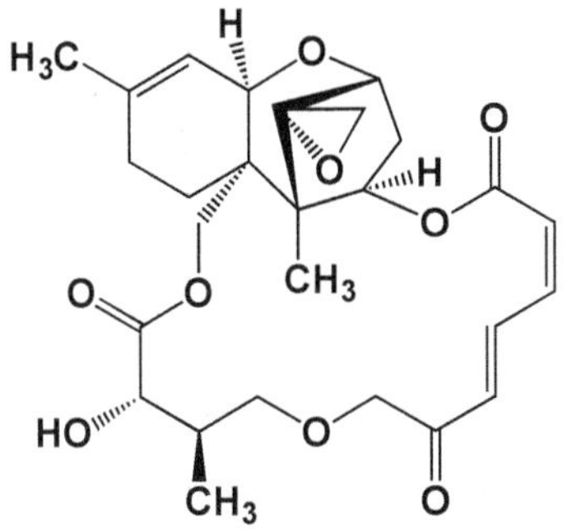

This macrocyclic trichothecene toxin (FW = 502.56 g/mol; CAS 3148-09-2), also called muconomycin A, is produced by the soil fungi *Myrothecium verrucaria*. Verrucarin's phytotoxicity and cytotoxicity in mammalian cell lines and potent antifungal action arise from its inhibition of eukaryotic protein biosynthesis at the initiation stage, interacting with the large ribosomal subunit (1-3). **Caution:** All verrucarins are extremely toxic and should be handled with care. **1**. Hernandez & Cannon (1982) *J. Antibiot.* **35**, 875. **2**. Fresno & Vázquez (1979) *Meth. Enzymol.* **60**, 566. **3**. Jiménez (1976) *Trends Biochem. Sci.* **1**, 28.

Verteporfin

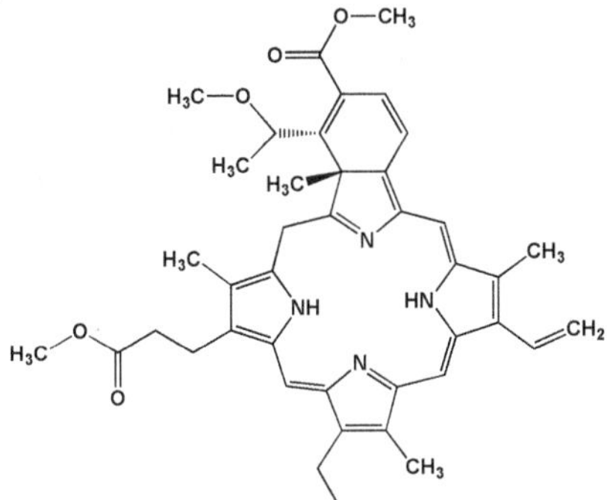

This benzoporphyrin-based photosensitizer (FW = 718.85 g/mol; CAS 129497-78-5), also known by the tradename Visudyne®, is used in the photodynamic therapy (PDT) to ablate abnormal blood vessels within the eye, most often associated with the wet form of macular degeneration. Verteporfin also inhibits drug- and starvation-induced autophagic degradation and the sequestration of cytoplasmic materials into autophagosomes; however, verteporfin does not inhibit LC3/Atg-8 processing or membrane recruitment in response to autophagic stimuli (1). Mitochondrial photodamage occurs at low verteporfin concentrations (≤ 1 µM) in murine hepatoma 1c1c7 cells (2). Significantly, apoptosis was not observed until several hours after irradiation of photosensitized cells.

Moreover, autophagy was cytoprotective, because PDT efficacy was significantly enhanced in a knockdown sub-line, in which the level of a critical autophagy protein (Atg7) was markedly reduced (2). Autophagy thus protects from phototoxicity, even when apoptosis is substantially delayed. Much higher concentrations (≥ 10 µM) of verteporfin had previously been shown to inhibit autophagosome formation. Other sensitizers (*e.g.*, porphycene (termed CPO) and mesochlorin) respectively localize in the endoplasmic reticulum and mitochondria (3). **1**. Donohue, Tovey, Vogl, *et al.* (2011) *J. Biol. Chem.* **286**, 7290. **2**. Andrzejak, Price & Kessel (2011) *Autophagy* **7**, 979. **3**. Kessel & Reiners (2007) *Photochem. Photobiol.* **83**, 1024.

Vesamicol

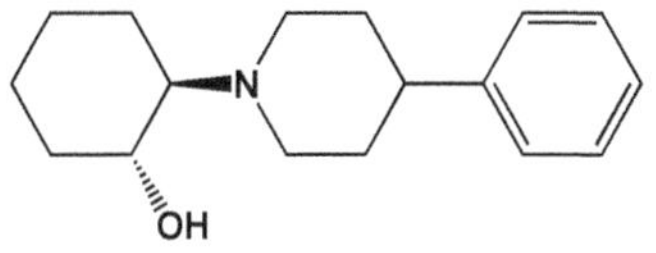

This substituted piperidine (FW = 259.39 g/mol; CAS 22232-64-0), also known as 2-(4-phenylpiperidino)cyclohexanol, allosterically inhibits the vesicular acetylcholine transporter (VAChT; TC 2.A.1.2.13), IC$_{50}$ ~50 nM, but is not itself transported (1). Its binding site is fundamentally different from the ACh binding site(s) on VAChT (2). Vesamicol exhibits monophasic, high-affinity binding characteristic of strong and specific contacts between a single, substantially preformed binding site and its ligand (2). **1**. Prior, Marshall & Parsons (1992) *Gen. Pharmacol.* **23**, 1017. **2**. Khare, Mulakaluri & Parsons (2010) *J. Neurochem.* **115**, 984.

VG, *See* Amiton

VI21497, *See* Merimepodib

Viagra™, *See* Sildenafil

Vibramycin, *See* Doxycycline

Vigabatrin

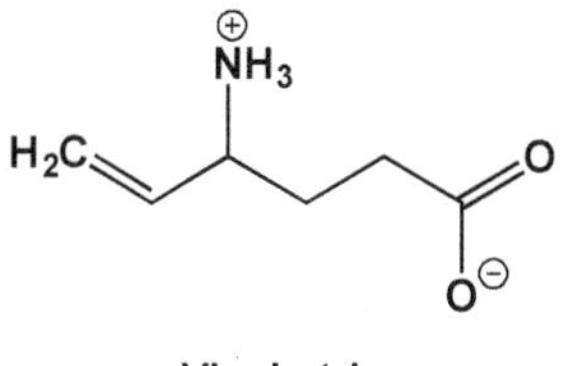

Vigabatrin

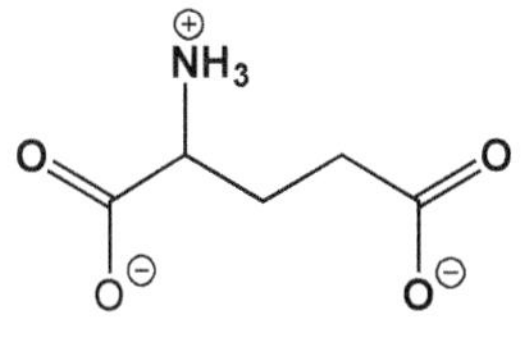

Glutamate

This synthetic γ-amino acid and first-in-class anticonvulsant (FW = 129.15 g/mol; CAS 60643-86-9), also known as vinyl-GABA, CPP-115, Sabril® and 4-amino-5-hexenoic acid, is a glutamate analogue and irreversible Michael-type mechanism-based inactivator of γ-aminobutyrate aminotransferase (1,2). The $S(+)$-stereoisomer is pharmacologically active. (*See also* cis-3-Aminocyclohex-4-ene-1-carboxylate) γ-Aminobutyrate aminotransferase inhibition consequentially increases the level of GABA, a neurotransmitter that inhibits dopamine release. In 2009, FDA approved Sabril as second-line antiepileptic therapy for adult patients with refractory complex partial seizures (CPS) who have already responded inadequately to alternative treatments. **Primary Mode of Action:** Inactivation proceeds by two divergent mechanisms (2). The major pathway involves removal of the γ-proton, tautomerization into the PLP ring, followed by a Michael addition of an active-site lysine residue at the conjugated vinyl group to give a stable covalent adduct with the protein. The minor inactivation mechanism also involves γ-proton loss, but tautomerization occurs through the vinyl group, followed by an enamine rearrangement that leads to attachment of the inactivator to the PLP bound to the protein. The X-ray structure of GABA transaminase complexed to vigabatrin suggests likely steps in the Michael-type addition/inhibition process (9). **Target(s):** 4-aminobutyrate transaminase (1-9); aspartate aminotransferase (8); glutamate 1-semialdehyde 2,1-aminomutase, *or* glutamate-1-semialdehyde aminotransferase (10); hypotaurine aminotransferase (11); and ornithine aminotransferase (8). **1**. Lippert, Metcalf, Jung & Casara (1977) *Eur. J. Biochem.* **74**, 441. **2**. Nanavati & Silverman (1991) *J. Am. Chem. Soc.* **113**, 9341. **3**. Sulaiman, Suliman & Barghouthi (2003) *J. Enzyme Inhib. Med. Chem.* **18**, 297. **4**. Ricci, Frosini, Gaggelli, *et al.* (2006) *Biochem. Pharmacol.* **71**, 1510. **5**. Lu & Silverman (2006) *J. Med. Chem.* **49**, 7404. **6**. Yuan & Silverman (2007) *Bioorg. Med. Chem. Lett.* **17**, 1651. **7**. Storici, De Biase, Bossa, *et al.* (2004) *J. Biol. Chem.* **279**, 363. **8**. John, Jones & Fowler (1979) *Biochem. J.* **177**, 721. **9**. Storici, De Biase, Bossa, *et al.* (2004) *J. Biol. Chem.* **279**, 363. **10**. Nair, Kannangara, Harwood & John (1990) *Biochem. Soc. Trans.* **18**, 656. **11**. Fellman (1987) *Meth. Enzymol.* **143**, 183.

Vilazodone

This serotonergic antidepressant (FW = 441.52 g/mol; CAS 163521-12-8; IUPAC: 5-(4-[4-(5-cyano-1*H*-indol-3-yl)butyl]piperazin-1-yl)benzofuran-2-carboxamide), also known by its codename EMD 68843 and its tradename Viibryd®, is an FDA-approved treatment of major depressive disorder as well as for generalized anxiety disorder and obsessive compulsive disorder. Vilazodone inhibits serotonin reuptake (IC$_{50}$ = 2.1 nM; K_i = 0.1 nM) and 5-HT$_{1A}$ receptor partial agonist (IC$_{50}$ = 0.2 nM; Agonist Efficacy = ~60–70%), with negligible affinity for serotonin 5-HT$_{1D}$, 5-HT$_{2A}$, and 5-HT$_{2C}$ receptors (1,2). Vilazodone also shows negligible inhibitory activity at the norepinephrine transporter, *or* NET (IC$_{50}$ = 56 nM) and dopamine transporter, *or* DAT (IC$_{50}$ = 56 nM). **1**. Page, Cryan, Sullivan, *et al.* (2002) *J. Pharmacol. Exp. Therap.* **302**, 1220. **2**. Hughes, Starr, Langmead, *et al.* (2005) *Europ. J. Pharmacol.* **510**, 49.

Vildagliptin

This oral antihyperglycemic drug (FW = 304.41 g/mol; CAS 274901-16-5), also known as NVP-LAF237 and 1-[[(3-hydroxy-1-adamantyl) amino]acetyl]-2-cyano-(*S*)-pyrrolidine, inhibits dipeptidyl-peptidase IV (IC$_{50}$ = 3.5 nM). By limiting proteolysis of the gastrointestinal hormones, glucagon-like peptides 1 and 2 (GLP-1 and GLP-2), vildagliptin is a so-called incretin enhancer that indirectly maintains insulin secretion and suppresses glucagon release (*See Incretin Hormone Enhancers; Saxagliptin*). Vildagliptin also reduces levels of reactive oxygen species as well as mitochondrial swelling, thereby providing cardioprotection by reducing the infarct size and ameliorating cardiac dysfunction during ischemia-reperfusion (4). **Target(s):** cytosol nonspecific dipeptidase (prolyl dipeptidase)1; dipeptidyl-peptidase IV (1-3). **1**. Jiang, Chen, Hsu, *et al.* (2005) *Bioorg. Med. Chem. Lett.* **15**, 687. **2**. Villhauer, Brinkman, Naderi, *et al.* (2003) *J. Med. Chem.* **46**, 2774. **3**. Brandt, Lambeir, Ketelslegers, *et al.* (2006) *Clin. Chem.* **52**, 82. **4**. Chinda, Sanit, Chattipakorn & Chattipakorn (2013) *Diab. Vasc. Dis. Res.* **11**, 75.

Vinblastine

This indole-indoline alkaloid and microtubule assembly/disassembly modulator (FW = 811 g/mol; CAS 865-21-4) from the periwinkle (*Catharanthus roseus*, formerly *Vinca rosea*) is a potent inhibitor of microtubule assembly and cellular processes depending on an organized microtubule cytoskeleton. Low concentrations (0.2–1 μM) of this antiproliferative agent do not induce microtubule disassembly. The alkaloid instead binds to tubulin to form a high-affinity tubulin·vinblastine complex that is incorporated by limited copolymerization with drug-free tubulin into the ends of assembling microtubules, where it kinetically suppresses microtubule dynamics by modulating the gain and loss of terminally bound tubulin·GTP or tubulin·GDP·P$_i$. By disrupting mitosis, vinblastine also triggers apoptosis in many cell lines, induces p53 expression, stimulates Raf-1 activation, and initiates phosphorylation of bcl-2 family proteins. Mounting evidence suggests that tumor cell apoptosis and checkpoint-linked apoptosis is synergistically enhanced by combined treatments with DNA- and cytoskeleton-damaging agents. When administered at a dose of 6 milligrams per square meter of body surface, vinblastine has a 1-day half-life in the bloodstream. Vinblastine has proven efficacy in the treatment of Hodgkin's disease, lymphocytic lymphoma, histiocytic lymphoma, advanced testicular cancer, advanced breast cancer, Kaposi's sarcoma, and Letterer-Siwe disease. Overexpression of Class III β-tubulin has emerged as a biomarker in a number of cancers, a prognostic indicator of more aggressive disease, and a predictor of resistance to taxanes and *Vinca* alkaloids (6). (*See also Vindesine; Vincristine; Vinorelbine*) **Target(s):** microtubule self-assembly and mitosis (1-6); xenobiotic-transporting ATPase, *or* multidrug-resistance protein (7-10); cadmium-transporting ATPase (11); CYP3A4, *or* cytochrome P450 3A4 (12); thromboxane synthetase (13). **1**. Jordan & Wilson (1998) *Meth. Enzymol.* **298**, 252. **2**. Jordan, Thrower & Wilson (1991) *Cancer Res.* **51**, 2212. **3**. Toso, Jordan, Farrell, Matsumoto & Wilson (1993) *Biochemistry* **32**, 1285. **4**. Scott & Tomkins (1975) *Meth. Enzymol.* **40**, 273. **5**. Rosenshine, Ruschkowski & Finlay (1994) *Meth. Enzymol.* **236**, 467. **6**. Powell, Kaizer, Koopmeiners, Iwamoto & Klein (2014) *Oncol Lett.* **7**, 405. **7**. Shapiro & Ling (1997) *Eur. J. Biochem.* **250**, 130. **8**. Garrigues, Escargueil & Orlowski (2002) *Proc. Natl. Acad. Sci. U.S.A.* **99**, 10347. **9**. Romsicki & Sharom (2001) *Biochemistry* **40**, 6937. **10**. Steinfels, Orelle, Fantino, *et al.* (2004) *Biochemistry* **43**, 7491. **11**. Li, Szczypka, Lu, Thiele & Rea (1996) *J. Biol. Chem.* **271**, 6509. **12**. Baumhakel, Kasel, Rao-Schymanski, *et al.* (2001) *Int. J. Clin. Pharmacol. Ther.* **39**, 517. **13**. Steinfels, Orelle, Fantino, *et al.* (2004) *Biochemistry* **43**, 7491.

Vinc-1 Head Fragment 840-849, Human

This decapeptide PDFPPPPPDL (MW = 1091.23 g/mol) corresponding to a proline-rich register (residues 840-849) of human vinculin inhibits actin-based membrane protrusion, actin-based propulsion of *Listeria monocytogenes* and *Shigella flexneri* (1) by disrupting binding interactions between filament (+)-end-tracking motors and their cargo (2), when microinjected at sub-μM concentrations. The analogous FEFPPPPPTDE register in *Listeria* ActA surface protein displaces human vasodilator-stimulated phosphoprotein (VASP) from FPPPP-containing proteins, suggesting that vinculin PDFPPPPPDL has the analogous mode of action (*See ActA Peptide FEFPPPPPTDE; ABM-1 & ABM-2 Sequences in Actin-Based Motors*). **1**. Laine, Zeile, Kang, Purich & Southwick (1997) *J. Cell Biol.* **138**, 1255. **2**. Purich (2010) *Enzyme Kinetics: Catalysis and Control*, Academic Press, New York.

Vincaleucoblastin, *See* Vinblastine

Vinclozolin

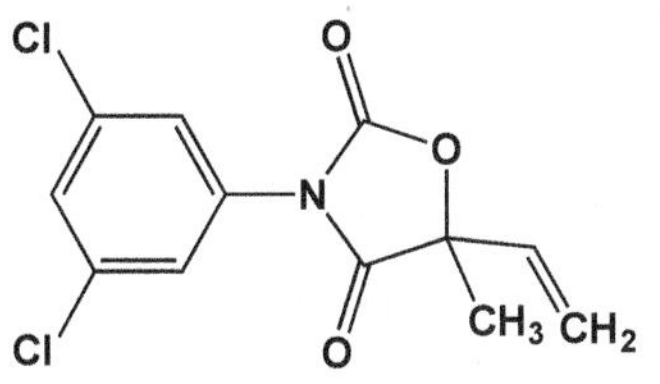

This dicarboximide-class fungicide (FW = 286.11 g/mol; CAS 50471-44-8), known also by its IUPAC name (RS)-3-(3,5-dichlorophenyl)-5-methyl-5-vinyloxazolidine-2,4-dione and the trade names Ronilan®, Curalan®, Vorlan®, and Touche®, is used to control blights, rots and molds in vineyards (especially *Botrytis*) and on raspberries, lettuce, kiwi, snap beans, and onions (1). Its precise mode of action remains unclear. Vinclozolin inhibits mycelial growth much more than spore germination (2).. Although chitin biosynthesis s inhibited by this fungicides, it was barely affected at the ED_{50} concentrations (the concentration required to reduce the growth or germination of the test species by 50%) and is thus unlikely to be the primary target (2). Moreover, it is without effect on respiration, membrane permeability or RNA production. Vinclozolin metabolites M_1 and M_2 are effective antagonists of the androgen receptor in rats, exhibiting K_i values of 92 and 9.7 μM respectively (3). **1.** Pothuluri, Freeman, Heinze, *et al.* (2000) *J. Agric. Food Chem.* **48**, 6138. **2.** Pappas & Fisher (1979) *Pesticide Sci.* **10**, 239. **3.** Kelce, Monosson, Gamcsik, *et al.* (1994) *Toxicol. Appl. Pharmacol.* **126**, 276.

Vincristine

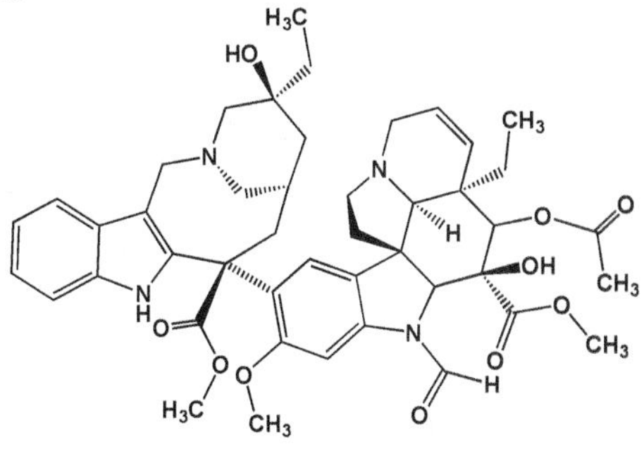

This indole-indoline alkaloid and microtubule assembly/disassembly modulator (FW = 824.97 g/mol; CAS 57-22-7), from the periwinkle (*Catharanthus minor*, formerly *Vinca minor*), is a potent inhibitor of microtubule self-assembly and cellular processes requiring a well-structured and properly organized microtubule cytoskeleton. **Mode of Action:** Vincristine is likely to exert effects that are similar to the antiproliferative action of vinblastine, which at low concentrations result from changes of spindle microtubule dynamics rather than from depolarization of the mitotic microtubules (1-3). (*See Vinblastine for additional comments on mode of action*) Marketed as Oncovin™ by Eli Lilly, vincristine (serum $t_{1/2}$ of ~85 hours) is used to treat acute leukemia, rhabdomyosarcoma, neuroblastoma, Hodgkin's disease and other lymphomas. Typical one-weekly therapeutic dose is 1.4 mg/m² body surface. Because neurons unavoidably require microtubules to maintain their anisometric structure, sensory and neurotransmission functions, and because vincririne penetrates neurons and blocks MT processes, significant peripheral vincristine neuropathy is often dose-limiting. (*See Vinblastine; Vindesine; Vinorelbine*) **Key Pharmacokinetic Parameters:** *See* Appendix II in Goodman & Gilman's THE PHARMACOLOGICAL BASIS OF THERAPEUTICS, 12th Edition (Brunton, Chabner & Knollmann, eds.) McGraw-Hill Medical, New York (2011). **Vincristine Resistance in Multiple Myeloma:** In vincristine-resistant RPMI8226/VCR multiple myeloma cells, multidrug resistance (MDR) involves the overexpression of MDR1 and survivin, exhibiting increased levels of activated ERK1/2, Akt, and NF-κB. By comparison, the levels of activated mTOR, p38MAPK, and JNK do not differ between RPMI8226/VCR cells and their vincristine-susceptible cells (4). Inhibition of ERK1/2, Akt, or NF-κB reverses drug-resistance of RPMI8226/VCR cells via the suppression of survivin expression, but does not affect MDR1 expression. RNA silencing of survivin expression completely reverses vincristine resistance, whereas MDR1 silencing only weakly suppresses vincristine resistance in RPMI8226/VCR cells (4). Such findings indicate that enhanced survivin expression via the activation of ERK1/2, Akt, and NF-κB plays a critical role in vincristine resistance in RPMI8226/VCR cells. **Target(s):**

CYP3A4, *or* cytochrome P450 3A4 (5); lactose synthase (6); thromboxane synthetase (7); xenobiotic-transporting ATPase, *or* multidrug-resistance protein (8). **1.** Jordan, Thrower & Wilson (1991) *Cancer Res.* **51**, 2212. **2.** Jordan, Thrower & Wilson (1992) *J. Cell. Sci.* **102**, 401. **3.** Toso, Jordan, Farrell, Matsumoto & Wilson (1993) *Biochemistry* **32**, 1285. **4.** Tsubaki, Takeda, Ogawa, *et al.* (2015 *Leuk. Res.* **39**, 445. **5.** Baumhakel, Kasel, Rao-Schymanski, *et al.* (2001) *Int. J. Clin. Pharmacol. Ther.* **39**, 517. **6.** West (1985) *Biochem. Soc. Trans.* **13**, 694. **6.** Maguire & Csonka-Khalifah (1987) *Biochim. Biophys. Acta* **921**, 426. **7.** Mao, Deeley & Cole (2000) *J. Biol. Chem.* **275**, 34166. **8.** Loe, Deeley & Cole (1998) *Cancer Res.* **58**, 5130.

Vindesine

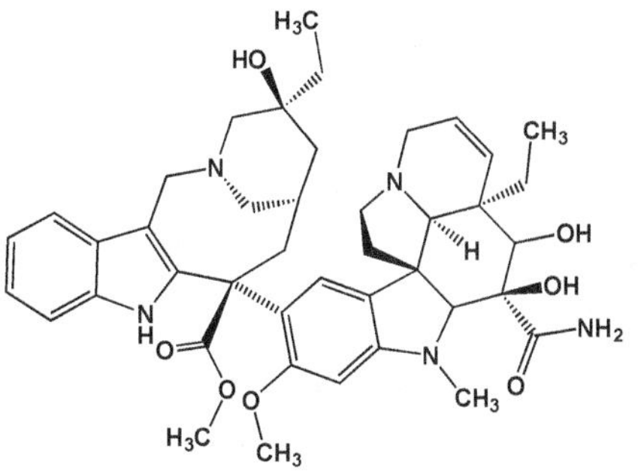

This semi-synthetic vinblastine derivative and microtubule assembly/disassembly modulator (FW = 753.94 g/mol; CAS 59917-39-4), marketed under the names Eldisine™ and Fildesin™, binds rapidly to tubulin and thereby undergoes limited incorporation into the ends of microtubules, thereby altering microtubule self-assembly and suppressing microtubule dynamics (1). Vindesine is mainly used to treat melanoma, lung cancers (carcinomas), and certain uterine cancers (*See also Vinblastine; Vincristine; Vinorelbine*). **1.** Jordan, Himes & Wilson (1985) *Cancer Res.* **45**, 2741.

Vineomycin A₁

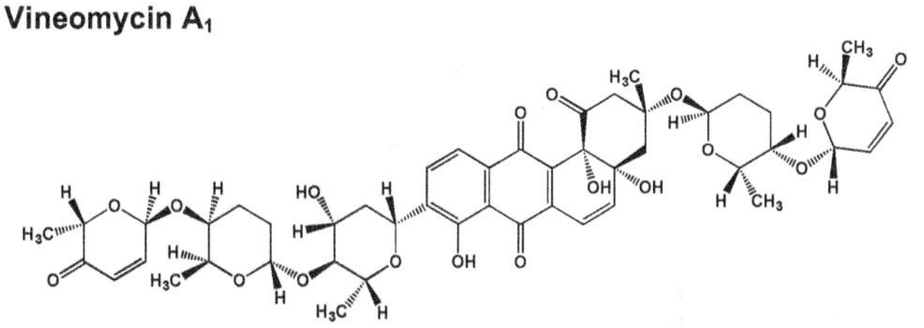

This *Streptomyces albogriseolus* natural product (FW = 934.99 g/mol; CAS 78164-00-8), also known as P-1894B, inhibits procollagen-proline 4-monooxygenase, *or* prolyl hydroxylase, K_i = 1.6 μM. Inasmuch as hydroxyprolyl residues are essential to stabilize collagen, vineomycin A₁ is likely to reduce T_m, the melting temperature for thermal denaturation of the tropocollagen triple helix, such that nonhydroxylated procollagen chains are degraded within the cell. Lacking suitable structural support, tissues, blood vessels, tendons, and skin become fragile. **1.** Ishimaru, Kanamaru, Takahashi, Ohta & Okazaki (1982) *Biochem. Pharmacol.* **31**, 915. **2.** Okazaki, Ohta, Kanamaru, Ishimaru & Kishi (1981) *J. Antibiot.* **34**, 1355.

Vinorelbine

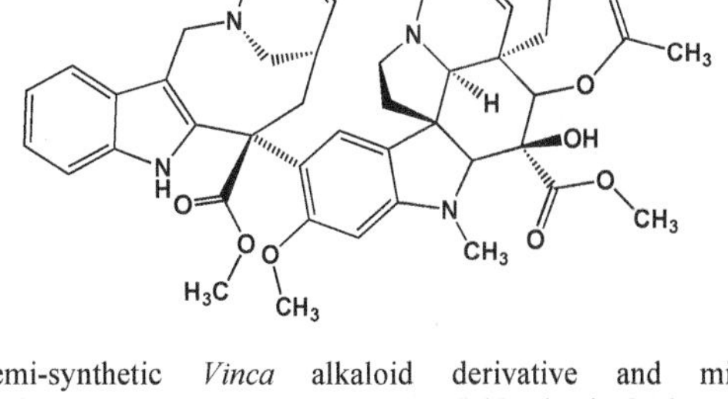

This semi-synthetic *Vinca* alkaloid derivative and microtubule assembly/disassembly modulator (FW = 778.95 g/mol; CAS 71486-22-1) binds rapidly to tubulin and thereby undergoes limited incorporation into the ends of microtubules, thereby altering microtubule self-assembly and suppressing microtubule dynamics (1,2). It inhibits proliferation of multiple human tumor cell lines (IC_{50} = 1.25 nM in HeLa cells) and blocks metaphase/anaphase transition by suppression of microtubule dynamics (IC_{50} = 3.8 nM). The most recent clinically approved *Vinca* alkaloid, Vinorelbine, shows improved efficacy and reduced toxicity and is effective in non–small-cell lung cancer, metastatic breast cancer, and ovarian cancer. It also shows promise in the treatment of lymphoma, esophageal cancer, and prostatic carcinoma. Vinorelbine's similarly acting difluoro analogue, **vinflunine** (FW = 816.92 g/mol; CAS 162652-95-1), contains a $-CF_2CH_3$ moiety in place of the ethyl group shown at the top of vinorelbine's structure. Vinflunine has 3x to 16x lower overall affinity for tubulin than vinorelbine, whereas vinorelbine has lower overall affinity than vinblastine under the conditions of the experiments, suggesting that their lower affinity for tubulin, may partly explain the high concentrations of drug required to block mitosis and cell proliferation (3,4). The lower potency and low tubulin binding affinity of vinflunine correlates well with the finding that it is the least cytotoxic of the *Vinca* alkaloids studied (5), and one cannot discount the possibility that drug clearance is strongly influenced by the off-rate constant for each of the *Vinca* alkaloid interactions with microtubules. Cellular uptake of *Vinca* alkaloids reaches peak intracellular drug concentrations (vinflunine, 4.2 µM; vinorelbine, 1.3 µM; vinblastine, 0.13 µM) that are considerably in excess of the medium concentrations corresponding to its mitotic IC_{50} value (2). Such findings suggest that each drug's efficacy may also depend on metabolism and efflux mechanisms. (*See also Vinblastine; Vincristine; Vindesine*) **Key Pharmacokinetic Parameters:** *See* Appendix II in Goodman & Gilman's *THE PHARMACOLOGICAL BASIS OF THERAPEUTICS*, 12th Edition (Brunton, Chabner & Knollmann, eds.) McGraw-Hill Medical, New York (2011). **1**. Jordan, Himes & Wilson (1985) *Cancer Res.* **45**, 2741. **2**. Ngan, Bellman, Hill, Wilson & Jordan (2001) *Mol. Pharmacol.* **60**, 225. **3**. Lobert, Vulevic, Correia (1996) *Biochemistry* **35**, 6806 **4**. Lobert, Ingram, Hill & Correia (1998) *Mol. Pharmacol.* **53**, 908. **5**. Kruczynski, Barret, Etievant, *et al.* (1998) *Biochemical Pharmacol.* **55**, 635.

Vinpocetine

This nootropic vincamine derivative (FW = 350.46 g/mol; CAS 42971-09-5; photosensitive) exhibits vasodilating activity by inhibiting 3',5'-cyclic-nucleotide phosphodiesterase (1-8), phosphodiesterase 1, IC_{50} = 20 µM (1-5,7), and phosphodiesterase 11, slightly inhibited (6). Vinpocetine inhibits retrograde axoplasmic transport of nerve growth factor (NGF) in the peripheral nerve, resulting in transganglionic degenerative atrophy (TDA) in the segmentally related ipsilateral superficial spinal dorsal horn (9). The latter is characterized by depletion of the marker enzymes fluoride-resistant acid phosphatase (FRAP) and thiamine monophosphatase (TMP). inhibition of retrograde axoplasmic transport of nerve growth factor (NGF) in the peripheral nerve. Blockade of retrograde transport of NGF results in transganglionic degenerative atrophy (TDA) in the segmentally related ipsilateral superficial spinal dorsal horn, which is characterized by depletion of the marker enzymes fluoride-resistant acid phosphatase (FRAP) and thiamine monophosphatase (TMP). Vinpocetine also inhibits the rise in sodium and calcium induced by 4-aminopyridine in synaptosomes (10). It produces a concentration- and state-dependent inhibition of $Na_V1.8$ sodium channel activity (11). Voltage-clamp experiments revealed a 3x increase in vinpocetine potency when whole-cell $Na_V1.8$ conductances were elicited from relatively depolarized potentials (–35 mV; IC_{50} = 3.5 µM) compared with hyperpolarized holding potentials (–90 mV; IC_{50} = 10.4 µM). Vinpocetine also produced an approximately 22 mV leftward shift in the voltage dependence of $Na_V1.8$ channel inactivation but did not affect the voltage range of channel activation (11). Vinpocetine-mediated stimulation of Ca^{2+}-activated K^+ current ($I_{K}Ca$) may result from the direct activation of large-conductance Ca^{2+}-activated K^+ (BK_{Ca}) channels and indirectly from elevated cytosolic Ca^{2+} (12). The main mechanism involved in the neuroprotective action of vinpocetine in the CNS is unlikely to be due to a direct inhibition of Ca^{2+} channels or PDE enzymes, but rather the inhibition of presynaptic Na^+ channel-activation unchained responses (13). **1**. Han, Werber, Surana, Fleischer & Michaeli (1999) *J. Biol. Chem.* **274**, 22337. **2**.

Hagiwara, Endo & Hidaka (1984) *Biochem. Pharmacol.* **33**, 453. **3**. Hidaka, Inagaki, Nishikawa & Tanaka (1988) *Meth. Enzymol.* **159**, 652. **4**. Sudo, Tachibana, Toga, *et al.* (2000) *Biochem. Pharmacol.* **59**, 347. **5**. Yu, Wolda, Frazier, *et al.* (1997) *Cell. Signal.* **9**, 519. **6**. Yuasa, Ohgaru, Asahina & Omori (2001) *Eur. J. Biochem.* **268**, 4440. **7**. Clapham & Wilderspin (1996) *Biochem. Soc. Trans.* **24**, 320. **8**. D'Angelo, Garzia, Andre, *et al.* (2004) *Cancer Cell* **5**, 137. **9**. Knyihar-Csillik, Vecsei, Mihaly, *et al.* (2007) *Ann. Anat.* **189**, 39. **10**. Sitges, Galván & Nekrassov (2005) *Neurochem. Int.* **46**, 533. **11**. Zhou, Dong, Crona, Maguire & Priestley (2003) *J. Pharmacol. Exp. Ther.* **306**, 498. **12**. Wu, Li & Chiang (2001) *Biochem. Pharmacol.* **61**, 877. **13**. Sitges & Nekrassov (1999) *Neurochem. Res.* **24**, 1585.

Vintafolide

This novel anticancer drug (FW = 1903.03 g/mol; CAS 742092-03-1), also known as MK-8109, is a disulfide-linked (see dashed box) conjugate consisting of vinblastine and folic acid, the former acting as an anti-mitotic agent (*See Vinblastine*) and the latter targeting the conjugate to cancer cells that overexpress folate receptors (FRs). Once within the target cell, vinblastine is released nonenzymatically through the action of intracellular glutathione and/or cysteine (1-3). FR-overexpressing cells include ~80-90% of ovarian tumors, gynecological cancers, pediatric ependymal brain tumors, and mesothelioma, as well as breast, colon, renal and lung tumors. Etarfolatide, a non-invasive companion diagnostic imaging agent, may be used to identify folate receptor-positive tumor cells that are likely to be susceptible to vintafolide therapy. Vintafolide is currently under development for treating late-stage ovarian cancer and adenocarcinoma of the lung. Combination therapy using vintafolide plus pegylated liposomal doxorubicin (PLD) offers improved outcome *versus* standard therapy in a randomized trial of patients with platinum-resistant ovarian cancer (4). **1**. Vlahov, Santhapuram, Kleindl, *et al.* (2006) *Bioorg. Med. Chem. Lett.* **16**, 5093. **2**. Reddy, Dorton, Westrick, *et al.* (2007) *Cancer Res.* **67**, 4434. **3**. Li, Sausville, Klein, *et al.* (2009) *J. Clin. Pharmacol.* **49**, 1467. **4**. Naumann, Coleman, Burger, *et al.* (2013) *J. Clin. Oncol.* **31**, 4400.

γ-Vinyl-GABA, *See Vigabatrin*

L-Vinylglycine

This alkenyl amino acid (FW = 101.11 g/mol; CAS, 52773-87-2), also known as 2-amino-3-butenoate, is a chiral mechanism-based enzyme inactivator of snake venom L-amino-acid oxidase by the L-stereoisomer (1,2) and D-alanine aminotransferase by the D-enantiomer. Vinylglycine (VG) is also an alternative substrate for threonine synthase (3). **Target(s):** alanine aminotransferase (4); alanine racemase (5); D-alanine aminotransferase, by the D-enantiomer (6); venom L-amino-acid oxidase (1,2,7); 1-aminocyclopropane-1-carboxylate synthase (8-10); aspartate aminotransferase (4,11,12); cysteine desulfurase (13); heart α-ketoglutarate dehydrogenase complex (14); threonine synthase (15); tryptophan 2-monooxygenase (16); UDP-N-acetylmuramate:L-alanine ligase, by the L-enantiomer (17,18). **1.** Walsh, Cromartie, Marcotte & Spencer (1978) *Meth. Enzymol.* **53**, 437. **2.** Maycock (1980) *Meth. Enzymol.* **66**, 294. **3.** Laber, Gerbling, Harde, *et al.* (1994) *Biochemistry* **33**, 3413. **4.** Cornell, Zuurendonk, Kerich & Straight (1984) *Biochem. J.* **220**, 707. **5.** Lacoste, Darriet, Neuzil & Le Goffic (1988) *Biochem. Soc. Trans.* **16**, 606. **6.** Soper, Manning, Marcotte & Walsh (1977) *J. Biol. Chem.* **252**, 1571. **7.** Marcotte & Walsh (1976) *Biochemistry* **15**, 3070. **8.** Jakubowicz (2002) *Acta Biochim. Pol.* **49**, 757. **9.** Feng & Kirsch (2000) *Biochemistry* **39**, 2436. **10.** McCarthy, Capitani, Feng, Gruetter & Kirsch (2001) *Biochemistry* **40**, 12276. **11.** Rando (1977) *Meth. Enzymol.* **46**, 28. **12.** Fitzpatrick, Cooper & Duffy (1983) *J. Neurochem.* **41**, 1370. **13.** Zheng, White, VCash & Dean (1994) *Biochemistry* **33**, 4714. **14.** Lai & Cooper (1986) *J. Neurochem.* **47**, 1376. **15.** Giovanelli, Veluthambi, Thompson, Mudd & Datko (1984) *Plant Physiol.* **76**, 285. **16.** Emanuele, Heasley & Fitzpatrick (1995) *Arch. Biochem. Biophys.* **316**, 241. **17.** Liger, Blanot & van Heijenoort (1991) *FEMS Microbiol. Lett.* **80**, 111. **18.** Liger, Masson, Blanot, van Heijenoort & Parquet (1995) *Eur. J. Biochem.* **230**, 80.

Vinylglycolate

This α-hydroxy acid (FW$_{free-acid}$ = 101.08 g/mol; CAS 6192-52-5), also known as vinylglycolic acid 2-hydroxy-3-butenoic acid, is an alternative substrate for lactate oxidase and lactate dehydrogenase. Mandelate racemase catalyzes its racemization (1). **Target(s):** L-α-hydroxyacid oxidase (2); phosphoenolpyruvate:protein phosphotransferase, *or* phosphoenol-pyruvate:phosphate transferase system, *Escherichia coli* (3). **1.** Garcia-Viloca, Gonzalez-Lafont & Lluch (2001) *J. Amer. Chem. Soc.* **123**, 709. **2.** Walsh, Cromartie, Marcotte & Spencer (1978) *Meth. Enzymol.* **53**, 437. **3.** Walsh & Kaback (1973) *J. Biol. Chem.* **248**, 5456.

4-Vinylpyridine

This substituted pyridine (FW = 105.14 g/mol; CAS 100-43-6), also known as 4-ethenylpyridine, reacts with sulfhydryl groups (1) and inactivates F$_O$ ATP synthase component (2); 6-pyruvoyltetrahydropterin synthase (3); thioredoxin reductase (4); and transducin (5). (**Caution:** Care should be exercised in its use as it is corrosive and a lachrymator. There are reports of adverse skin reactions with 4-vinylpyridine.) **1.** Cavins & Friedman (1970) *Anal. Biochem.* **35**, 489. **2.** Huang, Kantham & Sanadi (1987) *J. Biol. Chem.* **262**, 3007. **3.** Burgisser, Thöny, Redweik, *et al.* (1994) *Eur. J. Biochem.* **219**, 497. **4.** Nordberg, Zhong, Holmgren & Arner (1998) *J. Biol. Chem.* **273**, 10835. **5.** Ortiz & Bubis (2001) *Arch. Biochem. Biophys.* **387**, 233.

Viomycin

This cyclic peptide antibiotic (FW = 685.70 g/mol; CAS 32988-50-4), also called tuberactinomycin B, from *Streptomyces* sp., inhibits prokaryotic protein biosynthesis (1,2), and seems to bind at least to the small ribosomal subunit because it competes partially with the binding of streptomycin (3).

However, viomycin does not induce misreading like streptomycin1, but promotes association of the ribosomal subunits, stabilises 70S couples (4), and blocks trans-location (5,6). Interestingly, resistant mutants have been found containing either an altered 30S or an altered 50S subunit (7–9). Viomycin strengthens the association of the sensitive subunits, but does not affect the association of subunits derived from resistant mutants (4). **Target(s):** group I intron RNA splicing (10); GTP diphosphokinase (10); protein biosynthesis, ribosomal elongation and translocation (10-15); protein-synthesizing GTPase (elongation factor) (15). (**See also** *Tuberactinomycins*) **1.** Davies, Gorini & Davis (1965) *Molec. Pharmacol.* **1**, 93. **2.** Tanaka & Igusa (1968) *J. Antibiot.* **21**, 239. **3.** Masuda & Yamada (1976) *Biochim. Biophys. Acta* **453**, 333–339 (1976). **4.** Yamada & Nierhaus (1978) *Molec. Gen. Genet.* **161**, 261. **5.** Lion, Y. F. & Tanaka (1976) *Biochem. Biophys. Res. Commun.* **71**, 477. **6.** Modolell, J. & Vázquez, D. (1977) *Eur. J. Biochem.* **81**, 491. **7.** Yamada, Masuda, Shoji & Hori (1972) *J. Bact.* **112**, 1. **8.** Yamada, Masuda, Shoji & Hori (1974) *Antimicrob. Ag. Chemother.* **6**, 46. **9.** Yamada, Masuda, Mizugichi & Suga (1976) *Antimicrob. Ag. Chemother.* **9**, 817. **10.** Wank, Rogers, Davies & Schroeder (1994) *J. Mol. Biol.* **236**, 1001. **11.** Knutsson Jenvert & Holmberg Schiavone (2005) *FEBS J.* **272**, 685. **12.** Hausner, Geigenmuller & Nierhaus (1988) *J. Biol. Chem.* **263**, 13103. **13.** Misumi & Tanaka (1980) *Biochim. Biophys. Res. Commun.* **92**, 647. **14.** Modolell & Vazquez (1977) *Eur. J. Biochem.* **81**, 491. **15.** Campuzano & Modolell (1981) *Eur. J. Biochem.* **117**, 27. **IUPAC Name:** (S)-3,6-Diamino-N-((3S,9S,12S,15S,Z)-3((2R,4S)-6-amino-4-hydroxy-1,2,3,4-tetrahydropyridin-2-yl)-9,12-bis(hydroxymethyl)-2,5,8,11,14-pentaoxo-6-(ureidomethylene)-1,4,7,10,13-pentaazacyclohexadecan-15-yl)hexanamide disulfate

Vioxx, *See* Rofecoxib

VIR-576

LEAIPCSIPPEFLFGKPFVPLEAIPCSIPPEFLFGKPFVP

VIR-576

LEAIPMSIPPEVKFNKPFVP

VIRIP

This anti-retroviral (FW = 4483.33 g/mol) is based on "VIRIP", a naturally occurring peptide corresponding to a 20-residue segment near the C-terminus of α$_1$-antitrypsin (α$_1$-AT). VIR-576 is a conformationally restricted, disulfide-linked tandem repeat that recapitulates, albeit with somewhat higher affinity, the inhibitory action of VIRIP against Human Immunodeficiency Virus Type-1 (HIV-1) entry into the target cells. VIRIP and VIR-576 directly inhibit CD4$^+$ T cell activation *in vitro*. **1.** Forssmann, The, Stoll, *et al.* (2010) *Sci. Transl. Med.* **2**, 63re3.

Viracept, *See* Nelfinavir

Virazole 5'-Monophosphate, *See* Ribavirin

Virginiamycin M$_1$

This macrolactone antibiotic (FW = 529.59 g/mol; CAS 21411-53-0), also called vernamycin A, mikamycin A, ostreogrycin A, and streptogramin A, is produced by *Streptomyces virginiae* and inhibits protein biosynthesis (1-3). Virginiamycin M$_1$ inactivates the donor and acceptor sites of the peptidyltransferase (2). Greater inhibition is observed with ribosomes from Gram-positive microorganisms compared to those from Gram-negative bacteria (3). Virginiamycin is the active ingredient in veterinary medicine used to prevent laminitis, a painful hoof condition, in horses. Veterinary use of virginiamycin on poulty farms selectively controls the intestinal microflora in the upper small intestine, thereby preventing the bacterial breakdown and promoting assimilation of essential amino acids and glucose, significant reducing feed costs. Given the use patterns of virginiamycin in intensive animal farming practices, there is mounting concern about the spread of virginiamycin-resistant animal bacteria to humans and the transfer of antibiotic resistance genes from animal bacteria to human pathogens. The UK Veterinary Medicines Directorate, for example, announced in 2012 that importation of virginiamycin for veterinary use will be phased out, with a complete ban starting in late 2014. **Target(s):** peptidyltransferase (2); protein biosynthesis, initiation and elongation phases (1-5); and protein-synthesizing GTPase elongation factor (5). **1.** Pestka (1974) *Meth. Enzymol.* **30**, 261. **2.** Di Giambattista, Chinali & Cocito (1989) *J. Antimicrob. Chemother.* **24**, 485. **3.** Jiménez (1976) *Trends Biochem. Sci.* **1**, 28. **4.** Cundliffe (1969) *Biochemistry* **8**, 2063. **5.** Campuzano & Modolell (1981) *Eur. J. Biochem.* **117**, 27. **See also** *Pristinamycin IA*

Viridiofungins

Viridiofungin A

These tricarboxylic acids (FW$_{free-acid}$ = 591.7 g/mol for viridiofungin A) from *Trichoderma viride* are broad-spectrum antifungals that inhibit sphingolipid biosynthesis as well as other enzymes that act on di- and tricarboxylic acids (1). **Target(s):** geranylgeranyltransferase (1); serine palmitoyltransferase (1-3); and squalene synthase (1). **1.** Mandala & Harris (2000) *Meth. Enzymol.* **311**, 335. **2.** Dickson, Lester & Nagiec (2000) *Meth. Enzymol.* **311**, 3. **3.** Mandala, Thornton, Frommer, Dreikorn & Kurtz (1997) *J. Antibiot.* **50**, 339.

Viscumin

This toxic heterodimeric protein (MW = 60 kDa; consisting of Mr = 29,000 and 32,000 for its subunits) from mistletoe (*Viscum album*) potently inhibits protein synthesis in cell-free systems. The smaller subunit, designated the A-chain, is the inhibitory component, catalytically deactivating the 60 S ribosomal subunit. While the heterodimer inhibits ribosomes are inhibited at 1 ng/mL viscumin, the isolated A-chain is some 60x less active. Viscumin is a lectin that binds to Sepharose 4B and can be eluted with lactose. **1.** Olsnes, Stirpe, Sandvig & Pihl (1982) *J. Biol. Chem.* **257**, 13263.

Vismodegib

This novel, orally bioavailable cyclopamine-competitive antagonist (FW = 421.30 g/mol; CAS 879085-55-9), also known by the code name GDC-0449, the trade name Erivedge® and as 2-chloro-*N*-(4-chloro-3-pyridin-2-ylphenyl)-4-methylsulfonylbenzamide, is potent, and specific hedgehog inhibitor (IC$_{50}$ = 3 nM) that also inhibits P-gp (IC$_{50}$ = 3.0 μM). **Mode of Inhibitory Action:** Mainly as a way to regulate body patterning and organ development during embryogenesis, the Hedgehog (Hh) signaling pathways is largely quiescent in adults. Exceptions include Hh signaling in tissue maintenance/repair as well as several cancers. Aberrant constitutive activation of Hedgehog pathway signaling as well as uncontrolled cellular proliferation are associated with mutations in the cell surface receptors PTCH and SMO, which bind hedgehog ligands. Vismodegib inhibits ATP-binding cassette transporters, the overexpression of which confers a multidrug resistance phenotype. Vismodegib is the first FDA-approved Hedgehog (Hh) signaling pathway targeting agent for treatment of basal-cell carcinoma (BCC) and is also undergoing trials for use in metastatic colorectal cancer, small-cell lung cancer, advanced stomach cancer, pancreatic cancer, medulloblastoma and chondrosarcoma (1,2). The Smoothened receptor (Smo) mediates Hh pathway signaling that is involved in the pathogenesis, self-renewal of stem cells, and chemotherapeutic resistance of BCC (3). It is ≥95% protein bound in plasma at clinically relevant concentrations and has an approximately 200x longer single dose half-life in humans relative to rats. A strong linear relationship exists between plasma drug concentrations and α-1-Acid Glycoprotein (AAG), and vismodegib binds strongly to human AAG (K_d = 13 μM) and with lower affinity to albumin (K_d = 120 μM). Binding to rat AAG is ~20x weaker, whereas the binding affinity to rat and human albumin was similar (4). **Drug Interactions:** A clinical trial evaluating the effects of rabeprazole (a proton-pump inhibitor), itraconazole (a combined P-glycoprotein/ CYP3A4 inhibitor), and fluconazole (a CYP2C9 inhibitor) found that vismodegib can be administered along with these inhibitors absent risk of a clinically meaningful drug-drug interaction (5). ***Other Hedgehog inhibitors:*** *GANT61; BMS-833923; Purmorphamine; PF-5274857; LY2940680; Cyclopamine, LDE225, or NVP-LDE225, or Erismodegib; SANT-1* **1.** Scales & de Sauvage (2009) *Trends Pharmacol. Sci.* **30**, 303. **2.** Zhang Laterra, Pomper (2009) *Neoplasia* **11**, 96. **3.** Macha, Batra, Ganti (2013) *Cancer Manag. Res.* **31**, 197. **4.** Giannetti, Wong, Dijkgraaf, *et al.* (2011) *J. Med. Chem.* **54**, 2592. **5.** Malhi, Colburn, Williams, *et al.* (2016) *Cancer Chemother. Pharmacol.* **78**, 41.

Voglibose

This glucose analogue and oral antidiabetic agent (FW$_{free-base}$ = 267.3 g/mol; CAS 83480-29-9; M.P. = 162-163°C), also known as AO 128 and Basen, inhibits pig intestinal maltase and sucrase with IC$_{50}$ values of 15 and 4.6 nM, respectively. **Target(s):** α-glucosidase (1-3); maltase (3); and sucrase (3). **1.** Goke, Fuder, Wieckhorst, *et al.* (1995) *Digestion* **56**, 493. **2.** Oki, Matsui & Osajima (1999) *J. Agric. Food Chem.* **47**, 550. **3.** Horii, Fukase, Matsuo, *et al.* (1986) *J. Med. Chem.* **29**, 1038.

Volasertib

This ATP-competitive protein kinase inhibitor (FW$_{free-acid}$ = 618.81 g/mol; CAS 755038-65-4, 946161-17-7 (tri-HCl salt); Solubility: 50 mg/mL DMSO; <1 mg/mL Water), also known as BI6727 and systematically as *N*-((1*R*,4*R*)-4-(4-(cyclopropylmethyl)piperazin-1-yl)cyclohexyl)-4-((*R*)-7-ethyl-8-isopropyl-5-methyl-6-oxo-5,6,7,8-tetrahydropteridin-2-ylamino)-3-methoxybenzamide, targets the Polo-like kinase Plk1 (IC$_{50}$ = 0.87 nM), bringing about G$_2$/M cell-cycle arrest and cell death. This agent exhibits synergistic antitumor activity in combination with other chemotherapeutic agents. Low nanomolar concentrations of BI6727 display potent inhibitory activity against neuroblastoma. **1.** Rudolph, et al. (2009) *Clin. Cancer Res.* **15**, 3094. **2.** Grinshtein, *et al.* (2011) *Cancer Res.* **71**, 1385. **3.** Harris, et al. (2012) *BMC Cancer* **12**, 80.

Volinanserin

This highly selective serotonin antagonist (FW = 373.46 g/mol; CAS 139290-65-6), also known as (*R*)-(2,3-dimethoxyphenyl)-[1-[2-(4-fluorophenyl)ethyl]-4-piperidyl]methanol and MDL-100,907, targets HT$_{2A}$ receptors, showing favorable properties as an antidepressant, antipsychotic, and sleep aid. *See also* *Ketanserin; Pimavanserin; RH-34; Risperidone; Ritanserin; Setoperone.* **1**. Schmidt, Fadayel, Sullivan & Taylor (1992) *Eur. J. Pharmacol.* **223**, 65. **3**. Herth, Kramer, Piel, *et al.* (2009) *Bioorg. Medic. Chem.* **17**, 2989. **3**. Nic, Fox, Stutz, Rice & Cunningham (2009) *Behav. Neurosci.* **123**, 382.

Volkensin

This highly toxic heterodimeric protein (MW = 62 kDa; subunits of M_r 36 and 29 kDa) from the roots of *Adenia volkensii* is a galactose-specific lectin that inhibits protein synthesis in rabbit reticulocyte lysates and HeLa cells (1). In mice, the LD_{50} is 1.38 μg per kg body weight. **Use in Inducing Motor Neuron Loss:** In as early as four days after volkensin injection (5.0 ng), rat peroneal or sciatic nerves degenerate in the ventral horn of the lumbar spinal cord and L_4 and L_5 dorsal root ganglia. By two weeks, no retrogradely labeled motoneurons can be found in the treated peroneal pool. Moreover, loss of motor neurons is accompanied by severe loss of muscle mass. **1**. Stirpe, Barbieri, Abbondanza, *et al.* (1985) *J. Biol. Chem.* **260**, 14589. **2**. Nógrádi & Vrbová (1992) *Neuroscience* **50**, 975.

Vomitoxin

This trichothecene-class mycotoxin and probable carcinogen (FW = 296.32 g/mol; CAS 51481-10-8), also named deoxynivalenol (DON) and (3α,7α)-3,7,15-trihydroxy-12,13-epoxytrichothec-9-en-8-one, is a secondary metabolite of *Fusarium graminearum* and *F. culmorum*, common grain contaminants that cause head blight in wheat and also damage maize. Vomitoxin is a ribotoxin that interrupts protein synthesis by binding to ribosomes and triggering a Ribotoxic Stress Response (RSR), in which mitogen-activated protein kinases (MAPKs) are activated (1). In mononuclear phagocytes, DON induces p38 mobilization to the ribosome, attended by phosphorylation (2). Other effects include anorexia, suppressed immune response to pathogens, and superinduction of cytokines by T-helper cells. Vomitoxin is hazardous and, as its name implies, induces violent vomiting when administered at elevated concentrations. At lower concentrations, vomitoxin also stimulates tryptophan uptake in the CNS, increasing serotonin synthesis and most likely accounting for its anorexic effects. **1**. Rotter, Prelusky & Pestka (1996) *J. Toxicol. Environ. Health* **48**, 1. **2**. Bae & Pestka (2008) *Toxicol. Sci.* **105**, 59.

Vonoprazan

This **P**otassium-**C**ompetitive **A**cid **B**locker, *or* P-CAB (FW = 461.46; CAS 1260141-27-2; Solubility: 62 mg/mL DMSO; < 1 mg/mL H_2O), also known as TAK-438 and 5-(2-fluorophenyl)-*N*-methyl-1-(3-pyridinylsulfonyl)-1*H*-pyrrole-3-methanamine, reversibly inhibits the proton-pumping H^+/K^+, ATPase (IC_{50} = 19 nM at pH 6.5), controling gastric acid secretion (1). Its inhibitory potency is unaltered by ambient pH, whereas SCH28080 and lansoprazole inhibition is weaker at pH 7.5. The inhibition by vonoprazan and SCH28080 is reversible and achieved in a K^+-competitive manner, an action that is quite different from that by lansoprazole. When administered at a dose of 4 mg/kg (as the free base) orally, completely inhibited basal and 2-deoxy-D-glucose-stimulated gastric acid secretion in rats, and its effect on both was stronger than that of lansoprazole (1). The drug is approved in Japan for the treatment of acid-related diseases, including erosive oesophagitis, gastric ulcer, duodenal ulcer, peptic ulcer, gastric reflux, reflux esophagitis, and *Helicobacter pylori* eradication (2). **1**. Hori, Imanishi, Matsukawa, *et al.* (2010) *J. Pharmacol. Exp. Ther.* **335**, 231. **2**. Garnock-Jones (2015) *Drugs* **75**, 309.

Vorapaxar

This himbacine-based thrombin receptor antagonist (FW = 492.98 g/mol; CAS 618385-01-6), also known as SCH 530348, Zontivity®, and ethyl *N*-[(3*R*,3a*S*,4*S*,4a*R*,7*R*,8a*R*,9a*R*)-4-[(*E*)-2-[5-(3-fluorophenyl)-2-pyridyl]-vinyl]-3-methyl-1-oxo-3*a*,4,4*a*,5,6,7,8,8*a*,9,9*a*-decahydro-3*H*-benzo[*f*]iso-benzofuran-7-yl]carbamate, is an orally active and potent platelet-directed anticoagulant, binding with extreme affinity (1). Vorapaxar potently antagonizes the platelet thrombin receptor protease-activated receptor-1 (PAR-1), leaving intact thrombin's procoagulant function and demonstrating no effect on bleed time or coagulation parameters. high bioavailability and inhibited *ex vivo*. Thrombin Receptor-Activating Peptide- (*or* TRAP-) stimulates platelet aggregation in a potent and long-lasting manner (2). It is an FDA-approved treatment for acute coronary syndrome chest pain caused by coronary artery disease. Vorapaxar does not affect platelet aggregation, as induced by adenosine diphosphate (ADP), collagen or a thromboxane mimetic; nor does it alter coagulation parameters *ex vivo*. While PAR-1 receptors are also expressed in endothelial cells, neurons, and smooth muscle cells, vorapaxar's pharmacodynamic effects in these cell types have not been assessed. **1**. Chackalamannil, Wang, Greenlee, *et al.* (2008) *J. Med. Chem.* **51**, 3061. **2**. Oestreich (2009) *Curr. Opin. Investig. Drugs* **10**, 988.

Voriconazole

This triazole antifungal (FW = 349.31; CAS 137234-62-9, 173967-54-9; Solubility: 70 mg/mL DMSO; <1 mg/mL Water), also known as systematically as (2*R*,3*S*)-2-(2,4-difluorophenyl)-3-(5-fluoropyrimidin-4-yl)-1-(1*H*-1,2,4-triazol-1-yl)butan-2-ol, is used to control serious infections by invasive fungal infections (*e.g.,* invasive candidiasis, aspergillosis, and other emerging fungal pathogens) typically infecting immuno-compromized patients. **Mode of Action:** Voriconazole severely impairs growth by inhibiting sterol-14α-demethylase (cytochrome P450 CYP51), thereby blocking production of brassinosteroids, a recently recognized class of steroidal phytohormones (1). The natural voriconazole resistance of the woodland strawberry *Fragaria vesca* is conferred by the specific voriconazole-resistant CYP51 variant (1). Voriconazole also targets CYP3A4 and P-glycoprotein, *or* P-gp (2). **See also** *related compound,* **Fluconazole Key Pharmacokinetic Parameters:** *See* Appendix II in Goodman & Gilman's THE PHARMACOLOGICAL BASIS OF THERAPEUTICS, 12th Edition (Brunton, Chabner & Knollmann, eds.) McGraw-Hill Medical, New York (2011). **1**. Rozhon, Husar, Kalaivanan *et al.* (2013) *PLoS One* **8**, e53650. **2**. Park, Song, Kang, *et al.* (2012) *Clin. Nephrol.* **78**, 412.

Vorinostat

This histone deacetylase inhibitor (FW = 264.30 g/mol; CAS 149647-78-9; Solubility: 50 mg/mL DMSO; 1 mg/mL Water; *Abbreviation:* SAHA), also known as MK-0683, suberoylanilide hydroxamic acid, and Zolinza, targets HDAC1 and HDAC3 with IC_{50} values of 10 nM and 20 nM, respectively, resulting in a marked hyperacetylation of histone H4 (1). X-ray crystal structures of the uncomplexed HDAC as well as its binary complexes with trichostatin A and SAHA reveal an active site consisting of a tubular pocket, a Zn^{2+} binding site, as well as two Asp-His charge-relay systems, suggesting a likely mechanism for HDAC inhibition (2). Active-site residues contacting these inhibitors are conserved across the HDAC family, suggesting a histone deacetylation mechanism that should provide a framework for developing HDAC-directed inhibitors as antitumor agents. Vorinostat inhibits the prostate cancer cell growth (EC_{50} = 2.5-7.5 μM), inducing dose-dependent cell death (3). Vorinostat also inhibits the NF-κB pathway by suppressing the degradation of I-κBα and attenuating NF-κB p65 translocation to the nucleus (4). Transcriptome analysis of prostate cancer PC3 cells identified a subset of NF-κB target genes reversibly regulated by Vorinostat, as well as a group of interferon (IFN)-stimulated genes (ISGs) (5). Consistent with the induction of NF-κB target genes, vorinostat-mediated enhancement of vesicular stomatitis virus (VSV) oncolysis increases hyper-acetylation of NF-κB RELA/p65, a signaling event that also increases expression of several autophagy-related genes. VSV replication and cell killing was suppressed, when NF-κB signaling was inhibited. Inhibition of autophagy by 3-methyladenine enhanced expression of ISGs, indicating that vorinostat stimulates NF-κB activity in a reversible manner by modulating RELA/p65, inducing autophagy, suppressing IFN-mediated responses, and then enhancing VSV replication and apoptosis (5). **1.** Richon, *et al.* (1998) *Proc. Natl. Acad. Sci. USA* **95**, 3003. **2.** Finnin, Donigian, Cohen, *et al.* (1999) *Nature* **401**, 188. **3.** Butler, *et al.* (2000) *Cancer Res.* **60**, 5165. **4.** Zhong, Ding, Chen & Luo (2013) *Int. Immunopharmacol.* **17**, 329. **5.** Shulak, Beljanski, Chiang, *et al.* (2013) *J. Virol.* **88**, 2927.

Vorozole

This triazole derivative and orally available nonsteroidal aromatase inhibitor (FW = 324.8 g/mol; CAS 118949-22-7), also known by the code name R83842 and systematically as 6-[(4-chlorophenyl)(1,2,4-triazol-1-yl)methyl]-1-methylbenzotriazole, targets aromatase (IC_{50} = 1.4 nM, competitive inhibitor), significantly reducing plasma estradiol levels in postmenopausal women. (Note: Aromatase is an adrenal enzyme that converts androstenedione and estrone to estrogen.) Vorozole is highly effective in the treatment of these women with advanced breast cancer. In pregnant mare serum gonadotropin- (PMSG)- primed female rats, plasma estradiol levels at 2 hours after single oral administration of vorozole were significantly reduced by doses ≥1 μg/kg, with an ED_{50} of 3.4 μg/kg (2). An added feature is that vorozole displays >10000x selectivity for aromatase versus other cytochrome P450 enzymes (1). ***See also*** *Letrozole* **1.** Wouters, Snoeck & De Coster (1994) *Breast Cancer Res. Treat.* **30**, 89. **2.** Wouters, Van Ginckel, Krekels, Bowden & De Coster (1993) *J. Steroid Biochem. Mol. Biol.* **44**, 617.

Vortioxetine

This serotonin reuptake inhibitor ($FW_{free-base}$ = 298.45 g/mol; $FW_{hydrobromide}$ = 379.36 g/mol; CAS 508233-74-7), also known by its Code name Lu

AA21004, its trade name Brintellix®, and its systematic name 1-[2-(2,4-dimethylphenylsulfanyl)phenyl]piperazine, is an atypical antidepressant that targets the human 5-HT transporter (SERT) inhibitor (K_i = 1.6 nM); 5-HT$_{1A}$ receptor, with high-efficacy partial agonist action (K_i = 15 nM), the 5-HT$_{1B}$ receptor partial agonist (K_i = 33 nM), the 5-HT$_{1D}$ receptor antagonist (K_i = 54 nM), the 5-HT$_{3A}$ receptor antagonist (K_i = 3.7 nM), the 5-HT$_7$ receptor antagonist (K_i = 19 nM), and the norepinephrine transporter, *or* NET (K_i = 113 nM) (1,2). Vortioxetine likewise displays moderately high affinity for the β$_1$-adrenergic receptor (K_i = 46 nM). **Pharmacokinetics:** After oral administration, Lu AA21004 shows an extended absorption phase, a large volume of distribution, and a medium clearance rate, resulting in late t_{max} values and a mean elimination $t_{1/2}$ of 57 hr (3). See reference (4) for drug interactions, especially dose adjustment required when vortioxetine is co-administered with bupropion or rifampicin. **Metabolism:** Lu AA21004 was found *in vitro* to be oxidized to a 4-hydroxy-phenyl metabolite (by CYP2D6 with some contribution from CYP2C9), a sulfoxide (by CYP3A4/5 and CYP2A6), an N-hydroxylated piperazine (by CYP2C9 and CYP2C19), and a benzylic alcohol (by CYP2D6) that is further oxidized to 3-methyl-4-(2-piperazin-1-ylphenylsulfanyl)benzoic acid. Benzylic alcohol oxidation to the corresponding benzoic acid is primarily catalyzed by alcohol dehydrogenase and aldehyde dehydrogenase. **1.** Bang-Andersen, Ruhland, Jørgensen, *et al.* (2011) *J. Med. Chem.* **54**, 3206. **2.** Mørk, Pehrson, Brennum, *et al.* (2012) *J. Pharmacol. Exp. Ther.* **340**, 666. **3.** Areberg, Søgaard, Højer (2012) *Basic Clin. Pharmacol. Toxicol.* **111**, 198. **4.** Chen, Lee, Højer, *et al.* (2013) *Clin. Drug Investig.* **33**, 727.

Voruciclib

This potent flavone-based CDK inhibitor (FW = 487.90 g/mol; CAS 1000023-04-0; Soluble in DMSO), also named P1446A-05 and 2-[2-chloro-4-(trifluoromethyl)phenyl]-5,7-dihydroxy-8-[(2R,3S)-2-(hydroxymethyl)-1-methyl-3-pyrrolidinyl]-4H-chromen-4-one, selectively targets <u>C</u>yclin-<u>D</u>ependent <u>K</u>inases 4 and 6, with activity in multiple BRAF-mutant and wild type cell lines. The addition of a CDK 4/6 inhibitor to BRAF inhibitors is supported by extensive preclinical data. **1.** Diab, Martin, Simpson, *et al.* (2015) ASCO Annual Meeting, *J. Clin. Oncol.* **33**, *Suppl. Abstr.* 9076.

Voxtalisib

This mTOR/PI3K dual inhibitor (FW = 270.30 g/mol; CAS 934493-76-2; Solubility: 200 mM in DMSO; < 1 mg/mL in H_2O), also known as SAR245409, XL765, and 2-amino-8-ethyl-4-methyl-6-(1H-pyrazol-5-yl)pyrido[2,3-d]pyrimidin-7-one, targets p110γ (IC_{50} = 9 nM), but also inhibits DNA-PK and mTOR. Voxtalisib is pro-apoptotic to chronic lymphocytic leukemia (CLL) cells, irrespective of their ATM/p53 status, than PI3Kα or PI3Kδ isoform-selective inhibitors (2). SAR245409 also blocks CLL survival, adhesion and proliferation (2). Dual PI3K/mTOR inhibition results in increased anti-tumor activity, both *in vitro* and *in vivo*, in models of pancreatic adenocarcinoma (3). **1.** Papadopoulos, Egile, Ruiz-Soto, *et al.* (2015) *Leuk. Lymphoma* **56**, 1763. **2.** Thijssen, Ter Burg, van Bochove, *et al.* (2016) *Leukemia* **30**, 337. **3.** Mirzoeva, Hann, Hom, *et al.* (2011) *J. Mol. Med.* (Berlin). **89**, 877.

VP-16-213, *See Etoposide*

VPS34-IN1

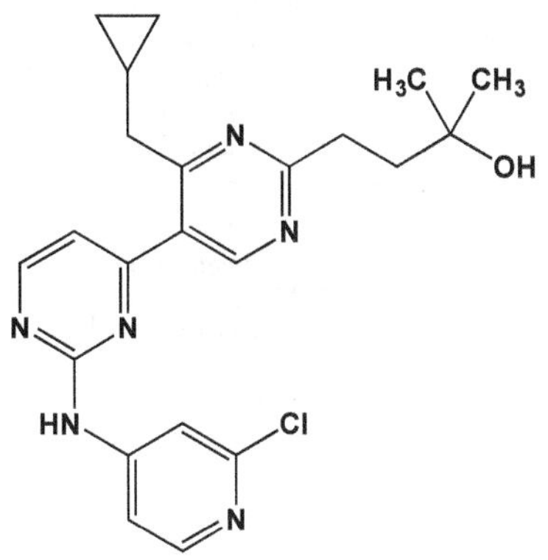

This first-in-class, cell-permeable inhibitor (FW = 424.93 g/mol) targets recombinant, insect cell-expressed class III PI3K isoform known as Vps34 (<u>V</u>acuolar <u>p</u>rotein <u>s</u>orting-<u>34</u>) with an IC_{50} ~25 nM, but does not significantly inhibit other lipid (or protein) kinases tested, including class I or class II phosphoinositide 3-kinases. PI3Ks phosphorylate the 3′-hydroxyl of the D-*myo*-inositol head group of phosphatidylinositol (PtdIns), thereby generating 3-phosphoinositides that orchestrate aspects of cell growth, proliferation and intracellular trafficking. Vps34 assists in controlling vesicular protein sorting, by forming a core complex with Vps15 (itself a serine/threonine protein kinase) and Beclin-1 (also known as Vps30 or ATG6) [24]. The resulting core complex then interacts with a growing list of proteins to form distinct complexes that regulate endosomal membrane trafficking processes, endosome-lysosome maturation, and autophagy. VPS34-IN1 also rapidly reduces endosomal PtdIns(3)P levels within 1 min of treatment. Use of VPS34-IN1 helped to demonstrate that Vps34 plays a role in regulating the phosphorylation and activity of the <u>S</u>erum- and <u>G</u>lucocorticoid-regulated protein <u>K</u>inase-<u>3</u> (SGK3), the only protein kinase known to possess a selective PtdIns(3)P-binding PX domain. That VPS34-IN1 suppresses Vps34 activity in cells is supported by the observed complete dispersal of PtdIns3P-binding GFP-labeled $2xFYVE_{Hrs}$ from endosomal membranes within 1-2 minutes. When present at 1 µM, VPS34-IN1 does not appreciably inhibit the activity of 340 protein kinases or 25 lipid kinases, including all isoforms of Class I and II PI3Ks. Use of VPS34-IN1 revealed phosphatidylinositol 3-phosphate binding SGK3 protein kinase as a downstream target of Class III PI-3 kinase. Other Vps34 inhibitors (***See*** *Wortmannin, 3-Methyadenine, LY294002, KU55933, Gö6946*) are not selective in their action. **1**. Bago, Malik, Munson, *et al*. (2014) *Biochem. J.* **463**, 413.

VR

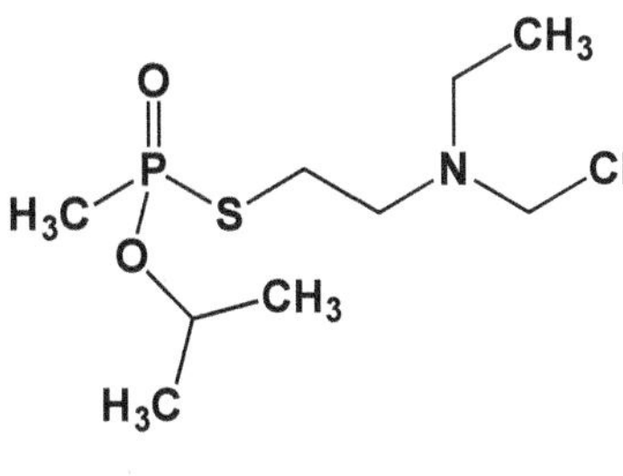

This membrane-penetrating chemical warfare (CW) agent (FW = 267.37 g/mol; CAS 159939-87-4); IUPAC Name: *S*-(2-diethylaminoethyl) isobutyl)methylphosphonothioate, also variously known as Russian VX, Soviet V-gas, Substance-33, and R-33, binds to and irreversibly inhibits acetylcholinesterase. VR has similar lethal dose (10–50 mg) levels to VX. Pre-treatment with pyridostigmine prior to exposure can reduce VR toxicity, but not death at high VR exposure. The oxime HI-6 is quite effective in reactivating acetylcholinesterase–VR adducts. The facile synthesis VR was conducive to production of so-called binary weapons, where less toxic precursors are combined while the CW projectile is in flight (4). **1**. Worek, Thiermann, Szinicz & Eyer (2004) *Biochem. Pharmacol.* **68**, 2237. **2**. Shih, Kan & McDonough (2005) *Chem. Biol. Interact.* **157-158**, 293. **3**. Aurbek, Thiermann, Szinicz, Eyer & Worek (2006) *Toxicology* **224**, 91. **4**. Mirzayanov (2009) *State Secrets: An Insider's Chronicle of the Russian Chemical Weapons Program*, p. 166. (ISBN 978-1-4327-2566-2).

VS-5584

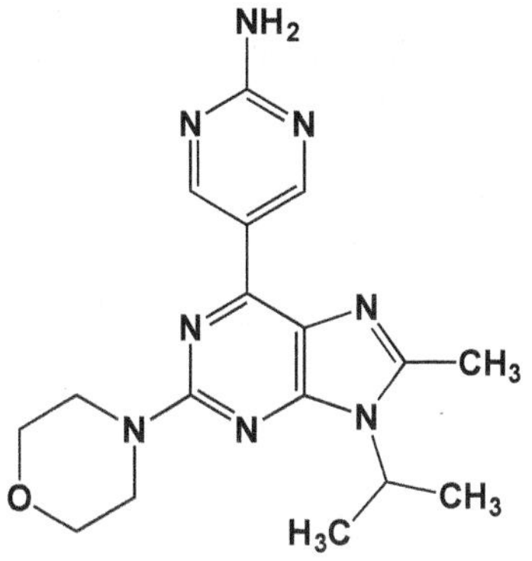

This potent ATP-competitive dual-PI3K/mTOR inhibitor FW = 345.32 g/mol; CAS 1246560-33-7; Solubility: 71 mg/mL DMSO), also known as SB2343 and 5-[8-methyl-9-(1-methylethyl)-2-(4-morpholinyl)-9*H*-purin-6-yl]-2-pyrimidinamine, selectively targets mTOR (IC_{50} = 3.4 nM), PI3Kα (IC_{50} =2.6 nM), PI3Kα-H1047 (IC_{50} = 3.3 nM), PI3Kβ (IC_{50} = 21 nM), PI3Kδ (IC_{50} = 2.7 nM), and PI3Kγ (IC_{50} = 3 nM), without relevant activity on 400 other lipid and protein kinases. VS-5584 likewise exhibits favorable pharmacokinetic properties after oral dosing in mice and is well tolerated. VS-5584 induces long-lasting and dose-dependent inhibition of PI3K/mTOR signaling in tumor tissue, leading to tumor growth inhibition in various rapalog-sensitive and -resistant human xenograft models. It also reduces the number of functional blood vessels in the tumor. **1**. Hart, Novotny-Diermayr, Goh, *et al*. (2013) *Mol. Cancer Ther.* **12**, 151.

VTP-27999

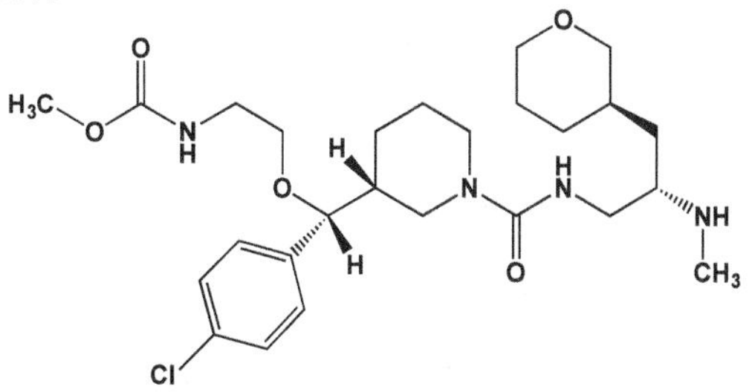

This renin inhibitor and antihypertensive (FW = 525.09 g/mol; CAS 942142-51-0) resembles aliskiren, showing a similar IC_{50} (0.3 nM) and $t_{1/2}$ (30 hours), but having a 7-8x higher bioavailability (~20%). VTP-27999 blocks renin, but also binds to prorenin, the latter inducing a conformational change (like that induced by acid), thereby allowing its recognition in a renin-specific assay. (***See*** *Aliskiren*). Notably, *in vitro* analysis demonstrated that VTP-27999 increases renin immunoreactivity for a given amount of renin by ≥ 30% without unfolding prorenin (1). It binds to acid-activated, intact prorenin, increasing immunoreactivity in a renin assay. No such increase in immunoreactivity is observed, when acid-activated prorenin bound to VTP-27999 is measured in a prosegment-directed assay (1). VTP-27999-induced increases in renin immunoreactivity is competitively prevented by aliskiren, and antibody displacement studies shows a higher affinity of the active site-directed antibodies in the presence of VTP-27999 (1). VTP-27999 accumulates at higher levels in HMC-1 cells than when on aliskiren, allowing this inhibitor to block intracellular renin at approximately five-fold lower medium levels (2). **1**. Krop, Lu, Verdonk, *et al*. (2013) *Hypertension* **61**, 1075. **2**. Lu, Krop, Batenburg, *et al*. (2014) *J. Hypertens.* **12**, 1255.

VU 0409106

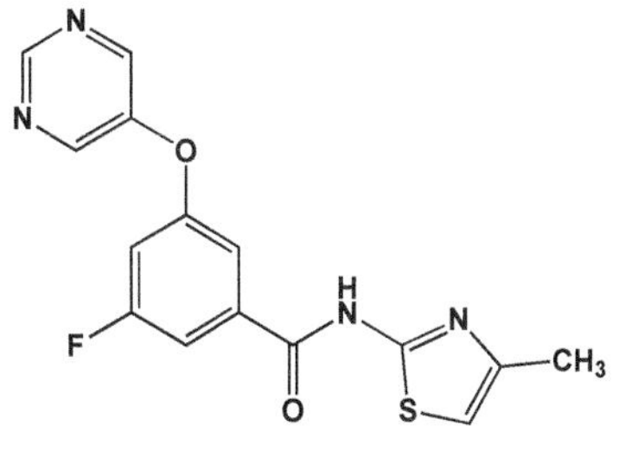

This potent and selective allosteric mGlu$_5$ inhibitor and anxiolytic (FW = 330.34 g/mol; CAS 1276617-62-9; Solubility: 100 mM in DMSO), also named 3-fluoro-N-(4-methyl-2-thiazolyl)-5-(5-pyrimidinyloxy)benzamide, targets the metabotropic glutamate receptor-5 (IC$_{50}$ = 49 nM), a G protein-coupled receptor that in humans is encoded by the GRM5 gene. VTP-27999 differs from aliskiren regarding its level of intracellular accumulation and its capacity to interfere with renin signaling via the (Pro)Renin Receptor in rat vascular smooth muscle cells, and the (P)RR determines prorenin-renin conversion and constitutive (but not regulated) (pro)renin release. 1. Felts, Rodriguez, Morrison, et al. (3013) Bioorg. Med. Chem. Lett. 23, 5779.

VUF 5574

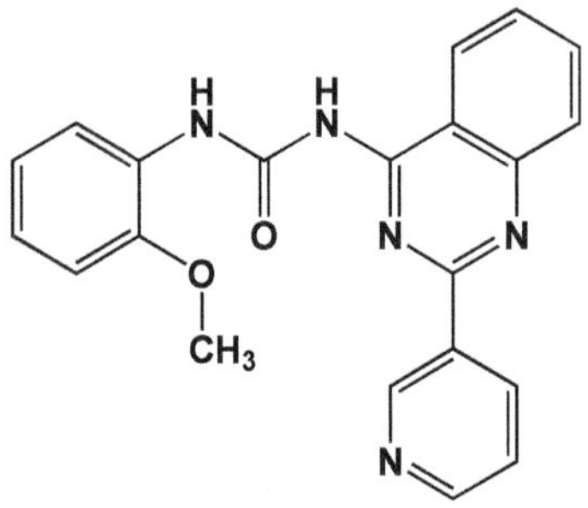

This potent A$_3$ antagonist (FW = 371.40 g/mol; CAS 280570-45-8; IUPAC: N-(2-Methoxyphenyl)-N'-[2-(3-pyridinyl)-4-quinazolinyl]urea) selectively targets the adenosine A$_3$ receptor (K_i = 4 nM), showing >2500-fold selectivity versus A$_1$ and A$_{2A}$ receptors. This GPCR couples to Gi/Gq and mediates a sustained cardioprotective action during cardiac ischemia. VUF 5574 competitively antagonizes the effect of an agonist in a functional A$_3$ receptor assay (i.e., inhibition of cAMP production in cells expressing the human adenosine A$_3$ receptor). A$_3$ receptor stimulation is also deleterious during ischaemia, and treatment with VUF 5574 appears to increase the resistance of the CA1 hippocampal region to ischaemic damage (3). 1. van Muijlwijk-Koezen, Timmerman, van der Goot, et al. (2000) J. Med. Chem. 43, 2227. 2. Baraldi & Borea (2000) Trends Pharmacol. Sci. 21, 456. 3. Pugliese, Coppi, Spalluto, Corradetti & Pedata (2006) Brit. J. Pharmacol. 147, 524.

VX

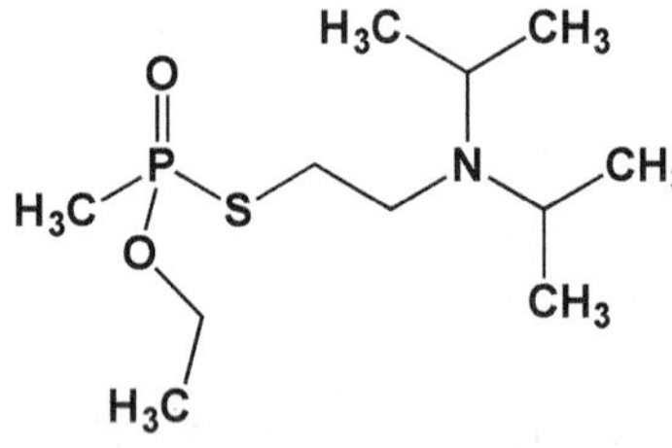

This chemical warfare agent (FW$_{free-base}$ = 267.37 g/mol; CAS 50782-69-9; 51848-47-6; 53800-40-1, and 65143-05-7), also named O-ethyl S-(2-diisopropylaminoethyl)methylphosphonothioate, is a colorless, odorless, and amber liquid with a low volatility and motor oil-like consistancy. Binary VX, also referred to as VX2, is formed by mixing O-(2-diisopropylaminoethyl) O'-ethyl methylphosphonite (Agent QL) with elemental sulfur (Agent NE) within chemical warfare munitions. VX was classified as a weapon of mass destruction by the UN Resolution 687, and its production and stockpiling are outlawed by the Chemical Weapons Convention of 1993. It is neutralized by reaction with aqueous sodium hydroxide and other nucleophiles, including hydroxylamine. **Caution:** The LD$_{50}$ of this irreversible cholinesterase inhibitor in rabbits is ~15 µg/kg subcutaneous and 7 µg/kg (intravenous) in rats. Its inhalation LC$_{50}$ is 30–

50 mg·min/m^3. Extreme care must be exercised, as a small drop contacting the skin is sufficient to kill a human. **Target(s):** acetylcholinesterase (1-7); cholinesterase, or butyrylcholinesterase (1,5-10); and Na$^+$/K$^+$ exchanging ATPase (11) 1. Duysen, Li, Xie, et al. (2001) J. Pharmacol. Exp. Ther. 299, 528. 2. Friboulet, Rieger, Goudou, Amitai & Taylor (1990) Biochemistry 29, 914. 3. Forsberg & Puu (1984) Eur. J. Biochem. 140, 153. 4. Worek, Thiermann, Szinicz & Eyer (2004) Biochem. Pharmacol. 68, 2237. 5. Aurbek, Thiermann, Szinicz, Eyer & Worek (2006) Toxicology 224, 91. 6. Bajgar, Kuca, Fusek, et al. (2007) J. Appl. Toxicol. 27, 458. 7. Schopfer, Voelker, Bartels, Thompson & Lockridge (2005) Chem. Res. Toxicol. 18, 747. 8. Gordon & Leadbeater (1977) Toxicol. Appl. Pharmacol. 40, 109. 9. Reiner, Simeon-Rudolf & Skrinjaric-Spoljar (1995) Toxicol. Lett. 82/83, 447. 10. Cerasoli, Griffiths, Doctor, et al. (2005) Chem. Biol. Interact. 157-158, 363. 11. Robineau, Leclercq, Gerbi, Berrebi-Bertrand & Lelievre (1991) FEBS Lett. 281, 145.

VX-478, *See Amprenavir*

VX-497, *See Merimepodib*

VX-509, *See Decernotinib*

VX-680, *See MK-0457*

VX-745

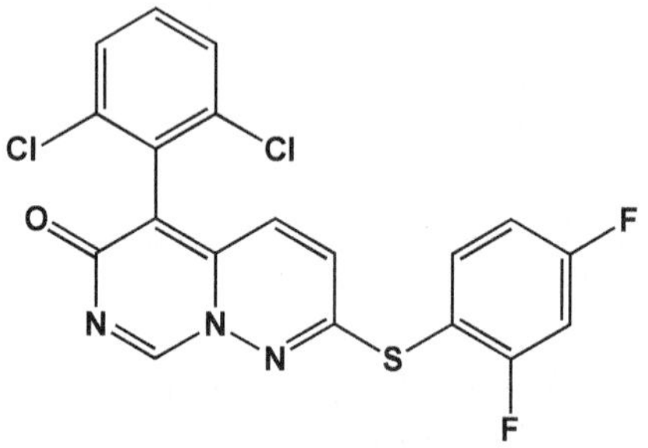

This anti-inflammatory agent (FW = 436.27 g/mol) is a small-molecule inhibitor of mitogen activated protein kinase (p38 MAP), exhibiting an IC$_{50}$ value of 0.8 nM. 1. Haddad (2001) Curr. Opin. Investig. Drugs 2, 1070. 2. Hideshima, Akiyama, Hayashi, et al. (2003) Blood 101, 703. 3. Stelmach, Liu, Patel, et al. (2003) Bioorg. Med. Chem. Lett. 13, 277. 4. Fitzgerald, Patel, Becker, et al. (2003) Nature Struct. Biol. 10, 764.

VX-765

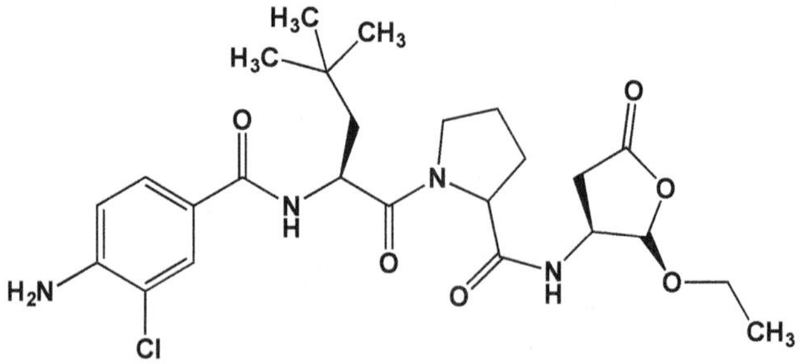

This oral interleukin (IL)-converting enzyme/caspase-1 inhibitor (FW = 508.99 g/mol; CAS 273404-37-8), also known as (S)-1-((R)-2-(4-amino-3-chlorobenzamido)-3,3-dimethylbutanoyl)-N-((2S,3S)-2-ethoxy-5-oxotetra-hydrofuran-3-yl)pyrrolidine-2-carboxamide, selectively targets caspase-1 (K_i = 0.8 nM) and exhibits potent anti-inflammatory activities by inhibiting the release of IL-1β and IL-18. 1. Wannamaker, Davies, Namchuk, et al. (2006) J. Pharmacol. Exp. Ther. 321, 509.

VX-950, *See Telaprevir*

– W –

W, *See Tryptophan; Tungsten and Tungsten Ions*

1400W

This slow, tight binding inhibitor (FW = 250.17 g/mol; CAS 214358-33-5), also named *N*-[[3-(aminomethyl)phenyl]methyl]ethanimidamide dihydrochloride, selectively targets inducible Nitric Oxide Synthase, *or* iNOS (K_i = 7 nM), with far weaker action against nNOS (K_d = 2 μM) and eNOS (K_d = 50 μM). This inhibitor is cell-permeable and active *in vivo*. The slow onset of inhibition by 1400W shows saturation kinetics with a maximal unimolecular rate constant of 0.028 s^{-1} and a binding constant of 2.0 μM (1). Inhibition is dependent on the cofactor NADPH. L-Arginine was a competitive inhibitor of 1400W binding with a K_s value of 3.0 μM. Inhibited enzyme did not recover activity after 2 hours. Thus, 1400W was either an irreversible inhibitor or an extremely slowly reversible inhibitor of human iNOS with a K_d value less than or equal to 7 nM (1). For the EMT6 murine mammary adenocarcinoma, in which iNOS is expressed in the tumor cells, continuous infusion of 1400W for 6 days at 10–12 mg/kg$^-$^{1}/hour^{-1} resulted in significant reduction in tumor weight (357 and 466 mg, respectively) compared with that of controls (2). **1**. Garvey, Oplinger, Furfine, *et al.* (1997) *J. Biol. Chem.* **272**, 4959. **2**. Thomsen, Scott, Topley, *et al.* (1997) *Cancer Res.* **57**, 3300.

W-5, See *N-(6-Aminohexyl)-1-naphthalenesulfonamide*

W-7, See *N-(6-Aminohexyl)-5-chloro-1-naphthalenesulfonamide*

W-12, See *N-(4-Aminobutyl)-1-naphthalenesulfonamide*

W-13, See *N-(4-Aminobutyl)-5-chloro-1-naphthalenesulfonamide*

W84

This synthetic allosteric modulator (FW = 708.54 g/mol; CAS 21093-51-6; Soluble to 10 mM in DMSO), also named hexamethylenebis[dimethyl-(3-phthalimidopropyl)ammonium] dibromide, increases the protective effect of atropine against organophosphate poisoning by stabilizing cholinergic antagonist-receptor complexes (1-3). It produces a much greater reduction in the affinity of carbachol than that of the competitive antagonists and as a consequence is said to exert a 'supra-additive' effect, when combined with a competitive antagonist (1). W84 acts equally well on the cardiac cholinoceptors of guinea pigs, rats, and pigs, respectively (3). **1**. Mitchelson (1975) *Eur. J. Pharmacol.* **33**, 237. **2**. Jepsen, Lüllmann, Mohr & Pfeffer (1988) *Pharmacol. Toxicol.* **63**, 163. **3**. Mohr, Staschen & Ziegenhagen (1992) *Pharmacol. Toxicol.* **70**, 198.

W146

This S1P$_1$R antagonist (MW = 342.37 g/mol; CAS 909725-61-7; Soluble to 20 mM as sodium salt), named systematically as *R*-3-amino-4-(3-hexylphenylamino)-4-oxobutylphosphonic acid, targets sphingosine 1-phosphate (subtype-1) receptors (K_i = 40 nM), with no effect at S1P$_2$, S1P$_3$ or S1P$_5$ receptors (1,2). W146 enhances capillary leakage and restores lymphocyte egress *in vivo*. **1**. Sanna, *et al.* (2006) *Nature Chem. Biol.* **2**, 434. **2**. Gonzalez-Cabrera, *et al.* (2007) *Biol. Chem.* 10th ed., **282**, 7254.

W-54011

This orally active nonpeptide C5a receptor antagonist (FW = 493.08 g/mol; CAS 405098-33-1), also named *N*-[(4-dimethylaminophenyl)methyl]-*N*-(4-isopropylphenyl)-7-methoxy-1,2,3,4-tetrahydronaphthalen-1-carboxamide, inhibits the binding of anaphylatoxin C5a, a potent for neutrophil/leukocyte chemotactic factor and inflammatory mediator, to human neutrophils (K_i = 2.2 nM). W-54011 also inhibits C5a-induced intracellular Ca^{2+} mobilization, chemotaxis, and generation of reactive super oxide species in human neutrophils with IC$_{50}$ values of 3.1, 2.7, and 1.6 nM, respectively. In C5a-induced intracellular Ca^{2+} mobilization assay with human neutrophils, W-54011 is not an agonist, even at 10 μM, and shifts the dose-response curves to C5a to the right without depressing the maximal responses. W-54011 inhibits C5a-induced intracellular Ca^{2+} mobilization in neutrophils from cynomolgus monkeys and gerbils, but not mice, rats, guinea pigs, rabbits, and dogs. **1**. Sumichika, Sakata, Sato, *et al.* (2002) *J. Biol. Chem.* **277**, 49403.

1400W, See *N-(3-Aminomethyl)benzylacetamidine*

Waldiomycin

This benz[*a*]anthraquinone antibiotic (FW = 702.71 g/mol) from the culture broth of *Streptomyces* sp. MK844-mF10, inhibits the histidine kinase WalK and the response regulator WalR, which comprise a two-component signal transduction system that is indispensable for the cell-wall metabolism of low-GC Gram-positive bacteria. Waldiomycin inhibits WalK from *Staphylococcus aureus* (MRSA) and *Bacillus subtilis* at IC$_{50}$ values of 8.8 and 10.2 μM, respectively, and shows antibacterial activity with MICs ranging from 4 to 8 μg mL^{-1} against MRSA and *B. subtilis*. Its structure was determined by two-dimensional NMR. Waldiomycin is structurally related to capoamycin and dioxamycin. **1**. Igarashi, Watanabe Hashida, *et al.* (2013) *J. Antibiot.* (Tokyo) **66**, 459.

Warburganal

This sulfhydryl-reactive, antifungal and antibiotic sesquiterpene dialdehyde (FW = 250.34 g/mol; CAS 62994-47-2) from the plants *Warburgia ugandensis* of East African is a potent inhibitor of numerous sulfhydryl-dependent enzymes, including alcohol dehydrogenase, which can be fully reactivated upon treatment with excess L-cysteine. Warburganal is an anti-feedant agent used to combat the African army worm. The minimal inhibitory concentrations for *Saccharomyces cerevisiae* and *Candida albicans* are both about 6 μg/mL. **1.** Taniguchi, Adachi, Haraguchi, Oi & Kubo (1983) *J. Biochem.* **94**, 149.

Warfarin

Warfarin

Vitamin K

This synthetic vitamin K mimic (FW = 308.33 g/mol; CAS= 81-81-2), also marketed under the tradenames Coumadin®, Jantoven®, Marevan®, Lawarin®, and Waran®, and systematically named 3-(α-acetonylbenzyl)-4-hydroxycoumarin, exhibits powerful anticoagulant action by inhibiting the reductase that converts vitamin K epoxide to its biologically active form, thereby creating a functional vitamin K deficiency. First prepared as a racemic mixture by Link and coworkers (1), warfarin is a derivative of the hemorrhagic agent dicoumarol (*See Dicoumarol*) and is a registered trademark of the Wisconsin Alumni Research Foundation. **Use as an Anticoagulant:** Warfarin and related coumarin anticoagulants act by inhibiting the posttranslational carboxylation of vitamin K-dependent coagulation factors II, VII, IX and X. These anticoagulants have no direct effect on an established thrombus, and they cannot reverse ischemic tissue damage. Instead, the aim of anticoagulation therapy is to prevent further clot progression as well as other thromboembolic complications. Upon oral dosing in humans, warfarin is completely absorbed, reaching peak concentrations within 4 hours. Its half-life is 20–60 hours, and its effects last for 2–5 days. Metabolism mainly occurs by liver CYP2C9. Warfarin dosing is complicated by many common medications as well as the presence of vitamin K in certain plant foodstuffs. Indeed, warfarin's anticoaguative effects are readily reversed by intravenous administration of (a) vitamin K, (b) prothrombin complex concentrate (PPC), which contains those factors formed in the absence of Warfarin, and/or (c) fresh frozen plasma (FFP), when available. Vitamin K1 is essential for sustaining the reversal achieved by PCC and FFP. Certain antibiotics, such as metronidazole, also retard Warfarin breakdown, and such effects also complicate its dosing. Use of Warfarin may also be compromised by off-target effects that suppress posttranslational γ-carboxylation of proteins (*e.g.*, osteocalcin, *or* matrix Gla protein) that are unrelated to blood clotting *per se*. (**See also** *Heparin, Clopidogrel Dabigatran, Rivaroxaban, and Apixaban*) **Use as a Rodenticide:** Because rats and mice experience life-threatening hemorrhage, after injesting sufficient quantities of Warfarin, this anticoagulant has been exploited as a reliable and powerful rodenticide sold under the tradenames R-Tox®, Co-Rax®, d-Con®, Dethmor®, Mar-Fin®, Rattunal®, Rax®, Rodex®, Rodex Blox®, Rosex®, Solfarin®, Tox-Hid®, Warf®, and Warfarat®. **Key Pharmacokinetic Parameters:** *See* Appendix II in Goodman & Gilman's THE PHARMACOLOGICAL BASIS OF THERAPEUTICS, 12th Edition (Brunton, Chabner & Knollmann, eds.) McGraw-Hill Medical, New York (2011). **Target(s):** Arylsulfatase (2,3); coagulation factor Xa (4,5); NAD(P)H dehydrogenase (quinone) (6); thyroxine 5'-deiodinase (7); vitamin K epoxide reductase, *primary anticoagulative target* (8-10); (*S*)-warfarin-7 hydroxylase (inhibited by the (*R*)-enantiomer) (11) **1.** Ikawa, Stahmann & Link (1944) *J. Amer. Chem. Soc.* **66**, 902. **2.** Ferrante, Messali, Ballabio & Meroni (2004) *Gene* **336**, 155. **3.** Franco, Meroni, Parenti, Levilliers, *et al.* (1995) *Cell* **81**, 15. **4.** Pejler, Lunderius & Tomasini-Johansson (2000) *Thromb. Haemost.* **84**, 429. **5.** Leadley, Jr. (2001) *Curr. Top. Med. Chem.* **1**, 151. **6.** Lind, Cadenas, Hochstein & Ernster (1990) *Meth. Enzymol.* **186**, 287. **7.** Goswami, Leonard & Rosenberg (1982) *Biochem. Biophys. Res. Commun.* **104**, 1231. **8.** Hildebrandt & Suttie (1982) *Biochemistry* **21**, 2406. **9.** Wallin & Guenthner (1997) *Meth. Enzymol.* **282**, 395. **10.** Suttie (1980) *Trends Biochem. Sci.* **5**, 302. **11.** Kunze, Eddy, Gibaldi & Trager (1991) *Chirality* **3**, 24.

WASP

This tetrameric cytoskeletal protein (MW = 210 kDa) NCBI Taxonomic Identifier = 9606), known as Wiskott-Aldrich Syndrome Protein, is a putative actin filament end-tracking protein that is activated by the Rho-family GTPase cdc42, thereby promoting actin-based motility. WASP inhibits cytosolic src protein-tyrosine kinase. **See also** *N-WASP inhibitor* **1.** Chong, Ia, Mulhern & Cheng (2005) *Biochim. Biophys. Acta* **1754**, 210.

(*S*)-WAY-100135

This systemically administered 5-HT$_{1A}$ receptor antagonist and anxiolytic agent (FW$_{free-base}$ = 395.55 g/mol; FW$_{di-HCl}$ = 468.47 g/mol; CAS 149007-54-5; Soluble to 100 mM in DMSO and to 5 mM in water, the latter with gentle warming), also named *N*-[2-[4-(2-methoxyphenyl)-1-piperazinyl]ethyl]-*N*-2-pyridinylcyclohexanecarboxamide, targets both presynaptic and postsynaptic serotonin type-1A receptors (IC$_{50}$ = 15 nM), with 100x selectivity *versus* other 5-HT subtypes as well as α$_1$, α$_2$ and D$_2$ receptors (IC$_{50}$ > 1000 nM). **1.** Cliffe et al (1993) *J. Med. Chem.* **36**, 1509. **2.** Fletcher et al (1993) *Eur. J. Pharmacol.* **237**, 283. **3.** Fletcher et al (1993) *Trends Pharmacol. Sci.* **14**, 41. **4.** Rodgers & Cole (1994) *Eur. J. Pharmacol.* **261**, 321.

WAY-100635

This 5-HT$_{1A}$ receptor antagonist (FW$_{free-base}$ = 422.57 g/mol; CAS 634908-75-1; Maleate salt is soluble to 25 mM in water and to 100 mM in DMSO), also named *N*-[2-[4-(2-methoxyphenyl)-1-piperazinyl]ethyl]-*N*-2-pyridinylcyclohexanecarboxamide, targets serotonin type-1A receptors (IC$_{50}$ = 2.2 nM and K_i = 0.84 nM for rat 5-HT$_{1A}$ receptors), with 100x selectivity *versus* other 5-HT subtypes. **1.** Zhuang, Kung & Kung (1994) *J. Med. Chem.* **37**, 1406. **2.** Forster, Cliffe, Bill, *et al.* (1995) *Eur. J. Pharmacol.* **281**, 81. **3.** Chemel, Roth, Armbruster, Watts & Nichols (2006) *Psychopharmacology* (Berlin) **188**, 244.

WAY-170523

This potent protease inhibitor (FW = 613.69 g/mol; CAS 307002-73-9), named systematically as benzofuran-2-carboxylic acid (2-{4-[benzyl(2-hydroxycarbamoyl-4,6-dimethylphenyl)sulfamoyl]phenoxy}ethyl)amide, inhibits matrix metalloproteinase-13 (collagenase 3), IC_{50} = 17 nM. **1.** Chen, Nelson, Levin, Mobilio, *et al.* (2000) *J. Amer. Chem. Soc.* **122**, 9648.

WAY-252623

This oral LXR agonist (FW = 422.78 g/mol; CAS 875787-07-8; Solubility: 100 mM in DMSO), also named 2-[(2-chloro-4-fluorophenyl)methyl]-3-(4-fluorophenyl)-7-(trifluoromethyl)-2H-indazole and LXR-623, targets the Liver X Receptors LXRβ and LXRα, with IC_{50} values of 24 and 179 nM, respectively. LXRs are ligand-inducible transcriptional activators that coordinately regulate cholesterol and lipid metabolism. The earlier discovery that oxysterols are potential endogenous ligands and the identification of first-generation of nonsteroidal LXR agonists helped to demonstrate that LXRs (a) mediate cholesterol efflux from macrophages via activation of ABCA1, ABCG1, and apoE, (b) promote HDL cholesterol elimination via the reverse cholesterol transport (RCT) process, and (c) modulate cholesterol absorption through intestinal activation of ABCA1 or ABCG5/G8 cholesterol transporters. Such considerations highlight the likelihood that LXRs may be druggable targets for controlling atherogenesis. In nonhuman primates with normal lipid levels, WAY-252623 significantly reduces total (50–55%) and LDL-cholesterol (LDLc) (70–77%) in a time- and dose-dependent manner as well as increased expression of the target genes ABCA1/G1 in peripheral blood cells. Statistically significant decreases in LDLc are noted as early as day-7, reaching a maximum by day-28, and exceeded reductions observed for simvastatin alone (20 mg/kg). Transient increases in circulating triglycerides and liver enzymes revert to baseline values over the course of the study. Complementary microarray analysis of duodenum and liver gene expression revealed differential activation of LXR target genes and suggested no direct activation of hepatic lipogenesis. **1.** Quinet, Basso, Halpern, *et al.* (2009) *J. Lipid. Res.* **50**, 2358. **2.** DiBlasio-Smith, Arai, Quinet, *et al.* (2008) *J. Transl. Med.* 6, 59.

WBA 8119, *See Brodifacoum*

WDR5 0103

This WDR antagonist (FW = 383.44 g/mol; CAS 890190-22-4; Soluble to 100 mM in DMSO), also named 3-[(3-methoxybenzoyl)amino]-4-(4-methyl-1-piperazinyl)benzoic acid methyl ester, targets WD-Repeat-containing protein-5, *or* WDR5 (K_d = 450 nM), a component of the mixed lineage leukemia (MLL) complex that methylates Lys-4 of histone H3 and was identified as a methylated Lys-4 histone H3-binding protein. WDR5-0103 disrupts WDR5 interactions with antagonist and inhibits MLL core complex methyltransferase activity *in vitro*. WDR5-0103 has no effect on a panel of seven other methyltransferases, including the histone-lysine *N*-methyltransferase SETD7. **1.** Senisterra, Wu, Allali-Hassani, *et al.* (2013) *Biochem. J.* 449, 151.

WEB 2086

This PAF receptor antagonist (FW = 455.96 g/mol; CAS 105219-56-5; Solubility: 100 mM in DMSO, 100 mM in Ethanol), also 4-[3-[4[(2-chlorophenyl)-9-methyl-6H-thieno[3,2-*f*][1,2,4]triazolo[4,3-*a*]diazepin-2-yl]-1-oxopropyl]morpholine, is a potent and selective platelet-activating factor receptor antagonist (K_i = 16.3 nM) that displays anti-inflammatory, antiangiogenic and anticancer activity. WEB 2086 inhibits growth and proliferation of MCF-7 breast cancer cells. **1.** Dent, Ukena, Chanez, *et al.* (1989) *FEBS Lett.* **244**, 365. **2.** Sariahmetoğlu, Cakici & Kanzik (1998) *Pharmacol. Res.* **38**, 173.

Wedelolactone

This ingredient in the Chinese herbal medicine *Han Lian Cao* (FW = 314.25 g/mol; CAS 524-12-9), also known as IKK Inhibitor II and 7-methoxy-5,11,12-trihydroxycoumestan, from the false daisy *Eclipta alba* suppresses LPS-induced caspase-11 expression by directly inhibiting the IKK complex. Wedelolactone is a cell-permeable, selective and irreversible inhibitor of IKKα and IKKβ kinase activity (IC_{50} < 10 μM), inhibiting NF-κB-mediated gene transcription by blocking the phosphorylation and degradation of IκBα. IKK Inhibitor II is without effect on p38 MAP kinase or Akt. Use of Han Lian Cao (墨旱蓮) is indicated for the treatment of coronary disease, dysentery, infantile urinary tract infections, digestive tract hemorrhage, acute nephritis, and shingles. **Target(s):** IKK kinase (1); 5-lipoxygenase (2) ; trypsin (3). **1.** Kobori, Yang, Gong, Heissmeyer, *et al.* (2004) *Cell Death Differ.* **11**, 123. **2.** Wagner & Fessler (1986) *Planta Med.* (5), 374. **3.** Syed, Deepak, Yogisha, Chandrashekar, *et al.* (2003) *Phytother Res.* **17**, 420.

WF-536, *See (R)-(+)-4-(1-Aminoethyl)-N-(4-pyridyl)benzamide*

WF-10129

This fungal metabolite (FW = 408.41 g/mol) is a potent inhibitor of peptidyl-dipeptidase A (angiotensin I-converting enzyme). WF-10129 inhibits the pressor response of angiotensin I, when administered intravenously at 0.3 mg/kg in rats. **1.** Mathey & Reuter (1989) *J. Basic Microbiol.* **29**, 623. **2.** Ando, Okada, Uchida, Hemmi, *et al.* (1987) *J. Antibiot.* **40**, 468.

WHI-P97

This quinazoline (FW$_{free-base}$ = 455.11 g/mol), also known as JAK3 inhibitor III and systematically named as 4-(3',5'-dibromo-4'-hydroxyphenyl)amino-6,7-dimethoxyquinazoline, is a cell-permeant inhibitor of JAK3 (Janus kinase 3) protein-tyrosine kinase, IC$_{50}$ = 11 μM. This inhibitor also reduces the invasiveness of EGFR-positive human cancer cells in a dose-dependent manner (K_i = 0.09 μM). **Target(s):** epidermal growth factor receptor protein-tyrosine kinase (1); JAK3 protein-tyrosine kinase (2). **1.** Ghosh, Narla, Zheng, Liu, *et al.* (1999) *Anticancer Drug Des.* **14**, 403. **2.** Sudbeck, Liu, Narla, Mahajan, *et al.* (1999) *Clin. Cancer Res.* **5**, 1569.

WHI-P131

This quinazoline (FW = 297.31 g/mol; CAS 202475-60-3), also known as 4-(4'-hydroxyphenyl)amino-6,7-dimethoxyquinazoline, JAK3 inhibitor I, and JANEX-1, is a potent and selective inhibitor of the Janus kinase 3 with an IC$_{50}$ value of 78 μM. WHI-P131 also acts as a potent inhibitor of glioblastoma cell adhesion and migration and blocks thrombin-induced platelet aggregation. **Target(s):** JAK3 (Jun N-terminal kinase; Janus kinase 3) protein-tyrosine kinase (1-3). **1.** Uckun, Ek, Liu & Chen (1999) *Clin. Cancer Res.* **5**, 2954. **2.** Sudbeck, Liu, Narla, Mahajan, *et al.* (1999) *Clin. Cancer Res.* **5**, 1569. **3.** Goodman, Niehoff & Uckun (1998) *J. Biol. Chem.* **273**, 17742.

WHI-P154

This cell-permeable quinazoline (FW = 376.21 g/mol; CAS 211555-04-3), also known as JAK3 inhibitor II and 4-[(3'-bromo-4'-hydroxyphenyl)amino]-6,7-dimethoxyquinazoline, is a potent inhibitor of Janus kinase 3 (JAK3) protein-tyrosine kinase (IC$_{50}$ = 5.6 μM). WHI-P154 also inhibits glioblastoma cell adhesion and migration. **Target(s):** epidermal growth factor receptor protein-tyrosine kinase (1); JAK3 (Janus kinase 3) protein-tyrosine kinase (1-3). **1.** Ghosh, Jennissen, Liu & Uckun (2001) *Acta Crystallogr. C* **57**, 76. **2.** Goodman, Niehoff & Uckun (1998) *J. Biol. Chem.* **273**, 17742. **3.** Sudbeck, Liu, Narla, Mahajan, *et al.* (1999) *Clin. Cancer Res.* **5**, 1569.

WIN Compounds

This class of antiviral drugs, originally developed at the Sterling-Winthrop Research Institute, bind within a hydrophobic pocket in the viral capsid, thereby replacing a hydrocarbon molecule that normally occupies the site. Lipid release from the capsid during viral entry permits the structural changes essential for uncoating and release of pathogen genome. WIN compounds are thought to block poliovirus and other picornovirus infectivity by preventing the transition to infective 135S particles. By replacing the lipid bound within the hydrophobic pocket, WIN compounds prevent structural transitions needed for virus uncoating. *See* WIN 51711

WIN 51711

This systemically active broad-spectrum antipicornavirus agent (FW = 342.44 g/mol; CAS 87495-31-6), also known as Disoxaril™ and named systematically as 5-[7-[4(4,5-dihydro-2-oxazolyl)phenoxy]heptyl]-3-methylisoxazole], exerts its antiviral effect through a direct interaction with picornaviral capsid protein(s), stabilizing the virion conformation, thereby preventing a thermally-induced conformational change needed for virus uncoating and replication (1). WIN 51711, which is not virucidal and is without effect on the kinetics of ^{3}H-uridine-labelled poliovirus-2, exhibits a broad spectrum of action against entero- and rhinovirus serotypes, inhibiting all 33 rhinovirus serotypes tested at concentrations ranging from 0.02-to-6 mg/L, and also inhibiting all echo-, Coxsackie A and B, and polio serotypes tested as well as EV-70 at concentrations ranging from 0.004 to 1 mg/L. *Note* The effectiveness of WIN 51711 may be enhanced when administered as a dimethyl-β-cyclodextrin inclusion complex. **1.** McKinlay (1985) *J. Antimicrob. Chemother.* **16**, 284.

WIN 13099, *See* N,N'-Bis(dichloroacetyl)-N,N'-diethyl-1,4-xylylenediamine

WIN 63759

This oral benzisothiazolone (FW$_{free-base}$ = 587.48 g/mol), also known as [6-methoxy-4-(1-methylethyl)-1-oxo-1,2-benzisothiazol-3(2H)-yl]methyl 2,6-dichloro-3-[2-(4-morpholinyl)ethoxy]benzoate *S,S* dioxide, is a potent inhibitor of human leukocyte elastase (K_i = 13 pM). **1.** Hlasta, Subramanyam, Bell, Carabateas, *et al.* (1995) *J. Med. Chem.* **38**, 739.

WIN 64733

This oral benzisothiazolone (FW$_{free-base}$ = 571.48 g/mol), also known as [6-methoxy-4-(1-methylethyl)-1-oxo-1,2-benzisothiazol-3(2H)-yl]methyl 2,6-dichloro-3-[2-(1-pyrrolidinyl)ethoxy]benzoate *S,S* dioxide, is a potent inhibitor of human leukocyte elastase (K_i = 14 pM). **1.** Hlasta, Subramanyam, Bell, Carabateas, *et al.* (1995) *J. Med. Chem.* **38**, 739.

WIPTIDE, *See* L-Threonyl-L-threonyl-L-tyrosyl-L-alanyl-L-aspartyl-L-phenylalanyl-L-isoleucyl-L-alanyl-L-serylglycyl-L-arginyl-L-threonylglycyl-L-arginyl-L-arginyl-L-asparaginyl-L-alanyl-L-isoleucinamide

Wiskostatin

This N-WASP inhibitor (FW = 426.15 g/mol; CAS 253449-04-6), also named 3,6-dibromo-α-[(dimethylamino)methyl]-9H-carbazole-9-ethanol, targets neural Wiskott-Aldrich Syndrome Protein, an actin filament end-tracking protein, by binding to its regulatory GTPase-binding domain, maintaining N-WASP in an inactive conformation and thereby interfering with its ability to activate Arp2/3 complex. Wiskostatin also inhibits PIP$_2$-induced actin polymerization (EC$_{50}$ ~4µM). (*See also WASP; N-WASP inhibitor*) **1.** Peterson et al (2004) Chemical inhibition of N-WASP by stabilization of a native autoinhibited conformation. *Nature Struct. Mol. Biol.* **11**, 747. **2.** Guerriero & Weisz (2006) *Am. J.Physiol. Cell Physiol.* **292**, C1562. **3.** Wegner et al (2008) *J. Biol. Chem.* **283**, 15912.

Withaferin A

This broadly active, epoxide-containing steroid lactone or withanolide (FW = 470.61 g/mol; CAS 5119-48-2; *Symbol*: WFA), also known as NSC-101088 and (4β,5β,6β,22R)-4,27-dihydroxy-5,6:22,26-diepoxyergosta-2,24-diene-1,26-dione, was first isolated from *Withania somnifer*, also known as Winter Cherry, a medicinal plant used traditionally in East Indian ayurvedic system of medicine to treat arthritis, immunological disorders, and bleeding conditions in women. Withaferin A binds to vimentin by covalently modifying a cysteine residue within its highly conserved α-helical coiled coil 2B domain of this intermediate filament protein (1). WFA induces vimentin filaments to aggregate *in vitro*, a property manifested *in vivo* as punctate cytoplasmic aggregates that colocalize vimentin and F-actin (1). WFA is also a proteasomal inhibitor that binds to specific catalytic β-subunit of the 20S proteasome (2). It exerts positive effect on osteoblast by increasing osteoblast proliferation and differentiation. WFA increases expression of osteoblast-specific transcription factor and mineralizing genes, promoted osteoblast survival and suppressed inflammatory cytokines. In osteoclasts, WFA treatment decreases osteoclast number directly by decreasing expression of tartarate-resistant acid phosphatase and receptor activator of NFκ-B and indirectly by decreasing osteoprotegrin/NFκ-B ligand ratio (2). At 1-3 µM, WFA also binds to Hsp90, inhibiting its action and leading to the degradation of its client proteins Akt and Cdk4 via a proteasome-dependent pathway in pancreatic cancer cells (3). Withaferin A also inhibits binding of Sp1 transcription factor to VEGF-gene promoter, thereby providing a mechanism for its antiangiogenic activity (4). While Withaferin A reduces cell survival in a dose-dependent manner (LD$_{50}$ = 16 µM), treatment with a 2.1-µM non-toxic dose significantly enhances radiation-induced cell killing, giving a sensitizer enhancement ratio of 1.5 for 37% survival and 1.4 for 10% survival. Cells accumulate at the G$_2$-M transition within 4 hour after a 1-hour treatment at 10.5 µM WFA (5). **1.** Bargagna-Mohan, Hamza, Kim, *et al.* (2007) *Chem. & Biol.* **14**, 623. **2.** Khedgikar, Kushwaha, Gautam, *et al.* (2013) *Cell Death Dis.* **4**, e778. **3.**

Gu, Yu, Gunaherath, *et al.* (2013) *Invest. New Drugs* **32**, 68. **4.** Kumar, Shilpa, and Salimath Bharati (2009) *Curr. Trends Biotechnol. Pharm.* **3**, 138. **5.** Devi, Akagi, Ostapenko, Tanaka & Sugahara (1996) *Int. J. Radiat. Biol.* **69**, 193.

Withania somnifera Glycoprotein

This glycoprotein (MW = 28 kDa; Symbol: WSG, from the folk medicinal plant *Withania somnifera* inhibits the phospholipase A$_2$ activity found in Indian cobra *Naja naja* venom. While unable to neutralize cobra venom toxicity, WSG prolongs the time to death in experimental mice by a factor of ten. **1.** Machiah & Gowda (2006) *Biochimie* **88**, 701.

Wnt-C59

This PORCN inhibitor (FW = 379.46 g/mol; CAS 1243243-89-1; Solubility: 76 mg/mL DMSO, <1 mg/mL H$_2$O), also known as C59 and 2-(4-(2-methylpyridin-4-yl)phenyl)-N-(4-(pyridin-3-yl)phenyl)acetamide, targets Wnt3A-mediated activation (IC$_{50}$ = 74 pM), thereby inhibiting both autocrine and paracrine signaling by sonic hedgehog, *or* Shh (1,2). Once-daily oral administration was sufficient to maintain blood concentrations well above the IC$_{50}$ (1). **Mode of Inhibitor Action:** The *PORCN* [*Official Name*: Porcupine homologue (*Drosophila*)] is a gene that codes for an endoplasmic reticulum transmembrane O-acyltransferase required for Wnt palmitoylation, secretion, and biologic activity in the processing of *Drosophila* wingless proteins. Knockdown of PORCN by multiple independent siRNAs results in a cell growth defect in a subset of epithelial cancer cell lines, suggesting that porcupine is a potentially useful druggable target in controlling cancer. Wnt-C59 inhibits PORCN activity *in vitro* at nanomolar concentrations, as assessed by inhibition of Wnt palmitoylation (1), Wnt interaction with the carrier protein Wntless/WLS, Wnt secretion, and Wnt activation of β-catenin reporter activity (1). Palmitoylation of Shh by Hedgehog acyltransferase (Hhat) also plays a critical role in regulating Shh's signaling potency of in cells. Porcupine inhibition by Wnt-C59 (100 nM) of serum-starved Wnt-3a cells for 3 hours blocked Wnt-3a palmitoyolation, whereas RU-SKI-43 (10 µM; *See also RU-SKI-43*) did not (3). WNT10B induces transcriptionally active β-catenin in highly aggressive metastatic triple-negative breast cancers (TNBC) and predicts survival-outcome of patients with both TNBC and basal-like tumors (4). **1.** Proffitt, *et al.* (2013) *Cancer Res.* **73**, 502. **2.** Petrova, et al. (2013) *Nature Chem. Biol.* **9**, 247. **3.** Wend, *et al.* (2013) *EMBO Mol. Med.* **5**, 264. **4.** Wend, Runke, Wend, *et al.* (2013) *EMBO Mol. Med.* **5**, 264.

Wogonin

This O-methylated flavone (FW = 284.07 g/mol; CAS 632-85-9) from the root of *Scutellaria baicalensis* (1), also known as Vogonin, Norwogonin 8-methyl ether, and 5,7-dihydroxy-8-methoxyflavone, as well as systematically as 5,7-dihydroxy-8-methoxy-2-phenyl-4H-chromen-4-one, is a potent sialidase inhibitor (1). **Target(s):** NAD(P)H:quinone acceptor oxidoreductase, *or* DT-diaphorase (2); xanthine oxidase (3); TPA-induced COX-2 expression (4); IL-6-induced angiogenesis by down-regulation of VEGF and VEGFR-1, not VEGFR-2 (5); TNF-alpha-induced MMP-9 expression by blocking the NFκB activation via MAPK signaling pathways in human aortic smooth muscle cells (6); microglial cell migration by suppression of NFκB activity (7); induction of G$_1$ phase arrest by inhibiting Cdk4 and cyclin D$_1$ concomitant with an elevation in p21Cip1 in human cervical carcinoma HeLa cells (8); P-glycoprotein multidrug resistance, thereby potentiating etoposide-induced apoptosis in cancer cells (9); human cytochrome P450 1A2 (10); indoleamine 3,5-dioxygenase (11,12). **1**. Nagai, Yamada & Otsuka (1989) *Planta Med.* **55**, 27. **2**. Liu, Liu, Iyanagi, *et al.* (1990) *Mol. Pharmacol.* **37**, 911. **3**. Chang, Lee, Lu & Chiang (1993) *Anticancer Res.* **13**, 2165. **4**. Park, Heo, Park & Kim (2001) *Eur. J. Pharmacol.* **425**, 153. **5**. Lin, Chang, Chen, Wu & Chiu (2006) *Planta Med.* **72**, 1305. **6**. Lee, Jeong, Yu, *et al.* (2006) *Biochem. Biophys. Res. Commun.* **351**, 118. **7**. Piao, Choi, Park, *et al.* (2008) *Int. Immunopharmacol.* **8**, 1658. **8**. Yang, Zhang, Hu, *et al.* (2009) *Biochem. Cell Biol.* **87**, 933. **9**. Lee, Enomoto, Koshiba & Hirano (2009) *Ann. N. Y. Acad. Sci.* **1171**, 132. **10**. Shao, Zhao, Li, *et al.* (2012) *Eur. Biophys. J.* **41**, 297. **11**. Chen, Corteling, Stevanato & Sinden (2012) *Biochem. Biophys. Res. Commun.* **429**, 117. **12**. Chen, Corteling, Stevanato & Sinden (2012) *Discov Med.* **14**, 327.

Wortmannin

This potent and cell-permeable furanosteroid antifungal agent (FW = 428.44 g/mol; CAS 19545-26-7), systematically named as (1α,11α)-11-(acetyloxy)-1-(methoxymethyl)-2-oxaandrosta-5,8-dieno[6,5,4-*bc*]furan-3,7,17-trione, from *Penicillium funiculosum* is a potent irreversible inhibitor of 1-phosphatidylinositol 3-kinase (IC$_{50}$ = 3 nM) and has a similar potency for other Class-I, -II, and -III PI3Ks *in vitro*. Wortmannin also inhibits a polo-type protein kinases, including mTOR and DNA-PK. Wortmannin is a widely used to block DNA repair, receptor-mediated endocytosis, transcytosis, vacuolar sorting, and cell proliferation. Wortmannin's short (10-min) intracellular half-life greatly limits its utility in experiments requiring sustained inhibitory action. Its derivative PX-866 is an orally active and stable PI3K inhibitor that is more suitable for sustained *in vivo* action (***See also** PX-866; as well as the highly selective Vps34 inhibitor, VPS34-IN1*). **Target(s):** DNA-dependent protein kinase (1); inositol-trisphosphate 3-kinase (2); inositol polyphosphate multi-kinase (3); [myosin light-chain] kinase (4-6); phosphatidylinositol-4,5 bisphosphate 3-kinase (7-9); 1-phosphatidylinositol 3-kinase (6,11-16); 1-phosphatidylinositol 4 kinase (26-30); phosphatidylinositol-4-phosphate 3-kinase (17-22); phospholipase D (23); polo kinase (24); protein-tyrosine kinase (25). **1**. Izzard, Jackson & Smith (1999) *Cancer Res.* **59**, 2581. **2**. Mayr, Windhorst & Hillemeier (2005) *J. Biol. Chem.* **280**, 13229. **3**. Saiardi, Resnick, Snowman, *et al.* (2005) *Proc. Natl. Acad. Sci. U.S.A.* **102**, 1911. **4**. Nakanishi, Kakita, Takahashi, *et al.* (1992) *J. Biol. Chem.* **267**, 2157. **5**. Grimm, Haas, B. Willipinski-Stapelfeldt, *et al.* (2005) *Cardiovasc. Res.* **65**, 211. **6**. S. P. Davies, H. Reddy, M. Caivano & Cohen (2000) *Biochem. J.* **351**, 95. **7**. Metzner, Heger, Hofmann, *et al.* (1997) *Biochem. Biophys. Res. Commun.* **232**, 719. **8**. Vanhaesebroeck, Welham, Kotani, *et al.* (1997) *Proc. Natl. Acad. Sci. U.S.A.* **94**, 4330. **9**. Stoyanov, Volinia, Hanck, *et al.* (1995) *Science* **269**, 690. **10**. Powis, Bonjouklian, Berggren, *et al.* (1994) *Cancer Res.* **54**, 2419. **11**. Linassier, MacDougall, Domin & Waterfield (1997) *Biochem. J.* **321**, 849. **12**. Yano, Agatsuma, Nakanishi, Saitoh, *et al.* (1995) *Biochem. J.* **312**, 145. **13**. Okada, Sakuma, Fukui, Hazeki & Ui (1994) *J. Biol. Chem.* **269**, 3563. **14**. Cataldi, Di Pietro, Centurione, *et al.* (2000) *Cell. Signal.* **12**, 667. **15**. Wang & Sul (1998) *J. Biol. Chem.* **273**, 25420. **16**. Huang, Dedousis, Bhatt & O'Doherty (2004) *J. Biol. Chem.* **279**, 21695. **17**. Wymann, Bulgarelli-Leva, Zvelebil, *et al.* (1996) *Mol. Cell. Biol.* **16**, 1722. **18**. Crljen, Volinia & Banfic (2002) *Biochem. J.* **365**, 791. **19**. Shepherd (2005) *Acta Physiol. Scand.* **183**, 3. **20**. Domin, Pages, Volinia, *et al.* (1997) *Biochem. J.* **326**, 139. **21**. Ono, Nakagawa, Saito, *et al.* (1998) *J. Biol. Chem.* **273**, 7731. **22**. Molz, Chen, Hirano & Williams (1996) *J. Biol. Chem.* **271**, 13892. **23**. Naccache, Caon, Gilbert, *et al.* (1993) *Lab. Invest.* **69**, 19. **24**. Johnson, Stewart, Woods, *et al.* (2007) *Biochemistry* **46**, 9551. **25**. Eriksson, Toivola, Sahlgren, *et al.* (1998) *Meth. Enzymol.* **298**, 542. **26**. Cutler, Heitman & Cardenas (1997) *J. Biol. Chem.* **272**, 27671. **27**. Zhao, Varnai, Tuymetova, *et al.* (2001) *J. Biol. Chem.* **276**, 40183. **28**. Jeganathan & Lee (2007) *J. Biol. Chem.* **282**, 372. **29**. Tóth, Balla, Ma, *et al.* (2006) *J. Biol. Chem.* **281**, 36369. **30**. Westergren, Ekblad, Jergil & Sommarin (1999) *Plant Physiol.* **121**, 507.

WP1066

This novel protein kinase inhibitor (FW = 356.22 g/mol; CAS 857064-38-1; Solubility = 44 mg/mL DMSO), also known as HY-15312 and (*S,E*)-3-(6-bromopyridin-2-yl)-2-cyano-*N*-(1-phenylethyl)acrylamide), is an AG490 analogue that targets JAK2^{V617F} (IC$_{50}$ = 2.3 μM), an activating somatic mutation in the Janus kinase-2 gene in patients with myeloproliferative disorder (1-4). WP1066 markedly and dose-dependently inhibits growth of HEL erythroid leukemia cells bearing the JAK2^{V617F} mutant, displaying IC$_{20}$, IC$_{50}$ and IC$_{80}$ values of 0.8 μM, 2.3 μM and 3.8 μM, respectively. **Other Targets:** WP1066 also inhibits STAT3 (IC$_{50}$ = 2.43 μM). At 0.5, 1.0, 2.0, 3.0, or 4.0 μM, WP1066 inhibits the phosphorylation of JAK2, STAT3, STAT5, and ERK1/2, respectively, without affecting the phosphorylation of JAK1 and JAK3 in HEL cells expressing JAK2^{V617F} (1). WP1066 (0.5-3.0 μM) also inhibits the proliferation of patient-derived AML colony-forming cells as well as AML cell lines OCIM2 and K562. When present at at concentrations of 0.5-4.0 μM, WP1066 dose-dependently decreases JAK2 and pJAK2 protein levels as well as downstream phosphorylation levels of STAT3, STAT5, and AKT in OCIM2 and K562 cells (1,2). WP1066 at concentrations of 2 μM inhibits OCIM2 cell proliferation by trapping cells at G$_0$-G$_1$. At 1-3 μM, it likewise induces apoptosis in both OCIM2 and K562 cells, activating both pro-caspase-3 and cleaving PARP (3). At 5 μM, WP1066 prevents the phosphorylation of STAT3, and at concentrations of 2.5μM WP1066 significantly inhibits cell survival and proliferation in Caki-1 and 786-O renal cancer cells (4). At 5 μM, WP1066 suppresses HIF1α and HIF2α expression as well as VEGF production in Caki-1 and 786-O renal cancer cells (4). **1**. Verstovsek, Manshouri, Quintás-Cardama, *et al.* (2008) *Clin. Cancer Res.* **3**, 788. **2**. Hatiboglu, Kong, Wei, *et al.* (2012) *Int. J. Cancer* **131**, 8. **3**. Ferrajoli, Faderl, Van, *et al.* (2007) *Cancer Res.* **67**, 11291. **4**. Horiguch, Asano, Kuroda, *et al.* (2010) *Br. J. Cancer* **102**, 1592.

WPE-III-31C

This cell-permeable (hydroxyethyl)urea peptidomimetic (FW = 640.82 g/mol), also known as γ-secretase inhibitor XVII, is a γ-secretase transition-state analogue (IC$_{50}$ = 300 nM), when assayed for its effect on A-β peptide formation (1-3). For cellular inhibition studies, 1 μM WPE-III-31C is a nominal concentration. Note also that incubation of cells with biotinylated-WPE-III-31C, followed by incubation with streptavidin, and later staining with anti-streptavidin antibody may be employed to localize the secretase within cells. **1**. Campbell, Iskandar, Reed & Xia (2002) *Biochemistry* **41**, 3372. **2**. Esler, Kimberly, Ostaszewski, *et al.* (2002) *Proc. Natl. Acad. Sci. U.S.A.* **99**, 2720. **3**. Kimberly, LaVoie, Ostaszewski, *et al.* (2002) *J. Biol. Chem.* **277**, 35113.

WT

This cationic peptide amide (MW = 2988 g/mol; *Sequence:* MLSLRQS IRFFKPATRTLCSSRYLL-NH$_2$; pI = 11.56, corresponding to a membrane-inserting pre-sequence for mitochondrial import of yeast cytochrome *c* oxidase subunit IV, inhibits bovine heart mitochondrial F$_1$ ATPase, IC$_{50}$ = 16 μM (1) When its single thiol group is fluorescently labeled, the resulting peptide is imported with first-order kinetics that the concentration of the lipid-bound peptide (2). **1**. Bullough, Ceccarelli, Roise & Allison (1989) *Biochim. Biophys. Acta* **975**, 377. **2**. Roise (1992) *Proc. Natl Acad. Sci. U.S.A.* **89**, 608.

WWL 123

This brain-penetrant ABHD6 inhibitor (FW = 436.51 g/mol; CAS 1338574-83-6; Soluble to 100 mM in DMSO), also named N-[[1,1'-biphenyl]-3-ylmethyl)-N-methylcarbamic acid 4'-(aminocarbonyl)[1,1'-biphenyl]-4-yl ester, targets α/β-hydrolase domain 6 (IC$_{50}$ = 0.43 μM), a serine hydrolyzing enzyme that possesses typical α/β-hydrolase family domains. Notably, ABHD6 produces ~4% of 2-arachidonoylglycerol (2-AG) in brain hydrolysis.[3] Moreover, monoacylglycerol lipase (MAGL), ABHD12, and ABHD6 control 99% of 2-AG signalling to the cannabinoid (CB) receptors in the brain . WWL-123 also blocks PTZ-induced seizures in healthy mice and spontaneous seizures in R6/2 mice. Its antiepileptic activity appears to be independent of CB$_1$ and CB$_2$ receptors, but is GABA$_A$ receptor dependent. **1**. Naydenov, Horne, Cheah, et al. (2014) Neuron **83**, 361.

Wy-50295

This oral anti-allergy agent (FW$_{free-acid}$ = 357.41 g/mol; CAS 133304-99-1) inhibits arachidonate 5-lipoxygenase (1,2). Although WY-50295T inhibits both in vitro and ex vivo rat whole blood leukocyte LTB4 formation (IC$_{50}$ = 40 μM; oral ED$_{50}$ of 18 mg/kg), it does not inhibit LTB4 production in calcium ionophore stimulated human whole blood at concentrations to 200 μM (3). Even when one seeks to reduce binding of WY-50295T to serum albumin (with 250 μM naphthalene sulfonic acid (> 99.9% binding to albumin primarily at the carboxylic site) and/or 250 μM sulfanilamide (binding to nonspecific sites), separately or in combination), WY-50295T still does not inhibit 5-LO and its free-drug blood levels remain unchanged. When purified human neutrophils are treated with WY-50295T in the presence of fatty acid saturated albumin (fraction V) was employed, WY-50295T' inhibition of 5-LO is also prevented, whereas 5 μM zileuton inhibits LTB4 production by 99%. Such results suggest that the high affinity binding of WY-50295T to human albumin and possibly the reduction of drug uptake (passive diffusion) using purified human versus rat neutrophils may account for the inactivity of WY-50295T in the human whole blood assay (3). **1**. Carlson, Kreft, Hartman, et al. (1999) Prostaglandins Leukot. Essent. Fatty Acids **60**, 31. **2**. Weichman, Berkenkopf, Chang, et al. (1991) Agents Actions Suppl. **34**, 201. **3**. Carlson, Kreft, Hartman, et al. (1999) Prostaglandins Leukot. Essent. Fatty Acids **60**, 31.

WZ4002

This potent orally bioavailable EGFR-PKI (FW = 494.98 g/mol; CAS 1213269-23-8; Soluble in DMSO), also named N-[3-[[5-chloro-2-[[2-methoxy-4-(4-methyl-1-piperazinyl)phenyl]amino]-4-pyrimidinyl]oxy] phenyl]-2-propenamide, targets Epidermal Growth Factor Receptor, including the Thr-790-Met mutation of the so-called Thr-790 "gatekeeper" residue (IC$_{50}$ = 8 nM) in gefitinib-resistant or erlotinib-resistant EGFR-mutant non-small-cell lung cancer patients. As a Michael-type inhibitor, WZ4002 is an irreversible inhibitor that undergoes 1,4-addition. WZ4002 is also effective against the EGFR Leu-758-Arg mutant (IC$_{50}$ = 8 nM), but not Thr-798-Ile of ERB2 (1-3). **1**. Zhou, Ercan, Chen, et al. (2009) Nature **462**, 1070. **2**. Sakuma, Yamazaki, Nakamura, et al. (2012) Lab Invest. **92**, 371. **3**. Zannetti, Iommelli, Speranza, Salvatore & Del Vecchio (2012) J. Nucl. Med. **53**, 443.

WZ4003

This highly specific protein kinase inhibitor (FW = 496.99 g/mol; CAS 1214265-58-3) targets the NUAK family SNF1-like kinases, NUAK1 (IC$_{50}$ = 20 nM) and NUAK2 (IC$_{50}$ = 100 nM). Note: NUAK1 is also known as AMPK-related protein kinase 5 (ARK5), an enzyme encoded by the human NUAK1 gene, and NUAK2 is also known as SNF1/AMP kinase-related kinase (SNARK), an enzyme encoded by the human NUAK2 gene. Both are activated by the LKB1 (liver kinase B1) tumour suppressor kinase. Notably, WZ4003 is without significant inhibition against 139 other kinases. Moreover, in all cell lines tested, WZ4003 inhibits phosphorylation of the well-characterized substrate, MYPT1 (or Myosin phosphate-targeting subunit 1), which is phosphorylated at Ser-445 by NUAK1 (1). **1**. Banerjee, Buhrlage, Huang, et al. (2014) Biochem. J. **457**, 215. **2**. Zhou, Ercan, Chen, et al. (2009) Nature **462**, 1070.

– X –

X, See *Xanthosine*

X-537A, See *Lasalosid A*

Xanthate, Potassium Ethyl, See *Potassium Xanthogenate*

Xanthidylate and Xanthidylic Acid, See *Xanthosine 5'-Monophosphate*

XAC

This xanthine amine congener and selective adenosine receptor antagonist (FW = 464.95 g/mol; CAS 96865-92-8; Solubility: 5 mM in DMSO), also known as *N*-(2-aminoethyl)-2-[4-(2,3,6,7-tetrahydro-2,6-dioxo-1,3-dipropyl-1*H*-purin-8-yl)phenoxy]acetamide hydrochloride, targets A_1 (IC_{50} = 1.8 nM) and A_2 (IC_{50} = 114 nM) adenosine receptors. XAC attenuates adenosine-induced vasodilation and exhibits proconvulsant activity *in vivo*. Pretreatment of animals with the adenosine receptor agonists 2-chloroadenosine, N6-cyclohexyladenosine or 5'-N-ethylcarboxamido-adenosine (1 mg/kg, i.p., 20 minutes prior to infusion) significantly decreased the seizure threshold of both XAC and caffeine (1). XAC rapidly equilibrates within the myocardial interstitial space and, as a result of blocking adenosine receptors, increases interstitial and venous adenosine concentrations (2). Increases in interstitial adenosine may partially overcome the adenosine receptor blockade by XAC, thereby reducing the effectiveness of XAC in attenuating the hypoxic vasodilatation. XAC attenuates intracoronary adenosine induced vasodilatation (mediated by endothelial adenosine receptors) much more effectively than it attenuates hypoxic vasodilatation (2). **1.** Morgan, Deckert, Jacobson, Marangos & Daly (1989) *Life Sci.* **45**, 719. **2.** Sawmiller, Linden & Berne (1994) *Cardiovasc.Res.* **28**, 604.

Xanthine

This naturally occurring purine (FW = 152.11 g/mol; CAS 69-89-6), symbolized by Xan; primarily lactam tautomer; λ_{max} = 267 nm, with ε = 10300 $M^{-1}cm^{-1}$). Xanthine has a pK_a value of 9.91 at 40°C. It was first prepared in the laboratory by Emil Fischer (1). **Target(s):** adenine phosphoribosyltransferase, weakly (2); cyclic-nucleotide 3' phosphodiesterase (3); hydroxy-acylglutathione hydrolase, *or* glyoxalase II (4); hypoxanthine(guanine) phosphoribosyltransferase (5-8); lactoylglutathione lyase, *or* glyoxalase I (4); methionine *S* adenosyltransferase (9); NADase, weakly (10,11); tRNA-guanine transglycosylase, weakly (12,13); urate-ribonucleotide phosphorylase (14); uricase, *or* urate oxidase (15-19). **1.** Fischer (1882) *Ann.* **215**, 309. **2.** Tuttle & Krenitsky (1980) *J. Biol. Chem.* **255**, 909. **3.** Beavo, Rogers, Crofford, *et al.* (1970) *Mol. Pharmacol.* **6**, 597. **4.** Oray & Norton (1980) *Biochem. Biophys. Res. Commun.* **95**, 624. **5.** Schimandle, Mole & Sherman (1987) *Mol. Biochem. Parasitol.* **23**, 39. **6.** Nussbaum & Caskey (1981) *Biochemistry* **20**, 4584. **7.** Doyle, Kanaani & Wang (1998) *Exp. Parasitol.* **89**, 9. **8.** Miller, Ramsey, Krenitsky & Elion (1972) *Biochemistry* **11**, 4723. **9.** Berger & Knodel (2003) *BMC Microbiol.* **3**, 12. **10.** Alivisatos, Kashket & Denstedt (1956) *Can. J. Biochem. Physiol.* **34**, 46. **11.** Webb (1966) *Enzyme and Metabolic Inhibitors*, vol. **2**, p. 492,

Academic Press, New York. **12.** Farkas, Jacobson & Katze (1984) *Biochim. Biophys. Acta* **781**, 64. **13.** Todorov & Garcia (2006) *Biochemistry* **45**, 617. **14.** Laster & Blair (1963) *J. Biol. Chem.* **238**, 3348. **15.** Van Pilsum (1953) *J. Biol. Chem.* **204**, 613. **16.** Bergmann, Kwietny-Govrin, Ungar-Waron, Kalmus & Tamari (1963) *Biochem. J.* **86**, 567. **17.** Baum, Hübscher & Mahler (1956) *Biochim. Biophys. Acta* **22**, 514. **18.** Mahler (1963) *The Enzymes*, 2nd ed. (Boyer, Lardy & Myrbäck, eds.), **8**, 285. **19.** Webb (1966) *Enzyme and Metabolic Inhibitors*, vol. 2, p. 284, Academic Press, New York. **See also** *Xanthosine; Xanthosine 5'-Monophosphate*

Xanthine Riboside, See *Xanthosine*

Xanthobilirubate

This pyrrole derivative ($FW_{free-acid}$ = 302.37 g/mol) is a more soluble, low-molecular-weight analogue of bilirubin that inhibits HIV-1 protease (K_i = 5 µM). At 100 µM, xanthobilirubic acid affected viral assembly, resulting in a 50% decrease in the generation of infectious particles, whereas the same concentrations of biliverdin or bilirubin exerts little or no effect on viral assembly. **1.** McPhee, Caldera, Bemis, *et al.* (1996) *Biochem. J.* **320**, 681.

Xanthophyll Dipalmitate Ester, See *Xanthophyll*

Xanthopterin

This pteridines (FW = 179.14 g/mol; CAS 119-44-8; Fluorescent: excitation λ_{max} = 380 and 400 nm; emission λ_{max} = 440 and 460 nm), also known as 2-amino-4,6-dihydroxypteridine, is a yellow pigment in butterfly wings. A metabolic end-product of the nonconjugated pteridine compound, xanthopterin is also found in a other insects, crabs, and even humans. Xanthopterin is an alternative substrate for xanthine oxidase and competitively inhibits urate formation (1). **Target(s):** dihydrofolate reductase, weakly (1); guanine deaminase (2); hypoxanthine(guanine) phosphoribosyltransferase (3); sepiapterin deaminase (4-6); xanthine oxidase, also alternative substrate (7-10). **1.** Bertino, Perkins & Johns (1965) *Biochemistry* **4**, 839. **2.** Dietrich & Shapiro (1953) *J. Biol. Chem.* **203**, 89. **3.** Miller, Ramsey, Krenitsky & Elion (1972) *Biochemistry* **11**, 4723. **4.** Tsusue (1971) *J. Biochem.* **69**, 781. **5.** Tsusue & Mazda (1977) *Experientia* **33**, 854. **6.** Tsusue (1967) *Experientia* **23**, 116. **7.** Wede, Altindag, Widner, Wachter & Fuchs (1998) *Free Rad. Res.* **29**, 331. **8.** Theorell (1951) *The Enzymes*, 1st ed. (Sumner & Myrbäck, eds.), **2** (part 1), 335. **9.** Webb (1966) *Enzyme and Metabolic Inhibitors*, vol. 2, p. 289, Academic Press, New York. **10.** Hofstee (1949) *J. Biol. Chem.* **179**, 633.

Xanthosine

This ribonucleoside (9-β-D-ribofuranosylxanthine; FW = 284.23 g/mol), abbreviated Xao, is produced by the deamination of guanosine. Xanthosine absorbs UV light strongly: at pH 2, ε = 8400 $M^{-1}cm^{-1}$ at 235 nm and 8950 $M^{-1}cm^{-1}$ at 263 nm; at pH 8, ε = 10200 $M^{-1}cm^{-1}$ at 248 nm; and at pH 11, ε = 8900 $M^{-1}cm^{-1}$ at 278 nm). **Target(s):** adenosine kinase, weakly (1);

dihydroorotase (2); guanine deaminase (3); inosine nucleosidase, as alternative substrate (4,5); purine nucleosidase, as alternative substrate for the enzyme from a number of sources (6,7). **1**. Kidder (1982) *Biochem. Biophys. Res. Commun.* **107**, 381. **2**. Bresnick & Blatchford (1964) *Biochim. Biophys. Acta* **81**, 150. **3**. Kimm, Park & Lee (1985) *Korean J. Biochem.* **17**, 139. **4**. Takagi & Horecker (1957) *J. Biol. Chem.* **225**, 77. **5**. Webb (1966) *Enzyme and Metabolic Inhibitors*, vol. 2, p. 471, Academic Press, New York. **6**. Koszalka & Krenitsky (1979) *J. Biol. Chem.* **254**, 8185. **7**. Atkins, Shelp & Storer (1989) *J. Plant Physiol.* **134**, 447.

Xanthosine 3',5'-Cyclic-Monophosphate

This cyclic nucleotide (FW$_{sodium-salt}$ = 368.18 g/mol) is a structural analogue of guanosine 3'5'-cyclic-monophosphate (cGMP) and often competes with cGMP in cGMP-dependent systems as well as with cAMP in cAMP-dependent reactions (1). cXMP is aalso an alternative substrate for cyclic-nucleoside-monophosphate phosphodiesterases, such as rat liver cGMP phosphodiesterase (2). **1**. Du Plooy, Michal, Weimann, Nelboeck & Paoletti (1971) *Biochim. Biophys. Acta* **230**, 30. **2**. Moss, Manganiello & Vaughan (1977) *J. Biol. Chem.* **252**, 5211.

Xanthosine 5'-Diphosphate

This GDP analogue (FW$_{free-acid}$ = 444.19 g/mol; CAS 10593-13-7; Abbreviation: XDP), produced nonenzymatically by the action of nitrous acid on GDP, often acts as an alternative substrate, and thus a competitive inhibitor, for GTP- and ATP-dependent enzymes. The metal-nucleotide complex formation constant for MgHXDP and MgXDP^{1-} should be similar to those the corresponding ADP complexes: 100 M^{-1} for MgHADP and 4000 M^{-1} for MgADP^{1-} is (1). Corresponding values for the manganese complexes would be 500 and 30000 M^{-1}, respectively, and for calcium ion complexes, 80 and 2000 M^{-1} (1). **Target(s)**: adenylosuccinate synthetase (2); amidophosphoribosyl-transferase (3); glycoprotein 3-α-L-fucosyltransferase (4); ribose-phosphate diphosphokinase, or phosphoribosyl pyrophosphate synthetase (5); xanthine phosphoribosyltransferase (6). **1**. O'Sullivan & Smithers (1979) *Meth. Enzymol.* **63**, 294. **2**. Van der Weyden & Kelley (1974) *J. Biol. Chem.* **249**, 7282. **3**. Holmes, McDonald, McCord, Wyngaarden & Kelley (1973) *J. Biol. Chem.* **248**, 144. **4**. Murray, Takayama, Schultz & Wong (1996) *Biochemistry* **35**, 11183. **5**. Fox & Kelley (1972) *J. Biol. Chem.* **247**, 2126. **6**. Miller, Adamczyk, Fyfe & Elion (1974) *Arch. Biochem. Biophys.* **165**, 349.

Xanthosine 5'-Monophosphate

This purine nucleotide (FW$_{monosodium-salt}$ = 365.18 g/mol; CAS 25899-70-1), abbreviated XMP, is the product of the IMP dehydrogenase reaction and, as the substrate for GMP synthase, it is the immediate biosynthetic precursor of GMP. XMP absorbs light strongly in the ultraviolet region of the spectrum (λ$_{max}$ = 278 nm (ε = 8900 M^{-1}cm^{-1}) at pH 10). **Target(s)**: adenylosuccinate synthetase (1-4); amido-phosphoribosyl-transferase (5-7);

hypoxanthine(guanine) phosphoribosyltransferase (8); IMP cyclohydrolase (9,10); orotate phiosphoribosyltransferase (11); orotidine-5'-phosphate decarboxylase (12-15). **1**. Spector, Jones & Elion (1979) *J. Biol. Chem.* **254**, 8422. **2**. Van der Weyden & Kelly (1974) *J. Biol. Chem.* **249**, 7282. **3**. Spector & Miller (1976) *Biochim. Biophys. Acta* **445**, 509. **4**. Matsuda, Shimura, Shiraki & Nakagawa (1980) *Biochim. Biophys. Acta* **616**, 340. **5**. Messenger & Zalkin (1979) *J. Biol. Chem.* **254**, 3382. **6**. Holmes, McDonald, McCord, Wyngaarden & Kelley (1973) *J. Biol. Chem.* **248**, 144. **7**. Reynolds, Blevins & Randall (1984) *Arch. Biochem. Biophys.* **229**, 623. **8**. Nagy & Ribet (1977) *Eur. J. Biochem.* **77**, 77. **9**. Szabados, Hindmarsh, Phillips, Duggleby & Christopherson (1994) *Biochemistry* **33**, 14237. **10**. Christopherson, Williams, Schoettle, *et al.* (1995) *Biochem. Soc. Trans.* **23**, 888. **11**. Ashihara (1978) *Z. Pflanzenphysiol.* **87**, 225. **12**. Silva & Hatfield (1978) *Meth. Enzymol.* **51**, 143. **13**. Porter & Short (2000) *Biochemistry* **39**, 11788. **14**. Fyfe, Miller & Krenitsky (1973) *J. Biol. Chem.* **248**, 3801. **15**. Miller, Butterfoss, Short & Wolfenden (2001) *Biochemistry* **40**, 6227. **See also** *3-N-Ribosylxanthine 5'-Monophosphate*

Xanthosine 5'-Triphosphate

This rare purine nucleotide, usually abbreviated XTP (FW$_{disodium-salt}$ = 568.13 g/mol; CAS 6253-56-1), which can be formed nonenzymatically by the action of nitrous acid on GTP, often acts as an alternative substrate (and thus competitively inhibits) for GTP- and ATP-dependent enzymes. XTP absorbs light strongly in the ultraviolet region of the spectrum (λ$_{max}$ = 278 nm [ε = 8900 M^{-1}cm^{-1}] at pH 10). **Target(s)**: adenylosuccinate synthetase (1); amidophosphoribosyltransferase (2); glucokinase, weak alternative substrate (3); GTP cyclohydrolase IIa (4); guanosine-triphosphate guanylyltransferase (5); ribose-phosphate diphosphokinase, or phosphoribosyl-pyrophosphate synthetase (6,7); xanthine phosphoribosyltransferase (8). **1**. Van der Weyden & Kelley (1974) *J. Biol. Chem.* **249**, 7282. **2**. Holmes, McDonald, McCord, Wyngaarden & Kelley (1973) *J. Biol. Chem.* **248**, 144. **3**. Porter, Chassy & Holmlund (1982) *Biochim. Biophys. Acta* **709**, 178. **4**. Graham, Xu & White (2002) *Biochemistry* **41**, 15074. **5**. Liu & McLennan (1994) *J. Biol. Chem.* **269**, 11787. **6**. Wong & Murray (1969) *Biochemistry* **8**, 1608. **7**. Fox & Kelley (1972) *J. Biol. Chem.* **247**, 2126. **8**. Miller, Adamczyk, Fyfe & Elion (1974) *Arch. Biochem. Biophys.* **165**, 349.

(+)-Xanthoxylol

This furofuran lignan (FW = 356.38 g/mol; CAS 111407-29-5) from *Fagora zanthoxyloides* exhibits anti-sickling properties by inhibiting sickle cell hemoglobin (HbS) polymerization (1,2). Based on NMR studies of related compunds, a revised structure has been proposed (3). **1**. Sofowora & Isaacs (1971) *Lloydia* **34**, 383. **2**. Dean & Schechter (1978) *N. Eng. J. Med.* **299**, 863. **3**. Pelter & Ward (1976) *Tetrahedroin* **32**, 2784.

Xanthurenate

This quinaldate (FW$_{sodium-salt}$ = 227.15 g/mol) *or* xanthurenic acid, which is formed in the enzymatic degradation of L-tryptophan, inhibits acetyl-CoA carboxylase (1), dihydrolipoamide dehydrogenase (2), kynurenine 3 hydroxylase (3), kynurenine:oxoglutarate aminotransferase, weakly (4); NAD⁺:ADP ribosyltransferase (poly(ADP-ribose) polymerase (5); IC50 = 0.19 mM); pyridoxal kinase (6,7). Xanthurenate is a also gametocyte-activating factor. It has also been reported to activate cell caspases and apoptosis (8). **1.** Tanabe, Nakanishi, Hashimoto, *et al.* (1981) *Meth. Enzymol.* **71**, 5. **2.** Furuta & Hashimoto (1982) *Meth. Enzymol.* **89**, 414. **3.** Shibata (1978) *Acta Vitaminol. Enzymol.* **32**, 195. **4.** Takeuchi, Otsuka & Shibata (1983) *Biochim. Biophys. Acta* **743**, 323. **5.** Banasik, Komura, Shimoyama & Ueda (1992) *J. Biol. Chem.* **267**, 1569. **6.** Takeuchi, Tsubouchi & Shibata (1985) *Biochem. J.* **227**, 537. **7.** Karawya, Mostafa & Osman (1981) *Biochim. Biophys. Acta* **657**, 153. **8.** Malina, Richter, Mehl & Hess (2001) *BMC Physiol.* **1**, 7.

2-(9-Xanthylmethyl)-2-(2'-carboxycyclopropyl)-glycine, See *LY 341495*

2-(9-Xanthylmethyl)-2-(2'-carboxy-3'-ethylcyclopropyl)glycine, See *2-(2'-Carboxy-3'-ethylcyclopropyl)-2-(9-xanthylmethyl)glycine*

XAV-939

This selective Wnt β-catenin-mediated transcription inhibitor (FW = 312.31 g/mol; CAS 284028-89-3; Solubility: 15 mg/mL DMSO; <1 mg/mL H₂O), known as 2-(4-(trifluoromethyl)phenyl)-7,8-dihydro-5*H*-thiopyrano [4,3-*d*]pyrimidin-4-ol, targets tankyrases TNKS1 and TNKS2, telomere-associated poly(adenosine diphosphate [ADP]-ribose) polymerasse (*or* PARPs), with IC₅₀ values of 11 nM and 4 nM, respectively. in both telomere function and the DNA damage response following exposure to ionizing radiation (l). In human cells, tankyrase 1 siRNA knockdown significantly elevates recombination specifically within telomeres, a phenotype with the potential of accelerating cellular senescence (2). Additionally, depletion of tankyrase 1 resulted in concomitant and rapid reduction of the nonhomologous end-joining protein DNA-PKcs, while Ku86 and ATM protein levels remained unchanged; DNA-PKcs mRNA levels were also unaffected (1). The requirement of tankyrase 1 for DNA-PKcs protein stability reflects the necessity of its PARP enzymatic activity, and depletion of tankyrase 1 resulted in proteasome-mediated DNA-PKcs degradation, explaining the associated defective damage response observed (*i.e.*, increased sensitivity to ionizing radiation-induced cell killing, mutagenesis, chromosome aberration and telomere fusion) (2). Dysregulated Wnt pathway activity has been implicated in many cancers, making this signal transduction pathway an attractive target for anticancer therapies. XAV939, which selectively inhibits β-catenin-mediated transcription. XAV939 stimulates β-catenin degradation by stabilizing axin, the concentration-limiting component of the destruction complex. Such findings suggest that tankyrase 1 is not only involved in telomeric DNA repair, but also has important implications for modifying cancer and aging (1,2). A high-resolution X-Ray crystal structure confirms that XAV-939 binds in the NAD+-donor site of tankyrase-1 (3). **XAV939-Prolongation of Anaphase in Normal & Tumor Cells:** In HeLa cells synchronized by a double-thymidine block, release into S-phase in the presence or absence of XAV939 shows no tendency to induce mitotic arrest. Staining with anti-centromere antibody (ACA), however, reveals a 3.5-fold increase in XAV939-treated mitotic cells possessing separated centromeres, as compared to untreated cells (4). Moreover, all cells with separated centromeres show a loss in cyclin B, indicating progression to anaphase. Cells isolated at the 10-hour time-point by mitotic shake-off, followed by Fluorescence In Situ Hybridization (using centromere (10cen)- and telomere (16ptelo)- specific probes) show that XAV939 induces loss of centromere cohesion, concomitant with persistent telomere cohesion, again indicating that the majority of mitotic cells were in anaphase. In further support of the conclusion that cells with cohered sister telomeres had initiated anaphase, XAV939-treated cells (displaying separated centromeres) show a 3.6-fold increase in cohered telomeres compared with control (4). **1.** Huang, Mishina, Liu, *et al.* (2009) *Nature* **461**, 614. **2.** Dregalla, Zhou, Idate, *et al.* (2010) *Aging* **2**, 691. **3.** Kirby, Cheung, Fazal, Shultz & Stams (2012) *Acta Crystallogr. Sect. F Struct. Biol. Cryst. Commun.* **68**, 115. **4.** Kim & Smith (2014) *Mol. Biol. Cell* **25**, 30.

XDP. See *Xanthosine 5'-Diphosphate*

XEN445

This potent and selective lipase inhibitor (FW = 366.34 g/mol; IUPAC Name: (*S*)-2-(3-(pyridin-2-ylmethoxy)pyrrolidin-1-yl)-5-(trifluoromethyl)-benzoic acid) targets Endothelial Lipase (EL), exhibiting good inhibition (IC₅₀ = 237 nM), absorption, distribution, metabolism, excretion properties in mice, suggesting likely *in vivo* efficacy in raising plasma HDLc concentrations. **Primary Mode of Inhibitory Action:** EL is secreted from vascular endothelial cells and hepatocytes and preferentially hydrolyses the *sn*-1 (*or* PLA₁) ester bond of phosphatidylcholine (PC) in HDL particles, releasing Lyso-PC and thereby reducting HDL particle size. Its role in HDL metabolism is further indicated by findings that EL knockout mice have HDLc concentrations that are 50–140% higher than in wild type mice. EL itself is a triglyceride lipase in the same protein family that also includes hepatic lipase (HL), lipoprotein lipase (LPL) and pancreatic lipase. While these lipases share a common serine protease/esterase catalytic triad, they differ in the lid region controlling substrate specificity. This property served as a druggable target for the development of XEN-445. XEN 445 demonstrated good *in vitro* potency against EL with high selectivity over LPL and HL, and, perhaps more importantly, showed *in vivo* efficacy in raising plasma HDLc in wild-type mice. **1.** Sun, Dean, Jia, *et al.* (2013) *Bioorg. Med. Chem.* **21**, 7724.

Xenon

This largely unreactive Period V element (Atomic Symbol: Xe; Atomic Weight = 131.29 g; Atomic Number = 54, with a full, noble gas octet: $1s^2 2s^2 2p^6 3s^2 3p^6 3d^{10} 4s^2 4p^6 4d^{10} 5s^2 5p^6$) is a heavy gas that has been employed as an anesthetic, and, with its extremely low blood-gas partition coefficient (0.115), xenon anesthesia is characterized by a rapid onset of action as well as prompt post-treatment recovery. The high volumes required for anesthesia (*e.g.*, patients must breathe 70% Xe & 20-30% O₂) and its very low natural abundance (~0.5 ppm in air) make its use somewhat expensive (1,2). Xe produces profound analgesia, inhibiting surgery-induced hemodynamic and catecholamine responses. Remarkably, Xe preconditioning (*i.e.*, a process whereby exposure to a stress/stimulus leads to decreased cellular damage/death upon later exposure to a greater or more sustained stress) exerts hypoxia-induced signaling (3,4), conferring neuroprotection in a mouse model of transient middle cerebral artery occlusion (5). Xenon acts to open plasmalemmal ATP-sensitive potassium channels to induce neuronal preconditioning that activates ERK, one of the downstream effectors of the MAPK signaling pathway (6). Xenon also inhibits glutamatergic NMDA receptors involved in learning and memory and affects synaptic plasticity within the amygdala and hippocampus, which play roles in fear conditioning models for post-traumatic stress disorder (PTSD). Xe substantially and persistently inhibits memory reconsolidation in a reactivation and time-dependent manner, suggesting its likely effectiveness as a therapy for PTSD (7). Xenon thus exerts its analgesic effects by inhibiting NMDA receptors, blocking of painful stimuli transmissions from peripheral tissues to the brain and it also avoids the development of pain hypersensitivity (7). Indeed, xenon exerts intra- and post-operative neuroprotective, cardioprotective, and renoprotective actions, mainly through the production of the hypoxia-inducible factor 1α (HIF-1α) and its downstream effector erythropoietin as well as noradrenalin reuptake inhibition. Because xenon activates the formation of erythropoietin (a well-known stimulator of erythrocyte formation), the World Anti-Doping Agency (WADA) banned it as performance-enhancing agent, effective September, 2014. Currently, the best method for measuring xenon in human plasma and blood uses gas chromatography/mass spectrometry, *or* GC/MS (8). **1.** Viatkin & Mizikov (2008) *Anesteziol. Reanimatol.* **5**, 103. **2.** Kennedy, Stokes & Downing (1992) *Anaesth Intensive Care* **20**, 66. **3.** **4.** Goetzenich, Hatam, Preuss, *et al.* (2014) *Interact. Cardiovasc. Thorac. Surg.* *18, 321.* **5.** Meloni, Gillis, Manoukian & Kaufman (2014) *PLoS One* **9**, e106189. **6.** Huang, Li, Li, Guo & Zou (2009) *Carcinogenesis* **30**, 737. **7.** Giacalone, Abramo, Giunta & Forfori (2013) *Clin. J. Pain.* **29**, 639. **8.** Thevis, Piper, Geyer *et al.* (2014) *Rapid Commun. Mass Spectrom.* **28**, 1501.

Xenopsin

This octapeptide (MW = 1030.25 g/mol; *Sequence:* pGlu-Gly-Lys-Arg-Pro-Trp-Ile-Leu, where pGlu indicates a cyclic pyroglutamate residue; Isoelectric Point = 8.85) from the clawed toad *Xenopus laevis* is structurally related to neurotensin and xenin. Xenopsin inhibits gastric acid secretion and also stimulates exocrine pancreatic secretion and growth. **Target(s):** gastric acid secretion(1); neurolysin, as an alternative substrate (2-4). **1.** Zinner, Kasher, Modlin & Jaffe (1982) *Amer. J. Physiol.* **243**, G195. **2.** Vincent, Dauch, Vincent & Checler (1997) *J. Neurochem.* **68**, 837. **3.** Barelli, Vincent & Checler (1993) *Eur. J. Biochem.* **211**, 79. **4.** Checler, Vincent & Kitabgi (1986) *J. Biol. Chem.* **261**, 11274.

Xerophthol, *See Retinol*

Xestospongin C

This membrane-permeable bis-1-oxaquinolizidine macrocycle (FW = 446.72 g/mol; CAS 524-95-8; Form: Solid (often as a film on container surface); Color: Off-white; Soluble in DMSO; Store at −20°C) from the Australian sponge *Xestospongia,* inhibits bradykinin- and carbamylcholine-stimulated Ca^{2+} efflux from the endoplasmic reticulum stores (1) and is likewise a highly potent, reversible inhibitor of IP_3-mediated Ca^{2+} release (IC_{50} = 358 nM). Does not interact with the IP_3 binding site and exhibits high selectivity over ryanodine receptors. **IUPAC Name:** (1*R*,4*aR*,11*R*,12*aS*,13*S*,16*aS*,23*R*,24*aS*)-eicosahydro-5*H*,17*H*-1,23:11,13-diethano-2*H*,14*H*-[1,11]dioxacycloeicosino[2,3-*b*:12,13-*b'*]dipyridine. **1.** Miyamoto, Izumi, Hori, *et al.* (2000) *Brit. J. Pharm.* **130**, 650. **2.** De Smet, Parys, Callewaert, *et al.* (1999) *Cell. Calcium* **26**, 9.

Xestospongin D

This membrane-permeable, bis-1-oxaquinolizidine macrocycle (FW = 462.72 g/mol), isolated to homogeneity from marine sponges, inhibits nitric-oxide synthase, showing competitive inhibition with respect to L-arginine and uncompetitive inhibition relative to NADPH and free Ca^{2+}. This inhibitory profile is similar to N^ω-nitro-L-arginine methylester (***See*** *L-NAME*). **1.** Venkateswara, Rao, Desaiah, Vig & Venkateswarlu (1998) *Toxicology* **129**, 103.

X-gal, *See 5-Bromo-4-chloro-3-indoyl-β-D-galactoside*

XIAP

This zinc finger-containing, X-chromosome-linked inducer of apotosis protein (MW_{human} = 56,000) is also known as hILP. Although a member of the IAP family (*i.e.,* Inhibitor-of-Apoptosis Protein family), XIAP is the only IAP family member that inhibits apoptosis as a consequence of its direct inhibitory action on caspases-3, -7, and -9 (1,2). XIAP is itself a target for inhibitors that, by blocking its inhibition of caspases-3, -7, and -9, should in turn promote apoptosis. **1.** Deveraux, Welsh & Reed (2000) *Meth. Enzymol.* **322**, 154. **2.** Deveraux, Roy, Stennicke, *et al.* (1998) *EMBO J.* **17**, 2215.

XK469

This quinoxaline derivative (FW_{free-acid} = 344.75 g/mol; CAS 157542-91-1), also known as NSC 697887 and 2-[4-(7-chloro-2-quinoxalinyl-

oxy)phenoxy]propionic acid, inhibits DNA topoisomerase IIβ (1) and induces reversible protein-DNA cross-links in mammalian cells. Both enantiomers of XK469 are active (1) XK469 also induces apoptosis in Waldenstrom's macroglobulinemia through multiple pathways (2). Note that the α-carbon of the lactate moiety is chiral: QK469R refers to the *R*-enantiomer, whereas QK469S refers to the *S*-enantiomer. **1.** Gao, Huang, Yamasaki, *et al.* (1999) *Proc. Natl. Acad. Sci. U.S.A.* **96**, 12168. **2.** Mensah-Osman, Al-Katib, Dandashi & Mohammad (2003) *Int. J. Oncol.* **23**, 1637.

XL019

This potent Janus kinase inhibitor (FW = 444.53 g/mol; CAS 945755-56-6), also named *N*-[4-[2-[[4-(4-morpholinyl)phenyl]amino]-4-pyrimidinyl]-phenyl]-(2*S*)-2-pyrrolidinecarboxamide, selectively targets JAK2 (IC_{50} = 2.2 nM), exhibiting lesser action against JAK1 (IC_{50} =134 nM), JAK3 (IC_{50} = 214 nM), and TYK2 (IC_{50} = 348 nM). **1.** Forsyth, Kearney, Kim, *et al.* (2012) *Bioorg. Med. Chem. Lett.* **22**, 7653.

XL147

This selective and reversible ATP-directed PI3K inhibitor (FW = 448.52 g/mol; CAS 956958-53-5; Solubility (25°C): 90 mg/mL DMSO, <1 mg/mL H_2O), also known as SAR245408 and *N*-(3-(benzo[*c*][1,2,5]thia-diazol-5-ylamino)quinoxalin-2-yl)-4-methylbenzenesulfonamide, targets both wild type (IC_{50} = 40 nM) and mutant p110α (IC_{50} = 40 nM), a Class-I phosphoinositide-3-kinases. **1.** Chakrabarty, Sánchez, Kuba, Rinehart, & Arteaga (2012) *Proc. Natl. Acad. Sci. U.S.A.* **109**, 2718.

XL184. *See Cabozantinib*

XL388

This highly potent and selective ATP-competitive inhibitor (FW = 455.50 g/mol; CAS 1251156-08-7), also known as [7-(6-amino-3-pyridinyl)-2,3-dihydro-1,4-benzoxazepin-4(5*H*)-yl][3-fluoro-2-methyl-4-(methylsulfonyl) phenyl]methanone, targets mTOR (IC_{50} = 9.9 nM), mTORC1, *or* p-p70S6K, pS6, and p-4E-BP1 (IC_{50} = 8 nM), and mTORC2, *or* pAKT and S473 (IC_{50} = 166 nM), displaying 1000x selectivity over the closely related PI3K kinases. XL388 displays good pharmacokinetics and oral exposure in multiple species with moderate bioavailability. **1.** Takeuchi, *et al.* (2013) *J. Med. Chem.* 2013, **56**, 2218.

XL765, *See Voxtalisib*

XL880, See *Foretinib*

XL888

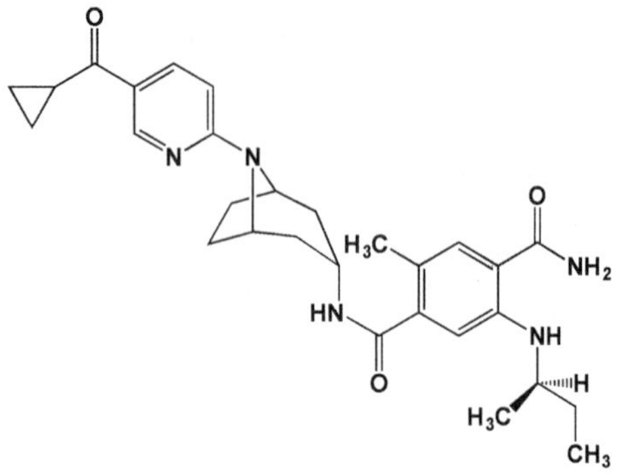

This ATP-competitive inhibitor (FW = 503.64 g/mol; CAS 1149705-71-4), also known as N^1-[(3-endo)-8-[5-(cyclopropylcarbonyl)-2-pyridinyl]-8-azabicyclo[3.2.1]oct-3-yl]-2-methyl-5-[[(1R)-1-methylpropyl]amino]-1,4-benzenedicarboxamide, targets HSP90 (IC$_{50}$ = 24 nM), blocking folding and conformational integrity of client proteins, of which many are steroid hormone receptors and signal transduction protein kinases. XL888 potently inhibits cell growth, induces apoptosis, and prevents the growth of vemurafenib-resistant melanoma cell lines in 3-dimensional cell culture, long-term colony formation assays, and human melanoma mouse xenografts. Reversal of the resistance phenotype by XL888 is associated with the degradation of PDGFRβ, COT, IGFR1, CRAF, ARAF, S6, cyclin D$_1$, and Akt, which in turn lead to the nuclear accumulation of FOXO3a, increases expression of BIM (Bcl-2-Interacting Mediator of cell death), and down-regulation of Mcl-1. **1**. Bussenius, Blazey, Aay, *et al.* (2012) *Bioorg. Med. Chem. Lett.* **22**, 5396. **2**. Paraiso, Haarberg, Wood, *et al.* (2012) *Clin. Cancer Res.* **18**, 2502.

XRP-6258, See *Cabazitaxel*

Xylan Polysulfate, See *Xylans; Pentosan Polysulfate*

Xylarate (Xylaric Acid)

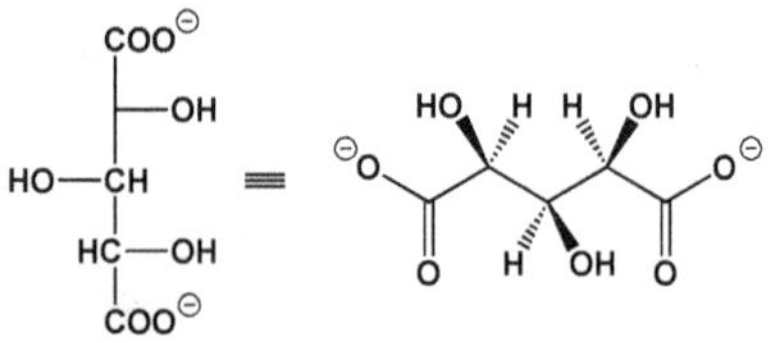

This achiral aldaric acid (FW$_{disodium-salt}$ = 224.08 g/mol; CAS 10158-64-2), which is formed from either D-xylose or L-xylose, inhibits 2-dehydro-3-deoxyglucarate aldolase. **1**. Fish & Blumenthal (1966) *Meth. Enzymol.* **9**, 529.

Xylarohydroxamate

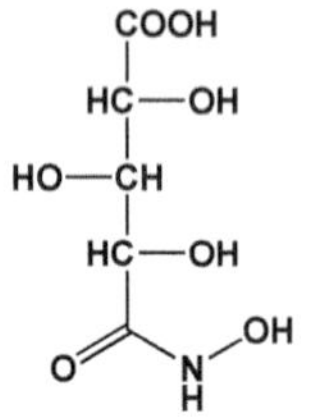

This aldaric acid monohydroxamate (FW$_{free-acid}$ = 195.13 g/mol) competitively inhibits *Escherichia coli* glucarate dehydratase (K_i = 0.8 mM). **1**. Gulick, Hubbard, Gerlt & Rayment (2000) *Biochemistry* **39**, 4590.

Xylazine

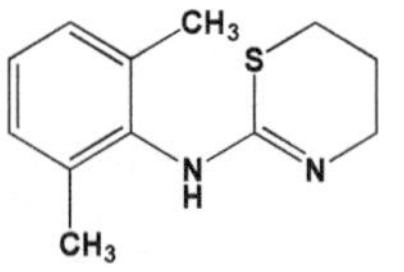

This sedative, anesthetic, muscle relaxant, and analgesic (FW = 220.33 g/mol; CAS 7361-61-7), also known as *N*-(2,6-dimethylphenyl)-5,6-dihydro-4*H*-1,3-thiazin-2-amine, is frequently employed in veterinary medicine, especially with horses and cattle. Clinically effective doses of xylazine markedly decrease basal insulin levels and completely abolish the rise in insulin produced by intravenous glucose. These changes in insulin levels lead to elevated fasting plasma glucose levels and glucose intolerance (1). Yohimbine (an α$_2$-adrenergic blocker) and phentol-amine (an α$_1$/ α$_2$-adrenergic blocker) reduces/abolishes xylazine-induced hyperglycemia and hypoinsulinemia (2). Prazosin and phenoxybenzamine (an α$_1$-adrenergic blockers) does not exert such antagonism, and α-adrenergic blocking agents alone does not change either plasma glucose or insulin concentrations. These findings suggest that xylazine-induced hyperglycemia and hypoinsulinemia are mediated by α$_2$-adrenergic receptors, possibly in β-cells of pancreatic islets that inhibit insulin release (2). **1**. Goldfine & Arieff (1979) *Endocrinology* **105**, 920. **2**. Hsu & Hummel (1981) *Endocrinology* **109**, 825.

3,4-Xylidine

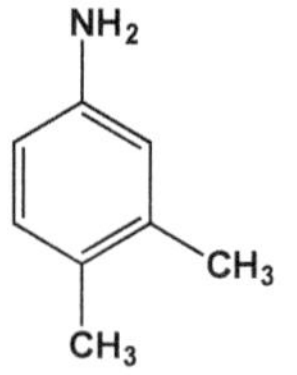

This aromatic base (FW = 121.18 g/mol; CAS 95-64-7), also known as 3,4-dimethylaniline, is often used in the production of dyes. The free base is sparingly soluble in water; however, it readily forms salts with mineral acids and these salts are considerably more soluble. The free base has a melting point of 49-51°C and a B.P. of 226-228°C. 3,4-Xylidine is *highly toxic* and care should be exercised in its handling. **Target:** β-fructofuranosidase, *or* invertase. **1**. Myrbäck (1960) *The Enzymes*, 2nd ed. (Boyer, Lardy & Myrbäck, eds.), **4**, 379.

Xylitol

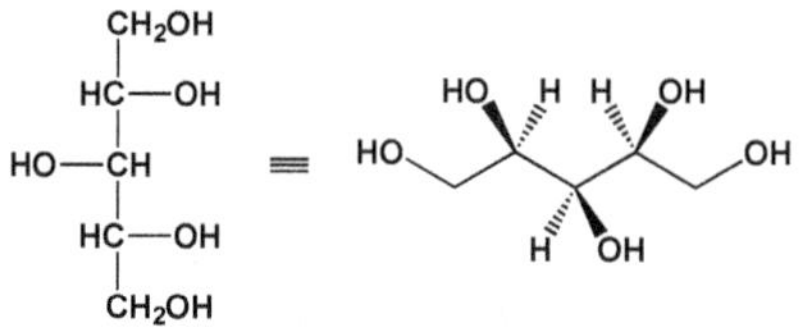

This achiral sugar alcohol (FW = 152.15 g/mol; CAS 87-99-0) is related to D- and L-xylose and can be produced by the action of either D- or L-xylulose reductase. It is also a byproduct in wood saccharification. Xylitol has a sweetness roughly equal to that of sucrose and is often used as an artificial sweetener. In fact, xylitol is the sweetest of all the common sugar alcohols. **Target(s):** aldose 1-epimerase, *or* mutarotase (1,2); α-amylase (3); D-arabinose isomerase (4,5); L-arabinose isomerase (6-10); L-fucose isomerase (2,5); xylose isomerase (7,11-16). **1**. Bailey, Fishman, Kusiak, Mulhern & Pentchev (1975) *Meth. Enzymol.* **41**, 471. **2**. Webb (1966) *Enzyme and Metabolic Inhibitors*, vol. **2**, p. 413, Academic Press, New York. **3**. Ali & Abdel-Moneim (1989) *Zentralbl. Mikrobiol.* **144**, 615. **4**. Yamanaka & Izumori (1975) *Meth. Enzymol.* **41**, 462. **5**. Izumori & Yamanaka (1974) *Agric. Biol. Chem.* **38**, 267. **6**. Yamanaka (1975) *Meth. Enzymol.* **41**, 458. **7**. Noltmann (1972) *The Enzymes*, 3rd ed. (Boyer, ed.), **6**, 271. **8**. Kim & Oh (2005) *J. Biotechnol.* **120**, 162. **9**. Nakamatu & Yamanaka (1969) *Biochim. Biophys. Acta* **178**, 156. **10**. Yamanaka & Izumori (1973) *Agric. Biol. Chem.* **37**, 521. **11**. Yamanaka (1966) *Meth. Enzymol.* **9**, 588 and (1975) *Meth. Enzymol.* **41**, 466. **12**. Smith, Rangarajan & Hartley (1991) *Biochem. J.* **277**, 255. **13**. Lehmacher & Bisswanger (1990) *Biol. Chem. Hoppe-Seyler* **371**, 527. **14**. Henrick, Collyer & Blow (1989) *J. Mol. Biol.* **208**, 129. **15**. Fenn, Ringe & Petsko (2004) *Biochemistry* **43**, 6464. **16**. Lonn, Gardonyi, van Zyl, Hahn-Hagerdal & Otero (2002) *Eur. J. Biochem.* **269**, 157.

Xylobiose

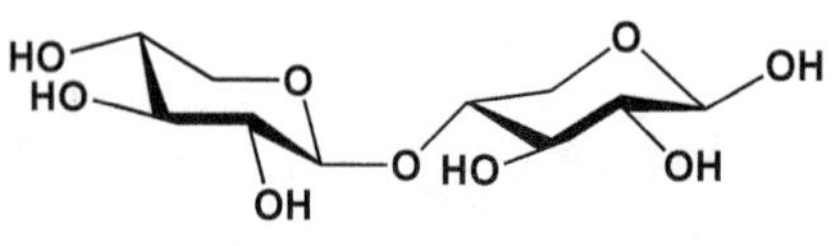

This reducing disaccharide (FW = 282.25 g/mol; CAS 6860-47-5), consisting of two β1,4-linked D-xylopyranosyl residues, is a structural subunit of most xylans and a substrate for a number of β glucosidases and β-xylosidases. **Target(s):** endo-1,4-β-xylanase (1); xylan 1,4-β-xylosidase (2). **1**. Kitpreechavanich, Hayashi & Nagai (1984) *J. Ferment. Technol.* **62**, 415. **2**. La Grange, Pretorius, Claeyssens & van Zyl (2001) *Appl. Environ. Microbiol.* **67**, 5512.

Xylocaine, *See Lidocaine*

L-*xylo*-Hexulose, *See L-Sorbose*

***p*-Xylohydroquinone,** *See 2,5-Dimethylhydroquinone*

D-Xyloketose, *See D-Xylulose*

L-Xyloketose, *See L-Xylulose*

Xylonojirimycin, *See 5-Amino-5-deoxy-D-xylopyranose*

***p*-Xyloquinol,** *See 2,5-Dimethylhydroquinone*

***p*-Xyloquinone**

This quinone, also called 2,5-dimethyl-1,4-benzoquinone and 2,5-dimethylquinone (FW = 136.15 g/mol; CAS 137-18-8), has been reported in a South American spider and inhibits NADH oxidation and bacterial luminescence. It can also serve as an electron acceptor in photosystem II. **Target(s):** α-amylase (1); β-amylase (1); α-glycerophosphatase (2); papain (3); pyruvate decarboxylase (4); succinate dehydrogenase (5); succinate oxidase (5,6); urease (7). **1**. Owens (1953) *Contrib. Boyce Thompson Inst.* **17**, 273. **2**. Hoffmann-Ostenhof & Putz (1948) *Monatsh. Chem.* **79**, 421. **3**. Hoffmann-Ostenhof & Biach (1946) *Experentia* **2**, 405. **4**. Kuhn & Beinert (1947) *Ber.* **80**, 101. **5**. Herz (1954) *Biochem. Z.* **325**, 83. **6**. Redfern & Whittaker (1962) *Biochim. Biophys. Acta* **56**, 440. **7**. Zaborska, Kot & Superata (2002) *J. Enzyme Inhib. Med. Chem.* **17**, 247.

D-Xylosamine

This D-glucosamine analogue, also known as β-D-xylopyranosylamine (FW$_{free\ base}$ = 135.14 g/mol; CAS 22738-07-4), can act as an alternative substrate and/or competitive inhibitor of a number of glucosamine-dependent systems. It is a potent, time-dependent inhibitor of xylan 1,4-β-xylosidase, *or* β-xylosidase, K_i = 40 nM (1) **1** Deleyn, Claeyssens & De Bruyne (1982) *Meth. Enzymol.* **83**, 639.

D-Xylose

α-D-Xylose   Fischer   β-D-Xylose
Projection

This D-aldopentose, also known as wood alcohol (FW = 150.13 g/mol; CAS 58-86-6) has the following relative compositions at 31°C in D$_2$O: 36.5% α-D-xylopyranose, 63% β-D-xylopyranose, and <1% furanose). **Targets:** aldose 1-epimerase, *or* mutarotase (1-5); α-amylase (6); α-L-arabinofuranosidase (2, 8); arabinogalactan endo-1,4-β-galactosidase (7); cellulose 1,4-β-cellobiosidase (8); Cu,Zn-superoxide dismutase (9); endo-1,4-β-xylanase (10,41); α-galactosidase (35-37); β-galactosidase, weakly (4,11-15); glucan 1,4-α-glucosidase, *or* glucoamylase (42,43); glucokinase (16-18,46); α-glucosidase (4,19,20,40); β-glucosidase (K_i = 495 mM for the sweet almond enzyme) (38,39); α-glucuronidase (21); hexokinase (22,51); isoamylase (23); lactose synthase (24); macrocellulase complex (25);

polyphosphate:glucose phosphotransferase (44,45); superoxide dismutase (9); thioglucosidase *or* myrosinase, weakly (26); α,α-trehalose phosphorylase (configuration-retaining) (47); xylan 1,4-β-xylosidase (27,29-34). **1**. Bentley (1962) *Meth. Enzymol.* **5**, 219. **2**. Bailey, Fishman, Kusiak, *et al.* (1975) *Meth. Enzymol.* **41**, 471. **3**. Fishman, Pentchev & Bailey (1975) *Meth. Enzymol.* **41**, 484. **4**. Webb (1966) *Enzyme and Metabolic Inhibitors,* vol. 2, Academic Press, New York. **5**. Bailey & Pentchev (1964) *Biochem. Biophys. Res. Commun.* **14**, 161. **6**. Sawai (1960) *J. Biochem.* **48**, 382. **7**. Nakano, Takenishi & Watanabe (1985) *Agric. Biol. Chem.* **49**, 3445. **6**. Mishra, Vaidya, Rao & Deshpande (1983) *Enzyme Microb. Technol.* **5**, 430. **7**. Ukeda, Hasegawa, Ishi & Sawamura (1997) *Biosci. Biotechnol. Biochem.* **61**, 2039. **10**. Lachke (1988) *Meth. Enzymol.* **160**, 679. **11**. Lester & Bonner (1952) *J. Bacteriol.* **63**, 759. **12**. Choi, Kim, Lee & Lee (1995) *Biotechnol. Appl. Biochem.* **22**, 191. **13**. Itoh, Suzuki & Adachi (1982) *Agric. Biol. Chem.* **46**, 899. **14**. Cowan, Daniel, Martin & Morgan (1984) *Biotechnol. Bioeng.* **26**, 1141. **15**. Levin & Mahoney (1981) *Antonie Leeuwenhoek* **47**, 53. **16**. Walker & Parry (1966) *Meth. Enzymol.* **9**, 381. **17**. Reeves, Montalvo & Sillero (1967) *Biochemistry* **6**, 1752. **18**. Kamel, Allison & Anderson (1966) *J. Biol. Chem.* **241**, 690. **19**. Kato, Matsushima & Akabori (1960) *J. Biochem.* **48**, 199. **20**. Halvorson & Ellias (1958) *Biochem. Biophys. Acta* **30**, 28. **21**. Nagy, Nurizzo, Davies, *et al.* (2003) *J. Biol. Chem.* **278**, 20286. **22**. Colowick (1973) *The Enzymes,* 3rd ed. (Boyer, ed.), **9**, 1. **23**. Kitagawa, Amemura & Harada (1975) *Agric. Biol. Chem.* **39**, 989. **24**. Ebner (1973) *The Enzymes,* 3rd ed. (Boyer, ed.), **9**, 363. **25**. Ljungdahl, Coughlan, Mayer, *et al.* (1988) *Meth. Enzymol.* **160**, 483. **26**. Tsuruo & Hata (1968) *Agric. Biol. Chem.* **32**, 1420. **27**. John & Schmidt (1988) *Meth. Enzymol.* **160**, 662. **28**. Hirano, Tsumuraya & Hashimoto (1994) *Physiol. Plant.* **92**, 286. **29**. Mujer & Miller (1991) *Physiol. Plant.* **82**, 367. **30**. Chinen, Oouchi, Tamaki & Fukuda (1982) *J. Biochem.* **92**, 1873. **31**. Vocadlo, Wicki, Rupitz & Withers (2002) *Biochemistry* **41**, 9736. **32**. Buttner & Bode (1992) *J. Basic Microbiol.* **32**, 159. **33**. Gomez, Isorna, Rojo & Estrada (2001) *Biochimie* **83**, 961. **34**. Bernier, Desrochers, Paice & Yaguchi (1987) *J. Gen. Appl. Microbiol.* **33**, 409. **35**. Oishi & Aida (1975) *Agric. Biol. Chem.* **39**, 2129. **36**. Soh, Ali & Lazan (2006) *Phytochemistry* **67**, 242. **37**. Sripuan, Aoki, Yamamoto, Tongkao & Kumagai (2003) *Biosci. Biotechnol. Biochem.* **67**, 1485. **38**. Dale, Ensley, Kern, Sastry & Byers (1985) *Biochemistry* **24**, 3530. **39**. Seidle, Marten, Shoseyov & Huber (2004) *Protein J.* **23**, 11. **40**. Thirunavukkarasu & Priest (1984) *J. Gen. Microbiol.* **130**, 3135. **41**. Rapp & Wagner (1986) *Appl. Environ. Microbiol.* **51**, 746. **42**. Buettner, Bode & Birnbaum (1987) *J. Basic Microbiol.* **27**, 299. **43**. Wong, Batt, Lee, Wagschal & Robertson (2005) *Protein J.* **24**, 455. **44**. Hsieh, Kowalczyk & Phillips (1996) *Biochemistry* **35**, 9772. **45**. Kowalczyk, Horn, Pan & Phillips (1996) *Biochemistry* **35**, 6777. **46**. Fernández, Herrero & Moreno (1985) *J. Gen. Microbiol.* **131**, 2705. **47**. Eis & Nidetzky (2002) *Biochem. J.* **363**, 335.

L-Xylose

α-L-Xylose   Fischer   β-L-Xylose
Projection

This uncommon L-aldopentose enantiomer (FW = 150.13 g/mol; CAS 609-06-3) inhibits aldose 1-epimerase (1,2), β-glucosidase (K_i = 120 mM for the sweet almond enzyme) (3), and lactose synthase, weakly (4). **1**. Bailey, Fishman, Kusiak, Mulhern & Pentchev (1975) *Meth. Enzymol.* **41**, 471. **2**. Fishman, Pentchev & Bailey (1975) *Meth. Enzymol.* **41**, 484. **3**. Dale, Ensley, Kern, Sastry & Byers (1985) *Biochemistry* **24**, 3530. **4**. Ebner (1973) *The Enzymes,* 3rd ed. (Boyer, ed.), **9**, 363. ***See also D-Xylose***

α-D-Xylose 1-Phosphate

This phosphorylated pentose analogue (FW$_{monosodium\text{-}salt}$ = 252.09 g/mol) of α-D-glucose 1-phosphate inhibits the *Schizophyllum commune* α,α-trehalose phosphorylase, a component of the α-D-glucopyranosyl α-D-glucopyranoside (α,α-trehalose)-degrading enzyme system in fungi and catalyzes glucosyl transfer from α,α-trehalose to phosphate with overall net retention of the anomeric configuration. **1**. Eis & Nidetzky (2002) *Biochem. J.* **363**, 335.

Xylosylzeatin, *See* Zeatin Xyloside

D-Xylulose

α-D-Xylulose　　　　Fischer　　　　β-D-Xylulose
　　　　　　　　　　Projection

This D-ketopentose (FW = 150.13 g/mol; CAS 551-84-8), an intermediate in the pentose phosphate shunt inhibits ribokinase (1) and ketohexokinase (*or* hepatic fructokinase), acting as an alternative substrate for the latter (2). **1**. Ogbunude, Lamour & Barrett (2007) *Acta Biochim. Biophys. Sin. (Shanghai)* **39**, 462. **2**. Raushel & Cleland (1977) *Biochemistry* **16**, 2169.

L-Xylulose

α-L-Xylulose　　　　Fischer　　　　β-L-Xylulose
　　　　　　　　　　Projection

This L-ketopentose intermediate (FW = 150.13 g/mol; CAS 527-50-4) of the glucuronate pathway inhibits arylglycosidases and glucosidase. The inhibition of yeast α-glucosidase by L-xylulose was of a competitive nature (IC$_{50}$ = 1 x 10^{-5} M). Both L-xylulose and L-fructose also inhibited the purified soybean glucosidase I (IC$_{50}$ = 1 x 10^{-4} M), but showed no inhibitory activity against soybean glucosidase II. Moreover, when influenza virus-infected MDCK cells are cultured in the presence of L-xylulose, there was a dose-dependent inhibition in the formation of complex types of oligosacc-harides on the viral glycoproteins consistent with the inhibition of the processing glucosidase I. This inhibition resulted in the accumulation of Glc$_3$Man$_9$(GlcNAc)$_2$ oligosaccharides on the viral glycoprotein. L-Fructose also inhibited glycoprotein processing in cell culture, and the inhibition resulted in the formation of similar oligosaccharides to those seen with L-xylulose. **1**. Muniruzzaman, Pan, Zeng, *et al.* (1996) *Glycobiology* **6**, 795.

D-Xylulose 1,5-Bisphosphate

This D-ribulose 1,5-bisphosphate analogue (FW$_{disodium-salt}$ = 354.05 g/mol) inhibits the two main reactions of ribulose-1,5-bisphosphate carboxylase, namely the reaction of carbon dioxide with D-ribulose 1,5-bisphosphate to produce two molecules of 3-phospho-D-glycerate as well as the oxidase activity, in which 3-phospho-D-glycerate and 2-phosphoglycolate are generated (1-3). **1**. McCurry & Tolbert (1977) *J. Biol. Chem.* **252**, 8344. **2**. Taylor, Fothergill & Andersson (1996) *J. Biol. Chem.* **271**, 32894. **3**. McCurry, Pierce, Tolbert & Orme-Johnson (1981) *J. Biol. Chem.* **256**, 6623.

D-Xylulose 5-Phosphate

This phosphorylated pentose-phosphate pathway intermediate (FW$_{monosodium-salt}$ = 252.09 g/mol) and it inhibits glucose-6-phosphate isomerase with K_i = 0.7 mM (1), acetyl-CoA synthetase, *or* acetate:CoA ligase (2), and ribose-5-phosphate isomerase (K_i = 4 mM) (3) **1**. Noltmann (1972) *The Enzymes*, 3rd ed. (Boyer, ed.), **6**, 271. **2**. O'Sullivan & Ettlinger (1976) *Biochim. Biophys. Acta* **450**, 410. **3**. Jung, Hartman, Lu & Larimer (2000) *Arch. Biochem. Biophys.* **373**, 409.

Y, See *Tyrosine; Yttrium and Yttrium Ions*

Y1

This synthetic peptide, corresponding to residues 13-33 of cholecystokinin, potently inhibits Yapsin 1 (K_i = 65 nM), the basic residue-specific yeast aspartyl protease also known as Sap9p from *Candida albicans*. Inhibitor-coupled agarose beads were successfully used to purify yapsin 1 to apparent homogeneity from conditioned medium of a yeast expression system. Cawley, Chino, Maldonado, Rodriguez, Loh & Ellman (2003) *J. Biol. Chem.* **278**, 5523.

Y 11

This FAK inhibitor (FW = 265.15 g/mol; CAS 1086639-59-9; Solubility: 100 mM in H₂O; 100 mM in DMSO; IUPAC Name: 1-(2-hydroxyethyl)-3,5,7-triaza-1-azaniatricyclo[3.3.1.1³,⁷]decane bromide, prevents focal adhesion kinase autophosphorylation at Tyr-397 (IC₅₀ ~ 50 nM, *in vitro*), with 20x–30x selectivity for FAK versus other rotein kinases. Y-11 decreases cell viability in a many cancer cell lines. **1.** Golubovskaya, Nyberg, Zheng, *et al.* (2008) *J. Med. Chem.* **51**, 7405.

Y 134

This selective estrogen receptor (ER) modulator, *or* SERM (FW = 472.60 g/mol; CAS 849662-80-2; Soluble to 100 mM in DMSO and to 100 mM in Ethanol), also named 6-hydroxy-2-(4-hydroxyphenyl)benzo[*b*]thien-3-yl]-[4-[4-(1-methylethyl)-1-piperazinyl]phenyl]methanone, ERα (K_i = 0.09 nM) over ERβ (K_i = 11.3 nM), showing no affinity for mineralocorticoid, glucocorticoid, androgen and progesterone receptors. Y-134 acts as an agonist in the bone and antagonist in reproductive tissue and uppresses estrogen-stimulated proliferation of ER-positive human breast cancer cells. **1.** Yang, Xu, Li, *et al.* (2005) *Bioorg. Med. Chem. Lett.* **15**, 1505. PMID: 15713417. **2.** Ning, Zhou, Weng, *et al.* (2007) *Brit. J. Pharmacol.* **150**, 19. PMID: 17115070.

Y-25130

This potent 5-HT₃ antagonist (FW = 386.30 g/mol; CAS 123040-16-4), also known as *N*-(1-azabicyclo[2.2.2]oct-3-yl)-6-chloro-4-methyl-3-oxo-3,4-dihydro-2*H*-1,4-benzoxazine-8-carboxamide hydrochloride, selectively targets serotonin receptors, but is free of dopamine receptor blocking activity (1). In an *in vitro* binding assay, Y-25130 inhibited the specific

binding of [³H]-quipazine to 5-HT3 receptors at the synaptic membranes of the rat cerebral cortex with a K_i value of 2.9 nM (2). Y-25130 is an orally active anti-emetic compound against cisplatin and doxorubicin- and cyclophosphamide-induced vomiting. **1.** Fukuda, Setoguchi, Inaba, Shoji & Tahara (1991) *Eur. J. Pharmacol.* **196**, 299. **2.** Sato, Sakamori, Haga, Takehara & Setoguch (1992) *Jpn. J. Pharmacol.* **59**, 443. **3.** Haga, Inaba, Shoji, *et al.* (1993) *Jpn. J. Pharmacol.* **63**, 377.

Y-26763

This active Y-27152 metabolite (FW = 276.29 g/mol; CAS 127408-31-5; Soluble to 100 mM when combined with 1 equivalent NaOH; IUPAC: *N*-[(3*S*,4*R*)-6-cyano-3,4-dihydro-3-hydroxy-2,2-dimethyl-2*H*-1-benzopyran-4-yl]-*N*-hydroxyacetamide, is a long-lasting Kᵢᵣ6 (K_ATP) channel opener that relaxes contracted rat aortic rings, IC₅₀ = 27 μM (1) and inhibits glucose-induced insulin secretion in isolated human pancreatic β-cells *in vitro* (2). (**See** *Y-27152*) Y-26763 also induces hypotension in spontaneously hypertensive rats and displays cardioprotective properties in dogs following systemic administration *in vivo*. **1.** Fukunari, Miyai, Shinagawa, Kawahara & Nakajima (1997) *Eur. J. Pharmacol.* **323**, 197. **2.** Cosgrove, Straub, Barnes, *et al.* (2004) *Eur. J. Pharmacol.* **486**, 133.

Y-27632

This cell-permeable protein kinase inhibitor (FW_free-base = 246.35 g/mol; FW_mono-HCl = 282.81 g/mol; FW_di-HCl = 320.27 g/mol; CAS 129830-38-2 and 146986-50-7; Solubility: 15 mg/mL DMSO; Freely soluble in H₂O as the d--HCl salt), also known systematically as (1*R*,4*r*)-4-((*R*)-1-aminoethyl)-*N*-(pyridin-4-yl)cyclohexanecarboxamide, is a vasodilator that targets ROCK1 (K_i = 0.14 μM), a RhoA/Rho kinase that plays important roles in mediating vasoconstriction and vascular remodeling in the pathogenesis of pulmonary hypertension. Rho-associated coiled coil-forming protein kinase (ROCK) is thought to regulate actomyosin-based contractility in many types of cells by phosphorylation of ROCK targets. Y-27632 also inhibits ROCK-II, while displaying little activity against PKC, PKA, and myosin light-chain kinase (MLCK), with K_i values of 26 μM, 25 μM and >250 μM, respectively, and >200-fold selectivity over other kinases. Y-27632 suppresses Rho-induced, p160 ROCK-mediated formation of actin stress fibres in cultured cells and dramatically corrects hypertension in several hypertensive rat models (1). Y-27632 also induces disassembly of actin stress fibers, attended by rapid dissociation of VASP and zyxin with a lifetime of 7-8 min, talin, paxillin and ILK with a lifetime of ~16 min, and then FAK, vinculin and kindlin-2 with a lifetime of 25-28 min (5). **See** *Y-39983 and (R)-(+)-trans-4-(1-Aminoethyl)-N-pyridyl)cyclohexanecarboxamide* **1.** Uehata, *et al.* (1997) *Nature* **389**, 990. **2.** Itoh, *et al.* (1999) *Nat. Med.* **5**, 221. **3.** Bito, *et al.* (2000) *Neuron* **26**, 431. **4.** Watanabe, *et al.* (2007) *Nat. Biotechnol.* **25**, 681. **5.** Lavelin, Wolfenson, Patla, *et al.* (2013) *PLoS* **8**:e73549.

Y-29794 Oxalate

This orally active and brain-penetrant non-peptide serine peptidase inhibitor (FW = 508.69 g/mol; CAS 146794-84-5; Soluble to 15 mM in Ethanol with gentle warming; 15 mM in DMSO; IUPAC Name: [2-[[8-(dimethyl-amino)octyl]thio]-6-(1-methylethyl)-3-pyridinyl]-2-thienylmethanone oxalate, targets rat brain prolyl endopeptidase (K_i = 0.95 nM) *in vitro* (1) and prevents amyloid β-peptide deposition *in vivo* within senescence-accelerated mice (2)   **1**. Nakajima, Ono, Kato, Maeda & Ohe (1992) *Neurosci. Lett.* **141**, 156. **2**. Kato, Fukunari, Sakai & Nakajima (1997) *J. Pharmacol. Exp. Ther.* **283**, 328.

Y-32885, See *(R)-(+)-4-(1-Aminoethyl)-N-(4-pyridyl)benzamide*

Y-39983

This cell-permeable protein kinase inhibitor and vasodilator ($FW_{free\text{-}base}$ = 280.33 g/mol; FW_{HCl} = 316.79 g/mol; CAS 471843-75-1), also known systematically as, selectively targets RhoA/Rho kinases that play important roles in mediating vasoconstriction and vascular remodeling in the pathogenesis of pulmonary hypertension. Aqueous outflow in the eye is regulated by the contraction and relaxation of the ciliary muscle (CM) and the trabecular meshwork (TM), and Rho-associated coiled coil-forming protein kinase (ROCK) is thought to regulate actomyosin-based contractility by phosphorylation of ROCK targets. Y-39983 also significantly enhances the survival and axonal regeneration of retinal ganglion cells (RGCs) in a rat optic nerve crush (ONC) model (2). Y-39983-mediated downregulation of active-RhoA, ROCK1 and ROCK2 expression coincided with the appearance of larger numbers of regenerating axons. *See Y-27632 and (R)-(+)-trans-4-(1-Aminoethyl)-N-pyridyl)cyclohexane carboxamide* 1. Nakajima, Nakajima, Minagawa, Shearer & Azuma (2005) *J. Pharm. Sci.* **94**, 701. **2**. Yang, Wang, Liu, *et al.* (2013) *Oncol. Rep.* **29**, 1140.

YC-1

This nitric oxide-independent activator of soluble guanylyl cyclase (FW = 304.34 g/mol; CAS 170632-47-0), also named 3-(5'-hydroxymethyl-2'-furyl)-1-benzylindazole, elevates cell levels of 3',5'-cyclic-GMP levels and inhibits collagen-stimulated aggregation of washed platelets (IC_{50} = 14.6 μM). YC-1 also displays antiproliferative activity *in vitro* and *in vivo* by inducing G_1 cell-cycle arrest in two human hepatocellular carcinoma (HCC) cell lines, and in HCC xenografts in athymic SCID mice. YC-1 exhibits low cytotoxicity in non-malignant cells. **1**. Ko, Wu, Kuo, Lee & Teng (1994) *Blood* **84**, 4226. **2**. Martin, Lee & Murad (2001) *Proc. Natl. Sci. U.S.A.* **98**, 12938. **3**. Wang, Pan & Guh (2005) *J. Pharmacol. Exp. Ther.* **312**, 917.

YCZ-18

This brassinosteroid biosynthesis inhibitor (FW = 415.87 g/mol) binds targets recombinant CYP90D1, a steroid C23-hydroxylase that is required in brassinosteroid (BR) biosynthesis, inducing a typical Type II binding spectrum (K_d = 0.79 μM). BRs are polyhydroxylated steroids that play critical hormone roles in regulating broad aspects of plant growth and development. Their structural diversity is generated by the action of several CYP's (*or* P450s). YCZ-18 down-regulates gene expression required for BRs in *Arabidopsis*. (*See Brassinazole*) 1. Oh, Matsumoto, Yamagami, *et al.* (2015) *PLoS One* **10**, :e0120812.

YE 120

This GPR35 agonist (FW = 330.17 g/mol; CAS 383124-82-1; Solubility: 100 mM in DMSO), also named 2-[3-cyano-5-(3,4-dichlorophenyl)-4,5-dimethyl-2(5H)-furanylidene]propanedinitrile, is a GTP Protein-Coupled Receptor 35, *or*, agonist (1) in dynamic mass redistribution (DMR) assays using resonant waveguide grating (RWG) biosensors to characterize stimulator-mediated cell responses including signaling, detecting redistribution of cellular contents in both directions that are perpendicular and parallel to the sensor surface (2). YE-120 also displays partial agonist activity in a β-arrestin translocation assay (EC_{50} = 10.2 μM). **1**. Deng, Hu, He, *et al.* (2011) *J. Med. Chem.* **54**, 7385. **2**. Fang, Ferrie, Fontaine, Mauro & Balakrishnan (2006) *Biophys. J.* **91**, 1925.

YH1885, See *Revaprazan*

YIC-C8-434, See *N-(3,5-Dimethoxy-4-n-octyloxy-cinnamoyl)-N'-(3,4-dimethylphenyl)piperazine*

YIL 781

This GHS-R_{1a} antagonist (FW = 409.51 g/mol; CAS 875258-85-8; Soluble to 100 mM in DMSO), also named 6-(4-fluorophenoxy)-2-methyl-3-[[(3S)-1-(1-methylethyl)-3-piperidinyl]methyl]-4(3H)-quinazolinone, targets the Ghrelin Receptor-1a (K_i = 17 nM), but displays much lower affinity for the motilin receptor (K_i = 6 μM). YIL 781 blocks the effects of ghrelin on insulin secretion both *in vivo* and *in vitro*, thereby improving glucose homeostasis, suppressing appetite, and promoting weight loss. Although this GHS-R_{1a} antagonist modestly delayed gastric emptying at the highest dose tested (10 mg/kg), delayed gastric emptying does not appear to be a requirement for weight loss because lower doses produced weight loss without an effect on gastric emptying. **1**. Esler, Rudolph, Claus, *et al.* (2007) *Endocrinology* **148**, 5175.

YJA20379-1, See *2-Amino-4,5-dihydro-8-phenylimidazole[2,1-b]thiazolo[5,4-g]benzothiazole*

YK 3-237

This SIRT1 activator (FW = 372.18 g/mol; CAS 1215281-19-8; Solubility: 100 mM in DMSO, systematically named *B*-[2-methoxy-5-[(1E)-3-oxo-3-(3,4,5-trimethoxyphenyl)-1-propen-1-yl]phenyl]boronic acid, activates the NAD^+-dependent deacetylase Sirtuin-1 *in vitro*, with deacetylation of both WTp53 (wild-type p53 acetylated at K382) as well as mtp53. In humans, the SIRT1 interactome includes some 150 proteins. Deacetylation of mtp53 resulted in depletion of mtp53 protein level and up-regulated the expression of WTp53-target genes, PUMA and NOXA. YK-3-237 also induces PARP-dependent apoptotic cell death and arrests the cell cycle at G_2/M phase of mtp53 TNBC cells. *See also SRT1720 Note*: Although YK-3-237 is originated as a boronic acid chalcone analog of combretastatin A-4 (CA-4), YK-3-237 does not bind to tubulin *in vitro* and exhibits potent antiproliferative activity toward a broad range of NCI cancer cell lines. **1.** Yi, Kang, Kim, *et al.* (2013) *Oncotarget* **4**, 984.

YK 4-279

This RNA helicase A inhibitor (FW = 366.20 g/mol; CAS 1037184-44-3; Soluble to 100 mM in DMSO), also named 4,7-dichloro-1,3-dihydro-3-hydroxy-3-[2-(4-methoxyphenyl)-2-oxoethyl]-2*H*-indol-2-one, binds to the oncogenic transciption factor EWS-FLI1, inhibiting Ewing's sarcoma family tumor (ESFT) cell growth and inducing apoptosis. Notably, Ewing's tumors contain a t(11;22) translocation, leading to expression of the oncogenic fusion protein EWS-FLI1, a disordered protein that currently defies standard crystallographic/multi-D-NMR structure-based small-molecule inhibitor design. EWS-FLI1 binding to RNA helicase A (RHA) is important for its oncogenic function. In a cell-free assay, NSC635437 reduced the direct binding of GST-RHA(647-1075) to full-length recombinant EWS-FLI1 (K_i = 9.5 μM). **1.** Erkizan, Kong, Merchant, et al (2009) *Nature Med.* **15**, 750.

YM155, *See Sepantronium bromide*

YM 175, *See Incadronate*

YM466

This orally active naphthylamidine and anticlotting agent (FW = 535.67 g/mol), also known as [*N*-[4-[(1-acetimidoyl-4-piperidyl)oxy]phenyl]-*N*-[(7-amidino-2-naphthyl)methyl]sulfamoyl]acetic acid and YM-60828, is a potent inhibitor of coagulation factor Xa (K_i = 1.3 nM), but a significantly weaker inhibitor of thrombin (K_i > 100 μM). YM466 also inhibits platelet aggregation induced by various agonists, reduces the incidence of occlusion, and exerts antithrombotic effects in venous thrombosis and arteriovenous shunt models. **1.** Taniuchi, Sakai, Hisamichi, Kayama, *et al.* (1998) *Thromb. Haemost.* **79**, 543. **2.** Leadley, Jr. (2001) *Curr. Top. Med. Chem.* **1**, 151.

YM 511

This aromatase inhibitor (FW = 354.21 g/mol; CAS 148869-05-0; Solubility: 100 mM in DMSO), also named 4-[[(4-bromophenyl)methyl]-4*H*-1,2,4-triazol-4-ylamino]benzonitrile, targets rat ovary (IC$_{50}$ = 0.4 nM) and human placenta (IC$_{50}$ = 0.12 nM), respectively, reducing plasma estrogen levels into ranges induced by ovariectomy and inhibiting testosterone-induced breast cancer cell growth *in vitro* (IC$_{50}$ = 0.13 nM). **1.** Kudoh, *et al.* (1995) *J. Steroid Biochem. Molec. Biol.* **54**, 265. **2.** Kudoh, *et al.* (1996) *J. Steroid Biochem. Molec. Biol.* **58**, 189. **3.** Okada, *et al.* (1996) *Chem. Pharm. Bull.* **44**, 1871.

YM-529, *See Minodronate*

YM 750

This anti-oxidative ACAT inhibitor (FW = 452.64 g/mol; CAS 138046-43-2; Solubility: 100 mM in DMSO; 10 mM in Ethanol), also known as *N*-cycloheptyl-*N*-(9*H*-fluoren-2-ylmethyl)-*N*'-(2,4,6-trimethylphenyl)urea, targets acyl-CoA:cholesterol acyltransferase (IC$_{50}$ = 0.18 μM), exhibiting both hypocholesterolemic and diet-induced atherosclerosis formation in mice (1-3). **1.** Nagata, Yonemoto, Iwasawa, *et al.* (1995) *Biochem. Pharmacol.* **49**, 643. **2.** Kamiya, Shirahase, Yoshimi, *et al.* (2000) *Chem. Pharm. Bull.* **48** 817. **3.** Miike, Shirahase, Jino, *et al* (2008) *Life Sci.* **82**, 79.

YM-09151-2, *See Nemonapride*

YM-60828, *See YM-466*

YM 90709

This novel selective inhibitor (FW = 359.43 g/mol; CAS 163769-88-8; Solubility: to 25 mM in Ethanol, to 50 mM in DMSO), also known as 5,6-dihydro-2,3-dimethoxy-6,6-dimethylbenz[7,8]indolizino[2,3-*b*]quinoxaline, targets the interleukin-5 (IL-5) receptor (IC$_{50}$ = 0.45-1.00 mM), blocking IL-5-mediated prolongation of eosinophil survival and IL-5-induced tyrosine phosphorylation of JAK2, without inhibiting GM-CSF-mediated effects. YM 90709 also inhibits antigen-induced eosinophil and lymphocyte recruitment in rat airways *in vivo*, without affecting peripheral blood or bone marrow leukocytes. **1.** Morokata, *et al.* (2002) *Int. Immunopharmacol.* **2**, 1693. **2.** Morokata, *et al.* (2004) **4**, 873. **3.** Morokata, *et al.* (2005) *Immunol. Lett.* **98**, 161.

YM-231146

This orally bioavailable antiangiogenesis agent (FW = 389.46 g/mol; IUPAC name: (3*Z*)-3-[6-(2-morpholin-4-ylethoxy)quinolin-2(1*H*)-ylidene]-1,3-dihydro-*H*-indol-2-one) inhibits VEGF receptor-2 tyrosine kinase, IC$_{50}$ = 80 nM, with much weaker (IC$_{50}$ > 1 μM) action against other kinases. YM-231146 inhibits VEGF-stimulated proliferation, VEGF-R2 autophosphorylation, and vessel sprout formation of human vascular endothelial cells at concentrations between 0.15-0.30 μM. YM-231146 did not inhibit cancer cell proliferation at these concentrations (IC$_{50}$ > 5 μM). Once-daily YM-231146 dosing (3-100 mg/kg) to human cancer xenografts elicited antitumor activity. **1.** Amino, Ideyama, Yamano, *et al.* (2005) *Biol. Pharm. Bull.* **28**, 2096.

YM-359445

This orally bioavailable VEGF receptor-2 tyrosine kinase inhibitor and anti-angiogenesis agent (FW$_{free-base}$ = 512.63 g/mol; often formulated as the mono-l-tartrate), also known as (3*Z*)-3-[6-[(4-methylpiperazin-1-yl)methyl]quinolin-2(1*H*)-ylidene]-2-oxoindoline-6-carbaldehyde *O*-(1,3-thiazol-4-ylmethyl)oxime, exhibits highly potent antitumor activity against established tumors. At 1 nM, YM-359445 also inhibits tyrosine kinases other than VEGFR2, including Flt-1, Flt-3, Flt-4, PDGFRβ, c-Fms, and c-Kit by 66%, 73%, 84%, 25%, 71%, and 78%, respectively. That YM-359445 is specific for tyrosine kinase, however, is indicated by its total inhibitory effect on VEGFR2 autophosphorylation at 3 nM, compared to IC$_{50}$ values of >1 μM for protein kinase A, protein kinase C$_\alpha$, phosphoinositide-dependent kinase-1, serum-/glucocorticoid-inducible kinase-1, as well as c-Jun NH$_2$-terminal kinase-3. The IC$_{50}$ against HUVEC proliferation induced by VEGF was only 1.5 nM, showing extremely potent antitumor action versus various human cancer xenografts. Moreover, because these tyrosine kinases would also be involved in tumor growth, multi-targeted tyrosine kinase inhibitors may be more effective at inhibiting tumor growth than agents that target a single angiogenic factor. **1.** Amino, Ideyama, Yamano, *et al.* (2006) *Clin. Cancer Res.* **12**, 1630.

YO-01027

This aspartate protease inhibitor (FW = 463.48 g/mol; CAS 209984-56-5; Solubility: 90 mg/mL DMSO; <1 mg/mL H$_2$O; Formulation: Suspend in 0.5% (w/v) hydroxypropylmethylcellulose (Methocel E4M) and 0.1% (w/v) Tween 80 in water), also known as Dibenzazepine (DBZ) and systematically as 7-(*S*)-[*N*'(3,5-difluorophenylacetyl)-L-alaninyl]amino-5-methyl-5,7-dihydro-6*H*-dibenz[*b,d*]azepin-6-one, targets γ-secretase (GS), thereby inhibiting cleavage of numerous type-I integral membrane proteins, including amyloid precursor protein (APP) and Notch. While APP cleavage contributes to the generation of toxic amyloid beta peptides in Alzheimer's disease, cleavage of the Notch receptor is required for normal physiological signaling between differentiating cells. **1.** Groth, *et al.* (2010) *Mol. Pharmacol.* **77**, 567.

Yohimbine

This alkaloid and α$_2$-adrenoceptor antagonist (FW$_{free-acid}$ = 354.45 g/mol; CAS 146-48-5; FW$_{HCl-salt}$ = 390.91 g/mol; CAS 65-19-0), also named 17α-hydroxyyohimban-16α-carboxylic acid methyl ester hydrochloride, from the bark of *Corynanthe yohimbe* and *Rauwolfia serpintina* is an α$_2$-adrenergic antagonist (1) with vasodilatory properties that have been exploited to treat atherosclerosis. Yohimbine binds to human α$_{2A}$ (pK_i = 8.52), α$_{2B}$ (pK_i = 8.00), and α$_{2C}$ (pK_i = 9.17) receptors. Yohimbine relaxes tissues previously made to contract by application of serotonin. When administered first, yohimbine likewise prevents serotonin-induced contraction. Such action is in marked contrast to the well known antagonism observed between yohimbine and adrenaline, which is effective only when applied first. Other important diastereoisomers are α yohimbine *or* rauwolscine (with all three E-ring hydrogens projecting backward and its methylacetoxy substituent projecting forward) and *allo*-yohimbine (with all three E-ring hydrogens projecting backward). Yohimbine's diastereomer is Rauwolscine (*See also* Rauwolscine). **Target(s):** α$_2$-adrenergic receptor (1); [β-adrenergic-receptor] kinase (2). **1.** Goldberg & Robertson (1983) *Pharmacol. Rev.* **35**, 143. **2.** Benovic, Regan, Matsui, Mayor, *et al.* (1987) *J. Biol. Chem.* **262**, 17251.

YopJ *Yersinia* Virulence Factor

This 33-kDa *Yersinia* virulence factor inhibits several signal transduction protein kinases, including c-Jun amino-terminal kinase, [mitogen-activated protein kinase] kinase, nuclear factor-κB signaling pathways. **1.** Orth, Palmer, Bao, Stewart, *et al.* (1999) *Science* **285**, 1920.

YS 035

This verapamil analogue (FW$_{HCl-Salt}$ = 395.93 g/mol; CAS 89805-39-0), also named *N,N*-bis-(3,4-dimethoxyphenethyl)-*N*-methylamine, inhibits Ca^{2+} uptake in many cell types, *e.g.* by muscle cells (inhibitor range 10-30 μM), by rat brain synaptosomes and less so by baby-hamster kidney cells (1). YS 035 partially inhibits the slow Ca^{2+} release induced by Ruthenium Red and the rapid Na$^+$-dependent efflux from heart mitochondria. The inhibition of the Na$^+$/Ca^{2+} exchange is noncompetitive (K_i = 28 μM). It also totally inhibits Ruthenium Red-induced Ca^{2+} efflux from liver mitochondria, but does not affect Ca^{2+} release induced by uncoupler, respiratory inhibitor or chelator, nor the mitochondrial ATP synthesis and membrane potential (1). YS 035 is likely to be especially useful for studies on mitochondrial Ca^{2+} transport. Electrophysiologic measurements on YS 035 effects on Purking cells have been reported (2). **1.** Deana, Panato, Cancellotti, Quadro & Galzigna (1984) *Biochem. J.* **218**, 899. **2.** Berger, Borchard, Hafner, Kammer & Weis (1991) *Naunyn. Schmiedebergs. Arch. Pharmacol.* **344**, 653.

Ytterbium and Ytterbium Ions

This lanthanide (Yb; Atomic number = 70; Atomic weight = 173.04) has two ionic forms, Yb^{2+} and Yb^{3+}, with respective ionic radii of 0.93 Å and 0.858 Å. Yb^{3+} is paramagnetic, making it a very useful NMR-shift reagent. The tree *Carya* spp. reportedly accumulates ionic ytterbium. **Target(s):** calcium channels (1,2); coagulation factor XIII (via Yb^{3+}) (3); phosphoprotein phosphatase (4). **1.** Mliner & Enyeart (1993) *J. Physiol.* **469**, 639. **2.** Beedle, Hamid & Zamponi (2002) *J. Membr. Biol.* **187**, 225. **3.** Fox, Yee, Pedersen, Le Trong, *et al.* (1999) *J. Biol. Chem.* **274**, 4917. **4.** Zhou, Clemens, Hakes, Barford & Dixon (1993) *J. Biol. Chem.* **268**, 17754. ***See also*** *Ytterbium-ATP Complex*

Ytterbium(III)-ATP Complex

The Yb(III)-ATP complex inhibits various ATP-dependent enzymes, most often by binding in place of MgATP^{2-} or CaATP^{2-}. **Target(s):** hexokinase (slow-binding)(1-3); phosphoglycerate kinase (4). **1.** Morrison (1982) *Trends Biochem. Sci.* **7**, 102. **2.** Viola, Morrison & WCleland (1980) *Biochemistry* **19**, 3131. **3.** Furumo & Viola (1989) *Inorg. Chem.* **28**, 820. **4.** Tanswell, Westhead & Williams (1974) *FEBS Lett.* **48**, 60.

Yttrium and Yttrium(III) Ion

This group 3 (*or* IIIA) transition metal element (Y; Atomic number = 39; Atomic weight = 88.90585; ionic radius = 0.893 Å) forms both oxygen exchange-active and -inert complexes with various anions and metabolites. The Poriferan *Melithoea* spp. and *Carya* spp. trees accumulate yttrium, as do the calcareous tissues in mammals have also been reported to be yttrium accumulators. **Target(s):** acetylcholinesterase (1,2); calcium channels (3,4); enolase (5); micrococcal nuclease (6). **1.** Marquis & Lerrick (1982) *Biochem. Pharmacol.* **31**, 1437. **2.** Marquis & Black (1985) *Biochem. Pharmacol.* **34**, 533. **3.** Mlinar & Enyeart (1993) *J. Physiol.* **469**, 639. **4.** Beedle, Hamid & Zamponi (2002) *J. Membr. Biol.* **187**, 225. **5.** Malmström (1961) *The Enzymes*, 2nd ed. (Boyer, Lardy & Myrbäck, eds.), **5**, 471. **6.** Cuatrecasas, Fuchs & Anfinsen (1967) *J. Biol. Chem.* **242**, 1541.

YZ9

This small-molecule inhibitor (FW = 234.21 g/mol; CAS 6093-71-6; Solubility: 50 mM in DMSO), also known as ethyl 7-hydroxy-2-oxo-2*H*-1-benzopyran-3-carboxylate, targets PFKFB3 (IC$_{50}$ = 0.183 μM), the hypoxia-inducible isoform of the 6-phosphofructo-2-kinase/fructose-2,6-bisphosphatases, which catalyze Fru-2,6-BP synthesis at >10x higher rates than other isoforms. YZ9 inhibition of PFKFB3 reduces the levels of Fru-2,6-BP in HeLa cells by 40%, resulting in a greater than 30% decrease in lactate, eventually resulting in apoptosis and necrosis. **1.** Seo, Kim, Neau, Sehgal & Lee (2011) *PLoS One* **6**, e24179.

– Z –

Zabofloxacin

This investigational fluoroquinolone-class antibiotic (FW = 401.39 g/mol; CAS 219680-11-2), also named DW-224a and 1-cyclopropyl-6-fluoro-7-[(8E)-8-(methoxyimino)-2,6-diazaspiro[3.4]oct-6-yl]-4-oxo-1,4-dihydro-1,8-naphthyridine-3-carboxylate, is a broad-spectrum systemic antibacterial agent that targets Type II DNA topoisomerases (gyrases), which are required for bacterial replication and transcription. Zabofloxacin has activity against *Neisseria gonorrhoeae*, including strains that are resistant to other quinolone antibiotics. The potency of zabofloxacin (MIC$_{50}$ = 16 ng/mL) was generally comparable with azithromycin, but 8-fold superior to ciprofloxacin. (For the prototypical member of this antibiotic class, **See** *Ciprofloxacin*) **1**. Jones, Biedenbach, Ambrose & Wikler (2008) *Diagn. Microbiol. Infect. Dis.* **62**, 110.

Zacopride

This anti-emetic and anxiolytic (FW = 346.26 g/mol; CAS 101303-98-4; Solubility: 100 mM H$_2$O; 100 mM DMSO), systematically named ($\pm$)-4-amino-N-1-azabicyclo[2.2.2]oct-3-yl-5-chloro-2-methoxybenzamide, is a serotonin receptor antagonist that targets the 5-HT$_3$ receptor antagonist (K_i = 0.38 nM), but a 5-HT$_4$ receptor agonist (K_i = 370 nM). [^{3}H]-Zacopride displays saturable binding to homogenates of the rat entorhinal cortex, with K_d = 0.76 nM, B$_{max}$ = 77.5 fmol/mg protein, n_{Hill} ~1 (1). [^{125}I]-Iodo-zacopride (K_d = 4.3 nM) has been to map 5-HT3 receptors in the rat central nervous system (2). **1**. Barnes, Costall & Naylor (1988) *J. Pharm. Pharmacol.* **40**, 548. **2**. Koscielniak, Ponchant, Laporte, *et al.* (1990) *Comptes rendus de l'Académie des sciences. Série III, Sciences de la vie* **311**, 231.

Zafirlukast

This leukotriene D$_4$ receptor antagonist (FW = 575.69 g/mol; CAS 107753-78-6), also known as ICI 204,219, Accolate®, Accoleit®, Vanticon®, and 4-(5-cyclopentyloxycarbonylamino-1-methylindol-3-ylmethyl)-3-methoxy-N-(*o*-tolylsulfonyl) benzamide, exerts its antiasthmatic activity by blocking the action of cysteinyl leukotrienes at CysLT1 receptors and thereby reducing airway constriction (1-4). **Other Target:** cytochrome P450, including CYP1A2, CYP3A, CYP2C9, CYP2C19, and CYP2D6 (5). **1**. Krell,

Aharony, Buckner, Keith, *et al.* (1990) *Amer. Rev. Respir. Dis.* **141**, 978. **2**. Creticos (2003) *Drugs* **63** (Suppl. 2), 1. **3**. Murata & Sugimoto (2002) *Nippon Yakurigaku Zasshi* **119**, 247. **4**. Bernstein (1998) *Amer. J. Respir. Crit. Care Med.* **157**, S220. **5**. Shader, Granda, von Moltke, Giancarlo & Greenblatt (1999) *Biopharm. Drug Dispos.* **20**, 385.

Zalcitabine, *See 2'-3'-dideoxycytidine*

Zaleplon

This benzodiazepine GABA$_A$ full agonist and anxiolytic (FW = 305.34 g/mol; CAS 151319-34-5), tradenamed Sonata®, Starnoc®, and Andante®, and systematically as N-(3-(3-cyanopyrazolo[1,5-*a*]pyrimidin-7-yl)phenyl)-N-ethylacetamide, selectively targets the benzodiazepine site (ω_1) on the α_1 sub-receptor (site associated with its anxiolytic effects), with lesser action at α_2 sub-receptor (site associated with muscle relaxant effects), α_3 sub-receptor (site associated with muscle relaxant effects), and α_5 sub-receptor (site associated with anticonvulsant effects). Its absorption is rapid, as is the onset of its therapeutic effects within 5–15 minutes following ingestion. Moreover, 1-1.5-hour half-life commends its use, when there are concerns about the residual effects of many benzodiazepine agents on performance-related skills (driving and piloting). Because Zaleplon is metabolized by aldehyde oxidase, its half-life can be prolonged by substances that inhibit aldehyde oxidase or shortened by substances that induce aldehyde oxidase. **1**. Weitzel, Wickman, Augustin & Strom (2000) *Clin. Ther.* **22**, 1254.

Zamifenacin

This muscarinic receptor antagonist (FW$_{free-base}$ = 415.53 g/mol; FW$_{fumarate-salt}$ = 415.53 g/mol; CAS 127308-98-9; Solubility: 100 mM in DMSO), also named UK 76654 and (3R)-1-[2-(1-,3-benzodioxol-5-yl)ethyl]-3-(diphenylmethoxy)piperidine, selectively targets M$_3$ receptors (pK$_i$ = 8.52), with about 5x lower affinity for M$_2$, M$_1$ and M$_4$ receptors (1). Zamifenacin was rapidly metabolized *in vitro* by liver microsomes from rat, dog, and man. Following oral administration to animals, metabolic clearance resulted in decreased bioavailability due to first-pass metabolism in rat and mouse (2). The major route of clearance of zamifenacin with the primary metabolic step resulting in opening of the methylenedioxy ring to yield the catechol. In man, this metabolite was excreted as the glucuronide conjugate (2). **1**. Watson, Reddy, Stefanich & Eglen (1995) *Eur. J. Pharmacol.* **285**, 135. **2**. Beaumont, Causey, Coates & Smith (1996) *Xenobiotica* **26**, 459.

Zanamivir

This antiviral (FW = 332.31 g/mol; CAS 139110-80-8), also known as Relenza™ and 4-guanidino-2,4-dideoxy-2,3-dehydro-N-acetylneuraminate, inhibits influenza virus A and B sialidase (*i.e.*, neuraminidase) at sub-nM concentrations *in vitro* (1-4). As a transition-state analogue of sialic acid, it forms a high-affinity complex with neuraminidase. Zanamivir's inhibitory action against human sialidases is minimal (IC$_{50}$ in the micromolar range), greatly contrasting from its action on viral sialidases. However, zanamivir is orally unavailable and is instead delivered directly to the respiratory tract by means of an inhaler (Diskhaler®) that holds small pouches of the powdered drug. This mode of delivery offers great advantage for preventing influenza infection, and, while overall whole-body bioavailability is low (~2 percent), zanamivir becomes highly concentrated in the respiratory tract, reaching levels that are 1000x its IC$_{50}$ within seconds after inhalation. For this reason, one cannot rule out inhibition of human sialidases in the respiratory tract. Analysis data for patients (all ages) admitted to hospital worldwide with laboratory-confirmed or clinically diagnosed infection with the pandemic influenza A H1N1pdm09 virus indicates that early treatment in adults admitted to hospital with suspected/proven influenza infection can be effective in reducing mortality (5). **1**. Pegg & von Itzstein (1994) *Biochem. Mol. Biol. Int.* **32**, 851. **2**. Woods, Bethell, Coates, Healy, *et al.* (1993) *Antimicrob. Agents Chemother.* **37**, 1473. **3**. Taylor & von Itzstein (1994) *J. Med. Chem.* **37**, 616. **4**. von Itzstein, Wu & Jin (1994) *Carbohydr. Res.* **259**, 301. **5**. Muthuri, Venkatesan, Myles, *et al.* (2014) *Lancet Respir. Med.* **2**, 395.

Zankiren

This peptide analogue (FW$_{\text{free-base}}$ = 705.98 g/mol; CAS 138742-43-5), also known as A-72517, is the first peptidic renin inhibitor with significant oral absorption. It inhibits human plasma renin at pH 7.4 with IC$_{50}$ = 3.1 nM (1-3). **1**. Kleinert, Stein, Boyd, Fung, *et al.* (1992) *Hypertension* **20**, 768. **2**. Rosenberg, Spina, Condon, Polakowski, *et al.* (1993) *J. Med. Chem.* **36**, 460. **3**. Kleinert, Rosenberg, Baker, *et al.* (1992) *Science* **257**, 1940.

Zantac, See Ranitidine

Zaprinast

This vasodilator and lead molecule for the development of Viagara (FW = 271.28 g/mol; CAS 37762-06-4; stability = 5 years at +20°C; Soluble to 100 mM in DMSO; stock solutions stable for up to 6 months at –20°C), also known as M&B 22,948 and systematically as 1,4-dihydro-5-(2-propoxyphenyl)-7*H*-1,2,3-triazolo[4,5-*d*]pyrimidine-7-one, inhibits 3′,5′-cGMP phosphodiesterase (1-10), with highest affinity toward type-5 and type-6 isozymes (*e.g.*, PDE-6, IC$_{50}$ = 0.2μM; PDE-5, IC$_{50}$ = 0.2-0.8μM; PDE-1, IC$_{50}$ = 2.6μM; PDE-2, IC$_{50}$ > 10μM; PDE-3, IC$_{50}$ > 10μM; and PDE-4, IC$_{50}$ > 10μM). Zaprinast also selectively increases cerebral blood flow. In actions that are unrelated to inhibition of cGMP phosphodiesterase, Zaprinast is a potent inhibitor of the mitochondrial pyruvate carrier (MPC), causing aspartate to accumulate at the expense of glutamate (11). (*See also UK5099*) The effects on both glutamate and aspartate are detectable at 30 minutes after addition of Zaprinast, becoming more apparent at 1 hour and remaining constant for 2–6 hours. Zaprinast also blocks the entry of glucose-derived pyruvate into the TCA cycle, and it inhibits pyruvate-driven O$_2$ consumption in brain mitochondria, actions consistent with its ability to block liver mitochondria MPC. Moreover, inactivation of the aspartate glutamate carrier in retina does not attenuate the metabolic effect of Zaprinast. Importantly, Zaprinast does not inhibit pyruvate dehydrogenase (PDH) activity. **1**. Thompson (1991) *Pharmacol. Ther.* **51**,

13. **2**. Burns, Rodger & Pyne (1992) *Biochem. J.* **283**, 487. **3**. Ni, Safai, Gardner & Humphreys (2001) *Kidney Int.* **59**, 1264. **4**. Coste & Grondin (1995) *Biochem. Pharmacol.* **50**, 1577. **5**. Zhang, Kuvelkar, Wu, Egan, Billah & Wang (2004) *Biochem. Pharmacol.* **68**, 867. **6**. Thomas, Francis & Corbin (1990) *J. Biol. Chem.* **265**, 14964. **7**. Wang, Wu, Myers, Stamford, Egan & Billah (2001) *Life Sci.* **68**, 1977. **8**. Wang, Wu, Egan & Billah (2003) *Gene* **314**, 15. **9**. Su & Vacquier (2006) *Mol. Biol. Cell* **17**, 114. **10**. Liu, Underwood, Li, Pamukcu & Thompson (2002) *Cell. Signal.* **14**, 45. **11**. Du, Cleghorn, Contreras, *et al.* (2013) *J. Biol. Chem.* **288**, 36129.

Zaragozic Acid A

This complex bicyclo-tricarboxylic acid (FW$_{\text{free-acid}}$ = 705.78 g/mol; CAS 142561-96-4), known simply as squalestatin 1 and systematically as (1*R*,2*S*,3*S*,5*S*,6*R*,7*R*)-5-[(4*S*,5*R*)-4-acetoxy-5-methyl-3-methylene-6-phenyl-hexyl]-7-[(*E*,4*S*,6*S*)-4,6-dimethyl-1-oxooct-2-enoxy]-2,6-dihydroxy-4,8-di-oxabicyclo[3.2.1]octane-1,2,3-tricarboxylic acid, inhibits protein farnesyltransferase, IC$_{50}$ = 216 nM. With squalene synthase, zaragozic acid A is a mechanism-based irreversible inactivation. *Note*: The terms zaragozic acid and zaragosate refer to the acid and conjugate base forms of a class of related squalestatins, so named for their capacity to inhibit squalene synthase. **Target(s):** phytoene synthase (1,2); protein farnesyltransferase (3-5); squalene synthase (3,6-14). **1**. Neudert, Martínez-Férez, Fraser & Sandmann (1998) *Biochim. Biophys. Acta* **1392**, 51. **2**. Fraser, Schuch & Bramley (2000) *Planta* **211**, 361. **3**. Nadin & Nicolaou (1996) *Angew. Chem. Int. Ed. Engl.* **35**, 1622. **4**. Gibbs, Pompliano, Mosser, Rands, *et al.* (1993) *J. Biol. Chem.* **268**, 7617. **5**. Gibbs (2001) *The Enzymes*, 3rd ed. (Tamanoi & Sigman, eds.), **21**, 81. **6**. Wentzinger, Bach & Hartmann (2002) *Plant Physiol.* **130**, 334. **7**. Kocarek, Kraniak & Reddy (1998) *Mol. Pharmacol.* **54**, 474. **8**. Keller (1996) *Biochim. Biophys. Acta* **1303**, 169. **9**. Procopiou, Cox, Kirk, Lester, *et al.* (1996) *J. Med. Chem.* **39**, 1413. **10**. Lindsey & Harwood (1995) *J. Biol. Chem.* **270**, 9083. **11**. Baxter, Fitzgerald, Hutson, McCarthy, *et al.* (1992) *J. Biol. Chem.* **267**, 11705. **12**. Thompson, Danley, Mazzalupo, Milos, Lira & Harwood (1998) *Arch. Biochem. Biophys.* **350**, 283. **13**. Soltis, McMahon, Caplan, Dudas, *et al.* (1995) *Arch. Biochem. Biophys.* **316**, 713. **1**. Brusselmans, Timmermans, Van de Sande, Van Veldhoven, *et al.* (2007) *J. Biol. Chem.* **282**, 18777.

Zardaverine

This pyridazinone heterocycle (FW = 268.21 g/mol; CAS 101975-10-4), also named 6-(4-difluoromethoxy-3-methoxyphenyl)-3(2*H*)-pyridazinone, is a phosphodiesterase III and IV inhibitor, IC$_5$ ≈ 1.5-1.8 μM (1). **Other Target(s):** Unlike other PDE III/IV inhibitors, zardaverine also induces G$_0$/G$_1$ cell-cycle arrest and apoptosis in sensitive Bel-7402 (IC$_{50}$ = 137 nM), Bel-7404 (IC$_{50}$ = 288 nM), QGY-7701 (IC$_{50}$ = 51 nM), and SMMC-7721 (IC$_{50}$ = 37 nM) hepatocellular carcinoma (HCC) cells by dysregulating cell cycle-associated proteins by means of a PDE III/IV-independent mechanism (2). The secondary target is signal transduction mediated by Rb (the retinoblastoma product of the well-characterized tumor suppressor gene *Rb*), which functions to prevent the cell from entering S phase. Phosphorylated or mutated forms of the Rb protein are incapable of arresting the cell in G$_1$ phase. Treatment with zardaverine dose-dependently decreases the total protein level of Rb and suppresses its phosphorylation at Ser780 in sensitive cells, whereas there are no such changes in insensitive cells, even at >30,000 nM (2). Low expression of Rb is thus a likely biomarker for those cancer cells that are sensitive to zardaverine's growth-inhibiting effects. **1**. Schudt, Winder, Muller & Ukena (1991) *Biochem. Pharmacol.* **42**, 153. **2**. Sun, Quan, Xie (2014) *PLoS One* **9**, e90627.

ZCL 278

This protein kinase inhibitor (FW = 584.89 g/mol; CAS 587841-73-4), also named 2-(4-bromo-2-chlorophenoxy)-*N*-[[[4-[[(4,6-dimethyl-2-pyrimidinyl)amino]sulfonyl]phenyl]amino]thioxomethyl]acetamide, selectively targets <u>C</u>ell <u>d</u>ivision <u>c</u>ontrol protein <u>42</u>, or Cdc42, binding (K_d = 6.4 µM) at the site for intersectin, its guanine nucleotide exchange factor. ZCL 278 inhibits Cdc42-mediated cellular effects, including microspike formation in 3T3 fibroblasts and neuronal branching in primary neonatal cortical neurons. It also suppresses cell motility and migration in PC3 cells, without cytotoxic effects. **1.** Friesland, Zhao, Chen, *et al.* (2013) *Proc. Natl. Acad. Sci U.S.A.* **110**, 1261.

ZD6126

This vascular disrupting agent and antineoplastic pro-drug (FW = 437.38 g/mol; CAS 219923-05-4), also known systematically as phosphoric acid mono-(5-acetylamino-9,10,11-trimethoxy-6,7-dihydro-5*H*-dibenzo[*a,c*]cyclohepten-3-yl) ester, binds reversibly with tubulin to prevent microtubule assembly (1-3), selectively disrupting the abnormal vasculature associated with disease processes such as cancer and macular degeneration. Such action has been attributed to the ability of these agents to selectively destroy the central regions of tumors, areas widely believed to contain cell populations resistant to cytotoxic therapies (3). Note that the cell-permeant form is formed by hydrolysis of the phosphate monoester. ZD6126 induces the apoptosis of CD31-positive vascular endothelial cells in tumors but not in the normal lung parenchyma. ZD6126 differs from its parent compound, colchicine, which binds extraordinarily tightly to tubulin. Cardiotoxicity limits the clinical use of ZD6126 (4). **1.** Goto, Yano, Zhang, *et al.* (2002) *Cancer Research* **62**, 3711. **2.** LoRusso, Gadgeel, Wozniak, *et al.* (2011) *Invest. New Drugs* **26**, 159. **3.** Chaplin, Horsman & Siemann (2006) *Curr. Opin. Investig. Drugs* **7**, 522. **4.** Lippert (2007) *Bioorg. & Med. Chem.* **15**, 605.

ZD 8321

This trifluoromethyl ketone (FW = 423.43 g/mol), also known as *N*-methoxycarbonyl-L-valyl-L-prolyl-L-valine trifluoromethyl ketone, inhibits porcine pancreatic elastase, K_i = 149 nM (1) and human leukocyte elastase, K_i = 13 nM (1,2). **1.** Veale, Bernstein, Bohnert, Brown, *et al.* (1997) *J. Med. Chem.* **40**, 3173. **2.** Kapui, Varga, Urban-Szabó, Mikus, *et al.* (2003) *J. Pharmacol. Exp. Ther.* **305**, 451.

ZD9238, See *Fulvestrant*

ZDP, See *5-Aminoimidazole-4-carboxamide-1-β-ribofuranosyl 5'-Diphosphate*

Zearalenone

This uterotropic mycotoxin (FW = 318.37 g/mol; CAS 17924-92-4), also known as F-2 mycotoxin and 2,4-dihydroxy-6-(10-dihydroxy-6-oxo-*trans*-1-undecenyl)benzoic acid *m*-lactone, is produced by several species of *Fusarium* (*e.g.*, *F. roseum*, now referred to as *F. semitectum*). Zearalenone is reduced by rat liver preparations to a more active estrogenic metabolite, α-zearalenol. Microsomal NAD(P)H-dependent zearalenone reductase is most active at pH 4-4.5, whereas the cytosolic reductase is active at neutral pH. The former enzyme reduced zearalenone only to α -zeaalenol and the latter reduced both α- and β-zearalenol. Zearalenone increases membrane permeability, promotes uterine synthesis of DNA, RNA, and protein, and inhibits cyclooxygenase and 7β-hydroxysteroid dehydrogenase. Zearalenone exerts its estrogen effects by binding to estrogen receptors and influencing transcription withinin the nucleus. It is also used in veterinary medicine as an estrogen-like anabolic agent. Zearalenone induces sister chromatid exchange and chromosomal aberration in CHO cells. **Target(s):** cyclooxygenase (1); 17β-hydroxysteroid dehydrogenase (2,3). **1.** Degen (1990) *J. Steroid Biochem.* **35**, 473. **2.** Krazeisen, Breitling, Moller & Adamski (2001) *Mol. Cell. Endocrinol.* **171**, 151. **3.** Krazeisen, Breitling, Moller & Adamski (2002) *Adv. Exp. Med. Biol.* **505**, 151.

Zeatin

trans-Zeatin   cis-Zeatin

The *trans*- and *cis*-isomers of N^6-isopentenyl-adenine (FW = 219.25 g/mol; CAS: 1637-39-4), also called 6-(4-hydroxy-3-methyl-*trans*-2-butenylamino)purine and (*E*)-2-methyl-4-(1*H*-purin-6-ylamino)-2-buten-1-ol, isolated from immature maize seeds (1). The *cis*-zeatin, also called N^6-*cis*-γ-methyl-γ-hydroxymethylallyladenine, is found in plant tRNA, whereas *trans*-zeatin is a competitive inhibitor of adenine and cytosine transporters and is reportedly an uncompetitive inhibitor of xanthine oxidase. **Target(s):** adenine transport (2); cytosine transport (2); indole-3-acetate β-glucosyltransferase (3); xanthine oxidase (4); *cis*-zeatin *O*-β-D-glucosyltransferase, by *trans*-zeatin (5). **1.** Letham (1963) *Life Sci.* **41**, 569. **2.** Gillissen, Burkle, Andre, Kuhn, Rentsch, Brandl & Frommer (2000) *Plant Cell* **12**, 291. **3.** Leznicki & Bandurski (1988) *Plant Physiol.* **88**, 1481. **4.** Sheu & Chiang (1996) *Anticancer Res.* **16**, 311. **5.** Veach, Martin, Mok, Malbeck, Vankova & Mok (2003) *Plant Physiol.* **131**, 1374.

Zeatin Riboside

β-D-ribosyl-*trans*-Zeatin   β-D-ribosyl-*cis*-Zeatin

This N^6-alkenyl-adenosine derivative (FW = 351.38 g/mol), also known as ribosylzeatin, is a plant cytokinin that is also a component of plant tRNA

(1). **Target(s):** adenosine Nucleosidase (2,3); barley β-glucosidase (1); GMP synthetase (4); tRNA methyltransferases (5). **1.** Dietz, Sauter, Wichert, Messdaghi & Hartung (2000) *J. Exp. Bot.* **51**, 937. **2.** Chen & Kristopeit (1981) *Plant Physiol.* **68**, 1020. **3.** Burch & Stuchbury (1986) *J. Plant Physiol.* **125**, 267. **4.** Spector & Beecham III (1975) *J. Biol. Chem.* **250**, 3101. **5.** Kerr & Borek (1973) *The Enzymes*, 3rd ed. (Boyer, ed.), **9**, 167.

Zeaxanthin

This plant pigment (FW = 568.88 g/mol; CAS 144-68-3), also called zeaxanthol and 3,3′-dihydroxy-β-carotene, is a member of the xanthophyll class of carotenes, provides the familiar yellow-orange color. Zeaxanthine is often found as the mono- and diester of fatty acids (most often palmitatic acid). Zeaxanthin noncompetitively inhibits β-carotene dioxygenase (1) and also inhibits β-glucuronidase (2). **1.** Grolier, Duszka, Borel, Alexandre-Gouabau & Azais-Braesco (1997) *Arch. Biochem. Biophys.* **348**, 233. **2.** Tappel & Dillard (1967) *J. Biol. Chem.* **242**, 2463.

Zebularine

This potent cytidine deaminase and DNA methylase inhibitor (FW = 228.20 g/mol; CAS 3690-10-6; *Symbol*: ZB), also known as 1-(β-D-ribofuranosyl)dihydropyrimidin-2-one or 1-[(2R,3R,4S,5R)-3,4-dihydroxy-5-(hydroxymethyl)oxolan-2-yl]pyrimidin-2-one, is a cytostatic for many cancer cells. It is metabolically phosphorylated to zebularine 5′-monophosphate (FW = 290.17 g/mol), also known as 2′-β-D-deoxyribose-pyrimidin-2-one 5′-monophosphate, which then inhibits dCMP deaminase, K_i = 12 nM (1). ZMP also elevates intracellular dCTP and TTP levels in whole cells (2-6). As a DNA methyltransferase (DNMT) inhibitor, ZB is similar in action to aza-CR and 5-Aza-CdR: all three are incorporated into DNA, forming a covalent irreversible adduct with DNMTs, thereby preventing the enzyme from methylating position-5 of cytosines clustered in regulatory CpG islands. ZB also sustains the demethylation state of the 5′-region of the tumor suppressor gene CDKN2A/p16 and other methylated genes in T24, HCT15, CFPAC-1, SW48, and HT-29 cells (6,7). Zebularine also inhibits growth of cancer cell lines but not normal cells (8), and, when used as an adjuvant in methotrexate therapy, ZB decreased gene expression of the three DNMTs and induced AhR gene promoter demethylation and its re-expression (9). It also inhibits tumorigenesis and stemness of colorectal cancer via p53-dependent endoplasmic reticulum stress (unfolded protein response) (10). **1.** Maley, Lobo & Maley (1993) *Biochim. Biophys. Acta* **1162**, 161. **2.** Barchi, Cooney, Hao, Weinberg, Taft, Marquez, Ford (1995) *J. Enzyme Inhib.* **9**, 147. **3.** Frick, Yang, Marquez & Wolfenden (1989) *Biochemistry* **28**, 9423. **4.** Carlow & Wolfenden (1998) *Biochemistry* (1998) **37**, 11873. **5.** Carlow, Carter, Mejlhede, Neuhard & Wolfenden (1999) *Biochemistry* **38**, 12258. **6.** Holy, Ludzisa, Votruba & Sediva (1985) *Collect. Czech. Chem. Commun.* **50**, 393. **7.** Yoo, Cheng & Jones (2004) *Biochem. Soc. Trans.* **32**, 910. **7.** Cheng, Yoo, Weisenberger, *et al.* (2004) *Cancer Cell* **6**,151. **8.** Cheng, Weisenberger, Gonzales, *et al.* (2004) *Mol. Cell Biol.* **24**, 1270. **9.** Andrade, Borges, Castro-Gamero, *et al.* (2013) *Anticancer Drugs* **25**, 72. **10.** Yang, Lin, Shun, *et al.* (2013) *Sci. Rep.* **3**, 3219.

Zenarestat

This aldose reductase inhibitor ($FW_{free-acid}$ = 441.64 g/mol; CAS 112733-06-9), also known as [3-(4-bromo-2-fluorobenzyl)-7-chloro-2,4-dioxo-1,2,3,4-tetrahydroquinazolin-1-yl]acetic acid, is used to treat diabetic neuropathy and cataracts by preventing sorbitol accumulation. During episodes of diabetic hyperglycemia, high levels of glucose and fructose are transported into the retina and lens cells, where they are reduced to sorbitol, a nontransported polyol that causes osmotic shock, inflicting great damage. Inhibiting the metabolism of glucose by the polyol pathway using aldose reductase inhibitors is a potential mechanism to slow or reverse the progression of polyneuropathy, a common complication of diabetes mellitus, causing pain and sensory and motor deficits in the limbs. **1.** Ao, Shingu, Kikuchi, Takano, *et al.* (1991) *Metabolism* **40**, 77.

Zerumbone

This naturally occurring monocyclic sesquiterpene (FW = 218.30 g/mol; CAS 471-05-6; IUPAC Name: (2E,6E)-2,6,9,9-tetramethyl-2,6,10E-cycloundecatrien-1-one; Solubility: ≥10 mg/mL DMSO), first isolated from rhizomes of the wild ginger (*Zingiber zerumbet*), potently inhibits (IC_{50} = 140 nM) phorbol ester activation of Epstein-Barr virus (1). Zerumbone markedly suppresses free radical generation, pro-inflammatory protein production, and cancer cell proliferation (2). Both VEGF expression and NF-κB activity in AGS cells (gastric adenocarcinoma) were reduced by treatment with zerumbone, thereby inhibiting angiogenesis (3). Zerumbone may therefore be a new anti-angiogenic and antitumor drug for the treatment of gastric cancer. ZER is selectively cytotoxic to Jurkat cells, showing dose- and time-dependencies, with IC_{50} values of 12, 9 and 5 µg/mL at 24, 48 and 72 hours of treatment, respectively, but does not produce an adverse effect on normal human peripheral blood mononuclear cells (4). ZER arrests Jurkat cells at the G_2/M cell-cycle phase. Its anti-proliferative effect on Jurkat cells occurs through the apoptotic intrinsic pathway via the activation of Caspase-3 and Caspase-9 (4). Zerumbone also increases oxidative stress in a thiol-dependent ROS-independent manner to increase DNA damage and sensitize colorectal cancer cells to radiation (5). **1.** Murakami, Takahashi, Jiwajinda, Koshimizu & Ohigashi (1999) *Biosci. Biotechnol. Biochem.* **63**, 1811. **2.** Murakami, Takahashi, Kinoshita, *et al.* (2002) *Carcinogenesis* **23**, 795. **3.** Tsuboi, Matsuo, Shamoto, Shibata, *et al.* (2014) *Oncol. Rep.* **31**, 57. **4.** Rahman, Rasedee, Chartrand, *et al.* (2014) *Nat. Prod. Commun.* **9**, 1237. **5.** Deorukhkar, Ahuja, Mercado, *et al.* (2015) *Cancer Med.* **4**, 278.

Zeta Inhibitory Peptide, Myristoylated

This synthetic myristoylated peptide (*Sequence*: Myr-SIYRRGARRWRKL; FW = 1928.40 g/mol; CAS 863987-12-6; Solubility: 1 mg/mL in ethanol-water (20%, vol/vol); *Symbol*: mZIP), also known as Cell-permeable z-Pseudosubstrate Inhibitory Peptide, is a 13-residue, cell-permeable inhibitor that mimics the endogenous pseudosubstrate region of protein kinase Mζ, *or* PKMζ (1). The latter is a constitutively active and atypical PKC isozyme implicated in the maintenance of Long-Term Potentiation (LTP). ZIP blocks PKMζ-induced synaptic potentiation in hippocampal slices *in vitro*. It also reverses late-phase LTP (IC_{50} = 1-2.5 µM), producing persistent loss of 1-day-old spatial memory following central administration *in vivo*. **Rationale:** The PKC family is comprised of three subfamilies: classical (α, $β_1$, $β_2$, γ), novel (ε, δ, η, θ), and atypical (λ/ι and ζ). All are inactive in the absence of an intramolecular interaction between a short N-terminal substrate-like peptide (called the pseudosubstrate) and the actual substrate binding region located within the catalytic domain. Peptides based on the pseudosubstrate sequence of PKC isoenzymes are selective *in vitro* inhibitors of specific isoforms of PKC. However, the unmodified peptide is plasma membrane-impermeable, requiring microinjection, expression in cells, or use of a membrane-solubilizing detergent, such as saponin. Peptides can also be fused with membrane-translocated peptides derived from the penetratin sequences in *Drosophila* Antennapedia protein or HIV TAT protein for micropinocytotic delivery directly into cells. Although originally designed with the goal of creating a cell-permeable ZIP derivative, Krotova, Hu, Xia, *et al.* (1) convincingly demonstrate that, at least in endothelial cells, the action of cell-penetrating myristoylated-ZIP is not limited to inhibitng PKCζ activity; instead, mZIP activates eNOS (endothelial nitric oxide synthase) via Akt phosphorylation, as well as a variety of NO-regulated enzymes. For this reason, the likely pluripotency of mZIP must be considered in all experiments relying on its use. **Control Peptides:** As in any experiment using an inhibitory peptide, one must conduct control experiments using a sequence-scrambled peptide to control for nonspecific actions of the test peptide. The control peptide is Scrambled Zeta Inhibitory Peptide (*Sequence*: Myr-RLYRKRIWRSAGR; FW = 1928.40 g/mol; CAS 908012-18-0; Solubility: 1 mg/mL acetonitrile/water (20%, vol/vol); *Symbol*: scr-ZIP). Another useful reagent is Biotinylated Zeta Inhibitory Peptide (FW = 2154.69 g/mol; Sequence: Myr-SIYRRGARRWRK(biotin)L; Solubility: 0.67 mg/mL acetonitrile/water (30%, vol/vol); *Symbol*: bZIP), the subcellular location of which which can be evaluated with fluorescently labeled streptavidin. **1.** Krotova, Hu, Xia, *et al.* (2006) *Brit. J. Pharmacol.* **148**, 732.

Zetekitoxin AB

This saxitoxin-like poison from the skin of the Panamanian golden frog *Atelopus zeteki* (FW = 552.47 g/mol; CAS 139158-99-9; IUPAC: [(3*R*,5*S*, 6*S*,11*R*,12*S*,14*Z*,16*S*,17*Z*)-14,17-diamino-19,19-dihydroxy-6-(hydroxy-methyl)-10-oxo-3-(sulfooxy)-8-oxa-1,9,13,15,18-pentaazapentacyclo[9.5.2. 1^{3,16}.0^{5,9}.0^{12,16}]nonadeca-14,17-dien-13-yl]methyl hydroxycarbamate) from the skin of the Panamanian golden frog *Atelopus zetek* (1), is a potent blocker of voltage-dependent sodium ion channels, acting on human heart channels (IC$_{50}$ = 280 pM), rat brain IIa channels (IC$_{50}$ = 6.1 pM), and rat skeletal muscle channels (IC$_{50}$ = 65 pM) (2). **1.** Kim, Brown & Mosher (1975) *Science* **189**, 151. **2.** Yotsu-Yamashita, Kim, Dudley, Choudhary Pfahnl, Oshima & Daly (2004) *Proc. Natl. Acad. Sci. U.S.A.* **101**, 4346.

Zileuton

This antiasthmatic hydroxyurea derivative (FW = 236.29 g/mol; CAS 111406-87-2), also known as Zyflo® and (±)-*N*-(1-benzo[β]thien-2-ylethyl)-*N*-hydroxyurea, is a potent inhibitor of 5-hydroxyeicosatetraenoic acid synthesis by 5-lipoxygenase (IC$_{50}$ = 0.5–0.9 μM), thereby blocking leukotriene production. It also inhibited leukotriene LT$_{B4}$ biosynthesis by rat PMNL (IC$_{50}$ = 0.4 μM), human PMNL (IC$_{50}$ = 0.4 μM) and human whole blood (IC$_{50}$ = 0.9 μM). Inhibition of human PMNL LT$_{B4}$ biosynthesis was removed readily by a simple wash procedure, indicating that zileuton's action is completely reversible. **1.** Carter, Young, Albert, Bouska, *et al.* (1991) *J. Pharmacol. Exp. Ther.* **256**, 929.

Zimelidine

This selective serotonin reuptake inhibitor, *or* SSRI (FW = 317.23 g/mol; CAS 56775-88-3), also known as (*Z*)-3-(4'-bromophenyl)-3-(3''-

pyridyl)dimethylallylamine or Zelmid™, was the first-in class antidepressant, but was withdrawn in 1983 in light of a five-times greater risk of Guillain-Barré syndrome. **Mode of Action:** The primary pharmacological action of SSRIs is presynaptic inhibition of serotonin (5-HT) reuptake, thereby increasing the concentration of this amine in the synaptic cleft. Serotonin exerts its effects through interactions with any of seven distinct 5-HT receptor families, and interaction between serotonin and post-synaptic receptors mediates a wide range of its actions. **Target(s):** amine oxidase (1); serotonin uptake (2,3) **1.** Obata & Yamanaka (2000) *Neurosci. Lett.* **286**, 131. **2.** Fuller (1995) *Prog. Drug Res.* **45**, 167. **3.** Ross, Ögren & Renuy (1976) *Acta Pharmacol. Toxicol.* **39**, 152.

Zinc and Zinc Ions

This Group 12 (or IIB) metal (Symbol = Zn; Atomic number = 30; Atomic weight = 65.409), which is located directly above cadmium in the Periodic Table, has a neutral-atom ground-state electronic configuration of $1s^2 2s^2 2p^6 3s^2 3p^6 4s^2 3d^{10}$. The radioisotope ^{65}Zn has half-life of 243.8 days. Zn^{2+} (ionic radius = 0.74 Å) is an essential metal ion in many enzymes (*e.g.*, alcohol dehydrogenase, NADPH:quinone reductase, methionine *S*-methyltransferase, alkaline phosphatase, phospholipase C, leucyl aminopeptidase, carboxypeptidases A, B, and H, interstitial collagenase, pitrilysin, mitochondrial intermediate peptidase, carbonic anhydrase, porphobilinogen synthase, mannose-6-phosphate isomerase, dopachrome δ-isomerase, and pyruvate carboxylase). The likely intracellular concentration of uncomplexed zinc has been put at 0.1–0.4 nM. Higher concentrations of Zn^{2+} inhibits over 1000 enzymes by nonselectively replacing active-site and/or structure-stabilizing divalent cations. Zinc ion is also transported into insulin-containing vesicles of β-cells by isoform-8 zinc transporters, where it acts as a counterion and stabilizes insulin and induces its oligomerization, polymerization, and crystallization. **1.** Vallee & Galdes (1984) *Adv Enzymol Relat Areas Mol Biol.* **56**, 283. **2.** Vallee & Falchuk (1993) *Physiol. Rev.* **73**, 79. **3.** Maret (2013) *Adv. Nutrition* **4**, 82.

Zincon

This chelator (FW$_{free-acid}$ = 440.06 g/mol; CAS 59600-76-9), also known as 2-carboxy-2'-hydroxy-5'-sulfoformazylbenzene, is used as a metal indicator in EDTA titrations and as a colorimetric/photometric indicator of zinc, copper, mercury, and gallium ions (1). A weakly alkaline aqueous solution of zincon is yellow or yellow-orange in color. The solution turns blue in the presence of Zn^{2+} or Cu^{2+} ions (λ$_{max}$ = 620 nm). Metal ion concentrations as low as 0.05 mg/mL can be detected using atomic absorption spectroscopy. The presence of calcium or magnesium ions does not affect the physiologic binding of zinc ions to Zn^{2+}-dependent proteins and enzymes, mainly as a result of zinc ion preference for nitrogen-containing ligands. Zincon has also been reported to inhibit a number of Zn^{2+}-dependent enzymes: for example, liver alcohol dehydrogenase (2) and formate dehydrogenase (3). **1.** Hilario, Romero & Celis (1990) *J. Biochem. Biophys. Methods* **21**, 197. **2.** McFarland, Watters & Petersen (1975) *Biochemistry* **14**, 624. **3.** Kanamori & Suzuki (1968) *Enzymologia* **35**, 185.

Zincov Inhibitor

This Zn(II) chelator (FW = 302.33 g/mol; CAS 71431-46-4), also known as Zincov™ and 2-(*N*-hydroxycarboxamido)-4-methylpentanoyl-L-alanyl-glycinamide, inhibits numerous zinc-containing enzymes (1), including fibrolase (2); fragilysin (3,4); hemagglutinin/protease (5,6); matrilysin, *or* matrix metalloproteinase 7 (7); meprin A (8); neprilysin (9); pitrilysin, *or* protease Pi (10); pseudolysin, *or Pseudomonas aeruginosa* elastase (6,11); thermolysin (11,12). **1**. Hudgin, Charleson, Zimmerman, Mumford & Wood (1981) *Life Sci.* **29**, 2593. **2**. Guan, Retzios, Henderson & Markland, Jr. (1991) *Arch. Biochem. Biophys.* **289**, 197. **3**. Obiso & Wilkins (1998) in *Handb. Proteolytic Enzymes* (Barrett, Rawlings & Woessner, eds.), p. 1211, Academic Press, San Diego. **4**. Moncrief, Obisco, Barroso, Kling, *et al.* (1995) *Infect. Immun.* **63**, 175. **5**. K. Yamamoto (1998) in *Handb. Proteolytic Enzymes* (Barrett, Rawlings & Woessner, eds.), p. 1056, Academic Press, San Diego. **6**. Häse & Finkelstein (1990) *Infect. Immun.* **58**, 4011. **7**. Oneda & Inouye (2001) *J. Biochem.* **129**, 429. **8**. Sterchi, Naim, Lentze, Hauri & Fransen (1988) *Arch. Biochem. Biophys.* **265**, 105. **9**. Sonnenberg, Sakane, Jeng, Koehn, *et al.* (1988) *Peptides* **9**, 173. **10**. Anastasi, Knight & Barrett (1993) *Biochem. J.* **290**, 601. **11**. Maeda & Morihara (1995) *Meth. Enzymol.* **248**, 395. **12**. Kuzuya & Inouye (2001) *J. Biochem.* **130**, 783.

Zinc-Protoporphyrin IX

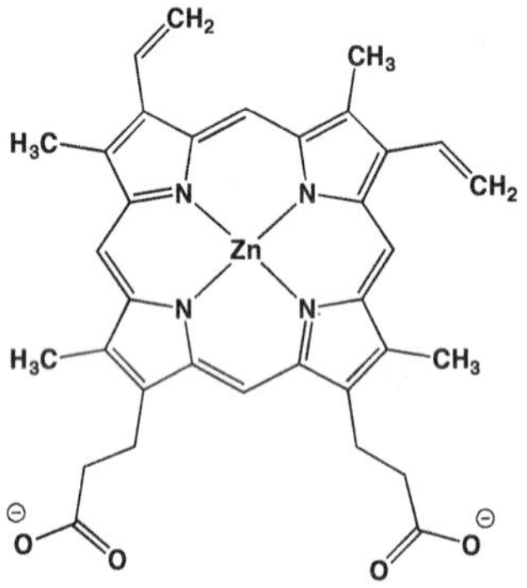

This porphyrin derivative (FW = 628.06 g/mol; CAS 553-12-8) inhibits ferrochelatase (1,2); guanylate cyclase (3); heme oxygenase (4,5); nitric-oxide synthase (6). **1**. Dailey, Jones & Karr (1989) *Biochim. Biophys. Acta* **999**, 7. **2**. Little & Jones (1976) *Biochem. J.* **156**, 309. **3**. Ignarro, Ballot & Wood (1984) *J. Biol. Chem.* **259**, 6201. **4**. Leon, Le Foll, Charriault-Marlangue, *et al.* (2003) *J. Neurochem.* **84**, 459. **6**. Maines (1981) *Biochim. Biophys. Acta* **673**, 339. **7**. Wolff, Naddelman, Lubeskie & Saks (1996) *Arch. Biochem. Biophys.* **333**, 27.

Zinc Sulfate

This inorganic salt ZnSO₄ (FW = 161.45 g/mol; CAS 7446-19-7) strongly inhibits human gingival matrix metalloproteinase-2 (IC_{50} = 15 nM) and matrix metalloproteinase-9 (IC_{50} = 40 nM). Both the monohydrate and heptahydrate are soluble in water and insoluble in ethanol. **Target(s):** dTTPase (1); ethylmorphine *N*-demethylase, *or* cytochrome P450 (2); gelatinase A, *or* matrix metalloproteinse-2 (3); gelatinase B, *or* matrix metalloproteinase-9 (3); glucan 1,3-β glucosidase (4); ribonuclease (5); steroid 5α-reductase (6). **1**. Schultes, Fischbach & Dahlmann (1992) *Biol. Chem. Hoppe-Seyler* **373**, 237. **2**. Arinc (1985) *Comp. Biochem. Physiol. B* **80**, 389. **3**. de Souza, Gerlach & Line (2000) *Dent. Mater.* **16**, 103. **4**. Nagasaki, Saito & Yamamoto (1977) *Agric. Biol. Chem.* **41**, 493. **5**. Mujica, Romero & Hernandez-Montes (1976) *Int. J. Fertil.* (2), 109. **6**. Stamatiadis, Bulteau-Portois & Mowszowicz (1988) *Brit. J. Dermatol.* **119**, 627. **See also** Zinc and Zinc Ions

Zinostatin

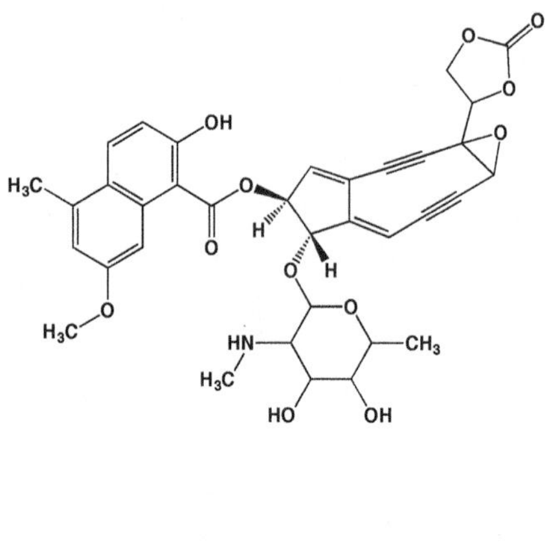

This antitumor antibiotic (FW = 659.65 g/mol; CAS 79633-18-4), also known as NCS and neocarzinostatin, consists of a 10.7 kDa protein and a labile chromophore, shown above. This second component reportedly is a potent inhibitor of casein kinase II. Zinostatin also intercalates DNA and is a strand scission agent. **1**. Harada, Haneda, Maekawa, *et al.* (1999) *Biol. Pharm. Bull.* **22**, 1122.

Ziprasidone

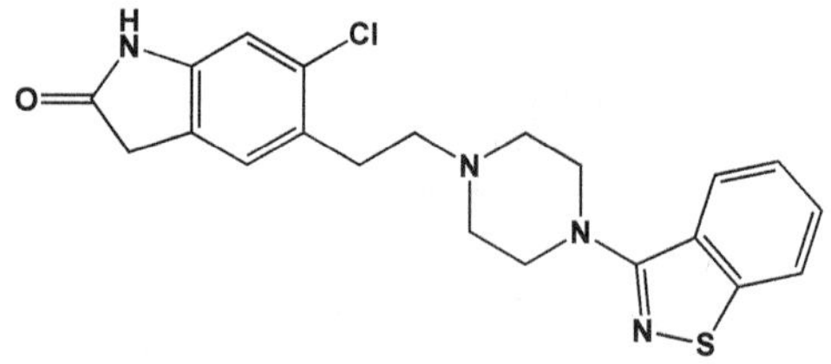

This broadly acting dopamine D_2 and serotonin 5-HT₂ antagonist (FW = 412.94 g/mol; CAS 146939-27-7), also known as Geodon®, Zeldox®, Zipwell®, and 5-{2-[4-(1,2-benzisothiazol-3-yl)-1-piperazinyl]ethyl}-6-chloro-1,3-dihydro-2*H*-indol-2-one, is an atypical antipsychotic pharmaceutical that has is been used in the treatment of Tourette syndrome, depression, and bipolar disorder. Like other atypical agents, ziprasidone exhibits extrapyramidal side-effects similar to that of placebo, without causing significant elevations in prolactin levels. In contrast to existing atypical agents, however, ziprasidone has a low propensity for causing weight gain. (**See also** *Amisulpride; Aripiprazole; Olanzapine; Quetiapine; Remoxipride; Risperidone; Sertindole; Zotepine*) **Pharmacokinetics:** Systemic bioavailability is 100%, when administered intramuscularly and 60%, when administered orally with food. After a single dose intramuscular administration, the peak serum concentration occurs within 1 hour. Steady-state plasma concentrations are achieved within 1–3 days. Ziprasidone undergoes liver metabolized by aldehyde oxidase, with minor action by CYP3A4. *See also* Appendix II in Goodman & Gilman's THE PHARMACOLOGICAL BASIS OF THERAPEUTICS, 12ᵗʰ Edition (Brunton, Chabner & Knollmann, eds.) McGraw-Hill Medical, New York (2011). **Targets:** Serotonin 5-HT₁ₐ receptor, as partial agonist (K_i = 12 nM); 5-HT₁ʙ receptor, as antagonist or inverse-agonist (K_i = 4.0 nM) (partial agonist); 5-HT₁ᴅ receptor, as antagonist or inverse-agonist (K_i = 2.3 nM); 5-HT₁ᴇ receptor, as antagonist or inverse-agonist (K_i = 360 nM); 5-HT₁ꜰ receptor (K_i > 10 μM); 5-HT₂ₐ receptor, as antagonist or inverse-agonist (K_i = 0.6 nM); 5-HT₂ʙ receptor, as antagonist or inverse agonist (K_i = 27 nM); 5-HT₂ᴄ receptor, as antagonist or inverse agonist (K_i = 13 nM); 5-HT₃ receptor (K_i > 10 μM); 5-HT₄ receptor, weakly (K_i = > 10 μM); 5-HT₅ₐ receptor, as antagonist or inverse-agonist (K_i = 291 nM); 5-HT₆ receptor, as antagonist or inverse-agonist (K_i = 61 nM); 5-HT7 receptor, as antagonist or iinverse-agonist (K_i = 6 nM); Dopamine D₁ receptor, as antagonist or inverse-agonist (K_i = 30 nM); D₂ receptor, as antagonist or inverse-agonist (K_i = 6.8 nM); D₃ receptor, as antagonist or inverse-agonist (K_i = 7.2 nM); D₄ receptor, as antagonist or inverse-agonist (K_i = 39 nM); D₅ receptor, as antagonist or inverse-agonist (K_i = 152 nM); α₁ₐ-adrenergic receptor, as antagonist or inverse-agonist (K_i = 18 nM); α₂ₐ-adrenergic receptor, as antagonist or inverse-agonist (K_i = 160 nM); Histamine H₁ receptor, as antagonist or inverse-agonist (K_i = 63 nM); muscarinic acetyl-choline receptor, weakly (K_i = >10 μM); Serotonin 5-HT transporter, as inhibitor (K_i = 53 nM); Norepinephrine NE transporter, as inhibitor (K_i = 48 nM); Dopamine DA transporter, as weak iinhibitor (K_i > 10 μM). **1**. Carnahan, Lund & Perry (2001) *Pharmacotherapy* **21**, 717. **2**. Akiskal & Tohen (2011). *Bipolar Psychopharmacotherapy: Caring for the Patient*. John Wiley & Sons. p. 209. **3**. Seeger, Seymour PA, Schmidt, *et al.* (1995) *J. Pharmacol.* Exper, Therapeut. **275**, 101. **4**. Schotte, Janssen, Gommeren, *et al.* (1996) *Psychopharmacol.* **124**, 57.

Ziram

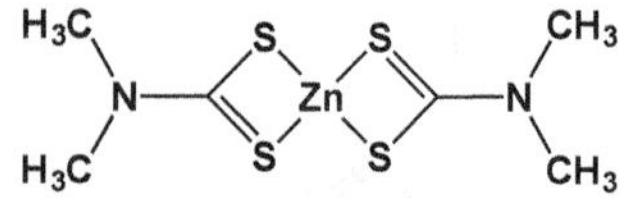

This toxic fungicide (FW = 305.83 g/mol; CAS 137-30-4) inhibits thyroid peroxidase as well as hepatic epoxide hydrolase and glutathione *S*-transferase. **Target(s):** dopamine b-monooxygenase (1); epoxide hydrolase (2); glutathione *S*-transferase (2); thyroid peroxidase (3). **1**. Weppelman (1987) *Meth. Enzymol.* **142**, 608. **2**. Schreiner & Freundt (1986) *Bull.*

Environ. Contam. Toxicol. **37**, 53. **3**. Marinovich, Guizzetti, Ghilardi, Viviani, *et al.* (1997) *Arch. Toxicol.* **71**, 508.

Zirconium and Zirconium Ions

This Group 4 (*or* IVA) transition metal (Symbol = Zr; Atomic Number = 40; Atomic Weight = 91.224) has a ground-state electronic configuration of $1s^2 2s^2 2p^6 3s^2 3p^6 4s^2 3d^{10} 4p^6 5s^2 4d^2$ (1,2). The ionic radii for Zr^{4+} is 0.079 nm. Naturally-occurring zirconium consists of five stable isotopes: ^{90}Zr at 51.45% natural abundance, ^{91}Zr at 11.22%, ^{92}Zr at 17.15%, ^{94}Zr at 17.38%, and ^{96}Zr at 2.80%. Of the stable isotopes, ^{91}Zr has a nuclear spin of 5/2. There are at least twenty known radioisotopes of which ^{93}Zr, a β^- emitter, is long-lived ($t_{1/2} = 1.5 \times 10^6$ years). ^{89}Zr is a β^+ emitter ($t_{1/2} = 3.27$ days) is used to label a protein or monoclonal antibody for positron emission tomography experiments. Tetravalent zirconium cations have been shown to induce iron acquisition in certain cells (1). Zr(II)O inhibits Ferroxidase (2) and D-arginase (3). **1**. Klaproth (1795) *Beiträge zur chemischen Kenntniss der Mineralkörper*, **1**, 117. **3**. Olakammi, Stokes, Pathan & Britigan (1997) *J. Biol. Chem.* **272**, 2599. **2**. Huber & Frieden (1970) *J. Biol. Chem.* **245**, 3979. **3**. Nadai (1958) *J. Biochem.* **45**, 1011.

Zizanin B, *See Cochliobolin B*

ZK 62711, *See Rolipram*

ZK 807191

This oral coagulation Factor Xa inhibitor (FW = 525.51 g/mol; IUPAC: 3-(5-carbamimidoyl-2-hydroxy-phenoxy)-2,6-difluoro-5-[3-(1-methyl-4,5-dihydro-1*H*-imidazol-2-yl)phenoxy]phenyl}methylamino)acetate, potently inhibits factor Xa (K_i = 0.1 nM) and has promise as a potential parenteral treatment of unstable angina pectoris. **1**. Vacca (1998) *Ann. Rep. Med. Chem.* **33**, 81. **2**. Lewis (2000) *Drugs* **3**, 525.

ZK 222584, *See Vatalanib*

ZK 807834, *See CI-1031*

ZK 816042

This amidine/imidazoline-containing antithrombotic agent (FW = 525.54 g/mol) is a potent inhibitor of coagulation factor Xa. NMR experiments performed on a lyophilized FXa-inhibitor complex indicate that the inhibitor binds with a distribution of orientations of the imidazoline ring. **1**. Studelska, McDowell, Adler, O'Connor, *et al.* (2003) *Biochemistry* **42**, 7942.

ZLN 024

This orally active allosteric activator ($FW_{HCl-Salt}$ = 361.69 g/mol; Soluble to 100 mM in DMSO) targets 5'-AMP-activated protein kinase (AMPK), increasing the activity of the $\alpha_1\beta_1\gamma_1$ heterotrimer by 1.5 times (EC_{50} = 0.42

μM) and $\alpha_2\beta_1\gamma_1$ by 1.7 times (EC_{50} = 0.95 μM). ZLN 024 also inhibits dephosphorylation at AMPK Thr-172 by PP2Cα. ZLN024 activated AMPK in myotubes and hepatocytes without increasing the ADP/ATP ratio. It stimulates glucose uptake and fatty acid oxidation, again without increasing the ADP/ATP ratio. ZLN 024 decreases fatty acid synthesis and glucose output in primary hepatocytes, and it mproves glucose tolerance and decreases cholesterol content in *db/db* mice. The *db/db* mouse is a model of obesity, diabetes, and dyslipidemia, wherein leptin receptor activity is deficient due to a point mutation in the gene for the leptin receptor. **1**. Li, *et al.* (2013) *Toxicol. Appl. Pharmacol.* **273**, 325. **2**. Zhang, *et al.* (2013) *PLoS One* **8**, e72092.

ZM 447439, *See ZM1*

ZMP, *See 5-Aminoimidazole-4-carboxamide-1-b-ribofuranosyl 5'-Monophosphate*

Zoanthoxanthin

This naturally fluorescent pigment (FW = 256.31 g/mol; CAS 40451-47-6), also known as 1,3,5,7-tetrazacyclopent[*f*]azulene, first isolated from a anemone-like anthozoan (*Parazoanthus* sp.) and containing a previously unknown heterocyclic system, inhibits electron transport and succinate dehydrogenase in assays conducted with fractionated mitochondria (1). Such molecules often trigger the degradation of mitochondria, affecting maturation and involution as well as serving as a line of defense against invaders (2). (*See also Parazoanthoxanthin*) **1**. Cariello & Tota (1974) *Experentia* **30**, 244. **2**. Rapoport & Schewe (1977) *Trends Biochem. Sci.* **2**, 186.

Zofenapril

This antihypertensive (FW = 429.55 g/mol; CAS 81872-10-8; (2*S*,4*R*)-1-((*S*)-3-(benzoylthio)-2-methylpropanoyl)-4-(phenylthio)pyrrolidine-2-carboxylic acid) target angiotensinogen converting enzyme and has profound and clinically useful cardioprotective properties. Upon oral administration, zofenopril is metabolically hydrolyzed to the active drug *S*-zofenoprilat. (*See Zofenapriat*) **Mode of Inhibitory Action:** *S*-Zofenoprilat is more potent than captoprilat as a rabbit lung ACE inhibitor, with the former having an IC_{50} value of 8 nM, compared to 23 nM for the latter. It is also a potent inhibitor of angiotensin I-induced contractions (EC_{50} = 3 nM) as well as a potentiator of bradykinin-induced contractions (EC_{50} = 1 nM) in isolated guinea pig ileum. It is without effect on the inotropic effects of angiotensin II, $BaCl_2$, PGE_1, histamine, serotonin, or acetycholine in the same tissue, suggesting that the pro-drug zofenopril is a specific inhibitor of ACE (1). Significantly, however, *S*-zofenoprilat releases H_2S in cell-free assays and directly relaxes vessels *in vitro* in a concentration-dependent manner. Administration *in vivo* of the *R*-zofenoprilat diastereoisomer, which does not inhibit ACE, does not modify blood pressure, but retains beneficial effects on vascular function and restores plasma/tissue H_2S levels (2). **Other Target(s):** With respect to interactions with the renal glycylsarcosine transport, zofenopril (IC_{50} = 81 μM) and fosinopril (IC_{50} = 55 μM) are potent inhibitors, whereas other ACE inhibitors exhibited low-affinity (3). Zofenopril also increases sarcoplasmic reticulum (SR) calcium cycling and stimulates active calcium uptake into the SR (4). Long-term administration of zofenopril promotes lowering of free radical oxidation products, with activation of protective antioxidant enzymes, and improvement in the total antioxidant activity of blood serum (5). **1**. DeForrest, Waldron, Krapcho, *et al.* (1989) *J. Cardiovasc. Pharmacol.* **13**, 887. **2**. Bucci, Vellecco, Cantalupo, *et al.* (2014) *Cardiovasc. Res.* **102**, 138. **3**. Lin, Akarawut & Smith (1999) *Pharm. Res.* **16**, 609. **4**. Frascarelli, Carnicelli, Ghelardoni, *et al.* (2009) *J. Cardiovasc. Pharmacol.* **54**, 456. **5**. Zdionchenko, Leksina, Timofeeva, *et al.* (2009) *Kardiologiia* **49**, 32.

Zofenoprilat

This angiotensin-converting enzyme inhibitor (FW$_{acid}$ = 325.44 g/mol; CAS 75176-37-3; also known as SQ 26703; IUPAC Name: (2S,4S)-1-[(2S)-2-methyl-3-sulfanylpropanoyl]-4-(phenylsulfanyl)pyrrolidine-2-carboxylic acid), which is generated from the prodrug zofenopril (1), reportedly mimics the effects of glutathione in certain systems (2). (**See Zofenapril**) **Target(s):** leucyl aminopeptidase3; peptidyl-dipeptidase A, *or* angiotensin I-converting enzyme, IC$_{50}$ = 0.4 nM (1); Xaa-Trp aminopeptidase, *or* aminopeptidase W (4-6). **1.** Ranadive, Chen & Serajuddin (1992) *Pharm. Res.* **9**, 1480. **2.** Del Corso, Dal Monte, Vilardo, Cecconi, *et al.* (1998) *Arch. Biochem. Biophys.* **350**, 245. **3.** Cappiello, Alterio, Amodeo, Del Corso, *et al.* (2006) *Biochemistry* **45**, 3226. **4.** Hooper (1998) in *Handb. Proteolytic Enzymes* (Barrett, Rawlings & Woessner, eds.), p. 1513, Academic Press, San Diego. **5.** Tieku & Hooper (1992) *Biochem. Pharmacol.* **44**, 1725. **6.** Tieku & Hooper (1993) *Biochem. Soc. Trans.* **21**, 250S.

Zolantidine

This potent brain-penetrating antihistamine (FW = 381.54 g/mol; CAS 118072-93-8), also known as SKF 95282 and systematically named (1-hydroxy-2-(1H-imidazol-1-yl)ethylidene)bisphosphonic acid, is a brain-penetrating histamine H$_2$ receptor antagonist (1,2). **1.** Young, Mitcell, Brown, *et al.* (1988) *J. Med. Chem.* **31**, 656. **2.** Calcutt, Ganellin, Griffiths, Leigh, *et al.* (1988) *Brit. J. Pharmacol.* **93**, 69.

Zoledronate

This aminobisphosphonate-class bone resorption inhibitor (FW$_{free-acid}$ = 272.09 g/mol; CAS, 118072-93-8), also known by its trade names Zometa®, Zomera®, Aclasta® and Reclast®, as well as its systematic name 2-(imidazol-1-yl)-1-hydroxyethane-1,1-bisphosphonate, inhibits osteoclast-mediated decalcification and is also an effective apoptotic inducer in tumor cells. **Mode of Inhibitory Action in Osteoporosis:** Zoledronate binds preferntially to crystalline hydroxyapatite within the bone matrix. Upon arrival of osteoclasts at the bone surface, zoledronate is released, whereupon it inhibits osteoclast farnesyl pyrophosphate (diphosphate) synthase, preventing protein prenylation within the mevalonate pathway. Inhibition of isoprenoid metabolite formation leads to dysregulation of downstream signaling required for osteoclast function, inducing apoptosis, thereby promoting osteoclast death. By preventing osteoclast-mediated bone resorption, zoledronate decreases bone turnover and stabilizes the bone matrix. Zoledronate also inhibits human vesicular endothelial cell adhesion, survival, migration and actin stress fiber formation by interfering with G-protein prenylation. **Potential Role in Cancer Chemotherapy:** The mevalonate pathway is also relevant to controlling cancer. Indeed, this pathway's final products, are cholesterol as well as various isoprenoids, including isopentenyl pyrophosphate (IPP), farnesyl pyrophosphate (FPP) and geranylgeranyl pyrophosphate (GGPP). The last three are critical for the post-translational modification of proteins needed for cell proliferation and differentiation, such as the small GTP-binding proteins RAS and RHOA. The mevalonate pathway also affects multidrug resistance (MDR), which often poses a major challenge to durable chemotherapy. Finally, because statins and aminobisphonates inhibit HMG-CoA reductase and FPP synthase, they are liely to limit the synthesis of steroids needed for the proliferation and survival of hormone-sensitive cell targets. **Target(s):** alkaline phosphatase (1); collagenase 3, *or* matrix metalloproteinase 13 (2); dimethylallyl-*trans*transferase, *or* geranyl-diphosphate synthase (3); enamelysin, *or* matrix metalloproteinase 20 (2); farnesyl*trans*transferase, *or* geranylgeranyl-diphosphate synthase (4,5); gelatinase A, *or* matrix metalloproteinase 2 (2); gelatinase B, *or* matrix metalloprotreinase 9 (2); geranyl *trans*transferase, *or* farnesyl-diphosphate synthase (5-10); *trans*-hexaprenyl *trans*transferase, *or* geranylgeranyl-diphosphate synthase (11); interstitial collagenases, *or* matrix metalloproteinase 1 (2,12); macrophage elastase, *or* matrix metalloproteinase 12 (2); neutrophil collagenases, *or* matrix metalloproteinase 8 (2); pyrophosphate-dependent phosphofructo-kinase, *or* diphosphate:fructose-6 phosphate 1-phosphotransferase (13); stromelysin 1, *or* matrix metalloproteinase 3 (2). **1.** Vaisman, McCarthy & Cortizo (2005) *Biol. Trace Elem. Res.* **104**, 131. **2.** Heikkila, Teronen, Moilanen, Konttinen, *et al.* (2002) *Anticancer Drugs* **13**, 245. **3.** Burke, Klettke & Croteau (2004) *Arch. Biochem. Biophys.* **422**, 52. **4.** Wiemer, Tong, Swanson & Hohl (2007) *Biochem. Biophys. Res. Commun.* **353**, 921. **5.** Ling, Li, Miranda. Oldfield & Moreno (2007) *J. Biol. Chem.* **282**, 30804. **6.** Sanders, Song, Chan, Zhang, Jennings, *et al.* (2005) *J. Med. Chem.* **48**, 2957. **7.** Dunford, Thompson, Coxon, Luckman, *et al.* (2001) *J. Pharmacol. Exp. Ther.* **296**, 235. **8.** Glickman & Schmid (2007) *Assay Drug Dev. Technol.* **5**, 205. **9.** J. Dunford, Kwaasi, Rogers, Barnett, *et al.* (2008) *J. Med. Chem.* **51**, 2187. **10.** Lühe, Künkele, Haiker, Schad, *et al.* (2008) *Toxicol. In Vitro* **22**, 899. **11.** Guo, Cao, Liang, Ko, *et al.* (2007) *Proc. Natl. Acad. Sci. U.S.A.* **104**, 10022. **12.** Derenne, Amiot, Barille, Collette, *et al.* (1999) *J. Bone Miner. Res.* **14**, 2048. **13.** Bruchhaus, Jacobs, Denart & E. Tannich (1996) *Biochem. J.* **316**, 57.

Zolmitriptan

This orally bioavailable potent serotonergic agonist (FW = 287.36 g/mol; CAS 139264-17-8; Solubility: 100 mM in DMSO; 100 mM in ethanol), also named (4S)-4-((3-(2-(dimethylamino)ethyl)-1H-indol-5-yl)methyl)-2-oxazolidinone, targets h5-HT$_{1D}$ receptors (K_i = 0.63 nM), h5-HT$_{1B}$ receptors (K_i = 5.0 nM), and h5-HT$_{1F}$ receptors (K_i = 63.1 nM), inhibiting trigeminal nerve stimulation-mediated neurogenic plasma protein extravasation in guinea pigs and blocking long excitatory postsynaptic potentials (EPSPs) in a rat model of spinal cord injury (1-3). Zolmitriptan has a dual action, peripherally inhibiting dilatation and inflammation of cranial vessel and also exhibiting a central nociceptive action in the brainstem nuclei (4). This dual action on migraine pain also shows beneficial effects on nausea and vomiting, due to its binding to the nucleus of the tractus solitarius, the vomiting control center. **1.** Murray, *et al* (2011) *J. Neurophysiol.* **106**, 925. **2.** Martin, *et al* (1997) *Brit. J. Pharmacol.* **121**, 157. **3.** Pauwels, *et al* (1997) *Neuropharmacol.* **36** 499. **4.** Pascual (1998) *Neurologia* **2**, 9.

Zolpidem

This widely prescribed and short-acting imidazopyridine-class hypnotic (FW = 307.39 g/mol; CAS 82626-48-0), also known by the trade names Ambien®, AmbienCR®, Intermezzo®, Stilnox®, and Sublinox®), and systematically named *N,N*-dimethyl-2-(6-methyl-2-*p*-tolylimidazo[1,2-

a]pyridin-3-yl)acetamide, potentiates the actions of the inhibitory neurotransmitter, γ-aminobutyric acid (GABA), by binding exclusively to the GABA$_A$ receptor, also known as the Benzodiazepine Omega-1 Receptor (1-3). Therefore, while zolpidem's direct action is not itself inhibitory, it has the overall effect of inhibiting neurotransmission by diminishing the chance of a successful action potential occurring. Indeed, α$_5$-subunit-containing GABA$_A$ receptors exhibit lower affinities for zolpidem (30-fold) than measured previously using recombinantly expressed type II receptors containing either α$_2$- or α$_3$-subunits (4). Zolpidem appears to have minimal next-day effects on cognition and psychomotor performance when administered at bedtime (5) and, as such, has become one of the most common GABA-potentiating sleeping medications. Paradoxically, caffeine appears, at least in some cases, to enhance zolpidem sedation. Ciprofloxacin also interacts with zolpidem in healthy volunteers, increasing its bioavailability by nearly 50%. **Key Pharmacokinetic Parameters:** *See* Appendix II in Goodman & Gilman's *THE PHARMACOLOGICAL BASIS OF THERAPEUTICS*, 12th Edition (Brunton, Chabner & Knollmann, eds.) McGraw-Hill Medical, New York (2011). **1.** Depoortere, Zivkovic, Lloyd, *et al.* (1986) *J. Pharmacol. Exp. Ther.* **237**, 649. **2.** von Moltke, Greenblat, Granda, *et al.* (1999) *Br. J. Clin. Pharmacol.* **48**, 89. **3.** Sanger, Griebel, Perrault, Claustre & Schoemaker (1999) *Pharmacol. Biochem. Behav.* 64, 269. **4.** Pritchett & Seeburg (1990) *J. Neurochem.* 54, 1802. **5.** Holm & Goa (2000) *Drugs* 59, 865.

Zomepirac

This analgesic and anti-inflammatory agent (FW = 277.70 g/mol; CAS 64092-48-4), also known as 5-(*p*-chlorobenzoyl)-1,4-dimethyl-pyrrole-2-acetic acid, inhibits prostaglandin synthesis. Zomepirac is converted metabolically to an acyl glucuronide conjugate which can form covalent adducts with proteins (for example, dipeptidyl peptidase IV) (1). Modification of tubulin inhibits microtubule assembly. **Target(s):** cyclooxygenase (2,3); tubulin polymerization (4). **1.** Wang, Gorrell, McGaughan & Dickinson (2001) *Life Sci.* **68**, 785. **2.** Thomas & Knoop (1982) *J. Infect. Dis.* **145**, 141. **3.** Llorens, Perez, Palomer & Mauleon (2002) *J. Mol. Graph. Model.* **20**, 359. **4.** Bailey, Worrall, de Jersey & Dickinson (1998) *Chem. Biol. Interact.* **115**, 153.

Zonisamide

This antiepileptic (FW = 212.22 g/mol; CAS 68291-97-4; Solubility: 10 mM in water; 100 mM in DMSO; Symbol: ZNS), 1,2-benzisoxazole-3-methanesulfonamide, blocks voltage-sensitive Na$^+$ and T-type Ca^{2+} channels, stimulates BK$_{Ca}$ channels, and modulates GABA, glutamate and monoamine neurotransmission. It inhibits lipid peroxidation and scavenges hydroxyl and nitric oxide free radicals. Zonisamide has a broad spectrum anticonvulsant, neuroprotective, and anti-Parkinson Disorder activities. ZNS (30 μM) reversibly increases the amplitude of K$^+$ outward currents, and paxilline (1 μM) is effective in suppressing the ZNS-induced increase of K$^+$ outward currents. In inside-out configuration, ZNS (30 μM), when applied to the intracellular face of the membrane, does not alter single-channel conductance; however, it did enhance the activity of large-conductance Ca^{2+}-activated K$^+$ (BK$_{Ca}$) channels primarily by decreasing mean closed time. In addition, the EC50 value for ZNS-stimulated BK$_{Ca}$ channels was 34 μM. Zonisamide causes a leftward shift in the activation curve of BK BK$_{Ca}$ channels, with no change in the gating charge of these channels. Moreover, ZNS at a concentration greater than 100 μM also reduced the amplitude of A-type K$^+$ current in these cells. **Key Pharmacokinetic Parameters:** *See* Appendix II in Goodman & Gilman's *THE*

PHARMACOLOGICAL BASIS OF THERAPEUTICS, 12th Edition (Brunton, Chabner & Knollmann, eds.) McGraw-Hill Medical, New York (2011). **1.** Huang, Huang & Wu (2007) *J. Pharmacol. Exp. Ther.* **321**, 98.

Zopiclone

This cyclopyrrolone class hypnotic (FW = 388.81 g/mol; CAS 43200-80-2), also named 4-methyl-1-piperazinecarboxylic acid 6-(5-chloro-2-pyridinyl)-6,7-dihydro-7-oxo-5*H*-pyrrolo[3,4-*b*]   pyrazin-5-yl   ester,   is   a benzodiazepine receptor (BZR) agonist, with little difference in affinity for the BZ$_1$ and BZ$_2$ receptor subtypes; however, its CNS effects differ from those of other BZR ligands, due to its interaction with a different subunit. ZPC shows high-affinity interactions for BZ receptors, as measured by its ability to displace [^{3}H]-flunitrazepam binding: K_i = 24 nM in cerebral cortex; K_i = 31 nM in cerebellum; and K_i = 36 nM in hippocampus (1). No other brain receptors such as γ-aminobutyric acid receptor, dopamine receptor, serotonin and noradrenergic receptors are reached by ZPC. ZD is rapidly and efficiently absorbed: its oral bioavailability was greater than 75% in all species except rats, where a first-pass effect of about 65% was recorded. Plasma protein binding is about 45% (2). **1.** Blanchard, Boireau & Julou (1983) *Pharmacology* 27(Suppl 2), 59. **2.** Gaillot, Heusse, Hougton, Marc Aurele & Dreyfus (1982) *Int Pharmacopsych.* **17**(Suppl 2), 76.

Zopolrestat

This benzothiazole derivative (FW = 419.38 g/mol) is an inhibitor of aldose reductase, acting primarily by binding to the enzyme complexed with NADP$^+$ to form a ternary dead-end complex that prevents turnover in the steady-state (1). **Target(s):** aldose reductase, *or* aldehyde reductase (1,2). **1.** Bohren & Grimshaw (2000) *Biochemistry* **39**, 9967. **2.** El-Kabbani, Rogniaux, Barth, Chung, *et al.* (2000) *Proteins* **41**, 407.

Zorubicin

This semi-synthetic glycoside antibiotic, also known as rubidazone (FW$_{free-base}$ = 631.69 g/mol; CAS 36508-71-1), is a similar to daunorubicin (1,2) It is used as an antineoplastic agent. **Target(s):** DNA polymerase (1); RNA polymerase (3,4). **1.** Sartiano, Lynch & Bullington (1979) *J. Antibiot. (Tokyo)* **32**, 1038. **2.** Young, Ozols & Myers (1981) *N. Engl. J. Med.* **305**, 139. **3.** Gabbay, Grier, Fingerle, Reimer, *et al.* (1976) *Biochemistry* **15**, 2062. **4.** Gabbay (1976) in *Inter. J. Quantum Chem., Quantum Biology Symposium No. 3*, p. 217.

Zosuquidar

This very potent drug efflux inhibitor (MW$_{\text{free-base}}$ = 527.62 g/mol; MW$_{\text{tri-HCl}}$ = 636.99 g/mol; CAS 167465-36-3; Solubility: 100 mg/mL DMSO; 23 mg/mL H$_2$O), also known as LY335979 and systematically as 4-[(1aα,6α,10bα)-1,1-difluoro-1,1a,6,10b-tetrahydrodibenzo-[a,e]cyclopropa-[c]cyclohepten-6-yl]-α-[(5-quinolinyloxy)methyl]-1-piperazineethanol tri-HCl, targets ABC transporters (P-glycoprotein), reversing multidrug resistance (K_i = 60 nM), when used at 0.1-0.5 μM in vitro. By blocking blocking P-gp-mediated efflux, GF120918 increases the intracellular concentration of a co-administered drug, often potentiating its cytotoxicity and effectiveness as an anti-tumor agent (1). LY335979 has been used both in vitro and in vivo as a tool inhibitor of P-glycoprotein to investigate the role of drug efflux transporters in the disposition of a test drug and to guide preclinical doses of candidate drugs. Treatment of mice bearing P388/ADR murine leukemia cells with LY335979 in combination with Dox or etoposide gave a significant increase in life span with no apparent alteration of pharmacokinetics (1). LY335979 also enhanced the antitumor activity of paclitaxel (Taxol) in a multidrug-resistant human non-small cell lung carcinoma nude mouse xenograft model (2). Zosuquidar either completely or partially restored drug sensitivity in all P-gp-expressing leukemia cell lines tested and enhanced the cytotoxicity of anthracyclines (daunorubicin, idarubicin, mitoxantrone) and gemtuzumab ozogamicin (Mylotarg) in primary AML blasts with active P-gp (2). In addition, P-gp inhibition by zosuquidar was found to be more potent than cyclosporine A in cells with highly active P-gp (2). 1. Dantzig, Shepard, Cao, et al. (1996) Cancer Res. 56, 4171. 2. Tang, Faussat, Perrot, et al. (2008) BMC Cancer 8, 51.

Zotarolimus

This mTOR inhibitor (FW = 966.21 g/mol; CAS 221877-54-9), also called ABT-578 and 42-deoxy-42-(1H-tetrazol-1-yl)-(42S)-rapamycin, targets smooth muscle cell proliferation (IC$_{50}$ = 2.9 nM) and endothelial cell proliferation (IC$_{50}$ = 2.6 nM). Zotarolimus is mechanistically similar to sirolimus (*See mTOR Inhibitors*), showing high-affinity binding to the immunophilin FKBP12 and comparable potency for inhibiting in vitro proliferation of both human and rat T cells (1). In intravenously dosed rats, pharmacokinetic studies demonstrated an terminal elimination half-lives of 9.4 hours for zotarolimus, compared to a value of 14 hours for sirolimus.

When given orally, $t_{1/2}$ values were 8 hours and 33 hours, respectively. Consistent with its shorter in vivo duration, zotarolimus has a statistically significant 4x reduction in potency for systemic immunosuppression in rat disease models (1). When applied to PC-coated stents, zotarolimus reduces neointima in the swine coronary model after 28 days, suggesting its application for coronary stenting (2). 1. Chen, Smith, Sheets, et al. (2007) J. Cardiovasc. Pharmacol. 49, 228. 2. Garcia-Touchard, Burke, Toner, Cromack & Schwartz (2006) Eur. Heart J. 27, 988.

Zotepine

This psychotropic/neurotropic tranquilizer (FW = 331.87 g/mol; CAS 26615-21-4; abbreviation = ZTP), also known as Nipolept™, Losizopilon™, Lodopin™, and Setous™, and named systematically as 2-(8-chlorobenzo[b][1]benzothiepin-6-yl)oxy-N,N-dimethylethanamine, exhibits atypical antipsychotic effects and also antidepressive effects in schizophrenia patients. ZTP is a high-affinity antagonist of dopamine (D$_1$ and D$_2$) receptors and serotonin (5-HT$_{2A}$, 5-HT$_{2C}$, 5-HT$_6$ and 5-HT$_7$) receptors, and is an especially effective inverse agonist of 5-HT$_{2A}$ receptors (1). Zotepine also inhibits norepinephrine re-uptake (2). At a higher concentrations (0.5 mM), zotepine inhibits the mitochondrial electron transport system Complex I (3). Given that norZTP (or N-desmethylzotepine; IUPAC: 2-[(8-chlorodibenzo[b,f]thiepin-10-yl)oxy]-N-methylethan-1-amine) is the major metabolite observed in humans, norZTP may contribute to the unique clinical profiles observed with the mother compound, ZTP. 1. Herrick-Davis, Grinde & Teitler (2000) J. Pharmacol. Exp. Ther. 295, 226. 2. Shobo, Kondo, Yamada, et al. (2010) J. Pharmacol. Exp. Ther. 333, 772. 3. Maurer & Moller (1997) Mol. Cell. Biochem. 174, 255.

Zovirax, See Acyclovir

Zoxamide

This haloketone-based antimitotic agent and fungicide (FW = 336.64 g/mol; CAS 156052-68-5), also known by its code name RH-7281, its proprietary names Gavel® an Electis®, as well as its systematic name (RS)-3,5-dichloro-N-(3-chloro-1-ethyl-1-methyl-2-oxopropyl)-p-toluamide, is a benzimid-azole-class fungicide that targets tubulin (most likely irreversibly), blocking microtubule self-assembly, while exerting minimal systemic cytotoxic action to treated plants or secondarily animals. With far fewer microtubules involved in mitosis than higher plants and animals, fungi are more susceptible spindle poisons, thus accounting for zoxamide's favorable therapeutic index. In animals, deactivating effects of metabolic conjugation by glutathione as well as glucuronide formation provide a rational for the observed absence of typically mitosis inhibition-associated toxicities of zoxamide in mammals in vivo (1). As formulated for use in limiting fungal infections, zoxamide is exceptionally resistant to wash-off by rain or irrigation, due to its penetration and binding into the plant cuticle. oxamide has also demonstrated control of the following pathogens: *Pseudoperonospora cubensis*, or downy mildew of cucurbits; *Bremia lactuae*, or downy mildew of lettuce; *Perenospora viciae*, or downy mildew of peas; *Phytophthora nicotianae*, or downy mildew of tobacco; *Plasmopara helianthi*, or downy mildew of sunflower, or *Phytophthora capsici*, or downy mildew of pepper and eggplants; and *Perenospora parasitica*, or downy mildew of crucifers. Failure to isolate mutants resistant to zoxamide results from the diploid nature of *Oomycete* fungi and the likelihood that target-site mutations would produce a recessive phenotype (2). Recent work has demonstrated that *Phytophthora capsici* has developed a novel non-target-site-based zoxamide resistance, involving two recessive genes and allowing resistance to emerge, if two compatible mating types co-exist in the same field (3). **1**. Oesch, Metzler, Fabian (2010) *Xenobiotica* **40**, 72. **2**. Young, Spiewak & Slawecki (2001) *Pest Manag Sci.* **57**, 1081. **3**. Bi, Chen, Cai, *et al.* (2014) *PLoS One* **9**, e89336.

ZPBK, See *N-Benzyloxycarbonyl-L-phenylalanine Bromomethyl Ketone*

ZPCK, See *N-Benzyloxycarbonyl-L-phenylalanine Chloromethyl Ketone*

ZR-448, See *Methyl (E,E)-11-Methoxy-3,7,11-trimethyldodeca-2,4-dienoate*

ZSTK474

This potent, ATP-site competitive, pan-Class I phosphoinositide-3-kinase PI3K inhibitor (FW = 417.41 g/mol; CAS 475110-96-4; Solubility (25°C): 21 mg/mL DMSO, <1 mg/mL Water), inhibits the activities of recombinant p110β (IC$_{50}$ = 7 nM); p110γ (IC$_{50}$ = 37 nM); and p110δ (IC$_{50}$ = 6 nM) (1,2), with little effect on mammalian target of rapamycin (mTOR) at 100 µM (3). Proliferation assays revealed that BEZ235 (dual PI3K/mTOR inhibitor) or ZSTK474 (pan-PI3K inhibitor), when combined with temsirolimus, inhibited cell growth compared to cells treated with any of the agents alone (4). Addition of a PI3K inhibitor synergistically overcomes cellular resistance to mTORC1 inhibitors regardless of PTEN status, substantially expanding the molecular phenotype of tumors likely to respond (4). ZSTK474 also demonstrated prophylactic efficacy in a rat model of rheumatoid arthritis through the inhibition of T cell and FLS functions (5).

1. Yaguchi, Fukui, Koshimizu, *et al.* (2006) *J. Natl. Cancer Inst.* **98**, 545. **2**. Kong & Yamori (2007) *Cancer Sci.* **98**, 1638. **3**. Kong, Okamura, Yoshimi & Yamori (2009) *Eur. J. Cancer* **45**, 857. **4**. Yang , Xiao, Meng & Leslie (2011) *PLoS One* **6**, e26343. **6**. Haruta, Mori, Tamura, *et al.* (2012) *Inflamm. Res.* **61**, 551.

α-Zurine 2G, See *Patent Blue Violet*

Zyklon-B

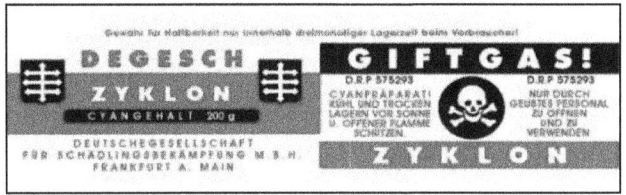

This notorious rodenticide, insecticide and fumigant consisted of pressurized hydrogen cyanide in combination with an adsorbent porous carrier (*i.e.*, diatomaceous earth or processed wood pulp), usually stored within a rupture-resistant metal container. (**See** *Cyanide*) The considerable surface area of the solid support facilitated maximal concentration of gaseous HCN, greatly enhancing its killing potential. Because gaseous HCN is odorless, ethyl bromoacetate (itself a volatile but noxious irritant) was included as a warning gas. Manufactured chiefly by Degesch (*or* Deutsche Gesellschaft für Schädlingsbekämpfung mbH, *or* German Corporation for Pest Control, Ltd.) and Tesch und Stabenow Internationale Gesellschaft für Schädlingsbekämpfung m.b.H., *or* Tesch & Stabenow, International Society for Pest Control, Ltd.), Zyklon (*translated*: Cyclone) was invented to prevent food loss in granaries and warehouses. During World War II, however, it was first employed to delouse the cramped prisoner quarters in concentration camps. Later, with warning gas omitted, Zyklon-B was employed in the systematic killing of one million Jews and "Social Deviants" (addicts, alcoholics, dissidents, draft dodgers, pacifists, prostitutes, vagrants, and common criminals) at Auschwitz-Birkenau, Majdanek, Mauthausen, Ravensbrück, and elsewhere. In a compendium wholly devoted to Inhibitor Science and the great good that can come of it, this final entry reminds us that we must never forget those lost in the *Shoah* or the great ill can come of even well-intentioned inventions.